CHILTON'S

AUTO REPAIR MANUAL 1988-1992

Publisher	Kerry A. Freeman, S.A.E.
Editor-In-Chief	Dean F. Morgantini, S.A.E.
Managing Editor	David H. Lee, A.S.E., S.A.E.
Senior Editors	Richard J. Rivele, S.A.E.
	Nick D'Andrea
	Ron Webb
Project Managers	Peter M. Conti, Jr., Ken Grabowski, A.S.E.
	Martin J. Gunther, Richard T. Smith
Editorial Staff	Robert E. Doughten, Jeff H. Fisher, A.S.E.
	Jacques Gordon, Michael L. Grady
	Ben Greisler, S.A.E., Jeffrey M. Hoffman
	Steve Horner, Neil Leonard, A.S.E.
	James R. Marotta, Robert McAnally
	Michael W. Parks, John H. Rutter
	Don Schnell, James B. Steele
	Larry E. Stiles, Jim Taylor
	Anthony Tortorici, A.S.E., S.A.E.
Manager of Manufacturing	John J. Cantwell
Production Manager	W. Calvin Settle, Jr.
Assistant Production Manager	Andrea M. Steiger
Mechanical Artists	Lisa Gressen, Marsha Park Herman
	Lorraine Martinelli, Kim Tansey
Special Projects	Peter Kaprielyan

OFFICERS

President	Gary R. Ingersoll
Sr. Vice President	Ronald A. Hoxter

Cover designed by Donna Monturo for DM Design

CHILTON BOOK COMPANY

ONE OF THE **DIVERSIFIED PUBLISHING COMPANIES,** A PART OF **CAPITAL CITIES/ABC, INC.**

Manufactured in USA
© 1991 Chilton Book Company
Chilton Way Radnor, Pa. 19089
ISBN 0-8019-7906-4
ISSN 0069-3634

1234567890   0987654321

CAR MODELS

TABLE OF CONTENTS

Car Sections

Unit Repair Sections

HOW TO USE THIS MANUAL

This manual is arranged in two sections:

Car Section

Car sections are grouped by manufacturer and arranged in alphabetical order. The text and illustrations that comprise the service procedures in each Car Section are arranged in the following order of systems and components: Engine Mechanical, Engine Lubrication, Engine Cooling, Engine Electrical, Emission Controls, Fuel System, Drive Axle, Manual Transmission/Transaxle, Clutch, Automatic Transmission/Transaxle, Front Suspension, Rear Suspension, Steering, Brakes, Chassis Electrical.

Specification charts are always located at the front of each section. All illustrations are located as close as possible to the pertinent text. Procedures are for all models in the particular section unless specifically noted otherwise.

Unit Repair Section

The Unit Repair Section contains troubleshooting and overhaul procedures for the major components and systems of your car. This portion of the book is intended to be used in conjunction with the Car Sections.

Every major Unit Repair Section contains an Identification or Application chart to correlate the information contained in that section. The sections are usually arranged by brands, manufacturers or types of components rather than models of cars. All overhaul procedures in the Unit Repair Section begin with the component removed from the car. The reason for this division of material is an economic one. The steps involved in overhauling an engine are virtually the same for all engines. However, the operation of removing the engine from the car varies greatly from model to model. By combining where possible, and separating where necessary, we are able to publish the maximum amount of information.

Locating Information

The Table of Contents, at the front of the book, lists the beginning of each Car and Unit Repair Section in the manual.

To find where a particular Car Section is located in the book, you need only look in the Table of Contents. Once you have found the proper section, you may wish to find where specific procedures are located in that section. Turn to the Index at the front of the section. At the upper left-hand side is a listing of the main topics within the section and the page number they will be found on. Following the main topics is an alphabetical listing of all the procedures within the section and their page numbers.

Safety Notice

Proper service and repair procedures are vital to the safe, reliable operation of all motor vehicles, as well as the personal safety of those performing repairs. This manual outlines procedures for servicing and repairing vehicles using safe effective methods. The procedures contain many NOTES and CAUTIONS which should be followed along with standard safety procedures to eliminate the possibility of personal injury or improper service which could damage the vehicle or compromise its safety.

It is important to note that repair procedures and techniques, tools and parts for servicing motor vehicles, as well as the skill and experience of the individual performing the work vary widely. It is not possible to anticipate all of the conceivable ways or conditions under which vehicles may be serviced, or to provide cautions as to all of the possible hazards that may result. Standard and accepted safety precautions and equipment should be used when handling toxic or flammable fluids, and safety goggles or other protection should be used during cutting, grinding, chiseling, prying, or any other process that can cause material removal or projectiles.

Some procedures require the use of tools specially designed for a specific purpose. Before substituting another tool or procedure, you must be completely satisfied that neither your personal safety, nor the performance of the vehicle will be endangered.

Part Numbers

Part numbers listed in this book are not recommendations by Chilton for any product by brand name. They are references that can be used with interchange manuals and aftermarket supplier catalogs to locate each brand supplier's discrete part number.

Although information in this manual is based on industry sources and is as complete as possible at the time of publication, the possibility exists that some car manufacturers made later changes which could not be included here. Information on very late models may not be available in some circumstances. While striving for total accuracy, Chilton Book Company cannot assume responsibility for any errors, changes, or omissions that may occur in the compilation of this data.

Copyright Notice

Chrysler/Eagle
Front Wheel Drive
Dodge—Stealth **Eagle**—Summit, Talon
Plymouth—Laser

SPECIFICATIONS

VEHICLE IDENTIFICATION CHART

It is important for servicing and ordering parts to be certain of the vehicle and engine identification. The VIN (vehicle identification number) is a 17 digit number visible through the windshield on the driver's side of the dash and contains the vehicle and engine identification codes. The tenth digit indicates model year and the eighth digit indicates engine code. It can be interpreted as follows:

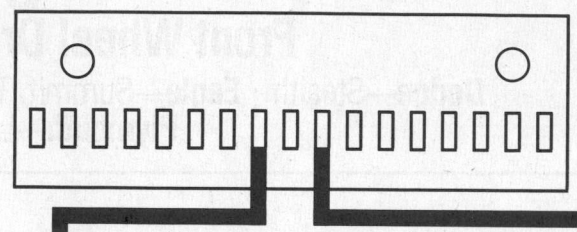

Engine Code						Model Year	
Code	Cu. In.	Liters	Cyl.	Fuel Sys.	Eng. Mfg.	Code	Year
T	107	1.8	4	MPI	Mitsubishi	K	1989
R	122	2.0	4	MPI	Mitsubishi	L	1990
U	122	2.0	4	MPI-Turbo	Mitsubishi	M	1991
X	92	1.5	4	MPI	Mitsubishi	N	1992
Y	98	1.6	4	MPI	Mitsubishi		
A	92	1.5	4	MPI	Mitsubishi		
S	181	3.0	6	MPI	Mitsubishi		
B	181	3.0	6	MPI	Mitsubishi		
C	181	3.0	6	Turbo	Mitsubishi		

MPI—Multi-port injection

ENGINE IDENTIFICATION

Year	Model	Engine Displacement cu. in. (liter)	Engine Series Identification (VIN)	No. of Cylinders	Engine Type
1989	Eagle Summit	92 (1.5)	X	4	SOHC
	Eagle Summit	98 (1.6)	Y	4	DOHC
1990	Eagle Summit	92 (1.5)	X	4	SOHC
	Eagle Summit	98 (1.6)	Y	4	DOHC
	Plymouth Laser	107 (1.8)	T	4	SOHC
	Plymouth Laser	122 (2.0)	R	4	DOHC
	Plymouth Laser	122 (2.0)	U	4	DOHC w/Turbo
	Eagle Talon	122 (2.0)	R	4	DOHC
	Eagle Talon	122 (2.0)	U	4	DOHC w/Turbo
1991-92	Eagle Summit	92 (1.5)	A	4	SOHC
	Plymouth Laser	107 (1.8)	T	4	SOHC
	Plymouth Laser	122 (2.0)	R	4	DOHC
	Plymouth Laser	122 (2.0)	U	4	DOHC w/Turbo
	Eagle Talon	122 (2.0)	R	4	DOHC
	Eagle Talon	122 (2.0)	U	4	DOHC w/Turbo

ENGINE IDENTIFICATION

Year	Model	Engine Displacement cu. in. (liter)	Engine Series Identification (VIN)	No. of Cylinders	Engine Type
1991-92	Dodge Stealth	181 (3.0)	S	6	SOHC
	Dodge Stealth	181 (3.0)	B	6	DOHC
	Dodge Stealth	181 (3.0)	C	6	DOHC w/Turbo

SOHC Single Overhead Cam Engine
DOHC Double Overhead Cam Engine

GENERAL ENGINE SPECIFICATIONS

Year	VIN	No. Cylinder Displacement cu. in. (liter)	Fuel System Type	Net Horsepower @ rpm	Net Torque @ rpm (ft. lbs.)	Bore × Stroke (in.)	Compression Ratio	Oil Pressure @ rpm
1989	X	4-92 (1.5)	MPI	81 @ 5500	91 @ 3000	2.972 × 3.228	9.4:1	54 @ 2000
	Y	4-98 (1.6)	MPI	113 @ 6500	99 @ 5000	3.243 × 2.955	9.2:1	54 @ 2000
1990	X	4-92 (1.5)	MPI	81 @ 5500	91 @ 3000	2.972 × 3.228	9.4:1	54 @ 2000
	Y	4-98 (1.6)	MPI	113 @ 6500	99 @ 5000	3.243 × 2.955	9.2:1	54 @ 2000
	T	4-107 (1.8)	MPI	92 @ 5000	105 @ 3500	3.172 × 3.388	9.0:1	41 @ 2000
	R	4-122 (2.0)	MPI	135 @ 6000	125 @ 5000	3.349 × 3.467	9.0:1	41 @ 2000
	U	4-122 (2.0)	Turbo	190 @ 6000	203 @ 3000	3.349 × 3.467	7.8:1	41 @ 2000
1991-92	A	4-92 (1.5)	MPI	92 @ 6000	93 @ 3000	2.97 × 3.23	9.2:1	54 @ 2000
	T	4-107 (1.8)	MPI	92 @ 5000	105 @ 3500	3.17 × 3.39	9.0:1	41 @ 2000
	R	4-122 (2.0)	MPI	135 @ 6000	125 @ 5000	3.35 × 3.47	9.0:1	41 @ 2000
	U	4-122 (2.0)	Turbo	190 @ 6000	203 @ 3000	3.35 × 3.47	7.8:1	41 @ 2000
	S	6-181 (3.0)	MPI	164 @ 5500	185 @ 4000	3.58 × 2.99	8.9:1	30-80 @ 2000
	B	6-181 (3.0)	MPI	222 @ 6000	201 @ 4500	3.58 × 2.99	10.0:1	30-80 @ 2000
	C	6-181 (3.0)	Turbo	300 @ 6000	307 @ 2500	3.58 × 2.99	8.0:1	30-80 @ 2000

MPI—Multi-port Injection

GASOLINE ENGINE TUNE-UP SPECIFICATIONS

Year	VIN	No. Cylinder Displacement cu. in. (liter)	Spark Plugs Type	Spark Plugs Gap (in.)	Ignition Timing (deg.) MT	Ignition Timing (deg.) AT	Compression Pressure (psi)	Fuel Pump (psi)	Idle Speed (rpm) MT	Idle Speed (rpm) AT	Valve Clearance (in.) In.	Valve Clearance (in.) Ex.
1988	F	4-132 (2.2)	RS9YC	0.035	①	①	②	34–36	800	700	0.006	0.008
	Z	4-150 (2.5)	RC12LYC	0.035	—	①	②	14–15	—	750	Hyd.	Hyd.
	J	6-182 (3.0)	RS9YC	0.035	—	①	②	36–37	—	800	Hyd.	Hyd.
1989	F	4-132 (2.2)	RS9YC	0.035	①	①	②	34–36	800	700	0.006	0.008
	H	4-150 (2.5)	RN12LYC	0.035	—	①	②	14–15	—	①	Hyd.	Hyd.
	U	6-182 (3.0)	RS9YC	0.035	—	①	②	28–30	—	①	Hyd.	Hyd.
1990	U	6-182 (3.0)	RS9YCX	0.035	—	①	②	28–30	—	①	Hyd.	Hyd.
1991	U	6-182 (3.0)	RS9YCX	0.035	—	①	②	③	—	①	Hyd.	Hyd.
1992		SEE UNDERHOOD SPECIFICATION STICKER										

Hyd.—Hydraulic
① Refer to Underhood Sticker
② The lowest reading should be no less than 75% of the highest reading.
③ Vehicles built before 10/9/91—28–30 psi
Vehicles built on or after 10/9/91—43 psi

FIRING ORDERS

NOTE: To avoid confusion, always replace spark plugs and wires one at a time.

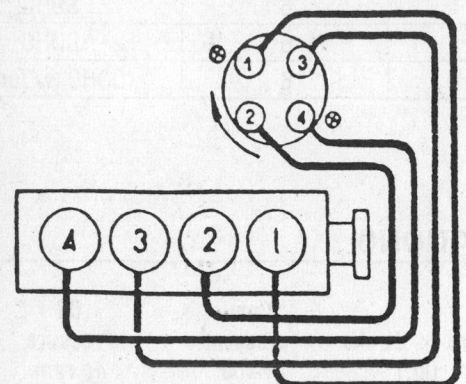

1.8L and 1989–90 1.5L Engines
Engine Firing Order: 1–3–4–2
Distributor Rotation: Clockwise

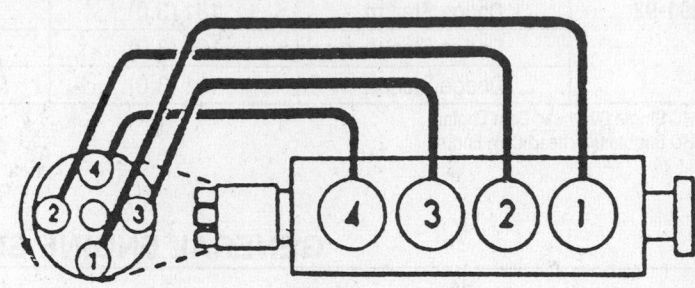

1991–92 1.5L Engine
Engine Firing Order: 1–3–4–2
Distributor Rotation: Counterclockwise

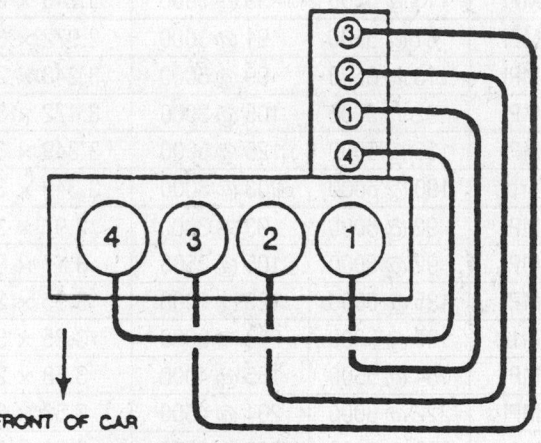

FRONT OF CAR

1.6L and 2.0L Engines
Engine Firing Order: 1–3–4–2
Distributorless Ignition System

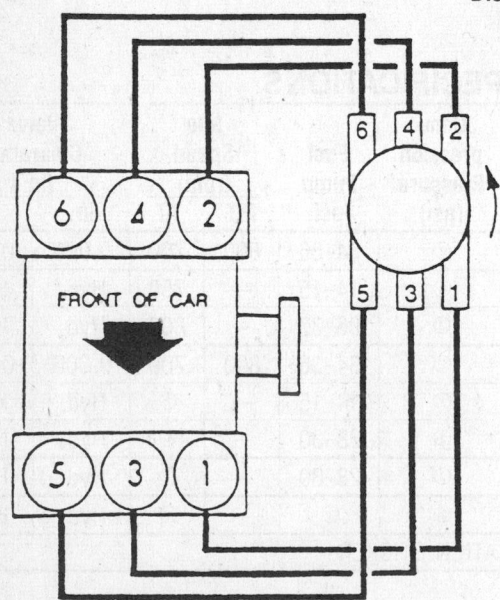

FRONT OF CAR

3.0L SOHC Engine
Engine Firing Order: 1–2–3–4–5–6
Distributor Rotation: Counterclockwise

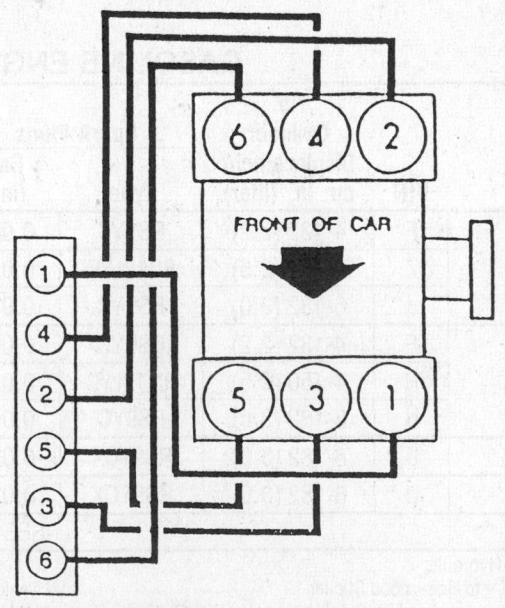

FRONT OF CAR

3.0L DOHC Engine
Engine Firing Order: 1–2–3–4–5–6

CAPACITIES

Year	Model	VIN	No. Cylinder Displacement cu. in. (liter)	Engine Crankcase (qts.) with Filter	Engine Crankcase (qts.) without Filter	Transmission (pts.) 4-Spd	Transmission (pts.) 5-Spd	Transmission (pts.) Auto.	Drive Axle (pts.)	Fuel Tank (gal.)	Cooling System (qts.)
1989	Summit	X	4-92 (1.5)	3.6①	3.0①	3.6	3.8②	13③	—	13.2	5.3
	Summit	Y	4-98 (1.6)	4.6①	4.1①	3.6	3.8	13③	—	13.2	5.3
1990	Summit	X	4-92 (1.5)	3.6①	3.0①	3.6	3.8②	13③	—	13.2	5.3
	Summit	Y	4-98 (1.6)	4.6①	4.2①	3.6	3.8	13③	—	13.2	5.3
	Laser	T	4-107 (1.8)	4.1①	3.6①	—	3.8	13③	—	16.0	6.6
	Laser	R	4-122 (2.0)	4.6①	4.1①	—	4.6⑤	13③	—	16.0	7.6
	Laser	U	4-122 (2.0)	4.6①	4.1①	—	4.8⑥	13③	⑦	16.0	7.6
	Talon	R	4-122 (2.0)	4.6①	4.1①	—	4.6⑤	13③	—	16.0	7.6
	Talon	U	4-122 (2.0)	4.6①	4.1①	—	4.8⑥	13③	⑦	16.0	7.6
1991–92	Summit	A	4-92 (1.5)	3.6⑧	3.0⑧	3.6	3.8	13③	—	13.2	5.3
	Laser	T	4-107 (1.8)	4.1⑧	3.6⑧	—	3.8	13③	—	16.0	6.6
	Laser	R	4-122 (2.0)	4.6⑧	4.1⑧	—	4.6⑤	13③	—	16.0	7.6
	Laser	U	4-122 (2.0)	4.6⑧	4.1⑧	—	4.8⑥	13③	⑦	16.0	7.6
	Talon	R	4-122 (2.0)	4.6⑧	4.1⑧	—	4.6⑤	13③	—	16.0	7.6
	Talon	U	4-122 (2.0)	4.6⑧	4.1⑧	—	4.8⑥	13③	⑦	16.0	7.6
	Stealth	S	6-181 (3.0)	4.7⑧	4.2⑧	—	5	15.8③	—	19.8	8.5
	Stealth	B	6-181 (3.0)	4.7⑧	4.2⑧	—	5	15.8③	—	19.8	8.5
	Stealth	C	6-181 (3.0)	5.2⑧	4.7⑧	—	5	—	⑨	19.8	8.5

① Use API class SF or SF/CC engine oil
② 3.8 pts. for transaxle models KM201, KM206.
   4.4 pts. for transaxle model KM210.
   See Vehicle Information Code Plate on firewall
   for transaxle number. Manual transaxles,
   use API class GL-4 or higher Hypoid Gear Oil
③ Automatic—Use Dexron II type fluid. Quantity
   shown includes converter. Check when hot.
④ Add 0.5 qt. for oil cooler on Turbo models.
⑤ 2WD-Turbo
⑥ 4WD-Turbo
⑦ Rear axle—with 4WD, capacity 0.75 qt. plus
   0.63 qt. in transfer case.
⑧ Use API class SG or SG/CD engine oil.
⑨ Rear axle with 4WD capacity 1.16 qts. plus
   0.29 qt. in transfer case.

CAMSHAFT SPECIFICATIONS

All measurements given in inches

Year	VIN	No. Cylinder Displacement cu. in. (liter)	Journal Diameter 1	Journal Diameter 2	Journal Diameter 3	Journal Diameter 4	Journal Diameter 5	Lobe Lift In.	Lobe Lift Ex.	Bearing Clearance	Camshaft End Play
1989	X	4-92 (1.5)	1.8110	1.8110	1.8110	—	—	1.5318	1.5344	0.0015–0.0031	0.002–0.008
	Y	4-98 (1.6)	1.020	1.020	1.020	1.020	1.020①	1.3858	1.3743	0.002–0.0035	0.004–0.008
1990	X	4-92 (1.5)	1.8110	1.8110	1.8110	—	—	1.5318	1.5344	0.0015–0.0031	0.002–0.008
	Y	4-98 (1.6)	1.020	1.020	1.020	1.020	1.020①	1.3858	1.3743	0.0020–0.0035	0.004–0.008

CAMSHAFT SPECIFICATIONS

All measurements given in inches

Year	VIN	No. Cylinder Displacement cu. in. (liter)	Journal Diameter					Lobe Lift		Bearing Clearance	Camshaft End Play
			1	2	3	4	5	In.	Ex.		
1990	T	4-107 (1.8)	1.3360–1.3366	1.3360–1.3366	1.3360–1.3366	1.3360–1.3366	1.336–1.337	1.4138	1.4138	0.0020–0.0035	0.004–0.008
	R	4-122 (2.0)	1.0217–1.0224	1.0217–1.0224	1.0217–1.0224	1.0217–1.0224	1.020①	1.3974	1.3858	0.0020–0.0035	0.040–0.008
	U	4-122 (2.0)	1.0217–1.0224	1.0217–1.0224	1.0217–1.0224	1.0217–1.0224	1.020①	1.3974	1.3858	0.0020–0.0035	0.004–0.008
1991–92	A	4-92 (1.5)	1.8110	1.8110	1.8110	1.8110	—	1.5059–1.5256	1.5197–1.5394	0.0024–0.0055	NA
	T	4-107 (1.8)	1.3360–1.3366	1.3360–1.3366	1.3360–1.3366	1.3360–1.3366	1.336–1.337	1.4138	1.4138	0.0020–0.0035	0.004–0.008
	R	4-122 (2.0)	1.0217–1.0224	1.0217–1.0224	1.0217–1.0224	1.0217–1.0224	1.020①	1.3974	1.3858	0.0020–0.0035	0.040–0.008
	U	4-122 (2.0)	1.0217–1.0224	1.0217–1.0224	1.0217–1.0224	1.0217–1.0224	1.020①	1.3974	1.3858	0.0020–0.0035	0.004–0.008
	S	6-181 (3.0)	1.34	1.34	1.34	1.34	—	1.6430–1.6440	1.6430–1.6440	0.0020–0.0035	0.004–0.008
	B	6-181 (3.0)	1.02	1.02	1.02	1.02	1.02	1.3776–1.3972	1.3661–1.3858	0.0020–0.0035	0.004–0.008
	C	6-181 (3.0)	1.02	1.02	1.02	1.02	1.02	1.3776–1.3972	1.3661–1.3858	0.0020–0.0035	0.004–0.008

① Six journals are used. All are 1.020 diameter.
  Bearing caps Nos. 2–5 are the same shape.
  "L" or "R" is stamped on No. 1 bearing cap.
  L=intake side, R=exhaust side. Bearing caps
  should be reinstalled at their original locations.
NA—Not available

CRANKSHAFT AND CONNECTING ROD SPECIFICATIONS

All measurements are given in inches.

Year	VIN	No. Cylinder Displacement cu. in. (liter)	Crankshaft				Connecting Rod		
			Main Brg. Journal Dia.	Main Brg. Oil Clearance	Shaft End-play	Thrust on No.	Journal Diameter	Oil Clearance	Side Clearance
1989	X	4-92 (1.5)	1.890	0.0008–0.0018	0.0020–0.0071	3	1.6500	0.0006–0.0017	0.0039–0.0098
	Y	4-98 (1.6)	2.240	0.0008–0.0020	0.0020–0.0071	3	1.7700	0.0008–0.0020	0.0039–0.0098
1990	X	4-92 (1.5)	1.890	0.0008–0.0018	0.0020–0.0071	3	1.6500	0.0006–0.0017	0.0039–0.0098
	Y	4-98 (1.6)	2.240	0.0008–0.0020	0.0020–0.0071	3	1.7700	0.0008–0.0020	0.0039–0.0098
	T	4-107 (1.8)	2.240	0.0008–0.0020	0.0020–0.0070	3	1.7700	0.0008–0.0020	0.0039–0.0098
	R	4-122 (2.0)	2.243–2.244	0.0008–0.0020	0.0020–0.0070	3	1.7709–1.7715	0.0008–0.0020	0.0040–0.0098
	U	4-122 (2.0)	2.243–2.244	0.0008–0.0020	0.0020–0.0070	3	1.7709–1.7715	0.0008–0.0020	0.0040–0.0098

CRANKSHAFT AND CONNECTING ROD SPECIFICATIONS

All measurements are given in inches.

| Year | VIN | No. Cylinder Displacement cu. in. (liter) | Crankshaft | | | | Connecting Rod | | |
			Main Brg. Journal Dia.	Main Brg. Oil Clearance	Shaft End-play	Thrust on No.	Journal Diameter	Oil Clearance	Side Clearance
1991–92	A	4-92 (1.5)	1.89	0.0008–0.0028	0.0020–0.0071	3	1.6500	0.0008–0.0024	0.0039–0.0098
	T	4-107 (1.8)	2.240	0.0008–0.0020	0.0020–0.0070	3	1.7700	0.0008–0.0020	0.0039–0.0098
	R	4-122 (2.0)	2.243–2.244	0.0008–0.0020	0.0020–0.0070	3	1.7709–1.7715	0.0008–0.0020	0.0040–0.0098
	U	4-122 (2.0)	2.243–2.244	0.0008–0.0020	0.0020–0.0070	3	1.7709–1.7715	0.0008–0.0020	0.0040–0.0098
	S	6-181 (3.0)	2.358	0.0008–0.0019	0.0020–0.0098	3	1.965	0.0006–0.0018	0.0040–0.0098
	B	6-181 (3.0)	2.358	0.0007–0.0017	0.0020–0.0098	3	1.965	0.0006–0.0018	0.0040–0.0098
	C	6-181 (3.0)	2.358	0.0007–0.0017	0.0020–0.0098	3	1.965	0.0006–0.0018	0.0040–0.0098

VALVE SPECIFICATIONS

| Year | VIN | No. Cylinder Displacement cu. in. (liter) | Seat Angle (deg.) | Face Angle (deg.) | Spring Test Pressure (lbs.) | Spring Installed Height (in.) | Stem-to-Guide Clearance (in.) | | Stem Diameter (in.) | |
							Intake	Exhaust	Intake	Exhaust
1989	X	4-92 (1.5)	44–44.5	45–45.5	53②	1.756①	0.0008–0.0020	0.0020–0.0035	0.2600	0.2600
	Y	4-98 (1.6)	44–44.5	45–45.5	53②	1.803③	0.0008–0.0019	0.0020–0.0033	0.2585–0.2586	0.2571–0.2579
1990	X	4-92 (1.5)	44–44.5	45–45.5	53②	1.756①	0.0008–0.0020	0.0020–0.0035	0.2600	0.2600
	Y	4-98 (1.6)	44–44.5	45–45.5	66②	1.902④	0.0008–0.0019	0.0020–0.0033	0.2585–0.2586	0.2571–0.2579
	T	4-107 (1.8)	44–44.5	45–45.5	62②	1.937⑦	0.0012–0.0024⑤	0.0020–0.0035⑥	0.3100	0.3100
	R	4-122 (2.0)	44–44.5	45–45.5	66②	1.902⑧	0.0008–0.0019⑤	0.0020–0.0033⑥	0.2585–0.2591	0.2571–0.2579
	U	4-122 (2.0)	44–44.5	45–45.5	66②	1.902⑧	0.0008–0.0019⑤	0.0020⑥	0.2585–0.2591	0.2571–0.2579
1991–92	A	4-92 (1.5)	44–44.5	45–45.5	② ⑨	⑩	0.0008–0.0020	0.0020–0.0035	0.2585–0.2591	0.2571–0.2579
	T	4-107 (1.8)	44–44.5	45–45.5	62②	1.937⑦	0.0012–0.0024⑤	0.0020–0.0035⑥	0.3100	0.3100
	R	4-122 (2.0)	44–44.5	45–45.5	66②	1.902⑧	0.0008–0.0019⑤	0.0020–0.0033⑧	0.2585–0.2591	0.2571–0.2579
	U	4-122 (2.0)	44–44.5	45–45.5	66②	1.902⑧	0.0008–0.0019⑤	0.0020–0.0033⑥	0.2585–0.2591	0.2571–0.2579
	S	6-181 (3.0)	44–44.5	45–45.5	74②	1.60–1.63	0.0012–0.0039	0.0020–0.0059	0.314	0.314
	B	6-181 (3.0)	44–44.5	45–45.5	62②	1.50–1.53	0.0008–0.0039	0.0020–0.0047	0.260	0.260

VALVE SPECIFICATIONS

Year	VIN	No. Cylinder Displacement cu. in. (liter)	Seat Angle (deg.)	Face Angle (deg.)	Spring Test Pressure (lbs.)	Spring Installed Height (in.)	Stem-to-Guide Clearance (in.) Intake	Stem-to-Guide Clearance (in.) Exhaust	Stem Diameter (in.) Intake	Stem Diameter (in.) Exhaust
1991–92	C	6-181 (3.0)	44–44.5	45–45.5	62 ②	1.50–1.53	0.0008–0.0039	0.0020–0.0047	0.260	0.260

NA—Not Available
① Free length, not installed height
   Used limit = 1.717
② At installed height
③ Free length, not installed height
   Used limit = 1.768
④ Free length, not installed height
   Used limit = 1.862
⑤ Used limit = 0.004
⑥ Used limit = 0.006
⑦ Free length, not installed height
   Used limit = 1.898
⑧ Free length, not installed height
   Used limit = 1.862
⑨ Intake: 51
   Exhaust: 64
⑩ Free length, not installed height
   Intake: 1.776–1.815
   Exhaust: 1.803–1.843

PISTON AND RING SPECIFICATIONS

All measurements are given in inches.

Year	VIN	No. Cylinder Displacement cu. in. (liter)	Piston Clearance	Ring Gap Top Compression	Ring Gap Bottom Compression	Ring Gap Oil Control	Ring Side Clearance Top Compression	Ring Side Clearance Bottom Compression	Ring Side Clearance Oil Control
1989	X	4-92 (1.5)	0.0008–0.0016	0.0079–0.0138	0.0079–0.0138	0.0079–0.0276	0.0012–0.0028	0.0008–0.0024	NA
	Y	4-98 (1.6)	0.0008–0.0016	0.0098–0.0157	0.0138–0.0197	0.0079–0.0276	0.0012–0.0028	0.0012–0.0028	NA
1990	X	4-92 (1.5)	0.0008–0.0016	0.0079–0.0138	0.0079–0.0138	0.0079–0.0276	0.0012–0.0028	0.0008–0.0024	NA
	Y	4-98 (1.6)	0.0008–0.0016	0.0098–0.0157	0.0138–0.0197	0.0079–0.0276	0.0012–0.0028	0.0012–0.0028	NA
	T	4-107 (1.8)	0.0004–0.0012	0.0118–0.0177	0.0079–0.0138	0.0080–0.0280	0.0018–0.0033	0.0008–0.0024	NA
	R	4-122 (2.0)	0.0008–0.0016	0.0098–0.0177	0.0138–0.0197	0.0079–0.0276	0.0012–0.0028	0.0012–0.0028	NA
	U	4-122 (2.0)	0.0012–0.0020	0.0098–0.0177	0.0138–0.0197	0.0079–0.0276	0.0012–0.0028	0.0012–0.0028	NA
1991–92	A	4-92 (1.5)	0.0008–0.0016	0.0079–0.0157	0.0079–0.0138	0.0079–0.0276	0.0012–0.0028	0.0008–0.0024	NA
	T	4-107 (1.8)	0.0004–0.0012	0.0118–0.0177	0.0079–0.0138	0.0080–0.0280	0.0018–0.0033	0.0008–0.0024	NA
	R	4-122 (2.0)	0.0008–0.0016	0.0098–0.0177	0.0138–0.0197	0.0079–0.0276	0.0012–0.0028	0.0012–0.0028	NA
	U	4-122 (2.0)	0.0012–0.0020	0.0098–0.0177	0.0138–0.0197	0.0079–0.0276	0.0012–0.0028	0.0012–0.0028	NA

PISTON AND RING SPECIFICATIONS

All measurements are given in inches.

| Year | VIN | No. Cylinder Displacement cu. in. (liter) | Piston Clearance | Ring Gap | | | Ring Side Clearance | | Oil Control |
				Top Compression	Bottom Compression	Oil Control	Top Compression	Bottom Compression	
1991–92	S	6-181 (3.0)	0.0012–0.0020	0.0118–0.0177	0.0098–0.0157	0.0118–0.0154	0.0020–0.0035	0.0008–0.0024	NA
	B	6-181 (3.0)	0.0012–0.0020	0.0118–0.0177	0.0177–0.0236	0.0079–0.0236	0.0012–0.0028	0.0008–0.0024	NA
	C	6-181 (3.0)	0.0012–0.0020	0.0118–0.0177	0.0177–0.0236	0.0079–0.0236	0.0012–0.0028	0.0008–0.0024	NA

NA—Not available

TORQUE SPECIFICATIONS

All readings in ft. lbs.

| Year | VIN | No. Cylinder Displacement cu. in. (liter) | Cylinder Head Bolts | Main Bearing Bolts | Rod Bearing Bolts | Crankshaft Pulley Bolts | Flywheel Bolts | Manifold | | Spark Plugs |
								Intake	Exhaust	
1989	X	4-92 (1.5)	①	36–40	23–25	51–72②	94–101	11–14	11–14	15–21③
	Y	4-98 (1.6)	④	47–51	36–38	80–94⑤	94–101	18–22	18–22	15–21③
1990	X	4-92 (1.5)	①	36–40	⑥	51–72②	94–101	11–14	11–14	15–21③
	Y	4-98 (1.6)	④	47–51	36–38	80–94⑤	94–101	18–22	18–22	15–21③
	T	4-107 (1.8)	51–54	37–39	24–25	80–94	94–101	13–18	18–22	15–21③
	R	4-122 (2.0)	65–72	47–51	36–38	80–94	94–101	18–22	18–22	15–21③
	U	4-122 (2.0)	65–72	47–51	36–38	80–94	94–101	18–22	18–22	15–21③
1991–92	A	4-92 (1.5)	①	47–51	36–38	51–72②	94–101	11–14	11–14	15–21③
	T	4-107 (1.8)	51–54	37–39	24–25	80–94	94–101	13–18	18–22	15–21③
	R	4-122 (2.0)	65–72	47–51	36–38	80–94	94–101	18–22	18–22	15–21③
	U	4-122 (2.0)	65–72	47–51	36–38	80–94	94–101	18–22	18–22	15–21③
	S	6-181 (3.0)	76–83	58	38	108–116	55	13	13	18③
	B	6-181 (3.0)	76–83	58	38	130–137	55	14	33	18③
	C	6-181 (3.0)	87–94	58	38	130–137	55	9–11	22⑦	18③

① 51–54 COLD
  58–61 HOT
② Pulley to Crankshaft Sprocket—9–11
③ Spark plugs used in aluminum heads
  should always have lubricated threads
④ 65–72 COLD
  72–80 HOT
⑤ Pulley to crankshaft sprocket—14–22
⑥ Torque to 14.5 ft. lbs., back off,
  torque again to 14.5 ft. lbs., then
  turn additional ¼ turn
⑦ See text for special sequence

BRAKE SPECIFICATIONS

All measurements in inches unless noted

Year	Model		Lug Nut Torque (ft. lbs.)	Master Cylinder Bore	Brake Disc Minimum Thickness	Brake Disc Maximum Runout	Standard Brake Drum Diameter	Minimum Lining Thickness Front	Minimum Lining Thickness Rear
1989	Summit ①		65–80	13/16	0.449	0.006	7.10	0.080	0.040
	Summit ②	front	65–80	7/8	0.882	0.006	—	0.080	0.080
		rear	65–80	—	0.331	0.006	—	0.080	0.080
1990	Summit ①		65–80	13/16	0.449	0.006	7.10	0.080	0.040
	Summit ②	front	65–80	7/8	0.882	0.006	—	0.080	0.080
		rear	65–80	—	0.331	0.006	—	0.080	0.080
	Laser	front	87–101	③	0.882	0.003	—	0.080	0.080
		rear	87–101	—	0.331	0.003	—	0.080	0.080
	Talon	front	87–101	③	0.882	0.003	—	0.080	0.080
		rear	87–101	—	0.331	0.003	—	0.080	0.080
1991–92	Summit ④		65–80	13/16	0.449	0.006	7.10	0.080	0.040
	Summit ⑤		65–80	7/8	0.646	0.006	7.10	0.080	0.040
	Laser	front	87–101	⑥	0.882	0.003	—	0.080	0.080
		rear	87–101	—	0.331	0.003	—	0.080	0.080
	Talon	front	87–101	⑥	0.882	0.003	—	0.080	0.080
		rear	87–101	—	0.331	0.003	—	0.080	0.080
	Stealth ⑦	front	87–101	⑨	0.880	0.003	—	0.080	0.080
		rear	87–101	—	0.650	0.003	—	0.080	0.080
	Stealth ⑧	front	87–101	1 1/16	1.12	0.003	—	0.080	0.080
		rear	87–101	—	0.720	0.003	—	0.080	0.080

① 1.5L engine
② 1.6L engine
③ Non-turbocharged engine: 7/8
Turbocharged engine: 15/16
④ Hatchback
⑤ Sedan
⑥ Non-turbocharged without ABS: 7/8
Non-turbocharged with ABS: 15/16
Turbocharged with FWD: 15/16
Turbocharged with AWD: 1
⑦ FWD—Front Wheel Drive
⑧ AWD—All Wheel Drive
⑨ Without ABS: 1
With ABS: 1 1/16

WHEEL ALIGNMENT

Year	Model		Caster Range (deg.)	Caster Preferred Setting (deg.)	Camber Range (deg.)	Camber Preferred Setting (deg.)	Toe-in (in.)	Steering Axis Inclination (deg.)
1989	Summit	front	2P–3P	2 1/3P	1/2N–1/2P	0	0	—
		rear	—	—	1N–0	2/3N	0	—
1990	Summit	front	2P–3P	2 1/3P	1/2N–1/2P	0	0	—
		rear	—	—	1N–0	2/3N	0	—
	Laser ①	front	1 5/6P–2 5/6P	2 1/3P	4/15N–11/15P	7/30P	0	—
		rear	—	—	1 1/4N–1/4N	3/4N	0	—

WHEEL ALIGNMENT

Year	Model		Caster Range (deg.)	Caster Preferred Setting (deg.)	Camber Range (deg.)	Camber Preferred Setting (deg.)	Toe-in (in.)	Steering Axis Inclination (deg.)
1990	Laser ②	front	$1\frac{9}{10}$P–$2\frac{9}{10}$P	$2\frac{2}{5}$P	$\frac{5}{12}$N–$\frac{7}{12}$P	$\frac{1}{12}$P	0	—
		rear	—	—	$1\frac{1}{4}$N–$\frac{1}{4}$N	$\frac{3}{4}$N	0	—
	Laser ③	front	$1\frac{4}{5}$P–$2\frac{4}{5}$P	$2\frac{3}{10}$P	$\frac{1}{3}$N–$\frac{2}{3}$P	$\frac{1}{6}$P	0	—
		rear	—	—	$2\frac{1}{20}$N–$1\frac{1}{20}$N	$1\frac{11}{20}$N	0.14	—
	Talon ②	front	$1\frac{9}{10}$P–$2\frac{9}{10}$P	$2\frac{2}{5}$P	$\frac{5}{12}$N–$\frac{7}{12}$P	$\frac{1}{12}$P	0	—
		rear	—	—	$1\frac{1}{4}$N–$\frac{1}{4}$P	$\frac{3}{4}$N	0	—
	Talon ③	front	$1\frac{4}{5}$P–$2\frac{4}{5}$P	$2\frac{3}{10}$P	$\frac{1}{3}$N–$\frac{2}{3}$P	$\frac{1}{6}$P	0	—
		rear	—	—	$2\frac{1}{20}$N–$1\frac{1}{20}$P	$1\frac{11}{20}$N	0.14	—
1991–92	Summit	front	$1\frac{5}{6}$P–$2\frac{5}{6}$P	$2\frac{1}{3}$P	$\frac{1}{2}$N–$\frac{1}{2}$P	0	0	—
		rear	—	—	$1\frac{1}{6}$N–$\frac{1}{6}$N	$\frac{2}{3}$N	0	—
	Laser ①	front	$1\frac{5}{6}$P–$2\frac{5}{6}$P	$2\frac{1}{3}$P	$\frac{4}{15}$N–$\frac{11}{15}$P	$\frac{7}{30}$P	0	—
		rear	—	—	$1\frac{1}{4}$N–$\frac{1}{4}$N	$\frac{3}{4}$N	0	—
	Laser ②	front	$1\frac{9}{10}$P–$2\frac{9}{10}$P	$2\frac{2}{5}$P	$\frac{5}{12}$N–$\frac{7}{12}$P	$\frac{1}{12}$P	0	—
		rear	—	—	$1\frac{1}{4}$N–$\frac{1}{4}$N	$\frac{3}{4}$N	0	—
	Laser ③	front	$1\frac{4}{5}$P–$2\frac{4}{5}$P	$2\frac{3}{10}$P	$\frac{1}{3}$N–$\frac{2}{3}$P	$\frac{1}{6}$P	0	—
		rear	—	—	$2\frac{1}{20}$N–$1\frac{1}{20}$N	$1\frac{11}{20}$N	0.14	—
	Talon ②	front	$1\frac{9}{10}$P–$2\frac{9}{10}$P	$2\frac{2}{5}$P	$\frac{5}{12}$N–$\frac{7}{12}$P	$\frac{1}{12}$P	0	—
		rear	—	—	$1\frac{1}{4}$N–$\frac{1}{4}$P	$\frac{3}{4}$N	0	—
	Talon ③	front	$1\frac{4}{5}$P–$2\frac{4}{5}$P	$2\frac{3}{10}$P	$\frac{1}{3}$N–$\frac{2}{3}$P	$\frac{1}{6}$P	0	—
		rear	—	—	$2\frac{1}{20}$N–$1\frac{1}{20}$N	$1\frac{11}{20}$N	0.14	—
	Stealth	front	$3\frac{5}{12}$P–$4\frac{5}{1}$P	$3\frac{11}{12}$P	$\frac{1}{2}$N–$\frac{1}{2}$P	0	0.12	—
		rear	—	—	④	⑤	0.01	—

N—Negative
P—Positive
① 1.8L Engine
② 2.0L Engine with FWD
③ AWD—All Wheel Drive
④ FWD: $\frac{1}{2}$N–$\frac{1}{2}$P
   AWD: $\frac{2}{3}$N–$\frac{1}{3}$P
⑤ FWD: 0
   AWD: $\frac{1}{6}$N

ENGINE MECHANICAL

NOTE: Disconnecting the negative battery cable on some vehicles may interfere with the functions of the on board computer systems and may require the computer to undergo a relearning process, once the negative battery cable is reconnected.

Engine Assembly

REMOVAL & INSTALLATION

The following procedure can be used on all vehicles. Slight variations may occur due to extra connections, etc., but the basic procedure should cover all models.

1. Relieve fuel system pressure.
2. Disconnect the negative battery cable.
3. Matchmark the hood and hinges and remove the hood assembly. Remove the air cleaner assembly and all adjoining air intake duct work.
4. Drain the engine coolant and remove the radiator assembly and intercooler.
5. Remove the transaxle.
6. Disconnect and tag for assembly reference the connections for the accelerator cable, heater hoses, brake vacuum hose, connection for vacuum hoses, high pressure fuel line, fuel return line, oxygen sensor connection, coolant temperature gauge connection, coolant temperature sensor connector, connection for thermo switch sensor, if equipped with automatic transaxle, the connection for the idle speed control, the motor position sensor connector, the throttle position sensor connector, the EGR temperature sensor connection (California vehicles), the fuel injector connectors, the power transistor connector, the ignition coil connector, the condenser and noise filter connector, the distributor and control harness, the connections for the alternator and oil pressure switch wires.
7. Remove the air conditioner drive belt and the air conditioning compressor. Leave the hoses attached. Do not

discharge the system. Wire the compressor aside.

8. Remove the power steering pump and wire aside.

9. Remove the exhaust manifold to head pipe nuts. Discard the gasket.

10. Attach a hoist to the engine and take up the engine weight. Remove the engine mount bracket. Remove any torque control brackets (roll stoppers). Note that some engine mount pieces have arrows on them for proper assembly. Double check that all cables, hoses, harness connectors, etc., are disconnected from the engine. Lift the engine slowly from the engine compartment.

To install:

11. Install the engine and secure all control brackets.

12. Install the exhaust pipe, power steering pump and air conditioning compressor.

13. Checking the tags installed at removal, reconnect all electrical and vacuum connections.

14. Install the transaxle.

15. Install the radiator assembly and intercooler.

16. Install the air cleaner assembly.

17. Fill the engine with the proper amount of engine oil. Connect the negative battery cable.

18. Refill the cooling system. Start the engine, allow it to reach normal operating temperature. Check for leaks.

19. Check the ignition timing and adjust if necessary.

20. Install the hood.

21. Road test the vehicle and check all functions for proper operation.

Engine Mounts

REMOVAL & INSTALLATION

1. Disconnect the negative battery cable. Remove the air cleaner and all necessary air duct work.

2. Raise and safely support the engine so it is not resting on the engine mount. One suggested way is a block of wood between a floor jack and the oil pan. Use care not to bend or damage any components.

3. Remove the engine mount bracket and body connection through bolt. Take note of the position of the arrow on the oval shaped mounting stopper plate. This is important.

4. Remove the engine mounting bracket and stopper plate.

5. Lower mounts (roll stoppers) are removed by removing the through bolt, then the frame bolts. On Stealth, the condenser and fan assembly and front catalytic converter must first be removed to gain access to the front mount.

6. The installation is the reverse of the removal procedure. Note the arrows on the stopper plates and make sure they are installed properly.

7. The front lower mount through bolt nut should not be tightened until the full weight of the engine is on the mount. Torque specifications are as follows:

Summit

Upper mount to engine nuts and bolts—36–47 ft. lbs. (50–65 Nm)

Upper mount through bolt nut— 65–80 ft. lbs. (90–110 Nm)

Lower mount through bolt nut— 33–43 ft. lbs. (45–60 Nm)

Laser and Talon

Upper mount to engine nuts and bolts—36–47 ft. lbs. (50–65 Nm)

Upper mount through bolt nut— 43–58 ft. lbs. (60–80 Nm)

Lower mount through bolt nut— 33–43 ft. lbs. (45–60 Nm)

Stealth

Upper mount to engine nuts and bolts—72–87 ft. lbs. (100–120 Nm)

Upper mount through bolt nut— 45–60 ft. lbs. (33–43 Nm)

Lower mount through bolt nut— 36–43 ft. lbs. (47–60 Nm).

Cylinder Head

REMOVAL & INSTALLATION

1.5L and 1.8L Engines

1. Relieve the fuel system pressure. Disconnect the negative battery cable.

2. Drain the cooling system.

3. Remove the air intake hose and the breather hose.

4. Disconnect the accelerator cable. There will be 2 cables if equipped with cruise-control.

5. Disconnect the high pressure fuel line.

6. Remove the upper radiator hose, the water breather hose, the water bypass hose and the heater hose.

7. Disconnect the PCV hose.

8. Remove the spark plug cables.

9. Disconnect and plug the fuel return line.

10. Disconnect the vacuum line for the brake booster.

11. Disconnect the electrical connections for the oxygen sensor, engine coolant temperature gauge unit and the water temperature sensor.

12. Disconnect the electrical connections for the ISC motor, throttle position sensor, distributor, MPS, fuel injectors, EGR temperature sensor (California vehicles), power transistor, condenser and ground cable.

13. Disconnect the engine control wiring harness.

14. Remove the clamp that holds the power steering pressure hose to the engine mounting bracket.

15. Place a jack and wood block under the oil pan and carefully lift just enough to take the weight off the engine mounting bracket. Then remove the bracket.

16. Remove the valve cover, gasket and half-round seal.

17. Remove the timing belt front upper cover.

18. If possible, rotate the crankshaft clockwise until the timing marks on the cam sprocket and belt align. Remove the sprocket bolt and remove the sprocket with the timing belt attached. On 1.8L engine, place on the timing belt front lower cover. On 1.5L engine, attach a flexible cord to the hood and suspend the sprocket so it cannot turn. Remove the timing belt rear upper cover.

19. Remove the exhaust pipe self-locking nuts and separate the exhaust pipe from the exhaust manifold. Discard the gasket.

20. Loosen the cylinder head mounting bolts in 3 steps, starting from the outside and working inward. Lift off the cylinder head assembly and remove the head gasket.

To install:

21. Thoroughly clean and dry the mating surfaces of the head and block. Check the cylinder head for cracks, damage or engine coolant leakage. Remove scale, sealing compound and carbon. Clean oil passages throughly. Check the head for flatness. End to end, the head should be within 0.002 in. normally with 0.008 in. the maximum allowed out of true. The total thickness allowed to be removed from the head and block is 0.008 in. maximum.

22. Place a new head gasket on the cylinder block with the identification marks facing upward. Do not use sealer on the gasket.

23. Carefully install the cylinder head on the block. Using 3 even steps, torque the head bolts in sequence, to 51–54 ft. lbs. (70–75 Nm).

24. Install a new exhaust pipe gasket and connect the exhaust pipe to the manifold. Install the upper rear timing cover.

25. Align the timing marks and install the cam sprocket. Torque the retaining bolt to 47–54 ft. lbs. (65–75 Nm) on 1.5L engine or 58–72 ft. lbs. (80–100 Nm) on 1.8L engine. Check the belt tension and adjust if necessary. Install the outer timing cover.

26. Apply sealer to the perimeter of the half-round seal. Install a new valve cover gasket. Install the valve cover.

27. Install the engine mount bracket. Once secure, remove the jack.

28. Install the clamp that holds the power steering pressure hose to the engine mounting bracket.

29. Connect or install all previously disconnected hoses, cables and electrical connections. Adjust the throttle cable(s).

30. Replace the O-rings and connect the fuel lines.

31. Install the air intake hose. Connect the breather hose.

32. Change the engine oil.

33. Fill the system with coolant.

34. Connect the negative battery cable, run the vehicle until the thermostat opens, fill the radiator completely.

35. Check and adjust the idle speed and ignition timing.

36. Once the vehicle has cooled, recheck the coolant level.

1.6L and 2.0L Engines

1. Relieve fuel system pressure. Disconnect the negative battery cable.

2. Drain the cooling system.

3. Disconnect the accelerator cable. There will be 2 cables if equipped with cruise-control.

4. Remove the air cleaner with the air intake hose.

5. Disconnect the oxygen sensor, engine coolant temperature sensor, the engine coolant temperature gauge unit and the engine coolant temperature switch on vehicles with air conditioning.

6. Disconnect the ISC motor, throttle position sensor, crankshaft angle sensor, fuel injectors, ignition coil, power transistor, noise filter, knock sensor on turbocharged engines, EGR temperature sensor (California vehicles), ground cable and engine control wiring harness.

7. Remove the upper radiator hose and the overflow tube.

8. Remove the spark plug cable center cover and remove the spark plug cables.

9. Disconnect and plug the high pressure fuel line.

10. Disconnect the small vacuum hoses.

11. Remove the heater hose and water bypass hose.

12. Remove the PCV hose.

13. If turbocharged, remove the vacuum hoses, water line and eyebolt connection for the oil line for the turbo.

14. Disconnect and plug the fuel return hose.

15. Disconnect the brake booster vacuum hose.

16. Remove the timing belt.

17. Remove the valve cover and the half-round seal.

18. On non-turbocharged engines, remove the exhaust pipe self-locking nuts and separate the exhaust pipe from the exhaust manifold. Discard the gasket.

19. On turbocharged engines, remove the sheet metal heat protector and remove the bolts that attach the turbocharger to the exhaust manifold.

20. Loosen the cylinder head mounting bolts in 3 steps, starting from the outside and working inward. Lift off the cylinder head assembly and remove the head gasket.

To install:

21. Thoroughly clean and dry the mating surfaces of the head and block. Check the cylinder head for cracks, damage or engine coolant leakage. Remove scale, sealing compound and carbon. Clean oil passages throughly. Check the head for flatness. End to end, the head should be within 0.002 in. normally with 0.008 in. the maximum allowed out of true. The total thickness allowed to be removed from the head and block is 0.008 in. maximum.

22. Place a new head gasket on the cylinder block with the identification marks facing upward. Do not use sealer on the gasket. Replace the turbo gasket and ring, if equipped.

23. Carefully install the cylinder head on the block. Using 3 even steps, torque the head bolts in sequence, to 65–72 ft. lbs. (90–100 Nm).

24. On turbocharged engine, install the heat shield. On non-turbocharged engine, install a new exhaust pipe gasket and connect the exhaust pipe to the manifold.

25. Apply sealer to the perimeter of the half-round seal and to the lower edges of the half-round portions of the belt-side of the new gasket. Install the valve cover.

26. Install the timing belt and all related items.

27. Connect or install all previously disconnected hoses, cables and electrical connections. Adjust the throttle cable(s).

28. Install the spark plug cable center cover.

29. Replace the O-rings and connect the fuel lines.

30. Install the air cleaner and intake hose. Connect the breather hose.

31. Change the engine oil.

32. Fill the system with coolant.

33. Connect the negative battery cable, run the vehicle until the thermostat opens, fill the radiator completely.

34. Check and adjust the idle speed and ignition timing.

35. Once the vehicle has cooled, recheck the coolant level.

3.0L SOHC Engine

1. Relieve fuel system pressure. Disconnect the negative battery cable.

2. Drain the cooling system.

3. Remove the air intake hose.

4. Remove the exhaust manifold.

5. Remove the air intake plenum and intake manifold.

6. Remove the timing belt.

7. Remove the camshaft sprocket and rear timing belt cover.

8. Remove the power steering pump bracket. If removing the rear (right) side head, remove the alternator brace.

9. Disconnect the water inlet pipe.

10. Remove the purge pipe assembly.

11. Remove the valve cover.

12. Loosen the cylinder head mounting bolts in 3 steps, starting from the outside and working inward. Lift off the cylinder head assembly and remove the head gasket.

To install:

13. Thoroughly clean and dry the mating surfaces of the head and block. Check the cylinder head for cracks, damage or engine coolant leakage. Remove scale, sealing compound and carbon. Clean oil passages throughly. Check the head for flatness. End to end, the head should be within 0.002 in. normally with 0.008 in. the maximum allowed out of true. The total thickness allowed to be removed from the head and block is 0.008 in. maximum.

14. Place a new head gasket on the cylinder block with the identification marks facing upward. Do not use sealer on the gasket.

15. Carefully install the cylinder head on the block. Make sure the head bolt washers are installed with the chamfered edge upward. Using 3 even steps, torque the head bolts in sequence, to 76–83 ft. lbs. (105–115 Nm).

16. Apply sealer to the lower edges of the half-round portions of the belt-side of the new gasket and install the valve cover.

17. Install the purge pipe assembly.

18. Connect the water inlet pipe.

19. Install the power steering pump bracket and alternator brace.

20. Install the rear timing belt cover and cam sprocket. Torque the retaining bolt to 60–70 ft. lbs. (81–95 Nm).

21. Install the timing belt and all related items.

22. Using all new gaskets, install the intake manifold, air intake plenum and exhaust manifold, following the proper torque sequences.

23. Install the air intake hose.

24. Change the engine oil.

25. Fill the system with coolant.

26. Connect the negative battery cable, run the vehicle until the thermostat opens, fill the radiator completely.

27. Check and adjust the idle speed and ignition timing.

28. Once the vehicle has cooled, recheck the coolant level.

3.0L DOHC Engine

1. Reieve fuel system pressure. Disconnect the negative battery cable.
2. Drain the cooling system.
3. Remove the air intake hoses.
4. Remove air intake plenum and intake manifold.
5. Remove the turbocharger if equipped, and exhaust manifold.
6. Remove the timing belt.
7. Remove the triple pipe assembly across the top of the engine.
8. Remove the breather hose.
9. Remove the spark plug cable center cover and remove the spark plug cables.
10. When removing the valve cover, note that bolts for the front head are black and bolts for the rear head are green. Also, all bolts are 10mm long except the 1 closest to the sprockets on the rear head which is 20mm long.
11. To remove the intake camshaft sprocket, hold the camshaft with a wrench on the hexagon near the end of the camshaft and remove the bolt.
12. Remove the center rear timing belt cover.
13. Remove the ignition coil.
14. Disconnect all water hoses from the thermostat housing and remove the housing.
15. Disconnect the water inlet from the front head.
16. Loosen the cylinder head mounting bolts in 3 steps, starting from the outside and working inward. Lift off the cylinder head assembly and remove the head gasket.

To install:

17. Thoroughly clean and dry the mating surfaces of the head and block. Check the cylinder head for cracks,

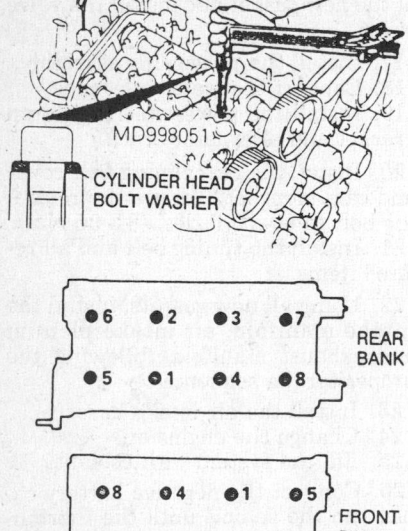

MD998051
CYLINDER HEAD BOLT WASHER

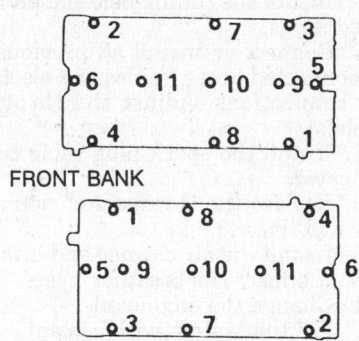

REAR BANK

FRONT BANK

Cylinder head bolt installation sequence—3.0L engine

damage or engine coolant leakage. Remove scale, sealing compound and carbon. Clean oil passages throughly. Check the head for flatness. End to end, the head should be within 0.002 in. normally with 0.008 in. the maximum allowed out of true. The total thickness allowed to be removed from the head and block is 0.008 in. maximum.

18. Place a new head gasket on the cylinder block with the identification marks facing upward. Do not use sealer on the gasket.
19. Carefully install the cylinder head on the block. Make sure the head bolt washers are installed with the chamfered edge upward. Using 3 even steps, torque the head bolts in sequence, to 76–83 ft. lbs. (105–115 Nm) for non-turbocharged engine or 87–94 ft. lbs. (120–130 Nm) for turbocharged engine.
20. Connect the water inlet to the front head.
21. Replace the gaskets and install the thermostat housing and connect the hoses.
22. Install the ignition coil and center rear timing belt cover.
23. Using the same procedure as in removal, install the intake camshaft sprocket. Torque the retaining bolt to 60–70 ft. lbs. (81–95 Nm).
24. Apply sealer to the lower edges of the half-round portions of the belt-side of the new gasket and install the valve cover. Make sure green bolts are installed on the rear head and black

REAR BANK

2 7 3
6 11 10 9 5
4 8 1

FRONT BANK

1 8 4
5 9 10 11 6
3 7 2

Valve cover bolt installation sequence—3.0L DOHC engine

bolts are installed on the front head. Also, make sure the longest bolt is installed in its proper location closest to the sprockets on the rear head. Tighten the bolts in the proper sequence to 26 inch lbs. Then retighten bolts 1–6 to 36 inch lbs.

25. Connect the spark plug cables and install the center cover.
26. Install the breather hose.
27. Install the triple pipe assembly across the top of the engine.
28. Install the timing belt and all related items.

29. Using all new gaskets, install the intake manifold, air intake plenum, turbocharger and exhaust manifold, following the proper torque sequences.
30. Install the air intake hoses.
31. Change the engine oil.
32. Fill the system with coolant.
34. Connect the negative battery cable, run the vehicle until the thermostat opens, fill the radiator completely.
35. Check and adjust the idle speed and ignition timing.
36. Once the vehicle has cooled, recheck the coolant level.

Valve Lash

ADJUSTMENT

1.5L Engine

NOTE: Incorrect valve clearances will cause unsteady engine operation, excessive noise and reduced engine output. Check the valve clearances and adjust as required while the engine is hot.

1. Warm the engine to operating temperature, turn OFF and disconnect the negative battery cable.
2. Remove all spark plugs so engine can be easily turned by hand.
3. Remove the valve cover.
4. Turn the crankshaft clockwise until the notch on the pulley is aligned with the T mark on the timing belt lower cover. This brings both No. 1 and No. 4 cylinder pistons to Top Dead Center (TDC).
5. Wiggle the rocker arms on No. 1 and No. 4 cylinders up and down to determine which cylinder is at TDC on the compression stroke. Both rocker arms should move if the piston in that cylinder is at TDC on the compression stroke.
6. Measure the valve clearance with a feeler gauge. When the No. 1 piston is at TDC on the compression stroke, check No. 1 intake and exhaust, No. 2 intake and No. 3 exhaust. Then turn the crankshaft clockwise 1 turn to bring No. 4 to TDC on its compression stroke. With No. 4 on TDC, compression stroke, check No. 2 exhaust, No. 3 intake and No. 4 intake and exhaust.
7. Valve lash specifications are: Exhaust–0.0098 in. hot or 0.0067 in. cold; Intake–0.0059 in. hot or 0.0028 in. cold. If the valve clearances are out of specification, loosen the rocker arm locknut and adjust the clearance using a feeler gauge while turning the adjusting screw. Be sure to hold the screw to prevent it from turning when tightening the locknut.
8. After adjusting the valves, install the valve cover and spark plugs, and connect the negative battery cable.

Rocker Arms/Shafts

REMOVAL & INSTALLATION

1.5L Engine

1. Disconnect the negative battery cable.

2. Remove the valve cover and discard the gasket.

3. Remove the rocker shaft hold-down bolts gradually and evenly and remove the rocker shaft/arm assemblies.

4. If disassembly is required, keep all parts in the exact order of removal.

Inspect the roller surfaces of the rockers. Replace if there are any signs of damage or if the roller does not turn smoothly. Check the inside bore of the rockers and the adjuster tip for wear.

To install:

5. Lubricate the rocker shaft with

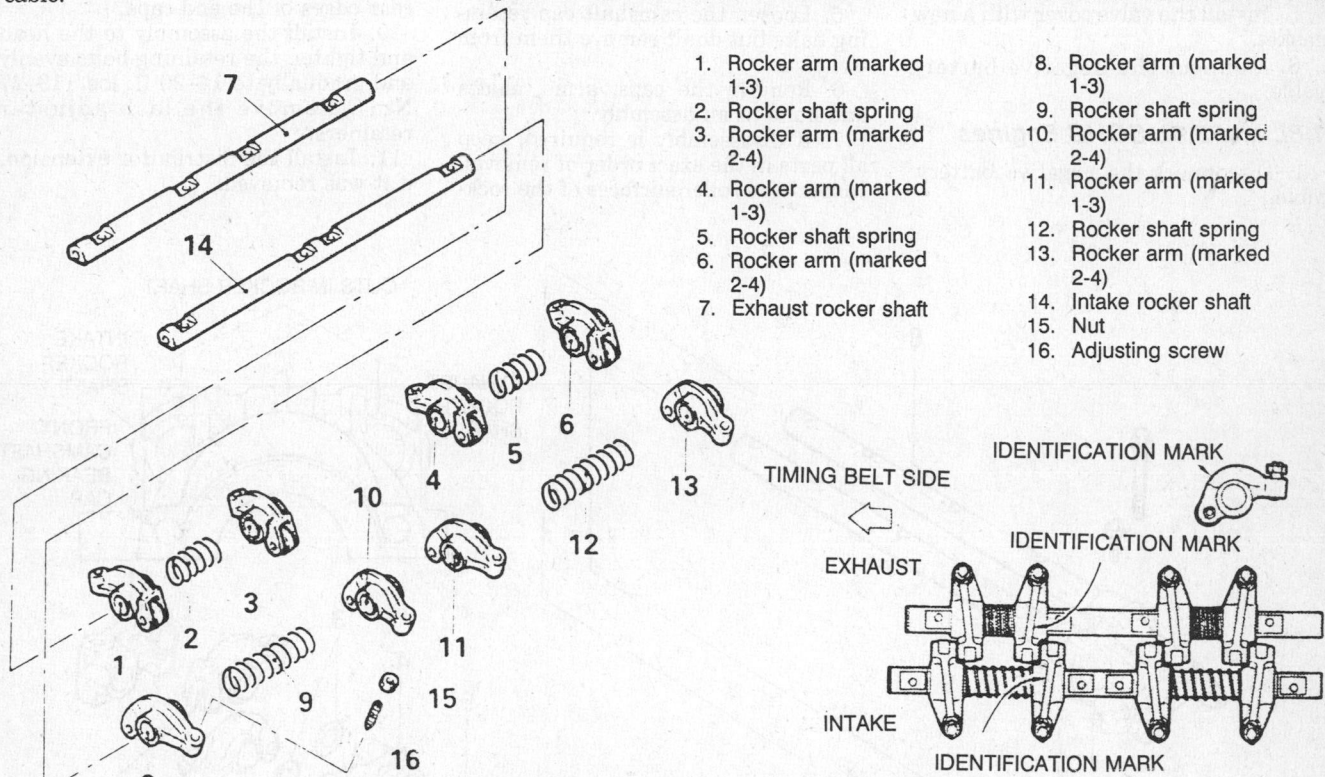

1. Rocker arm (marked 1-3)
2. Rocker shaft spring
3. Rocker arm (marked 2-4)
4. Rocker arm (marked 1-3)
5. Rocker shaft spring
6. Rocker arm (marked 2-4)
7. Exhaust rocker shaft
8. Rocker arm (marked 1-3)
9. Rocker shaft spring
10. Rocker arm (marked 2-4)
11. Rocker arm (marked 1-3)
12. Rocker shaft spring
13. Rocker arm (marked 2-4)
14. Intake rocker shaft
15. Nut
16. Adjusting screw

Rocker arm and shaft assembly—1989–90 1.5L engine

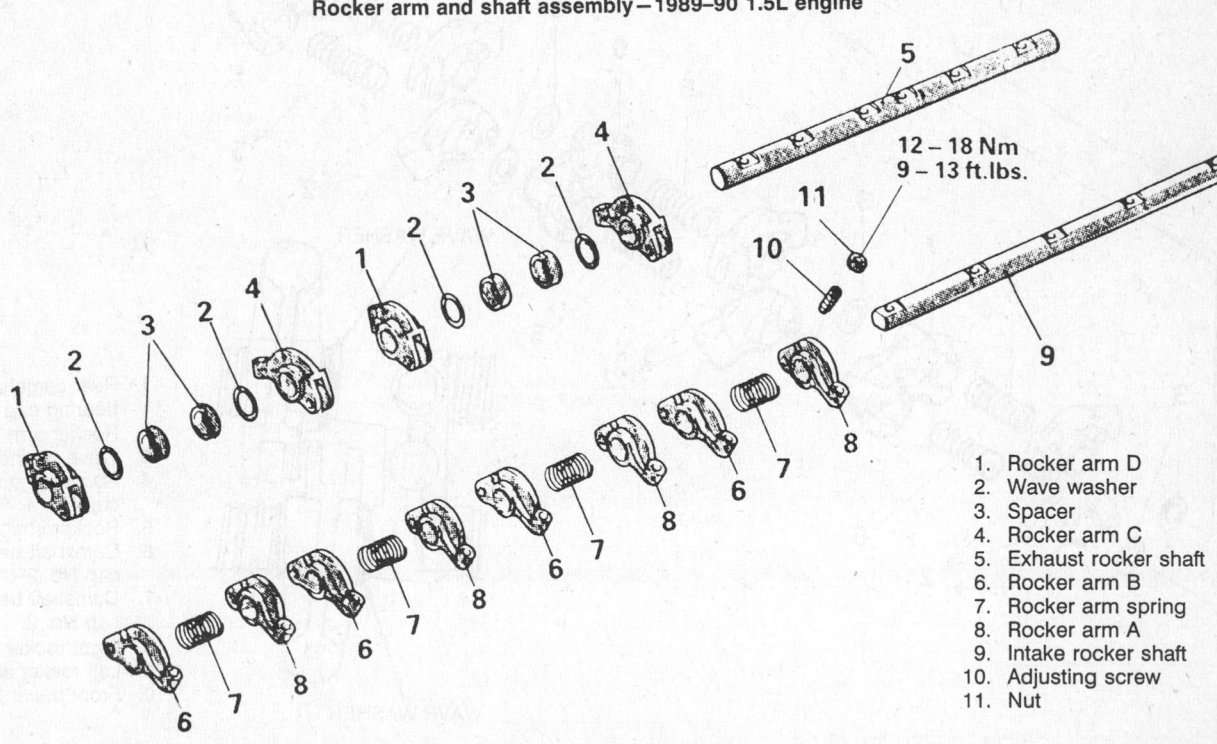

1. Rocker arm D
2. Wave washer
3. Spacer
4. Rocker arm C
5. Exhaust rocker shaft
6. Rocker arm B
7. Rocker arm spring
8. Rocker arm A
9. Intake rocker shaft
10. Adjusting screw
11. Nut

12 – 18 Nm
9 – 13 ft.lbs.

Rocker arm and shaft assembly—1991–92 1.5L engine

clean engine oil and install the rockers and springs in their proper places.

6. Install the rocker shaft assemblies on the engine and tighten the bolts gradually and evenly. Torque to 14–20 ft. lbs. (20–27 Nm) on 1989–90 engines or 21–25 ft. lbs. (29–35 Nm) on 1991–92 engines.

7. Install the valve cover with a new gasket.

8. Connect the negative battery cable.

1.8L and 3.0L SOHC Engines

1. Disconnect the negative battery cable.

2. Remove the valve cover. Install lash adjuster retainer tools MD998443 or equivalent to the rocker arms.

3. Remove the distributor extension if necessary.

4. Have a helper hold the rear of the camshaft down. If not, the belt will dislodge and valve timing will be lost.

5. Loosen the camshaft cap retaining bolts but don't remove them from the caps.

6. Remove the caps, arms, shafts and bolts as an assembly.

7. If disassembly is required, keep all parts in the exact order of removal. Inspect the roller surfaces of the rock-

ers. Replace if there are any signs of damage or if the roller does not turn smoothly. Check the inside bore of the rockers and lifter for wear.

To install:

8. Lubricate and assemble all parts.

9. Apply a drop of sealant to the rear edges of the end caps.

10. Install the assembly to the head and tighten the retaining bolts evenly and gradually to 14–20 ft. lbs. (19–27 Nm). Remove the lash adjuster retainers.

11. Install the distributor extension, if it was removed.

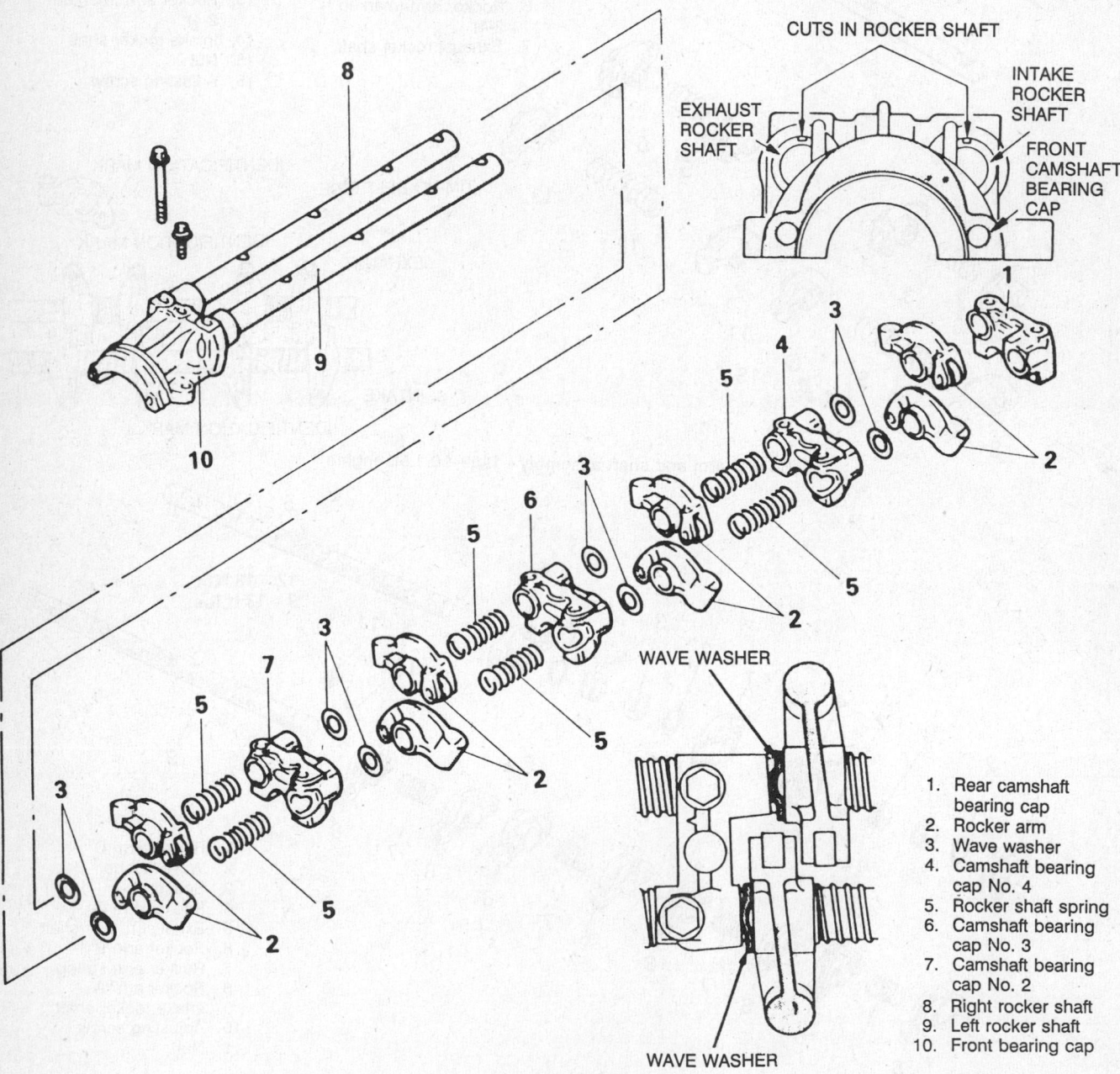

1. Rear camshaft bearing cap
2. Rocker arm
3. Wave washer
4. Camshaft bearing cap No. 4
5. Rocker shaft spring
6. Camshaft bearing cap No. 3
7. Camshaft bearing cap No. 2
8. Right rocker shaft
9. Left rocker shaft
10. Front bearing cap

Rocker arm and shaft assembly—1.8L engine

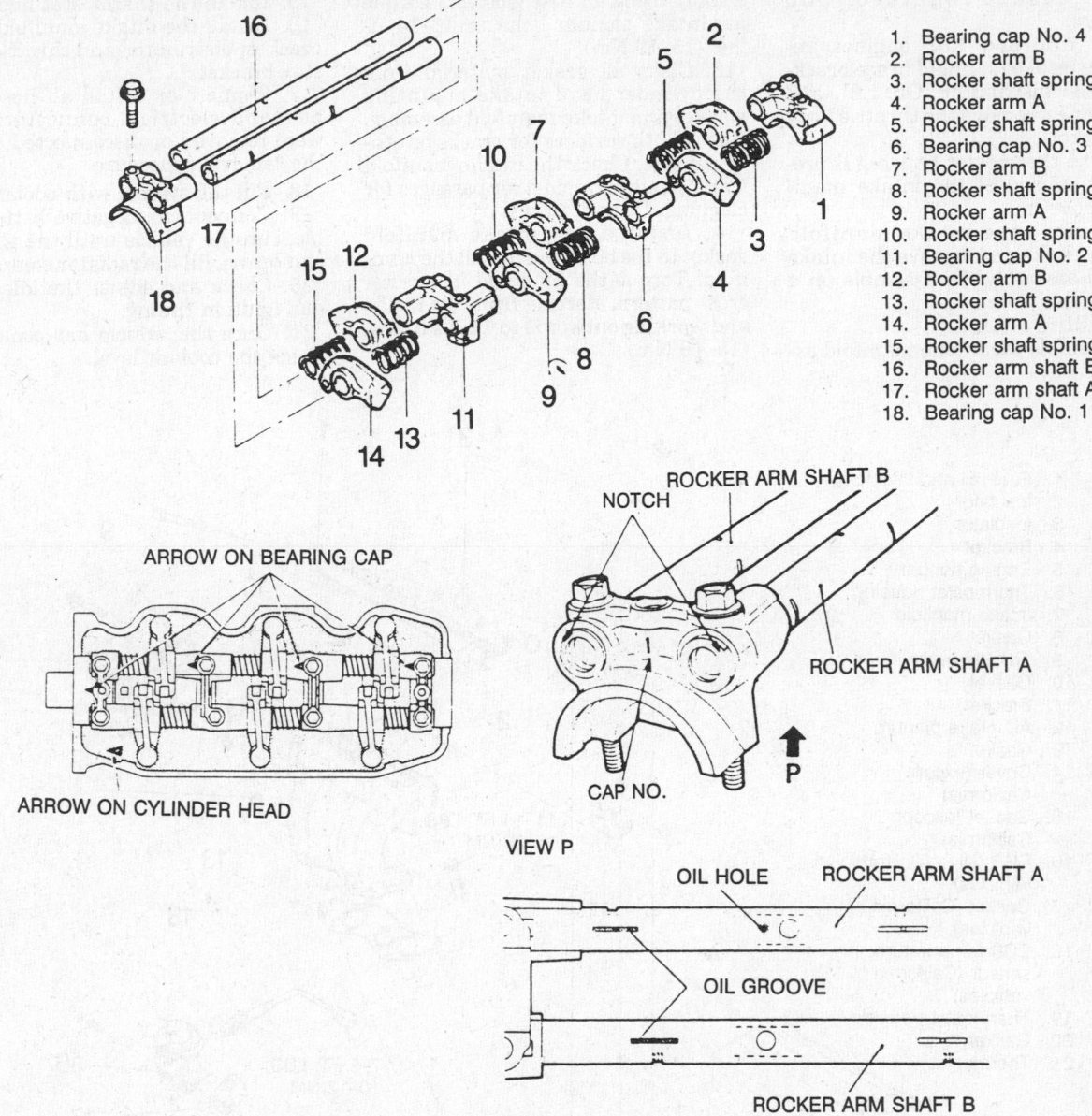

1. Bearing cap No. 4
2. Rocker arm B
3. Rocker shaft spring
4. Rocker arm A
5. Rocker shaft spring
6. Bearing cap No. 3
7. Rocker arm B
8. Rocker shaft spring
9. Rocker arm A
10. Rocker shaft spring
11. Bearing cap No. 2
12. Rocker arm B
13. Rocker shaft spring
14. Rocker arm A
15. Rocker shaft spring
16. Rocker arm shaft B
17. Rocker arm shaft A
18. Bearing cap No. 1

Rocker arm and shaft assembly—3.0L SOHC engine

12. Install the valve cover with a new gasket.

13. Connect the negative battery cable.

1.6L, 2.0L and 3.0L DOHC Engines

These engines do not use rocker shafts. The valves are actuated directly by the rocker arm. To remove, the camshaft must first be removed. It's recommended that all rocker arms and lash adjusters are replaced together.

Air Intake Plenum and Intake Manifold

REMOVAL & INSTALLATION

Except 3.0L Engine

1. Relieve the fuel system pressure.

2. Disconnect battery negative cable and drain the cooling system.

3. Disconnect the accelerator cable, breather hose and air intake hose.

4. Disconnect the upper radiator hose, heater hose and water bypass hose.

5. Remove all vacuum hoses and pipes as necessary, including the brake booster vacuum line.

6. Disconnect the high pressure fuel line, fuel return hose and throttle control cable brackets.

7. Remove and tag all electrical connectors that may interfere with the removal procedure, including spark plug wires.

8. Remove the fuel rail, fuel injec-

tors, pressure regulator and insulators.

9. On 1.5L and 1.8L engines, remove the intake manifold brace bracket and the distributor. On 1.6L and 2.0L engines, remove the throttle body stay bracket.

10. If the thermostat housing is preventing removal of the intake manifold, remove it.

11. Remove the intake manifold mounting bolts and remove the intake manifold assembly. Disassemble on a work bench.

To install:

12. Assemble the intake manifold assembly using all new gaskets. Torque air intake plenum bolts to 11–14 ft. lbs. (15–19 Nm).

13. Clean all gasket material from the cylinder head intake mounting surface and intake manifold assembly. Check both surfaces for cracks or other damage. Check the intake manifold water passages and jet air passages for clogging. Clean if necessary.

14. Install a new intake manifold gasket to the head and install the manifold. Torque the manifold in a criss-cross pattern, starting from the inside and working outwards to 11–14 ft. lbs. (15–19 Nm).

15. Install the thermostat housing.

16. Install the intake manifold brace bracket, distributor and throttle body stay bracket.

17. Connect or install all hoses, cables and electrical connectors that were removed or disconnected during the removal procedure.

18. Fill the system with coolant.

19. Connect the negative battery cable, run the vehicle until the thermostat opens, fill the radiator completely.

20. Check and adjust the idle speed and ignition timing.

21. Once the vehicle has cooled, recheck the coolant level.

1. Fuel rail and injectors
2. Insulator
3. Insulator
4. Bracket
5. Engine hanger
6. Thermostat housing
7. Intake manifold
8. Gasket
9. Throttle body
10. Gasket
11. Bracket
12. Air intake plenum
13. Gasket
14. Cover (except California)
15. Gasket (except California)
16. EGR valve (California vehicles)
17. Gasket (California vehicles)
18. EGR temperature sensor (California vehicles)
19. Thermostat housing
20. Gasket
21. Thermostat

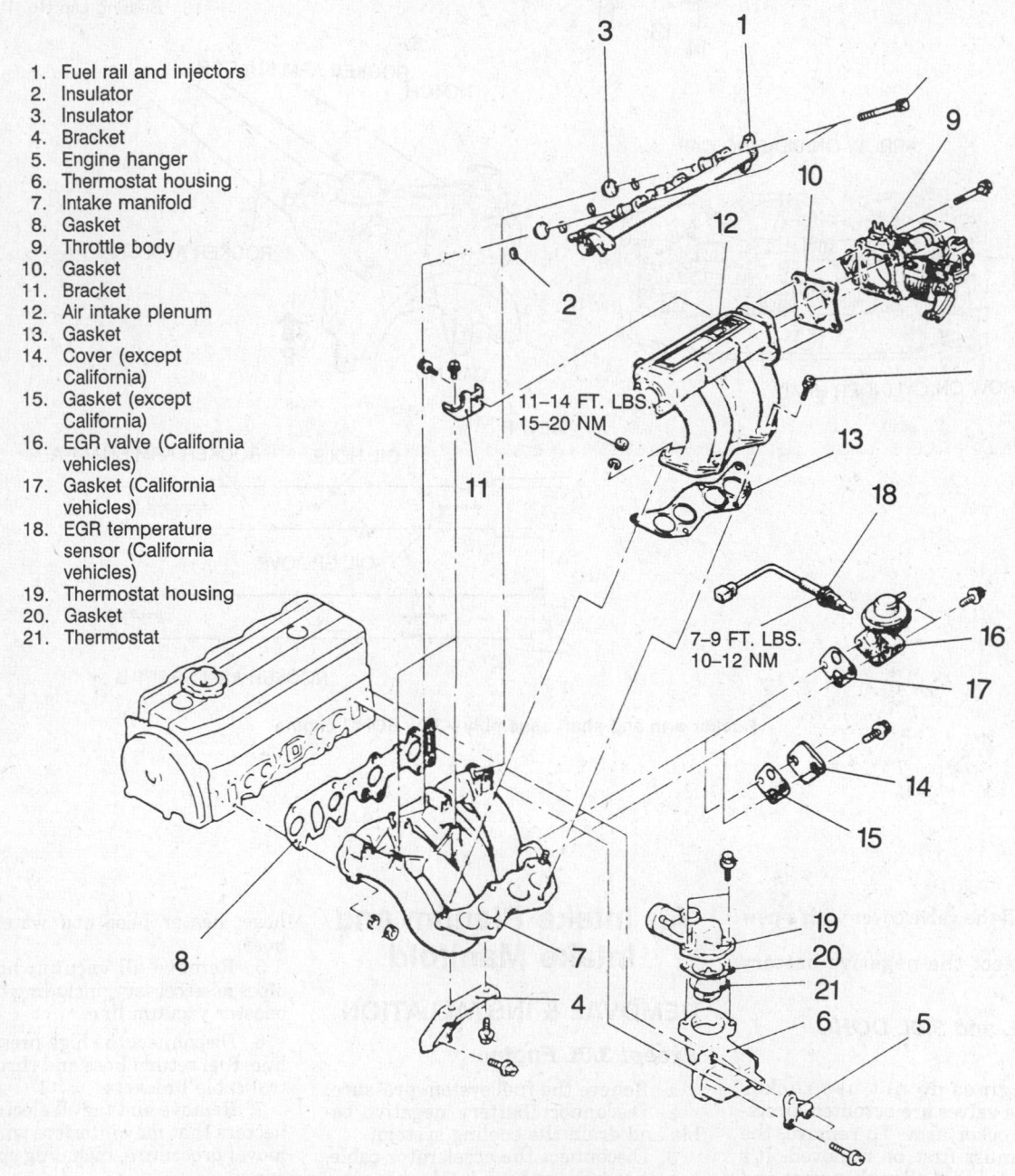

Air intake plenum and intake manifold—typical of all 4 cylinder engines; 1.8L engine shown

1. Air hose
2. Accelerator cable
3. Throttle body
4. Gasket
5. Air pipe
6. Vacuum hose
7. Brake booster vacuum hose
8. Wiring harness
9. Clutch booster vacuum hose
10. EGR temperature sensor
11. EGR valve
12. Gasket
13. Bolts
14. Gasket
15. Air intake plenum stay
16. Bolts
17. Nuts
18. Air intake plenum
19. Gasket

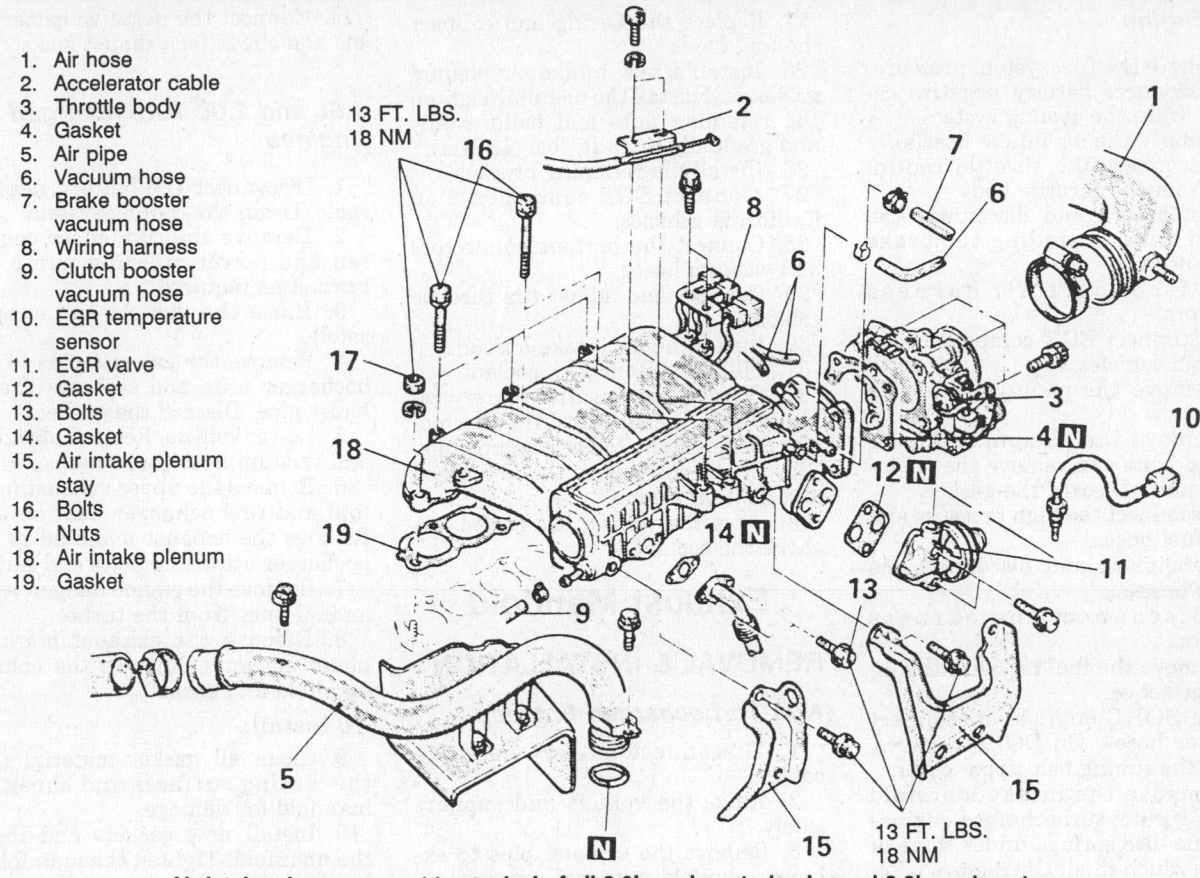

Air intake plenum assembly—typical of all 3.0L engine—turbocharged 3.0L engine shown

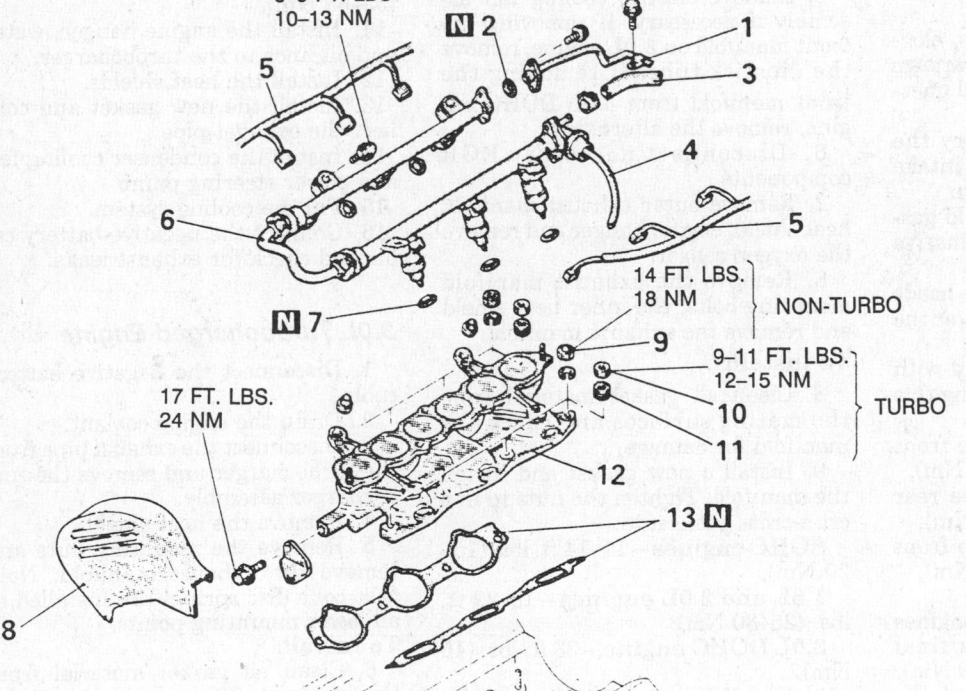

1. High pressure fuel hose
2. O-ring
3. Fuel return hose
4. Vacuum hoses
5. Injector copnnector
6. Fuel rail and injectors
7. Insulators
8. Timing belt upper cover
9. Mounting nut (Non-turbocharged engine
10. Mounting nut (turbocharged engine)
11. Cone disc spring (turbocharged engine)
12. Intake manifold
13. Gasket

Intake manifold and related parts—3.0L engine

3.0L Engine

1. Relieve the fuel system pressure.
2. Disconnect battery negative cable and drain the cooling system.
3. Remove the air intake hose(s).
4. Disconnect the throttle control cables from the throttle body.
5. Matchmark and disconnect the vacuum hoses including the brake booster hose.
6. Disconnect all harness connectors.
7. Disconnect EGR components on California vehicles.
8. Remove the plenum retaining bracket.
9. Remove the plenum retaining nuts and bolts and remove the air intake plenum. Discard the gasket.
10. Disconnect the high pressure and return fuel hoses.
11. Matchmark and disconnect the vacuum hoses.
12. Disconnect the harness connector.
13. Remove the fuel rail with the injectors attached.
14. On SOHC engines, disconnect the water hoses. On DOHC engines, remove the timing belt upper cover.
15. Remove the intake manifold mounting nuts; turbocharged engines have cone disc springs under some of the nuts which should be removed. Remove the intake manifold and discard the gaskets.

To install:

16. Check all items for cracks, clogging and warpage. Maximum warpage is 0.008 in. (0.2mm). Replace all questionable parts.
17. Thoroughly clean and dry the mating surfaces of the heads, intake manifold and air intake plenum.
18. Install new intake manifold gaskets to the heads with the adhesive side facing up.
19. Place the manifold on the heads install the cone disc springs and/or the lock washers.
20. Lubricate the studs lightly with oil, then install the nuts following this procedure:
   a. Tighten the nuts on the front bank to 26–43 inch lbs. (3–5 Nm).
   b. Tighten the nuts on the rear bank to 9–11 ft. lbs. (12–15 Nm).
   c. Tighten the nuts on the front bank to 9–11 ft. lbs. (12–15 Nm).
   d. Repeat Steps B and C.
   e. On non-turbocharged engines only, tighten the nuts to a final torque of 13–14 ft. lbs. (18–19 Nm).
21. On SOHC engines, connect the water hoses. On DOHC engines, install the timing belt upper cover.
22. Install the fuel rail assembly.
23. Connect the harness connector and vacuum hoses.

24. Replace the O-ring and connect the fuel hoses.
25. Install a new intake air plenum gasket and install the plenum. Tighten the retaining nuts and bolts evenly and gradually to 13 ft. lbs. (18 Nm).
26. Install the retaining bracket.
27. Connect EGR components on California vehicles.
28. Connect the harness connectors and vacuum hoses.
29. Connect and adjust the throttle cables.
30. Install the air intake hose(s).
31. Fill the system with coolant.
32. Connect the negative battery cable, run the vehicle until the thermostat opens, fill the radiator completely.
33. Check and adjust the idle speed and ignition timing.
34. Once the vehicle has cooled, recheck the coolant level.

Exhaust Manifold

REMOVAL & INSTALLATION

Non-Turbocharged Engines

1. Disconnect battery negative cable.
2. Raise the vehicle and support safely.
3. Remove the exhaust pipe to exhaust manifold nuts and separate exhaust pipe. Discard gasket.
4. Lower vehicle.
5. Remove electric cooling fan assembly if necessary. If removing the front manifold on 3.0L engine, remove the dipstick tube. If removing the front manifold from 3.0L DOHC engine, remove the alternator.
6. Disconnect necessary EGR components.
7. Remove outer exhaust manifold heat shield, engine hanger and remove the oxygen sensor.
8. Remove the exhaust manifold mounting bolts, the inner heat shield and remove the exhaust manifold.

To install:

8. Clean all gasket material from the mating surfaces and check the manifold for damage.
9. Install a new gasket and install the manifold. Tighten the nuts to in a criss-cross pattern to:
SOHC engines–11–14 ft. lbs. (15–20 Nm).
1.6L and 2.0L engines–18–22 ft. lbs. (25–30 Nm).
3.0L DOHC engine–33 ft. lbs. (45 Nm).
10. Install the heat shields.
11. Connect EGR components.
12. Install the electric cooling fan assembly, dipstick tube or alternator.
13. Install a new gasket and connect the exhaust pipe.

14. Connect the negative battery cable and check for exhaust leaks.

1.6L and 2.0L Turbocharged Engines

1. Disconnect the battery negative cable. Drain the cooling system.
2. Remove the condenser cooling fan and power steering pump and bracket as required.
3. Raise the vehicle and support safely.
4. Remove the exhaust pipe to turbocharger nuts and separate the exhaust pipe. Discard the gasket.
5. Lower vehicle. Remove air intake and vacuum hose connections.
6. Remove the upper exhaust manifold and turbocharger heat shields. Remove the exhaust manifold to turbocharger attaching bolts and nut.
7. Remove the engine hanger, water and oil lines from the turbo.
8. Remove the exhaust manifold mounting nuts. Remove the exhaust manifold and gasket.

To install:

9. Clean all gasket material from the mating surfaces and check the manifold for damage.
10. Install new gaskets and install the manifold. Tighten the manifold to head nuts in a criss-cross pattern to 18–22 ft. lbs. Tighten the manifold to turbo nut and bolts to 40–47 ft. lbs. (55–65 Nm).
11. Install the engine hanger, water and oil lines to the turbocharger.
12. Install the heat shields.
13. Install the new gasket and connect the exhaust pipe.
14. Install the condenser cooling fan and power steering pump.
15. Fill the cooling system.
16. Connect the negative battery cable and check for exhaust leaks.

3.0L Turbocharged Engine

1. Disconnect the negative battery cable.
2. Drain the engine coolant.
3. Disconnect the exhaust pipe from the turbocharger and remove the turbocharger assembly.
4. Remove the heat shield.
5. Remove the mounting nuts and remove the exhaust manifold. Note that cone disc springs are installed at all lower mounting points.
To install:
6. Clean all gasket material from the mating surfaces and check the manifold for damage.
7. Install new gaskets and install the manifold. Make sure all cone disc springs are in their original locations with the grooved side facing the nut.

Tighten the manifold nuts using the following procedure:

    a. Tighten all but the outer 2 nuts to 22 ft. lbs. (30 Nm).

    b. Tighten the outer 2 nuts to 34–38 ft. lbs. (47–53 Nm).

    c. Loosen the outer 2 nuts, then torque them to 22 ft. lbs. (30 Nm).

8. Install the heat shield.

9. Install the turbocharger assembly.

10. Fill the cooling system.

11. Connect the negative battery cable and check for exhaust leaks.

Turbocharger

Many turbocharger failures are due to oil supply problems. Heat soak after hot shutdown can cause the engine oil in the turbocharger and oil lines to "coke." Often the oil feed lines will become partially or completely blocked with hardened particles of carbon, blocking oil flow. Check the oil feed pipe and oil return line for clogging. Clean these tubes well. Always use new gaskets above and below the oil feed eyebolt fitting. Do not allow particles of dirt or old gasket material to enter the oil passage hole and that no portion of the new gasket blocks the passage.

REMOVAL & INSTALLATION

1.6L and 2.0L Engines

1. Disconnect the negative battery cable.

2. Drain the engine oil, cooling system and remove the radiator. On Laser and Talon with air conditioning, remove the condenser fan assembly with the radiator.

3. Disconnect the oxygen sensor connector and remove the sensor.

4. Remove the oil dipstick and tube on Laser and Talon.

5. Remove the air intake bellows hose, the wastegate vacuum hose, the connections for the air outlet hose, and the upper and lower heat shields.

6. On Laser and Talon, unbolt the power steering pump and bracket assembly and leaving the hoses connected, wire it aside.

7. Remove the self-locking exhaust manifold nuts, the triangular engine hanger bracket, the eyebolt and gaskets that connect the oil feed line to the turbo center section, and the water cooling lines. The water line under the turbo has a threaded connection.

8. Remove the exhaust pipe nuts and gasket and lift off the exhaust manifold. Discard the gasket.

9. Remove the 2 through bolts and 2 nuts that hold the exhaust manifold to the turbocharger.

10. Remove the 2 capscrews from the oil return line (under the turbo). Discard the gasket. Separate the turbo from the exhaust manifold. The 2 water pipes and oil feed line can still be attached.

11. Visually check the turbine wheel (hot side) and compressor wheel (cold side) for cracking or other damage. Check whether the turbine wheel and the compressor wheel can be easily turned by hand. Check for oil leakage. Check whether or not the wastegate valve remains open. If any problem is found, replace the part.

12. The wastegate can be checked with a pressure tester. Apply approximately 9 psi to the actuator and make sure the rod moves. Do not apply more than 10.3 psi or the diaphragm in the wastegate may be damaged. Do not attempt to adjust the wastegate valve.

To install:

13. Prime the oil return line with clean engine oil. Replace all locking nuts. Before installing the threaded connection for the water inlet pipe, apply light oil to the inner surface of the pipe flange. Assemble the turbocharger and exhaust manifold.

14. Install the exhaust manifold using a new gasket.

15. Connect the water cooling lines, oil feed line, and engine hanger.

16. If removed, install the power steering pump and bracket.

17. Install the heat shields, air outlet hose, wastegate hose and air intake bellows.

18. Install the oil dipstick tube and dipstick. Install the oxygen sensor.

19. Install the radiator assembly.

20. Fill the engine with oil, fill the cooling system and reconnect the negative battery cable.

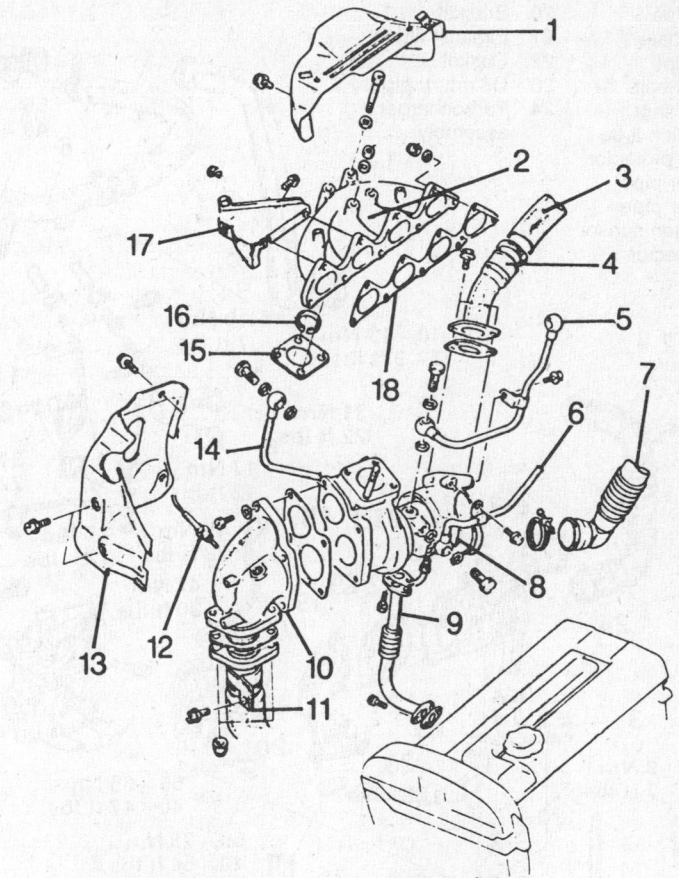

1. Upper heat shield	10. Exhaust fitting
2. Exhaust manifold	11. Exhaust pipe
3. Air hose connector	12. Oxygen sensor
4. Air inlet fitting	13. Lower heat shield
5. Oil feed pipe	14. Water line
6. Water line	15. Gasket
7. Connection–air intake	16. Ring
8. Turbocharger assembly	17. Brace/bracket
9. Oil drainback line	18. Manifold gasket

Turbocharger assembly–1.6L and 2.0L engines

3.0L Engine

RIGHT SIDE (FRONT) TURBOCHARGER

1. Disconnect the negative battery cable.
2. Remove the radiator.
3. Remove the right side transaxle bracket.
4. Remove the front exhaust pipe.
5. Carefully matchmark, diagram or photograph all air intake hoses and pipes along the front of the engine. It is imperative that all of these pieces are installed in the exact same positions when assembling. Remove the hoses and pipes and keep covered in a clean area.

6. Remove the alternator.
7. Remove the oil dipstick tube.
8. Remove the turbocharger heat protector.
9. Remove the water feed pipes.
10. Remove the oxygen sensor.
11. Remove the oil return line.
12. Remove the exhaust extension fitting and bracket.
13. Remove all air conditioning components preventing removal of the turbocharger.
14. Remove the oil feed tube.
15. Remove the turbocharger to exhaust manifold bolts and remove the turbocharger assembly.

To install:
16. Visually check the turbine wheel

(hot side) and compressor wheel (cold side) for cracking or other damage. Check whether the turbine wheel and the compressor wheel can be easily turned by hand. Check for oil leakage. Check whether or not the wastegate valve remains open. If any problem is found, replace the part.

17. Clean all mating surfaces. Pour clean engine oil through the oil pipe feed hole in the turbocharger.

18. Install a new gasket and ring a install the turbocharger to the manifold. Torque the bolts to 40–47 ft. lbs. (55–65 Nm).

19. Replace the eye-bolt rings and install the oil feed pipe.

1. Air hose	16. Turbocharger and fitting assembly	25. Air conditioner compressor
2. Air intake hose	17. Gasket	26. Tensioner pulley bracket
3. Air hose	18. Ring	27. Compressor bracket
4. Air hose	19. Oxygen sensor	28. Oil pipe
5. Air hose	20. Bracket	
6. Air pipe	21. Exhaust fitting	
7. Air hose	22. Gasket	
8. Air pipe	23. Oil return pipe	
9. Drive belt	24. Turbocharger assembly	
10. Alternator		
11. Dipstick tube		
12. Heat protector		
13. Water pipe		
14. Water pipe		
15. Oxygen sensor connector		

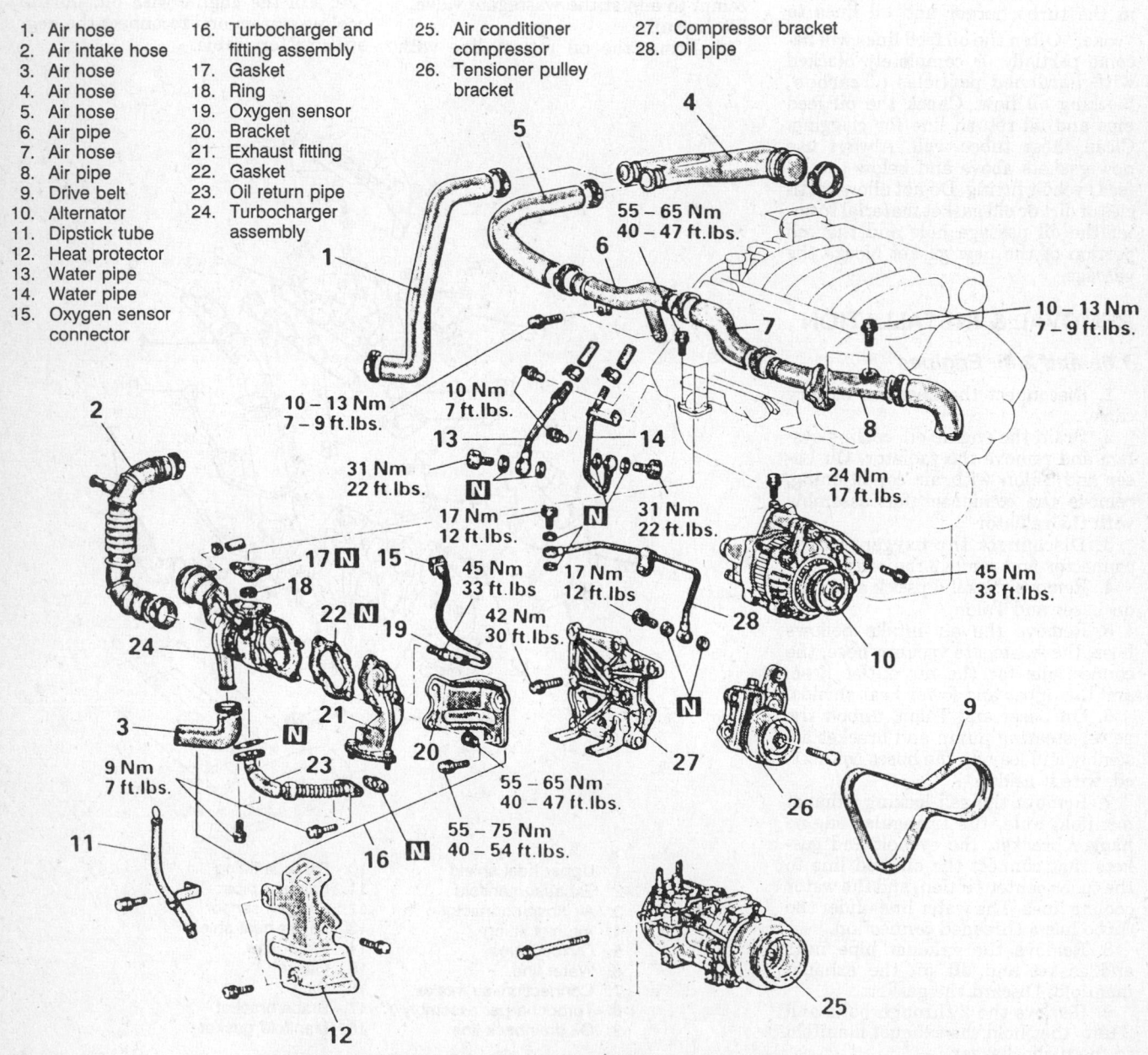

Right side (front) turbocharger and related parts—Stealth

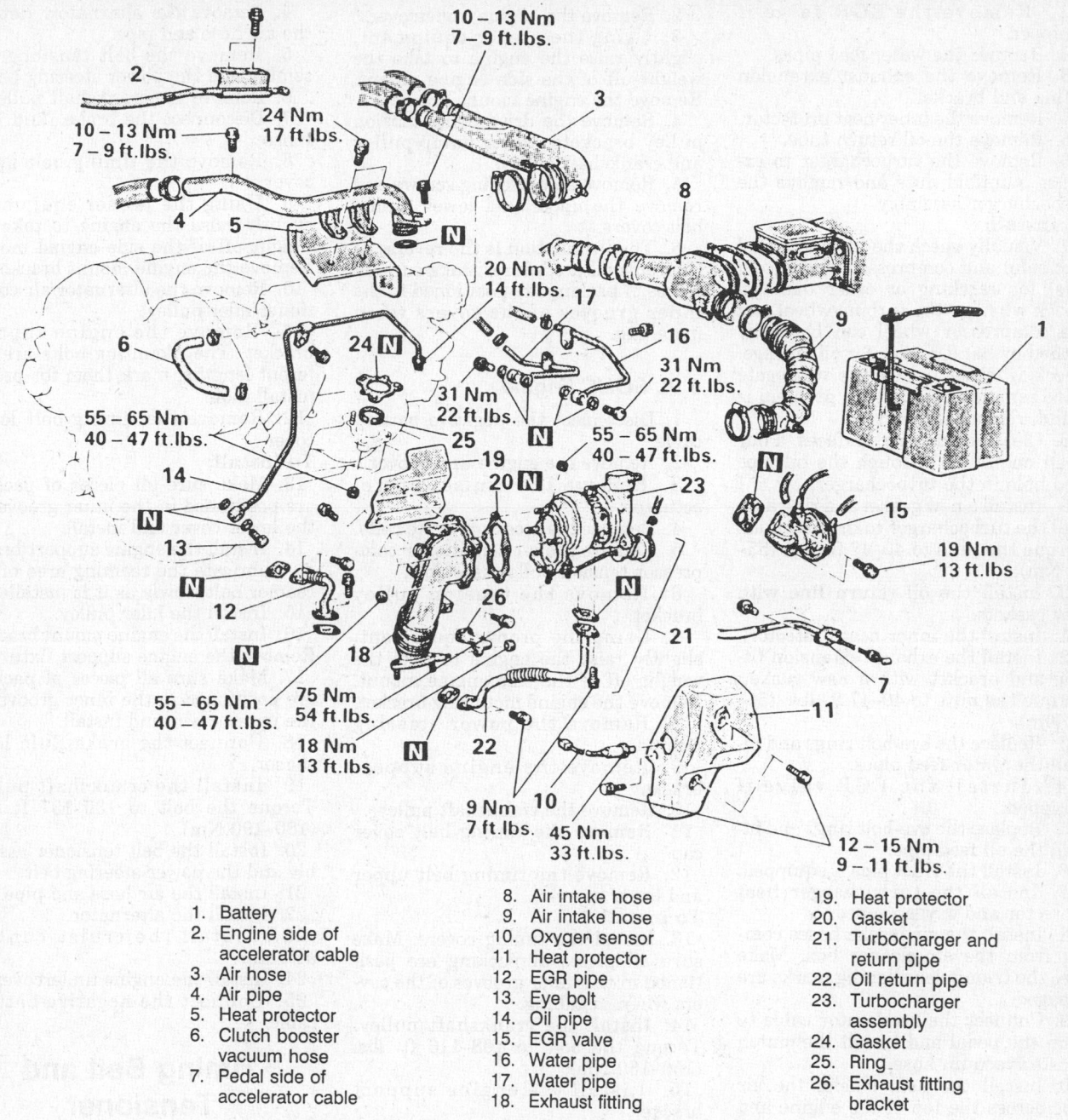

10 – 13 Nm
7 – 9 ft.lbs.

2

10 – 13 Nm
7 – 9 ft.lbs.

24 Nm
17 ft.lbs.

3

4   **5**

8

20 Nm
14 ft.lbs.   **17**   **16**

6   **24**   **9**   **1**

31 Nm
22 ft.lbs.   **31 Nm**
22 ft.lbs.

55 – 65 Nm
40 – 47 ft.lbs.   **25**   **55 – 65 Nm**
40 – 47 ft.lbs.

19 Nm
13 ft.lbs

14   **19**   **23**   **15**

20   **13**   **26**

12   **21**   **7**

18   **11**

55 – 65 Nm
40 – 47 ft.lbs.   **75 Nm**
54 ft.lbs.   **22**   **12 – 15 Nm**
9 – 11 ft.lbs.

18 Nm
13 ft.lbs.

9 Nm
7 ft.lbs.   **10**

45 Nm
33 ft.lbs.

1. Battery	8. Air intake hose
2. Engine side of accelerator cable	9. Air intake hose
3. Air hose	10. Oxygen sensor
4. Air pipe	11. Heat protector
5. Heat protector	12. EGR pipe
6. Clutch booster vacuum hose	13. Eye bolt
7. Pedal side of accelerator cable	14. Oil pipe
	15. EGR valve
	16. Water pipe
	17. Water pipe
	18. Exhaust fitting

19. Heat protector
20. Gasket
21. Turbocharger and return pipe
22. Oil return pipe
23. Turbocharger assembly
24. Gasket
25. Ring
26. Exhaust fitting bracket

Left side (rear) turbocharger and related parts—Stealth

20. Install the removed air conditioning components.
21. Install the exhaust extension fitting and bracket with a new gasket. Torque the nuts to 40–47 ft. lbs. (55–65 Nm).
22. Install the oil return line with new gaskets.
23. Install the oxygen sensor.
24. Replace the eye-bolt rings and install the water feed pipes.
25. Install the turbocharger heat protector.
26. Install the dipstick tube.
27. Install the alternator.
28. Install all air intake hoses and

pipes along the front of the engine. Make sure all are in their proper positions.
29. Install a new gasket and connect the front exhaust pipe.
30. Install the right side transaxle bracket.
31. Install the radiator.
32. Fill the system with coolant.
33. Connect the negative battery cable and check for exhaust leaks.

LEFT SIDE (REAR) TURBOCHARGER

1. Remove the battery.
2. Drain the coolant.

3. Remove the front exhaust pipe.
4. Disconnect the accelerator cable from the throttle body.
5. Remove the intake air hose, the air pipe across the top of the engine and its heat shield.
6. Remove the clutch booster vacuum hose and disconnect the accelerator cable from the pedal.
7. Remove the air intake hoses coming from the air cleaner box.
8. Remove the oxygen sensor and the turbocharger heat protector.
9. Remove the EGR pipe if equipped.
10. Remove the oil feed pipe.

11. Remove the EGR valve if equipped.

12. Remove the water feed pipes.

13. Remove the exhaust extension fitting and bracket.

14. Remove the inner heat protector.

15. Remove the oil return tube.

16. Remove the turbocharger to exhaust manifold nuts and remove the turbocharger assembly.

To install:

17. Visually check the turbine wheel (hot side) and compressor wheel (cold side) for cracking or other damage. Check whether the turbine wheel and the compressor wheel can be easily turned by hand. Check for oil leakage. Check whether or not the wastegate valve remains open. If any problem is found, replace the part.

18. Clean all mating surfaces. Pour clean engine oil through the oil pipe feed hole in the turbocharger.

19. Install a new gasket and ring a install the turbocharger to the manifold. Torque the nuts to 40–47 ft. lbs. (55–65 Nm).

20. Install the oil return line with new gaskets.

21. Install the inner heat protector.

22. Install the exhaust extension fitting and bracket with a new gasket. Torque the nuts to 40–47 ft. lbs. (55–65 Nm).

23. Replace the eye-bolt rings and install the water feed pipes.

24. Install the EGR valve if equipped.

25. Replace the eye-bolt rings and install the oil feed pipe.

26. Install the EGR pipe if equipped.

27. Install the turbocharger heat protector and oxygen sensor.

28. Install the air intake hoses coming from the air cleaner box. Make sure the triangular aligning marks are engaged.

29. Connect the accelerator cable to from the pedal and install the clutch booster vacuum hose.

30. Install the heat shield, the air pipe across the top of the engine and the air intake hose.

31. Connect the accelerator cable to the throttle body.

32. Install a new gasket and connect the front exhaust pipe.

33. Fill the system with coolant.

34. Install the battery.

35. Connect the negative battery cable and check for exhaust leaks.

Timing Belt Front Cover

REMOVAL & INSTALLATION

Except 3.0L Engine

1. Disconnect the negative battery cable.

2. Remove the engine undercover.

3. Using the proper equipment, slightly raise the engine to take the weight off of the side engine mount. Remove the engine mount bracket.

4. Remove the drive belts, tension pulley brackets, water pump pulley and crankshaft pulley.

5. Remove all attaching screws and remove the upper and lower timing belt covers.

6. The installation is the reverse of the removal procedure. Make sure all pieces of packing are positioned in the inner grooves of the covers when installing.

3.0L SOHC Engine

1. Disconnect the negative battery cable.

2. Remove the engine undercover.

3. Remove the cruise control actuator.

4. Remove the accessory drive belts.

5. Remove the air conditioner compressor tension pulley assembly.

6. Remove the tension pulley bracket.

7. Using the proper equipment, slightly raise the engine to take the weight off of the side engine mount. Remove the engine mounting bracket.

8. Remove the power steering pump.

9. Remove the engine support bracket.

10. Remove the crankshaft pulley.

11. Remove the timing belt cover cap.

12. Remove the timing belt upper and lower covers.

To install:

13. Install the timing covers. Make sure all pieces of packing are positioned in the inner grooves of the covers when installing.

14. Install the crankshaft pulley. Torque the bolt to 108–116 ft. lbs. (150–160 Nm).

15. Install the engine support bracket.

16. Install the power steering pump.

17. Install the engine mounting bracket and remove the engine support fixture.

18. Install the tension pulleys and drive belts.

19. Install the cruise control actuator.

20. Install the engine undercover.

21. Connect the negative battery cable.

3.0L DOHC Engine

1. Disconnect the negative battery cable.

2. Remove the engine undercover.

3. Remove the cruise control actuator.

4. Remove the alternator. Remove the air hose and pipe.

5. Remove the belt tensioner assembly and the power steering belt.

6. Remove the crankshaft pulley.

7. Disconnect the brake fluid level sensor.

8. Remove the timing belt upper cover.

9. Using the proper equipment, slightly raise the engine to take the weight off of the side engine mount. Remove the engine mount bracket.

10. Remove the alternator/air conditioner idler pulley.

11. Remove the engine support bracket. The mounting bolts are different lengths; mark them for proper installation.

12. Remove the timing belt lower cover.

To install:

13. Make sure all pieces of packing are positioned in the inner grooves of the lower cover and install.

14. Install the engine support bracket. Lubricate the reaming area of the reamer bolt slowly as it is installed.

15. Install the idler pulley.

16. Install the engine mount bracket. Remove the engine support fixture.

17. Make sure all pieces of packing are positioned in the inner grooves of the upper cover and install.

18. Connect the brake fluid level sensor.

19. Install the crankshaft pullet. Torque the bolt to 130–137 ft. lbs. (180–190 Nm).

20. Install the belt tensioner assembly and the power steering belt.

21. Install the air hose and pipe.

22. Install the alternator.

23. Install the cruise control actuator.

24. Install the engine undercover.

25. Connect the negative battery cable.

Timing Belt and Tensioner

ADJUSTMENT

1.5L and 1.8L Engines

1. Disconnect the negative battery cable.

2. Remove the timing belt covers.

3. On 1.8L engine, adjust the silent shaft (inner) belt tension first. Loosen the idler pulley center bolt so the pulley can be moved.

4. Move the pulley by hand so the long side of the belt deflects about ¼ in.

5. Hold the pulley tightly so the pulley cannot rotate when the bolt is tightened. Tighten the bolt to 15 ft. lbs. (20 Nm) and recheck the deflection amount.

6. To adjust the timing (outer) belt, first loosen the pivot side tensioner bolt and then the slot side bolt. Allow the spring to take up the slack.

7. Tighten the slot side tensioner bolt and then the pivot side bolt. If the pivot side bolt is tightened first, the tensioner could turn with bolt, causing over tension.

8. Turn the crankshaft clockwise. Loosen the pivot side tensioner bolt and then the slot side bolt. Tighten the slot bolt and then the pivot side bolt.

9. Check the belt tension on 1.5L engine by holding the tensioner and timing belt together by hand and give the belt a slight thumb pressure at a point level with tensioner center. Make sure the belt cog crest comes as deep as about ¼ of the width of the slot side tensioner bolt head. On 1.8L engine, the deflection of the longest span of the belt should be about 0.40 in. Do not manually overtighten the belt or it will howl.

10. Install the timing belt covers and all related items.

11. Connect the negative battery cable.

1.6L and 2.0L Engines

1. Disconnect the negative battery cable.

2. Remove the timing belt covers.

3. Adjust the silent shaft (inner) belt tension first. Loosen the idler pulley center bolt so the pulley can be moved.

4. Move the pulley by hand so the long side of the belt deflects about ¼ in.

5. Hold the pulley tightly so the pulley cannot rotate when the bolt is tightened. Tighten the bolt to 15 ft. lbs. (20 Nm) and recheck the deflection amount.

6. To adjust the timing (outer) belt, turn the crankshaft ¼ turn counterclockwise, then turn it clockwise to move No. 1 cylinder to TDC.

7. Loosen the center bolt. Using tool MD998738 or equivalent and a torque wrench, apply a torque of 1.88–2.03 ft. lbs. (2.6–2.8 Nm). Tighten the center bolt.

8. Screw the special tool into the engine left support bracket until its end makes contact with the tensioner arm. At this point, screw the special tool in some more and remove the set wire attached to the auto tensioner, if the wire was not previously removed. Then remove the special tool.

9. Rotate the crankshaft 2 complete turns clockwise and let it sit for approximately 15 minutes. Then, measure the auto tensioner protrusion (the distance between the tensioner arm and auto tensioner body) to ensure that it is within 0.15–0.18 in.

(3.8–4.5mm). If out of specification, repeat Step 1–4 until the specified value is obtained.

10. If the timing belt tension adjustment is being performed with the engine mounted in the vehicle, and clearance between the tensioner arm and the auto tensioner body cannot be measured, the following alternative method can be used:

  a. Screw in special tool MD998738 or equivalent, until its end makes contact with the tensioner arm.

  b. After the special tool makes contact with the arm, screw it in some more to retract the auto tensioner pushrod while counting the number of turns the tool makes until the tensioner arm is brought into contact with the auto tensioner body. Make sure the number of turns the special tool makes conforms with the standard value of 2½–3 turns.

  c. Install the rubber plug to the timing belt rear cover.

11. Install the timing belt covers and all related items.

12. Connect the negative battery cable.

3.0L SOHC Engine

1. Disconnect the negative battery cable.

2. Remove the timing belt covers.

3. Loosen the bolt that holds the tensioner in place and allow the spring to automatically apply tension to the belt.

4. Rotate the crankshaft 2 turns clockwise. Tighten the tensioner bolt to 20 ft. lbs. (25 Nm).

5. Measure the belt tension between the rear camshaft sprocket and the crankshaft with belt tension gauge. The specification is 46–68 lbs. (210–310 N).

6. Install the timing belt covers and all related items.

7. Connect the negative battery cable.

3.0L DOHC Engine

1. Disconnect the negative battery cable.

2. Remove the timing belt covers.

3. Turn the crankshaft ¼ turn counterclockwise, then turn it clockwise until all timing marks are aligned.

4. Loosen the center bolt on the tensioner pulley. Using tool MD998767 or equivalent and a torque wrench, apply a torque of 7.2 ft. lbs. (10 Nm). Tighten the tensioner bolt; make sure the tensioner doesn't rotate with the bolt.

5. Remove the set wire attached to the auto tensioner, if the wire was not previously removed.

6. Rotate the crankshaft 2 complete turns clockwise and let it sit for approximately 5 minutes. Then, make sure the set pin can easily be inserted and removed from the hole in the tensioner.

7. Measure the auto tensioner protrusion (the distance between the tensioner arm and auto tensioner body) to ensure that it is within 0.15–0.18 in. (3.8–4.5mm). If out of specification, repeat Step 1–4 until the specified value is obtained.

8. Install the timing belt covers and all related items.

9. Connect the negative battery cable.

REMOVAL & INSTALLATION

1.5L Engine

1. Disconnect the negative battery cable.

2. Remove the front engine mount bracket and accessory drive belts.

3. Remove timing belt upper and lower covers.

4. Make a mark on the back of the timing belt indicating the direction of rotation so it may be reassembled in the same direction if it is to be reused. Remove the timing belt.

NOTE: If coolant or engine oil comes in contact with the timing belt, they will drastically shorten its life. Also, do not allow engine oil or coolant to contact the timing belt sprockets or tensioner assembly.

5. Remove the tensioner spacer, tensioner spring and tensioner assembly.

6. Inspect the timing belt for cracks on back surface, sides, bottom and check for separated canvas. Check the tensioner pulley for smooth rotation.
To install:

7. Position the tensioner, tensioner spring and tensioner spacer on engine block.

8. Align the timing marks on the camshaft sprocket and crankshaft sprocket. This will position No. 1 piston on TDC on the compression stroke.

9. Position the timing belt on the crankshaft sprocket and keeping the tension side of the belt tight, set it on the camshaft sprocket.

10. Apply counterclockwise force to the camshaft sprocket to give tension to the belt and make sure all timing marks are aligned.

11. Loosen the pivot side tensioner bolt and the slot side bolt. Allow the spring to take up the slack.

12. Tighten the slot side tensioner bolt and then the pivot side bolt. If the pivot side bolt is tightened first, the

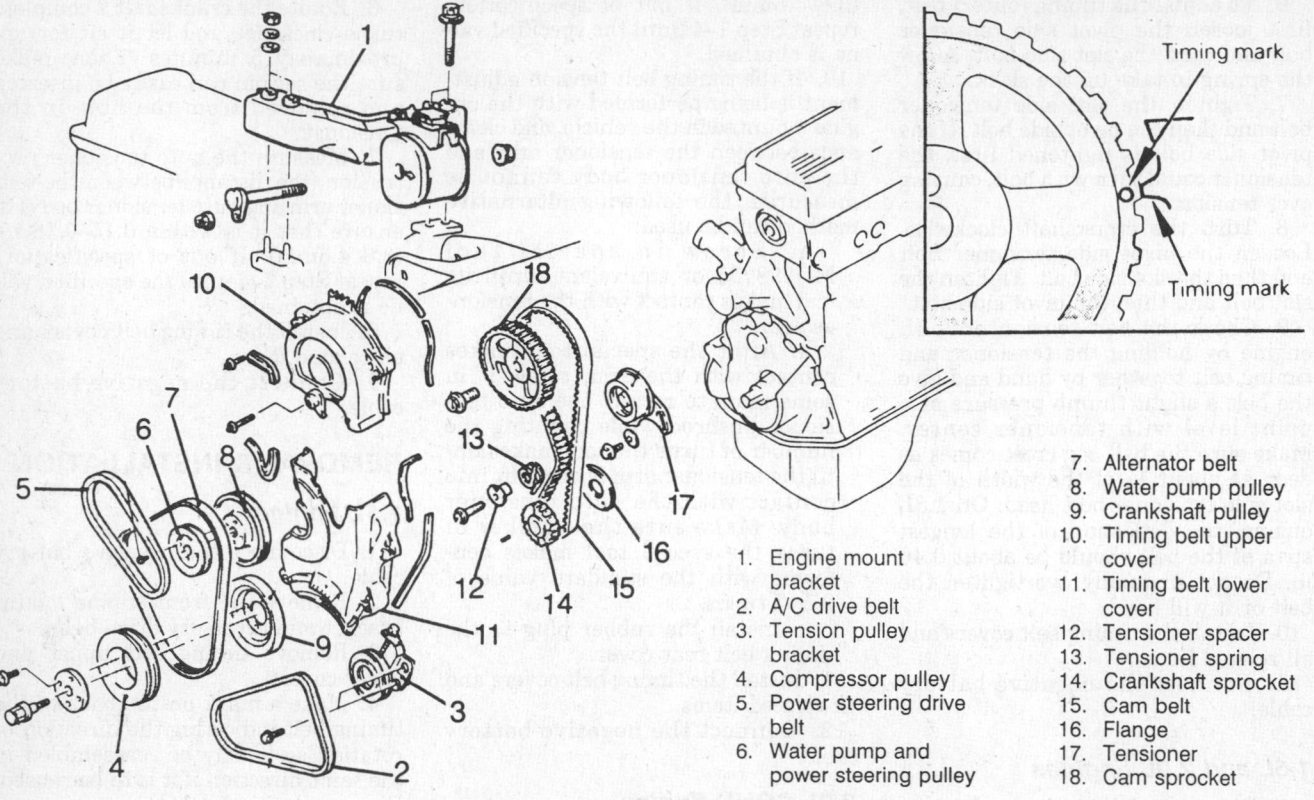

Timing belt and related parts—1.5L engine

1. Engine mount bracket
2. A/C drive belt
3. Tension pulley bracket
4. Compressor pulley
5. Power steering drive belt
6. Water pump and power steering pulley
7. Alternator belt
8. Water pump pulley
9. Crankshaft pulley
10. Timing belt upper cover
11. Timing belt lower cover
12. Tensioner spacer
13. Tensioner spring
14. Crankshaft sprocket
15. Cam belt
16. Flange
17. Tensioner
18. Cam sprocket

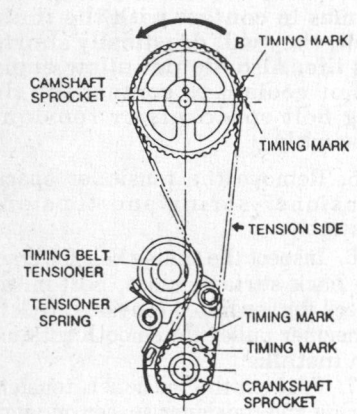

Timing marks alignment—1.5L engine

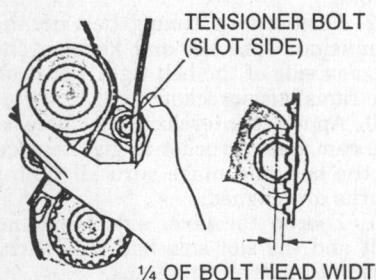

Checking timing belt tension— 1.5L engine

tensioner could turn with bolt, causing over tension.

13. Turn the crankshaft clockwise. Loosen the pivot side tensioner bolt and then the slot side bolt to allow the spring to take up any remaining slack. Tighten the slot bolt and then the pivot side bolt to 14–20 ft. lbs. (20–27 Nm).

14. Check the belt tension by holding the tensioner and timing belt together by hand and give the belt a slight thumb pressure at a point level with tensioner center. Make sure the belt cog crest comes as deep as about ¼ of the width of the slot side tensioner bolt head. Do not manually overtighten the belt or it will howl.

15. Install the timing belt covers and all related items.

16. Connect the negative battery cable.

1.6L and 2.0L Engines

NOTE: The 1.6L engine is not equipped with silent shafts. Disregard all instructions pertaining to silent shafts if working on that engine.

1. Disconnect the negative battery cable.

2. Remove the timing belt upper and lower covers.

3. Rotate the crankshaft clockwise and align the timing marks so No. 1

piston will be at TDC of the compression stroke. At this time the timing marks on the camshaft sprocket and the upper surface of the cylinder head should coincide, and the dowel pin of the camshaft sprocket should be at the upper side.

NOTE: Always rotate the crankshaft in a clockwise direction. Make a mark on the back of the timing belt indicating the direction of rotation so it may be reassembled in the same direction if it is to be reused.

4. Remove the auto tensioner and remove the timing belt.

5. Remove the timing belt tensioner pulley, tensioner arm, idler pulley, oil pump sprocket, special washer, flange and spacer.

6. Remove the silent shaft (inner) belt tensioner and remove the belt.

To install:

7. Align the timing marks of the silent shaft sprockets and the crankshaft sprocket with the timing marks on the front case. Wrap the timing belt around the sprockets so there is no slack in the upper span of the belt and the timing marks are still aligned.

8. Install the tensioner pulley and move the pulley by hand so the long side of the belt deflects about ¼ in.

9. Hold the pulley tightly so the pulley cannot rotate when the bolt is

tightened. Tighten the bolt to 15 ft. lbs. (20 Nm) and recheck the deflection amount.

10. Carefully push the auto tensioner rod in until the set hole in the rod aligned up with the hole in the cylinder. Place a wire into the hole to retain the rod.

11. Install the tensioner pulley onto the tensioner arm. Locate the pinhole in the tensioner pulley shaft to the left of the center bolt. Then, tighten the center bolt finger-tight.

12. When installing the timing belt, turn the 2 camshaft sprockets so their dowel pins are located on top. Align the timing marks facing each other with the top surface of the cylinder head. When you let go of the exhaust camshaft sprocket, it will rotate 1 tooth in the counter-clockwise direction. This should be taken into account when installing the timing belts on the sprocket.

NOTE: Both camshaft sprockets are used for the intake and exhaust camshafts and are provided with 2 timing marks. When the sprocket is mounted on the exhaust camshaft, use the timing mark on the right with the dowel pin hole on top. For the intake camshaft sprocket, use the 1 on the left with the dowel pin hole on top.

13. Align the crankshaft sprocket and oil pump sprocket timing marks. Install the timing belt as follows:

    a. Install the timing belt around the intake camshaft sprocket and retain it with 2 spring clips or binder clips.

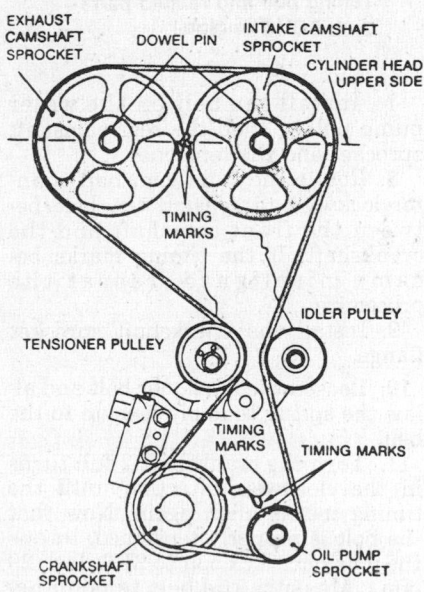

Timing belt and related parts — 1.6L and 2.0L engines

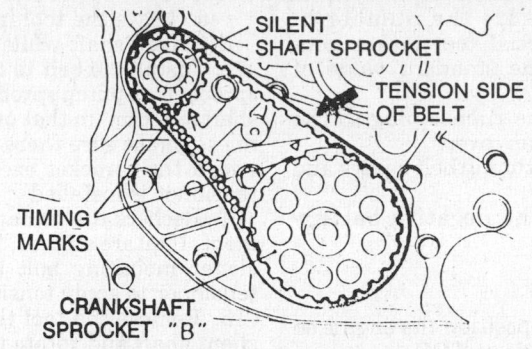

Silent shaft belt timing marks alignment — 1.8L and 2.0L engines

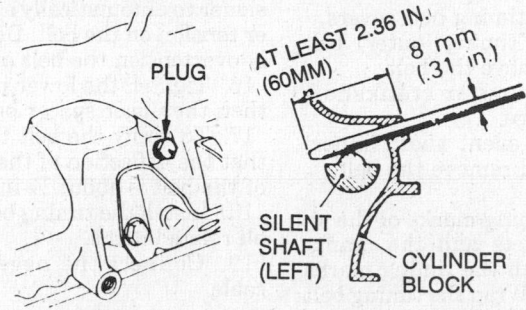

Checking the rear silent shaft for proper positioning

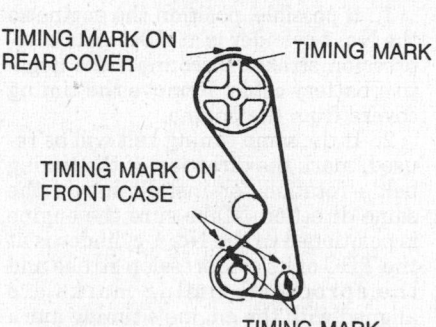

Timing belt timing marks alignment — 1.8L engine

    b. Install the timing belt around the exhaust sprocket, aligning the timing marks with the cylinder head top surface using 2 wrenches. Retain the belt with 2 spring clips.

    c. Install the timing belt around the idler pulley, oil pump sprocket, crankshaft sprocket and the tensioner pulley. Remove the 2 spring clips.

    d. Lift upward on the tensioner pulley in a clockwise direction and tighten the center bolt. Make sure all timing marks are aligned.

    e. Rotate the crankshaft a ¼ turn counterclockwise. Then, turn in clockwise until the timing marks are aligned again.

14. To adjust the timing (outer) belt, turn the crankshaft ¼ turn counterclockwise, then turn it clockwise to move No. 1 cylinder to TDC.

15. Loosen the center bolt. Using

tool MD998738 or equivalent and a torque wrench, apply a torque of 1.88–2.03 ft. lbs. (2.6–2.8 Nm). Tighten the center bolt.

16. Screw the special tool into the engine left support bracket until its end makes contact with the tensioner arm. At this point, screw the special tool in some more and remove the set wire attached to the auto tensioner, if the wire was not previously removed. Then remove the special tool.

17. Rotate the crankshaft 2 complete turns clockwise and let it sit for approximately 15 minutes. Then, measure the auto tensioner protrusion (the distance between the tensioner arm and auto tensioner body) to ensure that it is within 0.15–0.18 in. (3.8–4.5mm). If out of specification, repeat Step 1–4 until the specified value is obtained.

18. If the timing belt tension adjustment is being performed with the engine mounted in the vehicle, and clearance between the tensioner arm and the auto tensioner body cannot be measured, the following alternative method can be used:

    a. Screw in special tool MD998738 or equivalent, until its end makes contact with the tensioner arm.

    b. After the special tool makes contact with the arm, screw it in some more to retract the auto tensioner pushrod while counting the number of turns the tool makes until the tensioner arm is brought into contact with the auto tensioner body. Make sure the number of turns the special tool makes conforms with the standard value of 2½–3 turns.

   c. Install the rubber plug to the timing belt rear cover.

19. Install the timing belt covers and all related items.

20. Connect the negative battery cable.

1.8L Engine

1. If possible, position the engine so the No. 1 piston is at TDC.

2. Disconnect the negative battery cable.

3. Remove the timing belt covers.

4. Remove the timing (outer) belt tensioner and remove the belt.

5. Remove the outer crankshaft sprocket and flange.

6. Remove the silent shaft (inner) belt tensioner and remove the belt.

To install:

7. Align the timing marks of the silent shaft sprockets and the crankshaft sprocket with the timing marks on the front case. Wrap the timing belt around the sprockets so there is no slack in the upper span of the belt and the timing marks are still aligned.

8. Install the tensioner pulley and move the pulley by hand so the long side of the belt deflects about ¼ in.

9. Hold the pulley tightly so the pulley cannot rotate when the bolt is tightened. Tighten the bolt to 15 ft. lbs. (20 Nm) and recheck the deflection amount.

10. Install the timing belt tensioner fully toward the water pump and tighten the bolts. Place the upper end of the spring against the water pump body.

11. Align the timing marks of the camshaft, crankshaft and oil pump sprockets with their corresponding marks on the front case or rear cover.

NOTE: If the following step is not followed exactly, there is a 50 percent chance that the silent shaft alignment will be 180 degrees off. This will cause a noticeable vibration in the engine and the entire procedure will have to be repeated.

12. Before installing the timing belt, ensure that the left side (rear) silent shaft is in the correct position.:

   a. Remove the plug from the rear of the block and insert a suitable shaft.

   b. With the timing marks still aligned, the shaft must be able to go in at least 2⅓ in. If it can only go in about 1 in., turn the oil pump sprocket 1 complete revolution.

   c. Recheck and realign the timing marks.

   d. Leave the tool in place to hold the silent shaft while continuing.

13. Install the belt to the crankshaft sprocket, oil pump sprocket, then camshaft sprocket, in that order. While doing so, make sure there is no slack between the sprocket except where the tensioner is installed.

14. Recheck the timing marks' alignment. If all are aligned, loosen the tensioner mounting bolt and allow the tensioner to apply tension to the belt.

15. Remove the tool that is hold the silent shaft and rotate the crankshaft a distance equal to 2 teeth on the camshaft sprocket. This will allow the tensioner to automatically apply the proper tension on the belt. Do not manually overtension the belt or it will howl.

16. Tighten the lower mounting bolt then the upper spacer bolt.

17. To verify the belt tension, check that the deflection of the longest span of the belt is about ½ in.

18. Install the timing belt covers and all related items.

19. Connect the negative battery cable.

3.0L SOHC Engine

1. If possible, position the engine so the No. 1 cylinder is at TDC of its compression stroke. Disconnect the negative battery cable. Remove the timing covers from the engine.

2. If the same timing belt will be reused, mark the direction of the timing belt's rotation for installation in the same direction. Make sure the engine is positioned so the No. 1 cylinder is at the TDC of its compression stroke and the sprockets' timing marks are aligned with the engine's timing mark indicators.

3. Loosen the timing belt tensioner bolt and remove the belt. If the tensioner is not being removed, position it as far away from the center of the engine as possible and tighten the bolt.

4. If the tensioner is being removed, paint the outside of the spring to ensure that it is not installed backwards. Unbolt the tensioner and remove it along with the spring.

To install:

5. Install the tensioner, if removed, and hook the upper end of the spring to the water pump pin and the lower end to the tensioner in exactly the same position as originally installed. If not already done, position both camshafts so the marks align with those on the rear. Rotate the crankshaft so the timing mark aligns with the mark on the oil pump.

6. Install the timing belt on the crankshaft sprocket and while keeping the belt tight on the tension side, install the belt on the front camshaft sprocket.

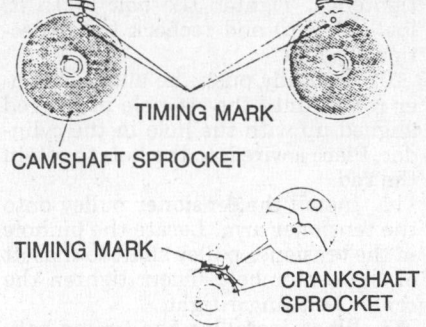

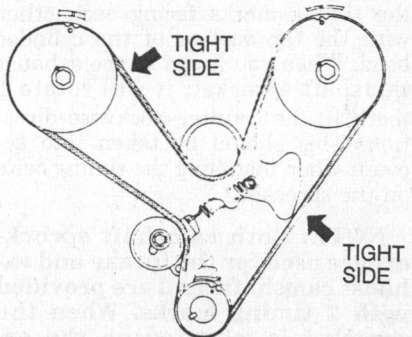

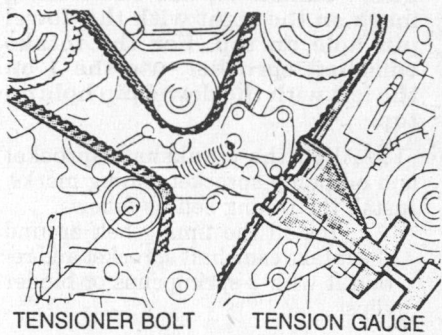

Timing belt and related parts—3.0L SOHC engine

7. Install the belt on the water pump pulley, then the rear camshaft sprocket and the tensioner.

8. Rotate the front camshaft counterclockwise to tension the belt between the front camshaft and the crankshaft. If the timing marks became misaligned, repeat the procedure.

9. Install the crankshaft sprocket flange.

10. Loosen the tensioner bolt and allow the spring to apply tension to the belt.

11. Turn the crankshaft 2 full turns in the clockwise direction until the timing marks align again. Now that the belt is properly tensioned, torque the tensioner lock bolt to 21 ft. lbs. (29 Nm). Measure the belt tension between the rear camshaft sprocket and the crankshaft with belt tension

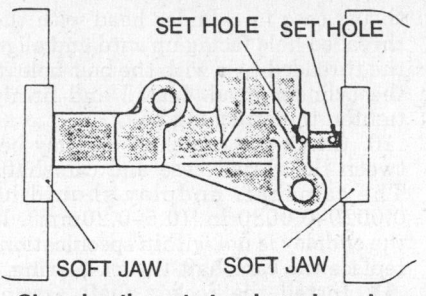

Clamping the auto-tensioner in a vice

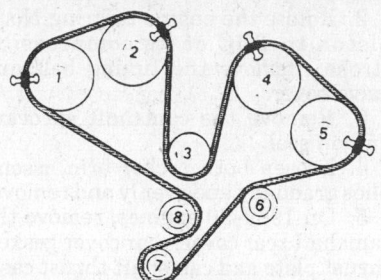

Timing belt installation— 3.0L DOHC engine

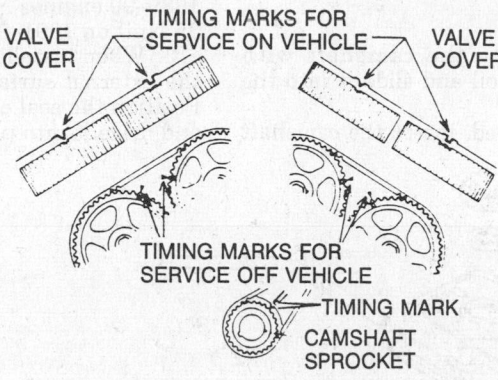

Aligning the timing marks—3.0L DOHC engine

gauge. The specification is 46–68 lbs. (210–310 N).

12. Install the timing belt covers and all related parts.

13. Connect the negative battery cable and road test the vehicle.

3.0L DOHC Engine

1. If possible, position the engine so the No. 1 cylinder is at TDC of its compression stroke. Disconnect the negative battery cable. Remove the timing covers from the engine.

2. If the same timing belt will be reused, mark the direction of the timing belt's rotation for installation in the same direction. Make sure the engine is positioned so the No. 1 cylinder is at the TDC of its compression stroke and the sprockets' timing marks are aligned with the engine's timing mark indicators on the valve covers or head.

3. Loosen the timing belt tensioner bolt and remove the belt.

4. Remove the tensioner assembly.

To install:

5. If the auto tensioner rod is fully extended, reset it as follows:

    a. Clamp the tensioner in a soft-jaw vice in level position.

    b. Slowly push the rod in with the vice until the set hole in the rod is aligned with the hole in the cylinder.

    c. Insert a stiff wire into the set holes to retain the position.

    d. Remove the assembly from the vice.

6. Leave the retaining wire in the

tension and install to the engine. Torque the retaining bolts to 17 ft. lbs. (24 Nm).

7. If the timing marks of the camshaft sprockets and crankshaft sprocket are not aligned at this point, proceed as follows:

NOTE: Keep fingers out from in between the camshaft sprockets. The sprockets may move unexpectedly because of valve spring pressure and could pinch fingers.

    a. Align the mark on the crankshaft sprocket with the mark on the front case. Then move the sprocket 3 teeth clockwise to lower the piston so the valve can't touch the piston when the camshaft are being moved.

    b. Turn each camshaft sprocket 1 at a time to align the timing marks with the mark on the valve cover or head. If the intake and exhaust valves of the same cylinder are opened simultaneously, they could interfere with each other. Therefore, if any resistance is felt, turn the other camshaft to move the valve.

    c. Align the timing mark of the crankshaft sprocket, then continue 1 tooth farther in the counterclockwise direction to facilitate belt installation.

8. Using 4 spring loaded paper clips to hold the belt on the cam sprockets,

install the belt to the sprockets in the following order:

    1st—exhuast camshaft sprocket for the front head

    2nd—intake camshaft sprocket for the front head

    3rd—water pump pulley

    4th—intake camshaft sprocket for the rear head

    5th—exhuast camshaft sprocket for the rear head

    6th—idler pulley

    7th—crankshaft sprocket

    8th—tensioner pulley

9. Turn the tensioner pulley so its pin holes are located above the center bolt. Then press the tensioner pulley against the timing belt and simultaneously tighten the center bolt.

10. Make certain that all timing marks are still aligned. If so, remove the 4 clips.

11. Turn the crankshaft ¼ turn counterclockwise, then turn it clockwise until all timing marks are aligned.

12. Loosen the center bolt on the tensioner pulley. Using tool MD998767 or equivalent and a torque wrench, apply a torque of 7.2 ft. lbs. (10 Nm). Tighten the tensioner bolt; make sure the tensioner doesn't rotate with the bolt.

13. Remove the set wire attached to the auto tensioner, if the wire was not previously removed.

14. Rotate the crankshaft 2 complete turns clockwise and let it sit for approximately 5 minutes. Then, make sure the set pin can easily be inserted and removed from the hole in the tensioner.

15. Measure the auto tensioner protrusion (the distance between the tensioner arm and auto tensioner body) to ensure that it is within 0.15–0.18 in. (3.8–4.5mm). If out of specification, repeat Step 1–4 until the specified value is obtained.

16. Install the timing belt covers and all related items.

17. Connect the negative battery cable.

Timing Sprockets and Oil Seals

REMOVAL & INSTALLATION

1. Disconnect the negative battery cable.

2. Remove the valve cover(s) and timing belt(s).

3. Use a suitable puller and remove the crankshaft sprocket.

7. Hold the camshaft using the hexagon cast into the shaft and remove the sprocket.

5. Pry the seals from the bores and replace using the proper installation tools.

6. Install the sprockets and torque the camshaft sprocket retaining bolts to 47–54 ft. lbs. (65–75 Nm) on 1.5L engine and 56–72 ft. lbs. (80–100 Nm) on all other engines. Torque the crankshaft sprocket retaining bolt, if equipped, to 80–94 ft. lbs. (110–130 Nm).

7. Install the timing belt(s) and valve cover(s).

8. Connect the negative battery cable and check for leaks.

Camshaft

REMOVAL & INSTALLATION

1.5L Engine

1. Disconnect the negative battery cable.

2. Rotate the engine to bring No. 1 piston to TDC of its compression stroke. Remove the timing belt and valve cover.

3. Remove the camshaft sprocket and oil seal.

4. Loosen both rocker arm assemblies gradually and evenly and remove.

5. On 1989–90 engines, remove the camshaft rear cover, rear cover gasket, thrust plate and camshaft thrust case.

6. Remove the camshaft from the head.

7. Carefully check all parts for damage and wear.

To install:

8. Lubricate the camshaft with heavy engine oil and slide it into the head.

9. If equipped, insert the camshaft thrust case in cylinder head with the threaded hole facing upward and align the threaded hole with the bolt hole in the cylinder head. Install and firmly tighten the attaching bolt.

10. Check the camshaft endplay between the thrust case and camshaft. The camshaft endplay should be 0.0020–0.0080 in. (0.5–0.20mm). If the endplay is not within specification, replace the camshaft thrust bearing.

11. Install the rocker shaft assemblies. Torque the bolts gradually and evenly to 14–20 ft. lbs. (20–27 Nm) on 1989–90 engines or 21–25 ft. lbs. (29–35 Nm) on 1991–92 engines.

12. When installing the oil seal, coat the external surface with engine oil. Position the seal on the camshaft end and drive it into place.

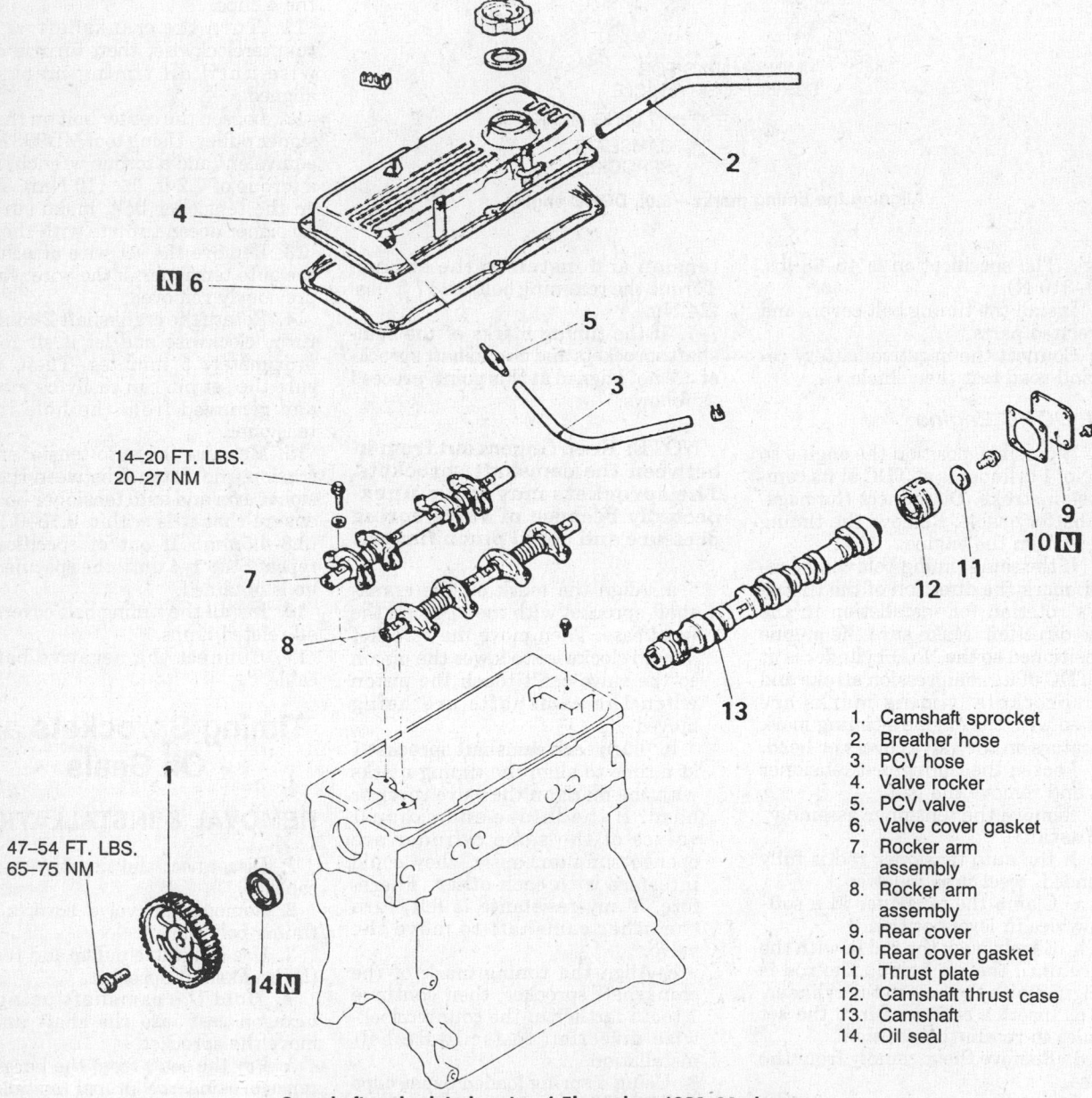

14–20 FT. LBS.
20–27 NM

47–54 FT. LBS.
65–75 NM

1. Camshaft sprocket
2. Breather hose
3. PCV hose
4. Valve rocker
5. PCV valve
6. Valve cover gasket
7. Rocker arm assembly
8. Rocker arm assembly
9. Rear cover
10. Rear cover gasket
11. Thrust plate
12. Camshaft thrust case
13. Camshaft
14. Oil seal

Camshaft and related parts—1.5L engine; 1989–90 shown

13. Install the camshaft sprocket, timing belt and valve cover with new gasket.

14. Connect the negative battery cable and check for leaks.

1.6L and 2.0L Engines

1. Relieve the fuel system pressure.
2. Disconnect battery negative cable.
3. Disconnect the accelerator cable.
4. Remove the timing belt cover and timing belt.
5. Remove the center cover, breather and PCV hoses, and spark plug cables.
6. Remove the rocker cover, semicircular packing, throttle body stay, crankshaft angle sensor, both camshaft sprockets, and oil seals.
7. Loosen the bearing cap bolts in 2–3 steps. Label and remove all camshaft bearing caps.

NOTE: If the bearing caps are difficult to remove, use a plastic hammer to gently tap the rear part of the camshaft.

8. Remove the intake and exhaust camshafts.
9. Check the camshaft journals for wear or damage. Check the cam lobes for damage. Also, check the cylinder head oil holes for clogging.
To install:
10. To install, lubricate the camshafts with heavy engine oil and position the camshafts on the cylinder head.

NOTE: Do not confuse the intake camshaft with the exhaust camshaft. The intake camshaft has a split on its rear end for driving the crank angle sensor.

11. Make sure the dowel pin on both camshaft sprocket ends are located on the top.
12. Install the bearing caps. Tighten the caps in sequence and in 2 or 3 steps. No. 2 and 5 caps are of the same shape. Check the markings on the caps to identify the cap number and intake/exhaust symbol. Only **L** (intake) or **R**

(exhaust) is stamped on No. 1 bearing cap. Also, make sure the rocker arm is correctly mounted on the lash adjuster and the valve stem end. Torque the retaining bolts to 15 ft. lbs. (20 Nm).

13. Apply a coating of engine oil to the oil seal. Using tool MD998307 or equivalent, press-fit the seal into the cylinder head.
14. Align the punch mark on the crank angle sensor housing with the notch in the plate. With the dowel pin on the sprocket side of the intake camshaft at top, install the crank angle sensor on the cylinder head.

NOTE: Do not position the crank angle sensor with the punch mark positioned opposite the notch; this position will result in incorrect fuel injection and ignition timing.

15. Install the timing belt, valve cover and all related parts.
16. Connect the negative battery cable and check for leaks.

1.8L and 3.0L SOHC Engines

1. Disconnect the negative battery cable. Remove the valve covers and timing belt.
2. Install auto lash adjuster retainer tools MD998443 or equivalent, on the rocker arms.
3. If removing the right side (front) camshaft on 3.0L engine, remove the distributor extension.
4. Remove the camshaft bearing caps but do not remove the bolts from the caps.
5. Remove the rocker arms, rocker shafts and bearing caps, as an assembly.
6. Remove the camshaft from the cylinder head.
7. Inspect the bearing journals on the camshaft, cylinder head, and bearing caps.
To install:
8. Lubricate the camshaft journals and camshaft with clean engine oil and install the camshaft in the cylinder head.
9. Align the camshaft bearing caps

with the arrow mark depending on cylinder numbers and install in numerical order.

10. Apply sealer at the ends of the bearing caps and install the assembly.
11. Torque the bearing cap bolts in the following sequence: No. 3, No. 2, No. 1 and No. 4 to 85 inch lbs. (10 Nm).
12. Repeat the sequence increasing the torque to 15 ft. lbs. (20 Nm).
13. Install the distributor extension if it was removed.
14. Install the timing belt, valve cover and all related parts.
15. Connect the negative battery cable and check for leaks.

3.0L DOHC Engine

1. Relieve the fuel system pressure.
2. Disconnect battery negative cable.
3. Remove the timing belt cover and timing belt.
4. Remove the center cover, breather and PCV hoses, and spark plug cables.
5. Remove the rocker cover, semicircular packing, throttle body stay, crankshaft angle sensor, both camshaft sprockets, and oil seals.
6. Remove the crank angle sensor and adaptor.
7. Loosen the bearing cap bolts in 2–3 steps. Label and remove all camshaft bearing caps.

NOTE: If the bearing caps are difficult to remove, use a plastic hammer to gently tap the rear part of the camshaft.

8. Remove the intake and exhaust camshafts.
9. Check the camshaft journals for wear or damage. Check the cam lobes for damage. Also, check the cylinder head oil holes for clogging.
To install:
10. To install, lubricate the camshafts with heavy engine oil and position the camshafts on the cylinder head.

NOTE: Do not confuse the intake camshaft with the exhaust camshaft. The intake camshaft has a V stamped on the hexagon and the exhaust camshaft has a C.

11. Make sure the dowel pin on both camshaft sprocket ends are located as shown.
12. Install the bearing caps. Tighten the caps in sequence and in 2 or 3 steps. Caps 2, 3 and 4 have a front mark. Install with the mark aligned with the front mark on the cylinder head. Intake caps have **I** stamped on the cap and exhaust caps have **E**. Also, make sure the rocker arm is correctly mounted on the lash adjuster and the

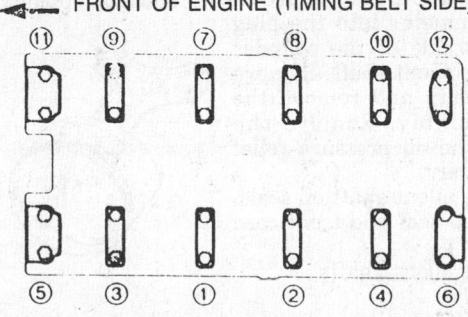

◄—— FRONT OF ENGINE (TIMING BELT SIDE)

Bearing cap tightening sequence—1.6L and 2.0L engines

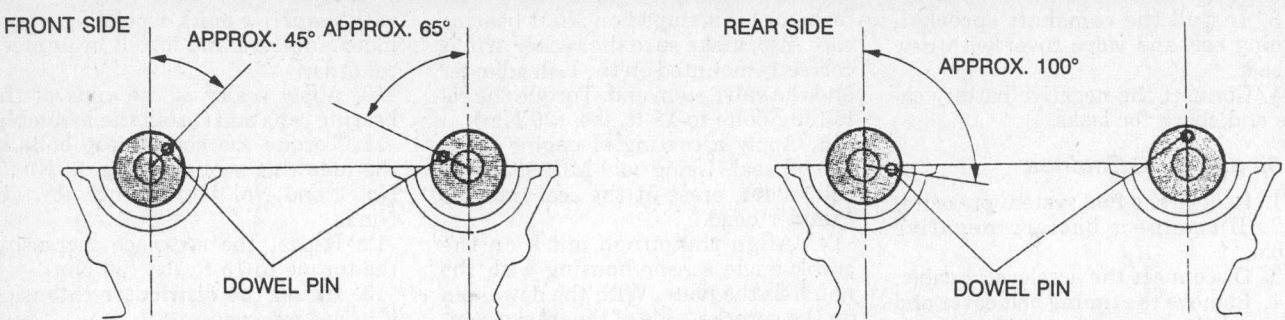

When installing the camshafts, position them so the dowel pins are as shown—3.0L DOHC engine

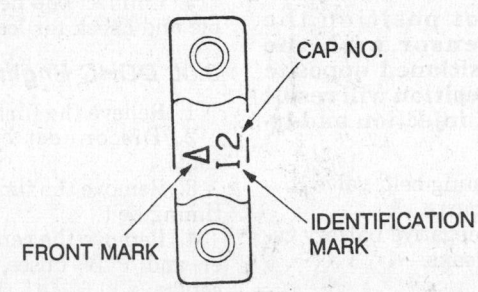

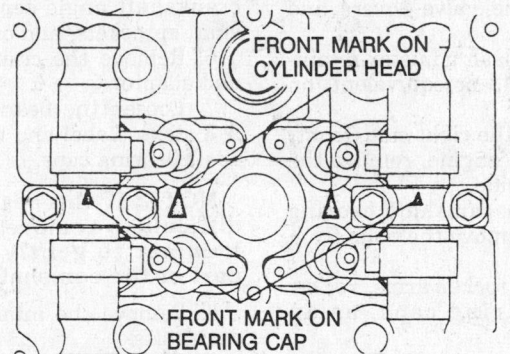

Align the marks on the caps and head as shown—3.0L DOHC engine

valve stem end. Torque the retaining bolts to 15 ft. lbs. (20 Nm).

13. Apply a coating of engine oil to the oil seals and install.

14. Install the timing belt, valve cover and all related parts.

15. Connect the negative battery cable and check for leaks.

Silent Shaft

REMOVAL & INSTALLATION

1.8L Engine

1. Disconnect the negative battery cable. Remove the timing belts.

2. Remove the oil filter, oil pressure switch, oil gauge sending unit and oil filter mounting bracket and gasket.

3. Drain engine oil. Remove engine oil pan, oil screen and gasket.

4. Remove the front engine cover which is also the oil pump cover. Different length bolts are used. Take note of their locations. If the cover sticks to the block, look for a special slot provided and pry with a suitable tool. Discard the shaft seal and gasket.

5. Remove the oil pump driven gear flange bolt. When loosening this bolt, first insert a suitable tool approximately ⅜ in. diameter into the plug hole on the left side of the cylinder block to hold the silent shaft. Remove the oil pump gears and remove the front case assembly. Remove the threaded plug, the oil pressure relief spring and plunger.

6. Remove the silent shaft oil seals, the crankshaft oil seal and front case gasket.

7. Remove the silent shafts.

To install:

8. Carefully install the silent shafts to the block.

9. Install the oil pump components.

10. Install new seals and install the front case with a new gasket.

11. Install the timing belts and all related items.

12. Install the oil pan, oil filter mounting bracket, oil switches and oil filter.

13. Connect the negative battery cable and check for leaks.

Piston and Connecting Rod

POSITIONING

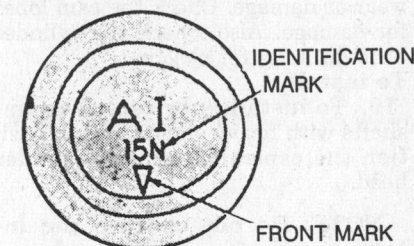

Piston identification marks—1.5L engine

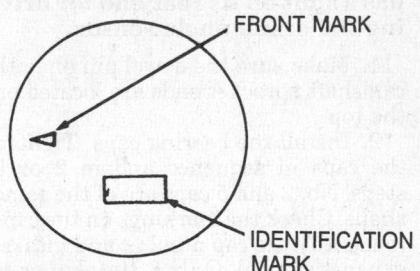

Piston identification marks—1.6L, 1.8L and 2.0L engines

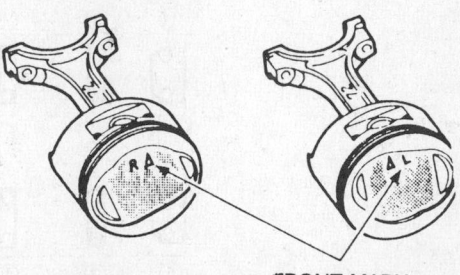

Piston identification marks—3.0L engine

ENGINE LUBRICATION SYSTEM

Oil Pan

REMOVAL & INSTALLATION

1. Disconnect the negative battery cable.

2. Raise the vehicle and support safely.

3. Remove the oil pan drain plug and drain the engine oil. On 1.6L engine equipped with turbocharger, remove the oil return pipe and gasket.

4. On 1.8L engine, disconnect and lower the exhaust pipe.

5. On 2.0L engine, remove the crossmember, disconnect and lower the exhaust pipe and on turbocharged engines, disconnect the return pipe for the turbocharger from the side of the oil pan.

6. Remove the oil pan mounting bolts, separate and remove the engine oil pan.

To install:

7. Thoroughly clean and dry the oil pan, cylinder block bolts and bolt holes.

8. Apply a thin bead of sealer around the surface of the oil pan.

9. Assemble the oil pan to the cylinder block within 15 minutes after applying the sealant.

10. Install the oil pan mounting bolts and torque to 4–6 ft. lbs. (6–8 Nm). On 1.6L engine equipped with turbocharger, install the oil return pipe using a new gasket.

11. Fill the engine with the proper amount of oil.

12. Connect the negative battery cable and check for leaks.

1. Oil filter
2. Oil pressure switch
3. Oil pressure gauge unit
4. Oil filter bracket
5. Gasket
6. Drain plug
7. Drain plug gasket
8. Oil pan
9. Oil screen
10. Gasket
11. Oil pump cover
12. Oil seal
13. Gasket
14. Flange bolt
15. Pump driven gear
16. Pump drive gear
17. Front case
18. Plug
19. Relief spring
20. Relief plunger
21. Silent shaft oil seal
22. Crankshaft front seal
23. Gasket
24. Right side silent shaft
25. Left side silent shaft
26. Silent shaft front bearing
27. Silent shaft rear bearing

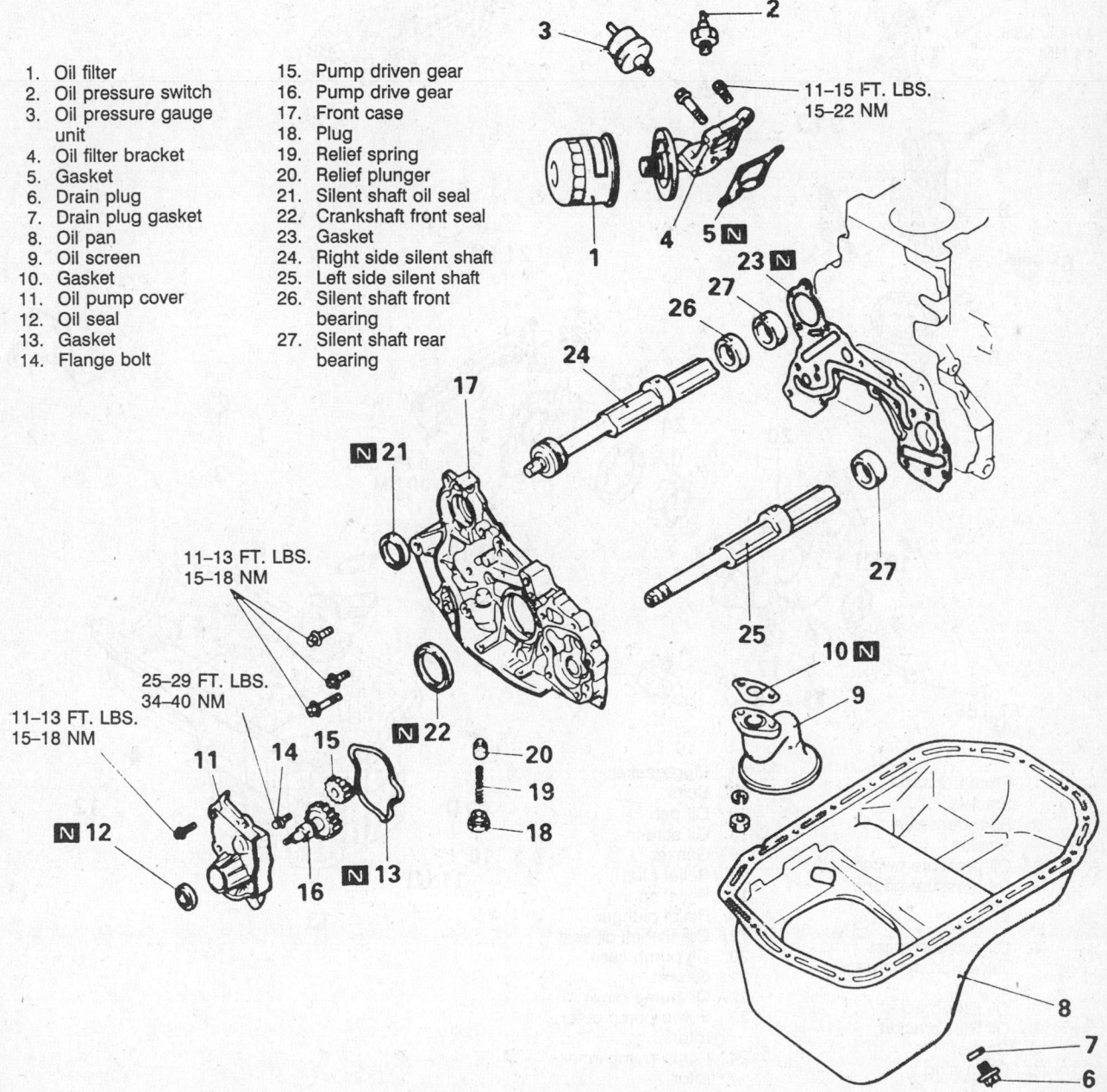

11–15 FT. LBS.
15–22 NM

11–13 FT. LBS.
15–18 NM

25–29 FT. LBS.
34–40 NM

11–13 FT. LBS.
15–18 NM

Front case, oil pump and silent shafts—1.8L engine—1.6L and 2.0L engines are similar

Oil Pump

REMOVAL & INSTALLATION

NOTE: Whenever the oil pump is disassembled or the cover removed, the gear cavity must be filled with petroleum jelly for priming purposes. Do not use grease.

1. Disconnect the negative battery cable.

2. Remove the front engine mount bracket and accessory drive belts.

3. Remove timing belt upper and lower covers.

4. Remove the timing belt and crankshaft sprocket.

5. Remove the oil pan.

6. Remove the oil screen and gasket.

7. Remove and tag the front cover mounting bolts. Note the lengths of the mounting bolts as they are removed for proper installation.

8. On 1.6L engine, remove the plug cap using tool MD998162 or equivalent, and remove the oil pressure switch.

9. Remove the front case cover and oil pump assembly. If necessary, the

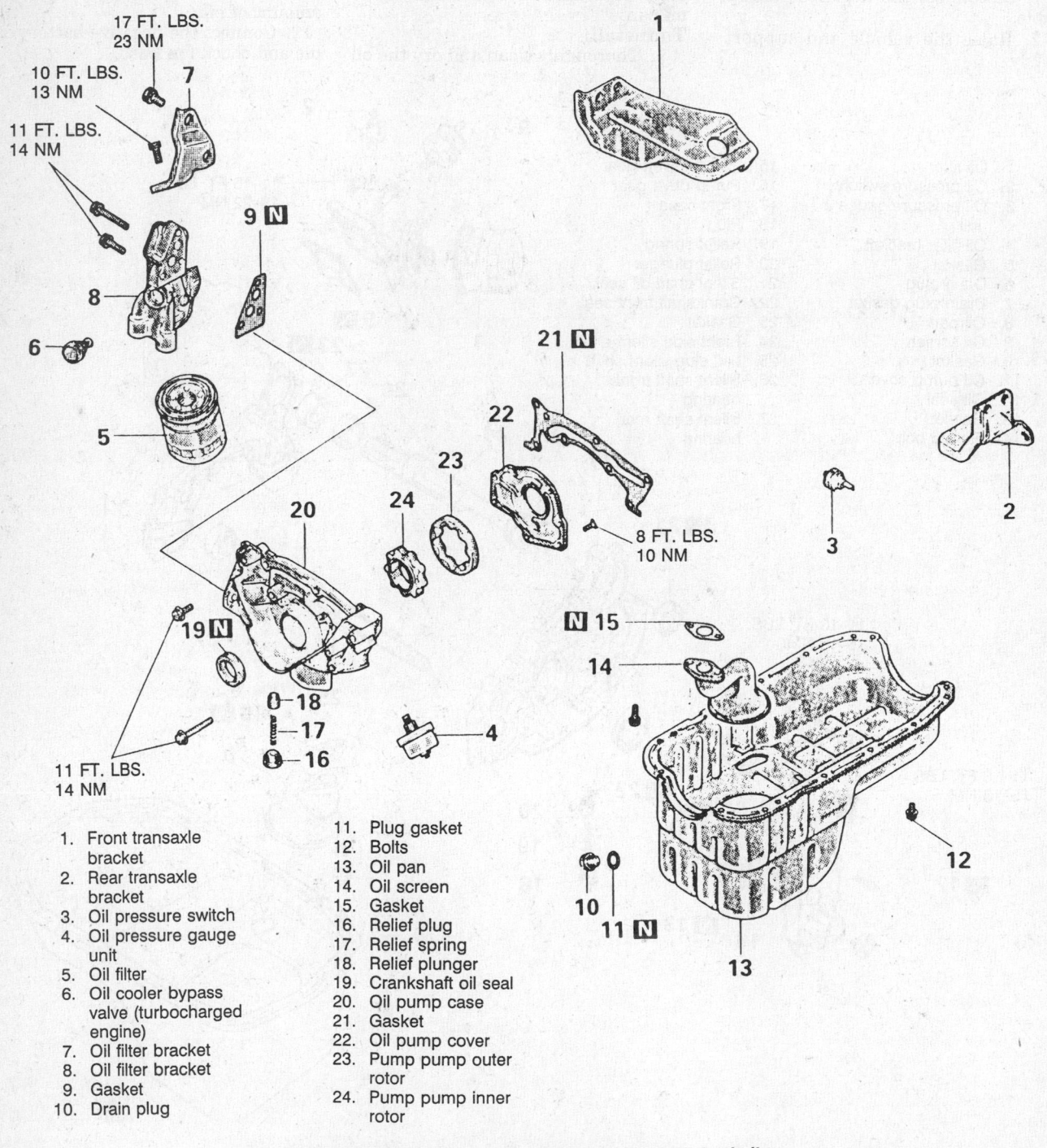

1. Front transaxle bracket
2. Rear transaxle bracket
3. Oil pressure switch
4. Oil pressure gauge unit
5. Oil filter
6. Oil cooler bypass valve (turbocharged engine)
7. Oil filter bracket
8. Oil filter bracket
9. Gasket
10. Drain plug
11. Plug gasket
12. Bolts
13. Oil pan
14. Oil screen
15. Gasket
16. Relief plug
17. Relief spring
18. Relief plunger
19. Crankshaft oil seal
20. Oil pump case
21. Gasket
22. Oil pump cover
23. Pump pump outer rotor
24. Pump pump inner rotor

Oil pump and related parts—3.0L engine—1.5L engine is similar

silent shaft can come out with the assembly. Disassemble as required.

NOTE: On 1.5L engine, the outer gear does not have any marks indicating its installed direction. Make a mark on the reverse side of the outer gear so it can be reinstalled in its proper position.

To install:

10. Thoroughly clean all gasket material from all mounting surfaces.

11. Apply engine oil to the entire surface of the gears or rotors. On 1.5L engine, make sure the outer gear is installed in the same direction as before according to the mark made at the time of removal.

12. On engines with silent shafts, install the drive/driven gears with the 2 timing marks aligned.

13. Assemble the front case cover and oil pump assembly to the engine block using a new gasket. On 1.6L engine, assemble the front case cover and oil pump assembly using tool MD998285 or equivalent, on the front end of the crankshaft.

14. Install the oil screen with new gasket.

15. Install the oil pan and timing belts.

16. Connect the negative battery cable and check for adequate oil pressure.

Rear Main Bearing Oil Seal

REMOVAL & INSTALLATION

1. Disconnect the negative battery cable.

2. Remove the transaxle from the vehicle.

3. Remove the flywheel/ring gear assembly.

4. If the crankshaft rear oil seal case is leaking, remove it. Otherwise, just remove the oil seal. Some engines have a separator that should also be removed.

To install:

5. Install the separator. Lubricate the inner diameter of the new seal with clean engine oil.

6. Install the oil seal in the crankshaft rear oil seal case using tool MD998376 or equivalent. Press the seal all the way in without tilting it. Force the oil separator into the oil seal case so the oil hole in the separator is at 7 o'clock position.

7. Install the seal case with a new gasket.

8. Install the flywheel and transaxle.

9. Connect the negative battery cable and check for leaks.

ENGINE COOLING

Radiator

REMOVAL & INSTALLATION

1. Disconnect the negative battery cable.

2. Drain the cooling system.

3. Disconnect the overflow tube. Some vehicles may also require removal of the overflow tank.

4. Disconnect upper and lower radiator hoses.

5. Disconnect electrical connectors for cooling fan and air conditioning condenser fan, if equipped. Remove the fan assembly if necessary.

6. Disconnect thermo sensor wires.

7. Disconnect and plug automatic transaxle cooler lines, if used.

8. Remove the upper radiator mounts and lift out the radiator/fan assembly.

9. Service the lower mounts as required.

To install:

10. Install the radiator and fan assembly, if removed as an assembly.

11. Connect the automatic transaxle cooler lines, if disconnected.

12. Connect the thermo wires.

13. Install the fan if removed separately.

14. Install the radiator hoses.

15. Install the overflow tube and reservoir, if removed.

16. Fill the system with coolant.

17. Connect the negative battery cable, run the vehicle until the thermo-

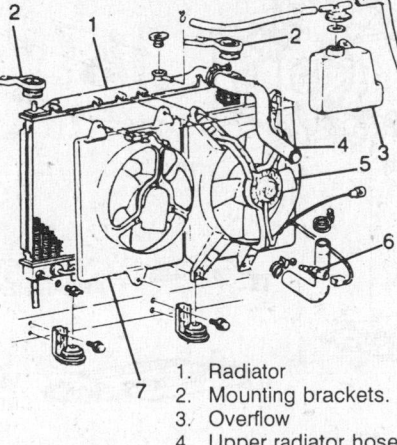

1. Radiator
2. Mounting brackets.
3. Overflow
4. Upper radiator hose
5. Cooling fan
6. Thermo switch
7. A/C condenser fan

Representative electric cooling fan and radiator assembly

stat opens, fill the radiator completely and check the automatic transaxle fluid level, if equipped.

18. Once the vehicle has cooled, recheck the coolant level.

Heater Core

REMOVAL & INSTALLATION

Summit

1. Disconnect the negative battery cable.

2. Drain the cooling system and disconnect the heater hoses.

3. Remove the front seats by removing the covers over the anchor bolts, the underseat tray, the seat belt guide ring, the seat mounting nuts and bolts and disconnect the seat belt switch wiring harness from under the seat. Then lift out the seats.

4. Remove the floor console by first taking out the coin holder and the console box tray. Remove the remote control mirror switch or cover. All of these items require only a plastic trim tool to carefully pry them out.

5. Remove the rear half of the console.

6. Remove the shift lever knob on manual transaxle vehicles.

7. Remove the front console box assembly.

8. A number of the instrument panel pieces may be retained by pin type fasteners. They may be removed using the following procedure:

   a. This type of clip is removed by pressing down on the center pin with a suitable blunt pointed tool. Press down a little more than $\frac{1}{16}$ in. (2mm); this releases the clip. Pull the clip outward to remove it.

   b. Do not push the pin inward more than necessary because it may damage the grommet or the pin may fall in if pushed in too far. Once the clips are removed, use a plastic trim stick to pry the piece loose.

9. Remove both lower cowl trim panels (kick panels).

10. Remove the ashtray.

11. Remove the center panel around the radio.

12. Remove the sunglass pocket at the upper left side of panel and the side panel into which it mounts.

13. Remove the driver's side knee protector and the hood release handle.

14. Remove the steering column top and bottom covers.

15. Remove the radio.

16. Remove the glove box striker and box assembly.

17. Remove the instrument panel lower cover, 2 small pieces in the center, by pulling forward.

18. Remove the heater control assembly screw.

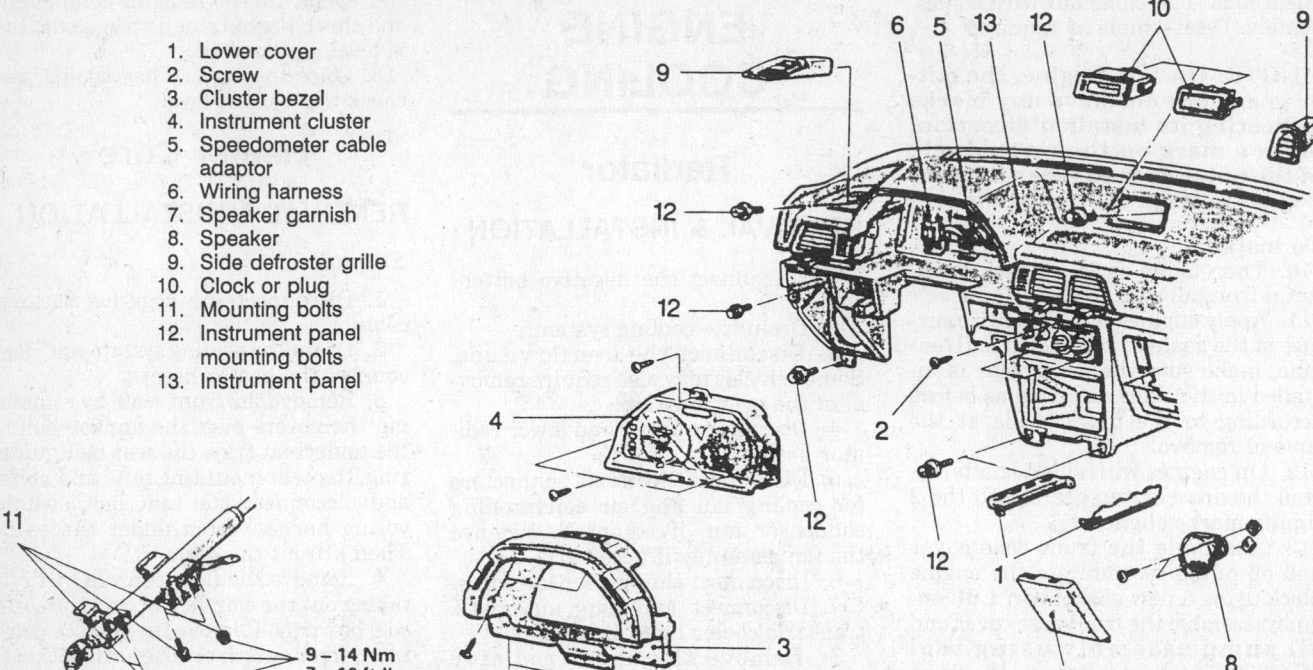

1. Lower cover
2. Screw
3. Cluster bezel
4. Instrument cluster
5. Speedometer cable adaptor
6. Wiring harness
7. Speaker garnish
8. Speaker
9. Side defroster grille
10. Clock or plug
11. Mounting bolts
12. Instrument panel mounting bolts
13. Instrument panel

9 – 14 Nm
7 – 10 ft.lbs.

Instrument panel and related parts—Summit

VEHICLES WITHOUT
REAR HEATER DUCTS

VEHICLES WITH
REAR HEATER DUCTS

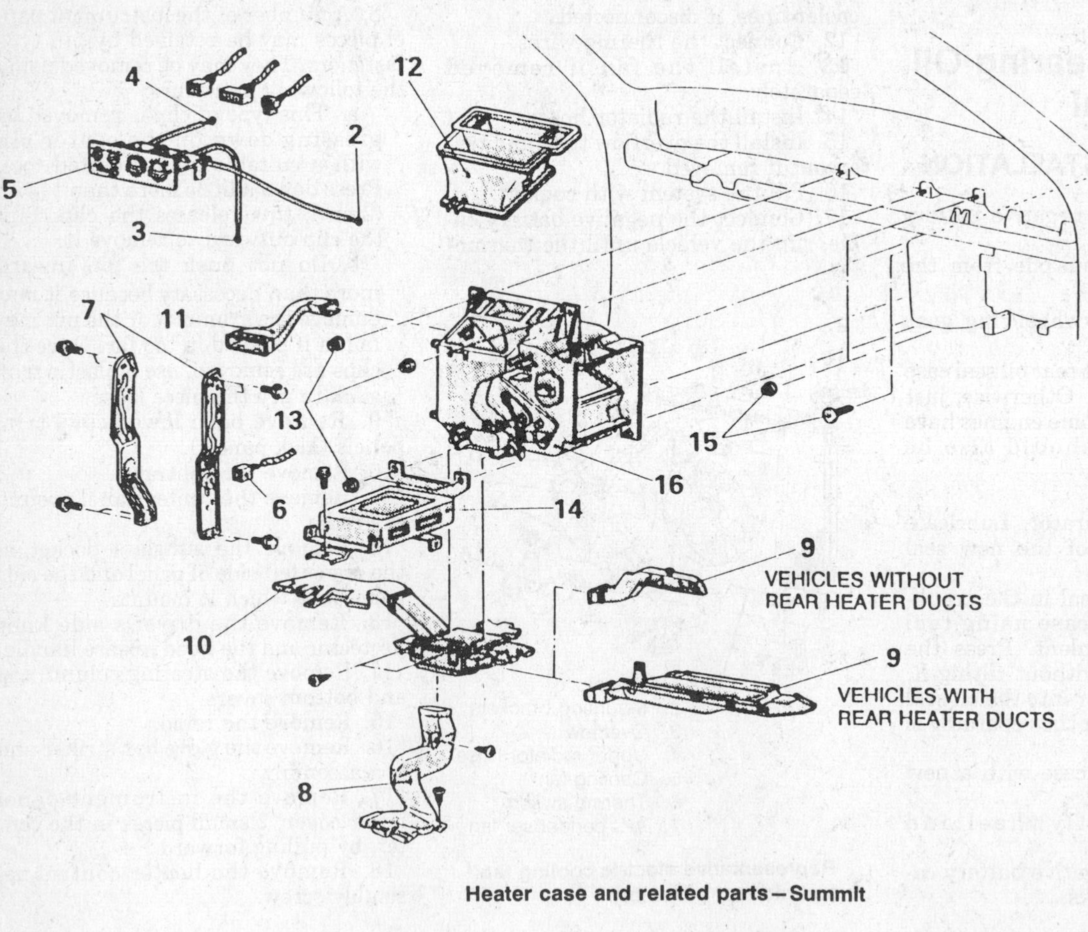

1. Heater hoses
2. Air selection control cable
3. Temperature control cable
4. Mode selection cable
5. Control head
6. ECI control relay connector
7. Center stay
8. Rear heater duct
9. Lap heater duct
10. Foot duct
11. Lap duct
12. Center vent duct
13. Mounting nuts
14. Automatic transaxle control unit
15. Evaporator mounting nuts and clips
16. Heater unit

Heater case and related parts—Summit

19. Remove the instrument cluster bezel and pull out the gauge assembly.

20. Remove the speedometer adapter by disconnecting the speedometer cable at the transaxle pulling the cable sightly towards the vehicle interior and giving a slight twist on the adapter to release it.

21. Insert a small flat-tipped tool to open the tab on the gauge cluster connector. Remove the harness connectors.

22. Remove, by prying with a plastic trim tool, the right side speaker cover and the speaker, the upper side defroster grilles and the clock or plug to gain access to some of the instrument panel mounting bolts.

23. Lower the steering column by removing the bolt and nut.

24. Remove the instrument panel bolts and the instrument panel.

25. Disconnect the air selection, temperature and mode selection control cables from the heater box and remove the heater control assembly.

26. Remove the connector for the ECI control relay.

27. Remove both stamped steel instrument panel supports.

28. Remove the heater duct work.

29. Remove the heater box mounting nuts.

30. Remove the automatic transaxle ELC control box.

31. Remove the evaporator mounting nuts and clips.

32. With the evaporator pulled toward the vehicle interior, remove the heater unit. Be careful not to damage the heater tubes or to spill coolant.

33. Remove the cover plate around the heater tubes and the core fastener clips. Pull the heater core from the heater box, being careful not to damage the fins or tank ends.

To install:

34. Thoroughly clean and dry the inside of the case. Install the heater core to the heater box. Install the clips and cover.

35. Install the evaporator and the automatic transaxle ELC box.

36. Install the heater box and connect the duct work.

37. Connect all wires and control cables.

38. Install the instrument panel assembly and the console by reversing their removal procedures.

39. Install the seats.

40. Refill the cooling system.

41. Evacuate and recharge the air conditioning system. Add 2 oz. of refrigerant oil during the recharge if the evaporator was replaced.

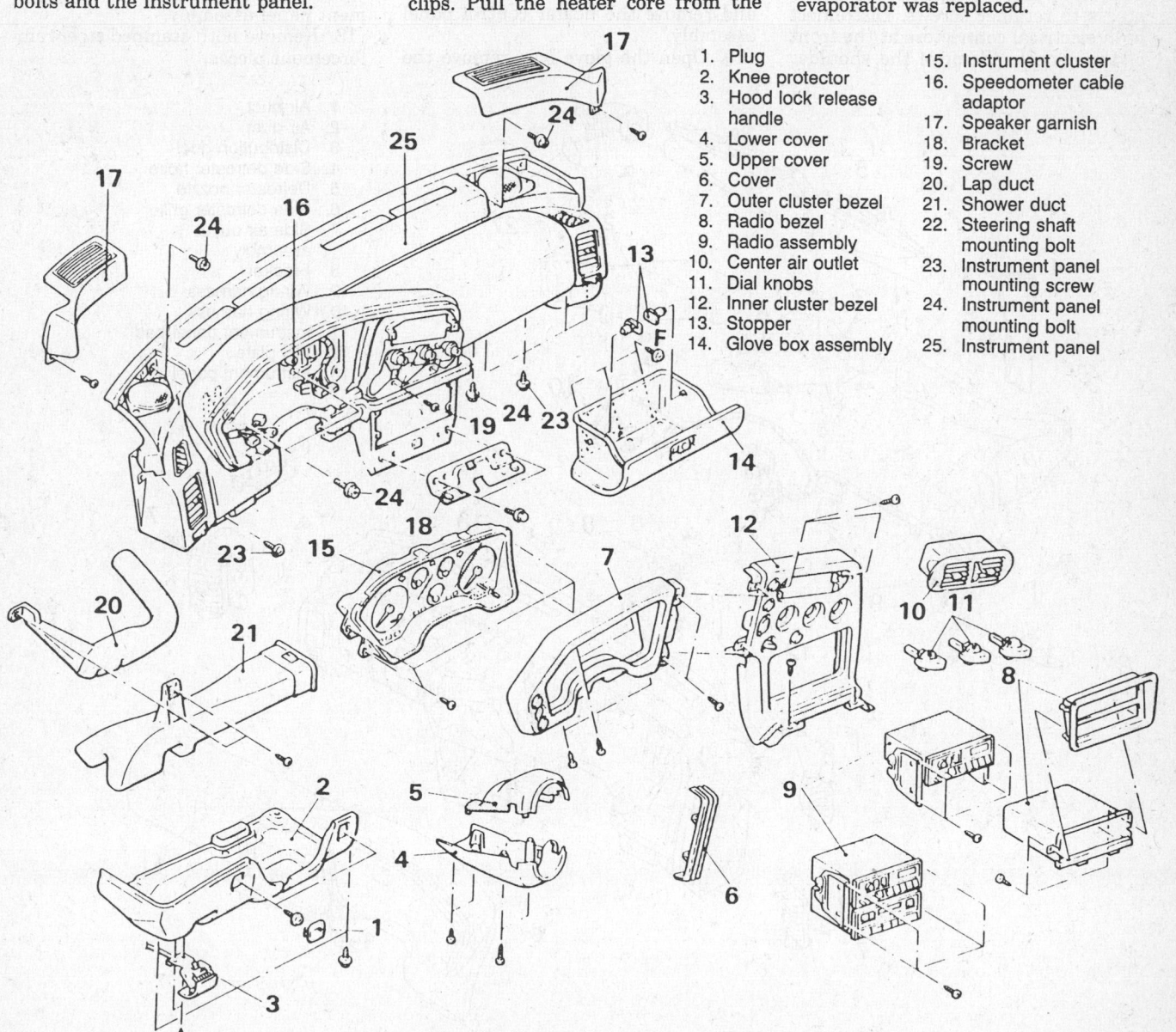

1. Plug
2. Knee protector
3. Hood lock release handle
4. Lower cover
5. Upper cover
6. Cover
7. Outer cluster bezel
8. Radio bezel
9. Radio assembly
10. Center air outlet
11. Dial knobs
12. Inner cluster bezel
13. Stopper
14. Glove box assembly
15. Instrument cluster
16. Speedometer cable adaptor
17. Speaker garnish
18. Bracket
19. Screw
20. Lap duct
21. Shower duct
22. Steering shaft mounting bolt
23. Instrument panel mounting screw
24. Instrument panel mounting bolt
25. Instrument panel

Instrument panel and related parts—Laser and Talon

42. Connect the negative battery cable and check the entire climate control system for proper operation. Check the system for leaks.

Laser and Talon

1. Disconnect the negative battery cable.

2. Drain the cooling system and properly discharge the air conditioning system and disconnect the refrigerant lines from the evaporator, if equipped. Cover the exposed ends of the lines to minimize contamination.

3. Remove the floor console by first removing the plugs, then the screws retaining the side covers and the small cover piece in front of the shifter. Remove the shifter knob, manual transmission, and the cup holder. Remove both small pieces of upholstery to gain access to retainer screws. Disconnect both electrical connectors at the front of the console. Remove the shoulder harness guide plates and the console assembly.

4. Locate the rectangular plugs in the knee protector on either side of the steering column. Pry these plugs out and remove the screws. Remove the screws from the hood lock release lever and the knee protector.

5. Remove the upper and lower column covers.

6. Remove the narrow panel covering the instrument cluster cover screws, and remove the cover.

7. Remove the radio panel and remove the radio.

8. Remove the center air outlet assembly by reaching through the grille and pushing the side clips out with a small flat-tipped tool while carefully prying the outlet free.

9. Pull the heater control knobs off and remove the heater control panel assembly.

10. Open the glove box, remove the plugs from the sides and the glove box assembly.

11. Remove the instrument gauge cluster and the speedometer adapter by disconnecting the speedometer cable at the transaxle, pulling the cable sightly towards the vehicle interior, then giving a slight twist on the adapter to release it.

12. Remove the left and right speaker covers from the top of the instrument panel.

13. Remove the center plate below the heater controls.

14. Remove the heater control assembly installation screws.

15. Remove the lower air ducts.

16. Drop the steering column by removing the bolts.

17. Remove the instrument panel mounting screws, bolts and the instrument panel assembly.

18. Remove both stamped steel reinforcement pieces.

1. Air duct
2. Air duct
3. Distribution duct
4. Side defroster hose
5. Defroster nozzle
6. Side defroster grille
7. Side air outlet assembly
8. Bracket
9. Wiring harness
10. Wiring harness
11. Instrument panel pad
12. VIN plate
13. Instrument panel

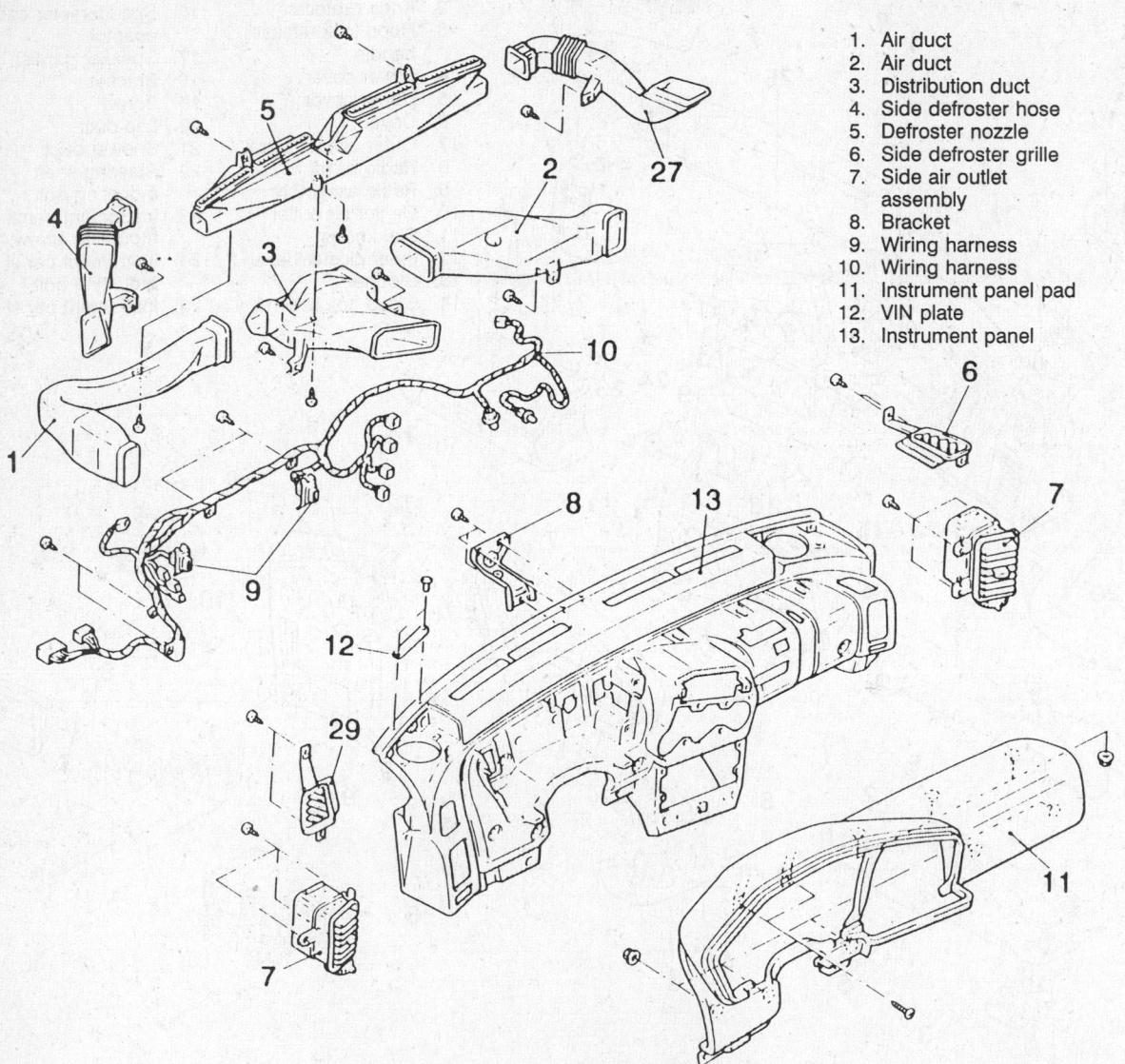

Instrument panel and duct work—Laser and Talon

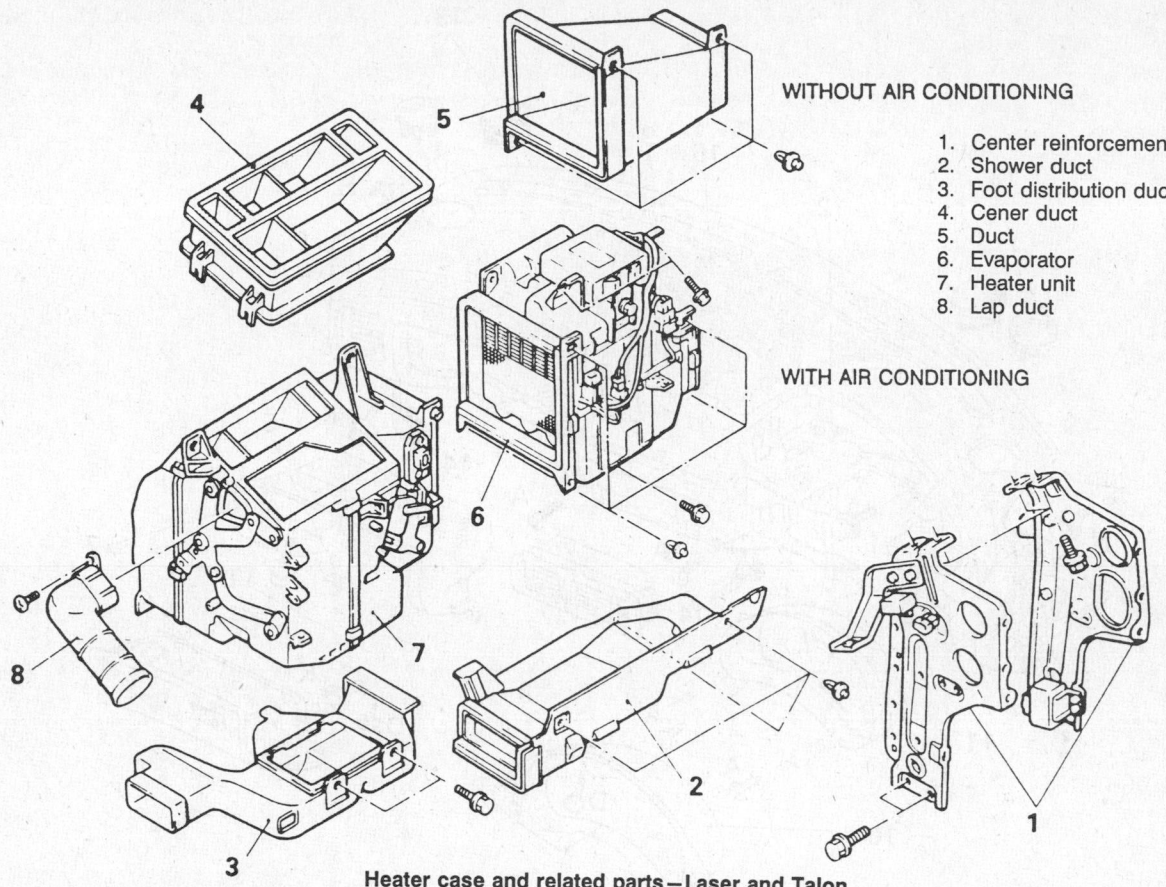

WITHOUT AIR CONDITIONING

1. Center reinforcement
2. Shower duct
3. Foot distribution duct
4. Cener duct
5. Duct
6. Evaporator
7. Heater unit
8. Lap duct

WITH AIR CONDITIONING

Heater case and related parts—Laser and Talon

19. Remove the lower duct work from the heater box.

20. Remove the upper center duct.

21. Vehicles without air conditioning will have a square duct in place of the evaporator; remove this duct if present. If equipped with air conditioning, remove the evaporator assembly:

   a. Remove the wiring harness connectors and the electronic control unit.

   b. Remove the drain hose and lift out the evaporator unit.

   c. If servicing the assembly, disassemble the housing and remove the expansion valve and evaporator.

22. With the evaporator removed, remove the heater unit. To prevent bolts from falling inside the blower assembly, set the inside/outside air-selection damper to the position that permits outside air introduction.

23. Remove the cover plate around the heater tubes and remove the core fastener clips. Pull the heater core from the heater box, being careful not to damage the fins or tank ends.

To install:

24. Thoroughly clean and dry the inside of the case. Install the heater core to the heater box. Install the clips and cover.

25. Install the heater box and connect the duct work.

26. Assemble the housing, evaporator and expansion valve, making sure the gaskets are in good condition. Install the evaporator housing.

27. Using new lubricated O-rings, connect the refrigerant lines to the evaporator.

28. Install the electronic transaxle ELC box. Connect all wires and control cables.

29. Install the instrument panel assembly and the console by reversing their removal procedures.

30. Evacuate and recharge the air conditioning system. If the evaporator was replaced, add 2 oz. of refrigerant oil during the recharge.

31. Connect the negative battery cable and check the entire climate control system for proper operation. Check the system for leaks.

Stealth

1. Disconnect the negative battery cable.

NOTE: If equipped with an air bag, be sure to disarm it before entering the vehicle.

2. Drain the coolant and disconnect

the heater hoses from the core tubes.

3. To remove the console, perform the following:

   a. Remove the cup holder and console plug.

   b. Remove the rear console.

   c. Remove the radio bezels and radio.

   d. Remove the switch bezel.

   e. Remove the side covers and front console garnish.

   f. If equipped with a manual transaxle, remove the shifter knob.

   g. Remove the mounting screws and remove the console assembly.

4. Remove the hood lock release handle from the instrument panel.

5. Remove the interior and dash lights rheostat and switch bezel to its right.

6. Remove the driver's knee protector. Remove the steering column covers.

7. Remove the glove box and cover.

8. Remove the center air outlet assembly.

9. Remove the climate control switch assembly.

10. Remove the instrument cluster bezel and cluster.

11. If equipped with front speakers, remove them. If not, remove the plug in their place.

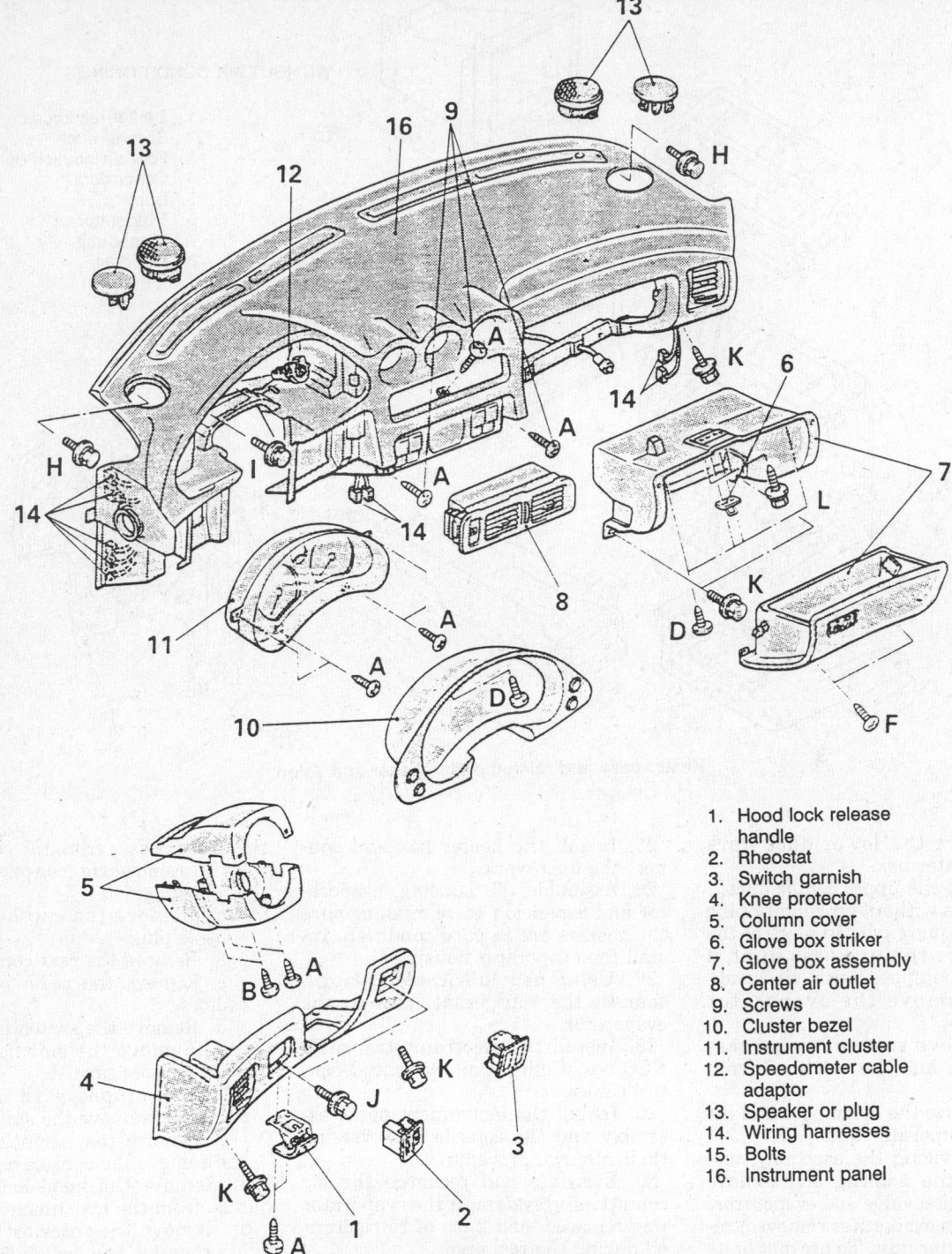

1. Hood lock release handle
2. Rheostat
3. Switch garnish
4. Knee protector
5. Column cover
6. Glove box striker
7. Glove box assembly
8. Center air outlet
9. Screws
10. Cluster bezel
11. Instrument cluster
12. Speedometer cable adaptor
13. Speaker or plug
14. Wiring harnesses
15. Bolts
16. Instrument panel

Instrument panel and related parts—Stealth

12. Disconnect the wiring harnesses on the right side of the instrument panel.

13. Remove the steering shaft support bolts and lower the steering column.

14. Remove the instrument panel mounting hardware and remove the instrument panel from the vehicle.

15. Remove the center reinforcement.

16. Remove the foot warmer ducts and lap duct.

17. If equipped with air conditioning, remove the evaporator case mounting bolt and nut to allow clearance for heater unit removal.

18. Remove the center duct above the heater unit.

19. Remove the heater unit and disassemble on a workbench. Remove the heater core from the heater case.

To install:

20. Thoroughly clean and dry the inside of the case and install the heater core and all related parts.

21. Install the heater unit to the vehicle and install the mounting screws.

Name	Symbol	Size mm (in.) (D x L)	Color	Shape
Tapping screw	A	5 x 16 (.20 x .63)	–	
	B	5 x 30 (.20 x 1.2)	–	
	C	4 x 12 (.16 x .47)	Black	
	D	5 x 16 (.20 x .63)	Black	
	E	4 x 16 (.16 x .63)	–	
Washer assembled screw	F	5 x 16 (.20 x .63)	–	
	G	4 x 12 (.16 x .47)	–	
Washer assembled bolt	H	6 x 16 (.24 x .63)	–	
	I	6 x 16 (.24 x .63)	–	
	J	6 x 20 (.24 x .79)	–	
	K	6 x 20 (.24 x .79)	Black	
	L	6 x 25 (.24 x .98)	Black	

Instrument panel fastener identification—Stealth

1. Heater hoses
2. Center reinforcement
3. undercover
4. Foot distribution duct
5. Foot shower duct
6. Lap duct
7. Evaporator mounting bolt and nut
8. Center duct
9. Heater unit
10. Plate
11. Heater core

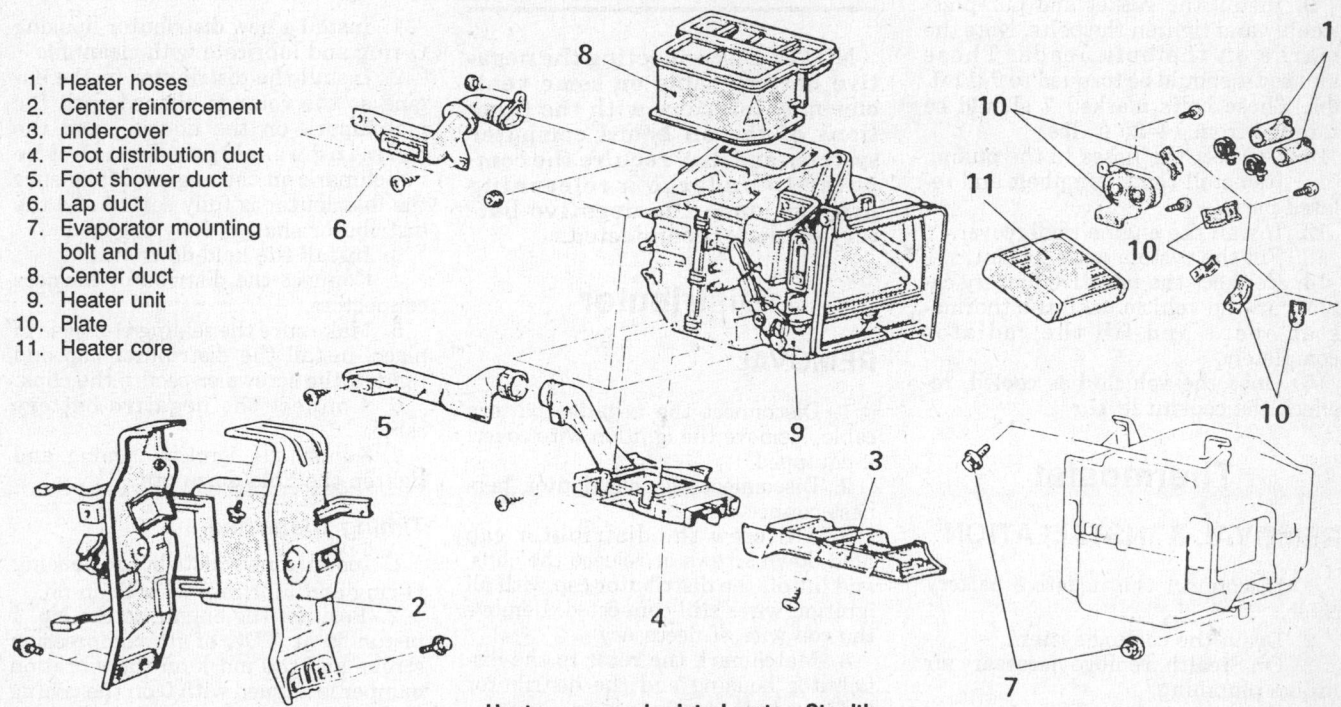

Heater case and related parts—Stealth

22. Install the center duct above the unit.

23. Secure the evaporator case with the bolt and nut.

24. Install the lap duct and foot warmer ducts.

25. Install the center reinforcement.

26. Install the instrument panel by reversing its removal procedure.

27. Install the hood lock release cable handle.

28. Install the console.

29. Fill the cooling system.

30. Connect the negative battery cable and check the entire climate control system for proper operation and leaks.

Water Pump

REMOVAL & INSTALLATION

1. Disconnect the negative battery cable.

2. Drain the cooling system.

3. Remove the engine undercover.

4. Remove the timing belt.

5. Disconnect the hoses from the pump, if equipped. Remove the alternator bracket.

6. Remove the water pump, gasket and O-ring where the water inlet pipe joins the pump.

To install:

7. Thoroughly clean and dry both gasket surfaces of the water pump and block.

8. Install a new O-ring into the groove on the front end of the water inlet pipe. Do not apply oils or grease to the O-ring. Wet with water only.

9. Install the gasket and pump assembly and tighten the bolts. Note the marks on the bolt heads. Those marked **4** should be torqued to 9–11 ft. lbs. Those bolts marked **7** should be torqued from 14–20 ft. lbs.

10. Connect the hoses to the pump.

11. Reinstall the timing belt and related parts.

12. Install the engine undercover.

13. Fill the system with coolant.

14. Connect the negative battery cable, run the vehicle until the thermostat opens and fill the radiator completely.

15. Once the vehicle has cooled, recheck the coolant level.

Thermostat

REMOVAL & INSTALLATION

1. Disconnect the negative battery cable.

2. Drain the cooling system.

3. On Stealth, remove necessary air intake plumbing.

4. Disconnect the radiator hose and overflow hose from the thermostat housing.

5. Remove the thermostat housing and gasket.

6. Remove the thermostat.

To install:

7. Install the thermostat so its flange seats tightly in the machined groove in the intake manifold or thermostat case.

8. Use a new gasket and reinstall the thermostat housing. Fill the system with coolant.

9. Install removed air intake plumbing.

10. Connect the negative battery cable, run the vehicle until the thermostat opens and fill the radiator completely.

11. Once the vehicle has cooled, recheck the coolant level.

Cooling System Bleeding

All vehicles are equipped with a self-bleeding thermostat. Slowly fill the cooling system in the conventional manner; air will vent through the jiggle valve in the thermostat. Run the vehicle until the thermostat has opened and continue filling the radiator. Recheck the coolant level after the vehicle has cooled.

ENGINE ELECTRICAL

NOTE: Disconnecting the negative battery cable on some vehicles may interfere with the functions of the on board computer systems and may require the computer to undergo a relearning process, once the negative battery cable is reconnected.

Distributor

REMOVAL

1. Disconnect the negative battery cable. Remove the ignition wire cover, if equipped.

2. Disconnect the distributor harness connectors.

3. Unscrew the distributor cap hold-down screws or release the clips, and lift off the distributor cap with all ignition wires still connected. Remove the coil wire, if necessary.

4. Matchmark the rotor to the distributor housing and the distributor housing to the engine.

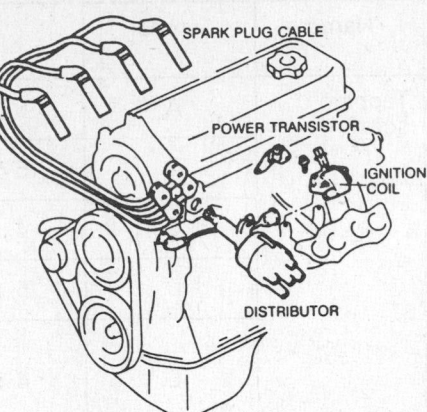

Ignition system—1.8L and 1989–90 1.5L engines

NOTE: Do not crank the engine during this procedure. If the engine is cranked, the matchmark must be disregarded.

5. Remove the hold-down nut.

6. Carefully remove the distributor from the engine.

INSTALLATION

NOTE: Some engines may be sensitive to the routing of the distributor sensor wires. If routed near the high-voltage coil wire or the spark plug wires, the electromagnetic field surrounding the high voltage wires could generate an occasional disruption of the ignition system operation.

Timing Not Disturbed

1. Install a new distributor housing O-ring and lubricate with clean oil.

2. Install the distributor in the engine so the rotor is aligned with the matchmark on the housing and the housing is aligned with the matchmark on the engine. Make sure the distributor is fully seated and the distributor shaft is fully engaged.

3. Install the hold-down nut.

4. Connect the distributor harness connectors.

5. Make sure the sealing O-ring is in place, install the distributor cap and tighten the screws or secure the clips.

6. Connect the negative battery cable.

7. Adjust the ignition timing and tighten the hold-down nut.

Timing Disturbed

1. Install a new distributor housing O-ring and lubricate with clean oil.

2. Position the engine so the No. 1 piston is at TDC of its compression stroke and the mark on the vibration damper is aligned with **0** on the timing indicator.

3. Install the distributor in the engine so the rotor is aligned with the position of the No. 1 ignition wire in the distributor cap and the housing is aligned with the matchmark on the engine. Make sure the distributor is fully seated and the distributor shaft is fully engaged.

NOTE: Make sure the rotor is pointing to where the No. 1 runner originates inside the cap, if equipped, and not where the No. 1 ignition wire plugs into the cap.

4. Install the hold-down nut.
5. Connect the distributor harness connectors.
6. Make sure the sealing O-ring is in place, install the distributor cap and tighten the screws or secure the clips.
7. Connect the negative battery cable.
8. Adjust the ignition timing and tighten the hold-down bolt.

Distributorless Ignition

REMOVAL & INSTALLATION

Crank Angle Sensor

1. Disconnect the negative battery cable.
2. Disconnect the sensor harness connector.
3. Unscrew the cap hold-down screws and lift off the cap.
4. Matchmark the coupling to the sensor housing and the housing to the engine.

NOTE: Do not crank the engine during this procedure. If the engine is cranked, the matchmark must be disregarded.

5. Remove the hold-down nut.
6. Carefully remove the crank angle sensor assembly from the engine.
To install:
7. If the timing is not disturbed, perform the following procedures:
   a. Install a new housing O-ring and lubricate with clean oil.
   b. Install the assembly in the engine so the coupling is aligned with the matchmark on the housing and the housing is aligned with the matchmark on the engine. Make sure the sensor assembly is fully seated and the shaft is fully engaged.
   c. Install the hold-down nut.
   d. Connect the harness connector.
   e. Make sure the sealing O-ring is in place, install the cap and tighten the screws.
   f. Connect the negative battery cable.

g. Adjust the ignition timing and tighten the hold-down nut.
8. If the timing is disturbed, perform the following procedures:
   a. Install a new housing O-ring and lubricate with clean oil.
   b. Position the engine so the No. 1 piston is at TDC of its compression stroke and the mark on the vibration damper is aligned with **0** on the timing indicator.
   c. Install the sensor in the engine so the factory matchmark on the coupling (notch) is aligned with the matchmark on the housing (punch mark) and the housing is aligned

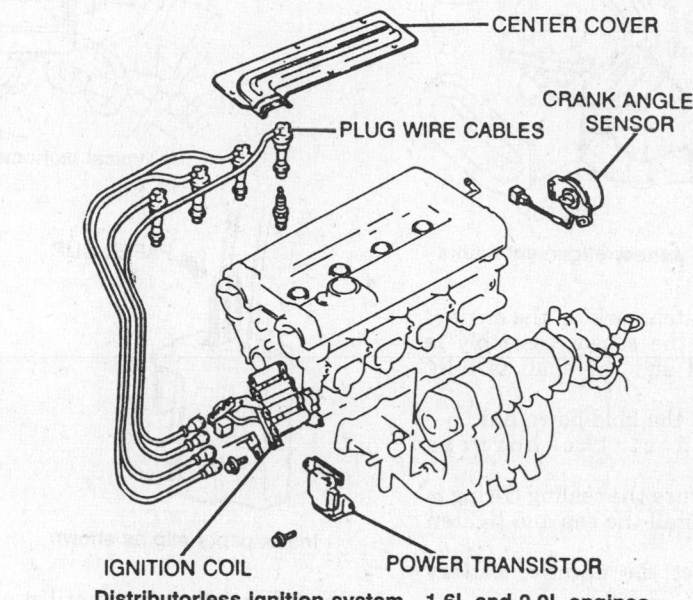

Distributorless Ignition system—1.6L and 2.0L engines

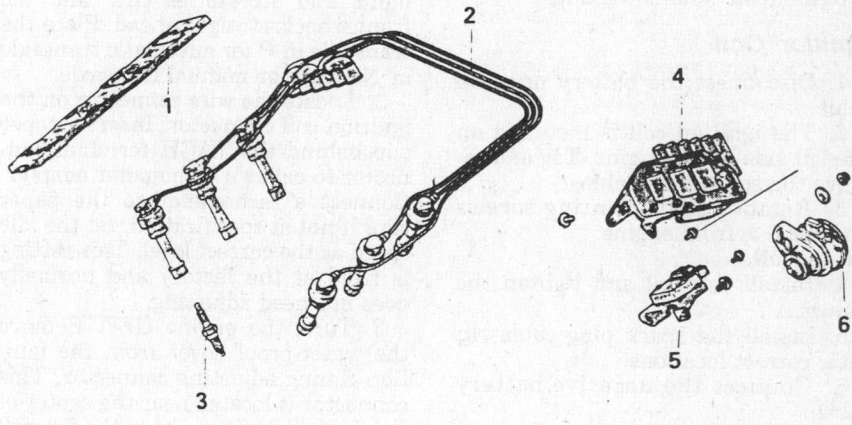

1. Center cover
2. Spark plug cable
3. Spark plug
4. Ignition coil
5. Power transistor
6. Crank angle sensor

Distributorless Ignition system—3.0L DOHC engine

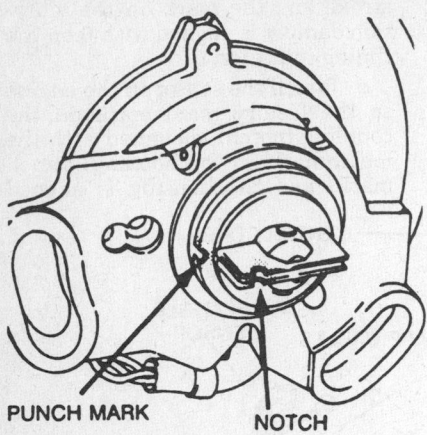

PUNCH MARK   NOTCH

Crank angle sensor alignment marks

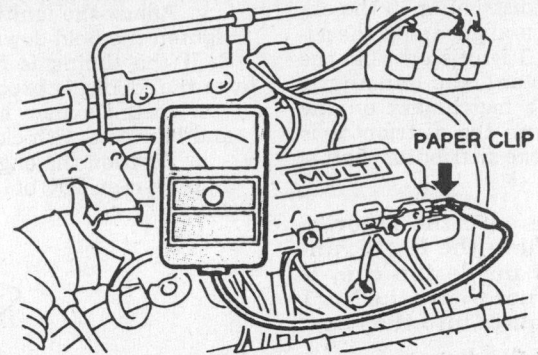

PAPER CLIP

Typical tachometer connector location

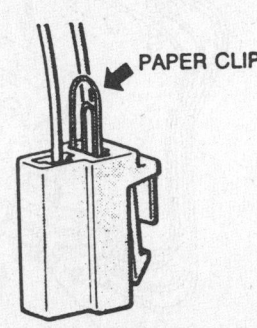

PAPER CLIP

Insert paper clip as shown

with the matchmark on the engine. Make sure the sensor assembly is fully seated and the shaft is fully engaged.

d. Install the hold-down nut.

e. Connect the harness connector.

f. Make sure the sealing O-ring is in place, install the cap and tighten the screws.

g. Connect the negative battery cable.

h. Adjust the ignition timing and tighten the hold-down nut.

Ignition Coil

1. Disconnect the battery negative cable.

2. The ignition coil is mounted on the left side of the engine. Tag and remove the spark plug cables.

3. Remove the mounting screws and remove from engine.

To install:

4. Install the coil and tighten the screws.

5. Install the spark plug cables in their correct locations.

6. Connect the negative battery cable.

Power Transistor

1. Disconnect the battery negative cable.

2. The power transistor is mounted on the left side of the engine near the coil. Remove and retaining screw and disconnect the wires to remove.

To install:

3. Install the power transistor and tighten the screws.

4. Connect the negative battery cable.

Ignition Timing

ADJUSTMENT

1. Start the engine, set the parking brake and run the engine until at normal operating temperature. Keep all lights and accessories OFF and the front wheels straight ahead. Place the transaxle in **P** for automatic transaxle or Neutral for manual transaxle.

2. Locate the wire connector on the ignition coil connector. Insert a paper clip behind the TACH terminal connector to act as a tachometer adapter. Connect a tachometer to the paper clip. If not at specification, set the idle speed at the correct level. This setting is fixed at the factory and normally does not need adjusting.

3. Turn the engine OFF. Remove the water-proof cover from the ignition timing adjusting connector. This connector is located near the center of the firewall on Summit, on the firewall just behind the battery on Laser, Talon and Stealth. Connect a jumper wire from this terminal to a good ground.

4. Connect a conventional power timing light to the No. 1 cylinder spark plug wire. Start the engine and run at idle.

5. Aim the timing light at the timing scale located near the crankshaft pulley.

6. Loosen the distributor or crank angle sensor hold-down nut just enough so the housing can be rotated.

7. Turn the housing in the proper direction until the specified timing is reached. Tighten the hold-down nut and recheck the timing. Turn the engine OFF.

8. Remove the jumper wire from the ignition timing adjusting terminal and install the water-proof cover.

9. Start the engine and check the actual timing (the timing without the terminal grounded). This reading should be 5 degrees more than the basic timing. This value may increase according to altitude. As long as the basic timing is correct, the engine is timed correctly. Also, actual timing may fluctuate because of slight variation accomplished by the ECU. The basic timing, though, should remain steady.

10. Turn the engine OFF and disconnect the timing apparatus and tachometer.

Alternator

PRECAUTIONS

Several precautions must be observed with alternator-equipped vehicles to avoid damage to the unit.

• If the battery is removed for any reason, make sure it is reconnected with the correct polarity. Reversing the battery connections may result in damage to the 1-way rectifiers.

• When utilizing a booster battery as a starting aid, always connect the positive to positive terminals and the negative terminal from the booster battery to a good engine ground on the vehicle being started.

• Never use a fast charger as a booster to start vehicles.

• Disconnect the battery cables when charging the battery with a fast charger.

• Never attempt to polarize the alternator.

• Do not use test lamps of more than 12 volts when checking diode continuity.

• Do not short across or ground any of the alternator terminals.

• The polarity of the battery, alternator and regulator must be matched and considered before making any electrical connections within the system.

• Never separate the alternator on

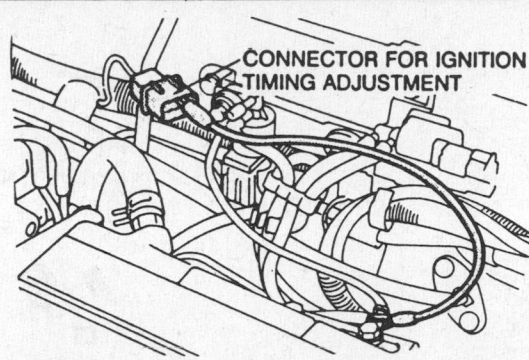

Ignition timing adjustment connector—1.5L engine

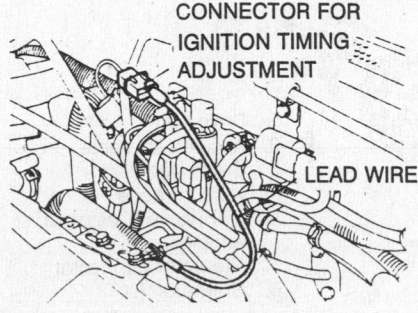

Ignition timing adjustment connector—1.6L engine

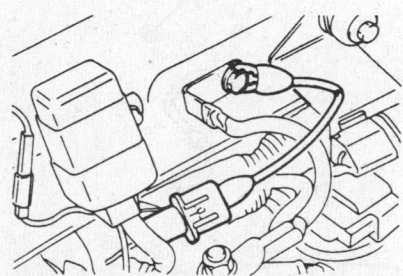

Ignition timing adjustment connector—Laser and Talon

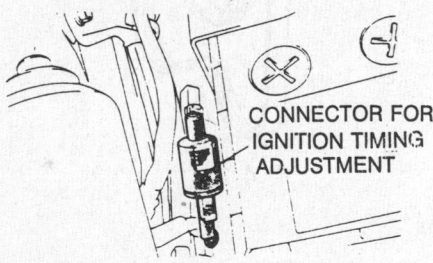

Ignition timing adjustment connector—Stealth

an open circuit. Make sure all connections within the circuit are clean and tight.

● Disconnect the battery ground terminal when performing any service on electrical components.

● Disconnect the battery if arc welding is to be done on the vehicle.

BELT TENSION ADJUSTMENT

1. Place a straight-edge along the belt between 2 pulleys.
2. Measure the deflection with a force of about 22 lbs. applied midway between the 2 pulleys. Deflection should be:

   a. Summit: 0.217–0.354 in. (5.5–9.0mm)

   b. Laser and Talon with 1.8L engine: 0.315–0.433 in. (8.0–11.0mm)

   c. Laser and Talon with 2.0L engine: 0.354–0.453 in. (9.0–11.5mm)

   d. Stealth with 3.0L SOHC engine: 0.236–0.354 in. (6.0–9.0mm)

   e. Stealth with 3.0L DOHC engine: 0.157–0.216 in. (4.0–5.5mm)

3. Belt tension can also be checked with a tension gauge. The value should be 55–110 lbs. (250–500 N).

4. Loosen the adjusting bolt or fixing bolt locknut on the alternator, alternator bracket or tension pulley. Then move the alternator or turn the adjusting bolt to adjust belt tension. Tighten the bolt or locknut when finished.

REMOVAL & INSTALLATION

Summit

1.5L ENGINE

1. Disconnect the negative battery cable.
2. Remove the left side cover panel under the vehicle.
3. Remove the drive belts.
4. Remove the water pump pulleys.
5. Remove the alternator upper bracket/brace.
6. Disconnect the alternator electrical connectors and remove the alternator.
7. The installation is the reverse of the removal procedure.
8. Connect the negative battery cable and check for proper operation.

1.6L ENGINE

1. Disconnect the negative battery cable.

SOHC ENGINE

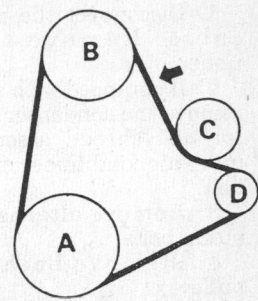

DOHC ENGINE WITHOUT AIR CONDITIONING

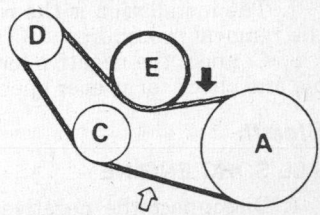

DOHC ENGINE WITH AIR CONDITIONING

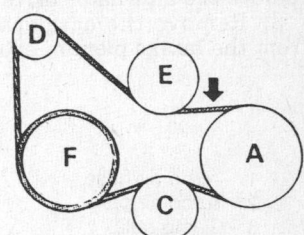

  A. Crankshaft pulley
  B. Power steering pulley
  C. Tensioner pulley
  D. Alternator pulley
  E. Idler pulley
  F. Air conditioner compressor pulley

Serpentine belt layout—Stealth

2. Remove the left side cover panel under the vehicle.
3. Remove the alternator and power steering drive belts and both water pump pulleys.
4. Remove the alternator adjuster brace.
5. Disconnect the alternator electrical connection.
6. Remove the the battery, windshield washer tank and battery tray.
7. Remove the attaching bolts at the top of the radiator and lift up the radiator. Do not disconnect the radiator hoses.
8. Remove the alternator.
9. The installation is the reverse of the removal procedure.
10. Connect the negative battery cable and check for proper operation.

Laser and Talon

1. Disconnect the negative battery cable. Remove the left side undercover.
2. If equipped with air conditioning, remove the condenser electric fan motor and shroud assembly. Then remove air conditioner compressor drive belt.
3. Remove alternator and water pump belts.
4. Remove both water pump pulleys.
5. Remove the alternator top brace.
6. Disconnect the alternator wiring and remove the alternator.
7. The installation is the reverse of the removal procedure.
8. Connect the negative battery cable and check for proper operation.

Stealth

3.0L SOHC ENGINE

1. Disconnect the negative battery cable.
2. Loosen the tensioner pulley and remove the alternator drive belt.
3. Remove the accelerator cable from the intake plenum extension.

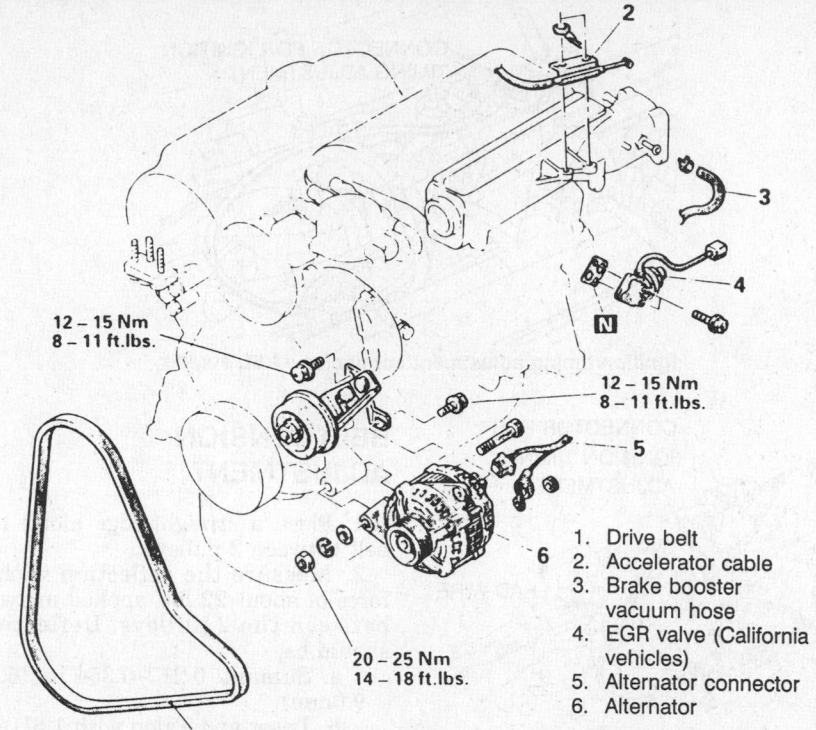

2
3
4
N
12 – 15 Nm
8 – 11 ft.lbs.
12 – 15 Nm
8 – 11 ft.lbs.
5
6
20 – 25 Nm
14 – 18 ft.lbs.
1
1. Drive belt
2. Accelerator cable
3. Brake booster vacuum hose
4. EGR valve (California vehicles)
5. Alternator connector
6. Alternator

Alternator and related parts – 3.0L SOHC engine

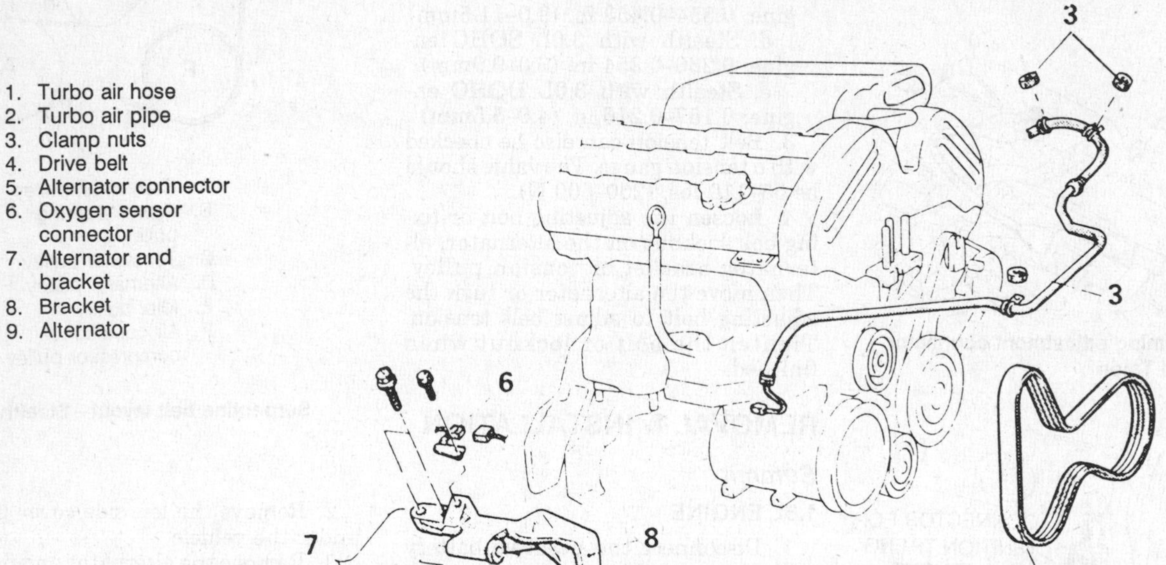

1. Turbo air hose
2. Turbo air pipe
3. Clamp nuts
4. Drive belt
5. Alternator connector
6. Oxygen sensor connector
7. Alternator and bracket
8. Bracket
9. Alternator
3
3
4
6
7
8
45 Nm
33 ft.lbs.
24 Nm
17 ft.lbs.
20 – 25 Nm
14 – 18 ft.lbs.
5
9
1
2

Alternator and related parts – 3.0L DOHC engine

4. Remove the brake booster vacuum hose.

5. If equipped with an EGR valve, remove it.

6. Disconnect the alternator connectors, remove the mounting bolts and remove the alternator.

7. The installation is the reverse of the removal procedure.

8. Connect the negative battery cable and check for proper operation.

3.0L DOHC ENGINE

1. Disconnect the negative battery cable. Remove the surge tank.

2. Remove the necessary air delivery hoses to gain access to the alternator.

3. Separate the air conditioning suction hose and suspend from the hood.

4. Loosen the tensioner pulley and remove the alternator drive belt.

5. Disconnect the oxygen sensor connector.

6. Disconnect the alternator wiring, remove the alternator bracket mounting bolts and remove the bracket and alternator as an assembly. Separate on a workbench.

7. The installation is the reverse of the removal procedure.

8. Connect the negative battery cable and check for proper operation.

Starter

REMOVAL & INSTALLATION

Summit

1. Disconnect the negative battery cable.

2. Disconnect the air-flow sensor assembly connector and remove the breather hose. Remove the resonator retaining nuts and remove the air intake hose and resonator assembly.

NOTE: Use care when removing the air cleaner cover because the air-flow sensor is attached and is a sensitive component.

3. Remove the heat shield from beneath the intake manifold on the 1.5L engine.

4. Disconnect the starter motor electrical connections.

5. Remove the starter motor mounting bolts and remove the starter.

6. The installation is the reverse of the removal procedure.

7. Connect the negative battery cable and check the starter for proper operation.

Laser and Talon

1. Remove the battery and battery tray.

2. Disconnect the speedometer cable on the transaxle end.

3. Remove the intake manifold brace on the 1.8L engine.

4. Disconnect the starter motor electrical connections.

5. Remove the starter motor mounting bolts and remove the starter.

6. The installation is the reverse of the removal procedure.

7. Connect the negative battery cable and check the starter for proper operation.

Stealth

1. Disconnect the negative battery cable.

2. Raise the vehicle and support safely.

3. Remove the engine undercovers.

4. Disconnect the wiring from the starter.

5. Remove the mounting bolts and remove the starter.

6. The installation is the reverse of the removal procedure.

7. Connect the negative battery cable and check the starter for proper operation.

FUEL SYSTEM

Fuel System Service Precaution

Safety is the most important factor when performing any type of maintenance, especially fuel system maintenance. Failure to conduct maintenance and repairs in a safe manner may result in serious personal injury or death. Maintenance and testing of the vehicle's fuel system components can be accomplished safely and effectively by adhering to the following rules and guidelines.

● To avoid the possibility of fire and personal injury, always disconnect the negative battery cable unless the repair or test procedure requires that battery voltage be applied.

● Always relieve the fuel system pressure prior to disconnecting any fuel system component (injector, fuel rail, pressure regulator, etc.), fitting or fuel line connection. Exercise extreme caution whenever relieving fuel system pressure to avoid exposing skin, face and eyes to fuel spray. Please be advised that fuel under pressure may penetrate the skin or any part of the body that it contacts.

● Always place a shop towel or cloth around the fitting or connection prior to loosening to absorb any excess fuel due to spillage. Ensure that all fuel spillage (should it occur) is quickly removed from engine surfaces. Ensure that all fuel soaked cloths or towels are deposited into a suitable waste container.

● Always keep a dry chemical (Class B) fire extinguisher near the work area.

● Do not allow fuel spray or fuel vapors to come into contact with a spark or open flame.

● Always use a backup wrench when loosening and tightening fuel line connection fittings. This will prevent unnecessary stress and torsion to fuel line piping. Always follow the proper torque specifications.

● Always replace worn fuel fitting O-rings with new. Do not substitute fuel hose or equivalent where fuel pipe is installed.

RELIEVING FUEL SYSTEM PRESSURE

1. Loosen the fuel filler cap to release fuel tank pressure.

2. Disconnect the fuel pump harness connector:

   a. Summit—remove the rear seat cushion to gain access to the connector.

   b. Laser and Talon—located at the rear of the fuel tank.

   c. Stealth—remove the fuel system access cover in the luggage compartment.

3. Start the vehicle and allow it to run until it stalls from lack of fuel. Turn the key to the **OFF** position.

4. Disconnect the negative battery cable, then reconnect the fuel pump connector.

5. Wrap shop towels around the fitting that is being disconnected to absorb residual fuel in the lines.

Fuel Tank

REMOVAL & INSTALLATION

1. Relieve fuel system pressure.

2. Disconnect the negative battery cable.

3. Raise the vehicle and support safely.

4. Drain the fuel from the fuel tank into an approved container.

5. Disconnect the return hose, high pressure hose and all other hoses and connectors connected to the pump/sending unit.

6. Disconnect the filler and vent hoses. Place a suitable support under the tank and remove the retaining nuts.

7. Lower the tank from the vehicle.

To install:

8. Install the fuel tank and all related items to the vehicle. Secure all tank retaining nuts.

9. Connect the negative battery cable and check the entire system for proper operation and leaks.

Fuel Filter

REMOVAL & INSTALLATION

1. Relieve the fuel pressure.

2. Disconnect the negative battery cable.

3. The filter is located in the engine compartment, mounted either on the firewall or inner fender panel.

4. Where applicable, remove the air cleaner assembly, battery and tray.

5. Hold the fuel filter nut securely with a backup wrench. Cover the hoses with shop towels and remove the eye bolt. Discard the gaskets.

6. On Stealth, the high pressure hose connection is accomplished with another eye bolt connection; first separate the flare nut connection at the line, then repeat Step 5. Otherwise, separate the flare nut connection at the filter. Discard the gaskets.

7. Remove the mounting bolts and remove the fuel hose from the vehicle.

To install:

8. If equipped with the flare fitting, install a new O-ring and tighten the fitting by hand before installing the filter to the vehicle.

9. Install the filter only finger-tight to its bracket temporarily.

10. Install new O-rings and connect the high pressure hose and eye bolt, then the main pipe and eye bolt. Tighten the eye bolts to 22 ft. lbs. (30 Nm).

11. Tighten the mounting bolts fully.

12. Install the air cleaner assembly, battery and tray, if removed.

13. Connect the negative battery cable, install the fuel filler cap, turn the key to the **ON** position to pressurize the fuel system and check for leaks.

Electric Fuel Pump

PRESSURE TESTING

1. Relieve fuel system pressure. Disconnect the battery negative cable.

2. Disconnect the fuel high pressure hose at the delivery pipe side.

3. Connect a fuel pressure gauge to tools MD998709 and MD998742 or exact equivalent, with appropriate adaptors, seals and/or gaskets to prevent leaks during the test. Install the gauge and adapter between the delivery pipe and high pressure hose.

4. Connect the negative battery cable.

5. Apply battery voltage to the terminal for fuel pump activation located in the engine compartment, to run the fuel pump and check for leaks.

6. Start the engine and run at curb idle speed.

7. Measure the fuel pressure.

8. Locate and disconnect the vacuum hose running to the fuel pressure regulator. Plug the end of the hose and measure the fuel pressure again. It should have increased about 10 psi.

9. Race the engine 2–3 times and check that the fuel pressure does not fall when the engine is running at idle.

10. Check to be sure there is fuel

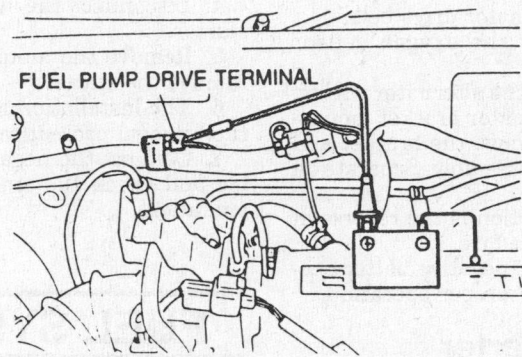

Fuel pump test terminal—Summit

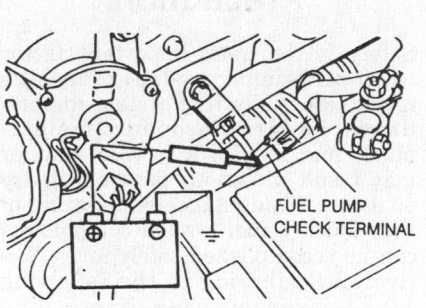

Fuel pump test terminal—Laser and Talon

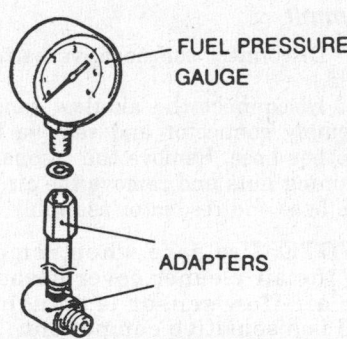

FUEL PRESSURE GAUGE

ADAPTERS

38 PSI AT IDLE

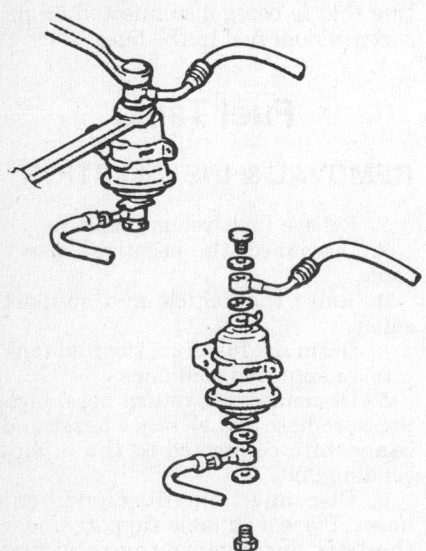

Fuel filter and hoses

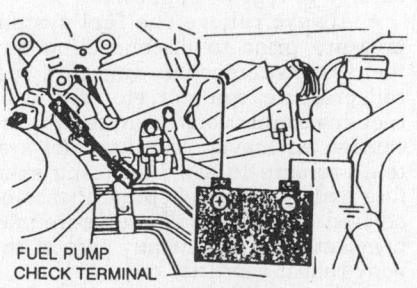

Fuel pump test terminal—Stealth

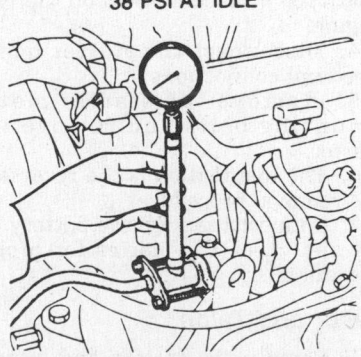

Fuel system pressure testing

1. Fuel pump
2. In-tank filter
3. Fuel gauge tank unit
4. Two-way valve
5. Vapor hose
6. Check valve
7. High pressure hose
8. Return hose
9. Tank drain plug

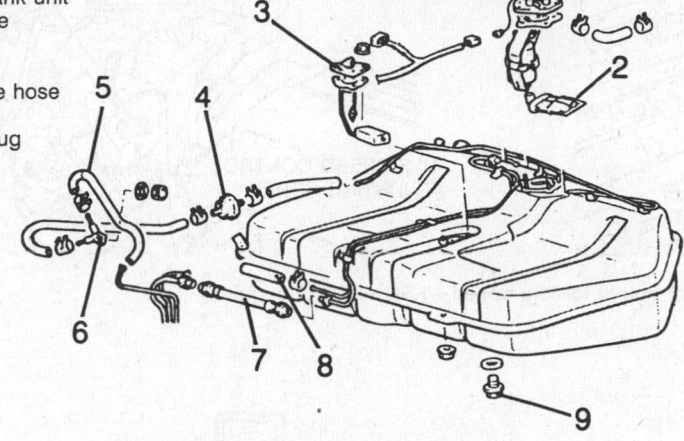

Fuel tank assembly and related parts

pressure in the return hose by gently pressing the fuel return hose with fingers while racing the engine. There will be no fuel pressure in the return hose when the volume of fuel flow is low.

11. If fuel pressure is too low, check for a clogged fuel filter, a defective fuel pressure regulator or a defective fuel pump, any of which will require replacement. If fuel pressure is too high, the fuel pressure regulator is defective and will have to be replaced or the fuel return is bent or clogged. If the fuel pressure reading does not change when the vacuum hose is disconnected, the hose is clogged or the valve is stuck in the fuel pressure regulator and it will have to be replaced.

12. Stop the engine and check for changes in the fuel pressure gauge. It should not drop. If the gauge reading does drop, watch the rate of drop. If fuel pressure drops slowly, the likely cause is a leaking injector which will require replacement. If the fuel pressure drops immediately after the engine is stopped, the check valve in the fuel pump isn't closing and the fuel pump will have to be replaced.

13. Relieve fuel system pressure.

14. Disconnect the high pressure hose and remove the fuel pressure gauge from the delivery pipe.

15. Install a new O-ring in the groove of the high pressure hose. Connect the hose to the delivery pipe and tighten the screws. After installation, apply battery voltage to the terminal for fuel pump activation to run the fuel pump. Check for leaks.

REMOVAL & INSTALLATION

1. Relieve fuel system pressure.
2. Disconnect the negative battery cable.

3. Raise the vehicle and support safely.

4. Drain the fuel from the fuel tank.

5. Disconnect the return hose, high pressure hose and all other hoses and connectors connected to the pump/sending unit.

6. Disconnect the filler and vent hoses. Place a suitable support under the tank and remove the retaining nuts. Lower the tank from the vehicle.

7. Remove the retaining nuts and

remove the fuel pump/sending unit assembly from the tank.

To install:

8. Install the replacement pump using a new gasket. Be certain the pump is installed in the same location, facing the same direction as before.

9. Install the fuel tank and all related items to the vehicle. Secure all tank retaining nuts.

10. Connect the negative battery cable and check the entire system for proper operation and leaks.

Fuel Injection

IDLE SPEED ADJUSTMENT

NOTE: The idle speed is controlled electronically and adjustment is usually unnecessary. However, the idle speed may be checked using the following procedures.

1.5L and 1.8L Engines

1. Warm the engine to operating temperature, leave lights, electric cooling fan and accessories **OFF**. The transaxle should be in **N** or **P** for automatic transaxle. The steering wheel in a neutral position for vehicles with power steering.

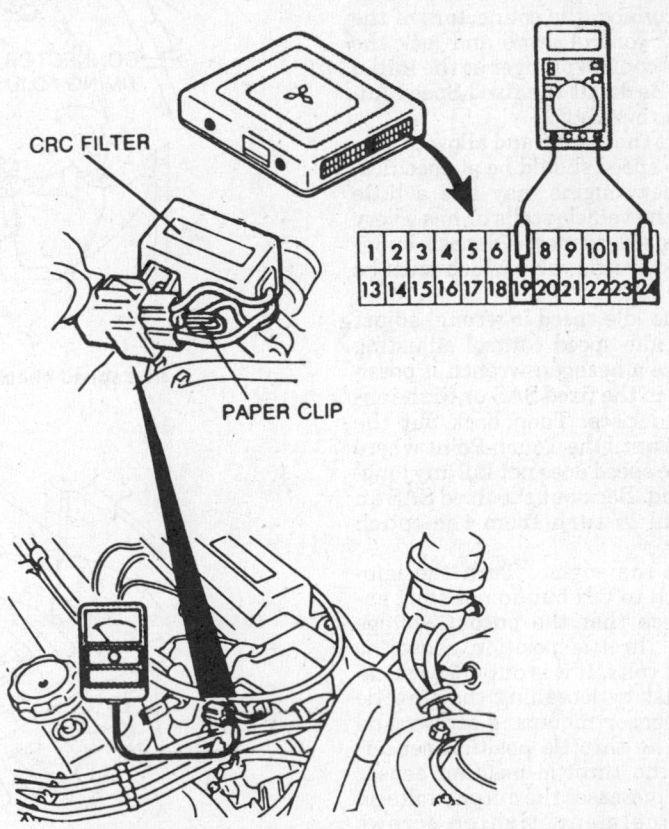

Idle speed check connections—1.5L and 1.8L engines

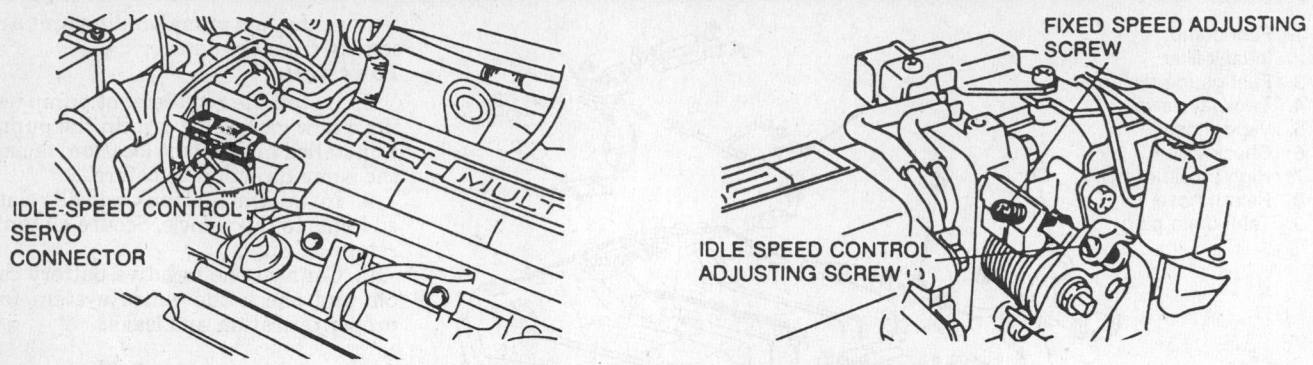

Adjustment points for idle speed—1.5L and 1.8L engines

2. Check the ignition timing and adjust, if necessary.

3. Connect a tachometer to the CRC filter connector. Use a paper clip for a tach adapter.

4. Run the engine for more than 5 seconds at 2000–3000 rpm. Allow the engine to idle for 2 minutes. Check the idle rpm. Curb idle should be 700 ± 100 rpm.

5. If adjustment is required, slacken the accelerator cable.

6. Connect a digital voltmeter between terminal **19** throttle position sensor output voltage) of the engine control unit and terminal **24** (ground).

7. Set the ignition switch to **ON**, without starting the engine, and hold it in that position for 15 seconds or more. Turn the ignition switch **OFF**.

8. Disconnect the connectors of the idle speed control servo and lock the idle speed control plunger at the initial position. Back out the fixed Speed Adjusting Screw (SAS).

9. Start the engine and allow to idle. Basic idle speed should be at specification. A new engine may idle a little lower. If the vehicle stalls or has a very low idle speed, suspect a deposit build-up on the throttle valve which must be cleaned.

10. If the idle speed is wrong, adjust with the idle speed control adjusting screw. Use a hexagon wrench if possible. Turn in the fixed SAS until the engine speed rises. Then back out the fixed SAS until the Touch Point where the engine speed does not fall any longer, is found. Back out the fixed SAS an additional ½ turn from the touch point.

11. Stop the engine. Turn the ignition switch to **ON** but do not start engine. Check that the output voltage from the throttle position sensor is 0.48–0.52 volts. If it is out of specification, adjust by loosening the throttle position sensor mounting screws and rotating the throttle position sensor. Turning the throttle position sensor clockwise increases the output voltage. After adjustment, tighten screws firmly.

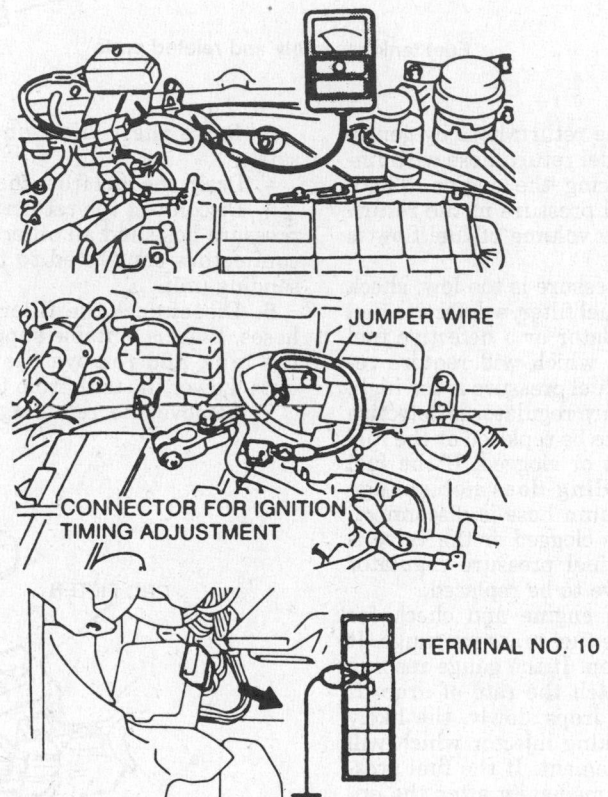

Idle speed check connections—1.6L and 2.0L engines

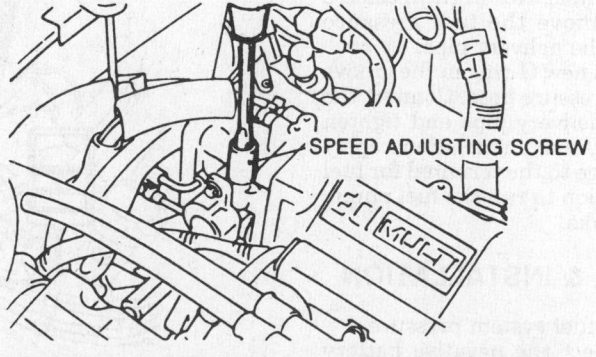

Adjustment point for idle speed—1.6L and 2.0L engines

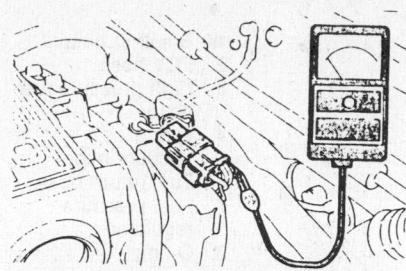

Idle speed check connector—3.0L SOHC engine

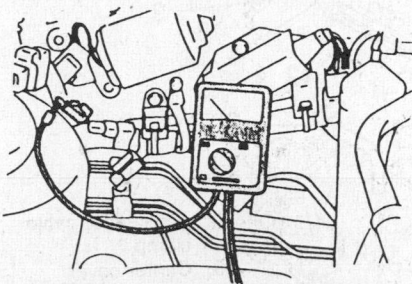

Idle speed check connector—3.0L DOHC engine

12. Turn the ignition switch **OFF**.

13. Adjust the free-play of the accelerator cable, reconnect the connectors of the idle speed control servo and remove the voltmeter.

14. Start the engine and check the curb idle. It should be 700 ± 100 rpm.

15. Turn the ignition switch to **OFF**, disconnect the negative battery cable for more than 10 seconds and reconnect. This clears any trouble codes introduced during testing.

16. Restart the engine, allow to run for 5 minutes and check for good idle quality.

1.6L, 2.0L and 3.0L Engines

1. Warm the engine to operating temperature, leave lights, electric cooling fan and accessories **OFF**. The transaxle should be in **N**. The steering wheel in a neutral position for vehicles with power steering.

2. Check the ignition timing and adjust, if necessary.

3. Connect a tachometer to the special terminal under the hood.

4. Run the engine for more than 5 seconds at 2000–3000 rpm. Allow the engine to idle for 2 minutes. Check the idle rpm. Curb idle should be 750 ± 100 rpm.

5. If adjustment is required, disconnect the waterproof female connector used for ignition timing adjustment. Connect this terminal to ground using a jumper wire.

6. Locate the self-diagnosis terminal under the dashboard and connect terminal No. **10** to ground with a jumper wire.

7. Start the engine and allow to idle. Check that the basic idle speed is at specification. On Stealth, the tachometer reading will be ⅓ of the actual engine speed. Multiply the reading by 3 to figure the actual engine speed. If the idle speed deviates from this speed, check the following:

a. A new engine will idle more slowly. Break-in should take approximately 300 miles.

b. If the vehicle stalls or has a very low idle speed, suspect a deposit buildup on the throttle valve which must be cleaned.

c. If the idle speed is high even though the speed adjusting screw is fully closed, check that the idle position switch (fixed speed adjusting screw) position has changed. If so, adjust the idle position switch.

d. If after all these checks the idle is still out of specification, it is probable that there is leakage resulting from deterioration of the Fast-Idle Air Valve (FIAV) and the throttle body will need to be replaced.

8. Turn the ignition switch **OFF** and stop the engine. Disconnect the jumper wire from the diagnosis connector, disconnect the jumper wire from from the ignition timing connector and reconnect the waterproof connector. Disconnect the tachometer.

9. Restart the engine, allow to run for 5 minutes and check for good idle quality.

Fuel Injector

REMOVAL & INSTALLATION

1.5L and 1.8L Engines

1. Relieve the fuel system pressure.

2. Disconnect the negative battery cable.

3. Wrap the connection with a shop towel and disconnect the high pressure fuel line at the fuel rail.

4. Disconnect the fuel return hose.

5. Disconnect the connectors from each connector.

6. Remove the injector rail retaining bolts. Make sure the rubber mounting bushings do not get lost.

7. Lift the rail assembly up and away from the engine.

8. Remove the injectors from the rail by pulling gently. Discard the lower insulator. Check the resistance through the injector. The specification is 13–15 ohms at 70°F (20°C).

To install:

9. Install a new grommet and O-ring to the injector. Coat the O-ring with light oil.

10. Install the injector to the fuel rail.

11. Replace the seats in the intake

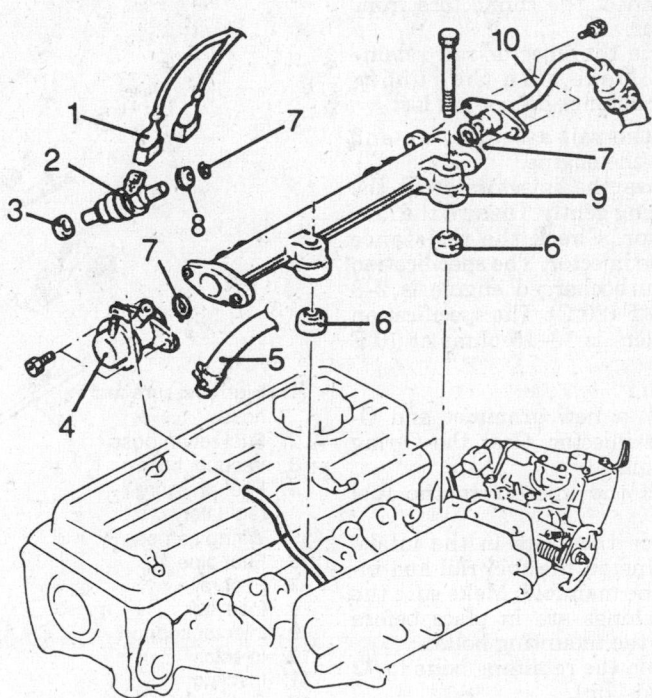

1. Injector Harness	6. Insulator
2. Injector	7. O-ring
3. Insulator	8. Injector grommet
4. Pressure regulator	9. Delivery pipe
5. Return hose	10. Pressure line

Fuel rail, injectors and related parts—1.5L and 1.8L engines

manifold. Install the fuel rail and injectors to the manifold. Make sure the rubber bushings are in place before tightening the mounting bolts.

12. Tighten the retaining bolts to 72 inch lbs. (11 Nm).

13. Connect the connectors to the injectors.

14. Replace the O-ring, lightly lubricate it and connect the fuel pressure regulator.

15. Connect the fuel return hose.

16. Replace the O-ring, lightly lubricate it and connect the high pressure fuel line.

17. Connect the negative battery cable and check the entire system for proper operation and leaks.

1.6L and 2.0L DOHC Engines

1. Relieve the fuel system pressure.

2. Disconnect the negative battery cable.

3. Wrap the connection with a shop towel and disconnect the high pressure fuel line at the fuel rail.

4. Disconnect the fuel return hose and remove the O-ring.

5. Disconnect the vacuum hose from the fuel pressure regulator. Remove the fuel pressure regulator and O-ring.

6. Disconnect the PCV hose. On Laser and Talon, remove the center cover.

7. Disconnect the connectors from each injector.

8. Remove the injector rail retaining bolts. Make sure the rubber mounting bushings do not get lost.

9. Lift the rail assembly up and away from the engine.

10. Remove the injectors from the rail by pulling gently. Discard the lower insulator. Check the resistance through the injector. The specification for 2.0L turbocharged engine is 2–3 ohms at 70°F (20°C). The specification for the others is 13–15 ohms at 70°F (20°C).

To install:

11. Install a new grommet and O-ring to the injector. Coat the O-ring with light oil.

12. Install the injector to the fuel rail.

13. Replace the seats in the intake manifold. Install the fuel rail and injectors to the manifold. Make sure the rubber bushings are in place before tightening the mounting bolts.

14. Tighten the retaining bolts to 72 inch lbs. (11 Nm).

15. Connect the connectors to the injectors and install the center cover. Connect the PCV hose.

16. Replace the O-ring, lightly lubricate it and connect the fuel pressure regulator.

17. Connect the fuel return hose.

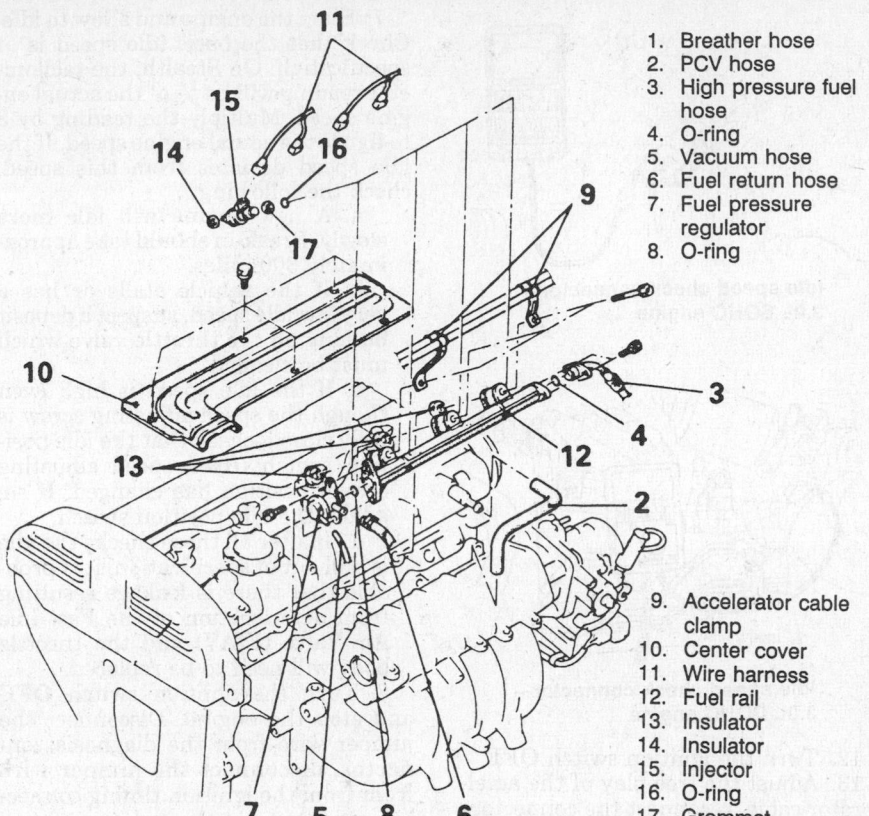

1. Breather hose
2. PCV hose
3. High pressure fuel hose
4. O-ring
5. Vacuum hose
6. Fuel return hose
7. Fuel pressure regulator
8. O-ring
9. Accelerator cable clamp
10. Center cover
11. Wire harness
12. Fuel rail
13. Insulator
14. Insulator
15. Injector
16. O-ring
17. Grommet

Fuel rail, injectors and related parts—1.6L and 2.0L engines

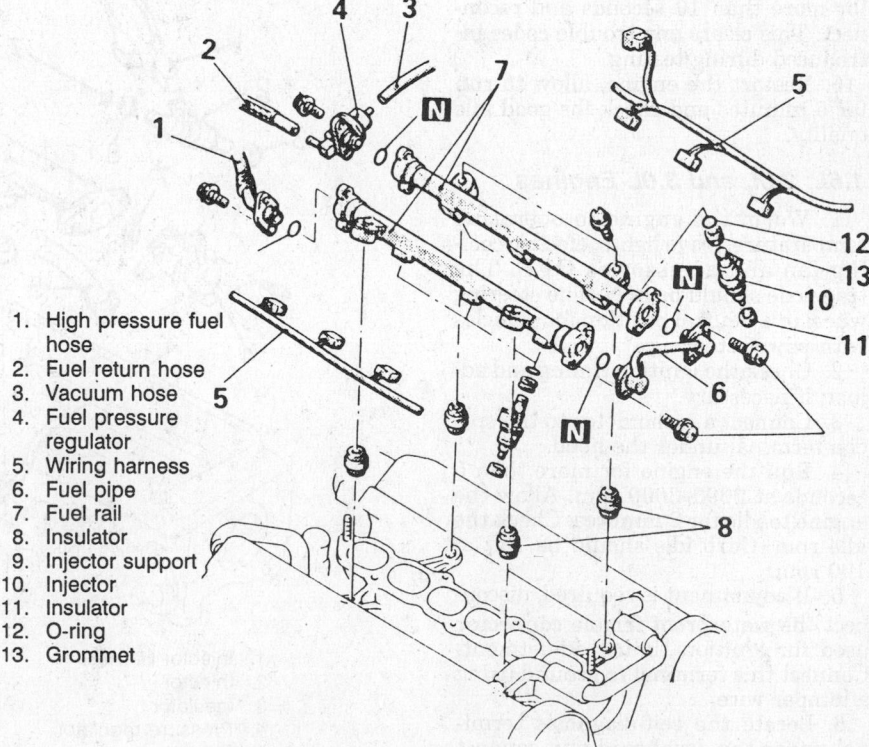

1. High pressure fuel hose
2. Fuel return hose
3. Vacuum hose
4. Fuel pressure regulator
5. Wiring harness
6. Fuel pipe
7. Fuel rail
8. Insulator
9. Injector support
10. Injector
11. Insulator
12. O-ring
13. Grommet

Fuel rail, injectors and related parts—3.0L engine

18. Replace the O-ring, lightly lubricate it and connect the high pressure fuel line.

19. Connect the negative battery cable and check the entire system for proper operation and leaks.

3.0L Engine

1. Relieve the fuel system pressure.
2. Disconnect the negative battery cable.
3. Drain the coolant.
4. Disconnect all components from the air intake plenum and remove the plenum from the intake manifold. Discard the gaskets.
5. Wrap the connection with a shop towel and disconnect the high pressure fuel line at the fuel rail.
6. Disconnect the fuel return hose and remove the O-ring.
7. Disconnect the vacuum hose from the fuel pressure regulator. Remove the fuel pressure regulator and O-ring.
8. Disconnect the connectors from each injector.
9. Remove the fuel pipe connecting the fuel rails. Remove the injector rail retaining bolts. Make sure the rubber mounting bushings do not get lost.
10. Lift the rail assemblies up and away from the engine.
11. Remove the injectors from the rail by pulling gently. Discard the lower insulator. Check the resistance through the injector. The specification for 2.0L turbocharged engine is 2–3 ohms at 70°F (20°C). The specification for the others is 13–15 ohms at 70°F (20°C).

To install:

12. Install a new grommet and O-ring to the injector. Coat the O-ring with light oil.
13. Install the injector to the fuel rail.
14. Replace the seats in the intake manifold. Install the fuel rails and injectors to the manifold. Make sure the rubber bushings are in place before tightening the mounting bolts.
15. Tighten the retaining bolts to 72 inch lbs. (11 Nm). Install the fuel pipe with new gasket.
16. Connect the connectors to the injectors.
17. Replace the O-ring, lightly lubricate it and connect the fuel pressure regulator. Connect the fuel return hose.
18. Replace the O-ring, lightly lubricate it and connect the high pressure fuel line.
19. Using new gaskets, install the intake plenum and all related items. Torque the plenum mounting bolts to 13 ft. lbs. (18 Nm).
20. Fill the cooling system.
21. Connect the negative battery ca-

ble and check the entire system for proper operation and leaks.

DRIVE AXLE

Halfshaft

REMOVAL & INSTALLATION

NOTE: If the vehicle is going to be rolled while the halfshafts are out of the vehicle, obtain 2 outer CV-joints or proper equivalent tools and install to the hubs. If the vehicle is rolled without the proper torque applied to the front wheel bearings, the bearings will no longer be usable.

1. Disconnect the negative battery cable.
2. Remove the cotter pin, halfshaft nut and washer. It is recommended that the halfshaft nut is removed while the vehicle is on the floor with the brakes applied.
3. Raise the vehicle and support safely. Remove the lower ball joint and the tie rod end from the steering knuckle.
4. On vehicles with an inner shaft, remove the center support bearing bracket bolts and washers.
5. On vehicles with an inner shaft, remove the halfshaft by setting up a puller on the outside wheel hub and pushing the halfshaft from the front hub. Then tap the joint case with a plastic hammer to remove the halfshaft shaft and inner shaft from the transaxle.
6. On vehicles without an inner shaft, remove the halfshaft by setting up a puller on the outside wheel hub and pushing the halfshaft from the front hub. After pressing the outer shaft, insert a prybar between the transaxle case and the halfshaft and pry the shaft from the transaxle. Do not pull on the shaft; doing so damages the inboard joint. Do not insert the prybar too far or the oil seal in the case may be damaged.

To install:

7. Inspect the halfshaft boot for damage or deterioration. Check the ball joints and splines for wear.
8. Replace the circlips on the ends of the halfshafts.
9. Insert the halfshaft into the transaxle. Make sure it is fully seated.
10. Pull the strut assembly out and install the other end to the hub.
11. Install the center bearing bracket bolts and tighten to 33 ft. lbs. (45 Nm).
12. Install the washer so the cham-

fered edge faces outward. Install the nut and tighten temporarily.

13. Install the tie rod end and ball joint.
14. Install the wheel and lower the vehicle to the floor. Tighten the axle nut with the brakes applied. Tighten the nut to a maximum torque of 188 ft. lbs. (260 Nm) maximum. Install the cotter pin and bend it securely.

CV-Boot

The vehicles use several different types of joints. Engine size, transaxle type, whether the joint is an inboard or outboard joint, even which side of the vehicle is being serviced could make a difference in joint type. Be sure to properly identify the joint before attempting joint or boot replacement. Look for identification numbers at the large end of the boots and/or on the end of the metal retainer bands.

The 4 types of joints used are the Birfield Joint, (B.J.), the Tripod Joint (T.J.), the Double Offset Joint (D.O.J.) and the Rzeppa Joint (R.J.). In addition, some left side shafts will have a round dynamic damper installed on the shaft. Special grease is generally used with these joints and is often supplied with the replacement joint and/or boot. Do not use regular chassis grease.

In most cases, a specification is called out for the distance between the large and small boot bands. This is so the boot will not be installed either too loose or too tight, which could cause early wear and cracking, allowing the grease to get out and water and dirt in, leading to early joint failure.

REMOVAL & INSTALLATION

Except Double Offset Joint

Although joint types vary, the basic procedures are the same, with the exception of the Double Offset Joint. The following is a general procedure which should apply to most applications.

1. Disconnect the negative battery cable. Remove the halfshaft.
2. Remove the snapring next to the tripod joint spider from the halfshaft with snapring pliers and remove the spider from the shaft. Do not disassemble the spider and use care in handling.
3. Side cutter pliers can be used to cut the metal retaining bands.
4. If the boot is reused, wrap vinyl tape around the spline part of the shaft so the boot will not be damaged when removed. Remove the dynamic damper, if used, and boots from the shaft.

To install:

5. Double check that the correct re-

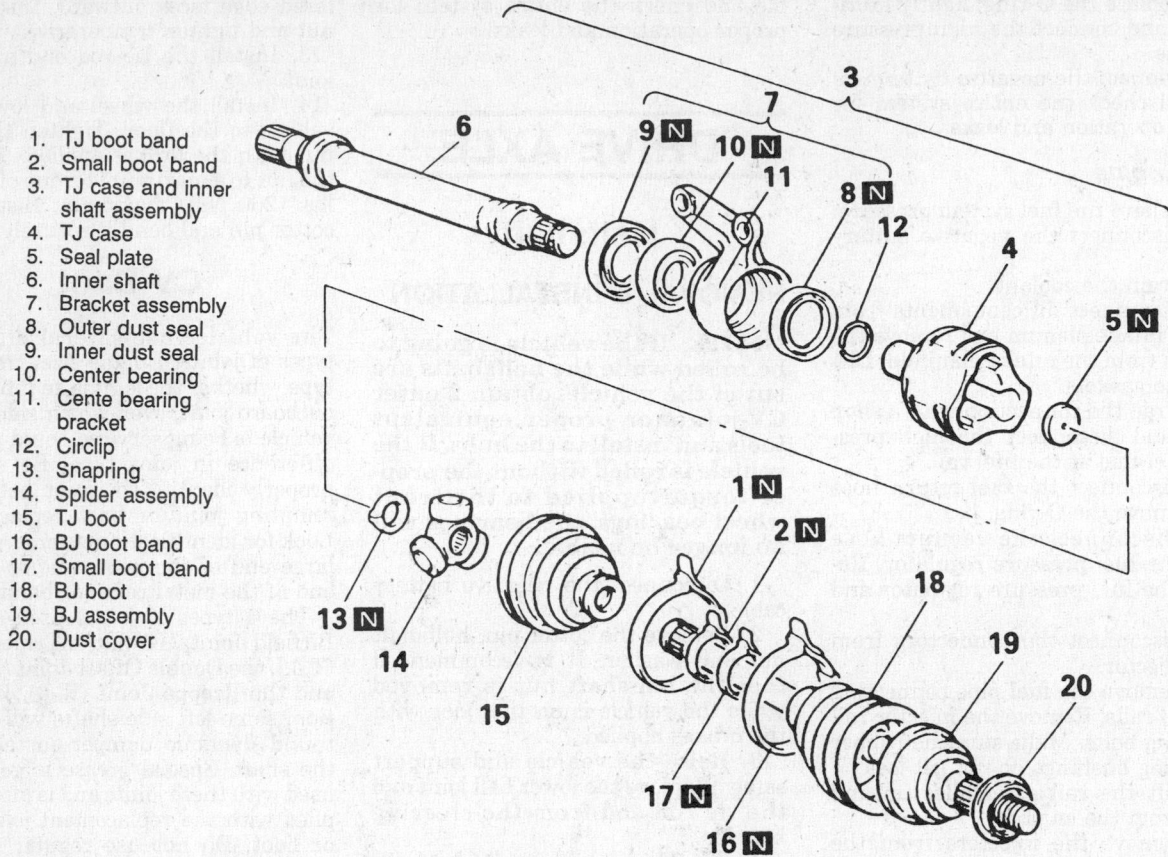

1. TJ boot band
2. Small boot band
3. TJ case and inner shaft assembly
4. TJ case
5. Seal plate
6. Inner shaft
7. Bracket assembly
8. Outer dust seal
9. Inner dust seal
10. Center bearing
11. Cente bearing bracket
12. Circlip
13. Snapring
14. Spider assembly
15. TJ boot
16. BJ boot band
17. Small boot band
18. BJ boot
19. BJ assembly
20. Dust cover

Representative halfshaft and CV-joint assemblies

CV-JOINT INSTALLED LENGTHS

DISTANCE A—SUMMIT

TRIPOD JOINTS

1.5L and 1.6L Engine w/AT,	LH Shaft—3.15 in. ±.12 in. (80mm ±3mm)
	RH Shaft—3.35 in. ±.12 in. (85mm ±3mm)
1.6L Engine w/MT,	—3.35 in. ±.12 in. (80mm ±3mm)

DOUBLE OFFSET JOINTS

1.6L Engine Non Turbo	—2.92 in. ±.12 in. (75mm ±3mm)
1.6L Engine Turbo	—3.15 in. ±.12 in. (80mm ±3mm)

DISTANCE A—LASER AND TALON

1.8L Engine up to 4-89	LH Shaft—3.15 in. ±.12 in. (80mm ±3mm)
1.8L Engine from 5-89	LH Shaft—2.95 in. ±.12 in. (75mm ±3mm)
1.8L Engine up to 4-89	RH Shaft—3.15 in. ±.12 in. (80mm ±3mm)
1.8L Engine from 5-89	RH Shaft—3.35 in. ±.12 in. (85mm ±3mm)
2.0L Engine up to 4-89	2WD-LH Shaft—2.95 in. ±.12 in. (75mm ±3mm)
2.0L Engine from 5-89	2WD-LH Shaft—3.15 in. ±.12 in. (80mm ±3mm)
2.0L Engine Turbo	2WD-LH Shaft—3.15 in. ±.12 in. (80mm ±3mm)
2.0L Engine Non Turbo	2WD-RH Shaft—3.15 in. ±.12 in. (80mm ±3mm)
2.0L Engine Turbo	2WD-RH Shaft—3.15 in. ±.12 in. (80mm ±3mm)
2.0L Engine Non Turbo	4WD-LH Shaft—3.35 in. ±.12 in. (85mm ±3mm)
2.0L Engine Non Turbo	4WD-RH Shaft—3.35 in. ±.12 in. (85mm ±3mm)

① Automatic transaxle
② Manual transaxle

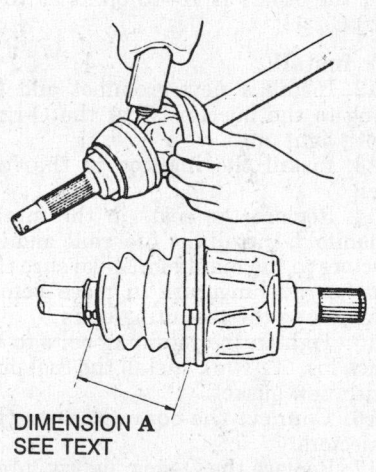

DIMENSION **A**
SEE TEXT

**CV-joint and boot assembly
with distance A indicated**

placement parts are being installed. Wrap vinyl tape around the splines to protect the boot and install the boots and damper, if used, in the correct order.

6. Fill the inside of the boot with the specified grease. Often the grease supplied in the replacement parts kit is meant to be divided in half, with half

being used to lubricate the joint and half being used inside the boot. Keep grease off the rubber part of the dynamic damper (if used).

7. Secure the boot bands with the halfshaft in a horizontal position. Make sure Distance "A" is set properly according to the chart. T.J. joints on Stealth should have Distance "A" set to 3.35 in. ± 0.12 in (85mm ± 3mm).

8. Install the halfshaft.

Double Offset Joint

1. Remove the halfshaft. The Double Offset Joint (D.O.J.) is bigger than other joints and in these applications, is only used as an inboard joint.

2. Side cutter pliers can be used to cut the metal retaining bands.

3. Locate and remove the large circlip at the base of the joint. Remove the outer race (the body of the joint).

4. Matchmark the shaft, D.O.J. inner race and cage. Remove the joint balls and the small snapring from the shaft. With a brass drift pin, tap lightly and evenly around the inner race to remove the race and the inner cage from the shaft.

5. If the boot is to be reused, wipe the grease from the splines and wrap the splines in vinyl tape before sliding the boot from the shaft.

To install:

6. Be sure to tape the shaft splines before installing the boots. Fill the inside of the boot with the specified grease. Often the grease supplied in the replacement parts kit is meant to be divided in half, with half being used to lubricate the joint and half being used inside the boot.

7. Install the cage onto the halfshaft so the small diameter side of the cage is installed first. Align the matchmarks made at disassembly on the inner race and shaft. With a brass drift pin, tap lightly and evenly around the inner race to install the race until it comes into contact with the rib of the shaft. Apply the specified grease to the inner race and cage and fit them together aligning the matchmarks. Insert the balls into the cage.

8. Install the outer race (the body of the joint) after filling with the specified grease. The outer race should be filled with this grease.

9. Tighten the boot bands securely. Make sure Distance "A" is set properly according to the chart.

10. Install the halfshaft.

Driveshaft and U-Joints

REMOVAL & INSTALLATION

Laser, Talon and Stealth With AWD

1. Disconnect the negative battery cable. Raise the vehicle and support safely.

2. The rear driveshaft is a 3-piece unit, with a front, center and rear propeller shaft. Remove the nuts and insulators from the center support bearing. Work carefully. There will be a number of spacers which will differ from vehicle to vehicle. Check the number of spacers and write down their locations for reference during reassembly.

3. Matchmark the rear differential companion flange and the rear driveshaft flange yoke. Remove the companion shaft bolts and remove the driveshaft, keeping it as straight as possible so as to ensure that the boot is not damaged or pinched. Use care to keep from damaging the oil seal in the output housing of the transfer case.

NOTE: Damage to the boot can be avoided and work will be easier if a piece of cloth or similar material is inserted in the boot.

4. Do not lower the rear of the vehicle or oil will flow from the transfer case. Cover the opening to keep dirt out.

To install:

5. Install the driveshaft to the vehicle and align the matchmarks at the rear yoke. Install the bolts and torque to 22–25 ft. lbs. (30–35 Nm) on Laser and Talon or 36–43 ft. lbs. (50–60 Nm) on Stealth.

6. Install the center support bearing with all spacers in place. Torque the retaining nuts to 22–25 ft. lbs. (30–35 Nm).

7. Check the fluid levels in the transfer case and rear differential case.

Rear Axle Shaft, Bearing and Seal

REMOVAL & INSTALLATION

Laser, Talon and Stealth With AWD

1. Disconnect the negative battery

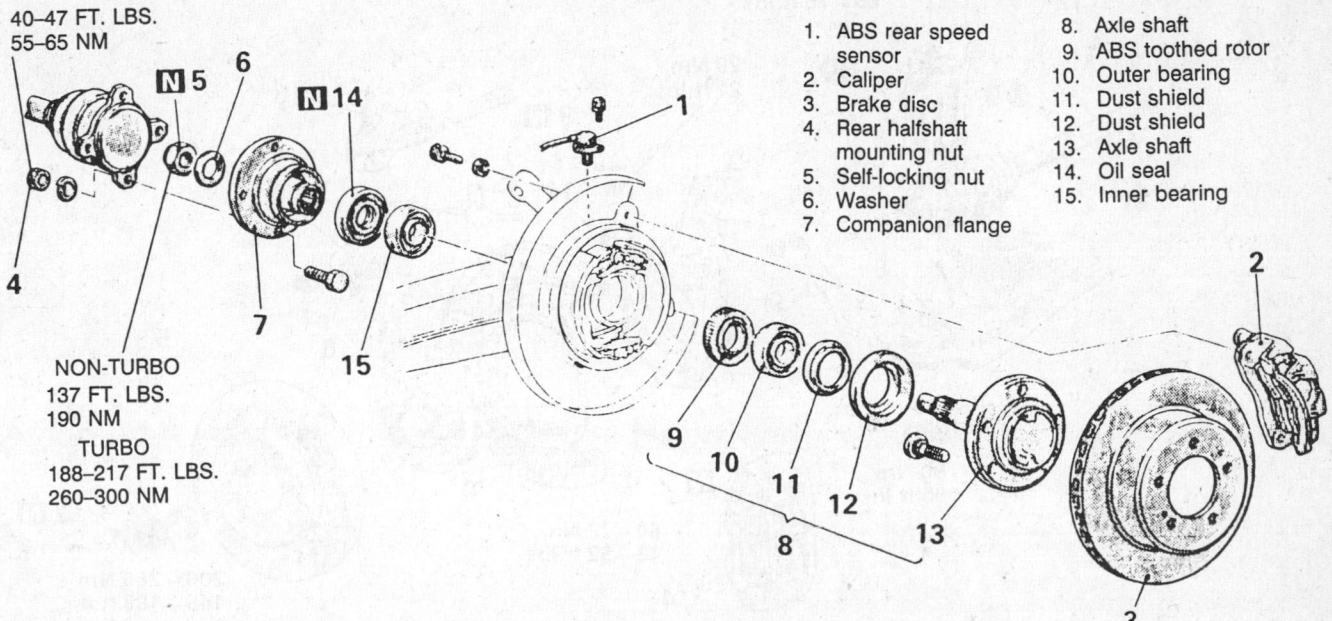

1. ABS rear speed sensor
2. Caliper
3. Brake disc
4. Rear halfshaft mounting nut
5. Self-locking nut
6. Washer
7. Companion flange
8. Axle shaft
9. ABS toothed rotor
10. Outer bearing
11. Dust shield
12. Dust shield
13. Axle shaft
14. Oil seal
15. Inner bearing

40–47 FT. LBS.
55–65 NM

NON-TURBO
137 FT. LBS.
190 NM

TURBO
188–217 FT. LBS.
260–300 NM

Rear axle shaft and related parts—Laser, Talon and Stealth with AWD

cable. Raise the vehicle and support safely.

2. Remove the bolts that attach the rear halfshaft to the companion flange.

3. Use a prybar to pry the inner shaft out of the differential case. Don't insert the prybar too far or the seal could be damage.

4. Remove the rear halfshaft from the vehicle.

5. If equipped with ABS, remove the rear wheel speed sensor.

6. Remove the caliper, pads and brake rotor.

7. Hold the axle shaft stationary and remove the axle shaft self-locking nut and washer.

8. Using a slide hammer, separate the axle shaft from the companion flange and remove.

9. Use a vice and gear puller tool to disassemble the axle shaft and companion flange assemblies.

To install:

10. Assemble the axle shaft and com-

panion shaft assemblies using new parts as required.

11. Install the axle shaft to the housing and slide the axle shaft over it. Install the washer and new self-locking nut. Hold the axle shaft stationary and torque the nut to 116–159 ft. lbs. (160–220 Nm) for Laser, Talon and non-turbocharged Stealth. Torque to 188–217 ft. lbs. (260–300 Nm) for turbocharged Stealth.

12. Install the brake rotor, pads and caliper.

1. ABS speed sensor connector
2. Cotter pin
3. Halfshaft end nut
4. Caliper
5. Brake dusc
6. Front hub bearing unit
7. Dust shield
8. Lower ball joint
9. Cotter pin
10. Tie rod end
11. Halfshaft
12. MacPhersonstrut mounting bolt
13. Hub and knuckle
14. Hub

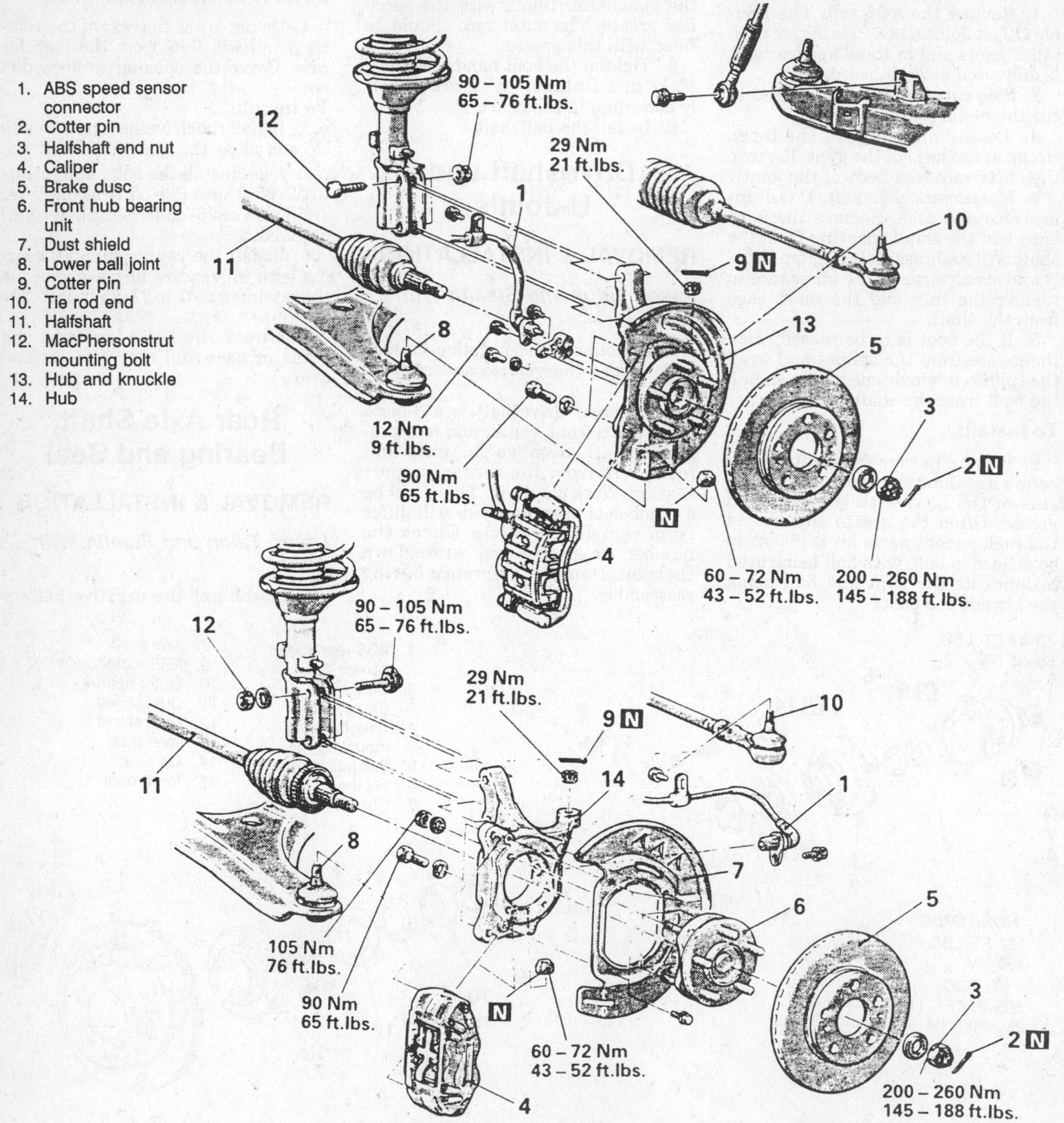

Front wheel hub, knuckle land bearing. FWD is on top and AWD in on the bottom.

13. Install the ABS rear wheel speed sensor.

14. Replace the circlip and install the rear halfshaft to the differential case. Make sure it snaps in place. Torque the companion flange bolts to 40–47 ft. lbs. (55–65 Nm).

15. Check the fluid level in the rear differential.

Front Wheel Hub, Knuckle and Bearing

REMOVAL & INSTALLATION

1. Disconnect the negative battery cable.

2. Remove the cotter pin, halfshaft nut and washer. It is recommended that the halfshaft nut is removed while the vehicle is on the floor with the brakes applied.

3. Raise the vehicle and support safely. If equipped with ABS, remove the front wheel speed sensor. Remove the ball joint and tie rod end from the steering knuckle.

4. Remove the caliper and pads and suspend with a wire.

5. On vehicles with an inner shaft, remove the center support bearing bracket bolts and washers. Remove the halfshaft by setting up a puller on the outside wheel hub and pushing the halfshaft from the front hub. Then tap the joint case with a plastic hammer to remove the halfshaft shaft and inner shaft from the transaxle.

6. On vehicles without an inner shaft, remove the halfshaft by setting up a puller on the outside wheel hub and pushing the halfshaft from the front hub. After pressing the outer shaft, insert a prybar between the transaxle case and the halfshaft and pry the shaft from the transaxle.

7. On Stealth with AWD, the front hub/bearing assembly can be serviced at this point as a unit. If the knuckle is being removed, proceed. All others require knuckle removal.

8. Unbolt the lower end of the strut and remove the hub and steering knuckle assembly.

9. Set up a puller with the knuckle/hub in a vise and pull the hub from the knuckle. Do not use a hammer to accomplish this or the bearing will be damaged.

10. Once the hub and outer bearing inner race are removed with a puller, the bearing outer races can be re-

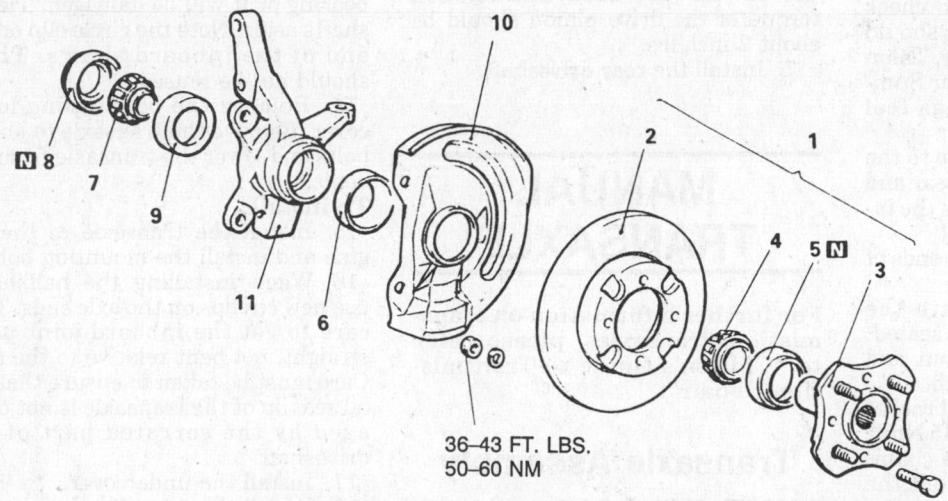

36–43 FT. LBS.
50–60 NM

1. Hub and brake disc
2. Barke disc
3. Front hub
4. Outer bearing inner race
5. Hub side oil seal
6. Outer bearing outer race
7. Inner bearing inner race
8. Halfhsaft side oil seal
9. Inner bearing outer race
10. Dust cover
11. Knuckle

Exploded view of the front wheel hub, knuckle and bearings—Summit

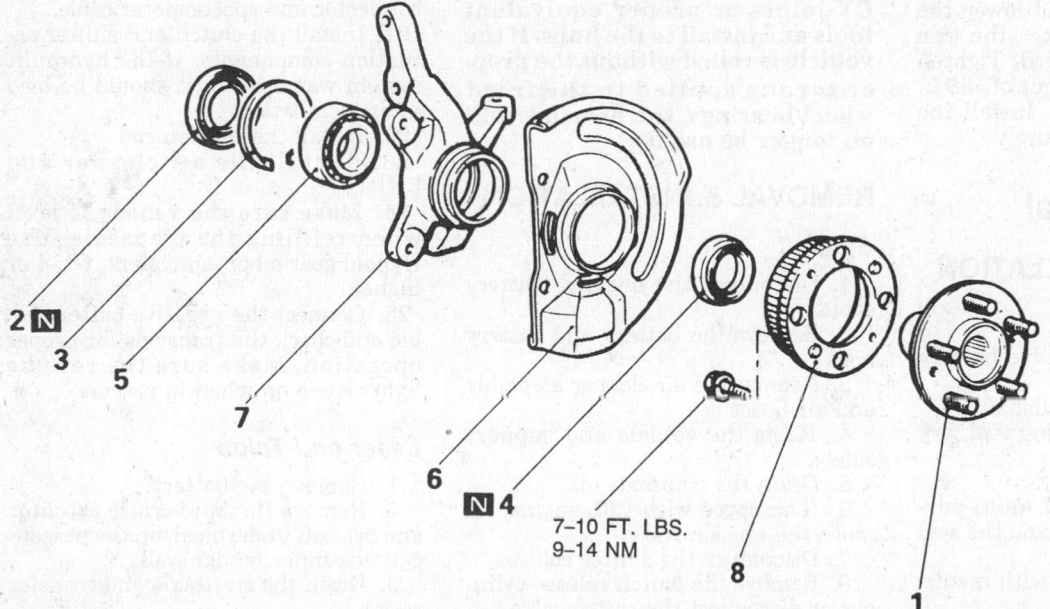

7–10 FT. LBS.
9–14 NM

1. Front hub
2. Halfshaft side oil seal
3. Snapring
4. Hub side oil seal
5. Wheel bearing
6. Dust shield
7. Knuckle
8. ABS toothed rotor

Exploded view of the front wheel hub, knuckle and bearings—except Summit

moved by tapping out with a brass drift pin and a hammer.

To install:

11. Assemble the hub/knuckle assembly with pressing tools, using new parts as required.

12. Install the knuckle assembly to the vehicle and install the strut bolts.

13. On AWD Stealth, torque the front hub/bearing assembly nuts to 76 ft. lbs. (105 Nm).

14. Apply a thin coat of grease to the outside of the outer races and install into the hub with a bearing driver.

15. Apply multi-purpose grease to the bearings, inside surface of the hub and the lip of the grease seal. Place the outside bearing into the knuckle and install the seal with a driver.

16. The hub is assembled to the knuckle with a puller. Draw the parts together firmly to seat the bearings. Use a small torque wrench to check the bearing turning torque. It should be 16 inch lbs. or less. for Laser, Talon and Stealth or 11 lbs. or less for Summit. Check that the bearings feel smooth when rotated.

17. Apply a thin coat of grease to the lip of the halfshaft side axle seal and drive into place until it contacts the inner bearing outer race.

18. Replace the circlips on the ends of the halfshafts.

19. Insert the halfshaft into the transaxle. Make sure it is fully seated.

20. Pull the strut assembly out and install the other end to the hub.

21. Install the center bearing bracket bolts and tighten to 33 ft. lbs. (45 Nm).

22. Install the washer so the chamfered edge faces outward. Install the nut and tighten temporarily.

23. Install the tie rod end and ball joint.

24. Install the wheel and lower the vehicle to the floor. Tighten the axle nut with the brakes applied. Tighten the nut to a maximum torque of 188 ft. lbs. (260 Nm) maximum. Install the cotter pin and bend it securely.

Pinion Seal

REMOVAL & INSTALLATION

Front Differential

1. Disconnect the negative battery cable.

2. Remove the front halfshaft.

3. Using a suitable prying tool, pry the seal from the case.

To install:

4. Apply a thin coat of multi-purpose grease to the seal lip and the seal contact surface.

5. Install the new seal with a suitable driver.

6. Install the front halfshaft.

Rear Differential

1. Raise the vehicle and support safely.

2. Matchmark the rear driveshaft and companion flange and remove the shaft. Don't let it hang from the transaxle. Tie it up to the underbody.

3. Hold the companion flange stationary and remove the large self-locking nut in the center of the companion flange.

4. With a suitable puller, remove the flange. Pry the old seal out.

To install:

5. Apply a thin coat of multi-purpose grease to the seal lip and the companion flange seal contacting surface. Install the new seal with a suitable driver.

6. Install the companion flange. Install a new locknut and torque to 116–160 ft. lbs. (157–220). The rotation torque of the drive pinion should be about 2 inch lbs.

7. Install the rear driveshaft.

MANUAL TRANSAXLE

For further information on transmissions/transaxles, please refer to "Chilton's Guide to Transmission Repair".

Transaxle Assembly

NOTE: If the vehicle is going to be rolled while the halfshafts are out of the vehicle, obtain 2 outer CV-joints or proper equivalent tools and install to the hubs. If the vehicle is rolled without the proper torque applied to the front wheel bearings, the bearings will no longer be usable.

REMOVAL & INSTALLATION

Summit

1. Disconnect the negative battery cable.

2. Remove the battery and battery tray.

3. Remove the air cleaner assembly and air hoses.

4. Raise the vehicle and support safely.

5. Drain the transaxle oil.

6. If equipped with 1.6L engine, remove the tension rod.

7. Disconnect the shifter cables.

8. Remove the clutch release cylinder or disconnect the clutch cable..

9. Disconnect the backup lamp

switch connector, speedometer cable connection and remove the starter motor.

10. Remove the transaxle mounting bolts and bracket.

11. Remove the sheet metal undercover.

12. Disconnect the tie rod ends and the lower ball joint connections.

13. Remove the halfshafts by inserting a prybar between the transaxle case and the driveshaft and prying the shaft from the transaxle. Do not pull on the driveshaft. Doing so damages the inboard joint. Use the prybar. Do not insert the prybar so far the oil seal in the case is damaged. Remove the right side shaft as just described. The left side shaft can be removed by tapping with a plastic hammer. Remove the shaft with the hub and knuckle as an assembly. Don't tap on the center bearing or it will be damaged. Tie the shafts aside. Note the circle clip on the end of the inboard shafts. These should not be reused.

14. Remove the bellhousing lower cover. Remove the transaxle to engine bolts and lower the transaxle from the vehicle.

To install:

15. Install the transaxle to the engine and install the mounting bolts.

16. When installing the halfshafts, use new circlips on the axle ends. Take care to get the inboard joint parts straight, not bent relative to the axle. Care must be taken to ensure that the oil seal lip of the transaxle is not damaged by the serrated part of the driveshaft.

17. Install the undercover.

18. Install the mounting brackets.

19. Install the starter.

20. Connect the backup light switch connector and speedometer cable.

21. Install the clutch and shifter actuation components. If the hydraulic system was opened, it should be bled after installation.

22. Install the tension rod.

23. Install the air cleaner and battery.

24. Make sure the vehicle is level when refilling the transaxle. Use Hypoid gear oil or equivalent, GL-4 or higher.

25. Connect the negative battery cable and check the transaxle for proper operation. Make sure the reverse lights come on when in reverse.

Laser and Talon

1. Remove the battery.

2. Remove the auto-cruise actuator and bracket underhood, on the passenger side inner fender wall.

3. Drain the transaxle and transfer case.

4. Remove the air intake hose.

5. Remove the cotter pin securing the select and shift cables and remove the cable ends from the transaxle.

6. Remove the connection for the clutch release cylinder and without disconnecting the hydraulic line, secure aside.

7. Disconnect the backup light switch and the speedometer cable.

8. Disconnect the starter electrical connections and remove the starter motor.

9. Remove the transaxle mount bracket.

10. Raise the vehicle and support safely. Remove the undercover.

11. Remove the cotter pin and disconnect the tie rod end from the steering knuckle.

12. Remove the self-locking nut and remove the lower arm ball joint.

13. Remove the halfshafts by inserting a prybar between the transaxle case and the driveshaft and prying the shaft from the transaxle. Do not pull on the driveshaft. Doing so damages the inboard joint. Use the prybar. Do not insert the prybar so far the oil seal in the case is damaged. On AWD, remove the right side shaft as just described. The left side shaft can be removed by tapping with a plastic hammer. Remove the shaft with the hub and knuckle as an assembly. Don't tap on the center bearing or it will be damaged. Tie the shafts aside. Note the circle clip on the end of the inboard shafts. These should not be reused.

14. On AWD vehicle, disconnect the front exhaust pipe.

15. On AWD vehicle, remove the transfer case by removing the attaching bolts, moving the transfer case to the left and lowering the front side. Remove it from the rear driveshaft. Be careful of the oil seal. Do not allow the prop shaft to hang; tie it up. Cover the transfer case openings to keep out dirt.

16. Remove the underpan from the transaxle bellhousing. On AWD, also remove the crossmember and the triangular gusset.

17. Remove the transaxle lower coupling bolt. It is just above the halfshaft opening on 2WD or transfer case opening on AWD.

18. Remove the transaxle assembly. On turbocharged vehicle, take care to prevent damaging the lower radiator hose with the transaxle housing. Wind tape around the lower hose and put tape on the transaxle housing. Support the transaxle assembly using the proper jack, move the transaxle to the right and lower it.

To install:

19. Install the transaxle to the engine and install the mounting bolts.

20. Install the transaxle lower coupling bolt.

21. Install the underpan, crossmember and the triangular gusset.

22. Install the transfer case on AWD vehicles and connect the exhaust pipe.

23. When installing the halfshafts, use new circlips on the axle ends. Take care to get the inboard joint parts straight, not bent relative to the axle. Care must be taken to ensure that the oil seal lip of the transaxle is not damaged by the serrated part of the driveshaft.

24. Connect the tie rod and ball joint to the steering knuckle.

25. Install the transaxle mount bracket.

26. Install the starter motor.

27. Connect the backup light switch and the speedometer cable.

28. Install the clutch release cylinder.

29. Connect the select and shift cables and install new cotter pins.

30. Install the air intake hose.

31. Install the auto-cruise actuator and bracket underhood, on the passenger side inner fender wall.

32. Install the battery.

33. Make sure the vehicle is level when refilling the transaxle. Use Hypoid gear oil or equivalent, GL-4 or higher.

34. Connect the negative battery cable and check the transaxle and transfer case for proper operation. Make sure the reverse lights come on when in reverse.

Stealth

1. Remove the battery and battery tray. Raise the vehicle and support safely. Drain the transaxle oil and remove the transfer case if equipped.

2. Remove the left side splash shield.

3. Remove the air cleaner assembly and all adjoining duct work.

4. Disconnect the shifter control cables and speedometer connector.

5. Remove the clutch release cylinder.

6. Disconnect the reverse light switch.

7. Support the weight of the transaxle and remove the transaxle mount through bolt. Remove the access plug, remove the bolts for the bracket and remove the brackets.

8. Disconnect the transaxle ground cable.

9. Disconnect the tie rod end and ball joint from the steering knuckle.

10. Remove the right frame member.

11. Remove the starter motor.

12. Remove the halfshafts by inserting a prybar between the transaxle case and the driveshaft and prying the shaft from the transaxle. Do not pull on the driveshaft. Doing so damages the inboard joint. Use the prybar. Do not insert the prybar so far the oil seal in the case is damaged. On AWD, remove the right side shaft as just described. The left side shaft can be removed by tapping with a plastic hammer. Remove the shaft with the hub and knuckle as an assembly. Don't tap on the center bearing or it will be damaged. Tie the shafts aside. Note the circle clip on the end of the inboard shafts. These should not be reused.

13. Remove the transaxle brackets.

14. Remove the transaxle assembly. On turbocharged vehicles, take care to prevent damaging the lower radiator hose with the transaxle housing. Wind tape around the lower hose and put tape on the transaxle housing. Support the transaxle assembly using the proper jack, move the transaxle away from the engine and lower it.

To install:

15. Install the transaxle to the engine and install the mounting bolts.

16. When installing the halfshafts, use new circlips on the axle ends. Take care to get the inboard joint parts straight, not bent relative to the axle. Care must be taken to ensure that the oil seal lip of the transaxle is not damaged by the serrated part of the driveshaft.

17. Install the starter motor and cover.

18. Install the right side frame member.

19. Install the ball joint and tie rod to the steering knuckle.

20. Connect the transaxle ground cable.

21. Install the side mount brackets and install the access plug.

22. Connect the reverse light switch.

23. Install the clutch release cylinder.

24. Connect the shifter control cables and speedometer connector.

25. Install the transfer case on AWD vehicles.

26. Install the air cleaner assembly and all adjoining duct work.

27. Install the left side splash shield.

28. Install the battery tray and battery.

29. Make sure the vehicle is level when refilling the transaxle. Use Hypoid gear oil or equivalent, GL-4 or higher.

30. Connect the negative battery cable and check the transaxle and transfer case for proper operation. Make sure the reverse lamps come on when in reverse.

LINKAGE ADJUSTMENT

There are 2 cables, the select cable and the shift cable.

1. On the transaxle, put select lever in **N** and move the transaxle shift lever

to put it in **4th** gear. Depress the clutch, if necessary, to shift.

2. Move the shift lever in the vehicle to the **4th** gear position until it contacts the stop.

3. Turn the adjuster turn buckle so the shift cable eye aligns with the eye in the gear shift lever. When installing the cable eye, make sure the flange side of the plastic bushing at the shift cable end is on the cotter pin side.

4. The cables should be adjusted so the clearance between the shift lever and the 2 stoppers are equal when the shift lever is moved to 3rd and 4th gear. Move the shift lever to each position and check that the shifting is smooth.

CLUTCH

Clutch Assembly

REMOVAL & INSTALLATION

1. Disconnect the negative battery cable.

2. Remove the transaxle assembly from the vehicle.

3. Remove the pressure plate attaching bolts. If the pressure plate is to be reused, loosen the bolts in succession, 1 or 2 turns at a time to prevent warping the the cover flange.

4. Remove the pressure plate release bearing assembly and the clutch disc. Do not use solvent to clean the bearing.

5. Inspect the condition of the clutch components and replace any worn parts.

To install:

6. Inspect the flywheel for heat damage or cracks. Use new bolts and replace if necessary.

7. Using the proper alignment tool, install the clutch disc to the flywheel. Install the pressure plate assembly and tighten the pressure plate bolts evenly to 11–16 ft. lbs. (15–22 Nm). Remove the alignment tool.

8. Apply a very light coat of high temperature grease to the clutch fork at the ball pivot and where the fork contacts the bearing. Also a little bit of grease can be applied to end of the release cylinder's pushrod and to the pushrod hole on the fork. Apply a light coat of grease on the transaxle input shaft splines.

9. Install a new clutch release bearing. Pack its inner surface with grease.

10. Install the transaxle assembly and check for proper clutch operation.

PEDAL HEIGHT/FREE-PLAY ADJUSTMENT

1. Measure the clutch pedal height from the face of the pedal pad to the firewall. If the pedal height is not within 6.60–6.89 in. (168–175mm) for Summit or 6.93–7.17 in (176–182mm) for Laser, Talon and Stealth, adjustment is necessary.

2. Measure the clutch pedal clevis pin play at the face of the pedal pad. If the clutch pedal clevis pin play is not within 0.04–0.12 in. (1–3mm) for Summit, Laser and Talon or 0.24–0.51 in. (6–13mm) for Stealth, adjustment is necessary.

3. If the clutch pedal height or clevis pin play are not within the standard value, adjust as follows:

a. For vehicles without cruise control, turn and adjust the bolt so the pedal height is the standard value, then tighten the locknut.

b. Vehicles with auto-cruise control system, disconnect the clutch switch connector and turn the switch to obtain the standard clutch pedal height. Then, lock with the locknut.

c. Turn the pushrod to adjust the clutch pedal clevis pin play to agree with the standard value and secure the pushrod with the locknut.

NOTE: When adjusting the clutch pedal height or the clutch pedal clevis pin play, be careful not to push the pushrod toward the master cylinder.

d. Check that when the clutch pedal is depressed all the way, the interlock switch switches over from **ON** to **OFF**.

Clutch Cable

ADJUSTMENT

To adjust the clutch cable, turn the adjusting wheel at the firewall to obtain the proper freeplay of about 1 inch.

REMOVAL & INSTALLATION

1. Disconnect the negative battery cable.

2. Remove the cable retaining clamps.

3. Remove the cotter pin from the clutch actuating arm at the transaxle.

4. Rotate the adjusting wheel counterclockwise to loosen the cable and remove the cable from the vehicle.

5. The installation is the reverse of the removal procedure.

6. Lubricate all pivot points. Adjust the cable.

Clutch Master Cylinder

REMOVAL & INSTALLATION

1. Disconnect the negative battery cable.

2. Remove necessary underhood components in order to gain access to the clutch master cylinder.

2. Loosen the line at the cylinder and allow the fluid to drain. Use care; brake fluid damages paint.

3. On Summit, Laser, Talon and FWD Stealth, remove the cotter pin at the clutch pedal and remove the washer and clevis pin. AWD Stealth has a clutch pedal booster which directly activates the master cylinder.

4. Remove the 2 nuts and pull the cylinder from the firewall. A seal should be between the mounting flange and firewall. This seal should be replaced.

5. The installation is the reverse of the removal procedure.

6. Lubricate all pivot points with grease.

7. Bleed the system at the slave cylinder using DOT 3 brake fluid.

Clutch Slave Cylinder

REMOVAL & INSTALLATION

1. Disconnect the negative battery cable. Remove necessary underhood components in order to gain access to the clutch slave cylinder.

2. Remove the hydraulic line and allow the system to drain.

3. Remove the bolts and pull the cylinder from the transaxle housing. On some 1.5L engines, instead of a pushrod bearing against the clutch arm, a clevis pin and yoke is used. Simply remove the circlip, pull out the clevis pin and remove the cylinder.

4. The installation is the reverse of the removal procedure.

5. Lubricate all pivot points with grease.

6. Bleed the system using DOT 3 brake fluid.

Hydraulic Clutch System Bleeding

1. Fill the reservoir with brake fluid.

2. Loosen the bleed screw, have the clutch pedal pressed to the floor.

3. Tighten the bleed screw and release the clutch pedal.

4. Repeat the bleeding operation until the fluid is free of air bubbles.

NOTE: It is suggested to attach a hose to the bleeder and place the other end into a container at least ½ full of brake fluid during the bleeding operation. Do not allow the reservoir to run out of fluid during the bleeding operation.

AUTOMATIC TRANSAXLE

For further information on transmissions/transaxles, please refer to "Chilton's Guide to Transmission Repair".

Transaxle Assembly

NOTE: If the vehicle is going to be rolled while the halfshafts are out of the vehicle, obtain 2 outer CV-joints or proper equivalent tools and install to the hubs. If the vehicle is rolled without the proper torque applied to the front wheel bearings, the bearings will no longer be usable.

REMOVAL & INSTALLATION

Summit

1. Disconnect the negative battery cable.
2. Remove the battery and battery tray.
3. Remove the air pipe and air hose.
4. Raise the vehicle and support safely.
5. Drain the transaxle oil.
6. If equipped with 1.6L engine, remove the tension rod.
7. Disconnect the control cable and cooler lines.
8. Disconnect the throttle control cable on 3 speed transaxle.
9. Disconnect the shift control solenoid valve connector on 4 speed transaxle.
10. Disconnect the inhibitor switch and kickdown servo switch on 4 speed transaxle.
11. Disconnect the pulse generator and oil temperature sensor on 4 speed transaxle.
12. Disconnect the speedometer cable and remove the starter.
13. Remove the transaxle mounting bolts and bracket.
14. Remove the under guard pan.
15. Disconnect the steering tie rod end and the ball joint from the steering arm.
16. Remove the halfshafts at the inboard side from the transaxle. Tie the joint aside.

17. Remove the bell housing cover and remove the driveplate bolts.
18. Remove the transaxle assembly lower connecting bolt, located just over the halfshaft opening.
19. Properly support the transaxle assembly and lower it moving it to the right for clearance.
To install:
20. After the torque converter has been mounted on the transaxle, install the transaxle assembly on the engine. Tighten the driveplate bolts to 34–38 ft. lbs. 46–53 Nm). Install the bell housing cover.
21. Replace the circlips and install the halfshafts to the transaxle.
22. Install the tie rods and ball joint to the steering arm.
23. Install the underguard and the mounting brackets.
24. Install the starter.
25. Connect the speedometer cable.
26. Connect the control cables, oil cooler lines and electrical connections.
27. Install the tension rod.
28. Install the air pipe and hose, battery tray and battery.
29. Refill with Dexron II, Mopar ATF Plus type 7176, or equivalent automatic transaxle fluid.
30. Start the engine and allow to idle for 2 minutes. Apply parking brake and move selector through each gear position, ending in **N**. Recheck fluid level and add if necessary. Fluid level should be between the marks in the **HOT** range.

Laser and Talon

1. Remove the battery and battery tray.
2. If equipped with auto-cruise, remove the control actuator and bracket.
3. Drain the transaxle fluid.
4. Remove the air cleaner assembly, intercooler and air hose.
5. Remove the adjusting nut and disconnect the shift cable.
6. Disconnect and tag the electrical connectors for the solenoid, neutral safety switch (inhibitor switch), the pulse generator kickdown servo switch and oil temperature sensor.
7. Disconnect the speedometer cable and oil cooler lines.
8. Disconnect the wires to the starter motor and remove the starter.
9. Remove the upper transaxle to engine bolts.
10. Support the transaxle and remove the transaxle mounting bracket.
11. Raise the vehicle and support safely. Remove the sheet metal under guard.
12. Remove the tie rod ends and the ball joints from the steering knuckle.
13. Remove the halfshafts by inserting a prybar between the transaxle case and the driveshaft and prying the

shaft from the transaxle. Do not pull on the driveshaft. Doing so damages the inboard joint. Use the prybar. Do not insert the prybar so far the oil seal in the case is damaged. Tie the halfshafts aside.
14. On AWD, disconnect the exhaust pipe, remove the frame pieces, and remove the transfer case.
15. Remove the lower bellhousing cover and remove the special bolts holding the flexplate to the torque converter. To remove, turn the engine crankshaft with a box wrench and bring the bolts into position 1 at a time. After removing the bolts, push the torque converter toward the transaxle so it doesn't stay on the engine side and allow oil to pour out the converter hub.
16. Remove the lower transaxle to engine bolts and remove the transaxle assembly.
To install:
17. After the torque converter has been mounted on the transaxle, install the transaxle assembly on the engine. Tighten the drive plate bolts to 34–38 ft. lbs. (46–53 Nm). Install the bell housing cover.
18. On AWD, install the transfer case and frame pieces. Connect the exhaust pipe using a new gasket.
19. Replace the circlips and install the halfshafts to the transaxle.
20. Install the tie rods and ball joint to the steering arm.
21. Install the transaxle mounting bracket.
22. Install the under guard.
23. Install the starter.
24. Connect the speedometer cable and oil cooler lines.
25. Connect the solenoid, neutral safety switch (inhibitor switch), the pulse generator kickdown servo switch and oil temperature sensor.
26. Install the shift control cable.
27. Install the air hose, intercooler and air cleaner assembly.
28. If equipped with auto-cruise, install the control actuator and bracket.
29. Refill with Dexron II, Mopar ATF Plus type 7176, or equivalent automatic transaxle fluid.
30. Start the engine and allow to idle for 2 minutes. Apply parking brake and move selector through each gear position, ending in **N**. Recheck fluid level and add if necessary. Fluid level should be between the marks in the **HOT** range.

Stealth

1. Remove the battery, battery tray and washer tank.
2. Remove the air cleaner assembly and adjoining duct work.
3. Disconnect the shifter control cable.

4. Disconnect and plug the oil cooler hoses.

5. Disconnect the inhibitor switch, kickdown servo switch, pulse generator, oil temperature sensor, shift control solenoid valve, and ground cable.

6. Disconnect the speedometer cable.

7. Raise the vehicle and support safely. Remove the undercovers.

8. Support the weight of the transaxle and remove the mount bracket. Remove the upper bellhousing bolts.

9. Disconnect the tie rod end and ball joint from the steering knuckle.

10. Remove the right frame member.

11. Remove the starter.

12. Remove the halfshafts by inserting a prybar between the transaxle case and the driveshaft and prying the shaft from the transaxle. Do not pull on the driveshaft. Doing so damages the inboard joint. Use the prybar. Do not insert the prybar so far the oil seal in the case is damaged. Tie the halfshafts aside.

13. Remove the remaining mounting brackets.

14. Remove the bellhousing cover plate.

15. Remove the special bolts holding the flexplate to the torque converter.

16. After removing the bolts, push the torque converter toward the transaxle so it doesn't stay on the engine side and allow oil to pour out the converter hub.

17. Remove the lower transaxle to engine bolts and remove the transaxle assembly.

To install:

18. After the torque converter has been mounted on the transaxle, install the transaxle assembly on the engine. Tighten the driveplate bolts to 34–38 ft. lbs. (46–53 Nm). Install the bell housing cover.

19. Install the mounting brackets.

20. Replace the circlips and install the halfshafts to the transaxle.

21. Install the starter and frame member.

22. Install the tie rods and ball joint to the steering arm.

23. Install the upper bellhousing bolts.

24. Install the transaxle mounting bracket.

25. Install the undercovers.

26. Connect the speedometer cable.

27. Connect the inhibitor switch, kickdown servo switch, pulse generator, oil temperature sensor, shift control solenoid valve, and ground cable.

28. Connect the oil cooler hoses.

29. Connect the shifter control cable.

30. Install the air cleaner assembly and adjoining duct work.

31. Install the washer tank, battery tray and battery.

32. Refill with Dexron II, Mopar ATF Plus type 7176, or equivalent automatic transaxle fluid.

33. Start the engine and allow to idle for 2 minutes. Apply parking brake and move selector through each gear position, ending in **N**. Recheck fluid level and add if necessary. Fluid level should be between the marks in the **HOT** range.

SHIFTER CONTROL CABLE ADJUSTMENT

1. The shifter cable adjustment is done at the neutral safety switch (inhibitor switch). Locate the switch on the transaxle and note the alignment holes in the arm and the body of the switch. Place the selector lever in **N**. Place the manual lever of the transaxle in the neutral position.

2. Align the holes on the switch.

3. If the cable needs to be adjusted, loosen the nut on the cable end and pull the cable end by hand until the alignment holes match. Tighten the nut. Check that the transaxle shifts and conforms to the positions of the selector lever.

THROTTLE CONTROL CABLE ADJUSTMENT

Some vehicles do not use a throttle linkage. Instead, the throttle position sensor provides an electric signal to the transaxle, so no linkage adjustment is required.

1. Check that the throttle lever is in the curb idle position, with the engine **OFF** but at normal operating temperature.

2. At the lower cable bracket, raise the cone shaped cover to uncover a small fitting on the cable. By loosening the locknut and adjuster nut, make the distance between the fitting on the cable and the lower collar is 0.020–0.060 in.

3. With the throttle in the wide open position, check that the cable does not bind.

FRONT SUSPENSION

MacPherson Strut

REMOVAL & INSTALLATION

1. Disconnect the negative battery cable. Raise and safely support vehicle.

2. Remove the brake hose and tube bracket. Do not pry the brake hose and tube clamp away when removing it.

3. Support the lower arm and remove the strut to knuckle bolts. Use a piece of wire to suspend the knuckle to keep the weight off the brake hose.

4. If equipped with ECS, disconnect the ECS connector at the top of the strut.

5. Before removing the top bolts, make matchmarks on the body and the strut insulator for proper reassembly. If this plate is installed improperly, the wheel alignment will be wrong. Remove the strut upper bolts and remove the strut assembly from the vehicle.

To install:

6. Install the strut to the vehicle and install the top bolts.

7. Connect the ECS connector.

8. Install to the knuckle and install the bolts.

9. Install the brake hose bracket.

10. Perform a front end alignment.

Lower Ball Joints

INSPECTION

The lower ball joints on these vehicles are not serviceable. If defective, the entire lower arm must be replaced. The ball joints can be checked using the following procedure:

1. Wiggle the ball joint a few times to make sure it is free.

2. Double-nut the stud and use a torque wrench to measure how much torque is required to turn it. Starting torque should be:

    a. Summit: 48 inch lbs. (5.5 Nm) or less.

    b. Laser and Talon: 26–87 inch lbs. (3–10 Nm).

    c. Stealth: 86–191 inch lbs. (10–22 Nm).

3. If the stud has more resistance than specified, replace the lower arm assembly. If the resistance is less, it may still be reused unless it has excessive play.

4. A new grease boot can be installed using a large socket for a driver.

Lower Control Arm

REMOVAL & INSTALLATION

1. Disconnect the negative battery cable.

2. Raise the vehicle and support safely.

3. Remove the sway bar and links.

4. Disconnect the ball joint stud from the steering knuckle.

5. Remove the inner mounting frame-through bolt and nut.

6. Remove the rear mount bolts. Remove the clamp if equipped.

7. Remove the rear rod bushing if servicing.

To install:

8. Assemble the control arm and bushing.

9. Install the control arm to the vehicle and install the through bolt. Replace the nut and snug temporarily.

10. Install the rear mount clamp, bolts and replacement nuts. Torque the bolts to 43–58 ft. lbs. (60–80 Nm) on Summit or to 70 ft. lbs. (95 Nm) on Laser, Talon and Stealth. The nut is torqued to 30 ft. lbs. (41 Nm).

11. Connect the ball joint stud to the knuckle. Install a new nut and torque to 43–52 ft. lbs. (60–72 Nm).

12. Install the sway bar and links.

13. Lower the vehicle to the floor for the final torquing of the frame mount through bolt.

14. Once the full weight of the vehicle is on the floor, torque the nuts to 75–90 ft. lbs. (102–122 Nm).

15. Connect the negative battery cable.

Sway Bar

REMOVAL & INSTALLATION

Summit, Laser and Talon

1. Disconnect the negative battery cable.

2. Raise and safely support vehicle. Remove the front exhaust pipe if necessary.

3. Remove the tie rod end from the steering knuckle.

4. Remove the center crossmember rear bolts.

5. Remove the stabilizer link bolts. On the ball stud type, hold ball stud with a hex wrench and remove the self-locking nut with a box wrench.

6. Remove the stabilizer bar mounts and remove the bar from the vehicle.

7. The installation is the reverse of the removal procedure. Lubricate all rubber parts when installing. Note that the bar brackets are marked left and right.

8. Tighten link bolts with rubber bushings just until the bushings are squashed to the width of the washer.

Stealth

1. Disconnect the negative battery cable.

2. Raise the vehicle and support safely.

3. Remove the front exhaust pipe and engine undercover.

4. Remove the left and right frame members.

5. On AWD vehicles with automatic transaxle, remove the transfer case bracket and transfer case.

6. Remove the sway bar link.

7. Remove the sway bar brackets and remove the sway bar from the vehicle.

To install:

8. Note that the bar brackets are marked left and right. Lubricate all rubber parts and install the bushings, the sway bar and brackets.

9. Install the sway bar link.

10. Install the transfer case and bracket.

11. Install the frame members.

12. Install the engine undercover and exhaust pipe.

13. Connect the negative battery cable.

Front Wheel Bearings

All vehicles are all front or all-wheel drive. Please refer to the Drive Axle section for bearing information.

REAR SUSPENSION

MacPherson Strut

REMOVAL & INSTALLATION

1. Disconnect the negative battery cable. Remove the trim panel inside the trunk or hatch area for access to the top mounting nuts.

2. Remove the top cap and mounting nuts. Disconnect the ECS connector if equipped.

3. Raise and safely support vehicle.

4. Remove the brake tube bracket bolt if necessary, then remove the strut lower mounting bolt.

5. Remove the rear strut assembly from the vehicle.

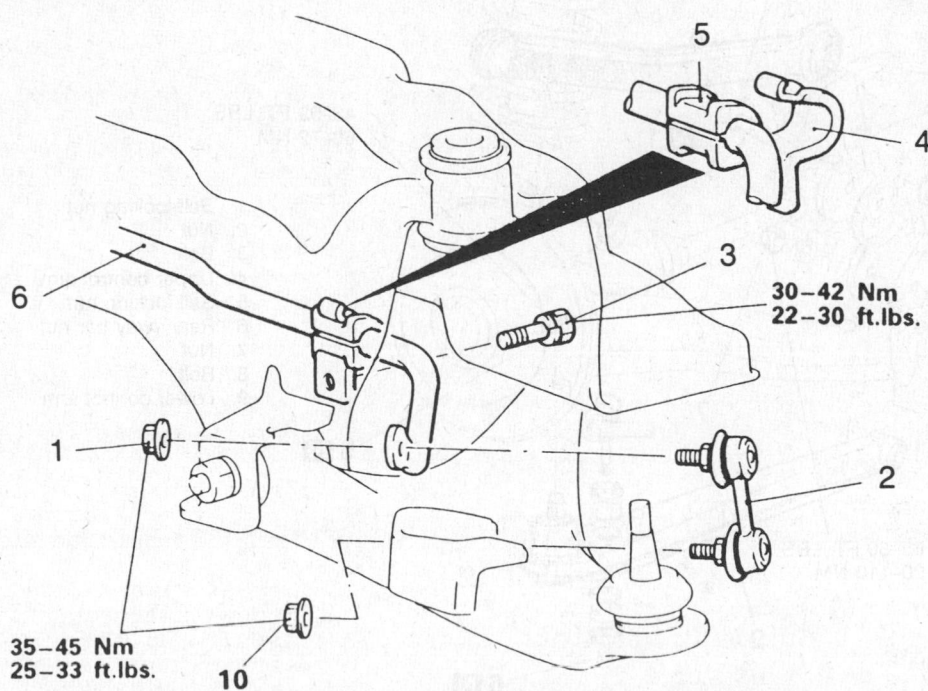

30–42 Nm
22–30 ft.lbs.

35–45 Nm
25–33 ft.lbs.

1. Link mounting nuts
2. Sway bar ball-type link
3. Bracket mount bolt
4. Bracket
5. Bushing
6. Sway bar

Typical sway bar and ball type link

6. Installation is the reverse of the removal procedure. Do not tighten the lower mounting nut until the full weight of the vehicle is on the ground.

Rear Control Arms

REMOVAL & INSTALLATION

Laser and Talon With AWD and Stealth

1. Disconnect the negative battery cable. On FWD Stealth, remove the rear strut assembly. Raise and safely support vehicle. Remove the brake line clamp bolt.

2. Remove the ball joint(s) from the rear trailing arm/steering knuckle.

3. If removing the lower arm, disconnect the sway bar link from the arm.

4. Matchmark and remove the inboard lower arm pivot bolt, if necessary, and remove the arm from the vehicle.

5. Installation is the reverse of the removal procedure. Replace all self-locking nuts. Do not torque the inboard pivot nuts until the full weight of the vehicle is on the ground.

6. On Laser and Talon, torque the lower inboard pivot nut to 65–80 ft. lbs. (90–110 Nm). All other nuts are torqued to 101–116 ft. lbs. (140–160 Nm).

7. Perform a rear wheel alignment.

Rear Trailing Arm

REMOVAL & INSTALLATION

Laser and Talon With AWD and Stealth

1. Disconnect the negative battery cable. Raise and safely support vehicle.

2. Remove the rear caliper from the brake disc and suspend with a wire. Remove the brake disc. Disconnect the parking brake cable and remove the mounting bolts along the trailing arm.

3. Remove the bolt(s) holding the speed sensor bracket to the knuckle and remove the assembly from the vehicle.

NOTE: The speed sensor has a pole piece projecting from it. This exposed tip must be protected from impact or scratches. Do not allow the pole piece to contact the toothed wheel during removal or installation.

4. On AWD, remove the rear axle to companion flange bolts and nuts and separate the axle from the companion flange. Remove the self-locking nut and remove the axle hub and companion flange. Remove the dust shield.

5. On FWD Stealth, remove the axle hub unit, parking brake shoes and backing plate. Remove the sway bar link bolt.

6. Remove the lower strut mounting bolt.

7. Remove the control arms from the trailing arm.

8. Remove the trailing arm front mounting nuts and bolts and remove the trailing arm from the vehicle. On AWD, remove the connecting rod at the front of the arm using tool MB991254 or equivalent.

To install:

9. Assemble the trailing arm and connecting rod. Install the trailing arm to the vehicle and install the install the front mounting nuts and bolts. Complete the final tightening of these when the full weight of the vehicle is on the ground.

10. Install the control arms to the trailing arm, using new self-locking nuts.

11. Install the lower strut bolt.

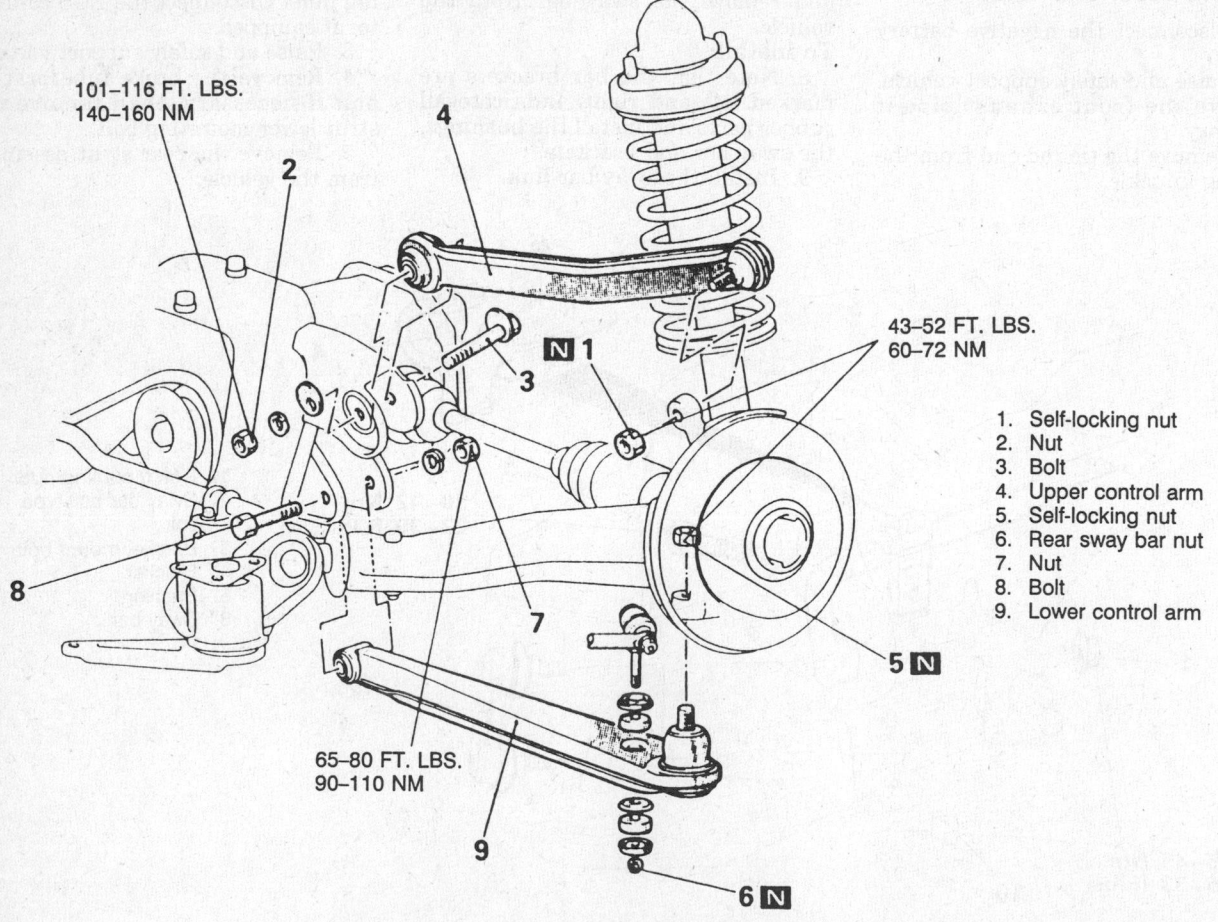

101–116 FT. LBS.
140–160 NM

43–52 FT. LBS.
60–72 NM

65–80 FT. LBS.
90–110 NM

1. Self-locking nut
2. Nut
3. Bolt
4. Upper control arm
5. Self-locking nut
6. Rear sway bar nut
7. Nut
8. Bolt
9. Lower control arm

Rear control arms—Laser and Talon with AWD and Stealth

12. On FWD Stealth, install the sway bar link. Install the parking brake parts and axle hub unit.

13. On AWD, install the dust shield, axle hub and companion flange with a new self-locking nut. Connect the rear axle to the companion flange.

14. Temporarily install the speed sensor to the knuckle; tighten the bolts only finger-tight.

15. Route the cable correctly and loosely install the clips and retainers. All clips must be in their original position and the sensor cable must not be twisted. Improper installation may cause cable damage and system failure.

NOTE: The wiring in the harness is easily damaged by twisting and flexing. Use the white stripe on the outer insulation to keep the sensor harness properly placed.

16. Use a brass or other non-magnetic feeler gauge to check the air gap between the tip of the pole piece and the toothed wheel. Correct gap is 0.012–0.035 inch (0.3–0.9mm). Tighten the 2 sensor bracket bolts to 10 ft. lbs. (14 Nm) with the sensor located so the gap is the same at several points on the toothed wheel. If the gap is incorrect, it is likely that the toothed wheel is worn or improperly installed.

17. Install the brake disc, caliper and connect the parking brake cable, if not already done. Install the mounting clamps bolts.

18. Double check everything for correct routing and installation. Lower the vehicle so its full weight is on the floor.

19. On Laser, Talon and FWD Stealth, torque the front mount nuts to 101–116 ft. lbs. (140–160 Nm). On AWD Stealth, tighten to 145–174 ft. lbs. (200–240 Nm).

20. Perform a rear wheel alignment.

Rear Wheel Bearings

REMOVAL & INSTALLATION

Summit

1. Raise the vehicle and support safely.
2. Remove the tire and wheel assembly.
3. If equipped with rear disc brakes, remove the caliper from the disc and remove the brake disc.
4. Remove the dust cap and bearing nut. Do not use an air gun to remove the nut.
5. Remove the outer wheel bearing.
6. Remove the drum and/or axle hub with the inner wheel bearing and the grease seal.

7. Remove the grease seal and remove the inner bearing.
To install:
8. Lubricate the inner bearing and install to the drum or hub.
9. Install a new grease seal.
10. To determine if the self-locking nut is reusable:
    a. Screw in the self-locking nut until about $1/10$ in. of the spindle is showing.
    b. Measure the torque required to turn the self-locking nut counterclockwise.
    c. The lowest allowable torque is 48 inch lbs. (5.5 Nm). If the measured torque is less than the specification, replace the nut.
11. Install the hub and/or hub to the vehicle.
12. Lubricate and install the outer wheel bearing to the spindle.
13. Torque the self-locking nut to 108–145 ft. lbs. (150–200 Nm).
14. Set up a dial indicator and measure the endplay while moving the hub or drum in and out. If the endplay exceeds 0.008 in. (0.002mm) , retorque the nut. If still beyond the limit, replace the bearings.
15. Install the grease cap and wheel.

Laser, Talon and Stealth

1. Raise the vehicle and support safely.
2. Remove the tire and wheel assembly.
3. Remove the bolt(s) holding the speed sensor bracket to the knuckle and remove the assembly from the vehicle.

NOTE: The speed sensor has a pole piece projecting from it. This exposed tip must be protected from impact or scratches. Do not allow the pole piece to contact the toothed wheel during removal or installation.

4. Remove the caliper from the brake disc and suspend with a wire.
5. Remove the brake disc.
6. Remove the grease cap, self-locking nut and tounged washer.
7. Remove the rear hub assembly. This assembly is not serviceable and must be replaced as a unit.

To install:
8. Install the hub assembly.
9. Install the tounged washer and a new self-locking nut. Torque the nut to 144–188 ft. lbs. (200–260 Nm), align with the indentation in the spindle, and crimp.
10. Set up a dial indicator and measure the endplay while moving the hub in and out. If the endplay exceeds 0.004 in. (0.01mm) for Laser and Talon or 0.002 in. (0.005mm) for Stealth,

retorque the nut. If still beyond the limit, replace the hub unit.
11. Install the grease cap and brake parts.
12. Temporarily install the speed sensor to the knuckle; tighten the bolts only finger-tight.
13. Route the cable correctly and loosely install the clips and retainers. All clips must be in their original position and the sensor cable must not be twisted. Improper installation may cause cable damage and system failure.

NOTE: The wiring in the harness is easily damaged by twisting and flexing. Use the white stripe on the outer insulation to keep the sensor harness properly placed.

14. Use a brass or other non-magnetic feeler gauge to check the air gap between the tip of the pole piece and the toothed wheel. Correct gap is 0.012–0.035 in. (0.3–0.9mm). Tighten the 2 sensor bracket bolts to 10 ft. lbs. (14 Nm) with the sensor located so the gap is the same at several points on the toothed wheel. If the gap is incorrect, it is likely that the toothed wheel is worn or improperly installed.
15. Install the wheel.

Rear Axle Assembly

REMOVAL & INSTALLATION

1. Raise the vehicle and support safely.
2. Remove the tire and wheel assembly.
3. If equipped with ABS, remove the bolts holding the speed sensor bracket to the trailing arm and remove the sensor assembly from the vehicle.

NOTE: The speed sensor has a pole piece projecting from it. This exposed tip must be protected from impact or scratches. Do not allow the pole piece to contact the toothed wheel during removal or installation.

4. If equipped with rear disc brakes, remove the caliper from the disc and remove the brake disc.
5. Remove the dust cap and bearing nut. Do not use an air gun to remove the nut.
6. Remove the outer wheel bearing.
7. Remove the drum and/or axle hub with the inner wheel bearing and the grease seal.
8. Remove the parking brake cable, brake hose, tube bracket and brake shoes with backing plate from the axle.
9. Remove the lateral rod mounting bolt and nut and secure the lateral rod to the axle beam with a piece of wire.

10. Using the proper equipment, slightly raise the torsion axle and arm assembly. Remove lower strut mounting bolt.

11. Remove the front trailing arm mount bolts and remove the rear axle assembly.

To install:

12. Install the rear axle assembly to the vehicle and install the strut mounting bolts. Install the front mount bolts and lateral rod bolts. Do not tighten these until the full weight of the vehicle is on the ground.

13. Install the backing plate, brake shoes, cable and hose.

14. On Summit, to determine if the self-locking nut is reusable:

    a. Screw in the self-locking nut until about $\frac{1}{10}$ in. of the spindle is showing.

    b. Measure the torque required to turn the self-locking nut counterclockwise.

    c. The lowest allowable torque is 48 inch lbs. (5.5 Nm). If the measured torque is less than the specification, replace the nut.

15. Remove the drum and/or axle hub. On Summit, lubricate and install the outer wheel bearing to the spindle. Torque the self-locking nut to 108–145 ft. lbs. (150–200 Nm).

16. On Laser and Talon, install the tounged washer and a new self-locking nut. Torque the nut to 144–188 ft. lbs. (200–260 Nm), align with the indentation in the spindle, and crimp.

17. Install the grease cap and brake parts.

18. Temporarily install the speed sensor to the knuckle; tighten the bolts only finger-tight.

19. Route the cable correctly and loosely install the clips and retainers. All clips must be in their original position and the sensor cable must not be twisted. Improper installation may cause cable damage and system failure.

NOTE: The wiring in the harness is easily damaged by twisting and flexing. Use the white stripe on the outer insulation to keep the sensor harness properly placed.

20. Use a brass or other non-magnetic feeler gauge to check the air gap between the tip of the pole piece and the toothed wheel. Correct gap is 0.012–0.035 in. (0.3–0.9mm). Tighten the 2 sensor bracket bolts to 10 ft. lbs. (14 Nm) with the sensor located so the gap is the same at several points on the toothed wheel. If the gap is incorrect, it is likely that the toothed wheel is worn or improperly installed.

21. Install the wheel.

22. Lower the vehicle so the full weight of the vehicle is on the floor.

23. On Summit, torque the front trailing arm bolt to 94–108 ft. lbs. (130–150 Nm). On Laser and Talon, torque the trailing arm bolt to 72–87 ft. lbs. ((100–120 Nm).

24. Torque the lateral rod nut to 58–72 ft. lbs. (80–100 Nm).

STEERING

Steering Wheel

NOTE: If equipped with an air bag, be sure to disarm it before entering the vehicle.

REMOVAL & INSTALLATION

Summit, Laser and Talon

1. Disconnect the negative battery cable.

2. Remove the horn pad and disconnect horn button connector.

3. Remove steering wheel retaining nut.

4. Matchmark the steering wheel to the shaft.

5. Use a steering wheel puller to remove the steering wheel. Do not hammer on steering wheel to remove it. The collapsible column mechanism may be damaged.

To install:

6. Line up the matchmarks and install the steering wheel. Torque the retaining nut to 29 ft. lbs. (40 Nm).

7. Install the steering wheel attaching nut and torque to 33 ft. lbs. (45 Nm).

8. Reconnect the horn connector and install the horn pad.

Stealth

1. Disconnect the negative battery cable.

2. Remove the air bag module mounting nut from behind the steering wheel.

3. Carefully disconnect the module connector.

4. Store the air bag module in a clean, dry place with the pad cover facing up.

5. Remove the steering wheel retaining nut. Matchmark the steering wheel to the shaft. Use a steering wheel puller to remove the wheel. Do not use a hammer or the collapsible mechanism in the column could be damaged.
from the steering column.

To install:

6. Center the clock spring by aligning the **NEUTRAL** mark on the clock spring with the mating mark on the casing.

7. Line up the matchmarks and install the steering wheel. Torque the retaining nut to 29 ft. lbs. (40 Nm).

Steering Column

NOTE: If equipped with an air bag, be sure to disarm it before entering the vehicle.

REMOVAL & INSTALLATION

1. Disconnect the negative battery cable.

2. Remove the instrument panel undercover or knee protector.

3. Remove the trim clip, foot shower duct and lap shower duct.

4. Remove the steering wheel and column upper and lower cover. Disconnect the key interlock cable if equipped.

5. Disconnect all connector to column-mounted items.

6. Remove the band from the steering joint cover and remove the joint assembly and gear box pinch bolt.

7. Remove the screws that attach the rubber seal to the firewall.

8. Remove the lower and upper column mounting bolts.

9. Remove the steering column assembly.

To install:

10. Install the column so the splines are inserted around the rack input shaft. Install the pinch bolt.

11. Install the mounting bolts.

12. Install the rubber seal screws.

13. Connect the connectors and interlock cable.

14. Install the column covers.

15. Install the remaining interior pieces.

16. Connect the negative battery cable and check all column-mounted switches for proper operation.

Manual Steering Rack

ADJUSTMENT

1. Remove the rack and pinion assembly.

2. Mount the rack in a vise and with a small torque wrench and an adapter to connect to the input shaft, position the rack at its center. Tighten the rack support cover, the bottom plug, to 11 ft. lbs. In the neutral position, rotate the shaft clockwise 1 turn in 4–6 seconds. Return the rack support cover 30–60 degrees and adjust the total pinion torque to 5–11 inch lbs.

3. When adjusting, set to the higher side of the specification. Make sure there is no ratcheting or catching when operating the rack. If the rack

cannot be adjusted to specification, check the rack support cover components or replace. After adjusting, lock the rack support cover with the locking nut.

REMOVAL & INSTALLATION

Summit

1. Disconnect the battery negative cable. Raise the vehicle and support safely.
2. Remove the pinch bolt holding the lower steering column joint to the rack and pinion input shaft.
3. Remove the cotter pins and disconnect the tie rod ends.
4. Remove the rack and pinion steering assembly and its rubber mounts.
5. The installation is the reverse of the removal procedure.
6. Perform a front end alignment.

Laser and Talon

1. Disconnect the negative battery cable. Raise the vehicle and support safely.
2. Remove the bolt holding lower steering column joint to the rack and pinion input shaft.
3. Remove the cotter pins and disconnect the tie rod ends.
4. Locate the triangular brace near the stabilizer bar brackets on the crossmember and remove both the brace and the stabilizer bar brackets.
5. Remove the through bolt from the round roll stopper and remove the rear bolts from the center crossmember.
6. Disconnect the front exhaust pipe.
7. Remove the rack and pinion steering assembly and its rubber mounts. Move the rack to the right to remove from the crossmember. Use caution to avoid damaging the boots.
To install:
8. Install the rack and install the mounting bolts. When installing the rubber rack mounts, align the projection of the mounting rubber with the indentation in the crossmember. Install the pinch bolt.
9. Connect the exhaust pipe.
10. Install the center member mounting bolts and roll stopper through bolt.
11. Install the stabilizer bar brackets and brace.
12. Connect the tie rod ends.
13. Perform a front end alignment.

Power Steering Rack

ADJUSTMENT

1. Disconnect the negative battery cable.

2. Raise the vehicle and support safely.
3. Remove the steering rack assembly from the vehicle.
4. Secure the steering rack assembly in a vise. Do not clamp the vise jaws on the steering housing tubes. Clamp the vise jaws only on the housing cast metal.
5. Remove the steering gear housing end plug from the steering gear shaft bore using tool 6103 or equivalent.
6. Remove the preload adjustment cap locknut from the steering gear housing bore using tool 6097 or equivalent.
7. In the neutral position, rotate the shaft clockwise 1 turn in 4–6 seconds. Return the rack support cover 30–60 degrees and adjust the total pinion torque to 5–11 inch lbs.
8. Secure the preload adjustment cap with a new locknut using tool 6097 or equivalent. Do not allow the adjustment cap to rotate when tightening the locknut.
9. Install the end plug using tool 6103 or equivalent.

REMOVAL & INSTALLATION

Summit

1. Disconnect the battery negative cable. Raise the vehicle and support safely.
2. Remove the pinch bolt holding the lower steering column joint to the rack and pinion input shaft.
3. Remove the cotter pins and disconnect the tie rod ends.
4. Disconnect the power steering fluid pressure pipe and return hose from the rack fittings.
5. Remove the rack and pinion steering assembly and its rubber mounts.
6. The installation is the reverse of the removal procedure.
7. Refill the reservoir and bleed the system.
8. Perform a front end alignment.

Laser and Talon

1. Disconnect the negative battery cable. Raise the vehicle and support safely.
2. Remove the bolt holding lower steering column joint to the rack and pinion input shaft.
3. Remove the cotter pins and disconnect the tie rod ends.
4. Locate the triangular brace near the stabilizer bar brackets on the crossmember and remove both the brace and the stabilizer bar brackets.
5. Remove the through bolt from the round roll stopper and remove the

rear bolts from the center crossmember.
6. Disconnect the front exhaust pipe.
7. Disconnect the power steering fluid pressure pipe and return hose from the rack fittings.
8. Remove the rack and pinion steering assembly and its rubber mounts. Move the rack to the right to remove from the crossmember. Use caution to avoid damaging the boots.
To install:
9. Install the rack and install the mounting bolts. When installing the rubber rack mounts, align the projection of the mounting rubber with the indentation in the crossmember. Install the pinch bolt.
10. Connect the power steering fluid lines to the rack.
11. Connect the exhaust pipe.
12. Install the center member mounting bolts and roll stopper through bolt.
13. Install the stabilizer bar brackets and brace.
14. Connect the tie rod ends.
15. Refill the reservoir and bleed the system.
16. Peform a front end alignment.

Stealth

1. Disconnect the negative battery cable.
2. Disconnect the front exhaust pipe.
3. If equipped with AWD, remove the transfer case assembly.
4. Remove the bolt holding lower steering column joint to the rack and pinion input shaft.
5. Remove the cotter pins and disconnect the tie rod ends.
6. Remove the left and right frame members.
7. Remove the stabilizer bar bracket.
8. If equipped with 4 wheel steering, disconnect the lines going to the rear pump.
9. Remove the rack and pinion steering assembly and its rubber mounts. Move the rack to the right to remove from the crossmember. Use caution to avoid damaging the boots.
To install:
10. Install the rack and install the mounting bolts. When installing the rubber rack mounts, align the projection of the mounting rubber with the indentation in the crossmember. Install the pinch bolt.
11. Connect the lines going to the 4 wheel steering rear pump and to the rack itself.
12. Install the frame members and torque the bolts to 50 ft. lbs. (68 Nm).
13. Connect the tie rods and instal new cotter pins.
14. Install the transfer case and

front exhaust pipe.

15. Refill the reservoir and bleed the system.

16. Peform a front end alignment.

Power Steering Pump

REMOVAL & INSTALLATION

Front

1. Disconnect the battery negative cable.

2. Remove the pressure switch connector from the side of the pump.

3. If the alternator is located under the oil pump, cover it with a shop towel to protect it from oil.

4. Disconnect the return fluid line. Remove the reservoir cap and allow the return line to drain the fluid from the reservoir. If the fluid is contaminated, disconnect the ignition high tension cable and crank the engine several times to drain the fluid from the gearbox.

5. Disconnect the pressure line.

6. Remove the pump drive belt and unbolt the pump from its bracket.

To install:

7. Install the pump, wrap the belt around the pulley and tighten the bolts.

8. Replace the O-rings and connect the pressure line. Connect the pressure line so the notch in the fitting aligns and contacts the pump's guide bracket.

9. Connect the return line.

10. Connect the pressure switch connector.

11. Adjust the belt tension and tighten the adjusting bolts.

12. Refill the reservoir and bleed the system.

Rear

STEALTH WITH FOUR WHEEL STEERING

1. Disconnect the negative battery cable. Raise the vehicle and support safely.

2. Drain the power steering fluid.

3. Remove the main muffler assembly.

4. Remove the rear shock absorber lower mounting bolts.

5. Remove the 2 small crossmember brackets.

6. Using the proper equipment, support the weight of the rear differential. Remove the large self-locking crossmember mounting nuts on the differential side.

7. Disconnect the pressure and suction hoses from the fittings on the pump.

8. Remove the pump retaining bolt and remove the pump from the rear differential assembly. Do not attempt

to disassemble the pump; it is not serviceable.

To install:

9. Replace the O-ring and install the pump assembly to the differential. Make sure the housing is fully seated and the gear is fully engaged. Install the retaining bolt.

10. Replace the O-ring and connect the fluid lines to the pump.

11. Install the large self-locking crossmember mounting nuts on the differential side. Torque to 80–94 ft. lbs. (110–130 Nm). Remove the support equipment.

12. Install the 2 small crossmember brackets.

13. Install the shock mounting bolts.

14. Install the muffler assembly.

15. Refill the reservoir and bleed the system.

16. To check and see if the system is functioning:

   a. Raise the vehicle safely so all 4 wheels turn freely.

   b. Run the vehicle at 50 mph.

   c. Turn the steering wheel quickly to the left and right and make sure the rear wheels steer in the same direction as the front wheels.

BELT ADJUSTMENT

1. Press the belt in about the center between the power steering pump pulley and the pulley it shares, usually water pump pulley. With reasonable pressure applied (about 22 lbs.) the belt should deflect about ¼–⅜ in.

2. Adjustment can be made by loosening the 3 bolts that hold the pump. Place a suitable bar or lever between the body of the pump and gently pry to get the desired tension.

3. Retighten the 3 bolts and check again.

SYSTEM BLEEDING

Front

1. Raise the vehicle and support safely.

2. Manually turn the pump pulley a few times.

3. Turn the steering wheel all the way to the left and to the right 5 or 6 times.

4. Disconnect the ignition high tension cable and, while operating the starter motor intermittently, turn the steering wheel all the way to the left and right 5–6 times for 15–20 seconds. During bleeding, make sure the fluid in the reservoir never falls below the lower position of the filter. If bleeding is attempted with the engine running, the air will be absorbed in the fluid. Bleed only while cranking.

5. Connect ignition high tension cable, start engine and allow to idle.

6. Turn the steering wheel left and right until there are no air bubbles in the reservoir. Confirm that the fluid is not milky and the level is up to the specified position on the gauge. Confirm that there is is very little change in the fluid level when the steering wheel is turned. If the fluid level

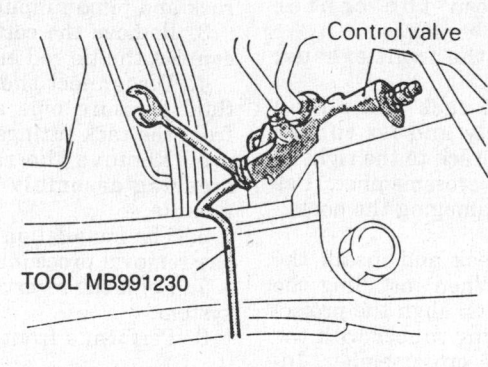

Control valve

TOOL MB991230

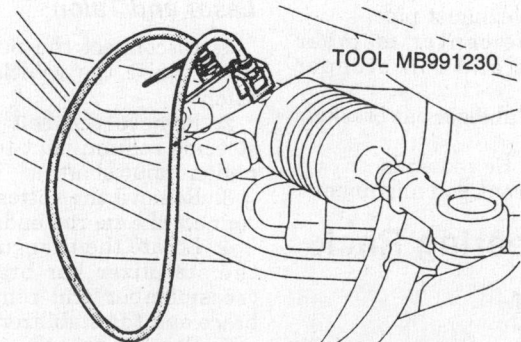

TOOL MB991230

Bleeding the rear steering system

changes more than 0.2 in. the air has not been completely bled. Repeat the process.

Rear

1. Bleed the front system as described above.
2. Start the engine and let it idle.
3. Loosen the bleeder screw on the left side of the control valve and install special tool MB991230 to the bleeder.
4. Turn the steering wheel all the way to the left, then immediately turn it half way back. Confirm that air has discharged with the fluid.
5. Repeat Step 5 two or three times as required, to remove all air from the rear system. Stop the engine.
6. Loosen the power cylinder (rear steering gear) bleeder screw about ⅛ turn and install the same special tool with the rotation prevention metal fixtures to prevent the bleeder from opening more.
7. Start the engine and run to 50 mph to circulate the fluid.
8. Maintain a speed of 20 mph and turn the steering wheel back and forth. Air should be discharged through the tube of the special tool and into the oil reservoir.
9. Repeat until all air is removed from the power cylinder.

Tie Rod Ends

REMOVAL & INSTALLATION

1. Disconnect the battery negative cable.
2. Raise the vehicle and support safely.
3. Wire brush the threads on the tie rod shaft and lubricate with penetrating oil. Loosen the locknut.
4. Remove the cotter pin and nut and press the tie rod end from the steering knuckle.
5. Hold the tie rod shaft with locking pliers and turn the tie rod end off, counting the number of turns for installation.
6. The installation is the reverse of the removal procedure. Install the tie rod end the same number of turns that it took to remove the old 1.
7. Perform a front end alignment.

BRAKES

For all brake system repair and service procedures not detailed below, please refer to "Brakes" in the Unit Repair section.

Master Cylinder

REMOVAL & INSTALLATION

1. Disconnect the negative battery cable.
2. Disconnect the fluid level sensor connector.
3. Disconnect the brake lines from the master cylinder. On Laser and Talon, a separate reservoir is used. Plug the lines to prevent drainage.
4. On Stealth, disconnect the low pressure hose.
5. Remove the 2 nuts securing the master cylinder and lift off.
To install:
6. Bench bleed the master cylinder.
7. Install to the studs and install the nuts.
8. Install the brake lines to the master cylinder.
9. Connect the negative battery cable and check the brakes for proper operation.

Proportioning Valve

REMOVAL & INSTALLATION

1. Disconnect the negative battery cable.
2. Locate the proportioning valve, usually below the master cylinder.
3. Tag and disconnect the brake lines from the valve.
4. Remove the proportioning valve from the engine compartment.
5. The installation is the reverse of the removal procedure.
6. Bleed the brakes in the following order:
Summit
   a. Left rear wheel cylinder or caliper
   b. Right front cylinder
   c. Right rear wheel cylinder or caliper
   d. Left front caliper
Laser and Talon
   a. Right rear caliper
   b. Left front caliper
   c. Left rear caliper
   d. Right front caliper
7. Connect the negative battery cable and check the brakes for proper operation.

Power Brake Booster

REMOVAL & INSTALLATION

1. Disconnect the negative battery cable.
2. Disconnect the vacuum hose from the booster. Pull it straight off. Prying off the vacuum hose could damage the check valve installed in the brake booster.

3. Remove the nuts attaching the master cylinder to the booster and remove the master cylinder.
4. From inside the passenger compartment, remove the cotter pin and clevis pin that secures the booster pushrod to the brake pedal.
5. Remove the nuts that attach the booster to the dash panel and remove it from the vehicle.
6. The installation is the reverse of the removal procedure.
7. Connect the negative battery cable, bleed the brakes and check for proper operation.

Brake Caliper

REMOVAL & INSTALLATION

Front Brakes

1. Disconnect the negative battery cable.
2. Raise the vehicle and support safely. Remove appropriate wheel assembly.
3. To disconnect the front brake hose, hold the nut on the brake hose side and loosen the flared brake line nut.
4. Remove the caliper lock pins and remove the caliper.
5. The installation is the reverse of the removal procedure. Make sure the brake hose is not twisted after installation. Refill the brake fluid as required and bleed the brakes.

Rear Brakes

SUMMIT, LASER AND TALON

1. Disconnect the negative battery cable.
2. Raise the vehicle and support safely. Remove appropriate wheel assembly.
3. Disconnect the parking brake cable from the actuator on the caliper.
4. To disconnect the brake hose, hold the nut on the brake hose side and loosen the flared brake line nut.
5. Remove the retaining bolts and remove rear caliper assembly.
6. The installation is the reverse of the removal procedure. Make sure the brake hose is not twisted after installation.
7. Refill the brake fluid as required and bleed the brakes.

STEALTH

1. Disconnect the negative battery cable.
2. Raise the vehicle and support safely. Remove appropriate wheel assembly.
3. To disconnect the brake hose, hold the nut on the brake hose side and loosen the flared brake line nut.

4. Remove the caliper lock pins and remove the caliper.

5. The installation is the reverse of the removal procedure. Make sure the brake hose is not twisted after installation. Refill the brake fluid as required and bleed the brakes.

Disc Brake Pads

REMOVAL & INSTALLATION

1. Disconnect battery negative cable.

2. Raise the vehicle and support safely.

3. Remove appropriate wheel assembly.

4. On the front of AWD Stealth, remove the pad retaining pins and pull the pads out of the caliper body.

5. On others, remove the caliper from its adaptor but do not allow the caliper to hang by the brake line. On some vehicles, the caliper can be flipped up by leaving the upper pin in place and using it as a pivot point. Take note of the clips, pins, antisqueal shims and other parts for reference at assembly.

6. On vehicles with rear disc brakes, it may help to loosen the parking brake cable from inside the car and disconnect the parking brake end from the rear caliper.

To install:

7. Use a large C-clamp to compress the piston(s) back into the caliper bore. On rear disc brakes with the parking brake mechanism incorporated into the caliper, a special tool is needed to turn the piston back into the bore.

8. Install the pads and all other small parts. Note that rear disc pads on calipers with the parking brake mechanism incorporated into the caliper should have a projection on the back side of the shoe that fits into the rear caliper piston.

9. Install the caliper. Make sure the brake hose is not twisted after installation. Connect the parking brake cable if disconnected.

Brake Rotor

REMOVAL & INSTALLATION

Summit Front Rotor

1. Loosen the large driveshaft nut while the vehicle is still on the ground with the brakes applied. Then raise and safely support vehicle. Remove appropriate wheel assembly.

2. Remove the axle end nut and lock washer.

3. Remove the caliper from its bracket. Do not allow the caliper to hang by the brake line. Remove the brake pads.

4. Remove the ball joint from the lower control arm.

5. Use and puller to push the halfshaft through the rotor/hub assembly.

6. Remove the bolts and separate the rotor from the hub.

To install:

7. Assemble the rotor and hub. Tighten the nuts to 40 ft. lbs. (54 Nm).

8. Install the assembly to the vehicle.

9. Install the washer so the chamfered edge faces outward. Install the nut and tighten temporarily.

10. Install the ball joint.

11. Install the brake components.

12. Install the wheel and lower the vehicle to the floor. Tighten the axle nut with the brakes applied. Tighten the nut to a maximum torque of 188 ft. lbs. (260 Nm) maximum. Install the cotter pin and bend it securely.

Except Summit Front Rotor

1. Raise the vehicle and support safely. Remove appropriate wheel assembly.

2. Remove the caliper and brake pads.

3. The rotor is held to the hub by 2 small threaded screws. Remove the bolts and pull off the rotor.

4. Installation is the reverse of the removal process.

Brake Drum

REMOVAL & INSTALLATION

1. Raise the vehicle and support safely.

2. Remove the wheel and tire assembly.

3. Remove the dust cap.

4. Remove the self-locking nut.

5. Remove the outer wheel bearing.

6. Remove the drum with the inner wheel bearing from the spindle. Remove the grease seal.

To install:

7. To determine if the self-locking nut is reusable:

a. Screw in the self-locking nut until about $\frac{1}{10}$ in. of the spindle is showing.

b. Measure the torque required to turn the self-locking nut counterclockwise.

c. The lowest allowable torque is 48 inch lbs. (5.5 Nm). If the measured torque is less than the specification, replace the nut.

8. Lubricate and install the inner wheel bearing. Install a new grease seal.

9. Install the drum to the spindle.

10. Lubricate and install the outer wheel bearing.

11. Torque the self-locking nut to 108–145 ft. lbs. (150–200 Nm).

12. Install the grease cap.

Brake Shoes

REMOVAL & INSTALLATION

1. Raise the vehicle and support safely. Remove appropriate wheel assembly.

2. Remove the brake drum. Remove the shoe to shoe spring.

3. Take note of the springs and clips for proper reassembly. Remove the shoe hold-down clips and remove the shoes.

To install:

4. Thoroughly clean and dry the backing plate. To prepare the backing plate, lubricate the bosses, anchor pin and parking brake actuating lever pivot surface lightly with lithium-based grease.

5. Remove, clean and dry all parts still on the old shoes. Lubricate the star wheel shaft threads with antisieze lubricant and transfer all parts to their proper locations on the new shoes.

6. Install shoes to the vehicle.

7. Connect the parking brake cable.

8. Adjust the star wheel.

9. To determine if the self-locking nut is reusable:

a. Screw in the self-locking nut until about $\frac{1}{10}$ in. of the spindle is showing.

b. Measure the torque required to turn the self-locking nut counterclockwise.

c. The lowest allowable torque is 48 inch lbs. (5.5 Nm). If the measured torque is less than the specification, replace the nut.

10. Remove any grease from the linings and install the drum to the spindle.

11. Lubricate and install the outer wheel bearing.

12. Torque the self-locking nut to 108–145 ft. lbs. (150–200 Nm).

13. Install the grease cap.

Wheel Cylinder

REMOVAL & INSTALLATION

1. Raise the vehicle and support safely.

2. Remove the wheel, drum and brake shoes.

3. Remove and plug the brake line from the wheel cylinder.

4. Remove the wheel cylinder re-

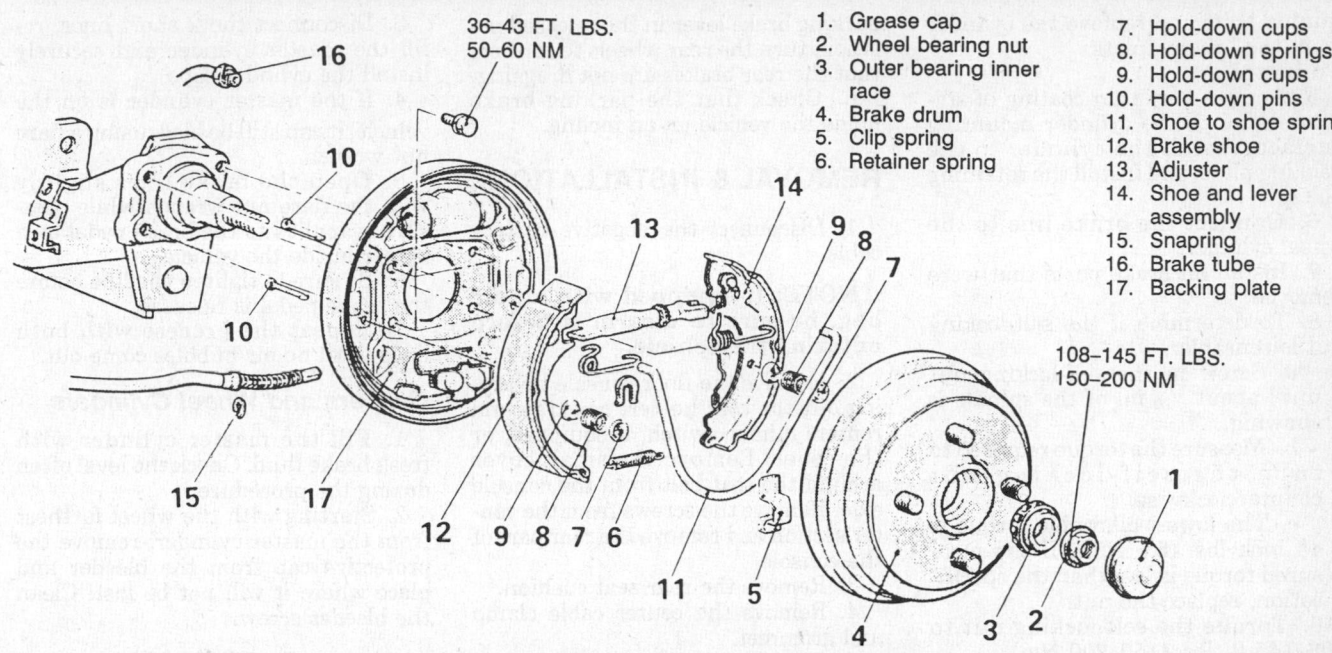

36–43 FT. LBS.
50–60 NM

1. Grease cap
2. Wheel bearing nut
3. Outer bearing inner race
4. Brake drum
5. Clip spring
6. Retainer spring
7. Hold-down cups
8. Hold-down springs
9. Hold-down cups
10. Hold-down pins
11. Shoe to shoe spring
12. Brake shoe
13. Adjuster
14. Shoe and lever assembly
15. Snapring
16. Brake tube
17. Backing plate

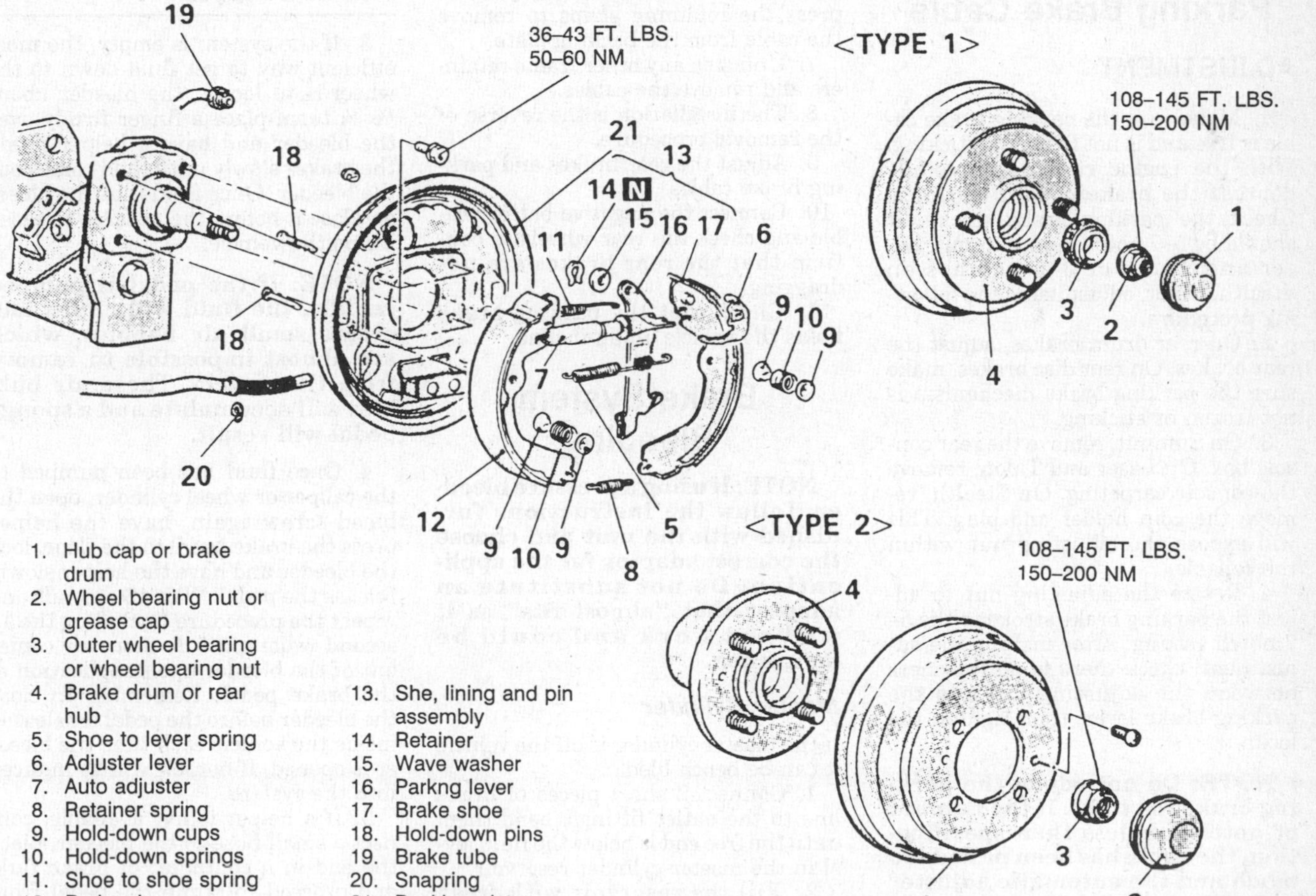

108–145 FT. LBS.
150–200 NM

Exploded view of the rear brakes—1989–90 Summit

36–43 FT. LBS.
50–60 NM

<TYPE 1>

108–145 FT. LBS.
150–200 NM

<TYPE 2>

108–145 FT. LBS.
150–200 NM

1. Hub cap or brake drum
2. Wheel bearing nut or grease cap
3. Outer wheel bearing or wheel bearing nut
4. Brake drum or rear hub
5. Shoe to lever spring
6. Adjuster lever
7. Auto adjuster
8. Retainer spring
9. Hold-down cups
10. Hold-down springs
11. Shoe to shoe spring
12. Brake shoe
13. She, lining and pin assembly
14. Retainer
15. Wave washer
16. Parkng lever
17. Brake shoe
18. Hold-down pins
19. Brake tube
20. Snapring
21. Backing plate

Exploded view of the rear brakes—1991–92 Summit

1–71

taining bolts and remove the cylinder from the backing plate.

To install:

5. Apply a very thin coating of silicone sealer to the cylinder mounting surface, install the cylinder to the backing plate, and install the retaining bolts.

6. Connect the brake line to the wheel cylinder.

7. Install all brake parts that were removed.

8. To determine if the self-locking nut is reusable:

   a. Screw in the self-locking nut until about $\frac{1}{10}$ in. of the spindle is showing.

   b. Measure the torque required to turn the self-locking nut counterclockwise.

   c. The lowest allowable torque is 48 inch lbs. (5.5 Nm). If the measured torque is less than the specification, replace the nut.

9. Torque the self-locking nut to 108–145 ft. lbs. (150–200 Nm).

10. Install the grease cap.

11. Install the tire and wheel assembly.

12. Bleed the rear brakes.

Parking Brake Cable

ADJUSTMENT

1. Make sure the parking brake cable is free and is not frozen or sticking. With the engine running, forcefully depress the brake pedal 5–6 times. Check the parking brake stroke. It should be 5–7 notches on Summit, Laser and Talon or 3–5 notches on Stealth. If not, adjust using the following procedure.

2. On rear drum brakes, adjust the rear brakes. On rear disc brakes, make sure the parking brake mechanism is not frozen or sticking.

3. On Summit, remove the rear console box. On Laser and Talon, remove the console carpeting. On Stealth, remove the coin holder and plug. This will expose the adjusting nut within the console.

4. Rotate the adjusting nut to adjust the parking brake stroke to the 5–7 notch setting. After making the adjustment, check there is no looseness between the adjusting nut and the parking brake lever, then tighten the locknut.

NOTE: Do not adjust the parking brake too tight. If the number of notches is less than specification, the cable has been pulled too much and the automatic adjuster will fail or the brakes will drag.

5. After adjusting the lever stroke, raise the rear of the vehicle. With the

parking brake lever in the released position, turn the rear wheels to confirm that the rear brakes are not dragging.

6. Check that the parking brake holds the vehicle on an incline.

REMOVAL & INSTALLATION

1. Disconnect the negative battery cable.

NOTE: If equipped with an air bag, be sure to disarm it before entering the vehicle.

2. Remove the floor console by prying out the coin holder, box tray and remote mirror switch, if equipped, or the cover. Remove the small cover around the seat belt from the console side. Remove the screws from the center section and remove the rear part of the console.

3. Remove the rear seat cushion.

4. Remove the center cable clamp and grommet.

5. Raise the vehicle and support safely.

6. At the rear wheel, remove the brake drum or disc and disconnect the cable end from the parking brake strut lever or actuator. If necessary, compress the retaining strips to remove the cable from the backing plate.

7. Unfasten any other frame retainers and remove the cables.

8. The installation is the reverse of the removal procedure.

9. Adjust the rear brakes and parking brake cables.

10. Connect the negative battery cable and check the rear wheels to confirm that the rear brakes are not dragging.

11. Check that the parking brake holds the vehicle on an incline.

Brake System Bleeding

NOTE: If using a pressure bleeder, follow the instructions furnished with the unit and choose the correct adaptor for the application. Do not substitute an adapter that "almost fits" as it will not work and could be dangerous.

Master Cylinder

If the master cylinder is off the vehicle it can be bench bled.

1. Connect 2 short pieces of brake line to the outlet fittings, bend them until the free end is below the fluid level in the master cylinder reservoir.

2. Fill the reservoir with fresh brake fluid. Pump the piston slowly until no more air bubbles appear in the reservoirs.

3. Disconnect the 2 short lines, refill the master cylinder and securely install the cylinder caps.

4. If the master cylinder is on the vehicle, it can still be bled, using a flare nut wrench.

5. Open the brake lines slightly with the flare nut wrench while pressure is applied to the brake pedal by a helper inside the vehicle.

6. Be sure to tighten the line before the brake pedal is released.

7. Repeat the process with both lines until no air bubbles come out.

Calipers and Wheel Cylinders

1. Fill the master cylinder with fresh brake fluid. Check the level often during the procedure.

2. Starting with the wheel farthest from the master cylinder, remove the protective cap from the bleeder and place where it will not be lost. Clean the bleeder screw.

─────── **CAUTION** ───────
When bleeding the brakes, keep face away from the brake area. Spewing fluid may cause facial and/or visual damage. Do not allow brake fluid to spill on the car's finish; it will remove the paint.

3. If the system is empty, the most efficient way to get fluid down to the wheel is to loosen the bleeder about $\frac{1}{2}$–$\frac{3}{4}$ turn, place a finger firmly over the bleeder and have a helper pump the brakes slowly until fluid comes out the bleeder. Once fluid is at the bleeder, close it before the pedal is released inside the vehicle.

NOTE: If the pedal is pumped rapidly, the fluid will churn and create small air bubbles, which are almost impossible to remove from the system. These air bubbles will accumulate and a spongy pedal will result.

4. Once fluid has been pumped to the caliper or wheel cylinder, open the bleed screw again, have the helper press the brake pedal to the floor, lock the bleeder and have the helper slowly release the pedal. Wait 15 seconds and repeat the procedure (including the 15 second wait) until no more air comes out of the bleeder upon application of the brake pedal. Remember to close the bleeder before the pedal is released inside the vehicle each time the bleeder is opened. If not, air will be induced into the system.

5. If a helper is not available, connect a small hose to the bleeder, place the end in a container of brake fluid and proceed to pump the pedal from inside the vehicle until no more air comes out the bleeder. The hose will prevent air from entering the system.

6. Repeat the procedure on remaining wheel cylinders in order:

Summit
a. Left rear wheel cylinder or caliper
b. Right front cylinder
c. Right rear wheel cylinder or caliper
d. Left front caliper

Laser, Talon and Stealth
a. Right rear caliper
b. Left front caliper
c. Left rear caliper
d. Right front caliper

7. Hydraulic brake systems must be totally flushed if the fluid becomes contaminated with water, dirt or other corrosive chemicals. To flush, bleed the entire system until all fluid has been replaced with the correct type of new fluid.

8. Install the bleeder cap on the bleeder to keep dirt out. Always road test the vehicle after brake work of any kind is done.

Anti-lock Brake System Service

PRECAUTIONS

● Certain components within the ABS system are not intended to be serviced or repaired individually. Only those components with removal and installation procedures should be serviced.

● Do not use rubber hoses or other parts not specifically specified for the ABS system. When using repair kits, replace all parts included in the kit. Partial or incorrect repair may lead to functional problems and require the replacement of components.

● Lubricate rubber parts with clean, fresh brake fluid to ease assembly. Do not use lubricated shop air to clean parts; damage to rubber components may result.

● Use only DOT 3 brake fluid from an unopened container.

● If any hydraulic component or line is removed or replaced, it may be necessary to bleed the entire system.

● A clean repair area is essential. Always clean the reservoir and cap thoroughly before removing the cap. The slightest amount of dirt in the fluid may plug an orifice and impair the system function. Perform repairs after components have been thoroughly cleaned; use only denatured alcohol to clean components. Do not allow ABS components to come into contact with any substance containing mineral oil; this includes used shop rags.

● The Anti-Lock control unit is a microprocessor similar to other computer units in the vehicle. Ensure that the ignition switch is **OFF** before removing or installing controller harnesses. Avoid static electricity discharge at or near the controller.

● If any arc welding is to be done on the vehicle, the ALCU connectors should be disconnected before welding operations begin.

Hydraulic Unit

REMOVAL & INSTALLATION

Laser and Talon

1. Disconnect the negative battery cable. Use a syringe or similar device to remove as much fluid as possible from the reservoir. Some fluid will be spilled from lines during removal of the hydraulic unit; protect adjacent painted surfaces.

2. On turbocharged engine, remove the center intercooler duct. Loosen the clamps and remove the bolts holding the duct to the air cleaner.

3. Disconnect the brake lines from the hydraulic unit. Correct reassembly is critical. Label or identify the lines before removal. Plug each line immediately after removal.

4. Remove the cover from the relay box. Disconnect the electrical harness to the hydraulic unit.

5. Disconnect the hydraulic unit ground strap from the chassis.

6. Remove the 3 nuts holding the hydraulic unit. Remove the unit upwards.

NOTE: The hydraulic unit is heavy; use care when removing it. The unit must remain in the upright position at all times and be protected from impact and shock.

7. Set the unit upright supported by blocks on the workbench. The hydraulic unit must not be tilted or turned upside down. No component of the hydraulic unit should be loosened or disassembled.

8. The bracket assemblies and relays may be removed if desired.

To install:

9. Install the relays and brackets if they were removed.

10. Install the hydraulic unit into the vehicle, keeping it upright at all times.

11. Install the retaining nuts and tighten them.

12. Connect the ground strap to the chassis bracket. Connect the hydraulic unit wiring harness.

13. Install the cover on the relay box.

14. Connect each brake line loosely to the correct port and double check the placement. Tighten each line to 10 ft. lbs. (13.5 Nm).

15. Fill the reservoir to the MAX line with brake fluid.

16. Bleed the master cylinder, then bleed the brake lines.

17. If equipped, install the intercooler air duct.

Stealth

1. Disconnect the negative battery cable. Remove the splash shield from beneath the car.

2. Use a syringe or similar device to remove as much fluid as possible from the reservoir. Some fluid will be spilled from lines during removal of the hydraulic unit; protect adjacent painted surfaces.

3. Lift the relay box with the harness attached and position it aside.

4. Remove the air intake duct.

5. Disconnect the brake lines from the hydraulic unit. Correct reassembly is critical. Label or identify the lines before removal. Plug each line immediately after removal. It will be necessary to hold the relay box aside to allow wrench access.

6. Disconnect the wiring harness connections at the hydraulic unit.

7. Disconnect the hydraulic unit ground strap from the chassis.

8. Remove the 3 bolts holding the hydraulic unit bracket. Remove the unit and the bracket.

NOTE: The hydraulic unit is heavy; use care when removing it. The unit must remain in the upright position at all times and be protected from impact and shock.

9. Set the unit upright supported by blocks on the workbench. The hydraulic unit must not be tilted or turned upside down. No component of the hydraulic unit should be loosened or disassembled.

10. Loosen the nut holding the bracket to the hydraulic unit and remove the bracket.

11. Disconnect the external ground wire from the bracket.

To install:

12. Install the bracket if was removed. Connect the ground wire to the bracket.

13. Install the hydraulic unit into the vehicle, keeping it upright at all times.

14. Install the retaining nuts and tighten them.

15. Connect the hydraulic unit wiring harness.

16. Connect each brake line loosely to the correct port and double check the placement. Tighten each line to 11 ft. lbs. (15 Nm).

17. Fill the reservoir to the MAX line with brake fluid.

18. Bleed the master cylinder, then bleed the brake lines.

19. Secure the relay box in position and install the air duct.

20. Install the splash shield.

Anti-Lock Control Unit

REMOVAL & INSTALLATION

1. Ensure that the ignition switch is **OFF** throughout the procedure.
2. Remove the interior right rear quarter trim panel and rear seat back and/or cushion.
3. Release the lock on the bottom of the connector; disconnect the multi-pin connector from the control unit. On Laser and Talon, access may be easier if the external ground is disconnected from the bracket.
4. Remove the retaining nuts and remove the control unit from its bracket. The bracket may be removed if desired.

To install:

5. Place the bracket in position. Install the controller and tighten the retaining nuts.
6. Connect the ground wire to the bracket if it was removed. Ensure a proper, tight connection. The ground must be connected before the multipin harness is connected.
7. Connect the multi-pin connector and secure the lock.
8. Install the rear quarter trim panel and seat.

G-Sensor

The G-Sensor is found only on all wheel drive vehicles.

REMOVAL & INSTALLATION

Laser and Talon

1. Ensure that the ignition switch is **OFF** throughout the procedure.
2. Remove the rear seat cushion.
3. Disconnect the wiring harness to G-sensor.
4. Remove the retaining bolts and remove the sensor.
5. To install, position the sensor, tighten the retaining bolts and connect the harness.
6. Install the rear seat cushion.

Stealth

1. Disconnect the negative battery cable. Remove the rearmost console assembly.
2. Remove the front console assembly.
3. Disconnect the G-sensor wiring harness.
4. Remove the G-sensor from the bracket. Remove the bracket if desired.

To install:

5. Reinstall the bracket. Tighten the bolts to 4 ft. lbs. (5 Nm.)
6. Install the G-sensor and connect the wiring harness.

7. Install the front and rear console assemblies.

Wheel Speed Sensors
— CAUTION —

Vehicles equipped with air bag systems will have wiring and system components in the fender or wheel well area. The ABS components must be correctly identified before beginning repairs. Improper work procedures may cause impaired function of the ABS and/or SRS systems

REMOVAL & INSTALLATION

1. Disconnect the negative battery cable. Raise and safely support the vehicle.
2. Remove the wheel and tire.
3. Remove the inner fender or splash shield.
4. Beginning at the sensor end, carefully disconnect or release each clip and retainer along the sensor wire. Take careful note of the exact position of each clip; they must be reinstalled in the identical position. Rear wheel sensor harnesses will be held by plastic wire ties; these may be cut away but must be replaced at reassembly.
5. Disconnect the sensor connector at the end of the harness.
6. Remove the 2 bolts holding the speed sensor bracket to the knuckle and remove the assembly from the vehicle.

NOTE: The speed sensor has a pole piece projecting from it. This exposed tip must be protected from impact or scratches. Do not allow the pole piece to contact the toothed wheel during removal or installation.

7. Remove the sensor from the bracket.

To install:

8. Assemble the sensor onto the bracket and tighten the bolt to 10 ft. lbs. (14 Nm). Note that the brackets are different for the left and right front wheels. Each bracket has identifying letters stamped on it.
9. Temporarily install the speed sensor to the knuckle; tighten the bolts only finger-tight.
10. Route the cable correctly and loosely install the clips and retainers. All clips must be in their original position and the sensor cable must not be twisted. Improper installation may cause cable damage and system failure.

NOTE: The wiring in the harness is easily damaged by twisting and flexing. Use the white stripe on the outer insulation to keep the sensor harness properly placed.

11. Use a brass or other non-magnetic feeler gauge to check the air gap between the tip of the pole piece and the toothed wheel. Correct gap is 0.012–0.035 in. (0.3–0.9mm). Tighten the 2 sensor bracket bolts to 10 ft. lbs. (14 Nm) with the sensor located so the gap is the same at several points on the toothed wheel. If the gap is incorrect, it is likely that the toothed wheel is worn or improperly installed.
12. Tighten the screws and bolts for the cable retaining clips.
13. Install the inner fender or splash shield.
14. Install the wheel and tire. Lower the vehicle to the ground.

Front Toothed Wheel Rings

REMOVAL & INSTALLATION

1. Disconnect the negative battery cable. Raise and safely support the vehicle.
2. Remove the wheel and tire.
3. Remove the wheel speed sensor and disconnect sufficient harness clips to allow the sensor and wiring to be moved out of the work area.

NOTE: The speed sensor has a pole piece projecting from it. This exposed tip must be protected from impact or scratches. Do not allow the pole piece to contact the toothed wheel during removal or installation.

4. Remove the front hub and knuckle assembly.
5. Remove the hub from the knuckle.
6. Support the hub in a vise with protected jaws. Remove the retaining bolts from the toothed wheel and remove the toothed wheel.

To install:

7. Fit the new toothed wheel onto the hub and tighten the retaining bolts to 7 ft. lbs. (10 Nm).
8. Assemble the hub to the knuckle.
9. Install the hub and knuckle assembly to the vehicle.
10. Install the wheel speed sensor.
11. Install the wheel and tire.
12. Lower the vehicle to the ground.

Rear Toothed Wheel Rings

REMOVAL & INSTALLATION

Front Wheel Drive

1. Disconnect the negative battery cable. Raise and safely support the vehicle.
2. Remove the wheel and tire.

3. Remove the wheel speed sensor and disconnect sufficient harness clips to allow the sensor and wiring to be moved out of the work area.

NOTE: The speed sensor has a pole piece projecting from it. This exposed tip must be protected from impact or scratches. Do not allow the pole piece to contact the toothed wheel during removal or installation.

4. Remove the hub assembly.
5. Support the hub in a vise with protected jaws. Remove the retaining bolts from the toothed wheel and remove the toothed wheel.

To install:

6. Fit the new toothed wheel onto the hub and tighten the retaining bolts to 7 ft. lbs. (10 Nm).
7. Install the hub assembly to the vehicle.
8. Install the tounged washer and hub nut. Tighten to a maximun of 188 ft. lbs. (260 Nm), crimp at the indentation and install the grease cap.
9. Install the wheel speed sensor.
10. Install the wheel and tire.
11. Lower the vehicle to the ground.

All Wheel Drive

1. Disconnect the negative battery cable. Raise and safely support the vehicle.
2. Remove the wheel and tire.
3. Disconnect the parking brake cable at the caliper or shoes.
4. Remove the speed sensor and its O-ring. Disconnect sufficient clamps and wire ties to allow the sensor to be moved well out of the work area.

NOTE: The speed sensor has a pole piece projecting from it. This exposed tip must be protected from impact or scratches. Do not allow the pole piece to contact the toothed wheel during removal or installation.

5. Remove the brake caliper and brake disc.
6. Remove the 3 retaining nuts and bolts holding the outer end of the driveshaft to the companion flange. Swing the axle shaft away and support it with stiff wire. Do not overextend the joint in the axle; do not allow it to hang of its own weight.
7. Remove the retaining nut and washer on the back of the driveshaft. Use special tool MB990767 or equivalent, to counterhold the hub.
8. Remove the companion flange from the knuckle.
9. Using an axle puller which bolts to the wheel lugs, remove the axle shaft assembly.
10. Fit the shaft assembly in a press with the toothed wheel completely

supported by a bearing plate such as special tool MB990560 or equivalent.
11. Press the toothed wheel off the axle shaft.

To install:

12. Press the new toothed wheel onto the shaft with the groove facing the axle shaft flange.
13. Install the axle shaft to the knuckle and fit the companion flange in place.
14. Install the lock washer and a new self-locking nut on the axle shaft. Hold the axle shaft stationary and torque the nut to 116–159 ft. lbs. (160–220 Nm) for Laser, Talon and non-turbocharged Stealth. Torque to 188–217 ft. lbs. (260–300 Nm) for turbocharged Stealth.
15. Swing the axle assembly into place and install the nuts and bolts. Tighten each to 45 ft. lbs. (61 Nm).
16. Install the brake disc and caliper.
17. Install the wheel speed sensor. Always use a new O-ring.
18. Connect the parking brake cable to the caliper.
19. Install the wheel and tire; lower the vehicle to the ground.

CHASSIS ELECTRICAL

Air Bag

DISARMING

1. Position the front wheels in the straight ahead position and place the key in the **LOCK** position.

1. Glove box assembly
2. Speaker cover
3. Right kickpanel
4. Right knee protector
5. Glove box frame
6. Lap heater duct
7. Electrical connector
8. Hose
9. MPI control unit
10. Blower motor assembly
11. Blower case
12. Packing seal
13. Fan
14. Blower motor

2. Disconnect the negative battery cable and insulate the cable end with high-quality electrical tape or similar non-conductive wrapping.
3. Wait at least 1 minute before working on the vehicle. The air bag system is designed to retain enough voltage to deploy for a short period of time even after the battery has been disconnected.
4. If necessary, enter the vehicle from the passenger side and turn the key to unlock the steering column.

Heater Blower Motor

REMOVAL & INSTALLATION

Summit

1. Disconnect the negative battery cable.
2. Remove the glove box assembly and pry off the speaker cover to the lower right of the glove box.
3. Remove the passenger side lower cowl side trim kick panel.
4. Remove the passenger side knee protector, which is the panel surrounding in the glove box opening.
5. Remove the glove frame along top of glove box opening.
6. Remove the lap heater duct. This is a small piece on vehicles without a rear heater and much larger on vehicles with a rear heater.
7. Disconnect the electrical connector from the blower motor.
8. Remove the cooling tube from the blower assembly.
9. Remove the Multi-Point Injection computer from the lower side of the cowl.
10. Remove the blower motor assembly.

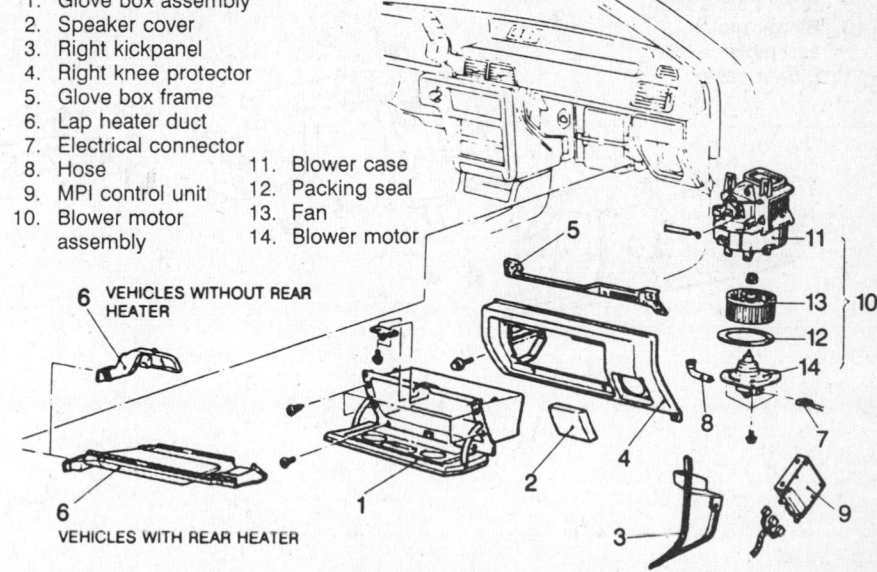

VEHICLES WITHOUT REAR HEATER

VEHICLES WITH REAR HEATER

Blower motor assembly—Summit

11. Separate the blower assembly case and packing seal from the blower motor flange.

12. Remove the fan retaining nut and fan in order to renew the motor.

To install:

13. Check that the blower motor shaft is not bent and that the packing and blower case are in good condition.

14. Assemble the fan and motor.

15. Install the blower assembly and connect the wiring and cooling tube.

16. Install the MPI computer.

17. Install the lap heater duct.

18. Install the glove box frame, interior trim pieces and glove box assembly.

19. Connect the negative battery cable and check the entire climate control system for proper operation.

Laser, Talon and Stealth

1. Disconnect the negative battery cable.

NOTE: If equipped with an air bag, be sure to disarm it before entering the vehicle.

2. On Laser and Talon, remove the right side duct, if equipped. On Stealth, remove the instrument panel undercover.

3. Remove the cooling tube from the blower assembly.

4. Remove the blower motor assembly.

5. Remove the packing seal.

6. Remove the fan retaining nut and fan in order to renew the motor.

To install:

7. Check that the blower motor shaft is not bent and that the packing is in good condition. Clean all parts of dust, etc.

8. Assemble the motor and fan. Install the blower motor and connect the wiring.

1. Duct, if so equipped
2. Molded hose
3. Blower motor assembly
4. Packing seal
5. Fan

Blower motor assembly—Laser and Talon

9. Install the cooling tube.

10. Install the duct or undercover.

11. Connect the negative battery cable and check the entire climate control system for proper operation.

Windshield Wiper Motor

REMOVAL & INSTALLATION
Summit

FRONT

1. Disconnect the negative battery cable.

2. Remove the windshield wiper arms by unscrewing the cap nuts and lifting the arms from the linkage posts.

3. Remove the front garnish panel.

4. Remove both windshield holders.

5. Remove the clips that hold the deck cover. If they are the pin type, they may be removed using the following procedure:

   a. Remove the clip by pressing down on the center pin with a suitable blunt pointed tool. Press down a little more than $\frac{1}{16}$ in. (2mm). This releases the clip. Pull the clip outward to remove it.

   b. Do not push the pin inward more than necessary because it may

1. Stopper
2. Blove box
3. Outer case
4. Undercover
5. Lower frame
6. Evaporator mounting bolt and nut
7. Air selection cable
8. Side frame
9. Blower assembly
10. Blower motor assembly
11. Blower case

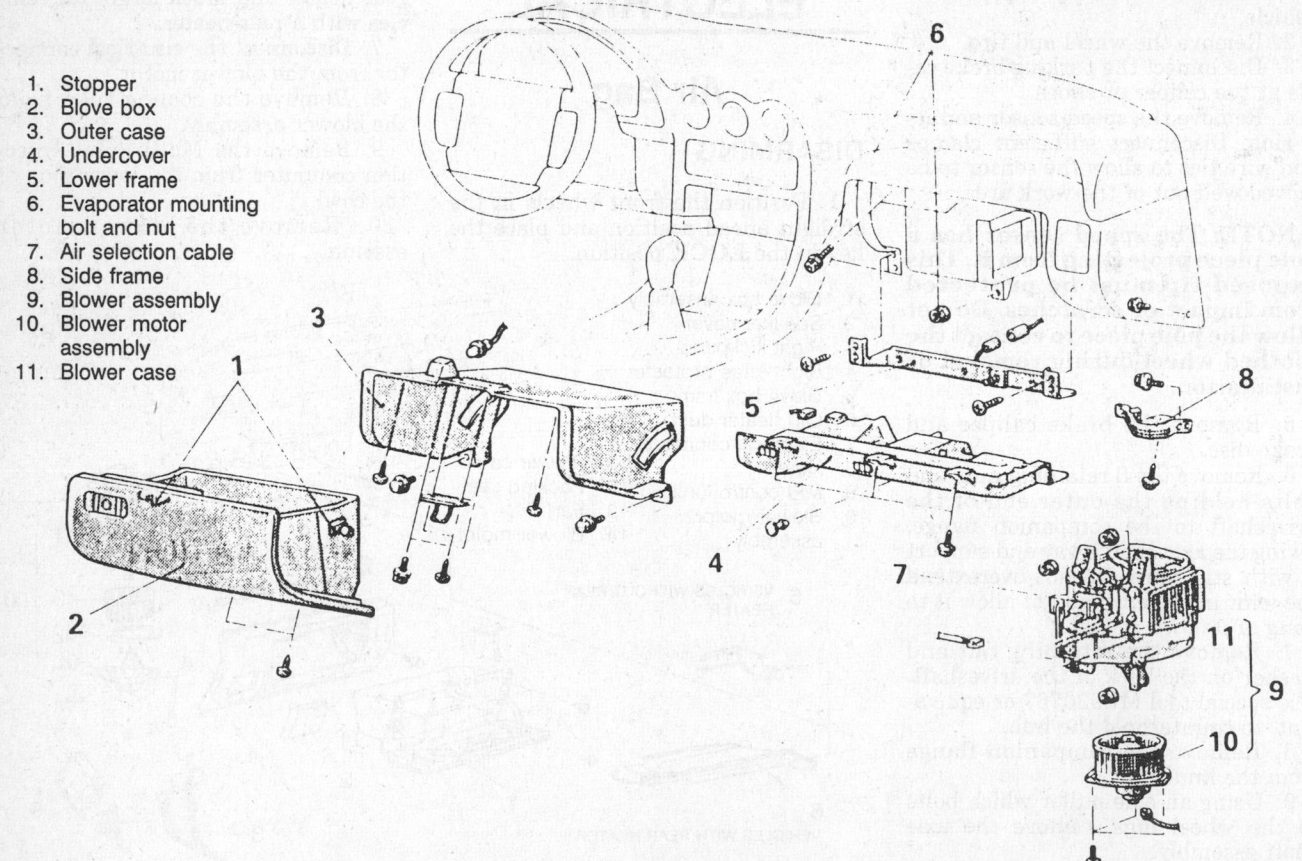

Blower motor assembly—Stealth

damage the grommet, or if pushed too far, the pin may fall in. Once the clips are removed, use a plastic trim stick to pry the deck cover loose.

6. Remove the air intake screen.

7. Loosen the wiper motor assembly mounting bolts and remove the windshield wiper motor. Disconnect the linkage and the motor assembly. If necessary, remove the linkage.

NOTE: The installation angle of the crank arm and motor has been factory set, do not remove them unless it is necessary to do so. If they must be removed, remove them only after marking their mounting positions.

To install:

8. Install the windshield wiper motor and connect the linkage.

9. When installing the trim and garnish pieces and reusing pin type clips, use the following procedure:

   a. With the pin pulled out, insert the trim clip into the hole in the trim.

   b. Push the pin inward until the pin's head is flush with the grommet.

   c. Check that the trim is secure.

10. Install the wiper arms.

11. Connect the negative battery cable and check the wiper system for proper operation.

REAR

1. Disconnect the negative battery cable.

2. Remove the hatchback wiper arm by removing the cap nut cover, unscrewing the cap nut and lifting the arm from the linkage post.

3. Remove the large interior trim panel. Use a plastic trim stick to unhook the trim clips of the liftgate trim. There will be a row of metal liftgate clips across the top. There will be 2 rows of trim clips that retain the rest of the panel.

4. Remove the rear wiper assembly. Do not loosen the grommet for the wiper post.

To install:

5. Install the motor and grommet. Mount the grommet so the arrow on the grommet is pointing downward.

6. Install the wiper arm.

7. Connect the negative battery cable and check rear wiper system for proper operation.

8. If operation is satisfactory, fit the tabs on the upper part of the liftgate trim into the liftgate clips and secure the liftgate trim.

Laser and Talon
FRONT

1. Disconnect the negative battery cable.

2. Remove the windshield wiper arms by unscrewing the cap nuts and lifting the arms from the linkage posts.

3. Remove the front garnish panel.

4. Remove the air inlet trim pieces.

5. Remove the hole cover.

6. Remove the wiper motor by loosening the mounting bolts, removing the motor assembly, then disconnecting the linkage.

NOTE: The installation angle of the crank arm and motor has been factory set; do not remove them unless it is necessary to do so. If they must be removed, remove them only after marking their mounting positions.

To install:

7. Install the windshield wiper motor and connect the linkage.

8. Reinstall all the trim pieces.

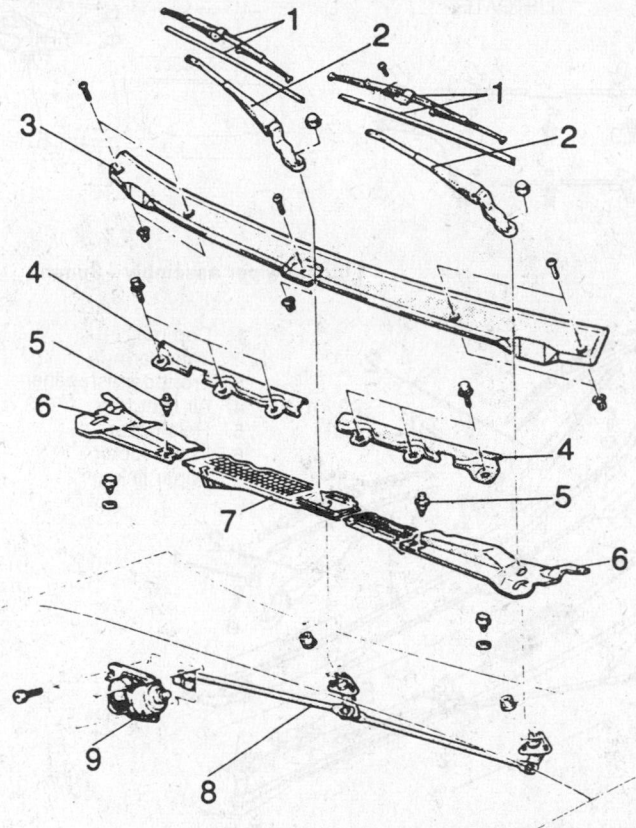

1. Wiper blades
2. Wiper arms
3. Front deck garnish
4. Windshield holder
5. Pin type trim clip
6. Deck cover
7. Air intake screen
8. Wiper linkage
9. Wiper motor

Windshield wiper assembly—Summit

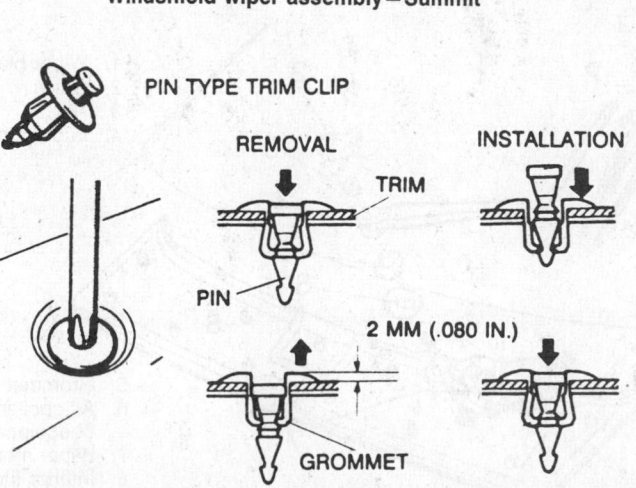

PIN TYPE TRIM CLIP

REMOVAL

INSTALLATION

TRIM

PIN

2 MM (.080 IN.)

GROMMET

Remove the pin clips with care so they can be reused

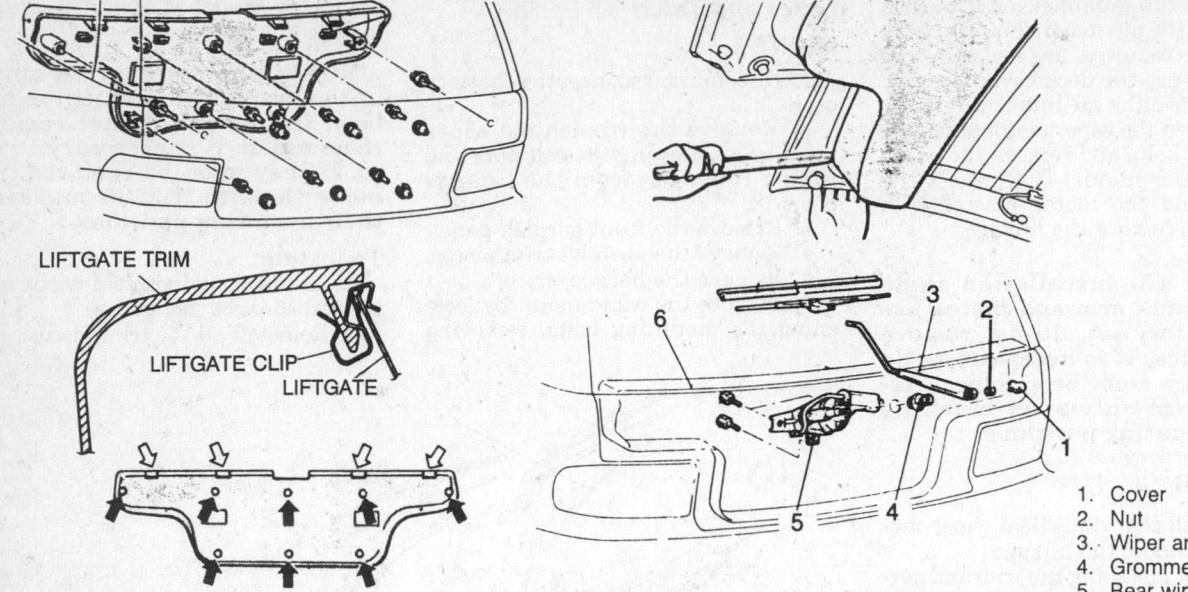

LIFTGATE TRIM

LIFTGATE CLIP

LIFTGATE

← **TRIM CLIPS**
⇐ **LIFTGATE CLIPS**

Liftgate wiper assembly—Summit

1. Cover
2. Nut
3. Wiper arm
4. Grommet
5. Rear wiper motor
6. Liftgate trim panel

1. Cap nut
2. Wiper arm
3. Front garnish panel
4. Air inlet trim
5. Hole cover
6. Wiper motor
7. Wiper linkage

Windshield wiper assembly—Laser and Talon

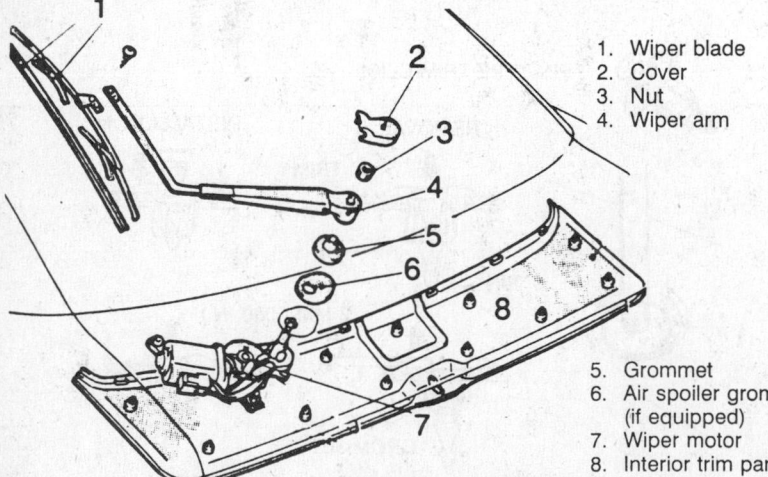

1. Wiper blade
2. Cover
3. Nut
4. Wiper arm

5. Grommet
6. Air spoiler grommet (if equipped)
7. Wiper motor
8. Interior trim panel

Liftgate wiper assembly—Laser and Talon

9. Reinstall the wiper blades. Note that the driver's side wiper arm should be marked **D** and the passenger's side wiper arm should be marked **A**. The identification marks should be located at the base of the arm, near the pivot. Install the arms so the blades are 1 inch from the garnish molding when parked.

10. Connect the negative battery cable and check the wiper system for proper operation.

REAR

1. Disconnect the negative battery cable.

2. Remove the rear wiper arm by removing the cover, unscrewing the nut and lifting the arm from the linkage post.

3. Remove the large interior trim panel. Use a plastic trim stick to unhook the trim clips of the liftgate trim.

4. If equipped with rear air spoiler, remove the grommet.

5. Remove the rear wiper assembly. Do not loosen the grommet for the wiper post.

To install:

6. Install the motor and grommet. Mount the grommet so the arrow on the grommet is pointing upward.

7. Install the wiper arm.

8. Connect the negative battery cable and check the rear wiper for proper operation.

9. If operation is satisfactory, fit the tabs on the upper part of the liftgate trim into the liftgate clips and secure the liftgate trim.

Stealth

FRONT

1. Disconnect the negative battery cable.

2. Remove the windshield wiper arms by unscrewing the cap nuts and lifting the arms from the linkage posts.

3. Remove the access hole cover.

4. Remove the wiper motor mounting bolts.

5. Detach the motor crank arm from the wiper linkage and remove the motor.

NOTE: The installation angle of the crank arm and motor has been factory set; do not remove them unless it is necessary to do so. If they must be removed, remove them only after marking their mounting positions.

To install:

6. Install the windshield wiper motor and connect the linkage.

7. Install the access hole cover.

8. Reinstall the wiper blades. Note that the driver's side wiper arm should be marked **D** and the passenger's side wiper arm should be marked **A**. The identification marks should be located at the base of the arm, near the pivot. Install the arms so the blades are parallel to the garnish molding when parked.

9. Connect the negative battery cable and check the wiper system for proper operation.

REAR

1. Disconnect the negative battery cable.

2. Remove the liftgate lower trim. Remove the clips that hold the trim by using the following procedure:

a. Remove the clip by pressing down on the center pin with a suitable blunt pointed tool. Press down a little more than $\frac{1}{16}$ in. (2mm). This releases the clip. Pull the clip outward to remove it.

b. Do not push the pin inward more than necessary because it may damage the grommet, or if pushed too far, the pin may fall in. Once the clips are removed, use a plastic trim stick to pry the trim cover loose.

3. Remove the rear spoiler, center brace and center brake light.

4. Lift the small cover, remove the retaining nut and remove the wiper arm and spacer.

5. Remove the mounting bolts and remove the wiper motor.

To install:

6. Install the motor and install the retaining bolts.

7. Install the spacer, wiper arm and retaining nut. The arm should be positioned so the upper tip points to the upper left corner of the rear window when parked. Connect the battery and check the operation of the motor before proceeding. If satisfactory, disconnect the cable and proceed.

8. Install the rear spoiler and related parts.

1. Wiper blade
2. Wiper arm
3. Deck garnish
4. Right side air inlet garnish
5. Hole cover
6. Wiper cover
7. Linkage
8. Battery
9. Battery tay
10. Washer tank
11. Washer motor
12. Level sensor
13. Washer nozzle
14. Washer tube

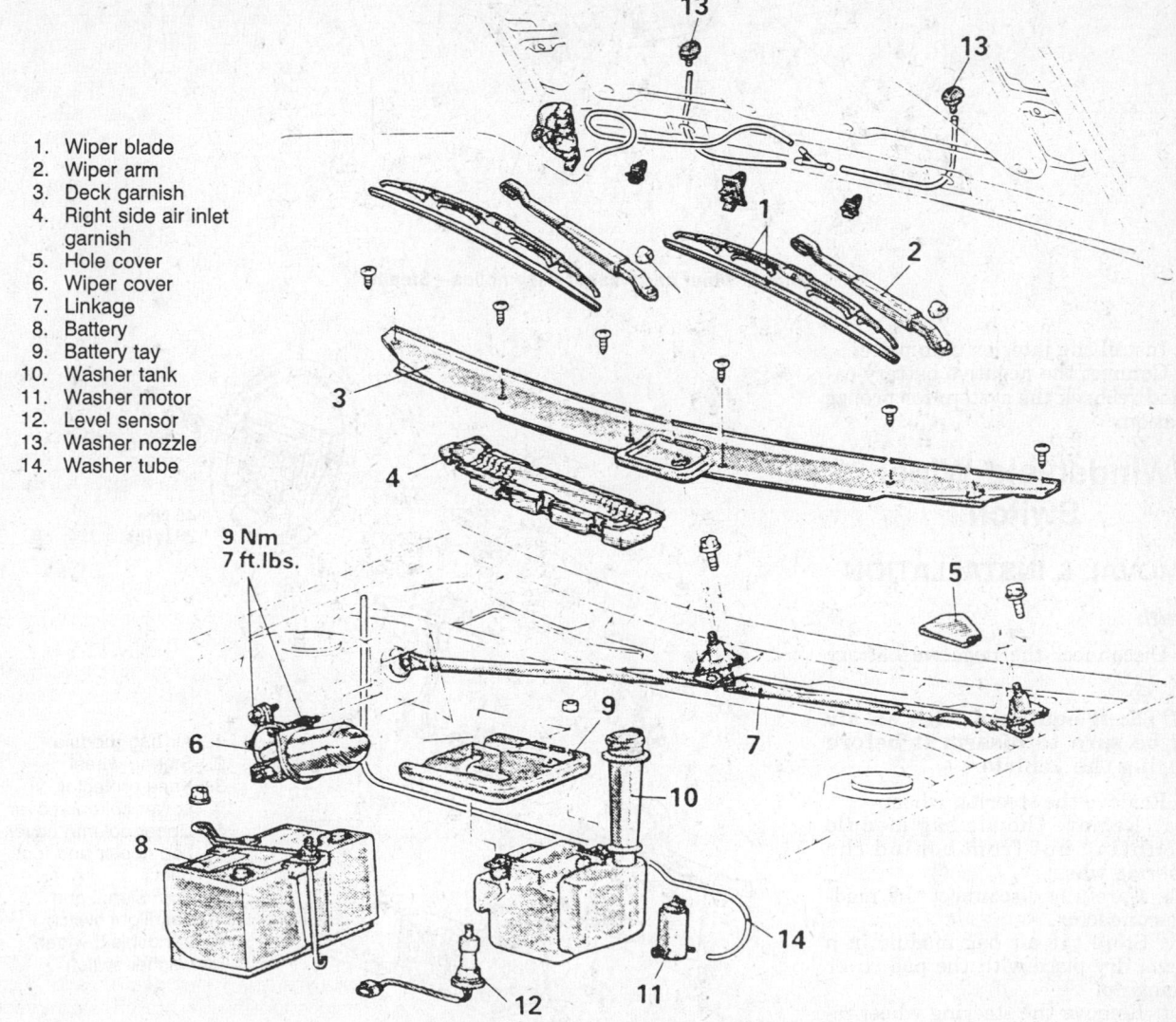

Windshield wiper assembly—Stealth

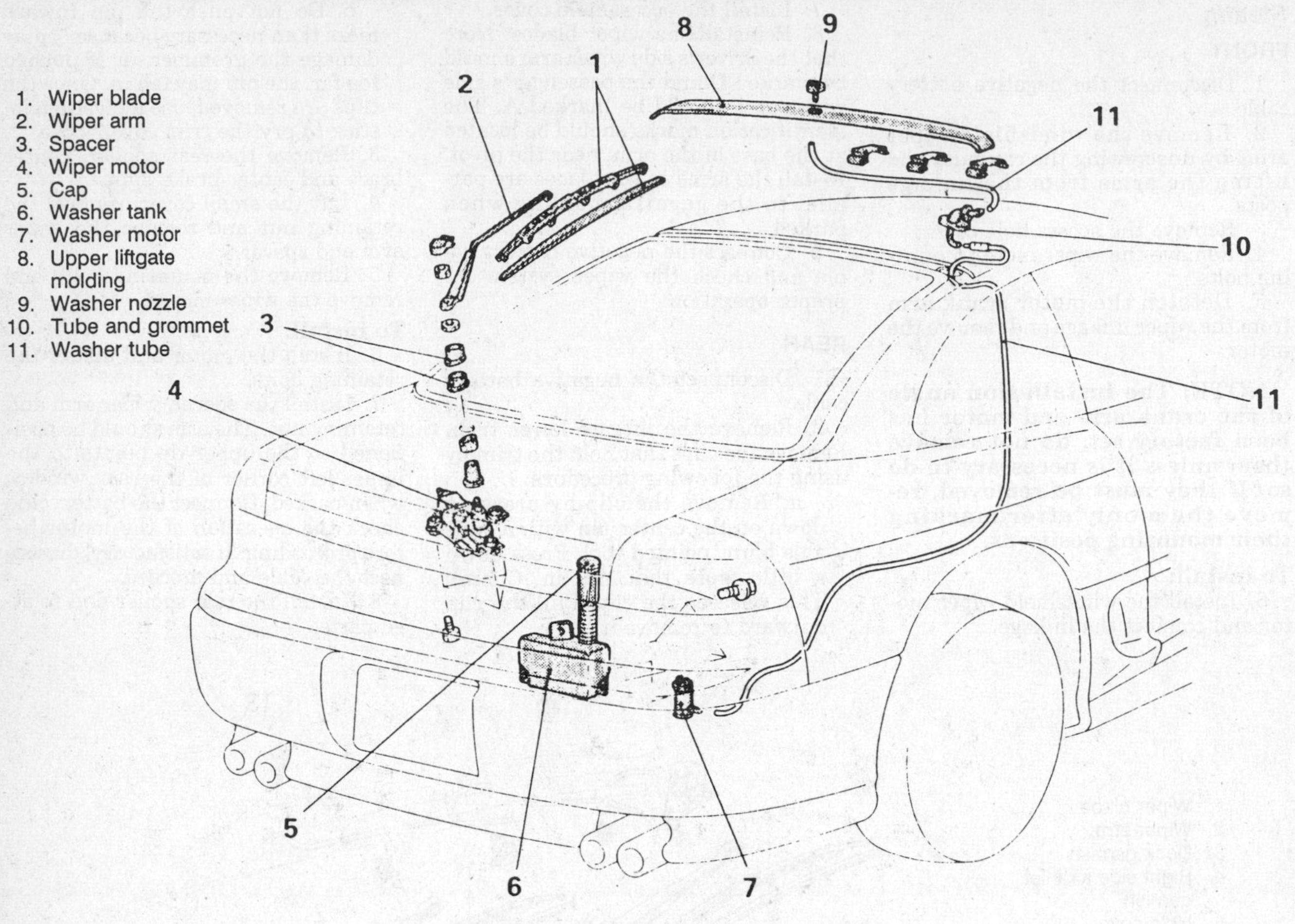

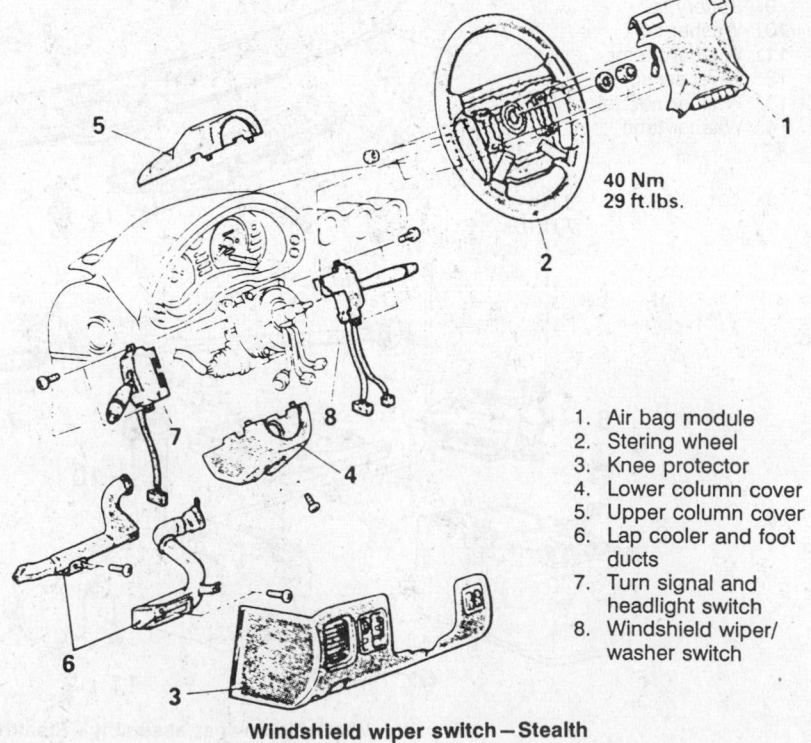

1. Wiper blade
2. Wiper arm
3. Spacer
4. Wiper motor
5. Cap
6. Washer tank
7. Washer motor
8. Upper liftgate molding
9. Washer nozzle
10. Tube and grommet
11. Washer tube

Liftgate wiper and washer assemblies — Stealth

9. Install the interior trim piece.
10. Connect the negative battery cable and recheck the system for proper operation.

Windshield Wiper Switch

REMOVAL & INSTALLATION

Stealth

1. Disconnect the negative battery cable.

NOTE: If equipped with an air bag, be sure to disarm it before entering the vehicle.

2. Remove the steering wheel:
   a. Remove the air bag module mounting nut from behind the steering wheel.
   b. Carefully disconnect the module connector.
   c. Store the air bag module in a clean, dry place with the pad cover facing up.
   d. Remove the steering wheel retaining nut and use a steering wheel

40 Nm
29 ft.lbs.

1. Air bag module
2. Stering wheel
3. Knee protector
4. Lower column cover
5. Upper column cover
6. Lap cooler and foot ducts
7. Turn signal and headlight switch
8. Windshield wiper/ washer switch

Windshield wiper switch — Stealth

puller to remove the wheel. Do not use a hammer or the collapsible mechanism in the column could be damaged.

3. Remove the hood lock release handle.

4. Remove the switches from the knee protector below the steering column, and remove the exposed retaining screws. Then remove the knee protector.

5. Remove the column covers.

6. Remove necessary duct work and disconnect the windshield wiper switch connectors.

7. Remove the retaining screws and remove the windshield wiper switch assembly from the steering column.

To install:

8. Install the wiper switch to the steering column and connect the connectors.

9. Install any removed duct work.

10. Install the column covers.

11. Install the knee protector and switches.

12. Install the hood release handle.

13. Center the clock spring by aligning the **NEUTRAL** mark on the clock spring with the mating mark on the casing. Then install the steering wheel and torque the retaining nut to 29 ft. lbs. (40 Nm).

14. Connect the negative battery cable and check the windshield wiper and washer for proper operation.

Instrument Cluster

REMOVAL & INSTALLATION

Summit

1. Disconnect the negative battery cable.

2. Remove the center panel.

3. Remove the knee protector. If pin type clips are used, they may be removed using the following procedure:

   a. This type of clip is removed by pressing down on the center pin with a suitable blunt pointed tool. Press down a little more than $1/16$ in. (2mm). This releases the clip. Pull the clip outward to remove it.

   b. Do not push the pin inward more than necessary because it may damage the grommet or the pin may fall in, if pushed in too far. Once the clips are removed, use a plastic trim stick if necessary to pry the knee protector loose.

4. Remove the instrument cluster bezel.

5. Remove the instrument cluster. Disassemble and remove gauges or the speedometer as required.

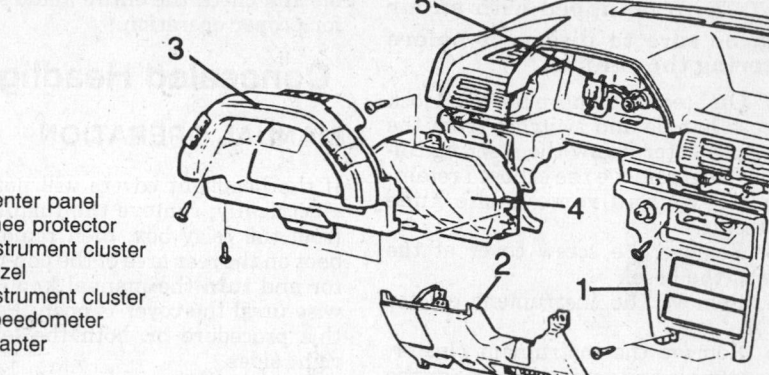

1. Center panel
2. Knee protector
3. Instrument cluster bezel
4. Instrument cluster
5. Speedometer adapter

Instrument cluster assembly—Summit

NOTE: If the speedometer cable adapter must be serviced, disconnect the cable at the transaxle end. Pull the cable slightly toward the vehicle interior, release the lock by turning the adapter to the right or left and remove the adapter.

6. The installation is the reverse of the removal procedure. Use care not to damage the printed circuit board or any gauge components.

7. Connect the negative battery cable and check all cluster-related items for proper operation.

Laser, Talon and Stealth

1. Disconnect the negative battery cable.

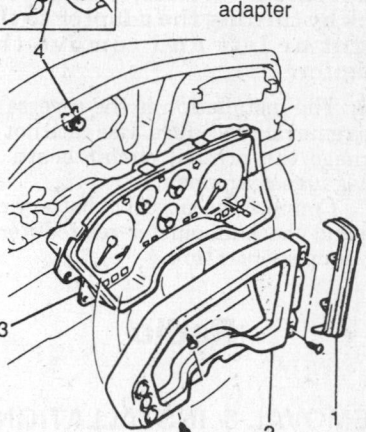

1. Screw cover
2. Instrument cluster bezel
3. Instrument cluster
4. Speedometer adapter

Instrument cluster assembly—Laser and Talon

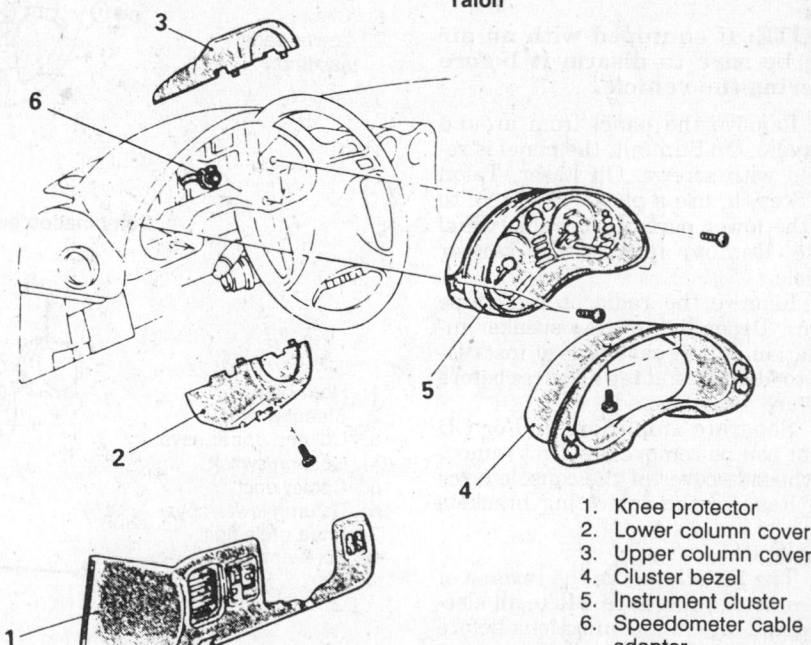

1. Knee protector
2. Lower column cover
3. Upper column cover
4. Cluster bezel
5. Instrument cluster
6. Speedometer cable adapter

Instrument cluster assembly—Stealth

NOTE: If equipped with an air bag, be sure to disarm it before entering the vehicle.

2. On Stealth, remove the hood lock release handle and switches from the knee protector below the steering column. Then remove the exposed retaining screws and remove the knee protector.

3. Remove the screw cover at the side of the bezel.

4. Remove the instrument cluster bezel.

5. Remove the instrument cluster. Disassemble and remove gauges or the speedometer as required.

NOTE: If the speedometer cable adapter must be serviced, disconnect the cable at the transaxle end. Pull the cable slightly toward the vehicle interior, release the lock by turning the adapter to the right or left and remove the adapter.

6. The installation is the reverse of the removal procedure. Use care not to damage the printed circuit board or any gauge components.

7. Connect the negative battery cable and check all cluster-related items for proper operation.

Radio

REMOVAL & INSTALLATION

1. Disconnect battery negative cable.

NOTE: If equipped with an air bag, be sure to disarm it before entering the vehicle.

2. Remove the panel from around the radio. On Summit, the panel is retained with screws. On Laser, Talon and Stealth, use a plastic trim tool to pry the lower part of the radio panel loose. Remove it from the center console.

3. Remove the radio or radio/tape player. Depending on the speaker installation, it may save time at installation to identify and tag all wires before they are disconnected.

4. Separate amplifiers and/or CD player can be removed by first removing the side cover of the console box.

5. Remove the mounting brackets from the radio.
To install:

6. The installation is the reverse of the removal procedure. Make all electrical and antenna connections before fastening the radio assembly in place.

7. Install the center panel.

8. Connect the negative battery ca-

ble and check the entire audio system for proper operation.

Concealed Headlights

MANUAL OPERATION

If the headlight covers will not raise electrically, remove the fusible link from the relay box, then remove the boot on the rear area of the pop-up motor and turn the manual knob clockwise until the cover is open. Perform this procedure on both the left and right sides.

Combination Switch

REMOVAL & INSTALLATION

Summit

NOTE: **The headlights, turn signals, dimmer switch, windshield/washer and, on some models, the cruise control function are all built into 1 multi-function combination switch that is mounted on the steering column.**

1. Disconnect the negative battery cable.

2. Remove the knee protector panel under the steering column, then the upper and lower column covers.

3. Remove the horn pad by pulling the lower end.

4. Matchmark and remove the steering wheel with a steering wheel puller. Do not hammer on the steering wheel to remove it or the collapsible mechanism may be damaged.

5. Disconnect all connectors, remove the wiring clip and remove the column switch assembly.

To install:

6. Install the switch assembly and secure the clip. Make sure no wires are pinched or out of place.

7. Install the steering wheel. Torque the steering wheel-to-column nut to 29 ft. lbs. (40 Nm).

8. Install the column covers and knee protector.

9. Connect the negative battery cable and check all functions of the combination switch for proper operation.

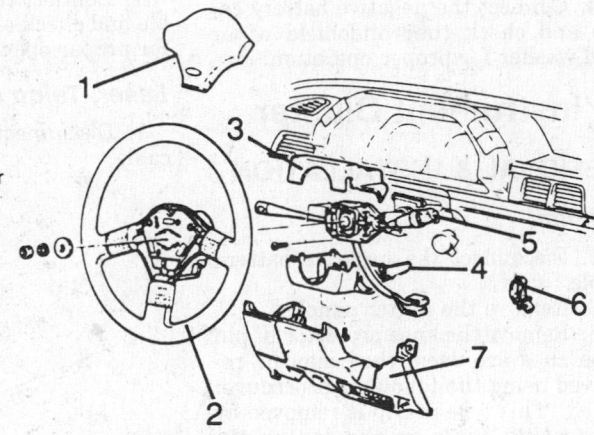

1. Horn pad
2. Steering wheel
3. Column upper cover
4. Column lower cover
5. Column switch
6. Clip
7. Lower panel assembly

Combination switch assembly—Summit

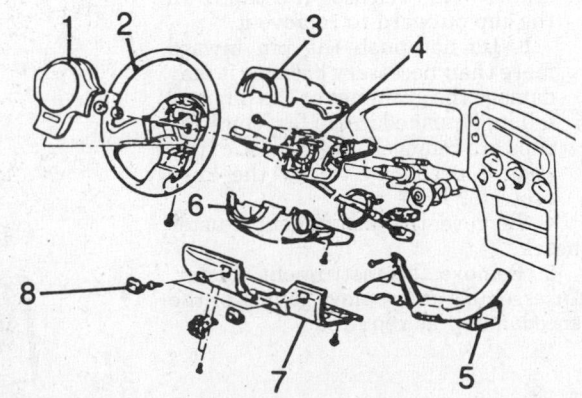

1. Horn pad
2. Steering wheel
3. Column upper cover
4. Column switch
5. Cooler duct
6. Column lower cover
7. Knee protector
8. Screw plugs

Combination switch assembly—Laser and Talon

Laser and Talon

NOTE: The headlights, turn signals, dimmer switch, windshield/washer and, on some models, the cruise control function are all built into 1 multi-function combination switch that is mounted on the steering column.

1. Disconnect the negative battery cable.
2. Remove the knee protector panel under the steering column, then the upper and lower column covers.
3. Remove the horn pad by pulling the lower end.
4. Matchmark and remove the steering wheel with a steering wheel puller. Do not hammer on the steering wheel to remove it or the collapsible mechanism may be damaged.
5. Locate the rectangular plugs in the knee protector on either side of the steering column. Pry these plugs out and remove the screws. Remove the screws from the hood lock release lever and remove the knee protector.
6. Remove the upper and lower column covers.
7. Remove the lap cooler ducts.
8. Remove the band retaining the switch wiring.
9. Disconnect all connectors, remove the wiring clip and remove the column switch assembly.

To install:
10. Install the switch assembly and secure the clip. Make sure no wires are pinched or out of place.
11. Install the lap cooler ducts.
12. Install the column covers and knee protector.
13. Install the steering wheel. Torque the steering wheel-to-column nut to 29 ft. lbs. (40 Nm).
14. Connect the negative battery cable and check all functions of the combination switch for proper operation.

Stealth

NOTE: The headlights, turn signals and dimmer switch are all built into 1 multi-function combination switch that is mounted on the left side of the steering column.

1. Disconnect the negative battery cable.

NOTE: If equipped with an air bag, be sure to disarm it before entering the vehicle.

2. Remove the steering wheel:
   a. Remove the air bag module mounting nut from behind the steering wheel.
   b. Carefully disconnect the module connector.
   c. Store the air bag module in a

clean, dry place with the pad cover facing up.
   d. Remove the steering wheel retaining nut and use a steering wheel puller to remove the wheel. Do not use a hammer or the collapsible mechanism in the column could be damaged.
3. Remove the hood lock release handle.
4. Remove the switches from the knee protector below the steering column, and remove the exposed retaining screws. Then remove the knee protector.
5. Remove the column covers.
6. Remove necessary duct work and disconnect the combination switch connectors.
7. Remove the retaining screws and remove the combination switch assembly from the steering column.

To install:
8. Install the switch to the steering column, and connect the connectors.
9. Install any removed duct work.
10. Install the column covers.
11. Install the knee protector and switches.
12. Install the hood release handle.
13. Center the clock spring by aligning the **NEUTRAL** mark on the clock spring with the mating mark on the casing. Then install the steering wheel and torque the retaining nut to 29 ft. lbs. (40 Nm).
14. Connect the negative battery cable and check all functions of the combination switch for proper operation.

Ignition Lock/Switch
REMOVAL & INSTALLATION
Summit, Laser and Talon

1. Disconnect the negative battery cable.
2. Remove the lower instrument panel knee protector.
3. Remove the lower steering column cover.
4. Remove the clip that holds the wiring against the steering column.
5. Unplug the ignition switch from the steering lock cylinder.
6. Insert the key into the steering lock cylinder and turn to the **ACC** position.
7. With a small pointed tool, push the lock pin of the steering lock cylinder inward and pull the lock out.

NOTE: When equipped with automatic transaxle, Laser and Talon have safety-lock systems and will have a key interlock cable installed in a slide lever on the side of the key lock.

To install:
8. Install the lock cylinder; make sure the lock pin snaps into place.

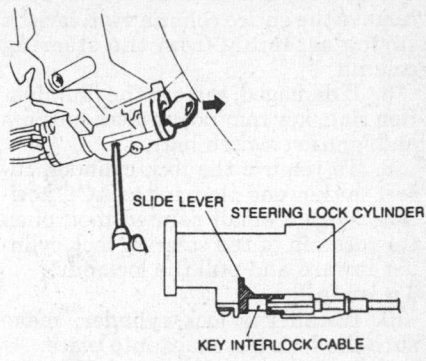
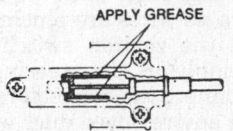

Ignition lock and optional automatic transaxle safety interlock cable—Summit, Laser and Talon

9. Install the ignition switch plug carefully and make sure no wires are pinched.
10. Install the wiring clip.
11. Install the steering column covers.
12. Install the knee protector.
13. Connect the negative battery cable and check the ignition switch and lock for proper operation.

Stealth

1. Disconnect the negative battery cable.

NOTE: If equipped with an air bag, be sure to disarm it before entering the vehicle.

2. Remove the steering wheel:
   a. Remove the air bag module mounting nut from behind the steering wheel.
   b. Carefully disconnect the module connector.
   c. Store the air bag module in a clean, dry place with the pad cover facing up.
   d. Remove the steering wheel retaining nut and use a steering wheel puller to remove the wheel. Do not use a hammer or the collapsible mechanism in the column could be damaged.
3. Remove the hood lock release handle.
4. Remove the switches from the knee protector below the steering column, and remove the exposed retaining screws. Then remove the knee protector.
5. Remove the column covers.
6. Remove necessary duct work and disconnect the windshield wiper and combination switch connectors.
7. Remove the retaining screws and

remove the entire column switch/clock spring assembly from the steering column.

8. If damaged, remove the illumination ring, key reminder switch harness and ignition switch harness.

9. To remove the lock cylinder, insert the key and place in the **ACC** position. With a small pointed tool, push the lock pin of the steering lock cylinder inward and pull the lock out.

To install:

10. Install the lock cylinder; make sure the lock pin snaps into place.

11. Install any other removed items, making sure no wires are pinched.

12. Install the column switch/clock spring assembly to the steering column and connect the connectors.

13. Install any removed duct work.

14. Install the column covers.

15. Install the knee protector and switches.

16. Install the hood release handle.

17. Center the clock spring by aligning the **NEUTRAL** mark on the clock spring with the mating mark on the casing. Then install the steering wheel and torque the retaining nut to 29 ft. lbs. (40 Nm).

18. Connect the negative battery cable and check all functions of column-mounted switches and the ignition switch for proper operation.

Stoplight Switch

ADJUSTMENT

1. Disconnect the negative battery cable.

NOTE: If equipped with an air bag, be sure to disarm it before entering the vehicle.

2. The stoplight switch works off the brake pedal lever. To adjust, disconnect the electrical connection and loosen the switch locknut.

3. Screw the switch inward until it contacts the stop on the brake pedal arm. Back out the switch ½–1 full turn. The gap between the switch plunger and the brake lever stop should be 0.020–0.040 in. (0.5–1.0mm).

4. Tighten the locknut and connect the wires.

5. Connect the negative battery cable.

6. Make sure the stoplights come on when the brake pedal is depressed and go out when the pedal is released. Also, make sure the cruise control system operates properly.

REMOVAL & INSTALLATION

1. Disconnect the negative battery cable.

NOTE: If equipped with an air bag, be sure to disarm it before entering the vehicle.

2. Locate the stoplight switch above the brake pedal lever.

3. Disconnect the wiring connectors from the switch and unscrew the switch.

4. The installation is the reverse of the removal procedure. Install the replacement switch and adjust to 0.020–0.040 in. clearance.

5. Connect the stoplight wires.

6. Connect the negative battery cable.

7. Make sure the stoplights come on when the brake pedal is depressed and go out when the pedal is released. Also, make sure the cruise control system operates properly.

Clutch Switch

ADJUSTMENT

The clutch interlock switch is located at the top of the clutch pedal arm. Note that there may be 2 switches; 1 will be a cruise control cut-out switch.

1. Clutch interlock switch adjustment is made with the pedal fully depressed.

2. Measure the gap between the switch plunger and the arm stop. The gap should be 0.140 in. (3.5mm).

3. If adjustment is necessary, loosen the locknut, turn and adjust.

4. After completing the adjustment, check that the pedal free-play, measured at the face of the pedal pad is 0.240–0.510 in. (6–13mm). The distance between the pedal pad and the firewall when the clutch is disengaged should be 2.20 in. or more for Summit and Stealth, 2.80 in. or more for Laser and Talon. If these dimensions are not right, the hydraulic clutch system may need further servicing.

REMOVAL & INSTALLATION

1. Disconnect the negative battery cable.

NOTE: If equipped with an air bag, be sure to disarm it before entering the vehicle.

2. Locate the interlock switch above the clutch pedal lever.

3. Disconnect the wiring connectors from the switch and unscrew the switch.

4. Installation is the reverse of the removal procedure. Install the replacement switch and adjust to 0.140 in. (3.5mm) clearance.

5. Reconnect the interlock wires.

6. Make sure the engine will not start unless the clutch pedal is depressed. Also, make sure the cruise control system operates properly.

Neutral Safety Switch

ADJUSTMENT

1. Locate the neutral safety switch on the top of the transaxle. Note that several different cable attaching methods have been used. The procedure here can be used as a general guide for all.

2. Place the selector lever in **N**.

3. Loosen the 2 adjusting nuts to free up the cable and lever.

4. Place the safety switch manual control lever in **N**.

5. Note that 1 end of the safety switch manual control lever has a 12mm wide square end. There is also a 12mm wide tab on the switch body flange. Loosen both retaining bolts and turn the safety switch until these portions align. Tighten the bolts, making sure the switch doesn't move.

6. Loosen the adjuster nuts and gently pull the cable to remove any slack. Gently tighten adjusting nut until it just starts to contact the adjuster. Secure adjusting nut with its locknut then turn nut to lock.

7. Verify that the switch lever moves to positions corresponding to each position of the selector lever.

8. Make sure the engine only starts in **P** and **N**. Also make sure the reverse lights come on in **R**.

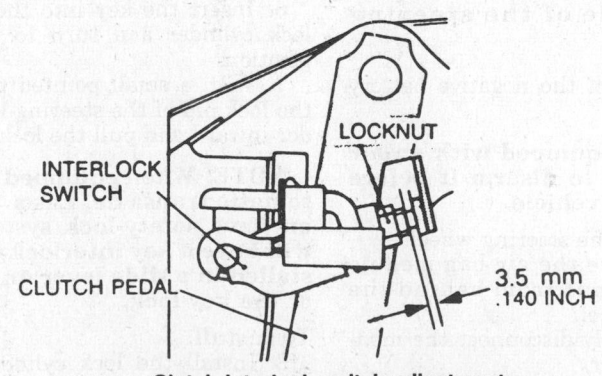

Clutch interlock switch adjustment

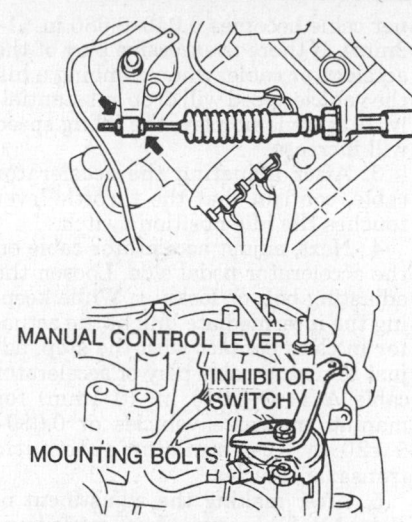

MANUAL CONTROL LEVER

INHIBITOR SWITCH

MOUNTING BOLTS

Automatic transaxle neutral safety (inhibitor) switch and adjustment

REMOVAL & INSTALLATION

1. Disconnect the negative battery cable.
2. Disconnect the selector cable from the lever.
3. Remove the 2 retaining screws and lift off the switch.
4. The installation is the reverse of the removal procedure. Do not tighten the bolts until the switch is adjusted.
5. Make sure the engine only starts in **P** and **N**. Also make sure the reverse lights come on in **R**.

Fuses, Fusible Links and Relays

LOCATION

Summit

FUSES AND FUSIBLE LINKS

The Summit has several fuse panels. One fuse panel is located on the passenger side, under the hood, just behind the battery. It shares the panel with a bank of relays. This panel also contains several fusible links. Another fuse panel is on the driver's side, under the hood, up front behind the headlight. It controls air conditioning functions. A third fuse panel is the multi-purpose fuse block located inside the vehicle, on the left side behind the driver's knee protector.

RELAYS

The Summit uses a number of relays. The headlight relay, power window relay, radiator fan motor relay and alternator relay share a panel with fuses and fusible links. It is located on the passenger side, under the hood, just behind the battery. Another panel is on the driver's side, under the hood, up front behind the headlight. This panel also shares space with air conditioning system fuses. These relays are the air conditioner compressor relay, the condenser fan motor relay and the condenser fan motor control relay.

The intermittent wiper relay is incorporated into the column switch. The seat belt warning timer relay is located behind the instrument panel to the right of the center air conditioning outlets. The Multi-Point Injection control relay is mounted behind the forward part of the console, on the left side, while the starter relay is on the right side. The defogger relay is located under the driver's left side knee protector. The door lock relay is behind the driver's side kick panel, at the bottom.

The multi-purpose fuse panel located under the driver's left side knee protector also contains the heater relay, the turn signal and hazard flasher unit and the defogger timer. The automatic seatbelt motor relay is located in the driver's side windshield post in Summit Hatchbacks, and inside the trim panel on the driver's side rear quarter panel, just behind the door post in Summit Sedans.

Laser and Talon

FUSES AND FUSIBLE LINKS

The Laser and Talon has several fuse panels. There are 3 main fusible links the MPI circuit—20 amp, the radiator fan motor circuit—30 amp and the ignition switch circuit—30 amp. They are found under the hood in a centralized junction with the battery positive cable clamp. Another fuse panel is located on the passenger side, under the hood, just forward of the strut tower. It shares the panel with a bank of relays. This panel contains fuses and several fusible links. Another fuse panel is on the driver's side, under the hood, back against the firewall. A fourth fuse panel is the multi-purpose fuse block located inside the vehicle, on the left side behind the driver's knee protector.

RELAYS

The Laser and Talon uses a number of relays. A centralized fuse/relay panel on the passenger side, under the hood, just forward of the strut tower contains the taillight relay, headlight relay, radiator fan motor relay, pop-up (retractable light) motor relay, power window relay, alternator relay and fog light relay. Also in the engine compartment on the driver's side a panel are 2 air conditioning condenser fan relays and the air conditioning compressor clutch relay. Inside the vehicle, the interior relay box contains the door lock relay, starter relay, defogger timer and room of other relays as required (use in Canada, etc).

Stealth

FUSES AND FUSIBLE LINKS

The Stealth has several fuse panels. One fuse panel is located on the passenger side, under the hood, just forward of the air flow box. It shares the panel with a bank of relays. This panel also contains several fusible links. Another fuse panel is the multi-purpose fuse block located under the instrument panel, on the left side behind the driver's knee protector.

RELAYS

The Stealth uses a number of relays. A centralized fuse/relay panel on the driver side, under the hood, just forward of the strut tower contains the radiator fan relay, air conditioning system relays and others. Also in the engine compartment in front of the air flow box is a relay bank containing the the taillight relay, headlight relay, pop-up (retractable light) motor relay, horn relay, alternator relay and fog light relay. Inside the vehicle, relays above the fuse box are for the blower motor and theft alarm horn.

Computers

LOCATION

Summit

Multi-Point Injection (MPI) control unit—is located under the instrument panel at the top of the passenger side kick panel, next to the blower motor.

Air conditioning control unit—is mounted behind the glove box.

Automatic transaxle control unit—is mounted on the floor at the very front of the console.

Cruise control unit—is under the instrument panel behind the driver's side knee protector.

Electric door lock control unit—is fastened to the body structure behind the driver's side kick panel.

Automatic seat belt control unit—is under the console next to hand brake handle.

Laser and Talon

Multi-Point Injection (MPI) control unit—is located under the instrument panel at the front of the center console.

Air conditioning control unit—is mounted behind the glove box.

Automatic transaxle control unit—is mounted on the floor at the very front of the console.

Cruise control unit—is mounted at top of instrument panel structure near where the dash pad and windshield meet.

Electric door lock control unit or theft-alarm control unit—is fastened to the body structure behind the passenger's side kick panel.

Automatic seat belt control unit—is fastened to the body structure under the trim panel at the base of the driver's side door latch pillar.

Anti Lock Brake control unit—is mounted behind the right side rear quarter trim panel.

Stealth

Multi-Point Injection (MPI) control unit—is located under the instrument panel at the front of the center console.

Automatic transaxle control unit—is mounted on the floor at the very front of the console.

Air conditioning control unit—is mounted at the front of the center console just above the MPI control unit.

Air conditioner compressor lock controller—is mounted on the bottom of the heater core housing under the right side of the instrument panel.

Cruise control unit—is located behind the right side kick panel.

Electronic Timing and Control System (ETACS) unit—is located just to the left of the the steering column.

Air bag diagnosis unit—is located under the arm rest in the console.

Electronic Suspension Control (ECS) control unit—is mounted behind the right rear trim panel behind an access door.

Anti Lock Brake control unit—is mounted behind the right side rear quarter trim panel.

Active exhaust control unit—is located in the rear luggage compartment, behind the left side trim panel.

Flashers

LOCATION

Summit, Laser and Talon

The turn signal and hazard flasher unit is located in the multi-purpose fuse panel located under the driver's left side knee protector.

Stealth

The turn signal and hazard flasher unit is mounted to the lower portion of the sheet metal behind the left side kick panel.

Cruise Control

ADJUSTMENT

Before starting adjustments, turn air conditioner and lights **OFF**. Warm engine until the idle is stable and the rpm is correct. Stop engine, ignition switch **OFF**. On 1.5L and 1.8L engines, turn the ignition switch to the **ON** position, without starting the engine. Leave in the position for approximately 15 seconds. Confirm there are no sharp bends in the accelerator cables. Check the inner cables for correct slack. If too loose or too tight, adjust with the following procedure:

1. Remove the air cleaner. If vehicle is equipped with a protective cover over the actuator, remove it.

2. First, adjust the accelerator cable on the throttle valve side. After loosening the adjustment bolts at the air intake plenum side and freeing the inner cable, use the adjusting bolts that secure the plate so the freeplay of the inner cable becomes 0.040–0.080 in. (1–2mm). If there is excessive play of the accelerator cable, when climbing a hill the vehicle speed will drop substantially. If there is no play, the idling speed will increase.

3. After adjusting the accelerator cable, confirm that the throttle lever touches the idle position switch.

4. Next, adjust accelerator cable on the accelerator pedal side. Loosen the adjusting bolt or locknut. While keeping the intermediate link of the actuator in close contact with the stop, adjust the inner cable play of accelerator cable **A** to 0–0.040 in. (0–1mm) for manual transaxle vehicles or 0.080–0.120 in. (2–3mm) for automatic transaxle vehicles.

5. After making the adjustment of the cable, make sure the throttle lever at the engine side moves 0.040–0.080 in. (1–2mm) when the actuator link is turned.

6. Confirm that the throttle valve fully opens and closes by operating the accelerator pedal.

7. Install the air cleaner.

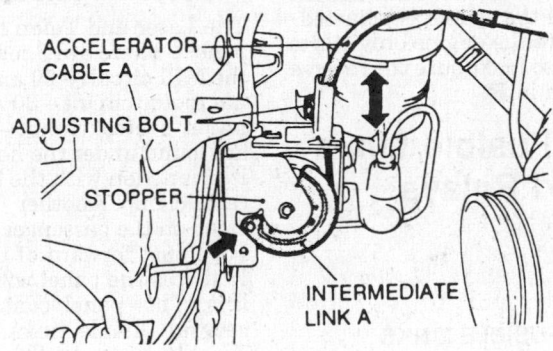

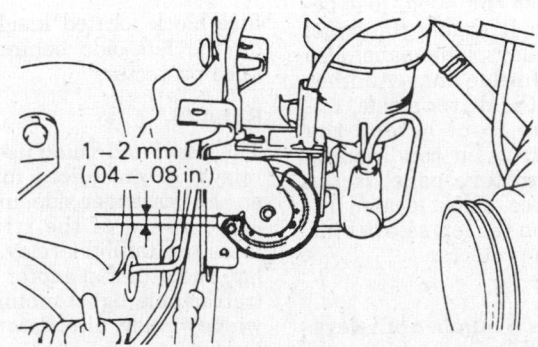

Typical cruise control adjustment points

Chrysler/Eagle
Front Wheel Drive
Eagle—Medallion, Premier
Dodge—Monaco

SPECIFICATIONS

1988

Engine Code						Model Year	
Code	Cu. In.	Liters	Cyl.	Fuel Sys.	Eng. Mfg.	Code	Year
F	132	2.2	4	MPI	Renault	J	1988
Z	150	2.5	4	TBI	AMC		
J	182	3.0	6	MPI	Renault		

MPI—Multi Port Injection
TBI—Throttle Body Injection

1989-92

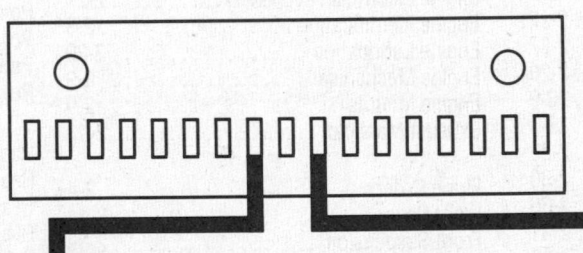

Engine Code						Model Year	
Code	Cu. In.	Liters	Cyl.	Fuel Sys.	Eng. Mfg.	Code	Year
F	132	2.2	4	MPI	Renault	K	1989
H	150	2.5	4	TBI	AMC	L	1990
U	182	3.0	6	MPI	Renault	M	1991
						N	1992

MPI—Multi Port Injection
TBI—Throttle Body Injection

ENGINE IDENTIFICATION

Year	Model	Engine Displacement cu. in. (liter)	Engine Series Identification (VIN)	No. of Cylinders	Engine Type
1988	Medallion	132 (2.2)	F	4	SOHC
	Premier	150 (2.5)	Z	4	OHV
	Premier	182 (3.0)	J	6	SOHC
1989	Medallion	132 (2.2)	F	4	SOHC
	Premier	150 (2.5)	H	4	OHV
	Premier	182 (3.0)	U	6	SOHC

ENGINE IDENTIFICATION

Year	Model	Engine Displacement cu. in. (liter)	Engine Series Identification (VIN)	No. of Cylinders	Engine Type
1990	Premier	182 (3.0)	U	6	SOHC
	Monaco	182 (3.0)	U	6	SOHC
1991–92	Premier	182 (3.0)	U	6	SOHC
	Monaco	182 (3.0)	U	6	SOHC

OHV Overhead Valve Engine
SOHC Single Overhead Cam Engine

GENERAL ENGINE SPECIFICATIONS

Year	VIN	No. Cylinder Displacement cu. in. (liter)	Fuel System Type	Net Horsepower @ rpm	Net Torque @ rpm (ft. lbs.)	Bore × Stroke (in.)	Compression Ratio	Oil Pressure @ rpm
1988	F	4-132 (2.2)	MPI	103 @ 5000	124 @ 2500	3.46 × 3.50	9.2:1	44 @ 3000
	Z	4-150 (2.5)	TBI	111 @ 4750	142 @ 2500	3.87 × 3.18	9.2:1	55 @ 3500
	J	6-182 (3.0)	MPI	150 @ 5000	171 @ 3750	3.66 × 2.87	9.3:1	60 @ 4000
1989	F	4-132 (2.2)	MPI	103 @ 5000	124 @ 2500	3.46 × 3.50	9.2:1	44 @ 3000
	H	4-150 (2.5)	TBI	111 @ 4750	142 @ 2500	3.88 × 3.19	9.2:1	37–75 @ 1600
	U	6-182 (3.0)	MPI	150 @ 5000	171 @ 3750	3.66 × 2.87	9.3:1	60 @ 4000
1990	U	6-182 (3.0)	MPI	150 @ 5000	171 @ 3750	3.66 × 2.87	9.3:1	60 @ 4000
1991–92	U	6-182 (3.0)	MPI	150 @ 5000	171 @ 3750	3.66 × 2.87	9.3:1	60 @ 4000

MPI Multiport Injection
TBI Throttle Body Injection

GASOLINE ENGINE TUNE-UP SPECIFICATIONS

Year	VIN	No. Cylinder Displacement cu. in. (liter)	Spark Plugs Type	Spark Plugs Gap (in.)	Ignition Timing (deg.) MT	Ignition Timing (deg.) AT	Compression Pressure (psi)	Fuel Pump (psi)	Idle Speed (rpm) MT	Idle Speed (rpm) AT	Valve Clearance (in.) In.	Valve Clearance (in.) Ex.
1988	F	4-132 (2.2)	RS9YC	0.035	①	①	②	34–36	800	700	0.006	0.008
	Z	4-150 (2.5)	RC12LYC	0.035	—	①	②	14–15	—	750	Hyd.	Hyd.
	J	6-182 (3.0)	RS9YC	0.035	—	①	②	36–37	—	800	Hyd.	Hyd.
1989	F	4-132 (2.2)	RS9YC	0.035	①	①	②	34–36	800	700	0.006	0.008
	H	4-150 (2.5)	RN12LYC	0.035	—	①	②	14–15	—	①	Hyd.	Hyd.
	U	6-182 (3.0)	RS9YC	0.035	—	①	②	28–30	—	①	Hyd.	Hyd.
1990	U	6-182 (3.0)	RS9YCX	0.035	—	①	②	28–30	—	①	Hyd.	Hyd.
1991	U	6-182 (3.0)	RS9YCX	0.035	—	①	②	③	—	①	Hyd.	Hyd.
1992			SEE UNDERHOOD SPECIFICATION STICKER									

Hyd.—Hydraulic
① Refer to Underhood Sticker
② The lowest reading should be no less than 75%
  of the highest reading.
③ Vehicles built before 10/9/91—28–30 psi
  Vehicles built on or after 10/9/91—43 psi

FIRING ORDERS

NOTE: To avoid confusion, always replace spark plugs and wires one at a time.

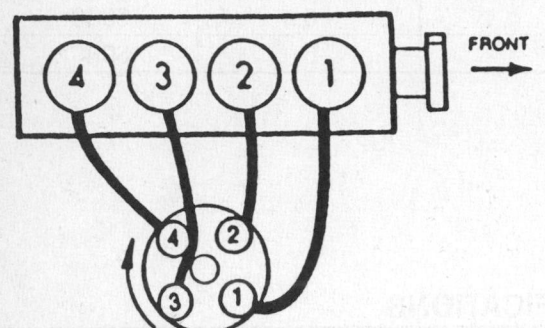

2.5L Engine
Engine Firing Order: 1–3–4–2
Distributor Rotation: Clockwise

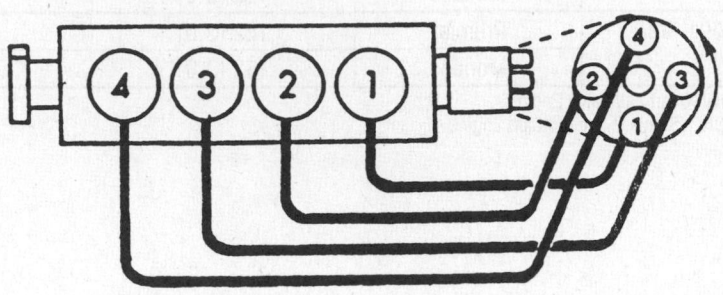

2.2L Engine
Engine Firing Order: 1–3–4–2
Distributor Rotation: Counterclockwise

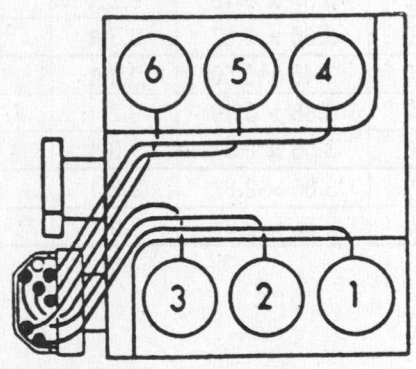

3.0L Engine (Vehicles Built Before Oct. 9, 1991)
Engine Firing Order: 1–6–3–5–2–4
Distributor Rotation: Counterclockwise

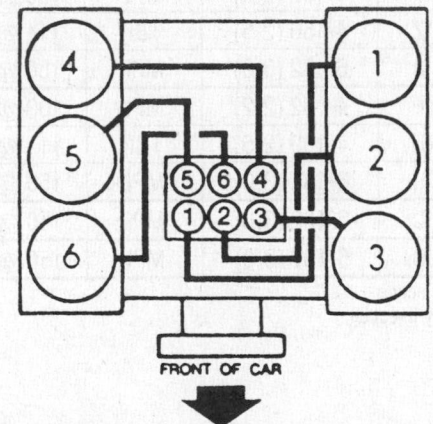

3.0L Engine (Vehicles Built On Or After Oct. 9, 1991)
Engine Firing Order: 1–6–3–5–2–4
Distributorless Ignition System

CAPACITIES

Year	Model	VIN	No. Cylinder Displacement cu. in. (liter)	Engine Crankcase (qts.) with Filter	without Filter	Transmission (pts.) 4-Spd	5-Spd	Auto.	Drive Axle (pts.)	Fuel Tank (gal.)	Cooling System (qts.)
1988	Medallion	F	4-132 (2.2)	5.25	4.75	4.8	—	12.8②	—	17	7.0
	Premier	Z	4-150 (2.5)	5.0	4.5	—	—	14.8②	1.32①	17	8.6
	Premier	J	6-182 (3.0)	6.0	5.5	—	—	14.8②	1.32①	17	8.6
1989	Medallion	F	4-132 (2.2)	5.0	4.5	4.8	—	12.8②	—	17	7.0
	Premier	H	4-150 (2.5)	5.0	4.5	—	—	14.8②	1.32①	17	8.6
	Premier	U	6-182 (3.0)	6.0	5.5	—	—	14.8②	1.32①	17	8.6

CAPACITIES

Year	Model	VIN	No. Cylinder Displacement cu. in. (liter)	Engine Crankcase (qts.) with Filter	without Filter	Transmission (pts.) 4-Spd	5-Spd	Auto.	Drive Axle (pts.)	Fuel Tank (gal.)	Cooling System (qts.)
1990	Premier	U	6-182 (3.0)	6.0	5.5	—	—	② ③	1.32 ①	17	8.6
	Monaco	U	6-182 (3.0)	6.0	5.5	—	—	② ③	1.32 ①	17	8.6
1991–92	Premier	U	6-182 (3.0)	6.0	5.5	—	—	② ③	1.32 ①	17	8.6
	Monaco	U	6-182 (3.0)	6.0	5.5	—	—	② ③	1.32 ①	17	8.6

① Differential requires a synthetic-type SAE grade 75W-140 gear lubricant. It is the only lubricant recommended. It is factory filled and designed to last the life of the differential under normal conditions. Periodic lubricant changes are not required.

② Use Mopar Mercon automatic transmission fluid only. This is the only fluid to be used in the ZF-4 transmission. Do not substitute any other type of fluid.

③ 1990–92 ZF-4 four speed automatic:
14.75 pts. Refill
5.5 pts. after fluid/filter change

CAMSHAFT SPECIFICATIONS

All measurements given in inches.

Year	VIN	No. Cylinder Displacement cu. in. (liter)	Journal Diameter 1	2	3	4	5	Lobe Lift In.	Ex.	Bearing Clearance	Camshaft End Play
1988	F	4-132 (2.2)	NA	NA	NA	NA	—	NA	NA	NA	0.002-0.005
	Z	4-150 (2.5)	2.0290-2.0300	2.0190-2.0200	2.0090-2.0100	1.9990-2.0000	—	0.240	0.250	0.001-0.003	—
	J	6-182 (3.0)	NA	NA	NA	NA	—	NA	NA	NA	0.003-0.0055
1989	F	4-132 (2.2)	NA	NA	NA	NA	—	NA	NA	NA	0.002-0.005
	H	4-150 (2.5)	2.0290-2.0300	2.0190-2.0200	2.0090-2.0100	1.9990-2.0000	—	0.240	0.250	0.001-0.003	—
	U	6-182 (3.0)	NA	NA	NA	NA	—	NA	NA	NA	0.003-0.0055
1990	U	6-182 (3.0)	NA	NA	NA	NA	—	NA	NA	NA	0.003-0.0055
1991–92	U	6-182 (3.0)	NA	NA	NA	NA	—	NA	NA	NA	0.003-0.0055

NA Not Available

CRANKSHAFT AND CONNECTING ROD SPECIFICATIONS

All measurements are given in inches.

Year	VIN	No. Cylinder Displacement cu. in. (liter)	Crankshaft				Connecting Rod		
			Main Brg. Journal Dia.	Main Brg. Oil Clearance	Shaft End-play	Thrust on No.	Journal Diameter	Oil Clearance	Side Clearance
1988	F	4-132 (2.2)	2.4660–2.4760	0.0015–0.0035	0.002–0.009	1	2.206–2.216	0.0008–0.0030	0.012–0.022
	Z	4-150 (2.5)	2.4996–2.5001	0.0020	0.002–0.007	2	2.0934–2.0955	0.0015–0.0020	0.010–0.019
	J	6-182 (3.0)	2.7576–2.7583	0.0015–0.0035	0.003–0.010	1	2.3611–2.3618	0.0008–0.0030	0.008–0.015
1989	F	4-132 (2.2)	2.4760	0.0015–0.0035	0.005–0.011	1	2.206–2.216	0.0008–0.0030	0.012–0.022
	H	4-150 (2.5)	2.4996–2.5001	0.0020	0.002–0.007	2	2.0934–2.0955	0.0015–0.0020	0.010–0.019
	U	6-182 (3.0)	2.7576–2.7583	0.0015–0.0035	0.002–0.007	1	2.3611–2.3618	0.0008–0.0030	0.008–0.015
1990	U	6-182 (3.0)	2.7576–2.7583	0.0015–0.0035	0.002–0.007	1	2.3611–2.3618	0.0008–0.0030	0.008–0.015
1991–92	U	6-182 (3.0)	2.7576–2.7583	0.0015–0.0035	0.002–0.007	1	2.3611–2.3618	0.0008–0.0030	0.008–0.015

VALVE SPECIFICATIONS

Year	VIN	No. Cylinder Displacement cu. in. (liter)	Seat Angle (deg.)	Face Angle (deg.)	Spring Test Pressure (lbs.)	Spring Installed Height (in.)	Stem-to-Guide Clearance (in.)		Stem Diameter (in.)	
							Intake	Exhaust	Intake	Exhaust
1988	F	4-132 (2.2)	②	②	NA	NA	0.004	0.004	0.315	0.315
	Z	4-150 (2.5)	45	45	200 @ 1.216	$1\frac{11}{16}$	0.001–0.003	0.001–0.003	0.312	0.312
	J	6-182 (3.0)	45	45	155 @ 1.220	$1\frac{13}{16}$	NA	NA	0.315	0.315
1989	F	4-132 (2.2)	②	②	NA	NA	0.004	0.004	0.315	0.315
	H	4-150 (2.5)	①	45	200 @ 1.21	$1\frac{11}{16}$	0.001–0.003	0.001–0.003	0.311–0.312	0.311–0.312
	U	6-182 (3.0)	45	45	155 @ 1.220	$1\frac{13}{16}$	NA	NA	0.315	0.315
1990	U	6-182 (3.0)	45	45	155 @ 1.220	$1\frac{13}{16}$	NA	NA	0.315	0.315
1991–92	U	6-182 (3.0)	45	45	155 @ 1.220	$1\frac{13}{16}$	NA	NA	0.315	0.315

NA Not Available
① Intake—44.5
　Exhaust—40.5
② Intake—60
　Exhaust—45

PISTON AND RING SPECIFICATIONS

All measurements are given in inches.

Year	VIN	No. Cylinder Displacement cu. in. (liter)	Piston Clearance	Ring Gap			Ring Side Clearance		
				Top Compression	Bottom Compression	Oil Control	Top Compression	Bottom Compression	Oil Control
1988	F	4-132 (2.2)	NA	①	①	①	NA	NA	NA
	Z	4-150 (2.5)	0.0013–0.0021	0.010–0.020	0.010–0.020	0.015–0.055	0.0010–0.0032	0.0010–0.0032	0.0010–0.0095
	J	6-182 (3.0)	NA	0.016–0.022	0.016–0.022	—	0.0010–0.0020	0.0010–0.0020	0.0015–0.0035
1989	F	4-132 (2.2)	NA	①	①	①	NA	NA	NA
	H	4-150 (2.5)	0.0013–0.0021	0.010–0.020	0.010–0.020	0.015–0.055	0.0010–0.0032	0.0010–0.0032	0.0010–0.0085
1990	U	6-182 (3.0)	0.0013–0.0021	0.016–0.022	0.016–0.022	NA	0.0010–0.0020	0.0010–0.0020	0.0015–0.0035
1991–92	U	6-182 (3.0)	0.0013–0.0021	0.016–0.022	0.016–0.022	NA	0.0010–0.0020	0.0010–0.0020	0.0015–0.0035

NA—Not Available
① The factory specifies only 1 type of ring for this engine. The ring gap is pre-adjusted.

TORQUE SPECIFICATIONS

All readings in ft. lbs.

Year	VIN	No. Cylinder Displacement cu. in. (liter)	Cylinder Head Bolts	Main Bearing Bolts	Rod Bearing Bolts	Crankshaft Pulley Bolts	Flywheel Bolts	Manifold		Spark Plugs
								Intake	Exhaust	
1988	F	4-132 (2.2)	①	69	46	96	44	11	13	11
	Z	4-150 (2.5)	②	80	33	80	48–54	23	23	28
	J	6-182 (3.0)	②	③	37	133	48–54	11	13	11
1989	F	4-132 (2.2)	①	69	46	96	44	11	13	11
	H	4-150 (2.5)	②	80	33	80	48–54	④	④	22
	U	6-182 (3.0)	②	③	35	133	48–54	11	13	11
1990	U	6-182 (3.0)	②	③	35	133	48–54	11	13	11
1991–92	U	6-182 (3.0)	②	③	35	133	48–54	11	13	11

① Torque in 3 steps, in sequence:
  1st—37 ft. lbs.
  2nd—59 ft. lbs.
  3rd—69 ft. lbs.
  Run engine for 15 minutes, shut off and allow to cool for 6 hours and recheck, should be 65–72 ft. lbs.
② See text
③ Tighten in 2 steps, in sequence:
  1st—20 ft. lbs.
  2nd—Angular torque 75 degrees
④ Torque bolts 1, 6, 7, 8 30 ft. lbs.
  2, 3, 4, 5 23 ft. lbs.
  9, 10 14 ft. lbs.

BRAKE SPECIFICATIONS
All measurements in inches unless noted

Year	Model		Lug Nut Torque (ft. lbs.)	Master Cylinder Bore	Brake Disc Minimum Thickness	Brake Disc Maximum Runout	Standard Brake Drum Diameter	Minimum Lining Thickness Front	Minimum Lining Thickness Rear
1988	Medallion	—	66	0.810	0.697②	0.002	9.00	0.06	0.06
	Premier	—	63①	0.945	0.807	0.003	8.92	0.06	0.06
1989	Medallion	—	66	0.810	0.697②	0.002	9.00	0.06	0.06
	Premier	—	63①	0.945	0.807	0.003	8.92	0.06	0.06
1990	Premier	front	63	0.945	0.807	0.003	—	0.06	0.06
		rear	63	—	0.380	0.003	8.92	0.06	0.06
	Monaco	front	63	0.945	0.807	0.003	—	0.06	0.06
		rear	63	—	0.380	0.003	8.92	0.06	0.06
1991–92	Premier	front	63	0.945	0.890	0.003	—	0.06	0.06
		rear	63	—	0.380	0.003	8.92	0.06	0.06
	Monaco	front	63	0.945	0.890	0.003	—	0.06	0.06
		rear	63	—	0.380	0.003	8.92	0.06	0.06

① Specification given is for steel wheels. Tighten lug nuts on aluminum wheels to 90 ft. lbs.
② Do not cut these rotors. Replace if beyond specification.

WHEEL ALIGNMENT

Year	Model	Caster Range (deg.)	Caster Preferred Setting (deg.)	Camber Range (deg.)	Camber Preferred Setting (deg.)	Toe-in (in.)	Steering Axis Inclination (deg.)
1988	Medallion	1½P–3½P	2½P	1/16P–13/16P	7/16P	5/64①	12¾
	Premier	15/16P–213/16P	2⅛P	9/16N–1/16N	5/16N	1/8①	NA
1989	Medallion	1½P–3½P	2½P	1/16P–13/16P	7/16P	5/64①	12¾
	Premier	1½P–3½P	2⅛P	9/16N–1/16N	5/16N	1/8①	NA
1990	Premier	1½P–2½P	2P	9/16N–1/16N	5/16N	0	NA
	Monaco	1½P–2½P	2P	9/16N–1/16N	5/16N	0	NA
1991–92	Premier	1½P–2½P	2P	9/16N–1/16N	5/16N	0	NA
	Monaco	1½P–2½P	2P	9/16N–1/16N	5/16N	0	NA

P Positive
N Negative
NA Not Applicable
① Toe-out

ENGINE MECHANICAL

NOTE: **Disconnecting the negative battery cable on some vehicles may interfere with the functions of the on-board computer systems and may require the computer to undergo a relearning process, once the negative battery cable is reconnected.**

Engine Assembly

REMOVAL & INSTALLATION

1. Disconnect the negative battery cable. Matchmark the hood to the hinges and remove the hood.
2. Disconnect the coil wire all vacuum hoses and fuel lines.
3. Disconnect the lower radiator hose and drain the coolant. Remove the air cleaner.
4. Remove the grille. Remove the screws retaining the front facia panel and radiator support and remove the panel and support.
5. Remove the radiator and cooling fan. If equipped with air conditioning, safely discharge the air conditioning system and remove the condenser and the radiator as an assembly.
6. Disconnect the electrical leads to the unit ECU or SBEC.
7. Remove the accelerator cable from the brackets on the valve cover.
8. Remove the bolts that attach the exhaust head pipes to the exhaust manifold. Remove the heater hoses and on automatic transaxle equipped vehicles, remove the cooler lines.
9. Raise the vehicle and safely support. Remove the underbody splash shield.
10. Remove the power steering pump mounting bolts and support the pump to the side. Remove the header pipe to converter bolts and remove the converter.
11. If equipped with automatic transaxle, disconnect the shifter linkages. On manual transaxle vehicles, disconnect the clutch cable at the transaxle.
12. Remove the wheel assemblies and remove the front stabilizer bar. Remove the brake calipers and support aside. Disconnect the tie rod ends from the steering knuckle. Remove the halfshaft retaining pin and remove the halfshaft assembly. Remove the strut-to-steering knuckle bolts.
13. Loosen the upper strut mounting bolts and swing the axle/strut assembly aside. Support the axles safely.
14. Disconnect the speedometer cable. Disconnect the vapor canister and remove it.
15. Loosen the bolts attaching the transmission support to the engine cradle. Remove the bolts attaching the left and right halves of the crossmember to the transaxle. Lower the vehicle.
16. Attach a suitable lifting device to the engine lifting eyes and lift the engine slightly. Remove the engine support bolts and remove the engine/transaxle assembly. Lift the engine out at an angle, make sure the transaxle clears the engine compartment. Separate the engine from the transaxle.

To install:

17. Position the engine/transaxle assembly in the vehicle and align the engine mounts with the engine cradle.
18. Install the engine mount bolts and remove the lifting device. Install the left and right sections of the crossmember.
19. Position and install the halfshafts to the transaxle, use new retaining pins. Install the shock absorber to steering knuckle bolts and attach the tie rod ends. Attach the front stabilizer bar.
20. Install the brake calipers and install the front wheels.
21. Install the converter to the header pipe. Install the power steering pump and adjust the belt tension. Connect the shift linkage and throttle cables.
22. Reconnect all electrical and vacuum leads. Install the canister and the air cleaner assemblies. Reconnect the fuel lines and coolant hoses.
23. Install the radiator and fan assemblies. Attach the front facia and support assembly. Install the grille.
24. Install the hood.
25. Check and fill all fluid levels properly.
26. Connect the negative battery cable and road test the vehicle.

Engine Mounts

REMOVAL & INSTALLATION

Except 3.0L Engine

1. Disconnect the negative battery cable.
2. Remove the engine mount upper attaching bolt.
3. Remove the engine pitch restrictor and bracket.
4. Raise the vehicle and support it safely.
5. Remove the engine mount bottom attaching bolt.
6. Carefully raise the engine and remove the engine mount.

7. Installation is the reverse of the removal procedure.

3.0L Engine

FRONT MOUNTS

1. Disconnect the negative battery cable.
2. Raise the vehicle and support safely.
3. Remove the engine mount stud locknut.
4. Disconnect the front engine damper and raise the engine enough to remove the mount.
5. Remove the mount through bolt and remove the mount.
6. The installation is the reverse of the removal procedure.
7. Torque mount retaining nuts to 48 ft. lbs. (65 Nm).

REAR MOUNT/CUSHION

1. Disconnect the negative battery cable.
2. Raise the vehicle and support safely.
3. Remove the underbody splash shield.
4. Using the proper equipment, support the weight of the transaxle.
5. Remove the nuts that attach the crossmember to the engine cradle.
6. Remove the large bolt that attaches the rear cushion to the support bracket.
7. Remove the bolt that attaches the exhaust pipe bracket to rear cushion.
8. Remove the crossmember and the rear cushion.
9. Remove the support bracket bolts and remove the bracket from the transaxle.

To install:

10. Install the support bracket to the transaxle and torque to 30 ft. lbs. (40 Nm).
11. Position the exhaust bracket on the rear cushion.
12. Install the rear cushion and the crossmember. Align the cushion to bracket bolt and nut. Leave it loose at this point.
13. Install, but don't tighten the exhaust bracket bolts.
14. Tighten the rear cushion to support bracket bolt to 49 ft. lbs. (67 Nm).
15. Tighten the exhaust bracket to rear cushion bolts to 23 ft. lbs. (31 Nm).
16. Install the bolts and nuts that attach the crossmember to the studs on the engine cradle. Torque to 44 ft. lbs. (60 Nm).
17. Remove the support apparatus from the transaxle.
18. Lower the vehicle. Connect the negative battery cable.

Cylinder Head

REMOVAL & INSTALLATION

2.2L Engine

1. Relieve the fuel system pressure.
2. Disconnect the negative battery cable and drain the cooling system. Remove the air inlet tube from the throttle body.
3. Remove the accessory drive belts. Remove the timing belt cover.
4. Loosen the bolts on the timing belt tensioner and remove the timing belt. Remove the spark plugs and wires.
5. Remove any hoses attached to the rocker cover and remove the rocker arm cover. Remove the distributor from the rear of the head.
6. Remove all of the cylinder head bolts except for the bolt at position No. 10 in the tightening sequence. Loosen the bolt at position No. 10 and pivot the cylinder head on that bolt. This can be done by tapping the opposite end of the head with an block of wood. This is necessary to free the cylinder head from the cylinder liners.

NOTE: When the cylinder head has been removed, retain the cylinder liners in the block with a liner hold-down clamp. This tool is designed to prevent the cylinder liners from being knocked out of position. If this happens, the sealing rings will dislodge or rip.

7. Once the head is free, remove the last bolt and remove the cylinder head. Clean all gasket material from mating surfaces.

To install:

8. Place the new cylinder head gasket on the block using the alignment dowel on the block to hold it in place.
9. Position the cylinder head on the block and insert the cylinder head bolts. Tighten the bolts in sequence to 37 ft. lbs. (51 Nm), then to 59 ft. lbs. (80 Nm), then to 69 ft. lbs. (94 Nm).
10. Install the distributor to the head. Install the rocker arm cover, using a new gasket. Tighten the rocker cover bolts to 35 inch lbs.
11. Install the timing belt and adjust the belt tension. Install the timing belt cover. Install the spark plugs and wires.

12. Install the accessory drive belts and reconnect all hoses that were disconnected.
13. Install the air inlet tube. Fill the cooling system.
14. Connect the negative battery cable. Run the engine and bleed the cooling system. Check for leaks.

2.5L Engine

1. Relieve the fuel system pressure.
2. Disconnect the negative battery cable and drain the cooling system.
3. Loosen the accessory drive belt and remove.
4. Remove the bolts attaching the air conditioning compressor and without disconnecting the pressure lines, move the compressor aside.
5. Disconnect the upper radiator hose and the heater hoses.
6. Remove the rocker arm cover. Remove the rocker arm assembly, keeping all of the valve train components in their original order for installation.
7. Remove the intake and exhaust manifolds.
8. Remove the cylinder head bolts and remove the cylinder head.
9. Clean all gasket material from mating surfaces.

To install:

10. To install, place the new cylinder head gasket on the block with the numbers facing UP.

NOTE: The cylinder head gasket used on this engine is a composite gasket and does not require the use of any sealing compound.

11. Place the cylinder head on the block and install the bolts. Following the proper sequence, tighten the bolts in the following steps:

   a. Torque all 10 bolts to 22 ft. lbs. (30 Nm)
   b. Torque all 10 bolts to 45 ft. lbs. (61 Nm)
   c. Retorque all 10 bolts to 45 ft. lbs. (61 Nm)
   d. Torque bolts No. 1–6 to 110 ft. lbs. (149 Nm)

   e. Torque bolt No. 7 to 100 ft. lbs. (136 Nm)
   f. Torque bolts No. 8–10 to 110 ft. lbs. (149 Nm)
   g. Retorque each bolt to its proper torque. Pay special attention to bolt No. 7; it should be overtorqued.
12. Install the valve train components in their original sequence. Place a new gasket on the cylinder head and install the rocker cover.
13. Connect all of the hoses removed and install the air conditioning compressor, tighten the mounting bolts to 20 ft. lbs. (27 Nm). Route the accessory drive belt and adjust the tension.
14. Connect the battery cable and fill the cooling system. Run the engine and bleed the cooling system. Check for leaks.

3.0L Engine

1. Relieve the fuel system pressure.
2. Disconnect the negative battery cable and drain the cooling system.
3. Remove the accessory drive belt and remove the air conditioning compressor from the cylinder head cover.
4. Remove the intake and exhaust manifolds.
5. Remove the spark plug wires. Remove the rocker arm cover.
6. Remove the alternator mounting bracket and remove the top timing case bolts that thread into the cylinder head.

NOTE: The timing sprocket and chain must be supported in place and not allowed to drop into the timing case. If the chain and sprocket slip into the case the timing case will have to be removed.

7. Turn the crankshaft until the camshaft sprocket dowel is straight up. A special tool is available called a timing chain support bracket. This support bracket and dummy bearing attaches to the timing case cover. On the left cylinder head, remove the distributor assembly.
8. Remove the threaded plug on the front of the timing case cover to gain access to the camshaft sprocket bolt.
9. Remove the cylinder head bolts. Remove the rocker shaft assembly.
10. Remove the rear camshaft cover and gasket at the rear of the cylinder head.
11. Loosen the camshaft thrust plate screw, located behind the timing sprocket, and move the thrust plate up. This will allow the camshaft to move back in the head as the sprocket bolt is removed.
12. Loosen the camshaft sprocket bolt and pull the camshaft back until the bolt is free from the camshaft, the bolt will stay in the sprocket. Use an

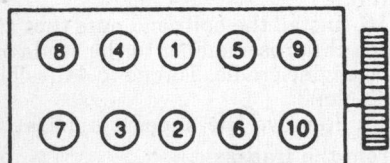

Head bolt torque sequence—2.2L engine

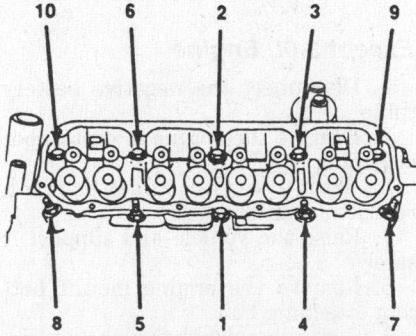

Head bolt torque sequence—2.5L engine

old pushrod or a long thin drift punch as a tool and insert it into the front and rear cylinder head bolt holes on the exhaust manifold side of the head. Tap the dowel down below the head gasket. The reason for this is that the cylinder head is not to be removed by pulling straight upward which would pull the cylinder liners loose from the block. It is to be bumped sideways to break the seal.

NOTE: Do not pull straight up on the cylinder head to remove it. This will cause the cylinder liners to come out of the block.

13. Position a block of wood on the intake manifold side of the head and strike it with a hammer, do the same on the exhaust manifold side of the head. Repeat this until the cylinder head is loose. Remove the cylinder head.

NOTE: When the cylinder head has been removed, retain the cylinder liners in the block with a liner hold-down clamp. This tool is designed to prevent the cylinder liners from being knocked out of position. If this happens, the sealing rings will dislodge or rip. Do not rotate the engine with the liner clamps in place.

14. Remove the cylinder head gasket and clean all gasket material from mating surfaces. Remove the cylinder head locating dowels. Check that the cylinder liners protrude between 0.002–0.005 in.

To install:

15. When installing, cut the gasket flush with the cylinder head gasket face at the back of the timing case cover and remove the pieces. Clean the back of the timing cover. Cut sections of new gasket to replace the pieces removed and attach them with adhesive. Install a small punch into the hole in the block below the locating dowel bolt holes. This will act as a stop for the dowel. Push the dowel into the block until it contacts the punch.

16. Install a new cylinder head gasket over the alignment dowels on the head. Place a small bead of RTV or equivalent, at the point where the head gasket meets the timing case cover.

17. Place the cylinder head on the block and install the top timing case cover-to-cylinder head bolts finger-tight.

18. Remove the timing sprocket support tool. Position the camshaft into the sprocket and align the dowel to the slot in the camshaft. Install the sprocket bolt and lightly tighten it. Slide the thrust plate into position and tighten the thrust plate bolt to 4 ft. lbs. (5.4 Nm).

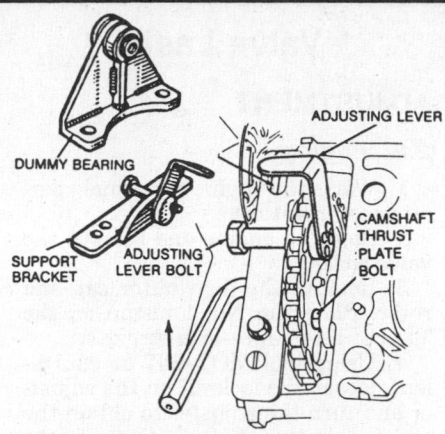

Special tools are recommended to retain cam drive when head is removed—3.0L engine

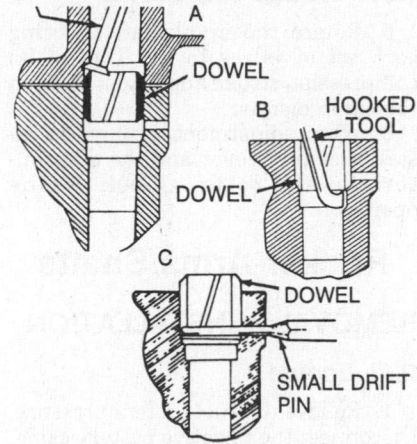

A—Push locating dowel below head gasket level so head can be bumped sideways to remove;

B—Pull dowel out with hook shaped tool; C—Use pin punch to gauge depth of dowel at installation—3.0L engine

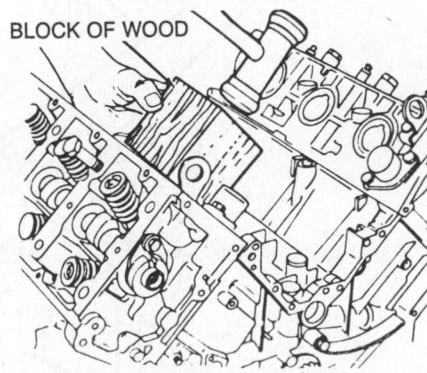

Lifting the head could dislodge the cylinder liner on 3.0L engine; instead, bump sideways as shown

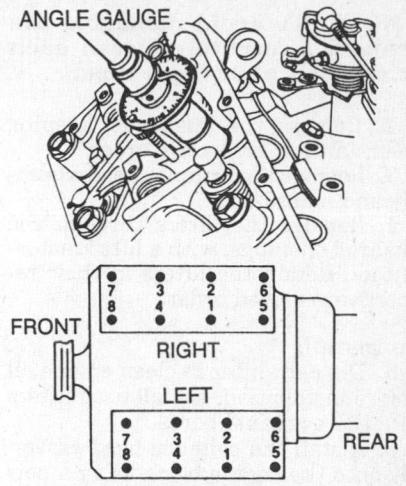

Head bolt torque sequence and torque angle gauge—3.0L engine

19. Install the rocker shaft assembly and install new head bolts. Using the proper sequence, tighten the cylinder head bolts in the following order:

a. Starting with bolt No. 1, torque all bolts to 44 ft. lbs. (60 Nm).

b. The following is performed on all bolts, one at a time. Starting with bolt No. 1, loosen the bolt completely, then tighten to 30 ft. lbs. (41 Nm).

c. Place an angle adapter on the torque wrench between the socket and wrench and angle tighten each bolt, in sequence, an additional 180 degrees ± 20 degrees.

d. Repeat Step C to ensure proper torque.

e. Check each bolt for at least 52 ft. lbs. (70 Nm) of torque.

20. Install the rocker covers, intake and exhaust manifolds.

21. Install the timing case plug, spark plug wires and the air conditioning compressor. Reconnect all hoses and fill the cooling system.

22. Install the distributor assembly on the left cylinder head.

23. Install the accessory drive belt and adjust the tension. Connect the negative battery terminal. Start the engine and bleed the cooling system. Check for leaks.

Valve Lifters

REMOVAL & INSTALLATION

2.5L Engine

1. Disconnect the negative battery cable. Remove the valve cover, the bridge and pivot assemblies, and the rocker arms.

NOTE: To avoid damaging the bridges, alternately loosen each bridge bolt a turn at a time.

2. Remove the pushrods, keeping them in their respective order.

3. Remove the cylinder head assembly and manifolds.

4. Remove the lifters through the pushrod openings, with a lifter removal tool. Retain the lifters in their respective removed order.

To install:

5. Dip each lifter in clean engine oil before installation. Install used lifters into their original bores.

6. Install the cylinder head assembly onto the engine block using a new head gasket. Tighten in sequence to the proper torque specification. Install the manifolds, if removed separately.

7. Install the pushrods into their original positions and install the rocker arms and bridges and the pivot assemblies. Tighten the bridge bolts a turn at a time, alternately, to avoid damaging the bridges.

8. Pour the remaining oil supplement over the valve train.

9. Install the valve cover and related parts.

10. Connect the negative battery cable and check for leaks.

3.0L Engine

1. Disconnect the negative battery cable.

2. Remove the rocker cover and the rocker shaft assembly.

3. On the rocker shaft, remove the retaining screw from the end of the shaft and carefully disassemble the rocker shaft components.

4. Remove the lifter and the lifter thrust washer from the rocker arm. Check the lifters for excess wear and check the rocker arm for blocked oil passages.

To install:

5. Lightly coat the lifter and thrust washer with clean engine oil. Install the lifter in the rocker arm.

NOTE: The lifter may tend to fall from the rocker arm. To prevent this use masking tape or wire to hold the tappet in place until the shaft assembly is installed.

6. Assemble the rocker shaft components in the order they were disassembled. Install the rocker shaft assembly on the cylinder head.

7. Install the rocker arm cover.

8. Connect the negative battery cable and check for leaks.

Valve Lash

ADJUSTMENT

2.2L Engine

1. Warm the engine to normal operating temperature.

2. Stop the engine and remove the valve cover.

3. Remove the distributor cap and rotor. Place the No. 1 piston on the TDC of its compression stroke.

4. Using tool MOT–647 or equivalent, loosen the locknut on the adjuster and turn the adjuster to obtain the proper clearance. Specification are 0.006 in. for intake valves and 0.008 for exhaust valves.

NOTE: Check the adjuster to be sure it is aligned evenly with the valve stem. If it is not aligned, valve damage could occur.

5. Rotate the crankshaft to bring each set of valves to the TDC of its compression stroke and adjust them in the same manner.

6. When adjustment is complete, install the valve cover and the distributor cap and rotor. Check engine operation.

Rocker Arms/Shafts

REMOVAL & INSTALLATION

2.2L Engine

1. Relieve the fuel system pressure. Disconnect the negative battery cable.

2. Remove the rocker arm cover retaining bolts and remove the rocker cover.

3. Remove the bolts retaining the rocker arm shaft to the cylinder head.

To install:

4. Position the rocker shaft assembly on cylinder head and tighten the rocker shaft retaining bolts to 66 inch lbs. on the 2.2L engine.

5. Install the rocker cover using a new gasket.

6. Connect the negative battery cable and check engine operation.

2.5L Engine

1. Disconnect the negative battery cable.

2. Disconnect and mark all vacuum hoses and electrical connections as required.

3. Remove the vacuum switch and bracket assembly. Remove the diverter valve and bracket assembly.

4. Remove the valve cover.

5. Remove the bolts at each bridge and pivot assembly. Alternately loosen each bolt a turn at a time to avoid damaging the bridges.

6. Remove the bridges, pivots and corresponding pairs of rocker arms and keep them in the order of removal.

7. Installation is the reverse of the removal procedure. Lubricate all parts prior to installation and install all components into their original positions.

8. At each bridge, tighten the bolts alternately a turn at a time to avoid damage to the bridge. Tighten the bolts to 19 ft. lbs. (26 Nm)

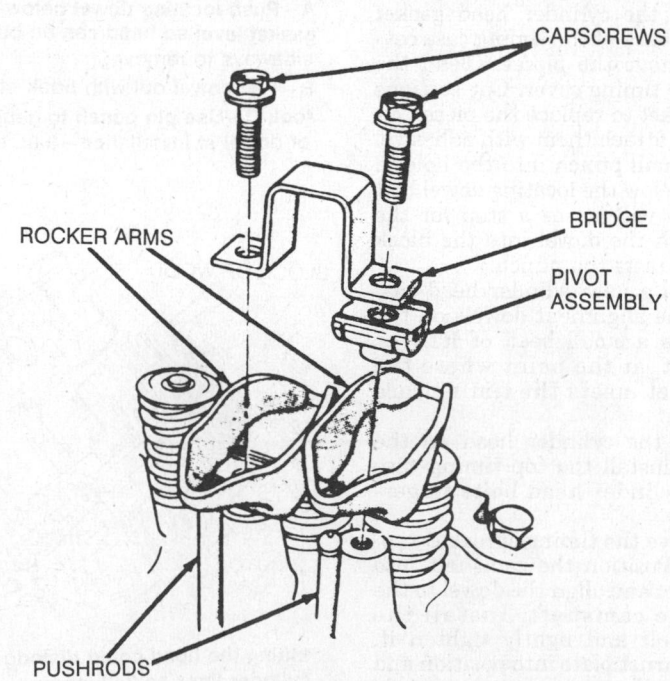

Rocker arm and bridge—2.5L engine

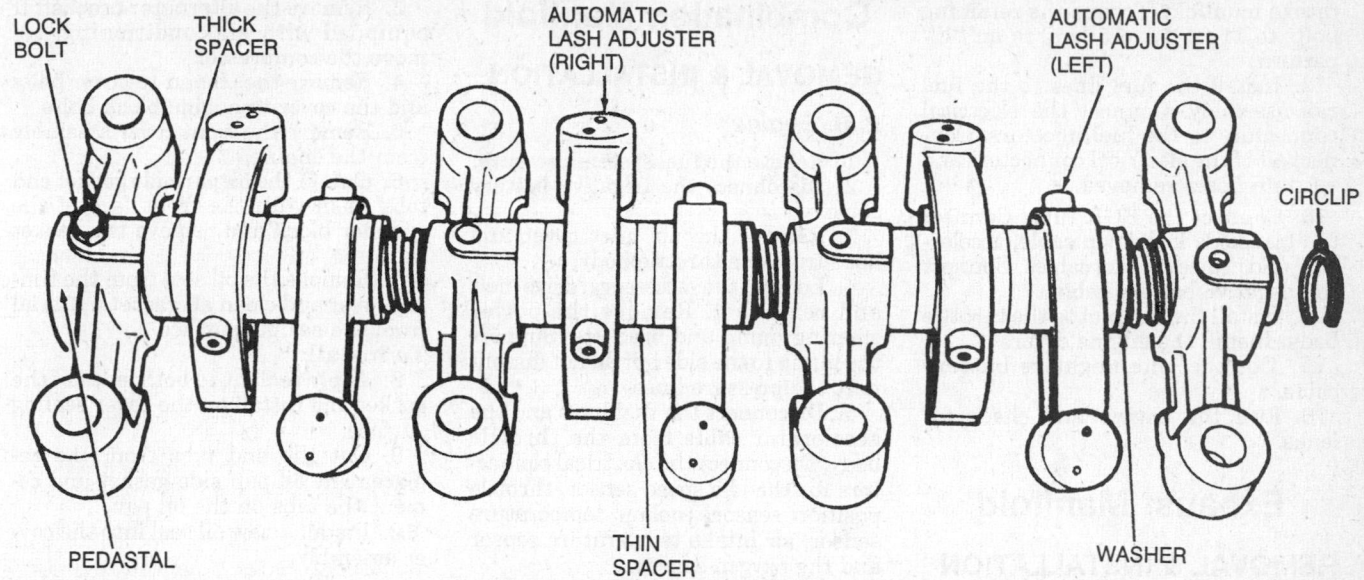

Rocker shaft assembly—3.0L engine

Labels on figure: LOCK BOLT, THICK SPACER, AUTOMATIC LASH ADJUSTER (RIGHT), AUTOMATIC LASH ADJUSTER (LEFT), CIRCLIP, PEDESTAL, THIN SPACER, WASHER

3.0L Engine

1. Relieve the fuel system pressure. Disconnect the negative battery cable.
2. Remove the engine cover mounting bolts and remove the engine cover.
3. Disconnect the vacuum hoses and electrical connectors, as required.
4. Remove the spark plug wire holder and loosen the accessory drive belt.
5. Remove the air conditioning compressor mounting bolts and position the compressor aside, if required.
6. Remove the power steering reservoir, idle speed regulator bracket, accelerator cable and bracket.
7. Remove the rocker arm cover attaching bolts and remove the rocker cover.
8. Label the rocker shaft assemblies. Remove the rocker arm shaft attaching bolts and remove the shaft assembly.

NOTE: Both left and right rocker shaft assemblies are identical and can be interchanged between cylinder heads. Always install them on the same cylinder head that they were removed from.

To install:

9. Lightly coat the rocker shaft assembly with clean engine oil and position on the cylinder head. Tighten the attaching bolts to 53 inch lbs. (6 Nm).
10. Before installing the rocker cover, apply a light coating of sealer to the top of the timing case cover at cylinder head joints area.
11. Install the accelerator cable and bracket, power steering reservoir, and idle speed regulator bracket.
12. Install the air conditioning compressor.

13. Install the spark plug wire holder.
14. Adjust the accessory drive belt.
15. Connect the vacuum hoses and electrical connectors.
16. Install the engine cover.
17. Connect the negative battery cable.

Intake Manifold

REMOVAL & INSTALLATION

2.2L Engine

1. Disconnect the negative battery cable.
2. Drain the cooling system.
3. Relieve the fuel system pressure.
4. Remove the air inlet/filter housing and tube.
5. Disconnect the fuel lines at the injector rail. Disconnect the vacuum lines at the intake manifold.
6. Disconnect the throttle linkage at the throttle body. Remove the electrical connectors from the injectors.
7. Remove the intake manifold retaining bolts and remove the intake manifold.
8. Clean the gasket mating surfaces.

To install:

9. Position the intake manifold on the head using a new gasket and insert the bolts. Torque the manifold bolts to 11 ft. lbs. (15 Nm) in a criss-cross pattern starting from the middle and working outward.
10. Connect the electrical leads to the injectors and the fuel lines to the fuel rail.
11. Connect the vacuum lines at the

manifold and the throttle linkage at the throttle body.
12. Attach the air inlet to the throttle body. Fill the cooling system and connect the negative battery cable.
13. Run the engine, bleed the cooling system and check for leaks.

3.0L Engine

1. Relieve the fuel system pressure.
2. Disconnect the negative battery cable.
3. Remove the engine cover retaining bolts and remove the cover.
4. Remove the air inlet cover from the throttle body.
5. Disconnect the transaxle kickdown cable, accelerator cable and cruise control cable from the throttle body. Remove the vacuum hoses from the intake manifold.
6. Remove the electrical connector from the throttle position sensor. Disconnect and tag the electrical connectors from the fuel injectors and lay the harness aside.
7. Remove the EGR tube. Remove the wire from the air temperature sensor.
8. Remove the fuel lines from the injector rails.
9. Remove the 4 bolts retaining the intake manifold and remove the manifold. Remove the O-rings from the cylinder heads and discard them.

NOTE: When the intake manifold has been removed the O-rings in the cylinder heads must be replaced.

10. Clean all gasket mating surfaces.
To install:
11. Use new O-rings and install the

intake manifold. Torque the retaining bolts to 11 ft. lbs. (15 Nm) in an "X" pattern.

12. Install the fuel lines to the fuel rail assembly. Connect the electrical connectors to the fuel injectors. Connect all of the electrical connectors and vacuum hoses removed.

13. Connect the EGR tube. Connect the transaxle kickdown cable, accelerator and cruise control cables. Connect the negative battery cable.

14. Install the air inlet to the throttle body. Install the engine cover.

15. Connect the negative battery cable.

16. Run the engine and check for leaks.

Exhaust Manifold

REMOVAL & INSTALLATION

2.2L Engine

1. Disconnect the negative battery cable.

2. Remove the exhaust manifold heat shield and hot air tube.

3. Remove the EGR tube from the manifold.

4. Remove the bolts retaining the header pipe to the manifold.

5. Remove the manifold mounting nuts. Remove the manifold and gaskets.

To install:

6. To install, place the manifold gaskets and the manifold on the block. Tighten the mounting nuts to 13 ft. lbs. (18 Nm) in a criss-cross pattern starting from the middle and working outward.

7. Install the heat shield and the EGR tube.

8. Connect the negative battery cable and check for exhaust leaks.

3.0L Engine

1. Disconnect the negative battery cable.

2. Disconnect the EGR tube from the right side manifold.

3. Raise the vehicle and support safely.

4. Remove the nuts retaining the header pipe to the manifolds.

5. On the right manifold, remove the nuts securing the dipstick tube to the manifold. On the left manifold, remove the starter heat shield and the heat stove.

6. Lower the vehicle.

7. Remove the manifold mounting nuts and remove the manifolds.

8. The installation is the reverse of the removal procedure. Tighten the nuts to 13 ft. lbs. (18 Nm).

9. Connect the negative battery cable and check for exhaust leaks.

Combination Manifold

REMOVAL & INSTALLATION

2.5L Engine

1. Relieve the fuel system pressure.

2. Disconnect the negative battery cable.

3. Remove the air inlet cover and hose from the throttle body.

4. Loosen the accessory drive belt and remove it. Remove the power steering pump and brackets. Support the pump to the side but do not disconnect the pressure lines.

5. Disconnect the fuel lines and the accelerator cable from the throttle body. Disconnect the electrical connectors for the idle speed sensor, throttle position sensor, coolant temperature sensor, air intake temperature sensor and the oxygen sensor.

6. Disconnect the electrical plug from the fuel injector. Disconnect the vacuum lines at the intake manifold.

7. Remove the bolts supporting the EGR tube to the exhaust manifold. Remove the heater hoses from the intake manifold.

8. Remove the intake/exhaust manifold mounting bolts and remove the manifolds from the engine.

To install:

9. Clean all of the gasket mounting surfaces.

10. To install, position the new intake manifold gasket and the new exhaust manifold spacers over the locating dowels and install the manifold to the head. Tighten the bolts in sequence and to the specified torque.

11. Install the EGR tube to the exhaust manifold. Connect the heater and vacuum hoses. Attach the fuel lines to the throttle body.

12. Reconnect all electrical connectors. Install the power steering pump and brackets.

13. Connect the accelerator cable. Install the accessory drive belt and adjust the tension. Install the air inlet tube and cover.

14. Connect the negative battery cable and fill the cooling system.

15. Run the engine and bleed the cooling system. Check for leaks.

Timing Chain Front Cover

REMOVAL & INSTALLATION

2.5L Engine

1. Disconnect the negative battery cable.

2. Remove the drive belts, engine fan and hub assembly, vibration damper, pulley and key.

3. Remove the alternator bracket. If equipped with air conditioning, remove the compressor.

4. Remove the oil pan-to-cover bolts and the cover-to-engine block bolts.

5. Remove the front cover assembly from the engine.

6. Cut off the oil pan side gasket end tabs flush with the front face of the cylinder block and remove the gasket tabs.

7. Remove the oil seal from the timing cover and clean all gasket material from the sealing surface.

To install:

8. Apply sealant to both sides of the gasket and install on the cover sealing surface.

9. Cut the end tabs from the replacement oil pan side gasket and cement the tabs on the oil pan.

10. Install a new oil seal into the cover assembly.

NOTE: The oil seal can be installed after the cover has been installed on the engine block, depending upon whether the cover aligning tools are available.

11. Coat the front cover seal end tab recesses with RTV sealant and position the seal on the cover bottom.

12. Position the cover on the engine block and position an alignment tool into the crankshaft opening.

NOTE: Two different types of alignment tools are available, without seal in housing or with seal in housing.

13. Install the cover-to-engine block bolts and the oil pan-to-cover bolts. Tighten the cover-to-engine block bolts to 5 ft. lbs. (7 Nm) and the oil pan-to-cover bolts to 11 ft. lbs. (15 Nm).

14. If not already done, install the seal.

15. Install the vibration damper.

16. Install the compressor and alternator bracket.

17. Install the engine fan and hub assembly. Install the belts and adjust.

18. Connect the negative battery cable and check for leaks.

3.0L Engine

1. Disconnect the negative battery cable.

2. Remove the rocker covers. Hold the camshaft sprocket in place and remove the distributor drive/camshaft sprocket bolt. Remove the distributor assembly.

3. Remove the accessory drive belt. Remove the nuts retaining the front engine vibration damper to the engine and move it toward the radiator.

4. Remove the crankshaft pulley nut and remove the crankshaft pulley.

NOTE: The crankshaft pulley nut is put on with a threaded lock installed with the nut. It may be necessary to strike the pulley with a brass hammer to loosen it.

5. Remove the timing cover mounting bolts. Place a suitable prying tool between the cylinder block and a special boss on the front cover and gently pry off the cover. Discard the gaskets.

6. Remove the oil seal from the cover.

To install:

7. To prevent the key from falling into the oil pan, rotate the crankshaft so the keyway points upward.

8. Apply a bead of RTV sealer to the points where the cylinder heads meet the block and the lower case meets the block.

9. Install the cover with new gasket over the alignment dowels. Tighten the bolts to 9 ft. lbs. (12 Nm).

10. Install the distributor assembly and install the rocker covers.

11. Install the crankshaft pulley, apply thread locking compound to the threads of the pulley nut and tighten to 133 ft. lbs. (180 Nm).

12. Install the accessory drive belt and adjust the belt tension.

NOTE: It is very important the accessory drive belt is routed correctly. If it is incorrectly routed the water pump could be driven in the wrong direction, causing the engine to overheat.

13. Install the engine vibration damper.

14. Connect the negative battery cable and check for leaks.

Timing Chain and Sprockets

REMOVAL & INSTALLATION

2.5L Engine

1. Disconnect the negative battery cable.

2. Remove the fan shroud assembly, accessory drive belts, water pump pulley, crankshaft vibration damper and timing case cover.

NOTE: It is a good practice to either remove the radiator or cover the radiator core area when working around the radiator, as damage can result to the radiator core.

3. Rotate the crankshaft until the **0** timing mark on the crankshaft sprocket aligns with the timing mark on the camshaft sprocket.

4. Remove the oil slinger from the crankshaft.

5. Remove the camshaft retaining

bolt, the sprocket and chain assembly.

6. If the timing chain tensioner is to be replaced, the oil pan must also be removed.

To install:

7. Turn the tensioner lever to the **UNLOCK** position and pull the tensioner block toward the tensioner lever to compress the spring. Hold the block and turn the tensioner lever to the lock **UP** position.

8. Install the crankshaft/camshaft sprockets and timing chain. Make sure the timing marks are aligned as indicated in Step 3.

9. Install the camshaft sprocket retaining bolt and washer. Torque the bolt to 80 ft. lbs. (108 Nm).

NOTE: To verify correct installation of the timing chain, rotate the crankshaft until the camshaft sprocket timing mark is approximately at the 1 o'clock position. There should be 20 pins (2 per link) between the marks.

10. Install the oil slinger.

11. Install the timing case cover and all related parts.

12. Connect the negative battery cable and check engine for proper operation.

3.0L Engine

1. Disconnect the negative battery cable. Remove the cylinder head cover.

2. Inspect the chain and sprocket for wear by pulling on the top of the chain. This will produce a gap between the bottom of the timing chain and the bottom of the area between the 2 sprocket teeth. The maximum gap is 0.067 in. (0.17mm). This must not be exceeded. This gap corresponds to a travel of 0.866 in. (2.19mm) by the timing chain tensioner plunger.

3. Use the solid end of a No. 51 drill bit (0.067 in. diameter) to gauge the gap. If the solid end of the drill bit fits into the gap between the timing chain and the 2 sprocket teeth, then the following parts must be replaced: timing chain shoes, tensioners, guides, sprockets, tensioner shoes. Use the following procedure.

4. Disconnect the negative battery cable.

5. Remove the front cover assembly. Turn the crankshaft until piston No. 1 is at TDC of the compression stroke.

NOTE: Keep all of the components from each side together. This will aid in assembly.

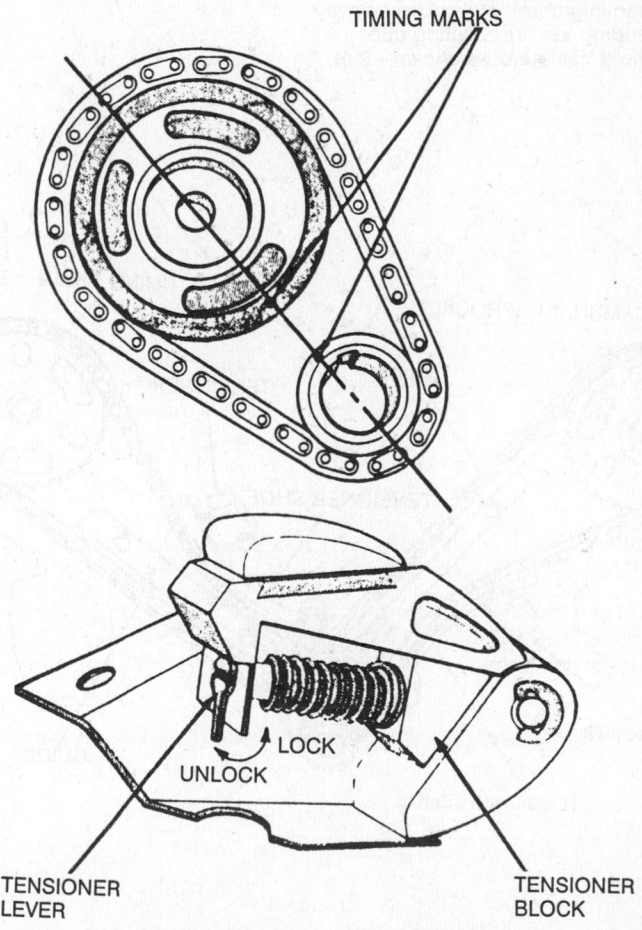

Timing mark alignment—2.5L engine

6. Remove the oil pump sprocket retaining bolts and remove the sprocket/chain assembly.

7. Remove the bolt attaching the right side camshaft sprocket to the camshaft. Remove the right side tensioner and let the tensioner shoe hang down.

8. Remove the right side timing chain and sprocket. Remove the right side chain guide and tensioner shoe.

9. Remove the bolt attaching the left side camshaft sprocket to the camshaft. Remove the left side tensioner and let the tensioner shoe hang down.

10. Remove the left side timing chain and sprocket. Remove the left side chain guide and tensioner shoe.

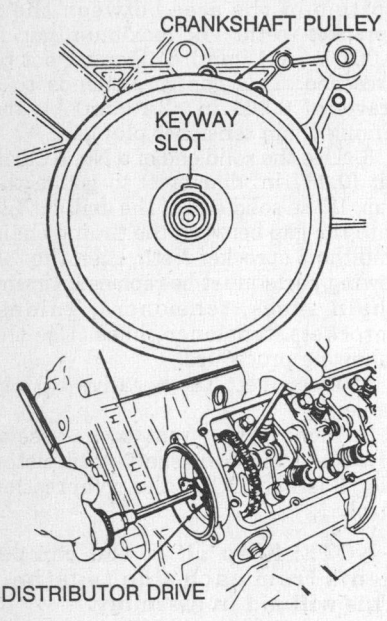

When removing crank pully, turn keyway to top keeping key from falling into engine; hold cam gear as shown—3.0L engine

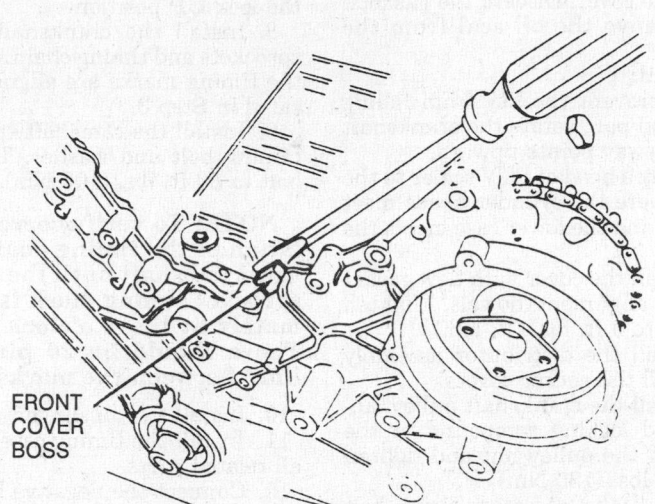

Cylinder block and front cover boss—3.0L engine

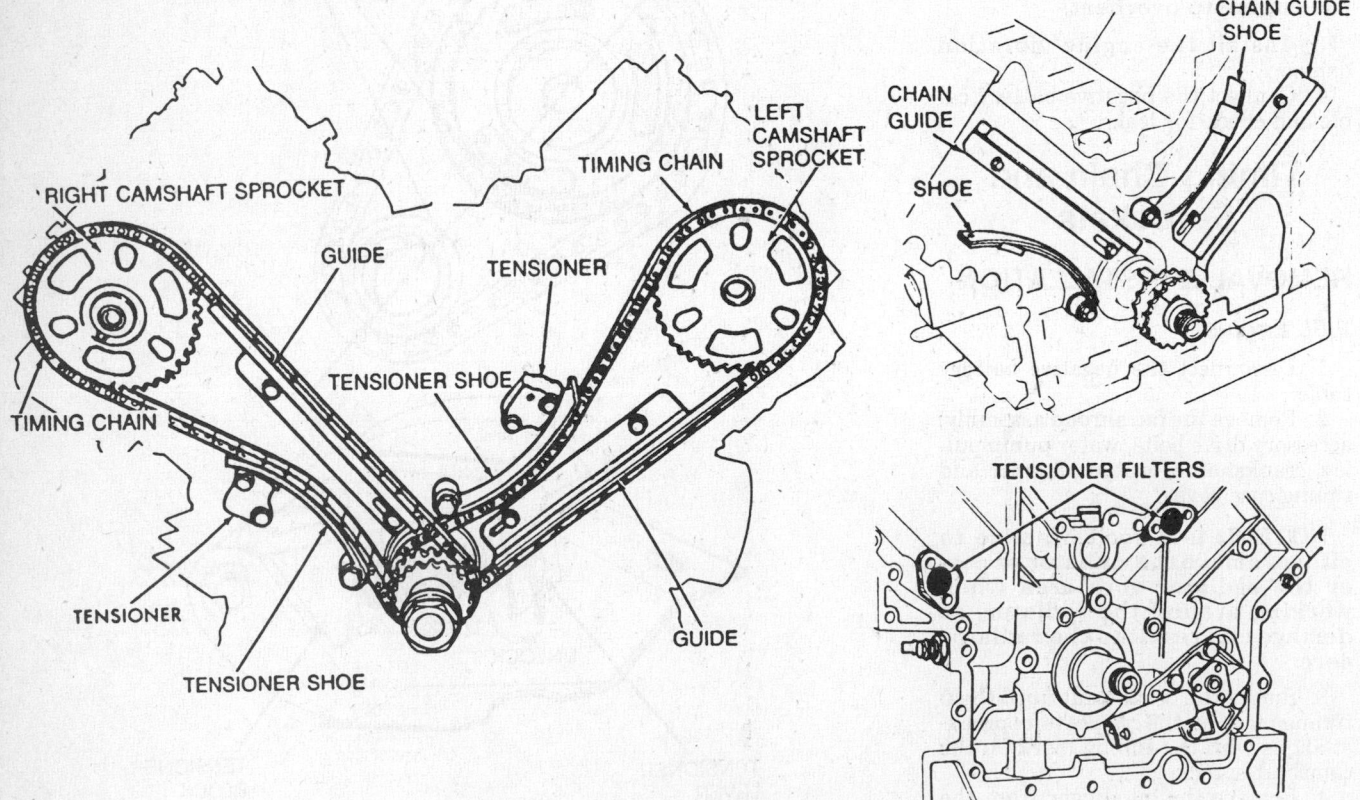

Timing chain and tensioner layout—3.0L engine

To install:

NOTE: **Inspect the timing chain tensioner. An opening shows the tensioner lock inside. This tensioner lock should not be removed. The lock is held in place by a spring that pushes a steel ball against the lock finger. If the lock is removed accidentally, replace the tensioner assembly because there is no way of checking the position of the lock finger in relation to the steel ball. When installing a tensioner use a thin blade tool to turn the ratchet counterclockwise. Then push the tensioner arm in. Position the tensioner over the filter and the tensioner shoe into the arm.**

11. To install, place the left and right chain guides into position and tighten the bolts to 48 inch lbs. (5.4 Nm). Install the tensioner shoes and tighten the mounting bolts to 9 ft. lbs.

12. Turn the left camshaft until the keyway slot is in the 11 o'clock position. Turn the right camshaft so the keyway is in the 8 o'clock position.

13. Turn the crankshaft until the keyway is aligned with the centerline of the left cylinder head.

NOTE: **The crankshaft has 3 sprockets on it. A sprocket each for the left and right timing chains and 1 for the oil pump drive. The timing mark is located on the center sprocket.**

14. Install the left camshaft sprocket. Install the left timing chain on the crankshaft. Position the single painted link of the timing chain on the tooth of the rear sprocket which is directly be-

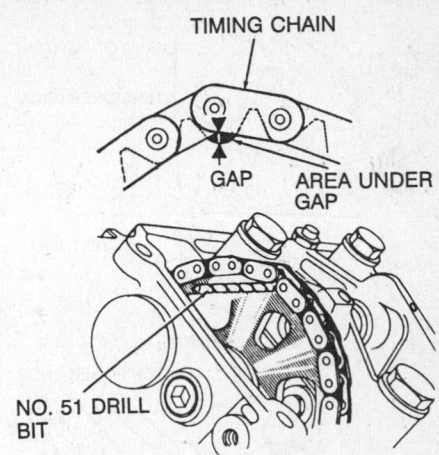

Measuring chain wear using a No. 51 drill as gauge—3.0L engine

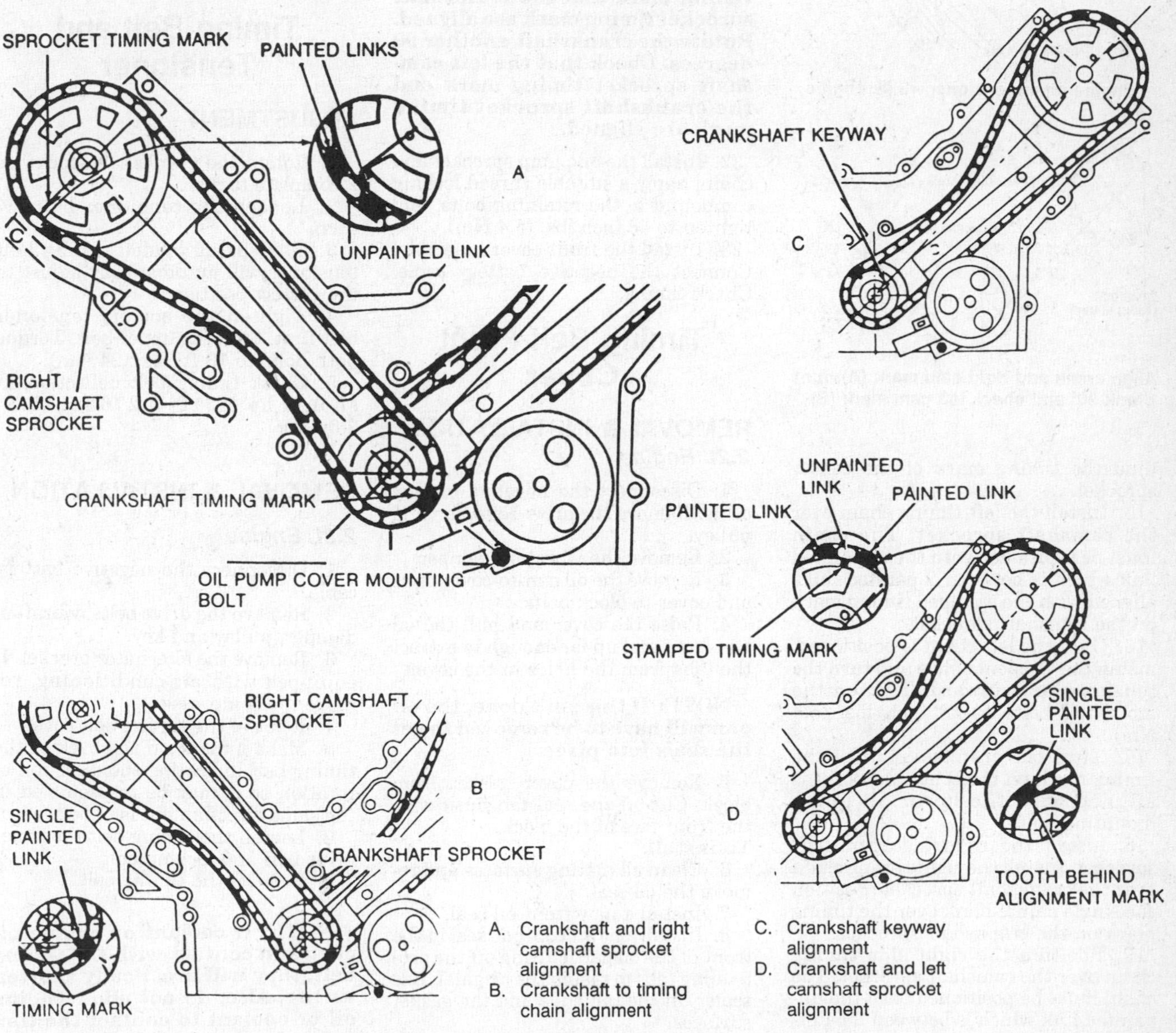

A. Crankshaft and right camshaft sprocket alignment
B. Crankshaft to timing chain alignment
C. Crankshaft keyway alignment
D. Crankshaft and left camshaft sprocket alignment

3.0L engine timing chains and marks are complex; locate and note marks before removal

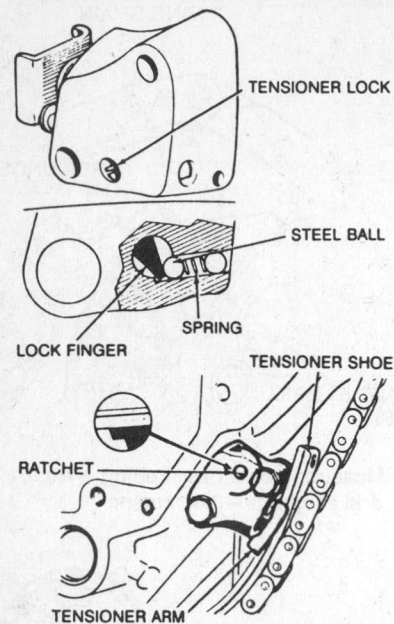

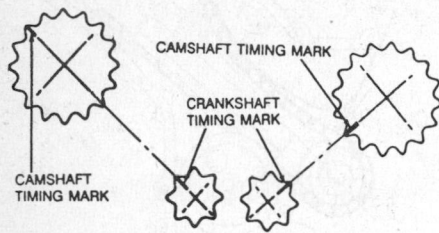

Engine chain tensioner—3.0L engine

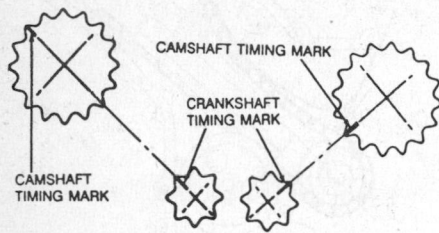

Align crank and right cam mark (A); turn crank 90° and check left cam mark (B)

hind the timing mark of the center sprocket.

15. Install the left timing chain over the camshaft sprocket. The chain must be positioned with the unpainted link which is between 2 painted links aligned with the stamped timing mark on the camshaft sprocket.

16. Once the left chain is positioned, install the tensioner shoe and turn the tensioner arm inward. Tighten the mounting bolts to 48 inch lbs. (5.4 Nm).

17. Turn the crankshaft until the timing mark on the center sprocket is aligned with the lower oil pump mounting bolt.

18. Install the right side camshaft sprocket. Install the right timing chain over the crankshaft sprocket. Position the single painted link over the timing mark on the crankshaft sprocket.

19. Position the right side timing chain over the camshaft sprocket. The chain must be positioned with the unpainted link which is between 2 painted links aligned with the stamped timing mark on the camshaft sprocket.

20. Once the right side chain is positioned, install the tensioner shoe and turn the tensioner arm inward. Tighten the mounting bolts to 48 inch lbs. (5.4 Nm).

21. Install the right camshaft sprocket bolt and tighten to 59 ft. lbs. (80 Nm). Push both of the chain tensioner shoes in to release them; this will adjust the chain tension.

NOTE: Once the crankshaft has been rotated, the painted marks on the chain will no longer align with the timing marks. When checking valve timing, it is the relation of the timing marks to each other that is used, not the position of the paint marks on the chains. To check, rotate the crankshaft 180 degrees. Check that the right camshaft sprocket timing mark and the crankshaft sprocket timing mark are aligned. Rotate the crankshaft another 90 degrees. Check that the left camshaft sprocket timing mark and the crankshaft sprocket timing mark are aligned.

22. Install the oil pump sprocket and chain, apply a suitable thread locking compound to the retaining bolts, and tighten to 48 inch lbs. (5.4 Nm).

23. Install the front cover assembly. Connect the negative battery cable. Check timing.

Timing Belt Front Cover

REMOVAL & INSTALLATION
2.2L Engine

1. Disconnect the negative battery cable. Remove the drive belts, fan and pulley.

2. Remove the vibration damper.

3. Remove the oil pan-to-cover bolts and cover-to-block bolts.

4. Raise the cover and pull the oil pan front seal up far enough to extract the tabs from the holes in the cover.

NOTE: If this isn't done, the oil pan will have to be removed to get the seals into place.

5. Remove the cover gasket from block. Cut off the seal tab flush with the front face of the block.
To install:

6. Clean all mating surfaces and remove the oil seal.

7. Install a new front oil seal.

8. Install a new neoprene seal in the front of the oil pan, cutting off the protruding tabs to match the original. Use sealer on the tab ends and the gasket surfaces.

9. Position the cover on the block and install the bolts. Tighten the cover

bolts to 4–6 ft. lbs. (5–8 Nm). The 4 lower bolts are tightened to 10–12 ft. lbs. (14–16 Nm).

10. Install the vibration damper and belts.

11. Connect the negative battery cable.

OIL SEAL REPLACEMENT

1. Disconnect the negative battery cable. Remove the timing belt cover.

2. Thoroughly clean and dry the cover.

3. Drive out the old seal. Install a new seal with a round driver.

4. Reinstall timing belt cover.

NOTE: The front oil seal can be installed with the cover in place only if the proper tool or equivalent is available.

Timing Belt and Tensioner

ADJUSTMENT

1. Rotate the crankshaft clockwise 2 complete turns.

2. Loosen the tensioner bolts ¼ turn.

3. The spring loaded timing belt tensioner will automatically adjust to the correct position.

4. Tighten the bottom tensioner bolt first, then the upper bolt. Torque both bolts to 18 ft. lbs. (24 Nm).

5. Check timing belt deflection. It should be 0.216–0.276 in. (5.5–7.0mm).

REMOVAL & INSTALLATION

2.2L Engine

1. Disconnect the negative battery cable.

2. Remove the drive belts, vibration damper, pulley and key.

3. Remove the alternator bracket. If equipped with air conditioning, remove the compressor.

4. Remove the timing belt cover.

5. Make a mark on the back of the timing belt indicating the direction of rotation so it may be reassembled in the same direction if it is to be reused.

6. Loosen the timing belt tensioner pivot bolt and locking bolt.

7. Remove the timing belt.

NOTE: If coolant or engine oil comes in contact with the timing belt, they will drastically shorten its life. Also, do not allow engine oil or coolant to contact the timing belt sprockets or tensioner assembly.

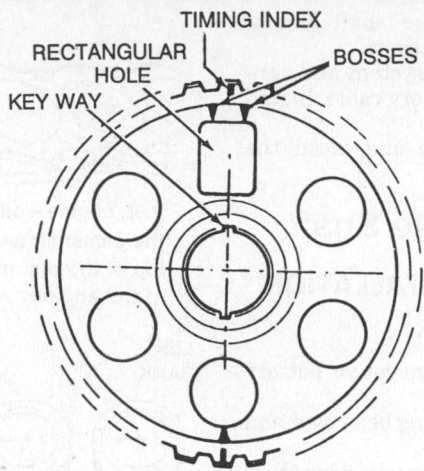

Camshaft sprocket timing marks alignment—2.2L engine

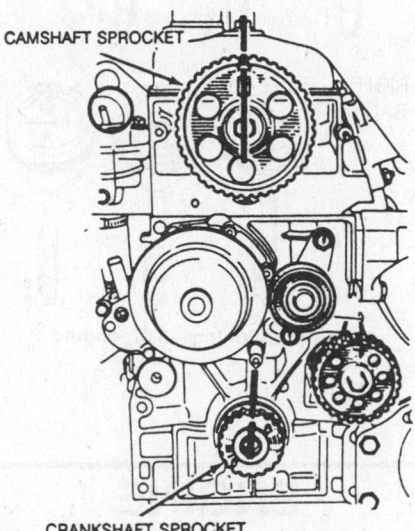

Timing index and timing marks alignment—2.2L engine

To install:

8. Inspect all parts for damage and wear. If any of the following is found, replacement is necessary:

   a. Timing belt—cracks on back surface, sides, bottom and separated canvas.

   b. Tensioner pulleys—turn the pulleys and check for binding, excessive play, unusual noise or if there is a grease leak.

9. Position the camshaft sprocket timing index in line with the static timing mark.

10. Position the crankshaft so No. 1 piston is at TDC on the compression stroke.

11. Remove the access hole plug in the cylinder block and insert the spe-cial tool used to apply pressure to the tensioner.

12. Loosen the timing belt tensioner bolts. Push the tensioner pulley towards the water pump to compress the tensioner spring. Tighten the tensioner bolts. This allows for easier installation of the timing belt.

13. Install the timing belt on the sprockets. If the original timing belt is being reused, install the timing belt with the arrow previously made pointing in the proper direction of rotation.

14. Loosen the tensioner bolts and allow the spring loaded tensioner to contact the belt. This will automatically tension the belt. Then, tighten the tensioner retaining bolts.

15. Position the timing belt cover over the sprockets and check the position of the camshaft sprocket timing mark with the index on the cover.

16. Install cylinder block plug, check the timing belt tension adjustment.

17. Install the compressor and alternator bracket.

18. Install vibration damper and drive belts.

19. Connect the negative battery cable and check the engine for proper operation.

Timing Sprockets

REMOVAL & INSTALLATION

1. Disconnect battery negative cable.

2. Turn crankshaft until piston No. 1 is at TDC of the compression stroke.

3. Remove the accessory drive belts.

4. Remove crankshaft vibration damper.

5. Remove the timing belt cover.

6. Remove the timing belt.

7. Make sure the camshaft sprocket has the rectangular hole upwards and timing mark is at the 12 o'clock position.

8. Loosen the camshaft sprocket bolt and gently tap the sprocket from the rear to remove.

To install:

9. Install the sprockets. Make sure the timing marks are aligned.

10. Install and adjust the timing belt.

11. Install the timing cover using a new seal, if required.

12. Install accessory belts, connect the negative battery cable, and check the timing.

Camshaft

REMOVAL & INSTALLATION

2.2L Engine

1. Relieve the fuel system pressure.

2. Disconnect the negative battery cable. Drain the cooling system.

3. Remove the intake and exhaust manifolds.

4. Remove the rocker cover and remove the rocker shaft assembly.

5. Remove the accessory drive belt. Remove the timing belt cover.

6. Remove the timing belt. Remove the cylinder head retaining bolts and remove the cylinder head.

7. Remove the camshaft sprocket and the bolts retaining the camshaft thrust plate.

8. Pry the oil seal out from around the camshaft and slide the camshaft from the head. Use care not to damage the camshaft lobes or the bearings.

To install:

9. Lubricate the camshaft with heavy oil and slide it into the head.

10. Install the camshaft thrust plate. Install a new camshaft oil seal using tool MOT–791–10 or equivalent. Install the camshaft sprocket and tighten the retaining bolt to 37 ft. lbs. (51 Nm).

11. Replace the gasket and install the cylinder head.

12. Install the timing belt and adjust the tension. Install the timing belt cover.

13. Install the rocker shaft assembly. Install the rocker cover, intake and exhaust manifolds.

14. Install the accessory drive belt and fill the cooling system.

15. Connect the negative battery cable. Run the engine and bleed the cooling system.

2.5L Engine

1. Relieve the fuel system pressure.

2. Disconnect the negative battery cable. Drain the cooling system.

3. Remove the radiator.

4. Remove the fan and water pump pulley.

5. Remove the grille, if necessary for clearance.

6. Remove the rocker cover, rocker arms and pushrods.

7. Remove the distributor, spark plugs and fuel pump.

8. Remove the lifters.

9. Remove the crankshaft hub and timing gear cover.

10. Remove the 2 camshaft thrust plate screws by working through the holes in the gear.

11. Remove the camshaft and gear assembly by pulling it through the front of the block. Be careful not to damage the bearings.

To install:

12. Lubricate the camshaft with heavy oil and install it into the block.

13. Install the timing chain and sprockets. Install the timing case cover.

14. Install the valve lifters and related components. Install the rocker cover.

15. Install the crankshaft hub and the water pump pulley. Install the accessory drive belts.

16. Position the distributor and tighten the hold-down bolt. Install the spark plugs.

17. Install the grille.

18. Connect the negative battery cable and check the timing.

3.0L Engine

The camshafts used in this engine are removed from the rear of the cylinder heads after the cylinder heads have been removed.

1. Relieve the fuel system pressure.

2. Disconnect the negative battery cable.

3. Drain the cooling system.

4. Remove the cylinder head(s).

5. Remove the camshaft cover at the rear of the cylinder head. Loosen the camshaft retainer bolt and slide the retainer away from the camshaft.

6. Slide the camshaft out of the head. Use care not to damage the camshaft lobes or bearings.

To install:

7. Coat the camshaft with heavy oil and slide it into the head. Position the retainer in the grove of the camshaft and tighten the mounting bolt to 9 ft. lbs. (12 Nm).

8. Push the camshaft to the front and check the camshaft endplay by inserting a feeler gauge between the retainer and the front of the camshaft.

9. Install the camshaft cover using a new gasket. Tighten the bolts to 48 inch lbs. (5.4 Nm).

10. Install the cylinder heads and all related parts.

11. Install the timing chains and sprockets.

12. Install the front cover assembly and the accessory drive belt.

13. Install the rocker shaft assemblies and the rocker covers.

14. Fill the cooling system and connect the negative battery cable. Install the air inlet tube.

15. Run the engine and bleed the cooling system.

Intermediate Shaft

REMOVAL & INSTALLATION

2.2L Engine

1. Disconnect the negative battery cable.

2. Remove the timing belt cover and the timing belt.

3. Remove the oil pump driveshaft cover, located on the side of the block.

4. Screw a piece of threaded rod into the top of oil pump driveshaft and remove it.

5. Remove the bolt retaining the intermediate shaft sprocket and remove the intermediate shaft sprocket.

6. Remove the bolts from the intermediate shaft cover. Remove the cover and gasket. Remove the bolt from the intermediate shaft retainer and pivot the retainer. Remove the intermediate shaft by pulling it from the block.

To install:

7. Coat the shaft with heavy oil and slide it into the block. Pivot the retainer into position and tighten the bolt. Install the shaft cover and loosely install the retaining bolts.

8. Install the shaft oil seal and align the cover using tool MOT-790 or equivalent. Tighten the cover retaining bolts.

9. Install the sprocket and bolt. Tighten the bolt to 37 ft. lbs. (51 Nm).

10. Install the oil pump driveshaft and cover.

11. Install the timing belt and cover, and check the belt tension.

12. Connect the negative battery cable.

Piston and Connecting Rod

POSITIONING

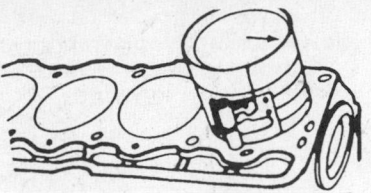

2.5L engine—oil squirt holes face the camshaft and the arrow on the top of the piston must face the front of the engine

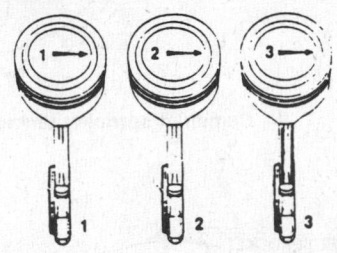

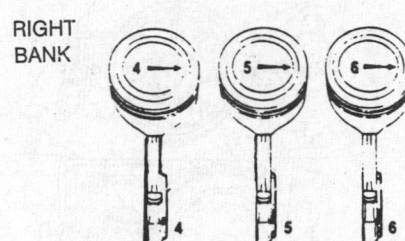

Piston positioning—3.0L engine

ENGINE LUBRICATION

Oil Pan

REMOVAL & INSTALLATION

2.2L Engine

1. Disconnect the negative battery cable.

2. Raise and safely support the vehicle.

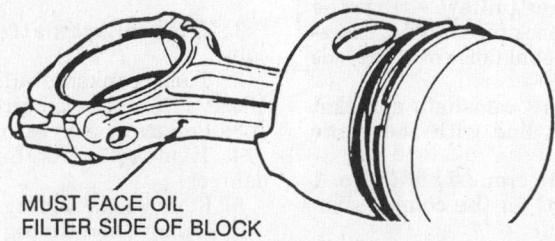

MUST FACE OIL FILTER SIDE OF BLOCK

Piston positioning—2.2L engine

3. Remove the underbody splash shield and drain the engine oil.

4. Remove the engine mount cushion nuts.

5. Lower the vehicle and position engine support tool MS–1900 or equivalent, on the inner fender flanges.

6. Tighten the support tool until the engine is raised enough to remove the oil pan.

NOTE: There are 3 sizes of bolts used to retain the oil pan on this engine. Note the location of each bolt when it is removed.

7. Raise and safely support the vehicle. Remove the oil pan bolts and remove the oil pan.

To install:

8. Thoroughly clean and dry the mating surfaces of the pan and block.

9. Install the oil pan to the engine block using a new gasket. Do not use any sealer on the gasket; it must be installed dry.

10. Tighten the oil pan bolts attaching the pan to the clutch/converter housing first, then tighten the remaining bolts. Tighten all of the bolts to 88 inch lbs. (10 Nm).

11. Install the splash shield and lower the vehicle.

12. Remove the support tool and install the engine mount nuts.

13. Fill the engine with the required amount of oil.

14. Connect the negative battery cable and check for leaks.

2.5L and 3.0L Engines

1. Disconnect the negative battery cable.

2. Raise and safely support the vehicle. Drain the oil.

3. Remove the front anti-sway bar retaining bolts and remove the sway bar.

4. Loosen the engine mount stud and nut assemblies. Remove the front tires.

5. Remove the lower ball joint retaining bolts and disengage the lower ball joints from the steering knuckles.

6. Remove the nuts at the center of the transaxle crossmember securing the rear of the transaxle to the crossmember.

7. Lower the vehicle and attach engine support tool MS–1900 to the engine.

8. With the vehicle lowered, loosen the 4 sub-frame attaching nuts. Remove the front 2 first, allowing the sub-frame to pivot to the ground. Support the rear of the sub-frame and remove the 2 rear nuts. Lower the sub-frame away from the vehicle.

9. Raise and support the vehicle. Remove the oil pan retaining bolts and remove the oil pan.

To install:

10. Thoroughly clean and dry the mating surfaces of the pan and block.

11. Install the oil pan to the engine block using a new gasket. Do not use any sealer on the gasket; it must be installed dry. Tighten all of the retaining bolts to 9 ft. lbs. (12 Nm).

12. Install the sub-frame assembly and tighten the mounting nuts to 92 ft. lbs. (125 Nm).

13. Connect the lower ball joints and tighten the attaching nut to 77 ft. lbs. (104 Nm). Tighten the transaxle-to-crossmember bolts to 20 ft. lbs. (27 Nm).

14. Remove the engine support tool. Attach the anti-sway bar and install the front wheels.

15. Lower the vehicle. Fill the crankcase with the appropriate quantity and grade of oil.

16. Connect the negative battery cable and check for leaks.

Oil Pump

REMOVAL & INSTALLATION

2.2L Engine

1. Disconnect the negative battery cable.

2. Remove the oil pump drive cover plate bolts and remove the cover.

3. Using a threaded rod, thread it into the top of the pump driveshaft. Remove the pump driveshaft by pulling it out of the block.

4. Raise and safely support the vehicle. Drain the oil.

5. Remove the oil pan. Remove the oil pump mounting bolts and remove the pump.

To install:

6. Install the oil pump using a new gasket. Tighten the mounting bolts to 33 ft. lbs. (45 Nm).

7. Install the oil pan. Fill the crankcase with the correct grade and quantity of oil.

8. Install the oil pump driveshaft and cover.

9. Connect the negative battery cable and check for leaks. Check for sufficient oil pressure.

2.5L Engine

1. Disconnect the negative battery cable.

2. Raise the vehicle and support safely. Drain the engine oil and remove the oil pan.

3. Remove the oil pump retaining bolts, the oil pump and gasket.

NOTE: The oil pump removal and installation will not affect the distributor timing because the distributor drive gear remains meshed with the camshaft. Do not disturb the position of the oil inlet tube and strainer assembly in the pump body. If the tube is moved in the body, a replacement tube and strainer must be installed to assure an airtight seal.

4. To ensure self priming, fill the gear cavity with petroleum jelly before installing the cover.

To install:

5. Install the pump with a new gas-

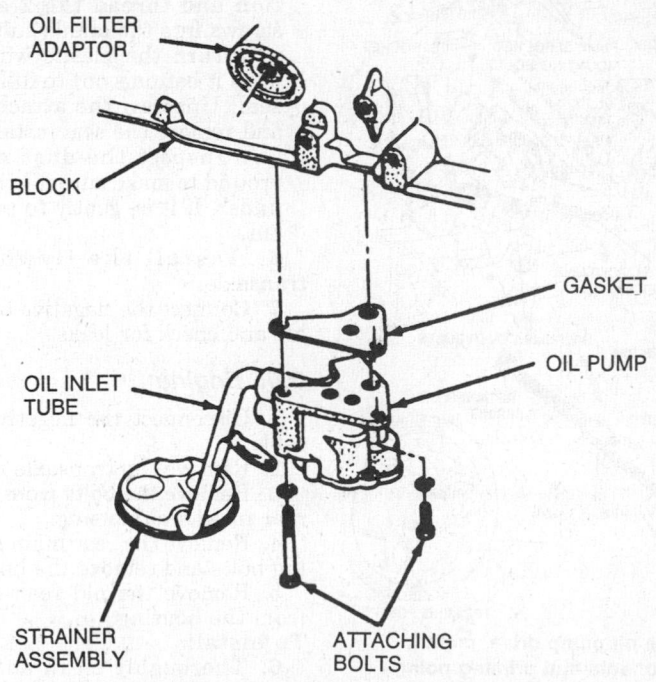

Oil pump and related parts—2.5L engine

ket. Tighten the short bolts to 10 ft. lbs. (14 Nm) and the long bolts to 17 ft. lbs. (23 Nm).

6. Install the oil pan and related parts using new gaskets and seals.

7. Fill the crankcase with the correct grade and quantity of oil.

8. Connect the negative battery cable and check for leaks. Check for sufficient oil pressure.

3.0L Engine

1. Disconnect the negative battery cable.

2. Remove the timing chain cover assembly.

3. Remove the bolts retaining the oil pump drive sprocket. Remove the sprocket and the oil pump drive chain.

4. Remove the oil pump mounting bolts and remove the oil pump.

To install:

5. Thoroughly clean and dry the mating surfaces of the pump and block. Install the oil pump to the block using a new gasket. Tighten the bolts to 9 ft. lbs. (12 Nm).

6. Install the oil pump drive sprocket and chain. Coat the threads of the sprocket bolts with a thread locking compound and torque them to 48 inch lbs. (5.4 Nm).

7. Install the timing chain cover.

8. Connect the negative battery cable and check for leaks. Check for sufficient oil pressure.

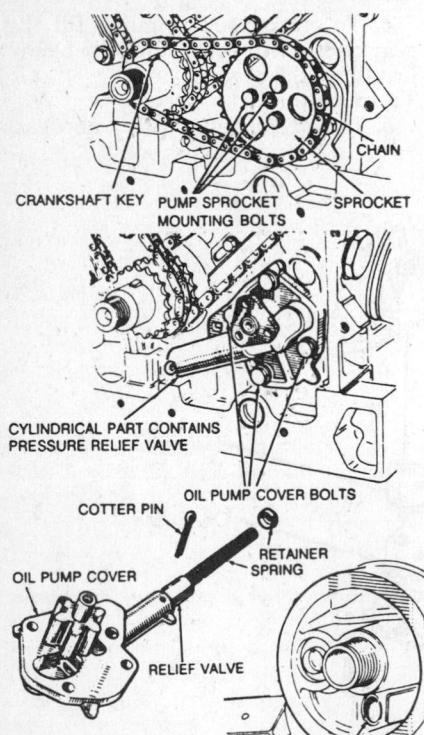

CRANKSHAFT KEY PUMP SPROCKET SPROCKET
MOUNTING BOLTS

CHAIN

CYLINDRICAL PART CONTAINS
PRESSURE RELIEF VALVE

COTTER PIN

OIL PUMP COVER BOLTS

RETAINER
SPRING

OIL PUMP COVER

RELIEF VALVE

PRIMING POINT

Engine oil pump drive, mount, components and priming point — 3.0L engine

Rear Main Oil Seal

REMOVAL & INSTALLATION

2.2L and 2.5L Engines

The rear main oil seal is a single unit and may be removed without removing the oil pan or crankshaft.

1. Disconnect the negative battery cable.

2. Remove the transaxle and flywheel.

3. Remove the rear main oil seal with a small prying tool. Be extremely careful not to scratch the crankshaft.

To install:

4. Lubricate the lips of the new seal with clean engine oil. Install the new seal by hand onto the rear crankshaft flange. The helical lip side of the seal should face the engine. Make sure the seal is firmly and evenly installed.

5. The new seal is installed with a special installer. Use the tool as follows:

   a. Back the plastic wing nut off until it contacts the cap nut on the end of the shaft.

   b. Lightly lubricate both the inside and outside edges of the seal.

   c. Install the seal on the tool with the dust shield facing toward the plastic wing nut.

   d. Fit the tool pilot in the center of the front surface of the installer into the pilot hole in the back of the crankshaft; the small dowel at the top of the front surface of the tool must fit into the corresponding small hole in the crankshaft at the same time. Hold the tool in this position and thread the 2 attaching screws into the crankshaft.

   e. Turn the plastic wing nut in until it bottoms out to fully seat the seal. Unscrew the attaching nuts and remove the seal installer.

   f. Inspect the dust shield all around to make sure it is not curled under. If it is, gently to pull the lip out.

6. Install the flywheel and transaxle.

7. Connect the negative battery cable and check for leaks.

3.0L Engine

1. Disconnect the negative battery cable.

2. Remove the transaxle assembly.

3. Remove the bolts from the lower rear main seal housing.

4. Remove the rear main seal housing bolts and remove the housing.

5. Remove the old rear main seal from the housing.

To install:

6. Thoroughly clean and dry the gasket mating surfaces.

7. Using a new gasket, install the rear main seal housing on the block.

8. Tighten the seal housing-to-block bolts first, then tighten the lower bolts. Torque all bolts to 9 ft. lbs. (12 Nm).

9. Install the new rear seal to tool MOT-259-01 or equivalent, and lightly coat the inner edges of the seal with oil. Install the seal to the seal housing by lightly tapping on the installation tool.

10. Remove the installation tool.

11. Install the transaxle assembly.

12. Connect the negative battery cable and check for leaks.

ENGINE COOLING

Radiator

REMOVAL & INSTALLATION

NOTE: Keep coolant off of the accessory drive belt and pulleys. Cover the belt and pulley with shop cloths prior to working on them. If coolant contacts the belts or pulleys, flush with water. Do not remove radiator cap or block drains when the system is hot.

Monaco and Premier

1. Disconnect the battery negative cable.

2. Remove the electric cooling fan assembly.

3. Attach one end of a ¼ in. hose about 3 feet long to the end of the radiator drain, the other end in a clean container. Open the drain and remove the radiator cap to drain the system.

4. Disconnect upper and lower hoses from the radiator.

5. Remove the front grille then disconnect the radiator from the air conditioning condenser by removing the top and bottom attaching screws. Lift out the radiator.

6. The installation is the reverse of the removal installation. Note that the radiator is equipped with alignment dowels on the bottom that fit into holes in the body crossmembers. Align these dowels when at installing.

7. Fill the radiator with coolant and bleed the system.

8. Connect the negative battery cable and check for leaks.

Medallion

1. Disconnect the negative battery cable.

2. Drain the coolant.

3. Matchmark and remove the hood.

4. If equipped with air conditioning, properly discharge the system.

5. Remove the 5 grille mounting screws and remove the grille.

6. Remove the radiator support and facia panel.

7. Disconnect hoses from the radiator.

8. Disconnect the connectors from the cooling fans and thermo switch.

9. If equipped with air conditioning, disconnect the refrigerant lines from the condenser. Cover the exposed ends of the lines to minimize contamination.

10. Remove any remaining mounting hardware and lift the radiator, condenser and cooling fans from the engine compartment as an assembly. Separate the components as required.

To install:

11. Assemble the components and position in the engine compartment. Install mounting hardware.

12. Replace the O-rings and connect the refrigerant lines to the condenser, if equipped.

13. Connect the electrical connectors.

14. Connect the hoses to the radiator.

15. Install the facia panel, radiator support, grille and hood.

16. Fill the radiator with coolant and bleed the system.

17. Evacuate and recharge the air conditioning system.

18. Connect the negative battery cable and check the entire climate control system for proper operation.

Electric Cooling Fan

TESTING

─────── **CAUTION** ───────
Make sure the key is in the OFF position when checking the electric cooling fan. If not, the fan could turn ON at any time, causing serious personal injury.

Monaco and Premier

All vehicles are equipped with an electric cooling fan systems designed to operate automatically under different conditions. The system reacts with the changes in engine temperature. When the engine coolant temperature reaches 198°F for 1988–89 or 188°F for 1990–92 , but below 212°F, the cooling fan switch low speed contacts close. This activates the cooling fan in low speed operation. When the coolant temperature exceeds 212°F, the fan switch activates the fan in high speed operation. If equipped with air conditioning, the coolant fan switch auto-

matically turns the cooling fan **ON** while the air conditioning is activated.

If the cooling fan does not operate with the air conditioning on but the compressor operates, repair an open between terminal **5** of the cooling fan relay connector and the compressor clutch relay.

COOLING FAN TEMPERATURE SWITCH

1. Turn ignition switch to **RUN**. Connect a jumper wire between coolant temperature switch terminals **A** and **B**. The cooling fan should operate.

2. If the cooling fan operates, replace the cooling fan temperature switch.

3. If the cooling fan does not operate, check the cooling fan relay.

COOLING FAN RELAY

1. Turn ignition switch to **RUN**, with the relay plugged in. Note that the cooling fan relay can be found in the relay panel in the engine compartment. Look for a panel on the driver's side fender wall area, with **4** relays. The front one should be the starter relay, next behind it is the ignition relay, then the radiator fan relay, with the back relay the air conditioning clutch relay. Connect a jumper wire between coolant temperature switch terminals **A** and **B**. Check for voltage at the cooling fan relay connector terminal **5**. If no voltage is present, repair an open between terminal **5** and the ignition switch.

2. If 12 volts is present, test for voltage at the cooling fan relay connector terminal **4**. If no voltage is present, repair an open between terminal **4** and fusible link **G**. The multiple fusible link connection is the main electrical feed near the battery positive cable.

3. If 12 volts is present at terminal **4**, check for voltage at the cooling fan relay connector terminal **2**. If voltage is present, some power is getting through, repair an open to ground since the system is trying to ground through the test meter. If zero voltage is indicated, check for voltage at connector terminal **1**.

4. If no voltage is present at connector terminal **1**, replace the cooling fan relay. If 12 volts is present, check and clean the connections at the cooling fan motor. If connections are okay, replace the cooling fan.

Medallion

1. Unplug the fan connector.

2. Using a jumper wire, connect the ground terminal of the fan connector to the negative battery terminal.

3. The fan should turn ON when the hot terminal is connected to the positive battery terminal.

4. If not, the fan is defective and should be replaced.

REMOVAL & INSTALLATION

Monaco and Premier

1. Disconnect the negative battery cable.

2. Remove the radiator support bracket screws. Remove the vibration cushion nuts.

3. Remove the upper radiator crossmember mounting screws and the crossmember.

4. Disconnect the electrical connectors from the fan. Remove the cooling fan and shroud mounting bolts and the fan by lifting upwards.

To install:

5. Install the fan into position and install the mounting bolts.

6. Install the radiator crossmember and support bracket.

7. Connect the negative battery cable and check the fan for proper operation.

Medallion

1. Disconnect the negative battery cable.

2. Unplug the connector.

3. Remove the mounting screws.

4. Remove the fan assembly.

5. The installation is the reverse of the removal procedure.

6. Connect the negative battery cable and check the fan for proper operation.

Heater Core

REMOVAL & INSTALLATION

Monaco and Premier

1. Disconnect the negative battery cable.

2. Drain the coolant. Properly discharge the air conditioning system, if equipped.

3. Remove the instrument panel lower trim cover, which is retained by 3 screws.

4. Remove the instrument panel support rod. Remove the screw attaching the steering column wiring harness bulkhead connector.

5. Disconnect the automatic transaxle shift cable from the lever.

   a. Compress the cable retainer tangs with pliers and slide the cable from the column mounting bracket.

   b. Loosen the screw that holds the anchoring bracket in place, move the bracket to the keyhole position and remove it from its mounting bracket.

6. Lift the indicator wire off of the pulley.

7. Pull the plastic sleeve down to ex-

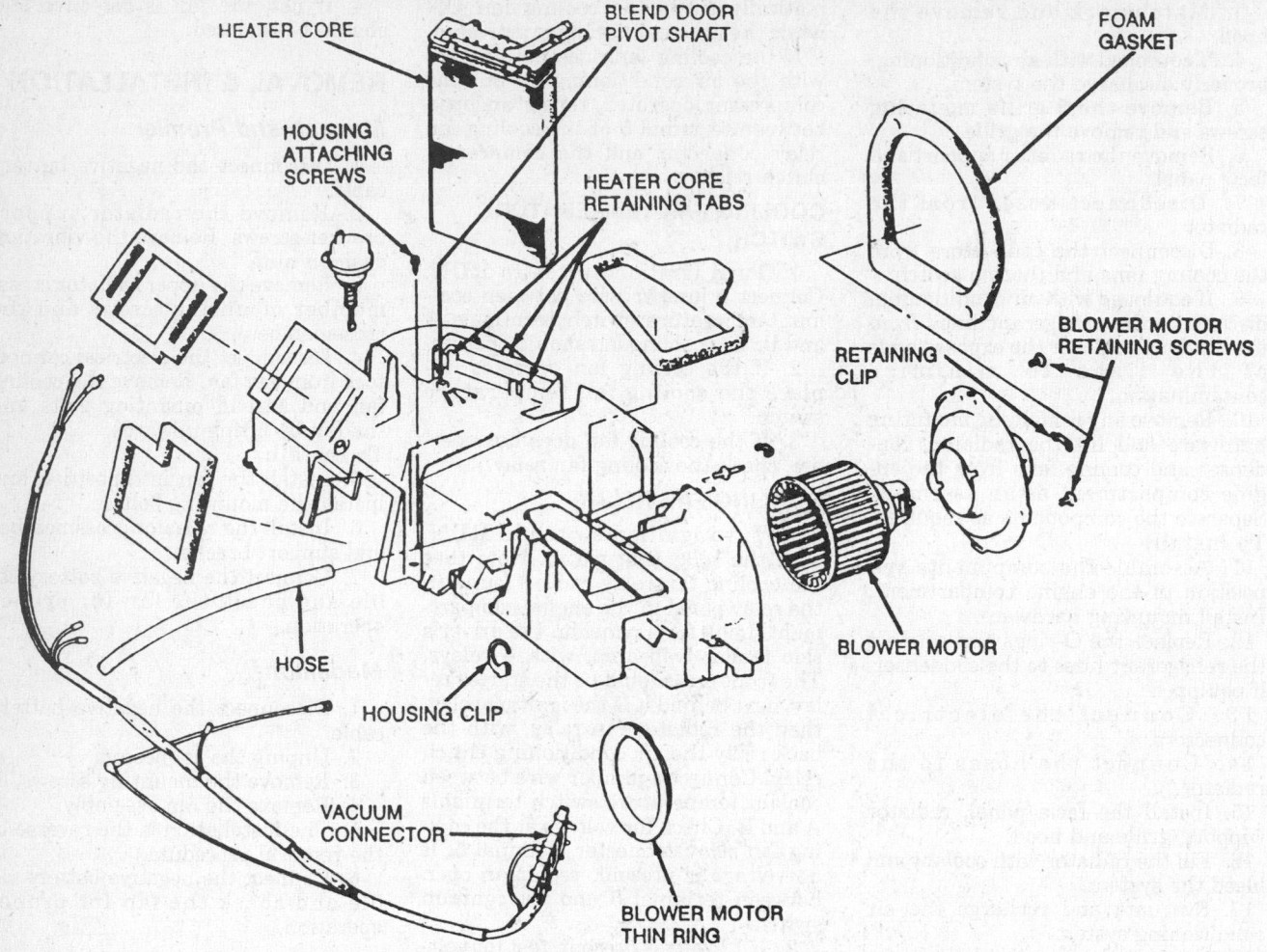

Heater core and upper air conditioning housing parts—Monaco and Premier

pose the steering column universal joint.

8. Make a reference mark on the steering column shaft and intermediate shaft.

9. Remove the bolt from the intermediate shaft.

10. Remove the 4 bolts and nuts that hold the steering column to the instrument panel and carefully lower to the vehicle floor.

11. Separate the steering column shaft from the intermediate shaft and remove the steering column assembly from the vehicle.

12. Remove the defroster grille from the top of the instrument panel.

13. Loosen but do not remove the nut located near the parking brake release handle and the nut which is located on the passenger side kick panel.

14. Remove the screws and lower the parking brake release handle.

15. Remove the ashtray.

16. Disconnect the cigarette lighter connectors.

17. Remove the screw from the ashtray cavity.

18. Disconnect all electrical connections.

19. Remove the bolts that hold the instrument panel to the center floor bracket.

20. Disconnect the interior temperature sensor.

21. Remove the floor duct extension.

NOTE: The heater core inlet and outlet tubes are made of plastic and may break if too much pressure is applied.

22. Remove the heater hoses from the heater core spouts.

23. Disconnect the coolant level switch connector.

24. Remove the coolant reservoir.

25. Disconnect the blower motor connector.

26. Disconnect the vacuum hoses.

27. Disconnect the refrigerant lines at the dash panel, if equipped.

28. Remove the retaining nuts from inside and outside of the vehicle and carefully pull the heater/air conditioning housing rearward to remove it.

29. Release the plastic tabs and re-

move the heater core from the housing.

To install:

30. Carefully insert the heater core into the housing and push until it snaps in place.

31. Before installing the housing, make sure the housing seals are in place and in good condition.

32. Position the heater/air conditioning housing to the dash panel. Make sure the drain tube extends through its opening in the upper floor and the blower motor connector and the vacuum line extends through the dash panel. Ensure that the ECU connectors are to the right of the drain tube.

33. Install new housing retaining nuts. Install the floor duct extension.

34. Install new O-rings on the refrigerant lines and lubricate with clean refrigerant oil. Press each line into its connector until it snaps into place.

35. Connect the vacuum hose and the blower electrical connector.

36. Install the coolant reservoir bracket and reservoir.

37. Reconnect the coolant level switch connector.

38. Carefully reconnect the heater hoses to the core.

39. Place the instrument panel into position so the mounting brackets engage the studs on the kick panels. Make sure the wiring harness is behind the center mounting bracket and connect all electrical connections.

40. Install the bolt to the brake support.

41. Install the screw into the ashtray cavity.

42. Install the 2 bolts to the center support bracket.

43. Connect the cigarette lighter connectors and the ashtray.

44. Tighten the nut located near the parking brake release handle and the nut located on the passenger side kick panel.

45. Install the parking brake release handle.

46. Install the bolts under the defroster grille, then install the grille.

47. Position and install the steering column shaft in the intermediate steering shaft U-joint. Align the 2 shafts using the reference marks made during removal. Install but do not tighten the U-joint bolt.

48. Attach the steering column to the instrument panel and tighten the bolts/nuts to 35 ft. lbs.

49. Tighten the bolt in the intermediate steering shaft U-joint and move the plastic sleeve into position.

50. Snap the shift cable into the mounting bracket.

51. Snap the shift cable head onto the mounting ball in the shift arm.

52. Loop the shift indicator wire over the pulley. Position the anchoring bracket over the screw.

53. Move the gearshift lever into **N** and check the position of the shift indicator. If the pointer is not aligned with the **N** mark on the display, slide the bracket forward/rearward to align the indicator. Tighten the screw.

54. Install the bulkhead connector and install the connector attaching screw.

55. Install the instrument panel support rod securely.

56. Install the instrument panel lower trim cover.

57. Evacuate and recharge the air conditioning system, if equipped.

58. Connect the negative battery cable and check the entire climate control system for proper operation. Check the system for leaks.

Medallion

1. Disconnect the negative battery cable.

2. Drain the cooling system. If equipped with air conditioning, properly discharge the system.

3. Remove the left and right rocker trim panels.

4. Disconnect the instrument panel wiring at the A-pillars.

5. Disconnect the ground cables at the rocker sills.

6. Disconnect the fuse panel and door buzzer.

7. Remove the lower instrument panel cover.

8. Open the glove box door and pull the edge of the console out to free it from the instrument panel.

9. On manual transmission equipped vehicles, pry off the boot shifter cover.

10. If equipped with automatic transmission remove the following:

   a. Remove the shift indicator plate by prying off with a suitable tool.

   b. Remove the shift lever knob by pulling straight off.

   c. Remove the shift indicator cover plate.

11. Remove the screws to free the console from the support. Pull the lower section of the console straight back and lift it up to remove it. Pull the upper section down and from the instrument panel.

12. Remove the radio bezel retaining screws.

13. Drill out the rivets that retain the radio.

14. Remove the radio bracket.

15. Remove the retaining screw from the heater control.

16. Remove the heater control knobs by pulling straight up.

17. Lower the heater control panel and disconnect the 2 cables and all the electrical connections.

18. Remove the upper and lower steering column covers.

19. Remove the bolt and nut at the steering joint connection under the dash.

20. Remove the 4 hex head bolts and 1 large Torx® head bolt holding the steering column in place.

21. Pull the steering column forward slightly and it will drop down. Disconnect the instrument panel wiring and remove the steering column.

22. Remove the speaker covers at the upper corners of the dash.

23. Remove the dash attaching bolts at each corner and remove the dash assembly.

24. Disconnect all electrical connections.

25. Disconnect and plug the heater hoses from the core.

26. On vehicles without air conditioning, remove the 3 remaining screws which retain the heater blower housing to the cowl panel and the housing by pulling it rearward.

27. On vehicles with air conditioning, perform the following:

   a. Disconnect the refrigerant lines from the expansion valve.

   b. Remove the heater evaporator housing retaining screws from inside the passenger compartment.

   c. Remove the vacuum reservoir from the bracket on the engine compartment side of the dash panel.

   d. Remove the 2 heater/evaporator housing retaining nuts in the engine compartment and the housing assembly.

28. Remove the screws that retain the heater core to the blower housing.

29. Spread the 4 retaining clips.

30. Remove the heater core by pulling straight up. Be careful of the capillary tube.

To install:

31. Install the heater core with the foam strips in place into the blower housing.

32. Make sure the 4 tabs clip into place.

33. Install the retaining screws.

34. On vehicles without air conditioning, position the heater housing against the cowl panel with the seals in place, then install the 3 heater housing retaining screws. On vehicles with air conditioning, mount and install the heater/evaporator housing into the vehicle and connect the heater hoses to the core and the air conditioning hoses to the evaporator.

35. Position the dashboard on the centering device and install the bolts in the corner of the dash.

36. Connect the wiring at the A-pillars.

37. Install the ground cables at the rocker sills.

38. Connect and adjust the heater control cables.

39. Install the control panel with the retaining screw.

40. Install the heater control knobs into the control panel.

41. Install the radio bracket, then secure the radio with rivets.

42. Install the lower console assembly.

43. Install the radio bezel and retaining screws.

44. Assemble the steering joint to the steering column.

45. Connect the speedometer cable.

46. Install the lower instrument panel cover.

47. Install the fuse panel and door buzzer.

48. Install the rocker trim panels.

49. On the vehicles without air conditioning, connect the heater hoses to the heater core.

50. Fill and bleed the cooling system.

51. Evacuate and recharge the air conditioning system, if equipped.

52. Connect the negative battery ca-

ble and check the entire climate control system for proper operation. Check the system for leaks.

Water Pump

REMOVAL & INSTALLATION

2.2L Engine

1. Disconnect the negative battery cable.
2. Drain the cooling system.
3. Remove the accessory drive belts.
4. Remove the timing belt cover.
5. Remove the water pump pulley bolt and remove the pulley.
6. Remove the timing belt and tensioner. Remove the hoses from the pump.
7. Remove the water pump attaching bolts and remove the water pump.
8. Clean the gasket mating surfaces.

To install:

9. Position the pump on the engine block using a new gasket. Tighten the bolts to 20 ft. lbs.
10. Install the timing belt and tensioner. Adjust the timing belt tension.
11. Install the timing belt cover. Install the water pump pulley and the accessory drive belts. Install the hoses.
12. Fill the cooling system and connect the negative battery cable. Bleed the cooling system.

2.5L Engine

1. Disconnect the negative battery cable.
2. Drain the cooling system. Remove the serpentine drive belt.
3. Disconnect the hoses from the engine. Remove the water pump pulley mounting bolts and remove the pulley.
4. Remove the water pump mounting bolts and remove the pump.
5. Clean the gasket mating surfaces.

To install:

6. Install the water pump using a new gasket. Tighten the bolts to 13 ft. lbs.
7. Install the hoses and the water pump pulley. Tighten the pulley retaining bolts to 20 ft. lbs.
8. Install the accessory drive belt.

NOTE: It is important that the serpentine belt is installed correctly. If it is incorrectly routed, the water pump could be rotated in the wrong direction, causing the engine to overheat.

9. Fill the cooling system. Connect the negative battery cable. Start the engine and bleed the cooling system.

3.0L Engine

1. Disconnect the negative battery cable.
2. Drain the cooling system.
3. Remove the spark plug wire holder from the top of the thermostat housing. Remove the nuts holding the engine damper to the engine.
4. Remove the accessory drive belt. Remove the upper and lower radiator hoses from the radiator.
5. Disconnect the electrical lead to the coolant temperature sensor.
6. At the back of the water pump, disconnect the hoses to the cylinder heads and the heater hoses.
7. Remove the water pump mounting bolts and remove the water pump.

To install:

8. Position the water pump to the block and tighten the mounting bolts to 13 ft. lbs.
9. Connect all of the hoses to the water pump, making sure they are not kinked. Connect the electrical lead to the coolant temperature sensor.
10. Install the accessory drive belt and the engine damper. Adjust the drive belt tension.

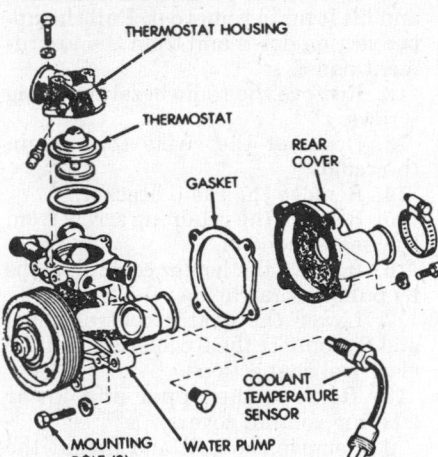

Water pump and thermostat assembly— 3.0L engine

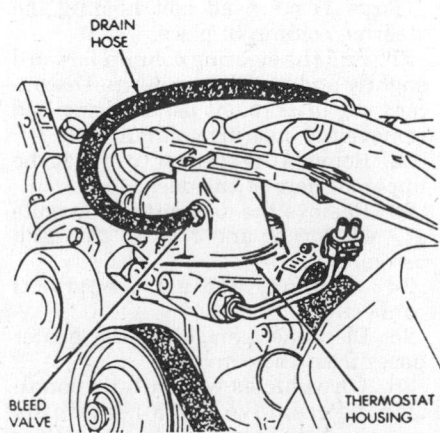

Cooling system bleed valve— 3.0L engine

11. Install the spark plug wire holder to the thermostat housing.
12. Fill the cooling system.
13. Connect the negative battery cable. Start the engine and bleed the cooling system. Check for leaks.

Thermostat

REMOVAL & INSTALLATION

1. Disconnect negative battery cable.
2. Drain the coolant until below the level of the thermostat housing. If useable, save it in a clean container.
3. Disconnect the coolant temperature sensor wire connector from the sensor.
4. Some engines have a spark plug wire holder on the top of the housing which should be moved.
5. Place shop towels on the serpentine belt. Chemicals deteriorate the synthetic materials in the belt. Always protect the belt and pulleys with clean shop towels.
6. Remove the thermostat housing and pull out the thermostat. Leave the radiator hose connected.

To install:

7. Clean gasket surfaces well. Inspect housing ports for blockage. Install a new thermostat and gasket.
8. Install the thermostat housing.
9. Install the spark plug wire holder on top of the housing.
10. Connect the coolant temperature sensor wire.
11. Fill the cooling system.
12. Connect the negative battery cable. Start the engine and bleed the cooling system. Check for leaks.

Cooling System Bleeding

3.0L Engine

NOTE: This procedure should be followed after any cooling system component has been replaced or removed and installed. It is essential that coolant does not contact the accessory drive belt or pulleys. Chemicals deteriorate the synthetic materials in the belt. Always protect the the the serpentine belt and pulleys with clean shop towels. When installing the drain hose to the air bleed valve on the thermostat housing, route the hose away from the belt, pulleys and cooling fan.

1. Attach one end of a 4-foot-long ¼ in. hose to the air bleed on the thermostat housing. Route the hose away from the drive belt and pulleys. Place the other end of the hose in a clean

container. The purpose of this hose is to keep coolant away from the belt and pulleys.

2. Open the bleed valve.

3. Slowly fill the coolant pressure bottle until a steady stream of coolant flows from the hose attached to the bleed valve. Close the bleed valve and continue filling to the full mark on the bottle. The full mark is the top of the post inside the bottle. Install the cap tightly on the coolant pressure bottle.

4. Remove the hose from the bleed valve, start and run the engine until the upper radiator hose is warm to the touch.

5. Turn the engine **OFF**. Reattach the drain hose to the bleed valve. Be sure to route the hose away from the belt and pulleys. Open the bleed valve until a steady stream of coolant flows from the hose. Close the bleed valve and remove the hose.

6. Check that the coolant pressure bottle is at or slightly above the full mark, at the top of the post inside the coolant pressure bottle. The full mark on the coolant pressure bottle is the correct coolant level for a cold engine. A hot engine will normally have a coolant level higher than the full mark.

ENGINE ELECTRICAL

NOTE: Disconnecting the negative battery cable on some vehicles may interfere with the functions of the on board computer systems and may require the computer to undergo a relearning process, once the negative battery cable is reconnected.

Distributor

REMOVAL

2.2L and 2.5L Engines

1. Disconnect the negative battery cable. Remove the distributor cap with the wires attached.

2. Matchmark the position of the rotor tip to the distributor housing and engine.

3. Disconnect and tag the distributor wiring and vacuum hose(s).

4. Remove the distributor hold-down bolt and pull the distributor up out of the engine. Note the position of the rotor in relation to the engine as the rotor stops rotating.

5. Do not rotate the engine with the distributor removed.

3.0L Engine

1. Disconnect the negative battery cable.

2. Remove accessory drive belt, if required.

3. Remove the timing belt cover.

4. Remove the spark plug wires from the spark plugs.

5. Remove the screws retaining the distributor cap.

6. Remove the screw that attach the distributor drive to the rotor and remove the rotor. Remove the dust shield from inside the housing.

7. Separate the distributor drive front and rear sections. Remove the distributor housing attaching bolts and remove the housing and seal.

INSTALLATION

NOTE: Some engines may be sensitive to the routing of the distributor sensor wires. If routed near the high-voltage coil wire or spark plug wires, the electromagnetic field surrounding the high-voltage wires could generate an occasional disruption of the ignition system operation.

TIMING NOT DISTURBED

2.2L AND 2.5L ENGINES

1. Install the distributor with the rotor pointing to the matchmark on the engine.

2. Rotate the housing and align the rotor-to-housing matchmark.

3. Install the distributor cap, hold-down clamp and nut. Tighten the nut hand-tight.

4. Connect the distributor wiring and negative battery cable. Run the engine and adjust the ignition timing.

5. With the timing correct, tighten the hold-down nut and recheck the timing.

3.0L ENGINE

1. Lightly coat the seal lips with clean engine oil and install the seal and housing.

2. Install the distributor drive rear section through the back of the cover, past the seal and into the housing.

3. Align the dowel in the top of the rear section with the dowel hole in the bottom of the distributor drive front section and tap them together.

4. Place the dust shield inside the distributor housing and the rotor on the the distributor drive. Install the retaining screw and tighten to 26 inch lbs.

5. Install the distributor cap and tighten the cap retaining bolts to 35 inch lbs.

6. Connect the spark plug wires.

TIMING DISTURBED

2.2L ENGINE

1. Remove the spark plug from the No. 1 cylinder and position a compression gauge or a thumb over the spark plug hole.

2. Slowly crank the engine until compression pressure starts to build up.

3. Continue cranking the engine so the timing mark or pointer aligns with the TDC mark.

4. Install the distributor with its drive meshed, so the rotor points to the No. 1 terminal on the distributor cap with No. 1 cylinder piston at TDC.

5. Install the distributor cap, hold-down clamp and nut. Tighten the nut hand-tight.

6. Connect the distributor wiring and negative battery cable. Run the engine and adjust the ignition timing.

7. With the timing correct, tighten the hold-down nut and recheck the timing.

2.5L ENGINE

1. Rotate the engine until the No. 1 piston is at TDC of the compression stroke.

2. Using an appropriate tool inserted in the distributor hole, rotate the oil pump gear so the slot in the oil pump shaft is slightly past the 3 o'clock position, relative to the length of the engine block.

3. With the distributor cap removed, install the distributor with the rotor at the 5 o'clock position, relative to the oil pump gear shaft slot. When the distributor is completely in place, the rotor should be at the 6 o'clock position. If not, remove the distributor and perform the entire procedure again.

4. Check the timing and tighten the lock bolt.

3.0L ENGINE

1. Remove the spark plug from the No. 1 cylinder and position a compression gauge or a thumb over the spark plug hole.

2. Slowly crank the engine until compression pressure starts to build up.

3. Continue cranking the engine so the timing mark or pointer aligns with the TDC mark.

4. Install the distributor drive rear section through the back of the cover, past the seal and into the housing.

5. Align the dowel in the top of the rear section with the dowel hole in the bottom of the distributor drive front section and tap them together.

6. Place the dust shield inside the distributor housing and the rotor on the the distributor drive. Install the

retaining screw and tighten to 26 inch lbs.

7. Install the distributor cap and tighten the cap retaining bolts to 35 inch lbs. Attach the spark plug wires.

Distributorless Ignition

All distributorless ignition components pertain to Monaco and Premier built on or after October 9, 1991 only.

REMOVAL & INSTALLATION

Crankshaft Timing Sensor

1. Disconnect the negative battery cable.
2. Disconnect the crankshaft timing sensor pick-up lead at the wiring harness connector.
3. Remove the sensor retaining bolts.
4. Remove the sensor from the transaxle bellhousing.
5. The installation is the reverse of the removal procedure. Tighten the retaining bolt to 105 inch lbs. (12 Nm).

Camshaft Position Sensor

1. Disconnect the negative battery cable.
2. Disconnect the camshaft position sensor lead at the wiring harness.
3. Remove the retaining bolt and remove the sensor from the cylinder head.

To install:

4. Install the sensor with the tab facing the rear of the engine.
5. Install the retaining bolt and tighten to 105 inch lbs. (12 Nm).
6. Connect the camshaft position sensor lead at the wiring harness.

Ignition Coil

1. Disconnect the negative battery cable.
2. Remove the spark plug wires from the coil by gripping the boot and not the cable.
3. Disconnect the electrical connector.
4. Remove the coil fasteners.
5. Remove the coil from the vehicle.
6. The installation is the reverse of the removal procedure.

Ignition Timing

ADJUSTMENT

Ignition timing is adjusted by the vehicle's Electronic Control Unit (ECU). The ECU uses input from sensors in the engine to determine various conditions during operation such as, manifold pressure, engine speed, manifold air temperature and coolant temperature. These inputs allow the ECU to adjust the timing under a variety of engine conditions. Therefore no timing adjustment is needed for normal vehicle service.

Alternator

PRECAUTIONS

Several precautions must be observed with alternator equipped vehicles to avoid damage to the unit.

- If the battery is removed for any reason, make sure it is reconnected with the correct polarity. Reversing the battery connections may result in damage to the one-way rectifiers.
- When utilizing a booster battery as a starting aid, always connect the positive to positive terminals and the negative terminal from the booster battery to a good engine ground on the vehicle being started.
- Never use a fast charger as a booster to start vehicles.
- Disconnect the battery cables when charging the battery with a fast charger.
- Never attempt to polarize the alternator.
- Do not use test lamps of more than 12 volts when checking diode continuity.
- Do not short across or ground any of the alternator terminals.
- The polarity of the battery, alternator and regulator must be matched and considered before making any electrical connections within the system.
- Never separate the alternator on an open circuit. Make sure all connections within the circuit are clean and tight.
- Disconnect the battery ground terminal when performing any service on electrical components.
- Disconnect the battery if arc welding is to be done on the vehicle.

BELT TENSION ADJUSTMENT

A single serpentine belt is used to drive all engine accessories. On the 2.2L engine, the drive belt tension is adjusted with the belt tension adjuster bolt, next to the alternator. On the 2.5L engine, the drive belt tension is adjusted with the power steering pump. On Premier the 3.0L engine, the drive belt tension is adjusted with the alternator.

When checking and adjusting belt tension, place the tension gauge on the longest belt span between pulleys. Specifications are: 180–200 lbs. for a new belt or 140–160 lbs. for a used belt.

REMOVAL & INSTALLATION

2.2L Engine

1. Disconnect the negative battery cable. Raise and support the vehicle safely.
2. Remove the lower splash shield.
3. Loosen but do not remove the locking bolt and adjusting nut from the drive belt tension adjuster.
4. Remove the lower alternator mounting nut. The nut is also used by the top tensioner mount.
5. Loosen the top alternator mounting bolt. Remove the tensioner from the alternator.

NOTE: Never use a sharp instrument to remove the drive belt from the pulley. The belt is made of synthetic material and may be damaged.

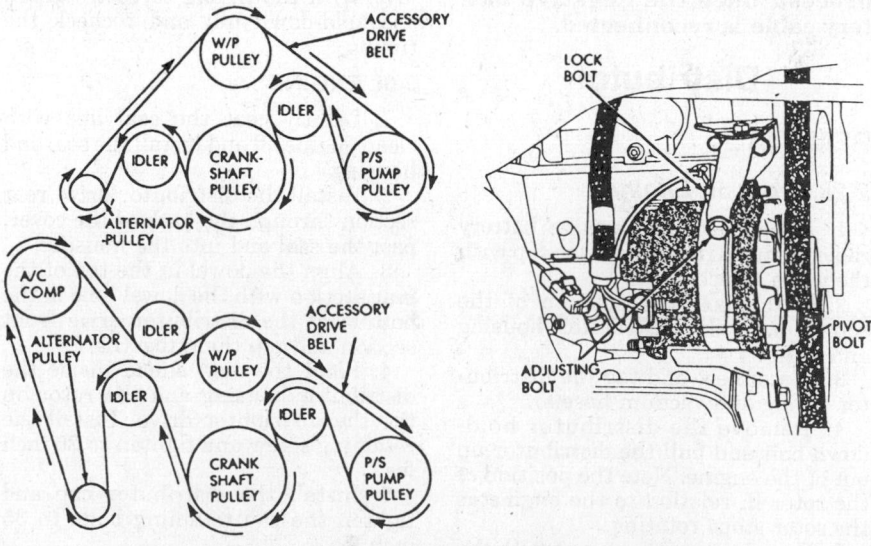

Serpentine belt and alternator adjust points – 3.0L engine

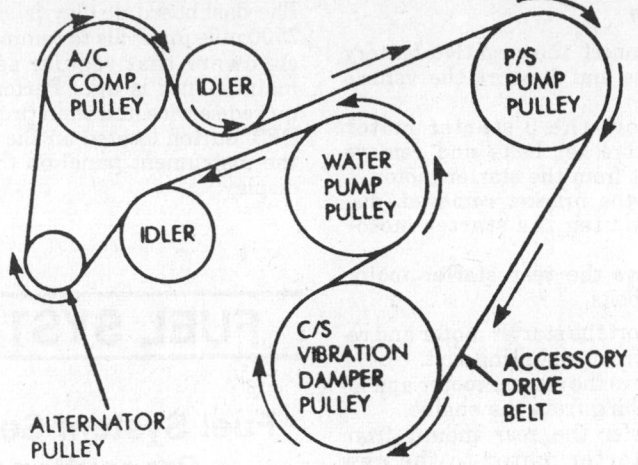

With A/C

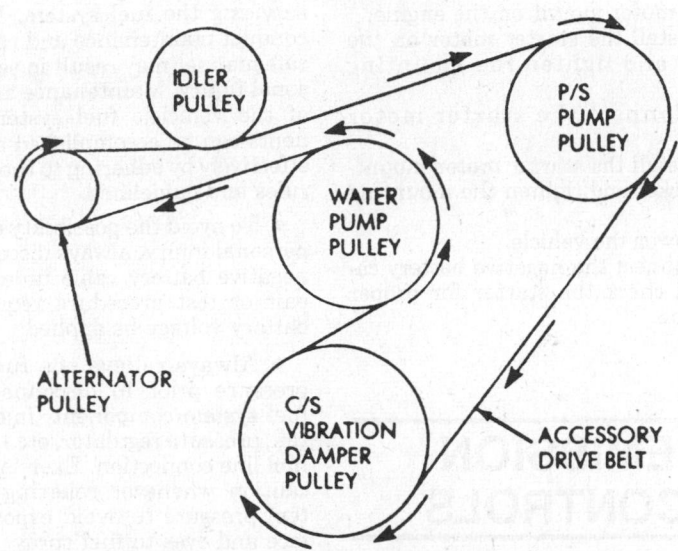

Without A/C
Serpentine belt arrangement—2.5L engine

6. Remove the serpentine drive belt from the alternator pulley.

7. Disconnect and tag the alternator wiring. Remove the top alternator mounting bolt.

8. Remove the alternator from the engine.

To install:

9. Position the alternator and install the top mounting bolt.

10. Install alternator lower mounting nut and tensioner. Tighten mounting nut finger-tight.

11. Install electrical connectors. Install serpentine belt.

12. Tighten the tensioner nut finger-tight and tensioner adjuster nut to obtain proper belt tension.

13. Tighten all mounting bolts and nuts.

14. Install splash shield and lower vehicle.

15. Connect the negative battery cable and check alternator operation.

2.5L Engine

1. Disconnect battery negative cable.

2. Remove the power steering pump locking nut from mounting bracket.

3. Loosen pivot bolt, adjusting bolt and 2 bolts located at rear of power steering pump.

4. Remove the drive belt and disconnect all electrical connectors from alternator.

5. Remove the pivot bolt and mounting bolts. Remove the alternator assembly.

To install:

6. Position the alternator on the engine and install the pivot bolt and mounting bolts finger-tight.

7. Reconnect the electrical connectors and install the drive belt.

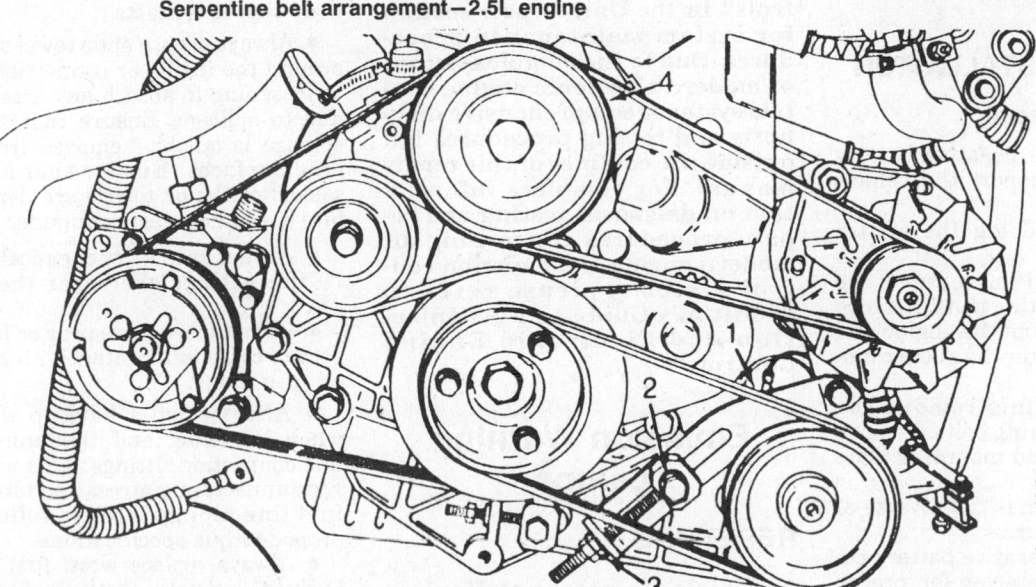

1. Power steering belt
2. Serpentine belt
3. Serpentine belt adjustment nut
4. Water pump

Accessory belts—2.2L engine

8. Install a belt tension gauge and tighten adjusting bolt to obtain the proper belt tension.

9. Torque pivot bolt to 30 ft. lbs. (40 Nm).

10. Torque the 2 rear power steering mounting bolts to 20 ft. lbs. (27 Nm).

11. Torque the locking nut to 20 ft. lbs. (27 Nm).

12. Connect the negative battery cable and check the alternator for proper operation.

3.0L Engine

1. Disconnect the negative battery cable.

2. Raise and safely support the vehicle.

3. Remove the lower splash shield.

4. Loosen the alternator adjusting bolt to relieve the belt tension. Remove alternator drive belt, mounting bolt and pivot bolt.

5. Disconnect all electrical connectors from the alternator and remove the alternator assembly.

To install:

6. Position the alternator on the engine and install pivot bolt and mounting bolts finger-tight.

7. Reconnect the electrical connectors and install the drive belt.

8. Torque the pivot bolt to 37 ft. lbs. (50 Nm).

9. Tighten the adjusting bolt to obtain the proper belt tension.

10. Torque the mounting bolt to 20 ft. lbs. (27 Nm).

11. Install the lower splash shield and lower the vehicle.

12. Connect the negative battery cable and check the alternator for proper operation.

Starter

REMOVAL & INSTALLATION

Premier and Monaco

1. Disconnect the negative battery cable. Raise and support the vehicle safely.

2. Disconnect and tag the starter motor wiring.

3. On 2.5L engine, remove the starter motor mounting bolts. Remove the starter motor from the engine and remove the bushing in the starter motor mounting plate.

4. On 3.0L engine, remove the starter motor mounting bolts. Remove the starter motor and mounting plate from the engine.

5. The installation is the reverse of the removal procedure.

6. Connect the negative battery cable and check the starter for proper operation.

Medallion

1. Disconnect the negative battery cable. Raise and support the vehicle safely.

2. Remove the 3 starter motor mounting bracket bolts and remove the bracket from the starter motor.

3. With the bracket removed, disconnect and tag the starter motor wiring.

4. Remove the rear starter motor mounting bolts.

5. Support the starter motor and remove the front mounting bolt.

6. Remove the starter motor and locating bushing from the engine.

7. Transfer the rear mount from the old starter motor to the new motor.

To install:

8. Place the bushing in the front starter motor mount on the engine.

9. Install the starter motor on the engine and tighten the mounting bolts.

10. Connect the starter motor wiring.

11. Install the starter motor mounting bracket and tighten the mounting bolts.

12. Lower the vehicle.

13. Connect the negative battery cable and check the starter for proper operation.

EMISSION CONTROLS

Please refer to "Emission Controls" in the Unit Repair section for system maintenance procedures. Due to the complex nature of modern electronic engine control systems, comprehensive diagnosis and testing procedures fall outside the confines of this repair manual. For complete information on diagnosis, testing and repair procedures concerning all modern engine and emission control systems, please refer to "Chilton's Guide to Fuel Injection and Electronic Engine Controls".

Emission Warning Lamps

RESETTING

A Vehicle Maintenance Monitor (VMM) is installed on most vehicles.

The dashboard display is activated at 7500 mile intervals to remind the vehicle owner that regular service and maintenance is due. Perform the required service and then press the **RESET** button located on the left side of the instrument panel on the monitor display.

FUEL SYSTEM

Fuel System Service Precautions

Safety is an important factor when servicing the fuel system. Failure to conduct maintenance and repairs in a safe manner may result in serious personal injury. Maintenance and testing of the vehicle's fuel system components can be accomplished safely and effectively by adhering to the following rules and guidelines.

● To avoid the possibility of fire and personal injury, always disconnect the negative battery cable unless the repair or test procedure requires that battery voltage be applied.

● Always relieve the fuel system pressure prior to disconnecting any fuel system component (injector, fuel rail, pressure regulator, etc.), fitting or fuel line connection. Exercise extreme caution whenever relieving fuel system pressure to avoid exposing skin, face and eyes to fuel spray. Please be advised that fuel under pressure may penetrate the skin or any part of the body that it contacts.

● Always place a shop towel or cloth around the fitting or connection prior to loosening to absorb any excess fuel due to spillage. Ensure that all fuel spillage is quickly removed from engine surfaces. Ensure that all fuel soaked cloths or towels are deposited into a suitable waste container.

● Always keep a dry chemical (Class B) fire extinguisher near the work area.

● Do not allow fuel spray or fuel vapors to come into contact with a spark or open flame.

● Always use a backup wrench when loosening and tightening fuel line connection fittings. This will prevent unnecessary stress and torsion to fuel line piping. Always follow the proper torque specifications.

● Always replace worn fuel fitting O-rings. Do not substitute fuel hose where fuel pipe is installed.

FUEL TUBES AND QUICK-CONNECT FITTINGS

Monaco and Premier

The fuel system in these vehicles utilize plastic fuel tubes with quick-connect fittings that have sealed O-rings; these O-rings do not have to be replaced when the fittings are disconnected. The quick-connect fitting consists of the O-rings, a retainer and the casing. When the fuel tube nipple is inserted into the quick-connect fitting, the shoulder of the nipple is locked in place by the retainer, and the O-rings seal the tube. The fuel tube nipples must first be lubricated with clean 30 weight engine oil prior to reconnecting the quick-connect fitting.

When the fittings are disconnected, the retainer will stay on the nipple of the component that the tube is being disconnected from. A fuel tube should never be inserted into a quick-connect fitting without the retainer being either on the tube or already in the quick-connect fitting. In either case, care must be taken to ensure that the retainer is locked securely into the quick-connect fitting.

If the quick-connect fitting has windows in the side of the casing, the retainer locking ears and the shoulder (stop bead) on the tube must be visible in the windows, or the retainer is not properly installed. After connecting a quick-connect fitting, the connection should be verified by pulling on the lines to ensure that the lock is secure.

There is a factory tool that can be used at the 3.0L engine fuel rail and fuel pressure regulator to remove the quick-connect fitting and retainer as an assembly. The retainer will remain in the fitting in the correct position. To install the fuel tube, push it over the nipple until a click is heard. Pull back on the tube to ensure that the connector is locked in place.

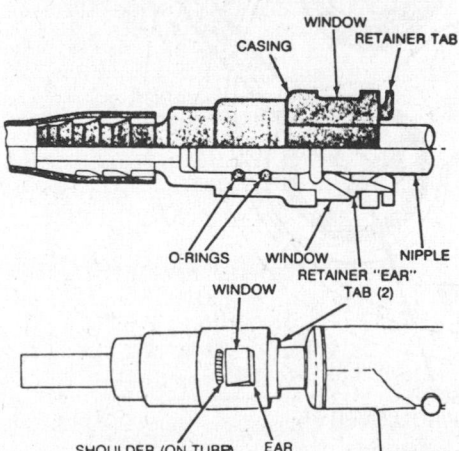

Special quick-connect fittings

RELIEVING FUEL SYSTEM PRESSURE

NOTE: Always wear eye protection when servicing the fuel system. Do not smoke or allow open flame near the fuel system or components during fuel system service.

1. Loosen the fuel filler cap to release fuel tank pressure.
2. Disconnect the fuel pump harness connector located in the rear of the vehicle near the fuel tank.
3. Start the vehicle and allow it to run until it stalls from lack of fuel. Turn the key to the **OFF** position.
4. Disconnect the negative battery cable, then reconnect the fuel pump connector.
5. Wrap shop towels around the fitting that is being disconnected to absorb residual fuel in the lines.

Fuel Tank

REMOVAL & INSTALLATION

1. Relieve the fuel system pressure.
2. Disconnect the negative battery cable.
3. Drain the fuel from the fuel tank.
4. Raise and safely support the vehicle. Remove the right rear wheel and inner fender splash shield.
5. Disconnect the fuel lines at the fuel filter and the electrical connectors from the tank. Disconnect the fuel tank vent tube from the filler neck. Disconnect the ground wire from the body.
6. Place a suitable support under the tank and remove the retaining straps. Lower the tank from the vehicle.
To install:
7. Raise the tank into position and install the tank straps.
8. Connect all hoses and wires to tank-mounted components.
9. Install the splash shield and wheel.
10. Connect the negative battery cable and check for leaks.

Fuel Filter

REMOVAL & INSTALLATION

1. Relieve the fuel pressure.
2. Disconnect the negative battery cable.
3. The filter is located on the frame rail near the rear of the vehicle.
4. Depress, or squeeze the retainer tabs together and slowly pull the connectors from the fuel filter. Note the retainer tabs stay on the fuel filter nipples.

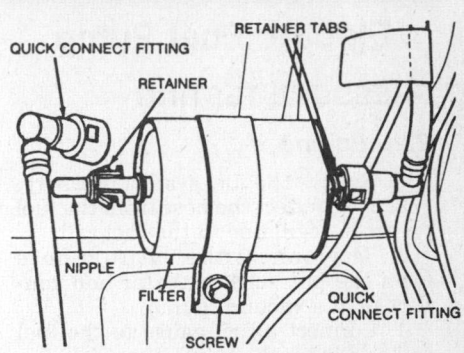

Fuel filter with special fittings

5. Remove the screw holding the fuel filter in place and remove the filter.
To install:
6. Carefully remove the retainers from the fuel filter nipples with a thin straight blade tool. Insert the tool between the filter nipple and the wedge portion of the retainer that seats against the shoulder of the nipple. Press the wedge back and slip the wedge over the nipple shoulder. Repeat this on the other side of the retainer and then pull the retainer off the nipple.
7. Push the retainers back into the fuel line quick-connect fittings. Ensure that the locking ears and the shoulder (stop bead) on the fuel tube are completely visible in the windows on the side of the quick-connect fitting.
8. The fuel filter may be marked **IN** and **OUT** at the nipple ends. The side marked **IN** is connected to the fuel line from the fuel tank. The side of the filter marked **OUT** is connected to the fuel line that runs to the engine. After determining the proper direction, install the fuel filter with the attaching screw.
9. Use a clean cloth to wipe the tube ends clean and lightly lubricate the fuel tube ends with clean 30 weight motor oil. The connectors contain O-rings which do not have to be replaced when the fittings are disconnected. Push the quick-connect fitting over the fuel tube until a click is heard. If the quick-connect fitting is type that has windows on the side, ensure that the locking ears on the retainer and the shoulder (stop bead) on the fuel tube are completely visible in the windows. Do not rely on the audible click to confirm that a secure connection has been made. Pull back on the quick-connect fitting to further ensure that the connection is complete and the connector is locked in place.
10. Connect the negative battery cable, install the fuel filler cap, turn the key to the **ON** position to pressurize the fuel system and check for leaks.

Electric Fuel Pump

PRESSURE TESTING

2.2L Engine

1. Relieve the fuel system pressure.
2. Disconnect the hose from the fuel pressure regulator to the fuel rail.
3. Disconnect the vacuum hose from the pressure regulator and connect it to a vacuum pump.
4. Connect a fuel gauge to the fuel rail and start the engine.
5. Check the fuel pressure readings and compare against the specification.
6. Apply 15 inches of vacuum to the pressure regulator. The pressure should drop about 3 psi.
7. Turn the ignition **OFF**. Remove the fuel gauge from fuel rail.
8. Reconnect the vacuum hose to the pressure regulator and hose from the pressure regulator to the fuel rail.

2.5L Engine

NOTE: The throttle body has 2 port plugs on it. The test port is located on the side of the fuel pressure regulator next to the fuel return tube connection.

1. Allow the engine to cool down before removing the test port.
2. Relieve the fuel system pressure.
3. Place a shop towel over the test port to catch fuel and slowly remove the test port plug from the throttle body.
4. Install fuel pressure test adapter along with a 0–30 psi (0–207 kPa) gauge into test port.
5. Start the engine and let it idle. Check the fuel pressure reading. If the pressure is not within specifications adjust the fuel pressure regulator as followed:
   a. Locate the fuel pressure regulator adjusting screw behind the aluminum plug in the nose of the fuel pressure regulator casing.
   b. Lightly tap the plug with a small punch and hammer until it pops out.
   c. Run the engine at 750–800 rpm, then turn the adjustment screw until the fuel pressure is within specifications.
6. Turn the ignition switch **OFF**. Disconnect the fuel gauge and pressure test adapter.
7. Install the plug in test port. Install the aluminum plug in front of the regulator adjusting screw.
8. Replace the fuel tank filler cap.

3.0L Engine

1. Relieve the fuel system pressure.
2. Remove the black fuel supply tube from the fuel rail using tool 6182 or equivalent. Slide the tool over the nipple and up into the connector until the handle fits the connector. Pull the fuel supply tube off the fuel rail.
3. Install fuel tube adapter 6175 or equivalent and a 0–60 psi gauge. Push the adapter female end with the quick connect fitting over the fuel rail. Push the male end with the nipple into the black fuel supply tube.
4. Start the engine and check the fuel pressure against the specification.
5. If the fuel pressure is not within specifications, check items such as a restricted fuel return hose, pressure regulator vacuum hose for leaks, faulty fuel pump or a faulty pressure regulator.
6. Remove the fuel tube adapter 6175 or equivalent.
7. Lightly lubricate the ends of the fuel supply tube with clean engine oil. Install the black fuel supply tube to fuel rail and grey fuel return tube to the pressure regulator.

REMOVAL & INSTALLATION

2.2L Engine

The fuel pump used on the 2.2L engine is mounted on a plate located under the vehicle in front of the rear axle assembly.

1. Release the fuel system pressure.
2. Disconnect the battery negative cable.
3. Raise and support the vehicle safely.
4. Disconnect the electrical connectors from pump.
5. Plug the pump inlet and outlet hoses to prevent fuel flow.
6. Disconnect fuel pump hoses. Wrap a shop towel around the hoses and remove them from the fuel pump.
7. Remove the pump retaining strap and remove the fuel pump.
8. Installation is the reverse of the removal procedure.

NOTE: The pump terminals are different sizes to ensure the pump rotates in the correct direction.

9. Connect the negative battery cable and check for proper operation.

2.5L and 3.0L Engines

1. Relieve the fuel system pressure.
2. Disconnect the negative battery cable.
3. Drain the fuel from the fuel tank.
4. Raise and safely support the vehicle. Remove the right rear wheel and inner fender splash shield.
5. Disconnect the fuel lines at the fuel filter and the electrical connectors from the tank. Disconnect the fuel tank vent tube from the filler neck. Disconnect the ground wire from the body.
6. Place a suitable support under the tank and remove the retaining straps. Lower the tank from the vehicle.
7. On vehicles built before October 9, 1991, remove the bolts holding the tank sending unit to the tank. On vehicles built on or after October 9, 1991, remove the fuel pump module retaining clamp. Pull the sending unit/pump from the tank, noting the position of the gasket.
8. Late 1991–92 modules cannot be

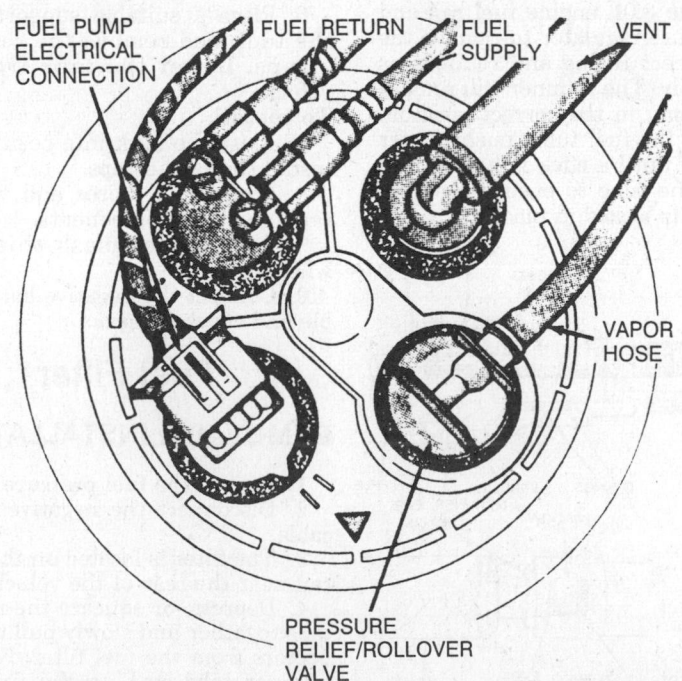

Top view of the fuel pump unit—Late 1991-92 Monaco and Premier

FUEL PUMP ELECTRICAL CONNECTION • FUEL RETURN • FUEL SUPPLY • VENT • VAPOR HOSE • PRESSURE RELIEF/ROLLOVER VALVE

disassembled and must be replaced as an assembly. On 1990 and early 1991 vehicles, disconnect the electrical connectors from the terminals on fuel pump and remove the pump holding bracket.

9. Disconnect the hose clamp at inlet port. Unscrew hose clamp and remove fuel pump.

NOTE: On some vehicles, there may be a tray in the bottom of the fuel tank that is contoured to hold the fuel filter. When installing the pump/sending unit. make sure the filter correctly fits into the tray.

To install:

10. Position the gasket so the holes in the gasket align with bolt holes in the fuel tank.

11. Install the sending unit/pump into the fuel tank and install the retaining bolts or clamp.

12. Install the fuel tank and all related items to the vehicle.

13. Connect the negative battery cable and check the entire system for proper operation and leaks.

Fuel Injection

IDLE SPEED ADJUSTMENT

The idle speed on fuel injected vehicles is controlled by the ECU through the use of an Idle Speed Control motor (ISC) or an idle speed regulator. The ISC motor does not require periodic adjustment.

On 2.5L engine, if the ISC is removed or replaced, it should be adjusted to establish the initial position of the plunger. To adjust the ISC, use the following procedure:

1. Start the engine and allow it to reach normal operating temperature.

2. Disconnect the ISC motor wire connector.

3. Locate the diagnostic terminals on the right side inner fender well. Connect a tachometer to terminals **D1–1** and **D1–3** of the diagnostic connector.

4. An adapter may be required to connect to the ISC motor. Fully extend the ISC motor plunger.

5. Adjust the plunger screw until the engine is running at 3500 rpm.

6. Remove adapter and reconnect the idle speed motor electrical connector. Idle speed should automatically return to normal.

Fuel Injector

REMOVAL & INSTALLATION

2.2L Engine

1. Relieve the fuel system pressure.

2. Disconnect the negative battery cable.

3. Disconnect the fuel lines from the fuel rail assembly.

4. Disconnect and tag the electrical leads from the fuel injectors and lay the harness aside.

5. Disconnect the accelerator cable from the the throttle body. On the 3.0L engine, remove the 4 screws attaching the engine cover and remove the cover.

6. Remove the fuel rail mounting bolts. Pull the fuel rail and injectors from the engine, using a back and forth twisting motion.

To install:

7. Install the fuel rail and injectors to the engine, be careful not to damage the O-rings on the injectors. Install the fuel rail hold-down bolts and connect the fuel lines.

8. Connect the electrical leads to the injectors. Connect the throttle cable. On the 3.0L engine install the engine cover plate.

9. Connect the negative battery cable. Turn the ignition to the **ON** position to pressurize the fuel system. Check for leaks.

2.5L Engine

1. Relieve the fuel system pressure.

2. Disconnect the negative battery cable.

3. Remove the air inlet tube from the throttle body. Disconnect the electrical lead from the fuel injector.

4. Remove the screws attaching the injector hold-down plate and remove the hold-down plate.

5. Using an appropriate tool, grasp the top of the injector and pull the injector out of the throttle body.

NOTE: The pintle at the bottom of the injector must be kept clean and undamaged. If the injector is dropped on the pintle, do not reuse the injector.

6. Remove the upper O-ring, injector alignment washer and the lower O-ring. Discard the O-rings.

To install:

7. Install new O-rings and install the alignment washer on the injector. Install the injector into the throttle body by pushing down on the injector.

8. Install the injector hold-down plate. Connect the electrical connector.

9. Install the air inlet tube and connect the negative battery cable.

3.0L Engine

1. Relieve the fuel system pressure.

2. Disconnect the negative battery cable.

3. Disconnect the fuel lines from the fuel rail assembly.

4. Disconnect and tag the electrical leads from the fuel injectors and lay the harness aside.

5. Disconnect the cruise and accelerator cables from the the throttle body.

6. Remove the 4 screws attaching the engine cover and remove the cover.

7. Remove the fuel rail mounting bolts and disconnect the vacuum line from the fuel pressure regulator. Pull the fuel rail and injectors from the engine, using a back and forth twisting motion.

8. Remove the retaining clip and separate the injectors from the fuel rail.

To install:

9. Assemble the fuel rail. Install the fuel rail and injectors to the engine, be careful not to damage the O-rings on the injectors. Install the fuel rail hold-down bolts and connect the fuel lines.

10. Connect the electrical leads to the injectors. Connect the throttle cable. Install the engine cover plate.

11. Connect the negative battery cable. Turn the ignition to the **ON** position to pressurize the fuel system and check for leaks.

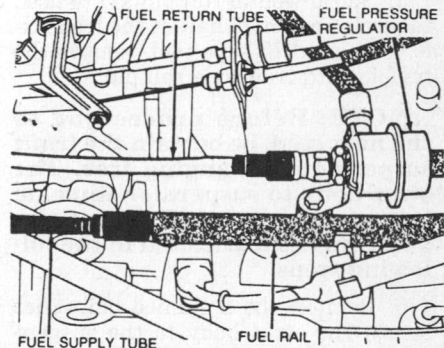

Return hose and regulator layout—3.0L engine

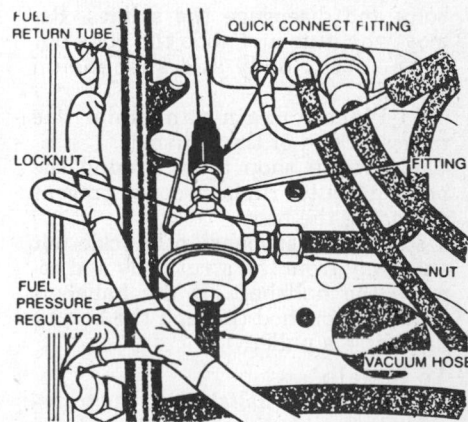

Fuel lines and routing—3.0L engine

DRIVE AXLE

Halfshaft

REMOVAL & INSTALLATION

The halfshafts are comprised of an inner CV-joint, an interconnecting shaft and an outer Rzeppa CV-joint with a stub shaft. The inner tripod CV-joint can be disassembled but must be replaced as a unit. The outer Rzeppa joint CV-joint cannot be disassembled and must be replaced as a unit. The protective rubber boots and clamps that cover each CV-joint are replaceable components.

1. Disconnect the negative battery cable. Raise the vehicle and support safely.
2. Remove the wheels.
3. Remove the brake caliper but do not disconnect the brake hose from the caliper. Wire it aside. Do not allow the brake hose to support the weight of the caliper.
4. Remove the halfshaft hub nut. A holding fixture may be required to hold the wheel hub/rotor when removing the nut.
5. Spiral wound roll pin(s) are used to retain each halfshaft at the transaxle. Using a drift type tool, remove the halfshaft-to-transaxle roll pin(s).

NOTE: **Before proceeding to the next step, be certain the front suspension is hanging free. The strut body to suspension knuckle bolts are splined. Remove the bolts only as instructed in the following steps.**

6. Remove the 2 splined bolts that attach the strut body to the suspension knuckle. Do this by first loosening and turning the nuts (do not turn the bolt heads) until they are almost at the end of the bolt threads. Tap the nuts with a brass hammer to loosen the bolts and disengage the splines. Remove the nuts and slide the bolts out of the strut body and suspension knuckle.
7. Place a drain pan under the transaxle end of the halfshaft.
8. Wrap a shop towel around the halfshaft outer rubber boot to prevent damaging the boot.
9. Tilt the suspension knuckle out and away from the strut body and remove the halfshaft. If the halfshaft cannot be pushed through the hub by hand, use a puller.

To install:

10. Install the halfshaft to the transaxle shaft and align the roll pin holes in each shaft.

NOTE: **One side of the roll pin hole in the transaxle shaft is beveled. Align the beveled side of that hole with the side of the hole in the CV-joint housing that is located in the housing "valley."**

11. Insert the roll pin(s) and seat with a hammer and a drift type tool.
12. Insert the halfshaft end through the hub.
13. Tilt the knuckle back into position and install the bolts. Hold the bolt heads with a wrench to prevent them from turning and tighten the nuts to 123 ft. lbs. (167 Nm).
14. Install the halfshaft end nut and tighten to 181 ft. lbs. (245 Nm), while holding the rotor and hub in place.
15. Install the brake caliper.
16. Check the differential fluid level. Fill with recommended synthetic gear oil.
17. Install the wheels.
18. Connect the negative battery cable.

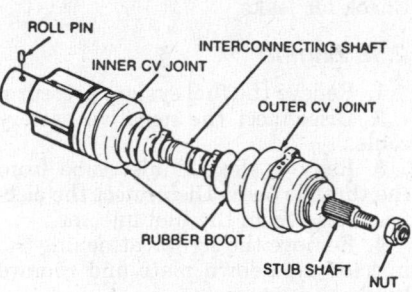

Typical halfshaft assembly

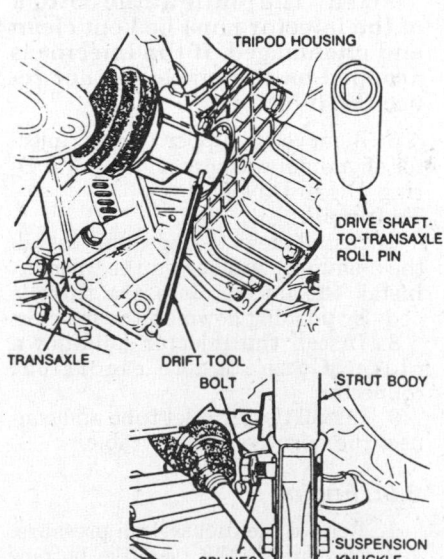

Remove halfshaft by driving out roll pin and removing from knuckle

CV-Joint Boot

The protective rubber boots and clamps that cover each CV-joint are replaceable components. The applicable CV-joint must be removed from the interconnecting shaft to replace a rubber boot.

REMOVAL & INSTALLATION

Outer CV-Joint

1. Disconnect the negative battery cable. Remove the halfshaft.
2. Side cutter pliers can be used to cut the metal retaining bands. If the rubber boot is re-useable, use care when cutting the clamps. Do not accidentally cut the boot when cutting the clamps.
3. Slide the rubber boot off of the CV housing for access to the plastic retainer.
4. Spread the plastic retainer at the seam with snapring pliers.
5. Tap the outer CV-joint with a plastic mallet to disengage the interconnecting shaft from the retainer.
6. Separate the outer CV-joint from the interconnecting shaft. Slide off the rubber boot. Inspect the boot for damage. If damaged, replace the boot.

To install:

7. Inspect the plastic retainer in the

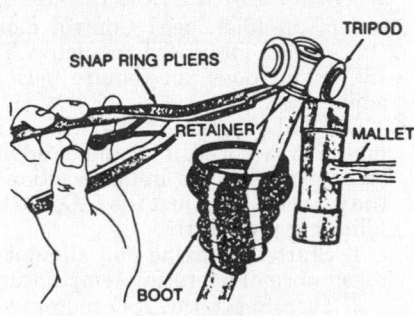

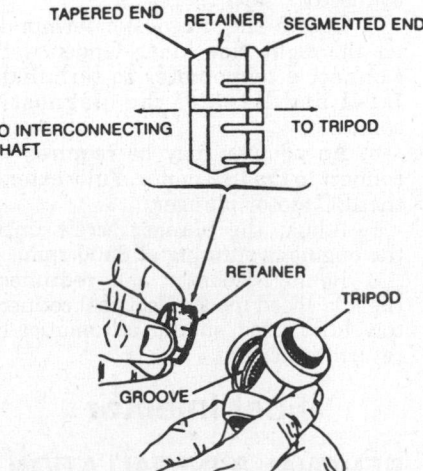

Typical joint removal and retainer orientation

outer CV-joint. Replace the retainer if it is damaged or defective. If a replacement retainer must be installed, ensure that it is installed correctly. The tapered end mates with the shaft and the segmented end mates with the CV-joint.

8. Install the rubber boot on the shaft. Thoroughly lubricate the CV-joint and the inside of the boot with the proper lubricant which is usually supplied with the service kit.

9. Align the shaft with the plastic retainer and the CV-joint, then tap the CV-joint onto the shaft with a plastic mallet. Continue tapping the CV-joint until the segmented end of the retainer snaps into position on the shaft.

10. Position the rubber boot on the clamp grooves machined in the CV-joint and in the shaft. Install the replacement clamps and crimp with the special tool.

11. Install the halfshaft in the vehicle.

Inner CV-Joint

1. Disconnect the negative battery cable. Remove the halfshaft.

2. Side cutter pliers can be used to cut the metal retaining bands. If the rubber boot is reuseable, use care when cutting the clamps. Do not accidentally cut the boot when cutting the clamps.

3. Slide the rubber boot off the CV-joint housing. Remove the inner CV-joint housing by pulling it straight away from the tripod.

4. Spread the plastic retainer with snapring pliers and tap the tripod with a plastic mallet to remove it from the shaft. If necessary, cut the retainer to remove it. Slide the rubber boot from the shaft. Inspect the boot for damage. If damaged, replace the boot.
To install:

5. Replace the complete CV-joint if any of its components are worn or damaged.

6. Ensure that the retainer is installed correctly. The segmented end mates with the tripod and the tapered end mates with the shaft. Install the plastic retainer in the tripod. Insert and force the segmented end of the retainer into the tripod until it snaps into the groove.

7. Install the rubber boot on the shaft. Align the tripod with the shaft and tap it with a plastic mallet until the retainer seats in the shaft groove. Thoroughly lubricate the CV-joint housing, tripod and the inside of the boot with the proper lubricant which is usually supplied with the service kit.

8. Position the CV-joint housing over the tripod/bearings. Position the seat of the rubber boot in the grooves in the housing and in the shaft.

9. Use caution in this step. The air must be allowed to vent from the boot. Insert a smooth rod between the rubber boot and the housing to allow the air pressure to equalize. Use care not to damage the rubber boot with the rod. Use a rod free from burrs or rough edges with the end chamferred. When the air pressure is equalized, remove the rod. Install the replacement clamps and crimp with the special crimping tool.

10. Install the halfshaft in the vehicle.

Front Wheel Hub and Bearings

REMOVAL & INSTALLATION

The front wheel hub can be removed without removing the bearing from the steering knuckle. However, the wheel hub must be removed before the bearing can be removed from the knuckle. The wheel hub and bearing are independently replaceable.

When servicing the steering knuckle, the wheel hub and bearing can and should be removed as a unit. Although the wheel bearing components can be disassembled for inspection, the bearing must be replaced as a unit only. If any of the bearing components are worn, damaged or defective, the complete bearing must be replaced.

1. Disconnect the negative battery cable. Raise and safely support the vehicle.

2. Remove the halfshaft end nut. Push the halfshaft inward and disengage the shaft splines from the wheel hub splines. If it does not push out easily, use a screw type puller to press the shaft from the hub.

3. Install a puller plate that can be

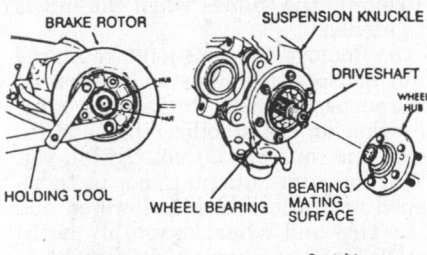

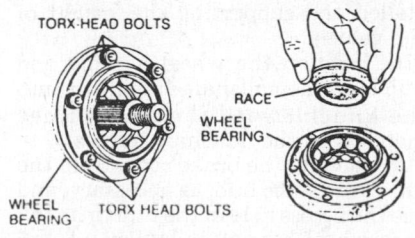

Front wheel bearing removal—Monaco and Premier

used with a slide hammer and pull the rotor/hub assembly from the steering knuckle. Use care to keep dirt and debris from the bearing as the hub assembly is removed.

4. If necessary, the rotor and hub can be separated by removing the rotor safety nuts. If these safety nuts are damaged by removal, replace with new ones.

5. Remove the Torx® bolts that attach the wheel bearing to the knuckle. Remove the bearing.

6. If the wheel hub and/or bearing are being replaced without the other, remove the outer race from the hub with a shop press and the appropriate adapters.
To install:

7. If a new wheel hub is being installed, force the original bearing outer race on the replacement hub with a shop press and a suitable length of steel pipe that has the correct inside diameter to fit around the hub. If a new bearing is being installed, do the same with the new race.

8. If the original wheel bearing is being installed, pack the bearing and lubricate both races (inner and outer) with an extreme pressure type wheel bearing lubricant.

9. If a replacement wheel bearing is being installed, prepare the bearing as follows:

   a. Remove and discard the plastic protective covers.

   b. Locate, remove and discard the plastic protective sleeve from the replacement bearing bore.

   c. Remove the inner and outer bearing races from the bearing.

   d. Pack the bearing with lubricant which may be supplied with the replacement bearing.

   e. Insert the bearing inner race in the bearing and force the bearing outer race on the wheel hub with a press and suitable length of steel pipe.

10. Install the bearing to the knuckle and install the bolts. Torque the bolts to 11 ft. lbs. (15 Nm).

11. Lubricate the bearing mating surface on the wheel hub bearing outer race with an extreme pressure type wheel bearing lubricant.

12. Position the wheel hub on the halfshaft and insert the hub into the wheel bearing. Tap the wheel hub with a brass hammer until 3 or 4 of the halfshaft threads extend beyond the hub.

13. If removed, install the brake rotor on the wheel hub, using new safety nuts if required.

14. Install the halfshaft-to-wheel hub nut. Use an appropriate holding tool to keep the hub from rotating while tightening the nut to 181 ft. lbs. (245 Nm) torque.

NOTE: Do not use an impact wrench to tighten the driveshaft-to-wheel hub nut. Use a torque wrench only to tighten the nut. It is also essential that the halfhaft-to-wheel hub nut be tighten to the specified torque. In addition to retaining the wheel hub on the halfshaft, the specified torque also establishes the wheel bearing preload.

15. Install the brake caliper, tire and wheel assembly and lower the vehicle.

16. Connect the negative battery cable. Depress the brake pedal several times to seat the brake pads before moving the vehicle.

Steering Knuckle

REMOVAL & INSTALLATION

The front wheel hub and bearing must be removed as a unit before the suspension knuckle can be removed. This is the only service situation where the wheel hub and bearing are removed together.

1. Disconnect the negative battery cable. Raise and safely support the vehicle.

2. Remove the tire and wheel assembly. Wrap a heavy shop cloth or towel around the outer CV-joint boot to protect it.

3. Remove the brake caliper. Do not disconnect the brake hose from the caliper. Wire it aside. Do not allow the brake hose to support the caliper weight.

4. Remove the halfshaft hub nut. A holding fixture may be required to hold the wheel hub/rotor when removing the nut.

5. Push the halfshaft inward, to disengage the shaft splines from the wheel hub splines. If it does not push out easily, use a screw type puller to press the shaft from the hub.

6. Install a puller plate that can be used with a slide hammer and pull the rotor/hub assembly from the suspension knuckle. Use care to keep dirt and debris from the bearing as the hub assembly is removed.

7. Remove the rotor from the hub by removing the rotor safety nuts. If these safety nuts are damaged by removal, always replace with new ones.

8. Rotate the wheel hub as necessary and use the access hole in the hub to remove each wheel bearing-to-suspension knuckle Torx® head bolt. Reinstall the brake rotor, attach a puller plate to it that can be used with a slide hammer and pull the rotor/hub assembly from the suspension knuckle. Use care to keep dirt and debris from the bearing as the hub assembly is removed.

9. Loosen but do not remove the stabilizer bar inner bracket retaining bolts at the engine cradle. Remove the stabilizer bar outer bracket retaining nuts at the suspension arm and remove the bracket from the retaining bolts. Note that the stabilizer bar outer bracket retaining nuts also retain the ball joint to the suspension arm. Move the stabilizer bar away from the suspension arm and reinstall one of the nuts on either of the ball joint retaining bolts.

10. Loosen but do remove the nuts and bolts at the bushings that attach the suspension arm to the engine cradle.

11. Remove the ball joint stud pinch bolt and disengage the ball joint stud from the suspension knuckle.

12. Note that the 2 bolts that hold the bottom of the MacPherson strut to the suspension knuckle have splines under the bolt head. This keeps the bolt from rotating. Turn the nuts (not the bolt heads) until they are almost at the end of the bolt threads. Tap the nuts with a brass hammer to loosen the bolts and disengage the splines. Remove the nuts and pull out the bolts. Remove the knuckle from the halfshaft.

To install:

13. Position the steering knuckle over the halfshaft and insert the ball joint stud into the knuckle. Install the pinch bolt. Note that there is a recess, groove or keyway machined into the ball joint stud. The pinch bolt must be seated in this groove. Torque to 77 ft. lbs. (104 Nm).

14. Position the knuckle to the strut and install the through bolts. Note that the splines under the bolt heads must be properly aligned in the strut hole. Tap the bolts in place and install the nuts. Use a wrench on the bolt head to keep the bolt from turning and stripping the splines when the nut is tightened.

15. Remove the ball joint retaining nut and position the stabilizer bar at the suspension arm. Position the outer bracket on the stabilizer bar and install the nuts but do not tighten yet. The ball joint nuts must not be tightened until the vehicle is lowered and the tire and wheel assembly is installed and supporting the weight of the vehicle.

16. Position the wheel bearing and hub over the halfshaft and insert into the knuckle with the hub splines mated with the halfshaft splines.

17. Remove the brake rotor from the hub. Rotate the hub, as necessary, and use the access hole in the hub to install each wheel bearing-to-knuckle Torx® head bolt. Tighten each bolt to 11 ft. lbs. (15 Nm). Install the brake rotor.

18. Install the halfshaft-to-wheel hub nut. Use an appropriate holding tool to keep the hub from rotating while tightening the nut to 181 ft. lbs. (245 Nm) torque.

NOTE: Do not use an impact wrench to tighten the halfshaft-to-wheel hub nut. Use a torque wrench only to tighten the nut. It is also essential that the halfshaft-to-wheel hub nut be tighten to the specified torque. In addition to retaining the wheel hub on the halfshaft, the specified torque also establishes the wheel bearing preload.

19. Install the brake caliper, tire and wheel assembly and lower the vehicle.

20. With the vehicle weight being supported by the tire and wheels, torque the suspension arms-to-engine cradle nuts and bolts at the bushings to 103 ft. lbs. (140 Nm), torque the stabilizer bar outer bracket retaining nuts to 60 ft. lbs. (81 Nm) and tighten the stabilizer bar inner bracket retaining bolts to 21 ft. lbs. (29 Nm).

21. Connect the negative battery cable. Depress the brake pedal several times to seat the brake pads before moving the vehicle.

Transaxle Output Shaft Seal

REMOVAL & INSTALLATION

1. Raise the vehicle and support it safely.

2. Remove the tires and wheels.

3. Remove the halfshaft.

4. Remove the differential drain plug and drain the lubricant into a clean container for reuse.

5. With a suitable tool, pry out the dust cover from the output shaft.

6. Loosen the shaft bolt in the center of the output shaft and pull the short shaft and bearing out of the transaxle case. Pry out the shaft seal.

7. Installation is the reverse of removal. Torque the short output shaft center bolt to 18 ft. lbs. (24 Nm).

8. Refill the differential with synthetic-type 75W–140 hypoid gear lubricant. Add oil until it starts to flow out of the fill plug opening.

MANUAL TRANSAXLE

For further information on transmissions/transaxles, please refer to "Chilton's Guide to Transmission Repair".

REMOVAL & INSTALLATION

1. Disconnect the negative battery cable.

2. Disconnect and remove the flexible heat tube from the engine.

3. Remove the TDC sensor retaining bolt and remove the sensor.

4. Remove the bolts retaining the steering bracket and remove the bracket.

5. Remove the bolts attaching the crossmember to the side sill and body.

6. Raise and safely support the vehicle.

7. Remove the front wheels. Disconnect and remove the passenger side tie rod.

8. Loosen the bolt retaining the coolant expansion tank and move the tank aside.

9. Attach an engine support tool to the engine and take up the engine weight.

10. Remove the bolts attaching the exhaust head pipe to the manifold. Remove the bolts attaching the exhaust head pipe to the converter and remove the head pipe.

11. Remove the crossmember by turning it and taking it out through the passenger side wheel well.

12. Disengage the clutch cable. Remove the upper steering knuckle mounting bolt and loosen the lower bolt.

13. Remove the halfshaft retaining pin. Swing each rotor and steering knuckle outward and remove the halfshafts from the transaxle.

14. Disconnect the reverse lockout cable and disconnect the shift rod from the lever. Disconnect the speedometer cable. Disconnect the ground strap at the transaxle.

15. Support the transaxle. Remove the transaxle support cushion nuts. Remove the bolts that attach the 2 transaxle mounting brackets to the transaxle.

16. Disconnect the wiring harness connector and remove the starter. Remove the bolts attaching the clutch housing to the engine.

17. Pull the transaxle straight back until the clutch shaft is clear of the engine and lower the transaxle.

To install:

18. Raise and position the transaxle in the vehicle. Align the release bearing and the release fork.

19. Install the transaxle-to-engine mounting bolts. Tighten to 37 ft. lbs. (51 Nm).

20. Install the starter and connect the electrical connectors. Slightly raise the transaxle and install the mounting brackets. Align the transaxle support cushion bolts and install the retaining nuts.

21. Connect the speedometer cable and the shift rods. Connect the clutch cable.

22. Install the halfshafts by tilting the steering knuckle in, then install the upper bolt and tighten both bolts to 148 ft. lbs. (200 Nm).

23. Install the axle retaining pins. Connect the ground strap to the case.

24. Install the crossmember through the wheel well opening and position it on the side sills. Install and tighten the bolts.

25. Connect the tie rods to the steering bracket and tighten the mounting bolts to 25 ft. lbs. (34 Nm). Connect the steering gear bracket to the steering rack and tighten the bolts to 30 ft. lbs. (41 Nm).

26. Install the front wheels. Install the TDC sensor and the heat tube. Connect the exhaust header pipe to the converter and manifold.

27. Check and fill the transaxle.

28. Remove the engine support tool.

29. Connect the negative battery cable and check the transaxle for proper operation. Make sure the reverse lights come on when in **R**.

CLUTCH

Clutch Assembly

REMOVAL & INSTALLATION

1. Disconnect the negative battery cable.

2. Remove the transaxle assembly from the vehicle.

3. Remove the pressure plate attaching bolts.

4. Remove the pressure plate release bearing assembly and the clutch disc.

To install:

5. Inspect the condition of the clutch components and replace any worn parts.

NOTE: The release bearing and pressure plate are not serviced separately. The bearing is permanently attached to the pressure plate diaphragm fingers. The pressure plate and bearing must be serviced as an assembly.

6. Inspect the flywheel for heat damage or cracks. Replace if damaged.

7. Install the clutch disc to the flywheel using alignment tool EMB-786-01 or equivalent. Install the pressure plate assembly and tighten the pressure plate bolts evenly to 18 ft. lbs. (25 Nm). Remove the alignment tool.

8. Install the transaxle assembly and check the clutch operation.

Clutch Cable

ADJUSTMENT

The Medallion is equipped with a cable-operated self-adjusting clutch mechanism. The adjustment is automatically set during operation by a quadrant mechanism on the clutch pedal assembly.

AUTOMATIC TRANSAXLE

For further information on transmissions/transaxles, please refer to "Chilton's Guide to Transmission Repair".

Transaxle Assembly

REMOVAL & INSTALLATION

Premier and Monaco

AR-4 TRANSAXLE

1. Disconnect the negative battery cable. Remove the windshield washer reservoir.

2. Disconnect the connectors at the control unit.

3. Disconnect and plug the fluid cooler hoses.

4. Remove the engine timing sensor.

5. Raise the vehicle and support safely. Remove the front wheels.

6. Remove the pins attaching the halfshafts to the transaxle output shafts.

7. Remove the upper steering knuckle mounting nut, remove the bolt, and loosen the lower bolt. Tilt the knuckles outward and slide the halfshafts from the transaxle.

8. Remove the transaxle splash shield.

9. Remove the brackets, retainers, tie straps or clips securing the transaxle electrical wiring to the vehicle body. Leave all components in place for transaxle removal.

10. Remove the starter and heat shield.

11. Remove the converter housing access plug and remove the converter bolts.

12. Remove the exhaust pipe clamp and bracket.

13. Using the proper equipment, support the weight of the transaxle.

14. Remove the bolts and nuts that transaxle crossmember to the engine cradle. Remove the bolt attaching the rear mount to the transaxle bracket.

Remove the crossmember and mount as an assembly.

15. Disconnect the shift cable from the bellcrank and remove the brace rod.

16. Move the bellcrank, link rod and bracket, and shift cable and bracket aside for clearance.

17. Remove the transaxle mount bracket.

18. Remove the transaxle to engine bolts. Note that 2 of these bolts are reversed.

19. Pull the transaxle back and away from the engine and remove from the engine.

To install:

NOTE: When installing, make sure the dowel pins are seated in the converter housing before tightening any bolts. Also, make sure the converter is aligned in the driveplate timing wheel. The timing wheel timing segments can be damaged if misaligned.

20. Lubricate the hub bore of the driveplate with with chassis grease. Install the converter.

21. Mount the transaxle on a suitable transaxle jack and raise into position.

22. Align the converter housing with the driveplate and engine dowel pins and slide the transaxle onto the engine.

23. Install the transaxle to engine bolts.

24. Coat the threads with Loctite® and install the converter bolts. Tighten to 24 ft. lbs. (33 Nm). Install the access plug.

25. Install the transaxle mount bracket and tighten the bolts to 29 ft. lbs. (40 Nm).

26. Install the bellcrank and bracket. Connect the bellcrank link rod to the shift lever.

27. Position the shift cable bracket on the case but do not tighten the bolts yet.

28. Attach the brace rod to the bellcrank and shift cable brackets. Tighten all bolts.

29. Snap the shift cable onto the bellcrank.

30. Install the transaxle crossmember but do not tighten the bolts yet.

31. Attach the transaxle bracket to the rear mount.

32. Tighten the transaxle bracket-to-rear mount bolt to 49 ft. lbs. (67 Nm). Tighten the crossmember bolts to 31 ft. lbs. (43 Nm). Remove the support equipment.

33. Install the exhaust pipe bracket and clamp. Install the starter and heat shield.

34. Verify that the transaxle output shaft O-rings are in position and are not damaged. Replace if necessary. In-

stall the halfshafts and install the retaining pins.

35. Install the halfshafts. Tilt the steering knuckles in and install the top bolts. Tighten the nuts to 148 ft. lbs. (200 Nm). Install the front wheels.

36. Install the engine timing sensor.

37. Connect the transaxle cooling hoses. Fill the differential section of the unit with synthetic type 75W–140 gear oil.

38. Route the wire harness properly and secure. Install the splash shield.

39. Fill the transaxle with Mopar Mercon™ transaxle fluid. Connect the control unit harness.

40. Connect the negative battery cable and check the transaxle for proper operation.

ZF-4 TRANSAXLE

1. Disconnect the negative battery cable.

2. Loosen the throttle valve cable adjusting nut and remove the cable from the engine bracket.

3. Remove the upper steering knuckle mounting nut, remove the bolt, and loosen the lower bolt.

4. Remove the halfshaft retaining pin. Swing each rotor and steering knuckle outward and slide the halfshafts from the transaxle.

5. Remove the underbody splash shield. Loosen the nut attaching the fill tube to the pan and drain the transaxle. When fluid has drained, tighten the nut.

6. Remove the converter housing covers. Remove the converter-to-flexplate bolts. Support the transaxle.

7. Remove the nuts attaching the crossmember to the side sills. Remove the large bolt and nut that attach the rear cushion to the support bracket.

8. Remove the support bracket and rear cushion.

9. Disconnect the header pipes from the exhaust manifold and the catalytic converter.

10. Loosen the engine cradle bolts only until there is ½–⅞ in. clearance between the cradle and the side sill.

11. Remove the front exhaust pipe. Remove the starter, plate and dowel.

12. Disconnect the shift cable from the transaxle lever. Remove the cable bracket bolts and separate the bracket from the case. Remove the brace rod.

13. Disconnect and remove the TDC sensor, speedometer sensor, and engine speed sensor. Disconnect and plug the transaxle cooling lines.

14. Using a suitable transaxle jack, support the weight of the transaxle. Remove the transaxle-to-engine bolts, then pull the transaxle back and away from the engine.

To install:

15. Position the transaxle to the engine. Install the transaxle-to-engine

bolts and tighten to 31 ft. lbs. 42 Nm).

16. Install removed sensors and connect all electrical leads. Connect the transaxle cooler lines. Install the brace rod.

17. Attach the shift bracket to the case and tighten the bolts to 125 inch lbs. (14 Nm). Install the shift cable to the bracket.

18. Install the starter. Connect the exhaust head pipes to the manifolds and he converter.

19. Install the rear support and cushion, install the mounting bolts and tighten to 49 ft. lbs. (66 Nm).

20. Tighten the engine cradle bolts to 92 ft. lbs. (125 Nm). Install the halfshafts.

21. Coat the threads with Loctite® and install the converter-to-flexplate bolts. Tighten to 24 ft. lbs. (33 Nm). Install the converter housing covers.

22. Install the halfshafts. Tilt the steering knuckles in and install the top bolts. Tighten the nuts to 148 ft. lbs. (200 Nm).

23. Install the front wheels. Install the under body splash shield. Attach the throttle valve cable.

24. Fill the transaxle with Mopar Mercon™ transaxle fluid. Check the differential fluid level. If it is low, fill with synthetic type 75W–140 gear oil.

25. Connect the negative battery cable and check the transaxle for proper operation.

Medallion

1. Disconnect the negative battery cable.

2. Disconnect and remove the flexible heat tube from the engine.

3. Remove the TDC sensor retaining bolt and remove the sensor.

4. Remove the bolts retaining the steering bracket and remove the bracket.

5. Remove the bolts attaching the crossmember to the side sill and body.

6. Raise and safely support the vehicle.

7. Remove the front wheels. Disconnect and remove the passenger side tie rod.

8. Loosen the bolt retaining the coolant expansion tank and move the tank aside.

9. Attach an engine support tool to take up the weight of the engine.

10. Remove the bolts attaching the exhaust head pipe to the manifold. Remove the bolts attaching the exhaust head pipe to the converter and remove the head pipe. Disconnect the coolant lines to the heat exchanger.

11. Remove the crossmember by turning it and taking out through the passenger side wheel well.

12. Disengage the shift cable and support it to the side. Remove the up-

per steering knuckle mounting bolt and loosen the lower bolt.

13. Remove the halfshaft retaining pin. Swing each rotor and steering knuckle outward and slide the halfshafts from the transaxle.

14. Disconnect the speedometer cable. Disconnect the ground strap at the transaxle. Disconnect the BVA module harness.

15. Support the transaxle. Remove the transaxle support cushion nuts. Remove the bolts that attach the 2 transaxle mounting brackets to the transaxle.

16. Disconnect the wiring harness connector and remove the starter. Remove the converter-to-flywheel bolts. Remove the transaxle-to-engine bolts.

17. Pull the transaxle straight back until the converter is clear of the engine and lower the transaxle. Install converter retainer BVI–465 or equivalent, to keep the converter from falling out.

To install:

18. Raise and position the transaxle into the vehicle. Apply a small amount of grease to the torque converter pilot. Align the painted marks on the converter with the painted marks on the flywheel. Coat the threads with Loctite® and install the converter to flywheel bolts. Tighten to 34 ft. lbs. (46 Nm).

19. Install the transaxle-to-engine mounting bolts.

20. Install the starter and connect the electrical connectors. Slightly raise the transaxle and install the mounting brackets. Align the transaxle support cushion bolts and install the retaining nuts.

21. Connect the speedometer cable and the shift cable.

22. Install the halfshafts by tilting the steering knuckle in, install the upper bolt and tighten both the nuts to 148 ft. lbs. (200 Nm).

23. Install the axle retaining pins. Connect the ground strap to the case.

24. Install the crossmember through the wheel well opening and position it on the side sills. Install and tighten the bolts.

25. Connect the tie rods to the steering bracket and tighten the mounting bolts to 25 ft. lbs. (34 Nm). Connect the steering gear bracket to the steering rack and tighten the bolts to 30 ft. lbs. (41 Nm). Connect the cooling lines.

26. Install the front wheels. Install the TDC sensor and the heat tube. Connect the exhaust header pipe to the converter and manifold. Connect the BVA wiring.

27. Check and fill the transaxle fluid. Remove the engine support tool and connect the negative battery cable.

SHIFT LINKAGE ADJUSTMENT

Premier and Monaco

1. Disconnect the negative battery cable.

2. Shift into the **P** detent.

3. Remove the shifter cover and locate the shift cable cross-lock where the cable meets the shifter bracket. Release the shift cable cross-lock by pulling it upward.

4. Move the transaxle shift lever all the way rearward into the **P** detent. Be sure the lever is centered in the detent. Verify positive engagement of the park lock by attempting to rotate the halfshafts. The shafts cannot be turned if the park lock is properly engaged.

5. On 2.5L engine, adjust by performing the following:

a. Grasp the shift cable and pull it rearward until the distance between the back of the lock tab and the seat (exposed threads) is 0.30 in. (7.62mm).

b. If adjustment is not correct, and the column shifter was biased toward the **D** gate, decrease the clearance by 0.040 in. (1mm) increments until correct.

c. If adjustment is not correct, and the column shifter was biased toward the **N** gate, increase the clearance by 0.040 in. (1mm) increments until correct.

6. On 3.0L engine, adjust by performing the following:

a. Verify that the cable is properly routed and secured and the cable grommet is fully seated in the floor pan. Press the cable cross-lock downward until it snaps in place.

b. Position the cable self-adjusting unit in the fork of the lower cable mounting bracket at the transaxle. Use the index key to properly index and seat the cable within the bracket.

c. Seat the cable core end fitting onto the transaxle operating lever pin.

d. At the transaxle end fitting, push the core-adjust slider mechanism until it snaps into a locked position. This will properly adjust and lock the gearshift cable.

7. Check the shift cable adjustment. The engine should start in **P** and **N** only.

THROTTLE VALVE CABLE ADJUSTMENT

1. Disconnect negative battery cable.

2. Loosen the cable locknuts and lift the threaded shank of the cable out of

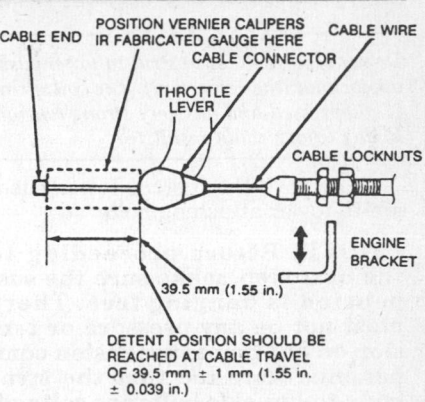

Throttle valve cable adjustment—Premier and Monaco

the engine bracket.

3. Place the throttle lever in the curb idle position.

4. An accurate measurement must now be made. Vernier calipers are suggested. If accurate calipers are not available, fabricate a cable adjustment gauge from a small piece of sheet stock or other material that can be shipped over the throttle cable wire. The gauge must be 1.55 in. (39.5mm) long.

5. Pull the cable wire forward and position the vernier calipers or fabricated gauge on the wire between the cable connector and cable end.

6. Pull the cable shank rearward to the detent position but not to the wide open throttle position; the detent position feels similar to a stop when reached.

7. Hold the cable shank at the detent position, then insert the cable shank into the cable bracket and tighten the cable locknuts.

8. Remove the vernier calipers or gauge and verify the adjustment. The cable detent position should be reached when the cable wire travels 1.55 in. ± 0.039 in. (39.5mm ± 1mm).

FRONT SUSPENSION

MacPherson Strut

REMOVAL & INSTALLATION

1. Raise and safely support vehicle. Do not support vehicle by placing supports under the suspension arms.

2. Remove the wheel and tire assemblies.

3. Remove the outer tie rod ends with a screw type puller.

---— **CAUTION** ——---

Do not remove the strut strut-to-tower cushion locknut (the center nut). The coil spring is compressed and has very strong tension. Bodily injury could result.

4. Remove the 3 strut tower cushion-to-tower attaching bolts.

NOTE: Before proceeding to the next step, make sure the suspension is hanging free. There must not be any pressure or tension on any front suspension components. Note too that the strut body-to-knuckle bolts are splined. Do not try to turn the bolt head. Turn the nuts only. Follow the procedure below.

Also, make sure brake hoses and/or ABS wiring will not be damaged.

5. Remove the splined bolts by loosening the nuts until they are almost at the end of the bolt threads. Tap the nuts with a brass hammer to loosen the bolts and disengage the splines. Remove the nuts, then the bolts.
6. For protection, wrap the halfshaft boot with heavy shop towels. Then press down on the suspension arm and pull the strut out of the wheel well.

To install:

7. Carefully route the strut into place and install the 3 upper strut tower cushion-to-body bolts finger-tight. Make sure the splines are aligned on the bolts and tap into place. Tighten the nuts only. Do not allow the bolt heads to turn or the splines will strip. Hold the bolt heads with a wrench

while the nuts are tightened. Torque the nuts to 123 ft. lbs. (167 Nm).
8. Torque the 3 upper bolts to 17 ft. lbs. (23 Nm).
9. Install the tie rod end and install the wheel.
10. Perform a front end alignment.

Lower Ball Joints

REMOVAL & INSTALLATION

1. Disconnect the negative battery cable. Raise and safely support vehicle.
2. Remove the wheel and tire assemblies.
3. Wrap a heavy shop cloth or towel around the halfshaft outer boot to protect it.
4. Loosen but do not remove the stabilizer bar inner bracket retaining bolts at the engine cradle. Remove the stabilizer bar outer bracket nuts at the suspension arm and remove the bracket. Note that the outer bracket nuts and bolts also fasten the ball joint to the suspension arm.
5. Remove the ball joint pinch bolt from the suspension knuckle. Loosen but do not remove the nuts and bolts at the bushings that attach the suspension arm to the engine cradle.
6. Disengage the ball joint stud from the suspension knuckle and remove the plastic washer from the stud. Remove the ball joint from the suspension arm by removing the bolts and tapping upward on it with a brass hammer.

To install:

7. Install the replacement ball joint assembly to the suspension arm.

Crimp the sleeves, but do not tighten the nuts yet. Install a new plastic washer on the stud.
8. Install to the knuckle. When installing the pinch bolt that holds the ball joint stud to the knuckle, make sure the bolt aligns with the groove in the stud. Torque to 77 ft. lbs. (105 Nm).
9. Do not tighten the stabilizer bar nuts or bolts until the vehicle is lowered and the tires are supporting the weight of the vehicle. At that time, torque the suspension arm-to-engine cradle nuts and bolts to 103 ft. lbs. (140 Nm), the stabilizer outer bracket nuts to 60 ft. lbs. (81 Nm), and the stabilizer bar inner bracket nuts to 21 ft. lbs. (29 Nm).
10. Perform a front end alignment.

Lower Control Arms

REMOVAL & INSTALLATION

1. Raise and safely support vehicle.
2. Remove the wheel and tire assemblies.
3. Wrap a heavy shop cloth or towel around the halfshaft outer boot to protect it.
4. Loosen but do not remove the stabilizer bar inner bracket retaining bolts at the engine cradle. Remove the stabilizer bar outer bracket nuts at the suspension arm and remove the bracket. Note that the outer bracket nuts and bolts also fasten the ball joint to the suspension arm. Reinstall a nut to keep the ball joint from separating from the arm.
5. Remove the ball joint pinch bolt

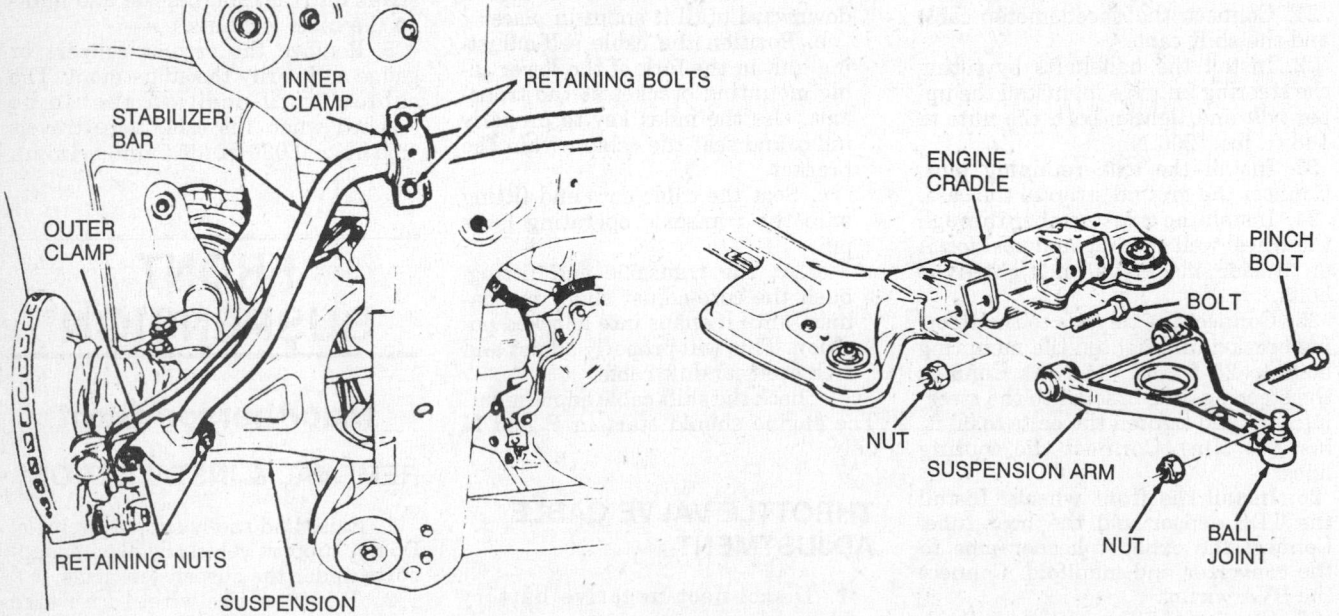

Control arm and sway bar assemblies

from the suspension knuckle. Remove the nuts and bolts at the bushings that attach the suspension arm to the engine cradle.

6. Disengage the ball joint stud from the suspension knuckle, remove the plastic washer from the stud and remove the arm from the vehicle.

To install:

7. If reuseable, transfer the ball joint. Inspect the suspension arm bushings and replace if necessary. Install the cradle bolts, but do not tighten until the full weight of the vehicle is on the ground.

8. Install the ball joint stud to the knuckle. When installing the pinch bolt that holds the ball joint stud to the knuckle, make sure the bolt aligns with the groove in the stud. Torque to 77 ft. lbs. (105 Nm).

9. Do not tighten the stabilizer bar nuts or bolts until the vehicle is lowered and the tires are supporting the weight of the vehicle. At that time, torque the suspension arm-to-engine cradle nuts and bolts to 103 ft. lbs. (140 Nm), the stabilizer outer bracket nuts to 60 ft. lbs. (81 Nm) and the stabilizer bar inner bracket nuts to 21 ft. lbs. (29 Nm).

Sway Bar

REMOVAL & INSTALLATION

1. For ease of stabilizer bar (sway bar) removal and installation, do not raise the vehicle. Leave the vehicle's weight on the tires.

2. Remove the bolts that hold the stabilizer bar inner brackets to the engine cradle.

3. Remove the retaining nuts from the stabilizer bar outer bracket bolts at the suspension arms. Note that the outer bracket nuts and bolts also fasten the ball joint to the suspension arm. Remove the stabilizer bar and brackets from the vehicle. Reinstall a nut to keep the ball joint from separating from the arm.

To install:

4. Inspect the stabilizer bushings and replace, if necessary.

5. Tighten stabilizer bar nuts or bolts only finger-tight until all the fasteners are in place. Then tighten the stabilizer outer bracket nuts to 60 ft. lbs. (81 Nm) and the stabilizer bar inner bracket nuts to 21 ft. lbs. (29 Nm).

REAR SUSPENSION

Shock Absorbers

REMOVAL & INSTALLATION

Never raise a Premier or Monaco with a lift positioned under the V-shaped rear crossmember. Never let the hoist arms come into contact with the lower edge of the rocker panel. If necessary, place a small block of wood between the hoist pad and the body lifting points so the vehicle does not rest on the rocker panels. The manufacturer does not recommend use of a twin post under-the-vehicle hoist.

1. Raise and safely support the rear of the vehicle.

2. Lift upward on the trailing arm to relieve the weight from the shock absorber.

3. Remove the top and bottom bolts and remove the shock absorber.

4. Installation is the reverse of removal. Torque the top bolt to 60 ft. lbs. (81 Nm) and the bottom bolt to 85 ft. lbs. (115 Nm).

Rear Wheel Bearings

REMOVAL & INSTALLATION

The rear wheel bearings and hubs are replaced as assemblies only. They are non-adjustable. The maximum allowable bearing endplay is 0.001 in. If the endplay exceeds this, the bearing/hub assembly should be replaced.

1. Raise and safely support the rear of the vehicle. Remove the wheel.

2. Remove the brake drum from the axle shaft hub.

3. Remove the axle shaft hub nut

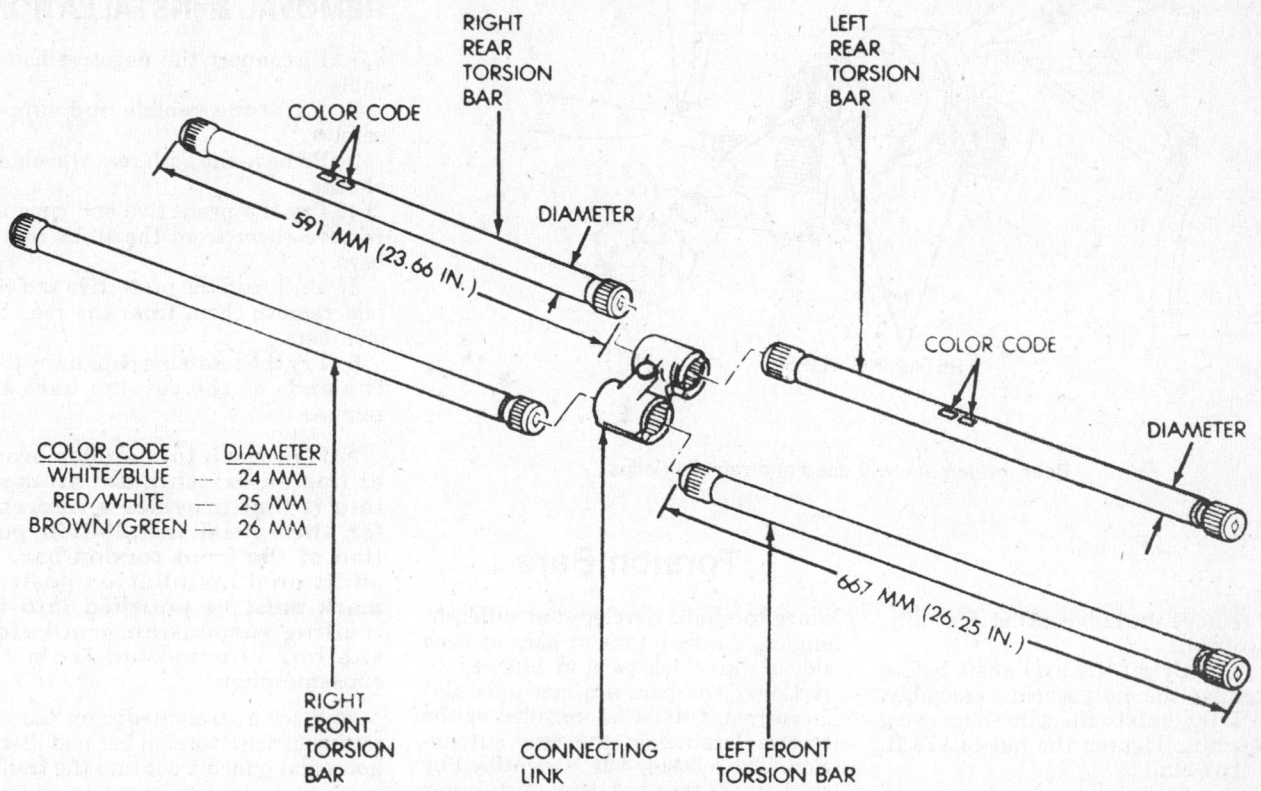

COLOR CODE	DIAMETER
WHITE/BLUE	24 MM
RED/WHITE	25 MM
BROWN/GREEN	26 MM

Rear torsion bars and identifying codes

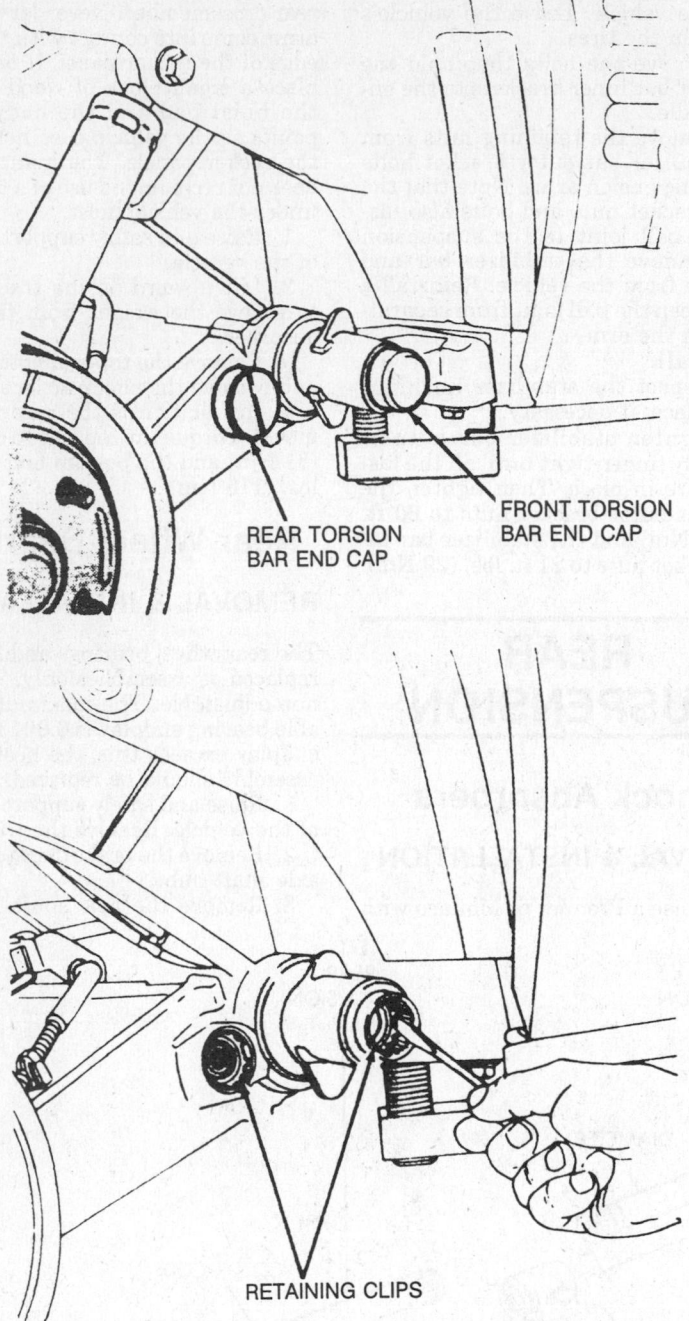

Rear torsion bar end caps and retaining clips

the letter **D** stamped twice on its outer end. In addition, the outer end of each torsion bar has 31 splines and the inner end has 30 splines. The torsion bars have a dot stamped onto their outer ends to assist in end identification and installation reference. This end with the dot must always be installed facing outward.

The front and rear torsion bars serve different purposes on Premier and Monaco. The front torsion bar is the actual suspension component, while the rear is used mostly as an anti-sway bar. The front bar is 26.25 in. (667mm) in length. The rear bar is 23.26 in. (591mm) in length. Also, the rear torsion bar in a vehicle is dependent upon the suspension package originally installed on the vehicle. The bars are color-coded to identify its diameter. The color codes are as follows:

- White/blue—0.96 in. (24mm)
- Red/white—1 in. (25mm)
- Brown/green—1.04 in. (26mm)

Never raise a Premier or Monaco with a lift positioned under the V-shaped rear crossmember. Never let the hoist arms come into contact with the lower edge of the rocker panel. If necessary, place a small block of wood between the hoist pad and the body lifting points so the vehicle does not rest on the rocker panels. The manufacturer does not recommend use of a twin post under-the-vehicle hoist.

REMOVAL & INSTALLATION

1. Disconnect the negative battery cable.
2. Raise the vehicle and support safely.
3. Remove the both rear wheels and shocks.
4. Pry the protective end caps and remove them from the front torsion bars.
5. Unthread the protective end caps and remove them from the rear torsion bars.
6. Pry the retaining clips away from the ends of the torsion bars and remove.

NOTE: Each torsion bar bracket has an existing dot stamped into it that provides a reference for the initial installation position of the front torsion bar. An additional installation position mark must be punched into the trailing suspension arm before the bar is removed from the crossmember.

7. Place a straightedge on the centerline of the 2 torsion bar installation holes and punch a dot into the trailing suspension arm adjacent to the rear torsion bar spline groove.

and remove the hub/bearing assembly.
To install:
4. Lightly oil the axle shaft before installing the hub/bearing assembly. Install the hub to the axle shaft using a new nut. Tighten the nut to 123 ft. lbs. (167 Nm).
5. Install the brake drum and wheel.

Torsion Bars

Since torque is developed at different angles, the rear torsion bars at each side of the vehicle twist in different directions. The bars are machined differently and must be installed at the correct location in the rear suspension. The left side bar is identified by the letter **G** stamped twice on its outer end. The right side bar is identified by

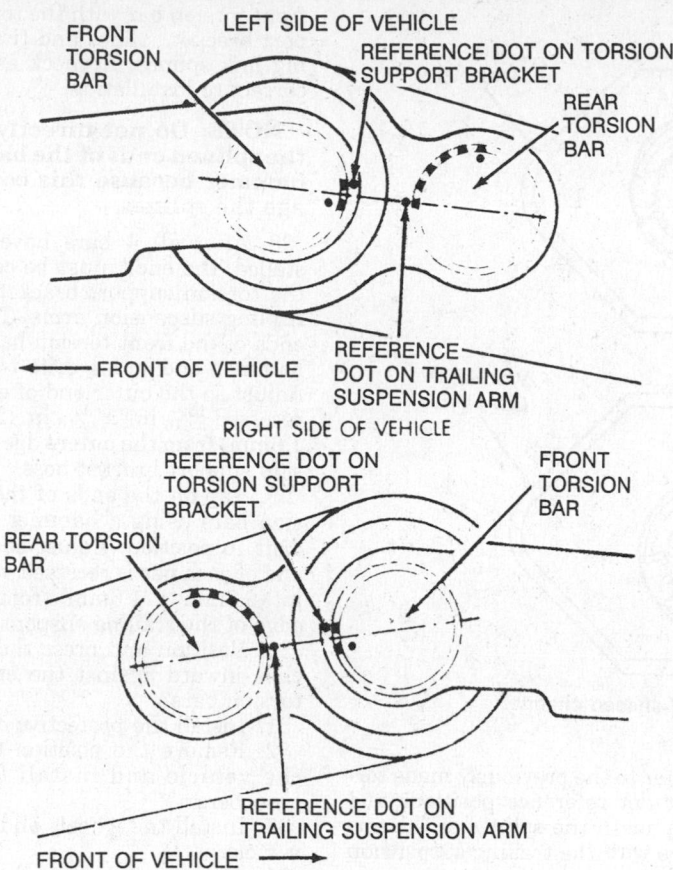

FRONT TORSION BAR

LEFT SIDE OF VEHICLE

REFERENCE DOT ON TORSION SUPPORT BRACKET

REAR TORSION BAR

← FRONT OF VEHICLE

REFERENCE DOT ON TRAILING SUSPENSION ARM

RIGHT SIDE OF VEHICLE

REFERENCE DOT ON TORSION SUPPORT BRACKET

FRONT TORSION BAR

REAR TORSION BAR

REFERENCE DOT ON TRAILING SUSPENSION ARM

FRONT OF VEHICLE →

Example of rear torsion bar installation positions

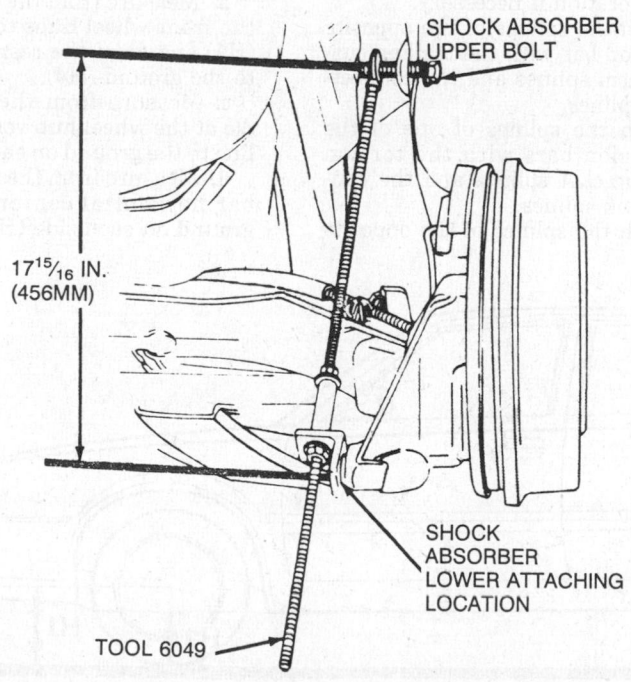

SHOCK ABSORBER UPPER BOLT

17¹⁵/₁₆ IN. (456MM)

SHOCK ABSORBER LOWER ATTACHING LOCATION

TOOL 6049

Special threaded tool installation

8. Note and record the relative positions of the installation reference dots on the ends of the torsion bars in respect to the dots on the torsion bar support bracket and the trailing suspension arm. In other words, count the number of splines between the dot on the bar and the dot made on the bracket or trailing arm. These dots will ensure that the proper "initial twist" is applied to the torsion bar upon installation, according to the suspension package originally installed in the vehicle.

9. Loosen and extract the front torsion bars using a slide hammer. Pull the bars out far enough to disengage the splines from the connecting link and the torsion support brackets.

10. Loosen and extract the rear torsion bars using a slide hammer. Pull the bars out far enough to disengage the splines from the connecting link and the trailing suspension arms.

11. The crossmember must be lowered before the torsion bars can be removed from it. Using the proper equipment, support the weight of the rear crossmember. Do not apply lifting force to the crossmember.

12. Loosen both torsion support bracket front bolts about 4 turns. Do not remove these bolts.

13. Loosen both torsion support bracket rear bolts about 10 turns. Do not remove these bolts.

14. Slowly allow the torsion support bracket and crossmember to lower about 1 inch.

15. Remove the torsion bars and connecting link from the crossmember.

To install:

16. Make sure the correct torsion bar will be installed to the correct side—**G** is stamped on left side bars; **D** is stamped on right side bars.

17. Insert the torsion bars into the rear crossmember. At this time, do not insert it so far as to engage the torsion bar splines with the splines in the support brackets or suspension arms.

18. Raise the crossmember and torsion support bracket and tighten the attaching bolts to 68 ft. lbs. (92 Nm).

NOTE: Before the torsion bar splines are meshed with the connecting link splines, the torsion support bracket splines and the trailing suspension arm splines must be positioned at the correct location in relation to the vehicle chassis.

19. Position the trailing suspension arms in the correct location on each side of the vehicle using 2 sets of special tool 6049 (threaded rod). Two spacers from tool set 7466 must also be used. Adjust each positioning tool so the distance between the center of the rod eyelet and the center of the

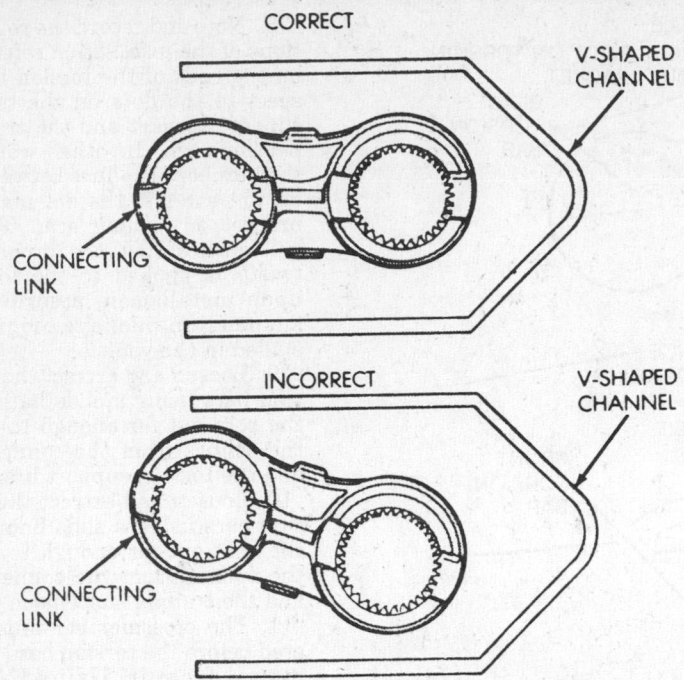

Center the connecting link in the V-shaped channel

hub is on the adjusting bracket is 17¹⁵⁄₁₆ in. (456mm).

20. After the distance has been properly set, install the positioning tools where the rear shocks are installed.

21. Insert the upper attaching bolt through the eyelet and loosely tighten it. Insert the spacer into the lower shock attaching bolt hole and insert the adjustable bracket hub into the spacer.

22. Liberally apply all-purpose lubricant to all torsion bar splines.

NOTE: If the bars are not being replaced, they must be installed in their exact original locations. The dot reference positions must be the same on both sides of the vehicle to prevent added stress from being applied to the bars.

23. Refer to the previously made torsion bar dot reference positions and correctly mesh the splines of the torsion bars with the trailing suspension arm splines.

24. Mesh the connecting link splines with the splines of the previously positioned rear torsion bar so the connecting link is correctly centered within the V-shaped channel in the rear crossmember. Use large pliers to assist in this operation if necessary.

25. Mesh the splines of the opposite rear torsion bar with the trailing suspension arm splines and the connecting link splines.

26. Mesh the splines of one of the front torsion bars with the torsion support bracket splines and the connecting link splines.

27. Mesh the splines of the opposite front torsion bar with the torsion support bracket splines and the connecting link splines. Recheck all bars for correct re-installation.

NOTE: Do not directly contact the splined ends of the bars with a hammer because this could damage the splines.

28. After all 4 bars have been installed, the ends must be centered in the torsion support brackets and the trailing suspension arms. Tap on the ends of the front torsion bars using a hammer and brass drift to position. Adjust so the outer end of each bar is recessed ¹³⁄₁₆ in. ± ¹⁄₁₆ in. (20.6mm ± 1.6mm) from the outer edge of the torsion support bracket boss.

29. Tap on the ends of the rear torsion bars using a hammer and brass drift to position. Adjust so the outer end of each bar is recessed ¼ in. ± ¹⁄₁₆ in. (6.3mm ± 1.6mm) from the outer edge of the trailing suspension arm.

30. Position and press the retaining clips inward against the ends of the torsion bars.

31. Install the protective caps.

32. Remove the position tools from the vehicle and install the shock absorbers.

33. Install the wheels and lower the vehicle.

34. Measure the vehicle height using the following procedure:

a. The vehicle should be unloaded, on a flat level surface, with a full tank of fuel, and with the tires all adjusted to the same proper pressure.

b. Measure from the centerline of the front wheel hubs to the ground (H1) and from the rear wheel hubs to the ground (H4).

c. Measure from the engine cradle at the wheel hub vertical centerline to the ground on each side (H2).

d. Measure from the front torsion bar horizontal center line to the ground on each side (H3).

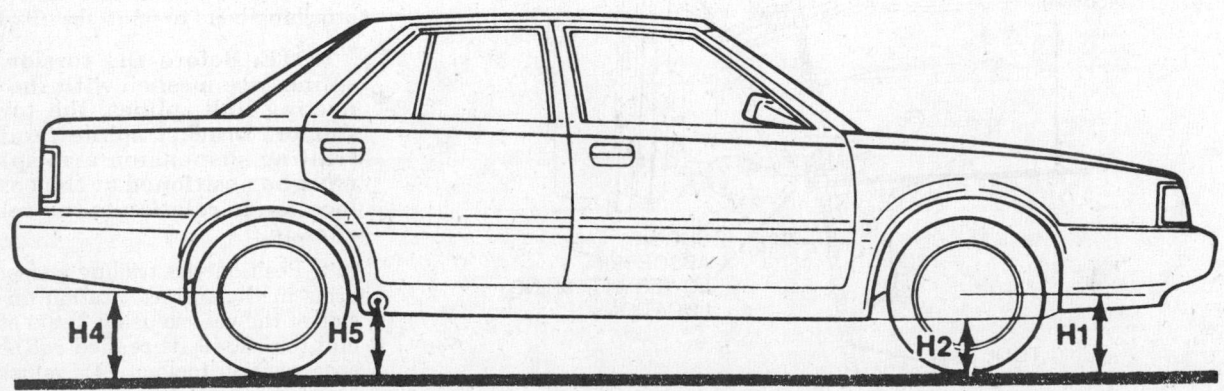

Various vehicle height measurement points

e. Subtract H2 from H1. This value should be 3.36–3.98 in. (85–101mm). If this height is not within specification, replace worn front end parts.

f. Subtract H3 from H4. This value should be 1.25–1.87 in. (31.5–47.5mm).

35. If the rear vehicle height is not within specification, reposition the front torsion bar(s). This is accomplished by adjusting the length of special tool 6049 (threaded rod) when re-installing the bar (Step 19 in the procedure). Do not attempt to adjust vehicle height with the rear torsion bars.

36. Connect the negative battery cable and road test the vehicle.

Rear Axle Assembly

REMOVAL & INSTALLATION

1. Raise and safely support the vehicle.
2. Remove the rear wheels.
3. Remove the parking brake cables from the body support.
4. Disconnect and plug the brake hoses at the axle. Remove the shock absorbers.
5. Support the axle assembly and remove the support bracket bolts. Lower the axle assembly and remove.

To install:
6. Installation is the reverse of removal. Position the axle under the vehicle and raise it into place. Install and tighten the support bracket bolts, tighten to 68 ft. lbs. (92 Nm).
7. Connect the brake hoses at the axle. Connect the parking brake cables. Install the shock absorbers, tighten the upper shock bolt to 60 ft. lbs. (81 Nm) and the lower bolt to 85 ft. lbs. (115 Nm).
8. Install the rear wheels, bleed the brake system and adjust the parking brake cable.

STEERING

Steering Wheel

REMOVAL & INSTALLATION

1. Disconnect the negative battery cable.
2. Unsnap the horn button and disconnect the wires. Remove the horn button.
3. Note the position of the reference mark on the end of the steering shaft. Matchmark the steering wheel to the shaft to emphisize the alignment.

4. Remove the nut and slide the wheel off the shaft. If required, use a suitable steering wheel puller. Disconnect the speed control wire connector, if equipped.

To install:
5. Install the electrical connector. Align the pin on the turn signal cam with the pin bore in the steering wheel and slide the wheel into place.
6. Align the wheel with the reference mark on the steering shaft and install the nut. Tighten the nut to 52 ft. lbs. (71 Nm).
7. Connect the negative battery cable.

Steering Column

REMOVAL & INSTALLATION

1. Disconnect battery negative cable.
2. Remove the screws that attach the instrument panel lower trim cover to the instrument panel. Remove the cover.
3. Remove the screws that attach the instrument panel support rod to the instrument panel. Remove the rod.
4. Disconnect the steering column wire harness connector.
5. Remove the screw that attaches the dash panel wire harness connector to the dash panel.
6. If equipped with a steering column gearshift mechanism, perform the following:

a. Disconnect the automatic transaxle shift cable receptacle from the shift lever ball joint with a small prybar.

b. Disconnect the shift cable retainer from the steering column bracket by depressing the cable retainer lock tabs with pliers.

c. Slide the shift cable retainer out of the steering column bracket.

d. Remove the screw that attaches the shift position indicator bracket to the steering column and remove the indicator wire from the pivot pin.

7. Detach the steering column boot from the steering column. Slide the upper half of the 2-piece boot downward over the lower half of the boot for access to the steering column shaft and intermediate shaft U-joint.
8. Matchmark the steering column shaft and the intermediate shaft U-joint coupling for installation alignment reference.
9. Remove the bolt from the intermediate steering shaft U-joint coupling clamp.
10. Remove the bolts and nuts that attach the steering column to the instrument panel.
11. Carefully lower the steering col-

umn to the vehicle floor. Disconnect the steering shaft from the intermediate shaft and remove the column from the vehicle.

To install:
12. Align the reference marks and insert the steering shaft into the intermediate shaft U-joint coupling clamp.
13. Install but do not tighten the intermediate shaft U-joint clamp bolt.
14. Place the steering column into position, install the nuts and bolts and torque to 33 ft. lbs. (45 Nm). Tighten the intermediate shaft U-bolt clamp bolt to 30 ft. lbs. (41 Nm).
15. Reassemble the shift mechanism. Check shift indicator alignment. Place the gearshift in **N** and observe the position of the shift indicator pointer. If the pointer is not aligned, loosen the shift position indicator bracket screw and move the bracket forward or backward to correctly align the pointer with the **N** on the quadrant. Tighten the screw.
16. Align the 2 halves of the steering shaft boot. Rotate the upper half of the boot until the **X** mark on the lower half is centered in the oval alignment cutout (window) in the upper half of the boot. Verify that the alignment mark on the metal boot flange is at the 6 o'clock position.
17. Install instrument panel components.
18. Connect the negative battery cable and check all column-mounted components for proper operation.

Power Steering Rack and Pinion

REMOVAL & INSTALLATION

1. Disconnect the negative battery cable. Remove the instrument panel lower cover.
2. Unsnap the steering shaft boot flange from the dash panel opening and slide the boot upward.
3. Remove the U-joint coupling clamp bolt. Matchmark the steering column shaft and the intermediate shaft U-joint coupling for installation alignment reference.
4. In the engine compartment, remove the splash shield from the dash panel.
5. Remove the fluid lines from their retaining block slots. Position a drain pan under the tubes and disconnect them from the steering gear.
6. Remove the front attaching nut from the right side bracket.
7. Raise the vehicle and support safely. Remove the left front wheel.
8. Disconnect the tie rod ends from the strut body brackets.
9. Remove the 3 bolts that attach

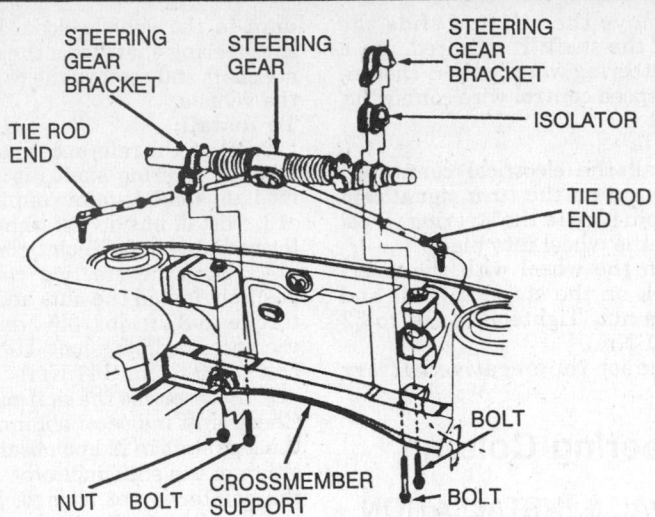

STEERING GEAR BRACKET
STEERING GEAR
STEERING GEAR BRACKET
ISOLATOR
TIE ROD END
TIE ROD END
BOLT
NUT BOLT CROSSMEMBER SUPPORT
BOLT

Rack and pinion steering gear and related parts

the steering gear brackets to the cross member support.

10. Tie the tie rods to the steering gear to keep them parallel to the rack. Remove the steering gear and tie rods through the left side of the vehicle.

To install:

11. Remove the tie rod ends from the steering gear to aid installation. Install the gear through the left opening. Position on the support and install the bracket attaching nut and bolts. Tighten to 40 ft. lbs. (55 Nm).

12. Attach the tie rod ends to the strut body brackets. Tighten the nuts to 35 ft. lbs. (48 Nm).

13. Connect the tie rods to the rack shaft spacer block and tighten the bolts to 55 ft. lbs. (75 Nm). Bend the lock tabs over the bolt flats.

14. Replace the O-rings and connect the fluid lines to the rack. Insert the lines into their slots.

15. Install the splash shield onto the dash panel.

16. Align the reference marks and insert the steering shaft into the intermediate shaft U-joint coupling clamp. Install the intermediate shaft U-joint clamp bolt. Tighten the intermediate shaft U-bolt clamp bolt to 30 ft. lbs. (41 Nm).

17. Lower the boot and install the instrument panel lower cover.

18. Fill the pump reservoir with power steering fluid.

19. Perform a front end alignment.

Power Steering Pump

REMOVAL & INSTALLATION

1. Disconnect the negative battery cable.

2. Raise and safely support the vehicle.

3. Remove the underbody splash shield. Loosen the accessory drive belt.

4. Disconnect and plug the power steering fluid lines.

5. Remove the pump mounting bolts and remove the pump.

To install:

6. Install the pump to the engine.

7. Replace the O-rings and connect the pressure lines.

8. Install the accessory drive belt and adjust the tension.

9. Lower the vehicle. Fill and bleed the system.

System Bleeding

1. With the wheels turned all the way to the left, add power steering fluid to the **COLD** mark on the fluid level indicator or until the reservoir is full.

2. Start the engine and run at fast idle momentarily, then shut the engine off and recheck fluid level. If necessary, add fluid.

3. Start the engine and bleed the system by turning the wheels from side to side without hitting the stops.

NOTE: Fluid with air in it has a red milky appearance.

4. Return the wheels to the center position and keep the engine running for 2-3 minutes.

5. Road test the vehicle and recheck the fluid level making sure it is at the **HOT** mark or the reservoir is full.

Tie Rod Ends

The rack and pinion system is mounted high in the body. The steering tie rods are connected to the steering rack shaft via a spacer block located at the center of the steering rack body. The tie rods are connected to strut body brackets.

REMOVAL & INSTALLATION

1. Raise and safely support vehicle.

2. Remove the wheel and tire assemblies.

3. Remove the nuts attaching the tie rod end ball studs to the strut body brackets.

4. Loosen the locknuts that secure the tie rod ends to the tie rods. A wire brush and solvent will help clear the threads of road debris.

5. Disconnect the tie rod ends with an appropriate press type tool.

6. Unscrew the tie rod ends. Counting the turns will make installation easier and get the front end alignment close.

7. Installation is the reverse of removal.

8. Torque the tie rod ends-to-strut bracket nuts to 35 ft. lbs. (48 Nm).

9. Perform a front end alignment.

BRAKES

For all brake system repair and service procedures not detailed below, please refer to "Brakes" in the Unit Repair section.

──── CAUTION ────

The ABS pump/motor assembly will keep the hydraulic accumulator charged to a pressure of 1600–200 psi (11,000–14,000 kPa) any time the ignition switch is in the ON position. The pump cannot run if the ignition switch is in the OFF position or if the negative battery cable is disconnected.

If the vehicle is equipped with ABS, depressurize the hydraulic accumulator before disassembling or disconnecting any part of the hydraulic system. Failure to do so could result in serious personal injury.

If the engine is hot, do not attempt brake system work inside the engine compartment. If brake fluid comes in contact with a hot engine, combustion is possible.

Master Cylinder

REMOVAL & INSTALLATION

Except ABS

1. Disconnect the negative battery cable.

2. Disconnect the fluid sensor electrical connector, if equipped.

3. Disconnect and plug the brake lines. Cover the master cylinder outlet ports and brake lines to prevent the entry of dirt.

4. If equipped with manual brake,

disconnect the master cylinder push-rod at the brake pedal.

5. Remove the master cylinder retaining bolts or nuts.

6. Remove the proportioning valve bracket, if required.

7. Remove the master cylinder from the vehicle.

8. Before installing the replacement master cylinder on the vehicle, bench bleed the master cylinder.

9. Installation is the reverse of the removal procedure.

10. The entire brake system must be bled after installing the master cylinder.

Proportioning Valve

REMOVAL & INSTALLATION

1. Disconnect the negative battery cable.

2. Disconnect and plug the brake lines from the proportioning valve.

3. Remove the bolt and nut attaching the valve to the bracket.

4. Remove the valve from the vehicle.

5. Installation is the reverse of removal.

6. Bleed the brakes after installation.

Power Brake Booster

REMOVAL & INSTALLATION

1. Disconnect the negative battery cable.

2. Disconnect the vacuum line from the booster.

3. Remove the clip retaining the throttle cables to the bracket on the booster. Remove the master cylinder.

4. Inside the vehicle, disconnect the connector from the brake light switch. Remove the pushrod from the brake pedal.

5. Remove the booster retaining nuts and remove the booster. Inspect the seal for damage.

To install:

6. Transfer parts to the replacement booster. Install the booster to the firewall and connect the pushrod to the brake pedal. Connect the brake light switch.

7. Install the master cylinder and clip the throttle cables in place.

8. Connect the negative battery cable and bleed the brake system.

Brake Caliper

REMOVAL & INSTALLATION

Front

1. If equipped with ABS, depressurize the hydraulic accumulator.

2. Raise and safely support vehicle. Remove wheel and tire assemblies.

3. Disconnect the brake hose from the caliper.

4. Remove caliper slide pins and lift caliper up and out of the bracket.

5. The installation is the reverse of removal. Replacement of the the copper washers that seal the brake hoses is necessary.

6. Bleed the brakes after installation. Pump brake pedal to seat the brakes before moving the vehicle.

Rear

NOTE: The rear calipers are not serviceable and must be replaced if defective.

1. If equipped with ABS, depressurize the hydraulic accumulator.

2. Raise and safely support vehicle. Remove wheel and tire assemblies.

3. Disconnect the brake hose from at the caliper. Retain the bolt but discard the 2 seal washers.

4. Unseat the operating lever return spring at the caliper.

5. Remove the operating lever attaching bolt and pry the lever off the drive disc.

6. Remove the lever return spring. Remove the operating lever from the parking brake cable.

7. Remove caliper slide pins and remove the caliper. Remove the brake shoe retaining pin and remove the brake shoes and anti-rattle spring.

To install:

8. A spanner tool may be required, along with an appropriate socket and extension, to turn the piston **clockwise** until it is fully seated in the bore.

9. Lubricate the caliper slide pins and bushings with silicone lubricant and install.

10. Replace the copper sealing washers and connect the hose to the caliper.

11. Install the operating lever and assemble the return spring to the brake operating lever.

12. Bleed the brakes after installation. Pump the brake pedal to seat the brakes before moving vehicle.

Disc Brake Pads

REMOVAL & INSTALLATION

1. Disconnect the negative battery cable. Raise the vehicle and safely support.

2. Remove the wheels.

3. On rear disc brakes, remove the retaining pin. Remove the caliper from the rotor.

4. Remove the disc brake pads.

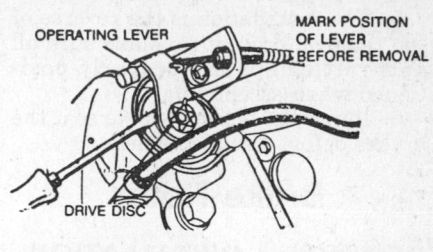

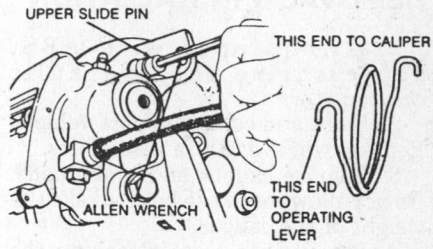

Rear disc brake caliper assembly—Monaco and Premier

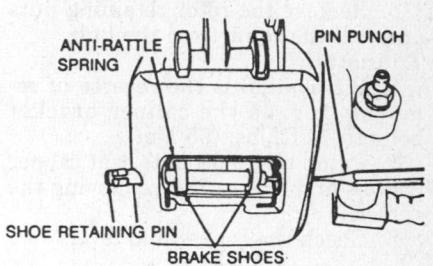

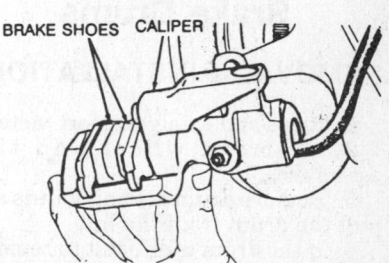

Rear caliper retaining pins—Monaco and Premier

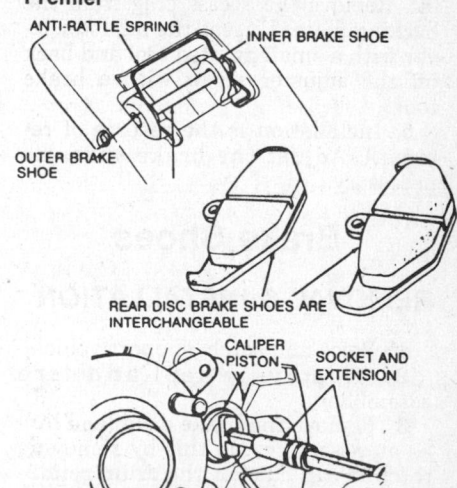

Rear brake pads with cranking piston back into the bore

5. The installation is the reverse of the removal procedure. Make sure all anti-rattle clips are properly positioned when assembling.

6. Pump the brake pedal to seat the brakes before moving vehicle.

Brake Rotor

REMOVAL & INSTALLATION

1. If equipped with ABS, depressurize the hydraulic accumulator.

2. Raise and safely support vehicle. Remove wheel and tire assemblies.

3. Remove caliper and brake pads. Do not allow the brake hose to take the weight of the caliper.

4. Remove the bolts attaching the caliper bracket to the steering knuckle.

5. Remove the rotor retaining nuts and pull the rotor from the hub.

To install:

6. Installation is the reverse of removal. Torque the caliper bracket bolts to 70 ft. lbs. (95 Nm).

7. Pump brake pedal to seat caliper pistons and brakes before moving the vehicle.

8. Check the brake fluid level.

Brake Drums

REMOVAL & INSTALLATION

1. Raise and safely support vehicle.

2. Remove wheel and tire assemblies.

3. Remove drum retaining nuts and pull the drum from the hub.

4. If the drum is difficult to remove, the brake shoes are probably holding the drum in place and must be backed off. Remove the access plug from the backing plate. Unseat the adjuster lever with a small pointed tool and back off the adjuster screw with a brake tool.

5. Installation is the reverse of removal. Adjust the brake shoes as necessary.

Brake Shoes

REMOVAL & INSTALLATION

1. Raise and safely support vehicle.

2. Remove wheel and tire assemblies.

3. Remove the brake drum and hub as an assembly. Do this by removing the hub cap but not the drum retaining nuts. Remove the large center hub nut and pull off the hub/drum.

4. If the drum is difficult to remove, the brake shoes are probably holding the drum in place and must be backed

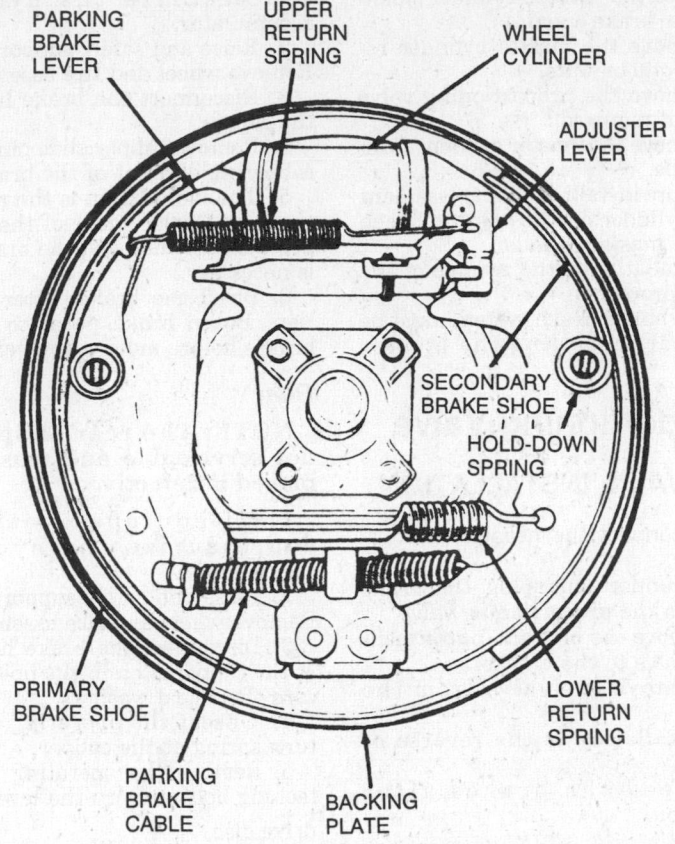

PARKING BRAKE LEVER · UPPER RETURN SPRING · WHEEL CYLINDER · ADJUSTER LEVER · SECONDARY BRAKE SHOE · HOLD-DOWN SPRING · LOWER RETURN SPRING · BACKING PLATE · PARKING BRAKE CABLE · PRIMARY BRAKE SHOE

Drum brake components

off. Remove the backing plate plug. Unseat the adjuster lever with a small pointed tool and back off the adjuster screw with a brake tool.

5. Remove the upper return spring with brake pliers. Install a wheel cylinder clamp to hold the piston in place. Remove the lower return spring.

6. Remove the parking brake adjuster cross lever (strut). Disengage the parking brake cable by moving the end of the cable away from the lever with expanding type snapring pliers and disengage the cable from the lever.

7. Remove the adjuster screw. Remove the secondary hold-down springs and the secondary brake shoe. Remove the horseshoe clip and remove the parking brake lever from the brake shoe.

8. Remove the primary brake shoe hold-down spring and remove the shoe.

To install:

9. Clean and lubricate the shoe contact surfaces on the backing plate, the parking brake lever pivot and the adjuster screw threads with moly grease.

10. Insert the parking brake lever pivot in the secondary shoe. Install the clip and crimp in place. Attach the parking brake lever to the lever.

11. Position the secondary shoe and lever on the backing plate and install the hold-down spring and pin.

12. Install the primary shoe and hold-down spring and pin.

13. When installing the adjuster screw, be sure the adjuster position is correct. The large notch in the adjuster screw goes to the brake shoe. The small notch goes to the adjuster lever. The long end goes to the secondary shoe.

14. Install the adjuster lever on the primary shoe pin and seat the lever in the adjuster screw. Make sure the adjuster lever is seated in the small notch of the adjuster screw.

15. Install the return springs. Sand the brakes clean.

16. Adjust the brake shoes and install the drum.

Wheel Cylinder

REMOVAL & INSTALLATION

1. Disconnect the negative battery cable. Raise and safely support vehicle.

2. Remove wheel and tire assemblies.

3. Remove drum retaining nuts and pull the drum from the hub.

4. Remove the brake shoe upper return spring and spread the shoes slightly to make room for cylinder removal.

5. Disconnect the brake line at the cylinder, remove the cylinder attaching bolts and remove the cylinder from the backing plate.

6. Installation is the reverse of removal. Torque the retaining bolts to 11 ft. lbs. (15 Nm).

7. Bleed the brake system. Adjust the brake shoes and parking brakes.

Parking Brake Cable

ADJUSTMENT

1. Adjust the rear brakes if necessary.

2. Apply and release the parking brake 5 times to center the shoes in the drums. Set the pedal on the first notch from the released position.

3. Raise and support the vehicle safely.

4. Tighten the cable at the equalizer so the wheels can just barely be turned forward, then loosen 1 turn. Be sure to hold the end of the cable screw to prevent the cable from turning.

5. Release the parking brake and check for rear brake drag. The wheels should rotate freely with the parking brake no applied.

REMOVAL & INSTALLATION

Front

1. Disconnect the negative battery cable.

2. Raise the vehicle and support safely.

3. Remove the jam nut and adjusting nut from the front cable. Lower the vehicle.

4. Remove the left side sill trim, driver's seat and lower dash trim cover.

5. Remove the rear seat cushion. Roll the carpeting aside for access to the cable.

6. Pull the cable through the holes in the floor pan and remove the cable from the actuating pedal.

7. The installation is the reverse of the removal procedure. Make sure the cable is routed properly before installing interior parts.

8. Adjust the rear brakes and cable.

Rear

1. Raise and safely support the vehicle.

2. Loosen the cable adjusting nut at the equalizer.

3. Remove the cotter pin for the cable to be replaced from the equalizer.

Then remove the retaining clip attaching the cable to the frame.

4. If only replacing the right side, remove the bolts attaching the cable to the rear axle housing. Disconnect the cable at the frame bracket.

5. If equipped with rear drums, remove the rear wheels and brake drum. Remove the shoes. Compress the locking tabs at the backing plate with a hose clamp or a suitably sized box wrench and remove the cable.

6. If equipped with rear disc brakes, disconnect the cable from the actuating arm and remove from the bracket.

7. Installation is the reverse of the removal process. Use a new cotter pin for the connection to the equalizer. Adjust the rear brakes and cable.

Brake System Bleeding

Except Anti-Lock Brakes

NOTE: If using a pressure bleeder, follow the instructions furnished with the unit and choose the correct adaptor for the application. Do not substitute an adapter that "almost fits" as it will not work and could be dangerous.

Master Cylinder

If the master cylinder is off the vehicle it can be bench bled.

1. Connect 2 short pieces of brake line to the outlet fittings, bend them until the free end is below the fluid level in the master cylinder reservoirs.

2. Fill the reservoir with fresh brake fluid. Pump the piston slowly until no more air bubbles appear in the reservoirs.

3. Disconnect the 2 short lines, refill the master cylinder and securely install the cylinder caps.

4. If the master cylinder is on the vehicle, it can still be bled, using a flare nut wrench.

5. Open the brake lines slightly with the flare nut wrench while pressure is applied to the brake pedal by a helper inside the vehicle.

6. Be sure to tighten the line before the brake pedal is released.

7. Repeat the process with both lines until no air bubbles come out.

Calipers and Wheel Cylinders

1. Fill the master cylinder with fresh brake fluid. Check the level often during the procedure.

2. Starting with the right rear wheel, remove the protective cap from the bleeder, if equipped, and place where it will not be lost. Clean the bleed screw.

CAUTION

When bleeding the brakes, keep face away from the brake area. Spewing fluid may cause facial and/or visual damage. Do not allow brake fluid to spill on the car's finish; it will remove the paint.

---

3. If the system is empty, the most efficient way to get fluid down to the wheel is to loosen the bleeder about ½–¾ turn, place a finger firmly over the bleeder and have a helper pump the brakes slowly until fluid comes out the bleeder. Once fluid is at the bleeder, close it before the pedal is released inside the vehicle.

NOTE: If the pedal is pumped rapidly, the fluid will churn and create small air bubbles, which are almost impossible to remove from the system. These air bubbles will eventually congregate and a spongy pedal will result.

4. Once fluid has been pumped to the caliper or wheel cylinder, open the bleed screw again, have the helper press the brake pedal to the floor, lock the bleeder and have the helper slowly release the pedal. Wait 15 seconds and repeat the procedure (including the 15 second wait) until no more air comes out of the bleeder upon application of the brake pedal. Remember to close the bleeder before the pedal is released inside the vehicle each time the bleeder is opened. If not, air will be induced into the system.

5. If a helper is not available, connect a small hose to the bleeder, place the end in a container of brake fluid and proceed to pump the pedal from inside the vehicle until no more air comes out the bleeder. The hose will prevent air from entering the system.

6. Repeat the procedure on remaining wheel cylinder and calipers in order:

   a. Left rear
   b. Right front
   c. Left front

7. Hydraulic brake systems must be totally flushed if the fluid becomes contaminated with water, dirt or other corrosive chemicals. To flush, bleed the entire system until all fluid has been replaced with the correct type of new fluid.

8. Install the bleeder cap(s) on the bleeder to keep dirt out. Always road test the vehicle after brake work of any kind is done.

Anti-Lock Brakes

PRESSURE BLEEDING

The brake lines may be pressure bled, using a standard diaphragm type pressure bleeder. Only diaphragm type pressure bleeding equipment should be used to bleed the system.

1. The ignition should be turned **OFF** and remain **OFF** throughout this procedure.

2. Depressurize the hydraulic accumulator.

—————— CAUTION ——————

Failure to depressurize the hydraulic accumulator prior to performing this operation, may result in personal injury and/or damage to the painted surfaces.

3. Remove the reservoir caps.

4. Install the pressure bleeder adapter.

5. Attach the bleeding equipment to the bleeder adapter. Charge the pressure bleeder to approximately 20 psi (138 kPa).

6. Connect a transparent hose to the caliper bleed screw. Submerge the free end of the hose in a clear glass container, which is partially filled with clean, fresh brake fluid.

7. With the pressure turned **ON**, open the caliper bleed screw ½–¾ turn and allow fluid to flow into the container. Leave the bleed screw open until clear, bubble-free fluid slows from the hose. If the reservoir has been drained or the hydraulic assembly removed from the vehicle prior to the bleeding operation, slowly pump the brake pedal 1–2 times while the bleed screw is open and fluid is flowing. This will help purge air from the hydraulic assembly. Tighten the bleeder screw to 7.5 ft. lbs. (10 Nm).

8. Repeat Step 7 at all calipers. Calipers should be bled in the following order:

    a. Left rear
    b. Right rear
    c. Left front
    d. Right front

9. After bleeding all 4 calipers, remove the pressure bleeding equipment and bleeder adapter by closing the pressure bleeder valve and slowly unscrewing the bleeder adapter from the hydraulic assembly reservoir. Failure to release pressure in the reservoir will cause spillage of brake fluid and could result in injury or damage to painted surfaces.

10. Using a syringe or equivalent method, remove excess fluid from the reservoir to bring the fluid level to full level.

11. Install the reservoir caps and connect the fluid level sensor connector. Turn the ignition **ON** and allow the pump to charge the accumulator.

MANUAL BLEEDING

1. Depressurize the hydraulic accumulator.

—————— CAUTION ——————

Failure to depressurize the hydraulic accumulator, prior to performing this operation may result in personal injury and/or damage to the painted surfaces.

2. Connect a transparent hose to the caliper bleed screw. Submerge the free end of the hose in a clear glass container, which is partially filled with clean, fresh brake fluid.

3. Slowly pump the brake pedal several times, using full strokes of the pedal and allowing approximately 5 seconds between pedal strokes. After 2 or 3 strokes, continue to hold pressure on the pedal, keeping it at the bottom of its travel.

4. With pressure on the pedal, open the bleed screw ½–¾ turn. Leave bleed screw open until fluid no longer flows from the hose. Tighten the bleed screw and release the pedal.

5. Repeat this procedure until clear, bubble-free fluid flows from the hose.

6. Repeat all steps at each of the calipers. Calipers should be bled in the following order:

    a. Left rear
    b. Right rear
    c. Left front
    d. Right front

Anti-Lock Brake System Service

PRECAUTIONS

Failure to observe the following precautions may result in system damage.

• Before performing electric arc welding on the vehicle, disconnect the ABS controller and the hydraulic modulator connectors.

• When performing painting work on the vehicle, do not expose the ABS controller to temperatures in excess of 185°F (85°C) for longer than 2 hrs. The system may be exposed to temperatures up to 200°F (95°C) for less than 15 min.

• Never disconnect or connect the ABS controller or hydraulic modulator connectors with the ignition switch ON.

• Never disassemble any component of the Anti-Lock Brake System (ABS) which is designated non-serviceable; the component must be replaced as an assembly.

• When filling the master cylinder, always use brake fluid which meets DOT-3 specifications; petroleum base fluid will destroy the rubber parts.

DEPRESSURIZING THE HYDRAULIC ACCUMULATOR

1. With the ignition **OFF**, pump the brake pedal a minimum of 40 times, using approximately 50 lbs. (222 N) pedal force. A noticeable change in pedal feel will occur when the accumulator is discharged.

2. When a definite increase in pedal effort is felt, stroke the pedal a few additional times. This should remove all hydraulic pressure from the system.

Pump/Motor Assembly

REMOVAL & INSTALLATION

1. Disconnect the negative battery cable. Depressurize the hydraulic accumulator.

—————— CAUTION ——————

Failure to depressurize the hydraulic accumulator, prior to performing this operation may result in personal injury and/or damage to the painted surfaces.

2. Disconnect all electrical connectors to the pump motor.

3. Disconnect and plug the high and low pressure hoses from the hydraulic assembly.

4. Remove the connector body from the engine mount.

5. Remove the heat shield.

6. Remove the retainer bolts that are used to mount the pump/motor.

7. Lift the pump/motor assembly off of the studs and out of the vehicle.

To install:

8. Position the assembly and install the retaining bolts.

9. Install the heat shield.

10. Install the connector body to the engine mount.

11. Connect the high and low pressure hoses to the hydraulic assembly.

12. Connect all electrical connectors to the pump motor.

13. Connect the negative battery cable and check the brakes for proper operation.

Hydraulic Assembly

REMOVAL & INSTALLATION

1. Disconnect the negative battery cable. Depressurize the hydraulic accumulator.

—————— CAUTION ——————

Failure to depressurize the hydraulic accumulator, prior to performing this operation may result in personal injury and/or damage to the painted surfaces.

2. Remove the fresh air intake ducts.

3. Disconnect all electrical connectors from the hydraulic unit and pump/motor.

4. Remove as much of the fluid as

possible from the reservoir on the hydraulic assembly.

5. Remove the pressure hose fitting from the hydraulic assembly.

6. Disconnect the return hose from the filter nipple. Cap the spigot on the reservoir.

7. Disconnect all brake tubes from the hydraulic assembly.

8. Remove the driver's side lower panel.

9. Disconnect the pushrod from the brake pedal.

10. Remove the 4 under-dash hydraulic assembly mounting nuts.

11. Remove the hydraulic assembly.

To install:

12. Have a helper position the hydraulic assembly on the vehicle and hold in place.

13. Install and torque the mounting nuts to 21 ft. lbs. (28 Nm).

14. Using Lubriplate® or equivalent, coat the bearing surface of the pedal pin.

15. Connect the pushrod to the pedal and install a new retainer clip.

16. Install the brake tubes. If the proportioning valves were removed from the hydraulic assembly, reinstall valves and tighten to 30 ft. lbs. (40 Nm).

17. Install the return hose to the nipple on the filter.

18. Install the pressure hose to the hydraulic assembly.

19. Fill the reservoir to the top of the screen.

20. Connect all electrical connectors to the hydraulic assembly.

21. Bleed the entire brake system.

22. Install the fresh air intake duct.

23. Connect the negative battery cable and check the assembly for proper operation.

Speed Sensors

REMOVAL & INSTALLATION

Front Sensor

1. Raise the vehicle and support safely. Remove the wheel and tire assembly.

2. Remove the screw from the clip that holds the sensor to the fender shield.

3. Carefully pull the sensor assembly grommet from the fender shield.

4. Unplug the connector from the harness. Remove the retainer clip from the strut damper bracket.

5. Remove the sensor mounting screw.

6. Carefully remove the sensor.

To install:

7. Coat the sensor with high temperature multi-purpose anti-corrosion compound before installing into the steering knuckle. Install the screw and tighten to 60 inch lbs. (7 Nm).

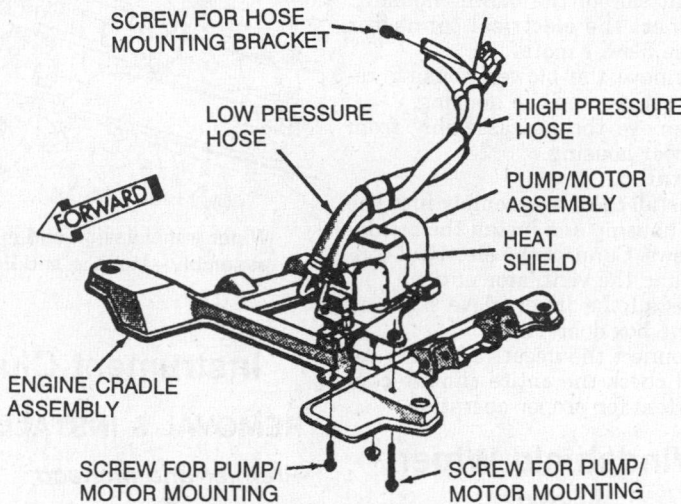

ABS pump motor assembly mounting

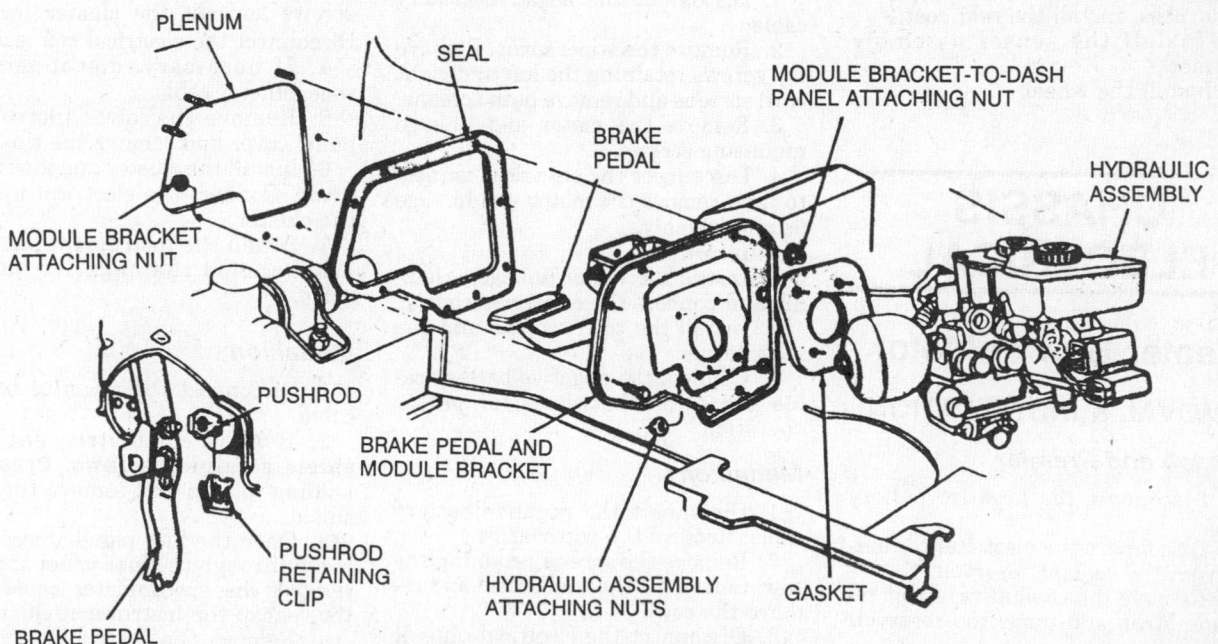

ABS hydraulic assembly mounting

8. Connect the sensor connector to the harness.

9. Install the sensor assembly grommet and attach the clip to the fender shield.

NOTE: Proper installation of the wheel speed sensor cables is critical to continued system operation. Be sure the cables are installed in retainers. Failure to install the cables in the retainers may result in contact with moving parts and/or over-extension of the cables, resulting in an open circuit.

10. Install the wheel.

Rear Sensor

1. Raise the vehicle and support safely. Remove the wheel and tire assembly.

2. Remove the rear seat and disconnect the sensor connector.

3. Carefully pull the sensor assembly grommet from the underbody and pull the harness through the hole.

4. Remove the screws that retain the wiring to the rear axle and floorpan.

5. Remove the bolt that retains the sensor to the rear bearing retainer.

6. Carefully remove the sensor.

To install:

7. Coat the sensor with high temperature multi-purpose anti-corrosion compound before installing into the bearing retainer. Install the screw and tighten to 60 inch lbs. (7 Nm).

8. Secure the wiring to the axle and floor pan.

9. Feed the sensor connector wire through the grommet and connect to the harness. Install the rear seat.

10. Install the sensor assembly grommet.

11. Install the wheel.

CHASSIS ELECTRICAL

Heater Blower Motor

REMOVAL & INSTALLATION

Monaco and Premier

1. Disconnect the negative battery cable.

2. Disconnect the electrical connector from the coolant reservoir.

3. Remove the coolant reservoir retaining strap and move the reservoir aside.

4. Remove the coolant reservoir mounting bracket. Disconnect the

electrical wires from the blower motor.

5. Remove the blower motor mounting bolts and the blower motor.

6. The installation is the reverse of the removal installation.

7. Connect the negative battery cable and check the entire climate control system for proper operation.

Medallion

1. Disconnect the negative battery cable.

2. Remove the glove box door straps and the glove box door. Remove the inner glove box.

3. Unclip the ventilator outlet from the right side of the blower housing. Disconnect the electrical connector from the blower motor.

4. Remove the blower housing retaining screws and the housing.

5. Remove the fan assembly from the blower housing.

To install:

6. Install the fan assembly into the blower housing and install the retaining screws. Connect the electrical connector and the ventilator outlet.

7. Install the inner glove box and the glove box door.

8. Connect the negative battery cable and check the entire climate control system for proper operation.

Windshield Wiper Motor

REMOVAL & INSTALLATION

Premier and Monaco

1. Disconnect the negative battery cable.

2. Remove the wiper arms. Remove the screws retaining the left and right cowl screens and remove both screens.

3. Remove the motor and linkage mounting screws.

4. Disconnect the electrical connector and remove the motor and linkage as an assembly.

To install:

5. Install the motor/linkage assembly and connect the electrical leads.

6. Install the cowl screens and the wiper arms.

7. Connect the negative battery cable and check the wipers for proper operation.

Medallion

1. Disconnect the negative battery cable. Remove the wiper arms.

2. Remove the screws retaining the cowl in front of the windshield and remove the cowl.

3. Disconnect the electrical plug at the wiper motor.

4. Remove the screws retaining the

wiper motor and wiper transmission and remove the assembly.

To install:

5. Install the wiper and transmission assembly. Connect the electrical plug to the wiper motor.

6. Install the cowl and the wiper arms.

7. Connect the negative battery cable and check the wipers for proper operation.

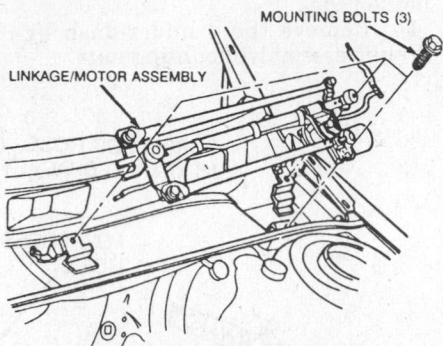

MOUNTING BOLTS (3)

LINKAGE/MOTOR ASSEMBLY

Wiper transmission and motor assembly—Monaco and Premier

Instrument Cluster

REMOVAL & INSTALLATION

Premier and Monaco

1. Disconnect the negative battery cable.

2. Remove the screws retaining the instrument cluster bezel and remove the bezel.

3. Remove the cluster retaining screws and tilt the cluster forward. Disconnect the electrical connectors.

4. If necessary, disconnect the speedometer cable.

5. Remove the lower instrument panel cover and remove the cluster.

6. Install the cluster and lower trim cover. Connect the electrical leads to the cluster.

7. Install the instrument panel bezel. Connect the negative battery cable.

Medallion

1. Disconnect the negative battery cable.

2. Remove the instrument glare shield retaining screws. Press the holding tabs in and remove the glare shield.

3. Open the fuse panel access door, reach through the fuse panel door and remove the speedometer cable from the rear of the instrument cluster.

4. Remove the instrument cluster mounting screws and pull the cluster forward. Disconnect the electrical wir-

ing and remove the cluster from the vehicle.

5. Install the cluster and connect the electrical wiring. Install the glare shield. Connect the speedometer cable and connect the negative battery cable.

Speedometer and Tachometer

REMOVAL & INSTALLATION

1. Disconnect the negative battery cable.
2. Remove the instrument cluster.

NOTE: Wear clean gloves when handling the cluster dial and gauge assembly. Finger prints and nails will mar the surface.

3. Remove the clear lens and black mask. Remove the mounting screws and remove the dial and gauge assembly from the cluster housing.
4. Grasp the pointer hub and slowly rotate the pointer assembly back and forth until the pointer contacts the trip reset shaft, while gently pulling upward of the hub away from the dial surface. Repeat this procedure as many times as it takes to remove the pointer from the shaft.
5. Remove the retaining screws and remove the speedometer from the faceplate.
To install:
6. Position the speedometer and secure with the screws.
7. Grasp the pointer by hand and gently place the bushing onto the movement shaft. The pointer tip should be indicating to approximately 90 mph.
8. Rotate the pointer assembly counterclockwise while gently pusing down on the hub toward the dial surface—a slight resistance should be felt. Clearance between the underside of the hub and the dial surface should be 0.020–0.060 in. (0.5–1.5mm) before the pointer tip is aligned with the zero horizontal graduation.
9. If the pointer is not properly aligned to the horizontal graduation, perform either of the following:
    a. Pointer too **high**—continue rotating the assembly counterclockwise until alignment is achieved.
    b. Pointer too **low**—Rotate the pointer clockwise until rotational resistance is felt—this does not refer to contact with the trip reset shaft. Continue rotating in the direction of resistance to compensate for initial misalignment. Release the hub, allowing the pointer to rotate back to its rest position.
    c. If alignment is not achieved, repeat the above as required.

10. Place the gauge faceplate and secure with the nuts. Install the retaining screws.
11. Position the black mask and clear lens and secure with screws.
12. Install the instrument cluster.
13. Connect the negative battery cable and check the speedometer for proper operation.

Radio

REMOVAL & INSTALLATION

1. Disconnect the battery negative cable.
2. Remove the instrument cluster bezel by removing the screws.
3. Remove the radio mounting screws.
4. Disconnect the electrical connector, the ground wire and unplug the antenna.
5. The installation is the reverse of the removal procedure.
6. Connect the negative battery cable and check all radio functions for proper operation.

Combination Switch

REMOVAL & INSTALLATION

Premier and Monaco

The windshield wiper, turn signal, headlight and dimmer switches are all combined in the combination switch on the left side of the steering column.

1. Disconnect the negative battery cable.
2. If not equipped with passive restraint, remove the lower instrument panel cover. If equipped with passive restraint, perform the following:
    a. Pull the ashtray from the receptacle, remove the receptacle and unplug the lighter.
    b. Remove the 2 screws fastening the console to the front bracket.
    c. Remove the armrest and the 2 screws fastening the console to the rear bracket.
    d. Reach inside the console and push out the seatbelt guides. Remove the console.
    e. Remove the bolts fastening the pivot bracket to the knee bolster. Loosen, but do not remove, the 2 bolts fastening the pivot bracket to the front console bracket.
    f. Remove the screw and the 2 Torx® screws that attach the bracket to the floor and slide the bracket back.
    g. Remove the screw located at the top of the knee bolster to the left of the steering column.

    h. Remove the screw attaching the air duct to the knee bolster.
    i. Remove the screw located at the bottom of the instrument panel holding the garnish penal.
    j. Remove both bolster end caps and the revealed nuts.
    k. Move the knee bolster toward the rear of the vehicle enough to gain access to the 2 screws holding the parking brake handle, then lower the handle.
    l. Remove the knee bolster.
3. Remove the screws, remove the support rod and pull the air duct aside.
4. Cut the plastic tie-wrap straps.
5. Loosen the hold-down nut in the center of the steering column electrical connector and separate the connector.
6. Separate the left side pod switch connector from the steering column connector by placing a flat blade tool between the connectors to disengage the locking tab. Push on the wire side of the left side pod switch connector and slide the connector out of the channels of the steering column connector.
7. Disconnect the electrical connector, then remove the bottom 2 screws (not the rivets) from the pod assembly.

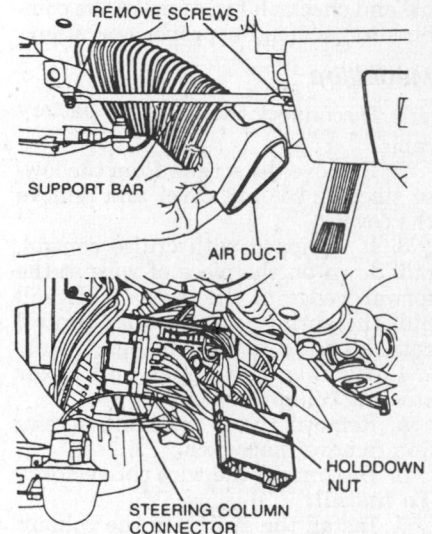

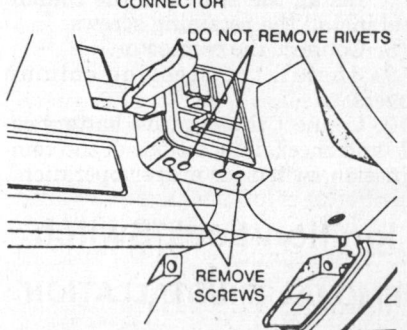

To remove combination switch, remove duct, electrical connection and pod screws

8. To gain access to the inside of the pod, remove the screws from the back of the left side switch pod assembly and remove the switch pod housing back cover.

NOTE: There are small retaining clips on the left side pod that may fall off when the switch is removed.

9. Carefully pull the switch pod far enough from the housing to expose the 2 screws, remove them and gently pull the switch forward and pull the harness out through the housing to remove the switch.

To install:

10. Route the switch assembly connector through the housing and along the underside of the steering column.

11. Connect the switch connector.

12. Connect the steering column connector and install the hold-down nut. Secure with a new tie.

13. Position the switch and secure with the screws.

14. Connect the air duct.

15. Install the lower support bar.

16. Install the knee bolster and console by reversing their removal procedure, if equipped, or install the lower instrument cover.

17. Connect the negative battery cable and check all functions of the combination switch for proper operation.

Medallion

1. Disconnect the negative battery cable.

2. Remove the screws from the lower steering column cover and remove the cover.

3. If equipped with cruise control, pull down on the piece of wire at the forward edge of the cover. This will pull the spring loaded cruise control commutator into its housing.

4. Remove the upper and lower steering column covers.

5. Remove the 2 retaining screws and remove the switch.

6. Disconnect the wire connectors.

To install:

7. Install the switch to the column and install the retaining screws.

8. Connect the connector.

9. Install the steering column covers.

10. Connect the negative battery cable and check all functions of the combination switch for proper operation.

Ignition Lock/Switch

REMOVAL & INSTALLATION

Medallion

1. Disconnect battery negative cable.

2. Remove 4 screws from lower steering column cover.

3. On cruise control equipped vehicles, pull down on the wire at the forward edge of the lower cover. This allow the spring loaded commutator brush to be pull into its housing.

NOTE: If the lower steering cover is removed before the commutator brush is pulled into its housing, the brush will be broken off the cover.

4. Remove the upper steering column cover and ignition switch cover. Remove the ignition switch mounting screw.

5. Remove the gray and black wire connectors and ignition switch mounting screw from beneath the key cylinder housing.

6. Insert the key and turn the key to the unmarked arrow on cylinder lock. Push in the locking tabs on the side of the housing with a punch and remove the switch.

7. Separate the switch from the wires by removing the screw retaining the connector. Feed the wiring harness through the lock cylinder hole.

8. Separate the tumbler by removing the 2 attaching screws.

To install:

9. Assemble the tumbler and connect the wiring to the switch.

10. Guide the wire harness through the lock cylinder hole and slide the switch into the hole. Press both locking tabs inward and slide the switch into place until it locks.

11. Install the ignition switch mounting screw at the bottom of cylinder housing and reconnect electrical connectors.

12. Install the ignition switch cover and steering column covers.

13. Connect the negative battery cable and check all functions of the ignition switch for proper operation.

Ignition Lock

REMOVAL & INSTALLATION

Premier and Monaco

1. Disconnect the negative battery cable. Remove the steering wheel.

2. Remove the turn signal cancelling cam, unlock the tabs, and slide the canceller off the steering shaft.

3. If equipped with a tilt wheel, remove the tilt control lever.

4. Remove the screws retaining the right and left switch pods. Remove the ignition switch trim ring.

5. Remove the screws from the pod housing/column cover. Remove the pod housing/column cover by pulling it up, then guide the pods through the cover and remove the cover.

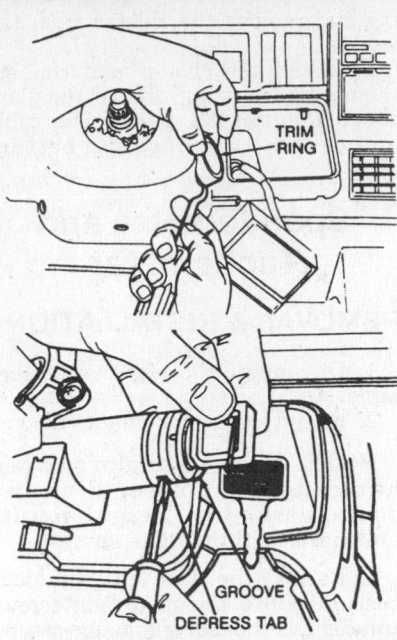

Ignition lock repair requires trim ring removal and pushing the lock tab

6. Insert the key into the ignition and align the key with groove in the lock cylinder housing. Push in the locking tab on the bottom of the housing, with a punch and remove the cylinder. Separate the switch from the wires by removing the screw retaining the connector.

To install:

7. Insert the key into the ignition and align key with groove in lock cylinder housing.

8. Depress the tab and install the lock cylinder.

9. Install the pod housing/column cover and install the pods.

10. Install the ignition switch trim caring.

NOTE: The retaining clips on the left and right switch pods must be in place when the switches are installed.

11. Install the tilt lever, if equipped. Install the turn signal cam with pin bore in the steering wheel. Reconnect the electrical connector.

12. Install the steering wheel, aligning the pin on the turn signal cam with the pin bore on the steering wheel.

13. Connect the negative battery cable and check for proper operation.

Ignition Switch

REMOVAL & INSTALLATION

Premier and Monaco

1. Disconnect battery negative cable.

2. Remove instrument panel lower cover attaching screws and remove cover.

3. Remove the horn pad. Disconnect the wires and remove the horn button.

4. Remove the steering wheel and the turn signal cancel cam.

5. If equipped with tilt wheel, use a small wrench to unscrew the tilt lever.

6. Remove the screws retaining the right and left switch pods. Remove the ignition switch trim ring.

7. Remove the screws from the pod housing/column cover. Remove the pod housing/column cover by pulling it up, guide the pods through the cover and remove the cover.

8. Remove the lower column shroud attaching screws and remove lower shroud.

9. Remove the upper column shroud attaching screws and remove upper shroud.

10. Remove the ignition switch retaining screws and separate the switch from cylinder housing.

11. Cut the tie straps and remove the harness anchor.

12. Loosen the retaining nut in the center of the steering column connector and separate the switch pod connector by disengaging locking tabs.

13. Remove the electrical harness from the channels of steering column connector and remove the ignition switch assembly.

To install:

14. Slide the switch pod connectors into the ignition switch connector and install the nut.

15. Install the ignition switch and secure with its retaining screws.

16. Route the wiring harness along the underside of steering column and secure with new tie straps. Install the harness anchor and secure with its retaining screw.

17. Install the pod housing/column cover and install the pods.

18. Install the ignition switch trim ring.

NOTE: The retaining clips on the left and right switch pods must be in place when the switches are installed.

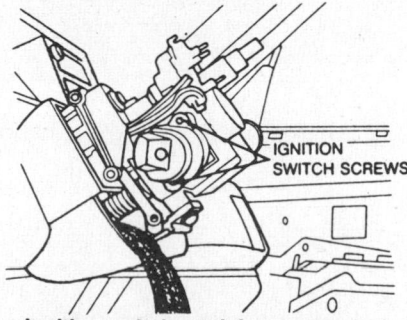

Ignition switch retaining screws— Monaco and Premier

19. Install the tilt lever, if equipped. Install the turn signal cam with pin bore in the steering wheel. Reconnect the electrical connector.

20. Install the steering wheel, aligning the pin on the turn signal cam with the pin bore on the steering wheel.

21. Connect the negative battery cable and check all functions of the ignition switch for proper operation.

Stoplight Switch

Two different switches are used. Without cruise control, the switch is attached to the brake pedal by the pushrod bolt. If equipped with cruise control, the switch is attached to a bracket on the brake support. Neither switch requires adjustment.

REMOVAL & INSTALLATION

1. Disconnect the negative battery cable.

2. Remove the bolt retaining the master cylinder pushrod to the brake pedal.

3. Disconnect the electrical wires from the stoplight switch and remove the switch.

To install:

4. Install the switch and the master cylinder pushrod to the brake pedal and install the retaining bolt.

5. Connect the electrical wires to the switch.

6. Connect the negative battery cable and check the operation of the brake lights.

Neutral Safety Switch

REMOVAL & INSTALLATION

Premier and Monaco

1. Disconnect battery negative cable.

2. Disconnect the neutral switch harness connector located in the engine compartment.

3. Raise and support the vehicle safely. Remove the splash shield.

4. Remove the bolt attaching the switch bracket to transaxle case and remove switch from case. Replace the O-ring.

5. The installation is the reverse of the removal procedure.

6. Connect the negative battery cable and check the switch for proper operation.

Medallion

A multi-function switch located on the transaxle assembly allows the vehicle to start only in **N** and **P** positions.

1. Disconnect the negative battery cable.

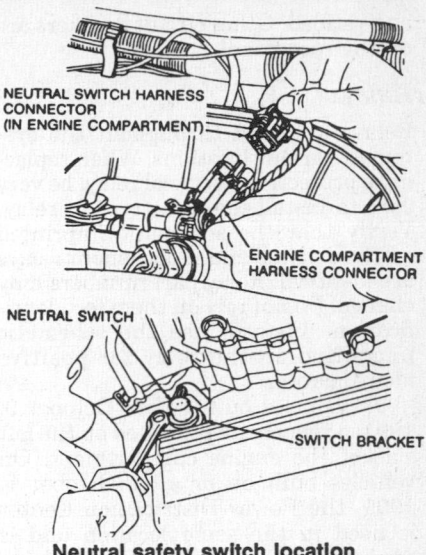

Neutral safety switch location

2. Remove the electrical connection from switch.

3. Remove the switch mounting screws and remove the switch.

4. The installation is the reverse of the removal procedure.

5. Connect the negative battery cable and check the switch for proper operation.

Fuses, Circuit Breakers and Relays

LOCATION

Fuses

The fuse panel is located above the parking brake release lever, under the instrument panel. On Monaco and Premier built on or after October 9, 1991, various cartridge fuses are in the Power Distribution Center, located on the left side of the engine compartment,

Fusible Links

Fusible links are used to prevent major wire harness damage in the event of a short circuit or an overload condition in the wiring circuits which are normally not fused, due to carrying high amperage loads or because of their locations within the wiring harness. Each fusible link is of a fixed value for a specific electrical load and should a link fail, the cause of the failure must be determined and repaired prior to installing a new fusible link of the same value.

Circuit Breakers

Circuit breakers are an integral part of the headlight switch, the wiper switch and the air conditioning circuit. They are used to protect each circuit from an overload. Other circuit breakers are on the fuse panel.

Relays

Relays are used throughout the system in various locations. When replacing a protective electrical relay, be very sure to install the same type of relay. Verify that the schematic imprinted on the original and replacement relays are identical. Relay part numbers may change. Do not rely on them for identification. Instead, use the schematic imprinted on the relay for positive identification.

On vehicles built before October 9, 1991, a relay bank is located on the left side of the engine compartment. On vehicles built on or after October 9, 1991, the Power Distribution Center is used in the same location and is equipped with additional fuses. Additional relay locations are as follows:

Power door lock relay — is located on the right side kick panel.

Passive restraint relays — are located under the seats.

Light outage module — is located behind the right side speaker in the trunk.

Passive restraint control module — is located on the left side of the trunk.

Headlight module — is located under the left side of the instrument panel.

Daytime running light module — is located in the right front area of the engine compartment.

Climate control relays and module — are located under the right side of the instrument panel.

Sun roof relay — is near the sun roof motor.

Computers

LOCATION

Engine Control Unit (except vehicles built after October 9, 1991) — is located under the right side of the instrument panel.

Single Board Engine Controller (vehicles built on or after October 9, 1991) — is located in the left front area of the engine compartment.

Transmission Controller — is located in the right front area of the engine compartment.

Anti Lock Brakes Controller — is located on the front of the right front strut tower in the engine compartment.

Flashers

LOCATION

The turn signal flasher is located behind the left side of the instrument panel. The hazard flasher is located behind the left side of the instrument panel.

Chrysler Corp.
Front Wheel Drive

"A" Body—Spirit, Acclaim, LeBaron Landau "E" Body—600, Caravelle, New Yorker "G" Body—Daytona "H" Body—Lancer, LeBaron GTS "J" Body—LeBaron "K" Body—Aries, Reliant, LeBaron "P" Body—Shadow, Sundance **TC** by Maserati

SPECIFICATIONS

VEHICLE IDENTIFICATION CHART
Except TC Maserati

It is important for servicing and ordering parts to be certain of the vehicle and engine identification. The VIN (vehicle identification number) is a 17 digit number visible through the windshield on the driver's side of the dash and contains the vehicle and engine identification codes. The tenth digit indicates model year and the eighth digit indicates engine code. It can be interpreted as follows:

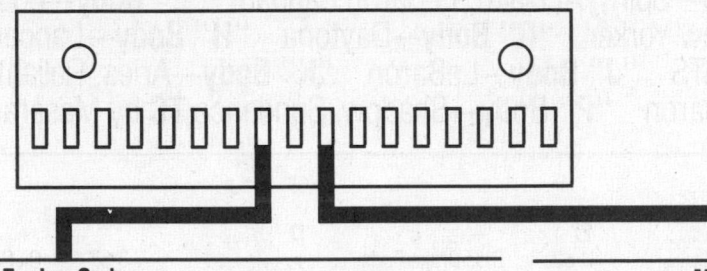

Engine Code						Model Year	
Code	Cu. In.	Liters	Cyl.	Fuel Sys.	Eng. Mfg.	Code	Year
A (1988–89)	135	2.2	4	Turbo II	Chrysler	J	1988
A (1991–92)	135	2.2	4	Turbo III	Chrysler	K	1989
C	135	2.2	4	Turbo IV	Chrysler	L	1990
D	135	2.2	4	EFI	Chrysler	M	1991
E	135	2.2	4	Turbo	Chrysler	N	1992
J	153	2.5	4	Turbo I	Chrysler		
K	153	2.5	4	EFI	Chrysler		
3	181	3.0	6	EFI	Mitsubishi		

EFI—Electronic Fuel Injection

VEHICLE IDENTIFICATION CHART
TC Maserati

It is important for servicing and ordering parts to be certain of the vehicle and engine identification. The VIN (vehicle identification number) is a 17 digit number visible through the windshield on the driver's side of the dash and contains the vehicle and engine identification codes. The tenth digit indicates model year and the fifth digit indicates engine code. It can be interpreted as follows:

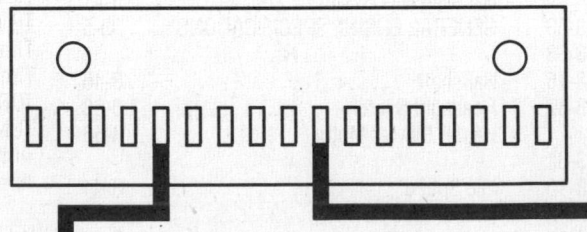

Engine Code						Model Year	
Code	Cu. In.	Liters	Cyl.	Fuel Sys.	Eng. Mfg.	Code	Year
A	135	2.2	4	Turbo	Chrysler	K	1989
R	135	2.2	4	Turbo	Chrysler ①	L	1990
S	181	3.0	6	EFI	Mitsubishi	M	1991
						N	1992

EFI—Electronic Fuel Injection
① Cylinder head and related parts by Maserati

ENGINE IDENTIFICATION

Year	Model	Engine Displacement cu. in. (liter)	Engine Series Identification (VIN)	No. of Cylinders	Engine Type
1988	Aries	135 (2.2)	D	4	OHV
	Aries	153 (2.5)	K	4	OHC
	Reliant	135 (2.2)	D	4	OHC
	Reliant	153 (2.5)	K	4	OHC
	Daytona	135 (2.2)	E	4	OHC
	Daytona	153 (2.5)	K	4	OHC
	600	135 (2.2)	D	4	OHC
	600	135 (2.2)	E	4	OHC
	600	153 (2.5)	K	4	OHC
	Caravelle	135 (2.2)	D	4	OHC
	Caravelle	135 (2.2)	E	4	OHC
	Caravelle	153 (2.5)	K	4	OHC
	LeBaron (P body)	135 (2.2)	D	4	OHC
	LeBaron (P body)	135 (2.2)	E	4	OHC
	LeBaron (P body)	153 (2.5)	K	4	OHC
	LeBaron (J body)	135 (2.2)	E	4	OHC
	LeBaron (J body)	153 (2.5)	K	4	OHC
	Lancer	135 (2.2)	E	4	OHC
	Lancer	153 (2.5)	K	4	OHC
	LeBaron GTS	135 (2.2)	E	4	OHC
	LeBaron GTS	153 (2.5)	K	4	OHC
	New Yorker	135 (2.2)	E	4	OHC
	Shadow	135 (2.2)	D	4	OHC
	Shadow	135 (2.2)	E	4	OHC
	Shadow	153 (2.5)	K	4	OHC
	Sundance	135 (2.2)	D	4	OHC
	Sundance	135 (2.2)	E	4	OHC
	Sundance	153 (2.5)	K	4	OHC
1989	Aries	135 (2.2)	D	4	OHC
	Aries	153 (2.5)	K	4	OHC
	Reliant	135 (2.2)	D	4	OHC
	Reliant	153 (2.5)	K	4	OHC
	Daytona	135 (2.2)	A	4	OHC
	Daytona	153 (2.5)	J	4	OHC
	Daytona	153 (2.5)	K	4	OHC
	LeBaron (J body)	135 (2.2)	A	4	OHC
	LeBaron (J body)	153 (2.5)	J	4	OHC
	LeBaron (J body)	153 (2.5)	K	4	OHC
	Lancer	135 (2.2)	D	4	OHC
	Lancer	135 (2.2)	A	4	OHC
	Lancer	153 (2.5)	J	4	OHC
	Lancer	153 (2.5)	K	4	OHC
	LeBaron GTS	135 (2.2)	D	4	OHC
	LeBaron GTS	135 (2.2)	A	4	OHC

ENGINE IDENTIFICATION

Year	Model	Engine Displacement cu. in. (liter)	Engine Series Identification (VIN)	No. of Cylinders	Engine Type
1989	LeBaron GTS	153 (2.5)	J	4	OHC
	LeBaron GTS	153 (2.5)	K	4	OHC
	Shadow	135 (2.2)	D	4	OHC
	Shadow	153 (2.5)	J	4	OHC
	Shadow	153 (2.5)	K	4	OHC
	Shadow	135 (2.2)	A	4	OHC
	Sundance	135 (2.2)	D	4	OHC
	Sundance	153 (2.5)	J	4	OHC
	Sundance	153 (2.5)	K	4	OHC
	Spirit	153 (2.5)	J	4	OHC
	Spirit	153 (2.5)	K	4	OHC
	Spirit	181 (3.0)	3	6	OHC
	Acclaim	153 (2.5)	J	4	OHC
	Acclaim	153 (2.5)	K	4	OHC
	Acclaim	181 (3.0)	3	6	OHC
	TC	135 (2.2)	A	4	OHC
	TC	135 (2.2)	R	4	DOHC
1990	Daytona	153 (2.5)	J	4	OHC
	Daytona	135 (2.2)	C	4	OHC
	Daytona	181 (3.0)	3	6	OHC
	LeBaron	153 (2.5)	K	4	OHC
	LeBaron	153 (2.5)	J	4	OHC
	LeBaron	135 (2.2)	C	4	OHC
	LeBaron	181 (3.0)	3	6	OHC
	LeBaron Landau	181 (3.0)	3	6	OHC
	Shadow	135 (2.2)	D	4	OHC
	Shadow	135 (2.2)	C	4	OHC
	Shadow	153 (2.5)	K	4	OHC
	Shadow	153 (2.5)	J	4	OHC
	Sundance	135 (2.2)	D	4	OHC
	Sundance	153 (2.5)	K	4	OHC
	Sundance	153 (2.5)	J	4	OHC
	Spirit	153 (2.5)	K	4	OHC
	Spirit	153 (2.5)	J	4	OHC
	Spirit	181 (3.0)	3	6	OHC
	Acclaim	153 (2.5)	K	4	OHC
	Acclaim	153 (2.5)	J	4	OHC
	Acclaim	181 (3.0)	3	6	OHC
	TC	181 (3.0)	S	6	OHC
	TC	135 (2.2)	R	4	DOHC
1991–92	Daytona	153 (2.5)	J	4	OHC
	Daytona	153 (2.5)	K	4	OHC
	Daytona	181 (3.0)	3	6	OHC
	LeBaron	153 (2.5)	K	4	OHC

ENGINE IDENTIFICATION

Year	Model	Engine Displacement cu. in. (liter)	Engine Series Identification (VIN)	No. of Cylinders	Engine Type
1991–92	LeBaron	153 (2.5)	J	4	OHC
	LeBaron	181 (3.0)	3	6	OHC
	LeBaron Landau	181 (3.0)	3	6	OHC
	Shadow	135 (2.2)	D	4	OHC
	Shadow	153 (2.5)	K	4	OHC
	Shadow	153 (2.5)	J	4	OHC
	Sundance	135 (2.2)	D	4	OHC
	Sundance	153 (2.5)	K	4	OHC
	Sundance	153 (2.5)	J	4	OHC
	Spirit	153 (2.5)	K	4	OHC
	Spirit	153 (2.5)	J	4	OHC
	Spirit	181 (3.0)	3	6	OHC
	Spirit R/T	135 (2.2)	A	4	DOHC
	Acclaim	153 (2.5)	K	4	OHC
	Acclaim	181 (3.0)	3	6	OHC
	TC	181 (3.0)	S	6	OHC

OHC—Overhead Cam
DOHC—Dual Overhead Cam
OHV—Overhead Valves

GENERAL ENGINE SPECIFICATIONS

Year	VIN	No. Cylinder Displacement cu. in. (liter)	Fuel System Type	Net Horsepower @ rpm	Net Torque @ rpm (ft. lbs.)	Bore × Stroke (in.)	Compression Ratio	Oil Pressure @ rpm
1988	D	4-135 (2.2)	EFI	99 @ 5600	121 @ 3200	3.44 × 3.62	9.5:1	30–80 @ 3000
	E	4-135 (2.2)	Turbo	146 @ 5200	170 @ 3600	3.44 × 3.62	8.0:1	30–80 @ 3000
	K	4-153 (2.5)	EFI	100 @ 4800	133 @ 2800	3.44 × 4.09	9.0:1	30–80 @ 3000
1989	A	4-135 (2.2)	Turbo	174 @ 5200	170 @ 3600	3.44 × 3.62	8.1:1	30–80 @ 3000
	D	4-135 (2.2)	EFI	99 @ 5600	121 @ 3200	3.44 × 3.62	9.5:1	30–80 @ 3000
	K	4-153 (2.5)	EFI	100 @ 4800	135 @ 2800	3.44 × 4.09	8.9:1	30–80 @ 3000
	J	4-153 (2.5)	Turbo	150 @ 4800	180 @ 2000	3.44 × 4.09	7.8:1	30–80 @ 3000
	3	6-181 (3.0)	EFI	141 @ 5000	171 @ 2000	3.59 × 2.99	8.6:1	30–80 @ 3000
	R	4-135 (2.2)	Turbo	200 @ 5500	220 @ 3400	3.44 × 3.62	7.4:1	30–80 @ 3000
1990	C	4-135 (2.2)	Turbo	174 @ 5200	210 @ 2400	3.44 × 3.62	8.0:1	30–80 @ 3000
	D	4-135 (2.2)	EFI	99 @ 4800	122 @ 3200	3.44 × 3.62	9.5:1	30–80 @ 3000
	K	4-153 (2.5)	EFI	100 @ 4800	135 @ 2800	3.44 × 4.09	8.9:1	30–80 @ 3000
	J	4-153 (2.5)	Turbo	150 @ 4800	180 @ 2000	3.44 × 4.09	7.8:1	30–80 @ 3000
	3	6-181 (3.0)	EFI	141 @ 5000	171 @ 2800	3.59 × 2.99	8.9:1	30–80 @ 3000
	S	6-181 (3.0)	EFI	141 @ 5000	170 @ 2800	3.59 × 2.99	8.9:1	30–80 @ 3000

GENERAL ENGINE SPECIFICATIONS

Year	VIN	No. Cylinder Displacement cu. in. (liter)	Fuel System Type	Net Horsepower @ rpm	Net Torque @ rpm (ft. lbs.)	Bore × Stroke (in.)	Compression Ratio	Oil Pressure @ rpm
1991–92	D	4-135 (2.2)	EFI	99 @ 4800	122 @ 3200	3.44 × 3.62	9.5:1	30–80 @ 3000
	K	4-153 (2.5)	EFI	100 @ 4800	135 @ 2800	3.44 × 4.09	8.9:1	30–80 @ 3000
	J	4-153 (2.5)	Turbo	150 @ 4800	180 @ 2000	3.44 × 4.09	7.8:1	30–80 @ 3000
	3	6-181 (3.0)	EFI	141 @ 5000	171 @ 2800	3.59 × 2.99	8.9:1	30–80 @ 3000
	S	6-181 (3.0)	EFI	141 @ 5000	170 @ 2800	3.59 × 2.99	8.9:1	30–80 @ 3000
	A	4-135 (2.2)	Turbo	224 @ 2800	217 @ 6000	3.44 × 3.62	8.5:1	30–80 @ 3000

EFI—Electronic Fuel Injection

ENGINE TUNE-UP SPECIFICATIONS

Year	VIN	No. Cylinder Displacement cu. in. (liter)	Spark Plugs Type	Spark Plugs Gap (in.)	Ignition Timing (deg.) MT	Ignition Timing (deg.) AT	Compression Pressure (psi)	Fuel Pump (psi)	Idle Speed (rpm) MT	Idle Speed (rpm) AT	Valve Clearance In.	Valve Clearance Ex.
1988	D	4-135 (2.2)	RN12YC	0.035	12B	12B	100①	15	850	850	Hyd.	Hyd.
	E	4-135 (2.2)	RN12YC	0.035	12B	12B	100①	55	900	900	Hyd.	Hyd.
	K	4-153 (2.5)	RN12YC	0.035	12B	12B	100①	15	850	850	Hyd.	Hyd.
1989	A	4-135 (2.2)	RN12YC	0.035	12B	—	100①	55	900	900	Hyd.	Hyd.
	D	4-135 (2.2)	RN12YC	0.035	12B	12B	100①	15	850	850	Hyd.	Hyd.
	J	4-153 (2.5)	RN12YC	0.035	12B	12B	100①	55	900	720	Hyd.	Hyd.
	K	4-153 (2.5)	RN12YC	0.035	12B	12B	100①	15	850	850	Hyd.	Hyd.
	3	6-181 (3.0)	RN11YC4	0.040	—	12B	178②	48	—	700	Hyd.	Hyd.
	R	4-135 (2.2)	DCPR-7E	0.030	12B	—	128	55	900	—	0.012	0.016
1990	C	4-135 (2.2)	RN12YC	0.035	12B	—	100①	55	900	—	Hyd.	Hyd.
	D	4-135 (2.2)	RN12YC	0.035	12B	12B	100①	15	850	850	Hyd.	Hyd.
	K	4-153 (2.5)	RN12YC	0.035	12B	12B	100①	15	850	850	Hyd.	Hyd.
	J	4-153 (2.5)	RN12YC	0.035	12B	12B	100①	55	900	850	Hyd.	Hyd.
	3	6-181 (3.0)	RN11YC4	0.040	—	12B	178②	48	—	700	Hyd.	Hyd.
	S	6-181 (3.0)	RN11YC4	0.040	—	12B	178②	48	—	700	Hyd.	Hyd.
	R	4-135 (2.2)	DCPR-7E	0.030	12B	—	128	55	900	—	0.012	0.016
1991	D	4-135 (2.2)	RN12YC	0.035	12B	12B	100①	39	850	850	Hyd.	Hyd.
	K	4-153 (2.5)	RN12YC	0.035	12B	12B	100①	39③	850	850	Hyd.	Hyd.
	J	4-153 (2.5)	RN12YC	0.035	12B	12B	100①	55	900	850	Hyd.	Hyd.
	3	6-181 (3.0)	RN11YC4	0.040	—	12B	178②	48	—	700	Hyd.	Hyd.
	S	6-181 (3.0)	RN11YC4	0.040	—	12B	178②	48	—	700	Hyd.	Hyd.
	A	4-135 (2.2)	RN9YC	0.035	NA	—	100①	55	850	—	Hyd.	Hyd.
1992		SEE UNDERHOOD SPECIFICATIONS STICKER										

① Minimum
② At 250 rpm
③ Early 1991 Shadow Convertible: 14.5 psi
NA—Not adjustable
Hyd.—Hydraulic

FIRING ORDERS

NOTE: To avoid confusion, always replace spark plug wires one at a time.

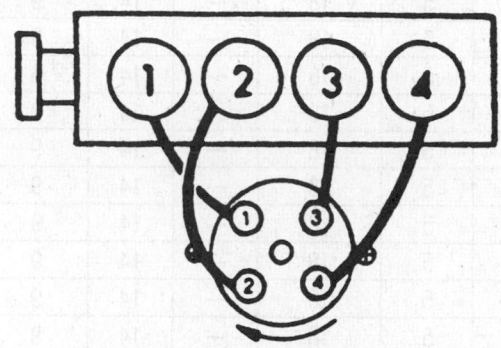

2.2L and 2.5L Engines (Except Turbo III)
Engine Firing Order: 1–3–4–2
Distributor Rotation: Clockwise

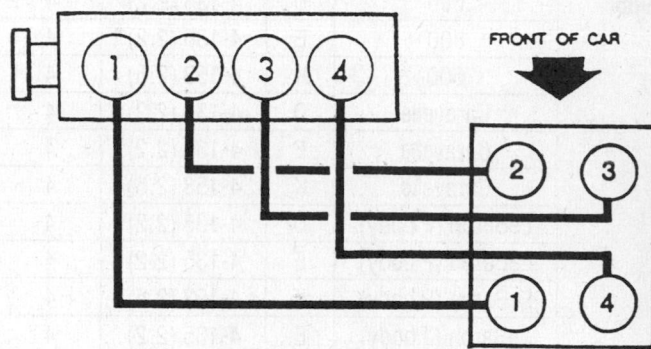

2.2L Turbo III Engine
Engine Firing Order: 1–3–4–2
Distributorless Ignition System

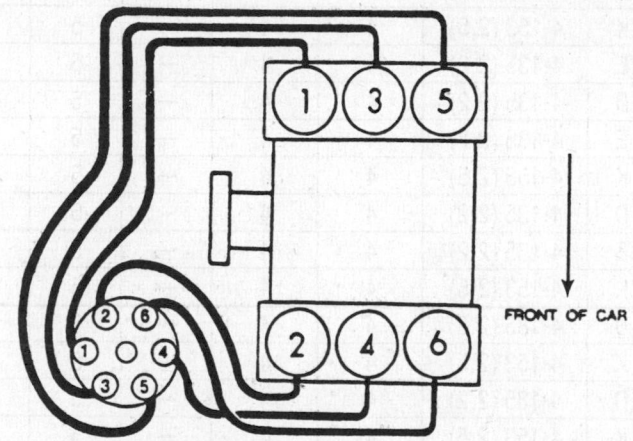

3.0L Engine
Engine Firing Order: 1–2–3–4–5–6
Distributor Rotation: Counterclockwise

CAPACITIES

Year	Model	VIN	No. Cylinder Displacement cu. in. (liter)	Engine Crankcase (qts.) with Filter	Engine Crankcase (qts.) without Filter	Transmission (pts.) 4-Spd	Transmission (pts.) 5-Spd	Transmission (pts.) Auto. ①	Drive Axle (pts.)	Fuel Tank (gal.)	Cooling System (qts.)
1988	Aries	D	4-135 (2.2)	4	4	—	5	18	—	14	9
	Aries	K	4-153 (2.5)	4	4	—	5	18	—	14	9
	Reliant	D	4-135 (2.2)	4	4	—	5	18	—	14	9
	Reliant	K	4-153 (2.5)	4	4	—	5	18	—	14	9
	Daytona	E	4-135 (2.2)	4	4	—	5	18	—	14	9
	Daytona	K	4-153 (2.5)	4	4	—	5	18	—	14	9

CAPACITIES

Year	Model	VIN	No. Cylinder Displacement cu. in. (liter)	Engine Crankcase (qts.) with Filter	without Filter	Transmission (pts.) 4-Spd	5-Spd	Auto. ①	Drive Axle (pts.)	Fuel Tank (gal.)	Cooling System (qts.)
1988	600	D	4-135 (2.2)	4	4	—	5	18	—	14	9
	600	E	4-135 (2.2)	4	4	—	5	18	—	14	9
	600	K	4-153 (2.5)	4	4	—	5	18	—	14	9
	Caravelle	D	4-135 (2.2)	4	4	—	5	18	—	14	9
	Caravelle	E	4-135 (2.2)	4	4	—	5	18	—	14	9
	Caravelle	K	4-153 (2.5)	4	4	—	5	18	—	14	9
	LeBaron (P body)	D	4-135 (2.2)	4	4	—	5	18	—	14	9
	LeBaron (P body)	E	4-135 (2.2)	4	4	—	5	18	—	14	9
	LeBaron (P body)	K	4-153 (2.5)	4	4	—	5	18	—	14	9
	LeBaron (J body)	E	4-135 (2.2)	4	4	—	5	18	—	14	9
	LeBaron (J body)	K	4-153 (2.5)	4	4	—	5	18	—	14	9
	Lancer	E	4-135 (2.2)	4	4	—	5	18	—	14	9
	Lancer	K	4-153 (2.5)	4	4	—	5	18	—	14	9
	LeBaron GTS	E	4-135 (2.2)	4	4	—	5	18	—	14	9
	LeBaron GTS	K	4-153 (2.5)	4	4	—	5	18	—	14	9
	New Yorker	E	4-135 (2.2)	4	4	—	5	18	—	14	9
	Shadow	D	4-135 (2.2)	4	4	—	5	18	—	14	9
	Shadow	E	4-135 (2.2)	4	4	—	5	18	—	14	9
	Shadow	K	4-153 (2.5)	4	4	—	5	18	—	14	9
	Sundance	D	4-135 (2.2)	4	4	—	5	18	—	14	9
	Sundance	E	4-135 (2.2)	4	4	—	5	18	—	14	9
	Sundance	K	4-153 (2.5)	4	4	—	5	18	—	14	9
1989	Aries	D	4-135 (2.2)	4	4	—	5	18	—	14	9
	Aries	K	4-153 (2.5)	4	4	—	5	18	—	14	9
	Reliant	D	4-135 (2.2)	4	4	—	5	18	—	14	9
	Reliant	K	4-153 (2.5)	4	4	—	5	18	—	14	9
	Daytona	A	4-135 (2.2)	4	4	—	5	—	—	14	9
	Daytona	J	4-153 (2.5)	4	4	—	5	18	—	14	9
	Daytona	K	4-153 (2.5)	4	4	—	5	18	—	14	9
	LeBaron (J body)	A	4-135 (2.2)	4	4	—	5	—	—	14	9
	LeBaron (J body)	J	4-153 (2.5)	4	4	—	5	18	—	14	9
	LeBaron (J body)	K	4-153 (2.5)	4	4	—	5	18	—	14	9
	Lancer	D	4-135 (2.2)	4	4	—	5	18	—	14	9
	Lancer	A	4-135 (2.2)	4	4	—	5	—	—	14	9
	Lancer	J	4-153 (2.5)	4	4	—	5	18	—	14	9
	Lancer	K	4-153 (2.5)	4	4	—	5	18	—	14	9
	LeBaron GTS	D	4-135 (2.2)	4	4	—	5	18	—	14	9
	LeBaron GTS	A	4-135 (2.2)	4	4	—	5	—	—	14	9
	LeBaron GTS	J	4-153 (2.5)	4	4	—	5	18	—	14	9
	LeBaron GTS	K	4-153 (2.5)	4	4	—	5	18	—	14	9
	Shadow	D	4-135 (2.2)	4	4	—	5	18	—	14	9
	Shadow	J	4-153 (2.5)	4	4	—	5	18	—	14	9
	Shadow	K	4-153 (2.5)	4	4	—	5	18	—	14	9

CAPACITIES

Year	Model	VIN	No. Cylinder Displacement cu. in. (liter)	Engine Crankcase (qts.) with Filter	Engine Crankcase (qts.) without Filter	Transmission (pts.) 4-Spd	Transmission (pts.) 5-Spd	Transmission (pts.) Auto. ①	Drive Axle (pts.)	Fuel Tank (gal.)	Cooling System (qts.)
1989	Shadow	A	4-135 (2.2)	4	4	—	5	—	—	14	9
	Sundance	D	4-135 (2.2)	4	4	—	5	18	—	14	9
	Sundance	J	4-153 (2.5)	4	4	—	5	18	—	14	9
	Sundance	K	4-153 (2.5)	4	4	—	5	18	—	14	9
	Spirit	J	4-153 (2.5)	4	4	—	5	18	—	14	9
	Spirit	K	4-153 (2.5)	4	4	—	5	18	—	14	9
	Spirit	3	6-181 (3.0)	4	4	—	—	18	—	14	9.5
	Acclaim	J	4-153 (2.5)	4	4	—	5	18	—	14	9
	Acclaim	K	4-153 (2.5)	4	4	—	5	18	—	14	9
	Acclaim	3	6-181 (3.0)	4	4	—	—	18	—	14	9.5
	TC	A	4-135 (2.2)	4	4	—	5	—	—	14	9
	TC	R	4-135 (2.2)	4	4	—	5	18	—	14	9
1990	Daytona	J	4-153 (2.5)	4	4	—	5	18	—	14	9
	Daytona	C	4-135 (2.2)	4	4	—	5	—	—	14	9
	Daytona	3	6-181 (3.0)	4	4	—	—	18	—	14	9.5
	LeBaron	K	4-153 (2.5)	4	4	—	5	18	—	14	9
	LeBaron	J	4-153 (2.5)	4	4	—	5	18	—	14	9
	LeBaron	C	4-135 (2.2)	4	4	—	5	—	—	14	9
	LeBaron	3	6-181 (3.0)	4	4	—	—	18	—	14	9.5
	LeBaron Landau	3	6-181 (3.0)	4	4	—	—	18	—	14	9.5
	Shadow	D	4-135 (2.2)	4	4	—	5	18	—	14	9
	Shadow	C	4-135 (2.2)	4	4	—	5	—	—	14	9
	Shadow	K	4-153 (2.5)	4	4	—	5	18	—	14	9
	Shadow	J	4-153 (2.5)	4	4	—	5	18	—	14	9
	Sundance	D	4-135 (2.2)	4	4	—	5	18	—	14	9
	Sundance	K	4-153 (2.5)	4	4	—	5	18	—	14	9
	Sundance	J	4-153 (2.5)	4	4	—	5	18	—	14	9
	Spirit	K	4-153 (2.5)	4	4	—	5	18	—	14	9
	Spirit	J	4-153 (2.5)	4	4	—	5	18	—	14	9
	Spirit	3	6-181 (3.0)	4	4	—	—	18	—	14	9.5
	Acclaim	K	4-153 (2.5)	4	4	—	5	18	—	14	9
	Acclaim	J	4-153 (2.5)	4	4	—	5	18	—	14	9
	Acclaim	3	6-181 (3.0)	4	4	—	—	18	—	14	9.5
	TC	S	6-181 (3.0)	4.5	4	—	—	18	—	14	9.5
	TC	R	4-135 (2.2)	4	4	—	5	18	—	14	9
1991–92	Daytona	J	4-153 (2.5)	4.5	4	—	5	18	—	14	9
	Daytona	K	4-135 (2.5)	4.5	4	—	5	—	—	14	9
	Daytona	3	6-181 (3.0)	4.5	4	—	—	18	—	14	9.5
	LeBaron	K	4-153 (2.5)	4.5	4	—	5	18	—	14	9
	LeBaron	J	4-153 (2.5)	4.5	4	—	5	18	—	14	9
	LeBaron	3	6-181 (3.0)	4.5	4	—	—	18	—	14	9.5
	LeBaron Landau	3	6-181 (3.0)	4.5	4	—	—	18	—	16	9.5
	Shadow	D	4-135 (2.2)	4.5	4	—	5	18	—	14	9

CAPACITIES

Year	Model	VIN	No. Cylinder Displacement cu. in. (liter)	Engine Crankcase (qts.) with Filter	Engine Crankcase (qts.) without Filter	Transmission (pts.) 4-Spd	Transmission (pts.) 5-Spd	Transmission (pts.) Auto. ①	Drive Axle (pts.)	Fuel Tank (gal.)	Cooling System (qts.)
1991–92	Shadow	K	4-153 (2.5)	4.5	4	—	5	18	—	14	9
	Shadow	J	4-153 (2.5)	4.5	4	—	5	18	—	14	9
	Sundance	D	4-135 (2.2)	4.5	4	—	5	18	—	14	9
	Sundance	K	4-153 (2.5)	4.5	4	—	5	18	—	14	9
	Sundance	J	4-153 (2.5)	4.5	4	—	5	18	—	14	9
	Spirit	K	4-153 (2.5)	4.5	4	—	5	18	—	16	9
	Spirit	J	4-153 (2.5)	4.5	4	—	5	18	—	16	9
	Spirit	3	6-181 (3.0)	4.5	4	—	—	18	—	16	9.5
	Spirit R/T	A	4-135 (2.2)	4.5	4	—	5	—		16	
	Acclaim	K	4-153 (2.5)	4.5	4	—	5	18	—	16	9
	Acclaim	J	4-153 (2.5)	4.5	4	—	5	18	—	16	9
	Acclaim	3	6-181 (3.0)	4.5	4	—	—	18	—	16	9.5
	TC	S	6-181 (3.0)	4.5	4	—	—	18	—	14	9.5

① 1989–91 A413 transaxle lockup—17 pts.

CAMSHAFT SPECIFICATIONS
All measurements given in inches.

Year	VIN	No. Cylinder Displacement cu. in. (liter)	Journal Diameter 1	2	3	4	5	Lobe Lift In.	Ex.	Bearing Clearance	Camshaft End Play
1988	D	4-135 (2.2)	1.375–1.376	1.375–1.376	1.375–1.376	1.375–1.376	1.375–1.376	NA	NA	—	0.005–0.020
	E	4-135 (2.2)	1.375–1.376	1.375–1.376	1.375–1.376	1.375–1.376	1.375–1.376	NA	NA	—	0.005–0.020
	K	4-153 (2.5)	1.375–1.376	1.375–1.376	1.375–1.376	1.375–1.376	1.375–1.376	NA	NA	—	0.005–0.020
1989	A	4-135 (2.2)	1.375–1.376	1.375–1.376	1.375–1.376	1.375–1.376	1.375–1.376	NA	NA	—	0.005–0.020
	D	4-135 (2.2)	1.375–1.376	1.375–1.376	1.375–1.376	1.375–1.376	1.375–1.376	NA	NA	—	0.005–0.020
	J	4-153 (2.5)	1.375–1.376	1.375–1.376	1.375–1.376	1.375–1.376	1.375–1.376	NA	NA	—	0.005–0.020
	K	4-153 (2.5)	1.375–1.376	1.375–1.376	1.375–1.376	1.375–1.376	1.375–1.376	NA	NA	—	0.005–0.020
	3	6-181 (3.0)	NA	NA	NA	NA	NA	①	①	—	NA
	R	4-135 (2.2)	NA	NA	NA	NA	NA	NA	NA	—	NA
1990	C	4-135 (2.2)	1.375–1.376	1.375–1.376	1.375–1.376	1.375–1.376	1.375–1.376	NA	NA	—	0.005–0.020
	D	4-135 (2.2)	1.375–1.376	1.375–1.376	1.375–1.376	1.375–1.376	1.375–1.376	NA	NA	—	0.005–0.020
	J	4-153 (2.5)	1.375–1.376	1.375–1.376	1.375–1.376	1.375–1.376	1.375–1.376	NA	NA	—	0.005–0.020
	K	4-153 (2.5)	1.375–1.376	1.375–1.376	1.375–1.376	1.375–1.376	1.375–1.376	NA	NA	—	0.005–0.020

CAMSHAFT SPECIFICATIONS

All measurements given in inches.

Year	VIN	No. Cylinder Displacement cu. in. (liter)	Journal Diameter 1	2	3	4	5	Lobe Lift In.	Ex.	Bearing Clearance	Camshaft End Play
1990	3	6-181 (3.0)	NA	NA	NA	NA	NA	①	①	—	NA
	S	6-181 (3.0)	NA	NA	NA	NA	NA	①	①	—	NA
	R	4-135 (2.2)	NA	NA	NA	NA	NA	NA	NA	—	NA
1991-92	D	4-135 (2.2)	1.375–1.376	1.375–1.376	1.375–1.376	1.375–1.376	1.375–1.376	NA	NA	—	0.005–0.020
	J	4-153 (2.5)	1.375–1.376	1.375–1.376	1.375–1.376	1.375–1.376	1.375–1.376	NA	NA	—	0.005–0.020
	K	4-153 (2.5)	1.375–1.376	1.375–1.376	1.375–1.376	1.375–1.376	1.375–1.376	NA	NA	—	0.005–0.020
	3	6-181 (3.0)	NA	NA	NA	NA	NA	①	①	—	NA
	S	6-181 (3.0)	NA	NA	NA	NA	NA	①	①	—	NA
	A	4-135 (2.2)	1.886–1.887	1.886–1.887	1.886–1.887	1.886–1.887	1.886–1.887	NA	NA	—	0.001–0.020

NA—Not available
① Height of cam lobe: 1.604–1.624 in.

CRANKSHAFT AND CONNECTING ROD SPECIFICATIONS

All measurements are given in inches.

Year	VIN	No. Cylinder Displacement cu. in. (liter)	Crankshaft Main Brg. Journal Dia.	Main Brg. Oil Clearance	Shaft End-play	Thrust on No.	Connecting Rod Journal Diameter	Oil Clearance	Side Clearance
1988	D	4-135 (2.2)	2.362–2.363	0.0004–0.0040	0.002–0.014	3	1.968–1.969	0.0008–0.0040	0.005–0.013
	E	4-135 (2.2)	2.362–2.363	0.0004–0.0040	0.002–0.014	3	1.968–1.969	0.0008–0.0040	0.005–0.013
	K	4-153 (2.5)	2.362–2.363	0.0004–0.0040	0.002–0.014	3	1.968–1.969	0.0008–0.0040	0.005–0.013
1989	A	4-135 (2.2)	2.362–2.363	0.0004–0.0040	0.002–0.014	3	1.968–1.969	0.0008–0.0040	0.005–0.013
	D	4-135 (2.2)	2.362–2.363	0.0004–0.0040	0.002–0.014	3	1.968–1.969	0.0008–0.0040	0.005–0.013
	J	4-153 (2.5)	2.362–2.363	0.0004–0.0040	0.002–0.014	3	1.968–1.969	0.0008–0.0040	0.005–0.013
	K	4-153 (2.5)	2.362–2.363	0.0004–0.0040	0.002–0.014	3	1.968–1.969	0.0008–0.0040	0.005–0.013
	3	6-181 (3.0)	2.361–2.362	0.0006–0.0020	0.002–0.010	3	1.968–1.969	0.0008–0.0028	0.004–0.010
	R	4-135 (2.2)	2.362–2.363	0.0011–0.0031	0.002–0.007	3	1.9695–1.9705	0.0006–0.0016	0.006–0.009
1990	C	4-135 (2.2)	2.362–2.363	0.0004–0.0040	0.002–0.014	3	1.968–1.969	0.0008–0.0040	0.005–0.013
	D	4-135 (2.2)	2.362–2.363	0.0004–0.0040	0.002–0.014	3	1.968–1.969	0.0008–0.0040	0.005–0.013
	J	4-153 (2.5)	2.362–2.363	0.0004–0.0040	0.002–0.014	3	1.968–1.969	0.0008–0.0040	0.005–0.013

CRANKSHAFT AND CONNECTING ROD SPECIFICATIONS

All measurements are given in inches.

Year	VIN	No. Cylinder Displacement cu. in. (liter)	Crankshaft Main Brg. Journal Dia.	Crankshaft Main Brg. Oil Clearance	Crankshaft Shaft End-play	Crankshaft Thrust on No.	Connecting Rod Journal Diameter	Connecting Rod Oil Clearance	Connecting Rod Side Clearance
1990	K	4-153 (2.5)	2.362–2.363	0.0004–0.0040	0.002–0.014	3	1.968–1.969	0.0008–0.0040	0.005–0.013
	3	6-181 (3.0)	2.361–2.362	0.0006–0.0020	0.002–0.010	3	1.968–1.969	0.0008–0.0028	0.004–0.010
	S	6-181 (3.0)	2.361–2.363	0.0006–0.0020	0.002–0.010	3	1.968–1.969	0.0008–0.0028	0.004–0.010
	R	4-135 (2.2)	2.362–2.363	0.0011–0.0031	0.002–0.007	3	1.9695–1.9705	0.0008–0.0016	0.006–0.009
1991–92	D	4-135 (2.2)	2.362–2.363	0.0004–0.0040	0.002–0.014		1.968–1.969	0.0008–0.0034	0.005–0.013
	J	4-153 (2.5)	2.362–2.363	0.0004–0.0040	0.002–0.014	3	1.968–1.969	0.0008–0.0034	0.005–0.013
	K	4-153 (2.5)	2.362–2.363	0.0004–0.0040	0.002–0.014	3	1.968–1.969	0.0008–0.0034	0.005–0.013
	3	6-181 (3.0)	2.361–2.362	0.0006–0.0020	0.002–0.010	3	1.968–1.969	0.0008–0.0028	0.004–0.010
	S	6-181 (3.0)	2.361–2.363	0.0006–0.0020	0.002–0.010	3	1.968–1.969	0.0008–0.0028	0.004–0.010
	A	4-135 (2.2)	2.362–2.363	0.0004–0.0040	0.002–0.014	3	1.968–1.969	0.0008–0.0034	0.005–0.013

VALVE SPECIFICATIONS

Year	VIN	No. Cylinder Displacement cu. in. (liter)	Seat Angle (deg.)	Face Angle (deg.)	Spring Test Pressure (lbs.)	Spring Installed Height (in.)	Stem-to-Guide Clearance (in.) Intake	Stem-to-Guide Clearance (in.) Exhaust	Stem Diameter ① (in.) Intake	Stem Diameter ① (in.) Exhaust
1988	D	4-135 (2.2)	45	45	114	1.65	0.001–0.003	0.0030–0.0047	0.3124	0.3103
	E	4-135 (2.2)	45	45	114	1.65	0.001–0.003	0.0030–0.0047	0.3124	0.3103
	K	4-153 (2.5)	45	45	114	1.65	0.001–0.003	0.0030–0.0047	0.3124	0.3103
1989	A	4-135 (2.2)	45	45	114	1.65	0.001–0.003	0.0030–0.0047	0.3124	0.3103
	D	4-135 (2.2)	45	45	114	1.65	0.001–0.003	0.0030–0.0047	0.3124	0.3103
	J	4-153 (2.5)	45	45	114	1.65	0.001–0.003	0.0030–0.0047	0.3124	0.3103
	K	4-153 (2.5)	45	45	114	1.65	0.001–0.003	0.0030–0.0047	0.3124	0.3103
	3	6-181 (3.0)	44.5	45.5	180	1.59	0.001–0.002	0.0020–0.0030	0.3130–0.3140	0.3120–0.3130
	R	4-135 (2.2)	NA	NA	NA	②	0.001–0.002	0.001–0.003	0.2750–0.2760	0.2750–0.2760

VALVE SPECIFICATIONS

Year	VIN	No. Cylinder Displacement cu. in. (liter)	Seat Angle (deg.)	Face Angle (deg.)	Spring Test Pressure (lbs.)	Spring Installed Height (in.)	Stem-to-Guide Clearance (in.) Intake	Exhaust	Stem Diameter ① (in.) Intake	Exhaust
1990	C	4-135 (2.2)	45	45	114	1.65	0.001–0.003	0.0030–0.0047	0.3124	0.3103
	D	4-135 (2.2)	45	45	114	1.65	0.001–0.003	0.0030–0.0047	0.3124	0.3103
	J	4-153 (2.5)	45	45	114	1.65	0.001–0.003	0.0030–0.0047	0.3124	0.3103
	K	4-153 (2.5)	45	45	114	1.65	0.001–0.003	0.0030–0.0047	0.3124	0.3103
	3	6-181 (3.0)	44.5	45.5	180	1.59	0.001–0.002	0.0020–0.0030	0.3130–0.3140	0.3120–0.3130
	S	6-181 (3.0)	44.5	45.5	180	1.59	0.001–0.002	0.0020–0.0030	0.3130–0.3140	0.3120–0.3130
	R	4-135 (2.2)	NA	NA	NA	②	0.001–0.002	0.0010–0.0030	0.2750–0.2760	0.2750–0.2760
1991–92	C	4-135 (2.2)	45	45	114	1.65	0.001–0.003	0.0030–0.0047	0.3124	0.3103
	D	4-135 (2.2)	45	45	114	1.65	0.001–0.003	0.0030–0.0047	0.3124	0.3103
	J	4-153 (2.5)	45	45	114	1.65	0.001–0.003	0.0030–0.0047	0.3124	0.3103
	K	4-153 (2.5)	45	45	114	1.65	0.001–0.003	0.0030–0.0047	0.3124	0.3103
	3	6-181 (3.0)	44.5	45.5	180	1.59	0.001–0.002	0.0020–0.0030	0.3130–0.3140	0.3120–0.3130
	S	6-181 (3.0)	44.5	45.5	180	1.59	0.001–0.002	0.0020–0.0030	0.3130–0.3140	0.3120–0.3130
	A	4-135 (2.2)	45	45	225	②	0.001–0.004	0.0020–0.0040	0.274	0.273

① If no range is given, the specification is the minimum allowable.
② Free length: 2.084–3.004 in.
NA—Not available

PISTON AND RING SPECIFICATIONS

All measurements are given in inches.

Year	VIN	No. Cylinder Displacement cu. in. (liter)	Piston Clearance	Ring Gap Top Compression	Bottom Compression	Oil Control	Ring Side Clearance Top Compression	Bottom Compression	Oil Control
1988	D	4-135 (2.2)	0.0005–0.0027	0.010–0.039	0.011–0.039	0.015–0.074	0.0015–0.0040	0.0015–0.0040	0.0002–0.0080
	E	4-135 (2.2)	0.0015–0.0037	0.010–0.039	0.011–0.039	0.015–0.074	0.0015–0.0040	0.0015–0.0040	0.0002–0.0080
	K	4-153 (2.5)	0.0010–0.0027	0.010–0.039	0.011–0.039	0.015–0.074	0.0015–0.0040	0.0015–0.0040	0.0002–0.0080

PISTON AND RING SPECIFICATIONS

All measurements are given in inches.

Year	VIN	No. Cylinder Displacement cu. in. (liter)	Piston Clearance	Ring Gap Top Compression	Ring Gap Bottom Compression	Ring Gap Oil Control	Ring Side Clearance Top Compression	Ring Side Clearance Bottom Compression	Ring Side Clearance Oil Control
1989	A	4-135 (2.2)	0.0005–0.0027	0.010–0.039	0.009–0.037	0.015–0.074	0.0016–0.0030	0.0016–0.0035	0.0002–0.0080
	D	4-135 (2.2)	0.0005–0.0027	0.010–0.039	0.011–0.039	0.015–0.074	0.0015–0.0040	0.0015–0.0040	0.0002–0.0080
	J	4-153 (2.5)	0.0006–0.0030	0.010–0.039	0.009–0.037	0.015–0.074	0.0016–0.0030	0.0016–0.0035	0.0002–0.0080
	K	4-153 (2.5)	0.0010–0.0027	0.010–0.039	0.011–0.039	0.015–0.074	0.0015–0.0040	0.0015–0.0040	0.0002–0.0080
	3	6-181 (3.0)	0.0008–0.0015	0.012–0.018	0.010–0.016	0.012–0.035	0.0020–0.0035	0.0008–0.0020	NA
	R	4-135 (2.2)	0.0005–0.0015	0.010–0.039	0.010–0.039	0.015–0.074	0.0015–0.0031	0.0015–0.0016	0.0002–0.0080
1990	C	4-135 (2.2)	0.0005–0.0027	0.010–0.039	0.009–0.037	0.015–0.074	0.0016–0.0030	0.0016–0.0035	0.0002–0.0080
	D	4-135 (2.2)	0.0005–0.0027	0.010–0.039	0.011–0.039	0.015–0.074	0.0015–0.0040	0.0015–0.0040	0.0002–0.0080
	J	4-153 (2.5)	0.0006–0.0030	0.010–0.039	0.009–0.037	0.015–0.074	0.0016–0.0030	0.0016–0.0035	0.0002–0.0080
	K	4-153 (2.5)	0.0010–0.0027	0.010–0.039	0.011–0.039	0.015–0.074	0.0015–0.0040	0.0015–0.0040	0.0002–0.0080
	3	6-181 (3.0)	0.0012–0.0020	0.012–0.018	0.010–0.016	0.012–0.035	0.0020–0.0035	0.0008–0.0020	NA
	S	6-181 (3.0)	0.0012–0.0020	0.012–0.018	0.010–0.016	0.012–0.035	0.0020–0.0035	0.0008–0.0020	NA
	R	4-135 (2.2)	0.0005–0.0015	0.010–0.039	0.010–0.039	0.012–0.035	0.0015–0.0031	0.0015–0.0016	0.0002–0.0080
1991–92	D	4-135 (2.2)	0.0005–0.0027	0.010–0.039	0.011–0.039	0.015–0.074	0.0015–0.0040	0.0015–0.0040	0.0002–0.0080
	J	4-153 (2.5)	0.0006–0.0030	0.010–0.039	0.009–0.037	0.015–0.074	0.0016–0.0030	0.0016–0.0035	0.0002–0.0080
	K	4-153 (2.5)	0.0010–0.0027	0.010–0.039	0.011–0.039	0.015–0.074	0.0015–0.0040	0.0015–0.0040	0.0002–0.0080
	3	6-181 (3.0)	0.0012–0.0020	0.012–0.018	0.010–0.016	0.012–0.035	0.0020–0.0035	0.0008–0.0020	NA
	S	6-181 (3.0)	0.0012–0.0020	0.012–0.018	0.010–0.016	0.012–0.035	0.0020–0.0035	0.0008–0.0020	NA
	A	4-135 (2.2)	0.0018–0.0039	0.014–0.039	0.014–0.039	0.001–0.039	0.0016–0.0040	0.0016–0.0040	0.0002–0.0040

NA—Not available

TORQUE SPECIFICATIONS

All readings in ft. lbs.

Year	VIN	No. Cylinder Displacement cu. in. (liter)	Cylinder Head Bolts	Main Bearing Bolts	Rod Bearing Bolts	Crankshaft Pulley Bolts	Flywheel Bolts	Manifold Intake	Manifold Exhaust	Spark Plugs
1988	D	4-135 (2.2)	①	30③	40③	85	70	17	17	26
	E	4-135 (2.2)	①	30③	40③	85	70	17	17	26
	K	4-153 (2.5)	①	30③	40③	85	70	17	17	26
1989	A	4-135 (2.2)	①	30③	40③	85	70	17	17	26
	D	4-135 (2.2)	①	30③	40③	85	70	17	17	26
	J	4-153 (2.5)	①	30③	40③	85	70	17	17	26
	K	4-153 (2.5)	①	30③	40③	85	70	17	17	26
	3	6-181 (3.0)	80	60	38	112	70	17	17	20
	R⑥	4-135 (2.2)	②	④	⑤	80	70	17	18	13
1990	C	4-135 (2.2)	①	30③	40③	85	70	17	17	26
	D	4-135 (2.2)	①	30③	40③	85	70	17	17	26
	J	4-153 (2.5)	①	30③	40③	85	70	17	17	26
	K	4-153 (2.5)	①	30③	40③	85	70	17	17	26
	3	6-181 (3.0)	80	60	38	112	70	17	17	20
	S	6-181 (3.0)	80	60	38	112	70	17	17	20
	R⑥	4-135 (2.2)	②	④	⑤	80	70	17	18	13
1991-92	D	4-135 (2.2)	①	30③	40③	85	70	17	17	26
	J	4-153 (2.5)	①	30③	40③	85	70	17	17	26
	K	4-153 (2.5)	①	30③	40③	85	70	17	17	26
	3	6-181 (3.0)	80	60	38	112	70	17	17	20
	S	6-181 (3.0)	80	60	38	112	70	17	17	20
	A	4-135 (2.2)	①	④	45–50	80	70	17	17	18

① Sequence: 45, 65, 65 plus ¼ turn
② Sequence: 32, 50, 65 plus ¼ turn
③ Plus ¼ turn
④ Sequence: 32, 43, 76
⑤ Sequence: 32, 47
⑥ TC Turbo

BRAKE SPECIFICATIONS

All measurements in inches unless noted.

Year	Model	Lug Nut Torque (ft. lbs.)	Master Cylinder Bore	Brake Disc Minimum Thickness	Brake Disc Maximum Runout	Standard Brake Drum Diameter	Minimum Lining Thickness Front	Minimum Lining Thickness Rear
1988	Aries	95	0.827	0.882	0.005	7.87	0.06	0.06
	Reliant	95	0.827	0.882	0.005	7.87	0.06	0.06
	600	95	0.827	0.882	0.005	8.66	0.06	0.06
	Caravelle	95	0.827	0.882	0.005	8.66	0.06	0.06
	LeBaron (K body)	95	0.827	0.882	0.005	7.87	0.06	0.06
	LeBaron (J body)	95	0.827	0.882	0.005	7.87	0.06	0.06
	Lancer	95	0.827	0.882	0.005	7.87	0.06	0.06
	LeBaron GTS	95	0.827	0.882	0.005	7.87	0.06	0.06
	New Yorker	95	0.827	0.882	0.005	8.66	0.06	0.06

BRAKE SPECIFICATIONS

All measurements in inches unless noted.

Year	Model	Lug Nut Torque (ft. lbs.)	Master Cylinder Bore	Brake Disc Minimum Thickness	Brake Disc Maximum Runout	Standard Brake Drum Diameter	Minimum Lining Thickness Front	Minimum Lining Thickness Rear
	Shadow	95	0.827	0.882	0.005	7.87	0.06	0.06
	Sundance	95	0.827	0.882	0.005	7.87	0.06	0.06
	Daytona	95	0.827	0.882	0.005	8.66	0.06	0.06
	rear disc	—	—	0.291	0.003	—	—	0.06
1989	Aries	95	0.827	0.882	0.005	7.87	0.06	0.06
	Reliant	95	0.827	0.882	0.005	7.87	0.06	0.06
	Lancer	95	0.827	0.882	0.005	7.87	0.06	0.06
	LeBaron GTS	95	0.827	0.882	0.005	7.87	0.06	0.06
	Shadow	95	0.827	0.882	0.005	7.87	0.06	0.06
	Sundance	95	0.827	0.882	0.005	7.87	0.06	0.06
	Spirit	95	0.827	0.882	0.005	8.66	0.06	0.06
	Acclaim	95	0.827	0.882	0.005	8.66	0.06	0.06
	Daytona	95	0.827	0.882	0.005	8.66	0.06	0.06
	solid rear disc	—	—	0.409	0.003	—	—	0.06
	vented rear disc	—	—	0.797	0.003	—	—	0.06
	LeBaron (J body)	95	0.827	0.882	0.005	8.66	0.06	0.06
	solid rear disc	—	—	0.409	0.003	—	—	0.06
	vented rear disc	—	—	0.797	0.003	—	—	0.06
	TC	95	NA	0.882	0.005	—	0.06	—
	rear disc	—	—	0.291	0.003	—	—	0.06
1990	LeBaron Landau	95	0.827	0.882	0.005	8.66	0.06	0.06
	Shadow	95	0.827	0.882	0.005	7.87	0.06	0.06
	Sundance	95	0.827	0.882	0.005	7.87	0.06	0.06
	Spirit	95	0.827	0.882	0.005	8.66	0.06	0.06
	Acclaim	95	0.827	0.882	0.005	8.66	0.06	0.06
	Daytona	95	0.827	0.882	0.005	8.66	0.06	0.06
	solid rear disc	—	—	0.409	0.003	—	—	0.06
	vented rear disc	—	—	0.797	0.003	—	—	0.06
	LeBaron (J body)	95	0.827	0.882	0.005	8.66	0.06	0.06
	solid rear disc	—	—	0.409	0.003	—	—	0.06
	vented rear disc	—	—	0.797	0.003	—	—	0.06
	TC	95	NA	0.882	0.005	—	0.06	—
	rear disc	—	—	0.291	0.003	—	—	0.06
1991-92	LeBaron Landau	95	0.827	0.882	0.005	8.66	0.06	0.06
	rear disc	—	—	0.797	0.003	—	—	0.06
	Shadow	95	0.827	0.882	0.005	7.87	0.06	0.06
	Sundance	95	0.827	0.882	0.005	7.87	0.06	0.06
	Spirit	95	0.827	0.882	0.005	8.66	0.06	0.06
	rear disc	—	—	0.797	0.003	—	—	0.06
	Acclaim	95	0.827	0.882	0.005	8.66	0.06	0.06
	rear disc	—	—	0.797	0.003	—	—	0.06

BRAKE SPECIFICATIONS

All measurements in inches unless noted.

Year	Model	Lug Nut Torque (ft. lbs.)	Master Cylinder Bore	Brake Disc		Standard Brake Drum Diameter	Minimum Lining Thickness	
				Minimum Thickness	Maximum Runout		Front	Rear
1991–92	Daytona	95	0.827	0.882	0.005	8.66	0.06	0.06
	solid rear disc	—	—	0.409	0.005	—	—	0.06
	vented rear disc	—	—	0.797	0.005	—	—	0.06
	LeBaron (J body)	95	0.827	0.882	0.005	8.66	0.06	0.06
	solid rear disc	—	—	0.409	0.005	—	—	0.06
	vented rear disc	—	—	0.797	0.005	—	—	0.06
	TC	95	NA	0.882	0.005	—	0.06	—
	rear disc	—	—	0.291	0.003	—	—	0.06

NA—Not available

WHEEL ALIGNMENT

Year	Model		Caster		Camber		Toe-in (in.)	Steering Axis Inclination (deg.)
			Range (deg.)	Preferred Setting (deg.)	Range (deg.)	Preferred Setting (deg.)		
1988	Aries	front	①	1³/₁₆P	¹/₄N–³/₄P	⁵/₁₆P	¹/₁₆	13⁵/₁₆
		rear	—	—	1¹/₄N–¹/₄N	¹/₂N	0	—
	Reliant	front	①	1³/₁₆P	¹/₄N–³/₄P	⁵/₁₆P	¹/₁₆	13⁵/₁₆
		rear	—	—	1¹/₄N–¹/₄N	¹/₂N	0	—
	Daytona	front	①	1³/₁₆P	¹/₄N–³/₄P	⁵/₁₆P	¹/₁₆	13⁵/₁₆
		rear	—	—	1¹/₄N–¹/₄N	¹/₂N	0	—
	600	front	①	1³/₁₆P	¹/₄N–³/₄P	⁵/₁₆P	¹/₁₆	13⁵/₁₆
		rear	—	—	1¹/₄N–¹/₄N	¹/₂N	0	—
	Caravelle	front	①	1³/₁₆P	¹/₄N–³/₄P	⁵/₁₆P	¹/₁₆	13⁵/₁₆
		rear	—	—	1¹/₄N–¹/₄N	¹/₂N	0	—
	LeBaron (P body)	front	①	1³/₁₆P	¹/₄N–³/₄P	⁵/₁₆P	¹/₁₆	13⁵/₁₆
		rear	—	—	1¹/₄N–¹/₄N	¹/₂N	0	—
	LeBaron (J body)	front	①	1³/₁₆P	¹/₄N–³/₄P	⁵/₁₆P	¹/₁₆	13⁵/₁₆
		rear	—	—	1¹/₄N–¹/₄N	¹/₂N	0	—
	Lancer	front	①	1³/₁₆P	¹/₄N–³/₄P	⁵/₁₆P	¹/₁₆	13⁵/₁₆
		rear	—	—	1¹/₄N–¹/₄N	¹/₂N	0	—
	LeBaron GTS	front	①	1³/₁₆P	¹/₄N–³/₄P	⁵/₁₆P	¹/₁₆	13⁵/₁₆
		rear	—	—	1¹/₄N–¹/₄N	¹/₂N	0	—
	New Yorker	front	①	1³/₁₆P	¹/₄N–³/₄P	⁵/₁₆P	¹/₁₆	13⁵/₁₆
		rear	—	—	1¹/₄N–¹/₄N	¹/₂N	0	—
	Shadow	front	①	1³/₁₆P	¹/₄N–³/₄P	⁵/₁₆P	¹/₁₆	13⁵/₁₆
		rear	—	—	1¹/₄N–¹/₄N	¹/₂N	0	—
	Sundance	front	①	1³/₁₆P	¹/₄N–³/₄P	⁵/₁₆P	¹/₁₆	13⁵/₁₆
		rear	—	—	1¹/₄N–¹/₄N	¹/₂N	0	—

WHEEL ALIGNMENT

Year	Model		Caster Range (deg.)	Caster Preferred Setting (deg.)	Camber Range (deg.)	Camber Preferred Setting (deg.)	Toe-in (in.)	Steering Axis Inclination (deg.)
1989	Aries	front	①	1³/₁₆P	¹/₄N–³/₄P	⁵/₁₆P	¹/₁₆	13⁵/₁₆
		rear	—	—	1¹/₄N–¹/₄N	¹/₂N	0	—
	Reliant	front	①	1³/₁₆P	¹/₄N–³/₄P	⁵/₁₆P	¹/₁₆	13⁵/₁₆
		rear	—	—	1¹/₄N–¹/₄N	¹/₂N	0	—
	Daytona	front	①	1³/₁₆P	¹/₄N–³/₄P	⁵/₁₆P	¹/₁₆	13⁵/₁₆
		rear	—	—	1¹/₄N–¹/₄N	¹/₂N	0	—
	LeBaron	front	①	1³/₁₆P	¹/₄N–³/₄P	⁵/₁₆P	¹/₁₆	13⁵/₁₆
		rear	—	—	1¹/₄N–¹/₄N	¹/₂N	0	—
	Lancer	front	①	1³/₁₆P	¹/₄N–³/₄P	⁵/₁₆P	¹/₁₆	13⁵/₁₆
		rear	—	—	1¹/₄N–¹/₄N	¹/₂N	0	—
	LeBaron GTS	front	①	1³/₁₆P	¹/₄N–³/₄P	⁵/₁₆P	¹/₁₆	13⁵/₁₆
		rear	—	—	1¹/₄N–¹/₄N	¹/₂N	0	—
	Shadow	front	①	1³/₁₆P	¹/₄N–³/₄P	⁵/₁₆P	¹/₁₆	13⁵/₁₆
		rear	—	—	1¹/₄N–¹/₄N	¹/₂N	0	—
	Sundance	front	①	1³/₁₆P	¹/₄N–³/₄P	⁵/₁₆P	¹/₁₆	13⁵/₁₆
		rear	—	—	1¹/₄N–¹/₄N	¹/₂N	0	—
	Spirit	front	①	1³/₁₆P	¹/₄N–³/₄P	⁵/₁₆P	¹/₁₆	13⁵/₁₆
		rear	—	—	1¹/₄N–¹/₄N	¹/₂N	0	—
	Acclaim	front	①	1³/₁₆P	¹/₄N–³/₄P	⁵/₁₆P	¹/₁₆	13⁵/₁₆
		rear	—	—	1¹/₄N–¹/₄N	¹/₂N	0	—
	TC	front	①	1³/₁₆P	¹/₄N–³/₄P	⁵/₁₆P	¹/₁₆	13⁵/₁₆
		rear	—	—	1¹/₄N–¹/₄N	¹/₂N	0	—
1990	Daytona	front	①	1³/₁₆P	¹/₄N–³/₄P	⁵/₁₆P	¹/₁₆	13⁵/₁₆
		rear	—	—	1¹/₄N–¹/₄N	¹/₂N	0	—
	LeBaron	front	①	1³/₁₆P	¹/₄N–³/₄P	⁵/₁₆P	¹/₁₆	13⁵/₁₆
		rear	—	—	1¹/₄N–¹/₄N	¹/₂N	0	—
	LeBaron Landau	front	①	1³/₁₆P	¹/₄N–³/₄P	⁵/₁₆P	¹/₁₆	13⁵/₁₆
		rear	—	—	1¹/₄N–¹/₄N	¹/₂N	0	—
	Shadow	front	①	1³/₁₆P	¹/₄N–³/₄P	⁵/₁₆P	¹/₁₆	13⁵/₁₆
		rear	—	—	1¹/₄N–¹/₄N	¹/₂N	0	—
	Sundance	front	①	1³/₁₆P	¹/₄N–³/₄P	⁵/₁₆P	¹/₁₆	13⁵/₁₆
		rear	—	—	1¹/₄N–¹/₄N	¹/₂N	0	—
	Spirit	front	①	1³/₁₆P	¹/₄N–³/₄P	⁵/₁₆P	¹/₁₆	13⁵/₁₆
		rear	—	—	1¹/₄N–¹/₄N	¹/₂N	0	—
	Acclaim	front	①	1³/₁₆P	¹/₄N–³/₄P	⁵/₁₆P	¹/₁₆	13⁵/₁₆
		rear	—	—	1¹/₄N–¹/₄N	¹/₂N	0	—
	TC	front	①	1³/₁₆P	¹/₄N–³/₄P	⁵/₁₆P	¹/₁₆	13⁵/₁₆
		rear	—	—	1¹/₄N–¹/₄N	¹/₂N	0	—
1991–92	Daytona	front	①	2³/₄P	¹/₄N–³/₄P	⁵/₁₆P	¹/₁₆	12¹/₂
		rear	—	—	1¹/₄N–¹/₄N	¹/₂N	0	—
	LeBaron	front	①	2³/₄P	¹/₄N–³/₄P	⁵/₁₆P	¹/₁₆	12¹/₂
		rear	—	—	1¹/₄N–¹/₄N	¹/₂N	0	—

WHEEL ALIGNMENT

Year	Model		Caster Range (deg.)	Caster Preferred Setting (deg.)	Camber Range (deg.)	Camber Preferred Setting (deg.)	Toe-in (in.)	Steering Axis Inclination (deg.)
1991–92	LeBaron Landau	front	①	2³/₄P	¼N–³/₄P	⁵/₁₆P	¹/₁₆	12½
		rear	—	—	1¼N–¼N	½N	0	—
	Shadow	front	①	2³/₄P	¼N–³/₄P	⁵/₁₆P	¹/₁₆	12½
		rear	—	—	1¼N–¼N	½N	0	—
	Sundance	front	①	2³/₄P	¼N–³/₄P	⁵/₁₆P	¹/₁₆	12½
		rear	—	—	1¼N–¼N	½N	0	—
	Spirit	front	①	2³/₄P	¼N–³/₄P	⁵/₁₆P	¹/₁₆	12½
		rear	—	—	1¼N–¼N	½N	0	—
	Acclaim	front	①	2³/₄P	¼N–³/₄P	⁵/₁₆P	¹/₁₆	12½
		rear	—	—	1¼N–¼N	½N	0	—
	TC	front	①	1³/₁₆P	¼N–³/₄P	⁵/₁₆P	¹/₁₆	13⁵/₁₆
		rear	—	—	1¼N–¼N	½N	0	—

N—Negative
P—Positive
① Not adjustable; variation between sides should
not exceed 1.5°

ENGINE MECHANICAL

NOTE: Disconnecting the negative battery cable on some vehicles may interfere with the functions of the on board computer systems and may require the computers to undergo a relearning process, once the negative battery cable is reconnected.

Engine Assembly

REMOVAL & INSTALLATION

2.2L and 2.5L Engines

1. Disconnect the negative battery cable and all engine ground straps. Relieve the fuel pressure.
2. Mark the hood hinge outline on the hood and remove the hood.
3. Drain the cooling system. Remove the radiator hoses, fan assembly, radiator and intercooler, if equipped.
4. Remove the air cleaner, duct hoses and oil filter.
5. Unbolt the air conditioning compressor from its mount, if equipped and position it aside.
6. Remove the power steering pump mounting bolts and position the pump aside, without disconnecting any fluid lines.
7. Label and disconnect all electrical connectors from the engine, alternator and fuel injection system.
8. Disconnect and plug the fuel lines and heater hoses.
9. Disconnect the throttle linkage.
10. Remove the alternator.
11. Raise the vehicle and support safely.
12. Disconnect the exhaust pipe from the manifold. Remove the right inner fender shield.
13. If equipped with a manual transaxle, remove the transaxle.
14. If equipped with an automatic transaxle, perform the following procedures:
   a. Remove the lower cover from the transaxle case.
   b. Remove the starter and set it aside.
   c. Matchmark the flexplate to the torque converter for installation purposes.
   d. Remove the torque converter bolts. Separate the converter from the flexplate. Remove the lower bellhousing bolts.
15. Lower the vehicle and support the transaxle, if still in the vehicle, with a floor jack or equivalent. Attach an engine lifting device to the engine.
16. Remove the remaining bellhousing bolts.

NOTE: If removing the insulator-to-rail screws, first mark the position of the insulator on the side rail to insure proper alignment during reinstallation.

17. Remove the front engine mount nut/bolt and the left insulator through bolt or the insulator bracket to transaxle bolts.
18. Lift the engine from the vehicle and remove.
To install:
19. Lower the engine into the engine compartment. Make sure the lifting device is supporting the full weight of the engine and loosely install all of the mounting bolts until all are threaded. Then tighten all bolts.
20. Remove the lifting device.
21. Raise the vehicle and support safely.
22. If equipped with a manual transaxle, install the transaxle.
23. If equipped with an automatic transaxle, install the torque converter bolts and torque to 55 ft. lbs. (75 Nm). Install the torque converter inspection plate and starter.
24. Connect the exhaust pipe. Lower the vehicle.
25. Install the alternator, power steering pump and air conditioning compressor, if equipped.
26. Connect the fuel lines and heater hoses.
27. Connect the throttle linkage.
28. Connect all remaining electrical connectors.
29. Install the air cleaner assembly and oil filter.
30. Install the radiator, fan assembly, hoses and intercooler, if equipped.
31. Fill the engine with the proper

amount of engine oil. Connect the negative battery cable.

32. Refill the cooling system. Start the engine, allow it to reach normal operating temperature. Check for leaks.

33. Check the ignition timing and adjust if necessary.

34. Install the hood.

3.0L Engine

1. Disconnect the negative battery cable. Relieve the fuel pressure.

2. Matchmark the hinge-to-hood position and remove the hood.

3. Drain the cooling system. Disconnect and label all engine electrical connections.

4. Remove the coolant hoses from the radiator and engine. Remove the radiator and cooling fan assembly.

5. Remove the air cleaner assembly. Disconnect the fuel lines from the engine. Disconnect the accelerator cable from the throttle body.

6. Raise the vehicle and support safely. Drain the engine oil.

7. Remove the air conditioning compressor mounting bolts, the drive belts and position the compressor aside. Disconnect the exhaust pipe from the exhaust manifold.

8. Remove the transaxle inspection cover, matchmark the converter to the flexplate, and remove the torque converter bolts.

9. Remove the power steering pump mounting bolts and set the pump aside, upright, with the fluid lines attached.

10. Remove the lower bellhousing bolts. Disconnect and label the starter motor wiring and remove the starter motor from the engine.

11. Lower the vehicle. Disconnect and label all electrical connectors from the engine, alternator and fuel injection system, vacuum hoses, and engine ground straps.

12. Support the transaxle with a floor jack or equivalent. Attach an engine lifting device to the engine.

13. Remove the upper transaxle-to-engine bolts.

14. To separate the engine mounts from the insulators, mark the right insulator-to-right frame support and remove the mounting bolts. Remove the front engine mount through bolt. Remove the left insulator through bolt from inside the wheel housing. Remove the insulator bracket-to-transaxle bolts.

15. Lift and remove the engine from the vehicle.

To install:

16. Lower the engine into the engine compartment. Align the engine mounts and install the bolts; do not tighten the bolts until all bolts have been installed. Torque the through bolts to 75 ft. lbs. (102 Nm).

17. Install the upper transaxle-to-engine mounting bolts and torque to 75 ft. lbs. (102 Nm). Remove the engine lifting fixture from the engine.

18. Raise the vehicle and support safely.

19. Align the converter marks, and install the torque converter bolts. Install the transaxle inspection cover.

20. Connect the exhaust pipe to the exhaust manifold. Install the starter motor and connect the wiring.

21. Install the power steering pump and air conditioning compressor. Adjust the drive belt tension, if necessary.

22. Lower the vehicle. Reconnect all vacuum hoses and electrical connections to the engine.

23. Connect the fuel lines and accelerator cable.

24. Install the radiator and fan assembly. Connect the fan motor wiring. Connect the radiator hoses and refill the cooling system.

25. Refill the engine with the proper oil to the correct level.

26. Connect the engine ground straps. Install the hood and align the matchmarks. Connect the battery.

27. Start and run the engine until it reaches normal operating temperatures and check for leaks. Adjust the transaxle linkage, if necessary.

Engine Mounts

REMOVAL & INSTALLATION

1. Disconnect the negative battery cable.

2. Matchmark the engine mount to its frame mounting location.

3. Raise the vehicle and support safely, if necessary. Using the proper equipment, support the weight of the engine.

4. Remove all bolts and nuts that attach the mount to the engine strut, transaxle or body and remove the mount assembly from the vehicle.

5. Remove the through bolt and separate the insulator from the yoke bracket as required.

6. The installation is the reverse of the removal procedure. Make sure the matchmarks are aligned before tightening bolts.

7. Tighten the lower yoke nut first, then the through bolt nut, then the body mounting bolts.

Cylinder Head

REMOVAL & INSTALLATION

2.2L and 2.5L Engines Except DOHC

1. Disconnect the negative battery cable and unbolt it from the head. Relieve the fuel pressure. Drain the cooling system. Remove the dipstick bracket nut from the thermostat housing and remove the ignition coil from the thermostat housing if it is installed there.

2. Remove the air cleaner assembly. Remove the upper radiator hose and disconnect the heater hoses.

3. Disconnect and label the vacuum lines, hoses and wiring connectors from the manifold(s), throttle body and from the cylinder head.

4. Disconnect all linkages and the fuel line from the throttle body. Unbolt the cable bracket. Remove the ground strap attaching screw from the firewall.

5. If equipped with air conditioning, remove the upper compressor mounting bolts. The cylinder head can be remove with the compressor and bracket still mounted. Remove the upper timing belt cover.

6. Raise the vehicle and support safely. Disconnect the exhaust pipe from the exhaust manifold. Disconnect the water hose and oil drain from the turbocharger, if equipped.

7. Rotate the engine by hand until the timing marks align. The No. 1 piston should be at TDC of its compression stroke. Lower the vehicle.

8. With the timing marks aligned, remove the camshaft sprocket. The camshaft sprocket can be suspended to keep the timing intact. Remove the spark plug wires from the spark plugs.

9. Remove the valve cover and curtain. Remove the cylinder head bolts and washers, starting from the outside and working inward.

10. Remove the cylinder head from the engine.

11. Clean the cylinder head gasket mating surfaces.

To install:

NOTE: Head bolt diameter is 11mm. These bolts are identified with the number "11" on the head of the bolt. The 10mm bolts used on previous vehicles will thread into an 11mm bolt hole, but will permanently damage the cylinder block. Make sure the correct bolts are used when replacing head bolts.

12. Using new gaskets and seals, install the head to the engine. Using new head bolts assembled with the old

washers, torque the cylinder head bolts in sequence, to 45 ft. lbs. (61 Nm). Repeating the sequence, torque the bolts to 65 ft. lbs. (88 Nm). With the bolts at 65 ft. lbs., turn each bolt an additional ¼ turn.

13. Install the timing belt.

14. Install or connect all items that were removed or disconnected during the removal procedure.

15. Refill the cooling system. Connect the negative battery cable. Start the engine and check for leaks using the DRB II to activate the fuel pump. Adjust the timing, as required.

2.2L DOHC Engine—TC

1. Disconnect the negative battery cable and unbolt it from the head. Relieve the fuel pressure. Drain the cooling system.

2. Remove the timing belt covers. Rotate the engine by hand until the timing marks align (No. 1 piston at TDC) and remove the timing belt.

3. Remove the air conditioning compressor and bracket from the cylinder head.

4. Disconnect the turbocharger coolant lines.

5. Remove the air cleaner assembly and separate the intake and exhaust manifolds from the cylinder head.

6. Disconnect and label all wiring connectors, hoses and ignition wires from the cylinder head.

7. Remove the cylinder head cover. Remove both camshafts to expose cylinder head bolts.

8. Remove the cylinder head bolts and washers, starting from the outside and working inward.

9. Remove the cylinder head and gasket from the engine.

10. Clean the cylinder head gasket mating surfaces.

To install:

11. Using new gaskets and seals, install the head to the engine. Using new head bolts assembled with the old washers, torque the cylinder head bolts in sequence, to 32 ft. lbs. (44 Nm). Repeating the sequence, torque the bolts to 50 ft. lbs. (69 Nm). Repeating the sequence a third time, torque the bolts to 65 ft. lbs. (88 Nm). With the bolts at 65 ft. lbs., turn each bolt an additional ¼ turn.

12. Install the camshafts and timing belt.

13. Install or connect all items that were removed or disconnected during the removal procedure.

14. Refill the cooling system. Connect the negative battery cable. Start the engine and check for leaks using the DRB II to activate the fuel pump. Adjust the timing as required.

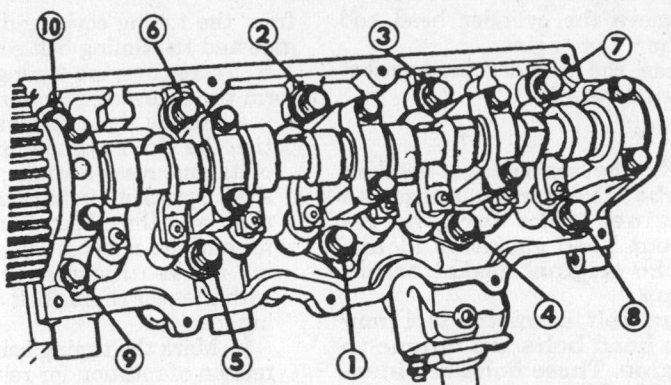

Cylinder head bolt torque sequence—2.2L and 2.5L SOHC engines

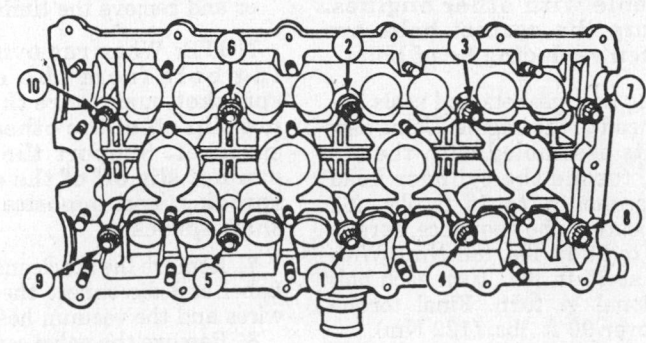

← TIMING BELT END

Cylinder head bolt torque sequence—2.2L DOHC engine—TC

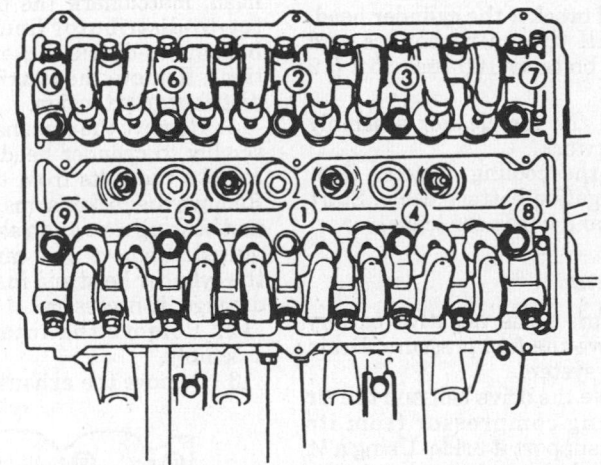

Cylinder head bolt torque sequence—2.2L Turbo III engine

2.2L Turbo III Engine

1. Disconnect the negative battery cable and unbolt it from the head. Relieve the fuel pressure. Drain the cooling system.

2. Remove the air cleaner assembly with all ductwork.

3. Remove the timing belt covers. Rotate the engine by hand until the timing marks align (No. 1 piston at TDC). Remove the timing belt.

4. Remove the air conditioning compressor and bracket from the cylinder head.

5. Disconnect the turbocharger coolant lines and separate the intake and exhaust manifolds from the cylinder head.

6. Remove the ignition cable cover and valve covers. Disconnect and label all wiring connectors, hoses and ignition wires from the cylinder head.

7. Remove the cylinder head and gasket from the engine.

8. Clean the cylinder head gasket mating surfaces.

To install:

NOTE: The head gasket used on the Turbo III engine is unique to the engine. Make sure the replacement head gasket is identical to the original gasket before installing.

Head bolt diameter is 11mm and the head bolts are unique to this engine. These bolts are identified with the number 11 on the head of the bolt and are not interchangeable with other engines. Make sure the correct bolts are used when replacing head bolts.

9. Using new gaskets and seals, install the head to the engine. Using new head bolts assembled with the old washers, torque the cylinder head bolts in sequence, to 45 ft. lbs. (61 Nm). Repeating the sequence, torque the bolts to 65 ft. lbs. (88 Nm). With the bolts at 65 ft. lbs., turn each bolt an additional ¼ turn. Final torque must be over 90 ft. lbs. (122 Nm).

10. Install the timing belt and all related items.

11. Install the intake and exhaust manifolds.

12. Install the air conditioning compressor and bracket the cylinder head.

13. Install the valve covers and torque the bolts to 105 inch lbs. (12 Nm).

14. Install the air cleaner assembly and all ductwork.

15. Refill the cooling system. Connect the negative battery cable. Start the engine and check for leaks.

3.0L Engine

1. Disconnect the negative battery cable. Relieve the fuel pressure. Drain the cooling system.

2. Remove the drive belt and the air conditioning compressor from its mount and support it aside. Using a ½ in. drive breaker bar, insert it into the square hole of the serpentine drive belt tensioner, rotate it counterclockwise to reduce the belt tension and remove the belt. Remove the alternator and power steering pump from the brackets and move them aside.

3. Raise the vehicle and support safely. Remove the right front wheel assembly and the right inner splash shield.

4. Remove the crankshaft pulleys and the torsional damper.

5. Lower the vehicle. Using a floor jack and a block of wood positioned under the oil pan, raise the engine slightly. Remove the engine mount bracket

from the timing cover end of the engine and the timing belt covers.

6. To remove the timing belt, perform the following procedures:

a. Rotate the crankshaft to position the No. 1 cylinder on the TDC of its compression stroke; the crankshaft sprocket timing mark should align with the oil pan timing indicator and the camshaft sprockets timing marks (triangles) should align with the rear timing belt covers timing marks.

b. Mark the timing belt in the direction of rotation for reinstallation purposes.

c. Loosen the timing belt tensioner and remove the timing belt.

NOTE: When removing the timing belt from the camshaft sprocket, make sure the belt does not slip off of the other camshaft sprocket. Support the belt so it can not slip off of the crankshaft sprocket and opposite side camshaft sprocket.

7. Remove the air cleaner assembly. Label and disconnect the spark plug wires and the vacuum hoses.

8. Remove the valve cover.

9. Install auto lash adjuster retainer tools MD998443 or equivalent, on the rocker arms.

10. If removing the front cylinder head, matchmark the distributor rotor-to-distributor housing and the housing-to-distributor extension locations. Remove the distributor and the distributor extension.

11. Remove the camshaft bearing assembly to cylinder head bolts (do not remove the bolts from the assembly). Remove the rocker arms, rocker shafts and bearing caps as an assembly, as required. Remove the camshafts from the cylinder head and inspect them for damage, if necessary.

12. Remove the intake manifold assembly.

13. Remove the exhaust manifold.

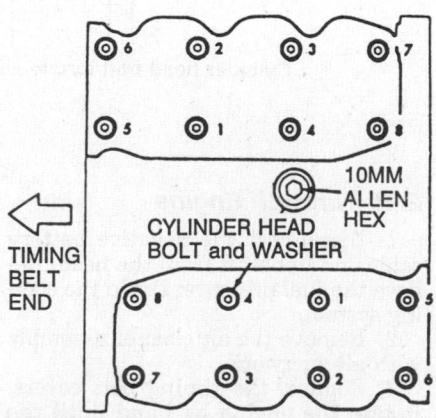

Cylinder head bolt torque sequence—3.0L engine

14. Remove the cylinder head bolts, starting from the outside and working inward. Remove the cylinder head from the engine.

15. Clean the gasket mounting surfaces and check the heads for warpage; the maximum warpage allowed is 0.008 in. (0.20mm).

To install:

16. Install the new cylinder head gaskets over the dowels on the engine block.

17. Install the cylinder heads on the engine and torque the cylinder head bolts in sequence using 3 even steps, to 70 ft. lbs. (95 Nm).

18. Install or connect all items that were removed or disconnected during the removal procedure.

19. When installing the timing belt over the camshaft sprocket, use care not to allow the belt to slip off the opposite camshaft sprocket.

20. Make sure the timing belt is installed on the camshaft sprocket in the same position as when removed.

21. Refill the cooling system. Connect the negative battery cable. Start the engine and check for leaks using the DRB II to activate the fuel pump. Adjust the timing as required.

Valve Lifters

REMOVAL & INSTALLATION

2.2L and 2.5L Engines Except DOHC

1. Disconnect the negative battery cable.

2. Remove the valve cover and curtain. If removing all lifters, remove the camshaft and rocker arms.

3. If only removing 1 lifter, rotate the crankshaft until the low point of the desired cam lobe is contacting the rocker arm.

4. Using the special valve spring compressor tool 4682 or equivalent, depress the valve spring without dislodging the keepers and slide the rocker arm out.

5. Remove the valve lifter(s) from the bore(s).

6. Lubricate the lifter(s) and their bore(s) with clean engine oil.

7. The installation is the reverse of the removal procedure.

8. Connect the negative battery cable.

2.2L Turbo III Engine

1. Disconnect the negative battery cable.

2. Remove the valve cover(s).

3. Remove the rocker arm shaft(s).

4. Slide the rocker arm(s) off the shaft and remove the lash adjuster from the rocker arm.

5. The installation is the reverse of the removal procedure.

6. Connect the negative battery cable.

3.0L Engine

1. Disconnect the negative battery cable. Remove the air cleaner assembly.

2. Remove the valve cover.

3. Using the valve lifter retainer tools MD998443 or equivalent, install them on the rocker arms to keep the lifters from falling out.

4. On the right side cylinder head, remove the distributor extension.

5. Have a helper hold the rear end of the camshaft down. If the rear of the camshaft cannot be held down, the belt will dislodge and the valve timing will be lost. Loosen the camshaft cap bolts but do not remove them from the caps. Remove the caps, arms, shafts and bolts all as an assembly.

6. Remove the lifter(s) from the rocker arm(s).

7. Lubricate the lifter(s) and their bore(s) with clean engine oil.

8. The installation is the reverse of the removal procedure.

Valve Lash

ADJUSTMENT

2.2L DOHC Engine

1. Disconnect the negative battery cable.

2. Remove the valve cover.

3. Check the clearance of all valves by inserting a feeler gauge between the camshaft and adjusting disc when the cam lobe is pointing straight up. Record all measurements.

4. The specifications are:
   Intake—0.012 in. (0.30mm)
   Exhaust—0.016 in. (0.40mm)

5. If not at specifications, remove the camshaft(s) and use the appropriate adjusting discs to bring clearance to specification.

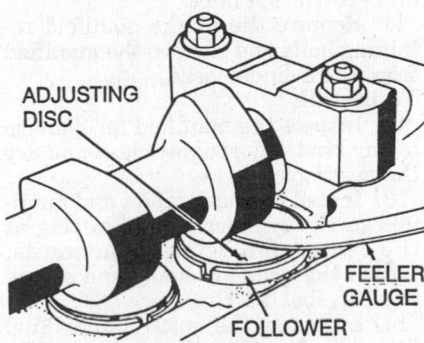

**Checking valve clearance—
2.2L DOHC engine**

Rocker Arms/Shafts

REMOVAL & INSTALLATION

2.2L and 2.5L Engines Except Turbo III

1. Disconnect the negative battery cable.

2. Remove the valve cover.

3. Rotate the crankshaft until the low point of the desired cam lobe is contacting the rocker arm.

4. Using the special valve spring compressor tool or equivalent, depress the valve spring without dislodging the keepers and slide the rocker arm out.

5. The installation is the reverse of the removal procedure.

2.2L Turbo III Engine

1. Disconnect the negative battery cable.

2. Remove the valve cover(s).

3. Remove the rocker arm retaining bolts in the proper removal sequence.

4. Remove the rocker shaft assembly from the cylinder head.

5. Keep all parts in order and disassemble as required. Inspect the lash adjusters carefully.

To install:

6. Lubricate and assemble the rocker arms to the shaft.

7. Make sure the lash adjusters are at least partially full of oil. This is indicated by little or no plunger travel when depressing. Install to the rocker arms.

8. Install the assembly and tighten the bolts in the proper sequence to 18 ft. lbs. (24 Nm).

9. Install the valve cover(s).

10. Connect the negative battery cable.

3.0L Engine

1. Disconnect the negative battery

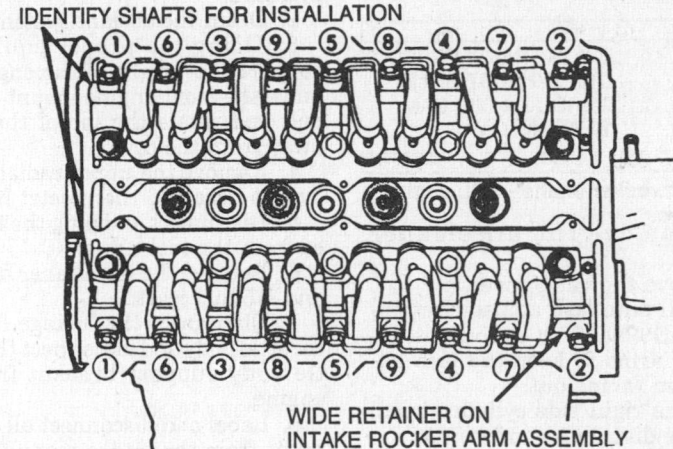

IDENTIFY SHAFTS FOR INSTALLATION

WIDE RETAINER ON INTAKE ROCKER ARM ASSEMBLY

Rocker shaft retaining bolt removal sequence—2.2L Turbo III engine

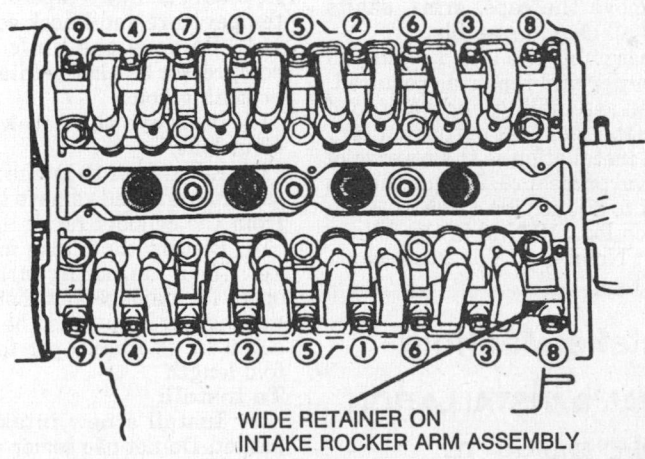

WIDE RETAINER ON INTAKE ROCKER ARM ASSEMBLY

Rocker shaft retaining bolt installation sequence—2.2L Turbo III engine

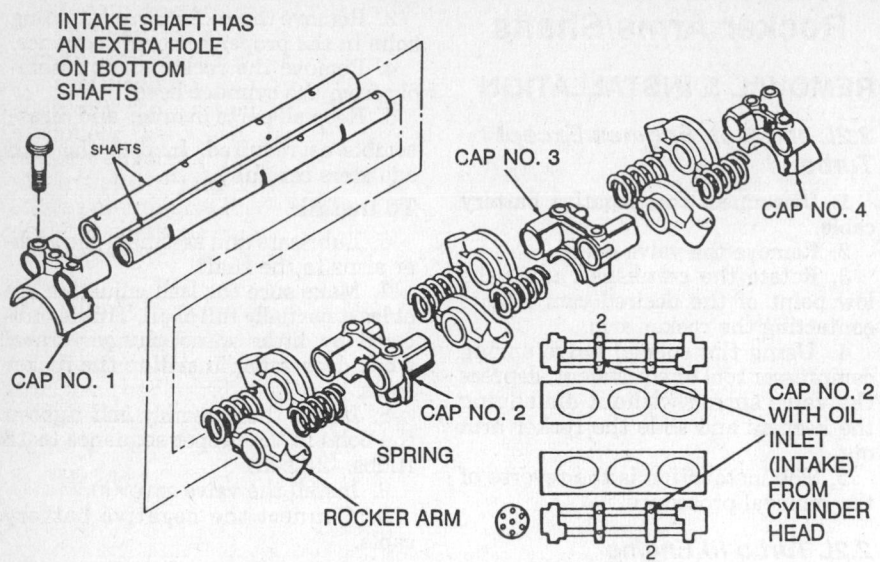

INTAKE SHAFT HAS
AN EXTRA HOLE
ON BOTTOM
SHAFTS

SHAFTS

CAP NO. 3

CAP NO. 4

CAP NO. 1

CAP NO. 2

SPRING

ROCKER ARM

CAP NO. 2
WITH OIL
INLET
(INTAKE)
FROM
CYLINDER
HEAD

Rocker shafts/arms assembly—3.0L engine

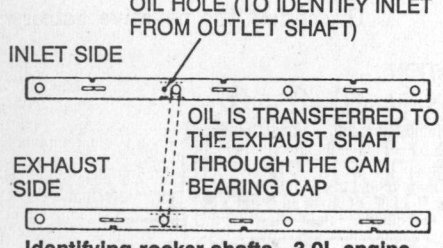

OIL HOLE (TO IDENTIFY INLET
FROM OUTLET SHAFT)

INLET SIDE

OIL IS TRANSFERRED TO
THE EXHAUST SHAFT
THROUGH THE CAM
BEARING CAP

EXHAUST
SIDE

Identifying rocker shafts—3.0L engine

cable. Remove the air cleaner assembly.

2. Remove the valve cover.

3. Install auto lash adjuster retainer tools MD998443 or equivalent, on the rocker arms to keep the lash adjusters from falling out.

4. On the right side cylinder head, remove the distributor extension.

5. Have a helper hold the rear end of the camshaft down. If the rear of the camshaft cannot be held down, the belt will dislodge and the valve timing will be lost. Loosen the camshaft cap bolts but do not remove them from the caps. Remove the caps, arms, shafts and bolts all as an assembly.

6. Disassemble the unit keeping all parts in order and repair as required.

7. When assembling, apply a drop of sealant to the rear edge of the rear cap.

8. The installation is the reverse of the removal procedure. Torque the cap bolts first to 85 inch lbs. (19 Nm), then to 180 inch lbs. (19 Nm) in the following order: No. 3 cap, No. 2 cap, No. 1 cap, No. 4 cap.

Intake Manifold

REMOVAL & INSTALLATION

2.2L DOHC Engine—TC

1. Disconnect the negative battery

cable. Relieve the fuel system pressure.

2. Drain the coolant system.

3. Using the proper equipment, support the weight of the engine. Remove the front engine mount through bolt and rotate the top of the engine away from the cowl.

4. Remove the upper radiator hose, bypass hose and thermostat housing.

5. Disconnect and plug the fuel hoses from the fuel tubes.

6. Remove the air cleaner assembly and all duct work.

7. Disconnect the linkage from the throttle body and disconnect the throttle body support bracket from the engine.

8. Label and disconnect all vacuum hoses from the intake manifold.

9. Disconnct the Throttle Position Sensor (TPS) and Automatic Idle Speed (AIS) motor wiring connectors from the throttle body.

10. Disconnect wiring from the fuel injectors, charge temperature, coolant temperature and knock sensors.

11. Remove the air conditioning compressor bracket to intake manifold attaching bolt.

12. Remove the intake manifold strut bolt.

13. Remove all 10 intake manifold attaching nuts and remove the manifold from the cylinder head.

14. Discard the intake manifold gasket. Clean the mating surfaces and inspect for damage and distortion. The mating surfaces must be flat within 0.006 in. (0.15mm) per foot of manifold length.

To install:

15. Install a new intake manifold gasket. Do not use sealer of any kind.

16. Position the manifold on the studs and install the retaining nuts.

Starting at the center and working outwards, torque the nuts gradually and evenly to 17 ft. lbs. (23 Nm).

17. Install the strut bolt and torque to 21 ft. lbs. (29 Nm). Install the air conditioning compressor bracket bolt and torque to 21 ft. lbs. (29 Nm).

18. Install the front engine mount through bolt.

19. Connect all hoses and wiring that was disconnected during the removal procedure.

20. Install the throttle body bracket and connect the linkage.

21. Install the upper radiator hose, bypass hose and thermostat housing.

22. Refill the cooling system. Connect the negative battery cable. Start the engine and check for leaks using the DRB II to activate the fuel pump.

2.2L Turbo III Engine

1. Disconnect the negative battery cable. Relieve the fuel system pressure. Drain the cooling system.

2. Remove the fresh air duct from the air filter housing. Remove the inlet hose from the intercooler.

3. Remove the radiator hose from the thermostat housing.

4. Remove the DIS ignition coil from the intake manifold.

5. Disconnect the throttle and speed control cables from the throttle body.

6. Disconnect the intercooler-to-throttle body outlet hose. Disconnect the vacuum hoses from the throttle body and carefully remove the harness.

7. Disconnect the AIS motor and TPS wiring connectors.

8. Remove the PCV breather/separator box and vacuum harness assembly. Remove the brake booster hose, vacuum vapor harness and fuel pressure regulator from the intake manifold.

9. Disconnect the fuel injector wiring harness and charge temperature sensor.

10. Wrap shop towels around the fittings and disconnect the fuel supply and return fuel lines.

11. Remove the intake manifold retaining bolts and remove the manifold from the cylinder head.

To install:

12. Inspect the manifold for damage of any kind. Thoroughly clean and dry the mating surfaces.

13. Install the new gasket and manifold to the cylinder head. Starting at the center and working outwards, torque the bolts gradually and evenly to 17 ft. lbs. (23 Nm).

14. Lubricate the quick connect fuel fittings with oil and connect to the chassis tubes. Ensure they are locked by pulling on them.

15. Install the PCV breather/separator box and vacuum harness assembly. Connect the brake booster hose, vacuum vapor harness and fuel pressure regulator to the intake manifold.

16. Connect the fuel injector wiring harness and charge temperature sensor. Connect the AIS motor and TPS wiring connectors.

17. Connect the vacuum hoses from the throttle body and carefully remove the harness. Connect the intercooler-to-throttle body outlet hose.

18. Connect the throttle and speed control cables from the throttle body.

19. Install the DIS ignition coil to the intake manifold.

20. Connect the radiator hose to the thermostat housing.

21. Install the inlet hose to the intercooler. Install the fresh air duct to the air filter housing.

22. Refill and bleed the cooling system. Connect the negative battery cable. Start the engine and check for leaks.

3.0L Engine

1. Disconnect the negative battery cable. Relieve the fuel system pressure.

2. Drain the cooling system.

3. Remove the throttle body to air cleaner hose.

4. Remove the throttle body and transaxle kickdown linkage.

5. Remove the AIS motor and TPS wiring connectors from the throttle body.

6. Remove and label the vacuum hose harness from the throttle body.

7. From the air intake plenum, remove the PCV and brake booster hoses and the EGR tube flange.

8. Disconnect and label the charge and temperature sensor wiring at the intake manifold.

9. Remove the vacuum connections from the air intake plenum vacuum connector.

10. Remove the fuel hoses from the fuel rail.

11. Remove the air intake plenum mounting bolts and remove the plenum.

12. Remove the vacuum hoses from the fuel rail and pressure regulator.

13. Disconnect the fuel injector wiring harness from the engine wiring harness.

14. Remove the fuel pressure regulator mounting bolts and remove the regulator from the fuel rail.

15. Remove the fuel rail mounting bolts and remove the fuel rail from the intake manifold.

16. Separate the radiator hose from the thermostat housing and heater hoses from the heater pipe.

17. Remove the intake manifold

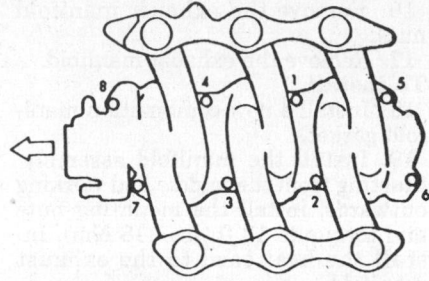

Intake manifold bolt torque sequence—3.0L engine

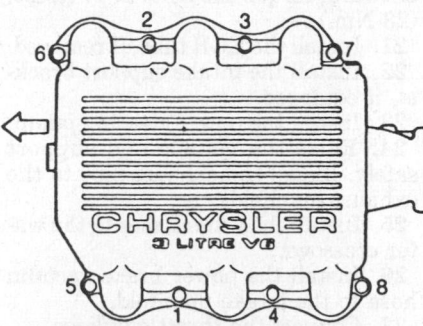

Air intake plenum bolt torque sequence—3.0L engine

mounting bolts and remove the manifold from the engine.

18. Clean the gasket mounting surfaces on the engine and intake manifold.

To install:

19. Using new gaskets, position the intake manifold on the engine and install the mounting nuts and washers.

20. Torque the mounting nuts gradually and evenly, in sequence, to 15 ft. lbs. (20 Nm).

21. Make sure the injector holes are clean. Lubricate the injector O-rings with a drop of clean engine oil and install the injector assembly onto the engine.

22. Install and torque the fuel rail mounting bolts to 10 ft. lbs. (14 Nm).

23. Install the fuel pressure regulator onto the fuel rail.

24. Install the fuel supply and return tube and the vacuum crossover holddown bolt.

25. Connect the fuel injection wiring harness to the engine wiring harness.

26. Connect the vacuum harness to the fuel pressure regulator and fuel rail assembly.

27. Remove the cover from the lower intake manifold and clean the mating surface.

28. Place the intake plenum gasket with the beaded sealant side up, on the intake manifold. Install the air intake plenum and torque the mounting bolts gradually and evenly, in sequence, to 10 ft. lbs. (14 Nm).

29. Connect or install all remaining items that were disconnected or re-

moved during the removal procedure.

30. Refill the cooling system. Connect the negative battery cable and check for leaks using the DRB II to activate the fuel pump.

Exhaust Manifold

REMOVAL & INSTALLATION

2.2L DOHC Engine—TC and 2.2L Turbo III Engine

1. Disconnect the negative battery cable.

2. Remove the turbocharger assembly.

3. Remove the coolant tube from the cylinder head.

3. Remove the exhaust manifold retaining nuts and remove the manifold.

4. Clean the gasket mounting surfaces. Inspect the manifolds for cracks, flatness and/or damage.

To install:

5. Install a new exhaust manifold gasket. Do not use sealer of any kind.

6. Position the manifold on the studs and install the retaining nuts. Starting at the center and working outwards, torque the nuts gradually and evenly to 17 ft. lbs. (23 Nm).

7. Using a new gasket, connect the coolant tube to the cylinder head.

8. Install the turbocharger assembly.

9. Start the engine and check for exhaust leaks.

3.0L Engine

1. Disconnect the negative battery cable. Raise the vehicle and safely support.

2. Disconnect the exhaust pipe from the rear exhaust manifold, at the articulated joint.

3. Disconnect the EGR tube from the rear manifold and disconnect the oxygen sensor wire.

4. Remove the crossover pipe to manifold bolts.

5. Remove the rear manifold to cylinder head nuts and the manifold.

6. Lower the vehicle and remove the heat shield from the manifold.

7. Remove the front manifold to cylinder head nuts and remove the manifold.

8. Clean the gasket mounting surfaces. Inspect the manifolds for cracks, flatness and/or damage.

To install:

9. When installing, the numbers 1–3–5 on the gaskets are used with the rear cylinders and 2–4–6 are on the gasket for the front cylinders. Torque the manifold to cylinder head nuts to 14 ft. lbs. (19 Nm).

10. Install the crossover pipe to the manifold.

11. Connect the EGR tube and oxygen sensor wire.

12. Connect the exhaust pipe to the rear exhaust manifold, at the articulated joint.

13. Connect the negative battery cable and check the manifolds for leaks.

Combination Manifold

REMOVAL & INSTALLATION

2.2L and 2.5L Engines Except DOHC

WITHOUT TURBOCHARGER

NOTE: On some vehicles, some of the manifold attaching bolts are not accessible or too heavily sealed from the factory and cannot be removed on the vehicle. Head removal would be necessary in these situations.

1. Disconnect the negative battery cable.
2. Relieve the fuel system pressure.
3. Drain the cooling system.
4. Remove the air cleaner and disconnect all vacuum lines, electrical wiring and fuel lines from the throttle body.
5. Disconnect the throttle linkage.
6. Loosen the power steering pump and remove the drive belt.
7. Remove the power brake vacuum hose from the intake manifold.
8. Remove the water hoses from the water crossover.
9. Raise and safely support the vehicle. Disconnect the exhaust pipe from the exhaust manifold.
10. Remove the power steering pump from its mounting bracket and set it aside.
11. Remove the intake manifold support bracket, if equipped.
12. Remove the EGR tube, if equipped.
13. Remove the intake manifold bolts.
14. Lower the vehicle.
15. Remove the intake manifold.

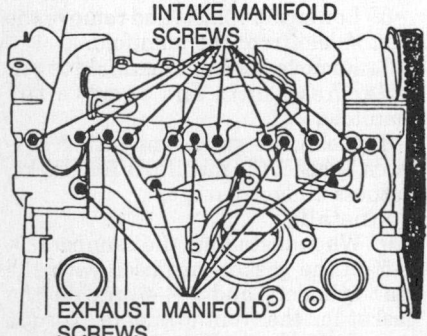

Combination manifold attaching nuts and bolts—2.2L and 2.5L non-turbocharged engines

16. Remove the exhaust manifold nuts.
17. Remove the exhaust manifold.

To install:

18. Install a new combination manifold gasket.
19. Install the manifold assembly. Starting from the middle and working outwards, install the mounting nuts and torque to 13 ft. lbs. (18 Nm). Install the heat cowl to the exhaust manifold.
20. Install the intake manifold. Starting from the middle and working outward, torque the bolts to 17 ft. lbs. (23 Nm.) .
21. Install the EGR tube, if removed.
22. Install the intake support bracket, if equipped.
23. Install the power steering pump.
24. Raise the vehicle and support safely. Install the exhaust pipe to the exhaust manifold.
25. Install the water hoses to the water crossover.
26. Install the power brake vacuum hose to the intake manifold.
27. Connect the throttle linkage.
28. Install all vacuum lines, electrical wiring and fuel lines to the carburetor or throttle body.
29. Install the air cleaner assembly.
30. Refill the cooling system.
31. Connect the negative battery cable and check the manifolds for leaks.

WITH TURBOCHARGER

NOTE: On some vehicles, some of the manifold attaching bolts are not accessible or too heavily sealed from the factory and cannot be removed on the vehicle. Head removal would be necessary in these situations.

1. Disconnect the negative battery cable. Drain the cooling system. Raise and safely support the vehicle.
2. Disconnect the exhaust pipe at the articulated joint. Disconnect the oxygen sensor at the electrical connection.
3. Remove the turbocharger to engine support bracket.
4. Loosen the oil drain back tube connector hose clamps. Move the tube down on the engine block fitting.
5. Disconnect the turbocharger coolant inlet tube from the engine block and disconnect the tube support bracket.
6. Remove the air cleaner assembly, including the throttle body adaptor, hose and air cleaner box with support bracket.
7. Disconnect the accelerator linkage, throttle body electrical connector and vacuum hoses.
8. Relocate the fuel rail assembly. Remove the bracket to intake manifold screws and the bracket to heat shield

clips. Lift and secure the fuel rail with injectors, wiring harness and fuel lines intact, up and aside.

9. Disconnect the turbocharger oil feed line at the oil sending unit tee fitting.
10. Disconnect the upper radiator hose from the thermostat housing.
11. Remove the cylinder head, manifolds and turbocharger as an assembly.
12. With the assembly on a workbench, loosen the upper turbocharger discharge hose end clamp.

NOTE: Do not disturb the center deswirler retaining clamp.

13. Remove the throttle body to intake manifold screws and throttle body assembly. Disconnect the turbocharger coolant return tube from the water box. Disconnect the retaining bracket on the cylinder head.
14. Remove the heat shield to intake manifold screws and the heat shield.
15. Remove the turbocharger to exhaust manifold nuts and the turbocharger assembly.
16. Remove the intake manifold bolts and the intake manifold.
17. Remove the exhaust manifold nuts and the exhaust manifold.

To install:

18. Place a new 2-sided Grafoil type intake/exhaust manifold gasket; do not use sealant.
19. Position the exhaust manifold on the cylinder head. Apply anti-seize compound to threads, install and torque the retaining nuts, starting at center and progressing outward in both directions, to 17 ft. lbs. (23 Nm). Repeat this procedure until all nuts are at 17 ft. lbs. (23 Nm).
20. Position the intake manifold on the cylinder head. Install and torque the retaining screws, starting at center and progressing outward in both directions, to 19 ft. lbs. (26 Nm). Repeat this procedure until all screws are at 19 ft. lbs. (26 Nm).
21. Connect the turbocharger outlet to the intake manifold inlet tube. Position the turbocharger on the exhaust manifold. Apply anti-seize compound to threads and torque the nuts to 30 ft. lbs. (41 Nm). Torque the connector tube clamps to 30 inch lbs. (41 Nm).
22. Install the tube support bracket to the cylinder head.
23. Install the heat shield on the intake manifold. Torque the screws to 105 inch lbs. (12 Nm).
24. Install the throttle body air horn into the turbocharger inlet tube. Install and torque the throttle body to intake manifold screws to 21 ft. lbs. (28 Nm). Torque the tube clamp to 30 inch lbs.
25. Install the cylinder head/mani-

folds/turbocharger assembly on the engine.

26. Reconnect the turbocharger oil feed line to the oil sending unit tee fitting and bearing housing, if disconnected. Torque the tube nuts to 10 ft. lbs. (14 Nm).

27. Install the air cleaner assembly. Connect the vacuum lines and accelerator cables.

28. Reposition the fuel rail. Install and torque the bracket screws to 21 ft. lbs. (28 Nm). Install the air shield to bracket clips.

29. Connect the turbocharger inlet coolant tube to the engine block. Torque the tube nut to 30 ft. lbs. (41 Nm). Install the tube support bracket.

30. Install the turbocharger housing-to-engine block support bracket and the screws hand tight. Torque the block screw 1st to 40 ft. lbs. (54 Nm). Torque the screw to the turbocharger housing to 20 ft. lbs. (27 Nm).

31. Reposition the drain back hose connector and tighten the hose clamps. Reconnect the exhaust pipe.

32. Connect the upper radiator hose to the thermostat housing.

33. Refill the cooling system.

34. Connect the negative battery cable and check the manifolds for leaks.

Turbocharger

REMOVAL & INSTALLATION

NOTE: On some vehicles, some of the turbocharger to exhaust manifold nuts are not accessible enough to loosen and cannot be removed on the vehicle. Head removal would be necessary in these situations.

1. Disconnect the negative battery cable. Drain the cooling system. Disconnect all air cleaner ducts from the turbocharger.

2. Disconnect the EGR valve tube at the EGR valve, if equipped. Disconnect the vacuum hose from the wastegate actuator.

3. Disconnect the turbocharger oil feed at the oil sending unit block or turbocharger and the coolant tube at the water box. Disconnect the oil/coolant support bracket from the cylinder head.

4. Remove the right intermediate shaft, bearing support bracket and outer driveshaft assemblies.

5. Remove the turbocharger to engine block support bracket.

6. Disconnect the exhaust pipe at the articulated joint. Disconnect the oxygen sensor electrical connection.

7. Loosen the oil drain-back tube connector clamps and move the tube hose down on the nipple.

8. Disconnect the coolant tube nut

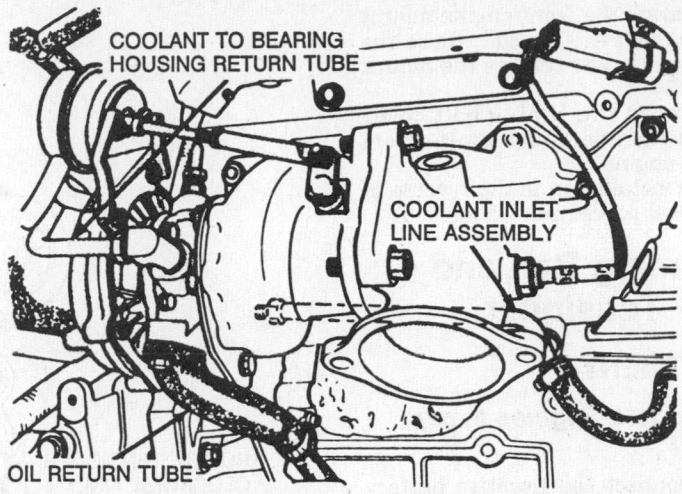

Coolant and oil tube connections on turbo—2.2L Turbo III engine; others similar

at the block outlet, below steering pump bracket, and the tube support bracket.

9. Remove the turbocharger to exhaust manifold nuts. Carefully route the oil and coolant lines and remove the assembly down and out of the vehicle.

To install:

NOTE: Before installing the turbocharger assembly, be sure it is first charged with oil. Failure to do this may cause damage to the turbocharger assembly.

10. Position the turbocharger on the exhaust manifold. Apply an anti-seize compound, Loctite® 771–64 or equivalent, to the threads and torque the retaining nuts to 40 ft. lbs. (54 Nm). Connect the vacuum hose.

11. Connect the coolant tubes using new gaskets where necessary.

12. Position the oil drain-back hose and torque the clamps to 30 inch lbs.

13. Install and torque the:
Turbocharger to engine support bracket block screw to 40 ft. lbs. (54 Nm).
Turbocharger housing screw to 20 ft. lbs. (27 Nm).
Articulated joint shoulder bolts to 21 ft. lbs. (28 Nm).

14. Install the right driveshaft assembly, the starter and the oil feed line.

15. Refill the cooling system. Connect the negative battery cable and check the turbocharger for proper operation.

Timing Belt Covers

REMOVAL & INSTALLATION

2.2L and 2.5L Engines

1. Disconnect the negative battery cable.

2. Remove the nuts and bolts that attach the upper cover to the valve cover, block or cylinder head.

3. Remove the bolt that attaches the upper cover to the lower cover.

4. Remove the upper cover.

5. Raise the vehicle and support safely. Remove the right wheel and side splash shield.

6. Remove the crankshaft pulley, water pump pulley and the belt(s).

7. Remove the lower cover attaching bolts.

8. Remove the lower cover.

9. The installation is the reverse of the removal procedure.

3.0L Engine

1. Disconnect the negative battery cable.

2. If equipped with air conditioning, loosen the adjustment pulley locknut, turn the screw counterclockwise to reduce the drive belt tension and remove the belt.

3. To remove the serpentine drive belt, insert a ½ in. breaker bar in to the square hole of the tensioner pulley, rotate it counterclockwise to reduce the drive belt tension and remove the belt.

4. Remove the air conditioning compressor and the air compressor bracket, if equipped, power steering pump and alternator from the mounts; support them aside. Remove power steering pump/alternator automatic belt tensioner bolt and the tensioner.

5. Raise and safely support the vehicle. Remove the right inner fender splash shield.

6. Remove the crankshaft pulley bolt and the pulley/damper assembly from the crankshaft.

7. Lower the vehicle and place a floor jack under the engine to support it.

8. Separate the front engine mount insulator from the bracket. Raise the engine slightly and remove the mount bracket.

9. Remove the timing belt cover bolts and the upper and lower covers from the engine.

10. The installation is the reverse of the removal procedure.

Timing Belt and Tensioner

ADJUSTMENT

2.2L and 2.5L Engines Except Turbo III

1. Disconnect the negative battery cable.

2. Raise the vehicle and support safely. Remove the right front inner splash shield.

3. Remove the tensioner cover.

4. Place the special tensioning tool C–4703 on the hex of the tensioner so the weight is at about the 10 o'clock position and loosen the bolt.

5. The tensioner should drop to the 9 o'clock position. Reposition the tool as required in order to have it end up at the 9 o'clock position (parallel to the ground, hanging toward the rear of the vehicle) ± 15 degrees.

6. Hold the tool in position and tighten the bolt. Do not pull the tool past the 9 o'clock position or the belt will be too tight and will cause howling or possible breakage.

7. Install the cover and the splash shield.

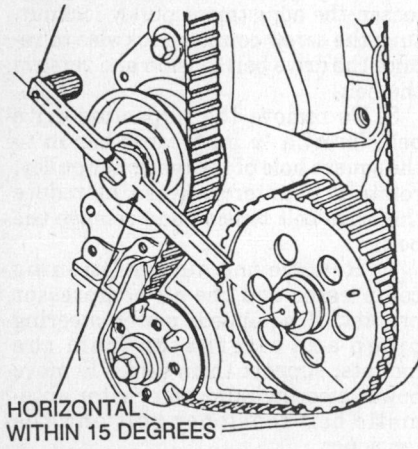

Adjusting the timing bolt tension— 2.2L and 2.5L engines except Turbo III

2.2L Turbo III Engine

1. Disconnect the negative battery cable.

2. Remove the timing covers.

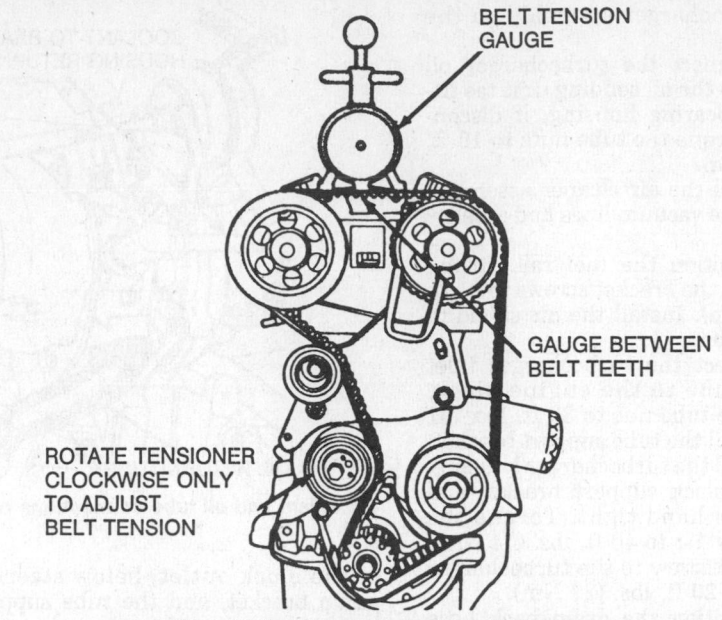

Checking the belt tension—2.2L Turbo III engine

3. Install a suitable belt tension gauge on the timing belt between the camshaft sprockets.

4. Rotate the tensioner clockwise to adjust the belt tension to 110 lbs. (445 N)

5. Rotate the crankshaft clockwise 2 revolutions and recheck the tension. Adjust as required.

6. Install the timing covers.

3.0L Engine

1. Disconnect the negative battery cable.

2. Remove the timing belt covers.

3. Loosen the bolt that holds the timing belt tensioner in place.

4. Allow the spring only to pull the tensioner in automatically. Do not manually move the tensioner or the belt will be too tight.

5. Tighten the tensioner locking bolt.

6. Install the timing belt covers and all related parts.

REMOVAL & INSTALLATION

2.2L and 2.5L Engines Except DOHC

1. If possible, position the engine so the No. 1 piston is at TDC of its compession stroke. Disconnect the negative battery cable.

2. Remove the timing belt covers. Remove the timing belt tensioner and allow the belt to hang free.

3. Place a floor jack under the engine and separate the right motor mount.

4. Remove the air conditioning compressor belt idler pulley, if equipped, and remove the mounting stud. Unbolt the compressor/alternator bracket and position it aside.

5. Remove the timing belt from the vehicle.

To install:

6. Turn the crankshaft sprocket

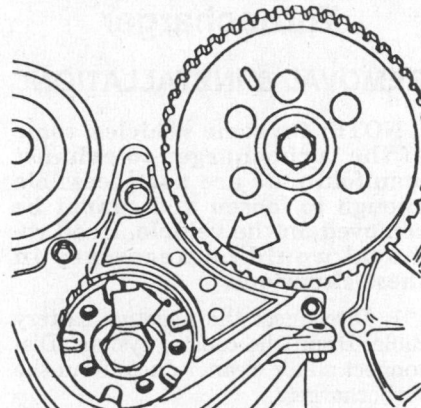

Alignment of the crankshaft sprocket and intermediate shaft sprocket—2.2L and 2.5L engines except Turbo III

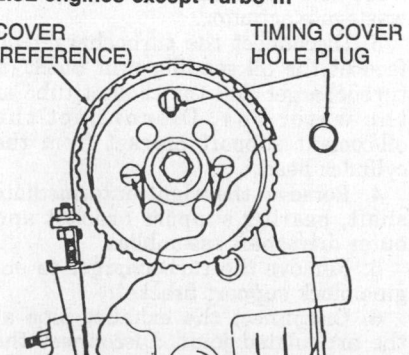

Alignment of arrows on the camshaft sprocket with the camshaft cap to cylinder head mounting line—2.2L and 2.5L SOHC engines

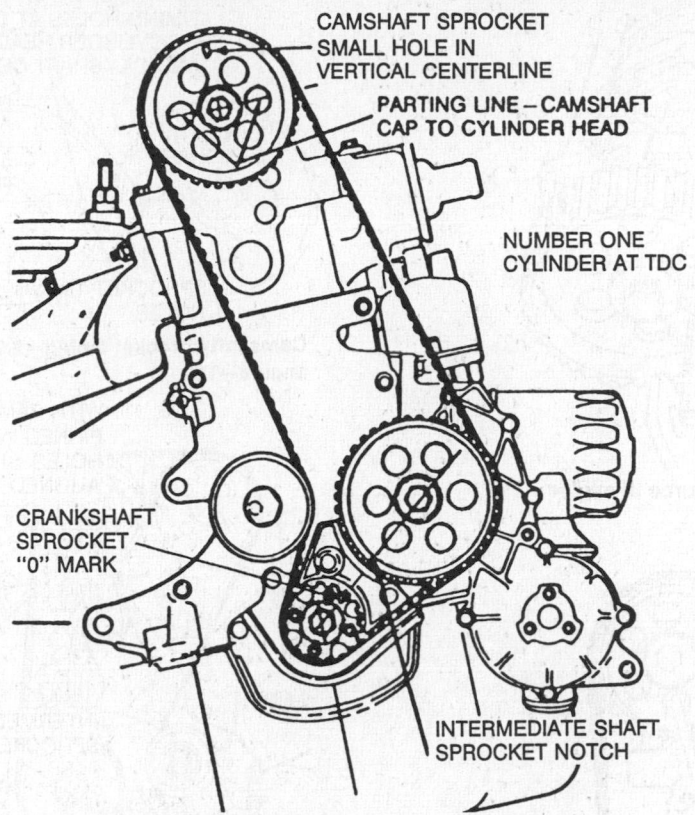

CAMSHAFT SPROCKET
SMALL HOLE IN
VERTICAL CENTERLINE

PARTING LINE—CAMSHAFT
CAP TO CYLINDER HEAD

NUMBER ONE
CYLINDER AT TDC

CRANKSHAFT
SPROCKET
"0" MARK

INTERMEDIATE SHAFT
SPROCKET NOTCH

Timing belt installation—2.2L and 2.5L SOHC engines

and intermediate shaft sprocket until the marks are in line. Use a straightedge from bolt to bolt to confirm alignment.

7. Turn the camshaft until the small hole in the sprocket is at the top and rows on the hub are in line with the camshaft cap to cylinder head mounting lines. Use a mirror to see the alignment so it is viewed straight on and not at an angle from above. Install the belt but let at hang free at this point.

8. Install the air conditioning compressor/alternator bracket, idler pulley and motor mount. Remove the floor jack. Raise the vehicle and support safely. Have the tensioner at an arm's reach because the timing belt will have to be held in position with one hand.

9. To properly install the timing belt, reach up and engage it with the camshaft sprocket. Turn the intermediate shaft counterclockwise slightly, then engage the belt with the intermediate shaft sprocket. Hold the belt against the intermediate shaft sprocket and turn clockwise to take up all tension; if the timing marks are out of alignment, repeat until alignment is correct.

10. Using a wrench, turn the crankshaft sprocket counterclockwise slightly and wrap the belt around it. Turn the sprocket clockwise so there is

no slack in the belt between sprockets; if the timing marks are out of alignment, repeat until alignment is correct.

NOTE: If the timing marks are in line but slack exists in the belt between either the camshaft and intermediate shaft sprockets or the intermediate and crankshaft sprockets, the timing will be incorrect when the belt is tensioned. All slack must be only between the crankshaft and camshaft sprockets.

11. Install the tensioner and install the mounting bolt loosely. Place the special tensioning tool C–4703 on the hex of the tensioner so the weight is at about the 9 o'clock position (parallel to the ground, hanging toward the rear of the vehicle) plus or minus 15 degrees.

12. Hold the tool in position and tighten the bolt to 45 ft. lbs. (61 Nm). Do not pull the tool past the 9 o'clock position; this will make the belt too tight and will cause it to howl or possibly break.

13. Lower the vehicle and recheck the camshaft sprocket positioning. If it is correct install the timing belt covers and all related parts.

14. Connect the negative battery cable and road test the vehicle.

2.2L Turbo III Engine

1. Disconnect the negative battery cable.

2. Remove the timing belt covers.

3. Place a floor jack under the engine and separate the right motor mount.

4. Raise the vehicle and support safely. Remove the lower accessory drive belt idler pulley bracket assembly.

5. Loosen the timing belt tensioner and remove the timing belt and idler pulley.

To install:

6. Remove the air cleaner fresh air duct, ignition cable cover, spark plugs and valve covers.

7. Loosen the rocker arm retaining bolts about 3 turns in the proper sequence. Check all lash adjusters and replace any that are damaged.

8. Align and pin both camshaft sprockets with $\frac{3}{16}$ in. drills or pin punches.

9. Install a dial indicator so the plunger is in the No. 1 spark plug hole. Rotate the crankshaft until the No. 1 piston is at TDC. Matchmark the crankshaft sprocket to the engine block for reference. Since there is no distributor, the intermediate shaft sprocket does not need to be timed.

10. Install the timing belt and idler pulley starting at the crankshaft and working counterclockwise. Make sure there is no slack between sprockets when installing.

11. Install a suitable belt tension gauge on the timing belt between the camshaft sprockets. Remove the pins from the camshaft sprockets.

12. Rotate the tensioner clockwise to adjust the belt tension to 110 lbs. (445 N). Torque the tensioner bolt 39 ft. lbs. (53 Nm).

13. Rotate the crankshaft clockwise 2 revolutions and recheck the timing and tension. Adjust as required.

14. Torque the rocker arm bolts in sequence to 18 ft. lbs. (24 Nm).

15. Install engine mount and timing belt covers.

16. Install the spark plugs, valve covers, ignition cable cover and air duct.

17. Connect the negative battery cable.

2.2L DOHC Engine—TC

1. If possible, position the engine so the No. 1 piston is at TDC of its compression stroke. Disconnect the negative battery cable.

2. Remove the timing belt covers.

3. Remove the timing belt tensioner and allow the belt to hang free.

4. Place a floor jack under the engine and separate the right motor mount.

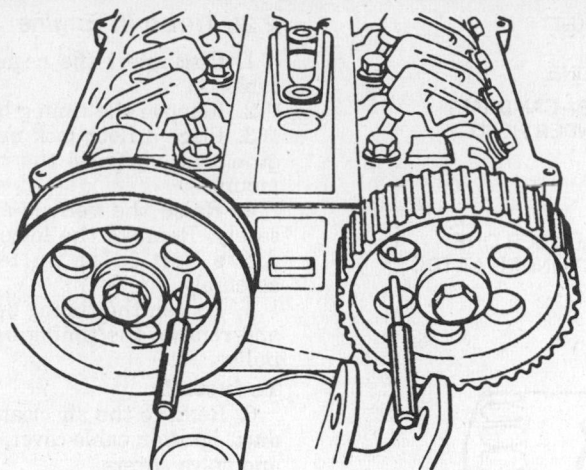

Camshaft pinned in position—2.2L Turbo III engine

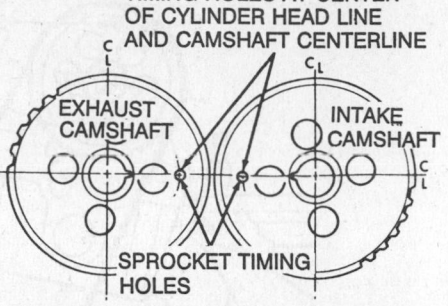

Camshaft sprocket timing—2.2L DOHC engine—TC

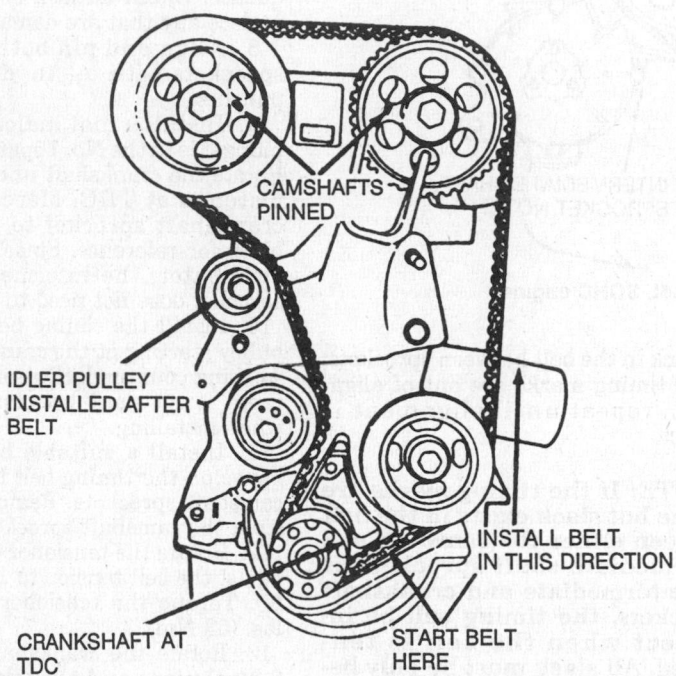

Timing belt properly installed—2.2L Turbo III engine

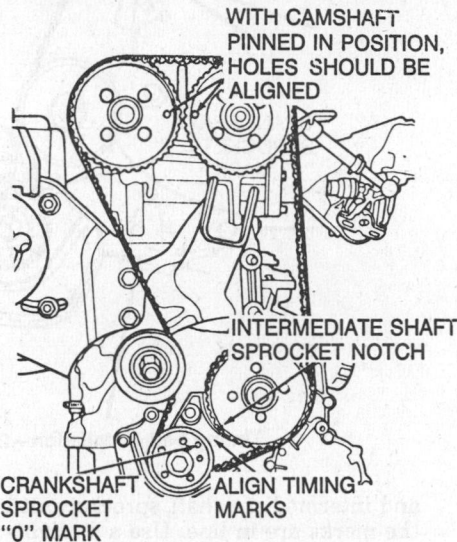

Timing belt installation—2.2L DOHC engine—TC

5. Remove the timing belt from the vehicle.

To install:

6. Turn the crankshaft sprocket and intermediate shaft sprocket until the marks are in line. Use a straight-edge from bolt to bolt to confirm alignment.

7. Nos. 1 and 6 camshaft journals have aligning pin holes to index with the blind holes in the camshaft. Turn the camshafts until the pin holes in the journals align with the aligning holes in the corresponding bearing caps. Install pin punches to secure this timing position. At this position, the sprocket timing holes on the camshaft sprockets should both be centered at the cylinder head mounting surface line.

8. Install the motor mount. Remove the floor jack. Raise the vehicle and support safely. Have the tensioner at an arm's reach because the timing belt will have to be held in position with one hand.

9. To properly install the timing belt, reach up and engage it with the camshaft sprockets, leaving no tension between sprockets. Turn the intermediate shaft counterclockwise slightly, then engage the belt with the intermediate shaft sprocket. Hold the belt against the intermediate shaft sprocket and turn clockwise to take up all tension; if the timing marks are out of alignment, repeat until alignment is correct.

10. Using a wrench, turn the crankshaft sprocket counterclockwise slightly and wrap the belt around it. Turn the sprocket clockwise so there is no slack in the belt between sprockets; if the timing marks are out of alignment, repeat until alignment is correct.

NOTE: If the timing marks are in line but slack exists in the belt anywhere except on the tensioner side, the timing will be incorrect when the belt is tensioned. All slack must be only between the crankshaft and exhaust camshaft sprockets.

11. Install the tensioner and install the mounting bolt loosely. Remove the pin punches from the camshafts. Place the special tensioning tool C–4703 on the hex of the tensioner so the weight is at about the 9 o'clock position (parallel to the ground, hanging toward the rear of the vehicle) ± 15 degrees.

12. Hold the tool in position and tighten the bolt to 45 ft. lbs. (61 Nm). Do not pull the tool past the 9 o'clock position; this will make the belt too tight and will cause it to howl or possibly break.

13. Rotate the crankshaft 2 full revolutions. With the No. 1 cylinder at TDC, all timing marks must be in line. Repeat the procedure if the timing is not correct.

14. Install the timing belt covers and all related parts.

15. Connect the negative battery cable and road test the vehicle.

3.0L Engine

1. If possible, position the engine so the No. 1 cylinder is at TDC of its compression stroke. Disconnect the negative battery cable. Remove the timing covers from the engine.

2. If the same timing belt will be reused, mark the direction of the timing belt's rotation for installation in the same direction. Make sure the engine is positioned so the No. 1 cylinder is at the TDC of its compression stroke and the sprockets timing marks are aligned with the engine's timing mark indicators.

3. Loosen the timing belt tensioner bolt and remove the belt. If not removing the tensioner, position it as far away from the center of the engine as possible and tighten the bolt.

4. If the tensioner is being removed, paint the outside of the spring to ensure it is not installed backwards. Unbolt the tensioner and remove it along with the spring.

To install:

5. Install the tensioner if removed, and hook the upper end of the spring to the water pump pin and the lower end to the tensioner in exactly the same position as originally installed. If not already done, position both camshafts so the marks align with those

on the alternator bracket (rear bank) and inner timing cover (front bank). Rotate the crankshaft so the timing mark aligns with the mark on the oil pump.

6. Install the timing belt on the crankshaft sprocket and while keeping the belt tight on the tension side (right side), install the belt on the front camshaft sprocket.

7. Install the belt on the water pump pulley, then the rear camshaft sprocket and the tensioner.

8. Rotate the front camshaft counterclockwise to tension the belt between the front camshaft and the crankshaft. If the timing marks came out of line, repeat the procedure.

9. Install the crankshaft sprocket flange.

10. Loosen the tensioner bolt and allow the spring to tension the belt.

11. Turn the crankshaft 2 full turns in the clockwise direction only until the timing marks are aligned and torque the tensioner lock bolt to 21 ft. lbs. (29 Nm).

12. Install the timing belt covers and all related parts.

13. Connect the negative battery cable and road test the vehicle.

Timing Sprockets

REMOVAL & INSTALLATION

2.2L and 2.5L Engines

1. Disconnect the negative battery cable. Remove the timing belt.

2. Remove the crankshaft sprocket bolt. Using the puller tool C–4685 or equivalent and the button from tool L–

4524 or equivalent, remove the crankshaft sprocket.

3. Using the tool C–4687 or equivalent, hold the camshaft and/or intermediate shaft sprocket, remove the center bolt and the sprocket(s).

4. Replace the seal(s) if they are leaking.

5. The installation is the reverse of the removal procedure. Torque the camshaft and intermediate sprocket bolts to 65 ft. lbs. (88 Nm) and the crankshaft sprocket bolt to 50 ft. lbs. (68 Nm).

3.0L Engine

1. Disconnect the negative battery cable.

2. Remove the timing belt.

3. To remove the camshaft sprocket, hold the sprocket with tool MB990775 or equivalent, and remove the retaining bolt and washer.

4. To remove the crankshaft sprocket, remove the bolt and remove the sprocket from the crankshaft. Replace any leaking seals.

5. The installation is the reverse of the removal procedure. Torque the camshaft sprocket bolt to 70 ft. lbs. (95 Nm) while holding the sprocket with the holding tool. Torque the crankshaft sprocket bolt. to 110 ft. lbs. (150 Nm).

Camshaft

REMOVAL & INSTALLATION

2.2L and 2.5L Engines Except DOHC

1. Disconnect the negative battery cable.

2. Turn the crankshaft so the No. 1 piston is at the TDC of the compression stroke. Remove the upper timing belt cover.

3. Remove the camshaft sprocket bolt and the sprocket and suspend tightly so the belt does not lose tension. If it does, the belt timing will have to be reset.

4. Remove the valve cover.

5. If the rocker arms are being reused, mark them for installation identification and loosen the camshaft bearing bolts, evenly and gradually.

6. Using a soft mallet, tap the rear of the camshaft a few times to break the bearing caps loose.

7. Remove the bolts, bearing caps and the camshaft with seals.

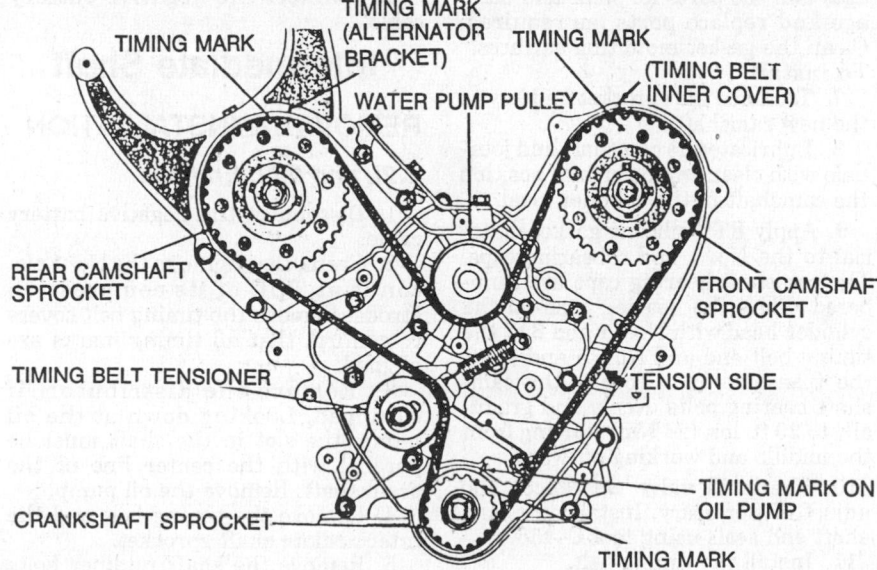

TIMING MARK (ALTERNATOR BRACKET)
TIMING MARK
WATER PUMP PULLEY
TIMING MARK (TIMING BELT INNER COVER)
REAR CAMSHAFT SPROCKET
FRONT CAMSHAFT SPROCKET
TIMING BELT TENSIONER
TENSION SIDE
TIMING MARK ON OIL PUMP
CRANKSHAFT SPROCKET
TIMING MARK

Timing belt Installation – 3.0L engine

NOTE: Take note of the color of the paint stripe on the rear camshaft seal. These stripes differentiate seal sizes. If a seal with a different color stripe is installed, a severe leak will develop if the seal is too small, or the cap will not be able to be fully installed if the seal is too big.

Also, oversized components can be identified as follows: the top of the bearing caps are painted green and "O/SJ" is stamped behind the oil galley plug on the end of the head. The barrel of an oversized camshaft is also painted green and "O/SJ" is stamped on the end of the shaft. If normal sized parts are installed in place of oversized ones, oil pressure will be significantly reduced.

8. Check the oil passages for blockages and the parts for wear and damage and replace parts, as required. Clean the gasket mounting surfaces.

To install:

9. Transfer the sprocket key to the new camshaft. New rocker arms and a new camshaft sprocket bolt are normally included with the camshaft package. Install the rocker arms, lubricate the camshaft and install with end seals installed.

10. Place the bearing caps with No. 1 at the timing belt end and No. 5 at the transaxle end. The camshaft bearing caps are numbered and have arrows facing forward. Torque the camshaft bearing bolts evenly and gradually to 18 ft. lbs. (24 Nm).

NOTE: Apply RTV silicone gasket material to the No. 1 and 5 bearing caps. Install the bearing caps before the seals are installed.

11. Mount a dial indicator to the front of the engine and check the camshaft endplay. Play should not exceed 0.020 in.

12. Install the camshaft sprocket and the new bolt.

13. Install the valve cover with a new gasket.

14. Connect the negative battery cable and check for leaks.

2.2L TURBO III Engine

1. Disconnect the negative battery cable.

2. Remove the cylinder head.

3. Remove the rocker shaft assemblies.

4. The thrust plates in the rear of the head are not interchangeable; the intake camshaft uses a wider plate. Identify the plates and remove them.

5. To remove the cam seal, push the cam toward the seal end and the seal will be pushed out of its bore in the head.

6. Carefully pull the camshaft from the head. The intake and exhaust camshafts are not interchangeable. If both are being removed, identify them for installation purposes.

To install:

7. Inspect the camshaft for wear and replace any parts that are damaged.

8. Lubricate the journals with fresh engine oil and insert the camshaft into the head.

9. Install the thrust plates and tighten the retaining nuts to 70 inch lbs. (8 Nm).

10. Install new camshaft seals flush with the head surface using installation tool C-4680.

11. Move the camshaft as far rearward as possible. Use a dial indicator and measure the endplay. Endplay specification is 0.001–0.008 in. (0.026–0.206mm).

12. Install the rocker shaft assemblies.

13. Install the cylinder head.

14. Connect the negative battery cable.

2.2L DOHC Engine—TC

1. Disconnect the negative battery cable.

2. Turn the crankshaft so the No. 1 piston is at the TDC of the compression stroke. With all timing marks aligned, remove the timing belt.

3. Remove the valve cover.

4. Remove the camshaft bearing caps nuts and washers.

5. Using a soft mallet, rap the camshaft caps a few times to break them loose.

6. Check the oil passages for blockages and the parts for wear and damage and replace parts, as required. Clean the gasket mounting surfaces.

To install:

7. Transfer the sprocket and key to the new camshaft.

8. Lubricate the camshaft and journals with clean engine oil and position the camshaft in the cylinder head.

9. Apply RTV silicone gasket material to the No. 1 and 6 bearing caps. The camshaft bearing caps are numbered. Place the bearing caps on the cylinder head with Nos. 1 and 6 at the timing belt end and Nos. 5 and 10 at the transaxle end. Torque the camshaft bearing bolts evenly and gradually to 20 ft. lbs. (24 Nm) starting from the middle and working outward.

10. Check all valve clearances and adjust, if necessary. Install new camshaft end seals using tool C-4680.

11. Install the timing belt.

12. Install the valve cover with new seals.

13. Connect the negative battery cable and check for leaks.

3.0L Engine

1. Disconnect the negative battery cable. Remove the air cleaner assembly and valve covers.

2. Install auto lash adjuster retainer tools MD998443 or equivalent, on the rocker arms.

3. If removing the right side (front) camshaft, remove the distributor extension.

4. Remove the camshaft bearing caps but do not remove the bolts from the caps.

5. Remove the rocker arms, rocker shafts and bearing caps, as an assembly.

6. Remove the camshaft from the cylinder head.

7. Inspect the bearing journals on the camshaft, cylinder head and bearing caps.

To install:

8. Lubricate the camshaft journals and camshaft with clean engine oil and install the camshaft in the cylinder head.

9. Align the camshaft bearing caps with the arrow mark (depending on cylinder numbers) and in numerical order.

10. Apply sealer at the ends of the bearing caps and install the assembly.

11. Torque the bearing cap bolts, in the following sequence: No. 3, No. 2, No. 1 and No. 4 to 85 inch lbs. (10 Nm).

12. Repeat the sequence increasing the torque to 175 inch lbs. (18 Nm).

13. Install the distributor extension, if it was removed.

14. Install the valve cover and all related parts.

15. Connect the negative battery cable.

Intermediate Shaft

REMOVAL & INSTALLATION

2.2L and 2.5L Engines

1. Disconnect the negative battery cable.

2. Crank the engine so the No. 1 piston is at TDC of its compression stroke. Remove the timing belt covers to confirm that all timing marks are aligned.

3. Remove the distributor, if equipped. Looking down at the oil pump, the slot in the shaft must be parallel with the center line of the crankshaft. Remove the oil pump.

4. Remove the timing belt and the intermediate shaft sprocket.

5. Remove the shaft retainer bolts and remove the retainer from the block.

Balance Shafts

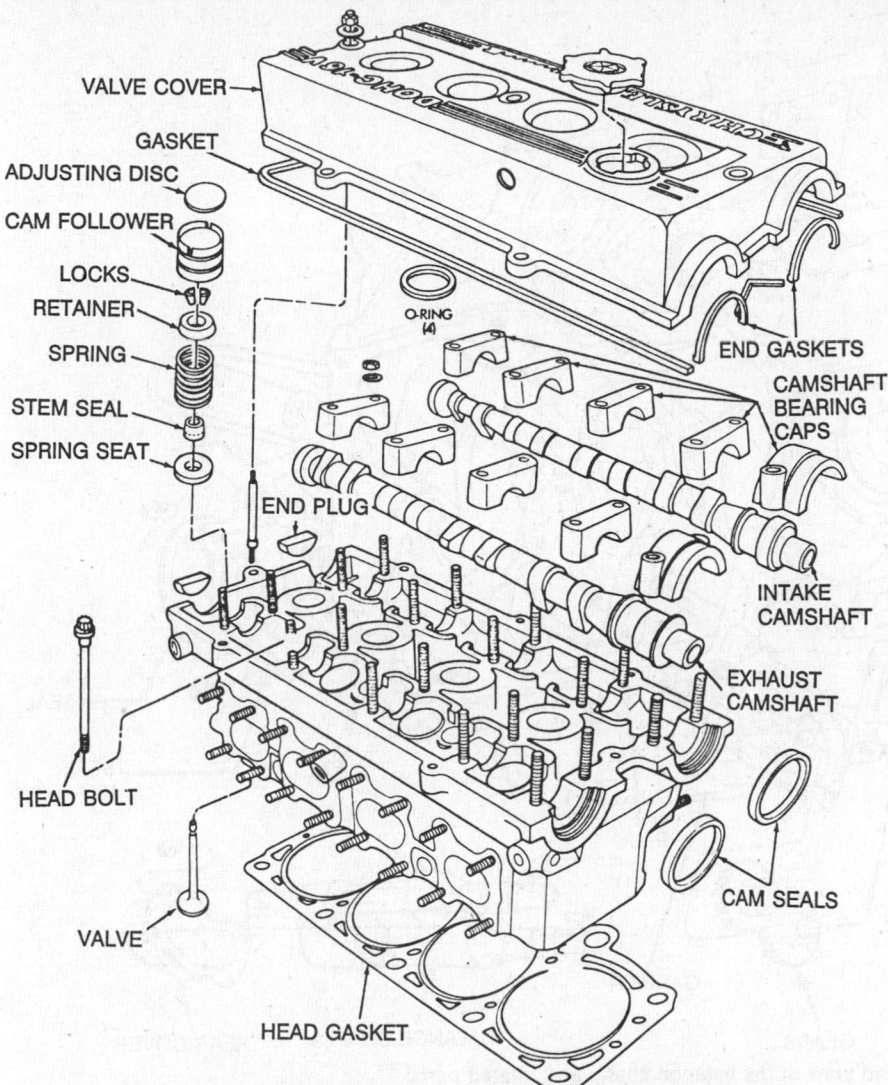

Cylinder head and valve assembly—2.2L DOHC engine—TC

Labels on diagram:
VALVE COVER, GASKET, ADJUSTING DISC, CAM FOLLOWER, LOCKS, RETAINER, SPRING, STEM SEAL, SPRING SEAT, END PLUG, HEAD BOLT, VALVE, HEAD GASKET, O-RING (4), END GASKETS, CAMSHAFT BEARING CAPS, INTAKE CAMSHAFT, EXHAUST CAMSHAFT, CAM SEALS

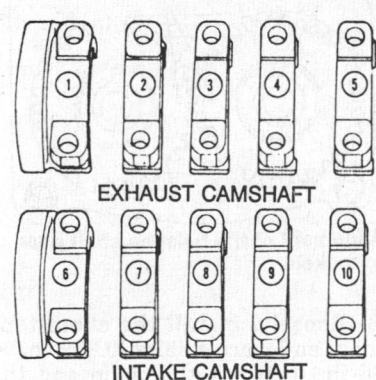

EXHAUST CAMSHAFT

INTAKE CAMSHAFT

◁ TIMING BELT END

Camshaft bearing caps installation— 2.2L DOHC engine

6. Remove the intermediate shaft from the engine.

7. If necessary, remove the front bushing using tool C–4697–2 and the rear bushing using tool C–4686–2.

To install:

8. Install the front bushing using tool C–4697–1 until the tool is flush with the block. Install the rear bushing using tool C–4686–1 until the tool is flush with the block.

9. Lubricate the distributor drive gear, if equipped, and install the intermediate shaft.

10. Replace the seal in the retainer and apply silicone sealer to the mating surface of the retainer. Install the retainer to the block and torque the bolts to 10 ft. lbs. (12 Nm).

11. Install the intermediate shaft sprocket and the timing belt.

12. With the timing belt properly installed, install the oil pump so the slot is parallel to the center line of the crankshaft. If equipped, install the distributor so the rotor is aligned with the No. 1 spark plug wire tower on the cap.

13. Connect the negative battery cable, check for leaks and adjust the ignition timing, as required.

REMOVAL & INSTALLATION

2.5L Engine and 2.2L Turbo III and IV Engines

1. Disconnect the negative battery cable. Raise the vehicle and support safely.

2. Remove the timing belt. Remove the oil pan, the oil pickup, the crankshaft belt sprocket and the front crankshaft oil seal retainer.

3. Remove the balance shaft chain cover, the guide and the tensioner.

4. Remove the balance shaft sprocket-to-shaft bolt, the gear cover to balance shaft bolt and the crankshaft sprocket-to-crankshaft bolts, then the sprockets with the balance shaft chain.

5. Remove the front gear cover-to-carrier housing stud, the gear cover and the balance shaft drive gears.

6. Remove the rear gear cover-to-carrier housing bolts, the rear cover and the balance shafts from the rear of the carrier.

7. If necessary, remove the carrier housing to crankcase bolts and the housing.

To install:

8. If the carrier housing is being installed, torque the carrier housing to crankcase bolts to 40 ft. lbs. (54 Nm).

9. Rotate the balance shafts until the keyways are facing upward, parallel to the vertical centerline of the engine.

10. Install the short hub gear on the sprocket driven shaft and the long hub gear on the gear driven shaft; make sure the gear timing marks are aligned (facing each other).

11. Install the front gear cover and torque the front gear cover to carrier housing stud bolt to 8.5 ft. lbs. (12 Nm).

12. Install the balance chain sprocket and torque the sprocket to crankshaft bolts to 11 ft. lbs. (13 Nm).

13. Rotate the crankshaft to position the No. 1 cylinder on the TDC of the compression stroke; the timing marks on the chain sprocket should align with the parting line on the left side of the No. 1 main bearing cap.

14. Position the balance shaft sprocket into the balance chain so the sprocket (yellow dot) timing mark mates with the yellow link on the chain.

15. Install the balance chain/sprocket assembly onto the crankshaft and the balance shaft. Torque the sprocket to shaft bolts to 21 ft. lbs. (28 Nm). If necessary to secure the crankshaft while tightening the bolts, place a block of wood between the crankcase and the crankshaft counterbalance.

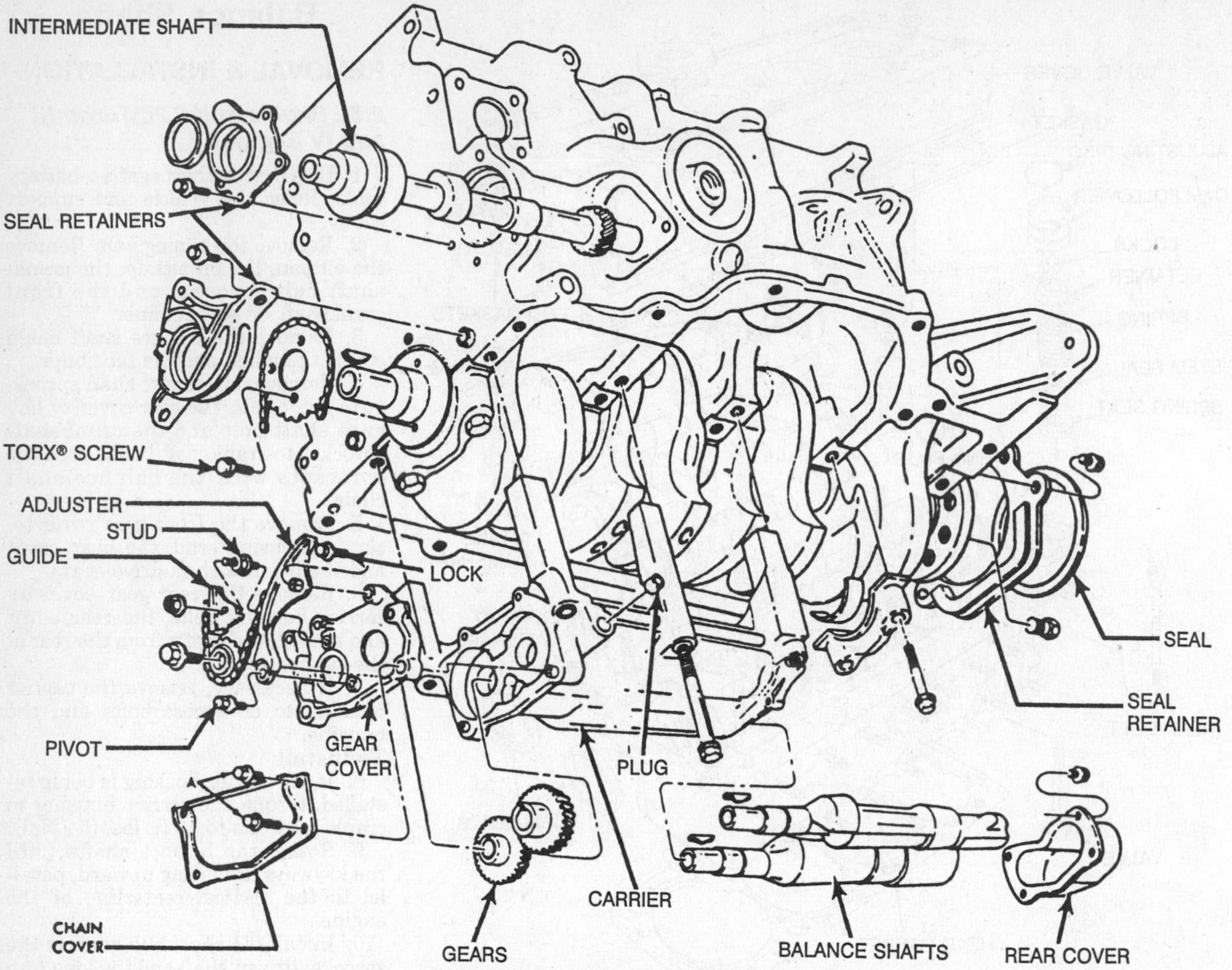

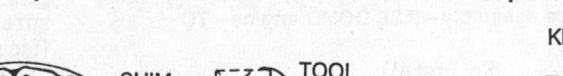

INTERMEDIATE SHAFT

SEAL RETAINERS

TORX® SCREW

ADJUSTER

STUD

GUIDE

LOCK

PIVOT

GEAR COVER

PLUG

SEAL

SEAL RETAINER

CHAIN COVER

GEARS

CARRIER

BALANCE SHAFTS

REAR COVER

Exploded view of the balance shafts and related parts

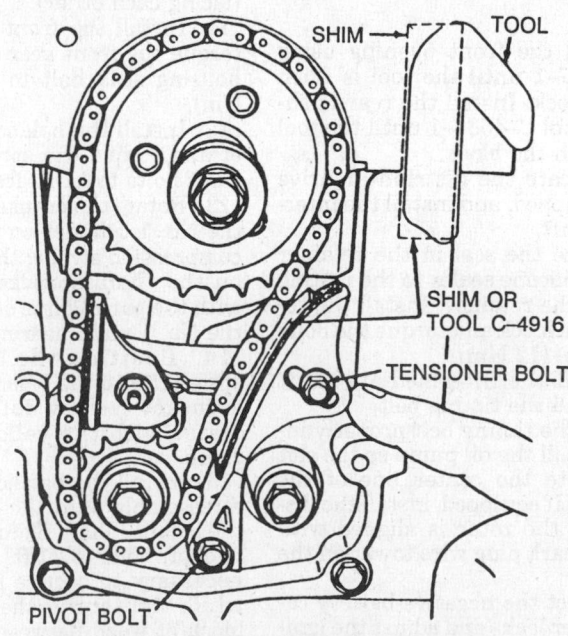

SHIM

TOOL

SHIM OR TOOL C–4916

TENSIONER BOLT

PIVOT BOLT

Adjusting the balance shaft chain tensioner

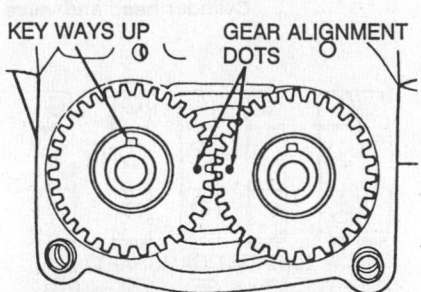

KEY WAYS UP

GEAR ALIGNMENT DOTS

Alignment of the balance shaft gear sprockets

16. Loosely, install the chain tensioners and place a shim (0.039 in. × 2.75 in.) between the chain and the tensioner. Apply firm pressure, to reduce the chain slack, to the tensioner shoe. Torque the tensioner to front gear cover bolts to 8.5 ft. lbs. (12 Nm).

17. Install the chain cover and the rear cover to the carrier housing and torque the bolts to 8.5 ft. lbs. (12 Nm).

18. Replace the crankshaft retainer seal, apply silicone sealer to the mating surface and install the retainer.

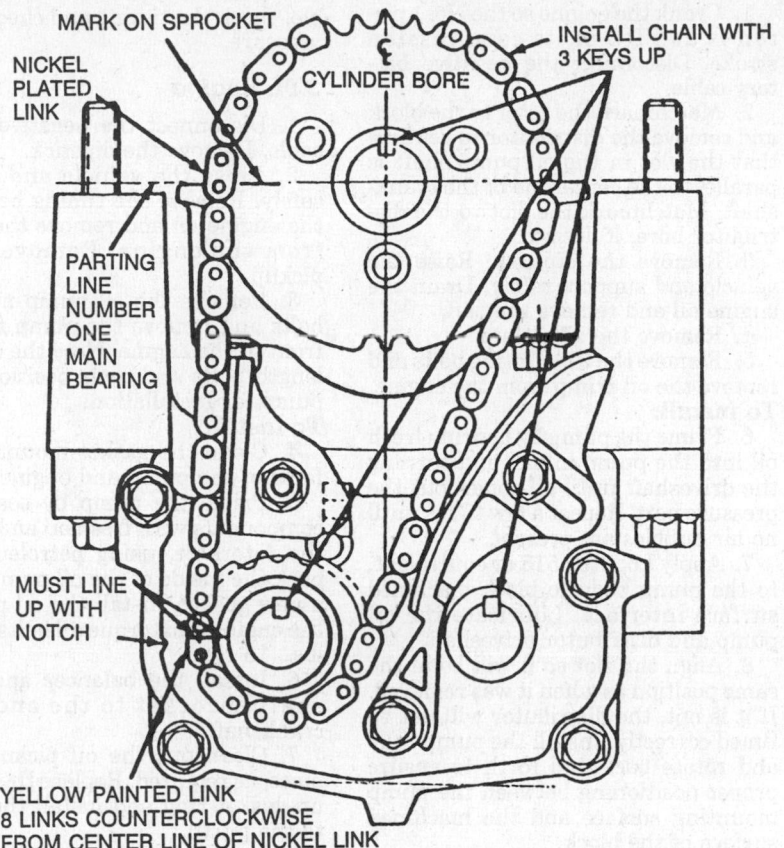

Balance shaft sprocket and crankshaft sprocket timing

MARK ON SPROCKET

NICKEL PLATED LINK

INSTALL CHAIN WITH 3 KEYS UP

CYLINDER BORE

PARTING LINE NUMBER ONE MAIN BEARING

MUST LINE UP WITH NOTCH

YELLOW PAINTED LINK 8 LINKS COUNTERCLOCKWISE FROM CENTER LINE OF NICKEL LINK

19. Install the oil pickup and oil pan.
20. Install the crankshaft sprocket and the timing belt.
21. Connect the negative battery cable and road test the vehicle.

Piston and Connecting Rod

POSITIONING

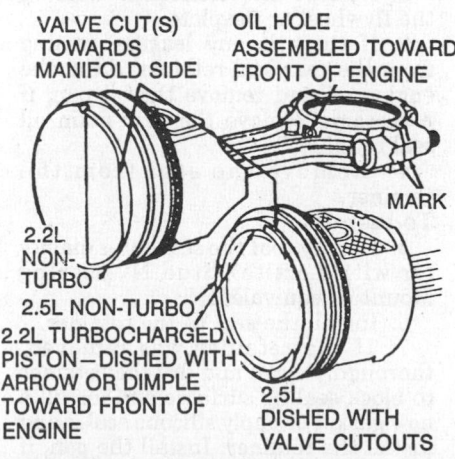

VALVE CUT(S) TOWARDS MANIFOLD SIDE

OIL HOLE ASSEMBLED TOWARD FRONT OF ENGINE

MARK

2.2L NON-TURBO

2.5L NON-TURBO

2.2L TURBOCHARGED PISTON—DISHED WITH ARROW OR DIMPLE TOWARD FRONT OF ENGINE

2.5L DISHED WITH VALVE CUTOUTS

Piston and rod positioning—2.2L and 2.5L non-turbocharged engines and 1988 2.2L turbocharged engine

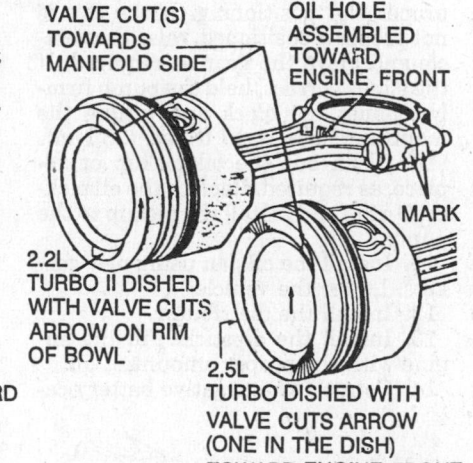

VALVE CUT(S) TOWARDS MANIFOLD SIDE

OIL HOLE ASSEMBLED TOWARD ENGINE FRONT

MARK

2.2L TURBO II DISHED WITH VALVE CUTS ARROW ON RIM OF BOWL

2.5L TURBO DISHED WITH VALVE CUTS ARROW (ONE IN THE DISH) TOWARD ENGINE FRONT

Piston and rod positioning—2.2L Turbo II and 1989 2.5L turbocharged engines

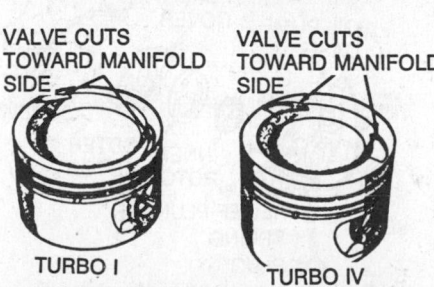

VALVE CUTS TOWARD MANIFOLD SIDE

VALVE CUTS TOWARD MANIFOLD SIDE

TURBO I

TURBO IV

Piston and rod positioning—1990-92 2.5L turbocharged and 2.2L Turbo IV engines

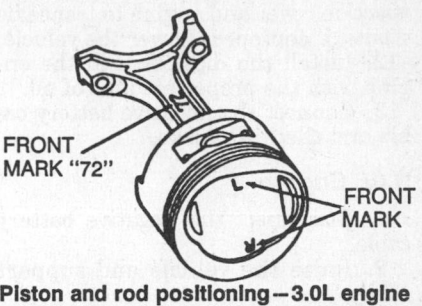

FRONT MARK "72"

FRONT MARK

Piston and rod positioning—3.0L engine

4 VALVE CUTS

OIL SLOT (BOTH SIDES)

MARK

PISTON DISHED WITH ARROW TOWARD FRONT OF ENGINE

Piston and rod positioning—2.2L DOHC engine—TC and 2.2L Turbo III engine

ENGINE LUBRICATION

Oil Pan

REMOVAL & INSTALLATION

2.2L and 2.5L Engines

1. Disconnect the negative battery cable. Remove the oil dipstick.
2. Raise the vehicle and support safely.
3. Drain the engine oil.
4. Remove the engine to transaxle struts, if equipped.
5. Remove the torque converter or clutch inspection cover.
6. Remove the oil pan retaining screws and remove the oil pan and the side seals.

To install:

7. Thoroughly clean and dry all sealing surfaces, bolts and bolt holes.
8. Apply silicone sealer to the 4 end seal-to-block corners and install the end seals making sure the corners are not twisted.
9. Apply silicone to the 4 pan-to-block corners. Install a new pan gasket or apply silicone sealer to the sealing surface of the pan and install to the engine making sure not to dislodge the end seals.
10. Install the retaining screws and torque to 17 ft. lbs. (23 Nm).
11. Install the torque converter in-

spection cover and engine to transaxle struts, if equipped. Lower the vehicle.

12. Install the dipstick. Fill the engine with the proper amount of oil.

13. Connect the negative battery cable and check for leaks.

3.0L Engine

1. Disconnect the negative battery cable.

2. Raise the vehicle and support safely.

3. Remove the torque converter bolt access cover.

4. Drain the engine oil.

5. Remove the oil pan retaining screws and remove the oil pan and gasket.

To install:

6. Thoroughly clean and dry all sealing surfaces, bolts and bolt holes.

7. Apply silicone sealer to the chain cover to block mating seam and the rear main seal retainer to block seam, if equipped.

8. Install a new pan gasket or apply silicone sealer to the sealing surface of the pan and install to the engine.

9. Install the retaining screws and torque to 50 inch lbs. (6 Nm).

10. Install the torque converter bolt access cover, if equipped. Lower the vehicle.

11. Install the dipstick. Fill the engine with the proper amount of oil.

12. Connect the negative battery cable and check for leaks.

Oil Pump

REMOVAL & INSTALLATION

2.2L and 2.5L Engines

NOTE: Many of the following steps pertain to engines with a distributor. Disregard these steps when working on Turbo III engine. Since that engine does not have a distributor, the oil pump can be installed without timing the distributor gear. The oil pump on all other engines must be properly timed.

1. Crank the engine so the No. 1 piston is at TDC of its compression stroke. Disconnect the negative battery cable.

2. Matchmark the rotor to the block and remove the distributor to confirm that the slot in the oil pump shaft is parallel to the centerline of the crankshaft. Matchmark the slot to the distributor bore, if desired.

3. Remove the dipstick. Raise vehicle and support safely. Drain the engine oil and remove the pan.

4. Remove the oil pickup.

5. Remove the 2 mounting bolts and remove the oil pump from the engine.

To install:

6. Prime the pump by pouring fresh oil into the pump intake and turning the driveshaft until oil comes out the pressure port. Repeat a few times until no air bubbles are present.

7. Apply Loctite® 515 or equivalent, to the pump body to block machined surface interface. Lubricate the oil pump and distributor driveshaft.

8. Align the slot so it will be in the same position as when it was removed. If it is not, the distributor will not be timed correctly. Install the pump fully and rotate back and forth to ensure proper positioning between the pump mounting surface and the machined surface of the block.

9. Install the mounting bolts finger-tight and lower the vehicle to confirm proper slot positioning. If the slot is not properly positioned, raise the vehicle and move the gear as required. If the slot is correct, hold the pump firmly against the block and torque the mounting bolts to 17 ft. lbs. (23 Nm).

10. Clean out the oil pickup or replace, as required. Replace the oil pickup O-ring and install the pickup to the pump.

11. Install the oil pan using new gaskets. Lower the vehicle.

12. Install the distributor.

13. Install the dipstick. Fill the engine with the proper amount of oil.

14. Connect the negative battery cable, check the timing and check the oil pressure.

3.0L Engine

1. Disconnect the negative battery cable. Remove the dipstick.

2. Raise the vehicle and support safely. Remove the timing belt, drain the engine oil and remove the oil pan from the engine. Remove the oil pickup.

3. Remove the oil pump mounting bolts and remove the pump from the front of the engine. Note the different length bolts and their position in the pump for installation.

To install:

4. Clean the gasket mounting surfaces of the pump and engine block.

5. Prime the pump by soaking its components with fresh oil and turning the rotors or, using petroleum jelly, pack the inside of the oil pump. Using a new gasket, install the oil pump on the engine and torque all bolts to 11 ft. lbs. (15 Nm).

6. Install the balancer and crankshaft sprocket to the end of the crankshaft.

7. Clean out the oil pickup or replace, as required. Replace the oil pickup gasket ring and install the pickup to the pump.

8. Install the timing belt, oil pan and all related parts.

9. Install the dipstick. Fill the engine with the proper amount of oil.

10. Connect the negative battery cable and check the oil pressure.

Rear Main Bearing Oil Seal

REMOVAL & INSTALLATION

1. Disconnect the negative battery cable.

2. Remove the transaxle. Remove the flywheel or flexplate.

3. If there is any leakage coming from the rear seal retainer, drain the engine oil and remove the oil pan, if necessary. Remove the rear main oil seal retainer.

4. Remove the seal from the retainer.

To install:

5. Lightly coat the seal outer diameter with Loctite® Stud N' Bearing Mount or equivalent.

6. Install the seal to the retainer.

7. If the retainer was removed, thoroughly clean and dry the retainer to block sealing surfaces and install a new gasket or apply silicone sealer and install the retainer. Install the pan, if it was removed.

8. Install the flywheel or flex plate and the transaxle.

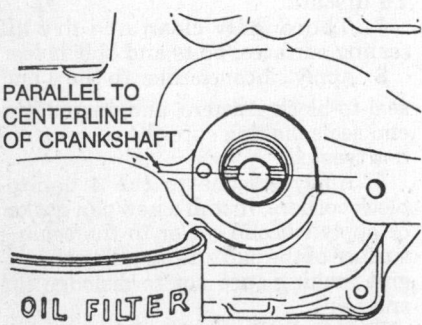

PARALLEL TO CENTERLINE OF CRANKSHAFT

OIL FILTER

Aligning the slot in the oil pump shaft— 2.2L and 2.5L engines

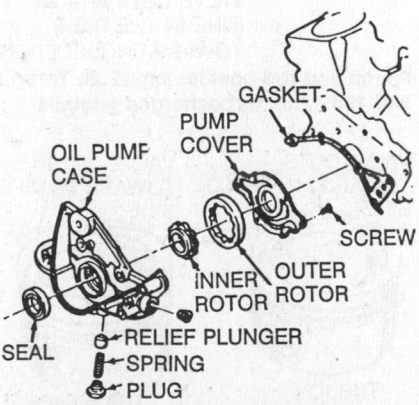

GASKET

PUMP COVER

OIL PUMP CASE

SCREW

INNER ROTOR   OUTER ROTOR

SEAL

RELIEF PLUNGER

SPRING

PLUG

Exploded view of the oil pump— 3.0L engine

9. Connect the negative battery cable and check for leaks.

ENGINE COOLING

Radiator

REMOVAL & INSTALLATION

1. Disconnect the negative battery cable.
2. Drain the coolant.
3. Remove the upper hose and coolant reserve tank hose from the radiator.
4. Remove the electric cooling fan.
5. Raise the vehicle and support safely. Remove the lower hose from the radiator.
6. Disconnect the automatic transaxle cooler hoses, if equipped, and plug them. Lower the vehicle.
7. Remove the mounting brackets and carefully lift the radiator out of the engine compartment.

To install:

8. Lower the radiator into position.
9. Install the mounting brackets.
10. Raise the vehicle, if necessary, and support safely. Connect the automatic transaxle cooler lines, if equipped.
11. Lower the vehicle and connect the lower hose.
12. Install the electric cooling fan.
13. Connect the upper hose and coolant reserve tank hose.
14. Fill the system with coolant.
15. Connect the negative battery cable, run the vehicle until the thermostat opens, fill the radiator completely and check the automatic transaxle fluid level, if equipped.
16. Once the vehicle has cooled, recheck the coolant level.

Heater Core

REMOVAL & INSTALLATION

Without Air Conditioning

1. Disconnect the negative battery cable. Drain the cooling system.
2. Clamp off the heater hoses near the heater core and remove the hoses from the core tubes. Plug the hose ends and the core tubes to prevent spillage of coolant.
3. Remove the glove box, right side kick and sill panels and all modules, relay panels and computer compo-
nents in the vacinity of the heater housing.
4. Remove the lower instrument panel silencers and reinforcements. Remove the radio and other dash-mounted optional equipment, as required.
5. Remove the floor console, if equipped. Remove the floor and defroster distribution ducts.
6. Remove the bolt holding the right side instrument panel to the right cowl.
7. Disconnect the blower motor wiring, antenna, resistor wiring and the temperature control cable.
8. On 1990-92 Daytona and LeBaron, using a suitable cutting device, cut the instrument panel along the indented line along the padded cover to the right of the glove box opening. Cut only plastic, not metal. Remove the reinforcement and the piece of instrument panel that is riveted to it.
9. Disconnect the demister hoses from the top of the housing, if equipped.
10. Disconnect the hanger strap from the package and rotate it aside.
11. Remove the retaining nuts from the package mounting studs at the firewall.
12. Fold the carpeting and insulation back to provide a little more working room and to prevent spillage from staining the carpeting. Pull the right side of the instrument panel out as far as possible.
13. Remove the heater housing from the dash panel and remove it from the passenger compartment. If the passenger seat is preventing removal, remove it.
14. To disassemble the housing assembly, remove the retaining screws from the cover and remove the cover.
15. Remove the retaining screw from the heater core and remove the core from the housing assembly.

To install:

16. Remove the temperature control door from the housing and clean the unit out with solvent. Lubricate the lower pivot rod and its well and install. Wrap the heater core with foam tape and place it in position. Secure it with its screw.
17. Assemble the housing, making sure all cover screws were used.
18. Connect the demister hoses. Install the nuts to the firewall and connect the hanger strap inside the passenger compartment.
19. Fold the carpeting back into position.
20. Install the bolt that attaches the right side of the instrument panel to the cowl.
21. Connect the blower motor wir-
ing, antenna, resistor wiring and the temperature control cable.
22. Install the air distribution ducts.
23. Install the floor console, if equipped.
24. Install the radio and all other dash mounted items that were removed during the disassembly procedure.
25. Install the lower instrument panel reinforcements and silencers.
26. Install all modules, relay panels and computer components that were removed during the disassembly procedure.
27. Install the glove box and right side kick and sill panels. Install the passenger seat.
28. Connect the heater hoses.
29. Fill the cooling system.
30. Connect the negative battery cable and check the entire climate control system for proper operation and leakage.

With Air Conditioning

1. Disconnect the negative battery cable. Properly discharge the air conditioning system. Drain the cooling system.
2. Clamp off the heater hoses near the heater core and remove the hoses from the core tubes. Plug the hose ends and the core tubes to prevent spillage of coolant.
3. Disconnect the H-valve connection at the valve and remove the H-valve. Remove the condensation tube.
4. Disconnect the vacuum lines at the brake booster and water valve.
5. Remove the glove box, right side kick and sill panels and all modules, relay panels and computer components in the vacinity of the housing.
6. Remove the lower instrument panel silencers and reinforcements. Remove the radio and other dash-mounted optional equipment, as required.
7. Remove the floor console, if equipped. Remove the floor and center distribution ducts.
8. Remove the bolt holding the right side instrument panel to the right cowl.
9. Disconnect the blower motor wiring, antenna, resistor wiring and the temperature control cable. Disconnect the vacuum harness at the connection at the top of the housing.
10. On 1990-92 Daytona and LeBaron, using a suitable cutting device, cut the instrument panel along the indented line along the padded cover to the right of the glove box opening. Cut only plastic, not metal. Remove the reinforcement and the piece of instrument panel that is riveted to it.
11. Disconnect the demister hoses

from the top of the housing, if equipped.

12. Disconnect the hanger strap from the package and rotate it aside.

13. Remove the retaining nuts from the package mounting studs at the firewall.

14. Fold the carpeting and insulation back to provide a little more working room and to prevent spillage from staining the carpeting. Pull the right side of the instrument panel out as far as possible.

15. Remove the entire housing assembly from the dash panel and remove it from the passenger compartment. Remove the passenger seat, if it is preventing removal.

16. To disassemble the housing assembly, remove the vacuum diaphragm and retaining screws from the cover and remove the cover.

17. Remove the retaining screw from the heater core and remove the core from the housing assembly.

To install:

18. Remove the temperature control door from the housing and clean the unit out with solvent. Lubricate the lower pivot rod and its well and install. Wrap the heater core with foam tape and place it in position. Secure it with its screw.

19. Assemble the housing, making sure all vacuum tubing is properly routed.

20. Feed the vacuum lines through the hole in the firewall and install the assembly to the vehicle. Connect the vacuum harness and demister hoses. Install the nuts to the firewall and connect the hanger strap inside the passenger compartment.

21. Fold the carpeting back into position.

22. Install the bolt that attaches the right side of the instrument panel to the cowl.

23. Connect the blower motor wiring, antenna, resistor wiring and the temperature control cable.

24. Install the center and floor distribution ducts.

25. Install the floor console, if equipped.

26. Install the radio and all other dash mounted items that were removed during the disassembly procedure.

27. Install the lower instrument panel reinforcements and silencers.

28. Install all modules, relay panels and computer components that were removed during the disassembly procedure.

29. Install the glove box and right side kick and sill panels. Install the passenger seat, if removed.

30. Connect the vacuum lines at the brake booster and water valve.

31. Using new gaskets, install the H-valve and condensation tube.

32. Connect the heater hoses.

33. Using the proper equipment, evacuate and recharge the air conditioning system.

34. Fill the cooling system.

35. Connect the negative battery cable and check the entire climate control system for proper operation and leakage.

Water Pump

REMOVAL & INSTALLATION

2.2L and 2.5L Engines

1. Disconnect the negative battery cable.

2. Drain the cooling system.

3. If the vehicle is equipped with air conditioning, remove the compressor from the bracket and position it aside.

4. Remove the alternator and bracket. Have a drain pan under the side mounting stud because the stud screws into a water jacket, and coolant

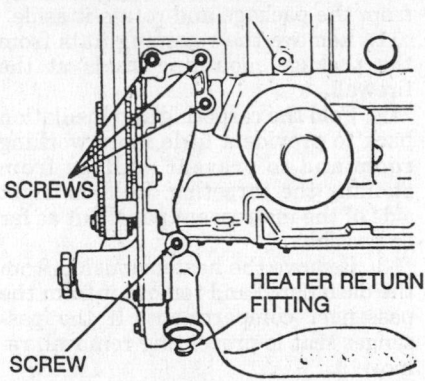

SCREWS

HEATER RETURN FITTING

SCREW

Water pump assembly—2.2L and 2.5L engines. Note differences on 2.2L Turbo III engine

will spill out when it is removed. Remove the pulley and belt from the water pump.

5. Disconnect the lower radiator hose and heater hose from the water pump.

6. Remove the water pump housing attaching screws and remove the assembly from the vehicle. Discard the O-ring. The 2.2L Turbo III engine is equipped with a spacer between the pump housing and block on the lower mounting stud.

7. Remove the water pump from the housing. The 2.2L Turbo III engine is equipped with a coolant deflector which an be re-used.

To install:

8. Using a new gasket or silicone sealer, install the water pump to the housing.

9. On 2.2L Turbo III engine, install the coolant deflector to the block and install the spacer to the lower stud. Install a new O-ring to the housing and install to the engine. Torque the 3 upper bolts to 21 ft. lbs. (30 Nm) and the lower nut to 50 ft. lbs. (68 Nm).

10. Install the water pump pulley and torque the bolts to 21 ft. lbs. (30 Nm). Connect the radiator hose and heater hose to the water pump.

11. Install all items removed to gain access to the water pump, then adjust the belts.

12. Remove the hex-head plug on the top of the thermostat housing. Fill the radiator with coolant until the coolant comes out the plug hole. Install the plug and continue to fill the radiator.

13. Connect the negative battery cable, run the vehicle until the thermostat opens, fill the radiator completely and check for leaks.

14. Once the vehicle has cooled, recheck the coolant level.

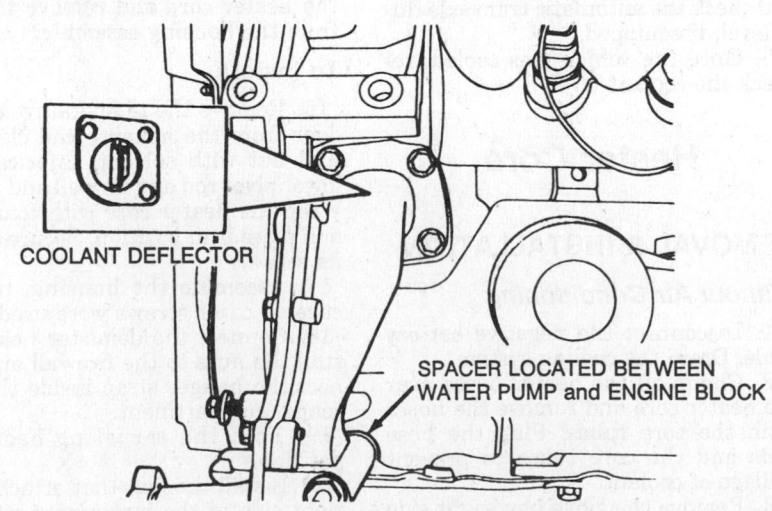

COOLANT DEFLECTOR

SPACER LOCATED BETWEEN WATER PUMP and ENGINE BLOCK

Water pump assembly—2.2L Turbo III engine

3.0L Engine

1. Disconnect the negative battery cable.

2. Drain the cooling system.

3. Remove the timing cover. If the same timing belt will be reused, mark the direction of the timing belt's rotation, for installation in the same direction. Make sure the engine is positioned so the No. 1 cylinder is at the TDC of its compression stroke and the sprockets timing marks are aligned with the engine's timing mark indicators.

4. Loosen the timing belt tensioner bolt and remove the belt. Position the tensioner as far away from the center of the engine as possible and tighten the bolt. Remove the water pump mounting bolts, separate the pump from the water inlet pipe and remove the pump from the engine.

To install:

5. Install the pump with a new gasket to the engine. Torque the water pump mounting bolts to 20 ft. lbs. (27 Nm).

6. If not already done, position both camshafts so the marks align with those on the alternator bracket (rear bank) and inner timing cover (front bank). Rotate the crankshaft so the timing mark aligns with the mark on the oil pump.

7. Install the timing belt on the crankshaft sprocket and while keeping the belt tight on the tension side (right side), install the belt on the front camshaft sprocket.

8. Install the belt on the water pump pulley, then the rear camshaft sprocket and the tensioner.

9. Rotate the front camshaft counterclockwise to tension the belt between the front camshaft and the crankshaft. If the timing marks are not aligned, repeat the procedure.

10. Install the crankshaft sprocket flange.

11. Loosen the tensioner bolt and allow the spring to tension the belt.

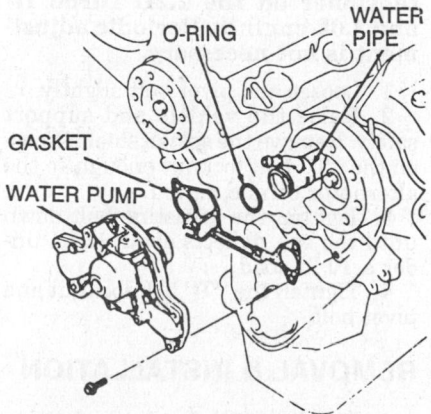

Water pump assembly—3.0L engine

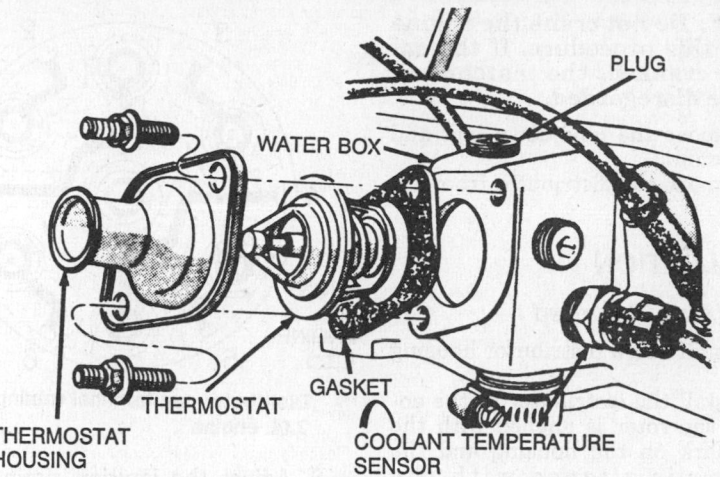

Thermostat and related items—2.2L and 2.5L SOHC engines

12. Turn the crankshaft 2 full turns in the clockwise direction only until the timing marks are aligned and torque the tensioner lock bolt to 21 ft. lbs. (29 Nm).

13. Refill the cooling system. This system uses a self-bleeding thermostat, so there is no need to bleed the system. Connect the negative battery cable, road test the vehicle and check for leaks.

Thermostat

REMOVAL & INSTALLATION

1. Disconnect the negative battery cable. Drain the coolant down to thermostat level or below.

2. Remove the thermostat housing.

3. Remove the thermostat and discard the gasket.

4. Clean the housing mating surfaces and use a new gasket.

5. The installation is the reverse of the removal procedure.

6. On 2.2L and 2.5L engines, remove the plug on top of the thermostat housing. On 2.2L Turbo III engine, remove the coolant temperature sensor on top of the housing. Fill the radiator with coolant until the coolant comes out the hole. Install the plug or sensor and continue to fill the radiator. The 3.0L engine thermostat is self-bleeding.

7. Connect the negative battery cable, run the vehicle until the thermostat opens, fill the radiator completely and check for leaks.

8. Once the vehicle has cooled, recheck the coolant level.

COOLING SYSTEM BLEEDING

To bleed air from the 2.2L and 2.5L engines, remove the plug or sensor on the top of the thermostat housing. Fill the radiator with coolant until the coolant comes out the hole. Install the plug and continue to fill the radiator. This will vent all trapped air from the engine.

The thermostat in the 3.0L engine is equipped with a small air vent valve that allows trapped air to bleed from the system during refilling. This valve negates the need for cooling system bleeding in those engines.

ENGINE ELECTRICAL

NOTE: Disconnecting the negative battery cable on some vehicles may interfere with the functions of the on board computer systems and may require the computers to undergo a relearning process, once the negative battery cable is reconnected.

Distributor

REMOVAL

1. Disconnect the negative battery cable.

2. Disconnect the distributor pickup lead wires. Remove the splash shield, if equipped.

3. Unscrew the distributor cap hold-down screws and lift off the distributor cap with all ignition wires still connected. Remove the coil wire, if necessary.

4. Matchmark the rotor to the distributor housing and the distributor housing to the engine.

NOTE: Do not crank the engine during this procedure. If the engine is cranked, the matchmark must be disregarded.

5. Remove the hold-down bolt and clamp or nut.

6. Remove the distributor from the engine.

INSTALLATION

Timing Not Disturbed

1. Install a new distributor housing O-ring.

2. Install the distributor in the engine so the rotor is aligned with the matchmark on the housing and the housing is aligned with the matchmark on the engine. Make sure the distributor is fully seated and the distributor shaft is fully engaged.

3. Install the hold-down clamp and snug the hold-down bolt or install the nut.

4. Connect the distributor pickup lead wires. Install the splash shield, if equipped.

5. Install the distributor cap and tighten the screws.

6. Connect the negative battery cable.

7. Adjust the ignition timing and tighten the hold-down bolt.

Timing Disturbed

1. Install a new distributor housing O-ring.

2. Position the engine so the No. 1 piston is at TDC of the compression stroke and the mark on the vibration damper is aligned with **0** on the timing indicator.

3. Install the distributor in the engine so the rotor is aligned with the position of the No. 1 ignition wire on the distributor cap and the housing is aligned with the matchmark on the engine. Make sure the distributor is fully seated and the distributor shaft is fully engaged.

NOTE: There are distributor cap runners inside the cap on 3.0L engine. Make sure the rotor is pointing to where the No. 1 runner originates inside the cap and not where the No. 1 ignition wire plugs into the cap.

4. Install the hold-down clamp and snug the hold-down bolt or install the nut.

5. Connect the distributor pickup lead wires. Install the splash shield, if equipped.

6. Install the distributor cap and tighten the screws.

7. Connect the negative battery cable.

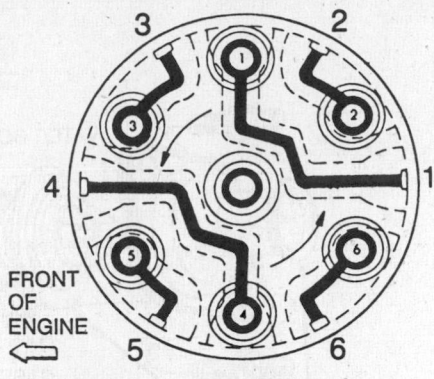

Distributor cap terminal routing— 3.0L engine

8. Adjust the ignition timing and tighten the hold-down bolt.

Ignition Timing

ADJUSTMENT

1. Start the engine, set the parking brake and run the engine until at normal operating temperature. Keep all lights and accessories **OFF**.

2. If a magnetic timing unit is available, insert the probe into the receptacle near the timing scale. The scale is located on the top of the bellhousing on 2.2L and 2.5L engines or near the crankshaft pulley on the 3.0L engine.

3. If a magnetic timing unit is not available, connect a conventional power timing light to the No. 1 cylinder spark plug wire.

4. If a Diagnostic Readout Box II (DRB II) is available, access the Basic Timing Mode.

5. If the DRB II is not available, disconnect the coolant sensor. This sensor is located on the side of the thermostat housing on 2.2L and 2.5L engines and between the distributor or thermostat housing on 3.0L engine. The Check Engine light on the instrument panel must be **ON**.

6. Aim the timing light at the timing scale or read the magnetic timing unit.

7. Loosen the distributor hold-down bolt or nut enough so the distributor can be rotated.

8. Turn the distributor in the proper direction until the specified timing according to the VECI label is reached. Tighten the hold-down bolt or nut and recheck the timing.

9. Turn the engine **OFF**. Connect the coolant sensor and check to make sure the Check Engine light does not come on when the vehicle is restarted. Disconnect the timing apparatus.

10. If the coolant temperature sensor was disconnected, erase the created fault code using the Erase Fault Code mode on the DRB II.

Alternator

PRECAUTIONS

Several precautions must be observed when working with the alternator to avoid damage to the unit.

• If the battery is removed for any reason, make sure it is reconnected with the correct polarity. Reversing the battery connections may result in damage to the one-way rectifiers.

• When utilizing a booster battery as a starting aid, always connect the positive to positive terminals and the negative terminal from the booster battery to a good engine ground on the vehicle being started.

• Never use a fast charger as a booster to start vehicles.

• Disconnect the battery cables when charging the battery with a fast charger.

• Never attempt to polarize the alternator.

• Do not use test lights of more than 12 volts when checking diode continuity.

• Do not short across or ground any of the alternator terminals.

• The polarity of the battery, alternator and regulator must be matched and considered before making any electrical connections within the system.

• Never separate the alternator on an open circuit. Make sure all connections within the circuit are clean and tight.

• Disconnect the battery ground terminal when performing any service on electrical components.

• Disconnect the battery if arc welding is to be done on the vehicle.

BELT TENSION ADJUSTMENT

NOTE: The belt tension is automatically adjusted by a dynamic tensioner on the 2.2L Turbo III and 3.0L engines. Periodic adjustment is not necessary.

1. Loosen the pivot bolt slightly.

2. Raise the vehicle and support safely. Remove the splash shield. Loosen the "T" bolt locknut enough so the alternator can be moved.

3. Tighten the adjusting bolt down until the belt deflects about ¼ in. under a 10 lb. load.

4. Tighten the "T" bolt locknut and pivot bolt.

REMOVAL & INSTALLATION

1. Disconnect the negative battery cable.

2. On 2.2L and 2.5L engines except Turbo III, remove the air conditioning compressor and position it aside, if equipped. Remove the oil filter to allow the alternator to be removed from above.

3. On 3.0L engine, release the dynamic belt tensioner using a ½ in. breaker bar and remove the belt. On other engines, loosen the mounting bolts and remove the drive belt.

4. Remove all mounting bolts, spacers and adjuster bolt, if equipped, and remove the alternator from the brackets. On Turbo III, remove the alternator bracket and separate the alternator from the bracket.

5. Remove the battery positive, field and ground terminals from the rear of the alternator. Remove the wire harness hold-down screw from the alternator, if equipped.

To install:

6. Connect all wiring to the proper terminals on the rear of the alternator and install the wire harness hold-down screw, if equipped.

7. Position the alternator in the mounting brackets.

8. Install the spacers, pivot bolt, adjuster slot bolt and adjuster bolt, if equipped. Install the belt.

9. Install the air conditioning compressor and oil filter, if they were removed.

10. Adjust the belt tension, if necessary.

11. Connect the negative battery cable and check for proper output.

Starter

REMOVAL & INSTALLATION

1. Disconnect the negative battery cable.

2. On 2.2L and 2.5L engines, remove the attaching nut and bolt at the top of the bellhousing. Raise the vehicle and support safely.

3. Remove the rear mount from the starter, if equipped. Remove the heat shield from the starter, if equipped.

4. Unbolt the starter and remove the starter from the vehicle.

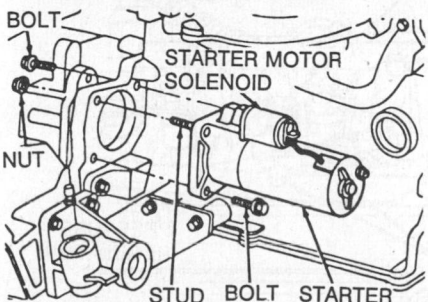

Removing or installing the starter—2.2L and 2.5L engines

5. Disconnect the solenoid lead wires from the starter.

To install:

6. Connect the solenoid lead wires and install the heat shield, if equipped.

7. On the 2.2L and 2.5L engines, install the lower bolt loosely, then lower the vehicle and install the nut and bolt from above and torque to 40 ft. lbs. (54 Nm).

8. Raise the vehicle again and torque the bottom bolt to the same value. Install the rear mount to the starter.

9. On the 3.0L engine, install all mounting bolts and torque to 40 ft. lbs. (54 Nm) evenly.

10. Connect the negative battery cable and check the starter for proper operation.

EMISSION CONTROLS

Please refer to "Emission Controls" in the Unit Repair section for system maintenance procedures. Due to the complex nature of modern electronic engine control systems, comprehensive diagnosis and testing procedures fall outside the confines of this repair manual. For complete information on diagnosis, testing and repair procedures concerning all modern engine and emission control systems, please refer to "Chilton's Guide to Fuel Injection and Electronic Engine Controls".

FUEL SYSTEM

Fuel System Service Precautions

Safety is the most important factor when performing not only fuel system maintenance but any type of maintenance. Failure to conduct maintenance and repairs in a safe manner may result in serious personal injury or death. Maintenance and testing of the vehicle's fuel system components can be accomplished safely and effectively by adhering to the following rules and guidelines.

• To avoid the possibility of fire and personal injury, always disconnect the negative battery cable unless the re-

pair or test procedure requires that battery voltage be applied.

• Always relieve the fuel system pressure prior to disconnecting any fuel system component (injector, fuel rail, pressure regulator, etc.), fitting or fuel line connection. Exercise extreme caution whenever relieving fuel system pressure to avoid exposing skin, face and eyes to fuel spray. Please be advised that fuel under pressure may penetrate the skin or any part of the body that it contacts.

• Always place a shop towel or cloth around the fitting or connection prior to loosening to absorb any excess fuel due to spillage. Ensure that all fuel spillage (should it occur) is quickly removed from engine surfaces. Ensure that all fuel soaked cloths or towels are deposited into a suitable waste container.

• Always keep a dry chemical (Class B) fire extinguisher near the work area.

• Do not allow fuel spray or fuel vapors to come into contact with a spark or open flame.

• Always use a backup wrench when loosening and tightening fuel line connection fittings. This will prevent unnecessary stress and torsion to fuel line piping. Always follow the proper torque specifications.

• Always replace worn fuel fitting O-rings with new. Do not substitute fuel hose or equivalent where fuel pipe is installed.

RELIEVING FUEL SYSTEM PRESSURE

1988

1. Disconnect the negative battery cable.

2. Loosen the fuel filler cap to release fuel tank pressure.

3. Remove the wiring harness connector from the (any) injector.

4. Using a jumper wire, ground either injector terminal.

5. Being careful not to allow contact between the jumper leads, connect a second jumper wire to the other terminal and touch the other end to the positive battery post for no longer that 10 seconds. This will relieve fuel pressure.

6. Remove the jumper wires and continue with fuel system service.

1989-92

1. Loosen the fuel filler cap to release fuel tank pressure.

2. Locate and disconnect the fuel injector harness connector.

3. Connect a jumper wire from terminal No. 1 of the appropriate connector to ground.

4. Being careful not to allow contact between the jumper leads, connect a jumper wire to terminal No. 2 of the connector and touch the other end of the jumper to the positive battery post for no longer than 5 seconds. This will relieve fuel pressure.

5. Remove the jumper wires and continue with fuel system service.

Fuel Tank

REMOVAL & INSTALLATION

1. Disconnect the negative battery cable.
2. Relieve the fuel pressure.
3. Raise the vehicle and support safely.
4. Using the proper equipment, drain the fuel tank.
5. Remove the screws that hold the filler neck to the quarter panel.
6. Disconnect the wiring and hoses from the tank.
7. Place a transmission jack or equivalent, under the center of the tank and apply slight pressure. Remove the tank straps.
8. Remove the filler tube from the tank.
9. Lower the tank and disconnect the vapor separator rollover valve hose. Remove the fuel tank from the vehicle.

To install:

10. Raise the tank into position and connect all harnesses and vacuum hoses.
11. Install the tank straps and tighten the retaining nuts.
12. Install the screws that hold the filler neck to the quarter panel.
13. Connect the negative battery cable, start the engine and check for leaks.

Fuel Filter

REMOVAL & INSTALLATION

─── CAUTION ───

Do not use conventional fuel filters, hoses or clamps when servicing this fuel system. They are not compatible with the injection system and could fail, causing personal injury or damage to the vehicle. Use only hoses and clamps specifically designed for fuel injection.

1. Disconnect the negative battery cable.
2. Relieve the fuel pressure.
3. The filter is located on the frame rail toward the rear of the vehicle. Raise the vehicle and support safely. Remove the filter retaining screw and remove the filter assembly from the mounting plate.

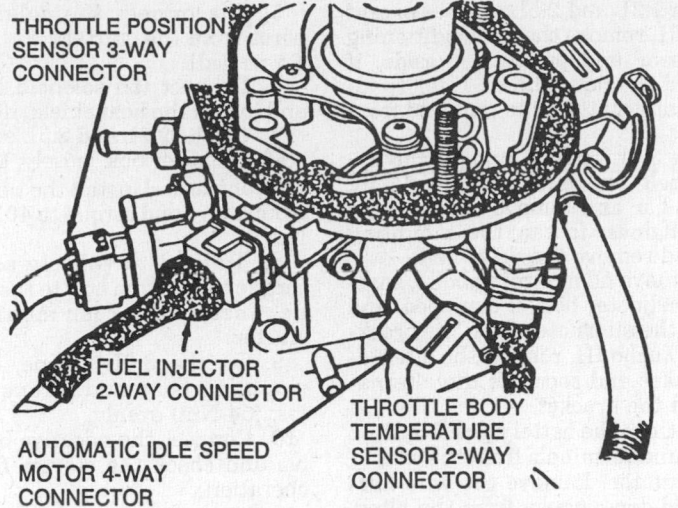

Fuel Injector harness location—2.2L and 2.5L non-turbocharged engine

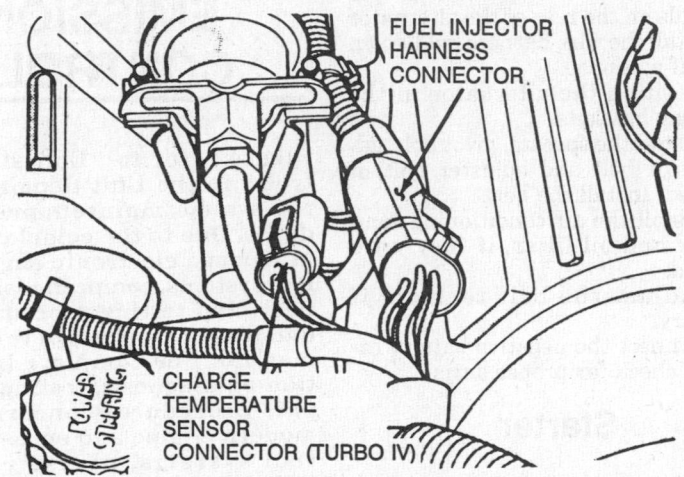

Fuel Injectors harness location—turbocharged engine, except TC and Turbo III. The connector may vary slightly between vehicles

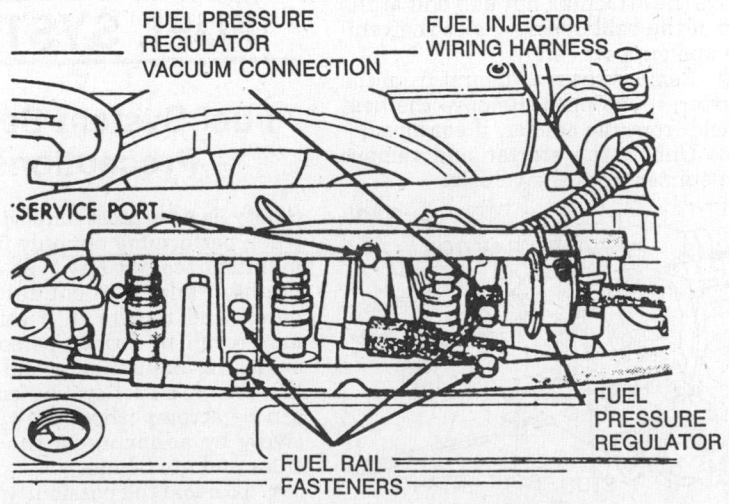

Fuel Injectors wiring harness—2.2L Turbo III engine

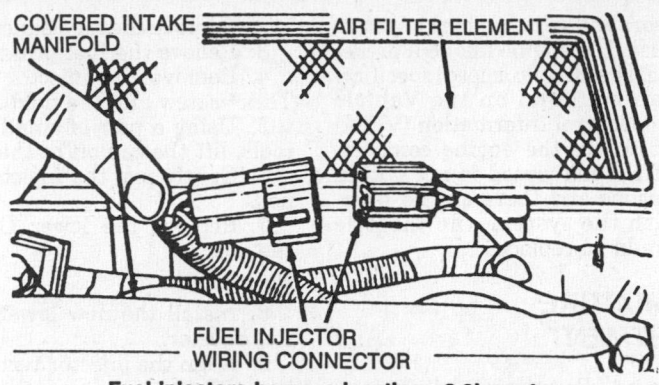

Fuel injectors harness location—3.0L engine

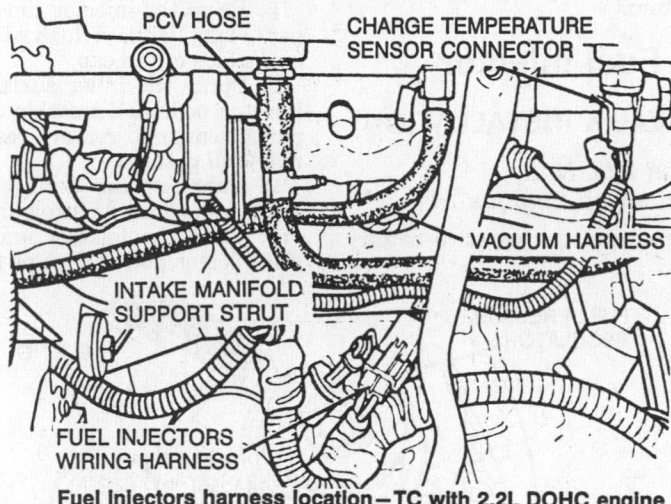

Fuel injectors harness location—TC with 2.2L DOHC engine

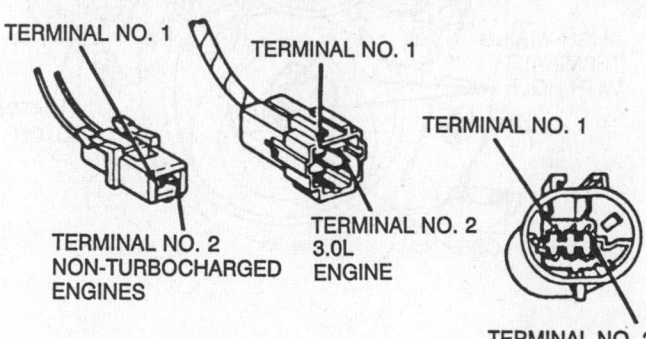

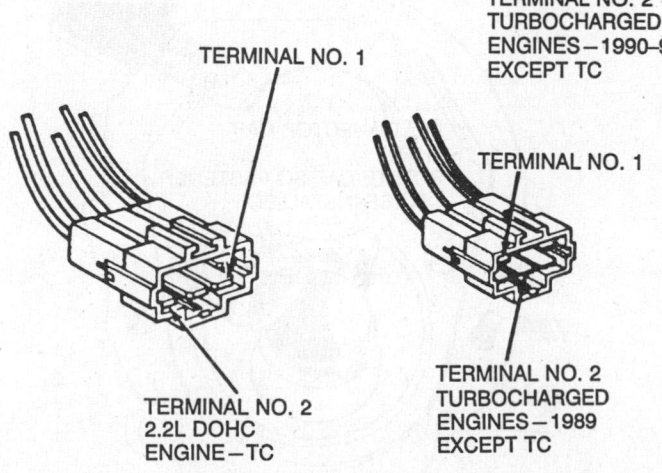

Fuel injector harness connector terminals

4. Loosen the outlet hose clamp on the filter and inlet hose clamp on the rear fuel tube.

5. Wrap a shop towel around the hoses to absorb fuel. Remove the hoses from the filter and fuel tube and discard the clamps and the filter.

To install:

6. Install the inlet hose on the fuel tube and tighten the new clamp to 10 inch lbs.

7. Install the outlet hose on the filter outlet fitting and tighten the new clamp to 10 inch lbs.

8. Position the filter assembly on the mounting plate and tighten the mounting screw to 75 inch lbs. (8 Nm).

9. Connect the negative battery cable, start the engine and check for leaks.

Electric Fuel Pump

PRESSURE TESTING

1. Relieve the fuel pressure.

2. Properly connect the fuel system pressure tester:

   a. Non-turbocharged engines—special tool C–4799 and adaptor 6539 or equivalent, is installed between the fuel supply hose and the engine fuel line assembly.

   b. Turbocharged engines—special tool C–4799 or equivalent, is installed to the fuel rail service valve.

3. With the key in the **RUN** position, put the DRB II in the activate auto shutdown relay mode; this will activate the fuel pump and pressurize the system.

4. If the pressure is within specifications, reinstall the fuel hose.

5. If fuel pressure is below specifications, install the tester with the adaptor in the fuel supply line between the tank and the filter and repeat the test.

6. If the pressure is 5 psi higher than in Step 5, replace the fuel filter. If no change is observed, squeeze the return hose. If pressure increases, replace the pressure regulator. If no change is observed, the problem is either a plugged in-tank sock filter or a defective pump.

7. If fuel pressure is above specifications, remove the fuel return line hose from the chassis line at the fuel tank and connect a 3 foot piece of fuel hose to the return line. Put the other end into a 2 gallon minimum capacity approved gasoline container. Repeat the test. If pressure is now correct, check the in-tank return hose for kinking. Replace the fuel pump assembly if the in-tank reservoir check valve or aspirator jet is obstructed.

8. If pressure is still above specifications, remove the fuel return hose from the throttle body. Connect a sub-

stitute hose to the throttle body return nipple and place the other end of the hose in a clean container. Repeat the test. If pressure is now correct, check for a restricted fuel return line. If no change is observed, replace the fuel pressure regulator.

REMOVAL & INSTALLATION

1. Disconnect the negative battery cable.
2. Relieve the fuel pressure.
3. Raise the vehicle and support safely.
4. Using the proper equipment, drain the fuel tank.
5. Remove the screws that hold the filler neck to the quarter panel.
6. Disconnect the wiring and hoses from the tank.
7. Place a transmission jack under the center of the tank and apply slight pressure. Remove the tank straps.
8. Remove the filler tube from the tank.
9. Lower the tank and disconnect the vapor separator rollover valve hose and remove the fuel tank from the vehicle.
10. Using a hammer and a brass drift, tap the lock ring counterclockwise to release the pump.
11. Partially pull the pump assembly out of the tank until the return line hose connection is visible at the of the pump assembly.
12. Disconnect the fuel fitting by pressing in on the ears.
13. Remove the pump from the tank with the O-ring. Discard the O-ring, pump inlet filter and inlet seal. Disassemble as required.
To install:
14. Install a new inlet seal and filter on the end of the pump.
15. Install a new O-ring to the pump.
16. Connect the reservoir hose to the pump assembly at the suction end of the pump. Press the female fitting onto the pump assembly male end until the ears snap in place.
17. Install the pump into the tank so the fuel return hose is not kinked.
18. Install the lock ring with a hammer and brass punch turning the ring clockwise.
19. Install the fuel tank.
20. Connect the negative battery cable, start the engine and check for leaks.

Fuel Injection

IDLE SPEED ADJUSTMENT

The idle speed is controlled by the Automatic Idle Speed motor (AIS). The AIS is controlled by the SMEC or SBEC, which receives data from vari-

ous sensors and switches in the system and adjusts the engine idle to a predetermined speed. Idle speed specifications can be found on the Vehicle Emission Control Information (VECI) label located in the engine compartment. If the idle speed is not within specifications and there are no problems with the system, the throttle body should be replaced.

IDLE MIXTURE ADJUSTMENT

There is no idle mixture adjustment provided with any Chrysler fuel injection system.

Fuel Injector

REMOVAL & INSTALLATION

2.2L and 2.5L Non-Turbocharged Engines

1. Disconnect the negative battery cable.

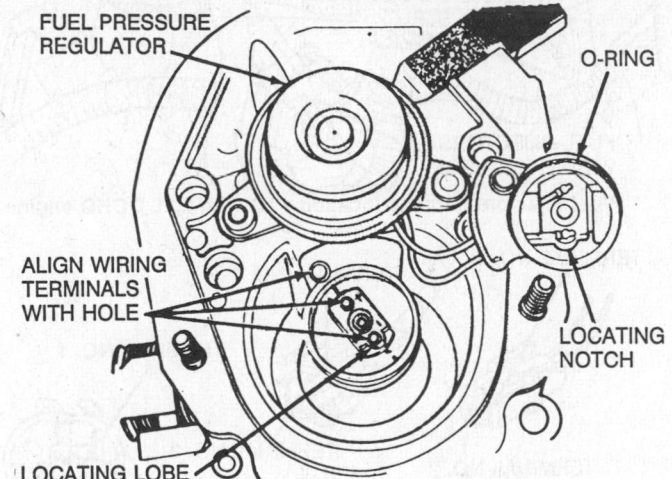

2. Remove the air cleaner assembly.
3. Relieve the fuel pressure.
4. Remove the injector hold-down Torx® screw and the hold-down.
5. Using a pair of small flat-tipped tools, lift the cap off of the injector.
6. Gently pry the injector from its pod.
7. Remove the lower O-ring from the pod.
To install:
8. Install the new lower O-ring on the injector.
9. Align the injector terminal housing with the locating socket in the injector cap.
10. Press the injector cap so the upper O-ring flange is flush with the lower surface of the cap.
11. Spray the inner surfaces of the injector pod with suitable carburetor parts cleaner to remove residual varnish and gasoline.
12. Lubricate the O-rings sparingly with unmedicated petroleum jelly.
13. Place the injector and cap into the injector pod and align the cap lo-

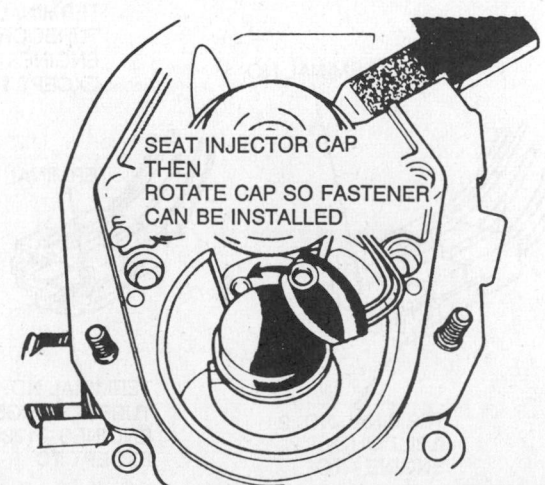

Installing the injector to the cap—2.2L and 2.5L non-turbocharged engines

cating pin with the locating hole in the casting.

14. Press firmly on the injector cap until it is flush with the casting surface.

15. Align the hole in the hold-down with the pin on the cap and install.

16. Push down on the cap, install the screw and torque to 35 inch lbs. (4 Nm).

17. Connect the negative battery cable and check for leaks using the DRB II to activate the fuel pump.

18. Install the air cleaner.

2.2L and 2.5L Turbocharged Engines

1. Disconnect the negative battery cable.

2. Relieve the fuel pressure.

3. Disconnect the injector wiring connector from the injector.

4. Unbolt the fuel rail from the rear of the engine. Position the fuel rail assembly so the fuel injectors are easily accessible. If necessary, disconnect the hoses from the fuel rail and remove it from the engine.

5. Remove the injector clip from the fuel rail and injector. Pull the injector straight out of the fuel rail receiver cup.

6. Check the injector O-ring for damage. If the O-ring is damaged, replace it. If the injector is being reused, install a protective cap on the injector tip to prevent damage.

7. Repeat the procedure for the remaining injectors.

To install:

8. Before installing an injector the rubber O-ring should be lubricated with a drop of clean engine oil to aid in installation.

9. Install injector top end into fuel rail receiver cup.

10. Install injector clip by sliding the open end into top slot of the injector and onto the receiver cup ridge into the side slots of clip.

11. Repeat the steps for the remaining injectors.

12. Install the fuel rail.

13. Connect the negative battery cable and check for leaks using the DRB II to activate the fuel pump.

3.0L Engine

1. Disconnect the negative battery cable.

2. Relieve the fuel pressure.

3. Remove the air cleaner to throttle body hose.

4. Disconnect the throttle cable from the throttle body and disconnect the kickdown linkage. Remove the throttle cable bracket attaching bolts.

5. Disconnect the connectors to the throttle body.

6. Matchmark and carefully remove the vacuum hoses from the throttle body.

7. Remove the PCV and brake booster hoses from the air intake plenum.

8. Remove the ignition coil from the intake plenum, if it is mounted there.

9. Remove the EGR tube flange from the intake plenum, if equipped.

10. Unplug the coolant temperature sensor and charge temperature sensor, if equipped.

11. Remove the vacuum connection from the air intake plenum vacuum connector.

12. Remove the fuel hoses from the fuel rail and plug them.

13. Remove the air intake plenum to intake manifold bolts and remove the plenum and gaskets. Cover the intake manifold openings.

14. Remove the vacuum hoses from the fuel rail.

15. Disconnect the fuel injector wiring harness.

16. Remove the fuel rail attaching bolts and remove the fuel rail with the wiring harness from the vehicle. Position the rail on the bench upside down so the injectors are easily accessible.

17. Remove the small connector retainer clip and unplug the injector. Remove the injector clip off the fuel rail and injector. Pull the injector straight out of the rail.

To install:

18. Lubricate the rubber O-ring with clean oil and install to the rail receiver cap. Install the injector clip to the **TOP** slot of the injector, plug in the connector and install the connector clip.

19. Install the fuel rail to the vehicle and plug in the injector harness. Connect the vacuum hoses to the fuel rail.

20. Install new intake plenum gaskets with the beaded sealer side up and install the intake plenum. Torque the attaching bolts and nuts to 115 inch lbs. (13 Nm).

21. Install the fuel hoses to the fuel rail.

22. Install or connect all items that were removed or disconnected from the intake plenum and throttle body.

23. Connect the negative battery cable and check for leaks using the DRB II to activate the fuel pump.

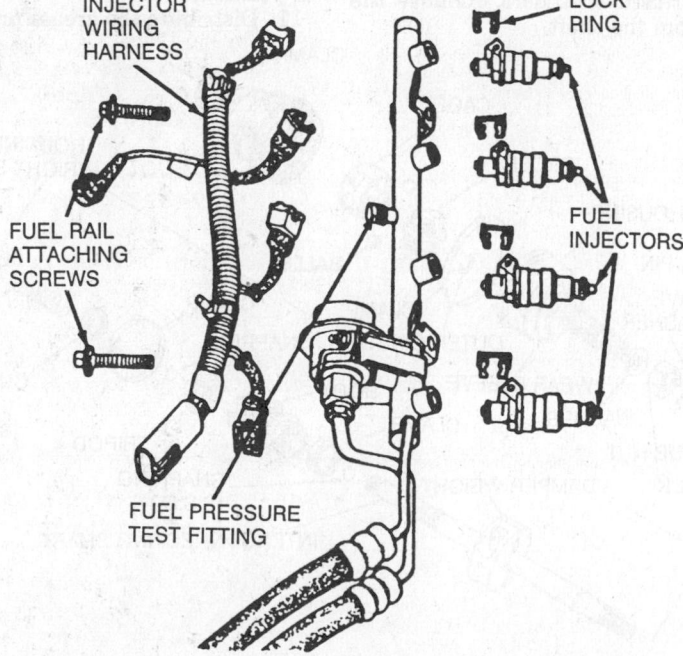

Fuel injector removal and installation—turbocharged engine

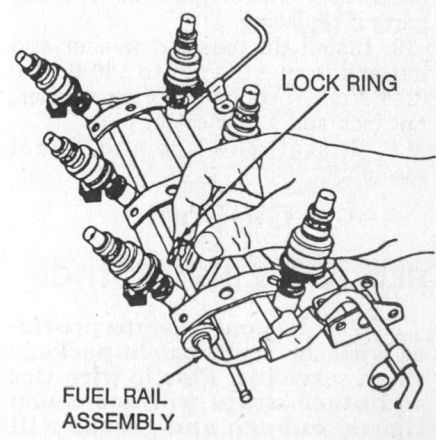

Fuel rail assembly—3.0L engine

DRIVE AXLE

Halfshaft

REMOVAL & INSTALLATION

1. Disconnect the negative battery cable.

2. Raise the vehicle and support safely.

3. Remove the tire and wheel assembly.

4. Remove the cotter pin from the end of the halfshaft. Remove the nut lock, spring washer, axle nut and washer.

5. Remove the ball joint retaining bolt and pry the control arm down to release the ball stud from the steering knuckle.

6. Position a drainpan under the transaxle where the halfshaft enters the differential or extension housing. Remove the halfshaft from the transaxle or center bearing. Unbolt the center bearing from the block and remove the intermediate shaft from the transaxle, if equipped.

To install:

7. Install the halfshaft or intermediate shaft to the transaxle, being careful not to damage the side seals. Make sure the inner joint clicks into place inside the differential. Install the center bearing retaining bolts if equipped, then install the outer shaft to the center bearing.

8. Pull the front strut out and insert the outer joint into the front hub.

9. If necessary, turn the ball joint stud to position the bolt retaining indent to the inside of the vehicle. Install the ball joint stud into the steering knuckle. Install the retaining bolt and nut and torque to 70 ft. lbs. (95 Nm). This nut and bolt combination is unique to this application and should not be replaced with conventional hardware. Use original equipment parts if replacing.

10. Install the axle nut washer and nut and torque the nut to 180 ft. lbs. (244 Nm). Install the spring washer, nut lock and a new cotter pin.

11. Install the tire and wheel assembly.

CV-Boot

REMOVAL & INSTALLATION

NOTE: Use only clamps provided with the replacement package when servicing. Plastic wire ties and other straps will not clamp tightly enough and grease will sling out, causing costly damage to the joint.

Inner Joint

1. Remove the halfshaft from the vehicle.

2. If cutting the boot away, mark and note the boot positioning on the shaft relative to the raised shoulders. Remove the boot clamps to gain access to the tripod retention system.

3. Separate the housing from the tripod according to the following:

NOTE: Hold the rollers in place when removing the housing from the tripod or the needle bearings may fall out.

a. A.C.I.—Has retaining tabs integral with the staked boot retaining collar. Hold the housing and lightly compress the CV-joint retention spring while bending the tabs back. Support the housing as the retention spring pushes it from the housing.

b. G.K.N.—Has retaining tabs integral with the housing cover. Hold the housing and lightly compress the CV-joint retention spring while bending the tabs back. Support the housing as the retention spring pushes it from the housing.

c. S.S.G.—Uses a wire ring tripod retainer which expands into a groove around the top of the housing. Use a suitable tool to pry the wire ring, without damaging it, out of the groove and slide the tripod from the housing.

4. Remove the snapring ring from the end of the shaft and remove the tripod.

5. If not already done, mark the boot positioning on the shaft relative to the raised shoulders. Remove the boot from the shaft.

6. Remove as much old grease as possible from the joint. Inspect all parts for wear or damage.

NOTE: Do not use petroleum based solvents on the joints, shaft or boot to clean; it will ruin hidden rubber seals within the joint. Use only chlorine based cleaner or hot soapy water to clean the joint, if necessary. Make sure the joint is completely dry before assembling.

To install:

7. On right inner joint of shafts of turbocharged vehicles, slide a new rubber washer seal over the stub shaft and down into the groove provided.

8. If the clamping device is not a staight strap, install it on the shaft first, then install the boot to the shaft in the proper position. Using the proper tool, C-4975 for crimping with plastic boot, C-4124 for crimping with rubber boot or C-4653 for clamping a strap, secure the clamp.

9. Slide the tripod on the shaft:

a. A.C.I.—Slide the tripod on the shaft with the non-chamfered edge facing the tripod retainer ring groove.

b. G.K.N—Slide the tripod on the shaft with the non-chamfered edge facing the tripod retainer ring groove.

c. S.S.G. Place the wire ring tripod retainer over the shaft, then slide the tripod. The tripod may installed either way; both ends are the same.

10. Install the snapring into its groove on the shaft to lock the tripod in position.

11. Distribute the grease provided in

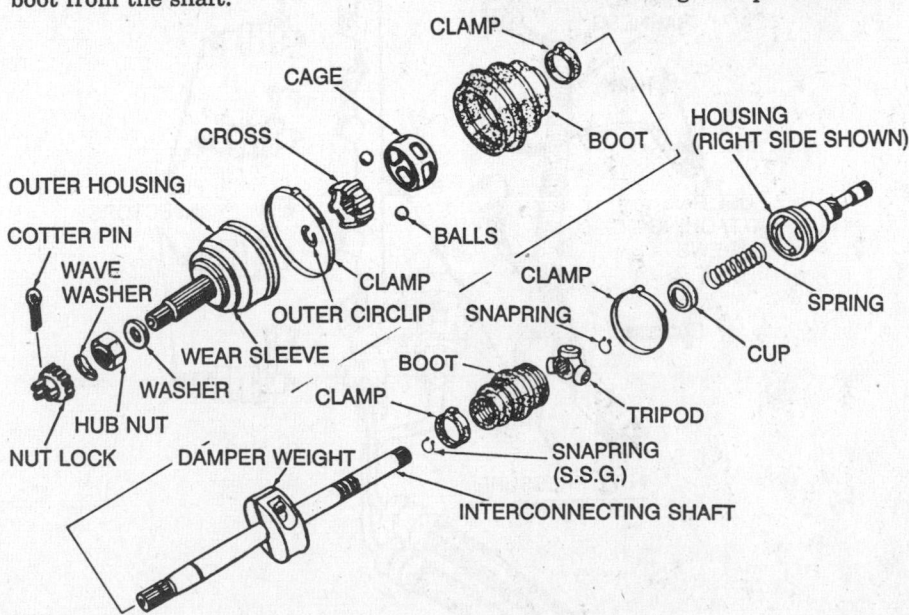

Typical driveshaft components

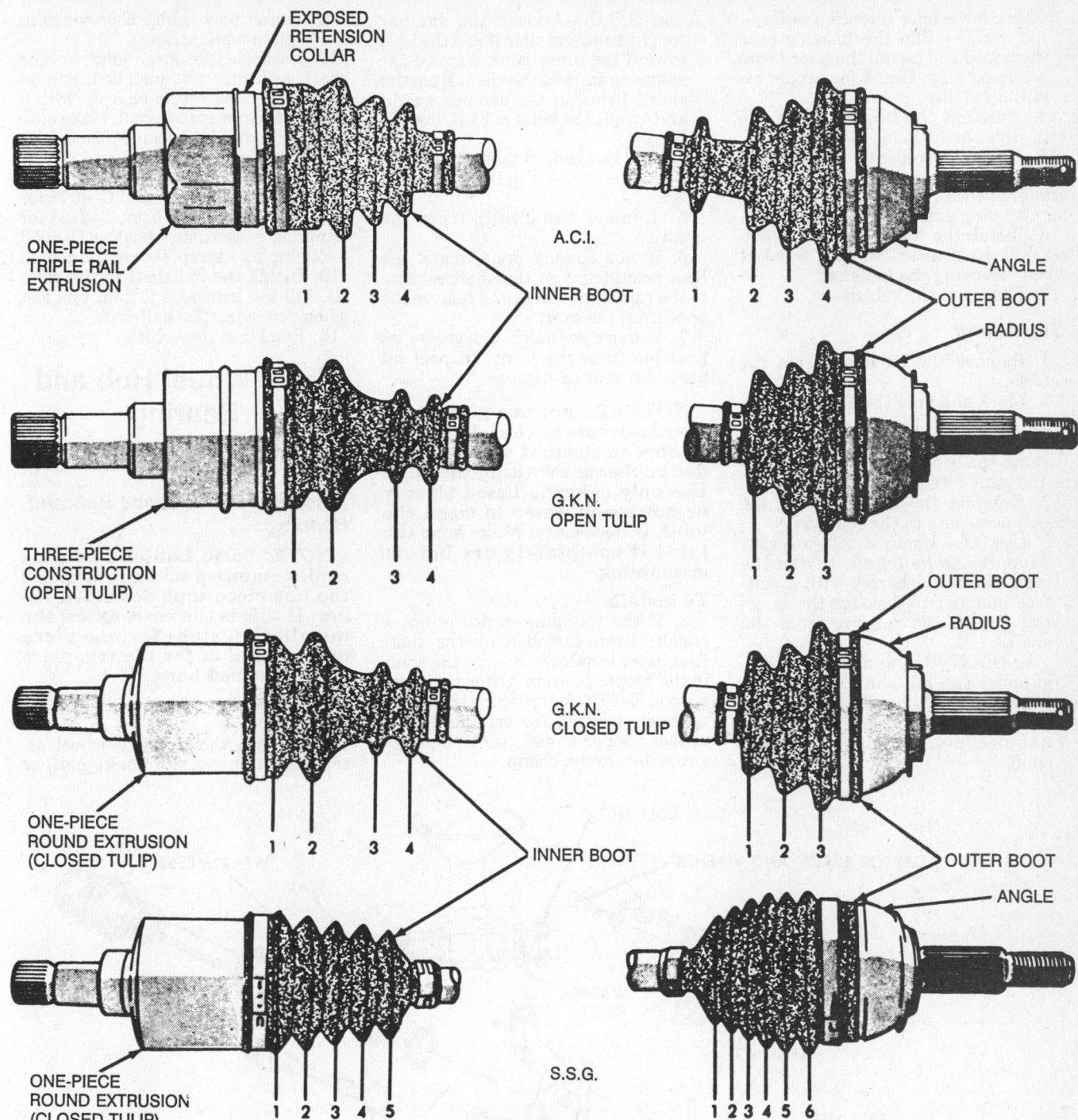

Halfshaft Identification.

the grease package as follows, or according to the instructions in the package:

a. A.C.I.—Distribute 1 of the 2 packets of grease into the boot and the remaining packet into the housing.

b. G.K.N—If equipped with 3 packets of grease, distribute 2 of the 3 packets into the boot and the remaining packet into the housing. Otherwise, distribute ½ of the pack-et of grease into the boot and the remaining amount into the housing.

c. S.S.G.—Distribute ½ of the packet of grease into the boot and the remaining amount into the housing.

12. Position the spring in the housing spring pocket with the spring cup attached to the exposed end of the spring. Place a dab of grease on the concave surface of the spring cup.

13. Keeping the spring centered, install the housing to the tripod as follows:

a. A.C.I.—Slip the housing onto the tripod. Do not bend the retaining tabs back into their original position. Instead, secure the boot to hold the housing. The tripod must be re-engaged to the housing with the shaft installed on the vehicle.

b. G.K.N—Slip the housing onto the tripod. Bend the retaining tabs back into their original positions.

Check for proper retention ability.

c. S.S.G.—Slip the housing onto the tripod and install the tripod wire retaining ring. Check for proper retention ability.

14. Position the larger end of the boot over the housing.

15. Using the proper tool, C–4975 for crimping with plastic boot, C–4124 for crimping with rubber boot or C–4653 for clamping a strap, secure the clamp.

16. Install the halfshaft to the vehicle. Fill the transaxle if fluid was lost when removing the halfshaft.

17. Road test the vehicle.

Outer Joint

1. Remove the halfshaft from the vehicle.

2. Mark and note the boot positioning on the shaft, relative to the raised shoulders, if cutting the boot away. Remove the boot clamps to gain access to the joint retention system.

3. Separate the housing from the tripod according to the following:

a. A.C.I.—Using a soft-jaw vise, support the halfshaft. Strike the joint assembly sharply with a soft-face hammer to dislodge the internal circlip and remove from the shaft.

b. G.K.N—Using a soft-jaw vise, support the halfshaft. Strike the joint assembly sharply with a soft-face hammer to dislodge the internal circlip and remove from the shaft.

c. S.S.G.—Loosen the damper weight bolts and slide it and the boot toward the inner joint. Expand the snapring and slide the joint from the shaft. Reinstall the damper weight and torque the bolts to 21 ft. lbs. (28 Nm).

4. If damaged, remove the wear sleeve from the CV-joint machined ledge.

5. Remove the circlip from the groove.

6. If not already done, mark the boot positioning on the shaft relative to the raised shoulders and remove the boot from the shaft.

7. Remove as much old grease as possible from the joint. Inspect all parts for wear or damage.

NOTE: Do not use petroleum based solvents on the joints, shaft or boot to clean; it will ruin hidden rubber seals within the joint. Use only chlorine based cleaner or hot soapy water to clean the joint, if necessary. Make sure the joint it completely dry before assembling.

To install:

8. If the clamping device is not a staight strap, install it on the shaft first, then install the boot to the shaft in the proper position. Using the proper tool, C–4975 for crimping with plastic boot, C–4124 for crimping with rubber boot or C–4653 for clamping a strap, secure the clamp.

9. Install new circlip if provided in the replacement package.

10. Position the outer joint on the shaft with hub nut installed, engage the splines and strike sharply with a soft-face hammer to install. Make sure the circlip did not become dislodged.

11. Position the larger end of the boot over the housing.

12. Using the proper tool C–4975 for crimping with plastic boot, C–4124 for crimping with rubber boot or C–4653 for clamping a strap, secure the clamp.

13. Install the halfshaft to the vehicle. Fill the transaxle if fluid was lost when removing the halfshaft.

14. Road test the vehicle.

Front Wheel Hub and Bearing

REMOVAL & INSTALLATION

Pressed In (Two-Piece Hub and Bearing)

NOTE: Some hub and bearing replacement packages include the one-piece unit described below. If this is the case, follow the installation steps for one-piece unit instead of for the two-piece unit described here.

1. Raise the vehicle and support safely.

2. Remove the tire and wheel assembly. Remove the brake caliper

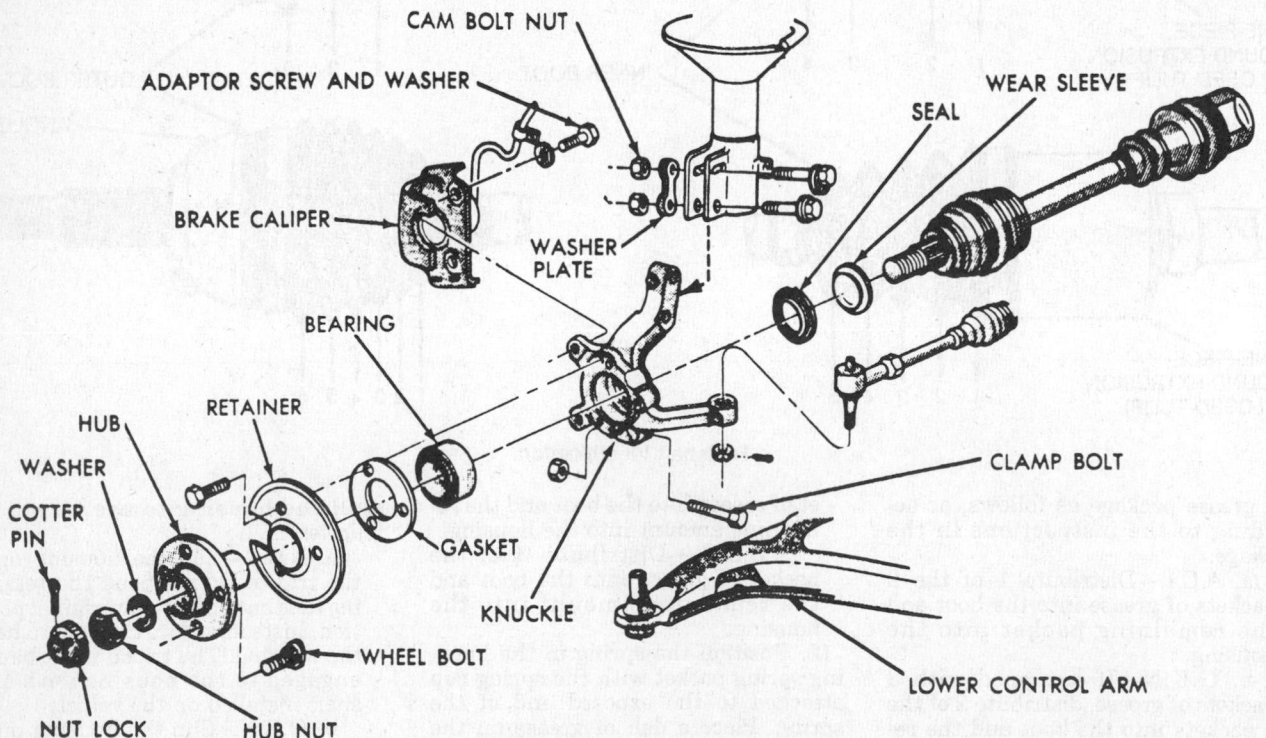

Front suspension components

from the adaptor and remove the adaptor. Remove the brake disc.

3. Remove the halfshaft.

NOTE: Knuckle removal is not necessary for bearing and hub replacement.

4. Disconnect the tie rod from the knuckle.

5. Matchmark the lower strut mount to the knuckle. Remove the 2 strut clamp bolts and remove the knuckle from the vehicle.

6. Attach the hub removal tool C–4811 or equivalent, and the triangular adapter, to the 3 rear threaded holes of the steering knuckle housing with the thrust button inside the hub bore.

7. Tighten the bolt in the center of the tool, to press the hub from the steering knuckle. Remove the removal tools.

8. Remove the bolts and bearing retainer from the outside of the steering knuckle.

9. Carefully pry the bearing seal from the machined recess of the steering knuckle and clean the recess.

10. Insert tool C–4811, or equivalent through the hub bearing and install bearing removal adapter to the outside of the steering knuckle. Tighten the tool to press the hub bearing from the steering knuckle. Discard the bearing and the seal.

To install:

11. Use tool C–4811 or equivalent, and the bearing installation adapter to press in the hub bearing into the steering knuckle.

12. Install a new seal, the bearing retainer and the bolts to the steering knuckle. Torque the bearing retainer bolts to 20 ft. lbs.

13. Use the tool C–4811 or equivalent, and the hub installation adapter, to press the hub into the hub bearing.

14. Using the bearing installation tool C–4698 or equivalent, drive the new dust seal into the rear of the steering the hub and bearing from the knuckle as required.

15. The installation of the knuckle and halfshaft is the reverse of the removal procedure. Torque the tie rod nut to 35 ft. lbs. (47 Nm).

16. Align the front end.

Bolt In (One-Piece Hub and Bearing)

NOTE: Knuckle removal is not necessary for bearing and hub replacement.

1. Raise the vehicle and support safely.

2. Remove the tire and wheel assembly. Remove the brake caliper from the adaptor and remove the adaptor. Remove the brake disc.

3. Remove the halfshaft.

4. Disconnect the tie rod from the knuckle.

5. Matchmark the lower strut mount to the knuckle. Remove the 2 strut clamp bolts and remove the knuckle from the vehicle.

6. Remove the 4 hub and bearing assembly mounting bolts from the rear of the knuckle and remove the assembly from the knuckle.

7. Carefully pry the bearing seal from the machined recess of the steering knuckle and clean the recess.

8. Thoroughly clean and dry the knuckle and bearing mating surfaces and the seal installation area.

To install:

9. Install the hub and bearing assembly to the knuckle and torque the bolts in a criss-cross pattern to 45 ft. lbs. (65 Nm).

10. Install a new seal and wear sleeve. Lubricate the circumferences of the seal and sleeve liberally with grease.

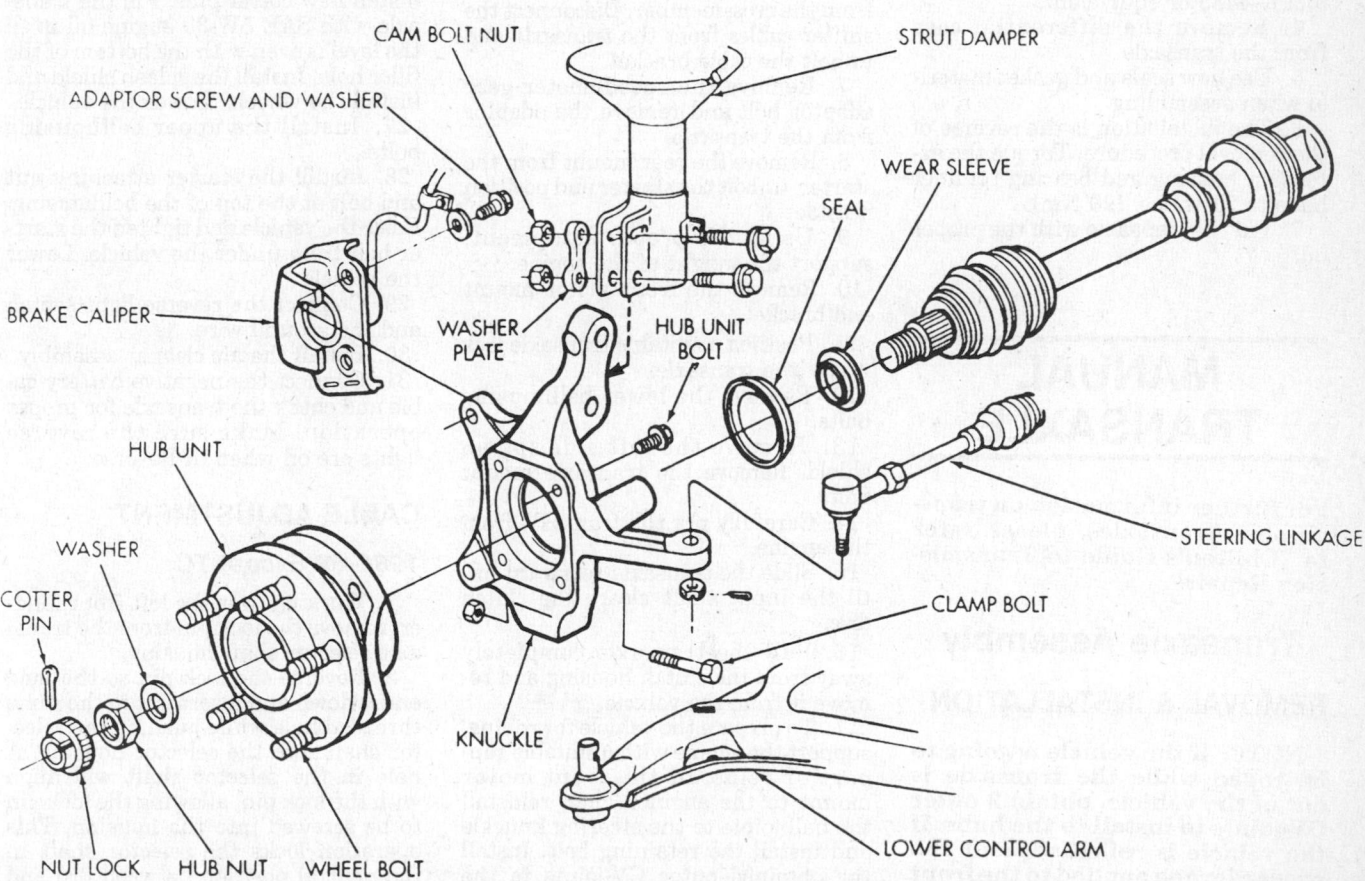

Bolt-In (one-piece) hub and bearing and related parts

11. The installation of the knuckle and halfshaft is the reverse of the removal procedure. Torque the tie rod nut to 35 ft. lbs. (47 Nm).

12. Align the front end.

Differential Case

REMOVAL & INSTALLATION

NOTE: The differential case can be removed from some vehicles with the transaxle installed. To do so, remove the halfshafts, remove the 2 K-frame mounting nuts and 2 bolts, and lower the K-frame to provide enough room to pull the differential case out of its housing and over the lowered frame.

Some individual parts in the Getrag manual transaxle are normally not serviceable; it must be replaced as an assembly.

1. Disconnect the negative battery cable. Raise the vehicle and support safely.

2. Remove the right side extension housing. Remove the differential cover.

3. Remove the bolts and the right side differential bearing retainer using tool L-4435 or equivalent.

4. Remove the differential case from the transaxle.

5. Use new seals and gasket material when assembling.

6. The installation is the reverse of the removal procedure. Torque the extension housing and bearing retainer bolts to 21 ft. lbs. (28 Nm).

7. Fill the transaxle with the proper oil.

MANUAL TRANSAXLE

For further information on transmissions/transaxles, please refer to "Chilton's Guide to Transmission Repair".

Transaxle Assembly

REMOVAL & INSTALLATION

NOTE: If the vehicle is going to be rolled while the transaxle is out of the vehicle, obtain 2 outer CV-joints to install to the hubs. If the vehicle is rolled without the proper torque applied to the front wheel bearings, the bearings will no longer be usable.

Different transaxles are used according to application. It is important to use the round identification tag screwed to the top of the case when obtaining parts for exact parts matching. The tag should be reinstalled for future reference.

1. Disconnect the negative battery cable.

2. Remove the air cleaner assembly with all ducts. Remove the upper bellhousing bolts. Disconnect the reverse light switch and the ground wire.

3. Remove the starter attaching nut and bolt at the top of the bellhousing.

4. Raise the vehicle and support safely. Remove the tire and wheel assemblies. Remove the axle end cotter pins, nut locks, spring washers and axle nuts.

5. Remove the ball joint retaining bolts and pry the control arm from the steering knuckle. Position a drainpan under the transaxle where the axles enter the differential or extension housing. Remove the axles from the transaxle or center bearing. Unbolt the center bearing and remove the intermediate axle from the transaxle, if equipped.

6. Remove the anti-rotation link from the crossmember. Disconnect the shifter cables from the transaxle and unbolt the cable bracket.

7. Remove the speedometer gear adaptor bolt and remove the adaptor from the transaxle.

8. Remove the rear mount from the starter, unbolt the starter and position it aside.

9. Using the proper equipment, support the weight of the engine.

10. Remove the front motor mount and bracket.

11. Position a suitable transaxle jack under the transaxle.

12. Remove the lower bellhousing bolts.

13. Remove the left side splash shield. Remove the transaxle mount bolts.

14. Carefully pry the transaxle from the engine.

15. Slide the transaxle rearward until the input shaft clears the clutch disc.

16. Pull the transaxle completely away from the clutch housing and remove it from the vehicle.

17. To prepare the vehicle for rolling, support the engine with a suitable support or reinstall the front motor mount to the engine. Then reinstall the ball joints to the steering knuckle and install the retaining bolt. Install the obtained outer CV-joints to the hubs, install the washers and torque the axle nuts to 180 ft. lbs. (244 Nm).

The vehicle may now be safely rolled.

To install:

18. Lubricate the pilot bushing and input shaft splines very lightly with high temperature lubricant.

19. Mount the transaxle securely on a suitable jack. Lift it in place until the input shaft is centered in the clutch housing opening. Roll the transaxle forward until the input shaft splines fully engage with the clutch disc and install the transaxle to clutch housing bolts.

20. Raise the transaxle and install the left side mount bolts.

21. Install the front motor mount and bracket.

22. Remove the engine and transaxle support fixtures.

23. Install the starter to the transaxle and install the lower bolt finger-tight.

24. Install a new O-ring to the speedometer cable adaptor and install to the extension housing; make sure it snaps in place. Install the retaining bolt.

25. Install the shift cable bracket and snap the cable ends in place. Install the anti-rotation link.

26. Install the axles and center bearing, if equipped. Install the ball joints to the steering knuckles. Torque the axle nuts to 180 ft. lbs. (244 Nm) and install new cotter pins. Fill the transaxle with SAE 5W-30 engine oil until the level is even with the bottom of the filler hole. Install the splash shield and install the wheels. Lower the vehicle.

27. Install the upper bellhousing bolts.

28. Install the starter attaching nut and bolt at the top of the bellhousing. Raise the vehicle and tighten the starter bolt from under the vehicle. Lower the vehicle.

29. Connect the reverse light switch and the ground wire.

30. Install the air cleaner assembly.

31. Connect the negative battery cable and check the transaxle for proper operation. Make sure the reverse lights are on when in Reverse.

CABLE ADJUSTMENT

1988–89 Except TC

1. Working over the left front fender, remove the lock pin from the transaxle selector shaft housing.

2. Reverse the lock pin so the long end is down and insert it into the same threaded hole while pushing the selector shaft into the selector housing. A hole in the selector shaft will align with the lock pin, allowing the lock pin to be screwed into the housing. This operation locks the selector shaft in the neutral position between 3rd and 4th gears.

3. Remove the gearshift knob, the

retaining nut and the pull-up ring from the gearshift lever.

4. If necessary, remove the shift lever boot and console to expose the gearshift linkage.

5. Fabricate 2 cable adjusting pins: $\frac{3}{16}$ in. diameter × 5 in. long with a $\frac{1}{2}$ in. 90 degree bend at one end.

6. Place a pin in the hole provided at the right side and the other in the hole provided at the rear side of the shifting mechanism; make sure the alignment holes match. Torque the selector (right side) and the crossover (left side) adjusting bolts to 4–5 ft. lbs.

7. Remove the lock pin from the selector shaft housing and reinstall the lock pin, with the long end up, in the selector shaft housing. Torque the lock pin to 10 ft. lbs. (12 Nm).

8. Check the first/reverse shifting and blockout into reverse.

9. Reinstall the console, boot, pull-up ring, retaining nut and knob.

1990–92 Except TC

1. Working over the left front fender, remove the lock pin from the transaxle selector shaft housing.

2. Reverse the lock pin so the long end is down and insert it into the same threaded hole while pushing the selector shaft into the selector housing. A hole in the selector shaft will align with the lock pin, allowing the lock pin to be screwed into the housing. This operation locks the selector shaft in the neutral position between 3rd and 4th gears.

3. Remove the gearshift knob, the retaining nut and the pull-up ring from the gearshift lever.

4. If necessary, remove the shift lever boot and console to expose the gearshift linkage. The selector cable is not adjustable.

5. Loosen the crossover cable adjusting screw and alow the cable to move in the slot. Tighten the screw to 70 inch lbs. (8 Nm).

6. Remove the lock pin from the selector shaft housing and reinstall the lock pin, with the long end up, in the selector shaft housing. Torque the lock pin to 10 ft. lbs. (12 Nm).

7. Check the first/reverse shifting and blockout into reverse.

8. Reinstall the console, boot, pull-up ring, retaining nut and knob.

1989–90 TC

1. Disconnect the negative battery cable.

2. Remove the console assembly to gain access to the cable ends.

3. Place the selector shaft in the neutral position.

4. Loosen the selector and crossover cables adjusting screws enough so the cables are free to move.

5. Install screw tool with tethered spacer block, which is taped to the shifter support bracket, to the support bracket.

6. Torque the cable adjusting screws to 70 inch lbs. (8 Nm).

7. Remove the tethered adjusting tool and attach it to the bracket for future use.

8. Install the console assembly.

9. Road test the vehicle and check for smooth shifting.

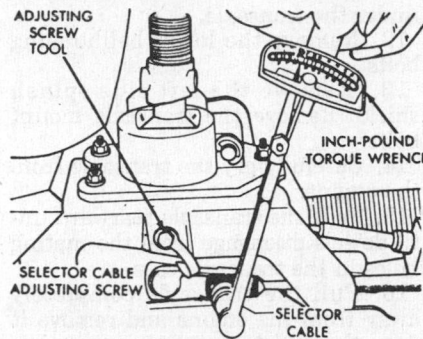

Adjusting the shifter cables — TC

CLUTCH

Clutch Assembly

REMOVAL & INSTALLATION

1. Disconnect the negative battery cable. Remove the transaxle.

2. Matchmark the clutch/pressure plate cover and flywheel. Insert a suitable clutch plate alignment tool into the clutch disc hub.

3. Loosen the flywheel to pressure plate bolts gradually and evenly to avoid warpage.

4. Remove the pressure plate/clutch assembly from the flywheel.

5. Sand the flywheel or replace it if it is scored, cracked or heat damaged.

6. Sparingly apply anti-sieze compound to the input shaft and clutch disc splines. Install a new release bearing.

To install:

7. Using a suitable clutch disc alignment tool, tighten the pressure plate bolts to center the disc.

8. Torque the pressure plate/clutch assembly mounting bolts to the flywheel gradually and evenly to 21 ft. lbs. (28 Nm).

9. Install the transaxle.

10. Connect the negative battery cable and check the clutch and reverse lights for proper operation.

PEDAL FREE-PLAY ADJUSTMENT

All vehicles are equipped with a self-adjusting cable operated mechanism and no adjustment is provided. The mechanism is located above the clutch pedal, where the cable and pivot points may be lubricated.

Clutch Cable

REMOVAL & INSTALLATION

1. Disconnect the negative battery cable.

2. Remove the clip from the cable mounting bracket on the shock tower and remove the cable from the bracket.

3. Remove the retainer from the clutch release lever on the transaxle.

4. Pry out the ball end of the cable

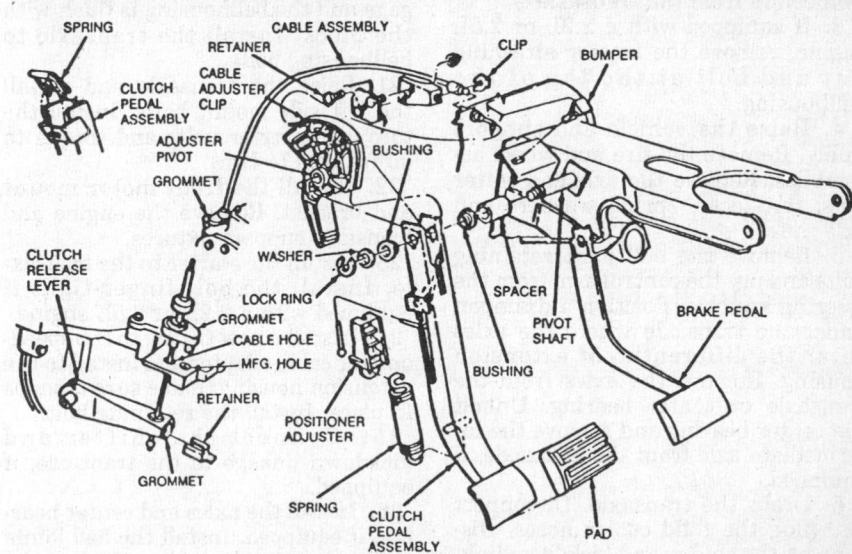

Clutch pedal, cable and related parts

from the position adjuster inside the pedal.

5. The installation is the reverse of the removal procedure. After installing, push the clutch pedal 2 or 3 times to allow the self-adjuster mechanism to function.

6. Connect the negative battery cable and check the clutch for proper operation.

AUTOMATIC TRANSAXLE

For further information on transmissions/transaxles, please refer to "Chilton's Guide to Transmission Repair".

Transaxle Assembly

REMOVAL & INSTALLATION

NOTE: If the vehicle is going to be rolled while the transaxle is out of the vehicle, obtain 2 outer CV-joints to install to the hubs. If the vehicle is rolled without the proper torque applied to the front wheel bearings, the bearings will no longer be usable.

1. Disconnect the negative battery cable. If equipped with 3.0L engine, drain the coolant. Remove the dipstick.

2. Remove the air cleaner assembly if it is preventing access to the upper bellhousing bolts. Remove the upper bellhousing bolts and water tube, where applicable. Unplug all electrical connectors from the transaxle.

3. If equipped with a 2.2L or 2.5L engine, remove the starter attaching nut and bolt at the top of the bellhousing.

4. Raise the vehicle and support safely. Remove the tire and wheel assemblies. Remove the axle end cotter pins, nut locks, spring washers and axle nuts.

5. Remove the ball joint retaining bolts and pry the control arm from the steering knuckle. Position a drainpan under the transaxle where the axles enter the differential or extension housing. Remove the axles from the transaxle or center bearing. Unbolt the center bearing and remove the intermediate axle from the transaxle, if equipped.

6. Drain the transaxle. Disconnect and plug the fluid cooler hoses. Disconnect the shifter and kickdown linkage from the transaxle, if equipped.

7. Remove the speedometer cable adaptor bolt and remove the adaptor from the transaxle.

8. Remove the starter. Remove the torque converter inspection cover, matchmark the torque converter to the flexplate and remove the torque converter bolts.

9. Using the proper equipment, support the weight of the engine.

10. Remove the front motor mount and bracket.

11. Position a suitable transaxle jack under the transaxle.

12. Remove the lower bellhousing bolts.

13. Remove the left side splash shield. Remove the transaxle mount bolts.

14. Carefully pry the transaxle from the engine.

15. Slide the transaxle rearward until dowels disengage from the mating holes in the transaxle case.

16. Pull the transaxle completely away from the engine and remove it from the vehicle.

17. To prepare the vehicle for rolling, support the engine with a suitable support or reinstall the front motor mount to the engine. Then reinstall the ball joints to the steering knuckle and install the retaining bolt. Install the obtained outer CV-joints to the hubs, install the washers and torque the axle nuts to 180 ft. lbs. (244 Nm). The vehicle may now be safely rolled.

To install:

18. Install the transmission securely on the transmission jack. Rotate the converter so it will align with the positioning of the flexplate.

19. Apply a coating of high temperature grease to the torque converter pilot hub.

20. Raise the transaxle into place and push it forward until the dowels engage and the bellhousing is flush with the block. Install the transaxle to bellhousing bolts.

21. Raise the transaxle and install the left side mount bolts. Install the torque converter bolts and torque to 55 ft. lbs. (74 Nm).

22. Install the front motor mount and bracket. Remove the engine and transaxle support fixtures.

23. Install the starter to the transaxle. Install the bolt finger-tight if equipped with a 2.2L or 2.5L engine.

24. Install a new O-ring to the speedometer cable adaptor and install to the extension housing; make sure it snaps in place. Install the retaining bolt.

25. Connect the shifter and kickdown linkage to the transaxle, if equipped.

26. Install the axles and center bearing, if equipped. Install the ball joints to the steering knuckles. Torque the axle nuts to 180 ft. lbs. (244 Nm) and

install new cotter pins. Install the splash shield and install the wheels. Lower the vehicle. Install the dipstick.

27. Install the upper bellhousing bolts and water pipe, if removed.

28. If equipped with 2.2L or 2.5L engine, install the starter attaching nut and bolt at the top of the bellhousing. Raise the vehicle again and tighten the starter bolt from under the vehicle. Lower the vehicle.

29. Connect all electrical wiring to the transaxle.

30. Install the air cleaner assembly, if removed. Fill the transaxle with the proper amount of Mopar ATF Plus Type 7176 or conventional Dexron®II.

31. Connect the negative battery cable and check the transaxle for proper operation.

UPSHIFT AND KICKDOWN LEARNING PROCEDURE

A-604 Ultradrive Transaxle

In 1989, the A-604 4 speed, electronic transaxle was introduced; it is the first to use fully adaptive controls. The controls perform their functions based on real time feedback sensor information. Although, the transaxle is conventional in design, functions are controlled by its ECM.

Since the A-604 is equipped with a learning function, each time the battery cable is disconnected, the ECM memory is lost. In operation, the transaxle must be shifted many times for the learned memory to be re-inputed to the ECM; during this period, the vehicle will experience rough operation. The transaxle must be at normal operating temperature when learning occurs.

1. Maintain constant throttle opening during shifts. Do not move the accelerator pedal during upshifts.

2. Accelerate the vehicle with the throttle ⅛-½ open.

3. Make fifteen to twenty 1/2, 2/3 and 3/4 upshifts. Accelerating from a full stop to 50 mph each time at the aforementioned throttle opening is sufficient.

4. With the vehicle speed below 25 mph, make 5-8 wide open throttle kickdowns to 1st gear from either 2nd or 3rd gear. Allow at least 5 seconds of operation in 2nd or 3rd gear prior to each kickdown.

5. With the vehicle speed greater than 25 mph, make 5 part throttle to wide open throttle kickdowns to either 3rd or 2nd gear from 4th gear. Allow at least 5 seconds of operation in 4th gear, preferably at road load throttle prior to performing the kickdown.

SHIFT LINKAGE ADJUSTMENT

1. Place the shifter in the **P** detent.
2. Loosen the clamp bolt on the gearshift cable bracket.
3. Pull the shift lever all the way to the front detent position and tighten the lock screw.
4. Check for proper neutral safety switch operation.

THROTTLE PRESSURE CABLE ADJUSTMENT

1. Run the engine until it reaches normal operating temperature.
2. Loosen the cable mounting bracket lock screw.
3. Position the bracket so both alignment tabs are touching the transaxle case surface and tighten the lock screws.
4. Release the cross lock on the cable assembly by pulling the cross lock up.
5. To ensure proper adjustment, the cable must be free to slide all the way toward the engine against its stop after the cross lock is released.
6. Move the transaxle throttle control lever fully clockwise and press the cross lock down until it snaps into position.
7. Road test the vehicle and check the shift points.

THROTTLE PRESSURE ROD ADJUSTMENT

1. Run the engine until it reaches normal operating temperature.
2. Loosen the adjustment swivel lock screw.
3. To ensure proper adjustment, the swivel must be free to slide along the flat end of the throttle rod. Disassembly, clean and lubricate as required.
4. Hold the transaxle throttle control lever firmly toward the engine and tighten the swivel screw.
5. Road test the vehicle and check the shift points.

FRONT SUSPENSION

MacPherson Strut

REMOVAL & INSTALLATION

1. Remove the 3 mounting nuts

from the shock tower under the hood.
2. Raise the vehicle and support safely.
3. Remove the brake hose bracket screw from the strut.
4. Matchmark the lower strut mount to the knuckle and remove the strut to knuckle bolts, nuts and nut plate.
5. The installation is the reverse of the removal procedure. Torque the upper mounting nuts to 20 ft. lbs. (27 Nm). Do not tighten the lower mounting bolts until the front end alignment has been completed.
6. Perform a front end alignment. Torque the strut to knuckle nuts to 75 ft. lbs. (100 Nm) plus ¼ turn.

Lower Ball Joints

INSPECTION

To inspect the ball joints, grasp the grease fitting by hand with the vehicle on the ground. If the grease fitting can be moved at all by hand, the ball joint should be replaced.

REMOVAL & INSTALLATION

The ball joints are welded to the lower control arms. This necessitates replacement of the control arm assembly. Do not attempt to replace ball joints that are welded to the control arm; replacement control arms are equipped with a new ball joint.

Lower Control Arms

REMOVAL & INSTALLATION

1. Raise the vehicle and support safely. Remove the tire and wheel assembly.
2. Remove the sway bar.
3. Remove the ball joint stud retaining bolt and nut.
4. Pry the lower control arm from the steering knuckle.
5. Remove the control arm to crossmember bolts, nuts bushings and retainers.
6. Remove the control arm from the vehicle.

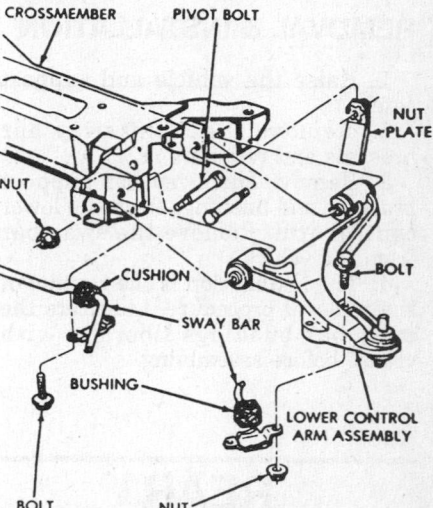

Lower control arm and related parts—1989–92 vehicles, except TC

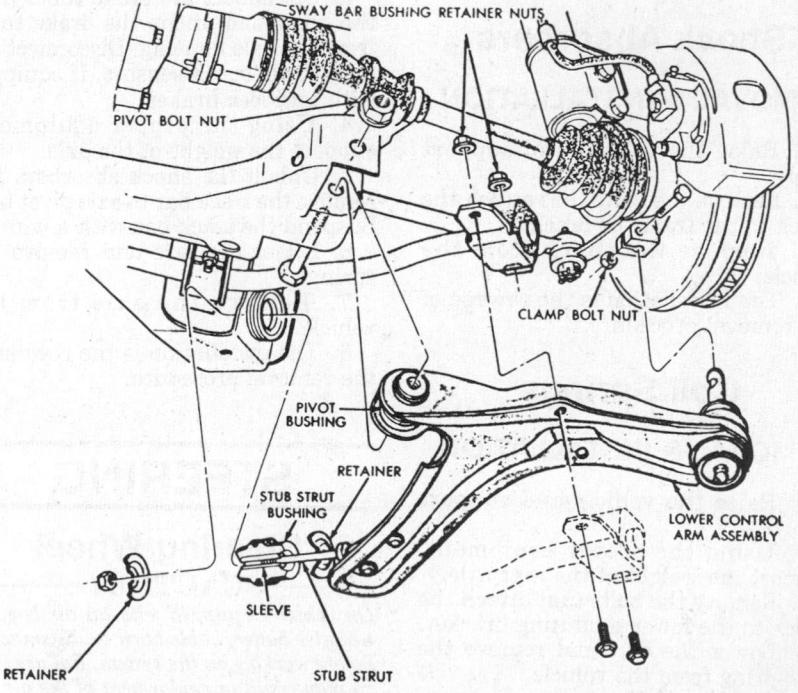

Lower control arm and related parts—Maserati TC and 1988 vehicles

7. Transfer all reusable parts to the new control arm and lubricate.

8. The installation is the reverse of the removal procedure.

9. Lower the vhicle so the full weight of the vehicle is on the ground.

10. On 1988 vehicles and TC, torque the pivot bolt to 120 ft. lbs. (163 Nm) and the stub strut nut to 70 ft. lbs. (95 Nm). On 1989–91 vehicles, torque both pivot bolts to 125 ft. lbs. (169 Nm).

11. Perform a front end alignment as required.

Sway Bar

REMOVAL & INSTALLATION

1. Raise the vehicle and support safely.

2. Remove the front sway bar brackets and retainers.

3. Remove the sway bar support brackets and bushings from the lower control arm. Remove the sway bar from the vehicle.

4. The installation is the reverse of the removal procedure. Lubricate the sway bar bushings liberally with grease before assembling.

REAR SUSPENSION

Shock Absorbers

REMOVAL & INSTALLATION

1. Raise the vehicle and support safely.

2. Remove the bolts that attach the shock to the frame or bracket.

3. Remove the shock from the vehicle.

4. The installation is the reverse of the removal procedure.

Coil Springs

REMOVAL & INSTALLATION

1. Raise the vehicle and support safely.

2. Using the proper equipment, support the weight of the rear axle.

3. Remove the bolts that attach the shock to the lower mounting bracket.

4. Lower the axle and remove the coil spring from the vehicle.

5. The installation is the reverse of the removal procedure.

Rear Wheel Bearings
REMOVAL & INSTALLATION

1. Raise the vehicle and support safely.

2. Remove the tire and wheel assembly.

3. Remove the dust cap.

4. Remove the cotter pin, nut lock and nut.

5. Remove the thrust washer and the outer wheel bearing.

6. Remove the drum with the inner wheel bearing and the grease seal.

7. Remove the grease seal and remove the inner bearing.

To install:

8. Lubricate the inner bearing and install to the drum.

9. Install a new grease seal.

10. Install the drum to the vehicle.

11. Lubricate and install the outer wheel bearing to the spindle.

12. Install the thrust washer.

13. Install and tighten the wheel bearng nut to 20–25 ft. lbs. (27–34 Nm) while rotating the drum.

14. Back off the adjusting nut ¼ turn then tighten it finger-tight.

15. Install the nut lock and a new cotter pin.

Rear Axle Assembly

REMOVAL & INSTALLATION

1. Raise the vehicle and support safely.

2. Disconnect the parking brake cable at the connection.

3. Disconnect the brake tubes from the hoses and unclip the brake tubes from the axle housing. Disconnect the rear wheel speed sensors, if equipped with anti-lock brakes.

4. Using the proper equipment, support the weight of the axle.

5. Unbolt the shock absorbers and remove the track bar to axle pivot bolt. Suspend the track bar with a wire.

6. Lower the axle and remove the springs.

7. Remove the axle from the vehicle.

8. The installation is the reverse of the removal procedure.

STEERING

Steering Wheel

CAUTION

On vehicles equipped with an air bag, the negative battery cable must be disconnected before working on the system. Failure to do so may result in deployment of the air bag and possible personal injury.

REMOVAL & INSTALLATION

Without Air Bag

1. Disconnect the negative battery cable.

2. Straighten the steering wheel so the front tires are pointing straight forward.

2. Remove the horn pad.

3. Remove the steering wheel hold-down nut and remove the damper, if equipped. Matchmark the steering wheel to the shaft.

4. Using a suitable steering wheel puller, pull the steering wheel off of the shaft.

5. The installation is the reverse of the removal procedure. Torque the hold-down nut to 45 ft. lbs. (60 nm).

With Air Bag

1. Disconnect the negative battery cable.

2. Straighten the steering wheel so the front tires are pointing straight forward.

3. Remove the 4 nuts located on the back side of the steering wheel that attach the airbag module to the steering wheel.

4. Lift the module and disconnect the connectors. Remove the speed control switch, if equipped.

NOTE: All columns except Acustar are equipped with a clockspring set screw held by a plastic tether on the steering wheel. Acustar-mounted clocksprings are auto-locking. If the steering column is not an Acustar and is lacking the set screw, obtain one before proceeding.

5. If equipped with the set screw, place it in the clockspring to ensure proper positioning when the steering wheel is removed.

6. Remove the steering wheel hold-down nut and remove the damper, if equipped. Matchmark the steering wheel to the shaft.

7. Using a suitable steering wheel puller, pull the steering wheel off of the shaft.

To install:

8. Position the steering wheel on the steering column. Make sure the flats on the hub of the steering wheel are aligned with the formations on the clockspring.

9. Pull the airbag and speed control connectors through the lower, larger hole in the steering wheel and pull the horn wire through the smaller hole at the top. Make sure the wires are not pinched anywhere.

10. Install the damper, if equipped.

11. Install the hold-down nut and torque to 45 ft. lbs. (60 nm).

12. If equipped with a clockspring set screw, remove the screw and place it in its storage location on the steering wheel.

13. Connect the horn wire.

14. Connect the speed control wire and install the speed control switch.

15. Connect the clockspring lead wire to the airbag module and install module to steering wheel.

NOTE: Do not allow anyone to enter the vehicle from this point on, until this procedure is completed.

16. Connect the DRB II to the Airbag System Diagnostic Module (ASDM) connector located to the right of the console.

17. From the passenger side of the vehicle, turn the key to the ON position.

18. Check to make sure nobody has entered the vehicle. Connect the negative battery cable.

19. Using the DRB II, read and record any active fault data or stored codes.

20. If any active fault codes are present, perform the proper diagnostic procedures before continuing.

21. If there are no active fault codes, erase the stored fault codes; if there are active codes, the stored codes will not erase.

22. From the passenger side of the vehicle, turn the key OFF, then ON and observe the instrument cluster airbag warning light. It should come on for 6–8 seconds, then go out, indicating the system is functioning normally. If the warning light either fails to come on, or stays lit, there is a system malfunction and further diagnostics are needed.

Steering Column

REMOVAL & INSTALLATION

1. Disconnect the negative battery cable.

2. Remove trim bezel, steering column cover and lower reinforcement, as required.

3. Disconnect all wiring connectors from below the instrument panel that lead up into the steering column.

4. Remove the nuts that attach the steering column assembly to the instrument panel support.

5. Firmly grasp the steering wheel and pull the steering column out, separating the stub shaft from the steering gear coupling.

6. The installation is the reverse of the removal procedure.

7. Connect the negative battery cable and check the steering column and all related components for proper operation.

Manual Rack and Pinion Steering Gear

REMOVAL & INSTALLATION

1. Disconnect the negative battery cable.

2. Raise the vehicle and support safely.

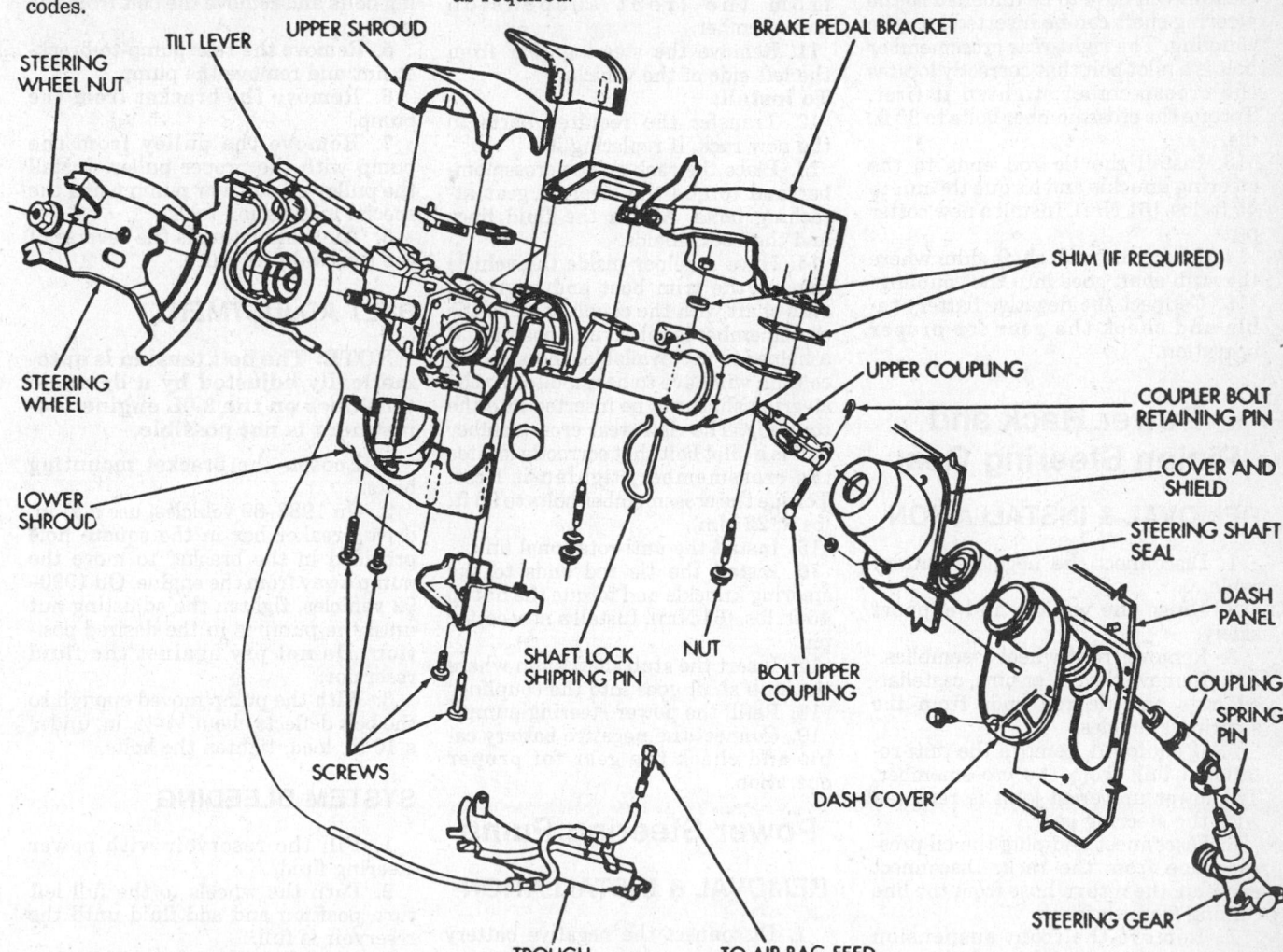

Exploded view of Acustar steering column

3. Remove front wheel assemblies.

4. Remove the cotter pins, castellated nuts and tie rod ends from the steering knuckles.

5. Remove the front suspension crossmember attaching bolts and nuts.

6. Lower the crossmember.

7. Remove the tie rod inner boot shields.

8. Remove the steering gear bolts from the front suspension crossmember.

9. Remove the steering gear from the left side of the vehicle.

To install:

10. Transfer the required parts to the new rack, if replacing it.

11. Place the rack on the crossmember and torque the steering gear attaching bolts to 21 ft. lbs. (29 Nm). Attach the boot shields.

12. Have a helper inside the vehicle remove the trim boot and align the stub shaft with the coupling while the crossmember is raised into position. If a helper is not available, the steering column will have to be unbolted so the steering shaft can be inserted into the coupling. The right rear crossmember bolt is a pilot bolt that correctly locates the crossmember; tighten it first. Torque the crossmember bolts to 90 ft. lbs.

13. Install the tie rod ends to the steering knuckle and torque the nut to 45 ft. lbs. (61 Nm). Install a new cotter pin.

14. Insert the stub shaft shim where the stub shaft goes into the coupling.

15. Connect the negative battery cable and check the gear for proper operation.

Power Rack and Pinion Steering Gear

REMOVAL & INSTALLATION

1. Disconnect the negative battery cable.

2. Raise the vehicle and support safely.

3. Remove front wheel assemblies.

4. Remove the cotter pins, castellated nuts and tie rod ends from the steering knuckles.

5. If equipped, remove the anti-rotational link from the crossmember. The lower universal joint is removed with the steering gear.

6. Disconnect and plug the oil pressure line from the rack. Disconnect and plug the return hose from the line coming from the rack.

7. Remove the front suspension crossmember attaching bolts and nuts.

8. Lower the crossmember.

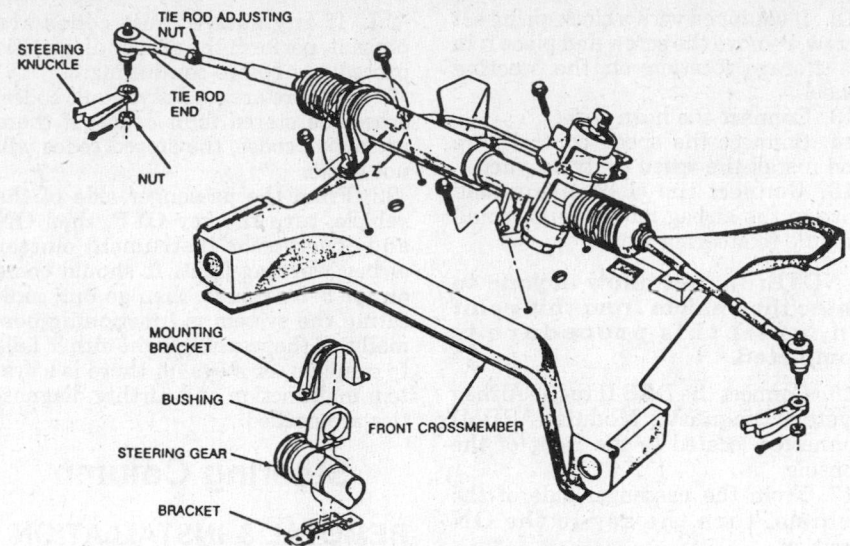

Rack and pinion steering gear mounting

9. Remove the tie rod inner boot shields.

10. Remove the steering gear bolts from the front suspension crossmember.

11. Remove the steering gear from the left side of the vehicle.

To install:

12. Transfer the required parts to the new rack, if replacing it.

13. Place the rack on the crossmember and torque the steering gear attaching bolts. Attach the fluid lines and the boot shields.

14. Have a helper inside the vehicle remove the trim boot and align the stub shaft with the coupling while the crossmember is raised into position. If a helper is not available, the steering column will have to be unbolted so the steering shaft can be inserted into the coupling. The right rear crossmember bolt is a pilot bolt that correctly locates the crossmember, tighten it first. Torque the crossmember bolts to 90 ft. lbs. (122 Nm).

15. Install the anti-rotational link.

16. Install the tie rod ends to the steering knuckle and torque the nut to 45 ft. lbs. (61 Nm). Install a new cotter pin.

17. Insert the stub shaft shim where the stub shaft goes into the coupling.

18. Refill the power steering pump.

19. Connect the negative battery cable and check the gear for proper operation.

Power Steering Pump

REMOVAL & INSTALLATION

1. Disconnect the negative battery cable.

2. Position a drain pan under the power steering pump.

3. Disconnect the fluid hoses from the pump and plug them.

4. Remove the front bracket attaching bolts and remove the belt from the pulley.

5. Remove the rear pump-to-bracket nut and remove the pump.

6. Remove the bracket from the pump.

7. Remove the pulley from the pump with the proper puller. Install the pulley on the new pump using the special installation tools.

8. The installation is the reverse of the removal procedure.

BELT ADJUSTMENT

NOTE: The belt tension is automatically adjusted by a dynamic tensioner on the 3.0L engine. Adjustment is not possible.

1. Loosen the bracket mounting bolts.

2. On 1988–89 vehicles, use a ½ in. drive breaker bar in the square hole provided in the bracket to move the pump away from the engine. On 1990–92 vehicles, tighten the adjusting nut until the pump is in the desired position. Do not pry against the fluid reservoir.

3. With the pump moved enough so the belt deflects about ¼–½ in. under a 10 lb. load, tighten the bolts.

SYSTEM BLEEDING

1. Fill the reservoir with power steering fluid.

2. Turn the wheels to the full left turn position and add fluid until the reservoir is full.

3. Start the engine and add fluid to bring the level to the correct level.

4. To purge the system of air, turn

the steering wheel from side to side without contacting the stops.

5. Return the wheel to the straight-ahead position and operate the engine for 2 minutes before road testing.

Tie Rod Ends

REMOVAL & INSTALLATION

1. Raise the vehicle and support safely.

2. Remove the cotter pin and nut from the tie rod end.

3. Using a suitable puller, remove the tie rod from the steering knuckle.

4. Loosen the sleeve clamp nut and bolt, if equipped, and unscrew the tie rod end from the sleeve or inner tie rod.

5. The installation is the reverse of the removal procedure. Torque the stud nuts to 45 ft. lbs. (61 Nm) and install a new cotter pin.

6. Perform a front end alignment as required.

BRAKES

For all brake system repair and service procedure not detained below, please refer to "Brakes" in the Unit Repair section.

Master Cylinder

REMOVAL & INSTALLATION

Except TC With Anti-Lock Brakes

1. Disconnect the negative battery cable.

2. Disconnect and plug the brake lines from the master cylinder.

3. Remove the nuts attaching the master cylinder to the power booster.

4. Remove the master cylinder from the mounting studs.

5. Remove the fluid reservoir from the cylinder.

To install:

6. Bench bleed the master cylinder.

7. Install to the studs and install the nuts.

8. Install the brake lines to the master cylinder.

9. Connect the negative battery cable and check the brakes for proper operation.

Combination Valve

REMOVAL & INSTALLATION

1. Disconnect the negative battery cable.

2. Raise the vehicle and support safely.

3. Tag and disconnect the brake lines from the valve.

4. Disconnect the wires to the pressure switch.

5. Remove the combination valve from the frame bracket.

6. The installation is the reverse of the removal procedure.

7. Bleed the brakes in the following order:

    a. Right rear wheel cylinder or caliper

    b. Left rear wheel cylinder or caliper

    c. Right front caliper

    d. Left front caliper

8. Connect the negative battery cable and check the brakes for proper operation.

Power Brake Booster

REMOVAL & INSTALLATION

1. Disconnect the negative battery cable. Disconnect the vacuum hose(s) from the booster.

2. Remove the nuts attaching the master cylinder to the booster and move the master cylinder aside.

3. From inside of the vehicle, remove the clip that secures the booster pushrod to the brake pedal.

4. Remove the nuts that attach the booster to the dash panel and remove it from the vehicle.

5. Transfer the check valve to the new booster.

6. The installation is the reverse of the removal procedure.

7. Connect the negative battery cable and check the brakes for proper operation.

Brake Caliper

REMOVAL & INSTALLATION

Except Rear Disc Brakes on 1988 Daytona and All TC With Anti-Lock Brakes

1. Raise the vehicle and support safely.

2. Remove the tire and wheel assembly.

3. Remove the caliper mounting pin(s).

4. Lift the caliper off of the rotor. Remove the outer pad from the caliper.

5. Remove the brake hose retaining bolt from the caliper.

To install:

6. Install the brake hose to the caliper using new copper washers.

7. Position the caliper over the rotor so the caliper engages the adaptor

correctly. Install the mounting pin(s). Install the hold-down spring, if equipped.

8. Fill the master cylinder and bleed the brakes.

1988 Daytona

1. Remove ⅔ of brake fluid from the master cylinder.

2. Remove access plug and insert a 4mm Allen wrench through hole.

3. Turn the retraction shaft counterclockwise a few turns to increase clearance between pads and rotor.

4. Remove the anti-rattle spring from outboard pad taking care not to damage it.

5. Back the caliper guide pins out just enough to free caliper from adapter.

6. Lift the caliper off of rotor and carefully suspend with wire.

7. Remove the brake pads.

8. Insert the Allen wrench through access hole and turn clockwise, if necessary, to retract piston further to increase clearance.

To install:

9. Lower the caliper over rotor and pads.

10. Install the guide pins and tighten to proper torque.

11. Insert the Allen wrench through the access hole and turn clockwise until snug (no clearance between pads and rotors) then back off ⅓ turn to obtain proper clearance.

12. Check the brake fluid level and add if necessary.

Disc Brake Pads

REMOVAL & INSTALLATION

Except Rear Disc Brakes on 1988 Daytona and All TC With Anti-Lock Brakes

1. Remove some of the fluid from the master cylinder.

2. Raise the vehicle and support safely. Remove the tire and wheel assemblies.

3. Remove the hold-down spring if necessary. Remove the caliper and remove the outer pad from the caliper.

4. Remove the inner pad from the adaptor.

To install:

5. Use a large C-clamp to compress the piston back into the caliper bore.

6. Install the inner pad to the adaptor.

7. Position the caliper over the rotor so the caliper engages the adaptor correctly and install the retainer pin(s).

8. Install the hold-down spring, if removed.

9. Refill the master cylinder.

1988 Daytona

1. Remove ⅔ of brake fluid from the master cylinder.
2. Remove access plug and insert a 4mm Allen wrench through hole.
3. Turn the retraction shaft counterclockwise a few turns to increase clearance between pads and rotor.
4. Remove the anti-rattle spring from outboard pad taking care not to damage it.
5. Back the caliper guide pins out just enough to free caliper from adapter.
6. Lift the caliper off of rotor and carefully suspend with wire.
7. Remove the brake pads.
8. Insert the Allen wrench through access hole and turn clockwise, if necessary, to retract piston further to increase clearance for new pads.

To install:

9. Install new inner and outer pads.

NOTE: The outboard pads are marked for right and left sides and must be properly installed.

10. Lower the caliper over rotor and pads.
11. Install the guide pins and tighten to proper torque.
12. Insert the Allen wrench through the access hole and turn clockwise until snug (no clearance between pads and rotors) then back off ⅓ turn to obtain proper clearance.
13. Check the brake fluid level and add, if necessary.

Brake Rotor

REMOVAL & INSTALLATION

1. Raise the vehicle and support safely. Remove the tire and wheel assembly.
2. Remove the caliper and brake pads.
3. Remove the factory installed clips, if equipped. It is not necessary to reinstall these clips.
4. Remove the rotor from the hub.
5. The installation is the reverse of the removal procedure.

Brake Drum

REMOVAL & INSTALLATION

1. Raise the vehicle and support safely.
2. Remove the wheel and tire assembly.
3. Remove the dust cap.
4. Remove the cotter pin and nut lock.
5. Remove the wheel bearing nut and washer from the spindle.
6. Remove the outer wheel bearing.

7. Remove the drum with the inner wheel bearing from the spindle. If the drum is difficult to remove, remove the plug from the rear of the backing plate and push the self adjuster lever away from the star wheel. Rotate the star wheel to retract the shoes. Remove the grease seal.

To install:

8. Lubricate and install the inner wheel bearing. Install a new grease seal.
9. Install the drum to the spindle.
10. Lubricate and install the outer wheel bearing, washer and nut. When the bearing preload is properly set, install the nut lock and a new cotter pin.
11. Install the grease cap.
12. Install the wheel and tire assembly. Adjust the rear brakes as required.

Brake Shoes

REMOVAL & INSTALLATION

1. Raise the vehicle and support safely. Remove the wheel and tire assemblies and the drums.
2. Remove the automatic adjuster spring and lever.
3. Rotate the automatic adjuster star wheel enough so both shoes move out far enough to be free of the wheel cylinder boots.
4. Disconnect the parking brake cable from the actuating lever.
5. Remove the lower shoe to shoe or shoe to anchor spring(s).
6. With the shoes held together by the upper shoe to shoe spring, remove them from the backing plate.

To install:

7. Thoroughly clean and dry the backing plate. To prepare the backing plate, lubricate the bosses, anchor pin and parking brake actuating lever pivot surface lightly with lithium based grease.
8. Remove, clean and dry all parts still on the old shoes. Lubricate the star wheel shaft threads with antisieze lubricant and transfer all parts to their proper locations on the new shoes.
9. Install the lower spring(s).
10. Connect the parking brake cable.
11. Install the automatic adjuster lever and spring.
12. Adjust the star wheel.
13. Remove any grease from the linings and install the drum.
14. Complete the brake adjustment with the wheels installed.

Wheel Cylinder

REMOVAL & INSTALLATION

1. Raise the vehicle and support safely.

2. Remove the wheel, drum and brake shoes.
3. Remove and plug the brake line from the wheel cylinder.
4. Remove the wheel cylinder bolts and remove the cylinder from the backing plate.

To install:

5. Apply a very thin coating of silicone sealer to the cylinder mounting surface, install the cylinder to the backing plate and install the retaining bolts.
6. Connect the brake line to the wheel cylinder.
7. Install all brake parts that were removed.
8. Install the tire and wheel assembly.
9. Bleed the brakes.

Parking Brake Cable

ADJUSTMENT

Except 1990–92 Daytona and LeBaron

1. Release the parking brakes fully.
2. Raise the vehicle and support safely.
3. Adjust the rear brakes.
4. Loosen the adjusting nut until there is slack in all the cables.
5. Rotate the rear wheels and tighten the cable adjusting nut until there is a slight drag at the wheels.
6. Continue to rotate the rear wheels and loosen the nut until all drag is eliminated.
7. Back off the nut an additional 2 turns.
8. Apply and release the parking brake several times. Upon the least release, verify there is no drag at the wheels.
9. To check the operation, make sure the parking brake holds on an incline.

1990–92 Daytona and LeBaron

The parking brake hand lever contains a self-adjusting loaded clockspring feature. Routine parking brake adjustment is not required.

REMOVAL & INSTALLATION

Front Cable

EXCEPT 1990–92 DAYTONA AND LEBARON

1. Disconnect the negative battery cable.
2. Loosen the adjusting nut and disengage the front cable from the equalizer bracket.
4. Lift the carpet and floor matting and remove the floor pan seal.

5. Pull the cable end forward and disconnect from the clevis.

6. Pull the cable through the hole and remove.

7. The installation is the reverse of the removal procedure.

8. Connect the negative battery cable and check the parking brakes for proper operation.

1990–92 DAYTONA AND LEBARON

--- **CAUTION** ---

The parking brake hand lever contains a self-adjusting loaded clockspring loaded to about 30 lbs. Care must be taken when handling components in the vicinity of the hand lever or serious personal injury may result.

1. Disconnect the negative battery cable.

2. Disengage the cable from the equalizer bracket in the console.

4. Lift the carpet and floor matting and remove the floor pan seal.

5. Separate the cable from the rear parking brake shoes lever.

6. Pull the cable through the hole and remove.

To install:

7. Install the cable and connect to the rear shoes and equalizer bracket. Install the floor pan seal and position the carpet.

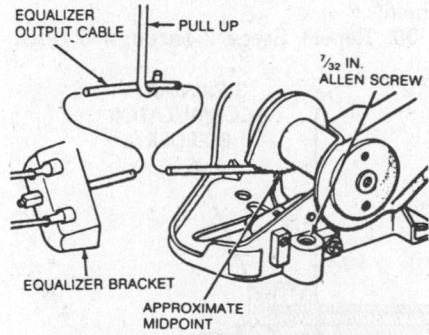

Self-adjusting parking brake lever assembly—1990–92 Daytona and LeBaron

8. To reload, lockout and adjust the system:

a. Pull on the equalizer output cable with at least 30 lbs. pressure to wind up the spring. Continue until the self-adjuster lockout pawl is positioned about midway between the self-adjuster sector.

b. Rotate the lockout pawl into the self-adjuster sector by turning the Allen screw clockwise. This action requires very little effort; do not force the screw.

c. Adjust the rear drum-in-hat parking brake shoes.

d. Turn the Allen screw counterclockwise about 15 degrees. When turning the lockout device, self-adjuster release is a snapping noise fol-

lowed by a detent that should be felt. Very light effort is required to seat the lockout device into the detent. Make sure to follow through into the detent.

e. Cycle the lever a few times to complete the adjustment. Thwe wheels should rotate freely.

9. Connect the negative battery cable and check the parking brakes for proper operation.

Rear Parking Brake Cable

EXCEPT TC

1. Disconnect the negative battery cable. Loosen the adjusting nut.

2. Raise the vehicle and support safely. Remove the wheels and drums. Disconnect the cable from the actuating lever on the rear brake shoe assembly.

3. Remove the retaining clip from the cable at the support bracket and pull the cable from the trailing arm assembly.

4. The installation is the reverse of the removal procedure.

5. Connect the negative battery cable and check the parking brakes for proper operation.

TC

1. Raise the vehicle and support safely. Remove the wheels.

2. Loosen the adjusting nut and disconnect the rear cable from the connector.

3. Remove the brake cable retaining clips from the hanger bracket and caliper.

4. Disconnect the cable from the parking brake lever on the caliper. Remove the cable guide from the trailing arm.

5. Pull the cable assembly from the hanger bracket.

6. The installation is the reverse of the removal procedure.

7. Connect the negative battery cable and check the parking brakes for proper operation.

Brake System Bleeding

Except Anti-Lock Brakes

NOTE: If using a pressure bleeder, follow the instructions furnished with the unit and choose the correct adaptor for the application. Do not substitute an adapter that "almost fits" as it will not work and could be dangerous.

MASTER CYLINDER

If the master cylinder is off the vehicle it can be bench bled.

1. Connect 2 short pieces of brake line to the outlet fittings, bend them until the free end is below the fluid level in the master cylinder reservoirs.

2. Fill the reservoir with fresh brake fluid. Pump the piston slowly until no more air bubbles appear in the reservoirs.

3. Disconnect the 2 short lines, refill the master cylinder and securely install the cylinder caps.

4. If the master cylinder is on the vehicle, it can still be bled, using a flare nut wrench.

5. Open the brake lines slightly with the flare nut wrench while pressure is applied to the brake pedal by a helper inside the vehicle.

6. Be sure to tighten the line before the brake pedal is released.

7. Repeat the procedure with both lines until no air bubbles come out.

CALIPERS AND WHEEL CYLINDERS

1. Fill the master cylinder with fresh brake fluid. Check the level often during the procedure.

2. Starting with the right rear wheel, remove the protective cap from the bleeder, if equipped, and place where it will not be lost. Clean the bleed screw.

--- **CAUTION** ---

When bleeding the brakes, keep face away from the brake area. Spewing fluid may cause facial and/or visual damage. Do not allow brake fluid to spill on the car's finish; it will remove the paint.

3. If the system is empty, the most efficient way to get fluid down to the wheel is to loosen the bleeder about ½–¾ turn, place a finger firmly over the bleeder and have a helper pump the brakes slowly until fluid comes out the bleeder. Once fluid is at the bleeder, close it before the pedal is released inside the vehicle.

NOTE: If the pedal is pumped rapidly, the fluid will churn and create small air bubbles, which are almost impossible to remove from the system. These air bubbles will eventually congregate and a spongy pedal will result.

4. Once fluid has been pumped to the caliper or wheel cylinder, open the bleed screw again, have the helper press the brake pedal to the floor, lock the bleeder and have the helper slowly release the pedal. Wait 15 seconds and repeat the procedure (including the 15 second wait) until no more air comes out of the bleeder upon application of the brake pedal. Remember to close the bleeder before the pedal is released inside the vehicle each time the bleed-

er is opened. If not, air will be induced into the system.

5. If a helper is not available, connect a small hose to the bleeder, place the end in a container of brake fluid and proceed to pump the pedal from inside the vehicle until no more air comes out the bleeder. The hose will prevent air from entering the system.

6. Repeat the procedure on remaining wheel cylinders in order:
   a. Left rear
   b. Right front
   c. Left front

7. Hydraulic brake systems must be totally flushed if the fluid becomes contaminated with water, dirt or other corrosive chemicals. To flush, bleed the entire system until all fluid has been replaced with the correct type of new fluid.

8. Install the bleeder cap(s), if equipped, on the bleeder to keep dirt out. Always road test the vehicle after brake work of any kind is done.

Daytona, LeBaron, Lebaron Landau, Spirit and Acclaim With Anti-Lock Brakes

The brake system must be bled any time air is permitted to enter the system through loosened or disconnected lines or hoses, or anytime the modulator is removed. Excessive air within the system will cause a soft or spongy feel in the brake pedal.

When bleeding any part of the system, the reservoir must remain close to **FULL** at all times. Check the level frequently and top off fluid as needed.

The Bendix Anti-lock 6 brake system must be bled as 2 separate brake systems. Proper procedures must be followed if the system is to work correctly. The normal portion of the brake system is bled in the usual fashion with either pressure or manual bleeding equipment and must be fully and properly bled before bleeding the modulator.

BLEEDING THE MODULATOR ASSEMBLY

To bleed the ABS unit, the battery must be relocated outside the vehicle and connected to the vehicle with jumper cables. This allows access to the 4 bleeder screws on top of the modulator assembly. Additionally, the DRB II must be connected to the diagnostic plug before bleeding begins; the DRB II is used to activate the system(s) during the procedure. The 4 components to be bled within the modulator are (in order) the secondary sump, the primary sump, the primary accumulator and the secondary accumulator. Use the following procedure to bleed the modulator assembly.

— CAUTION —

Wear eye protection when bleeding the modulator assembly and always use a hose on the bleed screw to direct the flow of fluid away from painted surfaces. Bleeding the modulator may result in the release of very high pressure fluid.

---

1. Connect a clear hose to the secondary sump bleeder screw and route the hose to a clear container.

2. Either install and pressurize the pressure bleeding equipment at the master cylinder or have an assistant provide light and constant pressure on the brake pedal.

3. Open the bleeder screw about ½–¾ turn. Use the DRB II to select the ACTUATE VALVES test; actuate the left front build/decay valve.

4. Bleed until the fluid flows free of air bubbles or until the brake pedal bottoms.

5. Tighten the bleeder screw and release the brake pedal if it was being held.

6. Repeat Steps 2 through 5 until the fluid is free of air bubbles. Remember to check the fluid reservoir level periodically.

7. Select and actuate the right rear build/decay valve and perform Steps 2–5 until the fluid flows without air bubbles.

8. Move the bleeder tube to the primary sump bleeder screw.

9. Pressurize the pressure bleeding equipment at the master cylinder or have an assistant provide light and constant pressure on the brake pedal.

10. Open the bleeder screw about ½–¾ turn. Using the DRB II, actuate the right front build/decay valve.

11. Bleed until the fluid flows free of air bubbles or until the brake pedal bottoms.

12. Tighten the bleeder screw and release the brake pedal if it was being held.

13. Repeat Steps 2 through 5 until the fluid is free of air bubbles. Remember to check the fluid reservoir level periodically.

14. Select and actuate the left rear build/decay valve. Perform Steps 2–5 until the fluid runs free of air bubbles.

15. Move the bleeder tube to the primary accumulator bleeder screw.

16. Pressurize the pressure bleeding equipment at the master cylinder or have an assistant provide light and constant pressure on the brake pedal.

17. Open the bleeder screw about ½–¾ turn. Using the DRB II, actuate the right front/left rear isolation valve.

18. Bleed until the fluid flows free of air bubbles or until the brake pedal bottoms.

19. Tighten the bleeder screw and release the brake pedal if it was being held.

20. Repeat Steps 2 through 5 until

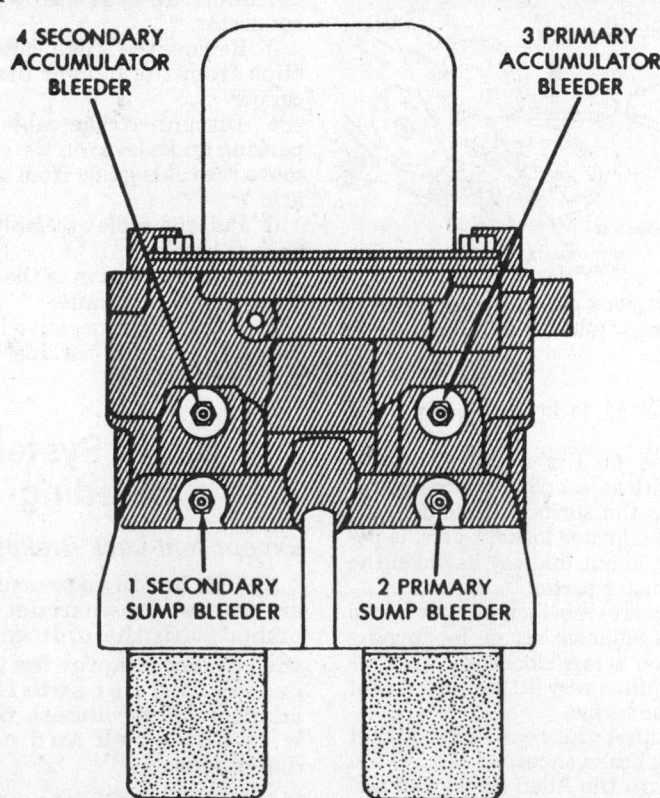

Bleeder locations for the ABS modulator assembly—Daytona, LeBaron, LeBaron Landau, Spirit and Acclaim

the fluid is free of air bubbles. Check the fluid reservoir level periodically.

21. Select and actuate the right front build/decay valve. Perform Steps 2–5 until the fluid runs free of air bubbles.

22. Move the bleeder tube to the secondary accumulator bleeder screw.

23. Pressurize the pressure bleeding equipment at the master cylinder or have an assistant provide light and constant pressure on the brake pedal.

24. Open the bleeder screw about ½–¾ turn. Using the DRB II, actuate the left front/right rear isolation valve.

25. Bleed until the fluid flows free of air bubbles or until the brake pedal bottoms.

26. Tighten the bleeder screw and release the brake pedal if it was being held.

27. Repeat Steps 2 through 5 until the fluid is free of air bubbles. Check the fluid reservoir level periodically.

28. Select and actuate the left front build/decay valve. Perform Steps 2–5 until the fluid runs free of air bubbles.

29. Remove the bleeding apparatus; fill the brake fluid reservoir to the correct level and install the cap.

TC With Anti-Lock Brakes
BOOSTER BLEEDING

1. Depressurize the hydraulic accumulator.

2. Connect all pump/motor and hydraulic assembly electrical connections. Be sure all brake lines and hose connections are tight.

3. Fill the reservoir to the full level.

4. Connect a transparent hose to the bleeder screw location on the right side of the hydraulic assembly. Place the other end of the hose into a clear container to receive brake fluid.

5. Open the bleeder screw ½–¾ of a turn.

6. Turn the ignition switch to the **ON** position. The pump/motor should run, discharging fluid into the container. After a good volume of fluid has been forced through the hose, an air-free flow in the plastic hose and container will indicate a good bleed.

7. Turn the ignition switch **OFF.**

NOTE: If the brake fluid does not flow, it may be due to a lack of prime to the pump/motor. Try shaking the return hose to break up air bubbles that may be present within the hose.

Should the brake fluid still not flow, turn the ignition switch OFF. Remove the return hose from the reservoir and cap nipple on the reservoir. Manually fill the return hose with brake fluid and connect to the reservoir. Repeat the bleeding process.

8. Remove the hose from the bleed-

er screw. Tighten the bleeder screw to 7.5 ft. lbs. (10 Nm). Do not overtighten.

9. Top off the reservoir to the correct fluid level.

10. Turn the ignition switch to the **ON** position. Allow the pump to charge the accumulator, which should stop after approximately 30 seconds.

PRESSURE BLEEDING

The brake lines may be pressure bled, using a standard diaphragm type pressure bleeder. Only diaphragm type pressure bleeding equipment should be used to bleed the system.

1. The ignition should be turned **OFF** and remain **OFF** throughout this procedure.

2. Depressurize the hydraulic accumulator.

——————— **CAUTION** ———————

Failure to depressurize the hydraulic accumulator, prior to performing this operation may result in personal injury and/or damage to the painted surfaces.

——————————————————

3. Remove the electrical connector from fluid level sensor on the reservoir cap and remove the reservoir cap.

4. Install the pressure bleeder adapter.

5. Attach the bleeding equipment to the bleeder adapter. Charge the pressure bleeder to approximately 20 psi (138 kPa).

6. Connect a transparent hose to the caliper bleed screw. Submerge the free end of the hose in a clear glass container, which is partially filled with clean, fresh brake fluid.

7. With the pressure turned **ON,** open the caliper bleed screw ½–¾ turn and allow fluid to flow into the container. Leave the bleed screw open until clear, bubble-free fluid slows from the hose. If the reservoir has been drained or the hydraulic assembly removed from the vehicle prior to the bleeding operation, slowly pump the brake pedal 1–2 times while the bleed screw is open and fluid is flowing. This will help purge air from the hydraulic assembly. Tighten the bleeder screw to 7.5 ft. lbs. (10 Nm).

8. Repeat Step 7 at all calipers. Calipers should be bled in the following order:
   a. Left rear
   b. Right rear
   c. Left front
   d. Right front

9. After bleeding all 4 calipers, remove the pressure bleeding equipment and bleeder adapter by closing the pressure bleeder valve and slowly unscrewing the bleeder adapter from the hydraulic assembly reservoir. Failure to release pressure in the reservoir will cause spillage of brake fluid and could

result in injury or damage to painted surfaces.

10. Using a syringe or equivalent method, remove excess fluid from the reservoir to bring the fluid level to full level.

11. Install the reservoir cap and connect the fluid level sensor connector. Turn the ignition **ON** and allow the pump to charge the accumulator.

MANUAL BLEEDING

1. Depressurize the hydraulic accumulator.

——————— **CAUTION** ———————

Failure to depressurize the hydraulic accumulator, prior to performing this operation may result in personal injury and/or damage to the painted surfaces.

——————————————————

2. Connect a transparent hose to the caliper bleed screw. Submerge the free end of the hose in a clear glass container, which is partially filled with clean, fresh brake fluid.

3. Slowly pump the brake pedal several times, using full strokes of the pedal and allowing approximately 5 seconds between pedal strokes. After 2–3 strokes, continue to hold pressure on the pedal, keeping it at the bottom of its travel.

4. With pressure on the pedal, open the bleed screw ½–¾ turn. Leave bleed screw open until fluid no longer flows from the hose. Tighten the bleed screw and release the pedal.

5. Repeat this procedure until clear, bubble-free fluid flows from the hose.

6. Repeat all steps at each of the calipers. Calipers should be bled in the following order:
   a. Left rear
   b. Right rear
   c. Left front
   d. Right front

Anti-Lock Brake System Service

PRECAUTIONS

Failure to observe the following precautions may result in system damage.

• Before performing electric arc welding on the vehicle, disconnect the control module and the hydraulic unit connectors.

• When performing painting work on the vehicle, do not expose the control module to temperatures in excess of 185°F (85°C) for longer than 2 hrs. The system may be exposed to temperatures up to 200°F (95°C) for less than 15 min.

• Never disconnect or connect the control module or hydraulic modulator connectors with the ignition switch **ON.**

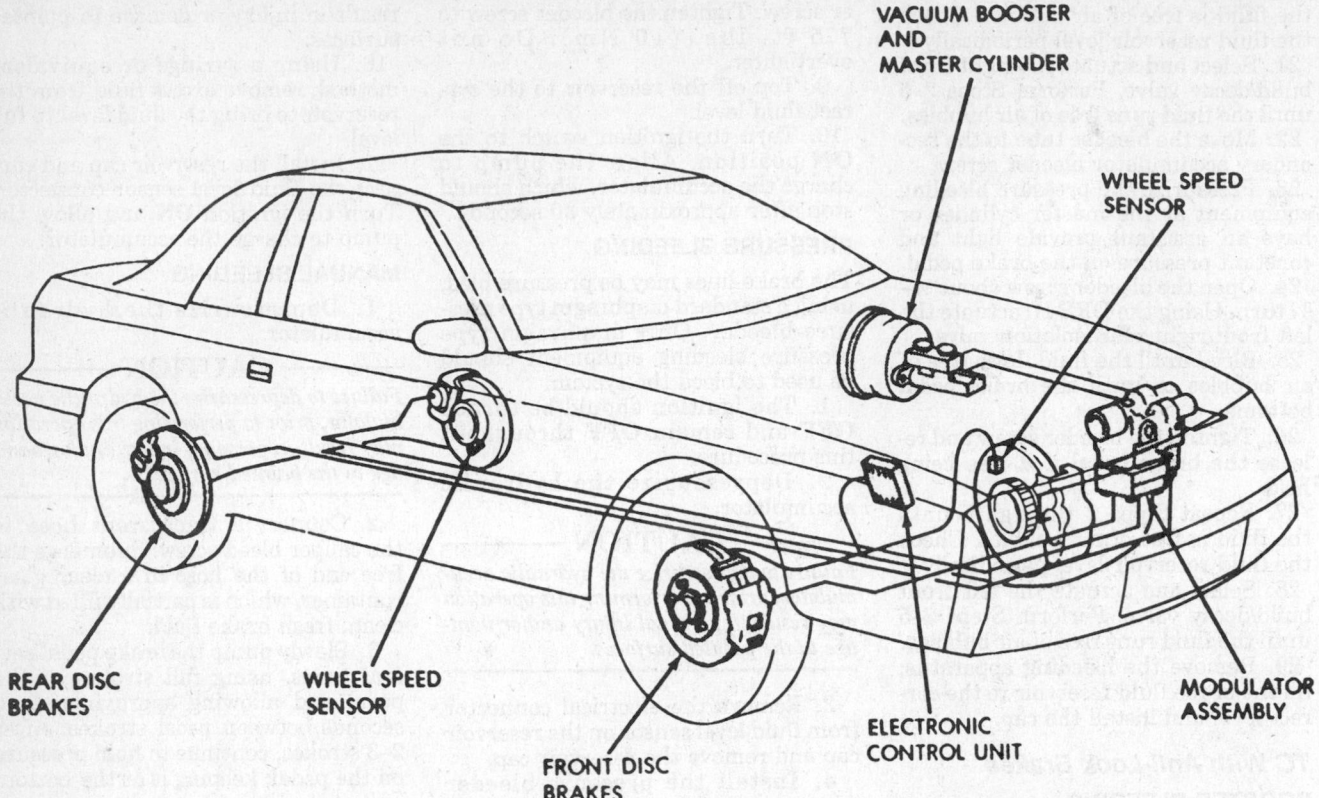

Anti-lock brake system components—Daytona, LeBaron, LeBaron Landau, Spirit and Acclaim

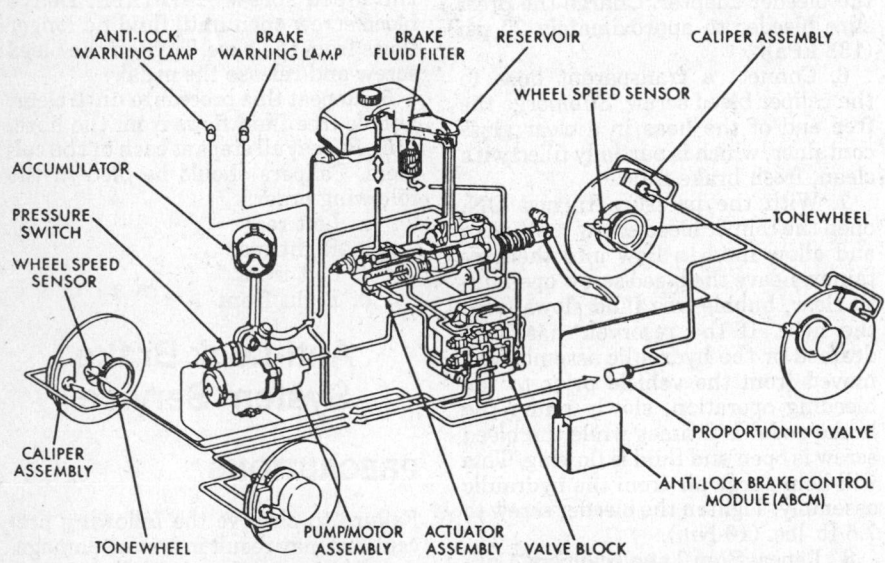

Anti-lock brake system components—TC

• Never disassemble any component of the Anti-Lock Brake System (ABS) which is designated non-servicable; the component must be replaced as an assembly.

• When filling the master cylinder, always use brake fluid which meets DOT-3 specifications; petroleum-based fluid will destroy the rubber parts.

DEPRESSURIZING THE HYDRAULIC ACCUMULATOR

TC

1. With the ignition **OFF**, pump the brake pedal a minimum of 25 times, using approximately 50 lbs. (222 N) pedal force. A noticeable change in pedal feel will occur when the accumulator is discharged.

2. When a definite increase in pedal effort is felt, stroke the pedal a few additional times. This should remove all hydraulic pressure from the system.

Rear Disc Brake Caliper and Pads

REMOVAL & INSTALLATION

TC

1. Depressurize the hydraulic accumulator. Remove ⅔ of brake fluid from the master cylinder.

2. Remove access plug and insert a 4mm Allen wrench through hole.

3. Turn the retraction shaft counterclockwise a few turns ro increase clearance between pads and rotor.

4. Remove the anti–rattle spring from outboard pad taking care not to damage it.

5. Back the caliper guide pins out just enough to free caliper from adapter.

6. Lift the caliper off of rotor and carefully suspend with wire.

7. Remove the brake pads.

8. Insert the Allen wrench through access hole and turn clockwise, if necessary to retract piston further to increase clearance for new pads.

To install:

9. Install new inner and outer pads.

NOTE: The outboard pads are marked for right and left hand sides and must be properly installed.

10. Lower the caliper over rotor and pads.

11. Install the guide pins and tighten to proper torque.

12. Insert the Allen wrench through the access hole and turn clockwise until snug, no clearance between pads and rotors, and back off ⅓ turn to obtain proper clearance.

13. Check the brake fluid level and add if necessary.

Pump/Motor Assembly

REMOVAL & INSTALLATION

Daytona, LeBaron, LeBaron Landau, Spirit and Acclaim

The pump and motor assembly used on the Bendix Anti-lock 6 brake system is not removable. If the pump or motor fails, the modulator assembly must be replaced.

TC

1. Depressurize the hydraulic accumulator.

———— CAUTION ————

Failure to depressurize the hydraulic accumulator prior to performing this operation may result in personal injury and/or damage to the painted surfaces.

2. Remove the fresh air intake ducts.

3. Disconnect all electrical connectors the pump motor.

4. Disconnect the high and low pressure hoses from the hydraulic assembly. Cap the spigot on the reservoir.

5. Disconnect the transmission shift selection cable bracket from the transaxle and move it aside.

6. Loosen the nuts on the 2 studs that position the pump/motor to the transaxle differential cover.

7. Remove the retainer bolts that are used to mount hose bracket and pump/motor. The engine inlet water extension pipe is also held in position by these bolts.

NOTE: Do not disturb the inlet water extension pipe, or engine coolant will leak out.

8. Disconnect the wiring harness retaining clip from the hose bracket.

9. Lift the pump/motor assembly off the studs and out of the vehicle.

10. Remove the heat shield from the pump/motor, if equipped, and discard.

To install:

11. Place a new heat shield to the pump/motor bracket, using fasteners provided.

12. Install the pump/motor assembly in the reverse order of the removal.

13. Readjust the gearshift linkage, if disturbed.

Modulator Assembly

REMOVAL & INSTALLATION

Daytona, LeBaron, LeBaron Landau, Spirit and Acclaim

1. Remove the battery, battery tray and the protective cover from the modulator.

2. Disconnect the electrical connector from the Delta P switch.

3. Remove the top bolt holding the modulator bracket to the fender shield.

4. Disconnect the 2 master cylinder supply tubes at the modulator. Loosen (but do not remove) the other end of the tubes at the master cylinder; swing the tubes aside without kinking them.

5. Elevate and safely support the vehicle.

6. From below, disconnect the modulator 10-pin electrical connector. Remove the remaining 4 brake tubes from the modulator assembly.

7. Remove the modulator bracket mounting bolt which is closest to the hydraulic junction block.

8. Loosen but do not fully remove the bracket mounting bolt closest to the radiator.

9. Lower the vehicle; lift the modulator assembly and bracket out of the vehicle.

To install:

10. Install the modulator and bracket into position. Use the protruding tab on the modulator to locate and hold the assembly. Make certain the bracket is held by the front mounting bolt.

11. Install but do not tighten the bolt holding the bracket to the fender shield.

12. Elevate and safely support the vehicle.

13. Install the bracket mounting bolt closest to the junction block. Tighten both lower mounting bracket bolts to 21 ft. lbs. (28 Nm).

14. Install the 4 hydraulic lines at the modulator; tighten the fittings to 12 ft. lbs. (16 Nm).

15. Reconnect the 10-pin electrical connector to the modulator.

16. Lower the vehicle. Connect the 2 supply tubes from the master cylinder to the modulator. Tighten the fittings at both ends of the tubes to 12 ft. lbs. (16 Nm).

17. Tighten the bolt holding the bracket to the fender shield (Step 11) to 21 ft. lbs. (28 Nm).

18. Bleed the base brake system in the usual fashion.

19. Bleed the modulator assembly following the correct sequences and procedure.

20. Install the protective cover on the modulator assembly.

21. Install the battery tray and battery. Connect the battery cables.

Hydraulic Assembly

REMOVAL & INSTALLATION

TC

1. Depressurize the hydraulic accumulator.

———— CAUTION ————

Failure to depressurize the hydraulic accumulator prior to performing this operation may result in personal injury and/or damage to the painted surfaces.

2. Remove the fresh air intake ducts.

3. Disconnect all electrical connectors from the hydraulic unit and pump/motor.

4. Remove as much of the fluid as possible from the reservoir on the hydraulic assembly.

5. Remove the pressure hose fitting (banjo bolt) from the hydraulic assembly. Use care not to drop the 2 washers used to seal the pressure hose fitting to the hydraulic assembly inlet.

6. Disconnect the return hose from the reservoir nipple. Cap the spigot on the reservoir.

7. Disconnect all brake tubes from the hydraulic assembly.

8. Remove the driver's side sound insulation panel.

9. Disconnect the pushrod from the brake pedal.

10. Remove the 4 underdash hydraulic assembly mounting nuts.

11. Remove the hydraulic assembly.

To install:

12. Position the hydraulic assembly on the vehicle.

13. Install and torque the mounting nuts to 21 ft. lbs. (28 Nm).

14. Using lubriplate or equivalent, coat the bearing surface of the pedal pin.

15. Connect the pushrod to the pedal and install a new retainer clip.

16. Install the brake tubes. If the proportioning valves were removed from the hydraulic assembly, reinstall valves and tighten to 20 ft. lbs. (27 Nm).

17. Install the return hose to the nipple on the reservoir.

18. Install the pressure hose to the hydraulic assembly; be sure the 2 washers are in there proper position.

Tighten the bango bolt to 13 ft. lbs. (18 Nm).

19. Fill the reservoir to the top of the screen.

20. Connect all electrical connectors to the hydraulic assembly.

21. Bleed the entire brake system.

22. Install the crosscar brace, if disturbed. Install the fresh air intake duct.

Sensor Block

REMOVAL & INSTALLATION

TC

1. Depressurize the hydraulic accumulator.

— CAUTION —

Failure to depressurize the hydraulic accumulator, prior to performing this operation may result in personal injury and/or damage to the painted surfaces.

2. Disconnect all electrical connectors from the reservoir on the hydraulic assembly.

3. Working from under the dash, disconnect the pushrod from the brake pedal.

4. Remove the driver's side sound insulator panel.

5. Remove the 4 hydraulic assembly mounting nuts.

6. Working from under the hood, pull the hydraulic assembly away from the dash panel and rotate the assembly enough to gain access to the sensor block cover.

NOTE: The brake lines should not be removed or deformed during this procedure.

7. Remove the sensor block cover retaining bolt and remove the sensor block cover. Care should be used not to damage the cover gasket during removal.

8. Disengage the locking tabs and disconnect the valve block connector (12 pin) from the sensor block.

9. Disengage the reed block connector, marked PUSH, by carefully pulling outward on the orange connector body. The connector is partially retained by a plastic clip and will only move outward approximately ½ in. (13mm).

10. Remove the 3 block retaining bolts.

11. Carefully disengage the sensor block pressure port from the hydraulic assembly and remove the sensor block from the vehicle. The sensor block pressure port is sealed with an O-ring and extra care should be taken to prevent damage to the seal.

12. Inspect the sensor block pressure port O-ring for damage. Replace the O-

ring if cut or damaged. Check the sensor block wiring for any mispositioning or damage. Correct any damage or replace the sensor block if damage cannot be corrected.

To install:

13. Pull the reed block connector (2 pin) outward to the disengage position prior to installing the sensor block on the hydraulic unit.

14. Throughly lubricate the sensor block pressure port O-ring with fresh, clean brake fluid. Carefully insert the pressure port into the hydraulic assembly's orifice, taking care not to cut or damage the O-ring. Position the sensor block for installation of the mounting bolts.

15. Install the sensor block mounting bolts. Tighten to 11 ft. lbs. (15 Nm).

16. Engage the reed block connector by pressing on the orange connector body marked PUSH.

17. Connect the valve block connector (12 pin) to the sensor block.

18. Install the sensor block cover, gasket and mounting bolt.

19. Connect the sensor block and control pressure switch connectors.

20. Install the hydraulic assembly by reversing the removal procedure.

Wheel Speed Sensors

REMOVAL & INSTALLATION

Daytona, LeBaron, LeBaron Landau, Spirit and Acclaim

FRONT WHEEL

1. Elevate and safely support the vehicle. Remove the wheel and tire.

2. Remove the clip holding the wiring grommet to the fender well.

3. Remove the screws holding the sensor wiring tube to the fender well.

4. Carefully remove the grommet from the fender shield.

5. Make certain the ignition switch is **OFF**. Disconnect the sensor wiring from the ABS harness.

6. Remove the triangular retaining clip from the bracket on the strut. Not all vehicles have this clip.

7. Remove the sensor wiring grommets from the bracket.

8. Remove the fastener holding the sensor head.

9. Carefully remove the sensor head from the steering knuckle. Do not use pliers on the sensor head; if it is seized in place, use a hammer and small punch to tap the edge of the sensor

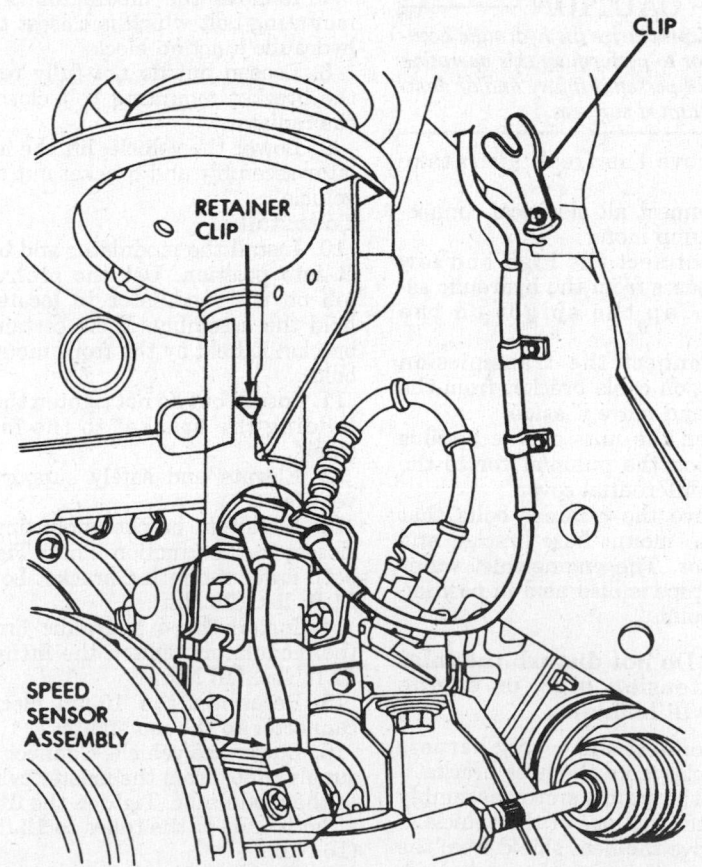

Front wheel speed sensor—Daytona, LeBaron, LeBaron Landau, Spirit and Acclaim

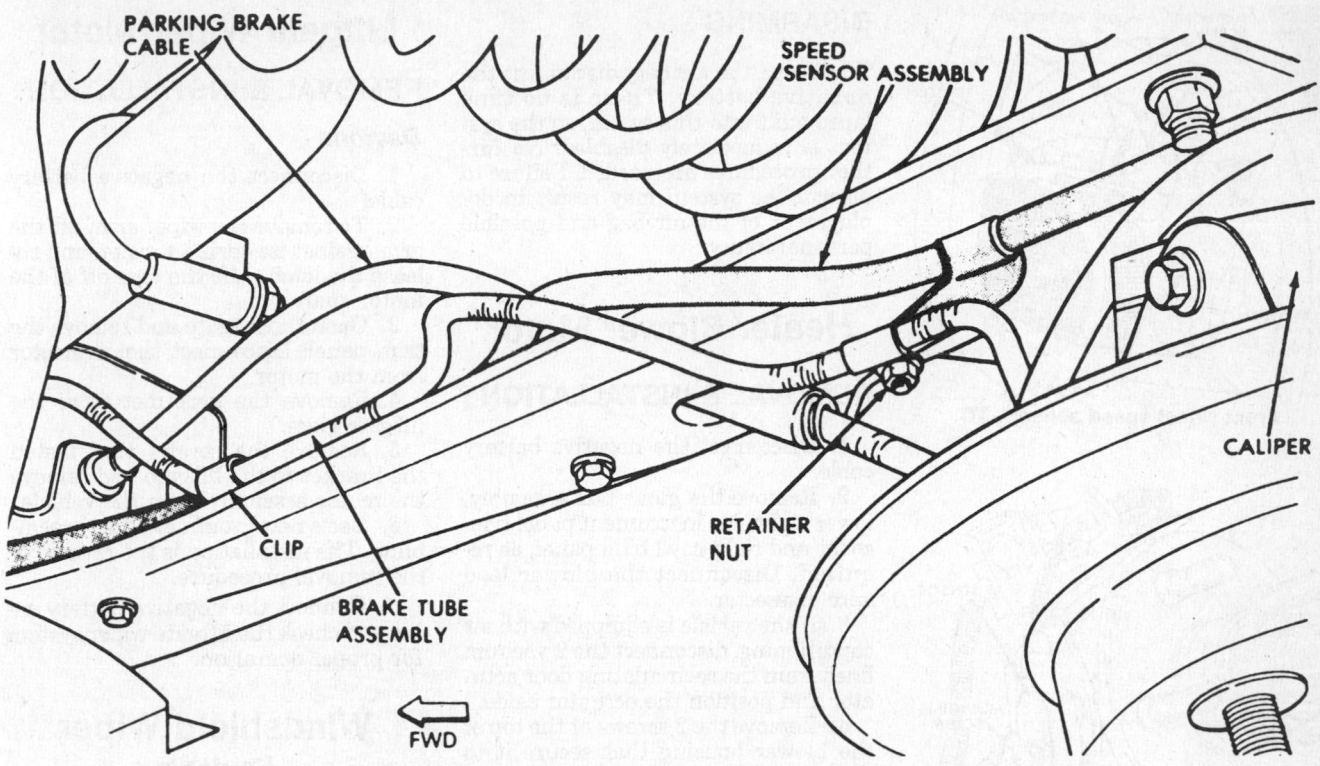

PARKING BRAKE CABLE

SPEED SENSOR ASSEMBLY

CALIPER

RETAINER NUT

CLIP

BRAKE TUBE ASSEMBLY

← FWD

Rear wheel speed sensor—Daytona, LeBaron, LeBaron Landau, Spirit and Acclaim

ear. The tapping and side-to-side motion will free the unit.

To install:

10. Connect the speed sensor to the ABS harness.

11. Push the sensor assembly grommet into the the hole in the fender shield. Install the retainer clip and screw.

12. Install the sensor wiring tube and tighten the retaining bolts to 35 inch lbs. (4 Nm).

13. Install the sensor grommets into the brackets on the fender shield and strut. Install the retainer clip at the strut.

14. Install the sensor to the knuckle. Install the retaining screw and tighten it to 60 inch lbs. (7 Nm).

NOTE: Proper installation of the sensor and its wiring is critical to system function. Make certain that wiring is installed in all retainers and clips. Wiring must be protected from moving parts and not be stretched during suspension movements.

15. Install the tire and wheel. Lower the vehicle to the ground.

REAR WHEEL

1. Elevate and safely support the vehicle. Remove the wheel and tire.

2. Remove the sensor assembly grommet from the underbody and pull the harness through the hole in the body.

3. Make certain the ignition switch is **OFF**. Disconnect the sensor wiring from the ABS harness.

4. Remove the clip retaining screw from the bracket just forward of the trailing arm bushing.

5. Remove the sensor and brake tube assembly clip from the inboard side of the trailing arm.

6. Remove the sensor wire retainer from the rear brake hose bracket.

7. Remove the outboard sensor assembly nut. This nut is also used to hold the brake tube clip.

8. Remove the fastener holding the sensor head.

9. Carefully remove the sensor head from the adapter assembly. Do not use pliers on the sensor head; if it is seized in place, use a hammer and small punch to tap the edge of the sensor ear. The tapping and side-to-side motion will free the unit.

To install:

10. Before installation, coat the sensor with high temperature multi-purpose grease.

11. Install the sensor; install the retaining screw and tighten it to 60 inch lbs. (7 Nm).

12. Install the outboard retaining nut.

13. Install the clips and nuts at and around the trailing arm.

14. Connect the sensor wiring to the ABS harness; make sure the connector lock is engaged.

15. Push the sensor assembly grommet into the the hole in the underbody.

NOTE: Proper installation of the sensor and its wiring is critical to system function. Make certain that wiring is installed in all retainers and clips. Wiring must be protected from moving parts and not be stretched during suspension movements.

16. Install the tire and wheel.

17. Lower the vehicle to the ground.

TC

1. Raise the vehicle and support safely. Remove the wheel and tire assembly.

2. Remove the sensor cable from the retainer clips.

3. Carefully pull the sensor assembly grommet from the floor pan.

4. Unplug the connector from the harness.

5. Remove the sensor mounting screw. Do not disturb the adjustment screw.

6. Carefully remove the sensor.

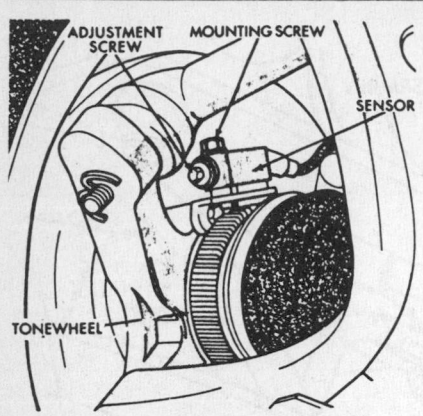

Front wheel speed sensor—TC

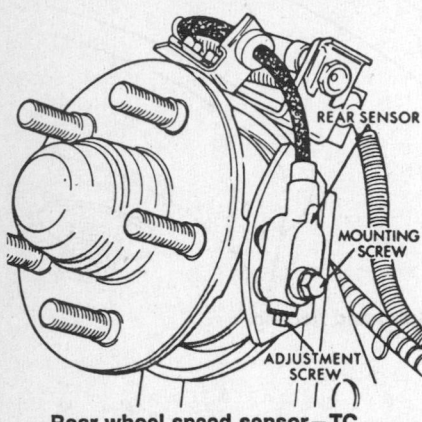

Rear wheel speed sensor—TC

To install:

7. To install, coat the sensor with high temperature multi-purpose anti-corrosion compound at all areas it contacts the bracket before installing into the steering knuckle. Install the screw and tighten to 85 inch lbs. (10 Nm).

8. Connect the sensor connector to the harness and install the sensor connector lock.

9. Install the sensor assembly grommet.

NOTE: Proper installation of the wheel speed sensor cables is critical to continued system operation. Be sure the cables are installed in retainers. Failure to install the cables in the retainers may result in contact with moving parts and/or over-extension of the cables, resulting in an open circuit.

CHASSIS ELECTRICAL

Air Bag

DISARMING

To disarm the air bag, disconnect the negative battery. There is no time lapse built into this sytem, so the system is immediately disabled. No further procedures are needed. Failure to disarm the system may result in deployment of the air bag and possible personal injury.

Heater Blower Motor

REMOVAL & INSTALLATION

1. Disconnect the negative battery cable.

2. Remove the glove box assembly, lower right side instrument panel trim cover and right cowl trim panel, as required. Disconnect the blower lead wire connector.

3. If the vehicle is equipped with air conditioning, disconnect the 2 vacuum lines from the recirculating door actuator and position the actuator aside.

4. Remove the 2 screws at the top of the blower housing that secure it to the unit cover.

5. Remove the 5 screws from around the blower housing and separate the blower housing from the unit.

6. Remove the 3 screws that secure the blower assembly to the heater or air conditioning housing and remove the assembly from the unit. Remove the fan from the blower motor.

7. The installation is the reverse of the removal procedure.

8. Connect the negative battery cable and check the blower motor for proper operation.

Windshield Wiper Motor

REMOVAL & INSTALLATION

1. Disconnect the negative battery cable.

2. If the cowl top plastic cover must be removed, remove the wiper arms and blades and remove the cover.

3. Remove the wiper motor cover and disconnect the motor wiring harness.

4. Disconnect the linkage drive crank from the motor crank arm.

5. Remove the motor mounting nuts and remove the wiper motor from the vehicle.

6. The installation is the reverse of the removal procedure.

7. Connect the negative battery cable and check the wiper motor for proper operation.

Liftgate Wiper Motor

REMOVAL & INSTALLATION

Daytona

1. Disconnect the negative battery cable.

2. To remove the wiper arm, lift the arm against its spring tension and release the latch. Lift the arm off of the motor shaft.

3. Open the liftgate and remove the trim panel. Disconnect the connector from the motor.

4. Remove the grommet from the liftgate glass.

5. Remove the screws that fasten the bracket to the liftgate and remove the motor assembly from the vehicle.

6. Use a new grommet when assembling. The installation is the reverse of the removal procedure.

7. Connect the negative battery cable and check the liftgate wiper system for proper operation.

Windshield Wiper Switch

REMOVAL & INSTALLATION

1988–89 Vehicles With Standard Column

1. Disconnect the negative battery cable.

2. Remove the lower steering column cover, if equipped.

3. Straighten the steering wheel so the tires are pointing straight-ahead.

NOTE: If equipped with an airbag, it is imperative that the steering wheel removal and installation procedure under Steering is followed.

4. Remove the steering wheel.

5. Remove the plastic wiring channel from the under the steering column.

6. Disconnect the wiper switch connector, intermittent wipe module connector and cruise control connector, if equipped.

7. Remove the side lock housing cover.

8. Remove the slotted hex-head screw that attaches the wiper switch to the turn signal switch, then remove the switch.

9. Remove the control knob from the end of the stalk. Pull the round nylon hider up the control stalk and remove the revealed screws that attach the control stalk sleeve to the wiper switch.

10. Rotate the control stalk shaft to the full clockwise position and remove

the shaft from the wiper switch by pulling it straight out.

To install:

11. Install the control shaft to the wiper switch, install the screws, the hider and the control knob.

12. Run the wiring through the opening and down the steering column, position the switch and install the hex-head screw. Make sure the dimmer switch rod is properly engaged.

13. Install the side lock housing cover.

14. Connect the wires and install the wiring channel.

15. Install the steering wheel torque the nut to 45 ft. lbs. (61 Nm).

16. Install the horn pad.

17. Connect the negative battery cable and check the wiper and washer, cruise control, turn signal switch and dimmer switch for proper operation.

18. Install the lower column cover, if equipped.

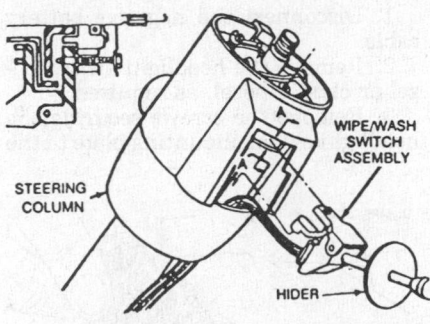

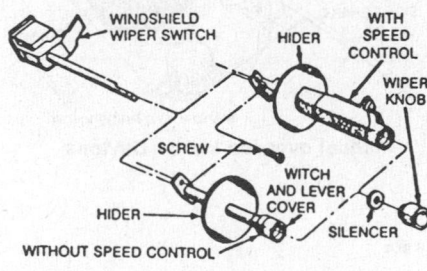

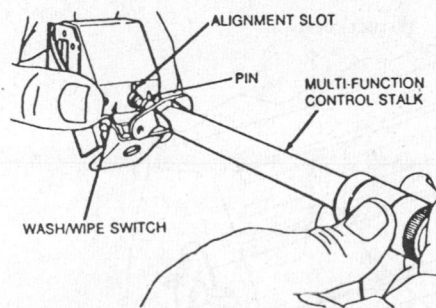

Removing the windshield wiper switch—standard column

Maserati TC and 1988–89 Vehicles With Tilt Wheel

1. Disconnect the negative battery cable.

2. Remove the lower steering column cover if equipped, and remove the plastic wiring channel from the under the steering column.

3. Straighten the steering wheel so the tires are pointing straight-ahead.

NOTE: If equipped with an airbag, it is imperative that the steering wheel removal and installation procedure under Steering is followed.

4. Remove the steering wheel.

5. Depress the lockplate with the proper depressing tool, remove the retaining ring from its groove and remove the tool, ring, lockplate, cancelling cam and spring.

6. Remove the switch stalk actuator screw and arm.

7. Remove the hazard switch knob.

8. Disconnect the turn signal switch, wiper switch, intermittent module and cruise control connectors, if equipped.

9. Remove the 3 screws and remove the turn signal switch. Tape the connector to the wires to aid in removal.

10. Remove the ignition key light.

11. Place the key in the **LOCK** position and remove the key. Insert a thin tool into the slot next to the switch mounting screw boss, depress the spring latch at the bottom of the slot releasing the lock. Remove the lock cylinder.

12. Remove the buzzer switch and wedge spring.

13. Remove the 3 housing cover screws and remove the housing cover.

14. Remove the wiper switch pivot pin with a punch and remove the switch.

15. Remove the control knob from the end of the stalk. Pull the round nylon hider up the control stalk and remove the revealed screws that attach the control stalk sleeve to the wiper switch.

16. Rotate the control stalk shaft to the full clockwise position and remove the shaft from the wiper switch by pulling it straight out.

To install:

17. Install the control shaft to the wiper switch, install the screws, the hider and the control knob.

18. Run the wiring through the opening and down the steering column, position the switch and install the wiper switch pivot pin.

19. Install the housing cover.

20. Install the buzzer switch and wedge spring.

21. Install the lock cylinder.

22. Install the ignition key light.

23. Install the turn signal switch, switch stalk actuator arm and hazard switch knob.

24. Install the spring, cancelling cam, lockplate and ring on the steering shaft. Depress the plate with the depressing tool and install the ring securely in the groove. Remove the tool slowly.

25. Connect the turn signal switch, wiper switch, intermittent module and cruise control connectors, if equipped. Install the channel.

26. Install the steering wheel and torque the nut to 45 ft. lbs. (61 Nm).

27. Install the horn pad.

28. Connect the negative battery cable and check the wiper and washer, cruise control, turn signal switch and dimmer switch for proper operation.

29. Install the lower column cover, if equipped.

1990–92 Daytona and LeBaron

1. Disconnect the negative battery cable.

2. Remove the panel vent grille above the switch pod assembly and remove the 2 revealed pod mounting screws.

3. Remove the 2 remaining screws under the pod and pull the pod out to disconnect the wiring harnesses. Remove the pod from the instrument panel.

4. Remove the inner panel from the pod. Disconnect the switch linkage from the buttons.

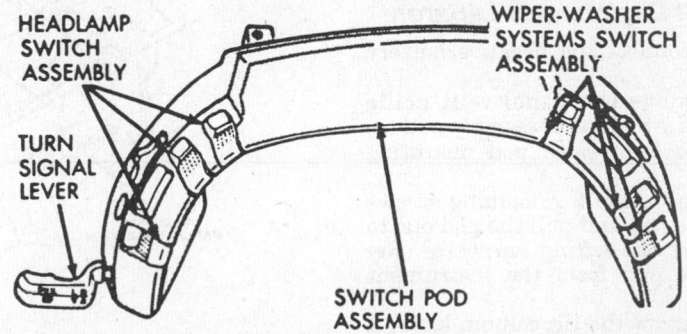

Instrument panel switch pod assembly—1990–92 Daytona and LeBaron

5. Remove the windshield wiper switch mounting screws and remove the entire switch assembly.

6. The installation is the reverse of the removal procedure.

7. Connect the negative battery cable and check the entire wiper system for proper operation.

Instrument Cluster

REMOVAL & INSTALLATION

Except 1990–92 Daytona and LeBaron

1. Disconnect the negative battery cable.

2. Remove the instrument cluster bezel. Cluster removal is not necessary if just removing gauges.

3. When only removing gauge(s) or the speedometer, remove the trip odometer reset knob, if necessary, remove the mask and lens assembly, and remove the desired gauge from the cluster. Disconnect the speedometer cable, if equipped, when removing the speedometer.

4. If equipped with automatic transaxle and column shift, remove the lower column cover and disconnect the gear indicator cable.

5. Remove the screws attaching the cluster to the instrument panel.

6. Pull the cluster out and disconnect all wiring harnesses and the speedometer cable, if equipped. Remove the cluster from the vehicle.
To install:

7. Position the cluster and feed the gear indicator cable through its slot.

8. Connect all wiring and install the speedometer cable to the speedometer, if removed; make sure the cable end is securely clicked in place.

9. Install the cluster retaining screws. Connect the gearshift indicator cable.

10. Install the cluster bezel.

11. Connect the negative battery cable, check all gauges and the speedometer for proper operation. Make sure the gearshift indicator is properly aligned.

1990–92 Daytona and LeBaron

1. Disconnect the negative battery cable.

2. Remove the panel vent grille above the switch pod assembly and remove the 2 revealed pod mounting screws.

3. Remove the 2 remaining screws under the pod and pull the pod out to disconnect the wiring harnesses. Remove the pod from the instrument panel.

4. Unscrew the tilt column lever, if equipped, remove the screws from un-

der the upper steering column shrouds and remove the shrouds.

5. Pull rearward to disengage the cluster trim bezel retaining clips and remove the bezel.

6. When only removing gauges or the speedometer, remove the mask and lens assembly and remove the desired assembly from the cluster.

7. Remove the screws attaching the cluster to the instrument panel.

8. Pull the cluster out and disconnect all wiring harnesses and the turbo gauge hose, if equipped. Remove the cluster from the vehicle.
To install:

9. Position the cluster and connect all wiring and the turbo hose, if it was disconnected.

10. Install the cluster mounting screws.

11. Install the cluster trim bezel.

12. Install the steering column shrouds and the tilt lever, if equipped.

13. Install the switch pod assembly and panel vent grille.

14. Connect the negative battery cable and check all gauges, switches and the speedometer for proper operation.

Radio

REMOVAL & INSTALLATION

NOTE: On vehicles equipped with a compact disc player, removal and installation procedures are the same as for the radio.

1. Disconnect the negative battery cable.

2. Remove the console or cluster bezel, as required.

3. Remove the screws that attach the radio to the instrument panel.

4. Pull the radio out, disconnect the connectors, ground cable and antenna, and remove the radio.

5. The installation is the reverse of the removal procedure.

6. Connect the negative battery cable and check all functions of the radio for proper operation.

Concealed Headlights

MANUAL OPERATION

1. Disconnect the negative battery cable.

2. Locate the manual override knob. On Daytona, they are located under access shields behind the bumper facia and under the center of the front bumper on LeBaron.

3. Remove the protective cover boot.

4. Rotate the manual override knob to raise the headlight cover(s).

5. Connect the negative battery cable.

Headlight Switch

REMOVAL & INSTALLATION

Except 1990–92 Daytona and LeBaron

1. Disconnect the negative battery cable.

2. Remove the headlight switch bezel or cluster bezel, as required.

3. Remove the screws securing the headlight switch mounting plate to the

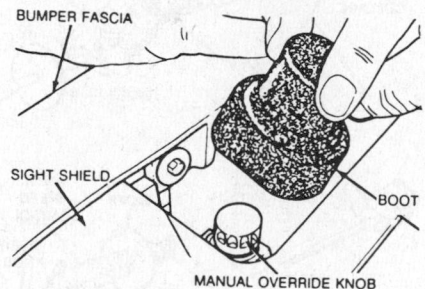

Manual override knob—Daytona

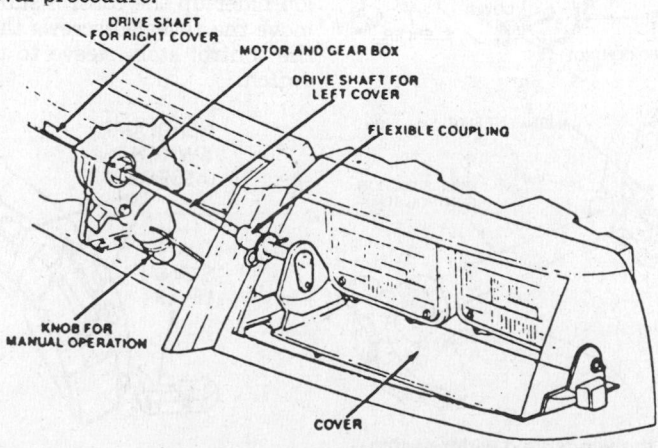

Manual override knob—LeBaron

instrument panel. Pull the assembly out and disconnect the connectors from the switch.

4. Depress the spring button and remove the headlight switch knob and stem.

5. Remove the escutcheon, if equipped, and remove the nut that attaches the switch to the mounting plate.

6. The installation is the reverse of the removal procedure.

7. Connect the negative battery cable and check the switch for proper operation.

1990–92 Daytona and LeBaron

1. Disconnect the negative battery cable.

2. Remove the panel vent grille above the switch pod assembly and remove the 2 revealed pod mounting screws.

3. Remove the 2 remaining screws under the pod and pull the pod out to disconnect the wiring harnesses. Remove the pod from the instrument panel.

4. Remove the turn signal switch lever by pulling it straight out of the pod.

5. Remove the inner panel from the pod. Remove the turn signal switch in order to gain access to the headlight switch.

6. Disconnect the switch linkage from the buttons.

7. Remove the switch mounting screws and remove the entire switch assembly.

8. The installation is the reverse of the removal procedure.

9. Connect the negative battery cable and check the system for proper operation.

Dimmer Switch

REMOVAL & INSTALLATION

Maserati TC and 1988–89 Vehicles

NOTE: The dimmer switch is incorporated into the combination switch on 1990–92 LeBaron Landau, Shadow, Sundance, Spirit and Acclaim. On 1990–92 Daytona and LeBaron, it is incorporated with the remote turn signal switch.

1. Disconnect the negative battery cable.

2. Remove the lower steering column cover, if equipped.

3. Unplug the switch, located on the lower portion of the steering column.

4. Holding the actuating rod against its upper seat, remove the

bolts that attach the switch to the column, and remove the switch.

5. The installation is the reverse of the removal procedure. Adjust the switch as required.

6. Connect the negative battery cable and check the switch for proper operation.

Turn Signal Switch

REMOVAL & INSTALLATION

1988–89 Vehicles With Standard Column

1. Disconnect the negative battery cable.

2. Remove the lower steering column cover, if equipped.

3. Straighten the steering wheel so the tires are pointing straight-ahead.

NOTE: If equipped with an airbag, it is imperative that the steering wheel removal and installation procedure under Steering is followed.

4. Remove the steering wheel.

5. Remove the plastic wiring channel from the under the steering column and disconnect the turn signal switch connector.

6. Remove the hazard switch knob. Remove the slotted hex-head screw that attaches the wiper switch to the turn signal switch.

7. Remove the 3 screws and pull the turn signal switch out of the column.
To install:

8. Run the wiring through the opening and down the steering column, position the switch and install the hex-head screw. Make sure the dimmer switch rod is properly engaged.

9. Install the 3 screws and the hazard switch knob.

10. Connect the wires and install the wiring channel.

11. Install the steering wheel and torque the nut to 45 ft. lbs. (61 Nm).

12. Install the horn pad.

13. Connect the negative battery cable and check the turn signal switch and dimmer switch for proper operation.

14. Install the lower column cover, if equipped.

Masratil TC and 1988–89 Vehicles With Tilt Wheel

1. Disconnect the negative battery cable.

2. Remove the lower steering column cover, if equipped and remove the plastic wiring channel from the under the steering column.

3. Straighten the steering wheel so the tires are pointing straight-ahead.

NOTE: If equipped with an airbag, it is imperative that the steering wheel removal and installation procedure under Steering is followed.

4. Remove the steering wheel.

5. Depress the lockplate with the proper depressing tool, remove the retaining ring from its groove and remove the tool, ring, lockplate, cancelling cam and spring.

6. Remove the stalk actuator screw and arm.

7. Remove the hazard switch knob.

8. Disconnect the turn signal switch connector.

9. Remove the 3 screws and remove the turn signal switch. Tape the connector to the wires to aid in removal.
To install:

10. Run the wiring through the opening and down the steering column, install the turn signal switch, switch stalk actuator arm and hazard switch knob.

11. Install the spring, cancelling cam, lockplate and ring on the steering shaft. Depress the plate with the depressing tool and install the ring securely in the groove. Remove the tool slowly.

12. Connect the turn signal switch connector and install the channel.

13. Install the steering wheel and torque the nut to 45 ft. lbs. (61 Nm).

14. Install the horn pad.

15. Connect the negative battery cable and check the turn signal switch and dimmer switch for proper operation.

16. Install the lower column cover, if equipped.

1990–92 Daytona and LeBaron

1. Disconnect the negative battery cable.

2. Remove the panel vent grille above the switch pod assembly and remove the 2 revealed pod mounting screws.

3. Remove the 2 remaining screws under the pod and pull the pod out to disconnect the wiring harnesses. Remove the pod from the instrument panel.

4. Remove the turn signal switch lever by pulling it straight out of the pod.

5. Remove the inner panel from the pod. Unplug the switch from the printed circuit board.

6. Remove the turn signal switch mounting screws and slide the switch out of the slot.

7. The installation is the reverse of the removal procedure.

8. Connect the negative battery cable and check the turn signal switch and dimmer function for proper operation.

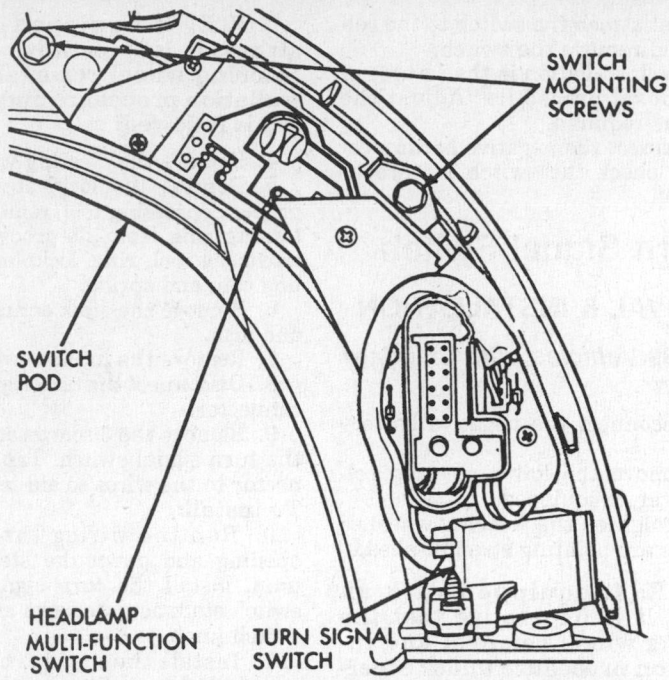

Turn signal switch location—1990–92 Daytona and LeBaron

Combination Switch

REMOVAL & INSTALLATION

1990–92 LeBaron Landau, Shadow, Sundance, Spirit and Acclaim

1. Disconnect the negative battery cable.
2. Remove the tilt lever, if equipped.
3. Remove the steering column covers.
4. Remove the combination switch tamper-proof mounting screws and pull the switch away from the steering column.
5. Loosen the connector screw; the screw will remain in the connector. Disconnect the connector from the switch.

6. The installation is the reverse of the removal procedure.
7. Connect the negative battery cable and check all functions of the combination switch for proper operation.

Ignition Lock

REMOVAL & INSTALLATION

Standard Column Except Acustar Column

NOTE: The Acustar column can be identified by the "halo" light around the ignition key cylinder.

1. Disconnect the negative battery cable.
2. Straighten the steering wheel so the tires are pointing straight-ahead.

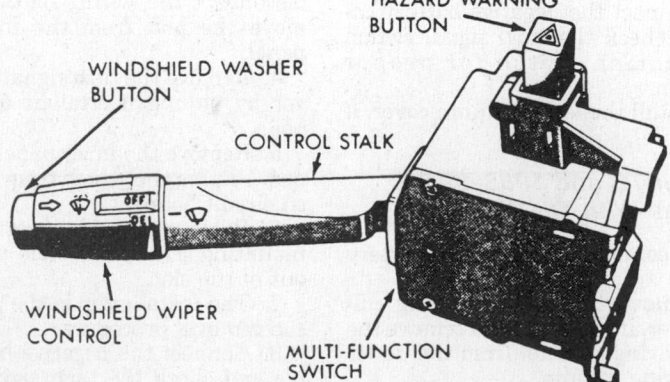

Combination switch—1990–92 LeBaron Landau, Shadow, Sundance, Spirit and Acclaim

NOTE: If equipped with an airbag, it is imperative that the steering wheel removal and installation procedure under Steering is followed.

3. Remove the steering wheel.
4. Remove the hazard switch knob. Remove the slotted hex-head screw that attaches the wiper switch to the turn signal switch.
5. Remove the 3 screws and pull the turn signal switch out of the column as far as it will go. For additional access, the switch can be unplugged from below.
6. Remove the ignition switch key light.
7. Place the key in the **LOCK** position and remove the key.
8. Insert 2 suitable small-diameter tools into both release holes and push inward to release the spring loaded lock retainers while simultaneously pulling the key lock cylinder out of its bore.

To install:

9. Install the key cylinder.
10. Install the ignition switch key light.
11. Install the turn signal switch and hazard switch knob, then connect all wiring.
12. Install the steering wheel and torque the nut to 45 ft. lbs. (61 Nm).
13. Install the horn pad.
14. Connect the negative battery cable and check the lock cylinder for proper operation.
15. Install the lower column cover, if equipped.

Saginaw Tilt Column

1. Disconnect the negative battery cable.
2. Straighten the steering wheel so the tires are pointing straight-ahead.

NOTE: If equipped with an airbag, it is imperative that the steering wheel removal and installation procedure under Steering is followed.

3. Remove the steering wheel.
4. Depress the lockplate with the proper depressing tool, remove the retaining ring from its groove, then remove the tool, ring, lockplate, cancelling cam and spring.
5. Remove the stalk actuator screw and arm.
6. Remove the hazard switch knob.
7. Remove the 3 screws and pull the turn signal switch out of the column as far as it will go. Unplug it below if necessary.
8. Remove the ignition key light.
9. Place the key in the **LOCK** position and remove the key. Insert a thin tool into the slot next to the switch

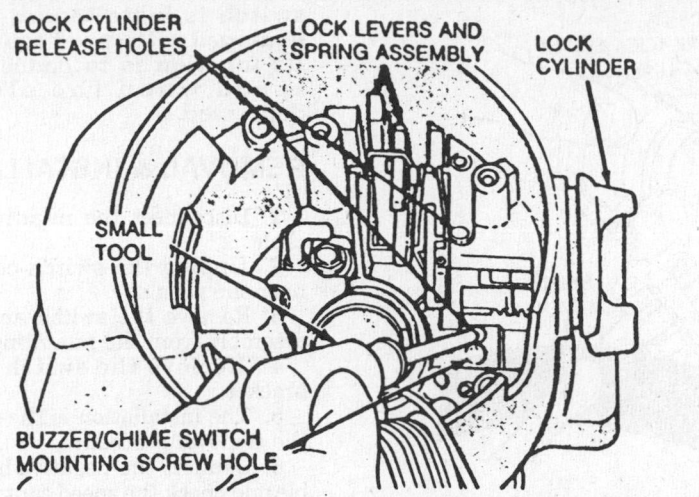

LOCK CYLINDER RELEASE HOLES

LOCK LEVERS AND SPRING ASSEMBLY

LOCK CYLINDER

SMALL TOOL

BUZZER/CHIME SWITCH MOUNTING SCREW HOLE

Removing the key lock cylinder—standard column, except Acustar steering column

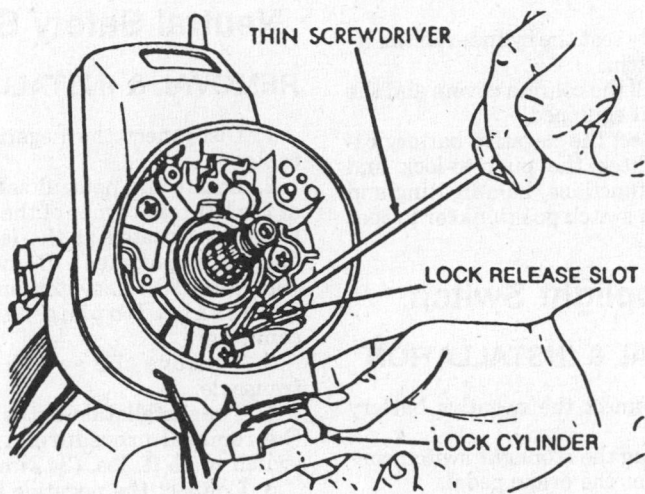

THIN SCREWDRIVER

LOCK RELEASE SLOT

LOCK CYLINDER

Removing the key lock cylinder—Saginaw steering column

mounting screw boss, depress the spring latch at the bottom of the slot releasing the lock and remove the lock cylinder.

To install:

10. Install the lock cylinder.
11. Install the ignition key light.
12. Install the turn signal switch, switch stalk actuator arm and hazard switch knob.
13. Install the spring, cancelling cam, lockplate and ring on the steering shaft. Depress the plate with the depressing tool and install the ring securely in the groove. Remove the tool slowly.
14. Connect the wires if they were disconnected.
15. Install the steering wheel and torque the nut to 45 ft. lbs. (61 Nm).
16. Install the horn pad.
17. Connect the negative battery cable and check the turn signal switch for proper operation.
18. Install the lower column cover, if removed.

Ignition Switch

REMOVAL & INSTALLATION

Except Acustar Steering Column

NOTE: The Acustar column can be identified by the "halo" light around the ignition key cylinder.

1. Disconnect the negative battery cable.
2. Remove the lower steering column cover.
3. Remove the steering column retaining nuts and allow the steering wheel to rest on the driver's seat.
4. Remove the 2 screws that attach the ignition switch to the column.
5. Rotate the switch 90 degrees and pull up to disengage it from the ignition switch rod.

To install:

6. Engage the switch with the rod, rotate the switch 90 degrees and push down until fully engaged.

7. Install the mounting screws finger-tight.
8. Place the key in the **LOCK** position and remove the key. Adjust the switch by pushing up gently on the switch to take up all slack in the rod.
9. Tighten the mounting screws and check the switch for proper operation in all positions.
10. Install the steering column and cover.

Ignition Lock/Switch

REMOVAL & INSTALLATION

Acustar Steering Column

NOTE: The Acustar column can be identified by the "halo" light around the ignition key cylinder.

1. Disconnect the negative battery cable.
2. Remove the tilt lever, if equipped.
3. Remove the upper and lower column covers.
4. Remove the 3 ignition switch Torx® screws; APEX 440–TX20H or equivalent required.
5. Pull the switch away from the column. Release the connector locks on the 2 wiring connectors and disconnect them from the switch.
6. Remove the key lock cylinder from the ignition switch:

   a. Insert the key and turn the switch to the **LOCK** position. Using

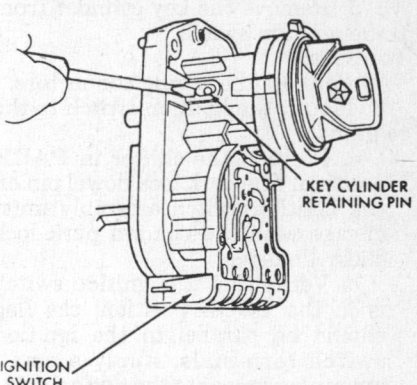

KEY CYLINDER RETAINING PIN

IGNITION SWITCH

Depressing the key cylinder retaining pin—Acustar column

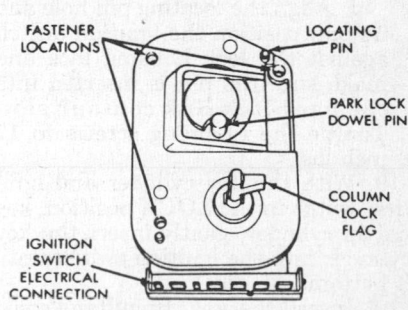

FASTENER LOCATIONS

LOCATING PIN

PARK LOCK DOWEL PIN

COLUMN LOCK FLAG

IGNITION SWITCH ELECTRICAL CONNECTION

Preparing the ignition switch for installation—Acustar column

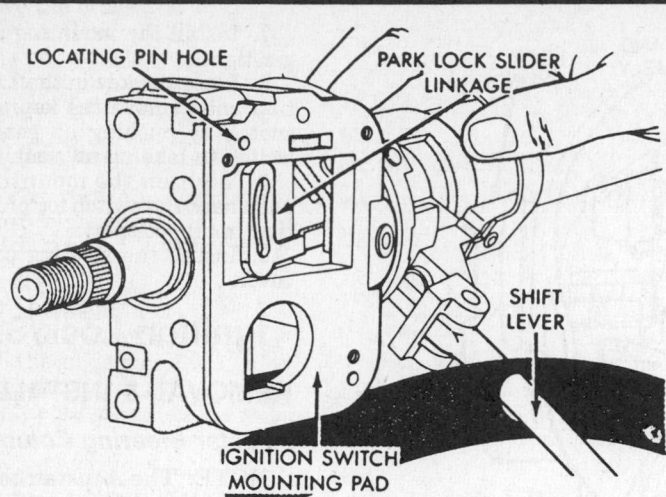

Ignition switch mounting pad—Acustar column

a suitable small tool, depress the key cylinder retaining pin flush with the key cylinder surface.

b. Rotate the key clockwise to the **OFF** position to unseat the key cylinder from the ignition switch assembly. The cylinder bezel should be about 1/8 in. above the ignition switch halo light ring. Do not attempt to remove the key cylinder yet.

c. With the key cylinder in the unseated position, rotate the key counterclockwise to the **LOCK** position and remove the key.

d. Remove the key cylinder from the ignition switch.

To install:

7. Connect the wiring connectors.

8. Mount the ignition switch to the column:

a. Position the shifter in **PARK** position. The park lock dowel pin on the ignition switch assembly must engage with the column park lock slider linkage.

b. Verify that the ignition switch is in the **LOCK** position; the flag should be parallel to the ignition switch terminals. Apply a small amount of grease to the flag and pin.

c. Position the park lock link to mid-travel.

d. Align the locating pin hole and its pin, position the ignition switch against the lock housing face and make sure the pin is inserted into the park lock link contour slot. Torque the retaining screws to 17 inch lbs.

9. With the key cylinder and ignition switch in the **LOCK** position, key not in cylinder, gently insert the key cylinder into the ignition switch until it bottoms.

10. Insert the key. Simultaneously, push in on the cylinder and rotate the key to the **RUN** position. This action should fully seat the cylinder in the ignition switch.

11. Install the column covers and the tilt lever, if equipped.

12. Connect the negative battery cable and check the push-to-lock and park lock functions, halo lighting and all ignition switch positions for proper operation.

Stoplight Switch

REMOVAL & INSTALLATION

1. Disconnect the negative battery cable.

2. Unplug the stoplight switch connectors near the brake pedal.

3. Remove the switch and bracket assembly from the brake pedal bracket.

4. Remove the switch from its bracket.

To install:

5. Install the switch and bracket assembly to the brake pedal bracket and push the switch forward as far as it will go; the brake pedal should move forward slightly.

6. Pull back on the brake pedal bringing the striker toward the switch until the pedal will not go back any farther.

7. This will cause the switch to ratchet backward into position and automatic adjustment is complete.

8. Connect the negative battery cable and check the switch for proper operation. Also, make sure the speed control system functions properly, if equipped.

Clutch Switch

NOTE: Some vehicles are installed with a clutch/starter interlock switch. Otherwise, a clutch switch is installed on vehicles equipped with speed control only. Its function is to cancel the set speed when the clutch is depressed.

REMOVAL & INSTALLATION

1. Disconnect the negative battery cable.

2. Unplug the switch connectors near the pedals.

3. Remove the switch and bracket assembly from the mounting bracket.

4. Remove the switch from its bracket.

5. The installation is the reverse of the removal procedure.

6. Connect the negative battery cable and check the speed control system for proper operation.

Neutral Safety Switch

REMOVAL & INSTALLATION

1. Disconnect the negative battery cable.

2. Locate the neutral safety switch at the left rear corner of the automatic transaxle, located at the left front of engine compartment. Do not confuse with the PRNDL switch on the A604 transaxle. Unplug the switch connector.

3. Remove the switch from the transaxle.

4. The installation is the reverse of the removal procedure. Torque the switch to 25 ft. lbs. (34 Nm).

5. Connect the negative battery cable and check the switch for proper operation.

Fuses, Circuit Breakers and Relays

LOCATION

Aries, Reliant, 600, Caravelle, LeBaron (K Body) and New Yorker

The fuse block is located behind a removeable access panel, below the steering column. The hazard and turn signal flashers along with the time delay and horn relays are also located behind the panel. Additional relays are mounted on the inner fender panel near the battery.

Spirit, Acclaim, Shadow and Sundance

The fuse block is located behind the steering column cover, accessible by removing the fuse access panel above the hood latch release lever. The relay and flasher module is located behind

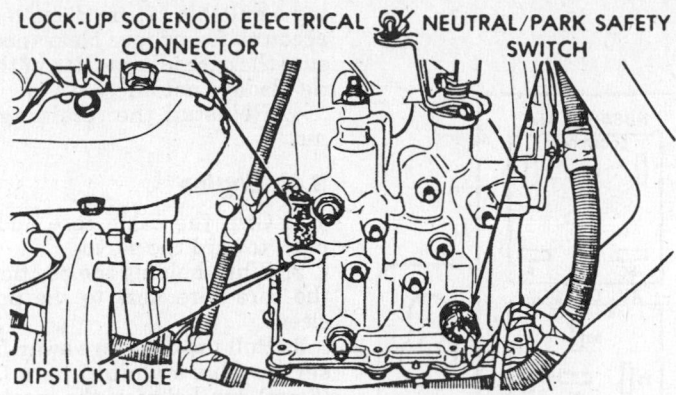

Neutral safety switch location—A413 automatic transaxle

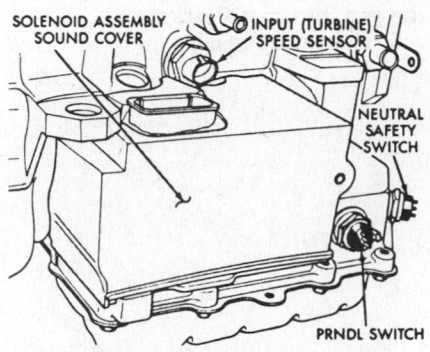

Neutral safety switch location—A604 automatic transaxle

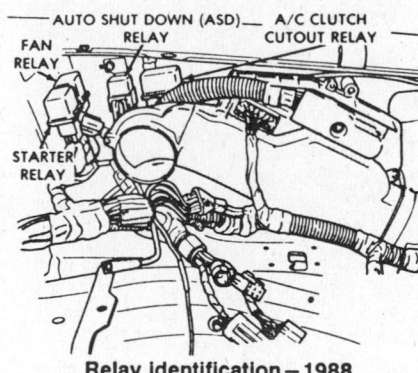

Relay identification—1988

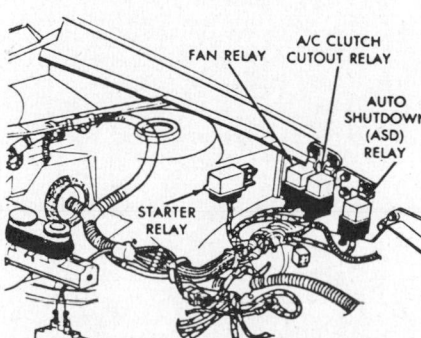

Relay identification—1989-90 vehicles and early 1991 Shadow Convertible

an access panel in the glovebox. Included in the module are the hazard and turn signal flashers along with the time delay and horn relays. Additional relays are mounted on the inner fender panel near the battery.

Lancer and LeBaron GTS

The fuse block is located behind the glove box door, accessible by removing the fuse access panel. The relay and flasher module is located behind the cupholder in the center of the instrument panel. The entire module can be removed by pushing it up and off of its mounting bracket. Included in the module are the hazard and turn signal flashers along with the time delay and horn relays. Additional relays are mounted on the inner fender panel near the battery.

Daytona, LeBaron (J Body) and TC

The fuse block is located behind a removeable access panel to the left of the lower portion of the steering column.

On TC and 1988–89 Daytona and LeBaron, the hazard and turn signal flashers along with the time delay and horn relays are also located behind the panel.

On 1990–92 Daytona and LeBaron, a relay bank is located on the left side kick panel. The Power Distribution Center, which contains additional relays and fuses, is located in the engine compartment behind the battery. Each item is identified on the cover.

Computers

LOCATION

1988: The Single Module Engine Controller (SMEC) is located in engine compartment, to the left of the battery.

1989–92: The Single Board Engine Controller (SBEC) is located in engine compartment, to the left of the battery.

If equipped with the A604 automatic transaxle, the transaxle controller is located in the right front of the engine compartment.

The body controller, if equipped, is located inside the passenger compartment, behind the right side kick panel.

Cruise Control

CABLE ADJUSTMENT

2.2L and 2.5L Engines

1. The clearance between the throttle stud and cable clevis should be $1/16$ in.

2. To adjust the cable, remove the retaining clip or loosen the retaining clamp nut at the throttle bracket.

3. Pull all slack out of the cable us-

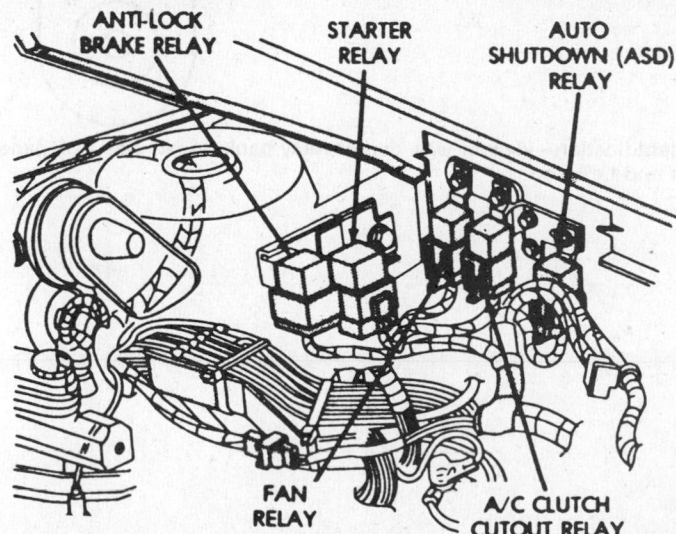

Relay identification—1991-92 vehicles without Power Distribution Center

ing a suitable $\frac{1}{16}$ in. diameter tool to account for proper clearance. Make sure the curb idle position of the throttle blade is not affected.

4. Reinstall the retaining clip or nut.

3.0L Engine

1. Grip the cable core and lightly push toward the servo.

2. While holding the position, mark the core wire next to the protective sleeve.

3. Pull the core wire away from the servo. There should be a 0.24 in. (6mm) gap between the mark on the core wire and the protective sleeve.

4. If the gap is not correct, remove the adjustment clip from the throttle bracket and move the sleeve to bring the gap into specification.

5. Reinstall the clip.

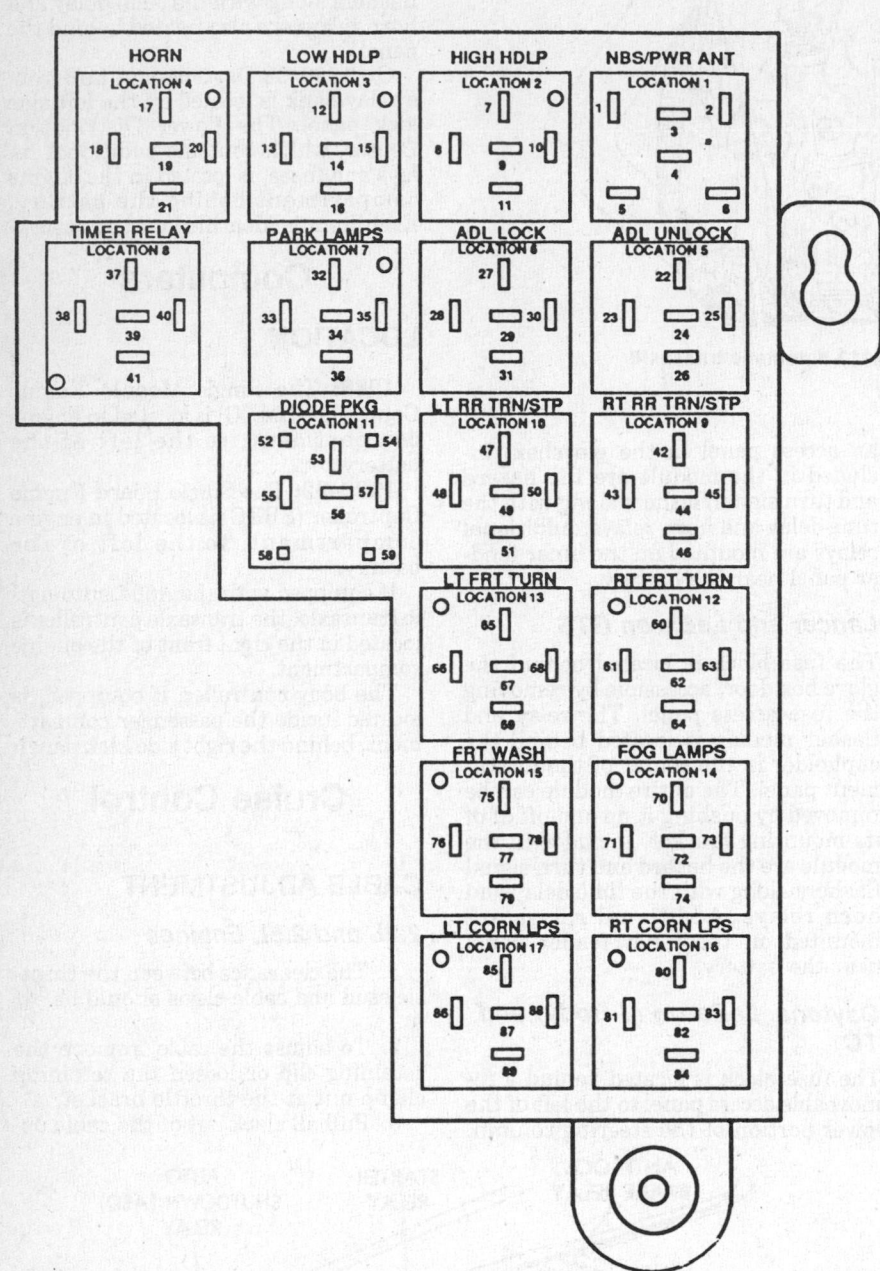

Relay identification—view of wire end of relay bank on left side kick panel—1990–92 Daytona and LeBaron

Chrysler Corp.
Front Wheel Drive
Chrysler—New Yorker, Imperial Dodge—Dynasty

SPECIFICATIONS

VEHICLE IDENTIFICATION CHART

It is important for servicing and ordering parts to be certain of the vehicle and engine identification. The VIN (vehicle identification number) is a 17 digit number visible through the windshield on the driver's side of the dash and contains the vehicle and engine identification codes. The tenth digit indicates model year and the eighth digit indicates engine code. It can be interpreted as follows:

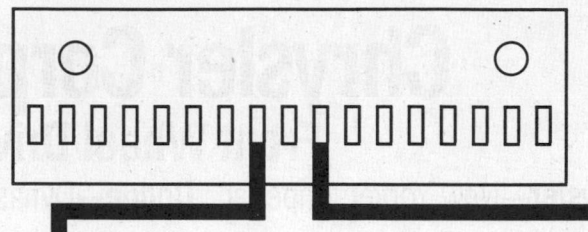

Engine Code					
Code	Cu. In.	Liters	Cyl.	Fuel Sys.	Eng. Mfg.
K	153	2.5	4	EFI	Chrysler Corporation
3	181	3.0	6	MPI	Mitsubishi
R	201	3.3	6	MPI	Chrysler Corporation
L	231	3.8	6	MPI	Chrysler Corporation

EFI—Electronic Fuel Injection
MPI—Multi Port Injection

Model Year	
Code	Year
J	1988
K	1989
L	1990
M	1991
N	1992

ENGINE IDENTIFICATION

Year	Model	Engine Displacement cu. in. (liter)	Engine Series Identification (VIN)	No. of Cylinders	Engine Type
1988	Dynasty	153 (2.5)	K	4	OHC
	Dynasty	181 (3.0)	3	6	OHC
	New Yorker Landau	181 (3.0)	3	6	OHC
1989	Dynasty	153 (2.5)	K	4	OHC
	Dynasty	181 (3.0)	3	6	OHC
	New Yorker Landau	181 (3.0)	3	6	OHC
1990	Dynasty	153 (2.5)	K	4	OHC
	Dynasty	181 (3.0)	3	6	OHC
	Dynasty	201 (3.3)	R	6	OHV
	New Yorker Landau	181 (3.0)	3	6	OHC
	New Yorker Landau	201 (3.3)	R	6	OHV
	New Yorker Salon	181 (3.0)	3	6	OHC
	New Yorker Salon	201 (3.3)	R	6	OHV
	New Yorker 5th Avenue	201 (3.3)	R	6	OHV
	Imperial	201 (3.3)	R	6	OHV

ENGINE IDENTIFICATION

Year	Model	Engine Displacement cu. in. (liter)	Engine Series Identification (VIN)	No. of Cylinders	Engine Type
1991–92	Dynasty	153 (2.5)	K	4	OHC
	Dynasty	181 (3.0)	3	6	OHC
	Dynasty	201 (3.3)	R	6	OHV
	New Yorker Salon	201 (3.3)	R	6	OHV
	New Yorker 5th Avenue	201 (3.3)	R	6	OHV
	New Yorker 5th Avenue	231 (3.8)	L	6	OHV
	Imperial	231 (3.8)	L	6	OHV

OHC—Overhead Cam
OHV—Overhead Valve

GENERAL ENGINE SPECIFICATIONS

Year	VIN	No. Cylinder Displacement cu. in. (liter)	Fuel System Type	Net Horsepower @ rpm	Net Torque @ rpm (ft. lbs.)	Bore × Stroke (in.)	Compression Ratio	Oil Pressure @ rpm
1988	K	4-153 (2.5)	EFI	100 @ 4800	135 @ 2800	3.44 × 4.09	8.9:1	30-80 @ 3000
	3	6-181 (3.0)	EFI	141 @ 5000	171 @ 2800	3.59 × 2.99	8.9:1	30-80 @ 3000
1989	K	4-153 (2.5)	EFI	100 @ 4800	135 @ 2800	3.44 × 4.09	8.9:1	30–80 @ 3000
	3	6-181 (3.0)	EFI	141 @ 5000	171 @ 2800	3.59 × 2.99	8.9:1	30–80 @ 3000
1990	K	4-153 (2.5)	EFI	100 @ 4800	135 @ 2800	3.44 × 4.09	8.9:1	30-80 @ 3000
	3	6-181 (3.0)	EFI	141 @ 5000	171 @ 2800	3.59 × 2.99	8.9:1	30–80 @ 3000
	R	6-201 (3.3)	EFI	183 @ 3600	183 @ 3600	3.66 × 3.19	8.9:1	30-80 @ 3000
1991–92	K	4-153 (2.5)	EFI	100 @ 4800	135 @ 2800	3.44 × 4.09	8.9:1	30-80 @ 3000
	3	6-181 (3.0)	EFI	141 @ 5000	171 @ 2800	3.59 × 2.99	8.9:1	30-80 @ 3000
	R	6-201 (3.3)	EFI	183 @ 3600	183 @ 3600	3.66 × 3.19	8.9:1	30-80 @ 3000
	L	6-231 (3.8)	EFI	150 @ 4400	203 @ 3200	3.78 × 3.42	9.0:1	30-80 @ 3000

EFI—Electronic Fuel Injection

GASOLINE ENGINE TUNE-UP SPECIFICATIONS

Year	VIN	No. Cylinder Displacement cu. in. (liter)	Spark Plugs Type	Gap (in.)	Ignition Timing (deg.) MT	Ignition Timing (deg.) AT	Compression Pressure (psi)	Fuel Pump (psi)	Idle Speed (rpm) MT	Idle Speed (rpm) AT	Valve Clearance In.	Valve Clearance Ex.
1988	K	4-153 (2.5)	RN12YC	0.035	—	12B	100①	15	—	850	Hyd.	Hyd.
	3	6-181 (3.0)	RN11YC4	0.040	—	12B	178②	48	—	700	Hyd.	Hyd.
1989	K	4-153 (2.5)	RN12YC	0.035	—	12B	100①	15	—	850	Hyd.	Hyd.
	3	6-181 (3.0)	RN11YC4	0.040	—	12B	178②	48	—	700	Hyd.	Hyd.
1990	K	4-153 (2.5)	RN12YC	0.035	—	12B	100①	15	—	850	Hyd.	Hyd.
	3	6-181 (3.0)	RN11YC4	0.040	—	12B	178②	48	—	700	Hyd.	Hyd.
	R	6-201 (3.3)	RN16YC5	0.050	—	12B	100①	48	—	750	Hyd.	Hyd.

GASOLINE ENGINE TUNE-UP SPECIFICATIONS

Year	VIN	No. Cylinder Displacement cu. in. (liter)	Spark Plugs		Ignition Timing (deg.)		Com- pression Pressure (psi)	Fuel Pump (psi)	Idle Speed (rpm)		Valve Clearance	
			Type	Gap (in.)	MT	AT			MT	AT	In.	Ex.
1991	K	4-153 (2.5)	RN12YC	0.035	—	12B	100①	15	—	850	Hyd.	Hyd.
	3	6-181 (3.0)	RN11YC4	0.040	—	12B	178②	48	—	700	Hyd.	Hyd.
	R	6-201 (3.3)	RN16YC5	0.050	—	12B	100①	48	—	750	Hyd.	Hyd.
	L	6-231 (3.8)	RN16YC5	0.050	—	12B	100①	48	—	750	Hyd.	Hyd.
1992		REFER TO UNDERHOOD STICKER										

① Minimum
② At 250 rpm
Hyd.—Hydraulic

FIRING ORDERS

NOTE: To avoid confusion, always replace spark plug wires one at a time.

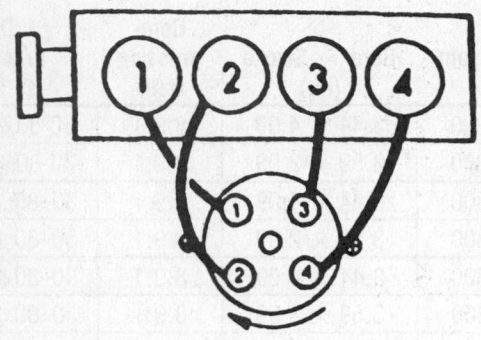

2.5L Engine
Engine Firing Order: 1–3–4–2
Distributor Rotation: Clockwise

3.0L Engine
Engine Firing Order: 1–2–3–4–5–6
Distributor Rotation: Counterclockwise

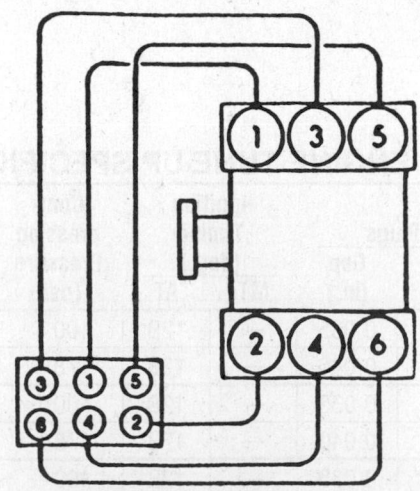

3.3L and 3.8L Engine
Engine Firing Order: 1–2–3–4–5–6
Distributorless Ignition System

CAPACITIES

Year	Model	VIN	No. Cylinder Displacement cu. in. (liter)	Engine Crankcase (qts.) with Filter	Engine Crankcase (qts.) without Filter	Transmission (pts.) 4-Spd.	Transmission (pts.) 5-Spd.	Transmission (pts.) Auto.	Drive Axle (pts.)	Fuel Tank (gal.)	Cooling System (qts.)
1988	Dynasty	K	4-153 (2.5)	4	4	—	—	18	—	16	9
	Dynasty	3	6-181 (3.0)	4	4	—	—	18	—	16	9.5
	New Yorker Landau	3	6-181 (3.0)	4	4	—	—	18	—	16	9.5
1989	Dynasty	K	4-153 (2.5)	4	4	—	—	17	—	16	9
	Dynasty	3	6-181 (3.0)	4	4	—	—	18	—	16	9.5
	New Yorker Landau	3	6-181 (3.0)	4	4	—	—	18	—	16	9.5
1990	Dynasty	K	4-153 (2.5)	4	4	—	—	17	—	16	9
	Dynasty	3	6-181 (3.0)	4	4	—	—	18	—	16	9.5
	Dynasty	R	6-201 (3.3)	4	4	—	—	18	—	16	9.5
	New Yorker Landau	3	6-181 (3.0)	4	4	—	—	18	—	16	9.5
	New Yorker Landau	R	6-201 (3.3)	4	4	—	—	18	—	16	9.5
	New Yorker Salon	3	6-181 (3.0)	4	4	—	—	18	—	16	9.5
	New Yorker Salon	R	6-201 (3.3)	4	4	—	—	18	—	16	9.5
	New Yorker 5th Avenue	R	6-201 (3.3)	4	4	—	—	18	—	16	9.5
	Imperial	R	6-201 (3.3)	4	4	—	—	18	—	16	9.5
1991–92	Dynasty	K	4-153 (2.5)	4	4	—	—	17	—	16	9
	Dynasty	3	6-181 (3.0)	4	4	—	—	18	—	16	9.5
	Dynasty	R	6-201 (3.3)	4	4	—	—	18	—	16	9.5
	New Yorker Salon	R	6-201 (3.3)	4	4	—	—	18	—	16	9.5
	New Yorker 5th Avenue	R	6-201 (3.3)	4	4	—	—	18	—	16	9.5
	New Yorker 5th Avenue	L	6-231 (3.8)	4	4	—	—	18	—	16	9.5
	Imperial	L	6-231 (3.8)	4	4	—	—	18	—	16	9.5

CAMSHAFT SPECIFICATIONS

All measurements given in inches.

Year	VIN	No. Cylinder Displacement cu. in. (liter)	Journal Diameter 1	Journal Diameter 2	Journal Diameter 3	Journal Diameter 4	Journal Diameter 5	Lobe Lift In.	Lobe Lift Ex.	Bearing Clearance	Camshaft End Play
1988	K	4-153 (2.5)	1.375– 1.376	1.375– 1.376	1.375– 1.376	1.375– 1.376	1.375– 1.376	NA	NA	—	0.005– 0.020
	3	6-181 (3.0)	NA	NA	NA	NA	NA	①	①	—	NA
1989	K	4-153 (2.5)	1.375– 1.376	1.375– 1.376	1.375– 1.376	1.375– 1.376	1.375– 1.376	NA	NA	—	0.005– 0.020
	3	6-181 (3.0)	NA	NA	NA	NA	NA	①	①	—	NA
1990	K	4-153 (2.5)	1.375– 1.376	1.375– 1.376	1.375– 1.376	1.375– 1.376	1.375– 1.376	NA	NA	—	0.005– 0.020
	3	6-181 (3.0)	NA	NA	NA	NA	NA	①	①	—	NA
	R	6-201 (3.3)	1.997– 1.999	1.980– 1.982	1.965– 1.967	1.949– 1.952	—	0.400	0.400	0.001– 0.005	0.005– 0.012

CAMSHAFT SPECIFICATIONS

All measurements given in inches.

Year	VIN	No. Cylinder Displacement cu. in. (liter)	Journal Diameter 1	2	3	4	5	Lobe Lift In.	Ex.	Bearing Clearance	Camshaft End Play
1991–92	K	4-153 (2.5)	1.375–1.376	1.375–1.376	1.375–1.376	1.375–1.376	1.375–1.376	NA	NA	—	0.005–0.020
	3	6-181 (3.0)	NA	NA	NA	NA	NA	①	①	—	NA
	R	6-201 (3.3)	1.997–1.999	1.980–1.982	1.965–1.967	1.949–1.952	—	0.400	0.400	0.001–0.005	0.005–0.012
	L	6-231 (3.8)	1.997–1.999	1.980–1.982	1.965–1.967	1.949–1.952	—	0.400	0.400	0.001–0.004	0.005–0.012

NA—Not Available
① Height of camshaft/lobe: 1.604–1.624 in.

CRANKSHAFT AND CONNECTING ROD SPECIFICATIONS

All measurements are given in inches.

Year	VIN	No. Cylinder Displacement cu. in. (liter)	Crankshaft Main Brg. Journal Dia.	Main Brg. Oil Clearance	Shaft End-play	Thrust on No.	Connecting Rod Journal Diameter	Oil Clearance	Side Clearance
1988	K	4-153 (2.5)	2.362–2.363	0.0002–0.0014	0.002–0.014	3	1.968–1.969	0.0008–0.0040	0.005–0.013
	3	6-181 (3.0)	2.361–2.362	0.0006–0.0020	0.002–0.010	3	1.968–1.969	0.0008–0.0028	0.004–0.010
1989	K	4-153 (2.5)	2.362–2.363	0.0002–0.0014	0.002–0.014	3	1.968–1.969	0.0008–0.0040	0.005–0.013
	3	6-181 (3.0)	2.361–2.362	0.0006–0.0020	0.002–0.010	3	1.968–1.969	0.0008–0.0028	0.004–0.010
1990	K	4-153 (2.5)	2.362–2.363	0.0002–0.0014	0.002–0.014	3	1.968–1.969	0.0008–0.0040	0.005–0.013
	3	6-181 (3.0)	2.361–2.362	0.0006–0.0020	0.002–0.010	3	1.968–1.969	0.0008–0.0028	0.004–0.010
	R	6-201 (3.3)	2.519	0.0007–0.0022	0.001–0.007	2	2.283	0.0008–0.0030	0.005–0.015
1991–92	K	4-153 (2.5)	2.362–2.363	0.0002–0.0014	0.002–0.014	3	1.968–1.969	0.0008–0.0040	0.005–0.013
	3	6-181 (3.0)	2.361–2.362	0.0006–0.0020	0.002–0.010	3	1.968–1.969	0.0008–0.0028	0.004–0.010
	R	6-201 (3.3)	2.519	0.0007–0.0022	0.003–0.009	2	2.283	0.0008–0.0030	0.005–0.015
	L	6-231 (3.8)	2.519	0.0007–0.0022	0.003–0.009	2	2.283	0.0008–0.0030	0.005–0.015

PISTON AND RING SPECIFICATIONS

All measurements are given in inches.

Year	VIN	No. Cylinder Displacement cu. in. (liter)	Piston Clearance	Ring Gap			Ring Side Clearance		
				Top Compression	Bottom Compression	Oil Control	Top Compression	Bottom Compression	Oil Control
1988	K	4-153 (2.5)	0.0006–0.0018	0.0010–0.0039	0.0011–0.0039	0.0015–0.0074	0.0015–0.0040	0.0015–0.0040	0.0002–0.0080
	3	6-181 (3.0)	0.0008–0.0015	0.0012–0.0018	0.0010–0.0016	0.0012–0.0035	0.0020–0.0035	0.0008–0.0020	NA
1989	K	4-153 (2.5)	0.0006–0.0018	0.0010–0.0039	0.0011–0.0039	0.0015–0.0074	0.0015–0.0040	0.0015–0.0040	0.0002–0.0080
	3	6-181 (3.0)	0.0008–0.0015	0.0012–0.0018	0.0010–0.0016	0.0012–0.0035	0.0020–0.0035	0.0008–0.0020	NA
1990	K	4-153 (2.5)	0.0006–0.0018	0.0010–0.0039	0.0011–0.0039	0.0015–0.0074	0.0015–0.0040	0.0015–0.0040	0.0002–0.0080
	3	6-181 (3.0)	0.0008–0.0015	0.0012–0.0018	0.0010–0.0016	0.0012–0.0035	0.0020–0.0035	0.0008–0.0020	NA
	R	6-201 (3.3)	0.0009–0.0022	0.0012–0.0022	0.0012–0.0022	0.0010–0.0040	0.0012–0.0037	0.0012–0.0037	0.0005–0.0089
1991–92	K	4-153 (2.5)	0.0006–0.0018	0.0010–0.0039	0.0011–0.0039	0.0015–0.0074	0.0015–0.0040	0.0015–0.0040	0.0002–0.0080
	3	6-181 (3.0)	0.0008–0.0015	0.0012–0.0018	0.0010–0.0016	0.0012–0.0035	0.0020–0.0035	0.0008–0.0020	NA
	R	6-201 (3.3)	0.0009–0.0022	0.0012–0.0022	0.0012–0.0022	0.0010–0.0040	0.0012–0.0037	0.0012–0.0037	0.0005–0.0089
	L	6-231 (3.8)	0.0009–0.0022	0.0012–0.0022	0.0012–0.0022	0.0010–0.0040	0.0012–0.0037	0.0012–0.0037	0.0005–0.0089

TORQUE SPECIFICATIONS

All readings in ft. lbs.

Year	VIN	No. Cylinder Displacement cu. in. (liter)	Cylinder Head Bolts	Main Bearing Bolts	Rod Bearing Bolts	Crankshaft Pulley Bolts	Flywheel Bolts	Manifold		Spark Plugs
								Intake	Exhaust	
1988	K	4-153 (2.5)	①	30③	40③	85	70	17	17	26
	3	6-181 (3.0)	80	60	38	112	70	17	17	20
1989	K	4-153 (2.5)	①	30③	40③	85	70	17	17	26
	3	6-181 (3.0)	80	60	38	112	70	17	17	20
1990	K	4-153 (2.5)	①	30③	40③	85	70	17	17	26
	3	6-181 (3.0)	80	60	38	112	70	17	17	20
	R	6-201 (3.3)	②	30③	40③	110	70	17	17	30
1991–92	K	4-153 (2.5)	①	30③	40③	85	70	17	17	26
	3	6-181 (3.0)	80	60	38	112	70	17	17	20
	R	6-201 (3.3)	②	30③	40③	110	70	17	17	30
	L	6-231 (3.8)	②	30③	40③	110	70	17	17	30

① Sequence: 45, 65, 65, plus ¼ turn
② Sequence: 45, 65, 65, plus ¼ turn
 Torque the small bolt in the rear of
 the head to 25 ft. lbs. last
③ Plus ¼ turn

BRAKE SPECIFICATIONS

All measurements in inches unless noted

Year	Model	Lug Nut Torque (ft. lbs.)	Master Cylinder Bore	Brake Disc Minimum Thickness	Brake Disc Maximum Runout	Standard Brake Drum Diameter	Minimum Lining Thickness Front	Minimum Lining Thickness Rear
1988	Dynasty	95	0.827	②	0.005	8.66	0.06	0.06
	New Yorker Landau	95	0.827	②	0.005	8.66	0.06	0.06
1989	Dynasty	95	0.827 ①	②	0.005	8.66	0.06	0.06
	New Yorker Landau	95	0.827 ①	②	0.005	8.66	0.06	0.06
1990	Dynasty	95	0.827 ①	②	0.005	8.66	0.06	0.06
	New Yorker Landau	95	0.827 ①	②	0.005	8.66	0.06	0.06
	New Yorker Salon	95	0.827 ①	②	0.005	8.66	0.06	0.06
	New Yorker 5th Ave.	95	0.827 ①	②	0.005	—	0.06	0.06
	Imperial	95	—	②	0.005	—	0.06	0.06
1991–92	Dynasty	95	0.827 ①	②	0.005	8.66	0.06	0.06
	New Yorker Landau	95	0.827 ①	②	0.005	8.66	0.06	0.06
	New Yorker Salon	95	0.827 ①	②	0.005	8.66	0.06	0.06
	New Yorker 5th Ave.	95	0.827 ①	②	0.005	—	0.06	0.06
	Imperial	95	—	②	0.005	—	0.06	0.06

① Except ABS
② Front: 0.882 in.
   Rear: 0.339 in.

WHEEL ALIGNMENT

Year	Model		Caster Range (deg.)	Caster Preferred Setting (deg.)	Camber Range (deg.)	Camber Preferred Setting (deg.)	Toe-in (in.)	Steering Axis Inclination (deg.)
1988	Dynasty	front	①	1³/₁₆	¼N–¾P	⁵/₁₆P	¹/₁₆	13⁵/₁₆
		rear	—	—	1¼N–¼N	½N	0	—
	New Yorker Landau	front	①	1³/₁₆	¼N–¾P	⁵/₁₆P	¹/₁₆	13⁵/₁₆
		rear	—	—	1¼N–¼N	½N	0	—
1989	Dynasty	front	①	1³/₁₆	¼N–¾P	⁵/₁₆P	¹/₁₆	13⁵/₁₆
		rear	—	—	1¼N–¼N	½N	0	—
	New Yorker Landau	front	①	1³/₁₆	¼N–¾P	⁵/₁₆P	¹/₁₆	13⁵/₁₆
		rear	—	—	1¼N–¼N	½N	0	—
1990	Dynasty	front	①	1³/₁₆	¼N–¾P	⁵/₁₆P	¹/₁₆	13⁵/₁₆
		rear	—	—	1¼N–¼N	½N	0	—
	New Yorker Landau	front	①	1³/₁₆	¼N–¾P	⁵/₁₆P	¹/₁₆	13⁵/₁₆
		rear	—	—	1¼N–¼N	½N	0	—
	New Yorker Salon	front	①	1³/₁₆	¼N–¾P	⁵/₁₆P	¹/₁₆	13⁵/₁₆
		rear	—	—	1¼N–¼N	½N	0	—
	New Yorker 5th Ave.	front	①	1³/₁₆	¼N–¾P	⁵/₁₆P	¹/₁₆	13⁵/₁₆
		rear	—	—	1¼N–¼N	½N	0	—
	Imperial	front	①	1³/₁₆	¼N–¾P	⁵/₁₆P	¹/₁₆	13⁵/₁₆
		rear	—	—	1¼N–¼N	½N	0	—

WHEEL ALIGNMENT

Year	Model		Caster Range (deg.)	Caster Preferred Setting (deg.)	Camber Range (deg.)	Camber Preferred Setting (deg.)	Toe-in (in.)	Steering Axis Inclination (deg.)
1991–92	Dynasty	front	①	2¾	¼N–¾P	⁵⁄₁₆P	¹⁄₁₆	12½
		rear	—	—	1¼N–¼N	½N	0	—
	New Yorker Salon	front	①	2¾	¼N–¾P	⁵⁄₁₆P	¹⁄₁₆	12½
		rear	—	—	1¼N–¼N	½N	0	—
	New Yorker 5th Ave.	front	①	2¾	¼N–¾P	⁵⁄₁₆P	¹⁄₁₆	12½
		rear	—	—	1¼N–¼N	½N	0	—
	Imperial	front	①	3	¼N–¾P	⁵⁄₁₆P ②	¹⁄₁₆	12½
		rear	—	—	1¼N–¼N	½N	0	—

① Not adjustable—variation between sides should not exceed 1½°
② With air suspension—¹⁄₁₀ in.

ENGINE MECHANICAL

NOTE: Disconnecting the negative battery cable on some vehicles may interfere with the functions of the on board computer systems and may require the computer to undergo a relearning process, once the negative battery cable is reconnected.

Engine Assembly

REMOVAL & INSTALLATION

2.5L Engine

1. Disconnect the negative battery cable and all engine ground straps. Relieve the fuel pressure.
2. Mark the hood hinge outline on the hood and remove the hood.
3. Drain the cooling system. Remove the radiator hoses, fan assembly and radiator.
4. Remove the air cleaner, duct hoses and oil filter.
5. Unbolt the air conditioning compressor from its mount, and position it aside.
6. Remove the power steering pump mounting bolts and position the pump aside, without disconnecting any fluid lines.
7. Label and disconnect all electrical connectors from the engine, alternator and fuel injection system.
8. Disconnect and plug the fuel lines and heater hoses.
9. Disconnect the throttle linkage.
10. Remove the alternator.
11. Raise the vehicle and support safely.

12. Disconnect the exhaust pipe from the manifold.
13. Remove the right inner fender shield. Remove the lower cover from the transaxle case.
14. Remove the starter and set it aside. Matchmark the flexplate to the torque converter for installation purposes. Remove the torque converter bolts. Separate the converter from the flexplate. Remove the lower bellhousing bolts.
0089Lower the vehicle and support the transaxle with a floor jack or equivalent. Attach an engine lifting device to the engine.
16.
Remove the remaining bellhousing bolts.

NOTE: If removing the insulator-to-rail screws, first mark the position of the insulator on the side rail to ensure proper alignment during reinstallation.

17. Remove the front engine mount nut/bolt and the left insulator through bolt or the insulator bracket to transaxle bolts.
18. Lift the engine from the vehicle and remove.
To install:
19. Lower the engine into the engine compartment. Make sure the lifting device is supporting the full weight of the engine and loosely install all of the mounting bolts until all are threaded. Then tighten all bolts.
20. Remove the lifting device.
21. Raise the vehicle and support safely.
22. If equipped with an automatic transaxle, install the torque converter bolts and torque to 55 ft. lbs. (75 Nm).
23. Install the torque converter inspection plate and starter.
24. Connect the exhaust pipe. Lower the vehicle.

25. Install the alternator, power steering pump and air conditioning compressor.
26. Connect the fuel lines and heater hoses.
27. Connect the throttle linkage.
28. Connect all remaining electrical connectors.
29. Install the air cleaner assembly and oil filter.
30. Install the radiator, fan assembly and hoses.
31. Fill the engine with the proper amount of engine oil. Connect the negative battery cable.
32. Refill the cooling system. Start the engine, allow it to reach normal operating temperature. Check for leaks.
33. Check the ignition timing and adjust if necessary.
34. Install the hood.

3.0L, 3.3L and 3.8L Engines

1. Disconnect the negative battery cable. Relieve the fuel pressure.
2. Matchmark the hinge-to-hood position and remove the hood.
3. Drain the cooling system. Disconnect and label all engine electrical connections.
4. Remove the coolant hoses from the radiator and engine. Remove the radiator and cooling fan assembly.
5. Remove the air cleaner assembly. Disconnect the fuel lines from the engine. Disconnect the accelerator cable from the engine.
6. Raise the vehicle and support safely. Drain the engine oil.
7. Remove the air conditioning compressor mounting bolts, the drive belts and position the compressor to the side. Disconnect the exhaust pipe from the exhaust manifold.
8. Remove the transaxle inspection cover, matchmark the converter to the

4-9

flexplate and remove the torque converter bolts.

9. Remove the power steering pump mounting bolts and set the pump aside, upright, with the fluid lines attached.

10. Remove the lower bellhousing bolts. Disconnect and label the starter motor wiring and remove the starter motor from the engine.

11. Lower the vehicle. Disconnect and label the vacuum hoses and engine ground straps.

12. Support the transaxle with a floor jack or equivalent. Attach an engine lifting device to the engine.

13. Remove the upper transaxle-to-engine bolts.

14. To separate the engine mounts from the insulators, mark the right insulator-to-right frame support and remove the mounting bolts. Remove the front engine mount through bolt. Remove the left insulator through bolt, from inside the wheel housing. Remove the insulator bracket-to-transaxle bolts.

15. Lift and remove the engine from the vehicle.

To install:

16. Lower the engine into the engine compartment. Align the engine mounts and install the bolts; do not tighten the bolts until all bolts have been installed. Torque the through bolts to 75 ft. lbs.

17. Install the upper transaxle-to-engine mounting bolts and torque to 75 ft. lbs. Remove the engine lifting fixture from the engine.

18. Raise the vehicle and support safely.

19. Align the converter marks, install the torque converter bolts. Install the transaxle inspection cover.

20. Connect the exhaust pipe to the exhaust manifold. Install the starter motor and connect the wiring.

21. Install the power steering pump and air conditioning compressor. Adjust the drive belt tension, if necessary.

22. Lower the vehicle. Reconnect all vacuum hoses and electrical connections to the engine.

23. Connect the fuel lines and accelerator cable.

24. Install the radiator and fan assembly. Connect the fan motor wiring. Connect the radiator hoses and refill the cooling system.

25. Refill the engine with the proper oil to the correct level.

26. Connect the engine ground straps. Install the hood and align the matchmarks. Connect the battery.

27. Start and run the engine until it reaches normal operating temperatures and check for leaks. Adjust the transaxle linkage, if necessary.

Engine Mounts

REMOVAL & INSTALLATION

1. Disconnect the negative battery cable.

2. Matchmark the engine mount to its frame mounting location.

3. Raise the vehicle and support safely, if necessary. Using the proper equipment, support the weight of the engine.

4. Remove all bolts and nuts that attach the mount to the engine strut, transaxle or body and remove the mount assembly from the vehicle.

5. Remove the through bolt and separate the insulator from the yoke bracket as required.

6. The installation is the reverse of the removal procedure. Make sure the matchmarks are aligned before tightening bolts.

7. Tighten the lower yoke nut first, then the through bolt nut, then the body mounting bolts.

Cylinder Head

REMOVAL & INSTALLATION

2.5L Engine

1. Relieve the fuel pressure. Disconnect the negative battery cable and unbolt it from the head. Drain the cooling system. Remove the dipstick bracket nut from the thermostat housing. Remove the ignition coil from the thermostat housing if it is mounted there.

2. Remove the air cleaner assembly. Remove the upper radiator hose and disconnect the heater hoses.

3. Disconnect and label the vacuum lines, hoses and wiring connectors from the manifolds, throttle body and from the cylinder head.

4. Disconnect the all linkages and the fuel line from the throttle body. Unbolt the cable bracket. Remove the ground strap attaching screw from the firewall.

5. Remove the upper air conditioning compressor mounting bolts. The cylinder head can be remove with the compressor and bracket still mounted. Remove the upper timing belt cover.

6. Raise the vehicle and support safely. Disconnect the exhaust pipe from the exhaust manifold.

7. Rotate the engine by hand, until the timing marks align. The No. 1 piston should be at TDC of its compression stroke. Lower the vehicle.

8. With the timing marks aligned, remove the camshaft sprocket. The camshaft sprocket can be suspended to keep the timing intact. Remove the spark plug wires from the spark plugs.

9. Remove the valve cover and cur-

tain. Remove the cylinder head bolts and washers, starting from the outside and working inward.

10. Remove the cylinder head from the engine.

11. Clean the cylinder head gasket mating surfaces.

To install:

12. Using new gaskets and seals, install the cylinder head to the engine block. Using new head bolts assembled with the old washers, torque the cylinder head bolts in sequence, to 45 ft. lbs. (61 Nm). Repeating the sequence, torque the bolts to 65 ft. lbs. (88 Nm). With the bolts at 65 ft. lbs., turn each bolt an additional ¼ turn.

13. Install the timing belt.

14. Install or connect all items that were removed or disconnected during the removal procedure.

15. Refill the cooling system. Connect the negative battery cable. Start the engine and check for leaks using the DRB I or II to activate the fuel pump.

16. Adjust the timing as required.

3.0L Engine

1. Relieve the fuel pressure. Disconnect the negative battery cable. Drain the cooling system.

2. Remove the compressor drive belt and the air conditioning compressor from its mount and support it aside. Using a ½ in. drive breaker bar, insert it into the square hole of the serpentine drive belt tensioner, rotate it counterclockwise to reduce the belt tension and remove the belt. Remove the alternator and power steering pump from the brackets and move them aside.

3. Raise the vehicle and support safely. Remove the right front wheel and the inner splash shield.

4. Remove the crankshaft pulleys and the torsional damper.

5. Lower the vehicle. Using a floor jack and a block of wood positioned under the oil pan, raise the engine slightly. Remove the engine mount bracket from the timing cover end of the engine and the timing belt covers.

6. To remove the timing belt, perform the following procedures:

   a. Rotate the crankshaft to position the No. 1 cylinder on the TDC of its compression stroke; the crankshaft sprocket timing mark should align with the oil pan timing indicator and the camshaft sprockets timing marks (triangles) should align with the timing marks on the rear timing belt covers.

   b. Mark the timing belt in the direction of rotation for reinstallation purposes.

   c. Loosen the timing belt tensioner and remove the timing belt.

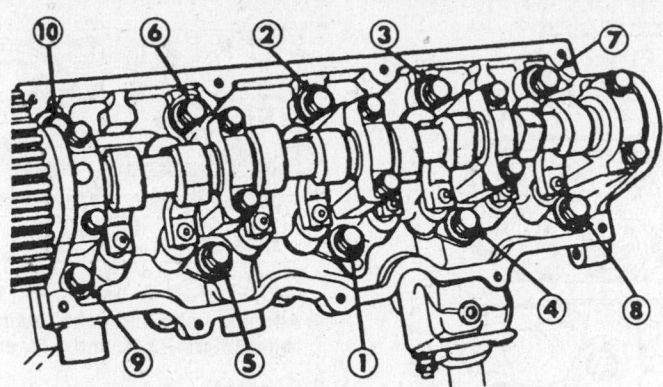

Cylinder head bolt torque sequence—2.5L engine

NOTE: When removing the timing belt from the camshaft sprocket, make sure the belt does not slip off of the other camshaft sprocket. Support the belt so it cannot slip off of the crankshaft sprocket and opposite side camshaft sprocket.

7. Remove the air cleaner assembly. Label and disconnect the spark plug wires and the vacuum hoses.

8. Remove the valve cover.

9. Install auto lash adjuster retainer tools MD998443 or equivalent, on the rocker arms.

10. If removing the front cylinder head, matchmark the distributor rotor to the distributor housing and the housing to distributor extension locations. Remove the distributor and the distributor extension.

11. Remove the camshaft bearing assembly to cylinder head bolts but do not remove the bolts from the assembly. Remove the rocker arms, rocker shafts and bearing caps as an assembly, as required. Remove the camshafts from the cylinder head and inspect them for damage.

12. Remove the intake manifold assembly.

13. Remove the exhaust manifold.

14. Remove the cylinder head bolts, starting from the outside and working inward. Remove the cylinder head from the engine.

15. Clean the gasket mounting surfaces and check the heads for warpage; maximum warpage is 0.008 in. (0.20mm).

To install:

16. Install the new cylinder head gasket(s) over the dowels on the engine block.

17. Install the cylinder head(s) on the engine and torque the cylinder head bolts in sequence using 3 even steps, to 70 ft. lbs. (95 Nm).

18. Install or connect all items that were removed or disconnected during the removal procedure.

19. When installing the timing belt over the camshaft sprocket, use care not to allow the belt to slip off the opposite camshaft sprocket.

20. Make sure the timing belt is installed on the camshaft sprocket in the same position as when removed.

21. Refill the cooling system. Connect the negative battery cable. Start the engine and check for leaks using the DRB II to activate the fuel pump.

22. Adjust the timing as required.

3.3L and 3.8L Engines

1. Relieve the fuel pressure. Disconnect the negative battery cable. Drain the cooling system.

2. Remove the intake manifold with the throttle body.

3. Disconnect the coil wires, coolant temperature sending unit wire, heater hoses and bypass hose.

4. Remove the closed ventilation system hoses, evaporation control system hoses and valve cover.

5. Remove the exhaust manifold.

6. Remove the rocker arm and shaft assemblies. Remove the pushrods and identify them in ensure installation in their original positions.

7. Remove the head bolts and remove the cylinder head from the block.

To install:

8. Clean the gasket mounting surfaces and install a new head gasket to the block.

9. Install the head to the block. Before installing the head bolts, inspect them for stretching. Hold a straightedge up to the threads. If the threads are not all on line, the bolt is stretched and should be replaced.

10. Torque the bolts in sequence to 45 ft. lbs. (61 Nm). Repeat the sequence and torque the bolts to 65 ft. lbs. (88 Nm). With the bolts at 65 ft. lbs., turn each bolt an additional ¼ turn.

11. Torque the lone head bolt in the rear of the head to 25 ft. lbs. (33 Nm) after the other 8 bolts have been properly torqued.

12. Install the pushrods, rocker arms and shafts and torque the bolts to 21 ft. lbs. (12 Nm).

13. Place a drop of silicone sealer onto each of the 4 manifold to cylinder head gasket corners.

— CAUTION —

The intake manifold gasket is composed of very thin and sharp metal. Handle this gasket with care or damage to the gasket or personal injury could result.

14. Install the intake manifold gasket and torque the end retainers to 105 inch lbs. (12 Nm).

15. Install the intake manifold and torque the bolts in sequence to 10 inch

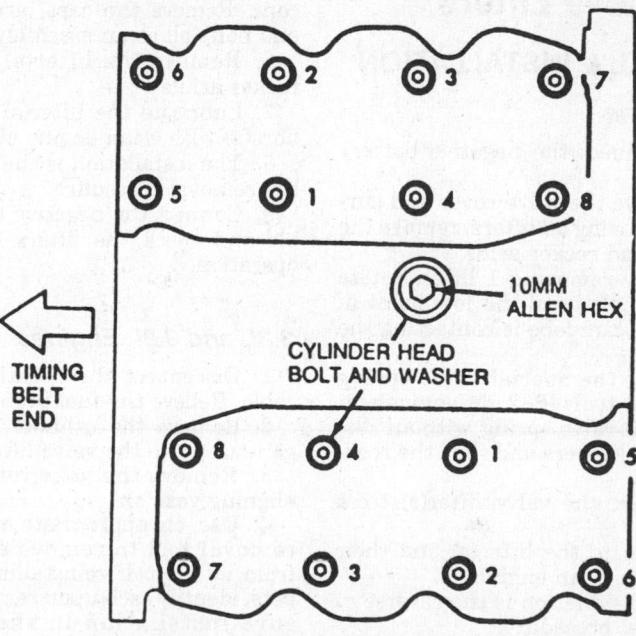

Cylinder head bolt torque sequence—3.0L engine

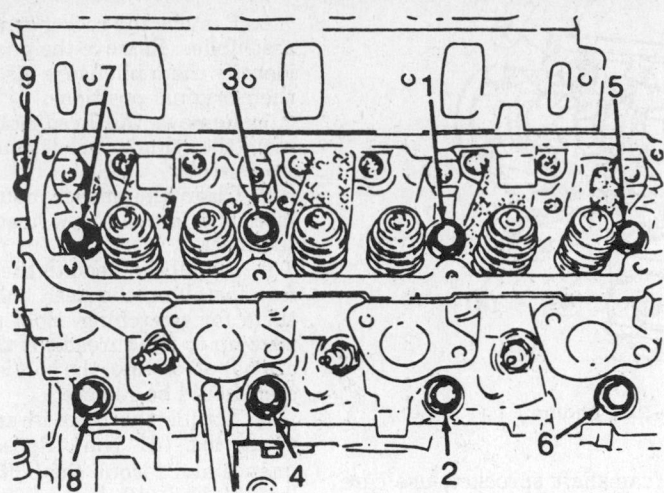

Cylinder head bolt torque sequence—3.3L and 3.8L engines

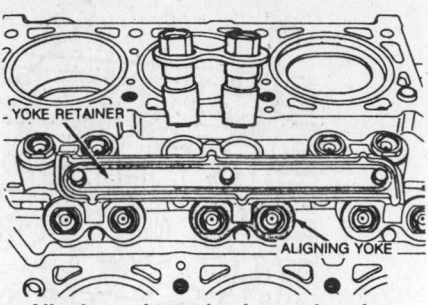

Aligning yoke and yoke retainer for roller lifters—3.3L and 3.8L engines

lbs. Repeat the sequence increasing the torque to 17 ft. lbs. (23 Nm) and recheck each bolt for 17 ft. lbs. After the bolts are torqued, inspect the seals to ensure that they have not become dislodged.

16. Lubricate the injector O-rings with clean oil and position the fuel rail in place. Install the rail mounting bolts.

17. Install the valve cover with a new gasket. Install the exhaust manifold.

18. Install or connect all remaining items that were removed or disconnected during the removal procedure.

19. Refill the cooling system. Connect the negative battery cable. Start the engine and check for leaks using the DRB I or II to activate the fuel pump.

Valve Lifters

REMOVAL & INSTALLATION

2.5L Engine

1. Disconnect the negative battery cable.

2. Remove the valve cover and curtain. If removing all lifters, remove the camshaft and rocker arms.

3. If only removing 1 lifter, rotate the crankshaft until the low point of the desired cam lobe is contacting the rocker arm.

4. Using the special valve spring compressor tool 4682 or equivalent, depress the valve spring without dislodging the keepers and slide the rocker arm out.

5. Remove the valve lifter(s) from the bore(s).

6. Lubricate the lifter(s) and their bore(s) with clean engine oil.

7. The installation is the reverse of the removal procedure.

8. Connect the negative battery ca-

ble and check the lifters for proper operation.

3.0L Engine

1. Disconnect the negative battery cable. Remove the air cleaner assembly.

2. Remove the valve cover.

3. Using the valve lifter retainer tools MD998443 or equivalent, install them on the rocker arms to keep the lifters from falling out.

4. On the right side cylinder head, remove the distributor extension.

5. Have a helper hold the rear end of the camshaft down. If the rear of the camshaft cannot be held down, the belt will dislodge and the valve timing will be lost. Loosen the camshaft cap bolts but do not remove them from the caps. Remove the caps, arms, shafts and bolts all as an assembly.

6. Remove the lifter(s) from the rocker arm(s).

7. Lubricate the lifter(s) and their bore(s) with clean engine oil.

8. The installation is the reverse of the removal procedure.

9. Connect the negative battery cable and check the lifters for proper operation.

3.3L and 3.8L Engines

1. Disconnect the negative battery cable. Relieve the fuel pressure.

2. Remove the cylinder head(s) to gain access to the valve lifter(s).

3. Remove the yoke retainer and aligning yoke(s).

4. Use an appropriate valve lifter removal tool to remove each lifter from its bore. If reinstalling the tappets, identify each upon removal to ensure installation in the original position.

To install:

5. Lubricate the lifter(s) and bore(s) and install.

6. Install aligning yoke(s).

7. Install the yoke retainer and torque the bolts to 105 inch lbs. (12 Nm).

8. Install the cylinder head(s) and all related components.

9. Connect the negative battery cable and check the lifters for proper operation.

Rocker Arms/Shafts

REMOVAL & INSTALLATION

2.5L Engine

1. Disconnect the negative battery cable.

2. Remove the valve cover.

3. Rotate the crankshaft until the low point of the desired cam lobe is contacting the rocker arm.

4. Use the special valve spring compressor tool or equivalent, depress the valve spring without dislodging the keepers, then slide the rocker arm out.

5. The installation is the reverse of the removal procedure.

3.0L Engine

1. Disconnect the negative battery cable. Remove the air cleaner assembly.

2. Remove the valve cover.

3. Using the auto lash adjuster retainer tools MD998443 or equivalent, install them on the rocker arms to keep the lash adjusters from falling out.

4. Remove the distributor extension, if necessary.

5. Have a helper hold the rear end of the camshaft down. If the rear of the camshaft cannot be held down, the belt will dislodge and the valve timing will be lost. Loosen the camshaft cap bolts but do not remove them from the caps. Remove the caps, arms, shafts and bolts all as an assembly.

6. Disassemble the unit, keeping all parts in order and repair as required.

7. When assembling, apply a drop of sealant to the rear edge of the rear cap.

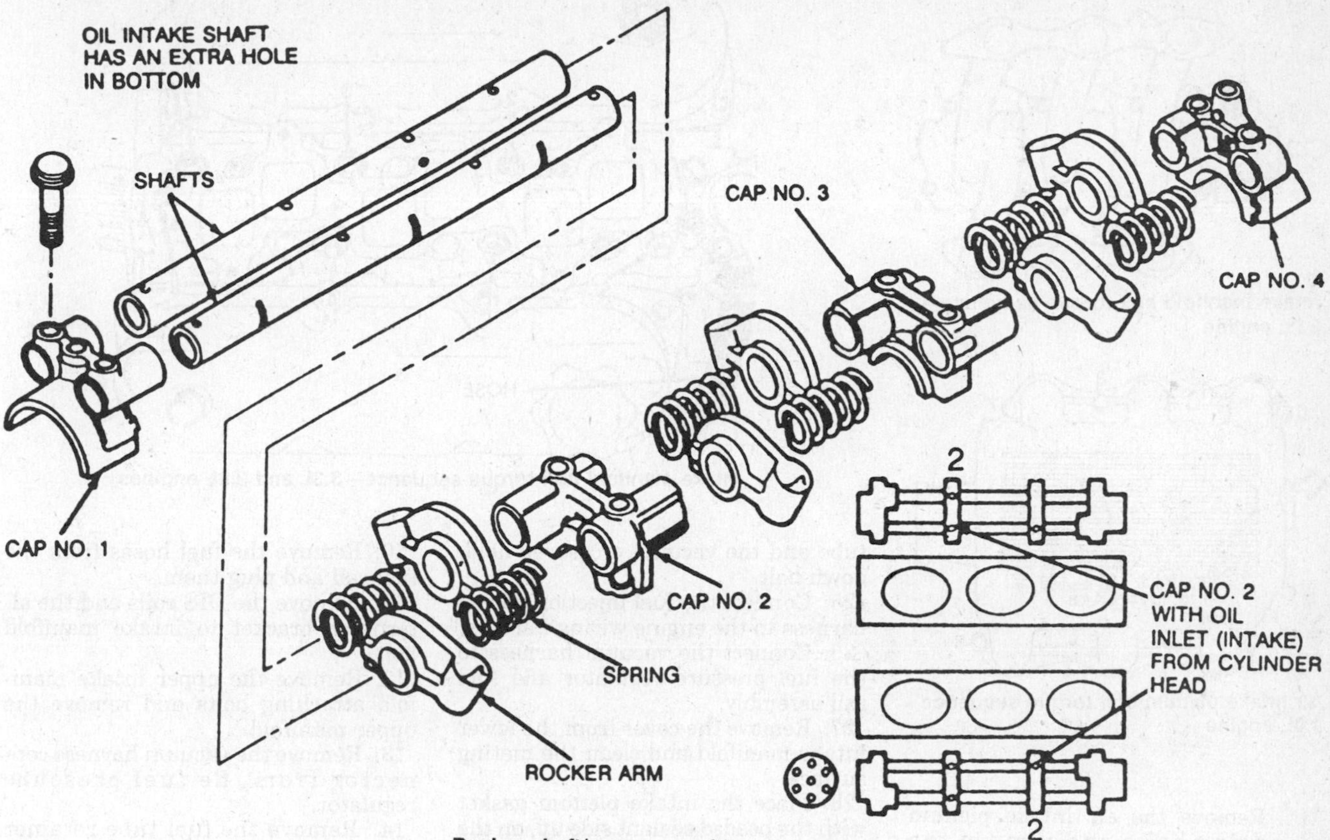

OIL INTAKE SHAFT
HAS AN EXTRA HOLE
IN BOTTOM

SHAFTS

CAP NO. 3

CAP NO. 4

CAP NO. 1

CAP NO. 2

SPRING

ROCKER ARM

2

2

CAP NO. 2
WITH OIL
INLET (INTAKE)
FROM CYLINDER
HEAD

Rocker shaft/arms assembly—3.0L engine

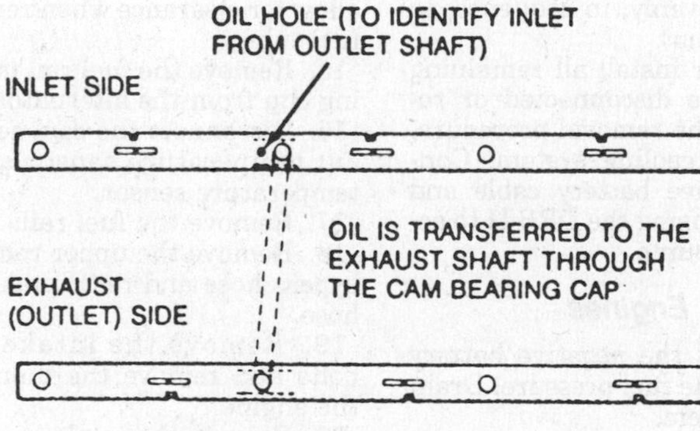

OIL HOLE (TO IDENTIFY INLET
FROM OUTLET SHAFT)

INLET SIDE

OIL IS TRANSFERRED TO THE
EXHAUST SHAFT THROUGH
THE CAM BEARING CAP

EXHAUST
(OUTLET) SIDE

Identifying the rocker shafts—3.0L engine

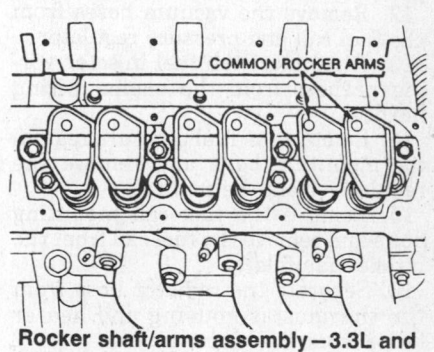

COMMON ROCKER ARMS

**Rocker shaft/arms assembly—3.3L and
3.8L engines**

8. The installation is the reverse of the removal procedure. Torque the cap bolts first to 85 inch lbs. (19 Nm), then to 180 inch lbs. (19 Nm) in the following order: No. 3 cap, No. 2 cap, No. 1 cap and No. 4 cap.

3.3L and 3.8L Engines

1. Disconnect the negative battery cable.
2. Remove the upper intake manifold assembly.
3. Remove the valve cover.
4. Remove the rocker shaft retaining bolts and retainers.
5. Remove the rocker shaft and arm assembly. Disassemble and repair as required.

6. The installation is the reverse of the removal procedure. Torque the retaining bolts gradually and evenly to 21 ft. lbs. (28 Nm).
7. Allow 20 minutes tappet-bleeddown time after rocker shaft installation before starting the engine.

Intake Manifold

REMOVAL & INSTALLATION

3.0L Engine

1. Disconnect the negative battery cable. Relieve the fuel system pressure.
2. Drain the cooling system.

3. Remove the throttle body to air cleaner hose.
4. Remove the throttle body and transaxle kickdown linkage.
5. Remove the AIS motor and TPS wiring connectors from the throttle body.
6. Remove and label the vacuum hose harness from the throttle body.
7. From the air intake plenum, remove the PCV and brake booster hoses and the EGR tube flange.
8. Disconnect and label the charge and temperature sensor wiring at the intake manifold.
9. Remove the vacuum connections from the air intake plenum vacuum connector.
10. Remove the fuel hoses from the fuel rail.

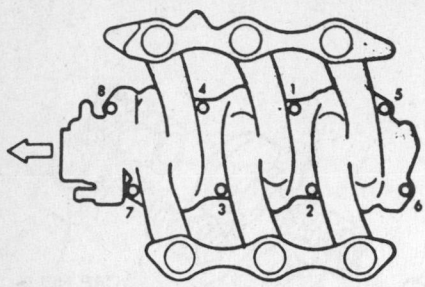

Intake manifold bolt torque sequence—
3.0L engine

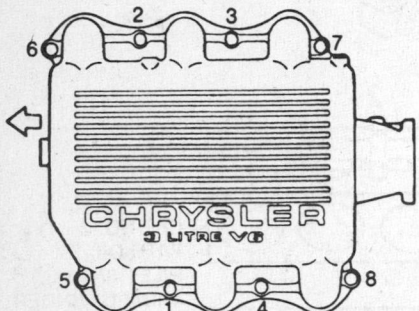

Air intake plenum bolt torque sequence—
3.0L engine

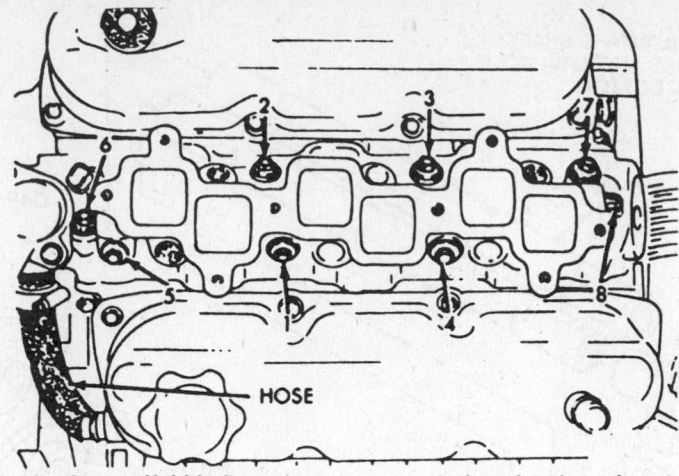

Intake manifold bolt torque sequence—3.3L and 3.8L engines

11. Remove the air intake plenum mounting bolts and remove the plenum.

12. Remove the vacuum hoses from the fuel rail and pressure regulator.

13. Disconnect the fuel injector wiring harness from the engine wiring harness.

14. Remove the fuel pressure regulator mounting bolts and remove the regulator from the fuel rail.

15. Remove the fuel rail mounting bolts and remove the fuel rail from the intake manifold.

16. Separate the radiator hose from the thermostat housing and heater hoses from the heater pipe.

17. Remove the intake manifold mounting bolts and remove the manifold from the engine.

18. Clean the gasket mounting surfaces on the engine and intake manifold.

To install:

19. Using new gaskets, position the intake manifold on the engine and install the mounting nuts and washers.

20. Torque the mounting nuts gradually and evenly, in sequence, to 15 ft. lbs. (20 Nm).

21. Make sure the injector holes are clean. Lubricate the injector O-rings with clean engine oil and install the injector assembly onto the engine.

22. Install and torque the fuel rail mounting bolts to 10 ft. lbs. (14 Nm).

23. Install the fuel pressure regulator onto the fuel rail.

24. Install the fuel supply and return

tube and the vacuum crossover hold-down bolt.

25. Connect the fuel injection wiring harness to the engine wiring harness.

26. Connect the vacuum harness to the fuel pressure regulator and fuel rail assembly.

27. Remove the cover from the lower intake manifold and clean the mating surface.

28. Place the intake plenum gasket with the beaded sealant side up, on the intake manifold. Install the air intake plenum and torque the mounting bolts gradually and evenly, in sequence, to 10 ft. lbs. (14 Nm).

29. Connect or install all remaining items that were disconnected or removed during the removal procedure.

30. Refill the cooling system. Connect the negative battery cable and check for leaks using the DRB II to activate the fuel pump.

3.3L and 3.8L Engines

1. Disconnect the negative battery cable. Relieve the fuel pressure. Drain the cooling system.

2. Remove the air cleaner to throttle body hose assembly.

3. Disconnect the throttle cable and remove the wiring harness from the bracket.

4. Remove AIS motor and TPS wiring connectors from the throttle body.

5. Remove the vacuum hose harness from the throttle body.

6. Remove the PCV and brake booster hoses from the air intake plenum.

7. Disconnect the charge temperature sensor electrical connector. Remove the vacuum harness connectors from the intake plenum.

8. Remove the cylinder head to the intake plenum strut.

9. Disconnect the MAP sensor and oxygen sensor connectors. Remove the engine mounted ground strap.

10. Remove the fuel hoses from the fuel rail and plug them.

11. Remove the DIS coils and the alternator bracket to intake manifold bolt.

12. Remove the upper intake manifold attaching bolts and remove the upper manifold.

13. Remove the vacuum harness connector from the fuel pressure regulator.

14. Remove the fuel tube retainer bracket screw and fuel rail attaching bolts. Spread the retainer bracket to allow for clearance when removing the fuel tube.

15. Remove the fuel rail injector wiring clip from the alternator bracket.

16. Disconnect the cam sensor, coolant temperature sensor and engine temperature sensor.

17. Remove the fuel rail.

18. Remove the upper radiator hose, bypass hose and rear intake manifold hose.

19. Remove the intake manifold bolts and remove the manifold from the engine.

20. Remove the intake manifold seal retaining screws and remove the manifold gasket.

21. Clean out clogged end water passages and fuel runners.

To install:

22. Clean and dry all gasket mating surfaces.

23. Place a drop of silicone sealer onto each of the 4 manifold-to-cylinder head gasket corners.

CAUTION

The intake manifold gasket is composed of very thin and sharp metal. Handle this gasket with care or damage to the gasket or personal injury could result.

24. Install the intake manifold gasket and torque the end retainers to 10 ft. lbs. (12 Nm).

25. Install the intake manifold and

torque the bolts in sequence to 10 inch lbs. Repeat the sequence increasing the torque to 17 ft. lbs. (23 Nm) and recheck each bolt for 17 ft. lbs. of torque. After the bolts are torqued, inspect the seals to ensure that they have not become dislodged.

26. Lubricate the injector O-rings with clean oil and position the fuel rail in place. Install the rail mounting bolts.

27. Connect the cam sensor, coolant temperature sensor and engine temperature sensor.

28. Install the fuel rail injector wiring clip to the alternator bracket.

29. Install the fuel rail attaching bolts and fuel tube retainer bracket screw.

30. Install the vacuum harness to the pressure regulator.

31. Install the upper intake manifold with a new gasket. Install the bolts only finger-tight. Install the alternator bracket to intake manifold bolt and the cylinder head to intake manifold strut and bolts. Torque the intake manifold mounting bolts to 21 ft. lbs. (28 Nm) starting from the middle and working outward. Torque the bracket and strut bolts to 40 ft. lbs. (54 Nm).

32. Install or connect all items that were removed or disconnected from the intake manifold and throttle body.

33. Connect the fuel hoses to the rail. Push the fittings in until they click in place.

34. Install the air cleaner assembly.

35. Connect the negative battery cable and check for leaks using the DRB II to activate the fuel pump.

Exhaust Manifold

REMOVAL & INSTALLATION

3.0L Engine

1. Disconnect the negative battery cable. Raise the vehicle and support safely.

2. Disconnect the exhaust pipe from the rear exhaust manifold at the articulated joint.

3. Disconnect the EGR tube from the rear manifold and unplug the oxygen sensor wire.

4. Remove the crossover pipe to manifold bolts.

5. Remove the rear manifold to cylinder head nuts and the manifold.

6. Lower the vehicle and remove the heat shield from the manifold.

7. Remove the front manifold to cylinder head nuts and the manifold.

8. Clean the gasket mounting surfaces. Inspect the manifolds for cracks, flatness and/or damage.

To install:

9. When installing, the numbers 1–3–5 on the gaskets are used with the

rear cylinders and 2–4–6 are on the gasket for the front cylinders. Torque the manifold to cylinder head nuts to 14 ft. lbs. (19 Nm).

10. Install the crossover pipe to the manifold.

11. Connect the EGR tube and oxygen sensor wire.

12. Connect the exhaust pipe to the rear exhaust manifold, at the articulated joint.

13. Connect the negative battery cable and check the manifolds for leaks.

3.3L and 3.8L Engines

1. Disconnect the negative battery cable.

2. If removing the rear manifold, raise the vehicle and support safely. Disconnect the exhaust pipe at the articulated joint from the rear exhaust manifold.

3. Separate the EGR tube from the rear manifold and disconnect the oxygen sensor wire.

4. Remove the alternator/power steering support strut.

5. Remove the bolts attaching the crossover pipe to the manifold.

6. Remove the bolts attaching the manifold to the head and remove the manifold.

7. If removing the front manifold, remove the heat shield, bolts attaching the crossover pipe to the manifold and the nuts attaching the manifold to the head.

8. Remove the manifold from the engine.

9. The installation is the reverse of the removal procedure. Torque all exhaust manifold attaching bolts to 17 ft. lbs. (23 Nm).

10. Start the engine and check for exhaust leaks.

Combination Manifold

REMOVAL & INSTALLATION

2.5L Engine

NOTE: On some cases, some of the manifold attaching bolts are not accessible or too heavily sealed from the factory and cannot be removed on the vehicle. Head removal would be necessary in these situations.

1. Disconnect the negative battery cable.

2. Relieve the fuel system pressure.

3. Drain the cooling system.

4. Remove the air cleaner and disconnect all vacuum lines, electrical wiring and fuel lines from the throttle body.

5. Disconnect the throttle linkage.

6. Loosen the power steering pump and remove the drive belt.

7. Remove the power brake vacuum hose from the intake manifold.

8. Remove the water hoses from the water crossover.

9. Raise the vehicle and support safely.

10. Disconnect the exhaust pipe from the exhaust manifold.

11. Remove the power steering pump from its mounting bracket and set it aside.

12. Remove the EGR tube.

13. Remove the intake manifold bolts.

14. Lower the vehicle.

15. Remove the intake manifold.

16. Remove the exhaust manifold nuts.

17. Remove the exhaust manifold.

To install:

18. Install a new combination manifold gasket.

19. Install the manifold assembly. Install the mounting nuts and torque to 13 ft. lbs. (18 Nm.) starting from the middle and working outward. Install the heat cowl to the exhaust manifold.

20. Install the intake manifold. Torque the bolts to 17 ft. lbs. (23 Nm.) starting from the middle and working outward.

21. Install the EGR tube, if removed.

22. Install the intake support bracket, if equipped.

23. Install the power steering pump.

24. Raise the vehicle and support safely. Install the exhaust pipe to the exhaust manifold.

25. Install the water hoses to the water crossover.

26. Install the power brake vacuum hose to the intake manifold.

27. Connect the throttle linkage.

28. Install all vacuum lines, electrical wiring and fuel lines to the throttle body.

29. Install the air cleaner assembly.

30. Refill the cooling system.

31. Connect the negative battery cable and check the manifolds for leaks.

Timing Chain Front Cover

REMOVAL & INSTALLATION

3.3L and 3.8L Engines

1. Disconnect the negative battery cable. Drain the cooling system.

2. Support the engine with a suitable engine support device and remove the right side motor mount.

3. Raise the vehicle and support safely. Drain the engine oil and remove the oil pan.

4. Remove the right wheel and splash shield.

5. Remove the drive belt.

6. Unbolt the air conditioning com-

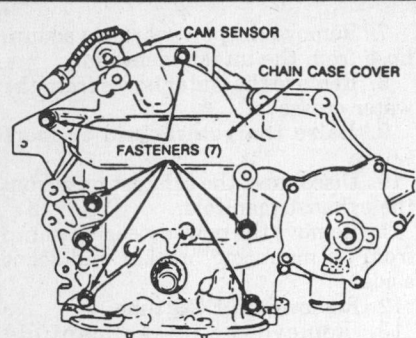

Timing chain cover—3.3L and 3.8L engines

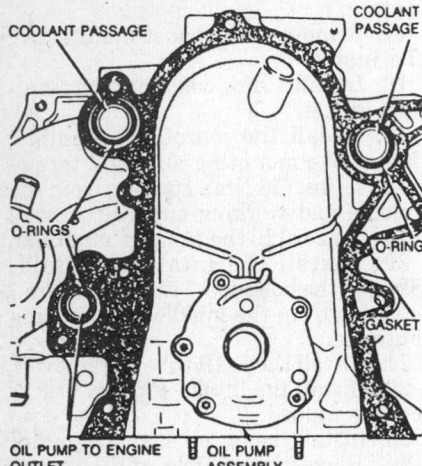

Timing cover removed—3.3L and 3.8L engines

pressor and position it to the side. Remove the compressor mounting bracket.

7. Remove the crankshaft pulley bolt and remove the pulley using a suitable puller.

8. Remove the idler pulley from the engine bracket and remove the bracket.

9. Remove the cam sensor from the timing chain cover.

10. Remove the cover mounting bolts and the cover from the engine. Make sure the oil pump inner rotor does not fall out. Remove the 3 O-rings from the coolant passages and the oil pump outlet.

To install:
11. Thoroughly clean and dry the gasket mating surfaces. Install new O-rings to the block.

12. Remove the crankshaft oil seal from the cover. The seal must be removed from the cover when installing to ensure proper oil pump engagement.

13. Using a new gasket, install the chain case cover to the engine.

14. Make certain that the oil pump is engaged onto the crankshaft before proceeding or there will be no oil pres-

sure. Install the attaching bolts and torque to 20 ft. lbs. (27 Nm).

15. Use tool C–4992 to install the crankshaft oil seal. Install the crankshaft pulley using a 5.9 in. suitable bolt used with thrust bearing and washer plate L–4524. Make sure the pulley bottoms out on the inner diameter of the crankshaft seal. Install the bolt and torque to 40 ft. lbs. (54 Nm).

16. Install the engine bracket and torque the bolts to 40 ft. lbs. (54 Nm). Install the idler pulley to the engine bracket.

17. To install the cam sensor, perform the following:

a. Clean off the old spacer from the sensor face completely. A new spacer must be attached to the cam sensor prior to installation; if a new spacer is not used, engine performance will be adversely affected.

b. Inspect the O-ring for damage and replace, if necessary. Lubricate the O-ring lightly with oil and push the sensor into its bore in the chain case cover until contact is made with the cam timing gear. Hold in this position and tighten the bolt to 9 ft. lbs. (12 Nm).

18. Install the air conditioning compressor and bracket.

19. Install the drive belt.

20. Install the inner splash shield and wheel.

21. Install the oil pan with a new gasket.

22. Install the motor mount.

23. Remove the engine temperature sensor and fill the cooling system until the level reaches the vacant sensor hole. Install the sensor and continue to fill the radiator. Fill the engine with the proper amount of oil.

24. Connect the negative battery cable and check for leaks.

Front Cover Oil Seal

REPLACEMENT

3.3L and 3.8L Engines

1. Disconnect the negative battery cable.

2. Raise the vehicle and support safely. Remove the right front wheel and the inner splash shield.

3. Remove the drive belt.

4. Remove the crankshaft bolt. Using a suitable puller, remove the crankshaft pulley.

5. Use tool C–4991 to remove the seal.

To install:
6. Clean out the bore. Place the seal with the spring toward the engine. Install the new seal using tool C–4992 until it is flush with the cover.

7. Install the crankshaft pulley using a suitable 5.9 in. bolt with thrust

bearing and washer plate L–4524. Make sure the pulley bottoms out on the inner diameter of the crankshaft seal. Install the bolt and torque to 40 ft. lbs. (54 Nm).

8. Install the drive belt.

9. Install the splash shield and wheel.

10. Connect the negative battery cable and check for leaks.

Timing Chain and Gears

REMOVAL & INSTALLATION

3.3L and 3.8L Engines

1. If possible, position the engine so the No. 1 piston is at TDC of its compression stroke. Disconnect the negative battery cable. Drain the coolant.

2. Remove the timing chain case cover.

3. Remove the camshaft gear attaching cup washer and remove the timing chain with both gears attached. Remove the timing chain snubber.

To install:
4. Assemble the timing chain and gears.

5. Turn the crankshaft and camshaft to line up with the keyway locations of the gears.

6. Slide both gears over their respective shafts and use a straight-edge to confirm alignment.

7. Install the cup washer and camshaft bolt. Torque the bolt to 35 ft. lbs. (47 Nm).

8. Check camshaft endplay. The specification with a new plate is 0.002–0.006 in. (0.051–0.052mm) and 0.002–0.010 in. (0.51–0.254mm) with a used plate. Replace the thrust plate if not within specifications.

9. Install the timing chain snubber.

10. Thoroughly clean and dry the gasket mating surfaces.

11. Install new O-rings to the block.

12. Remove the crankshaft oil seal from the cover. The seal must be removed from the cover when installing to ensure proper oil pump engagement.

13. Using a new gasket, install the chain case cover to the engine.

14. Make certain that the oil pump is engaged onto the crankshaft before proceeding or severe engine damage will result. Install the attaching bolts and torque to 20 ft. lbs. (27 Nm).

15. Use tool C–4992 to install the crankshaft oil seal. Install the crankshaft pulley using a 5.9 in. suitable bolt and thrust bearing and washer plate L–4524. Make sure the pulley bottoms out on the crankshaft seal diameter. Install the bolt and torque to 40 ft. lbs. (54 Nm).

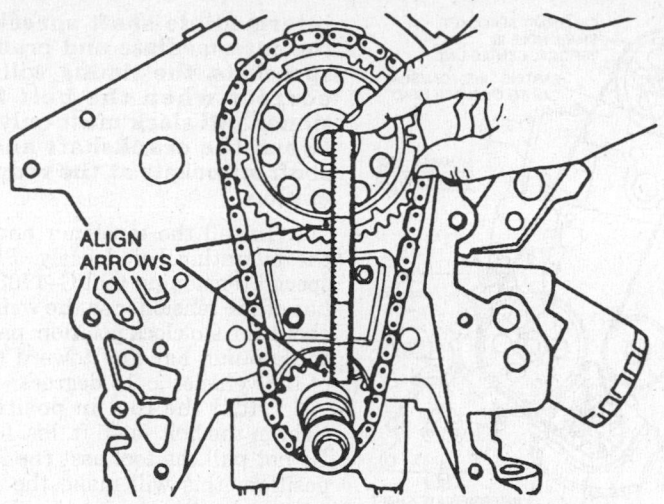

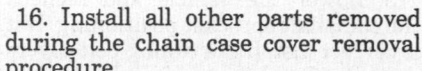

Aligning the timing marks—3.3L and 3.8L engines

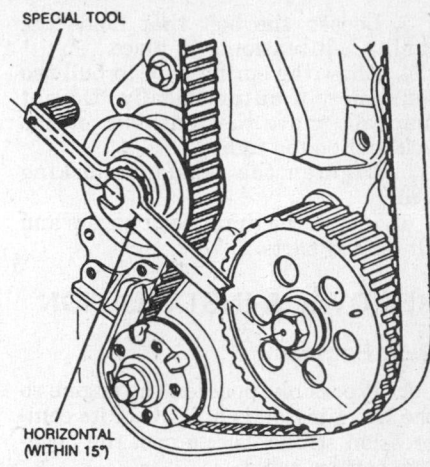

Adjusting the timing belt—2.5L engine

16. Install all other parts removed during the chain case cover removal procedure.

17. To install the cam sensor, first clean off the old spacer from the sensor face completely. Inspect the O-ring for damage and replace, if necessary. A new spacer must be attached to the cam sensor prior to installation; if a new spacer is not used, engine performance will be adversely affected. Oil the O-ring lightly and push the sensor into its bore in the chain case cover until contact is made with the cam timing gear. Hold in this position and tighten the bolt to 10 ft. lbs. (12 Nm).

18. Refill the cooling system and fill the engine with oil.

19. Connect the negative battery cable, road test the vehicle and check for leaks.

Timing Belt Front Cover

REMOVAL & INSTALLATION

2.5L Engine

1. Disconnect the negative battery cable.

2. Remove the nuts that attach the upper cover to the valve cover.

3. Remove the bolt that attaches the upper cover to the lower cover.

4. Remove the upper cover.

5. Raise the vehicle and support safely. Remove the right side splash shield.

6. Remove the crankshaft pulley, water pump pulley and belts.

7. Remove the lower cover attaching bolts.

8. Remove the lower cover.

9. The installation is the reverse of the removal procedure.

10. Connect the negative battery cable.

3.0L Engine

1. Disconnect the negative battery cable.

2. To remove the air conditioning compressor belt, loosen the adjustment pulley locknut, turn the screw counterclockwise to reduce the drive belt tension and remove the belt.

3. To remove the serpentine drive belt, insert a ½ in. breaker bar in to the square hole of the tensioner pulley, rotate it counterclockwise to reduce the drive belt tension and remove the belt.

4. Remove the air conditioning compressor and the air compressor bracket, power steering pump and alternator from the mounts and support them to the side. Remove power steering pump/alternator automatic belt tensioner bolt and the tensioner.

5. Raise the vehicle and support safely. Remove the right inner fender splash shield.

6. Remove the crankshaft pulley bolt and the pulley/damper assembly from the crankshaft.

7. Lower the vehicle and place a floor jack under the engine to support it.

8. Separate the front engine mount insulator from the bracket. Raise the engine slightly and remove the mount bracket.

9. Remove the timing belt cover bolts and the upper and lower covers from the engine.

To install:

10. Install the timing covers and bolts.

11. Install the engine mount bracket. The engine mount through bolt must be torqued to 75 ft. lbs. (102 Nm) on 1988 vehicles or 100 ft. lbs. (136 Nm) on 1989–91 vehicles with the engine support removed and the engine's weight on the mount.

12. Install the pulley damper assem-

bly to the crankshaft. Torque the bolt to 110 ft. lbs. (149 Nm). Install the splash shield.

13. Install the power steering pump/alternator automatic belt tensioner.

14. Install the air conditioning compressor bracket, compressor, power steering pump and alternator.

15. Install the belts.

16. Connect the negative battery cable and check all disturbed components for proper operation.

Timing Belt and Tensioner

ADJUSTMENT

2.5L Engine

1. Disconnect the negative battery cable.

2. Raise the vehicle and support safely. Remove the right front inner splash shield.

3. Remove the tensioner cover.

4. Place the special tensioning tool C–4703 on the hex of the tensioner so the weight is at about the 10 o'clock position, then loosen the bolt.

5. The tensioner should drop to the 9 o'clock position. Reposition the tool as required in order to have it end up at the 9 o'clock position, parallel to the ground, hanging toward the rear of the vehicle, ± 15 degrees.

6. Hold the tool in position and tighten the bolt. Do not pull the tool past the 9 o'clock position or the belt will be too tight and will cause howling or possible breakage.

7. Install the cover and the splash shield.

3.0L Engine

1. Disconnect the negative battery cable.

2. Remove the timing belt covers.

3. Loosen the bolt that holds the timing belt tensioner in place.

4. Allow the spring only to pull the tensioner in automatically. Do not manually move the tensioner or the belt will be too tight.

5. Tighten the tensioner locking bolt.

6. Install the timing belt covers and all related parts.

REMOVAL & INSTALLATION

2.5L Engine

1. If possible, position the engine so the No. 1 piston is at TDC of its compression stroke. Disconnect the negative battery cable.

2. Remove the timing belt covers. Remove the timing belt tensioner, and allow the belt to hang free.

3. Place a floor jack under the engine so the full weight of the engine is on the jack. Separate the right motor mount.

4. Remove the air conditioning compressor belt idler pulley, if equipped, and remove the mounting stud. Unbolt the compressor/alternator bracket and position it to the side.

5. Remove the timing belt from the vehicle.

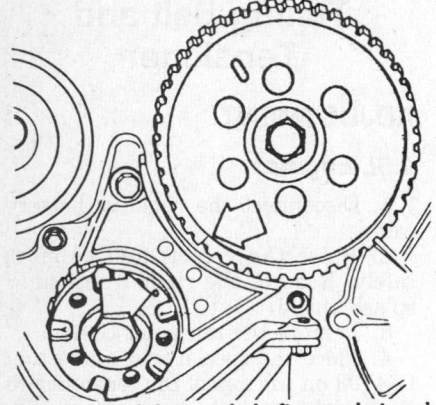

Alignment of the crankshaft sprocket and intermediate shaft sprocket—2.5L engine

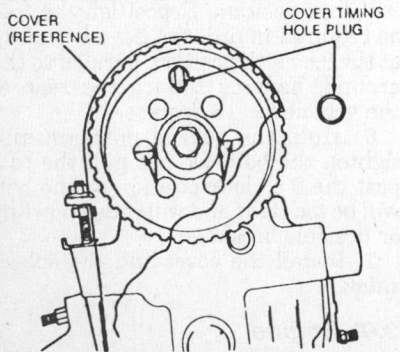

Alignment of the arrows on the camshaft sprocket with the camshaft cap-to-cylinder-mounting line—2.5L engine

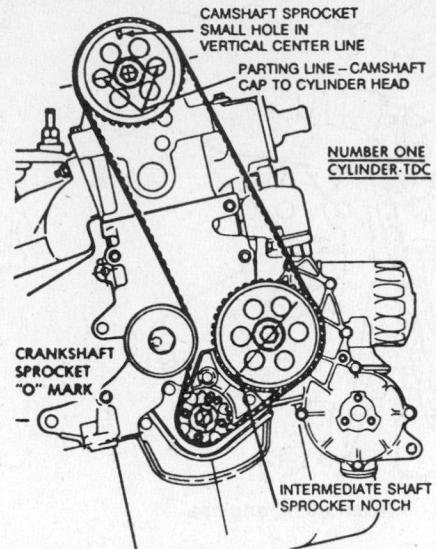

Proper timing belt positioning—2.5L engine

To install:

6. Turn the crankshaft sprocket and intermediate shaft sprocket until the marks are in line. Position a straight-edge from bolt to bolt to confirm alignment.

7. Turn the camshaft until the small hole in the sprocket is at the top and rows on the hub are in line with the camshaft cap to cylinder head mounting lines. Use a mirror to see the alignment so it is viewed straight on and not at an angle from above. Install the belt, but let at hang free at this point.

8. Install the air conditioning compressor/alternator bracket, idler pulley and motor mount. Remove the floor jack. Raise the vehicle and support safely. Have the tensioner at an arm's reach because the timing belt will have to be held in position with one hand.

9. To properly install the timing belt, reach up and engage it with the camshaft sprocket. Turn the intermediate shaft counterclockwise slightly, then engage the belt with the intermediate shaft sprocket. Hold the belt against the intermediate shaft sprocket and turn clockwise to take up all tension; if the timing marks are out of alignment, repeat until alignment is correct.

10. Using a wrench, turn the crankshaft sprocket counterclockwise slightly and wrap the belt around it. Turn the sprocket clockwise so there is no slack in the belt between sprockets; if the timing marks are out of alignment, repeat until alignment is correct.

NOTE: If the timing marks are in line but slack exists in the belt between either the camshaft and

intermediate shaft sprockets or the intermediate and crankshaft sprockets, the timing will be incorrect when the belt is tensioned. All slack must only be between the crankshaft and camshaft sprockets at the rear of the engine.

11. Install the tensioner and install the mounting bolt loosely. Place the special tensioning tool C–4703 on the hex of the tensioner so the weight is at about the 9 o'clock position, parallel to the ground, hanging toward the rear of the vehicle, ± 15 degrees.

12. Hold the tool in position and tighten the bolt to 45 ft. lbs. (61 Nm). Do not pull the tool past the 9 o'clock position; this will make the belt too tight and will cause it to howl or possibly break.

13. Lower the vehicle and recheck the camshaft sprocket positioning. If it is correct install the timing belt covers and all related parts.

14. Connect the negative battery cable and road test the vehicle.

3.0L Engine

1. If possible, position the engine so the No. 1 cylinder is at TDC of its compression stroke. Disconnect the negative battery cable. Remove the timing covers from the engine.

2. If the same timing belt will be reused, mark the direction of the timing belt's rotation, for installation in the same direction. Make sure the engine is positioned so the No. 1 cylinder is at the TDC of its compression stroke and the sprockets timing marks are aligned with the engine's timing mark indicators.

3. Loosen the timing belt tensioner bolt and remove the belt. If not removing the tensioner, position it as far away from the center of the engine as possible and tighten the bolt.

4. If the tensioner is being removed, paint the outside of the spring to ensure that it is not installed backwards. Unbolt the tensioner and remove it along with the spring.

To install:

5. Install the tensioner, if removed, and hook the upper end of the spring to the water pump pin and the lower end to the tensioner in exactly the same position as originally installed. If not already done, position both camshafts so the marks line up with those on the alternator bracket (rear bank) and inner timing cover (front bank). Rotate the crankshaft so the timing mark aligns with the mark on the oil pump.

6. Install the timing belt on the crankshaft sprocket and while keeping the belt tight on the tension side (right

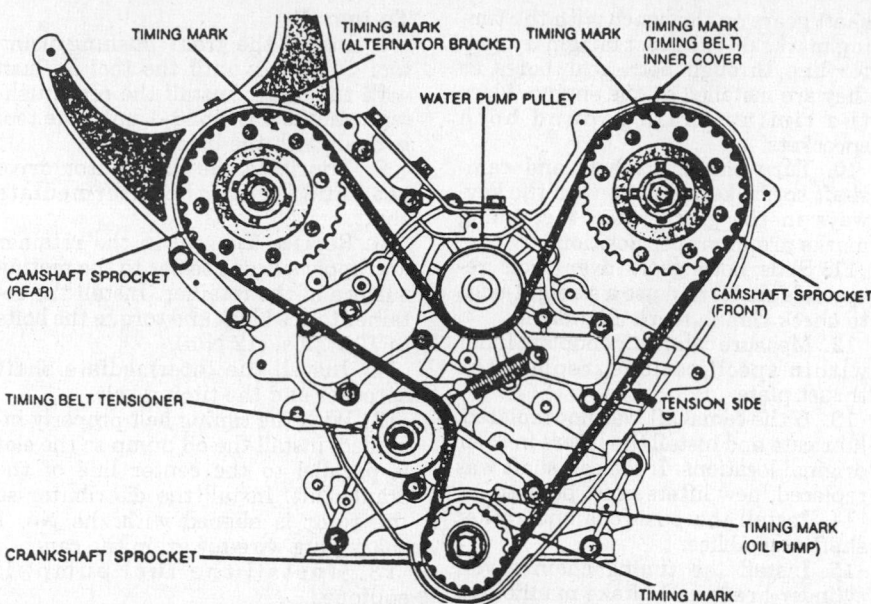

TIMING MARK • TIMING MARK (ALTERNATOR BRACKET) • TIMING MARK • TIMING MARK (TIMING BELT INNER COVER) • WATER PUMP PULLEY • CAMSHAFT SPROCKET (REAR) • CAMSHAFT SPROCKET (FRONT) • TIMING BELT TENSIONER • TENSION SIDE • CRANKSHAFT SPROCKET • TIMING MARK (OIL PUMP) • TIMING MARK

Proper timing belt positioning—3.0L engine

side), install the belt on the front camshaft sprocket.

7. Install the belt on the water pump pulley, then the rear camshaft sprocket and the tensioner.

8. Rotate the front camshaft counterclockwise to tension the belt between the front camshaft and the crankshaft. If the timing marks became misaligned, repeat the procedure.

9. Install the crankshaft sprocket flange.

10. Loosen the tensioner bolt and allow the spring to tension the belt.

11. Turn the crankshaft 2 full turns in the clockwise direction only until the timing marks align again. Now that the belt is properly tensioned, torque the tensioner lock bolt to 21 ft. lbs. (29 Nm).

12. Install the timing belt covers and all related parts.

13. Connect the negative battery cable and road test the vehicle.

Timing Sprockets

REMOVAL & INSTALLATION

2.5L Engine

1. Disconnect the negative battery cable. Remove the timing belt.

2. Remove the crankshaft sprocket bolt. Using the puller tool C–4685 or equivalent, and the button from tool L–4524 or equivalent, remove the crankshaft sprocket.

3. Using the tool C–4687 or equivalent, hold the camshaft and/or intermediate sprocket, remove the center bolt and the sprocket(s).

4. The installation is the reverse of

the removal procedure. Torque the camshaft and intermediate sprocket bolts to 65 ft. lbs. (88 Nm) and the crankshaft sprocket bolt to 50 ft. lbs. (68 Nm).

5. Connect the negative battery cable and road test the vehicle.

3.0L Engine

1. Disconnect the negative battery cable.

2. Remove the timing belt.

3. To remove the camshaft sprocket, hold the sprocket with tool MB990775 or equivalent, and remove the retaining bolt and washer.

4. To remove the crankshaft sprocket, remove the bolt and remove the sprocket from the crankshaft.

5. The installation is the reverse of the removal procedure. Torque the camshaft sprocket bolt to 70 ft. lbs. (95 Nm) while holding the sprocket with the holding tool. Torque the crankshaft sprocket bolt. to 110 ft. lbs. (150 Nm).

6. Connect the negative battery cable and road test the vehicle.

Camshaft

REMOVAL & INSTALLATION

2.5L Engine

1. Disconnect the negative battery cable.

2. Turn the crankshaft so the No. 1 piston is at the TDC of its compression stroke. Remove the upper timing belt cover.

3. Remove the camshaft sprocket bolt and the sprocket and suspend tightly so the belt does not lose ten-

sion. If it does, the belt timing will have to be reset.

4. Remove the valve cover.

5. If the rocker arms are being reused, mark them for installation identification and loosen the camshaft bearing bolts, evenly and gradually.

6. Using a soft mallet, tap the rear of the camshaft a few times to break the bearing caps loose.

7. Remove the bolts, bearing caps and the camshaft with seals.

NOTE: Take note of the color of the paint stripe on the rear camshaft seal. These stripes differentiate seal sizes. If a seal with a different color stripe is installed, a severe leak will develop if the seal is too small or the cap will not be able to be fully installed if the seal is too big.

Also, oversized components can be identified as follows: the top of the bearing caps are painted green and "O/SJ" is stamped behind the oil galley plug on the end of the head. The barrel of an oversized camshaft is also painted green and "O/SJ" is stamped on the end of the shaft. If normal sized parts are installed in place of oversized ones, oil pressure will be significantly reduced.

8. Check the oil passages for blockages and the parts for wear and damage and replace parts, as required. Clean the gasket mounting surfaces.
To install:

9. Transfer the sprocket key to the new camshaft. New rocker arms and a new camshaft sprocket bolt are normally included with the camshaft package. Install the rocker arms, lubricate the camshaft and install with end seals installed.

NOTE: Apply RTV silicone gasket material to the No. 1 and 5 bearing caps. Install the bearing caps before the seals are installed.

10. Place the bearing caps with No. 1 at the timing belt end and No. 5 at the transaxle end. The camshaft bearing caps are numbered and have arrows facing forward. Torque the camshaft bearing bolts evenly and gradually to 18 ft. lbs. (24 Nm).

11. Mount a dial indicator to the front of the engine and check the camshaft endplay. Play should not exceed 0.020 in.

12. Install the camshaft sprocket and the new bolt.

13. Install the valve cover with a new gasket.

14. Connect the negative battery cable and check for leaks.

3.0L Engine

1. Disconnect the negative battery cable. Remove the air cleaner assembly and valve covers.

2. Install auto lash adjuster retainer tools MD998443 or equivalent on the rocker arms.

3. If removing the right side (front) camshaft, remove the distributor extension.

4. Remove the camshaft bearing caps but do not remove the bolts from the caps.

5. Remove the rocker arms, rocker shafts and bearing caps, as an assembly.

6. Remove the camshaft from the cylinder head.

7. Inspect the bearing journals on the camshaft, cylinder head and bearing caps.

To install:

8. Lubricate the camshaft journals and camshaft with clean engine oil and install the camshaft in the cylinder head.

9. Align the camshaft bearing caps with the arrow mark depending on cylinder numbers and install in numerical order.

10. Apply sealer at the ends of the bearing caps and install the assembly.

11. Torque the bearing cap bolts, in the following sequence: No. 3, No. 2, No. 1 and No. 4 to 85 inch lbs. (10 Nm).

12. Repeat the sequence increasing the torque to 175 inch lbs. (18 Nm).

13. Install the distributor extension, if it was removed.

14. Install the valve cover and all related parts.

15. Connect the negative battery cable and road test the vehicle.

3.3L and 3.8L Engines

1. Relieve the fuel pressure. Disconnect the negative battery cable.

2. Remove the engine from the vehicle. Remove the intake manifold, cylinder heads, timing chain cover and timing chain from the engine.

3. Remove the rocker arm and shaft assemblies.

4. Label and remove the pusrods and lifters.

5. Remove the camshaft thrust plate.

6. Install a long bolt into the front of the camshaft to facilitate its removal. Remove the camshaft being careful not to damage the cam bearings with the cam lobes.

To install:

7. Install the camshaft to within 2 in. of its final installation position.

8. Install the camshaft thrust plate and 2 bolts and torque to 10 ft. lbs. (12 Nm).

9. Place both camshaft and crankshaft gears on the bench with the timing marks on the exact imaginary center line through both gear bores as they are installed on the engine. Place the timing chain around both sprockets.

10. Turn the crankshaft and camshaft so the keys line up with the keyways in the gears when the timing marks are in proper position.

11. Slide both gears over their respective shafts and use a straight-edge to check timing mark alignment.

12. Measure camshaft endplay. If not within specifications, replace the thrust plate.

13. If the camshaft was not replaced, lubricate and install the lifters in their original locations. If the camshaft was replaced, new lifters must be used.

14. Install the pushrods and rocker shaft assemblies.

15. Install the timing chain cover, cylinder heads and intake manifold.

16. Install the engine in the vehicle.

17. When everything is bolted in place, change the engine oil and replace the oil filter.

NOTE: If the camshaft or lifters have been replaced, add 1 pint of Mopar crankcase conditioner or equivalent, when replenishing the oil to aid in break in. This mixture should be left in the engine for a minimum of 500 miles and drained at the next normal oil change.

18. Fill the radiator with coolant.

19. Connect the negative battery cable, set all adjustments to specifications and check for leaks.

Intermediate Shaft

REMOVAL & INSTALLATION

2.5L Engine

1. Disconnect the negative battery cable.

2. Crank the engine around until the No. 1 piston is at TDC of its compression stroke. Remove the timing belt covers to confirm that all timing marks are lined up.

3. Remove the distributor. Looking down at the oil pump, the slot in the shaft must be parallel with the center line of the crankshaft. Remove the oil pump.

4. Remove the timing belt and the intermediate shaft sprocket.

5. Remove the shaft retainer bolts and remove the retainer from the block.

6. Remove the intermediate shaft from the engine.

7. If necessary, remove the front bushing using tool C–4697–2 and the rear bushing using tool C–4686–2.

To install:

8. Install the front bushing using tool C–4697–1 until the tool is flush with the block. Install the rear bushing using tool C–4686–1 until the tool is flush with the block.

9. Lubricate the distributor drive gear and install the intermediate shaft.

10. Replace the seal in the retainer and apply silicone sealer to the mating surface of the retainer. Install the retainer to the block and torque the bolts to 10 ft. lbs. (12 Nm).

11. Install the intermediate shaft sprocket and the timing belt.

12. With the timing belt properly installed, install the oil pump so the slot is parallel to the center line of the crankshaft. Install the distributor so the rotor is aligned with the No. 1 spark plug wire tower on the cap.

13. Install the fuel pump, if equipped.

14. Connect the negative battery cable and road test the vehicle.

Balance Shafts

REMOVAL & INSTALLATION

2.5L Engine

1. Disconnect the negative battery cable. Raise the vehicle and support safely.

2. Remove the timing belt. Remove the oil pan, the oil pickup, the crankshaft belt sprocket and the front crankshaft oil seal retainer.

3. Remove the balance shaft chain cover, the guide and the tensioner.

4. Remove the balance shaft sprocket-to-shaft bolt, the gear cover-to-balance shaft bolt and the crankshaft sprocket-to-crankshaft bolts, then the sprockets with the balance shaft chain.

5. Remove the front gear cover-to-carrier housing stud, the gear cover and the balance shaft drive gears.

6. Remove the rear gear cover-to-carrier housing bolts, the rear cover and the balance shafts from the rear of the carrier.

7. If necessary, remove the carrier housing-to-crankcase bolts and the housing.

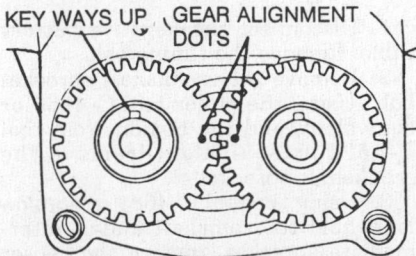

Aligning the balance shaft gear timing marks—2.5L engine

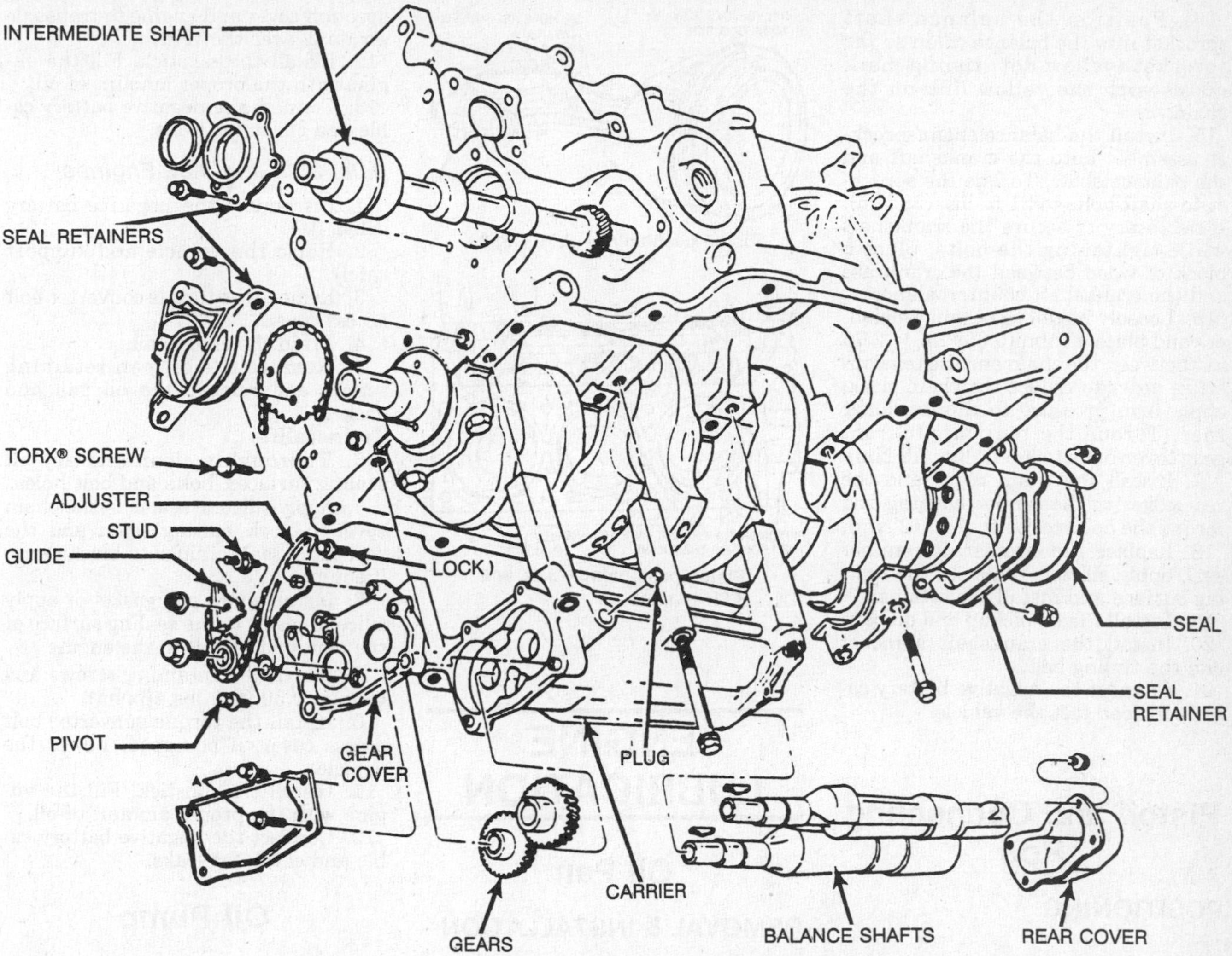

INTERMEDIATE SHAFT

SEAL RETAINERS

TORX® SCREW

ADJUSTER

STUD

GUIDE

LOCK

PIVOT

GEAR COVER

GEARS

CARRIER

PLUG

BALANCE SHAFTS

REAR COVER

SEAL

SEAL RETAINER

Exploded view of the balance shafts and related parts — 2.5L engine

To install:

8. If the carrier housing is being installed, torque the carrier housing-to-crankcase bolts to 40 ft. lbs. (54 Nm).

9. Rotate the balance shafts until the keyways are facing upward, parallel to the vertical centerline of the engine.

10. Install the short hub gear on the sprocket driven shaft and the long hub gear on the gear driven shaft; make sure the gear timing marks are aligned facing each other.

11. Install the front gear cover and torque the front gear cover-to-carrier housing stud bolt to 8.5 ft. lbs. (12 Nm).

12. Install the balance chain sprocket and torque the sprocket-to-crankshaft bolts to 11 ft. lbs. (13 Nm).

13. Rotate the crankshaft to position the No. 1 cylinder on the TDC of its compression stroke; the timing marks on the chain sprocket should align with the parting line on the left side of the No. 1 main bearing cap.

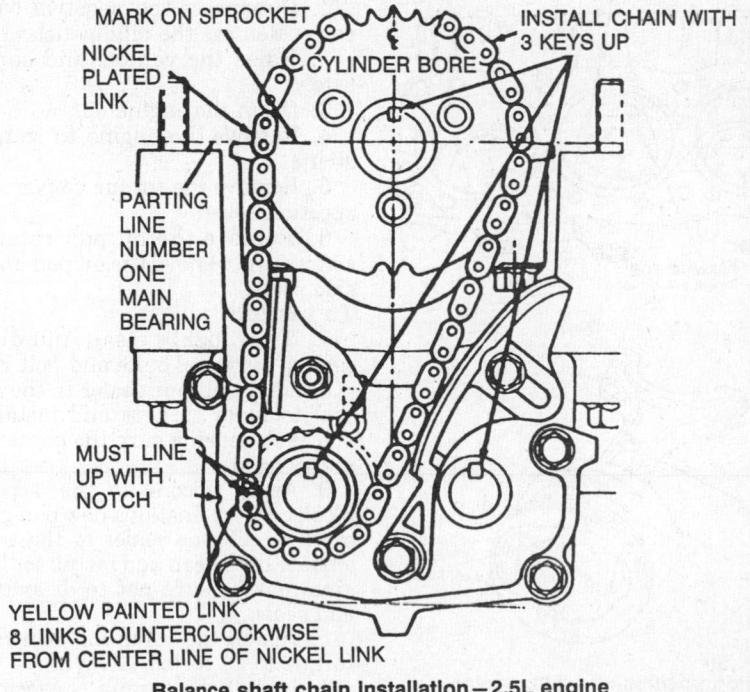

MARK ON SPROCKET

NICKEL PLATED LINK

CYLINDER BORE

INSTALL CHAIN WITH 3 KEYS UP

PARTING LINE NUMBER ONE MAIN BEARING

MUST LINE UP WITH NOTCH

YELLOW PAINTED LINK 8 LINKS COUNTERCLOCKWISE FROM CENTER LINE OF NICKEL LINK

Balance shaft chain installation — 2.5L engine

14. Position the balance shaft sprocket into the balance chain so the sprocket (yellow dot) timing mark mates with the yellow link on the chain.

15. Install the balance chain/sprocket assembly onto the crankshaft and the balance shaft. Torque the sprocket-to-shaft bolts to 21 ft. lbs. (28 Nm). If necessary to secure the crankshaft while tightening the bolts, place a block of wood between the crankcase and the crankshaft counterbalance.

16. Loosely install the chain tensioners and place a shim (0.039 in. × 2.75 in.) between the chain and the tensioner. In order to reduce the chain slack, apply firm pressure to the tensioner shoe. Torque the tensioner-to-front gear cover bolts to 8.5 ft. lbs. (12 Nm).

17. Install the chain cover and the rear cover to the carrier housing and torque the bolts to 8.5 ft. lbs. (12 Nm).

18. Replace the crankshaft retainer seal, apply silicone sealer to the mating surface and install the retainer.

19. Install the oil pickup and oil pan.

20. Install the crankshaft sprocket and the timing belt.

21. Connect the negative battery cable and road test the vehicle.

Piston and Connecting Rod

POSITIONING

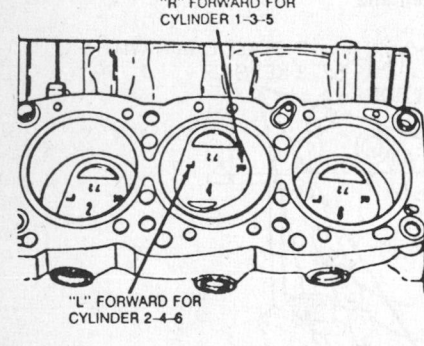

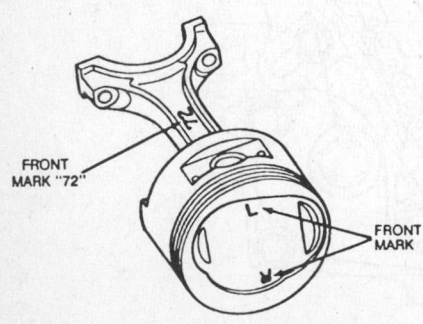

Piston positioning—3.0L engine

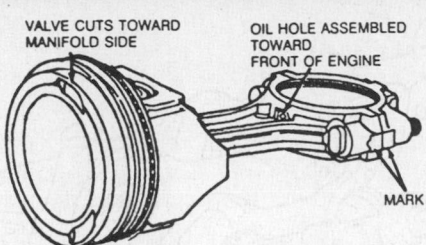

Piston positioning—2.5L engine

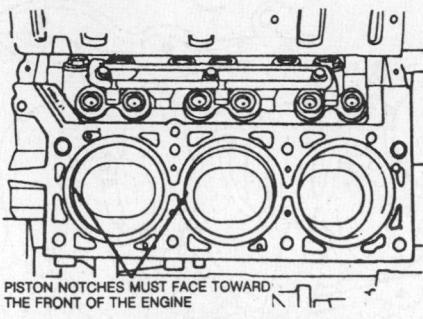

Piston positioning—3.3L and 3.8L engines

ENGINE LUBRICATION

Oil Pan

REMOVAL & INSTALLATION

2.5L Engine

1. Disconnect the negative battery cable. Remove the oil dipstick.

2. Raise the vehicle and support safely.

3. Drain the engine oil.

4. Remove the engine to transaxle struts.

5. Remove the torque converter inspection cover.

6. Remove the oil pan retaining screws and remove the oil pan and the side seals.

To install:

7. Thoroughly clean and dry all sealing surfaces, bolts and bolt holes.

8. Apply silicone sealer to the 4 end seal to block corners and install the end seals making sure the corners are not twisted.

9. Apply silicone to the 4 pan to block corners. Install a new pan gasket or apply silicone sealer to the sealing surface of the pan and install to the engine making sure not to dislodge the end seals.

10. Install the retaining screws and torque to 17 ft. lbs. (23 Nm).

11. Install the torque converter inspection cover and engine to transaxle struts. Lower the vehicle.

12. Install the dipstick. Fill the engine with the proper amount of oil.

13. Connect the negative battery cable and check for leaks.

3.0L, 3.3L and 3.8L Engines

1. Disconnect the negative battery cable.

2. Raise the vehicle and support safely.

3. Remove the torque converter bolt access cover.

4. Drain the engine oil.

5. Remove the oil pan retaining screws and remove the oil pan and gasket.

To install:

6. Thoroughly clean and dry all sealing surfaces, bolts and bolt holes.

7. Apply silicone sealer to the chain cover-to-block mating seam and the rear main seal retainer-to-block seam, if equipped.

8. Install a new pan gasket or apply silicone sealer to the sealing surface of the pan, and install to the engine.

9. Install the retaining screws and torque to 50 inch lbs. (6 Nm).

10. Install the torque converter bolt access cover, if equipped. Lower the vehicle.

11. Install the dipstick. Fill the engine with the proper amount of oil.

12. Connect the negative battery cable and check for leaks.

Oil Pump

REMOVAL & INSTALLATION

2.5L Engine

1. Crank the engine around so the No. 1 piston is at TDC of its compression stroke. Disconnect the negative battery cable.

2. Matchmark the rotor to the block and remove the distributor to confirm that the slot in the oil pump shaft is parallel to the centerline of the crankshaft. Matchmark the slot to the distributor bore, if desired.

3. Remove the dipstick. Raise the vehicle and support safely. Drain the engine oil and remove the pan.

4. Remove the oil pickup.

5. Remove the 2 mounting bolts and remove the oil pump from the engine.

To install:

6. Prime the pump by pouring fresh oil into the pump intake and turning the driveshaft until oil comes out the pressure port. Repeat a few times until no air bubbles are present.

7. Apply Loctite® 515 or equivalent, to the pump body-to-block machined surface interface. Lubricate the oil pump and distributor driveshaft.

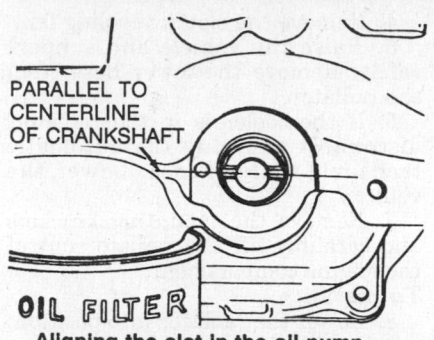

PARALLEL TO CENTERLINE OF CRANKSHAFT

OIL FILTER

Aligning the slot in the oil pump shaft—2.5L engine

8. Align the slot so it will be in the same position as when it was removed. If it is not, the distributor will not be timed correctly. Install the pump fully and rotate back and forth to ensure proper positioning between the pump mounting surface and the machined surface of the block.

9. Install the mounting bolts finger-tight and lower the vehicle to confirm proper slot positioning. If the slot is not properly positioned, raise the vehicle and move the gear, as required. If the slot is correct, hold the pump firmly against the block and torque the mounting bolts to 17 ft. lbs. (23 Nm).

10. Clean out the oil pickup or replace if necessary. Replace the oil pickup O-ring and install the pickup to the pump.

11. Install the oil pan using new gaskets. Lower the vehicle.

12. Install the distributor.

13. Install the dipstick. Fill the engine with the proper amount of oil.

14. Connect the negative battery cable, check the timing and check the oil pressure.

3.0L Engine

1. Disconnect the negative battery cable. Remove the dipstick.

2. Raise the vehicle and support safely. Remove the timing belt, drain the engine oil and remove the oil pan from the engine. Remove the oil pickup.

3. Remove the oil pump mounting bolts and remove the pump from the front of the engine. Note the different length bolts and their position in the pump for installation.

To install:

4. Clean the gasket mounting surfaces of the pump and engine block.

5. Prime the pump by pouring fresh oil into the pump and turning the rotors. Using a new gasket, install the oil pump on the engine and torque all bolts to 11 ft. lbs. (15 Nm).

6. Install the balancer and crankshaft sprocket to the end of the crankshaft.

7. Clean out the oil pickup or replace, if necessary. Replace the oil

pickup gasket ring and install the pickup to the pump.

8. Install the timing belt, oil pan and all related parts.

9. Install the dipstick. Fill the engine with the proper amount of oil.

10. Connect the negative battery cable and check the oil pressure.

3.3L and 3.8L Engines

1. Disconnect the negative battery cable. Remove the dipstick.

2. Raise the vehicle and support safely. Drain the oil and remove the oil pan.

3. Remove the oil pickup.

4. Remove the chain case cover.

5. Disassemble the oil pump and remove its components from the block.

To install:

6. Assemble the pump. Torque the cover screws to 10 ft. lbs. (12 Nm).

7. Prime the oil pump by filling the rotor cavity with fresh oil and turning

the rotors until oil comes out the pressure port. Repeat a few times until no air bubbles are present.

8. Install the chain case cover.

9. Clean out the oil pickup or replace, if necessary. Replace the oil pickup O-ring and install the pickup to the pump.

10. Install the oil pan.

11. Install the dipstick. Fill the engine with the proper amount of oil.

12. Connect the negative battery cable and check the oil pressure.

CHECKING

2.5L Engine

1. Remove the cover from the oil pump.

2. Check endplay of the inner rotor using a feeler gauge and a straight-edge placed across the pump body. The specification is 0.001–0.004 in. (0.03–0.09mm).

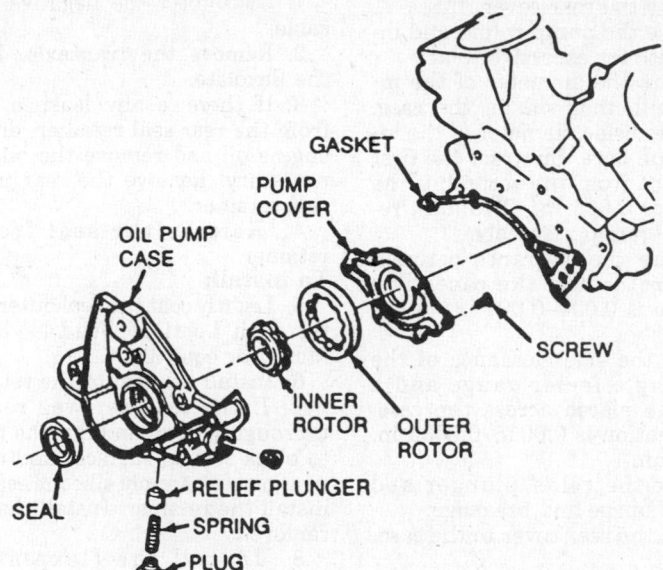

GASKET

PUMP COVER

OIL PUMP CASE

INNER ROTOR

OUTER ROTOR

SCREW

SEAL

RELIEF PLUNGER

SPRING

PLUG

Exploded view of the oil pump—3.0L engine

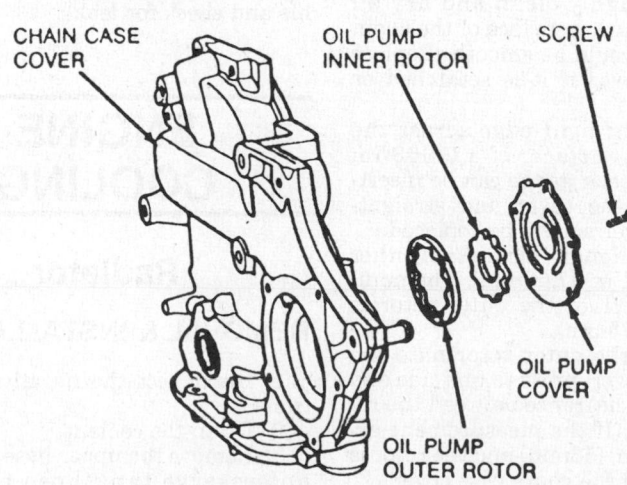

CHAIN CASE COVER

OIL PUMP INNER ROTOR

SCREW

OIL PUMP COVER

OIL PUMP OUTER ROTOR

Oil pump components—3.3L and 3.8L engines

3. Measure the clearance between the inner and outer rotors. The maximum clearance is 0.008 in. (0.20mm).

4. Measure the clearance between the outer rotor and the pump body. The maximum clearance is 0.014 in. (0.35mm).

5. The minimum thickness of the outer rotor is 0.944 in. (23.96mm). The minimum diameter of the outer rotor is 2.77 in. (62.70mm). The minimum thickness of the inner rotor is 0.943 in. (23.95mm).

6. Check the cover for warpage. The maximum allowable is 0.003 in. (0.076mm).

7. Check the pressure relief valve for damage. The spring's free length specification is 1.95 in. (49.50mm).

8. Assemble the outer rotor with the larger chamfered edge in the pump body. Torque the cover screws to 10 ft. lbs. (12 Nm).

3.0L Engine

1. Remove the rear cover.
2. Remove the pump rotors and inspect the case for excessive wear.
3. Measure the diameter of the inner rotor hub that sits in the case. Measure the inside diameter of the inner rotor hub bore. Subtract the first measurement from the second; if the result is over 0.006 in. (0.15mm), replace the oil pump assembly.
4. Measure the clearance between the outer rotor and the case. The specification is 0.004–0.007 in. (0.10–0.18mm).
5. Check the side clearance of the rotors using a feeler gauge and a straight-edge placed across the case. The specification is 0.0015–0.0035 in. (0.04–0.09mm).
6. Check the relief plunger and spring for damage and breakage.
7. Install the rear cover to the case.

3.3L and 3.8L Engines

1. Thoroughly clean and dry all parts. The mating surface of the chain case cover should be smooth. Replace the pump cover if it is scratched or grooved.
2. Lay a straight-edge across the pump cover surface. If a 0.003 in. (0.076mm) feeler gauge can be inserted between the cover and straight-edge, the cover should be replaced.
3. The minimum thickness of either rotor is 0.301 in. (7.63mm). The minimum diameter of the outer rotor is 3.14 in. (79.78mm).
4. Install the outer rotor onto the chain case cover, press to one side and measure the clearance between the rotor and case. If the measurement exceeds 0.022 in. (56mm) and the rotor is good, replace the chain case cover.
5. Install the inner rotor to the

chain case cover and measure the clearance between the rotors. If the clearance exceeds 0.008 in. (0.203mm), replace both rotors.

6. Place a straight-edge over the chain case cover between bolt holes. If a 0.004 in. (0.102mm) thick feeler gauge can be inserted under the straight-edge, replace the pump assembly.

7. Inspect the relief valve plunger for scoring and freedom of movement. Small marks may be removed with 400 grit wet or dry sandpaper.

8. The relief valve spring should have a free length of 1.95 in.

9. Assemble the pump using new parts where necessary.

Rear Main Bearing Oil Seal

REMOVAL & INSTALLATION

1. Disconnect the negative battery cable.
2. Remove the transaxle. Remove the flexplate.
3. If there is any leakage coming from the rear seal retainer, drain the engine oil and remove the oil pan, if necessary. Remove the rear main oil seal retainer.
4. Remove the seal from the retainer.

To install:

5. Lightly coat the seal outer diameter with Loctite® Stud N' Bearing Mount or equivalent.
6. Install the seal to the retainer.
7. If the retainer was removed, thoroughly clean and dry the retainer to block sealing surfaces and install a new gasket or apply silicone sealer and install the retainer. Install the pan, if removed.
8. Install the flexplate and transaxle.
9. Connect the negative battery cable and check for leaks.

ENGINE COOLING

Radiator

REMOVAL & INSTALLATION

1. Disconnect the negative battery cable.
2. Drain the coolant.
3. Remove the upper hose and coolant reserve tank hose from the radiator.

4. Remove the electric cooling fan.
5. Raise the vehicle and support safely. Remove the lower hose from the radiator.
6. If the cooler is in the radiator, disconnect and plug the automatic transaxle cooler hoses. Lower the vehicle.
7. Remove the mounting brackets and carefully lift the radiator out of the engine compartment.

To install:

8. Lower the radiator into position.
9. Install the mounting brackets.
10. Raise the vehicle and support safely. Connect the automatic transaxle cooler lines, if they were disconnected.
11. Connect the lower hose. Lower the vehicle.
12. Install the electric cooling fan.
13. Connect the upper hose and coolant reserve tank hose.
14. Fill the system with coolant and bleed the system.
15. Connect the negative battery cable, run the vehicle until the thermostat opens, fill the radiator completely and check the automatic transaxle fluid level.
16. Once the vehicle has cooled, recheck the coolant level.

Heater Core

REMOVAL & INSTALLATION

1. Disconnect the negative battery cable. Properly discharge the air conditioning system. Drain the cooling system.
2. Clamp off the heater hoses near the heater core and remove the hoses from the core tubes. Plug the hose ends and the core tubes to prevent spillage of coolant.
3. Disconnect the H-valve connection at the valve and remove the H-valve. Remove the condensation tube.
4. Disconnect the vacuum lines at the brake booster and water valve, if equipped.
5. Remove the right upper and lower under-panel silencers.
6. Remove the steering column cover and the ash tray.
7. Remove the left side under-panel silencer.
8. Remove the right side cowl trim piece.
9. Remove the glove box assembly and the right side instrument panel reinforcement.
10. Remove the center distribution and defroster adaptor ducts.
11. Disconnect the relay module, blower motor wiring and 25-way connector bracket and fuse block from the panel.

12. Disconnect the demister hoses from the top of the package.

13. Disconnect the temperature control cable and vacuum harness, if equipped. If equipped with Automatic Temperature Control (ATC), disconnect the instrument panel wiring from the rear of the ATC unit.

14. Disconnect the hanger strap from the package and rotate it out of the way.

15. Remove the retaining nuts from the package mounting studs at the firewall.

16. Fold the carpeting and insulation back to provide a little more working room and to prevent spillage from staining the carpeting.

17. Move the package rearward to clear the mounting studs and lower.

18. Pull the right side of the instrument panel out as far as possible. Rotate the package while removing it from under the instrument panel.

19. To disassemble the housing assembly, remove the vacuum diaphragm, if equipped. Then remove the retaining screws from the cover and remove the cover.

20. Remove the retaining screw from the heater core and remove the core from the housing assembly.

To install:

21. Remove the temperature control door from the housing and clean the unit out with solvent. Lubricate the lower pivot rod and its well and install. Wrap the heater core with foam tape and place it in position. Secure it with its screw.

22. Assemble the package, making sure all vacuum tubing is properly routed.

23. If equipped, feed the vacuum lines through the hole in the firewall and install the assembly to the vehicle. Connect the vacuum harness and demister hoses. Install the nuts to the firewall and connect the hanger strap inside the passenger compartment.

24. Fold the carpeting back into position.

25. Connect the wiring to the ATC unit, if equipped.

26. Install the fuse block. Connect the 25-way connector, relay module and blower motor wiring.

27. Install the center distribution and defroster adaptor ducts.

28. Install the right side instrument panel reinforcement and the glove box assembly.

29. Install the right side cowl trim piece, left side under-panel silencer, steering column cover, ash tray and right side under-panel silencers.

30. Connect the vacuum lines at the brake booster and water valve.

31. Using new gaskets, install the H-valve and condensation tube.

32. Connect the heater hoses.

33. Using the proper equipment, evacuate and recharge the air conditioning system.

34. Fill the cooling system.

35. Connect the negative battery cable and check the entire climate control system for proper operation and leakage.

Water Pump

REMOVAL & INSTALLATION

2.5L Engine

1. Disconnect the negative battery cable.

2. Drain the cooling system.

3. Remove the air conditioning compressor from the bracket and position it to the side.

4. Remove the alternator and bracket. Remove the pulley from the water pump.

5. Disconnect the lower radiator hose and heater hose from the water pump.

6. Remove the water pump housing attaching screws and remove the assembly from the vehicle. Discard the O-ring.

7. Remove the water pump from the housing.

To install:

8. Using a new gasket or silicone sealer, install the water pump to the housing.

9. Install a new O-ring to the housing and install to the engine. Torque the bolts to 21 ft. lbs. (30 Nm).

10. Install the water pump pulley and torque the bolts to 21 ft. lbs. (30 Nm). Connect the radiator hose and heater hose to the water pump.

11. Install all items removed to gain access to the water pump, then adjust the belts.

12. Remove the hex-head plug on the top of the thermostat housing. Fill the radiator with coolant until the coolant comes out the plug hole. Install the plug and continue to fill the radiator.

13. Connect the negative battery ca-

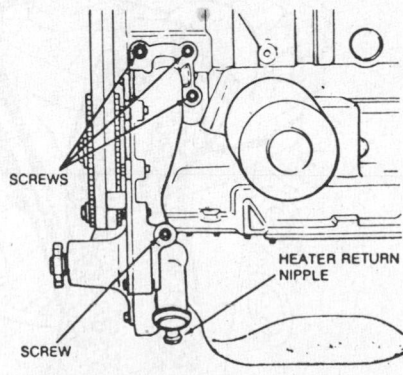

Water pump assembly—2.5L engine

ble, run the vehicle until the thermostat opens, fill the radiator completely and check for leaks.

14. Once the vehicle has cooled, recheck the coolant level.

3.0L Engine

1. Disconnect the negative battery cable.

2. Drain the cooling system.

3. Remove the timing cover. If the same timing belt will be reused, mark the direction of the timing belt's rotation, for installation in the same direction. Make sure the engine is positioned so the No. 1 cylinder is at the TDC of its compression stroke and the sprockets timing marks are aligned with the engine's timing mark indicators.

4. Loosen the timing belt tensioner bolt and remove the belt. Position the tensioner as far away from the center of the engine as possible and tighten the bolt. Remove the water pump mounting bolts, separate the pump from the water inlet pipe and remove the pump from the engine.

To install:

5. Install the pump with a new gasket to the engine. Torque the water pump mounting bolts to 20 ft. lbs. (27 Nm).

6. If not already done, position both camshafts so the marks line up with those on the alternator bracket (rear bank) and inner timing cover (front bank). Rotate the crankshaft so the timing mark aligns with the mark on the oil pump.

7. Install the timing belt on the crankshaft sprocket and while keeping the belt tight on the tension side (right side), install the belt on the front camshaft sprocket.

8. Install the belt on the water pump pulley, then the rear camshaft sprocket and the tensioner.

9. Rotate the front camshaft counterclockwise to tension the belt between the front camshaft and the crankshaft. If the timing marks became misaligned, repeat the procedure.

10. Install the crankshaft sprocket flange.

11. Loosen the tensioner bolt and allow the spring to tension the belt.

12. Turn the crankshaft 2 full turns in the clockwise direction only until the timing marks align again. Now that the belt is properly tensioned, torque the tensioner lock bolt to 21 ft. lbs. (29 Nm).

13. Refill the cooling system. This system uses a self-bleeding thermostat, so there is no need to bleed the system. Connect the negative battery cable and road test the vehicle.

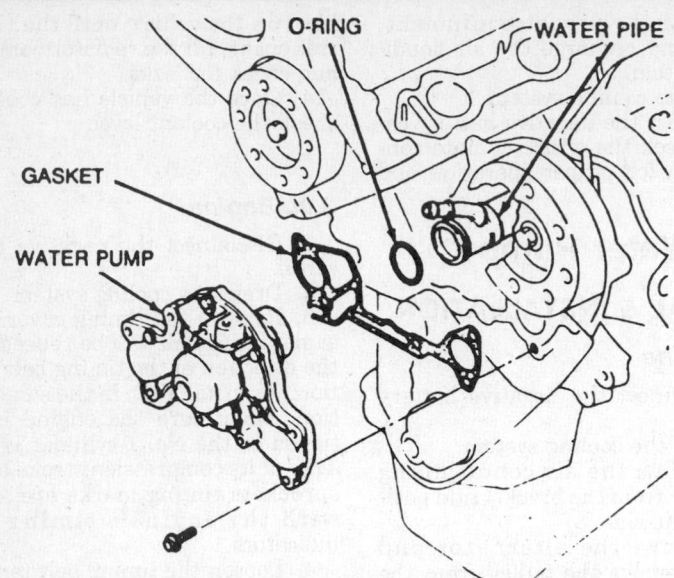

Water pump assembly—3.0L engine

3.3L and 3.8L Engines

1. Disconnect the negative battery cable.
2. Drain the cooling system.
3. Remove the serpentine belt.
4. Raise the vehicle and support safely. Remove the right front tire and wheel assembly and lower fender shield.
5. Remove the water pump pulley.
6. Remove the 5 mounting screws and remove the pump from the engine.
7. Discard the O-ring.

To install:

8. Using a new O-ring, install the pump to the engine. Torque the mounting bolts to 21 ft. lbs. (30 Nm).
9. Install the water pump pulley.
10. Install the fender shield and tire and wheel assembly. Lower the vehicle.
11. Install the serpentine belt.
12. Remove the engine temperature sending unit. Fill the radiator with coolant until the coolant comes out the sending unit hole. Install the sending unit and continue to fill the radiator.
13. Connect the negative battery cable, run the vehicle until the thermo-

stat opens, fill the radiator completely and check for leaks.
14. Once the vehicle has cooled, recheck the coolant level.

Thermostat

REMOVAL & INSTALLATION

1. Disconnect the negative battery cable. Drain the coolant down to thermostat level or below.
2. Remove the thermostat housing.
3. Remove the thermostat and discard the gasket.
4. Clean the housing mating surfaces and use a new gasket.
5. The installation is the reverse of the removal procedure.
6. To properly fill the system with coolant:
    a. On the 2.5L engine, remove the hex-head plug on the thermostat

housing. Fill the radiator with coolant until the coolant comes out the plug hole. Install the plug and continue to fill the radiator.
    b. The 3.0L engine is equipped with a self-bleeding thermostat; bleeding is not necessary.
    c. On 3.3L and 3.8L engines, remove the engine temperature sending unit. Fill the radiator with coolant until the coolant comes out the sending unit hole. Install the sending unit and continue to fill the radiator.
7. Connect the negative battery cable, run the vehicle until the thermostat opens, fill the radiator completely and check for leaks.
8. Once the vehicle has cooled, recheck the coolant level.

COOLING SYSTEM BLEEDING

To bleed air from the 2.5L engine, remove the plug on the top of the thermostat housing. Fill the radiator with coolant until the coolant comes out the hole. Install the plug and continue to fill the radiator. This will vent all trapped air from the engine.

The thermostat in the 3.0L engine is equipped with a small air vent valve

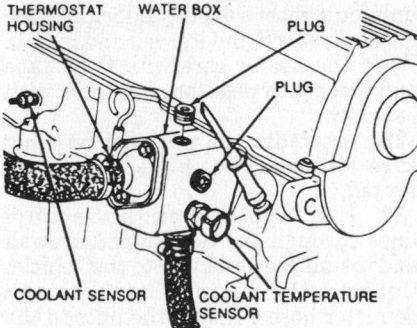

Cooling system bleed plug—2.5L engine

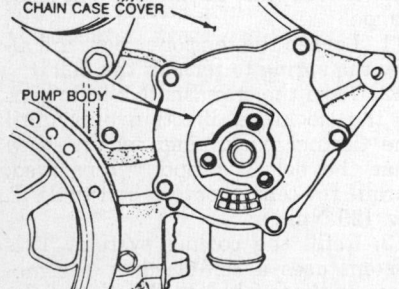

Water pump assembly—3.3L and 3.8L engines

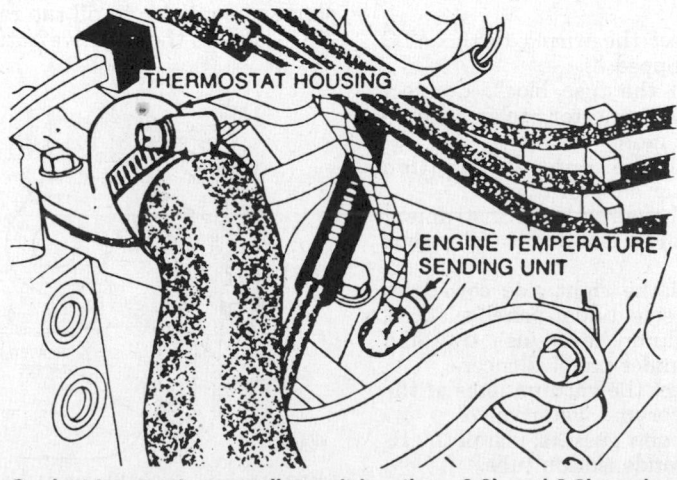

Coolant temperature sending unit location—3.3L and 3.8L engines

that allows trapped air to bleed from the system during refilling. This valve negates the need for cooling system bleeding in those engines.

On 3.3L and 3.8L engines, remove the engine temperature sending unit. Fill the radiator with coolant until the coolant comes out the hole. Install the switch and continue to fill the radiator. This will vent all trapped air from the engine.

ENGINE ELECTRICAL

NOTE: Disconnecting the negative battery cable on some vehicles may interfere with the functions of the on board computer systems and may require the computer to undergo a relearning process, once the negative battery cable is reconnected.

Distributor

REMOVAL

1. Disconnect the negative battery cable.
2. Disconnect the distributor pickup lead wires. Remove the splash shield, if equipped.
3. Unscrew the distributor cap hold-down screws and lift off the distributor cap with all ignition wires still connected. Remove the coil wire, if necessary.
4. Matchmark the rotor to the distributor housing and the distributor housing to the engine.

NOTE: Do not crank the engine during this procedure. If the engine is cranked, the matchmark must be disregarded.

5. Remove the hold-down bolt and clamp or nut.
6. Remove the distributor from the engine.

INSTALLATION

Timing Not Disturbed

1. Install a new distributor housing O-ring.
2. Install the distributor in the engine so the rotor is aligned with the matchmark on the housing and the housing is aligned with the matchmark on the engine. Make sure the distributor is fully seated and the distributor shaft is fully engaged.
3. Install the hold-down clamp and

snug the hold-down bolt or install the nut.
4. Connect the distributor pickup lead wires. Install the splash shield, if equipped.
5. Install the distributor cap and tighten the screws.
6. Connect the negative battery cable.
7. Adjust the ignition timing and tighten the hold-down bolt.

Timing Disturbed

1. Install a new distributor housing O-ring.
2. Position the engine so the No. 1 piston is at TDC of the compression stroke and the mark on the vibration damper is aligned with **0** on the timing indicator.
3. Install the distributor in the engine so the rotor is aligned with the position of the No. 1 ignition wire on the distributor cap and the housing is aligned with the matchmark on the engine. Make sure the distributor is fully seated and the distributor shaft is fully engaged.

NOTE: There are distributor cap runners inside the cap on 3.0L engine. Make sure the rotor is

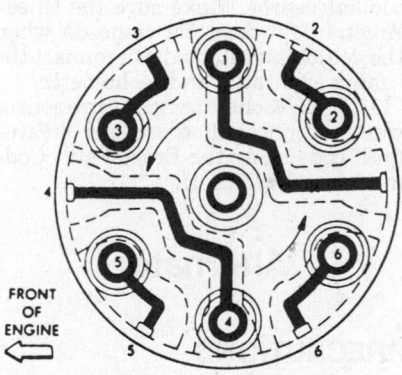

Distributor cap terminal routing— 3.0L engine

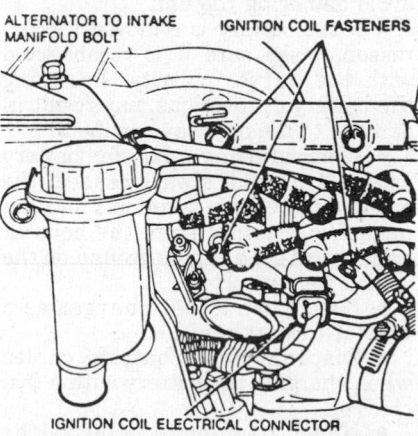

Ignition coil removal and installation— 3.3L and 3.8L engine

pointing to where the No. 1 runner originates inside the cap and not where the No. 1 ignition wire plugs into the cap.

4. Install the hold-down clamp and snug the hold-down bolt or install the nut.
5. Connect the distributor pickup lead wires. Install the splash shield, if equipped.
6. Install the distributor cap and tighten the screws.
7. Connect the negative battery cable.
8. Adjust the ignition timing and tighten the hold-down nut or bolt.

Distributorless Ignition

REMOVAL & INSTALLATION

Ignition Coil

1. Disconnect the negative battery cable.
2. Remove the spark plug wires from the coil.
3. Disconnect the electrical connector.
4. Remove the coil fasteners.
5. Remove the coil from the ignition module.
6. The installation is the reverse of the removal procedure.

Crankshaft Position Sensor

1. Disconnect the negative battery cable.
2. Disconnect the sensor lead at the harness connector.
3. Remove the sensor retaining bolt.
4. Pull the sensor straight up out the transaxle housing.
5. If the sensor is being reinstalled, remove any remains of the old spacer completely and attach a new spacer to the sensor. If a new spacer is not used, the sensor will not function properly. New sensors are packaged with a new spacer.

To install:

6. Install the sensor to the tranaxle

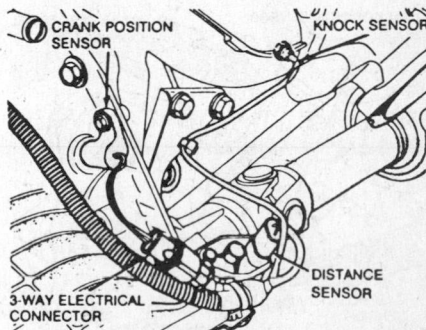

Crankshaft position sensor location— 3.3L and 3.8L engine

housing and push the sensor down until it contacts the drive plate.

7. Hold in this position and install the retaining bolt. Torque to 9 ft. lbs. (12 Nm).

8. Connect the sensor lead wire.

Camshaft Position Sensor

1. Disconnect the negative battery cable.

2. Disconnect the sensor lead at the harness connector.

3. Loosen the sensor retaining bolt sufficiently to allow the slotted mounting surface to slide past.

4. Pull the sensor straight up and out of the chain case cover. Resistance may be high due to the rubber O-ring.

5. If the sensor is being reinstalled, remove any remains of the old spacer completely and attach a new spacer to the sensor. If a new spacer is not used, the sensor will not function properly. New sensors are packaged with a new spacer.

To install:

6. Inspect the O-ring for damage and replace, if necessary.

7. Lubricate the O-ring with oil. Install the sensor to the chain case cover and push the sensor into its bore until contact is made with the cam timing gear.

8. Hold in this position and tighten the bolt to 9 ft. lbs. (12 Nm).

9. Connect the wire and rout it away from the accessory drive belt.

Ignition Timing

ADJUSTMENT

NOTE: The ignition timing on the distributorless 3.3L and 3.8L engines cannot be checked or changed.

1. Start the engine, set the parking brake and run the engine until at normal operating temperature. Keep all lights and accessories OFF.

2. If a magnetic timing unit is available, insert the probe into the receptacle near the timing scale. The scale is

located on the top of the bellhousing on the 2.5L engine and near the crankshaft pulley on the 3.0L engine.

3. If a magnetic timing unit is not available, connect a conventional power timing light to the No. 1 cylinder spark plug wire.

4. Connect the red lead of a tachometer to the negative primary terminal of the coil and connect the black lead to a good ground.

5. On 1988 vehicles, disconnect the coolant sensor wire.

6. On 1989–92 vehicles, connect the Diagnostic Readout Box II (DRB II) and access the Basic Timing Mode. If the DRB II is not available, disconnect the coolant sensor located near the thermostat housing. The Check Engine light on the instrument panel must be ON.

7. Aim the timing light at the timing scale or read the magnetic timing unit.

8. Loosen the distributor hold-down bolt just enough so the distributor can be rotated.

9. Turn the distributor in the proper direction until the specified timing according to the VECI label is reached. Tighten the hold-down bolt or nut and recheck the timing.

10. Turn the engine off. Connect the coolant sensor. Make sure the Check Engine light does not come on when the vehicle is restarted. Disconnect the timing apparatus and tachometer.

11. If the coolant temperature sensor was disconnected, erase the created fault code using the Erase Fault Code mode on the DRB II.

Alternator

PRECAUTIONS

Several precautions must be observed when working with the alternator to avoid damaging the unit.

- If the battery is removed for any reason, make sure it is reconnected with the correct polarity. Reversing the battery connections may result in damage to the one-way rectifiers.
- When utilizing a booster battery as a starting aid, always connect the positive to positive terminals and the negative terminal from the booster battery to a good engine ground on the vehicle being started.
- Never use a fast charger as a booster to start vehicles.
- Disconnect the battery cables when charging the battery with a fast charger.
- Never attempt to polarize the alternator.
- Do not use test lights of more

than 12 volts when checking diode continuity.

- Do not short across or ground any of the alternator terminals.
- The polarity of the battery, alternator and regulator must be matched and considered before making any electrical connections within the system.
- Never separate the alternator on an open circuit. Make sure all connections within the circuit are clean and tight.
- Disconnect the battery ground terminal when performing any service on electrical components.
- Disconnect the battery if arc welding is to be done on the vehicle.

BELT TENSION ADJUSTMENT

NOTE: The belt tension is automatically adjusted by a dynamic tensioner on the 3.0L, 3.3L and 3.8L engines. Periodic adjustment is not necessary.

1. Loosen the pivot bolt slightly.

2. Raise the vehicle and support safely. Remove the splash shield. Loosen the "T" bolt locknut enough so the alternator can be moved.

3. Tighten the adjusting bolt until the belt deflects about ¼ in. under a 10 lb. load.

4. Tighten the "T" bolt locknut and pivot bolt.

REMOVAL & INSTALLATION

1. Disconnect the negative battery cable.

2. On the 2.5L engine, remove the air conditioning compressor and position it to the side. Remove the oil filter to allow the alternator to be removed from above, if possible.

3. On 3.0L, 3.3L and 3.8L engines, release the dynamic belt tensioner and remove the belt. On all other engines, loosen the mounting bolts, move the alternator toward the engine and remove the drive belt(s).

4. Remove the mounting bolts and spacers and remove the alternator from the brackets.

5. Remove the battery positive, field and ground terminals from the rear of the alternator. Remove the wire harness hold-down screw from the alternator, if equipped.

To install:

6. Connect all wiring to the proper terminals on the rear of the alternator and install the wire harness hold-down screw, if equipped.

7. Position the alternator in the mounting brackets.

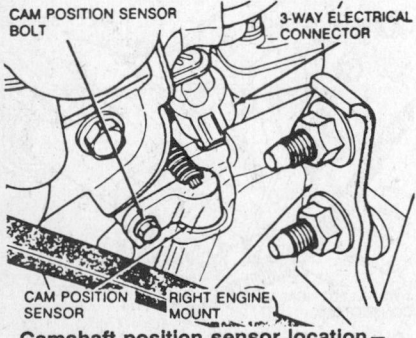

Camshaft position sensor location – 3.3L and 3.8L engines

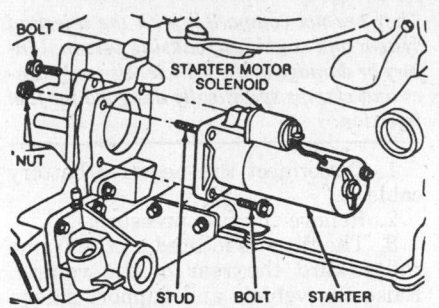

Starter mounting—2.5L engine

8. Install the spacers, pivot bolt and adjuster bolt. Install the belt.

9. Install the air conditioning compressor and oil filter, if they were removed.

10. Adjust the belt tension, if necessary.

11. Connect the negative battery cable.

Starter

REMOVAL & INSTALLATION

1. Disconnect the negative battery cable.

2. On the 2.5L engine, remove the attaching nut and bolt at the top of the bellhousing. Raise the vehicle and support safely.

3. Remove the rear mount and heat shield from the starter, if equipped.

4. Unbolt the starter and remove from the vehicle.

5. Disconnect the solenoid lead wires from the starter.

To install:

6. Connect the solenoid lead wires and install the heat shield, if equipped.

7. On the the 2.5L engine, install the lower bolt loosely, then lower the vehicle and install the nut and bolt from above and torque to 40 ft. lbs. (54 Nm).

8. Raise the vehicle and torque the bottom bolt to the same value. Install the rear mount to the starter.

9. On 3.0L, 3.3L and 3.8L engines, install all mounting bolts and torque to 40 ft. lbs. (54 Nm) evenly.

10. Connect the negative battery cable and check the starter for proper operation.

EMISSION CONTROLS

Please refer to "Emission Controls" in the Unit Repair section for system maintenance proce-dures. Due to the complex nature of modern electronic engine control systems, comprehensive diagnosis and testing procedures fall outside the confines of this repair manual. For complete information on diagnosis, testing and repair procedures concerning all modern engine and emission control systems, please refer to "Chilton's Guide to Fuel Injection and Electronic Engine Controls".

FUEL SYSTEM

Fuel System Service Precautions

Safety is the most important factor when performing not only fuel system maintenance but any type of maintenance. Failure to conduct maintenance and repairs in a safe manner may result in serious personal injury or death. Maintenance and testing of the vehicle's fuel system components can be accomplished safely and effectively by adhering to the following rules and guidelines.

• To avoid the possibility of fire and personal injury, always disconnect the negative battery cable unless the repair or test procedure requires that battery voltage be applied.

• Always relieve the fuel system pressure prior to disconnecting any fuel system component (injector, fuel rail, pressure regulator, etc.), fitting or fuel line connection. Exercise extreme caution whenever relieving fuel system pressure to avoid exposing skin, face and eyes to fuel spray. Please be advised that fuel under pressure may penetrate the skin or any part of the body that it contacts.

• Always place a shop towel or cloth around the fitting or connection prior to loosening to absorb any excess fuel due to spillage. Ensure that all fuel spillage (should it occur) is quickly removed from engine surfaces. Ensure that all fuel soaked cloths or towels are deposited into a suitable waste container.

• Always keep a dry chemical (Class B) fire extinguisher near the work area.

• Do not allow fuel spray or fuel vapors to come into contact with a spark or open flame.

• Always use a backup wrench when loosening and tightening fuel line connection fittings. This will pre-vent unnecessary stress and torsion to fuel line piping. Always follow the proper torque specifications.

• Always replace worn fuel fitting O-rings with new. Do not substitute fuel hose or equivalent where fuel pipe is installed.

RELIEVING FUEL SYSTEM PRESSURE

1988 Vehicles

1. Disconnect the negative battery cable.

2. Loosen the fuel filler cap to release fuel tank pressure.

3. Remove the wiring harness connector from the (any) injector.

4. Using a jumper wire, ground either injector terminal.

5. Being careful not to allow contact between the jumper leads, connect a second jumper wire to the other terminal and touch the other end to the positive battery post for no longer that 10 seconds. This will relieve fuel pressure.

6. Remove the jumper wires and continue with fuel system service.

1989–92 Vehicles

1. Loosen the fuel filler cap to release fuel tank pressure.

2. Locate and disconnect the fuel injector harness connector.

3. Connect a jumper wire from terminal No. 1 of the appropriate connector to ground.

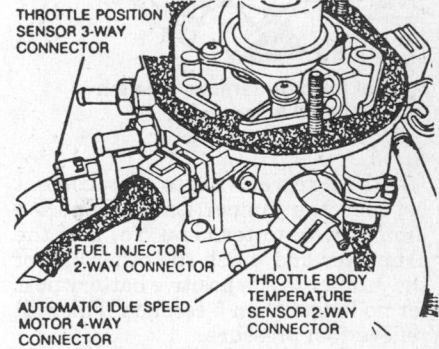

Fuel Injector harness location— 2.5L engine

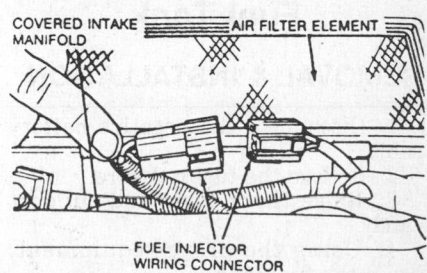

Fuel Injector harness location— 3.0L engine

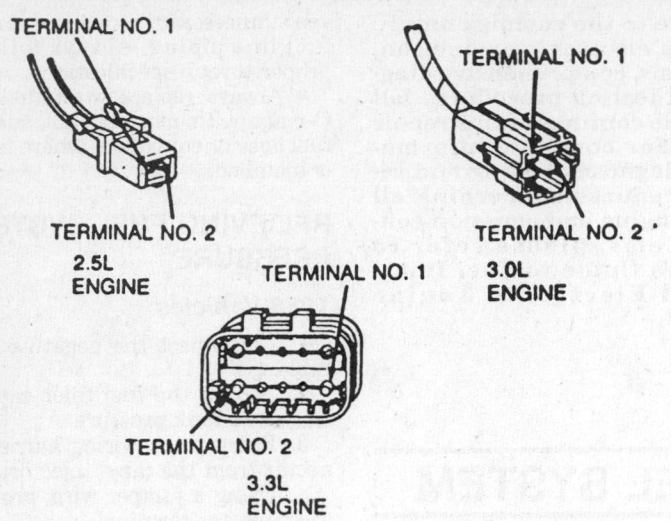

Fuel Injector harness connector terminal identification

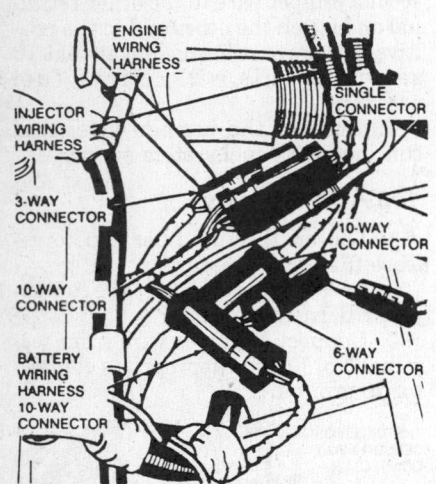

Fuel Injector harness location—
3.3L and 3.8L engines

4. Being careful not to allow contact between the jumper leads, connect a jumper wire to terminal No. 2 of the connector and touch the other end of the jumper to the positive battery post for no longer than 5 seconds. This will relieve fuel pressure.

5. Remove the jumper wires and continue with fuel system service.

Fuel Tank

REMOVAL & INSTALLATION

1. Disconnect the negative battery cable.
2. Relieve the fuel pressure.
3. Raise the vehicle and support safely.
4. Using the proper equipment, drain the fuel tank.
5. Remove the screws that hold the filler neck to the quarter panel.

6. Disconnect the wiring and hoses from the tank.
7. Place a suitable transmission jack or equivalent, under the center of the tank and apply slight pressure. Remove the tank straps.
8. Lower the tank and remove the filler tube from the tank.
9. Lower the tank farther and disconnect the vapor separator rollover valve hose and remove the fuel tank from the vehicle.

To install:
10. Raise the tank into position and connect all harnesses and vacuum hoses.
11. Install the tank straps and tighten the retaining nuts.
12. Install the screws that hold the filler neck to the quarter panel.
13. Connect the negative battery cable, start the engine and check for leaks.

Fuel Filter

REMOVAL & INSTALLATION

—— CAUTION ——
Do not use conventional fuel filters, hoses or clamps when servicing this fuel system.

They are not compatible with the injection system and could fail, causing personal injury or damage to the vehicle. Use only hoses and clamps specifically designed for fuel injection.

1. Disconnect the negative battery cable.
2. Relieve the fuel pressure.
3. The filter is located on the frame rail toward the rear of the vehicle. Raise the vehicle and support safely. Remove the filter retaining screw and remove the filter assembly from the mounting plate.
4. Loosen the outlet hose clamp on the filter and inlet hose clamp on the rear fuel tube.
5. Wrap a shop towel around the hoses to absorb fuel. Remove the hoses from the filter and fuel tube and discard the clamps and the filter.
To install:
6. Install the inlet hose on the fuel tube and tighten the new clamp to 10 inch lbs.
7. Install the outlet hose on the filter outlet fitting and tighten the new clamp to 10 inch lbs.
8. Position the filter assembly on the mounting plate and tighten the mounting screw to 75 inch lbs. (8 Nm).
9. Connect the negative battery cable, start the engine and check for leaks.

Electric Fuel Pump

PRESSURE TESTING

1. Relieve the fuel pressure.
2. Properly connect the fuel system pressure tester:
   a. 2.5L and 3.0L engines—special tool C-4799A or equivalent, is installed between the fuel supply hose and the engine fuel line assembly.
   b. 3.3L and 3.8L engines—special tool C-4799A or equivalent, is installed to the fuel rail service valve.
3. With the key in the **RUN** position, put the DRB II in the activate auto shutdown relay mode; this will activate the fuel pump and pressurize the system.

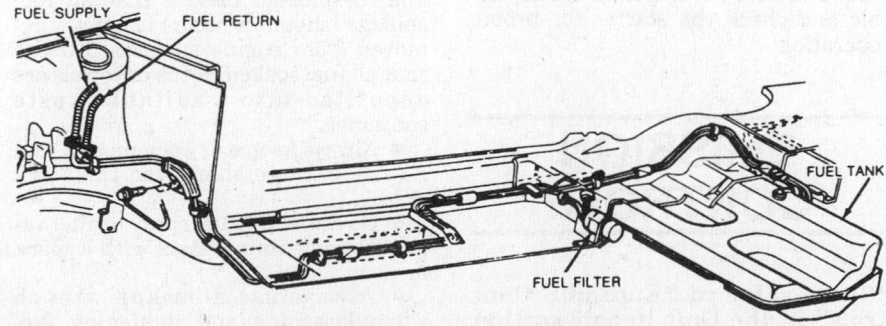

Fuel line layout and fuel filter location

4. If the pressure is within specifications, reinstall the fuel hose.

5. If fuel pressure is below specifications, install the tester in the fuel supply line between the tank and the filter and repeat the test.

6. If the pressure is 5 psi higher than in Step 5, replace the fuel filter. If no change is observed, squeeze the return hose. If pressure increases, replace the pressure regulator. If no change is observed, the problem is either a plugged in-tank sock filter or a defective pump.

7. If fuel pressure is above specifications, remove the fuel return line hose from the chassis line at the fuel tank and connect a 3 foot piece of fuel hose to the return line. Put the other end into a 2 gallon minimum capacity approved gasoline container. Repeat the test. If pressure is now correct, check the in-tank return hose for kinking. Replace the fuel pump assembly if the in-tank reservoir check valve or aspirator jet is obstructed.

8. If pressure is still above specifications, remove the fuel return hose from the throttle body. Connect a substitute hose to the throttle body return nipple and place the other end of the hose in a clean container. Repeat the test. If pressure is now correct, check for a restricted fuel return line. If no change is observed, replace the fuel pressure regulator.

REMOVAL & INSTALLATION

1. Disconnect the negative battery cable.

2. Relieve the fuel pressure.

3. Raise the vehicle and support safely.

4. Using the proper equipment, drain the fuel tank.

5. Remove the screws that hold the filler neck to the quarter panel.

6. Disconnect the wiring and hoses from the tank.

7. Place a suitable transmission jack, or equivalent, under the center of the tank and apply slight pressure. Remove the tank straps.

8. Lower the tank and remove the filler tube from the tank.

9. Lower the tank and disconnect the vapor separator rollover valve hose and remove the fuel tank from the vehicle.

10. Using a hammer and a brass drift, tap the lock ring counterclockwise to release the pump.

11. Partially pull the pump assembly, only 1 hose goes to the pump, which is bigger than the sending unit, out of the tank until the return line hose connection is visible at the bottom of the pump assembly.

12. Disconnect the fuel fitting.

13. Remove the pump from the tank

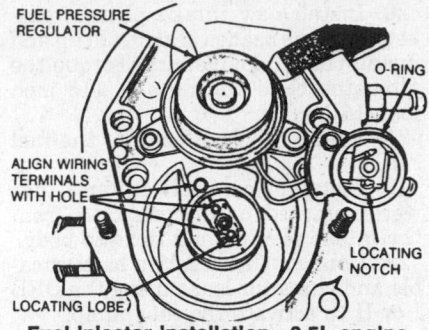

Fuel Injector Installation — 2.5L engine

with the O-ring. Discard the O-ring, pump inlet filter and inlet seal. Disassemble as required.

To install:

14. Install a new inlet seal and filter on the end of the pump.

15. Install a new O-ring to the pump.

16. Connect the reservoir hose to the pump assembly at the suction end of the pump. Press the female fitting onto the pump assembly male end until the ears snap in place.

17. Install the pump into the tank so the fuel return hose is not kinked.

18. Install the lock ring with a hammer and brass punch turning the ring clockwise.

19. Install the fuel tank.

20. Connect the negative battery cable, start the engine and check for leaks.

Fuel Injection

IDLE SPEED ADJUSTMENT

The idle speed is controlled by the Automatic Idle Speed (AIS) motor. The AIS motor is controlled by the SMEC or SBEC, which receives data from various sensors and switches in the system and adjusts the engine idle to a predetermined speed. Idle speed specifications can be found on the Vehicle Emission Control Information (VECI) label located in the engine compartment. If the idle speed is not within specifications and there are no problems with the system, the throttle body should be replaced.

IDLE MIXTURE ADJUSTMENT

There is no idle mixture adjustment provided with any Chrysler fuel injection system.

Fuel Injector

REMOVAL & INSTALLATION

2.5L Engine

1. Disconnect the negative battery cable.

2. Remove the air cleaner assembly.

3. Relieve the fuel pressure.

4. Remove the injector hold-down Torx® screw and the hold-down.

5. Using a small flat-tipped tool, lift the cap off of the injector.

6. Using the same tool, gently pry the injector from its pod.

7. Remove the lower O-ring from the pod.

To install:

8. Install the new lower O-ring on the injector.

9. Align the injector terminal housing with the locating socket in the injector cap.

10. Press the injector cap so the upper O-ring flange is flush with the lower surface of the cap.

11. Spray the inner surfaces of the injector pod with suitable carburetor parts cleaner to remove residual varnish and gasoline.

12. Lubricate the O-rings sparingly with clean oil.

13. Place the injector and cap into the injector pod and align the cap locating pin with the locating hole in the casting.

14. Press firmly on the injector cap until it is flush with the casting surface.

15. Align the hole in the hold-down with the pin on the cap and install.

16. Push down on the cap, install the screw and torque to 35 inch lbs. (4 Nm).

17. Connect the negative battery cable and check for leaks using the DRB II to activate the fuel pump.

18. Install the air cleaner.

3.0L Engine

1. Disconnect the negative battery cable.

2. Relieve the fuel pressure.

3. Remove the air cleaner to throttle body hose.

4. Disconnect the throttle cable from the throttle body and disconnect the kickdown linkage. Remove the throttle cable bracket attaching bolts.

5. Disconnect the connectors to the throttle body.

6. Matchmark and carefully remove the vacuum hoses from the throttle body.

7. Remove the PCV and brake booster hoses from the air intake plenum.

8. Remove the ignition coil from the intake plenum, if it is mounted there.

9. Remove the EGR tube flange from the intake plenum, if equipped.

10. Unplug the coolant temperature sensor and charge temperature sensor, if equipped.

11. Remove the vacuum connection from the air intake plenum vacuum connector.

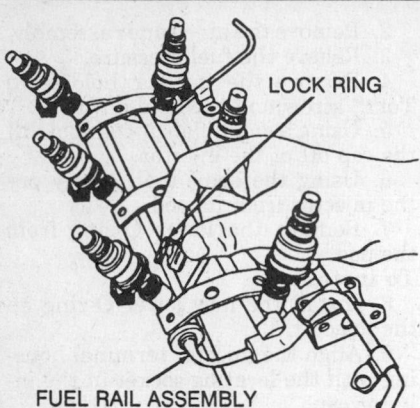

LOCK RING

FUEL RAIL ASSEMBLY

Fuel rail assembly—3.0L engine

12. Remove the fuel hoses from the fuel rail and plug them.

13. Remove the air intake plenum to intake manifold bolts and remove the plenum and gaskets. Cover the intake manifold openings.

14. Remove the vacuum hoses from the fuel rail.

15. Disconnect the fuel injector wiring harness.

16. Remove the fuel rail attaching bolts and remove the fuel rail with the wiring harness from the vehicle. Position the rail on the bench upside down so the injectors are easily accessible.

17. Remove the small connector retainer clip and unplug the injector. Remove the injector clip off the fuel rail and injector. Pull the injector straight out of the rail.

To install:

18. Lubricate the rubber O-ring with clean oil and install to the rail receiver cap. Install the injector clip to the **TOP** slot of the injector, plug in the connector and install the connector clip.

19. Install the fuel rail to the vehicle and plug in the injector harness. Connect the vacuum hoses to the fuel rail.

20. Install new intake plenum gaskets with the beaded sealer side up and install the intake plenum. Torque the attaching bolts and nuts to 115 inch lbs. (13 Nm).

21. Install the fuel hoses to the fuel rail.

22. Install or connect all items that were removed or disconnected from the intake plenum and throttle body.

23. Connect the negative battery cable and check for leaks using the DRB I or II to activate the fuel pump.

3.3L and 3.8L Engines

1. Disconnect the negative battery cable.

2. Relieve the fuel pressure.

3. Remove the air cleaner and hose assembly.

4. Disconnect the throttle cable. Remove the wiring harness from the throttle cable bracket and intake manifold water tube.

5. Remove the vacuum hose harness from the throttle body.

6. Remove the PCV and brake booster hoses from the air intake plenum.

7. Remove the EGR tube flange from the intake plenum, if equipped.

8. Unplug the charge temperature sensor and unplug all vacuum hoses from the intake plenum.

9. Remove the cylinder head to intake plenum strut.

10. Disconnect the MAP sensor and oxygen sensor connector. Remove the engine mounted ground strap.

11. Release the fuel hose quick disconnect fittings and remove the hoses from the fuel rail. Plug the hoses.

12. Remove the Direct Ignition System (DIS) coils and the alternator bracket-to-intake manifold bolt.

13. Remove the intake manifold bolts and rotate the manifold back over the rear valve cover. Cover the intake manifold.

14. Remove the vacuum harness from the pressure regulator.

15. Remove the fuel tube retainer bracket screw and fuel rail attaching bolts. Spread the retainer bracket to allow for clearance when removing the fuel tube.

16. Remove the fuel rail injector wiring clip from the alternator bracket.

17. Disconnect the cam sensor, coolant temperature sensor and engine temperature sensor.

18. Remove the fuel rail.

19. Position the rail on a work bench so the injectors are easily accessible.

20. Remove the small connector retainer clip and unplug the injector. Remove the injector clip off the fuel rail and injector. Pull the injector straight out of the rail.

To install:

21. Lubricate the rubber O-ring with clean oil and install to the rail receiver cap. Install the injector clip to the slot in the injector, plug in the connector and install the connector clip.

22. Install the fuel rail.

23. Connect the cam sensor, coolant temperature sensor and engine temperature sensor.

24. Install the fuel rail injector wiring clip to the alternator bracket.

25. Install the fuel rail attaching bolts and fuel tube retainer bracket screw.

26. Install the vacuum harness to the pressure regulator.

27. Install the intake manifold with a new gasket. Install the bolts only finger-tight. Install the alternator bracket to intake manifold bolt and the cylinder head to intake manifold strut and bolts. Torque the intake manifold mounting bolts to 21 ft. lbs. (28 Nm) starting from the middle and working outward. Torque the bracket and strut bolts to 40 ft. lbs. (54 Nm).

28. Install or connect all items that were removed or disconnected from the intake manifold and throttle body.

29. Connect the fuel hoses to the rail. Push the fittings in until they click in place.

30. Install the air cleaner assembly.

31. Connect the negative battery cable and check for leaks using the DRB II to activate the fuel pump.

FUEL RAIL RETAINER BRACKET SCREW

FUEL RAIL ATTACHING BOLTS (4)

LOWER INTAKE MANIFOLD SHOULD BE COVERED DURING SERVICE

Fuel rail assembly—3.3L and 3.8L engines

DRIVE AXLE

Halfshaft

REMOVAL & INSTALLATION

1. Disconnect the negative battery cable.

2. Raise the vehicle and support safely.

3. Remove the tire and wheel assembly.

4. Remove the cotter pin from the end of the halfshaft. Remove the nut lock, spring washer, axle nut and washer.

5. Remove the ball joint retaining bolt and pry the control arm down to release the ball stud from the steering knuckle.

6. Position a drain pan under the transaxle where the halfshaft enters the differential or extension housing.

7. Pull the strut assembly out—be careful of air suspension and ABS components if equipped—and remove the halfshaft from the hub and transaxle or center bearing. Unbolt the center bearing from the block and remove the intermediate shaft from the transaxle, if equipped.

To install:

7. Install the halfshaft or intermediate shaft to the transaxle, being careful not to damage the side seals. Make sure the inner joint clicks into pace inside the differential. Install the center bearing retaining bolts, if equipped. Install the outer shaft to the center bearing, if equipped.

8. Pull the front strut out—be careful of air suspension and ABS components if equipped—and insert the outer joint into the front hub.

9. If necessary, turn the ball joint stud to position the bolt retaining indent to the inside of the vehicle. Install the ball joint stud into the steering knuckle. Install the retaining bolt and nut.

10. Install the axle nut washer and nut and torque the nut to 180 ft. lbs. (244 Nm). Install the spring washer, nut lock and a new cotter pin.

11. Install the tire and wheel assembly.

CV-Boot

REMOVAL & INSTALLATION

NOTE: Use only clamps provided with the replacement package when servicing. Plastic wire ties and other straps will not clamp tightly enough and grease will sling out causing costly damage to the joint.

Inner Joint

1. Raise the vehicle and support

safely. Remove the halfshaft from the vehicle.

2. If cutting the boot away, mark and note the boot positioning on the shaft relative to the raised shoulders. Remove the boot clamps to gain access to the tripod retention system.

3. Separate the housing from the tripod according to the following:

NOTE: Always hold the rollers in place when removing the housing from the tripod or the needle bearing may fall out.

a. G.K.N.—Has retaining tabs integral with the housing cover. Hold the housing and lightly compress the CV-joint retention spring while bending the tabs back. Support the housing as the retention spring pushes it from the housing.

b. S.S.G.—Uses a wire ring tripod retainer which expands into a groove around the top of the housing. Use a suitable tool to pry the wire ring, without damaging it, out of the groove and slide the tripod from the housing.

4. Remove the snapring from the end of the shaft and remove the tripod.

5. If not already done, mark the

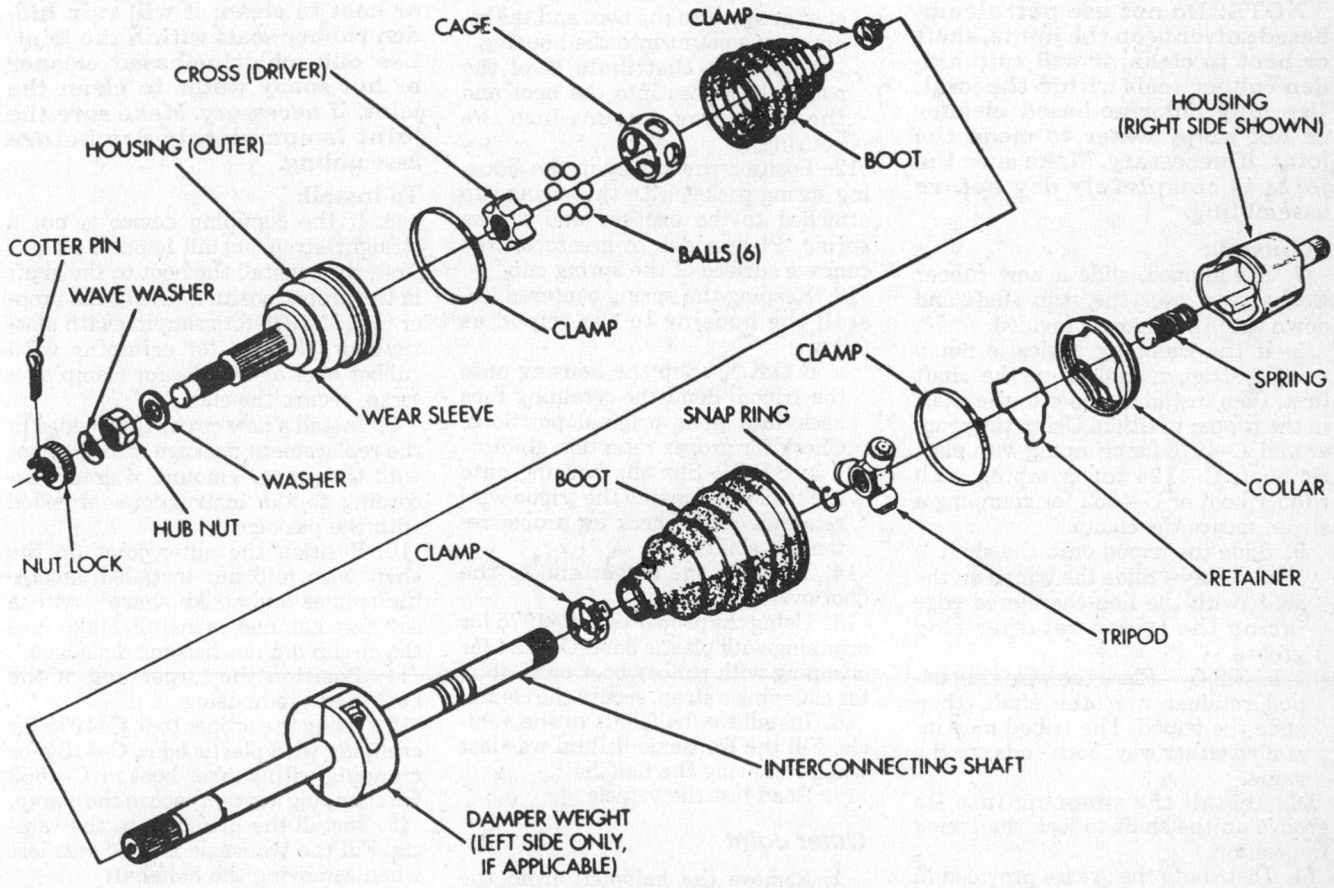

Exploded view of a typical halfshaft

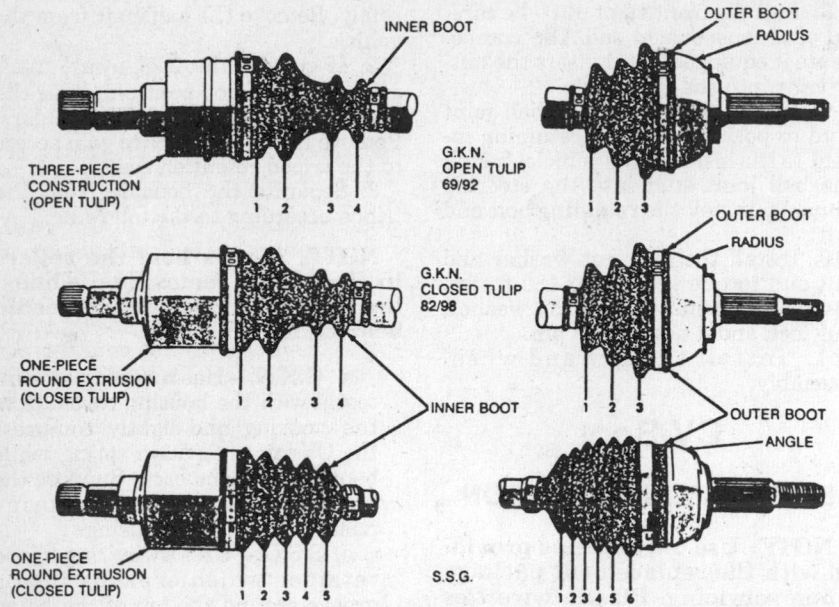

CV-joint boot identification

boot positioning on the shaft, relative to the raised shoulders and remove the boot from the shaft.

6. Remove as much old grease as possible from the joint. Inspect all parts for wear or damage.

NOTE: Do not use petroleum-based solvents on the joints, shaft or boot to clean; it will ruin hidden rubber seals within the joint. Use only chlorine-based cleaner or hot soapy water to clean the joint, if necessary. Make sure the joint is completely dry before assembling.

To install:

7. If equipped, slide a new rubber washer seal over the stub shaft and down into the groove provided.

8. If the clamping device is not a staight strap, install it on the shaft first, then install the boot to the shaft in the proper position. Using the proper tool, C–4975 for crimping with plastic boot, C–4124 for crimping with rubber boot or C–4653 for clamping a strap, secure the clamp.

9. Slide the tripod onto the shaft:

a. G.K.N—Slide the tripod on the shaft with the non-chamfered edge facing the tripod retainer ring groove.

b. S.S.G.—Place the wire ring tripod retainer over the shaft, then slide the tripod. The tripod may installed either way; both ends are the same.

10. Install the snapring into its groove on the shaft to lock the tripod in position.

11. Distribute the grease provided in the grease package as follows, or ac-

cording to the instructions in the package:

a. G.K.N—If equipped with 3 packets of grease, distribute 2 of the 3 packets into the boot and the remaining packet into the housing. Otherwise, distribute ½ of the packet of grease into the boot and the remaining amount into the housing.

b. S.S.G.—Distribute ½ of the packet of grease into the boot and the remaining amount into the housing.

12. Position the spring in the housing spring pocket with the spring cup attached to the exposed end of the spring. Place a dab of grease on the concave surface of the spring cup.

13. Keeping the spring centered, install the housing to the tripod as follows:

a. G.K.N—Slip the housing onto the tripod. Bend the retaining tabs back into their original positions. Check for proper retention ability.

b. S.S.G.—Slip the housing onto the tripod and install the tripod wire retaining ring. Check for proper retention ability.

14. Position the larger end of the boot over the housing.

15. Using the proper tool, C–4975 for crimping with plastic boot, C–4124 for crimping with rubber boot or C–4653 for clamping a strap, secure the clamp.

16. Install the halfshaft to the vehicle. Fill the transaxle if fluid was lost when removing the halfshaft.

17. Road test the vehicle.

Outer Joint

1. Remove the halfshaft from the vehicle.

2. If cutting the boot away, mark and note the boot positioning on the shaft relative to the raised shoulders. Remove the boot clamps to gain access to the joint retention system.

3. Separate the housing from the tripod according to the following:

a. G.K.N—Using a soft-jaw vise, support the halfshaft. Strike the joint assembly sharply with a soft-face hammer to dislodge the internal circlip and remove from the shaft.

b. S.S.G.—Loosen the damper weight bolts and slide it and the boot toward the inner joint. Expand the snapring and slide the joint from the shaft. Reinstall the damper weight and torque the bolts to 21 ft. lbs. (28 Nm).

4. If damaged, remove the wear sleeve from the CV-joint machined ledge.

5. Remove the circlip from the groove.

6. If not already done, mark the boot positioning on the shaft, relative to the raised shoulders and remove the boot from the shaft.

7. Remove as much old grease as possible from the joint. Inspect all parts for wear or damage.

NOTE: Do not use petroleum-based solvents on the joints, shaft or boot to clean; it will ruin hidden rubber seals within the joint. Use only chlorine-based cleaner or hot soapy water to clean the joint, if necessary. Make sure the joint is completely dry before assembling.

To install:

8. If the clamping device is not a straight strap, install it on the shaft first, then install the boot to the shaft in the proper position. Using the proper tool, C–4975 for crimping with plastic boot, C–4124 for crimping with rubber boot or C–4653 for clamping a strap, secure the clamp.

9. Install a new circlip if provided in the replacement package. Fill the boot with the proper amount of grease according to the instructions provided with the package.

10. Position the outer joint on the shaft with hub nut installed, engage the splines and strike sharply with a soft-face hammer to install. Make sure the circlip did not become dislodged.

11. Position the larger end of the boot over the housing.

12. Using the proper tool, C–4975 for crimping with plastic boot, C–4124 for crimping with rubber boot or C–4653 for clamping a strap, secure the clamp.

13. Install the halfshaft to the vehicle. Fill the transaxle if fluid was lost when removing the halfshaft.

14. Road test the vehicle.

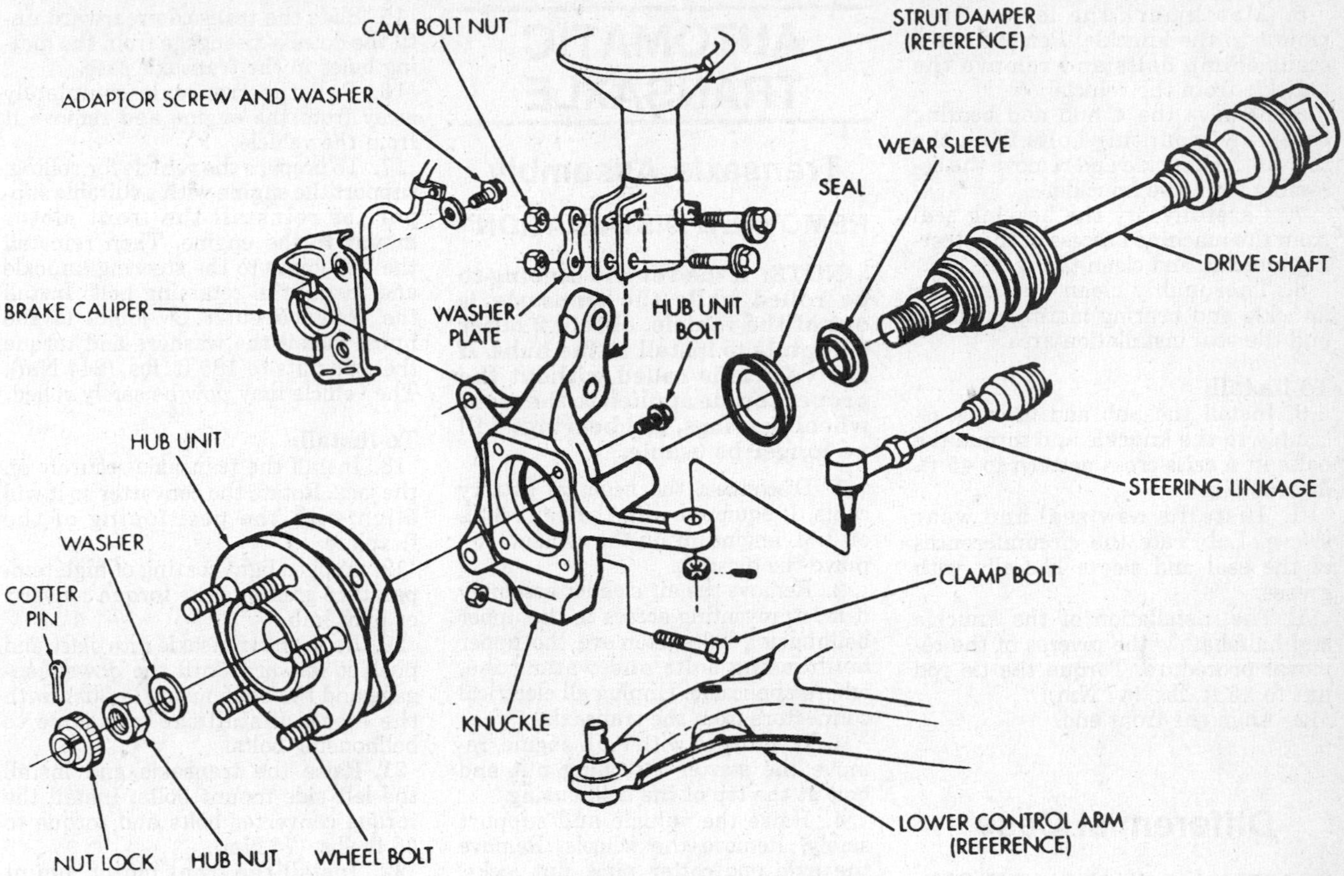

CAM BOLT NUT

STRUT DAMPER
(REFERENCE)

ADAPTOR SCREW AND WASHER

WEAR SLEEVE

SEAL

DRIVE SHAFT

BRAKE CALIPER

WASHER
PLATE

HUB UNIT
BOLT

STEERING LINKAGE

HUB UNIT

WASHER

COTTER
PIN

CLAMP BOLT

KNUCKLE

NUT LOCK   HUB NUT   WHEEL BOLT

LOWER CONTROL ARM
(REFERENCE)

Typical front suspension components

Front Wheel Hub and Bearing

REMOVAL & INSTALLATION

Pressed In (Two-Piece Hub and Bearing)

NOTE: Some hub and bearing replacement packages include the one-piece unit described below. If this is the case, follow the installation steps for one-piece unit instead of for the two-piece unit described here.

1. Raise the vehicle and support safely.
2. Remove the wheel. Remove the brake caliper from the adaptor and remove the adaptor. Remove the brake disc.
3. Remove the halfshaft.

NOTE: Knuckle removal is not necessary for bearing and hub replacement.

4. Disconnect the tie rod from the knuckle.
5. Matchmark the lower strut mount to the knuckle. Remove the 2 strut clamp bolts and remove the knuckle from the vehicle.
6. Attach the hub removal tool C–4811 or equivalent, and the triangular adapter to the 3 rear threaded holes of the steering knuckle housing with the thrust button inside the hub bore.
7. Tighten the bolt in the center of the tool, to press the hub from the steering knuckle. Remove the removal tools.
8. Remove the bolts and bearing retainer from the outside of the steering knuckle.
9. Carefully pry the bearing seal from the machined recess of the steering knuckle and clean the recess.
10. Insert the tool C–4811 or equivalent, through the hub bearing and install bearing removal adapter to the outside of the steering knuckle. Tighten the tool to press the hub bearing from the steering knuckle. Discard the bearing and the seal.

To install:

11. Use tool C–4811 or equivalent and the bearing installation adapter to press the hub bearing into the steering knuckle.
12. Install a new seal, the bearing retainer and the bolts to the steering knuckle. Torque the bearing retainer bolts to 20 ft. lbs.
13. Use the tool C–4811 or equivalent and the hub installation adapter, to press the hub into the hub bearing.
14. Using the bearing installation tool C–4698 or equivalent, drive the new dust seal into the rear of the steering the hub and bearing from the knuckle, as required.
15. The installation of the knuckle and halfshaft is the reverse of the removal procedure. Torque the tie rod nut to 35 ft. lbs. (47 Nm).
16. Align the front end.

Bolt In (One-Piece Hub and Bearing)

NOTE: Knuckle removal is not necessary for bearing and hub replacement.

1. Raise the vehicle and support safely.
2. Remove the wheel. Remove the brake caliper from the adaptor and remove the adaptor. Remove the brake disc.
3. Remove the halfshaft.
4. Disconnect the tie rod from the knuckle.

5. Matchmark the lower strut mount to the knuckle. Remove the 2 strut clamp bolts and remove the knuckle from the vehicle.

6. Remove the 4 hub and bearing assembly mounting bolts from the rear of the knuckle and remove the assembly from the knuckle.

7. Carefully pry the bearing seal from the machined recess of the steering knuckle and clean the recess.

8. Thoroughly clean and dry the knuckle and bearing mating surfaces and the seal installation area.

To install:

9. Install the hub and bearing assembly to the knuckle and torque the bolts in a criss-cross pattern to 45 ft. lbs. (65 Nm).

10. Install a new seal and wear sleeve. Lubricate the circumferences of the seal and sleeve liberally with grease.

11. The installation of the knuckle and halfshaft is the reverse of the removal procedure. Torque the tie rod nut to 35 ft. lbs. (47 Nm).

12. Align the front end.

Differential Case

REMOVAL & INSTALLATION

NOTE: The differential case can be removed from some vehicles with the transaxle installed. To do so, remove the halfshafts, remove the 2 K-frame mounting nuts and 2 bolts and lower the K-frame to provide enough room to pull the differential case out of its housing and over the lowered frame.

1. Disconnect the negative battery cable. Raise the vehicle and support safely.

2. Remove the right side extension housing. Remove the differential cover.

3. Remove the bolts and remove the right side differential bearing retainer using tool L–4435 or equivalent.

4. Remove the differential case from the transaxle.

5. Use new seals and gasket material when assembling.

6. The installation is the reverse of the removal procedure. Torque the extension housing and bearing retainer bolts to 21 ft. lbs. (28 Nm).

7. Fill the transaxle with the proper oil.

8. Connect the negative battery cable and check the differential for proper operation.

AUTOMATIC TRANSAXLE

Transaxle Assembly

REMOVAL & INSTALLATION

NOTE: If the vehicle is going to be rolled while the transaxle is out of the vehicle, obtain 2 outer CV-joints to install to the hubs. If the vehicle is rolled without the proper torque applied to the front wheel bearings, the bearings will no longer be usable.

1. Disconnect the negative battery cable. If equipped with the 3.0L, 3.3L or 3.8L engine, drain the coolant. Remove the dipstick.

2. Remove the air cleaner assembly if it is preventing access to the upper bellhousing bolts. Remove the upper bellhousing bolts and water tube, where applicable. Unplug all electrical connectors from the transaxle.

3. If equipped with 2.5L engine, remove the starter attaching nut and bolt at the top of the bellhousing.

4. Raise the vehicle and support safely. Remove the wheels. Remove the axle end cotter pins, nut locks, spring washers and axle nuts.

5. Remove the ball joint retaining bolts and pry the control arm from the steering knuckle. Position a drainpan under the transaxle where the axles enter the differential or extension housing. Remove the axles from the transaxle or center bearing. Unbolt the center bearing and remove the intermediate axle from the transaxle, if equipped.

6. Drain the transaxle. Disconnect and plug the fluid cooler hoses. Disconnect the shifter and kickdown linkage from the transaxle, if equipped.

7. Remove the speedometer cable adaptor bolt and remove the adaptor from the transaxle.

8. Remove the starter. Remove the torque converter inspection cover, matchmark the torque converter to the flexplate and remove the torque converter bolts.

9. Using the proper equipment, support the weight of the engine.

10. Remove the front motor mount and bracket.

11. Position a suitable transaxle jack under the transaxle.

12. Remove the lower bellhousing bolts.

13. Remove the left side splash shield. Remove the transaxle mount bolts.

14. Carefully pry the transaxle from the engine.

15. Slide the transaxle rearward until the dowels disengage from the mating holes in the transaxle case.

16. Pull the transaxle completely away from the engine and remove it from the vehicle.

17. To prepare the vehicle for rolling, support the engine with a suitable support or reinstall the front motor mount to the engine. Then reinstall the ball joints to the steering knuckle and install the retaining bolt. Install the obtained outer CV-joints to the hubs, install the washers and torque the axle nuts to 180 ft. lbs. (244 Nm). The vehicle may now be safely rolled.

To install:

18. Install the transaxle securely on the jack. Rotate the converter so it will align with the positioning of the flexplate.

19. Apply a light coating of high temperature grease to the torque converter pilot hub.

20. Raise the transaxle into place and push it forward until the dowels engage and the bellhousing is flush with the block. Install the transaxle to bellhousing bolts.

21. Raise the transaxle and install the left side mount bolts. Install the torque converter bolts and torque to 55 ft. lbs. (74 Nm).

22. Install the front motor mount and bracket. Remove the engine and transaxle support fixtures.

23. Install the starter to the transaxle. Install the bolt finger tight, if equipped with 2.5L engine.

24. Install a new O-ring to the speedometer cable adaptor and install to the extension housing; make sure it snaps in place. Install the retaining bolt.

25. Connect the shifter and kickdown linkage to the transaxle, if equipped.

26. Install the axles and center bearing, if equipped. Install the ball joints to the steering knuckles. Torque the axle nuts to 180 ft. lbs. (244 Nm) and install new cotter pins. Install the splash shield and install the wheels. Lower the vehicle. Install the dipstick.

27. Install the upper bellhousing bolts and water pipe, if removed.

28. If equipped with 2.5L engine, install the starter attaching nut and bolt at the top of the bellhousing. Raise the vehicle again and tighten the starter bolt from underneath the vehicle. Lower the vehicle.

29. Connect all electrical wiring to the transaxle.

30. Install the air cleaner assembly, if it was removed. Fill the transaxle with the proper amount of Mopar ATF Plus Type 7176. Conventional Dexron®II with Mercon may be used if Type 7176 is not available.

31. Connect the negative battery ca-

ble and check the transaxle for proper operation.

UPSHIFT AND KICKDOWN LEARNING PROCEDURE

A–604 Ultra-Drive Transaxle

In 1989, the A–604 4 speed, electronic transaxle was introduced; it is the first to use fully adaptive controls. The controls perform their functions based on real time feedback sensor information. Although, the transaxle is conventional in design, its functions are controlled by the ECM.

Since the A–604 is equipped with a learning function, each time the battery cable is disconnected, the ECM memory is lost. In operation, the transaxle must be shifted many times for the learned memory to be reinputed in the ECM; during this period, the vehicle will experience rough operation. The transaxle must be at normal operating temperature when learning occurs.

1. Maintain constant throttle opening during shifts. Do not move the accelerator pedal during upshifts.
2. Accelerate the vehicle with the throttle ⅛–½ open.
3. Make fifteen-to-twenty ½, ⅔ and ¾ upshifts. Accelerating from a full stop to 50 mph each time at the aforementioned throttle opening is sufficient.
4. With the vehicle speed below 25 mph, make 5–8 wide open throttle kickdowns to 1st gear from either 2nd or 3rd gear. Allow at least 5 seconds of operation in 2nd or 3rd gear prior to each kickdown.
5. With the vehicle speed greater than 25 mph, make 5 part throttle to wide open throttle kickdowns to either 3rd or 2nd gear from 4th gear. Allow at least 5 seconds of operation in 4th gear (preferably at road load throttle) prior to performing the kickdown.

SHIFT LINKAGE ADJUSTMENT

1. Place the shifter in the **P** detent.
2. Loosen the clamp bolt on the gearshift cable bracket.
3. Pull the shift lever all the way to the front detent position and tighten the lock screw.
4. Check for proper neutral safety switch operation.

THROTTLE PRESSURE CABLE ADJUSTMENT

1. Run the engine until it reaches normal operating temperature.
2. Loosen the cable mounting bracket lock screw.

3. Position the bracket so both alignment tabs are touching the transaxle case surface and tighten the lock screws.
4. Release the cross lock on the cable assembly by pulling the cross lock up.
5. To ensure proper adjustment, the cable must be free to slide all the way toward the engine against its stop after the cross lock is released.
6. Move the transaxle throttle control lever fully clockwise and press the cross lock down until it snaps into position.
7. Road test the vehicle and check the shift points.

FRONT SUSPENSION

MacPherson Strut

REMOVAL & INSTALLATION

Except with Automatic Air Suspension

1. Remove the 3 mounting nuts from the shock tower under the hood.
2. Raise the vehicle and support safely.
3. Remove the brake hose bracket screw from the strut.
4. Matchmark the lower strut mount to the knuckle and remove the strut to knuckle bolts, nuts and nut plate.
5. The installation is the reverse of the removal procedure. Torque the upper mounting nuts to 20 ft. lbs. (27 Nm). Do not tighten the lower mounting bolts until the front end alignment has been completed.
6. Perform a front end alignment. Torque the strut to knuckle nuts to 75 ft. lbs. (100 Nm) plus ¼ turn.

Air Suspension Strut

REMOVAL & INSTALLATION

1. Disconnect the negative battery cable.
2. Raise the vehicle and support safely. Remove the wheel and tire assembly.
3. To disconnect the air line, pull back on the plastic ring and pull the air line from the fitting.
4. Disconnect the electrical leads from the solenoid and the height sensor.
5. The solenoid has a molded square tang that fits into stepped notches in

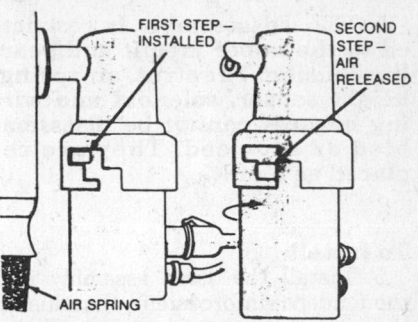

Air suspension spring solenoid positions

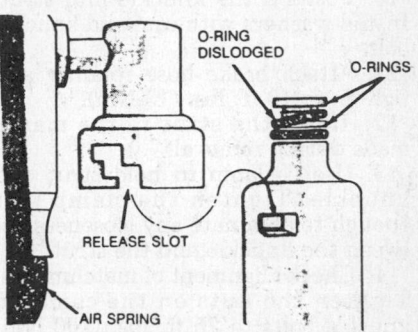

Air suspension solenoid removed

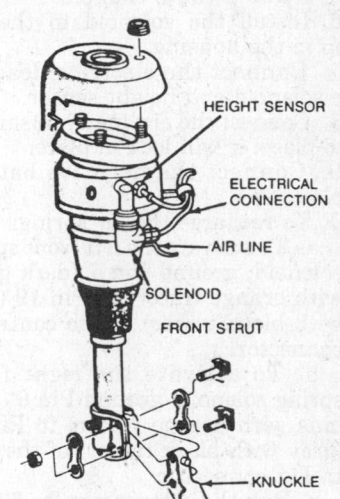

Air suspension strut assembly

the air spring housing to provide for exhaust and a retaining positions. To vent the air spring:

    a. Release the retaining clip.
    b. Rotate the solenoid to the first step in the housing and allow the air presure to vent.
    c. Rotate the solenoid farther to the release slot and remove it from the housing.

6. Matchmark the assembly to the knuckle.
7. Remove cam bolt, knuckle bolt, and washers. Disconnect the brake hose bracket retaining bolt.
8. Hold or support the strut. Remove the upper nuts from the shock tower. Remove the strut assembly.

NOTE: Disassembly is restricted to the upper mount and bearing housing. The strut, air spring, height sensor, solenoid and wiring harness cannot be disassembled or serviced. They are replaced as a unit.

To install:

9. Install the strut assembly into the fender reinforcement, then install the retaining nuts and washers. Tighten to 20 ft. lbs. (27 Nm).

10. Position the knuckle into strut. Install washers with cam and knuckle bolts.

11. Attach brake hose retainer and tighten to 10 ft. lbs. (13 Nm).

12. Index the strut to the marks made during removal.

13. Use C-clamp to hold strut and knuckle. Tighten the clamp just enough to eliminate any looseness between the knuckle and the strut.

14. Check alignment of matchmarks. Tighten the nuts on the cam and knuckle bolts to 75 ft. lbs. (100 Nm) plus ¼ turn.

15. Remove the C-clamp.

16. Install the solenoid to the top step in the housing.

17. Connect the electrical leads to the solenoid and height sensor.

18. Connect the air line by pushing it into place; it will lock in place.

19. Connect the negative battery cable.

20. To recharge the air spring:

   a. To activate the left front spring solenoid, ground Pin 7 (dark green with orange tracer) to Pin 19 (gray with black tracer) of the controller connector.

   b. To activate the right front spring solenoid, ground Pin 6 (dark blue with orange tracer) to Pin 19 (gray with black tracer) of the controller connector.

   c. Run the compressor for 60 seconds by jumping from pin No. 9 (black wire with red tracer) to pin No. 19 (gray wire with black tracer) of the controller connector.

   d. The air suspension controller is located behind the right side trunk trim panel.

21. Install the wheel and tire.

22. Check the system for proper operation.

Lower Ball Joints

INSPECTION

To inspect the ball joints, grasp the grease fitting by hand with the vehicle on the ground. If the grease fitting can be moved at all by hand, the ball joint should be replaced.

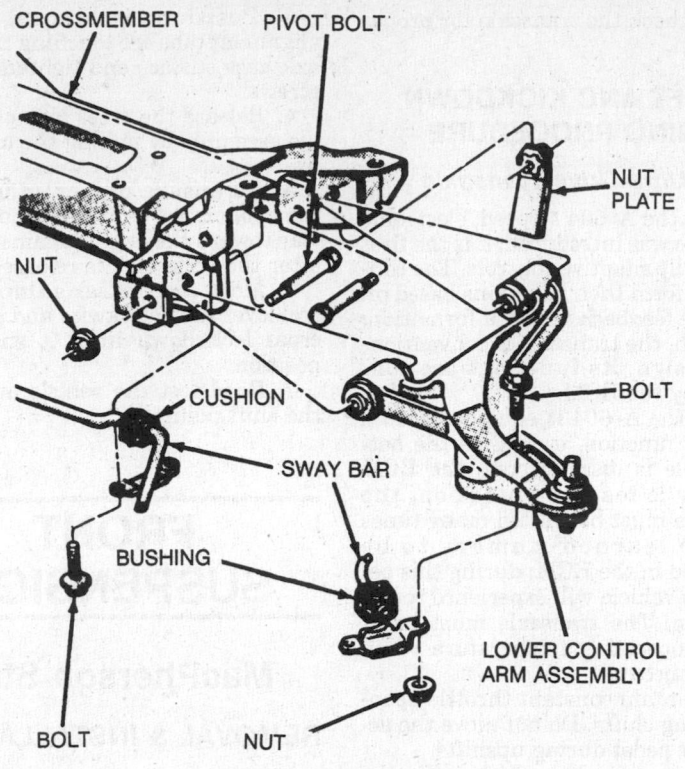

Lower control arm

REMOVAL & INSTALLATION

The ball joints are welded to the lower control arms. This necessitates replacement of the control arm assembly. Do not attempt to replace ball joints that are welded to the control arm; replacement control arms are equipped with a new ball joint.

Lower Control Arms

REMOVAL & INSTALLATION

1. Raise the vehicle and support safely. Remove the tire and wheel assembly.

2. Remove the sway bar.

3. Remove the ball joint stud retaining bolt and nut.

4. Pry the lower control arm from the steering knuckle.

5. Remove the control arm to crossmember bolts and nuts.

6. Remove the control arm from the vehicle.

7. The installation is the reverse of the removal procedure.

8. Lower the vehicle so the full weight of the vehicle is on the ground.

9. Torque the pivot bolts to 124 ft. lbs. (169 Nm).

10. Perform a front end alignment as required.

Sway Bar

REMOVAL & INSTALLATION

1. Raise the vehicle and support safely.

2. Remove the front sway bar brackets and retainers.

3. Remove the sway bar support brackets and bushings from the lower control arm. Remove the sway bar from the vehicle.

4. The installation is the reverse of the removal procedure. Lubricate the sway bar bushings liberally with grease before assembling.

REAR SUSPENSION

Shock Absorbers

REMOVAL & INSTALLATION

1. Raise the vehicle and support safely. Disconnect the height sensor and air line, if equipped. The air line is released by pulling back on the plastic retaining ring.

2. Remove the bolts that attach the shock to the frame or bracket.

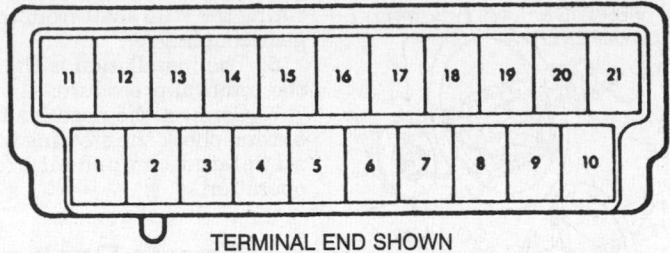

TERMINAL END SHOWN

Automatic air suspension controller connector terminals (terminal end)

3. Remove the shock from the vehicle.

4. The installation is the reverse of the removal procedure.

Coil Springs

REMOVAL & INSTALLATION

1. Raise the vehicle and support safely.

2. Using the proper equipment, support the weight of the rear axle.

3. Remove the bolts that attach the shock to the lower mounting bracket.

4. Lower the axle and remove the coil spring from the vehicle.

5. The installation is the reverse of the removal procedure.

Air Springs

REMOVAL & INSTALLATION

1. Disconnect the negative battery cable.

2. Raise the vehicle and support safely. Remove the wheel.

3. To disconnect the air line, pull back on the plastic ring and pull the air line from the fitting.

4. Disconnect the electrical leads from the solenoid and the height sensor.

5. The solenoid has a molded square tang that fits into stepped notches in the air spring housing to provide for exhaust and a retaining positions. To vent the air spring:

   a. Release the retaining clip.

   b. Rotate the solenoid to the first step in the housing and allow the air presure to vent.

   c. Rotate the solenoid farther to the release slot and remove it from the housing.

6. Release the upper air spring alignment/retaining clips.

7. Remove the nut that attaches the lower portion of the spring to the axle.

8. Pry the assembly down, pull the alignment studs through the retaining clips and remove the assembly from the vehicle.

To install:

9. Position the lower stud into its seat in the axle and the upper align-ment pins through the frame rail adaptor. Install the retaining clips.

10. Loosely install the lower mounting nut.

11. Install the solenoid to the top step in the housing.

12. Connect the electrical lead to the solenoid.

13. Connect the air line by pushing it into place; it will lock in place.

14. Connect the negative battery cable.

15. To partially recharge the air spring:

a.
To activate the right rear spring sole-noid, ground Pin 10 (light green with orange tracer) to Pin 19 (gray with black tracer) of the controller connector.

b.
Run the compressor for 60 seconds by jumping from pin No. 9 (black wire with red tracer) to pin No. 19 (gray wire with black tracer) of the controller connector.

c.
The air suspension controller is located behind the right side trunk trim panel.

15. When the air spring is properly inflated, torque the lower mounting nut to 50 ft. lbs. (68 Nm).

16. Install the wheel and tire.

17. Check the system for proper operation.

Rear Wheel Bearings

REMOVAL & INSTALLATION

1. Raise the vehicle and support safely.

2. Remove the tire and wheel assembly.

3. Remove the dust cap.

4. Remove the cotter pin, nut lock and nut.

5. Remove the thrust washer and the outer wheel bearing.

6. Remove the drum with the inner wheel bearing and the grease seal.

7. Remove the grease seal and re-move the inner bearing.

To install:

8. Lubricate the inner bearing and install to the drum.

9. Install a new grease seal.

10. Install the drum to the vehicle.

11. Lubricate and install the outer wheel bearing to the spindle.

12. Install the thrust washer.

13. Install and tighten the wheel bearng nut to 20–25 ft. lbs. (27–34 Nm) while rotating the drum.

14. Back off the adjusting nut ¼ turn then tighten it finger-tight.

15. Install the nut lock and a new cotter pin.

Rear Axle Assembly

REMOVAL & INSTALLATION

1. Raise the vehicle and support safely.

2. Disconnect the parking brake ca-ble at the connection.

3. Disconnect the brake tubes from the hoses and unclip the brake tubes from the axle housing. Disconnect the rear wheel speed sensors, if equipped with anti-lock brakes.

4. Disconnect the link from the sen-sor to the track bar used for automatic load leveling system, if equipped. Re-move the rear air spring, if equipped.

5. Using the proper equipment, support the weight of the axle.

6. Unbolt the shock absorbers and remove the track bar to axle pivot bolt. Suspend the track bar with a wire.

7. Lower the axle and remove the coil springs.

8. Remove the axle from the vehicle.

9. The installation is the reverse of the removal procedure.

STEERING

Steering Wheel

—— CAUTION ——

On vehicles equipped with an air bag, the negative battery cable must be disconnect-ed, before working on the system. Failure to do so may result in deployment of the air bag and possible personal injury.

REMOVAL & INSTALLATION

Without Air Bag

1. Disconnect the negative battery cable.

2. Straighten the steering wheel so the front tires are pointing straight forward.

3. Remove the horn pad retaining screws, unplug the connector and re-move the horn pad.

4. Remove the steering wheel hold-

down nut and remove the damper, if equipped. Matchmark the steering wheel to the shaft.

5. Using a suitable steering wheel puller, pull the steering wheel off of the shaft.

6. The installation is the reverse of the removal procedure. Torque the hold-down nut to 45 ft. lbs. (60 nm).

With Air Bag

1. Disconnect the negative battery cable.

2. Straighten the steering wheel so the front tires are pointing straight forward.

3. Remove the 4 nuts located on the back side of the steering wheel that attach the airbag module to the steering wheel.

4. Lift the module and disconnect the connectors. Remove the speed control switch, if equipped.

NOTE: All columns except Acustar are equipped with a clockspring set screw held by a plastic tether on the steering wheel. Acustar mounted clocksprings are auto-locking. If the steering column is not an Acustar and is lacking the set screw, obtain one before proceeding.

5. If equipped with the set screw, place it in the clockspring to ensure proper positioning when the steering wheel is removed.

6. Remove the steering wheel hold-down nut and damper, if equipped. Matchmark the steering wheel to the shaft.

7. Using a suitable steering wheel puller, pull the steering wheel off of the shaft.

To install:

8. Position the steering wheel on the steering column. Make sure the flats on the hub of the steering wheel are aligned with the formations on the clockspring.

9. Pull the air bag and speed control connectors through the lower, larger hole in the steering wheel and pull the horn wire through the smaller hole at the top. Make sure the wires are not pinched anywhere.

10. Install the damper, if equipped.

11. Install the hold-down nut and torque to 45 ft. lbs. (60 Nm).

12. If equipped with a clockspring set screw, remove the screw and place it in its storage location on the steering wheel.

13. Connect the horn wire.

14. Connect the speed control wire and install the speed control switch.

15. Connect the clockspring lead wire to the airbag module and install module to steering wheel.

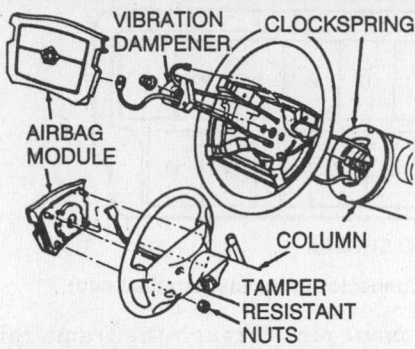

Air bag module and related components

NOTE: Do not allow anyone to enter the vehicle from this point on, until this procedure is completed.

16. Connect the DRB II to the Air bag System Diagnostic Module (ASDM) connector located to the right of the console.

17. From the passenger side of the vehicle, turn the key to the **ON** position.

18. Check to make sure nobody has entered the vehicle. Connect the negative battery cable.

19. Using the DRB II, read and record any active fault data or stored codes.

20. If any active fault codes are present, perform the proper diagnostic procedures before continuing.

21. If there are no active fault codes, erase the stored fault codes. If there are active codes, the stored codes will not erase.

22. From the passenger side of the vehicle, turn the key **OFF**, then **ON** and observe the instrument cluster airbag warning light. It should come on for 6–8 seconds, then go out, indicating the system is functioning normally. If the warning light either fails to come **ON** or stays lit, there is a system malfunction and the proper diagnostic procedures should be performed.

Steering Column

REMOVAL & INSTALLATION

1. Disconnect the negative battery cable.

2. Remove trim bezel, steering column cover and lower reinforcement, as required.

3. Disconnect all wiring connectors from below the instrument panel that lead up into the steering column.

4. Remove the nuts that attach the steering column assembly to the instrument panel support.

5. Firmly grasp the steering wheel and pull the steering column out, sepa-

rating the stub shaft from the steering gear coupling.

6. The installation is the reverse of the removal procedure.

7. Connect the negative battery cable and check the steering column and all related components for proper operation.

Power Rack and Pinion Steering Gear

REMOVAL & INSTALLATION

1. Disconnect the negative battery cable.

2. Raise the vehicle and support safely.

3. Remove the front wheels.

4. Remove the cotter pins, castellated nuts and tie rod ends from the steering knuckles.

5. The lower universal joint is removed with the steering gear.

6. Disconnect and plug the oil pressure line from the rack. Disconnect and plug the return hose from the line coming from the rack.

7. Remove the front suspension crossmember attaching bolts and nuts.

8. Lower the crossmember.

9. Remove the tie rod inner boot shields.

10. Remove the steering gear bolts from the front suspension crossmember.

11. Remove the steering gear from the left side of the vehicle.

To install:

12. Transfer the required parts to the new rack, if replacing it.

13. Place the rack on the crossmember and torque the steering gear attaching bolts to 21 ft. lbs. (29 Nm). Attach the fluid lines and the boot shields.

14. Have a helper inside the vehicle remove the trim boot and align the stub shaft with the coupling while the crossmember is raised into position. If a helper is not available, the steering column will have to be unbolted so the steering shaft can be inserted into the coupling. The right rear crossmember bolt is a pilot bolt that correctly locates the crossmember; tighten it first. Torque the crossmember bolts to 90 ft. lbs.

15. Install the tie rod ends to the steering knuckle and torque the nut to 45 ft. lbs. (61 Nm). Install a new cotter pin.

16. Insert the stub shaft shim where the stub shaft goes into the coupling.

17. Refill the power steering pump.

18. Connect the negative battery cable and check the gear for proper operation.

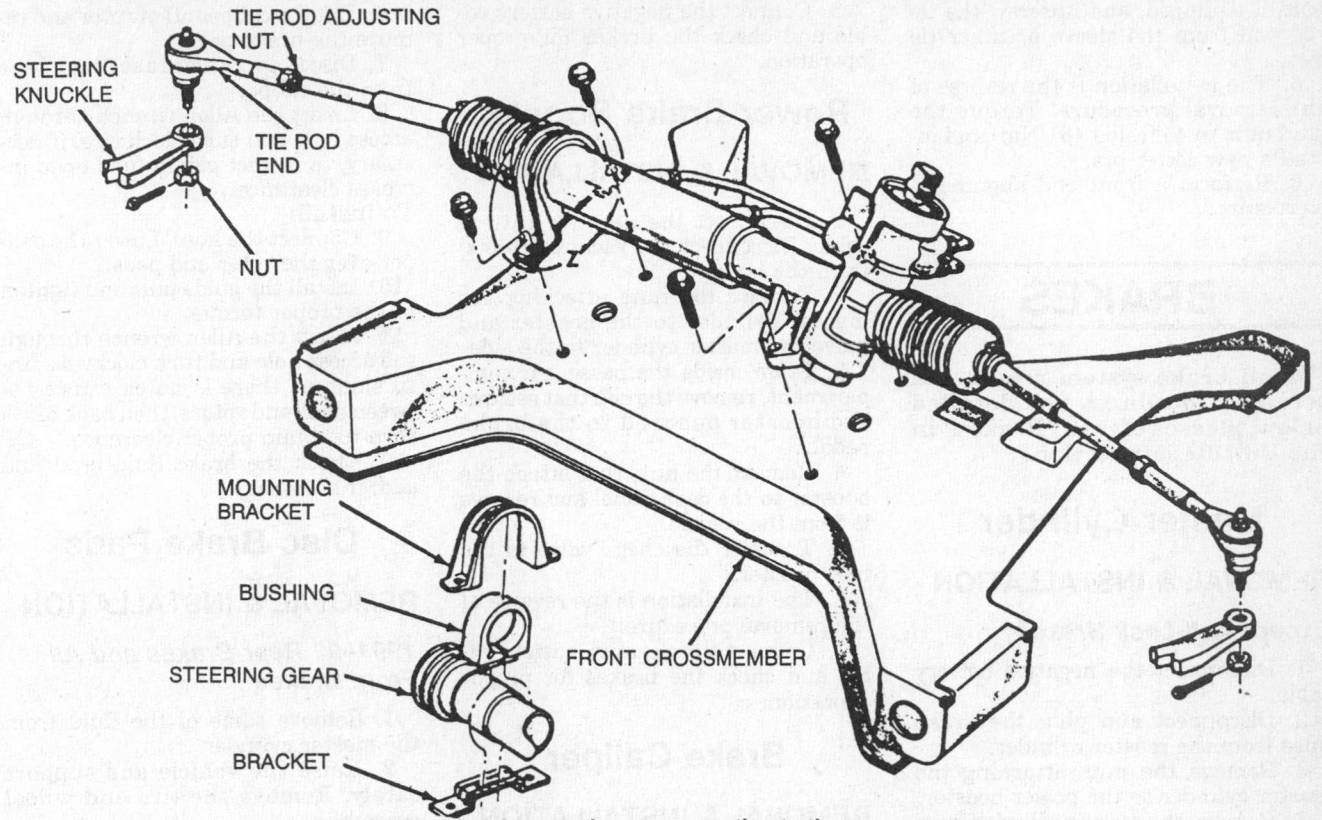

Rack and pinion steering gear mounting to the crossmember

Power Steering Pump

REMOVAL & INSTALLATION

1. Disconnect the negative battery cable.
2. Disconnect the vapor seperator hose from the throttle body.
3. Remove the drive belt from the pulley and remove bolts that are accessible from above.
4. Raise the vehicle and support safely. Remove the right front wheel and the splash shield, if necessary.
5. Position a drain pan under the return hose and disconnect the return hose from the line coming from the steering rack. Allow the fluid to drain and plug the hose.
6. Disconnect the pressure line from the pump.
7. Remove the lower mounting bolts and nuts.
8. Move the pump toward the firewall and remove the adjustment bracket.
9. Rotate the pump clockwise so the pump pulley faces the rear of the vehicle and pull the pump out of the engine compartment.
To install:
10. Position the pump and install the adjustment bracket.
11. Raise the vehicle and install the lower mounting bolts and nuts finger-tight.

12. Replace the O-rings on the pressure line and connect to the pump. Install the return hose and secure the clamps.
13. Wrap the drive belt around the pulley groove and adjust the belt tension, if possible. Tighten the mounting bolts to 30 ft. lbs. (41 Nm).
14. Install the splash shield, if removed.
15. Connect the vapor seperator to the throttle body.
16. Turn the wheels to the full left turn position and add fluid until the reservoir is full.
17. Start the engine and add fluid to bring the level to the correct level.
18. To purge the system of air, turn the steering wheel from side to side without contacting the stops.
19. Return the wheel to the straight-ahead position and operate the engine for 2 minutes before road testing.

BELT ADJUSTMENT

NOTE: The belt tension is automatically adjusted by a dynamic tensioner on 3.0L, 3.3L and 3.8L engines. Adjustment is not possible.

1. Loosen the bracket mounting bolts.
2. On 1988–89 vehicles, use a ½ in. drive breaker bar in the square hole provided in the bracket to move the

pump away from the engine. On 1990–91 vehicles, tighten the adjusting nut until the pump is in the desired position. Do not pry against the fluid reservoir.
3. With the pump moved enough so the belt deflects about ¼–½ in. under a 10 lb. load, tighten the bolts.

SYSTEM BLEEDING

1. Fill the reservoir with power steering fluid.
2. Turn the wheels to the full left turn position and add fluid until the reservoir is full.
3. Start the engine and add fluid to bring the level to the correct level.
4. To purge the system of air, turn the steering wheel from side to side without contacting the stops.
5. Return the wheel to the straight-ahead position and operate the engine for 2 minutes before road testing.

Tie Rod Ends

REMOVAL & INSTALLATION

1. Raise the vehicle and support safely.
2. Remove the cotter pin and nut from the tie rod end.
3. Using a suitable puller, remove the tie rod from the steering knuckle.
4. Loosen the sleeve clamp nut and

bolt, if equipped, and unscrew the tie rod end from the sleeve or inner tie rod.

5. The installation is the reverse of the removal procedure. Torque the stud nuts to 45 ft. lbs. (61 Nm) and install a new cotter pin.

6. Perform a front end alignment, as required.

BRAKES

For all brake system repair and service procedures not detailed below, please refer to "Brakes" in the Unit Repair section.

Master Cylinder

REMOVAL & INSTALLATION

Except Anti-Lock Brakes

1. Disconnect the negative battery cable.

2. Disconnect and plug the brake lines from the master cylinder.

3. Remove the nuts attaching the master cylinder to the power booster.

4. Remove the master cylinder from the mounting studs.

5. Remove the fluid reservoir from the cylinder.

To install:

6. Bench bleed the master cylinder.

7. Install to the studs and install the nuts.

8. Install the brake lines to the master cylinder.

9. Connect the negative battery cable and check the brakes for proper operation.

Combination Valve

REMOVAL & INSTALLATION

1. Disconnect the negative battery cable.

2. Raise the vehicle and support safely.

3. Tag and disconnect the brake lines from the valve.

4. Disconnect the wires to the pressure switch.

5. Remove the combination valve from the frame bracket.

6. The installation is the reverse of the removal procedure.

7. Bleed the brakes in the following order:

   a. Right rear wheel cylinder or caliper

   b. Left rear wheel cylinder or caliper

   c. Right front caliper

   d. Left front caliper

8. Connect the negative battery cable and check the brakes for proper operation.

Power Brake Booster

REMOVAL & INSTALLATION

1. Disconnect the negative battery cable. Disconnect the vacuum hose(s) from the booster.

2. Remove the nuts attaching the master cylinder to the booster and move the master cylinder to the side.

3. From inside the passenger compartment, remove the clip that secures the booster pushrod to the brake pedal.

4. Remove the nuts that attach the booster to the dash panel and remove it from the vehicle.

5. Transfer the check valve to the new booster.

6. The installation is the reverse of the removal procedure.

7. Connect the negative battery cable and check the brakes for proper operation.

Brake Caliper

REMOVAL & INSTALLATION

Front Brakes Except with ABS

1. Raise the vehicle and support safely.

2. Remove the tire and wheel assembly.

3. Remove the caliper mounting pins.

4. Lift the caliper off of the rotor. Remove the outer pad from the caliper.

5. Remove the brake hose retaining bolt from the caliper.

To install:

6. Install the brake hose to the caliper using new copper washers.

7. Position the caliper over the rotor so the caliper engages the adaptor correctly. Install the mounting pins.

8. Fill the master cylinder and bleed the brakes.

Rear Brakes

1988 VEHICLES

1. Remove ⅔ of brake fluid from the master cylinder.

2. Remove access plug and insert a 4mm Allen wrench through hole.

3. Turn the retraction shaft counterclockwise a few turns to increase clearance between pads and rotor.

4. Remove the anti-rattle spring from the outboard pad taking care not to damage it.

5. Back the caliper guide pins out just enough to free caliper from adapter.

6. Lift the caliper off of rotor and remove the brake pads.

7. Disconnect the brake fluid hose from the caliper.

8. Insert the Allen wrench through access hole and turn clockwise, if necessary, to retract piston further to increase clearance.

To install:

9. Connect the hose. Lower the caliper over the rotor and pads.

10. Install the guide pins and tighten to the proper torque.

11. Insert the Allen wrench through the access hole and turn clockwise until snug, so there is no clearance between pads and rotors, then back off ⅓ turn to obtain proper clearance.

12. Check the brake fluid level and add, if necessary.

Disc Brake Pads

REMOVAL & INSTALLATION

1991–92 Rear Brakes and All Front Brakes

1. Remove some of the fluid from the master cylinder.

2. Raise the vehicle and support safely. Remove the tire and wheel assemblies.

3. Remove the caliper and remove the outer pad from the caliper.

4. Remove the inner pad from the adaptor.

To install:

5. Use a large C-clamp to compress the piston back into the caliper bore.

6. Install the inner pad to the adaptor.

7. Position the caliper over the rotor so the caliper engages the adaptor correctly and install the retainer pin(s).

8. Install the hold-down spring, if it was removed.

9. Refill the master cylinder.

Rear Brakes

1988 VEHICLES

1. Remove ⅔ of brake fluid from the master cylinder.

2. Remove access plug and insert a 4mm Allen wrench through hole.

3. Turn the retraction shaft counterclockwise a few turns to increase clearance between pads and rotor.

4. Remove the anti-rattle spring from outboard pad taking care not to damage it.

5. Back the caliper guide pins out just enough to free caliper from adapter.

6. Lift the caliper off of rotor and carefully suspend with wire.

7. Remove the brake pads.

8. Insert the Allen wrench through access hole and turn clockwise, if nec-

essary to retract piston further to increase clearance for new pads.

To install:

9. Install new inner and outer pads.

NOTE: The outboard pads are marked for right and left hand sides and must be properly installed.

10. Lower the caliper over rotor and pads.

11. Install the guide pins and tighten to proper torque.

12. Insert the Allen wrench through the access hole and turn clockwise until snug, so there is no clearance between pads and rotors, then back off ⅓ turn to obtain proper clearance.

13. Check the brake fluid level and add, if necessary.

Brake Rotor

REMOVAL & INSTALLATION

1. Raise the vehicle and support safely. Remove the tire and wheel assembly.

2. Remove the caliper and brake pads.

3. Remove the factory installed clips, if equipped. It is not necessary to reinstall these clips.

4. Remove the adaptor, if necessary. Remove the rotor from the hub.

5. The installation is the reverse of the removal procedure.

Brake Drum

REMOVAL & INSTALLATION

1. Raise the vehicle and support safely.

2. Remove the wheel and tire assembly.

3. Remove the dust cap.

4. Remove the cotter pin and nut lock.

5. Remove the wheel bearing nut and washer from the spindle.

6. Remove the outer wheel bearing.

7. Remove the drum with the inner wheel bearing from the spindle. If the drum is difficult to remove, remove the plug from the rear of the backing plate and push the self adjuster lever away from the star wheel. Rotate the star wheel to retract the shoes. Remove the grease seal.

To install:

8. Lubricate and install the inner wheel bearing. Install a new grease seal.

9. Install the drum to the spindle.

10. Lubricate and install the outer wheel bearing, washer and nut. When the bearing preload is properly set, install the nut lock and a new cotter pin.

11. Install the grease cap.

12. Install the wheel and tire assembly. Adjust the rear brakes as required.

Brake Shoes

REMOVAL & INSTALLATION

1. Raise the vehicle and support safely. Remove the wheel and tire assemblies and the drums.

2. Remove the automatic adjuster spring and lever.

3. Rotate the automatic adjuster star wheel enough so both shoes move out far enough to be free of the wheel cylinder boots.

4. Disconnect the parking brake cable from the actuating lever.

5. Remove the lower shoe-to-shoe or shoe-to-anchor spring(s).

6. With the shoes held together by the upper shoe-to-shoe spring, remove them from the backing plate.

To install:

7. Thoroughly clean and dry the backing plate. To prepare the backing plate, lubricate the bosses, anchor pin and parking brake actuating lever pivot surface lightly with lithium based grease.

8. Remove, clean and dry all parts still on the old shoes. Lubricate the star wheel shaft threads with anti-sieze lubricant and transfer all parts to their proper locations on the new shoes.

9. Install the lower spring(s).

10. Connect the parking brake cable.

11. Install the automatic adjuster lever and spring.

12. Adjust the star wheel.

13. Remove any grease from the linings and install the drum.

14. Complete the brake adjustment with the wheels installed.

Wheel Cylinder

REMOVAL & INSTALLATION

1. Raise the vehicle and support safely.

2. Remove the wheel, drum and brake shoes.

3. Remove and plug the brake line from the wheel cylinder.

4. Remove the wheel cylinder bolts and remove the cylinder from the backing plate.

To install:

5. Apply a very thin coating of silicone sealer to the cylinder mounting surface, install the cylinder to the backing plate and install the retaining bolts.

6. Connect the brake line to the wheel cylinder.

7. Install all brake parts that were removed.

8. Install the tire and wheel assembly.

9. Bleed the brakes.

Parking Brake Cables

ADJUSTMENT

Except Anti-Lock Brakes

1. Release the parking brakes fully.

2. Raise the vehicle and support safely.

3. Adjust the rear brakes.

4. Loosen the adjusting nut until there is slack in all the cables.

5. Rotate the rear wheels and tighten the cable adjusting nut until there is a slight drag at the wheels.

6. Continue to rotate the rear wheels and loosen the nut until all drag is eliminated.

7. Back off the nut an additional 2 turns.

8. Apply and release the parking brake several times. Upon the least release, verify there is no drag at the wheels.

9. To check the operation, make sure the parking brake holds on an incline.

Anti-Lock Brakes

1. Fully release the parking brakes and pump the brakes several times. Raise the vehicle and support safely.

2. Tighten the cable adjusting nut until a very slight drag is felt at each rear wheel.

3. Loosen the adjusting nut 5 turns.

4. Actuate the parking brake lever on the rear calipers by manually pulling down and releasing each rear parking brake cable at the rear of the vehicle.

5. The parking brake lever should be touching the stop pin on both rear calipers. If not, loosen the adjusting nut 1 turn.

6. Repeat Steps 4 and 5 until the parking brake lever returns against the stop pin on both calipers.

7. When the adjustment is complete, the actuating levers on both calipers should return against the stop pins when the parking brakes are released and the wheels must rotate freely.

8. To confirm proper operation, make sure the parking brake holds on an incline.

REMOVAL & INSTALLATION

Front Cable

1. Disconnect the negative battery cable.

2. Loosen the adjusting nut and dis-

engage the front cable from the equalizer bracket.

4. Lift the carpet and floor matting and remove the floor pan seal.

5. Pull the cable end forward and disconnect from the clevis.

6. Pull the cable through the hole and remove.

7. The installation is the reverse of the removal procedure.

8. Adjust the cables, connect the negative battery cable and check the parking brakes for proper operation.

Rear Cable

DRUM BRAKES

1. Disconnect the negative battery cable.

2. Raise the vehicle and support safely.

3. Remove the rear wheels.

4. Back off the adjusting nut enough to provide slack in all cables. Disconnect the cables from the cable connectors.

5. Remove the drums. Disconnect the cable from the brake shoe lever.

6. Compress the retaining clips on the end of the cable housing and pull the cable from the backing plate.

7. Remove the retaining clip at the support bracket and remove the cable from the trailing arm assembly.

8. The installation is the reverse of the removal procedure.

9. Adjust the cables, connect the negative battery cable and check the parking brakes for proper operation.

DISC BRAKES

1. Disconnect the negative battery cable.

2. Raise the vehicle and support safely.

3. Remove the rear wheels.

4. Remove the brake cable retaining clips from the hanger bracket and caliper.

5. Disconnect the cable from the parking brake lever on the caliper.

6. Remove the cable guide attaching nut and screw.

7. Pull the cable assembly out from the hanger bracket and caliper.

8. The installation is the reverse of the removal procedure.

9. Adjust the cables, connect the negative battery cable and check the parking brakes for proper operation.

Brake System Bleeding

Except Anti-Lock Brakes

NOTE: If using a pressure bleeder, follow the instructions furnished with the unit and choose the correct adaptor for the application. Do not substitute an adapter that "almost fits" as it will not work and could be dangerous.

MASTER CYLINDER

If the master cylinder is off the vehicle it can be bench bled.

1. Connect 2 short pieces of brake line to the outlet fittings, bend them until the free end is below the fluid level in the master cylinder reservoirs.

2. Fill the reservoir with fresh brake fluid. Pump the piston slowly until no more air bubbles appear in the reservoirs.

3. Disconnect the 2 short lines, refill the master cylinder and securely install the cylinder caps.

4. If the master cylinder is on the vehicle, it can still be bled, using a flare nut wrench.

5. Open the brake lines slightly with the flare nut wrench while pressure is applied to the brake pedal by a helper inside the vehicle.

6. Be sure to tighten the line before the brake pedal is released.

7. Repeat the process with both lines until no air bubbles come out.

CALIPERS AND WHEEL CYLINDERS

1. Fill the master cylinder with fresh brake fluid. Check the level often during the procedure.

2. Starting with the right rear wheel, remove the protective cap from the bleeder, and place where it will not be lost. Clean the bleed screw.

CAUTION

When bleeding the brakes, keep face away from the brake area. Spewing fluid may cause facial and/or visual damage. Do not allow brake fluid to spill on the car's finish; it will remove the paint.

---

3. If the system is empty, the most effecient way to get fluid down to the wheel is to loosen the bleeder about ½–¾ turn, place a finger firmly over the bleeder and have a helper pump the brakes slowly until fluid comes out the bleeder. Once fluid is at the bleeder, close it before the pedal is released inside the vehicle.

NOTE: If the pedal is pumped rapidly, the fluid will churn and create small air bubbles, which are almost impossible to remove from the system. These air bubbles will eventually congregate and a spongy pedal will result.

4. Once fluid has been pumped to the caliper or wheel cylinder, open the bleed screw again, have the helper press the brake pedal to the floor, lock the bleeder and have the helper slowly release the pedal. Wait 15 seconds and repeat the procedure (including the 15 second wait) until no more air comes out of the bleeder upon application of the brake pedal. Remember to close the bleeder before the pedal is released inside the vehicle each time the bleeder is opened. If not, air will be induced into the system.

5. If a helper is not available, connect a small hose to the bleeder, place the end in a container of brake fluid and proceed to pump the pedal from inside the vehicle until no more air comes out the bleeder. The hose will prevent air from entering the system.

6. Repeat the procedure on remaining wheel cylinders in order:

   a. Left rear

   b. Right front

   c. Left front

7. Hydraulic brake systems must be totally flushed if the fluid becomes contaminated with water, dirt or other corrosive chemicals. To flush, bleed the entire system until all fluid has been replaced with new fluid.

8. Install the bleeder cap(s) on the bleeder to keep dirt out. Always road test the vehicle after brake work of any kind is done.

Anti-Lock Brakes

BOOSTER BLEEDING—BOSCH ABS III (1989–90)

1. The hydraulic accumulator must be depressurized.

2. Connect all pump/motor and hydraulic assembly electrical connections, if previously disconnected. Be sure that all brake lines and hose connections are tight.

3. Fill the reservoir to the full level.

4. Connect a transparent hose to the bleeder screw location on the right side of the hydraulic assembly. Place the other end of the hose into a clear container to receive brake fluid.

5. Open the bleeder screw ½–¾ of a turn.

6. Turn the ignition switch to the **ON** position. The pump/motor should run, discharging fluid into the container. After a good volume of fluid has been forced through the hose, an air-free flow in the plastic hose and container will indicate a good bleed.

7. Turn the ignition switch **OFF**.

NOTE: If the brake fluid does not flow, it may be due to a lack of prime to the pump/motor. Try shaking the return hose to break up air bubbles that may be present within the hose.

Should the brake fluid still not flow, turn the ignition switch to the OFF position. Remove the return hose from the reservoir and cap nipple on the reservoir. Manually fill the return hose with brake fluid and connect to the

reservoir. Repeat the bleeding process.

8. Remove the hose from the bleeder screw. Tighten the bleeder screw to 7.5 ft. lbs. (10 Nm). Do not overtighten.

9. Top off the reservoir to the correct fluid level.

10. Turn the ignition switch to the **ON** position. Allow the pump to charge the accumulator, which should stop after approximately 30 seconds.

Pressure Bleeding

The brake lines may be pressure bled, using a standard diaphragm type pressure bleeder. Only diaphragm type pressure bleeding equipment should be used to bleed the system.

1. The ignition should be turned **OFF** and remain **OFF** throughout this procedure.

2. Depressurize the hydraulic accumulator.

—— **CAUTION** ——

Failure to depressurize the hydraulic accumulator, prior to performing this operation may result in personal injury and/or damage to the painted surfaces.

3. Remove the electrical connector from fluid level sensor on the reservoir cap(s) and remove the reservoir cap(s).

4. Install the pressure bleeder adapter.

5. Attach the bleeding equipment to the bleeder adapter. Charge the pressure bleeder to approximately 20 psi (138 kPa).

6. Connect a transparent hose to the caliper bleed screw. Submerge the free end of the hose in a clear glass container, which is partially filled with clean, fresh brake fluid.

7. With the pressure turned **ON**, open the caliper bleed screw ½–¾ turn and allow fluid to flow into the container. Leave the bleed screw open until clear, bubble-free fluid slows from the hose. If the reservoir has been drained or the hydraulic assembly removed from the car prior to the bleeding operation, slowly pump the brake pedal 1–2 times while the bleed screw is open and fluid is flowing. This will help purge air from the hydraulic assembly. Tighten the bleeder screw to 7.5 ft. lbs. (10 N).

8. Repeat Step 7 at all calipers. Calipers should be bled in the following order:
   a. Left rear
   b. Right rear
   c. Left front
   d. Right front

9. After bleeding all 4 calipers, remove the pressure bleeding equipment and bleeder adapter by closing the pressure bleeder valve and slowly unscrewing the bleeder adapter from the hydraulic assembly reservoir. Failure to release pressure in the reservoir will cause spillage of brake fluid and could result in injury or damage to painted surfaces.

10. Using a syringe or equivalent method, remove excess fluid from the reservoir to bring the fluid level to full level.

11. Install the reservoir cap and connect the fluid level sensor connector. Turn the ignition **ON** and allow the pump to charge the accumulator.

Manual Bleeding

1. Depressurize the hydraulic accumulator.

—— **CAUTION** ——

Failure to depressurize the hydraulic accumulator, prior to performing this operation may result in personal injury and/or damage to the painted surfaces.

2. Connect a transparent hose to the caliper bleed screw. Submerge the free end of the hose in a clear glass container, which is partially filled with clean, fresh brake fluid.

3. Slowly pump the brake pedal several times, using full strokes of the pedal and allowing approximately 5 seconds between pedal strokes. After 2 or 3 strokes, continue to hold pressure on the pedal, keeping it at the bottom of its travel.

4. With pressure on the pedal, open the bleed screw ½–¾ turn. Leave the bleed screw open until fluid no longer flows from the hose. Tighten the bleed screw and release the pedal.

5. Repeat this procedure until clear, bubble-free fluid flows from the hose.

6. Repeat all steps at each of the calipers. Calipers should be bled in the following order:
   a. Left rear
   b. Right rear
   c. Left front
   d. Right front

Anti-Lock Brake System Service

PRECAUTIONS

Failure to observe the following precautions may result in system damage.

- Before performing electric arc welding on the vehicle, disconnect the Electronic Brake Control Module (EBCM) and the hydraulic modulator connectors.
- When performing painting work on the vehicle, do not expose the Electronic Brake Control Module (EBCM) to temperatures in excess of 185°F (85°C) for longer than 2 hrs. The system may be exposed to temperatures up to 200°F (95°C) for less than 15 min.
- Never disconnect or connect the Electronic Brake Control Module (EBCM) or hydraulic modulator connectors with the ignition switch ON.
- Never disassemble any component of the Anti-Lock Brake System (ABS) which is designated non-servicable; the component must be replaced as an assembly.
- When filling the master cylinder, always use brake fluid which meets DOT-3 specifications; petroleum-based fluid will destroy the rubber parts.

DEPRESSURIZING THE HYDRAULIC ACCUMULATOR

1. With the ignition **OFF**, pump the brake pedal a minimum of 40 times, using approximately 50 lbs. (222 N) pedal force. A noticeable change in pedal feel will occur when the accumulator is discharged.

2. When a definite increase in pedal effort is felt, stroke the pedal a few additional times. This should remove all hydraulic pressure from the system.

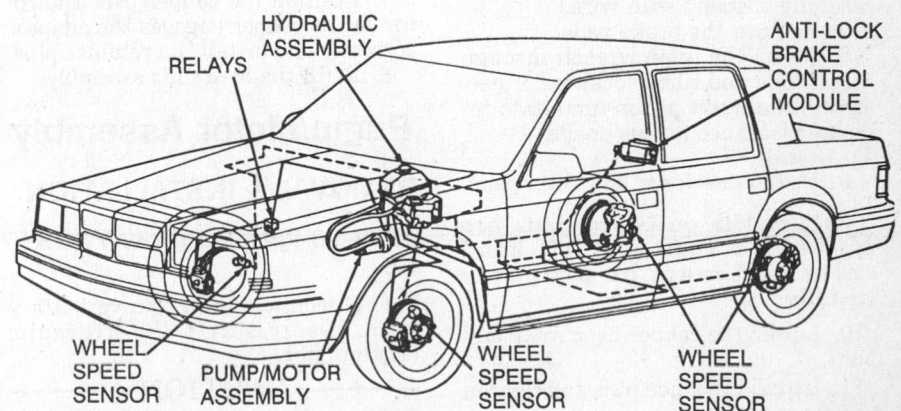

Bosch ABS III anti-lock brake system schematic – 1989–90 vehicles

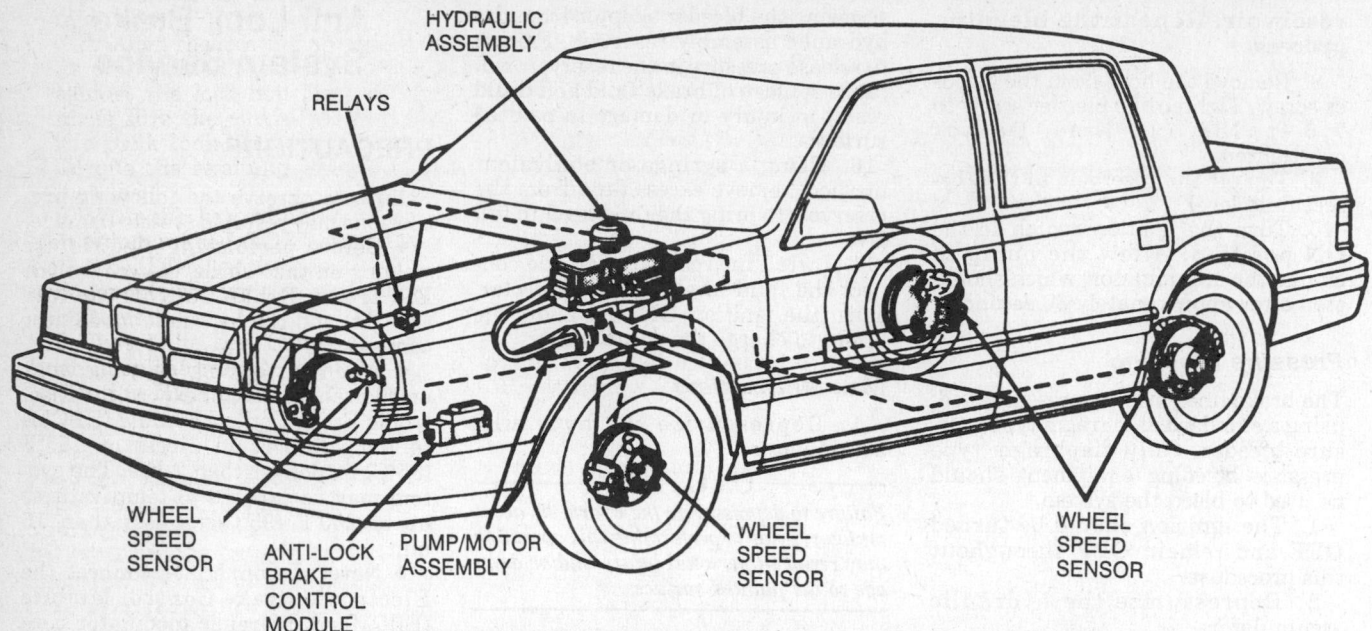

HYDRAULIC ASSEMBLY

RELAYS

WHEEL SPEED SENSOR

ANTI-LOCK BRAKE CONTROL MODULE

PUMP/MOTOR ASSEMBLY

WHEEL SPEED SENSOR

WHEEL SPEED SENSOR

Bendix System 10 anti-lock brake system schematic—1991–92 vehicles

Rear Disc Brake Caliper and Pads

REMOVAL & INSTALLATION

1989–90 Vehicles

1. Depressurize the hydraulic accumulator. Remove ⅔ of brake fluid from the hydraulic assembly fluid reservoir.
2. Remove the access plug and insert a 4mm Allen wrench through the hole.
3. Turn the retraction shaft counterclockwise a few turns to increase clearance between pads and rotor.
4. Remove the anti-rattle spring from outboard pad taking care not to damage it.
5. Back the caliper guide pins out just enough to free caliper from adapter.
6. Lift the caliper off of rotor and carefully suspend with wire.
7. Remove the brake pads.
8. Insert the Allen wrench through access hole and turn clockwise, if necessary, to retract piston further to increase clearance for new pads.
To install:
9. Install new inner and outer pads.

NOTE: **The outboard pads are marked for right and left hand sides and must be properly installed.**

10. Lower the caliper over rotor and pads.
11. Install the guide pins and tighten to proper torque.
12. Insert the Allen wrench through

the access hole and turn clockwise until snug, no clearance between pads and rotors, and back off ⅓ turn to obtain proper clearance.
13. Check the brake fluid level and add if necessary.

1991–92 Vehicles

1. Depressurize the hydraulic accumulator. Remove ⅔ of brake fluid from the hydraulic assembly fluid reservoir.
2. Raise the vehicle and support safely. Remove the tire and wheel assemblies.
3. Remove the caliper and outer pad from the caliper.
4. Remove the inner pad from the adaptor.
To install:
5. Use a large C-clamp to compress the piston back into the caliper bore.
6. Install the inner pad to the adaptor.
7. Position the caliper over the rotor so the caliper engages the adaptor correctly and install the retainer pins.
8. Refill the hydraulic assembly.

Pump/Motor Assembly

REMOVAL & INSTALLATION

1989–90 Vehicles (Bosch ABS III)

1. Disconnect the negative battery cable. Depressurize the hydraulic accumulator.

—— **CAUTION** ——
Failure to depressurize the hydraulic accumulator, prior to performing this operation

may result in personal injury and/or damage to the painted surfaces.

2. Remove the fresh air intake ducts.
3. Disconnect all electrical connectors to the pump motor.
4. Disconnect the high and low pressure hoses from the hydraulic assembly. Cap the spigot on the reservoir.
5. Disconnect the shift selection cable bracket from the transaxle and move it aside.
6. Loosen the nuts on the 2 studs that position the pump/motor to the transaxle differential cover.
7. Remove the retainer bolts that are used to mount hose bracket and pump/motor. The engine inlet water extension pipe is also held in position by these bolts.

NOTE: **Do not disturb the inlet water extension pipe, or engine coolant will leak out.**

8. Disconnect the wiring harness retaining clip from the hose bracket.
9. Lift the pump/motor assembly off of the studs and out of the vehicle.
10. Remove the heat shield from the pump/motor, if equipped and discard.
To install:
11. Place a new heat shield to the pump/motor bracket, using fasteners provided.
12. Install the pump/motor assembly in the reverse order of the removal.
13. Readjust the gearshift linkage, if it was disturbed.
14. Connect the negative battery cable and check the assembly for proper operation.

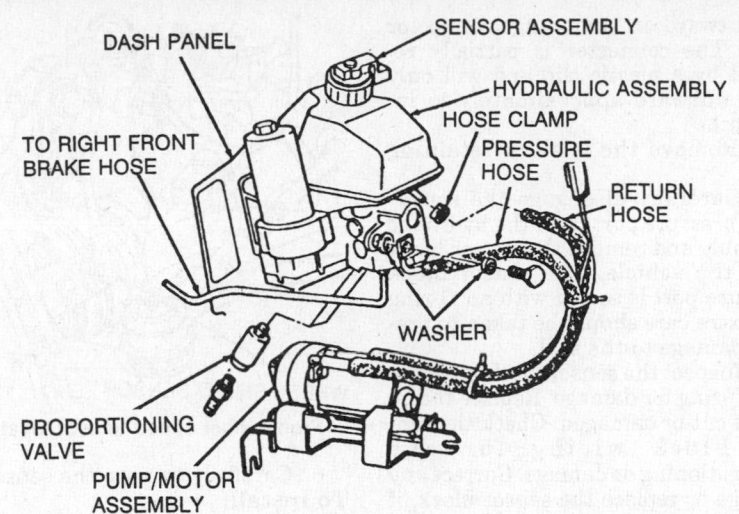

Hydraulic assembly, pump/motor assembly and related parts—1989–90 vehicles

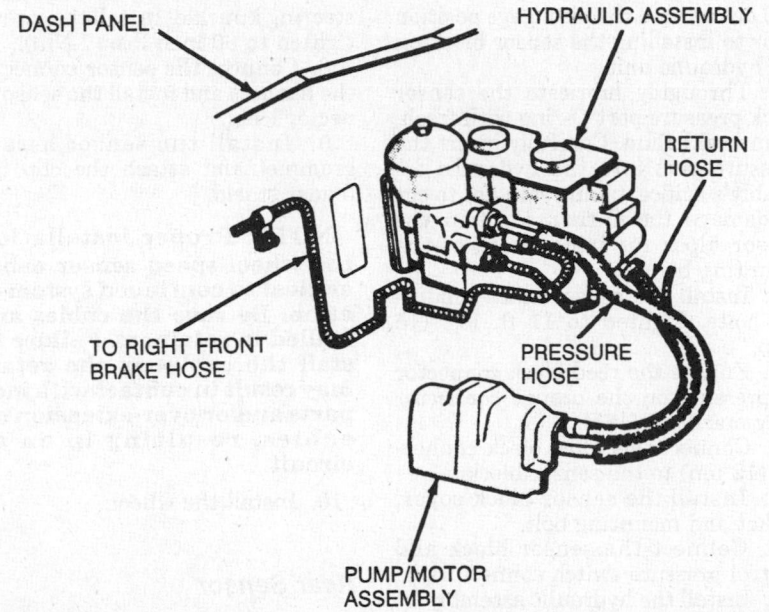

Hydraulic assembly, pump/motor assembly and related parts—1991–92 vehicles

1991–92 Vehicles (Bendix System 10)

1. Disconnect the negative battery cable.

— **CAUTION** —

Failure to depressurize the hydraulic accumulator, prior to performing this operation may result in personal injury and/or damage to the painted surfaces.

2. Depressurize the brake system.
3. Remove the fresh air intake ducts from the engine.
4. Remove the clip holding the high pressure line to the battery tray.
5. Disconnect the electrical connectors running across the engine compartment in the vicinity of the pump/motor high and low pressure hoses.

One of these connectors is the one for the pump/motor assembly.
6. Disconnect the high and low pressure hoses from the hydraulic assembly. Cap or plug the reservoir fitting.
7. Disconnect the pump/motor electrical connector from the engine mount.
8. Remove the heat shield bolt from the front of the pump bracket. Remove the heat shield.
9. Lift the pump/motor assembly from the bracket and out of the vehicle.

To install:
10. Fit the pump motor assembly onto the bracket; install the heat shield and its retaining bolt.

11. Install the pump/motor electrical connector to the engine mount.
12. Connect the high and low pressure hose to the hydraulic assembly. Tighten the high pressure line to 145 inch lbs. (16 Nm). Tighten the hose clamp on the low pressure hose to 10 inch lbs (1 Nm).
13. Connect the electrical connectors which were removed for access.
14. Install the high pressure line retaining clip to the battery tray if it was removed.
15. Install the fresh air intake ducts.
16. Bleed the brake system.

Hydraulic Assembly

REMOVAL & INSTALLATION

1. Disconnect the negative battery cable. Depressurize the hydraulic accumulator.

— **CAUTION** —

Failure to depressurize the hydraulic accumulator, prior to performing this operation may result in personal injury and/or damage to the painted surfaces.

2. Remove the fresh air intake ducts.
3. Disconnect all electrical connectors from the hydraulic unit and pump/motor.
4. Remove as much of the fluid as possible from the reservoir on the hydraulic assembly.
5. Remove the pressure hose fitting (banjo bolt) from the hydraulic assembly. Use care not to drop the 2 washers used to seal the pressure hose fitting to the hydraulic assembly inlet.
6. Disconnect the return hose from the reservoir nipple. Cap the spigot on the reservoir.
7. Disconnect all brake tubes from the hydraulic assembly.
8. Remove the driver's side sound insulation panel.
9. Disconnect the pushrod from the brake pedal by using a small, flat tool to release the retainer clip on the brake pedal pin. The center tang on the clip must be moved back enough to allow the lock tab to clear the pin. Disconnect the pushrod from the pedal pin.
10. Remove the 4 underdash hydraulic assembly mounting nuts.
11. Remove the hydraulic assembly.
To install:
12. Position the hydraulic assembly on the vehicle.
13. Install and torque the mounting nuts to 21 ft. lbs. (28 Nm).
14. Using Lubriplate® or equivalent, coat the bearing surface of the pedal pin.
15. Connect the pushrod to the pedal and install a new retainer clip.

16. Install the brake tubes. If the proportioning valves were removed from the hydraulic assembly, reinstall valves and tighten to 20 ft. lbs. (27 Nm).

17. Install the return hose to the nipple on the reservoir.

18. Install the pressure hose to the hydraulic assembly; be sure the 2 washers are in there proper position. Tighten the bango bolt to 13 ft. lbs. (18 Nm).

19. Fill the reservoir to the top of the screen.

20. Connect all electrical connectors to the hydraulic assembly.

21. Bleed the entire brake system.

22. Install the crosscar brace, if disturbed. Install the fresh air intake duct.

23. Connect the negative battery cable and check the assembly for proper operation.

Sensor Block

REMOVAL & INSTALLATION

1989–90 Vehicles (Bosch ABS III)

1. Disconnect the negative battery cable. Depressurize the hydraulic accumulator.

--- **CAUTION** ---

Failure to depressurize the hydraulic accumulator, prior to performing this operation may result in personal injury and/or damage to the painted surfaces.

2. Disconnect all electrical connectors from the reservoir on the hydraulic assembly.

3. Working from under the dash, disconnect the pushrod from the brake pedal.

4. Remove the driver's side sound insulator panel.

5. Remove the 4 hydraulic assembly mounting nuts.

6. Working from under the hood, pull the hydraulic assembly away from the dash panel and rotate the assembly enough to gain access to the sensor block cover.

NOTE: The brake lines should not be removed or deformed during this procedure.

7. Remove the sensor block cover retaining bolt and remove the sensor block cover. Care should be used not to damage the cover gasket during removal.

8. Disengage the locking tabs and disconnect the valve block connector (12 pin) from the sensor block.

9. Disengage the reed block connector, marked PUSH, by carefully pulling outward on the orange connector body. The connector is partially retained by a plastic clip and will only move outward approximately ½ in. (13mm).

10. Remove the 3 block retaining bolts.

11. Carefully disengage the sensor block pressure port from the hydraulic assembly and remove the sensor block from the vehicle. The sensor block pressure port is sealed with an O-ring and extra care should be taken to prevent damage to the seal.

12. Inspect the sensor block pressure port O-ring for damage. Replace the O-ring if cut or damaged. Check the sensor block wiring for any mispositioning or damage. Correct any damage or replace the sensor block, if damage cannot be corrected.

To install:

13. Pull the reed block connector (2 pin) outward to the disengage position prior to installing the sensor block on the hydraulic unit.

14. Throughly lubricate the sensor block pressure port O-ring with fresh, clean brake fluid. Carefully insert the pressure port into the hydraulic assembly's orifice, taking care not to cut or damage the O-ring. Position the sensor block for installation of the mounting bolts.

15. Install the sensor block mounting bolts. Tighten to 11 ft. lbs. (15 Nm).

16. Engage the reed block connector by pressing on the orange connector body marked **PUSH**.

17. Connect the valve block connector (12 pin) to the sensor block.

18. Install the sensor block cover, gasket and mounting bolt.

19. Connect the sensor block and control pressure switch connectors.

20. Install the hydraulic assembly by reversing the removal procedure.

21. Connect the negative battery cable and check the sensor block for proper operation.

Wheel Speed Sensors

REMOVAL & INSTALLATION

Front Sensor

1. Raise the vehicle and support safely. Remove the wheel and tire assembly.

2. Remove the screw from the clip that holds the sensor to the fender shield.

3. Carefully pull the sensor assembly grommet from the fender shield.

4. Unplug the connector from the harness. Remove the retaininer clip from the strut damper bracket.

5. Remove the sensor mounting screw.

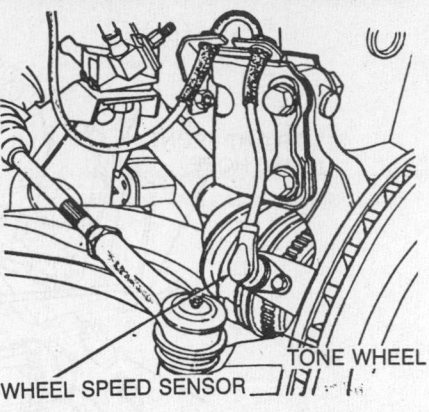

WHEEL SPEED SENSOR / TONE WHEEL

Front wheel speed sensor location

6. Carefully remove the sensor.

To install:

7. Coat the sensor with high temperature multi-purpose anti-corrosion compound before installing into the steering knuckle. Install the screw and tighten to 60 inch lbs. (7 Nm).

8. Connect the sensor connector to the harness and install the sensor connector lock.

9. Install the sensor assembly grommet and attach the clip to the fender shield.

NOTE: Proper installation of the wheel speed sensor cables is critical to continued system operation. Be sure the cables are installed in retainers. Failure to install the cables in the retainers may result in contact with moving parts and/or over-extension of the cables, resulting in an open circuit.

10. Install the wheel.

Rear Sensor

1. Raise the vehicle and support safely. Remove the wheel and tire assembly.

2. Carefully pull the sensor assembly grommet from the underbody and pull the harness through the hole.

3. Unplug the connector from the harness. Remove the retaininer clip from the strut damper bracket.

4. Remove the sensor spool grommet clip retaining screw from the body hose bracket, located in front of the inside of the trailing arm.

5. Remove the outboard sensor assembly retaining nut and sensor m009039g screw.

6. Carefully remove the sensor.

To install:

7. Coat the sensor with high temperature multi-purpose anti-corrosion compound before installing into the steering knuckle. Install the screw and tighten to 60 inch lbs. (7 Nm). Install the retaining nut.

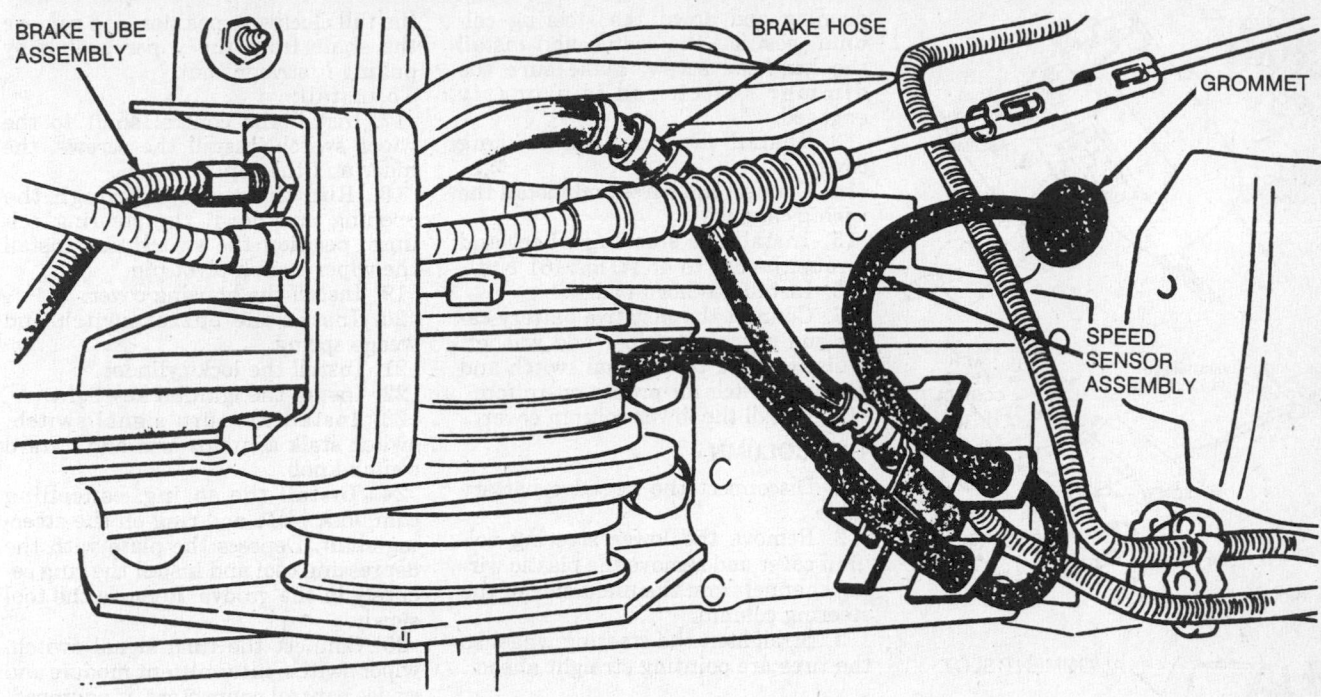

Rear speed sensor wiring routing along the body

8. Install the sensor spool grommet clip retaining screw.

9. Feed the sensor connector wire through the grommet and connect to the harness.

10. Install the sensor assembly grommet.

11. Install the wheel.

CHASSIS ELECTRICAL

Air Bag

DISARMING

To disarm the air bag, disconnect the negative battery. There is no time lapse built into this sytem, so the system is immediately disabled. No further procedures are needed. Failure to disarm the system may result in deployment of the air bag and possible personal injury.

Blower Motor

REMOVAL & INSTALLATION

1. Disconnect the negative battery cable.

2. Remove the glove box assembly, lower right side instrument panel trim cover and right cowl trim panel, as required. Disconnect the blower lead wire connector.

3. Disconnect the 2 vacuum lines from the recirculating door actuator and position the actuator to the side.

4. Remove the 2 screws at the top of the blower housing that secure it to the unit cover.

5. Remove the 5 screws from around the blower housing and separate the blower housing from the unit.

6. Remove the 3 screws that secure the blower assembly to the heater or air conditioning housing and remove the assembly from the unit. Remove the fan from the blower motor.

7. The installation is the reverse of the removal procedure.

8. Connect the negative battery cable and check the blower motor for proper operation.

Windshield Wiper Motor

REMOVAL & INSTALLATION

1. Disconnect the negative battery cable.

2. Remove the wiper arms and blades and remove the plastic cowl top cover.

3. Remove the attaching screws from each pivot assembly.

4. Remove the motor mounting bolts.

5. Remove the nut and disconnect the wiper linkage drive from the motor shaft.

6. Remove the wiper motor from the vehicle.

7. The installation is the reverse of the removal procedure.

8. Connect the negative battery cable and check the wiper motor for proper operation.

Windshield Wiper Switch

REMOVAL & INSTALLATION

NOTE: On 1990–92 vehicles, the windshield wiper switch is part of the combination switch.

1988–89 Vehicles
STANDARD COLUMN

1. Disconnect the negative battery cable.

2. Remove the lower steering column cover.

3. Straighten the steering wheel so the tires are pointing straight-ahead.

NOTE: For vehicles equipped with an airbag, it is imperative that the steering wheel removal and installation procedure under Steering is followed.

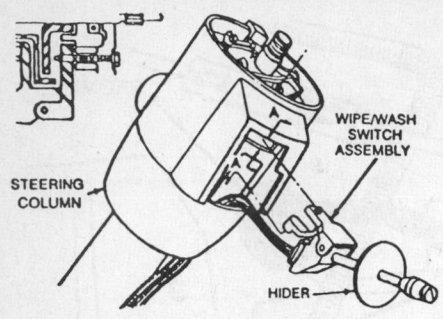

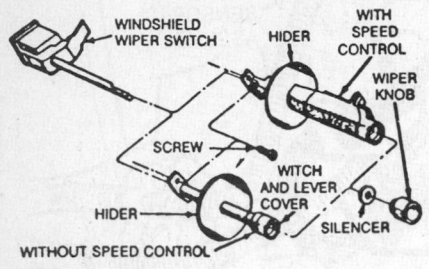

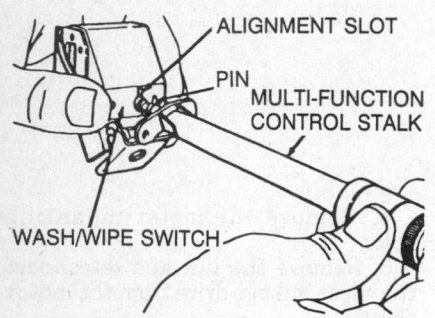

Removing the wiper switch—1988–89 standard column

4. Remove the steering wheel.

5. Remove the plastic wiring channel from the underside of the steering column.

6. Disconnect the wiper switch connector, intermittent wipe module connector and cruise control connector, if equipped.

7. Remove the side lock housing cover.

8. Remove the slotted hex-head screw that attaches the wiper switch to the turn signal switch and remove the switch.

9. Remove the control knob from the end of the stalk. Pull the round nylon hider up the control stalk and remove the revealed screws that attach the control stalk sleeve to the wiper switch.

10. Rotate the control stalk shaft to the full clockwise position and remove the shaft from the wiper switch by pulling it straight out.

To install:

11. Install the control shaft to the wiper switch, install the screws, the hider and the control knob.

12. Run the wiring through the opening and down the steering column, position the switch and install the hex-head screw. Make sure the dimmer switch rod is properly engaged.

13. Install the side lock housing cover.

14. Connect the wires and install the wiring channel.

15. Install the steering wheel and torque the nut to 45 ft. lbs. (61 Nm).

16. Install the horn pad.

17. Connect the negative battery cable and check the wiper and washer, cruise control, turn signal switch and dimmer switch for proper operation.

18. Install the lower column cover.

TILT COLUMN

1. Disconnect the negative battery cable.

2. Remove the lower steering column cover and remove the plastic wiring channel from the underside of the steering column.

3. Straighten the steering wheel so the tires are pointing straight-ahead.

NOTE: For vehicles equipped with an airbag, it is imperative that the steering wheel removal and installation procedure under Steering is followed.

4. Remove the steering wheel.

5. Depress the lock plate with the proper depressing tool, remove the retaining ring from its groove and remove the tool, ring, lock plate, cancelling cam and spring.

6. Remove the switch stalk actuator screw and arm.

7. Remove the hazard switch knob.

8. Disconnect the turn signal switch, wiper switch, intermittent module and cruise control connectors, if equipped.

9. Remove the 3 screws and remove the turn signal switch. Tape the connector to the wires to aid in removal.

10. Remove the ignition key light.

11. Place the key in the **LOCK** position and remove the key. Insert a thin tool into the slot next to the switch mounting screw boss, depress the spring latch at the bottom of the slot releasing the lock. Remove the lock cylinder.

12. Remove the buzzer switch and wedge spring.

13. Remove the 3 housing cover screws and remove the housing cover.

14. Remove the wiper switch pivot pin with a punch and remove the switch.

15. Remove the control knob from the end of the stalk. Pull the round nylon hider up the control stalk and remove the revealed screws that attach the control stalk sleeve to the wiper switch.

16. Rotate the control stalk shaft to the full clockwise position and remove the shaft from the wiper switch by pulling it straight out.

To install:

17. Install the control shaft to the wiper switch, install the screws, the hider and the control knob.

18. Run the wiring through the opening and down the steering column, position the switch and install the wiper switch pivot pin.

19. Install the housing cover.

20. Install the buzzer switch and wedge spring.

21. Install the lock cylinder.

22. Install the ignition key light.

23. Install the turn signal switch, switch stalk actuator arm and hazard switch knob.

24. Install the spring, cancelling cam, lock plate and ring on the steering shaft. Depress the plate with the depressing tool and install the ring securely in the groove. Remove the tool slowly.

25. Connect the turn signal switch, wiper switch, intermittent module and cruise control connectors, if equipped. Install the wiring channel.

26. Install the steering wheel and torque the nut to 45 ft. lbs. (61 Nm).

27. Install the horn pad.

28. Connect the negative battery cable and check the wiper and washer, cruise control, turn signal switch and dimmer switch for proper operation.

29. Install the lower column cover.

Instrument Cluster

REMOVAL & INSTALLATION

1. Disconnect the negative battery cable.

2. Remove the instrument cluster bezel. Cluster removal is not necessary if just removing gauges.

3. When only removing gauge(s) or the speedometer, remove the trip odometer reset knob, if necessary, remove the mask and lens assembly and remove the desired gauge from the cluster.

4. Remove the lower column cover and disconnect the gear indicator cable.

5. Remove the screws attaching the cluster to the instrument panel.

6. Pull the cluster out and disconnect all wiring harnesses. Remove the cluster from the vehicle.

To install:

7. Position the cluster and feed the gear indicator cable through its opening.

8. Connect all wiring in their proper positions.

9. Install the cluster retaining screws. Connect the gearshift indicator cable.

10. Install the cluster bezel.

11. Connect the negative battery cable and check all gauges and the speedometer for proper operation. Make sure the gearshift indicator is properly aligned.

Radio

REMOVAL & INSTALLATION

1. Disconnect the negative battery cable.

2. Remove the cluster bezel.

3. Remove the screws that attach the radio to the instrument panel.

4. Pull the radio out, disconnect the connectors, ground cable and antenna and remove the radio.

5. The installation is the reverse of the removal procedure.

6. Connect the negative battery cable and check the radio for proper operation.

Concealed Headlights

MANUAL OPERATION

1. Disconnect the negative battery cable.

2. Locate the manual override knob located under the center of the front bumper.

3. Rotate the manual override knob to raise the headlight cover(s).

4. Connect the negative battery cable.

Headlight Switch

REMOVAL & INSTALLATION

1. Disconnect the negative battery cable.

2. Remove the headlight cluster bezel.

3. Remove the screws securing the headlight and heated rear window

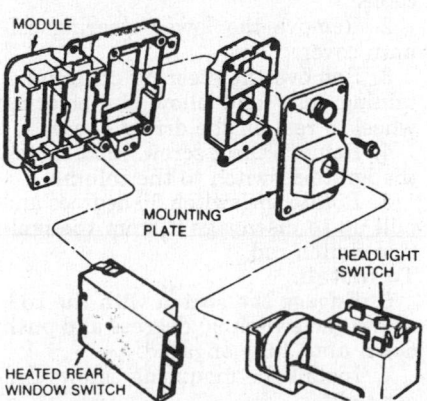

Headlight and heated rear window switches

switch module to the instrument panel. Pull the assembly out to disconnect the connectors from the switch.

4. Depress the spring button and remove the headlight switch knob and stem.

5. Remove the escutcheon and remove the nut that attaches the switch to the mounting plate.

6. The installation is the reverse of the removal procedure.

7. Connect the negative battery cable and check the switch for proper operation.

Dimmer Switch

REMOVAL & INSTALLATION

NOTE: On 1990–92 vehicles, the dimmer switch is part of the combination switch.

1. Disconnect the negative battery cable.

2. Remove the lower steering column cover, if equipped.

3. Unplug the switch, located on the lower portion of the steering column.

4. Holding the actuating rod against its upper seat, remove the bolts that attach the switch to the column and remove the switch.

5. The installation is the reverse of the removal procedure. Adjust the switch as required.

6. Connect the negative battery cable and check the switch for proper operation.

Turn Signal Switch

NOTE: On 1990–92 vehicles, the turn signal switch is part of the combination switch.

REMOVAL & INSTALLATION

1988–89 Vehicles

STANDARD COLUMN

1. Disconnect the negative battery cable.

2. Remove the lower steering column cover.

3. Straighten the steering wheel so the tires are pointing straight-ahead.

NOTE: For vehicles equipped with an airbag, it is imperative that the steering wheel removal and installation procedure under Steering is followed.

4. Remove the steering wheel.

5. Remove the plastic wiring channel from the underside of the steering column and disconnect the turn signal switch connector.

6. Remove the hazard switch knob. Remove the slotted hex-head screw

that attaches the wiper switch to the turn signal switch.

7. Remove the 3 screws and pull the turn signal switch out of the column.

To install:

8. Run the wiring through the opening and down the steering column, position the switch and install the hex-head screw. Make sure the dimmer switch rod is properly engaged.

9. Install the 3 screws and the hazard switch knob.

10. Connect the wires and install the wiring channel.

11. Install the steering wheel and torque the nut to 45 ft. lbs. (61 Nm).

12. Install the horn pad.

13. Connect the negative battery cable and check the turn signal switch and dimmer switch for proper operation.

14. Install the lower column cover.

TILT COLUMN

1. Disconnect the negative battery cable.

2. Remove the lower steering column cover and remove the plastic wiring channel from the underside of the steering column.

3. Straighten the steering wheel so the tires are pointing straight-ahead.

NOTE: For vehicles equipped with an airbag, it is imperative that the steering wheel removal and installation procedure under Steering is followed.

4. Remove the steering wheel.

5. Depress the lock plate with the proper depressing tool, remove the retaining ring from its groove and remove the tool, ring, lock plate, cancelling cam and spring.

6. Remove the stalk actuator screw and arm.

7. Remove the hazard switch knob.

8. Disconnect the turn signal switch connector.

9. Remove the 3 screws and remove the turn signal switch. Tape the connector to the wires to aid in removal.

To install:

10. Run the wiring through the opening and down the steering column, install the turn signal switch, switch stalk actuator arm and hazard switch knob.

11. Install the spring, cancelling cam, lock plate and ring on the steering shaft. Depress the plate with the depressing tool and install the ring securely in the groove. Remove the tool slowly.

12. Connect the turn signal switch connector and install the channel.

13. Install the steering wheel and torque the nut to 45 ft. lbs. (61 Nm).

14. Install the horn pad.

15. Connect the negative battery ca-

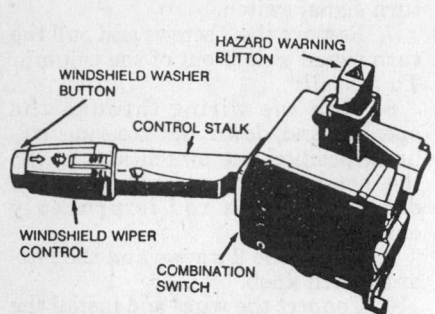

Combination switch—1990–92 vehicles

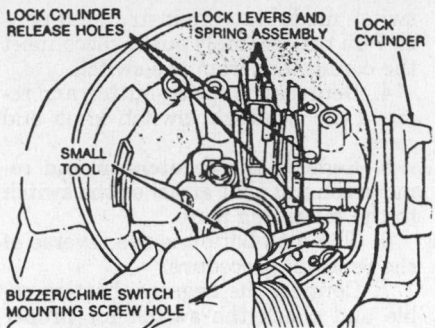

**Removing the key lock cylinder—
1988–89 standard column**

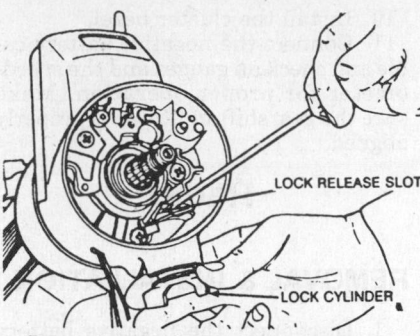

**Removing the key lock cylinder—
1988–89 tilt column**

ble and check the turn signal switch and dimmer switch for proper operation.

16. Install the lower column cover.

Combination Switch

REMOVAL & INSTALLATION

1990–92 Vehicles

1. Disconnect the negative battery cable.
2. Remove the tilt lever, if equipped.
3. Remove the steering column covers.
4. Remove the combination switch tamper-proof mounting screws and pull the switch away from the steering column.
5. Loosen the connector screw; the screw will remain in the connector.
6. Disconnect the connector from the switch.
7. The installation is the reverse of the removal procedure.
8. Connect the negative battery cable and check all functions of the combination switch for proper operation.

Ignition Lock

REMOVAL & INSTALLATION

1988–89 Vehicles

STANDARD COLUMN

1. Disconnect the negative battery cable.
2. Straighten the steering wheel so the tires are pointing straight-ahead.

NOTE: For vehicles equipped with an airbag, it is imperative that the steering wheel removal and installation procedure under Steering is followed.

3. Remove the steering wheel.
4. Remove the hazard switch knob. Remove the slotted hex-head screw that attaches the wiper switch to the turn signal switch.
5. Remove the 3 screws and pull the turn signal switch out of the column as

far as it will go. Unplug it below if necessary.

6. Remove the ignition switch key light.
7. Place the key in the **LOCK** position and remove the key.
8. Insert 2 suitable small diameter tools into both release holes and push inward to release the spring loaded lock retainers while simultaneously pulling the key lock cylinder out of its bore.

To install:

9. Install the key cylinder.
10. Install the ignition switch key light.
11. Install the turn signal switch and hazard switch knob. Connect the wires if they were disconnected.
12. Install the steering wheel and torque the nut to 45 ft. lbs. (61 Nm).
13. Install the horn pad.
14. Connect the negative battery cable and check the lock cylinder for proper operation.
15. Install the lower column cover, if equipped.

TILT COLUMN

1. Disconnect the negative battery cable.
2. Straighten the steering wheel so the tires are pointing straight-ahead.

NOTE: For vehicles equipped with an airbag, it is imperative that the steering wheel removal and installation procedure under Steering is followed.

3. Remove the steering wheel.
4. Depress the lock plate with the proper depressing tool, remove the retaining ring from its groove and remove the tool, ring, lock plate, cancelling cam and spring.
5. Remove the stalk actuator screw and arm.
6. Remove the hazard switch knob.
7. Remove the 3 screws and pull the turn signal switch out of the column as far as it will go. Unplug it below if necessary.
8. Remove the ignition key light.
9. Place the key in the **LOCK** posi-

tion and remove the key. Insert a thin tool into the slot next to the switch mounting screw boss, depress the spring latch at the bottom of the slot releasing the lock and remove the lock cylinder.

To install:

10. Install the lock cylinder.
11. Install the ignition key light.
12. Install the turn signal switch, switch stalk actuator arm and hazard switch knob.
13. Install the spring, cancelling cam, lock plate and ring on the steering shaft. Depress the plate with the depressing tool and install the ring securely in the groove. Remove the tool slowly.
14. Connect the wires if they were disconnected.
15. Install the steering wheel and torque the nut to 45 ft. lbs. (61 Nm).
16. Install the horn pad.
17. Connect the negative battery cable and check the turn signal switch for proper operation.
18. Install the lower column cover.

Ignition Switch

REMOVAL & INSTALLATION

1988–89 Vehicles

1. Disconnect the negative battery cable.
2. Remove the lower steering column cover.
3. Remove the steering column retaining nuts and allow the steering wheel to rest on the driver's seat.
4. Remove the 2 screws that attach the ignition switch to the column.
5. Rotate the switch 90 degrees and pull up to disengage it from the ignition switch rod.

To install:

6. Engage the switch with the rod, rotate the switch 90 degrees and push down until fully engaged.
7. Install the mounting screws finger tight.
8. Place the key in the **LOCK** position and remove the key. Adjust the

switch by pushing up gently on the switch to take up all slack in the rod.

9. Tighten the mounting screws and check the switch for proper operation in all positions.

10. Install the steering column and cover.

Ignition Lock/Switch

REMOVAL & INSTALLATION

1990–92 Vehicles

1. Disconnect the negative battery cable.

2. Remove the tilt lever, if equipped.

3. Remove the upper and lower column covers.

4. Remove the 3 ignition switch tamper-proof Torx® screws; APEX 440–TX20H or equivalent is required.

5. Pull the switch away from the column. Release the connector locks on the 2 wiring connectors and disconnect them from the switch.

6. Remove the key lock cylinder from the ignition switch by performing the following:

    a. Insert the key and turn the switch in the **LOCK** position. Using a suitable small tool, depress the key cylinder retaining pin flush with the key cylinder surface.

    b. Rotate the key clockwise to the **OFF** position to unseat the key cylinder from the ignition switch assembly. The cylinder bezel should be about ⅛ in. above the ignition switch halo light ring. Do not attempt to remove the key cylinder at this point.

    c. With the key cylinder in the unseated position, rotate the key counterclockwise to the **LOCK** position and remove the key.

    d. Remove the key cylinder from the ignition switch.

To install:

7. Connect the wiring connectors.

8. Mount ignition switch to the column by performing the following:

    a. Position the shifter in **PARK** position. The park lock dowel pin on the ignition switch assembly must engage with the column park lock slider linkage.

    b. Verify that the ignition switch is in the **LOCK** position. The flag should be parallel to the ignition switch terminals. Apply a small amount of grease to the flag and pin.

    c. Position the park lock link to mid-travel.

    d. Align the locating pin hole and its pin and position the ignition switch against the lock housing face, make sure the pin is inserted into the park lock link contour slot. Torque the retaining screws to 17 inch lbs.

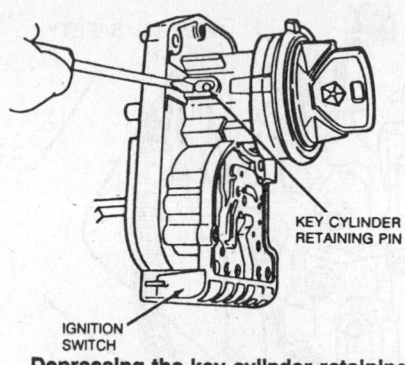

Depressing the key cylinder retaining pin

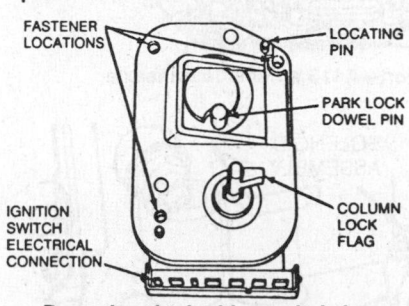

Preparing the ignition switch for installation

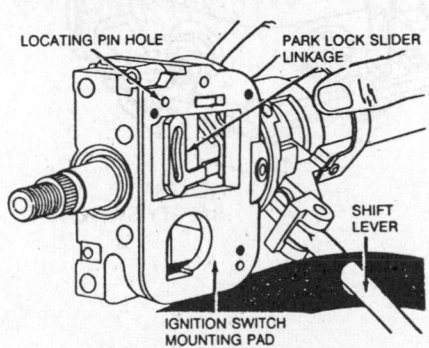

Ignition switch mounting pad

9. With the key cylinder and ignition switch in the **LOCK** position, key not in cylinder, gently insert the key cylinder into the ignition switch until it bottoms.

10. Insert the key. Simultaneously push in on the cylinder and rotate the key to the **RUN** position. This action should fully seat the cylinder in the ignition switch.

11. Install the column covers and the tilt lever, if equipped.

12. Connect the negative battery cable and check the push-to-lock and park lock functions, halo lighting and all ignition switch positions for proper operation.

Stoplight Switch

REMOVAL & INSTALLATION

1. Disconnect the negative battery cable.

2. Unplug the stoplight switch connectors near the brake pedal.

3. Remove the switch and bracket assembly from the brake pedal bracket.

4. Remove the switch from its bracket.

To install:

5. Install the switch and bracket assembly to the brake pedal bracket and push the switch forward as far as it will go; the brake pedal should move forward slightly.

6. Pull back on the brake pedal bringing the striker toward the switch until the pedal will not go back any farther.

7. This will cause the switch to ratchet backward into position and automatic adjustment is complete.

8. Connect the negative battery cable and check the switch for proper operation. Also, make sure the speed control system functions properly, if equipped.

Neutral Safety Switch

REMOVAL & INSTALLATION

1. Disconnect the negative battery cable.

2. Locate the neutral safety switch at the left rear corner of the automatic transaxle, in the left front of engine compartment. Do not confuse with the white PRNDL switch on the A604 automatic transaxle. Unplug the switch connector.

3. Remove the switch from the transaxle.

4. The installation is the reverse of the removal procedure. Torque the switch to 25 ft. lbs. (34 Nm).

5. Connect the negative battery cable and check the switch for proper operation.

Fuses, Circuit Breakers, Relays and Flashers

LOCATION

Fusible Links

On vehicles without a Power Distribution Center, fusible links are part of the the large wiring harness behind the battery. On vehicles with a Power Distribution Center, fusible links in the form of cartridge fuses, which resemble small relays but serve as fusible links, are located in the Center. Each item is identified on the cover of the Power Distribution Center.

Fuse Panels

The fuse panel, which contains fuses

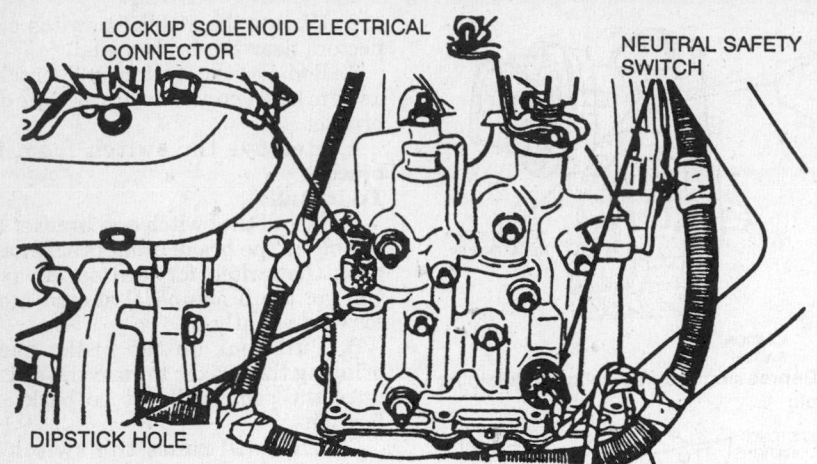

Neutral safety switch identification—A413 automatic transaxle

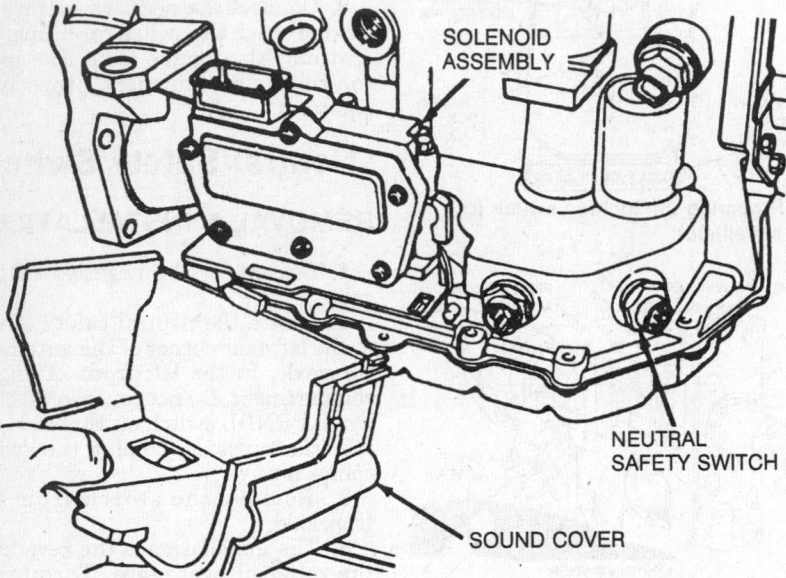

Neutral safety switch identification—A604 automatic transaxle

and circuit breakers, is located behind the glove box door. To remove the panel, pull it out from the bottom and slide the tabs out from the top. Additional fuses are in the Power Distribution Center located near the left side strut tower in the engine compartment. Each item is identified on the cover of the Power Distribution Center.

Relays, Flashers and Circuit Breakers

The relay and flasher module is located behind the cupholder, which also contains circuit breakers. The entire module can be removed by pushing it up and off of its mounting bracket. Additional relays are in the Power Distribution Center located near the left side strut tower in the engine compartment. Each item is identified on the cover of the Power Distribution Center.

Computers

LOCATION

1988

The Single Module Engine Controller (SMEC) is located in engine compartment, to the left of the battery.

1989–91

SINGLE BOARD ENGINE CONTROLLER (SBEC)

Located in the engine compartment, to the left of the battery.

TRANSAXLE CONTROLLER

If equipped with the A604 automatic transaxle, the transaxle controller is located in the right front of the engine compartment.

ANTI-LOCK BRAKES CONTROLLER

1989–90: The Bosch ABS 3 controller

is located behind the rear seat bulkhead trim panel in the trunk.
1991–92: The Bendix ABS 10 controller is located under the battery tray.

AIR SUSPENSION CONTROLLER

If equipped with automatic load leveling or automatic air suspension, the controller is located behind the right side trunk trim panel.

BODY CONTROLLER

The body controller, if equipped, is located inside the passenger compartment, behind the right side kick panel.

Cruise Control

ADJUSTMENT

2.5L Engine

1. The clearance between the throttle stud and cable clevis should be $\frac{1}{16}$ in.
2. To adjust the cable, remove the retaining clip or loosen the retaining clamp nut at the throttle bracket.
3. Pull all slack out of the cable using a suitable $\frac{1}{16}$ in. diameter tool to account for proper clearance. Make sure the curb idle position of the throttle blade is not affected.
4. Reinstall the retaining clip or nut.

3.0L, 3.3L and 3.8L Engines

1. Grip the cable core and lightly push toward the servo.
2. While holding the position, mark the core wire next to the protective sleeve.
3. Pull the core wire away from the servo. There should be a 0.24 in. (6mm) gap between the mark on the core wire and the protective sleeve.
4. If the gap is not correct, remove the adjustment clip from the throttle bracket and move the sleeve to bring the gap into specification.
5. Reinstall the clip.

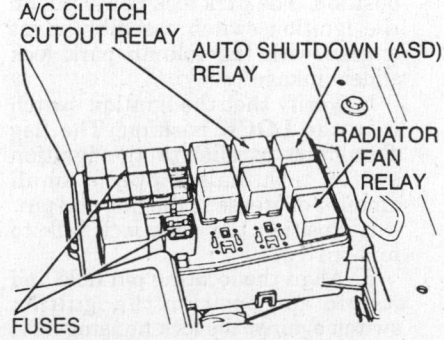

Relays and fuses in the Power Distribution Center—1990–92 vehicles. All are identified on its cover

Chrysler Corp.
Front Wheel Drive
Dodge—Omni Plymouth—Horizon

5

SPECIFICATIONS

VEHICLE IDENTIFICATION CHART

It is important for servicing and ordering parts to be certain of the vehicle and engine identification. The VIN (vehicle identification number) is a 17 digit number visible through the windshield on the driver's side of the dash and contains the vehicle and engine identification codes. The tenth digit indicates model year, and the eighth digit indicates engine code. It can be interpreted as follows:

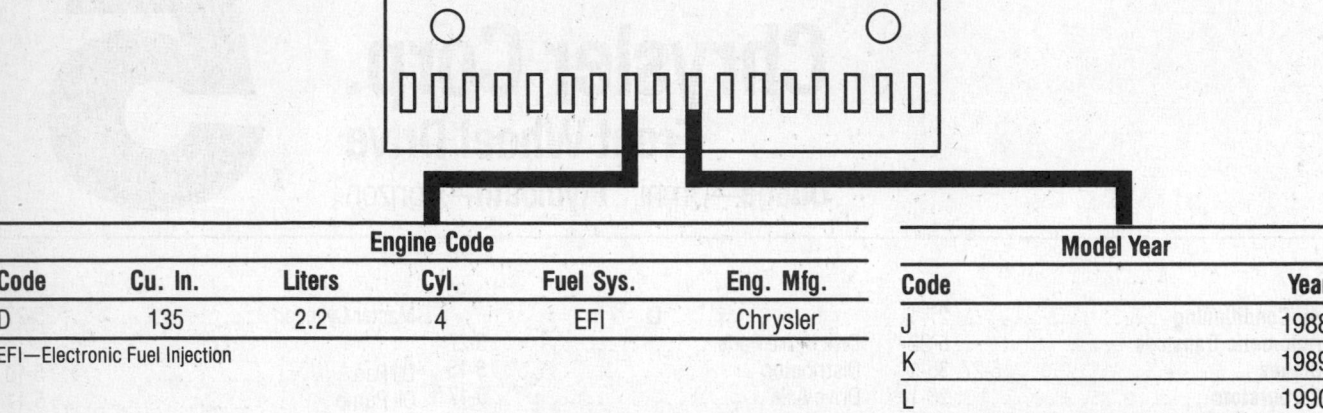

Engine Code

Code	Cu. In.	Liters	Cyl.	Fuel Sys.	Eng. Mfg.
D	135	2.2	4	EFI	Chrysler

EFI—Electronic Fuel Injection

Model Year

Code	Year
J	1988
K	1989
L	1990

ENGINE IDENTIFICATION

Year	Model	Engine Displacement cu. in. (liter)	Engine Series Identification (VIN)	No. of Cylinders	Engine Type
1988	Horizon	135 (2.2)	D	4	OHC
	Omni	135 (2.2)	D	4	OHC
1989	Horizon	135 (2.2)	D	4	OHC
	Omni	135 (2.2)	D	4	OHC
1990	Horizon	135 (2.2)	D	4	OHC
	Omni	135 (2.2)	D	4	OHC

OHC—Overhead Cam

GENERAL ENGINE SPECIFICATIONS

Year	VIN	No. Cylinder Displacement cu. in. (liter)	Fuel System Type	Net Horsepower @ rpm	Net Torque @ rpm (ft. lbs.)	Bore × Stroke (in.)	Compression Ratio	Oil Pressure @ rpm
1988	D	4-135 (2.2)	EFI	99 @ 5600	121 @ 3200	3.44 × 3.62	9.5:1	30–80 @ 3000
1989	D	4-135 (2.2)	EFI	99 @ 5600	121 @ 3200	3.44 × 3.62	9.5:1	30–80 @ 3000
1990	D	4-135 (2.2)	EFI	99 @ 5600	121 @ 3200	3.44 × 3.62	9.5:1	30–80 @ 3000

EFI Electronic Fuel Injection

GASOLINE ENGINE TUNE-UP SPECIFICATIONS

Year	VIN	No. Cylinder Displacement cu. in. (liter)	Spark Plugs Type	Spark Plugs Gap (in.)	Ignition Timing (deg.) MT	Ignition Timing (deg.) AT	Compression Pressure (psi)	Fuel Pump (psi)	Idle Speed (rpm) MT	Idle Speed (rpm) AT	Valve Clearance In.	Valve Clearance Ex.
1988	D	4-135 (2.2)	RN12YC	0.035	12B	12B	100①	15	850	850	Hyd.	Hyd.
1989	D	4-135 (2.2)	RN12YC	0.035	12B	12B	100①	15	850	850	Hyd.	Hyd.
1990	D	4-135 (2.2)	RN12YC	0.035	12B	12B	100①	15	850	850	Hyd.	Hyd.

① Minimum
Hyd.—Hydraulic

FIRING ORDERS

NOTE: To avoid confusion, always replace spark plug wires one at a time.

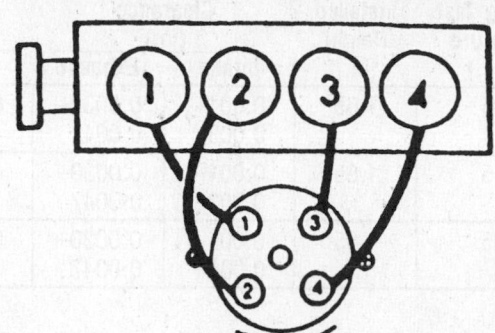

2.2L Engine
Engine Firing Order: 1–3–4–2
Distributor Rotation: Clockwise

CAPACITIES

Year	Model	VIN	No. Cylinder Displacement cu. in. (liter)	Engine Crankcase (qts.) with Filter	Engine Crankcase (qts.) without Filter	Transmission (pts.) 4-Spd	Transmission (pts.) 5-Spd	Transmission (pts.) Auto.	Drive Axle (pts.)	Fuel Tank (gal.)	Cooling System (qts.)
1988	Horizon	D	4-135 (2.2)	4	4	—	4.8	18	—	13	9
	Omni	D	4-135 (2.2)	4	4	—	4.8	18	—	13	9
1989	Horizon	D	4-135 (2.2)	4	4	—	4.8	18	—	13	9
	Omni	D	4-135 (2.2)	4	4	—	4.8	18	—	13	9
1990	Horizon	D	4-135 (2.2)	4	4	—	4.8	18	—	13	9
	Omni	D	4-135 (2.2)	4	4	—	4.8	18	—	13	9

CAMSHAFT SPECIFICATIONS

All measurements given in inches.

Year	VIN	No. Cylinder Displacement cu. in. (liter)	Journal Diameter 1	Journal Diameter 2	Journal Diameter 3	Journal Diameter 4	Journal Diameter 5	Lobe Lift In.	Lobe Lift Ex.	Bearing Clearance	Camshaft End Play
1988	D	4-135 (2.2)	1.375–1.376	1.375–1.376	1.375–1.376	1.375–1.376	1.375–1.376	NA NA	NA NA	—	0.005–0.020
1989	D	4-135 (2.2)	1.375–1.376	1.375–1.376	1.375–1.376	1.375–1.376	1.375–1.376	NA NA	NA NA	—	0.005–0.020
1990	D	4-135 (2.2)	1.375–1.376	1.375–1.376	1.375–1.376	1.375–1.376	1.375–1.376	NA NA	NA NA	—	0.005–0.020

NA—Not Available

CRANKSHAFT AND CONNECTING ROD SPECIFICATIONS

All measurements are given in inches.

Year	VIN	No. Cylinder Displacement cu. in. (liter)	Crankshaft Main Brg. Journal Dia.	Crankshaft Main Brg. Oil Clearance	Crankshaft Shaft End-play	Crankshaft Thrust on No.	Connecting Rod Journal Diameter	Connecting Rod Oil Clearance	Connecting Rod Side Clearance
1988	D	4-135 (2.2)	2.362–2.363	0.0003–0.0040	0.002–0.014	3	1.968–1.969	0.0008–0.0040	0.005–0.013
1989	D	4-135 (2.2)	2.362–2.363	0.0003–0.0040	0.002–0.014	3	1.968–1.969	0.0008–0.0040	0.005–0.013
1990	D	4-135 (2.2)	2.362–2.363	0.0003–0.0040	0.002–0.014	3	1.968–1.969	0.0008–0.0040	0.005–0.013

VALVE SPECIFICATIONS

Year	VIN	No. Cylinder Displacement cu. in. (liter)	Seat Angle (deg.)	Face Angle (deg.)	Spring Test Pressure (lbs.)	Spring Installed Height (in.)	Stem-to-Guide Clearance (in.)		Stem Diameter (in.)	
							Intake	Exhaust	Intake	Exhaust
1988	D	4-135 (2.2)	45	45	95	1.65	0.001–0.003	0.0030–0.0047	0.3124	0.3103
1989	D	4-135 (2.2)	45	45	95	1.65	0.001–0.003	0.0030–0.0047	0.3124	0.3103
1990	D	4-135 (2.2)	45	45	95	1.65	0.001–0.003	0.0030–0.0047	0.3124	0.3103

PISTON AND RING SPECIFICATIONS

All measurements are given in inches.

Year	VIN	No. Cylinder Displacement cu. in. (liter)	Piston Clearance	Ring Gap			Ring Side Clearance		
				Top Compression	Bottom Compression	Oil Control	Top Compression	Bottom Compression	Oil Control
1988	D	4-135 (2.2)	0.0005–0.0027	0.0100–0.0390	0.0110–0.0390	0.0150–0.0740	0.0015–0.0040	0.0015–0.0040	0.0002–0.0080
1989	D	4-135 (2.2)	0.0005–0.0027	0.0100–0.0390	0.0110–0.0390	0.0150–0.0740	0.0015–0.0040	0.0015–0.0040	0.0002–0.0080
1990	D	4-135 (2.2)	0.0005–0.0027	0.0100–0.0390	0.0110–0.0390	0.0150–0.0740	0.0015–0.0040	0.0015–0.0040	0.0002–0.0080

TORQUE SPECIFICATIONS

All readings in ft. lbs.

Year	VIN	No. Cylinder Displacement cu. in. (liter)	Cylinder Head Bolts	Main Bearing Bolts	Rod Bearing Bolts	Crankshaft Pulley Bolts	Flywheel Bolts	Manifold		Spark Plugs
								Intake	Exhaust	
1988	D	4-135 (2.2)	①	30②	40②	50	70	17	17	26
1989	D	4-135 (2.2)	①	30②	40②	50	70	17	17	26
1990	D	4-135 (2.2)	①	30②	40②	50	70	17	17	26

① Sequence:
    1st step 45 ft. lbs.
    2nd step 65 ft. lbs.
    3rd step plus ¼ turn
② Plus ¼ turn

BRAKE SPECIFICATIONS

All measurements in inches unless noted.

Year	Model	Lug Nut Torque (ft. lbs.)	Master Cylinder Bore	Brake Disc		Standard Brake Drum Diameter	Minimum Lining Thickness	
				Minimum Thickness	Maximum Runout		Front	Rear
1988	Horizon	95	0.827	0.431	0.005	7.87	0.06	0.06
	Omni	95	0.827	0.431	0.005	7.87	0.06	0.06
1989	Horizon	95	0.827	0.431	0.005	7.87	0.06	0.06
	Omni	95	0.827	0.431	0.005	7.87	0.06	0.06
1990	Horizon	95	0.827	0.431	0.005	7.87	0.06	0.06
	Omni	95	0.827	0.431	0.005	7.87	0.06	0.06

WHEEL ALIGNMENT

Year	Model		Caster Range (deg.)	Caster Preferred Setting (deg.)	Camber Range (deg.)	Camber Preferred Setting (deg.)	Toe-in (in.)	Steering Axis Inclination (deg.)
1988	Horizon	Front	①	1⁹/₁₀P	¼N–¾P	⁵/₁₆P	¹/₁₆	13³/₈
		Rear	—	—	1¼N–¼N	¾N	³/₃₂	13³/₈
	Omni	Front	①	1⁹/₁₀P	¼N–¾P	⁵/₁₆P	¹/₁₆	13³/₈
		Rear	—	—	1¼N–¼N	¾N	³/₃₂	13³/₈
1989	Horizon	Front	①	1⁹/₁₀P	¼N–¾P	⁵/₁₆P	¹/₁₆	13³/₈
		Rear	—	—	1¼N–¼N	¾N	³/₃₂	13³/₈
	Omni	Front	①	1⁹/₁₀P	¼N–¾P	⁵/₁₆P	¹/₁₆	13³/₈
		Rear	—	—	1¼N–¼N	¾N	³/₃₂	13³/₈
1990	Horizon	Front	①	1⁹/₁₀P	¼N–¾P	⁵/₁₆P	¹/₁₆	13³/₈
		Rear	—	—	1¼N–¼N	¾N	³/₃₂	13³/₈
	Omni	Front	①	1⁹/₁₀P	¼N–¾P	⁵/₁₆P	¹/₁₆	13³/₈
		Rear	—	—	1¼N–¼N	¾N	³/₃₂	13³/₈

① Variation between sides not to exceed 1½P
N—Negative
P—Positive

ENGINE MECHANICAL

NOTE: Disconnecting the negative battery cable on some vehicles may interfere with the functions of the on board computer systems and may require the computer to undergo a relearning process, once the negative battery cable is reconnected.

Engine Assembly

REMOVAL & INSTALLATION

1. Disconnect the negative battery cable and all engine ground straps. If equipped with fuel injection, relieve the fuel pressure.
2. Mark the hood hinge outline on the hood and remove the hood.
3. Drain the cooling system. Remove the radiator hoses, fan assembly and radiator.
4. Remove the air cleaner and air ducts.
5. Unbolt the air conditioning compressor from its mount, if equipped, and position it aside.
6. Remove the power steering pump mounting bolts and position the pump aside, without disconnecting any fluid lines.
7. Label and disconnect all electrical connectors and vacuum lines from the engine, alternator and fuel injection system.
8. Disconnect the fuel lines and heater hoses.
9. Disconnect the throttle linkage.
10. Remove the alternator.
11. Raise the vehicle and support safely.
12. Disconnect the exhaust pipe from the manifold. Remove the right inner fender shield.
13. If equipped with a manual transaxle, remove the transaxle.
14. If equipped with an automatic transaxle, perform the following procedures:
    a. Remove the lower cover from the transaxle case.
    b. Remove the starter and set it aside.
    c. Matchmark the flexplate to the torque converter, for installation purposes.
    d. Remove the torque converter bolts. Separate the converter from the flexplate. Remove the lower bellhousing bolts.
15. Lower the vehicle and support the transaxle, if still in the vehicle, with a floor jack or equivalent. Attach an engine lifting device to the engine.
16. Remove the remaining bellhousing bolts.

NOTE: If removing the insulator-to-rail screws, first mark the position of the insulator on the side rail to insure proper alignment during reinstallation.

17. Remove the front engine mount nut/bolt and the left insulator through bolt or the insulator bracket to transaxle bolts.
18. Lift the engine from the vehicle and remove.

To install:
19. Lower the engine into the engine compartment. Loosely install all of the mounting bolts. With all bolts installed, torque the:
    Engine to mount bolts to 40 ft. lbs.
    Engine to transaxle bolts to 70 ft. lbs.
    Torque converter bolts, if equipped, to 40 ft. lbs.
20. Remove the lifting device.
21. Raise the vehicle and support safely.
22. If equipped with a manual transaxle, install the transaxle.
23. If equipped with an automatic transaxle, install the torque converter inspection plate and starter.
24. Connect the exhaust pipe. Lower the vehicle.
25. Install the alternator, power steering pump and air conditioning compressor, if equipped.
26. Connect the fuel lines and heater hoses.
27. Connect the throttle linkage.
28. Connect all remaining electrical connectors and vacuum lines.
29. Install the air cleaner assembly and oil filter.
30. Install the radiator, fan assembly and hoses.

31. Fill the engine with the proper amount of engine oil. Connect the negative battery cable.

32. Refill the cooling system. Start the engine, allow it to reach normal operating temperature and check for leaks.

33. Check the ignition timing and adjust, if necessary.

34. Install the hood.

Engine Mounts

REMOVAL & INSTALLATION

1. Disconnect the negative battery cable.

2. Matchmark the engine mount to its frame mounting location.

3. Raise the vehicle and support safely, if necessary. Using the proper equipment, support the weight of the engine.

4. Remove all bolts and nuts that attach the mount to the engine strut, transaxle or body and remove the mount assembly from the vehicle.

5. Remove the through bolt and separate the insulator from the yoke bracket, as required.

6. The installation is the reverse of the removal procedure. Make sure that matchmarks are aligned before tightening bolts.

Cylinder Head

REMOVAL & INSTALLATION

1. Disconnect the negative battery cable and disconnect it from the head. Relieve the fuel pressure. Drain the cooling system. Remove the dipstick bracket nut from the thermostat housing.

2. Remove the air cleaner assembly. Remove the upper radiator hose and disconnect the heater hoses.

3. Disconnect and label the vacuum lines, hoses and wiring connectors from the manifolds, throttle body and cylinder head.

4. Disconnect the all linkages and fuel lines from the throttle body. Unbolt the cable bracket. Remove the ground strap attaching screw from the firewall.

5. Remove the upper air conditioning compressor mounting bolts, if equipped. The cylinder head can be remove with the compressor and bracket still mounted. Remove the upper timing belt cover.

6. Raise the vehicle and support safely. Disconnect the exhaust pipe from the exhaust manifold.

7. Rotate the engine by hand, until the timing marks align. The No. 1 piston should be at TDC of its compression stroke. Lower the vehicle.

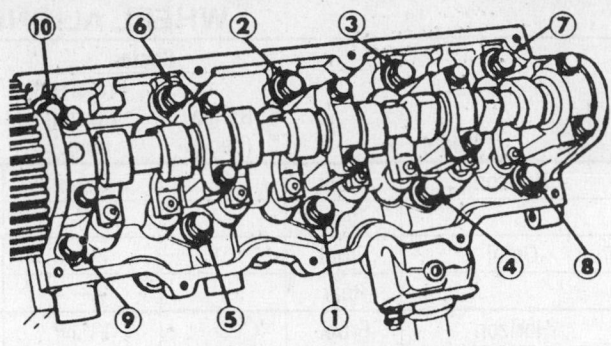

Cylinder head bolt torque sequence

8. With the timing marks aligned, remove the camshaft sprocket. The camshaft sprocket can be suspended to keep the timing intact. Remove the spark plug wires from the spark plugs.

9. Remove the valve cover and curtain. Remove the cylinder head bolts and washers, starting from the outside and working inward.

10. Remove the cylinder head from the engine.

11. Clean the cylinder head gasket mating surfaces.

To install:

12. Using a new gasket and seals, install the cylinder head to the engine block. Using new head bolts assembled with the old washers, torque the cylinder head bolts in sequence, to 45 ft. lbs. (61 Nm). Repeating the sequence, torque the bolts to 65 ft. lbs. (88 Nm). With the bolts at 65 ft. lbs., turn each bolt an additional ¼ turn.

NOTE: Head bolt diameter is 11mm. These bolts are identified with the number 11 on the head of the bolt. The 10mm bolts used on previous vehicles will thread into an 11mm bolt hole, but will permanently damage the cylinder block. Make sure the correct bolts are being used when replacing head bolts.

13. Install the timing belt.

14. Install or connect all items that were removed or disconnected during the removal procedure.

15. Refill the cooling system. Connect the negative battery cable. Start the engine and check for leaks using the DRB II to activate the fuel pump, if possible. Adjust the timing as required.

Valve Lifters

REMOVAL & INSTALLATION

1. Disconnect the negative battery cable.

2. Remove the valve cover and curtain. If removing all lifters, remove the camshaft and rocker arms.

3. If only removing 1 lifter, rotate the crankshaft until the low point of the desired cam lobe is contacting the rocker arm.

4. Using the special valve spring compressor tool 4682 or equivalent, depress the valve spring without dislodging the keepers and slide the rocker arm out.

5. Remove the valve lifter(s) from the bore(s).

6. Lubricate the lifter(s) and their bore(s) with clean engine oil.

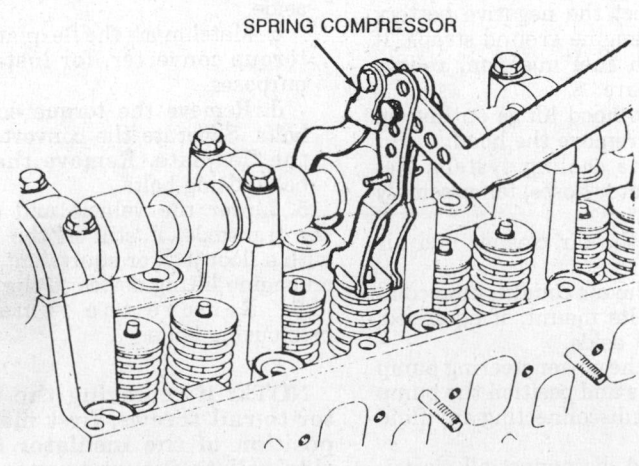

SPRING COMPRESSOR

Using the special tool to compress the valve spring

7. The installation is the reverse of the removal procedure.

Rocker Arms

REMOVAL & INSTALLATION

1. Disconnect the negative battery cable.
2. Remove the valve cover.
3. Rotate the crankshaft until the low point of the desired cam lobe is contacting the rocker arm.
4. Use the special valve spring compressor tool 4682 or equivalent, depress the valve spring without dislodging the keepers and slide the rocker arm out.
5. The installation is the reverse of the removal procedure.

Combination Manifold

REMOVAL & INSTALLATION

NOTE: On some vehicles, some of the manifold attaching bolts are not accessible or too heavily sealed from the factory and cannot be removed on the vehicle. Head removal would be necessary in these situations.

1. Disconnect the negative battery cable.
2. Relieve the fuel system pressure.
3. Drain the cooling system.
4. Remove the air cleaner and disconnect all vacuum lines, electrical wiring and fuel lines from the throttle body.
5. Disconnect the throttle linkage.
6. Loosen the power steering pump and remove the drive belt.
7. Remove the power brake vacuum hose from the intake manifold.
8. Remove the water hoses from the water crossover.

9. Raise and safely support the vehicle. Disconnect the exhaust pipe from the exhaust manifold.
10. Remove the power steering pump from its mounting bracket and set it aside.
11. Remove the intake manifold support bracket, if equipped.
12. Remove the EGR tube.
13. Remove the intake manifold bolts.
14. Lower the vehicle.
15. Remove the intake manifold.
16. Remove the exhaust manifold nuts.
17. Remove the exhaust manifold.

To install:
18. Install a new combination manifold gasket.
19. Install the manifold assembly. Starting from the middle and working outwards, install the mounting nuts and torque to 13 ft. lbs. (18 Nm). Install the heat cowl to the exhaust manifold.
20. Install the intake manifold. Starting from the middle and working outward, torque the bolts to 17 ft. lbs. (23 Nm).
21. Install the EGR tube.
22. Install the intake support bracket, if equipped.
23. Install the power steering pump.
24. Raise the vehicle and support safely. Install the exhaust pipe to the exhaust manifold.
25. Install the water hoses to the water crossover.
26. Install the power brake vacuum hose to the intake manifold.
27. Connect the throttle linkage.
28. Install all vacuum lines, electrical wiring and fuel lines to the throttle body.
29. Install the air cleaner assembly.
30. Refill the cooling system.
31. Connect the negative battery cable and check the manifolds for leaks.

Timing Belt Front Cover

REMOVAL & INSTALLATION

1. Disconnect the negative battery cable.
2. Remove the nuts that attach the upper cover to the valve cover.
3. Remove the bolt that attaches the upper cover to the lower cover and remove the upper cover.
4. Raise the vehicle and support safely. Remove the right side splash shield.
5. Remove the crankshaft pulley, water pump pulley and belts.
6. Remove the lower cover attaching bolts and remove the lower cover.
7. The installation is the reverse of the removal procedure.
8. Install the belts.

Timing Belt and Tensioner

ADJUSTMENT

1. Disconnect the negative battery cable.
2. Raise the vehicle and support safely. Remove the right front inner splash shield.
3. Remove the tensioner cover.
4. Place the special tensioning tool C–4703 on the hex of the tensioner so the weight is at about the 10 o'clock position and loosen the bolt.
5. The tensioner should drop to the 9 o'clock position. Reposition the tool, as required, in order to have it end up at the 9 o'clock position, parallel to the ground, hanging toward the rear of the vehicle, ± 15 degrees.
6. Hold the tool in position and tighten the bolt. Do not pull the tool past the 9 o'clock position or the belt will be too tight and will cause howling or possible breakage.
7. Install the cover and splash shield.
6. Install the timing belt covers and all related parts.

REMOVAL & INSTALLATION

1. If possible, position the engine so the No. 1 piston is at TDC of its compression stroke. Disconnect the negative battery cable.
2. Remove the timing belt covers. Remove the timing belt tensioner and allow the belt to hang free.
3. Place a floor jack under the engine and separate the right motor mount.
4. Remove the air conditioning

INTAKE MANIFOLD SCREWS

EXHAUST MANIFOLD SCREWS

Combination manifold attaching nuts and bolts

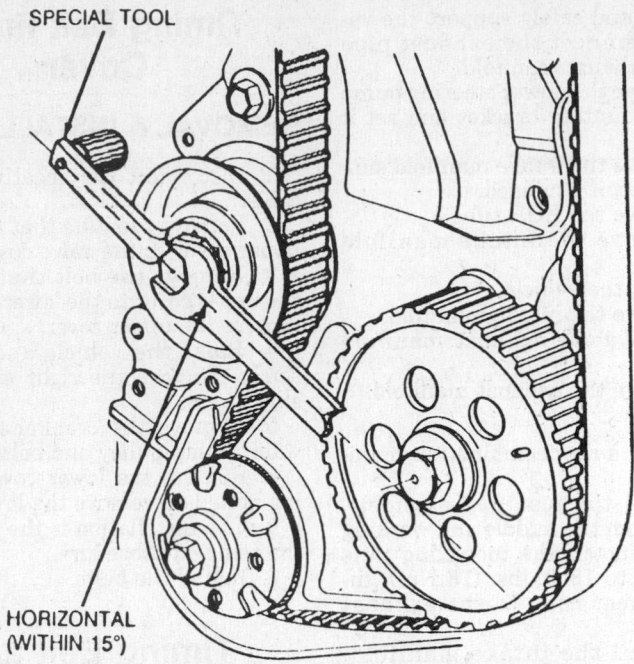

SPECIAL TOOL

HORIZONTAL
(WITHIN 15°)

Adjusting the timing belt

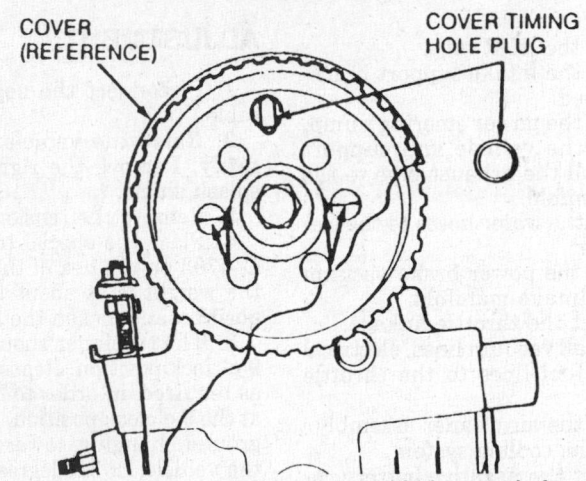

COVER
(REFERENCE)

COVER TIMING
HOLE PLUG

Alignment of the arrows on the camshaft sprocket with the camshaft cap to cylinder head mouting line

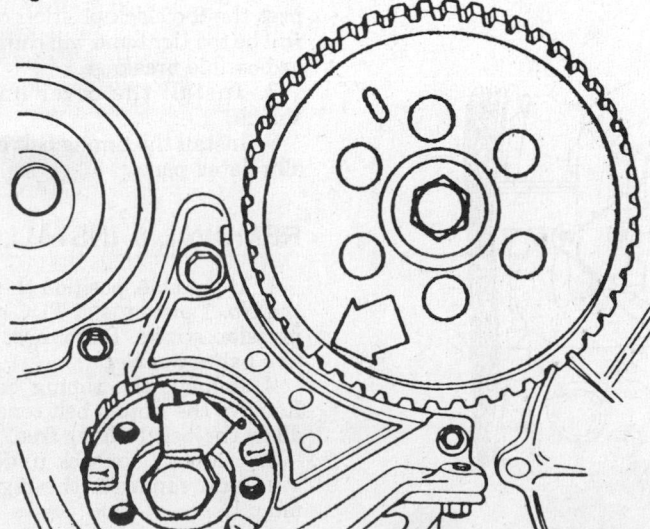

Alignment of the crankshaft and intermediate shaft sprockets

compressor belt idler pulley, if equipped, and remove the mounting stud. Unbolt the compressor/alternator bracket and position it aside.

5. Remove the timing belt from the vehicle.

To install:

6. Turn the crankshaft sprocket and intermediate shaft sprocket until the marks are in line. Position a straight-edge from bolt to bolt to confirm alignment.

7. Turn the camshaft until the small hole in the sprocket is at the top and rows on the hub are in line with the camshaft cap to cylinder head mounting lines. Use a mirror to see the alignment so it is viewed straight on and not at an angle from above. Install the belt but let it hang free at this point.

8. Install the air conditioning compressor/alternator bracket, idler pulley and motor mount. Remove the floor jack. Raise the vehicle and support safely. Have the tensioner at an arm's reach because the timing belt will have to be held in position with one hand.

9. To properly install the timing belt, reach up and engage it with the camshaft sprocket. Turn the intermediate shaft counterclockwise slightly, then engage the belt with the intermediate shaft sprocket. Hold the belt against the intermediate shaft sprocket and turn clockwise to take up all tension; if the timing marks are out of alignment, repeat until alignment is correct.

10. Using a wrench, turn the crankshaft sprocket counterclockwise slightly and wrap the belt around it. Turn the sprocket clockwise so there is no slack in the belt between sprockets; if the timing marks are out of alignment, repeat until alignment is correct.

NOTE: If the timing marks are in line but slack exists in the belt between either the camshaft and intermediate shaft sprockets or the intermediate and crankshaft sprockets, the timing will be incorrect when the belt is tensioned. All slack must be only between the crankshaft and camshaft sprockets.

11. Install the tensioner and install the mounting bolt loosely. Place the special tensioning tool C–4703 on the hex of the tensioner so the weight is at about the 9 o'clock position, parallel to the ground, hanging toward the rear of the vehicle, ± 15 degrees.

12. Hold the tool in position and tighten the bolt to 45 ft. lbs. (61 Nm). Do not pull the tool past the 9 o'clock position; this will make the belt too

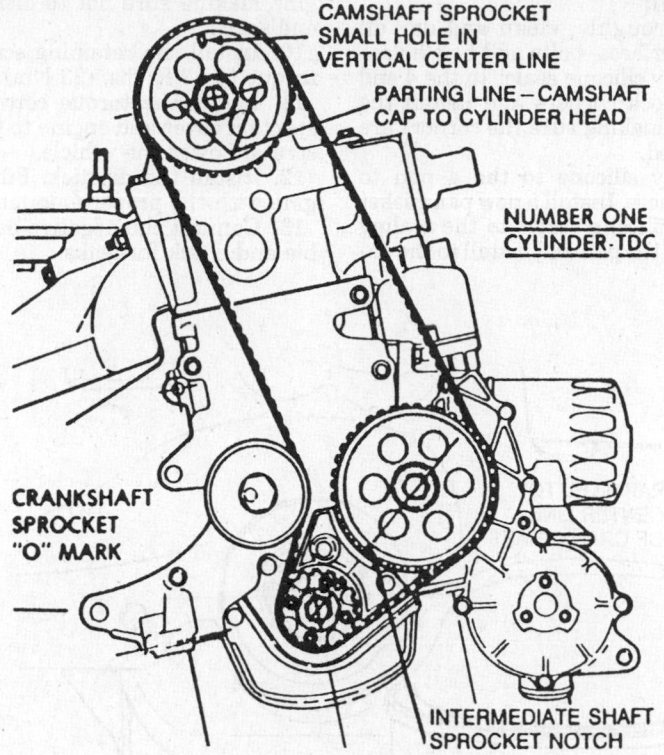

CAMSHAFT SPROCKET
SMALL HOLE IN
VERTICAL CENTER LINE

PARTING LINE – CAMSHAFT
CAP TO CYLINDER HEAD

NUMBER ONE
CYLINDER·TDC

CRANKSHAFT
SPROCKET
"O" MARK

INTERMEDIATE SHAFT
SPROCKET NOTCH

Timing belt properly Installed

tight and will cause it to howl or possibly break.

13. Lower the vehicle and recheck the camshaft sprocket positioning. If it is correct install the timing belt covers and all related parts.

14. Connect the negative battery cable and road test the vehicle.

Timing Sprockets

REMOVAL & INSTALLATION

1. Disconnect the negative battery cable. Remove the timing belt.

2. Remove the crankshaft sprocket bolt. Using the puller tool C–4685 or equivalent, and the button from tool L–4524 or equivalent, remove the crankshaft sprocket.

3. Using the tool C–4687 or equivalent, hold the camshaft and/or intermediate sprocket, remove the center bolt and the sprocket(s).

4. The installation is the reverse of the removal procedure. Torque the camshaft and intermediate sprocket bolts to 65 ft. lbs. (88 Nm) and the crankshaft sprocket bolt to 50 ft. lbs. (68 Nm).

Camshaft

REMOVAL & INSTALLATION

1. Disconnect the negative battery cable.

2. Turn the crankshaft so the No. 1 piston is at the TDC of its compression stroke. Remove the upper timing belt cover.

3. Remove the camshaft sprocket bolt and the sprocket and suspend tightly so the belt does not lose tension. If it does, the belt timing will have to be reset.

4. Remove the valve cover.

5. If the rocker arms are being reused, mark them for installation identification and loosen the camshaft bearing bolts, evenly and gradually.

6. Using a soft mallet, tap the rear of the camshaft a few times to break the bearing caps loose.

7. Remove the bolts, bearing caps and the camshaft with seals.

NOTE: Take note of the color of the paint stripe on the rear camshaft seal. These stripes differentiate seal sizes. If a seal with a different color stripe is installed, a severe leak will develop if the seal is too small or the cap will not be able to be fully installed if the seal is too big.

Also, oversized components can be identified as follows: the top of the bearing caps are painted green and "O/SJ" is stamped behind the oil galley plug on the end of the head. The barrel of an oversized camshaft is also painted green and "O/SJ" is stamped on the end of the shaft. If normal

sized parts are installed in place of oversized ones, oil pressure will be significantly reduced.

8. Check the oil passages for blockages, the parts for wear or damage and replace parts, as required. Clean the gasket mounting surfaces.

To install:

9. Transfer the sprocket key to the new camshaft. New rocker arms and a new camshaft sprocket bolt are normally included with the camshaft package. Install the rocker arms, lubricate the camshaft and install with end seals installed.

10. Place the bearing caps with No. 1 at the timing belt end and No. 5 at the transaxle end. The camshaft bearing caps are numbered and have arrows facing forward. Torque the camshaft bearing bolts evenly and gradually to 18 ft. lbs. (24 Nm).

NOTE: Apply RTV silicone gasket material to the No. 1 and 5 bearing caps. Install the bearing caps before the seals are installed.

11. Mount a dial indicator to the front of the engine and check the camshaft endplay. Play should not exceed 0.020 in.

12. Install the camshaft sprocket and the new bolt.

13. Install the valve cover with a new gasket.

14. Connect the negative battery cable and check for leaks.

Intermediate Shaft

REMOVAL & INSTALLATION

1. Disconnect the negative battery cable.

2. Crank the engine until the No. 1 piston is at TDC of its compression stroke. Remove the timing belt covers to confirm that all timing marks are in line.

3. Remove the fuel pump, if equipped. Remove the distributor. Looking down at the oil pump, the slot in the shaft must be parallel with the center line of the crankshaft. Remove the oil pump.

4. Remove the timing belt and the intermediate shaft sprocket.

5. Remove the shaft retainer bolts and remove the retainer from the block.

6. Remove the intermediate shaft from the engine.

7. If necessary, remove the front bushing using tool C–4697-2 and the rear bushing using tool C–4686-2.

To install:

8. Install the front bushing using tool C–4697-1 until the tool is flush

with the block. Install the rear bushing using tool C–4686–1 until the tool is flush with the block.

9. Lubricate the distributor drive gear and install the intermediate shaft.

10. Replace the seal in the retainer and apply silicone sealer to the mating surface of the retainer. Install the retainer to the block and torque the bolts to 10 ft. lbs. (12 Nm).

11. Install the intermediate shaft sprocket and the timing belt.

12. With the timing belt properly installed, install the oil pump so the slot is parallel to the center line of the crankshaft. Install the distributor so the rotor is aligned with the No. 1 spark plug wire tower on the cap.

13. Install the fuel pump, if equipped.

14. Connect the negative battery cable and road test the vehicle.

Piston and Connecting Rod

POSITIONING

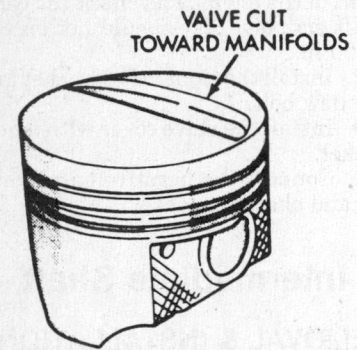

VALVE CUT TOWARD MANIFOLDS

ENGINE LUBRICATION

Oil Pan

REMOVAL & INSTALLATION

1. Disconnect the negative battery cable. Remove the oil dipstick.
2. Raise the vehicle and support safely.
3. Drain the engine oil.
4. Remove the engine to transaxle struts.
5. Remove the torque converter inspection cover.
6. Remove the oil pan retaining screws and remove the oil pan and the side seals.

To install:

7. Thoroughly, clean and dry all sealing surfaces, bolts and bolt holes.

8. Apply silicone sealer to the 4 end seal to block corners and install the end seals making sure the corners are not twisted.

9. Apply silicone to the 4 pan to block corners. Install a new pan gasket or apply silicone sealer to the sealing surface of the pan and install to the engine making sure not to dislodge the end seals.

10. Install the retaining screws and torque to 17 ft. lbs. (23 Nm).

11. Install the torque converter inspection cover and engine to transaxle struts. Lower the vehicle.

12. Install the dipstick. Fill the engine with the proper amount of oil.

13. Connect the negative battery cable and check for leaks.

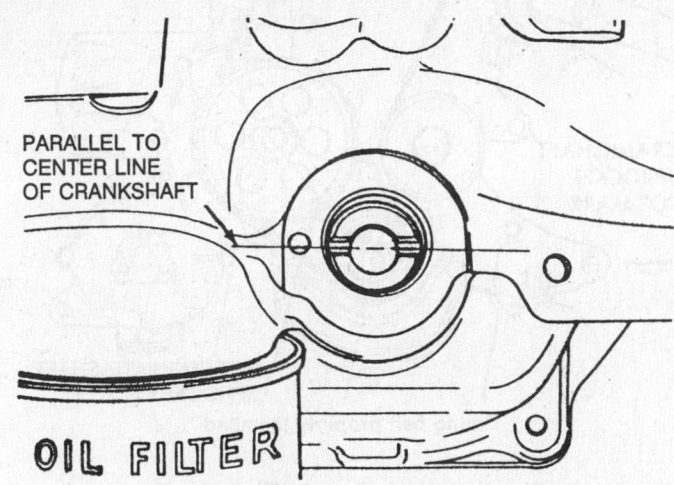

PARALLEL TO CENTER LINE OF CRANKSHAFT

OIL FILTER

Aligning the slot in the oil pump shaft

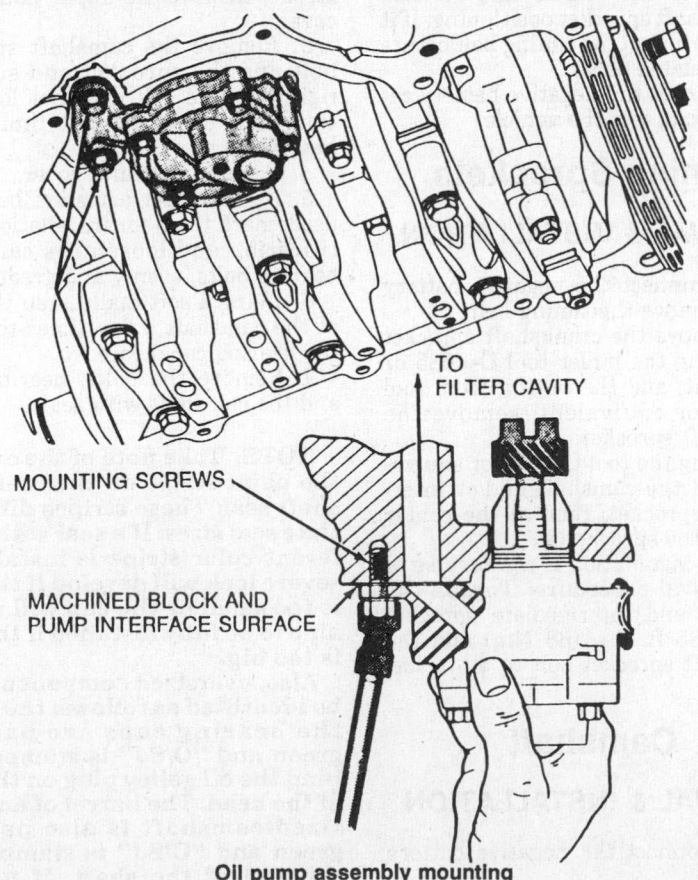

TO FILTER CAVITY

MOUNTING SCREWS

MACHINED BLOCK AND PUMP INTERFACE SURFACE

Oil pump assembly mounting

Oil Pump

REMOVAL & INSTALLATION

1. Crank the engine so the No. 1 piston is at TDC of its compression stroke. Disconnect the negative battery cable.

2. Matchmark the rotor to the block and remove the distributor to confirm that the slot in the oil pump shaft is parallel to the centerline of the crankshaft. Matchmark the slot to the distributor bore, if desired.

3. Remove the dipstick. Raise the vehicle and support safely. Drain the engine oil and remove the pan.

4. Remove the oil pickup.

5. Remove the 2 mounting bolts and remove the oil pump from the engine.

To install:

6. Prime the pump by pouring fresh oil into the pump intake and turning the driveshaft until oil comes out the pressure port. Repeat a few times until no air bubbles are present.

7. Apply Loctite® 515 sealant or equivalent, to the pump body to block machined surface interface. Lubricate the oil pump and distributor driveshaft.

8. Align the slot so it will be in the same position as when it was removed. If it is not, the distributor will not be timed correctly. Install the pump fully and rotate back and forth to ensure proper positioning between the pump mounting surface and the machined surface of the block.

9. Install the mounting bolts finger-tight and lower the vehicle to confirm proper slot positioning. If the slot is not properly positioned, raise the vehicle and move the gear, as required. If the slot is correct, hold the pump firmly against the block and torque the mounting bolts to 17 ft. lbs. (23 Nm).

10. Clean out the oil pickup or replace, as required. Replace the oil pickup O-ring and install the pickup to the pump.

11. Install the oil pan using new gaskets. Lower the vehicle.

12. Install the distributor.

13. Install the dipstick. Fill the engine with the proper amount of oil.

14. Connect the negative battery cable, check the timing and check the oil pressure.

Rear Main Bearing Oil Seal

REMOVAL & INSTALLATION

1. Disconnect the negative battery cable.

2. Remove the transaxle. Remove the flexplate or flywheel.

3. If there is any leakage coming from the rear seal retainer, drain the engine oil and remove the oil pan, if necessary. Remove the rear main oil seal retainer.

4. Remove the seal from the retainer.

To install:

5. Lightly coat the seal outer diameter with Loctite® Stud N' Bearing Mount or equivalent.

6. Install the seal to the retainer.

7. If the retainer was removed, thoroughly clean and dry the retainer to block sealing surfaces and install a new gasket or apply silicone sealer and install the retainer. Install the pan, if it was removed.

8. Install the flexplate or flywheel and transaxle.

9. Connect the negative battery cable and check for leaks.

ENGINE COOLING

Radiator

REMOVAL & INSTALLATION

1. Disconnect the negative battery cable.

2. Drain the coolant.

3. Remove the upper hose and coolant reserve tank hose from the radiator.

4. Remove the electric cooling fan.

5. Raise the vehicle and support safely. Remove the lower hose from the radiator.

6. Disconnect the automatic transaxle cooler hoses, if equipped, and plug them. Lower the vehicle.

7. Remove the mounting brackets and carefully lift the radiator out of the engine compartment.

To install:

8. Lower the radiator into position.

9. Install the mounting brackets.

10. Raise the vehicle and support safely. Connect the automatic transaxle cooler lines, if equipped.

11. Connect the lower hose. Lower the vehicle.

12. Install the electric cooling fan.

13. Connect the upper hose and coolant reserve tank hose.

14. Fill the system with coolant and bleed the system.

15. Connect the negative battery cable, run the vehicle until the thermostat opens, fill the radiator completely and check the automatic transaxle fluid level.

16. Once the vehicle has cooled, recheck the coolant level.

Heater Core

REMOVAL & INSTALLATION

Without Air Conditioning

1. Disconnect the negative battery cable.

2. Drain the coolant. Clamp off the heater hoses near the heater core and remove the hoses from the core tubes. Plug the hose ends and the core tubes to prevent spillage of coolant.

3. Disconnect the blower motor connector.

4. Remove the ash tray.

5. Depress the tab on the temperature control cable and pull the cable out of its housing on the heater assembly.

6. Remove the glove box assembly and unplug the resistor block.

7. Remove the 2 nuts fastening the heater assembly to the firewall and remove the screw attaching the heater support brace to the instrument panel.

8. Remove the heater support bracket nut. Disconnect the strap from the plenum stud and lower the assembly from under the instrument panel.

9. Depress the tab on the flag and pull the mode door control cable out of its housing on the heater assembly.

10. Move the assembly toward the right side of the vehicle and remove.

11. With the assembly on a workbench, remove the top cover and slide the heater out of its cavity in the assembly.

To install:

12. Clean the inside of the assembly and assemble.

13. Connect the mode control cable to the mode door crank and position the heater assembly under the instrument panel. Slide forward so the mounting studs and heater core tubes project through their holes in the firewall. Install the support bracket and brace to hold the assembly in place.

14. From the engine compartment, install the 2 nuts on the firewall and connect the heater hoses.

15. Connect the temperature control cable, blower motor wiring and resistor block connector.

16. Install the defroster duct.

17. Install the ash tray and glove box.

18. Fill the cooling system.

19. Connect the negative battery cable and check the heater for proper operation and leaks.

With Air Conditioning

1. Disconnect the negative battery cable. Properly discharge the air conditioning system. Drain the cooling system.

2. Clamp off the heater hoses near the heater core and remove the hoses from the core tubes. Plug the hose ends and the core tubes to prevent spillage of coolant.

3. Disconnect the H-valve connection at the valve and remove the H-valve. Remove the condensation tube. Disconnect the vacuum lines at the brake booster and water valve.

4. Disconnect the temperature door cable from evaporator/heater assembly.

5. Remove the glove box assembly.

6. On 1990 vehicles, remove the Air Bag Diagnostic Module (ABDM) from its mount and position aside. Remove the mounting bracket.

7. Disconnect the blower motor feed wire and vacuum harness.

8. Remove the central air duct cover from the central air distributor duct.

9. Remove the screws securing the central air conditioning air distributor duct. Remove the duct from under the dash panel.

10. From the engine compartment, remove the nuts that attach the unit to the firewall.

11. Remove the panel support bracket.

12. Remove the right side cowl lower panel and the top cover of the instrument panel.

13. Remove the instrument panel pivot bracket screw from the right side.

14. Remove the screws securing the lower instrument panel and steering column, if required.

15. Pull the carpet rearward as far as possible.

16. Remove the nut from the air conditioning to plenum mounting brace and blower motor ground cable. Support the unit and remove the brace from its stud.

17. Lift the unit and pull it rearward as far as possible to clear the dash panel and liner. Pull rearward on the lower instrument panel to gain enough clearance to remove the unit.

18. Slowly lower the unit to floor and slide out from the under dash panel.

19. With the unit on a workbench, remove the nut from the mode door actuator arm on the top cover. Remove the retaining clips from the front edge of the cover.

20. Remove the mode door actuator to cover screws and remove the actuator.

21. Remove the cover to heater evaporator assembly screws and remove the cover. Lift the mode door out of the unit.

22. Remove the heater core tube retaining bracket and screw. Lift the heater core out of the unit.

To install:

23. Remove the temperature control

door from the housing and clean the unit out with solvent. Lubricate the lower pivot rod and its well and install. Wrap the heater core with foam tape and place it in position. Secure it with its screw.

24. Assemble the package, making sure all vacuum tubing is properly routed.

25. Feed the vacuum lines through the hole in the firewall and install the assembly to the vehicle. Connect the vacuum harness and defroster duct adaptor. Install the nuts to the firewall and connect the hanger strap inside the passenger compartment.

26. Fold the carpeting back into position. Connect the blower motor wiring and install the ABDM.

27. Install the center distribution and defroster adaptor ducts.

28. Secure the lower instrument panel and steering column, as required.

29. Install the instrument panel pivot bracket screw at the right side.

30. Install the right side cowl lower panel and the top cover of the instrument panel. Install the panel support bracket.

31. Connect the vacuum lines at the brake booster and water valve. Using new gaskets, install the H-valve and condensation tube.

32. Connect the heater hoses.

33. Using the proper equipment, evacuate and recharge the air conditioning system.

34. Fill the cooling system.

35. Connect the negative battery cable and check the entire climate control system for proper operation and leakage.

Water Pump

REMOVAL & INSTALLATION

1. Disconnect the negative battery cable.

2. Drain the cooling system.

3. If equipped with air conditioning, remove the compressor from the bracket and position it aside.

4. Remove the alternator and bracket. Remove the pulley from the water pump.

5. Disconnect the lower radiator hose and heater hose from the water pump.

6. Remove the water pump housing attaching screws and remove the assembly from the vehicle. Discard the O-ring.

7. Remove the water pump from the housing.

To install:

8. Using a new gasket or silicone sealer, install the water pump to the housing.

9. Install a new O-ring to the housing and install to the engine. Torque the bolts to 21 ft. lbs. (30 Nm).

10. Install the water pump pulley and torque the bolts to 21 ft. lbs. (30 Nm). Connect the radiator hose and heater hose to the water pump.

11. Install all items removed to gain access to the water pump, then adjust the belts.

12. Remove the hex-head plug on the top of the thermostat housing. Fill the radiator with coolant until the coolant comes out the plug hole. Install the plug and continue to fill the radiator.

13. Connect the negative battery ca-

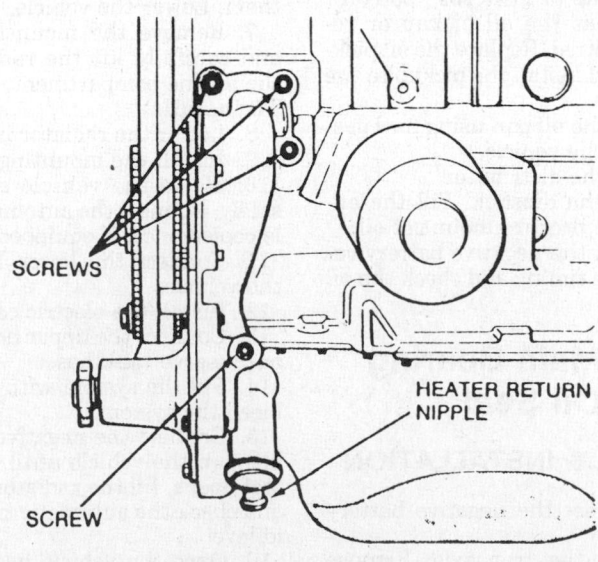

SCREWS

HEATER RETURN NIPPLE

SCREW

Water pump assembly

ble, run the vehicle until the thermostat opens, fill the radiator completely and check for leaks.

14. Once the vehicle has cooled, recheck the coolant level.

Thermostat

REMOVAL & INSTALLATION

1. Disconnect the negative battery cable. Drain the coolant down to thermostat level or below.
2. Remove the thermostat housing.
3. Remove the thermostat and discard the gasket.
4. Clean the housing mating surfaces and use a new gasket.
5. The installation is the reverse of the removal procedure.
6. Remove the hex-head plug on the thermostat housing. Fill the radiator with coolant until the coolant comes out the plug hole. Install the plug or valve and continue to fill the radiator.
7. Connect the negative battery cable, run the vehicle until the thermostat opens, fill the radiator completely and check for leaks.
8. Once the vehicle has cooled, recheck the coolant level.

COOLING SYSTEM BLEEDING

To bleed air from the engine, remove the hex-head plug on the top of the thermostat housing. Fill the radiator with coolant until the coolant comes out the hole. Install the switch or plug and continue to fill the radiator.

ENGINE ELECTRICAL

NOTE: Disconnecting the negative battery cable on some vehicles may interfere with the functions of the on board computer systems and may require the computer to undergo a relearning process, once the negative battery cable is reconnected.

Distributor

REMOVAL

1. Disconnect the negative battery cable.
2. Disconnect the distributor pickup lead wires. Remove the splash shield, if equipped.
3. Unscrew the distributor cap

hold-down screws and lift off the distributor cap with all ignition wires still connected.

4. Matchmark the rotor to the distributor housing and the distributor housing to the engine.

NOTE: Do not crank the engine during this procedure. If the engine is cranked, the matchmark must be disregarded.

5. Remove the hold-down bolt and clamp.
6. Remove the distributor from the engine.

INSTALLATION

Timing Not Disturbed

1. Install a new distributor housing O-ring.
2. Install the distributor in the engine so the rotor is aligned with the matchmark on the housing and the housing is aligned with the matchmark on the engine. Make sure the distributor is fully seated and the distributor shaft is fully engaged.
3. Install the hold-down clamp and snug the hold-down bolt.
4. Connect the distributor pickup lead wires. Install the splash shield, if equipped.
5. Install the distributor cap and tighten the screws.
6. Connect the negative battery cable.
7. Adjust the ignition timing and tighten the hold-down bolt.

Timing Disturbed

1. Install a new distributor housing O-ring.
2. Position the engine so the No. 1 piston is at TDC of its compression stroke and the mark on the vibration damper is aligned with **0** on the timing indicator.
3. Install the distributor in the engine so the rotor is aligned with the position of the No. 1 ignition wire on the distributor cap and the housing is aligned with the matchmark on the engine. Make sure the distributor is fully seated and that the distributor shaft is fully engaged.
4. Install the hold-down clamp and snug the hold-down bolt.
5. Connect the distributor pickup lead wires. Install the splash shield, if equipped.
6. Install the distributor cap and tighten the screws.
7. Connect the negative battery cable.
8. Adjust the ignition timing and tighten the hold-down bolt.

Ignition Timing

ADJUSTMENT

1. Start the engine, set the parking brake and run the engine until at normal operating temperature. Keep all lights and accessories off.
2. If a magnetic timing unit is available, insert the probe into the receptacle near the timing scale. The scale is located on the top of the bellhousing.
3. If a magnetic timing unit is not available, connect a conventional power timing light to the No. 1 cylinder spark plug wire.
4. Connect the red lead of a tachometer to the negative primary terminal of the coil and connect the black lead to a good ground.
5. Disconnect the coolant sensor located near the thermostat housing. The Check Engine light on the instrument panel must be on.
6. On 1989–90 vehicles, the Diagnostic Readout Box II (DRB II), if available, may be used in conjuction with the Basic Timing Mode.
7. Aim the timing light at the timing scale or read the magnetic timing unit.
8. Loosen the distributor hold-down bolt just enough so the distributor can be rotated.
9. Turn the distributor in the proper direction until the specified timing according to the VECI label is reached. Tighten the hold-down bolt and recheck the timing and idle speed.
10. Turn the engine **OFF**. Connect the vacuum hose or coolant sensor. Make sure the Check Engine light does not come ON when restarted.
11. Disconnect the timing apparatus and tachometer.
12. If the coolant temperature sensor was disconnected, erase the created fault code using the Erase Fault Code mode on the DRB II. The code can also be erased by disconnecting the battery. If the coolant sensor code is not erased at this point, it will disappear after 50–100 vehicle key ON/OFF cycles, providing there is no problem with that circuit.

Alternator

PRECAUTIONS

Several precautions must be observed when working with the alternator to avoid damaging the unit.

● If the battery is removed for any reason, make sure it is reconnected with the correct polarity. Reversing the battery connections may result in damage to the one-way rectifiers.

- When utilizing a booster battery as a starting aid, always connect the positive to positive terminals and the negative terminal from the booster battery to a good engine ground on the vehicle being started.
- Never use a fast charger as a booster to start vehicles.
- Disconnect the battery cables when charging the battery with a fast charger.
- Never attempt to polarize the alternator.
- Do not use test lights of more than 12 volts when checking diode continuity.
- Do not short across or ground any of the alternator terminals.
- The polarity of the battery, alternator and regulator must be matched and considered before making any electrical connections within the system.
- Never separate the alternator on an open circuit. Make sure all connections within the circuit are clean and tight.
- Disconnect the battery ground terminal when performing any service on electrical components.
- Disconnect the battery if arc welding is to be done on the vehicle.

BELT TENSION ADJUSTMENT

1. Loosen the pivot bolt slightly.
2. Raise the vehicle and support safely. Remove the splash shield. Loosen the adjuster slot bolt or "T" bolt locknut enough so the alternator can be moved. Loosen the adjusting bolt locknut, if equipped.
3. Tighten the adjusting bolt until the belt deflects about ¼ in. under a 10 lb. load.
4. Tighten the "T" bolt locknut or adjusting bolt locknut.

REMOVAL & INSTALLATION

1. Disconnect the negative battery cable.
2. Remove the air conditioning compressor and position it aside. Remove the oil filter to allow the alternator to be removed from above.
3. Loosen the mounting bolts, move the alternator toward the engine and remove the drive belts.
4. Remove the mounting bolts and spacers and remove the alternator from the brackets.
5. Remove the battery positive, field and ground terminals from the rear of the alternator. Remove the wire har-

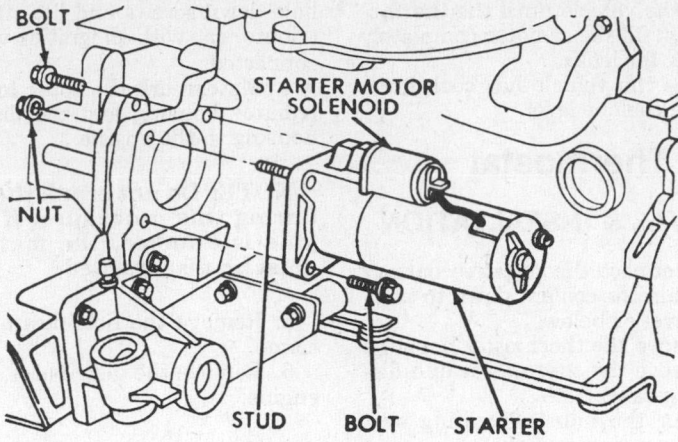

Starter motor mounting

ness hold-down screw from the alternator, if equipped.

To install:
6. Connect all wiring to the proper terminals on the rear of the alternator and install the wire harness hold-down screw, if equipped.
7. Position the alternator in the mounting brackets.
8. Install the spacers, pivot bolt and adjuster bolt. Install the belt.
9. Install the air conditioning compressor and oil filter.
10. Adjust the belt tension.
11. Connect the negative battery cable and check the alternator for proper operation.

Starter

REMOVAL & INSTALLATION

1. Disconnect the negative battery cable.
2. Remove the attaching nut and bolt at the top of the bellhousing. Raise the vehicle and support safely.
3. Remove the rear mount and heat shield from the starter.
4. Unbolt the starter and remove from the vehicle.
5. Disconnect the solenoid lead wires from the starter.

To install:
6. Connect the solenoid lead wires and install the heat shield.
7. Install the lower bolt loosely, then lower the vehicle and install the nut and bolt from above and torque to 40 ft. lbs. (54 Nm).
8. Raise the vehicle and torque the bottom bolt to the same value. Install the rear mount to the starter.
9. Connect the negative battery cable and check the starter for proper operation.

EMISSION CONTROLS

Please refer to "Emission Controls" in the Unit Repair section for system maintenance procedures. Due to the complex nature of modern electronic engine control systems, comprehensive diagnosis and testing procedures fall outside the confines of this repair manual. For complete information on diagnosis, testing and repair procedures concerning all modern engine and emission control systems, please refer to "Chilton's Guide to Fuel Injection and Electronic Engine Controls".

FUEL SYSTEM

Fuel System Service Precautions

Safety is the most important factor when performing not only fuel system maintenance but any type of maintenance. Failure to conduct maintenance and repairs in a safe manner may result in serious personal injury or death. Maintenance and testing of the vehicle's fuel system components can be accomplished safely and effectively by adhering to the following rules and guidelines.
- To avoid the possibility of fire and personal injury, always disconnect the negative battery cable unless the re-

pair or test procedure requires that battery voltage be applied.

● Always relieve the fuel system pressure prior to disconnecting any fuel system component (injector, fuel rail, pressure regulator, etc.), fitting or fuel line connection. Exercise extreme caution whenever relieving fuel system pressure to avoid exposing skin, face and eyes to fuel spray. Please be advised that fuel under pressure may penetrate the skin or any part of the body that it contacts.

● Always place a shop towel or cloth around the fitting or connection prior to loosening to absorb any excess fuel due to spillage. Ensure that all fuel spillage (should it occur) is quickly removed from engine surfaces. Ensure that all fuel soaked cloths or towels are deposited into a suitable waste container.

● Always keep a dry chemical (Class B) fire extinguisher near the work area.

● Do not allow fuel spray or fuel vapors to come into contact with a spark or open flame.

● Always use a backup wrench when loosening and tightening fuel line connection fittings. This will prevent unnecessary stress and torsion to fuel line piping. Always follow the proper torque specifications.

● Always replace worn fuel fitting O-rings with new. Do not substitute fuel hose or equivalent where fuel pipe is installed.

RELIEVING FUEL SYSTEM PRESSURE

1988 Vehicles

1. Disconnect the negative battery cable.
2. Loosen the fuel filler cap to release fuel tank pressure.
3. Remove the wiring harness connector from the (any) injector.
4. Using a jumper wire, ground either injector terminal.
5. Being careful not to allow contact between the jumper leads, connect a second jumper wire to the other terminal and touch the other end to the positive battery post for no longer that 10 seconds. This will relieve fuel pressure.
6. Remove the jumper wires and continue with fuel system service.

1989–90 Vehicles

1. Loosen the fuel filler cap to release fuel tank pressure.
2. Locate and disconnect the fuel injector harness connector on the throttle body.
3. Connect a jumper wire from ter-

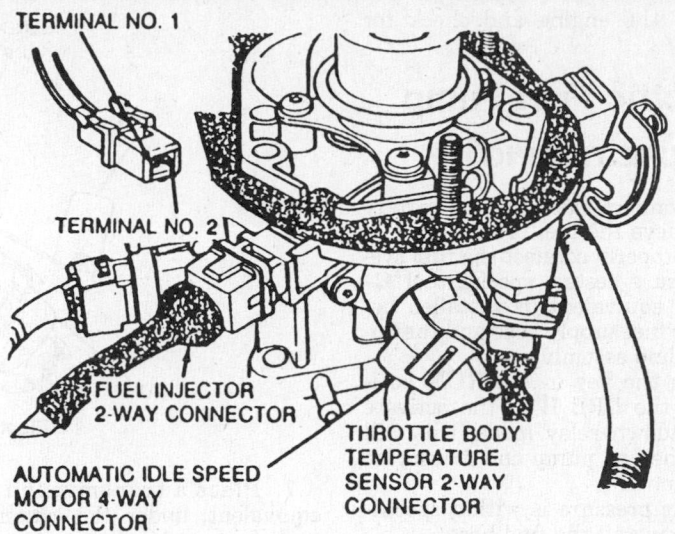

TERMINAL NO. 1

TERMINAL NO. 2

FUEL INJECTOR 2-WAY CONNECTOR

AUTOMATIC IDLE SPEED MOTOR 4-WAY CONNECTOR

THROTTLE BODY TEMPERATURE SENSOR 2-WAY CONNECTOR

Throttle body fuel injector harness connector

minal No. 1 of the connector to ground.
4. Being careful not to allow contact between the jumper leads, connect a jumper wire to terminal No. 2 of the connector and touch the other end of the jumper to the positive battery post for no longer than 5 seconds. This will relieve fuel pressure.
5. Remove the jumper wires and continue with fuel system service.

Fuel Tank

REMOVAL & INSTALLATION

1. Disconnect the negative battery cable.
2. Relieve the fuel pressure.
3. Raise the vehicle and support safely.
4. Using the proper equipment, drain the fuel tank.
5. Remove the screws that hold the filler neck to the quarter panel.
6. Disconnect the wiring and hoses from the tank.
7. Place a transmission jack or equivalent, under the center of the tank and apply slight pressure. Remove the tank straps.
8. Remove the filler tube from the tank.
9. Lower the tank and disconnect the vapor separator rollover valve hose. Remove the fuel tank from the vehicle.
To install:
10. Raise the tank into position and connect all harnesses and vacuum hoses.
11. Install the tank straps and tighten the retaining nuts.
12. Install the screws that hold the filler neck to the quarter panel.

13. Connect the negative battery cable, start the engine and check for leaks.

Fuel Filter

REMOVAL & INSTALLATION

——— CAUTION ———
Do not use conventional fuel filters, hoses or clamps when servicing this fuel system. They are not compatible with the injection system and could fail, causing personal injury or damage to the vehicle. Use only hoses and clamps specifically designed for fuel injection.

1. Disconnect the negative battery cable.
2. Relieve the fuel pressure.
3. The filter is located under the rear of the vehicle, in front of the fuel tank. Raise the vehicle and support safely. Remove the filter retaining screw and remove the filter assembly from the mounting plate.
4. Loosen the outlet hose clamp on the filter and inlet hose clamp on the rear fuel tube.
5. Wrap a shop towel around the hoses to absorb fuel. Remove the hoses from the filter and fuel tube and discard the clamps and the filter.
To install:
6. Install the inlet hose on the fuel tube and tighten the new clamp to 10 inch lbs.
7. Install the outlet hose on the filter outlet fitting and tighten the new clamp to 10 inch lbs.
8. Position the filter assembly on the mounting plate and tighten the mounting screw to 75 inch lbs. (8 Nm)
9. Connect the negative battery ca-

ble, start the engine and check for leaks.

Electric Fuel Pump

PRESSURE TESTING

1. Disconnect the negative battery cable. Relieve the fuel pressure.
2. To properly connect the fuel system pressure tester, special tool C–4799A or equivalent, is installed between the fuel supply hose and the engine fuel line assembly.
3. With the key in the **RUN** position, put the DRB II in the activate auto shutdown relay mode; this will activate the fuel pump and pressurize the system.
4. If the pressure is within specifications, reinstall the fuel hose.
5. If fuel pressure is below specifications, install the tester in the fuel supply line between the tank and the filter and repeat the test.
6. If the pressure is 5 psi higher than in Step 5, replace the fuel filter. If no change is observed, squeeze the return hose. If pressure increases, replace the pressure regulator. If no change is observed, the problem is either a plugged in-tank sock filter or a defective pump.
7. If fuel pressure is above specifications, remove the fuel return line hose from the chassis line at the fuel tank and connect a 3 foot piece of fuel hose to the return line. Put the other end into a 2 gallon minimum capacity approved gasoline container. Repeat the test. If pressure is now correct, check the in-tank return hose for kinking. Replace the fuel pump assembly if the in-tank reservoir check valve or aspirator jet is obstructed.
8. If pressure is still above specifications, remove the fuel return hose from the throttle body. Connect a substitute hose to the throttle body return nipple and place the other end of the hose in a clean container. Repeat the test. If pressure is now correct, check for a restricted fuel return line. If no change is observed, replace the fuel pressure regulator.

REMOVAL & INSTALLATION

1. Disconnect the negative battery cable.
2. Relieve the fuel pressure.
3. Raise the vehicle and support safely.
4. Using the proper equipment, drain the fuel tank.
5. Remove the screws that hold the filler neck to the quarter panel.
6. Disconnect the wiring and hoses from the tank.

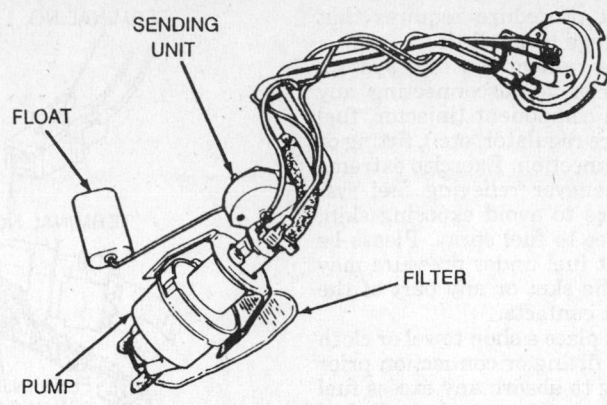

Electronic fuel pump assembly

7. Place a transmission jack or equivalent, under the center of the tank and apply slight pressure. Remove the tank straps.
8. Lower the tank and remove the filler tube from the tank.
9. Disconnect the vapor separator rollover valve hose and remove the fuel tank from the vehicle.
10. Using a hammer and a brass drift, tap the lock ring counterclockwise to release the pump.
11. Remove the pump from the tank with the O-ring. Discard the O-ring, pump inlet filter and inlet seal. Disassemble as required.
To install:
12. Install a new inlet seal and filter on the end of the pump.
13. Install a new O-ring to the pump.
14. Connect the reservoir hose to the pump assembly at the suction end of the pump. Press the female fitting onto the pump assembly male end until the ears snap in place.
15. Install the pump assembly into the tank.
16. Install the lock ring with a hammer and brass punch turning the ring clockwise.
17. Install the fuel tank.

18. Connect the negative battery cable, start the engine and check for leaks.

Fuel Injection

IDLE SPEED ADJUSTMENT

The idle speed is controlled by the Automatic Idle Speed (AIS) motor. The AIS motor is controlled by the SMEC or SBEC, which receives data from various sensors and switches in the system and adjusts the engine idle to a predetermined speed. Idle speed specifications can be found on the Vehicle Emission Control Information (VECI) label located in the engine compartment. If the idle speed is not within specifications and there are no problems with the system, the throttle body should be replaced.

IDLE MIXTURE ADJUSTMENT

There is no idle mixture adjustment provided with any Chrysler fuel injection system.

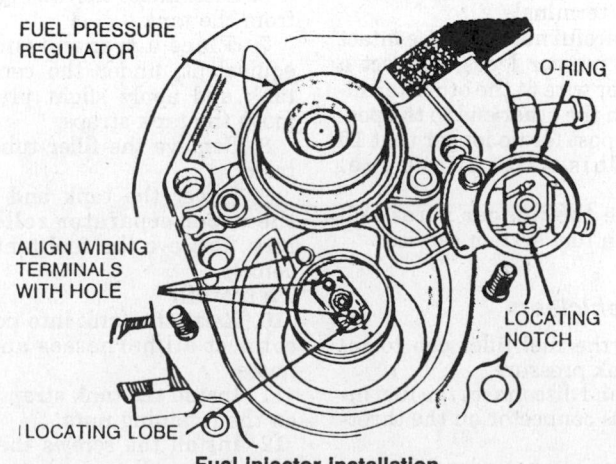

Fuel Injector Installation

Fuel Injector

REMOVAL & INSTALLATION

1. Disconnect the negative battery cable.
2. Remove the air cleaner assembly.
3. Relieve the fuel pressure.
4. Remove the injector hold-down Torx® screw and the hold-down.
5. Using a small flat-tipped tool, lift the cap off of the injector.
6. Using the same tool, gently pry the injector from its pod.
7. Remove the lower O-ring from the pod.
To install:
8. Install the new lower O-ring on the injector.
9. Align the injector terminal housing with the locating socket in the injector cap.
10. Press the injector cap so the upper O-ring flange is flush with the lower surface of the cap.
11. Spray the inner surfaces of the injector pod with suitable parts cleaner to remove residual varnish and gasoline.
12. Lubricate the O-rings sparingly with unmedicated petroleum jelly.
13. Place the injector and cap into the injector pod and align the cap lo-cating pin with the locating hole in the casting.
14. Press firmly on the injector cap until it is flush with the casting surface.
15. Align the hole in the hold-down with the pin on the cap and install.
16. Push down on the cap, install the screw and torque to 35 inch lbs. (4 Nm).
17. Connect the negative battery cable and check for leaks using the DRB II to activate the fuel pump.
18. Install the air cleaner.

DRIVE AXLE

Halfshaft

REMOVAL & INSTALLATION

1. Disconnect the negative battery cable.
2. Raise the vehicle and support safely.
3. Remove the tire and wheel assembly.
4. Remove the cotter pin from the end of the halfshaft. Remove the nut lock, spring washer, axle nut and washer.
5. Remove the ball joint retaining bolt and pry the control arm down to release the ball stud from the steering knuckle.
6. Position a drain pan under the transaxle where the halfshaft enters the differential or extension housing. Remove the halfshaft from the trans-axle or center bearing. Unbolt the center bearing from the block and remove the intermediate shaft from the trans-axle, if equipped.

To install:
7. Install the halfshaft or intermediate shaft to the transaxle, being careful not to damage the side seals. Make sure the inner joint clicks into pace inside the differential. Install the center bearing retaining bolts, if equipped. Install the outer shaft to the center bearing, if equipped.
8. Pull the front strut out and insert the outer joint into the front hub.
9. If necessary, turn the ball joint stud to position the bolt retaining indent to the inside of the vehicle. Install the ball joint stud into the steering knuckle. Install the retaining bolt and nut.
10. Install the axle nut washer and nut and torque the nut to 180 ft. lbs.

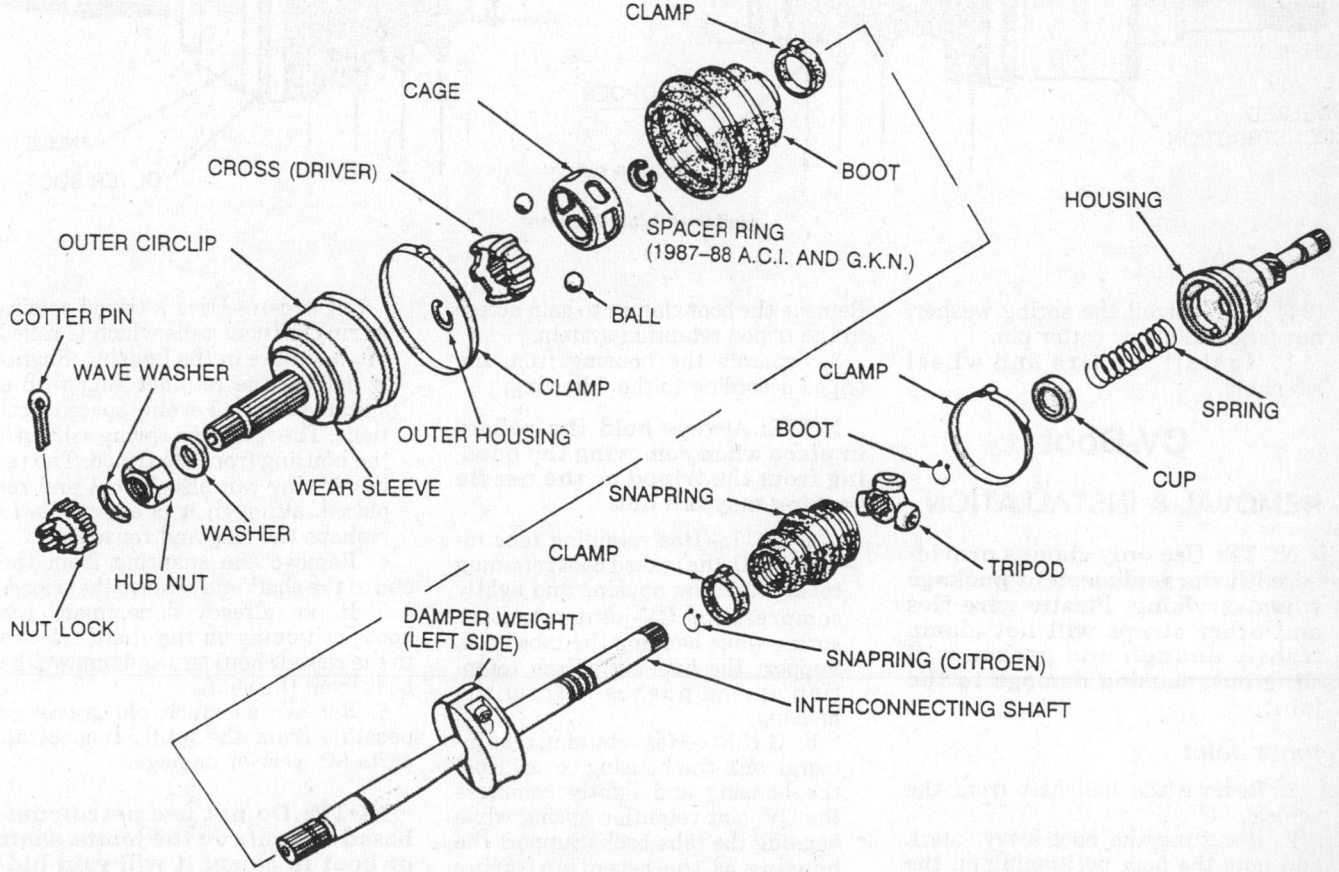

Exploded view of the halfshaft

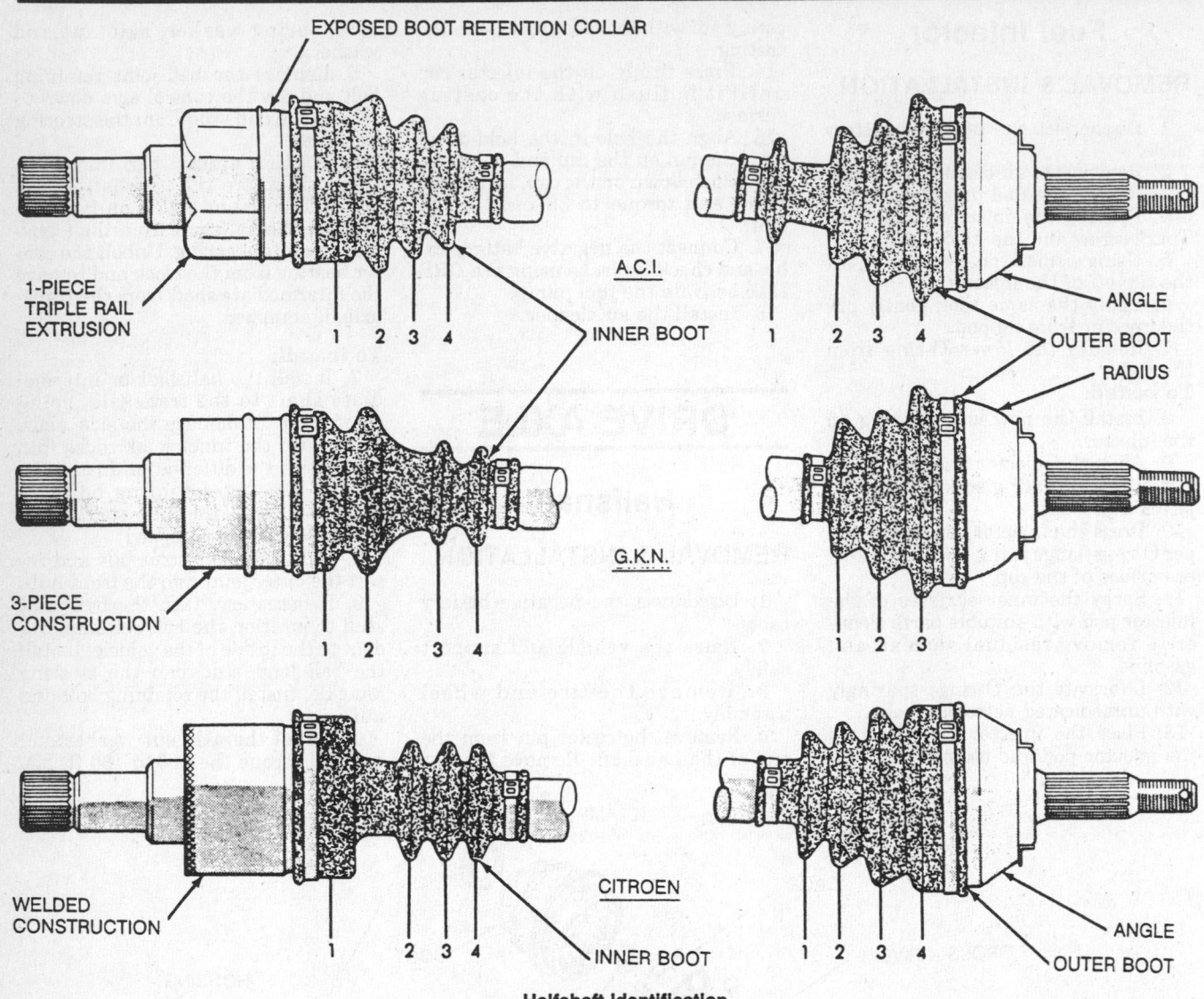

EXPOSED BOOT RETENTION COLLAR

1-PIECE
TRIPLE RAIL
EXTRUSION

3-PIECE
CONSTRUCTION

WELDED
CONSTRUCTION

INNER BOOT

A.C.I.

G.K.N.

CITROEN

INNER BOOT

ANGLE

OUTER BOOT

RADIUS

ANGLE

OUTER BOOT

Halfshaft Identification

(244 Nm). Install the spring washer, nut lock and a new cotter pin.

11. Install the tire and wheel assembly.

CV-Boot

REMOVAL & INSTALLATION

NOTE: Use only clamps provided with the replacement package when servicing. Plastic wire ties and other straps will not clamp tightly enough and grease will sling out, causing damage to the joint.

Inner Joint

1. Remove the halfshaft from the vehicle.

2. If cutting the boot away, mark and note the boot positioning on the shaft relative to the raised shoulders.

Remove the boot clamps to gain access to the tripod retention system.

3. Separate the housing from the tripod according to the following:

NOTE: Always hold the rollers in place when removing the housing from the tripod or the needle bearing may fall out.

a. A.C.I.—Has retaining tabs integral with the staked boot retaining collar. Hold the housing and lightly compress the CV-joint retention spring while bending the tabs back. Support the housing as the retention spring pushes it from the housing.

b. G.K.N.—Has retaining tabs integral with the housing cover. Hold the housing and lightly compress the CV-joint retention spring while bending the tabs back. Support the housing as the retention spring pushes it from the housing.

c. Citroen—Uses a tripod retainer ring without tabs which is rolled into a groove in the housing. Slightly deform the retainer ring with a suitable tool in 3 evenly spaced locations. The retention spring will push the housing from the tripod. The retainer ring can also be cut and replaced, although it is acceptable to reshape the ring and reuse it.

4. Remove the snapring from the end of the shaft and remove the tripod.

5. If not already done, mark the boot positioning on the shaft, relative to the raised shoulders and remove the boot from the shaft.

6. Remove as much old grease as possible from the joint. Inspect all parts for wear or damage.

NOTE: Do not use petroleum-based solvents on the joints, shaft or boot to clean; it will ruin hidden rubber seals within the joint.

Use only chlorine based cleaner or hot soapy water to clean the joint, if necessary. Make sure the joint it completely dry before assembling.

To install:

7. If equipped, slide a new rubber washer seal over the stub shaft and down into the groove provided.

8. If the clamping device is not a straight strap, install it on the shaft first, then install the boot to the shaft in the proper position. Using the proper tool, C–4124 for crimping with rubber boot or C–4653 for clamping a strap, secure the clamp.

9. Slide the tripod on the shaft:

   a. A.C.I.—Slide the tripod on the shaft with the non-chamfered edge facing the tripod retainer ring groove.

   b. G.K.N—Slide the tripod on the shaft with the non-chamfered edge facing the tripod retainer ring groove.

   c. Citroen—The tripod may installed either way; both ends are the same.

10. Install the snaping into its groove on the shaft to lock the tripod in position.

11. Distribute the grease provided in the grease package as follows, or according to the instructions in the package:

   a. A.C.I.—Distribute 1 of the 2 packets of grease into the boot and the remaining packet into the housing.

   b. G.K.N—Distribute 2 of the 3 packets of grease into the boot and the remaining packet into the housing.

   c. Citroen—Distribute $\frac{2}{3}$ of the packet of grease into the boot and the remaining amount into the housing.

12. Position the spring in the housing spring pocket with the spring cup attached to the exposed end of the spring. Place a dab of grease on the concave surface of the spring cup.

13. Keeping the spring centered, install the housing to the tripod as follows:

   a. A.C.I.—Slip the housing onto the tripod. Do not bend the retaining tabs back into their original position. Instead, secure the boot to hold the housing. The tripod must be reengaged to the housing with the shaft installed on the vehicle.

   b. G.K.N—Slip the housing onto the tripod. Bend the retaining tabs back into their original positions. Check for proper retention ability.

   c. Citroen—Slip the housing onto the tripod. Reform the retainer ring or install a new one. Stake in place using a punch. Check for proper retention ability.

14. Position the larger end of the boot over the housing.

15. Using the proper tool, C–4124 for crimping with rubber boot or C–4653 for clamping a strap, secure the clamp.

16. Install the halfshaft to the vehicle. Fill the transaxle if fluid was lost when removing the halfshaft.

17. Road test the vehicle.

Outer Joint

1. Remove the halfshaft from the vehicle.

2. If cutting the boot away, mark and note the boot positioning on the shaft relative to the raised shoulders. Remove the boot clamps to gain access to the joint retention system.

3. Separate the housing from the tripod according to the following:

   a. A.C.I.—Using a soft-jaw vise, support the halfshaft. Strike the joint assembly sharply with a soft-face hammer to dislodge the internal circlip and remove from the shaft.

   b. G.K.N—Using a soft-jaw vise, support the halfshaft. Strike the joint assembly sharply with a soft-face hammer to dislodge the internal circlip and remove from the shaft.

   c. Citroen—Using a soft-jaw vise, support the halfshaft. Strike the joint assembly sharply with a soft-face hammer to dislodge the internal circlip and remove from the shaft.

4. If damaged, remove the wear sleeve from the CV-joint machined ledge.

5. Remove the circlip from the groove.

6. If not already done, mark the boot positioning on the shaft, relative to the raised shoulders and remove the boot from the shaft.

7. Remove as much old grease as possible from the joint. Inspect all parts for wear or damage.

NOTE: Do not use petroleum based solvents on the joints, shaft or boot to clean; it will ruin hidden rubber seals within the joint. Use only chlorine based cleaner or hot soapy water to clean the joint, if necessary. Make sure the joint it completely dry before assembling.

To install:

8. If the clamping device is not a straight strap, install it on the shaft first, then install the boot to the shaft in the proper position. Using the proper tool, C–4124 for crimping with rubber boot or C–4653 for clamping a strap, secure the clamp. Install a new circlip if provided in the replacement package.

9. Fill the boot with the proper amount of grease according to the instructions provided with the package.

10. Position the outer joint on the shaft with hub nut installed, engage the splines and strike sharply with a soft-face hammer to install. Make sure the circlip did not become dislodged.

11. Position the larger end of the boot over the housing.

12. Using the proper tool, C–4124 for crimping with rubber boot or C–4653 for clamping a strap, secure the clamp.

13. Install the halfshaft to the vehicle. Fill the transaxle if fluid was lost when removing the halfshaft.

14. Road test the vehicle.

Front Wheel Hub and Bearing

REMOVAL & INSTALLATION

Pressed In (Two-Piece Hub and Bearing)

NOTE: Some hub and bearing replacement packages include the one-piece unit described below. If this is the case, follow the installation steps for one-piece unit instead of for the two-piece unit described here.

1. Raise the vehicle and support safely.

2. Remove the wheel. Remove the brake caliper from the adaptor and remove the adaptor. Remove the brake disc.

3. Remove the halfshaft.

NOTE: Knuckle removal is not necessary for bearing and hub replacement.

4. Disconnect the tie rod from the knuckle.

5. Matchmark the lower strut mount to the knuckle. Remove the 2 strut clamp bolts and remove the knuckle from the vehicle.

6. Attach the hub removal tool C–4811 or equivalent, and the triangular adapter, to the 3 rear threaded holes of the steering knuckle housing with the thrust button inside the hub bore.

7. Tighten the bolt in the center of the tool, to press the hub from the steering knuckle. Remove the removal tools.

8. Remove the bolts and bearing retainer from the outside of the steering knuckle.

9. Carefully pry the bearing seal from the machined recess of the steering knuckle and clean the recess.

10. Insert the tool C–4811 or equivalent, through the hub bearing and install bearing removal adapter to the outside of the steering knuckle. Tight-

en the tool to press the hub bearing from the steering knuckle. Discard the bearing and the seal.

To install:

11. Use tool C–4811 or equivalent, and the bearing installation adapter to press in the hub bearing into the steering knuckle.

12. Install a new seal, the bearing retainer and the bolts to the steering knuckle. Torque the bearing retainer bolts to 20 ft. lbs.

13. Use the tool C–4811 or equivalent, and the hub installation adapter, to press the hub into the hub bearing.

14. Using the bearing installation tool C–4698 or equivalent, drive the new dust seal into the rear of the steering the hub and bearing from the knuckle, as required.

15. The installation of the knuckle and halfshaft is the reverse of the removal procedure. Torque the tie rod nut to 35 ft. lbs. (47 Nm).

16. Align the front end.

Bolt In (One-Piece Hub and Bearing)

NOTE: Knuckle removal is not necessary for bearing and hub replacement.

1. Raise the vehicle and support safely.

2. Remove the wheel. Remove the brake caliper from the adaptor and remove the adaptor. Remove the brake disc.

3. Remove the halfshaft.

4. Disconnect the tie rod from the knuckle.

5. Matchmark the lower strut mount to the knuckle. Remove the 2 strut clamp bolts and remove the knuckle from the vehicle.

6. Remove the 4 hub and bearing assembly mounting bolts from the rear of the knuckle and remove the assembly from the knuckle.

7. Carefully pry the bearing seal from the machined recess of the steering knuckle and clean the recess.

8. Thoroughly clean and dry the knuckle and bearing mating surfaces and the seal installation area.

To install:

9. Install the hub and bearing assembly to the knuckle and torque the bolts in a criss-cross pattern to 45 ft. lbs. (65 Nm).

10. Install a new seal and wear sleeve. Lubricate the circumferences of the seal and sleeve liberally with grease.

11. The installation of the knuckle and halfshaft is the reverse of the removal procedure. Torque the tie rod nut to 35 ft. lbs. (47 Nm).

12. Align the front end.

Differential Case

REMOVAL & INSTALLATION

NOTE: The differential case can be removed from some vehicles with the transaxle installed. To do so, remove the halfshafts, remove the 2 K-frame mounting nuts and 2 bolts and lower the K-frame to provide enough room to pull the differential case out of its housing and over the lowered frame.

1. Disconnect the negative battery cable. Raise the vehicle and support safely.

2. Remove the right side extension housing. Remove the differential cover.

3. Remove the bolts and remove the right side differential bearing retainer using tool L–4435 or equivalent.

4. Remove the differential case from the transaxle.

5. Use new seals and gasket material when assembling.

6. The installation is the reverse of the removal procedure. Torque the extension housing and bearing retainer bolts to 21 ft. lbs. (28 Nm).

7. Fill the transaxle with the proper oil.

MANUAL TRANSAXLE

For further information on transmissions/transaxles, please refer to "Chilton's Guide to Transmission Repair".

Transaxle Assembly

REMOVAL & INSTALLATION

NOTE: If the vehicle is going to be rolled while the transaxle is removed, obtain 2 outer CV-joints to install to the hubs. If the vehicle is rolled without the proper torque applied to the front wheel bearings, the bearings will no longer be usable.

1. Disconnect the negative battery cable. Remove the air pump and mounting bracket, if equipped.

2. Remove the air cleaner assembly with all ducts. Remove the upper bellhousing bolts. Disconnect the reverse light switch and the ground wire to the transaxle case.

3. Remove the starter attaching nut and bolt at the top of the bellhousing.

4. Raise the vehicle and support safely. Remove the wheels. Remove the axle end cotter pins, nut locks, spring washers and axle nuts.

5. Remove the ball joint retaining bolts and pry the control arm from the steering knuckle. Position a drain pan under the transaxle where the axles enter the differential or extension housing. Remove the axles from the transaxle or center bearing. Unbolt the center bearing and remove the intermediate axle from the transaxle, if equipped.

6. Disconnect the shifter linkage from the transaxle.

7. Remove the speedometer gear adaptor bolt and remove the adaptor from the transaxle.

8. Remove the rear mount from the starter, unbolt the starter and position it aside.

9. Using the proper equipment, support the weight of the engine.

10. Remove the front motor mount and bracket.

11. Position a suitable transmission jack under the transaxle.

12. Remove the lower bellhousing bolts.

13. Remove the left side splash shield. Remove the transaxle mount bolts.

14. Carefully pry the transaxle from the engine.

15. Slide the transaxle rearward until the input shaft clears the clutch disc.

16. Pull the transaxle completely away from the clutch housing and remove it from the vehicle.

17. To prepare the vehicle for rolling, support the engine with a suitable support or reinstall the front motor mount to the engine. Then reinstall the ball joints to the steering knuckle and install the retaining bolt. Install the obtained outer CV-joints to the hubs, install the washers and torque the axle nuts to 180 ft. lbs. (244 Nm). The vehicle may now be safely rolled.

To install:

18. Lubricate the pilot bushing and input shaft splines very lightly with high temperature lubricant.

19. Mount the transaxle securely on a suitable jack. Lift it in place until the input shaft is centered in the clutch housing opening. Roll the transaxle forward until the input shaft splines fully engage with the clutch disc and install the transaxle to clutch housing bolts.

20. Raise the transaxle and install the left side mount bolts.

21. Install the front motor mount and bracket.

22. Remove the engine and transaxle support fixtures.

23. Install the starter to the transaxle and install the lower bolt finger-tight.

24. Install a new O-ring to the speedometer cable adaptor and install to the extension housing; make sure it snaps in place. Install the retaining bolt.

25. Install the shifter linkage and snap into place. Install the anti-rotation link, if equipped.

26. Install the axles and center bearing, if equipped. Install the ball joints to the steering knuckles. Torque the axle nuts to 180 ft. lbs. (244 Nm) and install new cotter pins. Fill the transaxle with SAE 5W-30 engine oil. Install the splash shield and install the wheels. Lower the vehicle.

27. Install the upper bellhousing bolts.

28. Install the starter attaching nut and bolt at the top of the bellhousing. Raise the vehicle and tighten the starter bolt from under the vehicle. Lower the vehicle.

29. Connect the reverse light switch and the ground wire.

30. Install the air cleaner assembly and air pump, if equipped.

31. Connect the negative battery cable and check the transaxle for proper operation.

LINKAGE ADJUSTMENT

1. Working over the left front fender, remove the lock pin from the transaxle selector shaft housing.

2. Reverse the lock pin so the long end is down and insert it into the same threaded hole while pushing the selector shaft into the selector housing. A hole in the selector shaft will align with the lock pin, allowing the lock pin to be screwed into the housing. This operation locks the selector shaft in the neutral position between 1st and 2nd gears.

3. Raise the vehicle and support safely.

4. Loosen the clamp bolt that secures the gear shift tube to the gear shift connector. Check to see that the connector slides and turns freely on the tube.

5. Position the shifter mechanism connector assembly so the isolator is just contacting the up standing flange and the rib on the isolator is pointing to the hole in the blockout bracket.

6. Hold the connector isolator in this position and tighten the clamp bolt on the gear shift tube to 14 ft. lbs. (19 Nm). Lower the vehicle.

7. Remove the lock pin from the selector shaft housing and reinstall the lock pin, with the long end up, in the selector shaft housing. Torque the lock pin to 10 ft. lbs. (12 Nm).

8. Check the first/reverse shifting and blockout into reverse.

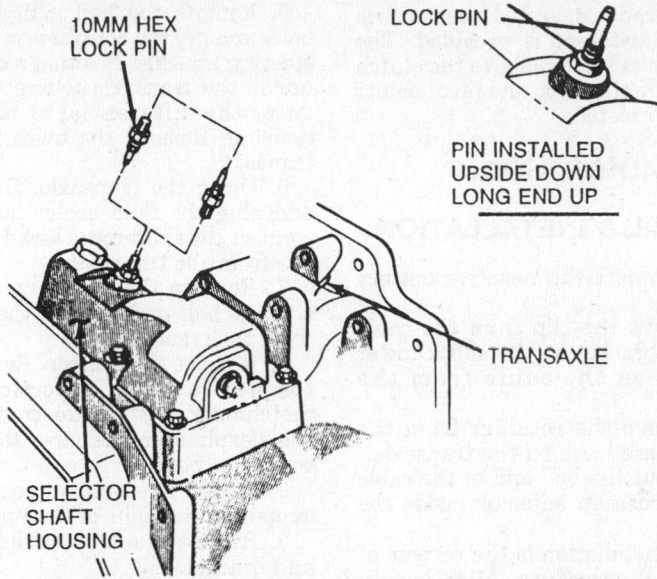

Manual transaxle pinned in the neutral position

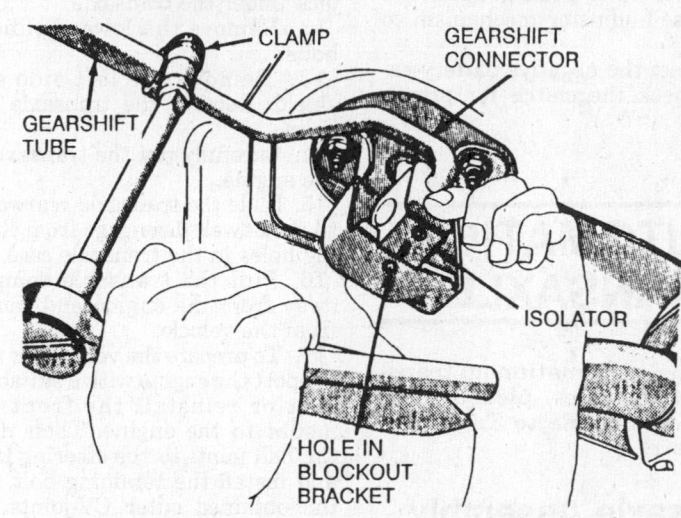

Adjusting the shifter linkage—manual transaxle

CLUTCH

Clutch Assembly

REMOVAL & INSTALLATION

1. Disconnect the negative battery cable. Remove the transaxle.

2. Matchmark the clutch/pressure plate cover and flywheel. Insert a suitable clutch plate alignment tool into the clutch disc hub.

3. Loosen the flywheel to pressure plate bolts gradually and evenly to avoid warpage.

4. Remove the pressure plate/clutch assembly from the flywheel.

5. Sand the flywheel or replace it, if scored, cracked or heat damaged.

6. Sparingly apply anti-sieze compound to the input shaft and clutch disc splines. Install a new release bearing.

To install:

7. Using a suitable clutch disc alignment tool, tighten the pressure plate bolts to center the disc.

8. Torque the pressure plate/clutch assembly mounting bolts to the flywheel gradually and evenly to 21 ft. lbs. (28 Nm).

9. Install the transaxle.

10. Connect the negative battery cable and check the clutch and reverse lights for proper operation.

PEDAL FREE-PLAY ADJUSTMENT

All vehicles are equipped with a self-adjusting cable operated mechanism and no adjustment is provided. The mechanism is located above the clutch pedal, where the cable and pivot points may be lubricated.

Clutch Cable

REMOVAL & INSTALLATION

1. Disconnect the negative battery cable.
2. Remove the clip from the cable mounting bracket on the shock tower and remove the cable from the bracket.
3. Remove the retainer from the clutch release lever on the transaxle.
4. Pry out the ball end of the cable from the position adjuster inside the pedal.
5. The installation is the reverse of the removal procedure. After installing, push the clutch pedal 2–3 times to allow the self-adjuster mechanism to function.
6. Connect the negative battery cable and check the clutch for proper operation.

AUTOMATIC TRANSAXLE

For further information on transmissions/transaxles, please refer to "Chilton's Guide to Transmission Repair".

Transaxle Assembly

REMOVAL & INSTALLATION

NOTE: If the vehicle is going to be rolled while the transaxle is removed, obtain 2 outer CV-joints to install to the hubs. If the vehicle is rolled without the proper torque applied to the front wheel bearings, the bearings will no longer be usable.

1. Disconnect the negative battery cable. Remove the air pump and mounting bracket, if equipped.
2. Remove the air cleaner assembly if it is preventing access to the upper bellhousing bolts. Remove the upper bellhousing bolts and unplug all electrical connectors from the transaxle.
3. Remove the starter attaching nut and bolt at the top of the bellhousing.
4. Raise the vehicle and support safely. Remove the wheels. Remove the axle end cotter pins, nut locks, spring washers and axle nuts.

5. Remove the ball joint retaining bolts and pry the control arm from the steering knuckle. Position a drain pan under the transaxle where the axles enter the differential or extension housing. Remove the axles from the transaxle.
6. Drain the transaxle. Disconnect and plug the fluid cooler hoses. Disconnect the shifter and kickdown linkage from the transaxle.
7. Remove the speedometer cable adaptor bolt and remove the adaptor from the transaxle.
8. Remove the starter. Remove the torque converter inspection cover, matchmark the torque converter to the flexplate and remove the torque converter bolts.
9. Using the proper equipment, support the weight of the engine.
10. Remove the front motor mount and bracket.
11. Position a suitable transmission jack under the transaxle.
12. Remove the lower bellhousing bolts.
13. Remove the left side splash shield. Remove the transaxle mount bolts.
14. Carefully pry the transaxle from the engine.
15. Slide the transaxle rearward until the dowels disengage from the mating holes in the transaxle case.
16. Pull the transaxle completely away from the engine and remove it from the vehicle.
17. To prepare the vehicle for rolling, support the engine with a suitable support or reinstall the front motor mount to the engine. Then reinstall the ball joints to the steering knuckle and install the retaining bolt. Install the obtained outer CV-joints to the hubs, install the washers and torque the axle nuts to 180 ft. lbs. (244 Nm). The vehicle may now be safely rolled.
To install:
18. Install the transaxle securely on the jack. Rotate the converter so it will align with the positioning of the flexplate.
19. Apply a light coating of high temperature grease to the torque converter pilot hub.
20. Raise the transaxle into place and push it forward until the dowels engage and the bellhousing is flush with the block. Install the transaxle to bellhousing bolts.
21. Raise the transaxle and install the left side mount bolts. Install the torque converter bolts and torque to 55 ft. lbs. (74 Nm).
22. Install the front motor mount and bracket. Remove the engine and transaxle support fixtures.
23. Install the starter to the transaxle. Install the bolt finger-tight.
24. Install a new O-ring to the speed-

ometer cable adaptor and install to the extension housing; make sure it snaps in place. Install the retaining bolt.
25. Connect the shifter and kickdown linkage to the transaxle.
26. Install the axles. Install the ball joints to the steering knuckles. Torque the axle nuts to 180 ft. lbs. (244 Nm) and install new cotter pins. Install the splash shield and install the wheels. Lower the vehicle.
27. Install the upper bellhousing bolts.
28. Install the starter attaching nut and bolt at the top of the bellhousing. Raise the vehicle again and tighten the starter bolt from under the vehicle. Lower the vehicle.
29. Connect all electrical wiring to the transaxle. Install the air pump, if equipped.
30. Install the air cleaner assembly, if it was removed. Fill the transaxle with the proper amount of Mopar ATF Plus Type 7176. Conventional Dexron®II may be used if Type 7176 is not available.
31. Connect the negative battery cable and check the transaxle for proper operation.

SHIFT LINKAGE ADJUSTMENT

1. Place the shifter in the **P** detent.
2. Loosen the clamp bolt on the gear shift cable bracket.
3. Pull the shift lever all the way to the front detent position and tighten the lock screw.
4. Check for proper neutral safety switch operation.

THROTTLE PRESSURE CABLE ADJUSTMENT

1. Run the engine until it reaches normal operating temperature.
2. Loosen the cable mounting bracket lock screw.
3. Position the bracket so both alignment tabs are touching the transaxle case surface and tighten the lock screws.
4. Release the cross lock on the cable assembly by pulling the cross lock up.
5. To ensure proper adjustment, the cable must be free to slide all the way toward the engine against its stop after the cross lock is released.
6. Move the transaxle throttle control lever fully clockwise and press the cross lock down until it snaps into position.
7. Road test the vehicle and check the shift points.

FRONT SUSPENSION

MacPherson Strut

REMOVAL & INSTALLATION

1. Remove the 3 mounting nuts from the strut tower under the hood.
2. Raise the vehicle and support safely.
3. Remove the brake hose bracket screw from the strut.
4. Matchmark the lower strut mount to the knuckle and remove the strut to knuckle bolts, nuts and nut plate.
5. The installation is the reverse of the removal procedure. Torque the upper mounting nuts to 20 ft. lbs. (27 Nm). Do not tighten the lower mounting bolts until the front end alignment has been completed.
6. Perform a front end alignment. Torque the strut to knuckle nuts to 75 ft. lbs. (100 Nm) plus ¼ turn.

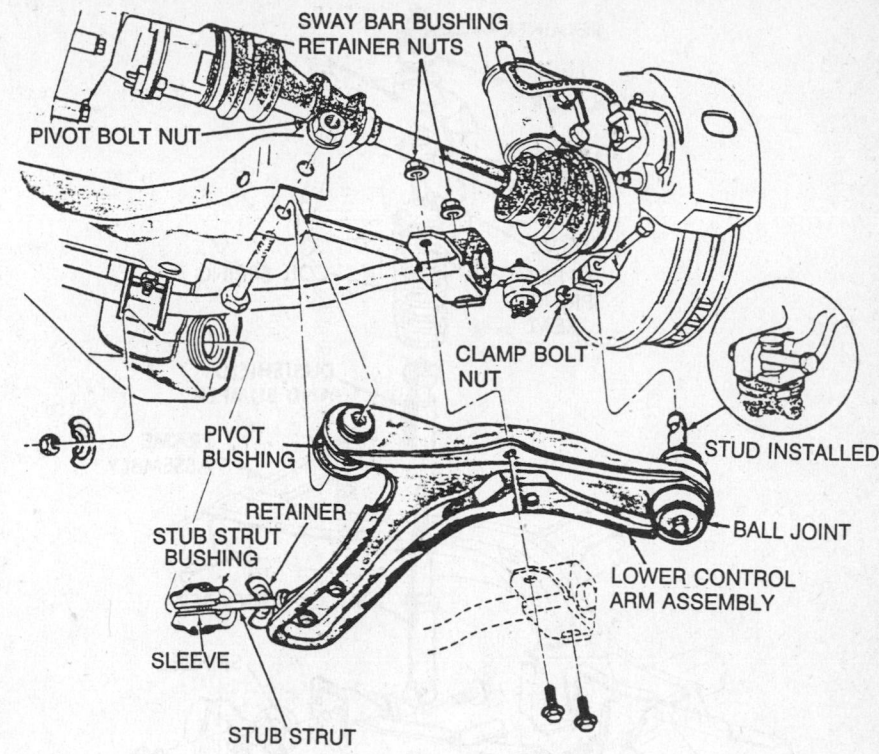

Lower control arm and related parts

Lower Ball Joints

INSPECTION

To inspect the ball joints, grasp the grease fitting by hand with the vehicle on the ground. If the grease fitting can be moved at all by hand, the ball joint should be replaced.

REMOVAL & INSTALLATION

The ball joints are welded to the lower control arms, necessitating replacement of the control arm assembly. Do not attempt to replace ball joints that are welded to the control arm; replacement control arms are equipped with a new ball joint.

Lower Control Arms

REMOVAL & INSTALLATION

1. Raise the vehicle and support safely. Remove the tire and wheel assembly.
2. Remove the sway bar.
3. Remove the ball joint stud retaining bolt and nut.
4. Pry the lower control arm from the steering knuckle.
5. Remove the control arm to cross-member bolts, nuts bushings and retainers.
6. Remove the control arm from the vehicle.
7. Transfer all reusable parts to the new control arm and lubricate.
8. The installation is the reverse of the removal procedure.
9. Lower the vehicle so it's full weight is on the ground.
10. Torque the pivot bolt to 120 ft. lbs. (163 Nm) and the stub strut nut to 70 ft. lbs. (95 Nm).
11. Perform a front end alignment, as required.

Sway Bar

REMOVAL & INSTALLATION

1. Raise the vehicle and support safely.
2. Remove the front sway bar brackets and retainers.
3. Remove the sway bar support brackets and bushings from the lower control arm. Remove the sway bar from the vehicle.
4. The installation is the reverse of the removal procedure. Lubricate the sway bar bushings liberally with grease before assembling.

REAR SUSPENSION

MacPherson Strut

REMOVAL & INSTALLATION

1. Remove interior rear trim panels, as required, to expose the upper strut mounting nut protective cap. Remove the cap.
2. Remove the upper mounting nut, isolator retainer and upper isolator.
3. Raise the vehicle and support safely.
4. Remove the lower strut mounting bolt.
5. Remove the strut from the vehicle.
6. The installation is the reverse of the removal procedure.

Rear Wheel Bearings

REMOVAL & INSTALLATION

1. Raise the vehicle and support safely.

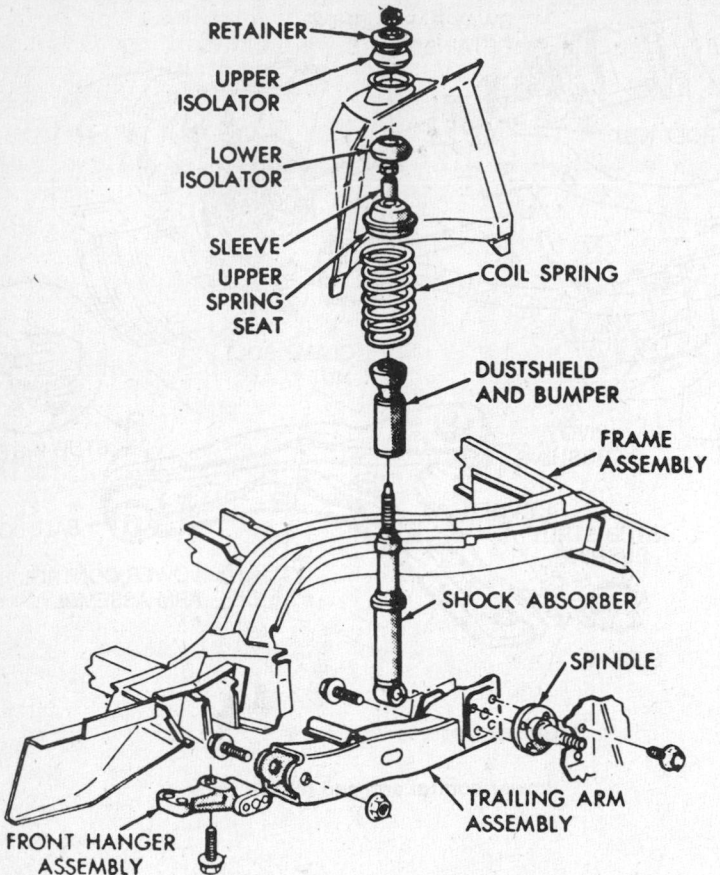

RETAINER

UPPER ISOLATOR

LOWER ISOLATOR

SLEEVE

UPPER SPRING SEAT

COIL SPRING

DUSTSHIELD AND BUMPER

FRAME ASSEMBLY

SHOCK ABSORBER

SPINDLE

TRAILING ARM ASSEMBLY

FRONT HANGER ASSEMBLY

Rear suspension components

2. Remove the tire and wheel assembly.

3. Remove the dust cap.

4. Remove the cotter pin, nut lock and nut.

5. Remove the thrust washer and the outer wheel bearing.

6. Remove the drum with the inner wheel bearing and the grease seal.

7. Remove the grease seal and remove the inner bearing.

To install:

8. Lubricate the inner bearing and install to the drum.

9. Install a new grease seal.

10. Install the drum to the vehicle.

11. Lubricate and install the outer wheel bearing to the spindle.

12. Install the thrust washer.

13. Install and tighten the wheel bearing nut to 20–25 ft. lbs. (27–34 Nm) while rotating the drum.

14. Back off the adjusting nut ¼ turn then tighten it finger-tight.

15. Install the nut lock and a new cotter pin.

Rear Axle Assembly

REMOVAL & INSTALLATION

1. Raise the vehicle and support safely. Remove the wheels.

2. Disconnect and plug the brake fittings and remove the retaining clips that hold the brake hoses.

3. Remove the parking brake cable adjusting nut.

4. Slip the ball ends of the cables through the connectors and release both parking cables from their brackets. Pull the cable through the brackets.

5. Using the proper equipment, safely support the weight of the rear axle assembly at 2 points.

6. Remove the strut lower mounting bolts.

7. Remove the trailing arm to hanger bracket mounting bolts.

8. Lower the axle assembly and remove from the vehicle.

To install:

9. Raise the assembly into position under the vehicle.

10. Install the trailing arm to hanger mounting bracket loosely.

11. Feed the parking brake cable through the bracket and slip the ball end through the brake connectors on the cable bracket. Install both retaining clips.

12. Install the parking brake adjusting nut and tighten until all slack is removed from the cabels.

13. Install the wheels and lower the

vehicle so all of its weight is on the ground.

14. With the vehicle on the ground, torque the trailing arm to hanger bracket mounting bolts and the lower strut mounting bolts to 40 ft. lbs. (55 Nm).

15. Raise the vehicle and bleed the brakes. Adjust the parking brakes.

STEERING

Steering Wheel
CAUTION

On vehicles equipped with an air bag, the negative battery cable must be disconnected, before working on the system. Failure to do so may result in deployment of the air bag and possible personal injury.

REMOVAL & INSTALLATION

Without Air Bag

1. Disconnect the negative battery cable.

2. Straighten the steering wheel so the front tires are pointing straight forward.

3. Remove the horn pad.

4. Remove the steering wheel hold-down nut. Matchmark the steering wheel to the shaft.

5. Using a suitable steering wheel puller, pull the steering wheel off of the shaft.

6. The installation is the reverse of the removal procedure. Torque the hold-down nut to 45 ft. lbs. (60 Nm).

With Air Bag

1. Disconnect the negative battery cable.

2. Straighten the steering wheel so the front tires are pointing straight forward.

3. Remove the 4 nuts located on the back side of the steering wheel that attach the air bag module to the steering wheel.

4. Lift the module and disconnect the connectors. Remove the speed control switch, if equipped.

NOTE: The column should be equipped with a clockspring set screw held by a plastic tether on the steering wheel. If the steering column is lacking the set screw, obtain one before proceeding.

5. Place the set screw in the clockspring to ensure proper positioning when the steering wheel is removed.

6. Remove the steering wheel hold-

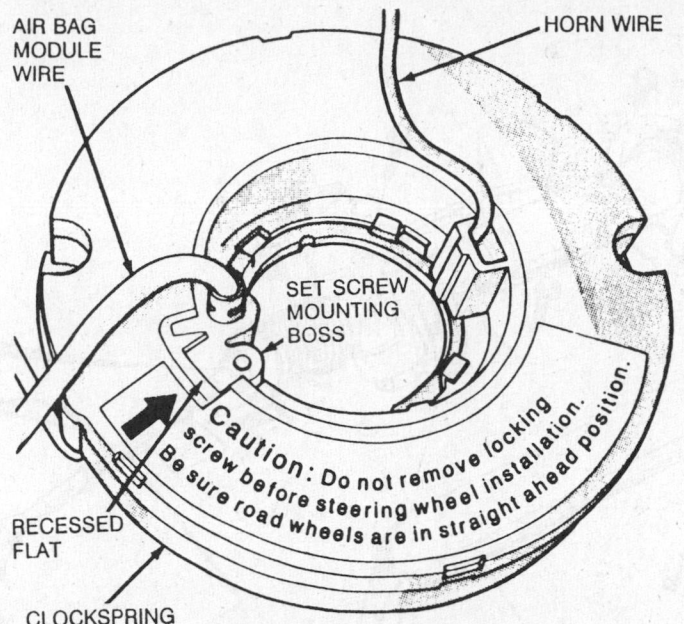

Air bag clockspring assembly

down nut. Matchmark the steering wheel to the shaft.

7. Using a suitable steering wheel puller, pull the steering wheel off of the shaft.

To install:

8. Position the steering wheel on the steering column. Make sure the flats on the hub of the steering wheel are aligned with the formations on the clockspring.

9. Pull the air bag and speed control connectors through the lower, larger hole in the steering wheel and pull the horn wire through the smaller hole at the top. Make sure the wires are not pinched anywhere.

10. Install the hold-down nut and torque to 45 ft. lbs. (60 nm).

11. Remove the clockspring set screw and place it in its storage location on the steering wheel.

12. Connect the horn wire.

13. Connect the speed control wire and install the speed control switch, if equipped.

14. Connect the clockspring lead wire to the air bag module and install module to steering wheel.

NOTE: Do not allow anyone to enter the vehicle from this point on, until this procedure is completed.

15. Connect the DRB II to the Airbag System Diagnostic Module (ASDM) connector located to the right of the console.

16. From the passenger side of the vehicle, turn the key to the **ON** position.

17. Check to make sure nobody has entered the vehicle. Connect the negative battery cable.

18. Using the DRB II, read and record any active fault data or stored codes.

19. If any active fault codes are present, perform the proper diagnostic procedures before continuing.

20. If there are no active fault codes, erase the stored fault codes. If there are active codes, the stored codes will not erase.

21. From the passenger side of the vehicle, turn the key **OFF**, then **ON** and observe the instrument cluster air bag warning light. It should come ON for 6–8 seconds, then go out, indicating the system is functioning normally. If the warning light either fails to come ON or stays lit, there is a system malfunction and the proper diagnostic procedures should be performed.

Steering Column

REMOVAL & INSTALLATION

1. Disconnect the negative battery cable.

2. Remove trim bezel, steering column cover and lower reinforcement, as required.

3. Disconnect all wiring connectors from below the instrument panel that lead up into the steering column.

4. Matchmark and remove the nuts that attach the steering column assembly to the instrument panel support.

NOTE: On 1988–89 vehicles, the upper coupling can be slipped out

of the lower coupling without removing the roll pin. On 1990 vehicles, the stub shaft is separated from the steering gear coupling.

5. Firmly grasp the steering wheel and pull the steering column out.

6. The installation is the reverse of the removal procedure.

7. Connect the negative battery cable and check the steering column and all related components for proper operation.

Manual Steering Rack and Pinion

REMOVAL & INSTALLATION

1. Disconnect the negative battery cable. On 1988–89 vehicles, drive out the upper roll pin, under the instrument panel, attaching the pinion shaft to the upper universal joint. Use a back-up device to steady the universal joint to prevent from damaging it.

2. Raise the vehicle and support safely.

3. Remove the front wheels.

4. Remove the cotter pins, castellated nuts and tie rod ends from the steering knuckles.

5. Remove the front suspension crossmember attaching bolts and nuts.

6. Lower the crossmember.

7. Remove the tie rod inner boot shields.

8. Remove the steering gear bolts from the front suspension crossmember.

9. Lower the rack to disengage it from the steering column stub shaft. Remove the steering gear from the left side of the vehicle.

To install:

10. Transfer the required parts to the new rack, if replacing it.

11. Place the rack on the crossmember and torque the steering gear attaching bolts to 21 ft. lbs. (29 Nm). Attach the boot shields.

12. Have a helper inside the vehicle remove the trim boot and align the stub shaft with the coupling while the crossmember is raised into position. If a helper is not available, the steering column will have to be unbolted so the steering shaft can be inserted into the coupling. Install the roll pin, if equipped.

13. The right rear crossmember bolt is a pilot bolt that correctly locates the crossmember; tighten it first. Torque the crossmember bolts to 90 ft. lbs.

14. Install the tie rod ends to the steering knuckle and torque the nut to 45 ft. lbs. (61 Nm). Install a new cotter pin.

15. Insert the stub shaft shim where

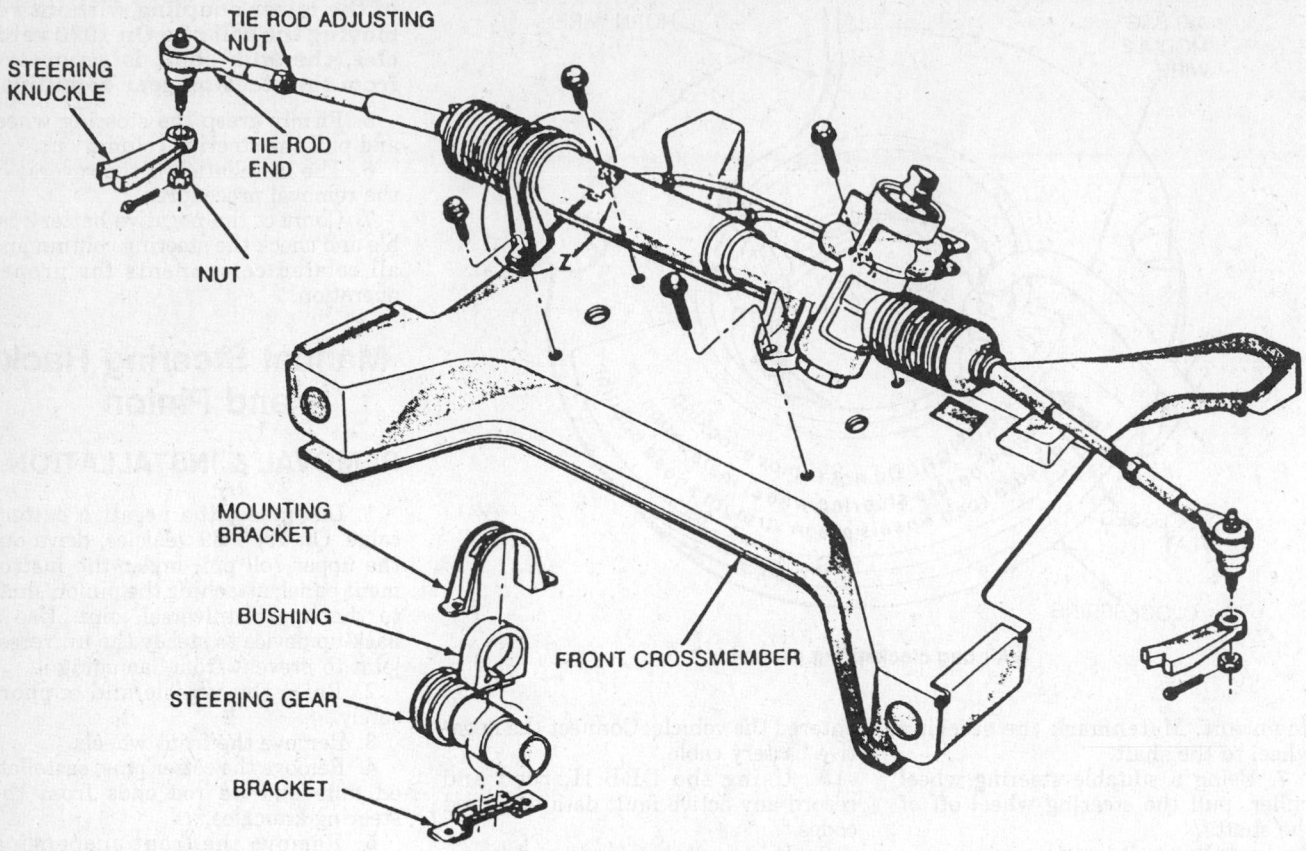

Rack and pinion steering gear mounting

the stub shaft goes into the coupling, if equipped.

16. Connect the negative battery cable and check the gear for proper operation.

Power Steering Rack and Pinion

REMOVAL & INSTALLATION

1. Disconnect the negative battery cable. On 1988–89 vehicles, drive out the upper roll pin, under the instrument panel, attaching the pinion shaft to the upper universal joint. Use a back-up device to steady the universal joint to prevent from damaging it.

2. Raise the vehicle and support safely.

3. Remove the front wheels.

4. Remove the cotter pins, castellated nuts and tie rod ends from the steering knuckles.

5. Disconnect and plug the oil pressure line from the rack.

6. Disconnect and plug the return hose from the line coming from the rack.

7. Remove the front suspension crossmember attaching bolts and nuts.

8. Lower the crossmember.

9. Remove the tie rod inner boot shields.

10. Remove the steering gear bolts from the front suspension crossmember.

11. Lower the rack to disengage it from the steering column stub shaft. Remove the steering gear from the left side of the vehicle.

To install:

12. Transfer the required parts to the new rack, if replacing it.

13. Place the rack on the crossmember and torque the steering gear attaching bolts to 21 ft. lbs. (29 Nm). Attach the fluid lines and the boot shields.

14. Have a helper inside the vehicle remove the trim boot and align the stub shaft with the coupling while the crossmember is raised into position. If a helper is not available, the steering column will have to be unbolted so the steering shaft can be inserted into the coupling. Install the roll pin, if equipped.

15. Install the tie rod ends to the steering knuckle and torque the nut to 45 ft. lbs. (61 Nm). Install a new cotter pin.

16. Insert the stub shaft shim where the stub shaft goes into the coupling, if equipped.

17. Refill the power steering pump.

18. Connect the negative battery cable and check the gear for proper operation.

Power Steering Pump

REMOVAL & INSTALLATION

1. Disconnect the negative battery cable.

2. Remove the drive belt from the pulley and remove bolts that are accessible from above.

3. Raise the vehicle and safely support.

4. Remove the right front wheel and the splash shield.

5. Position a drain pan under the return hose and disconnect the return hose from the line coming from the steering rack. Allow the fluid to drain, then plug the hose.

6. Disconnect the pressure line from the pump.

7. Remove the lower mounting bolts and nuts.

8. Move the pump toward the firewall and remove the adjustment bracket.

9. Rotate the pump clockwise so the pump pulley faces the rear of the vehicle and pull the pump out of the engine compartment.

To install:

10. Position the pump and install the adjustment bracket.

11. Raise the vehicle and install the lower mounting bolts and nuts finger-tight.

12. Replace the O-rings on the pressure line and connect to the pump. Install the return hose and secure the clamps.

13. Wrap the drive belt around the pulley groove and adjust the belt tension. Tighten the mounting bolts to 30 ft. lbs. (41 Nm).

14. Install the splash shield.

15. Turn the wheels to the full left turn position and add fluid until the reservoir is full.

16. Start the engine and add fluid to bring the level to the correct level.

17. To purge the system of air, turn the steering wheel from side to side without contacting the stops.

18. Return the wheel to the straight-ahead position and operate the engine for 2 minutes before road testing.

BELT ADJUSTMENT

1. Loosen the bracket mounting bolts.

2. Use a ½ in. drive breaker bar in the square hole provided in the bracket to move the pump away from the engine. Do not pry against the fluid reservoir.

3. With the pump moved enough so the belt deflects about ¼–½ in. under a 10 lb. load, tighten the bolts.

SYSTEM BLEEDING

1. Fill the reservoir with power steering fluid.

2. Turn the wheels to the full left turn position and add fluid until the reservoir is full.

3. Start the engine and add fluid to bring the level to the correct level.

4. To purge the system of air, turn the steering wheel from side to side without contacting the stops.

5. Return the wheel to the straight-ahead position and operate the engine for 2 minutes before road testing.

Tie Rod Ends

REMOVAL & INSTALLATION

1. Raise the vehicle and support safely.

2. Remove the cotter pin and nut from the tie rod end.

3. Using a suitable puller, remove the tie rod from the steering knuckle.

4. Loosen the sleeve clamp nut and bolt, if equipped, and unscrew the tie rod end from the sleeve or inner tie rod.

5. The installation is the reverse of the removal procedure. Torque the stud nuts to 45 ft. lbs. (61 Nm) and install a new cotter pin.

6. Perform a front end alignment as required.

BRAKES

For all brake system repair and service procedures not detailed below, please refer to "Brakes" in the Unit Repair section.

Master Cylinder

REMOVAL & INSTALLATION

1. Disconnect the negative battery cable.

2. Disconnect and plug the brake lines from the master cylinder.

3. Remove the nuts attaching the master cylinder to the power booster.

4. Remove the master cylinder from the mounting studs.

5. Remove the fluid reservoir from the cylinder.

To install:

6. Bench bleed the master cylinder.

7. Install to the studs and install the nuts.

8. Install the brake lines to the master cylinder.

9. Connect the negative battery cable and check the brakes for proper operation.

Combination Valve

REMOVAL & INSTALLATION

1. Disconnect the negative battery cable.

2. Raise the vehicle and support safely.

3. Tag and disconnect the brake lines from the valve.

4. Disconnect the wires to the pressure switch.

5. Remove the combination valve from the frame bracket.

6. The installation is the reverse of the removal procedure.

7. Bleed the brakes in the following order:

   a. Right rear wheel cylinder
   b. Left rear wheel cylinder
   c. Right front caliper
   d. Left front caliper

8. Connect the negative battery cable and check the brakes for proper operation.

Power Brake Booster

REMOVAL & INSTALLATION

1. Disconnect the negative battery cable. Disconnect the vacuum hose(s) from the booster.

2. Remove the nuts attaching the master cylinder to the booster and move the master cylinder aside.

3. From inside the passenger compartment, remove the clip that secures the booster pushrod to the brake pedal.

4. Remove the nuts that attach the booster to the dash panel and remove it from the vehicle.

5. Transfer the check valve to the new booster.

6. The installation is the reverse of the removal procedure.

7. Connect the negative battery cable and check the booster and brakes for proper operation.

Brake Caliper

REMOVAL & INSTALLATION

1. Raise the vehicle and support safely.

2. Remove the tire and wheel assembly.

3. Remove the caliper mounting pin.

4. Lift the caliper off of the pads, which should stay on the adaptor.

5. Remove the brake hose retaining bolt from the caliper.

To install:

6. Install the brake hose to the caliper using new copper washers.

7. Position the caliper over the rotor so the caliper engages the adaptor and pads correctly.

8. Install the caliper mounting pin.

9. Fill the master cylinder and bleed the brakes.

Disc Brake Pads

REMOVAL & INSTALLATION

1. Remove some of the fluid from the master cylinder.

2. Remove the wheels.

3. Remove the caliper and suspend securely.

4. Remove the inner and outer pads from the adaptor.

To install:

5. Use a large C-clamp to compress the piston back into the caliper bore.

6. Transfer the anti-rattle clips to the new brake pads, if necessary. Install the pads to the adaptor.

7. Position the caliper over the pads so the caliper engages the adaptor correctly.

8. Install the caliper mounting pin.
9. Refill the master cylinder.

Brake Rotor

REMOVAL & INSTALLATION

1. Raise the vehicle and support safely. Remove the tire and wheel assembly.
2. Remove the caliper and outer brake pad.
3. Remove the factory installed clips, if equipped. It is not necessary to reinstall these clips.
4. Remove the rotor from the hub.
5. The installation is the reverse of the removal procedure.

Brake Drums

REMOVAL & INSTALLATION

1. Raise the vehicle and support safely.
2. Remove the wheel and tire assembly.
3. Remove the dust cap.
4. Remove the cotter pin and nut lock.
5. Remove the wheel bearing nut and washer from the spindle.
6. Remove the outer wheel bearing.
7. Remove the drum with the inner wheel bearing from the spindle. If the drum is difficult to remove, remove the plug from the rear of the backing plate and push the self adjuster lever away from the star wheel. Rotate the star wheel to retract the shoes. Remove the grease seal.
To install:
8. Lubricate and install the inner wheel bearing. Install a new grease seal.
9. Install the drum to the spindle.
10. Lubricate and install the outer wheel bearing, washer and nut. When the bearing preload is properly set, install the nut lock and a new cotter pin.
11. Install the grease cap.
12. Install the wheel and tire assembly. Adjust the rear brakes as required.

Brake Shoes

REMOVAL & INSTALLATION

1. Raise the vehicle and support safely. Remove the wheels and the drums.
2. Remove the automatic adjuster spring and lever.
3. Rotate the automatic adjuster star wheel enough so both shoes move out far enough to be free of the wheel cylinder boots.

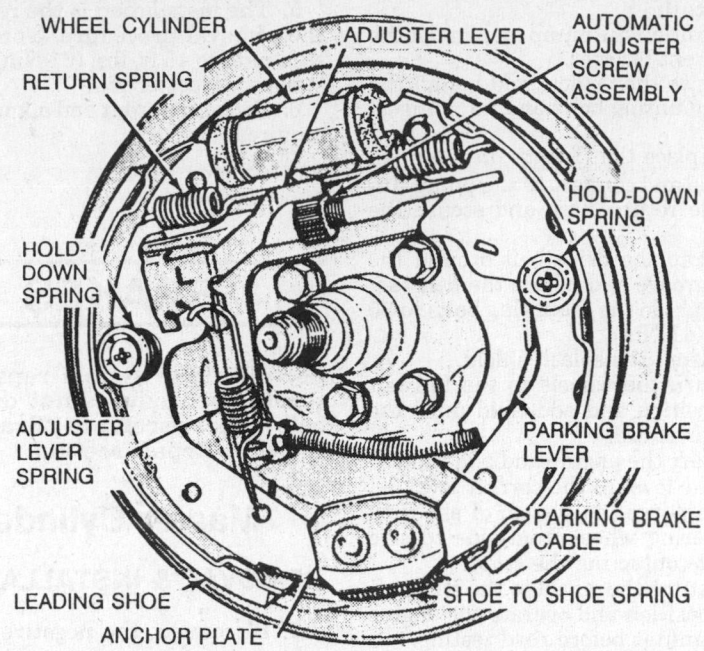

Rear drum brake components

4. Disconnect the parking brake cable from the actuating lever.
5. Remove the lower shoe to shoe spring.
6. With the shoes held together by the upper shoe to shoe spring, remove them from the backing plate.
To install:
7. Thoroughly clean and dry the backing plate. To prepare the backing plate, lubricate the bosses, anchor pin and parking brake actuating lever pivot surface lightly with lithium based grease.
8. Remove, clean and dry all parts still on the old shoes. Lubricate the star wheel shaft threads with anti-sieze lubricant and transfer all parts to their proper locations on the new shoes.
9. Install the lower spring.
10. Connect the parking brake cable.
11. Install the automatic adjuster lever and spring.
12. Adjust the star wheel.
13. Remove any grease from the linings and install the drum.
14. Complete the brake adjustment with the wheels installed.

Wheel Cylinder

REMOVAL & INSTALLATION

1. Raise the vehicle and support safely.
2. Remove the wheel, drum and brake shoes.
3. Remove and plug the brake line from the wheel cylinder.

4. Remove the wheel cylinder bolts and remove the cylinder from the backing plate.
To install:
5. Apply a very thin coating of silicone sealer to the cylinder mounting surface, install the cylinder to the backing plate and install the retaining bolts.
6. Connect the brake line to the wheel cylinder.
7. Install all brake parts that were removed.
8. Install the tire and wheel assembly.
9. Bleed the rear brakes.

Parking Brake Cable

ADJUSTMENT

1. Release the parking brakes fully.
2. Raise the vehicle and support safely.
3. Adjust the rear brakes.
4. Loosen the adjusting nut until there is slack in all the cables.
5. Rotate the rear wheels and tighten the cable adjusting nut until there is a slight drag at the wheels.
6. Continue to rotate the rear wheels and loosen the nut until all drag is eliminated.
7. Back off the nut an additional 2 turns.
8. Apply and release the parking brake several times. Upon the least release, verify there is no drag at the wheels.
9. To check the operation, make

sure the parking brake holds on an incline.

REMOVAL & INSTALLATION

Front Cable

1. Disconnect the negative battery cable.
2. Raise the vehicle and support safely.
3. Remove the cable adjusting nut and disengage the front cable from the connectors.
4. Loosen the heat shield for access and remove the cable housing to floor pan bracket and clips.
5. Remove the brake lever housing inside the passenger compartment and lift the floor mat for access to the floor pan. Pull the cable end forward and disconnect from the clevis.
6. Remove the floor seal panel.
7. Compress the cable housing retainer and push the cable out from the floor pan and remove from the vehicle.
To install:
8. Insert the cable housing retainer into the hole in the floor pan and install the seal panel.
9. Push the cable through the hole and engage the cable end to the clevis. Force the cable end retainer into the retainer until it is firmly seated.
10. Install the floor mat and brake lever housing.
11. Install the cable housing bracket and clips and install the heat shield.
12. Slide the adjuster end through the connectors and install the adjusting nut.
13. Adjust the cable, connect the negative battery cable and check the parking brakes for proper operation.

Rear Cable

1. Disconnect the negative battery cable.
2. Raise the vehicle and support safely.
3. Remove the rear wheels.
4. Remove the retaining clip from the brake cable bracket.
5. Remove the drums. Disconnect the cable from the brake shoe lever.
6. Compress the retaining clips on the end of the cable housing and pull the cable from the backing plate.
7. Remove the brake support plate from the axle.
8. The installation is the reverse of the removal procedure.
9. Adjust the cables, connect the negative battery cable and check the parking brakes for proper operation.

Brake System Bleeding

NOTE: If using a pressure bleed- er, follow the instructions furnished with the unit and choose the correct adaptor for the application. Do not substitute an adapter that "almost fits" as it will not work and could be dangerous.

Master Cylinder

If the master cylinder is off the vehicle it can be bench bled.
1. Connect 2 short pieces of brake line to the outlet fittings, bend them until the free end is below the fluid level in the master cylinder reservoirs.
2. Fill the reservoir with fresh brake fluid. Pump the piston slowly until no more air bubbles appear in the reservoirs.
3. Disconnect the 2 short lines, refill the master cylinder and securely install the cylinder caps.
4. If the master cylinder is on the vehicle, it can still be bled, using a flare nut wrench.
5. Open the brake lines slightly with the flare nut wrench while pressure is applied to the brake pedal by a helper inside the vehicle.
6. Be sure to tighten the line before the brake pedal is released.
7. Repeat the process with both lines until no air bubbles come out.

Calipers and Wheel Cylinders

1. Fill the master cylinder with fresh brake fluid. Check the level often during the procedure.
2. Starting with the right rear wheel, remove the protective cap from the bleeder, if equipped, and place where it will not be lost. Clean the bleed screw.

────── CAUTION ──────
When bleeding the brakes, keep face away from the brake area. Spewing fluid may cause facial and/or visual damage. Do not allow brake fluid to spill on the car's finish; it will remove the paint.

3. If the system is empty, the most effecient way to get fluid down to the wheel is to loosen the bleeder about ½-¾ turn, place a finger firmly over the bleeder and have a helper pump the brakes slowly until fluid comes out the bleeder. Once fluid is at the bleeder, close it before the pedal is released inside the vehicle.

NOTE: If the pedal is pumped rapidly, the fluid will churn and create small air bubbles, which are almost impossible to remove from the system. These air bubbles will eventually congregate and a spongy pedal will result.

4. Once fluid has been pumped to the caliper or wheel cylinder, open the bleed screw again, have the helper press the brake pedal to the floor, lock the bleeder and have the helper slowly release the pedal. Wait 15 seconds and repeat the procedure (including the 15 second wait) until no more air comes out of the bleeder upon application of the brake pedal. Remember to close the bleeder before the pedal is released inside the vehicle each time the bleeder is opened. If not, air will be induced into the system.
5. If a helper is not available, connect a small hose to the bleeder, place the end in a container of brake fluid and proceed to pump the pedal from inside the vehicle until no more air comes out the bleeder. The hose will prevent air from entering the system.
6. Repeat the procedure on remaining wheel cylinders in order:
   a. Left rear
   b. Right front
   c. Left front
7. Hydraulic brake systems must be totally flushed if the fluid becomes contaminated with water, dirt or other corrosive chemicals. To flush, bleed the entire system until all fluid has been replaced with the correct type of new fluid.
8. Install the bleeder cap(s) on the bleeder to keep dirt out. Always road test the vehicle after brake work of any kind is done.

CHASSIS ELECTRICAL

Air Bag

DISARMING

To disarm the air bag, disconnect the negative battery. There is no time lapse built into this sytem, so the system is immediately disabled. No further procedures are needed. Failure to disarm the system may result in deployment of the air bag and possible personal injury.

Heater Blower Motor

REMOVAL & INSTALLATION

1. Disconnect the negative battery cable.
2. Remove the glove box assembly, lower right side instrument panel trim cover and right cowl trim panel, as required. Disconnect the blower lead wire connector.
3. If not equipped with air condi-

tioning, remove the left side heater outlet duct. If equipped with air conditioning, disconnect the 2 vacuum lines from the recirculating door actuator and position the actuator aside.

4. Remove the 2 screws at the top of the blower housing that secure it to the unit cover.

5. Remove the 5 screws from around the blower housing and separate the blower housing from the unit.

6. Remove the 3 screws that secure the blower assembly to the heater or air conditioning housing and remove the assembly from the unit. Remove the fan from the blower motor.

7. The installation is the reverse of the removal procedure.

8. Connect the negative battery cable and check the blower motor for proper operation.

Windshield Wiper Motor

REMOVAL & INSTALLATION

Front

1. Disconnect the negative battery cable. Open the hood.

2. Remove the plastic wiper motor cover and washer hose attaching clip.

3. Disconnect the motor connector.

4. Remove the 3 bolts that attach the motor mounting bracket to the firewall.

5. Pull the motor out of its cavity and remove the drive crank mounting nut.

6. The installation is the reverse of the removal procedure.

7. Connect the negative battery ca-

ble and check the wiper motor for proper operation.

Rear

1. Disconnect the negative battery cable.

2. Remove the liftgate wiper using special tool C–3982.

NOTE: Prying the arm off of the shaft with a prying device could damage the arm permanently and possibly cause it to come off the shaft while in operation causing poor rear vision for the driver and/or a dangerous situation for the driver behind. Do not bend or push the spring clip at the base of the arm to release the arm; it is self-releasing. Use only the indicated tool.

3. Open the liftgate. Remove the plastic wiper motor cover.

4. Remove the pivot shaft nut, ring and seal assembly from the pivot shaft.

5. Disconnect the motor connector, remove the mounting screws and remove the wiper motor from the vehicle.

To install:

6. Install the motor to the liftgate and connect the connector.

7. Install the plastic cover.

8. Install the sealing washer about ½ in. down on the pivot shaft.

9. Install the seal, washer and nut on the pivot shaft and tighten the nut.

10. Install the wiper arm so it is parallel to the lower edge of the liftgate glass in the park position.

11. Connect the negative battery cable and check the motor for proper operation.

Windshield Wiper Switch

REMOVAL & INSTALLATION

1988–89 Vehicles

1. Disconnect the negative battery cable.

2. Disconnect the electrical connectors to the wiper switch and the turn signal switch.

3. Remove the column covers, horn pad and felt switch hider disc.

4. Turn the steering wheel so the access hole in the hub area of the steering wheel is in the 9 o'clock position, then loosen the turn signal pivot screw.

5. Disengage the dimmer switch pushrod from the wiper switch and remove the switch from the column.

To install:

6. Position the switch and engage the dimmer switch pushrod.

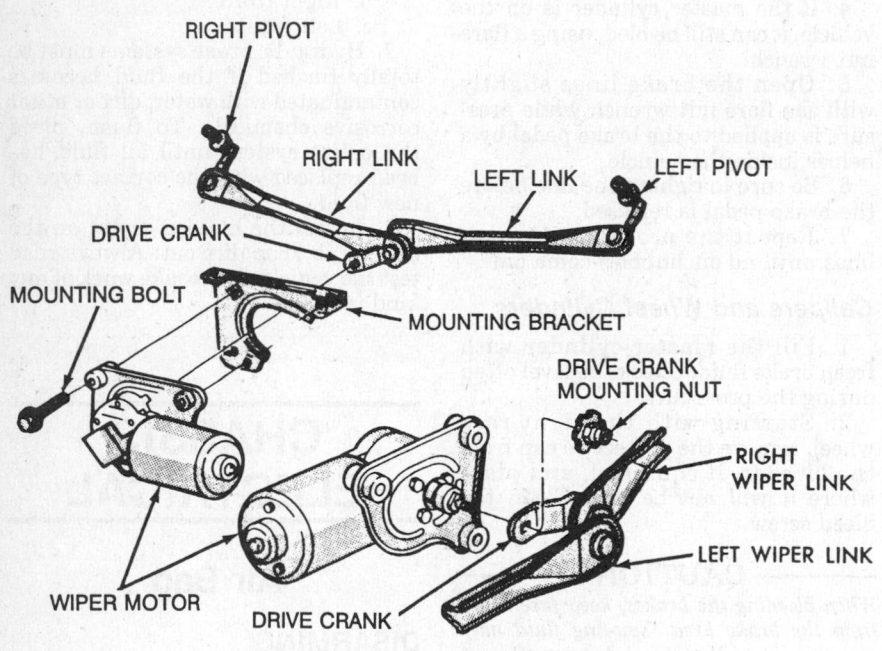

Wiper motor and linkage assembly

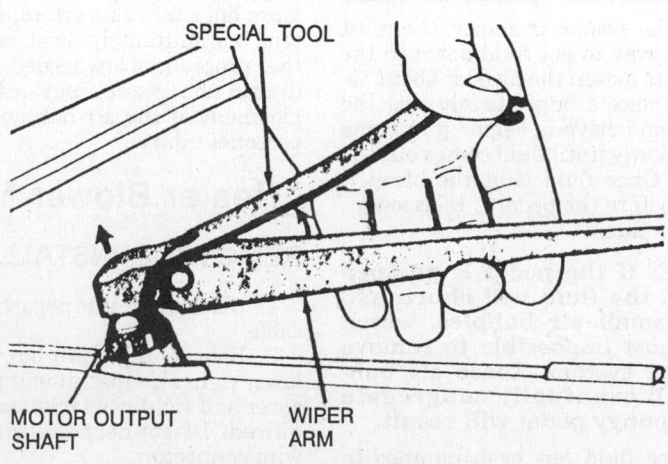

Removing the liftgate wiper arm

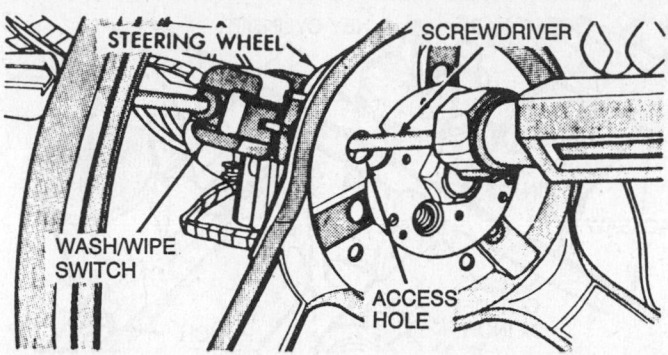

Access hole that reveals the turn signal lever screw—1988–89 vehicles

7. Secure the switch with the turn signal pivot screw.

8. Install the hider, column covers and horn pad.

9. Connect the connectors underneath the dash.

10. Connect the negative battery cable and check the switch for proper operation.

1990 Vehicles

1. Disconnect the negative battery cable.

2. Remove the lower steering column cover.

3. Straighten the steering wheel so the tires are pointing straight-ahead.

NOTE: Since the vehicle is equipped with an air bag, it is imperative that the steering wheel removal and installation procedure under Steering is followed.

4. Remove the steering wheel.

5. Remove the plastic wiring channel from the underside of the steering column.

6. Disconnect the wiper switch connector, intermittent wipe module connector and cruise control connector, if equipped.

7. Remove the side lock housing cover.

8. Remove the slotted hex-head screw that attaches the wiper switch to the turn signal switch and remove the switch.

9. Remove the control knob from the end of the stalk. Pull the round nylon hider up the control stalk and remove the revealed screws that attach the control stalk sleeve to the wiper switch.

10. Rotate the control stalk shaft to the full clockwise position and remove the shaft from the wiper switch by pulling it straight out.

To install:

11. Install the control shaft to the wiper switch, install the screws, the hider and the control knob.

12. Run the wiring through the opening and down the steering column, position the switch and install the hex-head screw. Make sure the dimmer switch rod is properly engaged.

13. Install the side lock housing cover.

14. Connect the wires and install the wiring channel and the column covers.

15. Install the steering wheel and torque the nut to 45 ft. lbs. (61 Nm).

16. Install the horn pad.

17. Connect the negative battery cable and check the wiper and washer, cruise control, turn signal switch and dimmer switch for proper operation.

18. Install the lower column cover.

Instrument Cluster

REMOVAL & INSTALLATION

1. Disconnect the negative battery cable.

2. Remove the instrument cluster bezel. Cluster removal is not necessary if just removing gauges.

3. When only removing gauges or the speedometer, remove the trip odometer reset knob, if necessary, remove the mask and lens assembly and remove the desired gauge from the cluster.

4. Remove the screws attaching the cluster to the instrument panel.

5. Pull the cluster out and disconnect the speedometer cable, if equipped, and all wiring harnesses.

6. Remove the cluster from the vehicle.

To install:

7. Position the cluster and connect the speedometer cable, if equipped.

8. Connect all wiring in their proper positions.

9. Install the cluster retaining screws.

10. Install the cluster bezel.

11. Connect the negative battery cable and check all gauges and the speedometer for proper operation.

Radio

REMOVAL & INSTALLATION

1. Disconnect the negative battery cable.

2. Remove the cluster or console bezel, as required.

3. Remove the screws that attach the radio to the instrument panel.

4. Pull the radio out, disconnect the connectors, ground cable and antenna and remove the radio.

5. The installation is the reverse of the removal procedure.

6. Connect the negative battery cable and check the radio for proper operation.

Headlight Switch

REMOVAL & INSTALLATION

1. Disconnect the negative battery cable.

2. Remove the left side trim bezel.

3. Remove the switch mounting screws, pull the switch assembly out and unplug the connector.

4. Remove the switch knob and shaft by depressing the release button on the switch.

5. Disengage the escutcheon from the mounting plate, unscrew the bracket retainer nut and remove the switch.

6. The installation is the reverse of the removal procedure.

7. Connect the negative battery cable and check the switch for proper operation.

Dimmer Switch

REMOVAL & INSTALLATION

1. Disconnect the negative battery cable.

2. Remove the lower steering column cover, if equipped.

3. Unplug the switch, located on the lower portion of the steering column.

4. Hold the actuating rod against its upper seat, remove the screws that attach the switch to the column and remove the switch.

5. The installation is the reverse of the removal procedure. Adjust the switch as required.

6. Connect the negative battery cable and check the switch for proper operation.

Turn Signal Switch

REMOVAL & INSTALLATION

1. Disconnect the negative battery cable.

2. Remove the lower steering column cover, if equipped.

3. Straighten the steering wheel so the tires are pointing straight-ahead.

NOTE: If equipped with an air bag, it is imperative that the steering wheel removal and installation procedure under Steering is followed.

4. Remove the steering wheel. On 1988–89 vehicles, remove the column covers.

5. Remove the plastic wiring channel from the underside of the steering column and disconnect the turn signal switch connector.

6. Remove the hazard switch knob, if necessary. Remove the slotted hex-head screw that attaches the wiper switch to the turn signal switch.

7. Remove the 3 screws and pull the turn signal switch out of the column.
To install:

8. Run the wiring through the opening and down the steering column, position the switch and install the hex-head screw. Make sure the dimmer switch rod is properly engaged.

9. Install the 3 screws and the hazard switch knob.

10. Connect the wires and install the wiring channel.

11. Install the steering wheel and torque the nut to 45 ft. lbs. (61 Nm). Install the column covers, if they were removed.

12. Install the horn pad.

13. Connect the negative battery cable and check the turn signal switch and dimmer switch for proper operation.

14. Install the lower column cover.

Ignition Lock

REMOVAL & INSTALLATION

1988–89 Vehicles

1. Disconnect the negative battery cable.

2. Place the key cylinder in the **LOCK** position and remove the key.

3. Remove the steering wheel, the column covers and the turn signal switch.

4. Using a hacksaw blade, cut the upper ¼ in. from the key cylinder retainer pin boss.

5. Using the proper size punch, drive the roll pin from the housing and remove the key cylinder.
To install:

6. Insert the cylinder into the housing, make sure it engages the lug on the ignition switch driver and install the roll pin.

7. Install the turn signal switch, column covers and the steering wheel.

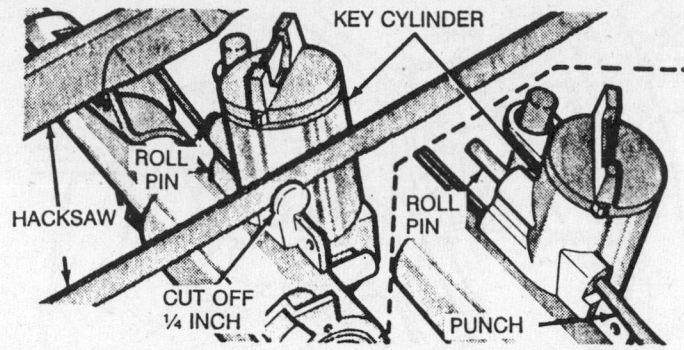

Cutting the key cylinder retainer pin boss—1988–89 vehicles

8. Connect the negative battery cable and check the lock cylinder for proper operation in all positions.

1990 Vehicles

1. Disconnect the negative battery cable.

2. Straighten the steering wheel so the tires are pointing straight-ahead.

NOTE: Since the vehicle is equipped with an air bag, it is imperative that the steering wheel removal and installation procedure under Steering is followed.

3. Remove the steering wheel.

4. Remove the hazard switch knob. Remove the slotted hex-head screw that attaches the wiper switch to the turn signal switch.

5. Remove the 3 screws and pull the turn signal switch out of the column as far as it will go. Unplug it below if necessary.

6. Remove the ignition switch key light.

7. Place the key in the **LOCK** position and remove the key.

8. Insert 2 suitable small diameter tools into both release holes and push inward to release the spring loaded lock retainers while simultaneously pulling the key lock cylinder out of its bore.
To install:

9. Install the key cylinder.

10. Install the ignition switch key light.

11. Install the turn signal switch and hazard switch knob. Connect the wires if they were disconnected.

12. Install the steering wheel and torque the nut to 45 ft. lbs. (61 Nm).

13. Install the horn pad.

14. Connect the negative battery cable and check the lock cylinder for proper operation in all positions.

Ignition Switch

REMOVAL & INSTALLATION

1. Disconnect the negative battery cable.

2. Remove the lower steering column cover, if equipped.

3. If necessary, remove the steering column retaining nuts and allow the steering wheel to rest on the driver's seat.

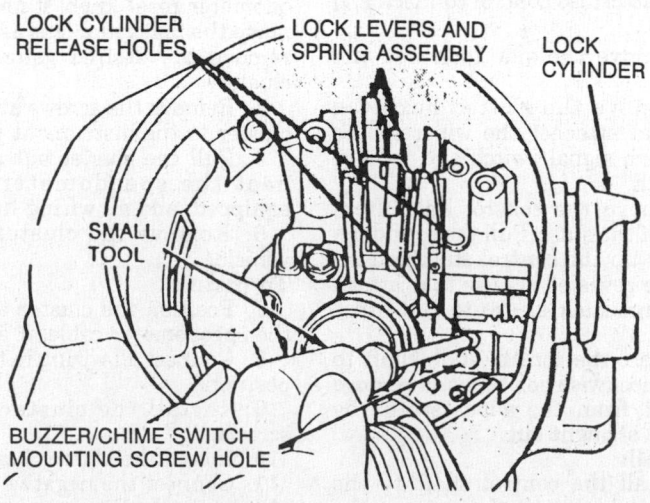

Removing the key lock cylinder—1990 vehicles

4. Remove the 2 screws that attach the ignition switch to the lower portion of the column.

5. Rotate the switch 90 degrees and pull up to disengage it from the ignition switch rod.

To install:

6. Engage the switch with the rod, rotate the switch 90 degrees and push down until fully engaged.

7. Install the mounting screws finger-tight.

8. Place the key in the **LOCK** position and remove the key. Adjust the switch by pushing up gently on the switch to take up all slack in the rod.

9. Tighten the mounting screws, connect the negative battery cable and check the switch for proper operation in all positions.

Stoplight Switch

REMOVAL & INSTALLATION

1. Disconnect the negative battery cable.

2. Unplug the stoplight switch connectors near the brake pedal.

3. Remove the switch and bracket assembly from the brake pedal bracket.

4. Remove the switch from its bracket.

To install:

5. Install the switch and bracket assembly to the brake pedal bracket and push the switch forward as far as it will go; the brake pedal should move forward slightly.

6. Pull back on the brake pedal, bringing the striker toward the switch, until the pedal will not go back any farther.

7. This will cause the switch to ratchet backward into position and automatic adjustment is complete.

8. Connect the negative battery cable and check the switch for proper operation. Also, make sure the speed control system functions properly, if equipped.

Neutral Safety Switch

REMOVAL & INSTALLATION

1. Disconnect the negative battery cable.

2. Locate the neutral safety switch at the left rear corner of the automatic transaxle, facing the front of the vehicle. Unplug the switch connector.

3. Remove the switch from the transaxle.

4. The installation is the reverse of the removal procedure. Torque the switch to 25 ft. lbs. (34 Nm).

5. Connect the negative battery cable and make sure the vehicle does not

start in any gear except **P** and **N**. Also make sure the reverse lights come ON when the shifter is placed in **R**.

Fuses, Circuit Breaker, Relays and Flashers

LOCATION

Fuses, Circuit Breaker and Flashers

The fuse block, which contains the fuses, circuit breaker and flashers, is located on the left side kick panel, below the left side of the instrument panel.

Relays

A/C Clutch Cutout Relay—located on the left front inner fender just in front of the strut tower.

Automatic Shutdown (ASD) Re-lay—located on the left front inner fender just in front of the strut tower.

Cooling Fan Motor Relay—located on the left front strut tower on 1988 vehicles and on the left front inner fender just in front of the strut tower on 1989–90 vehicles.

Horn Relay—located on the upper right side of the fuse block.

Rear Window Defroster Timer—part of the defroster switch.

Seatbelt Warning Buzzer—located on the fuse block.

Starter Relay—located on the left front strut tower.

Time Delay Relay—taped to the wiring harness near the fuse block.

Computers

LOCATION

1988—The Single Module Engine

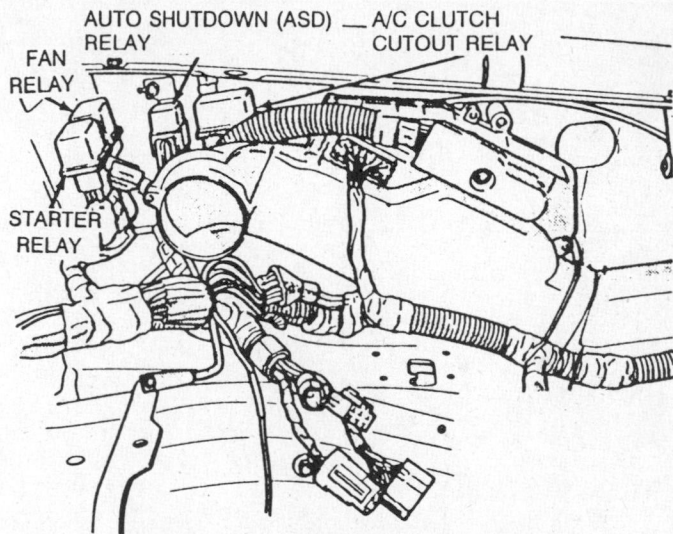

Underhood relay identification—1988 vehicles

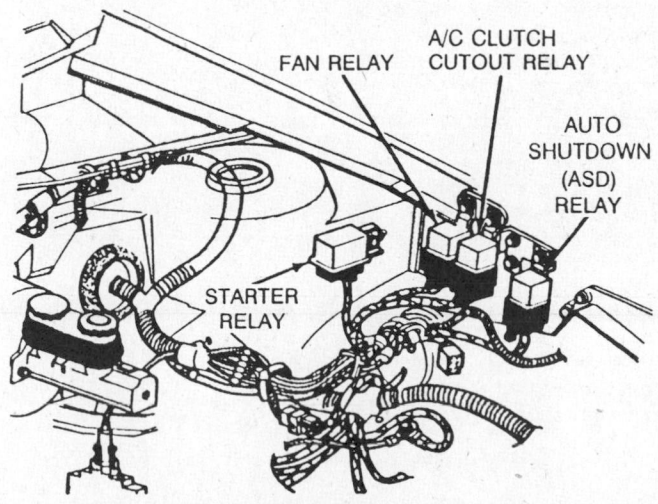

Underhood relay identification—1989–90 vehicles

Controller (SMEC) is located in the engine compartment, to the left of the battery.

1989–90 — The Single Board Engine Controller (SBEC) is located in the engine compartment, to the left of the battery.

The Airbag System Diagnostic Module (ASDM), if equipped, is located under the instrument panel, to the right of the front of the console.

Cruise Control

ADJUSTMENT

1. The clearance between the throttle stud and cable clevis should be $\frac{1}{16}$ in.

2. To adjust the cable, remove the retaining clip or loosen the retaining clamp nut at the throttle bracket.

3. Pull all slack out of the cable using a suitable $\frac{1}{16}$ in. diameter tool to account for proper clearance. Make sure the curb idle position of the throttle blade is not affected.

4. Reinstall the retaining clip or nut.

Chrysler Corp.
Rear Wheel Drive
Chrysler—Fifth Avenue, Newport **Dodge**—Diplomat
Plymouth—Caravelle, Grand Fury

SPECIFICATIONS
VEHICLE IDENTIFICATION CHART

It is important for servicing and ordering parts to be certain of the vehicle and engine identification. The VIN (vehicle identification number) is a 17 digit number visible through the windshield on the driver's side of the dash and contains the vehicle and engine identification codes. The tenth digit indicates model year, and the eighth digit indicates engine code. It can be interpreted as follows:

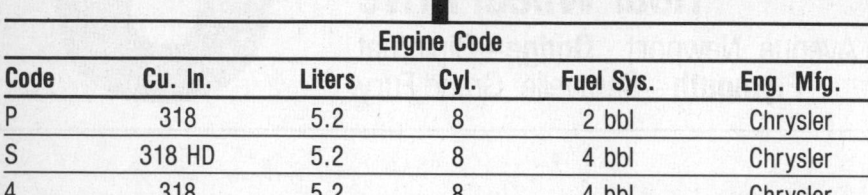

	Engine Code					Model Year	
Code	Cu. In.	Liters	Cyl.	Fuel Sys.	Eng. Mfg.	Code	Year
P	318	5.2	8	2 bbl	Chrysler	J	1988
S	318 HD	5.2	8	4 bbl	Chrysler	K	1989
4	318	5.2	8	4 bbl	Chrysler		

ENGINE IDENTIFICATION

Year	Model	Engine Displacement cu. in. (liter)	Engine Series Identification (VIN)	No. of Cylinders	Engine Type
1988	Diplomat	318 (5.2)	P③	8	OHV
	Gran Fury/Caravelle①	318 (5.2)	P③	8	OHV
	Gran Fury/Caravelle①	318 (5.2)	4④	8	OHV
	Gran Fury/Caravelle①	318 (5.2)	S②	8	OHV
	Fifth Avenue/Newport	318 (5.2)	P③	8	OHV
	Fifth Avenue/Newport	318 (5.2)	4④	8	OHV
	Fifth Avenue/Newport	318 (5.2)	S②	8	OHV
1989	Diplomat	318 (5.2)	P③	8	OHV
	Gran Fury/Caravelle①	318 (5.2)	P③	8	OHV
	Gran Fury/Caravelle①	318 (5.2)	4④	8	OHV
	Gran Fury/Caravelle①	318 (5.2)	S②	8	OHV
	Fifth Avenue/Newport	318 (5.2)	P③	8	OHV
	Fifth Avenue/Newport	318 (5.2)	4④	8	OHV
	Fifth Avenue/Newport	318 (5.2)	S②	8	OHV

① Caravelle—Canada oinly   ③ 318 2 bbl
② 318 4 bbl Heavy Duty   ④ 318 4 bbl Standard

GENERAL ENGINE SPECIFICATIONS

Year	VIN	No. Cylinder Displacement cu. in. (liter)	Fuel System Type	Net Horsepower @ rpm	Net Torque @ rpm (ft. lbs.)	Bore × Stroke (in.)	Compression Ratio	Oil Pressure @ rpm
1988	P	8-318 (5.2)	2 bbl	140 @ 3600	265 @ 1600	3.910 × 3.310	9.0:1	80 @ 3000
	4	8-318 (5.2)	4 bbl	165 @ 4000	240 @ 2000	3.910 × 3.310	8.6:1	80 @ 3000
	S	8-318 (5.2)HD	4 bbl	175 @ 4000	250 @ 3200	3.910 × 3.310	8.0:1	80 @ 3000
1989	P	8-318 (5.2)	2 bbl	140 @ 3600	265 @ 1600	3.910 × 3.310	9.0:1	80 @ 3000
	4	8-318 (5.2)	4 bbl	140 @ 3600	265 @ 2000	3.910 × 3.310	9.0:1	80 @ 3000
	S	8-318 (5.2)HD	4 bbl	175 @ 4000	250 @ 3200	3.910 × 3.310	8.0:1	80 @ 3000

HD Heavy Duty

TUNE-UP SPECIFICATIONS

Year	VIN	No. Cylinder Displacement cu. in. (liter)	Spark Plugs Type	Spark Plugs Gap (in.)	Ignition Timing (deg.) MT	Ignition Timing (deg.) AT	Compression Pressure (psi)	Fuel Pump (psi)	Idle Speed (rpm) MT	Idle Speed (rpm) AT	Valve Clearance In.	Valve Clearance Ex.
1988	P	8-318 (5.2)	RN12YC	0.035	—	7B	100	5.75–7.25	—	680	Hyd.	Hyd.
	4	8-318 (5.2)	RN12YC	0.035	—	16B	100	5.75–7.25	—	750	Hyd.	Hyd.
	S	8-318 (5.2)	RN12YC	0.035	—	16B	100	5.75–7.25	—	750	Hyd.	Hyd.
1989	P	8-318 (5.2)	RN12YC	0.035	—	7B	100	5.75–7.25	—	680	Hyd.	Hyd.
	4	8-318 (5.2)	RN12YC	0.035	—	16B	100	5.75–7.25	—	750	Hyd.	Hyd.
	S	8-318 (5.2)	RN12YC	0.035	—	16B	100	5.75–7.25	—	750	Hyd.	Hyd.

NOTE: The underhood specifications sticker often reflects tune-up specification changes made in production. Sticker figures must be used if they disagree with those in this chart.

Part numbers in this chart are not recommendations by Chilton for any product by brand name.

Hyd.—Hydraulic

FIRING ORDERS

NOTE: To avoid confusion, always replace spark plug wires one at a time.

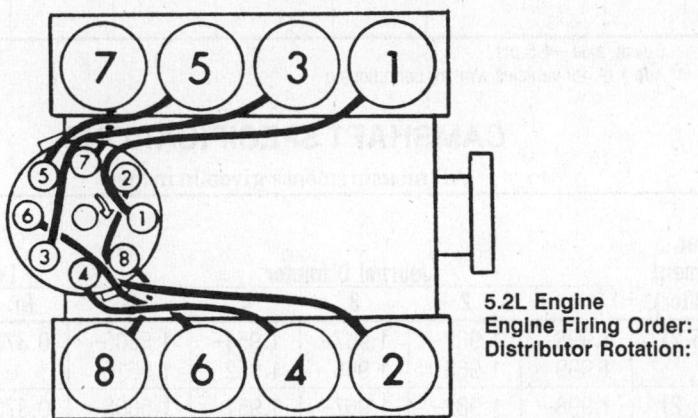

5.2L Engine
Engine Firing Order: 1–8–4–3–6–5–7–2
Distributor Rotation: Clockwise

CAPACITIES

Year	Model	VIN	No. Cylinder Displacement cu. in. (liter)	Engine Crankcase with Filter	Engine Crankcase without Filter	Transmission (pts.) 4-Spd	Transmission (pts.) 5-Spd	Transmission (pts.) Auto.	Drive Axle (pts.)	Fuel Tank (gals.)	Cooling System (qts.)
1988	Diplomat	P	8-318 (5.2)	5	4	—	—	16.4	②	18	15.5③
	Diplomat	4	8-318 (5.2)	5	4	—	—	16.3	16.4	18	15.5③
	Diplomat	S	8-318 (5.2)	5	4	—	—	16.3	16.4	18	15.5③
	Gran Fury/ Caravelle①	P	8-318 (5.2)	5	4	—	—	16.4	②	18	15.5③
	Gran Fury/ Caravelle①	4	8-318 (5.2)	5	4	—	—	16.3	16.4	18	15.5③
	Gran Fury/ Caravelle①	S	8-318 (5.2)	5	4	—	—	16.3	16.4	18	15.5③
	Fifth Avenue/ Newport	P	8-318 (5.2)	5	4	—	—	16.4	②	18	15.5③
	Fifth Avenue/ Newport	4	8-318 (5.2)	5	4	—	—	16.3	16.4	18	15.5③
	Fifth Avenue/ Newport	S	8-318 (5.2)	5	4	—	—	16.3	16.4	18	15.5③

CAPACITIES

Year	Model	VIN	No. Cylinder Displacement cu. in. (liter)	Engine Crankcase with Filter	Engine Crankcase without Filter	Transmission (pts.) 4-Spd	Transmission (pts.) 5-Spd	Transmission (pts.) Auto.	Drive Axle (pts.)	Fuel Tank (gals.)	Cooling System (qts.)
1989	Diplomat	P	8-318 (5.2)	5	4	—	—	16.4	②	18	15.5③
	Diplomat	4	8-318 (5.2)	5	4	—	—	16.3	16.4	18	15.5③
	Diplomat	S	8-318 (5.2)	5	4	—	—	16.3	16.4	18	15.5③
	Gran Fury/ Caravelle①	P	8-318 (5.2)	5	4	—	—	16.4	②	18	15.5③
	Gran Fury/ Caravelle①	4	8-318 (5.2)	5	4	—	—	16.3	16.4	18	15.5③
	Gran Fury/ Caravelle①	S	8-318 (5.2)	5	4	—	—	16.3	16.4	18	15.5③
	Fifth Avenue/ Newport	P	8-318 (5.2)	5	4	—	—	16.4	②	18	15.5③
	Fifth Avenue/ Newport	4	8-318 (5.2)	5	4	—	—	16.3	16.4	18	15.5③
	Fifth Avenue/ Newport	S	8-318 (5.2)	5	4	—	—	16.3	16.4	18	15.5③

① Caravelle-Canada only
② 7¼ in. axle—2.5 pts.
  8¼ in. axle—4.4 pts.
  9¼ in. axle—4.5 pts.
③ Add 1 qt. for vehicles with air conditioning

CAMSHAFT SPECIFICATIONS

All measurements given in inches.

Year	VIN	No. Cylinder Displacement cu. in. (liter)	Journal Diameter 1	Journal Diameter 2	Journal Diameter 3	Journal Diameter 4	Journal Diameter 5	Lobe Lift In.	Lobe Lift Ex.	Bearing Clearance	Camshaft End Play
1988	P	8-318 (5.2)	1.998–1.999	1.982–1.983	1.967–1.968	1.951–1.952	1.5605–1.5615	0.373	0.400	0.001–0.003	0.002–0.010
	4	8-318 (5.2)	1.998–1.999	1.982–1.983	1.967–1.968	1.951–1.952	1.5605–1.5615	0.373	0.400	0.001–0.003	0.002–0.010
	S	8-318 (5.2)	1.998–1.999	1.982–1.983	1.967–1.968	1.951–1.952	1.5605–1.5615	0.373	0.400	0.001–0.003	0.002–0.010
1989	P	8-318 (5.2)	1.998–1.999	1.982–1.983	1.967–1.968	1.951–1.952	1.5605–1.5615	0.373	0.400	0.001–0.003	0.002–0.010
	4	8-318 (5.2)	1.998–1.999	1.982–1.983	1.967–1.968	1.951–1.952	1.5605–1.5615	0.373	0.400	0.001–0.003	0.002–0.010
	S	8-318 (5.2)	1.998–1.999	1.982–1.983	1.967–1.968	1.951–1.952	1.5605–1.5615	0.373	0.400	0.001–0.003	0.002–0.010

CRANKSHAFT AND CONNECTING ROD SPECIFICATIONS

All measurements are given in inches.

Year	VIN	No. Cylinder Displacement cu. in. (liter)	Crankshaft Main Brg. Journal Dia.	Crankshaft Main Brg. Oil Clearance	Crankshaft Shaft End-play	Crankshaft Thrust on No.	Connecting Rod Journal Diameter	Connecting Rod Oil Clearance	Connecting Rod Side Clearance
1988	P	8-318 (5.2)	2.4995–2.5005	①	0.002–0.010	3	2.1240–2.1250	0.0005–0.0022	0.006–0.014
	4	8-318 (5.2)	2.4995–2.5005	①	0.002–0.010	3	2.1240–2.1250	0.0005–0.0022	0.006–0.014
	S	8-318 (5.2)	2.4995–2.5005	①	0.002–0.010	3	2.1240–2.1250	0.0005–0.0022	0.006–0.014

CRANKSHAFT AND CONNECTING ROD SPECIFICATIONS

All measurements are given in inches.

Year	VIN	No. Cylinder Displacement cu. in. (liter)	Crankshaft Main Brg. Journal Dia.	Crankshaft Main Brg. Oil Clearance	Crankshaft Shaft End-play	Crankshaft Thrust on No.	Connecting Rod Journal Diameter	Connecting Rod Oil Clearance	Connecting Rod Side Clearance
1989	P	8-318 (5.2)	2.4995–2.5005	①	0.002–0.010	3	2.1240–2.1250	0.0005–0.0022	0.006–0.014
	4	8-318 (5.2)	2.4995–2.5005	①	0.002–0.010	3	2.1240–2.1250	0.0005–0.0022	0.006–0.014
	S	8-318 (5.2)	2.4995–2.5005	①	0.002–0.010	3	2.1240–2.1250	0.0005–0.0022	0.006–0.014

① No. 1—0.0005–0.0015;
No. 2–5—0.0005–0.0020

VALVE SPECIFICATIONS

Year	VIN	No. Cylinder Displacement cu. in. (liter)	Seat Angle (deg.)	Face Angle (deg.)	Spring Test Pressure (lbs.)	Spring Installed Height (in.)	Stem-to-Guide Clearance (in.) Intake	Stem-to-Guide Clearance (in.) Exhaust	Stem Diameter (in.) Intake	Stem Diameter (in.) Exhaust
1988	P	8-318 (5.2)	45	45	177 @ 1.31	$1^{21}/_{32}$	0.0010–0.0030	0.0020–0.0040	0.3725	0.3715
	4	8-318 (5.2)	45	45	177 @ 1.31	$1^{21}/_{32}$	0.0010–0.0030	0.0020–0.0040	0.3725	0.3715
	S	8-318 (5.2)	45	45	193 @ 1.25	$1^{21}/_{32}$	0.0015–0.0035	0.0025–0.0045	0.3720	0.3710
1989	P	8-318 (5.2)	45	45	177 @ 1.31	$1^{21}/_{32}$	0.0010–0.0030	0.0020–0.0040	0.3725	0.3715
	4	8-318 (5.2)	45	45	177 @ 1.31	$1^{21}/_{32}$	0.0010–0.0030	0.0020–0.0040	0.3725	0.3715
	S	8-318 (5.2)	45	45	193 @ 1.25	$1^{21}/_{32}$	0.0015–0.0035	0.0025–0.0045	0.3720	0.3710

PISTON AND RING SPECIFICATIONS

All measurements are given in inches.

Year	VIN	No. Cylinder Displacement cu. in. (liter)	Piston Clearance	Ring Gap Top Compression	Ring Gap Bottom Compression	Ring Gap Oil Control	Ring Side Clearance Top Compression	Ring Side Clearance Bottom Compression	Ring Side Clearance Oil Control
1988	P	8-318 (5.2)	0.0005–0.0015 ①	0.0100–0.0200	0.0100–0.0200	0.0150–0.0550	0.0015–0.0030	0.0015–0.0030	0.0002–0.0050
	4	8-318 (5.2)	0.0005–0.0015 ①	0.0100–0.0200	0.0100–0.0200	0.0150–0.0550	0.0015–0.0030	0.0015–0.0030	0.0002–0.0050
	S	8-318 (5.2)	0.0005–0.0015 ①	0.0100–0.0200	0.0100–0.0200	0.0150–0.0550	0.0015–0.0030	0.0015–0.0030	0.0002–0.0050

PISTON AND RING SPECIFICATIONS

All measurements are given in inches.

Year	VIN	No. Cylinder Displacement cu. in. (liter)	Piston Clearance	Ring Gap			Ring Side Clearance		
				Top Compression	Bottom Compression	Oil Control	Top Compression	Bottom Compression	Oil Control
1989	P	8-318 (5.2)	0.0005–0.0015 ①	0.0100–0.0200	0.0100–0.0200	0.0150–0.0550	0.0015–0.0030	0.0015–0.0030	0.0002–0.0050
	4	8-318 (5.2)	0.0005–0.0015 ①	0.0100–0.0200	0.0100–0.0200	0.0150–0.0550	0.0015–0.0030	0.0015–0.0030	0.0002–0.0050
	S	8-318 (5.2)	0.0005–0.0015 ①	0.0100–0.0200	0.0100–0.0200	0.0150–0.0550	0.0015–0.0030	0.0015–0.0030	0.0002–0.0050

① High Performance engines—0.001–0.002

TORQUE SPECIFICATIONS

All readings in ft. lbs.

Year	VIN	No. Cylinder Displacement cu. in. (liter)	Cylinder Head Bolts	Main Bearing Bolts	Rod Bearing Bolts	Crankshaft Pulley Bolts	Flywheel Bolts	Manifold		Spark Plugs
								Intake	Exhaust	
1988	P	8-318 (5.2)	105	85	45	100	55	45	①	30
	4	8-318 (5.2)	105	85	45	100	55	45	①	30
	S	8-318 (5.2)	105	85	45	100	55	45	①	30
1989	P	8-318 (5.2)	105	85	45	100	55	45	①	30
	4	8-318 (5.2)	105	85	45	100	55	45	①	30
	S	8-318 (5.2)	105	85	45	100	55	45	①	30

① Nuts—15 ft. lbs., bolts—20 ft.lb.

BRAKE SPECIFICATIONS

All measurements in inches unless noted.

Year	Model	Lug Nut Torque (ft. lbs.)	Master Cylinder Bore	Brake Disc		Standard Brake Drum Diameter	Minimum Lining Thickness	
				Minimum Thickness	Maximum Runout		Front	Rear
1988	All	85	0.827①	0.940	0.004	10.000②	⅛	⅛
1989	All	85	0.827①	0.940	0.004	10.000②	⅛	⅛

① Heavy Duty—1.03
② Heavy Duty—11.000

WHEEL ALIGNMENT

Year	Model	Caster		Camber		Toe-in (in.)	Steering Axis Inclination (deg.)
		Range (deg.)	Preferred Setting (deg.)	Range (deg.)	Preferred Setting (deg.)		
1988	All	1¼P–3¾P	2½P	¼N–1¼P	½P	⅛	8
1989	All	1¼P–3¾P	2½P	¼N–1¼P	½P	⅛	8

N—Negative
P—Positive

ENGINE MECHANICAL

Engine Assembly

REMOVAL & INSTALLATION

1. Disconnect the negative battery cable. Matchmark the hood hinge positions and remove the hood.
2. Drain cooling system, remove the battery and carburetor air cleaner.
3. Remove the radiator/heater hoses and remove radiator. Set the fan shroud aside.
4. Remove the air conditioning compressor and set aside without removing lines.
5. Remove vacuum lines, distributor cap and wiring.
6. Remove the carburetor linkage, starter wires and oil pressure wire.
7. Remove the power steering hoses, if equipped.
8. Remove the starter motor, alternator, charcoal canister and horns.
9. Raise and support the vehicle safely.
10. Remove the exhaust pipe at the manifold.
11. Remove the bellhousing bolts and inspection plate.
12. Remove the torque converter-to-flexplate bolts from torque converter flexplate. Mark the converter and flexplate to aid in re-assembly.
13. Support the transmission with a transmission stand tool. Attach a C-clamp on front bottom of transmission torque converter housing. This will assure that the torque converter will be retained in proper position in the transmission housing.
14. Disconnect the engine from the torque converter flexplate.
15. Install engine lifting fixture. Attach a chain hoist to fixture eyebolt.
16. Remove engine front mount bolts.
17. Remove engine from engine compartment and support it safely on a engine repair stand.

To install:

18. After all repairs have been made, remove engine from repair stand and install in engine compartment.
19. Install bellhousing bolts and inspection plate. Remove stand from transmission.
20. Install torque converter-to-flexplate bolts and front end mounts. Remove C-clamp. Install inspection plate.
21. Remove engine lifting fixture and install carburetor and lines.
22. Install starter motor, alternator, charcoal canister and lines.

23. Install vacuum lines, distributor cap and wiring.
24. Install exhaust pipe. Torque to 24 ft. lbs. (33 Nm). Tighten nuts alternately so space between manifold flange and exhaust pipe flange is approximately equal.
25. Connect carburetor linkage and wiring to engine.
26. Install radiator, radiator hoses and heater hoses.
27. Install fan shroud. Fill cooling system.
28. Fill the engine crankcase with oil.
29. Install the battery and carburetor air cleaner. Connect vacuum hose and power steering hoses, if equipped.
30. Install air conditioning equipment, if equipped.
31. Run engine until full operating temperature is reached and adjust carburetor, as necessary.
32. Install hood.
33. Road test vehicle.

Engine Mounts

REMOVAL & INSTALLATION

1. Disconnect the negative battery cable.
2. Position the fan to clear the radiator hose and radiator top tank.
3. Disconnect throttle linkage at transmission and at carburetor. Raise and support the vehicle safely.
4. Remove torque nuts from insulator studs.
5. Raise the engine just enough to remove the engine front mount assembly.

To install:

6. Before installing the engine mount, identify whether the mount is right or left side.
7. Install the insulator to engine bracket and tighten.
8. Lower the engine and install washers and prevailing torque nuts to insulator studs; tighten the nuts.
9. Connect the throttle linkage at the transmission and carburetor.

Cylinder Head

REMOVAL & INSTALLATION

1. Disconnect the negative battery cable. Drain the cooling system.
2. Remove alternator, carburetor air cleaner and fuel line.
3. Disconnect the accelerator linkage.
4. Remove the vacuum control hose between the carburetor and distributor.
5. Remove the distributor cap and wires.
6. Disconnect the coil wires, heat in-

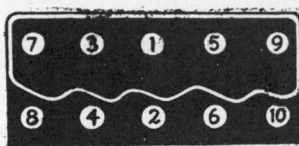

Cylinder head bolt tightening sequence

dicator sending unit wire, heater hoses and bypass hose.
7. Remove the closed ventilation system, evaporation control system and valve covers.
8. Remove the intake manifold, ignition coil and carburetor as an assembly.
9. Remove the exhaust manifolds.
10. Remove the rocker arm and shaft assemblies. Remove the pushrods and identify to ensure installation in original location. Inspect pushrods and replace worn or bent rods.
11. Remove the head bolts from each cylinder head and remove the cylinder heads.

To install:

12. Prior to installing the cylinder heads, clean all gasket surfaces of cylinder block and cylinder heads.
13. Inspect all surfaces with a straight-edge if there is any reason to suspect leakage. If out of flatness exceeds 0.004 in., either machine or replace the head.
14. Remove cylinder heads from holding fixtures, install gaskets and place heads on engine.
15. Clean pipe sealant from bolt threads and bolt holes. Apply Mopar Lock N' Seal® or equivalent, to bolt threads.
16. Install cylinder head bolts. Starting at top center, tighten all cylinder head bolts to 50 ft. lbs. (68 Nm) in sequence. Repeat procedure, retightening all cylinder head bolts to 105 ft. lbs. (142 Nm).
17. Install the pushrods, rocker arm and shaft assemblies with the notch on the end of rocker shaft pointing to centerline of engine and toward front of engine on the left bank and to the rear on right bank, making sure to install the long stamped steel retainers in the No. 2 and No. 4 positions, tighten to 200 inch lbs. (23 Nm).
18. Do not use any sealer on side composition gaskets.
19. Install side gaskets to cylinder head.
20. Clean the cylinder block front and rear gasket surfaces using an approved solvent.
21. Apply a thin, uniform coating of a quick dry cement to the intake manifold front and rear gaskets and cylinder block gasket surface. Allow to dry 4–5 minutes or until tack free.

NOTE: When installing gaskets,

the center hole in the gasket must engage the dowels in block. End holes in seals must be locked into tangs of head gasket.

22. Carefully install the front and rear intake manifold gaskets.

23. Place a drop, approximately ¼ in. diameter, of rubber sealer onto each of the 4 manifold-to-cylinder head gasket corners.

24. Carefully lower intake manifold into position on the cylinder block and cylinder heads. After the intake manifold is in place, inspect to make sure end seals are in place.

25. Install the finger-tight. Tighten the intake manifold bolts, in 3 stages, in sequence. The 1st stage to 25 ft. lbs. (34 Nm), the 2nd stage to 40 ft. lbs. (54 Nm) and 3rd stage to 45 ft. lbs. (61 Nm).

26. Install exhaust manifolds and tighten screws to 20 ft. lbs. (27 Nm) and nuts to 15 ft. lbs. (20 Nm).

27. Adjust spark plug gap and install the plugs, tightening to 30 ft. lbs. (41 Nm).

28. Install the ignition wires, heat indicator sending unit wire, heater hoses and bypass hose.

29. Install the vacuum control hoses between carburetor and distributor.

30. Install the throttle linkage and adjust, as necessary.

31. Install the distributor cap and wires.

32. Install the fuel line, alternator and drive belt. Tighten the alternator mounting bolt to 30 ft. lbs. (41 Nm) and adjusting strap bolt to 200 inch lbs. (23 Nm).

33. Be certain the valve covers are not distorted at screw holes; flatten if necessary. Thoroughly clean and dry the valve cover, mounting surface, bolts and bolt holes.

34. Place the new valve cover gaskets in position and install valve covers. Tighten to 80 inch lbs. (9 Nm) using load spreader fasteners.

35. Install the closed crankcase ventilation system and evaporation control system.

36. Fill the cooling system and install battery ground cable.

Valve Lifters

REMOVAL & INSTALLATION

Flat Lifters

1. Disconnect the negative battery cable.

2. Remove valve cover, rocker assembly and pushrods; identify pushrods to insure installation in original location.

3. Slide a lifter extractor tool through the opening in the cylinder head and seat the tool firmly in the head of lifter.

NOTE: Although it is possible to remove the valve lifters without removing the intake manifold, it is recommended the manifold be removed.

4. Pull the lifter out of the bore with a twisting motion. If all liters are to be removed, identify lifters to ensure installation in original locations. Work on 1 lifter at a time to avoid mixing of parts. Mixed parts are not compatible.

To install:

5. Lubricate lifters completely with engine oil.

6. Install lifters and pushrods in their original positions.

7. Install the rocker arm and shaft assembly.

8. Install the valve cover.

9. Start and operate engine. Warm up to normal operating temperature.

NOTE: To prevent damage to valve mechanism, the engine should not be run above fast idle until all hydraulic lifters have filled with oil and have become quiet.

Roller Lifters

1. Disconnect the negative battery cable.

2. Remove the valve cover. Remove the rocker assembly and pushrods. Identify the pushrods to insure proper installation.

3. Remove the intake manifold. Remove the valve lifter yoke retainer and aligning yokes.

4. Remove the valve lifters using a valve lifter removal tool. Identify the lifters to ensure proper installation.

5. Repair or replace the valve lifters as required.

6. Installation is the reverse order of the removal procedure.

7. When installing the aligning yokes, make sure the arrow points toward the camshaft. Torque the retaining bolt to 200 inch lbs. (23 Nm).

Rocker Arms/Shafts

REMOVAL & INSTALLATION

1. Disconnect the negative battery cable.

2. Disconnect the spark plug wires by pulling on the boot straight out in line with the plug.

3. Disconnect closed crankcase ventilation system and evaporation control system from valve cover.

4. Remove the valve cover and gasket.

5. Remove rocker shaft bolts and retainers.

6. Remove rocker arms and shaft assembly.

To install:

7. Before installing the rocker arm assemblies, check the oil drain holes for blockage.

8. Install the rocker arm and shaft assemblies with the notch of rocker shaft pointing to centerline of engine and toward front of engine on the left bank and to the rear on right bank, making sure to install the long stamped steel retainers in the No. 2 and No. 4 positions. Tighten bolts to 200 inch lbs. (23 Nm).

9. Clean the valve cover gasket surface. Inspect cover for distortion and flatten, if necessary.

10. Clean the head rail, if necessary. Install the valve cover and tighten bolts to 80 inch lbs. (9 Nm).

11. Install closed crankcase ventilation system and evaporation control system.

Intake Manifold

REMOVAL & INSTALLATION

1. Drain the cooling system. Disconnect the negative battery cable.

2. Remove the alternator, the air cleaner and disconnect the fuel line from the carburetor.

3. Disconnect all vacuum lines and throttle linkage attached to the carburetor and intake manifold.

4. Disconnect the spark plug wires from the plugs and remove the distributor cap and wires as an assembly.

5. Disconnect the wires from the coil and the temperature sending unit.

6. Disconnect the heater hose and bypass hose from the intake manifold.

7. Remove the intake manifold attaching bolts and remove the manifold, carburetor and coil from the engine as an assembly.

8. Clean all gasket mounting surfaces and firmly cement new gaskets to the engine.

9. Installation is the reverse order of the removal procedure. Torque

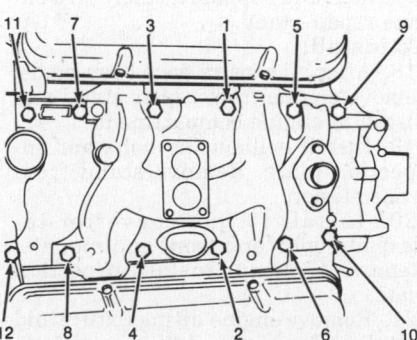

Intake manifold bolts tightening sequence

bolts to 45 ft. lbs. (61 Nm), in 3 passes, in sequence.

Exhaust Manifold

REMOVAL & INSTALLATION

1. Disconnect the negative battery cable.
2. Raise and support the vehicle safely.
3. Disconnect the exhaust manifold at the pipe flange. Access to these bolts is from underneath the vehicle.
4. If equipped, disconnect the air injection nozzles and carburetor heated air stove.
5. Disconnect any components of the EGR system which are in the way. Remove the exhaust manifold by removing the securing bolts and washers.
6. When the exhaust manifold is removed, sometimes the securing studs will come out with the nuts. If this occurs, studs must be replaced with the aid of sealing compound on the coarse thread ends. If this is not done, water leaks may develop at the studs.
7. Installation is the reverse order of the removal procedure. Torque the bolts to 20 ft. lbs. (27 Nm) and the nuts to 15 ft. lbs. (20 Nm).

NOTE: On the center branch of the manifold, no conical washers are used.

Timing Chain Front Cover

REMOVAL & INSTALLATION

1. Disconnect the negative battery cable.
2. Drain the cooling system.
3. Remove the water pump.
4. Remove the power steering pump.
5. Remove the pulley from the vibration damper and bolt and washer securing the vibration damper on the crankshaft.
6. Using a vibration damper pulling tool, remove the vibration damper from end of crankshaft.
7. Remove the fuel lines and fuel pump.
8. Loosen the oil pan bolts and remove the front bolt at each side.
9. Remove the chain case cover and gasket using extreme caution to avoid damaging the oil pan gasket.
To install:
10. Prior to installing the timing cover, be sure mating surfaces of chain case cover and cylinder block are clean and free from burrs.
11. Using a new cover gasket, care-fully install the chain case cover to avoid damaging oil pan gasket. A ⅛ in. diameter bead of sealer is recommended on the oil pan gasket. Do not tighten the chain case cover bolts at this time.
12. Lubricate the seal lip with lubriplate, position vibration damper hub slot on crankshaft. Damper will act as a pilot for the crankshaft seal.
13. Press the vibration damper on the crankshaft.
14. Tighten the chain case cover screws to 30 ft. lbs. (41 Nm) first, tighten the oil pan screws to 200 inch lbs. (23 Nm).
15. Install the vibration damper bolt with the washer and tighten to 135 ft. lbs. (183 Nm).
16. Position the pulley on the vibration damper and attach with bolts and lock washers. Tighten to 200 inch lbs. (23 Nm).
17. Install the fuel pump and fuel lines.
18. Install the water pump and housing assembly, using new gaskets. Tighten bolts to 30 ft. lbs. (41 Nm).
19. Install the power steering pump.
20. Install the fan/belt assembly, hoses and close drains.
21. Fill the cooling system. Connect the negative battery cable.

Front Cover Oil Seal

REPLACEMENT

1. Disconnect the negative battery cable.
2. Loosen and remove the belts from the crankshaft pulley.
3. Remove the radiator shroud screws and set the shroud back over the engine.
4. Remove the fan and shroud from the engine.
5. Remove the crankshaft pulley and vibration dampener bolt and washer from the end of the crankshaft.
6. Pull the vibration dampener from the end of crankshaft.
7. Using a seal removing tool behind the lips of the oil seal, pry outward, being careful not to damage the crankshaft seal surface of cover.
To install:
8. Install the new seal by installing the threaded shaft part of the installing tool into the threads of the crankshaft.
9. Place the seal into the opening, with the seal spring toward the inside of the engine.
10. Place the installing tool with the thrust bearing and nut on the shaft. Tighten nut until tool is flush with the timing chain cover.

11. Lubricate the dampener hub and install the vibration dampener.
12. Install the vibration dampener bolt and washer and torque to 135 ft. lbs. (183 Nm).
13. Install the pulley on the vibration dampener and torque to 200 inch lbs. (23 Nm).
14. Set the radiator shroud back over engine and install the fan and belts.
15. Install the radiator shroud to the radiator.
16. Connect the negative battery cable.

Timing Chain and Gears

REMOVAL & INSTALLATION

1. Position the engine at TDC on the compression stroke.
2. Disconnect the negative battery cable.
3. Remove the front timing cover.
4. Remove the camshaft sprocket attaching cup washer, fuel pump eccentric and remove timing chain with crankshaft and camshaft sprockets.
5. Place both camshaft sprocket and crankshaft sprocket on the bench with timing marks on exact imaginary center line through both camshaft and crankshaft bores.
6. Place the timing chain around both sprockets.
7. Turn the crankshaft and camshaft to line up with keyway location in crankshaft sprocket and in camshaft sprocket.
8. Lift the sprockets and chain, keep the sprockets tight against the chain in position.
9. Slide both sprockets evenly over

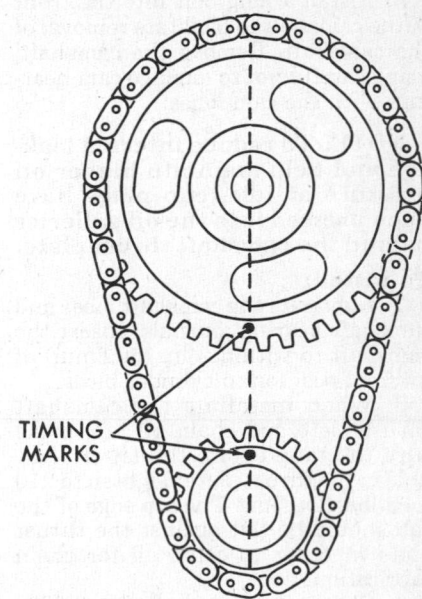

Alignment of timing marks

their respective shafts and use a straight-edge to check alignment of the timing marks.

10. Install the fuel pump eccentric, cup washer and camshaft bolt. Tighten bolt to 35 ft. lbs. (47 Nm).

11. Check the camshaft for 0.002–0.006 in. (0.051–0.0152mm) endplay with a new thrust plate and up to 0.010 in. (0.254mm) endplay with a used thrust plate. If not within these limits, install a new thrust plate.

12. Continue the installation in the reverse order of the removal procedure.

Camshaft

REMOVAL & INSTALLATION

1. Disconnect the negative battery cable. Drain the cooling system. Position the engine at TDC on the compression stroke.

2. Remove the air cleaner assembly. Remove the valve covers. Remove the rocker arm and shaft assemblies. Remove the distributor.

3. Remove the intake manifold assembly. Remove the pushrods and lifters. Be sure to identify the components so each part will be replaced in its original location.

4. Remove the radiator assembly. Remove the front cover assembly. Remove the camshaft gear and timing chain.

5. Properly discharge the air conditioning system and remove the air conditioning condenser.

NOTE: If necessary, remove the grille assembly to allow enough room to remove the camshaft.

6. Install a long bolt into the front of the camshaft to facilitate removal of the camshaft. Remove the camshaft, being careful not to damage cam bearings with the cam lobes.

NOTE: To reduce internal leakage and help maintain higher oil pressure at idle, cup plugs have been pressed into the oil galleries behind the camshaft thrust plate.

To install:

7. Lubricate the camshaft lobes and camshaft bearing journals. Insert the camshaft to within 2 in. (50.8mm) of its final position in cylinder block.

8. When installing the camshaft thrust plate and chain oil tab, make sure the tang enters the lower right hole in the thrust plate. Tighten to 210 inch lbs. (24 Nm). The top edge of the tab should be flat against the thrust plate in order to catch oil for chain lubrication.

9. Check the camshaft for 0.002–0.006 in. endplay with a new thrust

plate and up to 0.010 in. endplay with a used thrust plate. If not within limits install a new thrust plate.

10. Intall the intake manaifold and all related parts. Be sure to torque the intake manifold bolts to specification and in the proper sequence.

11. Connect the negative battery cable.

Piston and Connecting Rod

POSITIONING

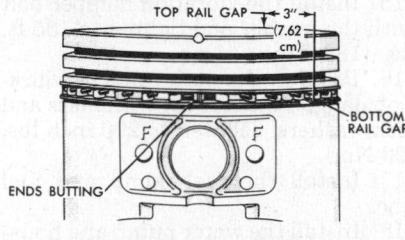

Oil ring installation

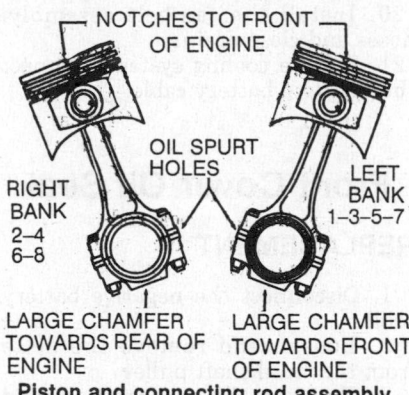

Piston and connecting rod assembly

ENGINE LUBRICATION

Oil Pan

REMOVAL & INSTALLATION

1. Disconnect the negative battery cable and remove dipstick.

2. Raise the vehicle, support it safely and drain the oil from the pan.

3. Remove the exhaust crossover pipe. Disconnect and lower center steering link.

4. Remove the starter nut and bolt and remove the starter.

5. Remove the torque converter inspection plate.

6. Remove the engine oil pan retaining bolts and remove the oil pan.

To install:

7. Inspect alignment of the oil strainer. The bottom of the strainer must be parallel with the machined surface of the cylinder block. The bottom of the strainer must touch the bottom of oil pan with $1/16$–$1/8$ in. (1.587–3.175mm) interference desirable.

8. Using a new pan gasket, add a drop of sealer at corners of rubber and cork.

9. Install the oil pan and torque the screws to 200 inch lbs. (23 Nm).

10. Install the torque converter inspection plate.

11. Install the starter and starter mounting nut and bolt.

12. Install the crossover pipe. Torque to 24 ft. lbs. (33 Nm).

13. Connect the center steering link.

14. Lower the vehicle, install dipstick and fill the engine with motor oil.

15. Connect the negative battery cable, start engine and check for leaks.

Oil Pump

REMOVAL & INSTALLATION

1. Disconnect the negative battery cable.

2. Raise the vehicle and support it safely.

3. Remove the oil pan.

4. Remove the oil pump from the rear main bearing cap.

5. Prime the oil pump before installation by filling the rotor cavity with engine oil and rotating the shaft.

6. Install the oil pump on the rear main bearing cap and tighten the retaining bolts to 30 ft. lbs. (41 Nm).

Rear Main Bearing Oil Seal

REMOVAL & INSTALLATION

Rope Type

1. Disconnect the negative battery cable.

2. Raise the vehicle and support it safely.

3. Remove the oil pan.

4. Remove the rear main bearing cap.

5. Remove the lower rope oil seal by prying from the side of the bearing cap with a small pry tool.

6. Install a new lower seal half in the cap. Tap the seal down into position with a rope seal installing tool.

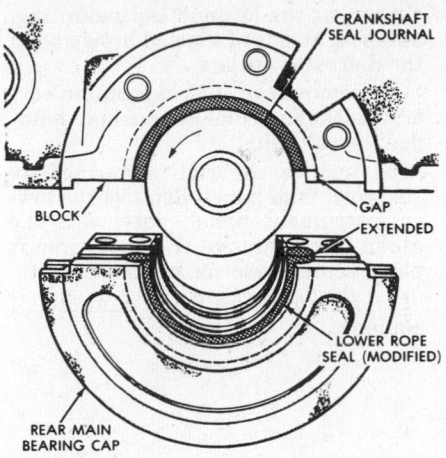

Modifying lower rope type rear main crankshaft seal

7. Cut the right bank seal end flush with the cap.

8. Remove the rope seal, rotate it end for end and re-install the seal back into the bearing cap with the cut end protruding above the surface so as to tightly fill the block half seal end compressed above the block/cap parting line.

9. Re-press the seal into the cap with the rope seal installing tool and cut the left bank side flush with the cap surface.

NOTE: This modification procedure ensures the protruding end is properly formed without a frayed end.

10. Lightly oil the lower rope seal half with engine oil. Install the side seals in the bearing cap. Be sure the side seal identified with yellow paint is installed on the right side.

11. Remove the upper seal half with a rope seal remover tool.

12. Screw the tool into the seal, being careful not to damage the crankshaft. Pull the seal out with the tool while rotating the crankshaft.

To install:

13. Lightly lubricate the new seal before installing it.

14. Install the upper rope seal using a rope seal installer tool. Carefully trim the upper seal after installation.

15. Install the rear main bearing cap, being careful not to crimp the extended side of the oil seal between the cap and the block.

16. Install the main bearing bolts and torque to 85 ft. lbs. (115 Nm).

17. Complete the assembly of the oil pump and oil pan assembly. Add sealer at the bearing cap to block joint to provide oil pan end sealing.

ENGINE COOLING

Radiator

REMOVAL & INSTALLATION

1. Disconnect the negative battery cable.

2. Place the heater temperature selector to FULL ON. Drain the cooling system by opening the drain cock at the bottom of the radiator. When the reserve tank is empty, remove the pressure cap.

3. Remove the oil cooler lines from the radiator.

4. Remove the upper and lower hose clamps and hoses. Remove the coolant reserve tank tube.

5. Remove the screws and position the shroud rearward to provide maximum clearance.

6. Loosen the retaining screws at the bottom of the radiator and remove the screws at the top.

7. Lift the radiator out of the engine compartment.

NOTE: Extreme care should be taken during removal not to damage the radiator cooling fins or water tubes.

8. Installation is the reverse order of the removal procedure. Fill the radiator to the top of neck and the reservoir tank to the MAX level. Warm the engine with the heater on and check the coolant level. Check the transmission fluid level after warm-up and add fluid, as necessary.

Heater Core

REMOVAL & INSTALLATION

Without Air Conditioning

1. Disconnect the negative battery cable.

2. Drain the radiator coolant.

3. Remove the air cleaner and disconnect the heater hoses. Plug the core tubes to prevent spillage.

4. Slide the front seat all the way back.

5. Remove the instrument cluster bezel assembly.

6. Remove the instrument panel upper cover by removing the mounting screws at the top inner surface of the glove box, above the instrument cluster, at the left end cap mounting, at the right side of the pad brow and in the defroster outlets.

7. Remove the steering column cover; it is the instrument panel piece under the column.

8. Remove the right intermediate side cowl trim panel. Remove the lower instrument panel; part with the glove box. Remove the instrument panel center to lower reinforcement.

9. Remove the floor console, if equipped.

10. Remove the right center air distribution duct. Detach the locking tab on the defroster duct.

11. Disconnect the temperature control cable from the housing. Disconnect the blower motor resistor block wiring.

12. Detach the vacuum lines from the water valve and tee in the engine compartment. Detach the wiring from the evaporator housing. Remove the vacuum lines from the inlet air housing and disconnect the vacuum harness coupling.

13. Remove the drain tube in the engine compartment. Remove the mounting nuts from the firewall.

14. Roll the heater unit back so the pipes clear and remove it.

15. Remove the blend air door lever from the shaft. Remove the screws and lift off the top cover. Lift the heater core out.

To install:

16. When installing the heater core, place the housing on the front floor under the instrument panel.

17. Tip the housing up under instrument panel and press mounting studs through the dash panel, making sure the defroster duct is properly seated on unit and gasket is installed properly. Connect the locking tab on the defroster duct.

18. While holding the housing in position, place the mounting bracket in position to the plenum stud and install the nut.

19. In engine compartment, install retaining nuts and tighten securely.

20. Connect electrical connectors to the resistor block and connect the control cable.

21. Connect vacuum lines in engine compartment, making sure the grommet is seated. Connect vacuum lines to inlet air housing and vacuum harness coupling.

22. Install right center air distribution duct.

23. Install instrument panel center to lower reinforcement.

24. Install lower instrument panel.

25. Install right intermediate side cowl trim panel.

26. Install steering column cover.

27. Install instrument panel upper cover.

28. Install cluster bezel assembly.

29. From engine compartment, remove plugs from core tubes and con-

nect hoses to heater. Install condensate tube and corbin clamp.

30. Fill cooling system and inspect for leaks.

31. Install air cleaner and connect battery negative cable.

With Air Conditioning

1. Disconnect the negative battery cable.

2. Discharge the air conditioning system.

3. Drain the radiator coolant.

4. Remove the air cleaner and disconnect the heater hoses. Plug the core tubes to prevent spillage.

5. Remove the H-type expansion valve.

6. Slide the front seat all the way back.

7. Remove the instrument cluster bezel assembly.

8. Remove the instrument panel upper cover by removing the mounting screws at the top inner surface of the glove box, above the instrument cluster, at the left end cap mounting, at the right side of the pad brow and in the defroster outlets.

9. Remove the steering column cover; it is the instrument panel piece under the column.

10. Remove the right intermediate side cowl trim panel. Remove the lower instrument panel; part with the glove box. Remove the instrument panel center to lower reinforcement.

11. Remove the floor console, if equipped.

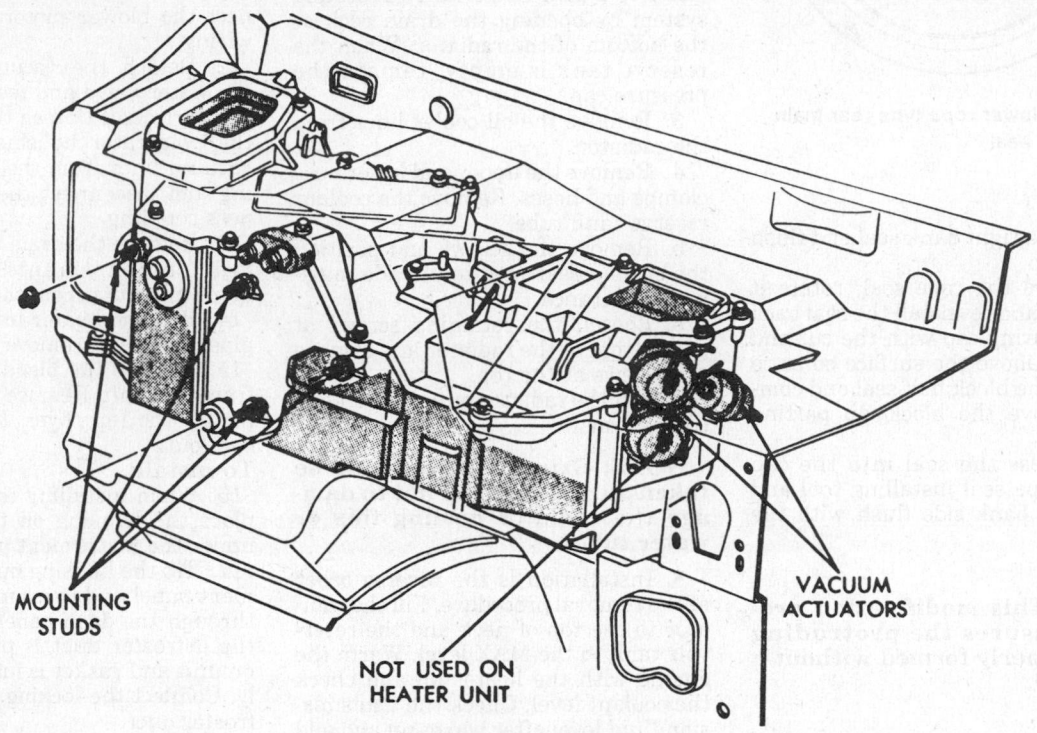

MOUNTING STUDS

NOT USED ON HEATER UNIT

VACUUM ACTUATORS

Front view of the heater-A/C housing

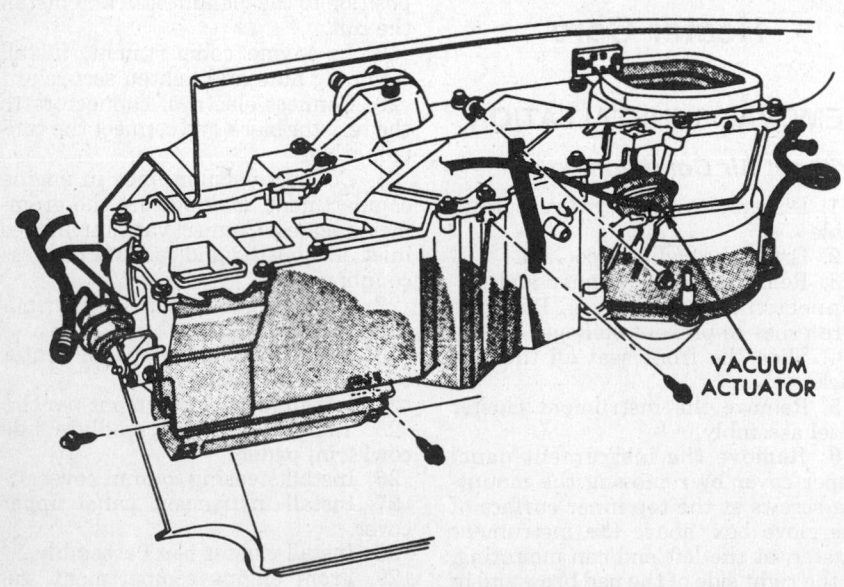

VACUUM ACTUATOR

Rear view of the heater-A/C housing

12. Remove the right center air distribution duct. Detach the locking tab on the defroster duct.

13. Disconnect the temperature control cable from the housing. Disconnect the blower motor resistor block wiring.

14. Detach the vacuum lines from the water valve and tee in the engine compartment. Detach the wiring from the evaporator housing. Remove the vacuum lines from the inlet air housing and disconnect the vacuum harness coupling.

15. Remove the drain tube in the engine compartment. Remove the mounting nuts from the firewall.

16. Remove the hanger strap from the rear of the evaporator and plenum stud.

17. Roll the heater/air conditioning unit back so the pipes clear and remove it.

18. Remove the blend air door lever

from the shaft. Remove the screws and lift off the top cover. Lift the heater core out.

To install:

19. When installing the heater core, place the evaporator housing on the front floor under the instrument panel.

20. Tip the evaporator housing up under instrument panel and press mounting studs through the dash panel, making sure the defroster duct and air conditioning distribution duct is properly seated on unit and gasket is installed properly. Connect the locking tab on the defroster duct.

21. While holding the housing in position, place the mounting bracket in position to the plenum stud and install the nut.

22. In engine compartment, install retaining nuts and tighten securely. Install condensate drain tube.

23. Connect electrical connectors to the resistor block and connect the control cable.

24. Connect vacuum lines in engine compartment, making sure the grommet is seated. Connect vacuum lines to inlet air housing and vacuum harness coupling.

25. Install right center air distribution duct.

26. Install instrument panel center to lower reinforcement.

27. Install lower instrument panel.

28. Install right intermediate side cowl trim panel.

29. Install steering column cover.

30. Install instrument panel upper cover.

31. Install cluster bezel assembly.

32. From engine compartment, remove plugs from core tubes and connect hoses to heater. Install condensate tube and corbin clamp.

33. Install H-valve and install refrigerant lines to valve. Replace gaskets.

34. Fill cooling system and inspect for leaks.

35. After the evaporator heater housing assembly is installed in the vehicle, evacuate and recharge the system with the proper amount of refrigerant. It is recommended, operation of all controls be tested and an overall performance test be made after the repair or replacement of the evaporator assembly.

36. Install air cleaner and connect battery negative cable.

Water Pump

REMOVAL & INSTALLATION

1. Disconnect the negative battery cable. Drain the cooling system.

2. Remove the fan shroud screws and move the shroud aside.

3. It may be necessary to remove the radiator on some vehicles to obtain the working clearance necessary to remove the water pump.

4. Loosen the alternator mounting bolts. Loosen the mounting bolts for the power steering pump, idler pulley, air conditioning compressor and air pump, if equipped. Remove all the accessory belts.

5. Remove the fan, spacer or fluid drive and the pulley.

NOTE: For fluid-coupled fan drives, do not set the drive unit down with its shaft pointing downward. Keep the unit in a vertical position as installed on the engine. This will prevent the silicone fluid from leaking out.

6. If necessary, remove the alternator or compressor mounting bracket bolts from the water pump to swing the alternator or compressor aside; keep the compressor in an upright position.

7. If necessary, unbolt the power steering pump and set it aside; leave the hoses connected. Also remove the air pump and brackets, if equipped.

8. Detach the hoses from the water pump. Remove the bolts which secure the water pump body to its engine block housing. Remove the water pump and discard the gasket.

To install:

9. Install the bypass hose to the pump with the second clamp temporarily in the center of the hose. Install the water pump with a new gasket, using sealer. Torque the bolts to 30 ft. lbs. (40.7 Nm).

10. Rotate the pump shaft by hand to be sure it rotates freely. Install the alternator or compressor mounting bracket to the pump if either was removed. Install the pulley, spacer or fluid drive and the fan. Torque the nuts to 15 ft. lbs. (20.3 Nm).

11. Reinstall all accessory drive belts. Adjust them to get about ½ in. of play under moderate thumb pressure on the longest run of belt between pulleys.

12. Install the radiator, if removed.

13. Install the fan shroud. Fill the cooling system to 1¼ in. below the filler neck with correct water and antifreeze mixture, without a coolant reserve tank. With a reserve tank, fill the radiator and fill the tank to the indicated level. Warm up the engine with the heater on and inspect the water pump for any leaks. Check the coolant level and add as required.

Thermostat

REMOVAL & INSTALLATION

1. Disconnect the negative battery cable.

2. Drain the cooling system to a level below the thermostat housing.

3. Remove the housing bolts and the thermostat housing and thermostat. Clean the gasket surfaces.

4. Installation is the reverse order of the removal procedure. Use a new gasket, dipped in water or sealant. Be sure the pellet end is facing the engine.

5. Refill the system, allow the engine to warm up with the heater on and check for leaks.

ENGINE ELECTRICAL

Distributor

REMOVAL

1. Disconnect the negative battery cable.

2. Remove the cap and wire assembly.

3. Disconnect the lead wire at the harness connector.

4. Mark the relative positions of the distributor and rotor on the engine block or distributor housing edge.

5. Remove the distributor mounting bolt and hold-down clamp and lift out the distributor. Should the distributor shaft rotate slightly during the removal, make a 2nd matchmark to indicated rotor positioning for installation.

INSTALLATION

Timing Not Disturbed

1. Align the distributor rotor-to-distributor housing matchmark, the distributor housing-to-engine matchmarks and the intermediate shaft with the oil pump drive.

2. Install the distributor into the engine until it seats.

3. Install the hold-down clamp and bolt; be sure to snug the hold-down bolt.

4. Connect the electrical harness connector.

5. Start the engine and check the ignition timing. Refer to the underhood vehicle emission information label for correct timing specifications.

Timing Disturbed

1. Rotate the crankshaft until No. 1 cylinder is at TDC.
2. The pointer on the timing chain case cover should be over the **0** mark on the crankshaft pulley.
3. The slot in the intermediate shaft which carries the gear that drives the oil pump and the distributor should be parallel to the crankshaft.
4. Hold the distributor over the mounting pad on the cylinder block so the distributor body flange coincides with the mounting pad and the rotor points to the No. 1 cylinder firing position.
5. Install the distributor while holding the rotor in position, allowing it to move only enough to engage the slot in the drive gear.
6. Install the cap, snug down the retaining bolt and check the ignition timing. Refer to the underhood vehicle emission information label for correct timing specifications.

Ignition Timing

ADJUSTMENT

The ignition timing test indicates correct timing of the engine only at idle and with the engine hot. Check timing as follows:

1. Connect tachometer and timing light. A magnetic timing probe receptacle is mounted to timing indicator and may be used.
2. Set parking brake and place transmission in **P** or **N** position. Start the engine and run until normal operating temperature is reached.
3. Connect jumper wire between carburetor switch and ground.
4. Adjust engine idle if necessary. Check ignition timing.
5. If timing is out of allowed specifications, loosen the hold-down bolt and rotate distributor housing.
6. Turn the distributor housing in the direction of rotor-rotation to retard the timing. Rotate the distributor housing against rotor rotation to advance the timing.
7. Tighten distributor hold-down bolt securely.
8. Recheck engine idle and remove test equipment.

Alternator

PRECAUTIONS

Several precautions must be observed with alternator equipped vehicles to avoid damage to the unit.

• If the battery is removed for any reason, make sure it is reconnected with the correct polarity. Reversing

the battery connections may result in damage to the one-way rectifiers.

• When utilizing a booster battery as a starting aid, always connect the positive to positive terminals and the negative terminal from the booster battery to a good engine ground on the vehicle being started.

• Never use a fast charger as a booster to start vehicles.

• Disconnect the battery cables when charging the battery with a fast charger.

• Never attempt to polarize the alternator.

• Do not use test lights of more than 12 volts when checking diode continuity.

• Do not short across or ground any of the alternator terminals.

• The polarity of the battery, alternator and regulator must be matched and considered before making any electrical connections within the system.

• Never separate the alternator on an open circuit. Make sure all connections within the circuit are clean and tight.

• Disconnect the battery ground terminal when performing any service on electrical components.

• Disconnect the battery if arc welding is to be done on the vehicle.

BELT TENSION

ADJUSTMENT

Satisfactory performance of the belt driven accessories depends on proper belt tension. The 2 tensioning methods are given in order of preference:

1. Belt tension gauge method
2. Torque equivalent method

Belt Tension Gauge Method

For this method, the belt is adjusted by measuring the tension of the belts with a belt tension gauge. Check belt tension in the middle of the span, between the 2 pulleys.

Torque Equivalent Method

Each adjustable accessory bracket is provided with a ½ in. (13mm) square hole for torque wrench use. Equivalent torque values for adjusting each accessory drive belt are specified.

REMOVAL & INSTALLATION

1. Disconnect the negative battery terminal.
2. Loosen the alternator mounting nut and bolt, the belt tensioner bracket bolt and remove the drive belt.
3. Remove the alternator mounting nut and bolt, the belt tensioner bracket bolt and spacer.

4. Disconnect the battery, field and ground leads from the alternator.
5. Disconnect the harness from the alternator and remove the alternator from the vehicle.
6. Position the harness to the alternator and reconnect the battery, field and ground leads to the alternator.
7. Position the alternator spacer between the end shields and install the mounting nut.
8. Install the adjustment bracket bolt and drive belt.
9. Adjust the drive belt tension.

Voltage Regulator

REMOVAL & INSTALLATION

1. Disconnect the negative battery cable. Disconnect the wiring harness plug from the voltage regulator.
2. Remove the mounting screws from the voltage regulator base and remove the regulator.
3. Prior to installing the voltage regulator, clean the mounting surface of any dirt or corrosion build-up; the regulator must have a good ground contact.
4. Position the voltage regulator to the mounting surface and install the mounting screws.
5. Connect the wiring harness plug to the voltage regulator. Test the system for proper operation.

Starter

REMOVAL & INSTALLATION

1. Disconnect the negative battery cable. Raise and support the vehicle safely.
2. Remove the cable from the starter and heat shield.
3. Disconnect the solenoid leads at the solenoid terminals.
4. Remove the starter securing bolts and remove the starter from the engine flywheel housing.
5. If fluid cooler tube bracket interferes with starter removal, remove the

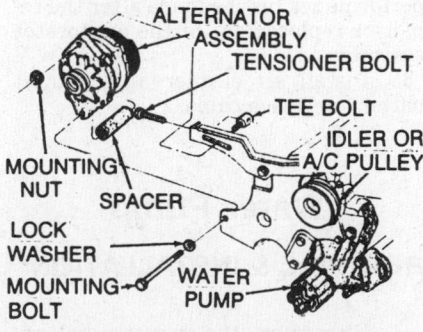

Alternator and mounting brackets

starter securing bolts, slide the cooler tube bracket off the stud and remove the starter.

To install:

6. Reverse order of the removal procedure. Be sure the starter and flywheel housing mating surfaces are free of dirt and oil.

7. When tightening the bolt and nut, hold the starter away from the engine to ensure proper alignment during its seating as the bolt is tightened. Do not damage the flywheel housing seal, if equipped.

EMISSION CONTROLS

Please refer to "Emission Controls" in the Unit Repair section for system maintenance procedures. Due to the complex nature of modern electronic engine control systems, comprehensive diagnosis and testing procedures fall outside the confines of this repair manual. For complete information on diagnosis, testing and repair procedures concerning all modern engine and emission control systems, please refer to "Chilton's Guide to Fuel Injection and Electronic Engine Controls".

FUEL SYSTEM

Fuel System Service Precautions

Safety is the most important factor when performing fuel system maintenance. Failure to conduct maintenance and repairs in a safe manner may result in serious personal injury or death. Maintenance and testing of the vehicle's fuel system components can be accomplished safely and effectively by adhering to the following rules and guidelines.

• To avoid the possibility of fire and personal injury, always disconnect the negative battery cable unless the repair or test procedure requires that battery voltage be applied.

• Always relieve the fuel system pressure prior to disconnecting any fuel system component (injector, fuel rail, pressure regulator, etc.), fitting or fuel line connection. Exercise extreme

caution whenever relieving fuel system pressure to avoid exposing skin, face and eyes to fuel spray. Please be advised that fuel under pressure may penetrate the skin or any part of the body that it contacts.

• Always place a shop towel or cloth around the fitting or connection prior to loosening to absorb any excess fuel due to spillage. Ensure that all fuel spillage (should it occur) is quickly removed from engine surfaces. Ensure that all fuel soaked cloths or towels are deposited into a suitable waste container.

• Always keep a dry chemical (Class B) fire extinguisher near the work area.

• Do not allow fuel spray or fuel vapors to come into contact with a spark or open flame.

• Always use a backup wrench when loosening and tightening fuel line connection fittings. This will prevent unnecessary stress and torsion to fuel line piping. Always follow the proper torque specifications.

• Always replace worn fuel fitting O-rings with new. Do not substitute fuel hose or equivalent where fuel pipe is installed.

RELIEVING FUEL SYSTEM PRESSURE

1. Loosen the fuel filler cap to release fuel tank pressure.

2. Place a container under the fuel inlet fitting to catch any fuel that may be trapped in the fuel line.

3. Relieve the fuel pressure by slowly loosening the fuel inlet line to the carburetor, using 2 wrenches to avoid twisting the line.

4. Fuel will spray slightly from the line into the container. Wrap a shop towel around the connection to avoid the spray of fuel.

5. When repairs have been completed, tighten the fuel lines and inspect for fuel leaks.

Fuel Tank

REMOVAL & INSTALLATION

1. Disconnect the negative battery cable.

2. Relieve the fuel pressure.

3. Raise the vehicle and support safely.

4. Using the proper equipment, drain the fuel tank.

5. Remove the screws that hold the filler neck to the quarter panel.

6. Disconnect the wiring and hoses from the tank.

7. Place a transmission jack or equivalent, under the center of the

tank and apply slight pressure. Remove the tank straps.

8. Remove the filler tube from the tank.

9. Lower the tank and disconnect the vapor separator rollover valve hose. Remove the fuel tank from the vehicle.

To install:

10. Raise the tank into position and connect all harnesses and vacuum hoses.

11. Install the tank straps and tighten the retaining nuts.

12. Install the screws that hold the filler neck to the quarter panel.

13. Connect the negative battery cable, start the engine and check for leaks.

Fuel Filter

REMOVAL & INSTALLATION

1. Locate the filter in the fuel line between the fuel pump and the carburetor.

2. Using hose-clamp pliers, remove the attaching clamps and pull the filter and accompanying hoses off.

3. Reverse this procedure for installation, using new hoses and clamps. Be sure the arrow on the filter is pointing toward the carburetor, in the direction of fuel flow.

Mechanical Fuel Pump

PRESSURE TESTING

1. Insert a tee fitting in fuel line at the carburetor.

2. Connect a 6 in. (152.4mm) piece of hose between the tee fitting and a fuel pressure gauge.

NOTE: The hose should not exceed 6 in. (152.4mm). The longer hose may collect fuel and additional weight of fuel would be added to pump pressure and result in an inaccurate reading.

3. Vent the pump for a few seconds to relieve air trapped in fuel chamber. If this is not done, the pump will not operate at full capacity and low pressure reading will result.

4. Connect a tachometer, then start engine and run at idle. The reading should be as shown in specifications, depending on the pump, and remain constant or return to 0 slowly, when the engine is stopped. An instant drop to 0 indicates a leaky outlet valve. If pressure is too low, a weak diaphragm main spring or improper assembly of diaphragm may be the cause. If pressure is too high, main spring is too strong or the air vent is plugged.

REMOVAL & INSTALLATION

1. Disconnect the negative battery cable.
2. Disconnect and plug the fuel lines from the fuel pump.
3. Remove the pump-to-block mounting bolts.
4. Remove the pump.
5. Remove the old gasket from the pump and replace with a new gasket during reinstallation.
6. Installation is the reverse order of removal procedure.

Carburetor

REMOVAL & INSTALLATION

1. Be sure the engine is cold before removing the carburetor from the engine. Disconnect the negative battery cable.
2. Remove the air cleaner.
3. Remove the fuel tank cap.
4. Place a container under the fuel inlet fitting to catch any be remaining in the fuel line.
5. Disconnect the fuel inlet line using a line wrench and a open end wrench to avoid twisting the line.
6. Disconnect the throttle linkage, choke linkage and all vacuum hoses.
7. Remove the carburetor mounting bolts or nuts and carefully remove the carburetor from the engine compartment. Hold the carburetor level to avoid spilling fuel from fuel bowl.

IDLE SPEED ADJUSTMENT

Holley 6280 Electronic Feedback Carburetor

NOTE: Before checking or adjusting the idle speed, check ignition and adjust ignition timing.

1. Disconnect and plug the vacuum hose at the EGR valve. Disconnect and plug the hose from the carburetor at the heated air temperature sensor. Remove air cleaner and disconnect and plug the canister purge hose at the canister and plug the vacuum hose at the ESA unit. Remove the PCV valve from the valve cover and allow the valve to draw underhood air. Install tachometer, start and run engine until normal operating temperature is reached. Turn OFF engine. Disconnect, then reconnect, fusible link at battery.
2. Ground the carburetor switch. Disconnect the engine harness lead from the oxygen sensor and ground the engine harness lead.

NOTE: Care should be exer-

cised so no pulling force is put on the wire attached to the oxygen sensor. The bullet connector to be disconnected is approximately 4 in. from the sensor. Use care in working around the sensor as the exhaust manifold is extremely hot.

3. Start the engine and allow it to reach normal operating temperatures.
4. Connect a jumper wire between the positive battery terminal and the solenoid idle stop lead wire. Be sure to attach the wire to the right solenoid or damage to the wiring harness will occur.
5. Open throttle slightly to allow solenoid plunger to extend. Remove solenoid outer screw and spring. Insert a ⅛ in. Allen wrench into solenoid and adjust solenoid idle speed.
6. Install the screw and spring. Turn the screw in until it lightly bottoms out. Remove jumper wire. Set the idle speed by turning out solenoid screw.
7. The solenoid rpm specification is 900 and the idle rpm specification is 680.
8. Remove the tachometer. Unplug and reconnect all hoses. Reinstall the PCV valve and the air cleaner.

Rochester Quadrajet Electronic Feedback Carburetor

1. Disconnect and plug the vacuum hose at the EGR valve.
2. Disconnect and plug the hose from the carburetor at the heated air temperature sensor.
3. Remove air cleaner and disconnect and plug the canister purge hose at the canister.
4. Remove the PCV valve from the valve cover and allow the valve to draw underhood air.
5. Install tachometer and start and run engine until normal operating temperatures are reached.
6. Disconnect carburetor electrical connector. Attach a jumper wire between the ground switch terminal of the wiring harness connector (violet wire) and a good ground.
7. Attach a jumper wire between solenoid coil terminal of the carburetor connector (red wire) and battery positive post. Open throttle slightly to allow solenoid plunger to extend.
8. Remove outer screw and spring from solenoid. Insert a ⅛ in. Allen wrench into solenoid and adjust solenoid idle speed.
9. The solenoid rpm specification is 800. The idle rpm specification is 750.
10. Install screw and spring and turn in the outer screw until it lightly bottoms out. Remove jumper wire from carburetor connector and battery.

Turn the outer solenoid screw until correct idle rpm is obtained.
11. Remove remaining jumper wire and reconnect carburetor connector. Remove tachometer, unplug and reconnect all hoses, reinstall the PCV valve and air cleaner.

SERVICE ADJUSTMENTS

For all carburetor service adjustment procedures and specifications, please refer to "Carburetor Service" in the Unit Repair section.

DRIVE AXLE

Driveshaft and U-Joints

The driveshaft is a one-piece tubular shaft with 2 universal joints, mounted at either end. The front joint yoke serves as a slip yoke on the transmission output shaft. The rear universal joint is the type that must be disassembled to be removed.

REMOVAL & INSTALLATION

1. Raise and support the vehicle safely.
2. Matchmark the driveshaft, U-joint and pinion flange before disassembly. These marks must be realigned during reassembly to maintain the balance of the driveline. Failure to align them may result in excessive vibration.
3. Remove both of the clamps from the differential pinion yoke and slide the driveshaft forward slightly to disengage the U-joint from the pinion yoke. Tape the 2 loose U-joint bearings together to prevent them from falling off.

NOTE: Do not disturb the bearing assembly retaining strap. Never allow the driveshaft to hang from either of the U-joints. Always support the unattached end of the shaft to prevent damage to the joints.

4. Lower the rear end of the driveshaft and gently slide the front yoke/driveshaft assembly rearward disengaging the assembly from the transmission output shaft. Be careful not to damage the splines or the surface with the output shaft seal rides on.
5. Check the transmission output shaft seal for sign of leakage.
6. Installation is the reverse order

of the removal procedure. Be sure to align the matchmarks. The torque for the clamp bolts is 14 ft. lbs. (19 Nm).

Rear Axle Shaft, Bearing and Seal

REMOVAL & INSTALLATION

NOTE: **Under no circumstances should rear axle bearing cones, cups, bores or journals be subjected to heating with a torch, hitting with a hammer or any other abnormal abuse, or permanent damage may result.**

1. Raise the vehicle and support it safely.
2. Remove the wheel cover and wheel and tire assembly. Remove the brake drum.
3. Loosen the housing cover and drain the lubricant from the rear axle. Remove the cover.
4. Turn the differential case to make the differential pinion shaft lock screw is accessible and remove the lock screw and pinion shaft.
5. Push the axle shafts toward the center of vehicle and remove the C-washers from the recessed groove of the axle shaft.
6. Remove the axle shaft from housing being careful not to damage the straight roller-type axle shaft bearing which will remain in the rear axle housing.
7. Inspect the axle shaft bearing surfaces for signs of imperfection, spalling or pitting. If any of these conditions are present both the shaft and the bearing should be replaced.
8. Remove the axle shaft seal from housing bore. Using a slide hammer motion, remove the axle shaft bearing. If the axle shaft and bearing show no signs of distress, they can be reinstalled along with a new axle shaft seal. Never reuse an axle shaft seal.

To install:

NOTE: **Remove all burrs from the bearing housing shoulder.**

9. Wipe the axle shaft bearing cavity of axle housing clean. The axle shaft oil seal bores at both ends of the housing should be smooth and free of rust and corrosion. This also applies to the brake support plate and housing flange face surface.
10. Insert the axle shaft bearing into cavity making sure it bottoms against the shoulder and it is not cocked in bore.

NOTE: **Do not use the seal to position or bottom the bearing in**

its bore as this would damage the seal.

11. Install the axle shaft bearing seal using bearing installer tool, until the outer flange of tool bottoms against housing flange face. This positions the seal to the proper depth beyond the end of the flange face.
12. Lubricate the bearing and seal area of the axle shaft, slide the axle shaft into place being careful the splines of the shaft do not damage the oil seal and properly engage with the splines of differential side gears.
13. With the axle shaft in place, install the C-washers in recessed grooves of axle shaft and pull outward on the shaft so the C-washers seat in the counterbore of the differential side gear.
14. Install the differential pinion shaft through the case and pinions, aligning the hole in shaft with the lock screw hole. Install the lock screw and tighten to 100 inch lbs. (11 Nm).
15. Clean up the mating surfaces and apply a $^{1}/_{16}$–$^{3}/_{32}$ in. bead of silicone rubber sealant along the bolt circle of the cover. Allow sealant to cure.

Pinion Seal

REMOVAL & INSTALLATION

1. Raise and safely support the vehicle.
2. Remove the driveshaft from the differential pinion flange and support it aside.
3. Remove the drive pinion nut and washer.
4. Using removal tool C–452 and holding tool C–3281 or equivalent, remove the drive pinion flange from the drive pinion.
5. Using a prybar and a hammer, remove the pinion seal from the housing and discard it; be careful not to scratch the pinion or the mounting surface.

To install:

6. Clean the pinion seal mounting surface.
7. Using seal installer tool C–4002 for 7¼ in. axle or tool C–4076 for 8¼ in. axle, drive the new seal into the housing until it seats.
8. Using installer tool C–3718 and holding tool C–3281 or equivalent, press the drive pinion flange onto the drive pinion.
9. Install the drive pinion washer and nut. Torque the drive pinion nut to 210 ft. lbs. (284 Nm).
10. Using an inch lb. torque wrench, check the pinion bearing preload; the preload should be 15–30 inch lbs. (1–3 Nm) for 7¼ in. axle or 20–35 inch lbs. (2–4 Nm) for 8¼ in. axle.

NOTE: **If the correct preload is not reached, continue tightening in very small increments until it is reached.**

11. Install the driveshaft to the drive pinion flange and torque the cap bolts to 10 ft. lbs. (14 Nm).
12. Check and/or refill the axle housing. Lower the vehicle.

Axle Housing

REMOVAL & INSTALLATION

1. Raise the vehicle and support it safely. Install suitable stands at the front of the rear springs.
2. Block the brake pedal in the UP position, using a wooden block or equivalent.
3. Drain the lubricant from differential housing.
4. Loosen and remove rear wheels. Do not removed drum retaining spring clips or brake drums.
5. Disconnect hydraulic brake lines at wheel cylinders and cap fittings to prevent loss of brake fluid.
6. Disconnect the parking brake cables.
7. Disconnect the driveshaft at differential pinion flange and secure in a near horizontal position to prevent damage to front universal joint.
8. Remove the shock absorbers from the spring plate studs and loosen rear spring U-bolt nuts and remove U-bolts.
9. Remove the axle assembly from vehicle.
10. Installation is the reverse order of the removal procedure.

AUTOMATIC TRANSMISSION

For further information on transmissions/transaxles, please refer to "Chilton's Guide to Transmission Repair".

Transmission Assembly

REMOVAL & INSTALLATION

1. Disconnect negative battery cable. Raise the vehicle and support it safely.
2. Remove the exhaust system.
3. Remove engine to transmission braces, if equipped.
4. Remove and plug the cooler lines at transmission.

5. Remove the starter motor and cooler line bracket.

6. Remove the torque converter access cover.

7. Drain the transmission oil, then reinstall the pan.

8. Mark torque converter and flexplate to aid in re-assembly. The crankshaft flange bolt circle, inner and outer circle of holes in the flexplate and the 4 tapped holes in front face of the torque converter all have 1 hole offset so these parts will be installed in the original position. This maintains balance of the engine and torque converter.

9. Rotate engine to position the bolts attaching torque converter to flexplate and remove bolts.

10. Mark parts for re-assembly, then disconnect propeller shaft at rear universal joint.

11. Carefully pull shaft assembly out of the extension housing.

12. Disconnect wire connector from the neutral safety switch.

13. Disconnect gearshift rod and torque shaft assembly from transmission.

14. When it is necessary to disassemble linkage rods from levers using plastic grommets as retainers, the grommets should be replaced with new grommets. Use a prying tool to force rod from grommet in lever, then remove the old grommet. Use pliers to snap new grommet into lever and rod into grommet.

15. Disconnect throttle rod from lever at the left side of transmission. Remove linkage bellcrank from transmission, if equipped.

16. Remove oil filler tube and speedometer cable.

17. Install engine support fixture with frame hooks or a suitable substitute, that will support rear of the engine.

18. Raise the transmission slightly with service jack to relieve load on the supports.

19. Remove bolts securing transmission mount to crossmember and crossmember to frame, remove the crossmember.

20. Remove all bellhousing bolts.

21. Carefully work the transmission and torque converter assembly rearward off engine block dowels and disengage converter hub from the end of the crankshaft. Attach a small C-clamp to the edge of the bellhousing to hold the torque converter in place during transmission removal.

22. Lower the transmission and remove the assembly from under the vehicle.

23. To remove the torque converter assembly, remove the C-clamp from the edge of the bellhousing, and carefully slide the assembly out of the transmission.

To install:

NOTE: The transmission and torque converter must be installed as an assembly; otherwise, the torque converter flexplate, pump bushing and oil seal will be damaged. The flexplate will not support a load; therefore, none of the weight of transmission should be allowed to rest on the plate during installation.

24. Rotate the pump gears with an alignment tool until the 2 small holes in handle are vertical.

25. Carefully slide the torque converter assembly over the input shaft and reaction shaft. Make sure the torque converter hub slots are also vertical and fully engage the pump inner gear lugs.

NOTE: Test for full engagement by placing a straight-edge on face of the case. The surface of torque converter front cover lug should be at least ½ in. to rear of straight-edge when torque converter is pushed all the way into transmission.

26. Maintain the small C-clamp to edge of the torque converter housing to hold the torque converter in place during transmission installation.

27. Inspect the torque converter flexplate for distortion or cracks and replace, if necessary. Torque the flexplate to crankshaft bolts to 55 ft. lbs. (75 Nm). When the flexplate replacement has been necessary, make sure both transmission dowel pins are in the engine block and they are protruding far enough to hold the transmission in alignment.

28. Coat the converter hub hole in the crankshaft with multi-purpose grease. Place transmission and torque converter assembly on a service jack and position assembly under vehicle for installation. Raise or tilt, as necessary, until the transmission is aligned with the engine.

29. Rotate the torque converter so the mark on the torque converter, made during removal, will align with the mark on the flexplate. The offset holes in plate are located next to ⅛ in. hole in the inner circle of plate. Carefully work the transmission assembly forward over the engine block dowels with the torque converter hub entering the crankshaft opening.

30. After the transmission is in position, install the converter housing bolts and tighten to 30 ft. lbs. (41 Nm). If equipped, install vibration damper weight on rear of the extension housing.

31. Install the crossmember to frame and lower transmission to install mount on extension to the crossmember. Tighten bolts.

32. The engine support fixture may now be removed.

33. Install the oil filler tube and speedometer cable.

34. Connect the throttle rod to the transmission lever.

35. Connect the gear shift rod and torque shaft assembly to the transmission lever and frame.

36. Place the wire connector on the combination back-up light and neutral/park starter switch.

37. Carefully guide the sliding yoke into the extension housing and on the output shaft splines. Align marks made at removal. Connect the propeller shaft to the rear axle pinion shaft yoke.

38. Rotate the crankshaft clockwise with socket wrench on the vibration dampener bolt, as needed to install the torque converter to flexplate bolts, matching marks made at removal. Tighten to 270 inch lbs. (31 Nm).

39. Install the torque converter access cover.

40. Install the starter motor and cooler line bracket.

41. Tighten the cooler lines to the transmission fittings.

42. Install the engine-to-transmission struts, if equipped. Tighten the bolts holding strut to transmission before the strut to engine bolts.

43. Replace the exhaust system, if it was disturbed and adjust for clearance.

44. Adjust the shift and throttle linkage.

45. Refill the transmission with Dexron®II type automatic transmission fluid.

SHIFT LINKAGE ADJUSTMENT

NOTE: When it is necessary to disassemble linkage rods from their levers which use plastic grommets for retainers, the grommets should be replaced with new ones.

1. Make sure all linkage is free, especially the adjustable slide on the shift rod, so the pre-load spring action is not reduced by friction. Disassemble, clean and lube, if necessary.

2. Put the shift lever in the **P** position.

3. With the adjustable swivel loose, move the shift lever all the way to the rear-most detent position, which is **P**.

4. Tighten swivel lock bolts to 90 inch lbs. (10 Nm).

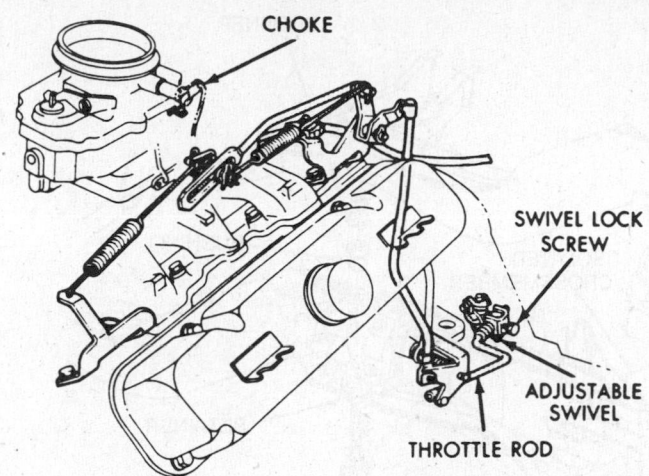

CHOKE

SWIVEL LOCK SCREW

ADJUSTABLE SWIVEL

THROTTLE ROD

Throttle rod and related parts

5. Verify the vehicle will only start in **P** or **N**.

THROTTLE LINKAGE ADJUSTMENT

1. Perform transmission throttle rod adjustment while engine is at normal operating temperature. Otherwise, make sure carburetor is not on fast idle cam.

2. Raise the vehicle and support it safely.

3. Loosen adjustment swivel lock screw.

4. To ensure proper adjustment, the swivel must be free to slide along the flat end of the throttle rod, this will insure the preload spring action is not restricted. Disassemble and clean or repair parts to assure free action, if necessary.

5. Hold transmission lever firmly forward against its internal stop and tighten swivel lock screw to 100 inch lbs. (11 Nm).

6. The adjustment is finished and linkage backlash was automatically removed by the preload spring.

7. Lower the vehicle, reconnect the choke, if disconnected, and test the linkage freedom of operation by moving the throttle rod rearward, slowly releasing it to confirm it will return fully forward.

FRONT SUSPENSION

Shock Absorbers

REMOVAL & INSTALLATION

1. Raise and safely support the vehicle.

2. Remove the front wheels.

3. Remove the nut and retainer from the shock absorber upper end.

4. Grip the shock absorber base, and remove the lower attaching nut, retainer and bushing.

5. Fully compress the shock absorber by pushing it upward, disengaging it from the lower control arm. Pull the shock absorber down firmly, and remove it from the vehicle.

6. Check the shock absorber bushings; if they are worn, cracked or scored, replace them. Remove and install the bushings with a press or using a drift and a hammer. To ease installation, lubricate with soapy water.

NOTE: Do not use oil to ease the installation.

7. Purge the new shock absorber by repeatedly extending it in the upright position and compressing it in the inverted position. It is normal to have more resistance to extend than to compress.

To install:

8. Fully compress the new shock absorber, insert the top end through the upper bushing, then install the retainer and nut. Torque the nut to 25 ft. lbs. (34 Nm).

NOTE: Be sure all the retainers are installed with the concave side in contact with the rubber.

9. Install the shock absorber to the lower control arm mount. Install the bolt, from the rear, or the retainer and nut finger-tight.

10. Lower the vehicle and torque the nut to 35 ft. lbs. (47 Nm), with the full weight of the vehicle on the wheels.

Torsion Bars

REMOVAL & INSTALLATION

1. Raise the safely support the vehicle so the front suspension is in full rebound position.

2. Release load on both torsion bars by turning anchor adjusting bolts in frame crossmember counterclockwise. Remove anchor adjusting bolt on torsion bar to be removed.

3. Raise lower control arms until clearance between crossmember ledge, at jounce bumper, and torsion bar end bushing is 2⅞ in. (63.0mm). Support lower control arms at this design height, equal to 3 passenger position with vehicle on ground. This is necessary to align sway bar and lower control arm attaching points for disassembly and component re-alignment and attachment during reassembly.

4. Remove sway bar to control arm attaching bolt and retainers.

5. Remove bolts attaching torsion bar end bushing to lower control arm.

6. Remove bolts attaching torsion bar pivot cushion bushing to crossmember and remove torsion bar and anchor assembly from crossmember.

7. Carefully separate anchor from torsion bar.

To install:

8. Carefully slide balloon seal over end of torsion bar, cupped end toward hex.

9. Coat hex end of torsion bar with waterproof grease.

10. Install torsion bar hex end into anchor bracket. With torsion bar in a horizontal position, the ears of the anchor bracket should be positioned nearly straight up. Position swivel into anchor bracket ears.

11. Place bushing end of bar into position on top of lower control arm. Then install anchor bracket assembly into crossmember anchor retainer and install anchor adjusting bearing and bolt.

12. Attach pivot cushion bushing to crossmember with the bolt and washer assemblies. Leave bolt and washer assemblies loose enough to install friction plates.

13. With lower control arms at design height install the bolt and nut assemblies attaching torsion bar bushing to lower control arm. Torque to 70 ft. lbs. (95 Nm).

14. Ensure that the torsion bar anchor bracket is fully seated in crossmember. Then install friction plates between crossmember and pivot cushion bushing with open end of slot to rear and bottomed out on mounting bolts. Tighten the cushion bushing bolts to 85 ft. lbs. (115 Nm).

15. Position balloon seal over anchor bracket.

16. Reinstall bolt, through sway bar, retainer cushions and sleeve and attach to lower control arm end bushing. Torque bolt to 50 ft. lbs. (68 Nm).

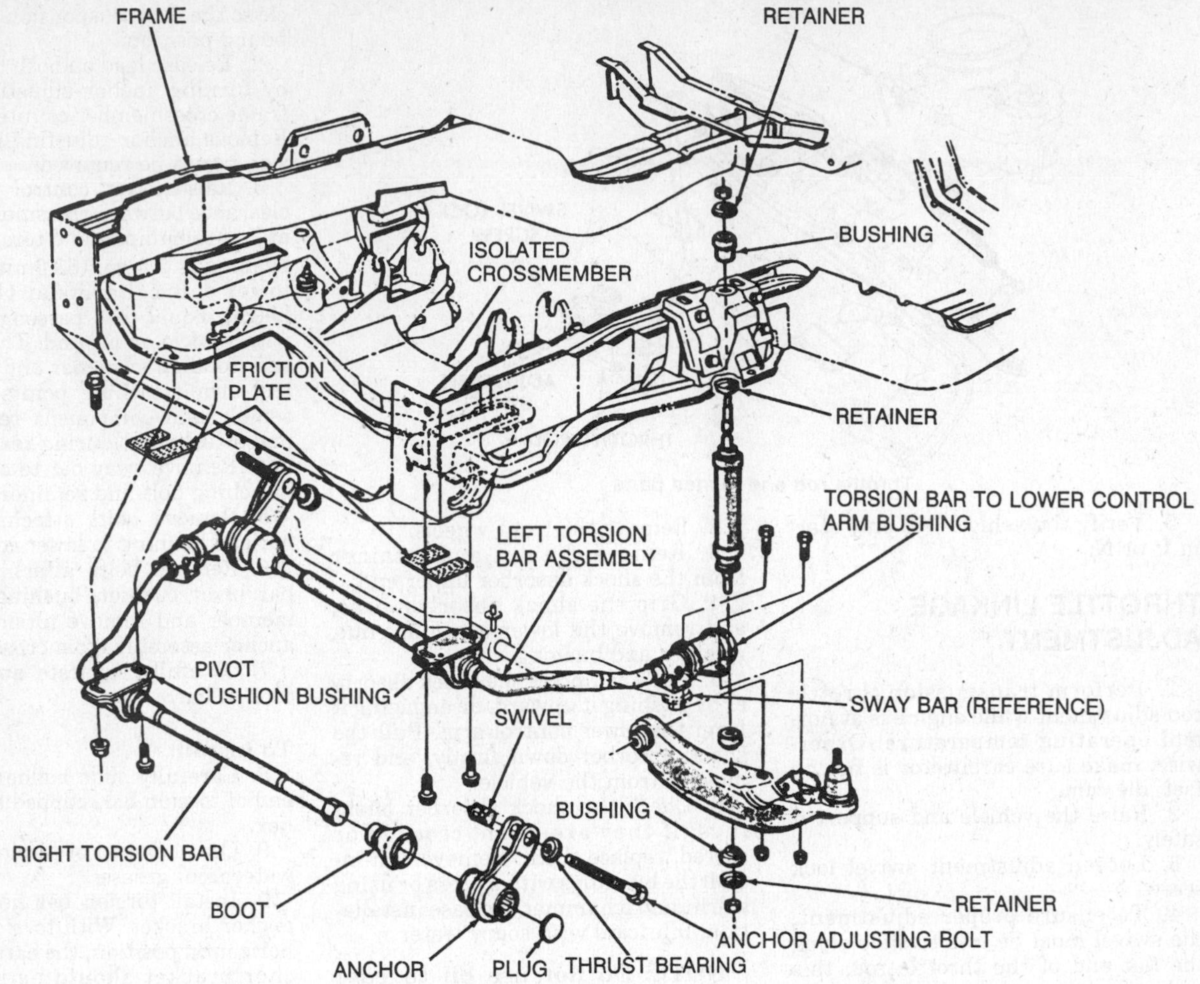

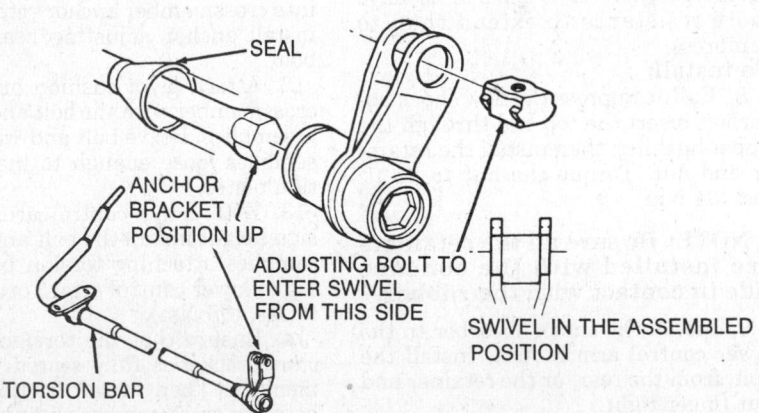

FRAME

ISOLATED CROSSMEMBER

RETAINER

BUSHING

FRICTION PLATE

RETAINER

TORSION BAR TO LOWER CONTROL ARM BUSHING

LEFT TORSION BAR ASSEMBLY

PIVOT CUSHION BUSHING

SWIVEL

SWAY BAR (REFERENCE)

BUSHING

RIGHT TORSION BAR

BOOT

ANCHOR   PLUG   THRUST BEARING

ANCHOR ADJUSTING BOLT

RETAINER

Exploded view of the torsion bar assemblies

SEAL

ANCHOR BRACKET POSITION UP

ADJUSTING BOLT TO ENTER SWIVEL FROM THIS SIDE

SWIVEL IN THE ASSEMBLED POSITION

TORSION BAR

Torsion bar anchor and swivel installation

17. Load torsion bar by turning anchor adjusting bolt clockwise.

18. Lower vehicle and adjust torsion bar height to specifications. Front vehicle height is measured from the head of the the front suspension front crossmember isolator bolt to the ground. To adjust turn torsion bar adjusting bolt clockwise to increase height and counterclockwise to reduce height. The height specification is 12½ in.

19. Align the front end.

Upper Ball Joints

INSPECTION

NOTE: Before making the inspection, verify that the wheel bearings are adjusted correctly and the control arm bushings are in good condition.

1. Place a jack under the lower control arm as close to the wheel as possible.
2. Raise the vehicle so the tire lightly contacts the floor.
3. Tighten the wheel bearing adjusting nut enough to remove all play.
4. Have an assistant try to move the top of the tire in and out while observing the upper ball joint. If there is any noticeable side play, replace the upper ball joint.
5. Correct the wheel bearing adjustment.

REMOVAL & INSTALLATION

———— **CAUTION** ————

The torsion bar remains under tension during this procedure.

NOTE: Turn the ignition key to the OFF or UNLOCKED position.

1. Raise and support the front of the vehicle. Place a jackstand under the lower control arm as close to the wheel as possible. Remove the wheel. The jackstand should not contact the brake splash shield and the rubber rebound bumper must not contact the frame.

2. Remove the cotter pin and nut that attaches the upper ball joint to the steering knuckle. Remove the cotter pin and nut from the lower ball joint, to enable the removal tool to be used.

3. Slide the ball joint removal tool onto the lower ball joint stud, allowing the tool to rest on the knuckle arm. Set the tool securely against the upper stud. Apply pressure to the upper stud by tightening the tool, and strike the knuckle sharply to loosen the stud. Do not strike the ball joint stud.

NOTE: The brake caliper may have to be removed for clearance.

4. After disengaging the ball joint, support the steering knuckle and brake assembly. Using tool, unscrew the upper ball joint from the upper control arm.

To install:

5. Position a new ball joint on the upper control arm and screw the joint into the arm. Be careful not to cross thread the joint in the arm. Torque to 125 ft. lbs. (162.5 Nm).

6. Position a new seal on the ball joint stud and install the seal in the ball joint making sure the seal is fully seated on the ball joint housing.

7. Position the ball joint stud in the steering knuckle and install the retaining nut. Torque the nut to 100 ft. lbs. (130 Nm). Install a new cotter pin.

8. To complete the installation, reverse the removal procedures. Lubricate the ball joint. Adjust the wheel alignment.

Lower Ball Joints

INSPECTION

NOTE: Before making the inspection, verify that the wheel bearings are adjusted correctly and the control arm bushings are in good condition.

1. Place a jack under the lower control arm as close to the wheel as possible.

2. Raise the vehicle until there is 1–2 in. (25.4–50.8mm) of clearance under the wheel.

3. Insert a bar under the wheel and pry upward. If the wheel raises noticeably the ball joint is worn.

4. Clamp a dial indicator to the lower control arm and measure the lower ball joint stud movement. The manufacturer's limit for lower ball joint play, measured at the joint and is 0.030 in. (0.76mm).

5. Correct the wheel bearing adjustment.

REMOVAL & INSTALLATION

NOTE: Turn the ignition key to the OFF or UNLOCKED position.

1. Raise the vehicle so the front suspension drops to the downward limit of its travel. Position jackstands beneath the front frame for extra support.

2. Remove the wheel and tire assembly.

3. Remove the brake caliper and tie it aside, so there is no strain on the flexible brake hose.

4. Remove the hub/rotor assembly and splash shield. Disconnect shock absorber from the lower control arm.

5. Unload the torsion bar by rotating the adjusting bolt counterclockwise.

6. Remove the cotter pin and nut from the upper and lower ball joints. Slide the ball joint removal tool over the upper stud, so it rests on the steering knuckle. Tighten the tool to place the lower ball joint under pressure. Using a hammer, strike the steering knuckle to loosen the ball joint.

7. Use the ball joint pressing tool to press the ball joint out of the lower control arm.

To install:

8. When installing the new ball joint, use a ball joint pressing tool to press it into the lower control arm.

9. Place a new seal over the ball joint. Press the retainer portion of the seal down over the ball joint housing using a pressing tool until it locks into position.

10. Insert the ball joint stud through the opening in the steering knuckle and install the stud retaining nut. Torque to 100 ft. lbs. (130 Nm). Install the cotter pin and lubricate the ball joint.

11. Load the torsion bar by rotating the adjusting bolt clockwise.

12. Install the shock absorber, the splash shield, hub/rotor assembly and brake caliper. Install the wheel and tire assembly.

13. Adjust the front wheel bearings. Remove the jackstands and lower the vehicle. Adjust the front suspension height.

Upper Control Arm

REMOVAL & INSTALLATION

1. Raise and support the front of the vehicle. Place a jackstand under the lower control arm as close to the wheel as possible. Remove the wheel. The jackstand should not contact the brake splash shield and the rubber rebound bumper must not contact the frame.

2. Remove the cotter pin and nut that attaches the upper ball joint to the steering knuckle. Remove the cotter pin and nut from the lower ball joint, to enable the removal tool to be used.

3. Slide the ball joint removal tool onto the lower ball joint stud, allowing the tool to rest on the knuckle arm. Set the tool securely against the upper stud. Apply pressure to the upper stud by tightening the tool and strike the knuckle sharply to loosen the stud. Never strike the ball joint stud.

4. Remove the rubber engine splash shield and the upper control arm pivot shaft nuts. It will be easier to reset the alignment if you mark the original pivot bar location. Remove the control arm and pivot shaft assembly.

5. Installation is the reverse order of removal procedure.

Lower Control Arm

REMOVAL & INSTALLATION

1. Separate the lower ball joint from the steering knuckle.

NOTE: Unload both bars, even if removing only 1 control arm, to reduce the sway bar reaction.

2. Raise the lower control arm until there is 2⅞ in. clearance between the crossmember ledge at the jounce bumper and the torsion bar bushing on the lower control arm. Unbolt the torsion bar busing from the control arm.

3. Remove the lower control arm pivot bolt and the control arm.

NOTE: If the control arm shaft bushings indicates wear or deterioration, replace them.

4. Installation is the reverse order of the removal procedure. Torque the control arm pivot bolt to 75 ft. lbs. (95.5 Nm) and the torsion bar end bushing-to-lower control arm to 50 ft. lbs. (65 Nm).

Sway Bar

REMOVAL & INSTALLATION

1. Raise the front of the vehicle and support it safely.

2. Release the load on both torsion bars by turning the anchor adjusting bolts counterclockwise.

3. Raise the lower control arms until there is 2⅞ in. clearance between the crossmember ledge, at the jounce bumper and the torsion bar bushing, on the lower control arm.

4. Support the lower control arms with a jackstands. Remove the sway bar-to-torsion bar bushing bolts, retainers, cushions and sleeves. Remove the retainer assembly strap and retainer bolts. Remove the sway bar.

NOTE: Inspect the cushions/bushings for wear or deterioration. Replace if necessary.

5. Installation is the reverse order of the removal procedure. Torque the sway bar-to-torsion bar to 50 ft. lbs. (65 Nm) and the sway bar retainer/strap bolts to 30 ft. lbs. (41 Nm). Load the torsion bar by turning the crossmember adjusting bolt clockwise. Lower the vehicle and adjust the torsion bar height.

Front Wheel Bearings

ADJUSTMENT

1. Raise the vehicle and support it safely.

2. Remove the grease cup, cotter pin and locknut.

3. Back off on the adjusting nut.

4. Check for free wheel rotation.

5. While rotating the wheel, tighten the wheel bearing adjustment nut to 240–300 inch lbs.

6. Loosen the nut ¼ turn (90 degrees). Retighten the nut so it is finger-tight.

7. Position the nut lock so 1 pair of the slots is in line with the cotter pin hole and install the cotter pin. This adjustment should give 0.001–0.003 in. endplay.

8. Install the rest of the components removed.

REMOVAL & INSTALLATION

1. Raise the vehicle and support it safely.

2. Remove the tire and wheel assembly.

3. Remove the brake caliper assembly and move to the side; do not disconnect the brake line.

NOTE: Avoid strain on the flexible brake hose.

4. Remove the grease cup, cotter pin and locknut.

5. Remove the adjusting nut and washer.

6. Remove the outer bearing and remove brake disc from spindle.

7. Remove the inner bearing by removing the bearing seal with a seal remover tool.

8. Installation is the reverse order of the removal procedure.

9. Adjust the wheel bearing and use a new cotter pin.

REAR SUSPENSION

Shock Absorbers

REMOVAL & INSTALLATION

1. Raise and safely support the vehicle under the axle assembly with jackstands, so as to relieve load from the shock absorbers.

2. Remove the nut, retainer and bushing, attaching the shock to the spring mounting plate. To avoid damage to the shock, grip the base of the shock below the base-to-reservoir weld while loosening the retaining nut.

3. At the upper mount, remove the shock attaching bolt/nut and the shock.

4. Purge the new shock of air by repeatedly extending it in its upright position and compressing it, in an inverted position. It is normal to have more resistance to extend than compress.

5. To install the shock, position it so the upper bolt or nut may be replaced, hand-tighten only. Align the shock with the spring mounting plate and install the bolt or nut, hand-tighten only.

6. Lower the vehicle and tighten the shock absorber mounting bolts. Torque the bottom bolt to 35 ft. lbs. (46 Nm) and the top bolt to 70 ft. lbs. (97.5 Nm).

Leaf Springs

REMOVAL & INSTALLATION

1. Raise and safely support the vehicle on jackstands; place the jackstands under the axle to relieve weight from the rear springs. Remove the wheels.

2. Disconnect the rear shock absorbers at the bottom. Lower the axle assembly to allow the rear springs to hang free. Disconnect the rear sway bar links, if equipped.

3. Remove the U-bolt nuts, bolts and spring plates. Remove the nuts securing the front spring hanger to the body mounting bracket.

4. Remove the rear spring hanger bolts and allow the spring to drop enough to allow the front spring hanger bolts to be removed.

5. Remove the front pivot bolt from the front spring hanger.

6. Remove the shackle nuts and shackle from the rear of the spring.

To install:

7. Assemble the shackle/bushings in the rear of the spring and hanger. Start the shackle bolt nut. Do not lubricate rubber bushings to ease installation or tighten the bolt nut.

8. Align the front spring hanger with the front spring eye and insert the pivot bolt and nut. Do not tighten.

9. Install the rear spring hanger-to-body bracket and torque the bolts to 35 ft. lbs. (45.5 Nm).

10. With the aid of a helper, raise the spring and insert the bolts in the spring hanger mounting bracket holes. Install the nuts and torque them to 35 ft. lbs. (45.5 Nm).

11. Position the axle assembly so it is correctly aligned with the spring center bolt.

12. Position the center bolt over the lower spring plate. Insert the U-bolt and nut. Tighten the U-bolt to 45 ft. lbs. (58.5 Nm). Connect the rear shock absorbers.

13. Lower the vehicle. Torque the pivot bolts to 105 ft. lbs. (136.5 Nm). and the shackle nuts to 35 ft. lbs. (45.5 Nm).

NOTE: Road test the vehicle and re-check the front suspension heights and correct, if necessary.

Rear Sway Bar

REMOVAL & INSTALLATION

1. Raise and safely support the vehicle.

2. Remove the sway bar link retaining nuts, retainers and insulators from the support.

3. Remove the nuts and retainer from each bracket fastened to each rail.

4. Remove the sway bar.

5. Installation is the reverse order of the removal procedure. Torque the retainer-to-bracket bolts to 17 ft. lbs. (23 Nm) and the link nuts to 8 ft. lbs. (11 Nm).

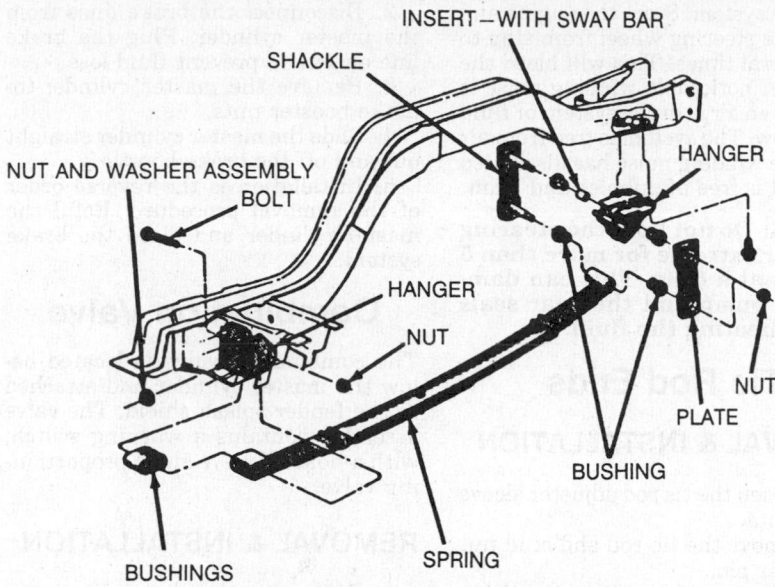

INSERT—WITH SWAY BAR

SHACKLE

HANGER

NUT AND WASHER ASSEMBLY

BOLT

HANGER

NUT

NUT

PLATE

BUSHING

BUSHINGS

SPRING

Exploded view of the rear spring assembly

STEERING

Steering Wheel

CAUTION

On vehicles equipped with an air bag, the negative battery cable must be disconnected before working on the system. Failure to do so may result in deployment of the air bag and possible personal injury.

REMOVAL & INSTALLATION

NOTE: All vehicles are equipped with collapsible steering columns. A sharp blow or excessive pressure on the column will cause it to collapse. Do not hammer on the steering wheel.

1. Disconnect the negative battery cable.
2. Remove the padded center assembly. This center assembly is often held on only by spring clips. There are usually holes in the back of the wheel so the pad can be pushed off. However, on some deluxe steering wheel pads, it is held on by screws behind the arms of the wheel. Remove the horn wire if necessary.
3. On tilt and telescoping steering column, remove the locking lever knob by releasing the clip on its underside. Remove the locking lever screws and the lever.
4. Remove the large center nut. Mark the steering wheel and steering shaft so the wheel may be replaced in its original position. In most cases, the wheel can only go on one way.

5. Using a puller, pull the steering wheel from the steering shaft.
6. Reverse the procedure to install the wheel. When placing the wheel on the shaft, make sure the front tires are in the straight-ahead position and the steering wheel and shaft are properly aligned. Tighten the retaining nut to 45 ft. lbs.

Steering Column

REMOVAL & INSTALLATION

1. Disconnect the negative battery cable.
2. Disconnect the column shift link by prying the shift rod out of the grommet in the shift lever.
3. Remove the steering shaft lower coupling-to-worm shaft roll pin.
4. Disconnect the wiring connectors at the steering column jacket.
5. Remove the steering wheel center pad assembly, and disconnect the horn switch if applicable.
6. Remove the steering wheel retaining nut and remove the steering wheel from the steering shaft.

NOTE: Do not bump or hammer on the steering shaft to remove the steering wheel.

7. Remove the floor plate to floor pan attaching screws.
8. To expose the steering column bracket retaining screws, remove instrument panel steering column cover and lower reinforcements.
9. Remove the nuts holding the steering column bracket to the instrument panel supports.

10. Carefully remove the lower coupler from the steering gear wormshaft, then remove the column assembly out through the passenger compartment.

NOTE: Do not damage the paint or trim during the removal procedure.

11. Should a new grommet be needed in the shift rod, install from the rod side of the lever.
12. The installation of the steering column is the reverse procedure of the removal.

Power Steering Gear

ADJUSTMENT

1. Disconnect the center link from the steering gear arm.
2. Start the engine and run at idle speed.
3. Turn the steering wheel gently from 1 stop to stop counting the number of turns. Then turn the wheel back exactly halfway, to center position.
4. Loosen the sector shaft adjusting screw until backlash is evident in steering gear arm. Feel backlash by holding the end of the steering gear arm between thumb and fore-finger with a light grip. Tighten the adjusting screw until backlash just disappears.
5. Continue to tighten to ⅜–½ turn from this position and tighten locknut to 28 ft. lbs. (38 Nm) to maintain this setting.

REMOVAL & INSTALLATION

1. Separate from the steering gear input shaft and remove the steering column.

NOTE: Chrysler Corporation recommends complete detachment from the floor and instrument panel of the steering column to avoid damage to the energy absorbing steering column components.

2. Remove the pressure and return fluid lines.
3. Raise the vehicle and support safely. Remove the retaining nut and washer from the steering gear arm sector shaft. With a puller tool, remove the steering gear arm.

NOTE: On some vehicles it may be necessary to remove the starter heat shield and remove the exhaust system.

4. Remove the steering gear assembly-to-frame bolts or nuts and remove the steering gear.
To install:
5. Center the sector shaft to its midpoint of travel.

6. Position the gear assembly on the frame and tighten the bolts or nuts.

7. Align the master serrations on the sector shaft to the splines in the steering arm, install and tighten the nut and washer.

8. Lower the vehicle and install the pressure and return fluid lines.

9. Install the steering column, fill the reservoir with fluid, start the engine and turn the steering wheel several times from stop to stop to bleed the system of air.

Power Steering Pump

REMOVAL & INSTALLATION

1. Back off the pump mounting and locking bolts. Remove the pump drive belt.

2. Disconnect and plug the pressure line and return hose.

3. Remove the pump bolts and pump with the bracket.

To install:

4. Place the pump in position and install the mounting bolts.

5. Install the pump drive belt and adjust. There should be no more than ½ in. of play, under moderate thumb pressure, on the longest run of belt. Some pump brackets have a ½ in. square hole for use in tensioning the belt. Torque the mounting bolts to 30 ft. lbs. (41 Nm).

6. Replace the pressure hose O-ring, and install the pressure line. Connect the return hose.

7. Fill the pump with power steering fluid.

8. Start the engine and rotate the steering wheel from stop to stop several times. This will bleed the system. Check the pump fluid level and fill as required.

9. Be certain the hoses are away from the exhaust manifolds and are not kinked or twisted.

BELT ADJUSTMENT

1. Disconnect the negative battery cable.

2. Install the pump drive belt and adjust.

3. There should be no more than ½ in. of play, under moderate thumb pressure, on the longest span of the belt.

4. If equipped with a ½ in. sq. hole on the adjusting bracket, use a ½ in. breaker bar to adjust the belt tension or torque wrench and adjust to specifications.

SYSTEM BLEEDING

Whenever the power steering system has been serviced, it is necessary to bleed the system. Start the engine and rotate the steering wheel from stop to stop several times. This will bleed the system. A noticeable winding noise is heard when air is in the system or fluid level is low. The system is free from air when the winding noise has dissipated and fluid is free of bubbles and foam.

NOTE: Do not hold the steering to either extreme for more than 5 seconds at a time. This can damage the pump and the gear seals by overheating the fluid.

Tie Rod Ends

REMOVAL & INSTALLATION

1. Loosen the tie rod adjuster sleeve clamp nuts.

2. Remove the tie rod end stud nut and cotter pin.

3. If the outer tie rod end is being removed, remove the stud from the steering knuckle. If the inner tie rod end is being removed, remove the stud from the center link. The studs on all the tie rod ends fit in a tapered hole, they can be removed with a ball joint removal tool.

NOTE: Use extreme care not to damage the rubber grease seals at the tie rod ends. If the seals become damaged they must be removed and the tie rod ends inspected.

4. Unscrew the tie rod end from the threaded sleeve and record the number of turns required to remove it; the threads may be left or right hand.

5. Installation is the reverse order of the removal procedure. Lubricate the tie rod threads. Screw in the tie rod end as many turns as were needed to remove it. This will give approximately correct toe-in. Torque the stud nuts to 40 ft. lbs. (52 Nm) and install new cotter pins.

6. Align the front end or adjust the toe.

BRAKES

For all brake system repair and service procedures not detailed below, please refer to "Brakes" in the Unit Repair section.

Master Cylinder

REMOVAL & INSTALLATION

1. Disconnect the negative battery cable.

2. Disconnect the brake lines from the master cylinder. Plug the brake line outlets to prevent fluid loss.

3. Remove the master cylinder-to-brake booster nuts.

4. Slide the master cylinder straight out and off the brake booster.

5. Installation is the reverse order of the removal procedure. Refill the master cylinder and bleed the brake system.

Combination Valve

The combination valve is located below the master cylinder and attached to the fender splash shield. The valve assembly contains a warning switch, with a hold off valve and a proportioning valve.

REMOVAL & INSTALLATION

1. Disconnect the electrical connector from the combination valve.

2. Disconnect and plug the brake tubes at the combination valve.

3. Remove the valve-to-fender splash shield bolts, then remove the combination valve from the vehicle.

4. Installation is the reverse order of the removal procedure. Bleed the brake system.

Power Brake Booster

REMOVAL & INSTALLATION

1. Disconnect the negative battery cable.

2. Remove the master cylinder-to-brake booster nuts and position the master cylinder aside without disconnecting the lines. Use care not to kink the brake lines.

3. Disconnect the vacuum hose from the brake booster.

4. Working under the dash, remove the nut and bolt or retainer clip attaching the brake booster pushrod to the brake pedal.

5. Remove the brake booster attaching nuts and washers.

6. Remove booster assembly from the vehicle.

7. Installation is the reverse order of the removal. Torque mounting nuts to 200–250 inch lbs. (22–28 Nm) and pushrod nut/bolt to 30 ft. lbs. (41 Nm).

Brake Caliper

REMOVAL & INSTALLATION

1. Raise and safely support the vehicle.

2. Remove the front wheel and tire assembly.

3. Using a C-clamp, force the piston into the caliper.

4. Remove the caliper retaining screws, clips and anti-rattle springs.

5. Remove the caliper from the disc by slowly sliding the caliper assembly out and away from the disc.

6. Disconnect the brake hose from caliper.

7. Remove the caliper from the vehicle.

To install:

8. Replace the copper washers. Connect the brake hose to the caliper.

9. Slowly slide the caliper assembly into position in the adapter and over the disc.

10. Align the caliper on the machined surfaces of the adapter. Be careful not to pull the dust boot from its groove as the piston and boot slide over the inboard shoe.

11. Tighten the retaining screws to 180 inch lbs. (20 Nm).

12. Bleed the brake system. Pump the brake pedal several times until a firm pedal has been obtained.

13. Check and refill the master cylinder reservoir, using approved DOT 3 brake fluid.

14. Install the wheel and tire assemblies. Tighten the stud nuts to 85 ft. lbs. (115 Nm).

15. Lower the vehicle and test the brakes.

Disc Brake Pads

REMOVAL & INSTALLATION

1. Raise and safely support the vehicle.

2. Remove the front wheel and tire assemblies.

3. Remove the caliper retaining screws, clips and anti-rattle springs.

4. Remove the caliper from the disc by slowly sliding the caliper assembly out and away from the disc.

5. Remove the outboard shoe assembly, flanges on the outboard shoe will retain the shoe to the caliper, by prying between pad and the caliper fingers.

6. Support the caliper to prevent damage to the flexible brake hose and remove the inboard pad. Do not allow the caliper to hang by the hose.

NOTE: Prior to installing the disc pads, check the caliper piston seal for leaks, evident by brake fluid in and around the boot area and inboard lining, and for ruptures of the piston dust boot. If the boot is damaged or fluid is evident, it will be necessary to disassemble the caliper assembly and overhaul or replace it. Check the mating surfaces of the abutments on the caliper and adapter. If corroded or rusty, clean surfaces with a wire brush.

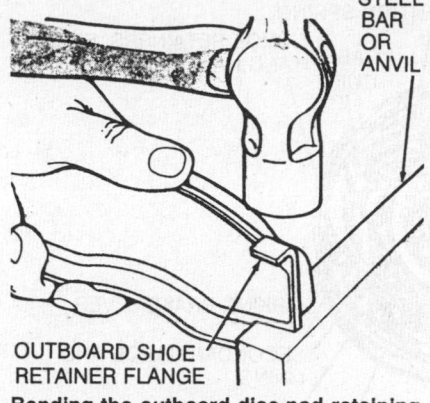

OUTBOARD SHOE
RETAINER FLANGE

Bending the outboard disc pad retaining flange to prevent brake rattle

To install:

7. Inspect the braking surfaces of the disc. Re-surface the disc if heavy scoring or warping is evident.

8. Slowly and carefully, push the piston back into the bore until it is bottomed. Watch for possible master cylinder reservoir overflow.

9. Slide the new outboard shoe assembly into the recess of the caliper.

NOTE: There should be no freeplay between the brake shoe flanges and the caliper fingers; this will cause disc brake rattle.

10. If freeplay is evident by vertical shoe movement after installation, perform the following:

a. Remove the shoe from the caliper and bend the flanges to create a slight interference fit to eliminate all vertical free-play when the shoe is installed.

b. Install the shoe after the above modification, if necessary, by snapping the shoe into place.

11. Position the inboard shoe into position on the adapter with the shoe flanges in the adapter ways.

12. Slowly slide the caliper assembly into position in the adapter and over the disc.

13. Align the caliper on the machined ways of the adapter. Be careful not to pull the dust boot from its groove as the piston and boot slide over the inboard shoe.

14. Install the anti-rattle springs and retaining clips. Tighten the retaining screws to 180 inch lbs. (20 Nm).

NOTE: The inboard shoe anti-rattle spring must be installed on top of the retainer spring plate.

15. Pump the brake pedal several times until a firm pedal has been obtained.

16. Check and refill the master cylinder reservoirs, if necessary, using an approved DOT 3 brake fluid. It should not be necessary to bleed the brake system after replacing the pads. However, if a firm pedal cannot be obtained, air may be in the the brake system.

17. Install the wheel and tire assemblies. Tighten the stud nuts to 85 ft. lbs. (115 Nm).

18. Lower the vehicle and test the brakes.

Brake Rotor

REMOVAL & INSTALLATION

1. Raise and safely support the vehicle.

2. Remove the front wheel and tire assemblies.

3. Remove the caliper retaining screws, clips and anti-rattle springs.

4. Remove the caliper from the rotor by slowly sliding the caliper assembly out and away from the rotor. Suspend the caliper on a wire.

5. Remove the grease cap, the cotter pin, the nut lock, the nut, the thrust washer and the outer wheel bearing.

6. Remove the rotor/hub assembly from the spindle.

To install:

7. Inspect the braking surfaces of the rotor. Re-surface the rotor if heavy scoring or warping is evident.

8. Install the rotor/hub assembly onto the spindle.

9. Install the outer bearing, the thrust washer and the adjusting nut.

10. Tighten the adjusting nut to 240–300 inch lbs. (27–34 Nm) while rotating the rotor/hub assembly.

11. Back off the adjusting nut to release the preload and retighten finger-tight.

12. Install the nut lock, a new cotter pin and the grease cup.

13. Install the caliper and the wheel/tire assembly.

14. Pump the brake pedal several times until a firm pedal has been obtained.

15. Check and refill the master cylinder reservoirs, using an approved DOT 3 brake fluid.

16. Lower the vehicle and test the brakes.

Brake Drums

REMOVAL & INSTALLATION

1. Raise and safely support the vehicle.

2. Remove the rear plug from the brake adjusting hole.

3. Insert a thin tool into the brake adjusting hole and hold the adjusting lever away from the notches of the adjusting screw.

4. Insert a brake adjusting tool into the brake adjusting hole and release

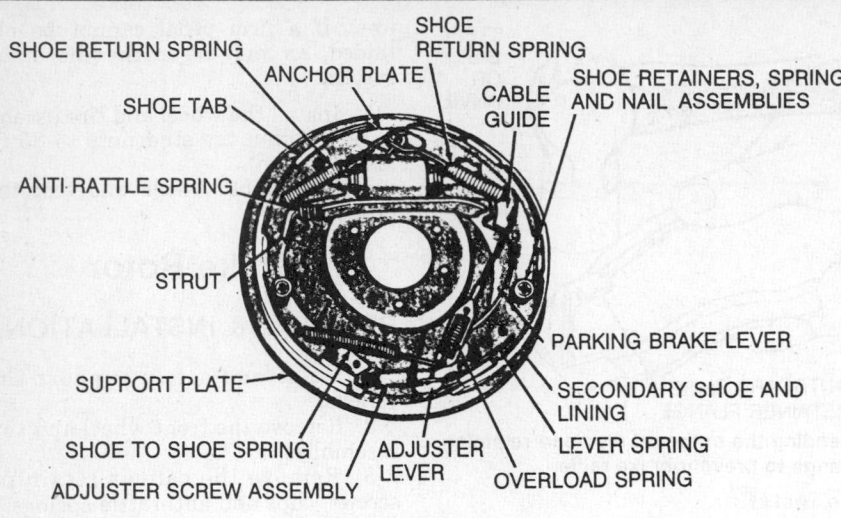

SHOE RETURN SPRING
SHOE RETURN SPRING
ANCHOR PLATE
SHOE TAB
CABLE GUIDE
SHOE RETAINERS, SPRING AND NAIL ASSEMBLIES
ANTI-RATTLE SPRING
STRUT
PARKING BRAKE LEVER
SUPPORT PLATE
SECONDARY SHOE AND LINING
SHOE TO SHOE SPRING
ADJUSTER LEVER
LEVER SPRING
ADJUSTER SCREW ASSEMBLY
OVERLOAD SPRING
LEFT REAR

Rear drum brake parts

the brake by adjusting the star adjuster downward.

5. Remove the rear wheels and the brake drums.

6. Install the brake drum and adjust the brakes. Install the wheel assemblies and lower the vehicle.

Brake Shoes

REMOVAL & INSTALLATION

1. Raise and safely support the vehicle.

2. Remove the rear wheels and the brake drums.

3. Inspect the brake lining for wear, shoe alignment, contamination from grease or brake fluid.

4. Using a brake return spring tool, remove the brake shoe return springs. Note how the secondary shoe return spring overlaps the primary shoe return spring.

5. Slide the eye of the automatic adjuster cable off the anchor and unhook it from the adjusting lever.

6. Remove the cable, overload spring, cable guide and the anchor plate.

7. Disengage the adjusting lever from the spring by sliding forward to clear the pivot, working out from under the spring.

8. Remove the spring from the pivot. Remove the shoe-to-shoe spring from the secondary shoe web and disengage the front primary shoe web. Remove the spring.

9. Disengage the primary and secondary shoes and remove the adjusting star wheel assembly from the shoes.

10. Remove the brake lever from the secondary shoes. Remove the shoes.

11. Disengage the parking brake lever from the parking brake cable.

To install:

12. Clean the backing plate with a suitable solvent, inspect for rough or rusted shoe contact areas. Clean and inspect the adjusting screws for damaged threads. Apply a thin film of lubricant to the threads, socket and washer.

13. Replace the adjuster screw assembly if corrosion of any part inhibits free operation.

14. Install new brake shoe return springs and hold-down springs where the old springs have been overheated or their strength is questionable.

NOTE: Spring paint discoloration or distorted end coils indicate a spring has been overheated.

15. Examine the brake drum for cracks, excessive scoring or excessive run-out. If drum run-out exceeds 0.006 in., the drum must be resurfaced. Do not exceed the suggested resurfacing amount that is marked on the drum.

16. Lubricate the shoe tab contact pads on the support plate with a thin film of a multi-purpose lubricant.

17. Engage the parking brake lever with the cable. Install the parking brake lever into the rectangular hole of the secondary brake shoe.

18. Slide the secondary shoe against the support plate, at the same time engage the shoe web with the pushrod and against the anchor.

19. Slide the parking brake strut behind the axle flange and into the slot in the parking brake lever. Slide the anti-rattle spring over the free end of the strut.

20. Slide the primary shoe into position and engage it with the pushrod, if equipped, and free end of the strut.

21. Install the anchor plate over the

anchor. Install the eye of the adjuster cable over the anchor.

22. Engage the primary shoe return spring into the web of the shoe and install the free end over the anchor using a brake return spring tool.

23. Insert the protruding hole rim of the cable guide into the hole in the secondary shoe web. Holding the guide in position, engage the secondary shoe return spring through both, hole in the guide and hole in the web.

24. Using a brake return spring tool, install the return spring over the anchor.

25. Using pliers, squeeze the ends of the spring loops, around the anchor, parallel.

26. Install the adjusting star wheel assembly between the primary and secondary shoes, with the star wheel next to the secondary shoe.

27. The left star wheel adjusting stud end is stamped **L**, indicating its position on the vehicle, and the star wheel is cadmium plated.

28. The right star wheel is black and the adjusting stud end is stamped **R**.

29. Install the shoe-to-shoe spring between the shoes. Engage the primary shoe first.

30. Vehicles with 11 in. brakes, install the shoe-to-shoe spring with the coil forward, opposite the adjuster lever.

31. Install the adjusting lever spring over the pivot pin on the shoe web. Install the adjusting lever slightly rearward to lock in position.

32. Using a brake spring hold-down tool, install the shoe retaining nails, retainers and springs.

33. Thread the adjuster cable over the guide and hook the end of the overload spring in the lever. Be sure the eye of the cable is pulled tight against the anchor and is in a straight line with the guide.

34. Install the brake drum and adjust the brakes. Install the wheel assemblies and lower the vehicle.

Wheel Cylinder

REMOVAL & INSTALLATION

1. Raise and safely support the vehicle.

2. Remove the wheel assembly and the brake drum.

3. Inspect the wheel cylinder for signs of leakage.

NOTE: If either cylinder shows signs of leakage, the cylinders should be replaced as a pair.

4. Remove the brake shoe assembly form the backing plate.

5. Disconnect and plug the brake line at the wheel cylinder.

6. Remove the wheel cylinder-to-backing plate bolts and separate the wheel cylinder from the backing plate.

7. Install a new wheel cylinder. Installation is the reverse order of the removal procedure. Torque the wheel cylinder-to-backing plate bolts to 75 inch lbs. (9 Nm). Bleed the brake system.

Parking Brake Cable

ADJUSTMENT

1. Raise the vehicle and support it safely.

2. Insert an adjusting tool through the rear brake adjusting hole and rotate the star wheel until a slight drag is felt while rotating the wheels. Back off the star wheel until no drag is felt with the aid of a welding rod type probe, to move the adjusting lever out of engagement with the star wheel.

3. Tighten the cable adjusting nut until a slight drag is felt in the rear wheels when the rear wheels are rotated. Loosen the cable adjusting nut until the rear wheels can be rotated freely. Back off the cable adjusting nut 2 additional turns.

4. Apply and release the parking brake several times and check to verify the rear wheels rotate freely, without any brake drag.

REMOVAL & INSTALLATION

Front Cable

1. Disengage the front parking brake cable from the left connector.

2. Using a suitable tool, force the cable housing and attaching clip forward out of the body crossmember.

3. Fold back the left front edge of the floor covering and remove the rubber cable cover from the floor pan.

4. Engage the parking brake and work the brake cable up and out of the clevis linkage.

5. Using a suitable tool, force the upper end of the cable housing and clip down out of the pedal assembly bracket.

6. Work the cable and housing assembly up through the floor pan.

To install:

7. Insert the parking brake cable through the floor pan.

8. Insert the retainer into the hole in the bottom of the parking brake pedal assembly bracket. Insert the cable through the hole and insert the cable end fitting into the linkage clevis. Force the upper cable housing into the retainer until firmly seated against the pedal assembly bracket.

9. Insert the retainer into the hole in the crossmember. Insert the cable through the hole in the crossmember and force the cable housing into the retainer until firmly seated.

10. Install the rubber cable cover and floor pan clip.

11. Attach the cable to the connector.

12. Adjust the service brakes and parking brake system.

13. Apply brakes several times and test for free wheel rotation.

Rear Cable

1. Raise and safely support the vehicle. Remove the rear wheels.

2. Disconnect the brake cable from the connector.

3. Remove the retaining clip from the brake cable bracket.

4. Remove the brake drum from the rear axle.

5. Remove the brake shoe return springs and adjuster mechanism.

6. Remove the brake shoe retaining springs.

7. Remove the brake shoe strut and spring and disconnect the brake cable from the operating arm.

8. Compress the retainers on the end of the brake cable housing and remove the cable from the support plate.

To install:

9. Insert the brake cable and housing into the brake support plate, making certain the housing retainers lock the housing firmly into place.

10. Holding the brake shoes in place on the support plate, engage the brake cable into the brake shoe operating lever. Install the parking brake strut and spring.

11. Install the brake shoe retaining springs, adjuster mechanism and brake shoe return springs.

12. Install the brake drum and wheel.

13. Insert the brake cable and housing into the cable bracket and install the retaining clip.

14. Insert the brake cable into the equalizer.

15. Adjust the service brake and the parking brake.

Brake System Bleeding

Master Cylinder

Complete bleeding will require a residual valve on outlet of each bleeder tube. Obtain an bleeder tool equipped with at least 1 residual valve on primary outlet tube. To modify secondary outlet tube use a flaring tool to flare the tube outlet and install the residual valve.

1. Clamp master cylinder in a vise and attach bleeding tubes.

2. Fill both reservoirs with approved brake fluid.

3. Using a brass rod or wood dowel depress pushrod slowly and allow the pistons to return under pressure of springs. Do this several times until all air bubbles are expelled.

4. Remove bleeding tubes from cylinder, plug outlets to prevent spillage and install caps.

5. Remove from vise and install master cylinder on vehicle.

Calipers and Wheel Cylinders

1. Fill the master cylinder with fresh brake fluid. Check the level often during the procedure.

2. Start with the wheel farthest from the master cylinder. Pop the cap off of the bleeder screw, if equipped and place where it will not be lost. Clean the bleed screw.

— CAUTION —
When bleeding the brakes, keep face away from the brake area. Spewing fluid may cause facial and/or visual damage. Do not allow brake fluid to spill on the car's finish, it will remove the paint.

3. If the system is empty, the most efficient way to get fluid down to the wheel is to loosen the bleeder about ½–¾ turn, place a finger firmly over the bleeder and have a helper pump the brakes slowly until fluid comes out the bleeder. Once fluid is at the bleeder, close it before the pedal is released inside the vehicle.

NOTE: If the pedal is pumped rapidly, the fluid will churn and create small air bubbles, which are almost impossible to remove from the system. These air bubbles will eventually congregate and a spongy pedal will result.

4. Once fluid has been pumped to the caliper or wheel cylinder, open the bleed screw again, have the helper press the brake pedal to the floor, lock the bleeder and have the helper slowly release the pedal. Wait 15 seconds and repeat the procedure, including the 15 second wait, until no more air comes out of the bleeder upon application of the brake pedal. Remember to close the bleeder before the pedal is released inside the vehicle each time the bleeder is opened. If not, air will be induced into the system.

5. If a helper is not available, connect a small hose to the bleeder, place the end in a container of brake fluid and proceed to pump the pedal from inside the vehicle until no more air comes out the bleeder. The hose will prevent air from entering the system.

6. Repeat the procedure on remaining wheel cylinders and calipers still working from the wheel farthest away from the master cylinder. The bleeding sequence is as follows: right rear, left rear, right front and left front.

7. Hydraulic brake systems must be totally flushed if the fluid becomes contaminated with water, dirt or other corrosive chemicals. To flush, simply bleed the entire system until all fluid has been replaced with new fluid.

8. Install the bleeder caps on the bleeders to keep dirt out.

9. Road test the vehicle after brake work of any kind is done.

CHASSIS ELECTRICAL

Air Bag

DISARMING

To disarm the air bag, disconnect the negative battery. There is no time lapse built into this sytem, so the system is immediately disabled. No further procedures are needed. Failure to disarm the system may result in deployment of the air bag and possible personal injury.

Heater Blower Motor

REMOVAL & INSTALLATION

1. Disconnect the battery ground cable.

2. Remove the blower motor feed and ground wires at the connector.

3. Remove the blower motor mounting nuts from the bottom of the recirculation housing or separate lower blower housing from upper housing.

4. Lower the blower motor assembly downward from under the instrument panel.

5. Remove the blower motor mounting plate screws.

6. Separate the blower motor housing from the blower motor and fan assembly housing.

To install:

7. Set the blower motor and fan assembly into the blower motor housing and install the housing to mounting plate screws.

8. Position the blower assembly into the recirculation housing and install the retaining nuts.

9. Install the blower motor feed and ground wire connector.

10. Install the battery ground cable.

11. Test the blower motor operation on all fan speeds.

Windshield Wiper Motor

REMOVAL & INSTALLATION

1. Disconnect the battery negative cable.

2. Remove the cowl screen.

3. Remove the crank-nut while holding the drive crank with a wrench to prevent overloading the gears.

4. Remove the drive crank from the motor and disconnect the electrical wiring connector.

5. Remove the nuts retaining the motor to the dash panel. Remove the motor carefully, so as not to lose the spacers and rubber grommet.

6. Installation is in the reverse order of the removal procedure. Be sure the wiper motor is correctly grounded by having the ground strap under 1 of the retaining nuts.

7. Tighten the crank nut to 95 inch lbs. (10.4mm).

Windshield Wiper Switch

REMOVAL & INSTALLATION

1. Disconnect the negative battery cable.

2. Remove the steering wheel.

3. If equipped with a tilt wheel, remove the lock plate cover and the lock plate.

4. Remove the lower instrument panel bezel.

5. If equipped with a tilt wheel, perform the following procedures:

   a. Remove the gear shift indicator.

   b. Remove the nuts retaining the column to the lower panel reinforcement.

   c. Remove the mounting bracket from the steering column after removing the retaining bolts.

6. Remove the wiring holder from the steering column by unsnapping the plastic retainers.

7. Remove the turn signal switch.

8. Remove the retaining screws and remove the lock housing cover.

9. Gently pull the wiper switch up from the column while guiding the wires through the column opening.

10. Installation is the reverse of the removal procedure.

Instrument Cluster

REMOVAL & INSTALLATION

1. Disconnect the negative battery cable.

2. Remove the instrument cluster bezel.

3. Loosen the shift pointer set screw and remove the pointer.

4. From under the dash, disconnect the speedometer cable.

NOTE: The speedometer cable is attached to the speedometer by a snap-on plastic ferrule, which attaches directly to the speedometer head and must be disconnected before the speedometer or cluster can be removed.

5. Remove the cluster retaining screws and pull the cluster away from the carrier. Disconnect the electrical wiring.

6. Remove the cluster assembly from the dash.

7. Installation is the reverse order of the removal procedure.

Speedometer

REMOVAL & INSTALLATION

NOTE: The disassembly of the cluster and speedometer will vary to a small degree from each vehicle line to another. It is most important to mask surfaces which may become scratched or damaged during the disassembly or assembly procedures. Extreme care should be exercised when handling the internal components of the speedometer/cluster assembly. The electronic units cannot be repaired but must be replaced.

1. Disconnect the negative battery cable. Remove the instrument cluster.

2. Remove the cluster and the printed circuit board assembly from the carrier.

3. Remove the trip odometer knob from the shaft.

4. Remove the plastic pins and pull the lens and mask assembly away from the cluster housing.

5. Remove the screws retaining the speedometer to the cluster housing and the remove speedometer.

6. Disconnect the connector-to-odometer switch, if equipped.

7. Installation is the reverse order of the removal procedure.

Radio

REMOVAL & INSTALLATION

1. Disconnect the negative battery cable.

2. Remove the center bezel and the radio-to-panel mounting screws.

3. Pull the radio out through the front face of the panel. Detach the an-

tenna lead, ground strap, power wire and speaker leads.

4. Installation is the reverse of the removal procedures.

Headlight Switch

REMOVAL & INSTALLATION

1. Disconnect negative battery cable.

2. Remove the instrument cluster bezel.

3. Remove the mounting screws from switch module and pull assembly away from panel.

4. To remove the knob and stem assembly, depress the headlight switch stem release button and pull assembly from switch.

5. Disconnect electrical wiring.

6. Remove the switch from the vehicle.

7. Installation is the reverse of the removal procedure.

Dimmer Switch

REMOVAL & INSTALLATION

1. Disconnect the negative battery cable.

2. Remove the steering column lower cover.

3. Disconnect the electrical connector from the switch.

4. Remove the dimmer switch retaining nuts. Disengage the switch from the actuating rod. Remove the switch from the vehicle.

5. Install the switch to its proper mounting. Insert two $^3/_{32}$ in. drill shanks through the alignment holes.

6. Install the actuator rod into the washer/wiper switch pocket. Once the switch is installed, remove the drill shanks.

Turn Signal Switch

REMOVAL & INSTALLATION

Standard Column

1. Disconnect the negative battery cable. Remove the steering wheel and the steering column cover.

2. Remove sound deadening insulation panel and lower instrument panel bezel.

3. Loosen the Allen screw on the gear shift housing and remove the gearshift indicator.

4. Place the shift lever in full clockwise position.

5. Pry out the plastic buttons retaining wiring the harness holder to the column. Remove the harness holder.

6. Disconnect wiring connector

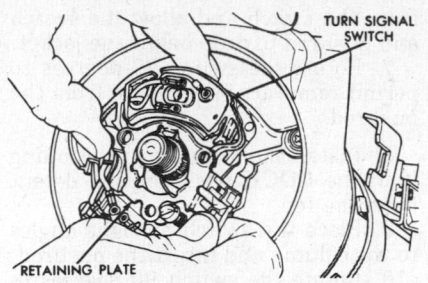

Turn signal switch retainer

from switch. Wrap the connector with tape to prevent snagging when removing the switch.

7. Remove the screw holding the turn signal lever assembly to the turn signal switch pivot. Leave the assembly in its installed location.

8. Remove the screws and bearing retainer fastening the turn signal switch to the upper bearing housing.

9. Remove the turn signal/hazard warning switch by gently pulling the switch up from the column while straightening and guiding wires up through column opening.

10. Installation is the reverse order of the removal procedure.

Tilt Column

1. Disconnect the negative battery cable.

2. Remove the lower steering column cover, if equipped and remove the plastic wiring channel from the under the steering column.

3. Straighten the steering wheel so the tires are pointing straight-ahead.

NOTE: If equipped with an airbag, it is imperative that the steering wheel removal and installation procedure under Steering is followed.

4. Remove the steering wheel.

5. Depress the lockplate with the proper depressing tool, remove the retaining ring from its groove and remove the tool, ring, lockplate, cancelling cam and spring.

6. Remove the stalk actuator screw and arm.

7. Remove the hazard switch knob.

8. Disconnect the turn signal switch connector.

9. Remove the 3 screws and remove the turn signal switch. Tape the connector to the wires to aid in removal.

To install:

10. Run the wiring through the opening and down the steering column, install the turn signal switch, switch stalk actuator arm and hazard switch knob.

11. Install the spring, cancelling cam, lockplate and ring on the steering shaft. Depress the plate with the depressing tool and install the ring securely in the groove. Remove the tool slowly.

12. Connect the turn signal switch connector and install the channel.

13. Install the steering wheel and torque the nut to 45 ft. lbs. (61 Nm).

14. Install the horn pad.

15. Connect the negative battery cable and check the turn signal switch and dimmer switch for proper operation.

16. Install the lower column cover, if equipped.

Ignition Lock

REMOVAL & INSTALLATION

Standard Column

1. Disconnect the negative battery cable. Remove the steering wheel and turn signal lever. Pull the turn signal switch aside.

2. Remove the retaining snapring

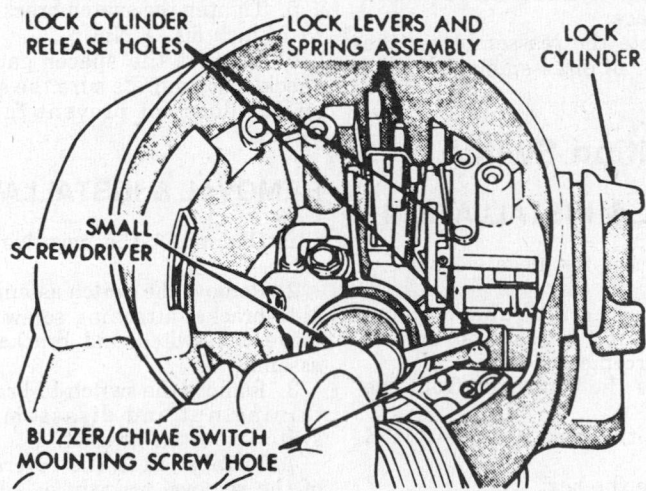

Releasing the ignition lock—standard column

and pry the upper bearing housing off the steering shaft.

3. Press out the pin attaching the lock plate to the steering shaft and remove the lock plate. Remove the lock lever guide plate. Remove the buzz/chime switch.

4. With the ignition lock cylinder in the **LOCK** position and the ignition key removed, insert 2 small diameter tools into the lock cylinder release openings to release the spring-loaded lock retainer. Pull the lock cylinder out of its housing.

5. To install, place the lock cylinder into the housing, positioning it in the **LOCK** position and remove the key. Insert the lock cylinder far enough into the housing to contact the switch actuator. Insert the key, press and turn until the retainer snaps into place.

6. Complete the reassembly in the reverse order of disassembly.

Tilt Column

1. Disconnect the negative battery cable. Remove the steering wheel, shaft lock cover, turn signal lever, tilt control lever and hazard warning knob.

2. Remove the lock plate, cancelling cam and spring, disconnect and pull the turn signal switch aside. Remove key light.

3. With the ignition lock cylinder in the **LOCK** position and the key removed, insert a thin suitable tool into the lock cylinder release opening to release the spring-loaded lock retainer. Pull the lock cylinder out of its housing.

4. To install, place the lock cylinder in its housing, positioning it in the **LOCK** position and remove the key. Insert the cylinder into the housing until it contacts the switch actuator. Move the switch actuator rod up and down to align the parts. When aligned, move the lock cylinder inward and snap into place.

5. Complete the reassembly in the reverse order of disassembly.

Ignition Switch

REMOVAL & INSTALLATION

1. Disconnect the negative battery cable.

2. Remove the instrument panel steering column bracket cover and lower reinforcement.

3. Remove the connector from the switch.

4. Place the key in the **LOCK** position.

5. Remove the key.

6. Remove the mounting screws

from the switch and allow the switch and pushrod to drop below the jacket.

7. Rotate the switch 90 degrees to permit removal of the switch from the pushrod.

8. Install the switch by positioning it in the **LOCK** position, 2nd detent from the top.

9. Place the switch at right angles to the column and insert the pushrod.

10. Rotate the switch 90 degrees to lock the actuator rod, align the switch on the bracket and install the screws.

11. With a light rearward load on the switch, tighten the screws. Check for proper operation.

Stoplight Switch

The stoplight switch or stoplight/speed control switch and mounting bracket are attached to the brake pedal bracket. If equipped with speed control, the stoplight switch is a combined unit.

ADJUSTMENT

1. Loosen the switch assembly pedal-to-bracket screw and slide the assembly away from the pedal blade or striker plate.

2. Depress the brake pedal and allow it to return to free position, do not pull the brake pedal back at any time.

3. Position a spacer gauge on the pedal striker plate. A clearance of 0.130–0.150 in. (3.302–3.810mm) is required for vehicles without speed control or 0.060–0.080 in. (1.542–2.032mm) is required for vehicles equipped with cruise control.

4. Slide the switch assembly toward the pedal striker until the switch plunger is fully depressed against the spacer gauge; on heavy duty or stoplight/speed control switches, depress the plunger until the switch body contacts the feeler gauge.

5. Tighten the switch bracket screw to 75 inch lbs. (8 Nm).

6. Remove the spacer gauge and check operation. Be sure the stoplight switch does not prevent full pedal return.

REMOVAL & INSTALLATION

1. Disconnect the negative battery cable.

2. Remove the switch assembly pedal-to-bracket attaching screw and remove the switch and bracket as an assembly.

3. Remove the switch-to-bracket retaining nut and disassemble the switch from the bracket.

4. Installation is the reverse order of the removal procedure. Adjust the switch.

Fuses, Circuit Breakers and Relays

LOCATION

Fusible Links

The fusible links are used to prevent major damage to wire harnesses in the event of a short circuit or an overload condition in the wiring circuits which normally are not fused, due to carrying high amperage loads or because of their locations within the wiring harness. Each fusible link is of a fixed value for a specific electrical load and should the link fail, the cause of the failure must be determined and repaired prior to installing a new fusible link of the same value.

When replacing fusible links connected to the battery terminal or starter relay, they should be serviced with the same type of prefabricated fusible link. All other fusible links can be replaced with fusible link wire cut from bulk rolls. Most fusible links are located near the large wiring harness rear of the battery.

NOTE: When replacing fusible links, use only rosin core solder. Do not use acid core solder.

Circuit Breakers

Circuit breakers are used in varied circuits to control amperage surges and if the circuit is opened, to re-set themselves as the heat from the current flow load has diminished. Should a continual interruption of power be experienced when operation of a controlled electrical component is attempted, repairs to the circuits/components or replacement of the component must be accomplished. Circuit breakers are located in the fuse panel and, if necessary, can be changed quickly by pulling the assembly from the fuse panel and inserting a new one in its place.

Fuse Panel

The relays and circuit breakers are located on the fuse panel. The fuse panel is located on the left side of the passenger compartment, either mounted to the underside of the dash panel or to the inner side of the firewall panel.

Various Relays

Accessory Power Relay—it supplies current to the dash bulk head and related accessories. It is located on the left side of the brake pedal support bracket.

Starter Relay—it supplies current directly from the battery to the starter solenoid. It is located in the upper left side of the firewall.

Ford Motor Co.
Front Wheel Drive
Ford—Escort, Tempo   Mercury—Topaz

SPECIFICATIONS

VEHICLE IDENTIFICATION CHART

It is important for servicing and ordering parts to be certain of the vehicle and engine identification. The VIN (vehicle identification number) is a 17 digit number visible through the windshield on the driver's side of the dash and contains the vehicle and engine identification codes. The tenth digit indicates model year and the eighth digit indicates engine code. It can be interpreted as follows:

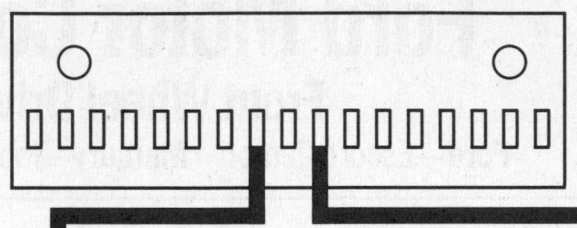

Engine Code						Model Year	
Code	Cu. In.	Liters	Cyl.	Fuel Sys.	Eng. Mfg.	Code	Year
8	112	1.8	4	EFI	Mazda	J	1988
9	116	1.9	4	CFI	Ford	K	1989
J	116	1.9	4	EFI	Ford	L	1990
X	140	2.3	4	EFI	Ford	M	1991
S	140	2.3	4	EFI	Ford	N	1992

CFI   Central Fuel Injection
EFI   Electronic Fuel Injection

ENGINE IDENTIFICATION

Year	Model	Engine Displacement cu. in. (liter)	Engine Series Identification (VIN)	No. of Cylinders	Engine Type
1988	Escort	116 (1.9)	9	4	OHC
	Escort	116 (1.9)	J	4	OHC
	Tempo	140 (2.3)	X	4	OHV
	Tempo	140 (2.3)	S	4	OHV
	Topaz	140 (2.3)	X	4	OHV
	Topaz	140 (2.3)	S	4	OHV
1989	Escort	116 (1.9)	9	4	OHC
	Escort	116 (1.9)	J	4	OHC
	Tempo	140 (2.3)	X	4	OHV
	Tempo	140 (2.3)	S	4	OHV
	Topaz	140 (2.3)	X	4	OHV
	Topaz	140 (2.3)	S	4	OHV
1990	Escort	116 (1.9)	9	4	OHC
	Escort	116 (1.9)	J	4	OHC
	Tempo	140 (2.3)	X	4	OHV
	Tempo	140 (2.3)	S	4	OHV
	Topaz	140 (2.3)	X	4	OHV
	Topaz	140 (2.3)	S	4	OHV

ENGINE IDENTIFICATION

Year	Model	Engine Displacement cu. in. (liter)	Engine Series Identification (VIN)	No. of Cylinders	Engine Type
1991–92	Escort	112 (1.8)	8	4	DOHC
	Escort	116 (1.9)	J	4	OHC
	Tempo	140 (2.3)	X	4	OHV
	Tempo	140 (2.3)	S	4	OHV
	Topaz	140 (2.3)	X	4	OHV
	Topaz	140 (2.3)	S	4	OHV

OHC Overhead Cam
OHV Overhead Valve
DOHC Dual Overhead Cam

GENERAL ENGINE SPECIFICATIONS

Year	VIN	No. Cylinder Displacement cu. in. (liter)	Fuel System Type	Net Horsepower @ rpm	Net Torque @ rpm (ft. lbs.)	Bore × Stroke (in.)	Compression Ratio	Oil Pressure @ rpm
1988	9	4-116 (1.9)	CFI	90 @ 4600	106 @ 3400	3.23 × 3.46	9.0:1	35–65 @ 2000
	J	4-116 (1.9)	EFI	110 @ 5400	115 @ 4200	3.23 × 3.46	9.0:1	35–65 @ 2000
	X	4-140 (2.3)	EFI	98 @ 4400	124 @ 2200	3.70 × 3.30	9.0:1	55–70 @ 2000
	S	4-140 (2.3)	EFI	100 @ 4400	130 @ 2600	3.70 × 3.30	9.0:1	55–70 @ 2000
1989	9	4-116 (1.9)	CFI	90 @ 4600	106 @ 3400	3.23 × 3.46	9.0:1	35–65 @ 2000
	J	4-116 (1.9)	EFI	110 @ 5400	115 @ 4200	3.23 × 3.46	9.0:1	35–65 @ 2000
	X	4-140 (2.3)	EFI	98 @ 4400	124 @ 2200	3.70 × 3.30	9.0:1	55–70 @ 2000
	S	4-140 (2.3)	EFI	100 @ 4400	130 @ 2600	3.70 × 3.30	9.0:1	55–70 @ 2000
1990	9	4-116 (1.9)	CFI	90 @ 4600	106 @ 3400	3.23 × 3.46	9.0:1	35–65 @ 2000
	J	4-116 (1.9)	EFI	110 @ 5400	115 @ 4200	3.23 × 3.46	9.0:1	35–65 @ 2000
	X	4-140 (2.3)	EFI	98 @ 4400	124 @ 2200	3.70 × 3.30	9.0:1	55–70 @ 2000
	S	4-140 (2.3)	EFI	100 @ 4400	130 @ 2600	3.70 × 3.30	9.0:1	55–70 @ 2000
1991–92	8	4-112 (1.8)	EFI	127 @ 6500	114 @ 4500	3.27 × 3.35	9.0:1	43–57 @ 3000
	J	4-116 (1.9)	EFI	88 @ 4400	108 @ 3800	3.23 × 3.46	9.0:1	35–65 @ 2000
	X	4-140 (2.3)	EFI	98 @ 4400	124 @ 2200	3.70 × 3.30	9.0:1	55–70 @ 2000
	S	4-140 (2.3)	EFI	100 @ 4400	130 @ 2600	3.70 × 3.30	9.0:1	55–70 @ 2000

CFI—Central Fuel Injection
EFI—Electronic Fuel Injection

GASOLINE ENGINE TUNE-UP SPECIFICATIONS

Year	VIN	No. Cylinder Displacement cu. in. (liter)	Spark Plugs Type	Gap (in.)	Ignition Timing (deg.) MT	Ignition Timing (deg.) AT	Compression Pressure (psi)	Fuel Pump (psi)	Idle Speed (rpm) MT	Idle Speed (rpm) AT	Valve Clearance In.	Valve Clearance Ex.
1988	9	4-116 (1.9)	AGSF-34C	0.044	10B	10B	①	13–17	760–840	760–840	Hyd.	Hyd.
	J	4-116 (1.9)	AGSF-24C	0.044	10B	10B	①	30–45	②	②	Hyd.	Hyd.
	X	4-140 (2.3)	AWSF-42C	0.054	15B	15B	①	45–60	820–880	690–750	Hyd.	Hyd.
	S	4-140 (2.3)	AWSF-42C	0.054	15B	15B	①	50–60	820–880	690–750	Hyd.	Hyd.
1989	9	4-116 (1.9)	AGSF-34C	0.044	10B	10B	①	13–17	760–840	760–840	Hyd.	Hyd.
	J	4-116 (1.9)	AGSF-24C	0.044	10B	10B	①	30–45	②	②	Hyd.	Hyd.
	X	4-140 (2.3)	AWSF-42C	0.054	15B	15B	①	45–60	820–880	690–750	Hyd.	Hyd.
	S	4-140 (2.3)	AWSF-42C	0.054	15B	15B	①	50–60	820–880	690–750	Hyd.	Hyd.

GASOLINE ENGINE TUNE-UP SPECIFICATIONS

Year	VIN	No. Cylinder Displacement cu. in. (liter)	Spark Plugs Type	Gap (in.)	Ignition Timing (deg.) MT	AT	Compression Pressure (psi)	Fuel Pump (psi)	Idle Speed (rpm) MT	AT	Valve Clearance In.	Ex.
1990	9	4-116 (1.9)	AGSF-34C	0.044	10B	10B	①	13–17	760–840	760–840	Hyd.	Hyd.
	J	4-116 (1.9)	AGSF-24C	0.044	10B	10B	①	35–45	②	②	Hyd.	Hyd.
	X	4-140 (2.3)	AWSF-42C	0.054	15B	15B	①	45–60	820–880	690–750	Hyd.	Hyd.
	S	4-140 (2.3)	AWSF-42C	0.054	15B	15B	①	50–60	820–880	690–750	Hyd.	Hyd.
1991	8	4-112 (1.8)	AGSP-32C	0.043	9-11B	9-11B	①	64–85	700–800	700–800	Hyd.	Hyd.
	J	4-116 (1.9)	AGSF-34C6	0.054	10B	10B	①	35–45	②	②	Hyd.	Hyd.
	X	4-140 (2.3)	AWSF-42C	0.054	②	②	①	45–60	②	②	Hyd.	Hyd.
	S	4-140 (2.3)	AWSF-42C	0.054	②	②	①	45–60	②	②	Hyd.	Hyd.
1992		SEE UNDERHOOD SPECIFICATIONS STICKER										

B Before Top Dead Center
① Lowest cylinder should be at least 75% of highest cylinder
② Refer to vehicle emission control information label

FIRING ORDERS

NOTE: To avoid confusion, al- ways replace spark plug wires one at a time.

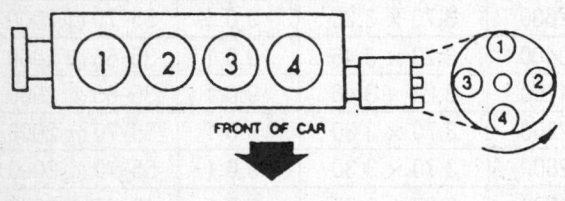

1.8L Engine
Engine Firing Order: 1–3–4–2
Distributor Rotation: Counterclockwise

1988–90 1.9L Engine
Engine Firing Order: 1–3–4–2
Distributor Rotation: Clockwise

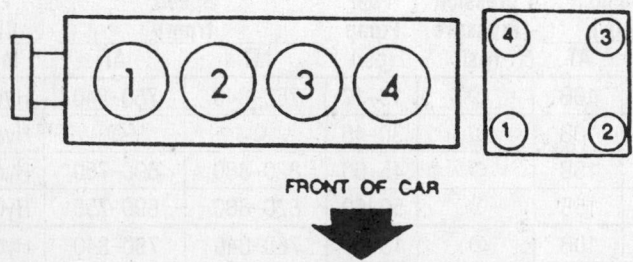

1991–92 1.9L Engine
Engine Firing Order: 1–3–4–2
Distributorless Ignition System

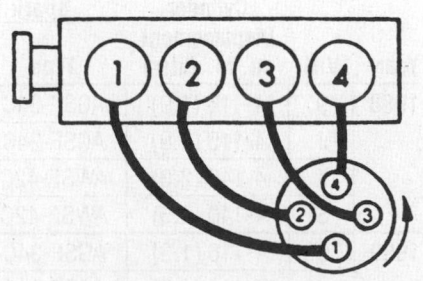

2.3L Engine
Engine Firing Order: 1–3–4–2
Distributor Rotation: Clockwise

CAPACITIES

Year	Model	VIN	No. Cylinder Displacement cu. in. (liter)	Engine Crankcase (qts.) with Filter	without Filter	Transmission (pts.) 4-Spd	5-Spd	Auto.	Drive Axle (pts.)	Fuel Tank (gal.)	Cooling System (qts.)
1988	Escort	9	4-116 (1.9)	4.0	3.5	6.1	6.1	16.6	①	13	②
	Escort	J	4-116 (1.9)	4.0	3.5	6.1	6.1	16.6	①	13	②
	Tempo	S	4-140 (2.3)	5.0	4.0	—	6.1	16.6	③	④	⑤
	Tempo	X	4-140 (2.3)	5.0	4.0	—	6.1	16.6	③	④	⑤
	Topaz	S	4-140 (2.3)	5.0	4.0	—	6.1	16.6	③	④	⑤
	Topaz	X	4-140 (2.3)	5.0	4.0	—	6.1	16.6	③	④	⑤
1989	Escort	9	4-116 (1.9)	4.0	3.5	6.1	6.1	16.6	①	13	②
	Escort	J	4-116 (1.9)	4.0	3.5	6.1	6.1	16.6	①	13	②
	Tempo	S	4-140 (2.3)	5.0	4.0	—	6.1	16.6	③	④	⑤
	Tempo	X	4-140 (2.3)	5.0	4.0	—	6.1	16.6	③	④	⑤
	Topaz	S	4-140 (2.3)	5.0	4.0	—	6.1	16.6	③	④	⑤
	Topaz	X	4-140 (2.3)	5.0	4.0	—	6.1	16.6	③	④	⑤
1990	Escort	9	4-116 (1.9)	4.0	3.5	6.1	6.1	16.6	①	13	②
	Escort	J	4-116 (1.9)	4.0	3.5	6.1	6.1	16.6	①	13	②
	Tempo	S	4.140 (2.3)	5.0	4.0	—	6.1	16.6	③	⑥	⑤
	Tempo	X	4.140 (2.3)	5.0	4.0	—	6.1	16.6	③	⑥	⑤
	Topaz	S	4.140 (2.3)	5.0	4.0	—	6.1	16.6	③	⑥	⑤
	Topaz	X	4.140 (2.3)	5.0	4.0	—	6.1	16.6	③	⑥	⑤
1991-92	Escort	8	4-112 (1.8)	4.1	3.6	—	7.2	13.4	①	13.2	⑦
	Escort	J	4-116 (1.9)	4.0	3.5	—	5.6	13.4	①	11.9	⑦
	Tempo	X	4-140 (2.3)	5.0	4.0	—	6.1	16.6	③	⑥	⑤
	Tempo	S	4-140 (2.3)	5.0	4.0	—	6.1	16.6	③	⑥	⑤
	Topaz	X	4-140 (2.3)	5.0	4.0	—	6.1	16.6	③	⑥	⑤
	Topaz	S	4-140 (2.3)	5.0	4.0	—	6.1	16.6	③	⑥	⑤

① Included in transaxle capacity
② Without air conditioning—8.3 qts.
Manual transaxle with air conditioning—6.8 qts.
Automatic transaxle with air conditioning—7.3 qts.
③ Included in transaxle capacity
All wheel drive rear capacity is 1.3 pts.
④ Without all wheel drive—15.4 gallons
With all wheel drive—14.2 gallons
⑤ With manual transaxle—7.3 qts.
With automatic transaxle—7.8 qts.
⑥ Without all wheel drive—15.9 gallons
With all wheel drive—14.2 gallons
⑦ With manual transaxle—5.3 qts.
With automatic transaxle—6.3 qts.

CAMSHAFT SPECIFICATIONS

All measurements given in inches.

Year	VIN	No. Cylinder Displacement cu. in. (liter)	Journal Diameter 1	2	3	4	5	Lobe Lift	Bearing Clearance	Camshaft End Play
1988	9	4-116 (1.9)	1.8007-1.8017	1.8007-1.8017	1.8007-1.8017	1.8007-1.8017	1.8007-1.8017	0.240	0.001-0.003	0.002-0.006
	J	4.116 (1.9)	1.8007-1.8017	1.8007-1.8017	1.8007-1.8017	1.8007-1.8017	1.8007-1.8017	0.240	0.001-0.003	0.002-0.006
	X	4.140 (2.3)	2.006-2.009	2.006-2.009	2.006-2.009	2.006-2.009	—	①	0.001-0.003	0.009
	S	4.140 (2.3)	2.006-2.009	2.006-2.009	2.006-2.009	2.006-2.009	—	0.262	0.001-0.003	0.009

CAMSHAFT SPECIFICATIONS

All measurements given in inches.

Year	VIN	No. Cylinder Displacement cu. in. (liter)	Journal Diameter 1	2	3	4	5	Lobe Lift	Bearing Clearance	Camshaft End Play
1989	9	4-116 (1.9)	1.8007–1.8017	1.8007–1.8017	1.8007–1.8017	1.8007–1.8017	1.8007–1.8017	0.240	0.001–0.003	0.002–0.006
	J	4-116 (1.9)	1.8007–1.8017	1.8007–1.8017	1.8007–1.8017	1.8007–1.8017	1.8007–1.8017	0.240	0.001–0.003	0.002–0.006
	X	4-140 (2.3)	2.006–2.009	2.006–2.009	2.006–2.009–	2.006–2.009	—	①	0.001–0.003	0.009
	S	4-140 (2.3)	2.006–2.009	2.006–2.009	2.006–2.009	2.006–2.009	—	0.262	0.001–0.003	0.009
1990	9	4-116 (1.9)	1.8007–1.8017	1.8007–1.8017	1.8007–1.8017	1.8007–1.8017	1.8007–1.8017	0.240	0.001–0.003	0.002–0.006
	J	4-116 (1.9)	1.8007–1.8017	1.8007–1.8017	1.8007–1.8017	1.8007–1.8017	1.8007–1.8017	0.240	0.001–0.003	0.002–0.006
	X	4-140 (2.3)	2.006–2.009	2.006–2.009	2.006–2.009	2.006–2.009	—	①	0.001–0.003	0.009
	S	4-140 (2.3)	2.006–2.009	2.006–2.009	2.006–2.009	2.006–2.009	—	0.262	0.001–0.003	0.009
1991–92	8	4-112 (1.8)	1.021–1.022	1.021–1.022	1.021–1.022	1.021–1.022	1.021–1.022	NA	0.001–0.003	0.003–0.007
	J	4-116 (1.9)	1.8007–1.8017	1.8007–1.8017	1.8007–1.8017	1.8007–1.8017	1.8007–1.8017	0.240	0.001–0.003	0.002–0.006
	X	4-140 (2.3)	2.006–2.009	2.006–2.009	2.006–2.009	2.006–2.009	—	①	0.001–0.003	0.009
	S	4-140 (2.3)	2.006–2.009	2.006–2.009	2.006–2.009	2.006–2.009	—	0.262	0.001–0.003	0.009

NA Not available   ① Intake—0.249 in.   Exhaust—0.239 in.

CRANKSHAFT AND CONNECTING ROD SPECIFICATIONS

All measurements are given in inches.

Year	VIN	No. Cylinder Displacement cu. in. (liter)	Crankshaft Main Brg. Journal Dia.	Main Brg. Oil Clearance	Shaft End-play	Thrust on No.	Connecting Rod Journal Diameter	Oil Clearance	Side Clearance
1988	9	4-116 (1.9)	2.2827–2.2835	①	0.004–0.008	3	1.7279–1.7287	0.0008–0.0015	0.004–0.011
	J	4-116 (1.9)	2.2827–2.2835	①	0.004–0.008	3	1.7279–1.7287	0.0008–0.0015	0.004–0.011
	X	4-140 (2.3)	2.2489–2.2490	0.0008–0.0015	0.004–0.008	3	2.1232–2.1240	0.0008–0.0015	0.004–0.011
	S	4-140 (2.3)	2.2489–2.2490	0.0008–0.0015	0.004–0.008	3	2.1232–2.1240	0.0008–0.0015	0.004–0.011
1989	9	4-116 (1.9)	2.2827–2.2835	①	0.004–0.008	3	1.7279–1.7287	0.0008–0.0015	0.004–0.011
	J	4-116 (1.9)	2.2827–2.2835	①	0.004–0.008	3	1.7279–1.7287	0.0008–0.0015	0.004–0.011
	X	4-140 (2.3)	2.2489–2.2490	0.0008–0.0015	0.004–0.008	3	2.1232–2.1240	0.0008–0.0015	0.004–0.011
	S	4-140 (2.3)	2.2489–2.2490	0.0008–0.0015	0.004–0.008	3	2.1232–2.1240	0.0008–0.0015	0.004–0.011

CRANKSHAFT AND CONNECTING ROD SPECIFICATIONS

All measurements are given in inches.

Year	VIN	No. Cylinder Displacement cu. in. (liter)	Crankshaft Main Brg. Journal Dia.	Crankshaft Main Brg. Oil Clearance	Crankshaft Shaft End-play	Crankshaft Thrust on No.	Connecting Rod Journal Diameter	Connecting Rod Oil Clearance	Connecting Rod Side Clearance
1990	9	4-116 (1.9)	2.2827–2.2835	①	0.004–0.008	3	1.7279–1.7287	0.0008–0.0015	0.004–0.011
	J	4-116 (1.9)	2.2827–2.2835	①	0.004–0.008	3	1.7279–1.7287	0.0008–0.0015	0.004–0.011
	X	4-140 (2.3)	2.2489–2.2490	0.0008–0.0015	0.004–0.008	3	2.1232–2.1240	0.0008–0.0015	0.004–0.011
	S	4-140 (2.3)	2.2489–2.2490	0.0008–0.0015	0.004–0.008	3	2.1232–2.1240	0.0008–0.0015	0.004–0.011
1991–92	8	4-112 (1.8)	1.9661–1.9668	0.0007–0.0014	0.003–0.011	4	1.7692–1.7699	0.0011–0.0027	NA
	J	4-116 (1.9)	2.2827–2.2835	①	0.004–0.008	3	1.7279–1.7287	0.0008–0.0015	0.004–0.011
	X	4-140 (2.3)	2.2489–2.2490	0.0008–0.0015	0.004–0.008	3	2.1232–2.1240	0.0008–0.0015	0.004–0.011
	S	4-140 (2.3)	2.2489–2.2490	0.0008–0.0015	0.004–0.008	3	2.1232–2.1240	0.0008–0.0015	0.004–0.011

① Without cylinder head: 0.0018–0.0026 in.
  With cylinder head: 0.0011–0.0019 in.

VALVE SPECIFICATIONS

Year	VIN	No. Cylinder Displacement cu. in. (liter)	Seat Angle (deg.)	Face Angle (deg.)	Spring Test Pressure (lbs.)	Spring Installed Height (in.)	Stem-to-Guide Clearance (in.) Intake	Stem-to-Guide Clearance (in.) Exhaust	Stem Diameter (in.) Intake	Stem Diameter (in.) Exhaust
1988	9	4-116 (1.9)	45	45.6	200 @ 1.09	1.44–1.48	0.0008–0.0027	0.0018–0.0037	0.3159–0.3167	0.3149–0.3156
	J	4-116 (1.9)	45	45.6	216 @ 1.016	1.44–1.48	0.0008–0.0027	0.0018–0.0037	0.3159–0.3167	0.3149–0.3156
	X	4-140 (2.3)	44–45	44–45	181 @ 1.08	1.49	0.0018	0.0023	0.3415–0.3422	0.3411–0.3418
	S	4-140 (2.3)	44–45	44–45	181 @ 1.08	1.49	0.0018	0.0023	0.3415–0.3422	0.3411–0.3418
1989	9	4-116 (1.9)	45	45.6	200 @ 1.09	1.44–1.48	0.0008–0.0027	0.0018–0.0037	0.3159–0.3167	0.3149–0.3156
	J	4-116 (1.9)	45	45.6	216 @ 1.016	1.44–1.48	0.0008–0.0027	0.0018–0.0037	0.3159–0.3167	0.3149–0.3156
	X	4-140 (2.3)	44–45	44–45	181 @ 1.08	1.49	0.0018	0.0023	0.3415–0.3422	0.3411–0.3418
	S	4-140 (2.3)	44–45	44–45	181 @ 1.08	1.49	0.0018	0.0023	0.3415–0.3422	0.3411–0.3418
1990	9	4-116 (1.9)	45	45.6	200 @ 1.09	1.44–1.48	0.0008–0.0027	0.0018–0.0037	0.3159–0.3167	0.3149–0.3156
	J	4-116 (1.9)	45	45.6	216 @ 1.016	1.44–1.48	0.0008–0.0027	0.0018–0.0037	0.3159–0.3167	0.3149–0.3156
	X	4-140 (2.3)	44–45	44–45	181 @ 1.08	1.49	0.0018	0.0023	0.3415–0.3422	0.3411–0.3418
	S	4-140 (2.3)	44–45	44–45	181 @ 1.08	1.49	0.0018	0.0023	0.3415–0.3422	0.3411–0.3418

VALVE SPECIFICATIONS

Year	VIN	No. Cylinder Displacement cu. in. (liter)	Seat Angle (deg.)	Face Angle (deg.)	Spring Test Pressure (lbs.)	Spring Installed Height (in.)	Stem-to-Guide Clearance (in.) Intake	Stem-to-Guide Clearance (in.) Exhaust	Stem Diameter (in.) Intake	Stem Diameter (in.) Exhaust
1991–92	8	4-112 (1.8)	45	45	NA	NA	0.0010–0.0024	0.0012–0.0026	0.2350–0.2356	0.2348–0.2354
	J	4-116 (1.9)	45	45.6	200 @ 1.09	1.44–1.48	0.0008–0.0027	0.0018–0.0037	0.3159–0.3167	0.3149–0.3156
	X	4-140 (2.3)	44–45	44–45	181 @ 1.08	1.49	0.0018	0.0023	0.3415–0.3422	0.3411–0.3418
	S	4-140 (2.3)	44–45	44–45	181 @ 1.08	1.49	0.0018	0.0023	0.3415–0.3422	0.3411–0.3418

NA—Not available

PISTON AND RING SPECIFICATIONS

All measurements are given in inches.

Year	VIN	No. Cylinder Displacement cu. in. (liter)	Piston Clearance	Ring Gap Top Compression	Ring Gap Bottom Compression	Ring Gap Oil Control	Ring Side Clearance Top Compression	Ring Side Clearance Bottom Compression	Ring Side Clearance Oil Control
1988	9	4-116 (1.9)	0.0016–0.0024	0.010–0.020	0.010–0.020	0.016–0.055	0.0015–0.0032	0.0015–0.0035	Snug
	J	4-116 (1.9)	0.0016–0.0024	0.010–0.020	0.010–0.020	0.016–0.055	0.0015–0.0032	0.0015–0.0035	Snug
	X	4-140 (2.3)	0.0012–0.0022	0.008–0.016	0.008–0.016	0.015–0.055	0.0020–0.0040	0.0020–0.0040	Snug
	S	4-140 (2.3)	0.0012–0.0022	0.008–0.016	0.008–0.016	0.015–0.055	0.0020–0.0040	0.0020–0.0040	Snug
1989	9	4-116 (1.9)	0.0016–0.0024	0.010–0.020	0.010–0.020	0.016–0.055	0.0015–0.0032	0.0015–0.0035	Snug
	J	4-116 (1.9)	0.0016–0.0024	0.010–0.020	0.010–0.020	0.016–0.055	0.0015–0.0032	0.0015–0.0035	Snug
	X	4-140 (2.3)	0.0012–0.0022	0.008–0.016	0.008–0.016	0.015–0.055	0.0020–0.0040	0.0020–0.0040	Snug
	S	4-140 (2.3)	0.0012–0.0022	0.008–0.016	0.008–0.016	0.015–0.055	0.0020–0.0040	0.0020–0.0040	Snug
1990	9	4-116 (1.9)	0.0016–0.0024	0.010–0.020	0.010–0.020	0.016–0.055	0.0015–0.0032	0.0015–0.0032	Snug
	J	4-116 (1.9)	0.0016–0.0024	0.010–0.020	0.010–0.020	0.016–0.055	0.0015–0.0032	0.0015–0.0032	Snug
	X	4-140 (2.3)	0.0012–0.0022	0.008–0.016	0.008–0.016	0.015–0.055	0.0020–0.0040	0.0020–0.0040	Snug
	S	4-140 (2.3)	0.0012–0.0022	0.008–0.016	0.008–0.016	0.015–0.055	0.0020–0.0040	0.0020–0.0040	Snug
1991–92	8	4-112 (1.8)	0.0015–0.0020	0.006–0.012	0.006–0.012	0.008–0.028	0.0012–0.0028	0.0012–0.0028	Snug
	J	4-116 (1.9)	0.0016–0.0024	0.010–0.020	0.010–0.020	0.016–0.055	0.0015–0.0032	0.0015–0.0035	Snug
	X	4-140 (2.3)	0.0012–0.0022	0.008–0.016	0.008–0.016	0.015–0.055	0.0020–0.0040	0.0020–0.0040	Snug
	S	4-140 (2.3)	0.0012–0.0022	0.008–0.016	0.008–0.016	0.015–0.055	0.0020–0.0040	0.0020–0.0040	Snug

TORQUE SPECIFICATIONS

All readings in ft. lbs.

Year	VIN	No. Cylinder Displacement cu. in. (liter)	Cylinder Head Bolts	Main Bearing Bolts	Rod Bearing Bolts	Crankshaft Pulley Bolts	Flywheel Bolts	Manifold Intake	Manifold Exhaust	Spark Plugs
1988	9	4-116 (1.9)	①	67–80	19–25	74–90	54–64	12–15	15–20	8–15
	J	4-116 (1.9)	①	67–80	19–25	74–90	54–64	12–15	15–20	8–15
	X	4-140 (2.3)	②	51–66	21–26	140–170	54–64	15–23	③	5–10
	S	4-140 (2.3)	②	51–66	21–26	140–170	54–64	15–23	③	5–10
1989	9	4-116 (1.9)	①	67–80	26–30	81–96	54–64	④	15–20	8–15
	J	4-116 (1.9)	①	67–80	26–30	81–96	54–64	④	15–20	8–15
	X	4-140 (2.3)	②	51–66	21–26	140–170	54–64	15–23	③	5–10
	S	4-140 (2.3)	②	51–66	21–26	140–170	54–64	15–23	③	5–10
1990	9	4-116 (1.9)	①	67–80	26–30	81–96	54–64	④	15–20	8–15
	J	4-116 (1.9)	①	67–80	26–30	81–96	54–64	④	15–20	8–15
	X	4-140 (2.3)	②	51–66	21–26	140–170	54–64	15–22	③	5–10
	S	4-140 (2.3)	②	51–66	21–26	140–170	54–64	15–22	③	5–10
1991–92	8	4-112 (1.8)	56–60	40–43	35–37	80–87	71–76	14–19	28–34	11–17
	J	4-116 (1.9)	①	67–80	26–30	81–96	54–67	12–15	16–19	8–15
	X	4-140 (2.3)	②	51–66	21–26	140–170	54–64	⑤	③	5–10
	S	4-140 (2.3)	②	51–66	21–26	140–170	54–64	⑤	③	5–10

① Tighten in sequence to 44 ft. lbs.
Loosen 2 turns
Retighten in sequence to 44 ft. lbs.
Turn all bolts 90 degrees
Turn all bolts additional 90 degrees

② Tighten, in sequence, in 2 steps: 52–59 ft. lbs.
70–76 ft. lbs.
③ Tighten in 2 stages—5–7 ft. lbs. then 20–30 ft. lbs.

④ Tighten to 12–15 ft. lbs. in the proper sequence
⑤ Tighten in 2 steps: 5–7 ft. lbs.
15–22 ft. lbs.

BRAKE SPECIFICATIONS

All measurements in inches unless noted

Year	Model	Lug Nut Torque (ft. lbs.)	Master Cylinder Bore	Brake Disc Minimum Thickness	Brake Disc Maximum Runout	Standard Brake Drum Diameter	Minimum Lining Thickness Front	Minimum Lining Thickness Rear
1988	Escort	80–105	①	0.882	0.003	7.145②	0.125	③
	Tempo	80–105	①	0.882	0.003	8.059	0.125	③
	Topaz	80–105	①	0.882	0.003	8.059	0.125	③
1989	Escort	80–105	①	0.882	0.003	7.145②	0.125	③
	Tempo	80–105	①	0.882	0.003	8.059	0.125	③
	Topaz	80–105	①	0.882	0.003	8.059	0.125	③
1990	Escort	85–105	①	0.882	0.003	7.145②	0.125	③
	Tempo	85–105	①	0.882	0.003	8.059	0.125	③
	Topaz	85–105	①	0.882	0.003	8.059	0.125	③
1991–92	Escort	65–87	0.87	④	0.004	9.00	0.08	0.04
	Tempo	85–105	①	0.882	0.003	8.059	0.125	0.059
	Topaz	85–105	①	0.882	0.003	8.059	0.125	0.059

① Primary bore—1.12 in.
Secondary bore—0.116 in.
② With styled steel wheels and/or 4 door hatchback, 4 door wagon—8.059 in.

③ Riveted linings—within 0.031 in. of rivet head
Bonded linings—0.059 in.
④ Front—0.79 in.
Rear—0.28 in.

WHEEL ALIGNMENT

Year	Model		Caster Range (deg.)	Caster Preferred Setting (deg.)	Camber Range (deg.)	Camber Preferred Setting (deg.)	Toe-in (in.)	Wheel Turning Angle (deg.)
1988	Escort	Front	1 5/8P-3 1/8P	2 3/8P	①	②	1/4N-0	③
		Rear			1 3/16N-1/2P	5/16N	3/16P-3/8P	
	Tempo	Front	1 11/16P-3 3/16P	2 7/16P	④	⑥	1/4N-0	③
		Rear					3/16N-3/16P	
	Topaz	Front	1 11/16P-3 3/16P	2 7/16P	④	⑥	1/4N-0	③
		Rear					3/16N-3/16P	
1989	Escort	Front	1 5/8P-3 1/8P	2 3/8P	①	②	1/4N-0	③
		Rear			1 3/16N-1/2P	5/16N	3/16P-3/8P	
	Tempo	Front	1 11/16P-3 3/16P	2 7/16P	④	⑥	1/4N-0	③
		Rear			⑤	⑦	3/16N-3/16P	
	Topaz	Front	1 11/16P-3 3/16P	2 7/16P	④	⑥	1/4N-0	③
		Rear			⑤	⑦	3/16N-3/16P	
1990	Escort	Front	1 5/8P-3 1/8P	2 3/8P	①	②	1/4N-0	③
		Rear			1 3/16N-1/2P	5/16N	3/16P-3/8P	
	Tempo	Front	1 11/16P-3 3/16P	2 7/16P	④	⑥	1/4N-0	③
		Rear			⑤	⑦	3/16N-3/16P	
	Topaz	Front	1 11/16P-3 3/16P	2 7/16P	④	⑥	1/4N-0	③
		Rear			⑤	⑦	3/16N-3/16P	
1991-92	Escort	Front	1P-2 7/8P	1 15/16P	3/4N-1/2P	1/16N	3/64N-13/64P	NA
		Rear			1N-1/2P	1/4N	3/64N-13/64P	
	Tempo	Front	1 11/16P-3 3/16P	2 7/16P	④	⑥	1/4N-0	③
		Rear			⑤	⑦	3/16N-3/16P	
	Topaz	Front	1 11/16P-3 3/16P	2 7/16P	④	⑥	1/4N-0	③
		Rear			⑤	⑦	3/16N-3/16P	

① Left wheel—3/8P-1 7/8P
Right wheel—0-1 1/2P
② Left wheel—1 1/8P
Right wheel—15/16P
③ Left wheel—14 21/32
Right wheel—15 3/32
④ Left wheel—21/32P-2 5/32P
Right wheel—7/32P-1 23/32P
⑤ With front wheel drive—29/32N-19/32P
With all wheel drive—13/13N-1 3/32P
⑥ Left wheel—1 13/32P
Right wheel—31/32P
⑦ With front wheel drive—5/32N
With all wheel drive—11/32P

ENGINE MECHANICAL

NOTE: Disconnecting the negative battery cable on some vehicles may interfere with the functions of the on board computer systems and may require the computer to undergo a relearning process, once the negative battery cable is reconnected.

Engine Assembly

REMOVAL & INSTALLATION

1.8L Engine

WITH AUTOMATIC TRANSAXLE

1.8L engine equipped with an automatic transaxle can be removed without removing the transaxle from the vehicle. The engine can be split from the transaxle and lifted out of the engine compartment.

1. Disconnect the negative battery cable.
2. Mark the position of the hood hinges and remove the hood.
3. If equipped with air conditioning, properly discharge the system.
4. Drain the cooling system.
5. Remove the air duct connecting the throttle body and resonance chamber.
6. Disconnect the power brake vacuum supply hose from the power booster.
7. If equipped with speed control, disconnect the necessary vacuum hoses from the intake plenum.
8. Disconnect the electrical connectors from the power steering pump, water thermoswitch, temperature sending unit, oil pressure switch, fuel injector wiring harness, exhaust gas oxygen sensor, throttle position sensor and distributor.

NOTE: Mark the position of the connectors prior to removal to ease reinstallation.

9. Disconnect all engine ground straps.
10. Disconnect the ignition coil high-tension lead from the distributor.
11. Disconnect the accelerator and kickdown cables from the throttle cam.
12. Remove the accelerator and kickdown cable bracket from the intake plenum and set the assembly aside.
13. Disconnect the heater core inlet and outlet hoses at the bulkhead.

14. Relieve the fuel system pressure.

15. Remove the necessary fuel line clips and disconnect the fuel pressure and return lines.

16. Remove the upper radiator hose.

17. Disconnect the electrical connectors from the cooling fan and the radiator thermoswitch.

18. Remove the starter motor.

19. Raise and safely support the vehicle.

20. Remove the right upper and both left and right lower splash shields.

21. Remove the radiator lower hose.

22. Disconnect the 2 transaxle cooling lines from the radiator and plug the lines.

23. Remove the air conditioner line routing bracket from the radiator and position the line aside.

24. Remove the halfshaft bearing support.

25. Remove the inspection plate from the oil pan, place a wrench on the crankshaft pulley, and rotate the crankshaft to gain access to the torque converter nuts. Remove the nuts.

26. Remove the power steering and air conditioner drive belt.

27. Remove the crankshaft pulley.

28. Remove the exhaust flex-pipe and mounting flange assembly from the exhaust manifold.

29. If equipped with air conditioning, remove the compressor.

30. Remove the power steering pump and bracket assembly with the hoses still connected. Suspend the pump with wire, aside of the work area.

31. Remove all accessible transaxle-to-engine bolts from the engine block.

32. Lower the vehicle.

33. Remove the radiator mounting brackets and the resonance duct.

34. Remove the radiator, fan and shroud assembly from the vehicle.

35. Remove the vaccum chamber canister located next to the intake plenum.

36. Remove the pressure regulator and bracket assembly and set it aside.

37. Remove the shutter valve actuator and bracket assembly and set it aside.

38. Remove the alternator and water pump drive belt and remove the alternator.

39. Install a suitable engine removal sling onto the engine lifting brackets. Place a suitable engine hoist into position and support the engine.

40. Remove the oil pan-to-transaxle attaching bolts and the remaining transaxle-to-engine bolts from the engine block.

41. Remove the engine vibration dampener.

42. Remove the engine mount.

43. Carefully separate the engine from the transaxle, then remove the engine from the vehicle.

44. Install the engine onto a suitable engine stand.

To install:

45. Install a suitable engine removal sling onto the engine lifting brackets.

46. Place a suitable engine hoist into position and install the engine sling. Remove the engine from the engine stand and lower it into the engine compartment.

47. Install the transaxle-to-engine upper right bolt and tighten to 41–59 ft. lbs. (55–80 Nm).

NOTE: Make sure the torque converter studs are properly seated in the flexplate mounting holes.

48. Install the engine mount. Tighten the bolt and nuts to 49–69 ft. lbs. (67–93 Nm).

50. Install the engine vibration dampener. Tighten the bolt and nuts to 41–50 ft. lbs. (55–80 Nm).

51. Remove the engine sling from the lifting brackets and remove the engine hoist.

52. Install the remaining transaxle-to-engine bolts and tighten to 41–59 ft. lbs. (55–80 Nm).

53. Install the alternator and the alternator and water pump drive belt.

54. Install the shutter valve actuator and bracket assembly.

55. Install the pressure regulator and bracket assembly.

56. Install the vacuum chamber canister located next to the intake plenum.

57. Place the power steering pump and bracket assembly into its mounting position.

58. Place the radiator, fan and shroud assembly into its mounting position.

59. Install the radiator mounting brackets along with the resonance duct. Tighten the mounting bolts to 69–95 inch lbs. (7.8–11.0 Nm).

60. Connect the cooling fan and radiator thermoswitch electrical connectors.

61. Raise and safely support the vehicle.

62. Install the oil pan-to-transaxle attaching bolts and tighten to 27–38 ft. lbs. (37–52 Nm).

63. Install the power steering pump and bracket assembly. Tighten the bolts to 27–38 ft. lbs. (37–52 Nm).

64. Install the lower radiator hose and clamps.

65. Connect the 2 transaxle cooling lines to the radiator.

66. If equipped, install the air conditioning compressor.

67. Install the air conditioning hose routing bracket to the radiator, if equipped. Tighten the bracket attaching nuts to 56–82 inch lbs. (6.4–9.3 Nm).

68. Install the crankshaft pulley and tighten the bolts to 109–152 inch lbs. (12–17 Nm).

69. Place a wrench on the crankshaft pulley and rotate the crankshaft to gain access to the torque converter studs. Install the torque converter nuts and tighten to 25–36 ft. lbs. (34–49 Nm). Install the transaxle inspection plate.

70. Install the power steering and air conditioning, if equipped, drive belt.

71. Install the halfshaft bearing support and tighten the bolts to 31–46 ft. lbs. (42–62 Nm).

72. Install the starter motor.

73. Connect the heater core inlet and outlet hoses at the bulkhead.

74. Install the exhaust flex-pipe, with a new gasket, to the exhaust manifold. Tighten the pipe-to-converter attaching nuts to 23–34 ft. lbs. (31–46 Nm).

75. Install the right and left lower splash shields and the right upper splash shield. Tighten the bolts to 69–95 inch lbs. (7.8–11.0 Nm).

76. Lower the vehicle.

77. Install the upper radiator hose and clamps.

78. Unplug the fuel pressure and return lines and connect them to the fuel rail. Install the necessary fuel line clips.

79. Install the accelerator and kickdown cable bracket onto the intake plenum. Tighten the bolts to 69–95 inch lbs. (7.8–11.0 Nm). Install the accelerator and kickdown cables onto the throttle cam.

80. Connect the power brake vacuum supply hose to the vacuum booster.

81. If equipped, connect the vehicle speed control vacuum hoses to the intake plenum.

82. Connect all engine ground straps.

83. Connect all remaining electrical connectors to their original locations, as marked during the removal procedure.

84. Connect the ignition coil high-tension lead into the distributor.

85. Install the air duct and resonance chamber assembly.

86. Fill the cooling system.

87. If equipped, recharge the air conditioning system according to the proper procedure.

88. Install the hood, aligning the marks that were made during the removal procedure.

89. Connect the negative battery cable.

90. Start the engine and check for leaks. Stop the engine and check the fluid levels.

WITH MANUAL TRANSAXLE

1.8L engine equipped with manual transaxles require the engine and transaxle to be removed as an assembly. Lift the assembly out of the engine compartment.

1. Disconnect the negative battery cable.

2. Mark the position of the hood hinges and remove the hood.

3. If equipped with air conditioning, properly discharge the system.

4. Drain the cooling system.

5. Remove the resonance duct and the air cleaner assembly.

6. Remove the battery and the battery tray.

7. Disconnect the accelerator cable from the throttle cam and remove the accelerator cable bracket from the intake plenum.

8. Remove the upper radiator hose and disconnect the radiator overflow hose from the radiator filler neck.

9. Disconnect the radiator thermoswitch and cooling fan electrical connectors.

10. Remove the attaching nuts to the radiator mounting brackets and remove the brackets.

11. Disconnect the alternator, oil pressure switch, throttle position sensor, idle speed control, manual lever position switch, fuel injector wiring harness, back-up light switch, water thermoswitch, oxygen sensor, power steering pump and distributor electrical connectors.

NOTE: Mark the position of the connectors prior to removal to ease reinstallation.

12. Disconnect all engine ground straps.

13. Disconnect the ignition coil high-tension lead from the distributor.

14. Properly relieve the fuel system pressure.

15. Disconnect the fuel pressure and return lines.

16. Disconnect the heater core inlet and outlet, power brake vacuum supply, purge control vacuum and, if equipped, speed control vacuum hoses.

NOTE: Mark the position of the hoses prior to removal to ease reinstallation.

17. Raise and safely support the vehicle.

18. Remove the right upper and lower splash shields.

19. Remove the clutch slave cylinder pipe bracket from the transaxle with the hose still connected. Position the slave cylinder aside.

NOTE: Be careful not to damage the pipe or the hose.

20. Disconnect the shift control rod and the extension bar from the transaxle.

21. Remove the battery duct.

22. Remove the radiator lower hose.

23. Remove the power steering and, if equipped, air conditioning compressor drive belt.

24. Remove the power steering pump and bracket assembly with the hoses still connected. Suspend the pump with wire aside of the work area.

25. Remove the air conditioning hose routing bracket, if equipped, from the transaxle crossmember and position the air conditioning hose aside.

26. If equipped, remove the air conditioning compressor with the hoses still connected. Suspend the compressor with wire aside of the work area.

27. Disconnect the speedometer cable from the transaxle.

28. Remove the exhaust pipe front mounting flange and support bracket from the exhaust manifold.

29. Mark the location and disconnect the wires from the starter motor.

30. Remove the stabilizer bar.

31. Remove the tie rod ends from the steering knuckles.

32. Remove the halfshafts from the transaxle.

33. Remove the transaxle front and rear mount attaching nuts from the crossmember.

34. Lower the vehicle.

35. Remove the radiator, fan and shroud assembly from the vehicle.

36. Install a suitable engine removal sling onto the engine lifting brackets.

37. Place a suitable engine hoist into position and support the engine.

38. Remove the engine vibration dampener.

39. Remove the engine mount, transaxle upper mount and the transaxle support bracket.

40. Remove the engine and transaxle assembly.

41. Remove the intake plenum support bracket.

42. Remove the starter motor.

43. Remove the transaxle front mount.

44. Remove all oil pan-to-transaxle bolts and transaxle-to-engine attaching bolts from the engine block and separate the transaxle from the engine.

45. Remove the clutch assembly from the engine.

46. Install the engine onto a suitable engine stand.

To install:

47. Install a suitable engine removal sling onto the engine lifting brackets. Place a suitable engine hoist into position and install the engine sling.

48. Remove the engine from the engine stand and lower the engine with the hoist still supporting it.

49. Install the clutch assembly.

50. Install the transaxle onto the engine.

51. Install the transaxle-to-engine bolts and tighten to 47–66 ft. lbs. (64–89 Nm).

52. Install the oil pan-to-transaxle attaching bolts and tighten to 27–38 ft. lbs. (37–52 Nm).

53. Position the transaxle front mount onto the transaxle and install the attaching bolts. Tighten the bolts to 27–38 ft. lbs. (37–52 Nm).

54. Position the starter motor into the transaxle housing and install the mounting bolts. Tighten the bolts to 27–38 ft. lbs. (37–52 Nm).

55. Install the intake plenum support bracket. Tighten the bolts to 27–38 ft. lbs. (37–52 Nm) and the nut to 14–19 ft. lbs. (19–25 Nm).

56. Using the engine hoist, position the engine and transaxle assembly into the engine compartment and align the engine mounting points with the engine mount and the mounting holes in the transaxle crossmember.

57. Install the attaching nuts to the transaxle front and rear mounts and the transaxle crossmember.

58. Position the engine mount into the vehicle.

59. Install the engine mount through-bolt and nut. Tighten them to 49–69 ft. lbs. (67–93 Nm).

60. Install the engine mount-to-engine attaching nuts. Tighten the nuts to 54–76 ft. lbs. (74–103 Nm).

61. Install the engine mount vibration dampener and attaching bolt and nut. Tighten the bolt and nut to 41–59 ft. lbs. (55–80 Nm).

62. Place the clutch slave cylinder and pipe assembly into its proper mounting position.

63. Install the engine support bracket and attaching bolts. Tighten the bolts to 41–59 ft. lbs. (55–80 Nm).

64. Install the transaxle upper mount and install the attaching bolts. Tighten the bolts to 32–45 ft. lbs. (43–61 Nm).

65. Install the transaxle upper mount attaching nuts. Tighten the nuts to 49–69 ft. lbs. (67–93 Nm).

66. Place the radiator, fan and shroud assembly into its mounting position.

67. Install the radiator mounting brackets and tighten the nuts to 69–95 inch lbs. (7.8–11.0 Nm).

68. Install the upper radiator hose and connect the expansion reservoir overflow tube to the radiator filler neck.

69. Connect the cooling fan and radiator thermoswitch electrical connectors.

70. Raise and safely support the vehicle.

71. Install the lower radiator hose.

72. Install the halfshafts.

73. Install the tie rod ends into the steering knuckle.

74. Install the stabilizer bar.

75. Connect the wires to the starter motor according to their positions as marked during the removal procedure.

76. Install the exhaust front mounting flange to the exhaust manifold while making sure to install a new gasket. Tighten the flange-to-manifold attaching nuts to 23–34 ft. lbs. (31–46 Nm).

77. Install the exhaust pipe support bracket. Tighten the bracket attaching bolts to 27–38 ft. lbs. (37–52 Nm).

78. Install the speedometer cable into the transaxle.

79. If equipped, install the air conditioning compressor. Tighten the mounting bolts to 15–22 ft. lbs. (20–30 Nm).

80. Install the air conditioning routing bracket, if equipped, to the transaxle crossmember. Tighten the bolt to 56–82 inch lbs. (6.4–9.3 Nm).

81. Install the power steering pump and bracket assembly. Tighten the pump mounting bolts to 27–38 ft. lbs. (37–52 Nm).

82. Install the power steering and air conditioning, if equipped, drive belt.

83. Install the battery duct and tighten the attaching bolts to 69–95 inch lbs. (7.8–11.0 Nm).

84. Install the extension bar to the transaxle and tighten the attaching nut to 23–34 ft. lbs. (31–46 Nm).

85. Connect the shift control rod to the transaxle and tighten the attaching nut to 12–17 ft. lbs. (16–23 Nm).

86. Install the clutch slave cylinder attaching bolts and tighten to 12–17 ft. lbs. (16–23 Nm).

87. Position the slave cylinder pipe and install the routing bracket and attaching bolt. Tighten the bolt to 12–17 ft. lbs. (16–23 Nm).

88. Install the right upper and lower splash shields. Tighten the bolts to 69–95 inch lbs. (7.8–11.0 Nm).

89. Lower the vehicle.

90. Connect the heater core and vacuum hoses according to their original positions as marked during the removal procedure.

91. Connect the fuel pressure and return lines.

92. Connect the ignition coil high tension lead into the distributor.

93. Connect all engine ground straps.

94. Connect all remaining electrical connectors according to the locations marked during the removal procedure.

95. Install the accelerator cable bracket to the intake plenum and connect the accelerator cable to the throttle cam.

96. Install the battery tray and the battery.

97. Install the air cleaner assembly and the resonance duct.

98. Fill the cooling system.

99. If equipped, recharge the air conditioning system.

100. Install the hood, aligning the marks that were made during the removal procedure.

101. Connect the negative battery cable.

102. Start the engine and check for leaks. Stop the engine and check the fluid levels.

1988–90 1.9L Engine

1. Mark position of hood hinges and remove hood.

2. Relieve the fuel system pressure. Remove air cleaner, air intake duct and heat tube.

3. Disconnect negative battery cable.

4. Drain the cooling system. Remove the secondary wire from the ignition coil.

5. Remove alternator air intake tube, if equipped. Remove the alternator drive belt. Remove alternator mounting bolts and lay alternator aside.

6. Disconnect and remove thermactor air pump, if equipped.

7. Disconnect radiator hoses and disconnect oil cooler lines, if equipped with automatic transaxle.

8. Remove radiator cooling fan and shroud as an assembly.

9. Remove the transaxle cooler line routing clip located at the radiator, if equipped with automatic transaxle. Remove the radiator and disconnect the heater at the metal tube.

10. Identify, tag and disconnect heater hoses, electrical connections and vacuum hoses as necessary. Disconnect the fuel pump supply and return lines. If equipped with power assist brakes, disconnect the power boost vacuum hose at the engine.

11. Disconnect kickdown rod at fuel charging assembly, if equipped with automatic transaxle.

12. Disconnect accelerator cable at the fuel charging assembly and remove the cable routing bracket attaching screws. Disconnect the vapor hose at the carbon canister tube.

13. Raise and safely support the vehicle.

14. Remove the clamp from the heater supply and return hose. Remove knee brace at front of starter motor and remove battery cable from starter. Remove the starter motor.

15. Disconnect exhaust inlet pipe at manifold.

16. Remove support bracket in front of converter cover, if equipped with automatic transaxle and remove cover. Remove inspection cover from manual transaxle.

17. Remove crankshaft pulley and damper.

18. Remove torque converter to flywheel nuts, if equipped with automatic transaxle.

19. Remove timing belt cover lower attaching bolts, if equipped with manual transaxle.

20. Remove converter housing, if equipped with automatic transaxle, or flywheel housing, if equipped with manual transaxle, attaching bolts.

21. Remove 2 oil pan-to-transaxle attaching bolts. Disconnect coolant bypass hose from intake manifold. Remove the bolt attaching the battery negative cable to the cylinder block.

22. Remove nut and bolt attaching insulator bracket to the engine bracket at front of engine.

23. Lower vehicle.

24. Install suitable lifting brackets on engine.

25. Use a suitable lifting device connected to the engine lifting brackets and raise engine just enough to remove the through bolt from the front engine insulator and remove insulator.

26. Remove the remaining timing belt cover bolts and remove the cover, if equipped with manual transaxle.

27. Remove insulator attaching bracket from engine.

28. Position a jack under the transaxle. Raise jack just enough to support the weight of the transaxle.

29. Remove the converter housing or flywheel housing upper attaching bolts.

30. Remove engine assembly from vehicle.

To install:

31. Carefully lower engine into the vehicle using a suitable lifting device.

32. Join the engine and the transaxle, making sure the alignment dowels on the back of the engine engage the transaxle housing.

NOTE: If equipped with an manual transaxle, make sure the transaxle input shaft engages the clutch disc. If equipped with an automatic transaxle, make sure the torque converter studs engage the flywheel.

33. Install the converter housing if automatic transaxle, flywheel housing if manual transaxle, upper attaching bolts.

34. Install 2 oil pan to transaxle attaching bolts and tighten to 30–40 ft. lbs. (40–54 Nm). Then loosen bolts ½ turn.

35. Remove jack from under the transaxle.

36. Position engine insulator attaching casting on the engine and install the attaching bolt and nut.

37. Attach the nuts to the insulator to casting.

38. Remove lifting device and the lifting brackets.
39. Connect electrical connectors, vacuum hoses and carbon canister vapor hose.
40. If equipped with an automatic transaxle, connect the kickdown rod.
41. Connect the heater hoses.
42. Connect the fuel supply and return lines at the throttle body.
43. If equipped with power brakes, connect the vacuum hose to the power booster.
44. Position the accelerator cable routing bracket and install the attaching bolts.
45. Connect the accelerator cable to the fuel charging assembly.
46. Connect the coolant bypass hose, if equipped with an manual transaxle.
47. Install radiator.
48. Install battery negative cable to cylinder block attaching bolt.
49. Attach the lower cooler line, if equipped with an automatic transaxle.
50. Connect the radiator lower hose.
51. Attach the upper cooler line, if equipped with an automatic transaxle.
52. Install radiator cooling fan and shroud assembly. Connect the cooling fan electrical connector.
53. Connect radiator upper hose.
54. Raise and safely support the vehicle.
55. Tighten the casting-to-insulator attaching nuts.
56. Install the torque converter-to-flywheel attaching bolts.
57. Install the crankshaft damper.
58. Install lower attaching bolts to converter housing, if equipped with automatic transaxle; flywheel housing if equipped with manual transaxle.
59. Install support bracket.
60. Install starter motor and connect battery cable.
61. Install starter brace at the front of the starter motor.
62. Connect the exhaust inlet pipe.
63. Install the cooler line routing bracket, if equipped with automatic transaxle.

64. Lower the vehicle.
65. Install the timing belt cover, if equipped with manual transaxle.
66. Install alternator and drive belt.
67. Connect negative battery cable.
68. Install alternator air intake tube.

69. Fill cooling system, overflow bottle and bleed the cooling system.
70. Fill the crankcase to the proper level.
71. Install the hood.

72. Start engine and check for coolant, oil and fuel leaks.
73. Install air cleaner assembly with the intake duct and heat tube and connect remaining vacuum hoses.

1991–92 1.9L Engine
WITH AUTOMATIC TRANSAXLE

On automatic transaxle vehicles, the 1.9L engine assembly is removed without the transaxle attached. The engine is lifted from the engine compartment with the transaxle assembly remaining in the vehicle, attached to the mounts.

1. Mark the position of the hood hinges and remove the hood.
2. Disconnect the negative battery cable.
3. Drain the cooling system.
4. Remove the air intake duct.
5. Remove the crankcase ventilation hose from the valve cover and the vacuum hose from the bottom side of the throttle body.
6. Disconnect the power brake booster supply hose.
7. Disconnect the following electrical connectors:
   a. Fuel charging harness, located at the right shock tower.
   b. Alternator harness, from the back side of the alternator.
   c. Oxygen sensor.
   d. Ignition coil.
   e. Radio suppressor, mounted on the coil bracket.
   f. Coolant temperature sensor, cooling fan sensor and temperature gauge sending unit, mounted on a common water tube near the thermostat housing.
   g. Radiator cooling fan.

NOTE: Mark the position of the electrical connectors to aid reinstallation.

8. Remove the idle air bypass valve.
9. Remove the ground strap from the stud on the left side of the cylinder head near the ignition coil.
10. Disconnect the accelerator cable and the transaxle kickdown cable from the throttle lever. Remove the cable bracket from the intake manifold and position aside.
11. Disconnect both heater hoses at the engine compartment bulkhead.
12. Properly relieve the fuel system pressure and disconnect the fuel supply and return hoses at the fuel supply manifold.
13. Remove the upper radiator hose.
14. Raise and safely support the vehicle.
15. Remove the right side and the right and left front splash shields.
16. Remove the lower radiator hose from the radiator.
17. Position a drain pan under the radiator and remove the lower transaxle oil cooler line.
18. Remove the 2 oil cooler line retaining bracket bolts from the bottom of the radiator.

19. Remove the radiator fan shroud lower mounting bolts.
20. Lower the vehicle.
21. Remove the radiator fan shroud upper mounting bolts and remove the fan and shroud assembly from the vehicle.
22. Remove the upper transaxle oil cooler line from the radiator and remove the radiator from the vehicle.
23. If equipped with air conditioning, properly discharge the system.
24. Disconnect the air conditioning suction line at the suction accumulator/drier. Plug or cap the openings to prevent the entrance of dirt and moisture.
25. Remove the accessory drive belt.
26. Remove the power steering return hose from the pump reservoir and the high-pressure hose from the power steering pump.
27. Remove the power steering and air conditioner line retainer bracket bolts from the alternator bracket. Position the hoses aside.
28. Remove the accessory drive belt automatic tensioner assembly.
29. Raise and safely support the vehicle.
30. Remove the drive belt idler pulley.
31. If equipped, remove the 4 air conditioning compressor mounting bolts. Remove the compressor assembly with the lines attached and position aside. Safety wire the compressor to the vehicle sub-frame.
32. Remove the catalytic converter inlet pipe.
33. Remove the transaxle kickdown cable support bracket from the back side of the engine block. Position the cable and the bracket aside.
34. Disconnect the oil pressure switch.
35. Disconnect the relay wire and the positive battery cable from the starter.
36. Remove the flywheel inspection shield.
37. Remove the 4 torque converter attaching nuts.
38. Remove the crankshaft dampener.
39. Remove the 5 engine-to-transaxle bolts.
40. Lower the vehicle.
41. Remove the 3 starter motor mounting bolts and remove the starter out of the top of the engine compartment.
42. Remove the 2 transaxle-to-engine mounting bolts.
43. Connect an engine removal sling to suitable engine lifting brackets. Position a suitable engine hoist and suport the engine.
44. Remove the right engine mount dampener and mount assembly.
45. With the engine assembly supported by the engine hoist, carefully

separate the assembly from the transaxle.

46. Lift the engine assembly out of the vehicle.

47. Install the engine onto a suitable engine stand.

To install:

48. Attach the engine removal sling to the engine lifting brackets and remove the engine from the stand with the engine hoist.

49. Carefully lower the engine into the vehicle and join the engine to the transaxle. Make sure the torque converter studs correctly engage the flywheel and the alignment dowels engage the transaxle housing.

50. Install the 2 transaxle-to-engine bolts, but do not fully tighten them at this time.

51. Install the right engine mount insulator and dampener.

52. Position the engine hoist aside and remove the sling from the engine lifting brackets.

53. Raise and safely support the vehicle.

54. Install the 5 engine-to-transaxle bolts, but do not fully tighten them at this time.

55. Install the crankshaft dampener and tighten the attaching bolt to 81–96 ft. lbs. (110–130 Nm).

56. Install the 4 torque converter attaching nuts and tighten to 25–36 ft. lbs. (34–49 Nm). Install the flywheel inspection plate.

57. Connect the oil pressure switch.

58. Install the kickdown cable support bracket.

59. If equipped, position the air conditioning compressor on the bracket and install the 4 mounting bolts. Tighten the bolts to 15–22 ft. lbs. (20–30 Nm).

60. Install the catalytic converter inlet pipe.

61. Lower the vehicle.

62. From above, position the starter motor and install the 3 mounting bolts. Connect the positive battery cable and the relay wire to the starter.

63. Tighten the 2 transaxle-to-engine bolts to 40–59 ft. lbs. (55–80 Nm).

64. Install the power steering high-pressure hose on the pump.

65. Install the accessory drive belt idler pulley and automatic tensioner.

66. Install the power steering return hose on the pump reservoir.

67. Install the power steering hose retainer bracket on the alternator bracket.

68. Install the accessory drive belt.

69. If equipped, connect the air conditioner suction line to the accumulator.

70. Install the radiator assembly.

71. Connect the upper transaxle oil cooler line at the radiator.

72. Position the cooling fan and shroud assembly and install the upper mounting bolts.

73. Raise and safely support the vehicle.

74. Install the lower shroud bolts and connect the lower transaxle oil cooler line.

75. Install the oil cooler line retaining bracket bolts.

76. Install the lower radiator hose.

77. Tighten the 5 engine-to-transaxle bolts to 27–38 ft. lbs. (37–52 Nm).

78. Install the left and right front splash shields and the right side splash shield.

79. Lower the vehicle.

80. Install the upper radiator hose.

81. Connect both heater hoses at the engine compartment bulkhead.

82. Install the accelerator cable bracket and attach the accelerator and kickdown cables to the throttle lever.

83. Install the idle air bypass valve.

84. Install the gound strap on the stud at the front left side of the cylinder head, near the ignition coil.

85. Connect the remaining electrical connectors according to the positions marked during the removal procedure.

86. Connect the fuel supply and return lines. Be sure to install the fuel line safety clips.

87. Connect the power brake suply hose, the vacuum hose on the bottom side of the throttle body, and the crankcase ventilation hose to the valve cover.

88. Install the air intake duct.

89. Connect the negative battery cable.

90. Fill the cooling system.

91. Install the hood, aligning the marks that were made during the removal procedure.

92. Start the engine and check for leaks. Stop the engine and check the fluid levels.

93. If equipped, evacuate and recharge the air conditioning system according to the proper procedure.

WITH MANUAL TRANSAXLE

On manual transaxle vehicles, the 1.9L engine assembly is removed with the transaxle attached. The engine is lifted out of the engine compartment.

1. Mark the position of the hood hinges and remove the hood.

2. Disconnect the battery cables and remove the battery and the battery tray.

3. Drain the cooling system.

4. Remove the air cleaner.

5. Disconnect the crankcase ventilation hose from the valve cover and the vacuum hose from the bottom side of the throttle body.

6. Remove the power brake supply hose.

7. Disconnect the following electrical connectors:

 a. Fuel charging harness, located at the right shock tower.

 b. Alternator harness, from the back side of the alternator.

 c. Oxygen sensor.

 d. Ignition coil.

 e. Radio suppressor, mounted on the coil bracket.

 f. Coolant temperature sensor, cooling fan sensor and temperature gauge sending unit, mounted on a common water tube near the thermostat housing.

 g. Radiator cooling fan.

NOTE: Mark the position of the electrical connectors to aid reinstallation.

8. Remove the idle air bypass valve.

9. Remove the ground strap from the stud on the left side of the cylinder head near the ignition coil.

10. Disconnect the accelerator cable from the throttle lever. Remove the cable bracket from the intake manifold and position aside.

11. Disconnect both heater hoses at the engine compartment bulkhead.

12. Properly relieve the fuel system pressure and disconnect the fuel supply and return hoses at the fuel supply manifold.

13. Remove the upper radiator hose.

14. If equipped, properly discharge the air conditioning system and disconnect the suction line at the accumulator.

15. Remove the accessory drive belt and the automatic tensioner and idler pulley.

16. Disconnect the power steering return hose from the pump reservoir and the high pressure hose from the pump.

17. Remove the power steering hose and air conditioning line retainer brackets from the alternator bracket.

18. Raise and safely support the vehicle.

19. Remove the right and left side and front splash shields.

20. Disconnect the lower radiator hose from the radiator and remove the radiator fan shroud lower mounting bolts.

21. If equipped, remove the 4 air conditioning compressor mounting bolts. Remove the compressor assembly with the lines attached and position aside. Safety wire the compressor to the vehicle sub-frame.

22. Remove the catalytic converter inlet pipe.

23. Disconnect the oil pressure switch.

24. Disconnect the relay wire and the positive battery cable at the starter.

25. Remove the transaxle extension bar and shift control rod.

26. Remove the crankshaft dampener.

27. Remove the front wheel and tire assemblies.

28. Remove both halfshaft assemblies.

29. Install suitable transaxle plugs into the differential side gears.

NOTE: Failure to install the transaxle plugs may allow the differential side gears to move out of position.

30. Disconnect the speedometer cable and the neutral switch on the transaxle.

31. Remove the clutch slave cylinder and line as an assembly from the transaxle and set it aside.

32. Remove the transaxle front and rear mount bolts.

33. Lower the vehicle.

34. Remove the radiator fan shroud upper mounting bolts and remove the fan shroud assembly from the vehicle.

35. Connect a suitable engine removal sling to the engine lifting brackets. Connect the sling to a suitable engine hoist, position the hoist and support the engine.

36. Remove the right engine mount dampener and mount assembly.

37. Remove the transaxle upper mount.

38. Lift the engine and transaxle assembly out of the vehicle and set it down on the floor.

To install:

39. Carefully lower the engine and transaxle assembly into the vehicle with the engine hoist.

40. Position the transaxle on its mounts and install the transaxle upper mount.

41. Install the right engine mount and mount dampener.

42. Remove the engine removal sling and the hoist.

43. Position the fan shroud assembly and install the upper mounting bolts.

44. Raise and safely support the vehicle.

45. Install the front and rear transaxle mount bolts.

46. Install the clutch slave cylinder and line assembly.

47. Connect the neutral switch and the speedometer cable.

48. Remove the transaxle plugs and install the halfshaft assemblies.

49. Install the crankshaft dampener and tighten the bolt to 81–96 ft. lbs. (110–130 Nm).

50. Install the transaxle extension bar and shift control rod.

51. Connect the relay wire and the positive battery cable to the starter.

52. Connect the oil pressure switch.

53. Install the catalytic converter inlet pipe.

54. If equipped, position the air conditioning compressor on its bracket and install the 4 mounting bolts.

55. Install the radiator fan shroud lower mounting bolts and install the lower radiator hose.

56. Install the left and right side and front splash shields.

57. Lower the vehicle.

58. Install the power steering hoses and install the power steering hose and air conditioner line retainer brackets.

59. Install the accessory drive belt idler pulley and automatic tensioner and install the accessory drive belt.

60. If equipped, connect the air conditioner suction line.

61. Install the upper radiator hose.

62. Connect the fuel supply and return hoses to the fuel supply manifold.

63. Connect both heater hoses.

64. Install the accelerator cable bracket on the intake manifold and connect the cable to the throttle lever.

65. Install the ground strap on the stud at the front left side of the cylinder head.

66. Install the idle air bypass valve.

67. Connect the remaining electrical connectors according to the positions marked during the removal procedure.

68. Connect the power brake supply hose, the crankcase ventilation hose and the vacuum line at the bottom of the throttle body.

69. Install the air cleaner assembly.

70. Install the battery tray and the battery. Connect the battery cables.

71. Fill the cooling system.

72. Install the hood, aligning the marks that were made during the removal procedure.

73. Start the engine and check for leaks. Stop the engine and check the fluid levels.

74. If equipped, evacuate and recharge the air conditioning system according to the proper procedure.

2.3L Engine

NOTE: This procedure describes the removal and installation of the engine and transaxle as an assembly.

1. Mark position of hood hinges and remove hood.

2. Remove negative ground cable from battery.

3. Relieve the fuel system pressure. Remove air cleaner.

4. Remove lower radiator hose to drain engine coolant.

5. Remove upper radiator hose and disconnect transaxle cooler lines at rubber hoses below radiator, if equipped with automatic transaxle.

6. Remove coil and disconnect coolant fan at electrical connection.

7. Remove radiator shroud and cooling fan as an assembly. Remove radiator.

8. Discharge air conditioning system, if equipped and remove pressure and suction lines from compressor.

CAUTION

Use extreme care when discharging air conditioning system, as the refrigerant is under high pressure and may cause personal injury.

---

9. Identify, tag and disconnect all electrical and vacuum lines as necessary.

10. Disconnect TV linkage or clutch cable at transaxle.

11. Disconnect accelerator linkage and fuel lines.

12. Remove coil and brackets assembly.

13. Disconnect power steering lines at pump and bracket at the cylinder head, if equipped.

14. Install 2 engine lifting eyes and install engine support tool to engine lifting eyes.

15. Raise and safely support the vehicle.

16. Remove battery cable from starter and remove hose from catalytic converter.

17. Remove bolt attaching exhaust pipe bracket to oil pan and 2 exhaust pipe to manifold attaching nuts.

18. Remove exhaust inlet pipe-to-exhaust manifold retaining nuts, pull exhaust system out of rubber insulating grommets and set aside.

19. Remove speedometer cable from transaxle.

20. Remove one heater hose from water pump inlet tube and the other from the steel tube on intake manifold.

21. Remove water pump inlet tube clamp bolt at engine block and clamp bolts at underside of oil pan. Remove inlet tube.

22. Remove bolts attaching control arms to body. Remove stabilizer bar brackets retaining bolts and remove brackets.

23. Halfshaft assemblies must be removed from transaxle at this time.

24. On manual transaxle equipped vehicles, remove roll restrictor nuts from transaxle. Pull roll restrictor from mounting bracket.

25. On manual transaxle equipped vehicles, remove shift stabilizer bar to transaxle attaching bolts. Remove shift mechanism to shift shaft attaching nut and bolt at transaxle.

26. On automatic transaxle equipped, disconnect manual shift cable clip from lever on transaxle. Remove manual shift linkage bracket bolts from transaxle and remove bracket.

27. Remove the left rear insulator mount bracket from body bracket.

28. Remove the left front insulator to transaxle mounting bolts.

29. Lower the vehicle. Install lifting equipment to the 2 lifting eyes on engine.

NOTE: Do not allow front wheels to touch floor.

30. Remove engine support tool.

31. Remove right No. 3A insulator intermediate bracket-to-engine bracket bolts, intermediate bracket-to-insulator attaching nuts and the nut on the bottom of the double ended stud which attaches the intermediate bracket-to-engine bracket. Remove bracket.

32. Carefully lower engine and transaxle assembly to the floor.

To install:

33. Raise and safely support the vehicle.

34. Position engine and transaxle assembly directly below engine compartment.

35. Slowly lower vehicle over engine and transaxle assembly.

NOTE: Do not allow the front wheels to touch the floor.

36. Install lifting equipment to both existing engine lifting eyes on engine.

37. Raise engine and transaxle assembly up through engine compartment and position accordingly.

38. Install right side No. 3A insulator intermediate attaching nuts to intermediate bracket. Tighten to 55–75 ft. lbs. (75–100 Nm). Attach intermediate bracket to engine bracket bolts. Tighten to 52–70 ft. lbs. (70–95 Nm). Install nut on bottom of double-ended stud that attaches the intermediate bracket-to-engine bracket. Tighten to 60–90 ft. lbs. (80–120 Nm).

39. Install engine support tool to engine lifting eye.

40. Remove lifting equipment.

41. Raise and safely support the vehicle.

42. Position transaxle jack under engine. Raise engine and transaxle assembly into mounted position.

43. Install insulator-to-bracket nut. Tighten to 45–65 ft. lbs. (61–68 Nm).

44. If equipped with manual transaxle, position roll restrictor onto starter studs. Install nuts attaching roll restrictor to transaxle. Tighten to 25–39 ft. lbs. (35–50 Nm).

45. Install starter cable to starter.

46. Install lower radiator hose and install retaining bracket and bolt.

47. If equipped with manual transaxle, install shift stabilizer bar-to-transaxle attaching bolt. Tighten to 23–35 ft. lbs. (31–47 Nm).

48. If equipped with manual transaxle, install shift mechanism-to-input shift shaft (on transaxle) bolt and nut. Tighten to 7–10 ft. lbs. (9–13 Nm).

49. If equipped with automatic transaxle, install manual shift linkage bracket bolts to transaxle. Install cable clip to lever on transaxle.

50. Install lower radiator hose to radiator.

51. Install speedometer cable to transaxle.

52. Position exhaust system up and into insulating rubber grommets located at rear of vehicle.

53. Install exhaust pipe-to-exhaust manifold studs. Install exhaust pipe bracket-to-oil pan bolt.

54. Connect pulse air hose to catalytic converter.

55. Place stabilizer bar and control arm assembly into position. Install control arm-to-body attaching bolts. Install stabilizer bar brackets and tighten all fasteners.

56. Halfshaft assemblies must be installed at this time.

57. Lower vehicle.

58. Remove engine support tool.

59. Connect any remaining electrical and vacuum lines.

60. Install heater hose.

61. Install air conditioning discharge and suction lines to compressor, if equipped. Do not charge at this time.

62. Connect fuel supply and return lines to engine.

63. Connect accelerator cable.

64. Install power steering pressure and return lines.

65. If equipped with automatic transaxle, connect TV linkage at transaxle.

66. If equipped with manual transaxle, connect clutch cable to shift lever on transaxle. Check clutch adjustment.

67. Install radiator shroud and coolant fan assembly.

68. If equipped with automatic transaxle, connect transaxle cooler lines to rubber hoses below radiator.

69. Fill cooling system.

70. Connect battery ground cable.

71. Install air cleaner assembly.

72. Install hood.

73. Charge air conditioning system, if equipped.

74. Check all fluid levels.

75. Start engine. Check for leaks.

Engine Mounts

REMOVAL & INSTALLATION

Tempo, Topaz and 1988–90 Escort

RIGHT ENGINE INSULATOR (NO. 3A)

1. Disconnect the negative battery cable. Place a floor jack and a block of wood under the engine oil pan. Raise the engine approximately ½ in. or enough to take the load off of the insulator.

2. Remove the lower support bracket attaching nut, bottom of the double ended stud. Remove the insulator-to-support bracket attaching nuts. Do not remove the nut on top of the double ended stud.

3. Remove the insulator support bracket from the vehicle. Remove the insulator attaching nuts through the right side front wheel opening.

4. Remove the insulator attaching bolts through the engine compartment. Work the insulator out of the body and remove it from the vehicle.

To install:

5. Work insulator into the body opening.

6. Position the insulator and install the attaching nuts and bolts. Tighten the nuts to 75–100 ft. lbs. (100–135 Nm) and tighten the bolts to 37–55 ft. lbs. (50–75 Nm).

7. Install insulator support casting on top of the insulator and engine support bracket. Make sure the double-edged stud is through the hole in the engine bracket.

8. Tighten the insulator support casting-to-insulator attaching nuts to 55–75 ft. lbs. (75–100 Nm). Install and tighten lower support bracket nut to 60–90 ft. lbs. (80–120 Nm).

9. Install the insulator casting-to-engine bracket bolt and tighten to 60–90 ft. lbs. (80–120 Nm).

10. Lower engine. Connect negative battery cable.

LEFT REAR ENGINE INSULATOR (NO. 4)

1. Disconnect the negative battery cable. Raise the vehicle and support safely. Place a transaxle jack and a block of wood under the transaxle.

2. Raise the transaxle approximately ½ in. or enough to take the load off of the insulator.

3. Remove the insulator attaching nuts from the support bracket. Remove the 2 through bolts and remove the insulator from the transaxle.

To install:

4. Install the insulator over the left rear transaxle housing and support bracket studs.

5. Install the 2 insulator through bolts and tighten to 30–45 ft. lbs. (41–61 Nm).

6. Install 2 insulator-to-support bracket attaching nuts. Tighten to 80–100 ft. lbs. (108–136 Nm).

7. Lower vehicle and remove floor jack. Connect negative battery cable.

NOTE: To remove the left rear support bracket, remove the left rear engine insulator No. 4. Then remove the support bracket attaching bolts. When installing the support bracket, torque the attaching bolts to 45–65 ft. lbs. (61–88 Nm).

LEFT FRONT ENGINE INSULATOR (NO. 1)

1. Disconnect the negative battery cable. Raise and the vehicle and support safely. Place a transaxle jack and a block of wood under the transaxle. Raise the transaxle approximately ½ in. or enough to take the load off of the insulator.

2. Remove the insulator-to-support bracket attaching nut. Remove the insulators and transaxle attaching bolts and remove the insulator from the vehicle.

3. Complete the installation of the insulator by reversing the removal procedure. Torque the insulator to transaxle attaching bolts to 25–37 ft. lbs. (35–50 Nm). Torque the insulator-to-support bracket nut to 80–100 ft. lbs. (108–136 Nm).

Cylinder Head

REMOVAL & INSTALLATION

1.8L Engine

1. Properly relieve the fuel system pressure.

2. Disconnect the negative battery cable.

3. Drain the cooling system.

4. Remove the bolts from the timing belt upper and middle covers. Remove the covers and gaskets.

5. Rotate the crankshaft by hand in the direction of normal engine rotation and align the timing marks located on the camshaft pulleys and seal plate.

6. Loosen the timing belt tensioner lock bolt and Temporarily secure the tensioner spring in the fully extended position.

7. Remove the timing belt from the camshaft pulleys and secure it aside to prevent damage during the removal and installation of the cylinder head.

NOTE: Do not allow the timing belt to become contaminated by oil or grease.

8. Tag and disconnect the vacuum hoses from the cylinder head cover.

9. Tag and disconnect the spark plug wires from the spark plugs and position the wires aside.

10. Remove the cylinder head cover and gasket.

11. Remove the air duct from the resonance chamber and throttle body.

12. Disconnect the accelerator cable and, if equipped with automatic transaxle, the kickdown cable from the throttle cam. Remove the cable bracket from the intake plenum.

13. Tag and disconnect all vacuum lines from the intake plenum.

14. Tag and disconnect all necessary electrical connectors from the cylinder head, exhaust manifold, intake plenum, and throttle body. Disconnect the ground straps.

15. Remove the upper radiator hose.

16. Remove the transaxle-to-engine block upper-right bolt.

17. Disconnect the fuel pressure and return lines and plug the lines.

18. Disconnect the ignition coil high-tension lead from the distributor.

19. Tag and disconnect the necessary hoses connected to the cylinder head and intake plenum.

20. Remove the 2 bolts from the transaxle vent tube routing brackets.

21. Raise and safely support the vehicle.

22. Remove the bolt from the water pump-to-cylinder head hose bracket.

23. Remove the exhaust front mounting flange and exhaust pipe support bracket from the exhaust manifold.

24. Remove the intake plenum support bracket.

25. Lower the vehicle.

26. Remove the cylinder head bolts in the proper sequence.

27. Remove the cylinder head assembly, with the intake plenum and exhaust manifold attached, from the vehicle.

28. Remove the intake plenum and exhaust manifold.

29. Inspect the cylinder head for damage, cracks, and leakage of water and oil. Measure the cylinder head for warpage in 6 directions. The maximum distortion allowable is 0.004 in. (0.10mm).

30. If the cylinder head distortion exceeds specification, machine the cylinder head surface. The cylinder head must be replaced if the cylinder head height is not within 5.268–5.276 in. (133.8–134.0mm).

31. Inspect the manifold contact surface distortion in 4 directions. The maximum distortion allowable is 0.006 in. (0.15mm). If the distortion exceeds specification, machine the manifold contact surface or replace the cylinder head, as necessary.

To install:

32. Remove all dirt, oil and old gasket material from all gasket contact surfaces.

33. Install the intake plenum and exhaust manifold.

34. Install a new head gasket onto the top of the engine block, using the dowel pins for reference.

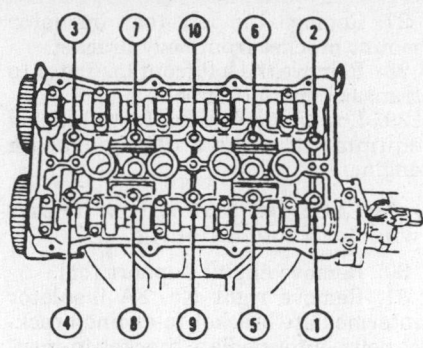

Cylinder head bolt removal sequence— 1.8L engine

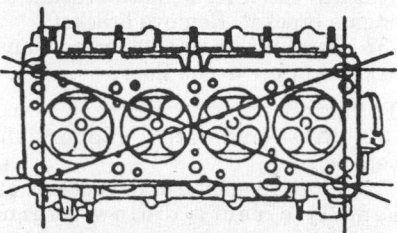

Cylinder head warpage measuring locations—1.8L engine

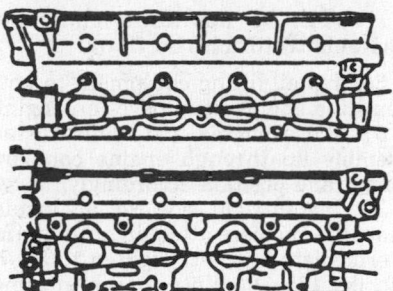

Manifold contact surface warpage measuring location—1.8L engine

35. Place the cylinder head into its monting position on top of the engine block.

36. Lubricate the cylinder head bolts with engine oil and install them finger-tight. Tighten the bolts in the proper sequence to 56–60 ft. lbs. (76–81 Nm).

37. Install the 2 bolts to the transaxle vent tube routing brackets.

38. Connect the heater hoses to the cylinder head and install the clamps.

39. Connect the ignition coil high-tension lead to the distributor.

40. Connect the fuel pressure and return lines to the fuel supply manifold and install the safety clips.

41. Install the transaxle-to-engine block upper-right bolt. If equipped with manual transaxle, tighten the bolt to 47–66 ft. lbs. (64–89 Nm). If equipped with automatic transaxle, tighten the bolt to 41–59 ft. lbs. (55–80 Nm).

42. Install the upper radiator hose and clamps.

43. Connect the ground straps and connect the electrical connectors that

were disconnected at the cylinder head, exhaust manifold, intake plenum, and throttle body.

44. Connect the vacuum lines to the intake plenum.

45. Install the accelerator and kickdown cable bracket onto the intake plenum and tighten the bolts to 69–95 inch lbs. (7.8–11.0 Nm). Connect the cable(s) to the throttle cam.

46. Install the cylinder head cover and gasket, then connect the hose running from the plenum to the cylinder head cover. Tighten the cover bolts to 43–78 inch lbs. (4.9–8.8 Nm).

47. Install the air duct to the resonance chamber and throttle body and tighten the clamps. Connect the hose going from the air duct to the cylinder head cover.

48. Install and connect the spark plug wires.

49. Raise and safely support the vehicle.

50. Install the intake plenum support bracket. Tighten the bolts to 27–38 ft. lbs. (37–52 Nm) and the nut to 14–19 ft. lbs. (19–25 Nm).

51. Install the bolt to the water pump-to-cylinder head hose bracket.

52. Install the exhaust front mounting flange with a new gasket to the exhaust manifold. Tighten the flange-to-manifold attaching nuts to 23–34 ft. lbs. (31–46 Nm).

53. Install the exhaust pipe support bracket. Tighten the bracket attaching bolts to 27–38 ft. lbs. (37–52 Nm).

54. Make sure the yellow ignition timing mark on the crankshaft pulley is aligned with the TDC mark on the timing belt cover.

55. Lower the vehicle.

56. Make sure the timing marks on the camshaft pulleys and seal plate are aligned. Install the timing belt so there is no looseness at the idler pulley side or between the 2 camshaft pulleys.

NOTE: Do not turn the crankshaft counterclockwise.

57. Turn the crankshaft 2 turns clockwise by hand and verify that the yellow ignition timing mark on the crankshaft pulley is aligned with the timing mark on the timing belt cover. Verify that the timing marks on the camshaft pulley and seal plate are aligned.

NOTE: If the timing marks are not aligned, remove the timing belt and repeat the procedure beginning with Step 54.

58. Turn the crankshaft $1^5/_6$ turns clockwise by hand and align the 4th tooth to the right of the **I** and **E** timing marks on the camshaft pulleys with the seal plate alignment marks.

59. Loosen the timing belt tensioner

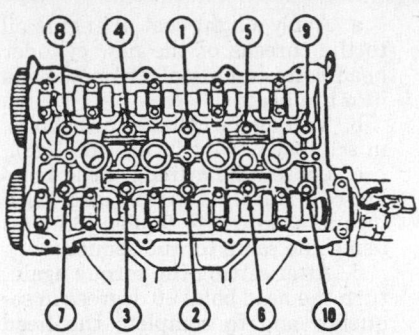

Cylinder head bolt torque sequence— 1.8L engine

lock bolt and apply tension to the timing belt. Tighten the tensioner lock bolt to 27–38 ft. lbs. (37–52 Nm).

60. Turn the crankshaft $2^1/_6$ turns clockwise and verify that the timing marks on the camshaft pulleys and the seal plate are aligned.

61. Install new gaskets onto the timing belt upper and middle covers and install the covers. Tighten the mounting bolts to 69–95 inch lbs. (8–11 Nm).

62. Fill the cooling system.

63. Connect the negative battery cable.

64. Start the engine and check for leaks.

1.9L Engine

1988–90

NOTE: The engine must be cold before removing the cylinder head, to reduce the possibility of warpage or distortion.

1. Disconnect the negative battery cable. Properly relieve the fuel system pressure.

2. Drain the cooling system and disconnect the heater hose at the fitting located under the intake manifold.

3. Disconnect the radiator upper hose at the cylinder head.

4. Disconnect the wiring terminal from the cooling fan switch.

5. Remove the air cleaner assembly.

6. Remove the PCV hose.

7. Identify, tag and disconnect the required vacuum hoses.

8. Remove the rocker arm cover.

9. Disconnect all accessory drive belts.

10. Remove the crankshaft pulley using the proper puller tool.

11. Remove the timing belt cover.

12. Set the engine on No. 1 cylinder to TDC prior to removing the timing belt.

13. Remove the distributor cap and spark plug wires as an assembly.

14. Loosen both belt tensioner attaching bolts using torque wrench adapter tool T81P-6254-A or equivalent.

15. Secure the belt tensioner as far left as possible.

16. Remove the timing belt.

17. Disconnect the EGR tube at the EGR valve.

18. Disconnect the fuel supply and return lines at the metal connectors, located on the right side of the engine, set rubber lines aside.

19. Disconnect the accelerator cable and, if equipped, the speed control cable.

20. Disconnect the alternator wiring harness.

21. Remove the alternator and its mounting bracket.

22. Raise and safely support the vehicle.

23. Disconnect the exhaust system at the exhaust pipe.

24. Lower the vehicle.

25. Remove the cylinder head bolts and washers. Discard the bolts.

NOTE: Do not reuse the cylinder head retaining bolts. Use new bolts when installing head.

26. Remove the cylinder head with the exhaust and intake manifolds attached. Discard the cylinder head gasket.

NOTE: Do not lay the cylinder head flat. Damage to the spark plug or gasket contact surfaces may result.

To install:

27. Clean all gasket material from the mating surfaces on the cylinder head and block and clean out the head bolt holes in the block.

28. Before final installation of the cylinder head to the engine, check the piston squish height as follows:

NOTE: Squish height is the clearance of the piston dome to the combustion chamber at piston TDC. No cylinder block deck machining or use of replacement crankshaft, piston or connecting rod causing the assembled squish height to be over or under tolerance specification, is permitted. If no parts other than the head gasket are replaced, the squish height should be within specification. If parts other than the head gasket are replaced, check the squish height. If the squish height is out of specification, replace the parts again and recheck the squish height.

a. Place a small amount of soft lead solder or shot of an appropriate thickness on the piston spherical areas shown.

b. Rotate the crankshaft to lower the piston in the bore and install the head gasket and cylinder head.

NOTE: A compressed (used) head gasket is preferred.

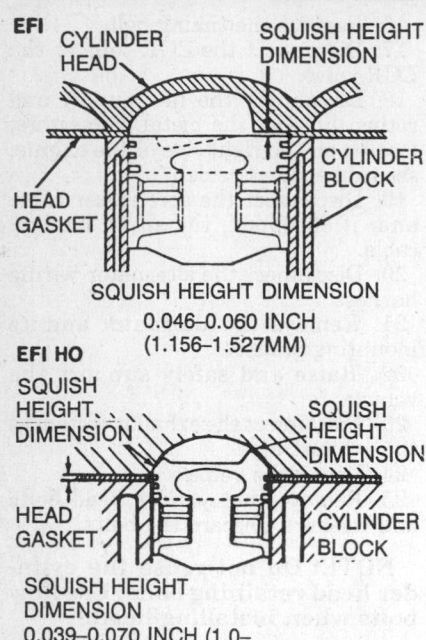

EFI

SQUISH HEIGHT DIMENSION
0.046–0.060 INCH
(1.156–1.527MM)

EFI HO

SQUISH HEIGHT DIMENSION
0.039–0.070 INCH (1.0–1.77MM)

Piston squish height—1.9L engine

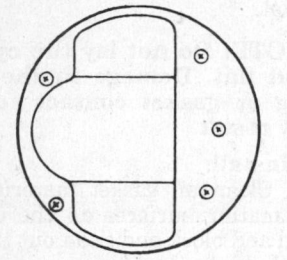

Solder locations to measure piston squish height

c. Install used head bolts and tighten the head bolts to 30–44 ft. lbs. (40–60 Nm) following proper sequence.

d. Rotate the crankshaft to move the piston through its TDC position.

e. Remove the cylinder head and measure the thickness of the compressed solder to determine squish height at TDC. The solder should be 0.039–0.070 in. (1.0–1.77mm) for EFI HO engines or 0.046–0.060 in. (1.156–1.527mm) for EFI engine.

29. If the camshaft has been turned or removed or if installing a replacement cylinder head, rotate the camshaft until the camshaft gear pointer is aligned with the timing mark on the cylinder head and the camshaft keyway is at the 6 o'clock.

30. Position the No. 1 piston 90 degrees BTDC, pulley keyway at 9 o'clock position, during the cylinder head installation.

31. Position the cylinder head gasket on the cylinder block.

32. Install the cylinder head and install new bolts and washers in the following order:

a. Apply a light coat of engine oil to the threads of the new cylinder head bolts and install the new bolts into the head.

b. Torque the cylinder head bolts, in sequence, to 44 ft. lbs. (60 Nm).

c. Loosen the cylinder head bolts approximately 2 turns and then torque again to 44 ft. lbs. (60 Nm) using the same torque sequence.

d. After setting the torque again, turn the head bolts 90 degrees in sequence and to complete the head bolt installation, turn the head bolts an additional 90 degrees in the same torque sequence.

NOTE: The cylinder head attaching bolts cannot be tightened to the specified torque more than once and must therefore be replaced when installing a cylinder head.

33. Raise and safely support the vehicle.

34. Connect the exhaust system at the exhaust pipe.

35. Lower the vehicle.

36. Install the alternator mounting bracket and the alternator. Connect the alternator wiring harness.

37. Connect the accelerator cable and, if equipped, the speed control cable.

38. Connect the fuel supply and return lines at the metal connector, located on the right side of the engine.

39. Connect the EGR tube to the EGR valve.

40. Rotate the crankshaft to bring No. 1 piston to TDC on its compression stroke. The crankshaft keyway should then be at the 12 o'clock position. The distributor rotor should be pointing toward the No. 1 spark plug firing position. Install the timing belt, the timing belt cover and the crankshaft pulley.

41. Install the distributor cap and spark plug wires.

42. Install the rocker arem cover.

43. Connect the required vacuum hoses.

44. Connect the wiring terminal to the cooling fan switch.

45. Connect the radiator upper hose at the cylinder head.

46. Connect the heater hose to the fitting located below the intake manifold.

47. Fill the cooling system to the proper level and connnect the negative battery cable.

48. Start the engine and check for leaks.

49. After engine has reached operating temperature, check and, if necessary, add coolant.

50. Adjust the ignition timing.

51. Install the PCV hose on the air cleaner assembly.

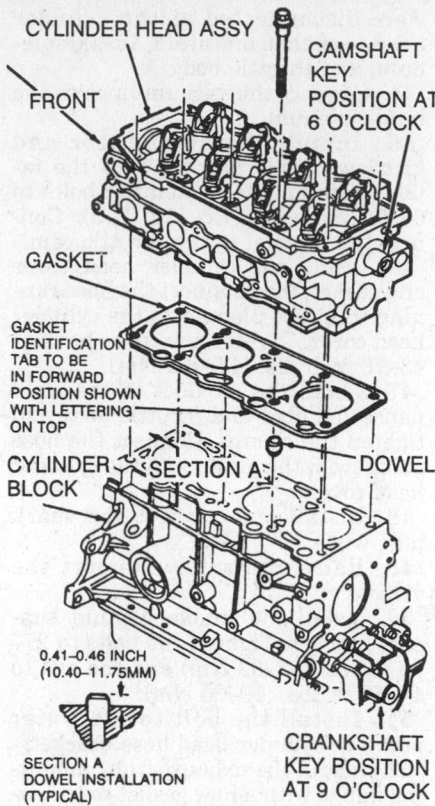

Cylinder head installation—1.9L engine

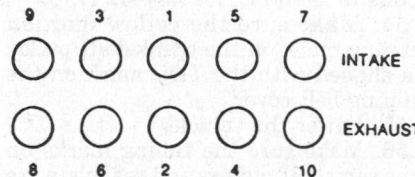

TIGHTENING SEQUENCE CYLINDER HEAD ATTACHING BOLTS

Cylinder head bolt torque sequence—1.9L engine

1991–92

1. Properly relieve the fuel system pressure.

2. Disconnect the negative battery cable.

3. Drain the cooling system.

4. Remove the air intake duct.

5. Remove the crankcase breather hose from the rocker arm cover and the vacuum hose from the bottom of the throttle body.

6. Remove the power brake supply hose.

7. Disconnect the electrical connectors at the following:

a. Fuel charging harness.

b. Alternator harness.

c. Crank angle sensor.

d. Oxygen sensor.

e. Ignition coil.

f. Radio suppressor.

g. Coolant temperature sensor, cooling fan sensor and temperature sending unit.

NOTE: Tag the connectors prior to removal to aid reinstallation.

8. Remove the ground strap from the stud on the left side of the cylinder head.

9. Disconnect the accelerator and the transaxle kickdown cables from the throttle lever and remove the cable bracket from the intake manifold.

10. Disconnect the heater hose containing the coolant temperature switches at the bulkhead.

11. Remove the upper radiator hose.

12. Disconnect the fuel supply and return lines.

13. Remove the oil level indicator tube mounting nut from the cylinder head stud.

14. Remove the power steering hose and the air conditioner line retainer bracket bolts from the alternator bracket.

15. Remove the accessory drive belt, alternator, and the drive belt automatic tensioner.

16. Raise and safely support the vehicle.

17. Remove the right side splash shield and remove the crankshaft dampener.

18. Remove the catalytic converter inlet pipe.

19. Remove the starter wiring harness from the retaining clip below the intake manifold.

20. Set the engine No. 1 cylinder on TDC.

21. Lower the vehicle.

22. Support the engine with a suitable floor jack.

23. Remove the right engine mount dampener and the right engine mount retaining bolts from the mount bracket on the engine. Loosen the right engine mount thru-bolt and roll the mount back aside.

24. Remove the timing belt cover.

25. Loosen the belt tensioner attaching bolt and pry the tensioner as far toward the rear of the engine as possible. Tighten the attaching bolt while in this position.

26. Remove the timing belt.

27. Roll the right engine mount back into position and install the mounting bolts. Lower the floor jack.

28. Remove the heater hose support bracket retaining bolt and the alternator bracket-to-cylinder head mounting bolt.

29. Remove the rocker arm cover.

30. Remove and discard the cylinder head bolts.

31. Remove the cylinder head with the exhaust and intake manifolds attached. Discard the cylinder head gasket.

NOTE: Do not lay the cylinder head flat. Damage to the spark plugs, valves or gasket surfaces may result.

To install:

32. Clean all gasket material from the mating surfaces on the cylinder head and block and clean out the head bolt holes in the block.

33. Before final installation of the cylinder head to the engine, check the piston squish height as follows:

NOTE: Squish height is the clearance of the piston dome to the combustion chamber at piston TDC. No cylinder block deck machining or use of replacement crankshaft, piston or connecting rod causing the assembled squish height to be over or under tolerance specification, is permitted. If no parts other than the head gasket are replaced, the squish height should be within specification. If parts other than the head gasket are replaced, check the squish height. If the squish height is out of specification, replace the parts again and recheck the squish height.

a. Place a small amount of soft lead solder or shot of an appropriate thickness on the piston spherical areas shown.

b. Rotate the crankshaft to lower the piston in the bore and install the head gasket and cylinder head.

NOTE: A compressed (used) head gasket is preferred.

c. Install used head bolts and tighten the head bolts to 30–44 ft. lbs. (40–60 Nm) following proper sequence.

d. Rotate the crankshaft to move the piston through its TDC position.

e. Remove the cylinder head and measure the thickness of the compressed solder to determine squish height at TDC. The solder should be 0.039–0.070 in. (1.0–1.77mm) for all engines.

34. Install the dowels in the cylinder block, if removed. Check the dowel height, it should be 0.41–0.46 in. (10.40–11.75mm) above the surface of the block. A dowel that is too long will not allow the cylinder head to sit properly.

35. Position the cylinder head gasket on the cylinder block.

36. Install the cylinder head and install new bolts and washers in the following order:

a. Apply a light coat of engine oil to the threads of the new cylinder head bolts and install the new bolts into the head.

b. Torque the cylinder head bolts in sequence to 44 ft. lbs. (60 Nm).

c. Loosen the cylinder head bolts approximately 2 turns and then torque again to 44 ft. lbs. (60 Nm) using the same torque sequence.

d. After setting the torque again, turn the head bolts 90 degrees in sequence and to complete the head bolt installation, turn the head bolts an additional 90 degrees in the same torque sequence.

NOTE: The cylinder head attaching bolts cannot be tightened to the specified torque more than once and must therefore be replaced when installing a cylinder head.

37. Install the rocker arm cover and the alternator bracket-to-cylinder head bolt.

38. Support the engine with a suitable floor jack.

39. Remove the right engine mount-to-mount bracket bolts. Roll the mount aside.

40. Make sure cylinder No. 1 is at TDC.

41. Install the timing belt and the timing belt cover.

42. Roll the right engine mount into place and install the 2 mounting bolts and the mount dampener. Remove the floor jack.

43. Raise and safely support the vehicle.

44. Install the crankshaft dampener.

45. Install the starter wiring harness on the retaining clip below the intake manifold.

46. Install the catalytic converter inlet pipe and the right side splash shield.

47. Lower the vehicle.

48. Install the alternator and the accessory drive belt automatic tensioner. Install the accessory drive belt.

49. Install both the power steering hose and air conditioner line retainer bracket bolts. Install the oil level indicator tube retainer bolt.

50. Connect the fuel supply and return lines.

51. Install the upper radiator hose and connect the heater hose at the engine compartment bulkhead. Install the heater hose support bracket retaining bolt.

52. Install the accelerator cable bracket on the intake manifold and connect the accelerator and kickdown cables to the throttle lever.

53. Install the ground strap at the left side of the cylinder head.

54. Connect all remaining electrical connectors according to their positions marked during the removal procedure.

55. Connect the power brake supply hose, crankcase breather hose and the vacuum line at the bottom of the throttle body.
56. Install the air intake duct.
57. Connect the negative battery cable.
58. Fill and bleed the cooling system.
59. Start the engine and check for leaks. Stop the engine and check the coolant level.

2.3L Engine

1. Disconnect the negative battery cable.
2. Disconnect the electric cooling fan switch at the plastic connector.
3. Drain the cooling system at the lower radiator hose.
4. Disconnect the heater hose at the heater inlet tube and disconnect the adapter hose at the water outlet connector.
5. Disconnect the upper radiator hose at the cylinder head.
6. Remove the air cleaner assembly.
7. Tag and disconnect the required electrical connectors and vacuum hoses.
8. Remove the distributor cap and spark plug wires as an assembly. Tag the spark plug wires prior to removal.
9. Disconnect all accessory drive belts.
10. Remove the rocker arm cover and gasket.
11. Remove the rocker arm fulcrum retaining bolts and remove the fulcrum, rocker arms and pushrods. Mark the location of each rocker arm, pushrod and fulcrum for reinstallation in its original position.
12. Properly relieve the fuel system pressure, then disconnect the fuel supply and return lines at the fuel rail.
13. Disconnect the accelerator cable and speed control cable, if equipped.
14. Raise and safely support the vehicle.
15. Disconnect the exhaust system at the exhaust pipe and the hose at the tube.
16. Lower the vehicle.
17. Remove the cylinder head bolts.
18. Remove the cylinder head and gasket with the exhaust and intake manifolds attached.

NOTE: Do not lay the cylinder head flat. Damage to spark plugs or gasket surfaces may result.

To install:
19. Clean all gasket material from the mating surfaces of the cylinder head and block.
20. Position the head gasket on the cylinder block.

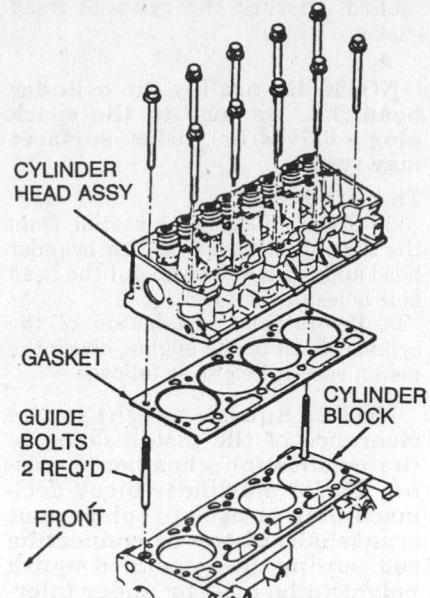

Cylinder head installation—2.3L engine

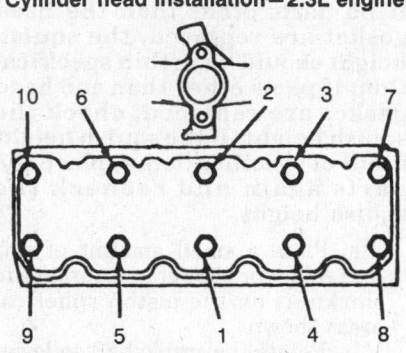

Cylinder head bolt torque sequence—2.3L engine

NOTE: Before installing the cylinder head, thread 2 cylinder head alignment studs T84P–6065–A or equivalent, into the block at opposite corners.

21. Install the cylinder head over the alignment studs onto the cylinder block. Start and run down several head bolts until snug. Remove the alignment studs and install the remaining head bolts. Tighten the bolts in sequence in 2 steps, first to 52–59 ft. lbs. (70–80 Nm) and then to 70–76 ft. lbs. (95–103 Nm).
22. Raise and safely support the vehicle.
23. Connect the exhaust system at the exhaust pipe and the hose to the metal tube.
24. Lower the vehicle.
25. Connect the accelerator cable and speed control cable, if equipped.
26. Connect the fuel supply and return lines.
27. Install the fulcrums, rocker arms and pushrods in their original positions. Tighten the fulcrum bolts to 19.5–26.5 ft. lbs. (26–38 Nm).

28. Install the rocker arm cover gasket and cover.
29. Install the distributor cap and spark plug wires as an assembly.
30. Connect the accessory drive belts.
31. Connect the required electrical connectors and vacuum hoses.
32. Install the air cleaner assembly.
33. Connect the cooling fan switch at the plastic connector.
34. Connect the upper radiator hose and the heater hose.
35. Fill the cooling system.
36. Connect the negative battery cable.
37. Start the engine and check for leaks.
38. After the engine has reached operating temperature, check and, if necessary, add coolant.

Valve Lifters

REMOVAL & INSTALLATION

1.8L Engine

NOTE: Hydraulic lash adjusters are used on the 1.8L engine between the valve stem and the camshaft to reduce noise and to provide maintenance-free valve clearance.

1. Disconnect the negative battery cable.
2. Remove the camshafts according to the proper procedure.
3. Mark the hydraulic lash adjusters and the cylinder head with alignment marks so the hydraulic lash adjusters can be installed in their original mounting positions.
4. Remove the hydraulic lash adjusters from the cylinder head.
To install:
5. Apply clean engine oil to the hydraulic lash adjuster friction surfaces.
6. If the hydraulic lash adjusters are being reused, install them in the positions from which they were removed.
7. Make sure the hydraulic lash adjusters move smoothly in their bores.
8. Install the camshafts.
9. Connect the negative battery cable.

1.9L Engine

1. Disconnect the negative battery cable.
2. Remove air cleaner assembly. Remove valve cover and gasket.
3. Remove rocker arms, lifter guides, lifter retainers and lifters.

NOTE: Always return lifters to the original bores unless they are being replaced.

To install:

4. Lubricate each lifter bore with engine oil.

5. If equipped with flat bottom lifters, install with oil hole in plunger upward. If equipped with roller lifters, install with plunger upward and position guide flats of lifters to be parallel with centerline of camshaft. Color orientation dots on lifters should be opposite the oil feed holes in cylinder head.

6. For roller lifters only, install lifter guide plates over tappet guide flats with notch toward exhaust side. For flat lifters, no guide plate is required.

7. Lubricate lifter plunger cap and valve tip with engine oil.

8. Install lifter guide plate retainers into rocker arm fulcrum slots, in both intake and exhaust side. Notch to be with exhaust valve lifter.

9. Install 4 rocker arms in lifter position No's 3, 6, 7 and 8.

10. Lubricate rocker arm surface that will contact fulcrum surface with engine oil.

11. Install 4 fulcrums. Fulcrums must be fully seated in slots of cylinder head.

12. Install 4 bolts. Tighten to 17–22 ft. lbs. (23–30 Nm).

13. Rotate the engine until the camshaft sprocket keyway is in the 6 o'clock position.

14. Repeat steps 9–12 in lifter position No's 1, 2, 4 and 5.

15. Install valve cover and gasket. Install air cleaner assembly.

16. Connect negative battery cable.

2.3L Engine

1. Disconnect the negative battery cable. Remove the cylinder head and related parts.

2. Using a magnet, remove the lifters. Identify, tag and place the lifters in a rack so they can be installed in the original positions.

3. If the lifters are stuck in their bores by excessive varnish or gum, it may be necessary to use a hydraulic lifter puller tool to remove the lifters. Rotate the lifters back and forth to loosen any gum and varnish which may have formed. Keep the assemblies intact until they are to be cleaned.

To install:

4. Install new or cleaned hydraulic lifters through the pushrod openings with a magnet.

5. Install the cylinder head and related parts.

6. Connect negative battery cable.

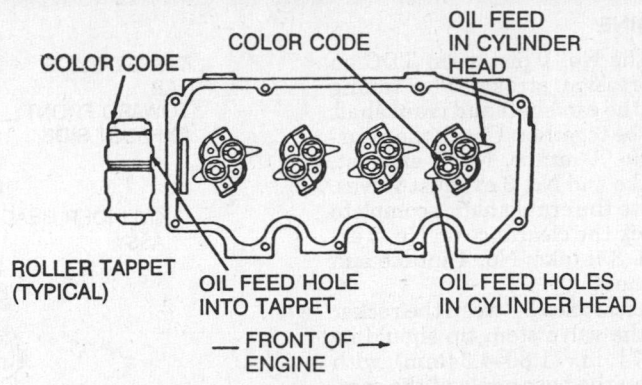

Roller tappet assembly installation—1.9L engine

Valve Lash

ADJUSTMENT

Collapsed Lifter Clearance

1.9L ENGINE

1. Connect an auxiliary starter switch in the starting circuit. Crank the engine with the ignition switch **OFF** until the No. 1 piston is on TDC of the compression stroke.

2. With the crankshaft in position, place the hydraulic lifter compressor tool T81P–6500–A or equivalent, on the rocker arm. Slowly apply pressure to bleed down the lifter until it completely bottoms. Hold the lifter in this position and check the available clearance between the rocker arm and the valve stem tip with a feeler gauge. The feeler gauge width must not exceed ⅜ in., in order to fit between the rails on the rocker arm.

3. The clearance should be as follows:

1988 CFI and 1989 EFI engines: 1.2–3.5mm.

1988 EFI and 1989 EFI-HO engines: 1.5–3.8mm.

1990 EFI with flat tappet: 0.7–5.2mm, 2.9mm normal.

1990–92 EFI with roller tappet: 0–4.5mm, 2.2mm normal.

1990 EFI-HO with flat tappet: 1.2–5.6mm, 3.4mm normal.

1990 EFI-HO with roller tappet: 0.5–4.9mm, 2.7mm normal.

4. If the clearance is less than specifications, check the fulcrum, lifter, camshaft lobe and valve tip for wear.

5. With the No. 1 piston on TDC at the end of the compression stroke check the following valves: No. 1 intake, No. 1 exhaust, No. 2 intake.

6. Rotate the crankshaft 180 degrees and check the following valves: No. 3 intake, No. 3 exhaust.

7. Rotate the crankshaft another 180 degrees TDC and check the following valves: No. 4 intake, No. 4 exhaust, No. 2 exhaust.

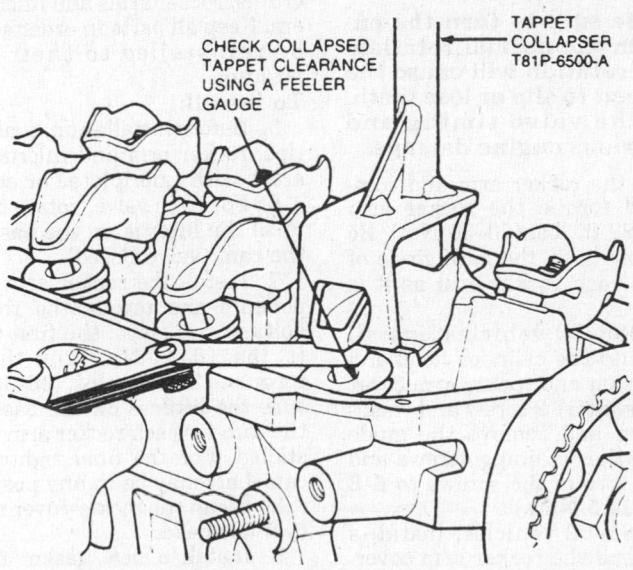

Checking collapsed lifter clearance—1.9L engine

2.3L ENGINE

1. Set the No. 1 piston on TDC on the compression stroke. The timing marks on the camshaft and crankshaft gears will be together. Check the clearance on No. 1 intake, No. 1 exhaust, No. 2 intake and No. 3 exhaust valves.

2. Rotate the crankshaft 1 complete turn. Check the clearance on No. 2 exhaust, No. 3 intake, No. 4 intake and No. 4 exhaust.

3. The clearance between the rocker arm and the valve stem tip should be 0.072–0.174 in. (1.80–4.34mm) with the lifter on the base circle of the cam.

Rocker Arms

REMOVAL & INSTALLATION

1.9L Engine

1. Disconnect the negative battery cable and remove the air cleaner assembly.

2. Remove and tag all necessary vacuum hoses from the rocker cover. Remove the spark plug wire retainers, if equipped. Remove the rocker cover from the cylinder head.

3. Remove the rocker cover and gasket from the engine.

4. Remove the rocker arm nuts, fulcrums, rocker arms and fulcrum washers. Keep all parts in order so they can be reinstalled to their original position.

To install:

5. Before installation, coat the valve tips, rocker arm and fulcrum contact areas with Lubriplate® or equivalent.

6. Rotate the engine until the lifter is on the base circle of the cam (valve closed).

NOTE: Be sure to turn the engine only in the normal rotation. Backward rotation will cause the camshaft belt to slip or lose teeth, altering the valve timing and causing serious engine damage.

7. Install the rocker arm and components and torque the rocker arm bolts to 17–22 ft. lbs. (23–30 Nm). Be sure the lifter is on the base circle of the cam for each rocker arm as it is installed.

8. On 1988–90 vehicles, install guide pins into the cylinder head and guide the gasket and rocker arm cover over the pins. Start 2 screw and washer assemblies and remove the guide pins. Install the retaining screws and washer and torque the screws to 6–8 ft. lbs. (8.0–11.5 Nm).

9. On 1991–92 vehicles, install a new gasket and the rocker arm cover. Install the 3 retaining bolts and tighten to 4–9 ft. lbs. (5–12 Nm).

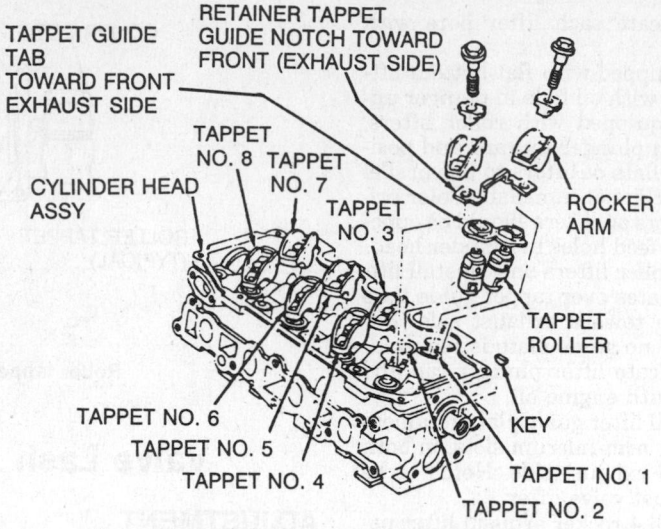

Rocker arm removal—1.9L engine

NOTE: Do not use any type of sealer with the rocker arm cover silicone gasket.

10. Connect all vacuum hoses and install the spark plug wire retainers, if equipped.

11. Connect the negative battery cable.

2.3L Engine

1. Disconnect the negative battery cable.

2. Remove and tag all necessary vacuum hoses from the rocker cover. Remove the oil fill cap and set it aside. Disconnect the PCV hose and set it aside.

3. Remove the rocker arm cover bolts. Remove the rocker cover and gasket from the engine.

4. Remove the rocker arm bolts, fulcrums, rocker arms and fulcrum washers. Keep all parts in order so they can be reinstalled to their original position.

To install:

5. Before installation, coat the valve tips, rocker arm and fulcrum contact areas with Lubriplate® or equivalent.

6. For each valve, rotate the engine until the lifter is on the base circle of the cam (valve closed).

7. Install the rocker arm and components and torque the rocker arm bolts in two steps: the first to 4.5–7.5 ft. lbs. (6–10 Nm) and the second torque to 20–26 ft. lbs. (26–38 Nm). Be sure the lifter is on the base circle of the cam for each rocker arm as it is installed. For the final tightening, the camshaft may be in any position.

8. Clean the rocker cover rail on the cylinder head.

9. Install a new gasket or, in the case of the moulded in place gasket, apply RTV sealer and install the rock-er cover. Install the retaining bolts and tighten to 5.9–8.5 ft. lbs. (8.0–11.5 Nm).

10. Install oil fill cap, all necessary vacuum hoses and the PCV hose.

11. Connect negative battery cable.

Intake Manifold

REMOVAL & INSTALLATION

1.8L Engine

1. Properly relieve the fuel system pressure.

2. Disconnect the negative battery cable.

3. Tag and disconnect the necessary vacuum hoses from the intake manifold and plenum.

4. Remove the vacuum chamber canister from the intake plenum.

5. Disconnect the idle speed control and bypass air hoses from the intake plenum.

6. Disconnect the accelerator cable and, if equipped with automatic transaxle, the kickdown cable from the throttle cam. Remove the cable bracket from the intake plenum.

7. Tag and disconnect the throttle body electrical connectors.

8. Disconnect the fuel pressure and return line spring lock couplings.

9. Disconnect the PCV hose from the intake plenum and cylinder head cover.

10. Disconnect the fuel pressure regulator vacuum hose and the fuel injector wiring harness electrical connectors.

11. Remove the fuel rail mounting bolts and remove the fuel rail.

12. Remove the 2 bolts from the transaxle vent tube and remove the vent tube from the intake plenum.

13. Remove the intake manifold upper mounting nuts.

14. Raise and safely support the vehicle.

15. Remove the intake plenum support bracket and the intake manifold lower mounting nuts.

16. Lower the vehicle.

17. Remove the intake manifold, intake plenum and throttle body as an assembly from the vehicle.

18. Remove the intake manifold gasket.

19. If necessary, separate the intake plenum and throttle body from the intake manifold.

20. Clean all gasket mating surfaces.

To install:

21. If necessary, install the throttle body and intake plenum onto the intake manifold.

22. Install the intake manifold gasket.

23. Install the intake manifold, intake plenum and throttle body assembly onto the intake manifold mounting studs.

24. Install the mounting nuts and tighten to 14–19 ft. lbs. (19–25 Nm) in the proper sequence.

25. Raise and safely support the vehicle.

26. Install the intake plenum support bracket and tighten the bolts to specification.

27. Lower the vehicle.

28. Place the fuel rail into position and install the mounting bolts. Tighten the bolts to 14–19 ft. lbs. (19–25 Nm).

29. Connect the fuel injector wiring harness electrical connectors and connect the vacuum hose to the pressure regulator.

30. Connect the PCV hose to the intake plenum and cylinder head cover.

31. Connect the fuel pressure and return line spring lock couplings.

32. Install the transaxle vent tube that bolts onto the intake plenum.

33. Connect the electrical connectors to the throttle body and the necessary vacuum hoses to the intake plenum and throttle body.

34. Connect the idle speed control and bypass air hoses to the intake plenum.

35. Install the cable bracket onto the intake plenum and connect the accelerator and, if equipped, kickdown cables to the throttle cam.

36. Install the inlet air duct that connects to the throttle body and the resonance chamber.

37. Connect the negative battery cable.

1988–90 1.9L CFI Engine

1. Raise and secure the hood in the open position.

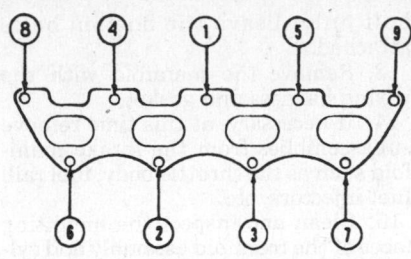

Intake manifold bolt torque sequence— 1.8L engine

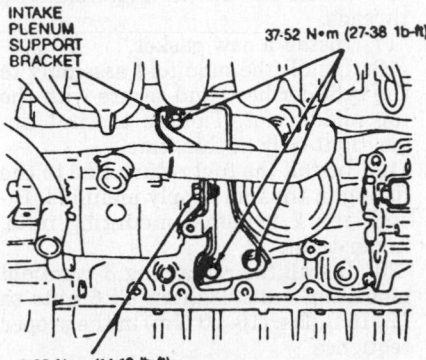

Intake plenum support bracket torque specifications—1.8L engine

2. Install protective fender covers.

3. Properly relieve the fuel system pressure. Disconnect the negative battery cable.

4. Partially drain the cooling system and disconnect the heater hose at the fitting located on the side of the intake manifold.

5. Remove air cleaner assembly.

6. Identify, tag and disconnect the vacuum hoses.

7. Identify, tag and disconnect wiring connectors at the following points:
   a. Coolant temperature sensor
   b. Air charge temperature sensor

8. Remove EGR supply tube.

9. Raise and safely support the vehicle.

10. Remove the PVS hose connectors. Label the connectors and set aside.

11. Remove the bottom 4 of the intake manifold retaining nuts, locations 2, 3, 6 and 7.

12. Lower the vehicle.

13. Disconnect fuel lines at the the throttle body.

14. Disconnect accelerator and, if equipped, the speed control cable.

15. Disconnect the throttle valve linkage at the throttle body and remove the cable bracket attaching bolts on vehicles equipped with automatic transaxle.

16. Remove the remaining 3 intake manifold attaching nuts, intake manifold and gasket.

NOTE: Do not lay the intake manifold flat as the gasket surfaces may be damaged.

To install:

17. Make sure the mating surfaces on the intake manifold and the cylinder head are clean and free of gasket material.

18. Install a new intake manifold gasket.

19. Position the intake manifold on the engine and install the attaching nuts. Tighten the nuts, in sequence, to 12–15 ft. lbs. (16–20 Nm).

20. Connect throttle valve linkage and install the cable bracket attaching bolts, if removed, on vehicles with automatic transaxle.

21. Connect accelerator cable and, if equipped, the speed control cable.

22. Connect fuel lines at the fuel charging assembly.

23. Raise and safely support the vehicle.

24. Connect heater hose to the fitting located on side of the intake manifold.

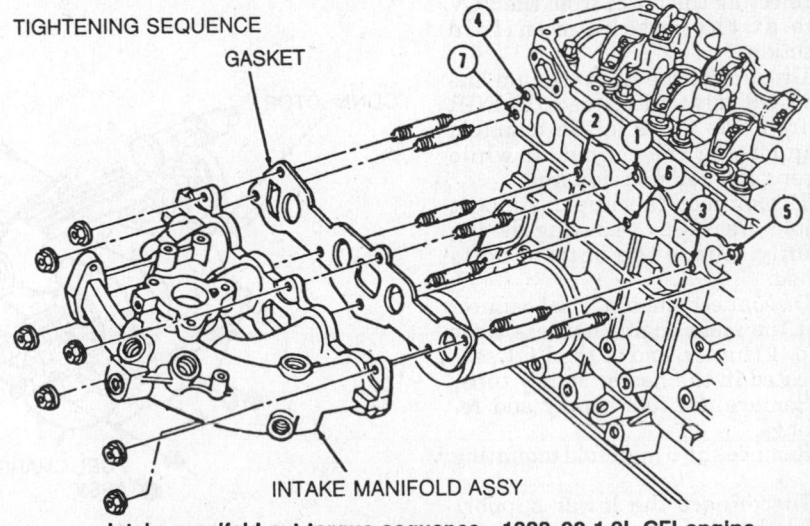

Intake manifold nut torque sequence—1988–90 1.9L CFI engine

25. Lower vehicle.
26. Connect EGR supply tube.
27. Connect the wiring connectors at the following points:
   a. Coolant temperature sensor
   b. Air charge temperature sensor
28. Connect vacuum hoses.
29. Install air cleaner assembly.
30. Fill cooling system to specified level.

NOTE: Because the cylinder head is an aluminum alloy, the cooling system must be filled only with the specified coolant, normally consisting of a 50/50 mix of water and permanent coolant/antifreeze. Test coolant to ensure that it meets the required low temperature protection for the area in which the vehicle will be operated.

31. Connect negative battery cable.
32. Start engine and check for fuel and coolant leaks. Bring engine to normal operating temperature and check again for coolant leaks.

1988–90 1.9L EFI-HO Engine

1. Disconnect the negative battery cable.
2. Properly relieve the fuel system pressure.
3. Remove the engine air cleaner outlet tube between the vane air meter and the air throttle body by loosening the 2 clamps.
4. Disconnect and remove the accelerator and speed control cables, if equipped, from the accelerator mounting bracket and throttle lever.
5. Disconnect the top manifold vacuum fitting connections by disconnecting the rear vacuum line to the dash panel vacuum tree and the vacuum line at the intake manifold tee.
6. Disconnect the PCV system by disconnecting the hoses from the PCV valve at the intake manifold connection.
7. Disconnect the EGR vacuum line at the EGR valve. Disconnect the EGR tube from the upper intake manifold by supporting the connector while loosening the compression nut.
8. Disconnect the upper support manifold bracket by removing the top bolt only. Leave the bottom bolts attached.
9. Disconnect the electrical connectors at the main engine harness, near the No. 1 runner, and at the ECT sensor located in the heater supply tube.
10. Remove the fuel supply and return lines.
11. Remove the 6 manifold mounting nuts.
12. Disconnect the lower support manifold bracket by removing the top

bolt only. Leave the bottom bolts attached.
13. Remove the manifold with the wiring harness and gasket.
14. If necessary, at this time remove subassemblies from the intake manifold such as the throttle body, fuel rail, fuel injectors, etc.
15. Clean and inspect the mounting faces of the manifold assembly and cylinder head. Both surfaces must be clean and flat.

To install:

16. Clean and oil the manifold stud threads.
17. Install a new gasket.
18. Install the manifold assembly to the cylinder head and secure with the top middle nut. Tighten the nut finger-tight only at this time.
19. Install the fuel return line to the fitting in the fuel supply manifold. Install the 2 manifold mounting nuts, finger-tight.
20. Install the remaining 3 manifold mounting nuts. Tighten all 6 nuts to 12–15 ft. lbs. (16–20 Nm) in the proper sequence.
21. Connect the upper and lower manifold support brackets and tighten the bolts to 15–22 ft. lbs. (20–30 Nm).
22. Install the EGR tube with oil-coated compression nut tightened to 29.5–40.5 ft. lbs. (40–55 Nm).
23. Connect the vacuum line to the throttle body port and connect the large PCV vacuum line to the upper manifold fitting.
24. Connect the rear manifold vacuum connections at the dash panel vacuum tree and connect the vacuum line(s) to the upper manifold.
25. Connect the accelerator and, if equipped, speed control cables.
26. Install the air supply tube. Tighten the clamps to 12–20 inch lbs. (1.4–2.3 Nm).

27. Connect the wiring harness at the ECT sensor in the heater supply tube and the main engine harness, near the No. 1 runner.
28. Connect the fuel supply hose from the fuel filter to the fuel rail and connect the fuel return line.
29. Reconnect the spring-lock coupling retaining clips on the fuel inlet and return fittings.
30. Fill the cooling system.
31. Connect the negative battery cable.
32. Start the engine and bring to normal operating temperature. Check for leaks. Stop the engine and check the coolant level.

1991–92 1.9L Engine

1. Properly relieve the fuel system pressure.
2. Disconnect the negative battery cable.
3. Partially drain the cooling system.
4. Remove the air intake tube.
5. Disconnect the fuel injector harness from the engine control harness at the right shock tower.
6. Disconnect the crankshaft position sensor.
7. Disconnect the fuel supply and return lines.
8. Remove the accelerator cable and, if equipped with automatic transaxle, kickdown cable from the throttle lever. Remove the cable bracket from the intake manifold and position the cables aside.
9. Remove the power brake supply hose, PCV line and the vacuum line from the bottom of the throttle body.
10. Remove the 7 attaching nuts from the intake manifold studs, slide the manifold assembly off of the studs and remove it from the cylinder head.

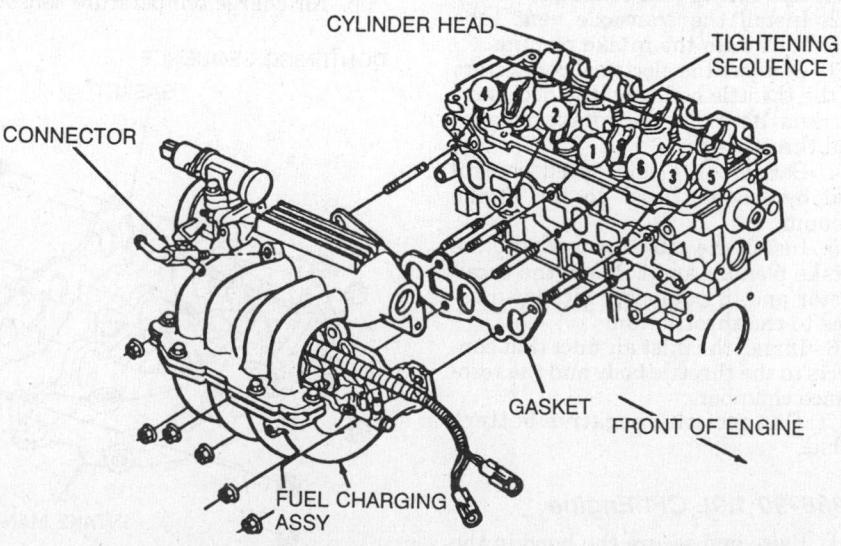

Intake manifold nut torque sequence—1988–90 1.9L EFI-HO engine

Remove and discard the intake manifold gasket.

To install:

11. Clean and inspect the mounting faces of the intake manifold and cylinder head. Both surfaces must be clean and flat.

12. Clean and oil the manifold studs and position a new gasket over them.

13. Install the intake manifold and the attaching nuts. Tighten the nuts to 12–15 ft. lbs. (16–20 Nm).

14. Install the vacuum line on the bottom of the throttle body, the power brake supply hose and the PCV line.

15. Install the accelerator cable bracket and connect the accelerator cable and, if equipped, kickdown cable on the throttle lever.

16. Connect the crankshaft position sensor electrical connector.

17. Connect the fuel supply and return lines. Install the fuel line retaining clips.

18. Connect the 2 fuel injector harness connectors to the engine control harness at the right shock tower.

19. Install the air intake tube.

20. Refill the cooling system.

21. Connect the negative battery cable.

22. Start the engine and bring to normal operating temperature. Check for leaks. Stop the engine and check the coolant level.

2.3L Engine

1. Disconnect the negative battery cable.

2. Properly relieve the fuel system pressure.

3. Drain the cooling system.

4. Remove accelerator cable.

5. Remove air cleaner assembly and heat stove tube at heat shield.

6. Remove required vacuum lines and electrical connectors.

7. Remove exhaust manifold heat shield. Disconnect the oxygen sensor wire at the connector.

8. Disconnect the throttle linkage.

9. Disconnect the speed control cable, if equipped.

10. Disconnect the fuel supply and return lines at the rubber connector.

11. Disconnect EGR tube at EGR valve.

12. Remove the intake manifold retaining bolts. Remove the intake manifold. Remove the gasket and clean the gasket contact surfaces.

To install:

13. Install intake manifold with gasket and retaining bolts. Torque the retaining bolts, in the proper sequence to 15–22 ft. lbs. (20–30 Nm).

14. Connect the oxygen sensor wire at their proper connector.

15. Connect EGR tube to EGR valve.

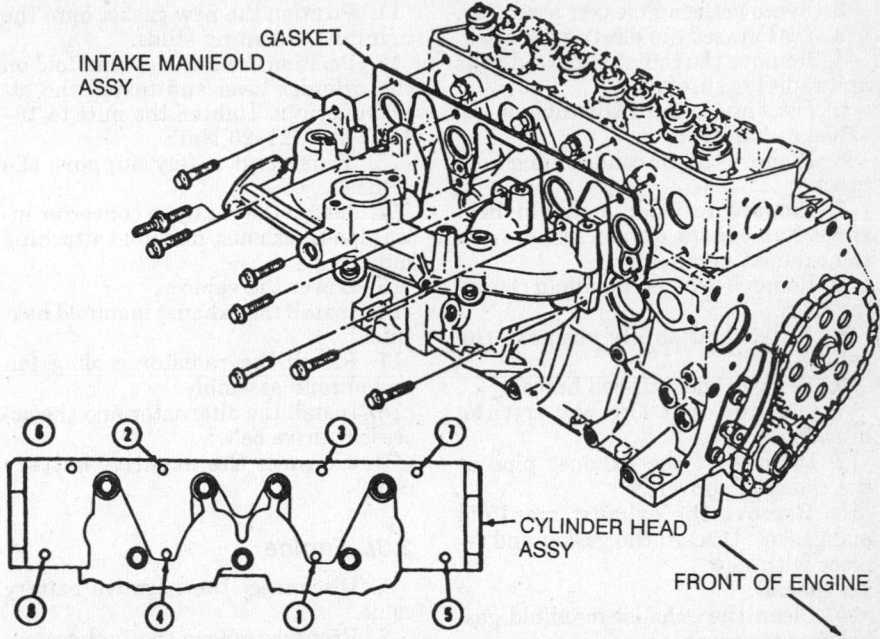

Intake manifold bolt torque sequence—2.3L engine

16. Connect the fuel supply and return lines.

17. Install vacuum lines.

18. Install air cleaner assembly and heat stove tube.

19. Install accelerator cable and speed control cable, if equipped.

20. Connect negative battery cable and fill the cooling system.

21. Start engine and bring to normal operating temperature. Check for leaks. Stop the engine and check the coolant level.

Exhaust Manifold

REMOVAL & INSTALLATION

1.8L Engine

1. Disconnect the negative battery cable.

2. Remove the resonance duct.

3. Partially drain the cooling system and disconnect the upper radiator hose.

4. Remove the cooling fan.

5. Raise and safely support the vehicle.

6. Remove the exhaust pipe from the exhaust manifold and remove the gasket.

7. Remove the 2 bolts from the exhaust pipe support bracket.

8. Remove the left lower splash shield.

9. Lower the vehicle.

10. Disconnect the oxygen sensor electrical connector.

11. Remove the exhaust manifold heat shield mounting bolts and remove the shield.

12. Remove the exhaust manifold mounting nuts and remove the assembly.

13. Remove all gasket material from the cylinder head and exhaust manifold.

To install:

14. Install a new gasket onto the exhaust manifold mounting studs.

15. Place the exhaust manifold onto the mounting studs and install the manifold mounting nuts. Tighten the nuts to 28–34 ft. lbs. (38–46 Nm).

16. Place the heat shield into its mounting position and install the shield mounting bolts. Tighten the bolts to 69–95 inch lbs. (7.8–11.0 Nm).

17. Connect the oxygen sensor electrical connector.

18. Install the cooling fan.

19. Connect the upper radiator hose.

20. Install the resonance duct.

21. Raise and safely support the vehicle.

22. Install the exhaust pipe support bracket.

23. Install a new gasket and install the exhaust pipe to the exhaust manifold. Tighten the attaching nuts to 23–34 ft. lbs. (31–46 Nm).

24. Install the left lower splash shield and tighten the bolts to 69–95 inch lbs. (7.8–11.0 Nm).

25. Lower the vehicle.

26. Refill the cooling system.

27. Connect the negative battery cable.

1.9L Engine

1988–90

1. Disconnect the negative battery cable.

2. Remove the air cleaner assembly.

3. Disconnect the electric fan wire.

4. Remove the radiator shroud bolts and radiator shroud.

5. Disconnect the EGR tube at the exhaust manifold.

6. Remove the air conditioning hose bracket.

7. Remove exhaust manifold heat stove. Remove the oxygen sensor from the exhaust manifold.

8. Remove exhaust manifold retaining nuts.

9. Raise and safely support the vehicle.

10. Remove the anti-roll brace.

11. Disconnect the water tube brackets.

12. Disconnect the exhaust pipe at the catalytic converter.

13. Remove the exhaust manifold and gasket. Discard the gasket and replace with new.

To install:

14. Clean the exhaust manifold gasket contact areas.

15. Position the gasket and exhaust manifold.

16. Install exhaust pipe to the catalyst.

17. Install anti-roll brace. Install water tube brackets.

18. Lower vehicle.

19. Install exhaust manifold retaining nuts. Tighten to 16–20 ft. lbs. (21–26 Nm).

20. Install exhaust manifold heat stove.

21. Install oxygen sensor in exhaust manifold. Tighten to 30–40 ft. lbs. (40–50 Nm).

22. Connect the EGR tube.

23. Install air conditioning hose brackets.

24. Position shroud and fan assembly on radiator and install bolts.

25. Connect electric fan wire.

26. Connect battery cable.

27. Install air cleaner assembly.

1991–92

1. Disconnect the negative battery cable.

2. Remove the accessory drive belt.

3. Remove the alternator.

4. Remove the radiator cooling fan and the shroud assembly.

5. Remove the exhaust manifold heat shield.

6. Raise and safely support the vehicle.

7. Remove the 2 catalytic converter inlet pipe-to-exhaust manifold attaching nuts.

8. Lower the vehicle.

9. Remove the 8 exhaust manifold attaching nuts and remove the exhaust manifold and gasket.

To install:

10. Clean the cylinder head and exhaust manifold gasket surfaces.

11. Position the new gasket onto the manifold mounting studs.

12. Position the exhaust manifold on the cylinder head and install the attaching nuts. Tighten the nuts to 16–19 ft. lbs. (21–26 Nm).

13. Raise and safely support the vehicle.

14. Install the catalytic converter inlet pipe-to-exhaust manifold attaching nuts.

15. Lower the vehicle.

16. Install the exhaust manifold heat shield.

17. Install the radiator cooling fan and shroud assembly.

18. Install the alternator and the accessory drive belt.

19. Connect the negative battery cable.

2.3L Engine

1. Disconnect the negative battery cable.

2. Properly relieve the fuel system pressure.

3. Drain the cooling system.

4. Remove the accelerator cable and position to the side.

5. Remove air cleaner assembly and heat stove tube at heat shield.

6. Identify, tag and disconnect all necessary vacuum lines.

7. Disconnect the exhaust pipe-to-exhaust manifold retaining nuts.

8. Remove exhaust manifold heat shield. Disconnect the oxygen sensor wire at the connector.

9. Disconnect the throttle linkage.

10. Disconnect the speed control cable, if equipped.

11. Disconnect the fuel supply and return lines at the rubber connector.

12. Disconnect EGR tube from the EGR valve.

13. Remove the intake manifold.

14. Remove the exhaust manifold retaining nuts. Remove the exhaust manifold from the vehicle.

To install:

15. Position exhaust manifold to the cylinder head using guide bolts in holes 2 and 3.

16. Install the attaching bolts in the remaining holes.

17. Tighten the attaching bolts until snug, then remove guide bolts and install the remaining attaching bolts.

18. Tighten all exhaust manifold bolts to specification using the following tightening procedure: torque retaining bolts in sequence to 5–7 ft. lbs. (7–10 Nm) then retorque, in sequence, to 20–30 ft. lbs. (27–41 Nm).

19. Install the intake manifold gasket and bolts. Torque the intake manifold retaining bolts, in the proper sequence to 15–22 ft. lbs. (20–30 Nm).

20. Connect the oxygen sensor wire at their proper connector.

21. Connect the EGR tube to EGR valve.

22. Install exhaust manifold studs.

23. Connect exhaust pipe to exhaust manifold.

24. Connect the fuel supply and return lines.

25. Install vacuum lines.

26. Install air cleaner assembly and heat stove tube.

27. Install accelerator cable and speed control cable, if equipped.

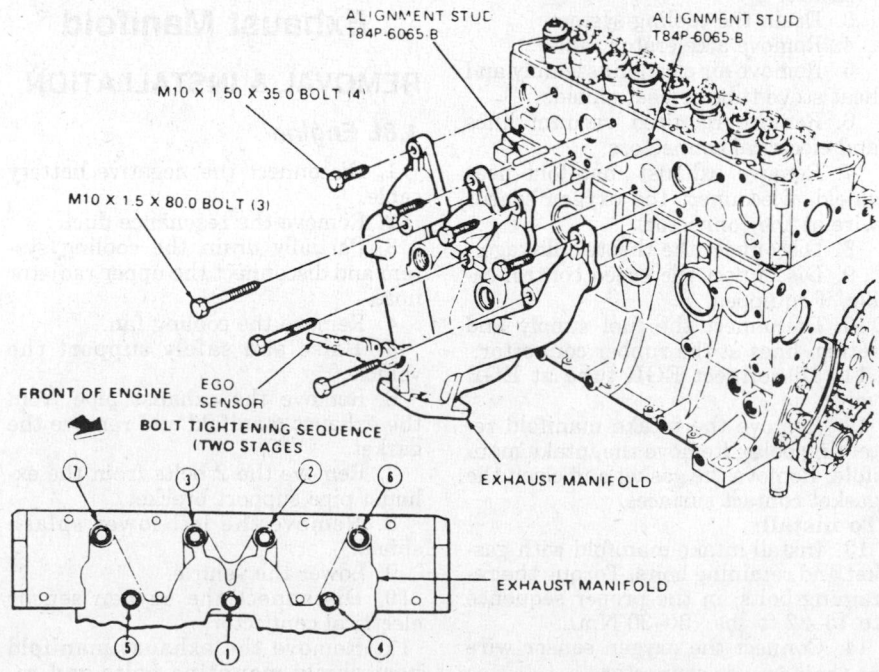

Exhaust manifold bolt torque sequence—2.3L engine

28. Connect the negative battery cable.
29. Fill the cooling system.
30. Start engine and check for leaks.

Timing Chain Front Cover

REMOVAL & INSTALLATION

2.3L Engine

1. Remove the engine and transaxle from the vehicle as an assembly and position in a suitable holding fixture. Remove the dipstick.

2. Remove accessory drive pulley, if equipped, Remove the crankshaft pulley attaching bolt and washer and remove pulley.

3. Remove front cover attaching bolts from front cover. Pry the top of the front cover away from the block.

4. Clean any gasket material from the surfaces.

5. Check timing chain and sprockets for excessive wear. If the timing chain and sprockets are worn, replace with new.

6. Check timing chain tensioner blade for wear depth. If the wear depth exceeds specification, replace tensioner.

7. Remove the oil pan.

NOTE: Oil pan removal is recommended to ensure proper sealing to front cover.

To install:

8. Clean and inspect all parts before installation. Clean the oil pan, cylinder block and front cover of gasket material and dirt.

9. Apply oil resistant sealer to a new front cover gasket and position gasket into front cover.

10. Remove the front cover oil seal and position the front cover on the engine.

11. Position front cover alignment tool T84P–6019–C or equivalent, onto the end of the crankshaft, ensuring the crank key is aligned with the keyway in the tool. Bolt the front cover to the engine and torque bolts to 6–9 ft. lbs. (8–12 Nm). Remove the front cover alignment tool.

12. Replace the front cover seal with new. Lubricate the hub of the crankshaft pulley with polyethylene grease to prevent damage to the seal during installation and initial engine start. Install crankshaft pulley.

13. Install the oil pan.

14. Install the accessory drive pulley, if equipped.

15. Install crankshaft pulley attaching bolt and washer. Tighten to 140–170 ft. lbs. (190–230 Nm).

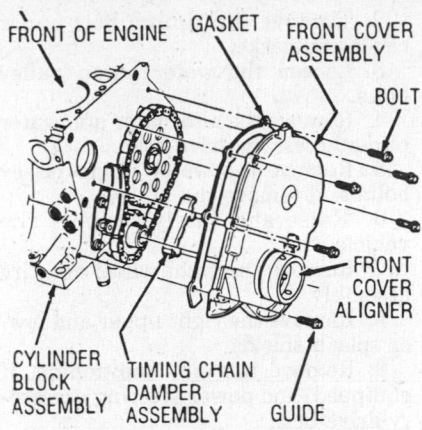

Front cover removal and installation— 2.3L engine

16. Remove engine from work stand and install in vehicle.

Front Cover Oil Seal

REPLACEMENT

2.3L Engine

NOTE: The removal and installation of the front cover oil seal on these engines can only be accomplished with the engine removed from the vehicle.

1. Remove the engine from the vehicle and position in a suitable holding fixture.

2. Remove bolt and washer at crankshaft pulley.

3. Remove the crankshaft pulley.

4. Using a suitable tool, remove the front cover oil seal.

To install:

5. Coat a new seal with grease. Using a suitable installation tool install the seal into the cover. Drive the seal in until it is fully seated. Check the seal after installation to be sure the spring is properly positioned in the seal.

6. Install crankshaft pulley, attaching bolt and washer. Torque the crankshaft pulley bolt to 140–170 ft. lbs. (190–230 Nm).

7. To complete the installation of the front cover oil seal, reverse the removal procedure.

Timing Chain and Sprockets

REMOVAL & INSTALLATION

2.3L Engine

1. Disconnect negative battery cable.

2. Remove engine and transaxle from vehicle as an assembly and position in a suitable holding fixture. Remove the dipstick.

3. Remove front cover from engine.

4. Check timing chain deflection as follows:

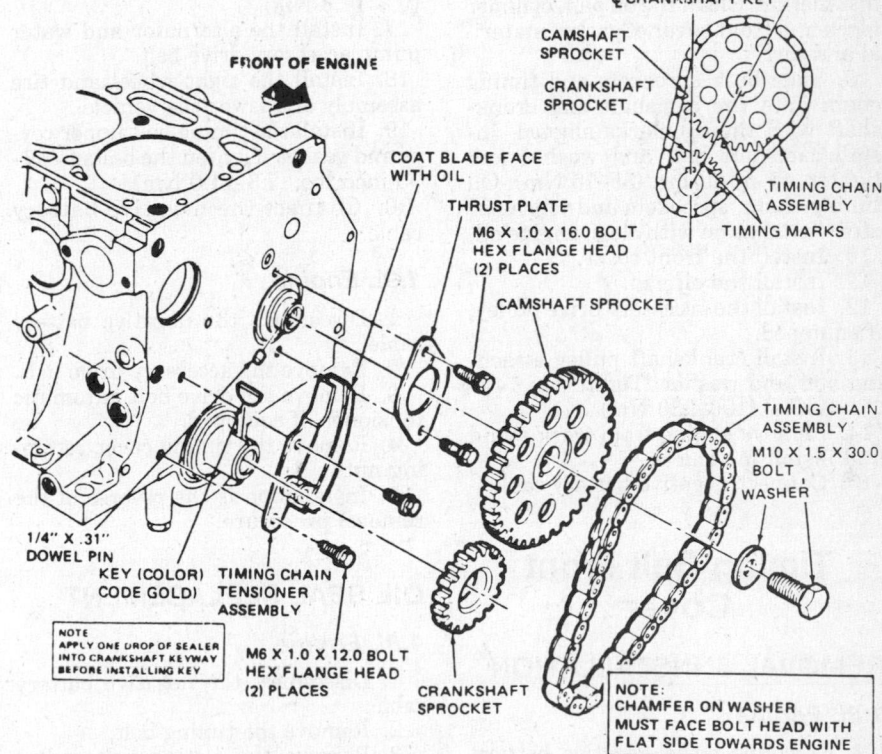

Timing chain tensioner, sprockets and timing chain installation—2.3L engine

a. Rotate crankshaft counter-clockwise, as viewed from the front of the engine, to take up slack on the left side of chain.

b. Make a reference mark on the block at approximately mid-point of chain. Measure from this point to chain.

c. Rotate crankshaft in opposite direction to take up slack on the right side of the chain. Force left side of chain out with fingers and measure distance between reference point and chain. The deflection is the difference between the 2 measurements.

d. If deflection measurement exceeds 0.5 in. (12.7mm), replace timing chain and sprockets. If wear on tensioner face exceeds 0.06 in. (1.5mm), replace tensioner.

5. Turn engine over until the timing marks are aligned. Remove camshaft sprocket attaching bolt and washer. Slide both sprockets and timing chain forward and remove as an assembly.

6. Check timing chain vibration damper for excessive wear and replace if necessary. The damper is located inside the front cover.

7. Remove the oil pan.

NOTE: Oil pan removal is recommended to ensure proper sealing to front cover upon installation.

To install:

8. Clean and inspect all parts before installation. Clean the oil pan, cylinder block and front cover of gasket material and dirt.

9. Slide both sprockets and timing chain onto the camshaft and crankshaft with timing marks aligned. Install camshaft bolt and washer and tighten 41–56 ft. lbs. (55–75 Nm). Oil timing chain, sprockets and tensioner after installation with clean engine oil.

10. Install the front cover.

11. Install the oil pan.

12. Install the accessory drive pulley, if equipped.

13. Install crankshaft pulley attaching bolt and washer. Tighten to 140–170 ft. lbs. (190–230 Nm).

14. Remove engine from work stand and install in vehicle.

15. Connect negative battery cable.

Timing Belt Front Cover

REMOVAL & INSTALLATION

1.8L Engine

1. Disconnect the negative battery cable.

2. Remove the timing belt upper cover and gasket.

3. Loosen the water pump pulley bolts.

4. Remove the alternator and water pump accessory drive belt.

5. Remove the water pump pulley bolts and remove the pulley.

6. Raise and safely support the vehicle.

7. Remove the right wheel and tire assembly.

8. Remove the right upper and lower splash shields.

9. Remove the air conditioning, if equipped, and power steering accessory drive belt.

10. Remove the crankshaft pulley, crankshaft pulley guide plate and timing belt outer and inner guide plates.

11. Remove the timing belt middle and lower covers along with the gaskets.

To install:

12. Install the timing belt middle and lower covers along with the gaskets.

13. Install the timing belt inner and outer guide plates, the crankshaft pulley and the crankshaft pulley guide plate. Tighten the bolts to 109–152 inch lbs. (12–17 Nm).

14. Install the air conditioning, if equipped, and power steering accessory drive belt.

15. Install the splash shields and tighten the bolts to 69–95 inch lbs. (7.8–11.0 Nm).

16. Install the water pump pulley and tighten the bolts to 69–95 inch lbs. (7.8–11.0 Nm).

17. Install the alternator and water pump accessory drive belt.

18. Install the right wheel and tire assembly and lower the vehicle.

19. Install the timing belt upper cover and gasket. Tighten the bolts to 69–95 inch lbs. (7.8–11.0 Nm).

20. Connect the negative battery cable.

1.9L Engine

1. Disconnect the negative battery cable.

2. Remove the accessory drive belt.

3. Remove the drive belt automatic tensioner, if equipped.

4. Remove the timing cover retaining nuts.

5. Installation is the reverse of the removal procedure.

OIL SEAL REPLACEMENT

1.8L Engine

1. Disconnect the negative battery cable.

2. Remove the timing belt.

3. Remove the timing belt pulley locking bolt.

4. Remove the timing belt pulley. If necessary, use a suitable puller.

5. Remove the Woodruff® key.

6. If necessary, cut the lip of the front oil seal to ease removal.

7. Use a suitable prying tool to remove the oil seal.

To install:

8. Lubricate the lip of the new oil seal with clean engine oil.

9. Using a suitable installation tool, install the seal evenly until it is flush with the edge of the oil pump body.

10. Install the timing belt pulley onto the shaft while making sure to match the alignment grooves.

11. Install the Woodruff® key with the tapered end facing the oil pump.

12. Install the timing belt pulley locking bolt. Tighten the locking bolt to 80–87 ft. lbs. (108–118 Nm).

13. Install the timing belt.

14. Connect the negative battery cable.

1.9L Engine

1988–90

1. Disconnect the negative battery cable.

2. Remove the accessory drive belts.

3. Remove the timing belt cover.

4. Remove the timing belt.

5. Remove the crankshaft damper.

6. Remove the crankshaft sprocket.

7. Remove the crankshaft front seal.

To install:

8. Coat the new seal with clean engine oil.

9. Install the crankshaft front seal using a suitable seal installer tool.

10. Install the crankshaft sprocket.

11. Install the crankshaft damper. Tighten the attaching bolt to 74–90 ft. lbs. (100–121 Nm).

12. Install the timing belt.

13. Install the timing belt cover.

14. Install the accessory drive belts.

15. Connect negative battery cable.

1991–92

1. Disconnect the negative battery cable.

2. Remove the accessory drive belt.

3. Raise and safely support the vehicle.

4. Remove the right side splash shield.

5. Remove the flywheel inspection shield.

6. Use a suitable tool to hold the flywheel in place.

7. Remove the crankshaft bolt and washer and remove the crankshaft dampener.

8. Remove the timing belt.

9. Remove the crankshaft sprocket and belt guide.

10. Using a suitable seal remover, remove the crankshaft seal from the oil pump body.

To install:

11. Lubricate the lip of the new seal with clean engine oil.

12. Install the new seal using a suitable installation tool.

13. Install the belt guide and crankshaft sprocket.

14. Install the timing belt.

15. Position the crankshaft dampener on the crankshaft. Install the attaching bolt and washer and tighten to 81–96 ft. lbs. (110–130 Nm).

16. Remove the flywheel holding tool and install the inspection shield.

17. Install the right splash shield and lower the vehicle.

18. Install the accessory drive belt.

19. Connect the negative battery cable.

20. Start the engine and check for leaks.

Timing Belt and Tensioner

ADJUSTMENT

1.8L Engine

1. Disconnect the negative battery cable.

2. Remove the timing belt upper and middle covers and gaskets.

3. Place a wrench onto the crankshaft pulley and rotate the crankshaft clockwise so the timing marks located on the camshaft pulley and the seal plate are aligned.

4. Rotate the crankshaft clockwise 2 complete revolutions and align the timing marks on the camshaft pulleys and seal plate.

5. Make sure the yellow ignition timing mark on the crankshaft pulley is aligned with the TDC mark on the timing belt cover.

6. Measure the timing belt deflection by applying 22 lbs. of pressure on the belt, at a point between the camshaft pulleys. The timing belt deflection should be 0.35–0.45 in. (9.0–11.5mm).

7. If the deflection is not within specification, loosen the tensioner lock bolt. Using a suitable prying tool to move the tensioner, tighten or loosen the belt, as required, so the deflection will meet specification. Tighten the tensioner lock bolt to 27–38 ft. lbs. (37–52 Nm) and recheck the timing belt deflection beginning with Step 3.

8. If the timing belt will not meet specification, it must be replaced.

9. Install new gaskets onto the timing belt covers and install. Tighten the bolts to 69–95 inch lbs. (7.8–11.0 Nm).

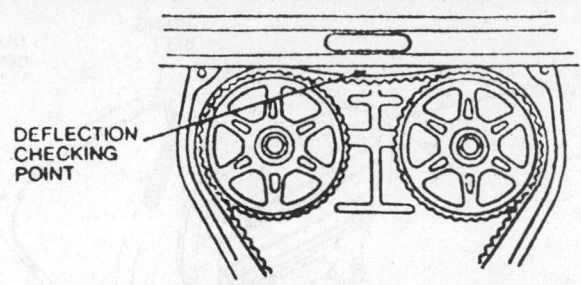

Timing belt deflection checking point—1.8L engine

1.9L Engine

The timing belt tensioner is spring-loaded, on the 1.9L engine. The spring automatically maintains the proper tension and periodic belt tension adjustments are not necessary.

REMOVAL & INSTALLATION

1.8L Engine

1. Disconnect the negative battery cable.

2. Remove the timing belt covers.

3. Rotate the crankshaft and align the timing marks located on the camshaft pulleys and the seal plate.

4. If the timing belt is to be reused, mark an arrow on the belt to indicate its rotational direction for installation reference.

5. Loosen the timing belt tensioner lock bolt and remove the timing belt.

To install:

6. Temporarily secure the timing belt tensioner in the far left position with the spring fully extended, then tighten the lock bolt.

7. Make sure the timing marks on the timing belt pulley and the engine block are aligned.

8. Make sure the timing marks on the camshaft pulleys and the seal plate are aligned.

9. Install the timing belt.

10. Loosen the tensioner lock bolt. Using a suitable prying tool, position the timing belt tensioner so the timing belt is taut, then tighten the tensioner lock bolt.

11. Turn the crankshaft 2 turns clockwise and align the timing belt pulley mark with the mark on the engine block.

12. Make sure the camshaft pulley marks are aligned with the seal plate marks.

NOTE: If the timing marks are not aligned, remove the belt and repeat the procedure.

13. Turn the crankshaft $1\frac{5}{6}$ turns clockwise and align the timing belt pulley mark with the tension set mark, at approximately the 10 o'clock position.

14. Apply tension to the timing belt tensioner and install the tensioner lock bolt. Tighten the bolt to 27–38 ft. lbs. (37–52 Nm).

15. Turn the crankshaft $2\frac{1}{6}$ turns clockwise and make sure the timing marks are aligned.

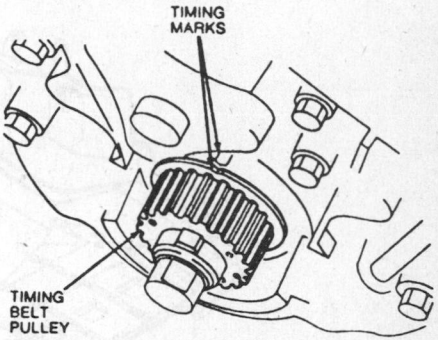

Timing belt pulley alignment marks—1.8L engine

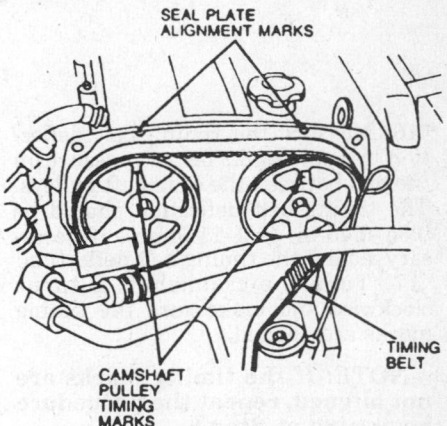

Camshaft pulley alignment marks—1.8L engine

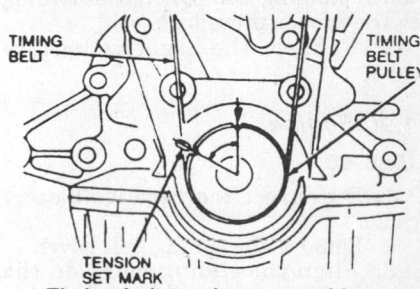

Timing belt tension set position—1.8L engine

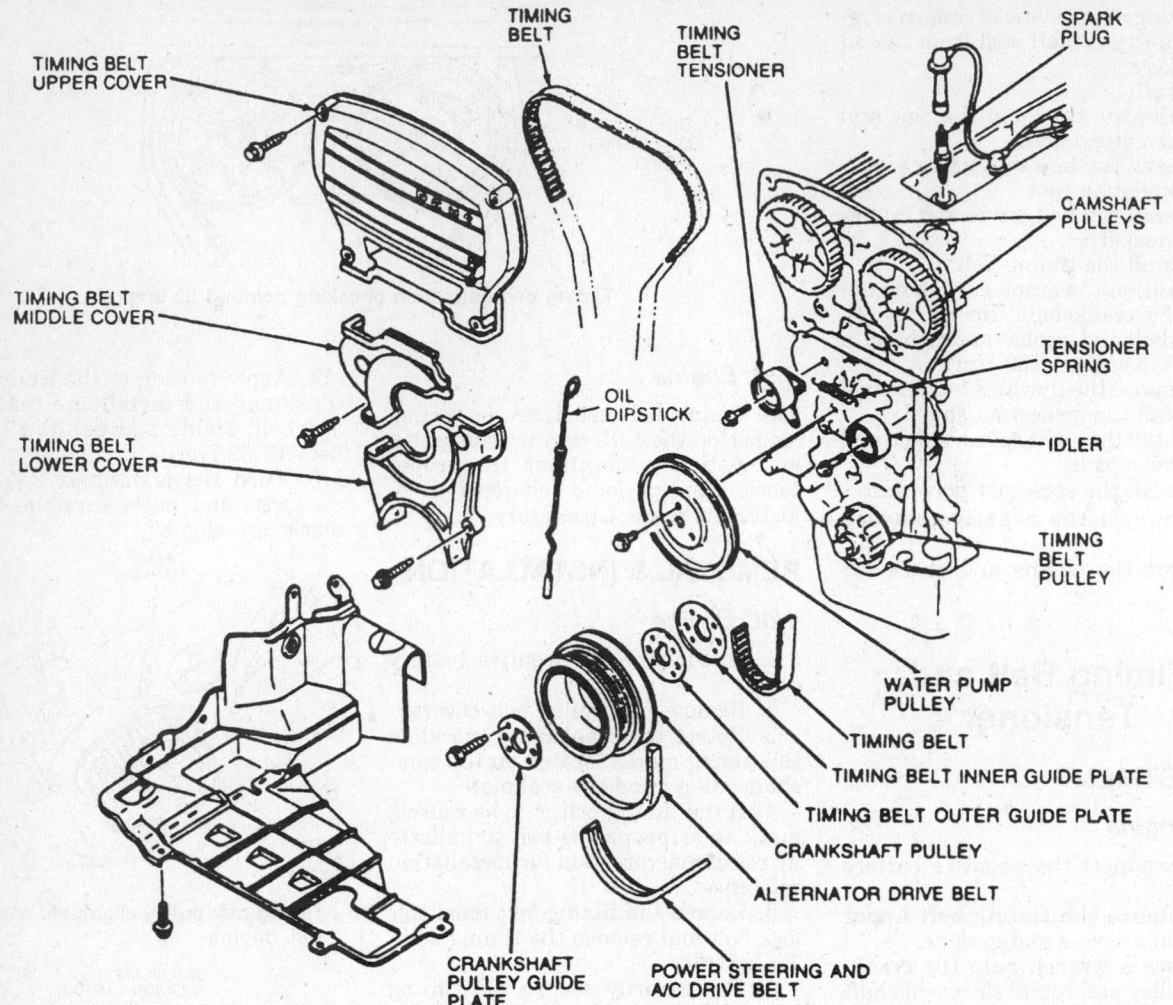

Timing belt removal and installation—1.8L engine

16. Measure the timing belt deflection by applying 22 lbs. of pressure on the belt between the camshaft pulleys. The timing belt deflection should be 0.35–0.45 in. (9.0–11.5mm). If necessary, adjust the timing belt deflection.

17. Turn the crankshaft 2 turns clockwise and make sure the timing marks are aligned.

NOTE: If the timing marks are not aligned, repeat the procedure beginning at Step 9.

18. Install the timing belt covers and the remaining components according to the proper procedure.

19. Connect the negative battery cable.

1.9L Engine

1988–90

1. Disconnect the negative battery cable.
2. Remove the timing belt cover.
3. Align the timing mark on the camshaft sprocket with the timing mark on the cylinder head.

4. Install the timing belt cover and confirm that the timing mark on the crankshaft pulley aligns with the **TDC** on the front cover.

5. Remove the timing belt cover.
6. Loosen both timing belt tensioner attaching bolts.
7. Pry the belt tensioner away from the belt as far as possible and tighten 1 of the tensioner attaching bolts.
8. Remove crankshaft pulley (damper) and remove the timing belt.

NOTE: With the timing belt removed and the No. 1 piston at TDC, Do not rotate the camshaft. If the camshaft must be rotated, align the crankshaft dampener 90 degrees BTDC.

To install:

9. Install the timing belt over the sprockets in a counterclockwise direction starting at the crankshaft. Keep the belt span from the crankshaft to the camshaft tight as the belt is installed over the remaining sprocket.

10. Loosen belt tensioner attaching bolts and allow the tensioner to snap against the belt.
11. Tighten 1 of the tensioner attaching bolts.
12. Install the crankshaft pulley, driveplate and pulley attaching bolt. Hold the crankshaft pulley stationary and tighten the pulley attaching bolt to 74–90 ft. lbs. (100–121 Nm).
13. To seat the belt on the sprocket teeth, complete the following:

    a. Connect the negative battery terminal.

    b. Crank engine several revolutions.

    c. Disconnect the negative battery terminal.

    d. Turn camshaft, as necessary, to align the timing pointer on the cam sprocket with the timing mark on the cylinder head.

NOTE: Do not turn the engine counterclockwise to align the timing marks.

    e. Position the timing belt cover on the engine and check to see that the timing mark on the crankshaft aligns with the TDC pointer on the cover. If the timing marks do not align, remove the belt, align the timing marks and return to Step 9.

14. Loosen the belt tensioner attaching bolt tightened in Step 11.

15. On 1988 vehicles, proceed as follows:

   a. Hold the crankshaft stationary and position a suitable torque wrench onto the camshaft sprocket bolt.

   b. Turn the camshaft sprocket counterclockwise. Tighten the belt tensioner attaching bolt when the torque wrench reads as follows:

   New belt — 27–32 ft. lbs.

   Used belt (30 days or more in service) — 10 ft. lbs.

16. On 1989–90 vehicles, the tensioner spring will apply the proper load on the belt. Tighten the belt tensioner bolt.

NOTE: The engine must be at room temperature. Do not set belt tension on a hot engine.

17. Install timing belt cover.
18. Install accessory drive belts.
19. Connect negative battery cable.

1991–92

1. Disconnect the negative battery cable.

2. Remove the accessory drive belt automatic tensioner and the accessory drive belt.

3. Remove the timing belt cover.

4. Align the timing mark on the camshaft sprocket with the timing mark on the cylinder head.

5. Confirm that the timing mark on the crankshaft sprocket is aligned with the timing mark on the oil pump housing.

6. Loosen the belt tensioner attaching bolt, pry the tensioner away from the timing belt and retighten the bolt.

7. Remove the spark plugs.

8. Raise and safely support the vehicle.

9. Remove the right side splash shield.

10. Remove the flywheel inspection shield.

11. Use a suitable tool to hold the flywheel in place.

12. Remove the crankshaft dampener bolt and washer and remove the dampener.

13. Remove the timing belt.

NOTE: With the timing belt removed and the No. 1 piston at TDC, Do not rotate the camshaft. If the camshaft must be rotated, align the crankshaft dampener 90 degrees BTDC.

To install:

14. Install the timing belt over the

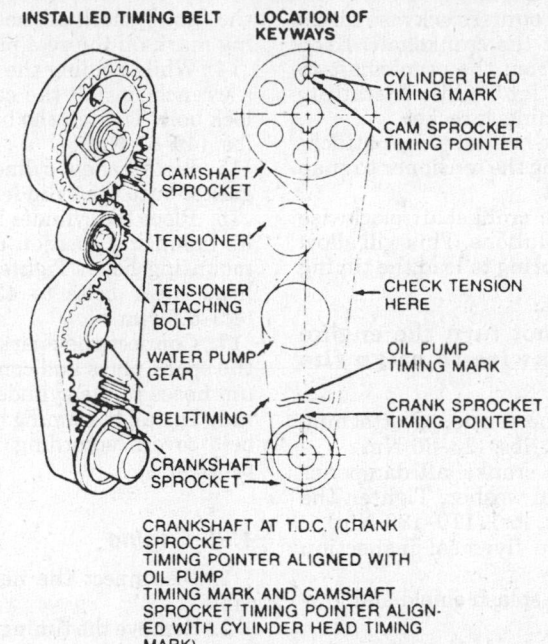

INSTALLED TIMING BELT   LOCATION OF KEYWAYS

CYLINDER HEAD TIMING MARK
CAM SPROCKET TIMING POINTER
CAMSHAFT SPROCKET
TENSIONER
TENSIONER ATTACHING BOLT
CHECK TENSION HERE
WATER PUMP GEAR
OIL PUMP TIMING MARK
BELT-TIMING
CRANK SPROCKET TIMING POINTER
CRANKSHAFT SPROCKET

CRANKSHAFT AT T.D.C. (CRANK SPROCKET TIMING POINTER ALIGNED WITH OIL PUMP TIMING MARK AND CAMSHAFT SPROCKET TIMING POINTER ALIGNED WITH CYLINDER HEAD TIMING MARK).

Timing belt pulley alignment — 1.9L engine

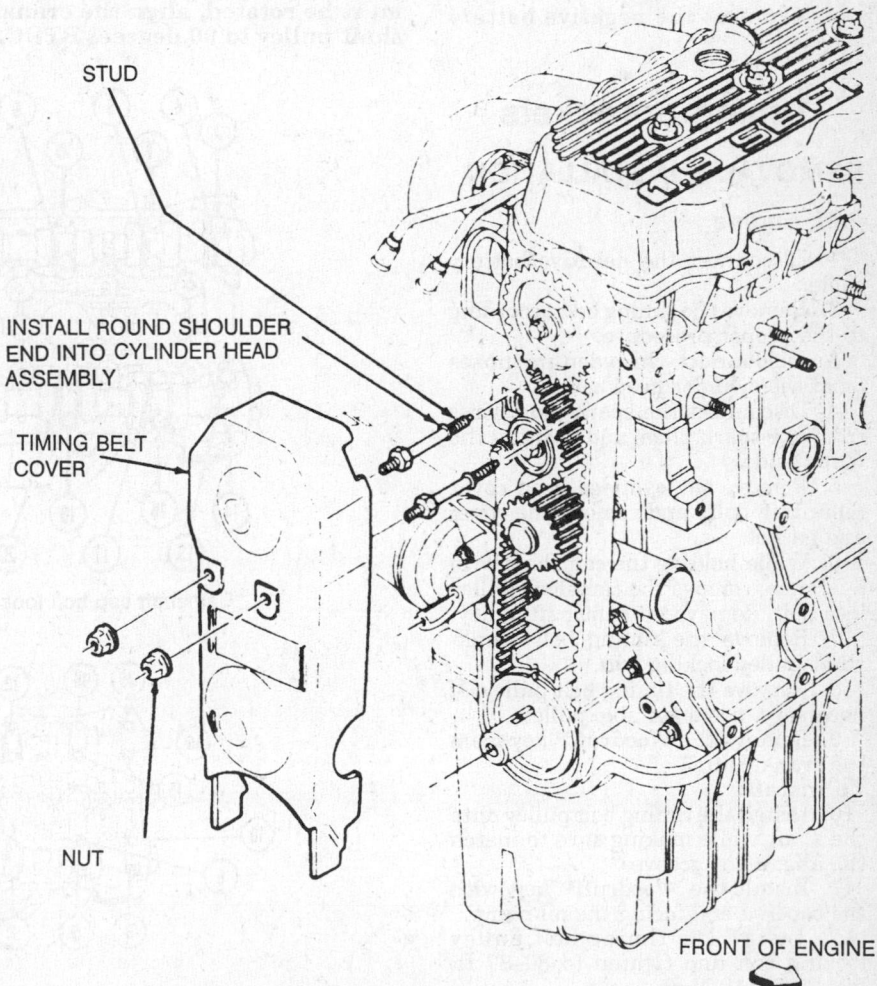

STUD

INSTALL ROUND SHOULDER END INTO CYLINDER HEAD ASSEMBLY

TIMING BELT COVER

NUT

1.9 SEFI

FRONT OF ENGINE

Timing belt cover removal and installation — 1991–92 1.9L engine

sprockets in a counterclockwise direction starting at the crankshaft. Keep the belt span from the crankshaft to the camshaft tight while installing over the remaining sprocket.

15. Loosen the belt tensioner attaching bolt, allowing the tensioner to snap against the belt.

16. Rotate the crankshaft clockwise 2 complete revolutions. This will allow the tensioner spring to load the timing belt.

NOTE: Do not turn the engine counterclockwise to align the timing marks.

17. Tighten the tensioner attaching bolt to 17–22 ft. lbs. (23–30 Nm).

18. Install the crankshaft dampener and the bolt and washer. Tighten the bolt to 81–96 ft. lbs. (110–130 Nm).

19. Install the flywheel inspection shield.

20. Install the splash shield and lower the vehicle.

21. Install the spark plugs.

22. Install the timing belt cover.

23. Install the accessory drive belt automatic tensioner and the accessory drive belt.

24. Connect the negative battery cable.

Timing Sprockets

REMOVAL & INSTALLATION

1.8L Engine

1. Disconnect the negative battery cable.

2. Remove the timing belt according to the proper procedure.

3. Disconnect the vacuum hoses from the cylinder head cover.

4. Disconnect the spark plug wires from the spark plugs and position the wires aside.

5. Remove the cylinder head cover mounting bolts and remove the cover and gasket.

6. While holding the camshaft with a wrench, remove the camshaft pulley lock bolt. Remove the camshaft pulley.

7. Remove the timing belt crankshaft pulley locking bolt.

8. Remove the timing belt pulley. If necessary, use a suitable puller.

9. Remove the Woodruff® key from the crankshaft.

To install:

10. Install the timing belt pulley onto the shaft while making sure to match the alignment grooves.

11. Install the Woodruff® key with the tapered end facing the oil pump.

12. Install the timing belt pulley locking bolt and tighten to 80–87 ft. lbs. (108–118 Nm).

13. Install the camshaft pulley with

the timing mark aligned with the timing mark on the seal plate.

14. While holding the camshaft with a wrench, install the camshaft pulley lock bolt. Tighten the bolt to 36–45 ft. lbs. (49–61 Nm).

15. Install a new cylinder head cover gasket onto the cylinder head.

16. Place the cylinder head cover into its mounting position and install the mounting bolts. Tighten the cylinder head cover bolts to 43–78 inch lbs. (4.9–8.8 Nm).

17. Connect the spark plug wires to the spark plugs and connect the vacuum hoses to the cylinder head cover.

18. Install the timing belt and timing belt covers according to the proper procedure.

1.9L Engine

1. Disconnect the negative battery cable.

2. Remove the timing belt cover and timing belt.

NOTE: With the timing belt removed and pistons at TDC, do not rotate the engine. If the camshaft must be rotated, align the crankshaft pulley to 90 degrees BTDC.

3. Remove the camshaft sprocket attaching bolt and washer and camshaft sprocket.

4. Remove the crankshaft sprocket.

5. Install the camshaft sprocket and attaching bolt and washer. Tighten to 37–46 ft. lbs. (50–62 Nm) on 1988 vehicles or 71–84 ft. lbs. (95–115 Nm) on 1989–92 vehicles.

6. Install the crankshaft sprocket.

7. Install the timing belt and cover.

8. Connect the negative battery cable.

Camshaft

REMOVAL & INSTALLATION

1.8L Engine

1. Disconnect the negative battery cable.

2. Remove the distributor assembly.

3. Remove the camshaft pulleys.

4. Remove the seal plate mounting bolts and remove the seal plate.

5. Loosen the camshaft cap bolts in the correct sequence.

6. Remove the camshaft caps and note their mounting locations for installation reference.

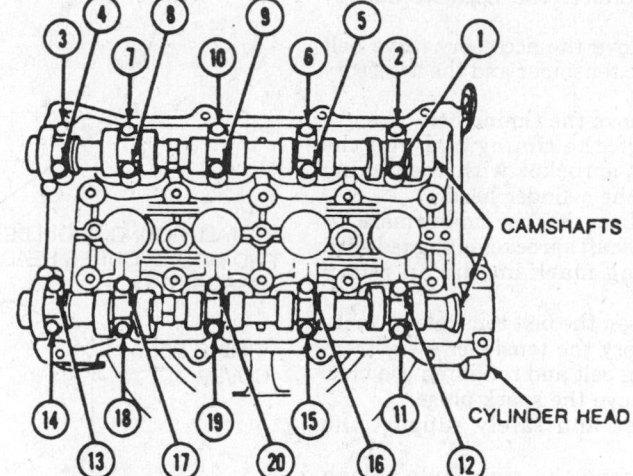

Camshaft cap bolt loosening sequence—1.8L engine

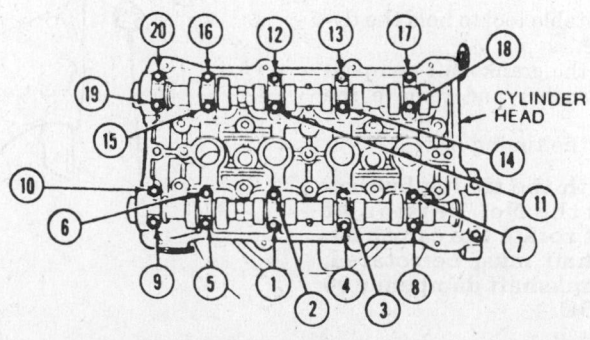

Camshaft cap bolt torque sequence—1.8L engine

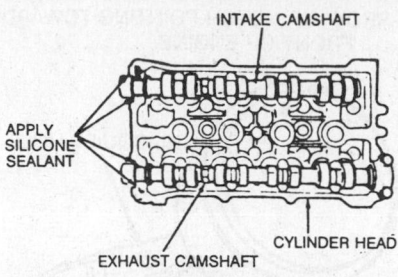

Silicone sealer application points— 1.8L engine

NOTE: The camshaft caps are numbered and have arrow marks for installation and direction reference.

7. Remove the camshaft and camshaft oil seal.

To install:

8. Apply clean engine oil to the camshaft journals and bearings.

9. Place the camshaft into its mounting position.

NOTE: The exhaust camshaft has a groove which must be installed into the distributor drive gear.

10. Apply silicone sealant to the required areas.

11. Install the camshaft caps according to the cap numbers and arrow marks.

12. Install the camshaft cap bolts and tighten them in the proper sequence to 100–126 inch lbs. (11.3–14.2 Nm).

13. Apply a small amount of clean engine oil to the lip of a new camshaft oil seal. Using a suitable installation tool, install the new seal.

14. Place the seal plate into its mounting position and install the mounting bolts. Tighten the bolts to 69–95 inch lbs. (7.8–11.0 Nm).

15. Install the camshaft pulleys and the distributor assembly.

16. Connect the negative battery cable.

1.9L Engine

1. Disconnect the negative battery cable.

2. Remove the air cleaner or air intake duct.

3. Remove the accessory drive belts and crankshaft pulley.

4. Remove the timing belt cover and valve cover.

5. Set the engine No. 1 cylinder at TDC prior to removing timing belt.

NOTE: Make sure the crankshaft is positioned at TDC and do not turn the crankshaft until the timing belt is installed.

6. Remove rocker arms and lifters as follows:

  a. Remove hex flange bolts.
  b. Remove fulcrums.
  c. Remove rocker arms.
  d. Remove lifter guide retainers.
  e. Remove lifters guides.
  f. Remove lifters.

7. Remove the distributor assembly on 1988–90 vehicles. On 1991–92 vehicles, remove the ignition coil assembly.

8. Remove timing belt.

9. Remove the camshaft sprocket and key.

10. Remove the camshaft thrust plate.

11. Remove the ignition coil and coil bracket on 1988–90 vehicles. On 1991–92 vehicles, remove the cup plug from the back of the cylinder head.

12. Remove the camshaft through the back of the head toward the transaxle.

13. Replace camshaft seal.

To install:

14. Thoroughly coat the camshaft bearing journals, cam lobe surfaces and thrust plate groove with a suitable lubricant.

15. Install camshaft through the rear of the cylinder head. Rotate camshaft during installation.

NOTE: Before installing the camshaft, apply a thin film of lubricant to the lip of the camshaft seal.

16. Install the camshaft thrust plate. Tighten attaching bolts to 6–9 ft. lbs. (8–13 Nm).

17. Align and install the cam sprocket over the cam key. Install attaching washer and bolt. While holding camshaft stationary, tighten the bolt to 37–46 ft. lbs. (50–62 Nm) on 1988 vehicles or 71–84 ft. lbs. (95–115 Nm) on 1989–92 vehicles.

18. On 1988–90 vehicles, install the ignition coil and coil bracket. On 1991–92 vehicles, install the cup plug using a suitable sealer. Use the sealer sparingly, as excess sealer may clog the oil holes in the camshaft.

19. Install the timing belt.

20. Install the timing belt cover.

21. Install the rocker arm assembly as follows:

NOTE: Lubricate all the parts with a heavy engine oil before installation.

  a. Install the lifters.
  b. Install the lifter guides.
  c. Install the lifter retainers.
  d. Install the rocker arms.
  e. Install the fulcrums.
  f. Install the rocker arm bolts. Tighten to 17–22 ft. lbs. (23–30 Nm).

22. Install the distributor assembly on 1988–90 vehicles. On 1991–92 vehicles, install the ignition coil assembly.

23. Install new valve cover gasket, if required.

NOTE: Make sure the surfaces on the cylinder head and valve cover are clean and free of sealant material.

24. On 1988–90 vehicles, install the rocker arm cover attaching bolts and studs. Tighten bolts and studs to 6–8 ft. lbs. (8–11.5 Nm). On 1991–92 vehicles, install the attaching bolts and tighten to 4–9 ft. lbs. (5–12 Nm).

25. Install the air intake duct or the air cleaner assembly.

26. Connect negative battery cable.

2.3L Engine

1. Disconnect the negative battery cable.

2. Drain the cooling system and crankcase. Properly relieve the fuel system pressure.

3. Remove the engine from the vehicle and position in a suitable holding fixture. Remove the engine oil dipstick.

4. Remove necessary drive belts and pulleys.

5. Remove the cylinder head.

6. Remove the distributor.

7. Using a magnet, remove the hydraulic lifters and label them so they can be installed in their original positions. If the lifters are stuck in the bores by excessive varnish, etc., use a suitable puller to remove them.

8. Remove the crankshaft pulley.

9. Remove the oil pan.

10. Remove the cylinder front cover and gasket.

11. Check the camshaft endplay as follows:

  a. Push the camshaft toward the rear of the engine and install a dial indicator tool, so the indicator point is on the camshaft sprocket attaching screw.

  b. Zero the dial indicator. Position a small prybar or equivalent, between the camshaft sprocket or gear and block.

  c. Pull the camshaft forward and release it. Compare the dial indicator reading with the camshaft endplay specification of 0.009 in.

  d. If the camshaft endplay is over the amount specified, replace the thrust plate.

12. Remove the timing chain, sprockets and timing chain tensioner.

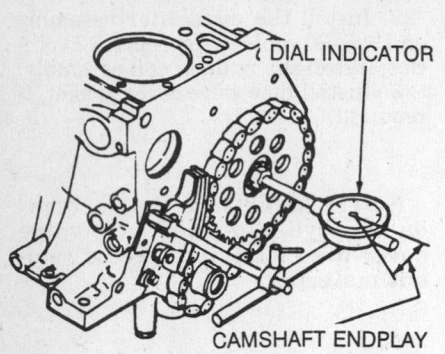

Checking camshaft endplay—2.3L engine

Piston and Connecting Rod

POSITIONING

NOTE: On 1.8L engine, the piston and rod assembly must be positioned in the engine block with the F mark facing the front of the engine.

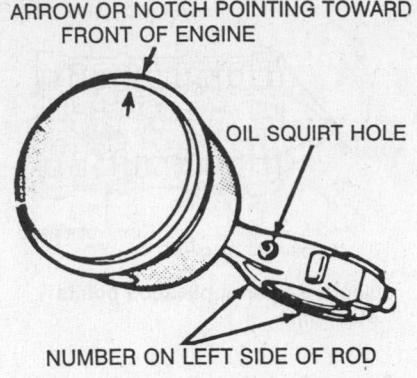

Piston and rod assembly—2.3L engine

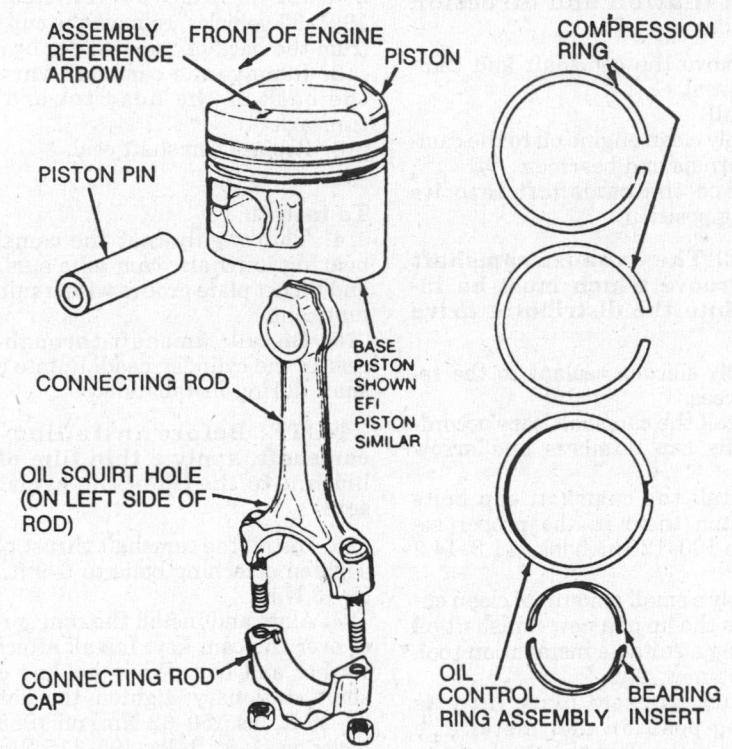

Piston and rod assembly—1.9L engine

13. Remove camshaft thrust plate. Carefully remove the camshaft by pulling it toward the front of the engine. Use caution to avoid damaging bearings, journals and lobes.

To install:

14. Clean and inspect all parts before installation.

15. Lubricate camshaft lobes and journals with heavy engine oil. Carefully slide the camshaft through the bearings in the cylinder block.

16. Install the thrust plate. Tighten attaching bolts to 6–9 ft. lbs (8–12 Nm).

17. Install the timing chain, sprockets and timing chain tensioner according to the proper procedure.

18. Install the cylinder front cover and crankshaft pulley.

19. Clean the oil pump inlet tube screen, oil pan and cylinder block gasket surfaces. Prime oil pump by filling the inlet opening with oil and rotate the pump shaft until oil emerges from the outlet tube. Install oil pump, oil pump inlet tube screen and oil pan.

20. Install the accessory drive belts and pulleys.

21. Lubricate the lifters and lifter bores with heavy engine oil. Install lifters into their original bores.

22. Install cylinder head.

23. Position No. 1 piston at TDC after the compression stroke. Position distributor in the block with the rotor at the No. 1 firing position. Install distributor retaining clamp.

24. Install engine in vehicle.

25. Connect engine temperature sending unit wire. Connect coil primary wire. Install distributor cap. Connect spark plug wires and the coil high tension lead.

26. Fill the cooling system and crankcase to the proper levels.

27. Connect negative battery cable.

28. Start the engine. Check and adjust ignition timing. Check for leaks.

ENGINE LUBRICATION

Oil Pan

REMOVAL & INSTALLATION

1.8L Engine

1. Disconnect the negative battery cable. Remove the oil filler cap.

2. Raise and safely support the vehicle.

3. Remove the drain plug and drain the engine oil into a suitable container.

4. Remove the right upper and right and left lower splash shields.

5. Remove the exhaust pipe front mounting flange and exhaust pipe support bracket from the exhaust manifold.

6. Remove the oil pan-to-transaxle attaching bolts.

7. Support the oil pan with a suitable jack stand.

8. Remove the oil pan-to-engine block attaching bolts.

NOTE: Do not force a prying tool between the engine block and the oil pan contact surface when trying to remove the oil pan. This may damage the oil pan contact surface and cause oil leakage.

9. Only at the most rearward points of the oil pan, next to the transaxle, use a suitable tool to carefully pry the

oil pan away from the engine block and remove the oil pan.

10. Use a suitable tool to pry the crankcase stiffeners away from the engine block and/or oil pan.

11. Remove the front and rear oil pan gaskets and end seals. Remove all sealant material from the engine block and oil pan.

NOTE: When removing the crankcase stiffeners and sealant material from the oil pan and engine block, be careful not to damage the oil pan and engine block contact surfaces.

To install:

12. Apply a bead of silicone sealant to the crankcase stiffeners along the inside of the bolt holes.

13. Install the crankcase stiffeners onto the oil pan.

14. Apply sealant to the proper areas of the end seals. Be sure to install the end seals with the projections in the notches.

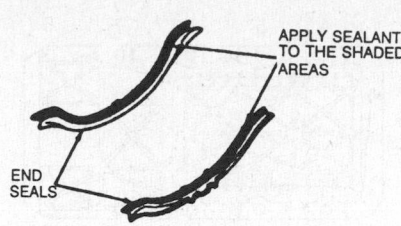

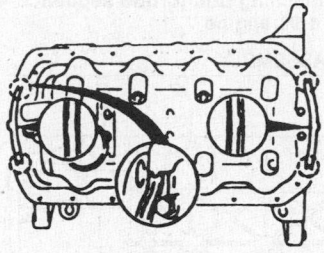

Sealant application areas and oil pan end seal installation—1.8L engine

15. Install the front and rear end seals onto the oil pan.

16. Apply a continuous bead of silicone sealant to the oil pan along the inside of the bolt holes. Overlap the sealant ends.

17. Place the oil pan into its mounting position and install the oil pan-to-engine block attaching bolts. Tighten the bolts to 69–95 inch lbs. (7.8–11.0 Nm).

NOTE: If the oil pan attaching bolts are to be resused, the old sealant must be removed from the bolt threads. Tightening the old attaching bolts with old sealant still on them may cause cracking inside the bolt holes.

18. Install the oil pan-to-transaxle attaching bolts and tighten to 27–38 ft. lbs. (37–52 Nm).

19. Install the oil drain plug and tighten to 22–30 ft. lbs. (29–41 Nm).

20. Install the exhaust front mounting flange to the exhaust manifold while making sure to install a new gasket. Tighten the mounting flange-to-exhaust manifold attaching nuts to 23–34 ft. lbs. (31–46 Nm).

21. Install the exhaust pipe support bracket and tighten the bolts to 27–38 ft. lbs. (37–52 Nm).

22. Install the splash shields. Tighten the bolts to 69–95 inch lbs. (7.8–11.0 Nm).

23. Lower the vehicle.

24. Fill the crankcase with the proper type and quantity of engine oil. Install the filler cap.

25. Connect the negative battery cable.

1.9L Engine

1. Disconnect negative cable at the battery.

2. Raise the vehicle and support safely.

3. Drain the crankcase.

4. On 1988–90 vehicles, disconnect cable at the starter, remove starter-brace located at the front of the starter and remove starter attaching bolts and starter.

5. Remove the 2 oil pan-to-transaxle bolts.

6. Disconnect the exhaust inlet pipe at the manifold and converter. Remove pipe.

7. Remove oil pan retaining bolts and oil pan.

8. Remove oil pan gasket and discard.

To install:

9. Clean the oil pan gasket surface and the mating surface on the cylinder block. Wipe the oil pan rail with a solvent-soaked cloth to remove oil traces.

10. Remove and clean the oil pump pick up tube and screen assembly. In-

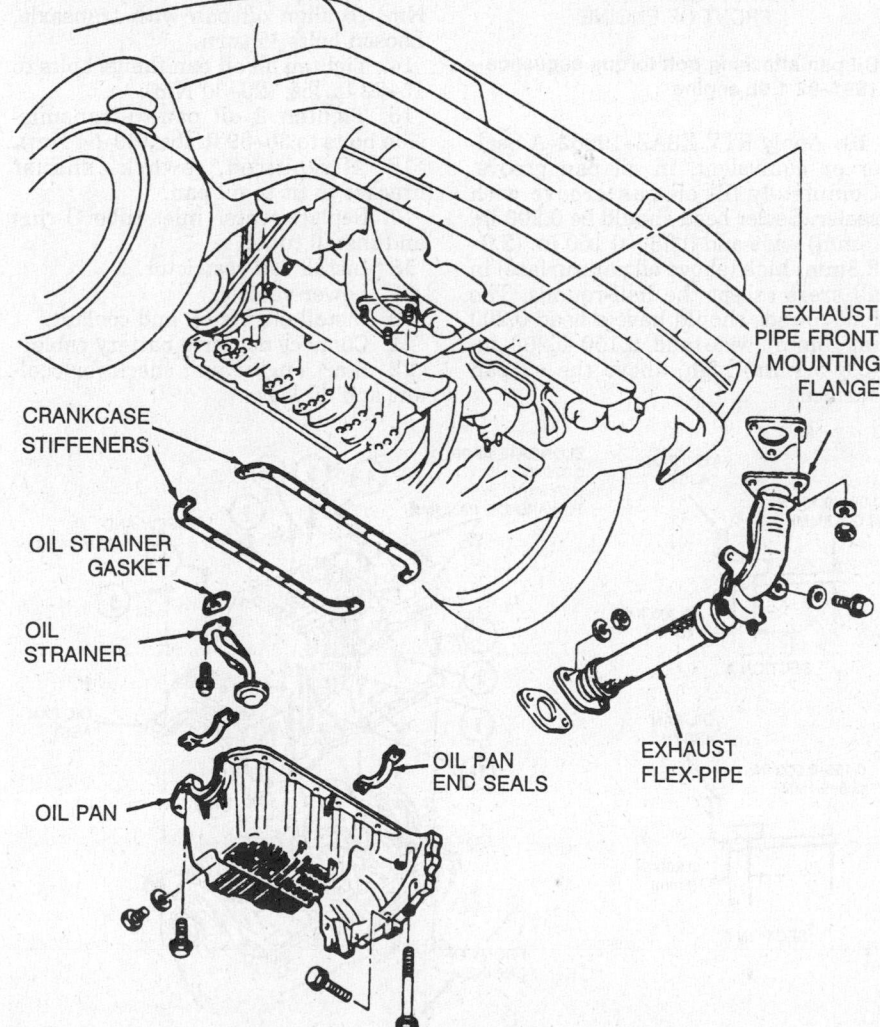

Oil pan removal and installation—1.8L engine

stall tube and screen assembly using a new gasket.

11. Apply a bead of suitable silicone rubber sealer at the corner of the oil pan front and rear seals and at the seating point of the oil pump to the block retainer joint.

12. Install the gasket in oil pan ensuring press fit tabs are fully engaged in oil pan gasket channel.

13. Install the oil pan and the attaching bolts. Tighten the bolts lightly until the 2 oil pan-to-transmission bolts can be installed.

NOTE: If the oil pan is installed on the engine outside of the vehicle, a transaxle case, or equivalent fixture must be bolted to the block to line the oil pan up, flush with the rear face of the block.

14. Tighten the 2 pan-to-transaxle bolts to 30–40 ft. lbs. (40–54 Nm), then loosen ½ turn.

15. Tighten the oil pan flange-to-cylinder block bolts to 15–22 ft. lbs. (20–30 Nm) in the proper sequence. Retighten the 2 oil pan-to-transaxle bolts to 30–40 ft. lbs. (40–55 Nm).

16. Install the transaxle inspection plate.

17. On 1988–90 vehicles, install the starter, starter brace at the starter and connect the starter cable.

18. Install the exhaust inlet pipe. Lower the vehicle and fill the crankcase.

19. Connect negative battery cable.

20. Start the engine and check for oil leaks.

2.3L Engine

1. Disconnect the negative battery cable. Raise the vehicle and support safely.

2. Drain the crankcase and drain the cooling system by removing the lower radiator hose.

3. Remove the roll restrictor on manual transaxle equipped vehicles.

4. Disconnect the starter cable.

5. Remove the starter.

6. Disconnect the exhaust pipe from oil pan.

7. Remove the engine coolant tube from the lower radiator hose, water pump and at the tabs on the oil pan. Position air conditioner line off to the side. emove the retaining bolts and remove the oil pan.

To install:

8. Clean both mating surfaces of oil pan and cylinder block making certain all traces of RTV sealant are removed. Ensure that the block rails, front cover and rear cover retainer are also clean.

9. Remove and clean oil pump pickup tube and screen assembly. After cleaning, install tube and screen assembly.

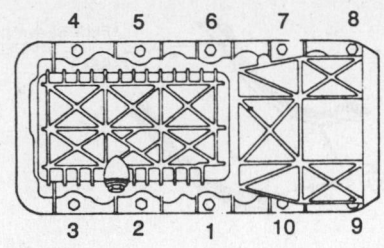

Oil pan attaching bolt torque sequence— 1988–90 1.9L engine

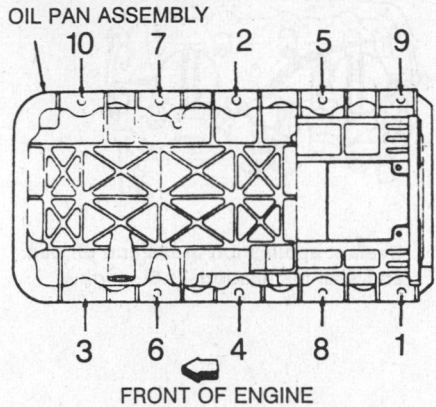

Oil pan attaching bolt torque sequence— 1991–92 1.9L engine

10. Apply RTV E8AZ–19562–A Sealer or equivalent, in oil pan groove. Completely fill oil pan groove with sealer. Sealer bead should be 0.200 in. (5mm) wide and 0.080–0.150 in. (2.0–3.8mm) high (above oil pan surface) in all areas except the half-rounds. The half-rounds should have a bead 0.200 in. (5mm) wide and 0.150–0.200 in. (3.8–5.1mm) high, above the oil pan surface.

NOTE: Applying RTV in excess of the specified amount will not improve the sealing of the oil pan, and could cause the oil pickup screen to become clogged with sealer. Use adequate ventilation when applying sealer.

11. Install oil pan to cylinder block within 5 minutes to prevent skinning over. RTV needs to cure completely before coming in contact with any engine oil, about 1 hour at ambient temperature between 65–75°F.

12. Install oil pan bolts lightly until the 2 oil pan-to-transmission bolts can be installed.

NOTE: If oil pan is installed on engine outside of vehicle, a transaxle case or equivalent, fixture must be bolted to the block to lin the oil pan up, flush with the rear face of block.

13. Install 2 oil pan-to-transaxle bolts. Tighten to 30–39 ft. lbs. (40–54 Nm) to align oil pan with transaxle. Loosen bolts ½ turn.

14. Tighten all oil pan flange bolts to 15–22 ft. lbs. (20–30 Nm).

15. Tighten 2 oil pan-to-transmission bolts to 30–39 ft. lbs. (40–54 Nm).

16. If required, rework exhaust bracket to fit to oil pan.

17. Replace water inlet tube O-ring and install tube.

18. Install roll restrictor.

19. Lower vehicle.

20. Install engine oil and coolant.

21. Connect negative battery cable.

22. Start engine and check for coolant and oil leaks.

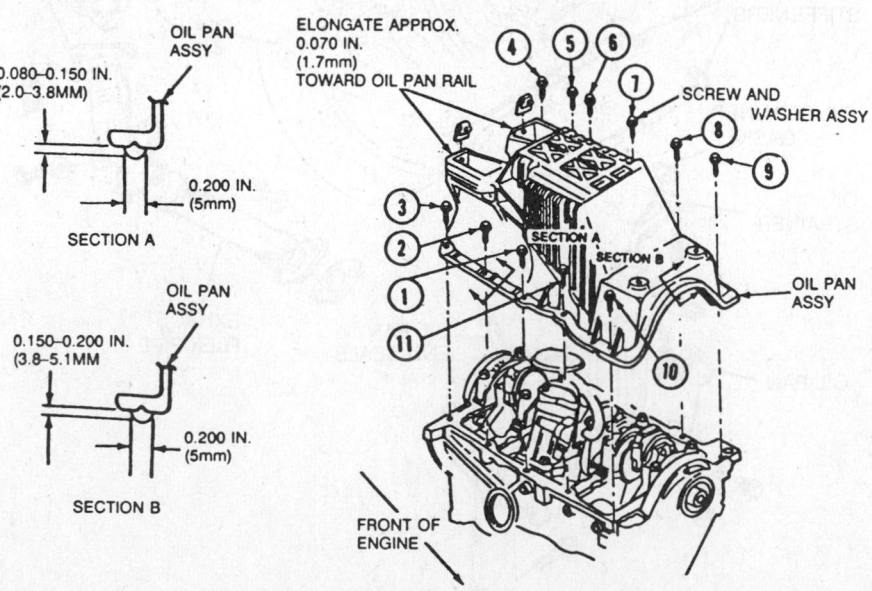

Oil pan installation—2.3L engine

Oil Pump

REMOVAL & INSTALLATION

1.8L Engine

1. Disconnect the negative battery cable.
2. Remove the timing belt and crankshaft pulley.
3. Remove the oil pan.
4. Remove the oil strainer mounting bolts and remove the oil strainer and gasket.
5. If equipped, remove the air conditioning compressor mounting bolts and position the compressor so it is free from the work area.
6. Remove the air conditioning compressor mounting bracket.
7. Remove the mounting bolt from the engine oil dipstick tube bracket and remove the alternator lower mounting bolt.
8. Remove all oil pump mounting bolts and remove the oil pump. Remove all gasket material from the oil pump.

To install:

9. Install a new gasket onto the oil pump.
10. Place the oil pump into its mounting position and install the pump mounting bolts. Tighten the bolts to 14–19 ft. lbs. (19–25 Nm).
11. Place the dipstick tube bracket bolt into its mounting position and install the mounting bolt.
12. Install the alternator lower mounting bolt and tighten to 27–38 ft. lbs. (37–52 Nm).
13. Install a new gasket onto the oil strainer, place the strainer into its mounting position and install the mounting bolts. Tighten to 69–95 inch lbs. (7.8–11.0 Nm).
14. Install the oil pan.
15. If equipped, place the air conditioning compressor bracket into its mounting position and install the mounting bolts. Tighten the bolts to 15–22 ft. lbs. (20–30 Nm).
16. Install the crankshaft pulley and timing belt.
17. Connect the negative battery cable.

1.9L Engine

1988–90

1. Disconnect the negative cable at the battery.
2. Loosen the alternator bolt on the alternator adjusting arm.
3. Lower the alternator to remove the accessory drive belt from the crankshaft pulley.

4. Remove the timing belt cover.
5. Set No. 1 cylinder at TDC. Loosen both belt tensioner attaching bolts. Using a prybar of other suitable tool, pry the tensioner away from the belt. While holding the tensioner away from the belt, tighten one of the tensioner attaching bolts.
6. Disengage timing belt from camshaft sprocket, water pump sprocket and crankshaft sprocket.
7. Raise and safely support the vehicle.
8. Drain the crankcase.
9. Remove the crankshaft pulley attaching bolt.
10. Remove the timing belt.
11. Remove the crankshaft driveplate assembly.
12. Remove the crankshaft pulley.
13. Remove the crankshaft sprocket.
14. Disconnect the starter cable at the starter.
15. Remove the starter-brace from the engine.
16. Remove the starter.
17. Remove 2 oil pan-to-transaxle bolts.
18. Remove oil pan retaining bolts and oil pan.
19. Remove 1 piece oil pan gasket.
20. Remove oil pump attaching bolts.
21. Remove oil pump and gasket.

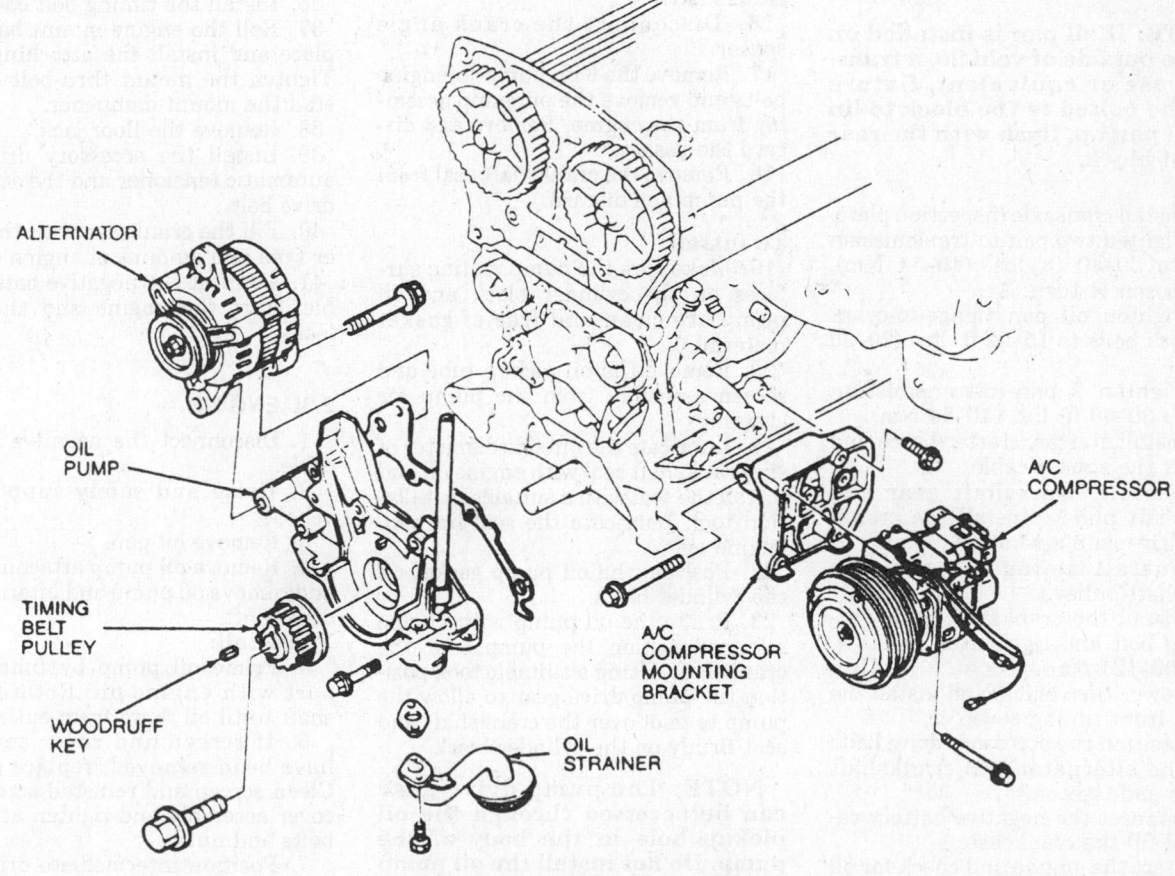

ALTERNATOR

OIL PUMP

TIMING BELT PULLEY

WOODRUFF KEY

A/C COMPRESSOR

A/C COMPRESSOR MOUNTING BRACKET

OIL STRAINER

Oil pump installation—1.8L engine

22. Remove oil pump seal.

To install:

23. Make sure the mating surfaces on the cylinder block and the oil pump are clean and free of gasket material.

24. Remove the oil pick-up tube and screen assembly from the pump for cleaning.

25. Lubricate the outside diameter of the oil pump seal with clean engine oil.

26. Install the oil pump seal using a suitable seal installer tool.

27. Install pick-up tube and screen assembly on the oil pump using a new gasket. Tighten attaching bolts to 6–9 ft. lbs. (8–12 Nm).

28. Lubricate the oil pump seal lip with light engine oil.

29. Position the oil pump gasket over the locating dowels.

30. Position the oil pump. Install attaching bolts and tighten to 5–7 ft. lbs. (7–9 Nm).

31. Apply a bead of sealer approximately ⅛ in. (3.0mm) wide at the corners of the block and at the seating point of the oil pump and the rear seal retainer joint.

32. Install gasket in oil pan ensuring press fit tabs are fully engaged in the oil pan gasket channel.

33. Position the oil pan on the cylinder block. Install oil pan attaching bolts. Tighten lightly until 2 oil pan-to-transmission bolts can be installed.

NOTE: If oil pan is installed on engine outside of vehicle, a transaxle case or equivalent, fixture must be bolted to the block to lin the oil pan up, flush with the rear face of block.

34. Install transaxle inspection plate.

35. Tighten two pan-to-transmission bolts to 30–40 ft. lbs. (40–54 Nm), then loosen ½ turn.

36. Tighten oil pan flange-to-cylinder block bolts to 15–22 ft. lbs. (20–30 Nm).

37. Tighten 2 pan-to-transmission bolts to 30–40 ft. lbs. (40–54 Nm).

38. Install starter, starter-brace and connect the starter cable.

39. Install crankshaft gear and crankshaft pulley. Install the crankshaft driveplate assembly.

40. Install timing belt over the crankshaft pulley.

41. Install the crankshaft pulley attaching bolt and tighten to 74–90 ft. lbs. (100–121 Nm).

42. Lower the vehicle and install the engine front timing cover.

43. Position the accessory drive belts over the alternator and crankshaft pulleys and tighten.

44. Connect the negative battery cable and fill the crankcase.

45. Start the engine and check for oil leaks.

1991–92

1. Disconnect the negative battery cable.

2. Remove the accessory drive belt and the automatic tensioner.

3. Support the engine with a suitable floor jack.

4. Remove the right engine mount dampener and remove the right engine mount bolts from the mount bracket.

5. Loosen the mount thru-bolt and roll the mount aside.

6. Remove the timing belt cover.

7. Make sure the No. 1 cylinder is at TDC.

8. Roll the engine mount back into place and install the 2 mount bolts. Remove the floor jack.

9. Loosen the belt tensioner attaching bolt and pry the tensioner to the rear of the engine. Tighten the attaching bolt.

10. Raise and safely support the vehicle.

11. Remove the right side splash shield.

12. Remove the catalytic converter inlet pipe.

13. Drain and remove the oil pan. Remove the oil filter.

14. Remove the crankshaft dampener and the timing belt.

15. Remove the crankshaft sprocket and the timing belt guide from the crankshaft.

16. Disconnect the crank angle sensor.

17. Remove the 6 oil pump-to-engine bolts and remove the oil pump assembly from the engine. Remove and discard the gasket.

18. Remove the crankshaft seal from the pump and discard.

To install:

19. Make sure the pump mating surfaces on the cylinder block and oil pump are clean and free of gasket material.

20. Remove the oil pickup tube and screen assembly from the pump for cleaning.

21. Lubricate the outside diameter of the crankshaft seal with engine oil and install the seal with a suitable installation tool. Lubricate the seal lip with engine oil.

22. Position the oil pump gasket on the cylinder block.

23. Prime the oil pump with engine oil and position the pump over the crankshaft. Using a suitable tool, position the pump drive gear to allow the pump to pilot over the crankshaft and seat firmly on the cylinder block.

NOTE: The pump drive gear can be accessed through the oil pickup hole in the body of the pump. Do not install the oil pump pickup tube and screen until the

pump has been correctly installed on the cylinder block.

24. Install the 6 oil pump bolts and tighten to 8–12 ft. lbs. (11–16 Nm).

NOTE: When the oil pump bolts are tightened, the gasket must not be below the cylinder block sealing surface.

25. Install the pickup tube and screen assembly on the oil pump using a new gasket. Tighten the attaching screws to 7–9 ft. lbs. (10–13 Nm).

26. Install the timing belt guide over the end of the crankshaft and install the crankshaft sprocket.

27. Make sure the No. 1 cylinder is at TDC.

28. Position the timing belt over the sprockets.

29. Connect the crank angle sensor.

30. Install the oil pan and the crankshaft dampener.

31. Install the catalytic converter inlet pipe.

32. Install the splash shield and lower the vehicle.

33. Install the timing belt. Tighten the tensioner attaching bolt to 17–22 ft. lbs. (23–30 Nm).

34. Support the engine with a suitable floor jack.

35. Remove the right engine mount bolts and roll the mount back.

36. Install the timing belt cover.

37. Roll the engine mount back into place and install the attaching bolts. Tighten the mount thru-bolt and install the mount dampener.

38. Remove the floor jack.

39. Install the accessory drive belt automatic tensioner and the accessory drive belt.

40. Fill the crankcase with the proper type and amount of engine oil.

41. Connect the negative battery cable, start the engine and check for leaks.

2.3L ENGINE

1. Disconnect the negative battery cable.

2. Raise and safely support the vehicle.

3. Remove oil pan.

4. Remove oil pump attaching bolts and remove oil pump and intermediate driveshaft.

To install:

5. Prime oil pump by filling inlet port with engine oil. Rotate pump shaft until oil flows from outlet port.

6. If screen and cover assembly have been removed, replace gasket. Clean screen and reinstall screen and cover assembly and tighten attaching bolts and nut.

7. Position intermediate driveshaft into distributor socket.

8. Insert intermediate driveshaft into oil pump. Install pump and shaft as an assembly.

NOTE: Do not attempt to force the pump into position if it will not seat. The shaft hex may be mis-aligned with the distributor shaft. To align, remove the oil pump and rotate the intermediate driveshaft into a new position.

9. Tighten the oil pump attaching bolts to 15–23 ft. lbs. (20–30 Nm).
10. Install oil pan with new gasket.
11. Connect negative battery cable.
12. Fill the crankcase. Start engine and check for leaks.

Rear Main Bearing Oil Seal

REMOVAL & INSTALLATION

1.8L Engine

1. Disconnect the negative battery cable.
2. Remove the transaxle assembly.
3. If equipped with manual transaxle, remove the clutch assembly and the flywheel. If equipped with automatic transaxle, remove the flexplate.
4. If necessary, remove the rear cover mounting bolts and remove the rear cover.
5. Using a suitable removal tool, remove the crankshaft rear oil seal.

To install:

6. If removed, install the rear cover and attaching bolts. Tighten the bolts to 69–95 inch lbs. (7.8–11.0 Nm).
7. Lubricate the lip of a new seal and install, using a suitable installation tool.
8. Install the flywheel and clutch assembly or flexplate.
9. Install the transaxle.
10. Connect the negative battery cable.

1.9L Engine

1. Disconnect the negative battery cable.
2. Raise the vehicle and support it safely. Remove the transaxle.
3. Remove flywheel and the engine cover plate.
4. With a suitable tool, remove the oil seal.

NOTE: Use caution to avoid damaging the oil seal surface.

To install:

5. Inspect the crankshaft seal area for any damage which may cause the seal to leak. If damage is evident, service or replace the crankshaft, as necessary.

6. Coat the crankshaft seal area and the seal lip with engine oil.
7. Using a suitable seal installer tool, install the seal.
8. Install the engine cover plate and the flywheel. Tighten the flywheel attaching bolts to 54–64 ft. lbs. (73–87 Nm).
9. Install the transaxle assembly.
10. Connect the negative battery cable, start the engine and check for leaks.

2.3L Engine

1. Disconnect the negative battery cable.
2. Remove transaxle.
3. Remove flywheel.
4. Remove rear cover plate.
5. Insert a suitable tool into seal cavity and pry out old seal.

NOTE: Use caution to avoid damaging the oil seal surface.

To install:

6. Inspect the crankshaft seal area for any damage which may cause the seal to leak. If damage is evident, service or replace the crankshaft, as necessary.
7. Coat the crankshaft seal area and the seal lip with engine oil.
8. Using a suitable seal installer tool, install the seal.
9. Install rear cover plate and 2 dowels.
10. Install the flywheel. Tighten attaching bolts to 54–64 ft. lbs. (73–87 Nm).
11. Install the transaxle assembly.
12. Connect the negative battery cable, start the engine and check for leaks.

ENGINE COOLING

Radiator

REMOVAL & INSTALLATION

Except 1991–92 Escort

1. Disconnect the negative battery cable.
2. Drain the coolant from the cooling system. Retain the coolant in a suitable container for reuse.
3. On the the Escort, remove the air intake tube from the radiator support.
4. Remove the upper hose from the radiator.
5. Remove the 2 fasteners retaining the upper end of the fan shroud to the radiator and sight shield.

NOTE: If equipped with air conditioning, remove the nut and screw retaining the upper end of the fan shroud to the radiator at the cross support and nut and screw at the inlet end of the tank.

6. Disconnect the electric cooling fan motor wires and air conditioning discharge line, if equipped, from the shroud and remove the fan shroud from the vehicle.
7. Loosen the hose clamp and disconnect the radiator lower hose from the radiator.
8. Disconnect the overflow hose from the radiator filler neck.
9. If equipped with an automatic transaxle, disconnect the oil cooler hoses at the transaxle using a quick-disconnect tool. Cap the oil tubes and plug the oil cooler hoses.
10. Remove the 2 nuts retaining the top of the radiator to the radiator support. If the stud loosens, make sure it is tightened before the radiator is installed. Tilt the top of the radiator rearward to allow clearance with the upper mounting stud and lift the radiator from the vehicle. Make sure the mounts do not stick to the radiator lower mounting brackets.

To install:

11. Make sure the lower radiator isomounts are installed over the bolts on the radiator support.
12. Position the radiator to the radiator support making sure the radiator lower brackets are positioned properly on the lower mounts.
13. Position the top of the radiator to the mounting studs on the radiator support and install 2 retaining nuts. Tighten to 5–7 ft. lbs. (7–9.5 Nm).
14. Connect the radiator lower hose to the engine water pump inlet tube. Install constant tension hose clamp between alignment marks on the hose.
15. Check to make sure the radiator lower hose is properly positioned on the outlet tank and install the constant tension hose clamp. The stripe on the lower hose should be indexed with the rib on the tank outlet.
16. Connect the oil cooler hoses to the automatic transaxle oil cooler lines, if equipped. Use an appropriate oil resistant sealer.
17. Position the fan shroud to the radiator lower mounting bosses. On vehicles with air conditioning, insert the lower edge of the shroud into the clip at the lower center of the radiator. Install 2 nuts and bolts retaining the upper end of the fan shroud to the radiator. Tighten the nuts on Tempo/Topaz to 35–41 inch lbs. (3.9–4.6 Nm). On Escort, tighten the nut to 23–33 inch lbs. (2.6–3.7 Nm). Do not overtighten.
18. Connect the electric cooling fan motor wires to the wire harness.

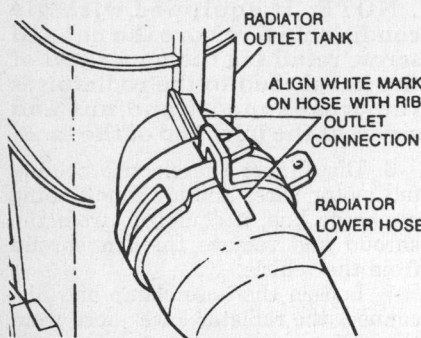

RADIATOR OUTLET TANK

ALIGN WHITE MARK ON HOSE WITH RIB OUTLET CONNECTION

RADIATOR LOWER HOSE

Radiator hose connection with alignment marks

19. Connect the upper hose to the radiator inlet tank fitting and install the constant tension hose clamp.

20. Connect the overflow hose to the nipple just below the radiator filler neck.

21. Install the air intake tube or sight shield.

22. Connect the negative battery cable.

23. Refill the cooling system. Start the engine and allow to come to normal operating temperature. Check for leaks. Confirm the operation of the electric cooling fan.

1991–92 Escort

1. Disconnect the negative battery cable.

2. Raise and safely support the vehicle. Drain the cooling system.

3. Remove the right side and front splash shields and remove the lower radiator hose.

4. If equipped with automatic transaxle, remove the lower oil cooler line from the radiator. Remove the oil cooler line brackets from the bottom of the radiator.

5. Lower the vehicle.

6. If equipped with automatic transaxle and air conditioning, remove the seal located between the radiator and fan shroud.

7. If equipped with automatic transaxle, remove the upper oil cooler line from the radiator.

8. If equipped with 1.8L engine, remove the resonance duct from the radiator isomounts.

9. Disconnect the cooling fan motor electrical connector and the cooling fan thermoswitch electrical connector.

10. Remove the 3 fan shroud attaching bolts and remove the shroud assembly by pulling it straight up.

11. Remove the upper radiator hose and the 2 upper radiator isomounts. Remove the radiator by lifting it straight up.

To install:

12. Make sure the radiator lower isomounts are installed over the bolts on the radiator support.

13. Position the radiator to the radiator support, making sure the radiator lower brackets are positioned properly on the lower isomounts.

14. Install the radiator upper isomounts, making sure the radiator locating pegs are positioned correctly. Install the upper radiator hose.

15. Lower the cooling fan shroud assembly into place and install the 3 shroud attaching bolts.

16. Connect the cooling fan motor electrical connector and thermoswitch electrical connector.

17. If equipped with 1.8L engine, install the resonance duct on the radiator isomounts.

18. Install the upper oil cooler line on the radiator.

19. If equipped with automatic transaxle and air conditioning, install the seal between the radiator and fan shroud.

20. Raise and safely support the vehicle. Install the lower oil cooler line on the radiator.

21. Install the lower radiator hose and install the right side and front splash shields.

22. Lower the vehicle and fill the cooling system.

23. Connect the negative battery cable, start the engine and check for coolant leaks.

Electric Cooling Fan

TESTING

Except 1991–92 Escort

1. Check the fuse or circuit breaker for power to the cooling fan motor.

2. Remove the connector(s) at the cooling fan motor(s). Connect a jumper wire and apply battery voltage to the positive terminal of the cooling fan motor.

3. Using an ohmmeter, check for continuity in the cooling fan motor.

NOTE: Remove the cooling fan connector at the fan motor before performing continuity checks. Perform continuity check of the motor windings only. The cooling fan control circuit is connected electrically to the ECM through the cooling fan relay center. Ohmmeter battery voltage must not be applied to the ECM.

4. Ensure proper continuity of the cooling fan motor ground circuit at the chassis ground connector.

1991–92 Escort

1. Make sure the ignition key is OFF.

2. Apply 12 volts to the **Y** wire at the cooling fan motor on all except

1.8L vehicles equipped with 4EAT automatic transaxle or 1.9L vehicles equipped with air conditioning. Replace the motor if it does not run.

3. On 1.8L vehicles equipped with 4EAT automatic transaxle or 1.9L vehicles equipped with air conditioning, apply 12 volts to the **BL** wire on the 1.8L engine or the **LG/Y** wire on the 1.9L engine at the cooling fan motor. Replace the motor if it does not run.

REMOVAL & INSTALLATION

Except 1991–92 Escort

1. Disconnect the negative battery cable.

2. Disconnect the wiring connector from the fan motor. Disconnect the wire loom from the clip on the shroud by pushing down on the lock fingers and pulling the connector from the motor end.

3. Remove the fasteners retaining the fan motor and shroud assembly and remove from the vehicle.

4. Remove the retaining clip from the motor shaft and remove the fan.

NOTE: A metal burr may be present on the motor shaft after the retaining clip has been removed. If necessary, remove burr to facilitate fan removal.

5. Unbolt and withdraw the fan motor from the shroud.

To install:

6. Install the fan motor in position in the fan shroud. Install the retaining nuts and washers or screws and tighten to 44–66 inch lbs. (5.0–7.5 Nm).

7. Position the fan assembly on the motor shaft and install the retaining clip.

8. Position the fan, motor and shroud as an assembly in the vehicle. Install the retaining nuts or screws and tighten nut to 35–41 inch lbs. (3.9–4.6 Nm) and screw to 23–33 inch lbs. (2.6–3.7 Nm) on Escort; 31–41 inch lbs. (3.5–4.6 Nm) on Tempo and Topaz.

9. Install the fan motor wire loom in the clip provided on the fan shroud. Connect the wiring connector to the fan motor. Be sure the lock fingers on the connector snap firmly into place.

10. Reconnect the battery cable.

11. Check the fan for proper operation.

1991–92 Escort

1. Disconnect the negative battery cable.

2. On 1.8L engine equipped vehicles, remove the resonance duct from the radiator isomounts.

3. Disconnect the cooling fan motor electrical connector.

4. Remove the 3 shroud attaching bolts and remove the cooling fan shroud assembly by pulling it straight up.

5. Working on a bench, remove the cooling fan retainer clip. Remove the cooling fan from the motor shaft.

6. Unclip the cooling fan motor electrical harness retainers and remove the harness from the retainers.

7. Remove the cooling fan motor attaching screws and remove the cooling fan motor from the shroud assembly.

To install:

8. Position the cooling fan motor on the shroud assembly and install the attaching screws.

9. Position the cooling fan motor electrical harness in the harness retainers and clip the retainers shut.

10. Install the cooling fan on the cooling motor shaft and install the retainer clip.

11. Carefully lower the cooling fan shroud assembly into place and install the attaching bolts. Connect the cooling fan motor electrical connector.

12. If equipped with 1.8L engine, install the resonance duct on the radiator isomounts.

13. Connect the negative battery cable.

Heater Core

REMOVAL & INSTALLATION

Tempo and Topaz

1. Disconnect the negative battery cable.

2. Drain the cooling system.

3. Disconnect the heater hoses from the heater core.

4. From inside the vehicle, remove the 2 screws retaining floor duct to the plenum. Remove one screw retaining floor duct to instrument panel. Remove floor duct.

5. Remove the 4 screws attaching the heater core cover to the heater case assembly.

6. Remove the heater core and cover from the plenum.

7. Complete the installation of the heater core by reversing the removal procedure. Check the system for proper operation.

1988–90 Escort

WITHOUT AIR CONDITIONING

1. Disconnect the negative battery cable.

2. Drain the cooling system into a clean container.

3. Loosen the heater hose clamps at the heater core tubes and disconnect the heater hoses from the heater core tubes. Cap the tubes to prevent spill-

ing coolant into the passenger compartment.

4. Remove the glove compartment door, liner and lower reinforcement.

5. Move the temperature control lever to the **WARM** position.

6. Remove the 4 screws attaching the heater core cover to the heater assembly and remove the cover.

7. Working in the engine compartment, loosen the 2 nuts attaching the heater case assembly to the dash panel.

8. Push the heater core tubes toward the passenger compartment to loosen the heater core from the heater case assembly.

9. Pull the heater core from the heater case assembly and remove the

heater core through the glove compartment opening.

To install:

10. Position the heater core in the core opening in the case assembly with the heater core tubes on the top side of the end tank.

11. Slide the heater core into the opening of the heater case assembly.

12. Position the heater core cover to the heater case assembly. Install the 4 attaching screws.

13. Tighten the 2 nuts attaching the heater case assembly to the dash panel.

14. Connect the heater hoses to the heater core tubes. Tighten the hose clamps.

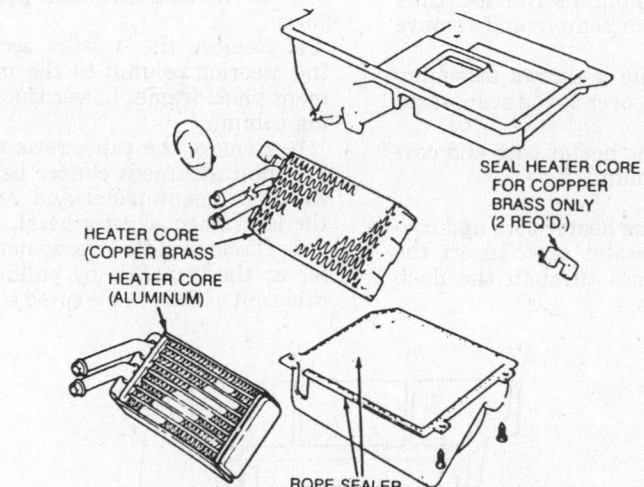

Heater core removal—Tempo and Topaz

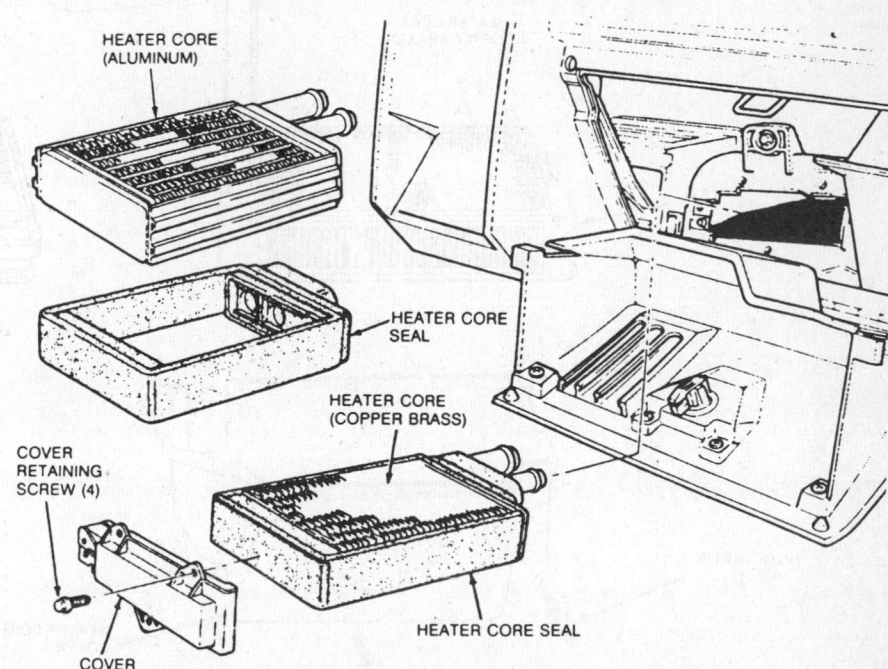

Heater core removal—1988–90 Escort without air conditioning

15. Fill the cooling system to the proper level with the removed coolant.

16. Install the glove compartment door, liner and hinge bar.

17. Check the heater for proper operation. Check the coolant level after the engine reaches normal operating temperature.

WITH AIR CONDITIONING

1. Disconnect the negative battery cable.

2. Drain cooling system into clean container.

3. Loosen the heater hose clamps at the heater core tubes and disconnect the heater hoses from the heater core tubes.

4. Working inside the vehicle, remove the screws attaching the floor duct to the plenum, instrument panel and evaporator assembly and remove the floor duct.

5. Remove the 4 screws attaching the heater core cover to the evaporator case.

6. Remove the heater core and cover from the plenum.

To install:

7. Position the heater core and cover to the evaporator case. Insert the heater core tubes through the dash panel seal holes.

8. Support the heater core and cover in place. Install the 4 attaching screws.

9. Position the floor duct to the evaporator case and instrument panel. Install the 4 attaching screws.

10. Connect the heater hoses to the heater core and tighten the hose clamps.

11. Fill the radiator with coolant to the proper level and check the system for proper operation.

1991–92 Escort

1. Disconnect the negative battery cable and drain the cooling system.

2. Disconnect the heater hoses at the bulkhead.

3. Remove the instrument panel as follows:

  a. Remove the 4 bolts securing the steering column to the instrument panel frame. Lower the steering column.

  b. Remove the cap screws securing the instrument cluster bezel to the instrument panel and remove the instrument cluster bezel.

  c. Disconnect the speedometer cable at the transaxle by pulling the cable out of the vehicle speed sensor.

  d. Remove the screws and bolts securing the instrument cluster to the instrument panel. Pull the instrument cluster out slightly and disconnect the electrical connectors from the rear of the instrument cluster.

  e. Disconnect the speedometer cable from the instrument cluster. Remove the instrument cluster from the instrument panel.

  f. Detach the hood release cable from the left lower dash trim panel. Carefully pry out both dash side panels.

  g. Remove the 4 retaining screws and the left lower dash trim panel. Disconnect all necessary electrical connectors.

  h. Remove the 2 hinge-to-instrument panel retaining screws and remove the glove compartment.

  i. Remove the climate control assembly and the ash tray.

  j. Remove the 7 accessory console retaining screws. Disconnect the radio antenna, radio wire connectors and cigarette lighter connector.

  k. Remove the retaining screws and the right lower dash trim panel. Disconnect the 3 amplifier wire connectors.

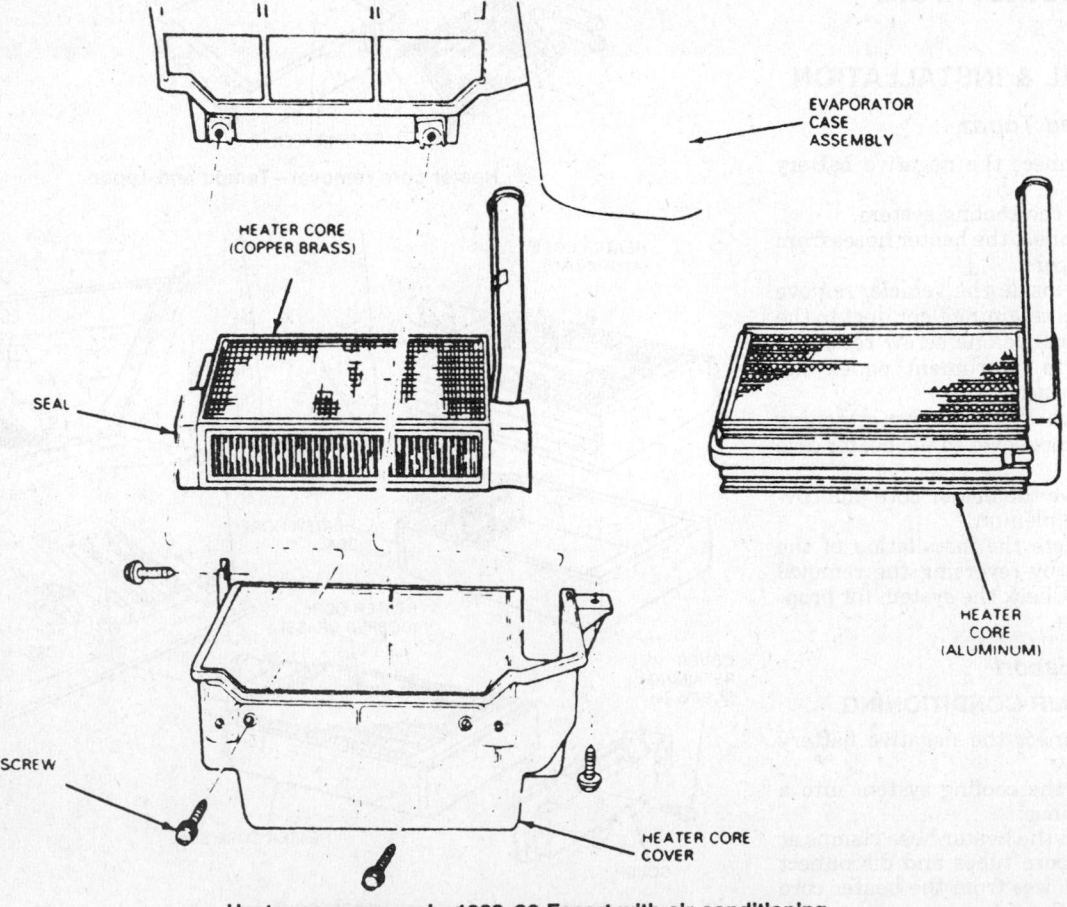

Heater core removal—1988-90 Escort with air conditioning

l. Remove the 4 bolts attaching the instrument panel frame to the floor pan. Remove the bolt from both of the lower instrument panel mounts.

m. Remove the 2 bolts from both of the upper instrument panel mounts. Remove the retaining screw and the defroster duct bezel.

n. Remove the 3 mounting bolts that attach the upper instrument panel to the cowl and remove the instrument panel from the vehicle.

NOTE: Use care to prevent any damage to the instrument panel or the surrounding interior trim.

4. Disconnect the mode selector and temperature control cables from the cams and retaining clips.

5. Remove the necessary defroster duct screws and loosen the capscrew that secures the heater-to-blower clamp.

6. Remove the 3 heater unit mounting nuts and disconnect the antenna lead from the retaining clip. Remove the heater unit.

7. Remove the insulator and the 4 brace capscrews. Remove the brace.

8. Remove the heater core from the heater unit.

To install:

9. Install the heater core into the heater unit and install the brace.

10. Install the brace capscrews and the insulator.

NOTE: If a new heater unit is being installed, save the keys that are found on the new unit for mode selector and temperature control cable adjustment.

11. Position the heater unit and attach the defroster and floor ducting. Install the heater unit mounting nuts.

12. Tighten the heater-to-blower clamp capscrew and install the defroster duct screws. Connect the antenna lead to the retaining clip.

13. Install the instrument panel by reversing the removal procedure.

14. Connect and adjust the mode selector and temperature control cables. Connect the heater hoses at the bulkhead.

15. Fill the cooling system and connect the negative battery cable. Start the engine and check for leaks. Check the coolant level and fill as necessary.

Water Pump

REMOVAL & INSTALLATION

1.8L Engine

1. Disconnect the negative battery cable.

2. Drain the cooling system.

3. Remove the timing belt according to the proper procedure.

4. Raise and safely support the vehicle.

5. Remove the engine oil dipstick tube bracket bolt from the water pump.

6. Remove the 2 bolts and the gasket from the water inlet pipe.

7. Remove all but the uppermost water pump mounting bolt.

8. Lower the vehicle.

9. Remove the remaining bolt and the water pump assembly.

10. If it is being reused, remove all gasket material from the water pump. Remove all gasket material from the engine block.

To install:

11. Install a new gasket onto the water pump.

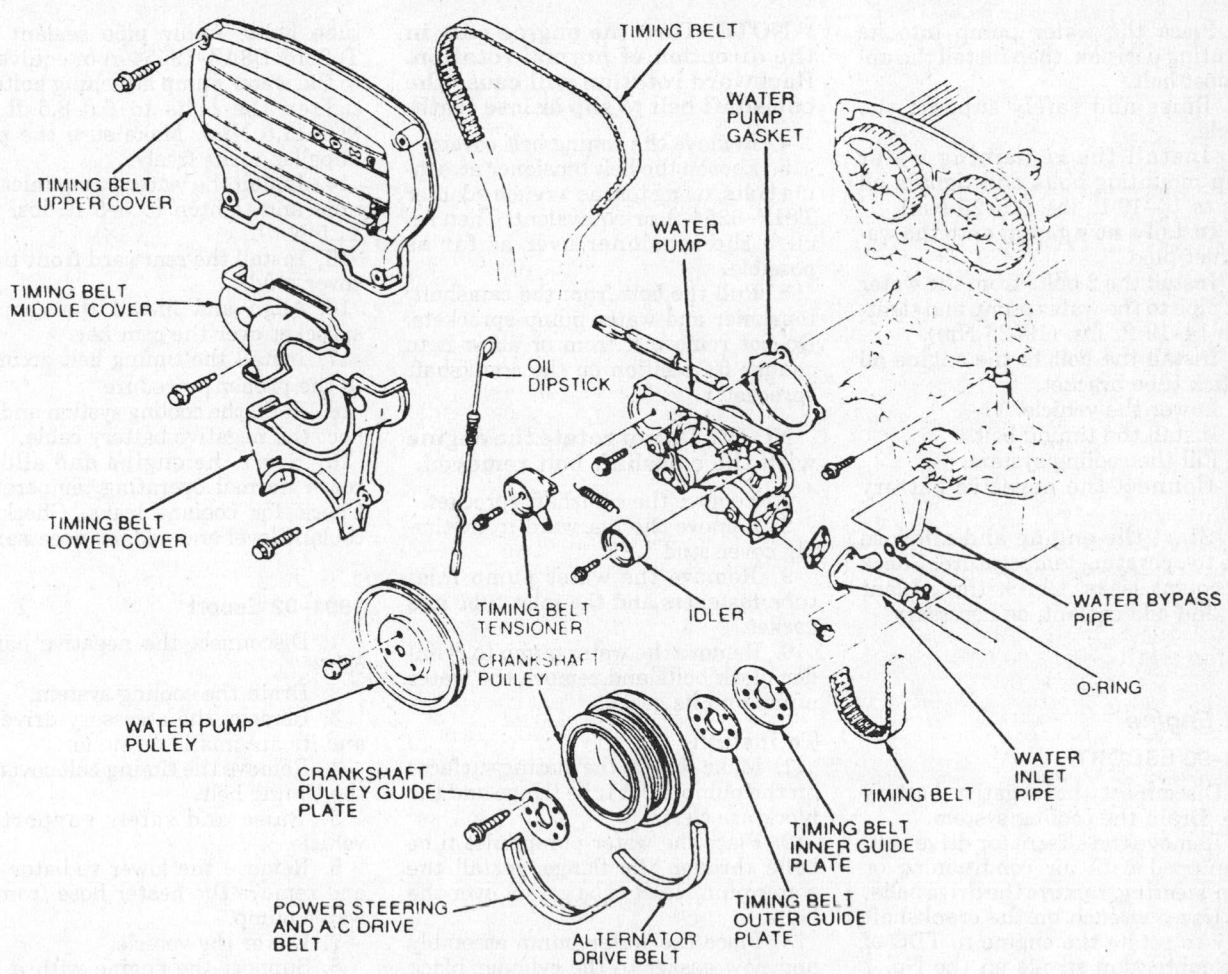

Water pump installation—1.8L engine

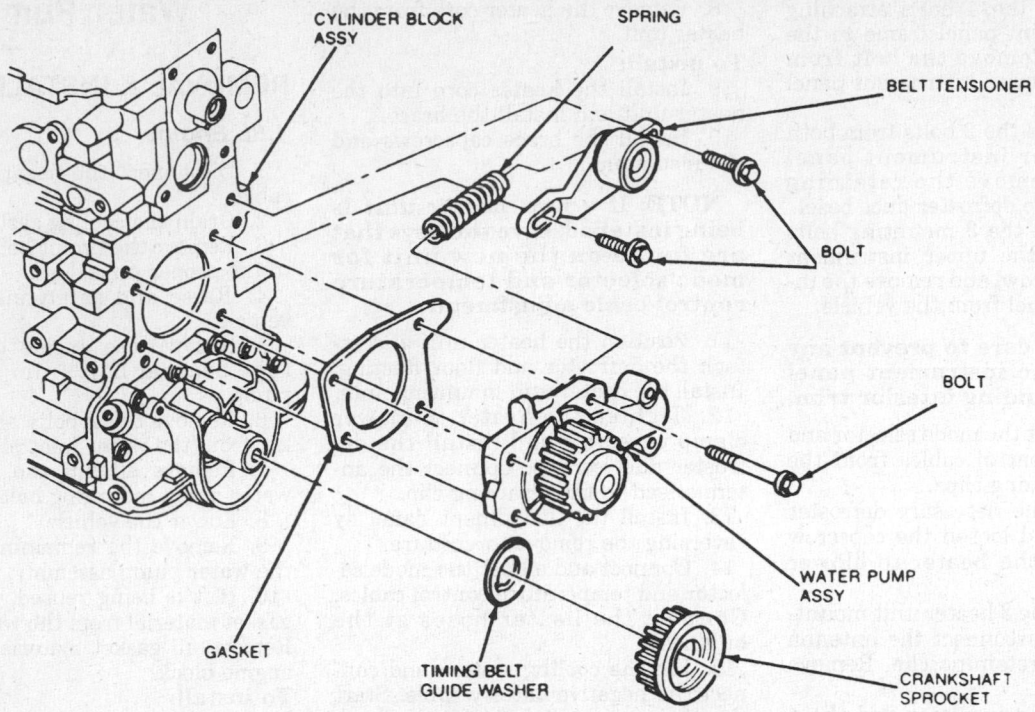

Water pump installation—1.9L engine, 1988-90 Escort

12. Place the water pump into its mounting position, then install the uppermost bolt.

13. Raise and safely support the vehicle.

14. Install the remaining water pump mounting bolts and tighten all bolts to 14–19 ft. lbs. (19–25 Nm).

15. Install a new gasket onto the water inlet pipe.

16. Install the 2 bolts from the water inlet pipe to the water pump and tighten to 14–19 ft. lbs. (19–25 Nm).

17. Install the bolt to the engine oil dipstick tube bracket.

18. Lower the vehicle.

19. Install the timing belt.

20. Fill the cooling system.

21. Connect the negative battery cable.

22. Start the engine and allow to come to operating temperature. Check for coolant leaks. Check the coolant level and add coolant, as necessary.

1.9L Engine

1988–90 ESCORT

1. Disconnect the negative battery cable. Drain the cooling system.

2. Remove the alternator drive belt. If equipped with air conditioning or power steering, remove the drive belts.

3. Use a wrench on the crankshaft pulley to rotate the engine to TDC of the compression stroke on the No. 1 cylinder.

NOTE: Turn the engine only in the direction of normal rotation. Backward rotation will cause the camshaft belt to slip or lose teeth.

4. Remove the timing belt cover.

5. Loosen the belt tensioner attaching bolts, using torque wrench adapter T81P–6254–A or equivalent. Then secure the tensioner over as far as possible.

6. Pull the belt from the camshaft, tensioner and water pump sprockets. Do not remove it from or allow it to change its position on the crankshaft sprocket.

NOTE: Do not rotate the engine with the camshaft belt removed.

7. Remove the camshaft sprocket.

8. Remove the rearward front timing cover stud.

9. Remove the water pump inlet tube fasteners and the inlet tube and gasket.

10. Remove the water pump to cylinder block bolts and remove the water pump and its gasket.

To install:

11. Make certain the mating surfaces on the pump, inlet tube flange and the block are clean.

12. Place the water pump inlet tube bolts through the flange. Install the water pump inlet tube gasket over the bolts.

13. Place the water pump assembly and new gasket to the cylinder block and through the water pump inlet tube bolts. Apply pipe sealant with Teflon® D8AZ–19554–A or equivalent, to the water pump attaching bolts and tighten the bolts to 6.0–8.5 ft. lbs. (8.0–11.5 Nm). Make sure the pump impeller turns freely.

14. Install the water pump inlet tube nuts and tighten to 6–8 ft. lbs. (8.0–11.5 Nm).

15. Install the rearward front timing cover stud.

16. Align and install the camshaft sprocket over the cam key.

17. Install the timing belt according to the proper procedure.

18. Refill the cooling system and connect the negative battery cable.

19. Start the engine and allow to reach normal operating temperature. Check for coolant leaks. Check the coolant level and add, as necessary.

1991–92 Escort

1. Disconnect the negative battery cable.

2. Drain the cooling system.

3. Remove the accessory drive belt and its automatic tensioner.

4. Remove the timing belt cover and the timing belt.

5. Raise and safely support the vehicle.

6. Remove the lower radiator hose and remove the heater hose from the water pump.

7. Lower the vehicle.

8. Support the engine with a suitable floor jack.

9. Remove the right engine mount attaching bolts and roll the engine mount aside.

10. Remove the water pump attaching bolts.

11. Using the floor jack, raise the engine enough to provide clearance for removing the water pump.

12. Remove the water pump and the gasket from the engine through the top of the engine compartment.

To install:

13. Make sure the mating surfaces of the cylinder block and water pump are clean and free of gasket material.

14. If the water pump is to be replaced, transfer the timing belt tensioner components to the new water pump.

15. With the engine supported and raised with a suitable floor jack, place the water pump and the gasket on the cylinder block and install the 4 attaching bolts. Tighten the bolts to 15–22 ft. lbs. (20–30 Nm).

16. Install the timing belt and cover.

17. Roll the right engine mount into position and install the mount bolts. Remove the floor jack.

18. Raise and safely support the vehicle.

19. Install the lower radiator hose and install the heater hose on the pump.

20. Install the crankshaft dampener and the splash shield.

21. Lower the vehicle.

22. Install the accessory drive belt automatic tensioner and the accessory drive belt.

23. Connect the negative battery cable.

24. Refill the cooling system.

25. Start the engine and allow to come to normal operating temperature. Check for coolant leaks. Check the coolant level and add as necessary.

2.3L Engine

1. Drain the cooling system.

2. Disconnect the negative battery cable.

3. If equipped with an air pump, remove it as follows:

    a. Loosen thermactor pump adjusting bolt and remove belt.

    b. Remove thermactor pump hose clamp located below the thermactor pump.

    c. Remove the thermactor pump bracket bolts.

    d. Remove thermactor pump and bracket as an assembly.

4. Loosen the water pump idler pulley and remove the belt from the water pump pulley.

5. Disconnect the heater hose at the water pump or the water pump inlet tube.

6. Disconnect the water pump inlet tube, if equipped.

7. Remove the 3 water pump retaining bolts and remove the water pump from its mounting.

To install:

8. Thoroughly clean both gasket mating surfaces on the water pump and cylinder block.

9. Coat the new gasket on both sides with a water resistant sealer and position on the cylinder block.

10. Install the water pump retaining bolts and tighten to 15–22 ft. lbs. (20–30 Nm).

11. Connect the water pump inlet tube, if equipped.

12. Connect the heater hose.

13. Install water pump belt on the pulley and adjust the tension. Install thermactor pump and bracket, if equipped.

14. Connect the negative battery cable.

15. Replace the engine coolant. Operate the engine until normal operating temperature is reached. Check for leaks and recheck the coolant level.

Thermostat

REMOVAL & INSTALLATION

Except 1991–92 Escort

1. Disconnect the negative battery cable.

2. Disconnect the wiring connector from the thermal switch in the thermostat housing.

3. Remove the radiator cap and drain the cooling system to a corresponding level just below water outlet connection. Drain the coolant into a clean receptacle for reuse.

4. Loosen the top hose clamp at the radiator, remove the water outlet connection retaining bolts, lift clear of the engine and remove the thermostat by rotating counterclockwise in the water outlet connection until the thermostat becomes free to move.

    NOTE: Do not pry the housing off.

To install:

5. Make sure the water outlet connection pocket and cylinder head mating surfaces are clean and free of gasket material.

6. Place the thermostat in position, fully inserted to compress the gasket, and rotate clockwise in the water outlet connection to secure. Position the water outlet connection to the cylinder head and tighten the bolts to 6–8 ft. lbs. (8–11 Nm) on the 1.9L engine or 12–18 ft. lbs. (16–24 Nm) on the 2.3L engine.

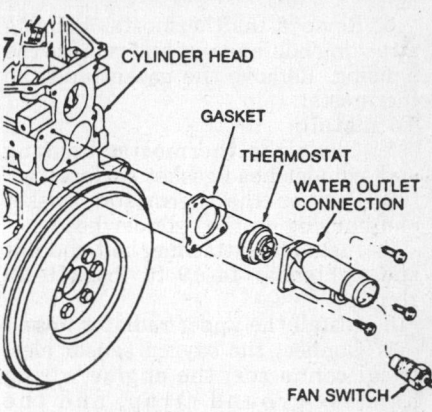

Thermostat installation—1988–90 1.9L engine

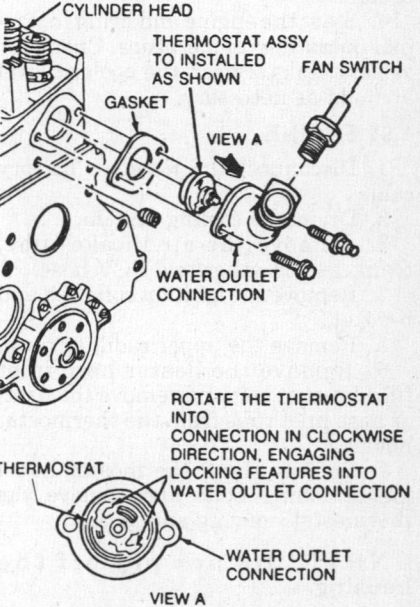

Thermostat installation—2.3L engine

7. Position the top hose to the radiator and tighten the clamps.

8. Refill the cooling system and connect the negative battery cable.

9. Start the engine and bring to normal operating temperature. Check for coolant leaks. Check the coolant level and add as necessary.

1991–92 Escort

1.8L ENGINE

1. Disconnect the negative battery cable.

2. Drain the cooling system.

3. Remove the air intake tube.

4. Disconnect the water thermoswitch connector, the engine wiring harness ground strap from the connector above the housing, and the exhaust gas oxygen sensor electrical connector.

5. Remove the upper radiator hose from the housing.

6. Remove the thermostat housing attaching bolt and nut and remove the housing. Remove the gasket and the thermostat.

To install:

7. Clean the thermostat housing and cylinder head gasket surfaces.

8. Position the thermostat, gasket and housing on the cylinder head.

9. Install the attaching bolt and nut and tighten to 14–19 ft. lbs. (19–26 Nm).

10. Install the upper radiator hose.

11. Connect the oxygen sensor electrical connector, the engine wiring harness ground strap, and the thermoswitch electrical connector.

12. Install the air intake tube.

13. Connect the negative battery cable.

14. Start the engine and bring to normal operating temperature. Check for coolant leaks. Check the coolant level and add as necessary.

1.9L ENGINE

1. Disconnect the negative battery cable.

2. Drain the cooling system.

3. Remove the air intake tube, crankcase breather and PCV hose.

4. Remove the ignition coil pack and bracket.

5. Remove the upper radiator hose.

6. Remove the heater hose inlet tube bracket bolt and remove the heater hose inlet tube from the thermostat housing.

7. Remove the 3 thermostat housing attaching bolts and remove the thermostat housing and gasket.

NOTE: Do not pry off the housing.

8. Remove the thermostat and the rubber seal from the housing.

To install:

9. Clean the thermostat housing pocket and cylinder head mating surfaces.

10. Place the thermostat into position, fully inserted to compress the rubber seal inside the housing.

NOTE: Make sure the thermostat tabs engage properly into the housing slots.

11. Position the thermostat housing and gasket on the cylinder head.

12. Install the 3 attaching bolts and tighten to 8.0–11.5 ft. lbs. (11–15 Nm).

13. Install the heater hose inlet pipe and the heater hose inlet pipe bracket bolt.

14. Install the upper radiator hose.

15. Install the ignition coil and bracket.

16. Install the crankcase breather, PCV hoses and the air intake tube.

17. Connect the negative battery cable.

18. Refill the cooling system.

19. Start the engine and bring to normal operating temperature. Check for coolant leaks. Check the coolant level and add as necessary.

Cooling System Bleeding

When the entire cooling system is drained, the following procedure should be used to ensure a complete fill.

1. Install the block drain plug, if removed, and close the draincock. With the engine off, add anti-freeze to the radiator to a level of 50 percent of the total cooling system capacity. Then add water until it reaches the radiator filler neck seat.

2. Install the radiator cap to the first notch to keep spillage to a minimum.

3. Start the engine and let it idle until the upper radiator hose is warm. This indicates that the thermostat is open and coolant is flowing through the entire system.

4. Carefully remove the radiator cap and top off the radiator with water. Install the cap on the radiator securely.

5. Fill the coolant recovery reservoir to the FULL COLD mark with anti-freeze, then add water to the FULL HOT mark. This will ensure that a proper mixture is in the coolant recovery bottle.

6. Check for leaks at the draincock and the block drain plug.

ENGINE ELECTRICAL

NOTE: Disconnecting the negative battery cable on some vehicles may interfere with the functions of the on board computer systems and may require the computer to undergo a relearning process, once the negative battery cable is reconnected.

Distributor

REMOVAL

1. Turn the engine over until No. 1 cylinder is at TDC of the compression stroke.

2. Mark the position of No. 1 cylinder wire on distributor base for reference when installing the distributor.

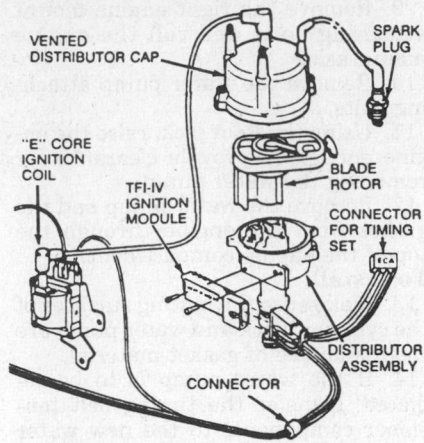

Ignition system components—typical

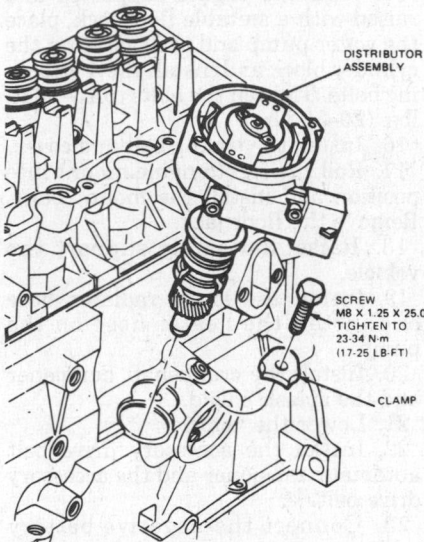

Distributor mounting—2.3L engine

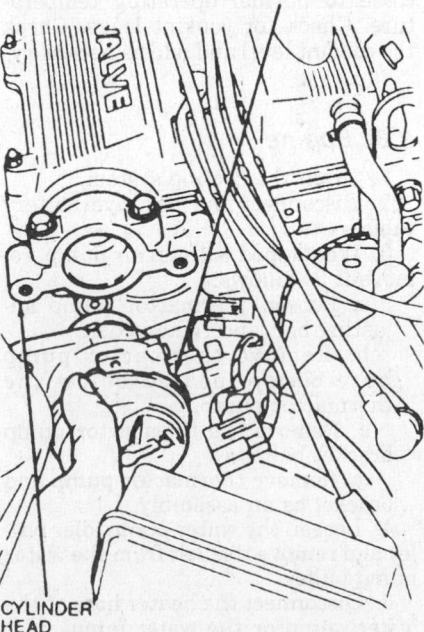

Distributor Installation—1.8L engine

3. Disconnect the negative battery cable.

4. Disconnect the primary wiring from the distributor. Disconnect the coil wire on the 1.8L engine.

5. Remove the cap screws and remove the distributor cap straight off to prevent damage to the rotor blade and spring. Remove rotor from distributor shaft and armature.

6. Scribe or paint an alignment mark on the distributor body, showing the position of the ignition rotor. Place another mark on the distributor body and cylinder head or block, showing the position of the body in relation to the head or block. These marks are used for reference when installing the distributor.

7. Remove the 2 distributor retaining bolts at the base of the distributor housing on 1.8L and 1.9L engines. Remove distributor hold-down bolt and clamp on the 2.3L engine.

8. Pull the distributor out of the head or block.

NOTE: Some engines are equipped with security type distributor retaining bolts and special tool T82L–12270–A or equivalent, must be used to remove these bolts.

INSTALLATION

Timing Not Disturbed

1.8L AND 1.9L ENGINES

1. Place the distributor in the cylinder head, seating the tang(s) of the drive coupling into the groove(s) at the end of the camshaft.

2. Install the distributor retaining bolts and tighten them so the distributor can just barely be moved.

3. Install the rotor and distributor cap and connect all wiring.

4. Set initial timing.

5. After timing has been set, tighten distributor hold-down bolts to 14–19 ft. lbs. (19–25 Nm) on the 1.8L engine or 44–62 inch lbs. (5–7 Nm) on the 1.9L engine.

2.3L ENGINE

1. Rotate distributor shaft so blade on rotor is pointing toward paint mark on distributor base made during removal.

2. Continue rotating rotor slightly so leading edge of the vane is centered in vane switch stator assembly.

3. Rotate distributor in block to align leading edge of vane and vane switch stator assembly. Verify rotor is pointing at No. 1 mark on distributor base.

NOTE: If vane and vane switch stator cannot be aligned by rotating distributor in cylinder block, remove distributor enough to just disengage distributor gear from camshaft gear. Rotate rotor enough to engage distributor gear on another tooth of camshaft gear.

4. Install the distributor retaining bolt and tighten so the distributor can just barely be moved.

5. Install the rotor, if removed, the distributor cap and all wiring. Tighten distributor cap to 18–23 inch lbs. (2.0–2.6 Nm).

6. Set initial timing according to procedures.

7. After timing has been set, tighten distributor hold-down bolt to 17–25 ft. lbs. (23–34 Nm).

Timing Disturbed

1.8L AND 1.9L ENGINES

1. If the crankshaft was rotated while the distributor was removed, the engine must be brought to TDC on the compression stroke of the No. 1 cylinder.

2. Remove the No. 1 spark plug. Place a finger over the hole and rotate the crankshaft slowly in the direction of normal rotation, until engine compression is felt.

NOTE: Turn the engine only in the direction of normal rotation. Backward rotation may cause the cam belt to slip or lose teeth, altering engine timing.

3. When engine compression is felt at the spark plug hole, indicating that the piston is approaching TDC, continue to turn the crankshaft until the timing mark on the pulley is aligned with the **0** mark on the engine front cover.

4. Turn the distributor shaft until the ignition rotor is at the No. 1 firing position.

5. Place the distributor in the cylinder head, seating the tang(s) of the drive coupling into the groove(s) at the end of the camshaft.

6. Install the distributor retaining bolts and tighten them so the distributor can just barely be moved.

7. Install the rotor and distributor cap and connect all wiring.

8. Set initial timing.

9. After timing has been set, tighten distributor hold-down bolts to 14–19 ft. lbs. (19–25 Nm) on the 1.8L engine or 44–62 inch lbs. (5–7 Nm) on the 1.9L engine.

2.3L ENGINE

1. If the crankshaft was rotated while the distributor was removed, the engine must be brought to TDC on the compression stroke of the No. 1 cylinder.

2. Remove the No. 1 spark plug. Place a finger over the hole and rotate the crankshaft slowly in the direction of normal rotation, until engine compression is felt.

3. When engine compression is felt at the spark plug hole, indicating that the piston is approaching TDC, continue to turn the crankshaft until the timing mark on the pulley is aligned with the **0** mark on the engine front cover.

4. Turn the distributor shaft until the ignition rotor is at the No. 1 firing position.

5. Rotate distributor shaft so blade on rotor is pointing toward paint mark on distributor base made during removal.

6. Continue rotating rotor slightly so leading edge of the vane is centered in vane switch stator assembly.

7. Rotate distributor in block to align leading edge of vane and vane switch stator assembly. Verify rotor is pointing at No. 1 mark on distributor base.

NOTE: If vane and vane switch stator cannot be aligned by rotating distributor in cylinder block, remove distributor enough to just disengage distributor gear from camshaft gear. Rotate rotor enough to engage distributor gear on another tooth of camshaft gear.

8. Install the distributor retaining bolt and tighten so the distributor can just barely be moved.

9. Install the rotor and distributor cap and connect all wiring. Tighten distributor cap to 18–23 inch lbs. (2.0–2.6 Nm).

10. Set initial timing according to procedures.

11. After timing has been set, tighten distributor hold-down bolt to 17–25 ft. lbs. (23–34 Nm).

Distributorless Ignition

Beginning in 1991, the 1.9L engine is equipped with a Distributorless Ignition System (DIS). The DIS consists of the following components: crankshaft sensor, ignition module, ignition coil pack, the spark angle portion of the ECU and the related wiring.

The crankshaft sensor is a variable reluctance-type sensor triggered by a 36-minus-1 tooth trigger wheel configuration pressed onto the rear of the crankshaft dampener. The signal generated by this sensor is called a Variable Reluctance Sensor (VRS) signal.

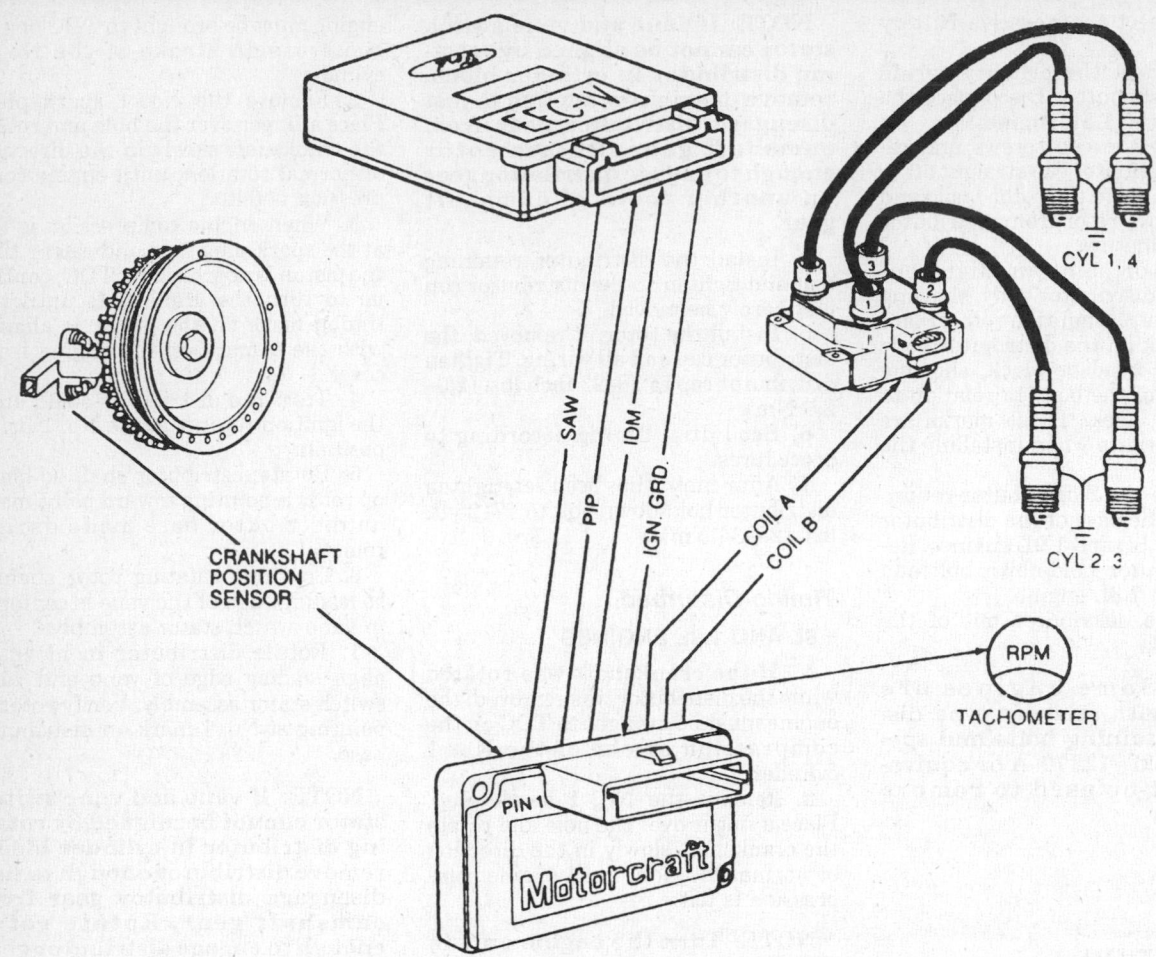

Distributorless ignition system—1991–92 1.9L engine

The VRS signal provides engine position and rpm information to the ignition module.

The ignition module is a micro-processor that recieves input from the crankshaft sensor in regards to engine position and engine speed and input from the ECU pertaining to spark advance. The ignition module uses this information to direct which coil to fire and to calculate the turn on and turn off times of the coils required to achieve the correct dwell and spark advance.

The ignition coil pack contains 2 separate ignition coils. Each ignition coil fires 2 cylinders simultaneously. When 1 cylinder is firing on the compression stroke, the other is firing on the exhaust stroke. During the next engine revolution, the reverse occurs. The spark plug fired on the exhaust stroke uses very little of the ignition coil's stored energy; the majority of the energy is used by the spark plug on the compression stroke. Since these 2 spark plugs are connected in series, the firing voltage of 1 plug will be negative with respect to ground, while the voltage of the other will be positive with respect to ground.

REMOVAL & INSTALLATION

Crankshaft Sensor

1. Disconnect the negative battery cable.
2. Raise and safely support the vehicle.
3. Remove the right side splash shield.
4. Disconnect the sensor electrical connector from the wiring harness.
5. Remove the crankshaft sensor mounting screws and remove the sensor.
6. Installation is the reverse of the removal procedure. Tighten the sensor attaching screws to 40–61 inch lbs. (5–7 Nm).

Ignition Module

NOTE: The ignition module is located on the left side of the engine compartment, in front of the left strut tower.

1. Disconnect the negative battery cable.
2. Remove the 3 module sub-bracket attaching nuts.

3. Gently pull the sub-bracket and module assembly straight up and disconnect the module electrical harness.
4. Remove the 2 module attaching screws from the sub-bracket. Remove the ignition module from the sub-bracket.
5. Installation is the reverse of the removal procedure. Tighten the module attaching screws to 24–35 inch lbs. (3–4 Nm). Tighten the sub-bracket attaching nuts to 62–88 inch lbs. (7–10 Nm).

Ignition Coil Pack

1. Disconnect the negative battery cable.
2. Disconnect the electrical connector from the coil pack.
3. Remove the spark plug wires by squeezing the locking tabs to release the coil boot retainers. Tag the wires and mark their position on the coil pack prior to removal.
4. Remove the coil pack attaching bolts and remove the coil pack.

NOTE: Save the capacitor for installation with the new coil pack.

5. Installation is the reverse of the removal procedure. Tighten the attaching bolts to 40–62 inch lbs. (5–7 Nm).

Ignition Timing

ADJUSTMENT

1.8L Engine

1. Apply the parking brake and make sure the vehicle is in **N** or **P**.
2. Start the engine and warm it up to normal operating temperature.
3. Turn off all electrical loads and accessories.
4. Connect a suitable timing light according to the manufacturers instructions.
5. Using a jumper wire, connect the **GROUND** terminal to the **TEN** terminal on the diagnosis connector.
6. Connect the positive lead of a suitable tachometer to the **IG** terminal on the diagnosis connector and connect the negative lead to the negative battery post.
7. Using the timing light, inspect the ignition timing. The ignition timing should be 9–11 degrees at 700–800 rpm. The yellow mark on the crankshaft pulley should be aligned with the corresponding mark on the timing belt cover.
8. If the marks are not aligned, loosen the distributor mounting bolts and turn the distributor until the ignition timing is within specification.
9. Tighten the distributor mounting bolts to 14–19 ft. lbs. (19–25 Nm).
10. Remove the jumper wire from the diagnosis connector and remove the timing light and tachometer.

1.9L Engine

NOTE: The initial timing on 1991–92 1.9L engine with distributorless ignition is fixed at 10 degrees ± 2 degrees and is not adjustable.

1. Ignition timing marks consist of a notch on the crankshaft pulley and a graduated scale molded into the camshaft belt cover. The number of degrees before or after TDC represented by each mark in the scale can be interpreted according to the decal affixed to the top of the belt cover.
2. With white paint or chalk, mark the notch in the crankshaft pulley and the appropriate mark in the degree scale. See the underhood emission control decal for timing specifications.
3. Warm the engine until it reaches normal operating temperature.
4. Shut the engine **OFF**. Make sure the transaxle is in **P** or **N**, apply the parking brake and block the wheels.

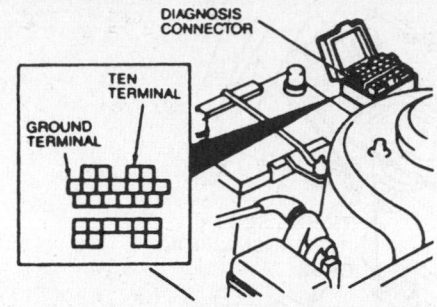

Diagnosis connector location—Escort with 1.8L engine

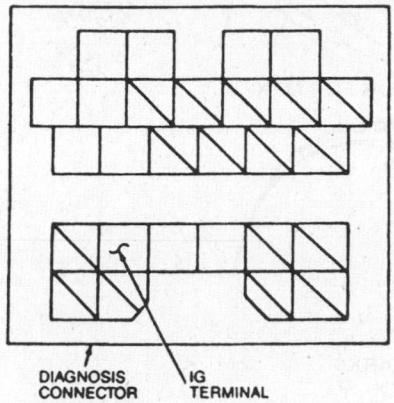

Diagnosis connector terminals—Escort with 1.8L engine

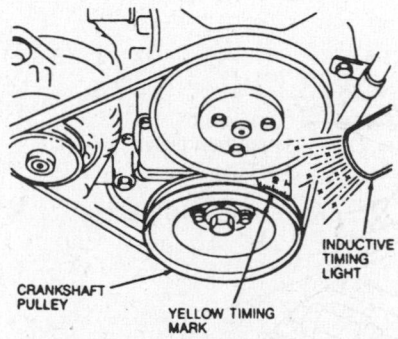

Checking ignition timing—Escort with 1.8L engine

Place the air conditioning/heater control switch in the **OFF** position.
5. Connect a suitable timing light to the engine. Connect a suitable tachometer to the engine.
6. Disconnect the single wire in-line spout connector or remove the shorting bar from the double wire spout connector.
7. Start the engine and allow to reach normal operating temperature. Aim the light at the marks. If they are not aligned, loosen the distributor clamp bolts slightly and rotate the distributor body until the marks are aligned under timing light illumination.

NOTE: To set timing correctly, a remote starter should not be used. Use the ignition key only to start the vehicle. Disconnecting the start wire at the starter relay will cause the TFI module to revert to start mode timing after the vehicle is started. Reconnecting the start wire after the vehicle is running will not correct the timing.

8. Tighten the distributor clamp bolts. Connect the spout connector and check the timing to verify distributor advance beyond initial setting.
9. Shut the engine **OFF** and remove all test equipment.

2.3L Engine

The timing marks are located on the flywheel and visible through an access hole in the transaxle case. On manual transaxle vehicles, the timing cover plate must be removed in order to view the timing marks and adjust the timing.

NOTE: Some distributor retaining bolts have a security type head and cannot be loosened to adjust timing, unless special tool T82L-12270-A or equivalent, is available.

1. Place the transaxle in the **P** or **N** position. Firmly apply the parking brake and block the wheels.
2. Open the hood, locate the timing marks and clean with a stiff brush or solvent. On vehicles with manual transaxle, it will be necessary to remove the cover plate which allows access to to the timing marks.
3. Using white chalk or paint, mark the specified timing mark and pointer.
4. Remove the in-line spout connector or remove the shorting bar from the double wire spout connector.
5. Connect a suitable inductive type timing light to the No. 1 spark plug wire. Do not puncture an ignition wire with any type of probing device.

NOTE: The high ignition coil charging currents generated in the EEC-IV ignition system may falsely trigger timing lights with capacitive or direct connect pickups. It is necessary that an inductive type timing light be used in this procedure.

6. Connect a suitable tachometer to the engine. The ignition coil connector allows a test lead with an alligator clip to be connected to the Distributor Electronic Control (DEC) terminal without removing the connector.
7. Start the engine and let it run until it reaches normal operating temperature.

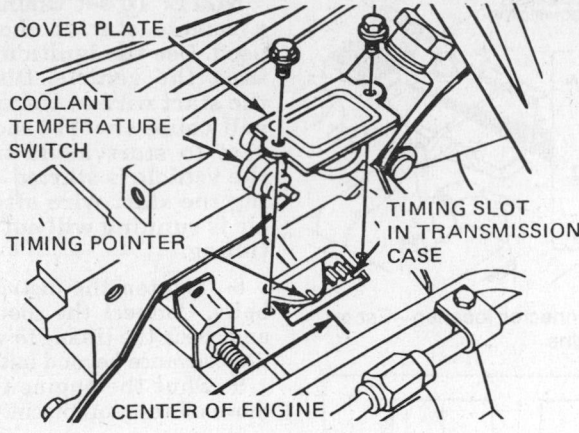

TIMING LOCATION FOR MTX

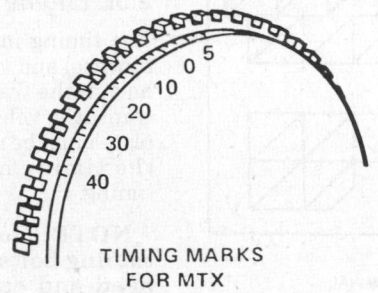

Timing marks—2.3L engine with manual transaxle

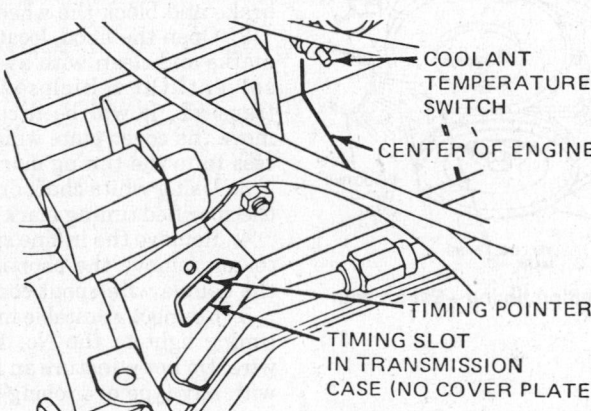

TIMING LOCATION FOR ATX

Timing marks—2.3L engine with automatic transaxle

8. Check the engine idle rpm if it is not within specifications, adjust as necessary. After the rpm has been adjusted or checked, aim the timing light at the timing marks. If they are not aligned, loosen the distributor clamp bolts slightly and rotate the distributor body until the marks are aligned under timing light illumination.

NOTE: To set timing correctly, a remote starter should not be used. Use the ignition key only to start the vehicle. Disconnecting the start wire at the starter relay will cause the TFI module to revert to start mode timing after the vehicle is started. Reconnecting the start wire after the vehicle is running will not correct the timing.

9. Tighten the distributor clamp bolts and recheck the ignition timing. Readjust the idle speed. Shut the engine **OFF**, remove all test equipment, reconnect the in-line spout connector and reinstall the cover plate on the manual transaxle vehicles.

Alternator

PRECAUTIONS

Several precautions must be observed with alternator equipped vehicles to avoid damage to the unit.

• If the battery is removed for any reason, make sure it is reconnected with the correct polarity. Reversing the battery connections may result in damage to the one-way rectifiers.

• When utilizing a booster battery as a starting aid, always connect the positive to positive terminals and the negative terminal from the booster battery to a good engine ground on the vehicle being started.

• Never use a fast charger as a booster to start vehicles.

• Disconnect the battery cables when charging the battery with a fast charger.

• Never attempt to polarize the alternator.

• Do not use test lights of more than 12 volts when checking diode continuity.

• Do not short across or ground any of the alternator terminals.

• The polarity of the battery, alternator and regulator must be matched and considered before making any electrical connections within the system.

• Never separate the alternator on an open circuit. Make sure all connections within the circuit are clean and tight.

- Disconnect the battery ground terminal when performing any service on electrical components.
- Disconnect the battery if arc welding is to be done on the vehicle.

BELT TENSION ADJUSTMENT

1.8L Engine

1. Loosen the alternator adjusting bolt.
2. Raise and safely support the vehicle.
3. Loosen the alternator mounting bolt.

NOTE: Do not pry against the stator frame. Position the prybar against a stronger point, such as the area around a case bolt.

4. Position a suitable belt tension gauge on the longest accessible span of belt and tighten the belt. Adjust the tension to 85.8–103.4 lbs. for a new belt or 68.2–85.8 lbs. for a used belt.
5. If a belt tension gauge is not available, adjust the tension to 0.31–0.35 in. (8–9mm) deflection for a new belt or 0.35–0.39 in. (9–10mm) deflection for a used belt.
6. Tighten the alternator adjusting bolt to 14–19 ft. lbs. (19–25 Nm).
7. Tighten the alternator mounting bolt to 27–38 ft. lbs. (37–52 Nm).
8. Lower the vehicle.

1.9L Engine

NOTE: An automatic tensioner maintains correct belt tension during operation on the 1991–92 1.9L engine. No adjustment is necessary.

1. Loosen the alternator pivot and adjustment bolts.
2. Install a ½ inch breaker bar or equivalent, to the support bracket that is located behind the alternator.
3. Apply tension to the belt using the breaker bar. Using a belt tension gauge, adjust the tesnsion to 140–180 lbs. for a new belt or 120–140 lbs. for a used belt. While maintaining proper belt tension, tighten the alternator adjustment bolt to 30 ft. lbs. (40 Nm).
4. Remove the belt tension gauge and breaker bar, start the engine and let it idle for 5 minutes.
5. Stop the engine and check the belt tension. If the tension is below 120 lbs., retension the belt to 120–140 lbs. and then tighten the adjustment bolt.
7. Tighten the alternator pivot bolt to 50 ft. lbs. (68 Nm) and the support bracket bolt to 35 ft. lbs. (47 Nm).

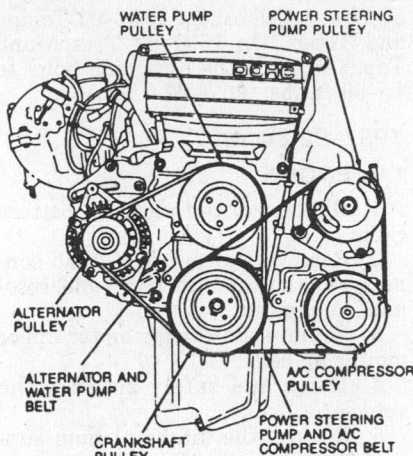

Drive belt arrangement—1.8L engine

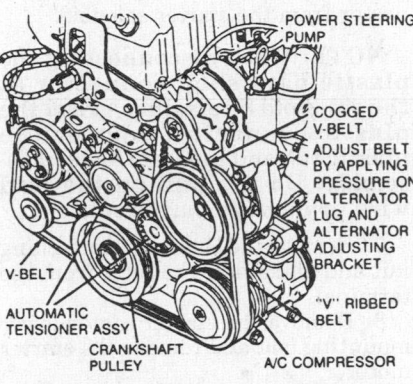

Drive belt arrangement—2.3L engine

2.3L Engine

1. Loosen the alternator pivot and adjustment bolts.
2. Using adjustable pliers or equivalent, apply tension to the belt. Place the bottom jaw of the pliers under the alternator adjustment boss and the top jaw in the notch at the top of the alternator adjustment bracket.
3. Squeeze the pliers together and, using a belt tension gauge, adjust the tension to 160 lbs. for a new belt or 140 lbs. for a used belt. While maintaining the proper belt tension, tighten the alternator adjustment bolt to 26 ft. lbs. (35 Nm).
4. Remove the belt tension gauge, start the engine and let it idle for 5 minutes.
5. Stop the engine and recheck the belt tension. Adjust to the proper specifications and retighten the adjustment bolt.
6. Tighten the alternator pivot bolt to 52 ft. lbs. (70 Nm).

REMOVAL & INSTALLATION

Except 1991–92 Escort

1. Disconnect the negative battery cable.

2. Disconnect the wire harness attachments to the integral alternator/regulator assembly. Pull the 2 connectors straight out.
3. Loosen the alternator pivot bolt. Remove the adjustment arm bolt from the alternator.
4. Disengage the alternator drive belt from the alternator pulley.
5. Remove the alternator pivot bolt and alternator/regulator assembly.
6. Remove the alternator fan shield, if equipped.
To install:
7. Position the integral alternator/regulator assembly on the engine.
8. Install the alternator pivot and adjuster arm bolts. Do not tighten the bolts until the belt is tensioned.
9. Install the drive belt over the alternator pulley.
10. Adjust the belt tension according to the proper procedure.
11. Connect wiring harness to the alternator/regulator assembly. Push both connectors straight in.
12. Attach the alternator fan shield to the alternator, if equipped.
13. Connect the negative battery cable.

1991–92 Escort

1.8L ENGINE

1. Disconnect the negative battery cable.
2. Remove the nut securing the wiring connector to the alternator.
3. Remove the field terminal wiring connector.
4. Remove the upper mounting bolt securing the alternator to the alternator bracket.
5. Loosen the lower alternator mounting bolt and pivot the alternator forward.
6. Remove the alternator belt from the pulley and position the belt aside.
7. Raise and safely support the vehicle.
8. Remove the lower splash shield located under the accessory belts.
9. Remove the alternator lower mounting bolt and remove the alternator from the vehicle.
To install:
10. Position the alternator into the vehicle and install the lower mounting bolt.
11. Install the lower splash shield.
12. Lower the vehicle.
13. Position the alternator belt onto the pulley and adjust the belt according to the proper procedure.
14. Tighten the upper mounting bolt to 14–19 ft. lbs. (19–25 Nm) and the lower mounting bolt to 27–38 ft. lbs. (37–52 Nm).
15. Install the field terminal wiring connector.

16. Position the wiring connector to the alternator and secure it with the nut.

17. Connect the negative battery cable.

1.9L ENGINE

1. Disconnect the negative battery cable.

2. Remove the accessory drive belt.

3. Remove the nut securing the wiring connector to the alternator.

4. Remove the 2 snap-in type wiring connectors at the alternator.

5. Remove the air conditioning hose bracket from the alternator bracket and position it aside.

6. Remove the alternator mounting bolts.

7. Remove the bolts securing the power steering reservoir and position it aside.

8. Remove the alternator from it's bracket.

To install:

9. Position the alternator onto its bracket.

10. Position the power steering reservoir and secure it with the bolts.

11. Install the alternator upper mounting bolt and tighten to 14–22 ft. lbs. (20–30 Nm). Install the alternator lower mounting bolt and tighten to 29–40 ft. lbs. (40–55 Nm).

12. Position the air conditioning hose bracket onto the alternator bracket and secure it with the bolts.

13. Install the 2 snap-in type wiring connectors into the alternator.

14. Position the wiring connector onto the alternator and secure it with the nut.

15. Install the accessory drive belt.

16. Connect the negative battery cable.

Starter

REMOVAL & INSTALLATION

Except 1991–92 Escort

1. Disconnect the negative battery cable.

2. Raise and safely support the vehicle.

3. Disconnect the starter cable at the starter terminal.

4. Remove the 2 bolts attaching the starter rear support bracket, if equipped. Remove the bracket.

5. If equipped with roll restrictor brace-to-starter studs on the transaxle housing, remove the nuts and remove the brace.

6. On Tempo and Topaz, remove the cable support.

7. For installation, reverse the removal procedure. Tighten the attaching studs or bolts to 30–40 ft. lbs. (41–

54 Nm) on all except 1991–92 Tempo and Topaz. On 1991–92 Tempo and Topaz, tighten the attaching bolts to 15–20 ft. lbs. (20.3–27.0 Nm).

1991–92 Escort

1.8L ENGINE

1. Disconnect the negative battery cable.

2. Remove the air duct that connects to the throttle body and resonance chamber.

3. Remove the starter motor upper mounting bolts.

4. Raise and safely support the vehicle.

5. Remove the intake plenum support bracket mounting bolts and remove the bracket.

6. Disconnect the S terminal connector from the starter solenoid.

NOTE: When disconnecting the plastic hard shell connector at the solenoid S terminal, grasp the plastic connector, depress the plastic tab and pull off the lead assembly. Do not pull on the lead wire or damage may result.

7. Remove the B terminal attaching nut and disconnect the cable from the terminal.

8. Remove the starter motor lower mounting bolt and remove the starter motor.

To install:

9. Place the starter motor into its mounting position and install the lower mounting bolt. Tighten the bolt to 15–20 ft. lbs. (20.3–27.0 Nm).

10. Connect the cable to the starter solenoid B terminal and install the attaching nut to the terminal. Tighten the nut to 80–120 inch lbs. (9.0–13.5 Nm).

11. Connect the electrical connector to the starter solenoid S terminal.

12. Install the intake plenum support bracket and tighten the attaching bolts to 27–38 ft. lbs. (37–52 Nm) and the attaching nut to 14–19 ft. lbs. (19–25 Nm).

13. Lower the vehicle.

14. Install the starter motor upper mounting bolts and tighten to 15–20 ft. lbs. (20.3–27.0 Nm).

15. Install the air duct that connects to the throttle body and resonance chamber.

16. Connect the negative battery cable.

1.9L ENGINE

1. Disconnect the negative battery cable.

2. If equipped with an automatic transaxle, remove the kickdown cable routing bracket from the engine block.

3. Disconnect the wire from the starter solenoid S terminal.

NOTE: When disconnecting the plastic hard shell connector at the solenoid S terminal, grasp the plastic connector, depress the plastic tab and pull off the lead assembly. Do not pull on the lead wire or damage may result.

4. Remove the attaching nut from the starter solenoid B terminal and disconnect the cable from the terminal.

5. Remove the starter motor mounting bolts and remove the starter motor.

To install:

6. Place the starter motor into its mounting position and install the mounting bolts. Tighten the bolts to 15–20 ft. lbs. (20.3–27.0 Nm).

7. Connect the cable to the starter solenoid B terminal and install the attaching nut. Tighten the nut to 80–120 inch lbs. (9–13.5 Nm).

8. Connect the wire to the starter solenoid S terminal.

9. If equipped with an automatic transaxle, install the kickdown cable routing bracket to the engine block.

10. Connect the negative battery cable.

EMISSION CONTROLS

Please refer to "Emission Controls" in the Unit Repair section for system maintenance procedures. Due to the complex nature of modern electronic engine control systems, comprehensive diagnosis and testing procedures fall outside the confines of this repair manual. For complete information on diagnosis, testing and repair procedures concerning all modern engine and emission control systems, please refer to "Chilton's Guide to Fuel Injection and Electronic Engine Controls".

Emission Warning Lamps

RESETTING

These vehicles have a "CHECK ENGINE" or "SERVICE ENGINE SOON" light that will light when there is a fault in the engine control system. Depending upon the system or sensor involved, the light may go out if the fault is intermittent. However, the fault code will remain stored in the

ECU until the system is serviced and the ECU memory cleared. When a fault is detected in certain systems or sensors, the light will remain lit until the system is serviced. When the system has been diagnosed, the problem corrected and the ECU memory cleared, the light will go out.

FUEL SYSTEM

Fuel System Service Precautions

Safety is the most important factor when performing not only fuel system maintenance but any type of maintenance. Failure to conduct maintenance and repairs in a safe manner may result in serious personal injury or death. Maintenance and testing of the vehicle's fuel system components can be accomplished safely and effectively by adhering to the following rules and guidelines.

● To avoid the possibility of fire and personal injury, always disconnect the negative battery cable unless the repair or test procedure requires that battery voltage be applied.

● Always relieve the fuel system pressure prior to disconnecting any fuel system component (injector, fuel rail, pressure regulator, etc.), fitting or fuel line connection. Exercise extreme caution whenever relieving fuel system pressure to avoid exposing skin, face and eyes to fuel spray. Please be advised that fuel under pressure may penetrate the skin or any part of the body that it contacts.

● Always place a shop towel or cloth around the fitting or connection prior to loosening to absorb any excess fuel due to spillage. Ensure that all fuel spillage (should it occur) is quickly removed from engine surfaces. Ensure that all fuel soaked cloths or towels are deposited into a suitable waste container.

● Always keep a dry chemical (Class B) fire extinguisher near the work area.

● Do not allow fuel spray or fuel vapors to come into contact with a spark or open flame.

● Always use a backup wrench when loosing and tightening fuel line connection fittings. This will prevent unnecessary stress and torsion to fuel line piping. Always follow the proper torque specifications.

● Always replace worn fuel fitting O-rings with new. Do not substitute fuel hose or equivalent, where fuel pipe is installed.

RELIEVING FUEL SYSTEM PRESSURE

The pressure in the fuel system must be released before disconnecting any fuel lines or components.

Except 1.9L Engine with Central Fuel Injection and 1991–92 Escort

1. Disconnect the negative battery cable.
2. A special valve is incorporated in the fuel rail assembly for the purpose of relieving the pressure in the fuel system.
3. Remove the fuel tank cap and remove the cap from the pressure relief valve.
4. Attach pressure gauge tool T80L-9974-B or equivalent, to the fuel pressure valve on the fuel rail assembly and release the pressure from the system into a suitable container.

1.9L Engine with Central Fuel Injection

1. Remove the electrical connector at the inertia switch located on the left side of the luggage compartment.
2. Release the fuel system pressure by cranking the engine for 15 seconds.

1991–92 Escort

1. Start the engine.
2. Remove the rear seat cushion and disconnect the fuel pump electrical connectors.
3. Wait for the engine to stall, then turn **OFF** the ignition switch.
4. Connect the fuel pump electrical connectors and install the rear seat cushion.

Fuel Line Push Connect Fittings

It is necessary to disconnect nylon push connect fittings to disconnect the fuel lines on most vehicles. The push connect fittings are designed with 2 different retaining clips. The fittings used with $5/16$ in. (7.9mm) diameter tubing use a hairpin clip. The fittings used with $1/4$ in. (6.35mm) and $1/2$ in. (12.7mm) diameter tubing use a duck bill clip. Each type of fitting requires different procedures for service.

REMOVAL & INSTALLATION

Hairpin Clip Fitting

1. Inspect the internal portion of the fitting for dirt accumulation. If more than a light coating of dust is present, clean the fitting before disassembly.

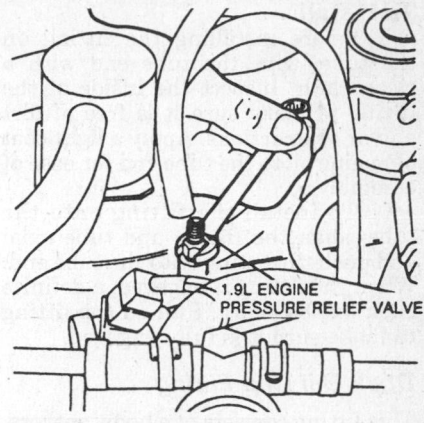

Pressure relief valve location—1988–90 1.9L EFI engine

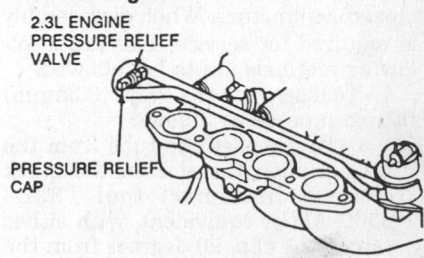

Pressure relief valve location—2.3L engine

2. Remove the hairpin clip from the fitting by first bending the shipping tab downward so it will clear the body. This is done, using hands only, by spreading the 2 clip legs about $1/8$ in. (3.2mm) each to disengage the body and pushing the legs into the fitting. Complete removal is accomplished by lightly pulling from the triangular end of the clip and working it clear of the tube and fitting.

NOTE: Do not use any tools when disconnecting the clip.

3. Grasp the fitting and hose assembly and pull in an axial direction to remove the fitting from the steel tube. Adhesion between sealing surfaces may occur. A slight twist of the fitting may be required to break this adhesion for easier removal.

NOTE: On 90 degree elbow connectors, excessive side loading could break the connector body.

4. When the fitting is removed from the tube end, inspect the clip to make sure it has not been damaged. If damaged, replace the clip. If undamaged, immediately install the clip to prevent loss or damage. To install the clip, insert the clip into any 2 adjacent openings with the triangular portion pointing away from the fitting opening. Install the clip to fully engage the body—legs of the hairpin clip locked on the outside of the body. Piloting with an index finger is necessary.

To install:

5. Before installing the fitting on the tube, wipe the tube end with a clean cloth. Inspect the inside of the fitting to make sure it is free of dirt and/or obstructions. Apply a light coat of engine oil to the tube end for ease of assembly.

6. To install the fitting onto the tube, align the fitting and tube axial and push the fitting onto the tube end. When the fitting is engaged, a definite click will be heard. Pull on the fitting to make sure it is fully engaged.

Duck Bill Clip Fitting

This fitting consists of a body, spacers, O-rings and a duck bill retaining clip. The clip maintains the fitting to the steel tube juncture. When disassembly is required for service, 1 of the 2 following methods are to be followed.

1. To disengage the ¼ in. (6.35mm) fitting, proceed as follows:

a. To disenage the tube from the fitting, align the slot on the quick connect/disconnect tool T82L–9500–AH or equivalent, with either tab on the clip, 90 degrees from the slots on the side of the fitting, and insert the tool.

b. This disengages the duck bill from the tube. Holding the tool and the tube with 1 hand, pull the fitting away from the tube.

NOTE: Only moderate effort is required if the tube has been properly disengaged. Use hands only.

c. After disassembly, inspect and clean the tube sealing surface. Also inspect the inside of the fitting for damage to the retaining clip. If the retaining clip appears to be damaged, replace it.

NOTE: Some fuel tubes have a secondary bead which aligns with the outer surface of the clip. These beads can make tool insertion difficult. If there is extreme difficulty, use the ½ in. (12.7mm) fitting method as an alternative.

2. To disengage the ½ in. (12.7mm) fitting, proceed as follows:

a. This method of disassembly disengages the retaining clip from the fitting body.

b. Use a pair of suitable narrow pliers such as 6 in. channel lock pliers. The pliers must have a jaw width of 0.2 in. (5.2mm) or less.

c. Align the jaws of the pliers with the openings in the side of the fitting case and compress the portion of the retaining clip that engages the fitting case. This disengages the retaining clip from the case. Often 1 side of the clip will disengage before

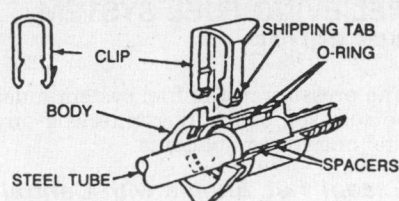

Hairpin clip fitting connector

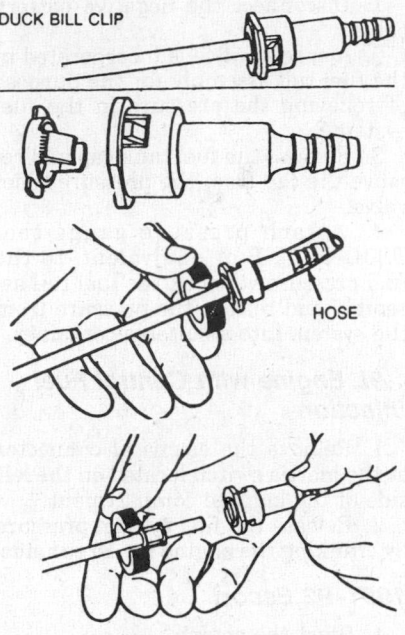

Duck bill clip fitting connector

the other. It is necessary to disengage the clip from both openings. Pull the fitting off of the tube.

NOTE: Only moderate effort is required if the retaining clip has been properly disengaged. Use hands only.

d. The retaining clip will remain on the tube. Disengage the clip from the tube bead and remove. Replace the retaining clip if it appears to be damaged.

NOTE: Slight ovality of the ring of the clip will usually occur. If there are no visible cracks and the ring will pinch back to its circular configuration, it is not damaged. If there is any doubt, replace the clip.

To install:

3. Install the clip into the body by inserting 1 of the retaining clip serrated edges on the duck bill portion into 1 of the window openings. Push on the other side until the clip snaps into place. Slide the fuel line back into the clip.

Fuel Tank

REMOVAL & INSTALLATION

Except 1991–92 Escort

1. Disconnect the negative battery cable.

2. Relieve the fuel system pressure.

3. Fuel should be drained from the tank as completely as possible prior to tank removal. This is accomplished by siphoning or pumping fuel out through the fuel filler neck.

NOTE: There are reservoirs inside the fuel tank to maintain fuel near the fuel pickup during vehicle cornering maneuvers and under low fuel operating conditions. These reservoirs could block siphon tubes or hoses from reaching the bottom of the fuel tank. This situation can be overcome with a few repeated attempts using different hose orientations.

4. Disconnect the fuel hoses and tubes. Disconnect the push connect fittings according to the proper procedure.

5. Disconnect the electrical hookup to the fuel tank sender unit. On some vehicles, the electrical connection is inaccessible on top of the tank and no intermediate connection point is provided. In these cases, the electrical connector must be disconnected from the fuel sender with the tank partially removed from the vehicle.

6. On Tempo and Topaz equipped with all-wheel drive, perform the following procedure to gain access to the fuel tank:

a. Remove the exhaust system.

b. Positiona a suitable jack under the rear axle.

c. Remove the bolts from the torque tube support bracket.

d. Remove 1 bolt each from the left and center differential support brackets.

e. Lower the differential approximately 6–8 in.

f. Remove 2 horizontal bolts retaining the axle pinion support crossmember.

7. On all vehicles, place a safety support under the fuel tank and remove the bolts or nuts from 1 end of the fuel tank straps. The straps are hinged at 1 end. Remove the bolts from the unhinged end and swing the straps aside.

8. Partially remove the tank and disconnect the fuel lines and the electrical connector from the fuel gauge sender, if required.

9. Remove the tank from the vehicle.

To install:

10. Before proceeding, check the following:

    a. Leak check the sender unit. If necessary, use fuel tank sender wrench T74P-9275-A or equivalent.

    b. Make sure the fuel vapor separator valve is installed completely on the tank top.

    c. Make all required fuel line, fuel return line, vapor vent and electrical connections which will be inaccessible after the tank is installed.

11. Position the fuel tank in the vehicle.

12. Bring the fuel tank straps around the tank and start attaching the nut or bolt. Align the tank with the straps.

13. Check the hoses and wiring mounted on the tank top, to make sure they are correctly routed and will not be pinched between tank and body.

14. Tighten the fuel tank strap attaching bolts to 21–29 ft. lbs. (28–40 Nm) on Escort or 25–39 ft. lbs. (34–54 Nm) on Tempo and Topaz.

15. Connect the fuel tank hoses and lines. Verify that the fuel supply, fuel return and vapor vent connections are made correctly.

16. Connect the electrical connections.

17. Install the fuel filler hoses that connect the fuel tank to the fuel filler pipe. Install new hose clamps and tighten.

18. On Tempo and Topaz with all-wheel drive, perform the following procedure:

    a. Position the crossmember axle pinion support and install 2 retaining bolts.

    b. Raise the differential assembly into position.

    c. Install the bolts retaining the left and center differential support brackets.

    d. Install the torque tube support bracket retaining bolts.

    e. Remove the jack.

    f. Install the exhaust system.

19. Replace the fuel that was drained from the tank.

20. Check all connections for leaks.

21. Connect the negative battery cable.

22. Turn the ignition key **ON** to run the fuel pump and pressurize the system. Check for fuel leaks.

1991–92 Escort

1. Relieve the fuel system pressure.

2. Disconnect the negative battery cable.

3. Completely drain the fuel tank by siphoning or pumping out the fuel through the fuel filler hose.

4. Remove the rear seat cushion.

5. Remove the ground strap retaining screw and the 3 remaining fuel pump assembly cover screws.

6. Remove the fuel pump assembly cover.

7. Remove the clips from the fuel hoses.

8. Disconnect the fuel hoses from the fuel pump assembly.

9. Raise and safely support the vehicle.

10. Loosen the filler neck clamp and disconnect the filler neck hose from the filler neck.

11. Loosen the clamp and disconnect the filler neck overflow hose from the overflow tube.

12. Disconnect the vapor hoses from the vapor tubes.

13. Remove the exhaust middle pipe heat shield.

14. Support the fuel tank with a suitable jack.

15. Remove the mounting bolts from the fuel tank straps, unclip the straps and remove them.

16. Remove the 3 fuel tank heat shield attaching bolts from the fuel tank and remove the fuel tank.

To install:

17. Place the fuel tank onto the fuel tank heat shield and install the heat shield attaching bolts into the fuel tank.

18. Clip the fuel tank straps into their mounting positions.

19. Install the fuel tank strap mounting bolts and tighten to 27–38 ft. lbs. (37–52 Nm).

20. Remove the support jack from under the fuel tank.

21. Install the exhaust middle pipe heat shield.

22. Connect the vapor hoses to the vapor tubes.

23. Connect the filler neck hose to the filler neck and install the attaching clamp.

24. Connect the filler neck overflow hose to the overflow tube and install the attaching clamp.

25. Lower the vehicle.

26. Connect the fuel hoses to the fuel pump assembly.

27. Install the clips onto the fuel hoses.

28. Position the fuel pump assembly cover and ground strap and install the retaining screws.

29. Connect the fuel pump assembly electrical connector.

30. Install the rear seat cushion.

31. Replace the fuel that was drained from the tank during the removal procedure. Check for leaks.

32. Connect the negative battery cable.

33. Turn the ignition switch **ON** to run the fuel pump and pressurize the system. Check for fuel leaks.

Fuel Filter

REMOVAL & INSTALLATION

Except 1991–92 Escort

1. Disconnect the negative battery cable.

2. Properly relieve the fuel system pressure.

3. Remove the push connect fittings according to the proper procedure. Install new retainer clips in each connector fitting.

NOTE: The flow arrow direction should be noted to ensure proper flow of fuel through the replacement filter.

4. Remove the filter from the bracket by loosening the filter retaining clamp enough to allow the filter to pass through.

To install:

5. Install the filter in the bracket, ensuring the proper direction of flow, as noted earlier. Tighten the clamp to 15–25 inch lbs. (1.7–2.8 Nm).

6. Install push connect fittings at both ends of the filter.

7. Connect the negative battery cable.

8. Start the engine and inspect for leaks.

1991–92 Escort

1. Properly relieve the fuel system pressure.

2. Disconnect the negative battery cable.

3. Position a suitable container below the fuel filter to collect any excess fuel that may leak from the filter and lines.

4. Remove the retaining clip from the fuel filter upper hose.

5. Disconnect the upper hose from the fuel filter and drain any excess fuel into the container. Plug the hose.

6. Loosen the fuel filter mounting clamp.

7. Raise and safely support the vehicle.

8. Remove the retaining clip from the fuel filter lower hose.

9. Disconnect the lower hose from the fuel filter and drain any excess fuel into the container. Plug the hose.

10. Lower the vehicle.

11. Remove the fuel filter.

To install:

12. Position the fuel filter and tighten the filter mounting clamp.

13. Connect the filter upper hose to the filter and install the upper hose retaining clip.

14. Raise and safely support the vehicle.

15. Connect the filter lower hose to the filter and install the lower hose retaining clip.

16. Lower the vehicle.
17. Connect the negative battery cable.
18. Start the engine and check for leaks.

Electric Fuel Pump

PRESSURE TESTING

Except 1.8L Engine

1. Make sure the ignition key is in the **OFF** position.
2. Properly relieve the fuel system pressure.
3. Disconnect the fuel pump output line and connect a suitable pressure tester.
4. Ground the fuel pump lead of the Self-Test connector through a jumper wire at the **FP** lead.
5. Turn the ignition key **ON**, to operate the fuel pump, and observe the fuel pressure.
6. The fuel pressure should be 13–17 psi on the 1.9L CFI engine, 35–45 psi on the 1.9L EFI engine or 45–60 psi on the 2.3L engine.
7. Turn the ignition key **OFF** and remove the jumper wire.
8. Properly relieve the fuel system pressure and remove the pressure tester.
9. Reconnect the fuel line.

1.8L Engine

1. Make sure the ignition key is in the **OFF** position.
2. Properly relieve the fuel system pressure.
3. Install a suitable fuel pressure tester in the fuel line between the fuel filter and the fuel rail.
4. Jump the fuel pump test connector terminals together, terminal **LG** and terminal **BK** at the Self-Test connector.
5. Turn the ignition key to **RUN**, to operate the fuel pump, and observe the fuel pressure.
6. The fuel pressure should be 64–85 psi.
7. Turn the ignition key **OFF** and remove the jumper wire.
8. Properly relieve the fuel system pressure and remove the pressure tester.

REMOVAL & INSTALLATION

NOTE: The fuel pump is mounted on the fuel sender assembly inside the fuel tank.

Except 1991–92 Escort

1. Properly relieve the fuel system pressure.
2. Disconnect the negative battery cable.

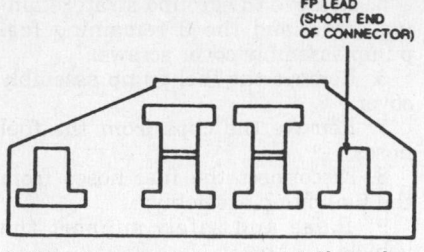

Self-Test connector—except 1.8L engine

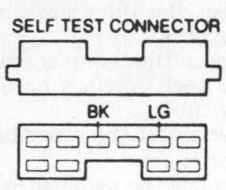

Fuel pump test connector terminals—1.8L engine

3. Remove the fuel tank.
4. Remove any dirt that has accumulated around the fuel pump retaining flange so it will not enter the fuel tank during removal and installation.
5. Turn the fuel pump locking ring counterclockwise using fuel tank sender wrench D84P–9257–A or equivalent, and remove the lock ring.
6. On all except all-wheel drive vehicles, remove the fuel pump and bracket assembly and remove the seal gasket and discard.
7. On all-wheel drive vehicles, proceed as follows:
   a. Partially lift up the sender unit and disconnect the jet pump line and the electrical connector to the resistor.
   b. Remove the fuel pump and bracket assembly and remove the seal gasket and discard.
   c. Remove the jet pump assembly attaching screw and remove the jet pump assembly.
To install:
8. Clean the fuel pump mounting flange and fuel tank mounting surface and seal ring groove.
9. Put a light coating of multi-purpose lubricant C1AZ–19590–BA or equivalent, on a new seal ring to hold it in place during assembly and install it in the fuel ring groove.
10. On all-wheel drive vehicles, install the jet pump assembly and attaching screw. Tighten the screw to 10–15 ft. lbs. (14–20 Nm).
11. Install the fuel pump and sender assembly carefully to ensure that the filter is not damaged. Make sure the locating keys are in the keyways and
12. On all-wheel drive vehicles, connect the jet pump line and the electrical connector to the resistor. Make sure the locating keyways and seal ring remain in place.

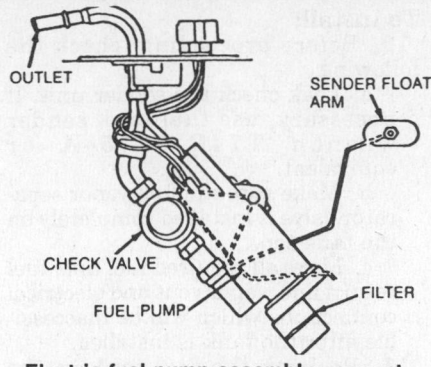

Electric fuel pump assembly—except 1991–92 Escort and Tempo/Topaz all-wheel drive vehicles.

13. Hold the assembly in place and install the locking ring finger-tight. Make sure all locking tabs are under the tank lock ring tabs.
14. Rotate the locking ring clockwise using fuel tank sender wrench D84P–9275–A or equivalent, until the ring stops against the stops.
15. Install the fuel tank into the vehicle according to the proper procedure.
16. Connect the negative battery cable.

1991–92 Escort

1. Properly relieve the fuel system pressure.
2. Disconnect the negative battery cable.
3. Remove the rear seat cushion. Disconnect the electrical connector at the fuel pump.
4. Remove the ground strap retaining screw and the 3 remaining fuel pump assembly cover screws.
5. Remove the fuel pump assembly cover.
6. Remove the clips from the fuel hoses.
7. Disconnect the fuel hoses from the fuel pump assembly.
8. Using a suitable removal tool, carefully remove and, if necessary, discard the fuel pump assembly spanner nut.
9. Remove the fuel pump assembly and discard the gasket.
To install:
10. Install a new gasket and position the fuel pump assembly in the tank.
11. Install the fuel pump assembly spanner nut.
12. Connect the fuel hoses to the fuel pump assembly and install the clips.
13. Position the fuel pump assembly cover and ground strap and install the retaining screws.
14. Connect the fuel pump assembly electrical connector.
15. Connect the negative battery cable.

16. Turn the ignition switch **ON** to run the fuel pump and pressurize the fuel system. Check for fuel leaks.

17. Install the rear seat cushion.

Fuel Injection

IDLE SPEED ADJUSTMENT

1.8L Engine

1. Apply the parking brake and make sure the vehicle is in **N** or **P**.

2. Start the engine and warm it up to normal operating temperature.

3. Turn off all electrical loads and accessories.

4. Using a jumper wire, connect the **GROUND** terminal to the **TEN** terminal on the diagnosis connector.

5. Connect the positive lead of a suitable tachometer to the **IG** terminal on the diagnosis connector and the tachometer negative lead to the negative battery terminal.

6. Check the vehicle idle speed when the electric cooling fan is not operating. The idle speed should be 700–800 rpm.

NOTE: When the parking brake is not applied, the idle speed for automatic transaxle vehicles is approximately 800 rpm.

7. If the idle speed is not within specification, adjust the idle speed by turning the idle speed adjusting screw until the idle speed is within specification.

8. Remove the jumper wire from the diagnosis connector and remove the tachometer.

1.9L CFI Engine

1. Engine **OFF**, remove air cleaner. Connect jumper wire between Self-Test Input (STI) and signal return pin on the self-test connector.

2. Turn ignition key **ON** but do not start engine. Wait for ISC plunger to retract approximately 10–15 seconds.

3. Disconnect vehicle harness from the ISC motor. Turn ignition key **OFF** and remove jumper wire.

4. Start engine, check idle rpm. If rpm is 600 ± 50, continue with Step 6. If not continue with Step 5.

5. Adjust throttle stop adjusting screw.

6. Shut engine OFF and reconnect vehicle harness to ISC motor. Reinstall air cleaner.

1.9L EFI Engine

1988–90

1. Disconnect idle speed control-air bypass solenoid.

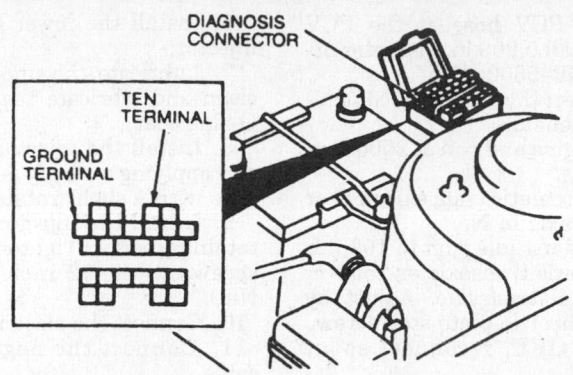

Diagnosis connector location—1991–92 Escort with 1.8L engine

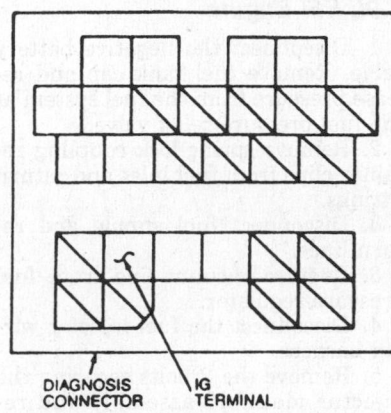

IG terminal location—1991–92 Escort with 1.8L engine

2. Start engine and run at 2000 rpm for 60 seconds.

3. Place transaxle in **N** or **P**.

4. Check/adjust idle rpm to 950 ± 50 on all except 1990 1.9L SEFI MA engine. On 1990 1.9L SEFI MA engine, check/adjust idle rpm to 650 ± 50. Adjust by turning the throttle plate stop screw.

5. Shut engine **OFF**, reconnect idle speed control-air bypass solenoid. Verify the throttle is not stuck in the bore and linkage is not preventing the throttle from closing.

6. Start engine and let stabilize for 2 minutes, then quickly open throttle and let return to idle, lightly depress and release accelerator. Check idle speed.

1991–92

1. Place the automatic transaxle in **P** or the manual transaxle in neutral.

2. Disconnect the negative battery cable for 5 minutes, then reconnect it.

3. Start the engine and let it stabilize for 2 minutes, then goose the engine and let it return to idle, lightly depress and release the accelerator and let the engine idle. If the engine does not idle properly, proceed to Step 4.

NOTE: If the electric fan comes on, wait until it turns itself off to check the engine idle.

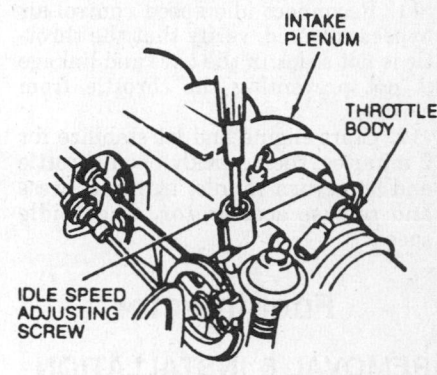

Idle speed adjusting screw location—1991–92 Escort with 1.8L engine

4. Stop the engine and disconnect the idle speed control air bypass solenoid.

5. Start the engine and run it at 2000 rpm for 60 seconds.

6. After the engine returns to idle, check the idle speed. It should be 600 ± 150 rpm.

7. If the idle rpm is incorrect, turn the plate stop screw until the idle speed is within specification.

8. Turn the engine **OFF** and reconnect the idle speed control-air bypass solenoid. Verify the throttle is not stuck in the bore and the linkage is not preventing the throttle from closing.

9. Start the engine and let it stabilize for 2 minutes, then goose the engine and let it return to idle; lightly depress and release the accelerator and let the engine idle. Check idle speed.

2.3L ENGINE

1. Apply the parking brake, block the drive wheels and place the vehicle in **P** or **N**.

2. Start the engine and let it run until it reaches normal operating temperature, then shut engine **OFF**.

3. Unplug spout line and verify that ignition timing is at base setting ± 2 degrees BTDC. Refer to decal under hood.

4. Remove PCV hose at the PCV valve and install 0.200 in. diameter orifice, tool T86P-9600-A.

5. Disconnect the idle speed control-air bypass solenoid.

6. Start engine and run at 2500 rpm for 30 seconds.

7. Place automatic transaxle in **D** or manual transaxle in **N**.

8. Check/adjust idle rpm to 1025 ± 50 for automatic transaxle or 1550 ± 50 for manual transaxle. Adjust by turning the throttle plate stop screw.

9. Engine **OFF**, reconnect spout line.

10. Remove orifice from PCV hose and reconnect to PCV valve.

11. Reconnect idle speed control-air bypass solenoid, verify that the throttle is not stuck in the bore and linkage is not preventing the throttle from closing.

12. Start engine and let stabilize for 2 minutes, then quickly open throttle and let return to idle, lightly depress and release accelerator. Check idle speed.

Fuel Injector

REMOVAL & INSTALLATION

1.8L Engine

1. Properly relieve the fuel system pressure.

2. Disconnect the negative battery cable.

3. Disconnect the fuel pressure and return lines from the fuel rail.

4. Disconnect the PCV hose from the intake plenum and cylinder head cover.

5. Disconnect the fuel pressure regulator vacuum hose and the fuel injector wiring harness electrical connectors.

6. Remove the fuel rail mounting bolts and remove the fuel rail.

7. Remove the fuel injectors, grommets and insulators.

8. Installation is the reverse of the removal procedure. Lubricate new O-rings with clean engine oil and install them on the fuel injectors prior to installation.

1.9L CFI Engine

1. Disconnect the negative battery cable.

2. Disconnect the electrical connector from the injector top.

3. Remove the fuel injector retaining screw and retainer.

4. Remove the injector and lower O-ring. Discard the O-ring.

To install:

5. Lubricate a new lower O-ring and the injector seat area with clean engine oil. Do not use transmission oil.

6. Install the lower O-ring on the injector.

7. Lubricate the upper O-ring and clean and lubricate the throttle body O-ring seat.

8. Install the injector by centering and applying a steady downward pressure with a slight rotational force.

9. Install the injector retainer and retaining screw. Tighten the retaining screws to 28–32 inch lbs. (3.2–3.6 Nm).

10. Connect the electrical connector.

11. Connect the negative battery cable.

1.9L EFI Engine

1. Disconnect the negative battery cable. Remove fuel tank cap and release pressure from the fuel system at the fuel pressure relief valve.

2. Remove spring-lock coupling retainer clips from fuel inlet and return fittings.

3. Disconnect fuel supply and return lines.

3. Remove vacuum line from fuel pressure regulator.

4. Disconnect the fuel injector wiring harness.

5. Remove the 2 bolts securing the injector manifold assembly and remove the assembly.

6. Carefully remove connectors from individual injectors(s) as required.

7. Grasping the injector's body, pull up while gently rocking the injector from side-to-side.

8. Inspect the injector O-rings (2 per injector) for signs of deterioration. Replace as required.

9. Inspect the injector "plastic hat" (covering the injector pintle) and washer for signs of deterioration. Replace as required. If hat is missing, look for it on intake manifold.

To install:

10. Use a light grade oil to lubricate new O-rings and install 2 on each injector.

11. Install the injector(s). Use a light, twisting, pushing motion to install the injector(s).

12. Carefully seat the fuel injector manifold assembly on the injectors and secure the manifold with the attaching bolts. Tighten to 15–22 ft. lbs. (20–30 Nm).

13. Connect the vacuum line to the fuel pressure regulator.

14. Connect fuel injector wiring harness.

15. Connect fuel supply and fuel return lines.

16. Reconnect spring-lock coupling retaining clips on fuel inlet and return fittings.

17. Check entire assembly for proper alignment and seating.

18. Connect negative battery cable.

2.3L Engine

1. Properly relieve the fuel system pressure.

2. Disconnect the engine air cleaner outlet tube from the air intake throttle body and the TP sensor from the wiring harness.

3. Disconnect the vacuum lines from the upper manifold and disconnect the EGR tube at the manifold connection.

4. Disconnect the air bypass valve connector, remove the accelerator and, if equipped, speed control cables and remove the manifold upper support bracket top bolt.

5. Remove the fuel supply manifold shield and the 4 upper manifold retaining bolts and 1 retaining shoulder stud.

6. Remove the upper manifold assembly and gasket and set it aside.

7. Disconnect the fuel supply and return lines and the vacuum line at the pressure regulator.

8. Disconnect the fuel injector wiring harness and disconnect the connectors from the injectors.

9. Remove the fuel supply manifold retaining bolts and remove the fuel supply manifold.

10. Grasping the injector body, pull up while gently rocking the injector from side-to-side.

11. Inspect the injector O-rings, the injector plastic hat and washer for signs of deterioration. Replace as necessary. If the hat is missing, look for it on the intake manifold.

12. Installation is the reverse of the removal procedure. Lubricate new O-rings with light engine oil and install on the injectors prior to installation. Tighten the fuel supply manifold retaining bolts and the upper intake manifold retaining bolts to 15–22 ft. lbs. (20–30 Nm).

DRIVE AXLE

Halfshaft

REMOVAL & INSTALLATION

Except 1991–92 Escort and All Wheel Drive Rear Halfshaft

NOTE: Halfshaft assembly removal and installation procedures are the same for automatic and manaual transaxles, except on the automatic transaxle, the right side halfshaft must be removed first. Differential rotator

tool T81P–4026–A or equivalent, is then inserted into the transaxle to drive the left side inboard CV-joint assembly from the transaxle. If only the left side halfshaft assembly is to be removed for service, remove the right side halfshaft assembly from the transaxle only. After removal, support it with a length of wire, then drive the left side halfshaft assembly from the transaxle.

1. Remove the cap from the hub and loosen the hub nut. Set the parking brake. The nut must be loosened without unstaking; the use of a chisel or similar tool may damage the spindle thread.

2. Raise and safely support the vehicle. Remove the wheel and tire assembly. Remove the hub nut/washer and discard the nut.

3. Remove the brake hose routing clip-to-strut bolt.

4. Remove the nut from the ball joint-to-steering knuckle bolt. Using a hammer and a punch, drive the bolt from the steering knuckle and discard the bolt/nut.

5. Using a prybar, separate the ball joint from the steering knuckle. Position the end of the prybar outside of the bushing pocket to avoid damage to the bushing; be careful not to damage the ball joint or CV-joint boot.

NOTE: The lower control arm ball joint fits into a pocket formed in the plastic disc brake rotor shield; bend the shield away from the ball joint while prying the ball joint from the steering knuckle.

6. Using a prybar, pry the halfshaft from the differential housing. Position the prybar between the differential housing and the CV-joint assembly. Be careful not to damage the differential oil seal, case, CV-joint boot or the transaxle.

NOTE: Shipping plugs T81P–1177–B or equivalent, must be installed in the differential housing after halfshaft removal. Failure to do so can result in dislocation of the differential side gears. Should the gears become misaligned, the differential will have to be removed from the transaxle to re-align the gears.

7. Using a piece of wire, support the end of the shaft from a convenient underbody component.

NOTE: Do not allow the shaft to hang unsupported, as damage to the outboard CV-joint may result.

8. Using a front hub removal tool, press the halfshaft's outboard CV-joint from the hub.

NOTE: Never use a hammer or separate the outboard CV-joint stub shaft from the hub. Damage to the CV-joint internal components may result.

To install:

9. Install a new circlip onto the inboard CV-joint stub shaft; the outboard CV-joint stub shaft does not have a circlip. To install the circlip properly, start one end in the groove and work the circlip over the stub shaft end and into the groove; this will avoid over expanding the circlip.

10. Carefully, align the splines of the inboard CV-joint stub shaft with the splines in the differential. Push the CV-joint into the differential until the circlip is seated in the differential side gear. Use care to prevent damage to the differential oil seal.

NOTE: A non-metallic mallet may be used to aid in seating the circlip into the differential side gear groove; if a mallet is necessary, tap only on the outboard CV-joint stub shaft.

11. Carefully, align the outboard CV-joint stub shaft splines with the hub splines and push the shaft into the hub, as far as possible; use the front hub replacer tool to firmly press the halfshaft into the hub.

12. Connect the control arm-to-steering knuckle and torque the new nut/bolt to 40–54 ft. lbs. (54–74 Nm). A new bolt and nut must be used.

13. Position the brake hose routing clip on the suspension strut and torque the bolt to 8 ft. lbs. (11 Nm).

14. Install the hub nut washer and a new hub nut.

15. Install the wheel/tire assembly and torque the lug nuts to 80–105 ft. lbs. (108–144 Nm). Lower the vehicle and torque the hub nut to 180–200 ft. lbs. (244–271 Nm).

16. Refill the transaxle and road test.

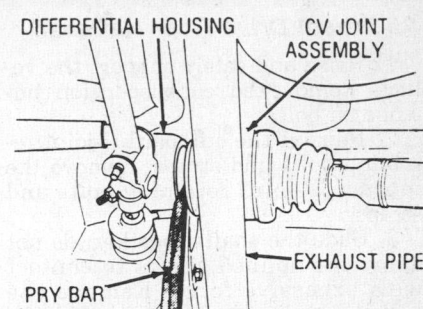

Removing halfshaft from transaxle

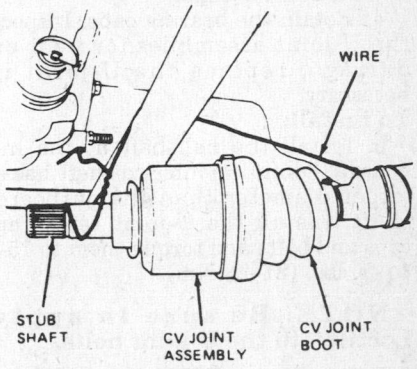

Support halfshaft by wiring to body

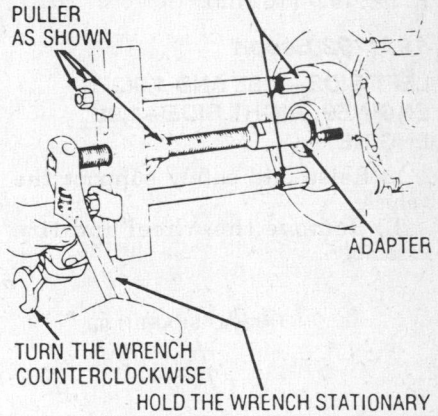

Removing hub from shaft assembly — except 1991–92 Escort

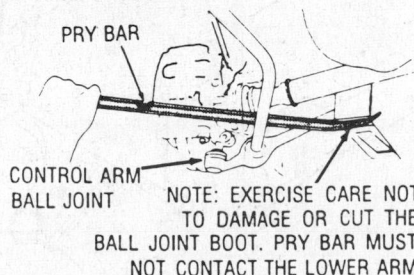

Separating lower ball joint from steering knuckle — except 1991–92 Escort

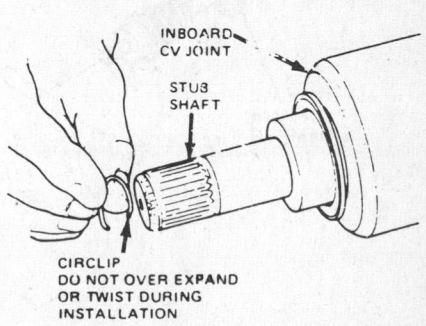

Stub shaft circlip installation

All Wheel Drive Rear Halfshaft

1. Raise and safely support the vehicle. Remove the rear suspension control arm bolt.

2. Remove the outboard U-joint retaining bolts and straps. Remove the inboard U-joint retaining bolts and straps.

3. Slide the shafts together; do not allow the splined shafts to contact with excessive force. Remove the halfshafts; do not drop the halfshafts as the impact may cause damage to the U-joint bearing cups.

4. Retain the bearing cups. Inspect the U-joint assemblies for wear or damage, replace the U-joint if necessary.

To install:

5. Install the halfshaft at the inboard U-joint; the inboard shaft has a larger diameter than the outboard shaft. Install the U-joint retaining caps and bolts and torque them to 15–17 ft. lbs. (21–23 Nm).

NOTE: Be sure to apply Loctite® to the U-joint bolts.

6. Install the halfshaft at the outboard U-joint. Install the U-joint retaining caps and bolts and torque them to 15–17 ft. lbs. (21–23 Nm).

7. Install the rear suspension control arm and torque the bolt to 60–86 ft. lbs. (82–116 Nm).

1991–92 Escort

LEFT SIDE—1.8L AND 1.9L ENGINES, RIGHT SIDE—1.9L ENGINE

1. Raise and safely support the vehicle.

2. Remove the wheel and tire assembly.

3. Remove the splash shield.

4. Carefully raise the staked portion of the halfshaft retaining nut using a suitable small chisel. Remove and discard the retaining nut.

5. Remove the cotter pin and nut from the tie rod end and remove the tie rod end from the steering knuckle using a suitable removal tool.

6. Remove the lower ball joint clamp bolt. Carefully pry down on the lower control arm to separate the ball joint from the steering knuckle.

7. Pull outward on the steering knuckle/brake assembly. Carefully pull the halfshaft from the steering knuckle and position it aside.

8. Removal of the left side halfshaft requires removal of the crossmember to allow access with a prybar. If the left side halfshaft is being removed, proceed as follows:

   a. Support the transaxle with a suitable transmission jack.

   b. Remove the 4 transaxle mount-to-crossmember attaching nuts.

   c. Remove the 2 crossmember attaching nuts at the rear of the crossmember.

   d. While supporting the rear of the crossmember, remove the 2 front mounting bolts. Remove the crossmember.

9. Position a drain pan under the transaxle.

10. Insert a prybar between the halfshaft and the transaxle case. Gently pry outward to release the halfshaft from the differential side gears. Be careful no to damage the transaxle case, oil seal, CV-joint or CV-joint boot.

11. Remove the halfshaft.

NOTE: Install suitable plugs after removing the halfshafts to prevent the differential side gears from becoming mispositioned. Should the gears become misaligned, the differential will have to be removed from the transaxle to align the gears.

To install:

12. Position the circlip on the inner CV-joint spline so the circlip gap is at the top. Lubricate the splines lightly with suitable grease.

13. Remove the plugs that were installed in the differential side gears.

14. Position the halfshaft so the CV-joint splines are aligned with the differential side gear splines. Push the halfshaft into the differential.

NOTE: When seated properly, the circlip can be felt as it snaps into the differential side gear groove.

15. Pull outward on the steering knuckle/brake assembly and insert the halfshaft into the steering knuckle.

16. Pry downward on the control arm and position the lower ball joint in the steering knuckle.

17. For left side halfshaft installation, proceed as follows:

   a. Position the crossmember in place.

   b. Install the 2 mounting bolts and the 2 attaching nuts. Tighten the nuts and bolts to 47–66 ft. lbs. (64–89 Nm).

   c. Install the 4 transaxle mount-to-crossmember attaching nuts. Tighten the nuts to 27–38 ft. lbs. (37–52 Nm).

   d. Remove the transmission jack.

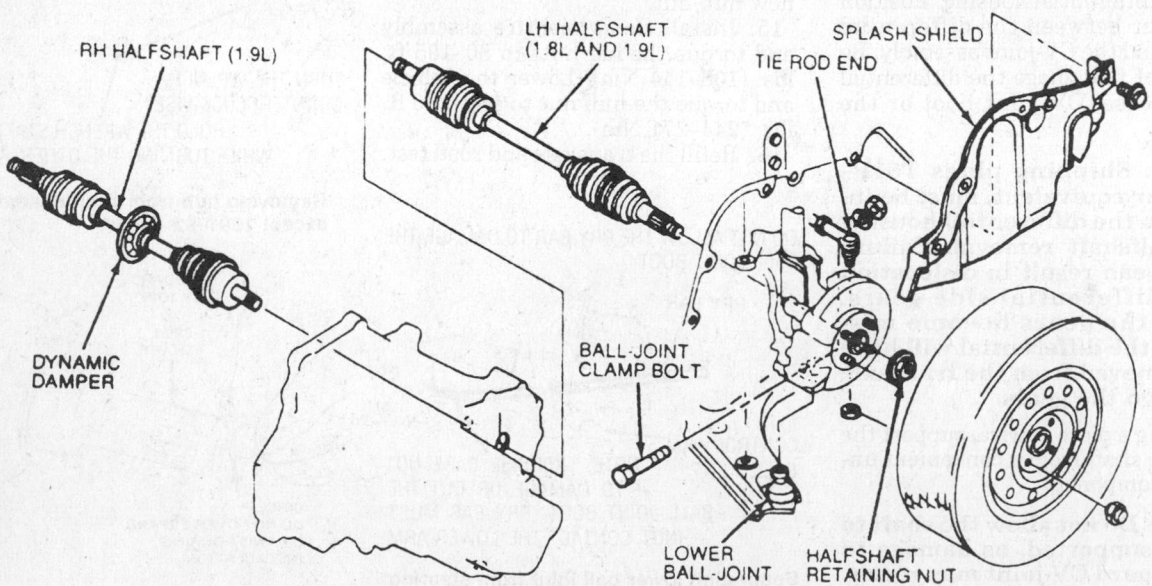

Halfshaft removal and installation—right side, 1.8L and 1.9L engines and left side, 1.8L engine

18. Install the lower ball joint clamp bolt and tighten to 32–43 ft. lbs. (43–59 Nm).

19. Install the tie rod end in the steering knuckle. Install the nut to the tie rod end and tighten to 31–42 ft. lbs. (42–57 Nm). Install a new cotter pin.

20. Install a new halfshaft retaining nut and tighten to 174–235 ft. lbs. (235–319 Nm). Stake the halfshaft retaining nut using a suitable chisel with a rounded cutting edge.

NOTE: If the nut splits or cracks after staking, replace it with a new nut.

21. Install the splash shield.

22. Install the wheel and tire assembly and lower the vehicle.

23. Check and refill the transaxle with the proper type and quantity of fluid.

RIGHT SIDE—1.8L ENGINE

NOTE: The right side halfshaft assembly is a 2 piece shaft with a bearing support bracket positioned between the 2 halves. The bearing support bracket is mounted on the cylinder block and must be unbolted if the entire halfshaft assembly is to be removed. If only the CV-joints/boots are to be serviced, the outboard shaft assembly may be removed, leaving the bearing support bracket mounted on the engine cylinder block.

1. Raise and safely support the vehicle.

2. Remove the right front wheel and tire assembly.

3. Remove the splash shield.

4. Carefully raise the staked portion of the halfshaft retaining nut using a suitable small chisel. Remove and discard the retaining nut.

5. Remove the cotter pin and nut from the tie rod end and remove the tie rod end from the steering knuckle using a suitable removal tool.

6. Remove the lower ball joint clamp bolt. Carefully pry down on the lower control arm to separate the ball joint from the steering knuckle.

7. Pull outward on the steering knuckle/brake assembly. Carefully pull the halfshaft from the steering knuckle and position it aside.

8. Position a drain pan under the transaxle.

9. Remove the 3 bearing support bracket mounting bolts.

10. Insert a prybar between the bearing support bracket and the starter bracket. Gently pry outward on the damper until the halfshaft disengages from the differential side gear.

11. Remove the halfshaft assembly. Install an appropriate differential plug in the differential side gear.

To install:

12. Position the circlip on the inner CV-joint spline so the circlip gap is at the top. Lubricate the splines lightly with a suitable grease.

13. Remove the differential plug from the side gear. Position the halfshaft assembly so the shaft splines are aligned with the differential side gear splines. Push the halfshaft into the differential.

NOTE: When seated properly, the circlip can be felt as it snaps into the differential side gear groove.

14. Pull outward on the steering knuckle/brake assembly and insert the halfshaft into the steering knuckle.

15. Pry downward on the control arm and position the lower ball joint in the steering knuckle. Install the lower ball joint clamp bolt and tighten to 32–43 ft. lbs. (43–59 Nm).

16. Install the tie rod end in the steering knuckle. Install the nut to the tie rod end and tighten to 31–42 ft. lbs. (42–57 Nm). Install a new cotter pin.

17. Position the bearing support bracket and install the 3 mounting bolts. Tighten the bolts in the proper sequence to 31–46 ft. lbs. (42–62 Nm).

18. Install a new halfshaft retaining nut and tighten to 174–235 ft. lbs. (235–319 Nm). Stake the retaining nut using a suitable chisel with the cutting edge rounded off.

NOTE: If the nut splits or cracks after staking, it must be replaced with a new nut.

19. Install the splash shield.

20. Install the right front wheel and lower the vehicle.

21. Check and refill the transaxle with the proper type and quantity of fluid.

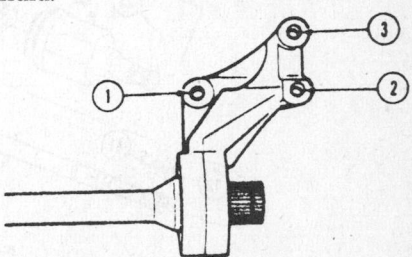

Bearing support bracket torque sequence

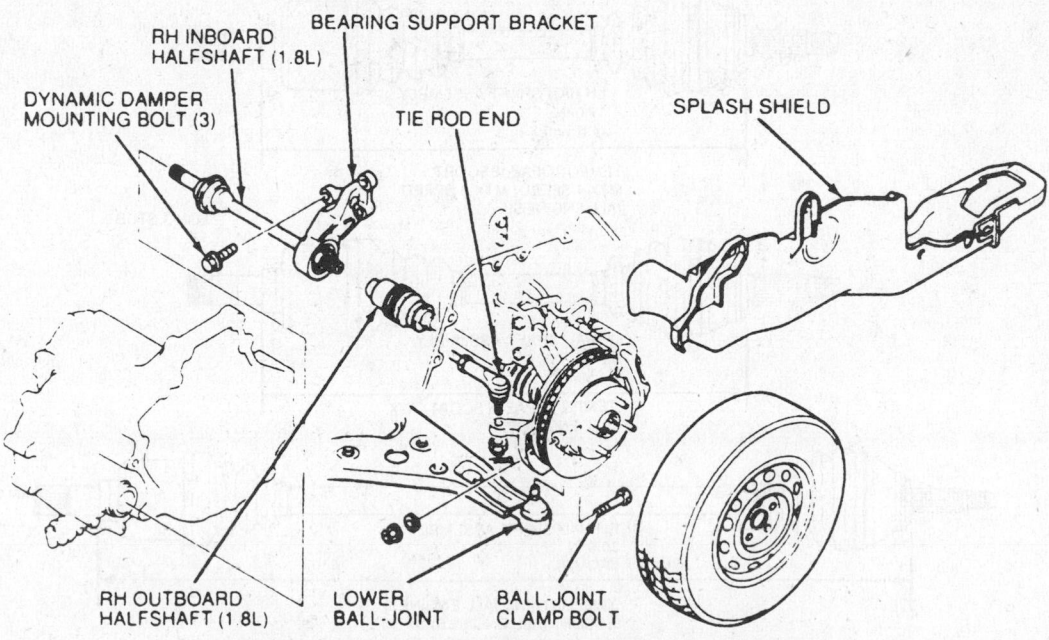

Halfshaft removal and installation—right side, 1.8L engine

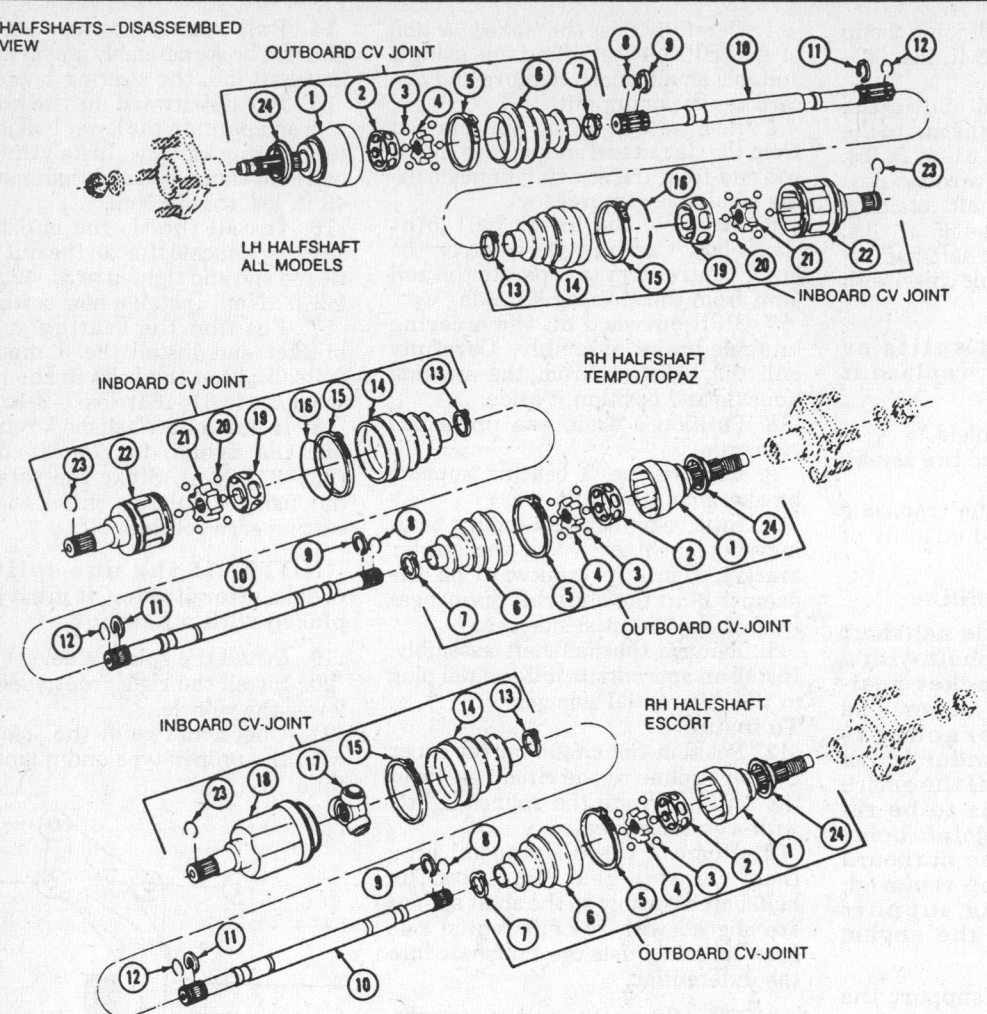

1. Outboard joint outer race and stub shaft
2. Ball cage
3. Balls (6)
4. Outboard joint inner race
5. Boot clamp (large)
6. Boot
7. Boot clamp (small)
8. Circlip
9. Stop ring
10. Interconnecting shaft
11. Stop ring
12. Circlip
13. Boot clamp (small)
14. Boot
15. Boot clamp (large)
16. Wire ring ball retainer
17. Tripod assy
18. Tripod outer race
19. Ball cage
20. Balls (6)
21. Inboard joint inner race
22. Inboard joint outer race and stub shaft
23. Circlip
24. Dust seal

Exploded view of halfshafts—except 1991-92 Escort

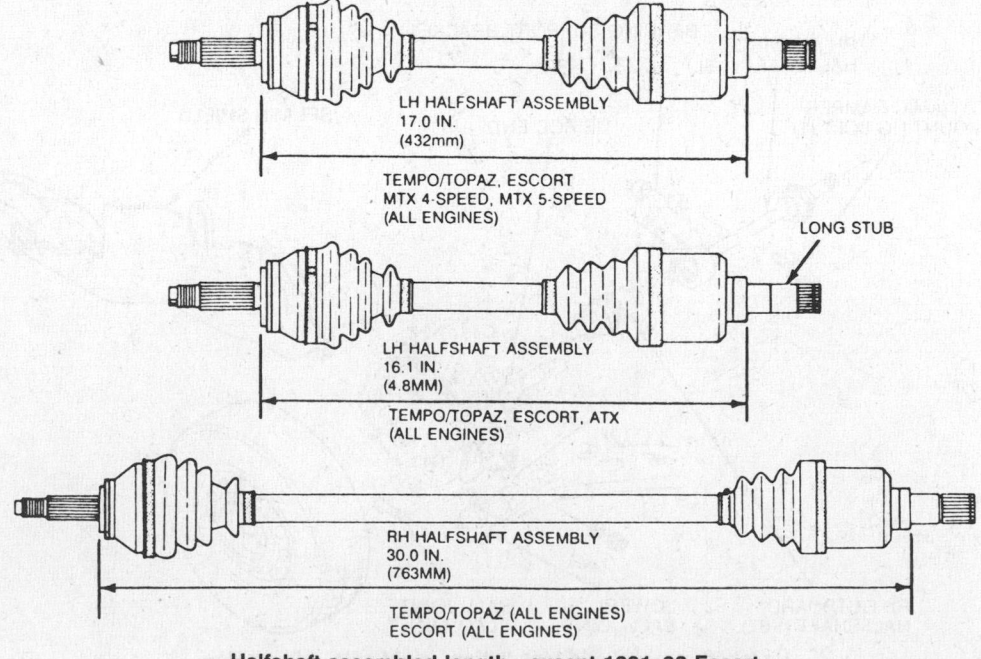

Halfshaft assembled length—except 1991-92 Escort

CV-Boot

REPLACEMENT

NOTE: When replacing a CV-boot, be aware of the transaxle type, transaxle ratio, engine size, CV-joint type, right or left side and inboard or outboard end.

Except 1991–92 Escort

NOTE: There are two different types of inboard CV-joints (Double Offset Joint and Tripod-Type) requiring different removal procedures.

Double Offset Joint Inboard CV-Joint Boot

1. Disconnect the negative battery cable.
2. Remove halfshaft assembly from vehicle. Place halfshaft in vise. Do not allow vice jaws to contact the boot or its clamp. The vise should be equipped with jaw caps to prevent damage to any machined surfaces.
3. Cut the large boot clamp using side cutters and peel away from the boot. After removing the clamp, roll boot back over shaft.
4. Remove wire ring ball retainer.
5. Remove outer race.
6. Pull inner race assembly out until it rests on the circlip. Using snapring pliers, spread stop ring and move it back on shaft.
7. Slide inner race assembly down the shaft to allow access to the circlip. Remove circlip.
8. Remove inner race assembly. Remove boot.

NOTE: Circlips must not be reused. Replace with new circlips before assembly.

9. When replacing damaged CV-boots, the grease should be checked for contamination. If the CV-joints were operating satisfactorily and the grease does not appear to be contaminated, add grease and replace the boot. If the lubricant appears contaminated, proceed with a complete CV-joint disassembly and inspection.
10. Remove balls by prying from cage.

NOTE: Exercise care to prevent scratching or other damage to the inner race or cage.

11. Rotate inner race to align lands with cage windows. Lift inner race out through the wider end of the cage.

To install:

12. Clean all parts (except boots) in a suitable solvent.

13. Inspect all CV-joint parts for excessive wear, looseness, pitting, rust and cracks.

NOTE: CV-joint components are matched during assembly. If inspection reveals damage or wear the entire joint must be replaced as an assembly. Do not replace a joint merely because the parts appear polished. Shiny areas in ball races and on the cage spheres are normal.

14. Install a new circlip, supplied with the service kit, in groove nearest end of shaft. Do not over-expand or twist circlip during installation.
15. Install inner race in the cage. The race is installed through the large end of the cage with the circlip counterbore facing the large end of the cage.
16. With the cage and inner race properly aligned, install the balls by pressing through the cage windows with the heel of the hand.
17. Assemble inner race and cage assembly in outer race.
18. Push the inner race and cage assembly by hand, into the outer race. Install with inner race chamfer facing out.
19. Install ball retainer into groove inside of outer race.
20. Install new CV-boot.
21. Tighten clamp securely but not to the point where the clamp bridge is cut or the boot is damaged.
22. Position stop ring and new circlip into grooves on shaft.
23. Fill CV-joint outer race with 3.2 oz. (90 grams) of grease, then spread 1.4 oz. (40 grams) of grease evenly inside boot for a total combined fill of 4.6 oz. (130 grams).
24. With boot peeled back, install CV-joint using soft tipped hammer. Ensure splines are aligned prior to installing CV-joint onto shaft.
25. Remove all excess grease from the CV-joint external surfaces.
26. Position boot over CV-joint. Before installing boot clamp, move CV-joint in or out, as necessary, to adjust to the proper length.

NOTE: Insert a suitable tool between the boot and outer bearing race and allow the trapped air to escape from the boot. The air should be released from the boot only after adjusting to the proper dimensions.

27. Ensure boot is seated in its groove and clamp in position.
28. Tighten clamp securely but not to the point where the clamp bridge is cut or the boot is damaged.
29. Install halfshaft assembly in vehicle.
30. Connect negative battery cable.

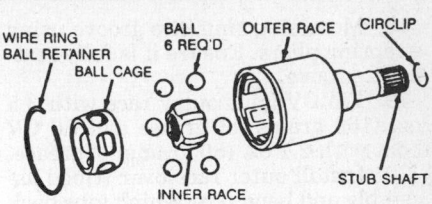

Double offset CV-joint

Tripod Inboard CV-Joint Boot

1. Disconnect the negative battery cable.
2. Remove halfshaft assembly from vehicle. Place halfshaft in vice. Do not allow vise jaws to contact the boot or its clamp. The vise should be equipped with jaw caps to prevent damage to any machined surfaces.
3. Cut the large boot clamp using side cutters and peel away from the boot. After removing the clamp, roll boot back over shaft.
4. Bend retaining tabs back slightly to allow for tripod removal.
5. Separate outer race from tripod.
6. Move stop ring back on shaft using snapring pliers.
7. Move tripod assembly back on shaft to allow access to circlip.
8. Remove circlip from shaft.
9. Remove tripod assembly from shaft. Remove boot.
10. When replacing damaged CV-boots, the grease should be checked for contamination. If the CV-joints were operating satisfactorily and the grease does not appear to be contaminated, add grease and replace the boot. If the lubricant appears contaminated, proceed with a complete CV-joint disassembly and inspection.

To install:

11. Clean all parts (except boots) in a suitable solvent.
12. Inspect all CV-joint parts for excessive wear, looseness, pitting, rust and cracks.

NOTE: CV-joint components are matched during assembly. If inspection reveals damage or wear the entire joint must be replaced as an assembly. Do not replace a joint merely because the parts appear polished. Shiny areas in ball races and on the cage spheres are normal.

13. Install new CV-boot.
14. Tighten clamp securely but not to the point where the clamp bridge is cut or the boot is damaged.
15. Install tripod assembly on shaft with chamfered side toward stop ring.
16. Install new circlip.
17. Compress circlip and slide tripod assembly forward over circlip to expose stop ring groove.

18. Move stop ring into groove using snapring pliers. Ensure it is fully seated in groove.

19. Fill CV-joint outer race with 3.5 oz. (100 grams) of grease and fill CV boot with 2.1 oz. (60 grams) of grease.

20. Install outer race over tripod assembly and bend 6 retaining tabs back into their original position.

21. Remove all excess grease from CV-joint external surfaces. Position boot over CV-joint. Move CV-joint in and out as necessary, to adjust to proper length.

NOTE: Insert a suitable tool between the boot and outer bearing race and allow the trapped air to escape from the boot. The air should be released from the boot only after adjusting to the proper dimensions.

22. Ensure boot is seated in its groove and clamp in position.

23. Tighten clamp securely but not to the point where the clamp bridge is cut or the boot is damaged.

24. Install a new circlip, supplied with service kit, in groove nearest end of shaft by starting one end in the groove and working clip over stub shaft end and into groove.

25. Install halfshaft assembly in vehicle.

26. Connect negative battery cable.

Outboard CV-Joint Boot

1. Disconnect the negative battery cable.

2. Remove halfshaft assembly from vehicle.

3. Place halfshaft in vice. Do not allow vise jaws to contact the boot or its clamp. The vise should be equipped with jaw caps to prevent damage to any machined surfaces.

4. Cut the large boot clamp using side cutters and peel away from the boot. After removing the clamp, roll boot back over shaft.

5. Support the interconnecting shaft in a soft jaw vise and angle the CV-joint to expose inner bearing race.

6. Using a brass drift and hammer, give a sharp tap to the inner bearing race to dislodge the internal circlip and separate the CV-joint from the interconnecting shaft. Take care not to drop the CV-joint at separation.

7. Remove the boot.

8. When replacing damaged CV-boots, the grease should be checked for contamination. If the CV-joints were operating satisfactorily and the grease does not appear to be contaminated, add grease and replace the boot. If the lubricant appears contaminated, proceed with a complete CV-joint disassembly and inspection.

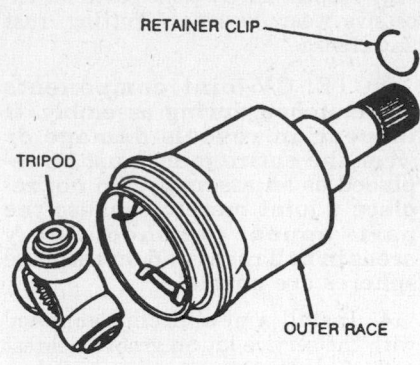

Tripod CV-Joint

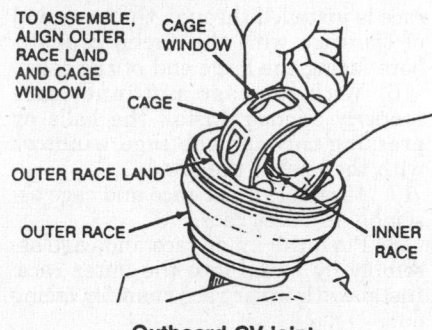

Outboard CV-Joint

9. Remove circlip located near the end of the shaft. Discard the circlip. Use new clip supplied with boot replacement kit and CV-joint overhaul kit.

10. Clamp CV-joint stub shaft in a vise with the outer face facing up. Care should be taken not to damage dust seal. The vise must be equipped with jaw caps to prevent damage to the shaft splines.

11. Press down on inner race until it tilts enough to allow removal of ball. A tight assembly can be tilted by tapping the inner race with wooden dowel and hammer. Do not hit the cage.

12. With cage sufficiently tilted, remove ball from cage. Remove all 6 balls in this manner.

13. Pivot cage and inner race assembly until it is straight up and down in outer race. Align cage windows with outer race lands while pivoting the bearing cage. With the cage pivoted and aligned, lift assembly from the outer race.

14. Rotate inner race up and out of the cage.

To install:

15. Clean all parts (except boots) in a suitable solvent.

16. Inspect all CV-joint parts for excessive wear, looseness, pitting, rust and cracks.

NOTE: CV-joint components are matched during assembly. If inspection reveals damage or

wear the entire joint must be replaced as an assembly. Do not replace a joint merely because the parts appear polished. Shiny areas in ball races and on the cage spheres are normal.

17. Apply a light coating of grease on inner and outer ball races. Install the inner race in cage.

18. Install inner race and cage assembly in the outer race.

19. Install the assembly vertically and pivot 90 degrees into position.

20. Align cage and inner race with outer race. Tilt inner race and cage and install one of the 6 balls. Repeat this process until the remaining balls are installed.

21. Install new CV-joint boot.

22. Tighten clamp securely but not to the point where the clamp bridge is cut or the boot is damaged.

23. Install the stop ring, if removed.

24. Install a new circlip, supplied with the service kit, in groove nearest the end of the shaft.

25. Pack CV-joint with grease. Any grease remaining in tube should be spread evenly inside boot.

26. With the boot "peeled" back, position CV-joint on shaft and tap into position using a plastic tipped hammer.

27. Remove all excess grease from the CV-joint external surfaces.

28. Position boot over CV-joint.

29. Ensure boot is seated in its groove and clamp into position.

30. Tighten clamp securely but not to the point where the clamp bridge is cut or the boot is damaged.

31. Install halfshaft assembly in vehicle.

32. Connect negative battery cable.

1991-92 Escort

1. Raise and safely support the vehicle.

2. Remove the halfshaft assembly from the vehicle.

3. Secure the halfshaft in a vise with protective jaw covers.

4. Using a suitable tool, pry up the locking tabs of the inner CV-boot bands. Remove the bands with pliers.

5. Slide the boot back to expose the tripod CV-joint. Mark the shaft and the CV-joint housing to ensure correct assembly.

6. Remove the retainer ring from the CV-joint housing and remove the CV-joint housing from the halfshaft.

7. Mark the tripod bearing and the shaft to ensure correct assembly. Using snapring pliers, remove the tripod snapring.

8. Using a soft-faced mallet, gently tap the tripod bearing from the shaft.

9. Wrap the shaft splines with tape to protect the CV-boot if the boot is to be reused.

10. Slide the inner CV-joint boot off the shaft. If the outer CV-joint boot is to be replaced, continue with the procedure.

11. On 1.9L right side halfshafts, pry up the rubber damper retaining band locking clip using a suitable tool. Remove the retaining band using pliers and remove the rubber damper from the shaft.

12. Using a suitable tool, pry up the outer CV-boot band locking tabs. Remove the bands with pliers.

13. Slide the outer CV-boot off the shaft.

NOTE:When replacing a damaged boot, check the grease for contamination by rubbing it between 2 fingers. Any gritty feeling indicates a contaminated CV-joint. A contaminated inner CV-joint must be completely disassembled, cleaned and inspected. The outer CV-joint is not serviceable and should be replaced as an assembly, if necessary. If the grease is not contaminated and the CV-joint has been operating satisfactorily, replace only the boot and add the required lubricant.

To install:

14. Cover the halfshaft splines with tape and install the outer CV-joint boot.

NOTE: The outer and inner CV-joint boots are different. Failure to correctly install the boot on the proper end of the halfshaft could lead to premature boot and/or CV-joint wear.

15. Fill the outer CV-joint housing with the proper type and amount of lubricant.

16. Position the CV-boot. Make sure the boot is fully seated in the shaft grooves and the CV-joint housing.

17. Insert a suitable tool between the boot and the CV-joint housing to allow trapped air to escape.

18. Position new bands on the outer CV-joint boot.

NOTE: Always use new bands. The bands should be mounted in the direction opposite the forward revolving direction of the halfshaft.

19. Wrap the bands around the boot in a clockwise direction, pull them

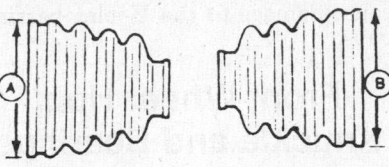

INNER CV BOOT   OUTER CV BOOT

Inner and outer CV-joint boots—1991 –92 Escort

tight with pliers and bend the locking tabs to secure the bands in position.

20. Work the CV-joint through it's full range of travel at various angles. The CV-joint should flex, extend and compress smoothly.

21. On 1.9L right side halfshafts, position the rubber damper on the halfshaft. Position a new band on the damper. Pull the band tight with pliers and fold it back. Lock the end of the band by bending the locking clip.

22. Position the inner CV-joint boot on the halfshaft.

23. Align the marks on the tripod bearing and the halfshaft. Install the tripod bearing on the halfshaft. If necessary, using a soft-faced mallet, tap the bearing into place.

24. Install the snapring.

25. Fill the inner CV-joint housing with the proper type and amount of lubricant. Coat the tripod bearing with the same lubricant.

CV-BOOT SPECIFICATIONS—1991–92 Escort

	1.9L Engine		1.8L Engine	
	Right Side	Left Side	Right Side	Left Side
Ⓐ	84.0 mm (3.31 in)	90.0 mm (3.54 in)	89.9 mm (3.54 in)	
Ⓑ	89.0 mm (3.50 in)		85.2 mm (3.35 in)	

CV-BOOT LUBRICANT SPECIFICATIONS— 1991–92 Escort

Halfshaft Assemblies	1.9L Engine		1.8L Engine	
	Right Side	Left Side	Right Side	Left Side
Differential Side	220 g (7.77 oz) Lt. Yellow	140 g (4.94 oz) Yellow	145 g (5.12 oz) Yellow	
Wheel Side	140 g (4.94 oz) Black		90 g (3.18 oz) Black	

HALFSHAFT LENGTH SPECIFICATIONS—1991–92 Escort

Item	Model	1.8L Engine	1.9L Engine
Halfshaft			
Length of joint (between center of joint)	Right side	631.2 mm (24.85 in)	918.7 mm (36.16 in)
	Left side	621.7 mm (24.48 in)	640.7 mm (25.22 in)
Shaft diameter	Right side	23.0 mm (0.91 in)	
	Left side	23.0 mm (0.91 in)	

26. Position the inner CV-joint housing over the tripod bearing, making sure to align the alignment marks. Install the retainer ring in the CV-joint housing.

27. Slide the inner CV-boot in place. Make sure the boot is fully seated in the shaft grooves and in the housing.

28. Insert a suitable tool between the boot and the CV-joint housing to allow trapped air to escape.

29. Position new bands on the inner CV-joint boot.

NOTE: Always use new bands. The bands should be mounted in the direction opposite the forward revolving direction of the halfshaft.

30. Wrap the bands around the boot in a clockwise direction, pull them tight with pliers and bend the locking tabs to secure the bands in position.

31. Work the CV-joint through it's full range of travel at various angles. The CV-joint should flex, extend and compress smoothly.

32. Measure the length of the assembled halfshaft. If the length is not as specified, check the CV-joints for freedom of movement to ensure that it was assembled correctly. Repair or replace any components as necessary.

Driveshaft and U-Joints

REMOVAL & INSTALLATION

Tempo and Topaz with All Wheel Drive

1. Raise the vehicle and support safely.

2. To maintain the driveshaft balance, mark the U-joints so they may be installed in their original position.

3. Remove the front U-joint retaining bolts and straps.

4. Support the driveshaft near the center bearing. Remove the driveshaft center bearing retaining bolts.

5. Slide the driveshaft toward the rear of the vehicle to disengage from the transfer case.

6. Remove the rear U-joint bolts and straps retaining the driveshaft, from the torque tube yoke flange.

7. Slide the driveshaft toward the front of the vehicle to disengage. Do not allow the splined shafts to contact with excessive force.

8. Remove the center bearing retaining bolts. Remove the driveshaft and retain the bearing cups with tape, if necessary.

9. Inspect the U-joint assemblies for wear and or damage, replace the U-joint, if necessary.

To install:

10. Install the driveshaft at the rear torque yoke flange. Ensure that the U-joint is in its original position.

11. Install the U-joint retaining bolts and caps. Torque them to 15–17 ft. lbs. (21–23 Nm). Position the front U-joint. Install the U-joint retaining caps and bolts. Torque them to 15–17 ft. lbs. (21–23 Nm).

12. Install the center bearing and retaining bolts. Torque to 23–30 ft. lbs. (31–41 Nm). Do not drop the assembled driveshafts as the impact may cause damage to the U-joint bearing cups.

Front Wheel Hub, Knuckle and Bearings

REMOVAL & INSTALLATION

Except 1991–92 Escort

1. Remove wheel cover/hub cover from wheel and tire assembly and loosen wheel nuts.

2. Remove hub nut retainer and washer by applying sufficient torque to nut to break locking tab and remove hub nut retainer. The hub nut retainer must be discarded after removal.

3. Raise and safely support the vehicle. Remove wheel and tire assembly.

4. Remove brake caliper by loosening caliper locating pins and rotating caliper off rotor starting from lower end of caliper and lifting upward. Do not remove caliper pins from caliper assembly. Lift caliper off rotor and hang it free of rotor. Do not allow caliper assembly to hang from brake hose. Support caliper assembly with a length of wire.

5. Remove rotor from hub by pulling it off hub bolts. If rotor is difficult to remove from hub, strike rotor sharply between studs with a rubber or plastic hammer. If rotor will not pull off, apply rust penetrator to inboard and outboard rotor hub mating surfaces. Install a 3 jaw puller and remove rotor by pulling on rotor outside diameter and pushing on hub center. If excessive force is required for removal, check rotor for lateral runout.

6. Lateral runout must be checked with nuts clamping hat section of rotor.

7. Remove rotor splash shield.

8. Disconnect lower control arm and tie rod from knuckle (leave strut attached.).

9. Loosen the 2 strut top mount-to-apron nuts.

10. Install a suitable hub removal tool and remove hub/bearing/knuckle assembly by pushing out CV-joint outer shaft until it is free of assembly.

11. Support knuckle with a length of wire, remove strut bolt and slide hub/knuckle assembly off strut.

12. Carefully remove support wire and transfer hub/bearing/knuckle assembly to bench.

13. Install a suitable front hub puller with jaws of puller on the knuckle bosses and remove hub.

NOTE: Ensure the shaft protector is centered, clears the bearing inside diameter and rests on the end face of the hub journal.

14. Remove snapring which retains bearing knuckle assembly and discard.

15. Using a hydraulic press, place a suitable front bearing spacer step side up on press plate and position knuckle on spacer with outboard side up. Install bearing removal tool on bearing inner race and press bearing out of knuckle.

16. Discard bearing.

17. Remove halfshaft.

18. Place halfshaft in vise. Remove bearing dust seal by uniformly tapping on outer edge with a light-duty hammer and screwdriver. Discard dust seal.

To install:

19. Place halfshaft in vise. Install a new dust seal using a suitable seal installer. Seal flange must face outboard.

20. Install halfshaft.

21. On bench, remove all foreign material from knuckle bearing bore and hub bearing journal to ensure correct seating of new bearing.

NOTE: If hub bearing journal is scored or damaged, replace hub. Do not attempt to service. The front wheel bearings are of a cartridge design and are pregreased, sealed and require no scheduled maintenance. The bearings are preset and cannot be adjusted. If a bearing is disassembled for any reason, it must be replaced as a unit. No individual service seals, rollers or races are available.

22. Place suitable bearing spacer step side down on hydraulic press plate and position knuckle on spacer with outboard side down. Position a new bearing in inboard side of knuckle. Install a suitable front bearing installer on bearing outer race face with undercut side facing bearing and press bearing into knuckle. Ensure that bearing seats completely against shoulder of knuckle bore.

NOTE: Ensure proper positioning of bearing installer during installation to prevent bearing damage.

23. Install a new snapring in knuckle groove using snapring pliers.

24. Place suitable front bearing spacer on arbor press plate and position hub on tool with lugs facing downward. Position knuckle assembly on hub barrel with outboard side down. Place a suitable front bearing installer on inner race of bearing and press down on tool until bearing is fully seated onto hub. Make sure hub rotates freely in knuckle after installation.

25. Suspend hub/knuckle/bearing assembly on vehicle with wire and attach strut loosely to knuckle. Lubricate CV-joint stub shaft splines with SAE 30 weight motor oil and insert shaft into hub splines as far as possible using hand pressure only. Check that spline are properly engaged.

26. On all except 1991–92 Tempo and Topaz, install suitable front hub installer and wheel bolt adapter to hub and stub shaft. Tighten hub installer tool to 120 ft. lbs. (162 Nm) to ensure that hub is fully seated. Remove tool and install washer and new hub nut retainer. Tighten hub nut retainer finger-tight.

27. On 1991–92 Tempo and Topaz, install washer and new hub nut. Rotate nut clockwise to seat CV-joint. Tighten hub nut to 188–254 ft. lbs. (255–345 Nm). Do not tighten with impact gun and do not move vehicle before retainer is tightened.

28. Complete installation of front suspension components.

29. Install disc brake rotor to hub assembly.

30. Install disc brake caliper over rotor.

31. Ensure outer brake shoe spring end is seated under upper arm of knuckle.

32. Install wheel and tire assembly, tightening wheel nuts finger-tight.

33. Lower vehicle and block wheels to prevent vehicle from rolling.

34. Tighten wheel nuts to 85–105 ft. lbs. (115–142 Nm).

35. On all except 1991–92 Tempo and Topaz, manually thread hub nut retainer assembly on constant velocity output shaft as far as possible using a 30mm or $1^3/_{16}$ in. socket, tighten retainer assembly to 180–200 ft. lbs.

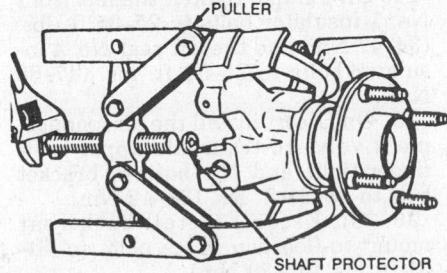

Front hub puller—except 1991–92 Escort

(245–270 Nm). Do not use power or impact tools to tighten the hub nut. Do not move the vehicle before retainer is tightened.

NOTE: During tightening, an audible click sound will indicate proper ratchet function of the hub nut retainer. As the hub nut retainer tightens, ensure that one of the 3 locking tabs is in the slot of the CV-joint shaft. If the hub nut retainer is damaged, or more than 1 locking tab is broken, replace the hub nut retainer.

36. Install wheelcover or hub cover and lower vehicle completely to ground.

37. Remove wheel blocks.

1991–92 Escort

1. Raise and safely support the vehicle.

2. Remove the front wheel and tire assembly, brake caliper and rotor.

3. Remove the nut securing the halfshaft to the hub.

4. Remove the outer tir rod end at the steering knuckle.

5. Remove the nuts and bolts and separate the shock/strut assembly from the steering knuckle.

6. Remove the nut and bolt and separate the lower ball joint from the steering knuckle.

7. Remove the front hub/steering knuckle assembly from the halfshaft.

8. Remove the oil seal from the rear of the hub/steering knuckle assembly.

9. Position the hub/steering knuckle assembly on a hydraulic press and press the front hub out of the steering knuckle using a suitable removal tool.

NOTE: If the bearing inner race remaons on the hub, use a grinder to grind a section of the bearing inner race until only 0.020 in. (0.5mm) remains. Remove the inner race with a suitable chisel.

10. Remove the E-clip from the steering knuckle.

11. Position the steering knuckle onto a hydraulic press and, using an appropriate bearing remover, press the bearing out of the steering knuckle.

NOTE: If the dust cover is removed, it must be replaced.

12. Scribe a mark in the dust cover and steering knuckle. Using a suitable chisel, remove the dust cover.

To install:

13. Scribe a mark on the new dust cover in the same position as on the previous mark. Align the marks on the steering knuckle to the mark on the dust cover and, using a suitable tool,

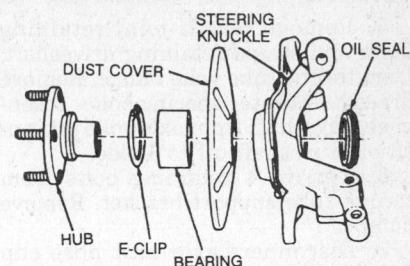

Front hub, knuckle and bearing assembly—1991–92 Escort

press the dust cover onto the steering knuckle.

14. Position the steering knuckle onto a press and press the bearing into the steering knuckle, using a suitable bearing installer.

15. Install the E-clip.

16. Position the hub onto the knuckle and press the hub into the bearing and the steering knuckle, using a suitable installation tool.

17. Using an appropriate seal installer, install a new oil seal onto the inboard side of the steering knuckle. Make sure the oil seal mounts flush with the steering knuckle.

18. Install the hub/steering knuckle assembly onto the ball joint and install the nut and bolt. Tighten to 32–43 ft. lbs. (43–59 Nm).

19. Install the outer tie rod.

20. Install the steering knuckle to the shock/strut assembly. Tighten the nuts and bolts to 69–93 ft. lbs. (93–127 Nm).

21. Install a new locknut securing the halfshaft to the front hub. Tighten the locknut to 174–235 ft. lbs. (235–319 Nm). Stake the locknut to prevent it from loosening.

22. Install the brake rotor, caliper and wheel and tire assembly.

23. Lower the vehicle.

Axle Housing

REMOVAL & INSTALLATION

Tempo and Topaz with All Wheel Drive

1. Disconnect the negative battery cable.

2. Raise and safely support the vehicle.

NOTE: Anytime a U-joint retaining bolt is removed, Loctite® or equivalent, must be applied to the retaining bolts prior to installation.

3. Position a hoist or jack under rear axle housing.

4. Remove muffler and exhaust system from catalytic converter back.

5. Remove rear U-joint retaining bolts and straps retaining driveshaft, from torque tube yoke flange. Remove driveshaft center bearing bolts. Disengage driveshaft from axle yoke and position driveshaft off to 1 side.

6. Remove 4 retaining bolts from torque tube support bracket. Remove damper.

7. Disconnect axle vent hose clip form body.

8. Remove axle retaining bolt from left side differential support bracket.

9. Remove axle retaining bolt from center differential support bracket.

10. Lower axle assembly and remove inboard U-joint retaining bolts and straps from each halfshaft. Remove and wire halfshaft assemblies aside.

11. Remove rear axle assembly.

To install:

12. Position rear axle assembly under vehicle. Raise axle far enough for U-joint and halfshaft assemblies to be installed.

13. Position each inboard U-joint to rear axle. Install U-joint straps and retaining bolts. Using a T-30 Torx® bit, tighten bolts to 15–17 ft. lbs. (21–23 Nm).

14. Raise into position being careful not to trap or pinch axle vent hose. Install bolts attaching differential housing to left side and center differential support bracket. Tighten to 70–80 ft. lbs. (95–108 Nm).

15. Attach axle vent hose clip to body.

16. Position torque tube and mounting bracket and damper to crossmember. Install 4 attaching bolts. Tighten to 28–35 ft. lbs. (38–47 Nm). Install driveshaft and retaining bolts to torque tube yoke flange. Using a T-30 Torx® bit, tighten to 15–17 ft. lbs. (21–23 Nm).

17. Install exhaust from catalytic converter back.

18. Check lubricant level in axle.

19. Lower vehicle.

MANUAL TRANSAXLE

For further information on transmissions/transaxles, please refer to "Chilton's Guide to Transmission Repair".

Transaxle Assembly

REMOVAL & INSTALLATION

Except 1991–92 Escort

1. Disconnect the negative battery cable. Wedge a 7 in. wooden block under the clutch pedal to hold the pedal up slightly beyond its normal position. Grasp the clutch cable, pull it forward and disconnect it from the clutch release shaft assembly. Remove the clutch casing from the rib on the top surface of the transaxle case.

2. Remove the upper 2 transaxle-to-engine bolts. Remove the air cleaner on Tempo and Topaz or the air management valve bracket-to-transaxle upper bolt on Escort.

3. Raise and safely support the vehicle.

4. On Tempo and Topaz, remove the front stabilizer bar-to-control arm nut and washer, on the driver's side and discard the nut. Remove both front stabilizer bar mounting brackets and discard the bolts.

5. Remove the lower control arm ball joint-to-steering knuckle nut/bolt and discard the nut/bolt; repeat this procedure on the opposite side.

6. Using a large prybar, pry the lower control arm from the steering knuckle; repeat this procedure on the opposite side.

NOTE: Be careful not to damage or cut the ball joint boot and do not contact the lower arm.

7. Using a large prybar, pry the left-side inboard CV-joint assembly from the transaxle.

NOTE: Plug the seal opening to prevent lubricant leakage.

8. Grasp the left-hand steering knuckle and swing it and the halfshaft outward from the transaxle; this will disconnect the inboard CV-joint from the transaxle.

NOTE: If the CV-joint assembly cannot be pried from the transaxle, insert a differential rotator tool through the left-side and tap the joint out; the tool can be used from either side of transaxle.

9. Using a wire, support the halfshaft in a near level position to prevent damage to the assembly during the remaining operations; repeat this procedure on the opposite side.

10. Disengage the locking tabs and remove the backup light switch connector from the transaxle backup light switch.

11. On Tempo and Topaz, remove the starter studs-to-engine roll restrictor bracket nuts and the engine roll restrictor. Remove the starter stud bolts.

12. On Escort, remove the starter bolts.

13. Remove the shift mechanism-to-shift shaft nut/bolt, the control selector indicator switch arm and the shift shaft.

14. Remove the shift mechanism stabilizer bar-to-transaxle bolt, control selector indicator switch and bracket assembly.

15. Using a crowfoot wrench, remove the speedometer cable from the transaxle.

16. On Tempo and Topaz, remove both oil pan-to-clutch housing bolts. On Escort, remove 2 stiffener brace retaining bolts.

17. Using a floor jack and a transaxle support, position it under the transaxle and secure the transaxle to it.

18. On Tempo and Topaz, remove the both left-hand rear No. 4 insulator-to-body bracket nuts and the left-hand front No. 1 insulator-to-body bracket bolts.

19. On Escort, remove both rear mount-to-floorpan bolts, loosen the nut at the bottom of the front mount and remove the front mount-to-transaxle bolts.

20. Lower the floor jack, until the transaxle clears the rear insulator. Support the engine by placing wood under the oil pan.

21. Remove the engine-to-transaxle bolts and lower the transaxle from the vehicle.

NOTE: On Tempo and Topaz, 1 of the engine-to-transaxle bolts attaches the ground strap and wiring loom stand off bracket.

To install:

22. Raise the transaxle into position and engage the input shaft with the clutch plate. Install the lower engine-to-transaxle bolts and torque to 28–31 ft. lbs. (38–42 Nm).

NOTE: Never attempt to start the engine prior to installing the CV-joints or differential side gear for dislocation and/or damage may occur.

23. On Escort, install the front mount-to-transaxle bolts and torque to 25–35 ft. lbs. (34–47 Nm); also, tighten the nut on the bottom of the front transaxle mount.

24. On Tempo, tighten the left front No. 1 insulator bolts to 25–35 ft. lbs. (34–47 Nm) and the left rear No. 4 insulator bolts to 35–50 ft. lbs. (47–68 Nm).

25. On Escort install the air management valve-to-transaxle upper bolt, finger-tight and the bottom bracket bolt to 28–31 ft. lbs. (38–42 Nm).

26. On Escort, install both rear mount-to-floorpan brace bolts to 40–51 ft. lbs. (55–70 Nm).

27. Remove the floor jack and adapter.

28. Using a crowfoot wrench, install the speedometer cable; be careful not to cross-thread the cable nut.

29. On Tempo and Topaz, install the oil pan-to-transaxle bolts and tighten to 28–38 ft. lbs. (38–51 Nm). On Escort, install the 2 stiffener brace bolts and tighten to 15–21 ft. lbs. (21–28 Nm).

30. Install the shifter stabilizer bar/control selector indicator switch-to-transaxle bolt and torque to 23–35 ft. lbs. (31–47 Nm).

31. Install the shift mechanism-to-shift shaft, the switch actuator bracket clamp and torque the bolt to 7–10 ft. lbs. (9–13 Nm); be sure to shift the transaxle into **4th** for 4-speed or **5th** for 5-speed and align the actuator.

32. On Escort, innstall the starter bolts and tighten to 30–40 ft. lbs. (41–54 Nm). On Tempo and Topaz, install the starter stud bolts and tighten to 30–40 ft. lbs. (41–54 Nm) and install the engine roll restrictor and the attaching nuts. Tighten the attaching nuts to 14–20 ft. lbs. (19–27 Nm).

33. Install the backup light switch connector to the transaxle switch.

34. Install the new circlip onto both inner joints of the halfshafts, insert the inner CV-joints into the transaxle and fully seat them; lightly, pry outward to confirm that the retaining rings are seated.

NOTE: When installing the halfshafts, be careful not to tear the oil seals.

35. Connect the lower ball joint to the steering knuckle, insert a new pinch bolt and torque the new nut to 37–44 ft. lbs. (50–60 Nm); be careful not to damage the boot.

36. Refill the transaxle and lower the vehicle.

37. On Escort, install the upper air management valve bracket-to-transaxle bolt and torque to 28–31 ft. lbs. (38–42 Nm). On Tempo and Topaz, install the air cleaner.

38. Install the both upper transaxle-to-engine bolts and torque to 28–31 ft. lbs. (38–42 Nm).

39. Connect the clutch cable to the clutch release shaft assembly and remove the wooden block from under the clutch pedal. Connect the negative battery cable.

NOTE: Prior to starting the engine, set the hand brake and pump the clutch pedal several times to ensure proper clutch adjustment.

1991–92 Escort

1. Disconnect the battery cables and remove the battery and the battery tray.

2. Remove the air hose and the resonance chamber.

3. Disconnect the speedometer cable at the transaxle.

4. Remove the retaining clip, then disconnect the slave cylinder line from the slave cylinder hose and plug the hose.

5. Disconnect the ground strap from the transaxle.

6. Remove the tie wrap and disconnect the 3 electrical connectors located above the transaxle. Remove the electrical connector support bracket.

7. Mount engine support bar D88L–6000–A or equivalent, and attach it to the engine hangers.

8. Remove the 3 nuts from the upper transaxle mount. Loosen the mount pivot nut and rotate the mount out of position. Remove the 3 bolts and the upper transaxle mount bracket.

9. Remove the 2 upper transaxle-to-engine bolts.

10. Raise and safely support the vehicle.

11. Remove the front wheel and tire assemblies.

12. Remove the inner fender splash shields.

13. Drain the transaxle fluid and install the drain plug.

14. Remove the halfshafts. Install 2 transaxle plugs between the differential side gears.

NOTE: Failure to install the transaxle plugs may cause the differential side gears to become improperly positioned.

15. Remove the plenum support bracket and remove the starter.

16. Remove the nut and the extension bar and the bolt and nut and shift control rod from the transaxle.

17. Remove both lower splash shields.

18. Remove the 2 transaxle mount-to-crossmember nuts and remove the lower crossmember and the front transaxle mount.

19. Position and secure a suitable jack under the transaxle.

20. Remove the 5 lower engine-to-transaxle bolts and lower the transaxle out of the vehicle.
To install:
21. Apply a thin coating of suitable grease to the spline of the input shaft.

22. Place the transaxle onto a suitable jack. Make sure the transaxle is secure.

23. Raise the transaxle into position on the engine.

24. Install the 5 lower engine-to-transaxle bolts and tighten to 27–38 ft. lbs. (37–52 Nm).

25. Install the front transaxle mount and tubing bracket. Tighten the bolts to 12–17 ft. lbs. (16–23 Nm).

26. Install the lower crossmember. Tighten the nuts and bolts to 47–66 ft. lbs. (64–89 Nm).

27. Install the 2 transaxle mount-to-crossmember nuts and tighten to 27–38 ft. lbs. (37–52 Nm).

28. Install both lower splash shields.

29. Install the shift control rod bolt and nut and tighten to 23–34 ft. lbs. (31–46 Nm).

30. Install the extension bar nut and tighten to 12–17 ft. lbs. (16–23 Nm).

31. Install the starter and the plenum support bracket.

32. Remove the transaxle plugs and install the halfshafts.

33. Install the inner fender splash shields.

34. Install the wheel and tire assemblies. Tighten the lug nuts to 65–87 ft. lbs. (88–118 Nm).

35. Lower the vehicle.

36. Install the 2 upper engine-to-transaxle bolts and tighten to 47–66 ft. lbs. (64–89 Nm).

37. Install the upper transaxle mount bracket and tighten the 3 bolts. Rotate the mount into position and tighten the pivot nut. Install and tighten the 3 upper mount nuts.

38. Remove the engine support bar.

39. Install the electrical connector support bracket. Connect the 3 electrical connectors and secure with the tie wrap.

40. Connect the ground strap to the transaxle.

41. Connect the slave cylinder line to the slave cylinder hose and install the retaining clip.

42. Add the proper type and amount of fluid to the transaxle.

43. Connect the speedometer cable.

44. Install the air hose and the resonance chamber.

45. Install the battery tray and the battery. Connect the battery cables.

46. Check for fluid leaks and proper operation.

CLUTCH

Clutch Assembly

REMOVAL & INSTALLATION

1. Disconnect the negative battery cable. Raise and safely support the vehicle. Remove the transaxle.

2. Matchmark the pressure plate assembly and the flywheel so they can be assembled in the same position.

3. Loosen the pressure plate-to-flywheel bolts 1 turn at a time, in sequence, until spring tension is relieved to prevent pressure plate cover distortion.

4. Support the pressure plate and remove the bolts. Remove the pressure plate and clutch disc from the flywheel.

5. Inspect the flywheel, clutch disc, pressure plate, throwout bearing and the clutch fork for wear; replace parts, as required.

NOTE: If the flywheel shows any signs of overheating (blue discoloration) or if it is badly grooved or scored, it should be refaced or replaced.

To install:

6. Clean the pressure plate and flywheel surfaces thoroughly. Position the clutch disc and pressure plate into the installed position, aligning the matchmarks made previously; support them with a dummy shaft or clutch aligning tool.

7. Install the pressure plate-to-flywheel bolts. Tighten them gradually in a criss-cross pattern to 12–24 ft. lbs. (17–32 Nm) on all except 1991–92 Escort, where the torque should be 13–20 ft. lbs. (18–26 Nm). Remove the alignment tool.

8. Lubricate the release bearing and install it in the fork.

9. To complete the installation, reverse the removal procedures. Lower the vehicle and connect the negative battery cable.

PEDAL HEIGHT/FREE-PLAY ADJUSTMENT

Except 1991–92 Escort

The pedal height and free-play are controlled by a self-adjusting feature.

1991–92 Escort

PEDAL HEIGHT

To determine if the pedal height requires adjustment, measure the distance from the bulkhead to the upper center of the pedal pad. The distance should be 7.72–8.03 in. (196–204mm). If and adjustment is necessary, proceed as follows:

1. Disconnect the clutch switch electrical connector.
2. Loosen the clutch switch locknut.
3. Turn the clutch switch until the correct height is achieved.
4. Tighten the locknut to 10–13 ft. lbs. (14–18 Nm).
5. Measure the pedal free-play.
6. Connect the electrical connector.

PEDAL FREE-PLAY

To determine if the pedal free-play requires adjustment, depress the clutch pedal by hand until clutch resistance is felt. Measure the distance between the upper pedal height and where the resistance is felt. Free-play should be

0.20–0.51 in. (5–13mm). If an adjustment is necessary, proceed as follows:

1. Loosen the pushrod locknut.
2. Turn the pushrod until the pedal free-play is within specification.
3. Check that the disengagement height is correct when the pedal is fully depressed. Minimum disengagement height is 1.6 in. (41mm).
4. Tighten the pushrod locknut to 9–12 ft. lbs. (12–17 Nm).

Clutch Cable

ADJUSTMENT

Tempo, Topaz and 1988–90 Escort

The free-play in the clutch is adjusted by a built in mechanism that allows the clutch controls to be self-adjusted during normal operation. The self-adjusting feature should be checked every 5000 miles. This is accomplished by insuring that the clutch pedal travels to the top of its upward position. Grasp the clutch pedal with hand or put foot under the clutch pedal, pull up on the pedal until it stops. Very little effort is required (about 10 lbs.). During the application of upward pressure, a click may be heard which means an adjustment was necessary and has been accomplished.

REMOVAL & INSTALLATION

Tempo, Topaz and 1988–90 Escort

1. Disconnect the negative battery cable.
2. Wedge a 7 in. wooden block under the clutch pedal to hold the pedal up slightly beyond its normal position.
3. Remove the air cleaner to gain access to the clutch cable.
4. Using a pair of pliers, grasp the clutch cable, pull it forward and disconnect it from the clutch release shaft assembly.

NOTE: Do not grasp the wire strand portion of the inner cable since it may cut the wires and cause cable failure.

5. Remove the clutch casing from the insulator which is located on the rib on the top of the transaxle case.
6. On the Tempo or Topaz, remove the panel from above the clutch pedal pad.
7. Remove the rear screw and move the clutch shield away from the brake pedal support bracket. Loosen the front retaining screw, located near the toe board, rotate the shield aside and snug the screw to retain the shield.

8. With the clutch pedal raised to release the pawl, rotate the gear quadrant forward, unhook the clutch cable and allow the quadrant to swing rearward; do not allow the quadrant to snap back.
9. Pull the cable through the recess between the clutch pedal and the gear quadrant and from the insulator of the pedal assembly.
10. Remove the cable from the engine compartment.

To install:

11. Lift the clutch pedal to disengage the adjusting mechanism.
12. Insert the clutch cable through the dash panel and the dash panel grommet.

NOTE: Be sure the clutch cable is routed under the brake lines and not trapped at the spring tower by the brake lines. If equipped with power steering, route the cable inboard of the power steering hose.

13. Push the clutch cable through the insulator on the stop bracket and through the recess between the pedal and the gear quadrant.
14. Lift the clutch pedal to release the pawl, rotate the gear quadrant forward and hook the cable into the gear quadrant.
15. Install the clutch shield on the brake pedal support bracket.
16. On the Tempo or Topaz, install the panel above the clutch pedal.
17. Using a piece of wire or tape, secure the pedal in the upmost position.
18. Insert the clutch cable through the insulator and connect the cable to the clutch release lever in the engine compartment.
19. Remove the wooden block from under the clutch pedal.
20. Depress the clutch pedal several times. Install the air cleaner and connect the negative battery cable.

Clutch Master Cylinder

REMOVAL & INSTALLATION

1991–92 Escort

1. Disconnect the battery cables and remove the battery and battery tray.
2. Disconnect the clutch pipe from the master cylinder using a suitable line wrench.
3. Disengage the clamp and remove the master cylinder hose from the clutch master cylinder. Prevent excess fluid loss by plugging the hose.
4. Remove the external mounting nut.
5. Remove the internal mounting nut and remove the master cylinder.

To install:

6. Align the pushrod and install the clutch master cylinder.

7. Install the external and internal mounting nuts and tighten to 14–19 ft. lbs. (19–25 Nm).

8. Connect the clutch pipe and tighten the nut to 10–16 ft. lbs. (13–22 Nm).

9. Install the hose and the clamp to the master cylinder.

10. Install the battery and battery tray.

11. Bleed the air from the system.

12. Test the system and make sure there is no leakage.

13. Connect the negative battery cable.

Clutch Slave Cylinder

ADJUSTMENT

1991–92 Escort

The clutch slave cylinder is not adjustable. The only adjustments necessary on the clutch control system are pedal height and pedal free-play.

REMOVAL & INSTALLATION

1991–92 Escort

1. Disconnect the pressure line. Plug the line to prevent leaking.

2. Remove the attaching bolts and remove the slave cylinder.

To install:

3. Install the slave cylinder.

4. Install the attaching bolts and tighten to 12–17 ft. lbs. (16–23 Nm).

5. Connect the pressure line and tighten the nut to 10–16 ft. lbs. (13–22 Nm).

6. Bleed the air from the system.

7. Press on the clutch pedal and make sure there is no leakage.

Hydraulic Clutch System Bleeding

NOTE: The fluid level in the reservoir must be maintained at the ¾ level or higher during air bleeding.

1. Remove the bleeder cap from the slave cylinder and attach a vinyl hose to the bleeder screw.

2. Place the other end of the hose in a container.

3. Slowly pump the clutch pedal several times.

4. With the clutch pedal depressed, loosen the bleeder screw to release the fluid and air.

5. Tighten the bleeder screw.

6. Repeat the last 3 steps until no air bubbles appear in the fluid.

AUTOMATIC TRANSAXLE

For further information on transmissions/transaxles, please refer to "Chilton's Guide to Transmission Repair".

Transaxle Assembly

REMOVAL & INSTALLATION

Except 1991–92 Escort

1. Disconnect the negative battery cable.

NOTE: Due to automatic transaxle case configuration, the right-side halfshaft assembly must be removed first. The differential rotator tool or equivalent, is then inserted into the transaxle to drive the left-side inboard CV-joint assembly from the transaxle.

2. Remove the air cleaner assembly.

3. Disconnect the electrical harness connector from the neutral safety switch.

4. Disconnect the throttle valve linkage and the manual lever cable from their levers.

NOTE: Failure to disconnect the linkage and allowing the transaxle to hang, will fracture the throttle valve cam shaft joint, which is located under the transaxle cover.

5. To prevent contamination, cover the timing window in the converter housing. If equipped, remove the bolts retaining the thermactor hoses.

6. If equipped, remove the ground strap, located above the upper engine mount, and the coil and bracket assembly.

7. Remove both transaxle-to-engine upper bolts; the bolts are located below and on both ides of the distributor. Raise and safely support the vehicle. Remove the front wheels.

8. Remove the control arm-to-steering knuckle nut, at the ball joint.

9. Using a hammer and a punch, drive the bolt from the steering knuckle; repeat this step on the other side. Discard the nut and bolt.

NOTE: Be careful not to damage or cut ball joint boot. The prybar must not contact lower arm.

10. Using a prybar, disengage the control arm from the steering knuckle; repeat this step on the other side.

NOTE: Do not hammer on the knuckle to remove the ball joints. The plastic shield installed behind the rotor contains a molded pocket into which the lower control arm ball joint fits. When disengaging the control arm from the knuckle, clearance for the ball joint can be provided by bending the shield back toward the rotor. Failure to provide clearance for the ball joint can result in damage to the shield.

11. Remove the stabilizer bar bracket-to-frame rail bolts and discard the bolts; repeat this step on the other side.

12. Remove the stabilizer bar-to-control arm nut/washer and discard the nut; repeat this step on the other side.

13. Pull the stabilizer bar from of the control arms.

14. Remove the brake hose routing clip-to-suspension strut bracket bolt; repeat this step on the other side.

15. Remove the steering gear tie rod-to-steering knuckle nut and disengage the tie rod from the steering knuckle; repeat this step on the other side.

16. Using a halfshaft removal tool, pry the halfshaft from the right side of the transaxle and support the end of the shaft with a wire.

NOTE: It is normal for some fluid to leak from the transaxle when the halfshaft is removed.

17. Using a differential rotator tool or equivalent, drive the left-side halfshaft from the differential side gear.

18. Pull the halfshaft from the transaxle and support the end of the shaft with a wire.

NOTE: Do not allow the shaft to hang unsupported, as damage to the outboard CV-joint may result.

19. Install seal plugs into the differential seals.

20. Remove the starter support bracket and disconnect the starter cable. Remove the starter bolts and the starter. If equipped with a throttle body, remove the hose and bracket bolts on the starter and a bolt at the converter and disconnect the hoses.

21. Remove the transaxle support bracket and the dust cover from the torque converter housing.

22. Remove the torque converter-to-flywheel nuts by turning the crankshaft pulley bolt to bring the nuts into position.

23. Position a suitable transmission jack under the transaxle and remove the rear support bracket nuts.

24. Remove the left front insulator-to-body bracket nuts, the bracket-to-body bolts and the bracket.

25. Disconnect the transaxle cooler lines.

26. Remove the manual lever bracket-to-transaxle case bolts.

27. Support the engine. Make sure the transaxle is supported and remove the remaining transaxle-to-engine bolts.

28. Make sure the torque converter studs will be clear the flywheel. Insert a prybar between the flywheel and the converter, then, pry the transaxle and converter away from the engine. When the converter studs are clear of the flywheel, lower the transaxle about 2–3 in. (51–76mm).

29. Disconnect the speedometer cable and lower the transaxle.

NOTE: When moving the transaxle away from the engine, watch the No. 1 insulator. If it contacts the body before the converter studs clear the flywheel, remove the insulator.

To install:

30. Raise the transaxle and align it with the engine and flywheel. Install the No. 1 insulator, if removed. Torque the transaxle-to-engine bolts to 25–33 ft. lbs. (34–45 Nm) and the torque converter-to-flywheel bolts to 23–39 ft. lbs. (31–53 Nm).

31. Install the manual lever bracket-to-transaxle case bolts and connect the transaxle cooler lines.

32. Install the left front insulator-to-body bracket nuts and torque the nuts to 40–50 ft. lbs. (55–70 Nm). Install the bracket-to-body and torque the bolts to 55–70 ft. lbs. (75–90 Nm).

33. Install the transaxle support bracket and the dust cover to the torque converter housing.

34. If equipped with a throttle body, install the hose and bracket bolts on the starter and a bolt to the converter and connect the hoses. Install the starter and the support bracket; torque the starter-to-engine bolts to 30–40 ft. lbs. (41–54 Nm). Connect the starter cable.

35. Remove the seal plugs from the differential seals and install the halfshaft by performing the following procedures:

   a. Prior to installing the halfshaft in the transaxle, install a new circlip onto the CV-joint stub.

   b. Install the halfshaft in the transaxle by carefully aligning the CV-joint splines with the differential side gears. Be sure to push the CV-joint into the differential until the circlip is felt to seat in the differential side gear. Use care to prevent damage to the differential oil seal.

   c. Attach the lower ball joint to the steering knuckle, taking care not to damage or cut the ball joint boot. Insert a new pinch bolt and a

new nut. While holding the bolt with a 2nd wrench, torque the nut to 40–54 ft. lbs. (54–74 Nm).

36. Engage the tie rod with the steering knuckle and torque the nut to 23–35 ft. lbs. (31–47 Nm).

37. Install the brake hose routing clip-to-suspension strut bracket and torque the bolt to 8 ft. lbs. (11 Nm).

38. Install the stabilizer bar to control arm and using a new nut, torque it to 98–125 ft. lbs. (133–169 Nm).

39. Install the stabilizer bar bracket-to-frame rail bolts and using new bolts, torque them to 60–70 ft. lbs. (81–95 Nm).

40. Install the wheels and lower the vehicle. Install the upper transaxle-to-engine bolts and torque to 25–33 ft. lbs. (34–45 Nm).

41. If equipped, install the ground strap, located above the upper engine mount, and the coil and bracket assembly.

42. If equipped, install the bolts retaining the thermactor hoses. Uncover the timing window in the converter housing.

43. Connect the throttle valve linkage and the manual lever cable to their levers.

44. Connect the electrical harness connector from the neutral safety switch.

45. Install the air cleaner assembly.

46. Connect the negative battery cable and road test the vehicle.

1991–92 Escort

1. Disconnect the battery cables and remove the battery and battery tray.

2. Disconnect the wiring harness retaining clip from the battery tray.

3. Remove the air cleaner assembly.

4. Disconnect the shift control cable from the manual lever.

5. Disconnect the speedometer cable from the transaxle by unsnapping the cable at the speedometer driven gear.

6. Disconnect the transaxle electronic control electrical connectors and separate the harness from the transaxle clips.

7. Remove the manual lever position switch wiring brackets and disconnect the ground cables from the top of the transaxle.

8. Remove the starter.

9. Disconnect the manual lever position switch wiring connectors.

10. Install engine support D88L–6000–A or equivalent, to support the engine.

11. Disconnect the kickdown cable at the throttle cam.

12. Place a suitable drain pan under the transaxle and disconnect the transaxle cooler lines at the transaxle.

13. Remove the upper transaxle mount bolts, the mount and the upper transaxle housing bolts.

14. Disconnect the oxygen sensor electrical connector, the transaxle vent hose, and the electrical connector at the vehicle speed sensor.

15. Raise and safely support the vehicle.

16. Remove the front wheel and tire assemblies.

17. Using a suitable hammer and a flat punch, straighten the detent in the halfshaft nut.

18. Remove the nuts securing the halfshafts to the steering knuckles and remove the nuts and bolts securing the lower ball joints to the steering knuckles. Separate the lower ball joints from the steering knuckles.

19. Disconnect the halfshaft mid-bearing bracket from the back of the engine.

20. Remove the halfshafts from both steering knuckles.

21. Remove the 3 engine/transaxle lower splash shields and the torque converter inspection plate. Remove the nuts securing the torque converter to the flexplate.

22. Disconnect the lower crossmember from the chassis and the transaxle mounts.

23. Remove the driver's side and then the passenger's side halfshafts. Install 2 transaxle plugs T88C–7025–AH or equivalent into the differential side gears.

NOTE: Failure to install the transaxle plugs may cause the differential side gears to become improperly positioned.

24. Position the drain pan and remove the drainplug from the transaxle. Drain the fluid from the differential cavity. Remove the transaxle pan and drain the transaxle fluid.

25. Position a suitable transmission jack under the transaxle. Secure the transaxle to the jack.

26. Remove the lower bolts securing the transaxle to the engine and carefully lower the transaxle out of the vehicle.

To install:

NOTE: A pin is used for securing the throttle cam in a fixed position on new and rebuilt transaxles. This pin must be removed to allow proper transaxle operation. If the pin is not removed, the throttle lever will remain in a fixed position. After removing the pin, apply sealant to the bolt from the previous transaxle. Install the bolt and tighten to 69–95 inch lbs. (8–11 Nm).

27. Secure the transaxle on the transmission jack.

28. Raise the transaxle into position and install the lower transaxle-to-engine bolts. Tighten the bolts to 41–59 ft. lbs. (55–80 Nm).

29. Position the torque converter to the flexplate and install the nuts. Tighten the nuts to 25–36 ft. lbs. (34–49 Nm).

30. Install the halfshafts.

31. Connect the crossmember to the transaxle mounts and the chassis. Tighten the crossmember-to-transaxle mount nuts to 27–38 ft. lbs. (37–52 Nm). Tighten the crossmember-to-chassis nuts and bolts to 47–66 ft. lbs. (64–89 Nm).

32. Install the engine/transaxle splash shields and the starter.

33. Position the lower ball joints into the steering knuckles and secure with the nuts and bolts. Tighten the nuts and bolts to 32–43 ft. lbs. (43–59 Nm).

34. Position the tie rod ends into the steering knuckles and install the nuts. Tighten to 31–42 ft. lbs. (42–57 Nm).

35. Install the wheel and tire assemblies. Tighten the lugs to 65–88 ft. lbs. (88–118 Nm).

36. Lower the vehicle.

37. Install the transaxle-to-engine bolts and tighten to 41–59 ft. lbs. (55–80 Nm).

38. Install the upper transaxle mount and tighten the nuts to 49–69 ft. lbs. (67–93 Nm).

39. Connect the transaxle vent hose, the electrical connector at the speed sensor, the speedometer cable and the oxygen sensor connector.

40. Connect the transaxle coller lines and connect the kickdown cable at the throttle body.

41. Remove the engine support.

42. Connect the ground wires to the transaxle and connect the manual lever position switch bracket and wiring connectors.

43. Connect the shift control cable to the cable bracket and to the selector lever.

44. Install the battery tray and battery.

45. Install the air cleaner assembly.

46. Connect the battery cables.

47. Add the proper type and quantity of transaxle fluid.

48. Check the transaxle for proper operation.

SHIFT LINKAGE ADJUSTMENT

Except 1991–92 Escort

1. Place the gear shift selector into **D**. The shift lever must be in the **D** position during linkage adjustment.

2. Working at the transaxle, loosen the transaxle lever-to-control cable nut.

3. Move the transaxle lever to the **D** position, 2nd detent from the most rearward position.

4. Torque the adjusting nut to 10–15 ft. lbs. (14–20 Nm).

5. Make sure all gears engage correctly and the vehicle will only start in **P** or **D**.

1991–92 Escort

1. Disconnect the negative battery cable. This will deactivate the shift-lock system.

2. Move the gear selector lever to **P**.

3. Remove the screw securing the gear selector knob to the gear selector lever. Remove the knob.

4. Remove the shift console as follows:

    a. Remove the rear seat ash tray and position both front seats to the rear-most position.

    b. Remove the 2 front retaining screws from the parking brake console and recline both front seats.

    c. Remove the 2 rear retaining screws from the parking brake console.

    d. With the parking brake engaged, remove the parking brake console.

    e. Remove the 2 front retaining screws from the shift console and remove the console.

5. Remove the position indicator mounting screws and disconnect the illumination bulb from the position indicator.

6. Disconnect the shift-lock servo and park range switch electrical connectors.

7. Remove the position indicator.

NOTE: Make sure the detent spring roller is in the P detent.

8. Loosen the shift control cable bracket mounting bolts.

9. Push the gear selector lever against the **P** range and hold it.

10. Tighten the shift control cable bracket mounting bolts to 69–95 inch lbs. (8–11 Nm).

11. Lightly press the gear selector pushrod and make sure the guide plate and guide pin clearances are within specifications.

12. Check that the guide plate and guide pin clearances are within the appropriate specifications when the selector lever is shifted to **N** and **OD**. If the clearances are not as specified, readjust the shift control cable.

13. Make sure the gear selector operates properly.

14. Connect the illumination bulb to the position indicator.

15. Connect the shift-lock servo and park range switch electrical connectors.

16. Install the position indicator and secure it with the mounting screws.

17. Install the shift console by reversing the removal procedure.

18. Position the gear selector knob onto the gear selector lever and secure the knob with the screw.

19. Connect the negative battery cable.

THROTTLE LINKAGE ADJUSTMENT

Tempo and Topaz

1. Disconnect the negative battery cable.

2. Remove the splash shield from the cable retainer bracket.

3. Loosen the trunnion bolt at the throttle valve rod.

4. Install a plastic clip to bottom the throttle valve rod; be sure the clip keeps the rod from telescoping.

5. Be sure the return spring is connected between the throttle valve rod

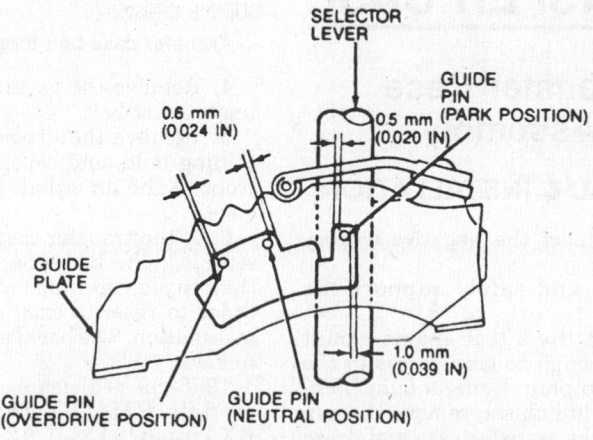

Shift control cable adjustment clearances—1991–92 Escort

and the retaining bracket to hold the transaxle throttle valve lever at it's idle position.

6. Make sure the throttle lever is resting on the throttle return control screw.

7. Tighten the throttle valve rod trunnion bolt and remove the plastic clip.

8. Install the splash shield. Connect the negative battery cable and check the vehicle's operation.

1988–90 Escort

1. Disconnect the negative battery cable.

2. Set the parking brake and place the transaxle shift lever into **P**.

3. Loosen the sliding trunnion block bolt, located on the throttle valve control rod assembly, a minimum of 1 turn.

4. Make sure the trunnion block slides freely on the control rod.

5. Using a jumper wire, connect it between the STI connector and the signal return ground on the self-test connector.

6. Turn the ignition switch to the **RUN** position but do not start the engine. The Idle Speed Control (ISC) plunger should retract; wait until the plunger is fully retracted, about 10 seconds.

7. Turn the ignition switch **OFF** and remove the jumper wire.

8. Using light force, pull the throttle valve rod upward to ensure the control lever is against the internal stop.

9. Allow the trunnion to slide on the rod to it's normal position.

10. Without relaxing the pressure on the throttle valve control lever, tighten the trunnion block bolt.

11. Connect negative battery cable.

TRANSFER CASE

Transfer Case Assembly

REMOVAL & INSTALLATION

1. Disconnect the negative battery cable.

2. Raise and safely support the vehicle.

3. Loosen the 2 rear engine mount bolts far enough to gain access to the transfer cup plug. Using a light hammer and a dull chisel, remove the cup plug from the transfer case and drain the oil.

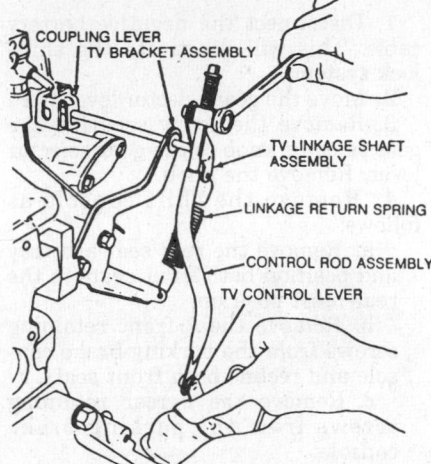

Throttle valve linkage adjustment— Tempo and Topaz

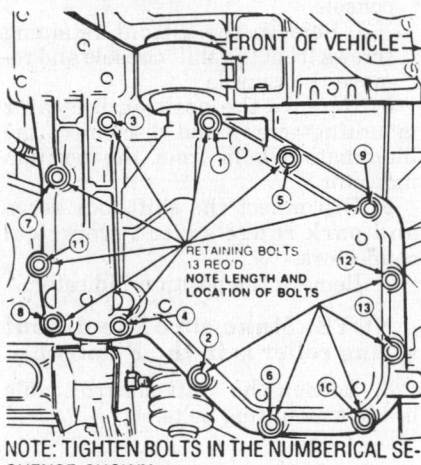

Throttle valve linkage adjustment— 1988–90 Escort

NOTE: TIGHTEN BOLTS IN THE NUMBERICAL SEQUENCE SHOWN

Transfer case bolt torque sequence

4. Remove the vacuum line retaining bracket bolt.

5. Remove the driveshaft front retaining bolts and caps; disengage the front of the driveshaft from the drive yoke.

6. If the transfer case is to be disassembled, check the backlash through the cup plug opening before removal in order to reset to existing backlash at installation. The backlash should be as follows:
1988–89 vehicles except transaxle models PMA-BX through PMA-BX10 – 0.012–0.024 in. (0.30– 0.60mm) on a 3 in. (76mm) radius.

1990–92 vehicles except transaxle models PMA-BX through PMA-BX10 – 0.012–0.047 in. (0.30– 1.20mm).

Vehicles with transaxle models PMA-BX through PMA-BX10 – 0.031– 0.066 in. (0.78–1.68mm).

7. Remove the vacuum motor shield bolts and the shield.

8. Remove the vacuum lines from the vacuum servo.

9. Remove the transfer case-to-transaxle bolts; note and record the length and locations of the bolts.

10. Remove the the transfer case from the vehicle.

To install:

11. Position the transfer case to the transaxle.

12. Install the transfer case bolts in the proper positions and torque the bolts, in sequence, to 15–19 ft. lbs. (21–25 Nm) for 1988–89 or 12–15 ft. lbs. (16–20 Nm) for 1990–92.

13. Install the vacuum motor supply hose connector, vacuum motor shield and torque the bolts to 7–12 ft. lbs. (9– 16 Nm).

14. Install the driveshaft to the drive yoke, lubricate the bolts with Loctite® and torque the bolts to 15–17 ft. lbs. (21–23 Nm). Install the vacuum line retaining bracket and torque the bolt to 7–12 ft. lbs. (9–16 Nm).

15. Refill the transaxle and lower the vehicle. Road test the vehicle and check the performance of the transfer case.

FRONT SUSPENSION

MacPherson Strut

REMOVAL & INSTALLATION

Except 1991–92 Escort

NOTE: All vehicles except 1988 Tempo with base suspension are equipped with gas pressurized shock absorbers which will extend unassisted. Do not apply heat or flame to the shock strut tube during removal.

1. Loosen but do not remove, 2 top mount-to-shock tower nuts.

2. Raise and safely support the vehicle. Raise vehicle to a point where it is possible to reach the 2 top mount-to-shock tower nuts and the strut-to-knuckle pinch bolt.

3. Remove wheel and tire assembly.

4. Remove brake flex line-to-strut bolt.

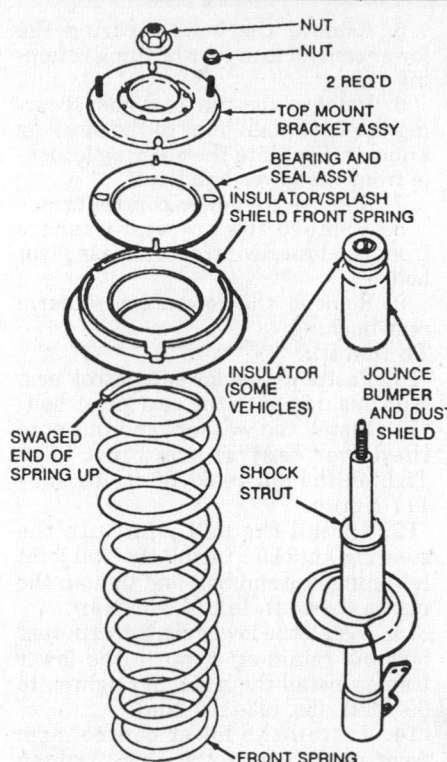

Front strut assembly—except 1991–92 Escort

5. Remove strut-to-knuckle pinch bolt.

6. Using a suitable tool, spread knuckle-to-strut pinch joint slightly.

7. Using a suitable bar, place top of bar under fender apron and pry down on knuckle until strut separates from knuckle. Be careful not to pinch brake hose.

NOTE: Do not pry against caliper or brake hose bracket.

8. Remove 2 top mount-to-shock tower nuts and remove strut from vehicle.

9. Install spring compressor in bench mount, install strut in compressor and compress spring.

10. Place deep 18mm socket on strut shaft nut. Insert an 8mm deep socket with ¼ in. drive wrench. Remove top shaft mounting nut from shaft while holding ¼ in. drive socket with a suitable extension.

NOTE: Do not attempt to remove shaft nut by turning shaft and holding nut. The nut must be turned and the shaft held to avoid possible damage to the shaft.

11. Loosen spring compressor tool and remove top mount bracket assembly, bearing, insulator and spring.
To install:
12. Install replacement strut in spring compressor.

NOTE: During reassembly of strut/spring assembly, be certain to follow correct sequence and proper positioning of bearing plate and seal assembly. If bearing and seal assembly are out of position, damage to the bearing will result.

13. Install spring, insulator, bearing and top mount bracket assembly.

14. Install top shaft mounting nut while holding shaft with ¼ drive 8mm deep socket and extension. Tighten nut to 35–50 ft. lbs. (48–68 Nm).

15. Install strut assembly in vehicle. Install 2 top mount-to-shock tower nuts. Tighten to 25–30 ft. lbs. (37–41 Nm).

16. Slide strut mounting flange onto knuckle.

17. Install strut-to-knuckle pinch bolt. Tighten to 68–80 ft. lbs. (92–110 Nm).

18. Install brake flex line-to-strut bolt.

19. Install wheel and tire assembly.

20. Lower vehicle.

21. Check alignment.

1991–92 Escort

1. Raise and safely support the vehicle.

2. Remove the front wheel and tire assembly.

3. Remove the clip securing the flexible brake hose to the strut assembly.

4. Remove the 2 nuts and 2 bolts securing the strut assembly to the steering knuckle.

5. Remove the upper mounting block nuts and remove the strut assembly from the vehicle.

6. Remove the cap from the top of the strut assembly.

7. Secure the strut assembly mounting block in a vise. Turn the piston rod nut 1 revolution to loosen.

8. Install an appropriate spring compressor onto the strut spring and compress the spring.

9. Remove the nut, mounting block, thrust bearing, upper spring seat, rubber spring seat, coil spring and bound stopper.

To install:

10. Position the bound stopper onto the strut piston rod.

11. With the coil spring compressed, position the spring onto the strut assembly.

12. Install the rubber spring seat, upper spring seat, thrust bearing, mounting block and piston rod nut. Tighten the piston rod nut to 58–81 ft. lbs. (79–110 Nm).

13. With the nut tightened to specification, carefully remove the spring compressor from the spring while making sure the spring is properly seated in the upper and lower spring seats.

14. Install the cap.

15. Position the strut assembly into the wheel housing. Make sure the direction indicator on the mounting block faces inboard.

16. Secure the upper mounting block to the strut tower with the 4 nuts.

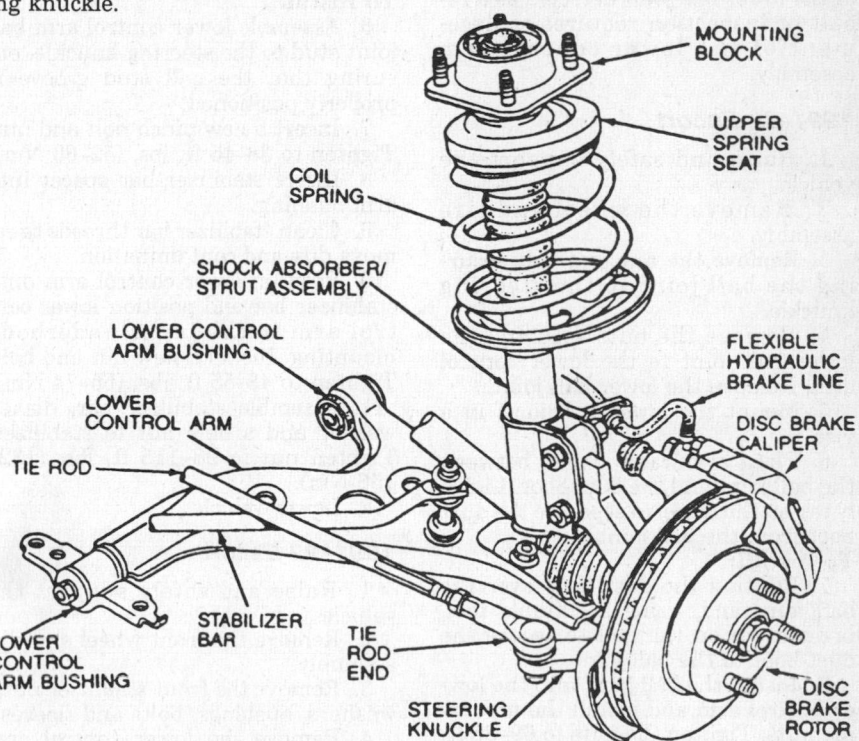

Front suspension assembly—1991–92 Escort

Tighten the nuts to 22–30 ft. lbs. (29–40 Nm).

17. Attach the strut assembly to the steering knuckle and install the bolts and nuts. Tighten to 69–93 ft. lbs. (93–127 Nm).

18. Position the flexible brake hose to the strut assembly and secure it with the brake hose clip.

19. Install the front wheel and tire assembly. Tighten the lug nuts to 65–87 ft. lbs. (88–118 Nm).

20. Lower the vehicle and check the front wheel alignment.

Lower Ball Joints

INSPECTION

1. Raise and safely support the vehicle so wheels are in the full-down position.

2. Have an assistant grasp lower edge of the tire and move wheel and tire assembly in and out.

3. As wheel is being moved in and out, observe lower end of knuckle and lower control arm. Any movement indicates abnormal ball joint wear.

4. If any movement is observed, install new lower control arm assembly.

REMOVAL & INSTALLATION

Excpet 1991–92 Escort

The lower ball joint is integral to the lower control assembly and cannot be serviced individually. Any movement of the lower ball joint detected as a result of inspection requires replacement of the lower control arm assembly.

1991–92 Escort

1. Raise and safely support the vehicle.

2. Remove the wheel and tire assembly.

3. Remove the nut and bolt securing the ball joint to the steering knuckle.

4. Remove the nuts securing the lower ball joint to the lower control arm. Remove the lower ball joint.

5. Mount the lower ball joint in a vise.

6. Place a suitable chisel between the ball joint and the dust boot. Lightly tap on the chisel to separate the dust boot from the ball joint.

To install:

7. Position the dust boot over the ball joint and, using a suitable tool, press down on the tool to secure the dust boot to the ball joint.

8. Install the ball joint into the lower control arm and install the mounting nuts. Tighten the nuts to 69–86 ft. lbs. (93–117 Nm).

9. Install the lower ball joint into the steering knuckle and secure it with the nut and bolt. Tighten the nut to 32–43 ft. lbs. (43–59 Nm).

10. Install the wheel and tire assembly and lower the vehicle.

Lower Control Arms

REMOVAL & INSTALLATION

Except 1991–92 Escort

1. Raise and safely support the vehicle.

2. Remove nut from stabilizer bar end. Pull off large dished washer.

3. Remove lower control arm inner pivot nut and bolt.

4. Remove lower control arm ball joint pinch bolt. Using a suitable tool, slightly spread knuckle pinch joint and separate control arm from steering knuckle. A drift punch may be used to remove bolt.

NOTE: Do not allow the steering knuckle/halfshaft to move outward. Over extension of the tripod CV-joint could result in separation of internal parts, causing failure of the joint.

5. Remove stabilizer bar spacer from the arm bushing.

NOTE: Make sure steering column is in unlocked position. Do not use a hammer to separate ball joint from knuckle.

To install:

6. Assemble lower control arm ball joint stud to the steering knuckle, ensuring that the ball stud groove is properly positioned.

7. Insert a new pinch bolt and nut. Tighten to 38–45 ft. lbs. (52–60 Nm).

8. Insert stabilizer bar spacer into arm bushing.

9. Clean stabilizer bar threads to remove dirt and contamination.

10. Position lower control arm onto stabilizer bar and position lower control arm to the inner underbody mounting. Install a new nut and bolt. Tighten to 48–55 ft. lbs. (65–74 Nm).

11. Assemble stabilizer bar, dished washer and a new nut to stabilizer. Tighten nut to 98–115 ft. lbs. (132–156 Nm).

12. Lower vehicle.

1991–92 Escort

1. Raise and safely support the vehicle.

2. Remove the front wheel and tire assembly.

3. Remove the front stabilizer nuts, washers, bushings, bolts and sleeves.

4. Remove the lower control arm front bushing bolt and washer.

5. Remove the bolts securing the lower control arm rear bushing retaining strap.

6. Remove the nut and bolt securing the lower ball joint to the steering knuckle. Separate the steering knuckle from the lower ball joint.

7. Remove the lower control arm.

8. Remove the nut and washers from the lower control arm rear pivot bolt.

9. Remove the lower control arm rear bushing.

To install:

10. Position the lower control arm rear bushing onto the rear pivot bolt.

11. Install the washers and nut onto the lower control arm pivot bolt. Tighten the nut to 69–86 ft. lbs. (93–117 Nm).

12. Install the ball joint into the steering knuckle. Install the ball joint retaining nut and bolt and tighten the nut to 32–43 ft. lbs. (43–59 Nm).

13. Install the lower control arm rear bushing retaining strap to the lower frame. Install the bolts and tighten to 69–86 ft. lbs. (93–117 Nm).

14. Install the lower control arm front pivot bolt and washer. Tighten the nut to 69–93 ft. lbs. (93–127 Nm).

15. Install the stabilizer bolts, washers, bushings, sleeves and nuts. Tighten the stabilizer nuts so 0.67–0.75 in. (17–19mm) of thread is exposed at the end of the bolt.

16. Install the wheel and tire assembly. Tighten the lug nuts to 65–87 ft. lbs. (88–118 Nm).

17. Lower the vehicle.

Stabilizer Bar

REMOVAL & INSTALLATION

Except 1991–92 Escort

1. Raise and safely support the vehicle.

2. Remove nut from stabilizer bar at each lower control arm and pull off large dished washer. Discard nuts.

3. Remove stabilizer bar insulator U-bracket bolts and U-brackets and remove stabilizer bar assembly. Discard bolts.

NOTE: Stabilizer bar U-bracket insulators can be serviced without removing the stabilizer bar assembly.

To install:

4. Slide new insulators onto the stabilizer bar and position them in the approximate location.

5. Clean stabilizer bar threads to remove dirt and contamination.

6. Install spacers into the control arm bushings from forward side of control arm so washer end of spacer

will seat against stabilizer bar machined shoulder and push mounting brackets over insulators.

7. Insert end of stabilizer bar into the lower control arms. Using new bolts, attach the stabilizer bar and the insulator U-brackets to the bracket assemblies. Hand start all 4 U-bracket bolts. Tighten all bolts halfway, then tighten bolts to 59–68 ft. lbs. (80–92 Nm) on Tempo/Topaz or 85–100 ft. lbs. (115–135 Nm) on Escort.

8. Using new nuts and the original dished washers (dished side away from bushing), attach the stabilizer bar to the lower control arm. Tighten nuts to 98–115 ft. lbs. (132–156 Nm).

9. Lower vehicle.

1991–92 Escort

1. Support the engine with engine support D88L–6000–A or equivalent.
2. Raise and safely support the vehicle.
3. Remove the wheel and tire assemblies.
4. Remove the nuts securing the steering gear mounting brackets and position the steering gear slightly forward.
5. Remove the stabilizer bar nuts, washers, bushings, sleeves and bolts from the lower control arm.
6. Remove the rear crossmember nuts from the rear transaxle mount and the vehicle frame.
7. Loosen the front crossmember bolts and nuts from the front transax-

Tightening Torque
A 37-52 N·m (27-38 LB-FT)
B 64-89 N·m (47-66 LB-FT)

Torque sequence and specifications for crossmember-to-frame and crossmember-to-transaxle mount bolts —1991–92 Escort

le mount and the vehicle frame. Lower the rear end of the crossmember.

8. Remove the nuts and bolts securing the chassis frame to the vehicle frame. Lower the chassis frame.

NOTE: The engine and transaxle mounts will support the chassis frame when unbolting the chassis frame from the vehicle frame.

9. Unbolt the stabilizer bar from the chassis frame and remove the stabilizer bar from the vehicle.
To install:
10. Position the stabilizer bar into the vehicle.
11. Secure the stabilizer bar to the chassis frame with the bolts. Tighten the bolts to 32–43 ft. lbs. (43–59 Nm).

12. Install the chassis frame to the vehicle frame with the bolts and nuts. Tighten the bolts and nuts to 69–93 ft. lbs. (93–127 Nm).
13. Position the crossmember to the vehicle frame and the transaxle mounts. Tighten the bolts and nuts to the specified torque.
14. Install the stabilizer bar bolts, sleeves, bushings, washers and nuts. Tighten the stabilizer bolts so 0.67–0.75 in. (17–19mm) of thread is exposed at the end of the bolt.
15. Position the steering gear and secure it with the brackets and nuts. Tighten the nuts to 28–38 ft. lbs. (37–52 Nm).
16. Install the wheel and tire assemblies. Tighten the lug nuts to 65–87 ft. lbs. (88–118 Nm).
17. Lower the vehicle.

REAR SUSPENSION

MacPherson Strut

REMOVAL & INSTALLATION

1988–90 Escort

1. Remove rear compartment access panels. On 4-door models, remove quarter trim panel.

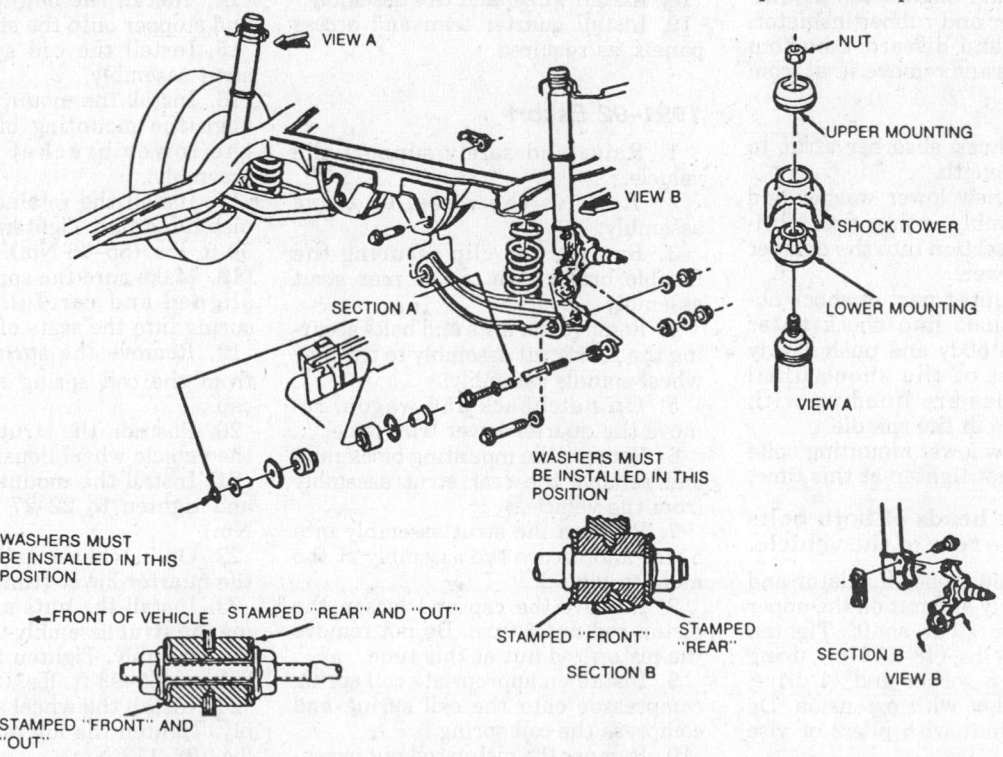

WASHERS MUST BE INSTALLED IN THIS POSITION

WASHERS MUST BE INSTALLED IN THIS POSITION

FRONT OF VEHICLE

STAMPED "REAR" AND "OUT"

STAMPED "FRONT" AND "OUT"

SECTION A

STAMPED "FRONT"   STAMPED "REAR"

SECTION B

SECTION B

VIEW B

NUT

UPPER MOUNTING

SHOCK TOWER

LOWER MOUNTING

VIEW A

Rear suspension—1988–90 Escort

NOTE: Do not attempt to remove shaft nut by turning shaft and holding nut. Nut must be turned and shaft held to avoid possible damage to shaft.

2. Loosen, but do not remove, top strut attaching nut using an 18mm deep socket while holding the strut rod with a ¼ drive, 8mm deep socket and suitable extension.

NOTE: If the strut is to be reused, do not grip the shaft with pliers or vise grips, as this will damage the shaft surface finish and result in severe oil leakage.

3. Raise and safely support the vehicle.
4. Remove tire and wheel assembly.

NOTE: If a frame contact lift is used, support the lower control arm with a floor jack. If a twin-post lift is used, support the body with floor jacks on lifting pads forward of the tie rod body bracket.

5. Remove stabilizer bar link from shock bracket, if equipped.
6. Remove clip retaining the brake line flexible hose to the rear shock strut and move aside.
7. Loosen and discard 2 nuts and bolts retaining strut to the spindle. Do not remove bolts at this time.
8. Remove and discard top mounting nut, washer and rubber insulator.
9. Remove and discard 2 bottom mounting bolts and remove strut from the vehicle.
To install:
10. Extend shock absorber strut to its maximum length.
11. Install a new lower washer and insulator assembly, using tire lubricant to ease insertion into the quarter panel shock tower.
12. Position upper part of shock absorber strut shaft into shock tower opening in the body and push slowly on lower part of the shock until mounting holes are lined up with mounting holes in the spindle.
13. Install new lower mounting bolts and nuts. Do not tighten at this time.

NOTE: The heads of both bolts must be to the rear of the vehicle.

14. Place a new upper insulator and washer assembly and nut on the upper shock absorber strut shaft. Tighten nut to 35–55 ft. lbs. (48–75 Nm), using the 18mm deep socket and ¼ drive, 8mm deep socket with extension. Do not grip the shaft with pliers or vise grips.
15. Tighten 2 lower mounting bolts to 70–96 ft. lbs. (95–130 Nm).

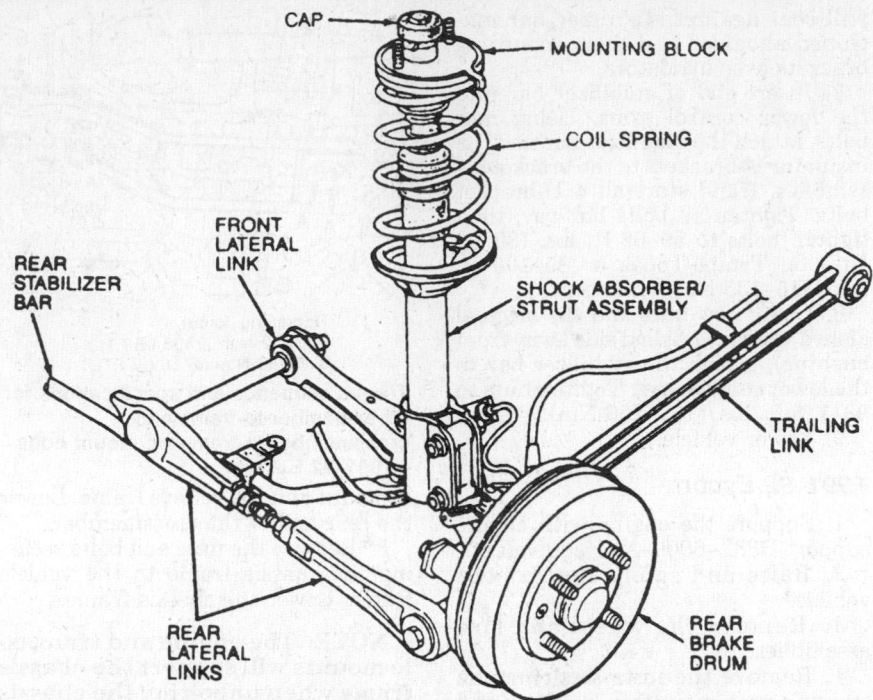

Rear suspension assembly—1991–92 Escort

16. Install stabilizer bar link to bracket on strut, if equipped. Tighten bolts to 40–55 ft. lbs. (55–75 Nm).
17. Install brake line flex hose and retaining clip.
18. Install wheel and tire assembly.
19. Install quarter trim and access panels, as required.

1991–92 Escort

1. Raise and safely support the vehicle.
2. Remove the wheel and tire assembly.
3. Remove the clip securing the flexible brake hose to the rear strut assembly.
4. Remove the nuts and bolts securing the rear strut assembly to the rear wheel spindle assembly.
5. On hatchback and wagon, remove the quarter lower trim panel.
6. Remove the mounting block nuts and remove the rear strut assembly from the vehicle.
7. Position the strut assembly into a vise and secure the assembly at the mounting block.
8. Remove the cap and loosen the piston rod nut 1 turn. Do not remove the piston rod nut at this time.
9. Install an appropriate coil spring compressor onto the coil spring and compress the coil spring.
10. Remove the piston rod nut, washer, retainer and mounting block.
11. Remove the coil spring.

12. Remove the bound stopper seat and stopper from the strut piston.
To install:
13. Position the strut assembly into a vise and secure.
14. Install the bound stopper seat and stopper onto the strut piston rod.
15. Install the coil spring onto the strut assembly.
16. Install the mounting block, then align the mounting block studs and the lower bracket of the strut assembly.
17. Install the retainer, washer and piston rod nut. Tighten the nut to 41–50 ft. lbs. (55–68 Nm).
18. Make sure the spring is properly aligned and carefully release the spring into the seats of the strut.
19. Remove the spring compressor from the coil spring and install the cap.
20. Position the strut assembly into the vehicle wheel housing.
21. Install the mounting block nuts and tighten to 22–27 ft. lbs. (29–40 Nm).
22. On hatchback and wagon, install the quarter lower trim panel.
23. Install the nuts and bolts securing the strut assembly to the rear spindle assembly. Tighten the lower strut bolts to 69–93 ft. lbs. (93–127 Nm).
24. Install the wheel and tire assembly. Tighten the lug nuts to 65–87 ft. lbs. (88–118 Nm).
25. Check the rear alignment and lower the vehicle.

Tempo and Topaz

NOTE: All Tempo and Topaz vehicles except 1988 Tempo with base suspension are equipped with gas-pressurized shock absorbers which will extend unassisted. Do not apply heat or flame to the shock strut during removal.

1. Open luggage compartment and loosen but do not remove, 2 nuts retaining the upper strut mount to body.
2. Raise and safely support the vehicle. Remove the wheel and tire assembly.
3. Place a jack stand under the control arms to support the suspension.

NOTE: Care should be taken when removing the strut that the rear brake flex hose is not stretched or the steel brake tube is not bent.

4. Remove bolt attaching brake hose bracket to strut and move it aside.
5. Remove 2 bolts attaching shock strut to spindle.
6. Remove 2 upper mount-to-body nuts.
7. Remove strut from vehicle.
8. Place strut, spring and upper mount assembly in spring compressor.

--- CAUTION ---

Attempting to remove the spring from the strut without first compressing the spring with a tool designed for that purpose could cause bodily injury.

NOTE: Do not attempt to remove shaft nut by turning shaft and holding nut. Nut must be turned and shaft held to avoid possible fracture of shaft at base of hex.

9. With the spring compressed, remove strut shaft-to-mount nut and then remove spring, strut and mount from compressor tool.
To install:
10. With spring compressed, install spring, spring insulator, top mount and upper washer on strut shaft.
11. Ensure spring is properly located in upper and lower spring seats. The spring end must be within 0.39 in. (10mm) of the step in the spring seat.
12. Tighten shaft nut to 35–50 ft. lbs. (48–68 Nm). Use 18mm deep socket to turn the nut and ¼ drive 8mm deep socket to hold shaft so it will not turn while tightening nut.
13. Insert 2 upper mount studs into strut tower and hand start 2 new nuts. Do not tighten at this time.
14. Position spindle into lower strut mount and install 2 new bolts. Tighten to 70–96 ft. lbs. (95–130 Nm).

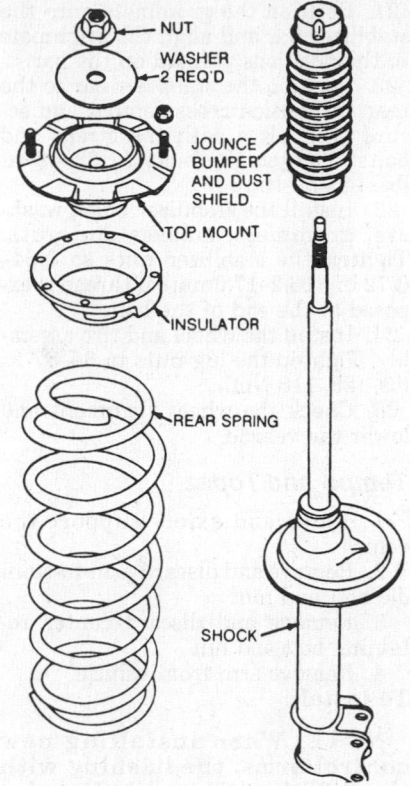

Rear strut assembly—Tempo and Topaz

NUT
WASHER 2 REQ'D
JOUNCE BUMPER AND DUST SHIELD
TOP MOUNT
INSULATOR
REAR SPRING
SHOCK

15. Install brake flex-hose bracket on the strut.
16. Install wheel and tire.
17. Remove jack stand and lower vehicle to the ground.
18. Tighten 2 top mount-to-body nuts to 25–30 ft. lbs. (27–41 Nm).

Coil Springs

REMOVAL & INSTALLATION

1988–90 Escort

1. Raise and safely support the vehicle.
2. Place floor jack under lower control arm. Raise lower control arm to curb position.

NOTE: If a twin-post lift is used, vehicle must be supported on jackstands place under jack pads of the underbody, forward of the tie rod bracket.

3. Remove tire and wheel assembly.
4. Remove and discard nut, bolt and washers retaining lower control arm to spindle.
5. Slowly lower control arm with floor jack until spring can be removed.

To install:
6. The spring insulator must be replaced when servicing spring.
7. Index the insulator on the spring and press insulator downward until it

snaps into place. Check again to ensure insulator is properly indexed against tip of the spring.
8. Install spring in control arm. Ensure spring is properly seated in control arm spring pocket.
9. Raise control arm and spring with floor jack. position spring in pocket on underbody.
10. Using a new bolt, nut and washers, attach control arm to spindle. Install bolt with the head toward front of the vehicle. Tighten to 70–96 ft. lbs. (95–130 Nm).
12. Install tire and wheel.
13. Remove floor jack and lower vehicle.

Rear Control Arms

REMOVAL & INSTALLATION

1988–90 Escort

1. Raise and safely support the vehicle.
2. Place floor jack under lower control arm. Raise lower control arm to curb position.

NOTE: If a twin-post lift is used, vehicle must be supported on jackstands placed under jack pads of the underbody forward of the tie rod bracket.

3. Remove tire and wheel assembly.
4. Remove nuts from control arm-to-body mounting and control arm-to-spindle mounting. Do not remove bolts at this time.
5. Remove and discard spindle end mounting bolt. Slowly lower control arm with floor jack until spring can be removed.
6. Remove and discard bolt from the body end and remove control arm from vehicle.
To install:
7. Attach lower control arm-to-body bracket using a new bolt and nut. Head of the bolt should face the front of the vehicle. Do not tighten at this time.
8. The spring insulator must be replaced when servicing spring.
9. Index the insulator on the spring and press insulator downward until it snaps into place. Place spring in spring pocket in lower control arm. Make sure spring is properly indexed.
10. Using a floor jack, raise lower control arm until it is in line with mounting hole in the spindle.
11. Install lower control arm to spindle using a new bolt, nut and washers. Do not tighten at this time. Bolt head should face the front of the vehicle.
12. Using the floor jack, raise lower control arm to curb height.
13. Tighten control arm-to-spindle bolt to 60–80 ft. lbs. (81–109 Nm).

14. Tighten control arm-to-body bolt to 52–74 ft. lbs. (70–100 Nm).
15. Install tire and wheel.
16. Remove floor jack and lower vehicle.

1991–92 Escort

1. Raise and safely support the vehicle.
2. Remove the wheel and tire assembly.
3. Remove the stabilizer nuts, washers, bushings, sleeves and bolts.
4. Remove the bolts securing the stabilizer bar brackets and grommets to the rear suspension crossmember.
5. Remove the stabilizer bar.
6. Remove the cap covering the front and rear lateral link pivot bolts.
7. Position a floor jack stand under the rear suspension crossmember.
8. Remove the bolts securing the rear suspension crossmember to the vehicle frame.
9. Lower the floor jack stand to allow the rear suspension crossmember to be lowered from the vehicle frame.
10. Remove the front and rear lateral link pivot nut, washer and bolt from the rear suspension crossmember.
11. Remove the front and rear lateral links from the rear suspension crossmember.
12. Remove the bolt, washers and nut securing the front and rear lateral links to the rear wheel spindle and remove the lateral links.
13. Remove the nuts securing the parking brake cable and cable bracket to the trailing link.
14. Remove the rear trailing link bolts and washers from the vehicle frame and rear wheel spindle. Remove the rear trailing link.

To install:

15. Position the rear trailing link and install the bolts and washers. Tighten the trailing link front bolt to 46–69 ft. lbs. (63–93 Nm) and the rear bolt to 69–93 ft. lbs. (93–127 Nm).
16. Position the parking brake cable and bracket to the trailing link and secure it with the nuts.
17. Position the front and rear lateral links to the rear wheel spindle and install the washers, bolt and nut. Tighten the front and rear lateral link nut at the rear wheel spindle to 63–86 ft. lbs. (85–117 Nm).
18. Position the front and rear lateral links to the rear suspension crossmember. Tighten the front and rear lateral link nut at the rear suspension crossmember to 50–70 ft. lbs. (68–95 Nm).
19. Install the cap.
20. Raise the floor jack stand to position the rear suspension crossmember to the vehicle frame. Install and tighten the bolts. Remove the floor jack stand from under the vehicle.

21. Position the grommets onto the stabilizer bar and align the grommets to the positions painted on the bar.
22. Position the stabilizer bar to the rear suspension crossmember and secure it in place with the straps and bolts. Tighten the bolts to 32–43 ft. lbs. (43–59 Nm).
23. Install the stabilizer bolts, washers, grommets, sleeves and nuts. Tighten the stabilizer nuts so 0.64–0.72 in. (16.2–17.0mm) of thread is exposed at the end of the bolt.
24. Install the wheel and tire assembly. Tighten the lug nuts to 65–87 ft. lbs. (88–118 Nm).
25. Check the wheel alignment and lower the vehicle.

Tempo and Topaz

1. Raise and safely support the vehicle.
2. Remove and discard arm-to-spindle bolt and nut.
3. Remove and discard center retaining bolt and nut.
4. Remove arm from vehicle.

To install:

NOTE: When installing new control arms, the bushing with the 0.39 in. (10mm) hole is installed to the center of the vehicle and the bushing with the 0.48 in. (12mm) hole is installed to the spindle. The offset on the arm must face up on the right side of the vehicle and down on the left side of the vehicle. The flange edge of the arm stamping must also face the rear of the vehicle.

5. Position arm at center of vehicle and insert new bolt and nut. Do not tighten at this time.
6. Move arm end up to spindle and insert new bolt, washer and nut. Ensure bolt engages both arms and spindle.
7. Tighten arm-to-body bolt to 30–40 ft. lbs. (40–54 Nm).
8. Tighten arm-to-spindle nut to 60–80 ft. lbs. (81–109 Nm).
9. Lower vehicle.

Rear Wheel Bearings

REMOVAL & INSTALLATION

Except Tempo and Topaz with AWD and 1991–92 Escort

1. Raise and safely support the vehicle.
2. Remove wheel and tire assembly. Remove grease cap from hub.
3. Remove cotter pin, nut retainer, adjusting nut and flatwasher from spindle. Discard cotter pin.

4. Pull hub and drum assembly off spindle being careful not to drop outer bearing assembly.
5. Remove outer bearing assembly.
6. Using seal remover, remove and discard grease seal. Remove inner bearing assembly from hub.
7. Wipe all lubricant from spindle and inside of hub. Cover spindle with a clean cloth and vacuum all loose dust and dirt from brake assembly. Carefully remove cloth to prevent dirt from falling on spindle.
8. Clean both bearing assemblies and cups using solvent. inspect bearing assemblies and cups for excessive wear, scratches, pits or other damage. Replace all worn or damaged parts as required.

NOTE: Allow solvent to dry before repacking bearings. Do not spin-dry bearings with air pressure.

9. If cups are replaced, remove them with wheel hub cup remover D80L–927–A and bearing cup puller T77F–1102–A or equivalent.

To install:

10. If inner or outer bearing cups were removed, install replacement cups using driver handle T80T–4000–W and bearing cup replacers T77F–1202–A and T73T–1217–A or equivalent. Support drum hub on wood block to prevent damage. Insure cups are properly seated in hub.

NOTE: Do not use cone and roller assembly to install cup as this will cause damage to bearing cup and cone and roller assembly.

11. Ensure all spindle and bearing surfaces are clean.
12. Using a bearing packer, pack bearing assemblies with a suitable wheel bearing grease. If a packer is not available, work in as much grease as possible between the rollers and the cages. Grease the cup surfaces.
13. Place inner bearing cone and roller assembly in inner cup. Apply light film of grease to lips of a new grease seal and install seal with rear hub seal replacer T81P–1249–A or equivalent. Ensure retainer flange is seated all around.
14. Apply light film of grease on spindle shaft bearing surfaces.
15. Install hub and drum assembly on spindle. Keep hub centered on spindle to prevent damage to grease seal and spindle threads.
16. Install outer bearing assembly and keyed flat washer on spindle. Install adjusting nut finger-tight. Adjust wheel bearings. Install a new cotter pin.
17. Install wheel and tire on drum.
18. Lower vehicle.

Tempo and Topaz with AWD

1. Raise and support the vehicle safely. Remove the tire and wheel assembly.

2. Remove the brake drum. Remove the parking brake cable from the brake backing plate.

3. Remove the brake line from the wheel cylinder. Remove the outboard U-joint retaining bolts. Remove the outboard end of the halfshaft from the wheel stub shaft yoke and wire it to the control arm.

4. Remove and discard the control arm to spindle bolt, washer and nut. Remove the tie rod nut, bushing and washer and discard the nut.

5. Remove and discard the 2 bolts retaining the spindle to the strut. Remove the spindle from the vehicle. Mount the spindle and backing plate assembly in a suitable vise.

6. Remove the cotter pin and nut attaching the stub shaft yoke to the stub shaft. Discard the cotter pin.

7. Remove the spindle and backing plate assembly from the vise. Remove the stub shaft yoke using a 2 jaw puller and shaft protector. After removing end yoke from spindle assembly, inspect the nylon bushing and replace, as necessary.

8. Position the spindle and backing plate assembly into a vise and remove the wheel stub shaft.

9. Remove the snapring retaining the bearing. Remove the bolts retaining the spindle to the backing plate and remove the backing plate.

10. Remove the spindle from the vise and mount it into a suitable press. With the spindle side facing upward, carefully press out the bearing from the spindle, using a driver handle and bearing cup driver. Discard the bearing after removal.

To install:

11. Mount the spindle in a press, spindle side facing down. Position a new bearing in the outboard side of the spindle and carefully press in the new bearing using a driver handle and bearing installer.

12. Remove the spindle from the press and mount it in a vise. Install the snapring retaining the bearing. Position the backing plate to the spindle and install the retaining bolts.

13. Install the wheel stub shaft. Install the stub shaft yoke and attaching nut. Torque the nut to 120–150 ft. lbs. install a new cotter pin.

14. Remove the spindle and backing plate assembly from the vise. Position the spindle onto the tie rod and then into the strut lower bracket. Insert 2 new strut-to-spindle bolts. Do not tighten at this time.

15. Install the tie rod bushing washer and new nut. Install the new control arm to spindle bolt, washers and nut. Do not tighten them at this time.

16. Install a jack stand to support the suspension at the normal curb height before tightening the fasteners.

17. Torque the spindle to strut bolts to 70–96 ft. lbs. Torque the tie rod nut to 52–74 ft. lbs. Torque the control arm to spindle nut to 60–86 ft. lbs.

18. Position the outboard end of the halfshaft to the wheel stub shaft yoke. Install the retaining caps and bolts and torque them to 15–17 ft. lbs.

19. Install the brake line to wheel cylinder. Install the parking brake cable and brake drum. Install the wheel assembly, torque the lugs nuts to 80–105 ft. lbs.

20. Lower the vehicle and bleed the brake system. Check and adjust the toe, if necessary.

1991–92 Escort

1. Raise and safely support the vehicle.

2. Remove the wheel and tire assembly.

3. Remove the brake drum or brake caliper and rotor, as necessary.

4. Remove the nut securing the rear wheel hub to the spindle and remove the hub and bearing assembly.

To install:

5. Install the rear wheel hub and bearing assembly onto the spindle.

6. Install the hub nut onto the spindle and tighten to 130–174 ft. lbs. (177–235 Nm).

7. Stake the hub nut and install the cap.

8. Install the brake drum or the brake caliper and rotor, as necessary.

9. Install the wheel and tire assembly. Tighten the lugnuts to 65–87 ft. lbs. (88–118 Nm).

10. Lower the vehicle.

ADJUSTMENT

Except Tempo and Topaz with AWD and 1991–92 Escort

1. Raise and safely support the vehicle.

2. Remove wheel cover or ornament and nut covers. Remove grease cap from hub.

3. Remove cotter pin and nut retainer. Discard cotter pin.

4. Back-off adjusting nut 1 full turn. Ensure nut turns freely on spindle threads. Correct any binding condition.

5. Tighten adjusting nut to 17–25 ft. lbs. (23–34 Nm) while rotating hub and drum assembly to seat bearings. Loosen adjusting nut ½ turn and

tighten adjusting nut to 24–28 inch lbs. (2.7–3.2 Nm) using inch lb. torque wrench.

6. Position adjusting nut retainer over adjusting nut so slots in nut retainer flange are in line with cotter pin hole in spline.

7. Install a new cotter pin and bend ends around retainer flange.

8. Check hub rotation. If hub rotates freely, install grease cap. If not, check bearings for damage and replace as necessary.

9. Install wheel and tire assembly, wheel cover or ornaments, and nut covers as required.

10. Lower vehicle.

Tempo and Topaz with AWD

Bearings on 4WD vehicles are not adjustable.

1991–92 Escort

1. Raise and safely support the vehicle.

2. Remove the wheel and tire assembly.

3. Remove the brake drum or the brake caliper and rotor, as necessary.

4. Position a dial indicator to the wheel hub.

5. By hand, push and pull the wheel hub in the axial direction and measure the wheel bearing play.

6. If the wheel bearing play exceeds 0.002 in. (0.05mm), check and adjust the locknut torque or replace the wheel bearing, if necessary.

7. Install the brake drum or brake caliper and rotor, as necessary.

8. Install the wheel and tire assembly and lower the vehicle.

STEERING

Steering Wheel

—— CAUTION ——

On vehicles equipped with an air bag, the negative battery cable and backup power supply must be disconnected, before working on the system. Failure to do so may result in deployment of the air bag and possible personal injury.

REMOVAL & INSTALLATION

Except 1991–92 Escort

1. Disconnect the negative battery cable. On air bag equipped vehicles, disconnect the backup power supply as follows:

a. Remove 2 screws retaining steering column opening cover to instrument panel and remove cover.

b. Remove 4 bolts retaining bolster and remove bolster.

c. Disconnect the connector from the backup power supply.

2. Remove the horn pad cover by removing the retaining screws from the steering wheel assembly.

NOTE: The emblem assembly is removed after the horn pad cover is removed, by pushing it out from the backside of the emblem.

3. Remove the energy absorbing foam from the wheel assembly, if equipped. Remember the energy absorbing foam must be installed when the steering wheel is assembled. Disconnect the horn pad wiring connector.

4. If equipped with air bag restraint system, remove the 4 nuts located on the back of the steering wheel holding the air bag module to the steering wheel.

5. Lift the air bag module from the wheel and disconnect the air bag module-to-clockspring connector.

6. Loosen the steering wheel retaining bolt 4–6 turns but do not remove. On air bag equipped vehicles, remove the bolt completely to remove the vibration damper, then reinstall the bolt loosely on the shaft.

7. Remove the steering wheel with a suitable puller. Do not use a knock-off type puller, because it will cause damage to the collapsible steering column. Loosen the retaining bolt, grasp the rim of the steering wheel and pull the steering wheel from the upper shaft.

To install:

8. Install the steering wheel assembly on the steering column, making sure the alignment marks are correct.

9. Install a new retaining bolt. Torque the bolt to 23–33 ft. lbs. (31–45 Nm). On air bag equipped vehicles, install the vibration damper before installing the bolt.

10. If equipped with air bag, connect the air bag module wire to Clockspring connector and place the module on the steering wheel with the 4 attaching nuts, torque the nuts to 35–53 inch lbs. (4–6 Nm).

11. On vehicles without air bag, connect the horn pad wiring connector and, if equipped, install the energy absorbing foam. Install the horn pad cover and torque the retaining screws to 8–10 inch lbs. (0.9–1.1 Nm).

12. On air bag equipped vehicles, connect the backup power supply connector and reinstall the bolster and steering column opening cover.

13. Reconnect the negative battery cable and check the steering wheel for proper operation.

1991–92 Escort

1. Disconnect the negative battery cable.

2. Remove the steering wheel cover retaining screws from the back side of the steering wheel and remove the cover.

NOTE: On 2-spoke steering wheels there are 2 retaining screws, and on 4-spoke steering wheels there are 4 retaining screws.

3. Disconnect the horn electrical connector and the speed control electrical connector, if equipped.

4. Remove the steering wheel mounting nut and remove the steering wheel with a suitable puller. Do not attempt to remove the steering wheel by hitting the column shaft with a hammer; the column may collapse.

To install:

5. Position the steering wheel and install the mounting nut. Tighten the nut to 29–36 ft. lbs. (39–49 Nm).

6. Connect the horn electrical connector and the speed control electrical connector, if equipped.

7. Position the steering wheel cover and install the retaining screws.

8. Connect the negative battery cable.

Steering Column

REMOVAL & INSTALLATION

Except 1991–92 Escort

NOTE: On air bag equipped vehicles, whenever the steering column is separated from the steering gear for any reason, the steering column must be locked to prevent the steering wheel from being rotated, which in turn will prevent damage to the air bag clockspring.

1. Disconnect the negative battery cable.

NOTE: Before disconnecting cable on air bag equipped vehicles, ensure wheels are in straight ahead position. Turn ignition switch to LOCK position and rotate steering wheel about 16 degrees counterclockwise until locked into position.

2. Remove steering column cover on lower portion of instrument panel (2 screws). On air bag equipped vehicles, remove the bolster and disconnect the backup power supply for the air bag module.

3. Remove speed control module, if equipped (2 screws).

4. Remove lower steering column shroud (5 screws).

5. Loosen, but do not remove, 2 nuts and 2 bolts retaining steering column to support bracket and remove upper shroud.

6. Disconnect all steering column electrical connections: ignition, wash/wipe, turn signal, key warning buzzer, speed control. On console shift automatic transaxle, remove interlock cable retaining screw and disconnect cable from steering column.

7. Loosen steering column to intermediate shaft clamp connection and remove bolt or nut.

8. Remove 2 nuts and 2 bolts retaining steering column to support bracket.

9. Pry open steering column shaft in area of clamp on each side of bolt groove with steering column locked. Open enough to disengage shafts with minimal effort. Do not use excessive force.

10. Inspect 2 steering column bracket clips for damage. If clips have been bent or excessively distorted, they must be replaced.

To install:

11. Engage lower steering shaft to intermediate shaft and hand start clamp bolt and nut.

12. Align 2 bolts on steering column support bracket assembly with outer tube mounting holes and hand start 2 nuts. Check for presence of 2 clips on outer bracket. The clips must be present to ensure adequate performance of vital parts and systems. Hand start 2 bolts through outer tube upper bracket and clip and into support bracket nuts. On console shift automatic transaxles, install interlock cable and retaining screw. Tighten to 30–38 inch lbs. (3.3–4.3 Nm).

13. Connect all quick-connect electrical connections: turn signal, wash/wipe, key warning buzzer, ignition, speed control and air bag clockspring connector, if equipped.

14. Install upper shroud.

15. Tighten steering column mounting nuts and bolts to 15–25 ft. lbs. (20–34 Nm).

16. On air bag equipped vehicles, unlock steering column and cycle steering wheel 1 turn left and 1 turn right to align intermediate shaft into column shaft. Power steering vehicles must have engine running.

17. Tighten steering shaft clamp nut to 20–30 ft. lbs. (27–40 Nm).

18. Install lower trim shroud with 5 screws.

19. Install speed control module, if equipped, with 2 screws.

20. On air bag equipped vehicles, connect the backup power supply and install the bolster.

21. Install steering column cover on instrument panel with 2 screws.
22. Connect battery ground cable.
23. Check steering column for proper operation.

1991–92 Escort

1. Disconnect the negative battery cable.
2. Remove the steering wheel.
3. Remove the combination switch and disconnect the ignition switch electrical connector.
4. Remove the shift-lock cable mounting bracket bolt and place the bracket and cable aside.
5. Remove the 4 steering column upper mounting bracket bolts and lower the column.
6. Remove the 5 set plate mounting nuts and remove the set plate.
7. Remove the intermediate shaft-to-pinion shaft bolt.
8. Remove the 2 steering column lower mounting bracket nuts and remove the column.

To install:

9. Position the steering column and install the 2 lower mounting bracket nuts.
10. Install the intermediate shaft-to-pinion shaft bolt and tighten to 30–36 ft. lbs. (40–50 Nm).
11. Position the set plate and install the 5 mounting nuts.
12. Install the 4 steering column upper mounting bracket bolts and tighten to 80–123 inch lbs. (9–14 Nm).
13. Position the shift-lock cable mounting bracket and install the bolt. Tighten the bolt to 37–55 inch lbs. (4–6 Nm).
14. Connect the ignition switch electrical connector and install the combination switch.
15. Install the steering wheel.
16. Connect the negative battery cable and inspect the shift-lock system.

Manual Rack and Pinion

REMOVAL & INSTALLATION

Except 1991–92 Escort

1. Disconnect the negative battery cable.
2. Turn the ignition key to the **RUN** position.
3. Remove the access trim panel from below the steering column.
4. Remove the intermediate shaft bolts at the rack and pinion input shaft and the steering column shaft.
5. Spread the slots enough to loosen the intermediate shaft at both ends. They cannot be separated at this time.
6. Raise the vehicle and support it safely.

7. Separate the tie rod ends from the steering knuckles, using a suitable tool. Turn the right wheel to the full left turn position.
8. Disconnect the speedometer cable at the transaxle on automatic transaxles only.
9. Disconnect the secondary air tube at the check valve. Disconnect the exhaust system at the manifold and remove the system.
10. Remove the gear mounting brackets and insulators. Keep separated as they are not interchangeable.
11. Turn the steering wheel full left so the tie rod will clear the shift linkage during removal.
12. Separate the gear intermediate shaft, with an assistant pulling upward on the shaft from the inside of the vehicle.

NOTE: Care should be taken during steering gear removal and installation to prevent tearing or damaging the steering gear bellows.

13. Rotate the gear forward and down to clear the input shaft through the dash panel opening.
14. With the gear in the full left turn position, move the gear through the right (passenger side) apron opening until the left tie rod clears the shift linkage and other parts so it may be lowered.
15. Lower the left side of the gear assembly and remove from the vehicle.

To install:

16. Rotate the input shaft to a full left turn stop. Position the right wheel to a full left turn.
17. Start the right side of the gear through the opening in the right apron. Move the gear in until the left tie rod clears all parts so it may be raised up to the left apron opening.
18. Raise the gear and insert the left side through the apron opening. Rotate the gear so the joint shaft enters the dash panel opening.
19. With an assistant guiding the intermediate shaft from the inside of the vehicle, insert the input shaft into the intermediate shaft coupling. Insert the intermediate shaft clamp bolts finger-tight. Do not tighten at this time.
20. Install the gear mounting insulators and brackets in their proper places. Ensure the flat in the left mounting area is parallel to the dash panel. Tighten the bracket bolts to 40–55 ft. lbs. (54–75 Nm) in the sequence as described below:
   a. Tighten the left (driver's side) upper bolt halfway.
   b. Tighten the left side lower bolt.
   c. Tighten the left side upper bolt.
   d. Tighten the right side bolts.

   e. Do not forget that the right and left side insulators and brackets are not interchangeable side to side.
21. Attach the tie rod ends to the steering knuckles. Tighten the castellated nuts to 27–32 ft. lbs. (36–43 Nm), then tighten the nuts until the slot aligns with the cotter pin hole. Insert a new cotter pin.
22. Install the exhaust system. Install the speedometer cable, if removed.
23. Tighten the gear input shaft to intermediate shaft coupling clamp bolt first. Then, tighten the upper intermediate shaft clamp bolt. Tighten both bolts to 20–37 ft. lbs. (28–50 Nm).
24. Install the access panel below the steering column. Turn the ignition key to the **OFF** position.
25. Check and adjust the toe. Tighten the tie rod end jam nuts, check for twisted bellows.

1991–92 Escort

1. Working inside the vehicle, remove the nuts securing the set plate and remove the set plate.
2. Remove the intermediate shaft-to-pinion shaft bolt from inside the vehicle.
3. Raise and safely support the vehicle.
4. Remove the front wheel and tire assemblies.
5. Remove the cotter pins and nuts securing the tie rod ends to the steering knuckles. Separate the tie rod ends from the steering knuckles using a suitable tool.
6. If equipped with manual transaxle, disconnect the extension bar.
7. Remove the nuts securing the steering gear brackets to the bulkhead. Remove the brackets.
8. Remove the steering gear from the vehicle.

To install:

9. Position the steering gear into its mounting position and install the brackets and nuts. Tighten the nuts to 27–38 ft. lbs. (37–52 Nm).
10. If equipped with a manual transaxle, connect the extension bar. Tighten the nut to 23–34 ft. lbs. (31–46 Nm).
11. Attach the tie rod ends to the steering knuckles. Install the nuts and tighten to 31–42 ft. lbs. (42–57 Nm). Install new cotter pins.
12. Install the front wheel and tire assemblies.
13. Lower the vehicle.
14. Install the intermediate shaft-to-pinion shaft bolt and tighten to 13–20 ft. lbs. (18–27 Nm).
15. Position the set plate and secure it with the nuts.

ADJUSTMENT

Except 1991–92 Escort

The yoke clearance is not adjustable except when overhauling the steering gear assembly. Pinion bearing preload is not adjustable because of the non-adjustable bearing usage. Tie rod articulation is preset and is not adjustable. If articulation is out of specification, replace the tie rod assembly. To check tie rod articulation, proceed as follows:

1. With the tie rod end disconnected from the steering knuckle, loop a piece of wire through the hole in the tie rod end stud.

2. Insert the hook of spring scale T74P-3504-Y or equivalent, through the wire loop. Effort to move the tie rod after initial breakaway should be 0.7–5.0 lbs.

NOTE: Do not damage tie rod neck.

3. Replace ball joint/tie rod assembly if effort falls outside this range. Save the tie rod end for use on the new tie rod assembly.

1991–92 Escort

RACK PRELOAD/SUPPORT YOKE ADJUSTMENT

1. Remove the rack and pinion assembly from the vehicle and mount it in a suitable vice.

2. Loosen the locknut.

3. Tighten the adjusting bolt using yoke adjustment adapter T90P-3504-JH in the yoke plug to 8.7 inch lbs. (1 Nm), then loosen the adjusting bolt 10–40 degrees from that position.

4. Measure the pinion turning torque using pinion shaft adapting tool T86P-3504-K. The correct torque at the neutral position ± 90 degrees should be 9–12 inch lbs. (1.0–1.3 Nm). At any other position the torque should be 14.7 inch lbs. (1.6 Nm) or less.

5. If the pinion torque is not within specification, re-adjust the adjusting bolt to achieve the correct pinion torque. Tighten the adjusting bolt locknut.

Power Rack and Pinion

ADJUSTMENT

Except 1991–92 Escort

The power rack and pinion steering gear provides for only rack yoke plug preload adjustment. This adjustment can be performed only with the gear

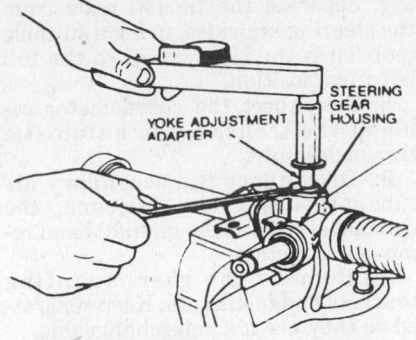

Tightening adjusting bolt on manual steering rack—1991–92 Escort

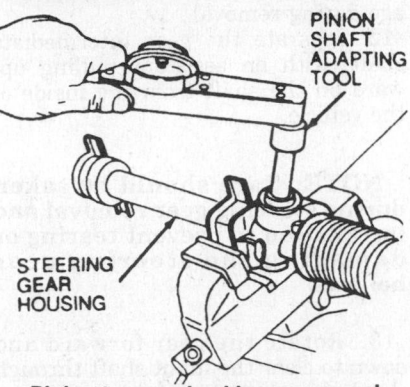

Pinion torque checking on manual steering rack—1991–92 Escort

out of the vehicle. To check rack yoke plug preload, proceed as follows:

RACK AND PINION WITH ONE-PIECE HOUSING

1. Disconnect the negative battery cable.

2. Raise and safely support the vehicle.

3. Remove power rack and pinion assembly from vehicle.

4. Clean exterior of steering gear thoroughly.

5. Mount steering gear in a suitable rack housing holding fixture.

NOTE: Do not mount gear in vice.

6. Do not remove external pressure lines, unless they are leaking or damaged. If these lines are removed, they must be replaced with new lines.

7. Drain power steering fluid by rotating unput shaft lock-to-lock twice using input shaft torque adapter T81P-3504-R or equivalent. Position adapter and wrench on input shaft.

8. Loosen yoke plug locknut with yoke locknut wrench T81P-3504-G or equivalent.

9. Loosen yoke plug using yoke plug adapter T87P-3504-G or equivalent.

10. With rack at center of travel, tighten yoke plug to 44–50 inch lbs. (5.0–5.7 Nm). Clean threads of yoke plug prior to tightening to prevent a false reading.

11. install yoke plug adapter T87P-3504-G or equivalent. Mark location of zero degree mark on housing. Back off adjuster so 48 degree mark lines up with zero degree mark.

12. Place yoke plug locknut wrench T81P-3504-G or equivalent, on yoke plug locknut. While holding yoke plug, tighten locknut to 40–50 ft. lbs. (54–68 Nm). Do not allow yoke plug to move while tightening or preload will be affected. Check input shaft torque after tightening locknut.

13. If external pressure lines were removed, the Teflon® seal rings must be replaced. Clean out Teflon® seal shreds from housing ports prior to installation of new lines.

14. Install power rack assembly in vehicle.

15. Lower vehicle.

16. Connect negative battery cable.

RACK AND PINION WITH TWO-PIECE HOUSING

1. Disconnect the negative battery cable.

2. Raise and safely support the vehicle.

3. Remove the power rack and pinion assembly from the vehicle.

4. Clean the exterior of the gear in the yoke plug area and mount the gear in a vise, gripping it near the center of the tube. Do not over-tighten.

5. Loosen and remove the yoke plug locknut.

6. Back off the yoke plug 1 turn.

7. Tighten the yoke plug to 45 inch lbs. (5.8 Nm) using yoke plug adapter T81P-3504-U or equivalent, and an inch-pound torque wrench with a full scale reading to 100 inch lbs. maximum.

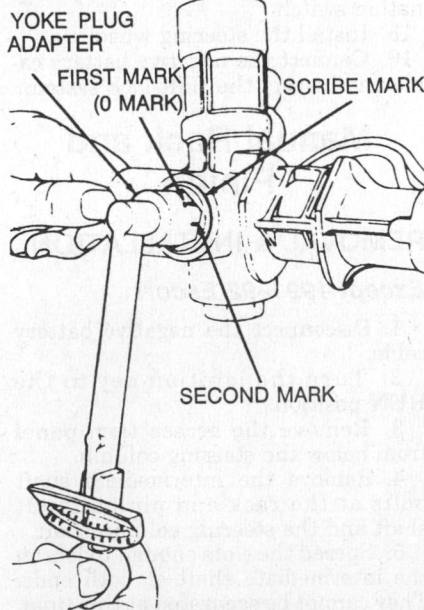

Yoke plug adjustment—power rack and pinion with 2-piece housing

8. Scribe the gear housing in line with the 0 mark on the yoke plug adapter tool.

9. Back off the yoke plug so the second mark on the yoke plug adapter tool aligns with the scribe mark on the gear housing.

10. Hold the plug, and install and tighten the locknut to 40–50 ft. lbs. (54–68 Nm) using yoke locknut wrench T81P–3504–G or equivalent.

REMOVAL & INSTALLATION

Except 1991–92 Escort

1. Disconnect the negative battery cable.

2. Turn the ignition key to the **RUN** position.

3. Remove access panel from dash below the steering column.

4. Remove screws from steering column boot at the dash panel and slide boot up intermediate shaft.

5. Remove intermediate shaft bolt at gear input shaft and loosen the bolt at the steering column shaft joint.

6. With a suitable tool, spread the slots enough to loosen intermediates shaft at both ends. The intermediate shaft and gear input shaft cannot be separated at this time.

7. Remove the air cleaner on Escort.

8. On Escort with air conditioning, wire the air conditioner liquid line above the dash panel opening. Doing so provides clearance for gear input shaft removal and installation.

9. Separate pressure and return lines at intermediate connections on Escort or at steering gear on Tempo and Topaz and drain fluid.

10. On Tempo and Topaz, remove the pressure switch.

11. Disconnect the exhaust secondary air tube at check valve. Raise the vehicle and support it safely. Disconnect exhaust system at exhaust manifold on Escort or at intermediate connection on Tempo and Topaz and remove exhaust system.

12. Separate tie rod ends from steering knuckles.

13. Remove left tie rod end from tie rod on manual transaxle vehicles. This will allow tie rod to clear the shift linkage.

NOTE: Mark location of rod end prior to removal.

14. Disconnect speedometer cable at transaxle, if equipped with automatic transaxle. Remove the vehicle speed sensor.

15. Remove transaxle shift cable assembly at transaxle on vehicles equipped with automatic transaxle.

16. Remove the driveshaft assembly on Tempo and Topaz with all wheel drive.

17. Turn steering wheel to full left turn stop for easier gear removal.

18. On Escort, remove screws holding the heater water tube to shake brace below the oil pan.

19. On Escort, remove nut from the lower of 2 bolts holding engine mount support bracket to transaxle housing. Tap bolt out as far as it will go.

20. Remove the gear mounting brackets and insulators.

21. Drape cloth towel over both apron opening edges to protect bellows during gear removal.

22. Separate gear from intermediate shaft by either pushing up on shaft with a bar from underneath the vehicle while pulling the gear down or with an assistant removing the shaft from inside the vehicle.

23. Rotate gear forward and down to clear the input shaft through the dash panel opening.

24. Make sure input shaft is in full left turn position. Move gear through the right (passenger) side apron opening until left tie rod clears left apron opening and other parts so it may be lowered. Guide the power steering hoses around the nearby components as the gear is being removed.

25. Lower the left side of the gear and remove the gear out of the vehicle. Use care not to tear the bellows.

To install:

26. Rotate the input shaft to a full left turn stop. Position the right road wheel to a full left turn.

27. Start the right side of the gear through the opening in the right apron. Move the gear in until the left tie rod clears all parts so it may be raised up to the left apron opening.

28. Raise the gear and insert the left side through the apron opening. Move the power steering hoses into their proper position at the same time. Rotate the gear so the joint shaft enters the dash panel opening.

29. With an assistant guiding the intermediate shaft from the inside of the vehicle, insert the input shaft into the intermediate shaft coupling. Insert the intermediate shaft clamp bolts finger-tight. Do not tighten at this time.

30. Install the gear mounting insulators and brackets in their proper places. Ensure the flat in the left mounting area is parallel to the dash panel. Tighten the bracket bolts to 40–55 ft. lbs. (54–75 Nm) in the sequence as described below:

   a. Tighten the left (driver's side) upper bolt halfway.

   b. Tighten the left side lower bolt.

   c. Tighten the left side upper bolt.

   d. Tighten the right side bolts.

   e. Do not forget that the right and left side insulators and brackets are not interchangeable side to side.

31. Attach the tie rod ends to the steering knuckles. Tighten the castellated nuts to 27–32 ft. lbs. (36–43 Nm), then tighten the nuts until the slot aligns with the cotter pin hole. Insert a new cotter pin.

32. On the Escort, install the engine mount nut.

33. On the Escort, install the heater water tube to the shake brace.

34. Install the exhaust system. Install the speedometer cable, if removed. Install the vehicle speed sensor and the transaxle shift cable.

35. On Tempo and Topaz with all wheel drive, install the driveshaft.

36. Connect the secondary air tube at the check valve. Connect the pressure and return lines at the intermediate connections or steering gear.

37. Install the pressure switch on Tempo and Topaz. On Escort, install the air cleaner.

38. Tighten the gear input shaft to intermediate shaft coupling clamp bolt first. Then, tighten the upper intermediate shaft clamp bolt. Tighten to 20–30 ft. lbs. (27–40 Nm).

39. Install the access panel below the steering column. Turn the ignition key to the **OFF** position.

40. Fill the system. Check and adjust the toe. Tighten the tie rod end jam nuts to 40–50 ft. lbs. (54–68 Nm), check for twisted bellows.

41. Connect negative battery cable.

1991–92 Escort

1. From inside the passenger compartment, remove the 5 set plate nuts and remove the set plate.

2. Remove the intermediate shaft-to-pinion shaft bolt.

3. Raise and safely support the vehicle.

4. Remove the front wheel and tire assemblies.

5. Remove the cotter pins and attaching nuts from the tie rod ends. Using a suitable tool, separate the tie rod ends from the steering knuckles.

6. If equipped with the 1.8L engine, remove the 2 screws from the power steering line retaining bracket and remove the bracket from the steering gear housing. If equipped with the 1.9L engine, remove the strap that holds the power steering lines to the steering gear housing and discard the strap.

7. Disconnect the high-pressure and return lines from the steering gear and plug the lines.

8. If equipped with manual transaxle, disconnect the extension bar and shift control rod from the transaxle.

9. Remove the nuts from the 2 steering gear mounting brackets.

10. Remove the splash shield from the left wheel well.

11. Remove the steering gear from the left side of the vehicle.

To install:

12. Position the steering gear in its mounting location and install the splash shield in the left wheel well.

13. Position the 2 steering gear mounting brackets and install the 2 nuts to each bracket. Tighten the nuts to 27–38 ft. lbs. (37–52 Nm).

14. If equipped with a manual transaxle, connect the extension bar and shift control rod. Tighten the extension bar nut to 23–34 ft. lbs. (31–46 Nm) and the shift control rod nut to 12–17 ft. lbs. (16–23 Nm).

15. Remove the plugs and connect the pressure and return lines to the steering gear. Tighten the flare nuts to 22–28 ft. lbs. (29–39 Nm).

16. If equipped with 1.8L engine, position the power steering line retaining bracket and install the 2 screws. If equipped with 1.9L engine, install a new strap to hold the power steering lines to the steering gear housing.

17. Position the tie rod ends in the steering knuckles and install the attaching nuts. Tighten the nuts to 31–42 ft. lbs. (42–57 Nm). Install new cotter pins.

18. Install the wheel and tire assemblies and lower the vehicle.

19. From inside the vehicle, install the intermediate shaft-to-pinion shaft bolt. Tighten the bolt to 13–20 ft. lbs. (18–27 Nm).

20. Position the set plate and install the 5 set plate nuts.

21. Fill the system with steering fluid.

Power Steering Pump

REMOVAL & INSTALLATION

1988–90 Escort

1. Disconnect the negative battery cable. Remove the air cleaner, thermactor air pump and belt. Remove the reservoir filler extension and cover the hole to prevent dirt from entering.

2. If equipped with EFI and remote reservoir, remove the reservoir supply hose at the pump, drain the fluid and plug or cap the opening at the pump to prevent entry of contaminants during removal.

3. From under the vehicle, loosen 1 pump adjusting bolt. Remove 1 pump to bracket mounting bolt and disconnect the fluid return line.

4. From above the vehicle, loosen 1 adjusting bolt and the pivot bolt. Remove the drive belt and the 2 remaining pump to bracket mounting bolts.

5. Remove the pump by passing the pulley through the adjusting bracket opening. Remove the pressure hose from the pump assembly.

To install:

6. From under the vehicle, connect the pressure hose to the pump. Pass the pulley through the opening in the adjusting bracket. Install the mounting bolts and tighten to 30–45 ft. lbs. (40–62 Nm).

7. If applicable, make sure the air pump belt is on the inner power steering pump pulley groove. Install the power steering pump belt and adjust. Tighten all bolts to 30–45 ft. lbs. (40–62 Nm).

NOTE: When adjusting belt tension, never pry on the pump or surrounding aluminum parts or brackets.

8. If not equipped with a remote reservoir, install the return line to the pump.

9. From above the vehicle, install the reservoir filler neck extension, if applicable. Install the air cleaner, if applicable.

10. Install remote reservoir supply to pump on EFI/remote vehicles.

11. Fill pump or remote reservoir with fluid and check operation.

1991–92 Escort

1.8L ENGINE

1. Disconnect the negative battery cable.

2. Loosen the power steering fluid reservoir-to-pump hose clamp and pull the hose from the reservoir. Plug the hose.

3. Remove the 2 reservoir mounting bolts and lift the reservoir from its mounting position.

4. Loosen the return hose clamp and pull the return hose from the reservoir. Plug the hose and remove the reservoir.

5. Disconnect the electrical connector from the power steering pressure switch.

6. Loosen the high-pressure line flare nut and disconnect the line from the pump. Plug the line.

7. Raise and safely support the vehicle.

8. Remove the 5 right front undercover bolts and remove the undercover.

9. Remove the belt tensioner adjustment bolt and remove the accessory drive belt from the pulley.

10. Lower the vehicle.

11. Remove the 3 pump mounting bracket bolts and remove the pump and the bracket.

12. Remove the bolt that attaches the pump to the mounting bracket.

13. Remove the nut and bolt that attaches the tensioner to the pump mounting bracket and remove the nut and bolt that attaches the tensioner to the pump.

To install:

14. Position the tensioner to the pump and install the bolt and nut. Tighten the nut to 14–19 ft. lbs. (19–25 Nm).

15. Position the tensioner to the pump mounting bracket and install the bolt and nut. Tighten the nut to 23–34 ft. lbs. (31–46 Nm).

16. Install the bolt that attaches the pump to the mounting bracket and tighten to 27–40 ft. lbs. (36–54 Nm).

17. Position the pump and bracket and install the 3 pump mounting bracket bolts. Tighten the bolts to 27–38 ft. lbs. (37–54 Nm).

18. Raise and safely support the vehicle.

19. Position the accessory drive belt on the pulley and install the belt tensioner adjustment bolt.

20. Position the right front undercover and install the 5 bolts.

21. Lower the vehicle.

22. Unplug the high-pressure line and connect the line to the pump. Tighten the flare nut to 12–17 ft. lbs. (16–24 Nm).

23. Connect the power steering pressure switch electrical connector.

24. Unplug the return hose and connect the hose to the reservoir. Tighten the clamp.

25. Position the reservoir and install the 2 mounting bolts.

26. Unplug the reservoir-to-pump hose and connect the hose to the reservoir. Tighten the clamp.

27. Fill the system with power steering fluid and adjust the accessory drive belt tension.

1.9L ENGINE

1. Disconnect the negative battery cable and drain the cooling system.

2. Loosen the belt tensioner and remove the drive belt from the pulley. Remove the belt tensioner bolt and remove the tensioner.

3. Support the engine with a suitable floor jack.

4. Remove the engine vibration dampener nut and bolt and remove the dampener.

5. Remove the 2 front engine mount nuts. Loosen the engine mount pivot bolt and nut and position the engine mount aside.

6. Raise the engine to gain access to the power steering pump pulley.

7. Hold the pulley in position with a suitable tool and remove the 3 pulley mounting bolts. Remove the pulley and lower the engine.

8. Position the engine mount and install the 2 nuts.

9. Loosen the clamp and disconnect the return line from the pump. Loosen the flare nut from the high-pressure line and disconnect the line from the pump.

10. Raise and safely support the vehicle.

11. Remove the 2 passenger side splash shields.

12. If equipped, remove the 4 compressor mounting bolts and position the air conditioning compressor aside.

13. Remove the lower radiator hose.

14. Remove the 3 power steering pump mounting bolts and remove the pump.

To install:

15. Position the power steering pump and install the 3 mounting bolts. Tighten the bolts to 30–45 ft. lbs. (40–62 Nm).

16. Install the lower radiator hose.

17. If equipped, position the air conditioning compressor and install the 4 mounting bolts. Tighten the bolts to 30–40 ft. lbs. (40–55 Nm).

18. Install the 2 passenger side splash shields and lower the vehicle.

19. Connect the high-pressure line to the power steering pump and tighten the nut. Connect the return line to the pump and position the clamp.

20. Support the engine with a suitable floor jack.

21. Remove the 2 front engine mount nuts and raise the engine to gain access to the pulley.

22. Position the pulley and, holding the pulley in place with a suitable tool, install the 3 pulley mounting bolts. Tighten the bolts to 15–22 ft. lbs. (20–30 Nm).

23. Lower the engine.

24. Position the engine mount and install the 2 nuts. Tighten the engine mount pivot bolt and nut.

25. Position the engine vibration dampener and install the bolt and nut.

26. Position the belt tensioner and install the bolt loosely. Position the accessory drive belt on the pulley and tighten the tensioner mounting bolt to 30–41 ft. lbs. (40–55 Nm).

27. Fill the cooling system.

28. Add the proper type and quantity of power steering fluid.

29. Connect the negative battery cable. Check that the pump operates properly and that there are no leaks.

Tempo and Topaz

1. Disconnect the negative battery cable. Loosen the alternator and remove the drive belt. Pivot the alternator to it most upright position or remove the alternator.

2. Remove the radiator overflow bottle. Loosen and remove the power steering pump drive belt. Mark the pulley and pump drive hub with paint or grease pencil for location reference.

3. Remove the pulleys from the pump shaft.

4. Remove the return line from the pump. Be prepared to catch any spilled fluid in a suitable container.

5. Back off the pressure line attaching nut completely. The line will separate from the pump connection when the pump is removed.

6. Remove the pump mounting bolts and remove the pump.

To install:

7. Install the pump on the mounting bracket. Guide the pressure line into the pump outlet fitting while installing the pump.

8. Install the pressure and return lines.

9. Install 2 pulleys on the hub by aligning the previously applied marks to maintain pulley balance.

10. Install the steering pump drive belt and alternator drive belt and adjust the tension.

11. Install the radiator overflow bottle.

12. Connect the negative battery cable. Fill the pump with fluid and check operation.

BELT ADJUSTMENT

1.8L Engine

1. Raise and safely support the vehicle.

2. Loosen the power steering pump mounting bolt and nuts.

3. Adjust the belt tension by turning the pump adjusting bolt.

4. Tighten the power steering pump mounting nut near the pump adjusting bolt.

5. Check the belt tension using either a belt tension gauge or using the deflection method.

6. If using a belt tension gauge, position the gauge on the longest accessible span of belt. The tension for a new belt should be 110–132 lbs. The tension for a used belt (more than 10 minutes running time) should be 95–110 lbs.

7. If using the deflection method, the deflection on a new belt should be 0.31–0.35 in. (8–9mm). The deflection on a used belt (more than 10 minutes running time) should be 0.35–0.39 in. (9–10mm).

8. Tighten the power steering pump mounting nut, located near the adjusting bolt, to 27–38 ft. lbs. (37–52 Nm).

9. Tighten the pump mounting bolt behind the pulley to 27–40 ft. lbs. (36–54 Nm) and the remaining pump mounting nut to 23–34 ft. lbs. (31–46 Nm).

10. Lower the vehicle.

1988–90 1.9L Engine

1. From engine compartment, loosen pivot bolt and upper adjustment bolt.

2. Raise and safely support the vehicle. From below vehicle:

   a. Loosen lower adjustment bolt.

   b. Apply pressure with ½ drive breaker bar and measure the belt tension with a suitable belt tension gauge. The belt tension should be 140 ± 20 lbs. for a new belt or 110 ± 10 lbs. for a used belt.

   c. Tighten lower adjustment bolt to 38 ft. lbs. (52 Nm).

3. Lower vehicle. From engine compartment, tighten pivot bolt and upper adjustment bolt to 38 ft. lbs. (52 Nm).

2.3L Engine and 1991–92 1.9L Enjgine

Belt tension is maintained by an automatic belt tensioner and does not require adjustment.

SYSTEM BLEEDING

If air bubbles are present in the power steering fluid, bleed the system by performing the following:

1. Fill the reservoir to the proper level.

2. Operate the engine until the fluid reaches normal operating temperature (165–175°F).

3. Turn the steering wheel all the way to the left then all the way to the right several times. Do not hold the steering wheel in the far left or far right position stops.

4. Check the fluid level and recheck the fluid for the presence of trapped air. If apparent that air is still in the system, fabricate or obtain a vacuum tester and purge the system as follows:

   a. Remove the pump dipstick cap assembly.

   b. Check and fill the pump reservoir with fluid to the **COLD FULL** mark on the dipstick.

   c. Disconnect the ignition coil wire or the coil pack electrical connector if equipped with distributorless ignition, and raise the front of the vehicle and support safely.

   d. Crank the engine with the starter and check the fluid level. Do not turn the steering wheel at this time.

   e. Fill the pump reservoir to the **COLD FULL** mark on the dipstick. Crank the engine with the starter while cycling the steering wheel lock-to-lock. Check the fluid level.

   f. Tightly insert a suitable size rubber stopper and air evacuator pump into the reservoir fill neck. Connect the ignition coil wire or coil pack electrical connector.

   g. With the engine idling, apply a 15 in. Hg vacuum to the reservoir for 3 minutes. As air is purged from the system, the vacuum will drop off. Maintain the vacuum on the system as required throughout the 3 minutes.

h. Remove the vacuum source. Fill the reservoir to the **COLD FULL** mark on the dipstick.

i. With the engine idling, re-apply 15 in. Hg vacuum source to the reservoir. Slowly cycle the steering wheel to lock-to-lock stops for approximately 5 minutes. Do not hold the steering wheel on the stops during cycling. Maintain the vacuum as required.

j. Release the vacuum and disconnect the vacuum source. Add fluid, as required.

k. Start the engine and cycle the wheel slowly and check for leaks at all connections.

l. Lower the front wheels.

5. In cases of severe aeration, repeat the procedure.

Tie Rod Ends

REMOVAL & INSTALLATION

1. Remove and discard cotter pin and nut from worn tie rod end ball stud.

2. Disconnect tie rod end from spindle, using tie rod end remover tool 3290–D and adapter T81P–3504–W or equivalent.

3. Holding tie rod end with a wrench, loosen tie rod jam nut.

4. Grip tie rod hex flats with a pair of suitable locking pliers, and remove tie rod end assembly from tie rod. Note depth to which tie rod was located, using jam nut as a marker.

To install:

5. Clean tie rod threads. Apply a light coating of disc brake caliper slide grease D7AZ–19590–A or equivalent, to tie rod threads. Thread new tie rod end on tie rod to same depth as removed tie rod end. Tighten jam nut.

6. Place tie rod end stud into steering spindle.

7. Install a new nut on tie rod end stud. Tighten nut to 27–32 ft. lbs. (36–43 Nm) on all except 1991–92 Escort where the torque is 31–42 ft. lbs. (42–57 Nm), and continue tightening nut to align next castellation with cotter pin hole in stud. Install a new cotter pin.

8. Set toe to specification and tighten jam nuts to 42–50 ft. lbs. (57–68 Nm) on all except 1991–92 Escort where the torque is 25–29 ft. lbs. (34–49 Nm). Do not twist bellows.

BRAKES

For all brake system repair and service procedures not detailed below, please refer to "Brakes" in the Unit Repair Section.

Master Cylinder

REMOVAL & INSTALLATION

Except 1991–92 Escort

1. Disconnect the negative battery cable.

2. Disconnect and plug the brake lines from the primary and secondary outlet ports of the master cylinder and pressure control valves.

3. Remove the nuts attaching the master cylinder to the brake booster assembly. Disconnect the brake warning light wire.

4. Slide the master cylinder forward and upward from the vehicle.

To install:

5. Before installation, bench bleed the new master cylinder as follows:

a. Mount the new master cylinder in a suitable holding fixture. Be careful not to damage the housing.

b. Fill the master cylinder reservoir with brake fluid.

c. Using a suitable tool inserted into the booster pushrod cavity, push the master cylinder piston in slowly. Place a suitable container under the master cylinder to catch the fluid being expelled from the outlet ports.

d. Place a finger tightly over each outlet port and allow the master cylinder piston to return.

e. Repeat the procedure until clear fluid only is expelled from the master cylinder. Plug the outlet ports and remove the master cylinder from the holding fixture.

6. Position the master cylinder over the booster pushrod and booster mounting studs. Install the nuts and tighten to 13–25 ft. lbs. (18–33 Nm).

7. Remove the plugs and connect the brake lines. Tighten the fittings.

8. Make sure the master cylinder reservoir is full. Have an assistant push down on the brake pedal. When the pedal is all the way down, crack open the brake line fittings, 1 at a time, to expel any remaining air in the master cylinder and brake lines. Tighten the fittings, then have the assistant allow the brake pedal to return.

9. Repeat Step 8 until all air is expelled from the master cylinder and brake lines. Final tighten the brake line fittings to 10–18 ft. lbs. (14–24 Nm).

10. Connect the brake warning indicator connector.

11. Make sure the master cylinder reservoir is full.

12. If necessary, bleed the brake system.

13. Connect the negative battery cable. Check for fluid leaks and check for proper operation.

1991–92 Escort

1. Disconnect the battery cables and remove the battery.

2. Disconnect the low fluid level sensor electrical connector.

3. Loosen the brake line fittings and disconnect the brake lines from the master cylinder.

4. If equipped with manual transaxle, remove the clamp and pull the clutch hose from the brake/clutch fluid reservoir.

5. Cap the lines and the master cylinder ports.

6. Remove the 2 mounting nuts and remove the master cylinder assembly.

To install:

7. Adjust the piston to pushrod clearance as follows:

a. Position master cylinder gauge T87C–2500–A or equivalent, on the end of the master cylinder. Loosen the set screw and push the gauge plunger against the bottom of the primary piston.

b. While holding the gauge in position, tighten the set screw.

c. Apply 19.7 in. Hg of vacuum to the power brake booster with a vacuum pump.

d. Invert the gauge and place it over the power brake pushrod.

e. Make sure there is no space between the end of the adjustment gauge and the power brake pushrod.

f. If there is space between the end of the adjustment gauge and the power brake pushrod, loosen the pushrod locknut and adjust the pushrod until there is no space.

NOTE: After the master cylinder is installed, the piston to pushrod clearance will be 0.016–0.024 in. (0.4–0.6mm) with no vacuum or 0.004–0.01 in. (0.1–0.3mm) with 500mm Hg of vacuum.

8. Before installation, bench bleed the new master cylinder as follows:

a. Mount the new master cylinder in a suitable holding fixture. Be careful not to damage the housing.

b. Fill the master cylinder reservoir with brake fluid.

c. Using a suitable tool inserted into the booster pushrod cavity, push the master cylinder piston in slowly. Place a suitable container under the master cylinder to catch

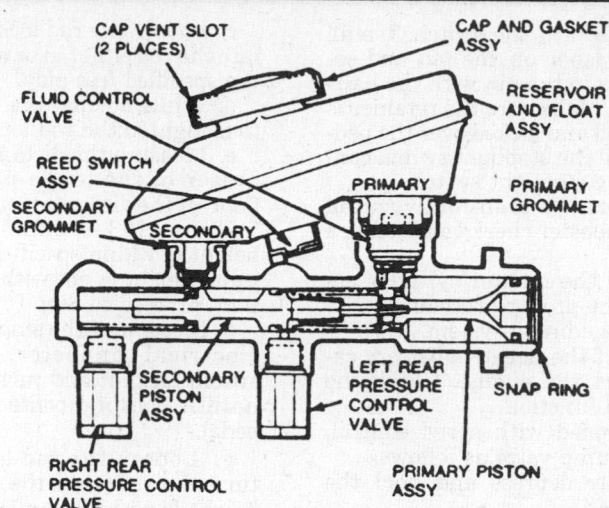

Master cylinder removal and installation — 1991–92 Escort

Master cylinder and pressure control valve assembly — except 1991–92 Escort

the fluid being expelled from the outlet ports.

d. Place a finger tightly over each outlet port and allow the master cylinder piston to return.

e. Repeat the procedure until clear fluid only is expelled from the master cylinder. Plug the outlet ports and remove the master cylinder from the holding fixture.

9. Position the master cylinder over the booster pushrod and booster mounting studs. Install the nuts and tighten to 8–12 ft. lbs. (10–16 Nm).

10. If equipped with manual transaxle, connect the clutch hose onto the brake/clutch fluid reservoir and install the clamp.

11. Remove the caps and connect the brake lines. Tighten the fittings.

12. Make sure the master cylinder reservoir is full. Have an assistant push down on the brake pedal. When the pedal is all the way down, crack open the brake line fittings, 1 at a time, to expel any remaining air in the master cylinder and brake lines. Tighten the fittings, then have the assistant allow the brake pedal to return.

13. Repeat Step 12 until all air is expelled from the master cylinder and brake lines. Tighten the brake line fittings to 10–16 ft. lbs. (13–22 Nm).

14. Connect the low fluid level sensor electrical connector.

15. Install the battery and connect the negative battery cable.

16. Make sure the master cylinder reservoir is full. Bleed the brakes, if necessary.

17. Check for brake fluid leaks and for proper brake operation.

Proportioning/ Combination Valve

REMOVAL & INSTALLATION

Except 1991–92 Escort

There are 2 pressure control valves housed in the master cylinder assembly. The valves reduce rear brake system hydraulic pressure when the pressure exceeds a preset value. The rear brake hydraulic pressure is limited in order to minimize rear wheel skidding during hard braking. Remove and install the pressure control valves as follows:

1. Disconnect the primary or secondary brake line, as necessary.

2. Loosen and remove the pressure control valve from the master cylinder housing.

To install:

3. Install the pressure control valve in the master cylinder housing port and tighten to 10–18 ft. lbs. (14–24 Nm).

4. Connect the brake line and tighten the fitting to 10–18 ft. lbs. (14–24 Nm).

5. Fill and bleed the brake system.

1991–92 Escort

A separate proportioning valve located on the underside of the vehicle regulates the hydraulic pressure to the rear brakes.

1. Raise and safely support the vehicle.

2. Loosen the brake line fittings and disconnect the brake lines from the proportioning valve.

3. Remove the 2 mounting bolts and remove the valve.

To install:

4. Position the valve and install the mounting bolts. Tighten to 14–17 ft. lbs. (19–23 Nm).

5. Connect the brake lines and tighten the fittings to 10–16 ft. lbs. (13–22 Nm).

6. Properly bleed the brake system.

7. Lower the vehicle.

Power Brake Booster

REMOVAL & INSTALLATION

Except 1991–92 Escort

1. Disconnect the battery ground cable and remove the brake lines from the master cylinder.

2. Remove the retaining nuts and remove the master cylinder.

3. From under the instrument panel, remove the stoplight switch wiring connector from the switch. Remove the pushrod retainer and outer nylon washer from the brake pin, slide the stoplight switch along the brake pedal pin, far enough for the outer hole to clear the pin.

4. Remove the switch by sliding it upward. Remove the booster to dash panel retaining nuts. Slide the booster pushrod and pushrod bushing off the brake pedal pin.

5. Disconnect the manifold vacuum hose from the booster check valve and move the booster forward until the booster studs clear the dash panel and remove the booster.

To install:

6. Align the pedal support and support spacer inside the vehicle and place the booster in position on the dash panel. Hand-start the retaining nuts.

7. Working inside the vehicle, install the pushrod and pushrod bushing on the brake pedal pin. Tighten the booster-to-dash panel retaining nuts to 13–25 ft. lbs. (18–33 Nm).

8. Position the stoplight switch so it straddles the booster pushrod with the stoplight switch slot toward the pedal blade and the hole just clearing the pin. Slide the stoplight switch down onto the pin. Slide the assembly toward the pedal arm, being careful not

to bend or deform the switch. Install the nylon washer on the pin and secure all parts to the pin with the hairpin retainer. Make sure the retainer is fully installed and locked over the pedal pin. Install the stoplight switch connector on the stoplight switch.

9. Connect the manifold vacuum hose to the booster check valve using a hose clamp.

10. Install the master cylinder according to the proper procedure.

11. Bleed the brake system.

12. Connect the negative battery cable and start the engine. Check the power brake function.

13. If equipped with speed control, adjust the dump valve as follows:

a. Firmly depress and hold the brake pedal.

b. Push in the dump valve until the valve collar bottoms against the retaining clip.

c. Place a 0.050–0.10 in. shim between the white button of the valve and the pad on the brake pedal.

d. Firmly pull the brake pedal rearward to its normal position, allowing the dump valve to ratchet backward in the retaining clip.

1991–92 Escort

1. Disconnect the negative battery cable.

2. Remove the master cylinder assembly.

3. Loosen the vacuum hose clamp and remove the hose from the power brake booster.

4. From inside the vehicle, remove the pin and discard.

5. Remove the clevis pin.

6. Remove the 4 booster mounting nuts and remove the booster. Remove and discard the gasket.

To install:

7. Install a new gasket over the studs and position the power brake booster.

8. From inside the vehicle, install the 4 mounting nuts and tighten to 14–19 ft. lbs. (19–25 Nm).

9. Lubricate the clevis pin with white lithium grease and install. Install a new pin.

10. Position the vacuum hose to the booster and install the clamp.

11. Install the master cylinder.

12. Adjust the brake pedal as follows:

a. Press the brake pedal several times to eliminate the vacuum in the booster.

b. Carefully press the pedal and measure the amount of free-play until resistance is felt. If the free-play is 0.16–0.28 in. (4–7mm), the pedal free-play is within specification. If the free-play is not within specification, proceed to Step c.

c. Loosen the rod locknut and rotate the rod either in or out to obtain the specified free-play.

d. While holding the rod in position, tighten the rod locknut.

e. Measure the distance from the center of the brake pedal to the floor. If the distance measures 7.60–7.72 in. (193–196mm), the pedal height is within specification. If the pedal height is not within specification, proceed to Step f.

f. Disconnect the stoplight switch electrical connector, loosen the switch locknut and turn the switch until it does not contact the brake pedal.

g. Loosen the rod locknut and turn the rod until the brake pedal height is within specification.

h. Turn the stoplight switch until it contacts the brake pedal, then turn it an additional ½ turn. Tighten the stoplight locknut and the rod locknut.

i. Connect the stoplight switch electrical connector and check the operation of the stoplights and brake system.

Brake Caliper

REMOVAL & INSTALLATION

Except 1991–92 Escort

1. Disconnect the negative battery cable.

2. Raise and safely support the vehicle.

3. Remove wheel and tire assembly from rotor mounting face.

4. Disconnect flexible brake hose from caliper. Remove hollow retaining bolt that connects hose fitting to caliper. Remove hose assembly from caliper and plug hose.

5. Remove caliper locating pins using torx drive bit D79P-2100-T40 or equivalent.

6. Lift caliper off rotor and integral knuckle and anchor plate using rotating motion.

NOTE: Do not pry directly against plastic piston or damage to piston will occur.

To install:

7. Position caliper assembly above rotor with anti-rattle spring under upper arm of knuckle. Install caliper over rotor with rotating motion. Ensure inner shoe is properly positioned.

NOTE: Ensure correct caliper assembly is installed on correct knuckle. The caliper bleed screw should be positioned on top of caliper when assembled on vehicle.

8. Lubricate locating pins and inside of insulators with silicone grease. Install locating pins through caliper insulators and into knuckle attaching holes. The caliper locating pins must be inserted and threads started by hand.

9. Using torx drive bit D79P-2100-T40 or equivalent, tighten caliper locating pins to 18–25 ft. lbs. (24–34 Nm).

10. Remove plug and install brake hose on caliper with new gasket on each side of fitting outlet. Insert attaching bolt through washers and fittings. Tighten bolts to 30–40 ft. lbs. (40–54 Nm).

11. Bleed brake system. Always replace rubber bleed screw cap after bleeding.

12. Fill master cylinder as required.

13. Install wheel and tire assembly. Tighten wheel nuts to 80–105 ft. lbs. (109–142 Nm).

14. Connect negative battery cable.

15. Pump brake pedal prior to moving vehicle to position brake linings.

16. Road test vehicle.

1991–92 Escort

FRONT CALIPER

1. Raise and safely support the vehicle. Remove the wheel and tire assembly.

2. Remove the brake pads.

3. Clamp the brake hose and remove the brake hose attaching bolt.

4. Disconnect the brake hose from the caliper and discard the 2 copper washers.

5. Remove the 2 caliper mounting bolts and remove the caliper.

To install:

6. Position the caliper and install the 2 caliper mounting bolts. Tighten the bolts to 29–36 ft. lbs. (39–49 Nm).

7. Install 2 new copper washers to the brake hose. Position the brake hose onto the caliper and install the attaching bolt. Tighten the bolt to 16–22 ft. lbs. (22–29 Nm).

8. Remove the clamp from the brake hose.

9. Install the brake pads.

10. Bleed the brake system.

11. Install the wheel and tire assembly and lower the vehicle.

REAR CALIPER

1. Raise and safely support the vehicle. Remove the wheel and tire assembly.

2. Remove the brake pads.

3. Remove the parking brake cable bracket bolt and position the bracket aside.

4. Remove the parking brake cable from the operating lever.

5. Clamp the brake hose, remove the brake line attaching bolt and re-

move the 2 washers. Discard the washers.

6. Disconnect the brake line and slide the caliper off the mounting bracket.

To install:

7. Position the caliper on the mounting bracket.

8. Install 2 new washers to the brake line. Position the brake line to the caliper and install the attaching bolt. Tighten the bolt to 16–22 ft. lbs. (22–29 Nm).

9. Remove the clamp from the brake hose.

10. Attach the parking brake cable to the operating lever and position the bracket and install the bracket bolt.

11. Install the brake pads.

12. Bleed the brake system.

13. Install the wheel and tire assembly and lower the vehicle.

Disc Brake Pads

REMOVAL & INSTALLATION

Except 1991–92 Escort

1. Remove master cylinder cap and check fluid level in reservoir. Remove brake fluid until reservoir is ½ full. Discard removed fluid.

2. Raise and safely support the vehicle.

3. Remove wheel and tire assembly.

4. Remove caliper locating pins.

5. Lift caliper assembly from integral knuckle and anchor plate and ro-

tor using rotating motion. Do not pry directly against plastic piston or damage will occur.

6. Remove outer shoe and lining assembly.

7. Remove inner shoe and lining assembly.

8. Inspect both rotor braking surfaces. Minor scoring or buildup of lining material does not require machining or replacement of rotor. Hand-sand glaze from both rotor braking surfaces using garnet paper 100-A (medium grit) or aluminum oxide 150-J (medium).

9. Suspend caliper inside fender housing with wire. Use care not to damage caliper or stretch brake hose.

To install:

10. Use a 4 in. C-clamp and wood block 2 ¾ in. x 1 in. (70mm x 25mm) and approximately ¾ in. (19mm) thick to seat caliper hydraulic piston in its bore.

NOTE: Extra care must be taken during this procedure to prevent damage to the plastic piston. Metal or sharp objects cannot come into direct contact with the piston surface or damage will result.

11. Remove all rust buildup from inside of caliper legs where the outer shoe makes contact.

12. Install inner shoe and lining assembly in caliper piston(s). Do not bend shoe clips during installation in piston.

13. Install correct outer shoe and lining assembly. Ensure clips are properly seated.

14. Install caliper over rotor.

15. Install wheel and tire assembly. Tighten wheel nuts to 80–105 ft. lbs. (109–142 Nm).

16. Pump brake pedal prior to moving vehicle to position brake linings. Check the fluid level in the master cylinder.

17. Connect negative battery cable.

18. Road test vehicle.

1991–92 Escort

FRONT DISC BRAKE PADS

1. Remove master cylinder cap and check fluid level in reservoir. Remove brake fluid until reservoir is ½ full. Discard removed fluid.

2. Raise and safely support the vehicle.

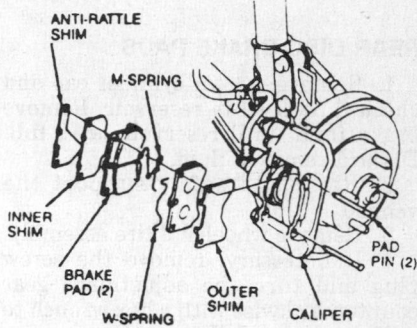

Front disc brake pad assembly—1991–92 Escort

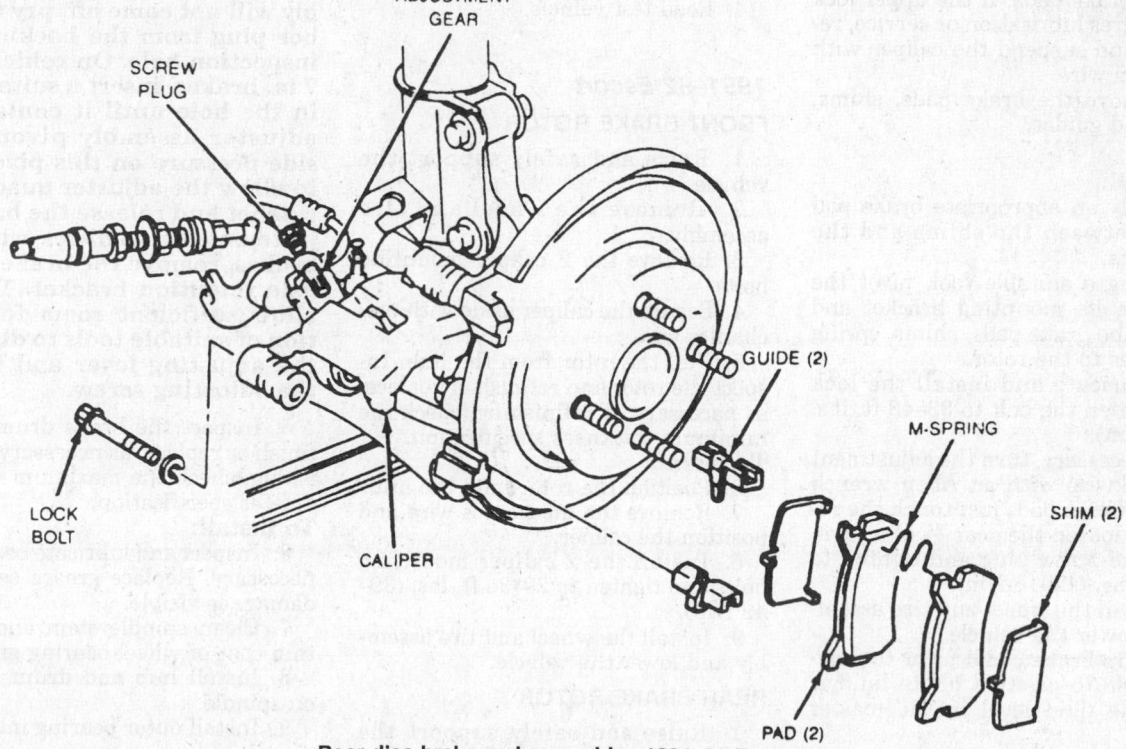

Rear disc brake pad assembly—1991–92 Escort

3. Remove wheel and tire assembly.

4. Remove the 2 brake pad pins and remove the M-spring and the W-spring.

5. Remove the brake pads and shims from the caliper.

To install:

6. Use a suitable tool to push the piston into the caliper bore.

7. Apply suitable grease between the shims and the brake pad guide plates and position the brake pads and shims into the caliper.

8. Install the W-spring and the M-spring. Install the 2 brake pad pins.

9. Install the wheel and tire assembly and lower the vehicle.

10. Pump brake pedal prior to moving vehicle to position brake linings. Check the fluid level in the master cylinder.

11. Road test vehicle.

REAR DISC BRAKE PADS

1. Remove master cylinder cap and check fluid level in reservoir. Remove brake fluid until reservoir is ½ full. Discard removed fluid.

2. Raise and safely support the vehicle.

3. Remove wheel and tire assembly.

4. If necessary, remove the screw plug and turn the adjustment gear counterclockwise with a hex wrench to pull the piston fully inward.

5. Remove the caliper lock bolt.

6. Using a suitable tool, pivot the caliper on its mounting bracket to access the brake pads. If the upper lock bolt requires lubrication or service, remove it and suspend the caliper with mechanics wire.

7. Remove the brake pads, shims, spring and guides.

To install:

8. Apply an appropriate brake pad grease between the shims and the brake pads.

9. Using a suitable tool, pivot the caliper on its mounting bracket and position the brake pads, shims, spring and guides to the rotor.

10. Lubricate and install the lock bolt. Tighten the bolt to 33–43 ft. lbs. (45–59 Nm).

11. If necessary, turn the adjustment gear clockwise with an Allen wrench until the brake pads just touch the rotor, then loosen the gear ⅓ of a turn. Install the screw plug and tighten to 9–12 ft. lbs. (12–16 Nm).

12. Install the wheel and tire assembly and lower the vehicle.

13. Pump brake pedal prior to moving vehicle to position brake linings. Check the fluid level in the master cylinder.

14. Road test vehicle.

Brake Rotor

REMOVAL & INSTALLATION

Except 1991–92 Escort

1. Disconnect the negative battery cable.

2. Raise and safely support the vehicle.

3. Remove wheel and tire assembly.

4. Remove caliper locating pins.

5. Lift caliper assembly from integral knuckle and anchor plate and rotor using rotating motion. Do not pry directly against plastic piston or damage will occur.

6. Position caliper aside and support it with a length of wire to avoid damaging caliper.

7. Remove rotor from hub assembly by pulling it off the hub studs. Inspect the rotor and refinish or replace, as necessary. If refinishing, check the minimum thickness specification.

To install:

8. If rotor is being replaced, remove protective coating from new rotor with carburetor degreaser. If original rotor is being installed, make sure rotor braking and mounting surfaces are clean.

9. Install rotor on hub assembly.

10. Install caliper assembly on rotor.

11. Install wheel and tire assembly. Tighten wheel nuts to 80–105 ft. lbs. (109–142 Nm).

12. Pump brake pedal prior to moving vehicle to position brake linings.

13. Connect negative battery cable.

14. Road test vehicle.

1991–92 Escort

FRONT BRAKE ROTOR

1. Raise and safely support the vehicle.

2. Remove the wheel and tire assembly.

3. Remove the 2 caliper mounting bolts.

4. Secure the caliper aside with mechanics wire.

5. Pull the rotor from the hub. Inspect the rotor and refinish or replace, as necessary. If refinishing, check the minimum thickness specification.

To install:

6. Position the rotor onto the hub.

7. Remove the mechanics wire and position the caliper.

8. Install the 2 caliper mounting bolts and tighten to 29–36 ft. lbs. (39–49 Nm).

9. Install the wheel and tire assembly and lower the vehicle.

REAR BRAKE ROTOR

1. Raise and safely support the vehicle.

2. Remove the wheel and tire assembly.

3. Remove the brake pads.

4. Remove the 2 rotor mounting screws.

5. Using a suitable tool, pivot the caliper on its mounting bracket and remove the rotor. Inspect the rotor and refinish or replace, as necessary. If refinishing, check the minimum thickness specification.

To install:

6. Using a suitable tool, pivot the caliper on its mounting bracket and position the rotor.

7. Install the 2 mounting screws.

8. Install the brake pads.

9. Install the wheel and tire assembly and lower the vehicle.

Brake Drums

REMOVAL & INSTALLATION

Except 1991–92 Escort and Tempo and Topaz with All Wheel Drive

1. Raise and safely support the vehicle.

2. Remove wheel and tire assembly.

3. Remove grease cap from hub. Remove cotter pin, nut lock, adjusting nut and keyed flat washer from spindle. Remove outer bearing.

4. Remove hub and drum assembly as a unit.

NOTE: If the hub/drum assembly will not come off, pry the rubber plug from the backing plate inspection hole. On vehicles with 7 in. brakes, insert a suitable tool in the hole until it contacts the adjuster assembly pivot. Apply side pressure on this pivot point to allow the adjuster quadrant to ratchet and release the brake adjustment. On vehicles with 8 in. brakes, remove the brake line-to-axle retention bracket. This will allow sufficient room for insertion of suitable tools to disengage the adjusting lever and back-off the adjusting screw.

5. Inspect the brake drum and refinish or replace, as necessary. If refinishing, check the maximum inside diameter specification.

To install:

6. Inspect and lubricate bearings, as necessary. Replace grease seal if any damage is visible.

7. Clean spindle stem and apply a thin coat of wheel bearing grease.

8. Install hub and drum assembly on spindle.

9. Install outer bearing into hub on spindle.

10. Install keyed flat washer and adjusting nut. Tighten nut finger-tight.

11. Adjust wheel bearing. Install nut retainer and a new cotter pin.

12. Install grease cap.

13. Install wheel and tire assembly. Tighten wheel nuts to 80–105 ft. lbs. (109–142 Nm).

14. Pump brake pedal prior to moving vehicle to position brake linings.

15. Connect negative battery cable.

16. Road test vehicle.

1991–92 Escort and Tempo and Topaz with All Wheel Drive

1. Raise and safely support the vehicle.

2. Remove wheel and tire assembly.

3. Remove the spring nut or attaching screws, if necessary.

4. Pull the brake drum from the hub. Inspect the drum and refinish or replace, as necessary. If refinishing, check the maximum inside diameter specification.

To install:

5. Position the brake drum on the hub.

6. Install the 2 drum attaching screws, if applicable.

7. Install the wheel and tire assembly and lower the vehicle.

Brake Shoes

REMOVAL & INSTALLATION

Except Tempo, Topaz, 1991–92 Escort and 1988–90 Escort with 8 Inch Brake Shoes

1. Raise and safely support the vehicle.

2. Remove wheel and tire assembly.

3. Remove hub and drum assembly.

4. Remove hold-down spring and pins.

5. Lift brake shoe and adjuster assembly up and away from anchor block and shoe guide. Do not damage the boots when rotating shoes off the wheel cylinder.

6. Remove lower shoe-to-shoe spring from leading and trailing shoe slots.

7. Hold brake shoe/adjuster assembly, remove leading shoe-to-adjuster strut retracting spring. This can be done by rotating shoe over adjuster quadrant until spring is slack and then disconnecting spring. The leading shoe should now be free.

8. Remove trailing shoe-to-parking brake strut retracting spring by pivoting strut downward until it disengages from trailing shoe.

9. Disassemble adjuster, if necessary, by pulling quadrant away from knurled pin in strut and rotating quadrant in either direction until quadrant teeth are no longer meshed with pin. Remove spring and slide quadrant out of slot. Do not overstress spring during disassembly.

10. Remove parking brake lever from trailing shoe and lining assembly by removing horeshoe retaining clip and spring washer, and lifting lever off pin on brake shoe.

To install:

11. Apply light coating of high temperature grease at points where brake shoes contact the backing plate.

12. Apply light uniform coating of multi-purpose lubricant to strut at contact surface between strut and adjuster quadrant.

13. Install adjuster quadrant pin into slot in strut and install adjuster spring. Pivot quadrant until it meshes with knurled pin in third and fourth notch of outboard end of quadrant.

14. Assemble parking brake lever to trailing shoe. Install spring washer and new horseshoe clip. Crimp clip until lever is securely fastened.

15. Install trailing shoe-to-parking brake strut retracting spring by attaching spring to slots in each part and

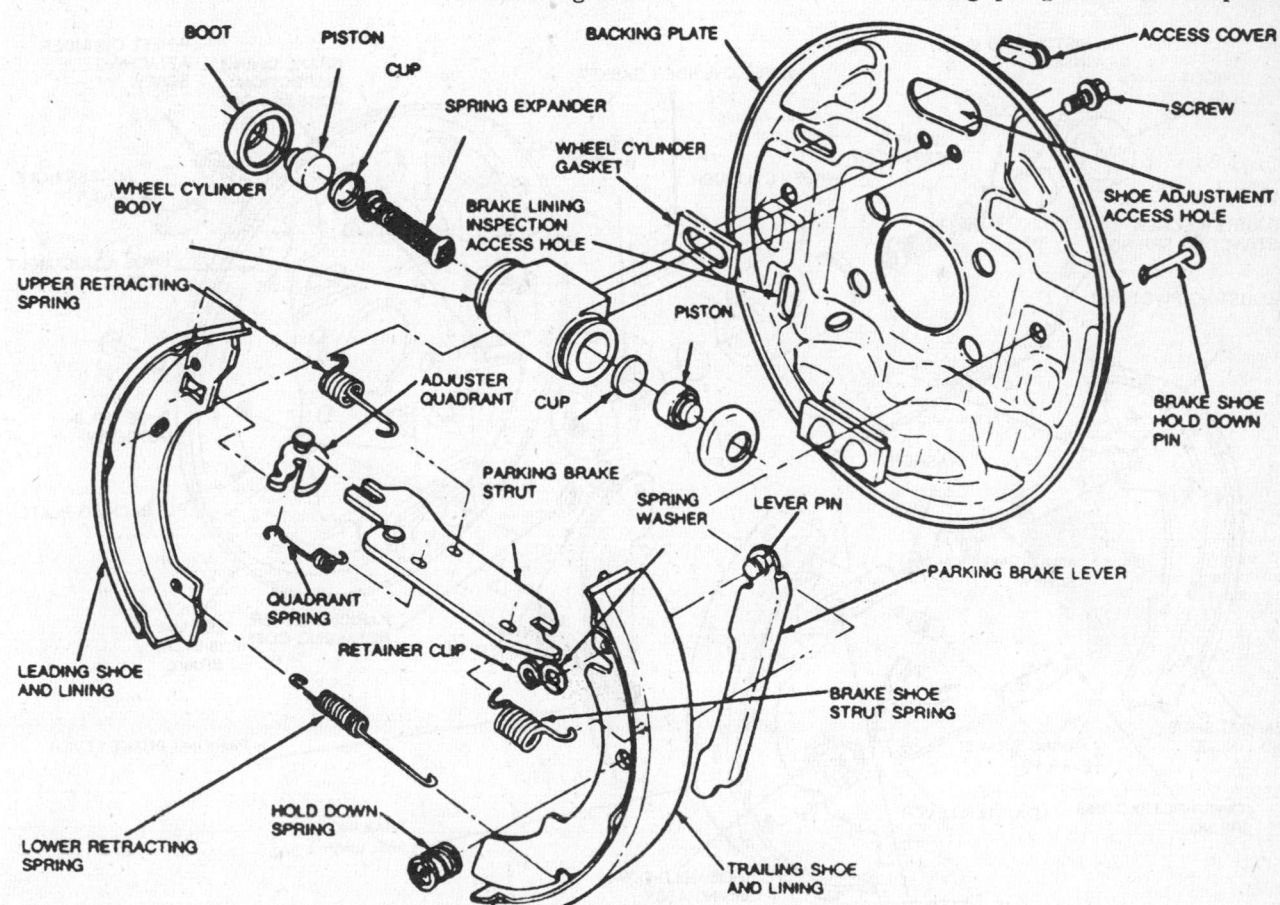

Rear brake assembly—1988–90 Escort with 7 inch brake shoes

pivoting strut into position to tension spring. Make sure the end of the spring, with the hook that is parallel to the center line of the coils, is installed in hole in shoe web. Installed spring should be flat against shoe web and parallel to strut.

16. Install lower shoe-to-shoe retracting spring between leading and trailing shoes. The spring hook with the longest straight section fits into hole in trailing shoe.

17. Install leading shoe-to-adjuster/strut retracting spring by installing spring to both parts and pivoting leading shoe over quadrant into position to tension spring.

18. Expand shoe and strut assembly to fit over anchor plate and wheel cylinder piston inserts.

19. Attach parking brake cable to parking brake lever.

20. Install hold-down pins and springs on each shoe and lining assembly.

21. Set brake shoe diameter using a suitable brake adjusting gauge.

22. Install hub/drum and wheel and tire assemblies.

23. Adjust wheel bearings.

24. Lower vehicle and check brake operation.

Tempo, Topaz and 1988–90 Escort with 8 Inch Brake Shoes

1. Raise and safely support the vehicle.

2. Remove the wheel, tire, and hub and drum assembly.

3. Remove 2 shoe hold-down springs and pins.

4. Lift the brake shoes, springs and adjuster assembly off backing plate and wheel cylinder assembly. Be careful not to bend adjusting lever during assembly removal.

5. Remove the parking brake cable from the parking brake lever.

6. Remove the retracting springs from the lower brake shoe attachments and upper shoe-to-adjusting lever attachment points. This will separate the brake shoes and disengage the adjuster mechanism.

7. Remove the horseshoe retaining clip and spring washer and slide the lever off the parking brake lever pin on the trailing shoe.

To install:

8. Apply a light coating of high temperature grease at the points where the brake shoes contact the backing plate.

9. Apply a light coating of lubricant to the adjuster screw threads and the socket end of the adjusting screw. Install the stainless steel washer over the socket end of the adjusting screw and install the socket. Turn the adjusting screw into the adjusting pivot nut to the limit of the threads and then back-off ½ turn.

10. Assemble the parking brake lever to the trailing shoe by installing the spring washer and a new horseshoe retaining clip. Crimp the clip until it retains the lever to the shoe securely.

11. Attach the parking brake cable to the parking brake lever.

12. Attach the lower shoe retracting spring to the leading and trailing shoe assemblies and install to backing plate. It will be necessary to stretch the retracting spring as the shoes are installed downward over the anchor plate to inside of shoe retaining plate.

13. Install the adjuster screw assembly between the leading shoe slot and the slot in the trailing shoe and parking brake lever. The adjuster socket end slot must fit into the trailing shoe and parking brake lever.

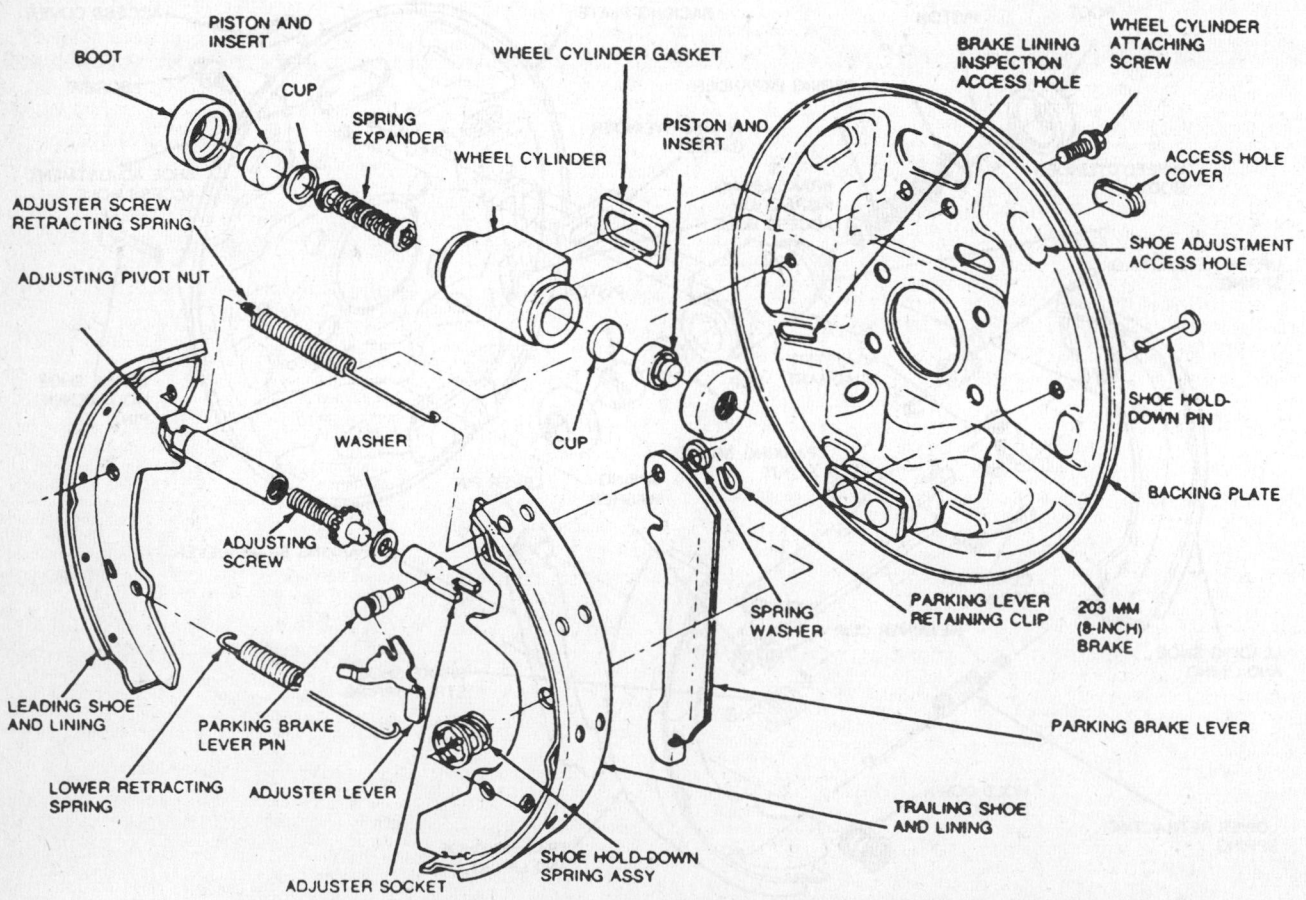

Rear brake assembly—Tempo, Topaz and 1988–90 Escort with 8 inch brake shoes

NOTE: The adjuster socket blade is marked R or L for the right or left brake assemblies. The R or L adjuster blade must be installed with the letter R or L in the upright position, facing the wheel cylinder, on the correct side to ensure that the deeper of the 2 slots in the adjuster sockets fits into the parking brake lever.

14. Assemble the adjuster lever in the groove located in the parking brake lever pin and into the slot of the adjuster socket that fits into the trailing shoe web.

15. Attach the upper retracting spring to the leading shoe slot. Using a suitable spring tool, stretch the other end of the spring into the notch on the adjuster lever. If the adjuster lever does not contact the star wheel after installing the spring, it is possible that the adjuster socket is installed incorrectly.

16. Set the brake shoe diameter using a suitable brake adjusting gauge.

17. Install the hub/drum and wheel/tire assemblies and adjust the wheel bearings.

18. Lower the vehicle and check brake operation.

1991–92 Escort

1. Raise and safely support the vehicle.

2. Remove the wheel and tire assembly and remove the brake drum.

3. Remove the 2 brake shoe return springs.

4. Remove the anti-rattle spring.

5. Push and turn the 2 brake shoe hold-down clips and remove the clips.

6. Remove the leading and trailing shoes from the backing plate.

To install:

7. Use a suitable high temperature grease to lubricate the brake shoe contact points on the backing plate.

8. Position the trailing brake shoe on the backing plate and install 1 of the brake shoe hold-down clips.

9. Position the leading brake shoe on the backing plate and install the other brake shoe hold-down clip.

10. Install the anti-rattle spring.

11. Install the 2 brake shoe return springs.

12. Press the brake pedal to verify operation of the automatic brake adjuster.

13. Install the brake drum and the wheel and tire assembly.

14. Lower the vehicle.

15. Check brake operation.

Wheel Cylinder

REMOVAL & INSTALLATION

Except 1991–92 Escort

1. Raise and safely support the vehicle. Remove wheel/tire and hub/drum assemblies.

2. Remove brake shoe assembly.

3. Disconnect brake tube from wheel cylinder.

4. Remove wheel cylinder attaching bolts and remove wheel cylinder.

NOTE: Use caution to prevent brake fluid from contacting brake linings and drum braking surface. Contaminated linings must be replaced.

To install:

5. Ensure ends of hydraulic fittings are free of foreign matter before making connections.

6. Position wheel cylinder and foam seal on backing plate and finger-tighten brake tube to cylinder.

7. Secure cylinder to backing plate by installing attaching bolts. Tighten bolts to 8–10 ft. lbs. (10–14 Nm).

8. Tighten tube nut fitting.

9. Install and adjust brakes.

10. Install hub/drum and wheel assembly.

11. Bleed brake system and lower the vehicle.

1991–92 Escort

1. Raise and safely support the vehicle.

2. Remove the wheel and tire assembly and remove the brake drum.

3. Remove the upper brake shoe return spring.

4. Clamp the wheel cylinder brake hose.

5. Using a suitable flare nut wrench, loosen the wheel cylinder-to-brake line flare nut.

6. Pull the clip from the brake hose retaining bracket and remove the brake hose from the retaining bracket.

7. Remove the brake line from the wheel cylinder.

8. Remove the 2 wheel cylinder mounting bolts and remove the wheel cylinder from the backing plate.

9. Remove and discard the wheel cylinder gasket.

To install:

10. Install a new wheel cylinder gasket onto the backing plate.

11. Position the wheel cylinder onto the backing plate and install the 2 mounting bolts. Tighten the bolts to 89–115 inch lbs. (10–13 Nm).

12. Position the brake line into the wheel cylinder fitting and tighten the wheel cylinder-to-brake line flare nut to 12–16 ft. lbs. (16–22 Nm).

13. Position the brake hose into the retaining bracket and install the clip. Remove the clamp from the wheel cylinder brake hose.

14. Install the brake shoe return spring.

15. Press the brake pedal to verify the operation of the automatic brake adjuster.

16. Install the brake drum and the wheel and tire assembly.

17. Bleed the brake system and lower the vehicle.

Parking Brake Cable

ADJUSTMENT

Except 1991–92 Escort

NOTE: The rear brake shoes should be properly adjusted before adjusting the parking brake.

1. With the engine running, apply approximately 100 lbs. pedal effort to the hydraulic service brake 3 times before adjusting the parking brake.

2. Block the front wheels and place the transaxle in **N**. Raise and safely support the rear of the vehicle just enough to rotate the wheels.

3. Place the parking brake control assembly in the 12th notch position, 2 notches from full application. Tighten the adjusting nut until approximately 1 in. (25mm) of threaded rod is exposed beyond the nut. Release the parking brake control and rotate the rear wheels by hand. There should be no brake drag.

4. If the brakes drag when the control assembly is fully released, or the handle travels too far on full apply, repeat the procedure and adjust the nut accordingly.

1991–92 Escort

1. Start the engine and place the transaxle in **R**.

2. With the vehicle moving in reverse, depress the brake pedal several times.

3. Stop the vehicle and place the transaxle in **P**. Stop the engine.

4. Remove the parking brake console as follows:

   a. Remove the rear seat ash tray.

   b. Position both front seats to the rear-most position.

   c. Remove the 2 front retaining screws from the parking brake console.

   d. Recline both front seats.

   e. Remove the 2 rear retaining screws and with the parking brake engaged, remove the parking brake console.

   f. Release the parking brake lever.

5. Turn the adjusting nut until the parking brake lever stroke is 5–7 notches when pulled with a force of 22 lbs.

6. Install the parking brake console by reversing the removal procedure.

REMOVAL & INSTALLATION

Except 1991–92 Escort

1. Place control assembly in seventh notch position and loosen adjusting nut. Completely release control assembly.

2. Raise vehicle. Remove rear parking brake cable from equalizer.

3. Remove hairpin clip holding cable to floor pan tunnel bracket.

4. Remove wire retainer holding cable to fuel tank mounting bracket. Remove cable from wire retainer. Remove cable and clip from the fuel pump bracket.

5. Remove screw holding cable retaining clip to rear sidemember. Remove cable from clip.

6. Remove the wheel and tire assembly and rear brake drum.

7. Disengage cable end from brake assembly parking brake lever. Depress cable prongs holding cable to backing plate. Remove cable through hole in backing plate.

To install:

8. Insert cable through hole in backing plate. Attach cable end to rear brake assembly parking brake lever.

9. Insert conduit end fitting into backing plate. Ensure retention prongs are locked into place.

10. Insert cable into rear attaching clip and attach clip to rear sidemember with screw.

11. Route cable through bracket in floorpan tunnel and install hairpin retaining clip.

12. Install cable end into equalizer.

13. Insert cable into wire retainer and snap retainer into hole in fuel tank mounting bracket. Insert cable and install clip into fuel pump bracket.

14. Install rear drum, wheel and tire assembly and wheel cover.

15. Lower vehicle.

16. Adjust parking brake.

1991–92 Escort

1. Remove the parking brake console as follows:

 a. Remove the rear seat ash tray.

 b. Position both front seats to the rear-most position.

 c. Remove the 2 front retaining screws from the parking brake console.

 d. Recline both front seats.

 e. Remove the 2 rear retaining screws and with the parking brake engaged, remove the parking brake console.

 f. Release the parking brake lever.

2. Remove the cable adjusting nut.

3. Raise and safely support the vehicle.

4. Remove the rear exhaust pipe and resonator heat shields.

5. Disconnect the equalizer return spring and remove the cables from the equalizer.

6. Remove the clip that attaches the cable to the retaining bracket located near the equalizer. Remove the cable from the bracket.

7. Remove the cable routing bracket bolt from the floorpan and remove the bracket.

8. Remove the 2 cable routing bracket nuts from the trailing link and remove the bracket.

9. Remove the 2 cable retaining bracket bolts from the backing plate and remove the bracket.

10. Remove the cable from the parking brake actuating lever.

To install:

11. Position the cable onto the parking brake actuating lever.

12. Position the parking brake cable bracket onto the backing plate and install the 2 bolts. Tighten the bolts to 14–19 ft. lbs. (19–25 Nm).

13. Position the cable routing bracket onto the trailing link and install the 2 nuts. Tighten the nuts to 12–17 ft. lbs. (16–23 Nm).

14. Position the cable routing bracket to the floor pan and install the mounting bolt. Tighten the bolt to 14–19 ft. lbs. (19–25 Nm).

15. Position the cable into the retaining bracket near the equalizer and install the clip.

16. Install the cables into the equalizer and install the cable return spring.

17. Install the rear exhaust pipe and resonator heat shields.

18. Lower the vehicle.

19. Install the adjusting nut.

20. Adjust the parking brake cable and install the parking brake console in the reverse order of removal.

Brake System Bleeding

1. Clean all the dirt from around the master cylinder filler cap.

2. Fill the reservoir with brake fluid. The reservoir must be at least ¾ full throughout the bleeding procedure.

3. If the master cylinder is known or suspected to have air in bore, it must be bled before any wheel cylinders or calipers.

4. To bleed the master cylinder, loosen 1 outlet fitting aproximately ¾ turn. Have an assistant push the brake pedal down slowly through full travel. Close the outlet fitting, then return the pedal slowly to the full released position. Wait 5 seconds, then repeat the operation until the air bubbles cease to appear.

5. Loosen the other outlet fitting approximately ¾ turn and repeat Step 4.

6. To continue to bleed the system, remove the rubber cap dust cap from the wheel cylinder bleeder fitting or caliper fitting. Check to make sure the bleeder fitting is positioned at the upper half on the front of the caliper, if not the caliper is located on the wrong side.

7. Attach a suitable length of rubber hose to the fitting. Submerge the free end of the hose in a container partially filled with clean brake fluid and loosen the bleeder fitting approximately ¾ of a turn.

8. Have the assistant push brake pedal down slowly through full travel. Close the bleeder fitting, then return the pedal to the full release position. Wait 5 seconds, then repeat this operation until the air bubbles cease to appear at the submerged end of the bleeder hose.

9. When the fluid is completely free of air bubbles, properly tighten the bleeder fitting and reinstall the rubber dust cap. Repeat this process on the opposite diagonal system. Refill the master cylinder reservoir after each wheel cylinder or caliper is bled and reinstall the master cylinder cap.

NOTE: If all wheels are to be bled, proceed as follows: right rear, left front, left rear and right front.

10. When the bleeding operation is completed, the fluid level should be filled to the maximum fill level indicated on the reservoir. Always ensure the disc brake pistons are returned to their normal positions by depressing the brake pedal several times until the normal pedal travel is established. Check the pedal feel. If the pedal feels spongy, repeat the bleeding procedure.

CHASSIS ELECTRICAL

Air Bag

A driver's side air bag can be installed as optional equipment on Tempo and Topaz vehicles.

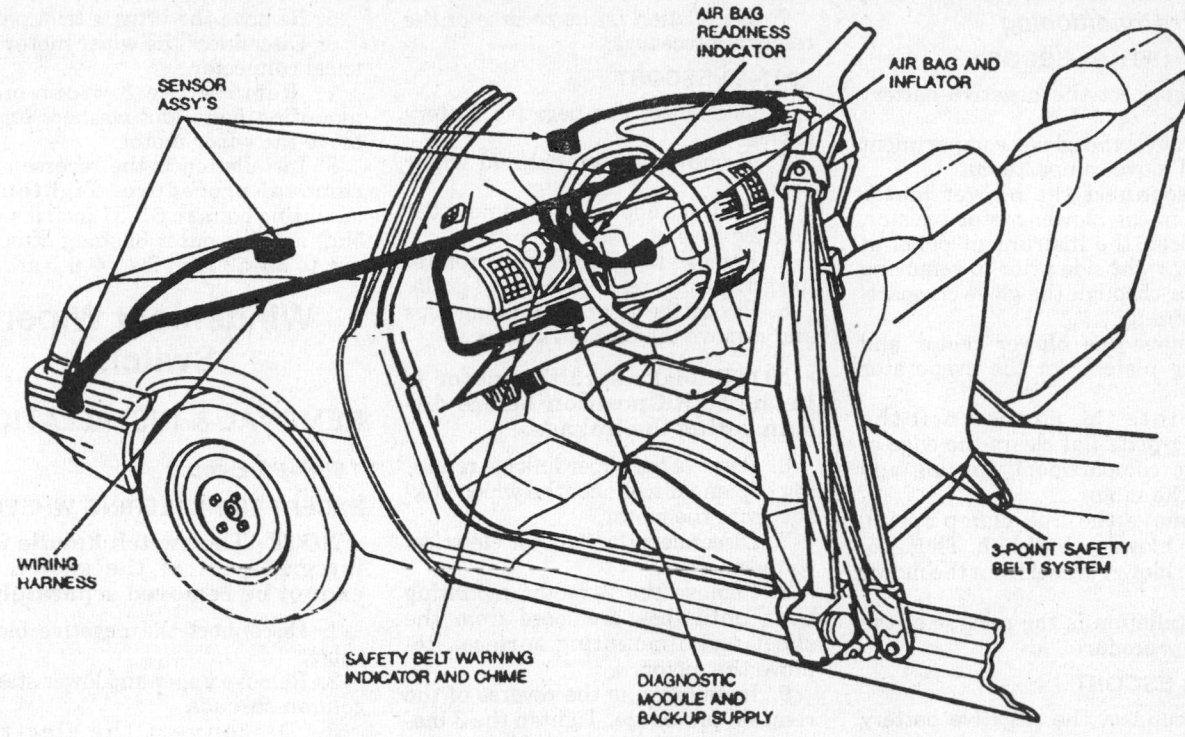

SENSOR ASSY'S

AIR BAG READINESS INDICATOR

AIR BAG AND INFLATOR

WIRING HARNESS

SAFETY BELT WARNING INDICATOR AND CHIME

DIAGNOSTIC MODULE AND BACK-UP SUPPLY

3-POINT SAFETY BELT SYSTEM

Air bag system component locations—Tempo and Topaz

DISARMING

1. Disconnect the negative battery cable.

NOTE: On most vehicles equipped with an air bag, a back-up power supply is included in the system to provide air bag deployment in the event the battery or cables are damaged in an accident before the sensors can close. The power supply is a capacitor that will leak down in approximately 15 minutes after the battery is disconnected or 1 minute if the battery positive cable is grounded. If the system is equipped with a backup power supply, it must be disconnected to disarm the system.

2. Remove the 2 screws retaining the steering column opening cover to the instrument panel and remove the cover.

3. Remove the 4 bolts retaining the bolster and remove the bolster.

4. Disconnect the backup power supply connector.

5. Remove the 4 nut and washer assemblies retaining the driver air bag to the steering wheel.

6. Disconnect the driver air bag connector.

7. Attach a jumper wire to the air bag terminals on the clockspring.

8. Connect the backup power supply and negative battery cable.

Heater Blower Motor

REMOVAL & INSTALLATION

Without Air Conditioning

1988–90 ESCORT

1. Disconnect the negative battery cable.

2. Remove the glove compartment door from the instrument panel.

3. Remove the 6 screws attaching the air inlet lower duct to the blower housing and upper inlet duct and remove the lower air inlet duct.

4. Remove the pushnut from the blower wheel hub.

5. Remove the blower motor flange attaching screws located inside the blower housing.

6. Pull the blower motor out from the blower housing and disconnect the blower motor wires from the motor.

7. Installation is the reverse of the removal procedure.

1991–92 ESCORT

1. Disconnect the negative battery cable.

2. Remove the trim panel below the glove compartment.

3. Remove the wiring bracket and bolt.

4. Disconnect the blower motor electrical connector.

5. Remove the 3 blower motor mounting bolts and remove the blower motor.

6. Remove the blower wheel retaining clip and remove the blower wheel from the blower motor.

7. Installation is the reverse of the removal procedure.

TEMPO AND TOPAZ

1. Disconnect the negative battery cable.

2. Remove the contents from the glove compartment and remove the glove compartment door.

3. Disconnect the blower motor wires from the blower motor resistor.

4. Loosen the instrument panel at the lower right side prior to removing the blower motor assembly through the glove compartment opening.

5. Remove the 4 screws retaining the blower motor and mounting plate to the heater case.

6. Rotate the motor until the mounting plate flats clear the edge of the glove compartment opening. Then, remove the motor and mounting plate from the vehicle.

7. Remove the blower motor and mounting plate seal from the mounting plate and make sure the mounting surface is clean.

8. Remove the pushnut from the blower wheel shaft and remove the blower wheel from the motor shaft.

9. Installation is the reverse of the removal procedure. Be sure to use a new mounting plate seal.

With Air Conditioning

EXCEPT 1991–92 ESCORT

1. Disconnect the negative battery cable.
2. Remove the glove compartment door and glove compartment.
3. Disconnect the blower motor wires from the blower motor resistor.
4. Loosen the instrument panel at the lower right side prior to removing the motor through the glove compartment opening.
5. Remove the blower motor and mounting plate from the evaporator case.
6. Rotate the motor until the mounting plate flat clears the edge of the glove compartment opening and remove the motor.
7. Remove the hub clamp spring from the blower wheel hub. Then, remove the blower wheel from the motor shaft.
8. Installation is the reverse of the removal procedure.

1991–92 ESCORT

1. Disconnect the negative battery cable.
2. Remove the trim panel below the glove compartment.
3. Remove the wiring bracket and bolt.
4. Disconnect the blower motor electrical connector.
5. Remove the 3 blower motor mounting bolts and remove the blower motor.
6. Remove the blower wheel retaining clip and remove the blower wheel from the blower motor.
7. Installation is the reverse of the removal procedure.

Windshield Wiper Motor

REMOVAL & INSTALLATION

Front

EXCEPT 1991–92 ESCORT

1. Disconnect the negative battery cable.
2. Lift the water shield cover from the cowl on the passenger side.
3. Disconnect the power lead from the motor.
4. Remove the linkage retaining clip from the operating arm on the motor by lifting locking tab up and pulling clip away from pin.
5. Remove the attaching screws from the motor and bracket assembly and remove.
6. Remove the operating arm from the motor. Unscrew the 3 bolts and separate the motor from the mounting bracket.

7. Installation is the reverse of the removal procedure.

1991–92 ESCORT

1. Disconnect the negative battery cable.
2. Remove the windshield wiper arms.
3. With the hood closed, remove the 7 screw covers.
4. Remove the 7 cowl grille retaining screws and remove the cowl grille.
5. Pry up the 4 baffle retaining clips and remove the baffle trim piece.

NOTE: Make sure the motor is in the PARK position before disconnecting the linkage.

6. Remove the wiper linkage retaining clip and disconnect the wiper linkage from the motor.
7. Disconnect the 2 motor electrical connectors.
8. Remove the 3 motor mounting bolts until they are loose from the sheet metal mounting surface. Remove the motor.
9. Installation is the reverse of the removal procedure. Tighten the 3 motor mounting bolts to 61–87 inch lbs. (7–9 Nm).

Rear

EXCEPT 1991–92 ESCORT

1. Disconnect the negative battery cable.
2. Remove wiper arm.
3. Remove pivot shaft attaching nut and spacers.
4. On Hatchback vehicles, remove liftgate inner trim panel. On the Station Wagon, remove the screws attaching the license plate housing. Disconnect license plate light wiring and remove housing.
5. Disconnect electrical connector to wiper motor.
6. On Hatchback vehicles, remove the 3 screws retaining the bracket to the door inner skin and remove complete motor, bracket and linkage assembly. On the Station Wagon, remove the motor and bracket assembly retaining screws and remove the motor and bracket assembly.
7. Installation is the reverse of the removal procedure.

1991–92 ESCORT

1. Disconnect the negative battery cable.
2. Remove the wiper arm by lifting the wiper arm attaching nut cover, removing the attaching nut and pulling the wiper arm from the pivot shaft.
3. Remove the shaft seal from the outer bushing attaching nut.
4. Remove the outer bushing attaching nut and remove the outer bushing.

5. Remove the liftgate trim panel.
6. Disconnect the wiper motor electrical connector.
7. Remove the 3 wiper motor mounting bolts and washers and remove the wiper motor.
8. Installation is the reverse of the removal procedure. Tighten the mounting bolts to 61–87 inch lbs. (7–9 Nm) and the outer bushing attaching nut to 35–52 inch lbs. (4–6 Nm).

Windshield Wiper Switch

REMOVAL & INSTALLATION

1988–90 Escort

EXCEPT TILT STEERING WHEEL

NOTE: The switch handle is an integral part of the switch and cannot be removed separately.

1. Disconnect the negative battery cable.
2. Remove upper and lower steering column shrouds.
3. Disconnect the electrical connector.

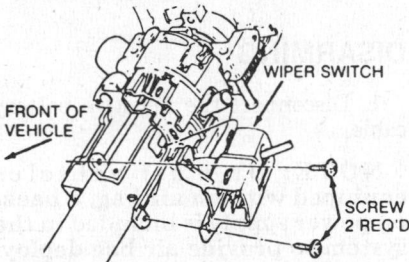

Wiper switch installation—1988–90 Escort without tilt steering wheel

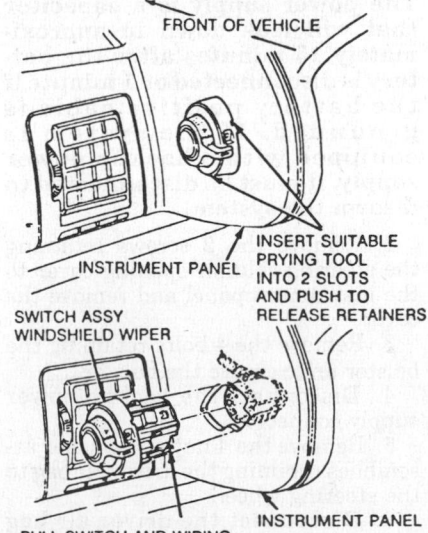

Wiper switch installation—Tempo and Topaz

4. Peel back the foam sight shield. Remove the 2 screws holding the switch and remove the wash/wiper switch.

5. Installation is the reverse of the removal procedure.

TILT STEERING WHEEL

1. Disconnect the negative battery cable.

2. Remove the steering column shroud.

3. Peel back the side shield and disconnect the switch wiring connector.

4. Remove the screw attaching the wiring retainer to the steering column.

5. Grasp the switch handle and pull straight out to disengage the wiper switch from the turn signal switch.

6. Installation is the reverse of the removal procedure.

Tempo and Topaz

1. Disconnect the negative battery cable.

2. Insert a suitable prying tool into the small slots on top and bottom of the switch bezel.

3. Push down on the tool to work the top of the switch away from the instrument panel.

4. Work the bottom portion of the switch from the panel and completely remove the switch from the panel opening. Hold the switch and pull the wiring at the rear of the switch until

the switch connector can be easily disconnected. Disconnect the connector and allow the wiring to hang from the switch mounting opening.

To install:

5. Connect the wiring connector to the new switch and route the wiring back into the mounting opening. Insert the switch into the opening so the graphics are properly aligned.

6. Push on the switch until the bezel seats against the instrument panel and the clips lock the switch into place.

7. Connect the negative battery cable.

Instrument Cluster

REMOVAL & INSTALLATION

Escort

1988–90

1. Disconnect the negative battery cable.

2. Remove the 2 retaining screws at the bottom of the steering column opening and snap the steering column cover out.

3. Remove the 10 cluster opening finish panel retainer screws and remove the finish panel.

4. Remove the 2 upper and lower screws retaining the cluster to the instrument panel.

5. Reach under the instrument panel and disconnect the speedometer cable by pressing down on the flat surface of the plastic connector (quick connect).

6. Pull the cluster away from the instrument panel. Disconnect the cluster feed plug from its receptacle in the printed circuit.

7. Installation is the reverse of the removal procedure.

1991–92

1. Disconnect the negative battery cable.

2. If equipped with a standard column, remove the 4 bolts securing the steering column to the instrument panel frame.

3. Lower the steering column.

4. Remove the cap screws securing the instrument cluster bezel to the instrument panel and remove the instrument cluster bezel.

5. Disconnect the speedometer cable at the transaxle by pulling the cable out of the vehicle speed sensor.

6. Remove the screws and bolts securing the instrument cluster to the instrument panel.

7. Pull the instrument cluster out slightly and disconnect the electrical connectors from the rear of the instrument cluster.

8. Disconnect the speedometer cable from the instrument cluster and remove the cluster from the instrument panel.

9. Installation is the reverse of the removal procedure. Make sure the instrument cluster is held in its forward most position while attaching the 2 upper screws.

Tempo and Topaz

1. Disconnect the negative battery cable.

2. Remove 2 retaining screws at the bottom of the steering column opening and snap the steering column cover out.

3. Remove the steering column trim shroud.

4. Remove the snap-in lower cluster finish panels.

5. Remove 4 cluster opening finish panel retaining screws and pull the panel rearward.

5. Disconnect the speedometer cable at the transaxle.

6. Remove the 4 screws retaining the instrument cluster and carefully pull rearward enough to disengage the speedometer cable.

7. Carefully pull the cluster away from the instrument panel. Disconnect the cluster feed plugs from the printed circuit.

8. Installation is the reverse of the removal procedure.

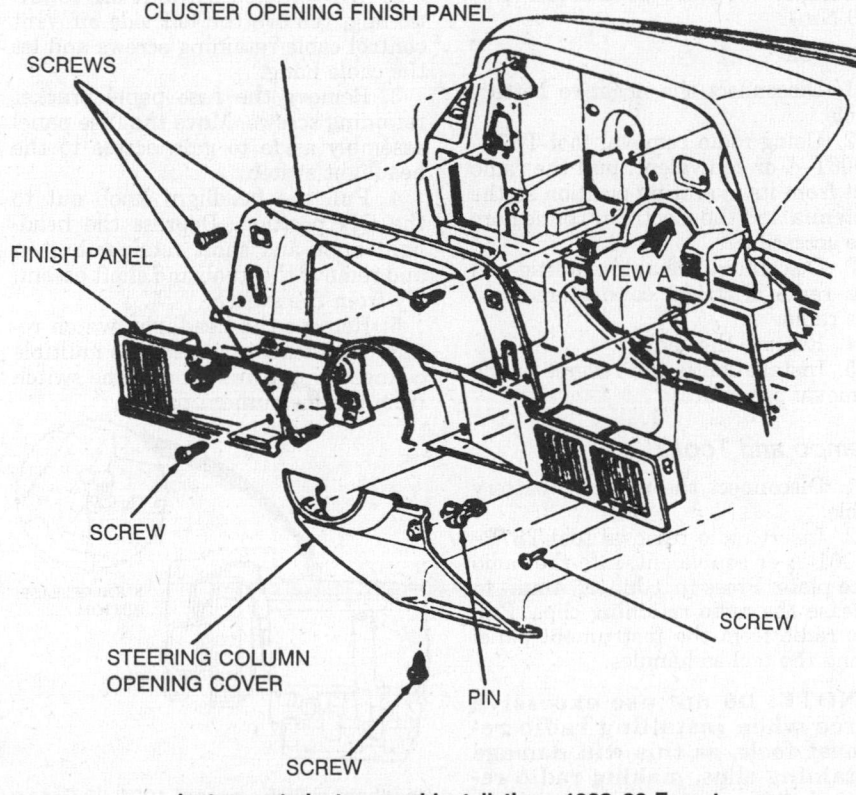

CLUSTER OPENING FINISH PANEL

SCREWS

FINISH PANEL

VIEW A

SCREW

STEERING COLUMN OPENING COVER

PIN

SCREW

SCREW

Instrument cluster panel Installation—1988–90 Escort

NOTE: If gauges are being removed from the cluster assembly, do not remove the gauge pointer because the magnetic gauges cannot be recalibrated.

Speedometer

REMOVAL & INSTALLATION

Except 1991–92 Escort

1. Disconnect the negative battery cable.
2. Remove the instrument cluster.
3. Remove the 7 screws that retain the lens and mask to the backplate.
4. Remove the 2 nuts retaining the fuel gauge assembly to the backplate. Remove the fuel gauge assembly and then remove the speedometer assembly.

To install:

5. Apply a $^3/_{16}$ in. ball of speedometer cable lubricant part D2AZ–19581–A or equivalent, in the drive hole of the speedometer head. Install speedometer head assembly into cluster.

NOTE: The speedometer is calibrated at the time of manufacture. Excessive rough handling of the speedometer may disturb the calibration.

6. Install the retaining screws to retain the lens and mask to the backplate.
7. Install the instrument cluster.
8. Connect the negative battery cable and check the operation of the speedometer.

1991–92 Escort

1. Disconnect the negative battery cable.
2. Remove the instrument cluster.
3. Remove the instrument cluster lens and shroud.
4. Remove the speedometer from the instrument cluster.
5. Installation is the reverse of the removal procedure. Be careful when handling the speedometer so as not to disturb the factory calibration.

Radio

REMOVAL & INSTALLATION

ESCORT

1988–90

1. Disconnect the negative battery cable.
2. Remove the center instrument trim panel.

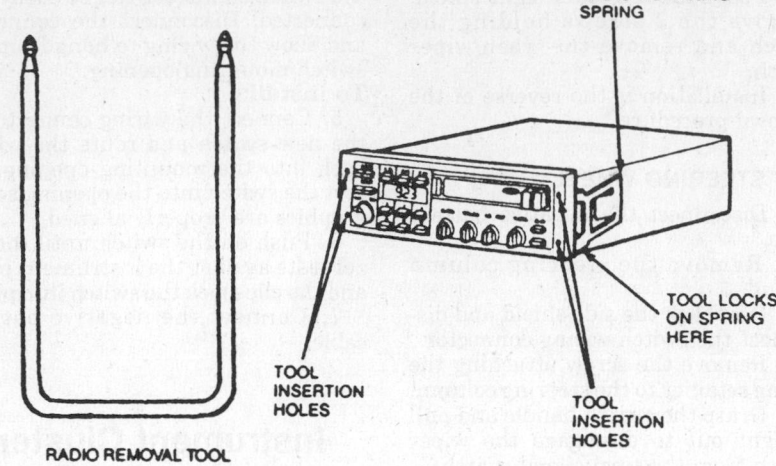

Radio removal procedure—Tempo, Topaz and 1991–92 Escort

3. Remove the 4 screws retaining the radio and mounting bracket to the instrument panel.
4. Pull the radio to the front and raise the back end of the radio slightly so the rear support bracket clears the clip in the instrument panel. Pull the radio out of the instrument panel slowly.
5. Disconnect the wiring connectors and antenna cable.
6. Transfer the mounting brackets to the new radio, if necessary.
7. Installation is the reverse of the removal procedure. Tighten the retaining screws to 14–16 inch lbs. (1.5–1.9 Nm).

1991–92

1. Disconnect the negative battery cable.
2. Using radio removal tool T87P–19061–A or equivalent, pull the radio out from its mounting position so the antenna and the electrical connectors are accessible.
3. Disconnect the antenna lead and the radio electrical connectors from the radio.
4. Remove the radio.
5. Installation is the reverse of the removal procedure.

Tempo and Topaz

1. Disconnect the negative battery cable.
2. Insert radio removal tool T87P–19061–A or equivalent, into the radio face plate. Press in 1 in. (25.4mm) to release the radio retaining clips. Pull the radio from the instrument panel using the tool as handles.

NOTE: Do not use excessive force when installing radio removal tools, as this will damage retaining clips, making radio removal difficult.

3. Disconnect the wiring connectors and antenna cable.
4. Transfer the rear mounting bracket to the new radio, if necessary.
5. Installation is the reverse of the removal procedure.

Headlight Switch

REMOVAL & INSTALLATION

Except 1991–92 Escort

1. Disconnect the negative battery cable.
2. On vehicles without air conditioning, remove the left side air vent control cable retaining screws and let the cable hang.
3. Remove the fuse panel bracket retaining screws. Move the fuse panel assembly aside to gain access to the headlight switch.
4. Pull the headlight knob out to the **ON** position. Depress the headlight knob and shaft retainer button and remove the knob and shaft assembly from the switch.
5. Remove the headlight switch retaining bezel. Disconnect the multiple connector plug and remove the switch from the instrument panel.

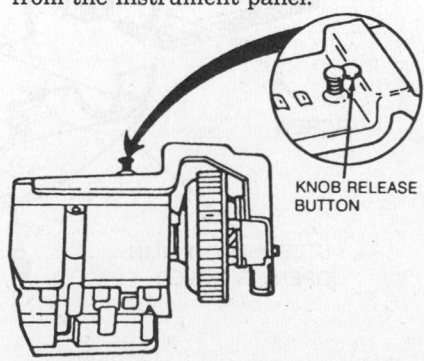

Headlight switch—except 1991–92 Escort

To install:

6. Install the headlight switch into the instrument panel. Connect the multiple connector and install the headlight switch retaining bezel.

7. Install the knob and shaft assembly by inserting the shaft into the switch and gently pushing until the shaft locks in position.

8. Move the fuse panel back into position and install the fuse panel bracket with the 2 retaining screws.

9. On vehicles without air conditioning, install the left side air vent control cable and bracket.

10. Connect the negative battery cable.

Combination Switch

On Tempo, Topaz and 1988–90 Escort, the combination switch assembly is a multi-function switch comprising turn signal, hazard, headlight dimmer and flash-to-pass functions. The switch lever on the left side of the upper steering column controls the turn signal, headlight dimmer and flash-to-pass functions. The hazard function is controlled by the actuating knob on the bottom part of the steering column.

On 1991–92 Escort, the combination switch assembly is a multi-function switch that controls the headlights, parking lights and tail lights, the turn signals, headlight dimmer and window wipers.

REMOVAL & INSTALLATION

Except 1991–92 Escort

1. Disconnect the negative battery cable.

2. Remove the 5 column shroud screws and remove the lower column shroud.

3. Loosen the 4 steering column attaching nuts enough to allow the removal of the upper trim shroud.

4. Remove the upper shroud.

5. Remove the turn signal switch lever by pulling the lever straight out from the switch. To make removal easier, work the outer end of the lever around with a slight rotary movement before pulling it out.

6. Peel back the foam sight shield from the turn signal switch.

7. Disconnect the turn signal switch electrical connectors.

8. Remove the 2 self-tapping screws that attach the turn signal switch to the lock cylinder housing and disengage the switch from the housing.

To install:

9. Align the turn signal switch mounting holes with the corresponding holes in the lock cylinder housing

and install 2 self-tapping screws until tight.

10. Apply the foam sight shield to the turn signal switch.

11. Install the turn signal switch lever into the switch by aligning the key on the lever with the keyway in the switch and pushing the lever toward the switch to full engagement.

12. Install the turn signal switch electrical connectors to full engagement.

13. Install the upper steering column trim shrouds.

14. Torque the steering column attaching nuts to 15–22 ft. lbs. (20–30 Nm).

15. Connect the negative battery cable.

16. Check the steering column and switch for proper operation.

1991–92 Escort

1. Disconnect the negative battery cable.

2. Remove the steering wheel cover retaining screws from the back side of the steering wheel and remove the cover.

3. Disconnect the horn electrical connector and the speed control electrical connectors, if equipped.

4. Remove the steering wheel mounting nut.

NOTE: Do not attempt to remove the steering wheel by hitting the column shaft with a hammer. The column may collapse.

5. Remove the steering wheel using a suitable puller.

6. Remove the 4 retaining screws from the steering column lower cover and remove the cover. Remove the upper cover.

7. Disconnect the 3 multi-function switch electrical connectors.

8. Remove the multi-function switch retaining screw, pull the electrical connectors from the retaining brackets and remove the switch.

9. Installation is the reverse of the removal procedure. Tighten the steering wheel mounting nut to 29–36 ft. lbs. (39–49 Nm).

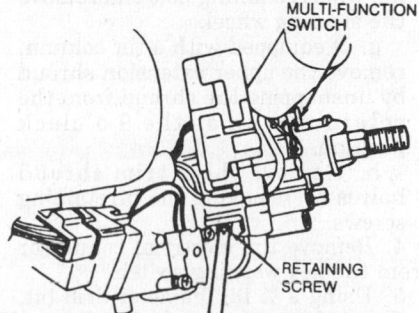

Combination switch removal—1991–92 Escort

Ignition Lock

REMOVAL & INSTALLATION

Except 1991–92 Escort
FUNCTIONAL LOCK

The following procedure pertains to vehicles that have functional lock cylinders. Lock cylinder keys are available for these vehicles or the lock cylinder key numbers are known and the proper key can be made.

1. Disconnect the negative battery cable.

2. If equipped with a tilt steering column, remove the upper extension shroud by unsnapping the shroud from the retaining clip at the 9 o'clock position.

3. Remove the steering column lower shroud on Escort. On Tempo and Topaz, remove the 5 attaching screws and the 2 trim shroud halves.

4. Disconnect the warning buzzer electrical connector. With the lock cylinder key, rotate the cylinder to the **RUN** position.

5. Take a ⅛ in. diameter pin or small wire punch and push on the cylinder retaining pin. The pin is visible through a hole in the mounting surrounding the key cylinder. Push on the pin and withdraw the lock cylinder from the housing.

To install:

6. Install the lock cylinder by turning it to the **RUN** position and depressing the retaining pin. Insert the lock cylinder into the housing. Be sure the lock cylinder is fully seated and aligned in the interlocking washer before turning the key to the **OFF** position. This action will permit the cylinder retaining pin to extend into the cylinder housing hole.

7. Rotate the lock cylinder, using the lock cylinder key, to ensure correct mechanical operation in all positions.

8. Install the electrical connector for the key warning buzzer.

9. Install the lower steering column shroud or trim shroud halves.

10. Connect the negative battery cable.

11. Check for proper start in **P** or **N**. Also, make certain the start circuit cannot be actuated in the **D** and **R** positions and that the column is locked in the **LOCK** position.

NON-FUNCTIONAL LOCK

The following procedure applies to vehicles in which the ignition lock is inoperative and the lock cylinder cannot be rotated due to a lost or broken lock cylinder key, the key number is not known or the lock cylinder cap is damaged and/or broken to the extent the lock cylinder cannot be rotated.

1. Make sure the wheels are in the straight ahead position. Disconnect the negative battery cable.

NOTE: On most vehicles equipped with an air bag, a backup power supply is included in the system to provide air bag deployment in the event the battery or cables are damaged in an accident before the sensors can close. The power supply is a capacitor that will leak down in approximately 15 minutes after the battery is disconnected or 1 minute if the battery positive cable is grounded. If the system is equipped with a backup power supply, it must be disconnected to disarm the system.

2. If equipped with an air bag, perform the following procedure:

a. Remove the 2 screws retaining the steering column opening cover to the instrument panel and remove the cover.

b. Remove the 4 bolts retaining the bolster and remove the bolster.

c. Disconnect the backup power supply connector.

d. Remove the 4 nut and washer assemblies retaining the driver air bag to the steering wheel.

e. Disconnect the driver air bag electrical connector from the contact assembly connectors and remove the air bag assembly.

CAUTION

When carrying a live air bag, make sure the bag and trim cover are pointed away from the body. In the unlikely event of an accidental deployment, the bag will then deploy with minimal chance of injury. In addition, when placing a live air bag on a bench or other surface, always face the bag and trim cover up, away from the surface. This will reduce the motion of the module if it is accidentally deployed.

---

f. Remove the steering wheel retaining bolt and remove the vibration damper, then reinstall the bolt loosely on the shaft.

g. Loosen the steering wheel on the shaft using a suitable puller.

NOTE: Do not use a knock-off type steering wheel puller or strike the retaining bolt with a hammer. This could cause damage to the steering shaft bearings.

h. Remove and discard the steering wheel retaining bolt and remove the steering wheel.

i. Remove the upper and lower shrouds.

j. Disconnect the air bag clockspring connector from the column harness.

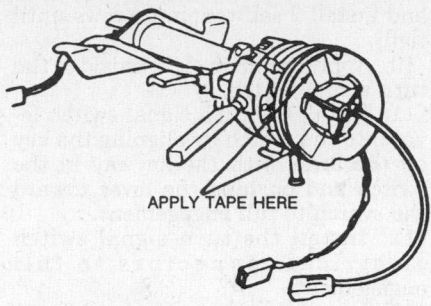

Air bag clockspring taping locations

NOTE: Before removing the air bag clockspring from the steering shaft, the clockspring must be taped to prevent the clockspring rotor from being turned accidentally and damaging the clockspring.

k. Remove the 2 screws that secure the clockspring to the retainer plate and remove the clockspring.

3. If not equipped with an air bag, perform the following procedure:

a. Remove the horn pad cover by removing 2 or 4 screws from the back of the steering wheel assembly.

NOTE: The emblem assembly is removed after the horn pad cover is removed, by pushing out from the backside of the emblem.

b. Remove the energy absorbing foam from the wheel assembly, if equipped. Remember to reinstall when the steering wheel is reassembled.

c. Disconnect the horn pad wiring connector.

d. Loosen the steering wheel retaining bolt 4–6 turns. Do not remove the bolt.

e. Loosen the steering wheel on the shaft using a suitable puller.

NOTE: Do not use a knock-off type steering wheel puller or strike the retaining bolt with a hammer. This could cause damage to the steering shaft bearings.

f. Remove and discard the steering wheel retaining bolt and remove the steering wheel.

g. If equipped with a tilt column, remove the upper extension shroud by unsnapping the shroud from the retaining clip at the 9 o'clock position.

h. Remove the 2 trim shroud halves by removing the 5 retaining screws.

4. Remove the electrical connector from the key warning switch.

5. Using a ⅛ in. diameter drill bit, drill out the retaining pin, being careful not to drill deeper than ½ in. (12.7mm).

6. Place a suitable chisel at the base of the ignition lock cylinder cap and, using a suitable hammer, strike the chisel with sharp blows to break the cap away from the lock cylinder.

7. Using a ⅜ in. diameter drill bit, drill down the middle of the ignition lock key slot approximately 1¾ in. (44mm) until the lock cylinder breaks loose from the breakaway base of the lock cylinder. Remove the lock cylinder and drill shavings from the lock cylinder housing.

8. Remove the retainer, washer and steering column lock gear. Thoroughly clean all drill shavings and other foreign materials from the casting.

9. Carefully inspect the lock cylinder housing for damage from the foregoing operation. If any damage is evident, the housing must be replaced.

To install:

10. Install the ignition lock drive gear, washer and retainer.

11. Install the ignition lock cylinder and check for smooth operation.

12. Connect the electrical connector to the key warning switch.

13. If equipped with an air bag, install the clockspring, steering wheel and air bag module as follows:

a. Place the clockspring onto the steering shaft. Install the 2 retaining screws that secure the clockspring to the retainer plate. Make sure the ground wire is secured with the lower retaining screw. Remove the tape that was installed during the removal procedure.

b. Connect the clockspring wire to the column harness.

c. Install the upper and lower shrouds.

d. Install the steering wheel on the steering column, making sure the alignment marks are correct. Install the vibration damper and a new retaining bolt. Tighten the bolt to 23–33 ft. lbs. (31–45 Nm).

e. Connect the air bag module wire to the clockspring connector and place the air bag module on the steering wheel. Install the 4 retaining nuts and tighten to 35–53 inch lbs. (4–6 Nm).

f. Connect the backup power supply connector and install the bolster and steering column opening cover.

g. Connect the negative battery cable and verify the air bag indicator.

14. If not equipped with an air bag, complete the installation as follows:

a. Install the trim shroud halves.

b. Install the steering wheel assembly on the steering column making sure the alignment marks are correct. Install a new retaining bolt and tighten to 23–33 ft. lbs. (31–45 Nm).

c. Connect the horn pad wiring connector. If equipped, install the energy absorbing foam.

d. Install the horn pad cover and 2 or 4 retaining screws. Make sure the wires are not pinched. Tighten the screws to 8–10 inch lbs. (0.9–1.1 Nm).

e. Connect the negative battery cable.

1991–92 Escort

1. Disconnect the negative battery cable.
2. Remove the steering wheel cover retaining screws from the back side of the steering wheel and remove the cover.
3. Disconnect the horn electrical connector and the speed control electrical connectors, if equipped.
4. Remove the steering wheel mounting nut.

NOTE: Do not attempt to remove the steering wheel by hitting the column shaft with a hammer. The column may collapse.

5. Remove the steering wheel using a suitable puller.
6. Remove the combination switch.
7. Disconnect the ignition switch electrical connector.
8. Remove the shift-lock cable mounting bracket bolt and position the bracket and cable aside.
9. Remove the 4 steering column upper mounting bracket bolts and lower the column.
10. Using a suitable hammer and chisel, make a groove in the head of each of the 2 column lock mounting bracket bolts.
11. Remove the bolts with a suitable screwdriver and discard the bolts.
12. Remove the steering column lock and mounting bracket.
To install:
13. Position the steering column lock and mounting bracket and install 2 new bolts, tightening them only enough to hold the column lock in position.
14. With the key in the ignition, verify the operation of the column lock. If necessary, reposition the column lock until it operates properly.
15. Tighten the mounting bracket bolts until the bolt heads break off.
16. Position the steering column and install the 4 upper mounting bracket bolts. Tighten the bolts to 80–123 inch lbs. (9–14 Nm).
17. If equipped with a tilt steering wheel, remove the upper mounting bracket retaining pin.
18. Position the shift-lock cable mounting bracket and install the bolt. Tighten to 37–55 inch lbs. (4–6 Nm).

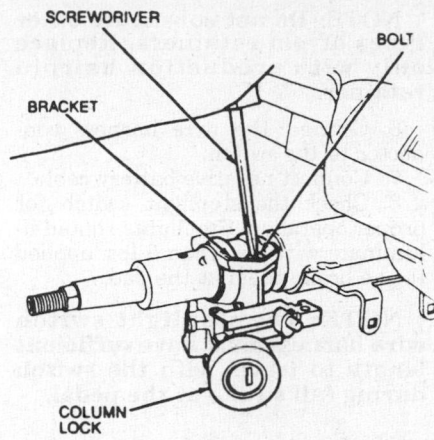

Ignition lock removal—1991–92 Escort

19. Connect the ignition switch electrical connector.
20. Install the combination switch.
21. Install the steering wheel and the mounting nut. Tighten to 29–36 ft. lbs. (39–49 Nm).
22. Connect the horn electrical connector and the speed control electrical connectors, if equipped.
23. Position the steering wheel cover and install the retaining screws.
24. Connect the negative battery cable.

Ignition Switch

REMOVAL & INSTALLATION

Except 1991–92 Escort

1. Disconnect the negative battery cable.
2. Remove the steering column upper and lower trim shroud by removing the self-tapping screws. The steering column attaching nuts may have to be loosened enough to allow removal of the upper shroud.
3. Remove 2 bolts and nuts holding the steering column assembly to the steering column bracket assembly and lower the steering column to the seat.
4. Remove the steering column shrouds.
5. Disconnect the electrical connector from the ignition switch.
6. Rotate ignition lock cylinder to the **RUN** position.
7. Remove 2 screws attaching the switch to the lock cylinder housing.
8. Disengage the ignition switch from the actuator pin.
To install:
9. Check to see that the actuator pin slot in the ignition switch is in the **RUN** position.

NOTE: A new switch assembly will be pre-set in the RUN position.

10. Make certain the ignition key lock cylinder is in approximately the **RUN** position to properly locate the lock actuator pin. The **RUN** position is achieved by rotating the key lock cylinder approximately 90 degrees from the **LOCK** position.
11. Install the ignition switch onto the actuator pin. It may be necessary to move the switch slightly back and fourth to align the switch mounting holes with the column lock housing threaded holes.
12. Install the new screws and tighten to 50–70 inch lbs. (5.6–7.9 Nm).
13. Connect the electrical connector to ignition switch.
14. Connect the negative battery cable.
15. Check the ignition switch for proper function including **START** and **ACC** positions. Also make certain the steering column is locked when in the **LOCK** position.
16. Position the top half of the shroud on the steering column.
17. Install the 2 bolts and nuts attaching the steering column assembly to the steering column bracket assembly. Tighten to 15–25 ft. lbs. (20–34 Nm).
18. Position lower shroud to upper shroud and install 5 self-tapping screws.

1991–92 Escort

1. Disconnect the negative battery cable.
2. Remove the combination switch.
3. Disconnect the ignition switch electrical connector.
4. Remove the 3 ignition switch mounting screws and remove the ignition switch.
5. Installation is the reverse of the removal procedure. Check the switch for proper operation.

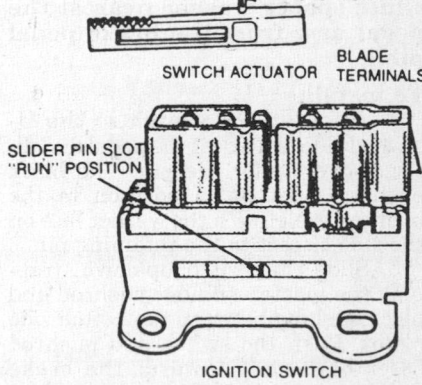

Ignition switch actuator position— except 1991–92 Escort

Stoplight Switch

ADJUSTMENT

1991–92 Escort

1. Measure the distance from the center of the brake pedal pad to the floor. The distance should be 7.60–7.72 in. (193–196mm). If not, proceed to Step 2.
2. Disconnect the stoplight switch electrical connector.
3. Loosen the stoplight locknut and turn the stoplight switch until it does not contact the brake pedal.
4. Loosen the rod locknut and turn the rod until the brake pedal height is within specification.
5. Turn the stoplight switch until it contacts the brake pedal, then turn it an additional ½ turn.
6. Tighten the stoplight locknut and the rod locknut.
7. Connect the stoplight switch electrical connector.
8. Check the operation of the stoplights.

REMOVAL & INSTALLATION

Except 1991–92 Escort

1. Disconnect the negative battery cable.
2. Disconnect the wire harness at the connector from the switch.

NOTE: The locking tab must be lifted before the connector can be removed.

3. Remove the hairpin retainer and white nylon washer. Slide the stoplight switch and the pushrod away from the pedal. Remove the switch by sliding the switch up/down.

NOTE: Since the switch side plate nearest the brake pedal is slotted, it is not necessary to remove the brake master cylinder pushrod black bushing and 1 white spacer washer nearest the pedal arm from the brake pedal pin.

To install:

4. Position the switch so the U-shaped side is nearest the pedal and directly over/under the pin. The black bushing must be in position in the pushrod eyelet with the washer face on the side closest to the retaining pin.
5. Slide the switch up/down, trapping the master cylinder pushrod and black bushing between the switch side plates. Push the switch and pushrod assembly firmly towards the brake pedal arm. Assemble the outside white plastic washer to the pin and install the hairpin retainer to trap the whole assembly.

NOTE: Do not substitute other types of pin retainers. Replace only with production hairpin retainer.

6. Connect the wire harness connector to the switch.
7. Connect negative battery cable.
8. Check the stoplight switch for proper operation. Stoplights should illuminate with less than 6 lbs. applied to the brake pedal at the pad.

NOTE: The stoplight switch wire harness must have sufficient length to travel with the switch during full stroke at the pedal.

1991–92 Escort

1. Disconnect the negative battery cable.
2. Disconnect the stoplight switch electrical connector.
3. Remove the stoplight switch retaining nuts and remove the stoplight switch.
4. Installation is the reverse of the removal procedure. Adjust the switch according to the proper procedure.

Clutch Switch

ADJUSTMENT

Except 1991–92 Escort

1. Remove panel above clutch pedal on Tempo and Topaz vehicles.
2. Disengage the wiring connector by flexing the retaining tab on the switch and withdrawing the connector.
3. Using a test light, check to see that the switch is open with the clutch pedal up (engaged) and closed at ap-

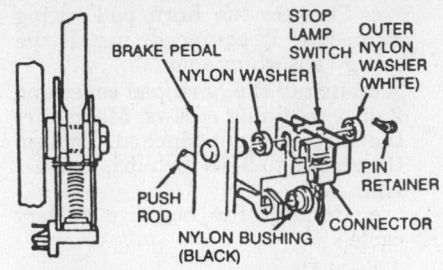

Stoplight switch installation—except 1991–92 Escort

proximately 1 in. (25.4mm) from the clutch pedal full down position (disengaged).
4. If the switch does not operate as outlined in Step 3, check to see if the self-adjusting clip is out of position on the rod. It should be near the end of the rod.
5. If the self-adjusting clip is out of position, remove and reposition the clip approximately 1 in. (25.4mm) from the end of the rod.
6. Reset the switch by pressing the clutch pedal to the floor. Repeat Step 3. If the switch is damaged or the clips do not remain in place replace the switch.

1991–92 Escort

1. Disconnect the negative battery cable.
2. Disconnect the clutch engage electrical connector.
3. Using a suitable ohmmeter, check the resistance between the connector terminals.
4. The ohmmeter should show continuity when the switch rod is pushed into the switch. The ohmmeter should show no continuity with the switch rod released.

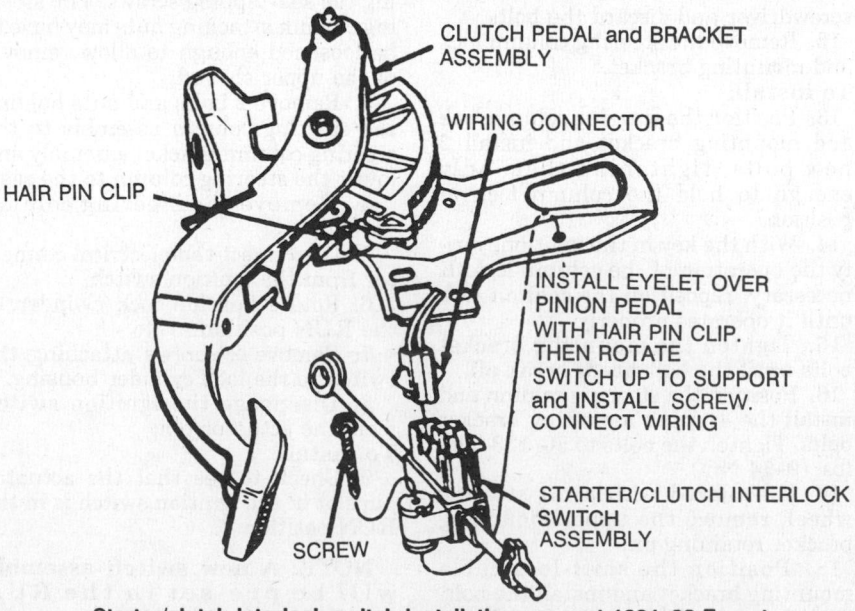

Starter/clutch interlock switch installation—except 1991–92 Escort

5. Replace the switch if it does not perform as specified.

REMOVAL & INSTALLATION

Except 1991–92 Escort

1. Disconnect the negative battery cable.
2. Remove panel above clutch pedal on Tempo and Topaz vehicles.
3. Disconnect the switch wiring connector.
4. Remove clutch interlock attaching screw and hairpin clip and remove the switch.
To install:

NOTE: **Always install the switch with the self-adjusting clip about 1 in. from the end of the rod. The clutch pedal must be fully up (clutch engaged). Otherwise, the switch may be misadjusted.**

5. Insert the eyelet end of the rod over the clutch pedal pin and secure it with the hairpin clip.
6. Swing the switch around to align the hole in the mounting boss with the corresponding hole in the bracket. Attach with the screw.
7. Reset the clutch interlock switch by pressing the clutch pedal to the floor.
8. Connect the wiring connector.
9. Install the panel above the clutch on Tempo and Topaz.
10. Connect the negative battery cable.

1991–92 Escort

1. Disconnect the negative battery cable.
2. Disconnect the electrical connector.
3. Remove the 2 retaining nuts.
4. Remove the clutch engage switch.
5. Installation is the reverse of the removal procedure.

Neutral Safety Switch

ADJUSTMENT

Except 1991–92 Escort

The mounting location of the neutral safety switch does not provide for adjustment of the switch position when installed. If the engine will not start in **P** or **N** or if it will start in **R** or any of the **D** ranges, check the control linkage adjustment and/or replace with a known good switch.

1991–92 Escort

The neutral safety switch function is performed by the Manual Lever Position Switch (MLPS). The MLPS is an adjustable switch that informs the automatic transaxle control unit of the position of the transaxle manual shaft. The MLPS will allow the vehicle to be started with the gear selector in the **P** or **N** positions when properly adjusted. The MLPS is located externally on the transaxle housing and is positioned on the manual shaft.

1. Remove the air cleaner assembly and air inlet tube.
2. Remove the nut securing the manual shaft lever to the transaxle manual shaft.
3. Remove the lever from the manual shaft.
4. Turn the manual shaft to the neutral position.
5. Loosen the MLPS mounting bolts.
6. Align the hole of the MLPS with the hole on the manual shaft lever by inserting a 0.079 in. (2.0mm) outside diameter pin.
7. Tighten the MLPS mounting bolts to 69–95 inch lbs. (8–11 Nm). Remove the pin.
8. Check the continuity of the switch as follows:
   a. Disconnect the switch connector.
   b. Using a suitable ohmmeter, check the switch for continuity at the connector.
   c. There should be continuity between the **BK/BL** and **BK/R** terminals with the shift lever in the **P** or **N** position.
9. If the continuity is incorrect, replace the MLPS.
10. Position the manual shaft lever to the manual shaft and install the nut. Tighten to 33–47 ft. lbs. (44–64 Nm).
11. Install the air cleaner assembly and air inlet tube.

REMOVAL & INSTALLATION

Except 1991–92 Escort

1. Set the parking brake.
2. Disconnect the negative battery cable.
3. Disconnect the wire connector from the neutral safety switch.
4. Remove the 2 retaining screws from the neutral start switch and remove the switch.

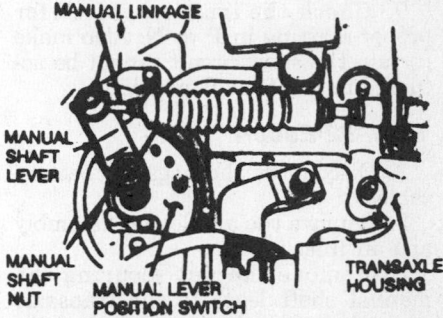

Manual lever position switch location— 1991–92 Escort

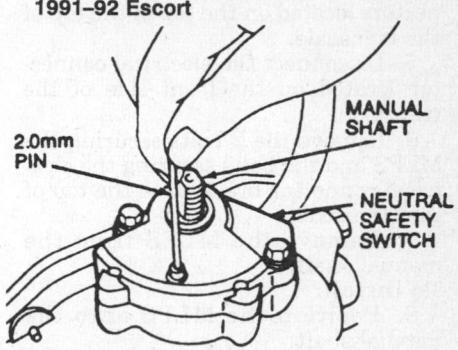

Manual lever position switch adjustment —1991–92 Escort

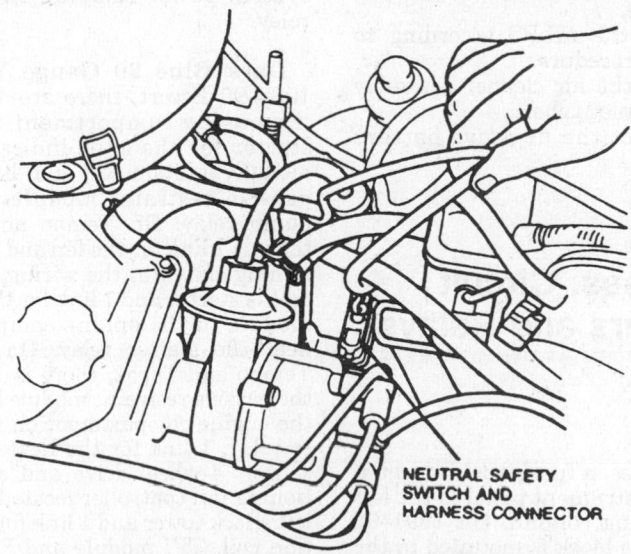

Neutral safety switch location—1988–90 Escort

To install:

5. Place the switch on the manual shift shaft and loosely install the retaining bolts.

6. Use a No. 43 (0.089 in.) drill and insert it into the hole provided in the switch.

7. Tighten the retaining bolts to 7–9 ft. lbs. (9–12 Nm) and remove the drill.

8. Connect the neutral start switch connector and connect negative battery cable.

9. Check the ignition switch for proper starting in **P** or **N**. Also make certain the start circuit cannot be actuated in the **D** or **R** position.

1991–92 Escort

1. Disconnect the negative battery cable.

2. Remove the air cleaner assembly and air inlet tube.

3. Remove the nut securing the manual shaft lever to the transaxle manual shaft and remove the lever.

4. Disconnect the 3 electrical connectors located on the top of the top of the transaxle.

5. Disconnect the electrical connector located on the front side of the transaxle.

6. Remove the 2 bolts securing the MLPS and the bolts securing the electrical connector brackets to the top of the transaxle.

7. Remove the MLPS from the manual shaft.

To install:

8. Position the MLPS onto the manual shaft.

9. Install the bolts securing the electrical connectors to the transaxle housing.

10. Install the bolts securing the MLPS, but do not tighten yet.

11. Connect the electrical connectors located on the top and on the side of the transaxle.

12. Adjust the MLPS according to the proper procedure.

13. Install the air cleaner assembly and the air inlet tube.

14. Connect the negative battery cable.

Fuses, Circuit Breakers and Relays

LOCATION

Fuses

On all vehicles, a fuse panel is located under the instrument panel to the left of the steering column. On 1991–92 Escort, a fuse block is mounted in the engine compartment.

Fusible Links

Fusible links are used to prevent major wire harness damage in the event of a short circuit or an overload condition in the wiring circuits that are normally not fused, due to carrying high amperage loads or because of their locations within the wiring harness. Each fusible link is of a fixed value for a specific electrical load and should a fusible link fail, the cause of the failure must be determine and repaired prior to installing a new fusible link of the same value. Please be advised that the color coding of replacement fusible links may vary from the production color coding that is outlined in the text that follows.

Dark Green 14 Gauge Wire – on Tempo, Topaz and 1988–90 Escort, there is 1 link located in the charging circuit near the starter motor relay.

Black 16 Gauge Wire – on 1988–90 Escort, there is 1 link for the rear window defogger located in the engine compartment on the starter relay. On 1988–90 Escort, Tempo and Topaz, there is 1 link for the headlight feed located in the engine compartment on the starter relay and 1 link for the ignition feed near the starter relay. On Tempo and Topaz, there is 1 link for the cooling fan relay located in the wiring assembly on the starter relay.

Brown 18 Gauge Wire – on Tempo and Topaz, there is 1 link used to protect the rear window defogger and the fuel door release. On 1991–92 Tempo and Topaz, there is 1 link in the charging circuit near the starter relay. On the 1988–90 Escort, there is 1 link used to protect the cooling fan motor circuit and 1 link for the EEC system power relay on the starter relay.

Dark Blue 20 Gauge Wire – on 1988–90 Escort, there are 4 links in the engine compartment near the starter for the shift indicator light module, ignition coil and distributor, passive restraint module and fuel pump relay. On Tempo and Topaz, there is 1 link for the fan and air conditioning clutch in the wiring assembly on the starter and 1 link for the air bag module in the engine compartment near the starter relay. On 1991–92 Tempo and Topaz, there is 1 link for the passive restraint module located in the engine compartment on the starter relay, 1 link for the heated oxygen sensor, 4-wheel drive and air conditioning fan controller located near the left shock tower and 1 link for the ignition coil, TFI module and ECA relay located near the left shock tower.

NOTE: Always disconnect the negative battery cable before servicing the high current fuses or serious personal injury may result.

Fuse Link Cartridge

Fuse link cartridges are used on 1991–92 Escort. Fuse link cartridges have a colored plastic housing with a clear "window" at the top. To check a fuse cartridge, look at the fuse element through the clear "window". The fuse link cartridges are located in the engine compartment fuse box. The following fuse link cartridges are listed according to their labels in the fuse box.

FUEL INJ – Pink 30 amp: to protect the electronic engine control circuit.

HEAD – Pink 30 amp: to protect the headlight circuit and the daytime running lights circuit.

MAIN – Black 80 amp for 1.8L engine or Dark Blue for 1.9L engine: to protect all cicuits, except starter and starter solenoid circuits.

BTN – Yellow 60 amp for 1.8L engine or Green 40 amp for 1.9L engine: to protect the courtesy lights, electronic automatic transaxle, electronic engine control, exterior lights, horn, interior lights, passive restraint, power door locks, radio, shift lock and warning chime circuits.

COOLING FAN – Pink 30 amp for 1.8L engine or Green 40 amp for 1.9L engine: to protect cooling fans circuit.

Circuit Breakers

Circuit breakers are used to protect the various components of the electrical system, such as headlights and windshield wipers. The circuit breakers are located either in the control switch or mounted on or near the fuse panel.

TEMPO AND TOPAZ

Headlights and Highbeam Indicator – one 22 amp circuit breaker incorporated in the lighting switch.

Alternator Voltage Sensing Circuit – one 18 amp circuit breaker located in engine compartment wiring assembly near starter relay on 1990–92 vehicles.

HEGO, All Wheel Drive Relays, Air Conditioning Fan Controller, Fan Tester and All Wheel Drive Switch – one 20 amp circuit breaker located in the engine compartment near the starter relay on 1990–92 vehicles.

Passive Restraint Module – one 20 amp circuit breaker located in the engine compartment near the starter relay on 1990–92 vehicles.

Power Windows, Power Seats, Power Door Locks and Power Lumbar—one 20 amp circuit breaker located in the fuse panel.

Windshield Wipers—one 8.25 amp circuit breaker located in the fuse panel.

ESCORT

Headlights and High Beam Indicator—one 22 amp circuit breaker incorporated in the lighting switch, on 1988–90 vehicles.

Liftgate Wiper—one 4.5 amp circuit breaker located in the instrument panel, to the left of the radio, on 1988–90 vehicles.

Windshield Wiper and Wiper Pump Circuit—one 8.25 amp circuit breaker located in the fuse panel, on 1988–90 vehicles.

Heater Blower Motor—one 30 amp circuit breaker located in the fuse panel under the dash, to the left of the steering column, on 1991–92 vehicles.

Various Relays

TEMPO AND TOPAZ

All Wheel Drive Relays—located behind the right side of the instrument panel.

Door Lock Control Relay—located below the left side of the instrument panel, near the fuse panel.

Cooling Fan Relay—located in the left front of the engine compartment.

Electronic Engine Control Power Relay—located behind the right side of the instrument panel.

Fuel Pump Relay—located behind the right side of the instrument panel.

Horn Relay—located behind the left side of the instrument panel, above the fuse panel.

Starter Relay—located on the left front fender apron in front of the strut tower.

Rear Window Defrost Relay—located behind the left side of the instrument panel, to the right of the steering column.

Shift Indicator Dimmer Relay—located behind the left side of the instrument panel, near the steering column.

Window Safety Relay—located behind the left side of the instrument panel, above the fuse panel.

1988–90 ESCORT

Cooling Fan Relay—located in the left front of the engine compartment, near the left headlight.

Electronic Engine Control (EEC) Power Relay—located behind the left side of the instrument panel.

Fuel Pump Relay—located behind the left side of the instrument panel.

Horn Relay—located behind the left side of the instrument panel, to the right of the steering column.

Starter Relay—located on the left side of the fender apron in front of the shock tower.

1991–92 ESCORT

Air Conditioning Relay—located on the rear of the right fender apron.

Cooling Fan Relay—located on the front of the left fender apron.

Door Lock Relay—located above the left cowl.

Daytime Running Lights Relay—located behind the right side of the instrument panel, near the blower motor.

Electronic Engine Control Power Relay—located behind the center of the instrument panel.

Fuel Pump Relay—located behind the center of the instrument panel.

Headlight Relay—located above the left cowl.

Horn Relay—located above the left cowl.

Parking Light Relay—located behind the left side of the instrument panel.

Vane Air Flow Meter Relay—located behind the center of the instrument panel.

Wide Open Throttle Cutout Relay—located on the rear of the right fender apron.

Computers

LOCATION

The Electronic Engine Control (EEC) module is located behind the left side of the instrument panel on all except 1991–92 Escort. On the 1991–92 Escort, the EEC module is located behind the center of the instrument panel.

Turn Signal/Hazard Warning Flashers

LOCATION

Except 1991–92 Escort

The turn signal flasher is located on the front side of the fuse panel. The hazard flasher is located on the rear of the fuse panel behind the turn signal flasher.

1991–92 Escort

The turn signal and hazard flasher switch use the same flasher unit. The flasher unit is located near the combination switch.

Cruise Control

ADJUSTMENT

Actuator Cable

EXCEPT 1991–92 ESCORT

1. With engine **OFF**, set the throttle linkage so the throttle plate is closed.
2. Remove the locking pin.
3. Pull the bead chain through the adjuster.
4. Insert the locking pin in the best hole of the adjuster to draw the bead chain tight without opening the throttle plate.

1991–92 ESCORT

1. Remove the cable adjusting clip from the cable housing.
2. Pull tightly on the cable until all of the slack is taken out.
3. Install the cable adjusting clip.

Vacuum Dump Valve

EXCEPT 1991–92 ESCORT

1. Firmly depress the brake pedal and hold in position.
2. Push in the dump valve until the valve collar bottoms against the retaining clip.
3. Place a 0.050–0.10 in. (1.27–2.54mm) shim between the white button of the valve and the pad on the brake pedal.
4. Firmly pull the brake pedal rearward to its normal position, allowing the dump valve to ratchet backwards in the retaining clip.

Clutch Switch

EXCEPT 1991–92 ESCORT

1. Prop the clutch pedal in the full-up position—pawl fully released from the sector.
2. Loosen the switch mounting screw.
3. Slide the switch forward toward the clutch pedal until the switch plunger cap is 0.030 in. (0.76mm) from contacting the switch housing. Tighten the attaching screw.
4. Remove the prop from the clutch pedal and test drive for clutch cancellation of a speed control.

Fuse Panel Reference Chart

Fuse	Fuse Rating	Color	Protected Component
(REAR WIPER)	10 amp	Red	Rear wiper and washer motors, WAC and A/C relays, DRL relay (Canada), climate control panel illumination
HAZARD	15 amp	Blue	Front and rear turn signal lamps, turn signal indicator light, hazard flasher
ROOM	10 amp	Red	Dome and map lamps, warning chime module, ignition key illumination, luggage lamp, vanity mirror lamp, radio (memory), door lock relay, shift-lock system, 4EAT control unit, ECA
ENGINE	15 amp	Blue	Cooling fan relays, IG relay, EEC relay
RADIO	15 amp	Blue	Radio, premium sound amplifier, remote control mirror, instrument panel dimmer switch
(DOOR LOCK)	30 amp	Light Green	Door lock motor
BELT	30 amp	Light Green	Passive restraint control module, passive restraint motors
(POWER WIND)	30 amp	Light Green	Power window motors
METER	15 amp	Blue	Instrument cluster, backup lamps, turn signal lamps, speed control system, shift-lock system, warning chime module, 4EAT control unit, passive restraint control module, hazard flasher ECA (1.8L Canada)
WIPER	20 amp	Yellow	Front wiper and washer motors, interval wiper
STOP	20 amp	Yellow	Stoplamp, horn, hi-mount stoplamp
TAIL	15 amp	Blue	Front side markers, rear side markers, tail lights, license lamps, engine compartment lamp, TNS relay
(MOON ROOF)	15 amp	Blue	Moon roof motor
(DEFOG)	20 amp	Yellow	Rear window defroster
HEATER	30 amp circuit breaker	—	Blower motor
(ON JOINT BOX SIDE) HEGO	10 amp	Red	HEGO sensor (1.9L engine)
(ON JOINT BOX SIDE) CIGAR	20 amp	Yellow	Cigar Lighter
REAR WIPER	10 amp	Red	Rear wiper and washer (wagon), Daytime Running Light (Canada), heater and air conditioner

Ford Motor Co.
Rear Wheel Drive
Ford—Crown Victoria, Mustang, Thunderbird
Lincoln—Mark VII, Town Car
Mercury—Cougar, Grand Marquis

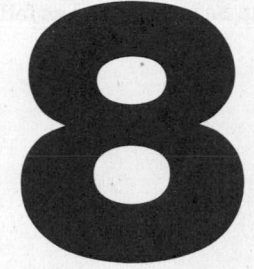

SPECIFICATIONS

VEHICLE IDENTIFICATION CHART

It is important for servicing and ordering parts to be certain of the vehicle and engine identification. The VIN (vehicle identification number) is a 17 digit number visible through the windshield on the driver's side of the dash and contains the vehicle and engine identification codes. The tenth digit indicates model year and the eighth digit indicates engine code. It can be interpreted as follows:

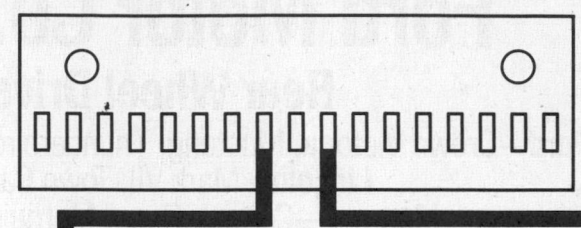

Engine Code							Model Year	
Code	Cu. In.	Liters	Cyl.	Fuel Sys.	Eng. Mfg.		Code	Year
A	140	2.3	4	EFI	Ford		J	1988
T	140	2.3	4 (Turbo)	EFI	Ford		K	1989
M	140	2.3	4	EFI	Ford		L	1990
4	232	3.8	6	EFI	Ford		M	1991
R	232 (SC)	3.8	6	EFI	Ford		N	1992
C	232 (SC)	3.8	6	EFI	Ford			
W	280	4.6	8	EFI	Ford			
F	302	5.0	8	EFI	Ford			
E	302 (HO)	5.0	8	EFI	Ford			
T	302 (HO)	5.0	8	EFI	Ford			
G	351 (HO)	5.8	8	VV	Ford			

EFI Electronic Fuel Injection
HO High Output
SC Supercharged
VV Variable Venturi

ENGINE IDENTIFICATION

Year	Model	Engine Displacement cu. in. (liter)	Engine Series Identification (VIN)	No. of Cylinders	Engine Type
1988	Mustang	140 (2.3)	A	4	OHC
	Mustang	302 (5.0) HO	E	8	OHV
	Thunderbird	140 (2.3)	T	4	OHC-Turbo
	Thunderbird	232 (3.8)	4	6	OHV
	Thunderbird	302 (5.0)	F	8	OHV
	Cougar	232 (3.8)	4	6	OHV
	Cougar	302 (5.0)	F	8	OHV
	Crown Victoria	302 (5.0)	F	8	OHV
	Crown Victoria	351 (5.8)	G	8	OHV
	Grand Marquis	302 (5.0)	F	8	OHV
	Grand Marquis	351 (5.8)	G	8	OHV
	Mark VII	302 (5.0) HO	E	8	OHV
	Town Car	302 (5.0)	F	8	OHV

ENGINE IDENTIFICATION

Year	Model	Engine Displacement cu. in. (liter)	Engine Series Identification (VIN)	No. of Cylinders	Engine Type
1989	Mustang	140 (2.3)	A	4	OHC
	Mustang	302 (5.0) HO	E	8	OHV
	Thunderbird	232 (3.8)	4	6	OHV
	Thunderbird	232 (3.8) SC	R①	6	OHV
	Cougar	232 (3.8)	4	6	OHV
	Cougar	232 (3.8) SC	R①	6	OHV
	Crown Victoria	302 (5.0)	F	8	OHV
	Crown Victoria	351 (5.8)	G	8	OHV
	Grand Marquis	302 (5.0)	F	8	OHV
	Grand Marquis	351 (5.8)	G	8	OHV
	Mark VII	302 (5.0) HO	E	8	OHV
	Town Car	302 (5.0)	F	8	OHV
1990	Mustang	140 (2.3)	A	4	OHC
	Mustang	302 (5.0) HO	E	8	OHV
	Thunderbird	232 (3.8)	4	6	OHV
	Thunderbird	232 (3.8) SC	R①	6	OHV
	Cougar	232 (3.8)	4	6	OHV
	Cougar	232 (3.8) SC	R①	6	OHV
	Crown Victoria	302 (5.0)	F	8	OHV
	Crown Victoria	351 (5.8)	G	8	OHV
	Grand Marquis	302 (5.0)	F	8	OHV
	Grand Marquis	351 (5.8)	G	8	OHV
	Mark VII	302 (5.0) HO	E	8	OHV
	Town Car	302 (5.0)	F	8	OHV
1991–92	Mustang	140 (2.3)	M	4	OHC
	Mustang	302 (5.0) HO	E	8	OHV
	Thunderbird	232 (3.8)	4	6	OHV
	Thunderbird	232 (3.8) SC	R①	6	OHV
	Thunderbird	302 (5.0) HO	T	8	OHV
	Cougar	232 (3.8)	4	6	OHV
	Cougar	302 (5.0) HO	T	8	OHV
	Crown Victoria	280 (4.6)	W	8	OHC
	Crown Victoria	302 (5.0)	F	8	OHV
	Crown Victoria	351 (5.8)	G	8	OHV
	Grand Marquis	280 (4.6)	W	8	OHC
	Grand Marquis	302 (5.0)	F	8	OHV
	Grand Marquis	351 (5.8)	G	8	OHV
	Mark VII	302 (5.0) HO	E	8	OHV
	Town Car	280 (4.6)	W	8	OHC

OHC Overhead Camshaft
OHV Overhead Valve
HO High Output
SC Supercharged
① Early Production could be Code C

GENERAL ENGINE SPECIFICATIONS

Year	VIN	No. Cylinder Displacement cu. in. (liter)	Fuel System Type	Net Horsepower @ rpm	Net Torque @ rpm (ft. lbs.)	Bore × Stroke (in.)	Compression Ratio	Oil Pressure @ rpm
1988	A	4-140 (2.3)	EFI	88 @ 4000	132 @ 2600	3.78 × 3.12	9.5:1	40–60 @ 2000 RPM
	T	4-140 (2.3) T	EFI	①	②	3.78 × 3.12	8.0:1	40–60 @ 2000 RPM
	4	6-232 (3.8)	EFI	140 @ 3800	215 @ 2400	3.81 × 3.39	9.0:1	40–60 @ 2000 RPM
	F	8-302 (5.0)	EFI	③	④	4.00 × 3.00	8.9:1	40–60 @ 2000 RPM
	E	8-302 (5.0) HO	EFI	225 @ 4200	300 @ 3200	4.00 × 3.00	9.0:1	40–60 @ 2000 RPM
	G	8-351 (5.8) HO	VV	180 @ 3600	285 @ 2400	4.00 × 3.50	8.3:1	40–60 @ 2000 RPM
1989	A	4-140 (2.3)	EFI	88 @ 4000	132 @ 2600	3.78 × 3.12	9.5:1	40–60 @ 2000 RPM
	4	6-232 (3.8)	EFI	140 @ 3800	215 @ 2400	3.81 × 3.39	9.0:1	40–60 @ 2000 RPM
	R	6-232 (3.8) SC	EFI	210 @ 4000	315 @ 2600	3.81 × 3.39	8.2:1	40–60 @ 2000 RPM
	C	6-232 (3.8) SC	EFI	210 @ 4000	315 @ 2600	3.81 × 3.39	8.2:1	40–60 @ 2000 RPM
	F	8-302 (5.0)	EFI	③	④	4.00 × 3.00	8.9:1	40–60 @ 2000 RPM
	E	8-302 (5.0) HO	EFI	225 @ 4200	300 @ 3200	4.00 × 3.00	9.0:1	40–60 @ 2000 RPM
	G	8-351 (5.8) HO	VV	180 @ 3600	285 @ 2400	4.00 × 3.50	8.3:1	40–60 @ 2000 RPM
1990	A	4-140 (2.3)	EFI	88 @ 4000	132 @ 2600	3.78 × 3.12	9.5:1	40–60 @ 2000 RPM
	4	6-232 (3.8)	EFI	140 @ 3800	215 @ 2400	3.81 × 3.39	9.0:1	40–60 @ 2000 RPM
	R	6-232 (3.8) SC	EFI	210 @ 4000	315 @ 2600	3.81 × 3.39	8.2:1	40–60 @ 2000 RPM
	C	6-232 (3.8) SC	EFI	210 @ 4000	315 @ 2600	3.81 × 3.39	8.2:1	40–60 @ 2000 RPM
	F	8-302 (5.0)	EFI	③	④	4.00 × 3.00	8.9:1	40–60 @ 2000 RPM
	E	8-302 (5.0) HO	EFI	225 @ 4200	300 @ 3200	4.00 × 3.00	9.0:1	40–60 @ 2000 RPM
	G	8-351 (5.8) HO	VV	180 @ 3600	285 @ 2400	4.00 × 3.50	8.3:1	40–60 @ 2000 RPM
1991–92	M	4-140 (2.3)	EFI	105 @ 4600	135 @ 2600	3.78 × 3.12	9.5:1	40–60 @ 2000 RPM
	4	6-232 (3.8)	EFI	140 @ 3800	215 @ 2400	3.81 × 3.39	9.0:1	40–60 @ 2000 RPM
	R	6-232 (3.8) SC	EFI	210 @ 4000	315 @ 2600	3.81 × 3.39	8.2:1	40–60 @ 2000 RPM
	C	6-232 (3.8) SC	EFI	210 @ 4000	315 @ 2600	3.81 × 3.39	8.2:1	40–60 @ 2000 RPM
	W	8-280 (4.6)	EFI	⑤	⑥	3.60 × 3.60	9.0:1	20–45 @ 1500 RPM
	F	8-302 (5.0)	EFI	③	④	4.00 × 3.00	8.9:1	40–60 @ 2000 RPM
	T	8-302 (5.0) HO	EFI	200 @ 4000	275 @ 3000	4.00 × 3.00	9.0:1	40–60 @ 2000 RPM
	E	8-302 (5.0) HO	EFI	225 @ 4200	300 @ 3200	4.00 × 3.00	9.0:1	40–60 @ 2000 RPM
	G	8-351 (5.8) HO	VV	180 @ 3600	285 @ 2400	4.00 × 3.50	8.3:1	40–60 @ 2000 RPM

■ Horsepower and torque are SAE net figures.
They are measured at the rear of the transmission with all accessories installed and operating. Since the figures vary when a given engine is installed in different models, some are representative rather than exact.
T Turbocharger
EFI Electronic Fuel Injection
HO High Output
VV Variable Venturi carburetor
SC Supercharged
① 150 @ 4400 with automatic transmission
190 @ 4600 with manual transmission
② 200 @ 3000 with automatic transmission
240 @ 3400 with manual transmission
③ 150 @ 3200 with single exhaust
160 @ 3400 with dual exhaust
④ 270 @ 2000 with single exhaust
280 @ 2200 with dual exhaust
⑤ 190 @ 4200 with single exhaust
210 @ 4600 with dual exhaust
⑥ 260 @ 3200 with single exhaust
270 @ 3400 with dual exhaust

GASOLINE ENGINE TUNE-UP SPECIFICATIONS

Year	VIN	No. Cylinder Displacement cu. in. (liter)	Spark Plugs Type	Gap (in.)	Ignition Timing (deg.) MT	AT	Compression Pressure (psi)	Fuel Pump (psi)	Idle Speed (rpm) MT	AT	Valve Clearance In.	Ex.
1987	A	4-140 (2.3)	AWSF-44C	0.044	①	①	NA	35	750	750	Hyd.	Hyd.
	W	4-140 (2.3) T	AWSF-320	0.034	①	①	NA	35	825–975	825–975	Hyd.	Hyd.
	3	6-232 (3.8)	AWSF-54	0.044	①	①	NA	35	—	550	Hyd.	Hyd.
	E	8-302 (5.0)	ASF-32C	0.044	①	①	NA	39	① ②	① ②	Hyd.	Hyd.
	F	8-302 (5.0)	ASF-32C	0.044	①	①	NA	35	①	①	Hyd.	Hyd.
	M	8-302 (5.0) HO	ASF42	0.044	①	①	NA	35	700	700	Hyd.	Hyd.
	G	8-351 (5.8)	ASF-32C	0.044	①	①	NA	6–8	650	650	Hyd.	Hyd.
1988	A	4-140 (2.3)	AWSF-44C	0.044	①	①	NA	35	750	750	Hyd.	Hyd.
	T	4-140 (2.3)	AWSF-32C	0.034	①	①	NA	35	825–975	825–975	Hyd.	Hyd.
	4	6-232 (3.8)	AWSF-54	0.044	①	①	NA	39	—	550	Hyd.	Hyd.
	F	8-302 (5.0)	ASF-32C	0.044	①	①	NA	39	①	①	Hyd.	Hyd.
	E	8-302 (5.0) HO	ASF42	0.044	①	①	NA	39	700	700	Hyd.	Hyd.
	G	8-351 (5.8)	ASF-32C	0.044	①	①	NA	6–8	650	650	Hyd.	Hyd.
1989	A	4-140 (2.3)	AWSF-44C	0.044	①	①	NA	35	750	750	Hyd.	Hyd.
	T	4-140 (2.3)	AWSF-32C	0.034	①	①	NA	35	825–975	825–975	Hyd.	Hyd.
	4	6-232 (3.8)	AWSF-54	0.044	①	①	NA	39	—	550	Hyd.	Hyd.
	R	6-232 (3.8) SC	AWSF-54	0.044	①	①	NA	39	650	550	Hyd.	Hyd.
	C	6-232 (3.8) SC	AWSF-54	0.044	①	①	NA	39	650	550	Hyd.	Hyd.
	F	8-302 (5.0)	ASF-32C	0.044	①	①	NA	39	①	①	Hyd.	Hyd.
	E	8-302 (5.0) HO	ASF42	0.044	①	①	NA	39	700	700	Hyd.	Hyd.
	G	8-351 (5.8)	ASF-32C	0.044	①	①	NA	6–8	650	650	Hyd.	Hyd.
	G	8-351 (5.8)	ASF-32C	0.044	①	①	NA	6–8	650	650	Hyd.	Hyd.
1990	A	4-140 (2.3)	AWSF-44C	0.044	①	①	NA	35	750	750	Hyd.	Hyd.
	4	6-232 (3.8)	AWSF-44C	0.054	①	①	NA	39	—	550	Hyd.	Hyd.
	R	6-232 (3.8) SC	AWSF-34PP	0.054	①	①	NA	39	650	550	Hyd.	Hyd.
	C	6-232 (3.8) SC	AWSF-34PP	0.054	①	①	NA	39	650	550	Hyd.	Hyd.
	F	8-302 (5.0)	AWSF-44C	0.054	①	①	NA	39	①	①	Hyd.	Hyd.
	E	8-302 (5.0) HO	ASF-42C	0.054	①	①	NA	39	700	700	Hyd.	Hyd.
	G	8-351 (5.8)	AWSF-44C	0.054	①	①	NA	6–8	650	650	Hyd.	Hyd.
1991						SEE UNDERHOOD SPECIFICATIONS STICKER						

NOTE: The underhood specifications sticker often reflects tune-up specifications changes made in production. Sticker figures must be used if they disagree with those in this chart.

T Turbocharger
B Before top dead center
HO High output
— Not applicable
HSC High Swirl Combustion
① Calibrations vary depending upon the model; refer to the underhood sticker
② The carbureted models use spark plug ASF-42 (.044) and the idle speed is 700 rpm

FIRING ORDERS

NOTE: To avoid confusion, always replace spark plug wires one at a time.

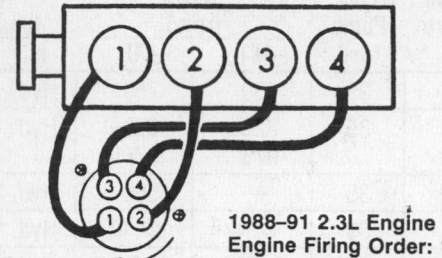

1988–91 2.3L Engine
Engine Firing Order: 1–3–4–2
Distributor Rotation: Clockwise

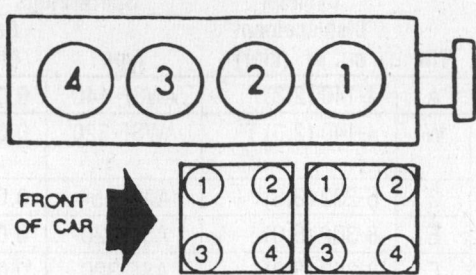

1991–92 2.3L Engine
Engine Firing Order: 1–3–4–2
Distributorless Ignition System

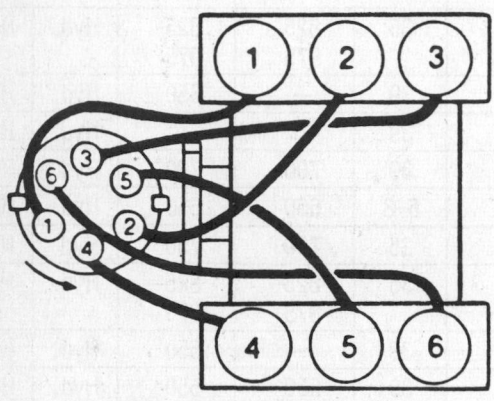

3.8L (except SC) Engine
Engine Firing Order: 1–4–2–5–3–6
Distributor Rotation: Counterclockwise

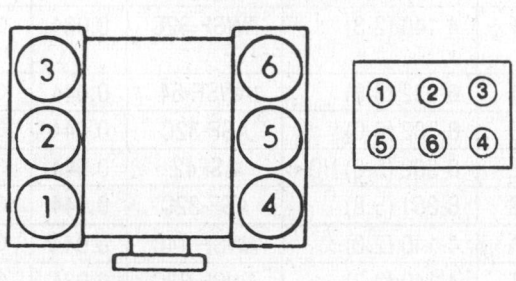

3.8L SC Engine
Engine Firing Order: 1–4–2–5–3–6
Distributorless Ignition System

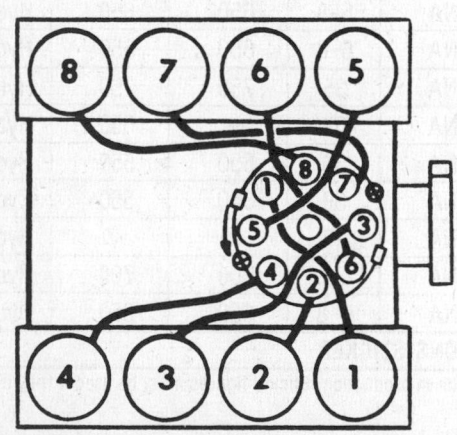

5.0L (except HO) Engine
Engine Firing Order: 1–5–4–2–6–3–7–8
Distributor Rotation: Counterclockwise

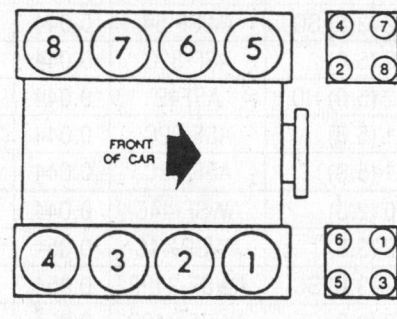

4.6L Engine
Engine Firing Order: 1–3–7–2–6–5–4–8
Distributorless Ignition System

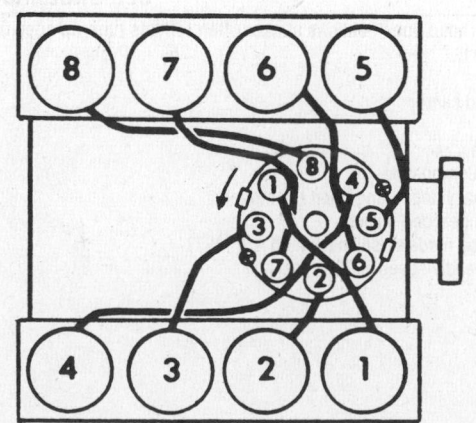

5.0L HO and 5.8L Engines
Engine Firing Order: 1–3–7–2–6–5–4–8
Distributor Rotation: Counterclockwise

CAPACITIES

Year	Model	VIN	No. Cylinder Displacement cu. in. (liter)	Engine Crankcase with Filter	Engine Crankcase without Filter	Transmission (pts.) 4-Spd	Transmission (pts.) 5-Spd	Transmission (pts.) Auto.	Drive Axle (pts.)	Fuel Tank (gal.)	Cooling System (qts.)
1988	Mustang	A	4-140 (2.3)	5	4	—	5.6	20	①	15.4	③
	Mustang	E	8-302 (5.0) HO	5	4	—	5.6	24.6	①	15.4	14.1
	T-Bird	T	4-140 (2.3) T	5	4	—	5.6	20	①	②	10.0
	T-Bird	4	6-232 (3.8)	5	4	—	—	24.6	①	18.8	11.8
	T-Bird	F	8-302 (5.0)	5	4	—	—	24.6	①	18.8	14.1
	Cougar	4	6-232 (3.8)	5	4	—	—	24.6	①	18.8	11.8
	Cougar	F	8-302 (5.0)	5	4	—	—	24.6	①	18.8	14.1
	Mark VII	E	8-302 (5.0) HO	5	4	—	—	24.6	①	22.3	14.1
	Crown Victoria	F	8-302 (5.0)	5	4	—	—	24.6	3.75	18.0	14.1
	Crown Victoria	G	8-351 (5.8)	5	4	—	—	24.6	3.75	20.0	14.1
	Grand Marquis	F	8-302 (5.0)	5	4	—	—	24.6	3.75	18.0	14.1
	Grand Marquis	G	8-351 (5.8)	5	4	—	—	24.6	3.75	20.0	14.1
	Town Car	F	8-302 (5.0)	5	4	—	—	24.6	3.75	22.3	14.1
1989	Mustang	A	4-140 (2.3)	5	4	—	5.6	20	①	15.4	③
	Mustang	E	8-302 (5.0)	5	4	—	5.6	24.6	①	15.4	14.1
	T-Bird	4	6-232 (3.8)	5	4	—	—	24.6	④	18.8	11.8
	T-Bird	R	6-232 (3.8) SC	5	4	—	6.3	24.6	④	18.8	11.9
	T-Bird	C	6-232 (3.8) SC	5	4	—	6.3	24.6	④	18.8	11.9
	Cougar	4	6-232 (3.8)	5	4	—	—	24.6	④	18.8	11.8
	Cougar	R	6-232 (3.8) SC	5	4	—	6.3	24.6	④	18.8	11.9
	Cougar	C	6-232 (3.8) SC	5	4	—	6.3	24.6	④	18.8	11.9
	Mark VII	E	8-302 (5.0) HO	5	4	—	—	24.6	3.75	22.1	14.1
	Crown Victoria	F	8-302 (5.0)	5	4	—	—	24.6	3.75	18.0	14.1
	Crown Victoria	G	8-351 (5.8)	5	4	—	—	24.6	3.75	20.0	14.1
	Grand Marquis	F	8-302 (5.0)	5	4	—	—	24.6	3.75	18.0	14.1
	Grand Marquis	G	8-351 (5.8)	5	4	—	—	24.6	3.75	20.0	14.1
	Town Car	F	8-302 (5.0)	5	4	—	—	24.6	3.75	22.3	14.1
1990	Mustang	A	4-140 (2.3)	5	4	—	5.6	19.4	①	15.4	③
	Mustang	E	8-302 (5.0) HO	5	4	—	5.6	24.6	①	15.4	14.1
	T-Bird	4	6-232 (3.8)	5	4	—	—	24.6	④	18.8	11.8
	T-Bird	R	6-232 (3.8) SC	5	4	—	6.3	24.6	④	18.8	11.9
	T-Bird	C	6-232 (3.8) SC	5	4	—	6.3	24.6	④	18.8	11.9
	Cougar	4	6-232 (3.8)	5	4	—	—	24.6	④	18.8	11.8
	Cougar	R	6-232 (3.8) SC	5	4	—	6.3	24.6	④	18.8	11.9
	Cougar	C	6-232 (3.8) SC	5	4	—	6.3	24.6	④	18.8	11.9
	Mark VII	E	8-302 (5.0) HO	5	4	—	—	24.6	3.75	22.1	14.1
	Crown Victoria	F	8-302 (5.0)	5	4	—	—	24.6	3.75	18.0	14.1
	Crown Victoria	G	8-351 (5.8)	5	4	—	—	24.6	3.75	20.0	14.1
	Grand Marquis	F	8-302 (5.0)	5	4	—	—	24.6	3.75	18.0	14.1
	Grand Marquis	G	8-351 (5.8)	5	4	—	—	24.6	3.75	20.0	14.1
	Town Car	F	8-302 (5.0)	5	4	—	—	24.6	3.75	18.0	14.1

CAPACITIES

Year	Model	VIN	No. Cylinder Displacement cu. in. (liter)	Engine Crankcase with Filter	Engine Crankcase without Filter	Transmission (pts.) 4-Spd	Transmission (pts.) 5-Spd	Transmission (pts.) Auto.	Drive Axle (pts.)	Fuel Tank (gal.)	Cooling System (qts.)
1991–92	Mustang	M	4-140 (2.3)	5	4	—	5.6	19.4	①	15.4	③
	Mustang	E	8-302 (5.0) HO	5	4	—	5.6	24.6	①	15.4	14.1
	T-Bird	4	6-232 (3.8)	5	4	—	—	24.6	④	19.0	11.8
	T-Bird	R	6-232 (3.8) SC	5	4	—	6.3	24.6	④	19.0	11.9
	T-Bird	C	6-232 (3.8) SC	5	4	—	6.3	24.6	④	19.0	11.9
	T-Bird	T	8-302 (5.0) HO	5	4	—	—	24.6	④	19.0	14.1
	Cougar	4	6-232 (3.8)	5	4	—	—	24.6	④	19.0	11.8
	Cougar	T	8-302 (5.0) HO	5	4	—	—	24.6	④	19.0	14.1
	Mark VII	E	8-302 (5.0) HO	5	4	—	—	24.6	3.75	22.1	14.1
	Crown Victoria	W	8-280 (4.6)	5	4	—	—	24.6	3.75	20.0	14.1
	Crown Victoria	F	8-302 (5.0)	5	4	—	—	24.6	3.75	18.0	14.1
	Crown Victoria	G	8-351 (5.8)	5	4	—	—	24.6	3.75	20.0	14.1
	Grand Marquis	W	8-280 (4.6)	5	4	—	—	24.6	3.75	20.0	14.1
	Grand Marquis	F	8-302 (5.0)	5	4	—	—	24.6	3.75	18.0	14.1
	Grand Marquis	G	8-351 (5.8)	5	4	—	—	24.6	3.75	20.0	14.1
	Town Car	W	8-280 (4.6)	5	4	—	—	24.6	3.75	20.0	14.1

① With 7.5 in. axle—3.5 pts.
  With 7.5 in. limited slip, 8.8 in. or 8.8 in. limited slip axles—3.75 pts.
② With automatic transmission—22.1 gal.
  With manual transmission—18.0 gal.
③ Equipped with manual transmission and air conditioning—9.7 qts.
  Except equipped with manual transmission and air conditioning—10.0 qts.
④ With 7.5 in. axle—3 pts.
  With 7.5 in. limited slip axle—2.75 pts.
  With 8.8 in. or 8.8 in. limited slip axles—3.25 pts.

CAMSHAFT SPECIFICATIONS

All measurements given in inches

Year	VIN	No. Cylinder Displacement cu. in. (liter)	Journal Diameter 1	Journal Diameter 2	Journal Diameter 3	Journal Diameter 4	Journal Diameter 5	Lobe Lift In.	Lobe Lift Ex.	Bearing Clearance	Camshaft End Play
1988	A	4-140 (2.3)	1.7713–1.7720	1.7713–1.7720	1.7713–1.7720	1.7713–1.7720	—	0.400	0.400	0.001–0.003	0.001–0.007
	T	4-140 (2.3)	1.7713–1.7720	1.7713–1.7720	1.7713–1.7720	1.7713–1.7720	—	0.400	0.400	0.001–0.003	0.001–0.007
	4	6-232 (3.8)	2.0515–2.0505	2.0515–2.0505	2.0515–2.0505	2.0515–2.0505	—	0.240	0.241	0.001–0.003	①
	F	8-302 (5.0)	2.0805–2.0815	2.0655–2.0665	2.0505–2.0515	2.0355–2.0365	2.0205–2.0215	0.2375	0.2474	0.001–0.003	0.0005–0.0055
	E	8-302 (5.0)	2.0805–2.0815	2.0655–2.0665	2.0505–2.0515	2.0355–2.0365	2.0205–2.0215	0.2780	0.2780	0.001–0.003	0.0005–0.0055
	G	8-351 (5.8)	2.0805–2.0815	2.0655–2.0665	2.0505–2.0515	2.0355–2.0365	2.0205–2.0215	0.2830	0.2830	0.001–0.003	0.001–0.007
1989	A	4-140 (2.3)	1.7713–1.7720	1.7713–1.7720	1.7713–1.7720	1.7713–1.7720	—	0.400	0.400	0.001–0.003	0.001–0.007
	4	6-232 (3.8)	2.0515–2.0505	2.0515–2.0505	2.0515–2.0505	2.0515–2.0505	—	0.245	0.259	0.001–0.003	①
	R	6-232 (3.8) SC	2.0515–2.0505	2.0515–2.0505	2.0515–2.0505	2.0515–2.0505	—	0.245	0.259	0.001–0.003	①

CAMSHAFT SPECIFICATIONS

All measurements given in inches

| Year | VIN | No. Cylinder Displacement cu. in. (liter) | Journal Diameter | | | | | Lobe Lift | | Bearing Clearance | Camshaft End Play |
			1	2	3	4	5	In.	Ex.		
	C	6-232 (3.8) SC	2.0515–2.0505	2.0515–2.0505	2.0515–2.0505	2.0515–2.0505	—	0.245	0.259	0.001–0.003	①
	F	8-302 (5.0)	2.0805–2.0815	2.0655–2.0665	2.0505–2.0515	2.0355–2.0365	2.0205–2.0215	0.2375	0.2474	0.001–0.003	0.001–0.007
	E	8-302 (5.0)	2.0805–2.0815	2.0655–2.0665	2.0505–2.0515	2.0355–2.0365	2.0205–2.0215	0.2780	0.2780	0.001–0.003	0.0055–0.0055
	G	8-351 (5.8)	2.0805–2.0815	2.0655–2.0665	2.0505–2.0515	2.0355–2.0365	2.0205–2.0215	0.2830	0.2830	0.001–0.003	0.001–0.007
1990	A	4-140 (2.3)	1.7713–1.7720	1.7713–1.7720	1.7713–1.7720	1.7713–1.7720	—	0.400	0.400	0.001–0.003	0.001–0.007
	4	6-232 (3.8)	2.0515–2.0505	2.0515–2.0505	2.0515–2.0505	2.0515–2.0505	—	0.245	0.259	0.001–0.003	①
	R	6-232 (3.8) SC	2.0515–2.0505	2.0515–2.0505	2.0515–2.0505	2.0515–2.0505	—	0.245	0.259	0.001–0.003	①
	C	6-232 (3.8) SC	2.0515–2.0505	2.0515–2.0505	2.0515–2.0505	2.0515–2.0505	—	0.245	0.259	0.001–0.003	①
	F	8-302 (5.0)	2.0805–2.0815	2.0655–2.0665	2.0505–2.0515	2.0355–2.0365	2.0205–2.0215	0.2375	0.2474	0.001–0.003	0.0005–0.0055
	E	8-302 (5.0) HO	2.0805–2.0815	2.0655–2.0665	2.0505–2.0515	2.0355–2.0365	2.0205–2.0215	0.2780	0.2780	0.001–0.003	0.0005–0.0055
	G	8-351 (5.8)	2.0805–2.0815	2.0655–2.0665	2.0505–2.0515	2.0355–2.0365	2.0205–2.0215	0.2780	0.2830	0.001–0.003	0.001–0.007
1991–92	M	4-140 (2.3)	1.7713–1.7720	1.7713–1.7720	1.7713–1.7720	1.7713–1.7720	—	0.2381	0.2381	0.001–0.003	0.001–0.007
	4	6-232 (3.8)	2.0515–2.0505	2.0515–2.0505	2.0515–2.0505	2.0515–2.0505	—	0.245	0.259	0.001–0.003	0.001–0.006
	R	6-232 (3.8) SC	2.0515–2.0505	2.0515–2.0505	2.0515–2.0505	2.0515–2.0505	—	0.245	0.259	0.001–0.003	0.001–0.006
	C	6-232 (3.8) SC	2.0515–2.0505	2.0515–2.0505	2.0515–2.0505	2.0515–2.0505	—	0.245	0.259	0.001–0.003	0.001–0.006
	W	8-280 (4.6)	1.0615–1.0605	1.0615–1.0605	1.0615–1.0605	1.0615–1.0605	1.0615–1.0605	0.259	0.259	0.001–0.003	0.001–0.006
	F	8-302 (5.0)	2.0805–2.0815	2.0655–2.0665	2.0505–2.0515	2.0355–2.0365	2.0205–2.0215	0.2375	0.2474	0.001–0.003	0.0005–0.0055
	T	8-302 (5.0) HO	2.0805–2.0815	2.0655–2.0665	2.0505–2.0515	2.0355–2.0365	2.0205–2.0215	0.278	0.278	0.001–0.003	0.0005–0.0055
	E	8-302 (5.0) HO	2.0805–2.0815	2.0655–2.0665	2.0505–2.0515	2.0355–2.0365	2.0205–2.0215	0.278	0.278	0.001–0.003	0.0005–0.0055
	G	8-351 (5.8)	2.0805–2.0815	2.0655–2.0665	2.0505–2.0515	2.0355–2.0365	2.0205–2.0215	0.278	0.283	0.001–0.003	0.0005–0.0055

① The endplay is controlled by the button and spring on the camshaft end

CRANKSHAFT AND CONNECTING ROD SPECIFICATIONS

All measurements are given in inches.

Year	VIN	No. Cylinder Displacement cu. in. (liter)	Crankshaft Main Brg. Journal Dia.	Crankshaft Main Brg. Oil Clearance	Crankshaft Shaft End-play	Thrust on No.	Connecting Rod Journal Diameter	Connecting Rod Oil Clearance	Connecting Rod Side Clearance
1988	A	4-140 (2.3)	2.3990–2.3982	0.0008–0.0015	0.004–0.008	3	2.0464–2.0472	0.0008–0.0015	0.0035–0.0105
	T	4-140 (2.3)	2.3990–2.3982	0.0008–0.0015	0.004–0.008	3	2.0464–2.0472	0.0008–0.0015	0.0035–0.0105
	4	6-232 (3.8)	2.5190–2.5198	0.0010–0.0014	0.004–0.008	3	2.3103–2.3111	0.0010–0.0014	0.0047–0.0114
	F	8-302 (5.0)	2.2482–2.2490	0.0004–0.0015	0.004–0.008	3	2.1228–2.1236	0.0008–0.0015	0.010–0.020
	E	8-302 (5.0) HO	2.2482–2.2490	0.0004–0.0015	0.004–0.008	3	2.1228–2.1236	0.0008–0.0015	0.010–0.020
	G	8-351 (5.8)	2.2994–3.0002	0.0008–0.0015	0.004–0.008	3	2.3103–2.3111	0.0008–0.0015	0.010–0.020
1989	A	4-140 (2.3)	2.3990–2.3982	0.0008–0.0015	0.004–0.008	3	2.0464–2.0472	0.0008–0.0015	0.0035–0.0105
	4	6-232 (3.8)	2.5190–2.5198	0.0010–0.0014	0.004–0.008	3	2.3103–2.3111	0.0010–0.0014	0.0047–0.0114
	R	6-232 (3.8) SC	①	②	0.004–0.008	3	2.3103–2.3111	0.0010–0.0014	0.0047–0.0114
	C	6-232 (3.8) SC	①	②	0.004–0.008	3	2.3103–2.3111	0.0010–0.0014	0.0047–0.0114
	F	8-302 (5.0)	2.2482–2.2490	0.0004–0.0015	0.004–0.008	3	2.1228–2.1236	0.0008–0.0015	0.010–0.020
	E	8-302 (5.0) HO	2.2482–2.2490	0.0004–0.0015	0.004–0.008	3	2.1228–2.1236	0.0008–0.0015	0.010–0.020
	G	8-351 (5.8)	2.2994–3.0002	0.0008–0.0015	0.004–0.008	3	2.3103–2.3111	0.0008–0.0015	0.010–0.020
1990	A	4-140 (2.3)	2.3990–2.3982	0.0008–0.0015	0.004–0.008	3	2.0465–2.0472	0.0008–0.0015	0.0035–0.0105
	4	6-232 (3.8)	2.5190–2.5198	0.0010–0.0014	0.004–0.008	3	2.3103–2.3111	0.0010–0.0014	0.0047–0.0114
	R	6-232 (3.8) SC	③	④	0.004–0.008	3	2.3103–2.3111	0.0010–0.0014	0.0047–0.0114
	C	6-232 (3.8) SC	③	④	0.004–0.008	3	2.3103–2.3111	0.0086–0.0027	0.0047–0.0114
	F	8-302 (5.0)	2.2482–2.2490	0.0004–0.0015	0.004–0.008	3	2.1228–2.1236	0.0008–0.0015	0.010–0.020
	E	8-302 (5.0) HO	2.2482–2.2490	0.0004–0.0015	0.004–0.008	3	2.1228–2.1236	0.0008–0.0015	0.010–0.020
	G	8-351 (5.8)	2.2994–3.0002	0.0008–0.0015	0.004–0.008	3	2.3103–2.3111	0.0008–0.0015	0.010–0.020
1991–92	M	4-140 (2.3)	2.2051–2.2059	0.0008–0.0015	0.003–0.008	3	2.0462–2.0472	0.0008–0.0015	0.0035–0.0105
	4	6-232 (3.8)	2.5190–2.5198	0.0010–0.0014	0.004–0.008	3	2.3103–2.3111	0.0010–0.0014	0.0047–0.0114
	R	6-232 (3.8) SC	③	④	0.004–0.008	3	2.3103–2.3111	0.0010–0.0014	0.0047–0.0114

CRANKSHAFT AND CONNECTING ROD SPECIFICATIONS

All measurements are given in inches.

Year	VIN	No. Cylinder Displacement cu. in. (liter)	Crankshaft Main Brg. Journal Dia.	Crankshaft Main Brg. Oil Clearance	Crankshaft Shaft End-play	Crankshaft Thrust on No.	Connecting Rod Journal Diameter	Connecting Rod Oil Clearance	Connecting Rod Side Clearance
	C	6-232 (3.8) SC	③	④	0.004–0.008	3	2.3103–2.3111	0.0010–0.0014	0.0047–0.0114
	W	8-280 (4.6)	2.6575	0.0011–0.0026	0.005–0.010	5	2.0866	0.0011–0.0027	0.0006–0.0177
	F	8-302 (5.0)	2.2482–2.2490	0.0004–0.0015	0.004–0.008	3	2.1228–2.1236	0.0008–0.0015	0.010–0.020
	T	8-302 (5.0) HO	2.2482–2.2490	0.0004–0.0015	0.004–0.008	3	2.1228–2.1236	0.0008–0.0015	0.010–0.020
	E	8-302 (5.0) HO	2.2482–2.2490	0.0004–0.0015	0.004–0.008	3	2.1228–2.1236	0.0008–0.0015	0.010–0.020
	G	8-351 (5.8)	2.2994–3.0002	0.0008–0.0015	0.004–0.008	3	2.3103–2.3111	0.0008–0.0015	0.010–0.020

① No. 1, 2, 3: 2.5194–2.5186 in.
  No. 4: 2.5100–2.5092 in.
② No. 1, 2, 3: 0.0009–0.0026 in.
  No. 4: 0.0014–0.0032 in.
③ No. 1, 2, 3: 2.5190–2.5198
  No. 4: 2.5104–2.5096
④ No. 1, 2, 3: 0.0005–0.0023
  No. 4: 0.0010–0.0028

VALVE SPECIFICATIONS

Year	VIN	No. Cylinder Displacement cu. in. (liter)	Seat Angle (deg.)	Face Angle (deg.)	Spring Test Pressure (lbs.)	Spring Installed Height (in.)	Stem-to-Guide Clearance (in.) Intake	Stem-to-Guide Clearance (in.) Exhaust	Stem Diameter (in.) Intake	Stem Diameter (in.) Exhaust
1988	A	4-140 (2.3)	45	44	128–141 @ 1.12	1.52	0.0010–0.0027	0.0015–0.0032	0.3416–0.3423	0.3411–0.3418
	T	4-140 (2.3)	45	44	152–166 @ 1.12	1.52	0.0010–0.0027	0.0015–0.0032	0.3416–0.3423	0.3411–0.3418
	4	6-232 (3.8)	44.5	45.8	190 @ 1.28	1.70	0.0010–0.0028	0.0015–0.0033	0.3415–0.3423	0.3410–0.3418
	F	8-302 (5.0)	45	44	①	③	0.0010–0.0027	0.0015–0.0032	0.3416–0.3423	0.3411–0.3418
	E	8-302 (5.0) HO	45	44	②	③	0.0010–0.0027	0.0015–0.0032	0.3416–0.3423	0.3411–0.3418
	G	8-351 (5.8)	45	44	195–215 @ 1.05	③	0.0010–0.0027	0.0015–0.0032	0.3416–0.3423	0.3411–0.3418
1989	A	4-140 (2.3)	45	44	128–141 @ 1.12	1.52	0.0010–0.0027	0.0015–0.0032	0.3416–0.3423	0.3411–0.3418
	4	6-232 (3.8)	44.5	45.8	220 @ 1.18	1.65	0.0010–0.0028	0.0015–0.0033	0.3415–0.3423	0.3410–0.3418
	R	6-232 (3.8) SC	44.5	45.8	220 @ 1.18	1.65	0.0010–0.0028	0.0015–0.0033	0.3415–0.3423	0.3410–0.3418
	C	6-232 (3.8) SC	44.5	45.8	220 @ 1.18	1.65	0.0010–0.0028	0.0015–0.0033	0.3415–0.3423	0.3410–0.3418
	F	8-302 (5.0)	45	44	①	③	0.0010–0.0027	0.0015–0.0032	0.3416–0.3423	0.3411–0.3418
	E	8-302 (5.0) HO	45	44	②	③	0.0010–0.0027	0.0015–0.0032	0.3416–0.3423	0.3411–0.3418
	G	8-351 (5.8)	45	44	195–215 @ 1.05	③	0.0010–0.0027	0.0015–0.0032	0.3416–0.3423	0.3411–0.3418

VALVE SPECIFICATIONS

Year	VIN	No. Cylinder Displacement cu. in. (liter)	Seat Angle (deg.)	Face Angle (deg.)	Spring Test Pressure (lbs.)	Spring Installed Height (in.)	Stem-to-Guide Clearance (in.)		Stem Diameter (in.)	
							Intake	Exhaust	Intake	Exhaust
1990	A	4-140 (2.3)	45	44	128–141 @ 1.12	1.52	0.0010–0.0027	0.0015–0.0032	0.3416–0.3423	0.3411–0.3418
	4	6-232 (3.8)	44.5	45.8	220 @ 1.18	1.65	0.0010–0.0028	0.0015–0.0033	0.3415–0.3423	0.3410–0.3418
	R	6-232 (3.8) SC	44.5	45.8	220 @ 1.18	1.65	0.0010–0.0028	0.0015–0.0033	0.3415–0.3423	0.3410–0.3418
	C	6-232 (3.8) SC	44.5	45.8	220 @ 1.18	1.65	0.0010–0.0028	0.0015–0.0033	0.3415–0.3423	0.3410–0.3418
	F	8-302 (5.0)	45	44	①	③	0.0010–0.0027	0.0015–0.0032	0.3416–0.3423	0.3411–0.3418
	E	8-302 (5.0) HO	45	44	②	③	0.0010–0.0027	0.0015–0.0032	0.3416–0.3423	0.3411–0.3418
	G	8-351 (5.8)	45	44	195–215 @ 1.05	③	0.0010–0.0027	0.0015–0.0032	0.3416–0.3423	0.3411–0.3418
1991–92	M	4-140 (2.3)	45	44	128–141 @ 1.12	1.52	0.0010–0.0027	0.0015–0.0032	0.3416–0.3423	0.3411–0.3418
	4	6-232 (3.8)	44.5	45.8	220 @ 1.18	1.65	0.0010–0.0028	0.0015–0.0033	0.3415–0.3423	0.3410–0.3418
	R	6-232 (3.8) SC	44.5	45.8	220 @ 1.18	1.65	0.0010–0.0028	0.0015–0.0033	0.3415–0.3423	0.3410–0.3418
	C	6-232 (3.8) SC	44.5	45.8	220 @ 1.18	1.65	0.0010–0.0028	0.0015–0.0033	0.3415–0.3423	0.3410–0.3418
	W	8-280 (4.6)	45	45.5	132 @ 1.10	1.57	0.0008–0.0027	0.0018–0.0037	0.2746–0.2754	0.2736–0.2744
	F	8-302 (5.0)	45	44	①	③	0.0010–0.0027	0.0015–0.0032	0.3416–0.3423	0.3411–0.3418
	T	8-302 (5.0) HO	45	44	②	③	0.0010–0.0027	0.0015–0.0032	0.3416–0.3423	0.3411–0.3418
	E	8-302 (5.0) HO	45	44	②	③	0.0010–0.0027	0.0015–0.0032	0.3416–0.3423	0.3411–0.3418
	G	8-351 (5.8)	45	44	195–215 @ 1.05	③	0.0010–0.0027	0.0015–0.0032	0.3416–0.3423	0.3411–0.3418

① Intake—194–214 @ 1.36 Exhaust—190–210 @ 1.20
② Intake—211–230 @ 1.33 Exhaust—200–226 @ 1.15
③ Intake—1.75–1.80 Exhaust—1.58–1.64

PISTON AND RING SPECIFICATIONS

All measurements are given in inches.

Year	VIN	No. Cylinder Displacement cu. in. (liter)	Piston Clearance	Ring Gap			Ring Side Clearance		
				Top Compression	Bottom Compression	Oil Control	Top Compression	Bottom Compression	Oil Control
1988	A	4-140 (2.3)	0.0030–0.0038	0.0100–0.0200	0.0100–0.0200	0.0150–0.0550	0.0020–0.0040	0.0020–0.0040	Snug
	T	4-140 (2.3)	0.0030–0.0038	0.0100–0.0200	0.0100–0.0200	0.0150–0.0550	0.0020–0.0040	0.0020–0.0040	Snug
	4	6-232 (3.8)	0.0014–0.0032	0.0100–0.0220	0.0100–0.0220	0.0150–0.0550	0.0016–0.0037	0.0016–0.0037	Snug
	F	8-302 (5.0)	0.0014–0.0022	0.0100–0.0200	0.0100–0.0200	0.0150–0.0550	0.0020–0.0040	0.0020–0.0040	Snug
	E	8-302 (5.0) HO	0.0030–0.0038	0.0100–0.0200	0.0100–0.0200	0.0150–0.0550	0.0020–0.0040	0.0020–0.0040	Snug

PISTON AND RING SPECIFICATIONS

All measurements are given in inches.

Year	VIN	No. Cylinder Displacement cu. in. (liter)	Piston Clearance	Ring Gap			Ring Side Clearance		
				Top Compression	Bottom Compression	Oil Control	Top Compression	Bottom Compression	Oil Control
	G	8-351 (5.8)	0.0018–0.0026	0.0100–0.0220	0.0100–0.0220	0.0150–0.0550	0.0020–0.0040	0.0020–0.0040	Snug
1989	A	4-140 (2.3)	0.0030–0.0038	0.0100–0.0200	0.0100–0.0200	0.0150–0.0550	0.0020–0.0040	0.0020–0.0040	Snug
	4	6-232 (3.8)	0.0014–0.0032	0.0110–0.0120	0.0090–0.0200	0.0150–0.0550	0.0016–0.0034	0.0016–0.0034	Snug
	R	6-232 (3.8) SC	0.0040–0.0045	0.0110–0.0120	0.0090–0.0200	0.0150–0.0580	0.0016–0.0034	0.0016–0.0034	Snug
	C	6-232 (3.8) SC	0.0040–0.0045	0.0110–0.0120	0.0900–0.0200	0.0150–0.0580	0.0016–0.0034	0.0016–0.0034	Snug
	F	8-302 (5.0)	0.0014–0.0022	0.0100–0.0200	0.0100–0.0200	0.0150–0.0550	0.0020–0.0040	0.0020–0.0040	Snug
	E	8-302 (5.0) HO	0.0030–0.0038	0.0100–0.0200	0.0100–0.0200	0.0150–0.0550	0.0020–0.0040	0.0020–0.0040	Snug
	G	8-351 (5.8)	0.0018–0.0026	0.0100–0.0220	0.0100–0.0220	0.0150–0.0550	0.0020–0.0040	0.0020–0.0040	Snug
1990	A	4-140 (2.3)	0.0030–0.0038	0.0100–0.0200	0.0100–0.0200	0.0150–0.0550	0.0020–0.0040	0.0020–0.0040	Snug
	4	6-232 (3.8)	0.0014–0.0032	0.0110–0.0120	0.0090–0.0200	0.0150–0.0550	0.0016–0.0034	0.0016–0.0034	Snug
	R	6-232 (3.8) SC	0.0035–0.0040	0.0110–0.0120	0.0090–0.0200	0.0150–0.0580	0.0016–0.0034	0.0016–0.0034	Snug
	C	6-232 (3.8) SC	0.0035–0.0040	0.0110–0.0120	0.0090–0.0200	0.0150–0.0580	0.0016–0.0034	0.0016–0.0034	Snug
	F	8-302 (5.0)	0.0014–0.0022	0.0100–0.0200	0.0100–0.0200	0.0150–0.0550	0.0020–0.0040	0.0020–0.0040	Snug
	E	8-302 (5.0) HO	0.0030–0.0038	0.0100–0.0200	0.0100–0.0200	0.0150–0.0550	0.0020–0.0040	0.0020–0.0040	Snug
	G	8-351 (5.8)	0.0018–0.0026	0.0100–0.0200	0.0100–0.0200	0.0150–0.0550	0.0020–0.0040	0.0020–0.0040	Snug
1991–92	M	4-140 (2.3)	0.0024–0.0033	0.0100–0.0200	0.0100–0.0200	0.0150–0.0490	0.0016–0.0033	0.0016–0.0033	Snug
	4	6-232 (3.8)	0.0014–0.0032	0.0110–0.0120	0.0090–0.0200	0.0150–0.0583	0.0016–0.0034	0.0016–0.0034	Snug
	R	6-232 (3.8) SC	0.0035–0.0040	0.0110–0.0120	0.0090–0.0200	0.0150–0.0583	0.0016–0.0034	0.0016–0.0034	Snug
	C	6-232 (3.8) SC	0.0035–0.0040	0.0110–0.0120	0.0090–0.0200	0.0150–0.0583	0.0016–0.0034	0.0016–0.0034	Snug
	W	8-280 (4.6)	0.0008–0.0018	0.0090–0.0193	0.0090–0.0193	0.0100–0.0300	0.0016–0.0035	0.0012–0.0031	Snug
	F	8-302 (5.0)	0.0014–0.0022	0.0100–0.0200	0.0100–0.0200	0.0150–0.0550	0.0020–0.0040	0.0020–0.0040	Snug
	T	8-302 (5.0) HO	0.0030–0.0038	0.0100–0.0200	0.0100–0.0200	0.0150–0.0550	0.0020–0.0040	0.0020–0.0040	Snug
	E	8-302 (5.0) HO	0.0030–0.0038	0.0100–0.0200	0.0100–0.0200	0.0150–0.0550	0.0020–0.0040	0.0020–0.0040	Snug
	G	8-351 (5.8)	0.0018–0.0026	0.0100–0.0200	0.0100–0.0200	0.0150–0.0550	0.0020–0.0040	0.0020–0.0040	Snug

TORQUE SPECIFICATIONS
All readings in ft. lbs.

Year	VIN	No. Cylinder Displacement cu. in. (liter)	Cylinder Head Bolts	Main Bearing Bolts	Rod Bearing Bolts	Crankshaft Pulley Bolts	Flywheel Bolts	Manifold Intake	Manifold Exhaust	Spark Plugs
1988	A	4-140 (2.3)	①	②	③	103–133	56–64	15–22	④	5–10
	T	4-140 (2.3)	①	②	③	103–133	56–64	15–22	④	5–10
	4	6-232 (3.8)	⑤	65–81	31–36	93–121	54–64	⑥	15–22	5–11
	F	8-302 (5.0)	⑦	60–70	19–24	70–90	75–85	23–25⑧	18–24	5–10
	E	8-302 (5.0) HO	⑦	60–70	19–24	70–90	75–85	23–25⑧	18–24	5–10
	G	8-351 (5.8)	⑨	90–105	40–45	70–90	75–85	23–25⑧	18–24	10–15
1989	A	4-140 (2.3)	①	②	③	103–133	56–64	20–29	④	5–10
	4	6-232 (3.8)	⑤	65–81	31–36	93–121	54–64	⑥	15–22	5–11
	R	6-232 (3.8) SC	⑤	65–81	31–36	93–121	54–64	⑩	15–22	5–11
	C	6-232 (3.8) SC	⑤	65–81	31–36	93–121	54–64	⑩	15–22	5–11
	F	8-302 (5.0)	⑦	60–70	19–24	70–90	75–85	15⑧	18–24	5–10
	E	8-302 (5.0) HO	⑦	60–70	19–24	70–90	75–85	⑪	18–24	5–10
	G	8-351 (5.8)	⑨	90–105	40–45	70–90	75–85	23–25⑧	18–24	10–15
1990	A	4-140 (2.3)	①	②	③	103–133	56–64	20–29	④	5–10
	4	6-232 (3.8)	⑤	65–81	31–36	103–132	54–64	⑥	15–22	5–11
	R	6-232 (3.8) SC	⑤	65–81	31–36	103–132	54–64	⑩	15–22	5–11
	C	6-232 (3.8) SC	⑤	65–81	31–36	103–132	54–64	⑩	15–22	5–11
	F	8-302 (5.0)	⑦	60–70	19–24	70–90	75–85	15⑧	18–24	5–10
	E	8-302 (5.0) HO	⑦	60–70	19–24	70–90	75–85	⑪	18–24	5–10
	G	8-351 (5.8)	⑨	90–105	40–45	70–90	75–85	23–25	18–24	10–15
1991–92	M	4-140 (2.3)	①	②	③	114–151	56–64	19–28	④	5–10
	4	6-232 (3.8)	⑫	65–81	31–36	103–132	54–64	⑥	15–22	5–11
	R	6-232 (3.8) SC	⑬	65–81	31–36	103–132	54–64	⑩	15–22	5–11
	C	6-232 (3.8) SC	⑬	65–81	31–36	103–132	54–64	⑩	15–22	5–11
	W	8-280 (4.6)	⑭	⑮	⑯	114–121		53–64⑧	15–22	6–7
	F	8-302 (5.0)	⑦	60–70	19–24	70–90	75–85	23–25⑧	18–24	5–10
	T	8-302 (5.0) HO	⑦	60–70	19–24	70–90	75–85	⑪	18–24	5–10
	E	8-302 (5.0) HO	⑦	60–70	19–24	70–90	75–85	⑪	18–24	5–10
	G	8-351 (5.8)	⑨	90–105	40–45	70–90	75–85	23–25⑧	18–24	10–15

① Torque in sequence in 2 steps:
Step 1—50–60 ft. lbs.
Step 2—80–90 ft. lbs.

② Tighten in 2 steps:
Step 1—50–60 ft. lbs.
Step 2—75–85 ft. lbs.

③ Tighten in 2 steps:
Step 1—25–30 ft. lbs.
Step 2—30–36 ft. lbs.

④ Tighten in 2 steps:
Step 1—178–204 inch lbs.
Step 2—20–30 ft. lbs.

⑤ A. Tighten in 4 steps:
Step 1—37 ft. lbs.
Step 2—45 ft. lbs.
Step 3—52 ft. lbs.
Step 4—50 ft. lbs.
B. Back off all bolts 2–3 turns
C. Repeat step A.

⑥ Tighten in 3 steps:
Step 1—7 ft. lbs.

Step 2—15 ft. lbs.
Step 3—24 ft. lbs.

⑦ Tighten in 2 steps:
Step 1—55–65 ft. lbs.
Step 2—65–72 ft. lbs.

⑧ After assembly, retorque with engine hot.

⑨ Tighten in 3 steps:
Step 1—85 ft. lbs.
Step 2—95 ft. lbs.
Step 3—105–112 ft. lbs.

⑩ Tighten in 2 steps:
Step 1—7.5 ft. lbs.
Step 2—11 ft. lbs.

⑪ Tighten in 2 steps:
Step 1—15–20 ft. lbs.
Step 2—23–25 ft. lbs.

⑫ A. Tighten in 4 steps:
Step 1—37 ft. lbs.
Step 2—45 ft. lbs.
Step 3—52 ft. lbs.
Step 4—59 ft. lbs.

B. Back off all bolts 2–3 turns
C. Tighten all bolts in sequence to 11–18 ft. lbs., then rotate long bolts an additional 85–105 degrees and short bolts an additional 65–85 degrees. Go to next bolt in sequence.

⑬ A. Tighten in 4 steps:
Step 1—37 ft. lbs.
Step 2—45 ft. lbs.
Step 3—52 ft. lbs.
Step 4—59 ft. lbs.
B. Back off all bolts 2–3 turns
C. Tighten all bolts in sequence to 48–55 ft. lbs., then rotate an additional 90–110 degrees. Go to next bolt in sequence.

⑭ Tighten in 3 steps:
Step 1—15–22 ft. lbs.
Step 2—Rotate 85–95 degrees
Step 3—Rotate 85–95 degrees

⑮ Tighten in 2 steps:
Step 1—22–25 ft. lbs.
Step 2—Rotate 85–95 degrees

⑯ Tighten in 2 steps:
Step 1—18–25 ft. lbs.
Step 2—Rotate 85–95 degrees

BRAKE SPECIFICATIONS
All measurements in inches unless noted

Year	Model	Lug Nut Torque (ft. lbs.)	Master Cylinder Bore	Brake Disc Minimum Thickness	Brake Disc Maximum Runout	Standard Brake Drum Diameter	Minimum Lining Thickness Front	Minimum Lining Thickness Rear
1988	Mustang	80–105	0.872	①	0.003	9.00	0.125	0.030
	T-Bird	80–105	0.872②	③	0.003	④	0.125	⑤
	Cougar	80–105	0.872②	③	0.003	④	0.125	⑤
	Mark VII	80–105	NA	⑥	0.003	—	0.125	0.125
	Crown Victoria	80–105	1.0	0.972	0.003	⑦	0.125	0.030
	Grand Marquis	80–105	1.0	0.972	0.003	⑦	0.125	0.030
	Town Car	80–105	1.0	0.972	0.003	10.00	0.125	0.030
1989	Mustang	80–105	0.872	①	0.003	9.00	0.125	0.030
	T-Bird	80–105	0.872②	⑧	0.003	9.80	0.125	⑤
	Cougar	80–105	0.872②	⑧	0.003	9.80	0.125	⑤
	Mark VII	80–105	NA	⑥	⑨	—	0.125	0.125
	Crown Victoria	80–105	1.0	0.972	0.003	⑦	0.125	0.030
	Grand Marquis	80–105	1.0	0.972	0.003	⑦	0.125	0.030
	Town Car	80–105	1.0	0.972	0.003	10.00	0.125	0.030
1990	Mustang	80–105	0.872	①	0.003	9.00	0.125	0.030
	T-Bird	80–105	0.938②	⑩	0.003	9.80	0.125	⑤
	Cougar	80–105	0.938②	⑩	0.003	9.80	0.125	⑤
	Mark VII	80–105	NA	⑥	⑨	—	0.125	0.125
	Crown Victoria	80–105	1.0	0.972	0.003	⑦	0.125	0.125
	Grand Marquis	80–105	1.0	0.972	0.003	⑦	0.125	0.125
	Town Car	80–105	1.0	0.972	0.003	10.00	0.125	0.125
1991–92	Mustang	80–105	0.872	①	0.003	9.00	0.125	0.030
	T-Bird	80–105	0.938②	⑩	0.003	9.80	0.125	⑤
	Cougar	80–105	0.938②	⑩	0.003	9.80	0.125	⑤
	Mark VII	80–105	NA	⑥	⑪	—	0.125	0.123
	Crown Victoria	80–105	1.0	⑫	⑬	⑦	0.125	⑭
	Grand Marquis	80–105	1.0	⑫	⑬	⑦	0.125	⑭
	Town Car	80–105	1.0	⑫	0.003	—	0.125	0.220

① 2.3L engine—0.810
   5.0L engine—0.972
② Except ABS equipped
③ Except Turbo Coupe—0.810
   Turbo Coupe Front—0.972
   Rear—0.895
④ Except heavy duty models—9.00
   Heavy duty models—10.00
⑤ With drum brakes—0.030
   With disc brakes—0.123
⑥ Front—0.972
   Rear—0.895

⑦ Except station wagon, police, taxi and trailer towing package—10.00
   Station wagon, police, taxi and trailer towing package—11.030
⑧ Front—0.935
   Rear—0.895
⑨ Front—0.003
   Rear—0.004
⑩ Front—0.935
   Rear—0.900

⑪ Front—0.003
   Rear—0.002
⑫ Front—0.972
   Rear—0.440
⑬ All except 1992 front—0.003
   1992 front—0.002
⑭ With drum brakes—0.030
   With disc brakes—0.220

WHEEL ALIGNMENT

Year	Model		Caster Range (deg.)	Caster Preferred Setting (deg.)	Camber Range (deg.)	Camber Preferred Setting (deg.)	Toe-in (in.)	Steering Axis Inclination (deg.)
1988	Mustang	Exc. 5.0L GT	$15\frac{5}{32}$P–$1\frac{29}{32}$P	$1\frac{3}{16}$P	$\frac{27}{32}$N–$\frac{21}{32}$P	$\frac{3}{32}$N	$\frac{1}{16}$–$\frac{5}{16}$	$15\frac{23}{32}$
		5.0L GT	$\frac{1}{2}$P–2P	$\frac{19}{32}$P	$\frac{5}{8}$N–$\frac{29}{32}$P	$\frac{5}{32}$P	$\frac{1}{16}$–$\frac{5}{16}$	$15\frac{23}{32}$
	Thunderbird	Exc. Turbo	$\frac{15}{32}$P–$1\frac{31}{32}$P	$1\frac{7}{32}$P	$1\frac{13}{32}$N–$\frac{3}{32}$P	$\frac{21}{32}$N	$\frac{1}{16}$–$\frac{5}{16}$	$15\frac{23}{32}$
		Turbo	$\frac{19}{32}$P–$2\frac{3}{32}$P	$1\frac{11}{32}$P	$\frac{3}{4}$N–$\frac{3}{4}$P	0	$\frac{1}{16}$–$\frac{5}{16}$	$15\frac{23}{32}$
	Cougar	—	$\frac{15}{32}$P–$1\frac{31}{32}$P	$1\frac{17}{32}$P	$1\frac{13}{32}$N–$\frac{3}{32}$P	$\frac{21}{32}$N	$\frac{1}{16}$–$\frac{5}{16}$	$15\frac{23}{32}$
	Mark VII	—	$\frac{5}{8}$P–$2\frac{3}{4}$P	$1\frac{1}{2}$P	$\frac{3}{4}$N–$\frac{3}{4}$P	0	0–$\frac{1}{4}$	11
	Crown Victoria	—	$3\frac{1}{4}$P–5P	4P	$1\frac{1}{4}$N–$\frac{1}{4}$P	$\frac{1}{2}$N	$\frac{1}{16}$–$\frac{3}{16}$	11
	Grand Marquis	—	$3\frac{1}{4}$P–5P	4P	$1\frac{1}{4}$N–$\frac{1}{4}$P	$\frac{1}{2}$N	$\frac{1}{16}$–$\frac{3}{16}$	11
	Town Car	—	$3\frac{1}{4}$P–5P	4P	$1\frac{1}{4}$N–$\frac{1}{4}$P	$\frac{1}{2}$N	$\frac{1}{16}$–$\frac{3}{16}$	11
1989	Mustang	Exc. 5.0L GT	$15\frac{5}{32}$P–$1\frac{29}{32}$P	$1\frac{3}{16}$P	$\frac{27}{32}$N–$\frac{21}{32}$P	$\frac{3}{32}$N	$\frac{1}{16}$–$\frac{5}{16}$	$15\frac{23}{32}$
		5.0L GT	$\frac{1}{2}$P–2P	$\frac{19}{32}$P	$\frac{5}{8}$N–$\frac{29}{32}$P	$\frac{5}{32}$P	$\frac{1}{16}$–$\frac{5}{16}$	$15\frac{23}{32}$
	Thunderbird	Front	$4\frac{3}{4}$P–$6\frac{1}{4}$P	$5\frac{1}{2}$P	$1\frac{1}{4}$N–$\frac{1}{4}$P	$\frac{1}{2}$N	0–$\frac{1}{4}$	$15\frac{23}{32}$
		Rear	—	—	1N–0	$\frac{1}{2}$N	$\frac{1}{16}$N–$\frac{3}{16}$P	—
	Cougar	Front	$4\frac{3}{4}$P–$6\frac{1}{4}$P	$5\frac{1}{2}$P	$1\frac{1}{4}$N–$\frac{1}{4}$P	$\frac{1}{2}$N	0–$\frac{1}{4}$	$15\frac{23}{32}$
		Rear	—	—	1N–0	$\frac{1}{2}$N	$\frac{1}{16}$N–$\frac{3}{16}$P	—
	Mark VII	—	$\frac{5}{8}$P–$2\frac{3}{4}$P	$1\frac{1}{2}$P	$\frac{3}{4}$N–$\frac{3}{4}$P	0	0–$\frac{1}{4}$	11
	Crown Victoria	—	$2\frac{1}{2}$P–$4\frac{1}{2}$P	$3\frac{1}{2}$P	$1\frac{1}{4}$N–$\frac{1}{4}$P	$\frac{1}{2}$N	$\frac{1}{16}$N–$\frac{3}{16}$P	11
	Grand Marquis	—	$2\frac{1}{2}$P–$4\frac{1}{2}$P	$3\frac{1}{2}$P	$1\frac{1}{4}$N–$\frac{1}{4}$P	$\frac{1}{2}$N	$\frac{1}{16}$N–$\frac{3}{16}$P	11
	Town Car	—	$2\frac{1}{2}$P–$4\frac{1}{2}$P	$3\frac{1}{2}$P	$1\frac{1}{4}$N–$\frac{1}{4}$P	$\frac{1}{2}$N	$\frac{1}{16}$N–$\frac{3}{16}$P	11
1990	Mustang	Exc. 5.0L GT	$15\frac{5}{32}$P–$2\frac{5}{8}$P	$1\frac{29}{32}$P	$\frac{1}{4}$N–$\frac{1}{4}$P	$\frac{1}{2}$N	$\frac{1}{4}$N–0	$15\frac{23}{32}$
		5.0L GT	$15\frac{5}{32}$P–$2\frac{5}{8}$P	$1\frac{29}{32}$P	$1\frac{3}{8}$N–$\frac{1}{8}$P	$\frac{5}{8}$N	$\frac{1}{4}$N–0	$15\frac{23}{32}$
	Thunderbird	Front	$4\frac{3}{4}$P–$6\frac{1}{4}$P	$5\frac{1}{2}$P	$1\frac{1}{4}$N–$\frac{1}{4}$P	$\frac{1}{2}$N	0–$\frac{1}{4}$	$15\frac{23}{32}$
		Rear	—	—	1N–0	$\frac{1}{2}$N	$\frac{1}{16}$N–$\frac{3}{16}$P	—
	Cougar	Front	$4\frac{3}{4}$P–$6\frac{1}{4}$P	$5\frac{1}{2}$P	$1\frac{1}{4}$N–$\frac{1}{4}$P	$\frac{1}{2}$N	0–$\frac{1}{4}$	$15\frac{23}{32}$
		Rear	—	—	1N–0	$\frac{1}{2}$N	$\frac{1}{16}$N–$\frac{3}{16}$P	—
	Mark VII	—	$\frac{5}{8}$P–$2\frac{3}{4}$P	$1\frac{1}{2}$P	$\frac{3}{4}$N–$\frac{3}{4}$P	0	0–$\frac{1}{4}$	11
	Crown Victoria	—	$2\frac{1}{2}$P–$4\frac{1}{2}$P	$3\frac{1}{2}$P	$1\frac{1}{4}$N–$\frac{1}{4}$P	$\frac{1}{2}$N	$\frac{1}{16}$N–$\frac{3}{16}$P	11
	Grand Marquis	—	$2\frac{1}{2}$P–$4\frac{1}{2}$P	$3\frac{1}{2}$P	$1\frac{1}{4}$N–$\frac{1}{4}$P	$\frac{1}{2}$N	$\frac{1}{16}$N–$\frac{3}{16}$P	11
	Town Car	—	$2\frac{1}{2}$P–$4\frac{1}{2}$P	$3\frac{1}{2}$P	$1\frac{1}{4}$N–$\frac{1}{4}$P	$\frac{1}{2}$N	$\frac{1}{16}$N–$\frac{3}{16}$P	11
1991–92	Mustang	Exc. 5.0L GT	$15\frac{5}{32}$P–$2\frac{5}{8}$P	$1\frac{29}{32}$P	$\frac{1}{4}$N–$\frac{1}{4}$P	$\frac{1}{2}$N	$\frac{1}{4}$N–0	$15\frac{23}{32}$
		5.0L GT	$15\frac{5}{32}$P–$2\frac{5}{8}$P	$1\frac{29}{32}$P	$1\frac{3}{8}$N–$\frac{1}{8}$P	$\frac{5}{8}$N	$\frac{1}{4}$N–0	$15\frac{23}{32}$
	Thunderbird	Front	$4\frac{3}{4}$P–$6\frac{1}{4}$P	$5\frac{1}{2}$P	$1\frac{1}{4}$N–$\frac{1}{4}$P	$\frac{1}{2}$N	0–$\frac{1}{4}$	$15\frac{23}{32}$
		Rear	—	—	1N–0	$\frac{1}{2}$N	$\frac{1}{16}$N–$\frac{3}{16}$P	—
	Cougar	Front	$4\frac{3}{4}$P–$6\frac{1}{4}$P	$5\frac{1}{2}$P	$1\frac{1}{4}$N–$\frac{1}{4}$P	$\frac{1}{2}$N	0–$\frac{1}{4}$	$15\frac{23}{32}$
		Rear	—	—	1N–0	$\frac{1}{2}$N	$\frac{1}{16}$N–$\frac{3}{16}$P	—
	Mark VII	—	$\frac{5}{8}$P–$2\frac{3}{4}$P	$1\frac{1}{2}$P	$\frac{3}{4}$N–$\frac{3}{4}$P	0	0–$\frac{1}{4}$	11
	Crown Victoria	—	$2\frac{1}{2}$P–$4\frac{1}{2}$P	$3\frac{1}{2}$P	$1\frac{1}{4}$N–$\frac{1}{4}$P	$\frac{1}{2}$N	$\frac{1}{16}$N–$\frac{3}{16}$P	11
	Grand Marquis	—	$2\frac{1}{2}$P–$4\frac{1}{2}$P	$3\frac{1}{2}$P	$1\frac{1}{4}$N–$\frac{1}{4}$P	$\frac{1}{2}$N	$\frac{1}{16}$N–$\frac{3}{16}$P	11
	Town Car	—	$4\frac{3}{4}$P–$6\frac{1}{4}$P	$5\frac{1}{2}$P	$1\frac{1}{4}$N–$\frac{1}{4}$P	$\frac{1}{2}$N	$\frac{1}{16}$N–$\frac{3}{16}$P	11

P—Positive
N—Negative

ENGINE MECHANICAL

NOTE: Disconnecting the negative battery cable on some vehicles may interfere with the functions of the on board computer systems and may require the computer to undergo a relearning process, once the negative battery cable is reconnected.

Engine Assembly

REMOVAL & Installation

2.3L Engine

1. Disconnect the negative battery cable and relieve the fuel system pressure.

2. Drain the cooling system and the crankcase.

3. Mark the position of the hood on the hinges and remove the hood.

4. On non-turbocharged engines, remove the air cleaner outlet hose. On turbocharged engines, disconnect the zip tube from the turbocharger inlet. Remove the ground strap on the turbocharger inlet elbow.

5. Remove the radiator upper and lower hoses. Disconnect the electrical connector to the cooling fan and remove the fan and shroud. If equipped with automatic transmission, disconnect the oil cooler lines from the radiator. Remove the radiator.

6. Disconnect the heater hose from the water pump. Tag and disconnect the wires from the alternator and starter. Disconnect the accelerator cable from the throttle body.

7. If equipped with air conditioning, remove the compressor from the mounting bracket and position it out of the way, leaving the refrigerant lines attached.

8. If equipped with power steering, remove the pump and position out of the way, leaving the hoses attached.

9. Disconnect the flexible fuel line at the fuel rail and plug the fuel line.

10. Disconnect the coil primary wire, the water temperature sending unit connector and the injector wiring harness connectors from the main wiring harness.

11. Remove the starter and remove the engine mount bolts.

12. Raise and safely support the vehicle. Remove the flywheel or converter housing upper retaining bolts.

13. Disconnect the muffler inlet pipe at the exhaust manifold. Disconnect the engine right and left mounts at the No. 2 crossmember pedestals. Remove

the flywheel or converter housing cover.

14. If equipped with a manual transmission, remove the flywheel housing lower retaining bolts. If equipped with an automatic transmission, disconnect the converter from the flywheel and disconnect the transmission oil cooler lines, if attached to the engine at the pan rail. Remove the converter housing lower retaining bolts.

15. Lower the vehicle. Support the transmission and flywheel or converter housing with a jack.

16. Attach suitable engine lifting equipment to the existing lifting brackets. Carefully lift the engine out of the engine compartment and install on a work stand.

To install:

17. Install the clutch, if removed.

18. Carefully lower the engine into the engine compartment. Make sure the studs on the exhaust manifold or turbocharger are aligned with the holes in the muffler inlet pipe.

19. If equipped with an automatic transmission, start the converter pilot into the crankshaft. If equipped with a manual transmission, start the transmission input shaft into the clutch disc. It may be necessary to adjust the position of the transmission in relation to the engine if the input shaft will not enter the clutch disc.

NOTE: If the engine hangs up after the shaft enters, turn the crankshaft slowly in a clockwise direction, with the transmission in gear, until the shaft splines mesh with the clutch disc splines.

20. Install the flywheel or converter housing upper retaining bolts. Remove the engine lifting equipment.

21. Remove the jack from the transmission. Raise and safely support the vehicle.

22. Install the flywheel or converter housing lower retaining bolts. If equipped with an automatic transmission, attach the converter to the flywheel and tighten the retaining nuts to 20–34 ft. lbs. (27–46 Nm).

23. Install the flywheel or converter housing dust cover. Install the left and right engine mounts to the No. 2 crossmember pedestal. Tighten the nuts and bolts on Thunderbird to 33–45 ft. lbs. (45–61 Nm) or Mustang to 80–106 ft. lbs.

24. Connect the muffler inlet pipe to the manifold or turbocharger. Connect the fuel line to the fuel rail.

25. Install the starter and connect the starter cable.

26. Lower the vehicle. Connect the oil pressure and water temperature sending unit connectors. Connect the coil and alternator wires. Connect the accelerator cable and the heater hoses.

27. If equipped with air conditioning, install the compressor in the mounting bracket. If equipped with power steering, install the pump. Install the drive belt.

28. Install the radiator, cooling fan and shroud. Connect the fan electrical connector. If equipped with automatic transmission, connect the oil cooler lines to the radiator. Install the upper and lower radiator hoses.

29. On non-turbocharged engines, install the air cleaner outlet hose. On turbocharged engines, connect the air inlet hose to the turbocharger. Connect the ground strap to the turbocharger inlet elbow.

30. Fill the crankcase with the proper type and quantity of oil. Fill and bleed the cooling system.

31. Connect the negative battery cable, start the engine and bring to normal operating temperature. Check for leaks. Check all fluid levels.

32. Align the hood on the hinges with the marks that were made during removal. Secure with the mounting bolts.

3.8L Engine

1. Disconnect the negative battery cable. Drain the cooling system and the crankcase.

2. Relieve the fuel system pressure and, if equipped, discharge the air conditioning system.

3. Disconnect the electrical connector to the underhood lamp. Mark the position of the hood on the hinges and remove the hood.

4. Remove the engine oil dipstick tube. Disconnect the alternator to voltage regulator wiring and remove the radiator upper sight shield.

5. Remove the air cleaner assembly. Remove the fan and the fan shroud.

6. Remove the upper radiator hose and disconnect the heater hoses. Disconnect the transmission oil cooler lines from the radiator.

7. Disconnect the power steering pressure hose assembly. Disconnect the air conditioning compressor clutch wire assembly and disconnect the compressor to condenser line.

8. Remove the coolant recovery reservoir and the wiring shield. Remove the accelerator cable mounting bracket and disconnect the fuel lines from the fuel rail.

9. Disconnect the power steering pump pressure and return tube bracket. Disconnect the engine control module wiring assembly and the ground wire assembly. Tag and disconnect the vacuum hoses.

10. Remove the duct assembly. Disconnect 1 end of the throttle control valve cable. Disconnect the bulkhead electrical connector and transmission pressure switches.

11. Remove the bolts retaining the transmission support assembly and remove the assembly from the vehicle.

12. Raise and safely support the vehicle. Remove the oil filter element.

13. Disconnect the oxygen sensor connector. Remove the accessory drive belts and remove the crankshaft pulley and drive belt tensioners.

14. Disconnect the lower radiator hose. Remove the engine-to-transmission bolts. Remove the water pump pulley and the water pump.

15. Remove the distributor cap and position out of the way. Remove the distributor rotor. Remove the starter.

16. Remove the converter housing assembly and the inlet pipe converter assembly. Remove the left and right front support insulator retaining nuts.

17. Remove the converter-to-flywheel nuts. Disconnect the oil level indicator sensor. Remove the exhaust manifold bolt lock retaining bolts.

18. Remove the thermactor air pump retaining bolts and remove the air pump. Disconnect the oil pressure engine unit gauge assembly.

19. Position engine lifting equipment and jacks. Raise the transmission assembly slightly. Remove the engine assembly from the vehicle and place on a workstand.

To install:

20. Remove the engine assembly from the workstand and position in the vehicle. Install the engine-to-transmission bolts. Remove the engine lifting equipment and jacks.

21. Connect the oil pressure engine unit gauge assembly. Place the air conditioning compressor in position and tighten the retaining bolts.

22. Connect the compressor to condenser discharge line and connect the compressor clutch wire assembly.

23. Connect the heater hoses, fuel lines and vacuum hoses. Connect the engine control module wiring assembly.

24. Install the radiator assembly and connect the transmission oil cooler lines. Partially raise the vehicle and support safely.

25. Install the converter-to-flywheel nuts and the converter housing cover. Install the left and right engine front support insulator nuts.

26. Install the starter and connect the lower radiator hose. Install the belt tensioners and the crankshaft pulley.

27. Install the converter assembly and connect the oxygen sensor. Install a new oil filter and lower the vehicle.

28. Position the thermactor air pump and install the retaining bolts. Install the pulley.

29. Install the wiring shield and install the distributor rotor and cap. Install the coolant recovery reservoir and connect the upper radiator hose.

30. Install the water pump and the pulley. Connect the alternator to voltage regulator wiring. Install the radiator upper sight shield. Partially raise and safely support the vehicle.

31. Connect the wiring assembly ground. Install the accelerator cable mounting bracket.

32. Connect the power steering pressure hose assembly. Install the fan and fan shroud and position the drive belts.

33. Position and install the transmission support assembly.

34. Fill the crankcase with the proper type and quantity of engine oil. Fill and bleed the cooling system.

35. Install the hood, aligning the marks that were made during removal. Connect the negative battery cable.

36. Start the engine and bring to normal operating temperature. Check for leaks. Check all fluid levels. Leak test, evacuate and charge the air conditioning system according to the proper procedure. Observe all safety precautions.

1989–92

1. Disconnect the negative battery cable. Drain the crankcase and the cooling system.

2. Relieve the fuel system pressure and discharge the air conditioning system.

3. Disconnect the electrical connector to the underhood lamp. Mark the position of the hood on the hinges and remove the hood.

4. Remove the left cowl vent screen and wiper module. On non-supercharged engines, disconnect the alternator to voltage regulator wiring assembly.

5. On supercharged engines, remove the upper intercooler tube at the supercharger and cooler assemblies. Remove the bolt retaining the cooler tube to the alternator bracket and remove the tube.

6. Remove the radiator upper sight shield. Release the belt tension and remove the drive and accessory/supercharger belts. Remove the air cleaner-to-throttle body tube.

7. On supercharged engines, disconnect the electric cooling fan electrical connector and remove the cooling fan/shroud assembly. On non-supercharged engines, remove the fan and shroud.

8. Remove the upper radiator hose and disconnect the heater hoses. If equipped with an automatic transmission, disconnect the oil cooler lines from the radiator.

9. Disconnect the lower radiator hose at the water pump. Remove the radiator. On supercharged engines it will also be necessary to remove the 2 push pins retaining the intercooler to the radiator assembly.

10. Disconnect the power steering pressure hose assembly. On non-supercharged engines, remove the power steering pump and bracket assembly and position aside.

11. Disconnect the air conditioner compressor clutch wire assembly. Disconnect and plug the refrigerant lines. Remove the compressor.

12. Remove the coolant recovery reservoir and remove the wiring shield. Remove the accelerator cable mounting bracket and position aside.

13. Disconnect the fuel lines from the fuel rail. Tag and disconnect the vacuum hoses.

14. On non-supercharged engines, disconnect the ground and coil wires. On supercharged engines, disconnect the DIS module wiring, remove the coil pack retaining bolts and position the coil pack aside.

15. On supercharged engines, remove the nuts retaining the lower intercooler tube to the supercharger elbow and lower intercooler tube bracket and remove the intercooler tube retaining bolt and nut at the alternator bracket.

16. On supercharged engines, remove the alternator bracket bolts, disconnect the alternator wiring and remove the alternator. Remove the power steering pump bracket assembly and position aside.

17. Disconnect the canister purge line and disconnect 1 end of the throttle control valve cable.

18. Raise and safely support the vehicle. Remove the oil filter element.

19. On supercharged engines, remove the 2 nuts retaining the lower intercooler tube to the intercooler and remove the intercooler and intercooler tube.

20. Remove the exhaust pipe-to-manifold nuts and remove the left exhaust shield. Disconnect the oxygen sensors.

21. If equipped with an automatic transmission, remove the inspection plug and remove the torque converter bolts.

22. Remove the engine-to-transmission bolts and remove the engine mount through bolts. On supercharged engines, remove the left mount retaining strap bolt.

23. Remove the crankshaft pulley assembly.

NOTE: If the crankshaft pulley and vibration damper have to be separated, mark the damper and pulley so they may be reassembled in the same relative position. This is important as the damper and pulley are initially balanced

as a unit. **If the crankshaft damper is being replaced, check if the original damper has balance pins installed. If so, new balance pins must be installed on the new damper in the same position as the original damper.**

24. Remove the starter. Remove the ground cable and remove the left and right starter harness retainers.

25. Disconnect the oil level indicator sensor and partially lower the vehicle. Disconnect the oil pressure sending unit gauge assembly.

26. Position a floor jack under the transmission and position suitable engine lifting equipment.

27. Remove the engine from the vehicle and position on a workstand.

To install:

28. Remove the engine assembly from the workstand and install engine lifting equipment.

29. Position the engine in the vehicle and install 2 engine-to-transmission bolts. Lower the engine onto the mounting seats, left side first, and remove the lifting equipment. Remove the jacks.

30. Tighten the 2 engine-to-transmission bolts to 40–50 ft. lbs. (55–68 Nm) and connect the oil pressure sending unit gauge assembly. Raise and safely support the vehicle.

31. Install the remaining engine-to-transmission bolts and tighten to 40–50 ft. lbs. (55–68 Nm).

32. Install the torque converter bolts and tighten to 20–34 ft. lbs. (27–46 Nm). Install the inspection plug.

33. Install and tighten the engine mount through bolts to 35–50 ft. lbs. (47–68 Nm). On supercharged engines, install the left mount retaining strap bolt and tighten to 33–45 ft. lbs. (45–61 Nm).

34. Install the starter. Install the starter harness retainer, ground cable and transmission oil cooler line bracket. Install the exhaust pipe-to-manifold nuts.

35. Install the crankshaft pulley assembly and tighten the bolts to 20–28 ft. lbs. (26–30 Nm).

36. Connect the oxygen sensors and the oil level indicator sensor. Install a new oil filter and lower the vehicle.

37. Connect the throttle control valve cable and the canister purge line.

38. On supercharged engines, perform the following:

a. Install the lower intercooler tube, intercooler and power steering pump bracket assembly.

b. Install the alternator, connect the wiring and install the alternator bracket bolts.

c. Install the intercooler tube bolts at the power steering bracket and install the nuts retaining the lower intercooler tube to the lower intercooler tube bracket and supercharger elbow.

d. Install the coil pack and retaining bolts.

39. Install the coolant recovery reservoir.

40. Connect the alternator-to-voltage regulator wiring assembly, the engine control module wiring assembly and engine feed harnesses. Connect the vacuum hoses.

41. On non-supercharged engines, connect the wiring assembly ground and coil wire. On supercharged engines, connect the DIS module wiring.

42. Connect the fuel lines to the fuel rail. Install the accelerator cable mounting bracket and the wiring shield.

43. Install the air conditioning compressor and retaining bolts. Tighten the bolts to 30–45 ft. lbs. (41–61 Nm).

44. Remove the plugs from the air conditioner compressor lines and connect the lines to the compressor. Connect the compressor clutch wire.

45. On non-supercharged engines, install the power steering pump bracket assembly. Connect the power steering hoses.

46. Install the radiator. On supercharged engines, install the intercooler to the radiator and install the retaining push pins.

47. Connect the lower radiator hose to the water pump and install the heater hoses. If equipped with an automatic transmission, install the cooler lines to the radiator.

48. Install the upper radiator hose and the fan and fan shroud. On supercharged engines, connect the electric cooling fan connector.

49. Position the drive belts and the accessory/supercharger belts. Install the radiator sight shield.

50. On supercharged engines, install the intercooler tube and bolts retaining the tube to the power steering bracket. Install the upper intercooler tube to the supercharger and cooler assemblies.

51. Install the cowl vent screen and wiper module. Install the hood, aligning the marks that were made during removal. Connect the underhood lamp wiring.

52. Fill the crankcase with the proper type and quantity of engine oil. Fill and bleed the cooling system.

53. Connect the negative battery cable, start the engine and bring to normal operating temperature. Check for leaks. Check all fluid levels.

54. Leak test, evacuate and charge the air conditioning system according to the proper procedure. Observe all safety precautions.

4.6L Engine

1. Disconnect the battery cables. Drain the crankcase and the cooling system.

2. Relieve the fuel system pressure and discharge the air conditioning system.

3. Mark the position of the hood on the hinges and remove the hood.

4. Remove the cooling fan, shroud and radiator.

5. Remove the wiper module and support bracket. Remove the air inlet tube.

6. Remove the 42-pin connector from the retaining bracket on the brake vacuum booster. Disconnect the 42-pin connector and transmission harness connector and position out of the way.

7. Disconnect the accelerator and cruise control cables. Disconnect the throttle valve cable.

8. Disconnect the electrical connector and vacuum hose from the purge solenoid. Disconnect the power supply from the power distribution box and starter relay.

9. Disconnect the vacuum supply hose from the throttle body adapter vacuum port. Disconnect the heater hoses.

10. Disconnect the alternator harness from the fender apron and junction block. Disconnect the air conditioning hoses from the compressor.

11. Disconnect the EVO sensor connector from the power steering pump and disconnect the body ground strap from the dash panel.

12. Raise and safely support the vehicle.

13. Disconnect the exhaust system from the exhaust manifolds and support with wire hung from the crossmember.

14. Remove the retaining nut from the transmission line bracket and remove the 3 bolts and stud retaining the engine to the transmission knee braces.

15. Remove the starter. Remove the 4 bolts retaining the power steering pump to the engine block and position out of the way.

16. Remove the plug from the engine block to access the torque converter retaining nuts. Rotate the crankshaft until each of the 4 nuts is accessible and remove the nuts.

17. Remove the 6 transmission-to-engine retaining bolts. Remove the engine insulator through bolts, 2 on the left insulator and 1 on the right insulator.

18. Lower the vehicle. Support the transmission with a floor jack and remove the bolt retaining the right engine insulator to the lower engine bracket.

19. Install an engine lifting bracket to the left cylinder head on the front and the right cylinder head on the rear. Connect engine lifting equipment to the lifting brackets.

20. Raise the engine slightly and carefully separate the engine from the transmission.

21. Carefully lift the engine out of the engine compartment and position on a workstand. Remove the engine lifting equipment.

To install:

22. Install engine lifting brackets as in Step 19. Connect engine lifting equipment to the brackets and remove the engine from the workstand.

23. Carefully lower the engine into the engine compartment. Start the converter pilot into the flexplate and align the paint marks on the flexplate and torque converter. Make sure the studs on the torque converter align with the holes in the flexplate.

24. Fully engage the engine to the transmission and lower onto the insulators. Remove the engine lifting equipment and brackets. Install the bolt retaining the right engine insulator to the frame.

25. Raise and safely support the vehicle. Install the 6 engine-to-transmission bolts and tighten to 30–44 ft. lbs. (40–60 Nm).

26. Install the engine insulator through bolts and tighten to 15–22 ft. lbs. (20–30 Nm). Install the 4 torque converter retaining nuts and tighten to 22–25 ft. lbs. (20–30 Nm). Install the plug into the access hole in the engine block.

27. Position the power steering pump on the engine block and install the 4 retaining nuts. Tighten to 15–22 ft. lbs. (20–30 Nm). Install the starter.

28. Position the engine to transmission braces and install the 3 bolts and 1 stud. Tighten the bolts and stud to 18–31 ft. lbs. (25–43 Nm).

29. Position the transmission line bracket to the knee brace stud and install the retaining nut. Tighten to 15–22 ft. lbs. (20–30 Nm).

30. Cut the wire and position the exhaust system to the manifolds. Install the 4 nuts and tighten to 20–30 ft. lbs. (27–41 Nm).

NOTE: Make sure the exhaust system clears the No. 3 crossmember. Adjust as necessary.

31. Lower the vehicle and connect the EVO sensor.

32. Connect the air conditioner lines to the compressor and connect the alternator harness from the fender apron and junction block.

33. Connect the heater hoses and connect the vacuum supply hose to the throttle body adapter vacuum port.

34. Connect the power supply to the power distribution box and starter relay. Connect the electrical connector and vacuum hose to the purge solenoid.

35. Connect and if necessary, adjust the throttle valve cable. Connect the accelerator and cruise control cables.

36. Connect the 42-pin engine harness connector and transmission harness connector. Install the 42-pin connector to the retaining bracket on the brake vacuum booster.

37. Install the wiper module and support bracket. Connect the fuel lines.

38. Install the radiator, cooling fan and shroud. Install the air inlet tube.

39. Fill the crankcase with the proper type and quantity of engine oil. Fill and bleed the cooling system.

40. Install the hood, aligning the marks that were made during removal. Connect the battery cables.

41. Start the engine and bring to operating temperature. Check for leaks. Check all fluid levels. Leak test, evacuate and charge the air conditioning system according to the proper procedure. Observe all safety precautions.

5.0L and 5.8L Engines

EXCEPT 1991–92 THUNDERBIRD AND COUGAR

1. Disconnect the battery cables. Drain the crankcase and the cooling system.

2. Relieve the fuel system pressure and discharge the air conditioning system.

3. Mark the position of the hood on the hinges and remove the hood. Disconnect the battery ground cables from the cylinder block.

4. Remove the air intake duct and the air cleaner, if engine mounted.

5. Disconnect the upper radiator hose from the thermostat housing and the lower hose from the water pump. If equipped with an automatic transmission, disconnect the oil cooler lines from the radiator.

6. Remove the bolts attaching the radiator fan shroud to the radiator. Remove the radiator. Remove the fan, belt pulley and shroud.

7. Remove the alternator bolts and position the alternator out of the way.

8. Disconnect the oil pressure sending unit wire from the sending unit. Disconnect the flexible fuel line from the fuel tank line. Plug the fuel tank line.

9. Disconnect the accelerator cable from the carburetor or throttle body. Disconnect the TV rod if equipped with an automatic transmission. Disconnect the cruise control cable, if equipped.

10. Disconnect the throttle valve vacuum line from the intake manifold, if equipped. Disconnect the transmis-

sion filler tube bracket from the cylinder block.

11. If equipped with air conditioning, disconnect the lines and electrical connectors at the compressor and remove the compressor. Plug the lines and the compressor fittings to prevent the entrance of dirt and moisture.

12. Disconnect the power steering pump bracket from the cylinder head. Remove the drive belt. Position the power steering pump out of the way in a position that will prevent the fluid from leaking.

13. Disconnect the power brake vacuum line from the intake manifold.

14. On 5.0L engines, disconnect the heater hoses from the heater tubes. On 5.8L engines, disconnect the heater hoses from the water pump and intake manifold. Disconnect the electrical connector from the coolant temperature sending unit.

15. Remove the converter housing-to-engine upper bolts.

16. On 5.8L engines, disconnect the primary wiring connector from the ignition coil. Disconnect the wiring to the solenoid on the left rocker cover. Remove the wire harness from the left rocker arm cover and position the wires out of the way. Disconnect the ground strap from the block.

17. On 5.0L engines, disconnect the wiring harness at the two 10-pin connectors.

18. Raise and safely support the vehicle. Disconnect the starter cable from the starter and remove the starter.

19. Disconnect the muffler inlet pipes from the exhaust manifolds. Disconnect the engine support insulators from the chassis. Disconnect the downstream thermactor tubing and check valve from the right exhaust manifold stud, if equipped.

20. Disconnect the transmission cooler lines from the retainer and remove the converter housing inspection cover. Disconnect the flywheel from the converter and secure the converter assembly in the housing. Remove the remaining converter housing-to-engine bolts.

21. Lower the vehicle and then support the transmission. Attach engine lifting equipment and hoist the engine.

22. Raise the engine slightly and carefully pull it from the transmission. Carefully lift the engine out of the engine compartment. Avoid bending or damaging the rear cover plate or other components. Install the engine on a workstand.

To install:

23. Attach the engine lifting equipment and remove the engine from the workstand.

24. Lower the engine carefully into

the engine compartment. Make sure the exhaust manifolds are properly aligned with the muffler inlet pipes.

25. Start the converter pilot into the crankshaft. Align the paint mark on the flywheel to the paint mark on the torque converter.

26. Install the converter housing upper bolts, making sure the dowels in the cylinder block engage the converter housing.

27. Install the engine support insulator-to-chassis attaching fasteners and remove the engine lifting equipment.

28. Raise and safely support the vehicle. Connect both muffler inlet pipes to the exhaust manifolds. Install the starter and connect the starter cable.

29. Remove the retainer holding the converter in the housing. Attach the converter to the flywheel. Install the converter housing inspection cover and install the remaining converter housing attaching bolts.

30. Remove the support from the transmission and lower the vehicle.

31. On 5.8L engines, connect the wiring harness to the left rocker arm cover and connect the coil wiring connector. On 5.0L engines, connect the wiring harness at the two 10-pin connectors.

32. Connect the coolant temperature sending unit wire and connect the heater hoses. Connect the wiring to the metal heater tubes, engine coolant temperature, air charge temperature and oxygen sensors.

33. Connect the transmission filler tube bracket. Connect the manual shift rod and the retracting spring. Connect the throttle valve vacuum line, if equipped.

34. Connect the acclerator cable and TV cable. Connect the cruise control cable, if equipped.

35. Remove the plug from the fuel tank line and connect the fuel line and the oil pressure sending unit wire.

36. Install the pulley, water pump belt and fan clutch assembly.

37. Position the alternator bracket and install the alternator bolts. Connect the alternator and ground cables. Adjust the drive belt tension.

38. Install the air conditioning compressor. Unplug and connect the refrigerant lines and connect the electrical connector to the compressor.

39. Install the power steering drive belt and power steering pump bracket. Connect the power brake vacuum line.

40. Install the fan on the water pump pulley. Place the shroud over the fan and install the radiator. Connect the radiator hoses and the transmission oil cooler lines. Position the shroud and install the bolts.

41. Connect the heater hoses to the heater tubes. Fill and bleed the cooling system. Fill the crankcase with the proper type and quantity of engine oil. Adjust the transmission throttle linkage.

42. Connect the negative battery cable. Start the engine and bring to normal operating temperature. Check for leaks. Check all fluid levels.

43. Install the air intake duct assembly. Install the hood, aligning the marks that were made during removal.

44. Leak test, evacuate and charge the air conditioning system according to the proper procedure. Observe all safety precautions.

1991–92 THUNDERBIRD AND COUGAR

1. Disconnect the negative battery cable. Drain the crankcase and the cooling system.

2. Relieve the fuel system pressure and discharge the air conditioning system.

3. Disconnect the electrical connector for the underhood lamp. Mark the position of the hood on the hinges and remove the hood.

4. Remove the oil dipstick. Disconnect and plug the refrigerant lines at the air conditioning compressor.

5. Disconnect the compressor clutch and power steering pressure switch electrical connectors. Disconnect the alternator wiring harness from the alternator and position the harness out of the way.

6. Remove the fan shroud and the fan. Remove the upper radiator hose.

7. Remove the air cleaner-to-throttle body tube. Disconnect and plug the transmission oil cooler lines at the radiator.

8. Disconnect the throttle and kickdown cables from the throttle body and remove the cable bracket retaining bolts. Position the cable and bracket assembly out of the way.

9. Tag and disconnect the vacuum lines at the upper intake manifold vacuum tree, air conditioning control panel vacuum supply hose, thermactor valve and EGR valve. Disconnect the electrical connector at the EGR valve.

10. Remove the upper intake manifold as follows:

a. Disconnect the spark plug wires from the spark plugs. Remove the wires and bracket assembly from the rocker arm cover attaching stud. Remove the distributor cap, adapter and spark plug wire assembly.

b. Disconnect the fuel lines from the fuel rail.

c. Disconnect the distributor wiring connector. Mark the position of the rotor in relation to the distributor housing and the distributor housing in relation the engine. Remove the distributor hold-down bolt and remove the distributor.

d. Disconnect the water temperature sending unit wire at the sending unit. Disconnect the hose from the intake manifold and the 2 throttle body cooler hoses.

e. Loosen the clamp on the water pump bypass hose at the thermostat housing and slide the hose off the housing. Tag and disconnect the wires at the engine coolant temperature, air charge temperature, throttle position and EGR sensors and the idle speed control solenoid. Tag and disconnect the injector wire connections and the fuel charging assembly wiring.

f. Pull the PCV valve out of the grommet at the rear of the lower intake manifold. Disconnect the fuel evaporative purge hose from the plastic connector at the front of the upper intake manifold.

g. Remove the retaining bolts and remove the upper intake manifold.

11. Disconnect the main engine wiring harness connectors at the right side of the dash panel. Position the engine wiring harness so it can be removed with the engine.

12. Disconnect the heater hoses at the engine. Disconnect the wiring harness from the coil and distributor and position the harness out of the way.

13. Disconnect the lower radiator hose from the water pump. Remove the radiator retaining bolts and remove the radiator.

14. Raise and safely support the vehicle. Remove the oil filter.

15. Remove the starter. Disconnect the oxygen sensors for the right and left catalytic converters. Disconnect the negative battery cable from the left side of the engine.

16. On the right side of the engine, disconnect the brackets for the transmission cooler lines, engine-to-body ground straps and the starter wiring harness.

17. Remove the torque converter inspection cover and mark 1 of the converter studs to the flywheel for alignment during reassembly. Remove the torque converter attaching nuts.

18. Remove the exhaust manifold heat shield at the left manifold flange and disconnect the exhaust pipe from the flange. Disconnect the right exhaust manifold flange.

19. Loosen the transmission mount retaining nut. Remove the converter housing to engine bolts and the motor mount through bolts.

20. Lower the vehicle and disconnect the power steering lines. Cap the lines to prevent contamination.

21. Support the transmission with a floor jack. Install suitable engine lift-

ing equipment on the engine lifting eyes.

22. Lift the engine assembly clear of the engine mounts and remove the engine from the vehicle. Place the engine on a workstand.

To install:

23. Install suitable engine lifting equipment on the engine lifting eyes and lift the engine from the workstand.

24. Carefully lower the engine into the engine compartment. Make sure the exhaust manifolds are properly aligned with the muffler inlet pipes.

25. Start the converter pilot into the crankshaft. Align the mark on the flywheel to the mark on the torque converter.

26. Position the retaining clip for the left oxygen sensor wiring near the left upper transmission-to-engine bolt. Install the converter housing upper bolts. Make sure the dowels in the cylinder block engage the converter housing.

27. Raise and safely support the vehicle. Install the remaining converter housing bolts and install the motor mount through bolts. Tighten the transmission mount retaining nut.

28. Connect the right exhaust manifold flange. Connect the exhaust pipe to the left exhaust manifold flange and install the heat shield.

29. Install the torque converter retaining nuts and the inspection cover.

30. On the right side of the engine, install the brackets for the transmission cooler lines, engine-to-body ground strap and the starter wiring harness.

31. Connect the negative battery cable to the left side of the engine. Connect the oxygane sensors for the catalytic converters.

32. Install the starter. Install a new oil filter and the oil pan drain plug. Lower the vehicle and connect the power steering lines.

33. Install the radiator. Connect the coolant overflow hose and the lower radiator hose.

34. Position and connect the wiring harness for the coil and distributor. Connect the heater hoses at the engine. Connect the main engine wiring harness connectors at the right side of the dash panel.

35. Install the upper intake manifold in the reverse order of removal. Be sure to use a new gasket.

36. Connect the vacuum lines at the upper intake manifold vacuum tee, air conditioning control panel vacuum supply hose, thermactor valve and EGR valve. Connect the electrical connector to the EGR valve.

37. Connect the throttle and kickdown cables to the throttle body

and install the cable bracket retaining bolts.

38. Connect the transmission oil cooler lines and the upper radiator hose. Install the fan shroud and the fan. Install the air cleaner-to-throttle body tube assembly.

39. Position and connect the wiring harness for the alternator. Connect the compressor clutch electrical connector and connect the refrigerant lines to the compressor.

40. Install the hood, aligning the marks that were made during removal. Connect the wiring connector for the underhood lamp.

41. Fill the engine with the proper type and quantity of engine oil. Fill and bleed the cooling system. Install the dipstick.

42. Fill the power steering system with the proper type and quantity of fluid. Connect the negative battery cable.

43. Start the engine and bring to normal operating temperature. Check for leaks. Check all fluid levels.

44. Leak test, evacuate and charge the air conditioning system according to the proper procedure. Observe all safety precautions.

Engine Mounts

REMOVAL & INSTALLATION

2.3L Engine

FRONT

1. Disconnect the negative battery cable. Raise and safely support the vehicle. Support the engine using a wood block and jack placed under the engine.

2. Remove the through bolts attaching both insulators to the No. 2 crossmember pedestal bracket. On Mustang convertible, remove nuts.

3. Disconnect shift linkage.

4. Raise the engine sufficiently to disengage the insulator from the crossmember pedestal bracket.

5. Remove the bolts attaching the insulator and bracket assembly to the engine. Remove the insulator and bracket assembly.

To install:

6. Position the insulator and bracket assembly to the engine. Install the attaching bolts. Tighten to 33–45 ft. lbs. (45–61 Nm).

7. Lower the engine into position making sure that the insulators are seated flat on the No. 2 crossmember. Hand start the bolts, lower the engine completely, then tighten the through bolts to 33–45 ft. lbs. (45–61 Nm).

8. On Mustang convertible, tighten the flange nut to 80–106 ft. lbs. (108–144 Nm).

9. Install fuel pump shield attach-

ing screw to left engine support, if equipped.

10. Install shift linkage. Lower the vehicle and connect the negative battery cable.

REAR

1. Disconnect the negative battery cable. Raise and safely support the vehicle.

2. Support the transmission with a jack and a wood block. Remove the nut(s) retaining the rear insulator to the crossmember.

3. Remove the 2 bolts and nuts retaining the crossmember to the body brackets. Remove the crossmember by raising the transmission slightly with the jack.

4. Remove the 2 bolts retaining the rear insulator to the transmission and remove the insulator and retainer. If equipped with automatic transmission, remove the 2 bolts retaining the rear insulator to the intermediate bracket.

To install:

5. Position the rear insulator and retainer on the transmission. Install the 2 retaining bolts and tighten to 50–70 ft. lbs. (68–95 Nm). If equipped with automatic transmission, tighten the 2 bolts to 33–45 ft. lbs. (46–61 Nm).

6. Install the crossmember to the body brackets. Tighten the retaining nuts and bolts to 35–50 ft. lbs. (48–68 Nm).

7. Lower the transmission and install the insulator to crossmember retaining nuts. Tighten to 25–35 ft. lbs. (34–47 Nm). If equipped with automatic transmission, tighten the nut to 65–85 ft. lbs. (85–115 Nm).

8. Lower the vehicle. Connect the negative battery cable.

3.8L Engine

FRONT

1. Disconnect the negative battery cable.

2. Remove fan shroud retaining screws. Remove the air tube to the remote air cleaner.

3. Raise and safely support the vehicle. Support engine using a jack and wood block placed under the engine.

4. Remove the engine mount through bolt. On supercharged engines, remove the retaining strap bolt.

5. Remove shift linkage.

6. Raise engine high enough to clear clevis brackets.

NOTE: Raise the engine carefully so as not to damage the lines and hoses at the rear of the engine.

7. Remove the accessories and oil

cooler line retaining clips from the engine support brackets.

8. Remove bolts retaining the engine mount and bracket assembly to engine. Remove insulator and bracket assembly.

NOTE: The left hand front engine mount removal on the supercharged engine may require lowering the front sub frame.

To install:

9. Position the engine mount and bracket assembly to the engine, install the retaining bolts and tighten to 25–35 ft. lbs. (34–47 Nm).

10. Install the accessories to the lower front engine mount support bracket stud.

11. Lower the engine into position and make sure the engine mounts are seated flat on the front sub frame. Install the through-bolt and tighten to 35–50 ft. lbs. (47–68 Nm). On supercharged engines, install the retaining strap bolt and tighten to 33–45 ft. lbs. (45–61 Nm).

12. Lower the vehicle and install the air tube and the fan shroud retaining screws. Connect the negative battery cable.

REAR

1. Disconnect the negative battery cable. Raise and safely support the vehicle.

2. Support the transmission with a jack and a wood block. Remove the rear nut(s) attaching the insulator-to-crossmember. Keep transmission weight on the mount during nut removal.

3. Remove the 2 bolts retaining the crossmember = to-body brackets. Remove the crossmember by raising the transmission slightly with the jack.

4. Remove the bolts retaining the rear engine mount to the transmission. Remove the insulator.

To install:

5. Position the engine mount and retainer on the transmission. Install the 2 retaining bolts and tighten to 50–70 ft. lbs. (68–95 Nm).

6. Install the crossmember-to-body brackets. Tighten the retaining bolts or nuts to 35–50 ft. lbs. (47–68 Nm).

7. Lower the transmission. Install the mount-to-crossmember nut(s). Tighten to 25–35 ft. lbs. (34–47 Nm) on 1988 vehicles or 65–85 ft. lbs. (88–115 Nm) on 1989–92 vehicles.

8. Lower the vehicle and connect the negative battery cable.

4.6L Engine
FRONT

1. Disconnect the battery cables. Drain the cooling system, relieve the

fuel system pressure and discharge the air conditioning system.

2. Remove the air inlet tube and the cooling fan and shroud. Remove the upper radiator hose.

3. Disconnect the fuel lines from the fuel rail. Remove the wiper module and support bracket.

4. Disconnect the air conditioning compressor outlet hose at the compressor and remove the bolt retaining the hose assembly to the right coil bracket.

5. Remove the 42-pin engine harness connector from the retaining bracket on the brake vacuum booster. Disconnect the 42-pin connector and transmission harness connector.

6. Disconnect the throttle valve cable from the throttle body. Disconnect the heater outlet hose.

7. Remove the upper stud and loosen the lower bolt retaining the heater outlet hose to the right cylinder head and position out of the way.

8. Remove the blower motor resistor. Remove the bolt retaining the right engine insulator to the lower engine bracket.

9. Disconnect the vacuum hoses from the EGR valve and EGR tube. Remove the 2 bolts retaining the EGR valve to the intake manifold. Disconnect both oxygen sensors.

10. Raise and safely support the vehicle. Remove the engine insulator through bolts, 2 from the left side and 1 from the right.

11. Remove the EGR tube line nut from the right exhaust manifold and remove the EGR valve and tube assembly.

12. Disconnect the exhaust pipes from the manifolds. Lower the exhaust and hang the pipes with wire from the crossmember.

13. Position a jack and a block of wood under the oil pan, rearward of the oil drain hole. Raise the engine approximately 4 in. (100mm).

14. Install a block of wood under the oil pan and lower the engine onto the wood block. Remove 3 retaining bolts each from the right and left engine insulators and remove the insulators.

To install:

15. Position the insulators on the engine block, install 3 retaining bolts and tighten to 45–60 (60–81 Nm). Raise the engine and remove the wood block.

16. Lower the engine onto the insulators. Position and connect the EGR valve and tube assembly to the exhaust manifold. Tighten the line nut to 26–33 ft. lbs.

NOTE: Loosen the line nut at the EGR valve prior to installing the assembly onto the vehicle. This will allow enough movement to align the EGR valve retaining bolts.

17. Install the engine insulator through bolts and tighten to 45–60 ft. lbs. (60–81 Nm) on 1991 vehicles or 15–22 ft. lbs. (20–30 Nm) on 1992 vehicles.

18. Cut the wire and position the exhaust manifolds. Install the 4 nuts and tighten to 20–30 ft. lbs. (27–41 Nm). Make sure the exhaust system clears the No. 3 crossmember, adjust as necessary.

19. Lower the vehicle and connect the oxygen sensors. Install the bolt retaining the right engine insulator to the frame. Tighten to 45–60 ft. lbs. (60–81 Nm).

20. Install a new gasket on the EGR valve and position to the intake manifold. Install the 2 EGR valve retaining bolts and tighten to 15–22 ft. lbs. (20–30 Nm).

21. Tighten the EGR tube line nut at the EGR valve to 26–33 ft. lbs. (35–45 Nm). Connect the vacuum hoses to the EGR valve and tube.

22. Install the blower motor resistor. Position the heater outlet hose. Install the upper stud and tighten the upper stud and lower bolt to 15–22 ft. lbs. (20–30 Nm). Install the gound strap onto the stud and tighten the nut to 15–22 ft. lbs. (20–30 Nm). Connect the heater outlet hose.

23. Connect and if necessary, adjust the throttle valve cable. Connect the 42-pin connector and transmission harness connector. Install the 42-pin engine harness connector to the retaining bracket on the brake vacuum booster.

24. Connect the air conditioning compressor outlet hose to the compressor and install the bolt retaining the hose assembly to the right coil bracket.

25. Install the upper radiator hose and connect the fuel lines. Install the wiper module and retaining bracket.

26. Install the cooling fan and shroud. Install the air inlet tube.

27. Fill and bleed the cooling system. Connect the battery cables, start the engine and check for leaks. Leak test, evacuate and charge the air conditioning system according to the proper procedure. Observe all safety precautions.

REAR

1. Disconnect the negative battery cable. Raise and safely support the vehicle.

2. Support the transmission with a jack and wood block. Remove the 2 nuts attaching the rear insulator to the crossmember.

3. Remove the 2 bolts attaching the support insulator to the transmission.

4. Raise the transmission with the jack and remove the insulator and retainer assembly.

To install:

5. Position the insulator and retainer assembly on the transmission. Install the 2 retaining bolts and tighten to 50–70 ft. lbs. (68–95 Nm).

6. Lower the transmission. Install the rear insulator-to-crossmember retaining nuts and tighten to 35–50 ft. lbs. (48–68 Nm).

7. Lower the vehicle and connect the negative battery cable.

5.0L and 5.8L Engines

CROWN VICTORIA, GRAND MARQUIS AND TOWN CAR— FRONT

1. Disconnect the negative battery cable. Remove fan shroud attaching screw.

2. Raise and safely support the vehicle. Support the engine using a jack and wood block placed under the engine.

3. Remove the through bolts attaching the insulators to the insulator support bracket.

4. Remove the bolts attaching the insulator assembly to the frame.

5. Raise the engine slightly with the jack and remove the insulator assembly.

To install:

6. Position the engine insulator assembly to the frame and install the attaching bolts. Tighten the bolts to 26–38 ft. lbs. (35–52 Nm).

7. Lower the engine into position and install the engine insulator assembly to insulator support bracket through bolts. Tighten the through bolts to 40–46 ft. lbs. (54–62 Nm) on 1988 vehicles or 45–65 ft. lbs. (61–88 Nm) on 1989–92 vehicles.

8. Lower the vehicle. Connect the negative battery cable and install the fan shroud attaching screws.

CROWN VICTORIA, GRAND MARQUIS AND TOWN CAR—REAR

1. Disconnect the negative battery cable. Raise and safely support the vehicle.

2. Support the transmission with a jack and wood block. Remove the 2 nuts attaching the rear insulator to the crossmember.

3. Remove the 2 bolts attaching the support insulator to the transmission.

4. Raise the transmission with the jack and remove the insulator and retainer assembly.

To install:

5. Position the insulator and retainer assembly on the transmission. Install the 2 retaining bolts and tighten to 50–70 ft. lbs. (68–95 Nm).

6. Lower the transmission. Install the rear insulator-to-crossmember retaining nuts and tighten to 35–50 ft. lbs. (48–68 Nm).

7. Lower the vehicle and connect the negative battery cable.

MARK VII, THUNDERBIRD, COUGAR AND MUSTANG—FRONT

1. Disconnect the negative battery cable. Remove fan shroud attaching screws.

2. Raise and safely support the vehicle. Support the engine using a jack and wood block placed under the engine.

3. Remove the nuts or bolts attaching the insulators to the No. 2 crossmember. On Thunderbird and Cougar, remove the through bolts.

4. Disconnect shift linkage on all except 1991–92 Thunderbird and Cougar.

5. Raise the engine sufficiently with the jack to disengage the insulator from the crossmember. If equipped, remove the transmission brace attached at the left or right engine mount bracket.

6. Remove the engine insulator and bracket assembly to the cylinder block attaching bolts. Remove the engine insulator assembly.

To install:

7. Position the insulator assembly on the engine and install the attaching bolts. Tighten the bolts to 35–60 ft. lbs. (48–81 Nm).

8. Attach the transmission brace to the right or left engine mount, if equipped. Tighten the nut to 35–60 ft. lbs. (48–81 Nm).

9. Lower the engine into position making sure that the insulators are seated flat on the No. 2 crossmember and the insulator studs are at the bottom of the slots.

NOTE: On 1991–92 Thunderbird and Cougar, the left mount, with the locating pin, must seat before the right mount.

10. Install and tighten the insulator nuts to 80–105 ft. lbs. (109–142 Nm). On Thunderbird and Cougar, install the through bolts and tighten to 35–45 ft. lbs. (45–61 Nm).

11. Lower the vehicle and install the fan shroud attaching screws. Connect the negative battery cable.

MARK VII—REAR

1. Disconnect the negative battery cable. Raise and safely support the vehicle.

2. Support the transmission with a jack and wood block. Remove the bolts that attach the rear mounts and crossmember to the transmission.

3. Remove the lower rebound insulator.

4. Raise the transmission slightly with the jack. Remove the upper insulators.

To install:

5. Position the upper insulator between the crossmember and the transmission.

6. Position the lower insulators and hand start the attaching bolts. Lower the transmission and tighten the bolts to 50–70 ft. lbs. (62–95 Nm).

7. Lower the vehicle and connect the negative battery cable.

MUSTANG AND 1988 THUNDERBIRD AND COUGAR— REAR

1. Disconnect the negative battery cable. Raise and safely support the vehicle.

2. Support the transmission with a jack and wood block. Remove the 2 nuts attaching the insulator to the crossmember.

3. Remove the 2 bolts and nuts attaching the crossmember to the body brackets and remove the crossmember by raising the transmission slightly with the jack.

4. Remove the 2 bolts attaching the rear insulator to the transmission and remove the insulator and retainer.

To install:

5. Position the rear insulator and retainer on the transmission. Install the 2 attaching bolts and tighten to 50–70 ft. lbs. (68–95 Nm).

6. Install the crossmember to the body brackets. Tighten the attaching nuts to 35–50 ft. lbs. (48–68 Nm).

7. Lower the transmission and install the insulator to crossmember attaching nuts. Tighten to 25–35 ft. lbs. (34–48 Nm).

8. Lower the vehicle and connect the negative battery cable.

1991–92 THUNDERBIRD AND COUGAR—REAR

1. Disconnect the negative battery cable. Raise and safely support the vehicle.

2. Remove the nut attaching the rear support to the No. 3 crossmember.

NOTE: This must be done while the transmission weight is still on the mount or the stud may twist out of the crossmember slot and damage the mount.

3. Support the transmission with a jack and a wood block. Remove the bolts that attach the rear mount and crossmember to the transmission bracket.

4. Raise the transmission slightly with the floor jack and remove the insulator.

To install:

5. Position the insulator between the crossmember and the transmission bracket. Install the bracket bolts

and tighten to 33–45 ft. lbs. (45–61 Nm).

6. Lower the transmission and install the nut. Tighten the nut to 65–85 ft. lbs. (88–115 Nm).

7. Lower the vehicle and connect the negative battery cable.

Cylinder Head

REMOVAL & Installation

2.3L Engine

1. Disconnect the negative battery cable. Drain the cooling system and relieve the fuel system pressure.

2. Remove the air cleaner on non-turbocharged engines. On turbocharged engines, remove the inlet tube between the turbocharger and throttle body.

3. If equipped, remove the heater hose retaining screw to the rocker cover.

4. Tag and disconnect the spark plug wires from the spark plugs. Remove the spark plug wires and, if equipped, the distributor cap. Remove the spark plugs.

5. Tag and disconnect the required vacuum hoses. Remove the dipstick and disconnect the dipstick tube from the bracket.

6. Remove the upper intake manifold and throttle body as follows:

    a. Tag and disconnect the electrical connectors and vacuum hoses.

    b. Disconnect the throttle linkage, cruise control and kickdown cable. Unbolt the accelerator cable from the bracket and position the cable out of the way.

    c. Disconnect the crankcase vent hose. Disconnect the PCV hose from the fitting on the underside of the upper intake manifold.

    d. Disconnect the EGR tube from the EGR valve. Remove the 4 upper intake manifold mounting bolts and the manifold.

7. Remove the rocker cover retaining bolts and remove the cover. Remove the intake manifold retaining bolts.

8. Remove the alternator belt and swing the alternator aside. Remove the bracket mounting bolts, if necessary.

9. Remove the upper radiator hose. Remove the timing belt cover retaining bolts and remove the cover.

10. Loosen the timing belt idler retaining bolts. Position the idler in the unloaded position and tighten the retaining bolts.

11. Remove the timing belt from the camshaft pulley and the auxiliary pulley.

12. Remove the 8 exhaust manifold retaining bolts. Remove the timing

belt idler and 2 bracket bolts. Remove the timing belt idler spring stop from the cylinder head.

13. Disconnect the oil sending unit wire, if necessary.

14. Remove the cylinder head bolts and the cylinder head. Clean all gasket mating surfaces and blow the oil out of the cylinder head bolt block holes.

15. Check the cylinder head for flatness using a straight edge and a feeler gauge. If the head gasket surface is warped greater than 0.006 in., it must be resurfaced. Do not grind more than 0.010 in. from the cylinder head.

To install:

16. Position the head gasket on the block. Position the camshaft with the pin approximateley 30 degrees to the right of the 6 o' clock position when facing the front of the cylinder head. The camshaft must be positioned this way to protect protruding valves.

17. Position the cylinder head on the block and install the cylinder head bolts. Tighten the bolts, in sequence, in 2 steps, first to 50–60 ft. lbs. (60–81 Nm) and then to 80–90 ft. lbs. (108–122 Nm).

18. Connect the oil sending unit wire, if necessary. Install the timing belt tensioner spring stop to the cylinder head.

19. Position the timing belt tensioner and tensioner spring to the cylinder head and install the retaining bolts. Rotate the tensioner against the spring with belt tensioner tool T74P–6254–A or equivalent, and temporarily tighten.

20. Install the 8 exhaust manifold retaining bolts. Tighten the bolts, in sequence, in 2 steps, first to 178–204 inch lbs. (20–23 Nm) and then to 20–30 ft. lbs. (27–40 Nm).

21. If equipped with a distributor, align the distributor rotor with the No. 1 plug location on the distributor cap. Align the camshaft sprocket with the pointer and align the crankshaft pulley with the pointer on the timing belt cover.

22. Install the timing belt over the sprockets. Loosen the tensioner retaining bolts, rotate the engine by hand 1 complete revolution and check the timing alignment.

23. Tighten the 10mm tensioner bolt to 28–40 ft. lbs. (38–54 Nm) and the 8mm bolt to 14–21 ft. lbs. (19–29 Nm).

24. Install the timing belt cover and tighten the retaining bolts to 6–9 ft. lbs. (8–12 Nm).

25. Install the rocker arm cover and tighten the retaining bolts to 62–97 inch lbs. (7–11 Nm).

26. Install the intake manifold. On 1988 vehicles, tighten the bolts, in sequence to 15–22 ft. lbs. (19–29 Nm). On 1989–92 vehicles, tighten the

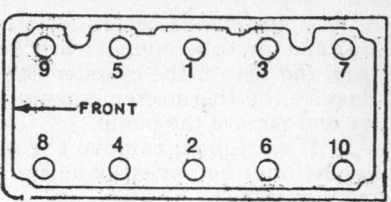

Cylinder head bolt torque sequence— 2.3L engine

bolts, in sequence, to 19–28 ft. lbs. (26–38 Nm).

27. Install the upper intake manifold and throttle body in the reverse order of removal. Tighten the upper intake-to-lower intake bolts to 15–22 ft. lbs. (20–30 Nm).

28. Position the alternator and install the drive belt. Install the upper radiator hose.

29. Install the dipstick and connect the necessary vacuum hoses. Install the spark plugs, spark plug wires and distributor cap, if equipped.

30. Position and connect the engine and alternator wiring harnesses. Install the hose from the air cleaner to the throttle body. If equipped, install the retaining heater hose screw to the rocker cover.

31. Fill and bleed the cooling system. Connect the negative battery cable, start the engine and bring to normal operating temperature. Check for leaks. If equipped with distributor ignition, check the ignition timing.

3.8L Engine

1. Disconnect the negative battery cable.

2. Relieve the fuel system pressure. Drain the cooling system.

3. Remove air cleaner assembly including the air intake duct and heat tube.

4. Loosen accessory drive belt idler. Remove drive belt.

5. If the left cylinder head is being removed, perform the following:

    a. On supercharged engines, remove the intercooler and intercooler tubes.

    b. Remove oil fill cap.

    c. Remove the power steering pump front mounting bracket attaching bolts.

    d. Remove the alternator assembly and accessory drive belt main idler.

    e. Remove the power steering/pump alternator bracket retaining bolts.

    f. Leaving the hoses connected, place the power steering pump/alternator bracket assembly aside in a position to prevent the fluid from leaking out.

6. If the right cylinder head is being removed, perform the following:

a. If equipped, disconnect the thermactor tube support bracket from the rear of the cylinder head. Remove the thermactor pump pulley and remove the pump.

b. If equipped, remove the air conditioner compressor belt and main drive belt.

c. If equipped, remove the compressor mounting bracket retaining bolts. Leave the hoses connected and position the compressor aside.

d.Remove the PCV valve.

7. Remove the upper intake manifold. On supercharged engines, remove the supercharger.

8. Remove valve rocker arm cover attaching screws. Remove the fuel rail and the lower intake manifold.

9. Remove the exhaust manifold(s).

10. Loosen rocker arm fulcrum attaching bolts enough to allow the rocker arm to be lifted off the pushrod and rotated to 1 side.

11. Remove the pushrods. Identify the position of each rod. The rods should be installed in their original position during assembly.

12. Remove the cylinder head attaching bolts and discard.

13. Remove the cylinder head(s). Clean all gasket mating surfaces.

14. Check the flatness of the cylinder head gasket surface using a straight edge and a feeler gauge. The allowable warpage is 0.003 in for every 6.0 inches. Do not machine more than 0.010 in.

To install:

NOTE: Lightly oil all bolt and stud bolt threads before installation except those specifying special sealant.

15. Position new head gasket(s) on the cylinder block using the dowels for alignment.

16. Position the cylinder head(s) on the block.

17. Install new cylinder head bolts. Apply a thin coating of pipe sealant to the threads of the short cylinder head bolts, nearest to the exhaust manifold. Do not apply sealant to the long bolts.

NOTE: Always use new cylinder head bolts to assure a leak-tight assembly. Torque retention with used bolts can vary, which may result in coolant or compression leakage at the cylinder head mating surface area.

18. On 1988 vehicles tighten the new cylinder head attaching bolts in sequence as follows:
    a. 37 ft. lbs. (50 Nm)
    b. 45 ft. lbs. (60 Nm)
    c. 52 ft. lbs. (70 Nm)
    d. 59 ft. lbs. (80 Nm)
    e. Back-off the attaching bolts 2–3 turns.

f. Repeat Steps a–d.

NOTE: When cylinder head attaching bolts have been tightened using multi-step torque procedure, it is not necessary to retighten the bolts after extended engine operation.

19. On 1989 vehicles tighten the new cylinder head attaching bolts in sequence as follows:
    a. 37 ft. lbs. (50 Nm)
    b. 45 ft. lbs. (60 Nm)
    c. 52 ft. lbs. (70 Nm)
    d. 59 ft. lbs. (80 Nm)
    e. Back-off the attaching bolts 2–3 turns
    f. In sequential order, tighten the bolts to 52 ft. lbs. (70 Nm)
    g. In sequential order, rotate the bolts an additional 90–110 degrees

NOTE: When cylinder head attaching bolts have been tightened using multi-step torque procedure, it is not necessary to retighten the bolts after extended engine operation.

20. On 1990–92 vehicles tighten the new cylinder head attaching bolts in numerical sequence as follows:
    a. 37 ft. lbs. (50 Nm)
    b. 45 ft. lbs. (60 Nm)
    c. 52 ft. lbs. (70 Nm)
    d. 59 ft. lbs. (80 Nm)
    e. Back-off the attaching bolts 2–3 turns
    f. On supercharged engines, tighten each long and short bolt to 48–55 ft. lbs. (65–75 Nm), rotate an additional 90–110 degrees, then go to the next bolt in sequence.
    g. On non-supercharged engines, tighten each long bolt to 11–18 ft. lbs. (15–25 Nm), rotate an additional 85–105 degrees, then go to the next bolt in sequence. Do the same for each short bolt except only rotate the short bolts 65–85 degrees.

NOTE: When cylinder head attaching bolts have been tightened using multi-step torque procedure, it is not necessary to retighten the bolts after extended engine operation.

21. Lubricate each pushrod with heavy engine oil and install, in their original positions.

22. For each valve rotate the crankshaft until the tappet rests on the base circle of the camshaft lobe, before tightening the fulcrum attaching bolts to 43 inch lbs. (5 Nm).

23. Lubricate the rocker arm assemblies with heavy engine oil and final tighten the fulcrum bolts to 19–25 ft. lbs. (25–35 Nm). Fulcrums must be fully seated in cylinder head and pushrods must be seated in rocker arm

sockets prior to final tightening. Final tightening can be done with the camshaft in any position.

NOTE: If the original valve train components are being installed, a valve clearance check is not required. If a component has been replaced, perform a valve clearance check.

24. Install the exhaust manifold(s).

25. Install the lower intake manifold and the fuel rail.

26. Position cover and new gasket on the cylinder head and install attaching bolts. Note the location of spark plug wire routing clip stud bolts. Tighten attaching bolts to 80–106 inch lbs. (9–12 Nm).

27. Install the upper intake manifold. On supercharged engines, install the supercharger.

28. Install the spark plugs, if removed.

29. Connect the spark plug wires to the spark plugs.

30. If the left cylinder head is being installed, perform the following:
    a. Install the oil filler cap.
    b. Install the alternator/power steering pump mounting bracket.
    c. Install the alternator assembly.
    d. Install the main accessory drive belt tensioner assembly.
    e. Install the power steering pump assembly.
    f. Install the power steering pump support bracket.
    g. On supercharged engines, install the intercooler tubes.

31. If the right cylinder head is being installed, perform the following:
    a. Install PCV valve.
    b. If equipped with air conditioning, install the compressor mounting and support brackets and install the compressor.
    c. If equipped, install the thermactor pump and pump pulley.
    d. If equipped, install the accessory drive belt idler pulley.
    e. If equipped, install the thermactor air control valve or air bypass valve hose. Tighten the clamps securely to the air pump assembly.

32. Install the accessory drive belt and tighten to specification. If equipped, attach the thermactor tube(s) support bracket to the rear of the cylinder head. Tighten attaching bolts to 30–40 ft. lbs. (40–55 Nm).

33. Connect cable to the battery negative terminal.

34. Fill and bleed the cooling system.

35. Start engine and check for coolant, fuel and oil leaks.

36. Check and, if necessary, adjust the curb idle speed.

37. Install the air cleaner assembly including the air intake duct and heat tube.

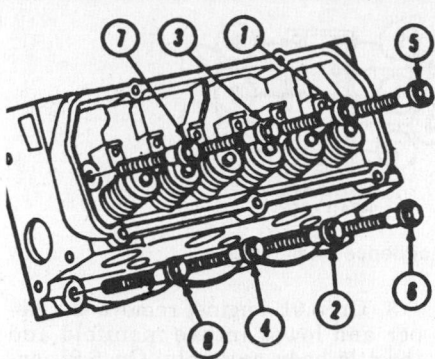

Cylinder head bolt torque sequence— 3.8L engine

4.6L Engine

1. Disconnect the negative battery cable.

2. Drain the cooling system and remove the cooling fan and shroud.

3. Relieve the fuel system pressure and disconnect the fuel lines.

4. Remove the air inlet tube and the wiper module. Release the belt tensioner and remove the accessory drive belt.

5. Tag and disconnect the ignition wires from the spark plugs. Disconnect the ignition wire brackets from the camshaft cover studs and remove the 2 bolts retaining the ignition wire tray to the coil brackets.

6. Remove the bolt retaining the air conditioner high pressure line to the right coil bracket. Disconnect both ignition coils and CID sensor.

7. Remove the nuts retaining the coil brackets to the front cover. Slide the ignition coil brackets and ignition wire assembly off the mounting studs and remove from the vehicle.

8. Remove the water pump pulley. Disconnect the alternator wiring harness from the junction block, fender apron and alternator. Disconnect the bolts retaining the alternator to the intake manifold and engine block and remove the alternator.

9. Disconnect the positive battery cable at the power distribution box. Remove the retaining bolt from the positive battery cable bracket located on the side of the right cylinder head.

10. Disconnect the vent hose from the canister purge solenoid and position the positive battery cable out of the way. Disconnect the canister purge solenoid vent hose from the PCV valve and remove the PCV valve from the camshaft cover.

11. Remove the 42-pin engine harness connector from the retaining bracket on the brake vacuum booster, disconnect and position out of the way.

12. Disconnect the HDR sensor, air conditioning compressor clutch and canister purge solenoid connectors.

13. Raise and safely support the vehicle.

14. Remove the bolts retaining the power steering pump to the engine block and front cover. The front lower bolt on the power steering pump will not come all the way out. Wire the power steering pump out of the way.

15. Remove the 4 bolts retaining the oil pan to the front cover. Remove the crankshaft damper retaining bolt and remove the damper, using a suitable puller.

16. Disconnect the EVO sensor and oil sending unit. Position the EVO sensor and oil pressure sending unit harness out of the way.

17. Disconnect the EGR tube from the right exhaust manifold. Disconnect the exhaust pipes from the exhaust manifolds. Lower the exhaust pipes and hang with wire from the crossmember.

18. Remove the bolt retaining the starter wiring harness to the rear of the right cylinder head. Lower the vehicle.

19. Remove the bolts and stud bolts retaining the camshaft covers to the cylinder heads and remove the covers.

20. Disconnect the accelerator, cruise control and throttle valve cables. Remove the accelerator cable bracket from the intake manifold and position out of the way.

21. Disconnect the vacuum hose from the throttle body elbow vacuum port, both oxygen sensors and the heater supply hose.

22. Remove the 2 bolts retaining the thermostat housing to the intake manifold and position the upper hose and thermostat housing out of the way.

NOTE: Two thermostat housing bolts also retain the intake manifold.

23. Remove the 9 bolts retaining the intake manifold to the cylinder heads and remove the intake manifold and gaskets.

24. Remove the 7 stud bolts and 4 bolts retaining the front cover to the engine and remove the front cover.

25. Remove the timing chains.

26. Remove the 10 bolts retaining the left cylinder head to the engine block and remove the head.

NOTE: The lower rear bolt cannot be removed due to interference with the brake vacuum booster. Use a rubber band to hold the bolt away from the engine block.

27. Remove the ground strap, 1 stud and 1 bolt retaining the heater return line to the right cylinder head.

28. Remove the 10 bolts retaining the right cylinder head to the engine block and remove the head.

NOTE: The lower rear bolt cannot be removed due to interference with the evaporator housing. Use a rubber band to hold the bolt away from the engine block.

29. Clean all gasket mating surfaces. Check the cylinder head and engine block for flatness. Check the cylinder head for scratches near the coolant passage and combustion chamber that could provide leak paths. Machine as necessary.

To install:

30. Rotate the crankshaft counterclockwise 45 degrees. The crankshaft keyway should be at the 9 o'clock position viewed from the front of the engine. This ensures that all pistons are below the top of the engine block deck face.

31. Rotate the camshaft to a stable position where the valves do not extend below the head face.

32. Position new head gaskets on the engine block. Install the lower rear bolts on both cylinder heads and retain with rubber bands as explained during the removal procedure.

33. Position the cylinder heads on the engine block dowels, being careful not to score the surface of the head face. Apply clean oil to the head bolts, remove the rubber band from the lower rear bolt and install all bolts hand-tight.

34. Tighten the head bolts as follows:
   a. Tighten the bolts, in sequence, to 15–22 ft. lbs. (20–30 Nm).
   b. Rotate each bolt, in sequence, 85–95 degrees.
   c. Rotate each bolt, in sequence, an additional 85–95 degrees.

35. Position the heater return hose and install the 2 bolts. Rotate the camshafts using the flats matched at the center of the camshaft until both are in time. Install cam positioning tools T91P–6256–A or equivalent, on the flats of the camshafts to keep them from rotating.

36. Rotate the crankshaft clockwise 45 degrees to position the crankshaft at TDC on No. 1 cylinder.

NOTE: The crankshaft must only be rotated in the clockwise direction and only as far as TDC.

37. Install the timing chains according to the proper procedure.

38. Install a new front cover seal and gasket. Apply silicone sealer to the lower corners of the cover where it meets the junction of the oil pan and cylinder block and to the points where the cover contacts the junction of the cylinder block and cylinder head.

39. Install the front cover and the stud bolts and bolts. Tighten to 15–22 ft. lbs. (20–30 Nm).

40. Position new intake manifold

gaskets on the cylinder heads. Make sure the alignment tabs on the gaskets are aligned with the holes in the cylinder heads.

NOTE: Before installing the intake manifold, inspect it for nicks and cuts that could provide leak paths.

41. Position the intake manifold on the cylinder heads and install the retaining bolts. Tighten the bolts, in sequence, to 15–22 ft. lbs. (20–30 Nm).

42. Install the thermostat and O-ring, then position the thermostat housing and upper hose and install the 2 bolts. Tighten to 15–22 ft. lbs. (20–30 Nm).

43. Connect the heater supply hose and both oxygen sensors. Connect the vacuum hose to the throttle body adapter vacuum port.

44. Connect and, if necessary, adjust the throttle valve cable. Install the accelerator cable bracket on the intake manifold and connect the accelerator and cruise control cables to the throttle body.

45. Apply silicone sealer to both places where the front cover meets the cylinder head. Install new gaskets on the camshaft covers.

46. Install the camshaft covers on the cylinder heads. Install the bolts and stud bolts and tighten to 6.0–8.8 ft. lbs. (8–12 Nm).

47. Raise and safely support the vehicle. Position the starter wiring harness to the right cylinder head and install the retaining bolt.

48. Cut the wire and position the exhaust pipes to the exhaust manifolds. Tighten the 4 nuts to 20–30 ft. lbs. (27–41 Nm).

NOTE: Make sure the exhaust system clears the No. 3 crossmember. Adjust as necessary.

49. Connect the EGR tube to the right exhaust manifold and tighten the line nut to 26–33 ft. lbs. (35–45 Nm). Connect the EVO sensor and oil sending unit.

50. Apply a small amount of silicone sealer in the rear of the keyway on the damper. Position the damper on the crankshaft, making sure the crankshaft key and keyway are aligned.

51. Using damper installer T74P-6316-B or equivalent, install the crankshaft damper. Install the damper bolt and washer and tighten to 114–121 ft. lbs. (155–165 Nm).

52. Install the 4 bolts retaining the oil pan to the front cover and tighten to 15–22 ft. lbs. (20–30 Nm).

53. Position the power steering pump on the engine and install the 4 retaining bolts. Tighten to 15–22 ft. lbs. (20–30 Nm). Lower the vehicle.

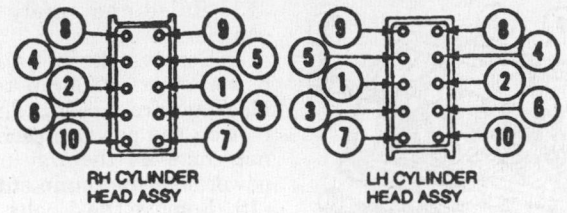

Cylinder head bolt torque sequence—4.6L engine

54. Connect the air conditioning compressor, HDR sensor and canister purge solenoid.

55. Connect the 42-pin engine harness connector and transmission harness connector. Install the 42-pin connector on the retaining bracket on the vacuum brake booster.

56. Install the PCV valve in the right camshaft cover and connect the canister purge solenoid vent hose.

57. Position the positive battery cable harness on the right cylinder head and install the bolt retaining the cable bracket to the cylinder head. Connect the positive battery cable at the power distribution box and battery.

58. Position the alternator and install the 2 retaining bolts. Tighten to 15–22 ft. lbs. (20–30 Nm). Install the 2 bolts retaining the alternator brace to the intake manifold and tighten to 6–8 ft. lbs. (8–12 Nm).

59. Install the water pump pulley and tighten the bolts to 15–22 ft. lbs. (20–30 Nm).

60. Position the ignition coil brackets and ignition wire assembly onto the mounting studs. Install the 7 nuts retaining the coil brackets to the front cover and tighten to 15–22 ft. lbs. (20–30 Nm).

61. Install the 2 bolts retaining the ignition wire tray to the coil bracket and tighten to 6.0–8.8 ft. lbs. (8–12 Nm). Connect both ignition coils and CID sensor.

62. Position the air conditioner high pressure line on the right coil bracket and install the bolt. Connect the ignition wires to the spark plugs and install the bracket onto the camshaft cover studs.

63. Install the accessory drive belt and the wiper module. Connect the fuel lines and install the cooling fan and shroud. Fill and bleed the cooling system.

64. Install the air inlet tube and connect the negative battery cable. Start the engine and bring to normal operating temperature. Check for leaks. Check all fluid levels.

5.0L and 5.8L Engines

1. Disconnect the negative battery cable.

2. Drain the cooling system and relieve the fuel system pressure.

3. On 5.0L engine, remove the upper and lower intake manifold and throttle body assembly. On 5.8L engine, remove the intake manifold and carburetor assembly.

4. If the air conditioning compressor is in the way of a cylinder head that is to be removed, proceed as follows:

a. Discharge the air conditioning system.

b. Disconnect and plug the refrigerant lines at the compressor. Cap the openings on the compressor.

c. Disconnect the electrical connector to the compressor.

d. Remove the compressor and the necessary mounting brackets.

5. If the left cylinder head is to be removed, disconnect the power steering pump bracket from the cylinder head and remove the drive belt from the pump pulley. Position the pump out of the way in a position that will prevent the oil from draining out.

6. Disconnect the oil level indicator tube bracket from the exhaust manifold stud, if necessary.

7. If the right cylinder head is to be removed, on some vehicles it is necessary to disconnect the alternator mounting bracket from the cylinder head.

8. Remove the thermactor crossover tube from the rear of the cylinder heads. If equipped, remove the fuel line from the clip at the front of the right cylinder head.

9. Raise and safely support the vehicle. Disconnect the exhaust manifolds from the muffler inlet pipes. Lower the vehicle.

10. Loosen the rocker arm fulcrum bolts so the rocker arms can be rotated to the side. Remove the pushrods in sequence so they may be installed in their original positions.

11. Remove the cylinder head attaching bolts and the cylinder heads. If necessary, remove the exhaust manifolds to gain access to the lower bolts. Remove and discard the head gaskets.

12. Clean all gasket mating surfaces. Check the flatness of the cylinder head using a straight edge and a feeler gauge. The cylinder head must not be warped any more than 0.003 in. in any 6.0 in. – 0.006 in. overall. Machine as necessary.

To install:

13. Position the new cylinder head gasket over the dowels on the block. Position the cylinder heads on the block and install the attaching bolts.

14. On 5.0L engine, tighten the bolts, in sequence, in 2 steps, first to 55–65 ft. lbs. (75–88 Nm), then to 65–72 ft. lbs. (88–97 Nm). On 5.8L engine, tighten the bolts, in sequence, in 2 steps, first to 95–105 ft. lbs. (129–142 Nm), then to 105–112 ft. lbs. (142–152 Nm).

NOTE: When the cylinder head bolts have been tightened following this procedure, it is not necessary to retighten the bolts after extended operation.

15. If removed, install the exhaust manifolds. Tighten the retaining bolts to 18–24 ft. lbs. (24–32 Nm).

16. Clean the pushrods, making sure the oil passages are clean. Check the ends of the pushrods for wear. Visually check the pushrods for straightness or check for runout using a dial indicator. Replace pushrods, as necessary.

17. Apply a suitable grease to the ends of the pushrods and install them in their original positions. Position the rocker arms over the pushrods and the valves.

18. Before tightening each fulcrum bolt, bring the lifter for the fulcrum bolt to be tightened onto the base circle of the camshaft by rotating the engine. When the lifter is on the base circle of the camshaft, tighten the fulcrum bolt to 18–25 ft. lbs. (24–34 Nm).

NOTE: If all the original valve train parts are reinstalled, a valve clearance check is not necessary. If any valve train components are replaced, a valve clearance check must be performed.

19. Install new rocker arm cover gaskets on the rocker arm covers and install the covers on the cylinder heads.

20. Raise and safely support the vehicle. Connect the exhaust manifolds to the muffler inlet pipes. Lower the vehicle.

21. If necessary, install the air conditioning compressor and brackets. Connect the refrigerant lines and electrical connector to the compressor.

22. If necessary, install the alternator bracket.

23. If the left cylinder head was removed, install the power steering pump.

24. Install the drive belts. Install the thermactor tube at the rear of the cylinder heads.

25. Install the intake manifold. Fill and bleed the cooling system.

26. Connect the negative battery cable, start the engine and bring to normal operating temperature. Check for leaks. Check all fluid levels.

27. If necessary, leak test, evacuate and charge the air conditioning system according to the proper procedure. Observe all safety precautions.

Valve Lifters

REMOVAL & INSTALLATION

2.3L Engine

The 2.3L engine is equipped with hydraulic lash adjusters which, while not being exactly the same as a conventional hydraulic lifter, perform the same function — maintain proper valve train clearance.

1. Disconnect the negative battery cable.

2. On turbocharged engine, proceed as follows:

   a. Disconnect the air intake tube at the throttle body by disconnecting the metal tube from the turbocharger and clamp from the throttle body end.

   b. Disconnect the PCV hose at the rear of the rocker cover.

   c. Remove the throttle body assembly.

3. If equipped with distributor ignition, tag and disconnect the spark plug wires from the spark plugs. Move the wires out of the way.

4. Remove the hose and the retaining bolts from the rocker arm cover and remove the cover.

5. Rotate the camshaft so the base circle of the cam is facing the cam follower to be removed.

6. Using valve spring compressor tool T88T–6565–BH or equivalent, compress the lash adjuster as required and/or depress the valve spring if necessary and slide the cam follower over the lash adjuster and out.

7. Lift out the hydraulic lash adjuster.

To install:

8. Rotate the camshaft so the base circle of the camshaft is facing the lash adjuster and cam follower to be installed. Place the hydraulic lash adjuster in position in the bore.

9. Using valve spring compressor tool T88T–6565–BH or equivalent, compress the lash adjuster, as necessary, to position the cam follower over the lash adjuster and the valve stem.

10. Before rotating the camshaft to the next position, make sure the lash adjuster just installed is fully compressed and released.

11. Clean the gasket mating surface of the rocker arm cover and cylinder head. Install a new gasket and the rocker arm cover. Install the mounting screws and tighten to 62–97 inch lbs. (7–11 Nm).

12. Install the remaining components in the reverse order of removal. Start the engine and check for oil leaks.

3.8L Engine

NOTE: Before replacing a tappet for noisy operation, be sure the noise is not caused by improper valve to rocker arm clearance or by worn rocker arms or pushrods.

1. Disconnect the negative battery cable. Disconnect the spark plug wires at the spark plugs.

2. Remove plug wire routing clips from the studs on the rocker arm cover attaching bolts. Lay the plug wires, with the routing clips toward the front of the engine.

3. Remove the upper intake manifold. On supercharged engine, remove the supercharger.

4. Remove the rocker arm covers. Remove the lower intake manifold.

5. Sufficiently loosen each rocker arm fulcrum attaching bolt to allow the rocker arm to be lifted off the pushrod and rotated to 1 side.

6. Remove the pushrods. The location of each pushrod should be identified. When the engine is assembled each rod should be installed in its original position.

7. If equipped with roller lifters, remove the 4 bolts holding the 2 guide plate retainers in place; the bolts are held captive in the retainers. Remove the 6 guide plates from the adjacent lifters.

8. Remove the lifters using a magnet. The location of each lifter should be identified. When the engine is assembled, each lifter should be installed in its original position.

NOTE: If the lifters are stuck in the bores due to excessive varnish or gum deposits, it may be necessary to use a claw-type tool to aid removal. When using a remover tool, rotate the lifter back and forth to loosen it from gum or varnish that may have formed on the lifter.

To install:

9. Clean the rocker arm cover and cylinder head mating surfaces.

10. Install each lifter in the bore from which it was removed. If new lifters are being installed, check the new lifters for free fit in the bores.

11. If equipped with roller lifters, align the flats on the side of the lifters and install the 6 guide plates between the adjacent lifters. Make sure the word **UP** is showing. Install the 2 guide plate retainers and tighten the 4 captive bolts to 7–10 ft. lbs. (9–14 Nm).

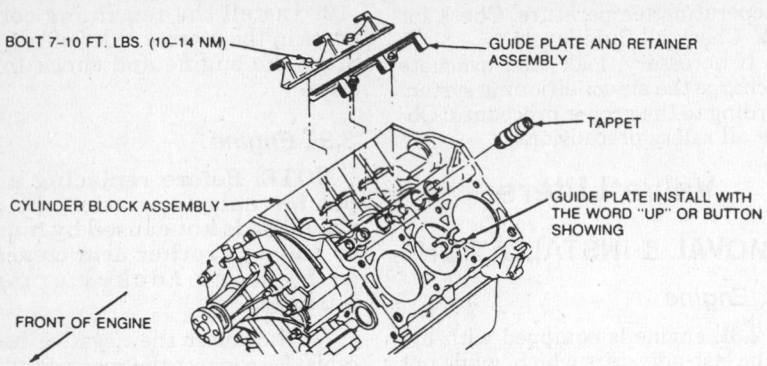

BOLT 7-10 FT. LBS. (10-14 NM)
GUIDE PLATE AND RETAINER ASSEMBLY
TAPPET
CYLINDER BLOCK ASSEMBLY
GUIDE PLATE INSTALL WITH THE WORD "UP" OR BUTTON SHOWING
FRONT OF ENGINE

Valve lifter installation—3.8L engine with roller lifters

12. Dip each pushrod in heavy engine oil and install in its original position.

13. For each valve, rotate the crankshaft until the lifter rests on the base circle of the camshaft lobe. Position the rocker arms over the pushrods. Install the fulcrums and tighten the bolts to 44 inch lbs. (5 Nm).

14. Lubricate all rocker arm assemblies with heavy engine oil. Final tighten the fulcrum bolts to 19-25 ft. lbs. (25-35 Nm). For final tightening the camshaft may be in any position.

NOTE: The fulcrums must be fully seated in the cylinder head and the pushrods must be seated in the rocker arm sockets prior to final tightening.

15. Install the lower intake manifold and the rocker arm covers. On non-supercharged engines, install the upper intake manifold. On supercharged engines, install the supercharger.

16. Install the spark plug wire routing clips and connect the wires to the spark plugs. Connect the negative battery cable, start the engine and check for oil and coolant leaks.

4.6L Engine

The 4.6L engine is equipped with hydraulic lash adjusters which, while not being exactly the same as a conventional hydraulic lifter, perform the same function—maintain proper valve train clearance.

1. Disconnect the negative battery cable.

2. Remove the right camshaft cover as follows:

   a. Disconnect the positive battery cable at the battery and at the power distribution box. Remove the retaining bolt from the positive battery cable bracket located on the side of the right cylinder head.

   b. Disconnect the High Data Rate (HDR) sensor, air conditioning compressor clutch and canister purge solenoid connectors. Position the harness out of the way.

   c. Disconnect the vent hose from

the purge solenoid and position the positive battery cable out of the way.

   d. Disconnect the ignition wires from the spark plugs. Remove the ignition wire brackets from the camshaft cover studs and position the wires out of the way.

   e. Remove the PCV valve from the camshaft cover grommet and position out of the way.

   f. Remove the bolts and stud bolts and remove the camshaft cover.

3. Remove the left camshaft cover as follows:

   a. Remove the air inlet tube. Relieve the fuel system pressure and disconnect the fuel lines.

   b. Raise and safely support the vehicle.

   c. Disconnect the EVO sensor and oil pressure sending unit and position the harness out of the way. Lower the vehicle.

   d. Remove the 42-pin engine harness connector from the retaining bracket on the brake vacuum booster. Disconnect and position out of the way.

   e. Remove the windshield wiper module.

   f. Disconnect the ignition wires from the spark plugs. Remove the ignition wire brackets from the studs and position the wires out of the way.

   g. Remove the bolts and stud bolts and remove the camshaft cover.

4. Position the piston of the cylinder being serviced at the bottom of its stroke and position the camshaft lobe on the base circle.

5. Install valve spring spacer tool T91P-6565-AH or equivalent, between the spring coils to prevent valve seal damage.

NOTE: If the valve spring spacer tool is not used, the retainer will hit the valve stem seal and damage the seal.

6. Install valve spring compressor

tool T91P-6565-A or equivalent, under the camshaft and on top of the valve spring retainer.

7. Compress the valve spring and remove the roller follower. Remove the valve spring compressor and spacer.

8. Remove the hydraulic lash adjuster.

To install:

9. Check the hydraulic lash adjusters. They must have no more than 1.5mm of plunger travel prior to installation.

10. Apply engine oil to the valve stem and tip, roller follower contact surfaces and lash adjuster bore. Install the lash adjusters.

11. Install valve spring spacer tool T91P-6565-AH or equivalent, between the spring coils. Compress the valve spring using valve spring compressor tool T91P-6565-A or equivalent, and install the roller follower.

NOTE: The piston must be at the bottom of its stroke and the camshaft at the base circle.

12. Remove the valve spring compressor and spacer.

13. Clean the sealing surfaces of the camshaft covers and cylinder heads. Apply silicone sealer to the places where the front cover meets the cylinder head.

14. Position new gaskets onto the camshaft covers and install the covers. Install the bolts and stud bolts and tighten to 6.0-8.8 ft. lbs. (8-12 Nm).

15. When installing the right camshaft cover, proceed as follows:

   a. Install the PCV into the camshaft cover grommet.

   b. Install the ignition wire brackets on the studs and connect the wires to the spark plugs.

   c. Position the harness and connect the canister purge solenoid, air conditioning compressor clutch and HDR sensor.

   d. Position the positive battery cable harness on the right cylinder head. Install the bolt retaining the cable bracket to the cylinder head.

   e. Connect the positive battery cable at the power distribution box and the battery.

16. When installing the left camshaft cover, proceed as follows:

   a. Install the ignition wire brackets on the studs and connect the wires to the spark plugs.

   b. Install the windshield wiper module.

   c. Connect the 42-pin connector and transmission harness connector. Install the connector on the retaining bracket.

   d. Raise and safely support the vehicle. Position and connect the

EVO sensor and oil pressure sending unit harness.

  e. Lower the vehicle. Connect the fuel lines.

17. Connect the negative battery cable. Start the engine and check for leaks.

5.0L and 5.8L Engines

1. Disconnect the negative battery cable. Remove the intake manifold and related parts.

2. Remove the crankcase ventilation hoses, PCV valve and elbows from the valve rocker arm covers.

3. Remove the valve rocker arm covers. Loosen the valve rocker arm fulcrum bolts and rotate the rocker arms to the side.

4. Remove the valve pushrods and identify them so that they can be installed in their original position.

5. If equipped with roller lifters, remove the lifter guide retainer bolts. Remove the retainer and lifter guide plates. Identify the guide plates so they may be reinstalled in their original positions.

6. Using a magnet, remove the lifters and place them in a rack so that they can be installed in their original bores.

NOTE: If the lifters are stuck in the bores due to excessive varnish or gum deposits, it may be necessary to use a claw-type tool to aid removal. When using a remover tool, rotate the lifter back and forth to loosen it from gum or varnish that may have formed on the lifter.

To install:

7. Lubricate the lifters and install them in their original bores. If new lifters are being installed, check them for free fit in their respective bores.

8. If equipped with roller lifters, install the lifter guide plates in their original positions, then install the guide plate retainer.

9. Install the pushrods in their original positions. Apply grease to the ends prior to installation.

10. Lubricate the rocker arms and fulcrum seats with heavy engine oil. Position the rocker arms over the pushrods and install the fulcrum bolts.

11. Before tightening each fulcrum bolt, rotate the crankshaft until the lifter is on the base circle of the cam. Tighten the fulcrum bolt to 18–25 ft. lbs. (24–34 Nm). Check the valve clearance.

12. Install the rocker arm covers and the intake manifold. Connect the negative battery cable, start the engine and check for leaks.

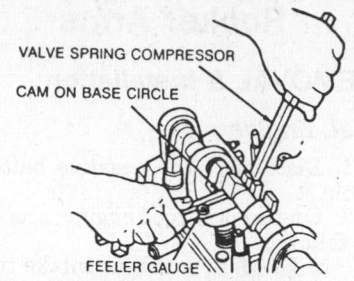

Checking collapsed lifter clearance— 2.3L engine

Valve Lash

ADJUSTMENT

2.3L Engnie

1. Disconnect the negative battery cable.

2. Remove the valve cover assembly.

3. Position the camshaft so that the base circle of the lobe is facing the cam follower of the valve to be checked.

4. Using valve spring compressor tool T88T-6565-BH or equivalent, slowly apply pressure to the cam follower until the the lash adjuster is completely collapsed.

5. With follower collapsed, insert a feeler gauge between the base circle of the camshaft and follower. The clearance should not be more than 0.040–0.050 in.

6. If the clearance is excessive, remove the cam follower and inspect for damage.

7. If the cam follower appears to be intact and not excessively worn, measure the valve spring assembled height to make sure the valve is not sticking.

8. If the valve spring assembled height is correct, check the camshaft for wear. If the camshaft dimensions are correct, replace the lash adjuster.

9. Install the valve cover and any other removed components.

3.8L Engine

The valve lash is not adjustable. If the collapsed lifter clearance is found to be incorrect, there are replacement pushrods available to compensate for excessive or insufficient clearance.

1. Disconnect the negative battery cable.

2. Remove the valve cover assembly on the side to be checked.

3. Turn the engine until the No. 1 piston is at TDC of it's compression stroke.

4. The following valves can be checked with the engine in this position:

  a. No. 1 intake — No. 1 exhaust
  b. No. 3 intake — No. 2 exhaust
  c. No. 6 intake — No. 4 exhaust

5. Rotate the engine 360 degrees and check the following valves:

  a. No. 2 intake — No. 3 exhaust
  b. No. 4 intake — No. 5 exhaust
  c. No. 5 intake — No. 6 exhaust

6. Check each of the lifters by placing a hydraulic lifter compressing tool on the rocker arm and slowly applying pressure to the lifter, until the lifter is collapsed.

7. Hold the lifter in this position and check the clearance between the rocker arm and the and the valve stem tip. The clearance should not exceed 0.375 in. (9.53mm) on 1988–89 engines. The clearance should be 0.09–0.19. (2.25–4.79mm) on 1990–92 engines.

8. Repeat this operation for each valve to be checked.

9. If the clearance is greater than specification, replace the pushrod with a longer one. If the clearance is less than specified, replace the pushrod with a shorter one.

5.0L and 5.8L Engines

The valve lash is not adjustable. If the collapsed lifter clearance is found to be incorrect, there are replacement pushrods available to compensate for excessive or insufficient clearance.

5.0L ENGINE EXCEPT HO

1. Install an auxiliary starter switch. Crank the engine with the ignition switch off until the No. 1 piston is at TDC on the compression stroke.

2. With the crankshaft in the positions designated in Steps 4, 5 and 6, position lifter bleed down wrench tool No. T71P-6513-B or equivalent, on the rocker arm. Slowly apply pressure to bleed down the lifter until the plunger is completely bottomed. Hold the lifter in this position and check the available clearance between the rocker arm and the valve stem tip with a feeler gauge.

3. The clearance should be 0.096–0.146 in. If the clearance is less than specification, install a shorter pushrod. If the clearance is greater than specification, install a longer pushrod.

4. The following valves can be checked with the engine in position 1, No. 1 piston at TDC on the compression stroke.

  a. No. 1 intake — No. 1 exhaust
  b. No. 7 intake — No. 5 exhaust
  c. No. 8 intake — No. 4 exhaust

5. Rotate the engine 360 degrees (1 revolution) from the 1st position and check the following valves:

  a. No. 5 intake — No. 2 exhaust
  b. No. 4 intake — No. 6 exhaust

6. Rotate the engine 90 degrees (¼ revolution) from the 2nd position and check the following valves:

  a. No. 2 intake — No. 7 exhaust
  b. No. 3 intake — No. 3 exhaust

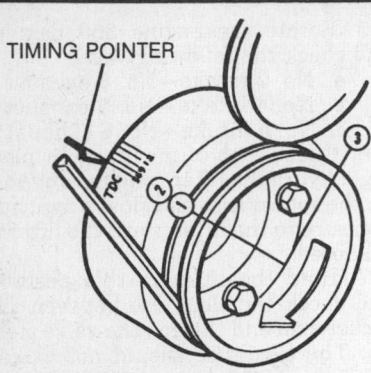

TIMING POINTER

POSITION 1—NO. 1 AT TDC ON
COMPRESSION STROKE
POSITION 2—ROTATE THE
CRANKSHAFT 360 DEGREES (1
REVOLUTION) CLOCKWISE FROM
POSITION 1
POSITION 3—ROTATE THE
CRANKSHAFT 90 DEGREES (¼
REVOLUTION) CLOCKWISE FROM
POSITION 2

**Engine valve adjusting positions—
5.0L and 5.8L engines**

   c. No. 6 intake—No. 8 exhaust

5.0L HO AND 5.8L ENGINES

1. Install an auxiliary starter switch. Crank the engine with the ignition switch off until the No. 1 piston is at TDC on the compression stroke.

2. With the crankshaft in the positions designated in Steps 4, 5 and 6, position lifter bleed down wrench tool No. T71P-6513-B or equivalent, on the rocker arm. Slowly apply pressure to bleed down the lifter until the plunger is completely bottomed. Hold the lifter in this position and check the available clearance between the rocker arm and the valve stem tip with a feeler gauge.

3. The clearance should be 0.123–0.146 in. on 5.0L HO engine and 0.096–0.146 in. on 5.8L engine. If the clearance is less than specification, install a shorter pushrod. If the clearance is greater than specification, install a longer pushrod.

4. The following valves can be checked with the engine in position 1, No. 1 piston at TDC on the compression stroke.

   a. No. 1 intake—No. 1 exhaust
   b. No. 4 intake—No. 3 exhaust
   c. No. 8 intake—No. 7 exhaust

5. Rotate the engine 360 degrees (1 revolution) from the 1st position and check the following valves:

   a. No. 3 intake—No. 2 exhaust
   b. No. 7 intake—No. 6 exhaust

6. Rotate the engine 90 degrees (¼ revolution) from the 2nd position and check the following valves:

   a. No. 2 intake—No. 4 exhaust
   b. No. 5 intake—No. 5 exhaust
   c. No. 6 intake—No. 8 exhaust

Rocker Arms

REMOVAL & Installation

2.3L Engine

1. Disconnect the negative battery cable.

2. On turbocharged engine, proceed as follows:

   a. Disconnect the air intake tube at the throttle body by disconnecting the metal tube from the turbocharger and clamp from the throttle body end.

   b. Disconnect the PCV hose at the rear of the rocker cover.

   c. Remove the throttle body assembly.

3. If equipped with distributor ignition, tag and disconnect the spark plug wires from the spark plugs. Move the wires out of the way.

4. Remove the hose and the retaining bolts from the rocker arm cover and remove the cover.

5. Rotate the camshaft so the base circle of the cam is facing the cam follower to be removed.

6. Using valve spring compressor tool T88T-6565-BH or equivalent, compress the lash adjuster as required and/or depress the valve spring if necessary and slide the cam follower over the lash adjuster and out.

To install:

7. Using valve spring compressor tool T88T-6565-BH or equivalent, compress the lash adjuster, as necessary, to position the cam follower over the lash adjuster and the valve stem.

8. Before rotating the camshaft to the next position, make sure the lash adjuster just installed is fully compressed and released.

9. Clean the gasket mating surface of the rocker arm cover and cylinder head. Install a new gasket and the rocker arm cover. Install the mounting screws and tighten to 62–97 inch lbs. (7–11 Nm).

10. Install the remaining components in the reverse order of removal. Start the engine and check for oil leaks.

3.8L Engine

1. Disconnect the negative battery cable.

2. Disconnect the spark plug wires from the spark plugs. Remove the spark plug wire routing clips from the rocker arm cover attaching bolt studs.

3. To remove the left rocker arm cover, proceed as follows:

   a. Remove the oil fill cap.

   b. On 1990–92 vehicles, remove the crankcase vent tube.

   c. On supercharged engines, remove the intercooler tubes and the oil cooler inlet tube.

4. To remove the right rocker arm cover, proceed as follows:

   a. Remove the PCV valve.

   b. Position the air cleaner assembly aside, if necessary.

   c. On supercharged engines, remove the air inlet tube and remove the throttle body assembly.

5. Remove the rocker arm cover attaching screws and remove the rocker arm covers.

6. Remove the rocker arm, fulcrum and bolt assemblies. Keep each assembly together and identify the assemblies so they may be reinstalled in their original positions.

To install:

7. Clean all gasket mating surfaces on the rocker arm covers and cylinder heads. Clean the rocker arms and fulcrums and inspect for wear or damage. Replace as necessary.

8. Apply grease to the pushrod tips and valve stem tips. Lubricate the fulcrums and rocker arms with heavy engine oil and install them over the pushrods and valve stems.

9. For each valve, rotate the crankshaft until the lifter is on the base circle of the camshaft. Install the fulcrum bolt and tighten to 18–25 ft. lbs. (25–35 Nm). Make sure the pushrod and fulcrum are fully seated prior to tightening.

10. Position new gaskets on the cylinder heads and install the rocker arm covers. Tighten the attaching bolts to 80–106 inch lbs. (9–12 Nm). Note the location of the spark plug wire routing clip stud bolts prior to installation.

11. After installing the left rocker arm cover, proceed as follows:

   a. Install the oil fill cap.

   b. On 1990–92 vehicles, install the crankcase vent tube.

   c. On supercharged engines, install the intercooler tubes and the oil cooler inlet tube.

12. After installing the right valve cover, proceed as follows:

   a. Install the PCV valve.

   b. Install the air cleaner assembly, if necessary.

   c. On supercharged engines, install the aie inlet tube and the throttle body assembly.

13. Install the spark plug wire routing clips and connect the wires to the spark plugs.

14. Connect the negative battery cable, start the engine and check for leaks.

4.6L Engine

1. Disconnect the negative battery cable.

2. Remove the right camshaft cover as follows:

   a. Disconnect the positive battery cable at the battery and at the power distribution box. Remove the retaining bolt from the positive battery cable bracket located on the side of the right cylinder head.

b. Disconnect the High Data Rate (HDR) sensor, air conditioning compressor clutch and canister purge solenoid connectors. Position the harness out of the way.

c. Disconnect the vent hose from the purge solenoid and position the positive battery cable out of the way.

d. Disconnect the ignition wires from the spark plugs. Remove the ignition wire brackets from the camshaft cover studs and position the wires out of the way.

e. Remove the PCV valve from the camshaft cover grommet and position out of the way.

f. Remove the bolts and stud bolts and remove the camshaft cover.

3. Remove the left camshaft cover as follows:

a. Remove the air inlet tube. Relieve the fuel system pressure and disconnect the fuel lines.

b. Raise and safely support the vehicle.

c. Disconnect the EVO sensor and oil pressure sending unit and position the harness out of the way. Lower the vehicle.

d. Remove the 42-pin engine harness connector from the retaining bracket on the brake vacuum booster. Disconnect and position out of the way.

e. Remove the windshield wiper module.

f. Disconnect the ignition wires from the spark plugs. Remove the ignition wire brackets from the studs and position the wires out of the way.

g. Remove the bolts and stud bolts and remove the camshaft cover.

4. Position the piston of the cylinder being serviced at the bottom of its stroke and position the camshaft lobe on the base circle.

5. Install valve spring spacer tool T91P–6565–AH or equivalent, between the spring coils to prevent valve seal damage.

NOTE: If the valve spring spacer tool is not used, the retainer will hit the valve stem seal and damage the seal.

6. Install valve spring compressor tool T91P–6565–A or equivalent, under the camshaft and on top of the valve spring retainer.

7. Compress the valve spring and remove the roller follower. Remove the valve spring compressor and spacer.
To install:
8. Apply engine oil to the valve stem

and tip and roller follower contact surfaces.

9. Install valve spring spacer tool T91P–6565–AH or equivalent, between the spring coils. Compress the valve spring using valve spring compressor tool T91P–6565–A or equivalent, and install the roller follower.

NOTE: The piston must be at the bottom of its stroke and the camshaft at the base circle.

10. Remove the valve spring compressor and spacer.

11. Clean the sealing surfaces of the camshaft covers and cylinder heads. Apply silicone sealer to the places where the front cover meets the cylinder head.

12. Position new gaskets onto the camshaft covers and install the covers. Install the bolts and stud bolts and tighten to 6.0–8.8 ft. lbs. (8–12 Nm).

13. When installing the right camshaft cover, proceed as follows:

a. Install the PCV into the camshaft cover grommet.

b. Install the ignition wire brackets on the studs and connect the wires to the spark plugs.

c. Position the harness and connect the canister purge solenoid, air conditioning compressor clutch and HDR sensor.

d. Position the positive battery cable harness on the right cylinder head. Install the bolt retaining the cable bracket to the cylinder head.

e. Connect the positive battery cable at the power distribution box and the battery.

14. When installing the left camshaft cover, proceed as follows:

a. Install the ignition wire brackets on the studs and connect the wires to the spark plugs.

b. Install the windshield wiper module.

c. Connect the 42-pin connector and transmission harness connec-

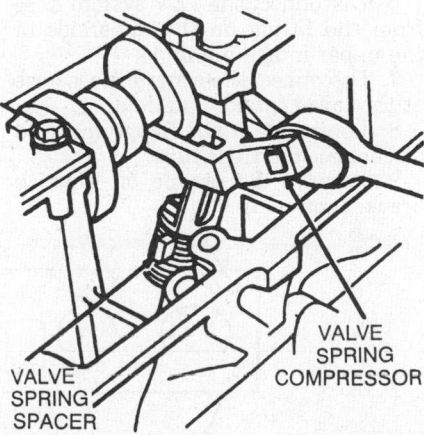

Camshaft follower removal—4.6L engine

tor. Install the connector on the retaining bracket.

d. Raise and safely support the vehicle. Position and connect the EVO sensor and oil pressure sending unit harness.

e. Lower the vehicle. Connect the fuel lines.

15. Connect the negative battery cable. Start the engine and check for leaks.

5.0L Engine

1. Disconnect the negative battery cable.

2. Before removing the right rocker arm cover, disconnect the PCV closure tube from the oil fill stand pipe at the rocker cover.

3. Remove the thermactor bypass valve and air supply hoses as necessary to provide clearance.

4. Disconnect the spark plug wires from the spark plugs. Remove the wires and bracket assembly from the rocker arm cover attaching stud and position the wires out of the way.

5. Remove the upper intake manifold.

6. Remove the attaching bolts and remove the covers.

7. Remove the rocker arm fulcrum bolt, fulcrum seat and rocker arm. Keep all rocker arm assemblies together. Identify each assembly so it may be reinstalled in its original position.
To install:
8. Clean all gasket mating surfaces of the rocker arm covers and cylinder heads. Clean and inspect the rocker arm assemblies for wear and/or damage. Replace as necessary.

9. Apply grease to the pushrod and valve stem tips and the underside of the fulcrum seats.

10. Rotate the crankshaft until the lifter is on the camshaft base circle and install the rocker, fulcrum seat and fulcrum bolt. Tighten the bolts to 18–25 ft. lbs. (24–34 Nm).

11. Position new rocker arm cover gaskets and install the rocker arm covers. Tighten the bolts to 6–9 ft. lbs. (8–12 Nm) on 1988 vehicles, 10–13 ft. lbs. (14–18 Nm) on 1989–92 vehicles, wait 2 minutes and tighten again to the same specification.

12. Install the crankcase ventilation tube in the right cover. Install the upper intake manifold.

13. Install the spark plug wires and bracket assembly on the rocker cover attaching stud. Connect the spark plug wires.

14. Install the air cleaner and intake duct assembly. Install the thermactor bypass valve and air supply hoses, if required.

15. Connect the negative battery cable, start the engine and check for leaks.

5.8L Engine

1. Disconnect the negative battery cable.

2. Before removing the right rocker arm cover, remove the cylinder head. Disconnect the automatic choke heat chamber air inlet hose from the inlet tube near the right rocker arm cover, if equipped.

3. Remove the crankcase ventilation fresh air tube from the rocker arm cover.

4. Remove the thermactor bypass valve and air supply hoses as necessary to provide clearance.

5. Disconnect the spark plug wires from the spark plugs. Remove the wires and bracket assembly from the rocker arm cover attaching stud and position the wires out of the way.

6. On the left side rocker arm cover, remove the wire harness from the retaining clips. Disconnect the wires at the solenoid mounted on the left rocker cover.

7. Remove the rocker arm cover attaching bolts and remove the rocker arm cover. Remove the rocker arm fulcrum bolt, fulcrum seat and rocker arm. Keep all rocker arm assemblies together. Identify each assembly so it may be reinstalled in its original position.

To install:

8. Clean all gasket mating surfaces of the rocker arm covers and cylinder heads. Clean and inspect the rocker arm assemblies for wear and/or damage. Replace as necessary.

9. Apply grease to the pushrod and valve stem tips and the underside of the fulcrum seats.

10. Rotate the crankshaft until the lifter is on the camshaft base circle and install the rocker, fulcrum seat and fulcrum bolt. Tighten the bolts to 18–25 ft. lbs. (24–34 Nm).

11. Position new rocker arm cover gaskets and install the rocker arm covers. Tighten the bolts to 3–5 ft. lbs. (4–

7 Nm) on 1988–90 vehicles, 10–13 ft. lbs. (14–18 Nm) on 1991–92 vehicles, wait 2 minutes and tighten again to the same specification.

12. Install the crankcase ventilation hoses on the rocker arm covers.

13. Install the spark plug wires and bracket assembly on the rocker arm cover attaching stud. Connect the spark plug wires.

14. Install the air cleaner, the thermactor bypass valve and air supply hoses.

15. Connect the negative battery cable, start the engine and check for leaks.

Intake Manifold

REMOVAL & INSTALLATION

2.3L Engine

1. Disconnect the negative battery cable.

2. Relieve the fuel system pressure and drain the cooling system.

3. Disconnect and label the electrical connectors at the following:
   a. air bypass valve
   b. throttle positioning sensor
   c. injector wiring harness
   d. air charge temperature sensor
   e. engine coolant temperature sensor
   f. EGR valve, if necessary
   g. fan switch, if necessary
   h. ignition control assembly, if equipped

3. Tag and disconnect the necessary vacuum lines.

4. Remove the throttle linkage shield. Disconnect the throttle linkage and if equipped, the cruise control and kickdown cables. Unbolt the accelerator cable from the bracket and position the cable out of the way.

5. If equipped with a turbocharger, remove the air charge intercooler and the throttle body. On all others, disconnect the air intake hose.

6. Disconnect the PCV system hose from the fitting on the underside of the upper intake manifold.

7. Disconnect the water bypass hose at the lower intake manifold.

8. Loosen the EGR flange nut and disconnect the EGR tube.

9. Remove the engine oil dipstick bracket retaining bolt.

10. Remove the upper intake manifold retaining bolts and/or studs and remove the upper intake manifold assembly.

11. Disconnect the fuel lines from the fuel supply manifold.

12. Disconnect the electrical connectors from the fuel injectors and move the harness aside.

13. Remove the fuel supply manifold retaining bolts and remove the manifold carefully. Injectors can be removed at this time by exerting a slight twisting/pulling motion.

14. Remove the lower intake manifold retaining bolts and remove the lower intake manifold. The front 2 bolts also secure an engine lift bracket.

To install:

15. Clean all gasket mating surfaces. Clean and oil the manifold bolt threads. Install a new intake manifold gasket.

16. Position the lower intake manifold to the head with the engine lift bracket. Install the manifold retaining bolts finger-tight.

17. On 1988 vehicles, tighten the manifold retaining bolts, in sequence, to 12–15 ft. lbs. (16–20 Nm). On 1989–90 vehicles, tighten the manifold retaining bolts, in sequence, in 2 steps, first to 5–7 ft. lbs. (7–10 Nm) and then to 20–29 ft. lbs. (26–38 Nm). On 1991–92 vehicles, tighten the manifold retaining bolts, in sequence, to 15–22 ft. lbs. (20–30 Nm).

18. Install the fuel supply manifold and injectors. Connect the electrical connectors to the injectors.

19. Install a new gasket and the upper intake manifold. Tighten the bolts to 15–22 ft. lbs. (20–30 Nm). Connect the fuel lines to the fuel supply manifold.

20. Install the engine oil dipstick and retaining bolt. Connect the EGR tube, water bypass line and PCV hose.

21. Connect the electrical connectors and vacuum lines to their original locations. Connect the throttle linkage. If equipped with a turbocharger, install the throttle body and air charge intercooler.

22. Fill and bleed the cooling system. Connect the negative battery cable, start the engine and check for leaks.

3.8L Engine

1. Disconnect the negative battery cable.

ATTACHING BOLT

FULCRUM

ROCKER ARM

FULCRUM GUIDE

THREADED PEDESTAL

Rocker arm assembly—5.0L and 5.8L engines

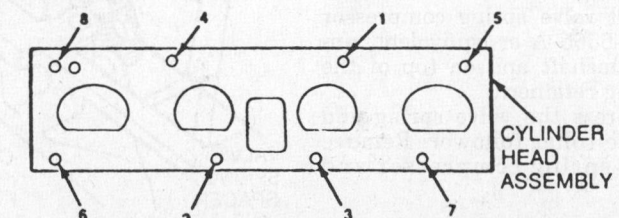

Intake manifold-to-cylinder head torque sequence—2.3L engine

CYLINDER HEAD ASSEMBLY

2. Drain the cooling system and relieve the fuel system pressure.

3. Remove the air cleaner assembly or air inlet tube.

4. Disconnect the accelerator cable at the throttle body. Disconnect the cruise control cable, if equipped.

5. If equipped with an automatic transmission, disconnect the transmission linkage at the upper intake manifold. Remove the retaining bolts from the accelerator cable mounting bracket and position the cables aside.

6. If equipped, disconnect the thermactor air supply hose at the check valve. The valve is located in the Y-pipe assembly.

7. Disconnect the fuel lines. If equipped, remove the supercharger.

8. Disconnect the radiator hose at the thermostat housing and the coolant bypass hose at the manifold.

9. Disconnect the heater tube at the intake manifold and remove the tube support bracket retaining nut. Remove the heater hose at the rear of the heater tube. Loosen the hose clamp at the heater elbow and remove the heater tube with the hose attached. Remove the heater tube with the lines attached and set the assembly aside.

10. Tag and disconnect the vacuum lines at the fuel rail assembly and intake manifold. Tag and disconnect the necessary electrical connectors.

11. If equipped with air conditioning, remove the compressor support bracket. Disconnect the 1 PCV line at the upper intake manifold and at the valve. Remove the second PCV line from the left rocker arm cover.

12. Remove the throttle body assembly. Remove the EGR valve assembly from the upper manifold.

13. Remove the retaining nut and remove the wiring retainer bracket located at the left front of the intake manifold and set aside with the spark plug wires.

14. Remove the upper intake manifold retaining bolts/studs and remove the upper intake manifold.

15. Remove the injectors and fuel rail assembly. Remove the heater water outlet hose.

16. Remove the lower intake manifold retaining bolts/studs and remove the lower intake manifold.

NOTE: The manifold is sealed at each end with RTV-type sealer. To break the seal, it may be necessary to pry on the front of the manifold with a small prybar. If it is necessary to pry on the manifold, use care to prevent damage to the machined surfaces.

To install:

17. Clean all gasket mating surfaces. Lightly oil all retaining bolt and stud threads.

18. Apply a dab of gasket adhesive to each cylinder head mating surface. Press new intake manifold gaskets in place, using location pins as necessary to aid in installation.

19. Apply a ⅛ in. bead of silicone sealer at each corner where the cylinder head joins the cylinder block. Install the front and rear intake manifold end seals.

20. Carefully lower the intake manifold into place on the cylinder heads and cylinder block. Use locating pins as necessary to guide the manifold.

21. Install the bolts and stud bolts in their original locations. On non-supercharged engines, tighten the bolts, in sequence, in 3 steps, first to 8 ft. lbs. (10 Nm), then to 15 ft. lbs. (20 Nm), and finally to 24 ft. lbs. (32 Nm).

22. On supercharged engines, tighten the bolts, in sequence, in 2 steps, first to 8 ft. lbs. (11 Nm) and then to 11 ft. lbs. (15 Nm).

23. Connect the rear PCV line to the upper intake tube. Install the front PCV tube so the mounting bracket sits over the lower intake manifold stud. Tighten the nut on the stud to 15–22 ft. lbs. (20–30 Nm).

24. Install the injectors and the fuel rail. On non-supercharged engines, install the upper intake manifold assembly. Install the bolts and stud bolts in their original locations. Tighten the 4 center bolts and then the end bolts in 3 steps, first to 8 ft. lbs. (10 Nm), then to 15 ft. lbs. (20 Nm), and finally to 24 ft. lbs. (32 Nm).

25. On supercharged engines, install the supercharger.

26. Install the EGR valve. Install the throttle body and cross-tighten the retaining nuts to 15–22 ft. lbs. (20–30 Nm).

27. Connect the rear PCV line at the PCV valve on the upper intake manifold. If equipped with air conditioning, install the compressor support bracket.

28. Connect the necessary electrical connectors and vacuum hoses. Connect the heater tube hose to the heater elbow and position the heater tube support bracket. Tighten the retaining nut to 15–22 ft. lbs. (20–30 Nm).

29. Connect the heater hose to the heater tube and connect the coolant bypass hose and radiator upper hose.

30. Connect the fuel lines. Position the accelerator cable mounting bracket and tighten the mounting bolts to 15–22 ft. lbs. (20–30 Nm).

31. Connect the transmission linkage at the upper intake manifold. If equipped, connect the cruise control cable.

32. Fill and bleed the cooling system. Connect the negative battery cable, start the engine and check for leaks.

33. Check and if necessary, adjust

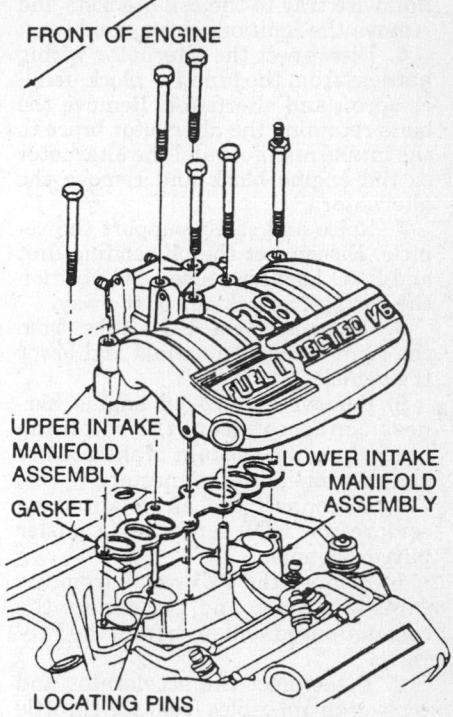

Upper intake manifold installation—3.8L engine

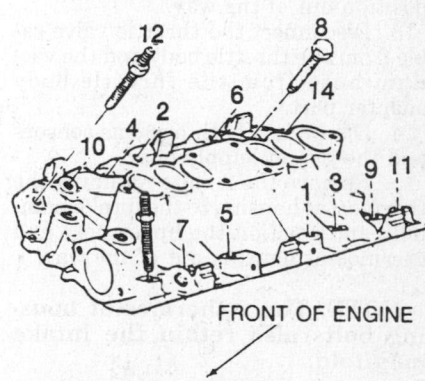

Lower intake manifold bolt torque sequence—3.8L engine

the engine idle speed, transmission throttle linkage and cruise control.

4.6L Engine

1. Disconnect the negative battery cable.

2. Drain the cooling system and relieve the fuel system pressure.

3. Remove the wiper module and the air inlet tube. Release the belt tensioner and remove the accessory drive belt.

4. Tag and disconnect the ignition wires from the spark plugs. Disconnect the ignition wire brackets from the camshaft cover studs.

5. Disconnect both ignition coils and CID sensor. Tag and disconnect all ignition wires from both ignition coils. Remove the 2 bolts retaining the igni-

tion wire tray to the coil brackets and remove the ignition wire assembly.

6. Disconnect the alternator wiring harness from the junction block, fender apron and alternator. Remove the bolts retaining the alternator brace to the intake manifold and the alternator to the engine block and remove the alternator.

7. Raise and safely support the vehicle. Disconnect the oil sending unit and EVO harness sensor and position the wiring harness out of the way.

8. Disconnect the EGR tube from the right exhaust manifold and lower the vehicle.

9. Remove the 42-pin engine harness connector from the retaining bracket on the vacuum brake booster and disconnect the connector.

10. Disconnect the air conditioning compressor, HDR sensor and canister purge solenoid.

11. Remove the PCV valve from the camshaft cover and disconnect the canister purge vent hose from the PCV valve.

12. Disconnect the accelerator and cruise control cables from the throttle body. Remove the accelerator cable bracket from the intake manifold and position out of the way.

13. Disconnect the throttle valve cable from the throttle body and the vacuum hose from the throttle body adapter port.

14. Disconnect both oxygens sensors and the heater supply hose.

15. Remove the 2 bolts retaining the thermostat housing to the intake manifold and position the upper hose and thermostat housing out of the way.

NOTE: The 2 thermostat housing bolts also retain the intake manifold.

16. Remove the bolts retaining the intake manifold to the cylinder heads and remove the intake manifold. Remove and discard the gaskets.

To install:

17. Clean all gasket mating surfaces. Position new intake manifold gaskets on the cylinder heads. Make sure the alignment tabs on the gaskets are aligned with the holes in the cylinder heads.

18. Install the intake manifold and the retaining bolts. Tighten the bolts, in sequence, to 15–22 ft. lbs. (20–30 Nm).

19. Inspect and if necessary, replace the O-ring seal on the thermostat housing. Position the housing and upper hose and install the 2 bolts. Tighten to 15–22 ft. lbs. (20–30 Nm).

20. Connect the heater supply hose and connect both oxygens sensors.

21. Connect the vacuum hose to the throttle body adapter vacuum port.

Connect and if necessary, adjust the throttle valve cable.

22. Install the accelerator cable bracket on the intake manifold and connect the accelerator and cruise control cables to the throttle body.

23. Install the PCV valve in the camshaft cover and connect the canister purge solenoid vent hose. Connect the air conditioning compressor, HDR sensor and canister purge solenoid.

24. Connect the 42-pin engine harness connector. Install the connector on the retaining bracket on the vacuum brake booster.

25. Raise and safely support the vehicle. Connect the EGR tube to the right exhaust manifold and tighten the line nut to 26–33 ft. lbs. (35–45 Nm).

26. Connect the EVO sensor and oil sending unit. Lower the vehicle.

27. Position the alternator and install the retaining bolts. Tighten to 15–22 ft. lbs. (20–30 Nm). Install the 2 bolts retaining the alternator brace to the intake manifold and tighten to 6.0–8.8 ft. lbs. (8–12 Nm).

28. Connect the ignition wires to the ignition coils in their proper positions. Connect the ignition wires to the spark plugs.

29. Connect the ignition wire brackets on the camshaft cover studs. Connect both ignition coils and CID sensor.

30. Install the accessory drive belt and the air inlet tube. Install the wiper module and connect the fuel lines.

31. Fill and bleed the cooling system. Connect the negative battery cable, start the engine and check for leaks.

5.0L Engine

1. Disconnect the negative battery cable.

2. Drain the cooling system and relieve the fuel system pressure.

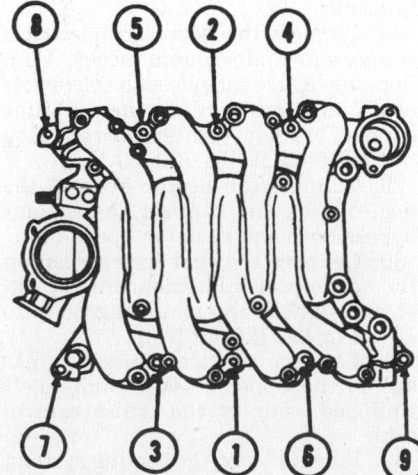

Intake manifold bolt torque sequence—4.6L engine

3. Disconnect the acclerator cable and cruise control linkage, if equipped, from the throttle body. Disconnect the TV cable, if equipped. Tag and disconnect the vacuum lines at the intake manifold fitting.

4. Tag and disconnect the spark plug wires from the spark plugs. Remove the wires and bracket assembly from the rocker arm cover attaching stud. Remove the distributor cap and wires assembly.

5. Disconnect the fuel lines and the distributor wiring connector. Mark the position of the rotor on the distributor housing and the distributor housing in the block. Remove the holddown bolt and remove the distributor.

6. Disconnect the upper radiator hose at the thermostat housing and the water temperature sending unit wire at the sending unit. Disconnect the heater hose from the intake manifold and disconnect the 2 throttle body cooler hoses.

7. Disconnect the water pump bypass hose from the thermostat housing. Tag and disconnect the connectors from the engine coolant temperature, air charge temperature, throttle position and EGR sensors and the idle speed control solenoid. Disconnect the injector wire connections and the fuel charging assembly wiring.

8. Remove the PCV valve from the grommet at the rear of the lower intake manifold. Disconnect the fuel evaporative purge hose from the plastic connector at the front of the upper intake manifold.

9. Remove the upper intake manifold cover plate and upper intake bolts. Remove the upper intake manifold.

10. If equipped, remove the heater tube assembly from the lower intake manifold studs. If necessary, remove the alternator and air conditioner braces from the intake studs. Disconnect the heater hose from the lower intake manifold.

11. Remove the lower intake manifold retaining bolts and remove the lower intake manifold.

NOTE: If it is necessary to pry the intake manifold away from the cylinder heads, be careful to avoid damaging the gasket sealing surfaces.

To install:

12. Clean all gasket mating surfaces. Apply a 1/8 in. bead of silicone sealer to the points where the cylinder block rails meet the cylinder heads.

13. Position new seals on the cylinder block and new gaskets on the cylinder heads with the gaskets interlocked with the seal tabs. Make sure the holes in the gaskets are aligned with the holes in the cylinder heads.

14. Apply a 1/16 in. bead of sealer to

the outer end of each intake manifold seal for the full width of the seal.

15. Using guide pins to ease installation, carefully lower the intake manifold into position on the cylinder block and cylinder heads.

NOTE: After the intake manifold is in place, run a finger around the seal area to make sure the seals are in place. If the seals are not in place, remove the intake manifold and position the seals.

16. Make sure the holes in the manifold gaskets and the manifold are in alignment. Remove the guide pins. Install the intake manifold attaching bolts and tighten, in sequence, to 23–25 ft. lbs. (31–34 Nm).

17. If required, install the heater tube assembly to the lower intake manifold studs.

18. Install the water pump bypass hose on the thermostat housing. Install the hoses to the heater tubes.

19. Install the distributor, aligning the housing and rotor with the marks that were made during removal. Install the distributor cap. Position the spark plug wires in the harness brackets on the rocker arm cover attaching stud and connect the wires to the spark plugs.

20. Install a new gasket and the upper intake manifold. Tighten the bolts to 12–18 ft. lbs. (16–24 Nm). Install the cover plate and connect the crankcase vent tube.

21. Connect the TV cable and cruise control cable, if equipped, to the throttle body. Connect the electrical connectors and vacuum lines.

22. Connect the coolant hoses to the EGR spacer. Fill and bleed the cooling system.

23. Connect the negative battery cable, start the engine and check for leaks. Check the ignition timing.

24. Operate the engine at fast idle. When engine temperatures have stabilized, tighten the intake manifold bolts to 23–25 ft. lbs. (31–34 Nm).

25. Connect the air intake duct and the crankcase vent hose.

5.8L Engine

1. Disconnect the negative battery cable and drain the cooling system.

2. Remove the air cleaner, crankcase ventilation hose and intake duct assembly. If equipped, disconnect the automatic choke heat tube.

3. Disconnect the accelerator cable and cruise control linkage, if equipped, from the carburetor. Disconnect the TV cable, if equipped, and remove the accelerator cable bracket.

4. Tag and disconnect the vacuum lines at the intake manifold and the wires from the coil.

5. Tag and disconnect the spark plug wires from the spark plugs. Remove the wires and bracket assembly from the rocker arm cover attaching stud. Remove the distributor cap and spark plug wires assembly.

6. Remove the carburetor fuel inlet line.

7. Disconnect the vacuum hoses and the wiring connector from the distributor. Mark the position of the rotor on the distributor housing and the distributor housing in the block. Re-

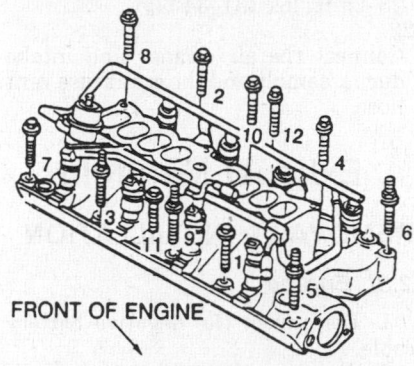

Lower intake manifold bolt torque sequence—5.0L engine

move the hold-down bolt and remove the distributor.

8. Disconnect the upper radiator hose at the thermostat housing and the water temperature sending unit wire at the sending unit. Disconnect the heater hose from the intake manifold. Disconnect the EGR cooler T-fitting from the heater return hose, if equipped.

9. Disconnect the water pump bypass hose at the thermostat housing. Disconnect the crankcase vent hose at the rocker arm cover. Disconnect the fuel evaporative purge hose, if equipped.

10. Remove the intake manifold and carburetor as an assembly.

NOTE: If it is necessary to pry the intake manifold away from the cylinder heads, be careful to avoid damaging the gasket sealing surfaces.

To install:

11. Clean all gasket mating surfaces. Apply a ⅛ in. bead of silicone sealer to the points where the cylinder block rails meet the cylinder heads.

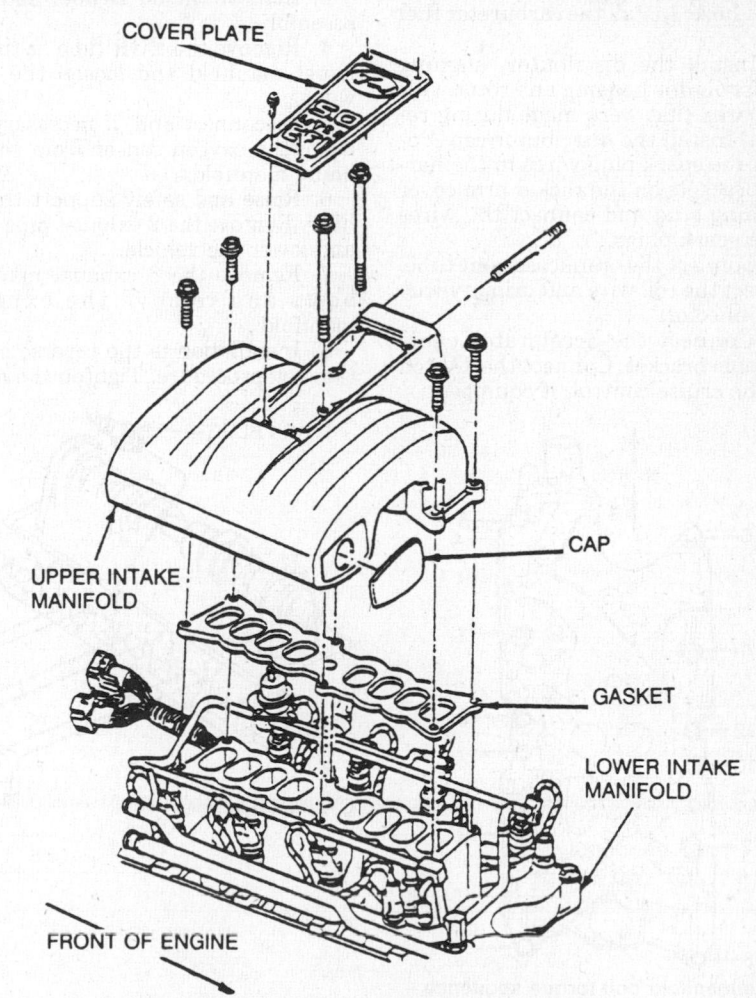

Upper intake manifold installation—5.0L engine

12. Position new seals on the cylinder block and new gaskets on the cylinder heads with the gaskets interlocked with the seal tabs. Make sure the holes in the gaskets are aligned with the holes in the cylinder heads.

13. Apply a $^1/_{16}$ in. bead of sealer to the outer end of each intake manifold seal for the full width of the seal.

14. Using guide pins to ease installation, carefully lower the intake manifold into position on the cylinder block and cylinder heads.

NOTE: After the intake manifold is in place, run a finger around the seal area to make sure the seals are in place. If the seals are not in place, remove the intake manifold and position the seals.

15. Make sure the holes in the manifold gaskets and the manifold are in alignment. Remove the guide pins. Install the intake manifold attaching bolts and tighten, in sequence, to 23–25 ft. lbs. (31–34 Nm).

16. Install the water pump bypass hose on the thermostat housing. Connect the upper radiator hose and the heater hose. Install the carburetor fuel line.

17. Install the distributor, aligning the distributor housing and rotor with the marks that were made during removal. Install the distributor cap. Position the spark plug wires in the harness brackets on the rocker arm cover attaching stud and connect the wires to the spark plugs.

18. Connect the crankcase vent tube. Connect the coil wire and primary wiring connector.

19. Connect the accelerator cable and cable bracket. Connect the TV rod and the cruise control, if equipped.

20. Connect all electrical connections and vacuum lines disconnected during removal. Fill and bleed the cooling system.

21. Connect the negative battery cable, start the engine and check for leaks. Adjust the ignition timing and connect the vacuum hoses to the distributor.

22.
Operate the engine at fast idle. When engine temperatures have stabilized, tighten the intake manifold bolts to 23–25 ft. lbs. (31–34 Nm).

23.
Connect the air cleaner and intake duct assembly and the crankcase vent hose.

Exhaust Manifold

REMOVAL & INSTALLATION

2.3L Engine

1. Disconnect the negative battery cable.

2. Remove the turbocharger, if equipped.

3. Remove the air cleaner and duct assembly.

4. Remove the EGR tube at the exhaust manifold and loosen the EGR valve.

5. Disconnect and, if necessary, remove the oxygen sensor from the exhaust manifold.

6. Raise and safely support the vehicle. Remove the 2 exhaust pipe bolts and lower the vehicle.

7. Remove the 8 exhaust manifold bolts and remove the exhaust manifold.

8. Installation is the reverse of the removal procedure. Tighten the manifold bolts, in sequence, in 2 steps, first to 15–17 ft. lbs. (20–30 Nm) and then to 20–30 ft. lbs. (27–41 Nm). Tighten the exhaust pipe bolts to 25–34 ft. lbs. (36–46 Nm).

3.8L Engine

LEFT SIDE

1. Disconnect the negative battery cable. Remove oil level dipstick tube support bracket.

2. Disconnect the oxygen sensor at the wiring connector.

3. Tag and disconnect the wires from the spark plugs.

4. Raise and safely support the vehicle.

5. Remove the manifold to exhaust pipe attaching nuts.

6. Lower the vehicle.

7. On supercharged engines, remove the intercooler tubes and remove the oil cooler tube and dipstick tube support brackets from the studs.

8. Remove exhaust manifold attaching bolts and manifold.

9. Installation is the reverse of removal procedure. Tighten the manifold retaining bolts to 15–22 ft. lbs. (20–30 Nm).

RIGHT SIDE

1. Disconnect the negative battery cable. On 1988 vehicles, remove the air cleaner assembly and heat tube. On supercharged engines, remove the air cleaner inlet tube.

2. On 1988 vehicles, disconnect the thermactor hose from the downstream air tube check valve.

3. On non-supercharged engines, disconnect the coil secondary wire from the coil. Tag and disconnect the wires from the spark plugs.

4. On non-supercharged engines,

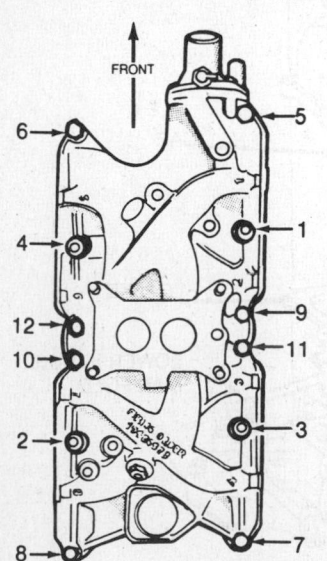

Intake manifold bolt torque sequence—5.8L engine

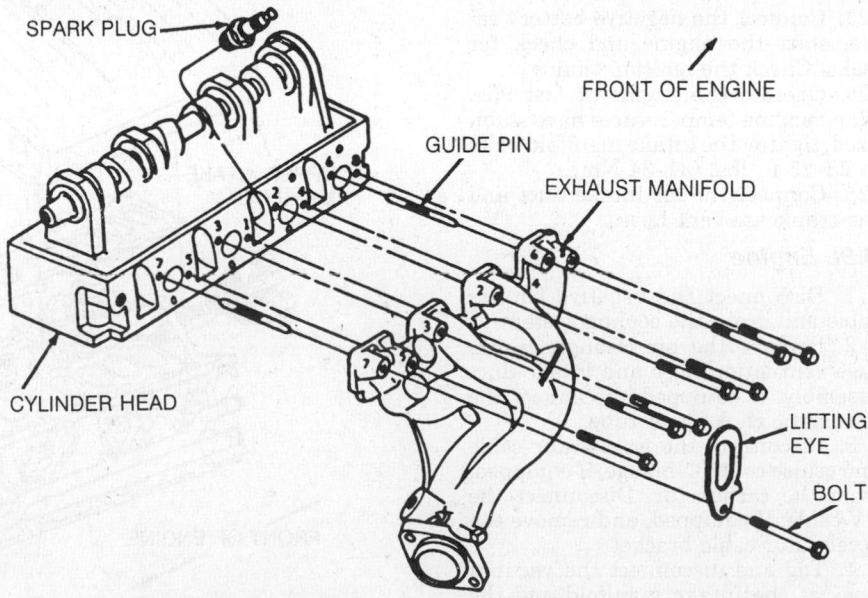

Exhaust manifold installation and torque sequence—2.3L engine

remove the spark plugs and the outer heat shield.

5. Raise and safely support the vehicle. Disconnect the EGR tube.

6. If equipped with automatic transmission, remove the dipstick tube.

7. On 1989–90 vehicles, remove the thermactor downstream air tube. Use cutters to cut the tube clamp at the underbody catalyst.

8. Remove the manifold-to-exhaust pipe retaining nuts and lower the vehicle.

9. Remove the exhaust manifold retaining bolts and remove the manifold.

10. Installation is the reverse of the removal procedure. Tighten the exhaust manifold retaining bolts to 15–22 ft. lbs. (20–30 Nm).

4.6L Engine

1. Disconnect the battery cables. Remove the air inlet tube.

2. Drain the cooling system and remove the cooling fan and shroud. Relieve the fuel system pressure and disconnect the fuel lines.

3. Remove the upper radiator hose. Remove the wiper module and support bracket.

4. Discharge the air conditioning system. Disconnect and plug the compressor outlet hose at the compressor and remove the bolt retaining the hose assembly to the right coil bracket.

5. Remove the 42-pin engine harness connector from the retaining bracket on the brake vacuum booster. Disconnect the connector.

6. Disconnect the throttle valve cable from the throttle body. Disconnect the heater outlet hose.

7. Remove the nut retaining the ground strap to the right cylinder head. Remove the upper stud and lower bolt retaining the heater outlet hose to the right cylinder head and position out of the way.

8. Remove the blower motor resistor and remove the bolt retaining the right engine insulator to the lower engine bracket. Disconnect both oxygen sensors.

9. Raise and safely support the vehicle. Remove the engine insulator through bolts.

10. Remove the EGR tube line nut from the right exhaust manifold.

11. Disconnect the exhaust pipes from the manifolds. Lower the exhaust system and hang it from the crossmember with wire.

12. To remove the left exhaust manifold, remove the engine insulator from the engine block and remove the 8 bolts retaining the exhaust manifold.

13. Position a jack and a block of wood under the oil pan, rearward of the oil drain hole. Raise the engine approximately 4 in. (100mm).

14. Remove the 8 bolts retaining the right exhaust manifold and remove the manifold.

To install:

15. If the exhaust manifolds are being replaced, transfer the oxygen sensors and tighten to 27–33 ft. lbs. (37–45 Nm). On the right manifold, transfer the EGR tube connector and tighten to 33–48 ft. lbs. (45–65 Nm).

16. Clean the mating surfaces of the exhaust manifolds and cylinder heads.

17. Position the exhaust manifolds to the cylinder heads and install the retaining bolts. Tighten, in sequence, to 15–22 ft. lbs. (20–30 Nm).

18. Position and connect the EGR valve and tube assembly to the exhaust manifold. Tighten the line nut to 26–33 ft. lbs. (35–45 Nm).

19. Install the left engine insulator and tighten the bolts to 15–22 ft. lbs. (20–30 Nm). Lower the engine onto the engine insulators and remove the jack. Install the engine insulator through bolts and tighten to 15–22 ft. lbs. (20–30 Nm).

20. Cut the wire and position the exhaust system. Tighten the nuts to 20–30 ft. lbs. (27–41 Nm).

NOTE: Make sure the exhaust system clears the No. 3 crossmember. Adjust as necessary.

21. Lower the vehicle. Connect both oxygen sensors and install the bolt retaining the right engine insulator to the frame. Tighten to 15–22 ft. lbs. (20–30 Nm).

22. Install the blower motor resistor. Position the heater outlet hoses. Install the upper stud and lower bolt and tighten to 15–22 ft. lbs. (20–30 Nm). Install the ground strap onto the stud and tighten the nut to 15–22 ft. lbs. (20–30 Nm).

23. Connect the heater outlet hose. Connect and if necessary, adjust the throttle valve cable.

24. Connect the 42-pin connector

NOTE: ENGINE REMOVED FOR CLARITY
NOTE: LEFT EXHAUST MANIFOLD SHOWN, RIGHT EXHAUST MANIFOLD TYPICAL

Exhaust manifold bolt torque sequence— 4.6L engine

and transmission harness connector. Install the connector to the retaining bracket on the brake vacuum booster.

25. Connect the air conditioning compressor outlet hose to the compressor and install the bolt retaining the hose assembly to the right coil bracket.

26. Install the upper radiator hose and connect the fuel lines. Install the wiper module and retaining bracket.

27. Install the cooling fan and shroud. Connect the battery cables, start the engine and check for leaks. Install the air inlet tube.

28. Leak test, evacuate and charge the air conditioning system according to the proper procedure. Observe all safety precautions.

5.0L and 5.8L Engines

1. Disconnect the negative battery cable.

2. Remove the thermactor hardware from the right exhaust manifold. Remove the air cleaner and inlet duct, if necessary.

3. Tag and disconnect the spark plug wires. Remove the spark plugs.

4. Disconnect the engine oil dipstick tube from the exhaust manifold stud.

5. Raise and safely support the vehicle. Disconnect the exhaust pipes from the exhaust manifolds.

6. Remove the engine oil dipstick tube by carefully tapping upward on the tube. Disconnect the oxygen sensor connector.

7. Lower the vehicle. If equipped, remove the nuts attaching the alternator rear brace to the right exhaust manifold and remove the brace.

8. Remove the attaching bolts and washers and remove the exhaust manifolds.

9. Installation is the reverse of the removal procedure. Working from the center to the ends, tighten the exhaust manifold attaching bolts to 18–24 ft. lbs. (24–32 Nm).

Timing Chain Front Cover

REMOVAL & INSTALLATION

3.8L Engine

1. Disconnect the negative battery cable and drain the cooling system.

2. Remove the air cleaner assembly and air intake duct.

3. On non-supercharged engines, remove the fan/clutch assembly and shroud. On supercharged engines, remove the electric cooling fan assembly.

4. Remove the accessory drive belt idlers, drive belts and the water pump pulley.

5. Remove the power steering pump

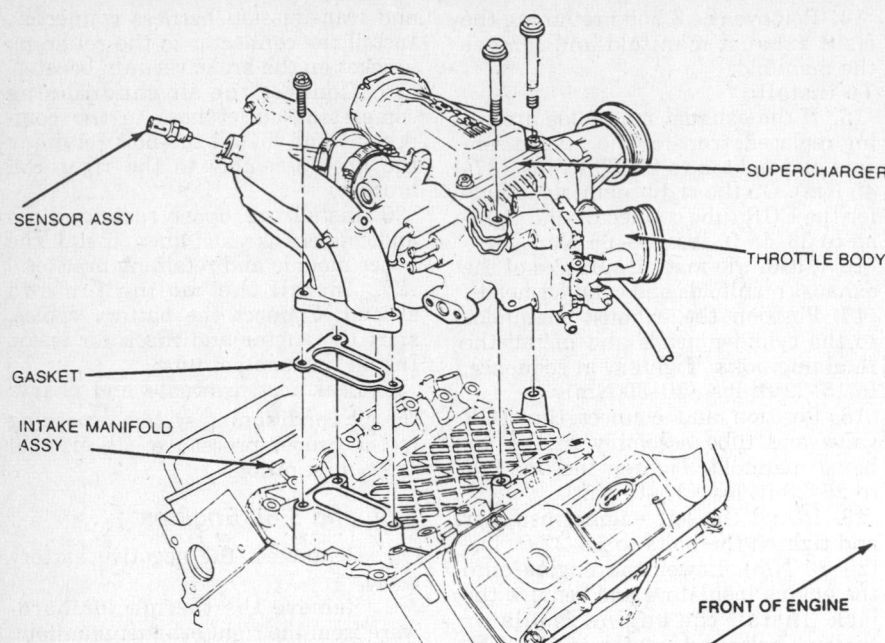

Removing the supercharger assembly—3.8L SC engine

SENSOR ASSY

SUPERCHARGER

THROTTLE BODY

GASKET

INTAKE MANIFOLD ASSY

FRONT OF ENGINE

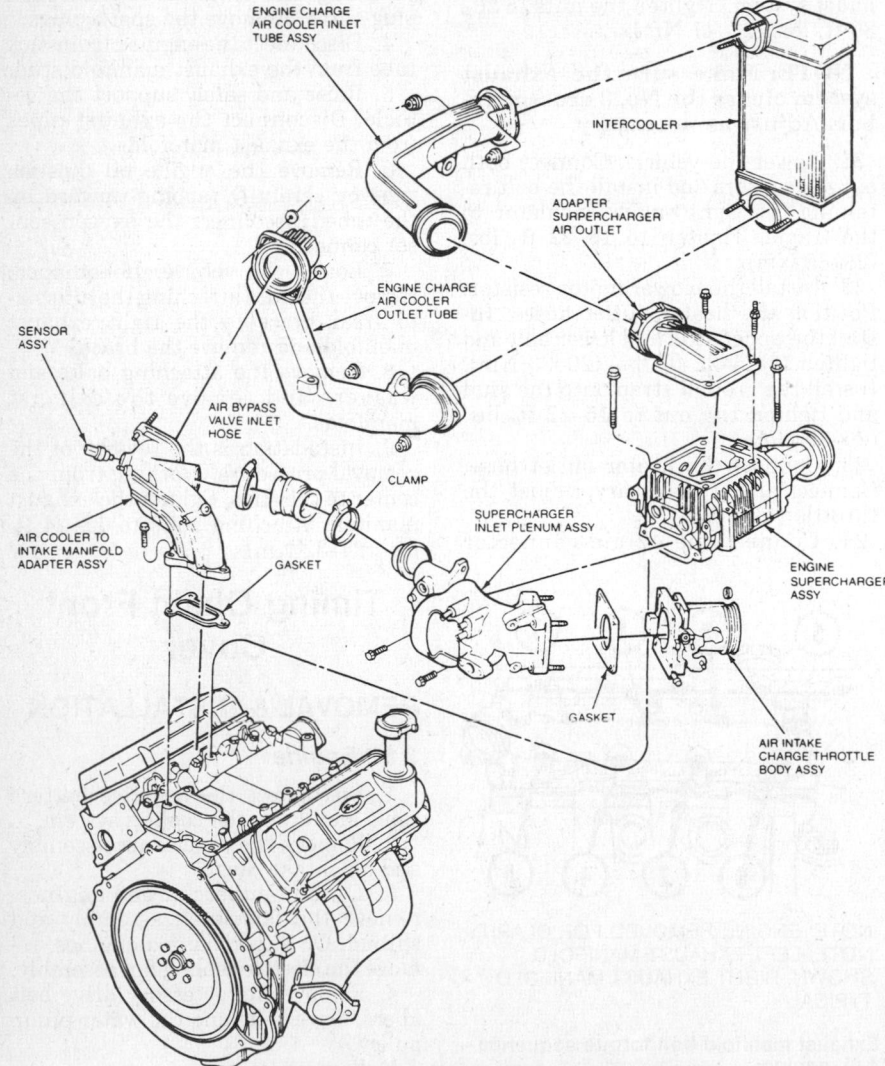

Supercharger system components—3.8L SC engine

ENGINE CHARGE AIR COOLER INLET TUBE ASSY

INTERCOOLER

ADAPTER SUPERCHARGER AIR OUTLET

ENGINE CHARGE AIR COOLER OUTLET TUBE

SENSOR ASSY

AIR BYPASS VALVE INLET HOSE

CLAMP

SUPERCHARGER INLET PLENUM ASSY

AIR COOLER TO INTAKE MANIFOLD ADAPTER ASSY

GASKET

ENGINE SUPERCHARGER ASSY

GASKET

AIR INTAKE CHARGE THROTTLE BODY ASSY

bracket retaining bolts. Leaving the hoses connected, place the pump/bracket assembly aside in a position to prevent fluid from leaking out.

6. If equipped with air conditioning, remove the compressor front support bracket but leave the compressor in place.

7. Disconnect the coolant bypass hose and heater hose at the water pump. Disconnect the upper radiator hose at the thermostat housing.

8. On non-supercharged engines, disconnect the coil wire from the distributor cap and remove the cap with the secondary wires attached.

9. On non-supercharged engines, remove the distributor hold-down clamp and lift the distributor out of the front cover.

10. On supercharged engines, remove the hold-down clamp and lift the camshaft synchronizer from the front cover.

11. Raise and safely support the vehicle. Remove the crankshaft damper and pulley using a puller.

NOTE: If the crankshaft pulley and vibration damper have to be separated, mark the damper and pulley so they may be reassembled in the same relative position. This is important as the damper and pulley are initially balanced as a unit. If the crankshaft damper is being replaced, check if the original damper has balance pins installed. If so, new balance pins must be installed on the new damper in the same position as the original damper. The crankshaft pulley, new or original, must also be installed in the same relative position as originally installed.

12. Remove the oil filter. On supercharged engines, remove the oil cooler.

13. Disconnect the lower radiator hose at the water pump. Remove the oil pan.

NOTE: The front cover cannot be removed without lowering the oil pan.

14. Remove the front cover retaining bolts. It is not necessary to remove the water pump.

NOTE: Do not overlook the cover retaining bolt located behind the oil filter adapter. The front cover will break if pried on and all retaining bolts are not removed.

15. Remove the front cover and water pump as an assembly. Remove and discard the cover gasket.

NOTE: The front cover contains the oil pump and water pump. If a new front cover is to be

installed, remove the water pump and oil pump from the old front cover.

To install:

16. Clean all gasket mating surfaces. If reusing the front cover, replace the front cover oil seal.

17. Position a new gasket on the cylinder block and install the front cover using dowels for proper alignment. Install the front cover retaining bolts and tighten to 15–22 ft. lbs. (20–30 Nm).

18. Raise and safely support the vehicle. Install the oil pan. Connect the lower radiator hose and install the oil filter.

19. Coat the crankshaft damper sealing surface with clean engine oil. Apply a small amount of silicone sealer to the crankshaft keyway.

20. Position the crankshaft pulley key in the crankshaft keyway and install the damper, using a suitable installation tool.

21. Install the damper washer and retaining bolt and tighten to 103–132 ft. lbs. (140–180 Nm). Install the crankshaft pulley and tighten the retaining bolts to 19–28 ft. lbs. (26–38 Nm).

22. Lower the vehicle. Connect the coolant bypass hose.

23. On non-supercharged engines, install the distributor with the rotor pointing at the No. 1 distributor cap tower. Install the distributor cap and coil wire. On supercharged engines, install the camshaft synchronizer.

24. Connect the upper radiator hose at the thermostat housing. Connect the heater hose.

25. If equipped with air conditioning, install the compressor and mounting brackets. Tighten retaining bolts to 30–45 ft. lbs. (41–61 Nm).

26. Install the power steering pump and mounting bracket. Tighten the retaining bolts to 30–45 ft. lbs. (41–61 Nm).

27. Install the water pump pulley. Position the accessory drive belts over the pulleys.

28. On non-supercharged engines, install the fan/clutch assembly and fan shroud. Cross-tighten the fan/clutch assembly retaining bolts to 12–18 ft. lbs. (16–24 Nm).

29. On supercharged engines, install the electric cooling fan assembly and connect the harness connector to the fan motor.

30. Fill and bleed the cooling system. Fill the crankcase with the proper type and quantity of engine oil. Connect the negative battery cable.

31. Start the engine and check for leaks. Check the ignition timing and curb idle speed and adjust, as necessary.

4.6L Engine

1. Disconnect the negative battery cable.

2. Remove the cooling fan and shroud. Loosen the water pump pulley bolts, remove the accessory drive belt and remove the water pump pulley.

3. Raise and safely support the vehicle.

4. Remove the bolts retaining the power steering pump to the engine block and cylinder front cover. The lower front bolt on the power steering pump will not come all the way out. Wire the power steering pump out of the way.

5. Remove the 4 bolts retaining the oil pan to the front cover. Remove the crankshaft damper retaining bolt and washer. Remove the damper using a puller.

6. Lower the vehicle. Remove the bolt retaining the air conditioner high pressure line to the right coil bracket.

7. Remove the front bolts and loosen the remaining bolts on the camshaft covers. Using plastic wedges or similar tools, prop up both camshaft covers. Disconnect both ignition coils and CID sensor.

8. Remove the 3 nuts retaining the right coil bracket to the front cover. Position the power steering hose out of the way.

9. Remove the 4 nuts retaining the left coil bracket to the front cover. Slide both coil brackets and ignition wires off the mounting studs and lay the assembly on top of the engine.

10. Disconnect the High Data Rate (HDR) sensor. Remove the 7 stud bolts and 4 bolts retaining the front cover to the engine and remove the front cover.

To install:

11. Inspect and replace the front cover seal as necessary and clean the sealing surfaces of the cylinder block. Apply silicone sealer to the oil pan where it meets the cylinder block and to the points where the cylinder head meets the cylinder block.

12. Install the front cover and the attachings studs and bolts. Tighten to 15–22 ft. lbs. (20–30 Nm). Connect the HDR sensor.

13. Position the coil brackets and ignition wires as an assembly onto the mounting studs. Position the power steering hose and install the 7 nuts retaining the coil brackets to the front cover. Tighten the nuts to 15–22 ft. lbs. (20–30 Nm). Connect both ignition coils and CID sensor.

14. Remove the plastic wedges holding up the camshaft covers. Apply silicone sealer where the front cover meets the cylinder head and make sure the camshaft cover gaskets are properly positioned. Install the front retaining bolts into the camshaft cover and tighten the bolts to 6.0–8.8 ft. lbs. (8–12 Nm).

15. Position the air conditioner high pressure line on the right coil bracket and install the bolt. Raise and safely support the vehicle.

16. Apply a small amount of silicone sealer in the rear of the keyway in the

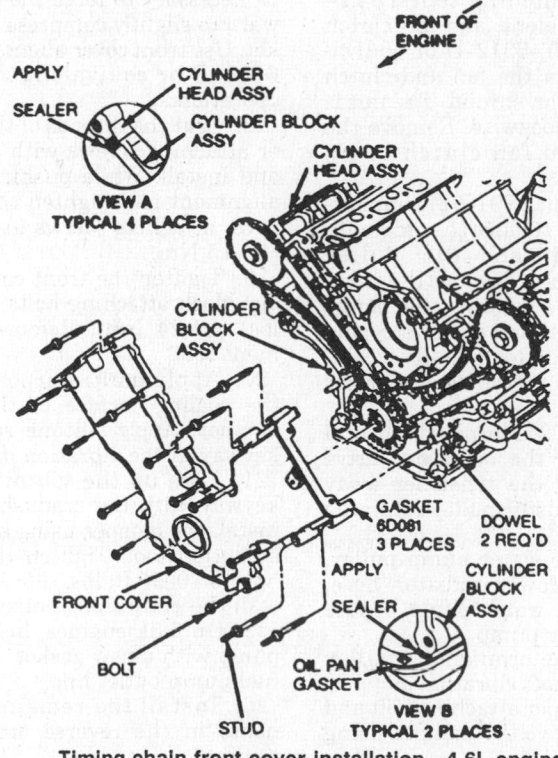

Timing chain front cover installation—4.6L engine

damper. Position the damper on the crankshaft and install, using a suitable installation tool. Install the damper bolt and washer and tighten to 114–121 ft. lbs. (155–165 Nm).

17. Install the 4 bolts retaining the oil pan to the front cover. Tighten to 15–22 ft. lbs. (20–30 Nm).

18. Position the power steering pump on the engine and install the 4 retaining bolts. Tighten to 15–22 ft. lbs. (20–30 Nm). Lower the vehicle.

19. Install the water pump pulley with the 4 bolts. Tighten to 15–22 ft. lbs. (20–30 Nm). Install the accessory drive belt and the cooling fan and shroud.

20. Connect the negative battery cable, start the engine and check for leaks.

5.0L and 5.8L Engines

1. Disconnect the negative battery cable.

2. Drain the cooling system and the crankcase. Remove the air inlet tube, if equipped.

3. On 1991–92 Thunderbird and Cougar, disconnect the upper radiator hose at the engine.

4. On all except 1991–92 Thunderbird and Cougar, remove the fan shroud attaching bolts and position the shroud over the fan. Remove the fan and clutch assembly from the water pump shaft and remove the shroud.

5. On 1991–92 Thunderbird and Cougar, remove the fan and clutch assembly from the water pump shaft using fan clutch holding tool T84T-6312-C or equivalent, and fan clutch nut wrench T84T-6312-D or equivalent, and position the fan and clutch assembly in the fan shroud. The nut is turned counterclockwise. Remove the fan shroud and fan/clutch as an assembly.

6. On all except 1991–92 Thunderbird and Cougar, remove the air conditioner drive belt and idler pulley bracket, if equipped. Remove the alternator drive belt. Remove the power steering drive belt and power steering pump, if equipped. Remove all accessory brackets that attach to the water pump.

7. On 1991–92 Thunderbird and Cougar, remove the accessory drive belt by rotating the tensioner away from the belt by using pulley retaining bolts only.

8. Remove the water pump pulley. Disconnect the lower radiator hose, heater hose and water pump bypass hose at the water pump.

9. Remove the crankshaft pulley from the crankshaft vibration damper. Remove the damper attaching bolt and washer and remove the damper using a puller.

10. On 5.8L engines, disconnect the fuel pump outlet line from the fuel pump. Remove the fuel pump attaching bolts and lay the pump to 1 side with the flexible fuel line still attached.

11. Remove the fuel line from the clip on the front cover, if equipped.

12. Remove the oil pan-to-front cover attaching bolts. Use a thin blade knife to cut the oil pan gasket flush with the cylinder block face prior to separating the cover from the cylinder block.

13. Remove the cylinder front cover and water pump as an assembly.

NOTE: Cover the front oil pan opening while the cover assembly is off to prevent foreign material from entering the pan.

To install:

14. If a new front cover is to be installed, remove the water pump from the old front cover and install it on the new front cover.

15. Clean all gasket mating surfaces. Pry the old oil seal from the front cover and install a new 1, using a seal installer.

16. Coat the gasket surface of the oil pan with sealer, cut and position the required sections of a new gasket on the oil pan and apply silicone sealer at the corners. Apply sealer to a new front cover gasket and install on the block.

17. Position the front cover on the cylinder block. Use care to avoid seal damage or gasket mislocation. It may be necessary to force the cover downward to slightly compress the pan gasket. Use front cover aligner tool T61P-6019-B or equivalent to assist the operation.

18. Coat the threads of the front cover attaching screws with pipe sealant and install. While pushing in on the alignment tool, tighten the oil pan to cover attaching screws to 9–11 ft. lbs. (12–15 Nm).

19. Tighten the front cover to cylinder block attaching bolts to 12–18 ft. lbs. (16–24 Nm). Remove the alignment tool.

20. Apply multi-purpose grease to the sealing surface of the vibration damper. Apply silicone sealer to the keyway of the vibration damper.

21. Line up the vibration damper keyway with the crankshaft key and install the damper using a suitable installation tool. Tighten the retaining bolt to 70–90 ft. lbs. (95–122 Nm). Install the crankshaft pulley.

22. On 5.8L engines, install the fuel pump with a new gasket. Connect the fuel pump outlet line.

23. Install the remaining components in the reverse order of their removal.

24. Fill the crankcase with the proper type and quantity of engine oil. Fill and bleed the cooling system.

25. Connect the negative battery cable, start the engine and check for leaks.

Front Cover Oil Seal

REPLACEMENT

3.8L Engine

1. Disconnect the negative battery cable.

2. On non-supercharged engines, remove the fan shroud and position it back over the fan. Remove the fan/clutch assembly and shroud.

3. On supercharged engines, disconnect the electric cooling fan connector and remove the fan assembly.

4. Loosen the accessory drive belt idlers. Raise and safely support the vehicle.

5. Disengage the drive belts and remove the crankshaft pulley. On supercharged engines, remove the upper and lower crankshaft shields.

6. Remove the crankshaft damper retaining bolt and remove the damper using a puller.

7. Using a small prybar, remove the seal from the front cover. Use care to prevent damage to the cover and crankshaft.

To install:

8. Inspect the front cover and crankshaft damper for damage, nicks, burrs or other roughness which may cause the seal to fail. Service or replace components as necessary.

9. Lubricate the seal lip using clean engine oil. Install the seal using a suitable seal installer.

10. Lubricate the seal surface on the damper with clean engine oil. Install the damper using a suitabel installation tool.

11. Install the damper retaining bolt and tighten to 103–132 ft. lbs. (140–180 Nm). Install the crankshaft pulley and tighten the retaining bolts to 20–28 ft. lbs. (26–38 Nm).

12. Install the remaining components in the reverse order of their removal. Connect the negative battery cable, start the engine and check for leaks.

4.6L Engine

1. Disconnect the negative battery cable.

2. Release the belt tensioner and remove the accessory drive belt.

3. Raise and safely support the vehicle.

4. Remove the crankshaft damper retaining bolt and washer. Remove the damper using a puller.

5. Using a small prybar, remove the front cover seal.

To install:

6. Lubricate the seal bore in the front cover and seal lip with clean engine oil. Install the seal, using a seal installer.

7. Apply a small amount of silicone sealer to the rear of the damper keyway. Using a damper installer, install the crankshaft damper. Be sure the key on the crankshaft aligns with the keyway in the damper.

8. Install the crankshaft damper retaining bolt and washer and tighten to 114–121 ft. lbs. (155–165 Nm).

9. Lower the vehicle and install the accessory drive belt.

10. Connect the negative battery cable, start the engine and check for leaks.

5.0L and 5.8L Engines

1. Disconnect the negative battery cable.

2. Remove the fan shroud and position it back over the fan. Remove the fan/clutch assembly and shroud.

3. Remove the accessory drive belts.

4. Remove the crankshaft pulley from the damper and remove the damper retaining bolt. Remove the damper using a puller.

5. Remove the seal using a seal removal tool.

To install:

6. Lubricate the seal lip with clean engine oil and install using a seal installer.

7. Apply clean engine oil to the sealing surface of the vibration damper. Line up the crankshaft damper keyway with the crankshaft key and install the damper using a damper installation tool.

8. Install the damper retaining bolt and tighten to 70–90 ft. lbs. (95–122 Nm).

9. Install the remaining components in the reverse order of their removal.

Timing Chain and Sprockets

REMOVAL & INSTALLATION

3.8L Engine

1. Disconnect the negative battery cable.

2. Remove the timing chain front cover.

3. Remove the camshaft bolt and washer from the end of the camshaft.

4. Remove the distributor drive gear, camshaft sprocket and timing chain.

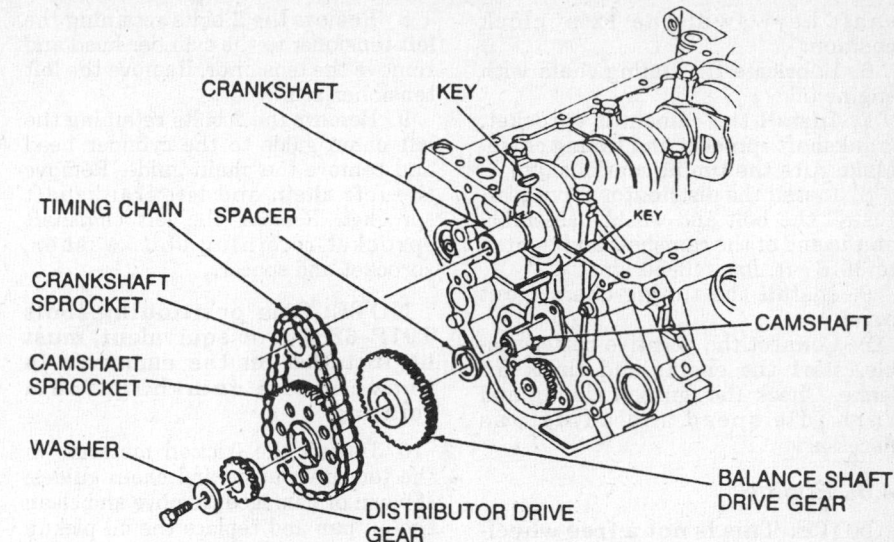

Timing chain and sprockets installation—1988 3.8L engine

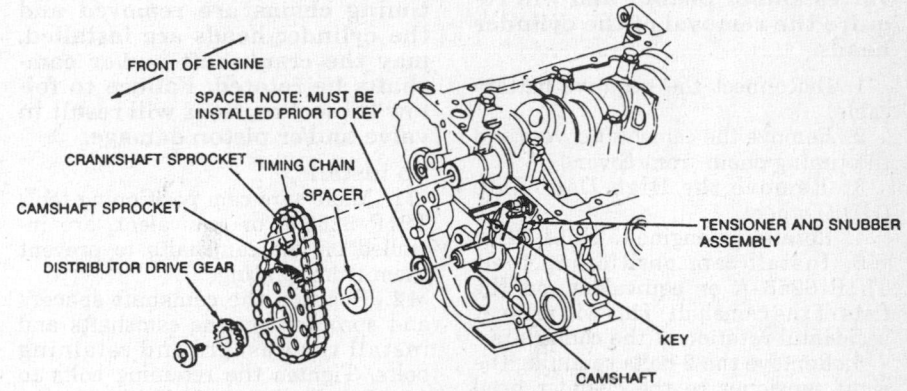

Timing chain and sprockets installation—1989–92 3.8L engines

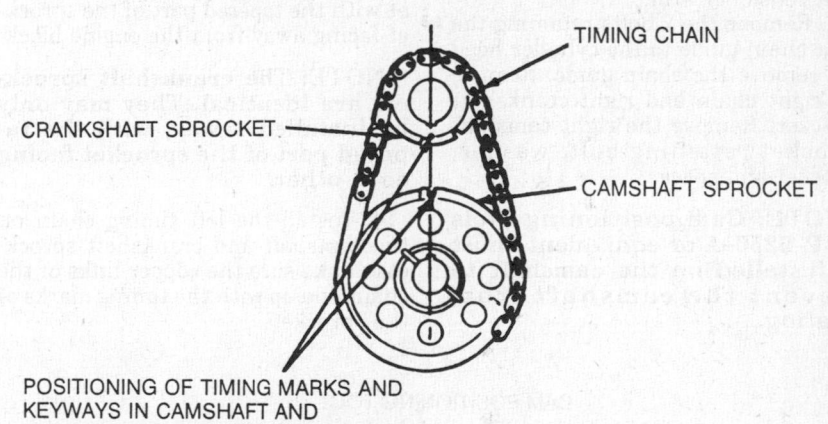

POSITIONING OF TIMING MARKS AND KEYWAYS IN CAMSHAFT AND CRANKSHAFT SPROCKETS MUST BE IN LINE AS SHOWN WITH NO. 1 PISTON AT TDC FIRING

Timing chain sprocket alignment—3.8L engine

NOTE: The engine in 1988 vehicles is equipped with an internal balance shaft. The balance shaft is driven off the camshaft, by a gear positioned behind the camshaft timing sprocket. When removing the timing chain and sprockets, care should be taken to keep this gear in its proper position.

To install:

5. Rotate the crankshaft to position the No. 1 piston at TDC and the crank-

shaft keyway at the 12 o' clock position.

6. Lubricate the timing chain with engine oil.

7. Install the camshaft sprocket, crankshaft sprocket and timing chain. Make sure the timing marks align.

8. Install the distributor drive gear. Install the bolt and washer assembly on the end of the camshaft and tighten to 30–37 ft. lbs. (40–50 Nm).

9. Install the timing chain front cover.

10. Connect the negative battery cable, start the engine and check for leaks. Check the ignition timing and curb idle speed and adjust, as necessary.

4.6L Engine

NOTE: This is not a free wheeling engine. If it has "jumped time," there will be damage to the valves and/or pistons and will require the removal of the cylinder heads.

1. Disconnect the negative battery cable.

2. Remove the camshaft covers and the timing chain front cover.

3. Remove the High Data Rate (HDR) wheel.

4. Rotate the engine to No. 1 TDC.

5. Install cam positioning tools T91P–6256–A or equivalent, on the flats of the camshaft. This will prevent accidental rotation of the camshafts.

6. Remove the 2 bolts retaining the right tensioner to the cylinder head and remove the tensioner. Remove the right tensioner arm.

7. Remove the 2 bolts retaining the right chain guide to the cylinder head and remove the chain guide. Remove the right chain and right crankshaft sprocket. Remove the right camshaft sprocket retaining bolt, washer, sprocket and spacer.

NOTE: Cam positioning tools T91P–6256–A or equivalent, must be installed on the camshaft to prevent the camshaft from rotating.

8. Remove the 2 bolts retaining the left tensioner to the cylinder head and remove the tensioner. Remove the left tensioner arm.

9. Remove the 2 bolts retaining the left chain guide to the cylinder head and remove the chain guide. Remove the left chain and left crankshaft sprocket. Remove the left camshaft sprocket retaining bolt, washer, sprocket and spacer.

NOTE: Cam positioning tools T91P–6256–A or equivalent, must be installed on the camshaft to prevent the camshaft from rotating.

10. Inspect the friction material on the tensioner arms and chain guides. If worn or damaged, remove and clean the oil pan and replace the oil pickup tube.

NOTE: At no time, when the timing chains are removed and the cylinder heads are installed, may the crankshaft and/or camshafts be rotated. Failure to follow these directions will result in valve and/or piston damage.

To install:

11. Make sure cam positioning tools T91P–6256–A or equivalent, are installed on the camshafts to prevent them from rotating.

12. Position the camshaft spacers and sprockets on the camshafts and install the washers and retaining bolts. Tighten the retaining bolts to 81–95 ft. lbs. (110–130 Nm).

13. Install the left crankshaft sprocket with the tapered part of the sprocket facing away from the engine block.

NOTE: The crankshaft sprockets are identical. They may only be installed 1 way, with the tapered part of the sprocket facing each other.

14. Install the left timing chain on the camshaft and crankshaft sprockets. Make sure the copper links of the chain line up with the timing marks of the sprockets.

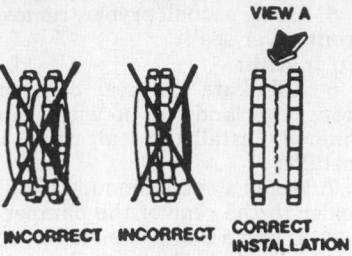
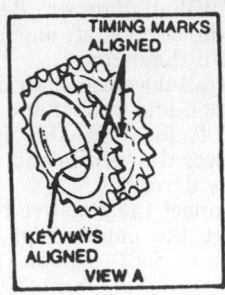

Crankshaft sprocket positioning— 4.6L engine

NOTE: If the copper links of the timing chain are not visible, pull the chain taught until the opposite sides of the chain contact 1 another and lay it on a flat surface. Mark the links at each end of the chain and use them in place of the copper links.

15. Install the right crankshaft sprocket with the tapered part of the sprocket facing the left crankshaft sprocket.

16. Install the right timing chain on the camshaft and crankshaft sprockets. Make sure the copper links of the chain line up with the timing marks of the sprockets.

17. It is necessary to bleed the timing chain tensioners before installation. Proceed as follows:

a. Position the timing chain tensioner in a soft-jawed vice.

b. Using a small pick or similar tool, hold the ratchet lock mechanism away from the ratchet stem and slowly compress the tensioner plunger by rotating the vise handle.

NOTE: The tensioner must be compressed slowly or damage to the internal seals will result.

c. Once the tensioner plunger bottoms in the tensioner bore, continue to hold the ratchet lock mechanism and push down on the ratchet stem until flush with the tensioner face.

d. While holding the ratchet stem flush to the tensioner face, release the ratchet lock mechanism and install a paper clip or similar tool in the tensioner body to lock the tensioner in the collapsed position.

e. The paper clip must not be removed until the timing chain, ten-

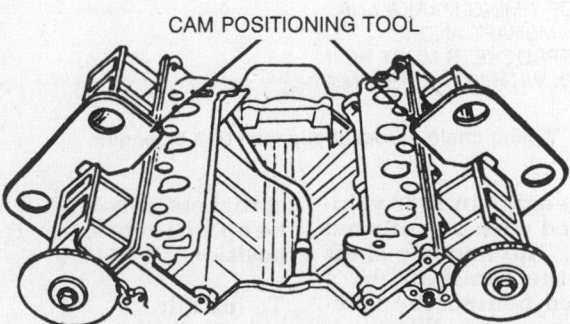

Camshaft positioning tools installation—4.6L engine

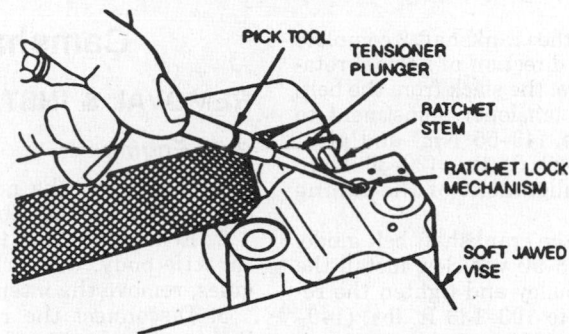

NOTE: WITH EITHER CHAIN POSITIONED AS SHOWN, MARK EACH END AND USE MARKS AS TIMING MARKS

Alternate timing chain marking procedure—4.6L engine

PICK TOOL
TENSIONER PLUNGER
RATCHET STEM
RATCHET LOCK MECHANISM
SOFT JAWED VISE

Timing chain tensioner bleeding procedure—4.6L engine

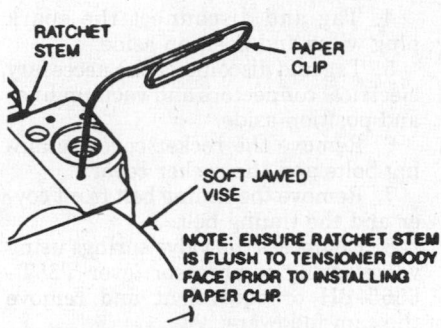

RATCHET STEM
PAPER CLIP
SOFT JAWED VISE

NOTE: ENSURE RATCHET STEM IS FLUSH TO TENSIONER BODY FACE PRIOR TO INSTALLING PAPER CLIP.

Timing chain tensioner locking procedure—4.6L engine

sioner, tensioner arm and timing chain guide are completely installed on the engine.

18. Lubricate the tensioner arm contact surfaces with engine oil and install the right and left tensioner arms on their dowels.

19. Install the right and left timing chain tensioners and secure with 2 bolts on each. Tighten the bolts to 15–22 ft. lbs. (20–30 Nm).

20. Install the right and left timing chain guides and secure with 2 bolts on each. Tighten the bolts to 6.0–8.8 ft. lbs. (8–12 Nm).

21. Remove the paper clips from the timing chain tensioners and make sure all timing marks are aligned.

22. Remove the camshaft positioning tools.

23. Installation of the remaining components is the reverse of removal.

24. Connect the negative battery cable, start the engine and check for leaks and proper operation.

5.0L and 5.8L Engines

1. Disconnect the negative battery cable and drain the cooling system.

2. Remove the timing chain front cover.

3. Rotate the crankshaft until the timing marks on the sprockets are aligned.

COPPER TIMING LINK RH CHAIN
COPPER TIMING LINK LH CHAIN
TIMING MARK RH CAM GEAR
TIMING MARK LH CAM GEAR
TDC
CRANKSHAFT GEAR KEYWAYS POSITIONED AT 315 DEGREES
CRANKSHAFT GEAR TIMING MARKS
COPPER TIMING LINKS RH AND LH CRANKSHAFT GEARS

Timing chain and sprockets alignment—4.6L engine

4. Remove the camshaft retaining bolt, washer and eccentric. Slide both sprockets and the timing chain forward and remove them as an assembly.

To install:

5. Position the sprockets and timing chain on the camshaft and crankshaft simultaneously. Make sure the timing marks on the sprockets are aligned.

6. Install the washer, eccentric and camshaft sprocket retaining bolt. Tighten the bolt to 40–45 ft. lbs. (54–61 Nm).

7. Install the timing chain front cover and remaining components.

8. Fill and bleed the cooling system. Connect the negative battery cable, start the engine and check for leaks.

9. Check and adjust the ignition timing and idle speed, as necessary.

Timing Belt Front Cover

REMOVAL & INSTALLATION

2.3L Engine

1. Disconnect the negative battery cable and drain the cooling system.

2. Remove the automatic belt tensioner and accessory drive belt. Remove the upper radiator hose.

3. Remove the crankshaft pulley bolt and pulley. Remove the thermostat housing and gasket.

4. Remove the timing belt outer cover retaining bolt(s). Release the cover interlocking tabs, if equipped, and remove the cover.

To install:

5. Position the timing belt front cover. Snap the interlocking tabs into place, if necessary. Install the timing belt outer cover retaining bolt(s) and tighten to 71–106 inch lbs. (8–12 Nm).

6. Install the thermostat housing and a new gasket. Install the upper radiator hose.

7. Install the crankshaft pulley and retaining bolt. Tighten to 103–133 ft. lbs. (140–180 Nm).

8. Install the water pump pulley and the automatic belt tensioner. Install the accessory drive belt.

9. Connect the negative battery cable, start the engine and check for leaks.

OIL SEAL REPLACEMENT

2.3L Engine

1. Disconnect the negative battery cable.

2. Remove the timing belt front cover and timing belt.

3. Use a suitable puller to remove the crankshaft, camshaft and auxiliary shaft sprockets.

4. Use seal remover tool T74P-6700–B or equivalent, to remove the crankshaft, camshaft and auxiliary shaft seals. Position the tool so the jaws are gripping the thin edge of the seal. Operate the jackscrew on the tool to remove the seal.

To install:

5. Lubricate the lips of the new seals with clean engine oil.

6. Use seal replacer tool T74P-6150–A or equivalent, to install the seals.

7. Install the crankshaft, camshaft and auxiliary shaft sprockets. Tighten the camshaft sprocket retaining bolt to 52–70 ft. lbs. (70–95 Nm) and the auxiliary sprocket retaining bolt to 28–40 ft. lbs. (38–54 Nm).

8. Install the timing belt and timing belt front cover.

9. Connect the negative battery cable, start the engine and check for leaks.

Timing Belt and Tensioner

REMOVAL & INSTALLATION

2.3L Engine

1. Disconnect the negative battery cable.

2. Remove the timing belt front cover.

3. Loosen the belt tensioner adjustment screw, position belt tensioner tool T74P-6254–A or equivalent, on the tension spring roll pin and release the belt tensioner. Tighten the adjustment screw to hold the tensioner in the released position.

4. On 1991–92 vehicles, remove the bolts holding the timing sensor in place and pull the sensor assembly free of the dowel pin.

5. Remove the crankshaft pulley, hub and belt guide. Remove the timing belt. If the belt is to be reused, mark the direction of rotation so it may be reinstalled in the same direction.

To install:

6. Position the crankshaft sprocket to align with the TDC mark and the camshaft sprocket to align with the camshaft timing pointer. On 1988–90 vehicles, remove the distributor cap and set the rotor to the No. 1 firing position by turning the auxiliary shaft.

7. Install the timing belt over the crankshaft sprocket and then counterclockwise over the auxiliary and camshaft sprockets. Align the belt fore-and-aft on the sprockets.

8. Loosen the tensioner adjustment bolt to allow the tensioner to move

against the belt. If the spring does not have enough tension to move the roller against the belt, it may be necessary to manually push the roller against the belt and tighten the bolt.

9. To make sure the belt does not jump time during rotation in Step 10, remove a spark plug from each cylinder.

10. Rotate the crankshaft 2 complete turns in the direction of normal rotation to remove the slack from the belt. Tighten the tensioner adjustment to 29–40 ft. lbs. (40–55 Nm) and pivot bolts to 14–22 ft. lbs. (20–30 Nm). Check the alignment of the timing marks.

11. Install the crankshaft belt guide.

12. On 1988–90 vehicles, install the crankshaft pulley and tighten the retaining bolt to 103–133 ft. lbs. (140–180 Nm). On 1991–92 vehicles, proceed as follows:

   a. Install the timing sensor onto the dowel pin and tighten the 2 longer bolts to 14–22 ft. lbs. (20–30 Nm).

   b. Rotate the crankshaft 45 degrees couterclockwise and install the crankshaft pulley and hub assembly. Tighten the bolt to 114–151 ft. lbs. (155–205 Nm).

   c. Rotate the crankshaft 90 degrees clockwise so the vane of the crankshaft pulley engages with timing sensor positioner tool T89P-6316–A or equivalent. Tighten the 2 shorter sensor bolts to 14–22 ft. lbs. (20–30 Nm).

   d. Rotate the crankshaft 90 degrees counterclockwise and remove the sensor positioner tool.

   e. Rotate the crankshaft 90 degrees clockwise and measure the outer vane to sensor air gap. The air gap must be 0.018–0.039 in. (0.458–0.996mm).

13. Install the timing belt front cover, spark plugs and remaining components.

14. Connect the negative battery cable, start the engine and check the ignition timing.

Timing Sprockets

REMOVAL & INSTALLATION

2.3L Engine

1. Disconnect the negative battery cable.

2. Remove the timing belt front cover and the timing belt.

3. Remove the camshaft and auxiliary shaft sprocket retaining bolts. Remove the crankshaft, camshaft and auxiliary shaft sprockets using suitable pullers.

To install:

4. Install the crankshaft, camshaft and auxiliary shaft sprockets. Tighten

the camshaft sprocket retaining bolt to 52–70 ft. lbs. (70–95 Nm) and the auxiliary sprocket retaining bolt to 28–40 ft. lbs. (38–54 Nm).

5. Install the timing belt and timing belt front cover.

6. Connect the negative battery cable.

Camshaft

REMOVAL & INSTALLATION

2.3L Engine

1. Disconnect the negative battery cable and drain the cooling system.

2. Remove the air intake and the throttle body. On turbocharged engines, remove the intercooler.

3. Disconnect the radiator hoses. Remove the cooling fan and shroud assembly.

4. Tag and disconnect the spark plug wires and position aside.

5. Tag and disconnect the necessary electrical connectors and vacuum lines and position aside.

6. Remove the rocker cover retaining bolts and the rocker cover.

7. Remove the timing belt front cover and the timing belt.

8. Compress the valve springs using valve spring compressor lever T88T-6565–BH or equivalent and remove the cam followers.

9. Remove the camshaft sprocket retaining bolt. Remove the camshaft sprocket using a suitable puller. Remove the camshaft seal using a seal removal tool.

10. Remove the 2 screws and the camshaft rear retainer.

11. Raise and safely support the vehicle. Remove the right and left engine support bolts and nuts.

12. Position a block of wood and a jack under the engine. Raise the engine as high as it will go. Place blocks of wood between the engine mounts and chassis brackets and remove the jack.

13. Lower the vehicle and remove the camshaft.

To install:

14. Make sure the threaded plug is in the rear of the camshaft. If not, remove the plug from the old camshaft and install.

15. Coat the camshaft lobes with multi-purpose grease and lubricate the journals with heavy engine oil before installation. Carefully slide the camshaft through the bearings.

16. Install the camshaft rear retainer and tighten the 2 screws to 6–9 ft. lbs. (8–12 Nm). Install a new camshaft seal using a suitable seal installer.

17. Install the camshaft sprocket and tighten the retaining bolt to 52–70 ft. lbs. (70–95 Nm).

18. Install the timing belt and timing belt front cover.

19. Raise and safely support the vehicle. Position a block of wood and a jack and raise the engine. Remove the blocks of wood, lower the engine and remove the jack.

20. Install the engine support bolts and nuts and lower the vehicle.

21. Install the remaining components in the reverse order of removal.

22. Connect the negative battery cable, start the engine and check for leaks. Check the ignition timing, if necessary.

3.8L, 5.0L and 5.8L Engine

1. Disconnect the negative battery cable and drain the cooling system.

2. Relieve the fuel system pressure and discharge the air conditioning system.

3. Remove the radiator. If equipped with air conditioning, remove the condenser.

4. Remove the grille.

5. Remove the intake manifolds and the lifters. On the 3.8L engine, remove the oil pan.

6. Remove the timing chain front cover, the timing chain and spacer. On 1988 3.8L engine, mark the relation of the balance shaft drive and driven gears. Remove the balance shaft drive gear.

7. Remove the thrust plate. Remove the camshaft, being careful not to damage the bearing surfaces.

To install:

8. Lubricate the cam lobes and journals with heavy engine oil. Install the camshaft, being careful not to damage the bearing surfaces while sliding into position.

9. Install the thrust plate. Tighten the bolts to 6–10 ft. lbs. (8–14 Nm) on the 3.8L engine and 9–12 ft. lbs. (12–16 Nm) on the 5.0L and 5.8L engines.

10. On 1988 3.8L engine, install the balance shaft drive gear, aligning the marks that were made during removal.

11. Install the timing chain and sprockets. Install the engine front cover.

12. Install the lifters and the intake manifolds. On 3.8L engine, install the oil pan.

13. Install the grille. If equipped with air conditioning, install the condenser.

14. Install the radiator. Fill and bleed the cooling system.

15. Connect the negative battery cable. Start the engine and check for leaks.

4.6L Engine

1. Disconnect the negative battery cable and drain the cooling system. Relieve the fuel system pressure.

2. Remove the right and left camshaft covers.

3. Remove the timing chain front cover. Remove the timing chains.

4. Rotate the crankshaft counterclockwise 45 degrees to make sure all pistons are below the top of the engine block deck face.

NOTE: The crankshaft must be in this position prior to rotating the camshafts or damage to the pistons and/or valve train will result.

5. Install valve spring compressor tool T91P-6565-A or equivalent, under the camshaft and on top of the valve spring retainer.

NOTE: Valve spring spacer tool T91P-6565-AH or equivalent, must be installed between the spring coils and the camshaft must be at the base circle before compressing the valve spring.

6. Compress the valve spring far enough to remove the roller follower. Repeat Steps 5 and 6 until all roller followers are removed.

7. Remove the bolts retaining the camshaft cap cluster assemblies to the cylinder heads. Tap upward on the camshaft caps at points near the upper bearing halves and gradually lift the camshaft clusters from the cylinder heads.

8. Remove the camshafts straight upward to avoid bearing damage.

To install:

9. Apply heavy engine oil to the camshaft journals and lobes. Position the camshafts on the cylinder heads.

10. Install and seat the camshaft cap cluster assemblies. Hand start the bolts.

11. Tighten the camshaft cluster retaining bolts in sequence to 6.0–8.8 ft. lbs. (8–12 Nm).

NOTE: Each camshaft cap cluster assembly is tightened individually.

12. Loosen the camshaft cap cluster retaining bolts approximately 2 turns or until the heads of the bolts are free. Retighten all bolts, in sequence, to 6.0–8.8 ft. lbs. (8–12 Nm).

NOTE: The camshafts should turn freely with a slight drag.

13. Install cam positioning tools T91P-6256-A or equivalent, on the flats of the camshafts and install the spacers and camshaft sprockets. Install the bolts and washers and tighten to 81–95 ft. lbs. (110–130 Nm).

14. Install valve spring compressor T91P-6565-A or equivalent, under the camshaft and on top of the valve spring retainer.

NOTE: Valve spring spacer tool T91P-6565-AH or equivalent, must be installed between the spring coils and the camshaft must be at the base circle before compressing the valve spring.

15. Compress the valve spring far enough to install the roller followers.

16. Repeat Steps 14 and 15 until all roller followers are installed.

17. Rotate the crankshaft clockwise 45 degrees to position the crankshaft at TDC.

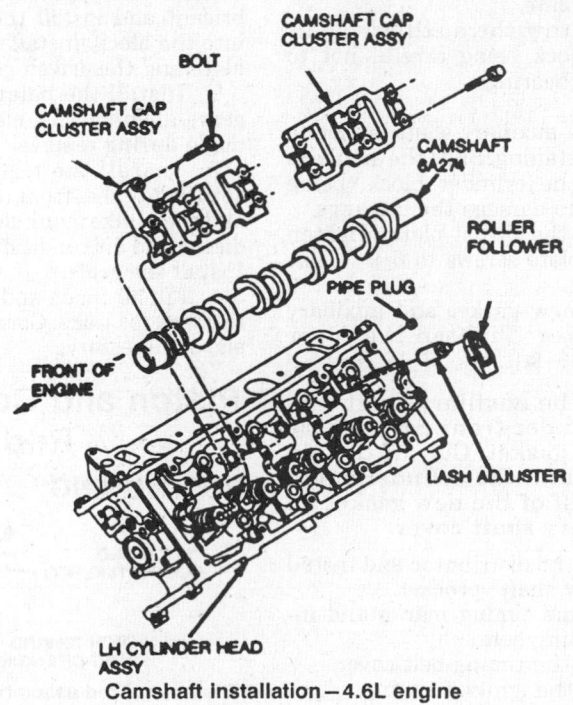

Camshaft Installation—4.6L engine

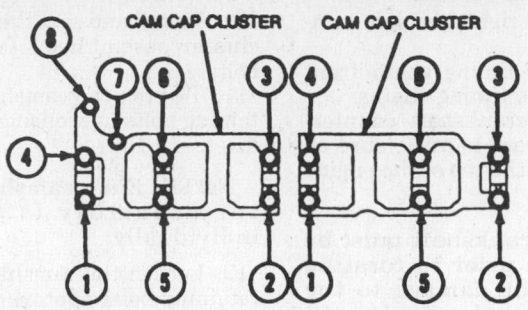

Camshaft cap cluster bolt torque sequence—4.6L engine

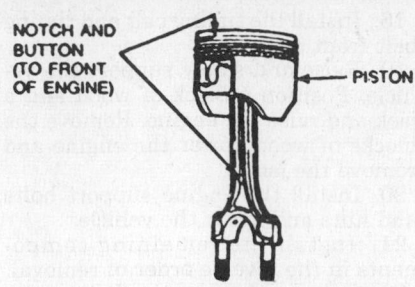

Piston and rod assembly—3.8L engine

NOTE: The crankshaft must only be rotated in the clockwise direction and only as far as TDC.

18. Install the timing chains and install the timing chain front cover. Install the camshaft covers.
19. Install the remaining components in the reverse order of removal.
20. Connect the negative battery cable. Start the engine and check for leaks.

Auxiliary Shaft

REMOVAL & INSTALLATION

2.3L Engine

1. Disconnect the negative battery cable. Remove the front timing belt cover.
2. Remove the timing belt. Remove the auxiliary shaft sprocket retaining bolt. Remove the sprocket using a puller.
3. On 1988–90 vehicles, remove the distributor.
4. Remove the auxiliary shaft cover and thrust plate.
5. Withdraw the auxiliary shaft from the block being careful not to damage the bearings.
To install:
6. Dip the auxiliary shaft in engine oil before installing. Slide the auxiliary shaft into the cylinder block, being careful not to damage the bearings.
7. Install the thrust plate. Tighten the thrust plate screws to 6–9 ft. lbs. (8–12 Nm).
8. Install a new gasket and auxiliary shaft cover. Tighten the cover screws to 6–9 ft. lbs. (8–12 Nm).

NOTE: The auxiliary shaft cover and cylinder front cover share a common gasket. Cut off the old gasket around the cylinder cover and use half of the new gasket on the auxiliary shaft cover.

9. Insert the distributor and install the auxiliary shaft sprocket.
10. Align the timing marks and install the timing belt.
11. Install the timing belt cover.
12. Check the ignition timing.

Balance Shaft

REMOVAL & INSTALLATION

1988 3.8L Engine

1. Disconnect the negative battery cable and drain the cooling system.
2. Relieve the fuel system pressure. Discharge the air conditioning system.
3. Remove the radiator. If equipped with air conditioning, remove the condenser. Remove the grille.
4. Remove the intake manifold assembly.
5. Remove the cylinder front cover assembly.
6. Remove the camshaft timing sprocket and the timing chain. Mark the relationship of the balance shaft driven and drive gears. Remove the drive gear.
7. Remove the balance shaft gear from the end of the balance shaft. Remove the balance shaft thrust plate and remove the shaft.
To install:
8. To install, lubricate the journals of the balance shaft with assembly lubricant and install the balance shaft into the block. Install the shaft thrust plate and the driven gear.
9. Install the balance shaft drive gear, aligning the marks that were made during removal.
10. Install the timing chain and sprockets, the front cover assembly and the intake manifold. Install the radiator and air conditioning condenser. Install the grille.
11. Fill all fluids and run the engine to check for leaks. Correct all fluid levels, as necessary.

Piston and Connecting Rod

POSITIONING

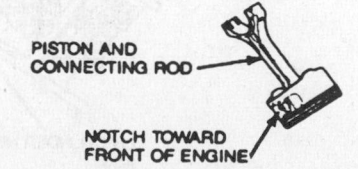

Piston and rod assembly—4.6L engine

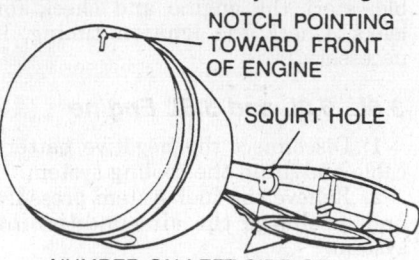

Piston and rod assembly—2.3L engine

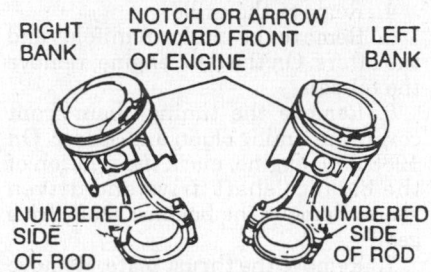

Piston and rod assembly—5.0L and 5.8L engines

ENGINE LUBRICATION

Oil Pan

REMOVAL & INSTALLATION

2.3L Engine
1988 THUNDERBIRD TURBO COUPE

1. Disconnect the negative battery cable. Raise and safely support the vehicle.
2. Drain the crankcase and the cooling system. Disconnect the low oil level sensor, if equipped.
3. Remove the right and left engine support through bolts. Using a jack and a block of wood, raise the engine as far as it will go. Place wood blocks between the mounts and the crossmember pedestal brackets. Remove the jack.

4. Remove the steering gear retaining nuts and bolts. Remove the bolt retaining the steering flex coupling to the steering gear. Position the steering gear forward and down.

5. Remove the shake brace. Disconnect the starter cable and remove the starter.

6. Remove the engine rear support to crossmember nuts. Position a jack under the transmission and raise.

7. Remove the oil pan retaining bolts and reinforcements. Remove the oil pump and lay it in the oil pan. Remove the oil pan.

To install:

8. Clean the oil pan and the gasket mating surfaces. Remove and clean the oil pickup tube and screen assembly. After cleaning, reinstall.

9. Position the oil pan gasket to the oil pan with gasket sealer.

10. Place the oil pump in the oil pan, then install the pump when the pan is in position. Position the oil pan to the cylinder block. Position the reinforcements and install the retaining bolts. Tighten to 90–120 inch lbs. (10–13 Nm).

11. Lower the jack under the transmission and install the crossmember nuts. Remove the jack. Do not tighten the crossmember nuts until the front bolts are tightened.

12. Install a new oil filter and connect the low oil level sensor, if equipped. Position the flex coupling to the steering gear and install the retaining bolt. Install the steering gear.

13. Install the shake brace. Install the starter and connect the starter cable.

14. Raise the engine enough to remove the wood blocks. Lower the engine and remove the jack. Install the engine support through bolts and tighten to 33–45 ft. lbs. (45–61 Nm).

15. Lower the vehicle. Fill the crankcase with the proper type and quantity of engine oil.

16. Connect the negative battery cable, start the engine and check for leaks.

1988–90 MUSTANG

1. Disconnect the negative battery cable and drain the cooling system.

2. Disconnect the lectrical connector to the cooling fan and remove the fan and shroud assembly. Disconnect the radiator hoses at the radiator. If equipped with an automatic transmission, disconnect the oil cooler lines at the radiator.

3. Raise and safely support the vehicle. Drain the crankcase and disconnect the low oil level sensor, if equipped.

4. Remove the right and left engine support through bolts, except convertible. On convertible, remove the nuts.

Using a jack and a block of wood, raise the engine as high as it will go. Place wood blocks between the mounts and the pedestal brackets. Remove the jack.

5. Remove the shake brace. Remove the sway bar retaining bolts and lower the sway bar.

6. Disconnect the cable at the starter and remove the starter. Remove the steering gear retaining bolts and lower the gear.

7. Remove the oil pan retaining bolts and allow the oil pan to drop to the crossmember. Rotate the crankshaft to position No. 4 piston up in the cylinder bore so the oil pan clears the crankshaft throw. Remove the oil pan.

To install:

8. Clean the oil pan and the gasket mating surfaces. Remove and clean the oil pickup tube and screen assembly. After cleaning, reinstall.

9. Apply silicone sealer to the points where the rear main bearing cap meets the cylinder block, to the corners of the engine front cover and to where the front cover meets the cylinder block. Position the oil pan gasket to the cylinder block.

10. Position the oil pan and pan reinforcements to the cylinder block and install the retaining bolts. Tighten to 71–106 inch lbs. (8–12 Nm).

11. Position the steering gear and install the bolts and nuts. Install the starter and connect the cable.

12. Raise the engine enough to remove the wood blocks. Lower the engine and remove the jack. Install the shake brace.

13. Install the right and left engine through bolts, except convertible. Tighten the bolts to 65–85 ft. lbs. (88–119 Nm). On convertible, install the right and left engine support nuts. Tighten the nuts to 80–106 ft. lbs. (108–144 Nm).

14. Install the sway bar. Connect the low oil level sensor, if equipped. Install a new oil filter.

15. Lower the vehicle. Install the cooling fan and shroud assembly. Connect the cooling fan electrical connector and the radiator hoses. If equipped with an automatic transmission, connect the oil cooler lines.

16. Fill the engine with the proper type and quantity of engine oil. Fill and bleed the cooling system.

17. Connect the negative battery cable, start the engine and check for leaks.

1991–92 MUSTANG

1. Disconnect the negative battery cable. Remove the air cleaner outlet tube at the throttle body.

2. Remove the engine oil dipstick. Remove the engine mount retaining nuts.

3. Remove the oil cooler lines at the radiator, if equipped. Disconnect the electrical connector to the cooling fan and remove the cooling fan and shroud.

4. Raise and safely support the vehicle. Drain the crankcase.

5. Remove the starter cable from the starter and remove the starter.

6. Disconnect the exhaust manifold tube to the inlet pipe bracket. Disconnect the catalytic converter at the inlet pipe.

7. Remove the insulator and retainer assembly at the transmission. Remove the transmission mount retaining nuts to the crossmember. If equipped with an automatic transmission, remove the oil cooler lines from the retainer at the block.

8. Position a jack and a wood block under the engine. Raise the engine approximately 2.5 in. (63.5mm) high and place wood blocks between the mounts and crossmember.

9. Remove the jack and position it under the transmission. Raise the jack slightly.

10. Remove the oil pan retaining bolts and lower the pan to the chassis. Remove the oil pump drive and pickup tube assembly. Remove the oil pan with the oil pump.

To install:

11. Clean the oil pan and the gasket mating surfaces. Clean the oil pump pickup tube screen.

12. Install the oil pan gasket in the groove in the oil pan. Position the oil pan assembly on the crossmember and install the oil pump drive and pickup tube assembly.

13. Apply silicone sealer to the points where the rear main bearing cap meets the cylinder block, to the corners of the engine front cover and to where the front cover meets the cylinder block.

14. Install the oil pan assembly. Install the oil pan flange bolts tight enough to compress the oil pan gasket to the point that the 2 transmission holes are aligned with the 2 tapped holes in the oil pan, but loose enough to allow movement of the pan, relative to the block.

15. Install the 2 oil pan/transmission bolts and tighten to 30–39 ft. lbs. (40–50 Nm) to align the oil pan with the transmission, then loosen the bolts ½ turn.

16. Tighten all oil pan flange bolts to 90–120 inch lbs. (10–13 Nm). Tighten the 2 oil pan/transmission bolts to 30–39 ft. lbs. (40–54 Nm).

17. Install a new oil filter. Position the jack and wood block under the engine, raise the engine and remove the wood blocks from under the mounts. Shift the engine/transmission backward to its original position.

18. Install the insulator/bracket assembly to the crossmember and lower the engine on the insulators. Raise the transmission with the jack and install the insulator.

19. Connect the automatic trnsmission oil coolerline to the engine, if equipped. Install the transmission mount retaining nuts at the crossmember.

20. Connect the rear exhaust pipe just behind the catalytic converter. Install the starter and starter cable and lower the vehicle.

21. Install the cooling fan and shroud on the radiator. Connect the electrical connector to the cooling fan. If equipped, connect the oil cooler lines to the radiator.

22. Install the engine mount retaining nuts. Install the dipstick and fill the crankcase with the proper type and quantity of engine oil.

23. Connect the negative battery cable, start the engine and check for leaks.

3.8L Engine

1988

1. Disconnect the negative battery cable.

2. Raise and safely support the vehicle. Drain the oil and remove the oil filter.

3. Remove the catalytic converter assembly from the exhaust manifold.

4. Remove the starter and remove the torque converter cover.

5. Remove the oil pan retaining bolts and remove the oil pan.

To install:

6. Clean all gasket mating surfaces.

7. Using a new gasket, apply a ⅛ in. bead of sealer to all gasket mating surfaces.

8. Install oil pan and tighten the retaining bolts to 80–106 inch lbs. (9–12 Nm).

9. Install the starter and torque converter cover.

10. Install the catalytic converter assembly. Lower the vehicle.

11. Fill the crankcase with oil and connect the negative battery cable.

12. Run the engine and check for leaks.

1989–92

1. Disconnect the negative battery cable and remove the air inlet tube.

2. Remove the 2 bolts retaining the sight shield and position aside. Remove the hood weather seal.

3. Remove the wipers. Remove the left cowl vent screen and the wiper module. On supercharged engines, remove the intercooler tubes.

4. Install a suitable engine support fixture. Raise and safely support the vehicle.

5. Remove the engine mount through bolts. On supercharged engine, remove the left mount retaining strap bolt.

6. Partially lower the vehicle and raise the engine with the support fixture.

7. Raise and safely support the vehicle. Remove the starter.

8. Drain the crankcase and remove the oil filter.

9. Remove the wire loom, ground strap and automatic transmission oil cooler lines, if equipped.

10. Remove the oil pan-to-bell housing bolts and the bolts at the crankshaft position sensor, if equipped. Remove the remaining oil pan retaining bolts.

11. Remove the steering shaft pinch bolts and separate the steering shaft. Position a jack under the front of the sub-frame.

12. Remove the 6 rearward bolts on the front of the sub-frame. Loosen the 2 front sub-frame bolts.

13. Remove the lower strut-to-control arm bolts and nuts and lower the sub-frame. Remove the oil pan.

To install:

14. Clean the gasket mating surfaces and the oil pan. Apply silicone sealer to the oil pan.

15. Fit the oil pan to the cylinder block. Make sure enough clearance has been provided to allow the oil pan to be installed without sealer being scraped off under the cylinder block.

16. Install the oil pan retaining bolts at the cylinder block and bell housing and install the lower crankshaft sensor shield, if equipped. Tighten the bolts to 80–106 inch lbs. (9–12 Nm).

17. Raise the sub-frame into position and install the lower strut mount-to-control arm bolts. Tighten to 103–144 ft. lbs. (140–195 Nm).

18. Install the 2 front sub-frame bolts and the 6 bolts at the rear of the front sub-frame member. Install a ¾ in. outside diameter pipe or equivalent, into both front left and right sub-frame and body alignment holes. Tighten 1 bolt at each corner. Remove the alignment tools and tighten the bolts to 70–95 ft. lbs. (95–130 Nm).

19. Connect the steering shaft and install the pinch bolt. Tighten to 30–42 ft. lbs. (41–57 Nm).

20. Install the transmission cooler lines, wire loom and ground strap. Install a new oil filter.

21. Install the starter and partially lower the vehicle.

22. Lower the engine with the support fixture. Seat the left side locating pin before the right. Partially raise the vehicle and support safely.

23. Install the engine mount through bolts and tighten to 35–50 ft. lbs. (47–68 Nm). On supercharged engine, install the left mount retaining strap bolt and tighten to 33–45 ft. lbs. (45–61 Nm). Lower the vehicle.

24. Remove the engine support fixture. On supercharged engine, install the intercooler tubes.

25. Install the wiper module and the left cowl vent screen. Install the wipers and the hood weather seal.

26. Install the sight shield and the 2 retaining bolts. Install the air duct assembly. Fill the crankcase with the proper type and quantity of engine oil.

27. Connect the negative battery cable, start the engine and check for leaks.

4.6L Engine

1. Disconnect the battery cables and remove the air inlet tube.

2. Drain the cooling system and remove the cooling fan and shroud. Releive the fuel system pressure and disconnect the fuel lines.

3. Remove the upper radiator hose. Remove the wiper module and support bracket.

4. Discharge the air conditioning system. Disconnect and plug the compressor outlet hose at the compressor and remove the bolt retaining the hose assembly to the right coil bracket.

5. Remove the 42-pin engine harness connector from the retaining bracket on the brake vacuum booster and disconnect the connector and transmission harness connector.

6. Disconnect the throttle valve cable from the throttle body and disconnect the heater outlet hose.

7. Remove the nut retaining the ground strap to the right cylinder head. Remove the upper stud and loosen the lower bolt retaining the heater outlet hose to the right cylinder head and position out of the way.

8. Remove the blower motor resistor. Remove the bolt retaining the right engine insulator to the lower engine bracket.

9. Disconnect the vacuum hoses from the EGR valve and tube. Remove the 2 bolts retaining the EGR valve to the intake manifold.

10. Raise and safely support the vehicle. Drain the crankcase and remove the engine insulator through bolts.

11. Remove the EGR tube line nut from the right exhaust manifold and remove the EGR valve and tube assembly.

12. Disconnect the exhaust from the exhaust manifolds. Lower the exhaust system and support it with wire from the crossmember.

13. Position a jack and a block of wood under the oil pan, rearward of the oil drain hole. Raise the engine approximately 4 in. and insert 2 wood blocks approximately 2½ in. thick un-

der each engine insulator. Lower the engine onto the wood blocks and remove the jack.

14. Remove the 16 bolts retaining the oil pan to the engine block and remove the oil pan.

NOTE: It may be necessary to loosen, but not remove, the 2 nuts on the rear tranmission insulator and with a jack, raise the transmission extension housing slightly to remove the pan.

To install:

15. Clean the oil pan and the gasket mating surfaces.

16. Position a new gasket on the oil pan. Apply silicone sealer to where the front cover meets the cylinder block and rear seal retainer meets the cylinder block. Position the oil pan on the engine and install the bolts. Tighten the bolts, in sequence, to 15–22 ft. lbs. (20–30 Nm).

17. Position the jack and wood block under the oil pan, rearward of the oil drain hole and raise the engine enough to remove the wood blocks. Lower the engine and remove the jack.

18. Install the engine insulator through bolts and tighten to 15–22 ft. lbs. (20–30 Nm).

19. Position the EGR valve and tube assembly in the vehicle and connect to the exhaust manifold. Tighten the line nut to 26–33 ft. lbs. (35–45 Nm).

NOTE: Loosen the line nut at the EGR valve prior to installing the assembly into the vehicle. This will allow enough movement to align the EGR valve retaining bolts.

20. Cut the wire and position the exhaust system to the manifolds. Install the 4 nuts and tighten to 20–30 ft. lbs. (27–41 Nm). Make sure the exhaust system clears the crossmember. Adjust as necessary.

21. Install a new oil filter and lower the vehicle.

22. Install the bolt retaining the right engine insulator to the lower engine bracket. Tighten to 15–22 ft. lbs. (20–30 Nm).

23. Install a new gasket on the EGR valve and position on the intake manifold. Install the 2 bolts retaining the EGR valve to the intake manifold and tighten to 15–22 ft. lbs. (20–30 Nm). Tighten the EGR tube line nut at the EGR valve to 26–33 ft. lbs. (35–45 Nm). Connect the vacuum hoses to the EGR valve and tube.

24. Install the blower motor resistor. Position the heater outlet hose, install the upper stud and tighten the upper and lower bolts to 15–22 ft. lbs. (20–30 Nm). Install the gound strap on the stud and tighten to 15–22 ft. lbs. (20–30 Nm).

25. Connect the heater outlet hose and the throttle valve cable. If necessary, adjust the throttle valve cable.

26. Connect the 42-pin connector and transmission harness connector. Install the harness connector on the brake vacuum booster.

27. Connect the air conditioning compressor outlet hose to the compressor and install the bolt retaining the hose to the right coil bracket.

28. Install the upper radiator hose and connect the fuel lines. Install the wiper module and retaining bracket.

29. Install the cooling fan and shroud and fill the cooling system. Fill the crankcase with the proper type and quantity of engine oil.

30. Connect the negative battery cable and install the air inlet tube. Start the engine and check for leaks.

5.0L and 5.8L Engines

CROWN VICTORIA, GRAND MARQUIS AND TOWN CAR

1. Disconnect the negative battery cable. Relieve the fuel system pressure.

2. On 5.8L engine, remove the air cleaner assembly and air ducts.

3. Disconnect the accelerator and TV cables at the throttle body or carburetor. On 5.8L engine, remove the accelerator mounting bracket retaining bolts and remove the bracket.

4. Remove the fan shroud attaching bolts, positioning the fan shroud back over the fan. Remove the dipstick and tube assembly.

5. Disconnect the wiper motor electrical connector and remove the wiper motor. Disconnect the windshield washer hose and remove the wiper motor mounting cover.

6. Remove the thermactor air dump tube on 5.0L engine. Remove the thermactor crossover tube at the rear of the vehicle.

7. Raise and safely support the vehicle. Drain the crankcase. Remove the filler tube from the oil pan and drain the transmission.

8. Disconnect the starter cable and remove the starter. Disconnect the fuel line.

9. Disconnect the exhaust system from the manifolds. Remove the oxygen sensors from the exhaust manifolds.

10. Remove the thermactor secondary air tube to torque converter housing clamps. Remove the converter inspection cover.

11. Disconnect the exhaust pipes to the catalytic converter outlet. Remove the catalytic converter secondary air tube and the inlet pipes to the exhaust manifold.

12. Loosen the rear engine mount attaching nuts and remove the engine mount through bolts. Remove the shift crossover bolts at the transmission.

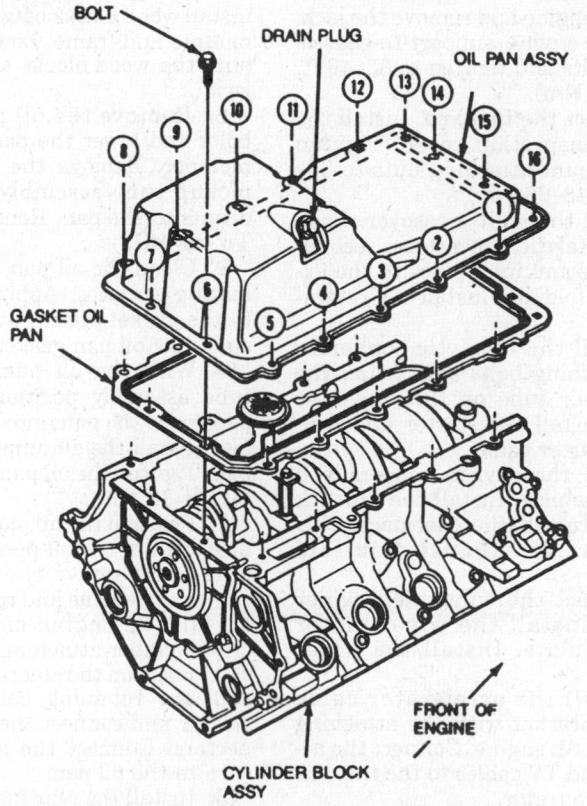

Oil pan bolt torque sequence—4.6L engine

13. Remove the brake line retainer from the front crossmember and disconnect the transmission kickdown rod.

14. Position a jack and wood block under the engine and raise the engine as high as it will go. Place wood blocks between the engine mounts and the chassis brackets, lower the engine and remove the jack.

15. Remove the oil pan retaining bolts and lower the oil pan. Remove the 2 bolts retaining the oil pump pickup tube and screen to the oil pump and the nut from the main bearing cap stud. Allow the pickup tube to drop into the oil pan.

16. Rotate the crankshaft, as required, for clearance and remove the oil pan from the vehicle.

To install:

17. Clean the oil pan and the gasket mating surfaces. Clean the oil pump pickup tube and screen assembly.

18. Install a new oil filter. Position a new oil pan gasket on the cylinder block. Place the oil pickup tube and screen in the oil pan and position the oil pan on the crossmember.

19. Install the pickup tube and screen with a new gasket. Install the bolts and tighten to 12–18 ft. lbs. (16–24 Nm). Position the oil pan and install the retaining bolts. Tighten to 7–10 ft. lbs. (9–14 Nm).

20. Position the jack and wood block under the engine and raise the engine enough to remove the wood blocks. Lower the engine and remove the jack. Install the engine support-to-chassis through bolts and tighten to 33–46 ft. lbs. (45–62 Nm).

21. Connect the fuel line. Install the converter inspection cover. Tighten the rear mount attaching nuts to 35–50 ft. lbs. (48–68 Nm).

22. Install the shift crossover. Position the catalytic converters, secondary air tube and inlet pipes to the exhaust manifold and install the retaining nuts.

23. Install the catalytic converter outlet attaching bolts and install the secondary air tube on the converter housing. Install the starter and connect the starter cable.

24. Install the oxygen sensors and lower the vehicle. Install the dipstick and tube. Install the thermactor air dump valve to exhaust manifold clamp.

25. Connect the windshield wiper hose and install the wiper motor mounting plate. Install the wiper motor.

26. Install the accelerator cable mounting bracket with the attaching screws on 5.8L engine. Connect the accelerator and TV cables to the throttle body or carburetor.

27. Position the shroud and install the retaining bolts. Install the thermactor tube to the rear of the engine. Install the air cleaner assembly and air ducts.

28. Fill the crankcase with the proper type and quantity of engine oil. Fill the transmission with the proper type and quantity of tranmission fluid.

29. Connect the negative battery cable. Start the engine and check for leaks.

MARK VII, MUSTANG AND 1988 THUNDERBIRD AND COUGAR

1. Disconnect the negative battery cable and remove the air cleaner tube.

2. Remove the oil level indicator from the left side of the cylinder block. Remove the fan shroud and position the shroud over the fan.

3. Raise and safely support the vehicle. Drain the crankcase and remove the oil level sensor wiring from the oil pan.

4. Disconnect the electrical connectors from the starter and remove the starter. Remove the catalytic converter and muffler inlet pipes.

5. Remove the engine mount-to-No. 2 crossmember attaching bolts or nuts. Support the transmission and remove the No. 3 crossmember and rear insulator support assemblies.

6. Remove the steering gear attaching bolts and position the steering gear forward out of the way.

7. Position a jack and wood block under the oil pan. Raise the engine and install wood blocks between the engine mounts and frame. Lowert the engine onto the wood blocks and remove the jack.

8. Remove the oil pan attaching bolts and lower the pan to the crossmember. Remove the oil pump and pickup tube assembly and allow to drop into the pan. Remove the pan.

To install:

9. Clean the oil pan and the gasket mating surfaces. Apply gasket sealer to the gasket mating surfaces and install new oil pan gaskets.

10. With the oil pump and pickup tube assembly positioned in the oil pan, raise the pan onto the crossmember. Install the oil pump and then the pan. Tighten the oil pan bolts to 6–9 ft. lbs. (8–12 Nm).

11. Position the oil pan and the wood block under the oil pan. Raise the engine and remove the wood blocks. Lower the engine and remove the jack. Install the engine mount-to-No. 2 crossmember attaching nuts or bolts.

12. Position the steering gear and install the retaining bolts. Install the starter and connect the electrical connectors. Connect the oil level sensor wire to the oil pan.

13. Install the rear insulator and the No. 3 crossmember. Install the catalytic converter and muffler inlet pipes. Lower the vehicle.

14. Install the fan shroud and install the oil level indicator to the side of the cylinder block. Install the air cleaner assembly.

15. Fill the crankcase with the proper type and quantity of engine oil. Connect the negative battery cable, start the engine and check for leaks.

1989–92 THUNDERBIRD AND COUGAR

1. Disconnect the negative battery cable and remove the oil level dipstick. Disconnect the air cleaner cover retaining clips to allow free movement when the engine is raised.

2. Remove the 2 bolts retaining the radiator shroud to the radiator and pull the shroud loose from the lower retaining clips.

3. Install a suitable engine support fixture. Raise and safely support the vehicle.

4. Drain the crankcase and remove the engine mount through bolts. Loosen the transmission mount nut to allow the mount to move when the engine is raised. Partially lower the vehicle.

5. Raise the engine approximately 2 in. using the support fixture. Raise and safely support the vehicle.

6. Remove the power steering cooler line retaining clips. Remove the bolt securing the transmission lines to the right side of the engine block.

7. Disconnect the electrical connector from the low oil level sensor located in the oil pan, if equipped. Remove the oil pan retaining bolts.

8. Remove the steering shaft pinch bolt and separate the steering shaft from the power steering rack assembly.

9. Position 2 jack stands under the engine support sub-frame. Remove the lower strut-to-control arm bolts and nuts from both sides.

10. While supporting the engine support sub-frame on jack stands, remove the 6 rearward bolts on the sub-frame. Loosen the 2 froward bolts on the sub-frame. Lower the sub-frame.

11. Remove the oil pump/pickup tube assembly and place it in the oil pan. Remove the pan.

To install:

12. Clean the oil pan and the gasket mating surfaces.

13. Apply a thin coat of silicone sealer to the engine block and to the engine block side of a new oil pan gasket. Allow the adhesive to set-up for approximately 5 minutes, before positioning the gasket to the engine.

14. Place the oil pump and pickup tube assembly in the oil pan and position the pan on the sub-frame. Install the oil pump/pickup tube and oil pump

drive to the engine. Tighten the oil pump retaining bolts to 22–32 ft. lbs. (30–43 Nm).

15. Position the oil pan to the engine and install all the pan bolts hand tight, then tighten the bolts evenly to 9–11 ft. lbs. (12–15 Nm). Connect the electrical connector to the low oil level sensor, if equipped.

16. Raise the sub-frame into position while supporting the sub-frame on the jackstands. Install a ¾ in. outside diameter pipe or equivalent, into both front left and right sub-frame and body alignment holes. Tighten 1 bolt at each corner. Remove the alignment tools and tighten the bolts to 70–95 ft. lbs. (95–130 Nm).

17. Install the lower strut-to-control arm bolts and nuts and tighten to 103–144 ft. lbs. (140–195 Nm). Remove the 2 jackstands used for installing the sub-frame.

18. Connect the steering shaft and install the steering shaft pinch bolt. Tighten the pinch bolt to 30–42 ft. lbs. (41–57 Nm). Install the bolt securing the transmission lines to the right side of the engine block.

19. Secure the power steering cooler line retaining clips and partially lower the vehicle. Lower the engine onto the engine mounts and remove the engine support fixture.

20. Raise and safely support the vehicle. Tighten the transmission mount nut to 65–85 ft. lbs. (88–115 Nm). Install the engine mount through bolts.

21. Install a new oil filter and lower the vehicle. Position the fan shroud into the lower retaining clips and install the 2 bolts. Connect the air filter cover retaining clips.

22. Fill the crankcase with the proper type and quantity of engine oil. Install the dipstick and connect the negative battery cable. Start the engine and check for leaks.

Oil Pump

REMOVAL &INSTALLATION

Except 3.8L and 4.6L Engines

1. Disconnect the negative battery cable. Remove the oil pan.
2. Remove the oil pump inlet tube and screen assembly.
3. Remove the oil pump attaching bolts and gasket. Remove the oil pump intermediate shaft.
To install:
4. Prime the oil pump by filling either the inlet or outlet ports with engine oil and rotating the pump shaft to distribute the oil within the pump body.
5. Position the intermediate driveshaft into the distributor socket. With the shaft firmly seated in the distribu-

tor socket, the stop on the shaft should touch the roof of the crankcase. Remove the shaft and position the stop, as necessary.

6. Position a new gasket on the pump body, insert the intermediate shaft into the oil pump and install the pump and shaft as an assembly.

NOTE: Do not attempt to force the pump into position if it will not seat readily. The driveshaft hex may be misaligned with the distributor shaft. To align, rotate the intermediate shaft into a new position.

7. Tighten the oil pump attaching screws to 14–21 ft. lbs. (19–29 Nm) on the 2.3L engine and 22–32 ft. lbs. (30–43 Nm) on the 5.0L and 5.8L engines.
8. Clean and install the oil pump inlet tube and screen assembly.
9. Install the oil pan and the remaining components.

3.8L Engine

NOTE: The timing chain front cover houses the oil pump on the 3.8L engine. If the oil pump housing is scored, worn or grooved, the entire front cover will have to be replaced.

1. Disconnect the negative battery cable. Raise and safely support the vehicle.
2. Remove the oil filter.
3. Remove the cover/filter mount assembly. On supercharged engines, remove the oil cooler assembly.
4. Lift the pump gears from their mounting pocket in the front cover.
5. Clean all gasket mounting surfaces.
6. Inspect the mounting pocket for wear. If excessive wear is present, complete timing cover assembly replacement is necessary.
7. Inspect the cover/filter mount gasket to timing cover surface for flatness. Place a straight edge across the flat and check clearance with a feeler gauge. If the measured clearance exceeds 0.0016 in. (0.04mm), replace the cover/filter mount.
8. Replace the pump gears if wear is excessive.
9. Remove the plug from the end of the pressure relief valve passage using a small drill and slide hammer. Use caution when drilling.
10. Remove the spring and valve from the bore. Clean all dirt and metal chips from the bore and valve. Inspect all parts for wear. Replace as necessary.
To install:
11. Install the valve and spring after lubricating them with engine oil. The end with the smaller diameter goes in first.

12. Install a new plug. The plug can be tapped into the bore using a plastic tipped hammer. Make sure the plug is 0–0.010 in. (0–0.25mm) below the machined surface.
13. Lightly pack the gear pocket with petroleum jelly. Install the gears in the cover pocket, making sure petroleum jelly fills all the voids between the gears and pockets.

NOTE: Failure to properly coat the oil pump gears may result in failure of the pump to prime when the engine is started.

14. Position the pump body O-ring seal and install the pump body to the front cover using alignment dowels on the front cover.
15. Tighten the pump body retaining bolts to 18–22 ft. lbs. (25–30 Nm) for M8 bolts and 30–40 ft. lbs. (40–55 Nm) for M10 bolts.
16. Install the oil cooler on supercharged engine. Install a new oil filter.
17. Connect the negative battery cable, start the engine and check for leaks and proper oil pressure.

4.6L Engine

1. Disconnect the negative battery cable.
2. Remove the camshaft covers, front cover, and oil pan.
3. Remove the timing chains.
4. Remove the 4 bolts retaining the oil pump to the cylinder block and remove the pump.
5. Remove the 2 bolts retaining the oil pickup tube to the oil pump and remove the bolt retaining the oil pickup tube to the main bearing stud spacer. Remove the pickup tube.
To install:
6. Clean the oil pickup tube and replace the O-ring.
7. Position the tube on the oil pump and hand-start the 2 bolts. Install the bolt retaining the pickup tube to the main bearing stud spacer hand tight.
8. Tighten the pickup tube-to-oil pump bolts to 6.0–8.8 ft. lbs. (8–12 Nm). Tighten the pickup tube to main bearing stud spacer bolt to 15–22 ft. lbs. (20–30 Nm).
9. Rotate the inner rotor of the oil pump to align with the flats on the crankshaft and install the oil pump flush with the cylinder block. Install the 4 retaining bolts and tighten to 6.0–8.8 ft. lbs. (8–12 Nm).
10. Install a new oil filter. Install the timing chains.
11. Install the oil pan, front cover and camshaft covers.
12. Fill the crankcase with the proper type and quantity of engine oil. Connect the negative battery cable, start the engine and check for leaks.

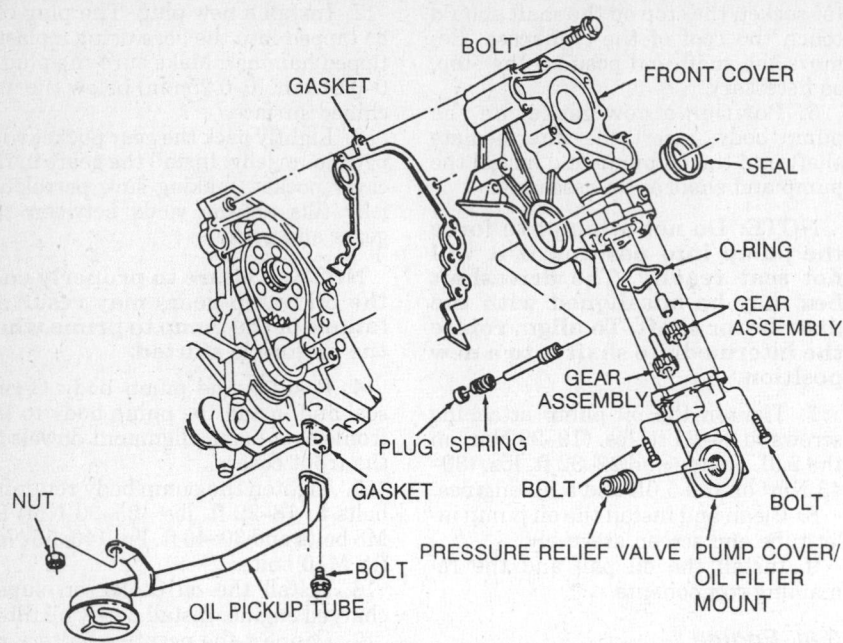

Oil pump and timing chain front cover exploded view—3.8L engine

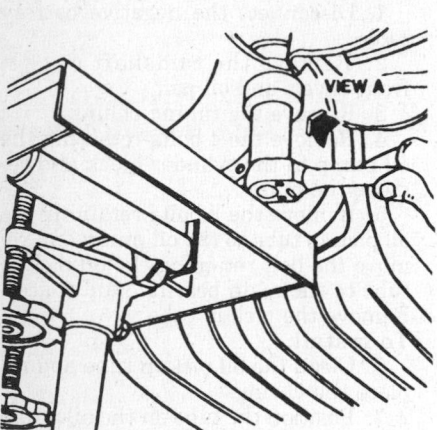

Rear main bearing oil seal installation

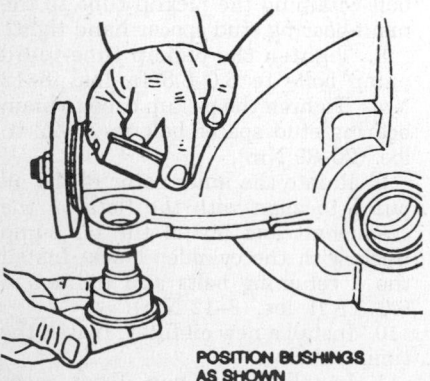

Front differential mount bushing installation—1989–92 Thunderbird and Cougar

Rear Main Bearing Oil Seal

REMOVAL & INSTALLATION

1. Disconnect the negative battery cable. Remove the transmission and on manual transmission equipped vehicles, remove the clutch and flywheel.

2. Punch 2 holes in the crankshaft rear oil seal on opposite sides of the crankshaft, just above the bearing cap to cylinder block split line. Install a sheet metal screw in each of the holes or use a small slide hammer and pry the crankshaft rear main oil seal from the block.

NOTE: Use extreme caution not to scratch the crankshaft oil seal surface.

3. Clean the oil seal recess in the cylinder block and main bearing cap.

4. Coat the seal and all of the seal mounting surfaces with oil. Position the seal on rear main seal installer T82L–6701–A or equivalent, and position the tool and seal to the rear of the engine.

5. Alternate bolt tightening to seat the seal properly. The rear face of the seal must be within 0.005 in. (0.127mm) of the rear face of the block.

ENGINE COOLING

Radiator

REMOVAL & INSTALLATION

Except 1989–92 3.8L SC Engine

1. Disconnect the negative battery cable. Drain the cooling system.

2. Disconnect the upper, lower and overflow hoses at the radiator.

3. If equipped with an automatic transmission, disconnect the fluid cooler lines at the radiator.

4. On Thunderbird and Mustang with 2.3L engine, remove the electric cooling fan/shroud assembly. On all other vehicles, remove the 2 upper fan shroud retaining bolts at the radiator support, lift the fan shroud sufficiently to disengage the lower retaining clips and lay the shroud back over the fan.

5. Remove the radiator upper support retaining bolts and remove the supports. Lift the radiator from the vehicle.

To install:

6. If a new radiator is to be installed, transfer the petcock from the old radiator to the new one. If equipped with automatic transmission, transfer the fluid cooler line fittings from the old radiator.

7. Position the radiator assembly into the vehicle. Install the upper supports and the retaining bolts. If equipped with automatic transmission, connect the fluid cooler lines.

8. On Thunderbird and Mustang with the 2.3L engine, install the electric cooling fan/shroud assembly. On all other vehicles, place the fan shroud into the clips on the lower radiator support and install the 2 upper shroud retaining bolts. Position the shroud to maintain approximately 1 in. (25mm) clearance between the fan blades and the shroud.

9. Connect the radiator hoses. Close the radiator petcock. Fill and bleed the cooling system.

10. Start the engine and bring to operating temperature. Check for coolant and transmission fluid leaks.

11. Check the coolant and transmission fluid levels.

1989–92 3.8L SC Engine

1. Disconnect the negative battery cable.

2. Remove the intercooler.

3. Drain the cooling system and disconnect the upper and lower radiator hoses and the overflow hose at the radiator.

4. If equipped with an automatic transmission, disconnect the fluid cooler lines at the radiator.

5. Remove the overflow hose from the clip on the fan shroud. Remove the 2 shroud upper retaining bolts at the radiator support and remove the Wiring harness retaining clip from the fan shroud. Lift the electric cooling fan/shroud assembly from the radiator, disengaging the shroud from the lower retaining clips.

6. Remove the 2 bolts retaining the top of the air duct to the intercooler and remove the upper 2 radiator retaining bolts. Tilt the radiator and support assembly toward the engine and lift the radiator from the vehicle.

To install:

7. If a new radiator is to be installed and the vehicle is equipped with automatic transmission, transfer the fluid cooler line fittings from the old radiator.

8. Position the radiator and support assembly in the vehicle and install the 2 upper retaining bolts.

9. Cut the retaining strap from the air duct. The duct should spring out from the support assembly. Lift the top of the duct and insert the tabs on the bottom of the duct into the clips at the bottom of the intercooler. Install the 2 bolts that retain the top of the duct to the intercooler.

10. Connect the fluid cooler lines to the radiator. Position the engine cooling fan and stud assembly into the radiator lower clips. Attach the top of the radiator to the top of the support with the 2 bolts.

11. Connect the radiator and overflow hoses to the radiator. Route the overflow hose through the retaining clip. Make sure the draincock is closed and fill the cooling system.

12. Install the intercooler. Connect the cooling fan electrical connector and install the harness clip to the fan shroud.

13. Start the engine and bring to operating temperature. Check for coolant and transmission fluid leaks.

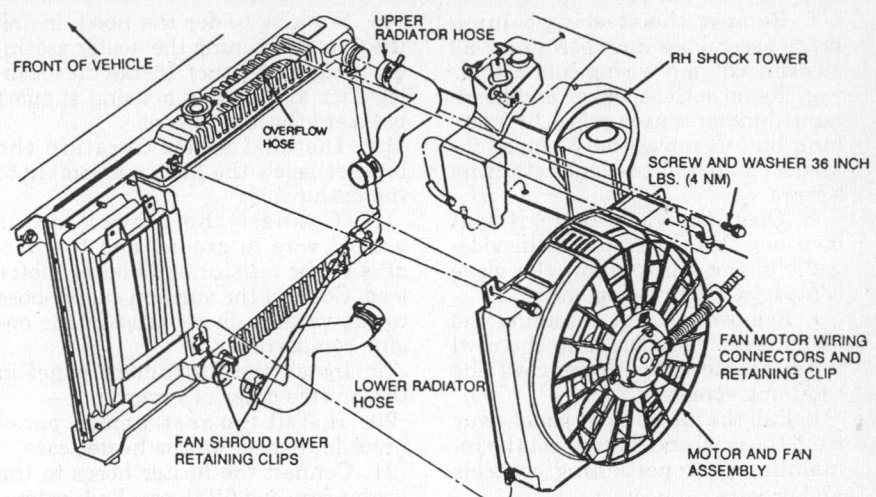

Electric cooling fan installation – Thunderbird and Cougar with 3.8L SC engine

14. Check the coolant and transmission fluid levels.

Electric Cooling Fan

REMOVAL & INSTALLATION

Mustang and Thunderbird with 2.3L Engine

1. Disconnect the negative battery cable.

2. Remove the fan wiring harness from the routing clip. Disconnect the wiring harness from the fan motor connector by pulling up on the single lock finger to separate the connectors.

3. Remove the 4 mounting bracket attaching screws and remove the fan assembly from the vehicle.

4. Remove the retaining clip from the end of the motor shaft and remove the fan.

NOTE: A metal burr may be present on the motor after the retaining clip is removed. Deburring of the shaft may be required to remove the fan.

5. Remove the nuts attaching the fan motor to the mounting bracket.

6. Installation is the reverse of the removal procedure.

Thunderbird and Cougar with 3.8L SC Engine

1. Disconnect the negative battery cable.

2. Disconnect the fan motor wiring connector at the side of the fan shroud. Remove the male terminal connector retaining clip from the shroud mounting tab.

3. Remove the overflow hose from the fan shroud retaining clip and remove the 2 shroud upper retaining bolts at the radiator support.

4. Lift the cooling fan module past the radiator, disengaging the shroud from the 2 lower retaining clips.

5. Installation is the reverse of the removal procedure. Tighten the shroud retaining bolts to 36 inch lbs. (4 Nm).

Heater Core

REMOVAL & INSTALLATION

Without Air Conditioning

1988 THUNDERBIRD AND COUGAR

1. Disconnect the negative battery cable.

2. Remove the instrument panel as follows:

   a. Disconnect all underhood wiring connectors from the main wiring harness. Disengage the rubber grommet seal from the dash panel and push the wiring harness and connectors into the passenger compartment.

   b. Remove the steering column opening cover. Remove 2 screws each from the lower left and right finish panels and remove the panels by carefully pulling on the upper portion of the panels to disengage the clips.

   c. Remove the right and left insulator panels and the right and left cowl side trim covers. Remove the 10 cluster finish panel retaining screws.

   d. If not equipped with a console, open the ash tray and remove the 2 outboard screws inside the ash tray opening. Remove the 4 screws retaining the center finish panel and remove the panel.

   e. If equipped with a console, remove the console front finish panel and remove the 4 screws retaining the center finish panel.

f. Remove the steering column brace assemblies and disconnect all steering column wiring connectors.

g. Remove the speaker and head-lamp dimmer sensor grilles by snapping out. Remove the 3 upper instrument panel-to-cowl retaining screws.

h. Open the glove compartment door and flex the stops on the sidewalls of the bin to allow the glove compartment to swing open.

i. Remove the nut retaining the instrument panel brace to the cowl top and remove the 2 lower cowl side retaining screws.

j. Pull the instrument panel away from the cowl and disconnect the remaining air conditioning controls and/or wire connectors.

3. Remove the right instrument panel brace located above the heater case and attached to the cowl.

4. Drain the cooling system and disconnect the hoses from the heater core. Plug the hoses and the core.

5. Remove the screw attaching the air inlet duct and blower housing assembly support bracket to the cowl top panel.

6. Disconnect the vacuum supply hose from the in-line vacuum check valve in the engine compartment and disconnect the blower motor harness from the resistor and motor.

7. Working under the hood, remove the 2 nuts retaining the heater assembly to the dash panel.

8. In the passenger compartment, remove the screw attaching the heater assembly support bracket to the cowl top panel. Remove the 1 screw retaining the bracket below the heater assembly to the dash panel.

9. Carefully pull the heater assembly away from the dash panel and remove the heater assembly from the vehicle.

10. Remove the 4 heater core access cover attaching screws and remove the access cover from the case.

11. Lift the heater core and seal from the case. Remove the seal from the heater core tubes.

To install:

12. Install the heater core tube seal on the heater core tubes. Inspect the heater core sealer in the heater case and replace, if necessary.

13. Install the heater core in the case with the seals on the outside of the case. Position the access cover on the case and install the attaching screws.

14. Position the heater assembly in the vehicle. Install the screw attaching the heater assembly support bracket to the cowl top panel.

15. Check the heater assembly drain tube to make sure it is through the dash panel and is not pinched or kinked.

16. Working under the hood, install the 2 nuts retaining the heater assembly to the dash panel. Install the air inlet duct and blower housing support bracket attaching screws.

17. Install 1 screw to retain the bracket below the heater assembly to the dash panel.

18. Connect the blower motor ground wire to ground and the harness to the resistor and blower motor lead. Connect the vacuum supply hose to the vacuum check valve in the engine compartment.

19. Install the instrument panel in the reverse order of removal.

20. Install the instrument panel brace located above the heater case.

21. Connect the heater hoses to the heater core and fill the cooling system. Check the heating system for proper operation.

1989–92 THUNDERBIRD AND COUGAR

1. Disconnect the negative battery cable.

2. Remove the instrument panel as follows:

a. Disconnect the underhood wiring at the left side of the dash panel.

b. Disengage the wiring connector from the dash panel and push the wiring harness into the passenger compartment.

c. Remove the steering column lower trim cover by removing the 3 screws at the bottom, 1 screw on the left side and pulling to disengage the 5 snap-in retainers across the top.

d. Remove the steering column lower opening reinforcement. 6 screws retain the reinforcement to the instrument panel.

e. Remove the steering column upper and lower shrouds and disconnect the wiring from the steering column.

f. Remove the shift interlock switch and disconnect the steering column lower universal joint.

g. Support the steering column and remove the 4 nuts retaining the column to the support. Remove the column from the vehicle.

h. Remove the 1 screw retaining the left side of the instrument panel to the parking brake bracket.

i. Install the steering column lower opening reinforcement using the 4 screws, 1 at each corner. This will prevent the instrument panel from twisting when being removed.

j. Remove the right and left cowl side trim panels.

k. Remove the console assembly and remove the 2 nuts retaining the center of the instrument panel to the floor.

l. Open the glove compartment, squeeze the sides of the bin and lower to the full open position. From under the instrument panel and through the glove compartment opening, disconnect the wiring, vacuum lines and control cables.

m. Remove 2 screws from the right side and 2 screws from the left side retaining the instrument panel to the cowl side.

n. Remove the right and left upper finish panels by pulling up to disengage the snap-in retainers. There are 3 on the right side, 4 on the left side.

o. Remove the 4 screws retaining the instrument panel to the cowl top. Remove the right and left roof rail trim panel. Remove the door frame weatherstrip.

p. Carefully pull the instrument panel away from the cowl and disconnect any remaining wiring or controls.

3. Remove the right instrument panel brace located above the heater case and attached to the cowl.

4. Drain the coolant from the cooling system and remove the hoses from the heater core. Plug the hoses and the core.

5. Disconnect the black vacuum supply hose from the in-line vacuum check valve in the engine compartment.

6. Disconnect the blower motor wire harness from the resistor and motor lead.

7. Working under the hood, remove the 3 nuts retaining the heater assembly to the dash panel.

8. In the passenger compartment, remove the screw attaching the heater assembly support bracket to the cowl top panel.

9. Remove the 1 screw retaining the bracket below the heater assembly to the dash panel.

10. Carefully pull the heater assembly away from the dash panel and remove the heater assembly from the vehicle.

11. Remove the 4 heater core access cover attaching screws and remove the access cover.

12. Remove the seal from the heater core tubes and pull the heater core from the case.

To install:

13. Inspect the heater core sealer in the case and replace, if necessary.

14. Install the heater core in the case with the seals on the outside of the case. Install the heater core tube seal on the heater core tubes.

15. Position the heater core access cover and seal on the case and install the 4 attaching screws.

16. Position the heater assembly in the vehicle. Install the screw attaching the heater assembly support bracket to the cowl top panel.

17. Working under the hood, install the 3 nuts retaining the heater assembly to the dash panel.

18. Install 1 screw to retain the bracket below the heater assembly to the dash panel.

19. Connect the blower motor and the harness to the resistor and blower motor lead.

20. Connect the black vacuum supply hose to the vacuum check valve in the engine compartment.

21. Install the right instrument panel brace and install the instrument panel by reversing the removal procedure.

22. Connect the heater hoses to the heater core and fill the cooling system. Check heater operation.

MUSTANG

1. Disconnect the negative battery cable.

2. Remove the floor console and instrument panel as follows:

   a. Remove the 2 access covers at the rear of the console by snapping them out. Remove the 4 armrest-to-floor bracket retaining bolts and remove the armrest assembly by snapping it out of the console.

   b. Remove the gear sgift lever opening finish panel by snapping out. If equipped with a manual transmission, the shift boot is attached to the bottom of the finish panel. Remove the shift knob and slide the boot and finish panel up the shift lever to remove.

   c. Pull up the emergency brake lever. Remove the 4 retaining screws and lift up the top finish panel. Disconnect the necessary wire connectors.

   d. Remove the 2 console-to-rear floor bracket retaining screws. Insert a small prybar into the 2 notches at the bottom of the front upper finish panel and snap it out.

   e. Remove the radio assembly. Open the glove compartment door and drop the glove compartment assembly down. Remove the 2 console-to-instrument panel retaining screws.

   f. Remove the 4 console-to-bracket retaining screws and remove the console.

   g. Disconnect all underhood wiring connectors from the main wiring harness. Disengage the rubber grommet seal from the dash panel and push the wiring harness and connectors into the passenger compartment.

   h. On 1988–89 vehicles, remove the 2 screws attaching the steering column shroud to the dash panel and remove the shroud. On 1990–92 vehicles, remove the 3 bolts attaching the steering column opening

cover and reinforcement panel. Remove the cover.

   i. On 1988–89 vehicles, remove the 3 screws and the steering column cover assembly. On 1990–92 vehicles, remove the steering column opening reinforcement by removing 2 bolts, remove the 2 bolts retaining the lower steering column opening reinforcement and remove the reinforcement.

   j. Remove the 6 steering column retaining nuts. 2 are retaining the hood release mechanism and 4 retain the column to the lower brake pedal support. Lower the steering column to the floor.

   k. On 1988–89 vehicles, remove the 4 screws retaining the steering column brace to the cowl side. On 1990–92 vehicles, remove the steering column upper and lower shrouds and disconnect the wiring from the multi-function switch.

   l. Remove the brake pedal support nut and snap out the defroster grille.

   m. Remove the screws from the speaker covers. Snap out the speaker covers. Remove the front screws retaining the right and left scuff plates at the cowl trim panel. Remove the right and left side cowl trim panels.

   n. Disconnect the wiring at the right and left cowl sides. Remove the cowl side retaining bolts, 1 on each side.

   o. Open the glove compartment door and flex the glove compartment bin tabs inward. Drop down the glove compartment door assembly.

   p. Remove the 5 cowl top screw attachments. Gently pull the instrument panel away from the cowl. Disconnect the speedometer cable and wire connectors.

3. Drain the coolant from the cooling system and remove the hoses from the heater core. Plug the hoses and the core.

4. Remove the screw attaching the air inlet duct and blower housing assembly support bracket to the cowl top panel.

5. Disconnect the black vacuum supply hose from the in-line vacuum check valve in the engine compartment.

6. Disconnect the blower motor wire harness from the resistor and motor head.

7. Working under the hood, remove the 2 nuts retaining the heater assembly to the dash panel.

8. In the passenger compartment, remove the screw attaching the heater assembly support bracket to the cowl top panel. Remove the 1 screw retaining the bracket below the heater assembly to the dash panel.

9. Carefully pull the heater assembly away from the dash panel and remove from the vehicle.

10. Remove the 4 heater core access cover attaching screws and remove the access cover from the case.

11. Lift the heater core and seal from the case. Remove the seal from the heater core tubes.

To install:

12. Install the heater core tube seal on the heater core tubes. Inspect the heater core sealer in the heater case and replace, if necessary.

13. Install the heater core in the case with the seals on the outside of the case. Position the heater core access cover on the case and install the 4 attaching screws.

14. Position the heater assembly in the vehicle. Install the screw attaching the heater assembly support bracket to the cowl top panel.

15. Check the heater assembly drain tube to ensure it is through the dash panel and is not pinched or kinked.

16. Working under the hood, install the 2 nuts retaining the heater assembly to the dash panel. Install the air inlet duct and blower housing support bracket attaching screw. Install 1 screw to the retainer bracket below the heater assembly to the dash panel.

17. Connect the blower motor ground wire to ground and the harness to the resistor and blower motor lead.

18. Connect the black vacuum supply hose to the vacuum check valve in the engine compartment.

19. Install the instrument panel and floor console by reversing the removal procedure.

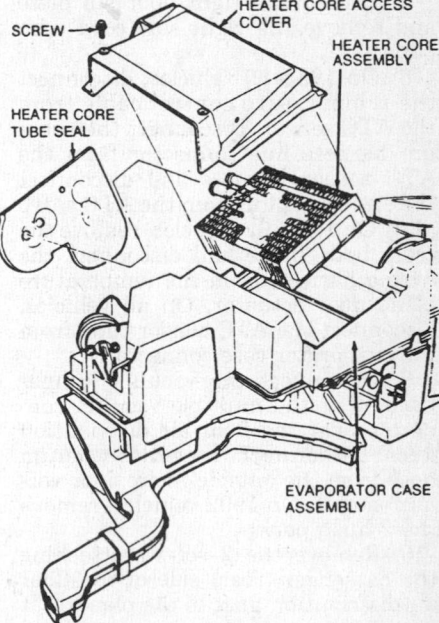

Heater core installation—Mustang

20. Connect the heater hoses to the heater core and fill the cooling system. Check the system for proper operation.

With Air Conditioning

CROWN VICTORIA, GRAND MARQUIS AND TOWN CAR WITH AUTOMATIC TEMPERATURE CONTROL

1. Disconnect the negative battery cable.

2. Drain the cooling system and disconnect the heater hoses from the heater core tubes. Plug the hoses and the heater core tubes.

3. On 1988 vehicles, remove the 1 bolt and on 1989–92 vehicles remove the 3 nuts located below the windshield wiper motor attaching the left end of the plenum to the dash panel. On all vehicles, remove the 1 nut retaining the upper left corner of the evaporator case to the dash panel.

4. Disconnect the vacuum supply hose(s) from the vacuum source. On 1989–92 vehicles, disconnect the vacuum harness from the thermal blower lockout switch. On all vehicles, push the grommet and vacuum supply hoses into the passenger compartment.

5. On 1989–92 vehicles, remove the right and left lower instrument panel insulators.

6. On 1988–89 vehicles, remove the glove box and the instrument panel pad. On 1990–92 vehicles, remove all instrument panel mounting screws and pull the instrument panel back as far as it will go without disconnecting the wiring harness. Make sure the nuts attaching the instrument panel braces to the dash panel are removed.

7. Loosen the right door sill plate and remove the right side cowl trim panel.

8. On 1988–89 vehicles, disconnect the temperature control cable from the ATC sensor. Disconnect the vacuum harness line connector from the ATC sensor harness and disconnect the electrical plug from the ATC servo plug. On 1990–92 vehicles, remove the cross body brace and disconnect the wiring harness from the temperature blend door actuator. On all vehicles, disconnect the ATC sensor tube from the evaporator case connector.

9. Disconnect the vacuum jumper harness at the multiple vacuum connector near the floor air distribution duct. Disconnect the white vacuum hose from the outside-recirc door vacuum motor. On 1992 vehicles, remove the 2 hush panels.

10. Remove the 2 screws attaching the passenger (rear) side of the floor air distribution duct to the plenum. It may be necessary to remove the 2 screws attaching the partial (lower)

panel door vacuum motor to the mounting bracket to gain access to the right screw.

11. Remove 1 plastic push fastener retaining the floor air distribution duct to the left end of the plenum and 2 screws on the rear face of the plenum and remove the floor air ditribution duct.

12. Remove 2 nuts from the 2 studs along the lower flange of the plenum. Carefully move the plenum rearward to allow the heater core tubes and the stud at the top of the plenum to clear the holes in the dash panel. Remove the plenum from the vehicle by rotating the top of the plenum forward, down and out from under the instrument panel. Carefully pull the lower edge of the instrument panel rearward, as necessary, while rolling the plenum from behind the instrument panel.

13. On 1988–89 vehicles, remove the ATC servo from the plenum. On all vehicles, remove the 4 retaining screws from the heater core cover and remove the cover from the plenum assembly. Pull the heater core and seal assembly from the plenum assembly.

To install:

14. Carefully install the heater core and seal assembly into the plenum assembly. Visually check to ensure the core seal is properly positioned. Position the heater core cover and install the 4 retaining screws. On 1988–89 vehicles, install the ATC servo on the plenum.

15. Position the plenum on the rear of the dash panel with the heater core tubes and the stud at the top of the plenum through the holes in the dash panel. Install the 2 nuts removed from the lower flange of the plenum.

16. Install the plastic push fastener retaining the left end of the floor air distribution duct to the left end of the plenum.

17. Install the screws that attach the rear of the floor air distribution duct to the plenum. If necessary, tighten the screws attaching the partial (lower) panel door vacuum motor to the mounting bracket.

18. Connect the white vacuum hose from the outside-recirc door vacuum motor. Connect the vacuum jumper harness at the multiple vacuum connector near the floor air distribution duct. On 1988–89 vehicles, route and connect the vacuum harness connector at the ATC sensor and connect the electrical plug to the ATC servo plug. Do not block the sensor aspirator exhaust port with the excess vacuum harness.

19. Connect the ATC sensor tube to the evaporator case connector. On 1990–92 vehicles, install the cross

body brace and connect the wiring harness to the blend door actuator.

20. Replace the right side cowl trim panel and tighten the screws in the right door sill plate.

21. On 1988–89 vehicles, install the glove compartment door and instrument panel pad. On 1990–91 vehicles, push the instrument panel back into position and install all instrument panel mounting screws. Install the right and left lower instrument panel insulators.

22. Push the vacuum supply hose(s) into the engine compartment and seat the grommet in the dash panel. Connect the vacuum supply hose(s) to the vacuum source and connect the vacuum harness to the thermal blower lockout switch.

23. Install the 1 nut retaining the upper left corner of the evaporator case to the dash panel. Install the 1 bolt on 1988 vehicles or 3 nuts on 1989–92 vehicles located below the windshield wiper motor, that attach the left end of the plenum to the dash panel. On 1992 vehicles, install the hush panels.

24. Unplug the heater core hoses and tubes and connect the heater hoses to the heater core tubes. Fill the cooling system.

25. Connect the negative battery cable and check the system for proper operation.

CROWN VICTORIA AND GRAND MARQUIS WITH MANUAL AIR CONDITIONING

1. Disconnect the negative battery cable.

2. Drain the cooling system and disconnect the heater hoses from the heater core tubes. Plug the hoses and the heater core tubes.

3. On 1988 vehicles, remove the 1 bolt and on 1989–92 vehicles remove the 3 nuts located below the windshield wiper motor attaching the left end of the plenum to the dash panel. Remove the 1 nut retaining the upper left corner of the evaporator case to the dash panel.

4. Disconnect the vacuum supply hose(s) from the vacuum source. Push the grommet and vacuum supply hoses into the passenger compartment.

5. Remove the right and left lower instrument panel insulators.

6. On 1988–89 vehicles, remove the glove compartment and the instrument panel pad. On 1990–92 vehicles, remove all instrument panel mounting screws and pull the instrument panel back as far as it will go without disconnecting the wiring harness. Make sure the nuts attaching the instrument panel braces to the dash panel are removed.

7. Loosen the right door sill plate and remove the right side cowl trim

panel. Remove the bolt attaching the lower right end of the instrument panel to the side cowl.

8. Disengage the temperature control cable housing from the bracket on top of the plenum. Disconnect the cable from the temperature blend door crank arm.

9. Disconnect the vacuum jumper harness at the multiple vacuum connector near the floor air distribution duct. Disconnect the white vacuum hose from the outside-recirculating door vacuum motor.

10. On 1989–92 vehicles, remove the 2 hush panels.

11. Remove 1 plastic push fastener retaining the floor air distribution duct to the left end of the plenum. Remove the left screw and loosen the right screw on the rear face of the plenum and remove the floor air distribution duct.

12. Remove the 2 nuts from the 2 studs along the lower flange of the plenum.

13. Carefully move the plenum rearward to allow the heater core tubes and the stud at the top of the plenum to clear the holes in the dash panel. Remove the plenum from the vehicle by rotating the top of the plenum forward, down and out from under the instrument panel. Carefully pull the lower edge of the instrument panel rearward, as necessary, while rolling the plenum from behind the instrument panel.

14. Remove the 4 retaining screws from the heater core cover and remove the cover from the plenum assembly. Pull the heater core and seal assembly from the plenum assembly.

To install:

15. Carefully install the heater core and seal assembly into the plenum assembly. Visually check to ensure that the core seal is properly positioned. Position the heater core cover and install the 4 retaining screws.

16. Route the vacuum supply hose through the dash panel and seat the grommet in the opening.

17. Position the plenum under the instrument panel with the register duct opening up and the heater core tubes down. Rotate the plenum up behind the instrument panel and position the plenum to the dash panel. Insert the heater core tubes and mounting studs through their respective holes in the dash panel and the evaporator case.

18. Install the 2 nuts on the studs along the lower flange of the plenum. Install the 1 bolt or 3 nuts below the windshield wiper motor to attach the left end of the plenum to the dash panel. Install 1 nut to retain the upper left corner of the evaporator case to the dash panel.

19. Position the floor air distribution duct on the plenum. Install the 2 screws and plastic push fastener. If removed, position the panel door vacuum motor to the mounting bracket and install the 2 attaching screws.

20. Connect the white vacuum hose to the outside-recirculating door vacuum motor. Connect the vacuum jumper harness to the plenum harness at the multiple vacuum connector near the floor air distribution duct. Install the floor duct.

21. Connect the temperature control cable housing to the bracket on top of the plenum and connect the temperature control cable to the temperature blend door crank arm. Adjust the temperature cable.

22. Install the bolt to attach the lower right end of the instrument panel to the side cowl. Install the right side cowl trim panel and tighten the right door sill plate attaching screws.

23. On 1988–89 vehicles, install the glove compartment and instrument panel pad. On 1990–91 vehicles, push the instrument panel back into position and install all instrument panel mounting screws. Install the right and left lower instrument panel insulators.

24. Push the vacuum supply hoses into the engine compartment and seat the grommet in the dash panel. Connect the vacuum supply hose(s) to the vacuum source.

25. Install the right and left lower instrument panel insulators and install the 2 hush panels.

26. Unplug the heater core tubes and the heater hoses and connect the heater hoses to the heater core tubes. Fill the cooling system.

27. Connect the negative battery cable and check the system for proper operation.

MARK VII

1. Disconnect the negative battery cable and drain the cooling system.

2. Discharge the refrigerant from the air conditioning system according to the proper procedure.

3. Remove the instrument panel as follows:

   a. Disconnect all underhood electrical connectors of the main wiring harness. Disengage the rubber grommet from the dash panel.

   b. Remove the right and left sound insulator assemblies from under the instrument panel. Remove the bulb and socket assemblies, if necessary.

   c. Remove the steering column opening trim cover and the steel reinforcement.

   d. If necessary, locate and remove the demister feed Y-connector from the driver's side floor duct outlet.

   e. Remove the left and right cowl trim panels. Remove the 2 screws attaching the hood release to the cowl panel before removing the left trim panel.

   f. Remove the 5 steering column trim shroud screws and remove the shrouds.

   g. Disconnect all electrical connector quick couplers from the steering column switches.

   h. Remove the 4 nuts attaching the steering column to the support. Lower the column to rest on the seat cushion.

   i. Remove the defroster opening grille panel and remove the screws attaching the floor console to the instrument panel and floor. Move the console rearward.

   j. Remove the screw(s) attaching the instrument panel to the floor. Remove the screws attaching the instrument panel to the cowls. Remove the bolt or nut attaching the instrument panel to the support bracket.

   k. Disconnect the main wiring harness behind the instrument panel, on the right side of the steering column support, at the blower motor and at the left and right cowl panels.

   l. Disconnect the radio antenna lead from the radio. Disconnect any vacuum hoses attached to the instrument panel.

   m. Remove the right and left A-pillar garnish mouldings.

   n. Remove the 3 screws attaching the instrument panel to the dash panel and pull/push the wiring harness and connectors into the passenger compartment. Remove the instrument panel.

4. Disconnect the heater hoses from the heater core. Plug the hoses and the core. Disconnect the wire harness connector from the clutch cycling pressure switch, located on top of the suction accumulator/drier.

5. Disconnect the liquid line and the accumulator/drier inlet tube from the evaporator core tubes. Use a backup wrench to prevent component damage. Cap all fittings to prevent the entrance of dirt and excessive moisture.

6. Working under the hood, remove the 2 nuts retaining the accumulator/drier bracket to the dash panel. Position the accumulator/drier and liquid line aside and remove the 2 evaporator assembly retaining nuts.

7. Disconnect the wiring harness connectors, as necessary. Disconnect the harness connectors from the blower motor wires and blower motor speed controller.

8. Disconnect the automatic temperature control sensor hose and elbow from the evaporator case. Discon-

nect the automatic temperature control harness at the control assembly.

9. Disconnect the rear seat duct adapter from the floor duct. Remove the 3 evaporator attaching screws and remove the evaporator assembly from the vehicle.

NOTE: Whenever an evaporator case is removed, it will be necessary to replace the suction accumulator/drier.

11. Remove the 5 heater core access cover attaching screws and remove the access cover from the evaporator case.

12. Lift the heater core and seal from the evaporator case. Remove the seal from the heater core tubes.
To install:

13. Install the heater core in the evaporator case.

14. Position the heater core access cover on the evaporator case and install the 5 attaching screws. Install the heater core seal.

15. Position the evaporator assembly to the dash panel. Install the 3 attaching screws located in the passenger compartment, but do not tighten at this time. Check the evaporator drain tube to be certain it is through the opening.

16. Working in the engine compartment, install 2 nuts to retain the evaporator assembly to the dash panel. Then, tighten the retaining nuts and attaching screws.

17. Position the instrument panel near the dash panel and connect the radio antenna, automatic temperature control harness to control assembly, harness to blower motor speed controller and blower motor wires. Also attach any additional wire harness connectors disconnected during removal.

18. Move the instrument panel into position and install the attaching screws. Connect the automatic temperature control sensor hose and elbow assembly.

NOTE: Make sure the air conditioning plenum (attached to the instrument panel) is correctly aligned and sealed at the evaporator outlet opening. Air leakage to the floor area will result if the plenum is not sealed at the evaporator case outlet.

19. Install the nut retaining the instrument panel to the brake pedal and steering column support. Position the steering column to the brake pedal and steering column support and install the retaining nuts.

20. Install the steering column opening reinforcement and shroud. Install the screws to attach the lower center of the instrument panel to the floor brace.

21. Install the defroster opening grille and the right and left side cowl trim panels. Install the right and left instrument panel sound insulators.

22. Install the rear seat duct adapter and the console assembly. Install the instrument panel right hand finish panel and the steering column opening cover.

23. Position the accumulator/drier mounting bracket over the studs on the dash panel and loosely install the 2 nuts.

24. Connect the accumulator/drier inlet tube to the evaporator core outlet tube using a new O-ring lubricated with clean refrigerant oil. Do not tighten the connection.

25. Connect the liquid line to the evaporator core inlet tube using a new O-ring lubricated with clean refrigerant oil. Do not tighten the connection.

26. Tighten the 2 nuts retaining the accumulator/drier to the dash panel. Tighten the 2 refrigerant line connections at the evaporator core. Use a backup wrench to prevent component damage.

27. Connect the harness connector to the clutch cycling pressure switch. Connect the heater hoses to the heater core and fill the cooling system.

28. Connect the negative battery cable. Leak test, evacuate and charge the refrigerant system according to the proper procedure. Observe all safety precautions.

29. Check the automatic temperature control system and all instrument panel functions for proper operation.

1988 THUNDERBIRD AND COUGAR WITH AUTOMATIC TEMPERATURE CONTROL

1. Disconnect the negative battery cable and drain the cooling system.

2. Discharge the refrigerant from the air conditioning system according to the proper procedure.

3. Disconnect and plug the heater hoses at the heater core. Plug the heater core tubes.

4. Disconnect the wire harness connector from the clutch cycling pressure switch located on top of the suction accumulator/drier. Disconnect the air conditioning lines from the evaporator core tubes. Cap the lines and the core to prevent the entrance of dirt and moisture.

5. Remove the 2 nuts retaining the accumulator/drier bracket to the dash panel. Position the accumulator/drier and liquid line aside and remove the 2 evaporator retaining nuts.

6. Working in the passenger compartment, remove the steering column opening cover from the instrument panel.

7. Remove the instrument panel right and left sound insulators and re-

move the right and left side cowl trim panels.

8. Remove the ash tray and remove the 2 screws attaching the lower center portion of the instrument panel to the floor brace.

9. Remove the shroud from the steering column and disconnect the transmission indicator cable from the steering column. Remove the instrument panel pad.

10. Remove the steering column-to-brake pedal support retaining nuts and lay the steering column down against the front seat.

11. Remove the nut attaching the instrument panel to the brake pedal and steering column support. Remove the instrument panel attaching screws and lay the instrument panel back on the front seat. Disconnect wiring harness connectors, as necessary.

12. Disconnect the harness connectors from the blower motor wires and the blower motor speed controller.

13. Disconnect the automatic temperature control sensor hose and elbow from the evaporator case. Disconnect the automatic temperature control harness at the control assembly.

14. Disconnect the antenna cable from the strap on the evaporator case. Using a sharp knife, carefully slit the carpet on the top of the transmission tunnel, then fold the carpet back to expose the rear seat heater duct.

15. Using a saw or a hot knife, cut the top and side from the rear seat heat duct. Remove the top from the floor duct.

16. Remove the 3 evaporator case attaching screws and remove the evaporator case.

NOTE: Whenever an evaporator case is removed, it will be necessary to replace the suction accumulator/drier.

17. Remove the 5 heater core access cover attaching screws and remove the access cover from the evaporator case.

18. Lift the heater core and seals from the evaporator case. Remove the 2 seals from the heater core tubes.
To install:

19. Install the heater core tube seals on the heater core tubes, thin seal first. Install the heater core in the evaporator case with the seals on the outside of the case.

20. Position the heater core access cover on the evaporator case and install the 5 attaching screws.

21. Position the evaporator assembly to the dash panel. Install the 3 attaching screws in the passenger compartment, but do not tighten at this time. Check the evaporator drain tube to ensure it is not pinched or kinked.

22. Working in the engine compartment, install the 2 nuts to retain the

evaporator case to the dash panel. Then, tighten the retaining nuts and attaching screws.

23. Place the top cut from the rear seat heat duct on the duct. Tape it in place with duct tape. Be sure to cover all seams to prevent air leakage. Then, reposition the carpet over the rear seat heat duct.

24. Position the instrument panel near the dash panel and connect the radio antenna, automatic temperature control harness to the control assembly, harness to the blower motor speed controller, blower motor wires and the automatic temperature control sensor hose and elbow assembly. Also, attach any additional wire harness connectors disconnected during removal.

25. Move the instrument panel into position and install the attaching screws.

NOTE: Make sure the air conditioning plenum, attached to the instrument panel, is correctly aligned and sealed at the evaporator outlet opening. Air leakage to the floor area will result if the plenum is not sealed at the evaporator case outlet.

26. Install the nut retaining the instrument panel to the brake pedal and steering column support. Position the steering column to the brake pedal and steering column support and install the retaining nuts.

27. Connect the transmission selector indicator to the steering column and adjust, as necessary. Install the steering column opening reinforcement and the shroud.

28. Install the 2 screws to attach the lower center portion of the instrument panel to the floor brace. Install the side cowl trim panels and instrument panel sound insulators.

29. Install the instrument panel pad and the steering column opening cover.

30. Position the accumulator/drier mounting bracket over the studs on the dash panel and loosely install the 2 nuts.

31. Connect the air conditioning lines using new O-rings lubricated with clean refrigerant oil. Do not tighten the connections at this time.

32. Tighten the 2 nuts retaining the accumulator/drier to the dash panel, then tighten the air conditioning lines.

33. Connect the harness connector to the clutch cycling pressure switch and connect the heater hoses to the heater core.

34. Fill the cooling system and connect the negative battery cable. Leak test, evacuate and charge the air conditioning system according to the proper procedure. Observe all safety precautions.

35. Check the heater and air conditioning systems for proper operation.

1988 THUNDERBIRD AND COPUGAR WITH MANUAL AIR CONDITIONING

1. Disconnect the negative battery cable and drain the cooling system.
2. Discharge the refrigerant from the air conditioning system according to the proper procedure.
3. Remove the instrument panel as follows:

a. Disconnect all underhood wiring connectors from the main wiring harness. Disengage the rubber grommet seal from the dash panel and push the wiring harness and connectors into the passenger compartment.

b. Remove the steering column opening cover. Remove 2 screws each from the lower left and right finish panels and remove the panels by carefully pulling on the upper portion of the panels to disengage the clips.

c. Remove the right and left insulator panels and the right and left cowl side trim covers. Remove the 10 cluster finish panel retaining screws.

d. If not equipped with a console, open the ash tray and remove the 2 outboard screws inside the ash tray opening. Remove the 4 screws retaining the center finish panel and remove the panel.

e. If equipped with a console, remove the console front finish panel and remove the 4 screws retaining the center finish panel.

f. Remove the steering column brace assemblies and disconnect all steering column wiring connectors.

g. Remove the speaker and headlamp dimmer sensor grilles by snapping out. Remove the 3 upper instrument panel-to-cowl retaining screws.

h. Open the glove compartment door and flex the stops on the sidewalls of the bin to allow the glove compartment to swing open.

i. Remove the nut retaining the instrument panel brace to the cowl top and remove the 2 lower cowl side retaining screws.

j. Pull the instrument panel away from the cowl and disconnect the remaining air conditioning controls and/or wire connectors.

4. Disconnect and plug the heater hoses at the heater core. Plug the heater core tubes.

5. Disconnect the air conditioning lines from the evaporator core tubes. Cap the lines and the core to prevent the entrance of dirt and moisture.

6. Remove the screw attaching the air inlet duct and blower housing assembly support brace to the cowl top panel.

7. Disconnect the vacuum supply hose from the in-line vacuum check valve in the engine compartment. Disconnect the blower motor wiring.

8. Working under the hood, remove the 2 nuts retaining the evaporator case to the dash panel. In the passenger compartment, remove the screw attaching the evaporator case support bracket to the cowl top panel.

9. Remove the 1 nut retaining the bracket below the evaporator case to the dash panel. Carefully pull the evaporator case away from the dash panel and remove the evaporator case from the vehicle.

NOTE: Whenever an evaporator case is removed, it will be necessary to replace the suction accumulator/drier.

10. Remove the 5 heater core access cover attaching screws and remove the access cover from the evaporator case.

11. Lift the heater core and seals from the evaporator case and remove the seals from the heater core tubes.

To install:
12. Install the heater core tube seals on the heater core tubes, thin seal first. Install the heater core in the evaporator case with the seals on the outside of the case.

13. Position the heater core access cover on the evaporator case and install the 5 attaching screws.

14. Position the evaporator case in the vehicle and install the screw attaching the evaporator case support bracket to the cowl top panel. Check the evaporator case drain tube to make sure it is through the dash panel and is not pinched or kinked.

15. Install 1 nut retaining the mounting bracket at the left end of the evaporator case to the dash panel and another nut to retain the bracket below the evaporator case to the dash panel.

16. Working under the hood, install 2 nuts retaining the evaporator case to the dash panel. Tighten the 4 nuts, 2 in the engine compartment and 2 in the passenger compartment, and the 1 support bracket attaching screw.

17. Connect the blower motor wiring and connect the vacuum supply hose to the vacuum check valve in the engine compartment.

18. Using new O-rings lubricated with clean refrigerant oil, connect the air conditioning lines to the evaporator core.

19. Install the instrument panel in the reverse order of removal.

20. Connect the heater hoses to the heater core and fill the cooling system. Leak test, evacuate and charge the air

conditioning system according to the proper procedure. Observe all safety precautions.

21. Check the heater and air conditioning system for proper operation.

1989–92 THUNDERBIRD AND COUGAR

1. Disconnect the negative battery cable and drain the cooling system.

2. Discharge the refrigerant from the air conditioning system according to the proper procedure.

3. Remove the instrument panel according to the following procedure:

a. Disconnect the underhood wiring at the left side of the dash panel.

b. Disengage the wiring connector from the dash panel and push the wiring harness into the passenger compartment.

c. Remove the steering column lower trim cover by removing the 3 screws at the bottom, 1 screw on the left side and pulling to disengage the 5 snap-in retainers across the top.

d. Remove the steering column lower opening reinforcement. 6 screws retain the reinforcement to the instrument panel.

e. Remove the steering column upper and lower shrouds and disconnect the wiring from the steering column.

f. Remove the shift interlock switch and disconnect the steering column lower universal joint.

g. Support the steering column and remove the 4 nuts retaining the column to the support. Remove the column from the vehicle.

h. Remove the 1 screw retaining the left side of the instrument panel to the parking brake bracket.

i. Install the steering column lower opening reinforcement using the 4 screws, 1 at each corner. This will prevent the instrument panel from twisting when being removed.

j. Remove the right and left cowl side trim panels.

k. Remove the console assembly and remove the 2 nuts retaining the center of the instrument panel to the floor.

l. Open the glove compartment, squeeze the sides of the bin and lower to the full open position. From under the instrument panel and through the glove compartment opening, disconnect the wiring, vacuum lines and control cables.

m. Remove 2 screws from the right side and 2 screws from the left side retaining the instrument panel to the cowl side.

n. Remove the right and left upper finish panels by pulling up to disengage the snap-in retainers. There are 3 on the right side, 4 on the left side.

o. Remove the 4 screws retaining the instrument panel to the cowl top. Remove the right and left roof rail trim panel. Remove the door frame weather-strip.

p. Carefully pull the instrument panel away from the cowl and disconnect any remaining wiring or controls.

4. Disconnect the liquid line and accumulator/drier inlet tube from the evaporator core at the dash panel. Cap the refrigerant lines and evaporator core to prevent the entrance of dirt and excessive moisture.

5. Disconnect the heater hoses from the heater core. Plug the hoses and heater core tubes.

6. Disconnect the black vacuum supply hose from the in-line vacuum check valve in the engine compartment. Disconnect the blower motor wiring.

7. Working under the hood, remove the 2 nuts retaining the evaporator case to the dash panel. In the passenger compartment, remove the screw attaching the evaporator case support bracket to the cowl top panel.

8. Remove 1 nut retaining the bracket below the evaporator case to the dash panel. Carefully pull the evaporator case away from the dash panel and remove the evaporator case assembly from the vehicle.

NOTE: Whenever an evaporator case is removed, it will be necessary to replace the suction accumulator/drier.

9. Remove the 4 heater core access cover attaching screws and remove the access cover from the evaporator case.

10. Remove the tube seal from the heater core tubes. Slide the heater core and seals from the evaporator case.

To install:

11. Install the heater core in the evaporator case with the tube seal on the outside of the case.

12. Position the heater core access cover on the evaporator case and install the 4 attaching screws.

13. Position the evaporator case assembly in the vehicle and install the screw attaching the evaporator case support bracket to the cowl top panel. Check the evaporator case drain tube to make sure it is through the dash panel and is not pinched or kinked.

14. Install 1 nut retaining the mounting bracket at the left end of the evaporator case to the dash panel and another nut to retain the bracket below the evaporator case to the dash panel.

15. Working under the hood, install the 2 nuts retaining the evaporator case to the dash panel. Tighten the 4 nuts, 2 in the engine compartment and 2 in the passenger comparment and the 1 support bracket attaching screw.

16. Connect the blower motor wiring. Connect the black vacuum supply hose to the vacuum check valve in the engine compartment.

17. Using new O-rings lubricated with clean refrigerant oil, connect the liquid line and suction accumulator inlet tube to the evaporator core.

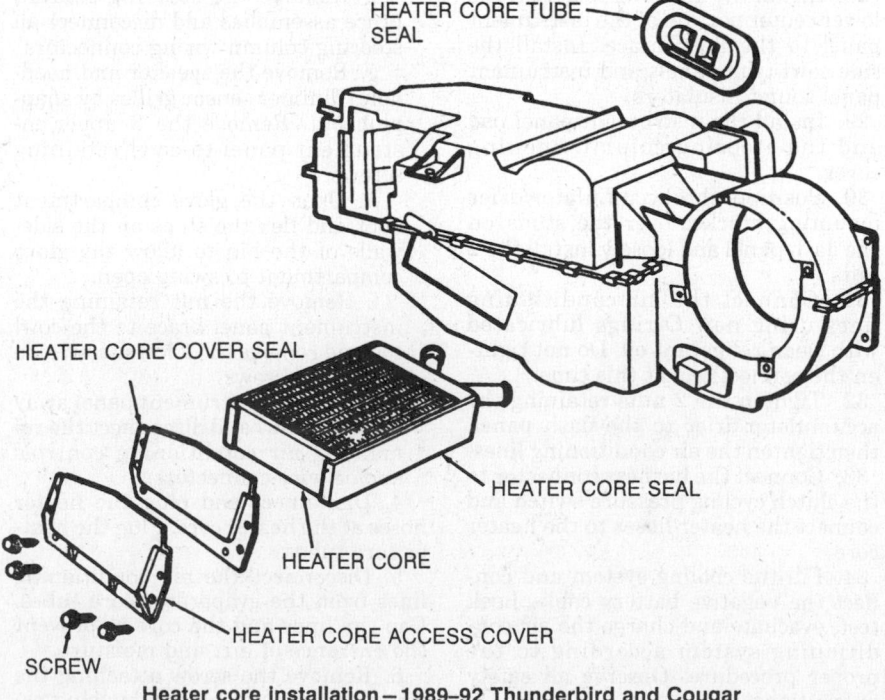

HEATER CORE TUBE SEAL

HEATER CORE COVER SEAL

HEATER CORE SEAL

HEATER CORE

HEATER CORE ACCESS COVER

SCREW

Heater core installation – 1989–92 Thunderbird and Cougar

18. Install the instrument panel by reversing the removal procedure.

19. Connect the heater hoses to the heater core and fill the cooling system.

20. Leak test, evacuate and charge the system according to the proper procedure. Observe all safety precautions.

21. Check the system for proper operation.

MUSTANG

1. Disconnect the negative battery cable and drain the cooling system.

2. Discharge the refrigerant from the air conditioning system according to the proper procedure.

3. Remove the instrument panel according to the following procedure:

a. Remove the 2 access covers at the rear of the console by snapping them out. Remove the 4 armrest-to-floor bracket retaining bolts and remove the armrest assembly by snapping it out of the console.

b. Remove the gear sgift lever opening finish panel by snapping out. If equipped with a manual transmission, the shift boot is attached to the bottom of the finish panel. Remove the shift knob and slide the boot and finish panel up the shift lever to remove.

c. Pull up the emergency brake lever. Remove the 4 retaining screws and lift up the top finish panel. Disconnect the necessary wire connectors.

d. Remove the 2 console-to-rear floor bracket retaining screws. Insert a small prybar into the 2 notches at the bottom of the front upper finish panel and snap it out.

e. Remove the radio assembly. Open the glove compartment door and drop the glove compartment assembly down. Remove the 2 console-to-instrument panel retaining screws.

f. Remove the 4 console-to-bracket retaining screws and remove the console.

g. Disconnect all underhood wiring connectors from the main wiring harness. Disengage the rubber grommet seal from the dash panel and push the wiring harness and connectors into the passenger compartment.

h. On 1988–89 vehicles, remove the 2 screws attaching the steering column shroud to the dash panel and remove the shroud. On 1990–92 vehicles, remove the 3 bolts attaching the steering column opening cover and reinforcement panel. Remove the cover.

i. On 1988–89 vehicles, remove the 3 screws and the steering column cover assembly. On 1990–92 vehicles, remove the steering col-

umn opening reinforcement by removing 2 bolts, remove the 2 bolts retaining the lower steering column opening reinforcement and remove the reinforcement.

j. Remove the 6 steering column retaining nuts. 2 are retaining the hood release mechanism and 4 retain the column to the lower brake pedal support. Lower the steering column to the floor.

k. On 1988–89 vehicles, remove the 4 screws retaining the steering column brace to the cowl side. On 1990–92 vehicles, remove the steering column upper and lower shrouds and disconnect the wiring from the multi-function switch.

l. Remove the brake pedal support nut and snap out the defroster grille.

m. Remove the screws from the speaker covers. Snap out the speaker covers. Remove the front screws retaining the right and left scuff plates at the cowl trim panel. Remove the right and left side cowl trim panels.

n. Disconnect the wiring at the right and left cowl sides. Remove the cowl side retaining bolts, 1 on each side.

o. Open the glove compartment door and flex the glove compartment bin tabs inward. Drop down the glove compartment door assembly.

p. Remove the 5 cowl top screw attachments. Gently pull the instrument panel away from the cowl. Disconnect the speedometer cable and wire connectors.

4. Disconnect the liquid line and the accumulator/drier inlet tube from the evaporator core at the dash panel. Cap the refrigerant lines and evaporator core tube to prevent the entrance of dirt and excessive moisture.

5. Disconnect the heater hoses from the heater core tubes and plug the hoses and tubes.

6. Remove the screw attaching the air inlet duct and blower housing assembly support brace to the cowl top panel.

7. Disconnect the black vacuum supply hose from the in-line vacuum check valve in the engine compartment. Disconnect the blower motor wires from the wire harness and disconnect the wire harness from the blower motor resistor.

8. Working under the hood, remove the 2 nuts retaining the evaporator case to the dash panel. Inside the passenger compartment, remove the 2 screws attaching the evaporator case support brackets to the cowl top panel.

9. Remove the 1 screw retaining the bracket below the evaporator case to the dash panel. Carefully pull the evaporator case away from the dash

panel and remove the evaporator case assembly from the vehicle.

NOTE: Whenever an evaporator case is replaced, it will be necessary to replace the suction accumulator/drier.

10. Remove the 4 heater core access cover attaching screws and remove the cover from the case.

11. Lift the heater core and seal from the case. Remove the seal from the heater core tubes.

To install:

12. Install the heater core tube seal on the heater core tubes.

13. Inspect the heater core sealer in the evaporator case. Replace with suitable caulking cord, if necessary.

14. Install the heater core in the case with the seals on the outside of the case. Position the heater core access cover on the case and install the 4 attaching screws.

15. Position the evaporator case assembly in the vehicle. Install the screws attaching the evaporator case support brackets to the cowl top panel. Check the evaporator case drain tube to ensure it is through the dash panel and is not pinched or kinked.

16. Install 1 screw retaining the bracket below the evaporator case to the dash panel. Working under the hood, install the 2 nuts retaining the evaporator case to the dash panel. Tighten the 4 nuts and 2 screws in the engine compartment. Tighten the 2 screws in the passenger compartment and the 2 support bracket attaching screws.

17. Connect the blower motor wire harness to the resistor and blower motor. Connect the black vacuum supply hose to the vacuum check valve in the engine compartment.

18. Using new O-rings lubricated with clean refrigerant oil, connect the liquid line and suction accumulator inlet to the evaporator core tubes. Tighten each connection using a backup wrench to prevent component damage.

19. Install the instrument panel by reversing the removal procedure.

20. Connect the heater hoses to the heater core and fill the cooling system.

21. Connect the negative battery cable. Leak test, evacuate and charge the refrigerant system according to the proper procedure. Observe all safety precautions.

22. Check the system for proper operation.

Water Pump
REMOVAL & INSTALLATION

2.3L Engine

1. Disconnect the negative battery cable and drain the cooling system.

2. Remove the 4 bolts retaining the pulley to the water pump shaft. Remove the fan and shroud.

3. Remove the air conditioning and power steering belts. Remove the water pump pulley.

4. Remove the heater hose to the water pump and the lower radiator hose.

5. Remove the timing belt outer cover bolt, release the interlocking tabs and remove the cover.

6. Remove the water pump retaining bolts and remove the water pump.

7. Installation is the reverse of the removal procedure. Clean all gasket mating surfaces prior to installation. Apply pipe sealant to the water pump bolts and tighten to 14–21 ft. lbs. (20–30 Nm). Tighten the pulley retaining bolts to 15–22 ft. lbs. (20–30 Nm).

8. Fill and bleed the cooling system. Operate the engine until normal operating temperatures have been reached and check for leaks.

3.8L Engine

1. Disconnect the negative battery cable and drain the cooling system.

2. On 1988 vehicles, remove the air cleaner assembly and air intake duct.

3. On all except supercharged engine, remove the fan/clutch assembly and shroud.

4. Rotate the main accessory drive belt tensioner. Remove the main drive belt and water pump pulley.

5. On 1988 vehicles, remove the power steering pump and mounting bracket attaching bolts. Loosen the air conditioner compressor drive belt idler and remove the air conditioner belt. Leaving the hoses connected, place the pump aside in a position to prevent fluid from leaking out.

6. On 1989–92 vehicles, remove the power steering pump pulley and remove the water pump to power steering pump brace.

NOTE: On supercharged engines, it may be necessary to remove the intercooler to gain access to the power steering pump pulley.

7. On 1988 vehicles with air conditioning, remove the compressor front support bracket but leave the compressor in place.

8. On all except supercharged engine, disconnect the coolant bypass hose(s) and the heater hose at the water pump. On supercharged engine, disconnect the oil cooler tube and bypass hose and remove the upper crankshaft sensor cover.

9. Disconnect the lower radiator hose. Remove the water pump retaining bolts and the pump. If a prybar is used to assist removal, be careful not to damage the mating surfaces.

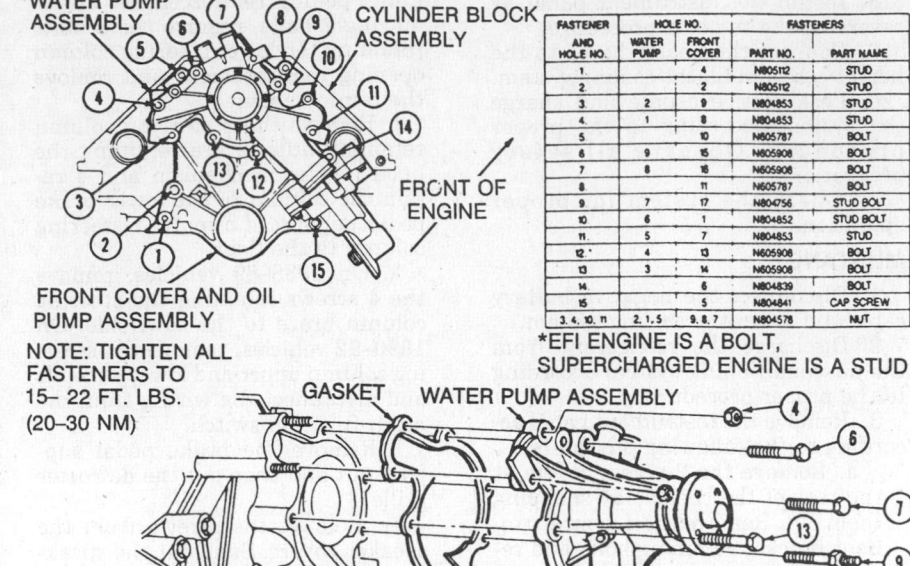

FASTENER AND HOLE NO.	HOLE NO.		FASTENERS	
	WATER PUMP	FRONT COVER	PART NO.	PART NAME
1		4	N805112	STUD
2		2	N805112	STUD
3	2	9	N804853	STUD
4	1	8	N804853	STUD
5		10	N605787	BOLT
6	9	15	N605908	BOLT
7	8	16	N605908	BOLT
8		11	N605787	BOLT
9	7	17	N804756	STUD BOLT
10	6	1	N804852	STUD BOLT
11	5	7	N804853	STUD
12*	4	13	N605908	BOLT
13	3	14	N605908	BOLT
14		6	N804839	BOLT
15		5	N804841	CAP SCREW
3, 4, 10, 11	2, 1, 5	9, 8, 7	N804578	NUT

*EFI ENGINE IS A BOLT, SUPERCHARGED ENGINE IS A STUD

NOTE: TIGHTEN ALL FASTENERS TO 15– 22 FT. LBS. (20–30 NM)

Water pump fastener and hole location—3.8L engine

10. Installation is the reverse of the removal procedure. Clean all gasket mating surfaces prior to installation. Tighten the water pump retaining bolts to 15–22 ft. lbs. (20–30 Nm). Fill and bleed the cooling system. Operate the engine until normal operating temperatures have been reached and check for leaks.

NOTE: The threads of the No. 1 water pump retaining bolt must be coated with pipe sealant before installing.

4.6L Engine

1. Disconnect the negative battery cable.

2. Drain the cooling system, remove the cooling fan and the shroud.

3. Release the belt tensioner and remove the accessory drive belt.

4. Remove the 4 bolts retaining the water pump pulley to the water pump and remove the pulley.

5. Remove the 4 bolts retaining the water pump to the engine assembly and remove the water pump.

6. Installation is the reverse of the removal procedure. Be sure to clean the sealing syrfaces of the water pump and block and use a new O-ring. Lubricate the O-ring with clean anti-freeze prior to installation.

7. Tighten the water pump-to-engine bolts and the pulley-to-water pump bolts to 15–22 ft. lbs. (20–30 Nm). Fill and bleed the cooling system. Operate the engine until normal oper-

ating temperatures have been reached and check for leaks.

5.0L and 5.8L Engines

1. Disconnect the negative battery cable.

2. Drain the cooling system. Remove the air inlet tube, if equipped.

3. On 1991–92 Thunderbird and Cougar, disconnect the upper radiator hose at the engine.

4. On all except 1991–92 Thunderbird and Cougar, remove the fan shroud attaching bolts and position the shroud over the fan. Remove the fan and clutch assembly from the water pump shaft and remove the shroud.

5. On 1991–92 Thunderbird and Cougar, remove the fan and clutch assembly from the water pump shaft using fan clutch holding tool T84T-6312–C or equivalent, and fan clutch nut wrench T84T–6312–D or equivalent, and position the fan and clutch assembly in the fan shroud. The nut is turned counterclockwise. Remove the fan shroud and fan/clutch as an assembly.

6. On all except 1991–92 Thunderbird and Cougar, remove the air conditioner drive belt and idler pulley bracket, if equipped. Remove the alternator drive belt. Remove the power steering drive belt and power steering pump, if equipped. Remove all accessory brackets that attach to the water pump.

7. On 1991–92 Thunderbird and

Cougar, remove the accessory drive belt by rotating the tensioner away from the belt by using pulley retaining bolts only.

8. Remove the water pump pulley. Disconnect the lower radiator hose, heater hose and water pump bypass hose at the water pump.

9. Remove the water pump attaching bolts and remove the water pump. Discard the gasket.

10. Installation is the reverse of the removal procedure. Clean all gasket mating surfaces prior to installation. Tighten the water pump attaching bolts to 12–18 ft. lbs. (16–24 Nm).

11. Fill and bleed the cooling system. Operate the engine until normal operating temperatures have been reached and check for leaks.

Thermostat

REMOVAL & INSTALLATION

2.3L Engine

1. Drain the cooling system to a level below the thermostat.

2. Remove the upper radiator hose and disconnect the heater hose at the thermostat housing located on the left front lower side of the engine.

3. Remove the thermostat housing retaining bolts and remove the housing. Remove the thermostat by rotating counterclockwise in the housing until the thermostat becomes free to remove. Do not pry out the thermostat.

4. Remove and discard the gasket.
To install:

5. Clean all gasket mating surfaces and position a new gasket on the cylinder head opening. The gasket must be positioned on the cylinder head, before the thermostat is installed.

6. Install the thermostat into the

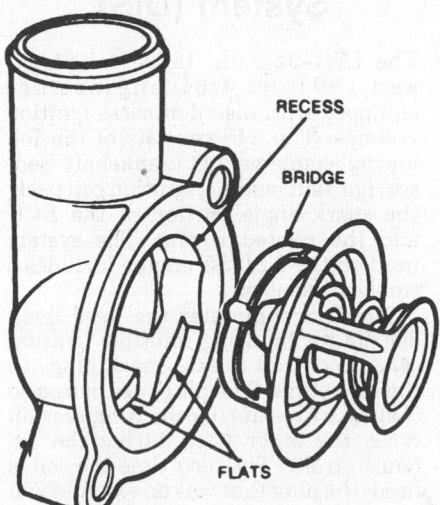

RECESS

BRIDGE

FLATS

Thermostat Installation—typical

housing with the bridge section in the housing. Turn the thermostat clockwise to lock it into position on the flats cast into the housing.

NOTE: It is important that the rubber thermostat gasket be pressed and the correct thermostat installation alignment be made to provide coolant flow to the heater. Insert and rotate the thermostat to the left or right until it stops in the thermostat housing. Visually check for full width of heater outlet tube opening to be visible within the thermostat port in assembly. This port alignment at assembly is required to provide maximum coolant flow to the heater.

7. Position the thermostat housing against the gasket on the cylinder head. Install and tighten the retaining bolts to 14–21 ft. lbs. (19–29 Nm).

8. Connect the upper radiator hose and the heater hose to the thermostat housing. Fill the cooling system. Start the engine and bring to normal operating temperature. Check for leaks.

3.8L and 4.6L Engines

1. Drain the cooling system to a level below the thermostat.

2. Disconnect the upper radiator hose at the thermostat housing.

3. Remove the 2 thermostat housing retaining bolts and remove the thermostat housing.

4. On 3.8L engine, remove the thermostat and gasket. On 4.6L engine, remove the thermostat and O-ring seal.

5. Installation is the reverse of the removal procedure. Make sure all mating surfaces are clean prior to installation. Use a new gasket on 3.8L engine or a new O-ring on 4.6L engine. On 3.8L engine, install the thermostat into the housing and turn clockwise to lock into position on the flats cast into the housing.

6. Tighten the thermostat housing retaining bolts to 15–22 ft. lbs. (20–30 Nm). Fill the cooling system. Start the engine and bring to normal operating temperature. Check for leaks.

5.0L and 5.8L Engines

1. Drain the cooling system to a level below the thermostat.

2. Disconnect the upper radiator hose and the bypass hose at the thermostat housing.

3. To gain access to the thermostat housing, either mark the location of the distributor, loosen the hold-down clamp and rotate the distributor, or remove the distributor cap and rotor.

4. Remove the thermostat housing retaining bolts and the housing and gasket. Remove the thermostat.

To install:

5. Clean the gasket mating surfaces. Position a new gasket on the intake manifold.

6. Install the thermostat in the housing, rotating slightly to lock the thermostat in place on the flats cast into the housing. Install the housing on the manifold and tighten the bolts to 12–18 ft. lbs. (16–24 Nm).

NOTE: If the thermostat has a bleeder valve, the thermostat should be positioned with the bleeder valve at the 12 o'clock position as viewed from the front of the engine.

7. Install the distributor cap and rotor, or reposition the distributor for correct ignition timing, as necessary. Tighten the hold-down bolt to 18–26 ft. lbs. (24–35 Nm).

8. Connect the bypass hose and the upper radiator hose to the thermostat housing. Fill the cooling system.

9. Start the engine and bring to normal operating temperature. Check for leaks.

Cooling System Bleeding

When the entire cooling system is drained, the following procedure should be used to ensure a complete fill.

1. Install the block drain plug, if removed and close the draincock. On 1991–92 3.8L engine, remove the vent plug on the intake manifold behind the thermostat housing. With the engine off, add anti-freeze to the radiator to a level of 50 percent of the total cooling system capacity. Then add water until it reaches the radiator filler neck seat.

NOTE: On Mustang equipped with the 2.3L engine, disconnect the heater hose at the connection on the thermostat housing. Fill the radiator until coolant is visible at the connection in the thermostat housing or the coolant level in the radiator reaches the radiator filler neck seat. Install the heater hose and tighten the hose clamps.

2. Install the radiator cap to the first notch to keep spillage to a minimum. On 1991–92 3.8L engine, install the vent plug.

3. Start the engine and let it idle until the upper radiator hose is warm. This indicates that the thermostat is open and coolant is flowing through the entire system.

4. Carefully remove the radiator cap and top off the radiator with water. Install the cap on the radiator securely.

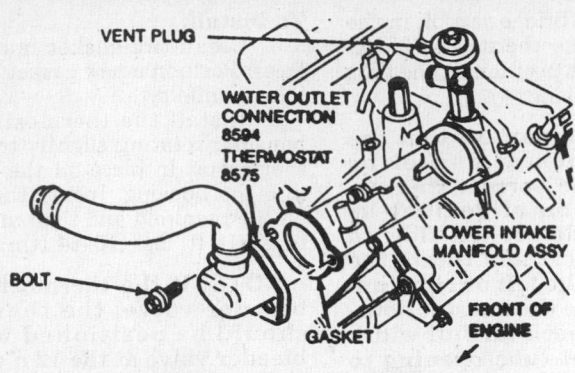

Vent plug location—1991–92 3.8L engine

5. Fill the coolant recovery reservoir to the FULL COLD mark with anti-freeze, then add water to the FULL HOT mark. This will ensure that a proper mixture is in the reservoir.

ENGINE ELECTRICAL

NOTE: Disconnecting the negative battery cable on some vehicles may interfere with the functions of the on board computer systems and may require the computer to undergo a relearning process, once the negative battery cable is reconnected.

Distributor

REMOVAL

1. Disconnect the negative battery cable.
2. Mark the position of the No. 1 cylinder wire tower on the distributor base. Remove the distributor cap and position the cap and ignition wires to the side.
3. Disconnect the wiring harness plug from the distributor connector. Disconnect and plug the vacuum hoses from the vacuum diaphragm assembly, if equipped.
4. Rotate the engine, in normal direction of rotation, until No. 1 piston is on Top Dead Center (TDC) of the compression stroke. The TDC mark on the crankshaft pulley and the pointer on the timing cover should align. The rotor tip should be pointing at the No. 1 spark plug wire position on the distributor cap.
5. Scribe a mark on the distributor body and the engine to indicate the position of the rotor tip and the position of the distributor in the engine.
6. Remove the hold-down bolt and

clamp located at the base of the distributor. Remove the distributor from the engine. Pay attention to the direction the rotor tip points if it moves from the No. 1 position when the drive gear disengages. For reinstallation purposes, the rotor should be at this point to insure proper gear mesh and timing.
7. Avoid turning the engine, if possible, while the distributor is removed. If the engine is turned from TDC position, TDC timing marks will have to be reset before the distributor is installed.

INSTALLATION

NOTE: Before installing, visually inspect the distributor. The drive gear should be free of nicks, cracks and excessive wear. The distributor drive shaft should move freely, without binding. If equipped with an O-ring, it should fit tightly and be free of cuts.

Timing Not Disturbed

1. Position the distributor in the engine with the rotor aligned to the marks made on the distributor or at the position the rotor pointed when the distributor was removed. Engage the oil pump intermediate shaft and insert the distributor until fully seated on the engine. If the distributor does not fully seat, turn the engine slightly to fully engage the intermediate shaft.
2. After the distributor has been fully seated into the block, recheck the timing mark and rotor alignment. Install the hold-down clamp and bolt. Snug the mounting bolt so the distributor can be turned for ignition timing purposes.
3. Install the distributor cap and connect the distributor to the wiring harness.
4. Connect the negative battery cable. Check and, if necessary, set the ignition timing. Tighten the distributor hold-down clamp bolt to 14–21 ft. lbs. (19–28 Nm) on the 2.3L and 3.8L engines or 17–25 ft. lbs. (23–34 Nm) on

the 5.0L and 5.8L engines. Recheck the ignition timing after tightening the bolt.
5. If equipped, connect the vacuum diaphragm hoses.

Timing Disturbed

If the engine was cranked with the distributor removed, the following procedure will enable the proper setting of the timing.

1. Disconnect the No. 1 spark plug wire and remove the No. 1 spark plug.
2. Place a finger over the spark plug hole and crank the engine slowly until compression is felt.
3. Align the TDC mark on the crankshaft pulley with the pointer on the timing cover. This places the No. 1 cylinder at TDC on the compression stroke.
4. Turn the distributor shaft until the rotor points to the No. 1 spark plug tower on the cap.
5. Install the distributor into the engine, aligning the marks made on the block and the distributor housing. Install the distributor hold-down clamp and bolt. Snug the bolt so the distributor housing can be moved for timing purposes.
6. Install the No. 1 spark plug and connect the spark plug wire. Install the distributor cap and connect the distributor to the wiring harness.
7. Connect the negative battery cable and set the ignition timing. Tighten the distributor hold-down clamp bolt to 14–21 ft. lbs. (19–28 Nm) on the 2.3L and 3.8L engines or 17–25 ft. lbs. (23–34 Nm) on the 5.0L and 5.8L engines. Recheck the ignition timing after tightening the bolt.
8. If equipped, connect the vacuum diaphragm hoses.

Distributorless Ignition System (DIS)

The 1991–92 2.3L, 1989–91 3.8L SC and 1991–92 4.6L engines are equipped with distributorless ignition systems. The DIS consists of the following components: crankshaft sensor, ignition module, ignition coil pack, the spark angle portion of the ECU and the related wiring. The system used on the 3.8L SC engine includes a camshaft sensor.

The DIS eliminates the need for a distributor by using multiple ignition coils. Each coil fires 2 spark plugs at the same time. The plugs are paired so that as 1 fires during the compression cycle, the other fires during the exhaust stroke. The next time the coil is fired, the plug that was on exhaust will be on compression and the 1 that was on compression will be on exhaust.

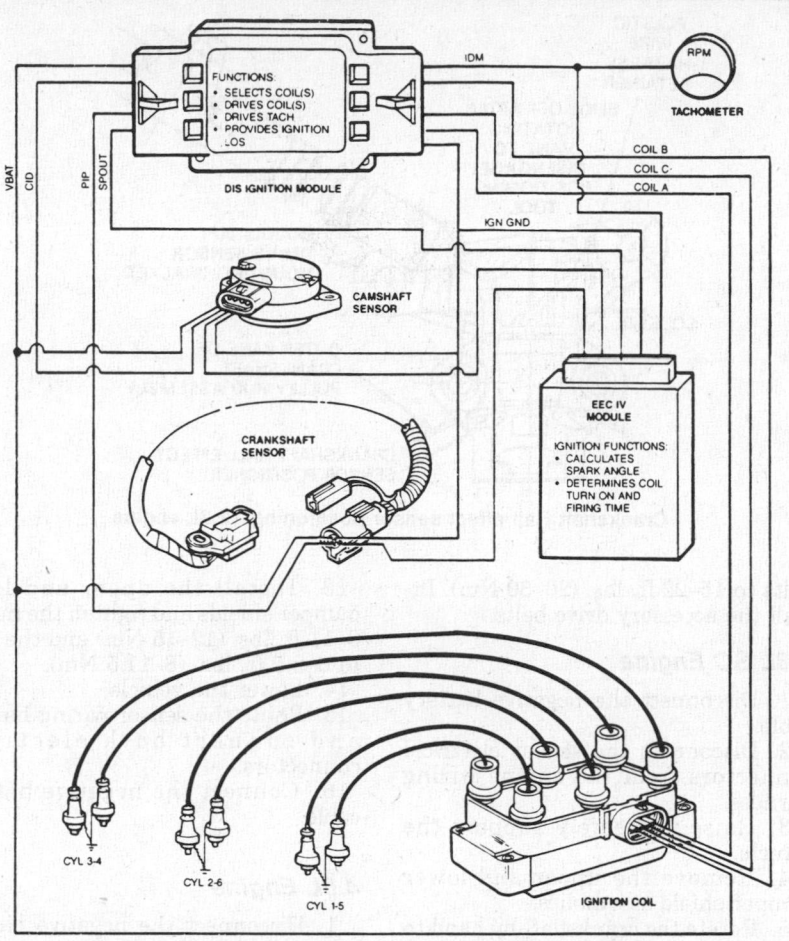

Distributorless ignition system—3.8L SC engine

provides engine position and rpm information to the ignition module.

The crankshaft sensor used on the 2.3L and 3.8L SC engines ia a Hall effect magnetic switch, activated by vanes on the crankshaft damper and pulley assembly. The signal generated by this sensor is called the Profile Ignition Pick-up (PIP). The PIP signal provides base timing and rpm information to the ECU. In addition, the crankshaft sensor on the 2.3L engine provides a Cylinder Identification (CID) signal. The CID signal is used to synchronize the ignition coils.

The camshaft sensor used on the 3.8L SC engine is a Hall effect magnetic switch, activated by a single vane which is driven by the camshaft. This sensor provides CID information for the ignition coil and for fuel system synchronization.

The ignition module is a microprocessor that receives input from the crankshaft and camshaft sensors in regards to engine position, base timing and engine speed and input from the ECU pertaining to spark advance. The ignition module uses this information to direct which coil to fire and to calculate the turn and turn off times of the coils required to achieve the correct dwell and spark advance.

REMOVAL & INSTALLATION

Crankshaft Sensor

2.3L ENGINE

1. Disconnect the negative battery cable.
2. Disconnect the crankshaft timing sensor assembly electrical connectors from the engine harness.
3. Remove the large electrical connector from the crankshaft timing sensor assembly by prying out the red

The spark in the exhaust cylinder is wasted but little of the coil energy is lost. The ignition coils are mounted together in coil packs. There are 2 coil packs used on the 2.3L and 4.6L engines, each containing 2 ignition coils. The 3.8L SC engine uses 1 coil pack containing 3 separate ignition coils.

The crankshaft sensor used on the 4.6L engine is a variable reluctance-type sensor triggered by a 36-minus-1 tooth trigger wheel configuration pressed onto the rear of the crankshaft damper. The signal generated by this sensor is called a Variable Reluctance Sensor (VRS) signal. The VRS signal

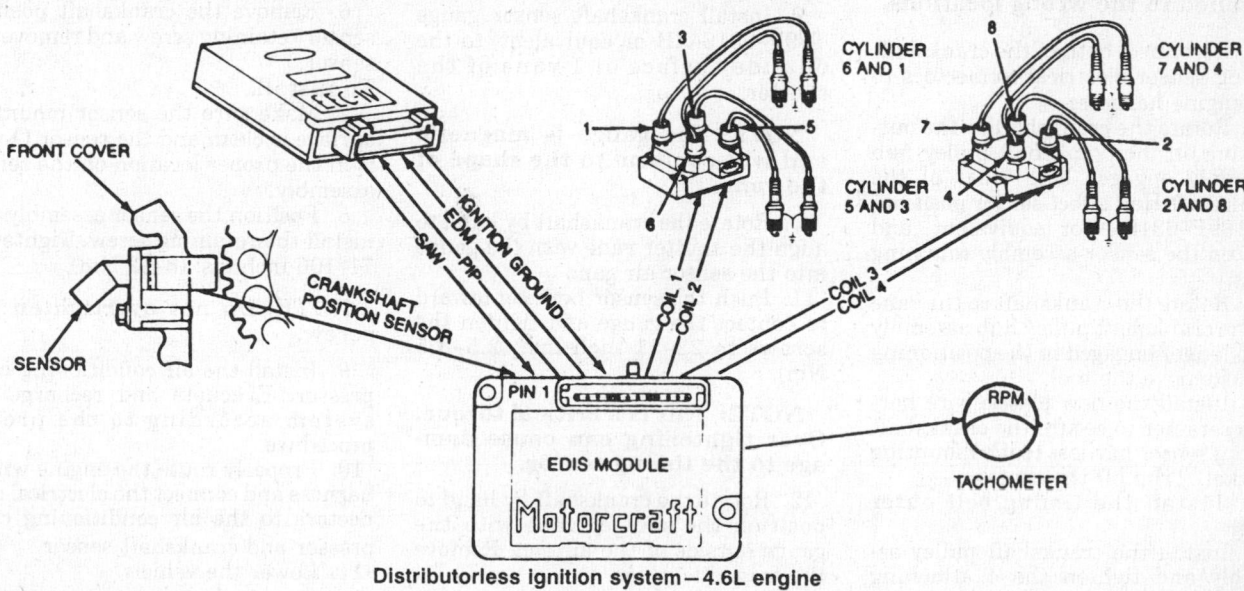

Distributorless ignition system—4.6L engine

retaining clip and removing the 4 wires.

4. Remove the crankshaft pulley assembly by removing the accessory drive belts and then the 4 bolts that retain it to the crankshaft pulley hub assembly.

5. Remove the timing belt outer cover.

6. Rotate the crankshaft so the keyway is at the 10 o'clock position. This will place the vane window of both inner and outer vane cups over the crankshaft timing sensor assembly.

NOTE: The vane cups are attached to the crankshaft pulley hub assembly.

7. Remove the 2 crankshaft timing sensor assembly retaining bolts and the plastic wire harness retainer which secures the crankshaft timing sensor harness to it's mounting bracket. Then remove the crankshaft timing sensor assembly, sliding the electrical wires out from behind the inner timing belt cover.

To install:

8. Remove the large electrical connector from the new crankshaft timing sensor assembly.

9. Position the crankshaft timing sensor assembly. First slide the electrical wires behind the inner timing belt cover. Then, hold the sensor assembly loosely in place with the retaining bolts, but do not tighten at this time.

10. Install the large electrical connector onto the crankshaft timing sensor assembly.

NOTE: Make sure the 4 wires to the large electrical connector are installed in the proper locations as indicated. The sensor will not function properly if the wires are installed in the wrong locations.

11. Reconnect both of the crankshaft timing sensor electrical connectors to the engine harness.

12. Rotate the crankshaft so the outer vane on the crankshaft pulley hub assembly engages both sides of the crankshaft Hall effect sensor positioner T89P-6316-A or equivalent, and tighten the sensor assembly retaining bolts.

13. Rotate the crankshaft so the vane on the crankshaft pulley hub assembly is no longer engaged in the positioning tool. Remove the tool.

14. Install the new plastic wire harness retainer to secure the crankshaft timing sensor harness to it's mounting bracket. Trim off the excess.

15. Install the timing belt outer cover.

16. Install the crankshaft pulley assembly and tighten the 4 attaching

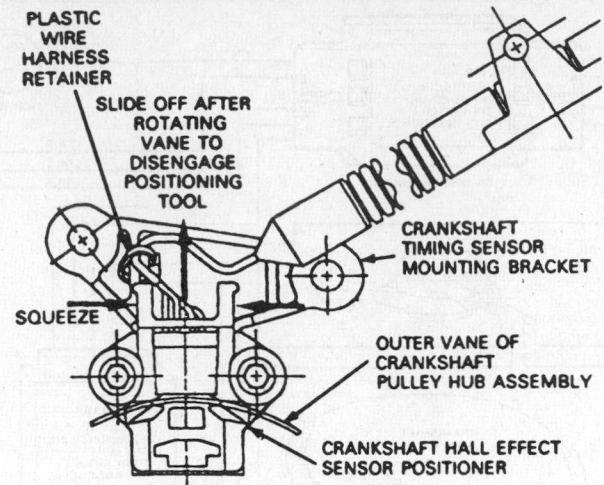

Crankshaft Hall effect sensor positioning—2.3L engine

bolts to 15–22 ft. lbs. (20–30 Nm). Install the accessory drive belts.

3.8L SC Engine

1. Disconnect the negative battery cable.

2. Disconnect the sensor electrical connectors from the engine wiring harness.

3. Raise and safely support the vehicle.

4. Remove the upper and lower damper shield assemblies.

5. Rotate the crankshaft by hand to position the metal vane of the shutter, attached to the rear of the damper, outside of the sensor air gap.

6. Remove the crankshaft sensor retaining screws and remove the sensor.

To install:

7. Position the crankshaft sensor assembly on the bracket.

8. Install 2 sensor retaining screws but do not tighten.

9. Install crankshaft sensor gauge T89P-6316-AH or equivalent, to the outside surface of 1 vane of the shutter.

NOTE: The gauge is magnetic and will conform to the shape of the vane.

10. Rotate the crankshaft by hand to align the shutter vane with the gauge into the sensor air gap.

11. Push the sensor housing inward to contact the gauge and tighten the screws to 22–31 inch lbs. (2.5–3.5 Nm).

NOTE: This is a critical torque. Over tightening can cause damage to the timing sensor.

12. Rotate the crankshaft by hand to position the shutter vane with the gauge outside of the air gap. Remove the magnetic gauge.

13. Install the upper and lower damper shields and tighten the nuts to 9–11 ft. lbs. (12–15 Nm) and the bolts to 6–8.5 ft. lbs. (8–11.5 Nm).

14. Lower the vehicle.

15. Route the sensor wiring harness and connect both electrical connectors.

16. Connect the negative battery cable.

4.6L Engine

1. Disconnect the negative battery cable.

2. Remove the serpentine belt.

3. Raise and safely support the vehicle.

4. Disconnect the crankshaft sensor and air conditioning compressor electrical connectors from the engine wiring harness.

5. Properly discharge the air conditioning system and remove the air conditioning compressor.

6. Remove the crankshaft position sensor retaining screw and remove the sensor.

To install:

7. Make sure the sensor mounting surface is clean and the sensor O-ring is in the proper location on the sensor assembly.

8. Position the sensor assembly and install the retaining screw. Tighten to 71–106 inch lbs. (8–12 Nm).

NOTE: Do not overtighten the screw.

9. Install the air conditioning compressor. Evacuate and recharge the system according to the proper procedure.

10. Properly route the engine wiring harness and connect the electrical connectors to the air conditioning compressor and crankshaft sensor.

11. Lower the vehicle.

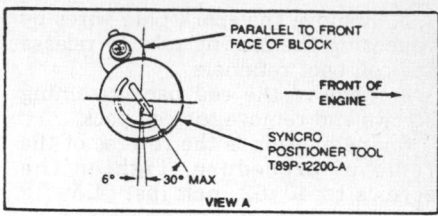

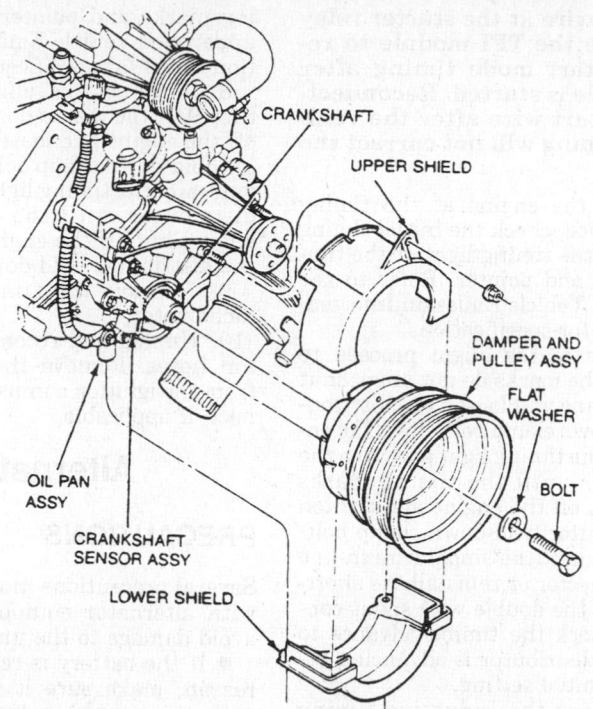

Crankshaft sensor removal and installation—3.8L SC engine

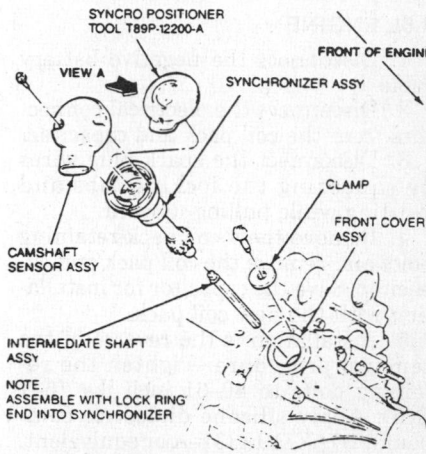

Synchronizer assembly installation— 3.8L SC engine

12. Install the serpentine belt and connect the negative battery cable.

Camshaft Sensor

3.8L SC ENGINE

1. Disconnect the negative battery cable.
2. Disconnect the camshaft sensor electrical connector.
3. Remove the camshaft sensor retaining screws and remove the sensor.
4. Installation is the reverse of the removal procedure. Tighten the retaining screws to 22–31 inch lbs. (2.5–3.5 Nm).

Synchronizer Assembly

3.8L SC ENGINE

The synchronizer assembly mounts in place of the distributor. It provides the mechanical link between the camshaft sensor and the camshaft.

NOTE: Before starting this procedure, set the No. 1 cylinder to 26 degrees after TDC of the compression stroke. Then note the position of the camshaft sensor electrical connector. The installation procedure requires that the connector be located in the same position.

1. Disconnect the negative battery cable.
2. Remove the camshaft sensor assembly.
3. Remove the syncrhonizer clamp, bolt and washer.
4. Remove the synchronizer from the front engine cover, by pulling it

out. The oil pump intermediate shaft will come out with the assembly.

NOTE: If the replacement synchronizer does not contain a plastic locator cover tool, a special service tool such as the synchro positioner T89P-12200-A or equivalent, must be used to install the synchronizer. Failure to use this special tool will cause the synchronizer timing to be out of adjustment, this could lead to engine damage.

5. To install the synchronizer, position the synchronizer so that gear engagement occurs when the arrow on the locator tool is pointed 30 degrees counterclockwise from the front face of the engine block. This will locate the camshaft sensor electrical connector to the position it was in before removal.
6. Install the synchronizer base clamp and tighten the mounting bolt to 15–22 ft. lbs. (20–30 Nm).
7. Remove the positioner tool and install the camshaft sensor. Connect the sensor electrical lead and connect the negative battery cable.

Ignition Module

1. Disconnect the negative battery cable.
2. Disconnect the electrical connectors at the module.
3. Remove the module retaining screws and remove the module.
To install:
4. On 2.3L and 3.8L SC engines, ap-

ply an even coating of silicone dielectric compound WA–10, D7AZ–19A331–A or equivalent to the mounting surface of the module.
5. Install the module and the retaining screws. Tighten the screws to 22–31 inch lbs. (2.5–3.5 Nm) on 2.3L and 3.8L engines and 24–35 inch lbs. (3–4 Nm) on 4.6L engines.
6. Connect the electrical connectors to the module and connect the negative battery cable.

Ignition Coil Pack

2.3L ENGINE

1. Disconnect the negative battery cable.
2. Squeeze the locking tabs of the coil wire retainer by hand and remove the spark plug wires with a twisting and pulling motion. Do not pull on the wire.
3. Disconnect the engine harness electrical connector from the ignition coil assembly.
4. Remove the ignition coil assembly by removing the 4 retaining screws.

NOTE: On vehicles equipped with power steering, it may be necessary to remove the intake coil and bracket as an assembly.

5. Installation is the reverse of the removal procedure.

3.8L SC ENGINE

1. Disconnect the negative battery cable.
2. Disconnect the electrical harness connector from the coil pack.

3. Remove the spark plug wires by squeezing the locking tabs to release the coil boot retainers.

4. Remove the coil pack retaining screws and remove the coil pack.

5. Installation is the reverse of the removal procedure. Tighten the screws to 40–62 inch lbs. (4.5–7.0 Nm).

4.6L ENGINE

1. Disconnect the negative battery cable.

2. Disconnect the electrical connectors from the coil pack and capacitor.

3. Disconnect the spark plug wires by squeezing the locking tabs and twisting while pulling upward.

4. Remove the 4 coil pack retaining bolts and remove the coil pack and capacitor. Save the capacitor for installation with the new coil pack.

5. Installation is the reverse of the removal procedure. Tighten the retaining bolts to 40–61 inch lbs. (5–7 Nm). Apply silicone dielectric compound D7AZ–19A331–A or equivalent, to all spark plug wire boots prior to installation.

Ignition Timing

ADJUSTMENT

NOTE: Always refer to the Vehicle Emission Information Label to verify the timing adjustment procedure.

Distributorless Ignition System

Base timing for distributorless engines is set from the factory at 10 degrees BTDC and is not adjustable.

Distributor Ignition System

EXCEPT 5.8L ENGINE

1. Locate the timing marks and pointer on the crankshaft pulley and the timing cover. Clean the marks so they will be visible with a timing light. Apply chalk or bright-colored paint, if necessary.

2. Place the transaxle in **P** or **N**. The air conditioning and heater controls should be in the **OFF** position.

3. Connect a suitable inductive timing light according to the manufacturer's instructions.

4. Disconnect the single wire in-line spout connector or remove the shorting bar from the double wire spout connector.

5. Start the engine and allow it to warm up to operating temperature.

NOTE: To set timing correctly, a remote starter should not be used. Use the ignition key only to start the vehicle. Disconnecting

the start wire at the starter relay will cause the TFI module to revert to start mode timing after the vehicle is started. Reconnecting the start wire after the vehicle is running will not correct the timing.

6. With the engine at the timing rpm specified, check the initial timing by aiming the timing light at the timing marks and pointer. Refer to the underhood Vehicle Emission Information Label for specifications.

7. If the marks align, proceed to Step 8. If the marks do not align, shut off the engine and loosen the distributor hold-down clamp bolt. Start the engine, aim the timing light and turn the distributor until the timing marks align. Shut off the engine and tighten the distributor hold-down clamp bolt.

8. Reconnect the single wire in-line spout connector or reinstall the shorting bar on the double wire spout connector. Check the timing advance to verify the distributor is advancing beyond the initial setting.

9. Remove the inductive timing light.

5.8L ENGINE

1. Locate the timing marks and pointer on the crankshaft pulley and the timing cover. Clean the marks so they will be visible with a timing light. Apply chalk or bright-colored paint, if necessary.

2. Place the transaxle in **P** or **N**. The air conditioning and heater controls should be in the **OFF** position.

3. Disconnect the vacuum hoses from the distributor vacuum advance connection at the distributor and plug the hoses.

4. Connect a suitable inductive timing light and a tachometer according to the manufacturer's instructions.

5. If equipped with a barometric pressure switch, disconnect it from the ignition module and place a jumper wire across the pins at the ignition module connector (yellow and black wires).

6. Start the engine and allow it to warm up to operating temperature.

NOTE: To set timing correctly, a remote starter should not be used. Use the ignition key only to start the vehicle. Disconnecting the start wire at the starter relay will cause the TFI module to revert to start mode timing after the vehicle is started. Reconnecting the start wire after the vehicle is running will not correct the timing.

7. With the engine at the timing rpm specified, check the initial timing by aiming the timing light at the tim-

ing marks and pointer. Refer to the underhood Vehicle Emission Information Label for specifications.

8. If the marks align, proceed to Step 9. If the marks do not align, shut off the engine and loosen the distributor hold-down clamp bolt. Start the engine, aim the timing light and turn the distributor until the timing marks align. Shut off the engine and tighten the distributor hold-down clamp bolt.

9. Remove the timing light and tachometer.

10. Unplug and reconnect the vacuum hoses. Remove the jumper wire from the ignition connector and reconnect, if applicable.

Alternator

PRECAUTIONS

Several precautions must be observed with alternator equipped vehicles to avoid damage to the unit.

• If the battery is removed for any reason, make sure it is reconnected with the correct polarity. Reversing the battery connections may result in damage to the one-way rectifiers.

• When utilizing a booster battery as a starting aid, always connect the positive to positive terminals and the negative terminal from the booster battery to a good engine ground on the vehicle being started.

• Never use a fast charger as a booster to start vehicles.

• Disconnect the battery cables when charging the battery with a fast charger.

• Never attempt to polarize the alternator.

• Do not use test lamps of more than 12V when checking diode continuity.

• Do not short across or ground any of the alternator terminals.

• The polarity of the battery, alternator and regulator must be matched and considered before making any electrical connections within the system.

• Never separate the alternator on an open circuit. Make sure all connections within the circuit are clean and tight.

• Disconnect the battery ground terminal when performing any service on electrical components.

• Disconnect the battery if arc welding is to be done on the vehicle.

BELT TENSION ADJUSTMENT

Except 1988–91 Crown Victoria and Grand Marquis

All vehicles except the 1988–91 Crown

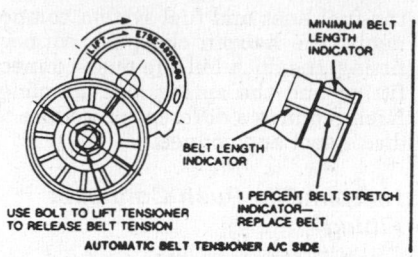 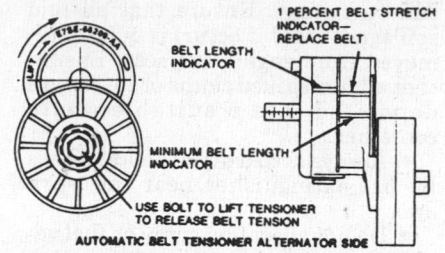

Belt wear indicator marks—2.3L engine

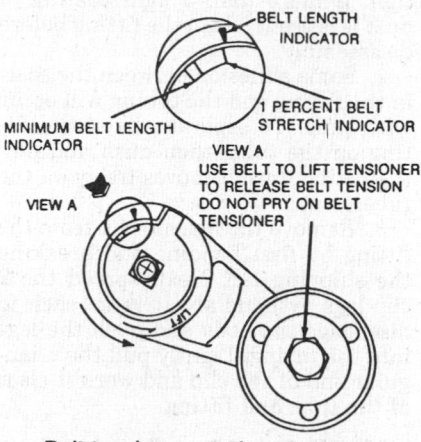

Belt tensioner—3.8L and 5.0L HO engines

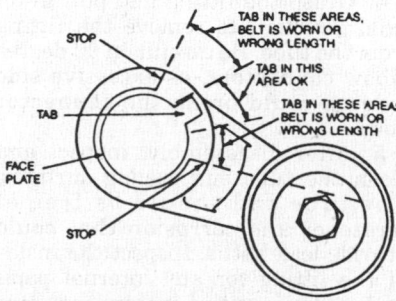

Belt replacement checking—3.8L and 5.0L HO engines

Victoria and Grand Marquis and 1988–90 Town Car are equipped with an automatic belt tensioner. No adjustment is necessary or possible. The belt tensioner is equipped with a belt wear indicator; when 1 percent belt stretch is indicated, the drive belt must be replaced. If the wear indicator is difficult to see on the 3.8L or 5.0L HO engines, locate the tab on the tensioner face plate. The tab should be approximately between the stops.

1988–91 Crown Victoria and Grand Marquis 1988–90 Town Car

1. Loosen the alternator pivot and adjustment bolts.
2. Install a suitable belt tension gauge midway between the pulleys on the longest accessible belt span. Install an open end wrench over the alterna-

tor adjustment boss, then apply tension to the belt, using the wrench.
3. Set the tension on a new belt to 170 lbs. or a used belt to 140 lbs. While maintaining the tension, tighten the alternator adjustment bolt to 29 ft. lbs. (39 Nm).
4. Remove the belt tension gauge, start the engine and let it idle for 5 minutes.
5. Shut off the engine and install the tension gauge. Apply tension with the open end wrench and slowly loosen the adjustment bolt to allow belt tension to increase to the used belt specification, 140 lbs. Tighten the adjustment bolt to 29 ft. lbs. (39 Nm).
6. Tighten the pivot bolt to 50 ft. lbs. (68 Nm).

REMOVAL & INSTALLATION

1. Disconnect the negative battery cable.
2. Tag and disconnect the wiring connectors from the rear of the alternator. To disconnect push-on type terminals, depress the lock tab and pull straight off.
3. Loosen the alternator pivot bolt and remove the adjusting bolt. Disengage the drive belt from the alternator pulley.
4. Remove the alternator pivot bolt and the alternator.
5. Installation is the reverse of the removal procedure.

Voltage Regulator

REMOVAL & INSTALLATION

1. Disconnect the negative battery cable. The regulator is located behind the battery on some models and it is necessary to remove the battery to remove the regulator.
2. Remove the regulator mounting screws, unlock the wire connectors and remove the regulator.

NOTE: Always disconnect the connector plug from the regulator before removing the mounting screws.

3. Installation is the reverse of the removal procedure.

Starter

REMOVAL & INSTALLATION

1. Disconnect the negative battery cable.
2. Raise the vehicle and support it safely.
3. Disconnect the starter cable and, if equipped, relay connector from the starter.
4. Remove the starter bolts and the starter.
5. Position the starter to the engine and tighten the mounting bolts to 15–20 ft. lbs. (20–27 Nm).
6. Reconnect the electrical leads. Connect the negative battery cable.

EMISSION CONTROLS

Please refer to "Emission Controls" in the Unit Repair section for system maintenance procedures. Due to the complex nature of modern electronic engine control systems, comprehensive diagnosis and testing procedures fall outside the confines of this repair manual. For complete information on diagnosis, testing and repair procedures concerning all modern engine and emission control systems, please refer to "Chilton's Guide to Fuel Injection and Electronic Engine Controls".

Emission Warning Lamps

RESETTING

These vehicles have an CHECK ENGINE lamp that will light when there is a fault in the engine control system. This light cannot be reset without diagnosing the fault in the system. When the system has been diagnosed and the problem corrected, the light will go out.

Service Lamp

1988 Thunderbird and Cougar

The SERVICE light will light after approximately 7500 miles of operation. To reset the interval light, simultaneously depress both the TRIP and TRIP RESET buttons on the instrument cluster. Hold until 3 short beeps verify the service reminder has been reset.

1989–92 THUNDERBIRD AND COUGAR

The optional Vehicle Maintenance Monitor (VMM) alerts the vehicle operator to when engine oil needs to be changed and when fuel, oil, coolant and washer fluids are low. To reset the VMM after an oil change, proceed as follows:

1. Turn the ignition key **OFF**, then turn it **ON**, but do not start the engine.

2. Within 16 seconds of turning the key to **ON**, stick a straightened paperclip into the reset switch hole and firmly push in the switch. The left side of the display will now flash.

NOTE: The reset switch is very small and is located to the left of the word "OK" on the VMM panel.

3. Keep pushing down on the reset switch with the paperclip until the left side of the display stops flashing. The VMM is now reset. Do not stop pushing in the switch until the display stops flashing, or the VMM will not be reset.

FUEL SYSTEM

Fuel System Service Precautions

Safety is the most important factor when performing not only fuel system maintenance but any type of maintenance. Failure to conduct maintenance and repairs in a safe manner may result in serious personal injury or death. Maintenance and testing of the vehicle's fuel system components can be accomplished safely and effectively by adhering to the following rules and guidelines.

● To avoid the possibility of fire and personal injury, always disconnect the negative battery cable unless the repair or test procedure requires that battery voltage be applied.

● Always relieve the fuel system pressure prior to disconnecting any fuel system component (injector, fuel rail, pressure regulator, etc.), fitting or fuel line connection. Exercise extreme caution whenever relieving fuel system pressure to avoid exposing skin, face and eyes to fuel spray. Please be advised that fuel under pressure may penetrate the skin or any part of the body that it contacts.

● Always place a shop towel or cloth around the fitting or connection prior to loosening to absorb any excess fuel

due to spillage. Ensure that all fuel spillage (should it occur) is quickly removed from engine surfaces. Ensure that all fuel soaked cloths or towels are deposited into a suitable waste container.

● Always keep a dry chemical (Class B) fire extinguisher near the work area.

● Do not allow fuel spray or fuel vapors to come into contact with a spark or open flame.

● Always use a backup wrench when loosening and tightening fuel line connection fittings. This will prevent unnecessary stress and torsion to fuel line piping. Always follow the proper torque specifications.

● Always replace worn fuel fitting O-rings with new. Do not substitute fuel hose or equivalent where fuel pipe is installed.

RELIEVING FUEL SYSTEM PRESSURE

Fuel supply lines on all fuel injected engines will remain pressurized for some period of time after the engine is shut OFF. This pressure must be relieved before servicing the fuel system. Pressure is relieved through the fuel pressure relief valve.

On the 1988 2.3L turbocharged engine this valve is located in the flexible fuel supply tube approximately 12 inches back from where it connects to the engine fuel rail. On all other engines, the valve is located on the fuel rail.

To relieve the fuel system pressure, first remove the fuel tank cap to relieve pressure in the tank, then remove the cap on the fuel pressure relief valve. Attach fuel pressure gauge T80L-9974-A or equivalent, and drain the system through the drain tube into a suitable container. Remove the fuel pressure gauge and replace the cap on the relief valve.

Fuel Line Couplings

REMOVAL & INSTALLATION

There are 3 methods in use to connect

the fuel lines and fuel system components, the hairpin clip push connect fitting, the duck bill clip push connect fitting and the spring lock coupling. Each requires a different procedure to disconnect and connect.

Hairpin Clip Push Connect Fitting

1. Inspect the visible internal portion of the fitting for dirt accumulation. If more than a light coating of dust is present, clean the fitting before disassembly.

2. Some adhesion between the seals in the fitting and the tubing will occur with time. To separate, twist the fitting on the tube, then push and pull the fitting until it moves freely on the tube.

3. Remove the hairpin clip from the fitting by first bending and breaking the shipping tab. Next, spread the 2 clip legs by hand about ⅛ in. each to disengage the body and push the legs into the fitting. Lightly pull the triangular end of the clip and work it clear of the tube and fitting.

NOTE: Do not use hand tools to complete this operation.

4. Grasp the fitting and pull in an axial direction to remove the fitting from the tube. Be careful on 90 degree elbow connectors, as excessive side loading could break the connector body.

5. After disassembly, inspect and clean the tube end sealing surfaces. The tube end should be free of scratches and corrosion that could provide leak paths. Inspect the inside of the fitting for any internal parts such as O-rings and spacers that may have been dislodged from the fitting. Replace any damaged connector.

To connect:

6. Install a new connector if damage was found. Insert a new clip into any 2 adjacent openings with the triangular portion pointing away from the fitting opening. Install the clip until the legs of the clip are locked on the outside of the body. Piloting with an index finger is necessary.

7. Before installing the fitting on

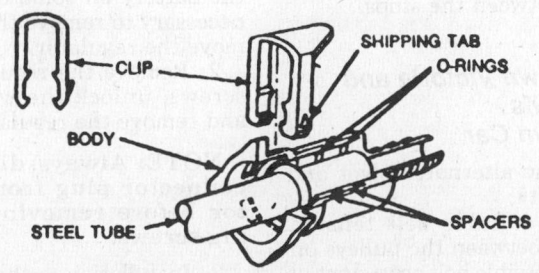

Hairpin clip push connect fitting

the tube, wipe the tube end with a clean cloth. Inspect the inside of the fitting to make sure it is free of dirt and/or obstructions.

8. Apply a light coating of engine oil to the tube end. Align the fitting and tube axially and push the fitting onto the tube end. When the fitting is engaged, a definite click will be heard. Pull on the fitting to make sure it is fully engaged.

Duck Bill Clip Push Connect Fitting

1. Inspect the visible internal portion of the fitting for dirt accumulation. If more than a light coating of dust is present, clean the fitting before disassembly.

2. Some adhesion between the seals in the fitting and the tubing will occur with time. To separate, twist the fitting on the tube, then push and pull the fitting until it moves freely on the tube.

3. Align the slot on push connect disassembly tool T90T-9550-B or T90T-9550-C or equivalent, with either tab on the clip, 90 degrees from the slots on the side of the fitting and insert the tool. This disengages the duck bill retainer from the tube.

4. Holding the tool and the tube with 1 hand, pull the fitting away from the tube.

NOTE: Use hands only. Only moderate effort is required if the tube has been properly disengaged.

5. After disassembly, inspect and clean the tube end sealing surfaces. The tube end should be free of scratches and corrosion that could provide leak paths. Inspect the inside of the fitting for any internal parts such as O-rings and spacers that may have been dislodged from the fitting. Replace any damaged connector.

6. Some fuel tubes have a secondary bead which aligns with the outer surface of the clip. These beads can make tool insertion difficult. If there is extreme difficulty, use the following disassembly method:

   a. Using pliers with a jaw width of 0.2 in. (5mm) or less, align the jaws with the openings in the side of the fitting case and compress the portion of the retaining clip that engages the fitting case. This disengages the retaining clip from the case. Often 1 side of the clip will disengage before the other. The clip must be disengaged from both openings.

   b. Pull the fitting off the tube by hand only. Only moderate effort is required if the retaining clip has been properly disengaged.

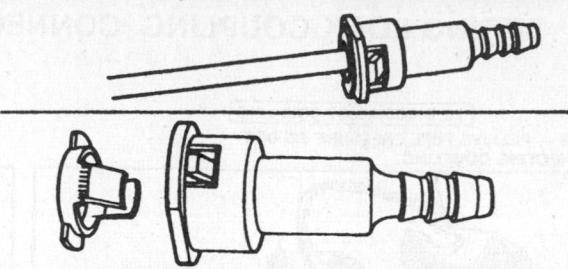

Duck bill clip push connect fitting

   c. After disassembly, inspect and clean the tube end sealing surfaces. The tube end should be free of scratches and corrosion that could provide leak paths. Inspect the inside of the fitting for any internal parts such as O-rings and spacers that may have been dislodged from the fitting. Replace any damaged connector.

   d. The retaining clip will remain on the tube. Disengage the clip from the tube bead and remove.

To connect:

7. Install a new connector if damage was found. Install the new replacement clip into the body by inserting 1 of the retaining clip serrated edges on the duck bill portion into 1 side of the window openings. Push on the other side until the clip snaps into place.

8. Before installing the fitting on the tube, wipe the tube end with a clean cloth. Inspect the inside of the fitting to make sure it is free of dirt and/or obstructions.

9. Apply a light coating of engine oil to the tube end. Align the fitting and tube axially and push the fitting onto the tube end. When the fitting is engaged, a definite click will be heard. Pull on the fitting to make sure it is fully engaged.

Spring Lock Coupling

The spring lock coupling is a fuel line coupling held together by a garter spring inside a circular cage. When the coupling is connected together, the flared end of the female fitting slips behind the garter spring inside the cage of the male fitting. The garter spring and cage then prevent the flared end of the female fitting from pulling out of the cage. As an additional locking feature, most vehicles have a horseshoe shaped retaining clip that improves the retaining reliability of the spring lock coupling.

Fuel Filter

REMOVAL & INSTALLATION

In-Line Filter

1. Disconnect the negative battery

cable and relieve the fuel system pressure.

2. Raise and safely support the vehicle.

3. Remove the push connect fittings at both ends of the filter. Install new retainer clips in each push connect fitting.

4. On all except Crown Victoria, Grand Marquis and Town Car, remove the fuel filter from the bracket by loosening the worm gear clamp. Note the direction of the flow arrow as installed in the bracket to ensure proper direction of fuel flow through the replacement filter.

5. On Crown Victoria, Grand Marquis and Town Car, remove the fuel filter and retainer from the metal bracket. Remove the filter from the retainer. Note that the direction of the flow arrow points to the open end of the retainer. Remove the rubber insulator rings.

To install:

6. On all except Crown Victoria, Grand Marquis and Town Car, install the fuel filter into the bracket, ensur-

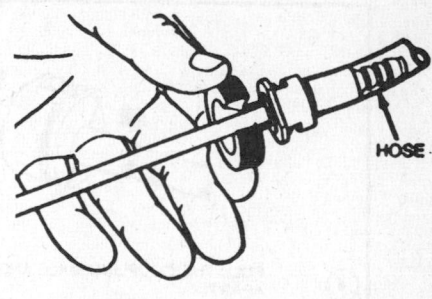

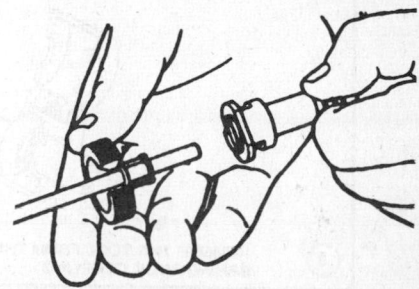

Duck bill clip push connect fitting disconnect tool

SPRING LOCK COUPLING CONNECT AND DISCONNECT PROCEDURE

TO DISCONNECT COUPLING

CAUTION — RELIEVE FUEL PRESSURE BEFORE DISCONNECTING COUPLING

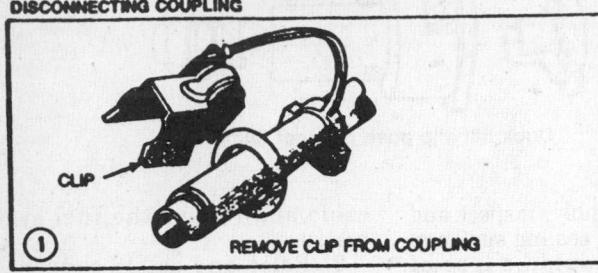

① CLIP
REMOVE CLIP FROM COUPLING

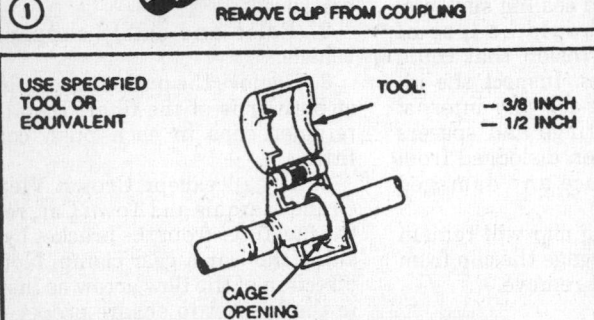

USE SPECIFIED TOOL OR EQUIVALENT

TOOL:
— 3/8 INCH
— 1/2 INCH

CAGE OPENING

② FIT TOOL TO COUPLING SO THAT TOOL CAN ENTER CAGE OPENING TO RELEASE THE GARTER SPRING.

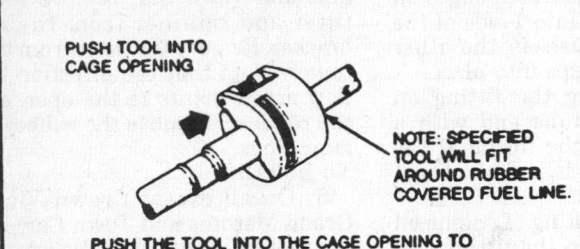

PUSH TOOL INTO CAGE OPENING

NOTE: SPECIFIED TOOL WILL FIT AROUND RUBBER COVERED FUEL LINE.

③ PUSH THE TOOL INTO THE CAGE OPENING TO RELEASE THE FEMALE FITTING FROM THE GARTER SPRING

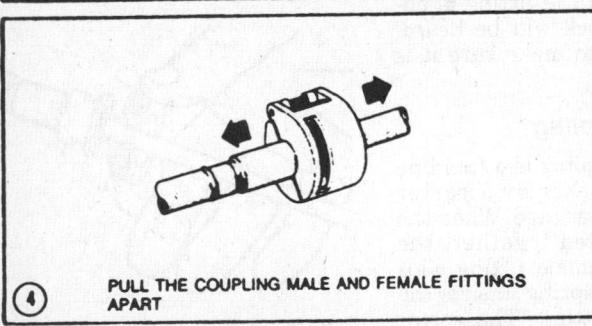

④ PULL THE COUPLING MALE AND FEMALE FITTINGS APART

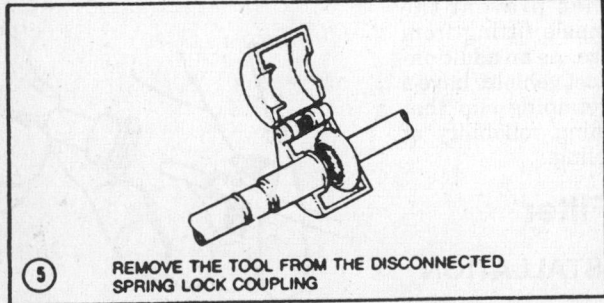

⑤ REMOVE THE TOOL FROM THE DISCONNECTED SPRING LOCK COUPLING

TO CONNECT COUPLING

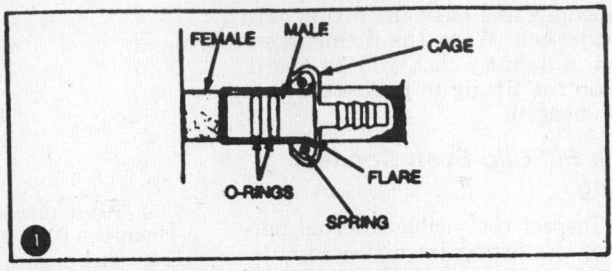

FEMALE   MALE   CAGE

O-RINGS   FLARE

SPRING

①

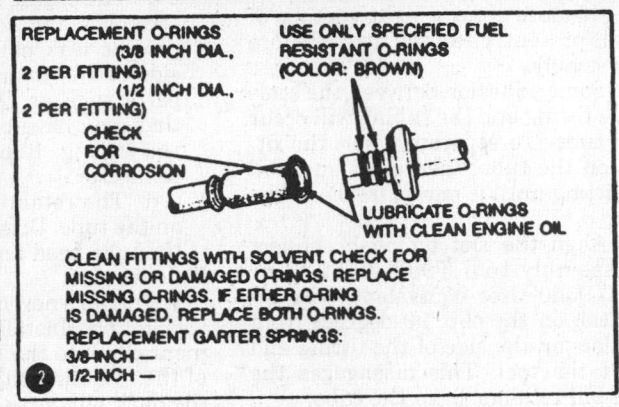

REPLACEMENT O-RINGS (3/8 INCH DIA., 2 PER FITTING) (1/2 INCH DIA., 2 PER FITTING)

USE ONLY SPECIFIED FUEL RESISTANT O-RINGS (COLOR: BROWN)

CHECK FOR CORROSION

LUBRICATE O-RINGS WITH CLEAN ENGINE OIL

CLEAN FITTINGS WITH SOLVENT. CHECK FOR MISSING OR DAMAGED O-RINGS. REPLACE MISSING O-RINGS. IF EITHER O-RING IS DAMAGED, REPLACE BOTH O-RINGS. REPLACEMENT GARTER SPRINGS:
3/8-INCH —
② 1/2-INCH —

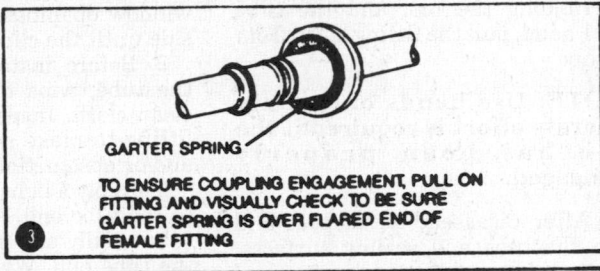

GARTER SPRING

TO ENSURE COUPLING ENGAGEMENT, PULL ON FITTING AND VISUALLY CHECK TO BE SURE GARTER SPRING IS OVER FLARED END OF FEMALE FITTING

③

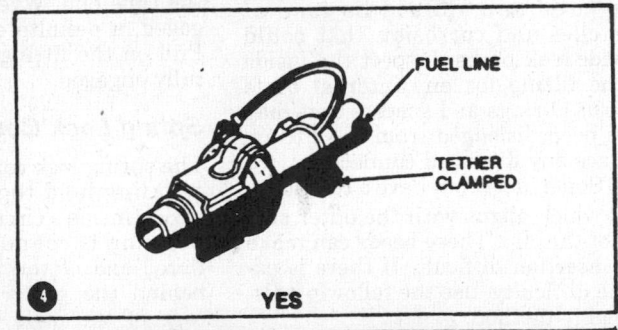

FUEL LINE

TETHER CLAMPED

④ YES

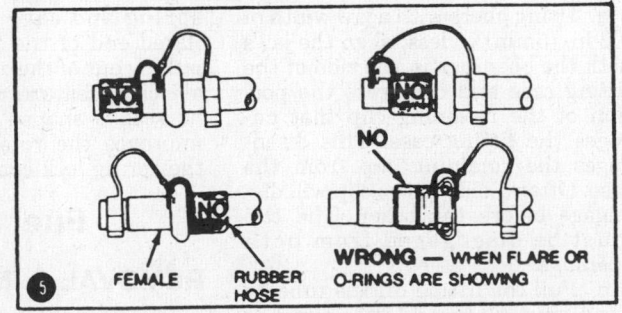

NO   NO

NO

FEMALE   RUBBER HOSE   NO

⑤ **WRONG** — WHEN FLARE OR O-RINGS ARE SHOWING

ing the proper direction of flow. Tighten the worm gear clamp to 15–25 inch lbs. (1.7–2.8 Nm).

7. On Crown Victoria, Grand Marquis and Town Car, install the rubber insulator rings, place the filter into the retainer with the flow arrow pointing out of the retainer open end, and install the retainer on the metal bracket. Tighten the retaining bolts to 27–44 inch lbs. (3–5 Nm).

8. Install the push connect fittings onto the filter ends. Start the engine and check for leaks.

9. Lower the vehicle.

Mechanical Fuel Pump

PRESSURE TESTING

1. Connect a suitable pressure gauge to the carburetor end of the fuel line.

2. Start the engine and read the pressure after 10 seconds. The engine should be able to run for over 30 seconds on the fuel in the carburetor bowl.

3. The fuel pump pressure should be 6–8 psi. If pump pressure is too low or too high, install a new fuel pump.

REMOVAL & INSTALLATION

NOTE: Before removing the pump, rotate the engine so that

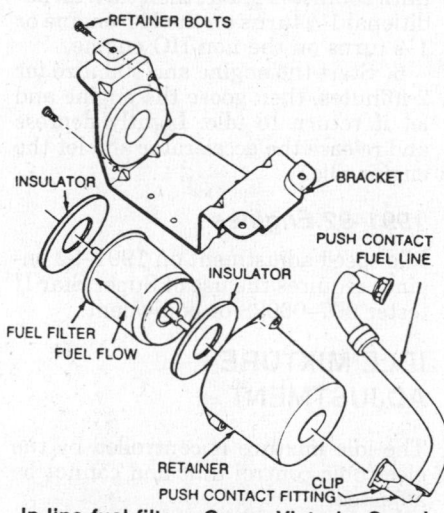

In-line fuel filter—Crown Victoria, Grand Marquis and Town Car

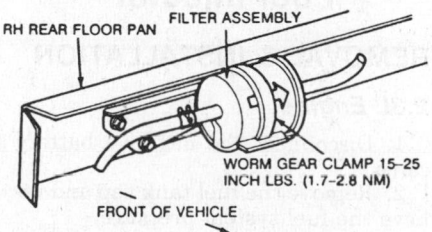

In-line fuel filter—Mustang

the low point of the cam lobe is against the pump arm. This can be determined by rotating the engine with the fuel pump mounting bolts loosened slightly; when tension (resistance) is removed from the arm, proceed with removal.

1. Disconnect the negative battery cable. Remove the inlet, outlet and vapor return, if equipped, lines from the pump.

2. Remove the fuel pump mounting bolts and remove the pump and gasket.

To install:

3. Clean all gasket material from the pump mounting surface on the engine and apply a coat of oil-resistant sealer to the new gasket.

4. Install the retaining bolts into the fuel pump and install the gaskets on the bolts. Position the pump on the engine. Turn the retaining bolts alternately and evenly and tighten to 19–27 ft. lbs. (26–37 Nm).

5. Reinstall the fuel lines, start the engine and check for leaks.

Electric Fuel Pump

PRESSURE TESTING

1. Relieve the fuel system pressure and connect a fuel pressure gauge to the valve on the fuel rail.

2. Ground the fuel pump lead of the self-test connector through a jumper wire at the **FP** lead.

3. Turn the ignition key to the **RUN** position to operate the fuel pump.

4. Observe the fuel pressure gauge. the indicated pressure should be 35–45 psi.

5. Remove the fuel pressure gauge and the jumper wire.

REMOVAL & INSTALLATION

1. Disconnect the negative battery cable and relieve the fuel system pressure.

2. Remove the fuel tank and place it on a bench.

3. Remove any dirt that has accumulated around the fuel pump retaining flange so it will not enter the tank during pump removal and installation.

4. Turn the fuel pump locking ring counterclockwise and remove the locking ring.

5. Remove the fuel pump and bracket assembly. Remove and discard the seal ring.

To install:

6. Clean the fuel pump mounting flange, fuel tank mounting surface and seal ring groove.

7. Apply a light coating of grease on a new seal ring to hold it in place during assembly and install in the seal ring groove.

8. Install the fuel pump and bracket assembly carefully to ensure the filter is not damaged. Make sure the locating keys are in the keyways and the seal ring remains in the groove.

9. Hold the pump assembly in place and install the locking ring fingertight. Make sure all the locking tabs are under the tank lock ring tabs.

10. Rotate the locking ring clockwise until the ring is against the stops.

11. Install the fuel tank in the vehicle. Add a minimum of 10 gallons of fuel to the tank and check for leaks.

12. Install a suitable fuel pressure gauge to the valve on the fuel rail.

13. Turn the ignition switch from **OFF** to **ON** for 3 seconds. Repeat this procedure 5–10 times until the pressure gauge shows at least 35 psi. Check foe fuel leaks.

14. Remove the pressure gauge, start the engine and check for leaks.

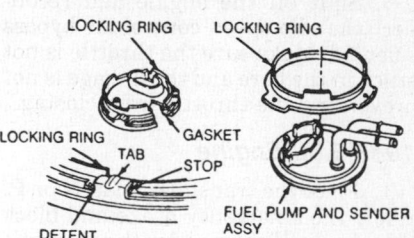

Typical in-tank fuel pump installation

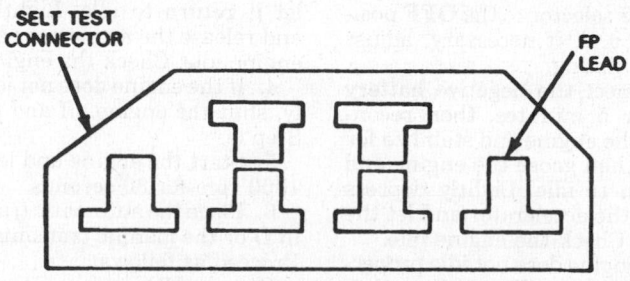

Self-test connector terminal location

Fuel Injection

IDLE SPEED ADJUSTMENT

1988–90 2.3L Engine

1. Place the transmission in **N** or **P**. Apply the emergency brake and block the wheels. If equipped with automatic brake release, disconnect the vacuum hose and plug it.

2. Bring the engine to normal operating temperature. Place the air conditioner/heater selector to the **OFF** position. Check and, if necessary, adjust the ignition timing.

3. Disconnect the negative battery terminal for 5 minutes, then reconnect. Start the engine and stabilize for 2 minutes, then goose the engine and let it return to idle. Lightly depress and release the accelerator and let the engine idle. Check the engine idle.

4. If the engine does not idle properly, shut the engine off and proceed to Step 5.

5. Disconnect the idle speed control-air bypass solenoid. Start the turbocharged engine and let it run at 2000 rpm for 2 minutes. Start the non-turbocharged engine and let it run at 1500 rpm for 30 seconds.

6. Turn the throttle plate stop screw and set the idle speed to the following specification:
1988 non-turbocharged engine—525 ± 25 rpm
1988 turbocharged engine—750 ± 50 rpm
1989–90 with automatic transmission—650 ± 25 rpm
1989–90 with manual transmission—600 ± 25 rpm

7. Shut off the engine and reconnect the idle speed control-air bypass solenoid. Make sure the throttle is not stuck in the bore and the linkage is not preventing the throttle from closing.

1988 3.8L Engine

1. Place the transmission in **N** or **P**. Apply the emergency brake and block the wheels. If equipped with automatic brake release, disconnect the vacuum hose and plug it.

2. Bring the engine to normal operating temperature. Place the air conditioner/heater selector to the **OFF** position. Check and, if necessary, adjust the ignition timing.

3. Disconnect the negative battery terminal for 5 minutes, then reconnect. Start the engine and stabilize for 2 minutes, then goose the engine and let it return to idle. Lightly depress and release the accelerator and let the engine idle. Check the engine idle.

4. If the engine does not idle properly, shut the engine off and proceed to Step 5.

5. Start the engine and let it run at 2500 rpm for 30 seconds.

6. Place the automatic transmission in **D** or the manual transmission in **N**. Back out the throttle stop screw to 550 ± 20 rpm, then back out an additional ½ turn.

1989–90 3.8L Engine

1. Place the transmission in **N** or **P**. Apply the emergency brake and block the wheels. If equipped with automatic brake release, disconnect the vacuum hose and plug it.

2. Bring the engine to normal operating temperature. Place the air conditioner/heater selector to the **OFF** position. Check and, if necessary, adjust the ignition timing.

3. Disconnect the negative battery terminal for 5 minutes, then reconnect. Start the engine and stabilize for 2 minutes, then goose the engine and let it return to idle. Lightly depress and release the accelerator and let the engine idle. Check the engine idle.

4. If the engine does not idle properly, shut the engine off and proceed to Step 5.

5. Back out the throttle plate stop screw clear off the throttle lever pad. With a 0.010 in. feeler gauge between the throttle plate stop screw and the throttle lever pad, turn the screw in until contact is made, then turn an additional 1½ turns.

6. Start the engine and stabilize for 2 minutes, then goose the engine and let it return to idle. Lightly depress and release the accelerator and let the engine idle.

1988 5.0L Engine

1. Place the transmission in **N** or **P**. Apply the emergency brake and block the wheels. If equipped with automatic brake release, disconnect the vacuum hose and plug it.

2. Bring the engine to normal operating temperature. Place the air conditioner/heater selector to the **OFF** position. Check and, if necessary, adjust the ignition timing.

3. Disconnect the negative battery terminal for 5 minutes, then reconnect. Start the engine and stabilize for 2 minutes, then goose the engine and let it return to idle. Lightly depress and release the accelerator and let the engine idle. Check the engine idle.

4. If the engine does not idle properly, shut the engine off and proceed to Step 5.

5. Start the engine and let it run at 1800 rpm for 30 seconds.

6. Place the automatic transmission in **D** or the manual transmission in **N**. Proceed as follows:
On non-HO engine with automatic transmission, back out the throttle stop screw to 575 ± 20 rpm, then back out an additional ½ turn.
On HO engine with automatic transmission, back out the throttle stop screw to 625 ± 20 rpm, then back out an additional ½ turn.
On HO engine with manual transmission, back out the throttle stop screw to 700 ± 20 rpm, then back out an additional ½ turn.

1989–90 5.0L Engine

1. Place the transmission in **N** or **P**. Apply the emergency brake and block the wheels. If equipped with automatic brake release, disconnect the vacuum hose and plug it.

2. Bring the engine to normal operating temperature. Place the air conditioner/heater selector to the **OFF** position. Check and, if necessary, adjust the ignition timing.

3. Disconnect the negative battery terminal for 5 minutes, then reconnect. Start the engine and stabilize for 2 minutes, then goose the engine and let it return to idle. Lightly depress and release the accelerator and let the engine idle. Check the engine idle.

4. If the engine does not idle properly, shut the engine off and proceed to Step 5.

5. Back out the throttle plate stop screw clear off the throttle lever pad. With a 0.010 in. feeler gauge between the throttle plate stop screw and the throttle lever pad, turn the screw in until contact is made, then turn an additional 1½ turns on the HO engine or 1⅞ turns on the non-HO engine.

6. Start the engine and stabilize for 2 minutes, then goose the engine and let it return to idle. Lightly depress and release the accelerator and let the engine idle.

1991–92 Engines

Idle speed adjustment on 1991–92 engines requires the use of Super Star II tester 007–00028 or equivalent.

IDLE MIXTURE ADJUSTMENT

The idle mixture is controlled by the electronic control unit and cannot be adjusted.

Fuel Injector

REMOVAL & INSTALLATION

2.3L Engine

1. Disconnect the negative battery cable.

2. Remove the fuel tank cap and relieve the fuel system pressure.

3. Disconnect the air intake, electrical connectors, throttle linkage, vacu-

um lines and EGR tube from the upper intake manifold and throttle body. Tag the electrical connectors and vacuum lines prior to removal for installation reference.

4. Remove the upper intake manifold retaining bolts and remove the upper intake manifold and throttle body assembly.

5. Disconnect the electrical connectors from the injectors.

6. Disconnect the fuel lines from the fuel supply manifold.

7. Remove the fuel supply manifold retaining bolts, carefully disengage the manifold and fuel injectors from the engine and remove the manifold and injectors.

8. Remove the fuel injectors from the manifold.

To install:

9. Lubricate new O-rings with clean light grade oil and install 2 on each injector.

NOTE: Never use silicone grease as it will clog the injectors.

10. Install the fuel injector supply manifold and injectors into the intake manifold. Push the fuel rail down to make sure all the fuel injector O-rings are fully seated in the fuel rail cups and intake manifold.

11. Install the fuel manifold assembly retaining bolts and tighten to 15–22 ft. lbs. (20–30 Nm) while holding the assembly down.

12. Connect the fuel lines to the manifold assembly.

13. After the fuel rail assembly has been installed and before the fuel injector wire connectors have been connected, connect the negative battery cable and turn the key to the **ON** position. This will cause the fuel pump to run for 2–3 seconds and pressurize the system.

14. Check for fuel leaks, especially where the fuel injector is installed into the fuel rail.

15. Disconnect the negative battery cable.

16. Install the upper intake manifold in the reverse order of removal.

17. Connect the fuel injector wire connectors.

18. Connect the negative battery cable. Start the engine and let it idle.

19. Turn the engine **OFF** and check for fuel leaks.

3.8L Engine Except Supercharged

1. Disconnect the negative battery cable.

2. Remove the fuel tank cap and relieve the fuel system pressure.

3. Disconnect the electrical connectors at the air bypass valve, TP sensor and EGR position sensor.

4. Disconnect the throttle linkage at the throttle ball and the transmission linkage from the throttle body.

5. Remove the 2 bolts securing the bracket to the intake manifold and position the bracket with the cables out of the way.

6. Disconnect all the vacuum lines from the upper intake manifold and throttle assembly. Tag all lines prior to removal for ease of reinstallation.

7. Remove the upper intake manifold retaining bolts and remove the upper intake manifold and throttle body assembly.

8. Disconnect the fuel lines from the fuel rail assembly.

9. Remove the fuel pressure regulator.

10. Disconnect the electrical connectors from the fuel injectors.

11. Remove the fuel rail retaining bolts. Carefully disengage the fuel rail from the fuel injectors and remove the fuel rail.

NOTE: It may be easier to remove the injectors with the fuel rail as an assembly.

12. Remove the injector retaining clips, as required.

13. Grasping the injector body, pull while gently rocking the injector from side-to-side to remove the injector from the fuel rail or the intake manifold.

14. Inspect the pintle protection cap and washer for signs of deterioration. Replace the complete injector, as required. If the cap is missing, look for it in the intake manifold.

NOTE: The pintle protection cap is not available as a separate part.

To install:

15. Lubricate new O-rings with light grade oil and install 2 on each injector.

NOTE: Never use silicone grease as it will clog the injectors.

16. Install the injectors in the intake manifold using a light, twisting pushing motion.

17. Reinstall the injector retaining clips, as required.

18. Install the fuel rail, pushing it down to ensure all injector O-rings are fully seated in the fuel rail cups and intake manifold.

19. Install the retaining bolts while holding the fuel rail down and tighten to 87 inch lbs. (10 Nm).

20. Install the fuel pressure regulator retaining bolt and tighten to 15–22 ft. lbs. (20–30 Nm).

21. Connect the fuel lines to the fuel rail.

22. With the injector wiring disconnected, connect the negative battery cable and turn the ignition to the **RUN** position to allow the fuel pump to pressurize the system. Check for fuel leaks.

23. Disconnect the negative battery cable.

24. Connect the fuel injector wiring harness.

25. Install the upper intake manifold and throttle body assembly by reversing the removal procedure. Tighten the upper intake manifold retaining bolts to 24 ft. lbs. (32 Nm).

26. Connect the negative battery cable, start the engine and check for fuel leaks.

3.8L Supercharged Engine

1. Disconnect the negative battery cable.

2. Remove the fuel tank cap and relieve the fuel system pressure.

3. Remove the supercharger assembly.

4. Disconnect the fuel lines from the fuel rail assembly.

5. Remove the 4 fuel rail assembly retaining bolts and remove the fuel pressure regulator bracket retaining bolt.

6. Disconnect the electrical connectors from the injectors.

7. Carefully disengage the fuel rail from the fuel injectors and remove the fuel rail.

NOTE: It may be easier to remove the injectors with the fuel rail as an assembly.

8. Grasping the injector body, remove the injector from the fuel rail or intake manifold by pulling while gently rocking the injector from side-to-side.

9. Inspect the pintle protection cap and washer for signs of deterioration. Replace the complete injector, as required. If the cap is missing, look for it in the intake manifold.

NOTE: The pintle protection cap is not available as a separate part.

To install:

10. Lubricate new O-rings with light grade oil and install 2 on each injector.

NOTE: Never use silicone grease as it will clog the injectors.

11. Install the injectors, using a light, twisting, pushing motion.

12. Place the fuel rail assembly over each of the injectors and seat the injectors into the fuel rail.

NOTE: It may be easier to seat the injectors in the fuel rail and then seat the entire assembly in the lower intake manifold.

13. Install the fuel rail assembly retaining bolts and tighten to 70–97 inch lbs. (8–11 Nm). Install the fuel pressure regulator bracket retaining bolt and tighten to 15–22 ft. lbs. (20–30 Nm).

14. Install the supercharger assembly.

15. Connect the negative battery cable. Turn the ignition from **OFF** to **ON** several times without starting the engine to check for fuel leaks. Check all connections at the fuel rail and injectors.

16. Start the engine and warm to operating temperature. Check for fuel or coolant leaks.

4.6L Engine

1. Disconnect the negative battery cable.

2. Remove the fuel tank cap and relieve the fuel system pressure.

3. Disconnect the vacuum line at the pressure regulator.

4. Disconnect the fuel lines from the fuel rail.

5. Disconnect the electrical connectors from the injectors.

6. Remove the fuel rail assembly retaining bolts.

7. Carefully disengage the fuel rail from the fuel injectors and remove the fuel rail.

NOTE: It may be easier to remove the injectors with the fuel rail as an assembly.

8. Grasping the injector body, pull while gently rocking the injector from side-to-side to remove the injector from the fuel rail or intake manifold.

9. Inspect the pintle protection cap and washer for signs of deterioration. Replace the complete injector, as required. If the cap is missing, look for it in the intake manifold.

NOTE: The pintle protection cap is not available as a separate part.

To install:

10. Lubricate new O-rings with light grade oil and install 2 on each injector.

NOTE: Never use silicone grease as it will clog the injectors.

11. Install the injectors using a light, twisting, pushing motion.

12. Install the fuel rail, pushing it down to ensure all injector O-rings are fully seated in the fuel rail cups and intake manifold.

13. Install the retaining bolts while holding the fuel rail down and tighten to 70–105 inch lbs. (8–12 Nm).

14. Connect the fuel lines to the fuel rail and the vacuum line to the pressure regulator.

15. With injector wiring disconnected, connect the negative battery cable and turn the ignition switch to the **RUN** position to allow the fuel pump to pressurize the system.

16. Check for fuel leaks.

17. Disconnect the negative battery cable.

18. Connect the electrical connectors to the fuel injectors.

19. Connect the negative battery cable and start the engine. Let it idle for 2 minutes.

20. Turn the engine **OFF** and check for leaks.

5.0L Engine

1. Disconnect the negative battery cable.

2. Remove the fuel tank cap and relieve the fuel system pressure.

3. Partially drain the cooling system into a suitable container.

4. Disconnect the electrical connectors at the air bypass valve, TP sensor and EGR sensor.

5. Disconnect the throttle linkage at the throttle ball and transmission linkage from the throttle body. Remove the 2 bolts securing the bracket the bracket to the intake manifold and position the bracket with the cables out of the way.

6. Disconnect the upper intake manifold vacuum fitting connections by disconnecting all vacuum lines to the vacuum tree, vacuum lines to the EGR valve, vacuum line to the fuel pressure regulator and canister purge line.

7. Disconnect the PCV system by disconnecting the hose from the fitting on the rear of the upper manifold and disconnect the PCV vent closure tube at the throttle body.

8. Remove the 2 EGR coolant lines from the fittings on the EGR spacer.

9. Remove the 6 upper intake manifold retaining bolts.

10. Remove the upper intake and throttle body as an assembly from the lower intake manifold.

11. Disconnect the fuel lines from the fuel rail.

12. Remove the 4 fuel rail assembly retaining bolts.

13. Disconnect the electrical connectors from the injectors.

14. Carefully disengage the fuel rail from the fuel injectors.

NOTE: It may be easier to remove the injectors with the fuel rail as an assembly.

15. Grasping the injector body, pull up while gently rocking the injector from side-to-side to remove the injector from the fuel rail or intake manifold.

16. Inspect the pintle protection cap and washer for signs of deterioration.

Replace the complete injector, as required. If the cap is missing, look for it in the intake manifold.

NOTE: The pintle protection cap is not available as a separate part.

To install:

17. Lubricate new O-rings with light grade oil and install 2 on each injector.

NOTE: Never use silicone grease as it will clog the injectors.

18. Install the injectors using a light, twisting, pushing motion.

19. Install the fuel rail, pushing it down to ensure all the injector O-rings are fully seated in the fuel rail cups and intake manifold.

20. Install the retaining bolts while holding the fuel rail down and tighten to 70–105 inch lbs. (8–12 Nm).

21. Connect the fuel lines to the fuel rail.

22. With the injector wiring disconnected, connect the negative battery cable and turn the ignition switch to the **RUN** position to allow the fuel pump to pressurize the system.

23. Check for fuel leaks.

24. Disconnect the negative battery cable.

25. Connect the electrical connectors to the injectors.

26. Install the upper intake manifold and throttle body assembly by reversing the removal procedure.

27. Refill the cooling system and connect the negative battery cable.

28. Start the engine and let it idle for 2 minutes. Turn the engine **OFF** and check for leaks.

DRIVE AXLE

Rear Halfshaft

REMOVAL & INSTALLATION

1989–92 Thunderbird and Cougar

NOTE: Before removing the rear halfshafts, new inboard CV-joint stub shaft circlips, new differential oil seals and new hub retainer nuts must be available for assembly.

1. Remove the wheelcover/hub cover and remove the hub retainer nut. Loosen the wheel nuts.

2. Raise and support the vehicle safely by the frame only. Remove the wheel nuts and remove the wheel and tire assembly.

3. If equipped with drum brakes, remove the brake drum.

4. If equipped with disc brakes, perform the following:

    a. Remove the anti-lock brake sensors, if equipped.

    b. Pull back on the parking brake cable release lever and at the same time, pull on the cable. This will relax the cable so the cable end can be removed from the brake caliper attachment.

    c. Remove the parking brake cable from the brake caliper.

    d. Remove the upper and lower caliper retaining bolts and remove the caliper. Support the caliper out of the way with a wire, do not allow it to hang from the brake hose.

    e. Remove the brake rotor.

5. Remove the upper control arm nuts and bolt. Wire the upper control arm to the top of the shock absorber, out of the way.

6. Using a paint marker, mark the position of the lower control arm in relation to the knuckle with the lower bushings in the relaxed position.

NOTE: Failure to mark this relationship will result in bushing wind-up on assembly and incorrect ride height, causing misalignment and premature tire wear.

7. Use a suitable puller to free the halfshaft from the hub.

8. Remove the lower control arm to knuckle attaching bolts. Remove the knuckle assembly while supporting the outboard CV-joint and boot. Carefully rest the halfshaft on the lower control arm.

9. If equipped with drum brakes, wire the knuckle assembly to the top of the shock. Do not allow the knuckle assembly to hang from the brake hose.

10. Remove the halfshaft from the differential using CV-joint remover tool T89P–3514–A or equivalent. Push the tool outward until the CV-joint is freed from the differential side gear.

NOTE: Be careful not to damage the differential oil seal, differential housing and/or CV-joint boot.

11. Remove the halfshaft from the vehicle. Insert plugs into the differential housing to prevent fluid loss.

To Install:

12. Remove the differential plugs and install new differential oil seals.

13. Install new circlips on the halfshaft, by sliding it into the groove on the splined end of the shaft.

14. Lightly lubricate the stub shaft splines and carefully align the splines on the shaft with the splines in the differential.

15. Push the halfshaft inward to seat the circlip in the differential side gear groove. Use care not to damage the seal.

16. Engage the hub splines with the outboard CV-joint splines.

17. Install the lower control arm bolts and nuts. Align the paint marks and tighten the bolts to 119–147 ft. lbs. (160–200mm).

18. Install a new hub retaining nut and pull the CV-joint into the hub as far as possible by hand.

19. Install the upper arm retaining bolt and nut and tighten to 119–147 ft. bs. (160–200 Nm).

20. If equipped with drum brakes, install the brake drum. If equipped with disc brakes, proceed as follows:

    a. Install the brake rotor.

    b. Install the brake caliper assembly to the rotor with the outer brake shoe against the rotor's braking surface. This prevents pinching the piston boot between the inner brake shoe and the piston.

    c. Install the upper and lower caliper retaining bolts and tighten to 80–99 ft. lbs. (108–135 Nm).

    d. Install the parking brake cable to the brake caliper. Operate the release lever to remove the slack in the cable.

    e. Install the anti-lock brake sensor, if equipped. Tighten the retaining bolts to 15–19 ft. lbs. (19–27 Nm).

21. Check inboard CV-joint circlip engagement by attempting to pull the inboard CV-joint from the axle. If the CV-joint circlip is not seated, push the CV-joint in until the circlip is fully engaged in the side gear.

22. Check the axle lube level and fill, as necessary.

23. Install the wheel and tire assembly and tighten the wheel nuts to 80–106 ft. lbs. (108–144 Nm). Lower the vehicle.

24. Tighten the hub nut to 250 ft. lbs. (340 Nm). Install the wheelcover/hub cover.

CV-Boot

REMOVAL & INSTALLATION

1989–92 Thunderbird and Cougar

1. Remove the halfshaft from the vehicle and clamp in a vise. Do not allow the vise jaws to contact the boot or its clamp.

NOTE: The vise should be equipped with jaw caps to prevent damage to any machined surfaces.

2. Cut and remove both boot clamps and slide the boot back on the shaft.

3. Slide the outer race off the tripod.

NOTE: When replacing damaged CV-joint boots, the grease should be checked for contamination or gritty feeling. If the CV-joints are operating satisfactory and the grease does not feel contaminated, add grease and replace the boot. If the grease appears contaminated the CV-joint should be disassembled and inspected.

4. Move the stop ring back on the shaft using snapring pliers.

5. Move the tripod back on the shaft to allow access to the circlip.

6. Remove the circlip from the shaft and remove the tripod from the shaft.

7. Remove the stop ring and remove the inboard CV-joint boot.

8. Reposition the halfshaft in the vise and remove the outboard CV-joint boot.

NOTE: The outboard CV-joint is permanently retained to the inter-connecting shaft and cannot be disassembled. Outboard CV-joints are serviced as an assembly, including the inter-connecting shaft, boot, clamps grease and circlips.

To install:

9. Slide the outboard boot on the shaft. Before positioning the boot over the CV-joint, pack the CV-joint and boot with grease. The total amount of grease required is is 7.05 ounces (200 grams) for vehicles without anti-lock brakes and 8.82 ounces (250 grams) for vehicles equipped with anti-lock brakes.

10. Position the boot on the CV-joint and install the boot clamps.

11. Slide the inboard CV-joint boot on the shaft.

12. With the stop ring installed past the splines, install the tripod assembly with the chamfered side toward the stop ring.

13. Start 1 end of a new circlip in the groove of the halfshaft and work the circlip over the stub shaft end and into the groove. This will avoid over-expanding the circlip.

14. Compress the circlip and slide the tripod assembly forward over the circlip to expose the stop ring groove.

15. Move the stop ring into the groove using snapring pliers and make sure it is fully seated in the groove.

16. Fill the CV-joint outer race and boot with grease. The total amount of grease required is is 9 ounces (250 grams) for vehicles without anti-lock brakes and 10.58 ounces (300 grams) for vehicles equipped with anti-lock brakes.

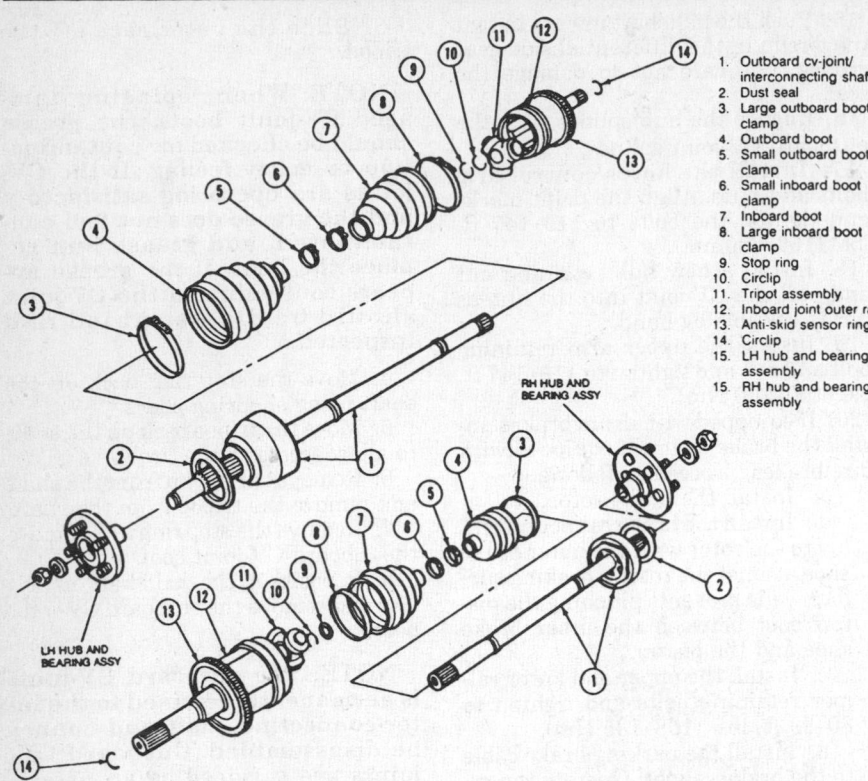

1. Outboard cv-joint/interconnecting shaft
2. Dust seal
3. Large outboard boot clamp
4. Outboard boot
5. Small outboard boot clamp
6. Small inboard boot clamp
7. Inboard boot
8. Large inboard boot clamp
9. Stop ring
10. Circlip
11. Tripot assembly
12. Inboard joint outer race
13. Anti-skid sensor ring
14. Circlip
15. LH hub and bearing assembly
15. RH hub and bearing assembly

Disassembled view of the halfshafts—1989–92 Thunderbird and Cougar

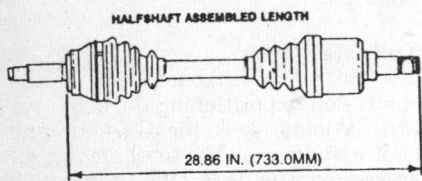

HALFSHAFT ASSEMBLED LENGTH

28.86 IN. (733.0MM)

Halfshaft assembled length—1989–92 Thunderbird and Cougar

17. Install the outer race on the tripod assembly.
18. Position the boot over the CV-joint. Move the CV-joint in and out, as necessary, to adjust to the proper length.
19. Release any air pressure by inserting a dulled tool between the boot and the outer bearing race.
20. Seat the boot in the groove and clamp in position without cutting the boot.

Driveshafts and U-Joints

REMOVAL & INSTALLATION

Except 1989–92 Thunderbird and Cougar

1. Raise and safely support the vehicle. Matchmark the rear driveshaft yoke and the companion flange so that the parts may be reassembled in the same way to maintain balance.

NOTE: Mark VII vehicles may have a balance weight attached to 1 of the flange bolts. This bolt should be reinstalled in it's original position.

2. Remove the flange bolts and disconnect the driveshaft from the axle companion flange.
3. Allow the rear of the driveshaft to drop down slightly. Pull the driveshaft and slip yoke out of the transmission extension housing.
4. Plug the transmission to prevent fluid leakage.

To install:

5. Lubricate the yoke splines and install the yoke into the transmission extension housing, aligning the splines. Be careful not to bottom the slip yoke hard against the transmission seal.
6. Rotate the pinion flange, as necessary, to align the matchmarks made during removal. Install the driveshaft yoke to the pinion flange. Install the bolts and tighten to 71–95 ft. lbs. (95–130 Nm).

1989–92 Thunderbird and Cougar

1. Drain the fuel tank.
2. Raise and safely support the vehicle by the frame.

3. Remove the crossmember on the forward side of the fuel tank.
4. Remove the exhaust pipe insulator from the left rocker panel hanger stud.
5. Remove the exhaust pipe rear insulator from the exhaust pipe hanger stud.
6. Remove the exhaust pipe at the muffler. Lower the pipe and support with a wire.
7. Remove the exhaust pipe heat shield.
8. Remove the driveshaft hoop on the rear side of the tank.
9. Remove the fuel tank filler tube retaining bolt from the right frame rail.
10. Carefully place a transmission jack under the fuel tank and remove the front heat shield.
11. Remove the support on the forward side of the fuel tank.
12. Remove the fuel tank support straps and lower the tank approximatelt 6 in.
13. Locate the original paint mark on the axle companion flange and mark the driveshaft flange in the same location. If the original mark is not visible matchmark both flanges.
14. Remove the driveshaft retaining bolts and separate the driveshaft from the axle companion flange. Pull the driveshaft rearward to remove. Install a plug in the extension housing to prevent fluid loss.

To install:

15. Lubricate the slip yoke splines and remove the plug from the transmission extension. Install the driveshaft assembly. Do not allow the slip yoke to bottom on the output shaft with excessive force.
16. Align the marks on the driveshaft with the axle companion flange. Install and tighten the bolts to 71–95 ft. lbs. (95–130 Nm).
17. Raise the fuel tank and install the support straps. Tighten the retaining bolts to 21–29 ft. lbs. (28–40 Nm).
18. Install the fuel tank filler tube retaining bolt. Tighten to 36–48 inch lbs. (4.0–5.5 Nm).
19. Install the driveshaft hoop and tighten the retaining bolts to 30–44 ft. lbs. (40–61 Nm).
20. Install the exhaust heat shield.
21. Install the support on the forward side of the fuel tank and tighten the bolts to 30–44 ft. lbs. (40–61 Nm).
22. Install the front heat shield.
23. Install the exhaust pipe to the muffler and tighten the bolts to 21–29 ft. lbs. (28–40 Nm).
24. Install the front and rear exhaust pipe insulator on the hanger stud.
25. Install the crossmember on the forward side of the fuel tank and tighten the bolts to 12–17 ft. lbs. (16–24 Nm).

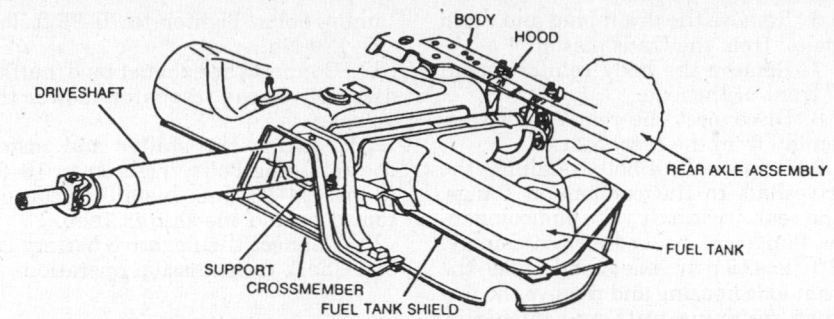

Driveshaft installation—1989–92 Thunderbird and Cougar

Rear Axle Shaft, Bearing and Seal

REMOVAL & INSTALLATION

Except 1989–92 Thunderbird and Cougar

1. Raise and safely support the vehicle. Remove wheel and tire assembly and remove brake drum or brake rotor.

2. If equipped, remove the anti-lock brake speed sensor.

3. Clean all dirt from the area of the carrier cover. Drain the axle lubricant by removing the housing cover.

4. Remove differential pinion shaft lock bolt and pinion shaft.

5. Push flanged end of axle shafts toward the center of the vehicle and remove the C-lock from button end of the axle shaft. Remove the axle shaft from the housing, being careful not to damage the oil seal.

6. Insert wheel bearing and seal replacer tool T85L-1225-AH or equivalent, in the bore and position it behind the bearing so the tangs on the tool engage the bearing outer race. Remove bearing and seal as a unit using an impact slide hammer.

To install:

7. Lubricate the new bearing with rear axle lubricant. Install the bearing into the housing bore using a suitable bearing installer.

8. Install a new axle seal using a seal installer.

NOTE: Check for the presence of an axle shaft O-ring on the spline end of the shaft and install, if not present.

9. Carefully slide the axle shaft into the axle housing, without damaging the bearing or seal assembly. Start the splines into the side gear and push firmly until the button end of the axle shaft can be seen in the differential case.

10. Install the C-lock on the button end of the axle shaft splines, then push the shaft outboard until the shaft splines engage and the C-lock seats in the counterbore of the differential side gear.

11. Insert the differential pinion shaft through the case and pinion gears, aligning the hole in the shaft with the lock bolt hole. Apply a suitable locking compound to the lock bolt and install in the case and pinion shaft. Tighten to 15–30 ft. lbs. (20–41 Nm).

12. Cover the inside of the differential case with a shop rag and clean the machined surface of the carrier and cover. Remove the shop rag.

13. Apply a bead of silicone sealer to the cover and install on the carrier. Tighten the bolts in a criss-cross pattern. Final torque the cover retaining bolts to 10–15 ft. lbs. (15–20 Nm).

14. Add rear axle lubricant to the carrier to a level ¼–⁹⁄₁₆ in. below the bottom of the fill hole. Install the filler plug and tighten to 15–30 ft. lbs. (20–41 Nm).

15. Install the anti-lock speed sensor, if equipped. Tighten the retaining bolt to 40–60 inch lbs. (4.5–6.8 Nm).

16. Install the brake calipers and rotors or the brake drums, as required. Install the wheel and tire assembly and lower the vehicle.

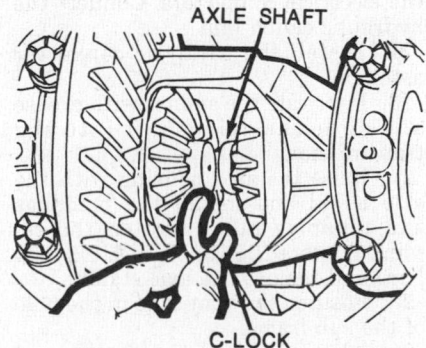

Removing the axle shaft C-locks

For further information on transmissions/transaxles, please refer to "Chilton's Guide to Transmission Repair".

Transmission Assembly

REMOVAL & INSTALLATION

Except 1989–92 Thunderbird and Cougar

1. Disconnect the negative battery cable.

2. Raise and support the vehicle safely.

3. Matchmark the driveshaft for reasembly. Disconnect the driveshaft from the rear U-joint flange. Slide the driveshaft off the transmission output shaft and install an extension housing seal installation tool into the extension housing to prevent lubricant from leaking.

4. Remove the bolts and remove the catalytic converter.

5. Remove the 2 nuts attaching the rear transmission support to the crossmember. Remove the bolts.

6. On the Thunderbird Turbo Coupe with the 2.3L turbocharged engine, remove the catalytic converter and inlet pipe.

7. Using a suitable jack, support the engine and tranmsmission.

8. Remove the 2 nuts from the crossmember bolts. Remove the bolts, raise the jack slightly and remove the crossmember.

9. Lower the transmission to expose the 2 bolts securing the shift handle to the shift tower. Remove the 2 nuts and bolts and remove the shift handle.

10. Disconnect the wiring harness from the backup lamp switch. On the 5.0L engine, disconnect the neutral sensing switch.

11. Remove the bolt from the speedometer cable retainer and remove the speedometer driven gear from the transmission.

12. Remove the 4 bolts that secure the transmission to the flywheel housing.

13. Remove the transmission and jack rearward until the transmission input shaft clears the flywheel housing. If necessary lower the engine enough to obtain clearance for removing the transmission.

NOTE: Do not depress the clutch while the transmission is removed.

To install:

14. Make sure the mounting surface of the transmission and flywheel housing are clean and free of dirt, paint and burrs.

15. Install 2 guide pins in the flywheel housing lower mounting bolt holes. Raise the transmission and move forward on the guide pins until the input shaft splines enter the clutch hub slpines and the case is positioned against the flywheel housing.

16. Install the 2 upper transmission-to-flywheel housing mounting bolts snug and remove the 2 guide pins. Install the 2 lower mounting bolts and tighten all the bolts to 45–65 ft. lbs. (61–88 Nm).

17. Raise the transmission with a jack until the shift handle can be secured to the shift tower. Install and tighten the attaching bolts and washers to 23–32 ft. lbs. (31–43 Nm).

18. Connect the speedometer cable to the extension housing and tighten the attaching screw to 36–54 inch lbs. (48–68 Nm).

19. Raise the rear of the transmission with the jack and install the transmission support. Install and tighten the attaching bolts to 36–50 ft. lbs. (48–68 Nm).

20. With the transmission extension housing resting on the engine rear support, install the attaching bolts and tighten to 25–35 ft. lbs. (38–48 Nm).

21. Connect the backup lamp switch wiring harness. On 5.0L engine, connect the neutral sensing switch to the wiring harness.

22. Install the catalytic converter. Tighten the attaching bolts to 20–30 ft. lbs. (27–41 Nm).

23. Remove the extension housing installation tool and slide the forward end of the driveshaft over the transmission output shaft. Connect the driveshaft to the rear U-joint flange. Make sure the matchmarks align. Tighten the U-bolt nuts to 42–57 ft. lbs. (56–77 Nm).

24. Fill the transmission with the proper type and quantity of fluid.

25. Lower the vehicle. Check the shift and crossover motion for full shift engagement and smooth crossover operation.

1989–92 Thunderbird and Cougar

1. Disconnect the negative battery cable.

2. Shift the transmission into the **N** position.

3. Remove the shift knob and the console top cover.

4. Remove the 2 shifter retaining bolts and remove the shifter.

5. Raise and support the vehicle safely.

6. Remove the drain plug and drain the oil from the transmission.

7. Remove the body reinforcement in front of the axle.

8. Disconnect the rear exhaust assembly from the resonator.

9. Remove the 4 bolts retaining the driveshaft to the companion flange. The rear driveshaft yoke and companion flange are marked for reassembly.

10. Position an axle stand under the front axle housing and remove the forward retaining nuts and bushings. Loosen the rear retaining nuts to allow the axle to tilt for driveshaft removal.

11. Pull the vent tube from the hole in the sub-frame.

12. Lower the front of the axle housing with the axle stand and slide the driveshaft out of the transmission above the axle housing. Let the driveshaft rest on the front driveshaft support and axle assembly.

13. Remove the catalytic converter assembly.

14. Disconnect the hydraulic clutch line.

15. Disconnect the electrical connectors and remove the starter.

16. Position a transmission jack under the transmission. Remove the crossmember and the bellhousing to engine bolts.

17. Move the transmission to the rear until the input shaft clears the flywheel and lower the transmission from the vehicle.

To install:

18. Install guide studs in the engine block and raise the transmission until the input shaft splines are aligned with the clutch disc splines.

19. Slide the transmission forward on the guide studs until it is against the bellhousing. Install the bellhousing-to-engine retaining bolts and tighten to 28–38 ft. lbs. (38–51 Nm).

20. Install the crossmember and tighten the bolts to 35–50 ft. lbs. (47–68 Nm). Remove the transmission jack.

21. Install the starter and connect the electrical connectors. Connect the hydraulic clutch line.

22. Install the catalytic converter assembly.

23. Lubricate the splines with grease and slide the driveshaft into the transmission.

24. Raise the axle housing with the axle stand and install the bushings and retaining nuts. Tighten the retaining nuts to 68–100 ft. lbs. (92–136 Nm) and remove the axle stand.

25. Position the vent tube in the hole of the sub-frame.

26. Align the driveshaft yoke and companion flange and install the retaining bolts. Tighten to 70–95 ft. lbs. (95–129 Nm).

27. Connect the exhaust pipe muffler assembly to the resonator. Lower the vehicle.

28. Position the shifter and install the retaining bolts. Tighten to 18–24 ft. lbs. (24–33 Nm). Install the console top cover and the shifter knob.

29. Connect the negative battery cable. Check transmission operation.

CLUTCH

Clutch Assembly

REMOVAL & INSTALLATION

Except 1989–92 Thunderbird and Cougar

1. Disconnect the negative battery cable. On Mustang, lift the clutch pedal to its uppermost position to disengage the pawl and quadrant. Push quadrant forward, unhook cable from quadrant and allow quadrant to slowly swing rearward.

2. Raise and safely support the vehicle. Remove the dust shield.

3. On Mustang, disconnect cable from the release lever. Remove the retaining clip and remove the clutch cable from the flywheel housing. On Thunderbird Turbo Coupe, remove the clutch slave cylinder.

4. Remove starter and bolts that secure engine rear plate to front lower part of flywheel housing.

5. Remove the transmission, then the flywheel housing.

6. Remove clutch release lever from housing by pulling it through the window in housing until retainer spring is disengaged from pivot. Remove release bearing from release lever.

7. Loosen the pressure plate cover attaching bolts evenly to release spring tension gradually and avoid distorting cover. If same pressure plate and cover are to be installed, mark cover and flywheel so that pressure plate can be installed in its original position.

8. Inspect the flywheel for scoring, cracks or other damage and machine or replace, as necessary. Inspect the pilot bearing for damage and free movement. Replace, as necessary.

To install:

9. If removed, install the flywheel. Make sure the mating surfaces of the flywheel and the crankshaft flange are clean prior to installation. Tighten the flywheel bolts to 56–64 ft. lbs. (73–87 Nm) on 2.3L engines and 75–85 ft. lbs. (102–115 Nm) on 5.0L engine.

10. Position the clutch disc and pressure plate assembly on the flywheel. The 3 dowel pins on the flywheel must be properly aligned with the pressure plate. Bent, damaged or missing dowels must be replaced. Start the pressure plate bolts but do not tighten them.

11. Align the clutch disc using a suitable alignment tool inserted in the pilot bearing. Alternately tighten the bolts a few turns at a time, until they are all tight. Final torque the bolts to 12–24 ft. lbs. (17–32 Nm). Remove the alignment tool.

12. Apply a light coating of multipurpose long-life grease to the release lever pivot pocket, release lever fork and flywheel housing pivot ball. Fill the grease groove of the release bearing hub with the same grease. Clean all excess grease from the inside bore of the bearing hub.

13. Install the release bearing on the release lever and install the lever in the flywheel housing.

14. Install the flywheel housing. Tighten the bolts to 28–38 ft. lbs. (38–52 Nm) on the 2.3L engine and 38–55 ft. lbs. (52–74 Nm) on the 5.0L engine.

15. Install the remaining components in the reverse order of removal.

1989–92 Thunderbird and Cougar

1. Disconnect the negative battery cable.

2. Disconnet the clutch hydraulic system master cylinder from the clutch pedal.

3. Raise and support the vehicle safely.

4. Remove the starter.

5. Disconnect the hydraulic coupling at the transmission with tool T88T-70522-A or equivalent, by sliding the white plastic sleeve toward the slave cylinder and applying a slight tug on the tube.

6. Remove the transmission.

7. Matchmark the assembled position of the pressure plate to the flywheel.

8. Loosen the pressure plate and cover attaching bolts evenly until the pressure plate springs are expanded, and remove the bolts.

9. Remove the pressure plate/cover assembly and the clutch disc from the flywheel.

10. Inspect the flywheel for scoring, cracks or other damage and machine or replace, as necessary. Inspect the pilot bearing for damage and free movement. Replace, as necessary.

To install:

11. If removed, install the flywheel. Make sure the mating surfaces of the flywheel and the crankshaft flange are clean prior to installation. Tighten the

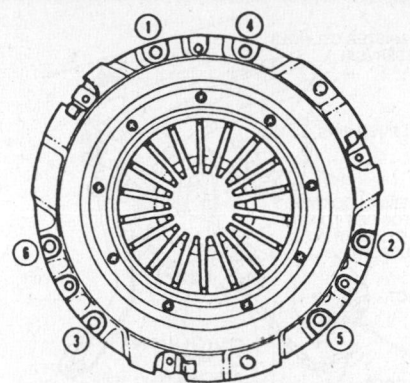

Pressure plate bolt torque sequence— 1989–92 Thunderbird and Cougar

flywheel bolts to 54–64 ft. lbs. (73–87 Nm).

12. Position the clutch disc on the flywheel so a suitable alignment tool can enter the clutch pilot bearing and align the disc.

13. If reinstalling the original pressure plate, align the matchmarks. Position the pressure plate on the flywheel and install the retaining bolts hand tight. Tighten the bolts, in sequence, to 20–28 ft. lbs. (27–39 Nm). Remove the alignment tool.

14. Install the remaining components in the reverse order of removal. Tighten the flywheel housing-to-engine bolts to 40–49 ft. lbs. (54–67 Nm).

NOTE: Reuse the aluminum washers under the attaching bolts to prevent galvanic corrosion.

Clutch Cable

REMOVAL & INSTALLATION

Mustang

NOTE: Whenever the clutch cable is disconnected, it is mandatory that the proper method for installing the clutch cable be followed.

1. Lift the clutch pedal to its upward most position to disengage the pawl and quadrant. Push the quadrant forward, unhook the cable from the quadrant and allow it to slowly swing rearward.

2. Remove the screw that holds the cable insulator to the dash panel and pull the cable through the dash panel and into the engine compartment.

3. Remove the cable bracket screw from the fender apron.

4. Raise and support the vehicle safely.

5. On 5.0L engine, remove the dust cover from the bell housing.

6. Remove the clip retainer holding the cable to the bell housing.

7. On the 5.0L engine, slide the ball on the end of the cable through the hole in the clutch release lever and remove the cable.

8. On the 2.3L engine, remove the hairpin clip, clevis pin and clevis from the end of the cable.

To install:

NOTE: The clutch pedal must be lifted to disengage the adjusting mechanism during cable installation. Failure to do so will cause damage the self-adjuster mechanism. A prying instrument should never be used to install the cable into the quadrant.

9. Slide the cable through the hole in the bell housing and through the hole in the the release lever. On the 5.0L engine, slide the ball on the end of the cable assembly into the cable ball pocket on the clutch release lever. On the 2.3L engine, place the cable ball into the clevis. Install the clevis and clevis pin onto the clutch release lever and into the clevis pin.

10. Install the clutch cable retaining clip on the bell housing.

11. On the 5.0L engine, install the dust shield on the bell housing.

12. Push the cable assembly into the engine compartment and lower the vehicle. Install the cable bracket screw in the fender apron.

13. Push the cable into the hole in the dash panel and secure the insulator with a screw.

14. Install the cable assembly by lifting the clutch pedal to disengage the pawl and quadrant. Then, pushing the quadrant forward, hook the end of the cable over the rear of the quadrant.

15. Depress the clutch pedal several times to adjust the cable.

Clutch Master Cylinder

REMOVAL & INSTALLATION

Thunderbird and Cougar

1988

1. Disconnect the negative battery cable.

2. Remove the clutch slave cylinder.

3. Remove master cylinder reservoir by removing the 2 nuts.

4. Disengage the pushrod from the clutch pedal.

5. Remove master cylinder by turning it 45 degrees clockwise, then gently pull the master cylinder out.

6. Remove the pressure line, if necessary.

7. Installation is the reverse of the removal procedure. Tighten the master cylinder reservoir nuts to 18–24 inch lbs. (2.0–2.7 Nm). Bleed the system if the pressure line was removed.

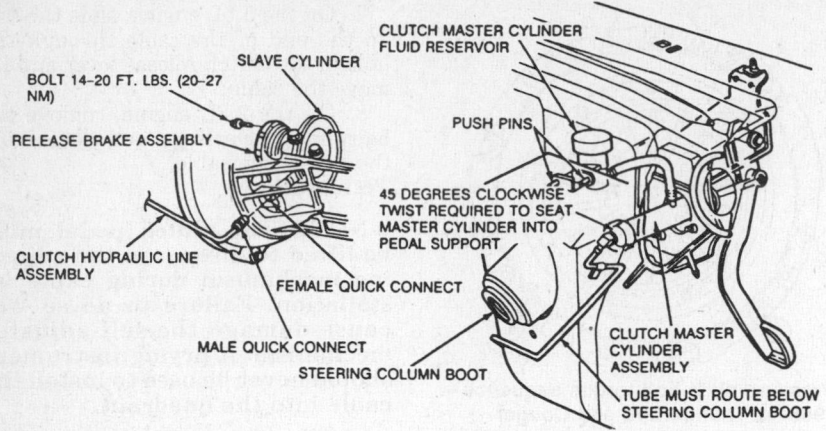

Clutch master cylinder installation—1989–92 Thunderbird and Cougar

1989–92

1. Disconnect the negative battery cable.

2. Disconnect the clutch pedal from the pushrod.

3. Disconnect the hydraulic line from the slave cylinder by depressing the white retainer bushing with tool T88T–70522–A or equivalent, while pulling slightly on the line.

4. Remove the 2 push pins retaining the clutch master cylinder reservoir to the left shock tower.

5. Rotate the master cylinder 45 degrees counterclockwise, then carefully pull the master cylinder through the dash panel, noting the routing of the hydraulic line to the slave cylinder.

6. If the master cylinder is to be replaced, position the master cylinder in a vise and drive out the roll pin using a drift. Remove the O-ring from the tube connection of the master cylinder.

To install:

7. Install a new O-ring onto the clutch tube and install the tube into the master cylinder. Install the roll pin.

8. Position the clutch master cylinder in the engine compartment and route the hydraulic line to the slave cylinder.

9. Install the master cylinder to the dash panel and install the clutch master cylinder fluid reservoir.

10. Install the hydraulic line to the slave cylinder and connect the pushrod to the clutch pedal.

11. Fill the reservoir and bleed the system.

Clutch Slave Cylinder

REMOVAL & INSTALLATION

Thunderbird and Cougar

1988

NOTE: Do not depress the clutch pedal while the slave cylinder is removed from the clutch housing.

1. Remove dust cover by removing the self tapping screw.

2. Unlatch slave cylinder from the transmission housing bracket.

3. Remove the pressure line, if necessary.

4. Installation is the reverse of the removal procedure. Bleed the system if the pressure line was removed.

1989–92

1. Disconnect the negative battery cable.

2. Raise and support the vehicle safely.

3. Disconnect the hydraulic line from the slave cylinder by depressing the white retainer bushing with tool T88T–70522–A or equivalent, while pulling slightly on the line.

4. Remove the transmission.

5. Remove the clutch release bearing by rotating the assembly against the spring tension until the spring pushes the bearing off the slave cylinder.

6. Remove the clutch slave cylinder retaining bolts and remove the slave cylinder.

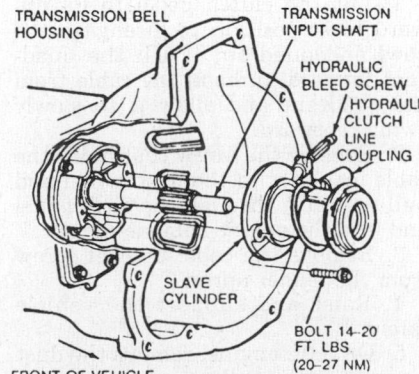

Clutch slave cylinder installation— 1989–92 Thunderbird and Cougar

To install:

7. Position the slave cylinder over the input shaft aligning the bleeder screw and line coupling with holes in the transmission housing.

8. Install the slave cylinder retaining bolts and tighten to 15–19 ft. lbs. (20–27 Nm).

9. Install the release bearing and transmission.

10. Connect the hydraulic cylinder to the slave cylinder and bleed the system.

Hydraulic Clutch System Bleeding

Thunderbird and Cougar

NOTE: On 1989–92 vehicles, be sure to pump the clutch at least 30 times to make sure that air is in the system. If the slave cylinder is pushed off the clutch plate, a similar pedal feel may occur. Pumping the clutch pushes fluid from the clutch reservoir into the slave cylinder, pushing it out to meet the clutch plate.

1. Clean all dirt and grease from the cap to make sure that no foreign substances enter the system.

2. Remove the cap and diaphragm and fill the reservoir to the top with the approved fluid.

3. Raise and support the vehicle safely.

4. Attach a hose to the bleeder valve at the slave cylinder.

NOTE: Keep the clutch fluid reservoir full at all times to prevent air from being pulled into the system.

5. While the clutch pedal is being depressed, slightly open the bleeder valve and observe air bubbles in the clutch fluid at the end of the hose.

6. Close the bleeder valve before releasing the clutch pedal.

7. Repeat Steps 5 and 6, as necessary, until no air bubbles are observed.

8. Lower the vehicle and fill the reservoir. Road test the vehicle.

AUTOMATIC TRANSMISSION

For further information on transmissions/transaxles, please refer to "Chilton's Guide to Transmission Repair".

Transmission Assembly

REMOVAL & INSTALLATION

1. Disconnect the negative battery cable. Raise the vehicle and support safely.

2. Drain the fluid from the transmission by removing all oil pan bolts except the 2 at the front. Loosen the 2 at the front and drop the oil pan at the rear to allow the fluid to drain into a container. When drained, reinstall a few of the bolts to hold the pan in place.

3. Remove the converter bottom cover and remove the converter drain plug, if equipped, to allow the converter to drain. After the converter has drained, reinstall the drain plug and tighten. Remove the converter to flywheel nuts by turning the converter to expose the bolts.

NOTE: Crank the engine over with a wrench on the crankshaft pulley attaching bolt. On belt driven OHC engines, never rotate the pulley in a counterclockwise direction as viewed from the front.

4. Matchmark and disconnect the driveshaft assembly. On 1989–92 Thunderbird and Cougar, proceed as follows:

    a. Loosen the differential housing assembly rear mounting nuts approximately ¼ in.

    b. Position an axle stand under the front of the differential housing and remove the forward mounting nuts and bushings. Pull the vent tube from the hole in the sub-frame.

    c. Lower the front of the differential housing with the axle stand and slide the driveshaft out of the transmission above the axle housing. Let the drive shaft rest on the front driveshaft support and axle assembly.

5. Remove the speedometer cable or sensor from the extension housing.

6. Disconnect the manual control shift rod or cable and the downshift rod or cable from the transmission control levers.

7. Remove the starter cable and remove the starter.

8. Remove the electrical wires and vacuum lines, as required from the transmission assembly. Remove the bellcrank bracket, if equipped, from the converter housing.

9. Place a support under the transmission and slightly raise it. It may be necessary to raise the engine hood and loosen the fan shroud.

10. Remove the rear crossmember and engine rear support. Disconnect the necessary exhaust components.

11. Lower the transmission to expose the oil cooler line fittings. Disconnect the lines from the transmission.

12. Support the engine and remove the dipstick tube and all the bell housing retaining bolts except for the top 2.

13. Chain the transmission to the jack or support unit for safety.

14. Remove the 2 top bolts from the converter housing and move the transmission rearward and down from under the vehicle. Hold the converter in place to avoid having it drop from the transmission.

To install:

15. Tighten the converter drain plug to 8–28 ft. lbs. (11–38 Nm).

16. Position the converter to the transmission and rotate into position to make sure the drive flats are fully engaged in the pump gear.

NOTE: Lubricate the pilot with chassis grease.

17. Raise the converter and transmission assembly into position. Rotate the converter until the studs and drain plug are in alignment with the holes in the flywheel. Align the orange balancing marks on the converter stud and flywheel bolt hole if balancing marks are present.

NOTE: The converter face must rest squarely against the flywheel. This indicates that the converter pilot is not binding in the engine crankshaft. To ensure the converter is properly seated, grasp a converter stud. It should move freely back and forth in the flywheel hole. If the converter will not move, the transmission must be removed and the converter repositioned so the impeller hub is properly engaged in the pump gear.

18. Install the transmission-to-engine attaching bolts. Tighten the bolts to 40–50 ft. lbs. (55–68 Nm) on all except 2.3L engine. On 2.3L engine, tighten the bolts to 28–38 ft. lbs. (38–51 Nm).

19. Remove the safety chain from around the transmission.

20. Install a new O-ring on the lower end of the transmission filler tube and install the tube to the transmission case.

21. Connect the speedometer cable to the transmission case, if equipped.

22. Connect the oil cooler lines to the right side of the transmission case.

23. Position the crossmember on the side supports. Position the rear mount on the crossmember and install the attaching bolts and/or nuts.

24. Secure the engine rear support to the extension housing.

25. Install any exhaust system components, if removed.

26. Lower the transmission and remove the jack.

27. Secure the crossmember to the side supports with the attaching bolts.

28. Connect the TV linkage rod or cable and the manual linkage rod.

29. Install the converter to flywheel attaching nuts and tighten to 20–34 ft. lbs. (27–46 Nm). Install the converter housing cover.

30. Secure the starter motor in place and connect all electrical connections.

31. Install the driveshaft, making sure the index marks are aligned. On 1989–92 Thunderbird, proceed as follows:

    a. Raise the differential housing with the axle stand and install the bushings and retaining nuts. Tighten to 68–100 ft. lbs. (92–136 Nm). Remove the axle stand.

    b. Position the vent tube in the hole of the sub-frame.

    c. Align the driveshaft yoke and companion flange and install the retaining bolts. Tighten to 70–95 ft. lbs. (95–129 Nm).

32. Lower the vehicle. Fill the transmission with the proper type and quantity of fluid, start the engine and check the transmission for leakage. Adjust the linkage as required.

MANUAL LINKAGE ADJUSTMENT

Crown Victoria, Grand Marquis and Town Car

1988–89

1. Place the selector lever in the **OVERDRIVE** position, tight against the overdrive stop. An 8 lb. weight should be hung on the selector lever to ensure the lever remains against the overdrive stop during the linkage adjustment.

2. Loosen the shift rod adjusting bolt.

3. Shift the transmission into **OVERDRIVE** by pushing the column shift rod downward to the lowest position and pulling up 3 detents.

4. Make sure the selector lever has not moved from the overdrive stop. Tighten the bolt to 14–23 ft. lbs. (19–31 Nm).

5. Check the transmission operation for all selector lever detent positions.

1990–92

1. Loosen the adjusting stud nut at the transmission lever.

2. From the passenger compartment, place the steering column selector lever in **OVERDRIVE** and hold the selector lever in position by placing a 3 lb. weight on the lever.

3. Rotate the transmission lever clockwise to low and return it 2 detent positions counterclockwise to the **OVERDRIVE** position.

4. Align the flats of the adjusting stud with the flats of the cable slot and install the cable on the stud.

NOTE: Do not push or pull on the rod while assembling the rod to the stud.

5. Tighten the adjusting stud nut and washer assembly to 10–18 ft. lbs. (13–25 Nm).

6. Check the shift lever for proper operation.

Mark VII, Thunderbird, Cougar and Mustang

1. Position the transmission selector lever in the **OVERDRIVE** position.

NOTE: The shift lever should be held against the rearward OVERDRIVE stop when the linkage is adjusted.

2. Raise and safely support the vehicle. Loosen the manual lever shift cable or rod retaining nut. Move the transmission manual lever to the **OVERDRIVE** position.

3. With the transmission selector lever and manual lever in the **OVERDRIVE** position, tighten the retaining nut to 10–18 ft. lbs. (13–25 Nm).

4. Check the operation of the transmission in each selector lver position.

THROTTLE VALVE CABLE ADJUSTMENT

Automatic Overdrive Transmission

1. Set the parking brake and place the shift selector in **N**.

2. Remove the air cleaner cover and inlet tube from the throttle body inlet to access the throttle lever and cable.

3. Using a small pry bar, pry the grooved pin on the cable assembly out of the grommet on the throttle body. Then push out the white locking tab.

3. Check the plastic block with pin and tab, it should slide freely on the notched rod. If not, the white tab may not be pushed out far enough.

4. While holding the throttle lever firmly against the idle stop, push the grooved pin into the grommet on the throttle lever as far as it will go.

5. Make sure that the throttle lever does not move while pushing the pin into the grommet.

6. Install the air cleaner cover and inlet tube.

FRONT SUSPENSION

NOTE: If equipped with the level ride air suspension, power to the air system must be shut OFF before servicing the suspension. The switch is located in the luggage compartment, on the drivers side rear fender well.

Shock Absorbers

REMOVAL & INSTALLATION

NOTE: Purge a new shock of air by repeatedly extending it in its normal position and compressing it while inverted.

1. Remove the nut, washer and bushing from the upper end of the shock absorber.

2. Raise and safely support the vehicle by the frame rails allowing the front wheels to hang.

3. Remove the 2 bolts securing the shock absorber to the lower control arm and remove the shock absorber.
To install:

4. Install a new bushing and washer on the top of the shock absorber and position the unit inside the front spring. Install the 2 lower attaching bolts and torque them to 12–18 ft. lbs. (16–24 Nm).

5. Lower the vehicle.

6. Place a new bushing and washer on the shock absorber top stud and install a new attaching nut. Tighten to 22–26 ft. lbs. (30–41 Nm).

MacPherson Strut

REMOVAL & INSTALLATION

Mark VII

1. Turn the air suspension switch **OFF**.

2. Turn the ignition switch to the **UNLOCKED** position to allow freee movement of the front wheels.

3. From inside the engine compartment, loosen but do not remove, the 1 strut-to-upper mount retaining nut. A suitable tool positioned in the slot will hold the rod stationary while loosening the nut.

4. Raise the vehicle and position safety stands under the lower control arms as far outboard as possible, verifying the lower sensor mounting bracket is clear. Lower the vehicle until vehicle weight is supported by the lower arms.

5. Remove the wheel and tire assembly. Remove the brake caliper and

---

support with a length of wire. Do not let the caliper hang by the brake hose.

6. Remove and discard the strut-to-upper mount retaining nut and then the 2 lower nuts and bolts attaching the strut to the spindle.

NOTE: The strut should be held firmly during removal of the last bolt since the gas pressure will cause the strut to fully extend when removed.

7. Lift the strut up from the spindle to compress the rod and then remove the strut. Remove the jounce bumper.
To install:

8. Prime the new strut by extending and compressing the strut 5 times. Install the jounce bumper.

9. Place the strut rod through the upper mount and hand start a new nut. Tighten the nut to 55–92 ft. lbs. (75–125 Nm).

10. Compress the strut and position onto the spindle. Install 2 new lower retaining bolts and hand start the nuts.

11. Raise the vehicle to remove the vehicle load from the lower control arms. Tighten the lower retaining nuts to 140–200 ft. lbs. (190–271 Nm).

12. Install the brake caliper and the wheel and tire assembly. Remove the safety stand and lower the vehicle to the ground.

13. Turn the air suspension switch **ON**. Check the front end alignment.

1989–92 Thunderbird and Cougar

1. Remove the plastic cover at the upper strut mount, if equipped. If equipped with automatic ride control, remove the actuator assembly as follows:

   a. Make sure the vehicle is level. Turn the ignition switch **OFF**.

   b. Disconnect the actuator connector from the wiring harness connector. Remove the actuator cover by snapping off.

   c. Slide the actuator connector off the cover by inserting a suitable tool between the connector and white Christmas tree track to separate the 2 parts prior to sliding the connector off.

   d. Squeeze the 2 actuator retaining tabs firmly inward with 1 hand and lift the actuator off the mounting bracket with the other hand.

   e. Grasp the piston rod end at the 9mm hex with a socket wrench.

   f. Loosen the nut retaining the actuator mounting bracket to the strut with a 19mm box wrench while holding the socket wrench.

   g. Remove the nut and mounting bracket.

2. Remove the 3 upper strut retain-

ing nuts and collar plate from the mounting studs in the engine compartment.

3. Raise and safely support the vehicle. Remove the wheel and tire assembly.

4. Remove the lower strut mounting bolt and nut and remove the nut at the stabilizer link upper mounting stud. Separate the link from the spindle using a suitable joint separator tool.

5. Support the lower control arm with a jack. Raise the control arm and spindle with the jack until the stabilizer link can be completely separated from the spindle. Position the link aside.

6. Remove the spindle to upper control arm retaining nut and bolt. Lower the jack to separate the spindle from the upper control arm. Support the spindle with a length of wire; do not let it hang free.

7. Lower the support for the lower control arm and remove the strut assembly from the vehicle.

To install:

8. Position the strut over the lower arm. Insert the lower strut bolt into the control arm.

9. Using a jack, raise the control arm and strut into position. Align the upper strut mounting studs with the holes.

10. Remove the wire supporting the spindle and position the spindle to the upper control arm. Raise the lower control arm using the jack and attach the spindle to the upper control arm.

11. Install a new spindle retaining bolt from the front of the vehicle and install the nut. Tighten to 59–66 ft. lbs. (80–90 Nm).

12. Position the stabilizer bar link and lower the spindle assembly until the link can be installed. Install the nut on the link stud and tighten to 48–55 ft. lbs. (65–75 Nm).

13. Remove the jack from the lower arm. Install the shock nut, but do not tighten at this time.

14. Install the wheel and tire assembly and lower the vehicle. Make sure the upper strut mounting studs are aligned with the holes.

15. Install the collar plate and 3 nuts to the upper mounting studs. Tighten to 16–23 ft. lbs. (22–31 Nm).

16. Install the washer, nut and automatic ride actuator, if equipped. Install the plastic cover, if equipped.

17. Neutralize the front suspension bushings by pushing down and releasing on the front of the vehicle. Then tighten the lower strut nut to 140–162 ft. lbs. (190–220 Nm).

NOTE: The lower strut nut must be tightened with the vehicle weight on the wheels.

Mustang and 1988 Thunderbird and Cougar

1. Disconnect the negative battery cable. If equipped with programmed ride control on 1988 Thunderbird, remove the actuator assembly as follows:

a. Make sure the vehicle is on level area. Turn the ignition switch to the **LOCK** position.

b. Disconnect the actuator connector from the wiring harness connector.

c. Remove the actuator cover by snapping off. Slide the actuator connector off the cover.

d. Remove the 2 screws retaining the actuator to the mounting bracket. Remove the actuator by lifting off.

e. Grasp the actuator mounting bracket with water pump pliers and hold firmly. Loosen the nut retaining the mounting bracket to the strut and remove the nut and mounting bracket.

2. Place the ignition switch in the **UNLOCKED** position to permit free movement of the front wheels.

3. Raise the vehicle by the lower control arms until the wheels are just off the ground. From the engine compartment, remove and discard the 3 upper mount retaining nuts. Do not remove the pop-rivet holding the camber plate position.

NOTE: If the upper mount on 1988 Thunderbird and Cougar is to be replaced, remove the strut rod nut at this time.

4. Continue to raise the front of the vehicle by the lower control arms and position safety stands under the frame jacking pads, rearward of the wheels.

5. Remove the wheel and tire assembly and remove the brake caliper. Support the caliper with a length of wire; do not let the caliper hang by the brake hose.

6. Remove the 2 lower nuts that attach the strut to the spindle, leaving the bolts in place. Carefully remove both spindle-to-strut bolts, push the bracket free of the spindle and remove the strut.

7. Compress the strut to clear the upper mount of the body mounting pad. Remove the upper mount and jounce bumper, if necessary.

To install:

8. Install the upper mount and jounce bumper, if removed.

9. Position the 3 upper mount studs into the body mounting pad and camber plate and start 3 new nuts.

10. Compress the strut and position into the spindle. Install 2 new lower retaining bolts and hand start the nuts. Remove the suspension load from the control arms by lowering the

vehicle. Tighten the lower retaining nuts to 140–200 ft. lbs. (190–271 Nm).

11. Raise the suspension control arms and tighten the upper mount retaining nuts to 40–55 ft. lbs. (54–75 Nm).

12. Install the brake caliper and the wheel and tire assembly.

13. Lower the vehicle to the ground and check the front end alignment.

Coil Springs

REMOVAL & INSTALLATION

Crown Victoria, Grand Marquis and Town Car

1. Raise and safely support the vehicle. Remove the wheel and tire assembly.

2. Disconnect the stabilizer bar link from the lower arm. Remove the shock absorber.

3. Remove the steering link from the pitman arm.

4. Using spring compressor tool D78P-5310-A or equivalent, install 1 plate with the pivot ball seat facing downward into the coils of the spring. Rotate the plate, so it is flush with the upper surface of the lower arm.

5. Install the other plate with the pivot ball seat facing upward into the coils of the spring. Insert the upper ball nut through the coils of the spring, so the nut rests in the upper plate.

6. Insert the compression rod into the opening in the lower arm, through the upper and lower plate and upper ball nut. Insert the securing pin through the upper ball nut and compression rod.

NOTE: This pin can only be inserted 1 way into the upper ball nut because of a stepped hole design.

7. With the upper ball nut secured, turn the upper plate so it walks up the coil until it contacts the upper spring seat. Then back-off ½ turn.

8. Install the lower ball nut and thrust washer on the compression rod and screw on the forcing nut. Tighten the forcing nut until the spring is compressed enough so it is free in its seat.

9. Remove the 2 lower arm pivot bolts, disengage the lower arm from the frame crossmember and remove the spring.

10. If a new spring is to be installed, perform the following:

a. Mark the position of the upper and lower plates on the spring with chalk.

b. With an assistant, compress a new spring for installation and measure the compressed length and the amount of curvature of the old spring.

11. Loosen the forcing nut to relieve the spring tension and remove the tools from the spring.

To install:

12. Assemble the spring compressor and locate in the same position as indicated in Step 10a.

13. Before compressing the coil spring, make sure the upper ball nut securing the pin is inserted properly.

14. Compress the coil spring until the spring height reaches the dimension obtained in Step 10b.

15. Position the coil spring assembly into the lower arm and reverse the removal procedure.

1989–92 Thunderbird and Cougar

1. Remove the strut assembly from the vehicle.

NOTE: The upper strut mount cannot be rotated when the strut and spring are assembled. Mark the position of the upper mount to the coil spring with chalk or paint, prior to disassembly. If the upper mount is not properly positioned during assembly, it will not install in the vehicle.

2. Position the strut assembly in spring compressor tool 086–00029 or equivalent.

3. Compress the spring. Remove the strut nut and washer and remove the upper mount.

4. Release the spring compressor to remove the coil spring.

To install:

5. If installing a new spring or upper mount, transfer the reference marks from the removed part to the new part.

6. Position the strut and the spring in the spring compressor tool and compress the spring to install the upper mount.

7. Install the upper mount, aligning the refernce marks. Install the washer and nut and tighten to 37–52 ft. lbs. (50–71 Nm).

8. Release the spring compressor, making sure the spring is properly seated at top and bottom.

9. Install the strut assembly in the vehicle.

Mustang and 1988 Thunderbird and Cougar

1. Raise and safely support the vehicle, allowing the control arms to hang free.

2. Remove the wheel and tire assembly and the brake caliper. Suspend the caliper with a length of wire; do not let the caliper hang by the brake hose.

3. Disconnect the tie rod end from the steering spindle and disconnect the stabilizer link from the lower arm.

4. Remove the steering gear bolts, if necessary and position the gear so the suspension arm bolt can be removed.

5. On all vehicles ecept Mustang with 5.0L engine, use spring compressor tool T82P–5310–A or equivalent to place the upper plate in position into the spring pocket cavity on the crossmember. The hooks on the plate should be facing the center of the vehicle.

6. On Mustang with 5.0L engine, use spring compressor tool D78P–5310–A or equivalent, to install a plate between the coils near the toe of the spring. Mark the location of the upper plate on the coils for installation.

7. Install the compression rod into the lower arm spring pocket hole, through the coil spring and into the upper plate.

8. Install the lower plate, lower ball nut, thrust washer and bearing and forcing nut onto the compression rod. Tighten the forcing nut until a drag on the nut is felt.

9. Remove the suspension arm-to-crossmember nuts and bolts. The compressor tool forcing nut may have to be tightened or loosened for easy bolt removal.

10. Loosen the compression rod forcing nut until spring tension is relieved and remove the forcing nut. Remove the compression rod and coil spring.

To install:

11. Place the insulator on top of the spring. Position the spring into the lower arm pocket. Make sure the spring pigtail is positioned between the 2 holes in the lower arm spring pocket.

12. Position the spring into the upper spring seat in the crossmember.

13. On all except Mustang with 5.0L engine, insert the compression rod through the control arm and spring, then hook it to the upper plate. The upper plate is installed with the hooks facing the center of the vehicle.

14. On Mustang with 5.0L engine, install the upper plate between the coils in the location marked during removal.

15. Install the lower plate, ball nut, thrust washer and bearing and forcing nut onto the compression rod.

16. Tighten the forcing nut, position the lower arm into the crossmember and install new lower arm-to-crossmember bolts and nuts. Do not tighten at this time.

17. Remove the spring compressor tool from the vehicle. Raise the suspension arm to a normal attitude position with a jack. Tighten the lower arm-to-crossmember attaching nuts to

110–150 ft. lbs. (149–203 Nm). Remove the jack.

18. Install the steering gear-to-crossmember bolts and nuts, if removed. Hold the bolts and tighten the nuts to 90–100 ft. lbs. (122–135 Nm).

19. Connect the stabilizer bar link to the lower suspension arm. Tighten the attaching nut to 8–12 ft. lbs. (12–16 Nm).

20. Position the tie rod into the steering spindle and install the retaining nut. Tighten the nut to 35 ft. lbs. (47 Nm) and continue tightening the nut to align the next castellation with the hole in the stud. Install a new cotter pin.

21. Install the brake caliper and the wheel and tire assembly. Lower the vehicle.

Upper Ball Joints

INSPECTION

Except 1989–92 Thunderbird and Cougar

1. Raise the vehicle and place floor jacks beneath the lower control arms.

2. Have an assistant grasp the bottom of the tire and move the wheel in and out.

3. As the wheel is being moved, observe the upper control arm where the spindle attaches to it. Any movement between the upper part of the spindle and the upper ball joint indicates a bad ball joint which must be replaced.

1989–92 Thunderbird and Cougar

1. Raise the vehicle and place jacks under the sub-frame. This will minimize the load on the ball joints.

2. Attach a dial indicator in such a way as to measure the lateral movement between the spindle and the arm.

3. Grasp the tire at the top and bottom and slowly move the tire in and out. Note the reading on the dial indicator. If the reading exceeds 0.015 in. (0.4mm), replace the ball joint.

Removal and Installation

1991–92 TOWN CAR AND 1992 CROWN VICTORIA AND GRAND MARQUIS

1. Raise and safely support the vehicle with safety stands behind the lower arm. Remove the wheel and tire assembly.

2. Position a floor jack under the lower arm at the lower ball joint area. The floor jack will support the spring load on the lower arm.

3. Remove the retaining nut and pinch bolt from the upper ball joint stud.

4. Mark the position of the alignment cams. When replacing the ball joint this will approximate the current alignment.

5. Remove the 2 nuts retaining the ball joint to the upper arm. Remove the ball joint and spread the slot with a suitable pry bar to separate the ball joint stud from the spindle.

To install:

NOTE: The upper ball joints differ from side to side. Be sure to use the proper ball joint on each side.

6. Position the ball joint on the upper arm and insert the ball stud into the spindle.

7. Install the pinch bolt and retaining nut. Tighten to 51–67 ft. lbs. (68–92 Nm).

8. Install the alignment cams to the approximate position at removal. If not marked, position to neutral position.

9. Install the 2 nuts attaching the ball joint to the arm. Hold the cams and tighten the nuts to 90–109 ft. lbs. (122–149 Nm).

10. Remove the floor jack from the lower arm and install the wheel and tire assembly. Remove the safety stands and lower the vehicle.

11. Check and adjust the front end alignment.

1988–90 Town Car and 1988–91 Crown Victoria and Grand Marquis

NOTE: Ford Motor Company recommends replacement of the control arm and ball joint as an assembly. However, aftermarket replacement parts are available, which can be installed using the following procedure.

1. Raise the vehicle and support on frame points so that the front wheels fall to their full down position. Remove the wheel and tire assembly.

2. Drill a ⅛ in. hole completely through each ball joint attaching rivet.

3. Using a large chisel, cut off the head of each rivet and drive them from the arm.

4. Place a jack under the lower arm and raise to compress the coil spring.

5. Remove the cotter pin and attaching nut from the ball joint stud.

6. Using a ball joint removal tool, loosen the ball joint stud from the spindle and remove the ball joint from the arm.

To install:

7. Clean all metal burrs from the arm and install the new ball joint, using the service part nuts and bolts to attach the ball joint. Do not attempt to rivet the ball joint once it has been removed.

8. Install the ball joint stud into the spindle. Tighten the ball joint-to-upper spindle nut to 60–90 ft. lbs. (81–122 Nm). Continue to tighten until the slot for the cotter pin is aligned. Install a new cotter pin.

9. Install the wheel and tire assembly and lower the vehicle. Check front end alignment.

1989–92 Thunderbird and Cougar

The ball joint is an integral part of the upper control arm. If the ball joint is defective, the entire upper control arm must be replaced.

Lower Ball Joints

INSPECTION

Except 1989–92 Thunderbird and Cougar

1. Support the vehicle in normal driving position with ball joints loaded.

2. Wipe the wear indicator and ball joint cover checking surface clean.

3. The checking surface should project outside the cover. If the checking surface is inside the cover, replace the lower arm assembly.

1989–92 Thunderbird and Cougar

1. Raise the vehicle and place jacks under the sub-frame. This will minimize the load on the ball joints.

2. Attach a dial indicator in such a way as to measure the lateral movement between the spindle and the arm.

3. Grasp the tire at the top and bottom and slowly move the tire in and out. Note the reading on the dial indicator. If the reading exceeds 0.015 in. (0.4mm), replace the ball joint.

REMOVAL & INSTALLATION

Except 1989–92 Thunderbird and Cougar

The ball joint is an integral part of the lower control arm. If the ball joint is defective, the entire lower control arm must be replaced.

1989–92 Thunderbird and Cougar

1. Remove the lower control arm.
2. Remove and discard the joint boot seal.
3. Press out the ball joint using ball joint remover tool D89P–3010–A and cup tool D84P–3395–A4 or equivalent, and a suitable press.

To install:

4. Install the ball joint with ball joint replacer tool D89P–3010–B, cup tool D84P–3395–A4 or equivalent, and a suitable press.

5. Make sure the ball joint is fully seated in the control arm and the ball joint seal is free of cuts or tears.

6. Install the lower control arm. Check the front end alignment.

Upper Control Arms

REMOVAL & INSTALLATION

1991–92 Town Car and 1992 Crown Victoria and Grand Marquis

1. Raise and safely support the vehicle on safety stands positioned on the frame just behind the lower arm.

2. Remove the wheel and tire assembly and position a floor jack under the lower arm.

3. Remove the retaining nut from the upper ball joint stud to spindle pinch bolt. Tap the pinch bolt to remove from the spindle.

4. Using a suitabel pry bar, spread the slot to allow the ball joint stud to release out of the spindle.

5. Remove the upper arm retaining bolts and the upper arm.

To install:

6. Transfer the rebound bumper from the old arm to the new arm, or replace the bumper if worn or damaged.

7. Use reference marks from the camber and caster cams as initial settings.

8. Position the upper arm shaft to the frame bracket. Install the 2 retaining bolts and washers. Position the arm in the center of the slot adjustment range and tighten to 100–140 ft. lbs. (136–190 Nm).

9. Connect the upper ball joint stud to the spindle and install the retaining pinch bolt and nut. Tighten the nut to 52–66 ft. lbs. (70–90 Nm).

10. Install the wheel and tire assembly and lower the vehicle. Check the front end alignment.

1988–90 Town Car and 1988–91 Crown Victoria and Grand Marquis

1. Raise and safely support the vehicle on safety stands positioned on the frame just behind the lower arm. Remove the wheel and tire assembly.

2. Remove the cotter pin from the upper ball joint stud nut. Loosen the nut a few turns but do not remove.

3. Install a ball joint press T57P–3006–B or equivalent between the upper and lower ball joint studs with the adapter screw on top.

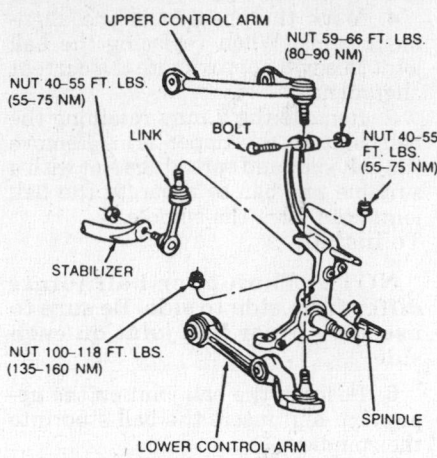

Front suspension assembly—1989-92
Thunderbird and Cougar

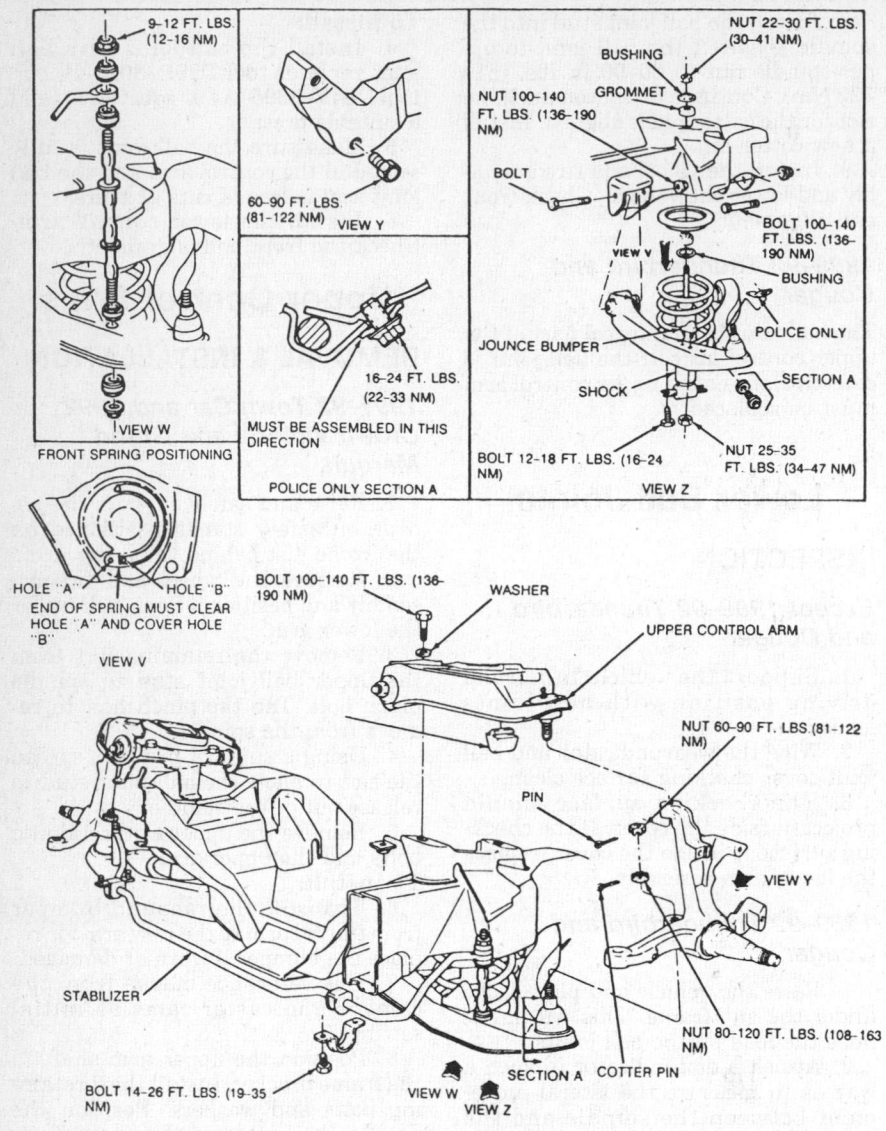

Front suspension assembly—1988-90 Town Car and 1988-91 Crown Victoria and Grand Marquis

NOTE: This tool should be seated firmly against the ends of both studs, not against the nuts or lower stud cotter pin.

4. With a wrench, turn the adapter screw until the tool places the stud under compression. Tap the spindle near the upper stud with a hammer to loosen the stud in the spindle.

NOTE: Do not loosen the stud from the spindle with tool pressure only. Do not contact the boot seal with the hammer.

5. Remove the tool from between the ball joint studs and place a floor jack under the lower arm.
6. Remove the upper arm attaching bolts and the upper arm.
To install:
7. Transfer the rebound bumper

from the old arm to the new arm, or replace the bumper if worn or damaged.
8. Position the upper arm shaft to the frame bracket. Install the 2 attaching bolts and washers. Tighten to 100–140 ft. lbs. (136–190 Nm).
9. Connect the upper ball joint stud to the spindle and install the attaching nut. Tighten the nut to 60–90 ft. lbs. (81–122 Nm). Continue to tighten the nut until the slot for the cotter pin is aligned. Install a new cotter pin.
10. Install the wheel and tire assembly and lower the vehicle. Check the front end alignment.

1989–92 Thunderbird and Cougar

1. Raise and safely support the vehicle. Remove the wheel and tire assembly.

2. Remove and discard the upper spindle-to-ball joint bolt and nut. Slightly spread the spindle at the slot and remove the ball joint.
3. Lower the vehicle. Break off the flags on the upper control arm pivot bolt heads.
4. Remove the upper control arm bolts and the control arm.
To install:
5. Position the upper control arm and install new bolts without the flags and nuts.
6. Hold the upper control arm at a horizontal position and tighten the nuts to 72–88 ft. lbs. (98–120 Nm).

NOTE: If it is necessary to tighten the bolts, due to nut access, tighten the bolts to 81–88 ft. lbs. (110–120 Nm).

7. Raise the vehicle. Attach the spindle to the upper control arm. Install a new bolt and nut from the front of the vehicle and tighten to 59–66 ft. lbs. (80–90 Nm).
8. Install the wheel and tire assembly and lower the vehicle. Check the front end alignment.

Lower Control Arms

REMOVAL & INSTALLATION

Crown Victoria, Grand Marquis and Town Car

1. Raise the front of the vehicle and position safety stands on the frame behind the lower control arms. Remove the wheel and tire assembly.
2. Remove the brake caliper and suspend with a length of wire; do not let the caliper hang by the brake hose. Remove the brake rotor and dust shield. Remove the anti-lock brake sensor, if equipped.
3. Remove the jounce bumper, inspect and save for installation if in

good condition. Remove the shock absorber.

4. Disconnect the stabilizer link from the lower arm , if necessary. Disconnect the steering center link from the pitman arm.

5. Remove the cotter pin and loosen the lower ball joint stud nut 1–2 turns.

NOTE: Do not remove the nut at this time.

6. Install a suitable ball joint press tool to place the ball joint stud under compression. With the stud under compression, tap the spindle sharply with a hammer to loosen the stud in the spindle. Remove the ball joint press tool.

7. Place a floor jack under the lower arm and install a suitable spring compression tool.

8. Remove the coil spring, the ball joint nut and remove the coil spring.

To install:

9. Position the arm assembly ball joint stud into the spindle and install the nut. Tighten to 80–120 ft. lbs. (108–163 Nm). Continue to tighten until the slot for the cotter pin is aligned. Install a new cotter pin.

10. Position the coil spring into the upper spring pocket and raise the lower arm, aligning the holes in the arm with the holes in the crossmember. Install the bolts and nuts with the washer installed on the front bushing. Do not tighten at this time.

NOTE: Make sure the pigtail of the lower coil of the spring is in the proper location of the seat on the lower arm.

11. Remove the spring compressor tool.

12. Connect the steering center link at the pitman arm and install the nut. Tighten to 43–47 ft. lbs. (59–63 Nm). Continue to tighten until the slot for the cotter pin is aligned. Install a new cotter pin.

13. Install the shock absorber and the jounce bumper.

14. Install the dust shield, rotor and caliper. Install the anti-lock brake sensor, if equipped.

15. Position the stabilizer link to the lower control arm and install the link, bushing and retaining nut. Tighten to 9–15 ft. lbs. (12–20 Nm).

16. Install the wheel and tire assembly and lower the vehicle. With the vehicle supported on the wheels and tires at normal curb height, tighten the lower control arm-to-crossmember bolts to 100–140 ft. lbs. (136–190 Nm).

17. Check the front end alignment.

Mark VII, Mustang and 1988 Thunderbird and Cougar

1. On Mark VII, turn the air suspension switch **OFF**.

2. Raise and safely support the vehicle. Allow the control arms to hang free. Remove the wheel and tire assembly.

3. If necessary, remove the brake caliper and suspend with a length of wire; do not let the caliper hang by the brake hose. Remove the brake rotor and dust shield.

4. Disconnect the tie rod end from the steering spindle. Disconnect the stabilizer bar link from the lower arm.

5. Remove the steering gear bolts and lower the gear out of the way to provide clearance, if necessary, for suspension arm bolt removal.

6. On Mark VII, disconnect the lower end of the height sensor from the lower control arm sensor mounting stud. Remove the sensor mounting stud and screw from the lower arm, noting the position of the stud on the lower arm bracket.

7. Remove the cotter pin and loosen the lower ball joint stud nut 1–2 turns. Do not remove the nut at this time. Tap the spindle boss sharply to relieve the stud pressure.

8. On all except Mark VII, install a suitable spring compressor and compress the spring so it is free in the seat. On Mark VII, proceed as follows:

a. Disconnect the electrical connector and then disconnect the air line.

b. Remove the solenoid clip.

c. Rotate the solenoid counterclockwise to the first stop.

d. Pull the solenoid straight out slowly to the second stop to bleed the air from the system.

e. Push in, then rotate clockwise to the first stop.

f. Install the solenoid clip. Inspect the wire harness connector and ensure the rubber gasket is in place at the bottom of the connector cavity.

g. Remove and discard the air spring-to-lower arm fastener clip.

9. Remove and discard the ball joint nut and raise he entire strut and spindle assembly. Wire out of the way to obtain working room.

10. Remove and discard the suspension arm-to-crossmember nuts and bolts. Remove the lower control arm and, except Mark VII, remove the coil spring.

To install:

11. On all except Mark VII, position the coil spring into the lower arm pocket. Make sure the spring pigtail is positioned between the 2 holes in the pocket.

12. Position the lower arm to the crossmember and install new arm-to-crossmember bolts and nuts. Do not tighten at this time.

13. On all except Mark VII, raise the control arm with a jack to a normal attitude position and remove the spring compressor.

14. On Mark VII, position the air spring in the arm and install the new fastener. Install the sensor mounting stud and screw to the lower arm in the same position as on the replaced arm. Connect the lower end of the sensor to the lower arm mounting stud. Raise the control arm to curb height with a jack.

15. With the jack in place, tighten the lower arm-to-crossmember attaching nuts to 110–150 ft. lbs. (149–203 Nm).

16. Tighten the ball joint stud nut to 100–120 ft. lbs. (136–163 Nm) and install a new cotter pin. Remove the jack.

17. Install the dust shield, rotor and brake caliper, if removed. Install the steering gear-to-crossmember bolts and nuts, if removed. Hold the bolts and tighten the nuts to 90–100 ft. lbs. (122–136 Nm).

18. Position the tie rod into the steering spindle and install the retaining nut. Tighten the nut to 35 ft. lbs. (47 Nm) and continue tightening the nut to align the next castellation with the hole in the stud. Install a new cotter pin.

19. Connect the stabilizer bar link to the lower control arm. Tighten the retaining nut to 9–12 ft. lbs. (12–16 Nm).

20. On all except Mark VII, install the wheel and tire assembly and lower the vehicle. Check the front end alignment.

21. On Mark VII, refill the air spring as follows:

a. Connect the air line and then the electrical connector to the air spring solenoid.

b. Turn the air suspension switch **ON**. The diagnostic pigtail must be ungrounded.

c. Connect a battery charger to reduce battery drain.

d. Cycle the ignition from the **OFF** to the **RUN** position, hold in the **RUN** position for a minimum of 5 seconds, then return to the **OFF** position. The drivers door is open with all other doors shut.

e. Ground the diagnostic pigtail by connecting a lead from the pigtail to vehicle ground. The pigtail must remain grounded during the spring fill sequence.

f. While applying the brakes, turn the ignition switch to the **RUN** position. The door must be open; do not start the vehicle. The warning indicator will blink continuously once every 2 seconds to indicate the spring pump sequence has been entered.

g. To fill the front spring(s), close and open the door twice. After a 6

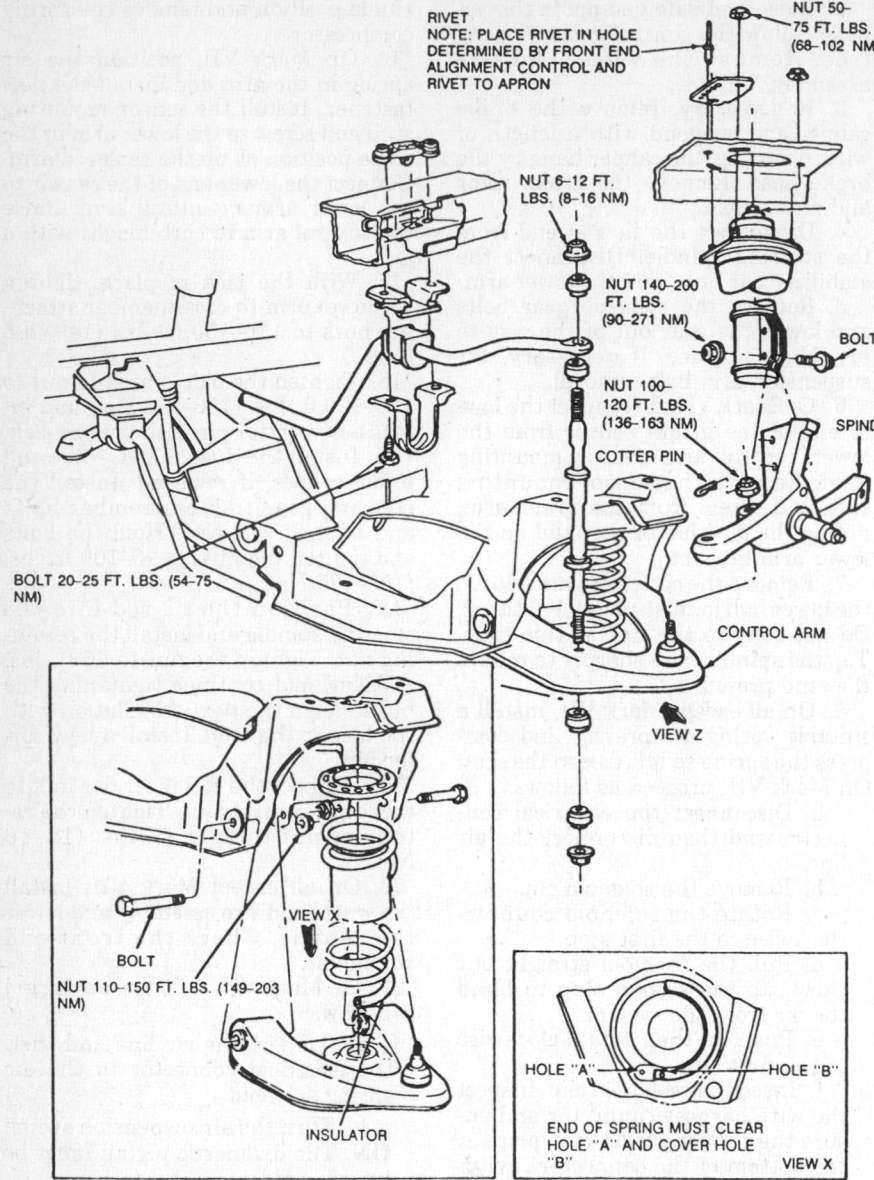

RIVET
NOTE: PLACE RIVET IN HOLE DETERMINED BY FRONT END ALIGNMENT CONTROL AND RIVET TO APRON

NUT 50–75 FT. LBS. (68–102 NM)

NUT 6–12 FT. LBS. (8–16 NM)

NUT 140–200 FT. LBS. (90–271 NM)

BOLT

NUT 100–120 FT. LBS. (136–163 NM)

COTTER PIN

SPINDLE

BOLT 20–25 FT. LBS. (54–75 NM)

CONTROL ARM

VIEW Z

VIEW X

BOLT

NUT 110–150 FT. LBS. (149–203 NM)

INSULATOR

HOLE "A" — HOLE "B"

END OF SPRING MUST CLEAR HOLE "A" AND COVER HOLE "B"

VIEW X

Front suspension assembly—Mustang and 1988 Thunderbird and Cougar

second delay, the front spring will be filled for 60 seconds.

h. Immediately after completion of the air spring fill, turn the air suspension switch **OFF** to prevent deflation of the air springs while the vehicle is raised. Inspect all springs for proper inflation; no folds or creases.

22. Install the wheel and tire assembly and lower the vehicle. Turn the air suspension switch **ON**. Check the front end alignment.

NOTE: Any further vehicle leveling will be done when the vehicle is in normal operation on the ground.

1989–92 Thunderbird and Cougar

1. Raise and safely support the ve-

hicle. Remove the wheel and tire assembly.

2. Loosen the ball joint nut 3–4 turns. Rap the spindle to separate the ball joint. Leave the nut attached.

3. Support the spindle with a wire. Mark the position of the camber adjusting cam. Remove the nut attaching the tension strut to the control arm.

4. Remove the lower strut bolt and remove the pivot bolt.

5. Remove the ball joint nut and remove the arm.

To install:

6. Position the control arm in the vehicle and loosely install the pivot bolt.

7. Install the tension strut washer and insulators and loosely install the strut to control arm attaching nut.

8. Install a new ball joint nut. In-

stall a new lower strut bolt and nut, but do not tighten at this time.

9. Tighten the ball joint nut to 81–118 ft. lbs. (110–160 Nm). Tighten the tension strut to control arm nut to 103–118 ft. lbs. (140–160 Nm).

10. Remove the wire holding the spindle. Install the wheel and tire assembly and lower the vehicle.

11. Push down on the front of the vehicle and release to neutralize the suspension. Tighten the lower strut nut to 140–162 ft. lbs. (190–220 Nm).

NOTE: The lower strut nut must be tightened with the vehicle weight on the wheels.

12. Align the camber marks at the pivot bolt and tighten the nut to 98–114 ft. lbs. (135–155 Nm).

13. Check the front end alignment.

Stabilizer Bar

REMOVAL & INSTALLATION

Except 1989–92 Thunderbird and Cougar

1. Raise the front of the vehicle and place jackstands under the lower control arms.

2. Disconnect the stabilizer bar from the links and the insulator mounting clamps. Remove the stabilizer bar.

3. Cut the worn insulators from the stabilizer bar.

4. Installation is the reverse of the removal procedure. Coat the necessary parts of the stabilizer bar with rubber lubricant prior to installation.

1989–92 Thunderbird and Cougar

1. Remove the air inlet tube. Remove the stabilizer bar retaining bracket bolts and brackets.

2. Remove the serpentine drive belt. Raise and safely support the vehicle.

3. Remove the wheel and tire assemblies. Remove the crankshaft vibration damper.

4. Remove the cotter pins and nuts from the tie rod ends. Separate the tie rod ends from the spindles.

5. Remove the transmission oil cooler line bracket. Remove the stabilizer bar to lower link retaining nuts.

6. Remove the stabilizewr bar link from the stabilizer bar using joint separator tool D88L-3006-A or equivalent. Be careful not to damage the ball joint seal.

7. Remove the stabilizer bar through the right wheel opening. Remove the bushings from the stabilizer bar.

To install:

8. Install the bushings onto the sta-

bilizer bar and position the bar in the vehicle.

9. Attach the stabilizer links to the bar and tighten the retaining nuts to 48–55 ft. lbs. (65–75 Nm).

10. Install the stabilizer bar bracket and retaining bolts. Tighten to 48–55 ft. lbs. (65–75 Nm).

11. Install the transmission oil cooler lines. Install the tie rod ends to the spindles. Tighten the nuts to 39–54 ft. lbs. (53–73 Nm) and install new cotter pins.

12. Install the crankshaft damper. Install the wheel and tire assemblies and lower the vehicle.

13. Install the serpentine drive belt and the air inlet tube.

Front Wheel Bearings

ADJUSTMENT

Except 1991–92 Town Car, 1992 Crown Victoria and Grand Marquis and 1989–92 Thunderbird and Cougar

1. Raise and safely support the front of the vehicle.

2. Remove the wheel cover and grease cap.

3. Remove the cotter pin and nut retainer.

4. Loosen the adjusting nut 3 turns and rock the wheel back and forth a few times to release the brake pads from the rotor.

5. While rotating the wheel and hub assembly in a counterclockwise direction, tighten the adjusting nut to 17–25 ft. lbs. (23–34 Nm).

6. Back off the adjusting nut ½ turn, then retighten to 10–28 inch lbs.(1.1–3.2 Nm).

7. Install the nut retainer and a new cotter pin. Check the wheel rotation. If it is noisy or rough, the bearings either need to be cleaned and repacked or re-

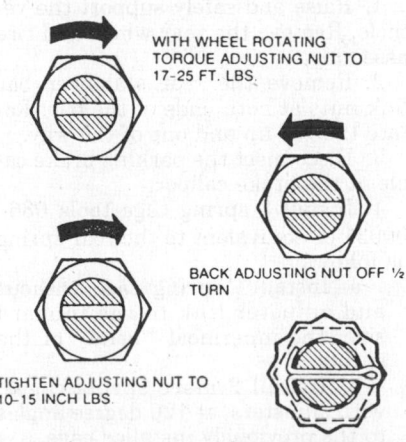

WITH WHEEL ROTATING TORQUE ADJUSTING NUT TO 17-25 FT. LBS.

BACK ADJUSTING NUT OFF ½ TURN

TIGHTEN ADJUSTING NUT TO 10-15 INCH LBS.

INSTALL THE LOCK AND NEW COTTER PIN

Wheel bearing adjustment procedure

placed. After adjustment is completed, replace the grease cap.

8. Lower the vehicle. Before driving the vehicle, pump the brake pedal several times to restore normal brake pedal travel.

1991–92 Town Car, 1992 Crown Victoria and Grand Marquis and 1989–92 Thunderbird and Cougar

The front wheel bearings are of a hub unit design and are pregreased, sealed and require no maintenance. The bearings are preset and cannot be adjusted.

REMOVAL & INSTALLATION

Except 1991–92 Town Car, 1992 Crown Victoria and Grand Marquis and 1989–92 Thunderbird and Cougar

1. Raise and support the vehicle safely. Remove the wheel and tire assembly and the caliper. Suspend the caliper with a length of wire; do not let it hang from the brake hose.

2. Pry off the dust cap. Tap out and discard the cotter pin. Remove the nut retainer.

3. Being careful not to drop the outer bearing, pull off the brake disc and wheel hub assembly.

4. Remove the inner grease seal using a prybar. Remove the inner wheel bearing.

5. Clean the wheel bearings with solvent and inspect them for pits, scratches and excessive wear. Wipe all the old grease from the hub and inspect the bearing races. If either bearings or races are damaged, the bearing races must be removed and the bearings and races replaced as an assembly.

6. If the bearings are to be replaced, drive out the races from the hub using a brass drift.

7. Make sure the spindle, hub and bearing assemblies are clean prior to installation.

To install:

8. If the bearing races were removed, install new ones using a suitable bearing race installer. Pack the bearings with a bearing packer. If a packer is not available, work as much grease as possible between the rollers and cages.

9. Coat the inner surface of the hub and bearing races with grease.

10. Install the inner bearing in the hub. Being careful not to distort it, install the oil seal with its lip facing the bearing. Drive the seal in until its outer edge is even with the edge of the

hub. Lubricate the lip of the seal with grease.

11. Install the hub/disc assembly on the spindle, being careful not to damage the oil seal.

12. Install the outer bearing, washer and spindle nut. Install the caliper and the wheel and tire assembly and adjust the bearings.

1991–92 Town Car, 1992 Crown Victoria and Grand Marquis and 1989–92 Thunderbird and Cougar

1. Raise and safely support the vehicle. Remove the wheel and tire assembly.

2. Remove and discard the grease cap from the hub.

3. Remove the brake caliper. Suspend the caliper with a length of wire; do not let it hang from the brake hose.

4. Remove the rotor. Remove and discard the wheel hub nut.

5. Remove the hub and bearing assembly.

To install:

6. Install the hub and bearing assembly. Install a new wheel hub nut and tighten to 238 ft. lbs. (322 Nm).

7. Install the rotor ans a new grease cap. Install the brake caliper.

8. Install the wheel and tire assembly and lower the vehicle.

REAR SUSPENSION

Shock Absorbers

REMOVAL & INSTALLATION

Crown Victoria, Grand Marquis and Town Car

WITHOUT AUTOMATIC LEVELING

1. If equipped with air suspension, turn the air suspension switch **OFF**.

2. Raise and safely support the vehicle. Make sure the rear axle is supported.

3. Remove the shock absorber retaining nut, washer and insulator from the stud on the upper side of the frame. Discard the nut. Compress the shock to clear the hole in the frame and remove the inner insulator and washer from the upper retaining stud.

NOTE: All vehicles are equipped with gas pressurized shock absorbers which will extend unassisted.

4. Remove the self-locking retaining nut and disconnect the shock absorber lower stud from the mounting bracket on the rear axle.

To install:

5. Prime the new shock absorber as follows:

    a. With the shock absorber right side up, extend it fully.

    b. Turn the shock upside down and fully compress it.

    c. Repeat the previous 2 steps at least 3 times to make sure any trapped air has been expelled.

6. Place the inner washer and insulator on the upper retaining stud and position the shock absorber with the stud through the hole in the frame.

7. While holding the shock absorber in position, install the outer insulator, washer and a new stud nut on the upper side of the frame. Tighten the nut to 14–26 ft. lbs. (19–35 Nm).

8. Extend the shock absorber and place the lower stud in the mounting bracket hole on the rear axle housing. Install a new self-locking nut and tighten to 52–85 ft. lbs. (70–115 Nm).

9. Lower the vehicle and, if equipped, turn the air suspension switch **ON**.

WITH AUTOMATIC LEVELING

NOTE: Disconnect the height sensor connector link before allowing the rear axle to hang free. Then, raise the vehicle on a hoist so the suspension arms hang free with the ignition switch in the OFF position. The rear shock absorbers will vent air through the compressor and a hissing noise will be heard. When the noise stops, the air lines can be disconnected. A residual pressure of 8–24 psi will remain in the air lines.

1. Disconnect the air line by pushing in on the retainer ring and pulling the line out.

2. Remove the top retaining nut, washer and bushing.

3. Remove the bottom retaining nut and washer. Remove the shock absorber.

To install:

4. Position the shock absorber and install the bottom retaining washer and nut. Tighten to 52–85 ft. lbs. (70–115 Nm).

5. Install the top bushing, washer and retaining nut. Tighten to 14–26 ft. lbs. (19–35 Nm).

NOTE: Check the rubber sleeve on the shock absorber to be sure it is not wrapped up. To assist in identifying wrap-up during installation, a white stripe is on the rubber sleeve and on the shock absorber body. The stripes should align. To correct a wrap-up condition, loosen the upper shock retaining nut and turn the shock to align the stripes. Retaighten the retaining nut.

6. Connect the air line to the shock absorber by pushing in on the retainer ring and installing the air line.

7. Connect the height sensor connecting link and lower the vehicle.

Mark VII, Thunderbird, Cougar and Mustang

1. On Mark VII, turn the air suspension switch **OFF**. On Thunderbird and Cougar, proceed as follows to remove the actuation assembly, if equipped with automatic ride control:

    a. Make sure the vehicle is on a flat surface and the ignition is in the **OFF** position.

    b. Remove the luggage compartment side trim panel and disconnect the actuator wiring connector.

    c. Squeeze the 2 actuator retaining tabs firmly inward with 1 hand and lift the actuator off the mounting bracket with the other hand.

    d. Grasp the actuator mounting bracket with water pump pliers and hold firmly. While holding the bracket, loosen the bracket retaining nut. Remove the bracket.

2. Raise the vehicle and support it by the rear axle housing. Open the luggage compartment. On Mustang 3-door, open the hatch back door.

3. Remove the trim panels, as necessary, to gain access to the shock absorber. Remove the shock absorber retaining nut washer and insulator.

4. Remove the shock absorber bolt washer and nut at the lower arm and remove the shock absorber.

NOTE: Vehicles are equipped with gas pressurized shock absorbers which will extend unassisted.

To install:

5. Prime the new shock absorber as follows:

    a. With the shock absorber right side up, extend it fully.

    b. Turn the shock upside down and fully compress it.

    c. Repeat the previous 2 steps at least 3 times to make sure any trapped air has been expelled.

6. Place the inner washer and insulator on the upper retaining stud and position the stud through the shock tower mounting hole.

7. Attach the lower end of the shock absorber with the retaining bolt and nut. Tighten the bolt to 45–60 ft. lbs. (61–81 Nm) on Mark VII. 1988 Thunderbird and Cougar and Mustang with handling package and 55–70 ft. lbs. (75–95 Nm) on Mustang without handling package. Tighten the nut to 110–120 ft. lbs. (150–162 Nm) on 1989–92 Thunderbird and Cougar.

8. Install the upper insulator, washer and retaining nut and tighten to 27 ft. lbs. (37–47 Nm).

9. On Thunderbird and Cougar, install the shock actuator, if necessary.

10. Lower the vehicle. On Mark VII, turn the air suspension switch **ON**.

Coil Springs

REMOVAL & INSTALLATION

Crown Victoria, Grand Marquis and Town Car

1. Raise the vehicle and support the rear axle housing. Place jack stands under the frame side rails.

2. Remove the rear stabilizer bar, if equipped.

3. Disconnect the lower studs of both rear shock absorbers from the mounting brackets on the axle tube.

4. Unsnap the right parking brake cable from the right upper arm retainer before lowering the axle.

5. Lower the axle housing until the coil springs are released. Remove the springs and insulators.

To install:

6. Position the spring in the upper and lower seats with an insulator between the upper end of the spring and frame seat.

7. Raise the axle and connect the shock absorbers to the mounting brackets. Install new retaining nuts and tighten to 52–85 ft. lbs. (70–115 Nm).

8. Snap the right parking cable into the upper arm retainer. Install the stabilizer bar, if equipped.

9. Remove the jack stands and lower the vehicle.

1989–92 Thunderbird and Cougar

1. Raise and safely support the vehicle. Remove the rear wheel and tire assembly.

2. Remove the rear stabilizer bar link nuts at both ends of the bar. Rotate the bar up and out of the way.

3. Disconnect the parking brake cable at the brake caliper.

4. Install 3 spring cage tools 086-00031 or equivalent to the rear spring as follows:

    a. Install 1 spring cage without and adjuster link to the inboard side, the innermost "bend" of the spring.

    b. Install 2 more spring cages, with adjusters, at 120 degree angles to the previously installed cage.

5. Place a jack under the lower rear control arm as far outboard as possible.

REAR WHEEL DRIVE FORD MOTOR CO. 8

6. Support the rear knuckle and caliper assembly by wiring the upper control arm to the body.

7. Remove the lower shock absorber mounting bolt and nut. Mark the toe adjustment cam-to-subframe position and loosen both inboard pivot bolts on the lower control arm.

NOTE: The control arm must not be lowered until the pivot bolts are loose. Do not attempt to remove the plastic cap on the front pivot nut.

8. Remove the 2 bolts and nuts attaching the lower control arm to the knuckle. Lower the control arm by lowering the jack. Make sure the spring cages properly seat on the spring as the control arm is dropped.

9. Remove the jack, pull the control arm down fully by hand and remove the rear spring with the cages in place. Remove the spring insulators, if necessary.

10. If the springs are to be replaced, use a suitable coil spring compressor to compress the spring and remove the spring cages.

To install:

11. If a new spring is to be installed, it first must be compressed and caged. Compress the spring to the length of the original spring. If replacing a broken spring, compress the spring to approximately 10½ in. (267mm).

12. Install the spring insulators, if removed. Install the spring, with the cages in place, onto the upper and lower control arm seats.

NOTE: The short cage, without the adjuster, must be inboard. the spring pigtails may be in any position.

13. Position 2 jack stands under the front bumper reinforcement to prevent the rear of the vehicle lifting off the hoist.

14. Position a jack under the lower control arm and raise the lower control arm up to the knuckle bores. Make sure the spring seats properly. Install the bolts and nuts attaching the lower control arm to the knuckle and tighten the bolts to 118–148 ft. lbs. (160–200 Nm).

15. Remove the wire supporting the knuckle, caliper and upper control arm. Install the lower shock absorber mount bolt and nut and tighten the nut to 110–120 ft. lbs. (150–162 Nm).

16. Remove the jack and the jack stands. Remove the spring cages.

17. Connect the parking brake cable to the caliper. Install the rear stabilizer bar links and retaining nuts.

18. Install the wheel and tire assembly and lower the vehicle.

19. Set the toe adjustment cam to the mark made at the time of removal.

Tighten the front lower control arm-to-sub-frame nut to 184–229 ft. lbs. (250–310 Nm). Tighten the rear lower control arm-to-sub-frame nut to 125–170 ft. lbs. (170–230 Nm).

20. Check the rear wheel toe setting and adjust as necessary.

Mustang and 1988 Thunderbird and Cougar

1. Raise and safely support the vehicle. Support the body at the rear body crossmember.

2. Remove the stabilizer bar, if equipped.

3. Support the axle with a suitable jack.

4. Place another jack under the lower arm axle pivot bolt. Remove and discard the bolt and nut. Lower the jack slowly until the coil spring load is relieved.

5. Remove the coil spring and insulator from the vehicle.

To install:

6. Place the upper spring insulator on top of the spring. Place the lower spring insulator on the lower arm.

7. Position the coil spring on the lower arm spring seat, so the pigtail on the lower arm is at the rear of the vehicle and pointing toward the left side of the vehicle.

8. Slowly raise the jack until the arm is in position. Insert a new rear pivot bolt and nut with the nut facing outward. Do not tighten at this time.

9. Raise the axle to curb height. Tighten the lower arm-to-axle pivot bolt to 70–100 ft. lbs. (95–135 Nm).

10. Install the stabilizer bar, if equipped. Remove the crossmember supports and lower the vehicle.

Rear Control Arms

REMOVAL & INSTALLATION

Crown Victoria, Grand Marquis and Town Car

UPPER ARM

NOTE: If both arms are to be replaced, remove and install 1 at a time to prevent the axle from rolling or slipping sideways.

1. If equipped, turn the air suspension switch **OFF**.

2. Raise the vehicle and support the frame side rails with jack stands.

3. Support the rear axle under the differential pinion nose as weel as under the axle.

4. Unsnap the parking brake cable from the upper arm retainer. If equipped, disconnect the height sensor from the ball stud on the left upper control arm.

5. Remove and discard the nut and

bolt retaining the upper arm to the axle housing. Disconnect the arm from the housing.

6. Remove and discard the nut and bolt retaining the upper arm to the frame bracket and remove the arm.

To install:

7. Hold the upper arm in place on the front arm bracket and install a new retaining bolt and self-locking nut. Do not tighten at this time.

8. Secure the upper arm to the axle housing with new retaining bolts and nuts. The bolts must be pointed toward the front of the vehicle.

9. Raise the suspension with a jack until the upper arm rear pivot hole is in position with the hole in the axle bushing. Install a new pivot bolt and nut with the nut facing inboard.

10. Tighten the upper arm-to-axle pivot bolts to 103–133 ft. lbs. (140–180 Nm) and upper arm-to-frame pivot bolts to 120–150 ft. lbs. (162–203 Nm).

11. Snap the parking brake cable into the upper arm retainer. Connect the height sensor to the ball stud on the left upper arm, if equipped.

12. Remove the supports from the frame and axle and lower the vehicle. If equipped, turn the air suspension switch **ON**.

LOWER ARM

1. If equipped, turn the air suspension switch **OFF**.

2. Raise the vehicle and support the frame side rails with jack stands.

3. Remove the stabilizer bar, if equipped.

4. Support the axle with jack stands under the differential pinion nose as well as under the axle.

5. Remove and discard the lower arm pivot bolts and nuts and remove the lower arm.

To install:

6. Position the lower arm to the frame bracket and axle. Install new bolts and nuts.

7. Raise the axle. Tighten the lower arm-to-axle pivot bolt to 103–133 ft. lbs. (140–180 Nm) and lower arm-to-frame pivot bolt to 120–150 ft. lbs. (162–203 Nm).

8. Install the stabilizer bar, if equipped.

9. Remove the jack stands and lower the vehicle. If equipped, turn the air suspension switch **ON**.

Mark VII

UPPER ARM

NOTE: If 1 arm needs to be replaced, replace the other arm also.

1. Turn the air suspension switch **OFF**.

2. Raise and safely support the ve-

hicle. Allow the suspension to be at full rebound.

3. On the right side, disconnect the rear height sensor from the side arm. Note position of the sensor adjustment bracket on the upper arm.

4. Remove and discard the upper arm pivot bolts and nuts and remove the upper arm.

To install:

5. Place the upper arm into position and install new pivot bolts and nuts. At the body bracket, the nut must face outboard. At the axle, the nut must face inboard. Do not tighten at this time.

6. Connect the rear height sensor to the arm. Set the adjustment bracket to the same position as on the replaced arm and tighten the nut to 6–10 ft. lbs. (8–14 Nm).

7. Using a suitable jack, raise the axle to curb height. Tighten the front upper arm bolt to 80–105 ft. lbs. (108–142 Nm) and the rear upper arm bolt to 70–100 ft. lbs. (95–135 Nm).

8. Remove the supports and lower the vehicle. Turn the air suspension switch **ON**.

LOWER ARM

NOTE: If 1 arm needs to be replaced, replace the other arm also.

1. Turn the air suspension switch **OFF**.

2. Raise and safely support the vehicle. Allow the suspension to be at full rebound.

3. Remove the wheel and tire assembly.

4. Vent the air springs to atmospheric pressure as follows:

   a. Disconnect the electrical connector and then disconnect the air line.

   b. Remove the solenoid clip.

   c. Rotate the solenoid counterclockwise to the first stop.

   d. Pull the solenoid straight out slowly to the second stop to bleed the air from the system.

   e. Push in, then rotate clockwise to the first stop.

   f. Install the solenoid clip. Inspect the wire harness connector and ensure the rubber gasket is in place at the bottom of the connector cavity.

   g. Reconnect the air line and electrical connector.

5. Remove and discard the 2 air spring-to-lower control arm bolts and remove the air spring from the lower arm.

6. Remove and discard the bolts and remove the arm from the vehicle.

To install:

6. Position the control arm and install new pivot bolts and nuts with the nuts facing outward. Do not tighten at this time.

7. Install 2 new air spring-to-arm bolts, but do not tighten at this time.

8. Using a jack, raise the axle to curb height. Tighten the lower arm front bolt to 80–105 ft. lbs. (108–142 Nm) and the rear bolt to 70–100 ft. lbs. (95–135 Nm).

9. Tighten the air spring-to-arm bolt to 25–35 ft. lbs. (34–48 Nm). Make sure the air sprin piston is flat on the lower arm. Remove the jack.

10. Refill the air spring as follows:

   a. Turn the air suspension switch **ON**. The diagnostic pigtail must be ungrounded.

   b. Connect a battery charger to reduce battery drain.

   c. Cycle the ignition from the **OFF** to the **RUN** position, hold in the **RUN** position for a minimum of 5 seconds, then return to the **OFF** position. The drivers door is open with all other doors shut.

   d. Ground the diagnostic pigtail by connecting a lead from the pigtail to vehicle ground. The pigtail must remain grounded during the spring fill sequence.

   e. While applying the brakes, turn the ignition switch to the **RUN** position. The door must be open; do not start the vehicle. The warning indicator will blink continuously once every 2 seconds to indicate the spring pump sequence has been entered.

   f. To fill the rear spring(s), close and open the door once. After a 6 second delay, the rear spring will be filled for 60 seconds.

   g. Immediately after completion of the air spring fill, turn the air suspension switch **OFF** to prevent deflation of the air springs while the vehicle is raised. Inspect all springs for proper inflation; no folds or creases.

11. Install the wheel and tire assembly and lower the vehicle. Turn the air suspension switch **ON**.

NOTE: Any further vehicle leveling will be done when the vehicle is in normal operation on the ground.

1989–92 Thunderbird and Cougar

UPPER ARM

1. Raise and safely support the vehicle. Remove the rear wheel and tire assembly.

2. Support the knuckle and hub assembly so it cannot swing outward.

3. Remove the inner and outer pivot bolts and nuts at the upper control arm and remove the arm.

To install:

4. Install the upper arm. Loosely install the bolts and nuts.

NOTE: The inner pivot bolt used for camber adjustment has a specially shaped washer under the bolt head. Make sure fasteners are used in the correct locations.

5. Install the wheel and tire assembly and lower the vehicle.

6. Tighten the outboard nut to 118–148 ft. lbs. (160–200 Nm).

7. Set the camber and tighten the inner pivot nut to 81–98 ft. lbs. (110–133 Nm).

LOWER ARM

1. Remove the coil spring.

2. Remove the inner control arm pivot bolts and nuts and remove the arm.

NOTE: Do not attempt to remove the plastic cap on the front pivot nut.

3. Remove the toe compensating link from the control arm.

To install:

4. Inspect the large nut used at the inner front arm attachment fro condition of plastic cap. Use a new nut if the cap is cracked, loose or missing.

5. Install the toe compensating link on the arm.

6. Install the lower control arm to the sub-frame and loosely install the pivot bolts and nuts.

7. Tighten the toe compensating link nut to 118–148 ft. lbs. (160–200 Nm).

8. Install the spring and reattach the control arm at the knuckle.

9. Check and adjust the rear toe.

Mustang and 1988 Thunderbird and Cougar

UPPER ARM

NOTE: If 1 arm needs to be replaced, replace the other arm also.

1. Raise and safely support the vehicle at the rear crossmember.

2. Remove and discard the upper arm pivot bolts and nuts and remove the control arm.

To install:

3. Place the upper arm into the bracket of the body side rail. Install a new pivot bolt and nut with the nut facing outboard. Do not tighten at this time.

4. Using a jack, raise the suspension until the upper arm-to-axle pivot hole is in position with the hole in the axle bushing. Install a new pivot bolt and nut with the nut facing inboard.

5. Raise the suspension to curb height. Tighten the front upper arm bolt to 70–100 ft. lbs. (95–135 Nm) and the rear upper arm bolt to 80–105 ft. lbs. (108–142 Nm).

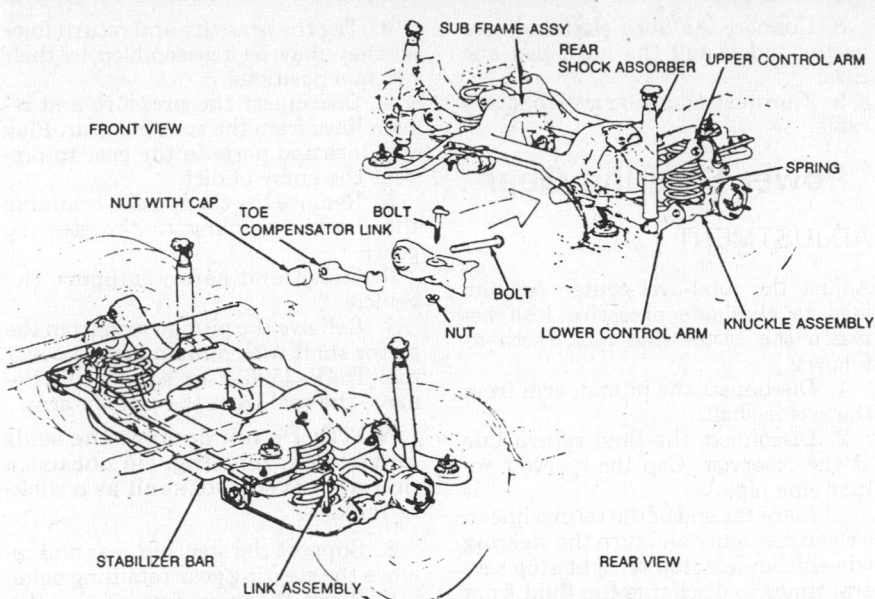

Rear suspension assembly—1989–92 Thunderbird and Cougar

6. Remove the supports and lower the vehicle.

LOWER ARM

NOTE: If 1 arm needs to be replaced, replace the other arm also.

1. Raise and safely support the vehicle at the rear crossmember.
2. Remove the stabilizer bar, if equipped.
3. Place a jack under the lower arm-to-axle pivot bolt. Remove and discard the bolt and nut. Lower the jack slowly until the coil spring can be removed.
4. Remove and discard the lower arm-to-frame pivot bolt and nut. Remove the lower arm.

To install:

5. Position the lower arm assembly into the front arm bracket. Install a new pivot bolt and nut with the nut facing outwards. Do not tighten at this time.
6. Position the coil spring on the lower arm spring seat, so the pigtail on the lower arm is at the rear of the vehicle and pointing toward the left side of the vehicle.
7. Slowly raise the jack until the arm is in position. Insert a new rear pivot bolt and nut with the nut facing outward. Do not tighten at this time.
8. Raise the axle to curb height. Tighten the lower arm front bolt to 80–105 ft. lbs. (108–142 Nm) and the rear bolt to 70–100 ft. lbs. (95–135 Nm).
9. Install the stabilizer bar, if equipped. Remove the crossmember supports and lower the vehicle.

STEERING

Steering Wheel
——— CAUTION ———

If equipped with an air bag, the negative battery cable and air bag backup power supply must be disconnected, before working on the system. Failure to do so may result in deployment of the air bag and possible personal injury.

REMOVAL & INSTALLATION

With Air Bag

1. Center the front wheels to the straight ahead position.

2. Disconnect the negative battery cable and air bag backup power supply.

NOTE: The backup power supply allows air bag deployment if the battery or battery cables are damaged in an accident before the crash sensors close. The power supply is a capacitor that will leak down in approximately 15 minutes after the battery is disconnected or in 1 minute if the battery positive cable is grounded. The backup power supply must be disconnected before any air bag related service is performed.

3. Remove the 4 air bag module retaining nuts and lift the module off the steering wheel. Disconnect the electrical connector from the air bag module and remove the module.

——— CAUTION ———

When carrying a live air bag, make sure the bag and trim cover are pointed away from the body. In the unlikely event of an accidental deployment, the bag will then deploy with minimal chance of injury. In addition, when placing a live air bag on a bench or other surface, always face the bag and trim cover up, away from the surface. This will reduce the motion of the air bag if it is accidentally deployed.

4. Disconnect the cruise control wire harness from the steering wheel, if equipped.
5. Remove and discard the steering wheel bolt. Remove the steering wheel using a suitable puller. Route the contact assembly wire harness through the steering wheel as the wheel is lifted off the shaft.

NOTE: Do not use a knock-off type steering wheel puller or strike the retaining bolt with a hammer. This could cause damage to the steering shaft bearing.

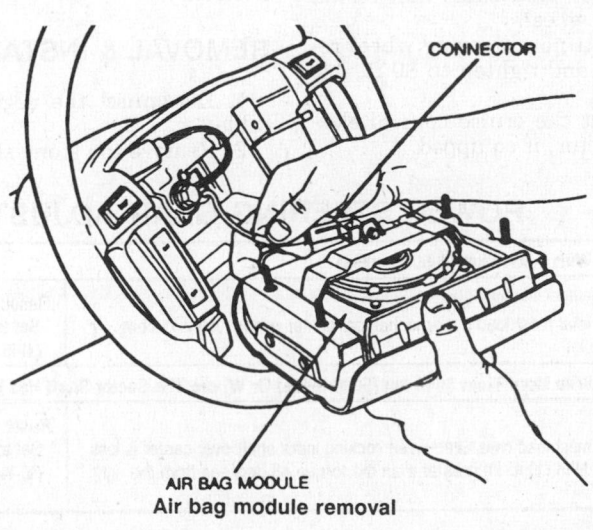

Air bag module removal

To install:

6. Make sure the front wheels are in the straight ahead position.

7. Route the contact assembly wire harness through the steering wheel opening at the 3 o'clock position and install the steering wheel on the steering shaft. The steering wheel and shaft alignment marks should be aligned. Make sure the air bag contact wire is not pinched.

8. Install a new steering wheel retaining bolt and tighten to 23–33 ft. lbs. (31–48 Nm).

9. If equipped, connect the cruise control wire harness to the wheel and snap the connector assembly into the steering wheel clip. Make sure the wiring does not get trapped between the steering wheel and contact assembly.

10. Connect the air bag wire harness to the air bag module and install the module to the steering wheel. Tighten the module retaining nuts to 3–4 ft. lbs. (4–6 Nm).

11. Connect the air bag backup power supply and negative battery cable. Verify the air bag warning indicator.

Without Air Bag

1. Disconnect the negative battery cable.

2. Remove the horn pad and cover assembly. Disconnect the horn electrical connector.

3. Disconnect the cruise control switch electrical connector, if equipped.

4. Remove and discard the steering wheel bolt. Remove the steering wheel using a suitable puller.

NOTE: Do not use a knock-off type steering wheel puller or strike the retaining bolt with a hammer. This could cause damage to the steering shaft bearing.

To install:

5. Align the index marks on the steering wheel and shaft and install the steering wheel.

6. Install a new steering wheel retaining bolt and tighten to 30 ft. lbs. (41 Nm).

7. Connect the cruise control electrical connector, if equipped.

8. Connect the horn electrical connector and install the horn pad and cover.

9. Connect the negative battery cable.

Power Steering Gear

ADJUSTMENT

Adjust the total-over-center position load to eliminate excessive lash between the sector and rack teeth as follows:

1. Disconnect the pitman arm from the sector shaft.

2. Disconnect the fluid return line at the reservoir. Cap the rservoir return line pipe.

3. Place the end of the return line in a clean container and turn the steering wheel from left stop to right stop several times to discharge the fluid from the gear.

4. Turn the steering wheel to 45 degrees from the left stop.

5. Using a feet pound torque wrench on the steering wheel nut, determine the torque required to rotate the shaft slowly approximately ¼ turn from the 45 degree position. If equipped with tilt column, place the steering wheel in the center tilt position.

6. Turn the steering wheel back to center and determine the torque required to rotate the shaft back and forth across the center position. If the reading is not to specification, loosen the nut and turn the adjuster screw until the reading is to specification. Tighten the wheel nut while holding the screw in place.

7. Check the readings and replace the pitman arm and steering wheel hub cover.

8. Connect the fluid return line to the reservoir and fill the reservoir. Check the belt tension and adjust, if necessary.

REMOVAL & INSTALLATION

1. Disconnect the negative battery cable.

2. Remove the stone shield.

3. Tag the pressure and return lines so they may be reassembled in their original positions.

4. Disconnect the pressure and return lines from the steering gear. Plug the lines and ports in the gear to prevent the entry of dirt.

5. Remove the clamp bolts retaining the flexible coupling to the steering gear.

6. Raise and safely support the vehicle.

7. Remove the pitman arm from the sector shaft with pitman arm remover tool T64P–3590–F or equivalent. Remove the tool from the pitman arm.

NOTE: Do not damage the seals and/or gear housing. Do not use a non-approved tool such as a pickle fork.

8. Support the steering gear and remove the steering gear retaining bolts.

9. Work the gear free of the flex coupling and remove the gear.

10. If the flex coupling did not come off with the gear, lift it off the shaft.

To install:

11. Turn the steering wheel to the straight ahead position.

12. Center the steering gear input shaft with the indexing flat facing downward on all except 1991–92 Town Car and 1992 Crown Victoria and Grand Marquis. On 1991–92 Town Car and 1992 Crown Victoria and Grand Marquis, center the steering gear input shaft with the centerline of the 2 indexing flats at 4 o'clock.

13. Slide the steering gear input shaft into the flex coupling and into place on the frame side rail. Install the retaining bolts and tighten to 50–65 ft. lbs. (68–88 Nm).

14. Make sure the wheels are in the straight ahead position. Install the pitman arm on the sector shaft and install the lockwasher and nut. Tighten the nut to 200–250 ft. lbs. (271–339 Nm). Install and tighten the sector shaft and retaining bolts.

15. Move the flex coupling into place on the steering gear input shaft. Install the retaining bolt and tighten to 20–30 ft. lbs. (27–41 Nm).

16. Connect the prssure and return lines to the steering gear and tighten

POWER STEERING GEAR ADJUSTMENT SPECIFICATIONS

Vehicles With 0-8045 km (0-5000 miles)	
Checking: Reset if total meshload over mechanical center is not 1.69-2.71 N-m (15-24 lb-in)	Reset: Set torque measured rocking across center to a value 1.24-1.69 N-m (11-15 lb-in) greater than that measured 45 degrees from the right stop.
Vehicles With More Than 8045 km (5000 miles) Or Where The Sector Shaft Has Been Replaced	
Checking: Reset if meshload measured while rocking input shaft over center is less than 1.2 N-m (10 lb-in) greater than the torque 45 degrees from the right stop.	Reset: Set torque measured rocking across center to a value 1.13-1.58 N-m (10-14 lb-in) greater than that measured 45 degrees from the right stop.

the lines. Fill the reservoir and turn the steering wheel from stop-to-stop to distribute the fluid. Check the fluid level and add fluid, if necessary.

17. Start the engine and turn the steering wheel from left to right. Check for leaks. Install the stone shield.

Power Steering Rack and Pinion

ADJUSTMENT

Rack Yoke Plug Clearance

The rack yoke plug clearance adjustment is not a normal service adjustment. It is only required when the input shaft and valve assembly is removed.

1. Clean the exterior of the steering rack thoroughly.

2. Install 2 long bolts and washers through the bushings and attach the rack to bench mounted holding fixture T57L–500–B or equivalent.

3. Do not remove the external pressure lines, unless they are leaking or damaged. If the lines are removed, install new seals. If the lines are damaged, they must be replaced.

4. Drain the power steering fluid by rotating the input shaft lock-to-lock twice using pinion shaft torque adapter tool T74P–3504–R or equivalent. Cover the ports on the valve housing with a shop cloth while draining the gear to avoid possible oil sparay.

5. Insert an inch pound torque wrench with a maximum capacity of 30–60 inch lbs. (3.39–6.77 Nm) into the pinion shaft torque adapter tool. Position the adapter and wrench on the input shaft splines.

6. Loosen the yoke plug locknut with pinion housing locknut wrench T78P–3504–H or equivalent. Loosen the yoke plug with a ¾ in. socket wrench.

7. With the rack at the center of travel, tighten the yoke plug to 45–50 inch lbs. (5.0–5.6 Nm). Clean the threads of the yoke plug prior to tightening to prevent a false reading.

8. Back off the yoke plug approximately ⅛ turn, 44 degrees minimum to 54 degrees maximum until the torque required to initiate and sustain rotation of the input shaft is 7–18 inch lbs. (0.79–2.03 Nm) for base power steering or 7–24 inch lbs. (0.79–2.71 Nm) for handling package.

9. Place pinion housing yoke locknut wrench T78P–3504–H or equivalent, on the yoke plug locknut. While holding the yoke plug, tighten the locknut to 44–66 ft. lbs. (60–89 Nm).

NOTE: Do not allow the yoke

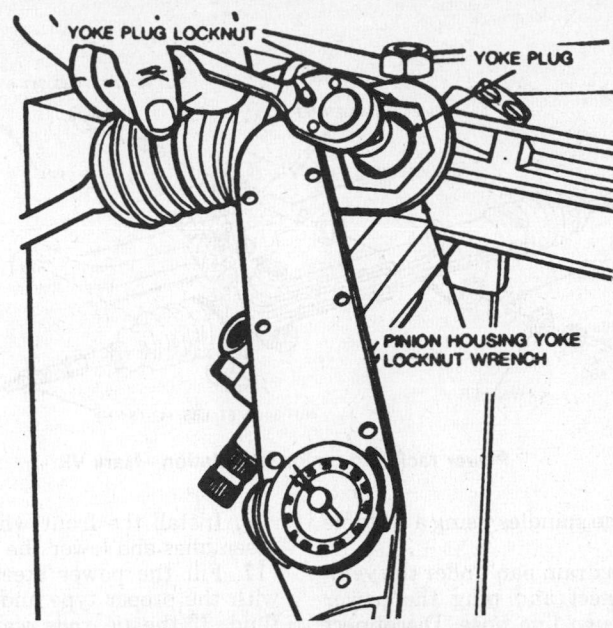

Rack yoke plug clearance adjustment

plug to move while tightening or the preload will be affected.

10. Install the steering rack in the vehicle.

REMOVAL & INSTALLATION

Except 1989–92 Thunderbird and Cougar

1. Disconnect the negative battery cable. Turn the ignition switch to the **RUN** position.

2. On Mark VII, turn **OFF** the air suspension switch, located in the luggage compartment.

3. Raise and safely support the vehicle. Position a drain pan to catch the fluid from the power steering lines.

4. Remove the 1 bolt retaining the flexible coupling to the input shaft.

5. Remove the front wheel and tire assemblies. Remove the cotter pins and nuts from the tie rod ends and separate the tie rod studs from the spindles.

6. Remove the 2 nuts, insulator washers and bolts retaining the steering rack to the crossmember. Remove the front rubber insulators.

7. Position the rack to allow access to the hydraulic lines and disconnect the lines.

8. Remove the steering rack.

To install:

9. Install new plastic seals on the hydraulic line fittings.

10. Install the rack on the mounting spikes and install the hydraulic lines. Tighten the fittings to 10–15 ft. lbs. (14–20 Nm) on 1988–89 vehicles and 20–25 ft. lbs. (27–33 Nm) on 1990–92 vehicles.

NOTE: The hoses are designed to swivel when properly tightened. Do not attempt to eliminate looseness by over-tightening the fittings.

11. Install the front rubber insulators. Make sure all rubber insulators are pushed completely inside the gear housing before installing the mounting bolts.

12. Insert the input shaft into the flexible coupling. Install the mounting bolts, insulator washers and nuts. Tighten the nuts to 30–40 ft. lbs. (41–54 Nm) while holding the bolts. Install and tighten the flexible coupling bolt to 20–30 ft. lbs. (28–40 Nm).

13. Connect the tie rod ends to the spindle arms and install the retaining nuts. Tighten to 35–47 ft. lbs. (48–63 Nm). After tightening, tighten the nuts to their nearest cotter pin castellation and install 2 new cotter pins.

14. Lower the vehicle. Turn the ignition switch to **OFF** and connect the negative battery cable. On Mark VII, turn the air suspension switch **ON**.

15. Fill the power steering system with the proper type and quantity of fluid. If the tie rod ends were loosened, check and adjust the front end alignment.

1989–92 Thunderbird and Cougar

1. Disconnect the negative battery cable. Raise and safely support the vehicle.

2. Remove the front wheel and tire assemblies.

3. Remove the cotter pins and nuts from the tie rod ends. Separate the tie

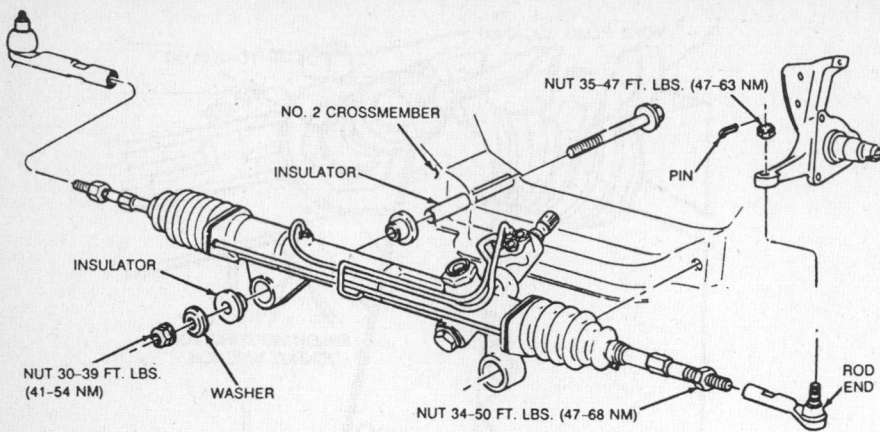

Power rack and pinion installation—Mark VII

Labels in diagram:
- NO. 2 CROSSMEMBER
- INSULATOR
- NUT 35–47 FT. LBS. (47–63 NM)
- PIN
- INSULATOR
- NUT 30–39 FT. LBS. (41–54 NM)
- WASHER
- NUT 34–50 FT. LBS. (47–68 NM)
- ROD END

rods from the spindles using a suitable tool.

4. Place a drain pan under the vehicle. Disconnect and plug the power steering return line hose. Disconnect the power steering pressure line at the intermediate fitting and position out of the way.

5. Remove the steering shaft retaining bolt. Remove the rack-to-subframe bolts and nuts. The nuts are accessed through the hole in the front crossmember.

6. Lower the rack as necessary to remove the pressure line inlet tube. Remove and discard the plastic seal on the inlet tube. Cut the tie strap securing the pressure line to seach tube.

7. Remove the steering rack from the vehicle.

To install:

8. Install a new seal on the pressure line inlet tube.

9. Install the insulators from the rear side of the rack housing making sure they are fully seated. Use a suitable rubber lubricant to aid in installation.

10. Install and position the rack to the front crossmember. Install the pressure line inlet tube to the rack.

11. Align the steering shaft to allow the rack to completely seat on the crossmember. Install the steering rack retaining bolts and nuts. Tighten the bolts to 100–144 ft. lbs. (135–195 Nm).

12. Install the steering shaft retaining bolt and tighten to 20–30 ft. lbs. (28–40 Nm).

13. Secure the pressure line to the rack tube with a new tie strap. Connect the power steering pressure line.

14. Connect the power steering return line and tighten the clamp to 12–18 inch lbs. (1.4–2.0 Nm).

15. Install the outer tie rod ends to the spindles. Install the nuts and tighten to 39 ft. lbs. (53 Nm). Continue to tighten the nuts until the castellations line up with the stud bores, then install new cotter pins.

16. Install the front wheel and tire assemblies and lower the vehicle.

17. Fill the power steering system with the proper type and quantity of fluid. If the tie rods were loosened, check and adjust the front end alignment.

Power Steering Pump

REMOVAL & INSTALLATION

NOTE: On the 3.8L SC engine, the intercooler and intercooler tubes must be removed to gain access to the power steering pump.

1. Disconnect the negative battery cable.

2. Disconnect the fluid return hose at the reservoir and drain the fluid into a container.

3. Remove the pressure hose from the pump and, if necessary, drain the fluid into a container. Do not remove the fitting from the pump.

4. On all except 4.6L engine, remove the pump mounting bracket. On 4.6L engine, remove the 4 pump retaining bolts. Disconnect the belt from the pulley and remove the pump.

5. On engines with the fixed pump system, remove the pulley before removing the pump.

To install:

6. On non-fixed pump systems, install the pulley on the pump, if removed.

7. On all except 4.6L engine, place the pump on the mounting bracket and install the bolts at the front of the pump. Tighten to 30–45 ft. lbs. (40–62 Nm).

8. On 4.6L engine, place the pump on the mounting bosses of the engine block and install the bolts at the side of the pump. Tighten to 15–22 ft. lbs. (20–30 Nm).

9. On fixed pump systems, install the pulley.

10. Place the belt on the pump pulley and adjust the tension, if necessary.

11. Install the pressure hose to the pump fitting. Tighten the tube nut with a tube nut wrench rather than with an open-end wrench. Tighten to 20–25 ft. lbs. (27–34 Nm).

NOTE: Do not overtighten this fitting. Swivel and/or end play of the fitting is normal and does not indicate a loose fitting. Overtightening the tube nut can collapse the tube nut wall, resulting in a leak and requiring replacement of the entire pressure hose assembly. Use of an open-end wrench to tighten the nut can deform the tube nut hex which may result in improper torque and may make further servicing of the system difficult.

12. Connect the return hose to the pump and tighten the clamp. Fill the reservoir with the proper type and quantity of fluid.

BELT ADJUSTMENT

Except 1988–91 Crown Victoria and Grand Marquis

All vehicles except the 1988–91 Crown Victoria and Grand Marquis and 1988–90 Town Car are equipped with an automatic belt tensioner. No adjustment is necessary or possible. The belt tensioner is equipped with a belt wear indicator; when 1 percent belt stretch is indicated, the drive belt must be replaced. If the wear indicator is difficult to see on the 3.8L or 5.0L HO engines, locate the tab on the tensioner face plate. The tab should be approximately between the stops.

1988–91 Crown Victoria and Grand Marquis 1988–90 Town Car

NOTE: The power steering pump and alternator are driven by the same belt. Belt adjustment is made using the alternator.

1. Loosen the alternator pivot and adjustment bolts.

2. Install a suitable belt tension gauge midway between the pulleys on the longest accessible belt span. Install an open end wrench over the alternator adjustment boss, then apply tension to the belt, using the wrench.

3. Set the tension on a new belt to 170 lbs. or a used belt to 140 lbs. While maintaining the tension, tighten the alternator adjustment bolt to 29 ft. lbs. (39 Nm).

4. Remove the belt tension gauge, start the engine and let it idle for 5 minutes.

5. Shut off the engine and install the tension gauge. Apply tension with the open end wrench and slowly loosen the adjustment bolt to allow belt tension to increase to the used belt specification, 140 lbs. Tighten the adjustment bolt to 29 ft. lbs. (39 Nm).

6. Tighten the pivot bolt to 50 ft. lbs. (68 Nm).

SYSTEM BLEEDING

1. Disconnect the ignition coil and raise the front wheels off the floor.
2. Fill the power steering fluid reservoir.
3. Crank the engine with the starter and add fluid until the level remains constant.
4. While cranking the engine, rotate the steering wheel from lock-to-lock.

NOTE: The front wheels must be off the floor during lock-to-lock rotation of the steering wheel.

5. Check the fluid level and add fluid, if necessary.
6. Connect the ignition coil wire. Start the engine and allow it to run for several minutes.
7. Rotate the steering wheel from lock-to-lock.
8. Shut off the engine and check the fluid level. Add fluid, if necessary.
9. If air is still present in the system, purge the system of air using power steering pump air evacuator tool 021-00014 or equivalent, as follows:

    a. Make sure the power steering pump reservoir is full to the COLD FULL mark on the dipstick.

    b. Tightly insert the rubber stopper of the air evacuator assembly into the pump reservoir fill neck.

    c. Apply 15 in. Hg maximum vacuum on the pump reservoir for a minimum of 3 minutes with the engine idling. As air purges from the system, vacuum will fall off. Maintain adequate vacuum with the vacuum source.

    d. Release the vacuum and remove the vacuum source. Fill the reservoir to the COLD FULL mark.

    e. With the engine idling, apply 15 in. Hg vacuum to the pump reservoir. Slowly cycle the steering wheel from lock-to-lock every 30 seconds for approximately 5 minutes. Do not hold the steering wheel on the stops while cycling. Maintain adequate vacuum with the vacuum source as the air purges.

    f. Release the vacuum and remove the vacuum source. Fill the reservoir to the COLD FULL mark.

    g. Start the engine and cycle the steering wheel. Check for oil leaks at all connections. In severe cases of aeration, it may be necessary to repeat Steps 9b–9f.

Tie Rod Ends

REMOVAL & INSTALLATION

Crown Victoria, Grand Marquis and Town Car

1. Raise and support the vehicle safely.
2. Remove the cotter pin and nut from the tie rod end ball stud.
3. Loosen the tie rod adjusting sleeve clamp bolts and remove the rod end from the spindle arm or center link, using ball stud remover tool 3290–D or equivalent.
4. Remove the tie rod end from the sleeve, counting the exact number of turns required to do so.
To install:
5. Install the new tie rod end into the sleeve, using the exact number of turns it took to remove the old one.
6. Install the stud and stud nut. Tighten to 43–47 ft. lbs. (59–63 Nm), then continue tightening the nut to align its next catellation with the cotter pin hole in the stud. Install a new cotter pin.
7. Check the toe and adjust if necessary. Loosen the clamps from the sleeve and oil the sleeve, clamps, bolts and nuts. Position the adjusting sleeve clamps and tighten the clamp nuts to 20–22 ft. lbs. (27–29 Nm).

Mark VII, Thunderbird, Cougar and Mustang

1. Raise and safely support the vehicle.
2. Remove the cotter pin and nut from the tie rod end ball stud. Disconnect the tie rod end from the spindle using ball stud remover tool 3290–D or equivalent.
3. Holding the tie rod end with a wrench, loosen the tie rod jam nut. Grip the tie rod end with pliers and remove the assembly from the tie rod, but first note the depth to which the tie rod was located by using the jam nut as a marker.
To install:
4. Clean the tie rod threads.
5. Thread the new tie rod end onto the tie rod to the same depth as the removed tie rod end.
6. Place the tie rod end ball stud into the spindle and install the nut. Make sure the front wheels are in the straight ahead position.
7. Tighten the nut to 35 ft. lbs. (48 Nm) and continue tightening the nut to align the next castellation of the nut with the cotter pin hole in the stud. Install a new cotter pin.
8. Set the toe to specification.

Tighten the jam nut to 35–50 ft. lbs. (48–68 Nm).

BRAKES

For all brake system repair and service procedures not detailed below, please refer to "Brakes" in the Unit Repair section.

Master Cylinder

REMOVAL & INSTALLATION

Except Mark VII, Thunderbird and Cougar with Anti-Lock Brakes

NOTE: The master cylinder on Mark VII, Thunderbird and Cougar with anti-lock brakes is part of the hydraulic actuation assembly and cannot be removed separately.

1. Disconnect the negative battery cable.
2. On Crown Victoria, Grand Marquis and Town Car with anti-lock brakes, depress the brake pedal several times to exhaust all vacuum in the system.
3. Remove the brake lines from the primary and secondary outlet ports of the master cylinder.
4. Disconnect the brake warning indicator connector.
5. On Crown Victoria, Grand Marquis and Town Car with anti-lock brakes, disconnect the Hydraulic Control Unit (HCU) supply hose at the master cylinder and secure in a position to prevent loss of brake fluid.
6. Remove the nuts attaching master cylinder to the brake booster assembly.
7. Slide the master cylinder forward and upward from the vehicle.
To install:
8. Before installation, bench bleed the new master cylinder as follows:

    a. Mount the new master cylinder in a holding fixture. Be careful not to damage the housing.

    b. Fill the master cylinder reservoir with brake fluid.

    c. Using a suitable tool inserted into the booster pushrod cavity, push the master cylinder piston in slowly. Place a suitable container under the master cylinder to catch the fluid being expelled from the outlet ports.

    d. Place a finger tightly over each outlet port and allow the master cylinder piston to return.

    e. Repeat the procedure until

clear fluid only is expelled from the master cylinder. Plug the outlet ports and remove the master cylinder from the holding fixture.

9. Mount the master cylinder on the booster. Install a new seal in the groove in the master cylinder mounting face on Crown Victoria, Grand Marquis and Town Car equipped with anti-lock brakes.

10. Attach the brake fluid lines to the master cylinder. On Crown Victoria, Grand Marquis and Town Car equipped with anti-lock brakes, install the HCU supply hose to the master cylinder.

11. Connect the brake warning indicator switch connector.

12. Bleed the system. Operate the brakes several times, then check for external hydraulic leaks.

Brake Control Valves

There are several types of valves in use. Not all valves perform the same function nor are they located in the same place on every vehicle.

Town Car — 1988–89 vehicles use a 3-way brake control valve containing a pressure differential valve, metering valve and proportioning valve located on the frame. 1990–92 vehicles use a brake pressure control valve that contains twin brake proportioning valves located on the frame.

Crown Victoria and Grand Marquis — a pressure control valve is screwed into the master cylinder. In addition, some vehicles have a metering valve located on the frame. 1992 vehicles with anti-lock brakes have a proportioning valve located on the frame.

Mark VII — a proportioning valve is located in the rear outlet port of the hydraulic actuator.

Thunderbird and Cougar — 1988 vehicles without anti-lock brakes have a brake control valve which contains a proportioning valve and shuttle valve located on the fender apron. 1989–92 vehicles without anti-lock brakes have 2 pressure control valves housed in the master cylinder. 1988 vehicles with anti-lock brakes have a proportioning valve located in the rear outlet port of the hydraulic actuator. 1989–92 vehicles with anti-lock brakes have a control valve which is located on a bracket below the hydraulic actuation unit.

Mustang — a brake control valve which contains a proportioning valve and shuttle valve located on the fender apron.

REMOVAL & INSTALLATION

1. Disconnect the negative battery cable.

2. Disconnect the brake line(s) from the valve. Disconnect the electrical connector, if equipped.

3. Unscrew the valve from the master cylinder or remove the mounting screw from the frame or fender apron, as required.

4. Installation is the reverse of the removal procedure. Bleed the brake system.

Power Brake Booster

REMOVAL & INSTALLATION

1. Disconnect the negative battery cable. On Thunderbird, Cougar and Mustang, remove the air cleaner.

2. On Crown Victoria, Grand Marquis and Town Car equipped with anti-lock brakes, pump the brake pedal several times until all vacuum is removed from the booster. Remove the cruise control actuator cable and cruise control servo.

3. On Mustang and 1988 Thunderbird equipped with the 2.3L engine, perform the following:

a. Relieve the fuel system pressure.

b. Disconnect the accelerator cable from the throttle body. Remove the screw that secures the accelerator cable to the accelerator shaft bracket and remove the cable from the bracket.

c. Remove the screws that secure the accelerator shaft bracket to the manifold and rotate the bracket toward the engine. Remove the horn.

d. Disconnect the 2 manifold injector connectors located near the oil dipstick retaining bracket. Disconnect the 2 fuel hoses to the fuel supply manifold.

e. Remove the 3 bolts holding the oil dipstick bracket to the upper intake manifold. Remove the dipstick and bracket.

f. Remove the windshield wiper motor and remove the vacuum hoses directly over the brake booster at the dash panel vacuum tee.

g. Remove the bolt holding the clutch cable stand, move the bracket to the side rail at the fender inner panel.

h. If equipped with cruise control, move the cruise control cable aside to clear the booster.

4. Disconnect the manifold vacuum hose from the booster check valve.

5. Disconnect the brake lines from the master cylinder, remove the master cylinder-to-booster retaining nuts and remove the master cylinder. On Crown Victoria, Grand Marquis and Town Car equipped with anti-lock brakes, disconnect the hydraulic control unit supply hose.

NOTE: On Crown Victoria, Grand Marquis and Town Car without anti-lock brakes, it may be possible to remove the master cylinder and move aside without disconnecting the brake lines. Be careful not to kink the brake lines.

6. Working inside the vehicle below the instrument panel, remove the stoplight switch connector. Remove the switch retaining pin and slide the switch off the brake pedal pin just far enough for the outer arm to clear the pin, then remove the switch. Be careful not to damage the switch.

7. Remove the booster-to-dash panel attaching nuts. If necessary, remove the cowl top intrusion bolt.

8. On Mustang and 1988 Thunderbird and Cougar equipped with cruise control, remove and set aside the control amplifier which is mounted to the lower outboard booster stud.

9. Slide the booster pushrod, washers and bushing off the brake pedal pin. Remove the booster.

10. Installation is the reverse of the removal procedure. Tighten the booster-to-dash panel attaching nuts and the master cylinder attaching nuts to 13–25 ft. lbs. (18–34 Nm). If the brake lines were disconnected, bleed the brake system.

Brake Caliper

REMOVAL & INSTALLATION

Front

1. Raise and safely support the vehicle. Remove the front wheel and tire assembly.

2. If equipped with a hollow brake hose retaining bolt, remove the bolt and plug the brake hose.

3. If the brake line screws into the caliper, loosen the brake line fitting that connects the brake hose to the brake line at the frame bracket. Remove the retaining clip from the hose and bracket and disengage the hose from the bracket. Unscrew the hose from the caliper.

4. Remove the caliper locating pins and remove the caliper. If removing both calipers, mark the right and left sides so they may be reinstalled correctly.

To install:

5. Install the caliper over the rotor with the outer brake shoe against the rotor's braking surface. On 1989–92 Thunderbird and Cougar, make sure the ant-rattle spring is under the arm of the knuckle.

6. Lubricate the inside of the locating pin insulators with silicone dialectric grease. Install the caliper lo-

cating pins and tighten to 45–65 ft. lbs. (61–88 Nm) on all except 1989–92 Thunderbird and Cougar. On 1989–92 Thunderbird and Cougar, tighten the locating pins to 18–25 ft. lbs. (25–34 Nm).

7. If equipped with a hollow brake hose retaining bolt, install new copper washers on each side of the brake hose fitting outlet and install the bolt, through the hose fitting and into the caliper. Tighten the bolt to 30–44 ft. lbs. (40–60 Nm) on 1989–92 Thunderbird and Cougar. On all ther vehicles, tighten the bolt to 25 ft. lbs. (34 Nm).

8. If the brake hose screws into the caliper, thread the brake hose into the caliper and tighten to 20–30 ft. lbs. (28–41 Nm).

NOTE: This is a special self-sealing fitting that does not require a gasket. When the hose is correctly tightened, there should be 1 or 2 threads of the fitting still showing at the caliper. It is not necessary for the hose fitting to be flush with the caliper for sealing, so do not over-tighten.

9. Bleed the brake system, install the wheel and tire assembly and lower the vehicle.

10. To position the brake pads, aplly the brake pedal several times before moving the vehicle.

Rear

CROWN VICTORIA, GRAND MARQUIS AND TOWN CAR

1. Raise and safely support the vehicle. Remove the rear wheel and tire assembly.

2. Disconnect the flexible brake hose from the caliper and remove the brake fitting retaining bolt from the caliper. Plug the hose and the caliper fitting.

3. Remove the caliper locating pins. Lift the caliper off the rotor and anchor plate using a rotating motion.

NOTE: Do not pry directly against the plastic piston or damage to the piston will occur.

4. Position the caliper assembly above the rotor with the anti-rattle spring located on the lower adapter support arm. Install the caliper over the rotor with a rotating motion. Make sure the inner shoe is properly positioned.

5. Install the caliper locating pins and start them in the threads by hand. Tighten them to 19–25 ft. lbs. (26–34 Nm).

6. Install the brake hose on the caliper with a new gasket on each side of the fitting outlet. Insert the retaining bolt and tighten to 30–40 ft. lbs. (40–54 Nm).

7. Bleed the brake system, install the wheel and tire assembly and lower the vehicle.

8. Pump the brake pedal prior to moving the vehicle to position the linings.

THUNDERBIRD, COUGAR AND 1991–92 MARK VII

1. Raise and safely support the vehicle. Remove the rear wheel and tire assembly.

2. Remove the brake hose from the caliper.

3. Release the parking brake cable tension, if necessary. Remove the cable retaining clip and disconnect the cable end from the lever.

4. Hold the slider pin hex heads with an open end wrench and remove the pinch bolts.

5. Lift the caliper assembly away from the anchor plate. Remove the slider pins and boots from the anchor plate.

To install:

6. Apply silicone dialectric compound to the inside of the slider pin boots and to the slider pins.

7. Position the slider pins and boots in the anchor plate. Position the caliper assembly on the anchor plate. Make sure the brake shoes and anti-rattle springs are installed correctly.

8. Remove the residue from the pinch bolt threads and apply locking compound. Install the pinch bolts and tighten to 23–26 ft. lbs. (31–35 Nm) while holding the slider pins with an open end wrench.

9. Attach the cable end to the parking brake lever and install the cable retaining clip. Adjust the parking brake.

10. Using new washers, connect the brake flex hose to the caliper. Tighten the retaining bolt to 30–45 ft. lbs. (40–60 Nm).

11. Bleed the brake system, install the wheel and tire assembly and lower the vehicle.

12. Pump the brake pedal prior to moving the vehicle to position the linings.

1988–90 MARK VII

1. Raise and safely support the vehicle. Remove the rear wheel and tire assembly.

2. Disconnect the parking brake cable from the lever and bracket. Be careful to avoid kinking or cutting the cable or return spring.

3. Remove the caliper locating pins. Lift the caliper away from the anchor plate by pushing the caliper upward toward the anchor plate and then rotate the lower end out of the anchor plate.

4. If insufficient clearance between the caliper and the brake pads prevents removal of the caliper, proceed as follows:

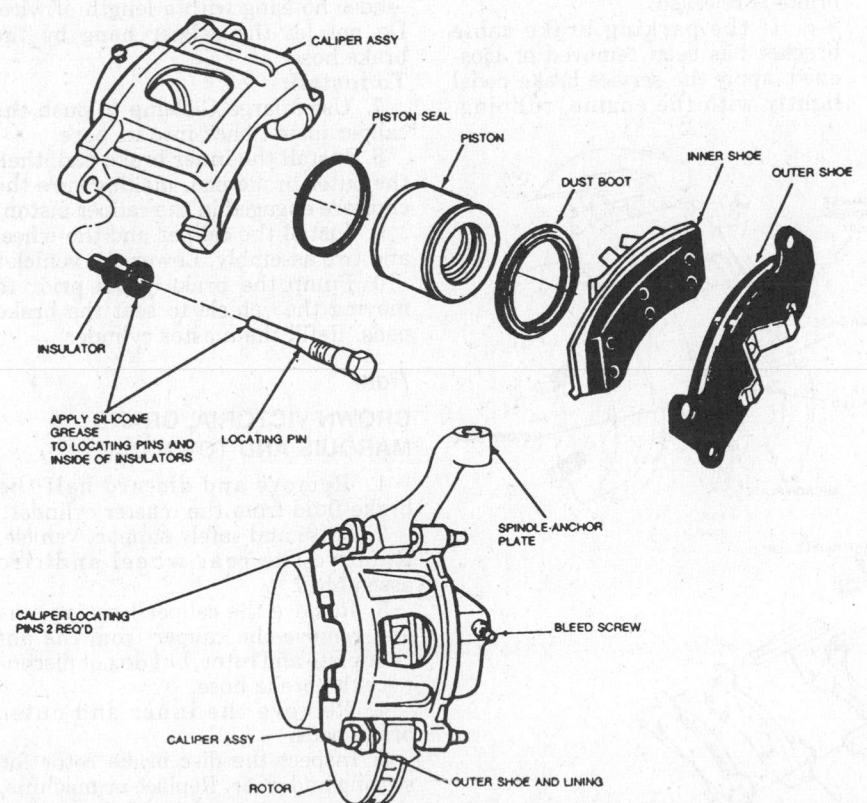

Front disc brake assembly—except 1989–92 Thunderbird and Cougar

a. Lower the vehicle and remove ½ of the brake fluid from the master cylinder.

b. Raise and safely support the vehicle. Loosen the caliper parking brake end retainer ½ turn, maximum, to allow the piston to be forced back into its bore.

c. To loosen the end retainer, remove the parking brake lever, then mark the end retainer and caliper housing to make sure the end retainer is not loosened more than ½ turn. Force the piston back in its bore and then remove the caliper.

NOTE: If the retainer must be loosened more than ½ turn, the seal between the thrust screw and the housing may be broken and brake fluid may leak into the parking brake mechanism chamber. In this case, the end retainer must be removed and the caliper overhauled.

To install:

5. If the end retainer has been loosened only ½ turn, install the caliper in the anchor plate without the brake pads. Tighten the end retainer to 75–96 ft. lbs. (101–130 Nm).

6. Install the parking brake lever on its keyed spline. The lever arm must point down and rearward. The parking brake cable will then pass freely under the axle. Tighten the retainer screw to 16–22 ft. lbs. (22–29 Nm). The parking brake lever must rotate freely after tightening the retainer screw. Remove the caliper from the anchor plate.

7. Install the brake pads.

8. Install the flexible hose by placing a new washer on each side of the fitting outlet and install the attaching bolt through the washers and fitting. Tighten to 20–30 ft. lbs. (27–40 Nm).

9. Position the upper tab of the caliper housing on the anchor plate upper abutment surface. Rotate the caliper housing until it is completely over the rotor. Be careful not to damage the piston boot.

10. Lubricate the locating pins and inside of the insulator with silicone dielectric compound. Apply threadlocking compound to the locating pin threads.

11. Hand start the locating pins through the caliper insulators and into the anchor plate. Tighten to 29–37 ft. lbs. (40–50 Nm).

12. Connect the parking brake cable to the bracket and the lever on the caliper. Bleed the brake system and fill the master cylinder.

13. Adjust the caliper as follows:

a. With the engine running, pump the service brake lightly about 40 times. Allow at least 1 second between pedal applications.

b. Check the parking brake for excessive travel or very light effort. In either case, repeat pumping the service brake pedal or check the parking brake cable for proper tension. The calipers must return to the OFF position when the parking brake is released.

c. If the parking brake cable bracket has been removed or loosened, apply the service brake pedal lightly with the engine, running.

While the service brake pedal is being applied, rotate the parking brake bracket until the bracket lever stop contacts the actuating lever. Hold the parking brake bracket in this position while tightening the bolts to 30–37 ft. lbs. (40–50 Nm).

14. Install the wheel and tire assembly and lower the vehicle. Make sure a firm brake pedal application is obtained.

Disc Brake Pads

REMOVAL & INSTALLATION

Front

1. Remove and discard half the brake fluid from the master cylinder.

2. Raise and safely support vehicle. Remove the front wheel and tire assemblies.

3. Remove the caliper locating pins and remove the caliper from the anchor plate and rotor, but do not disconnect the brake hose.

4. Remove the outer brake pad from the caliper assembly and remove the inner brake pad from the caliper piston.

5. Inspect the disc brake rotor for scoring and wear. Replace or machine, as necessary.

6. Suspend the caliper inside the fender housing with a length of wire. Do not let the caliper hang by the brake hose.

To install:

7. Use a large C-clamp to push the caliper piston back into its bore.

8. Install the inner brake pad, then the outer brake pad, making sure the clips are engaged in the caliper piston.

9. Install the caliper and the wheel and tire assembly. Lower the vehicle.

10. Pump the brake pedal prior to moving the vehicle to seat the brake pads. Refill the master cylinder.

Rear

CROWN VICTORIA, GRAND MARQUIS AND TOWN CAR

1. Remove and discard half the brake fluid from the master cylinder.

2. Raise and safely support vehicle. Remove the rear wheel and tire assemblies.

3. Remove the caliper locating pins and remove the caliper from the anchor plate and rotor, but do not disconnect the brake hose.

4. Remove the inner and outer brake pads.

5. Inspect the disc brake rotor for scoring and wear. Replace or machine, as necessary.

6. Suspend the caliper inside the fender housing with a length of wire.

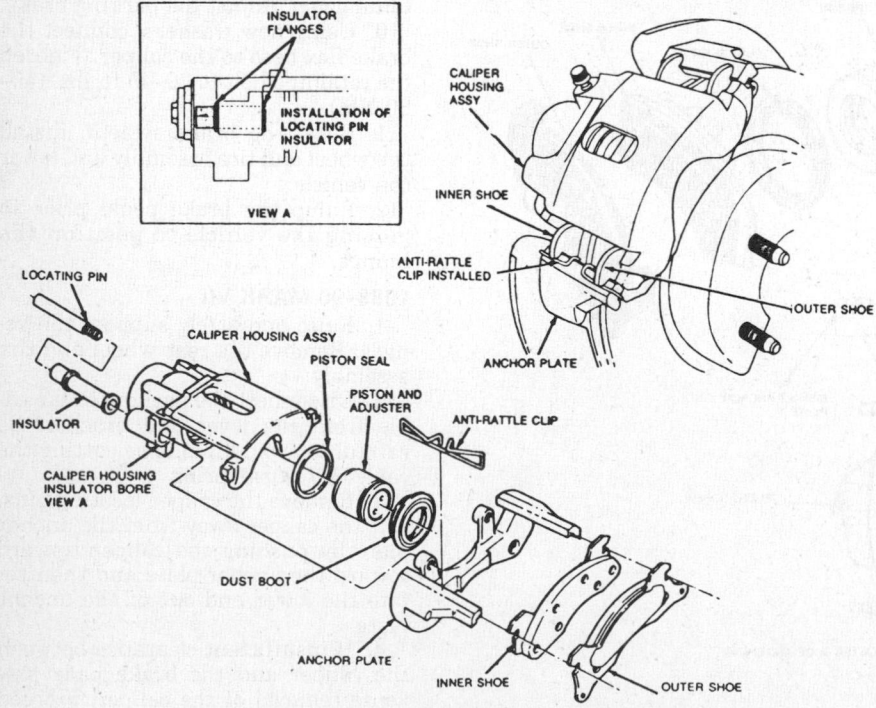

Exploded view of the rear disc brakes—1988–90 Mark VII

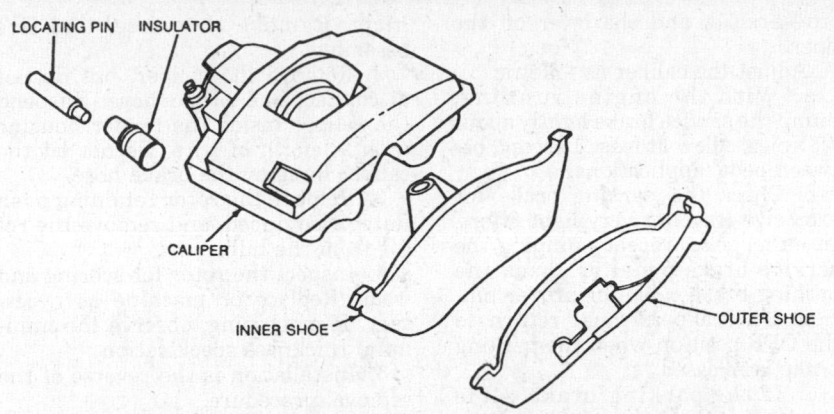

LOCATING PIN   INSULATOR

CALIPER

INNER SHOE

OUTER SHOE

Front disc brake assembly—1989–92 Thunderbird and Cougar

Do not let the caliper hang by the brake hose.

To install:

7. Use a large C-clamp to push the caliper piston back into its bore.

8. Install the inner brake pad, then the outer brake pad, making sure the clips are engaged in the caliper piston.

9. Install the caliper and the wheel and tire assembly. Lower the vehicle.

10. Pump the brake pedal prior to moving the vehicle to seat the brake pads. Refill the master cylinder.

THUNDERBIRD, COUGAR AND 1991–92 MARK VII

1. Remove and discard half the brake fluid from the master cylinder.

2. Raise and safely support vehicle. Remove the rear wheel and tire assemblies.

3. Remove the caliper from the anchor plate and rotor, but do not disconnect the brake hose. Suspend the caliper inside the fender housing with a length of wire. Do not let the caliper hang by the brake hose.

4. Remove the brake pads from the anchor plate.

5. Inspect the disc brake rotor for scoring and wear. Replace or machine, as necessary.

To install:

6. Using brake piston turning tool T87P-2588-A or equivalent, rotate the caliper piston clockwise until it is fully seated. Make sure that 1 of the 2 slots in the piston face is positioned so it will engage the nib on the brake pad.

7. Install the brake pads on the anchor plate. Install the caliper and wheel and tire assembly and lower the vehicle.

8. Pump the brake pedal prior to moving the vehicle to seat the brake pads. Refill the master cylinder.

1988–90 MARK VII

1. Remove and discard half the brake fluid from the master cylinder.

2. Raise and safely support vehicle. Remove the rear wheel and tire assemblies.

3. Remove the caliper from the anchor plate and rotor, but do not disconnect the brake hose. Suspend the caliper inside the fender housing with a length of wire. Do not let the caliper hang by the brake hose.

4. Remove and discard the caliper pin insulators. Remove the disc brake pads and the disc brake rotor.

5. Inspect the disc brake rotor for scoring and wear. Replace or machine, as necessary.

To install:

6. Install new pin insulators in the caliper housing. Make sure both insulator flanges straddle the housing holes.

7. Install the caliper on the anchor plate, without the disc brake pads or

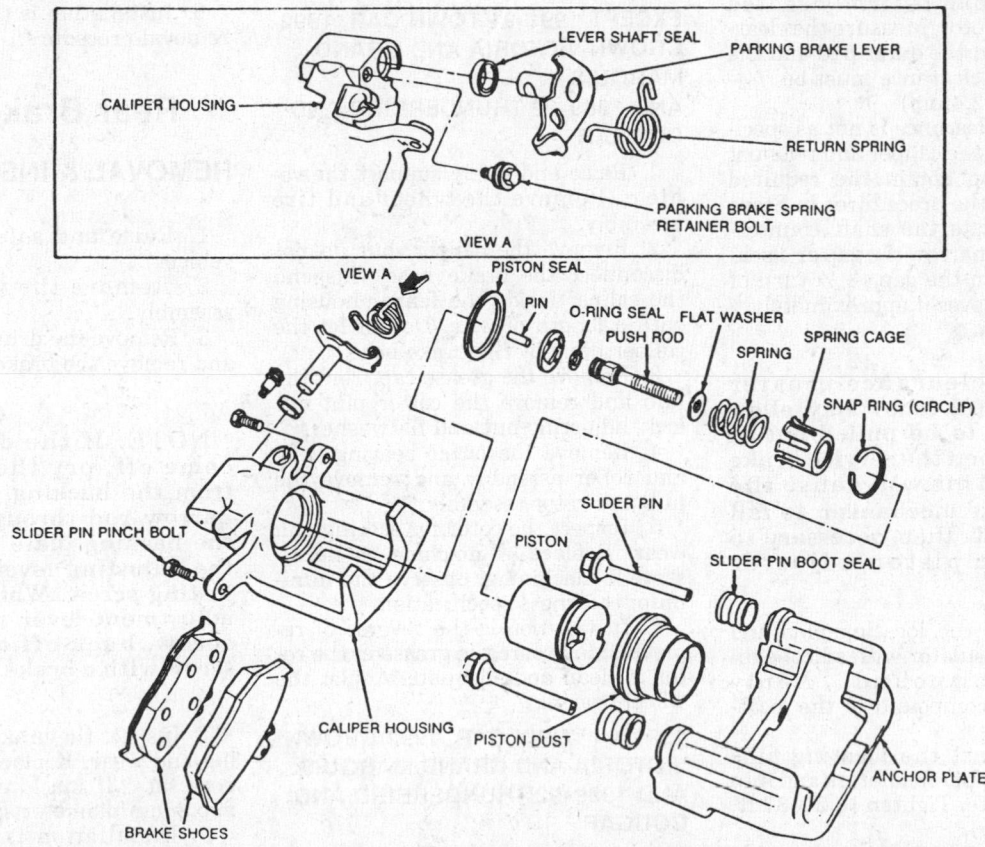

CALIPER HOUSING

LEVER SHAFT SEAL

PARKING BRAKE LEVER

RETURN SPRING

PARKING BRAKE SPRING RETAINER BOLT

VIEW A

VIEW A

PISTON SEAL

PIN

O-RING SEAL   FLAT WASHER

PUSH ROD

SPRING

SPRING CAGE

SNAP RING (CIRCLIP)

SLIDER PIN PINCH BOLT

PISTON

SLIDER PIN

SLIDER PIN BOOT SEAL

CALIPER HOUSING

PISTON DUST

ANCHOR PLATE

BRAKE SHOES

Exploded view of the rear disc brakes—1989–92 Thunderbird and Cougar and 1991–92 Mark VII

rotor in position. Install tool T75P-2588-B or equivalent on the caliper.

8. While holding the tool shaft, rotate the tool handle counterclockwise until the tool is firmly seated against the piston.

9. Loosen the tool handle ¼ turn. While holding the handle, rotate the tool shaft clockwise until the piston is fully bottomed in its bore. The piston will continue to turn even after it is bottomed. When there is no further inward movement of the piston and the tool handle is rotated until there is firm seating force, the piston is bottomed. Remove the tool and the caliper from the anchor plate.

10. Lubricate the anchor plate sliding ways with disc brake caliper slide grease. Do not get any lubricant on the braking surfaces.

11. Install the anti-rattle clip on the lower rail of the anchor plate. Install the inner brake pad on the anchor plate with the lining toward the rotor. Install the rotor.

12. Install the outer brake pad with the lining toward the rotor. Position the upper tab of the caliper housing on the anchor plate upper abutment surface. Rotate the caliper housing until it is completely over the rotor. Be careful not to damage the piston boot.

13. Adjust the piston position as follows:

a. Pull the caliper outboard until the inner pad is firmly seated against the rotor. Measure the clearance between the outer pad and the caliper. The clearance must be $\frac{1}{32}$–$\frac{3}{32}$ in. (0.8–2.4mm).

b. If the clearance is not as specified, remove the caliper and readjust the piston to obtain the required gap. Follow the procedures in Steps 7–9 and rotate the shaft counterclockwise to narrow the gap or clockwise to widen the gap. A ¼ turn of the piston moves it approximately $\frac{1}{16}$ in. (1.6mm).

NOTE: A clearance greater than $\frac{3}{32}$ in. (2.4mm) may allow the adjuster to be pulled out of the piston when the service brake is applied. This will cause the parking brake mechanism to fail to adjust. It is then necessary to replace the piston/adjuster assembly.

14. Lubricate the locating pins and inside of the insulator with silicone dielectric compound. Apply threadlocking compound to the locating pin threads.

15. Hand start the locating pins through the caliper insulators and into the anchor plate. Tighten to 29–37 ft. lbs. (40–50 Nm).

16. Connect the parking brake cable

to the bracket and the lever on the caliper.

17. Adjust the caliper as follows:

a. With the engine running, pump the service brake lightly about 40 times. Allow at least 1 second between pedal applications.

b. Check the parking brake for excessive travel or very light effort. In either case, repeat pumping the service brake pedal or check the parking brake cable for proper tension. The calipers must return to the OFF position when the parking brake is released.

c. If the parking brake cable bracket has been removed or loosened, apply the service brake pedal lightly with the engine, running. While the service brake pedal is being applied, rotate the parking brake bracket until the bracket lever stop contacts the actuating lever. Hold the parking brake bracket in this position while tightening the bolts to 30–37 ft. lbs. (40–50 Nm).

18. Install the wheel and tire assembly and lower the vehicle. Make sure a firm brake pedal application is obtained.

Brake Rotor

REMOVAL & INSTALLATION

Front

EXCEPT 1991–91 TOWN CAR, 1992 CROWN VICTORIA AND GRAND MARQUIS AND 1989–92 THUNDERBIRD AND COUGAR

1. Raise and safely support the vehicle. Remove the wheel and tire assembly.

2. Remove the caliper, but do not disconnect the brake hose. Suspend the caliper inside the fender housing with a length of wire. Do not let the caliper hang by the brake hose.

3. Remove the grease cap from the hub and remove the cotter pin, nut lock, adjusting nut and flatwasher.

4. Remove the outer bearing cone and roller assembly and remove the hub and rotor assembly.

5. Inspect the rotor for scoring and wear. Replace or machine as necessary. If machining, observe the minimum thickness specification.

6. Installation is the reverse of removal. Make sure the grease in the rotor is clean and adequate. Adjust the wheel bearings.

1991–91 TOWN CAR, 1992 CROWN VICTORIA AND GRAND MARQUIS AND 1989–92 THUNDERBIRD AND COUGAR

1. Raise and safely support the ve-

hicle. Remove the wheel and tire assembly.

2. Remove the caliper, but do not disconnect the brake hose. Suspend the caliper inside the fender housing with a length of wire. Do not let the caliper hang by the brake hose.

3. Remove the rotor retaining push nuts, if equipped, and remove the rotor from the hub.

4. Inspect the rotor for scoring and wear. Replace or machine as necessary. If machining, observe the minimum thickness specification.

5. Installation is the reverse of the removal procedure.

Rear

1. Raise and safely support the vehicle. Remove the wheel and tire assembly.

2. Remove the caliper, but do not disconnect the brake hose. Suspend the caliper inside the fender housing with a length of wire. Do not let the caliper hang by the brake hose.

3. On Mark VII, Thunderbird and Cougar, remove the anchor plate.

4. Remove the rotor retaining push nuts, if equipped, and remove the rotor from the hub.

5. Inspect the rotor for scoring and wear. Replace or machine as necessary. If machining, observe the minimum thickness specification.

6. Installation is the reverse of the removal procedure.

Rear Brake Drum

REMOVAL & INSTALLATION

1. Raise and safely support the vehicle.

2. Remove the wheel and tire assembly.

3. Remove the drum retaining nuts and remove the brake drum.

NOTE: If the drum will not come off, pry the rubber plug from the backing plate. Insert a narrow rod through the hole in the backing plate and disengage the adjusting lever from the adjusting screw. While holding the adjustment lever away from the screw, back off the adjusting screw with a brake adjusting tool.

4. Inspect the brake drum for scoring and wear. Replace or machine as necessary. If machining, observe the maximum diameter specification.

5. Installation is the reverse of removal.

Rear Brake Shoes

REMOVAL & INSTALLATION

Except 1989–92 Thunderbird and Cougar

1. Raise and safely support the vehicle. Remove the rear wheel and tire assemblies. Remove the brake drum.
2. Remove the shoe-to-anchor springs and unhook the cable eye from the anchor pin. Remove the anchor pin plate.
3. Remove the shoe hold-down springs, shoes, adjusting screw, pivot nut, socket and automatic adjustment parts.
4. Remove the parking brake link, spring and retainer. Disconnect the parking brake cablw from the parking brake lever.
5. After removing the rear brake secondary shoe, disassemble the parking brake lever from the shoe by removing the retaining clip and spring washer.

To install:

6. Before installing the rear brake shoes, assemble the parking brake lever to the secondary shoe and secure it with the spring washer and retaining clip.
7. Apply a light coating of caliper slide grease at the points where the brake shoes contact the backing plate. Be careful not to get any lubricant on the brake linings.
8. Position the brake shoes on the backing plate. The primary shoe with the short lining faces the front of the vehicle, the secondary to the rear. Secure the assembly with the hold-down springs. Install the parking brake link, spring and retainer. Back-off the parking brake adjustment, then connect the parking brake cable to the parking brake lever.
9. Install the anchor pin plate on the anchor pin. Place the cable eye over the anchor pin with the crimped side toward the drum. Install the primary shoe to the anchor pin.
10. Install the cable guide on the secindary shoe web with the flanged hole fitted into the hole in the secondary shoe web. Thread the cable around the cable guide groove.

NOTE: The cable must be positioned in the groove and not between the guide and the shoe web.

11. Install the secondary shoe-to-anchor spring. Make sure the cable eye is not cocked or binding on the anchor pin when installed. All parts should be flat on the anchor pin.
12. Apply a thin coat of lubricant to the threads and the socket end of the adjusting screw. Turn the adjusting screw into the adjusting pivot nut to the limit of the threads, then back-off ½ turn.

NOTE: Make sure the socket end of the adjusting screw is stamped with an R or L, indicating the right or left side of the vehicle. The adjusting screw assemblies must be installed on the correct side for proper brake shoe adjustment.

13. Place the adjusting socket on the screw and install the assembly between the shoe ends with the adjusting screw toothed wheel nearest the secondary shoe.
14. Hook the cable hook into the hole in the adjusting lever. The adjusting levers are stamped with an **R** or **L** to indicate their installation on the right or left side.
15. Position the hooked end of the adjuster spring completely into the large hole in the primary shoe web. Connect the loop end of the spring to the adjuster lever hole.
16. Pull the adjuster lever, cable and automatic adjuster spring down and toward the rear, engaging the pivot hook in the large hole of the secondary shoe web.
17. Make sure the upper ends of the brake shoes are seated against the anchor pin and the shoes are centered on the backing plate.
18. Adjust the brakes using brake adjustment gauge D81L–1103–A or equivalent.

19. Install the brake drum, wheel and tire assemblies and lower the vehicle.
20. Apply the brakes several times while backing up the vehicle. After each stop, the vehicle must be moved forward.

1989–92 Thunderbird and Cougar

1. Raise and safely support the vehicle. Remove the rear wheel and tire assemblies. Remove the brake drum.
2. Disconnect the parking brake cable from the parking brake lever.
3. Remove the 2 brake shoe hold-down retainers, springs and pins.
4. Spread the brake shoes over the piston shoe guide slots. Lift the brake shoes, springs and adjuster off the backing plate as an assembly. Be careful not to bend the adjusting lever.
5. Remove the adjuster spring. To separate the shoes, remove the retracting springs.
6. Remove the parking brake lever retaining clip and spring washer. Remove the lever from the pin.

To install:

7. Apply a light coating of caliper slide grease to the backing plate brake shoe contact areas.
8. Apply a light coat of lubricant to the threaded areas of the adjuster screw and socket. Assemble the brake adjuster with the stainless steel washer. Turn the socket all the way down on the screw, then back off ½ turn.

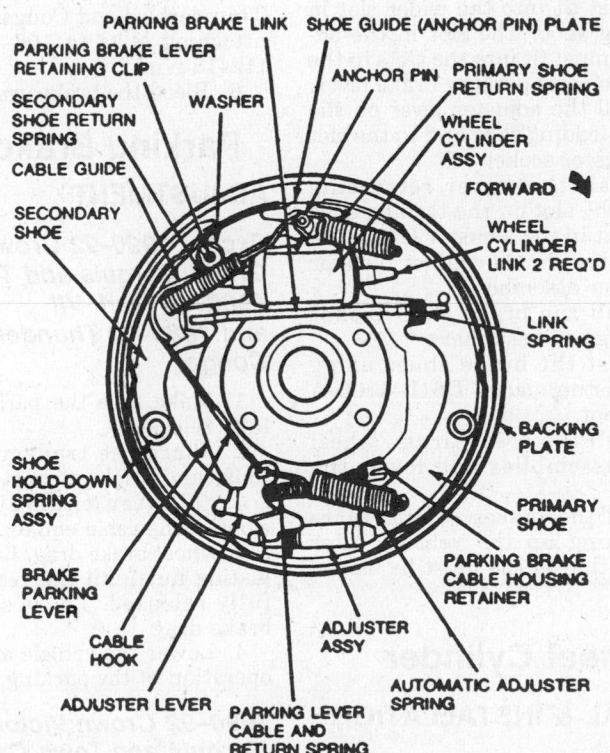

Brake shoe assembly—except 1989–92 Thunderbird and Cougar

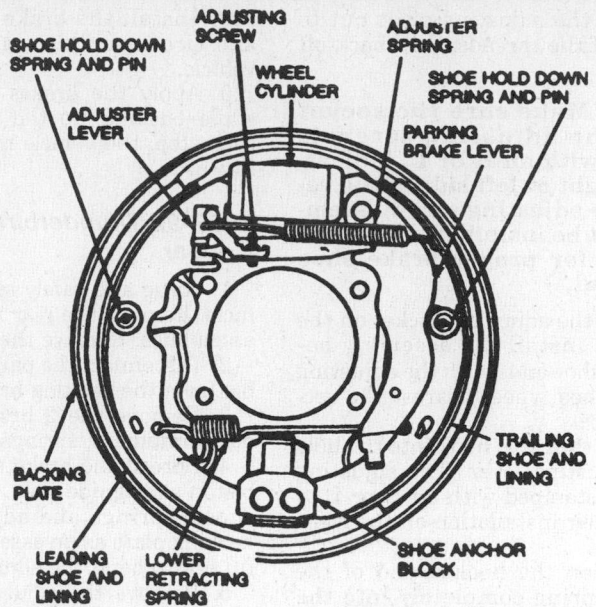

Brake shoe assembly—1989–92 Thunderbird and Cougar

9. Install the parking brake lever to the trailing shoe with the spring washer and new retaining clip. Crimp the clip to securely retain the lever.

10. Position the trailing shoe on the backing plate and attach the parking brake cable. Position the leading shoe on the backing plate and attach the lower retracting spring to the brake shoes.

11. Install the adjuster assembly to the slots in the brake shoes. The socket end must fit into the wider slot in the leading shoe. The slot in the adjuster nut must fit into the slots in the trailing shoe and parking brake lever.

12. Install the adjuster lever on the pin on the leading shoe and to the slot in the adjuster socket.

13. Install the upper retracting spring in the slot on the trailing shoe and the slot in the adjuster lever. The adjuster lever should contact the star and adjuster assembly.

14. Install the brake shoe anchor pins, springs and retainers.

15. Adjust the brake shoes using brake adjusting gauge D81L–1103–A or equivalent.

16. Install the brake drum, wheel and tire assemblies and lower the vehicle.

17. Apply the brakes several times while backing up the vehicle. After each stop, the vehicle must be moved forward.

Wheel Cylinder

REMOVAL & INSTALLATION

1. Remove the wheel and tire assembly and the brake drum.

2. Remove the brake shoe assembly.

3. Disconnect the brake line from the wheel cylinder at the backing plate.

4. Remove the wheel cylinder attaching bolts and remove the wheel cylinder.

5. Installation is the reverse of the removal procedure. Tighten the wheel cylinder attaching bolts to 10–20 ft. lbs. (14–28 Nm) except on 1989–92 Thunderbird and Cougar. On 1989–92 Thunderbird and Cougar, tighten the attaching bolts to 106–160 inch lbs. (12–18 Nm).

6. Bleed the brake system.

Parking Brake Cable

ADJUSTMENT

Except 1990–92 Crown Victoria, Grand Marquis and Town Car, 1988–90 Mark VII and 1989–92 Thunderbird and Cougar

1. Make sure the parking brake is fully released.

2. Place the transmission in **N**. Raise and safely support the vehicle.

3. Tighten the adjusting nut against the cable equalizer, causing a rear wheel brake drag. Loosen the adjusting nut until the rear brakes are fully released. There should be no brake drag.

4. Lower the vehicle and check the operation of the parking brake.

1990–92 Crown Victoria, Grand Marquis and Town Car

NOTE: The following proce-

dure is to be used only if a new parking brake control assembly is installed. All components of the parking brake system must be installed prior to the adjustment procedure. The parking brake control with automatic tensioning is preset by means of a shipping clip. The foolowing procedure must be followed in sequence and must be done with the vehicle weight on the axle.

1. Verify removal of the shipping clip. The take up reel will apply tension to the system.

2. Depress the parking brake control to the 8th notch.

3. Push the parking brake control pedal to release.

4. Check function as follows:

   a. Apply the parking brake with a full stroke, to the 9th or 10th notch.

   b. Release the parking brake by shifting the vehicle into a forward gear with the engine running. The control must release.

   c. Apply the parking brake with a full stroke, to the 9th or 10th notch.

   d. Manually release the parking brake with the push to release feature.

NOTE: With the control in the OFF position, the rear brakes must not drag. Check for movement of the rear cables from their conduits when the intermediate cable is deflected with a force of 10–15 lbs.

1988–90 Mark VII

1. Raise and safely support the vehicle.

2. Back off the parking brake cable adjusting nut until the cables are loose.

3. Make sure each rear disc brake is properly adjusted by movin the caliper lever in the applied direction. If the lever moves more than 20 degrees, using hand pressure of approximately 50 lbs. maximum, adjust the brake pads.

4. Tighten the parking brake cable adjusting nut until 1 or both of the parking brake actuating levers just begin to move.

5. Apply and release the parking brake control. Check the parking brake levers on the calipers to determine if they are fully returned to the stop position by attempting to pull them rearward.

NOTE: If the lever does not contact the parking brake caliper lever stop, the cable adjustment is too tight, repeat the adjustment procedure.

1989–92 Thunderbird and Cougar

1. Apply the parking brake control fully on. Release the parking brake control. Repeat the application and release.

2. Place the transmission in **N**. Raise and safely support the vehicle by the axles.

3. On 1989–90 vehicles, proceed as follows:

   a. With the parking brake control in the **OFF** position, release the tensioner by rotating the locking lever away from the threaded rod.

   b. The tensioner spring will take up the cable slack and preload the cables.

   c. Do not pull down on the locking lever for it will pull the cables down and cause the cables to have low tension.

NOTE: Lock the tensioner by releasing the locking lever. Make sure the locking lever is secure by rotating it toward the threaded rod.

4. On 1991–92 vehicles, proceed as follows:

   a. With the parking brake control in the **OFF** position, graps the tensioner around the housing, then using a hook tool, hook the end into the rounded end of the clip between the clip and the housing.

   b. Unlock the clip by pulling downward with the tool and support tensioner; the tensioner spring will take up cable slack and preload the cables.

   c. While holding the tensioner, lock the clip by pushing up on the bottom of the clip. If the clip does not slide up, move the assembly slightly to align the closest groove on the adjuster rod to the clip.

5. Examine the tensioner for remaining cable take up capability. If none is present, check all cables, parking brake control and brackets for possible damage or deflection.

Brake System Bleeding

Without Anti-Lock Brakes

1. Clean all dirt from the master cylinder filler cap.

2. If the master cylinder is known or suspected to have air in the bore, it must be bled before any of the wheel cylinders or calipers. To bleed the master cylinder, loosen the upper secondary left front outlet fitting approximately ¾ turn. Have an assistant depress the brake pedal slowly through it's full travel. Close the outlet fitting and let the pedal return slowly to the fully released position. Wait 5 seconds and then repeat the operation until all air bubbles disappear.

3. Repeat Step 2 with the right-hand front outlet fitting.

4. Continue to bleed the brake system by removing the rubber dust cap from the wheel cylinder bleeder fitting or caliper fitting at the right-hand rear of the vehicle. Place a suitable box wrench on the bleeder fitting and attach a rubber drain tube to the fitting. The end of the tube should fit snugly around the bleeder fitting. Submerge the other end of the tube in a container partially filled with clean brake fluid and loosen the fitting ¾ turn.

5. Have an assistant push the brake pedal down slowly through it's full travel. Close the bleeder fitting and allow the pedal to slowly return to it's full release position. Wait 5 seconds and repeat the procedure until no bubbles appear at the submerged end of the bleeder tube. Secure the bleeder fitting and remove the bleeder tube. Install the rubber dust cap on the bleeder fitting.

6. Repeat the procedure in Steps 4 and 5 in the following sequence: left front, left rear and right front. Refill the master cylinder reservoir after each wheel cylinder or caliper has been bled and install the master cylinder cover and gasket. When brake bleeding is completed, the fluid level should be filled to the maximum level indicated on the reservoir.

7. Always make sure the disc brake pistons are returned to their normal positions by depressing the brake pedal several times until normal pedal travel is established. If the pedal feels spongy, repeat the bleeding procedure.

With Anti-Lock Brakes
CROWN VICTORIA, GRAND MARQUIS AND TOWN CAR

The anti-lock brake system must be bled in 2 steps.

1. The master cylinder and hydraulic control unit must be bled using the Rotunda Anti-Lock Brake Breakout Box/Bleeding Adapter tool No. T90P-50–ALA or equivalent. If this procedure is not followed, air will be trapped in the hydraulic control unit which will eventually lead to a spongy brake pedal. To bleed the master cylinder and the hydraulic control unit, disconnect the 55-pin plug from the electronic control unit and install the Anti-Lock Brake Breakout Box/Bleeding Adapter to the wire harness 55-pin plug.

   a. Place the Bleed/Harness switch in the **BLEED** position.

   b. Turn the ignition to the **ON** position. At this point the red off light should come ON.

   c. Push the motor button on the adapter down to start the pump motor. The red OFF light will turn OFF and the green ON light will turn ON. The pump motor will run for 60 seconds after the motor button is pushed. If the pump motor is to be turned off for any reason before the 60 seconds has elapsed, push the abort button to turn the pump motor off.

   d. After 20 seconds of pump motor operation, push and hold the valve button down. Hold the valve button down for 20 seconds and then release it.

   e. The pump motor will continue to run for an additional 20 seconds after the valve button is released.

2. The brake lines can now be bled in the normal fashion. Bleed the brake system by removing the rubber dust cap from the caliper fitting at the right-hand rear of the vehicle. Place a suitable box wrench on the bleeder fitting and attach a rubber drain tube to the fitting. The end of the tube should fit snugly around the bleeder fitting. Submerge the other end of the tube in a container partially filled with clean brake fluid and loosen the fitting ¾ turn.

3. Have an assistant push the brake pedal down slowly through it's full travel. Close the bleeder fitting and allow the pedal to slowly return to it's full release position. Wait 5 seconds and repeat the procedure until no bubbles appear at the submerged end of the bleeder tube. Secure the bleeder fitting and remove the bleeder tube. Install the rubber dust cap on the bleeder fitting.

4. Repeat the bleeding procedure at the left front, left rear and right front in that order. Refill the master cylinder reservoir after each caliper has been bled and install the master cylinder and gasket. When brake bleeding is completed, the fluid level should be filled to the maximum level indicated on the reservoir.

5. Always make sure the disc brake pistons are returned to their normal positions by depressing the brake pedal several times until normal pedal travel is established. If the pedal feels spongy, repeat the bleeding procedure.

With Anti-Lock Brakes
MARK VII, THUNDERBIRD AND COUGAR

The front brakes can be bled in the same manner as a vehicle without anti-lock brakes or they can be bled with a pressure bleeder. The rear brakes must be bled with a pressure bleeder or with a fully charged accumulator.

Pressure Bleeding

1. Clean all dirt from the reservoir filler cap area. Attach a suitable pressure bleeder to the reservoir cap opening.
2. Maintain 35 psi pressure on the system through the pressure bleeder.
3. Remove the dust cap from the right front caliper bleeder fitting. Attach a rubber drain tube to the fitting, making sure the tube fits snugly.
4. With the ignition switch in the **OFF** position and the brake pedal in the fully released position, open the bleeder fitting for 10 seconds at a time until an air-free stream of brake fluid flow is observed.
5. Repeat the procedure at the left front, right rear and left rear calipers, in that order.
6. Place the ignition switch in the **RUN** position and pump the brake pedal several times to complete the bleeding procedure and to fully charge the accumulator.
7. Turn the ignition switch to the **OFF** position and remove the pressure bleeder. Siphon off the excess fluid in the reservoir to adjust the level to the MAX mark with a fully charged accumulator.

Rear Brake Bleeding with a Fully Charged Accumulator

1. Remove the dust cap from the right rear caliper bleeder fitting. Attach a rubber drain tube to the fitting, making sure the tube fits snugly.
2. Turn the ignition switch to the **RUN** position. This will turn on the electric pump to charge the accumulator, as required.
3. Have an assistant hold the brake pedal in the applied position. Open the bleeder fitting for 10 seconds at a time until an air-free stream of brake fluid flow is observed.

——— **CAUTION** ———

To prevent possible injury, care must be used when opening the bleeder screws due to the high pressures available from a fully charged accumulator.

4. Repeat the procedure at the left rear caliper.
5. Pump the brake pedal several times to complete the bleeding procedure.
6. Adjust the fluid level in the reservoir to the MAX mark with a fully charged accumulator.

NOTE: If the pump motor is allowed to run continuously for approximately 20 minutes, a thermal safety switch inside the motor may shut the motor off to prevent it from overheating. If that happens, a 2–10 minute cool down period is typically required

before normal operation can resume.

Anti-Lock Brake System Service

PRECAUTIONS

- Before servicing any high pressure component, discharge the hydraulic pressure from the system.
- Do not allow brake fluid to contact any electrical connections.
- Use care when opening the bleeder screws due to the high system pressure from the accumulator.

Relieving Anti-Lock Brake System Pressure

——— **CAUTION** ———

Before servicing any component which contains high pressure, it is mandatory that the hydraulic pressure in the system be discharged or personal injury could result.

To discharge the system, turn the ignition **OFF** and pump the brake pedal a minimum of 20 times until an increase in pedal force is clearly felt.

Hydraulic Control Unit (HCU)

REMOVAL & INSTALLATION

Except Mark VII, Thunderbird and Cougar

1. Disconnect the negative battery cable.

2. Remove the aie cleaner and air outlet tube.
3. Disconnect the 19-pin connector from the HCU to the wiring harness and disconnect the 4-pin connector from the HCU to the pump motor relay.
4. Remove the 2 lines from the inlet ports and the 4 lines from the outlet ports of the HCU. Plug each port to prevent brake fluid from spilling onto the paint and wiring.
5. Remove the 3 nuts retaining the HCU assembly to the mounting bracket and remove the assembly from the vehicle. The nut on the front of the HCU also retains the relay mounting bracket.
6. Install in the reverse order of removal. Tighten the 3 retaining nuts to 12–18 ft. lbs. (16–24 Nm) and the brake lines to 10–18 ft. lbs. (14–24 Nm). Bleed the brake system and check for fluid leaks.

Mark VII, Thunderbird and Cougar

The hydraulic actuation assembly contains all the anti-lock brake hydraulic components: master cylinder and fluid reservoir, hydraulic pump motor and accumulator and solenoid valve block assembly.

1. Discharge the hydraulic pressure in the system.
2. Disconnect the negative battery cable.
3. On Thunderbird and Cougar, remove the air cleaner housing and duct assembly.
4. Tag and disconnect the electrical connectors from the fluid level indicator, main valve, solenoid valve block,

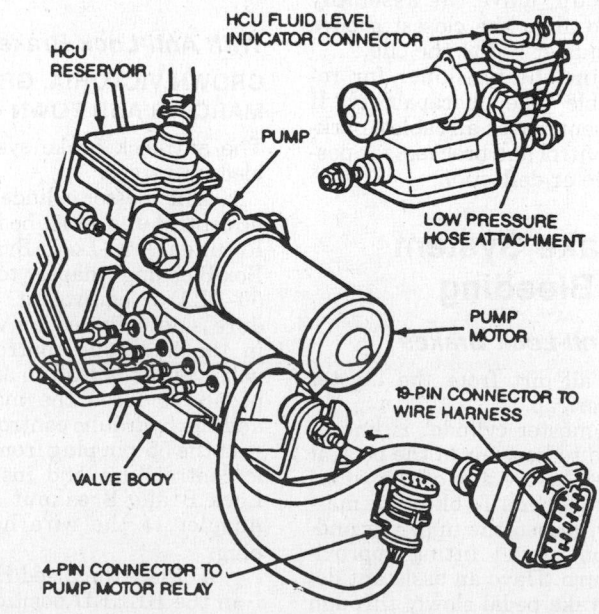

Hydraulic control unit—Crown Victoria, Grand Marquis and Town Car

pressure warning switch, hydraulic pump motor and ground connector from the master cylinder portion of the assembly.

5. Disconnect the brake line fittings. Immediately plug each port to prevent fluid loss and contamination.

NOTE: Do not allow brake fluid to come in contact with any electrical connectors.

6. On Mark VII, remove the accumulator. On Thunderbird and Cougar, remove the trim panel under the steering column.

7. Disconnect the actuation assembly pushrod from the brake pedal by removing the hairpin connector next to the stoplight switch. Slide the switch, pushrod and plastic bushings off the pedal pin.

8. Remove the 4 retaining nuts that hold the actuation assembly to the brake pedal support bracket.

9. Remove the actuation assembly.

To install:

10. Mount the actuation assembly with the rubber boot and foam gasket to the engine side of the dash panel with the 4 mounting studs and pushrod inserted in the proper holes.

11. Working in the passenger compartment, loosely start 4 retaining locknuts attaching the actuation assembly to the pedal support bracket.

12. Connect the pushrod to the brake pedal pin by sliding the flanged plastic bushing, pushrod and washer onto the brake pedal pin. Position the stoplight switch so the slot on the switch bracket straddles the pushrod on the brake pedal pin, with the hole on the opposite leg of the switch bracket just clearing the pin. Slide the switch onto the pedal pin until it bottoms. Install the outer nylon bushing and secure the assembly with the hairpin retainer.

13. Tighten the 4 locknuts to 13–25 ft. lbs. (18–34 Nm).

14. From the engine compartment, connect the brake tubes and tighten the nuts to 10–18 ft. lbs. (14–24 Nm). Connect the electrical connectors.

15. On Mark VII, screw in the accumulator, making sure the O-ring is in place. Tighten to 30–34 ft. lbs. (40–46 Nm). On Thunderbird and Cougar, install the air cleaner and duct assembly.

16. Connect the negative battery cable and bleed the brake system.

Front Wheel Speed Sensor

REMOVAL & INSTALLATION

Crown Victoria, Grand Marquis and Town Car

1. Disconnect the negative battery cable.

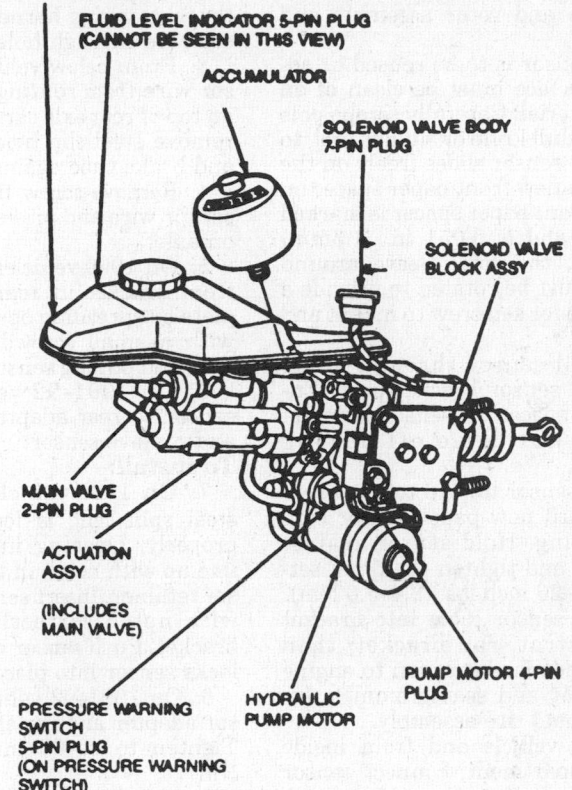

Hydraulic actuation assembly—Thunderbird and Cougar

2. From inside engine compartment, disconnect sensor assembly 2-pin connector from the wiring harness.

3. Remove the steel routing clip attaching the sensor wire to the tube bundle on the left sensor or remove the plastic routing clip attaching the sensor wire to the frame on the right sensor.

4. Remove the rubber coated spring steel clip holding the sensor wire to the frame.

5. Remove the sensor wire from the steel routing clip on the frame and from the dust shield.

6. Remove the sensor attaching bolt from the front spindle and slide the sensor out of the mounting hole.

To install:

7. Install the sensor into the mounting hole in the front spindle and attach with the mounting bolt. Tighten to 40–60 inch lbs. (4.5-6.8 Nm).

8. Insert the sensor routing grommets into the dust shield and steel bracket on the frame. Route the wire into the engine compartment.

9. Install the rubber coated steel clip that holds the sensor wire to the frame into hole in frame.

10. Install the steel clip that holds sensor wire to tube bundle on left side, or plastic clip that holds sensor to frame on right side.

11. Connect 2-pin connector to wire

harness. Connect the negative battery cable.

Mark VII and 1988 Thunderbird

1. Disconnect the negative battery cable.

2. From inside engine compartment, disconnect sensor electrical connector for right or left front sensor.

3. Raise and safely support the vehicle. Disengage wire grommet at right or left shock tower and pull sensor cable connector through hole. Use care not to damage connector.

4. Remove sensor wire from bracket on shock strut and side rail.

5. Remove wheel and tire assembly.

6. Loosen 5mm setscrew holding sensor to sensor bracket post. Remove sensor through hole in disc brake splash shield.

7. To remove sensor bracket or sensor bracket post in case of damage, the caliper and hub and rotor assembly must be removed. After removing the hub and rotor assembly, remove 2 brake splash shield attaching bolts which attach sensor bracket.

NOTE: Replace the toothed sensor ring, if damaged.

To Install:

8. Install sensor bracket with sensor bracket post, if removed. Tighten post retaining bolt to 40–60 inch lb.

(4.5–6.8 Nm) and splash shield attaching bolts to 10–15 ft. lbs. (13–20 Nm). Install hub and rotor assembly and caliper.

9. If a sensor is to be reused or adjusted, pole face must be clean of all foreign material. Carefully scrape pole face with a dull knife or similar tool, to ensure that sensor slides freely on the post. Glue a new front paper spacer on pole face, front paper spacer is marked with an **F** and is 0.051 in. (1.3mm) thick. Also, the steel sleeve around post bolt must be rotated to provide a new surface for setscrew to indent and lock into.

10. Install sensor through brake shield onto sensor bracket post. Ensure paper spacer on sensor is intact and does not come off during installation.

11. Push sensor toward toothed sensor ring until new paper sensor contacts the ring. Hold sensor against sensor ring and tighten the 5mm setscrew to 21–26 inch lbs. (2.4–3.0 Nm).

12. Insert sensor cable into bracket on shock strut, rail bracket; then through inner fender apron to engine compartment and seat grommet. Install wheel and tire assembly.

13. Lower vehicle and from inside engine compartment, connect sensor electrical connection. Connect the negative battery cable.

1989–92 Thunderbird and Cougar

1. Disconnect the negative battery cable. Raise and safely support the vehicle.

2. From underside of vehicle, up front near radiator support, disconnect sensor electrical connector for right or left front sensor.

3. Remove routing clips along wiring harness.

4. Remove Torx® head screw securing sensor to front spindle.

To install:

5. Install sensor into hole in spindle. No adjustment is necessary. Install Torx® head screw and tighten to 40–60 inch lbs. (4.5–6.8 Nm).

6. Route wiring using clips previously removed. Ensure wiring is routed properly as shown.

7. Connect sensor wiring connector to harness connector.

Rear Wheel Speed Sensor

REMOVAL & INSTALLATION

Crown Victoria, Grand Marquis and Town Car

1. Disconnect the negative battery cable.

2. From inside luggage compartment disconnect 2-pin sensor connector from wiring harness and push sensor wire through hole in floor.

3. From below vehicle, remove sensor wire from routing bracket located on top of rear axle carrier housing, and remove steel clip holding sensor wire and brake tube against axle housing.

4. Remove screw from clip holding sensor wire and brake tube to bracket on axle.

5. On 1990 vehicles, remove sensor from bracket in rear brake backing plate by spreading open steel split ring with a small screwdriver or similar tool, and pulling sensor out of bracket.

6. On 1991–92 vehicles, remove sensor to rear adapter retaining bolt and remove sensor.

To install:

7. On 1990 vehicles, ensure that steel split ring is located in groove properly. Opening in ring must not line up with notch in tube shaped sensor retainer. Insert sensor into bracket with notch correctly aligned with bracket. Push sensor in until split ring locks sensor into place.

8. On 1991–92 vehicles, insert sensor adapter and install retaining bolt. Tighten to 40–60 inch lbs. (4.5–6.8 Nm).

9. Attach clip holding sensor and brake tube to bracket on axle housing and secure with screw. Tighten to 40–60 inch lbs. (4.5–6.8 Nm).

10. Install steel clip around axle tube that holds sensor wire and brake tube against axle tube and push spool-shaped grommet into clip located on top of axle carrier housing.

11. Push sensor wire connector up through hole in floor and seat large round grommet into hole.

12. Connect sensor 2-pin connector to wiring harness inside luggage compartment.

1988–90 Mark VII

1. Disconnect the negative battery cable.

2. From inside the luggage compartment, disconnect the wheel sensor electrical connector located behind the forward luggage compartment trim panel.

3. Lift the luggage compartment carpet and push the sensor wire grommet through the hole in the luggage compartment floor.

4. Raise the vehicle and remove the appropriate wheel and tire assembly.

5. Carefully remove the wheel sensor wiring from the axle shaft housing. The wiring harness has 3 different types of retainers. The inboard retainer is a clip located on top of the differential housing. Bend the clip out of the way enough to remove the wiring harness. The second retainer is a C-clip located in the center of the axle shaft housing. Pull rearward on the clip to disengage the clip from the axle housing.

NOTE: Do not bend the clip open beyond the amount necessary to remove the clip from the axle housing. The third clip is at the connection between the rear wheel brake tube and the flexible hose. Remove the hold-down bolt and open the clip to remove the wiring harness.

6. Remove the rear wheel caliper and rotor assemblies.

7. Remove the wheel sensor E8 Torx® head retaining bolt. Slip the grommet out of the rear brake splash shield and pull the sensor wire outward through the hole.

8. Inspect the sensor bracket for possible damage. If damaged, remove the two 6mm self-tapping screws attaching the bracket to the axle adapter and remove the bracket.

NOTE: Replace the toothed sensor ring, if damaged.

To install:

9. Install the sensor bracket, if removed. Tighten the screws to 11–15 ft. lbs. (15–20 Nm)

10. Loosen the 5mm setscrew on the sensor and ensure that the sensor slides freely on the sensor bracket post.

11. If a sensor is to be reused or adjusted, the pole face must be clean of all foreign material. Carefully scrape the pole face with a dull knife or similar tool. Glue a new rear paper spacer on the pole face. Rear paper spacer is marked with a **R** and is 0.043-inch thick. If desired, a feeler gauge may be used instead of a paper spacer. If used, remove paper spacer prior to adjusting. Also, the steel sleeve around the post bolt must be rotated to provide a new surface for the setscrew to indent and lock into.

12. Insert the sensor into large hole in the sensor bracket and install the E8 Torx® head retaining bolt into the snesor bracket post. Tighten the bolt to 40–60 inch lbs. (4.5–6.8 Nm).

13. Push the sensor toward the toothed ring until the new paper sensor makes contact with the sensor ring. Hold the sensor against the toothed ring and tighten the 5mm setscrew to 21–26 inch lbs. (2.4–3.0 Nm).

14. Install the caliper and rotor.

15. Push the wire and connector through the splash shield hole and engage the grommet into the shield eyelet. Install the sensor wire in the retainers along the axle housing.

16. Push the connector through the

hole in the luggage compartment and seat the grommet in the luggage compartment floorpan.

17. From inside the luggage compartment, connect the cable electrical connector. Install the carpet as necessary.

18. Check the function of the sensor by driving the vehicle and observing the "Check Anti-Lock Brakes" lamp in the overhead console.

1991–92 Mark VII and 1988 Thunderbird

1. Disconnect the negative battery cable.

2. From inside the luggage compartment, disconnect the wheel sensor electrical connector located behind the forward luggage compartment trim panel.

3. Lift the luggage compartment carpet and push the sensor wire grommet through the hole in the luggage compartment floor.

4. Raise the vehicle and remove the appropriate wheel and tire assembly.

5. Carefully remove the wheel sensor wiring from the axle shaft housing. The wiring harness has 3 different types of retainers. The inboard retainer is a clip located on top of the differential housing. Bend the clip out of the way enough to remove the wiring harness. The second retainer is a C-clip located in the center of the axle shaft housing. Pull rearward on the clip to disengage the clip from the axle housing.

NOTE: Do not bend the clip open beyond the amount necessary to remove the clip from the axle housing. The third clip is at the connection between the rear wheel brake tube and the flexible hose. Remove the hold-down bolt and open the clip to remove the wiring harness.

6. Remove the sensor retaining bolt and remove the sensor. On 1988 Thunderbird, inspect the sensor bracket for possible damage. If damaged, remove the bracket to disc brake adapter attaching screws and remove the bracket.

NOTE: Replace the toothed sensor ring, if damaged.

To install:

7. On 1988 Thunderbird, insert the sensor into the large hole in the adapter and install the retaining bolt. Tighten to 40–60 inch lbs. (4.5–6.7 Nm). On 1991–92 Mark VII, install the sensor in the rear brake adapter and tighten the retaining screws to 11–15 ft. lbs. (15–20 Nm).

8. Install the sensor wire in the retainers along the axle housing. Push

the connector through the hole in the luggage compartment and seat the grommet in the luggage compartment floorpan.

9. Install the wheel and tire assembly and lower the vehicle.

10. From inside the luggage compartment, connect the cable electrical connector. Install the carpet, as necessary.

11. Check the sensor function by driving the vehicle and observing the "Check Anti-Lock Brakes" indicator.

1989–92 Thunderbird and Cougar

1. Disconnect the negative battery cable.

2. From inside luggage compartment, disconnect wheel sensor electrical connector located rearward of wheel well, behind carpeting on sides of luggage compartment.

3. Lift luggage compartment carpet and push sensor wire grommet through hole in luggage compartment floor.

4. Raise and safely support the vehicle.

5. Remove plastic clip holding sensor wire to axle carrier housing.

6. Remove wheel sensor retaining bolt using a ½ in. socket.

To install:

7. Align sensor locating tab and bolt hole with axle housing and push into position.

8. Install sensor retaining bolt and tighten to 14–20 ft. lbs. (19–27 Nm).

9. Install plastic clip retaining sensor wire to axle carrier housing and push electrical connector through hole in floor into luggage compartment. Ensure that rubber grommet is properly seated in hole in floor.

10. Lower the vehicle. Connect sensor electrical connector to connector on harness.

CHASSIS ELECTRICAL

Air Bag

DISARMING

1. Disconnect the negative battery cable and the backup power supply.

NOTE: The backup power supply allows air bag deployment if the battery or battery cables are damaged in an accident before the crash sensors close. The power supply is a capacitor that will

leak down in approximately 15 minutes after the battery is disconnected or in 1 minute if the battery positive cable is grounded. The backup power supply must be disconnected before any air bag related service is performed.

2. Remove the 4 nut and washer assemblies retaining the driver air bag module to the steering wheel.

3. Disconnect the driver air bag module connector and attach a jumper wire to the air bag terminals on the clockspring.

4. If equipped with a passenger air bag, open the glove compartment and rotate all the way down, past the stops. Disconnect the passenger air bag connector and attach a jumper wire to the air bag terminals on the wiring harness side of the passenger air bag module connector.

Heater Blower Motor

REMOVAL & INSTALLATION

Grand Marquis, Crown Victoria and Town Car

1. Disconnect the negative battery cable.

2. Disconnect the blower motor lead connector from the wiring harness connector.

3. Remove the blower motor cooling tube from the blower motor.

4. Remove the 4 retaining screws.

5. Turn the motor and wheel assembly slightly to the right so that the bottom edge of the mounting plate follows the contour of the wheel well splash panel. Lift up on the blower and remove it from the blower housing.

6. Installation is the reverse of removal. Connect the negative battery cable.

Mark VII, 1988 Thunderbird and Cougar

1. Disconnect the negative battery cable.

2. Remove the recirculation duct assembly and disconnect the blower electrical connector.

3. Remove the 4 retaining screws and remove the blower motor and wheel assembly from the blower housing.

4. Remove the pushnut from the blower motor shaft and remove the blower wheel from the shaft.

5. Installation is the reverse of removal. Connect the negative battery cable.

1989–92 Thunderbird and Cougar

1. Disconnect the negative battery cable.

2. Remove the glove compartment liner to gain access to the blower motor mounting screws.

3. Remove the 4 retaining screws and remove the blower motor and wheel assembly from the blower housing.

4. Remove the pushnut from the blower motor shaft and remove the blower wheel from the shaft.

5. Installation is the reverse of removal. Connect the negative battery cable.

Mustang

1. Disconnect the negative battery cable. Loosen glove compartment assembly by squeezing the sides together to disengage the retainer tabs.

2. Let the glove compartment and door hang down in front of instrument panel and remove blower motor cooling hose.

3. Disconnect electrical wiring harness. Remove 4 screws attaching motor to housing. Pull motor and wheel out of housing.

4. Installation is the reverse of the removal procedure. Connect the negative battery cable.

Windshield Wiper Motor

REMOVAL & INSTALLATION

1988–91 Crown Victoria and Grand Marquis, 1988–89 Town Car

1. Disconnect the negative battery cable.

2. Disconnect the 2 push-on wire connectors from the motor.

3. Remove the hood seal. Remove the right wiper arm and blade assembly from the pivot shaft.

4. Remove the windshield wiper linkage cover by removing the 2 attaching screws and hose clip.

5. Remove the linkage retaining clip from the operating arm on the motor by lifting the locking tab up and pulling the clip away from the pin.

6. Remove the 3 bolts that retain the motor to the dash panel extension and remove the motor.

7. Installation is the reverse of the removal procedure.

1992 Crown Victoria and Grand Marquis, 1990–92 Town Car

1. Disconnect the negative battery cable.

2. Remove the rear hood seal and the wiper arm assemblies.

3. Remove the cowl vent screws and disconnect the washer hoses from the washer jets.

4. Disconnect the electrical connectors from the wiper motor.

5. Remove the wiper assembly attaching screws, lift the assembly out and disconnect the washer hose.

6. Unsnap and remove the linkage cover.

7. Remove the linkage retaining clip from the motor operating arm by lifting the locking tab and pulling the clip away from the pin.

8. Remove the motor retaining screws and remove the motor from the vehicle.

9. Installation is the reverse of removal. Connect the negative battery cable.

Mark VII and 1988 Thunderbird and Cougar

1. Turn the wipers on until they reach full travel on the windshield then turn the key **OFF**.

2. Disconnect the negative battery cable and remove the arm and blade assemblies.

3. Remove the left hand cowl vent screen.

4. Disconnect the linkage drive arm from the motor crankpin after removing the clip.

5. Disconnect the electrical connector from the wiper motor, remove the 3 retaining bolts and remove the motor from the vehicle.

6. Installation is the reverse of removal. Tighten the motor mounting screws to 60–85 inch lbs. (7–9 Nm).

1989–92 Thunderbird and Cougar

1. Disconnect the negative battery cable.

2. With the wipers in the park position remove the arm and blade assemblies.

3. Remove the left hand cowl vent screen.

4. Remove the vacuum manifolds from the wiper module and disconnect the electrical connectors.

5. Remove the 5 screws and 1 nut from the wiper module and remove the wiper module.

6. Disconnect the linkage drive arm from the motor crankpin after removing the clip.

7. Remove the 3 wiper motor retaining screws and pull the motor from the opening.

8. Installation is the reverse of removal. Before installing the arm and blade assembly to the pivot shaft, cycle the motor and turn off the switch to ensure it is in the park position.

Mustang

1. Disconnect the negative battery cable.

2. Remove the right hand wiper arm assembly.

3. Remove the cowl top grille retaining screws and grille.

4. Disconnect the linkage drive arm from the motor crankpin after removing the clip.

5. Disconnect the electrical connector from the wiper motor, remove the 3 retaining bolts and remove the motor from the vehicle.

6. Installation is the reverse of removal. Before installing the arm and blade assembly to the pivot shaft, make sure the motor is in the park position.

Wiper Switch

REMOVAL & INSTALLATION

1988–89 Crown Victoria, Grand Marquis, Town Car and Mark VII, 1988 Thunderbird and Cougar

NOTE: The wiper switch function is a component of the combination switch on all other vehicles.

1. Disconnect the negative battery cable.

2. Remove the split steering column cover retaining screws.

3. Separate the halves and remove the wiper switch retaining screws.

4. Disconnect the electrical connector and remove the wiper switch.

5. The installation of the wiper switch is the reverse of the removal procedure.

Instrument Cluster

REMOVAL & INSTALLATION

Standard Cluster

CROWN VICTORIA, GRAND MARQUIS AND 1988–89 TOWN CAR

1. Disconnect the negative battery cable.

2. On 1988–89 vehicles, disconnect the speedometer cable. On 1988–89 Crown Victoria and Grand Marquis, remove the headlight switch knob and shaft assembly.

3. Remove the instrument cluster trim cover attaching screws and remove the trim cover.

4. Except 1992 Grand Marquis, remove the lower steering column cover retaining screws and remove the lower cover. On 1992 Grand Marquis, remove the knee bolster retaining screws and remove the knee bolster.

5. On all except 1992 Grand Marquis, remove the lower half of the steering column shroud.

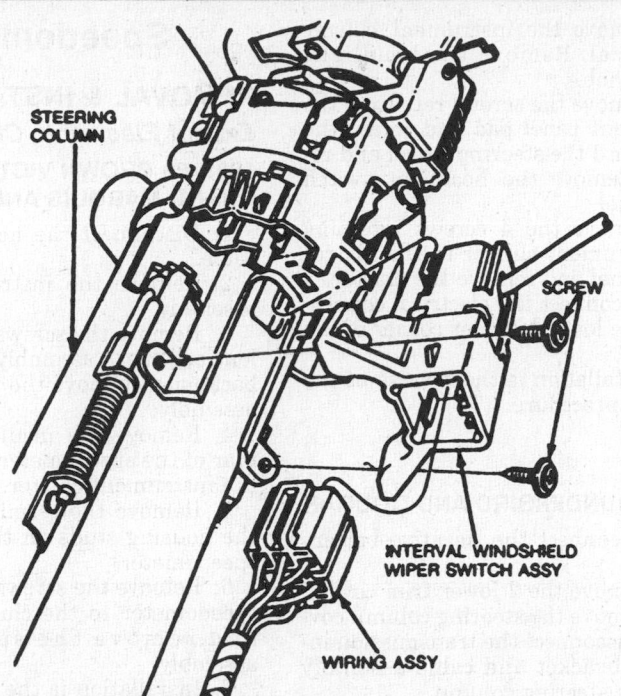

Wiper switch installation—1988–89 Mark VII

6. Remove the screw holding the transmission indicator column bracket to the steering column. Detach the cable loop from the pin and cane shift lever. Remove the column bracket from the column.

7. Remove the 4 cluster retaining screws. Disconnect the cluster feed plugs from the receptacle and remove the cluster assembly.

8. Installation is the reverse of the removal procedure.

MARK VII LSC

1. Disconnect the negative battery cable.

2. Remove the instrument cluster finish panel, disconnecting the warning indicator module connectors.

3. Remove the instrument panel binnacle moulding.

4. Remove the 5 mask to backplate retaining screws. Do not remove the 3 top screws securing the lens to the mask. Remove the mask and lens assembly.

5. Lift the main dial assembly from the backplate. Some effort may be required to pull the quick connect terminals from the clips.

6. Installation is the reverse of the removal procedure.

1988 THUNDERBIRD AND COUGAR, EXCEPT TURBO COUPE OR WITH 5.0L ENGINE

1. Disconnect the negative battery cable.

2. Remove the 2 lower trim covers. Remove the steering column cover and disconnect the transmission indicator bracket and cable assembly from the steering column.

3. Move the shift lever and remove the cluster trim panel. Remove the 4 cluster mounting screws.

4. Pull the bottom of the cluster toward the steering wheel. Reaching behind and under the cluster, unplug the 2 connectors.

5. Swing the bottom of the cluster out to clear the crash pad and remove.

6. Installation is the reverse of the removal procedure.

1988 THUNDERBIRD TURBO COUPE AND THUNDERBIRD AND COUGAR WITH 5.0L ENGINE

1. Disconnect the negative battery cable.

2. Remove the steering column shroud, if equipped. Remove the 10 retaining screws and the cluster trim cover.

3. Remove the 4 screws retaining the instrument cluster to the instrument panel.

4. Pull the cluster away from the instrument panel and disconnect the feed plugs from the printed circuit. Disconnect the light sockets and, if equipped, pull the boost gauge rubber tube from the gauge nipple.

5. Remove the cluster.

6. Installation is the reverse of the removal procedure.

1989–92 THUNDERBIRD AND COUGAR

1. Disconnect the negative battery cable.

2. Remove the 2 retaining screws and remove the cluster trim panel.

3. Remove the 4 cluster mounting screws.

4. Pull the bottom of the cluster towards the steering wheel.

5. Reach behind and under the cluster and unplug the 2 connectors.

6. If equipped, disconnect the vacuum line for the boost gauge.

7. Remove the cluster by swinging the bottom of the cluster out to clear the top of the steering column shroud.

8. Installation is the reverse of removal.

MUSTANG

1. Disconnect the negative battery cable.

2. Remove the switch assembly on the right and left sides of the cluster assembly.

3. Remove the 2 upper and 3 lower retaining screws from the instrument cluster trim cover and remove the trim cover.

4. Remove the 4 retaining screws from the instrument cluster to panel.

5. Pull the cluster away from the instrument panel. Reach behind the instrument cluster to disconnect the speedometer cable by pressing on the flat surface of the plastic connector (quick disconnect).

6. Pull the cluster further away from the instrument panel and disconnect the cluster printed circuit connectors from their receptacles in the cluster backplate.

7. Remove the cluster.

8. Installation is the reverse of the removal procedure.

Electronic Cluster

1992 CROWN VICTORIA AND GRAND MARQUIS, 1990–92 TOWN CAR

1. Disconnect the negative battery cable and set the parking brake.

2. Unsnap the center moulding on the left and right sides of the instrument panel. Remove the steering column cover and column shroud.

3. Remove the knobs from the auto dim and auto lamp switches, if equipped. Remove the 13 screws retaining the instrument panel and pull the panel out.

4. Move the shift lever to the 1 position, if required, for easier access.

5. Disconnect the electrical connectors from the warning lamp module, switch module and center panel switches, if equipped.

6. Remove the instrument cluster carefully so as not to scratch the cluster lens. Disconnect the electrical connector from the front of the cluster.

7. Disconnect the transmission indicator assembly from the cluster by

carefully bending the bottom tab down and pulling the indicator assembly forward.

8. Pull the cluster out and disconnect the electrical connectors on the rear of the cluster. Remove the instrument cluster.

9. Installation is the reverse of the removal procedure.

1988–89 TOWN CAR

1. Disconnect the negative battery cable.

2. Remove the steering column cover and lower instrument panel trim cover. Remove the keyboard trim panel and trim panel on left side of column.

3. Remove the 10 retaining instrument cluster trim cover screws and remove trim cover.

4. Remove the speedometer cable from the clip at the accelerator bracket stud.

5. Remove the 4 screws retaining the instrument cluster to the instrument panel and pull cluster forward. Reach behind the cluster, disconnect both feed plugs and ground wire from their receptacles in the cluster backplate.

6. Disconnect the speedometer cable by pressing on the flat surface of the plastic connector (quick disconnect).

7. Remove the attaching screw from the transmission indicator cable bracket to the steering column. Detach the cable loop from the pin on the shift cane lever of the steering column.

8. Remove the plastic clamp from around steering column. Remove the cluster.

9. Installation is the reverse order of the removal procedure.

1988–90 MARK VII

1. Disconnect the negative battery cable. Remove the screws retaining instrument finish panel and rotate top of panel toward steering wheel. Disconnect electrical and air sensor connectors at right hand portion of finish panel. Remove panel.

2. Remove 6 screws retaining instrument panel pad and rotate pad toward steering wheel and remove.

3. Remove 4 screws retaining instrument cluster to instrument panel and remove.

4. Disconnect electrical connector at lower left rear corner of cluster.

5. Installation is the reverse order of the removal procedure.

1991–92 MARK VII

1. Disconnect the negative battery cable.

2. Remove the instrument cluster trim bezel. Remove the headlight switch knob.

3. Remove the screws retaining the instrument panel pad and rotate the pad toward the steering wheel and remove. Remove the headlight switch trim panel.

4. Remove the 4 screws retaining the instrument cluster to the instrument panel and remove the cluster.

5. Disconnect the electrical connector at the lower left rear corner of the cluster.

6. Installation is the reverse of the removal procedure.

1988 THUNDERBIRD AND COUGAR

1. Disconnect the negative battery cable.

2. Remove the 2 lower trim covers.

3. Remove the steering column cover and disconnect the transmission indicator bracket and cable assembly from the steering column.

4. Move the shift lever and remove the cluster trim panel. Remove the 4 cluster mounting screws.

5. Pull the bottom of the cluster toward the steering wheel. Reach behind and under the cluster and unplug the 2 connectors.

6. Swing the bottom of the cluster out to clear the crash pad and remove the cluster.

7. Installation is the reverse of the removal procedure.

1989–92 THUNDERBIRD AND COUGAR

1. Disconnect the negative battery cable.

2. Remove the headlight switch knob.

3. Remove the cluster finish panel by removing 2 screws on the upper inside surface. Carefully pull away the finish panel while detaching the spring clips surrounding the finish panel.

4. Unplug the connector on the rear of the switch assembly. If equipped, disconnect the autolamp module.

5. Place a clean, soft cloth over the steering column shroud to prevent scratching or damage. Remove the 4 cluster retaining screws.

6. Pull the bottom of the cluster toward the steering wheel. Place a clean, soft cloth over the lens to prevent potential scratches.

7. Reaching behind and under the cluster, unplug the 2 connectors. Swing the bottom of the cluster out to clear the crash pad and remove the cluster.

8. Installation is the reverse of the removal procedure.

Speedometer

REMOVAL & INSTALLATION
Except Electronic Cluster
1988–89 CROWN VICTORIA, GRAND MARQUIS AND TOWN CAR

1. Disconnect the negative battery cable.

2. Remove the instrument cluster assembly.

3. Remove the screws attaching the lens and mask assembly to the cluster backplate. Remove the lens and mask assembly.

4. Remove the insulator from the rear of the speedometer at the back of the instrument cluster.

5. Remove the terminal nuts from the housing studs on the back of the speedometer.

6. Remove the screws attaching the speedometer to the cluster backplate and remove the speedometer assembly.

7. Installation is the reverse of the removal procedure.

1990–92 CROWN VICTORIA AND GRAND MARQUIS

1. Disconnect the negative battery cable.

2. Remove the instrument cluster assembly.

3. Keeping the cluster face up, remove the lens and mask retaining screws.

4. Remove the lens and mask assembly. Use caution handling the mask to prevent scratches.

5. Remove the transmission indicator assembly. Lift the temperature gauge and fuel gauge from the cluster. Set the face up to avoid damage.

6. Lift the speedometer assembly out of the cluster.

7. Installation is the reverse of the removal procedure.

MARK VII LSC

1. Disconnect the negative battery cable.

2. Remove the instrument cluster finish panel, disconnecting the warning indicator module connectors.

3. Remove the instrument panel binnacle moulding.

4. Remove the 5 mask to backplate retaining screws. Do not remove the 3 top screws securing the lens to the mask. Remove the mask and lens assembly.

5. Lift the main dial assembly from the backplate. Some effort may be required to pull the quick connect terminals from the clips.

6. Remove the screws retaining the fuel gauge, temperature gauge and tachometer. The speedometer is integral with the main dial.

7. Installation is the reverse of the removal procedure.

THUNDERBIRD, COUGAR AND MUSTANG

1. Disconnect the negative battery cable.

2. Remove the instrument cluster assembly.

3. Remove the 7 screws retaining the mask and lens assembly.

4. Remove the 2 speedometer head assembly retaining screws and remove the speedometer.

5. Installation is the reverse of the removal procedure.

Electronic Cluster

The speedometer is an integral part of the electronic cluster and cannot be removed separately.

Headlight Switch

REMOVAL & INSTALLATION

1988–89 Town Car

1. Disconnect the negative battery cable.

2. Remove the headlight switch knob.

3. Remove the auto dimmer bezel and the autolamp delay bezel, if equipped.

4. Remove the steering column lower shroud.

5. Remove the lower left instrument panel trim bezel.

6. Remove the 5 screws that retain the headlight switch mounting bracket to the instrument panel.

7. Carefully pull the switch and bracket from the instrument panel and disconnect the wiring connector(s) from the headlight switch.

8. Remove the locknut and screw that retain the headlight switch to the switch bracket.

9. Installation is the reverse of the removal procedure.

1990–92 Town Car

1. Disconnect the negative battery cable.

2. Remove the headlight switch knob and auto dimmer knob, if equipped.

3. Remove the right and left mouldings from the instrument panel by pulling away from the instrument panel and snapping out of the retainers.

4. Remove 12 screws retaining the finish panel and remove the panel.

5. Remove the 2 headlight switch bracket retaining screws and pull the bracket and switch from the instrument panel.

6. Remove the nut retaining the switch to the bracket, disconnect the connector and remove the switch.

7. Installation is the reverse of the removal procedure.

1988–89 Crown Victoria and Grand Marquis

1. Disconnect the negative battery cable.

2. Pull the headlight switch shaft out to the headlight **ON** position.

3. From under the instrument panel, depress the headlight switch knob and shaft retainer button on the headlight switch. Hold the button in and pull the knob and shaft assembly straight out.

4. Remove the autolamp control bezel and remove the locknut.

5. From under the instrument panel, move the switch toward the front of the vehicle while tilting it downward.

6. Disconnect the wiring from the switch and remove the switch from the vehicle.

7. Installation is the reverse of removal.

1990–92 Crown Victoria and Grand Marquis

1. Disconnect the negative battery cable.

2. Remove the right and left mouldings from the instrument panel by pulling up and snapping out of the retainers.

3. Remove the screws retaining the finish panel to the instrument panel.

4. Remove the headlight switch knob from the shaft and remove the finish panel.

5. Remove the 2 headlight bracket retaining screws and pull the bracket and switch from from the instrument panel.

6. Remove the nut retaining the switch to the bracket.

7. Disconnect the electrical connector and remove the switch.

8. Installation is the reverse of removal.

1988–89 Mark VII and 1988 Thunderbird and Cougar

1. Disconnect the negative battery cable. Remove the lens assembly attaching screws and then the lens assembly.

2. Remove the screws securing the switch to the instrument panel and pull the switch out from the panel.

NOTE: If equipped with auto lamp/auto dimmer system, remove this control first.

3. Disconnect the electrical connec-

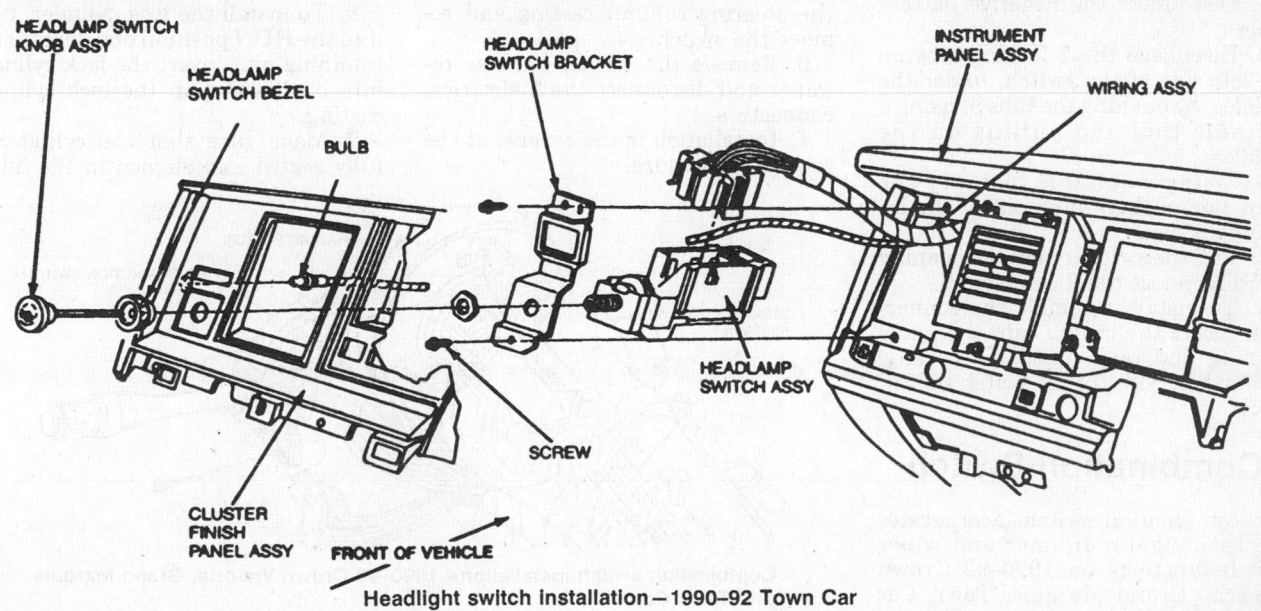

Headlight switch installation—1990–92 Town Car

tor and remove the switch from the vehicle.

4. Installation is the reverse order of the removal procedure.

1990–92 Mark VII

1. Disconnect the negative battery cable.

2. Remove the center moulding and the headlight switch knob.

3. Remove the 5 screws retaining the cluster finish panel and snap out the headlight switch lens.

4. Remove the 2 screws retaining the headlight switch. Remove the switch from the instrument panel and disconnect the electrical connector.

5. Installation is the reverse of the removal procedure.

1989–92 Thunderbird and Cougar

1. Disconnect the negative battery cable.

2. Remove the 2 cluster finish panel retaining screws.

3. Pull the headlight switch knob off.

4. Unsnap the cluster finish panel.

5. Disconnect the electrical connector to the headlight dimmer sensor assembly.

6. Through the opening in the instrument panel, depress the shaft release button on the switch and remove the shaft. The switch must be in the full ON position to release the shaft.

7. Remove the headlight switch retaining nut, pull the switch through the opening to disconnect the wiring connector.

8. Installation is the reverse of removal.

1988–92 Mustang

1. Disconnect the negative battery cable.

2. Disengage the 2 locking tabs on the left side of the switch, under the paddles, by pushing the tabs in using a suitable tool and pulling on the paddles.

3. Using a suitable tool, pry the right side of the switch out of the instrument panel.

4. Pull the switch out of the opening and disconnect the 2 connectors.

5. To install, assemble the connectors, insert the switch into the panel opening and push until the locking tabs on both sides of the switch slide in place.

Combination Switch

The combination switch incorporates the turn signal, dimmer and wiper switch functions on 1990–92 Crown Victoria, Grand Marquis, Town Car and Mark VII, 1989–92 Thunderbird and Cougar and 1988–92 Mustang. The combination switch incorporates only the turn signal and dimmer function on all other vehicles.

REMOVAL & INSTALLATION

1988–89 Crown Victoria, Grnad Marquis, Town Car and Mark VII, 1988 Thunderbird and Cougar

1. Disconnect the negative battery cable.

2. Remove the switch lever by grasping and pulling straight out.

3. Remove the steering column cover retaining screws and remove the cover.

4. On Crown Victoria, Grand Marquis and Town Car, remove the shroud retaining screws and remove the shroud.

5. With the wiring connectors exposed, carefully lift the connector retainer tabs and disconnect the connectors.

6. Remove the switch retaining screws and lift up the switch assembly.

7. Installation is the reverse of the removal procedure.

1990–92 Crown Victoria, Grand Marquis, Town Car and Mark VII

1. Disconnect the negative battery cable.

2. If equipped with tilt column, move to the lowest position and remove the tilt lever.

3. Remove the ignition lock cylinder.

4. Remove the shroud screws and remove the upper and lower shrouds.

5. Remove the 2 self-tapping screws attaching the combination switch to the steering column casting and remove the switch.

6. Remove the wiring harness retainer and disconnect the 2 electrical connectors.

7. Installation is the reverse of the removal procedure.

Mustang and 1989–92 Thunderbird and Cougar

1. Disconnect the negative battery cable.

2. Remove the shroud retaining screws and remove the upper and lower shrouds.

3. Remove the switch retaining screws and lift up the switch assembly.

4. With the wiring connectors exposed, carefully lift the connector retainer tabs and disconnect the connectors.

5. Installation is the reverse of the removal procedure.

Ignition Lock
REMOVAL & INSTALLATION

1. Disconnect the negative battery cable and, if equipped, the air bag backup power supply.

2. On Thunderbird, Cougar and 1988–89 Mark VII and Mustang equipped with tilt column, remove the upper extension shroud by unsnapping the shroud retaining clip at the 9 o'clock position.

3. On all except 1990–92 Crown Victoria, Grand Marquis, Town Car and Mark VII, remove the trim shroud halves by removing the attaching screws. Remove the electrical connector from the key warning switch.

4. Place the gear shift lever in **P**, for column shift only, and turn the igntion to the **RUN** position.

5. Place a ⅛ in. diameter wire pin or small drift punch in the hole in the casting surrounding the lock cylinder and depress the retaining pin while pulling out on the lock cylinder to remove it from the column housing.

To install:

6. To install the lock cylinder, turn it to the **RUN** position and depress the retaining pin. Insert the lock cylinder into its housing in the lock cylinder casting.

7. Make sure that the cylinder is fully seated and aligned in the inter-

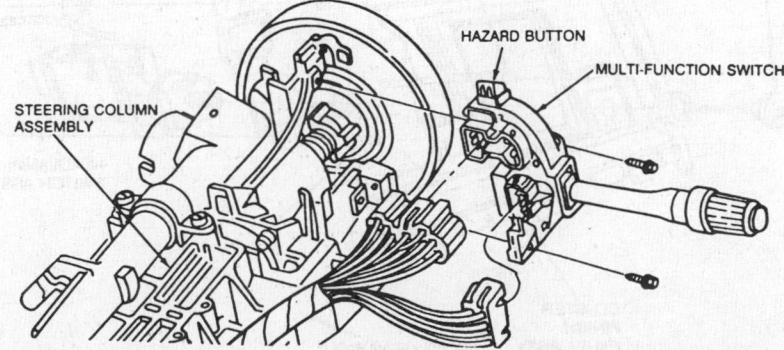

Combination switch installation – 1990–92 Crown Victoria, Grand Marquis and Town Car

locking washer before turning the key to the **OFF** position. This action will permit the cylinder retaining pin to extend into the hole in the lock cylinder housing.

8. Using the ignition key, rotate the cylinder to ensure the correct mechanical operation in all positions.

9. Check for proper start in **P** or **N**. Also make sure that the start circuit cannot be actuated in **D** or **R** positions and that the column is locked in the **LOCK** position.

10. Connect the key warning buzzer electrical connector and install the trim shrouds, if required.

Ignition Switch

REMOVAL & INSTALLATION

1. Disconnect the negative battery cable.

2. On all 1988–89 vehicles with tilt column, except Thunderbird and Cougar, remove the upper extension shroud by unsnapping the shroud from the retaining clips at the 9 o'clock position.

3. On all except 1989 Thunderbird and Cougar, remove the steering column shroud. On 1989 Thunderbird and Cougar, proceed as follows:

 a. Remove the 3 screws and remove the lower shroud.

 b. Remove the 4 nuts holding the steering column to the column mounting bracket.

 c. Lower the column until the ignition switch retaining screws are accessible.

 d. Remove the upper shroud.

4. On 1990–92 Crown Victoria, Grand Marquis and Town Car, remove the instrument panel lower steering column cover. On 1990–92 Mark VII, remove the steering column opening trim cover.

5. Disconnect the electrical connector from the ignition switch.

6. Rotate the ignition key lock cylinder to the **RUN** position.

7. Remove the 2 screws attaching the ignition switch.

8. Disengage the ignition switch from the actuator pin and remove the switch.

To install:

9. Adjust the new ignition switch by sliding the carrier to the **RUN** position.

10. Check to ensure that the ignition key lock cylinder is in the **RUN** position. The **RUN** position is achieved by rotating the key lock cylinder approximately 90 degrees from the **LOCK** position.

11. Install the ignition switch onto the actuator pin.

12. Align the switch mounting holes

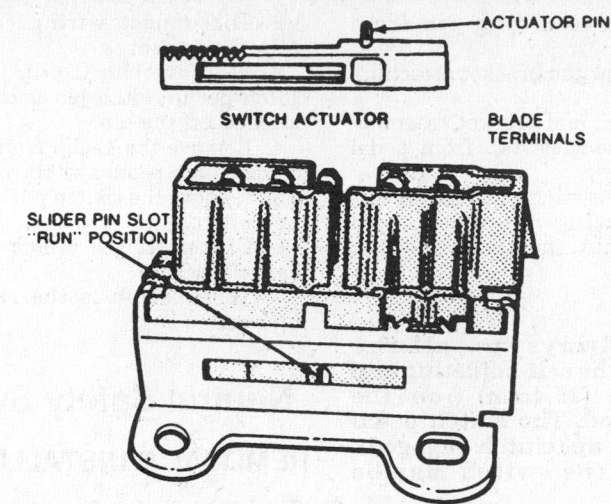

Ignition switch assembly

and install the attaching screws. Tighten the screws to 50–69 inch lbs. (5.6–7.9 Nm).

13. Connect the electrical connector to the ignition switch.

14. Connect the negative battery cable. Check the ignition switch for proper function in **START** and **ACC** positions. Make sure the column is locked in the **LOCK** position.

15. Install the remaining components in the reverse order of removal.

Stoplight Switch

REMOVAL & INSTALLATION

1. Disconnect the negative battery cable.

2. Disconnect the wire harness at the connector from the switch. The locking tab on the connector must be lifted before the connector can be removed.

3. Remove the hairpin retainer, slide the stoplight switch, the pushrod and the nylon washers and bushings away from the pedal and remove the switch.

NOTE: Since the switch side plate nearest the brake pedal is slotted, it is not necessary to remove the brake master cylinder pushrod and 1 washer from the brake pedal pin.

To install:

4. Position the switch so the U-shaped side is nearest the pedal and directly over/under the pin. Then slide the switch down/up trapping the master cylinder pushrod and black bushing between the switch side plates. Push the switch and pushrod assembly firmly toward the brake pedal arm. Assemble the outside white plastic washer to the pin and install the hairpin retainer to trap the whole assembly.

5. Assemble the wire harness connector to the switch. Check the switch for proper operation.

NOTE: The stoplight switch wire harness must be long enough to travel with the switch during full pedal stroke. If wire length is insufficient, reroute the harness or service, as required.

Clutch Switch

ADJUSTMENT

Except 1988 Thunderbird

The clutch switch on all vehicles except 1988 Thunderbird is self-adjusting.

1988 Thunderbird

1. If the adjusting clip is out of position on the rod, remove both halves of the clip.

2. Position both halves of the clip closer to the switch and snap the clips together on the rod.

3. Depress the clutch pedal to the floor to adjust the switch.

REMOVAL & INSTALLATION

Mustang

1. Disconnect the negative battery cable. Disconnect wiring connector.

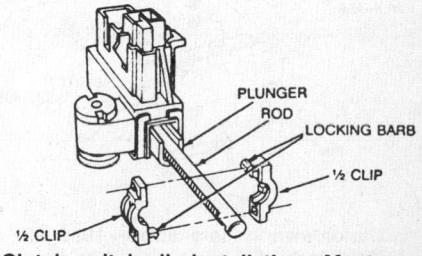

Clutch switch clip Installation–Mustang

2. Remove retaining pin from clutch pedal.

3. Remove switch bracket attaching screw.

4. Lift switch and bracket assembly upward to disengage tab from pedal support.

5. Move the switch outward to disengage actuating rod eyelet from clutch pedal pin and remove switch from vehicle.

To install:

NOTE: **Always install the switch with the self-adjusting clip about 1.0 in. (25.4mm) from the end of the rod. The clutch pedal must be fully up (clutch engaged). Otherwise, the switch may be misadjusted.**

6. Place eyelet end of rod onto pivot pin.

7. Swing switch assembly around to line up hole in mounting boss with hole in bracket.

8. Install attaching screw.

9. Replace retaining pin on clutch pedal.

10. Connect wiring connector.

Thunderbird and Cougar

1988

1. Disconnect the negative battery cable.

2. Remove the switch bracket mounting nuts.

3. Disconnect the electrical connector.

4. Remove the switch and bracket assembly.

5. Installation is the reverse of the removal procedure.

1989–92

1. Disconnect the negative battery cable.

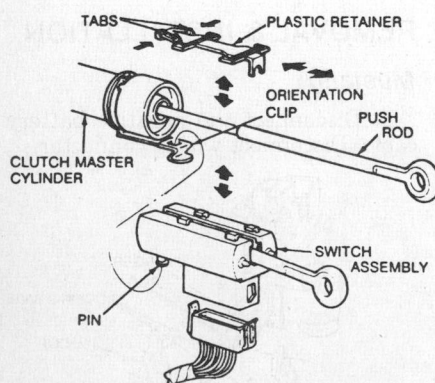

Clutch switch installation – 1989–92 Thunderbird

2. Disconnect wiring connector from the switch.

3. Remove the C-clip from the clutch pedal switch pin and slide the pushrod off the pin.

4. Remove the C-clip from the end of the clutch pedal switch rod.

5. Remove the switch pushrod from the switch.

6. Disconnect the switch from the plastic bracket.

7. Installation is the reverse of removal.

Neutral Safety Switch

REMOVAL & INSTALLATION

Crown Victoria, Grand Marquis and Town Car

1. Set the parking brake.
2. Place the selector lever in the manual **L** position.
3. Remove the air cleaner assembly.
4. Disconnect the negative battery cable.
5. Disconnect the neutral start switch electrical harness from the switch by lifting the harness straight up off the switch without side-to-side motion.
6. Reach in the area of the left hand dash panel, using a 24 inch extension, universal adapter and socket tool T74P–77247–A or equivalent, and remove the neutral start switch and O-ring.

NOTE: **Use of different tools could crush or puncture the walls of the switch.**

To install:

7. Install the neutral start switch and new O-ring using socket tool T74P–77247–A or equivalent.
8. Tighten the switch to 8–11 ft. lbs. (11–15 Nm).
9. Connect the neutral start switch to the wiring harness.
10. Connect the negative battery cable.
11. Check that the vehicle starts in the **N** or **P** position.

Mustang with 5.0L Engine, Thunderbird, Cougar and Mark VII

1. Place the selector lever in the manual **L** position.
2. Disconnect the negative battery cable.
3. Raise and support the vehicle safely.

4. Disconnect the neutral start switch electrical harness from the switch by pushing the harness straight up off the switch using a suitable long-bladed tool under the rubber plug section of the harness.
5. Install socket tool T74P–77247–A or equivalent, and rachet on the neutral start switch. Once the ratchet and socket tool are over the switch, reach from the rear of the transmission over the extension housing area and remove the neutral start switch and O-ring.

NOTE: **Use of different tools could crush or puncture the walls of the switch.**

To install:

6. Install the neutral start switch and new O-ring using socket tool T74P–77247–A or equivalent.
7. Tighten the switch to 8–11 ft. lbs. (11–15 Nm).
8. Connect the neutral start switch to the wiring harness.
9. Lower the vehicle and connect the negative battery cable.
10. Check that the vehicle starts in the **N** or **P** position.

Fuses, Circuit Breakers and Relays

LOCATION

Fuses

All vehicles are equipped with a fuse panel located on the left side of the lower instrument panel. In addition, 1990–92 Town Car and 1992 Crown Victoria and Grand Marquis are equipped with a fuse box located in the right front of the engine compartment. 1989–92 Thunderbird and Cougar are equipped with a high-current fuse box located in the engine compartment on the left fender apron.

Fusible Links

Fusible links are used to protect the main wiring harness and selected branches from complete burn-out, should a short circuit or electrical overload occur.

Circuit Breakers

Circuit breakers are used on certain electrical components requiring high amperage. The advantage of the circuit breaker is its ability to open and close the electrical circuit as the load demands, rather than the necessity of a part replacement.

Ford Motor Co.
Front Wheel Drive
Festiva

SPECIFICATIONS

VEHICLE IDENTIFICATION CHART

It is important for servicing and ordering parts to be certain of the vehicle and engine identification. The VIN (vehicle identification number) is a 20 digit number visible through the windshield on the driver's side of the dash and contains the vehicle and engine identification codes. The tenth digit indicates model year and the eighth digit indicates engine code. It can be interpreted as follows:

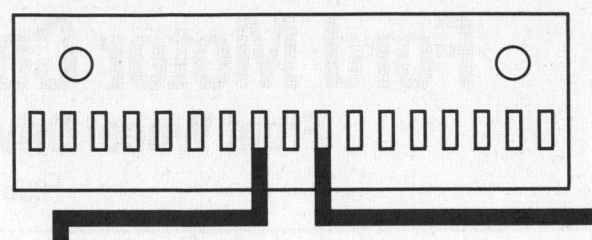

Engine Code

Code	Cu. In.	Liters	Cyl.	Fuel Sys.	Eng. Mfg.
K	81	1.3	4	2 bbl	Kia Motors
H	81	1.3	4	EFI	Kia Motors

EFI—Electronic Fuel Injection

Model Year

Code	Year
J	1988
K	1989
L	1990
M	1991
N	1992

ENGINE IDENTIFICATION

Year	Model	Engine Displacement cu. in. (liter)	Engine Series Identification (VIN)	No. of Cylinders	Engine Type
1988	Festiva	81 (1.3)	K	4	OHC
1989	Festiva	81 (1.3)	K	4	OHC
	Festiva	81 (1.3)	H	4	OHC
1990	Festiva	81 (1.3)	H	4	OHC
1991-92	Festiva	81 (1.3)	H	4	OHC

OHC—Overhead Cam

GENERAL ENGINE SPECIFICATIONS

Year	VIN	No. Cylinder Displacement cu. in. (liter)	Fuel System Type	Net Horsepower @ rpm	Net Torque @ rpm (ft. lbs.)	Bore × Stroke (in.)	Compression Ratio	Oil Pressure @ rpm
1988	K	4-81 (1.3)	2 bbl	58 @ 5000	73 @ 3500	2.79 × 3.29	9.7:1	50–64 @ 3000
1989	K	4-81 (1.3)	2 bbl	58 @ 5000	73 @ 3500	2.79 × 3.29	9.7:1	50–64 @ 3000
	H	4-81 (1.3)	EFI	63 @ 5000	73 @ 3000	2.79 × 3.29	9.7:1	50–64 @ 3000

GENERAL ENGINE SPECIFICATIONS

Year	VIN	No. Cylinder Displacement cu. in. (liter)	Fuel System Type	Net Horsepower @ rpm	Net Torque @ rpm (ft. lbs.)	Bore × Stroke (in.)	Compression Ratio	Oil Pressure @ rpm
1990	H	4-81 (1.3)	EFI	63 @ 5000	73 @ 3000	2.79 × 3.29	9.7:1	50–64 @ 3000
1991–92	H	4-81 (1.3)	EFI	63 @ 5000	73 @ 3000	2.79 × 3.29	9.7:1	50–64 @ 3000

EFI—Electronic Fuel Injection

GASOLINE ENGINE TUNE-UP SPECIFICATIONS

Year	VIN	No. Cylinder Displacement cu. in. (liter)	Spark Plugs Type	Gap (in.)	Ignition Timing (deg.) MT	Ignition Timing (deg.) AT	Compression Pressure (psi)	Fuel Pump (psi)	Idle Speed (rpm) MT	Idle Speed (rpm) AT	Valve Clearance In.	Valve Clearance Ex.
1988	K	4-81 (1.3)	AGS32C	0.040	TDC	—	①	3–6	700–750	–	0.012	0.012
1989	K	4-81 (1.3)	AGS32C	0.040	TDC	—	①	3–6	700–750	700–750	0.012	0.012
	H	4-81 (1.3)	AGS32C	0.040	2	2	①	64–85	800–900	800–900	Hyd.	Hyd.
1990	H	4-81 (1.3)	AGS32C	0.040	10	10	①	64–85	800–900	800–900	Hyd.	Hyd.
1991	H	4-81 (1.3)	AGS32C	0.040	10	10	①	64–85	680–720	830–870	Hyd.	Hyd.
1992	SEE UNDERHOOD SPECIFICATIONS											

Hyd.—Hydraulic
① The lowest cylinder pressure should be within
   75% of the highest cylinder pressure reading

FIRING ORDERS

NOTE: To avoid confusion, always replace spark plug wires one at a time.

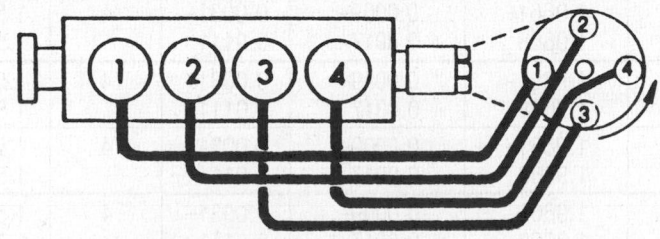

1.3L Engine
Engine Firing Order: 1–3–4–2
Distributor Rotation: Counterclockwise

CAPACITIES

Year	Model	VIN	No. Cylinder Displacement cu. in. (liter)	Engine Crankcase with Filter	Engine Crankcase without Filter	Transmission (pts.) 4-Spd	Transmission (pts.) 5-Spd	Transmission (pts.) Auto.	Drive Axle (pts.)	Fuel Tank (gal.)	Cooling System (qts.)
1988	Festiva	K	4-81 (1.3)	3.9	3.6	5.2	5.2	—	—	10	5.3
1989	Festiva	K	4-81 (1.3)	3.9	3.6	5.2	5.2	—	—	10	5.3
	Festiva	H	4-81 (1.3)	3.9	3.6	—	—	11.2	—	10	5.3
1990	Festiva	H	4-81 (1.3)	3.9	3.6	—	5.2	11.2	—	10	5.3
1991-92	Festiva	H	4-81 (1.3)	3.9	3.6	—	5.2	11.2	—	10	5.3

CAMSHAFT SPECIFICATIONS

All measurements given in inches.

Year	VIN	No. Cylinder Displacement cu. in. (liter)	Journal Diameter 1	Journal Diameter 2	Journal Diameter 3	Journal Diameter 4	Journal Diameter 5	Lobe Lift In.	Lobe Lift Ex.	Bearing Clearance	Camshaft End Play
1988	K	4-81 (1.3)	1.7103–1.7112	1.7091–1.7100	1.7103–1.7112	—	—	1.4185–1.4224	1.4185–1.4224	②	0.002–0.007
1989	K	4-81 (1.3)	1.7103–1.7112	1.7091–1.7100	1.7103–1.7112	—	—	1.4185–1.4224	1.4185–1.4224	②	0.002–0.007
	H	4-81 (1.3)	1.7103–1.7112	1.7091–1.7100	1.7103–1.7112	—	—	1.4331–1.4371	1.4331–1.4371	②	0.002–0.007
1990	H	4-81 (1.3)	1.7103–1.7112	1.7091–1.7100	1.7103–1.7112	—	—	1.4331–1.4371	1.4331–1.4371	②	0.002–0.007
1991-92	H	4-81 (1.3)	1.7103–1.7112	1.7091–1.7100	1.7103–1.7112	—	—	1.4331–1.4371	1.4331–1.4371	②	0.002–0.007

① Figure shown indicates total lobe height
② Front and Rear Bearing—0.0014–0.0033 in.
   Center Bearing— 0.0026–0.0045

CRANKSHAFT AND CONNECTING ROD SPECIFICATIONS

All measurements are given in inches.

Year	VIN	No. Cylinder Displacement cu. in. (liter)	Crankshaft Main Brg. Journal Dia.	Crankshaft Main Brg. Oil Clearance	Crankshaft Shaft End-play	Crankshaft Thrust on No.	Connecting Rod Journal Diameter	Connecting Rod Oil Clearance	Connecting Rod Side Clearance
1988	K	4-81 (1.3)	1.9661–1.9668	0.0009–0.0017	0.0031–0.0111	4	1.5724–1.5731	0.0011–0.0027	0.0043–0.0103
1989	K	4-81 (1.3)	1.9661–1.9668	0.0009–0.0017	0.0031–0.0111	4	1.5724–1.5731	0.0011–0.0027	0.0043–0.0103
	H	4-81 (1.3)	1.9661–1.9668	0.0009–0.0017	0.0031–0.0111	4	1.5724–1.5731	0.0011–0.0027	0.0043–0.0103
1990	H	4-81 (1.3)	1.9661–1.9668	0.0009–0.0017	0.0031–0.0111	4	1.5724–1.7531	0.0011–0.0027	0.0043–0.0103
1991-92	H	4-81 (1.3)	1.9661–1.9668	0.0009–0.0017	0.0031–0.0111	4	1.5724–1.5731	0.0011–0.0027	0.0043–0.0103

VALVE SPECIFICATIONS

Year	VIN	No. Cylinder Displacement cu. in. (liter)	Seat Angle (deg.)	Face Angle (deg.)	Spring Test Pressure (lbs.)	Spring Installed Height (in.)	Stem-to-Guide Clearance (in.) Intake	Exhaust	Stem Diameter (in.) Intake	Exhaust
1988	K	4-81 (1.3)	45	45	NA	NA	0.008	0.008	0.2744–0.2750	0.2742–0.2748
1989	K	4-81 (1.3)	45	45	NA	NA	0.008	0.008	0.2744–0.2750	0.2742–0.2748
	H	4-81 (1.3)	45	45	NA	NA	0.008	0.008	0.2744–0.2750	0.2742–0.2748
1990	H	4-81 (1.3)	45	45	NA	NA	0.008	0.008	0.2744–0.2750	0.2742–0.2748
1991–92	H	4-81 (1.3)	45	45	NA	NA	0.008	0.008	0.2744–0.2750	0.2742–0.2748

NA—Not Available

PISTON AND RING SPECIFICATIONS

All measurements are given in inches.

Year	VIN	No. Cylinder Displacement cu. in. (liter)	Piston Clearance	Ring Gap Top Compression	Bottom Compression	Oil Control	Ring Side Clearance Top Compression	Bottom Compression	Oil Control
1988	K	4-81 (1.3)	①	0.006–0.012	0.006–0.012	0.008–0.028	0.001–0.003	0.001–0.003	snug
1989	K	4-81 (1.3)	①	0.006–0.012	0.006–0.012	0.008–0.028	0.001–0.003	0.001–0.003	snug
	H	4-81 (1.3)	①	0.006–0.012	0.006–0.012	0.008–0.028	0.001–0.003	0.001–0.003	snug
1990	H	4-81 (1.3)	①	0.006–0.012	0.006–0.012	0.008–0.028	0.001–0.003	0.001–0.003	snug
1991–92	H	4-81 (1.3)	①	0.006–0.012	0.006–0.012	0.008–0.028	0.001–0.003	0.001–0.003	snug

① Optimum—0.0015–0.0020 in.
   Limit—0.006 in.

TORQUE SPECIFICATIONS

All readings in ft. lbs.

Year	VIN	No. Cylinder Displacement cu. in. (liter)	Cylinder Head Bolts	Main Bearing Bolts	Rod Bearing Bolts	Crankshaft Pulley Bolts	Flywheel Bolts	Manifold Intake	Exhaust	Spark Plugs
1988	K	4-81 (1.3)	①	40–43	②	③	71–76	14–20	12–17	10–17
1989	K	4-81 (1.3)	①	40–43	②	③	71–76	14–20	12–17	10–17
	H	4-81 (1.3)	①	40–43	②	③	71–76	14–20	12–17	10–17
1990	H	4-81 (1.3)	①	40–43	②	③	71–76	14–20	12–17	10–17

TORQUE SPECIFICATIONS
All readings in ft. lbs.

Year	VIN	No. Cylinder Displacement cu. in. (liter)	Cylinder Head Bolts	Main Bearing Bolts	Rod Bearing Bolts	Crankshaft Pulley Bolts	Flywheel Bolts	Manifold Intake	Manifold Exhaust	Spark Plugs
1991–92	H	4-81 (1.3)	①	40–43	②	③	71–76	14–20	12–17	10–17

① Tighten in sequence in 2 steps:
    Step 1—35–40 ft. lbs.
    Step 2—56–60 ft. lbs.
② Tighten in 2 steps:
    Step 1—11–13 ft. lbs.
    Step 2—22–25 ft. lbs.
③ Pulley bolts—109–152 inch lbs.
    Sprocket bolt—80–87 ft. lbs.

BRAKE SPECIFICATIONS
All measurements in inches unless noted

Year	Model	Lug Nut Torque (ft. lbs.)	Master Cylinder Bore	Brake Disc Minimum Thickness	Brake Disc Maximum Runout	Standard Brake Drum Diameter	Minimum Lining Thickness Front	Minimum Lining Thickness Rear
1988	Festiva	65–87	0.75/0.59	0.43	0.003	6.69	0.125	0.040
1989	Festiva	65–87	0.75/0.59	0.43	0.003	6.69	0.125	0.040
1990	Festiva	65–87	0.75/0.59	0.43	0.003	6.69	0.125	0.040
1991–92	Festiva	65–87	0.75/0.59	0.43	0.003	6.69	0.125	0.040

WHEEL ALIGNMENT

Year	Model	Caster Range (deg.)	Caster Preferred Setting (deg.)	Camber Range (deg.)	Camber Preferred Setting (deg.)	Toe-in (in.)	Steering Axis Inclination (deg.)
1988	Festiva	1⁵/₁₆P–1¹³/₁₆P	1⁹/₁₆P	¼N–1⁹/₁₆P	1¹/₁₆P	¹/₃₂P–¹/₄P	14³/₁₆
1989	Festiva	1⁵/₁₆P–1¹³/₁₆P	1⁹/₁₆P	¼N–1⁹/₁₆P	1¹/₁₆P	¹/₃₂P–¹/₄P	14³/₁₆
1990	Festiva	1⁵/₁₆P–1¹³/₁₆P	1⁹/₁₆P	¼N–1⁹/₁₆P	1¹/₁₆P	¹/₃₂P–¹/₄P	14³/₁₆
1991–92	Festiva	1⁵/₁₆P–1¹³/₁₆P	1⁹/₁₆P	¼N–1⁹/₁₆P	1¹/₁₆P	¹/₃₂P–¹/₄P	14³/₁₆

P—Positive
N—Negative

ENGINE MECHANICAL

NOTE: Disconnecting the negative battery cable on some vehicles may interfere with the functions of the on board computer systems and may require the computer to undergo a relearning process, once the negative battery cable is reconnected.

Engine Assembly

REMOVAL & INSTALLATION

NOTE: The engine and transaxle are removed as an assembly.

1. Properly relieve the fuel system pressure on EFI equipped vehicle.
2. Disconnect the battery cables. Remove the battery and battery tray.
3. Mark the hinge location and remove the hood.
4. Drain the radiator coolant, engine oil, transaxle fluid and, if equipped, the power steering fluid into suitable containers.
5. Properly discharge the air conditioning system, if equipped.
6. If carburetor equipped, remove the air cleaner assembly. On EFI engine, disconnect the vane airflow meter connector. Remove the vane airflow meter and hose.
7. Remove the radiator and cooling fan as an assembly.
8. Disconnect the accelerator cable from the mounting bracket and throttle lever.
9. Disconnect the speedometer cable from the transaxle.
10. Disconnect the fuel hoses. Plug or cover the hose openings to prevent dirt from entering and to avoid fuel leakage.
11. Disconnect the heater hoses and the brake booster vacuum hose.
12. On carbureted engine, disconnect the carburetor-to-chassis hoses and the wide open throttle vacuum switch connector. On EFI engine, disconnect the vacuum hose at the throttle body.
13. If equipped with automatic transaxle, disconnect the transaxle vacuum hose.
14. Tag and disconnect the carbon canister hoses and the engine harness connectors and grounds. Disconnect the distributor wiring at the coil.
15. Disconnect the power steering lines, if equipped. Disconnect the air conditioning lines and the air conditioning electrical connector, if equipped.
16. On automatic transaxles, remove the nut that connects the shift lever to

the manual shaft assembly. Remove the shift cable from the transaxle. On manual transaxles, disconnect the clutch control cable from the transaxle.
17. Raise and safely support the vehicle. Remove the front wheel and tire assemblies.
18. Remove the stabilizer mounting nuts and brackets.
19. Remove the lower arm clamp bolts and nuts. Pull the lower arms downward, separating the lower arms from the knuckles.
20. Remove the halfshafts and install differential plugs T87C-7025-C or equivalent, between the differential side gears.
21. If equipped with a manual transaxle, disconnect the shift control rod and stabilizer bar from the transaxle.
22. Remove the catalytic converter inlet pipe.
23. Support the engine using 3-bar engine support D88L-6000-A or equivalent.
24. Remove the crossmember attaching bolts.
25. Remove the front and rear engine mount-to-crossmember attaching nuts and remove the crossmember.
26. Lower the vehicle.
27. Remove the attaching bolt, nut and washer from the side mount. Remove the side mount-to-engine attaching nuts.
28. Position a suitable jack or hoist and attach it to the engine. Carefully remove the engine and transaxle as an assembly through the bottom of the vehicle.
29. Remove the gusset plates, starter motor and flywheel cover.
30. If equipped with automatic transaxle, remove the torque converter bolts.
31. Remove the engine-to-transaxle bolts and separate the transaxle from the engine.

To install:
32. Mount the transaxle to the engine. Install the engine-to-transaxle bolts and tighten to 41-59 ft. lbs. (55-80 Nm).
33. If equipped with automatic transaxle, install the torque converter bolts and tighten to 26-36 ft. lbs. (34-49 Nm).
34. Install the flywheel cover and tighten the bolts to 61-87 inch lbs. (7-10 Nm). Install the starter motor.
35. Install the gusset plates and tighten the bolts to 27-38 ft. lbs. (37-52 Nm).
36. Raise the engine and transaxle assembly into position in the engine compartment.
37. Install the side mount and tighten the side mount-to-engine attaching nuts to 29-40 ft. lbs. (39-54 Nm). Install the attaching bolt, washer and

nut to the side mount and tighten to 29-40 ft. lbs. (39-54 Nm).
38. Raise and safely support the vehicle.
39. Position the crossmember and install the attaching bolts. Tighten the bolts to 47-66 ft. lbs. (64-89 Nm).
40. Install the front and rear engine mount-to-crossmember attaching nuts. Tighten the front nuts to 32-38 ft. lbs. (43-52 Nm) and the rear nut to 21-34 ft. lbs. (28-46 Nm).
41. Install the catalytic converter inlet pipe and tighten the nuts to 23-34 ft. lbs. (31-46 Nm).
42. If equipped with manual transaxle, install the shift control rod and stabilizer bar.
43. Remove the differential plugs and install the halfshafts. Install the lower arm ball joint to the knuckle and tighten the clamp nut and bolt to 32-40 ft. lbs. (43-54 Nm).
44. Install the stabilizer bracket and mounting nuts. Tighten the mounting nuts to 40-50 ft. lbs. (54-68 Nm).
45. Install the front wheel and tire assemblies and lower the vehicle.
46. If equipped with manual transaxle, connect the clutch cable. If equipped with automatic transaxle, install the shift lever on the manual shaft assembly and tighten the nut to 34-57 ft. lbs. (44-64 Nm). Attach the shift cable to the transaxle.
47. Connect the distributor wiring to the coil and connect all engine harness connectors and grounds.
48. If equipped with air conditioning, connect the lines and the electrical connector.
49. If equipped with automatic transaxle, connect the transaxle vacuum hose.
50. On EFI engine, connect the vacuum hose at the throttle body.
51. Connect the brake booster vacuum hose, the carbon canister hoses, the heater hoses and the fuel lines.
52. On carbureted engine, connect the carburetor-to-chassis hoses and the wide open throttle vacuum switch connector.
53. Connect the speedometer cable and connect the accelerator cable to the throttle lever and mounting bracket.
54. If equipped, connect the power steering lines.
55. Install the radiator and cooling fan.
56. On EFI engine, install the vane airflow meter and hose. Connect the vane airflow meter connector.
57. On carburetor equipped engine, install the air cleaner assembly.
58. Install the hood, aligning the marks that were made during the removal procedure.
59. Install the battery carrier and the battery. Connect the battery cables.

60. Add the proper types and quantities of engine oil, transaxle fluid and coolant.

61. If equipped, add power steering fluid to the reservoir.

62. If equipped, charge the air conditioning system.

63. Start the engine. Check for leaks and proper fluid levels. Road test.

Engine Mounts

REMOVAL & INSTALLATION

Front Mount

1. Disconnect the negative battery cable. Remove the front mount through bolt attaching nut.

2. Properly support the engine.

3. Raise and support the vehicle safely.

4. Remove the front mount to cross-member attaching nuts.

5. Raise the vehicle, as required, to gain sufficient clearance to remove the front mount. Remove the front mount from the crossmember. Note and record the position of the mount to ensure proper installation.

To install:

6. Install the engine mount onto the crossmember in the original installation position.

7. Secure the mount to the crossmember with the attaching nuts. Torque the attaching nuts to 32–38 ft. lbs. (43–52 Nm).

8. Lower the vehicle.

9. Move the engine as necessary until the holes in the mount align with the holes in the engine bracket. Install the through bolt and attaching nut. Torque the nut to 29–40 ft. lbs. (39–54 Nm).

10. Remove the engine support.

Rear Mount

1. Disconnect the negative battery cable. Raise the vehicle and support safely.

2. Properly support the engine.

3. Remove the mount-to-crossmember attaching nut.

4. Remove the mount-to-engine attaching bolts.

5. If necessary, raise the engine to gain access to the rear mount. Remove the mount from the crossmember.

To install:

6. Position the mount onto the rear engine bracket.

7. Install the mount to engine bracket bolts. Torque the bolts to 29–40 ft. lbs. (39–54 Nm).

8. Lower the engine and mount onto the crossmember.

9. Install the attaching nut and torque to 21–34 ft. lbs. (28–46 Nm).

10. Remove the engine support.

Side Mount

1. Disconnect the negative battery cable. Properly support the engine.

2. Remove the through bolt, nut and washer.

3. Remove the bracket-to-engine attaching nuts.

4. Remove the side mount and bracket as an assembly.

To install:

5. Position the engine mount and bracket onto the engine.

6. Install the engine-to-bracket attaching nuts. Torque the nuts to 29–40 ft. lbs. (39–54 Nm).

7. Position the washer against the mount. Install the through bolt and nut. Torque the nut and bolt to 29–40 ft. lbs. (39–54 Nm).

8. Remove the engine support.

Cylinder Head

REMOVAL & INSTALLATION

1. Disconnect the negative battery cable. Drain the cooling system.

2. Position the engine at TDC on the compression stroke.

3. Remove the valve cover. Remove the timing belt cover and timing belt.

4. Remove the exhaust manifold. Remove the intake manifold.

5. Remove the spark plug wires and spark plugs. Remove the distributor.

6. Remove the front and rear engine lift hangers. Remove the engine ground wire.

7. Remove the wiring harness connector. Remove the upper radiator hose. Remove the bypass hose and bracket.

8. Remove the cylinder head retaining bolts. Remove the cylinder head from the engine. Discard the gasket.

9. Clean all mating surfaces of dirt and old gasket material.

To install:

10. Position the cylinder head gasket on the engine block. Install the cylinder head and tighten the bolts, in sequence, in 2 equal steps. The final torque should be 56–60 ft. lbs. (75–81 Nm).

11. Connect the radiator hose, bypass hose, wiring harness connectors,

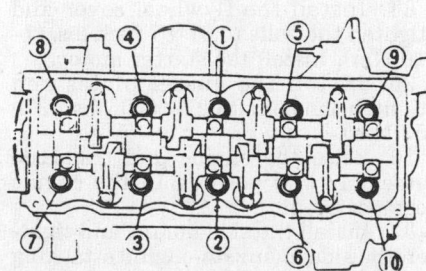

Cylinder head bolt torque sequence

engine lift hangers and engine ground wire.

12. Install the distributor, spark plugs and wires.

13. Install the intake and exhaust manifolds. Install the timing belt and cover.

14. Install the valve cover.

15. Fill the cooling system.

16. Connect the negative battery cable. Start the engine and check for leaks.

Valve Lifters

REMOVAL & INSTALLATION

All EFI engines are equipped with hydraulic lash adjusters that automatically maintain valve lash.

1. Disconnect the negative battery cable.

2. Remove the valve cover and the rocker arm shaft assemblies.

3. Remove the hydraulic lash adjuster from the rocker arm.

To install:

4. Pour engine oil into the oil reservoir in the rocker arm. Apply engine oil to the new hydraulic lash adjuster.

5. Install the hydraulic lash adjuster into the rocker arm.

NOTE: Be careful not to damage the O-ring when installing the hydraulic lash adjuster.

6. Install the rocker arm shaft assemblies and install the valve cover.

7. Connect the negative battery cable.

Valve Lash

ADJUSTMENT

Carbureted Engine

1. Start the engine and allow to reach normal operating temperature. Stop the engine.

2. Remove the valve cover.

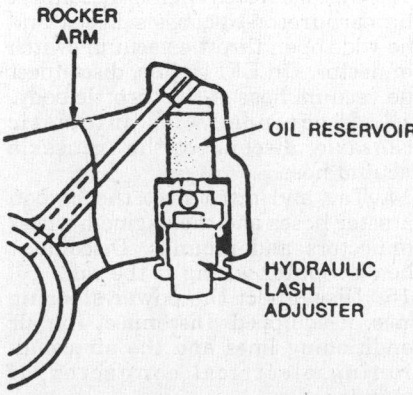

Rocker arm and hydraulic lash adjuster assembly—EFI engine

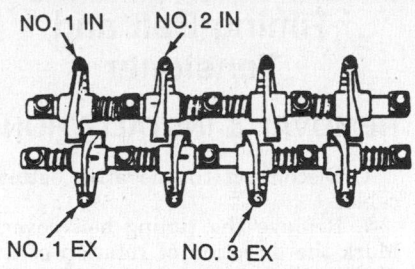

NO. 1 IN NO. 2 IN

NO. 1 EX NO. 3 EX

Intake and exhaust valve arrangement

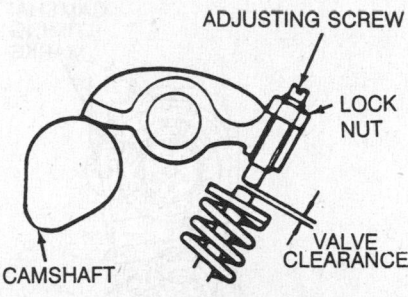

ADJUSTING SCREW

LOCK NUT

CAMSHAFT

VALVE CLEARANCE

Valve lash adjustment

3. Turn the crankshaft by hand, in the direction of normal rotation, until the piston of the No. 1 cylinder reaches TDC on the compression stroke.

4. Adjust the No. 1 and No. 2 intake valves and the No. 1 and No. 3 exhaust valves. The valve lash should be 0.012 in. (0.30mm).

5. Rotate the crankshaft 360 degrees so that the No. 4 piston is at TDC of the compression stroke. Adjust the remaining valves.

6. Install the valve cover using a new gasket. Tighten the valve cover attaching bolts to 44–79 inch lbs. (5–9 Nm).

EFI Engine

Inspect hydraulic lash adjuster operation by pushing down each rocker arm by hand. If a rocker arm moves down, replace the hydraulic lash adjuster.

Rocker Arms/Shafts

REMOVAL & INSTALLATION

1. Disconnect the negative battery cable.

2. On carbureted engine, remove the air cleaner assembly. If equipped with fuel injection, remove the air hose and the resonance chamber.

3. Disconnect the accelerator cable from the throttle lever and routing bracket. Remove the PCV valve.

4. Remove the spark plug wires from the routing clips. Remove the upper timing belt cover.

5. Remove the valve cover retaining bolts. Remove the valve cover. Discard the gasket.

6. Remove the rocker arm shaft retaining bolts. Remove the rocker arms/shafts assemblies from the engine.

7. Installation is the reverse of the removal procedure. Torque the rocker arm bolts in sequence to 16–21 ft. lbs. (22–28 Nm). Check the valve lash adjustment.

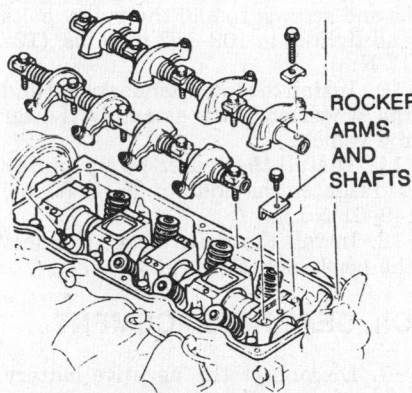

ROCKER ARMS AND SHAFTS

Rocker arm/shaft assembly removal and installation

Intake Manifold

REMOVAL & INSTALLATION

1. Disconnect the negative battery cable. Drain the cooling system.

2. Remove the air cleaner assembly, if equipped with a carburetor. If equipped with fuel injection, remove the intake manifold bracket.

3. Disconnect the accelerator cable.

4. Identify, tag and disconnect the necessary vacuum hoses and electrical connectors.

5. Support the intake manifold by hand and remove the retaining bolts. Remove the intake manifold from the cylinder head.

6. Remove the old gasket material and thoroughly clean the intake manifold and cylinder head surfaces.

To install:

7. Apply a new gasket to the cylinder head surface and hold in place.

8. Position the intake manifold onto the new gasket and install the retaining bolts. Torque the retaining bolts to 14–20 ft. lbs. (19–26 Nm) in a criss-cross pattern, from the inside out.

9. Connect the vacuum hoses and electrical wiring to their respective connections. Install the accelerator cable.

10. Install the air cleaner assembly on carbureted engine. Install the intake manifold bracket on EFI engine and tighten to 22–34 ft. lbs. (31–46 Nm).

11. Refill the cooling system to the proper level. Connect the negative battery cable.

Exhaust Manifold

REMOVAL & INSTALLATION

1. Disconnect the negative battery cable.

2. Raise and safely support the vehicle.

3. Disconnect the catalytic converter inlet pipe from the exhaust manifold by removing the 3 attaching nuts.

4. If equipped, remove the pulse air tube to catalytic converter inlet pipe attaching nuts.

5. Unbolt the catalytic converter support bracket.

6. Lower the vehicle.

7. Remove the air cleaner assembly on carbureted engine. On fuel injected vehicle, remove the throttle body-to-air cleaner hose.

8. Remove the exhaust manifold heat shroud.

9. Separate the oxygen sensor wiring connector from the routing bracket and disconnect the electrical connector.

10. If equipped, unbolt the pulse air routing bracket clamp. Remove the pulse air tube and gaskets. Discard the gaskets.

11. Support the exhaust manifold by hand and remove the attaching nuts and bolts. Separate the exhaust manifold from the cylinder head and inlet pipe. Remove the inlet pipe and exhaust manifold gaskets and discard.

12. If necessary, remove the oxygen sensor. Inspect the sensor gasket for damage and replace if necessary.

To install:

13. Remove all existing gasket material from the exhaust manifold, cylinder head inlet pipe and, if equipped, the pulse air tube flange surfaces. Clean all threaded surfaces.

14. If removed, position the gasket onto the oxygen sensor and install into the exhaust manifold.

15. Apply a new gasket onto the cylinder head studs and position the exhaust manifold onto the gasket. Install the attaching nuts and torque to 12–17 ft. lbs. (16–23 Nm).

16. Install the heat shroud.

17. If equipped, install the pulse air tube and mounting bracket clamp. On fuel injected vehicle, install the air hose. On carbureted engine, install the air cleaner assembly.

18. Connect the oxygen sensor electrical connector and secure the connector in the routing bracket.

19. Raise the vehicle and support it safely.

20. Position a new gasket over the exhaust manifold studs and, if equipped, 2 new gaskets onto the pulse air tube studs.

21. Raise the catalytic converter inlet

pipe into position on the exhaust manifold and pulse air tube studs and support by hand. Install the attaching nuts and torque to 23–34 ft. lbs. (31–46 Nm).

22. Install the catalytic converter inlet pipe support bracket.

23. Lower the vehicle and connect the negative battery cable.

24. Start the engine and inspect for exhaust gas leaks.

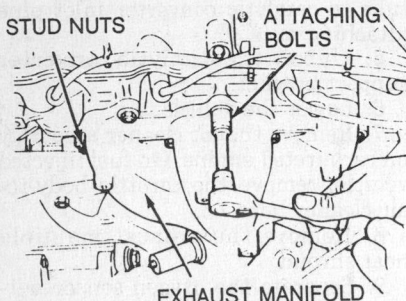

STUD NUTS   ATTACHING BOLTS

EXHAUST MANIFOLD

Exhaust manifold mounting bolt and nut locations

Timing Belt Front Cover

REMOVAL & INSTALLATION

1. Disconnect the negative battery cable. Remove the drive belts.

2. Remove the 3 water pump pulley attaching bolts and remove the water pump pulley.

3. Raise and safely support the vehicle.

4. Remove the right front wheel and tire assembly and the right inner fender panel.

5. Remove the 4 attaching bolts and the screws from the crankshaft pulley. Remove the spacer and outer pulley, if equipped. Remove the inner spacer, inner pulley and the baffle or guide plates, as required.

6. Remove the attaching bolts and the upper and lower covers.

To install:

7. Install the upper and lower cov-

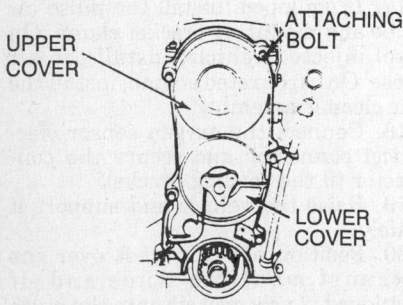

UPPER COVER   ATTACHING BOLT

LOWER COVER

Upper and lower timing belt covers with attaching bolts

ers. Install the attaching bolts and tighten to 69–95 inch lbs. (8–11 Nm).

8. Install the crankshaft pulley baffle with the curved lip facing outward or install the large guide plate and then the small guide plate, as required.

9. Install the inner pulley with the deep recess facing outward. Install the spacer and then the outer pulley, spacer and screws. Install the pulley bolts and tighten to 109–152 inch lbs. (12–17 Nm).

10. Install the inner fender panel and the wheel and tire assembly. Lower the vehicle.

11. Install the water pump pulley and tighten the bolts to 36–45 ft. lbs. (49–61 Nm).

12. Install the drive belts. Connect the negative battery cable.

OIL SEAL REPLACEMENT

1. Disconnect the negative battery cable.

2. Remove the timing belt covers, timing belt and crankshaft sprocket.

3. Use a suitable tool to pry the crankshaft seal from the oil pump housing.

To install:

4. Lubricate the lip of the new seal with engine oil.

5. Use a suitable tool to install the seal into the oil pump housing.

6. Install the crankshaft sprocket, timing belt and timing belt covers.

7. Connect the negative battery cable.

Timing Belt and Tensioner

REMOVAL & INSTALLATION

1. Disconnect the negative battery cable.

2. Remove the timing belt covers. Mark the direction of rotation of the timing belt.

3. Remove the timing belt tensioner

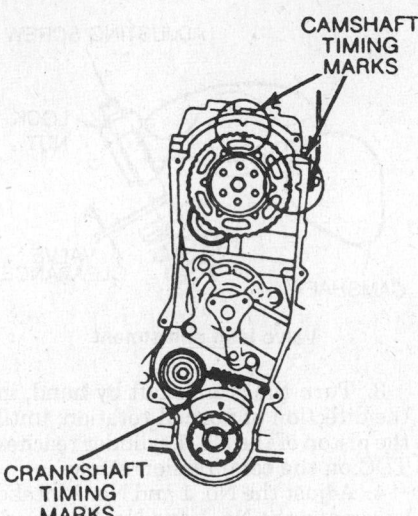

CAMSHAFT TIMING MARKS

CRANKSHAFT TIMING MARKS

Camshaft and crankshaft timing mark alignment

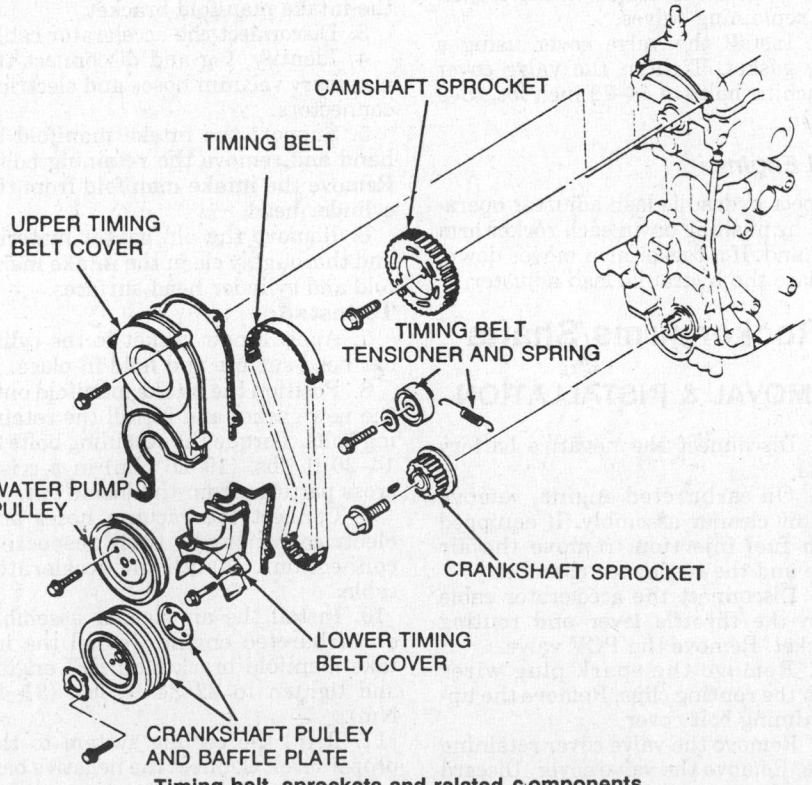

CAMSHAFT SPROCKET

TIMING BELT

UPPER TIMING BELT COVER

TIMING BELT TENSIONER AND SPRING

WATER PUMP PULLEY

LOWER TIMING BELT COVER

CRANKSHAFT SPROCKET

CRANKSHAFT PULLEY AND BAFFLE PLATE

Timing belt, sprockets and related components

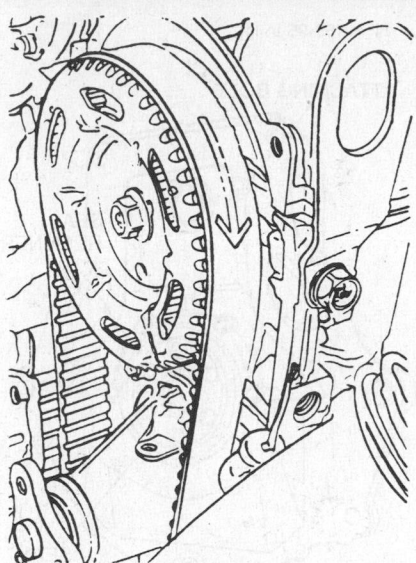

Direction of timing belt rotation

spring and retaining bolt. Remove the timing belt.

To install:

4. Align the camshaft and crankshaft timing marks with the marks located on the cylinder head and oil pump housing.

5. If reusing the original timing belt, install the timing belt with the mark made for the direction of rotation.

6. Install the timing belt tensioner spring and cover on the pulley. Position the tensioner and spring assembly on the engine and install the attaching bolt. Do not tighten the bolt at this time.

7. Reconnect the free end of the spring to the spring anchor. Torque the tensioner bolt to 14–19 ft. lbs. (19–26 Nm).

8. Install the timing belt covers and connect the negative battery cable.

Timing Sprockets

REMOVAL & INSTALLATION

Camshaft Sprocket

1. Disconnect the negative battery cable.

2. Remove the timing belt and timing belt tensioner.

3. With the appropriate size open end wrench or medium prybar, hold the camshaft stationary and remove the camshaft sprocket retaining bolt.

4. Pull the camshaft sprocket with dowel pin from the camshaft. Take care not to lose the dowel pin.

To install:

5. Install the camshaft sprocket, dowel pin and retaining bolt.

6. Hold the camshaft stationary and

torque the retaining bolt to 36–45 ft. lbs. (49–61 Nm).

7. Install the timing belt and tensioner.

8. Connect the negative battery cable.

Crankshaft Sprocket

1. Disconnect the negative battery cable.

2. Remove the timing belt and timing belt tensioner.

3. If equipped with manual transaxle, place the shift lever in **4th** gear and apply the parking brake. On automatic transaxle vehicle, use a suitable flywheel holding tool.

4. Remove the crankshaft sprocket retaining bolt.

5. Pull the crankshaft sprocket and key from the crankshaft. Make certain not to lose the key when removing the crankshaft sprocket. Replace the key if worn or damaged.

To install:

6. Position the crankshaft sprocket onto the crankshaft and align the keyways. Install the key.

7. Coat the threads of the retaining bolt with a suitable non-hardening compound. Install the retaining bolt and torque to 80–94 ft. lbs. (108–128 Nm).

8. Remove the flywheel holding tool, if necessary.

9. Install the timing belt.

10. Connect the negative battery cable.

Camshaft

REMOVAL & INSTALLATION

1. Disconnect the negative battery cable. Drain the cooling system.

2. Remove the cylinder head from the engine.

3. Position the cylinder head in a

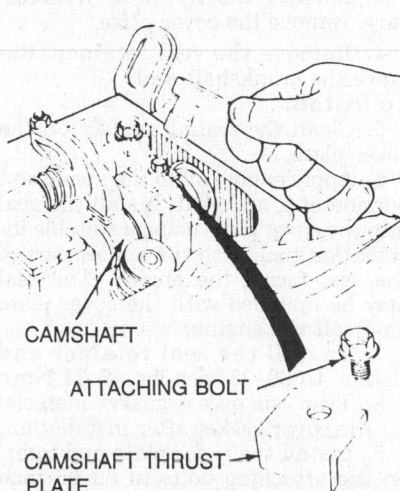

Camshaft and camshaft thrust plate

suitable holding fixture. Remove the camshaft sprocket. Remove the rocker arm/shaft assemblies.

4. Remove the camshaft thrust plate and the camshaft from the cylinder head.

5. Remove the camshaft seal.

To install:

6. Install the camshaft and the camshaft thrust plate.

7. Lubricate the lip of the new camshaft seal with engine oil and install, using a suitable installation tool.

8. Install the rocker arm/shaft assemblies and the camshaft sprocket.

9. Install the cylinder head. Check the valve lash.

10. Fill the cooling system and connect the negative battery cable.

Piston and Connecting Rod

POSITIONING

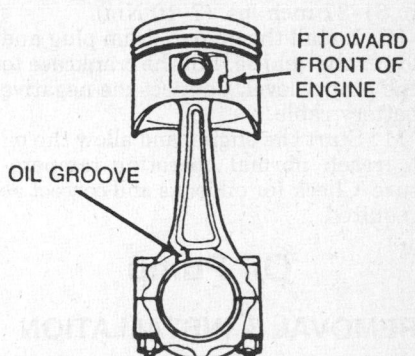

Piston and connecting rod positioning

ENGINE LUBRICATION

Oil Pan

REMOVAL & INSTALLATION

1. Disconnect the negative battery cable. Raise and support the vehicle safely. Drain the engine oil.

2. Remove the flywheel dust cover retaining bolts and remove the cover.

NOTE: Depending on the position of the crankshaft, the oil pan may encounter interference during removal from the crankshaft counterweights or connecting rods. If necessary, rotate the crankshaft retaining bolt until the oil pan can be removed without crankshaft interference.

3. Support the oil pan and remove the oil pan to cylinder block bolts, nuts and stiffeners. Lower the oil pan. Discard the oil pan gasket.

4. As required, remove the baffle plate from the oil pan on vehicles equipped with fuel injection.

To install:

5. Clean the oil pan and cylinder block sealing surfaces to remove all traces of existing gasket material. From beneath the engine, apply a suitable oil resistant sealant to the joint line formed at the cylinder block and front and rear engine covers.

6. On EFI engine, install the baffle plate.

7. Apply the new rubber gasket to the oil pan.

8. Raise the oil pan and gasket against the cylinder block. Install the stiffeners, bolts and nuts. Torque the oil pan bolts in an alternate pattern to 69–78 inch lbs. (8–9 Nm).

9. Install the flywheel dust cover and attaching bolts. Torque the bolts to 61–87 inch lbs. (7–10 Nm).

10. Install the oil pan drain plug and lower the vehicle. Fill the crankcase to the proper level. Connect the negative battery cable.

11. Start the engine and allow the oil to reach normal operating temperature. Check for oil leaks and correct as required.

Oil Pump

REMOVAL & INSTALLATION

1. Disconnect the negative battery cable. Raise and support the vehicle safely. Remove the crankshaft sprocket.

2. Drain the engine oil. Remove the oil pan.

3. Remove the oil pump assembly retaining bolts. Remove the oil pump assembly and gasket from the engine. Discard the gasket.

4. Remove the pickup tube and screen.

5. Remove the screws from the oil pump cover. Remove the cover. Remove the oil pump gears.

6. Remove the front seal from the pump assembly. Remove the cotter pin, spring and relief valve from the oil pump body.

To install:

7. Clean the oil pump housing and components with a suitable solvent and allow to dry.

8. Lubricate the oil pump relief valve and install into the bore. Install the spring, retainer and cotter pin.

9. Lubricate the lip of the new crankshaft seal and install the seal into the pump, using a suitable installation tool.

10. Lubricate and install the gears in the pump body and install the pump body cover. Coat the screws with a suitable locking compound and tighten.

11. Clean the cylinder block contact surface to remove the old gasket material and sealant. Thoroughly coat both sides of the new oil pump gasket with a suitable sealant compound. Apply the gasket to the oil pump and remove any excess sealant.

NOTE: Do not allow the sealant compound to enter the oil pump discharge opening once the gasket is in place. This opening must be free and clear before the oil pump is installed onto the cylinder block.

12. Position the oil pump against the cylinder block surface and install the retaining bolts. Torque the bolts to 14–19 ft. lbs. (19–25 Nm).

13. Install a new gasket onto the oil pump inlet and bolt the pickup tube to the oil pump. Torque the bolts to 69–95 inch lbs. (8–11 Nm).

14. Install the oil pan and the crankshaft sprocket.

15. Lower the vehicle. Fill the crankcase to the proper level with engine oil. Connect the negative battery cable.

16. Start the engine and allow the oil to reach normal operating temperature. Check for leaks and correct as required.

Rear Main Bearing Oil Seal

REMOVAL & INSTALLATION

1. Disconnect the negative battery cable.

2. Remove the transaxle from the vehicle.

3. Remove the flywheel. If necessary, remove the cover plate.

4. Remove the seal retainer. Remove the crankshaft seal.

To install:

5. Clean the sealing surface on the cover plate.

6. Apply engine oil to the inside and outside of a new seal. Install the seal into the cover plate using a suitable installation tool, with the hollow part of the seal facing the engine. The seal may be installed with the cover plate on or off the engine.

7. Install the seal retainer and tighten to 69–95 inch lbs. (8–11 Nm).

8. Trim the excess gasket material off the cover gasket after installation.

9. Install the cover plate and tighten the attaching bolts to 69–95 inch lbs. (8–11 Nm).

10. Install the flywheel and tighten

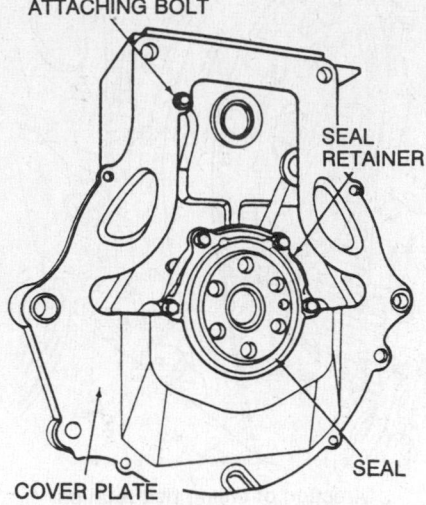

N·m (69–95 in-lb).

ATTACHING BOLT

SEAL RETAINER

SEAL

COVER PLATE

Rear crankshaft oil seal

the bolts to 71–76 ft. lbs. (96–103 Nm).

11. Install the transaxle. Connect the negative battery cable.

ENGINE COOLING

Radiator

REMOVAL & INSTALLATION

1. Disconnect the negative battery cable.

2. Disconnect the coolant recovery hose from the filler neck.

3. Loosen the retaining clamp and disconnect the upper radiator hose from the radiator.

4. Disconnect the cooling fan wiring harness connector. Disengage the wiring harness from the routing clamps on the cooling fan shroud.

5. Remove the radiator cap from the filler neck. Raise the vehicle and support safely.

6. Position a fluid catch pan under the radiator. Open the drain valve and drain the cooling system.

7. Disconnect the radiator temperature switch wires, if necessary.

8. Loosen the retaining clamp and disconnect the lower radiator hose.

9. Lower the vehicle.

10. Support the radiator by hand and remove the 4 bolts attaching the radiator support brackets to the vehicle body. Raise the radiator/cooling fan/shroud assembly from the crossmem-

ber mounting insulator supports and remove from vehicle. Disconnect the cooling fan and shroud from the radiator as required.

To install:

11. Lower the radiator into the normal operating position making certain the mounting insulators engage with their supports. Attach the radiator to the support brackets with the 4 bolts.

12. Connect the coolant recovery hose, upper radiator hose and the cooling fan wiring. Raise the vehicle and connect the lower radiator hose and temperature switch wires, if necessary.

13. Close the drain valve and lower the vehicle. Connect the negative battery cable. Fill the cooling system to the proper level.

14. Start the engine and allow to reach normal operating temperature. Inspect for coolant leaks and correct as required.

Heater Core

REMOVAL & INSTALLATION

1. Disconnect the negative battery cable.

2. Remove the instrument panel as follows:

   a. Remove the steering wheel, steering column covers and the combination switch.

   b. Remove the screws securing the instrument cluster bezel and move the bezel rearward.

   c. Disconnect the electrical connectors from the switches on the instrument cluster bezel and remove the bezel.

   d. Remove the left and right heater ducts.

   e. Disconnect the speedometer cable at the transaxle.

   f. Remove the 4 screws securing the instrument cluster and move the cluster toward the rear of the vehicle.

   g. Disconnect the instrument cluster electrical connectors and the speedometer cable from the instrument cluster. Remove the instrument cluster.

   h. On 1988–89 vehicles, remove the spacer brace bolts under the steering column and remove the spacer brace.

   i. On 1990–92 vehicles, remove the 4 shield nuts and the shield and remove the 2 shield bracket bolts and the shield bracket.

   j. Remove the screws securing the glove box hinges to the glove box and remove the glove box.

   k. Open the fuse panel cover, remove the fuse panel attaching

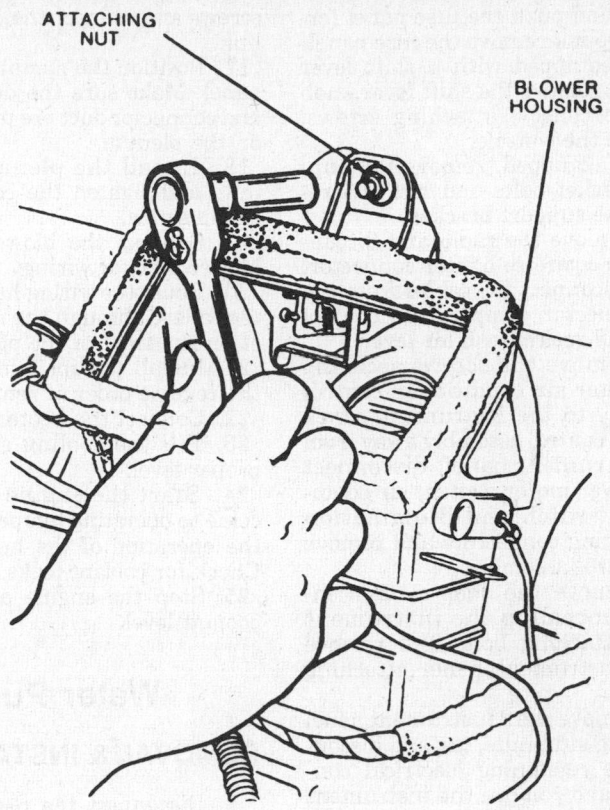

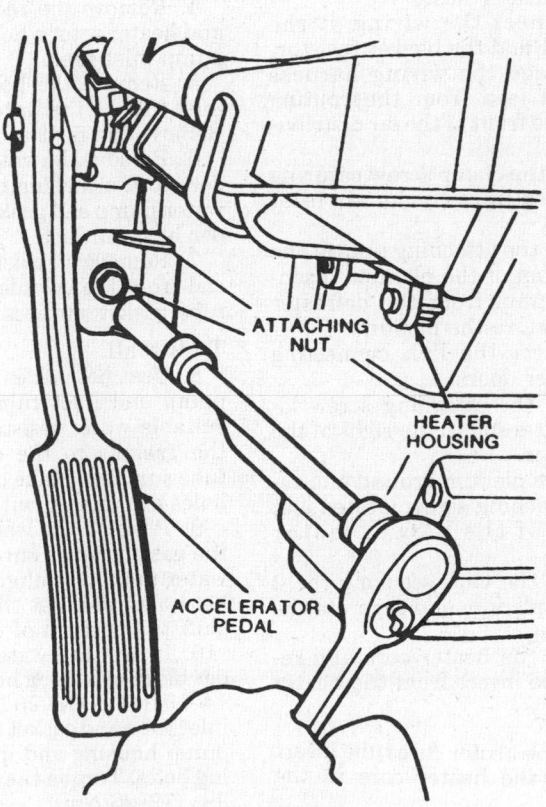

Plenum attaching nut locations

screws and push the fuse panel forward. Do not remove the fuse panel.

l. If equipped with a shift lever console, remove the shift lever knob and the console attaching screws. Remove the console.

m. If equipped, remove the support bracket bolts and nut and remove the support bracket.

n. Remove the radio and disconnect the cigarette lighter connector.

o. Disconnect the cables from the mode selector, temperature control lever and recirc/fresh air lever.

p. Remove the screws securing the heater/air conditioner control assembly to the instrument panel. Pull the control assembly away from the instrument panel, disconnect the blower motor switch, air conditioning switch and illumination light wiring connectors and remove the control assembly.

q. Remove the snap-in trim inserts concealing the instrument panel attaching bolts and remove the 7 instrument panel attaching bolts.

r. Remove the 2 instrument panel attaching stud nuts, tag and disconnect the remaining electrical connectors and remove the instrument panel.

3. Drain the cooling system.
4. In the engine compartment, disconnect the heater hoses.
5. Disconnect the wiring at the blower motor and the blower resistor.
6. Disengage the wiring harness and antenna lead from the routing bracket on the front of the air distribution housing.
7. Loosen the clamp screw securing the connector duct to the air inlet housing.
8. Remove the attaching nuts at the top and bottom of the plenum, disengage the plenum from the defroster ducts and remove the plenum.
9. Disconnect the link connecting the 2 defroster doors.
10. Remove the attaching screw located just above and to the right of the blower resistor.
11. Turn the plenum around and remove the attaching screw located just to the left of the blower motor opening.
12. Remove the clips securing the 2 halves of the plenum and separate the plenum halves.
13. Remove the heater core and remove the tube insert from the heater core.

To install:
14. Install the heater core tube insert and position the heater core in the plenum.
15. Install the remaining plenum half and the plenum retaining clips.
16. Install the plenum attaching

screws and connect the defroster door link.

17. Position the plenum on the dash panel. Make sure the defroster ducts and connector duct are properly seated on the plenum.
18. Install the plenum attaching nuts and tighten the connector duct clamp screw.
19. Connect the blower motor and blower resistor wiring.
20. Route the wiring harness and antenna lead through the routing bracket on the front of the plenum.
21. Install the instrument panel in the reverse order of removal.
22. Connect the heater hoses.
23. Fill the cooling system to the proper level.
24. Start the engine and allow to come to operating temperature. Check the operation of the heating system. Check for coolant leaks.
25. Stop the engine and check the coolant level.

Water Pump

REMOVAL & INSTALLATION

1. Disconnect the negative battery cable.
2. Remove the timing belt.
3. Drain the cooling system.
4. Remove the radiator lower hose and heater return hose from the water pump inlet tube.
5. Remove the bolts attaching the inlet tube to the water pump housing. Remove the inlet tube and gasket.
6. Remove the water pump-to-cylinder block attaching bolts. Remove the water pump and gasket from the cylinder block surface.
7. Remove all existing gasket material from the cylinder block and inlet tube gasket surfaces.

To install:
8. Coat both sides of the new water pump and inlet tube gaskets with a suitable water resistant sealer. Apply the gaskets to the engine and inlet tube surfaces. Make certain the gasket holes are aligned with the bolt holes.
9. Position the water pump against the gasket. Make sure the holes in the water pump are aligned with the gasket holes and that the pump does not shift the position of the gasket.
10. Install the water pump-to-cylinder block attaching bolts and torque to 14–19 ft. lbs. (19–26 Nm). Position the inlet tube and gasket against the water pump housing and install the attaching bolts. Torque the bolts to 14–19 ft. lbs. (19–26 Nm).
11. Connect the inlet tube hoses and install the timing belt.
12. Fill the cooling system to the

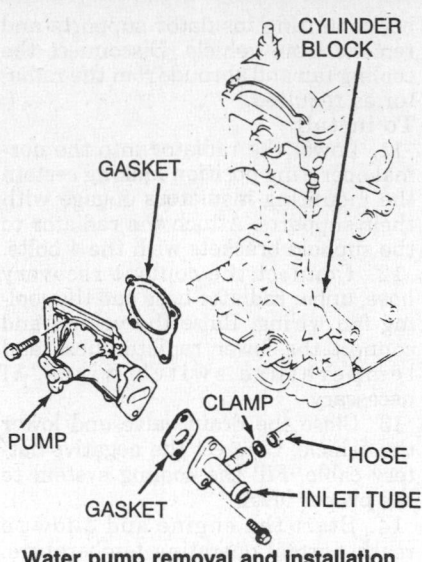

Water pump removal and installation

proper level. Connect the negative battery cable.

13. Start the engine and allow to reach normal operating temperature. Check for coolant leaks.

Thermostat

REMOVAL & INSTALLATION

1. Disconnect the negative battery cable.
2. Disconnect the cooling fan temperature switch wire.
3. Remove the radiator cap and drain the cooling system to a level below the radiator upper hose. Disconnect the radiator upper hose from the thermostat housing.
4. Remove the thermostat housing-to-cylinder head attaching bolts. Remove the thermostat housing and housing gasket. Withdraw the thermostat from the cylinder head.
5. Remove all gasket material from the thermostat housing and cylinder block surfaces.

To install:
6. Install the thermostat in the cylinder head, with the valve end first and the sub valve at the top.
7. Coat a new gasket with a suitable water resistant sealer. Apply the gasket to the cylinder block surface making sure the gasket and cylinder block holes are evenly aligned.
8. Position the thermostat housing onto the cylinder head making sure the bolt holes are aligned and the gasket does not shift. Install the housing attaching bolts but do not tighten at this time. Before tightening the bolts, ensure that the thermostat flange is properly seated against the recess of the housing. Torque the bolts to 14–19 ft. lbs. (19–26 Nm).

9. Connect the radiator upper hose to the thermostat housing. Fill the cooling system to the proper level and install the radiator cap. Connect the cooling fan temperature switch wire and the negative battery cable.

10. Start the engine and allow to reach normal operating temperature. Inspect for leaks.

Cooling System Bleeding

When the entire cooling system is drained, the following procedure should be used to ensure a complete fill.

1. Install the block drain plug, if removed and close the draincock. With the engine off, add anti-freeze to the radiator to a level of 50 percent of the total cooling system capacity. Then add water until it reaches the radiator filler neck seat.

2. Install the radiator cap to the first notch to keep spillage to a minimum.

3. Start the engine and let it idle until the upper radiator hose is warm. This indicates that the thermostat is open and coolant is flowing through the entire system.

4. Carefully remove the radiator cap and top off the radiator with water. Install the cap on the radiator securely.

5. Fill the coolant recovery reservoir to the FULL COLD mark with anti-freeze, then add water to the FULL HOT mark. This will ensure that a proper mixture is in the coolant recovery bottle.

6. Check for leaks at the draincock and block plug.

ENGINE ELECTRICAL

NOTE: Disconnecting the negative battery cable on some vehicles may interfere with the functions of the on board computer systems and may require the computer to undergo a relearning process, once the negative battery cable is reconnected.

Distributor

REMOVAL

1. Disconnect the negative battery cable.

2. Disconnect the coil wire from the distributor.

3. Remove the distributor cap attaching screws, pull off the distributor cap and swing it aside.

NOTE: If replacing the distributor cap, mark the distributor cap towers with the cylinder numbers before removing the spark plug wires, to aid reinstallation.

4. If equipped with a vacuum advance unit, tag the location and disconnect the vacuum hoses.

5. On 1988–89 vehicles, disconnect the white altitude connector from the distributor wiring harness. Remove the coil positive terminal nut and disconnect the distributor harness connector and suppression capacitor wire. Pull the distributor connector off the coil ground terminal tab. Open the harness routing clip and free the distributor primary circuit wires.

6. On 1990–92 vehicles, disconnect the distributor electrical connector.

7. Scribe a timing reference mark across the distributor mounting flange and cylinder head surface to ensure that the distributor will be installed without altering the timing.

8. Remove the base flange mounting bolts and remove the distributor assembly from its mounting bore.

9. Remove the flange base O-ring and inspect for damage. Replace the O-ring as required. Coat the O-ring with clean engine oil and install into the flange base.

INSTALLATION

Timing Not Disturbed

1. Insert the distributor assembly into the cylinder head mounting bore. Rotate the distributor until the offset drive tang aligns and engages with the camshaft slot.

2. After the distributor is engaged with the camshaft, align the timing reference marks scribed across the flange base and cylinder head. When the timing marks are aligned, install and tighten the mounting bolts to 14–18 ft. lbs. (19–25 Nm).

3. On 1988–89 vehicles, position the distributor-to-coil primary harness and the supression capacitor lead in the harness routing clip and close the clip. Connect the harness to the coil primary terminals. Connect the supression capacitor and battery leads to the positive terminal. Connect the white altitude connector to the distributor wiring harness.

4. On 1990–92 vehicles, connect the distributor electrical connector.

5. Install the distributor cap and connect the coil wire. If the spark plug wires were removed, connect them to the proper distributor cap towers, as marked during the removal procedure.

6. If equipped with a vacuum advance unit, connect the vacuum lines.

7. Connect the negative battery cable.

Timing Disturbed

1. If the crankshaft was rotated while the distributor was removed, the engine must be brought to TDC on the compression stroke of the No. 1 cylinder.

2. Remove the No. 1 spark plug. Place a finger over the hole and rotate the crankshaft slowly in the direction of normal rotation, until engine compression is felt.

NOTE: Turn the engine only in the direction of normal rotation. Backward rotation may cause the cam belt to slip or lose teeth, altering engine timing.

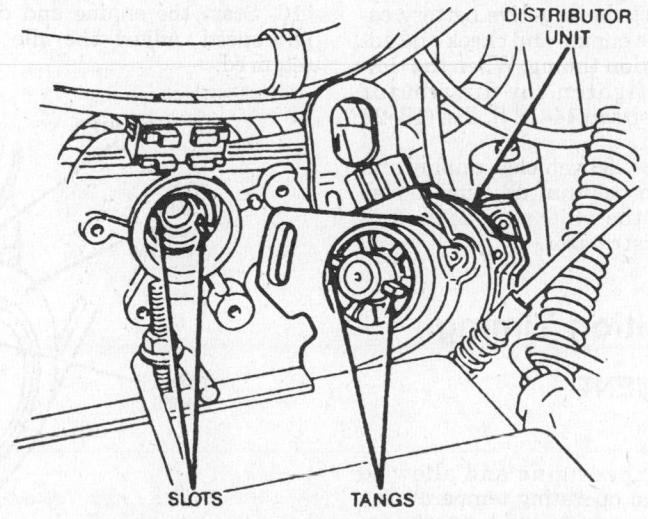

Distributor offset drive tangs and camshaft slots

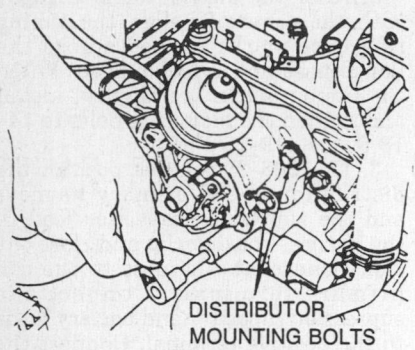

Distributor mounting bolt location

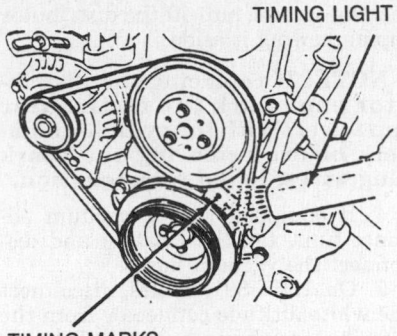

Ignition timing marks location—1988–89 vehicles

3. When engine compression is felt at the spark plug hole, indicating that the piston is approaching TDC, continue to turn the crankshaft until the TDC timing mark on the pulley is aligned with the **0** mark on the engine front cover.

4. Insert the distributor assembly into the cylinder head mounting bore. Rotate the distributor until the offset drive tang aligns and engages with the camshaft slot. Install the mounting bolts, leaving them loose enough that the distributor can be moved by hand.

5. On 1988–89 vehicles, position the distributor-to-coil primary harness and the supression capacitor lead in the harness routing clip and close the clip. Connect the harness to the coil primary terminals. Connect the supression capacitor and battery leads to the positive terminal. Plug the vacuum advance hoses.

6. On 1990–92 vehicles, connect the distributor electrical connector.

7. Install the distributor cap and connect the coil wire. If the spark plug wires were removed, connect them to the proper distributor cap towers, as marked during the removal procedure.

8. Connect the negative battery cable. Start the engine and check and adjust the ignition timing. When the timing is set, tighten the distributor mounting bolts to 14–18 ft. lbs. (19–25 Nm).

9. On 1988–89 vehicles, unplug and connect the vacuum advance hoses and connect the white altitude connector at the distributor.

Ignition Timing

ADJUSTMENT

1988–89

1. Start the engine and allow to reach normal operating temperature.

2. Stop the engine and connect a tachometer. Start the engine and check

the idle speed. Adjust the idle speed, if necessary.

3. Disconnect the vacuum hoses from the vacuum advance unit and plug the hose openings. Disconnect the white altitude connector at the distributor.

4. Turn off all electrical accessories.

5. Connect a timing light according to the manufacturers instructions. Start the engine.

6. With the timing light, observe the timing marks on the crankshaft pulley and timing case. The correct spark timing at idle is TDC $\pm$ 1 degree on carbureted engine or 2 degrees $\pm$ 1 degree on EFI engine.

7. If the timing is not as specified, loosen the distributor mounting bolts and rotate the distributor clockwise to advance the timing or counterclockwise to retard the timing.

8. When the timing is adjusted to specification, tighten the distributor mounting bolts.

9. Stop the engine. Remove the timing light. Unplug the vacuum hoses and connect them to the vacuum advance unit. Connect the white altitude connector.

10. Start the engine and check the idle speed. Adjust the idle speed as required.

1990–92

1. Place the transaxle in **P** or **N**, then make sure the air conditioner and heater fan is **OFF**.

2. Connect an inductive timing light to the No. 1 spark plug wire. Connect a tachometer.

3. Start the engine and allow it to warm up to normal operating temperature.

4. Ground the black 1-pin STI self-test connector located near the brake master cylinder.

5. Check and adjust the idle speed, if necessary.

6. Check the base ignition timing. The white ignition timing mark on the crankshaft pulley should align with the white pointer on the timing belt cover.

7. If the white timing mark and the white pointer do not line up, loosen distributor mounting bolts and rotate the distributor until the timing marks are properly aligned.

8. Tighten the distributor mounting bolts to 14–18 ft. lbs. (19–25 Nm).

9. Remove the jumper wire connecting the STI connector to ground.

10. Increase the engine rpm and check the timing marks to be sure the ignition timing changes.

11. Remove the timing light and tachometer.

Alternator

PRECAUTIONS

Several precautions must be observed with to avoid damage to the alternator.

• If the battery is removed for any reason, make sure it is reconnected with the correct polarity. Reversing the battery connections may result in damage to the one-way rectifiers.

• When utilizing a booster battery as a starting aid, always connect the

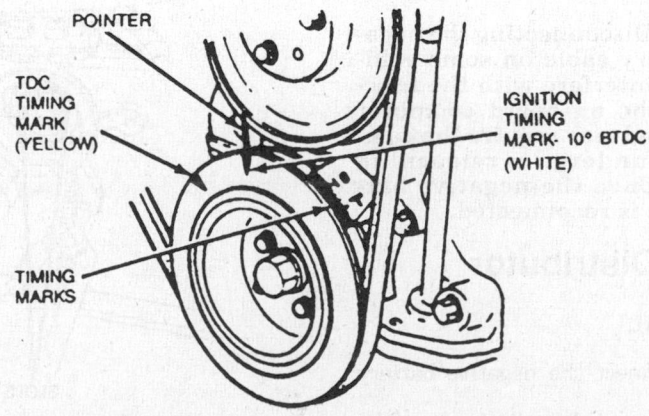

Ignition timing marks location—1990–92 vehicles

positive to positive terminals and the negative terminal from the booster battery to a good engine ground on the vehicle being started.

• Never use a fast charger as a booster to start vehicle. Disconnect the battery cables when charging the battery with a fast charger.

• Never attempt to polarize the alternator.

• Do not use test lamps of more than 12 volts when checking diode continuity.

• Do not short across or ground any of the alternator terminals.

• The polarity of the battery, alternator and regulator must be matched and considered before making any electrical connections within the system.

• Never separate the alternator on an open circuit. Make sure all connections within the circuit are clean and tight.

• Disconnect the battery ground terminal when performing any service on electrical components.

• Disconnect the battery if arc welding is to be done on the vehicle.

BELT TENSION ADJUSTMENT

1. Inspect the condition of the drive belt prior to adjustment. If the inspection reveals a severely glazed, frayed, oil contaminated or cracked belt, the belt must be replaced.

2. Loosen the alternator adjustment bolt.

3. Raise the vehicle and support it safely.

4. Loosen the lower alternator mounting/pivot bolt.

5. Lower the vehicle.

6. Position a suitable prybar between the engine and the alternator. Position the bar against the alternator in an area around a case bolt. Do not pry on the stator frame.

7. Adjust the belt tension by prying on the bar. Measure the belt tension using a belt tension gauge or by using the deflection method.

8. If using a belt tension gauge, position the gauge on the longest accessible belt span. The belt tension should be 110–132 lbs. for a new belt or 95–110 lbs. for a used belt (more than 10 minutes running time).

9. If using the deflection method, apply approximately 22 lbs. of pressure to the middle of the longest accessible belt span. The deflection should be 0.31–0.35 in. (8–9mm) for a new belt or 0.35–0.39 in. (9–10mm) for a used belt (more than 10 minutes running time).

10. When the belt tension is as speci-

fied, tighten the adjustment bolt to 14–19 ft. lbs. (19–25 Nm).

11. Raise and safely support the vehicle.

12. Tighten the alternator mounting/pivot bolt to 27–46 ft. lbs. (37–52 Nm).

13. Lower the vehicle.

REMOVAL & INSTALLATION

1. Disconnect the negative battery cable.

2. Pull the rubber boot away from the **B** terminal to expose the terminal nut. Remove the nut and electrical lead from the terminal post.

3. Remove the alternator adjusting bracket bolt.

4. Disconnect the electrical connector(s) from the alternator housing.

5. Raise and safely support the vehicle. Remove the alternator mounting/pivot bolt.

6. Disconnect the drive belt from the alternator.

7. Remove the alternator. If necessary, bend the catalytic converter shield brace to allow proper clearance.

8. Installation is the reverse of the removal procedure. Adjust the belt tension and tighten the adjustment bolt to 14–19 ft. lbs. (19–25 Nm) and the mounting/pivot bolt to 27–46 ft. lbs. (37–52 Nm).

Starter

REMOVAL & INSTALLATION

Automatic Transaxle

1. Disconnect the negative battery cable.

2. Remove the 2 upper starter motor mounting bolts.

3. Raise and safely support the vehicle.

4. Remove the 2 bolts that secure the manifold-to-cylinder block bracket, then remove the bracket.

5. Remove the bolt that secures the mounting bracket to the support bracket and remove the support bracket.

6. Remove the 2 nuts and washers that secure the mounting bracket to the starter motor and remove the mounting bracket.

7. Disconnect the **B** and **S** terminal connectors at the starter solenoid.

8. Remove the lower starter motor mounting bolt and remove the starter motor.

9. Installation is the reverse of the removal procedure. Tighten the starter motor mounting bolts to 23–34 ft. lbs. (31–46 Nm).

Manual Transaxle

1. Disconnect the negative battery cable.

2. Disconnect the **B** and **S** terminal connectors at the starter solenoid.

3. Remove the 2 bolts that secure the starter motor support bracket to the transaxle.

4. Remove the starter motor mounting bolts and remove the starter motor.

5. Installation is the reverse of the removal procedure. Tighten the starter motor mounting bolts to 23–34 ft. lbs. (31–46 Nm).

EMISSION CONTROLS

Please refer to "Emission Controls" in the Unit Repair section for system maintenance procedures. Due to the complex nature of modern electronic engine control systems, comprehensive diagnosis and testing procedures fall outside the confines of this repair manual. For complete information on diagnosis, testing and repair procedures concerning all modern engine and emission control systems, please refer to "Chilton's Guide to Fuel Injection and Electronic Engine Controls".

Emission Warning Lamps

RESETTING

The Malfunction Indicator Lamp (MIL) is a dual function light that informs the driver of possible engine malfunctions and emission system failure. The MIL is controlled by the ECU. The ECU functions to monitor engine, ignition and emission related components and signals the driver when the engine is running improperly or emissions are unsatisfactory. If the MIL illuminates during vehicle operation, the cause of the fault or malfunction must be determined and corrected.

FUEL SYSTEM

Fuel System Service Precautions

Safety is the most important factor

when performing not only fuel system maintenance but any type of maintenance. Failure to conduct maintenance and repairs in a safe manner may result in serious personal injury. Maintenance and testing of the vehicle's fuel system components can be accomplished safely and effectively by adhering to the following rules and guidelines.

• To avoid the possibility of fire and personal injury, always disconnect the negative battery cable unless the repair or test procedure requires that battery voltage be applied.

• Always relieve the fuel system pressure prior to disconnecting any fuel system component (injector, fuel rail, pressure regulator, etc.), fitting or fuel line connection. Exercise extreme caution whenever relieving fuel system pressure to avoid exposing skin, face and eyes to fuel spray. Please be advised that fuel under pressure may penetrate the skin or any part of the body that it contacts.

• Always place a shop towel or cloth around the fitting or connection prior to loosening to absorb any excess fuel. Ensure that all fuel spillage (should it occur) is quickly removed from engine surfaces. Ensure that all fuel soaked cloths or towels are deposited into a suitable waste container.

• Always keep a dry chemical (Class B) fire extinguisher near the work area.

• Do not allow fuel spray or fuel vapors to come into contact with a spark or open flame.

• Always use a backup wrench when loosening and tightening fuel line connection fittings. Always follow the proper torque specifications.

• Always replace worn fuel fitting O-rings with new. Do not substitute fuel hose or equivalent where fuel pipe is installed.

RELIEVING FUEL SYSTEM PRESSURE

EFI Engine

1. Remove the rear seat cushion.
2. Disconnect the electrical connector from the fuel pump/sending unit.
3. Start the engine and let it run until it stalls. Turn the ignition key OFF.
4. Reconnect the electrical lead.

Fuel Tank

REMOVAL & INSTALLATION

1. Remove the rear seat as follows:
    a. Remove the right and left front attaching bolts.
    b. Fold the rear seat forward.
    c. Remove the right and left anchor nuts on the rear side of the seat and remove the seat.
2. Remove the screw and retainers and remove the left rear quarter trim panel.
3. If EFI equipped, start the engine and disconnect the fuel pump/sending unit connector. After the engine stalls, turn the ignition key OFF.
4. Disconnect the negative battery cable.
5. Drain the fuel from the tank as completely as possible. This is accomplished by siphoning or pumping the fuel out through the fuel filler neck.
6. Remove the rear carpet holddown pins using a suitable tool. Fold the carpet forward until the sending unit access plate is uncovered.
7. Remove the sending unit access plate attaching screws, lift the access plate and disconnect the sending unit wiring.
8. Disconnect the fuel supply line at the sending unit and the fuel return line from the top of the fuel tank.
9. Remove the fuel tank cover plate.
10. Disconnect the filler neck hose, overflow hose and the 2 vapor separator hoses from the fuel tank.
11. Raise and safely support the vehicle.
12. Disconnect the vapor hose from the vapor line.
13. Position a suitable jack under the fuel tank and remove the 4 attaching bolts.
14. Move the fuel tank toward the left and lower it from the vehicle.

To install:

15. Raise the fuel tank and slide it into position from the left side of the vehicle. Install the attaching bolts.
16. Connect the vapor hose to the vapor line and lower the vehicle.
17. Connect the vapor separator hoses, overflow hose and fuel filler hose to the fuel tank.
18. Connect the fuel return hose to the fitting on the top of the fuel tank and the fuel supply hose to the fitting on the fuel sending unit.
19. Add fuel to the tank and check for leaks.
20. Connect the negative battery cable and the fuel sending unit wiring. Start the engine and check for leaks. Stop the engine.
21. Install the fuel line cover plate and the fuel sender access plate.
22. Position the rear carpet and secure it in position with the retainers.
23. Install the left rear quarter panel and the rear seat.

Fuel Filter

REMOVAL & INSTALLATION

The fuel filter is located in the rear left corner of the engine compartment next to the carbon canister.

EFI Engine

1. Properly relieve the fuel system pressure.
2. Disconnect the negative battery cable.
3. Remove the clamp and line at the inlet of the fuel filter. Plug the end to prevent spillage.
4. Remove the attaching bolts from the outlet of the fuel filter.
5. Remove the fuel filter from it's brace.

To install:

6. Install the fuel filter into it's brace.
7. Install the line onto the filter outlet with the attaching bolts. Tighten the bolts to 18–25 ft. lbs. (25–34 Nm).
8. Unplug and install the supply line onto the fuel filter inlet and secure with the clamp.
9. Connect the fuel pump connector and install the rear seat cushion. Connect the negative battery cable.
10. Run the engine and check for leaks.

Carbureted Engine

1. Disconnect the negative battery cable.
2. Remove the clamp and line at the inlet of the filter. Plug the end to prevent spillage.
3. Remove the clamp and line at the filter outlet and remove the filter from it's brace.
4. Installation is the reverse of the removal procedure. Be sure to install the filter with the arrow pointing in the direction of fuel flow.

Mechanical Fuel Pump

The fuel pump is located on the firewall side of the cylinder head, near the distributor.

PRESSURE TESTING

1. Disconnect the negative battery cable.
2. Disconnect the fuel hose at the carburetor and attach a suitable fuel pressure gauge.
3. Remove the fuel return hose at the fuel pump and plug or cap the exposed port.
4. Connect the negative battery cable.
5. With the engine idling or cranking normally, check the fuel pressure.

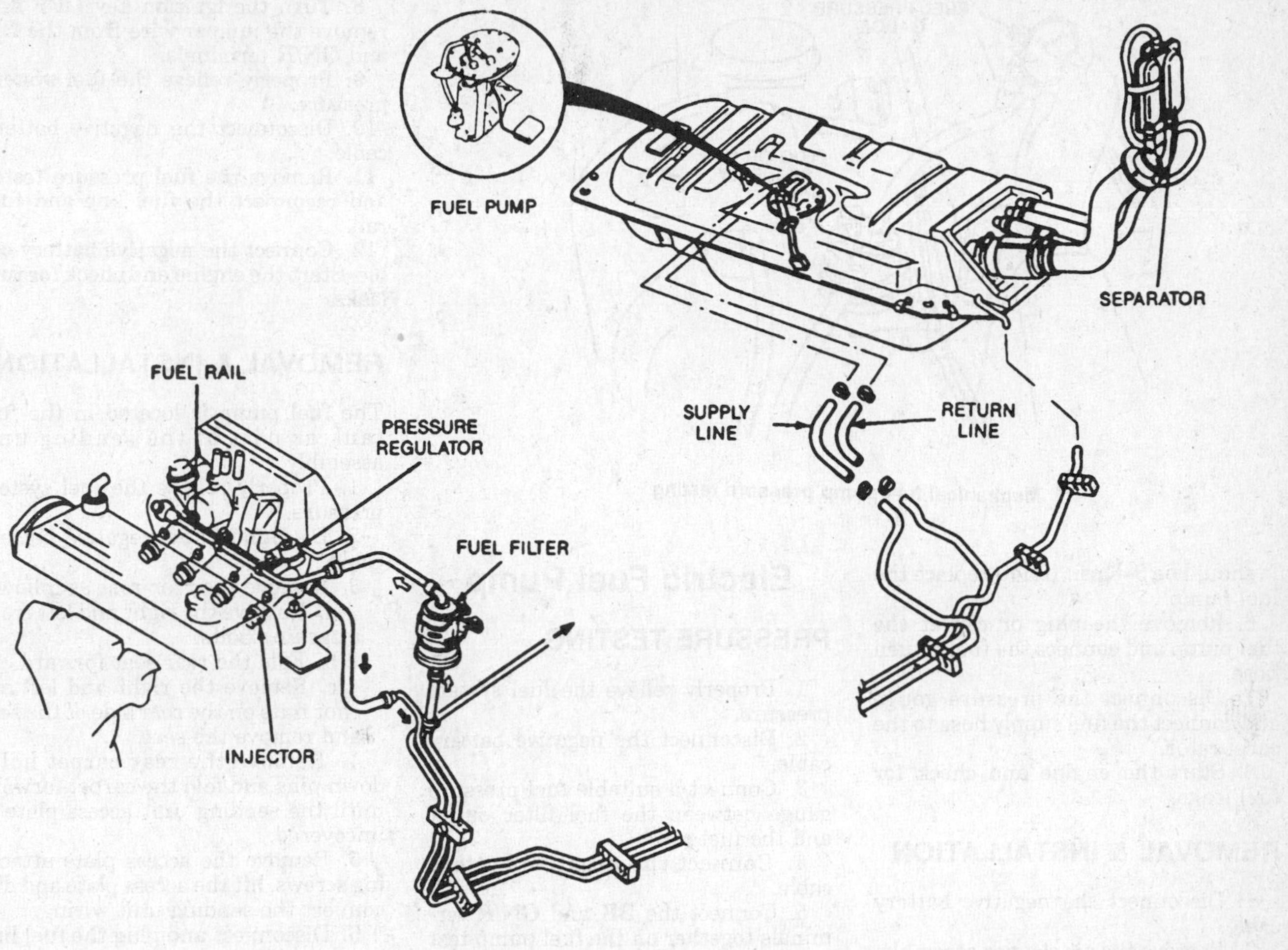

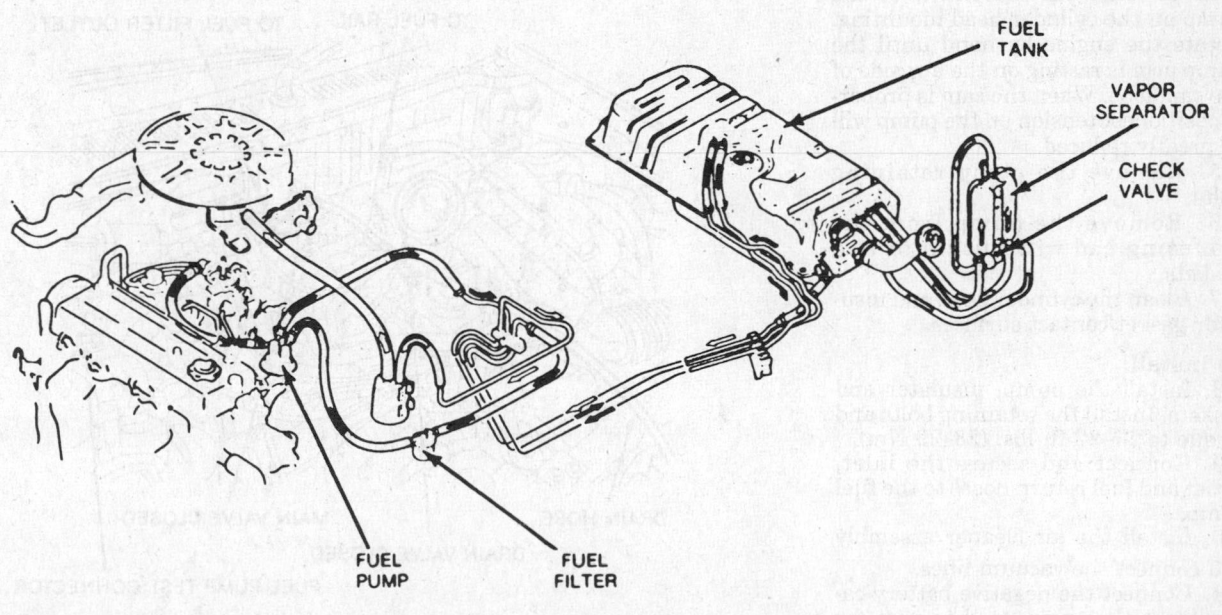

FUEL PUMP

SEPARATOR

FUEL RAIL

PRESSURE REGULATOR

FUEL FILTER

INJECTOR

SUPPLY LINE

RETURN LINE

EFI engine fuel supply system

FUEL TANK

VAPOR SEPARATOR

CHECK VALVE

FUEL PUMP

FUEL FILTER

Carbureted engine fuel supply system

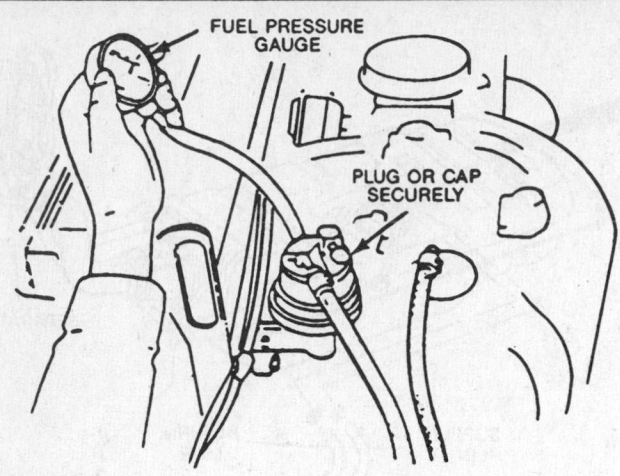

FUEL PRESSURE GAUGE

PLUG OR CAP SECURELY

Mechanical fuel pump pressure testing

It should be 3–6 psi. If not, replace the fuel pump.

6. Remove the plug or cap at the fuel pump and connect the fuel return hose.

7. Disconnect the pressure gauge and connect the fuel supply hose to the carburetor.

8. Start the engine and check for fuel leaks.

REMOVAL & INSTALLATION

1. Disconnect the negative battery cable.

2. Remove the air cleaner assembly. Identify and tag all vacuum hoses as required.

3. Tag and disconnect the fuel pump inlet, outlet and return hoses.

4. Loosen the fuel pump retaining bolts to allow for movement of the pump on the cylinder head mounting. Rotate the engine by hand until the pump arm is resting on the low side of the cam lobe. When the cam is properly positioned, tension on the pump will be greatly reduced.

5. Remove the pump retaining bolts.

6. Remove the pump from the mounting pad with insulator and gaskets.

7. Clean the cylinder head and insulator gasket contact surfaces.

To install:

8. Install the pump, insulator and gaskets. Install the retaining bolts and torque to 17–22 ft. lbs. (23–29 Nm).

9. Connect and secure the inlet, outlet and fuel return hoses to the fuel pump.

10. Install the air cleaner assembly and connect the vacuum lines.

11. Connect the negative battery cable. Start the engine and inspect for fuel leaks.

Electric Fuel Pump

PRESSURE TESTING

1. Properly relieve the fuel system pressure.

2. Disconnect the negative battery cable.

3. Connect a suitable fuel pressure gauge between the fuel filter outlet and the fuel rail.

4. Connect the negative battery cable.

5. Connect the **BK** and **GN/R** terminals together on the fuel pump test connector.

6. Turn the ignition key **ON** but do not start the engine.

7. The fuel pressure reading should be 64–85 psi.

8. Turn the ignition key **OFF** and remove the jumper wire from the **BK** and **GN/R** terminals.

9. Properly relieve the fuel system pressure.

10. Disconnect the negative battery cable.

11. Remove the fuel pressure tester and reconnect the fuel line and fuel rail.

12. Connect the negative battery cable. Start the engine and check for fuel leaks.

REMOVAL & INSTALLATION

The fuel pump is located in the fuel tank as part of the sending unit assembly.

1. Properly relieve the fuel system pressure.

2. Disconnect the negative battery cable.

3. Remove the rear seat as follows:

   a. Remove the right and left front attaching bolts.

   b. Fold the rear seat forward.

   c. Remove the right and left anchor nuts on the rear side of the seat and remove the seat.

4. Remove the rear carpet holddown pins and fold the carpet forward until the sending unit access plate is uncovered.

5. Remove the access plate attaching screws, lift the access plate and disconnect the sending unit wiring.

6. Disconnect and plug the fuel line at the sending unit.

7. Remove the sending unit retaining screws and remove the sending unit. Discard the gasket.

8. Remove the fuel filter from the

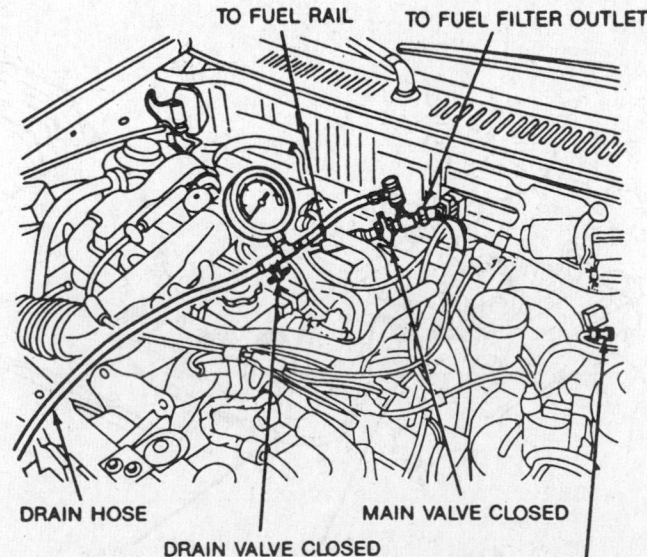

TO FUEL RAIL   TO FUEL FILTER OUTLET

DRAIN HOSE

DRAIN VALVE CLOSED

MAIN VALVE CLOSED

FUEL PUMP TEST CONNECTOR

Electric fuel pump pressure testing

pump. Remove the fuel pump wires from the sending unit.

9. Remove the retaining clamp screw and the pump outlet hose clamp. Remove the fuel pump from the sending unit.

To install:

10. Install the fuel pump to the sending unit bracket and secure with the retaining clamp.

11. Install the pump outlet hose and secure with the clamp.

12. Connect the fuel pump wires to the sending unit and install the fuel pump filter.

13. Position a new gasket on the fuel tank and install the sending unit with the attaching screws.

14. Connect the fuel line and the sending unit wiring.

15. Connect the negative battery cable. Start the engine and check for leaks at the fuel line connections. Stop the engine.

16. Install the access cover with the attaching screws.

17. Position the carpet and install the retaining pins.

18. Install the rear seat.

Carburetor

REMOVAL & INSTALLATION

1. Disconnect the negative battery cable.

2. Remove the air cleaner assembly.

3. Loosen the retaining clamp and disconnect the fuel supply line. Plug the hose opening to prevent contamination and the entry of foreign matter.

4. Disconnect the vacuum hoses from the carburetor. Identify each hose with its respective opening to ensure proper installation.

5. Disconnect the carburetor wiring connectors.

6. Disconnect the choke heater wire at the choke cap.

7. Move the throttle to the wide open position and disengage the throttle cable from the throttle lever.

8. Remove the carburetor retaining nuts and washers. Lift the carburetor upward from the intake manifold studs. Disconnect the throttle kicker diaphragm link from the carburetor linkage. If the EFE heater sticks to the carburetor base, gently remove it. Discard the carburetor flange gaskets and replace with new.

To install:

9. Thoroughly clean the carburetor, EFE heater and intake manifold gasket contact surfaces and install new gaskets.

10. Position the carburetor over the intake manifold mounting studs and support by hand. While supporting the carburetor, connect the throttle kicker

diaphragm link to the carburetor linkage. Install and tighten the mounting nuts and washers.

11. Move the throttle to the wide open position and connect the throttle cable to the throttle lever.

12. Connect the choke heater wire.

13. Connect the carburetor wires to their respective connectors.

14. Connect the vacuum hoses to their original openings.

15. Connect the the fuel supply line and install the retaining clamp.

16. Install the air cleaner assembly and connect the negative battery cable.

17. Start the engine and adjust the idle speed, if necessary.

IDLE SPEED ADJUSTMENT

1. Disconnect the cooling fan electrical connector. Check the ignition timing and adjust, if necessary. Adjust the idle mixture, as required.

2. Place the transmission selector lever in **N** and firmly apply the parking brake. Make certain the air conditioning system is **OFF**. Be sure all electrical accessories are **OFF**.

3. Connect a tachometer to the engine.

4. Start the engine and allow to reach normal operating temperature. Make certain the choke is fully open.

5. Allow the engine to remain at idle and observe the idle speed reading. The idle speed should be 700–750 rpm.

6. If the idle speed is not within specifications, rotate the idle speed adjusting screw, located on the right side of the carburetor, as required until the correct idle speed is obtained.

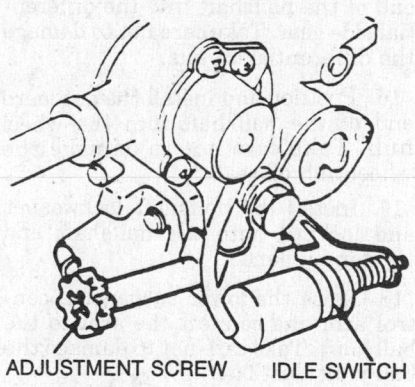

ADJUSTMENT SCREW   IDLE SWITCH

Carburetor curb idle adjustment screw location

SERVICE ADJUSTMENTS

For all carburetor service adjustment procedures and specifications, please refer to "Carburetor Service" in the Unit Repair section.

Fuel Injection

IDLE SPEED ADJUSTMENT

NOTE: The test connector is located on the passenger side of the engine compartment, near the shock tower.

1. Disconnect the cooling fan electrical connector. Check the ignition timing and adjust if necessary.

2. Apply the parking brake. Make certain the air conditioning system is **OFF**. Be sure all electrical accessories are **OFF**.

3. Connect a tachometer to the check connector (clear, pin No. 1). On 1989–90 vehicles, connect a jumper wire between the test connector (black, pin No. 1) and ground. On 1991–92 vehicles, connect a jumper wire between the black test connector (1 pin, Y/BL wire) and ground.

4. Check the idle speed on the tachometer. The idle speed should be 800–900 rpm on 1989 vehicles in **P**, 680–720 rpm on 1990–92 manual transaxle vehicles in **N** or 830–870 on 1990–92 automatic transaxle vehicles in **P**.

5. If necessary, turn the idle air adjust screw to obtain the correct idle speed.

6. After adjustment, remove the jumper wire and the tachometer.

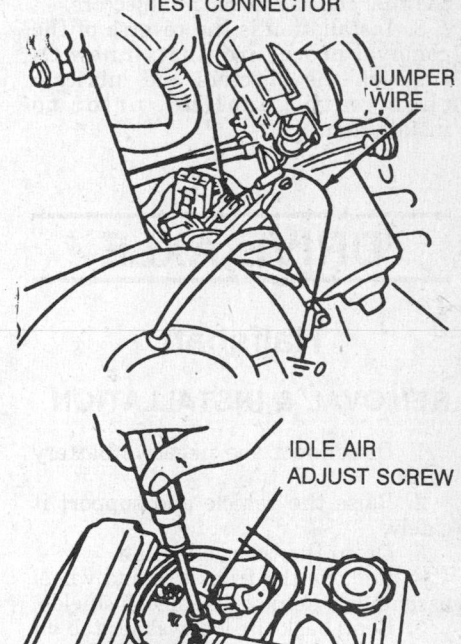

TEST CONNECTOR

JUMPER WIRE

IDLE AIR ADJUST SCREW

EFI engine test connector and idle air adjuster screw locations

IDLE MIXTURE ADJUSTMENT

The idle mixture screw is preset and sealed at the factory. Idle mixture cannot be adjusted.

Fuel Injector

REMOVAL & INSTALLATION

1. Properly relieve the fuel system pressure.
2. Disconnect the negative battery cable.
3. Remove the intake plenum as follows:
   a. Drain the cooling system.
   b. Disconnect the throttle cable and the air duct from the throttle body.
   c. Mark all vacuum and coolant hoses for ease of reassembly and remove the hoses from the throttle body.
   d. Disconnect the electrical connector at the throttle position sensor.
   e. Remove the intake plenum retaining bolts and/or nuts and remove the intake plenum and gasket.
4. Remove the fuel inlet and return lines from the fuel rail.
5. Remove the electrical connectors at the injectors.
6. Remove the pressure regulator.
7. Remove the attaching bolts and the fuel rail. Remove the injectors.
8. Installation is the reverse of the removal procedure. Install new O-rings on the injectors and lubricate them with gasoline, prior to installation.

DRIVE AXLE

Halfshaft

REMOVAL & INSTALLATION

1. Disconnect the negative battery cable.
2. Raise the vehicle and support it safely.
3. Drain the transaxle fluid.
4. Remove the front tire and wheel assemblies. Remove the splash shields.
5. Bend back the lockwasher tab on the halfshaft locknut slot. Lock the brakes and loosen but do not remove, the halfshaft locknut.
6. Remove the stabilizer mounting nuts and brackets.
7. Remove the clamp bolt and nut from the lower suspension control arm. With a suitable prybar, pry the lower suspension control arm downward to disconnect the ball joint. Be careful not to tear or puncture the dust boot when disconnecting the ball joint.
8. With the proper tool, separate the halfshaft from the transaxle.

NOTE: The halfshaft must be separated from the transaxle gradually. If the halfshaft is pulled or jerked suddenly, the differential and wheel hub oil seals may be damaged. If necessary, use a suitable puller to push the driveshaft from the wheel hub.

9. Remove and discard the halfshaft locking nut.
10. Withdraw the halfshaft from the wheel hub and the transaxle. Wrap tape around the inboard and outboard splines to prevent damage.
11. Install differential plug tool T87C–7025–C or equivalent, to prevent the side gear from moving.

To install:

12. Inspect the differential and wheel hub oil seals for damage and replace the seals as required.
13. Remove the protective tape. Remove the circlips from the inboard halfshaft spline ends and replace with new. Coat the inboard and outboard halfshaft spline ends with grease.
14. Remove the differential gear holding plugs.
15. Position and install the inboard end of the halfshaft into the differential side gear. Take care not to damage the differential oil seal.
16. Position and install the outboard end of the halfshaft into the wheel hub. Take care not to damage the wheel hub oil seal.
17. Install the halfshaft lockwasher and locknut onto the halfshaft and tighten by hand.
18. Raise the lower suspension control arm and connect the arm to the ball joint. Take care not to damage the ball joint dust boot.
19. Install the lower suspension arm clamp nut and bolt. Hold the bolt stationary and torque the nut to 32–40 ft. lbs. (43–54 Nm).
20. Make certain the brakes are still locked and torque the outboard halfshaft locknut to 116–174 ft. lbs. (157–235 Nm). Bend a tab of the the lockwasher into a slot in the locknut with the proper tool.

NOTE: Do not stake the locking tab with a pointed tool. Make sure the locking tab is depressed at least 0.16 in. (4mm) into the lock-nut slot to ensure proper locking capabilty. After the lockwasher is locked into place, grasp the wheel hub and pull to ensure that the halfshaft is installed properly. Rotate the wheel hub by hand to ensure that the wheel hub turns smoothly.

21. Install the stabilizer brackets and mounting bracket nuts. Tighten the nuts to 40–50 ft. lbs. (54–68 Nm).
22. Install the splash shields. Install the tire and wheel assemblies. Install and tighten the transaxle drain plug.
23. Fill the transaxle with the proper grade and type fluid to specification. Lower the vehicle.

CV-Boot

REMOVAL & INSTALLATION

Except 1988–89 Manual Transaxle

1. Raise and support the vehicle safely.
2. Remove the halfshaft from the vehicle. Support the assembly in a vise with protective jaws.

NOTE: During disassembly and assembly, do not allow dust or similar foreign matter to enter the halfshaft joints.

3. Remove the large boot clamp from the inboard CV-joint. Roll the boot back over the shaft. Remove the wire ring bearing retainer. Remove the outer race.

NOTE: Before removing the outer race, matchmark the outer race and tripot bearing for reassembly.

4. Matchmark the tripot bearing and the shaft. Remove the tripot bearing snapring. Using a suitable tool, remove the tripot bearing from the shaft.
5. Remove the small clamp and the CV-joint boot from the halfshaft.
6. If replacing the outboard CV-joint boot, remove the dynamic damper, then remove the clamps and slide the boot off of the shaft from the inboard side.
To install:
7. Cover the halfshaft splines with tape.
8. If replacing the outboard CV-joint boot, slide the boot onto the halfshaft and onto the ouboard CV-joint. Install the dynamic damper onto the halfshaft at a distance of 18.99–19.27 in. (482.5–489.5mm) from the

outboard end of the halshaft with the outboard halfshaft fully pushed onto the halfshaft.

9. Install the inboard CV-joint boot onto the halfshaft.

10. Install the tripot assembly on the halfshaft using a suitable tool. Observe the alignment marks that were made during the removal procedure.

11. Install the tripot assembly retaining ring.

12. Fill the CV-joint outer race with 3.5 oz. of high temperature CV-joint grease. Install the outer race over the tripot joint and install the wire ring bearing retainer. Observe the alignment marks that were made during the removal procedure.

13. Position the CV-joint boot. Make sure the boot is fully seated in the grooves in the shaft and outer race. Extend or compress the joint, as necessary, until the distance between the CV-joint boot clamp grooves measures 3.5 in. (90mm).

14. Insert a suitable tool between the boot and the outer bearing race to allow trapped air to escape from the boot. Install the boot clamps, wrapping them around the boots in the opposite direction of halfshaft rotation. Pull the clamps tight with a suitable tool and bend the locking tabs to secure in position.

15. Install the halfshaft in the vehicle. Lower the vehicle.

1988–89 Manual Transaxle

1. Raise and safely support the vehicle.

2. Remove the halfshaft assembly and clamp it in a vise with protected jaws.

NOTE: Do not allow dust or other foreign matter to enter the halfshaft joints during disassembly and assembly.

3. Remove the boot bands and slide the inboard CV-joint boot off of the inboard CV-joint.

4. Mark the outer ring and halfshaft with paint for proper positioning during assembly. Do not use a punch to make the mark.

5. Use a suitable tool to remove the large circlip that secures the ball joint in the outer ring. Withdraw the ball joint out of the outer ring.

6. Mark the halfshaft and ball joint inner ring with paint for proper positioning during assembly. Do not use a punch to make the mark.

7. Remove the snaping securing the halfshaft in the ball joint inner ring. Remove the ball joint assembly from the halfshaft.

8. Mark the inner ring and cage with paint for proper alignment at assembly. Insert a suitable tool between

the ball cage and inner ring to remove the balls. Be careful not to lose the balls. Turn the cage approximately 30 degrees to separate it from the inner ring.

9. Remove the inboard CV-joint boot.

10. If replacing the outboard CV-joint boot, remove the clamps and remove the CV-joint boot from the inboard end of the halfshaft. On the right side halfshaft, the dynamic damper must be removed before removing the boot.

To install:

11. Place tape on the halfshaft splines.

12. If replacing the outboard CV-joint boot, slide the boot onto the halfshaft from the inboard end.

13. If replacing the outboard boot on the right side halfshaft, install the dynamic damper onto the halfshaft at a distance of 18.99–19.27 in. (482.5–489.5mm) from the ouboard end of the halfshaft when the outboard CV-joint is fully pushed onto the halfshaft.

14. Slide the inboard CV-joint boot onto the halfshaft.

15. Observing the alignment marks made during the removal procedure, reassemble the ball cage, balls and inner ring. Apply molydenum disulfide grease to the ball joint.

16. Observing the alignment marks made during the removal procedure, replace the ball joint onto the halfshaft, Secure the ball joint onto the halfshaft with the snapring.

17. Observing the alignment marks made during the removal procedure, replace the halfshaft and ball joint into the outer ring. Secure the ball joint in the outer ring with a new large circlip.

18. Carefully fit the boots in their grooves on the joints. Wrap new boot bands around the boots in the opposite direction of halfshaft forward rotation. Use a suitable tool to apply tension while installing the boot bands. Bend the locking tabs down to secure the boot bands.

19. Remove the tape from the halfshaft splines and install the halfshafts. Lower the vehicle.

Front Wheel Hub, Knuckle and Bearings

REMOVAL & INSTALLATION

1. Disconnect the negative battery cable.

2. Raise the vehicle and support it safely.

3. Unbolt and remove front wheel from the hub assembly.

4. With a suitable tool, straighten

the staked edge of the halfshaft attaching nut. Take care not to damage the halfshaft threads.

5. Remove and discard the halfshaft attaching nut.

6. Remove the retaining clip securing the caliper hose to the strut bracket.

7. Remove the cotter pin and tie rod end attaching nut. Discard the cotter pin and set the nut aside. Inspect the nut for damage and replace as required.

8. Using a tie rod end separator tool, release the tie rod end from the steering knuckle arm. If the tie rod appears to be siezed, strike the knuckle sharply with a soft-tipped hammer to acheive separation.

9. Support the brake caliper and remove the brake caliper attaching bolts. Lift the caliper assembly from the steering knuckle.

NOTE: After the caliper assembly is lifted from the steering knuckle, do not allow it to be suspended by the brake hose. Support the caliper by a length of rope or wire attached to the MacPherson strut.

10. Remove the clamp bolt and nut at the point where the lower control arm ball joint connects to the steering knuckle. With a medium prybar, release the lower ball joint from the steering knuckle by prying downward on the lower control arm.

11. Remove the 2 bolts that position the steering knuckle between the MacPherson strut bracket flanges.

12. Slide the knuckle/hub assembly from the end of the halfshaft. If binding occurs, tap the end of the shaft with a soft-tipped hammer. If the wheel hub is rusted to the halfshaft, use either a 2 jaw or a hub puller to achieve separation.

13. Remove the wheel hub/rotor assembly from the steering knuckle/dust shield assembly using a suitable puller.

14. Remove the bearing preload spacer from the hub.

NOTE: The spacer is preselected to yield the correct bearing preload. Save the removed spacer for use during assembly.

15. Clamp the hub/rotor assembly in a vise with protective jaw caps. Scribe aligning marks on the hub and rotor for use during assembly. Remove the attaching bolts and the rotor.

16. Remove the outer bearing from the wheel hub using a suitable bearing splitter, driver and press. Remove the outer and inner grease seals and discard.

17. Remove the races from the steer-

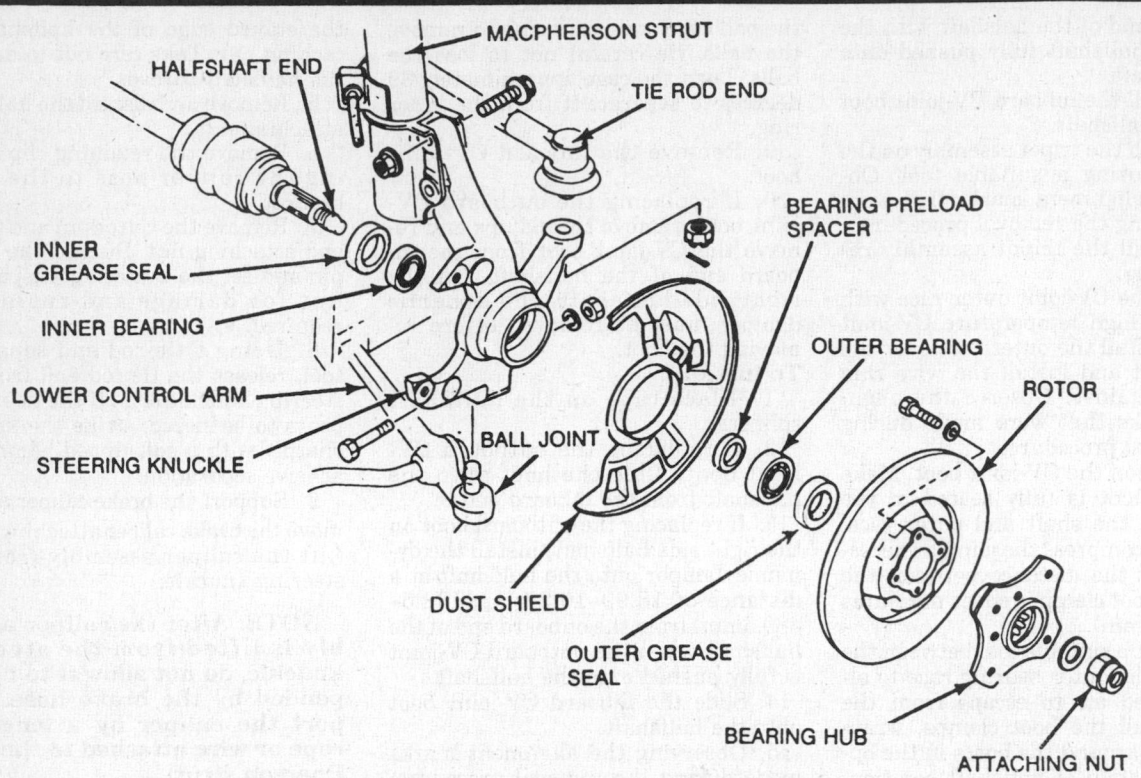

MACPHERSON STRUT

HALFSHAFT END

TIE ROD END

INNER GREASE SEAL

INNER BEARING

LOWER CONTROL ARM

STEERING KNUCKLE

BALL JOINT

DUST SHIELD

BEARING PRELOAD SPACER

OUTER BEARING

ROTOR

OUTER GREASE SEAL

BEARING HUB

ATTACHING NUT

Front wheel hub, knuckle and bearing assembly

ing knuckle using a suitable puller and slide hammer. If necessary, remove the dust shield from the steering knuckle.

To install:

18. Clean and inspect all components that will be reused. Check the hub, knuckle and rotor dust shield for cracks, scoring, rusting, etc.

19. If the brake rotor dust shield was removed, install a new 1 using a suitable installation tool.

20. If the bearings or knuckle are being replaced, bearing preload must be checked as follows before assembly.

   a. Install the outer bearing races in the steering knuckle using suitable tools.

   b. Lubricate the bearing races and bearing with a thin film of clean engine oil. Install the bearings in the steering knuckle.

   c. Install spacer selection tool T87C-1104-B or equivalent, and clamp the bolt head in a vise.

   d. Tighten the center bolt in increments, to 36, 72, 108 and 145 ft. lbs. (49, 98, 147 and 196 Nm). After tightening the center bolt to a specified increment, seat the bearings by rotating the steering knuckle.

   e. Remove the tool/steering knuckle from the vise. Remount the assembly in the vise, clamping it where the MacPherson strut mounts.

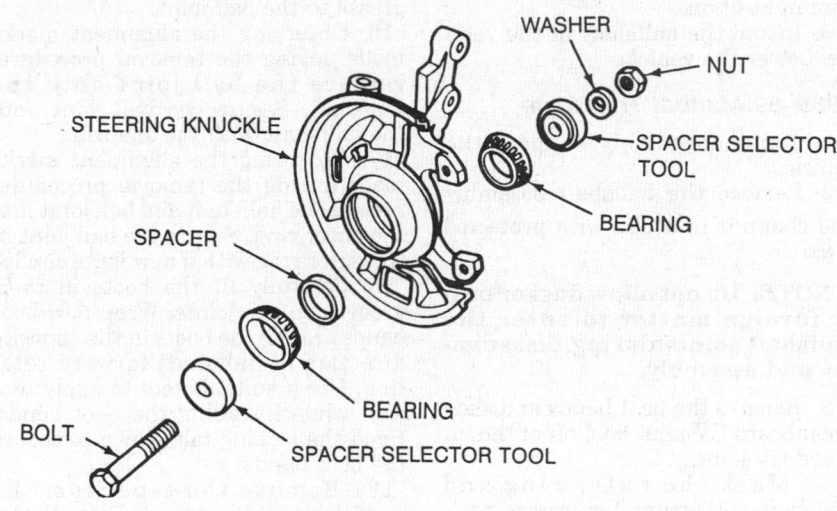

STEERING KNUCKLE

SPACER

BOLT

WASHER

NUT

SPACER SELECTOR TOOL

BEARING

BEARING

SPACER SELECTOR TOOL

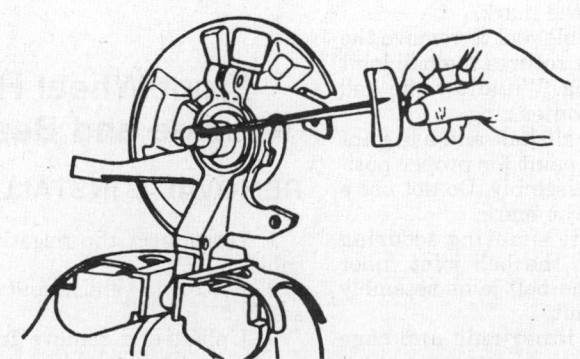

Front wheel bearing preload checking procedure

f. Measure the amount of torque required to rotate the spacer selector tool, using an inch pound torque wrench. The torque wrench reading must be taken just as the tool starts to rotate.

g. If the torque wrench indicates 2.2–10.4 inch lbs. (0.25–1.8 Nm), the spacer is the correct thickness. If the torque wrench indicates less than 2.2 inch lbs. (0.25 Nm), a thinner spacer must be installed. If the torque wrench indicates more than 10.4 inch lbs. (1.8 Nm), a thicker spacer must be installed.

h. Each bearing spacer has a numerical code that identifies it's thickness, stamped onto the outer diameter of the spacer. The numbers range from 1–21, with 1 being the thinnest spacer. If the number stamped on the spacer is not legible, measure the spacer with a micrometer and compare it to the spacer thickness chart to determine the number.

i. Changing the spacer thickness by 1 number, either higher or lower, will change the bearing preload by 1.7–3.5 inch lbs. (0.2–0.4 Nm).

21. If the bearings or knuckle are not being replaced, install the races in the steering knuckle using suitable tools.

22. Pack the bearings and the hub area with a suitable high temperature wheel bearing grease. Place the inner bearing into the steering knuckle bore.

23. Lubricate the lip of the new inner grease seal with the bearing grease. Form the lubricant into a strip, concentrated along the edges of the seal lip. Install the inner seal into the bore, using a suitable installation tool.

24. Place the original bearing preload spacer or the spacer selected from the bearing preload check procedure, in the steering knuckle bore. Position the bearing removed from the wheel hub in the steering knuckle bore.

25. Lubricate the lip of the new outer grease seal with the bearing grease. Form the lubricant into a strip, concentrated along the edges of the seal lip. Install the outer seal into the bore, using a suitable installation tool.

26. Position the rotor on the hub, observing the original aligning marks, and install the attaching bolts. Tighten the attaching bolts to 33–40 ft. lbs. (44–54 Nm).

27. Position the hub/rotor assembly in the steering knuckle bore and press it into position using a suitable driver.

28. Clean the halfshaft spline end and lubricate with a coating of wheel bearing grease. Apply a thin film of clean SAE 30 weight oil to the steering knuckle/rotor hub assembly up to the point where the uppermost arm of the steering knuckle seats into the MacPherson strut bracket. Guide the steering knuckle/rotor/hub assembly onto the halfshaft and the Macpherson strut.

29. Install the strut-to-steering knuckle bolts and attaching nuts. Tighten the nuts to 69–86 ft. lbs. (93–117 Nm).

30. Position the lower control arm ball joint in the steering knuckle. Install the lower control arm pinch bolt and attaching nut. Tighten the nut to 32–40 ft. lbs. (43–54 Nm).

31. Position the caliper on the steering knuckle and install the attaching bolts. Tighten the bolts to 29–36 ft. lbs. (39–49 Nm). Position the caliper hose in the strut routing bracket and install the retaining clip.

32. Install a new halfshaft attaching nut and tighten to 116–174 ft. lbs. (157–235 Nm). After installation, the wheel hub assembly must rotate freely by hand. Stake the halfshaft attaching nut into the shaft groove.

NOTE: Do not use a pointed tool to stake the nut. If the nut cracks even slightly during staking, replace it with another new one.

33. Connect the tie rod end to the steering knuckle and install the attaching nut. Tighten the attaching nut to 22–33 ft. lbs. (29–44 Nm). Install a new cotter pin through the nut and ball stud. If the openings in the nut and the hole in the ball stud are not aligned, tighten the nut slightly, just to the point of alignment. Never loosen the nut.

34. Install the wheel and tire assembly. Tighten the attaching bolts to 65–87 ft. lbs. (88–118 Nm). Lower the vehicle.

FRONT WHEEL BEARING PRELOAD SPACER THICKNESS

Stamped Mark	Thickness In. (mm)
1	0.2474 (6.285)
2	0.2490 (6.325)
3	0.2506 (6.365)
4	0.2522 (6.405)
5	0.2538 (6.445)
6	0.2554 (6.485)
7	0.2570 (6.525)
8	0.2586 (6.565)
9	0.2602 (6.605)
10	0.2618 (6.645)
11	0.2634 (6.685)
12	0.2650 (6.725)
13	0.2666 (6.765)
14	0.2682 (6.805)
15	0.2698 (6.845)
16	0.2714 (6.885)
17	0.2730 (6.925)
18	0.2746 (6.965)
19	0.2762 (7.005)
20	0.2778 (7.045)
21	0.2794 (7.085)

MANUAL TRANSAXLE

For further information on transmissions/transaxles, please refer to "Chilton's Guide to Transmission Repair".

Transaxle Assembly

REMOVAL & INSTALLATION

1. Disconnect the negative battery cable.
2. Disconnect the back-up switch wiring connector.
3. Disconnect the neutral switch wiring connector.

4. Loosen the clutch cable adjusting nut and disengage the cable from the release lever.

5. Remove the starter.

6. Disconnect the speedometer cable.

7. Remove the 2 bolts from the top of the clutch housing.

8. Install an engine support bar tool. Raise and support the vehicle safely.

9. Remove the nut and bolt attaching the shift rod to the input shift rail.

10. Remove the nuts and bolts attaching the lower control arms to the steering knuckles.

11. Disengage the halfshafts from the differential side gears.

12. Install a differential side gear plug tool to prevent the side gears from moving.

13. Remove the mounting bracket attaching bolts.

14. Remove the crossmember.

15. Position a suitable transmission jack under the transaxle and secure the jack with a safety chain.

16. Remove the remaining lower transaxle attaching bolts. Pull the transaxle away from the engine and lower it from the vehicle.

To install:

17. Raise the transaxle into position and seat against the rear of the engine.

18. Install the lower transaxle attaching bolts. Torque the bolts to 47–66 ft. lbs. (64–89 Nm).

19. Install the mounting brackets and remove the transmission jack.

20. Install the crossmember and remove the differential plugs.

21. Remove and discard the old halfshaft circlips. Install new circlips and engage the halfshafts with the differential side gears.

22. Connect the lower control arms to the steering knuckles. Install the lower control arm attaching bolts and nuts.

23. Position the shift rod on the input shift rail and install the attaching bolt and nut.

24. Lower the vehicle and remove the engine support bar.

25. Install the 2 bolts at the top of the clutch housing. Torque the bolts 47–66 ft. lbs. (64–89 Nm).

26. Install the starter.

27. Connect the clutch cable to the release lever. Connect the neutral and back-up switch wiring connectors.

28. Remove the speedometer gear and sleeve assembly from the transaxle case bore. With a clean rag, wipe the assembly and reinsert the sleeve into the transaxle. Remove the sleeve and check the oil level. The oil level should be between the **F** and **L** marks on the gear sleeve. If the level is not within the normal operating range, add oil through the speedometer bore as required.

29. Install the speedometer sleeve and gear assembly and connect the speedometer cable.

30. Connect the negative battery cable.

31. Adjust the clutch pedal free-play.

CLUTCH

Clutch Assembly

REMOVAL & INSTALLATION

1. Disconnect the negative battery cable.

2. Remove the transaxle assembly.

NOTE: During the removal procedure, do not allow oil or grease to come in contact with the clutch disc facing if the disc is to be reused. Handle the disc with clean rags wrapped around the edges and do not touch the disc facing. Even a small amount of dirt or grease may cause the clutch to grab or slip.

3. If the pressure plate is to be reused, paint or scribe alignment marks on the pressure plate and flywheel for assembly reference.

4. Install an appropriate locking tool to prevent the flywheel from turning.

5. Loosen the pressure plate attaching bolts in an alternate pattern 1 turn at a time. This will relieve the pressure plate spring tension evenly and prevent distortion of the pressure plate. Remove the pressure plate and clutch disc after the bolts are removed. Replace all clutch components as required.

6. Inspect the flywheel for scoring, cracks and heat checks. Resurface or replace the flywheel, as necessary.

7. Inspect the pilot bearing for damage. Make sure the bearing turns easily. If replacement is necessary, remove the flywheel and remove the pilot bearing.

To install:

8. If necessary, install a new pilot bearing using a suitable installation tool. Use only a driver tool that contacts the bearing outer race. A driver tool that contacts the inner race or the bearing area is unsuitable.

9. If the flywheel was removed, clean the sealant from the flywheel attaching bolts. Coat the bolt threads with a suitable sealer compound.

10. Make sure the crankshaft flange and the back of the flywheel are clean.

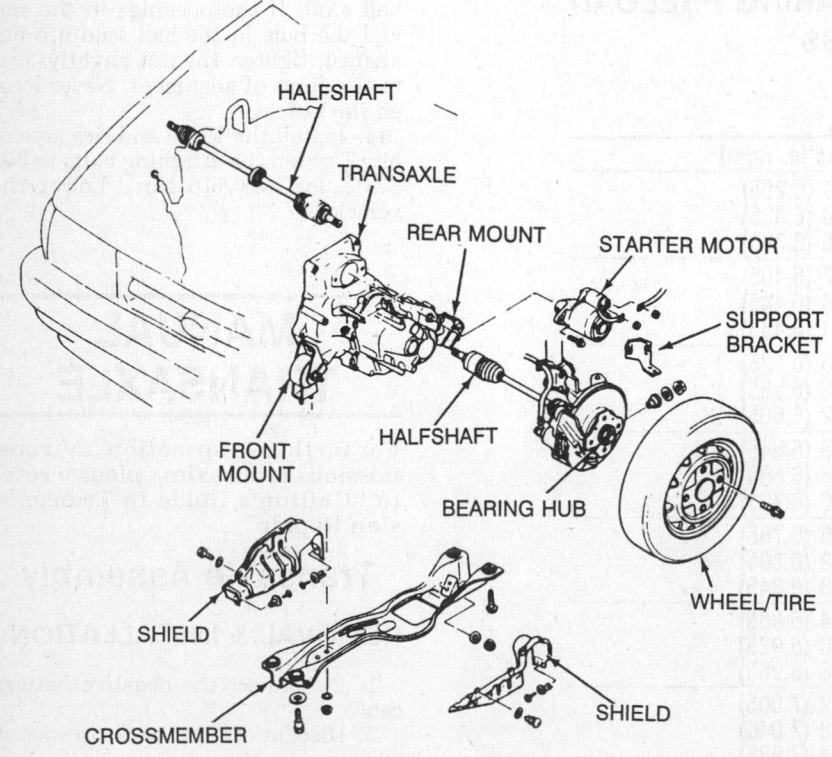

Manual transaxle removal and installation

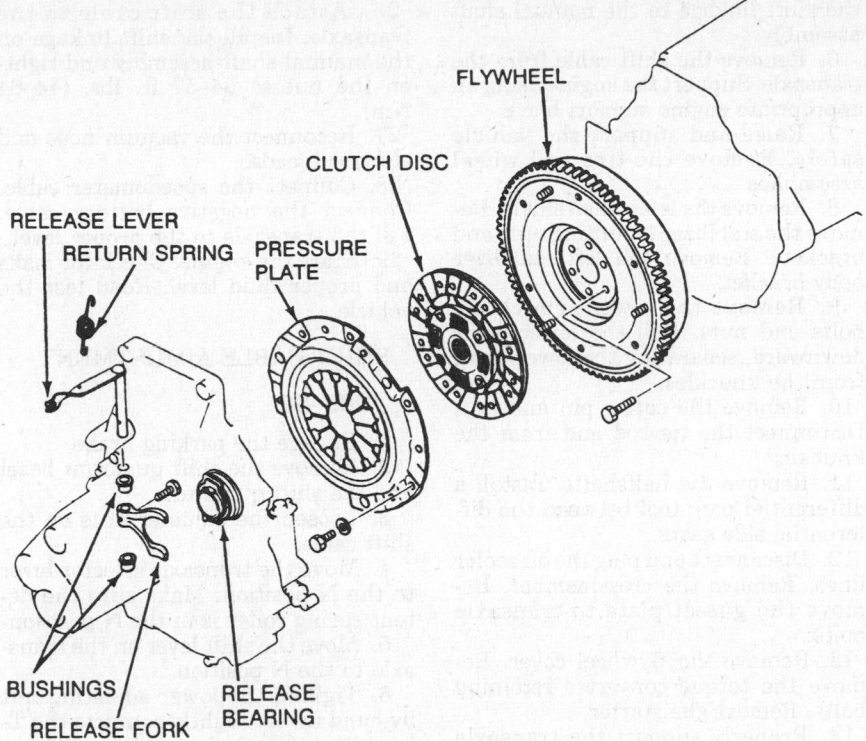

FLYWHEEL

CLUTCH DISC

RELEASE LEVER
RETURN SPRING   PRESSURE PLATE

BUSHINGS
RELEASE FORK   RELEASE BEARING

Exploded view of the clutch assembly

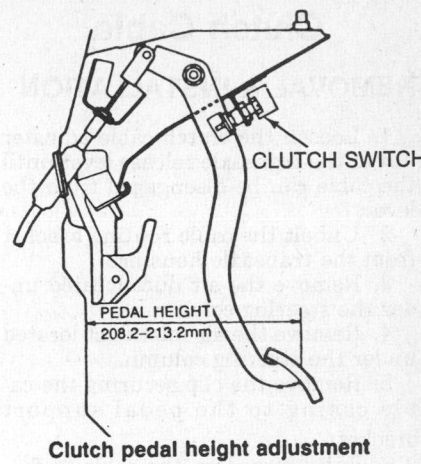

CLUTCH SWITCH

PEDAL HEIGHT
208.2–213.2mm

Clutch pedal height adjustment

CLUTCH SWITCH

PEDAL HEIGHT

PEDAL FREE-PLAY   9–15mm
(0.35–0.59 IN)

DISENGAGEMENT HEIGHT

Clutch pedal free-play adjustment

Position the flywheel on the crankshaft and install the attaching bolts. Tighten the bolts to 71–76 ft. lbs. (96–103 Nm).

11. Position the clutch disc on the flywheel and install a clutch alignment tool to hold the disc in place.

NOTE: When installing the clutch disc, make sure the disc dampener springs are facing away from the flywheel. A new disc will be stamped FLYWHEEL to indicate the correct installation postion.

12. Align the reference marks, if present, and position the pressure plate on the flywheel and install the attaching bolts. Torque the bolts evenly, in an alternate pattern, to 13–20 ft. lbs. (18–26 Nm). The bolts must be tightened in this manner to prevent distortion of the pressure plate.

13. Remove the clutch alignment tool.

14. Clean the clutch disc splines on the input shaft with a dry rag and coat the spline surfaces with clutch grease.

15. Install the transaxle.

16. Connect the negative battery cable.

17. Adjust the clutch pedal free-play.

PEDAL HEIGHT ADJUSTMENT

NOTE: The clutch pedal height is controlled by the clutch switch stop bolt.

1. To eliminate the possibility that the clutch cable is affecting the pedal height, disconnect it at the transaxle release lever. Move the floor carpet and insulation out of the way of the dash panel to ensure an accurate measurement.

2. Measure the distance from the upper center of the pedal to the cowl panel. The pedal height should be from 8.2–8.4 in. (208.2–213.2mm). If the pedal height is within this range, no adjustment is necessary. If the pedal height is not within specification, inspect the clutch pedal mounting for damaged, worn or missing parts. If the mounting appears to be satisfactory, proceed as follows:

   a. Remove the air duct located under the steering column.

   b. Locate the clutch switch and loosen the attaching nuts. Thread the switch in or out until the pedal height is within specification. Tighten the attaching nuts when the correct height is obtained.

   c. Connect the clutch cable to the

transaxle release lever and adjust the pedal free-play.

   d. If the pedal height changes after connecting the clutch cable, check for binding along the cable route.

   e. Install the air duct. Place the insulation and floor carpet in their original positions.

FREE-PLAY ADJUSTMENT

1. Carefully move the clutch pedal back and forth and measure the amount of travel. If the clutch pedal free-play is 0.35–0.59 in. (9–15mm), no adjustment is necessary. If the free-play is not within specification, proceed to Step 2.

2. Pull back the transaxle release lever and measure the clearance between the lever and the cable pin. Thread the adjuster in or out until the clearance between the pin and the lever is 0.06–0.10 in. (1.5–2.5mm).

3. Check the free-play at the clutch. If it is not within specification, inspect the clutch release components for a problem.

Clutch Cable

REMOVAL & INSTALLATION

1. Loosen the clutch cable adjuster nut at the transaxle release lever until the cable can be disengaged from the lever.

2. Unbolt the cable routing bracket from the transaxle housing.

3. Remove the air duct located under the steering column.

4. Remove the air inlet duct located under the steering column.

5. Remove the clip securing the cable casing to the pedal support bracket.

6. Pull upward on the cable to disengage it from the hook on the pedal.

7. If necessary, loosen the attaching nut and remove the routing bracket from the cable. Withdraw the cable through the hole in the bulkhead.

To install:

8. If necessary, position the routing bracket on the cable casing and tighten the attaching nuts.

9. Install the cable. Make sure the instrument panel grommet is properly seated.

10. Pull upward on the cable and hook it over the top of the clutch pedal.

11. Install the cable casing retaining clip.

12. Install the air duct.

13. Connect the cable to the release lever. Check the clutch pedal free-play and adjust if necessary.

AUTOMATIC TRANSAXLE

For further information on transmissions/transaxles, please refer to "Chilton's Guide to Transmission Repair".

Transaxle Assembly

REMOVAL & INSTALLATION

1. Disconnect the negative battery cable. Loosen the front wheel bolts.

2. Drain the transaxle fluid. Disconnect the speedometer cable from the transaxle.

3. Disconnect the transaxle electrical connectors, which are located next to the governor.

4. Disconnect the transaxle ground wire. Disconnect the transaxle vacuum hose.

5. Remove the nut which connects the shift linkage to the manual shaft assembly.

6. Remove the shift cable from the transaxle. Support the engine using an appropriate engine support bar.

7. Raise and support the vehicle safely. Remove the tire and wheel assemblies.

8. Remove the left splash shield. Remove the stabilizer mounting nuts and brackets. Remove the left stabilizer body bracket.

9. Remove the lower arm clamp bolts and nuts. Pull the lower arms downward, separating the lower arms from the knuckles.

10. Remove the cotter pin and nut. Disconnect the tie rod end from the knuckle.

11. Remove the halfshafts. Install a differential plug tool between the differential side gears.

12. Disconnect and plug the oil cooler lines. Remove the crossmember. Remove the gusset plate to transaxle bolts.

13. Remove the flywheel cover. Remove the torque converter retaining bolts. Remove the starter.

14. Properly support the transaxle assembly.

15. Remove the engine-to-transaxle retaining bolts. Carefully remove the transaxle from the vehicle.

To install:

16. Position the transaxle under the vehicle. Install the engine-to-transaxle bolts. Tighten to 41–59 ft. lbs. (55–80 Nm).

17. Install the starter. Install the torque converter bolts and tighten to 26–36 ft. lbs. (34–49 Nm).

18. Install the flywheel cover and tighten the bolts to 61–87 inch lbs. (7–10 Nm).

19. Install the crossmember and tighten the bolts to 47–66 ft. lbs. (64–89 Nm). Install the front engine mount-to-crossmember attaching nuts and tighten to 32–38 ft. lbs. (43–52 Nm). Install the rear engine mount-to-crossmember attaching nut and tighten to 21–34 ft. lbs. (28–46 Nm).

20. Install the halfshafts. Connect the oil cooler lines.

21. Connect the tie rod ends to the steering knuckles and tighten the attaching nuts to 26–30 ft. lbs. (35–40 Nm). Install new cotter pins.

22. Attach the lower arm ball joints to the knuckles. Tighten the lower arm clamp nuts and bolts to 32–40 ft. lbs. (43–54 Nm).

23. Install the stabilizer body bracket and mounting nuts. Tighten the nuts to 40–50 ft. lbs. (54–68 Nm).

24. Install the splash shield and the front wheels.

25. Lower the vehicle. Remove the engine support tool.

26. Attach the shift cable to the transaxle. Install the shift linkage on the manual shaft assembly and tighten the nut to 34–57 ft. lbs. (44–64 Nm).

27. Reconnect the vacuum hose and electrical leads.

28. Connect the speedometer cable. Connect the negative battery cable. Fill the transaxle to the proper level.

29. Start the engine. Check for leaks and proper fluid level. Road test the vehicle.

SHIFT CABLE ADJUSTMENT

1988–89

1. Engage the parking brake.

2. Remove the shift quadrant bezel and the shift quadrant.

3. Loosen the adjuster nuts on the shift cable.

4. Move the transaxle selector lever to the **N** position. Make sure the detent spring roller is in the **N** position.

5. Move the shift lever on the transaxle to the **N** position.

6. Tighten the lower adjusting nut by hand until it lightly contacts the T-joint, then loosen it a half turn. Tighten the upper adjuster nut to 69–95 inch lbs. (8–11 Nm).

7. Press the selector interlock button and push the selector lever toward **R** with a force of 4.4 lbs. Note the distance that the selector lever has moved. The distance of movement should be no more than 0.31 in. (8mm). Move the selector lever back to **N**.

8. Pull the selector lever toward **D** in the same way as in Step 7. Note the distance that the selector lever has moved. The distance of movement should be no more than 0.31 in. (8mm).

9. Compare the distances noted in Steps 7 and 8. If the distance recorded from **N** to **R** is larger than the distance recorded from **N** to **D**, loosen the upper adjuster nut and tighten the lower adjuster nut so the larger distance becomes smaller.

10. If the distance recorded from **N** to **D** is larger than the distance recorded from **N** to **R**, loosen the lower adjuster nut and tighten the upper adjuster nut so the larger distance becomes smaller.

11. Check the manual linkage operation. If the selector lever does not shift smoothly, set the selector lever to **P**. Loosen the attaching screws on the detent spring and roller assembly. Adjust the position of the detent spring roller.

12. If the position of the detent spring roller is adjusted, repeat Steps 3–10.

NOTE: Make sure the linkage adjustment has not affected operation of the neutral safety switch. With the parking brake and service brakes applied, try to start the engine in each gearshift position. The engine must crank only in the N and P positions. If the engine cranks in any other position, check the linkage adjustment and neutral safety switch operation.

13. Install the shift quadrant and the shift quadrant bezel.

1990–92

1. Disconnect the negative battery cable.
2. Remove the shift lever knob and the shift console attaching screws. Remove the shift console.
3. Shift the selector lever to **P**.
4. Remove the 4 shift quadrant attaching screws and the shift quadrant.

NOTE: Make sure the detent spring roller is in the **P** detent.

5. Loosen adjustment nuts "A" and "B" until they reach the ends of the cable thread.
6. Move the shift lever on the transaxle to the **P** position.
7. Tighten adjustment nut "A" by hand until it lightly contacts the T-joint, then tighten adjustment nut "B" to 80–97 inch lbs. (9–11 Nm).
8. Lightly press the selector pushrod and make sure the guide plate and guide pin clearances are within specification.
9. Check that the plate and pin

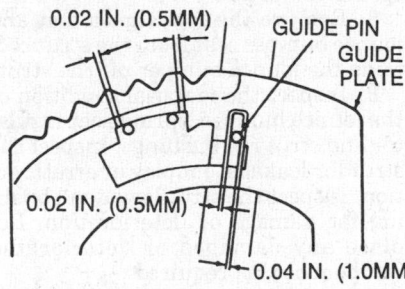

Guide plate and guide pin clearances— 1990–92 vehicles

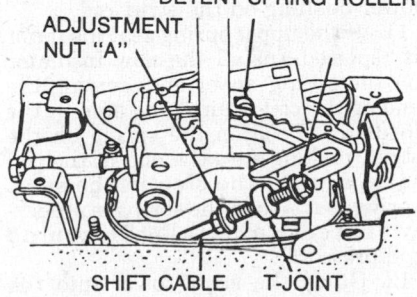

Shift linkage adjustment points

clearances are within the same specifications when the selector lever is shifted to **N** and **D**. If the clearances are not as specified, readjust the shift cable.
10. Make sure the selector lever operates properly.
11. Install the shift quadrant, shift console and selector lever knob.
12. Connect the negative battery cable.

FRONT SUSPENSION

MacPherson Strut

REMOVAL & INSTALLATION

1. Raise the vehicle and support it safely.
2. Remove the wheel and tire assembly.
3. Remove the brake line clip from the strut lower mounting bracket and disengage the brake line.
4. Remove the 2 nuts and bolts securing the strut lower bracket to the steering knuckle.
5. In the engine compartment, remove the 2 nuts securing the strut mounting block in the strut tower.
6. Disengage the strut lower bracket from the steering knuckle and lower the strut clear of the wheel well.
To install:
7. Place the strut assembly with spacer plate in the strut tower with the white alignment mark facing outward.
8. Install the upper mounting block stud nuts and torque to 22–27 ft. lbs. (29–36 Nm).
9. Engage the steering knuckle in the strut tower lower bracket and install the mounting bolts and nuts. Torque to 69–86 ft. lbs. (93–117 Nm).
10. Position the brake line into the strut lower mounting bracket cutout and install the retaining clip.
11. Install the wheel and tire assembly and lower the vehicle.

Lower Ball Joints

NOTE: The ball joint is an integral part of the control arm. If inspection proves the ball joint to be bad, the entire lower control arm must be replaced.

INSPECTION

Control Arm Installed

Check for ball joint wear by raising

and safely supporting the vehicle until the wheel and tire assembly is clear of the floor. Support the lower control arm so there is no load on the suspension strut. Try to rock the wheel top-to-bottom; if any wobble is felt, look for movement between the control arm and steering knuckle. If the ball joint appears tight, check and adjust the wheel bearing preload, then repeat the wobble check. Any movement still present is a sign of ball joint wear. Replace the lower control arm.

Control Arm Removed

Make sure the ball joint stud swivels freely but is not loose. Grip the ball joint stud with a suitable adapter and check the stud rotating torque with a low-reading torque wrench. It should be in the range of 16–27 inch lbs. (1.8–3.1 Nm).

Lower Control Arms

REMOVAL & INSTALLATION

1. Raise and support the vehicle safely. Remove the lower control arm pivot bolt at the frame bracket.
2. Remove the ball joint clamp bolt and and nut from the steering knuckle assembly.
3. Remove the stabilizer bar bushing retaining nut from the rear of the control arm and remove the rear bushing washer and bushing.
4. Lower the control arm, prying the ball joint stud out of the steering knuckle, if necessary. Disengage and remove the control arm from the stabilizer end.
5. Inspect the control arm for deformation or cracks and check the pivot bushing for deterioration. Verify that the ball joint swivels freely but is not loose. If the control arm pivot bushing is to be replaced, remove the old bushing with C-frame tool T74P-3044-A1, bushing tool T81P-5493-B2 and receiver cup tool T88C-5493E or equivalents. Center the new bushing in the center of the control arm eye and install using the removal tools. Replace the lower control arm/ball joint assembly as required.
6. If the ball joint boot is damaged or deteriorated, pry the boot off with a small cold chisel. Install the new boot onto the ball joint using a suitable adapter such as a ¾ in. socket to properly seat the boot.
To install:
7. Position the front bushing washer and bushing onto the stabilizer end. Engage the control arm with the stabilizer.
8. Raise the control arm inner end into the pivot bracket on the frame and start the pivot bolt to hold the con-

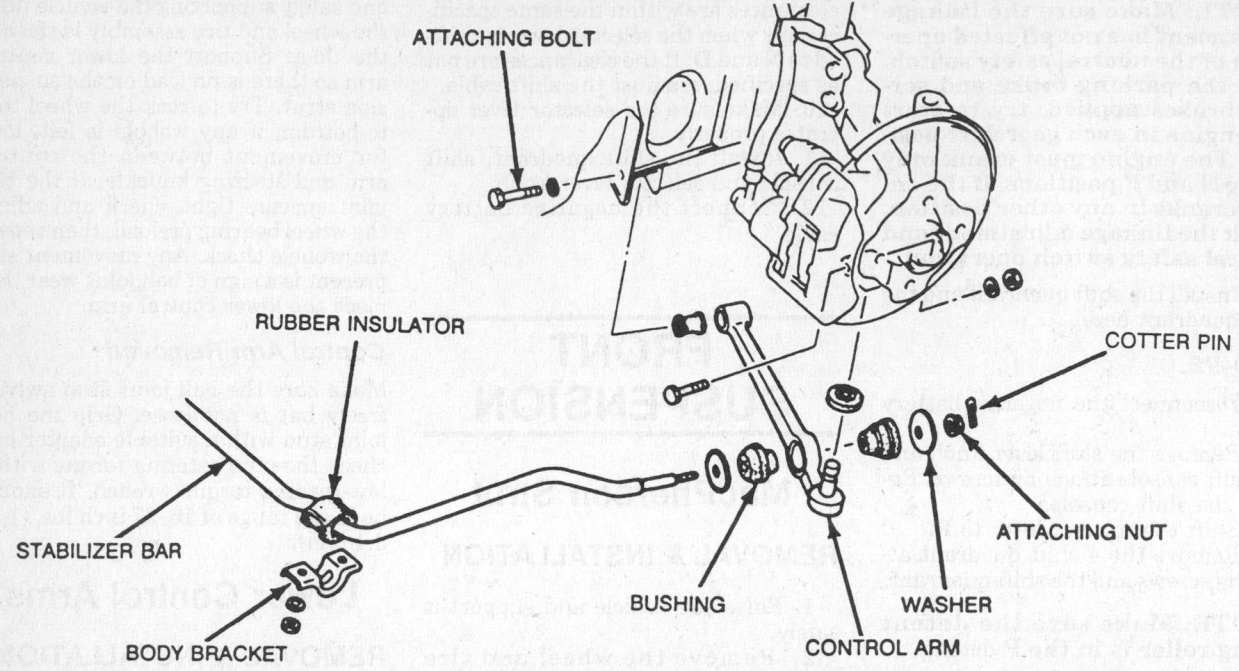

ATTACHING BOLT

RUBBER INSULATOR

COTTER PIN

STABILIZER BAR

ATTACHING NUT

BODY BRACKET

BUSHING

WASHER

CONTROL ARM

Lower control arm and stabilizer assembly

trol arm in place. Do not completely tighten the bolt at this time.

9. Engage the control arm ball joint stud with the clamp bore in the steering knuckle and install the clamp bolt and nut.

10. Install the stabilizer rear bushing and washer onto the stabilizer end with the retaining nut. Torque the retaining nut to 47–57 ft. lbs. (64–77 Nm).

11. Torque the pivot bolt at the control arm frame bracket to 32–40 ft. lbs. (43–54 Nm).

12. Hold the steering clamp bolt stationary and torque the clamp nut to 32–40 ft. lbs. (43–54 Nm).

13. Lower the vehicle.

Stabilizer Bar

REMOVAL & INSTALLATION

1. Raise and safely support the vehicle.

2. Remove the stabilizer mounting bracket nuts and mounting brackets.

3. Remove the split bushings from the stabilizer bar. Replace deteriorated or worn bushings as required.

4. Remove the stabilizer bushing nuts at the lower control arms and remove the rear washers and bushings.

5. Pull the stabilizer bar forward to disengage it from both lower control arms. Remove the bushings and washers. Replace deteriorated or worn bushings as required.

To install:

6. Install the control arm bushing

washers on the ends of the stabilizer bar and install the control arm front bushings.

7. Support the stabilizer bar by hand and insert the ends of the bar into the lower control arms. Install the control arm bushings and washers with the retaining nuts. Make the retaining nuts finger-tight.

8. Install the split bushings on the the stabilizer bar cross bar with the split side forward and position them next to the white alignment marks on the bar.

9. Install the stabilizer bar mounting brackets. Torque the bracket retaining nuts to 40–50 ft. lbs. (54–68 Nm).

10. Torque the control arm bushing retaining nuts to 47–57 ft. lbs. (64–77 Nm).

11. Lower the vehicle.

REAR SUSPENSION

MacPherson Strut

REMOVAL & INSTALLATION

1. Raise the vehicle and support it safely.

2. Remove the rear wheel and tire assembly.

3. Install a spring compressor tool

and release the strut spring tension.

4. From the cargo compartment, remove the rear quarter trim panel.

5. Remove the jam nut and flanged nut from the strut rod and remove the bushing washer and upper bushing.

6. Remove the strut lower end mounting bolt from the torsion beam.

7. Withdraw the strut assembly downward and separate it from the spring and seat insulator. Remove the spring compressor.

8. Remove the lower grommet and jounce bumper seat from the strut rod. Slide the jounce bumper off the strut.

9. Inspect the material condition of the jounce bumper, spring seat insulator and strut rod bushings. Inspect the strut for leakage, endplay or erratic action. Inspect the strut lower end bushing for damage or deterioration. Replace any damaged or deteriorated components, as required.

To install:

10. Slide the jounce bumper onto the strut rod. Install the bumper seat and lower bushing on the strut rod.

11. If the upper spring seat insulator is replaced, install the new insulator on the spring upper end, seating the end of the coil against the step in the insulator. Position the spring on the strut, making sure the end of the coil seats against the step in the strut spring seat. When the spring is properly seated, reinstall the spring compressor.

12. Guide the strut tower into the strut mounting hole through the wheel well.

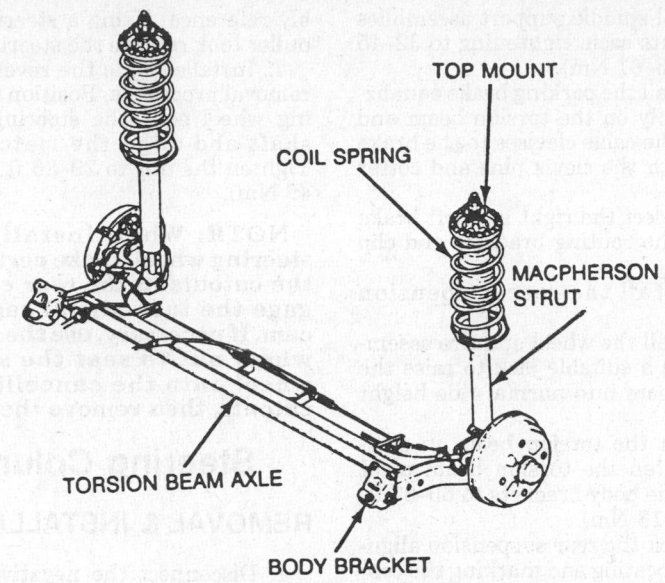

TOP MOUNT

COIL SPRING

MACPHERSON STRUT

TORSION BEAM AXLE

BODY BRACKET

Rear suspension assembly

13. Align the strut lower end with the mounting hole in the torsion beam. Start the mounting bolt in by hand to hold the strut in position.

14. From the cargo compartment, install the rod upper end bushing, bushing washer and flanged nut. Torque the flanged nut to 12–18 ft. lbs. (16–24 Nm). Hold the flanged nut stationary and tighten the locknut.

15. Torque the lower strut mounting bolt to 40–50 ft. lbs. (54–68 Nm).

16. Back off on the spring compressor slowly to release the spring tension. Remove the spring compressor.

17. Install the rear quarter trim panel.

18. Install the wheel and tire assembly. Lower the vehicle.

Rear Wheel Bearings

REMOVAL & INSTALLATION

1. Raise the vehicle and support it safely. Make sure the parking brake is fully released.

2. Remove the wheel and tire assembly.

3. Remove the grease/dust cap.

4. With a small cape chisel, carefully raise the staked portion of the locknut.

NOTE: The drum/hub locknuts are threaded left and right. The left hand threaded locknut is located on the right side of the vehicle. Turn this locknut clockwise to loosen. The right hand threaded locknut is located on the left side of the vehicle and is turned counterclockwise to loosen.

5. Remove the locknut and washer. Discard the locknut.

6. Pull the brake drum bearings and hub assembly away from the spindle shaft. Take care not to damage the spindle shaft threads.

7. With a small roll head prybar or equivalent, remove the bearing grease seal from the bearing hub. Discard the seal regardless of condition.

8. Remove the inner and outer bearings from the bearing hub. If the bearings are to be reused, identify and tag each bearing for installation reference. Replace worn or damaged bearings as required.

9. If the bearings are being replaced, remove the bearing races using a suitable tool.

To install:

10. If the bearings are being replaced, install the new bearing races in the hub using suitable installation tools.

11. Pack the bearings and the hub with high temperature wheel bearing grease.

12. Position the inner bearing in the hub. Install and seat a new grease seal with a suitable driving tool. Lubricate the lip of the seal with the wheel bearing grease.

13. Position the brake drum and hub assembly on the spindle. Keep the hub centered during positioning to prevent damage to the new grease seal and spindle threads.

14. Install the outer bearing, lockwasher and new locknut.

15. Adjust the bearing preload.

16. Install the grease cap, wheel and tire assembly.

17. Lower the vehicle and connect the negative battery cable.

ADJUSTMENT

A staked attaching nut and a flat washer are used to hold the bearings and hub in position on the spindle shaft. The attaching nuts are left and right hand thread. The left hand threaded nut, located on the right side of the vehicle, must be turned counterclockwise to tighten and the right hand threaded nut, located on the left side of the vehicle, must be turned clockwise to tighten.

1. Make sure the parking brake is fully released.

2. Raise the vehicle and support it safely. Remove the wheel and tire assembly.

3. Remove the grease cap. Rotate the brake drum to make sure there is no brake drag.

4. With a small cape chisel, carefully raise the staked portion of the locknut.

5. Remove the locknut and discard. Install a new locknut.

6. To seat the bearings, torque the locknut to 18–22 ft. lbs. (25–29 Nm). Rotate the brake drum by hand while tightening the locknut.

7. Loosen the locknut until it can be turned by hand.

8. Before the bearing preload can be set, the amount of seal drag must be measured and added to the the required preload. To measure the seal drag proceed as follows:

a. Install the proper size nut onto a wheel stud and rotate the brake drum until the stud is in the 12 o'clock position.

b. Place an inch pound torque wrench onto the nut to measure the amount of force required to rotate the break drum.

c. Pull the torque wrench and note and record the torque reading when rotation begins. This value will be used to calculate the bearing preload range.

9. The required preload range, without seal drag, is 1.3–4.3 inch lbs. (0.15–0.49 Nm). To calculate the preload, add the seal drag value obtained in Step 8c to the minimum and maximum preload specifications. For example, if the seal drag was 2.2 inch lbs. (0.25 Nm), then the minimum preload specification would be 1.3 inch lbs. (0.15 Nm) + 2.2 inch lbs. (0.25 Nm) = 3.5 inch lbs. (0.40 Nm) and the maximum preload specification would be 4.3 inch lbs. (0.49 Nm) + 2.2 inch lbs. (0.25 Nm) = 6.5 inch lbs. (0.74 Nm). Therefore, for a seal drag of 2.2 inch lbs. (0.25 Nm), the bearing preload should be within the range of 3.5–6.5 inch lbs. (0.40–0.74 Nm).

10. After the preload range is determined, tighten the locknut slightly.

11. Rotate the brake drum until the

nut and wheel are returned to the 12 o'clock position. Position the inch lb. torque wrench onto the nut and measure the amount of pull required to rotate the brake drum. Tighten the locknut until the torque shown on the torque wrench is within the range that was calculated in Step 9.

12. With the proper tool, stake the locknut in place.

NOTE: If the nut splits or cracks after staking, it must be replaced with a new nut.

13. Install the grease cap.
14. Install the wheel and tire assembly. Lower the vehicle.

Torsion Beam

REMOVAL & INSTALLATION

1. Raise and safely support the vehicle.
2. Remove the wheel and tire assemblies.
3. Remove the rear struts and disconnect the brake lines.
4. Disconnect the parking brake cable clevises at the brake backing plates.
5. Remove the parking brake equalizer and cables from the torsion beam.
6. Remove the 4 nuts from the back of each brake assembly to release the backing plates and wheel spindle supports.
7. Remove the torsion beam pivot bolts from the body brackets and carefully lower the torsion beam from the vehicle.

NOTE: If the torsion beam body brackets are not to be replaced, it may be desirable to leave them in place on the body. The bracket mounting holes are slotted to permit side-to-side adjustment of the torsion beam for true tracking of the rear suspension. If removed, they require alignment when the torsion beam is installed. However, if the torsion beam is repaired or replaced, the alignment must be checked at assembly.

To install:

8. If removed, install the torsion beam pivot brackets on the body with flatwashers, lockwashers and 3 bolts on each side. Do not tighten the mounting bolts at this time.
9. If installing a new torsion beam, install the pivot bushings in the beam arms.
10. Install the bushing flange washers and position the beam arms in the body brackets. Align the pivot bolt holes and install the bolts but do not tighten the nuts yet.
11. Install the brake backing plates

and wheel spindle support assemblies with 4 nuts each, tightening to 32–45 ft. lbs. (43–61 Nm).

12. Install the parking brake equalizer assembly on the torsion beam and connect the cable clevises to the brake levers with the clevis pins and cotter pins.
13. Connect the right and left brake lines at the routing brackets and clip in place.
14. Install the rear suspension struts.
15. Install the wheel and tire assemblies. Use a suitable jack to raise the torsion beam into normal ride height position.
16. With the torsion beam in position, tighten the torsion beam pivot bolts at the body brackets to 69–87 ft. lbs. (93–118 Nm).
17. Check the rear suspension alignment by locating and marking the center of the underbody, at a point equidistant from the right and left body bracket inboard mounting bolts. From this point, measure the distance to the centers of the strut lower mounting bolts, right and left. If these measurements are not within 0.2 in. (5mm), shift the torsion beam body brackets side-to-side to center the suspension.
18. When centered, tighten the body bracket mounting bolts, the upper bolts to 40–50 ft. lbs. (54–68 Nm) and the lower bolt to 69–87 ft. lbs. (93–118 Nm).
19. Bleed the rear brakes and lower the vehicle.

STEERING

Steering Wheel

REMOVAL & INSTALLATION

1. Disconnect the negative battery cable.
2. On 1988–89 vehicles, pry off the trim insert in the center of the steering wheel cover. Take care not to damage the cover.
3. On 1990–92 vehicles, remove the 2 screws from the back of the steering wheel. Disconnect the horn wire and remove the steering wheel cover.
4. Remove the steering wheel nut.
5. On 1988–89 vehicles, remove the attaching screws and washers locate to the left and right of the steering column stud. Remove the 2 screws from the back of the steering wheel spokes. Disconnect the horn wire and remove the cover assembly.
6. Matchmark the steering wheel and steering column shaft for assem-

bly reference. Using a steering wheel puller tool, remove the steering wheel.

7. Installation is the reverse of the removal procedure. Position the steering wheel onto the steering column shaft and align the matchmarks. Tighten the nut to 29–36 ft. lbs. (39–49 Nm).

NOTE: When installing the steering wheel, make certain that the cutouts in the rear cover engage the turn signal cancelling cam. If necessary, use the steering wheel nut to seat the steering wheel onto the cancelling cam cutouts, then remove the nut.

Steering Column

REMOVAL & INSTALLATION

1. Disconnect the negative battery cable.
2. Remove the steering wheel, combination switch and ignition switch.
3. On 1988–89 vehicles, remove the instrument panel spacer brace and air duct from below the steering column.
4. On 1990–92 vehicles, remove the steering column shield and the air duct from below the steering column.
5. Remove the 2 nuts securing the steering column upper mounting bracket to the instrument panel crossmember. When free, the upper end of the column may be lowered as needed for access to the intermediate shaft universal joint at the lower end.
6. With paint or marking pen, make an index mark at the juncture of the steering column shaft and the intermediate shaft upper universal joint to assure correct alignment during assembly. Remove the universal joint clamp screw.
7. Loosen the 2 nuts securing the steering column hinge bracket to the clutch/brake pedal support. Remove the steering column assembly by pulling to the rear, disengaging it from the universal joint. Remove the shim clips from the upper mounting bracket.

To install:

8. Install the joint clamp bolt but do not tighten it at this time as it may need to be shifted up or down on the shaft to line up with the steering column without binding.
9. Install the steering column, aligning the index marks on the column shaft and universal joint and engaging the column hinge bracket with the pedal support studs. Do not tighten the universal joint clamp bolt yet.
10. Tighten the hinge bracket nuts and raise the upper end of the column to seat under the instrument panel. Position the shim clips on the column upper bracket flanges.

11. Install the 2 steering column upper retaining nuts.

12. Turn the steering wheel lock-to-lock several times to align the universal joints, then tighten both universal joint clamp bolts.

13. Install the instrument panel brace or steering column shield, as necessary. Install the air duct.

14. Install the ignition switch, combination switch and the steering wheel.

Manual Steering Rack and Pinion

ADJUSTMENT

Only the rack preload is adjustable and only to a limited degree, since it is primarily determined by the yoke spring. Since adjustment requires removal of the steering gear, it should only be undertaken after a thorough inspection of front suspension and steering column components fails to reveal damage or binding elsewhere. If necessary, adjust the rack yoke preload as follows:

1988–89

1. Remove the steering rack from the vehicle.

2. Center the steering rack in a protected jaw vise, make sure there is equal left and right tie rod extension.

3. Measure the pinion operating torque with an inch lb. torque wrench and pinion torque adapter tool T87C–3504–C or equivalent. Within 90 degrees of the centered rack position, pinion torque should be 8–11.5 inch lbs. (0.9–1.3 Nm). Beyond 90 degrees, left or right, pinion torque should not exceed 13.3 inch lbs. (1.5 Nm).

4. If the pinion torque is not within the specified limits, tighten or loosen the rack adjusting screw to increase or decrease the rack preload.

NOTE: Do not loosen the adjusting screw so that it no longer makes contact with the yoke spacer. Any clearance at this point will allow the rack to deflect under load, resulting in reduced tooth engagement with the pinion.

5. When the pinion operating torque is within specification, tighten the jam nut on the adjusting screw. With a suitable adapter, torque the jam nut to 7.4–11 ft. lbs. (10–15 Nm) to retain the adjustment.

6. Install the steering rack in the vehicle.

1990–92

1. Remove the steering rack from the vehicle.

2. Center the steering rack in a protected jaw vise, make sure there is equal left and right tie rod extension.

3. Remove the locknut and the yoke plug and clean the yoke plug threads. Apply sealant to the yoke plug threads and install the yoke plug. Tighten to 78–95 inch lbs. (9–11 Nm).

4. Slowly cycle the rack back and forth through 90 percent of it's full stroke. Then center the rack so the tie rods are equally extended.

5. Loosen the yoke plug, then tighten it to 22–30 inch lbs. (2.5–3.4 Nm).

6. Use a spring scale to measure the force needed to turn the pinion 180 degrees from the rack center position.

7. Adjust the pinion to the position where the most force was needed to turn it.

8. Tighten the yoke plug to 48 inch lbs. (5.4 Nm). Install the locknut and tighten to 29–36 ft. lbs. (39–49 Nm).

REMOVAL & INSTALLATION

1. Disconnect the negative and positive battery cables and remove the battery from the vehicle.

2. Matchmark the steering column lower universal joint and steering rack pinion for assembly reference. Remove the steering column and intermediate shaft assembly from the vehicle.

3. Cut the plastic tie wrap securing the steering column boot to the steering rack.

4. Raise the vehicle and support safely. Remove the front tire and wheel assemblies.

5. Using the proper tool, separate both tie rod ends from the steering knuckles.

6. Remove the catalytic converter.

7. Remove the plastic tie rod splash shield from the right inner fender.

8. Remove the steering rack mounting bolts and lower the steering rack until it is free of the steering column boot. Slide the rack to the right, through the inner fender tie rod opening, until the left tie rod is clear of the left inner fender, then lower the left end until the steering rack assembly can be withdrawn from the left side of the vehicle.

NOTE: While maneuvering the tie rod boots in and out of the inner fender openings, guide the steering rack assembly carefully to avoid cutting or nicking the boots.

To install:

9. From under the vehicle, insert the right side tie rod through the right inner fender tie rod opening, far enough to allow raising the left end of the assembly to enter the left inner fender opening. Shift the assembly to the left taking care not to catch the boots.

10. Align the steering rack pinion shaft housing with the steering column boot. Raise the steering rack into the boot.

11. Install the steering rack mounting bolts from left to right. Torque the bolts to 23–34 ft. lbs. (31–46 Nm).

12. Connect the tie rod ends to the steering knuckles. If the tie rod ends are not properly aligned with the knuckle ends during installation, release the small end boot clips before rotating the tie rods. This is done to avoid twisting the boots.

13. Attach the right side tie rod splash shield on the right inner fender panel.

14. Install the catalytic converter.

15. Install the tire and wheel assemblies and lower the vehicle.

16. Secure the steering column boot to the steering rack housing with a new tie wrap.

17. Align the matchmarks made on the steering column lower universal joint and the steering rack pinion shaft. Install the steering column when the proper alignment is acheived.

18. Install the battery and connect the battery cables.

Power Steering Rack and Pinion

REMOVAL & INSTALLATION

1. Disconnect the negative battery cable.

2. Remove the intermediate shaft.

3. Disconnect and plug the high pressure and return lines.

4. Raise the vehicle and support safely. Remove the front tire and wheel assemblies.

5. Using the proper tool, separate both tie rod ends from the steering knuckles.

6. Remove the tie rod end splash shields and the right fender splash shield.

7. Remove the front catalytic converter nuts and separate the converter from the inlet pipe.

8. Place alignment marks on the right tie rod end to ease installation. Loosen the jam nut and remove the right tie rod end.

9. Remove the steering rack mounting bolts and lower the steering rack until it is free of the steering column boot. Slide the rack to the left and pull the right tie rod through the fender

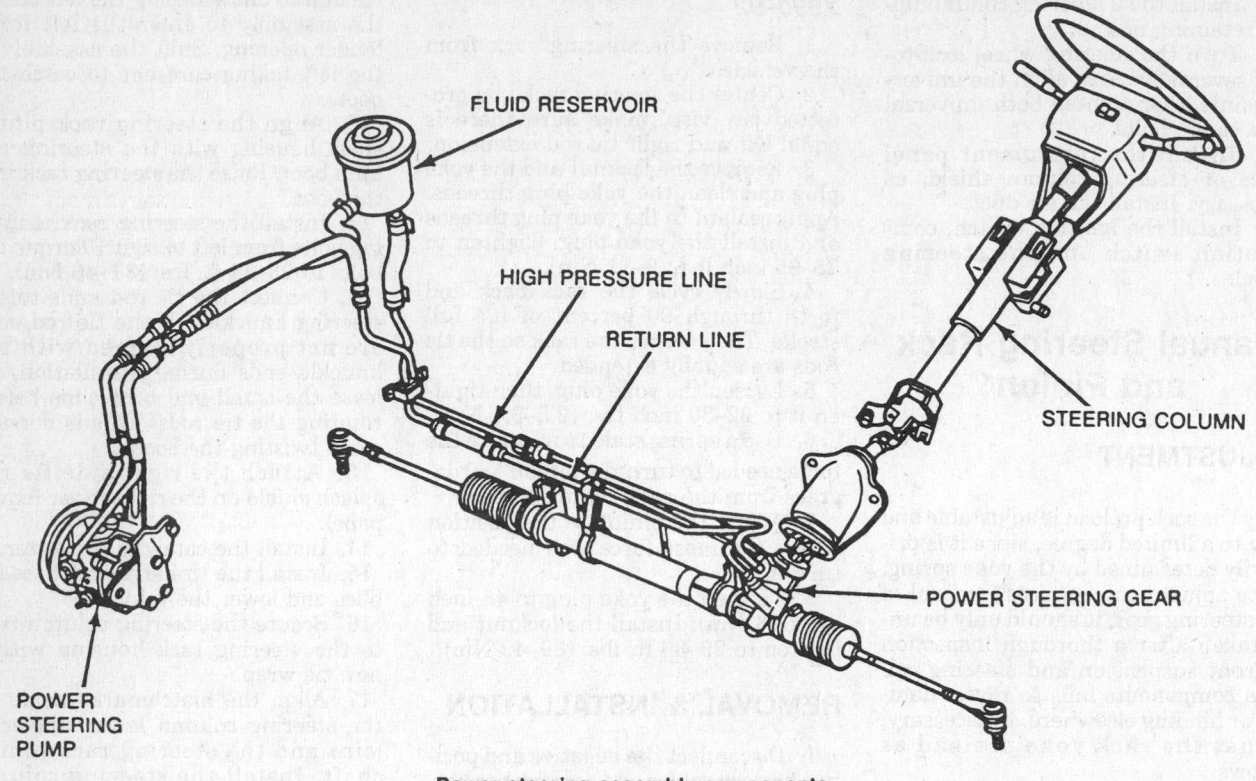

FLUID RESERVOIR

HIGH PRESSURE LINE

RETURN LINE

STEERING COLUMN

POWER STEERING GEAR

POWER STEERING PUMP

Power steering assembly components

opening. Remove the steering gear by sliding it to the right.

To install:

10. Position the steering rack in it's mounting location.

11. Attach the intermediate shaft to the steering gear pinion and tighten the clamp bolt to 13–20 ft. lbs. (18–26 Nm). Guide the intermediate shaft into the steering column hole.

12. Lower the vehicle.

13. With an assistant lifting the steering gear, align the intermediate shaft with the universal joint and install the clamp bolt.

14. Raise and safely support the vehicle.

15. Install the steering rack mounting bolts and tighten to 23–34 ft. lbs. (31–46 Nm).

16. Install the right tie rod end and attach the tie rod ends to the steering knuckles. Install the tie rod end nuts and tighten to 23–34 ft. lbs. (39–44 Nm). Install new tie rod end cotter pins.

17. Attach the catalytic converter to the inlet pipe and install the attaching nuts. Tighten the nuts to 23–34 ft. lbs. (31–46 Nm).

18. Install the tie rod end splash shields and the right fender splash shield.

19. Install the front wheel and tire assemblies and lower the vehicle.

20. Connect the high pressure and return lines.

21. Connect the negative battery cable.

22. Add power steering fluid and bleed the air from the system. Check for leaks.

Power Steering Pump

REMOVAL & INSTALLATION

1. Disconnect the negative battery cable.

2. Remove the air duct and air cleaner unit.

3. Disconnect the electrical connector from the fluid pressure switch.

4. Disconnect and plug the fluid lines.

5. Remove the adjustment bolt and the locknut, washer and bracket bolt.

6. Loosen the mounting bolt and disconnect the drive belt. Remove the mounting bolt and remove the power steering pump.

7. Installation is the reverse of the removal procedure. Tighten the high pressure line nuts to 29–36 ft. lbs. (39–49 Nm). Adjust the drive belt tension.

BELT TENSION ADJUSTMENT

1. Remove the air duct and air cleaner.

2. Loosen the pump mounting bolt. Loosen the adjusting locknut.

3. Using a belt tension gauge or the deflection method, adjust the tension at the adjusing bolt.

4. If using a belt tension gauge, set new belt tension to 110–132 lbs. or used belt (more than 10 minutes of run time) tension to 95–110 lbs.

5. If using the deflection method, apply approximately 22 lbs. of pressure to the middle of the longest accessible belt span. Adjust the tension to 0.31–0.35 in. (8–9mm) for a new belt or 0.35–0.39 in. (9–10mm) for a used belt (more than 10 minutes of run time).

6. Tighten the pump mounting bolt to 27–40 ft. lbs. (36–54 Nm) and the adjustment locknut to 27–38 ft. lbs. (37–52 Nm).

7. Install the air cleaner and air duct.

SYSTEM BLEEDING

1. Add power steering fluid to the **L** mark on the reservoir cap dipstick.

2. Run the engine until it reaches normal operating temperature.

3. Turn the steering wheel lock to lock approximately 10 times.

4. Shut the engine off with the wheels in the straight ahead position.

5. Check the fluid level, the level should be between the **L** and **H** marks on the reservoir cap dipstick. Repeat the procedure if needed.

Tie Rod Ends

REMOVAL & INSTALLATION

1. Raise the vehicle and support it safely.

2. Remove the wheel and tire assembly.

3. Remove the cotter pin and nut from the tie rod end stud. Discard the cotter pin. Examine the nut for damage and replace as required.

4. Separate the tie rod end from the steering knuckle using a suitable tool.

5. With paint or a suitable marker, mark the tie rod end, jam nut and tie rod to ease assembly without changing the toe-in setting.

6. Loosen the jam nut and unscrew the tie rod end counting the number of turns required for removal. Replace the tie rod end as required.

NOTE: If new tie rod ends are being installed, place the old and new ends side-by-side and place alignment marks in the new end that match as closely as possible to the marks on the old end. Please be advised that the existing jam nut may not seat in exactly the same position on the new end and the toe-in setting may have to be checked and/or readjusted as a precaution.

To install:

7. When replacing a tie rod end, install a new dust boot over the stud with a suitable adapter. A ¾ in. socket will accomplish the task simply and effectively.

8. Thread the jam nut and tie rod end onto the tie rod observing the alignment marks and the number of turns required for installation.

9. Install the tie rod end into the steering knuckle. If the tie rod is correctly aligned, the taper should seat without twisting the tie rod or boot. Torque the stud nut to 26–30 ft. lbs. (35–40 Nm) and install a new cotter pin. If the cotter pin does not align with stud bore, tighten (do not loosen) the nut until the castellations align with the pin bore.

10. Install the wheel and tire assembly. Lower the vehicle and connect the negative battery cable. Check the toe-in setting, if necessary.

BRAKES

For all brake system repair and service procedures not detailed below, please refer to "Brakes" in the Unit Repair section.

Master Cylinder

REMOVAL & INSTALLATION

1. Disconnect the negative battery cable. Disconnect the low fluid level sensor connector.

2. Disconnect the brake lines from the master cylinder connections. Plug or cover the line openings and master cylinder ports.

3. Remove the attaching nuts and washers and separate the master cylinder from the power booster mounting studs. Clean the master cylinder and power booster contact surfaces with a clean shop towel.

To install:

4. If a new master cylinder is being installed, check the pushrod length adjustment as follows:

a. Position master cylinder gauge T87C–2500–A or equivalent, on the end of the master cylinder, loosen the set screw and push the gauge plunger against the bottom of the primary piston.

b. While holding the gauge in position, tighten the set screw.

c. Invert the master cylinder gauge and place it over the brake booster pushrod.

d. If the clearance is not zero, loosen the pushrod locknut and adjust the pushrod.

NOTE: Proper pushrod length adjustment is critical. If the pushrod is adjusted too long, the brakes will drag. If the pushrod is adjusted too short, the brake pedal will be low.

5. Before installation, bench bleed a new master cylinder as follows:

a. Mount the new master cylinder in a suitable holding fixture. Be careful not to damage the housing.

b. Fill the master cylinder reservoir with brake fluid.

c. Using a suitable tool inserted into the booster pushrod cavity, push the master cylinder piston in slowly. Place a suitable container under the master cylinder to catch the fluid being expelled from the outlet ports.

d. Place a finger tightly over each outlet port and allow the master cylinder piston to return.

e. Repeat the procedure until clear fluid only is expelled from the master cylinder. Plug the outlet ports and remove the master cylinder from the holding fixture.

6. Position the master cylinder onto the power booster mounting studs.

7. Install the attaching washers and nuts. Torque the nuts to 7–12 ft. lbs. (10–16 Nm).

8. Connect the brake lines to master cylinder connections.

9. Make sure the master cylinder reservoir is full. Have an assistant slowly push down on the brake pedal. When the pedal is all the way down, crack open the brake line fittings, 1 at a time, to expel any remaining air in the master cylinder and brake lines. Tighten the fittings, then have the assistant allow the brake pedal to return.

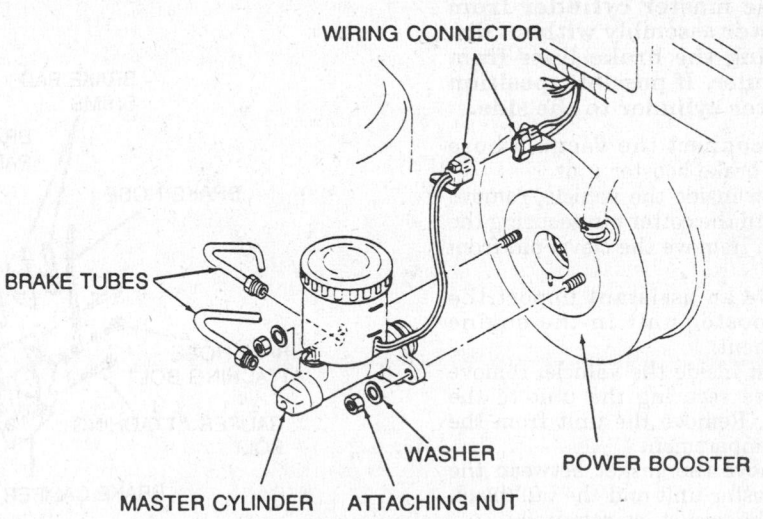

Master cylinder assembly

10. Repeat Step 9 until all air is expelled from the master cylinder and brake lines. Tighten the brake line fittings.

11. Connect the low fluid level sensor.

12. Make sure the master cylinder reservoir is full. If necessary, bleed the entire brake system.

13. Connect the negative battery cable.

Proportioning Valve

The proportioning valve is located in the engine compartment. It is mounted to the dash panel below and to the right of the brake booster. The valve is not repairable and must be replaced if determined to be faulty.

REMOVAL & INSTALLATION

1. Disconnect the negative battery cable. Loosen the connector nuts and disconnect the brake lines from the proportioning valve. Plug or cover the line openings to prevent the entry of dirt and grease.

2. Loosen the valve attaching bolts and remove the valve from the dash panel.

3. Installation is the reverse of the removal procedure.

4. Bleed the brake system.

5. Inspect for proper brake operation and inspect for leaks around the valve connections.

Power Brake Booster

REMOVAL & INSTALLATION

1. Disconnect the negative battery cable. Remove the master cylinder.

NOTE: It may be possible to remove the master cylinder from the booster assembly without disconnecting the brake lines from the cylinder. If possible, position the master cylinder to the side.

2. Disconnect the vacuum hose from the brake booster unit.

3. From inside the vehicle, remove and discard the cotter pin securing the clevis pin. Remove the clevis pin from the clevis.

4. Have an assistant upport the power booster unit in the engine compartment.

5. From inside the vehicle, remove the 4 nuts securing the unit to the bulkhead. Remove the unit from the engine compartment.

6. Remove the gasket between the power booster unit and the bulkhead. Replace the gasket, as required.

To install:

7. Position the gasket onto the power brake booster studs and have an assistant position the unit against the bulkhead.

8. From inside the vehicle, secure the power booster to the bulkhead with the 4 retaining nuts. Torque the retaining nuts to 12–17 ft. lbs. (16–23 Nm).

9. Lubricate the clevis with a coating of white lithium grease or equivalent. From inside the vehicle, attach the clevis to the brake pedal with the clevis pin. Secure the clevis pin with a new cotter pin.

10. Connect the vacuum to the power brake booster.

11. Install the master cylinder.

12. Bleed the brake system.

Brake Caliper

REMOVAL & INSTALLATION

1. Raise and safely support the vehicle.

2. Remove the wheel and tire assembly.

3. Remove the brake pads. Remove the brake hose attaching bolt and plug the hose end. Discard the seal washers.

4. Remove the caliper attaching bolts and the anti-squeak caps.

5. Remove the caliper from the vehicle.

6. Installation is the reverse of the removal procedure.

7. Tighten the caliper mounting bolts to 29–36 ft. lbs. (39–49 Nm). Use new seal washers on the brake hose and tighten the brake hose attaching bolt to 16–22 ft. lbs. (22–29 Nm). Bleed the brake system.

Disc Brake Pads

REMOVAL & INSTALLATION

1. Remove approximately ⅓ of the brake fluid from the master cylinder. Raise and support the vehicle safely.

2. Remove the tire and wheel assembly.

3. Remove the brake pad pin retainer. Disengage the anti-rattle spring from the brake pads.

4. Remove the brake pad pins and the anti-rattle spring.

5. Remove the brake pads and shims. Do not discard the shims found behind the brake pads.

To install:

6. Use a suitable tool to push the piston back into the caliper bore.

7. Apply the grease supplied with the brake pad set to both surfaces of the inner shim and to the back of the brake pad.

8. Install the brake pads, making sure the shims are installed.

9. Install the brake pad pins, anti-rattle spring and brake pad pin retainer.

10. Install the wheel and tire assembly and lower the vehicle.

11. Apply the brake several times to seat the pads. Check the brake fluid level in the master cylinder. Add fluid as necessary.

Brake Rotor

REMOVAL & INSTALLATION

1. Disconnect the negative battery cable.

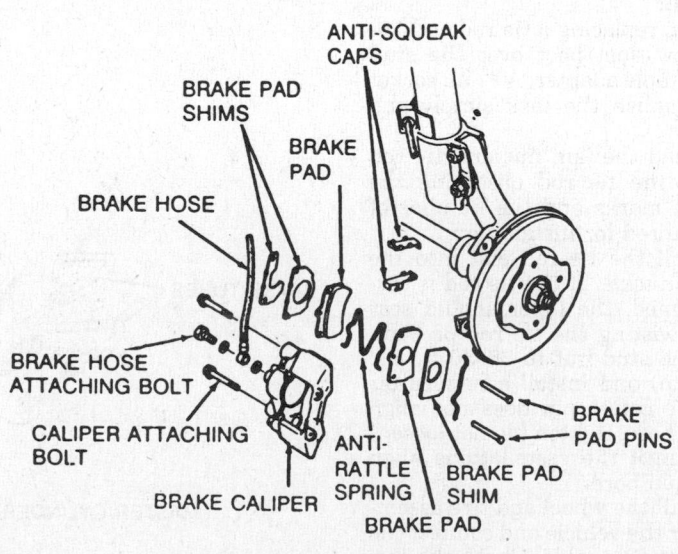

Front disc brake assembly

2. Raise the vehicle and support it safely.

3. Unbolt and remove front wheel from the hub assembly.

4. With a suitable tool, straighten the staked edge of the halfshaft attaching nut. Take care not to damage the halfshaft threads.

5. Remove and discard the halfshaft attaching nut.

6. Remove the retaining clip securing the caliper hose to the strut bracket.

7. Remove the cotter pin and tie rod end attaching nut. Discard the cotter pin and set the nut aside. Inspect the nut for damage and replace as required.

8. Using a tie rod end separator tool, release the tie rod end from the steering knuckle arm. If the tie rod appears to be siezed, strike the knuckle sharply with a soft-tipped hammer to acheive separation.

9. Support the brake caliper and remove the brake caliper attaching bolts. Lift the caliper assembly from the steering knuckle.

NOTE: After the caliper assembly is lifted from the steering knuckle, do not allow it to be suspended by the brake hose. Support the caliper by a length of rope or wire attached to the Mac-Pherson strut.

10. Remove the clamp bolt and nut at the point where the lower control arm ball joint connects to the steering knuckle. With a medium prybar, release the lower ball joint from the steering knuckle by prying downward on the lower control arm.

11. Remove the 2 bolts that position the steering knuckle between the Mac-Pherson strut bracket flanges.

12. Slide the knuckle/hub assembly from the end of the halfshaft. If binding occurs, tap the end of the shaft with a soft-tipped hammer. If the wheel hub is rusted to the halfshaft, use either a 2 jaw or a hub puller to

achieve separation.

13. Remove the wheel hub/rotor assembly from the steering knuckle/dust shield assembly using a suitable puller.

14. Remove the bearing preload spacer from the hub.

NOTE: The spacer is preselected to yield the correct bearing preload. Save the removed spacer for use during assembly.

15. Clamp the hub/rotor assembly in a vise with protective jaw caps. If the rotor is to be reused, scribe aligning marks on the hub and rotor for use during assembly. Remove the attaching bolts and the rotor.

To install:

16. Place the bearing preload spacer in the steering knuckle bore.

17. Position the rotor on the hub, observing the original aligning marks if the rotor is being resused and install the attaching bolts. Tighten the attaching bolts to 33–40 ft. lbs. (44–54 Nm).

18. Position the hub/rotor assembly in the steering knuckle bore and press it into position using a suitable driver.

19. Clean the halfshaft spline end and lubricate with a coating of wheel bearing grease. Apply a thin film of clean SAE 30 weight oil to the steering knuckle/rotor hub assembly up to the point where the uppermost arm of the steering knuckle seats into the Mac-Pherson strut bracket. Guide the steering knuckle/rotor/hub assembly onto the halfshaft and the Macpherson strut.

20. Install the strut-to-steering knuckle bolts and attaching nuts. Tighten the nuts to 69–86 ft. lbs. (93–117 Nm).

21. Position the lower control arm ball joint in the steering knuckle. Install the lower control arm pinch bolt and attaching nut. Tighten the nut to 32–40 ft. lbs. (43–54 Nm).

22. Position the caliper on the steering knuckle and install the attaching

bolts. Tighten the bolts to 29–36 ft. lbs. (39–49 Nm). Position the caliper hose in the strut routing bracket and install the retaining clip.

23. Install a new halfshaft attaching nut and tighten to 116–174 ft. lbs. (157–235 Nm). After installation, the wheel hub assembly must rotate freely by hand. Stake the halfshaft attaching nut into the shaft groove.

NOTE: Do not use a pointed tool to stake the nut. If the nut cracks even slightly during staking, replace it with another new one.

24. Connect the tie rod end to the steering knuckle and install the attaching nut. Tighten the attaching nut to 22–33 ft. lbs. (29–44 Nm). Install a new cotter pin through the nut and ball stud. If the openings in the nut and the hole in the ball stud are not aligned, tighten the nut slightly, just to the point of alignment. Never loosen the nut.

25. Install the wheel and tire assembly. Tighten the attaching bolts to 65–87 ft. lbs. (88–118 Nm). Lower the vehicle.

Brake Drums

REMOVAL & INSTALLATION

1. Raise and safely support the vehicle.

2. Remove the tire and wheel assembly.

3. Carefully raise the staked portion of the attaching nut using a suitable tool.

4. Remove and discard the locknut.

NOTE: The locknuts are right and left hand thread. The left hand threaded locknut is located on the right side of the vehicle. Turn this locknut clockwise to loosen. The right hand threaded locknut is located on the left side of the vehicle and is turned counterclockwise to loosen.

5. Remove the brake drum and bearings as an assembly. Be careful not to let the outer bearing fall out of the hub during removal.

To install:

6. Make sure the bearings and the hub contain adequate lubricant.

7. Position the brake drum, bearings and hub assembly on the spindle. Keep the drum centered on the spindle to prevent damage to the grease seal and spindle threads.

8. Install the outer bearing, washer

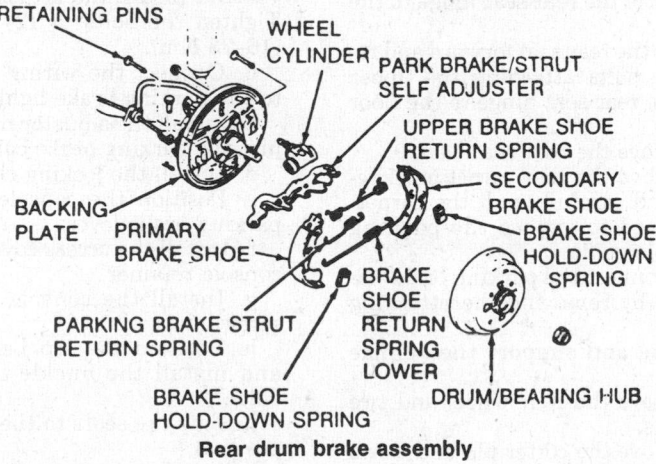

RETAINING PINS
WHEEL CYLINDER
PARK BRAKE/STRUT SELF ADJUSTER
UPPER BRAKE SHOE RETURN SPRING
SECONDARY BRAKE SHOE
BRAKE SHOE HOLD-DOWN SPRING
BACKING PLATE
PRIMARY BRAKE SHOE
PARKING BRAKE STRUT RETURN SPRING
BRAKE SHOE RETURN SPRING LOWER
DRUM/BEARING HUB
BRAKE SHOE HOLD-DOWN SPRING

Rear drum brake assembly

and a new locknut.

9. Properly adjust the wheel bearing preload.

Brake Shoes

REMOVAL & INSTALLATION

1. Raise and support the vehicle safely.

2. Remove the tire and wheel assembly. Remove the brake drum.

3. Remove the brake shoe hold-down springs and pins.

4. Remove the brake shoe return springs. Pull the brake shoes away from the backing plate and remove.

To install:

5. Lubricate the backing plate shoe pads with a suitable high temperature grease.

6. Install the brake shoe upper return spring on the primary brake shoe. Position the primary brake shoe on the backing plate and install the hold-down pin and spring.

7. Connect the upper return spring to the secondary brake shoe and position the shoe against the backing plate. Install the secondary brake shoe hold-down pin and spring.

8. Install the parking brake return spring and the lower brake shoe return spring.

9. Set the self adjuster to the fully released position. Place a suitable tool against the adjuster cam and push it to the released position.

10. Install the brake drum, wheel and tire assembly and lower the vehicle.

11. Push the brake pedal several times to set the self adjuster.

Wheel Cylinder

REMOVAL & INSTALLATION

1. Raise and support the vehicle safely.

2. Remove the rear brake shoes.

3. Disconnect the brake line from the wheel cylinder. Plug or cover the brake line opening to prevent the entry of dirt or grease.

4. Remove the 2 wheel cylinder attaching bolts and remove the wheel cylinder from the backing plate.

To install:

5. Position the wheel cylinder onto the backing plate and install the retaining bolts. Torque the retaining bolts to 7–9 ft. lbs. (10–13 Nm).

6. Connect the brake line to the wheel cylinder.

7. Install the rear brake shoes.

8. Bleed the brake system.

Parking Brake Cable

ADJUSTMENT

1. Make sure the parking brake is fully released.

2. Remove the parking brake console access cover.

3. Remove the locking clip from the cable adjuster nut.

4. Raise and support the vehicle safely.

5. Tighten the cable adjuster nut until there is a slight brake drag when the rear wheels are rotated.

6. Back off on the adjuster nut until the brake drag disappears.

7. Check the operation of the parking brake. The rear brakes should be fully applied when the brake lever is pulled upward 11–16 notches.

REMOVAL & INSTALLATION

1. Remove the parking brake console and parking brake lever as follows:

a. Slide both front seats all the way forward.

b. Remove the bolts that attach the lap belt buckles to their mounting brackets.

c. Remove the 2 console attaching screws.

d. Remove the retainer located at the front of the console.

e. Remove the access cover and remove the parking brake console.

f. Remove the locking clip from the cable adjuster nut and remove the cable adjuster nut.

g. Disconnect the wiring connector from the parking brake light switch.

h. Remove the attaching bolts and the parking brake lever.

2. Remove the attaching screws and parking brake console mounting bracket.

3. Remove the bolts attaching the lower half of the rear seat hinge to the floor pan.

4. Fold the rear seat forward and remove the bolts attaching the upper half of the rear seat hinge to the floor pan.

5. Remove the rear seat.

6. Remove the rear carpet push retainers and carefully pull the carpeting forward to expose the parking brake cable guide.

7. Disconnect the parking brake cable guide by removing the attaching screws.

8. Raise and support the vehicle safely.

9. Remove the rear wheel and tire assemblies.

10. Remove the cotter pin and clevis pin attaching the parking brake cable ends to the rear brake levers.

11. Remove the routing bracket retaining clips.

12. Disengage the parking brake routing sleeves from the torsion beam routing brackets.

13. Remove the nut and bolt attaching the parking brake routing bracket to the fuel tank.

14. Remove the parking brake cable equalizer attaching bolts.

15. Withdraw the lever end of the cable through the body opening and remove from the vehicle.

To install:

16. Position the lever end of the cable through the body opening.

17. Position the cable routing bracket on the fuel tank and install the attaching bolt and nut.

18. Make sure the cable seal is properly positioned in the floor pan.

19. Position the cable equalizer and install the attaching bolts. Make sure the equalizer spacers are in position before tightening the attaching bolts.

20. Route the cable ends through the body brackets and install the retaining clips.

21. Seat the cable sleeves in the torsion beam routing brackets.

22. Attach the cable ends to the brake levers using the clevis pins and new cotter pins.

23. Install the rear wheel and tire assemblies and lower the vehicle.

24. Route the end of the cable through the park brake lever.

25. Position the cable guide and secure with the attaching screws.

26. Position the carpet and install the push retainers. Install the rear seat and torque the retaining bolts to 28–38 ft. lbs. (38–51 Nm).

27. Position the console mounting bracket and install the attaching screws.

28. Install the parking brake lever and console as follows:

a. Position the parking brake lever and install the attaching bolts. Tighten the bolts to 12–17 ft. lbs. (16–23 Nm).

b. Connect the wiring connector to the parking brake light switch.

c. Install the adjuster nut and adjust the parking brake cable.

d. Install the locking clip.

e. Position the console over the parking brake lever.

f. Install the access cover and the console retainer.

g. Install the console attaching screws.

h. Position the lap belt buckles and install the buckle-to-bracket bolts.

i. Slide the seats to their original position.

Brake System Bleeding

When any part of the hydraulic system has been disconnected for service, air may enter the system and cause spongy pedal action. The bleeding procedure is used to remove air from the hydraulic circuits.

The brake hydraulic circuits form a split diagonal hydraulic system. The left front and right rear form 1 circuit while the right front and left rear form the other circuit. When bleeding 1 of these circuits, bleed the rear wheel first and then the front wheel at the opposite corner.

Never reuse brake fluid that has been drained from the hydraulic system or that has been allowed to stand in an open container for an extended period of time.

Bleed the brake system as follows: .

1. Clean all dirt from the master cylinder filler cap. Fill the master cylinder with DOT 3 brake fluid.

NOTE: Do not allow the master cylinder to run dry during the bleeding procedure.

2. If the master cylinder is known or suspected to contain air, it must be bled before the wheel cylinders or caliper. Bleed the master cylinder as follows:

   a. Loosen the front line fitting and have an assistant push the brake pedal slowly through it's full travel.

   b. While the assistant holds the pedal down, tighten the brake line fitting. After the line fitting is tightened, the assistant may release the brake pedal.

   c. Repeat the procedure on the rear brake line.

   d. Repeat the entire process several times to make sure all air has been removed from the master cylinder.

3. Remove the bleeder screw cap from the appropriate rear wheel cylinder. Position a box end wrench on the bleeder fitting.

4. Attach a rubber hose to the bleeder fitting. The hose must fit snugly around the bleeder fitting.

5. Submerge the other end of the hose in a container partially filled with brake fluid.

6. Loosen the bleeder fitting approximately ¾ turn. Have an assistant push the brake pedal slowly through it's full travel and hold it there.

7. Close the bleeder fitting, then have the assistant release the brake pedal.

8. Repeat Steps 6 and 7 until air bubbles no longer appear at the submerged end of the bleeder hose.

9. When the fluid entering the bottle is completely free of bubbles, tighten the bleeder screw, remove the hose and install the bleeder screw cap.

10. Repeat Steps 3–9 at the front caliper located diagonally to the wheel cylinder just completed.

11. If necessary, bleed the other diagonal circuit in the same manner.

12. Check the master cylinder fluid level and add, if necessary.

13. Check the pedal feel. If the pedal is still spongy, repeat the bleeding procedure.

CHASSIS ELECTRICAL

Heater Blower Motor

REMOVAL & INSTALLATION

1. Disconnect the negative battery cable.

2. Remove the airflow duct located below the steering column.

3. Disconnect the blower motor wiring.

4. Remove the attaching screws and the blower motor.

5. Remove the blower wheel attaching nut and remove the blower wheel and washer.

6. Installation is the reverse of the removal procedure.

Windshield Wiper Motor

REMOVAL & INSTALLATION

1. Disconnect the negative battery cable. Disconnect the wiring at the wiper motor.

2. Remove the wiper motor attaching bolts.

3. Remove the mounting plate attaching screws and pull the plate away from the dash panel.

4. Using a suitable tool, pry the linkage pivot off the output arm. Remove the wiper motor from the vehicle.

To install:

5. Position the motor on the mounting plate and connect the output arm to the linkage pivot.

6. Position the mounting plate and install the attaching screws.

7. Install the wiper motor attaching bolts and tighten to 61–87 inch lbs. (7–10 Nm). Make sure the ground wire is installed with the top left attaching bolt.

8. Connect the wiper motor wiring connector and the negative battery cable. Check the wiper motor for proper operation and linkage movement.

Instrument Cluster

REMOVAL & INSTALLATION

1. Disconnect the battery negative cable.

2. If equipped with tilt wheel, release the tilt lock and lower the steering column. If not equipped with tilt wheel, remove the steering column covers.

3. Remove the screws securing the instrument cluster bezel to the instrument panel.

4. Pull the instrument cluster bezel away from the instrument panel.

5. If equipped with rear window defogger, disconnect the wiring from the switch.

6. If equipped with rear window wiper, disconnect the wiring from the switch.

7. Remove the screws securing the instrument cluster in the instrument panel.

8. Pull the cluster away from the instrument panel.

9. Reach behind the cluster, press the lock tab and disconnect the speedometer cable.

10. Lift the lock tab and disconnect the 2 electrical connectors from the back of the instrument cluster.

11. Remove the instrument cluster from the vehicle.

To install:

12. Position the instrument cluster in the instrument panel opening.

13. Connect the electrical connectors to the back of the instrument cluster.

14. Slide the instrument cluster into the instrument panel.

15. Connect the speedometer cable.

16. Install and tighten the instrument cluster attaching screws.

17. Position the instrument cluster bezel in the instrument panel opening. If necessary, connect the rear defogger and rear wiper switch wiring.

18. Install and tighten the instrument cluster bezel attaching screws.

19. If equipped with tilt wheel, raise the steering column and lock in desired position. If not equipped with tilt wheel, install the steering column covers.

20. Connect the negative battery cable.

21. Check the operation of all instruments, gauges and indicator lights.

Speedometer

REMOVAL & INSTALLATION

1. Disconnect the negative battery cable. Remove the instrument cluster from the vehicle.
2. Remove the odometer reset button.
3. Remove the cluster illumination bar attaching screws, remove the screws attaching the illumination bar wiring to the cluster circuit board and remove the bar.
4. Press down on the lock tabs and remove the cluster lens.
5. Remove the circuit board attaching screws and remove the circuit board and gauges from the cluster housing.
6. Remove the speedometer.

NOTE: If equipped with a tachometer cluster, the speedometer and cluster are removed as an assembly. If the speedometer is being replaced, transfer the tachometer and gauges to the new cluster. On vehicles without a tachometer, the speedometer is a separate module that can be removed and installed without removing the gauges.

7. Installation is the reverse of the removal procedure. Check the speedometer for proper operation.

Radio

REMOVAL & INSTALLATION

1. Disconnect the negative battery cable.
2. Remove the bezel attaching screws and the trim bezel.
3. Remove the 2 screws retaining the radio in the instrument panel. Pull the radio out of the console.
4. Disconnect the electrical connectors and antenna wire from the radio.
5. Remove the rubber mounting insulator from the radio ground stud. Remove the nut and the ground wire from the stud.
6. Remove the radio.
7. Installation is the reverse of the removal procedure. Check the radio for proper operation.

Combination Switch

The combination switch controls the windshield wiper, turn signal and headlight operation.

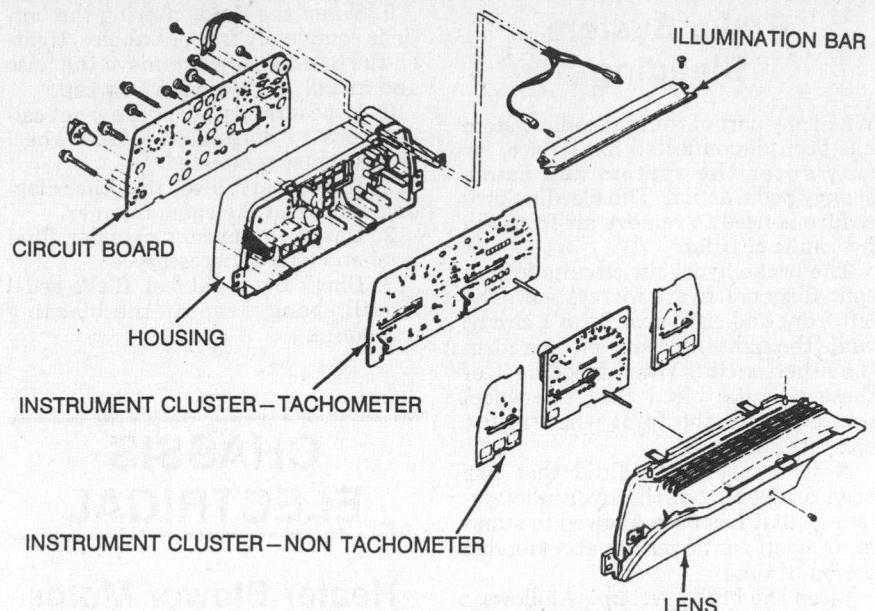

ILLUMINATION BAR

CIRCUIT BOARD

HOUSING

INSTRUMENT CLUSTER—TACHOMETER

INSTRUMENT CLUSTER—NON TACHOMETER

LENS

Instrument cluster exploded view

REMOVAL & INSTALLATION

1. Disconnect the negative battery cable.
2. Remove the steering wheel.
3. Remove the attaching screws and, if necessary, the clips from the lower steering column cover. Separate and remove the upper and lower steering column covers.
4. Release the wiring harness clip and unplug the 4 wiring harness connectors from the rear of the combination switch. From below the steering column, loosen the band clamp securing the switch hub to the steering column jacket.
5. Pull the switch assembly off the steering column.

To install:
6. Slide the combination switch assembly onto the steering column seating the switch against the column jacket. Make certain that the switch is level, then, tighten the band clamp on the switch hub to hold the switch assembly in place.
7. Plug the 4 wiring harness connectors into the rear of the switch and clip the harness in place.
8. Position the upper and lower steering column covers together and secure with the attaching screws.
9. Install the steering wheel. Connect the negative battery cable.

Ignition Lock

REMOVAL & INSTALLATION

1. Disconnect the negative battery

cable. Remove the steering wheel, combination switch and ignition switch.
2. If necessary, remove the shift-lock cable attaching bolt and disconnect the cable from the lock housing.
3. Using slim-nose locking pliers, grip and remove the round head mounting screws securing the steering lock housing and cap to the steering column jacket. Remove the lock housing. Discard the screws.

To install:
4. Position the steering lock housing onto the steering column jacket and install the mounting cap with new mounting screws. Tighten the screws enough to hold the lock in postion.
5. Using the ignition key, verify that the mechanism locks and unlocks positively and without binding. If necessary, reposition the lock housing until proper operation is obtained, then tighten the mounting screws until the heads break off.
6. If necessary, install the shift-lock cable and attaching bolt. Tighten the bolt to 37–54 inch lbs. (4–6 Nm).
7. Install the ignition switch, combination switch and steering wheel by reversing the removal procedures. Connect the negative battery cable.

Ignition Switch

REMOVAL & INSTALLATION

1988–90

1. Disconnect the negative battery cable.

2. Remove the attaching screws and, if necessary, the clips from the lower steering column cover. Remove the lower and upper covers.

3. Release the ignition switch wiring harness from the harness clip.

4. Remove the retaining screw above the switch to release the switch from the steering column lock housing. Remove the switch.

To install:

5. Push the ignition switch into the steering lock housing bore and engage the switch operating tang on the lock cylinder. If necessary, turn the lock cylinder with the ignition key until the tang aligns with the switch slot.

6. Install the retaining screw above the switch to secure the switch in the lock housing.

7. Position the switch wiring harness in the harness clip and close the clip.

8. Install the steering column covers with the screws and, if necessary, the clips.

9. Connect the negative battery cable.

1991–92

1. Disconnect the negative battery cable.

2. Remove the 4 screws from the upper half of the lower steering column cover, then remove the cover half.

3. Remove the upper steering column cover.

4. Remove the 5 clips from the lower half of the lower steering column cover, then remove the cover half.

5. Remove the 4 shield nuts and the steering column shield.

6. Remove the 2 shield bracket bolts and the shield bracket.

7. Remove the air discharge duct located below the steering column.

8. Remove the steering column attaching nuts and lower the steering column mounting bracket.

9. Remove the tie strap securing the key warning buzzer switch wires to the lock cylinder housing.

10. Remove the ignition switch attaching screw and remove the switch harness from the routing clip.

11. Separate the ignition switch wiring connectors and remove the switch.

To install:

12. Position the ignition switch in the lock cylinder housing and install the attaching screw.

13. Connect the switch wiring connectors. Position the switch wiring in the routing clip and close the clip.

14. Position the key warning buzzer switch wires and secure them to the lock cylinder housing with the tie strap.

15. Raise the steering column into position and install the attaching bolts. Tighten the bolts to 23–34 ft. lbs. (31–46 Nm).

16. Install the air discharge duct.

17. Install the shield bracket and the shield with the attaching bolts and nuts.

18. Install the steering column covers.

19. Connect the negative battery cable.

Stoplight Switch

ADJUSTMENT

1. Disconnect the negative battery cable. Disconnect the switch wiring connector.

2. Loosen the upper and lower attaching nuts enough to allow for rotation of the switch.

3. Connect an ohmmeter across the switch terminals.

4. Rotate the switch until the ohmmeter indicates continuity.

5. Slowly rotate the switch toward the brake pedal until the ohmmeter indicates that the switch is open (infinite resistance).

6. Rotate the switch toward the brake pedal ½ additional turn and tighten the attaching nuts to retain the adjustment.

7. Connect the switch wiring connector and check the switch for proper operation.

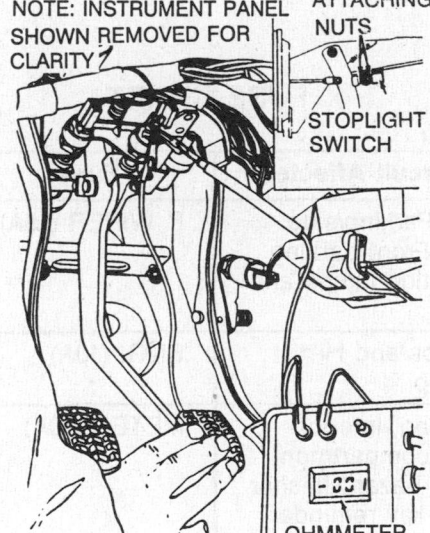

NOTE: INSTRUMENT PANEL SHOWN REMOVED FOR CLARITY

ATTACHING NUTS

STOPLIGHT SWITCH

OHMMETER

Stoplight switch adjustment

REMOVAL & INSTALLATION

1. Disconnect the negative battery cable. Disconnect the stoplight switch wiring connector.

2. Remove the upper attaching nut and lower the switch from the bracket.

3. Remove the lower attaching nut from the switch.

4. Installation is the reverse of the removal procedure. Adjust the switch after installation.

Clutch Switch

ADJUSTMENT

1. To eliminate the possibility that the clutch cable is affecting the pedal height, loosen the cable adjusting nut and disengage the cable pin from the transaxle release lever.

2. Move the floor carpet and insulation out of the way of the dash panel to gain sufficient room for an accurate measurement.

3. Measure the distance from the upper center of the pedal to the cowl panel. The pedal height should be 8.2–8.4 in. (208.2–213.2mm).

4. If the pedal height is within this range, no adjustment is necessary. If the pedal height is not within specification, proceed to Step 5.

5. Remove the air duct from under the instrument panel.

6. Locate the clutch switch and loosen the attaching nuts. Turn the switch in or out until the pedal height is within specification. Tighten the attaching nuts.

7. Connect the clutch cable to the transaxle release lever and adjust the pedal free-play.

8. Measure the clutch pedal height. If the pedal height has changed after connecting the clutch cable, check for binding along the cable route.

9. Install the air duct. Place the insulation and floor carpet in their original positions.

REMOVAL & INSTALLATION

1. Disconnect the negative battery cable. Move the floor carpet aside.

2. Remove the air duct located under the steering column.

3. Disconnect the clutch switch wiring connector.

4. Loosen the switch upper attaching nut and lower the switch from the mounting bracket. Remove the lower attaching nut.

5. Installation is the reverse of the removal procedure. Adjust the clutch pedal height after installation.

Neutral Safety Switch

The neutral safety switch is located in the lower right side of the transaxle.

REMOVAL & INSTALLATION

1. Disconnect the negative battery cable. Raise and support the vehicle safely.

2. Disconnect the neutral safety switch electrical wires.

3. Place a drain pan under the transaxle, to catch any excess transaxle fluid.

4. Remove the neutral safety switch from its mounting.

5. Installation is the reverse of the removal procedure. Be sure to replace any lost fluid.

Fuses, Circuit Breakers and Relays

LOCATION

Fuses

The fuse panel is located in the passenger compartment, to the left of the steering column. It is concealed behind an access panel that clips into position on the instrument panel. The fuses are the cartridge type that must be removed for inspection. When making replacements, install only cartridge type fuses with the same amperage rating as the fuse that was removed.

Fusible Links

The main fuse links are located in the engine compartment in the front of the left strut tower. The main fuse link panel contains 3 fusible links— **PTC** on carbureted engine or **EGI** on EFI engine, as well as **MAIN**, and **HEAD**. The ends of the fusible links are connected to the main fuse panel through standard push-on connectors. To remove a link, grasp the insulator and pull until the connector separates from the panel. Install the new link by reversing the removal procedure.

Relays

Air Conditioning Relays—located in the left front corner of the engine compartment, left of the cooling fan. There are 3 air conditioning relays, the main relay, the wide open throttle cut-off relay and the condenser fan relay.

Cooling Fan Relay—located on the fender apron, behind the left headlight.

Daytime Running Light Relay—located on the fender apron, behind the left headlight on Canadian vehicles only.

EFE Relay—located in the passenger compartment mounted on a bracket behind the left upper corner of the instrument panel, on carbureted vehicles only.

Fuel Pump Relay—located on the left side of the instrument panel, to the left of the electronic control unit, on EFI vehicles only.

Headlight Relay—located on the fender apron, behind the left headlight.

Main Relay—located in the left front corner of the engine compartment, attached to the fender apron.

Parking Light Relay—located in the right front corner of the engine compartment, on the fender apron.

Computers

LOCATION

The electronic control unit is located behind the instrument panel on the drivers side of the vehicle.

Flashers

LOCATION

The turn signal and hazard flashers are controlled by a single flasher unit. The flasher unit is located under the instrument panel, behind the electronic control unit.

FUSE CHART

Fuse	Item for Circuit Affected	Fuse	Item for Circuit Affected
TAIL (15A)	License lamp, Parking/side marker lamps, Front parking lamps, Illumination lamps and taillamps	F. WIPER (15A)	Front wiper and washer
STOP (15A)	Horn, Stoplamps and Hi-mount Stoplamp	ENG. (10A)	Charging system and Emission control system
HAZARD (15A)	Seat belt warning, Interior light, Luggage compartment lamp, Turn and hazard flasher lamps, Ignition key reminder buzzer and Radio system	METER (10A)	Seat belt warning, Turn and hazard flasher lamps, Cooling fan system, Backup lamp and cluster and warning lamps
CIGAR (15A)	Radio system, Cigar lighter and remote control mirror	R. DEF. (15A)	Rear window defroster
R. WIPER (15A)	Rear wiper and washer		
HEATER (15A)	Heater and Air conditioner		
FAN (15A)	Heater and Air conditioner and Cooling fan system		

Ford Motor Co.
Front Wheel Drive
Probe

SPECIFICATIONS

VEHICLE IDENTIFICATION CHART

It is important for servicing and ordering parts to be certain of the vehicle and engine identification. The VIN (vehicle identification number) is a 17 digit number visible through the windshield on the driver's side of the dash and contains the vehicle and engine identification codes. The tenth digit indicates model year and the eighth digit indicates engine code. It can be interpreted as follows:

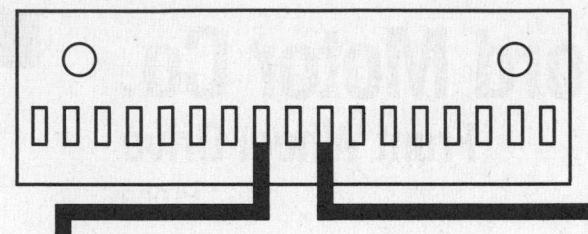

Engine Code

Code	Cu. In.	Liters	Cyl.	Fuel Sys.	Eng. Mfg.
C	133	2.2	4	MPFI	Mazda
L	133	2.2	4	MPFI ①	Mazda
U	182	3.0	6	MPFI	Mazda

MPFI—Multiport Fuel Injection
① Turbocharged

Model Year

Code	Year
K	1989
L	1990
M	1991
N	1992

ENGINE IDENTIFICATION

Year	Model	Engine Displacement cu. in. (liter)	Engine Series Identification (VIN)	No. of Cylinders	Engine Type
1989	Probe GL	133 (2.2)	C	4	OHC
	Probe LX	133 (2.2)	C	4	OHC
	Probe GT	133 (2.2)	L ①	4	OHC
1990	Probe GL	133 (2.2)	C	4	OHC
	Probe LX	182 (3.0)	U	6	OHC
	Probe GT	133 (2.2)	L ①	4	OHC
1991–92	Probe GL	133 (2.2)	C	4	OHC
	Probe LX	182 (3.0)	U	6	OHC
	Probe GT	133 (2.2)	L ①	4	OHC

OHC—Overhead Cam
① Turbocharged

GENERAL ENGINE SPECIFICATIONS

Year	VIN	No. Cylinder Displacement cu. in. (liter)	Fuel System Type	Net Horsepower @ rpm	Net Torque @ rpm (ft. lbs.)	Bore × Stroke (in.)	Compression Ratio	Oil Pressure @ rpm
1989	C	4-133 (2.2)	MPFI	110 @ 4700	130 @ 3000	3.39 × 3.70	8.5:1	57 @ 3000
	L	4-133 (2.2)	MPFI	145 @ 4300	190 @ 3500	3.39 × 3.70	7.8:1	57 @ 3000
1990	C	4-133 (2.2)	MPFI	110 @ 4700	130 @ 3000	3.39 × 3.70	8.5:1	57 @ 3000
	L	4-133 (2.2) ①	MPFI	145 @ 4300	190 @ 3500	3.39 × 3.70	7.8:1	57 @ 3000
	U	6-182 (3.0)	MPFI	140 @ 4800	160 @ 3000	3.50 × 3.14	9.3:1	60 @ 2500
1991–92	C	4-133 (2.2)	MPFI	110 @ 4700	130 @ 3000	3.39 × 3.70	8.5:1	57 @ 3000
	L	4-133 (2.2) ①	MPFI	145 @ 4300	190 @ 3500	3.39 × 3.70	7.8:1	57 @ 3000
	U	6-182 (3.0)	MPFI	140 @ 4800	160 @ 3000	3.50 × 3.14	9.3:1	60 @ 2500

MPFI—Multiport Fuel Injection
① Turbocharged

GASOLINE ENGINE TUNE-UP SPECIFICATIONS

Year	VIN	No. Cylinder Displacement cu. in. (liter)	Spark Plugs Type	Gap (in.)	Ignition Timing (deg.) MT	Ignition Timing (deg.) AT	Compression Pressure (psi)	Fuel Pump (psi)	Idle Speed (rpm) MT	Idle Speed (rpm) AT	Valve Clearance In.	Valve Clearance Ex.
1989	C	4-133 (2.2)	AGSP-33C	0.040	5–7①	5–7①②	④	64–85	725–775	725–775	Hyd.	Hyd.
	L	4-133 (2.2)	AGSP-33C	0.040	8–10	—	④	64–85	725–775	—	Hyd.	Hyd.
1990	C	4-133 (2.2)	AGSP-33C	0.040	5–7①	5–7①②	④	64–85	725–775	725–775	Hyd.	Hyd.
	L	4-133 (2.2)	AGSP-33C	0.040	8–10	—	④	64–85	725–775	725–775	Hyd.	Hyd.
	U	6-182 (3.0)	AWSF-32P	0.044	⑩	⑩	④	35–40	③	③	Hyd.	Hyd.
1991	C	4-133 (2.2)	AGSP-33C	0.040	5–7①	5–7①②	④	64–85	725–775	725–775	Hyd.	Hyd.
	L	4-133 (2.2)	AGSP-33C	0.040	8–10	8–10②	④	64–85	725–775	725–775	Hyd.	Hyd.
	U	6-182 (3.0)	AWSF-32P	0.044	③	③	④	35–40	③	③	Hyd.	Hyd.
1992		REFER TO UNDERHOOD SPECIFICATIONS										

Hyd.—Hydraulic
① Distributor vacuum hoses disconnected and plugged
② Transaxle in Park
③ Refer to vehicle information label
④ The compression pressure in the lowest cylinder should be at least 75% of the reading in the highest cylinder.

FIRING ORDERS

NOTE: To avoid confusion, always replace spark plug wires one at a time.

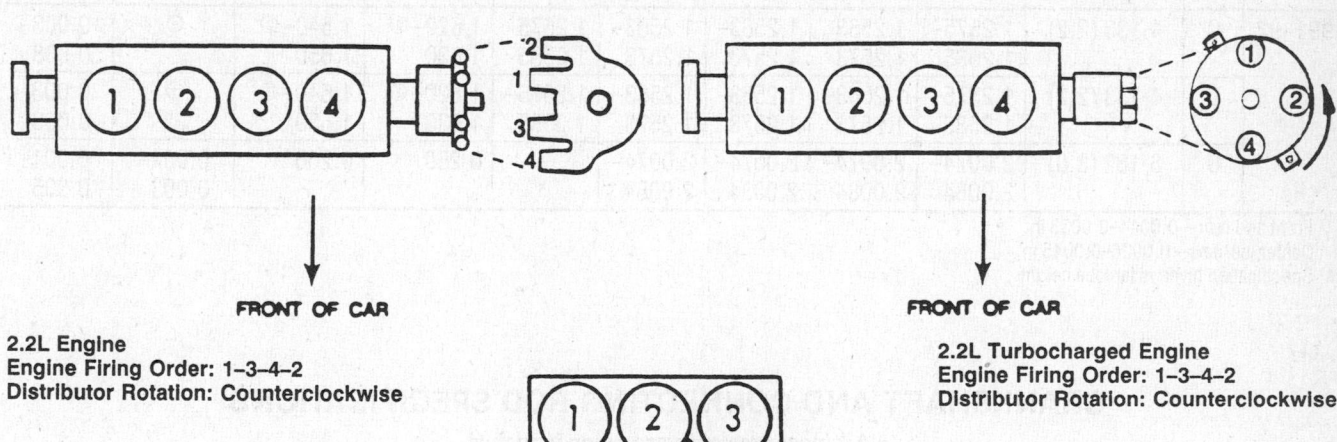

FRONT OF CAR

2.2L Engine
Engine Firing Order: 1–3–4–2
Distributor Rotation: Counterclockwise

FRONT OF CAR

2.2L Turbocharged Engine
Engine Firing Order: 1–3–4–2
Distributor Rotation: Counterclockwise

FRONT OF CAR

3.0L Engine
Engine Firing Order: 1–4–2–5–3–6
Distributor Rotation: Clockwise

CAPACITIES

Year	Model	VIN	No. Cylinder Displacement cu. in. (liter)	Engine Crankcase (qts.) with Filter	without Filter	Transmission (pts.) 4-Spd	5-Spd	Auto.	Drive Axle (pts.)	Fuel Tank (gal.)	Cooling System (qts.)
1989	Probe	C	4-133 (2.2)	4.4	4.1	—	7.2	14.4	—	15.1	7.9
	Probe	L	4-133 (2.2)	4.4	4.1	—	7.8	—	—	15.1	7.9
1990	Probe	C	4-133 (2.2)	4.4	4.1	—	7.2	14.4	—	15.1	7.9
	Probe	L	4-133 (2.2)	4.4	4.1	—	7.8	14.4	—	15.1	7.9
	Probe	U	6-182 (3.0)	4.5	4.0	—	7.8	14.4	—	15.1	11.0
1991-92	Probe	C	4-133 (2.2)	4.4	4.1	—	7.2	14.4	—	15.1	7.9
	Probe	L	4-133 (2.2)	4.4	4.1	—	7.8	14.4	—	15.1	7.9
	Probe	U	6-182 (3.0)	4.5	4.0	—	7.8	14.4	—	15.1	11.0

CAMSHAFT SPECIFICATIONS

All measurements given in inches.

Year	VIN	No. Cylinder Displacement cu. in. (liter)	Journal Diameter 1	2	3	4	5	Lobe Lift In.	Ex.	Bearing Clearance	Camshaft End Play
1989	C	4-133 (2.2)	1.2575-1.2585	1.2563-1.2573	1.2563-1.2573	1.2563-1.2573	1.2575-1.3585	1.620-②1.630	1.640-②1.650	①	0.003-0.008
	L	4-133 (2.2)	1.2575-1.2585	1.2563-1.2573	1.2563-1.2573	1.2563-1.2573	1.2575-1.2585	1.620-②1.630	1.640-②1.650	①	0.003-0.008
1990	C	4-133 (2.2)	1.2575-1.2585	1.2563-1.2573	1.2563-1.2573	1.2563-1.2573	1.2575-1.2585	1.620-②1.630	1.640-②1.650	①	0.003-0.008
	L	4-133 (2.2)	1.2575-1.2585	1.2563-1.2573	1.2563-1.2573	1.2563-1.2573	1.2575-1.2585	1.620-②1.630	1.640-②1.650	①	0.003-0.008
	U	6-182 (3.0)	2.0074-2.0084	2.0074-2.0084	2.0074-2.0084	2.0074-2.0084	—	0.260	0.260	0.001-0.003	0.001-0.005
1991-92	C	4-133 (2.2)	1.2575-1.2585	1.2563-1.2573	1.2563-1.2573	1.2563-1.2573	1.2575-1.2585	1.620-②1.630	1.640-②1.650	①	0.003-0.008
	L	4-133 (2.2)	1.2575-1.2585	1.2563-1.2573	1.2563-1.2573	1.2563-1.2573	1.2575-1.2585	1.620-②1.630	1.640-②1.650	①	0.003-0.008
	U	6-182 (3.0)	2.0074-2.0084	2.0074-2.0084	2.0074-2.0084	2.0074-2.0084	—	0.260	0.260	0.001-0.003	0.001-0.005

① Front and rear—0.0014-0.0033 in.
   Center journals—0.0026-0.0045 in.
② Specification given is for lobe height.

CRANKSHAFT AND CONNECTING ROD SPECIFICATIONS

All measurements are given in inches.

Year	VIN	No. Cylinder Displacement cu. in. (liter)	Crankshaft Main Brg. Journal Dia.	Main Brg. Oil Clearance	Shaft End-play	Thrust on No.	Connecting Rod Journal Diameter	Oil Clearance	Side Clearance
1989	C	4-133 (2.2)	2.3597-2.3604	①	0.003-0.007	3	2.0055-2.0061	0.0011-0.0026	0.0040-0.0103
	L	4-133 (2.2)	2.3597-2.3604	①	0.003-0.007	3	2.0055-2.0061	0.0011-0.0026	0.0040-0.0103

CRANKSHAFT AND CONNECTING ROD SPECIFICATIONS

All measurements are given in inches.

Year	VIN	No. Cylinder Displacement cu. in. (liter)	Crankshaft				Connecting Rod		
			Main Brg. Journal Dia.	Main Brg. Oil Clearance	Shaft End-play	Thrust on No.	Journal Diameter	Oil Clearance	Side Clearance
1990	C	4-133 (2.2)	2.3597–2.3604	①	0.003–0.007	3	2.0055–2.0061	0.0011–0.0026	0.0040–0.0103
	L	4-133 (2.2)	2.3597–2.3604	①	0.003–0.007	3	2.0055–2.0061	0.0011–0.0026	0.0040–0.0103
	U	6-182 (3.0)	2.5190–2.5198	0.0010–0.0014	0.004–0.008	3	2.1253–2.1261	0.0010–0.0014	0.0060–0.0140
1991–92	C	4-133 (2.2)	2.3597–2.3604	①	0.003–0.007	3	2.0055–2.0061	0.0011–0.0026	0.0040–0.0103
	L	4-133 (2.2)	2.3597–2.3604	①	0.003–0.007	3	2.0055–2.0061	0.0011–0.0026	0.0040–0.0103
	U	6-182 (3.0)	2.5190–2.5198	0.0010–0.0014	0.004–0.008	3	2.1253–2.1261	0.0010–0.0014	0.0060–0.0140

① No. 1, 2, 4 and 5—0.0010–0.0017 in.
  No. 3—0.0012–0.0019 in.

VALVE SPECIFICATIONS

Year	VIN	No. Cylinder Displacement cu. in. (liter)	Seat Angle (deg.)	Face Angle (deg.)	Spring Test Pressure (lbs.)	Spring Installed Height (in.)	Stem-to-Guide Clearance (in.)		Stem Diameter (in.)	
							Intake	Exhaust	Intake	Exhaust
1989	C	4-133 (2.2)	45	45	—	—	0.008	0.008	0.2744–0.2750	0.2742–0.2748
	L	4-133 (2.2)	45	45	—	—	0.008	0.008	0.2744–0.2750	0.2742–0.2748
1990	C	4-133 (2.2)	45	45	—	—	0.008	0.008	0.2744–0.2750	0.2742–0.2748
	L	4-133 (2.2)	45	45	—	—	0.008	0.008	0.2744–0.2750	0.2742–0.2748
	U	6-182 (3.0)	45	44	65 @ 1.58	1.58	0.0010–0.0028	0.0015–0.0033	0.3134–0.3126	0.3129–0.3121
1991–92	C	4-133 (2.2)	45	45	—	—	0.008	0.008	0.2744–0.2750	0.2742–0.2748
	L	4-133 (2.2)	45	45	—	—	0.008	0.008	0.2744–0.2750	0.2742–0.2748
	U	6-182 (3.0)	45	44	65 @ 1.58	1.58	0.0010–0.0028	0.0015–0.0033	0.3126–0.3134	0.3121–0.3129

PISTON AND RING SPECIFICATIONS

All measurements are given in inches.

Year	VIN	No. Cylinder Displacement cu. in. (liter)	Piston Clearance	Ring Gap			Ring Side Clearance		
				Top Compression	Bottom Compression	Oil Control	Top Compression	Bottom Compression	Oil Control
1989	C	4-133 (2.2)	0.0014–0.0030	0.008–0.014	0.006–0.012	0.012–0.035	0.001–0.003	0.001–0.003	NA
	L	4-133 (2.2)	0.0014–0.0030	0.008–0.014	0.006–0.012	0.006–0.014	0.001–0.003	0.001–0.003	NA

PISTON AND RING SPECIFICATIONS

All measurements are given in inches.

Year	VIN	No. Cylinder Displacement cu. in. (liter)	Piston Clearance	Ring Gap			Ring Side Clearance		
				Top Compression	Bottom Compression	Oil Control	Top Compression	Bottom Compression	Oil Control
1990	C	4-133 (2.2)	0.0014–0.0030	0.008–0.014	0.006–0.012	0.012–0.035	0.001–0.003	0.001–0.003	NA
	L	4-133 (2.2)	0.0014–0.0030	0.008–0.014	0.006–0.012	0.006–0.014	0.001–0.003	0.001–0.003	NA
	U	6-182 (3.0)	0.0014–0.0022	0.010–0.020	0.010–0.020	0.010–0.049	0.001–0.003	0.001–0.003	NA
1991–92	C	4-133 (2.2)	0.0014–0.0030	0.008–0.014	0.006–0.012	0.012–0.035	0.001–0.003	0.001–0.003	NA
	L	4-133 (2.2)	0.0014–0.0030	0.008–0.014	0.006–0.012	0.006–0.014	0.001–0.003	0.001–0.003	NA
	U	6-182 (3.0)	0.0014–0.0022	0.010–0.020	0.010–0.020	0.010–0.049	0.001–0.003	0.001–0.003	NA

NA—Not available

TORQUE SPECIFICATIONS

All readings in ft. lbs.

Year	VIN	No. Cylinder Displacement cu. in. (liter)	Cylinder Head Bolts	Main Bearing Bolts	Rod Bearing Bolts	Crankshaft Sprocket Bolts	Flywheel Bolts	Manifold		Spark Plugs
								Intake	Exhaust	
1989	C	4-133 (2.2)	59–64	61–65	48–51	108–116	71–76	14–22	16–21	11–17
	L	4-133 (2.2)	59–64	61–65	48–51	108–116	71–76	14–22	16–21	11–17
1990	C	4-133 (2.2)	59–64	61–65	48–51	108–116	71–76	14–22	16–21	11–17
	L	4-133 (2.2)	59–64	61–65	48–51	108–116	71–76	14–22	16–21	11–17
	U	6-182 (3.0)	①	55–63	23–39	92–122	54–64	②	15–22	7–15
1991–92	C	4-133 (2.2)	59–64	61–65	48–51	108–116	71–76	14–22	16–21	11–17
	L	4-133 (2.2)	59–64	61–65	48–51	108–116	71–76	14–22	16–21	11–17
	U	6-182 (3.0)	③	55–63	25	107	54–64	②	18	7–15

① Tighten in 2 steps
  A. 33–41 ft. lbs.
  B. 63–73 ft. lbs.

② Tighten in 2 steps
  A. 11 ft. lbs.
  B. 21 ft. lbs.

③ Tighten in 4 steps
  A. 59 ft. lbs.
  B. Back off all bolts a minimum of 360°
  C. 37 ft. lbs.
  D. 68 ft. lbs.

BRAKE SPECIFICATIONS

All measurements in inches unless noted

Year	Model	Lug Nut Torque (ft. lbs.)	Master Cylinder Bore	Brake Disc		Standard Brake Drum Diameter	Minimum Lining Thickness	
				Minimum Thickness	Maximum Runout		Front	Rear
1989	Probe GL	65–87	0.875	0.860	0.004	9.0	0.120	0.040
	Probe LX	65–87	0.875	0.860	0.004	9.0	0.120	0.040
	Probe GT	65–87	0.875	0.860 ①	0.004	—	0.120	0.040
1990	Probe GL	65–87	0.875	0.860	0.004	9.0	0.120	0.040
	Probe LX	65–87	0.875	0.860 ①	0.004	—	0.120	0.040
	Probe GT	65–87	0.875	0.860 ①	0.004	—	0.120	0.040
1991–92	Probe GL	65–87	0.875	0.860	0.004	9.0	0.120	0.040
	Probe LX	65–87	0.875	0.860 ①	0.004	—	0.120	0.040
	Probe GT	65–87	0.875	0.860 ①	0.004	—	0.120	0.040

① Rear disc—0.315 in.

WHEEL ALIGNMENT

Year	Model		Caster Range (deg.)	Caster Preferred Setting (deg.)	Camber Range (deg.)	Camber Preferred Setting (deg.)	Toe-in (in.)	Steering Axis Inclination (deg.)
1989	Probe GL	Front	0.47P–1.97P	1.22P	0.47N–1.03P	0.28P	0–0.24	12.78
		Rear	—	—	0.25N–1.25P	0.50P①	0.12N–0.12P	—
	Probe LX	Front	0.47P–1.97P	1.22P	0.47N–1.03P	0.28P	0–0.24	12.78
		Rear	—	—	0.25N–1.25P	0.50P①	0.12N–0.12P	—
	Probe GT	Front	0.47P–1.97P	1.22P	0.47N–1.03P	0.28P	0–0.24	12.78
		Rear	—	—	0.25N–1.25P	0.50P①	0.12N–0.12P	—
1990	Probe GL	Front	0.47P–1.97P	1.22P	0.47N–1.03P	0.28P	0–0.24	12.78
		Rear	—	—	0.25N–1.25P	0.50P①	0.12N–0.12P	—
	Probe LX	Front	0.47P–1.97P	1.22P	0.47N–1.03P	0.28P	0–0.24	12.78
		Rear	—	—	0.25N–1.25P	0.50P①	0.12N–0.12P	—
	Probe GT	Front	0.47P–1.97P	1.22P	0.47N–1.03P	0.28P	0–0.24	12.78
		Rear	—	—	0.25N–1.25P	0.50P①	0.12N–0.12P	—
1991–92	Probe GL	Front	0.47P–1.97P	1.22P	0.47N–1.03P	0.28P	0–0.24	12.78
		Rear	—	—	0.25N–1.25P	0.50P①	0.12N–0.12P	—
	Probe LX	Front	0.47P–1.97P	1.22P	0.47N–1.03P	0.28P	0–0.24	12.78
		Rear	—	—	0.25N–1.25P	0.50P①	0.12N–0.12P	—
	Probe GT	Front	0.47P–1.97P	1.22P	0.47N–1.03P	0.28P	0–0.24	12.78
		Rear	—	—	0.25N–1.25P	0.50P①	0.12N–0.12P	—

P—Positive
N—Negative
① Not Adjustable

ENGINE MECHANICAL

NOTE: Disconnecting the negative battery cable on some vehicles may interfere with the functions of the on board computer systems and may require the computer to undergo a relearning process, once the negative battery cable is reconnected.

Engine Assembly

REMOVAL & INSTALLATION

2.2L Engine

1. Relieve the fuel pressure and disconnect the negative battery cable.
2. Mark the hood hinge-to-hood location and remove the hood.
3. Drain the cooling system, the engine oil and the automatic transmission fluid.
4. Remove the battery, the battery carrier and the fuse holder.
5. Remove the air filter assembly and ducts. Disconnect the accelerator cable and the cruise control cable, if equipped.
6. Label and disconnect the electrical connectors from the electronic fuel injection system, the ignition coil, the thermostat housing sensors, the oxygen sensor, the radiator sensors and the cooling fan assembly.
7. If equipped with an automatic transaxle, disconnect and plug the

cooler lines from the radiator. Remove the radiator cooling fan assembly and the radiator.

8. If equipped with a manual transaxle, remove the clutch release cylinder and move it aside. Raise and safely support the vehicle.

9. Remove the front exhaust pipe-to-exhaust manifold nuts, the exhaust pipe-to-catalytic converter nuts and the front exhaust pipe. Lower the vehicle.

10. Discharge the air conditioning system and remove the air conditioning lines from the compressor. Disconnect the electrical connector from the compressor clutch.

11. Disconnect and plug the power steering lines from the power steering pump.

12. Disconnect the ground strap from the engine.

13. Disconnect and plug the heater hoses and the fuel lines.

14. Label and disconnect the vacuum lines from the brake booster chamber, the carbon canister, the bulkhead mounted solenoids and the distributor.

15. If equipped with an automatic transaxle, label and disconnect the electrical connectors from the transaxle.

16. Disconnect the speedometer cable from the transaxle.

17. If equipped with a turbocharger, disconnect the hoses and pipe and cover it with a clean rag.

18. Raise and safely support the vehicle. Remove the halfshafts from the transaxle.

19. Disconnect the shift control cable, if equipped with an automatic transaxle or rod, if equipped with a manual transaxle from the transaxle. Lower the vehicle.

20. Using an engine lifting device, attach it to the engine and support it's weight.

21. Disconnect the engine mount bolts and remove the engine/transaxle assembly from the vehicle.

22. If neccessary, remove the transaxle-to-engine bolts and support the engine on an engine stand.

To install:

23. If the transaxle was removed from the engine, install it and torque the bolts to 66–86 ft. lbs. (89–117 Nm).

24. Lower the engine/transaxle assembly into the vehicle and secure the engine mount bolts.

25. Install the halfshafts.

NOTE: When installing the halfshafts, hold the shafts to prevent damage to the seals, boots and joints caused by moving the joints through angles greater than 20 degrees.

26. Depending on which transaxle the vehicle is equipped with, connect the shift cable control or rod. If equipped with a manual transaxle, install the clutch release cylinder. If equipped with an automatic transaxle, connect the electrical connectors to the transaxle.

27. Connect the speedometer cable to the transaxle and the power steering lines to the power steering pump.

28. If equipped with air conditioning, use new O-rings and connect the high pressure and suction lines to the compressor. Reconnect the electrical connector to the compressor clutch.

29. Connect the engine ground strap. On a non-turbocharged vehicle, install the front exhaust pipe. If equipped with a turbocharger, connect the oil pipe and hoses to the turbocharger.

30. Install the radiator and the cooling fan assembly and reconnect the electrical connectors. If equipped with an automatic transaxle, reconnect the oil cooler lines to the radiator.

31. Connect the vacuum lines to the carbon canister, the bulkhead mounted solenoids, distributor and the brake booster.

32. Connect the heater hoses to the engine and the fuel lines to the fuel system. Connect the electrical connectors to the oxygen sensor, thermostat housing sensors, the coil and the electronic fuel injection assembly.

33. Install the accelerator cable and the cruise control cable, if equipped. Install the air filter and ducts.

34. Install the battery carrier, battery and the fuse holder. Connect the battery cables.

35. Refill the cooling system, the crankcase and the automatic transaxle. Charge the air conditioning system. Refill the power steering reservoir and bleed the system.

36. Start the engine, allow it to reach normal operating temperatures and check for leaks.

37. Install the hood, aligning the marks that were made during the removal procedure.

3.0L Engine

1. Relieve the fuel system pressure and disconnect the battery cables. Mark the position of the hood on it's hinges and remove the hood assembly.

2. Drain the cooling system and discharge the air conditioning system.

3. Remove the air cleaner assembly from the engine compartment and the vacuum valve assembly from the right side shock tower.

4. Disconnect and plug the fuel lines.

5. Remove the upper radiator hose.

6. Tag and disconnect the alternator, air conditioning compressor clutch, ignition coil and the engine coolant temperature sensor connectors.

7. Tag and disconnect the TFI module connector, injector wiring harness, air charge temperature sensor and the throttle position sensor.

8. Disconnect the oil pressure sending switch, ground straps on both sides of the engine and the block heater, if equipped.

9. Disconnect the knock sensor on the back side of engine, EGR sensor and the oil level sensor on the back side of the oil pan.

10. Disconnect and tag all vacuum lines, heater hoses and crankcase ventilation hoses.

11. Remove the high pressure and return lines from the power steering pump.

12. Discharge the air conditioning system and disconnect the air conditioning lines from the condenser and chassis, leaving the manifold lines attached to the compressor.

13. Disconnect the accelerator linkage, transmission throttle valve linkage and the speed control cable, if equipped.

14. Remove the battery, battery tray and the fuse box assembly.

15. Disconnect and set aside the speed control servo assembly and the transmission shift cable, if equipped with an automatic transaxle.

16. Disconnect all automatic transmission wiring connectors and the speedometer cable on conventional (analog) cluster vehicles.

17. Disconnect the Vehicle Speed Sensor (VSS) on electronic cluster vehicles.

18. Disconnect and plug the cooler lines at the transmission, if equipped.

19. Remove the clutch slave cylinder with the hose attached, if equipped with a manual transaxle.

20. Remove the radiator, cooling fan and shroud.

21. Raise and safely support the vehicle. Remove the front tires and wheels.

22. Remove the lower radiator hose, the front exhaust pipe and the starter motor.

NOTE: On vehicles with an automatic transaxle, it is advised that the torque converter nuts be removed at this time to facilitate the removal of the transaxle asembly from the engine after the engine/transaxle assembly is removed from the vehicle.

23. Remove the shift control rod and the extension bar on manual transaxle vehicles.

24. Remove the stabilizer links, tie rod ends and disconnect the lower ball

joints to disengage the control arms from the spindle.

25. Remove the dynamic damper mounting bolts on the right halfshaft assembly.

26. Disengage both halfshafts by pulling outward on both side brake and spindle assemblies.

27. Install 2 transaxle plugs into the differential side gears.

NOTE: Failure to install the transaxle plugs may allow the differential side gears to become misaligned.

28. Disconnect the lower transmission mount and safely, lower the vehicle.

29. Install and position the lifting devices. Disconnect the lower front engine mount.

30. Disconnect the right side upper engine mount at the timing cover and the left side upper engine mount at the transaxle case.

31. Carefully, lift the engine and the transaxle assembly out of the vehicle.

To install:

32. Lower the engine and the transaxle assembly into the vehicle.

33. Connect and tighten the upper and lower engine mounts. Remove the engine lifting devices.

34. Remove both transaxle plugs and install the halfshafts on both sides.

35. Install the dynamic damper mounting bolts on the right side halfshaft.

36. Engage the control arms and install the lower ball joints, tie rod ends and the stabilizer links.

37. Install the shift control rod and extension bar, if equipped with a manual transaxle.

38. Replace the torque converter nuts, if equipped with an automatic transaxle.

39. Install the starter, front exhaust pipe and the lower radiator hose.

40. Replace the front tires and wheels. Safely, lower the vehicle.

41. Install the cooling fan, shroud and the radiator.

42. Install the clutch release cable with the hose attached, if equipped.

43. Reconnect the cooler lines at the transmission, if equipped.

44. Connect the Vehicle Speed Sensor (VSS) on electronic cluster vehicles.

45. Connect all automatic transmission wiring connectors and the speedometer cable on conventional (analog) cluster vehicles.

46. Install the speed control servo assembly and the transmission shift cable, if equipped.

47. Replace the battery, battery tray and the fuse box assembly.

48. Connect the accelerator linkage, transmission throttle valve linkage and the speed control cable, if equipped.

49. Connect the air conditioning lines from the condensor and chassis and recharge the air conditioning system.

50. Install the high pressure and return lines to the power steering pump.

51. Reconnect all vacuum lines, heater hoses and crankcase ventilation hoses.

52. Reconnect the knock sensor on the back side of engine, EGR sensor and the oil level sensor on the back side of the oil pan.

53. Replace the oil pressure sending switch, ground straps on both sides of the engine and the block heater, if equipped.

54. Reconnect the TFI module connector, injector wiring harness, air charge temperature sensor and the throttle position sensor.

55. Install the alternator, air conditioning compressor clutch, ignition coil and the engine coolant temperature sensor.

56. Connect the fuel lines and replace the upper radiator hose.

57. Replace the air cleaner assembly to the engine compartment and the vacuum valve assembly on the right side shock tower.

58. Refill the cooling system.

59. Reconnect the battery cables and install the hood assembly, aligning the marks that were made during the removal procedure.

60. Check all fluid levels and refill if needed. Check for any leaks.

61. Recharge the air conditioning system.

Engine Mounts

REMOVAL & INSTALLATION

1. Disconnect the negative battery cable.

2. If necessary, raise and support the vehicle safely.

3. Using an engine lifting device, attach it to the engine and support it's weight.

4. Remove the engine mount(s)-to-engine bolts, the engine mount(s)-to-chassis bolts and the mounts.

5. To install, reverse the removal procedure. Remove the engine lift.

Cylinder Head

REMOVAL & INSTALLATION

2.2L Engine

1. Disconnect the negative battery cable. Remove the drive belts.

2. Remove the crankshaft pulley as follows:

a. Raise and safely support the vehicle.

b. Remove the right front wheel and tire assembly.

c. Remove the right inner fender panel.

d. Remove the 6 bolts, pulley and baffle plate.

e. Lower the vehicle.

3. Remove the timing belt covers and timing belt.

4. Remove the exhaust manifold, intake manifold and the distributor.

5. Remove rocker arm cover.

6. Drain the cooling system.

7. Remove the spark plug wires and the spark plugs.

8. Disconnect the electrical connectors from the thermostat housing sensors. Remove the upper radiator hose and the water bypass hose.

9. Remove the front and rear engine lifting eyes and the engine ground wire.

10. Remove the cylinder head bolts, a little at a time, in the reverse order of installation. Remove the cylinder head and discard the gasket.

11. Clean the gasket mounting surfaces. Check the cylinder head for cracks and/or other damage. Check the cylinder head gasket surface for warpage, using a straight-edge. The head gasket surface must be flat within 0.006 in. (0.15mm). Resurface, repair or replace the cylinder head, as necessary.

To install:

12. Position a new cylinder head gasket on the cylinder block. Install the cylinder head and torque the bolts, in sequence, to 59–64 ft. lbs. (80–86 Nm), in 2 equal steps.

13. Install the distributor and the front and rear engine lifting eyes.

14. Install the spark plugs and spark plug wires.

15. Install the intake and exhaust manifolds.

16. Install the rocker arm cover.

17. Install and adjust the timing belt. Replace the timing covers.

18. Install the crankshaft pulley and the drive belts.

19. Fill the cooling system and connect the negative battery cable.

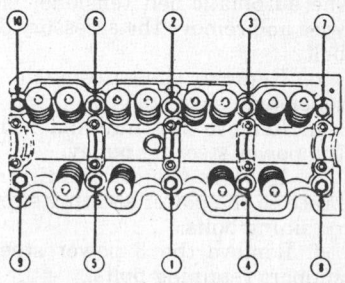

Cylinder head bolt torque sequence—2.2L engines

20. Run the engine and check for any leaks. Check the ignition timing.

3.0L Engine

1. Relieve the fuel system pressure and disconnect the negative battery cable.

2. Drain the cooling system.

3. Remove the air cleaner hoses from the throttle body and rocker arm cover.

4. Disconnect the fuel lines from the fuel supply manifold.

5. Tag and disconnect the vacuum lines from the throttle body.

6. Disconnect the air charge temperature sensor, throttle position sensor and air bypass solenoid electrical connectors.

7. Remove the EGR supply tube and the MAP sensor from the throttle body.

8. Disconnect the throttle cable and, if equipped with automatic transaxle, the throttle valve control cable from the throttle lever.

9. Remove the fuel rail bracket bolt from the throttle body, remove the 6 throttle body attaching bolts and remove the throttle body.

10. Disconnect the fuel injector harness stand-offs from the inboard rocker arm cover studs and each injector and remove from the engine.

11. Disconnect the upper radiator hose and heater hoses and move them aside.

12. Disconnect the engine coolant temperature sensor and the coolant temperature sending unit connectors.

13. Mark the distributor housing to block position, remove the distributor cap and mark the rotor position. Remove the distributor.

14. Tag and disconnect the spark plug wires from the spark plugs and remove the wires and the distributor cap.

15. Remove the ignition coil and bracket assembly from the left cylinder head and set aside.

16. If the left cylinder head is being removed, perform the foolowing:

    a. Remove the power steering protective shroud.

    b. Using a suitable tool, rotate the automatic belt tensioner clockwise and remove the accessory drive belt.

    c. Remove the automatic belt tensioner.

    d. Remove the nut and remove the power steering pulley.

    e. Remove the air conditioning brace to the power steering support retaining bolts.

    f. Remove the 3 power steering support retaining bolts.

    g. Remove the engine oil dipstick tube attaching nut from the exhaust manifold stud. Rotate or remove the tube from the manifold.

NOTE: The power steering support bracket may be pulled away from the engine with the alternator and power steering pump intact.

17. Remove the spark plugs.

18. Remove the exhaust manifold(s), heat shield(s) and inlet pipe(s).

19. Remove the rocker arm covers.

20. Loosen the rocker arm fulcrum retaining bolts enough to allow the rocker arm to be lifted off the pushrod and rotated to 1 side.

NOTE: The No. 3 intake valve pushrod must be removed to allow removal of the intake manifold, regardless of which cylinder head is being removed.

21. Remove the pushrods. Note the position of each so they may be reinstalled in their original positions.

22. Remove the intake manifold.

23. Remove the cylinder head retaining bolts and remove the cylinder head(s). Remove and discard the cylinder head gasket(s).

24. Clean all gasket mating surfaces. Check the cylinder head(s) for cracks or other damage. Check for warpage using a straight-edge. The cylinder head gasket surface must be flat within 0.007 in. (0.018mm). Resurface, repair or replace the cylinder head(s), as necessary.

To install:

25. Position new head gasket(s) on the cylinder block, using the dowel pins for alignment. Carefully position the cylinder head(s) on the block.

26. Lightly oil the threads and install the cylinder head bolts, finger tight. On 1990 vehicles, tighten the bolts, in sequence, in 2 steps; first to 37 ft. lbs. (50 Nm), and then to 68 ft. lbs. (92 Nm). On 1991–92 vehicles, tighten the bolts, in sequence, to 59 ft. lbs. (80 Nm), then back off all bolts a minimum of 1 full turn. Retighten the bolts, in sequence, in 2 steps; first to 37 ft. lbs. (50 Nm), and then to 68 ft. lbs. (92 Nm).

27. Install the intake manifold.

28. Dip each pushrod in heavy engine oil, then install in their original positions.

29. For each valve, rotate the crankshaft until the lifter rests on the base circle of the camshaft lobe, before tightening the fulcrum mounting bolts. Position the rocker arms over the pushrods and tighten the fulcrum mounting bolts to 24 ft. lbs. (32 Nm).

NOTE: If the original valve train components are being installed, a valve clearance check is not required. If a component has been replaced, perform a valve clearance check.

30. Install the exhaust manifold(s) and tighten the retaining bolts to 18 ft. lbs. (25 Nm). Install the inlet pipe retaining nuts and tighten to 30 ft. lbs. (41 Nm).

31. Install the dipstick tube into the cylinder block. Tighten the retaining nut to 13 ft. lbs. (18 Nm).

32. Install the spark plugs and tighten to 8 ft. lbs. (11 Nm).

33. Install the rocker arm covers.

34. Install the fuel injector electrical harness to the injectors and inboard rocker arm cover studs. Connect the engine harness to the main harness and secure with the retainers.

35. Install the distributor and the distributor cap. Connect the spark plug wires to the spark plugs.

36. Position a new gasket and the throttle body on the lower intake manifold. Install the attaching bolts and tighten to 15–22 ft. lbs. (20–30 Nm).

37. Install the fuel rail bracket bolt on the throttle body. Connect the throttle cable and, if equipped with automatic transaxle, the throttle valve control cable to the throttle lever.

38. Install the MAP sensor and the EGR supply tube to the throttle body.

39. Connect the electrical connectors for the air charge temperature sensor, throttle position sensor and air bypass solenoid.

40. Install the ignition coil and bracket. Tighten the bolts to 35 ft. lbs. (48 Nm).

41. If the left cylinder head was removed, perform the following:

    a. Install the power steering support bracket. Tighten the 3 retaining bolts to 35 ft. lbs. (48 Nm).

    b. Install the air conditioning brace to the power steering support bracket retaining bolt. Tighten the bolt to 18 ft. lbs. (25 Nm).

    c. Install the power steering pump pulley. Tighten the retaining nut to 47 ft. lbs. (64 Nm).

    d. Install the automatic belt tensioner. Tighten the retaining bolt to 35 ft. lbs. (48 Nm). Install the accessory drive belt.

    e. Install the power steering protective shroud. Tighten the 2 retaining bolts to 7 ft. lbs. (10 Nm).

42. Connect the fuel lines to the fuel supply rail.

43. Connect the upper radiator and heater hoses. Connect the vacuum lines to their original locations.

44. Drain and change the engine oil.

45. Install the air cleaner fresh air hose to the throttle body and air cleaner. Install the closure hose to the rocker arm cover.

46. Fill and bleed the cooling system.

47. Connect the negative battery cable.

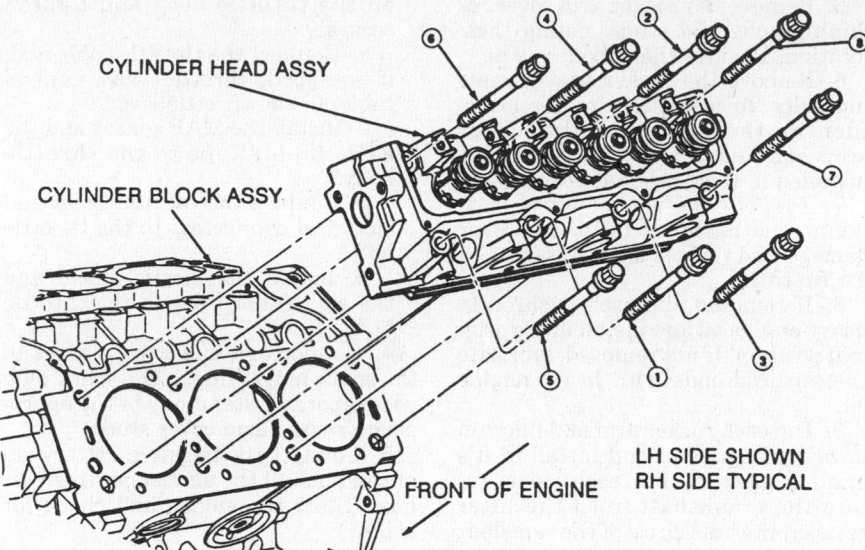

CYLINDER HEAD ASSY

CYLINDER BLOCK ASSY

FRONT OF ENGINE

LH SIDE SHOWN
RH SIDE TYPICAL

Cylinder head bolt torque sequence—3.0L engine

48. Start the engine and check for leaks. Check the ignition timing.

Valve Lifters

REMOVAL & INSTALLATION

2.2L Engine

The rocker arms are equipped with hydraulic lash adjusters, which ride directly on the camshaft.

3.0L Engine

NOTE: Before replacing a lifter for noisy operation, ensure that the noise is not caused by improper valve-to-rocker arm clearance, worn rocker arms or pushrods.

1. Disconnect the negative battery cable.
2. Drain the cooling system.
3. Remove the rocker arm covers and the throttle body and intake manifold assembly.
4. Loosen each rocker arm fulcrum mounting bolt to allow the rocker arm to be lifted off the pushrod and rotated to one side.
5. Remove the pushrods, marking the location of each pushrod to ensure the proper replacement in the original position.
6. Remove the lifter(s), using a magnet. Mark the location of each lifter to ensure the proper replacement in the original position.

NOTE: If the lifters are stuck in the bores, it may be neccessary to use a hydraulic lifter puller or equivalent.

To install:
7. Lubricate each lifter and the bore

with heavy engine oil and install the lifters into the bore.
8. Lubricate each pushrod with the heavy engine oil and insert in their original position.
9. For each valve, rotate the crankshaft until the lifter rests on the base circle of the camshaft lobe.
10. Place the rocker arms over the pushrods. Position the fulcrums and tighten the mounting bolts to 20–28 ft. lbs.
11. Lubricate all the rocker arm assemblies.

NOTE: Fulcrums must be fully seated in the cylinder head and the pushrods must be seated in the rocker arm sockets prior to final tightening.

12. Install the throttle body and intake manifold and the rocker arm covers.
13. Connect the negative battery cable.
14. Refill the cooling system and check for leaks.

Valve Lash

CHECKING

2.2L Engine

1. Warm up the engine to normal operating temperature.
2. Check the condition of the engine oil and check the oil pressure. The oil pressure should be 21–36 psi at 1000 rpm.
3. Stop the engine and remove the rocker arm cover.
4. Push down on the hydraulic lash adjuster side of the rocker arm to

make sure the hydraulic lash adjuster cannot be compressed.
5. If the hydraulic lash adjuster can be compressed, it must be replaced.

3.0L Engine

The valve stem-to-rocker arm clearance should be within specification with the valve lifter completely collapsed. To determine the rocker arm-to-valve stem clearance, proceed as follows:

1. Disconnect the negative battery cable.
2. Remove the rocker arm covers.
3. Rotate the engine until the No. 1 cylinder is at TDC of its compression stroke and check the clearance between the rocker arm and the following valves: No. 1 intake and exhaust, No. 2 exhaust, No. 3 intake, No. 4 exhaust and No. 6 intake.
4. Rotate the crankshaft 360 degrees and check the clearance between the rocker arm and the following valves: No. 2 intake, No. 3 exhaust, No. 4 intake, No. 5 intake and exhaust and No. 6 exhaust.
5. The clearance should be 0.09–0.19 in. (2.25–4.79mm).

Rocker Arms/Shafts

REMOVAL & INSTALLATION

2.2L Engine

1. Disconnect the negative battery cable. Remove the vent hose and the PCV valve from the rocker arm cover.
2. Remove the spark plug wire clips and move the wires aside.
3. Remove the rocker arm cover-to-cylinder head bolts, the rocker arm cover and the gasket.
4. Remove the rocker arm shaft assemblies-to-cylinder head bolts and the assemblies from the cylinder head.
5. If neccessary, separate the rocker arms and springs from the shafts; be sure to keep the parts in order for reinstallation purposes.
6. Clean the gasket mating surfaces. Inspect the rocker arms and shafts for wear.

To install:
7. Assemble the rocker arms and springs onto the shafts.

NOTE: When installing the rocker arm shafts onto the cylinder head, observe the notches at the ends of the shafts, they are different and cannot be interchanged.

8. Install the rocker arm/shaft assemblies onto the cylinder head and torque the rocker arm shaft-to-cylinder head bolts to 13–20 ft. lbs. (18–26 Nm), in 2 steps.

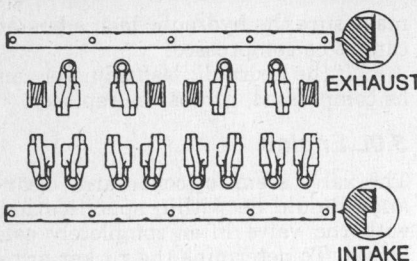

EXHAUST

INTAKE

Exploded view of the rocker arm/shaft assemblies—2.2L engines

9. Using a new gasket, position it onto the cylinder head, apply silicone sealant to the corners at both ends of the cylinder head and install the rocker arm cover. Torque the rocker arm cover-to-cylinder head bolts to 52–69 inch lbs. (6–8 Nm).

10. Install the spark plug retaining clips. Install the vent hose and the PCV valve onto the rocker arm cover.

11. Connect the negative battery cable.

12. Start the engine and check for leaks.

3.0L Engine

1. Disconnect the negative battery cable.

2. Tag and disconnect the spark plug wires from the spark plugs. Remove the spark plug wire/separator assembly from the rocker arm cover retaining studs and move aside.

3. If the left side (front) rocker arm cover is being removed, perform the following:

    a. Remove the fuel injector harness stand-offs from the inboard rocker arm cover studs. Move the harness aside.

    b. Remove the air cleaner closure hose from the oil fill adapter.

4. If the right side (rear) rocker arm cover is being removed, perform the following:

    a. Remove the air cleaner tube and the plastic shield from the throttle body.

    b. Remove the EGR supply tube and the MAP sensor from the throttle body.

    c. Tag and disconnect the vacuum hoses and electrical connectors from the throttle body.

    d. Disconnect the throttle cable and, if equipped with automatic transaxle, the throttle valve control cable from the throttle lever.

    e. Remove the fuel rail bracket bolt and the throttle body attaching bolts. Remove the throttle body.

    f. Remove the PCV valve.

    g. Remove the fuel injector electrical harness stand-offs from the inboard rocker arm cover studs. Move the harness aside.

5. Remove the rocker arm cover retaining bolts and studs, noting their location. Remove the rocker covers.

6. Remove the rocker arm mounting bolts, fulcrums and rocker arms. Identify the position of the rocker arms and fulcrums so they may be reinstalled in their original positions.

7. Inspect the rocker arms, fulcrums and pushrods for wear and/or damage and replace as necessary.

To install:

8. If removed, dip each pushrod in heavy engine oil and install in it's original position. If not removed, lubricate the pushrod ends with heavy engine oil.

9. Dip each rocker arm and fulcrum in heavy engine oil and install in it's original position. For each valve, rotate the crankshaft until the lifter rests on the base circle of the camshaft lobe, before tightening the fulcrum mounting bolts to 24 ft. lbs. (32 Nm).

NOTE: If the original valve train components are being installed, a valve clearance check is not required. If a component has been replaced, perform a valve clearance check.

10. Apply a 5mm bead of silicone sealant at the cylinder head to the intake manifold rail step.

11. Carefully position the cover on the cylinder head and install 1 retaining bolt and 7 studs on the left cover and/or 2 bolts and 6 studs on the right cover. Tighten to 9 ft. lbs. (12 Nm).

NOTE: When positioning the cover to the cylinder head, use a straight down approach to align the bolt holes. Once the cover is in contact with the silicone sealer, any adjustment for bolt alignment can roll the gasket from the cover channel, resulting in leaks.

12. If the left side (front) rocker arm cover is being installed, perform the following:

    a. Install the fuel injector harness stand-offs to the inboard rocker cover studs.

    b. Install the air cleaner closure hose to the oil fill adapter.

13. If the right side (rear) rocker arm cover is being installed, perform the following:

    a. Install the fuel injector harness stand-offs to the inboard rocker cover studs.

    b. Install the PCV valve and connect the hoses.

    c. Position a new gasket and the throttle body on the lower intake manifold. Install the 6 attaching bolts and tighten to 15–22 ft. lbs. (20–30 Nm).

    d. Install the fuel rail bracket bolt

on the throttle body and tighten securely.

    e. Connect the throttle cable and, if equipped, throttle valve control cable, to the throttle lever.

    f. Install the MAP sensor and the EGR supply tube to the throttle body.

    g. Connect the vacuum hoses and electrical connectors to the throttle body.

    h. Install the plastic shield and the air cleaner tube on the throttle body.

14. Connect the spark plug wires to the spark plugs. Install the spark plug wire separator stand-offs to the appropriate rocker arm cover studs.

15. Connect the negative battery cable and install the air cleaner fresh air hose. Start the engine and check for leaks.

Intake Manifold

REMOVAL & INSTALLATION

2.2L Engine

1. Relieve the fuel pressure and disconnect the negative battery cable.

2. Drain the cooling system.

3. From the bottom of the intake manifold, remove the water hose.

4. Disconnect the accelerator cables from the throttle body.

5. Remove the air duct from the throttle body.

6. Label and disconnect the vacuum lines and coolant hoses from the throttle body.

7. Disconnect the electrical connectors from the throttle position sensor, the idle switch and the bypass air control valve.

8. Remove the engine lifting bracket mounting bolts from the throttle body and the engine block.

9. Disconnect the coolant line/EGR hose bracket from the throttle body and the throttle cable brackets from the intake plenum.

10. Remove the wire loom bracket and the EGR back-pressure variable transducer bracket from the right side of the plenum.

11. Remove the PCV hose and the vacuum line assembly bracket from the intake plenum.

12. Label and disconnect the vacuum lines from the intake plenum.

13. Remove the plenum-to-intake manifold nuts/bolts, the plenum and the gasket.

14. Disconnect the electrical connectors from the fuel injectors. Carefully, bend the wire harness retainer brackets away from the wire harness and move the harness assembly away from the intake manifold.

15. Disconnect the fuel pressure and return lines at the fuel rail.

16. Disconnect the EGR pipe from the intake manifold. Label and disconnect any electrical connectors and hoses from the intake manifold.

17. Remove the intake manifold bracket-to-manifold nuts and the bracket. Remove the intake manifold-to-cylinder head nuts/bolts, the manifold and gasket.

18. If necessary, remove the fuel rail and fuel injectors from the intake manifold.

To install:

19. Clean all gasket mating surfaces.

20. Using a new gasket, position the intake manifold on the cylinder head studs and torque the nuts/bolts to 14–22 ft. lbs. (19–30 Nm).

21. Install the intake manifold bracket-to-manifold nuts to 14–22 ft. lbs. (19–30 Nm).

22. Connect the fuel lines to the fuel rail. Connect the electrical connectors to the fuel injectors.

23. Using a new gasket, install the intake plenum onto the intake manifold and torque the nuts/bolts to 14–19 ft. lbs. (19–25 Nm).

24. Connect the vacuum lines to the intake manifold. Install the retaining bolts on the vacuum line assembly bracket to the intake plenum.

25. Install the PCV hose to the intake plenum.

26. Install the wire loom bracket and the EGR variable transducer bracket to the right side of the plenum. Install the wire loom bracket mounting bolts.

27. Install the throttle cable bracket, engine lifting bracket mounting bolt and coolant line/EGR hose bracket to the intake plenum and throttle body.

28. Install the vacuum and coolant hoses to the throttle body.

29. Connect the throttle position sensor, idle switch and bypass air control valve connectors.

30. Install the air duct and the throttle cables to the throttle body.

31. Connect the EGR pipe and connect the water hose to the bottom of the intake manifold.

32. Connect the negative battery cable and fill the cooling system. Start the engine and check for leaks.

3.0L Engine

1. Relieve the fuel system pressure and disconnect the negative battery cable.

2. Drain the cooling system.

3. Remove the air cleaner hoses from the throttle body and rocker cover.

4. Disconnect the fuel lines from the fuel supply manifold. Cover the fuel line ends with clean shop rags to prevent dirt from entering.

5. Tag and disconnect the vacuum lines and electrical connectors from the throttle body.

6. Remove the plastic shield and the EGR supply tube from the throttle body.

7. Disconnect the throttle cable and, if equipped with automatic transaxle, the throttle valve control cable from the throttle lever.

8. Remove the fuel rail bracket bolt and the 6 throttle body mounting bolts. Remove the throttle body.

9. Disconnect the fuel injector harness stand-offs from the injector inboard rocker cover studs and each injector and remove from the engine.

10. Remove the brace from the fuel supply manifold and throttle body. Remove the fuel supply manifold and fuel injectors.

NOTE: The intake manifold assembly can be removed with the fuel supply manifold and fuel injectors in place.

11. Disconnect the upper radiator hose from the thermostat housing and disconnect the heater hoses.

12. Disconnect the engine coolant temperature sensor and coolant temperature sending unit connectors.

13. Tag and disconnect the spark plug wires from the spark plugs.

14. Remove the distributor cap. Mark the position of the rotor and the distributor in the engine and remove the distributor.

15. Remove the ignition coil and bracket assembly from the left side (front) cylinder head and set aside.

16. Remove the rocker arm covers.

17. Loosen the retaining bolt from the No. 3 intake valve and rotate the rocker arm fulcrum away from the valve retainer. Remove the pushrod.

18. Remove the intake manifold retaining bolts. Before attempting to remove the manifold, break the seal between the manifold and cylinder block. Place a suitable prybar between the manifold, near the thermostat, and the transaxle. Carefully pry upward to loosen the manifold.

19. Lift the intake manifold away from the engine. Place shop rags in the lifter valley to catch any dirt or gasket material. Clean all gasket mating surfaces.

To install:

20. Apply silicone sealer to the intersection of the cylinder block and cylinder head assembly at the 4 corners of the lifter valley.

21. Install the front and rear intake manifold seals. Install the intake manifold gaskets onto the cylinder heads and insert the locking tabs on the cylinder head gaskets.

NOTE: Make sure the side of the gasket marked TO INTAKE MANIFOLD is facing away from the cylinder head.

22. Carefully lower the intake manifold into position to prevent disturbing the silicone sealer. Install bolts No. 1, 2, 3 and 4 and snug. Install the remaining bolts and tighten, in sequence, to 11 ft. lbs. (15 Nm). Then tighten, in sequence, to 21 ft. lbs. (28 Nm).

23. Install the thermostat and housing, if removed, using a new gasket. Tighten the mounting bolts to 19 ft. lbs. (25 Nm).

24. If removed, lubricate and install new O-rings on the fuel injectors and install the fuel injectors in the fuel rail, using a light twisting-pushing motion. Install the fuel rail and injectors into the intake manifold, pushing down to seat the O-rings. While holding the fuel rail assembly in place, install the 4 retaining bolts and tighten to 7 ft. lbs. (10 Nm).

25. Install the No. 3 cylinder intake valve pushrod. Apply oil to the pushrod and fulcrum prior to installation. Rotate the crankshaft to place the lifter on the base circle of the camshaft and tighten the rocker arm bolt to 24 ft. lbs. (32 Nm).

26. Install the rocker arm covers and connect the fuel injector electrical harness.

27. Position a new gasket and the throttle body on the intake manifold. Install the mounting bolts and tighten to 15–22 ft. lbs. (20–30 Nm).

28. Install the fuel rail bracket bolt on the throttle body and connect the throttle cable and, if equipped, the throttle valve control cable to the throttle lever.

29. Installl the MAP sensor and the EGR tube to the throttle body.

30. Connect the vacuum hoses and the electrical connectors in their original positions on the throttle body. Install the plastic shield on the throttle body.

31. Install the fuel supply manifold brace. Tighten the retaining bolts to 7 ft. lbs. (10 Nm).

32. Connect the PCV hose at the PCV valve. Connect all remaining vacuum hoses.

33. Install the EGR tube and nut, if equipped. Tighten the nuts on both ends to 37 ft. lbs. (50 Nm).

34. Connect the fuel lines to the fuel rail.

35. Install the distributor assembly, aligning the housing and rotor with the marks that were made during the removal procedure. Install the distributor cap and connect the spark plug wires to the spark plugs. Install the

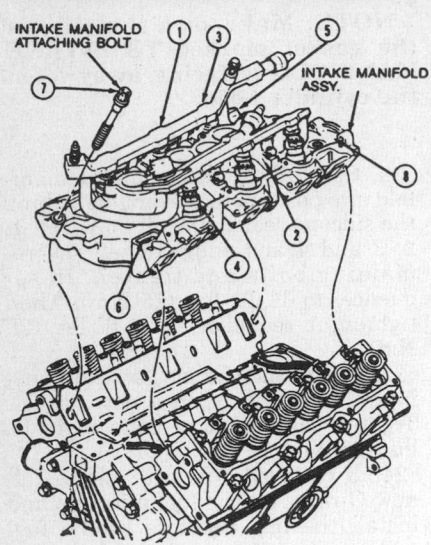

Intake manifold bolt torque sequence— 3.0L engine

wiring stand-offs to the rocker arm cover studs.

36. Install the ignition coil and bracket assembly. Tighten the mounting bolts to 35 ft. lbs. (48 Nm).

37. Connect the engine coolant temperature sensor and coolant temperature sending unit connectors.

38. Install the upper radiator and heater hoses. Fill and bleed the cooling system.

39. Drain and replace the engine oil.

40. Install the air cleaner hoses to the throttle body and rocker cover.

41. Connect the negative battery cable. Start the engine and check for coolant, oil, fuel and vacuum leaks. Check the ignition timing.

Exhaust Manifold

REMOVAL & INSTALLATION

2.2L Engine

1. Disconnect the negative battery cable and the oxygen sensor connector.

2. Remove the turbocharger assembly, if equipped.

3. Remove the oxygen sensor from the exhaust manifold on non-turbocharged vehicles.

4. Disconnect the exhaust pipe from the exhaust manifold and remove the heat shield.

5. Remove the exhaust manifold-to-cylinder head bolts and the exhaust manifold.

6. Remove the inner heat shield and the gaskets.

7. Clean the mating surfaces on the exhaust manifold and the cylinder head.

To install:

8. Position the inner heat shield on the studs

9. Install the exhaust manifold gaskets with the raised edge of the gasket facing the exhaust manifold.

10. Install the exhaust manifold and tighten the bolts to 16–21 ft. lbs. (22–28 Nm).

11. Install the outer heat shield and tighten the bolts to 14–22 ft. lbs. (19–30 Nm).

12. Install the exhaust gas oxygen sensor on non-turbocharged vehicles.

13. Install the turbocharger assembly, if equipped.

14. Connect the exhaust pipe to the exhaust manifold, using a new gasket. Tighten the bolts to 23–34 ft. lbs. (31–46 Nm).

15. Connect the exhaust oxygen sensor wire and connect the negative battery cable.

3.0L Engine

LEFT SIDE (FRONT) MANIFOLD

1. Remove the oil dipstick tube, support bracket and heat shield retaining nuts. Carefully rotate the tube away from the manifold stud.

2. Raise and safely support the vehicle.

3. Remove the exhaust manifold-to-front exhaust pipe attaching nuts.

4. Lower the vehicle and remove the exhaust manifold attaching bolts and the manifold.

To install:

5. Clean all gasket mating surfaces. Lightly oil the bolt and stud threads.

6. Install the exhaust manifold on the cylinder head with the attaching bolts. Tighten the bolts to 18 ft. lbs. (25 Nm).

7. Raise and safely support the vehicle.

8. Connect the exhaust pipe to the manifold and tighten the attaching nuts to 20 ft. lbs. (27 Nm). Lower the vehicle.

9. Rotate the oil dipstick tube and bracket into position over the manifold stud. Install the heat shield and retaining nuts. Tighten the nuts to 13 ft. lbs. (18 Nm).

RIGHT SIDE (REAR) MANIFOLD

1. Raise and safely support the vehicle.

2. Remove the EGR supply tube from the exhaust manifold, if equipped. Use a back-up wrench on the lower fitting adapter.

3. Remove the heat shield retaining nuts and the manifold-to-exhaust pipe retaining nuts.

4. Remove the exhaust manifold retaining bolts and the manifold.

To install:

5. Clean all gasket mating surfaces. Lightly oil the bolt and stud threads.

6. Install the exhaust manifold on the cylinder head with the attaching

bolts. Tighten the bolts to 15–22 ft. lbs. (20–30 Nm).

7. Connect the exhaust pipe to the manifold. Tighten the attaching nuts to 20 ft. lbs. (27 Nm).

8. Install the spark plug heat shield and retaining nuts. Tighten the nuts to 12–15 ft. lbs. (16–20 Nm).

9. Connect the EGR supply tube to the exhaust manifold. Tighten to 37 ft. lbs. (50 Nm). Lower the vehicle.

Turbocharger

REMOVAL & INSTALLATION

1. Disconnect the negative battery cable.

2. Drain the cooling system.

3. Remove the air inlet and outlet hoses from the turbocharger assembly.

4. Remove the heat shields from the exhaust manifold and turbocharger.

5. Disconnect the oxygen sensor electrical connector and place the wire over the front of the vehicle, away from the heat shield.

6. From the top of the turbocharger, remove the oil feed line. From the lower portion of the turbocharger, remove oil return line and gasket.

7. Disconnect the coolant inlet and outlet hoses from the turbocharger.

8. Remove the EGR tube from the exhaust manifold. Disconnect the turbo boost control solenoid valve electrical connector.

9. Remove the air tube from the turbo boost control solenoid valve at the turbocharger outlet air hose.

10. From under the turbocharger, remove the retaining bracket-to-turbocharger bolt.

11. Discharge the air conditioning system and remove the refrigerant line from the head of the compressor.

12. Remove the oxygen sensor from the exhaust manifold.

13. Disconnect the converter inlet pipe from the turbocharger joint pipe. Remove the exhaust manifold-to-cylinder head bolts, the exhaust manifold/turbocharger assembly and the gasket from the vehicle.

14. Remove the exhaust manifold-to-turbocharger nuts, the manifold and the gasket from the turbocharger. Remove the joint pipe-to-turbocharger nuts, the pipe and the gasket from the turbocharger.

15. Clean the gasket mounting surfaces.

To install:

16. Using a new gasket, install the joint pipe and heat shield assembly on the turbocharger and torque the nuts to 27–46 ft. lbs (37–63 Nm).

17. Using a new gasket, install the exhaust manifold on the turbocharger

and torque the nuts to 20–29 ft. lbs. (27–39 Nm).

18. Using a new gasket, position the exhaust manifold/turbocharger assembly onto the cylinder head and torque the nuts to 16–21 ft. lbs. (22–28 Nm).

19. Using a new gasket, install the converter inlet pipe to the joint pipe and torque the nuts to 26–36 ft. lbs. (34–49 Nm).

20. Install the mounting bolt to the retaining bracket under the turbocharger assembly.

21. Install the oil return line and coolant outlet hose to the turbocharger assembly.

22. Install the air tube to the turbocharger outlet air hose.

23. Connect the turbo boost control solenoid valve electrical connector.

24. Install the oxygen sensor. To ease the remaining installation procedure, leave the wiring assembly aside of the heat shield assembly.

25. Install the EGR tube onto the exhaust manifold and install the coolant inlet hose onto the turbocharger assembly.

26. Pour 0.85 oz. (25ml) of engine oil in the oil inlet fitting, then install the oil feed line onto the turbocharger assembly.

27. Install the heat shields onto the exhaust manifold and turbocharger assembly.

28. Install the inlet and outlet hoses on the turbocharger compressor housing.

29. Connect the oxygen sensor electrical connector and install the wiring assembly.

30. Connect the refrigerant line to the compressor.

31. Fill the cooling system and connect the negative battery cable.

32. After replacing the turbocharger, perform the following:

   a. Disconnect the electrical connector from the ignition coil.

   b. Crank the engine for approximately 20 seconds.

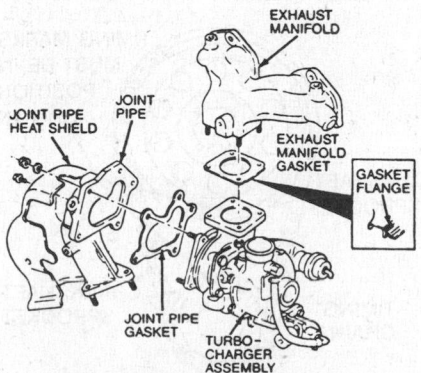

Turbocharger/exhaust manifold assembly—2.2L engine

   c. Reconnect the electrical connector to the ignition coil.

   d. Start the engine and operate it at idle for approximately 30 seconds.

   e. Stop the engine, disconnect the negative battery cable and depress the brake pedal for at least 5 seconds to cancel the malfunction code.

   f. Reconnect the negative battery cable.

33. Start the engine, allow it to reach normal operating temperatures and check for leaks and engine operation. Recharge the air conditioning system.

Timing Chain Front Cover

REMOVAL & INSTALLATION

3.0L Engine

1. Disconnect the negative battery cable.

2. Drain the cooling system.

3. Remove the 2 retaining bolts and remove the power steering protective shroud.

4. Using a suitable tool on the idler pulley tensioner, release the tension on the accessory drive belt and remove the belt. Remove the belt tensioner.

5. Raise and safely support the vehicle.

6. Remove the right front wheel and tire assembly and the plastic inner fender shield. Using a suitable tool, turn the water pump idler pulley tensioner clockwise to release the tension on the water pump belt. Remove the belt.

7. Lower the vehicle and support the engine with a floor jack.

8. If equipped with manual transaxle, remove the right engine mount from the water pump bracket.

9. Remove the 3 nuts that attach the right upper engine mount to the timing cover. Lower the floor jack carefully, allowing the engine to rest on the remaining mounts.

10. Raise and safely support the vehicle.

11. Remove the crankshaft damper bolt and flat washer. Using a suitable puller, remove the damper from the crankshaft.

12. Remove the 3 nuts and 1 bolt that attach the right side of the subframe to the body. Pull the subframe down slightly to remove the damper from the vehicle.

13. Disconnect the water pump-to-front cover hose from the water pump connection.

NOTE: The timing cover may be removed with the water pump hose attached.

14. Drain the engine oil and remove the oil pan. Discard the oil pan gasket.

15. Remove the 4 lowest timing cover retaining bolts and lower the vehicle.

16. Support the bottom of the engine using care to prevent damage to the crankshaft and oil pump assembly. Remove the 3 nuts and 1 bolt attaching the upper engine mount to the top of the front cover.

17. Remove the 6 remaining timing cover mounting bolts. Pry the timing cover away from the cylinder block. Carefully pull the cover over the end of the crankshaft and lower it through the bottom of the engine compartment.

To install:

18. Clean all gasket material and old silicone sealer from all gasket mating surfaces. Pry the old crankshaft seal from the timing cover.

19. Lubricate the lip of a new crankshaft seal and install in the timing cover, using a seal installation tool.

20. Install a new timing cover gasket over the cylinder block dowels and install the timing cover. Install the 6 upper mounting bolts, finger tight. Apply pipe sealant to bolt No. 5 prior to installation.

21. Raise and safely support the vehicle.

22. Install the 4 lower mounting bolts, finger tight. Apply pipe sealant to bolt No. 2 prior to installation.

23. Tighten the timing cover bolts, in sequence, to 18 ft. lbs. (25 Nm).

24. Lower the vehicle and install the upper engine mount.

25. Raise and safely support the vehicle.

26. Install the oil pan with a new gasket. Tighten the mounting bolts to 9 ft. lbs. (12 Nm).

27. Coat the crankshaft damper sealing surface with clean engine oil. Install the damper, using a suitable tool. Install the damper bolt and flat washer and tighten to 92–122 ft. lbs. (125–165 Nm).

28. Lower the vehicle. Using a suitable floor jack, raise the engine and install the right engine mount nuts. Tighten the nuts to 55–76 ft. lbs. (74–103 Nm).

29. Install the subframe nuts and bolt. Tighten the bolt to 27–40 ft. lbs. (36–54 Nm) and the nuts to 69–97 ft. lbs. (93–132 Nm).

30. Lower the engine and remove the floor jack. If equipped with manual transaxle, install the mount to the water pump bracket.

31. Raise and safely support the vehicle. Install the water pump belt, plastic shield and the right front wheel and tire assembly.

32. Connect the hose from the timing cover to the water pump.

33. Lower the vehicle.

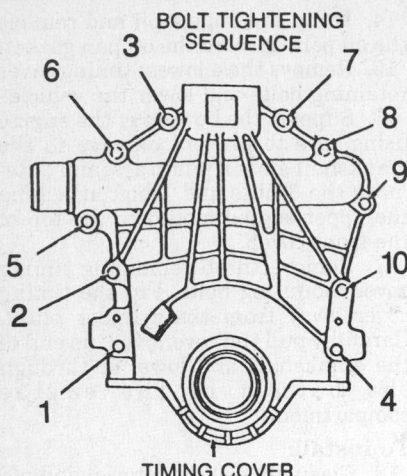

BOLT TIGHTENING SEQUENCE

TIMING COVER

Timing chain front cover bolt torque sequence—3.0L engine

34. Install the accessory drive belt.

35. Install the power steering protective shroud. Tighten the retaining bolts to 7 ft. lbs. (10 Nm).

36. Fill the crankcase with the proper type and quantity of engine oil. Fill the cooling system.

37. Connect the negative battery cable, start the engine and check for leaks.

Front Cover Oil Seal

REPLACEMENT

3.0L Engine

1. Remove the plastic belt shield from the power steering pump. Using a ½ inch drive breaker bar on the idler pulley tensioner, release the tension on the accessory drive belt and remove the belt.

2. Raise and safely support the vehicle.

3. Remove the right front wheel and tire assembly and the plastic inner fender shield.

4. Using a wrench, turn the water pump idler pulley tensioner clockwise to release tension on the water pump belt. Remove the belt.

5. Lower the vehicle and support the engine with a floor jack.

6. If equipped with manual transaxle, remove the right engine mount from the water pump bracket.

7. Remove the 3 nuts that attach the right upper engine mount to the timing cover. Lower the floor jack carefully, allowing the engine to rest on the remaining mounts.

8. Raise and safely support the vehicle.

9. Remove the crankshaft damper bolt and flat washer. Using a suitable puller, remove the damper from the crankshaft.

10. Remove the 3 nuts and 1 bolt that attach the right side of the subframe to the body. Pull the subframe down slightly to remove the damper from the vehicle.

11. Remove the front cover oil seal using a small prybar.

To install:

12. Inspect the timing cover and shaft seal surface of the crankshaft damper for damage, nicks, burrs or other roughness which may cause the new seal to fail. Service or replace components, as necessary.

13. Lubricate the seal lip with clean engine oil and install the seal in the timing cover, using a suitable installation tool.

14. Coat the crankshaft damper sealing surface with clean engine oil. Install the damper, using a suitable tool. Install the damper bolt and flat washer and tighten to 92–122 ft. lbs. (125–165 Nm).

15. Lower the vehicle. Using a suitable floor jack, raise the engine and install the right engine mount nuts. Tighten the nuts to 55–76 ft. lbs. (74–103 Nm).

16. Install the subframe nuts and bolt. Tighten the bolt to 27–40 ft. lbs. (36–54 Nm) and the nuts to 69–97 ft. lbs. (93–132 Nm).

17. Lower the engine and remove the floor jack. If equipped with manual transaxle, install the mount to the water pump bracket.

18. Raise and safely support the vehicle. Install the water pump belt, plastic shield and the right front wheel and tire assembly.

19. Lower the vehicle.

20. Install the accessory drive belt and the power steering pulley shield.

Timing Chain and Sprockets

REMOVAL & INSTALLATION

3.0L Engine

1. Disconnect the negative battery cable and drain the cooling system.

2. Remove the crankshaft pulley and damper. Remove the timing cover.

3. Rotate the crankshaft until the No.1 piston is at the Top Dead Center (TDC) and the timing marks are aligned.

4. Remove the camshaft sprocket retaining bolt and washer.

5. Slide the sprockets and the chain forward and remove as an assembly.

6. Clean and inspect all the parts prior to installation.

To install:

7. Slide the sprockets and the chain on as as assembly with the timing marks aligned.

8. Install the camshaft retaining bolt and washer. Tighten the retaining bolt to 41–51 ft. lbs. (55–70 Nm) and lubricate the chain and sprockets with engine oil.

NOTE: The camshaft retaining bolt has a drilled passage for timing chain lubrication. If damaged, do not replace with a standard bolt. Clean the oil passage with solvent prior to installation.

9. Position the timing cover gasket onto the cylinder block alignment dowels.

10. Install the timing cover onto the cylinder block, being careful not to damage the seal.

11. Install the oil pan using a new gasket.

12. Install the crankshaft damper and pulley.

13. Refill the crankcase and the cooling system. Connect the negative battery cable.

14. Start the engine and check for any leaks. Recheck the timing.

Timing Belt Front Cover

REMOVAL & INSTALLATION

2.2L Engine

1. Disconnect the negative battery cable.

2. Loosen the air conditioning compressor and alternator adjusting and pivot bolts, rotate the compressor and alternator toward the engine and remove the drive belts.

3. Raise and safely support the vehicle.

4. Remove the right front wheel and tire assembly and the right inner fender panel. Remove the 6 bolts, the crankshaft pulley and baffle plate.

5. Lower the vehicle.

6. Support the engine with a floor jack. Remove the 2 nuts and dowels from the right engine mount and remove the mount.

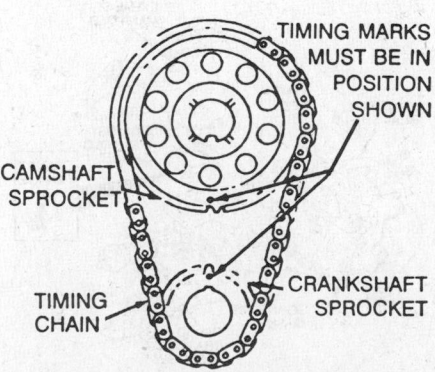

TIMING MARKS MUST BE IN POSITION SHOWN

CAMSHAFT SPROCKET

CRANKSHAFT SPROCKET

TIMING CHAIN

Timing mark alignment—3.0L engine

7. Remove the 7 bolts that retain the timing belt covers and remove the covers.

To install:

8. Install the lower cover gasket and the lower cover. Tighten the bolts to 61–87 inch lbs. (7–10 Nm).

9. Install the upper cover gasket and the upper cover. Tighten the bolts to 61–87 inch lbs. (7–10 Nm).

10. Position the engine mount on the engine and install the 2 nuts and dowels. Remove the floor jack.

11. Install the crankshaft sprocket baffle with the curved outer lip facing outward. Install the crankshaft pulley with the deep recess facing out and install the 6 bolts. Tighten the bolts to 109–152 inch lbs. (12–17 Nm).

12. Install the drive belts. Adjust the belt tension and tighten the adjusting and pivot bolts.

13. Install the right inner fender panel and connect the negative battery cable.

OIL SEAL REPLACEMENT

1. Disconnect the negative battery cable. Remove the timing belt.

2. If equipped with a manual transaxle, place the shift lever in the **4TH** gear and firmly apply the parking brake.

3. If equipped with an automatic transaxle, remove the lower flywheel cover and install a suitable flywheel locking tool, onto the flywheel ring.

4. Remove the crankshaft sprocket-to-crankshaft bolt, the sprocket and the key.

5. Using a small prybar, pry the oil seal from the engine block; be careful not to score the crankshaft or the seal seat.

To install:

6. Using an oil seal installation tool or equivalent, lubricate the seal lip with engine oil and drive the new into the engine until it seats.

7. Install the crankshaft key and sprocket. Torque the crankshaft sprocket-to-crankshaft bolt to 108–116 ft. lbs. (147–157 Nm).

8. If necessary, remove the flywheel locking tool.

9. Install the timing belt and connect the negative battery cable.

Timing Belt and Tensioner

REMOVAL & INSTALLATION

1. Disconnect the negative battery cable.

2. Remove the drive belts, crankshaft pulley and the timing belt covers.

3. Remove the timing belt tensioner spring and retaining bolt. Remove the idler pulley retaining bolt.

4. Mark the direction of rotation of the timing belt and remove the belt.

To install:

5. Align the camshaft and crankshaft sprockets with the marks on the cylinder head and the oil pump housing.

6. Install the timing belt. If reusing the old belt, observe the direction of rotation mark made during the removal procedure.

7. Place the timing belt tensioner and spring in position. Temporarily secure the tensioner with the spring fully extended. Make sure the timing belt is installed so there is no looseness at the water pump pulley at the idler side.

8. Loosen the idler bolt. Turn the crankshaft twice in the direction of rotation; align the timing marks.

NOTE: Always turn the crankshaft in the correct direction of rotation only. If the crankshaft is turned in the opposite direction, the timing belt may lose tension and correct belt timing may be lost.

9. Check to see that the timing marks are correctly aligned. Tighten the tensioner bolt to 27–38 ft. lbs. (37–52 Nm).

10. Measure the belt deflection between the crankshaft and camshaft pulleys. The correct deflection should be 0.30–0.33 (7.5–8.5mm) at 22 ft. lbs. (98 Nm) of pressure. If the deflection is not correct, repeat Steps 8 and 9.

11. Install the timing belt covers, crankshaft pulley and drive belts.

12. Connect the negative battery cable.

Timing Sprockets

REMOVAL & INSTALLATION

2.2L Engine

1. Disconnect the negative battery cable. Remove the timing belt.

2. Insert a proper tool through one of the camshaft sprocket holes to keep it from turning.

3. Remove the sprocket bolt and the sprocket from the camshaft.

4. If equipped with a manual transaxle, place the shift lever in **4th** gear and apply the parking brake. If equipped with an automatic transaxle, remove the flywheel dust cover and install a flywheel locking tool to hold the flywheel.

5. Remove the crankshaft sprocket bolt, sprocket and key.

To install:

6. Install the camshaft sprocket,

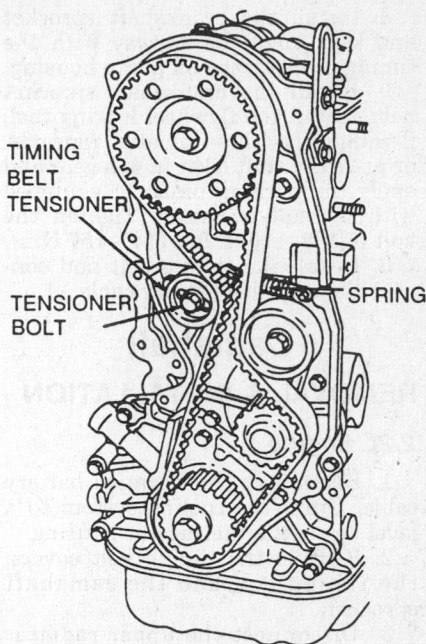

Timing belt installation – 2.2L engines

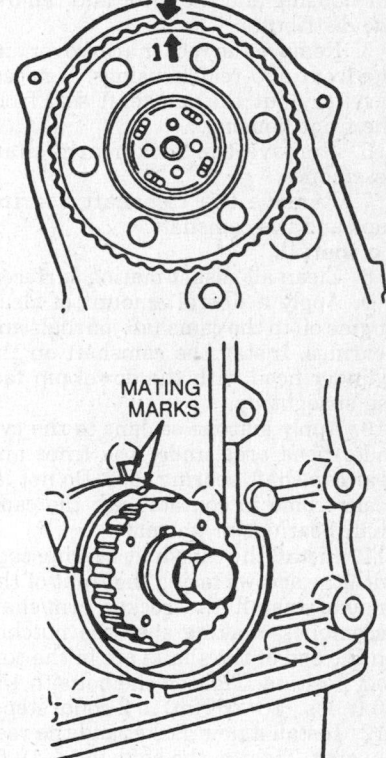

Timing mark locations – 2.2L engines

aligning the dowel with the number **1** mark.

7. Install the camshaft sprocket bolt. Hold the sprocket with a suitable tool and tighten the bolt to 35–48 ft. lbs. (47–65 Nm).

8. Install the crankshaft sprocket and key. Align the keyway with the timing mark on the oil pump housing.

9. Install the crankshaft sprocket bolt. Install the flywheel locking tool, if equipped with automatic transaxle, or place the shift lever in **4th** gear and apply the parking brake, if equipped with manual transaxle. Tighten the bolt to 108–116 ft. lbs. (147–157 Nm).

10. Install the timing belt and connect the negative battery cable.

Camshaft

REMOVAL & INSTALLATION

2.2L Engine

1. Disconnect the negative battery cable. Drain the cooling system to a level below the thermostat housing.

2. Remove the timing belt covers, the timing belt and the camshaft sprocket.

3. Disconnect the upper radiator hose and the electrical connectors from the thermostat housing.

4. Mark the position of the distributor housing and the rotor and remove the distributor.

5. Remove the rocker arm cover and the front and rear housings. If necessary, pry out the camshaft seal from the front housing.

6. Remove the rocker arm/shaft assemblies.

7. Remove the camshaft bearing caps and the camshaft.

To install:

8. Clean all gasket mating surfaces.

9. Apply a liberal amount of clean engine oil to the camshaft journals and bearings. Install the camshaft on the cylinder head with the dowel pin facing straight up.

10. Apply silicone sealant to the cylinder head area under the front and rear camshaft bearing caps. Do not let sealer come in contact with the camshaft bearings or journals.

11. Install the camshaft bearing caps with the arrows facing the front of the engine. Install the rocker arm/shaft assemblies, making sure the notches on the end of the shafts are in the correct position. Tighten the bolts to 13–20 ft. lbs. (18–26 Nm) in 2 equal steps.

12. Install a new gasket and the rear housing. Tighten the bolts to 14–19 ft. lbs. (19–25 Nm).

13. If the camshaft seal was removed, lubricate the lip of a new seal and install in the front housing, using a suitable installation tool. Install a new gasket and the front housing. Tighten the bolts to 14–19 ft. lbs. (19–25 Nm).

14. Install the rocker arm cover, tightening the retaining bolts to 52–69 inch lbs. (6–8 Nm).

15. Install the distributor, aligning the marks that were made during the removal procedure.

16. Connect the electrical connectors and the upper radiator hose.

17. Install the camshaft sprocket, the timing belt and the timing belt covers.

18. Fill the cooling system, connect the negative battery cable and start the engine. Check the ignition timing and check for leaks.

3.0L Engine

1. Disconnect the negative battery cable. Remove the engine assembly from the vehicle.

2. Remove the timing covers, rocker arm covers and the intake manifold.

3. Remove the hydraulic lifters using a magnet and keep them in order, so they may be reinstalled in their original positions. If the lifters are stuck in the bores, use a hydraulic lifter puller or equivalent, to remove them.

4. Check the camshaft endplay. If the endplay is excessive, replace the thrust plate.

5. Remove the timing chain and sprockets.

6. Remove the camshaft thrust plate. Remove the camshaft by pulling it toward the front of the engine.

NOTE: Use caution to avoid damaging the bearings, journals and lobes.

7. Clean and inspect all the parts prior to installation.

To install:

8. Lubricate the camshaft lobes and the journals with heavy engine oil. Carefully slide the camshaft through the bearings in the cylinder block.

9. Install the thrust plate and tighten the bolts to 6–8 ft. lbs. (8–12 Nm).

10. Install the timing chain and sprockets. Check the camshaft sprocket bolt for blockage of drilled oil passages.

11. Install the timing cover and the crankshaft damper.

12. Lubricate the lifters and lifter bores with heavy engine oil and install the lifters into their original bores.

13. Install the pushrods, rocker arms, rocker covers and intake manifold.

14. Install the engine assembly and replace the negative battery cable.

15. Fill the cooling system and the crankcase with the proper fluids. Run the engine and check for leaks.

Piston and Connecting Rod

POSITIONING

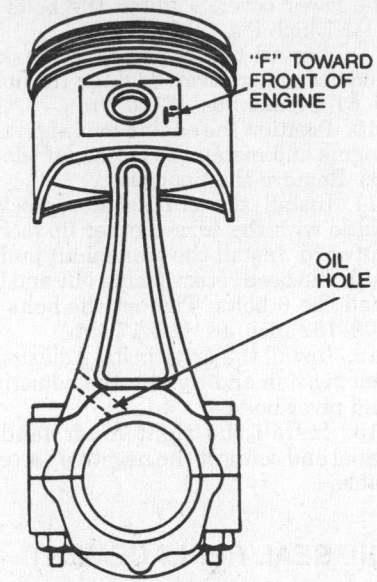

Piston and connecting rod—2.2L engine

Piston and connecting rod—3.0L engine

ENGINE LUBRICATION

Oil Pan

REMOVAL & INSTALLATION

2.2L Engine

1. Disconnect the negative battery cable.

2. Raise and support the vehicle, safely.

3. Remove the right wheel and tire assembly and the right inner splash shield.

4. Drain the crankcase.

5. Remove the engine-to-flywheel housing support bracket, the flywheel housing dust cover bolts and cover.

6. Remove the front exhaust pipe and the exhaust pipe support bracket.

7. Remove the oil pan-to-engine bolts, the oil pan, the oil pickup tube and the stiffener.

8. Clean the gasket mounting surfaces.

To install:

9. Using silicone sealant, apply a continuous bead on both sides of the stiffener, along the inside of the bolt holes.

10. Install the oil pan and stiffener and tighten the mounting bolts to 69–104 inch lbs. (8–12 Nm).

11. Install the flywheel housing dust cover and tighten the bolts to 49–95 inch lbs. (8–11 Nm).

12. Install the exhaust pipe support bracket and the front exhaust pipe.

13. Install the flywheel housing support bracket-to-flywheel housing and tighten the bolts to 27–38 ft. lbs. (37–52 Nm).

14. Tighten the flywheel housing-to-engine support bracket bolts to 27–38 ft. lbs. (37–52 Nm).

15. Install the oil pan drain plug. Install the oil temperature sending unit, if equipped.

16. Install the inner fender splash shield and the wheel and tire assembly. Lower the vehicle.

17. Add engine oil to the proper level.

18. Connect the negative battery cable, start the engine and check for leaks.

3.0L Engine

1. Disconnect the negative battery cable. Raise and safely support the vehicle.

2. Drain the engine oil and remove the starter motor.

3. Remove the front and rear transaxle-to-engine braces.

4. Disconnect the low oil level sensor connector from the dash panel side of the oil pan.

5. Remove the exhaust inlet pipe from the manifolds and position it aside.

6. Drain the cooling system and remove the water pump.

7. Remove the water pump bracket and idler pulley tensioner.

8. Remove the mounting bolts and nut from the front end of the right crossmember.

9. Loosen, but do not remove the bolts and nut from the rear end of the right crossmember.

NOTE: Allow the crossmember to drop as low as possible to allow the removal of the oil pan. If any attempt is made to remove the oil pan without lowering the crossmember first, damage to the baffle may occur. The oil pan must be pulled straight down without turning or prying it out.

10. Remove the oil pan mounting bolts and the oil pan.

To install:

11. Apply a $1/5$ in. (4–5mm) bead of silicone sealer to the junction of the rear main bearing cap and the cylinder block and the junction of the front cover assembly and the cylinder block.

12. Position the oil pan gasket on the oil pan with the bend against the pan

surface and secure with gasket adhesive.

13. Place the oil pan on the cylinder block and tighten the mounting bolts to 9 ft. lbs. (12 Nm).

14. Lift the right crossmember into place and tighten all the nuts and bolts.

15. Install the water pump mounting bracket and the idler pulley tensioner.

16. Install the water pump and the exhaust inlet pipe.

17. Connect the oil level sensor and install the transaxle-to-engine braces.

18. Install the starter motor.

19. Lower the vehicle. Refill the crankcase and the cooling system.

20. Connect the negative battery cable. Run the engine and check for leaks.

Oil Pump

REMOVAL & INSTALLATION

2.2L Engine

1. Disconnect the negative battery cable. Raise and safely support the vehicle.

2. Remove the crankshaft sprocket. Drain the engine oil and remove the oil pan.

3. Remove the oil pump pickup tube-to-oil pump bolts, the tube and gasket.

4. Remove the oil pump-to-cylinder block bolts, the pump and gasket.

5. If necessary, pry the oil seal from the pump and clean the seal bore.

6. Clean the gasket mounting surfaces. Inspect the pump and gears for wear; if necessary, replace the parts.

To install:

7. If necessary, press a new seal into the oil pump until it seats and lubricate the seal lip with engine oil. Install a new O-ring into the oil pump.

8. Apply a continuous bead of silicone sealer to the oil pump gasket surface.

NOTE: When using sealant, do not allow the sealant to squeeze into the pump's outlet hole in the pump or cylinder block.

9. Install the oil pump to the cylinder block; be careful not to cut the oil seal lip. Tighten the upper oil pump-to-cylinder block bolts to 14–19 ft. lbs. (19–25 Nm) and the lower oil pump-to-cylinder block bolts to 27–38 ft. lbs. (37–52 Nm).

10. Install the oil pump pickup tube using a new gasket.

11. Install the oil pan and the crankshaft sprocket.

12. Connect the negative battery cable and refill the crankcase. Start engine and check for leaks.

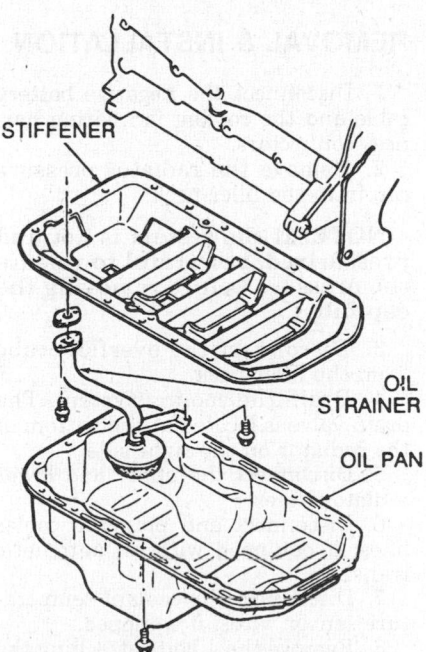

Oil pan, pickup tube and stiffener—2.2L engine

STIFFENER

OIL STRAINER

OIL PAN

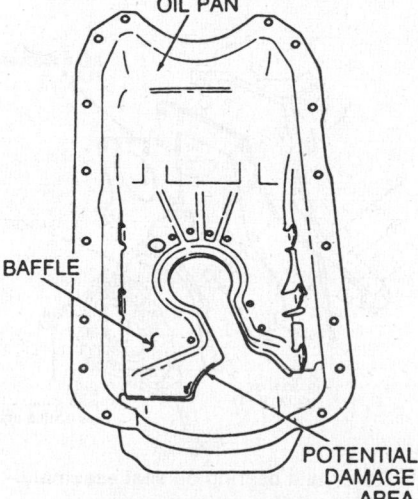

OIL PAN

BAFFLE

POTENTIAL DAMAGE AREA

Oil pan—3.0L engine

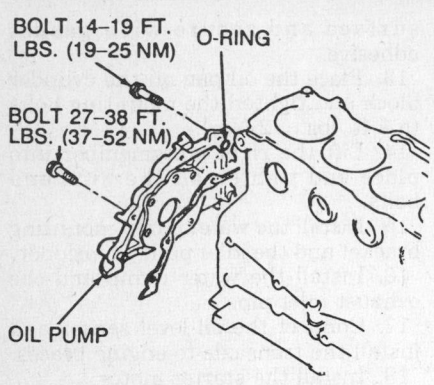

BOLT 14–19 FT. LBS. (19–25 NM)
O-RING
BOLT 27–38 FT. LBS. (37–52 NM)
OIL PUMP

Oil pump assembly—2.2L engine

3.0L Engine

1. Disconnect the negative battery cable. Raise and support the vehicle, safely. Drain the engine oil.
2. Remove the oil pan and the oil pump mounting bolt.
3. Remove the oil pump and the intermediate shaft from the rear main bearing cap.
4. Pull the intermediate shaft out of the oil pump.

To install:

5. Insert the pump intermediate shaft into the drive hole in the pump assembly until it clicks into place.
6. Pour a small amount of clean oil into the outlet hole in the body of the oil pump.
7. Lift the oil pump assembly into place guiding the intermediate shaft through the hole in the rear main bearing cap. Seat the pump securely on the locating dowels.
8. Install the pump mounting bolt and tighten to 35 ft. lbs. (48 Nm).
9. Install the oil pan.
10. Lower the vehicle and refill the crankcase. Connect the negative bat-

PUMP ATTACHING BOLT
OIL PUMP ASSY
OIL PUMP INTERMEDIATE SHAFT ASSY
CYLINDER BLOCK ASSY
DOWEL

Oil pump assembly—3.0L engine

tery cable, run the engine and check for leaks.

Rear Main Bearing Oil Seal

The rear main oil seal is a solid ring type. The transaxle must be removed to replace the seal.

REMOVAL & INSTALLATION

2.2L Engine

1. Disconnect the negative battery cable.
2. Raise and support the vehicle, safely. Remove the transaxle assembly.
3. If equipped with a manual transaxle, remove the clutch and flywheel assembly.
4. If equipped with an automatic transaxle, remove the flexplate-to-crankshaft bolts, the flexplate and shim plates.
5. If necessary, remove the rear engine plate.
6. Remove the rear oil seal housing mounting bolts, the housing and the gasket.
7. Using a small prybar, pry the oil seal from the oil seal housing. Clean the gasket mounting surfaces.

To install:

8. Using a new oil seal, face the seal's hollow side toward the engine and press it into the oil seal housing until it seats. Lubricate the seal lip with engine oil.
9. Apply sealant to the rear oil seal housing-to-oil pan surface.
10. Using a new gasket, install the rear oil seal housing and torque the seal housing-to-engine bolts to 69–104 inch lbs. (8–12 Nm).
11. Install the rear engine plate, if removed. Tighten the bolts to 14–22 ft. lbs. (19–30 Nm).

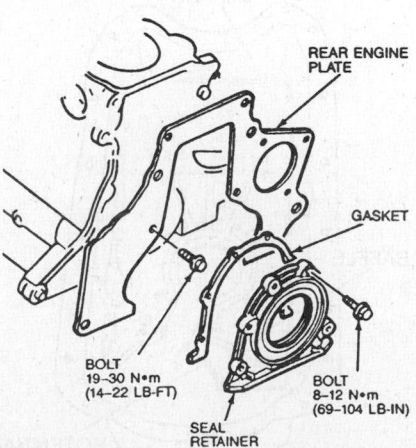

REAR ENGINE PLATE
GASKET
BOLT 19–30 N•m (14–22 LB-FT)
SEAL RETAINER
BOLT 8–12 N•m (69–104 LB-IN)

Rear main bearing oil seal assembly—2.2L engine

12. Install the clutch and flywheel assembly or the flexplate, as applicable. Tighten the flywheel or flexplate bolts to 71–76 ft. lbs. (96–103 Nm).
13. Install the transaxle, lower the vehicle and connect the negative battery cable.

3.0L Engine

1. Disconnect the negative battery cable.
2. Raise and safely support the vehicle. Remove the transaxle assembly.
3. Remove the clutch/flywheel assembly if equipped with manual transaxle or the flexplate if equipped with automatic transaxle.
4. Using a seal removal tool, remove the rear main oil seal. Use care to avoid damaging the oil seal surface.

To install:

5. Apply engine oil to the outer lips and inner seal edge. Using a suitable tool, install the new rear main oil seal. Make sure the seal is seated properly.
6. Install the clutch/flywheel assembly or flexplate, as applicable. Tighten the flywheel or flexplate bolts to 54–64 ft. lbs. (73–87 Nm).
7. Install the transaxle assembly and lower the vehicle. Connect the negative battery cable.

ENGINE COOLING

Radiator

REMOVAL & INSTALLATION

1. Disconnect the negative battery cable and the cooling fan wiring harness connectors.
2. Remove the radiator pressure cap from the filler neck.

NOTE. If the system is hot and pressurized, be careful to release the pressure before removing the cap fully.

3. Disconnect the overflow tube from the filler neck.
4. Drain the cooling system. The drain valve is located at the bottom of the radiator on the right side.
5. Disconnect the upper and lower radiator hoses.
6. Disconnect and plug the cooler lines, if equipped with an automatic transaxle.
7. Disconnect the coolant temperature sensor wires, if equipped.
8. Remove the 4 bolts attaching the radiator upper tank brackets to the radiator core support.

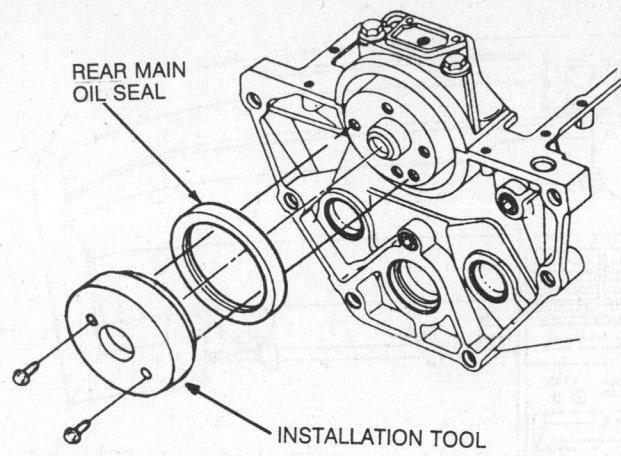

Rear main bearing oil seal—3.0L engine

9. Remove the radiator and the cooling fan as an assembly.

10. Remove the fan shroud mounting bolts.

11. Remove the fan and shroud assembly from the radiator.

To install:

12. Install the fan and shroud assembly. Tighten the mounting bolts to 61–87 inch lbs. (7–10 Nm).

13. Install the radiator, making sure the lower tank engages the insulators.

14. Install the upper radiator insulators and tighten the retaining bolts to 69–95 inch lbs. (8–11 Nm).

15. Unplug and connect the cooler lines, if required.

16. Reattach the wiring harness to the routing clips and install the upper and lower radiator hoses to the radiator.

17. Connect the overflow tube to the radiator and connect the cooling fan wiring connectors.

18. Close the radiator drain valve and fill the system with coolant.

19. Connect the negative battery cable, warm the engine to pressurize the system and check for leaks.

20. Recheck the coolant level and refill if neccessary.

Heater Core

REMOVAL & INSTALLATION

Without Air Conditioning

1. Disconnect the negative battery cable and drain the cooling system.

2. Remove the steering column.

3. Remove the instrument cluster.

4. Remove the floor console as follows:

   a. If equipped with automatic transaxle, remove the 2 transaxle selector lever knob mounting screws and remove the knob. Remove the selector trim piece. Remove the 4 selector bezel mounting screws. Lift the bezel and disconnect the illumination bulb and the shift control switch wiring harness. Remove the bezel.

   b. If equipped with manual transaxle, Slide the shifter boot down and remove the shift knob. Remove the trim piece and the boot.

   c. Remove the front ash tray and cigarette lighter.

   d. Remove the 4 screws located on the front of the console.

   e. Position the front seats to gain access to the console rear access hole covers.

   f. Remove the covers from each side of the console and remove the retaining screws. Position the front seats all the way to the rear.

   g. Lift the console from the rear and disconnect the power mirror adjust switch and programmed ride control switch electrical connectors.

   h. Apply the parking brake and carefully remove the floor console.

5. Remove the hood release handle.

6. Remove the left and right console kick panels and instrument panel dash side covers.

7. Remove the heater control panel as follows:

   a. Remove the bezel cover from the control assembly face.

   b. Remove the 4 attaching screws from the control assembly housing.

   c. Remove the left and right side sound deadening panels.

   d. Remove the REC/FRESH control cable at the REC/FRESH selector door assembly.

   e. Disconnect the blower switch and control assembly illumination electrical connectors.

   f. Remove the temperature control cable from the temperature blend door assembly at the right side of the heater case.

   g. Remove the function selector cable from the function control door assembly at the left side of the heater case.

   h. Remove the control assembly and control cables as an assembly. Note how the cables are routed while removing, to aid installation.

8. Remove the radio assembly.

9. If equipped, remove the trip computer.

10. Remove the instrument panel access cover to gain access to the center instrument panel mounting nut. Remove the mounting nut.

11. Remove the remaining instrument panel mounting bolts located at the ends of the instrument panel and at the sides of the center console.

12. Remove the A-B trim pillar, lift the instrument panel up and tilt it toward the rear of the vehicle.

13. Disconnect all remaining electrical connectors and remove the instrument panel from the vehicle.

14. Disconnect the heater hoses from the heater core extension tubes. Cap the tubes to prevent spilling coolant into the inside of the vehicle.

15. Remove the push pin from the blower case and heater case and remove the main air duct.

16. Remove the 3 heater case mounting nuts and remove the heater case by pulling it straight out. Be careful not to damage the heater core extension tubes.

17. Remove the 2 screws attaching the heater core tube braces to the heater case and remove the tube braces.

18. Remove the heater core by pulling it straight out.

To install:

19. Slide the heater core assembly into the heater case and install the tube braces.

20. Position the heater case onto its mounting studs, being careful not to damage the heater core extension tubes. Install the 3 mounting nuts.

21. Install the main air duct.

22. Make sure the rubber grommets for the extension tubes are still in place in the engine compartment.

23. Uncap the extension tubes, if necessary, and connect the heater hoses.

24. Install the instrument panel in the reverse order of removal.

25. Fill the cooling system.

26. Connect the negative battery cable. Check the operation of the heater system and check for coolant leaks.

With Air Conditioning

1. Disconnect the negative battery cable.

2. Drain the cooling system and discharge the refrigerant from the air conditioning system.

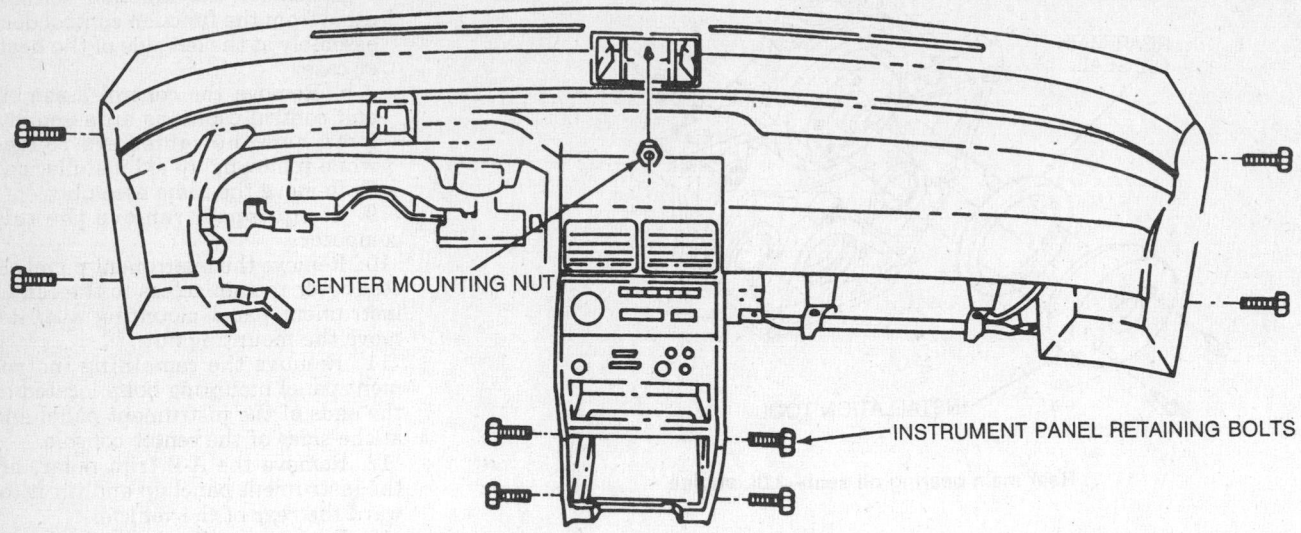

CENTER MOUNTING NUT

INSTRUMENT PANEL RETAINING BOLTS

Instrument panel mounting nut and bolt locations

3. Remove the steering column.

4. Remove the instrument cluster.

5. Remove the floor console as follows:

a. If equipped with automatic transaxle, remove the 2 transaxle selector lever knob mounting screws and remove the knob. Remove the selector trim piece. Remove the 4 selector bezel mounting screws. Lift the bezel and disconnect the illumination bulb and the shift control switch wiring harness. Remove the bezel.

b. If equipped with manual transaxle, Slide the shifter boot down and remove the shift knob. Remove the trim piece and the boot.

c. Remove the front ash tray and cigarette lighter.

d. Remove the 4 screws located on the front of the console.

e. Position the front seats to gain access to the console rear access hole covers.

f. Remove the covers from each side of the console and remove the retaining screws. Position the front seats all the way to the rear.

g. Lift the console from the rear and disconnect the power mirror adjust switch and programmed ride control switch electrical connectors.

h. Apply the parking brake and carefully remove the floor console.

6. Remove the hood release handle.

7. Remove the left and right console kick panels and instrument panel dash side covers.

8. If equipped with automatic temperature control, remove the control panel as follows:

a. Remove the bezel cover from the control assembly face.

b. Remove the 4 attaching screws from the control assembly housing.

c. Pull the control assembly out enough to gain access to the electrical connectors.

d. Disconnect the electrical connectors and remove the control assembly.

9. If equipped with manual air conditioning, remove the control panel as follows:

a. Remove the bezel cover from the control assembly face.

b. Remove the 4 attaching screws from the control assembly housing.

c. Remove the left and right side sound deadening panels.

d. Remove the REC/FRESH control cable at the REC/FRESH selector door assembly.

e. Disconnect the blower switch and control assembly illumination electrical connectors.

f. Remove the temperature control cable from the temperature blend door assembly at the right side of the heater case.

g. Remove the function selector cable from the function control door assembly at the left side of the heater case.

h. Remove the control assembly and control cables as an assembly. Note how the cables are routed while removing, to aid installation.

10. Remove the radio assembly.

11. If equipped, remove the trip computer.

12. Remove the instrument panel access cover to gain access to the center instrument panel mounting nut. Remove the mounting nut.

13. Remove the remaining instrument panel mounting bolts located at the ends of the instrument panel and at the sides of the center console.

14. Remove the A-B trim pillar, lift the instrument panel up and tilt it toward the rear of the vehicle.

15. Disconnect all remaining electrical connectors and remove the instrument panel from the vehicle.

16. Disconnect the heater hoses from the heater core extension tubes. Cap the tubes to prevent spilling coolant into the inside of the vehicle.

17. Remove the fuel evaporative system carbon canister.

18. Disconnect the air conditioning lines from the evaporator. On the inlet side, use spring coupling tool T81P–19623–G2 or equivalent. On the outlet side, use spring coupling tool T83P–19623–C or equivalent. Plug the line and evaporator tube openings to prevent the entrance of dirt and moisture into the system.

19. Disconnect the electrical connectors from the air conditioning relays at the top of the evaporator case.

20. Remove the air duct bands and the drain hose.

21. Remove the evaporator case attaching nuts and carefully remove the evaporator case from the vehicle.

22. Remove the 3 heater case mounting bolts.

23. If equipped with automatic temperature control, disconnect the electrical connectors from the function control and temperature blend door actuator motors.

24. Remove the heater case by pulling it straight out, being careful not to damage the heater core extension tubes.

25. Remove the 2 screws attaching the heater core tube braces to the

heater case and remove the tube braces.

26. Remove the heater core by pulling it straight out of the heater case.
To install:

27. Slide the heater core assembly into the heater case and install the tube braces.

28. Position the heater case onto its mounting studs, being careful not to damage the heater core extension tubes. Install the 3 mounting nuts.

29. Carefully position the evaporator case into the vehicle and install the attaching nuts.

30. Install the drain hose to the evaporator case and install the air duct bands. Connect the electrical connectors.

31. Unplug and connect the air conditioner lines and evaporator core tubes.

32. Install the carbon canister.

33. Make sure the rubber grommets for the heater core extension tubes are still in place in the engine compartment.

34. Uncap the extension tubes, if necessary, and connect the heater hoses.

35. Install the instrument panel in the reverse order of removal.

36. Fill the cooling system.

37. Connect the negative battery cable. Evacuate and charge the air conditioning system.

38. Check for proper operation of the heater and air conditioning system. Check for coolant leaks.

Water Pump

REMOVAL & INSTALLATION

2.2L Engine

1. Disconnect the negative battery cable.
2. Drain the cooling system.
3. Remove the timing belt.

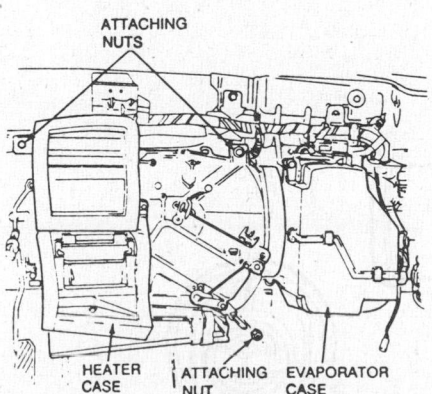

Heater case assembly—vehicles equipped with air conditioning

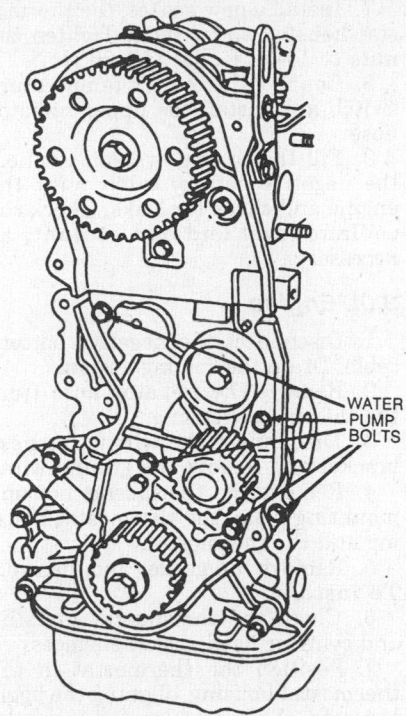

Water pump mounting bolt locations—2.2L engine

4. Remove the water pump-to-engine bolts, the water pump and the O-ring. Discard the O-ring.
To install:

5. Clean the mating surfaces of the water pump and the engine block.
6. Install a new O-ring onto the water pump.
7. Install the water pump and torque the bolts 14–19 ft. lbs. (19–25 Nm).
8. Install the timing belt.
9. Fill the cooling system.
10. Connect the negative battery cable, start the engine and check for leaks. Check the coolant level and add coolant, as necessary.

3.0L Engine

1. Disconnect the negative battery cable. Raise and safely support the vehicle.
2. Drain the cooling system and remove the water pump belt.

NOTE: The accessory drive belt may be left installed and the pump belt pulled aside. The accessory drive belt must be removed however, if the water pump belt is to be replaced.

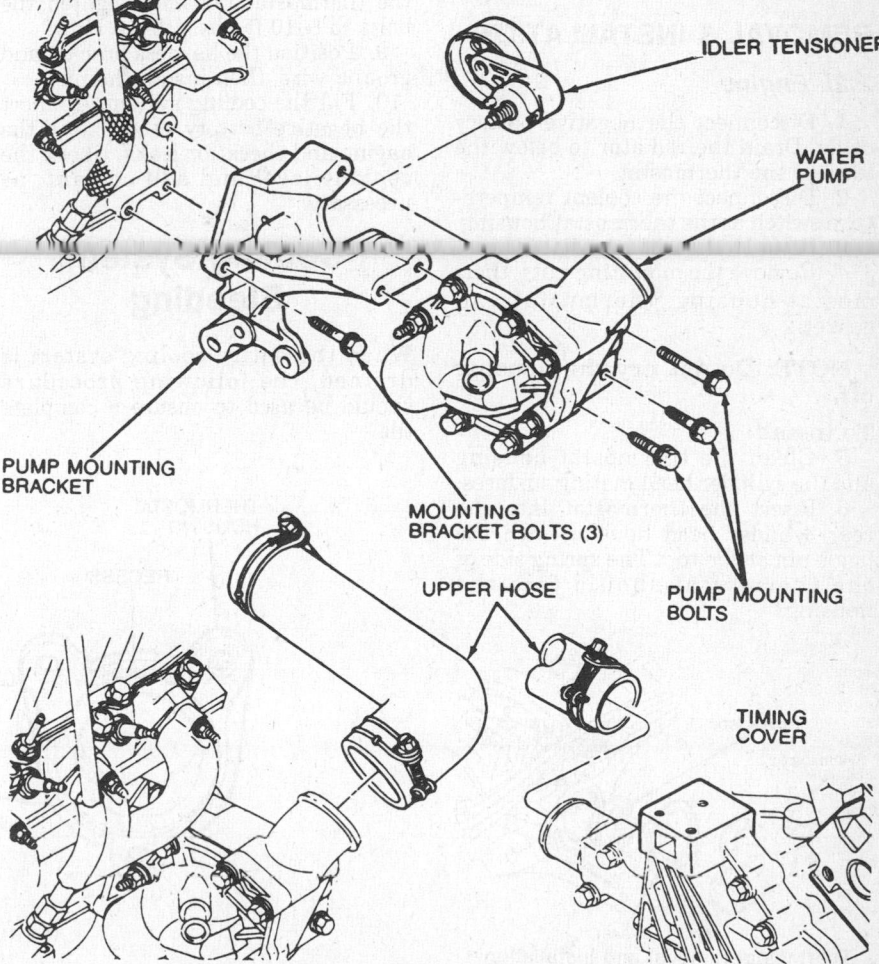

Water pump removal and installation—3.0L engine

10-23

3. Remove the upper water pump and heater hoses from the water pump.

4. Remove the lower radiator hose from the water pump steel tube.

5. Remove the steel tube brace bolt from the water pump mounting bracket.

6. Remove the water pump mounting bolts and remove the water pump.

To install:

7. Clean the mounting surfaces and install a new gasket.

8. Install the the water pump onto the mounting bracket and tighten the mounting bolts to 15–22 ft. lbs. (20–30 Nm).

9. Install the steel tube brace bolt. Install the lower radiator hose on the steel tube.

10. Install the heater and upper water pump hoses.

11. Install the water pump belt and lower the vehicle.

12. Connect the negative battery cable and fill the cooling system. Start the engine and check for leaks. Check the coolant level and add coolant, as necessary.

Thermostat

REMOVAL & INSTALLATION

2.2L Engine

1. Disconnect the negative battery cable. Drain the radiator to below the level of the thermostat.

2. Disconnect the coolant temperature switch at the thermostat housing.

3. Remove the upper radiator hose.

4. Remove the mounting nuts, thermostat housing, thermostat and gasket.

NOTE: Do not pry the housing off.

To install:

5. Clean the thermostat housing and the cylinder head mating surfaces.

6. Insert the thermostat into the rear cylinder head housing with the jiggle pin at the top. The spring side of the thermostat should face the housing.

7. Install a new gasket, the thermostat housing and 2 nuts. Tighten the nuts to 14–22 ft. lbs. (19–30 Nm).

8. Connect the coolant temperature switch and install the upper radiator hose.

9. Fill the cooling system. Connect the negative battery cable, start the engine and check for leaks. Check the coolant level and add coolant, as necessary.

3.0L Engine

1. Disconnect the negative battery cable. Drain the cooling system.

2. Remove the radiator hose from the thermosat housing.

3. Disconnect the wiring harness bracket and remove the ground wire.

4. Remove the thermostat housing mounting bolts, the thermostat housing and the thermostat.

5. Remove the gasket and discard.

To install:

6. Clean the thermostat housing and cylinder head gasket surfaces.

7. Position the thermostat in the thermostat housing, aligning the jiggle pin with the recess located near the top of the thermostat housing.

8. Position a new gasket and install the thermostat housing. Tighten the bolts to 8–10 ft. lbs. (10–14 Nm).

9. Position the harness bracket and ground wire, then install the nut.

10. Fill the cooling system. Connect the negative battery cable, start the engine and check for leaks. Check the coolant level and add coolant, as necessary.

Cooling System Bleeding

When the entire cooling system is drained, the following procedure should be used to ensure a complete fill.

1. Install the block drain plug, if removed and close the draincock. With the engine OFF, add anti-freeze to the radiator to a level of 50 percent of the total cooling system capacity. Then add water until it reaches the radiator filler neck seat.

2. Install the radiator cap to the first notch to keep spillage to a minimum.

3. Start the engine and let it idle until the upper radiator hose is warm. This indicates that the thermostat is open and coolant is flowing through the entire system.

4. Carefully remove the radiator cap and top off the radiator with water. Install the cap on the radiator securely.

5. Fill the coolant recovery reservoir to the FULL COLD mark with anti-freeze, then add water to the FULL HOT mark. This will ensure that a proper mixture is in the coolant recovery bottle.

6. Check for leaks at the draincock and at the block plug, if removed.

ENGINE ELECTRICAL

NOTE: Disconnecting the negative battery cable on some vehicles may interfere with the functions of the on board computer systems and may require the computer to undergo a relearning process, once the negative battery cable is reconnected.

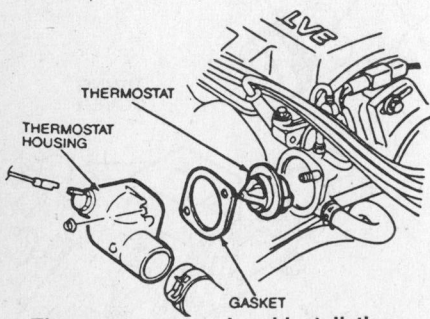

Thermostat removal and installation— 2.2L engine

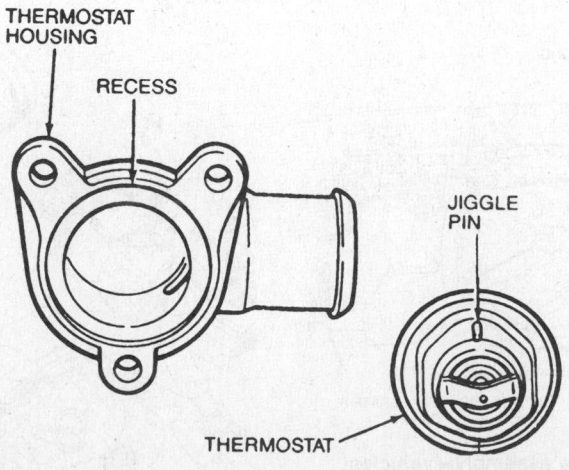

Thermostat jiggle pin and thermostat housing recess location—3.0L engine

Distributor

REMOVAL

1. Disconnect the negative battery cable.

2. Remove the distributor cap screws and cap, position the cap aside.

3. If not equipped with a turbocharger, perform the following procedures:

   a. Disconnect and label the vacuum lines from the distributor vacuum diaphragm.

   b. Disconnect the distributor electrical connectors from the ignition coil. On 3.0L engine disconnect the TFI-IV harness connector.

4. If equipped with a turbocharger, disconnect the distributor wiring harness connector located near the upper side of the distributor.

5. Using a wrench on the crankshaft pulley, rotate the crankshaft to position the No. 1 piston on TDC of the compression stroke; the crankshaft pulley notch should align with the timing plate indicator.

6. Using chalk or paint, mark the relationship of the distributor housing-to-cylinder head and the rotor-to-distributor housing; this will assist in installation.

7. Remove the distributor hold-down bolt(s) and the distributor.

NOTE: Some 3.0L engines may be equipped with a security type distributor hold-down bolt, use the proper tool to remove.

8. Inspect the O-ring and replace it, if necessary, on 2.2L engine.

INSTALLATION

Timing Not Disturbed

1. Using engine oil, lubricate the O-ring on 2.2L engine.

2. Align the rotor-to-distributor housing and the distributor housing-to-cylinder head.

3. Install the distributor. Make sure to engage the drive dog with the camshaft slot on 2.2L engine. Snug the distributor hold-down bolt(s).

4. Connect the electrical and the vacuum connections. Install the distributor cap.

5. Start the engine and check or adjust the ignition base timing.

6. Tighten the hold-down bolt(s) and recheck the timing, adjust if neccessary.

Timing Disturbed

1. Using engine oil, lubricate the O-ring on 2.2L engine.

2. Remove the spark plug from the

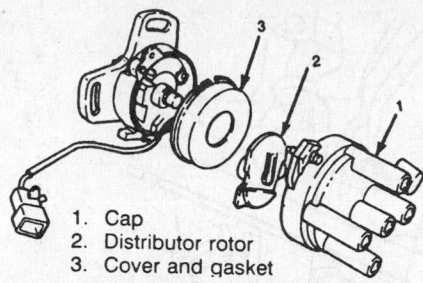

1. Cap
2. Distributor rotor
3. Cover and gasket

Exploded view of the distributor assembly—2.2L turbocharged engine

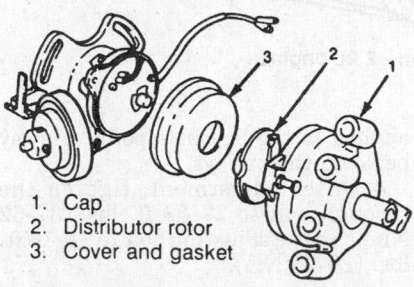

1. Cap
2. Distributor rotor
3. Cover and gasket

Exploded view of the distributor assembly—2.2L non-turbocharged engine

No. 1 cylinder and press a thumb over the spark plug hole.

3. Using a wrench on the crankshaft pulley, rotate the crankshaft to position the No. 1 piston on TDC of the compression stroke; pressure will be felt at the spark plug hole and the crankshaft pulley notch should align with the timing plate indicator.

4. Align the rotor to the No. 1 spark plug wire terminal of the distributor cap.

5. Install the distributor. Make sure to engage the drive dog with the camshaft slot on 2.2L engine. Snug the distributor hold-down bolt(s).

6. Connect the electrical and vacuum connections. Replace the distributor cap.

7. Start the engine and adjust the ignition base timing.

8. Tighten the hold-down bolt(s) and recheck the timing.

Ignition Timing

ADJUSTMENT

1. Turn **OFF** all of the accessories. Set the idle speed to specifications.

2. Firmly set the parking brake and position the gear shift selector in **P** for automatic transaxle or **N** for manual transaxle.

3. Start and operate the engine until normal operating temperatures are reached.

NOTE: A remote starter should not be used. Use the ignition key only to start the vehicle. On the 3.0L engine, disconnecting the start wire at the starter relay will cause the TFI module to revert to start mode timing after the vehicle is started. Reconnecting the start wire after the vehicle is running will not correct the timing.

4. On 2.2L non-turbocharged engine, disconnect and plug both vacuum hoses from the distributor vacuum diaphragm.

5. On 2.2L turbocharged engine or 1991–92 2.2L non-turbocharged engine, connect a jumper wire between the black 1-pin STI test connector and ground. On 3.0L engine, disconnect the single wire in-line spout connector or remove the shorting bar from the double wire spout connector.

6. Using a timing light, point it at the timing plate which is located at the crankshaft pulley, connect it to the engine and check the ignition timing; the timing should be 5–7 degrees BTDC for 2.2L non-turbocharged engine, 8–10 degrees BTDC for 2.2L turbocharged engine or 10 degrees BTDC on the 3.0L engine.

7. If the ignition timing is not within specifications, loosen the distributor hold-down bolt(s), rotate the distributor to align the timing marks and tighten the hold-down bolt(s).

8. Connect the vacuum lines, remove the jumper wire or connect the spout connector or shorting bar, as required. Remove the timing light.

Alternator

PRECAUTIONS

Several precautions must be observed with alternator equipped vehicles to avoid damage to the unit.

• If the battery is removed for any reason, make sure it is reconnected

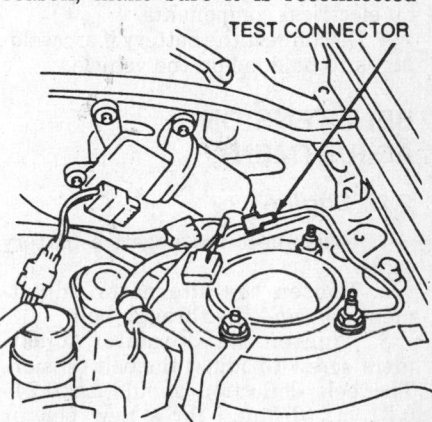

TEST CONNECTOR

STI test connector location—2.2L engine

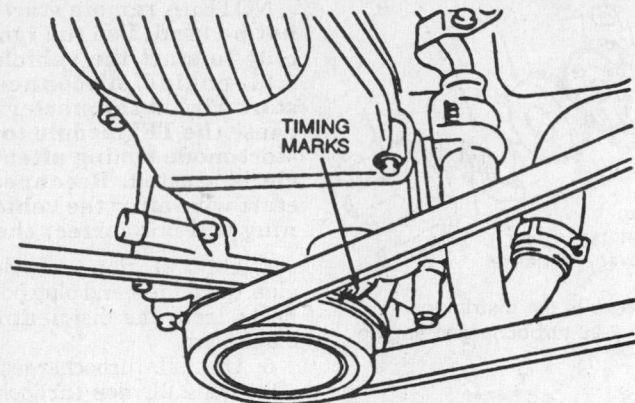

Timing mark location—2.2L engines

with the correct polarity. Reversing the battery connections may result in damage to the one-way rectifiers.

- When utilizing a booster battery as a starting aid, always connect the positive to positive terminals and the negative terminal from the booster battery to a good engine ground on the vehicle being started.
- Never use a fast charger as a booster to start vehicles.
- Disconnect the battery cables when charging the battery with a fast charger.
- Never attempt to polarize the alternator.
- Do not use test lights of more than 12 volts when checking diode continuity.
- Do not short across or ground any of the alternator terminals.
- The polarity of the battery, alternator and regulator must be matched and considered before making any electrical connections within the system.
- Never separate the alternator on an open circuit. Make sure all connections within the circuit are clean and tight.
- Disconnect the battery ground terminal when performing any service on electrical components.
- Disconnect the battery if arc welding is to be done on the vehicle.

BELT TENSION ADJUSTMENT

2.2L Engine

1. Disconnect the negative battery cable.
2. Loosen the alternator adjustment bolt and the through bolt.
3. Tighten the alternator adjustment screw to adjust the belt tension. The belt deflection should be 0.24–0.31 in. (6–8mm) for a new belt or 0.27–0.35 in. (7–9mm) for a used belt with a force of approximately 22 lbs.

applied to the belt at a point midway between the pulleys.
4. After adjustment, tighten the through bolt to 27–38 ft. lbs. (37–52 Nm) and the adjusting bolt to 13–18 ft. lbs. (18–25 Nm).

3.0L Engine

The 3.0L engine uses an automatic tensioner to maintain proper tension on the alternator drive belt. No adjustment is necessary.

REMOVAL & INSTALLATION

2.2L Engine

1. Disconnect the negative battery cable. Raise and support the vehicle, safely.
2. Remove the catalytic converter as follows:
   a. From both ends of the converter, remove the flange nuts and washers.
   b. Remove the resonator pipe-to-body insulator.
   c. Push the rear exhaust assembly rearward and remove the converter with the gaskets.
3. From the rear of the alternator, depress the lock tab(s) and disconnect the electrical connectors.
4. Loosen the alternator's adjusting and through bolts, tilt it and remove the drive belt.
5. Remove the alternator's adjusting bracket lock bolt and through bolt.
6. While supporting the alternator, slide (lower) it between the steering gear and the right halfshaft.
To install:
7. Reposition the alternator and install the mounting bolts finger tight.
8. Install the drive belt and adjust the belt tension.
9. After adjustment, tighten the through bolt to 27–38 ft. lbs. (37–52 Nm) and the adjusting lock bolt to 13–18 ft. lbs. (19–25 Nm).

10. Connect the electrical connections at the rear of the alternator.
11. Replace the converter and reposition the exhaust.
12. Lower the vehicle. Connect the negative battery cable.

3.0L Engine

1. Disconnect the negative battery cable.
2. Remove the accessory drive belt.
3. Remove and set aside the windshield washer reservoir.
4. Remove the power steering pump reservoir return hose from the pump body.
5. Remove the power steering pump high pressure hose from the pump assembly.
6. Remove the upper and middle accessory support bracket mounting bolts.
7. Pull back on the idler tensioner, using the proper tool and remove the lower accessory support bracket mounting bolt.
8. Remove the mounting bolt from the side of the accessory bracket at the air conditioning compressor brace.
9. Raise the alternator/accessory support bracket to clear the engine and set aside to remove the electrical connectors.
10. Disconnect the electrical connectors from the alternator and remove the alternator pivot bolt.
11. Remove the alternator mounting bolt from the back side of the alternator. Remove the alternator from the accessory support bracket.
To install:
12. Position the alternator on the accessory support bracket and install the alternator pivot bolt.
13. Install the alternator mounting bolt at the rear of the alternator.
14. Position the alternator/accessory support bracket in the proper position and connect the electrical connectors on the back side of the alternator.
15. Install the mounting bolt through the air conditioning compressor brace into the support bracket.
16. Install the middle accessory support bracket mounting bolt.
17. Pull back on the idler tensioner, using the proper tool and install the lower accessory support bracket mounting bolt.
18. Install the upper accessory support bracket mounting bolt.
19. Connect the power steering pressure and return hoses.
20. Install the accessory drive belt.
21. Replace the windshield washer resevoir and connect the negative battery cable.
22. Fill and bleed the power steering system.

Starter

REMOVAL & INSTALLATION

2.2L Engine

1. Disconnect the negative battery cable. Raise and support the vehicle, safely.

2. If equipped with a manual transaxle, remove the exhaust pipe bracket.

3. Remove the transaxle-to-engine bracket and intake manifold-to-engine bracket.

4. Disconnect the electrical connectors from the starter.

5. Remove the starter mounting bolts and the starter.

To install:

6. Install the starter and torque the bolts to 23–34 ft. lbs. (31–46 Nm).

7. Connect the electrical connectors to the starter.

8. Install the intake manifold-to-engine bracket and tighten the bolts to 14–22 ft. lbs. (19–30 Nm).

9. If equipped with an automatic transaxle, install the transaxle-to-engine bracket and torque the bellhousing bolt to 66–86 ft. lbs. (89–117 Nm) and the 3 other mounting bolts to 27–38 ft. lbs. (37–52 Nm).

10. If equipped with a manual transaxle, install the transaxle-to-engine bracket and connect the exhaust pipe bracket. Tighten the bracket bolts to 32–45 ft. lbs. (43–61 Nm).

11. Lower the vehicle.

12. Connect the negative battery cable and check the starter for proper operation.

3.0L Engine

1. Disconnect the negative battery cable.

2. Raise and safely support the vehicle.

3. If equipped with automatic transaxle, remove the kickdown cable routing bracket from the engine block.

4. Disconnect the electrical connections at the starter motor.

5. Remove the starter mounting bolts and remove the starter.

To install:

6. Position the starter and install the mounting bolts. Tighten the bolts to 15–20 ft. lbs. (20–27 Nm).

7. Connect the electrical connections at the starter motor.

8. If necessary, install the kickdown cable routing bracket to the engine block.

9. Lower the vehicle. Connect the negative battery cable and check the starter for proper operation.

EMISSION CONTROLS

Please refer to "Emission Controls" in the Unit Repair section for system maintenance procedures. Due to the complex nature of modern electronic engine control systems, comprehensive diagnosis and testing procedures fall outside the confines of this repair manual. For complete information on diagnosis, testing and repair procedures concerning all modern engine and emission control systems, please refer to "Chilton's Guide to Fuel Injection and Electronic Engine Controls".

Emission Warning Lamps

All vehicles are equipped with a **CHECK ENGINE** light. This light should come on briefly when the ignition key is turned **ON** but should turn off when the engine is started. If the light stays on after the engine is started or if it comes on at any time during engine operation, a problem in the electronic engine control system is indicated.

There are also 2 different service interval reminder systems, the Vehicle Maintenance Monitor or System Scanner. The Vehicle Maintenance Monitor is available only on vehicles with an analog instrument cluster. It will indicate a service interval check on a module located on the front center portion of the headliner, just above the rear view mirror. The System Scanner is found on vehicles with electronic instrument clusters. It will indicate a service interval check on a display, located in the lower left corner of the instrument cluster.

RESETTING

To reset the service interval light on vehicles equipped with the Vehicle Maintenance Monitor, insert a small pointed instrument into the service interval cancel switch. Depress the switch once.

To cancel the service check message on the System Scanner, press the **SERV** button on the keyboard, located to the right of the instrument cluster, and hold until 3 tones are sounded.

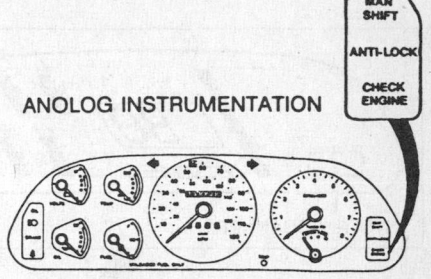

ANOLOG INSTRUMENTATION

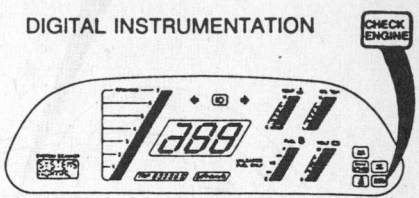

DIGITAL INSTRUMENTATION

Check engine light location

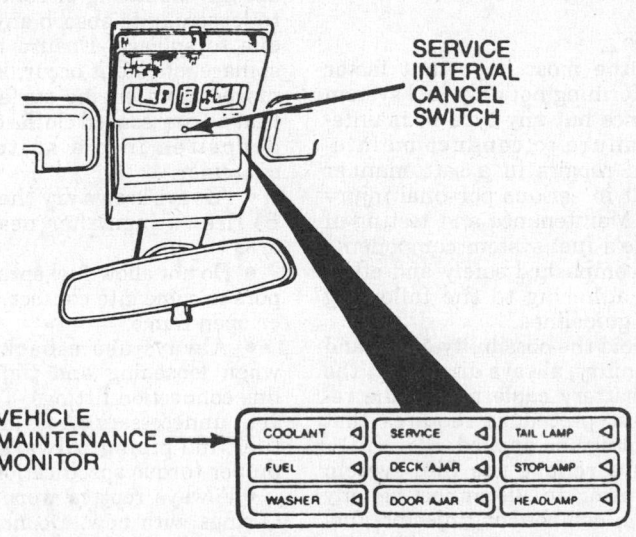

SERVICE INTERVAL CANCEL SWITCH

VEHICLE MAINTENANCE MONITOR

COOLANT	SERVICE	TAIL LAMP
FUEL	DECK AJAR	STOPLAMP
WASHER	DOOR AJAR	HEADLAMP

Vehicle Maintenance Monitor service interval cancel switch location

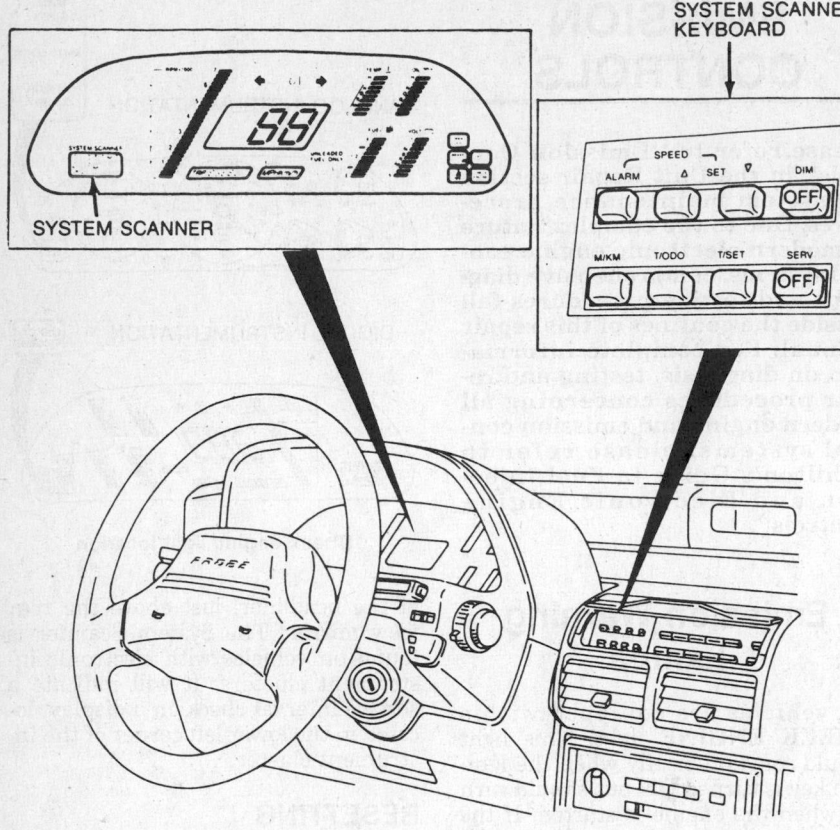

System Scanner display and keyboard location

RELIEVING FUEL SYSTEM PRESSURE

1. Start the engine.
2. From under the instrument panel, disconnect the fuel pump relay from the relay box.
3. After the engine stalls, turn the ignition switch **OFF** and reconnect the fuel pump relay to the relay box.

Fuel Tank

REMOVAL & INSTALLATION

1. Relieve the fuel system pressure.
2. Disconnect the negative battery cable.
3. Depress the clips on each end of the rear seat cushion and remove the cushion.
4. Disconnect the sending unit electrical connector, remove the 4 attaching screws and the sending unit access cover.
5. Disconnect the fuel supply and return lines from the sending unit.
6. Raise and safely support the vehicle.
7. Position a suitable container under the fuel tank drain plug. Remove the plug and drain the fuel tank.
8. Disconnect the vapor hoses and fuel filler neck hose at the fuel filler neck assembly.
9. Remove the 2 parking brake cable retaining brackets from the chassis to gain access to the fuel tank.
10. Remove the fuel tank mounting strap.
11. Support the fuel tank, remove the 3 attaching bolts and brackets and remove the fuel tank.

To install:

12. Position the fuel tank and install the 3 attaching bolts and brackets. Tighten the bolts to 16–22 ft. lbs. (22–30 Nm).
13. Install the fuel tank mounting strap. Tighten the mounting bolt to 32–45 ft. lbs. (43–61 Nm).
14. Install the 2 parking brake cable retaining brackets to the chassis.
15. Connect the fuel tank filler neck hose and the 3 vapor hoses to the fuel tank filler neck assembly.
16. Install the fuel tank drain plug and tighten to 9–13 ft. lbs. (12–18 Nm).
17. Lower the vehicle and connect the fuel supply and return hoses to the sending unit.
18. Install the sending unit access cover with the 4 attaching screws. Reconnect the sending unit electrical connector.
19. Position the rear seat cushion over the floor, making sure to align the retaining pins with the clips. Push down firmly until the 2 retaining pins

FUEL SYSTEM

Fuel System Service Precautions

Safety is the most important factor when performing not only fuel system maintenance but any type of maintenance. Failure to conduct maintenance and repairs in a safe manner may result in serious personal injury or death. Maintenance and testing of the vehicle's fuel system components can be accomplished safely and effectively by adhering to the following rules and guidelines.

• To avoid the possibility of fire and personal injury, always disconnect the negative battery cable unless the repair or test procedure requires that battery voltage be applied.

• Always relieve the fuel system pressure prior to disconnecting any fuel system component (injector, fuel rail, pressure regulator, etc.), fitting or fuel line connection. Exercise extreme caution whenever relieving fuel system pressure to avoid exposing skin, face and eyes to fuel spray. Please be advised that fuel under pressure may penetrate the skin or any part of the body that it contacts.

• Always place a shop towel or cloth around the fitting or connection prior to loosening to absorb any excess fuel due to spillage. Ensure that all fuel spillage, should it occur, is quickly removed from engine surfaces. Ensure that all fuel soaked cloths or towels are deposited into a suitable waste container.

• Always keep a dry chemical (Class B) fire extinguisher near the work area.

• Do not allow fuel spray or fuel vapors to come into contact with a spark or open flame.

• Always use a backup wrench when loosening and tightening fuel line connection fittings. This will prevent unnecessary stress and torsion to fuel line piping. Always follow the proper torque specifications.

• Always replace worn fuel fitting O-rings with new. Do not substitute fuel hose or equivalent where fuel pipe is installed.

are locked into the rear seat retaining clips.

20. Fill the fuel tank with fuel and check for leaks.

21. Connect the negative battery cable.

Fuel Filter

REMOVAL & INSTALLATION

1. Relieve the fuel system pressure.
2. Disconnect the fuel lines from both ends of the fuel filter. Plug the lines to prevent leakage.
3. Remove the in-line fuel filter from its mounting bracket.
4. Installation is the reverse of the removal procedure. Check for any fuel leaks.

Electric Fuel Pump

PRESSURE TESTING

1. Relieve the pressure in the fuel system, then disconnect the negative battery cable.
2. Install a suitable fuel pressure gauge between the fuel filter and the fuel rail.
3. On 2.2L engines, connect a jumper wire between the **BK** and **LG** terminals of the fuel pump test connector. On the 3.0L engine, ground the fuel pump lead of the self-test connector through a jumper wire at the **FP** lead.
4. Connect the negative battery cable, turn the ignition key **ON** and check the fuel pump pressure. The pressure should be 64–85 psi. on the 2.2L engines or 35–40 psi. on the 3.0L engine.
5. Remove the jumper wire, relieve the fuel system pressure and disconnect the negative battery cable.

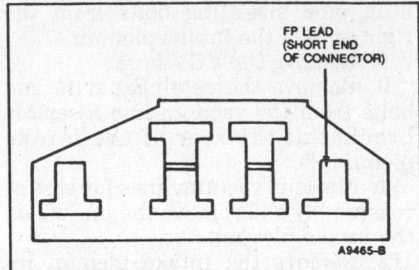

Self-test connector terminal locations— 3.0L engine

6. Remove the fuel pressure gauge and reconnect the fuel line.
7. Connect the negative battery cable.

REMOVAL & INSTALLATION

The fuel pump is mounted on the fuel sending unit assembly in the fuel tank.

1. Relieve the fuel pressure and disconnect the negative battery cable.
2. Depress the clips on each end of the rear seat cushion and remove the cushion.
3. Disconnect the electrical connector from the fuel pump/sending unit.
4. Remove the attaching screws from the fuel pump/sending unit access cover and remove the cover.
5. Disconnect the fuel supply and return hoses from the fuel pump/sending unit.
6. Remove the attaching screws and the fuel pump/sending unit from the fuel tank.
7. Disconnect the sending unit electrical connector, remove the sending unit attaching nuts and remove the sending unit from the fuel pump assembly.

To install:

8. Attach the sending unit to the fuel pump assembly and install the nuts. Connect the sending unit electrical connector.

9. Install the fuel pump/sending unit into the fuel tank and install the mounting screws.
10. Connect the fuel supply and return lines.
11. Install the access cover and the mounting screws.
12. Connect the sending unit electrical connector.
13. Position the rear seat cushion over the floor, making sure to align the retaining pins with the clips. Push down firmly until the 2 retaining pins are locked into the rear seat retaining clips.
14. Connect the negative battery cable, start the engine and check for proper system operation and for fuel leaks.

Fuel Injection

IDLE SPEED ADJUSTMENT

2.2L Engine

1. Check the ignition timing and adjust to specification, if necessary.
2. Turn off all lights and other unnecessary electrical loads. Idle speed adjustment must be done while the radiator cooling fan is not operating.
3. Set the parking brake and place the transaxle selector lever in **N** on manual transaxle vehicles or **P** on automatic transaxle vehicles. Warm up the engine and run it for 3 minutes at 2500–3000 rpm.
4. Ground the black 1-pin test connector located near the driver's side strut tower.
5. Attach a suitable tachometer according to the manufacturer's instructions.
6. Check the idle speed. It should be 750 rpm ± 25 rpm.
7. If the idle speed is not correct, remove the blind cap from the throttle body and adjust the idle speed by turning the idle air adjust screw.
8. After adjusting the idle speed, install the blind cap, disconnect the jumper wire from the test connector and remove the tachometer from the engine.

NOTE: Do not tamper with the adjustment screw located just to the left of the idle air adjust screw. Doing so may result in damage to the throttle body.

3.0L Engine

The curb idle speed is controlled by the EEC-IV processor and the idle rpm control device and cannot be adjusted.

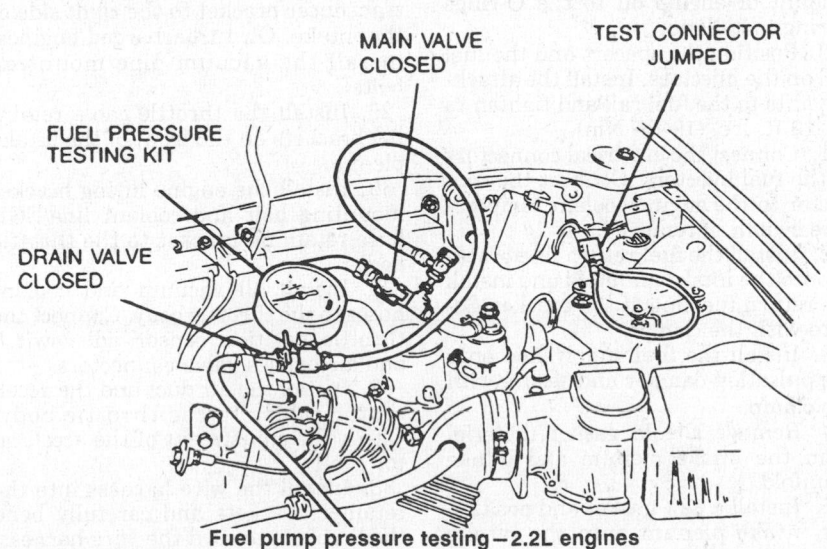

Fuel pump pressure testing—2.2L engines

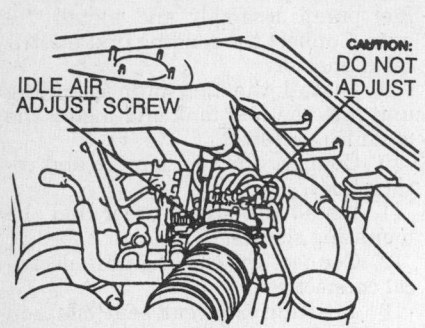

Idle speed adjustment—2.2L engines

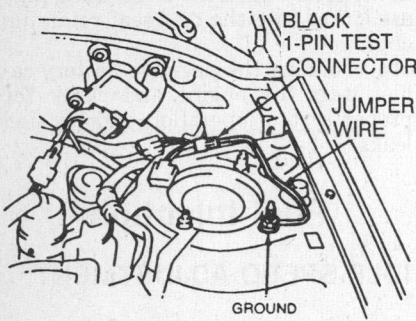

Idle speed adjustment test connector location—2.2L engines

IDLE MIXTURE ADJUSTMENT

The air/fuel mixture is controlled by the ECA and is not adjustable.

Fuel Injector

REMOVAL & INSTALLATION

2.2L Engine

1. Relieve the fuel system pressure.
2. Disconnect the negative battery cable and drain the cooling system.
3. Remove the accelerator cables and the air duct from the throttle body.
4. Mark all vacuum and coolant hoses for ease of reassembly and remove the hoses from the throttle body and engine block.
5. Disconnect the throttle position sensor, idle switch and air bypass valve connectors.
6. Remove the engine lifting bracket mounting bolt and coolant line/EGR hose retaining bracket from the throttle body.
7. Remove the throttle cable retaining brackets and wire loom bracket from the right side of the intake plenum.
8. On the turbocharged 2.2L engine, disconnect the vacuum pipe mounting bolts from the right side of the intake plenum. On non-turbocharged 2.2L engines, remove the vac-

uum pipe mounting bolts from the right side of the intake plenum.
9. Remove the PCV hose.
10. Remove the retaining nuts and bolts from the vacuum line assembly bracket at the rear of the intake plenum.
11. Mark all vacuum lines for ease of reassembly and remove the lines from the intake plenum.
12. Remove the intake plenum retaining bolts and nuts and remove the intake plenum and gasket.

NOTE: After removing the intake plenum, cover the intake manifold ports with a clean cloth to prevent dust and dirt from entering.

13. Remove the electrical connectors from the fuel injectors. On normally aspirated engines with automatic transmissions, remove the electrical connector from the engine coolant temperature switch.
14. Carefully bend the wire harness retainer brackets away from the wire harness and move the harness assembly away from the intake manifold.
15. Remove the fuel supply line from the pulsation damper.
16. Remove the fuel return line bracket from the intake manifold and remove the clamp and return fuel line at the bracket.
17. Remove the attaching bolts, spacers, insulators and the fuel rail with the injectors, pressure regulator and pulsation damper attached.
18. Remove the fuel injectors, grommets and O-rings from the fuel rail. Remove the O-rings from the fuel injectors.

To install:

19. Position the insulators and fuel injectors into the intake manifold. Position the grommets and new O-rings onto the fuel injectors. Apply a small amount of engine oil to the O-rings during installation.
20. Position the spacers and the fuel rail on the injectors. Install the attaching nuts to the fuel rail and tighten to 14–19 ft. lbs. (19–25 Nm).
21. Connect the electrical connectors to the fuel injectors. Connect the connector to the engine coolant temperature switch, if removed.
22. Install the fuel return line bracket onto the intake manifold and install the return fuel line at the bracket. Secure with the clamp.
23. Install the fuel supply line onto the pulsation damper and secure with the clamp.
24. Remove all old gasket material from the intake plenum and intake manifold.
25. Install a new gasket and position the intake plenum onto the intake

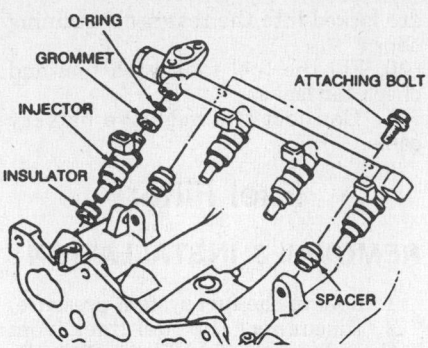

Fuel rail and fuel injector assembly—2.2L engines

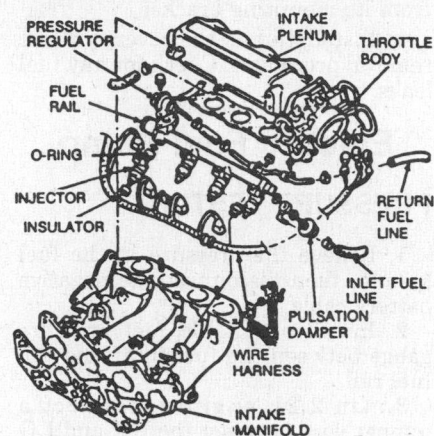

Air intake plenum and intake manifold assembly—2.2L engines

manifold. Install the attaching bolts and nuts and tighten to 14–19 ft. lbs. (19–25 Nm).
26. Connect the vacuum lines and install the retaining bolts on the vacuum line assembly bracket to the intake manifold.
27. Install the PCV hose and the wire loom bracket mounting bolts.
28. On non-turbocharged engines, install the EGR back-pressure variable transducer bracket to the right side of the intake. On turbocharged engines, install the vacuum pipe mounting bolts.
29. Install the throttle cable retaining brackets on the front of the intake plenum.
30. Install the engine lifting bracket mounting bolt and coolant line/EGR hose retaining bracket to the throttle body.
31. Install all vacuum and coolant hoses to the throttle body. Connect the throttle position sensor, idle switch and air bypass valve connectors.
32. Install the air duct and the accelerator cables to the throttle body. Check the adjustment of the accelerator cable.
33. Install the wire harness into the retainer brackets and carefully bend the brackets toward the wire harness.

34. Connect the negative battery cable and properly fill and bleed the cooling system.

3.0L Engine

1. Relieve the fuel system pressure.
2. Disconnect the negative battery cable.
3. Remove the air cleaner tube and plastic shield from the throttle body.
4. Remove the EGR supply tube and disconnect all the vacuum hoses from the air intake throttle body. Tag the hoses prior to removal for ease of installation.
5. Disconnect the air charge temperature sensor, idle speed control solenoid and throttle position sensor.
6. Remove the manifold absolute pressure sensor from the throttle body.
7. Disconnect the throttle cable and the throttle valve control cable, if equipped with an automatic transaxle, from the throttle lever.
8. Remove the fuel rail bracket bolt from the throttle body.
9. Remove the air intake throttle body mounting bolts and lift off the throttle body.
10. Disconnect the fuel supply and return lines.
11. Remove the fuel injector wiring harness.
12. Disconnect the vacuum line from the fuel pressure regulator.
13. Remove the fuel injector manifold mounting bolts.
14. Disengage the fuel rail assembly by lifting and rocking the rail.
15. Remove the injectors by lifting while gently rocking side to side.

NOTE: Handle the injectors and rail assembly with extreme care to prevent damage to the sealing areas and metering orifices.

To install:
16. Inspect the injector O-rings for wear or damage.
17. Lubricate new O-rings with a lightweight grade oil and install 2 on each injector.
18. Make sure the injector cups are free of contamination or damage.
19. Install the injectors in the fuel rail using a light twisting-pushing motion.
20. Install the rail assembly and the injectors carefully into the lower intake manifold, one side at a time.
21. Push down on the fuel rail to make sure the O-rings are seated.
22. Install the retaining bolts and tighten to 7 ft. lbs. (10 Nm) while holding the fuel rail in place.
23. Connect the fuel supply and return lines.
24. Before connecting the fuel injector harness, connect the negative battery cable and turn the key to the **ON** position. This will pressurize the fuel system.
25. Check for leaks where the injector is installed into the intake manifold and fuel rail.
26. Connect the fuel injector wiring harness.
27. Position the air intake throttle body and gasket on the lower intake manifold.

NOTE: Lightly oil all bolt threads before installation.

28. Install the mounting bolts and tighten to 15–22 ft. lbs. (20–30 Nm).
29. Install the fuel rail bracket on the throttle body and tighten securely.
30. Install the throttle cable and throttle valve control cable, if equipped with an automatic transaxle, on the throttle lever.
31. Install the manifold absolute pressure sensor on the throttle body.
32. Connect the throttle position sensor, idle speed control solenoid and the air charge temperature sensor.
33. Connect all vacuum hoses to the air intake throttle body.
34. Install the EGR supply tube. Install the plastic shield on the throttle body.
35. Install the air cleaner tube onto the throttle body and connect the negative battery cable.

DRIVE AXLE

Halfshaft

REMOVAL & INSTALLATION

1. Disconnect the negative battery cable. Raise and safely support the vehicle.
2. Remove the front wheel and tire assembly and the necessary inner fender splash guards.
3. Remove the stabilizer link assembly from the lower control arm.
4. Using a cape chisel and a hammer, raise the staked portion of the hub nut.
5. Using an assistant to depress the brake pedal, loosen, do not remove, the hub nut.
6. Remove the lower control arm ball joint clamp bolt. Using a prybar, pry the lower control arm downward to separate the ball joint from the steering knuckle.

NOTE: If removing the right halfshaft, remove the dynamic damper from the cylinder block.

7. Separate the halfshaft from the transaxle by positioning a prybar between the halfshaft and transaxle case. Pry out the halfshaft while pull-

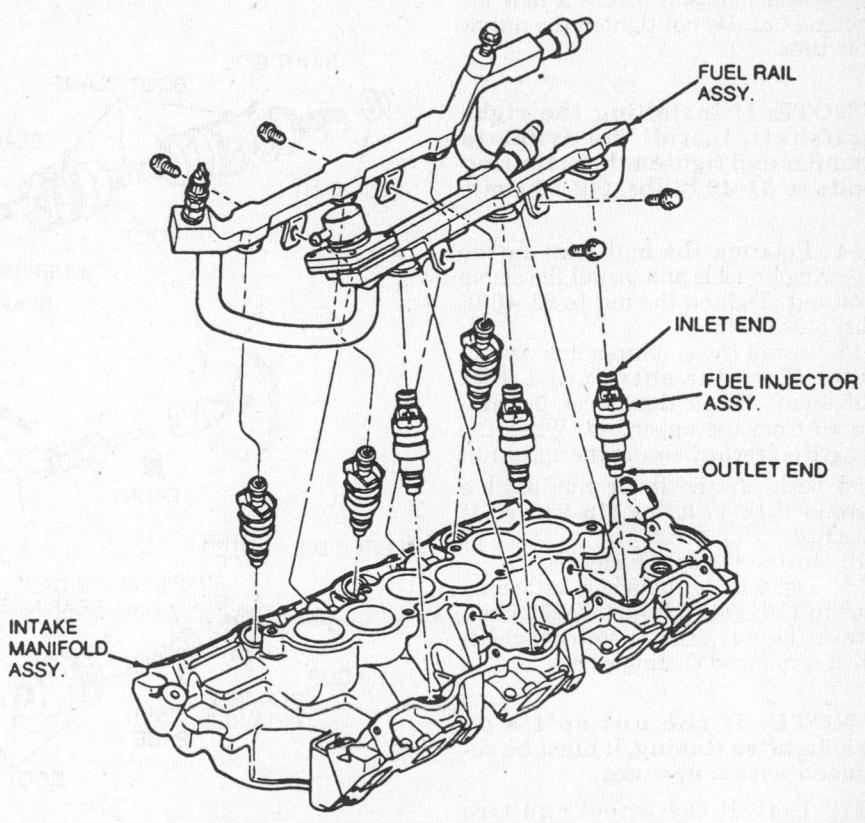

Fuel rail and fuel injector assembly—3.0L engine

FUEL RAIL ASSY.

INLET END

FUEL INJECTOR ASSY.

OUTLET END

INTAKE MANIFOLD ASSY.

ing out on the steering knuckle. Be careful not to damage the transaxle case, transaxle oil seal, CV-joint or CV-joint boot.

8. Remove and discard the hub nut. Pull the halfshaft out of the wheel hub.

NOTE: If the halfshaft binds in the hub splines, use a plastic hammer to tap it out or a wheel puller to press it out. Never use a metal hammer.

9. Using transaxle plugs or equivalent, install them into the halfshaft openings of the transaxle case; this will keep the differential side gears from becoming mispositioned.

To install:

10. On the end of each halfshaft, install a new retaining clip. Start 1 end of the clip in the groove and work the clip over the stub shaft end and into the groove. This will prevent over-expanding the clip.

11. Remove the transaxle plugs and inspect the transaxle oil seals. Replace, if necessary.

12. Lubricate the halfshaft splines with a suitable grease, align the splines with the differential side gears and push the halfshaft into the differential. Make sure the retaining clip is seated in the differential side gear groove.

13. Position the halfshaft through the wheel hub and install a new attaching nut. Do not tighten the nut at this time.

NOTE: If installing the right halfshaft, install the dynamic damper and tighten the mounting bolts to 31–46 ft. lbs. (42–62 Nm).

14. Position the ball joint in the steering knuckle and install the clamp bolt/nut. Tighten the nut to 32–40 ft. lbs. (43–54 Nm).

15. Install the stabilizer link assemblies. Turn the nuts until 1.0 in. (25.4mm) of bolt thread can be measured from the upper nut. When the length is reached, secure the upper nut and back off the lower nut until a torque of 12–17 ft. lbs. (16–23 Nm) is reached.

16. Install the splash shields.

17. Tighten the halfshaft attaching nut to 116–174 ft. lbs. (157–235 Nm). Stake the nut using a suitable chisel with a rounded cutting edge.

NOTE: If the nut splits or cracks after staking, it must be replaced with a new nut.

17. Install the wheel and tire assembly.

CV-Boot

REMOVAL & INSTALLATION

Inner Boot

MANUAL TRANSAXLE

1. Disconnect the negative battery cable. Raise and safely support the vehicle.

2. Remove the halfshaft and place it in a soft jawed vise, do not allow the vise to contact the boot.

3. Cut the inner joint boot clamps with the proper tool and move the boot rearward.

4. Using paint or chalk, matchmark the outer race to the halfshaft.

5. Using a small prybar, pry the wire ring bearing retainer from the joint and remove the outer race.

6. Using paint or chalk, matchmark the bearing assembly to the halfshaft. Using a pair of snapring pliers, remove the inner race snapring. Remove the inner race, cage and the ball bearings as an assembly.

7. Remove the boot.

8. Check the grease for grit; if necessary, clean the parts and regrease the assembly.

To install:

9. Install the new boot on the shaft; be careful not to cut the boot on the shaft splines.

10. Align the matchmarks and install the inner race, cage and ball bearings as an assembly. Install the snapring. Make sure the chamfer on the bearing cage faces the snapring.

11. Lubricate the outer race with 1.4–2.1 oz. of suitable CV-joint grease. Align the matchmarks and install the outer race and the large wire snapring.

12. Position the boot over the outer race so it is extended to 3.5 in. between the clamps. Using a dull blade prybar, lift the boot end to expell the trapped air.

13. Using new boot clamps, wrap them around the boots in a clockwise direction, pull them tight using a pair of pliers and bend the locking tabs to secure them in place.

14. Work the CV-joint through it's full range of travel at various angles; it should flex, extend and compress smoothly.

15. Install the halfshaft into the vehicle.

AUTOMATIC TRANSAXLE

1. Remove the halfshaft from the vehicle and position it in a soft jawed vise, do not allow the vise to contact the boot.

2. Using a pair of side cutters, cut inner joint boot clamps and move the boot rearward.

3. Using paint or chalk, matchmark the outer race to the tripot bearing.

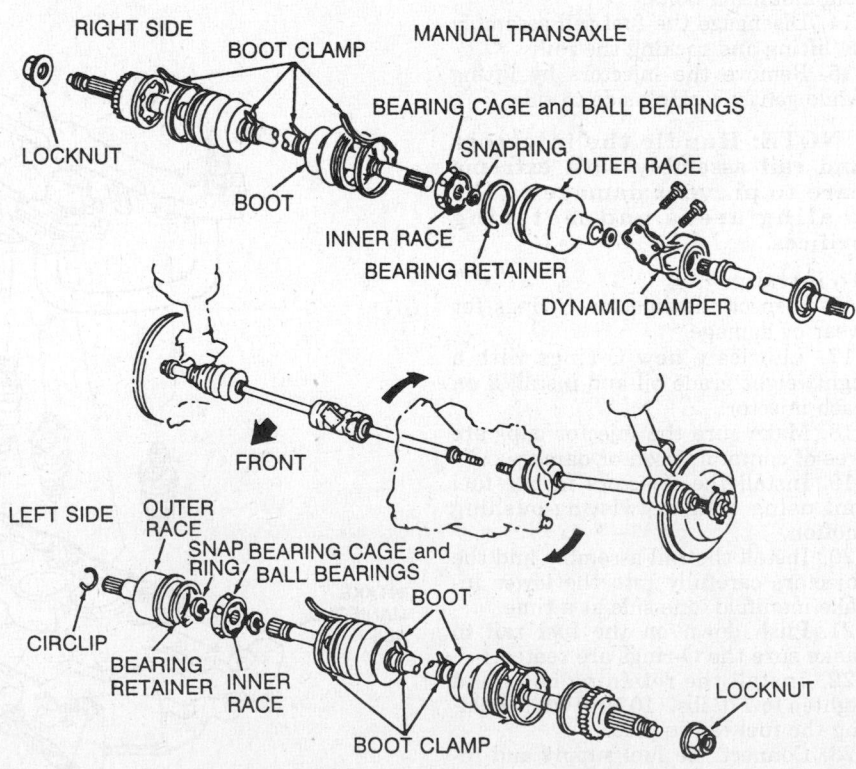

Exploded view of the manual transaxle halfshafts

4. Using a small prybar, pry the wire ring bearing retainer from the joint and remove the outer race.

5. Using paint or chalk, matchmark the tripot bearing to the halfshaft. Using a pair of snapring pliers, remove the tripot bearing snapring.

6. Using a brass drift and a hammer, drive the tripot bearing assembly from the halfshaft.

7. Wrap tape around the shaft splines and remove the boot.

8. Check the grease for grit; if necessary, clean the parts and regrease the assembly.

To install:

9. Install the new boot on the shaft; be careful not to cut the boot on the shaft splines.

10. Align the tripot bearing-to-halfshaft matchmarks. Using a brass drift and a hammer, drive the bearing assembly onto the halfshaft. Install the snapring.

11. Fill the CV-joint outer race with 3.5 oz. of grease.

12. Align the matchmarks and install the outer race and the large wire snapring.

13. Position the boot over the outer race so it is extended to 3.5 in. between the clamps. Using a dull blade prybar, lift the boot end to expell the trapped air.

14. Using new boot clamps, wrap them around the boots in a clockwise direction, pull them tight using a pair of pliers and bend the locking tabs to secure them in place.

15. Work the CV-joint through it's full range of travel at various angles; it should flex, extend and compress smoothly.

16. Install the halfshaft into the vehicle.

Outer Boot

The outer CV-joint is not serviceable and must be replaced with the shaft as an assembly.

1. Remove the inner joint and boot assembly from the halfshaft.

2. Using a pair of side cutters, cut the outer joint boot clamps and remove the boot from the halfshaft.

3. Position the outer boot over the outer race so it is extended to 3.5 in. between the clamps. Using a dull blade prybar, lift the boot end to expell the trapped air.

4. Using new boot clamps, wrap them around the boots in a clockwise direction, pull them tight using a pair of pliers and bend the locking tabs to secure them in place.

5. To complete the installation, reverse the removal procedures. Install the halfshaft into the vehicle.

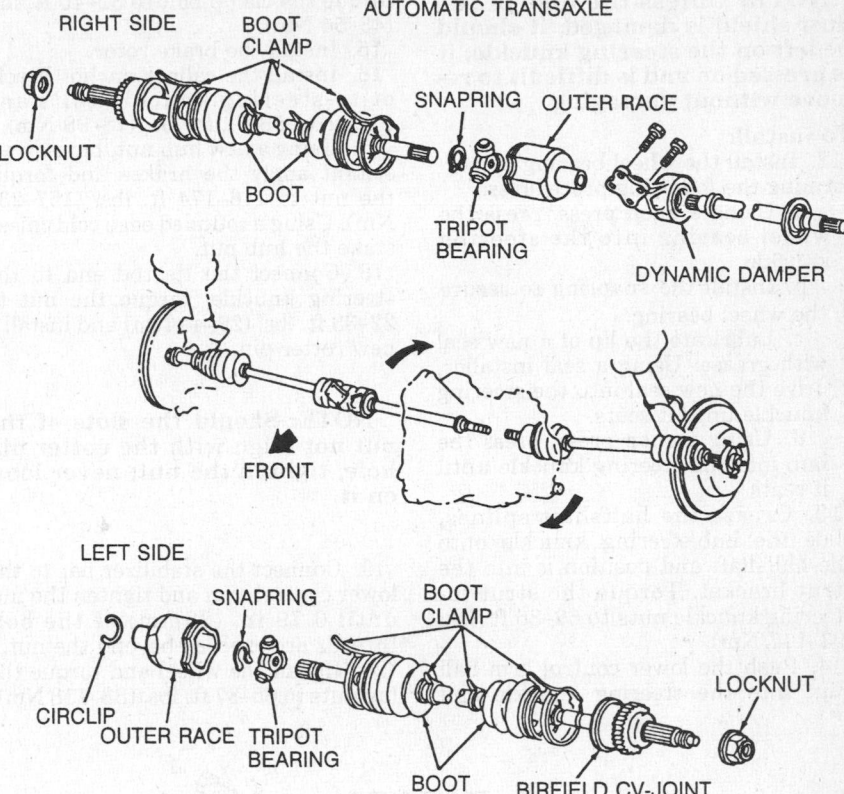

Exploded view of the automatic transaxle halfshafts

Front Wheel Hub, Knuckle/Spindle and Bearings

REMOVAL & INSTALLATION

1. Raise and safely support the vehicle. Remove the front wheel and tire assembly.

2. Using a small cape chisel and a hammer, raise the staked portion of the hub nut.

3. Have an assistant apply the brakes and remove the hub nut (discard it).

4. Remove the stabilizer bar-to-control arm, bolt, nut, washers and bushings.

5. At the tie rod end, remove the cotter pin and nut. Using a tie rod end separator tool or equivalent, separate the tie rod end from the steering knuckle.

6. Remove the caliper and anchor bracket and support the caliper assembly with mechanics wire.

7. Remove the brake disc rotor.

8. Remove the lower control arm ball joint clamp nut/bolt. Using a prybar, pry the lower control arm downward and separate the ball joint from the steering knuckle.

9. Remove the steering knuckle-to-strut attaching bolts and slide the steering knuckle assembly from the strut bracket.

10. Slide the steering knuckle assembly from the halfshaft and support the halfshaft with mechanics wire; be careful not to damage the seals. Should the wheel hub bind on the halfshaft, use a plastic hammer to jar it free.

NOTE: If the halfshaft splines bind in the hub, it may be necessary to use a 2-jawed wheel puller to separate them.

11. Remove the wheel bearing by performing the following procedures:

   a. Using a prybar, pry the grease seal from the hub/steering knuckle assembly.

   b. Using a shop press, press the hub from the steering knuckle.

NOTE: If the inner race remains on the hub, grind the inner race to approximately 0.020 in. (0.5mm) and use a chisel to remove it.

   c. Remove the snapring from the steering knuckle.

   d. Using a suitable puller, remove the wheel bearing from the steering knuckle.

NOTE: Unless the disc brake dust shield is damaged, it should be left on the steering knuckle; it is pressed on and is difficult to remove without damaging.

To install:

12. Install the wheel bearing by performing the following procedures:

   a. Using a shop press, press the wheel bearing into the steering knuckle.

   b. Install the snapring to secure the wheel bearing.

   c. Lubricate the lip of a new seal with grease. Using a seal installer, drive the new seal into the steering knuckle until it seats.

   d. Using a shop press, press the hub into the steering knuckle until it seats.

13. Grease the halfshaft splines, slide the hub/steering knuckle onto the halfshaft and position it into the strut bracket. Torque the strut-to-steering knuckle nuts to 69–86 ft. lbs. (93–117 Nm).

14. Push the lower control arm ball joint into the steering knuckle and torque the clamp bolt to 32–40 ft. lbs. (43–54 Nm).

15. Install the brake rotor.

16. Install the caliper anchor bracket-to-steering knuckle bolts and torque to 58–72 ft. lbs. (78–98 Nm).

17. Using a new hub nut, have an assistant apply the brakes and torque the nut to 116–174 ft. lbs. (157–235 Nm). Using a rounded edge cold chisel, stake the hub nut.

18. Connect the tie rod end to the steering knuckle, torque the nut to 22–33 ft. lbs. (29–44 Nm) and install a new cotter pin.

NOTE: Should the slots of the nut not align with the cotter pin hole, tighten the nut; never loosen it.

19. Connect the stabilizer bar to the lower control arm and tighten the nut until 0.79 in. (20mm) of the bolt threads are exposed beyond the nut.

20. Install the wheel and torque the lug nuts to 65–87 ft. lbs. (88–118 Nm).

MANUAL TRANSAXLE

For further information on transmissions/transaxles, please refer to "Chilton's Guide to Transmission Repair".

Transaxle Assembly

REMOVAL & INSTALLATION

1. Disconnect the battery cables, negative cable first. Remove the battery and the battery tray.

2. Disconnect the main fuse block and disconnect the coil wire from the distributor. Disconnect and mark the wiring assembly, as necessary.

3. Disconnect the electrical connector from the air flow meter and remove the air cleaner assembly.

4. On 2.2L EFI engine, remove the resonance chamber and bracket. On

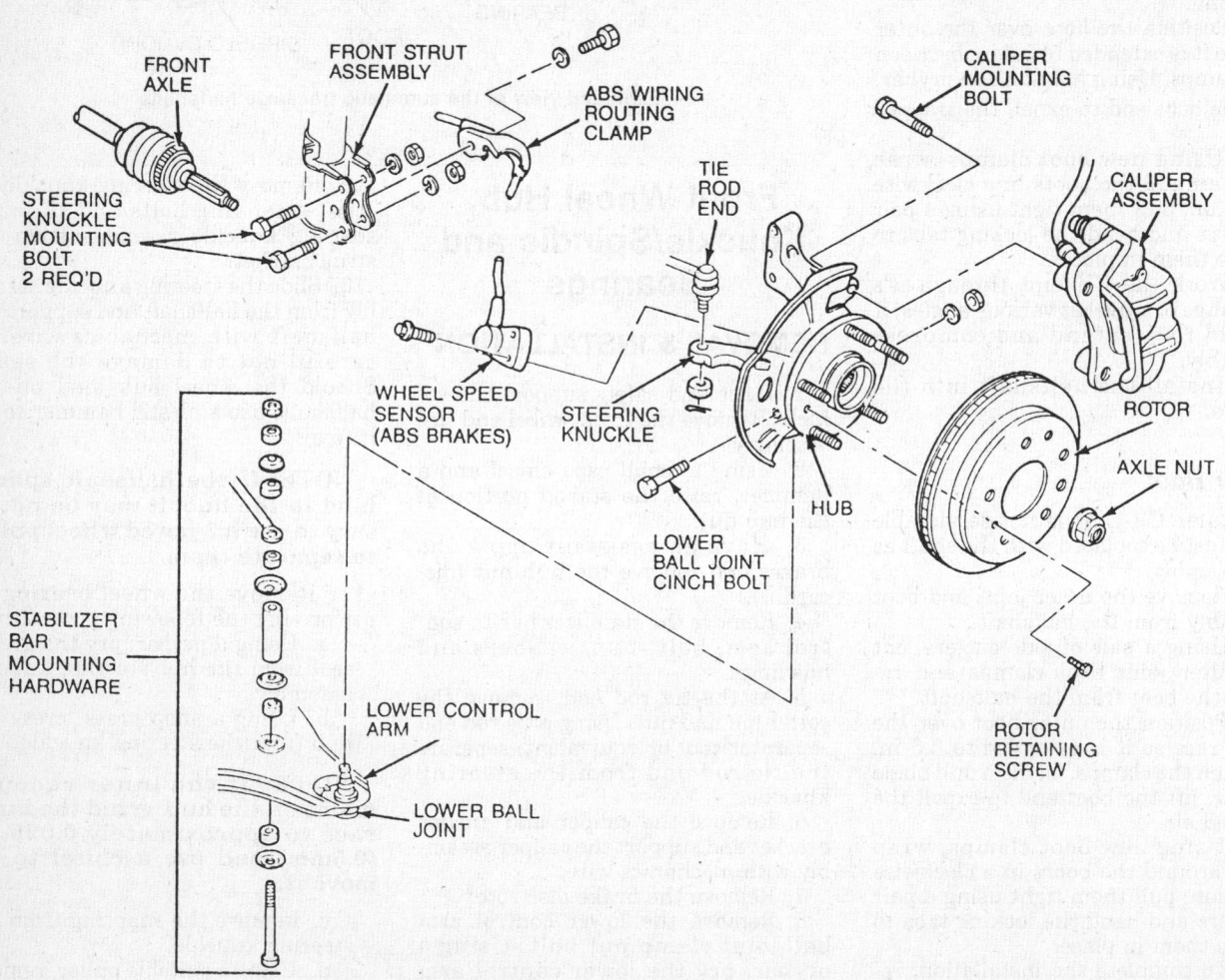

Front wheel hub/knuckle assembly and related components

2.2L turbocharged engine, remove the throttle body-to-intercooler air hose and the air cleaner-to-turbocharger air hose.

5. Disconnect the speedometer cable (analog) or harness (digital).

6. If equipped with the 3.0L engine, drain the engine coolant and close the drain valve. Remove the upper radiator hose.

7. Disconnect both ground wires from the transaxle. Raise and safely support the vehicle.

8. Remove the front wheel and tire assemblies and the splash shields. Drain the transaxle.

9. Remove the slave cylinder and move it aside.

10. Remove the tie rod ends-to-steering knuckle cotter pins and nuts. Disconnect the tie rod ends from the steering knuckle.

11. Remove the stabilizer link assemblies from the lower control arm.

12. Remove the lower control ball joint-to-steering knuckle nut/bolt. Using a prybar, pry the lower control arm downward to separate the ball joint from the steering knuckle.

13. Remove the right joint shaft bracket.

14. Position a prybar between the halfshaft and transaxle case; pry the halfshafts from the transaxle and suspend them on a wire.

15. Using transaxle plugs or equivalent, install them into the halfshaft openings of the transaxle case; this will keep the differential side gears from becoming mispositioned.

16. Remove the gusset plate-to-transaxle bolts on 2.2L engine. Disconnect the extension bar and shift control rod.

17. Remove the front exhaust pipe on the 3.0L engine.

18. Remove the flywheel-to-transaxle inspection plate on the 2.2L engine.

19. Remove the starter motor and the access brackets.

20. Using an engine support bar or equivalent, attach to the engine and support it's weight.

21. Remove the center transaxle mount/bracket, the left transaxle mount and the right transaxle mount-to-frame nut/bolt.

22. Remove the crossmember and the left lower arm as an assembly.

23. Using a transmission jack or equivalent, attach it to the transaxle.

24. Remove the transaxle-to-engine bolts, lower the transaxle and remove it from the vehicle.

To install:

25. Using clutch grease, apply a small amount to the input shaft splines.

26. Raise and position the transaxle. Install the transaxle-to-engine bolts

and torque to 66–86 ft. lbs. (89–117 Nm).

27. Install the center transaxle mount/bracket and torque the bolts to 27–40 ft. lbs. (36–54 Nm) and the nuts to 47–66 ft. lbs. (64–89 Nm); do not install the throttle air hose bracket nut.

28. Install the left transaxle mount and torque the left transaxle-to-mount bolts on the 2.2L EFI engine to 27–38 ft. lbs. (37–52 Nm) or on the 2.2L turbocharged engine and 3.0L engine to 49–69 ft. lbs. (67–93 Nm). Torque the mount-to-bracket nut and bolt to 49–69 ft. lbs. (67–93 Nm).

29. Install the crossmember and the left side arm as an assembly. Tighten the bolts to 27–40 ft. lbs. (36–54 Nm) and the nuts to 55–69 ft. lbs. (75–93 Nm).

30. Install the right transaxle mount bolt and nut and tighten to 63–86 ft. lbs. (85–117 Nm).

31. Install the starter motor and access brackets.

32. Install the flywheel inspection cover on 2.2L engine only. Tighten the bolts to 69–95 inch lbs. (8–11 Nm).

33. Connect the extension rod and control rod. Replace the front exhaust pipe on 3.0L engine.

34. Install the slave cylinder and tighten the bolts to 14–19 ft. lbs. (19–26 Nm).

35. Install the gusset plate-to-transaxle bolts and tighten to 27–38 ft. lbs. (37–52 Nm) on the 2.2L engine.

36. On the end of each halfshaft, install a new retaining clip.

37. Remove the transaxle plugs and install the halfshaft until the retaining clips snap into place. Attach the lower arm ball joints to the knuckles.

38. Install and torque the tie rod end-to-steering knuckle nut to 22–33 ft. lbs. (29–44 Nm) and the lower control arm ball joint-to-steering knuckle nut and bolt to 32–40 ft. lbs. (43–54 Nm).

39. Install the stabilizer link assembly-to-lower control arm. Turn the upper nuts (on each assembly) until 1.0 in. (25.4mm) of bolt thread can be measured above the nuts. When the length is reached, secure the upper nut and torque the lower nut to 12–17 ft. lbs. (16–23 Nm).

40. Install the splash shields and the front wheel and tire assemblies; torque the lug nuts to 65–87 ft. lbs. (88–118 Nm). Lower the vehicle.

41. Connect the ground wires to the transaxle case and tighten to 69–95 inch lbs. (8–11 Nm).

42. On 2.2L engine, install the resonance chamber and bracket; torque to 69–95 inch lbs. (8–11 Nm). On turbocharged engines, install the throttle body-to-intercooler air hose and torque the bracket-to-mount nut to 47–66 ft. lbs. (64–89 Nm).

43. On 3.0L engine, install the upper radiator hose and fill the cooling system.

44. Install the air cleaner assembly and tighten to 69–95 inch lbs. (8–11 Nm).

45. Connect the electrical connector to the air flow meter. Connect the previously marked wiring assembly, if disconnected.

46. Reconnect the main fuse block and connect the coil wire to the distributor.

47. Remove the engine support bracket.

48. Connect the speedometer cable or harness, as applicable.

49. Install the battery tray, battery and connect the battery cables.

50. Refill the transaxle assembly. Connect the negative battery cable, start the engine and check for leaks.

CLUTCH

Clutch Assembly

REMOVAL & INSTALLATION

1. Disconnect the negative battery cable. Raise and support the vehicle, safely. Remove the transaxle.

2. Using a clutch alignment tool or equivalent, position it through the pressure plate, clutch plate and into the pilot bushing; this will keep the assembly from dropping.

3. Using paint or chalk, matchmark the pressure plate-to-flywheel position.

4. Remove the pressure plate-to-flywheel bolts, a little at a time, evenly, to relieve the spring pressure.

5. Remove the pressure plate, clutch disc and alignment tool from the engine.

6. Inspect the pressure plate, clutch disc and release bearing for wear and/or damage and replace, as necessary.

7. Inspect the flywheel for scoring, cracks or other wear or damage. Remove the flywheel if machining or replacement is necessary. Inspect the pilot bearing for excessive wear or scoring and replace, if necessary.

To install:

8. If removed, install the flywheel. Make sure the crankshaft flange and flywheel mating surfaces are clean. Tighten the flywheel bolts, in sequence, to 71–76 ft. lbs. (96–103 Nm) on 2.2L engines or 54–64 ft. lbs. (73–87 Nm) on the 3.0L engine.

9. If removed, install a new pilot bearing using a suitable installation tool. The pilot bearing should be

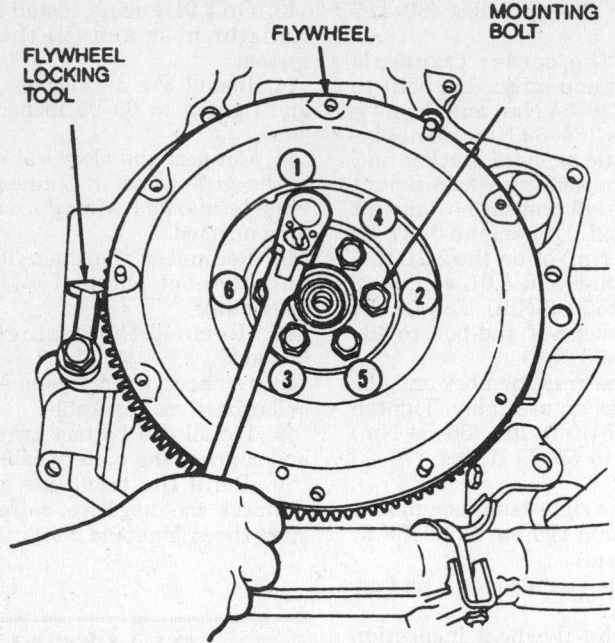

FLYWHEEL
LOCKING
TOOL

FLYWHEEL

MOUNTING
BOLT

Flywheel bolt torque sequence

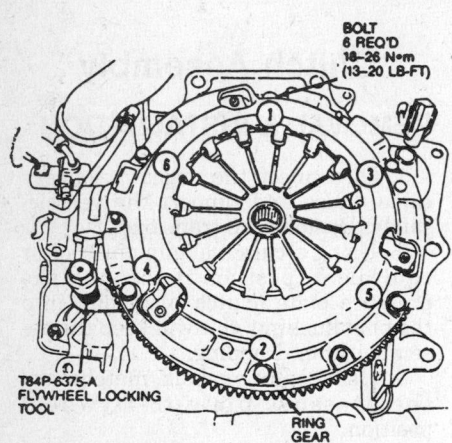

BOLT
6 REQ'D
18-26 N·m
(13-20 LB-FT)

T84P-6375-A
FLYWHEEL LOCKING
TOOL

RING
GEAR

Pressure plate bolt torque sequence

0.150–0.165 in. (3.8–4.2mm) below the surface of the crankshaft flange.

10. Using clutch grease, apply a small amount to the clutch disc and input shaft splines. Also apply a small amount of clutch grease to the clutch release bearing and release fork where they contact one another as well as the fork pivot.

11. Install the clutch plate and alignment tool.

12. Align the pressure plate to the flywheel matchmark. Install the pressure plate-to-flywheel bolts and torque, evenly, a little at a time, to 13–20 ft. lbs. (18–26 Nm) in the proper sequence.

13. Install the transaxle assembly and lower the vehicle.

14. Connect the negative battery cable. Check for proper clutch operation.

PEDAL HEIGHT/FREE-PLAY ADJUSTMENT

Pedal Height

1. To determine if the pedal height requires an adjustment, measure the distance from the bulkhead to the upper center of the pedal pad. The distance should be 8.524–8.720 in. (216.5–221.5mm).

2. To adjust, remove the lower dash panel and the air ducts.

3. Loosen the locknut and turn the

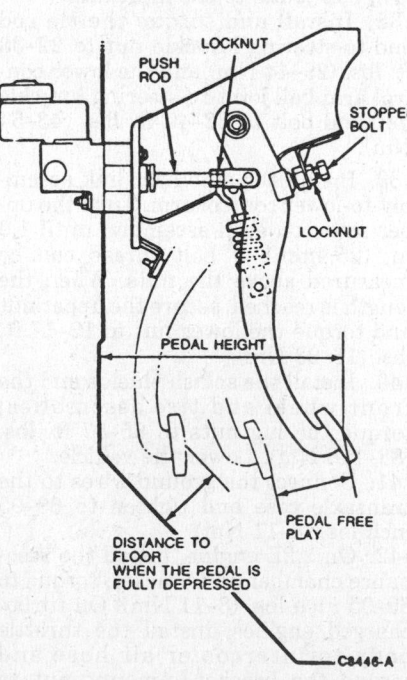

LOCKNUT

PUSH
ROD

STOPPER
BOLT

LOCKNUT

PEDAL HEIGHT

PEDAL FREE
PLAY

DISTANCE TO
FLOOR
WHEN THE PEDAL IS
FULLY DEPRESSED

C8446-A

Clutch pedal adjustment

stopper bolt until the desired pedal height is reached.

4. Tighten the locknut.

5. Install the ducts and the lower dash panel.

Pedal Free-Play

1. To determine if the pedal free-play needs adjustment, measure the free-play. The free-play should be 0.20–0.51 in. (5–13mm).

2. To adjust, remove the lower dash panel and the air ducts.

3. Loosen the locknut and turn the pushrod until the pedal play is within specifications.

4. Measure the distance from the floor to the center of the pedal pad when the pedal is fully depressed. The distance should be 2.7 in. (68mm) or more.

5. Tighten the locknut and replace the lower dash panel and the air ducts.

Clutch Master Cylinder

REMOVAL & INSTALLATION

1. Disconnect the negative battery cable. Remove the ABS relay box, if equipped.

2. Disconnect the pressure line to the cylinder, using the proper wrench.

3. Remove the mounting nuts and remove the clutch master cylinder.

To install:

4. Install the clutch master cylinder and tighten the mounting nuts to 14–19 ft. lbs. (19–26 Nm).

5. Connect the pressure line and tighten the nut securely.

6. Install the ABS relay box, if equipped.

7. Bleed the air from the clutch system, connect the negative battery cable and road test the vehicle.

Clutch Slave Cylinder

REMOVAL & INSTALLATION

1. Disconnect the negative battery cable.

2. Disconnect the pressure line at the slave cylinder. Plug the pressure line to prevent leaking.

3. Remove the slave cylinder mounting bolts and remove the slave cylinder.

To install:

4. Install the slave cylinder and tighten the mounting bolts to 12–17 ft. lbs. (16–23 Nm).

5. Connect the pressure line and tighten the nut to 10–16 ft. lbs. (13–22 Nm).

6. Connect the negative battery cable.

7. Bleed the air from the clutch system and road test the vehicle.

Hydraulic Clutch System Bleeding

NOTE: The fluid in the master cylinder reservoir must be maintained at the ¾ level or higher during air bleeding.

1. Remove the bleeder cap from the slave cylinder and attach a vinyl hose to the bleeder screw.
2. Place the other end of the hose in a container.
3. Slowly, pump the clutch pedal several times.
4. With the clutch pedal depressed, loosen the bleeder screw to release the fluid and air.
5. Tighten the bleeder screw. Repeat this procedure until there are no air bubbles in the fluid in the container.

AUTOMATIC TRANSAXLE

For further information on transmissions/transaxles, please refer to "Chilton's Guide to Transmission Repair".

Transaxle Assembly

REMOVAL & INSTALLATION

2.2L Engine

1. Disconnect the battery cables (negative cable first). Remove the battery and the battery tray.
2. Disconnect the main fuse block and disconnect the coil wire from the distributor.
3. Disconnect the electrical connector from the air flow meter and remove the air cleaner assembly.
4. Remove the resonance chamber and bracket.
5. Disconnect the speedometer cable (analog cluster) or harness (digital cluster).
6. Disconnect the transaxle electrical connectors and separate the harness from the transaxle clips.
7. Disconnect both ground wires, the range selector cable and the kickdown cable from the transaxle. Raise and safely support the vehicle.
8. Remove the front wheel and tire assemblies and the splash shields. Drain the transaxle.
9. Disconnect and plug the oil cooler hoses from the transaxle.
10. Remove the tie rod ends-to-steering knuckle cotter pins and nuts. Disconnect the tie rod ends from the steering knuckle.
11. Remove the stabilizer link assemblies from the lower control arm.
12. Remove the lower control ball joint-to-steering knuckle nut/bolt. Using a prybar, pry the lower control arm downward to separate the ball joint from the steering knuckle.
13. Remove the right joint shaft bracket.
14. Position a prybar between the halfshaft and transaxle case; pry the halfshafts from the transaxle and suspend them on a wire.
15. Using transaxle plugs or equivalent, install them into the halfshaft openings of the transaxle case; this will keep the differential side gears from becoming mispositioned.
16. Remove the gusset plate-to-transaxle bolts.
17. Remove the torque converter-to-transaxle cover, the starter and the access brackets.
18. Using paint or chalk, matchmark the torque converter-to-flexplate position and remove the mounting nuts.
19. Using an engine support bar or equivalent, attach to the engine and support it's weight.
20. Remove the center transaxle mount/bracket, the left transaxle mount and the right transaxle mount-to-frame nut/bolt.
21. Remove the crossmember and the left lower arm as an assembly.
22. Using a transmission jack or equivalent, attach it to the transaxle.
23. Using a prybar, position it between the torque converter and flexplate; pry the torque converter off the studs and move it into the transaxle.
24. Remove the transaxle-to-engine bolts, lower the transaxle and remove it from the vehicle.

To install:

25. Raise and position the transaxle, align the torque converter-to-flexplate matchmark and studs. Install the transaxle-to-engine bolts and torque to 66–86 ft. lbs. (89–117 Nm).
26. Install the center transaxle mount/bracket and torque the bolts to 27–40 ft. lbs. (36–54 Nm) and the nuts to 47–66 ft. lbs. (64–89 Nm).
27. Install the left transaxle mount. Tighten the transaxle-to-mount nut to 63–86 ft. lbs. (85–117 Nm). Tighten the mount-to-bracket bolt and nut to 49–69 ft. lbs. (67–93 Nm).
28. Install the crossmember and left lower arm as an assembly. Tighten the bolts to 27–40 ft. lbs. (36–54 Nm) and the nuts to 55–69 ft. lbs. (75–93 Nm).
29. Install the right transaxle mount bolt and nut. Tighten to 63–86 ft. lbs. (85–117 Nm).
30. Install the starter motor and access brackets.
31. Install the torque converter nuts and tighten to 32–45 ft. lbs. (43–61 Nm).
32. Install the torque converter cover and tighten the bolts to 69–95 inch lbs. (8–11 Nm).
33. Install the gusset plate-to-tranaxle bolts and tighten to 27–38 ft. lbs. (37–52 Nm).
34. On the end of each halfshaft, install a new retaining clip.
35. Remove the transaxle plugs and install the halfshaft until the retaining clips snap into place.
36. Attach the lower ball joints to the steering knuckle.
37. Install the tie rods and tighten to 22–33 ft. lbs. (29–44 Nm). Install new cotter pins.
38. Install the bolts and nuts to the lower arm ball joints and tighten to 32–40 ft. lbs. (43–54 Nm).
39. Install the stabilizer link assembly-to-lower control arm. Turn the upper nuts (on each assembly) until 1.0 inch (25.4mm) of bolt thread can be measured above the nuts. When the length is reached, secure the upper nut and torque the lower nut to 12–17 ft. lbs. (16–23 Nm).
40. Install the oil cooler hoses to the transaxle.
41. Install the splash shields and the front wheels; torque the lug nuts to 65–87 ft. lbs. (88–118 Nm).
42. Connect and adjust the kickdown cable. Connect the range selector cable and torque the bolt to 22–29 ft. lbs. (29–39 Nm).
43. Install the resonance chamber and bracket; torque to 69–95 inch lbs. (8–11 Nm).
44. Connect the electrical connectors and attach the harness to the transaxle clips. Connect the gound wires.
45. Connect the speedometer cable or harness, as necessary.
46. Install the air filter assemby and connect the air flow meter connector
47. Connect the center distributor terminal lead and main fuse block.
48. Install the battery carrier and the battery. Connect the battery cables.
49. Remove the engine support bracket.
50. Refill the transaxle and check for leaks.

3.0L Engine

1. Disconnect the battery cable and remove the battery and battery tray.
2. Disconnect the main fuse block.
3. Disconnect the air cleaner hose from the air cleaner, remove the bolt/nut/washer assemblies and remove the air cleaner.
4. Remove the cruise control actuator mounting bolts and nut and move the assembly aside.

5. Disconnect the speed sensor or speedometer cable from the transaxle.

6. Move the pinch clamps on the transaxle cooler lines aside, then disconnect and plug the lines.

7. Disconnect the transaxle electrical connectors, then disconnect the harness from the routing brackets. Remove the transaxle wiring harness bracket and disconnect the ground straps.

8. Disconnect the shift cable from the transaxle and the routing bracket. Disconnect the kickdown cable from the bracket and the throttle cam.

9. Install a suitable engine support device to support the engine and transaxle. Remove all accessible transaxle-to-engine bolts from the top of the engine compartment and remove the transaxle upper mount nuts.

10. Raise and safely support the vehicle.

11. Remove the front wheel and tire assemblies and the inner fender splash shields. Drain the transaxle.

12. Disconnect the stabilizer links from the lower control arms and the bolts/nuts from the ball joints. Separate the ball joints from the steering knuckles by prying downward on the lower control arm while pushing in on the rotor.

13. Remove the mounting bolts from the right halfshaft dynamic damper. Remove the halfshafts by inserting a prybar between the shaft and transaxle case and prying out.

14. Install suitable transaxle plugs in the transaxle to prevent the differential side gears from moving out of position.

15. Remove the starter and bracket and the transaxle support bracket.

16. Remove the torque converter inspection plate. Matchmark the converter and the flexplate and remove the attaching nuts. Use a prybar to move the converter away from the flexplate, disengaging the converter studs.

17. Position a transmission jack under the transaxle. Remove the rear lower mount bolts, front lower mount through-bolt. Remove the left front crossmember and lower control arm as an assembly.

18. Remove the remaining transaxle-to-engine bolts and lower the transaxle from the vehicle.

To install:

19. Raise the transaxle into position, aligning the matchmark and the torque converter studs with the flexplate. Install the transaxle-to-engine lower bolts and tighten to 66–86 ft. lbs. (89–117 Nm).

20. Install the left front crossmember and lower control arm assembly. Tighten the bolts to 27–40 ft. lbs. (36–54 Nm) and the nut to 55–69 ft. lbs. (75–93 Nm).

21. Install the front lower mount through bolt and tighten to 66–86 ft. lbs. (85–117 Nm). Install the rear lower mount bolts and tighten to 49–69 ft. lbs. (67–93 Nm).

22. Install the torque converter attaching nuts and tighten to 32–45 ft. lbs. (43–61 Nm). Install the inspection plate and mounting bolt.

23. Install the transaxle support bracket and the starter motor and bracket.

24. Install a new retaining clip on the end of each halfshaft. Remove the transaxle plugs and install the halfshaft, making sure the clips lock in place.

25. Install the mounting bolts to the right halfshaft dynamic damper. Tighten to 31–46 ft. lbs. (42–62 Nm).

26. Attach the ball joints to the steering knuckles. Install the bolts and nuts and tighten to 27–40 ft. lbs. (36–54 Nm).

27. Install the stabilizer link assemblies. Turn the nuts until 1.0 in. (25.4mm) of bolt thread can be measured from the upper nut. When the length is reached, secure the upper nut and back off the lower nut until a torque of 12–17 ft. lbs. (16–23 Nm) is reached.

28. Install the splash guards and the wheel and tire assemblies. Tighten the lug nuts to 65–87 ft. lbs. (88–118 Nm). Lower the vehicle.

29. Install the upper mount nuts and tighten to 47–66 ft. lbs. (64–89 Nm). Install the remaining transaxle-to-engine bolts and tighten to 66–86 ft. lbs. (89–117 Nm).

30. Remove the engine support bar.

31. Connect the kickdown cable to the throttle cam and the cable bracket. Tighten the adjusting nuts.

32. Connect the ground straps and install the wiring harness bracket.

33. Connect the shift cable to the routing bracket and the transaxle. Install the attaching nut and tighten to 33–47 ft. lbs. (44–64 Nm).

34. Connect the transaxle electrical connectors, then connect the harness routing brackets to the transaxle.

35. Unplug the transaxle cooler lines and connect them to the radiator. Install the pinch clamps.

36. Connect the speed sensor or speedometer cable to the transaxle.

37. Position the cruise control actuator and install the mounting bolts and nut.

38. Position the air cleaner assembly and install the bolt/nut/washer assemblies. Connect the air cleaner hose and install the clamp.

39. Connect the main fuse block.

40. Install the battery tray and battery. Connect the battery cables.

41. Refill the transaxle and check for leaks. Adjust the kickdown cable.

SHIFT CABLE ADJUSTMENT

1. Disconnect the negative battery cable. Shift the gear selector to the **P** detent.

2. Remove the selector lever mounting screws and remove the selector knob.

3. Remove the selector trim piece and the indicator mounting screws. Disconnect the illumination bulb.

4. Disconnect the shift control switch and programmed ride control switch wiring harnesses.

5. Remove the position indicator.

NOTE: Make sure the detent spring roller is in the P detent.

6. Loosen nuts **A** and **B** and the trunnion bolt.

7. Turn the transaxle-mounted shift lever clockwise to put the transaxle in the **P** position.

8. Tighten the nut **A** by hand until it contacts the spacer, then an additional ½ turn.

9. Tighten the trunnion bolt to 67–96 inch lbs. (8–11 Nm). Tighten nut **B** to 67–96 inch lbs. (8–11 Nm).

10. Lightly, press the selector pushrod and make sure the guide plate and guide pin clearances are within specifications.

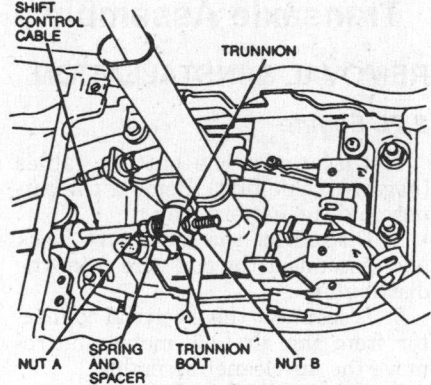

Shift cable adjustment

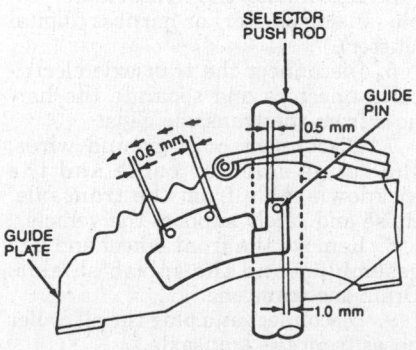

Guide plate and guide pin clearances

11. Check that the guide plate and guide pin clearances are within the specifications when the selector lever is shifted to **N** and **D**.

12. Connect the illumination bulb.

13. Connect the shift control switch and the programmed ride control switch wiring harnesses.

14. Install the position indicator and tighten the mounting screws.

15. Install the selector trim piece and position the selector knob. Tighten the knob screws.

16. Connect the negative battery cable.

THROTTLE LINKAGE ADJUSTMENT

1. Warm the engine to operating temperature and confirm that the engine is running within idle specifications.

2. Measure the free-play at the accelerator pedal. The free-play should be 0.04–0.12 in. (1–3mm).

3. To adjust, loosen the locknut and adjust at the cable housing bracket located near the throttle body.

4. Tighten the locknut after proper free-play is achieved.

5. Press the accelerator pedal all the way to the floor. Make sure the throttle plates in the throttle body are in the wide-open position.

6. If necessary, loosen the locknut and adjust at the pedal stop on the pedal bracket.

7. Tighten the locknut after adjustment is complete.

KICKDOWN CABLE ADJUSTMENT

1. From the left front wheel well, remove the splash shield.

2. At the transaxle, remove the square head plug, marked **L**, and install an adapter and a pressure gauge in the hole.

3. Rotate the kickdown cable locknuts to the furthest point from the

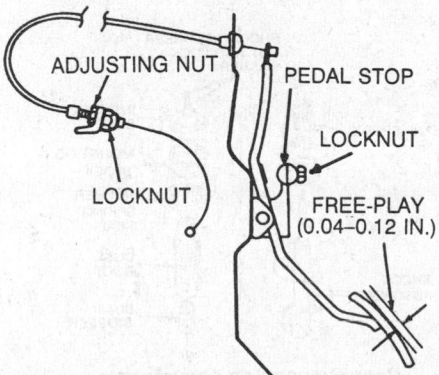

Throttle linkage adjustment

throttle cam to loosen the cable all the way.

4. Place the transaxle into the **P** position and warm the engine; the idle speed should be 700–800 rpm.

5. Rotate the locknuts toward the throttle cam until the line pressure exceeds 63–66 psi, rotate the locknuts away from the throttle cam until the line pressure is 63–66 psi and tighten the locknuts.

6. Turn the engine **OFF**, install the square head plug and torque to 43–87 inch lbs. (5–10 Nm).

NOTE: If installing a new kickdown cable, fully open the throttle valve, crimp the pin with the protector installed and remove the protector.

FRONT SUSPENSION

MacPherson Strut

REMOVAL & INSTALLATION

1. Raise and support the vehicle, safely.

2. Remove the wheel and tire assembly.

3. Remove the rubber cap from the strut mounting block. If equipped, disconnect the programmed ride control module connector.

4. At the inside of the strut mounting block and chassis strut tower, place an alignment mark.

5. If equipped, remove the programmed ride control module.

6. If equipped with anti-lock brakes, disconnect the electrical harness and remove the bracket.

7. Remove the brake caliper-to-steering knuckle bolts and suspend the caliper with mechanics wire; do not disconnect the pressure hose.

8. Remove the U-clip from the brake line hose and slide it out of the strut bracket.

9. Remove the strut-to-steering knuckle bolts.

10. Remove the vane airflow meter assembly and the ignition coil bracket.

11. Remove the strut-to-chassis nuts and remove the strut from the vehicle.

12. Place the strut assembly in a suitable holding fixture. Loosen, but do not remove the shock nut. Compress the spring with a suitable compressor tool, then remove the shock nut. Gradually release the spring compressor.

13. Remove the programmed ride control module bracket, if equipped, strut mounting block, spring seat, dust boot, bump stopper and the coil spring from the strut assembly.

To install:

14. Install the coil spring, bump stopper, dust boot and the upper spring seat on the strut assembly.

15. Install the strut mounting block and the programmed ride control module bracket, if equipped, making sure the notch on the mounting block is 180 degrees from the knuckle mounting bracket on the strut.

16. Compress the spring with the compressor tool and install the shock nut. Tighten the nut to 47–69 ft. lbs. (64–84 Nm). Gradually release the compressor tool and remove from the strut assembly.

17. Install the strut in the shock tower. Align the strut-to-chassis matchmark and torque the strut-to-chassis nuts to 34–46 ft. lbs. (46–63 Nm).

18. Install the vane airflow meter assembly and the ignition coil bracket.

19. If equipped, install the programmed ride control module and the connector.

20. Install the rubber cap on the strut tower.

21. Align the strut to the steering knuckle and torque the nuts/bolts to 69–86 ft. lbs. (93–117 Nm).

22. Install the brake caliper and the brake hose in its bracket.

23. Install the wheel and tire assembly. Tighten the lug nuts to 65–87 ft. lbs. (88–118 Nm). Lower the vehicle.

Lower Ball Joints

The ball joints can only be serviced by replacing the lower control arms.

Lower Control Arms

REMOVAL & INSTALLATION

1. Raise and safely support the vehicle under the frame away from the lower control arm. Remove the wheel and tire assembly.

2. Remove the brake caliper-to-steering knuckle bolts and support the caliper with mechanics wire.

3. Remove the stabilizer link assembly from the lower control arm.

4. Remove the ball-joint clamp bolt from the steering knuckle. Using a prybar, pry downward to separate the ball joint from the steering knuckle.

5. If equipped with an automatic transaxle, remove the harmonic damper from the chassis sub-frame; the damper is located on the left side near the lower control arm.

6. Remove the lower control arm-

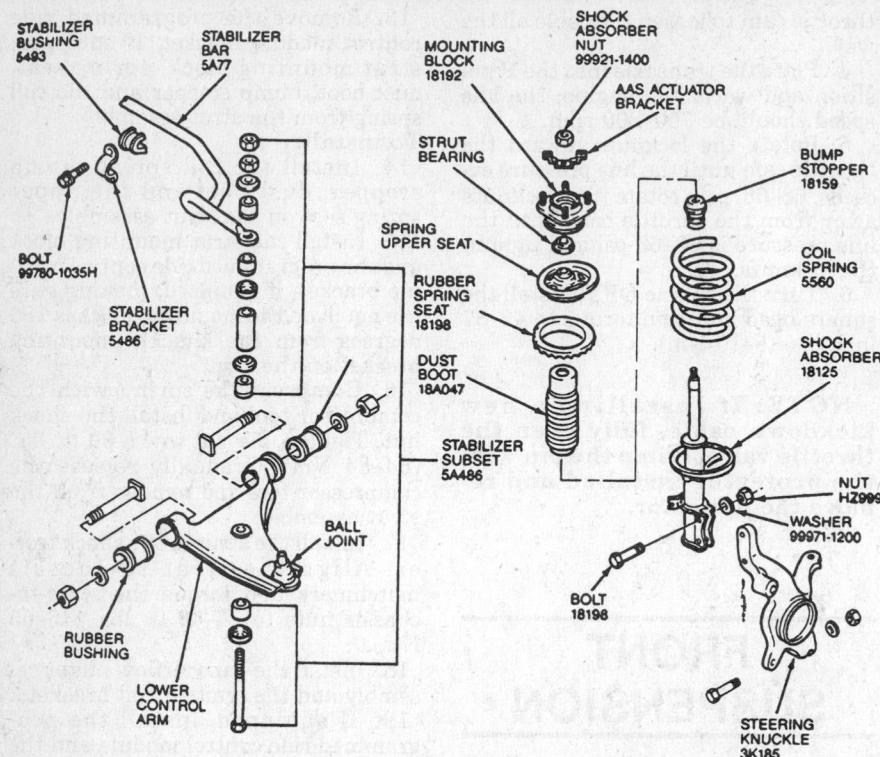

Front suspension components – exploded view

to-chassis nuts/bolts and the lower control arm.

To install:

7. Install the control arm and tighten the mounting bolts to 69–93 ft. lbs. (93–127 Nm).

8. Install the harmonic damper, if equipped.

9. Install the ball joint stud into the steering knuckle and tighten the bolt to 32–40 ft. lbs. (43–54 Nm).

10. Install the stabilizer bar link assembly. Install the brake caliper.

11. Install the wheel and tire assembly. Tighten the lug nuts to 65–87 ft. lbs. (88–118 Nm). Lower the vehicle.

12. Check and/or adjust the front wheel alignment.

Stabilizer Bar

REMOVAL & INSTALLATION

1. Raise and safely support the vehicle.

2. Remove the tire and wheel assemblies.

3. Remove the stabilizer bar link assembly mounting bolts from the lower control arm.

4. Remove the mounting bolt from the stabilizer bar bushing. Remove the stabilizer bar.

To install:

5. Install the stabilizer bar link assembly mounting bolts at the control arm. Hand tighten only.

6. Install the stabilizer bar bushing. Tighten the bushing bolt to 27–40 ft. lbs. (36–54 Nm).

7. Tighten the link nut until 0.79 in. (20mm) of thread remains above the nut.

8. Install the tire and wheel assemblies. Tighten the lug nuts to 65–87 ft. lbs. (88–118 Nm). Lower the vehicle.

REAR SUSPENSION

MacPherson Strut

REMOVAL & INSTALLATION

1. Raise and support the vehicle, safely. Remove the wheel and tire assembly.

2. Remove the upper trunk side garnish and lower trunk side trim to gain access to the strut assembly.

3. If equipped with programmed ride control, disconnect the programmed ride control module connector and removed the module.

4. If equipped with anti-lock brakes, remove the anti-lock brake harness and remove the bracket.

5. If equipped with drum brakes, remove the drum and backing plate assembly. If equipped with rear disc brakes, remove the rear disc brake caliper and rotor assembly.

6. Remove the brake line U-clip from the strut housing.

7. Loosen the trailing arm bolt and remove the spindle-to-strut bolts.

8. From inside the vehicle, remove the strut-to-chassis nuts. Remove the strut assembly.

To install:

9. Position the strut into the strut tower and torque the strut-to-chassis nuts to 34–46 ft. lbs. (46–63 Nm).

10. If equipped with programmed ride control, install the module and reconnect the connector.

11. Install the lower trunk side trim and the upper trunk side garnish.

12. Install the spindle-to-strut mounting bolts and tighten to 69–86 ft. lbs. (93–117 Nm). Tighten the trailing arm mounting bolt to 64–86 ft. lbs. (86–117 Nm).

13. Install the brake drum and backing plate or the caliper and rotor assembly, as applicable. Install the brake line U-clip onto the strut.

14. If equipped, install the ABS brake harness and bracket.

15. Install the wheel and tire assembly and tighten the lug nuts to 65–87 ft. lbs. (88–118 Nm). Lower the vehicle.

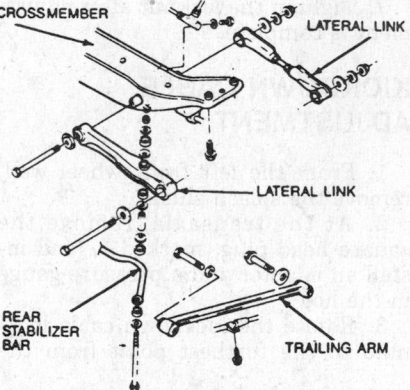

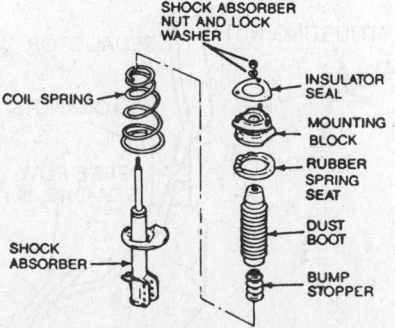

Rear suspension components – exploded view

Rear Control Arms

REMOVAL & INSTALLATION

1. Raise and support the vehicle, safely.
2. Remove the tire and wheel assembly.
3. Remove the brake drum and the backing plate assembly or the rear brake caliper and rotor assembly, if equipped.
4. Loosen, but do not remove the spindle to strut assembly mounting bolts.
5. Remove the common lateral link arm bolt and nut from the spindle.
6. Remove the trailing arm mounting bolt at the spindle and the spindle to strut assembly mounting bolts.
7. Remove the spindle from the strut assembly.
8. Remove the rear stabilize bar.
9. Remove the nut from the common lateral link mounting bolt at the rear crossmember and remove the rear lateral link.

NOTE: Because of lack of clearance between the fuel tank and the common lateral link mounting bolt, the bolt and the front lateral link cannot be removed at this time.

10. Remove the parking brake mounting bolts from the trailing arm assembly.
11. Remove the trailing arm mounting bolt and the trailing arm.
12. Remove the exhaust mounting bolts and the brake line retaining bracket from the rear crossmember. Remove the mounting bolts from the end of the crossmember.
13. Remove the rear crossmember and front lateral link as an assembly. Remove the common lateral link mounting bolt from the rear crossmember and remove the front lateral link from the crossmember.

To install:

14. Position the front lateral link on the crossmember and install the common lateral link mounting bolt.
15. Install the crossmember into the vehicle and install the mounting bolts, exhaust mounting bolts and the brake line retaining bracket bolt to the crossmember.
16. Tighten the crossmember mounting bolts to 27–40 ft. lbs. (36–54 Nm) and the brake line retaining bracket bolt to 13–20 ft. lbs. (18–26 Nm).
17. Position the trailing arm into the body mounting bracket and tighten the mounting bolt to 49–69 ft. lbs. (63–93 Nm).
18. Install the parking brake cable mounting bolts to the trailing arm.

19. Position the rear lateral link onto the common lateral link mounting bolt at the rear crossmember and install the nut to the bolt. Tighten the mounting bolt and nut at the rear crossmember to 64–86 ft. lbs. (86–117 Nm).
20. Install the rear stabilizer bar assembly.
21. Place the spindle onto the strut assembly mounting bracket and tighten the mounting bolts to 69–86 ft. lbs. (93–117 Nm).
22. Install the common lateral link arm bolt and nut through the spindle and tighten to 64–86 ft. lbs. (86–117 Nm).
23. Install the trailing arm mounting bolt and tighten to 64–86 ft. lbs. (86–117 Nm).
24. Install the brake drum and backing plate assembly or the brake caliper and rotor assembly, as applicable.
25. Install the wheel and tire assembly and tighten the lug nuts to 65–87 ft. lbs. (88–118 Nm). Lower the vehicle.

Rear Wheel Bearings

REMOVAL & INSTALLATION

1. Raise and support the vehicle, safely.
2. Remove the wheel and tire assembly and the grease cap.
3. Using a cape chisel and a hammer, raise the staked portion of the hub nut.
4. Remove and discard the hub nut.
5. Remove the brake drum or disc brake rotor assembly from the spindle.
6. Using a small prybar, pry the grease seal from the brake drum or rotor and discard it.
7. Remove the snapring. Using a

shop press, press the wheel bearing from the brake drum or rotor.

To install:

8. Using a shop press, press the new wheel bearing into the brake drum or rotor until it seats and install the snapring.
9. Lubricate the new seal lip with grease and install the seal, using a suitable installation tool.
10. Position the brake drum or rotor onto the wheel spindle.
11. Install a new locknut and tighten to 73–131 ft. lbs. (98–117 Nm).
12. Check the wheel bearing endplay.
13. Using a dull cold chisel, stake the locknut.

NOTE: If the nut splits or cracks after staking, it must be replaced with a new nut.

14. Install the grease cap and the wheel and tire assembly. Tighten the lug nuts to 65–87 ft. lbs. (88–118 Nm). Lower the vehicle.

ADJUSTMENT

1. Raise and safely support the vehicle. Make sure the parking brake is fully released.
2. Remove the wheel and tire assembly.
3. Rotate the drum or rotor to make sure there is no brake drag.
4. Install a suitable dial indicator and check the wheel bearing endplay. Endplay should not exceed 0.008 in. (0.2mm).
5. If the endplay exceeds specification, replace the wheel bearing.

STEERING

Steering Wheel

REMOVAL & INSTALLATION

1. Disconnect the negative battery cable.
2. Remove the steering wheel horn pad mounting screws from the back side of the steering wheel.
3. Remove the steering wheel cover pad and disconnect the horn wire from the cover pad.
4. Remove the steering wheel attaching nut.
5. Matchmark the steering wheel to the steering shaft.
6. On 1989–90 vehicles, remove the steering wheel, using a steering wheel puller. On 1991–92 vehicles, grasp the steering wheel and firmly pull the steering wheel to remove it.

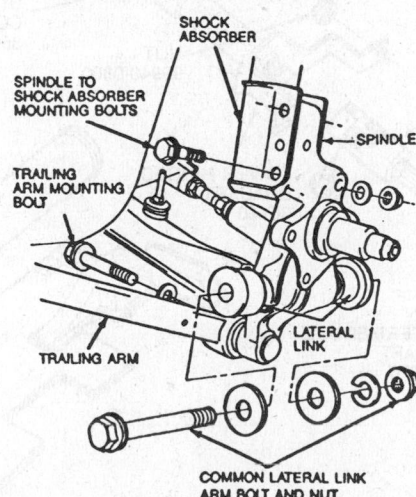

SHOCK ABSORBER

SPINDLE TO SHOCK ABSORBER MOUNTING BOLTS

SPINDLE

TRAILING ARM MOUNTING BOLT

LATERAL LINK

TRAILING ARM

COMMON LATERAL LINK ARM BOLT AND NUT

Rear spindle removal

NOTE: Do not use a steering wheel puller to remove the steering wheel on 1991–92 vehicles. The steering wheel puller will collapse the steering shaft, causing the steering wheel to rub the plastic trim behind the wheel.

To install:

7. Install the steering wheel over the steering shaft making sure the matchmarks are aligned.

8. Install the steering wheel mounting nut and tighten to 29–36 ft. lbs. (39–49 Nm).

9. Connect the horn wire. Replace the cover pad and replace the mounting screws. Connect the negative battery cable.

Steering Column

REMOVAL & INSTALLATION

1. Disconnect the negative battery cable.

2. Remove the steering wheel.

3. Remove the column cover screws and the cover.

4. Remove the attaching screws from the instrument cover. Carefully, pull the cover outward and disconnect the electrical connectors from the cover. Remove the ignition illumination bulb and the instrument cover.

5. Loosen the instrument cluster cover-to-hinge screws, remove the instrument cluster-to-dash screws and the instrument cluster cover.

6. Remove the lower panel, the lap duct and the defrost duct.

7. Disconnect the electrical connectors from the turn signal switch assembly.

8. Remove the U-joint cinch bolt from the lower end of the steering shaft.

9. Remove the mounting nuts from the hinge bracket.

10. Remove the cluster support nuts and the upper steering column brackets nuts/bolts and the steering shaft assembly.

11. At the steering rack, lift the boot from the intermediate shaft U-joint

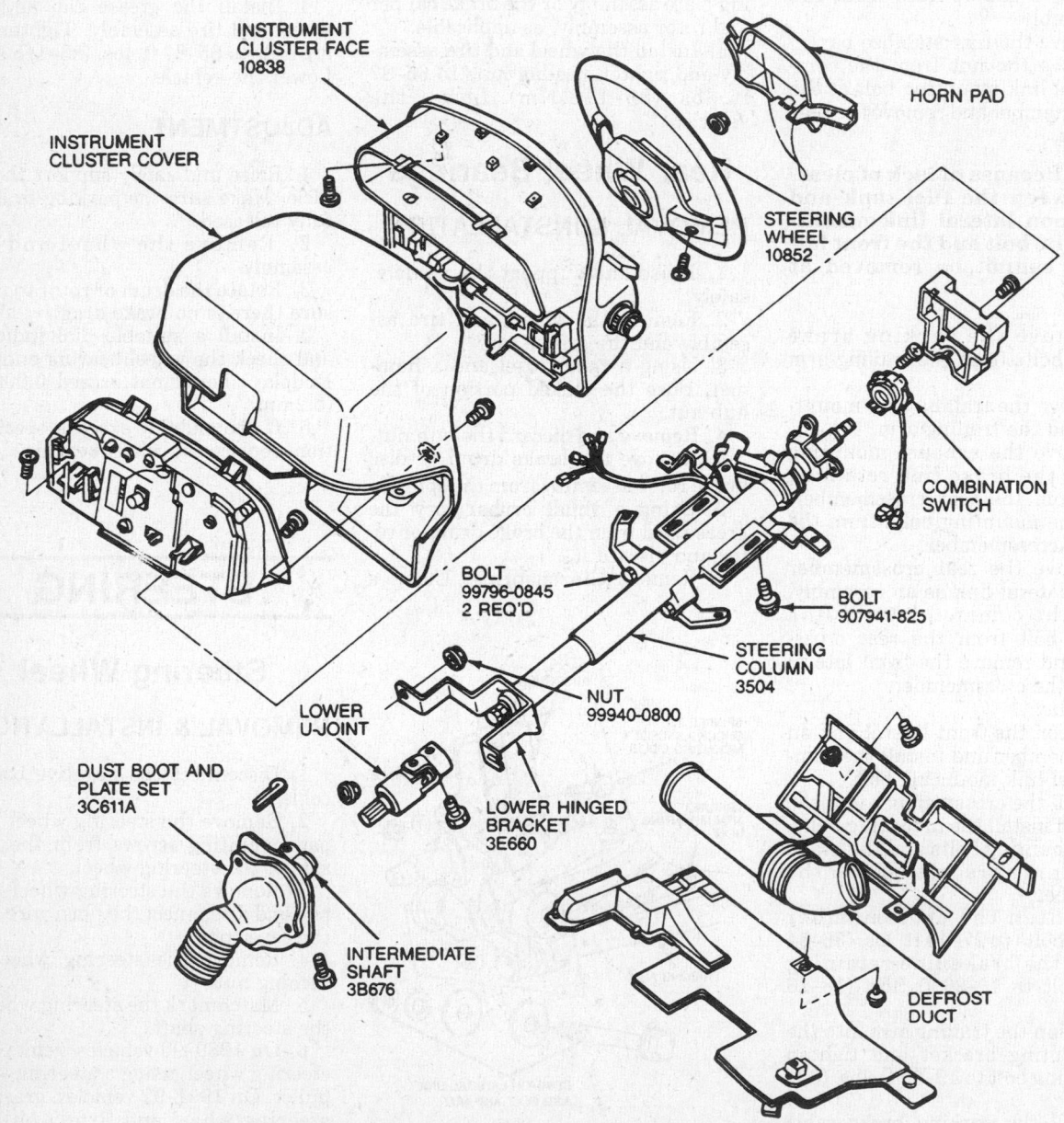

Steering column assembly and related components—1989–90 vehicles

and remove the lower U-joint cinch bolt.

12. Remove the intermediate shaft dust cover assembly nuts, the intermediate shaft and the dust cover assembly.

To install:

13. Using an assistant to support the intermediate shaft and dust cover assembly, guide the lower U-joint onto the steering rack pinion.

14. Install the lower intermediate shaft U-joint cinch bolt and torque it to 13–20 ft. lbs. (18–26 Nm). Install the dust cover nuts.

15. Using an assistant to support the steering column, guide the column into the upper intermediate U-joint; do not install the cinch bolt at this time.

16. Install the hinge bracket nuts; do not tighten at this time.

17. Install the upper cinch bolt into the intermediate U-joint and tighten to 13–20 ft. lbs. (18–26 Nm).

18. Install the upper column bracket bolts. Tighten the hinge bracket nuts to 12–17 ft. lbs. (16–23 Nm) and the upper bracket bolts to 6.5–10 ft. lbs. (8.8–14 Nm).

19. Install the cluster support nuts and tighten to 6.5–10 ft. lbs. (8.8–14 Nm).

20. Connect the electrical connectors at the ignition switch. Install the instrument cluster cover.

21. Connect the electrical connectors and install the ignition illumination bulb into the instrument cover and install the cover.

22. Install the steering wheel and connect the negative battery cable.

Power Rack and Pinion

ADJUSTMENT

Standard Rack Assembly

1. Disconnect the negative battery cable. Remove the steering rack assembly from the vehicle and place it in a holding fixture.

2. Using a pinion torque adapter tool and an inch pound torque wrench, check the pinion turning torque; it should be 89–124 inch lbs. (10–14 Nm).

3. If the torque is not to specifications, loosen the locknut.

4. Using a yoke torque gauge tool, torque the adjusting cover to 7.2 ft. lbs. (9.8 Nm), loosen the cover, retorque to 3.6 ft. lbs. (4.9 Nm) and loosen the cover 45 degrees.

5. Using a yoke locknut wrench tool, torque the locknut to 36–43 ft. lbs. (49–59 Nm).

6. Install the steering rack assem-

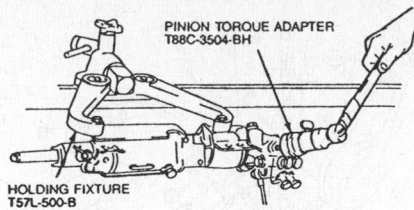

Measuring steering rack pinion torque

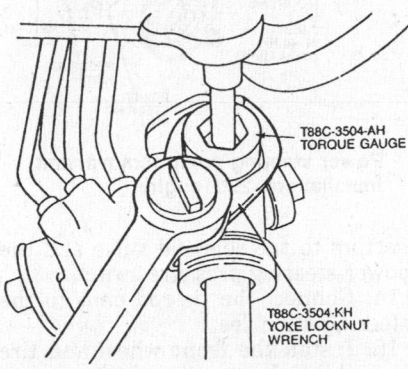

Tightening the steering rack locknut

bly. Refill the power steering reservoir. Start the engine and bleed the system. Test drive and check the steering operation.

Electronic Rack Assembly

1. Disconnect the negative battery cable. Remove the steering rack assembly from the vehicle and place it in a holding fixture.

2. Using a pinion torque adapter tool and an inch pound torque wrench, check the pinion turning torque; it should be 89–124 inch lbs. (10–14 Nm).

3. If the torque is not to specifications, loosen the locknut.

4. Using a yoke torque gauge tool, torque the adjusting cover to 39–48 inch lbs. (4.5–5.5 Nm), then loosen it 35 degrees.

5. Using a yoke locknut wrench tool, torque the locknut to 29–36 ft. lbs. (40–50 Nm).

6. Install the steering assembly. Refill the power steering reservoir. Connect the negative battery cable.

7. Start the engine and bleed the system. Test drive and check the steering operation.

REMOVAL & INSTALLATION

Standard Rack Assembly

The non-electronically controlled power rack and pinion steering gear has an integral valve and power assist system.

1. Disconnect the negative battery cable.

2. Raise and safely support the vehicle. Remove the front wheel and tire assemblies.

3. Disconnect the tie rod ends from the steering knuckles.

4. From both sides of the vehicle, remove the lower inner fender plastic splash shield.

5. At the steering assembly, pull back the steering column dust boot, turn the steering shaft until the clamp bolt is accessible and lock the steering column. Using paint, matchmark the steering column pinion shaft-to-intermediate shaft lower universal joint location.

6. Remove the clamp bolt from the intermediate shaft lower universal joint.

7. Disconnect and plug both hydraulic lines from the steering rack assembly; move the lines aside.

8. Remove the steering rack assembly-to-chassis bolts. Lower the steering assembly until it clears the bulkhead, slide it toward the right side until the left tie rod clears the left lower control arm, move it toward the left side and remove from the vehicle.

To install:

9. Move the steering rack assembly into position, so the pinion shaft is just below the intermediate shaft universal joint.

10. Raise the steering rack assembly into position, align the pinion shaft-to-intermediate universal joint matchmark and install the steering assembly-to-chassis bolts. Torque the steering rack assembly-to-chassis bolts to 27–40 ft. lbs. (36–54 Nm).

11. Install the pinion shaft-to-intermediate universal joint clamp bolt and torque to 13–20 ft. lbs. (18–26 Nm).

12. Connect the hydraulic lines to the steering rack assembly.

13. Connect the tie rod ends to the steering knuckles. Install the plastic splash shields.

14. Install the wheel and tire assemblies.

15. Lower the vehicle. Refill the power steering reservoir. Start the engine, bleed the power steering system and check for leaks.

Electronic Rack Assembly

The Variable Assist Power Steering (VAPS) system automatically adjusts the power steering pressure. It provides light steering effort during low speed and parking maneuvers and higher steering effort at higher speeds for improved road feel.

The completely automatic system (no driver controls) consists of: a VAPS control unit, a steering angle sensor, a vehicle speed sensor, a solenoid valve, a test connector and interconnecting wiring.

1. Disconnect the negative battery cable.

2. Raise and safely support the vehicle. Remove the front wheel and tire assemblies.

3. Disconnect the tie rod ends from the steering knuckles.

4. From both sides of the vehicle, remove the lower inner fender plastic dust shields.

5. At the steering assembly, pull back the steering column dust boot, turn the steering shaft until the clamp bolt is accessible and lock the steering column. Using paint, matchmark the steering column pinion shaft-to-intermediate shaft lower universal joint location.

6. Remove the clamp bolt from the intermediate shaft lower universal joint.

7. Disconnect the electrical connector from the solenoid valve and the power steering pressure switch.

8. Disconnect and plug the hydraulic lines from the steering rack assembly; discard both copper washers from each fitting and move the lines aside.

9. Remove the steering rack assembly-to-chassis bolts. Lower the steering rack assembly until it clears the bulkhead, slide it toward the right side until the left tie rod clears the left lower control arm, move it toward the left side and remove from the vehicle.

To install:

10. Move the steering rack assembly into position, so the pinion shaft is just below the intermediate shaft universal joint.

11. Raise the steering rack assembly into position, align the pinion shaft-to-intermediate universal joint matchmark and install the steering rack assembly-to-chassis bolts. Torque the steering assembly-to-chassis bolts to 27–40 ft. lbs. (36–54 Nm).

12. Install the pinion shaft-to-intermediate universal joint clamp bolt and torque to 13–20 ft. lbs. (18–26 Nm).

13. Using new copper washers, connect the hydraulic lines to the steering assembly. Connect the electrical con-

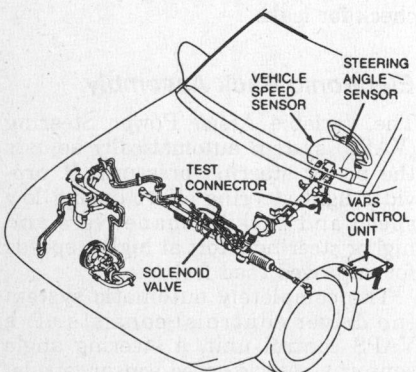

Electronic variable assist power steering system

nectors to the solenoid valve and the power steering pressure switch.

14. Connect the tie rod ends to the steering knuckles.

15. Install the front wheel and tire assemblies. Lower the vehicle.

16. Refill the power steering reservoir. Start the engine, bleed the power steering system and check for leaks.

Power Steering Pump

REMOVAL & INSTALLATION

2.2L Engine

1. Disconnect the negative battery cable.

2. At the right fender, remove the inner fender splash shield.

3. Loosen the power steering pump and remove the drive belt.

4. Disconnect and plug the pressure and return hoses from the pump

5. Remove the pump-to-bracket bolts and the pump; if necessary, remove the drive pulley from the pump.

To install:

6. Position the pump on the bracket and torque the bolts to 27–34 ft. lbs. (31–46 Nm).

7. Connect the pressure and return hoses to the pump.

8. Install the drive belt. Refill the power steering reservoir. Connect the negative battery cable, start the engine and bleed the system.

3.0L Engine

1. Disconnect the negative battery cable.

2. Remove the washer reservoir and place aside.

3. Remove the plastic shield and the accessory drive belt.

4. Remove the drive pulley from the pump.

5. Disconnect and plug both power steering hoses at the pump.

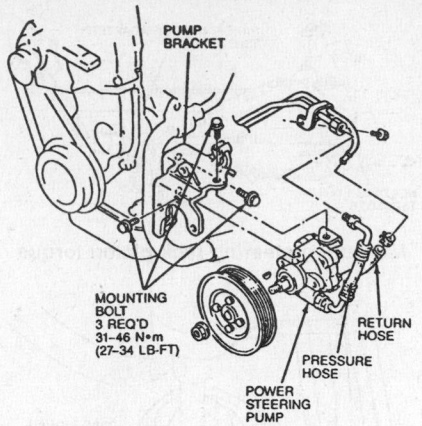

Power steering pump removal and Installation—2.2L engine

6. Remove the pump mounting bolts and lift the pump from the accessory support bracket.

To install:

7. Position the pump on the accessory support bracket.

8. Install the pump mounting bolts and tighten to 15–22 ft. lbs. (20–30 Nm).

9. Install the drive pulley and replace both power steering hoses.

10. Install the drive belt and the plastic shield. Install the washer reservoir and connect the negative battery cable.

11. Fill the pump reservoir and bleed the system.

BELT ADJUSTMENT

2.2L Engine

1. Loosen the air conditioning compressor pivot and adjustment lock bolts.

2. Check the belt deflection by applying approximately 22 lbs. pressure midway bwtween the compressor and crankshaft pulleys.

3. Turn the tension adjusting bolt until the belt deflection is 0.27–0.35 in. (7–9mm) for a new belt or 0.31–0.39 in. (8–10mm) for a used belt.

4. Tighten the pivot and adjustment lock bolts to 27–38 ft. lbs. (37–52 Nm).

3.0L Engine

Power steering belt tension is maintained by an automatic tensioner. No adjustment is necessary.

SYSTEM BLEEDING

1. Raise and support the vehicle, safely.

2. Disconnect the coil wire. Refill the power steering pump reservoir to the specified level.

3. Crank the engine. Check and refill the reservoir.

4. Crank the engine and rotate the steering wheel from lock-to-lock. Check and refill the power steering pump reservoir.

5. Connect the coil wire, start the engine and allow it to run for several minutes.

6. Rotate the steering wheel from lock-to-lock several times, until the air bubbles are eliminated from the fluid.

7. Turn the engine **OFF**. Check and/or refill the reservoir.

8. Disconnect the negative battery cable, depress the brake pedal for at least 5 seconds and reconnect the negative battery cable.

Tie Rod Ends

REMOVAL & INSTALLATION

1. Raise and support the vehicle, safely. Remove the tire and the wheel assembly.
2. Remove the cotter pin and the tie rod mounting nut.
3. Disconnect the tie rod from the steering knuckle, using the proper tool.
4. Matchmark the tie rod end jam nut to the tie rod end.
5. Loosen the jam nut on the tie rod and remove the tie rod end.

To install:

6. Thread the jam nut and the tie rod end onto to the tie rod.
7. Align the marks on the tie rod with the jam nut and tie rod end. Tighten the jam nut to 51–72 inch lbs. (69–98 Nm).
8. Reinstall the tie rod in the steering knuckle. Tighten the mounting bolt to 22–33 ft. lbs. (29–44 Nm).
9. Align the slots in the nut with the hole in the ball joint stud and install the cotter pin.

NOTE: If the slots in the nut do not align with the hole in the ball-joint stud, tighten the nut for proper alignment. Never loosen the nut.

10. Install the tire and wheel assembly and lower the vehicle.

BRAKES

For all brake system repair and service procedures not detailed below, please refer to "Brakes" in the Unit Repair section.

Master Cylinder

REMOVAL & INSTALLATION

1. Disconnect the negative battery cable and low fluid level sensor connector from the master cylinder.
2. Disconnect the brake lines from the master cylinder. Plug the lines and the master cylinder outlet ports.
3. Remove the master cylinder-to-power booster nuts and the master cylinder.

To install:

4. Check the adjustment of the master cylinder pushrod as follows:

a. Insert a pencil in the pushrod socket of the master cylinder. Mark the edge of the socket opening with a hack saw blade.
b. Measure the length of the pencil to the saw mark with a ruler.
c. With a ruler measure how far the master cylinder pushrod protrudes out of the booster assembly.
d. Measure the master cylinder boss with a ruler. Subtract the length of the boss from the length of the pencil. The difference in length between the master cylinder pushrod and the corrected pencil length is equal to the pushrod clearance.
e. Adjust the pushrod length to get the correct clearance. It should be 0.040 in. (1mm) shorter than the pushrod socket.

5. Before installation, bench bleed the new master cylinder as follows:

a. Mount the new master cylinder in a suitable holding fixture. Be careful not to damage the housing.
b. Fill the master cylinder reservoir with brake fluid.
c. Using a suitable tool inserted into the booster pushrod cavity, push the master cylinder piston in slowly. Place a suitable container under the master cylinder to catch the fluid being expelled from the outlet ports.
d. Place a finger tightly over each outlet port and allow the master cylinder piston to return.
e. Repeat the procedure until clear fluid only is expelled from the master cylinder. Plug the outlet ports and remove the master cylinder from the holding fixture.

6. Position the master cylinder over the booster pushrod and booster mounting studs. Install the mounting nuts.
7. Remove the plugs and connect the brake lines. Tighten the fittings.
8. Make sure the master cylinder reservoir is full. Have an assistant push down on the brake pedal. When the pedal is all the way down, crack open the brake line fittings, 1 at a time, to expel any remaining air in the master cylinder and brake lines. Tighten the fittings, then have the assistant allow the brake pedal to return.
9. Repeat Step 8 until all air is expelled from the master cylinder and brake lines. Final tighten the brake line fittings.
10. Connect the low fluid level sensor wiring.
11. Make sure the master cylinder reservoir is full.
12. If necessary, bleed the brake system.
13. Connect the negative battery cable. Check for fluid leaks and check for proper operation.

Proportioning Valve

REMOVAL & INSTALLATION

1. Disconnect the fuel filter bracket and position the fuel filter aside.
2. Disconnect the brake lines connected to the proportioning valve, using a tubing wrench.
3. Remove the brake lines between the proportioning valve and the master cylinder, using a tubing wrench.
4. Remove the proportioning valve mounting bolts and the proportioning valve.

To install:

5. Position the proportioning valve on the bulkhead and loosely install 1 of the attaching bolts.
6. Loosely install the brake lines between the valve and the master cylinder.

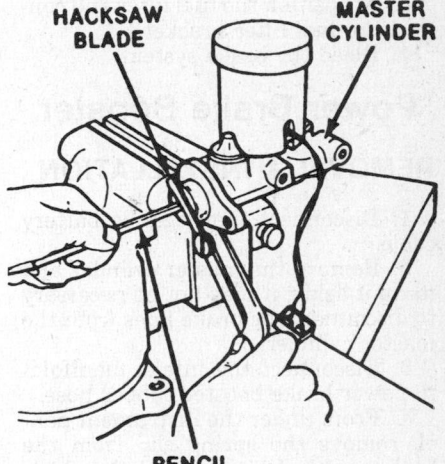

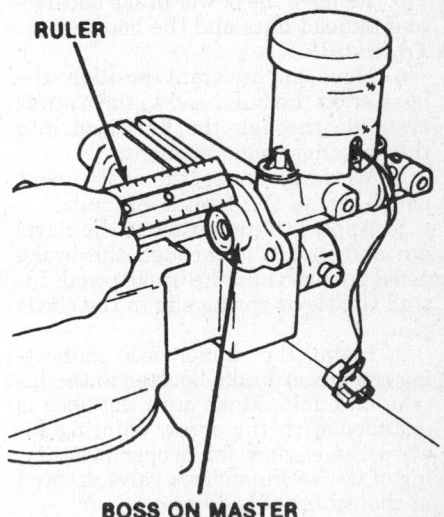

Master cylinder pushrod adjustment procedure

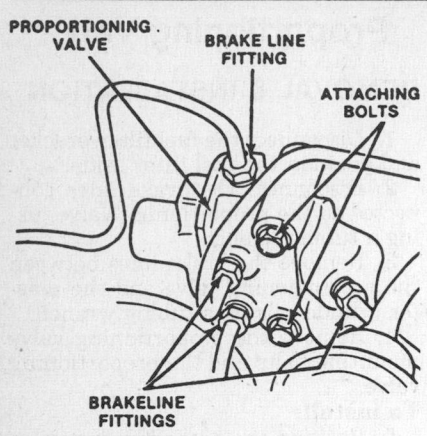

Proportioning valve location

7. Loosely install the remaining brake lines into the valve.

8. Install the other attaching bolt and tighten both bolts.

9. Tighten all the brake lines, using a tubing wrench.

10. Reposition the fuel filter and connect the fuel filter bracket.

11. Bleed the brake system.

Power Brake Booster

REMOVAL & INSTALLATION

1. Disconnect the negative battery cable.

2. Remove the master cylinder and move it aside; it may not be necessary to disconnect the brake lines from the master cylinder.

3. Disconnect the intake manifold-to-power brake booster rubber hose.

4. From under the instrument panel, remove the spring clip from the brake pedal clevis pin and the clevis pin.

5. Remove the power brake booster-to-bulkhead nuts and the booster.

To install:

6. Have an assistant position the booster on the bulkhead so the 4 studs protrude through the bulkhead into the passenger compartment.

7. Working under the instrument panel, install the 4 retaining nuts.

8. Apply lithium grease to the clevis pin and install it through the brake pedal pushrod and the brake pedal. Install the clevis spring clip in the clevis pin.

9. Install the rubber hose connecting the power brake booster to the intake manifold. Make sure the hose is installed with the arrow pointing toward the engine, for proper positioning of the vacuum check valve, located in the center of the hose.

10. Check the adjustment of the master cylinder pushrod as follows:

a. Insert a pencil in the pushrod socket of the master cylinder. Mark the edge of the socket opening with a hack saw blade.

b. Measure the length of the pencil to the saw mark with a ruler.

c. With a ruler measure how far the master cylinder pushrod protrudes out of the booster assembly.

d. Measure the master cylinder boss with a ruler. Subtract the length of the boss from the length of the pencil. The difference in length between the master cylinder pushrod and the corrected pencil length is equal to the pushrod clearance.

e. Adjust the pushrod length to get the correct clearance. It should be 0.040 in. (1mm) shorter than the pushrod socket.

11. Install the master cylinder. Bleed the brake system if the brake lines were disconnected.

Brake Caliper

REMOVAL & INSTALLATION

Front

1. Raise and support the vehicle, safely. Remove the front wheel and tire assemby.

2. Remove the banjo bolt mounting the brake flex hose to the caliper.

3. Remove and discard the copper washers that seal the flex hose banjo fitting.

4. Remove the caliper mounting bolt.

5. Pivot the caliper off the brake pads and slide the caliper off the guide pin.

6. Remove the guide pin dust boots and push out the caliper guide pin bushing.

To install:

7. Lubricate the guide pin bushings with a high temperature grease and install them in the caliper.

8. Install the guide pin bushing dust boots.

9. Position the caliper onto the guide pin. Pivot the caliper onto the brake pads. It may be neccessary to pull outward slightly on the caliper, to provide the necessary clearance.

10. Install the caliper mounting bolt and tighten to 23–30 ft. lbs. (31–41 Nm).

11. Install new copper washers and the banjo bolt on the flex hose banjo fitting.

12. Position the flex hose on the caliper and replace the banjo bolt. Tighten the bolt to 16–22 ft. lbs. (22–29 Nm).

13. Bleed the front brakes and install the wheel and tire assembly.

14. Lower the vehicle.

Rear

1. Raise and safely support the vehicle. Remove the wheel and tire assembly.

2. Loosen the parking brake cable housing adjustment nut. Remove the cable housing from the bracket and the parking brake lever.

3. Remove the banjo bolt mounting the brake flex hose to the caliper.

4. Remove and discard the copper washers from the banjo fitting.

5. Remove the caliper mounting bolt.

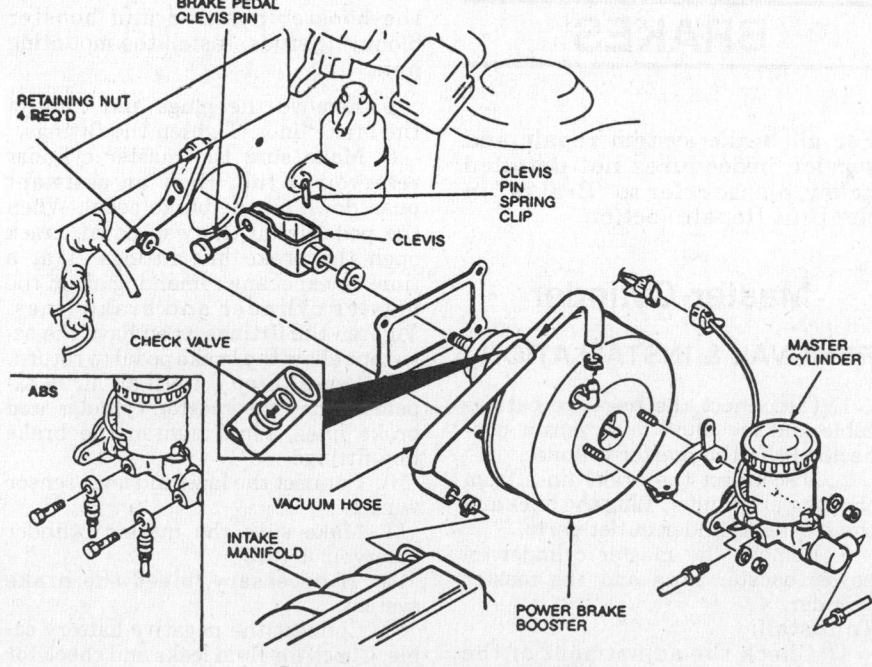

Power brake booster assembly

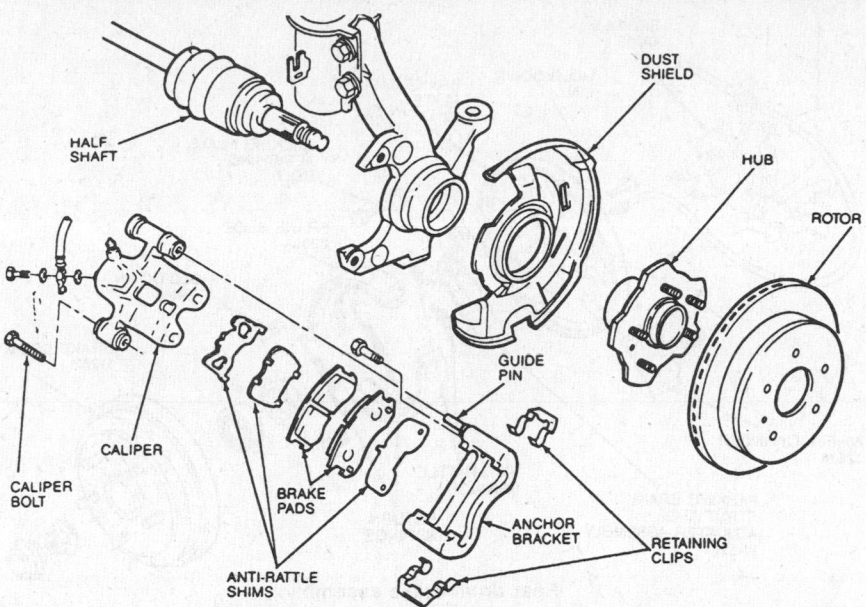

Front disc brake assembly

7. Remove the brake pads and retaining clips from the caliper anchor.
To install:

8. Install the retaining clips and install the brake pads into the caliper anchor; the pad with the wear indicator is the inboard pad.

9. Install the caliper onto the fixed guide pin and rotate the caliper downward over the brake pads. Install the caliper mounting bolt and torque to 23–30 ft. lbs. (31–34 Nm).

10. Install the wheel and tire assembly and tighten the lug nuts to 65–87 ft. lbs. (80–118 Nm). Lower the vehicle.

11. Pump the brake pedal several times to seat the pads. Check and/or add fluid to the brake master cylinder reservoir.

Rear

1. Remove approximately ⅔ of the brake fluid from the master cylinder reservoir.

2. Raise and support the vehicle, safely.

3. Remove the wheel and tire assembly.

4. Loosen the parking brake cable housing adjusting nut. Remove the cable housing from the bracket and the parking brake lever.

5. Remove the caliper mounting bolt and pivot the caliper to clear the brake shoes. If necessary, pry the caliper outward. Remove the caliper and support it with mechanics wire.

6. Remove the **V** spring from the disc pads. Remove the disc pads, the anti-rattle shims and retaining clips.

NOTE: If the anti-rattle shims are to be reused, they must be installed in their original positions.

6. Pivot the caliper off the brake pads and slide the caliper off the guide pin.

7. Remove the disc pads and the caliper guide pin from the anchor bracket.

8. Remove the bolts from the anchor bracket and remove the anchor bracket.
To install:

9. Install the anchor bracket and tighten the mounting bolts to 33–49 ft. lbs. (45–67 Nm).

10. Install the caliper guide pin and the brake pads and shims.

11. Lubricate the guide pin bushings with high temperature grease. Install the caliper onto the guide pin and pivot the caliper over the brake pads. Tighten the mounting bolts to 12–17 ft. lbs. (16–24 Nm).

12. Install new copper washers and the banjo bolt mounting the flex hose to the caliper. Tighten the banjo bolt to 16–20 ft. lbs. (22–26 Nm).

13. Position the parking brake cable into the parking brake lever and bracket.

14. Adjust the parking brake cable so there is no clearance between the cable end and the parking brake lever. Tighten the parking brake cable locknut to 14–21 ft. lbs. (20–28 Nm).

15. Bleed the rear brakes.

16. Install the wheel and tire assembly and lower the vehicle.

Disc Brake Pads

REMOVAL & INSTALLATION

Front

1. Remove approximately ⅔ of the

brake fluid from the master cylinder reservoir.

2. Raise and support the vehicle, safely.

3. Remove the wheel and tire assembly.

4. Using a prybar, insert it through the caliper opening and pry outwards to drive the caliper piston into it's bore.

5. Remove the caliper mounting bolt and rotate the caliper upward. Slide the caliper off the fixed guide pin and support it with mechanics wire.

6. Label the anti-rattle shims so they can be reinstalled in their original positions.

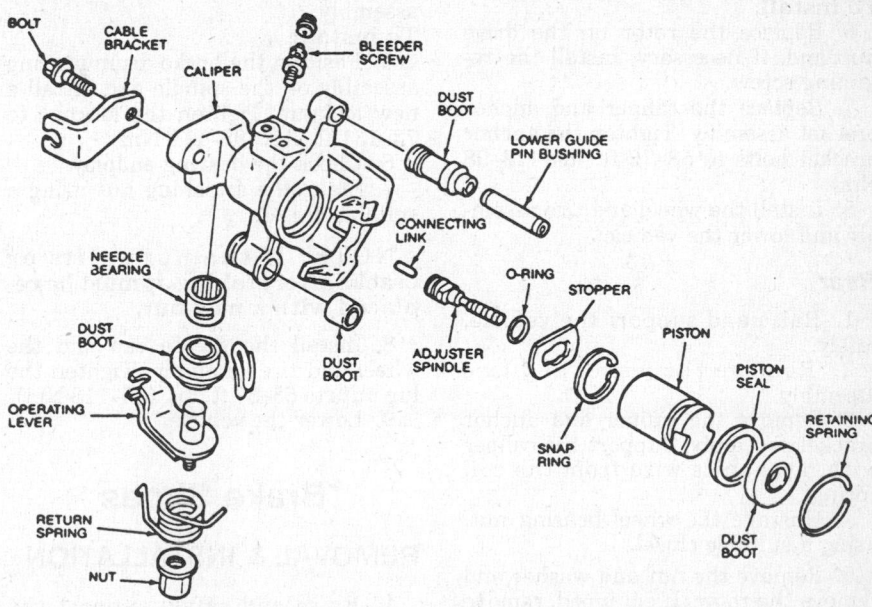

Exploded view of the rear disc brake caliper assembly

To install:

7. Install the retaining clips, the anti-rattle shims, the disc pads and the **V** spring.

8. Using high temperature grease, lubricate the guide pin busings.

9. Install the caliper onto the guide pin and pivot the caliper over the disc pads. Install the caliper mounting bolt and tighten to 12–17 ft. lbs. (16–24 Nm).

10. Install the parking brake cable, adjust the cable so there is no clearance between the cable end and the parking brake lever. Torque the parking brake cable locknut to 14–21 ft. lbs. (20–28 Nm).

11. Install the wheel and tire assembly. Lower the vehicle.

12. Pump the brake pedal several times to seat the pads. Check and/or add fluid to the brake master cylinder reservoir.

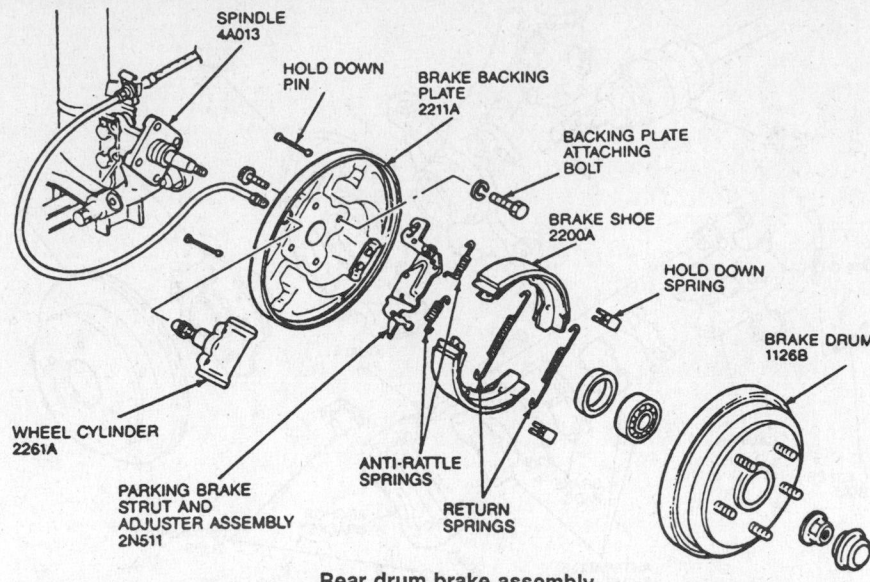

Rear drum brake assembly

Brake Rotor

REMOVAL & INSTALLATION

Front

1. Raise and support the vehicle, safely.

2. Remove the wheel and tire assembly.

3. Remove the caliper and the anchor bracket. Support the caliper assembly with mechanics wire from the coil spring.

4. Remove the rotor retaining screw, if necessary, and remove the rotor from the drive hub.

5. Inspect the rotor for scoring, wear and/or damage. Machine or replace, as necessary. If maching the rotor, observe the minimum thickness specification.

To install:

6. Replace the rotor on the drive hub and, if necessary, install the retaining screw.

7. Replace the caliper and anchor bracket assembly. Tighten the anchor bracket bolts to 58–72 ft. lbs. (78–98 Nm).

8. Install the wheel and tire assembly and lower the vehicle.

Rear

1. Raise and support the vehicle, safely.

2. Remove the wheel and tire assembly.

3. Remove the caliper and anchor bracket assembly. Support the caliper with mechanics wire from the coil spring.

4. Unstake the wheel bearing nut, using a suitable chisel.

5. Remove the nut and washer and remove the rotor. If equipped, remove the ABS signal ring from the rotor.

6. Inspect the rotor for scoring, wear and/or damage. Machine or replace, as necessary. If maching the rotor, observe the minimum thickness specification.

7. Installation is the reverse of the removal procedure.

Brake Drums

REMOVAL & INSTALLATION

1. Raise and safely support the vehicle.

2. Remove the wheel and tire assembly and remove the grease cap.

3. Carefully raise the staked portion of the attaching nut using a suitable chisel.

4. Remove and discard the hub nut. Remove the brake drum/bearing assembly.

To install:

5. Position the brake drum/bearing assembly on the spindle and install a new locknut. Tighten the locknut to 73–131 ft. lbs. (98–117 Nm).

6. Check the bearing endplay.

7. Stake the attaching nut using a suitable chisel.

NOTE: If the nut splits or cracks after staking, it must be replaced with a new nut.

8. Install the grease cap and the wheel and tire assembly. Tighten the lug nuts to 65–87 ft. lbs. (88–118 Nm).

9. Lower the vehicle.

Brake Shoes

REMOVAL & INSTALLATION

1. Raise and safely support the vehicle.

2. Remove the wheel and tire assembly and the brake drum.

3. Remove the brake shoe return springs and anti-rattle spring.

4. Remove the brake shoe hold-down springs; push the pin inward and twist the hold-down spring to disengage it from the hold-down pin.

5. Remove the rear shoe from the parking brake strut. Remove the front shoe.

NOTE: Unless they are broken or worn, leave the parking brake strut, adjuster mechanism and the adjuster spring in place.

To install:

6. Using high temperature grease, lubricate the 6 shoe contact pads and the adjuster mechanism.

NOTE: If new shoes are being installed, reface the brake drums.

7. Position the rear brake shoe in the parking brake strut and install the rear hold-down pin and spring.

8. Position the front brake shoe against the parking brake strut and backing plate and install the hold-down pin and spring.

9. Install the brake shoe return springs.

10. Using a small prybar between the knurled quadrant and the parking brake strut, twist the prybar until the quadrant just touches the backing plate.

11. Install the drum, wheel and tire assembly. Lower the vehicle.

12. Check the master cylinder reservoir fluid level. Firmly apply the brakes 2–3 times to adjust the rear brakes.

Wheel Cylinder

REMOVAL & INSTALLATION

1. Raise and safely support the vehicle. Remove the wheel and tire assembly.
2. Remove the brake drum and the brake shoes.
3. Using a tubing wrench, disconnect the brake line from the wheel cylinder.
4. Remove the wheel cylinder-to-backing plate bolts and the wheel cylinder.

To install:

5. install the wheel cylinder and loosely install the mounting bolts.
6. Connect the brake line to the wheel cylinder and tighten the fitting, using a tubing wrench.
7. Tighten the wheel cylinder mounting bolts to 7–9 ft. lbs. (10–13 Nm).
8. Install the brake shoes and the brake drum. Bleed the rear brakes.
9. Install the wheel and tire assembly and lower the vehicle.

Parking Brake Cable

ADJUSTMENT

The parking brake adjustment is performed at the parking brake lever.

1. Remove the center console-to-chassis screws and the console.
2. Adjust the parking brake adjusting nut so the brakes are fully applied when the parking brake lever can be lifted 7–10 notches.
3. Reinstall the center console.

REMOVAL & INSTALLATION

1. Raise and safely support the vehicle.
2. Using a pair of needle-nose pliers, remove the parking brake return spring from each backing plate; be careful not to stretch the spring.
3. If equipped with drum brakes, remove the attaching bolts from the parking brake cable housing and pull it away from the backing plate. Disconnect the cables from the backing plate parking brake levers.
4. If equipped with disc brakes, loosen the locknut and remove the parking brake cable from the lever.
5. Unbolt the cable housing clamps from the trailing arms and the body.
6. Disconnect the parking brake return spring from the parking brake equalizer.
7. Disconnect the parking brake cable from the equalizer and remove the cable from the vehicle.
8. Inspect the parking brake cable for free movement in the cable hous-ing; if necessary, lubricate it with grease.

To install:

9. Install the 2 cable ends in the equalizer. Apply a suitable grease to the cable clamps and the brake cable.
10. Install the cable housing clamps on the body and the trailing arms.
11. Connect the cables to the backing plate parking brake levers.

NOTE: On rear disc brakes, there must be no clearance between the cable end and the lever.

12. If euipped with drum brakes, position the parking brake cable against the backing plate and install the attaching bolts.
13. If equipped with disc brakes, tighten the locknut to 12–17 ft. lbs. (16–22 Nm).
14. Lower the vehicle and adjust the parking brake.

Brake System Bleeding

When any part of the hydraulic system has been disconnected for service, air may enter the system and cause spongy pedal action. The bleeding procedure is used to remove air from the hydraulic circuits.

The brake hydraulic circuits form a split diagonal hydraulic system. The left front and right rear form 1 circuit while the right front and left rear form the other circuit. When bleeding 1 of these circuits, bleed the rear wheel first and then the front wheel at the opposite corner.

Never reuse brake fluid that has been drained from the hydraulic system or that has been allowed to stand in an open container for an extended period of time.

Bleed the brake system as follows:

1. Clean all dirt from the master cylinder filler cap. Fill the master cylinder with DOT 3 brake fluid.

NOTE: Do not allow the master cylinder to run dry during the bleeding procedure.

2. If the master cylinder is known or suspected to contain air, it must be bled before the wheel cylinders or caliper. Bleed the master cylinder as follows:
   a. Loosen the front line fitting and have an assistant push the brake pedal slowly through it's full travel.
   b. While the assistant holds the pedal down, tighten the brake line fitting. After the line fitting is tightened, the assistant may release the brake pedal.
   c. Repeat the procedure on the rear brake line.
   d. Repeat the entire process several times to make sure all air has been removed from the master cylinder.
3. Remove the bleeder screw cap from the appropriate rear wheel cylinder. Position a box end wrench on the bleeder fitting.
4. Attach a rubber hose to the bleeder fitting. The hose must fit snugly around the bleeder fitting.
5. Submerge the other end of the hose in a container partially filled with brake fluid.
6. Loosen the bleeder fitting approximately ¾ turn. Have an assistant push the brake pedal slowly through it's full travel and hold it there.
7. Close the bleeder fitting, then have the assistant release the brake pedal.
8. Repeat Steps 6 and 7 until air bubbles no longer appear at the submerged end of the bleeder hose.
9. When the fluid entering the bottle is completely free of bubbles, tighten the bleeder screw, remove the hose and install the bleeder screw cap.
10. Repeat Steps 3–9 at the front caliper located diagonally to the wheel cylinder just completed.
11. If necessary, bleed the other diagonal circuit in the same manner.
12. Check the master cylinder fluid level and add, if necessary.
13. Check the pedal feel. If the pedal is still spongy, repeat the bleeding procedure.

Anti-Lock Brake System Service

PRECAUTION

Failure to observe the following precautions may result in system damage or personal injury.

- Before servicing any high pressure component, be sure to discharge the hydraulic pressure from the system.
- Do not allow the brake fluid to contact any of the electrical connectors.
- Use care when opening the bleeder screws due to the high system pressure from the accumulator.

RELIEVING ANTI-LOCK BRAKE SYSTEM PRESSURE

1. Turn the ignition key **OFF**.
2. Pump the brake pedal a minimum of 20 times until an increase in pedal force is clearly felt.

Hydraulic Actuation Unit

REMOVAL & INSTALLATION

1. Disconnect the negative battery cable. Relieve the anti-lock brake system pressure.

2. Remove the fuel filter and air filter assemblies.

3. Remove the coil and disconnect the wiring harness from the bottom of the coil and the fuel filter mounting bracket.

4. Remove the coil and fuel filter mounting bracket. Disconnect the electrical connectors.

5. Remove the banjo bolts and the copper washers from the brake lines at the actuation unit. Disconnect the brake lines between the master cylinder and the actuation unit.

NOTE: Observe the routing of the brake lines to ensure proper installation.

6. Remove the routing clip from the brake lines. Using an extension and a 6-point, 10mm crowfoot wrench, disconnect the 4 brake lines at the actuation unit.

NOTE: Observe the routing of the brake lines to ensure proper installation.

7. Remove the mounting nuts, lockwashers and washers. Lift the hydraulic actuation unit from the mounting bracket. Remove the mounting bushings from the actuation assembly. If neccessary, remove the mounting bracket.

To install:

8. Install the mounting bracket and bushings and the actuation unit. Tighten the mounting nuts to 14–19 ft. lbs. (19–25 Nm).

9. Connect the electrical connectors.

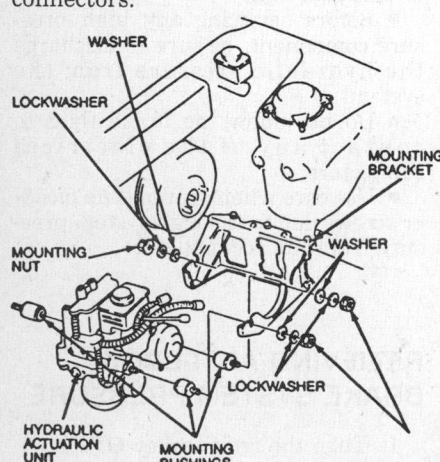

Hydraulic actuation unit removal and installation

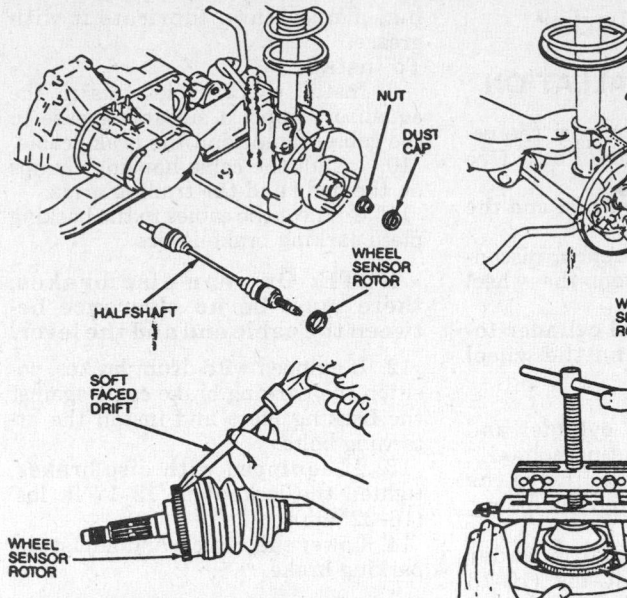

Front wheel sensor rotor removal

10. Using an extension and a 6-point, 10mm crowfoot wrench, install 4 brake lines to the actuation assembly. Tighten to 9–16 ft. lbs. (13–22 Nm).

11. Position the 2 brake lines between the actuation assembly and the master cylinder. Install the banjo bolts with new washers and tighten the banjo bolts to 14–23 ft. lbs. (20–29 Nm).

12. Install the coil and fuel filter bracket. Tighten th nuts to 34–46 ft. lbs. (56–63 Nm). Connect the wire harness to the bottom of the coil and fuel filter mounting bracket.

13. Install the coil, air filter and fuel filter assemblies.

14. Fill and bleed the brake system. Connect the negative battery cable.

Wheel Sensor Rotor

REMOVAL & INSTALLATION

Front

1. Raise and support the vehicle, safely. Remove the tire and wheel assembly.

2. Remove the halfshaft assembly.

3. Using a soft-faced drift, tap the sensor rotor from the outboard CV-joint.

To install:

4. Position the sensor rotor on the CV-joint with the chamfered edge facing the halfshaft.

5. Using a soft-faced drift, tap the sensor rotor onto the outboard CV-joint and install the halfshaft.

6. Install the wheel and tire assembly and lower the vehicle.

Rear

1. Raise and support the vehicle, safely.

Rear wheel sensor rotor removal

2. Remove the tire and wheel assembly.

3. Remove the caliper and anchor bracket. Remove the rotor assembly.

4. Remove the sensor rotor from the rotor, using a puller.

To install:

5. Position the rotor in a press with the wheel studs facing down.

6. Press the sensor onto the rotor, using a suitable installation tool.

7. Install the rotor, caliper and anchor bracket.

8. Install the tire and wheel assembly. Lower the vehicle.

Speed Sensor

REMOVAL & INSTALLATION

Front

1. Raise and support the vehicle, safely. Remove the front tire and wheel assembly.

2. Remove the retaining bolts and the speed sensor from steering knuckle.

3. Remove the routing bracket from the strut assembly.

4. Remove the routing bracket from the inner fender well and disconnect the wiring harness.

5. Remove the speed sensor.

NOTE: The speed sensors are not interchangeable. L and R are indicated on the bracket.

To install:

6. Route the sensor wiring harness in the vehicle and connect the wiring harness.

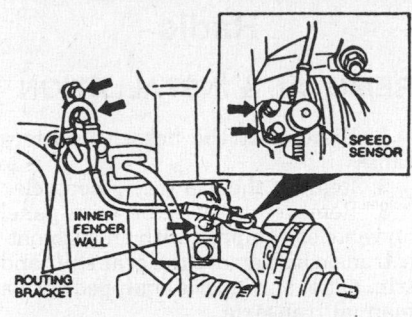

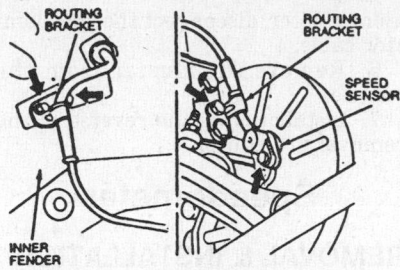

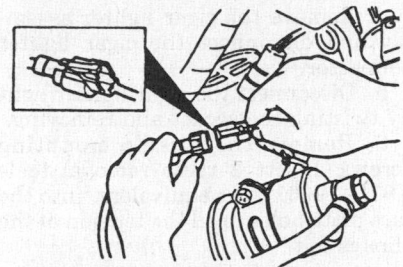

Front speed sensor location

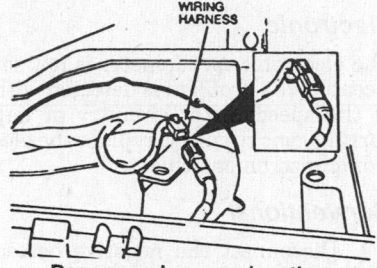

Rear speed sensor location

7. Install the routing bracket onto the inner fender well.

8. Install the routing bracket onto the strut assembly. Tighten the bolts to 12–17 ft. lbs. (16–23 Nm).

9. Install the speed sensor into the knuckle and tighten the bolts to 12–17 ft. lbs. (16–23 Nm).

10. Make sure the wiring harness will clear all suspension components.

11. Install the wheel and tire assembly. Lower the vehicle.

Rear

1. Raise and support the vehicle, safely. Remove the tire and wheel assembly.

2. Remove the retaining bolt and the speed sensor from steering knuckle.

3. Remove the routing bracket from the strut assembly.

4. Remove the routing bracket from the inner fender well.

5. Remove the interior panels as neccessary to gain access to the wiring harness.

6. Disconnect the wiring harness and remove the speed sensor.

NOTE: The speed sensors are not interchangeable. L and R are indicated on the bracket.

To install:

7. Route the sensor wiring harness in the vehicle and connect the wiring harness.

8. Install the routing bracket onto the inner fender well.

9. Install the routing bracket onto the strut assembly. Tighten the nut to 12–17 ft. lbs. (16–23 Nm).

10. Install the speed sensor into the

knuckle and tighten the bolt to 12–17 ft. lbs. (16–23 Nm).

11. Make sure the wiring harness will clear all suspension components.

12. Install any interior panels that were removed.

CHASSIS ELECTRICAL

Heater Blower Motor

REMOVAL & INSTALLATION

1. Disconnect the negative battery cable.

2. Remove the sound deadening panel from the passenger side.

3. Remove the glove box assembly and the brace.

4. Remove the cooling hose from the blower motor assembly.

5. Disconnect the electrical connector from the blower motor.

6. Remove the blower motor-to-blower motor housing screws and blower motor.

7. If necessary, remove the blower wheel-to-blower motor clip and the wheel.

8. To install, reverse the removal procedure and check the blower motor operation.

Windshield Wiper Motor

REMOVAL & INSTALLATION

Front

1. Disconnect the negative battery cable.

2. Unscrew the retaining cap and remove the wiper blade/arm assemblies.

3. Disconnect the hose from the washer jet.

4. Remove the lower moulding and wiper linkage cover.

5. Pull the wiper linkage off the wiper motor output arm.

6. Disconnect the electrical connectors from the wiper motor.

7. Remove the wiper motor mounting bolts and remove the motor from the vehicle.

8. To install, reverse the removal procedure. Check the wiper motor operation.

Rear

1. Disconnect the negative battery cable.

2. Lift the cover and remove the wiper blade/arm assembly retaining nut and the assembly.

3. From the pivot arm, remove the boot, the retaining nut and the mount.

4. From the inner side of the liftgate, pry off the trim panel.

5. Disconnect the electrical connector from the wiper motor.

6. Remove the wiper motor mounting bolts and the motor from the vehicle.

7. To install, reverse the removal procedure. Check the wiper motor operation.

Windshield Wiper Switch

REMOVAL & INSTALLATION

Front

1. Disconnect the negative battery cable.

2. Remove the instrument cluster module as follows:

   a. Remove the steering wheel.

   b. Remove the 2 column cover screws and remove the cover.

   c. Remove the 9 cluster module mounting screws.

   d. Carefully pull the cluster module outward and disconnect the 7 electrical connectors from the cover.

   e. Remove the ignition switch illumination bulb and remove the cluster module.

3. Gently, pull the washer/interval rate control switch knob and wiper control switch knob from the windshield wiper switch.

4. From the rear of the instrument cluster module cover, remove the windshield wiper switch housing screws and the switch.

5. To install, reverse the removal procedure. Check the windshield wiper switch operation.

Rear

1. Disconnect the negative battery cable.
2. Remove the instrument cluster module as follows:
   a. Remove the steering wheel.
   b. Remove the 2 column cover screws and remove the cover.
   c. Remove the 9 cluster module mounting screws.
   d. Carefully pull the cluster module outward and disconnect the 7 electrical connectors from the cover.
   e. Remove the ignition switch illumination bulb and remove the cluster module.
3. Gently, pull the front washer/interval rate control switch knob and the front wiper control switch knob from the windshield wiper switch.
4. From the rear of the instrument cluster module cover, remove the windshield wiper switch housing screws and the switch.
5. Remove the rear wiper/washer switch-to-instrument cluster module cover screws.
6. Remove the control switch button by releasing the tangs. Remove the rear wiper/washer switch from the rear of the instrument cover.
7. To install, reverse the removal procedure. Check the windshield wiper/washer switch and the rear wiper/washer switch operation.

Instrument Cluster

REMOVAL & INSTALLATION

1. Disconnect the negative battery cable.
2. Remove the instrument cluster module as follows:
   a. Remove the steering wheel.
   b. Remove the 2 column cover screws and remove the cover.
   c. Remove the 9 cluster module mounting screws.
   d. Carefully pull the cluster module outward and disconnect the 7 electrical connectors from the cover.
   e. Remove the ignition switch illumination bulb and remove the cluster module.
3. Loosen the 2 cover hinge screws and remove the 6 upper cluster cover screws. Remove the cover.

NOTE: During removal, be careful not to rip the rubber seal that joins the upper and lower portions of the cluster cover panels.

4. Remove the lower cluster cover panel and remove the 4 cluster mounting screws.
5. Disconnect the electrical connectors from the back of the cluster. If equipped with conventional instru-

ment cluster, disconnect the speedometer cable.
6. Remove the cluster from the vehicle.
7. Installation is the reverse of the removal procedure.

Speedometer

REMOVAL & INSTALLATION

Electronic

The electronic speedometer is not serviceable. If a problem is detected within the speedometer circuitry or supporting logic circuitry, replace the electronic instrument cluster.

Conventional

1. Disconnect the negative battery cable.
2. Remove the instrument cluster.
3. Remove the lens assembly from the instrument cluster.
4. Remove the speedometer subassembly-to-instrument cluster screws and the subassembly from the cluster.
5. To install, reverse the removal procedure.

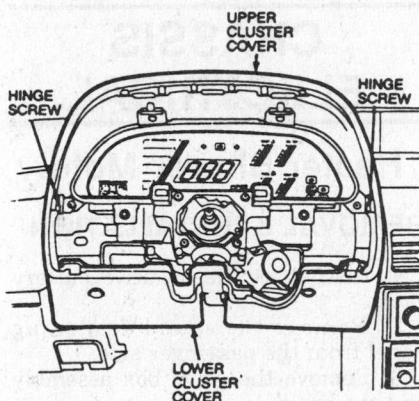

Electronic instrument cluster removal

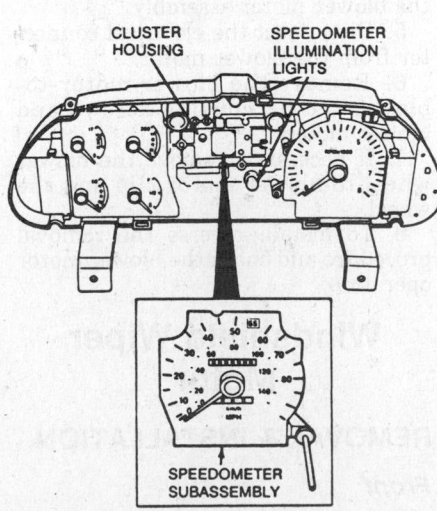

Speedometer subassembly removal

Radio

REMOVAL & INSTALLATION

1. Disconnect the negative battery cable.
2. Remove the ash tray receptacle.
3. Remove the selector trim panel on vehicles equipped with an automatic transaxle. Remove the gear shift and trim panel on vehicles equipped with a manual transaxle.
4. Remove the cigar lighter assembly and disconnect the cigar lighter connector.
5. Disconnect the cigar lighter light by twisting the socket and removing.
6. Remove the 2 radio mounting screws. Insert 2 radio removal tools T87P–19061–A or equivalent, into the face plate holes until the tension of the clips is felt.

NOTE: The first time the radio is removed requires the flashing in the bezel facing to be broken, which requires extra force.

7. Flex outward on both sides of the radio simultaneously and pull the radio from the instrument panel, using the tools as handles.
8. Disconnect the electrical connectors and antenna cable from the radio.
9. Installation is the reverse of the removal procedure.

Concealed Headlights

MANUAL OPERATION

A manual control knob is located under each headlight retractor motor and is accessible from under the front fascia. The retractor motor is mounted on a bracket which is attached to the radiator support.

Headlight Switch

REMOVAL & INSTALLATION

1. Disconnect the negative battery cable. Remove the turn signal switch.
2. Gently, pull the rotary knob from the headlight switch.
3. From the rear of the instrument cluster module cover, remove the rotary switch housing screws and the switch.
4. To install, reverse the removal procedures. Check the headlight switch operation.

Turn Signal Switch

The turn signal switch incorporates the headlight high beam switch.

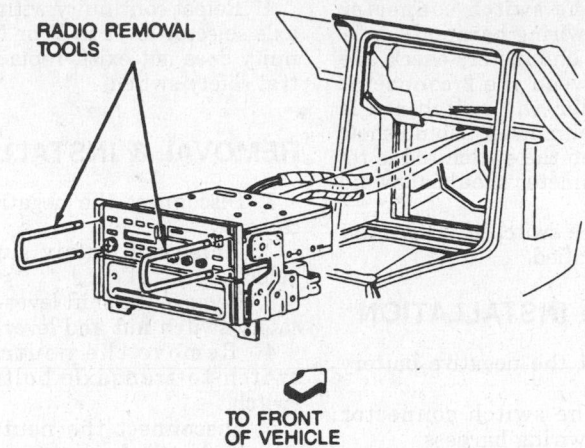

RADIO REMOVAL TOOLS

TO FRONT OF VEHICLE

Radio removal procedure

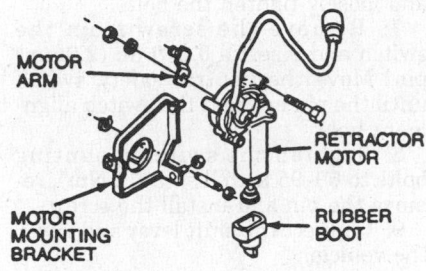

MOTOR ARM

RETRACTOR MOTOR

MOTOR MOUNTING BRACKET

RUBBER BOOT

Headlight retractor motor assembly

REMOVAL & INSTALLATION

1. Disconnect the negative battery cable. Remove the steering wheel.

2. Remove the center cover mounting screws and the cover.

3. Remove cluster module mounting screws, pull the cluster module away from the dash and disconnect the electrical connectors. Remove the cluster module.

4. Remove the turn signal lever-to-turn signal switch screw and the lever.

5. From the rear of the instrument cluster module, remove the turn signal switch screws and the switch.

6. To install, reverse the removal procedure. Check the turn signal switch operation.

Ignition Lock/Switch

REMOVAL & INSTALLATION

1989 Vehicles

1. Disconnect the negative battery cable. Remove the steering wheel.

2. Remove the steering column cover-to-instrument cover screws and the cover.

3. Remove the instrument cover-to-instrument cluster cover screws, pull the cover forward, disconnect the electrical connectors from the rear and remove the cover.

4. Remove the instrument cluster cover-to-dash screws and the cover.

5. Remove the lower panel, the lap duct and the defrost duct.

6. Remove the lower hinge bracket support nuts.

7. Remove the upper steering column-to-support bracket nuts/bolts and lower the steering column.

8. Remove the ignition switch-to-ignition switch housing screw.

9. At the left side of the steering column, disconnect the ignition switch snap connectors and the protective looming from the ignition switch wires.

10. Note the position of each wire in the 4-wire terminal connector. From the 4-wire connector, disconnect the key-in warning buzzer wires: green wire and the red wire/orange tracer.

NOTE: Use a paper clip or equivalent, to disengage the wire tangs from the 4-wire connector.

11. To install, reverse the removal procedure. Check the ignition switch operation.

1990 Vehicles

1. Disconnect the negative battery cable.

2. Remove the lower instrument cluster cover as follows:

  a. Remove the steering wheel.

  b. Remove the 2 column cover screws and the cover.

  c. Remove the 9 cluster module mounting screws.

  d. Carefully pull the cluster module outward and disconnect the electrical connectors.

  e. Remove the ignition switch illumination bulb and remove the cluster module.

  f. Loosen the 2 cover hinge screws and the 6 upper cluster cover screws. Remove the cover.

  g. Remove the lower cluster cover panel.

3. Loosen the shift cable nut retaining nut and allow the cable to hang down.

4. Remove the ignition switch mounting screw.

5. Disconnect the ignition switch wire connectors located to the left of the steering column.

6. Remove the protective looming from the ignition wires, note the position of each wire in the 4 terminal protector for replacing.

7. Disconnect the key-in warning buzzer wires (green wire and red wire/orange tracer) by disengaging the tang with a paper clip or suitable tool.

8. Remove the ignition switch. Remove the lock cylinder mounting screws and the lock cylinder.

9. To install, reverse the removal procedure. Check the operation of the ignition switch.

1991–92 Vehicles

1. Disconnect the negative battery cable.

2. Remove the instrument cluster trim cover and the instrument cluster lower panel.

3. If equipped with automatic transaxle, remove the shift-lock cable attaching bolt and disconnect the cable from the lock cylinder housing.

4. Remove the steering column lower panel and lower plate.

5. Remove the side register duct.

6. If equipped, remove the tilt lever through-bolt, tilt lever and attaching nuts. Remove the tilt lever housing attaching bolts and spring.

7. Disconnect the ignition switch electrical connectors and remove the ignition switch retaining screw.

8. Remove the retainers from the lock cylinder electrical connector and disconnect the connector from the lock cylinder housing.

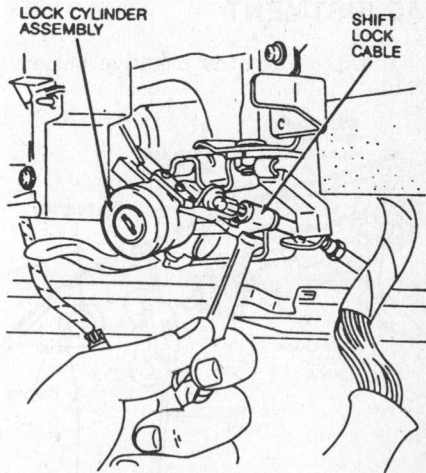

LOCK CYLINDER ASSEMBLY

SHIFT LOCK CABLE

Lock cylinder assembly removal— 1990–92 vehicles

9. Remove the ignition switch from the lock cylinder housing.

10. Remove the 2 lock cylinder assembly mounting screws and remove the lock cylinder assembly.

11. Installation is the reverse of the removal procedure.

Stoplight Switch

The stoplight switch is located at the top of the brake pedal and is used to adjust the brake pedal height.

ADJUSTMENT

1. Check the distance between the center of the pedal pad to the floor. The distance should be 8.74–8.94 in. (222–227mm).

2. If the distance is not as specified, disconnect the negative battery cable and the stoplight switch connector.

3. Loosen the stoplight switch locknut.

4. Rotate the switch until the pedal height is 8.54–8.74 in. (217–222mm).

5. Tighten the switch locknut and connect the stoplight switch connector.

6. Connect the negative battery cable.

REMOVAL & INSTALLATION

1. Disconnect the negative battery cable.

2. Disconnect the electrical connector from the stoplight switch.

3. Remove the stoplight switch locknut and the switch.

4. Installation is the reverse of the removal procedure. Adjust the switch.

Clutch Switch

The clutch engage switch is located next to the clutch pedal.

ADJUSTMENT

1. Disconnect the negative battery cable.

2. Unplug the switch connector from the main wiring harness.

3. Using an ohmmeter, check the resistance between the 2 connector terminals. When the switch rod is pushed in, the ohmmeter should show continuity; when the switch rod is released, the ohmmeter should show no continuity.

4. Replace the switch if it does not perform as specified.

REMOVAL & INSTALLATION

1. Disconnect the negative battery cable.

2. Unplug the switch connector from the main wiring harness.

3. Remove the switch mounting bolts and remove the switch.

4. Installation is the reverse of the removal procedure.

Neutral Safety Switch

The neutral safety switch is located on the top left side of the transaxle.

ADJUSTMENT

1. Unplug the 3-pronged neutral safety switch connector, located under the battery tray. Connect an ohmmeter between terminals **A** and **B**.

2. With the transaxle selector lever in **P** or **N**, there should be continuity between the terminals.

3. If continuity does not exist, adjust the switch as follows:

   a. Raise and safely support the vehicle.

   b. Turn the manual shaft to the **N** position and loosen the switch mounting bolts.

   c. Remove the screw from the switch and insert a 0.079 in. (2.0mm) pin.

   d. Move the neutral safety switch until the pin engages the switch alignment hole.

   e. Tighten the switch mounting bolts to 69–95 inch lbs. (8–11 Nm), remove the pin and install the screw.

   f. Lower the vehicle.

4. Retest continuity with the transaxle selector lever in **P** or **N**. If continuity does not exist, replace the neutral safety switch.

REMOVAL & INSTALLATION

1. Disconnect the negative battery cable.

2. Raise and safely support the vehicle.

3. Remove the shift lever-to-neutral safety switch nut and lever.

4. Remove the neutral safety switch-to-transaxle bolts and the switch.

5. Disconnect the neutral safety switch electrical connector and then remove the switch from the vehicle.

To install:

6. Install the neutral safety switch and loosely tighten the bolts.

7. Remove the screw from the switch and insert a 0.079 in. (2.0mm) pin. Move the neutral safety switch until the pin engages the switch alignment hole.

8. Tighten the switch mounting bolts to 69–95 inch lbs. (8–11 Nm), remove the pin and install the screw.

9. Connect the shift lever and lower the vehicle.

Fuses, Circuit Breakers and Relays

LOCATION

Fuses

The main fuse block is located in the left side of the engine compartment near the battery. The interior fuse block is located above the left side kick panel. The fuses are a plug in type and are color-coded by amp rating.

Circuit Breakers

A bimetal circuit breaker, used to protect the rear window defroster circuit, is located just above the interior fuse block.

Relays

The main relay box is located in the engine compartment on the upper left side of the bulkhead. There is also a relay box located inside the vehicle under the left side of the instrument panel, on the bulkhead.

  EFI Main Relays (2)—located in the main relay box

  Horn Relay—located in the main relay box

  Cooling Fan Relay No. 1—located in the main relay box

  Cooling Fan Relay No. 2—located in the main relay box

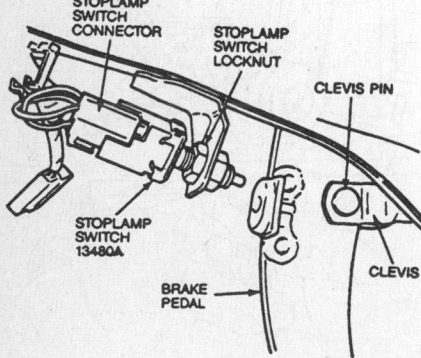

Stoplight switch location

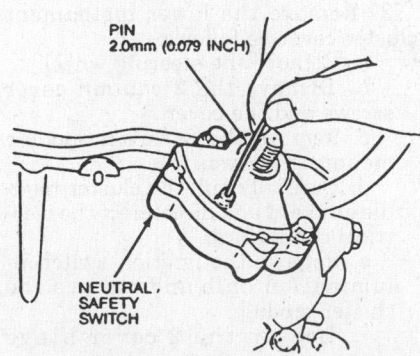

Neutral safety switch adjustment

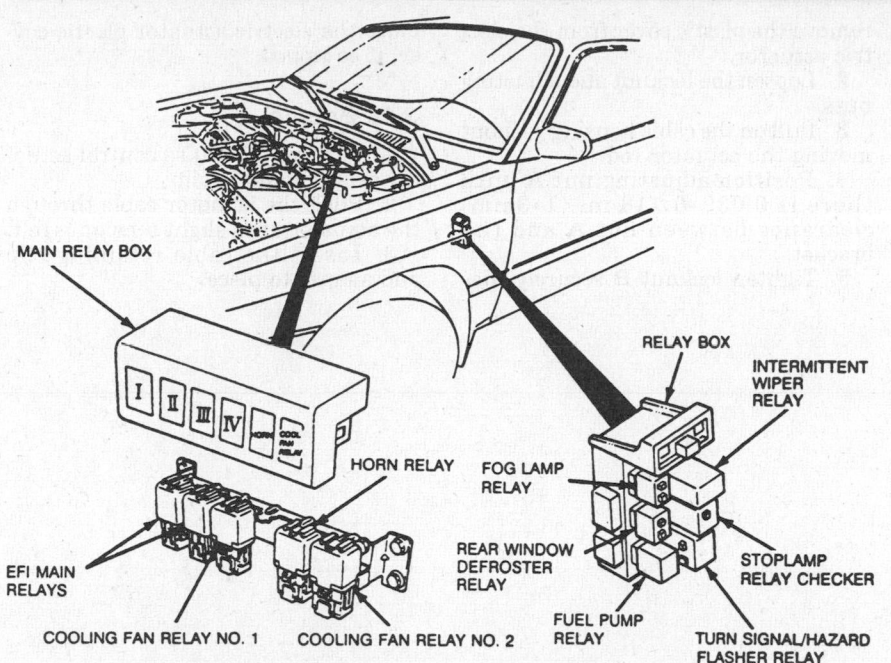

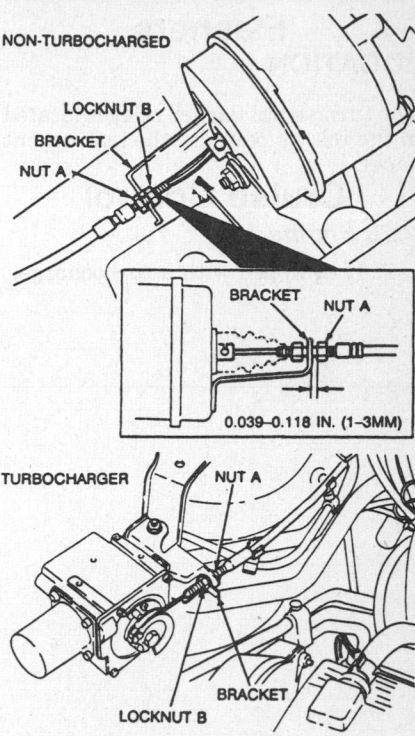

Main relay box and relay box contents and locations

Cruise control actuator cable adjustment—2.2L engines

Turn Signal/Hazard Flasher Relay—located in the relay box

Fuel pump Relay—located in the relay box

Rear Window Defroster Relay—located in the relay box

Intermittent Wiper Relay—located in the relay box

Stoplight/Tail Light Checker Relay—located in the relay box

Fog Light Relay—located in the relay box

ABS Relay—located in the engine compartment, near the master cylinder

Air Conditioning Relay—on Probe LX only, located behind the right side of the instrument panel, to the left of the blower motor

Blower Motor Relay—located in the engine compartment, forward of the battery

Condenser Fan Relay—on Probe GL and Probe GT only, located in the engine compartment, on right front of condenser

Dimmer Relay—located in the engine compartment, forward of the washer reservoir

Power Door Lock Relay—located behind the left interior rear quarter trim panel

NOTE: Cooling fan relay No. 1 is used only on vehicles equipped with the electronically controlled 4EAT automatic transaxle.

Computers

LOCATION

The computer controlling engine oper-
ation, the Electronic Control Assembly (ECA), is located behind the instrument panel, forward of the center console.

The control units for the anti-lock brake system and the variable assist power steering are both located under the driver's seat.

A central processing unit is located directly above the interior fuse block.

This microprocessor controls the key reminder alarm, lights-off reminder alarm, seat belt alarm, vehicle maintenance reminder alarm, turbocharger overboost alarm, seat belt timer, key illumination light timer and interior light timer.

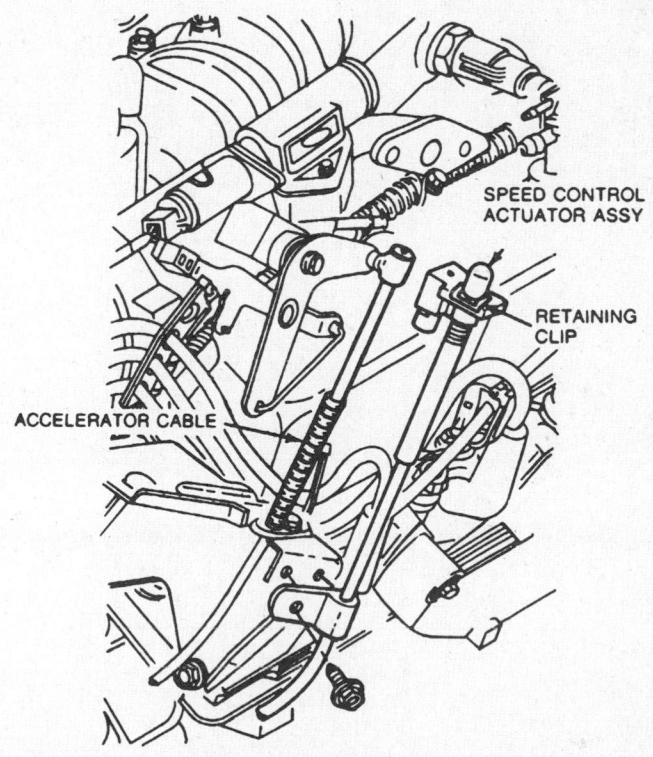

Cruise control actuator cable adjustment—3.0L engine

Flashers

LOCATION

The turn/signal flasher relay is located in the relay box under the instrument panel.

Cruise Control

2.2L Engine

1. If equipped with a turbocharger, remove the plastic cover from the electric actuator.
2. Loosen the locknut and adjusting nuts.
3. Pull on the cable housing without moving the actuator rod.
4. Position adjusting nut **A** until there is 0.039–0.118 in. (1–3mm) clearance between nut **A** and the bracket.
5. Tighten locknut **B** securely. Replace the electric actuator plastic cover, if equipped.

3.0L Engine

1. Remove the cruise control actuator cable retaining clip.
2. Push the actuator cable through the adjuster until slight tension is felt.
3. Insert the cable retaining clip and snap into place.

Ford Motor Co.
Front Wheel Drive
Ford—Taurus, Taurus SHO
Lincoln—Continental
Mercury—Sable

SPECIFICATIONS

VEHICLE IDENTIFICATION CHART

It is important for servicing and ordering parts to be certain of the vehicle and engine identification. The VIN (vehicle identification number) is a 17 digit number visible through the windshield on the driver's side of the dash and contains the vehicle and engine identification codes. The tenth digit indicates model year and the eighth digit indicates engine code. It can be interpreted as follows:

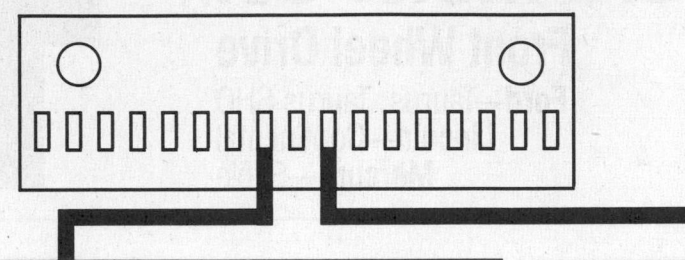

Engine Code							Model Year	
Code	Cu. In.	Liters	Cyl.	Fuel Sys.	Eng. Mfg.		Code	Year
D (1988–'90)	153	2.5 HSC	4	CFI	Ford		J	1988
D (1991–'92)	153	2.5 HSC	4	SEFI	Ford		K	1989
U (1988–'90)	182	3.0	6	EFI	Ford		L	1990
U (1991–'92)	182	3.0	6	SEFI	Ford		M	1991
Y	182	3.0 SHO	6	SEFI	Yamaha		N	1992
4 (1988)	232	3.8	6	EFI	Ford			
4 (1989–'92)	232	3.8	6	SEFI	Ford			

CFI   Central Fuel Injection
EFI   Electronic Fuel Injection
SEFI   Sequential Electronic Fuel Injection
HSC   High Swirl Combustion
SHO   Super High Output

ENGINE IDENTIFICATION

Year	Model	Engine Displacement cu. in. (liter)	Engine Series Identification (VIN)	No. of Cylinders	Engine Type
1988	Taurus	153 (2.5) HSC-CFI	D	4	OHV
	Taurus	182 (3.0) EFI	U	6	OHV
	Taurus	232 (3.8) EFI	4	6	OHV
	Sable	153 (2.5) HSC-CFI	D	4	OHV
	Sable	182 (3.0) EFI	U	6	OHV
	Sable	232 (3.8) EFI	4	6	OHV
	Continental	232 (3.8) EFI	4	6	OHV
1989	Taurus	153 (2.5) HSC-CFI	D	4	OHV
	Taurus	182 (3.0) EFI	U	6	OHV
	Taurus SHO	182 (3.0) SEFI	Y	6	DOHC
	Taurus	232 (3.8) SEFI	4	6	OHV
	Sable	182 (3.0) EFI	U	6	OHV
	Sable	232 (3.8) SEFI	4	6	OHV
	Continental	232 (3.8) SEFI	4	6	OHV
1990	Taurus	153 (2.5) HSC-CFI	D	4	OHV
	Taurus	182 (3.0) EFI	U	6	OHV
	Taurus SHO	182 (3.0) SEFI	Y	6	DOHC

ENGINE IDENTIFICATION

Year	Model	Engine Displacement cu. in. (liter)	Engine Series Identification (VIN)	No. of Cylinders	Engine Type
1990	Taurus	232 (3.8) SEFI	4	6	OHV
	Sable	182 (3.0) EFI	U	6	OHV
	Sable	232 (3.8) SEFI	4	6	OHV
	Continental	232 (3.8) SEFI	4	6	OHV
1991-92	Taurus	153 (2.5) HSC-SEFI	D	4	OHV
	Taurus	182 (3.0) SEFI	U	6	OHV
	Taurus SHO	182 (3.0) SEFI	Y	6	DOHC
	Taurus	232 (3.8) SEFI	4	6	OHV
	Sable	182 (3.0) SEFI	U	6	OHV
	Sable	232 (3.8) SEFI	4	6	OHV
	Continental	232 (3.8) SEFI	4	6	OHV

SEFI Sequential Electronic Fuel Injection
HSC High Swirl Combustion
EFI Electronic Fuel Injection
CFI Central Fuel Injection
SHO Super High Output
OHV Overhead Valve
DOHC Double Overhead Camshaft

GENERAL ENGINE SPECIFICATIONS

Year	VIN	No. Cylinder Displacement cu. in. (liter)	Fuel System Type	Net Horsepower @ rpm	Net Torque @ rpm (ft. lbs.)	Bore × Stroke (in.)	Compression Ratio	Oil Pressure @ rpm
1988	D	4-153 (2.5)	CFI	88 @ 4600	130 @ 2800	3.70 × 3.60	9.7:1	55–70 @ 2000
	U	6-182 (3.0)	EFI	140 @ 4800	160 @ 3000	3.50 × 3.10	9.3:1	55–70 @ 2000
	4	6-232 (3.8)	EFI	140 @ 3800	215 @ 2200	3.81 × 3.39	9.0:1	40–60 @ 2000
1989	D	4-153 (2.5)	CFI	88 @ 4600	130 @ 2800	3.70 × 3.60	9.7:1	55–70 @ 2000
	U	6-182 (3.0)	EFI	140 @ 4800	160 @ 3000	3.50 × 3.10	9.3:1	55–70 @ 2000
	Y	6-182 (3.0)	SEFI	220 @ 6200	200 @ 4800	3.50 × 3.15	9.8:1	13 @ 800
	4	6-232 (3.8)	SEFI	140 @ 3800	215 @ 2200	3.81 × 3.39	9.0:1	40–60 @ 2000
1990	D	4-153 (2.5)	CFI	90 @ 4400	130 @ 2600	3.70 × 3.60	9.0:1	55–70 @ 2000
	U	6-182 (3.0)	EFI	140 @ 4800	160 @ 3000	3.50 × 3.10	9.3:1	40–60 @ 2500
	Y	6-182 (3.0)	SEFI	220 @ 6200	200 @ 4800	3.50 × 3.15	9.8:1	13 @ 800
	4	6-232 (3.8)	SEFI	140 @ 3800	215 @ 2200	3.81 × 3.39	9.0:1	40–60 @ 2500
1991-92	D	4-153 (2.5)	SEFI	105 @ 4400	140 @ 2400	3.70 × 3.60	9.0:1	55–70 @ 2000
	U	6-182 (3.0)	SEFI	140 @ 4800	160 @ 3000	3.50 × 3.10	9.3:1	40–60 @ 2500

GENERAL ENGINE SPECIFICATIONS

Year	VIN	No. Cylinder Displacement cu. in. (liter)	Fuel System Type	Net Horsepower @ rpm	Net Torque @ rpm (ft. lbs.)	Bore × Stroke (in.)	Com-pression Ratio	Oil Pressure @ rpm
1991–92	Y	6-182 (3.0)	SEFI	220 @ 6200	200 @ 4800	3.50 × 3.15	9.8:1	13 @ 800
	4	6-232 (3.8)	SEFI	①	②	3.81 × 3.39	9.0:1	40–60 @ 2500

NA Not available
CFI Central Fuel Injection
EFI Electronic Fuel Injection
SEFI Sequential Electronic Fuel Injection
① Except Continental and Taurus Police—
   140 @ 3800
   Continental and Taurus Police—155 @ 4000
② Except Continental and Taurus Police—
   215 @ 2200
   Continental and Taurus Police—220 @ 2200

ENGINE TUNE-UP SPECIFICATIONS

Year	VIN	No. Cylinder Displacement cu. in. (liter)	Spark Plugs Type	Gap (in.)	Ignition Timing (deg.) MT	Ignition Timing (deg.) AT	Com-pression Pressure (psi)	Fuel Pump (psi)	Idle Speed (rpm)① MT	Idle Speed (rpm)① AT	Valve Clearance In.	Valve Clearance Ex.
1988	D	4-153 (2.5)	AWSF-32C	0.044	10B	10B	NA	13–17	725	650	Hyd.	Hyd.
	U	6-182 (3.0)	AWSF-32C	0.044	—	10B	NA	35–45	—	625	Hyd.	Hyd.
	4	6-232 (3.8)	AWSF-44C	0.056	①	①	NA	35–45	①	①	Hyd.	Hyd.
1989	D	4-153 (2.5)	AWSF-32C	0.044	10B	10B	NA	13–17	725	650	Hyd.	Hyd.
	U	6-182 (3.0)	AWSF-32C	0.044	—	10B	NA	35–45	—	625	Hyd.	Hyd.
	Y	6-182 (3.0)	AGSP-32P	0.044	10B	—	NA	36–39	800	—	0.008 ③	0.012 ③
	4	6-232 (3.8)	AWSF-44C	0.054	①	①	NA	35–45	①	①	Hyd.	Hyd.
1990	D	4-153 (2.5)	AWSF-32C	0.044	—	①	NA	13–17	—	①	Hyd.	Hyd.
	U	6-182 (3.0)	AWSF-32C	0.044	—	①	NA	35–45	—	①	Hyd.	Hyd.
	Y	6-182 (3.0)	AGSP-32P	0.044	10B	—	NA	35–45	①	—	0.008 ③	0.012 ③
	4	6-232 (3.8)	AWSF-44C	0.054	—	①	NA	35–45	—	①	Hyd.	Hyd.
1991	D	4-153 (2.5)	AWSF-32C	0.044	—	①	②	45–60	—	①	Hyd.	Hyd.
	U	6-182 (3.0)	AWSF-32C	0.044	—	①	②	35–45	—	①	Hyd.	Hyd.
	Y	6-182 (3.0)	AGSP-32P	0.044	10B	—	②	35–45	①	—	0.008 ③	0.012 ③
	4	6-232 (3.8)	AWSF-44C	0.054	—	①	②	35–45	—	①	Hyd.	Hyd.
1992		SEE UNDERHOOD SPECIFICATIONS STICKER										

NA Not available
B Before top dead center
Hyd. Hydraulic valve lash lifters.
① The Calibration levels vary from vehicle
   to vehicle. Refer to the Vehicle Emission
   Control Information label for ignition
   timing and idle speed specifications.
② Pressure in lowest cylinder must be at
   least 75% of pressure in highest cylinder.
③ Shim set bucket type valve lifter is used.

FIRING ORDERS

NOTE: To avoid confusion, al- ways replace spark plug wires one at a time.

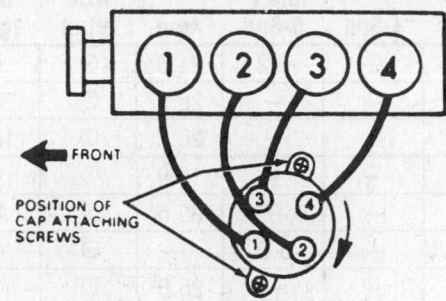

2.5L Engine
Engine Firing Order: 1–3–4–2
Distributor Rotation: Clockwise

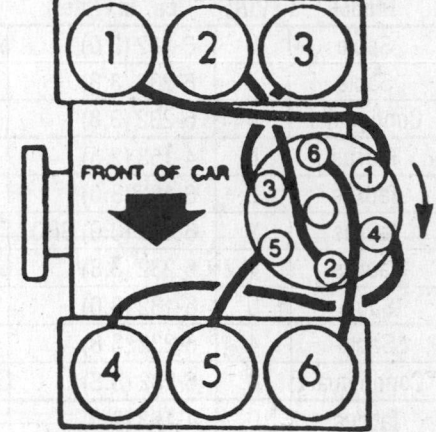

3.0L Engine
Engine Firing Order: 1–4–2–5–3–6
Distributor Rotation: Clockwise

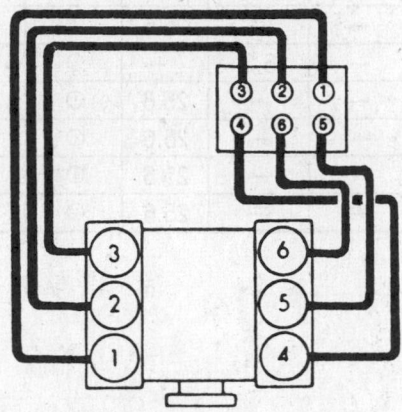

3.0L SHO Engine
Engine Firing Order: 1–4–2–5–3–6
Distributorless Ignition System

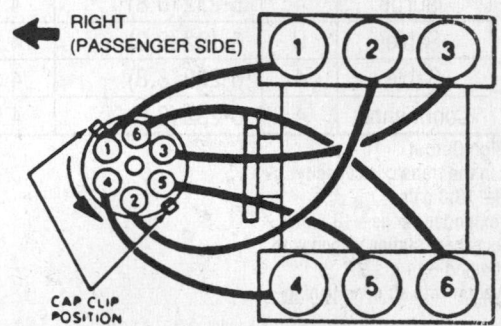

3.8L Engine
Engine Firing Order: 1–4–2–5–3–6
Distributor Rotation: Counterclockwise

CAPACITIES

Year	Model	VIN	No. Cylinder Displacement cu. in. (liter)	Engine Crankcase (qts.) with Filter	Engine Crankcase (qts.) without Filter	Transmission (pts.) 4-Spd	Transmission (pts.) 5-Spd	Transmission (pts.) Auto.	Drive Axle (pts.)	Fuel Tank (gal.)	Cooling System (qts.)
1988	Taurus	D	4-153 (2.5)	5.0	4.5	—	6.2	16.6	①	②	8.3
	Taurus	U	6-182 (3.0)	4.5	4.0	—	6.2	21.8	①	②	③
	Taurus	4	6-232 (3.8)	5.0	4.5	—	—	26.2	①	②	12.1
	Sable	D	4-153 (2.5)	5.0	4.5	—	6.2	16.6	①	②	8.3
	Sable	U	6-182 (3.0)	4.5	4.0	—	6.2	21.8	①	②	③
	Sable	4	6-232 (3.8)	5.0	4.5	—	—	26.2	①	②	12.1
	Continental	4	6-232 (3.8)	4.5	—	—	—	26.2	①	18.6	11.1
1989	Taurus	D	4-153 (2.5)	5.0	4.5	—	6.2	16.6	①	②	8.3
	Taurus	U	6-182 (3.0)	4.5	4.0	—	6.2	21.8	①	②	③
	Taurus	Y	6-182 (3.0) SHO	4.5	4.0	—	6.2	21.8	①	②	11.6
	Taurus	4	6-232 (3.8)	5.0	4.5	—	—	26.2	①	②	12.1
	Sable	D	4-153 (2.5)	5.0	4.5	—	6.2	16.6	①	②	8.3

CAPACITIES

Year	Model	VIN	No. Cylinder Displacement cu. in. (liter)	Engine Crankcase (qts.) with Filter	Engine Crankcase (qts.) without Filter	Transmission (pts.) 4-Spd	Transmission (pts.) 5-Spd	Transmission (pts.) Auto.	Drive Axle (pts.)	Fuel Tank (gal.)	Cooling System (qts.)
1989	Sable	U	6-182 (3.0)	4.5	4.0	—	6.2	21.8	①	②	③
	Sable	4	6-232 (3.8)	5.0	4.5	—	—	26.2	①	②	12.1
	Continental	4	6-232 (3.8)	4.5	4.0	—	—	26.2	①	18.6	12.1
1990	Taurus	D	4-153 (2.5)	5.0	4.5	—	—	16.0	①	②	8.3
	Taurus	U	6-182 (3.0)	4.5	4.0	—	—	25.6	①	②	③
	Taurus	Y	6-182 (3.0) SHO	5.0	4.5	—	6.2	—	①	②	11.6
	Taurus	4	6-232 (3.8)	4.5	4.0	—	—	25.6	①	②	12.1
	Sable	U	6-182 (3.0)	4.5	4.0	—	—	25.6	①	②	③
	Sable	4	6-232 (3.8)	4.5	4.0	—	—	25.6	①	②	12.1
	Continental	4	6-232 (3.8)	4.5	4.0	—	—	25.6	①	18.6	12.1
1991-92	Taurus	D	4-153 (2.5)	5.0	4.5	—	—	25.6	①	②	8.3
	Taurus	U	6-182 (3.0)	4.5	4.0	—	—	25.6	①	②	③
	Taurus	Y	6-182 (3.0) SHO	5.0	4.5	—	6.2	—	①	②	11.6
	Taurus	4	6-232 (3.8)	4.5	4.0	—	—	25.6	①	②	12.1
	Sable	U	6-182 (3.0)	4.5	4.0	—	—	25.6	①	②	③
	Sable	4	6-232 (3.8)	4.5	4.0	—	—	25.6	①	②	12.1
	Continental	4	6-232 (3.8)	4.5	4.0	—	—	25.6	①	18.6	12.1

SHO Super High Output
① Included in the transaxle capacity
② Standard—16.0 gals.
Optionnl extended range—18.6 gals.
③ All models except station wagon with air conditioning—11 qts.
Station Wagon with air conditioning—11.8 qts.

CAMSHAFT SPECIFICATIONS

All measurements given in inches

Year	VIN	No. Cylinder Displacement cu. in. (liter)	Journal Diameter 1	Journal Diameter 2	Journal Diameter 3	Journal Diameter 4	Journal Diameter 5	Lobe Lift In.	Lobe Lift Ex.	Bearing Clearance	Camshaft End Play
1988	D	4-153 (2.5)	2.006–2.008	2.006–2.008	2.006–2.008	2.006–2.008	2.006–2.008	0.249	0.239	0.001–0.003	0.009
	U	6-182 (3.0)	2.007–2.008	2.007–2.008	2.007–2.008	2.007–2.008	—	0.260	0.260	0.001–0.003	0.005
	4	6-232 (3.8)	2.050–2.052	2.050–2.052	2.050–2.052	2.050–2.052	—	0.240	0.241	0.001–0.003	①
1989	D	4-153 (2.5)	2.006–2.008	2.006–2.008	2.006–2.008	2.006–2.008	2.006–2.008	0.249	0.239	0.001–0.003	0.009
	U	6-182 (3.0)	2.007–2.008	2.007–2.008	2.007–2.008	2.007–2.008	—	0.260	0.260	0.001–0.003	0.001–0.005
	Y	6-182 (3.0)	1.2189–1.2195	1.2189–1.2195	1.2189–1.2195	1.2189–1.2195	1.2189–1.2195	0.335	0.315	0.001–0.003	0.012
	4	6-232 (3.8)	2.050–2.052	2.050–2.052	2.050–2.052	2.050–2.052	—	0.240	0.241	0.001–0.003	①
1990	D	4-153 (2.5)	2.006–2.008	2.006–2.008	2.006–2.008	2.006–2.008	2.006–2.008	0.249	0.239	0.001–0.003	0.009

CAMSHAFT SPECIFICATIONS

All measurements given in inches

Year	VIN	No. Cylinder Displacement cu. in. (liter)	Journal Diameter 1	2	3	4	5	Lobe Lift In.	Ex.	Bearing Clearance	Camshaft End Play
1990	U	6-182 (3.0)	2.007–2.008	2.007–2.008	2.007–2.008	2.007–2.008	—	0.260	0.260	0.001–0.003	0.001–0.005
	Y	6-182 (3.0)	1.2189–1.2195	1.2189–1.2195	1.2189–1.2195	1.2189–1.2195	1.2189–1.2195	0.335	0.315	0.001–0.003	0.012
	4	6-232 (3.8)	2.050–2.052	2.050–2.052	2.050–2.052	2.050–2.052	—	0.245	0.259	0.001–0.003	①
1991–92	D	4-153 (2.5)	2.006–2.008	2.006–2.008	2.006–2.008	2.006–2.008	2.006–2.008	0.249	0.239	0.001–0.003	0.009
	U	6-182 (3.0)	2.007–2.008	2.007–2.008	2.007–2.008	2.007–2.008	—	0.260	0.260	0.001–0.003	0.001–0.005
	Y	6-182 (3.0)	1.2189–1.2195	1.2189–1.2195	1.2189–1.2195	1.2189–1.2195	1.2189–1.2195	0.335	0.315	0.001–0.003	0.012
	4	6-232 (3.8)	2.050–2.052	2.050–2.052	2.050–2.052	2.050–2.052	—	0.245	0.259	0.001–0.003	①

① The camshaft is retained by a spring; there is no endplay.

CRANKSHAFT AND CONNECTING ROD SPECIFICATIONS

All measurements are given in inches.

Year	VIN	No. Cylinder Displacement cu. in. (liter)	Crankshaft Main Brg. Journal Dia.	Main Brg. Oil Clearance	Shaft End-play	Thrust on No.	Connecting Rod Journal Diameter	Oil Clearance	Side Clearance
1988	D	4-153 (2.5)	2.2489–2.2490	0.0008–0.0015	0.004–0.008	3	2.1232–2.1240	0.0008–0.0014	0.0035–0.0105
	U	6-182 (3.0)	2.5190–2.5198	0.0010–0.0014	0.004–0.008	3	2.1253–2.1261	0.0010–0.0014	0.006–0.014
	4	6-232 (3.8)	2.5190–2.5198	0.0010–0.0014	0.004–0.008	3	2.3103–2.3111	0.0010–0.0014	0.0047–0.0114
1989	D	4-153 (2.5)	2.2489–2.2490	0.0008–0.0015	0.004–0.008	3	2.1232–2.1240	0.0008–0.0014	0.0035–0.0105
	U	6-182 (3.0)	2.5190–2.5198	0.0010–0.0014	0.004–0.008	3	2.1253–2.1261	0.0010–0.0014	0.006–0.014
	Y	6-182 (3.0)	2.5187–2.5197	0.0011–0.0022	0.0008–0.0087	3	2.0463–2.0472	0.0009–0.0022	0.0063–0.0123
	4	6-232 (3.8)	2.5190–2.5198	0.0010–0.0014	0.004–0.008	3	2.3103–2.3111	0.0010–0.0014	0.0047–0.0114
1990	D	4-153 (2.5)	2.2489–2.2490	0.0008–0.0015	0.004–0.008	3	2.1232–2.1240	0.0008–0.0015	0.0035–0.0105
	U	6-182 (3.0)	2.5190–2.5198	0.0010–0.0014	0.004–0.008	3	2.1253–2.1261	0.0010–0.0014	0.006–0.014
	Y	6-182 (3.0)	2.5187–2.5197	0.0011–0.0022	0.0008–0.0087	3	2.0463–2.0472	0.0009–0.0022	0.0063–0.0123
	4	6-232 (3.8)	2.5190–2.5198	0.0010–0.0014	0.004–0.008	3	2.3103–2.3111	0.0010–0.0014	0.0047–0.0114

CRANKSHAFT AND CONNECTING ROD SPECIFICATIONS

All measurements are given in inches.

| Year | VIN | No. Cylinder Displacement cu. in. (liter) | Crankshaft | | | | Connecting Rod | | |
			Main Brg. Journal Dia.	Main Brg. Oil Clearance	Shaft End-play	Thrust on No.	Journal Diameter	Oil Clearance	Side Clearance
1991–92	D	4-153 (2.5)	2.2489–2.2490	0.0008–0.0015	0.004–0.008	3	2.1232–2.1240	0.0008–0.0015	0.0035–0.0105
	U	6-182 (3.0)	2.5190–2.5198	0.0010–0.0014	0.004–0.008	3	2.1253–2.1261	0.0010–0.0014	0.006–0.014
	Y	6-182 (3.0)	2.5187–2.5197	0.0011–0.0022	0.0008–0.0087	3	2.0463–2.0472	0.0009–0.0022	0.0063–0.0123
	4	6-232 (3.8)	2.5190–2.5198	0.0010–0.0014	0.004–0.008	3	2.3103–2.3111	0.0010–0.0014	0.0047–0.0114

VALVE SPECIFICATIONS

| Year | VIN | No. Cylinder Displacement cu. in. (liter) | Seat Angle (deg.) | Face Angle (deg.) | Spring Test Pressure (lbs.) | Spring Installed Height (in.) | Stem-to-Guide Clearance (in.) | | Stem Diameter (in.) | |
							Intake	Exhaust	Intake	Exhaust
1988	D	4-153 (2.5)	45	44	182 @ 1.13	1.49	0.0018	0.0023	0.3422	0.3418
	U	6-182 (3.0)	45	44	185 @ 1.11	1.58	0.0010–0.0028	0.0015–0.0032	0.3126	0.3121
	4	6-232 (3.8)	46	46	190 @ 1.28	1.70	0.0010–0.0028	0.0015–0.0033	0.3423–0.0033	0.3418–0.3410
1989	D	4-153 (2.5)	45	44	182 @ 1.13	1.49	0.0018	0.0023	0.3422	0.3418
	U	6-182 (3.0)	45	44	185 @ 1.11	1.58	0.0010–0.0028	0.0015–0.0032	0.3126	0.3121
	Y	6-182 (3.0)	45	45.5	120.8 @ 1.19	1.52	0.0010–0.0023	0.0012–0.0025	0.2346–0.2352	0.2344–0.2350
	4	6-232 (3.8)	46	46	220 @ 1.18	1.65	0.0010–0.0028	0.0015–0.0033	0.3423–0.3415	0.3418–0.3410
1990	D	4-153 (2.5)	45	44	182 @ 1.13	1.49	0.0018	0.0023	0.3415–0.3422	0.3411–0.3418
	U	6-182 (3.0)	45	44	180 @ 1.16	1.58	0.0010–0.0028	0.0015–0.0033	0.3134–0.3126	0.3129–0.3121
	Y	6-182 (3.0)	45	45.5	120.8 @ 1.19	1.52	0.0010–0.0023	0.0012–0.0025	0.2346–0.2352	0.2344–0.2350
	4	6-232 (3.8)	44.5	45.8	220 @ 1.18	1.65	0.0010–0.0028	0.0015–0.0033	0.3423–0.3415	0.3418–0.3410
1991–92	D	4-153 (2.5)	45	44	182 @ 1.13	1.49	0.0018	0.0023	0.3415–0.3422	0.3411–0.3418
	U	6-182 (3.0)	45	44	180 @ 1.16	1.58	0.0010–0.0028	0.0015–0.0033	0.3134–0.3126	0.3129–0.3121
	Y	6-182 (3.0)	45	45.5	120.8 @ 1.19	1.52	0.0010–0.0023	0.0012–0.0025	0.2346–0.2352	0.2344–0.2350
	4	6-232 (3.8)	44.5	45.8	220 @ 1.18	1.65	0.0010–0.0028	0.0015–0.0033	0.3423–0.3415	0.3418–0.3410

PISTON AND RING SPECIFICATIONS

All measurements are given in inches.

Year	VIN	No. Cylinder Displacement cu. in. (liter)	Piston Clearance	Ring Gap Top Compression	Ring Gap Bottom Compression	Ring Gap Oil Control	Ring Side Clearance Top Compression	Ring Side Clearance Bottom Compression	Ring Side Clearance Oil Control
1988	D	4-153 (2.5)	0.0012–0.0022	0.0080–0.0160	0.0080–0.0160	0.0150–0.0550	0.0020–0.0040	0.0020–0.0040	Snug
	U	6-182 (3.0)	0.0014–0.0022	0.0100–0.0200	0.0100–0.0200	0.0100–0.0490	0.0016–0.0037	0.0016–0.0037	Snug
	4	6-232 (3.8)	0.0014–0.0032	0.0100–0.0200	0.0100–0.0200	0.0150–0.0583	0.0016–0.0037	0.0016–0.0037	Snug
1989	D	4-153 (2.5)	0.0012–0.0022	0.0080–0.0160	0.0080–0.0160	0.0150–0.0550	0.0020–0.0040	0.0020–0.0040	Snug
	U	6-182 (3.0)	0.0014–0.0022	0.0100–0.0200	0.0100–0.0200	0.0100–0.0490	0.0016–0.0037	0.0016–0.0037	Snug
	Y	6-182 (3.0)	0.0012–0.0020	0.0120–0.0180	0.0120–0.0180	0.0080–0.0200	0.0008–0.0024	0.0006–0.0022	0.0024–0.0050
	4	6-232 (3.8)	0.0014–0.0032	0.011–0.022	0.0100–0.0200	0.0150–0.0583	0.0016–0.0037	0.0016–0.0037	Snug
1990	D	4-153 (2.5)	0.0012–0.0022	0.0080–0.0160	0.0080–0.0160	0.0150–0.0550	0.0020–0.0040	0.0020–0.0040	Snug
	U	6-182 (3.0)	0.0014–0.0022	0.0100–0.0200	0.0100–0.0200	0.0100–0.0490	0.0012–0.0031	0.0012–0.0031	Snug
	Y	6-182 (3.0)	0.0012–0.0020	0.0120–0.0180	0.0120–0.0180	0.0080–0.0200	0.0008–0.0024	0.0006–0.0022	0.0024–0.0059
	4	6-232 (3.8)	0.0014–0.0032	0.011–0.022	0.0100–0.0200	0.0150–0.0583	0.0016–0.0034	0.0016–0.0034	Snug
1991–92	D	4-153 (2.5)	0.0012–0.0022	0.0080–0.0160	0.0080–0.0160	0.0150–0.0550	0.0020–0.0040	0.0020–0.0040	Snug
	U	6-182 (3.0)	0.0014–0.0022	0.0100–0.0200	0.0100–0.0200	0.0100–0.0490	0.0012–0.0031	0.0012–0.0031	Snug
	Y	6-182 (3.0)	0.0012–0.0020	0.0120–0.0180	0.0120–0.0180	0.0080–0.0200	0.0008–0.0024	0.0006–0.0022	0.0024–0.0059
	4	6-232 (3.8)	0.0014–0.0032	0.011–0.022	0.0100–0.0200	0.0150–0.0583	0.0016–0.0034	0.0016–0.0034	Snug

TORQUE SPECIFICATIONS

All readings in ft. lbs.

Year	VIN	No. Cylinder Displacement cu. in. (liter)	Cylinder Head Bolts	Main Bearing Bolts	Rod Bearing Bolts	Crankshaft Pulley Bolts	Flywheel Bolts	Manifold Intake	Manifold Exhaust	Spark Plugs
1988	D	4-153 (2.5)	①	51–66	21–26	140–170	54–64	15–23	②	5–10
	U	6-182 (3.0)	③	65–81	26	141–169	54–64	④	19	5–10
	4	6-232 (3.8)	⑤	65–81	31–36	93–121	54–64	⑥	16–24	5–11
1989	D	4-153 (2.5)	①	51–66	21–26	140–170	54–64	15–23	②	5–10
	U	6-182 (3.0)	⑦	65–81	26	141–169	54–64	④	19	5–10
	Y	6-182 (3.0)	⑩	⑪	⑫	112–127	⑬	12–17	26–38	16–20
	4	6-232 (3.8)	⑤	65–81	31–36	93–121	54–64	⑥	16–24	5–11

TORQUE SPECIFICATIONS
All readings in ft. lbs.

Year	VIN	No. Cylinder Displacement cu. in. (liter)	Cylinder Head Bolts	Main Bearing Bolts	Rod Bearing Bolts	Crankshaft Pulley Bolts	Flywheel Bolts	Manifold Intake	Manifold Exhaust	Spark Plugs
1990	D	4-153 (2.5)	①	51–66	21–26	140–170	54–64	15–23	②	5–10
	U	6-182 (3.0)	⑦	63–69	26	107	54–64	⑨	19	5–11
	Y	6-182 (3.0)	⑩	⑪	⑫	112–127	⑬	12–17	26–38	16–20
	4	6-232 (3.8)	⑤	65–81	31–36	103–132	54–64	⑥	15–22	5–11
1991–92	D	4-153 (2.5)	①	52–66	21–26	140–170	54–64	⑧	②	5–10
	U	6-182 (3.0)	⑦	55–63	26	107	59	⑨	19	5–11
	Y	6-182 (3.0)	⑩	⑪	⑫	113–126	⑬	11–17	26–38	17–19
	4	6-232 (3.8)	⑭	65–81	31–36	103–132	54–64	⑥	15–22	5–11

① Tighten in 2 steps: 52–59 ft. lbs. and then the final torque of 70–76 ft. lbs.

② Tighten in 2 steps
Step 1: 5–7 ft. lbs.
Step 2: 20–30 ft. lbs.

③ Tighten in 2 steps: 48–54 ft. lbs. and then the final torque of 63–80 ft. lbs.

④ Tighten in 3 steps: 11, 18 and the final torque of 24 ft. lbs.

⑤ Tighten in 4 steps:
Step 1: 37 ft. lbs.
Step 2: 45 ft. lbs.
Step 3: 52 ft. lbs.
Step 4: 59 ft. lbs.
Back off all bolts 2–3 revolutions, then repeat steps 1–4.

⑥ Tighten in 3 steps:
Step 1: 7 ft. lbs.
Step 2: 15 ft. lbs.
Step 3: 24 ft. lbs.

⑦ Tighten in 2 steps:
Step 1: 37 ft. lbs.
Step 2: 68 ft. lbs.

⑧ Tighten in 2 steps:
Step 1: 5–7 ft. lbs.
Step 2: 15–22 ft. lbs.

⑨ Tighten in 2 steps:
Step 1: 11 ft. lbs.
Step 2: 21 ft. lbs.

⑩ Tighten in 2 steps:
Step 1: 37–50 ft. lbs.
Step 2: 62–68 ft. lbs.

⑪ Tighten in 2 steps:
Step 1: 34–50 ft. lbs.
Step 2: 58–65 ft. lbs.

⑫ Tighten in 2 steps:
Step 1: 22–26 ft. lbs.
Step 2: 33–36 ft. lbs.

⑬ Tighten in 2 steps:
Step 1: 29–43 ft. lbs.
Step 2: 51–58 ft. lbs.

⑭ A. Tighten in 4 steps:
Step 1: 37 ft. lbs.
Step 2: 45 ft. lbs.
Step 3: 52 ft. lbs.
Step 4: 59 ft. lbs.
B. In sequence, loosen each bolt 2–3 revolutions
C. Tighten long bolts to 11–18 ft. lbs., then an additional 85–105 degrees. Tighten short bolts to 11–18 ft. lbs., then an additional 65–85 degrees.

BRAKE SPECIFICATIONS
All measurements in inches unless noted

Year	Model	Lug Nut Torque (ft. lbs.)	Master Cylinder Bore	Brake Disc Minimum Thickness	Brake Disc Maximum Runout	Standard Brake Drum Diameter	Minimum Lining Thickness Front	Minimum Lining Thickness Rear
1988	Taurus	80–105	0.875	0.974	0.003	③	0.125	0.030
	Sable	80–105	0.875	0.974	0.003	③	0.125	0.030
	Continental	80–105	—	0.974①	0.002①	—	0.125	0.123
1989	Taurus	80–105	0.974	0.974	0.003	③	0.125	0.030
	Taurus SHO	80–105	0.875	0.974①	②	—	0.125	0.123
	Sable	80–105	0.875	0.974	0.003	③	0.125	0.030
	Continental	80–105	0.940	0.974①	0.002①	—	0.125	0.123
1990	Taurus	80–105	0.875	0.974①	0.003	③	0.125	0.030
	Taurus SHO	80–105	0.875	0.974①	②	—	0.125	0.123
	Sable	80–105	0.875	0.974①	0.003	③	0.125	0.030
	Continental	80–105	1.000	0.974①	0.002①	—	0.125	0.123
1991–92	Taurus	80–105	0.875	0.974	0.003	③	0.125	0.030
	Taurus SHO	80–105	0.875	0.974①	②	—	0.125	0.123
	Sable	80–105	0.875	0.974	0.003	③	0.125	0.030
	Continental	80–105	1.000	0.974①	0.002①	—	0.125	0.123

① Front and rear

② Front—0.003 in.
Rear—0.002 in.

③ Sedan—8.85
Wagon—9.84

WHEEL ALIGNMENT

Year	Model		Caster Range (deg.)	Caster Preferred Setting (deg.)	Camber Range (deg.)	Camber Preferred Setting (deg.)	Toe-in (in.)	Steering Axis Inclination (deg.)
1988	Taurus	Front	3P–6P	4P	$1^3/_{32}$N–$^3/_{32}$P	$^1/_2$N	$^7/_{32}$–$^1/_{64}$	$15^3/_8$
		Rear	—	—	$1^5/_8$N–$^1/_4$N	—	$^{13}/_{64}$N–$^{19}/_{64}$P②	—
	Sable	Front	3P–6P①	4P	$1^3/_{32}$N–$^3/_{32}$P	$^1/_2$N	$^7/_{32}$–$^1/_{64}$	$15^3/_8$
		Rear	—	—	$1^5/_8$N–$^1/_4$N	—	$^{13}/_{64}$N–$^{19}/_{64}$P②	—
	Continental	Front	4P–6P①	5P	$1^3/_{32}$N–$^3/_{32}$P	$^1/_2$N	$^7/_{32}$–$^1/_{64}$	$15^3/_8$
		Rear	—	—	$1^5/_8$N–$^1/_4$N	—	$^{13}/_{64}$N–$^{19}/_{64}$P②	—
1989	Taurus	Front	3P–6P①	4P	$1^3/_{32}$N–$^3/_{32}$P	$^1/_2$N	$^7/_{32}$–$^1/_{64}$	$15^3/_8$
		Rear	—	—	$1^5/_8$N–$^1/_4$N	—	$^{13}/_{64}$N–$^{19}/_{64}$P②	—
	Sable	Front	3P–6P①	4P	$1^3/_{32}$N–$^3/_{32}$P	$^1/_2$N	$^7/_{32}$–$^1/_{64}$	$15^3/_8$
		Rear	—	—	$1^5/_8$N–$^1/_4$N	—	$^{13}/_{64}$N–$^{19}/_{64}$P②	—
	Continental	Front	4P–6P	5P	$1^3/_{32}$N–$^3/_{32}$P	$^1/_2$N	$^7/_{32}$–$^1/_{64}$	$15^3/_8$
		Rear	—	—	$1^5/_8$N–$^1/_4$N	$1^{15}/_{16}$	$^{13}/_{64}$N–$^{19}/_{64}$P②	$15^1/_2$
1990	Taurus Sedan	Front	$2^{13}/_{16}$P–$5^{13}/_{16}$P①	$3^{13}/_{16}$P	$1^3/_{32}$N–$^3/_{32}$P	$^1/_2$N	$^7/_{32}$N–$^1/_{64}$P	$15^1/_2$
		Rear	—	—	$1^5/_8$N–$^1/_4$N	$^7/_8$N	$^1/_{32}$N–$^3/_{32}$P②	—
	Sable	Front	$2^{11}/_{16}$P–$5^{11}/_{16}$P①	$3^{11}/_{16}$P	$1^3/_{32}$N–$^3/_{32}$P	$^1/_2$N	$^7/_{32}$N–$^1/_{64}$P	$15^1/_2$
		Rear	—	—	$1^5/_8$N–$^1/_4$N	$^7/_8$N	$^1/_{32}$N–$^3/_{32}$P②	—
	Taurus-Sable Wagon	Front	$2^{11}/_{16}$P–$5^{11}/_{16}$P①	$3^{11}/_{16}$P	$1^3/_{64}$N–$^3/_{32}$P	$^1/_2$N	$^7/_{32}$N–$^1/_{64}$P	$15^1/_2$
		Rear	—	—	$1^5/_8$N–$^1/_4$N	$^7/_8$N	$^1/_{32}$N–$^3/_{32}$P②	—
	Continental	Front	$3^5/_8$P–$5^1/_8$P①	$4^3/_8$P	$1^{11}/_{16}$–$^1/_2$N	$1^1/_8$N	$^7/_{32}$N–$^1/_{32}$P	$15^1/_2$
		Rear	—	—	2N–$^5/_8$N	$1^5/_{16}$N	$^1/_{32}$N–$^7/_{32}$P②	—
1991–92	Taurus Sedan	Front	$2^{13}/_{16}$P–$5^{13}/_{16}$P①	$3^{13}/_{16}$P	$1^3/_{32}$N–$^3/_{32}$P	$^1/_2$N	$^7/_{32}$N–$^1/_{64}$P	$15^1/_2$
		Rear	—	—	$1^5/_8$N–$^1/_4$N	$^7/_8$N	$^1/_{32}$N–$^3/_{32}$P②	—
	Sable	Front	$2^{11}/_{16}$P–$5^{11}/_{16}$P①	$3^{11}/_{16}$P	$1^3/_{32}$N–$^3/_{32}$P	$^1/_2$N	$^7/_{32}$N–$^1/_{64}$P	$15^1/_2$
		Rear	—	—	$1^5/_8$N–$^1/_4$N	$^7/_8$N	$^1/_{32}$N–$^3/_{32}$P②	—
	Taurus-Sable Wagon	Front	$2^{11}/_{16}$P–$5^{11}/_{16}$P①	$3^{11}/_{16}$P	$1^3/_{64}$N–$^3/_{32}$P	$^1/_2$N	$^7/_{32}$N–$^1/_{64}$P	$15^1/_2$
		Rear	—	—	$1^5/_8$N–$^1/_4$N	$^7/_8$N	$^1/_{32}$N–$^3/_{32}$P②	—
	Continental	Front	$3^5/_8$P–$5^1/_8$P①	$4^3/_8$P	$1^{11}/_{16}$–$^1/_2$N	$1^1/_8$N	$^7/_{32}$N–$^1/_{32}$P	$15^1/_2$
		Rear	—	—	2N–$^5/_8$N	$1^5/_{16}$N	$^1/_{32}$N–$^7/_{32}$P②	—

N–Negative
P–Positive

① The caster measurements are made by turning each individual wheel through the prescribed angle of sweep

② Individual sides

ENGINE MECHANICAL

NOTE: Disconnecting the negative battery cable on some vehicles may interfere with the functions of the on board computer systems and may require the computer to undergo a relearning process, once the negative battery cable is reconnected.

Engine Assembly

REMOVAL & INSTALLATION

2.5L Engine

1. Disconnect the negative battery cable and relieve the fuel system pressure.

2. If equipped with automatic transaxle, remove the transaxle timing window cover and rotate the engine until the flywheel timing marker is aligned with the timing pointer.

3. Place a reference mark on the crankshaft pulley at the 12 o'clock position (TDC) then rotate the crankshaft pulley mark to the 6 o'clock postion (BTDC).

4. Mark the position of the hood hinges and remove the hood.

5. Remove the air cleaner assembly and drain the cooling system.

6. Disconnect the upper radiator hose at the engine.

7. Identify, tag and disconnect all electrical wiring and vacuum hoses as required.

8. Disconnect the crankcase ventilation hose at the valve cover and intake manifold.

9. Disconnect the fuel lines and heater hoses.

10. Disconnect the engine ground wire.

11. Disconnect the accelerator and throttle valve control cables at the throttle body.

12. Properly discharge the air conditioning system and remove the suction and discharge lines from the compressor, if equipped.

13. On manual transaxle equipped vehicles, remove the engine damper brace.

14. Remove the driver belt and water pump pulley.

15. Remove the air cleaner-to-canister hose.

16. Raise the vehicle and support safely.

17. Drain the engine oil and remove the oil filter.

18. Disconnect the starter cable and remove the starter motor.

19. On automatic transaxle equipped vehicles, remove the converter nuts and align the previously made reference mark as close to the 6 o'clock (BTDC) position as possible with the converter stud visible.

NOTE: The flywheel timing marker must be in the 6 o'clock (BDC) position for proper engine removal and installation.

20. Remove the engine insulator nuts.

21. Disconnect the exhaust pipe from the manifold.

22. Disconnect the canister and halfshaft brackets from the engine.

23. Remove the lower engine-to-transaxle retaining bolts.

24. Disconnect the lower radiator hose.

25. Lower the vehicle and position a floor jack under the transaxle.

26. Disconnect the power steering lines from the pump.

27. Install engine lifting eyes tool D81L–6001–D or equivalent and engine support tool T79P–6000–A or equivalent.

28. Connect suitable lifting equipment to support the engine and remove the upper engine-to-transaxle retaining bolts.

29. Remove the engine from the vehicle and support on a suitable holding fixture.

To install:

30. Make sure the timing marker is in the 6 o'clock (BDC) position.

31. Remove the engine from the stand and position it in the vehicle. Remove the lifting equipment.

32. Install the upper engine-to-transaxle bolts and tighten to 26–34 ft.

lbs. (34–47 Nm). Use a floor jack under the transaxle to aid alignment.

33. Connect the power steering lines to the pump.

34. Raise the vehicle and support it safely.

35. Connect the lower radiator hose to the tube.

36. Install the lower engine-to-transaxle attaching bolts and tighten to 26–34 ft. lbs. (34–47 Nm).

37. Connect the halfshaft bracket to the engine and the exhaust pipe to the manifold.

38. Install the engine insulator nuts and tighten to 40–55 ft. lbs. (54–75 Nm).

39. Position the marks on the crankshaft pulley as close to 6 o'clock position (BDC) as possible and install the converter nuts. Tighten the nuts to 20–33 ft. lbs. (27–46 Nm).

40. Install the starter and connect the starter cable.

41. Install the oil filter and make sure the oil drain plug is tight.

42. Lower the vehicle.

43. Install the air cleaner-to-canister hose and the water pump pulley and drive belt.

44. Connect the air conditioning lines to the compressor, if equipped.

45. Connect the accelerator cable and throttle valve control cable at the throttle body.

46. Connect the negative battery cable at the engine and connect the heater hoses and fuel lines.

47. Connect the crankcase ventilation hose at the valve cover and the intake manifold.

48. Connect the engine control sensor wiring assembly and vacuum lines.

49. Connect the upper radiator hose at the engine and install the air cleaner assembly.

50. Connect the negative battery cable.

51. Rotate the engine until the flywheel timing marker is aligned with the timing pointer. Install the timing window cover.

52. Connect the electrical connector at the inertia switch.

53. Fill the cooling system with the proper amount and type of coolant and fill the crankcase with the proper engine oil to the required level.

54. Install the hood.

55. Charge the air conditioning system, if equipped.

56. Check all fluid levels and start the vehicle. Check for leaks.

3.0L Engine

1. Disconnect the battery cables and drain the cooling system. Mark the position of the hood on the hinges and remove the hood.

2. Evacuate the air conditioning

system safely and properly. Relieve the fuel system pressure. Remove the air cleaner assembly. Remove the battery and the battery tray.

3. Remove the integrated relay controller, cooling fan and radiator with fan shroud. Remove the engine bounce damper bracket on the shock tower.

4. Remove the evaporative emission line, upper radiator hose, starter brace and lower radiator hose.

5. Remove the exhaust pipes from both exhaust manifolds. Remove and plug the power steering pump lines.

6. Remove the fuel lines and remove and tag all necessary vacuum lines.

7. Disconnect the ground strap, heater lines, accelerator cable linkage, throttle valve linkage and speed control cable.

8. Disconnect and label the following wiring connectors; alternator, air conditioning clutch, oxygen sensor, ignition coil, radio frequency supressor, cooling fan voltage resistor, engine coolant temperature sensor, Thick film ignition module, injector wiring harness, ISC motor wire, throttle position sensor, oil pressure sending switch, ground wire, block heater, if equipped, knock sensor, EGR sensor and oil level sensor.

9. Raise the vehicle and support it safely. Remove the engine mount bolts and engine mounts. Remove the transaxle to engine mounting bolts and transaxle brace assembly.

10. Lower the vehicle. Install a suitable engine lifting plate onto the engine and use a suitable engine hoist to remove the engine from the vehicle. Remove the main wiring harness from the engine.

To install:

11. Install the main wiring harness on the engine. Position the engine in the vehicle and remove the engine lifting plate.

12. Raise the vehicle and support it safely. Install the engine mounts and bolts and tighten to 40–55 ft. lbs. (54–75 Nm). Install the transaxle brace assembly and tighten the bolts to 40–55 ft. lbs. (54–75 Nm).

13. Connect all wiring connectors according to their labels.

14. Connect the ground strap, heater lines, accelerator cable linkage, throttle valve linkage and speed control cables.

15. Connect the power steering pump lines.

16. Connect the exhaust pipes to the exhaust manifolds.

17. Connect the fuel lines and vacuum lines.

18. Install the evaporative emission line, upper radiator hose, starter brace and lower radiator hose.

19. Install the integrated relay controller, cooling fan and radiator with

fan shroud. Install the engine bounce damper bracket on the shock tower.

20. Install the battery tray and the battery.

21. Install the air cleaner assembly and charge the air conditioning system.

22. Fill the cooling system with the proper type and quantity of coolant. Fill the crankcase with the correct type of motor oil to the required level.

23. Install the hood.

24. Connect the negative battery cable. Start the engine and check for leaks.

3.0L SHO Engine

1. Disconnect the battery cables and remove the battery and battery tray.

2. Drain the cooling system and relieve the fuel system pressure.

3. Disconnect the wiring connector retaining the under hood light, if equipped, mark the position of the hood hinges and remove the hood.

4. Remove the oil level indicator.

5. Disconnect the alternator and voltage regulator wiring assembly.

6. Remove the radiator upper sight shield.

7. Discharge the air conditioning system.

8. Remove the radiator coolant recovery reservoir assembly.

9. Remove the integrated relay controller, air cleaner hose assembly, upper radiator hose, electric fan and shroud assembly.

10. Remove the lower radiator hose and the radiator.

11. Disconnect the fuel inlet and return hose.

12. Remove the Barometric Air Pressure (BAP) sensor.

13. Remove the engine vibration damper and bracket assembly from the right side of the engine.

14. Remove the engine to damper bracket.

15. Remove the retaining bolt from the power steering reservoir and place the reservoir aside. Disconnect the hose to the power steering cooler at the pump.

16. Disconnect the throttle linkage and disconnect and tag the vacuum hoses.

17. Disconnect the heater hoses at the heater core.

18. Disconnect the electrical connectors from the harness on the rear of the engine.

19. Loosen the belt tensioner pulleys and remove the air conditioning compressor/alternator belt and the steering pump belt. Remove the lower tensioner pulley.

20. Disconnect the cycling switch on the top of the suction accumulator/drier.

21. Disconnect the air conditioning line at the dash panel and remove the accumulator and bracket assembly.

22. Remove the alternator assembly.

23. Disconnect the air conditioning discharge hose and remove the air conditioning compressor and bracket assembly.

24. Raise the vehicle and support it safely.

25. Place a drain pan under the oil pan and drain the motor oil and remove the filter element.

26. Remove the wheel and tire assemblies. Disconnect the oil level sensor switch.

27. Disconnect the right lower ball joint, tie rod end and stabilizer bar.

28. Disconnect the center support bearing bracket and right-hand CV-joint from the transaxle.

29. Disconnect the oxygen sensor assembly and the 4 exhaust catalyst to engine retaining bolts.

30. Remove the starter motor assembly.

31. Remove the lower transaxle to engine retaining bolts.

32. Remove the engine mount to sub-frame nuts.

33. Remove the crankshaft pulley assembly.

34. Lower the vehicle and remove the upper transaxle to engine retaining bolts.

35. Install engine lifting eyes.

36. Position a floor jack under the transaxle.

37. Position suitable engine lifting equipment, raise the transaxle assembly slightly and remove the engine from the vehicle.

To install:

38. Position the engine assembly in the vehicle.

39. Install the upper transaxle to engine bolts and remove the floor jack and engine lifting equipment. Remove the engine lifting eyes.

40. Raise the vehicle and support it safely.

41. Install the crankshaft pulley assembly. Tighten the retaining bolt to 113–126 ft. lbs. (152–172 Nm).

42. Install the engine mount to subframe nuts and the lower transaxle to engine retaining bolts. Tighten the bolts to 25–35 ft. lbs. (34–47 Nm).

43. Install the starter motor assembly.

44. Install the 4 exhaust catalyst to engine retaining nuts and tighten them to 19–34 ft. lbs. (27–47 Nm). Apply anti-seize compound to the threads, then install the oxygen sensor assembly. Tighten to 27–33 ft. lbs. (37–45 Nm).

45. Connect the center support bearing bracket and install the right-hand CV-joint.

46. Connect the right lower ball joint, tie rod end and stabilizer bar.

47. Connect the oil level sensor and install the wheel and tire assemblies.

48. Install the oil filter. Install the oil drain plug and tighten to 15–24 ft. lbs. (20–33 Nm).

49. Lower the vehicle.

50. Install the air conditioning compressor and bracket assembly, tighten to 27–40 ft. lbs. (36–55 Nm) and connect the air conditioning discharge hose.

51. Install the alternator assembly and tighten to 36–53 ft. lbs. (48–72 Nm).

52. Install the accumulator and bracket assembly and connect the cycling switch to the top of the accumulator.

53. Install the lower belt tensioner. Install the power steering and air conditioning compressor/alternator belts and tighten the tensioner pulleys.

54. Connect the electrical connectors from the harness on the rear of the engine.

55. Connect the heater hoses, vacuum hoses and throttle linkage.

56. Connect the hose from the power steering cooler at the pump and install the power steering reservoir.

57. Install the damper bracket to the engine and install the engine vibration damper and bracket assembly to the right side of the engine.

58. Install the BAP sensor.

59. Connect the fuel inlet and return hoses.

60. Install the radiator assembly and the lower radiator hose.

61. Install the electric fan and shroud assembly, upper radiator hose, air cleaner hose, integrated relay controller, radiator coolant recovery reservoir and radiator upper sight shield.

62. Connect the alternator and voltage regulator wiring.

63. Install the oil level indicator tube.

64. Install the hood and connect the under hood light wiring, if equipped.

65. Install the battery tray and the battery.

66. Install the negative battery cable.

67. Fill the cooling system with the proper type and quantity of coolant and fill the crankcase with the proper type of motor oil to the required level.

68. Drain, evacuate, pressure test and recharge the air conditioning system.

69. Start the engine and check for leaks.

3.8L Engine
TAURUS AND SABLE

1. Drain the cooling system and dis-

connect the battery ground cable. Properly relieve the fuel system pressure.

2. Disconnect the underhood light wiring connector. Mark position of hood hinges and remove hood.

3. Remove the oil level indicator tube.

4. Disconnect alternator to voltage regulator wiring assembly.

5. Remove the radiator upper sight shield. Remove the engine cooling fan motor relay retaining bolts and position cooling fan motor relay aside.

6. Remove the air cleaner assembly.

7. Disconnect the radiator electric fan and motor assembly. Remove fan shroud.

8. Remove upper radiator hose.

9. Disconnect the transaxle oil cooler inlet and outlet tubes and cover the openings to prevent the entry of dirt and grease. Disconnect the heater hoses.

10. Disconnect the power steering pressure hose assembly.

11. Disconnect the air conditioner compressor clutch wire assembly. Discharge the air conditioning system and disconnect the compressor-to-condenser line.

12. Remove the radiator coolant recovery reservoir assembly. Remove the wiring shield.

13. Remove accelerator cable mounting bracket.

14. Disconnect fuel inlet and return lines.

15. Disconnect power steering pump pressure and return tube brackets.

16. Disconnect the engine control sensor wiring assembly.

17. Identify, tag and disconnect all necessary vacuum hoses.

18. Disconnect the ground wire assembly. Remove the duct assembly.

19. Disconnect one end of the throttle control valve cable. Disconnect the bulkhead electrical connector and transaxle pressure switches.

20. Remove transaxle support assembly retaining bolts and remove transaxle and support assembly from vehicle.

21. Raise the vehicle and support safely. Remove the wheel and tire assemblies. Drain the engine oil and remove the filter.

22. Disconnect the oxygen sensor assembly.

23. Loosen and remove drive belt assembly. Remove the crankshaft pulley and drive belt tensioner assemblies.

24. Remove the starter motor assembly. Remove the converter housing assembly and remove the inlet pipe converter assembly.

25. Remove the engine left and right front support insulator retaining nuts.

26. Remove the converter-to-flywheel nuts.

27. Disconnect the oil level indicator sensor. Remove crankshaft pulley assembly.

28. Disconnect the lower radiator hose.

29. Remove the engine-to-transaxle bolts and partially lower engine. Remove the wheel assemblies.

30. Remove the water pump pulley retaining bolts and the water pump pulley.

31. Remove the distributor cap and position aside. Remove distributor rotor.

32. Remove the exhaust manifold bolt lock retaining bolts. Remove the thermactor air pump retaining bolts and the thermactor air pump.

33. Disconnect the oil pressure engine unit gauge assembly.

34. Install engine lifting eyes and connect suitable lifting equipment to the lifting eyes.

35. Position a suitable jack under the transaxle and raise the transaxle a small amount.

36. Remove the engine from the vehicle and position in a suitable holding fixture.

To install:

NOTE: Lightly oil all bolt and stud threads before installation except those specifying special sealant.

37. Remove the engine assembly from the work stand and position it in the vehicle.

38. Install the engine to transaxle bolts and remove the jack from under the transaxle and the engine lifting equipment. Remove the engine lifting eyes.

39. Tighten the engine to transaxle bolts to 41–50 ft. lbs. (55–68 Nm).

40. Connect the oil pressure engine unit gauge assembly.

41. Install the air conditioning compressor and tighten the retaining bolts to 30–45 ft. lbs. (41–61 Nm). Connect the compressor to condenser discharge line and the compressor clutch wire assembly.

42. Connect the heater hoses, vacuum hoses and the fuel tube hose and return line hose.

43. Connect the engine control module wiring assembly.

44. Connect the transaxle oil cooler inlet and outlet tubes.

45. Install the radiator assembly.

46. Partially raise the vehicle and support it safely.

47. Install the converter to flywheel bolts and tighten to 20–34 ft. lbs. (27–46 Nm).

48. Install the left and right transaxle and engine mount retaining nuts and install the converter housing cover.

49. Install the starter motor.

50. Connect the lower radiator hose.

51. Install the drive belt tensioner assembly and the crankshaft pulley assembly. Tighten the crankshaft pulley retaining bolts to 20–28 ft. lbs. (26–38 Nm).

52. Install the catalytic converter assembly and connect the heated exhaust gas oxygen sensor.

53. Install the oil filter and connect the oil level indicator sensor.

54. Lower the vehicle.

55. Position the thermactor air supply pump and install the retaining bolts.

56. Connect the vacuum pump and install the exhaust air supply pump pulley assembly.

57. Install the wiring shield.

58. Install the distributor cap and rotor.

59. Install the radiator coolant recovery reservoir assembly, upper radiator hose and water pump pulley.

60. Connect the alternator-to-voltage regulator wiring assembly and the engine control module wiring assembly.

61. Connect the wiring asembly ground.

62. Install the accelerator cable mounting bracket.

63. Connect the power steering pressure hose assembly and the power steering line.

64. Install the fan shroud.

65. Connect the radiator electric motor assembly and install the engine cooling fan motor relay assembly.

66. Install the drive belts.

67. Position and install the transaxle support assembly.

68. Install the radiator upper sight shield.

69. Partially raise the vehicle and support it safely. Install the wheel and tire assemblies.

70. Install the hood and connect the negative battery cable.

71. Fill the cooling system with the proper type and quantity of coolant and fill the crankcase with the proper type of motor oil to the required level.

72. Drain, evacuate, pressure test and recharge the air conditioning system.

73. Start the engine and check for leaks.

CONTINENTAL

1. Disconnect the negative battery cable.

2. Relieve the fuel system pressure, drain the cooling system and properly discharge the air conditioning system.

3. Tag and disconnect the alternator-to-voltage regulator, electric cooling fan, transaxle pressure switch, air conditioning compressor clutch, electronic engine control and ground wiring.

4. Disconnect the heater hoses, power steering hoses and brackets, air conditioning discharge hose and fuel lines.

5. Tag and disconnect the vacuum lines. Disconnect the throttle cable at the throttle valve.

6. Remove the electric cooling fan and motor assembly. Remove the fan shroud.

7. Remove the engine oil dipstick and the radiator sight shield. Remove the integrated controller relay and position aside.

8. Remove the air cleaner assembly.

9. Disconnect the upper radiator hose and remove the coolant recovery reservoir. Remove the wiring shield.

10. Remove the air suspension compressor and position aside. Remove the accelerator cable mounting bracket.

11. Remove the transaxle support assembly. Remove the air conditioning compressor.

12. Raise and safely support the vehicle.

13. Drain the engine oil and remove the oil filter. Disconnect the oxygen sensor.

14. Release the tension at the drive belts. Remove the crankshaft pulley and drive belt tensioner.

15. Remove the starter. Remove the catalytic converter housing cover and remove the converter and inlet pipe assembly from the engine.

16. Remove the nuts at the transaxle and engine mounts. Remove the torque converter-to-flywheel nuts.

17. Disconnect the oil level indicator sensor and the lower radiator hose.

18. Loosen the engine-to-transaxle bolts, leaving them loosely installed.

19. Partially lower the vehicle and remove the front wheel and tire assemblies.

20. Remove the drive belts and the water pump pulley. Remove the radiator assembly.

21. Remove the distributor cap and position aside. Remove the distributor rotor.

22. Remove the exhaust manifold lock bolts and the thermactor air pump. Disconnect the oil pressure sending unit.

23. Install suitable engine lifting equipment and position a transmission jack. Completely remove the engine-to-transaxle bolts.

24. Raise the transaxle assembly using the jack and lift the engine from the engine from the vehicle.

To install:

25. Position the engine assembly in the vehicle and align the engine-to-transaxle bolt bores. Install the engine-to-transaxle bolts that are accessible but do not tighten at this time.

26. Remove the transmission jack and the engine lifting equipment.

27. Install the oil pressure sending unit. Install the air conditioning compressor and tighten the retaining bolts to 30–45 ft. lbs. (41–61 Nm).

28. Connect the air conditioning compressor discharge hose and the clutch wiring to the compressor.

29. Connect the heater hoses and the fuel lines. Connect the vacuum hoses and routing clips.

30. Connect the transaxle oil cooler lines and the transaxle pressure switch wiring.

31. Install the radiator assembly. Raise and safely support the vehicle.

32. Install the remaining transaxle-to-engine bolts. Tighten all the bolts to 40–50 ft. lbs. (55–68 Nm).

33. Install the torque converter-to-flywheel bolts and tighten to 20–34 ft. lbs. (27–46 Nm). Install the converter housing cover.

34. Install the transaxle mount retaining nuts and tighten to 50–70 ft. lbs. (68–95 Nm).

35. Install the starter and the lower radiator hose.

36. Install the drive belt tensioner assembly and the crankshaft pulley assembly. Tighten the crankshaft pulley retaining bolts to 20–28 ft. lbs. (26–38 Nm).

37. Install the catalytic converter and inlet pipe assembly. Connect the oxygen sensor.

38. Install the oil filter and the oil drain plug. Connect the low oil level sensor.

39. Partially lower the vehicle.

40. Install the thermactor air pump and tighten the mounting bolts to 20–30 ft. lbs. (40–55 Nm). Install the vacuum hose at the pump and the air pump pulley.

41. Install the wiring shield. Install the distributor rotor and cap and connect the distributor wiring.

42. Install the coolant recovery reservoir and connect the top radiator hose.

43. Install the air suspension compressor and connect the wiring. Install the water pump pulley.

44. Connect the alternator to voltage regulator wiring, the electronic engine control wiring and the ground wires.

45. Connect the power steering hoses and the throttle cable at the throttle valve. Install the accelerator cable mounting bracket.

46. Install the fan shroud and connect the fan wiring. Install the cooling fan relay and position the drive belts.

47. Install the transaxle support assembly and the upper radiator sight shield. Partially raise the vehicle.

48. Install the exhaust air supply pump valve and hose assembly. Install the drive belts. Partially lower the vehicle.

49. Install the integrated controller relay and the engine oil dipstick. Install the front wheel and tire assemblies and tighten the lug nuts to 85–105 ft. lbs. (115–142 Nm).

50. Lower the vehicle. Install the hood, aligning the marks that were made during the removal procedure.

51. Install the air cleaner assembly and connect the negative battery cable.

52. Fill the engine with the proper type and quantity of engine oil and coolant. Leak test, evacuate and charge the air conditioning system. Observe all safety precautions.

53. Start the engine and check for leaks.

Cylinder Head

REMOVAL & INSTALLATION

2.5L Engine

1. Disconnect the negative battery cable. Drain the cooling system.

2. Remove the air cleaner assembly. Properly relieve the fuel system pressure.

3. On 1988–90 vehicles, disconnect the heater hose at the fitting located under the intake manifold. On 1991–92 vehicles, disconnect the heater hose at the heater inlet tube and disconnect the adapter hose at the water outlet connector.

4. Disconnect the upper radiator hose at the cylinder head and the electric cooling fan switch at the plastic connector.

5. Disconnect distributor cap and spark plug wire and remove as an assembly.

6. Remove spark plugs, if necessary.

7. Disconnect and tag required vacuum hoses. Disconnect the accessory drive belts.

8. Remove dipstick. Disconnect the choke cap wire.

9. Remove rocker cover retaining bolts and remove cover. Disconnect the EGR tube at the EGR valve.

10. Remove the rocker arm fulcrum bolts, the fulcrums, rocker arms and pushrods. Identify the location of each so they may be reinstalled in their original positions.

11. Disconnect the fuel supply and return lines at the rubber connections. Disconnect the accelerator and speed control cable, if equipped.

12. Raise the vehicle and support it safely. Disconnect the exhaust system at the exhaust pipe, hose and tube. Lower the vehicle.

13. Remove the cylinder head bolts. Remove the cylinder head and gasket with the exhaust manifold and intake manifold.

To install:

14. Clean all gasket material from

the mating surface of the cylinder head and block. Position the cylinder head gasket on the cylinder block, using a suitable sealer to retain the gasket.

15. Before installing the cylinder head, thread 2 cylinder head alignment studs through the head bolt holes in the gasket and into the block at opposite corners of the block.

16. Install the cylinder head and cylinder head bolts. Run down several head bolts and remove the 2 guide bolts. Replace them with the remaining head bolts. Torque the cylinder head bolts in 2 steps, first to 52–59 ft. lbs. (70–80 Nm) and then to 70–76 ft. lbs. (95–103 Nm).

17. Raise and support the vehicle safely. Connect the exhaust system at the exhaust pipe and hose to metal tube.

18. Lower the vehicle. Install the thermactor pump drive belt, if equipped. Connect the accelerator cable and speed control cable, if equipped.

19. Connect the fuel supply and return lines. Connect the choke cap wire, if equipped.

20. Install the pushrods, rocker arms, fulcrums and fulcrum bolts in their original positions. Install the rocker arm cover.

21. Connect the EGR tube at the EGR valve. Install the distributor cap and spark plug wires as an assembly. Install the spark plugs, if removed.

22. Connect all accessory drive belts.

23. Connect the required vacuum hoses. Install the air cleaner assembly. Connect the electric cooling fan switch at the connector.

24. Connect the upper radiator hose and the heater hose. Fill the cooling system. Connect the negative battery cable.

25. Start the engine and check for leaks. After the engine has reached normal operating temperature, check and if necessary add coolant.

3.0L Except SHO Engine

1. Disconnect the negative battery cable. Properly relieve the fuel system pressure. Drain the cooling system. Remove the air cleaner assembly.

2. Loosen the accessory drive belt idler pulley, remove the drive belt.

3. If the left cylinder head is being removed, perform the following:
    a. Disconnect the alternator electrical connectors.
    b. Rotate the tensioner clockwise and remove the accessory drive belt.
    c. Remove the automatic belt tensioner assembly.
    d. Remove the alternator.
    e. Remove the power steering mounting bracket retaining bolts.

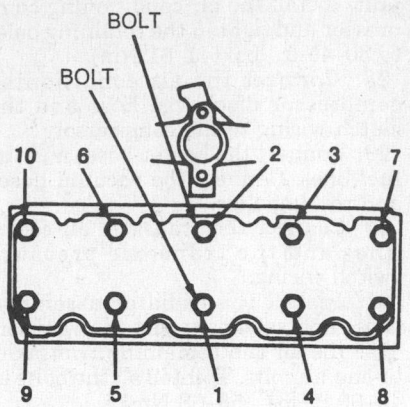

Cylinder head bolt torque sequence— 2.5L engine

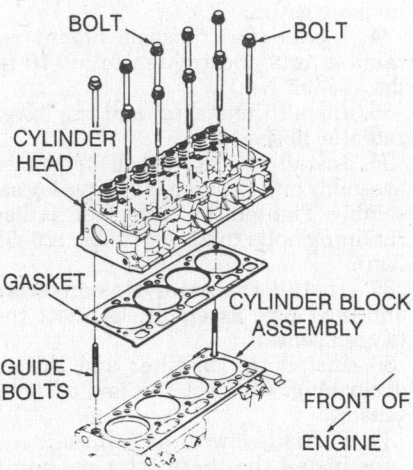

Cylinder head installation—2.5L engine

Leave the hoses connected and place the pump aside in a position to prevent fluid from leaking out.
    f. Remove the engine oil dipstick tube from the exhaust manifold.

4. If the right head is being removed, perform the following:
    a. Remove the alternator belt tensioner bracket.
    b. Remove the heater supply tube retaining brackets from the exhaust manifold.
    c. Remove the vehicle speed sensor cable retaining bolt and the EGR vacuum regulator sensor and bracket.

5. Remove the exhaust manifolds from both heads. Remove the PCV and the rocker arm covers. Loosen the rocker arm fulcrum attaching bolts enough to allow the rocker arm to be lifted off the pushrod and rotated to one side.

6. Remove the pushrods. Be sure to identify and label the position of each pushrod. The rods should be installed in their original position during reassembly.

7. Remove the intake manifold.

8. Remove the cylinder head attaching bolts and remove the cylinder heads from the engine. Remove and discard the old cylinder head gaskets.
To install:

9. Lightly oil all bolt and stud bolt threads before installation. Clean the cylinder head, intake manifold, rocker arm cover and cylinder head gasket contact surfaces. If the cylinder head was removed for a cylinder head gasket replacement, check the flatness of the cylinder head and block gasket surfaces.

NOTE: **If the flat surface of the cylinder head is warped, do not plane or grind off more than 0.010 in. If the head is machined past its resurface limit, the head will have to be replaced with a new one.**

10. Position new head gaskets on the cylinder block using the dowels in the engine block for alignment. If the dowels are damaged, they must be replaced.

11. Position the cylinder head on the cylinder block. Tighten the cylinder head attaching bolts in 2 steps following the proper torque sequence. The first step is 37 ft. lbs. (50 Nm) and the second step is 68 ft. lbs. (92 Nm).

NOTE: **When cylinder head attaching bolts have been tightened using the above procedure, it is not necessary to retighten the bolts after extended engine operation. The bolts can be rechecked for tightness if desired.**

12. Install the intake manifold. Connect the coolant temperature sending unit connectors.

13. Dip each pushrod end in oil conditioner or heavy engine oil. Install the pushrods in their original position.

14. Before installation, coat the valve tips, rocker arm and fulcrum contact areas with Lubriplate® or equivalent. Lightly oil all the bolt and stud threads before installation.

15. Rotate the engine until the lifter is on the base circle of the cam (valve closed).

16. Install the rocker arm and components and torque the rocker arm fulcrum bolts to 24 ft. lbs. (32 Nm). Be sure the lifter is on the base circle of the cam for each rocker arm as it is installed.

NOTE: **The fulcrums must be fully seated in the cylinder head and the pushrods must be seated in the rocker arm sockets prior to the final tightening.**

17. Install the exhaust manifolds, the oil dipstick tube. Install the remaining components by reversing the removal procedure.

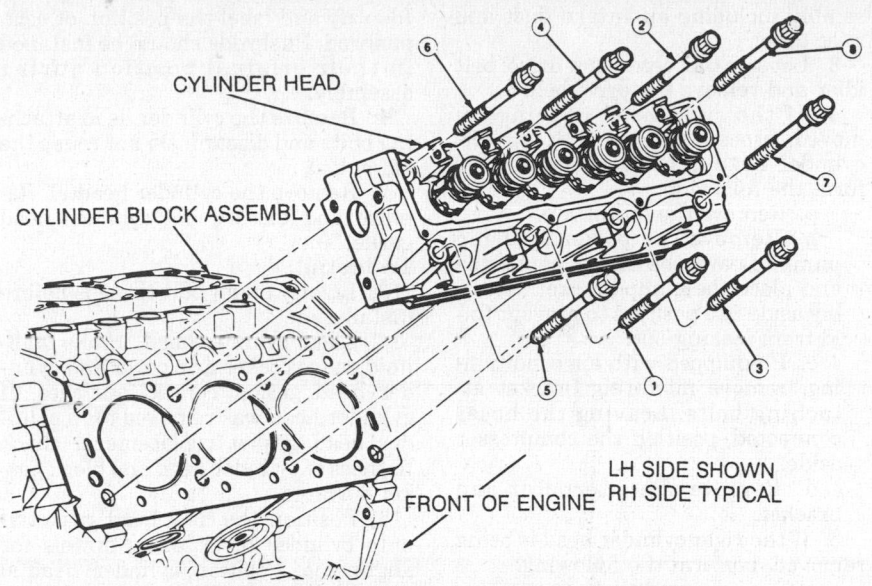

Cylinder head bolt torque sequence—3.0L engine

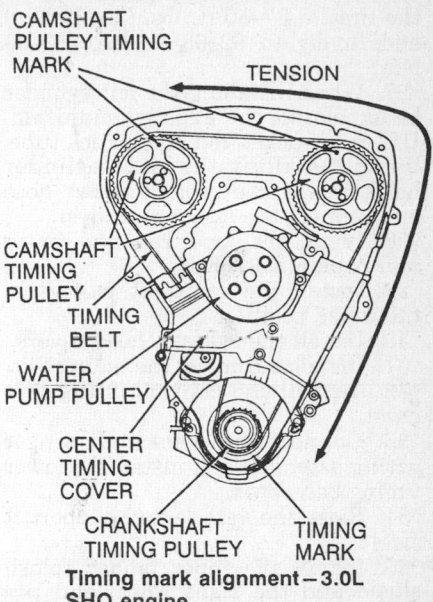

Timing mark alignment—3.0L SHO engine

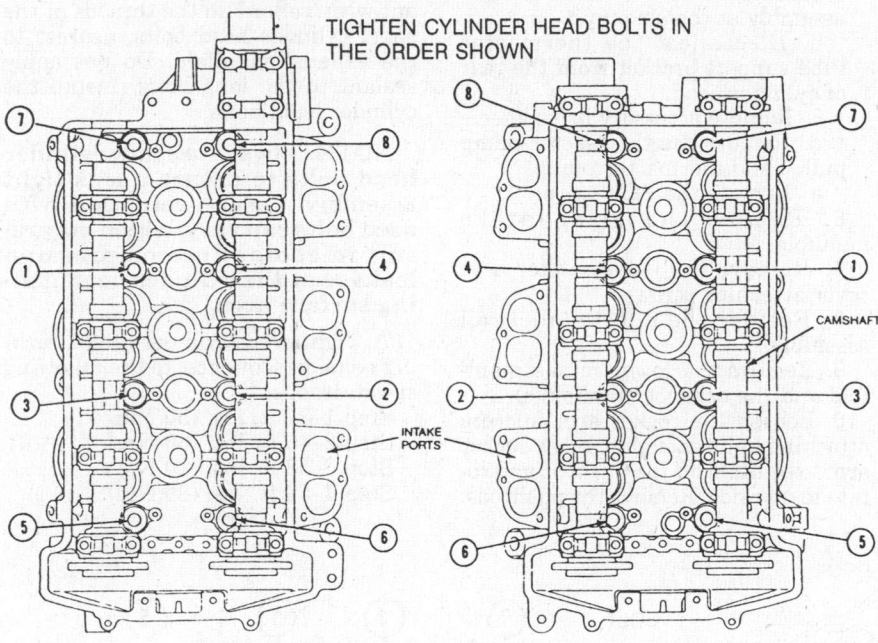

Cylinder head bolt torque sequence—3.0L SHO engine

18. Start the engine and check for leaks.

19. Check and if necessary, adjust the transaxle throttle linkage and speed control. Install the air cleaner outlet tube duct.

3.0L SHO Engine

1. Disconnect the negative battery cable.

2. Drain the cooling system. Properly relieve the fuel system pressure.

3. Remove the air cleaner outlet tube.

4. Remove the intake manifold.

5. Loosen the accessory drive belt idlers and remove the drive belts.

6. Remove the upper timing belt cover.

7. Remove the left idler pulley and bracket assembly.

8. Raise the vehicle and support it safely.

9. Remove the right wheel and inner fender splash shield.

10. Remove the crankshaft damper pulley.

11. Remove the lower timing belt cover.

12. Align both camshaft pulley timing marks with the index marks on the upper steel belt cover.

13. Release the tension on the belt by loosening the tensioner nut and rotat-

ing the tensioner with a hex head wrench. When tension is released, tighten the nut. This will hold the tensioner in place. Lower the vehicle until the wheels touch but keep the vehicle supported.

14. Disconnect the crankshaft sensor wiring assembly.

15. Remove the center cover assembly.

16. Remove the timing belt noting the location of the letters **KOA** on the belt. The belt must be installed in the same direction.

17. Remove the cylinder head covers.

18. Remove the camshaft timing pulleys.

19. Remove the upper rear and the center rear timing belt covers.

20. If the left cylinder head is being removed, remove the DIS coil bracket and the oil dipstick tube. If the right cylinder head is being removed, remove the coolant outlet hose.

21. Remove the exhaust manifold on the left cylinder head. On the right cylinder head the exhaust manifold must be removed with the head.

22. Remove the cylinder head to block retaining bolts.

23. Remove the cylinder head.

To install:

NOTE: Lightly oil all bolt and stud bolt threads before installation except those specifying special sealant.

24. Clean the cylinder head and engine block mating surfaces of all gasket material.

25. Position the cylinder head and gasket on the engine block and align with the dowel pins.

26. Install the cylinder head bolts and tighten, in sequence, in 2 steps,

the first to 37–50 ft. lbs. (49–69 Nm) and finally to 62–68 ft. lbs. (83–93 Nm).

27. When installing the left cylinder head, install the exhaust manifold, DIS coil bracket and oil dipstick tube. When installing the right cylinder head, install the coolant outlet hose and connect the exhaust catalyst.

28. Install the upper rear and center rear timing belt covers.

29. Install the camshaft pulleys in the timed position.

30. Install the cylinder head covers.

31. Install and adjust the timing belt.

32. Install the center timing belt cover.

33. Connect the crankshaft sensor wiring assembly and install the lower timing belt cover.

34. Raise the vehicle and support it safely.

35. Install the inner fender splash shield and the right wheel and tire assembly.

36. Install the left idler pulley and bracket.

37. Install the upper timing belt cover.

38. Install the accessory drive belts.

39. Install the intake manifold.

40. Install the air cleaner oulet tube.

41. Connect the negative battery cable.

42. Fill the engine cooling system with the proper type and quantity of coolant.

43. Start the engine and check for coolant, fuel or oil leaks.

3.8L Engine

1. Drain the cooling system and disconnect the negative battery cable.

2. Properly relieve the fuel system pressure. Remove the air cleaner as-sembly including air intake duct and heat tube.

3. Loosen the accessory drive belt idler and remove the drive belt.

4. If the right head is being re-moved, proceed to Step 5. If the left cylinder head is being removed, per-form the following:
   a. Remove the oil fill cap.
   b. Remove the power steering pump. Leave the hoses connected and place the pump/bracket assem-bly aside in a position to prevent flu-id from leaking out.
   c. If equipped with air condition-ing, remove mounting bracket at-taching bolts. Leaving the hoses connected, position the compressor aside.
   d. Remove the alternator and bracket.

5. If the right cylinder head is being removed, perform the following:
   a. Disconnect the thermactor air control valve or bypass valve hose assembly at the air pump.
   b. Disconnect the thermactor tube support bracket from the rear of cylinder head.
   c. Remove accessory drive idler.
   d. Remove the thermactor pump pulley and thermactor pump.
   e. Remove the PCV valve.

6. Remove the upper intake manifold.

7. Remove the valve rocker arm cover attaching screws.

8. Remove the injector fuel rail assembly.

9. Remove the lower intake mani-fold and the exhaust manifold(s).

10. Loosen the rocker arm fulcrum attaching bolts enough to allow rocker arm to be lifted off the pushrod and ro-tate to one side. Remove the pushrods.

Identify and label the position of each pushrod. Pushrods should be installed in their original position during assembly.

11. Remove the cylinder head attach-ing bolts and discard. Do not reuse the old bolts.

12. Remove the cylinder head(s). Re-move and discard old cylinder head gasket(s).

To install:

13. Lightly oil all bolt threads before installation.

14. Clean cylinder head, intake mani-fold, valve rocker arm cover and cylin-der head gasket contact surfaces. If cylinder head was removed for a cylin-der head gasket replacement, check flatness of cylinder head and block gas-ket surfaces.

15. Position the new head gasket(s) onto cylinder block using dowels for alignment. Position cylinder head(s) onto block.

16. Apply a thin coating of pipe seal-ant with Teflon® to the threads of the short cylinder head bolts, nearest to the exhaust manifold. Do not apply sealant to the long bolts. Install the cylinder head bolts.

NOTE: Always use new cylinder head bolts to ensure a leak-tight assembly. Torque retention with used bolts can vary, which may re-sult in coolant or compression leakage at the cylinder head mat-ing surface area.

17. Tighten the cylinder head attach-ing bolts, in sequence, to the following specifications:
   Step 1–37 ft. lbs. (50 Nm)
   Step 2–45 ft. lbs. (60 Nm)
   Step 3–52 ft. lbs. (70 Nm)
   Step 4–59 ft. lbs. (80 Nm)

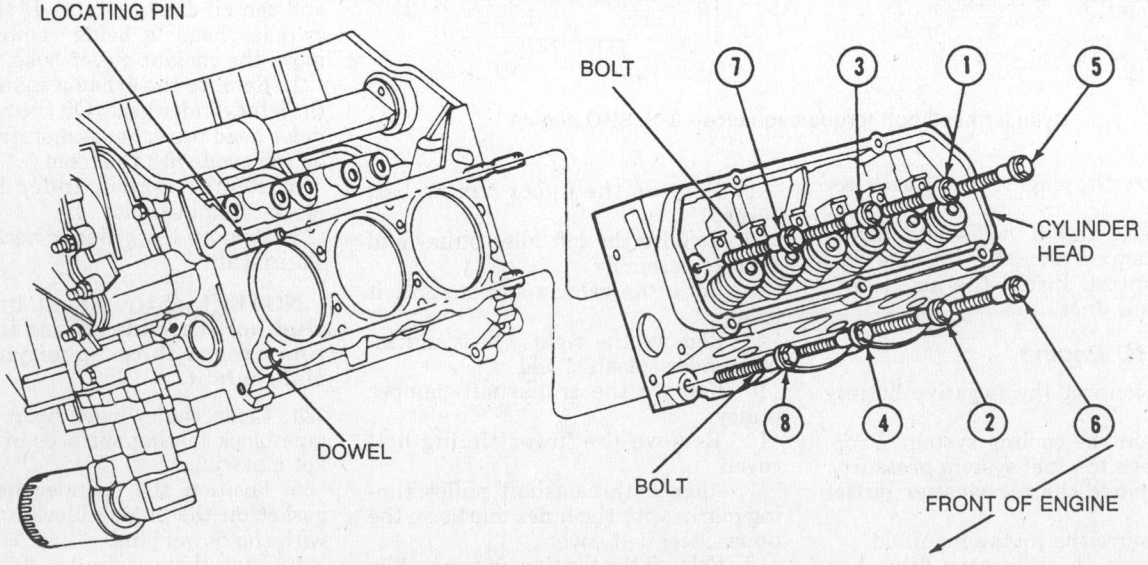

Cylinder head bolt torque sequence–3.8L engine

18. In sequence, retighten the cylinder head bolts 1 at a time in the following manner:

a. Long cylinder head bolts: Loosen the bolts and back them out 2–3 turns. Retighten to 11–18 ft. lbs. (15–25 Nm). Then tighten the bolt an additional 85–105 degrees and go to the next bolt in sequence.

b. Short cylinder head bolts: Loosen the bolts and back them out 2–3 turns. Retighten to 11–18 ft. lbs. (15–25 Nm). Then tighten the bolt an additional 65–85 degrees.

NOTE: When cylinder head attaching bolts have been tightened using the above procedure, it is not necessary to retighten bolts after extended engine operation. However, bolts can be checked for tightness if desired.

19. Dip each pushrod end in oil conditioner or heavy engine oil. Install pushrods in their original position.

20. For each valve, rotate crankshaft until the tappet rests on the heel (base circle) of the camshaft lobe. Torque the fulcrum attaching bolts to 43 inch lbs. maximum.

21. Lubricate all rocker arm assemblies with oil conditioner or heavy engine oil.

22. Tighten the fulcrum bolts a second time to 19–25 ft. lbs. (25–35 Nm). For final tightening, camshaft may be in any position.

NOTE: If original valve train components are being installed, a valve clearance check is not required. If a component has been replaced, perform a valve clearance check.

23. Install the exhaust manifold(s), lower intake manifold and injector fuel rail assembly.

24. Position the cover(s) and new gasket on cylinder head and install attaching bolts. Note location of spark plug wire routing clip stud bolts. Tighten attaching bolts to 80–106 inch lbs. (9–12 Nm).

25. Install the upper intake manifold and connect the secondary wires to the spark plugs.

26. If the left cylinder head is being installed, perform the following: install oil fill cap, compressor mounting and support brackets, power steering pump mounting and support brackets and the alternator/support bracket.

27. If the right cylinder head is being installed, perform the following: install the PCV valve, alternator bracket, thermactor pump and pump pulley, accessory drive idler, thermactor air control valve or air bypass valve hose.

28. Install the accessory drive belt. Attach the thermactor tube(s) support bracket to the rear of the cylinder head. Tighten the attaching bolts to 30–40 ft. lbs. (40–55 Nm).

29. Connect the negative battery cable and fill the cooling system.

30. Start the engine and check for leaks.

31. Check and, if necessary, adjust curb idle speed.

32. Install the air cleaner assembly including air intake duct and heat tube.

Valve Lifters

REMOVAL & INSTALLATION

2.5L Engine

1. Disconnect the negative battery cable. Remove the cylinder head.

2. Using a magnet, remove the lifters. Identify, tag and place the lifters in a rack so they can be installed in the original positions.

3. If the lifters are stuck in their bores by excessive varnish or gum, it may be necessary to use a hydraulic lifter puller tool to remove the lifters. Rotate the lifters back and forth to loosen any gum and varnish which may have formed. Keep the assemblies intact until the are to be cleaned.

4. Install the lifters through the pushrod openings with a magnet.

5. Install the cylinder head and related parts.

3.0L Engine

1. Disconnect the negative battery cable.

2. Drain the cooling system and relieve the fuel system pressure.

3. Disconnect the fuel lines from the fuel supply manifold and remove the throttle body.

4. Disconnect the spark plug wires from the spark plugs. Remove the ignition wire/separator assembly from the rocker cover retaining studs.

5. Mark the position of the distributor housing and rotor and remove the distributor.

6. Remove the rocker arm covers. Loosen the No. 3 intake valve rocker arm retaining bolt to allow the rocker arm to be rotated to 1 side.

7. Remove the intake manifold assembly.

8. Loosen the rocker arm fulcrum retaining bolt enough to allow the rocker arm to be lifted off the pushrod and rotated to 1 side.

9. Remove the pushrod(s). If more than 1 is removed, identify each pushrods location. The pushrods should be installed in their original position during reassembly.

10. Remove the lifter(s) using a magnet.

NOTE: If the lifter(s) are stuck in the bore(s) due to excessive varnish or gum deposits, it may be necessary to use a claw-type tool to aid removal. Rotate the lifter back and forth to loosen it from the gum or varnish that may have formed on the lifter.

To install:

11. Clean all gasket mating surfaces. Place a rag in the lifter valley to catch any stray gasket material.

12. Lubricate each lifter and bore with heavy engine oil. Install the lifter in the bore, checking for free fit.

13. Install the intake manifold and new gaskets.

14. Dip each pushrod end in oil conditioner and install in it's original position.

15. For each valve, rotate the crankshaft until the lifter rests on the base circle of the camshaft lobe. Position the rocker arms over the pushrod and valve. Tighten the retaining bolt to 8 ft. lbs. (11 Nm) to initially seat the fulcrum into the cylinder head and onto the pushrod. Final tighten the bolt to 24 ft. lbs. (32 Nm).

16. Install the rocker arm covers and the distributor.

17. Install the throttle body and connect the fuel lines to the fuel supply manifold. Install the safety clips.

18. Install the coolant hoses. Fill and bleed the cooling system. Drain and change the crankcase oil.

19. Connect the air cleaner hoses to the throttle body and rocker cover.

20. Connect the negative battery cable, start the engine and check for leaks. Check the ignition timing.

3.8L Engine

1. Disconnect the negative battery cable. Disconnect the secondary ignition wires at the spark plugs.

2. Remove the plug wire routing clips from mounting studs on the rocker arm cover attaching bolts. Lay plug wires with routing clips toward the front of engine.

3. Remove the upper intake manifold, rocker arm covers and lower intake manifold.

4. Sufficiently loosen each rocker arm fulcrum attaching bolt to allow the rocker arm to be lifted off the pushrod and rotated to one side.

5. Remove the pushrods. The location of each pushrod should be identified and labeled. When engine is assembled, each rod should be installed in its original position.

6. If equipped with roller lifters, remove the 2 tappet guide plate retainers and 6 guide plates.

7. Remove the lifters using a magnet. The location of each lifters should be identified and labeled. When engine

is assembled, each lifter should be installed in its original position.

NOTE: If lifters are stuck in bores due to excessive varnish or gum deposits, it may be necessary to use a hydraulic lifter puller tool to aid removal. When using a remover tool, rotate lifter back and forth to loosen it from gum or varnish that may have formed on the lifter.

To install:

8. Lightly oil all bolt and stud threads before installation. Using solvent, clean the cylinder head and valve rocker arm cover sealing surfaces.

9. Lubricate each lifter and bore with oil conditioner or heavy engine oil.

10. Install each lifter in bore from which it was removed. If a new tappet(s) is being installed, check new lifter for a free fit in bore.

11. If equipped with roller lifters, align the flats on the sides of the lifters and install the 6 guide plates between the adjacent lifters. Make sure the word "up" and/or button is showing. Install the 3 guide plate retainers and tighten the 4 bolts to 6–10 ft. lbs. (8–14 Nm).

12. Dip each pushrod end in oil conditioner or heavy engine oil. Install pushrods in their orignial positions.

13. For each valve, rotate crankshaft until lifter rests onto heel (base circle) of camshaft lobe. Position rocker arms over pushrods and install the fulcrums. Initially tighten the fulcrum attaching bolts to 44 inch lbs. maximum.

14. Lubricate all rocker arm assemblies with suitable heavy engine oil.

15. Finally tighten the fulcrum bolts to 19–25 ft. lbs. (25–35 Nm). For the final tightening, the camshaft may be in any position.

NOTE: Fulcrums must be fully seated in the cylinder head and pushrods must be seated in rocker arm sockets prior to the final tightening.

16. Complete the installation of the lower intake manifold, valve rocker arm covers and the upper intake manifold by reversing the removal procedure.

17. Install the plug wire routing clips and connect wires to the spark plugs.

18. Start the engine and check for oil or coolant leaks.

Valve Lash

CHECKING

The valve stem-to-rocker arm clearance for all engines except the 3.0L

SHO should be within specification with the valve lifter completely collapsed. To determine the rocker arm to valve lifter clearance, make the following checks.

2.5L Engine

1. Set the No. 1 piston on TDC on the compression stroke. The timing marks on the camshaft and crankshaft gears will be together. Check the clearance in No. 1 intake, No. 1 exhaust, No. 2 intake and No. 3 exhaust valves.

2. Rotate the crankshaft 1 complete turn, 180 degrees for the camshaft gear. Check the clearance in No. 2 exhaust, No. 3 intake, No. 4 intake and No. 4 exhaust.

3. The clearance between the rocker arm and the valve stem tip should be 0.072–0.174 in. (1.80–4.34mm) with the lifter on the base circle of the cam.

3.0L and 3.8L Engines, Except SHO

1. Rotate the engine until the No. 1 cylinder is at TDC of its compression stroke and check the clearance between the rocker arm and the following valves.
   a. No. 1 intake and No. 1 exhaust
   b. No. 3 intake and No. 2 exhaust
   c. No. 6 intake and No. 4 exhaust
2. Rotate the crankshaft 360 degrees and check the clearance between the rocker arm and the following valves.
   a. No. 2 intake and No. 3 exhaust
   b. No. 4 intake and No. 5 exhaust
   c. No. 5 intake and No. 6 exhaust
3. The clearance should be 0.09–0.19 in. (2.25–4.79mm).

3.0L SHO Engine

1. Remove the valve cover.
2. Remove the intake manifold assembly.
3. Insert a feeler gauge under the cam lobe at a 90 degree angle to the camshaft. Clearance for the intake valves should be 0.006–0.010 in. (0.15–0.25mm). Clearance for the exhaust valves should be 0.010–0.014 in. (0.25–0.35mm).

NOTE: The cam lobes must be directed 90 degrees or more away from the valve lifters.

ADJUSTMENT

3.0L SHO Engine

1. Disconnect the negative battery cable.
2. Remove the valve cover.
3. Remove the intake manifold assembly.
4. Install lifter compressor tool

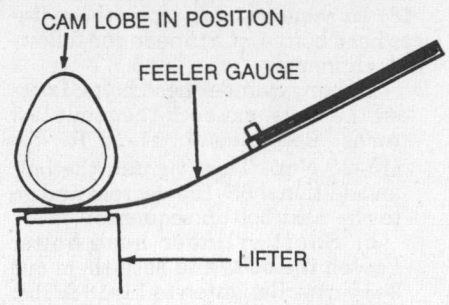

Checking valve clearance on the 3.0L SHO engine

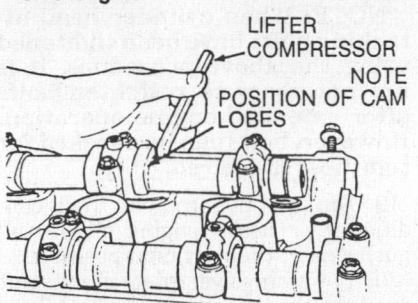

Valve lifter compressor tool—3.0L SHO engine

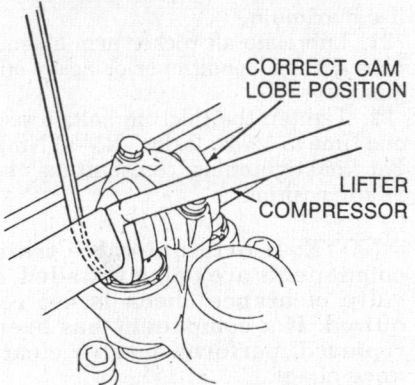

Valve lifter holding tool—3.0L SHO engine

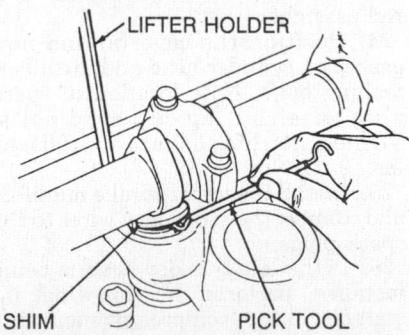

Removing the shim from the valve lifter —3.0L SHO engine

T89P-6500-A or equivalent, under the camshaft next to the lobe and rotate it downward to depress the valve lifter.

5. Install valve lifter holding tool T89P-6500-B or equivalent, and remove the compressor tool.

6. Using pick tool T71P-19703-C

or equivalent, lift the adjusting shim and remove the shim with a magnet.

7. Determine the size of the shim by the numbers on the bottom face of the shim or by measuring with a micrometer.

8. Install the replacement shim with the numbers down. Make sure the shim is properly seated.

9. Release the lifter holder tool by installing the compressor tool.

10. Repeat the procedure for each valve by rotating the crankshaft as necessary.

Rocker Arms

REMOVAL & INSTALLATION

2.5L Engine

1. Disconnect the negative battery cable.

2. Remove the oil fill cap and rocker arm filter and set aside. Disconnect the PCV hose and set it aside.

3. Disconnect the throttle linkage cable from the top of the rocker arm cover. Disconnect the speed control cable from the top of the rocker arm cover, if equipped.

4. Remove the rocker arm cover bolts. Remove the rocker cover and gasket from the engine.

5. Remove the rocker arm bolts, fulcrums, rocker arms and fulcrum washers. Keep all parts in order so they can be reinstalled to their original position.

To install:

6. Clean the cylinder head and rocker arm cover mating surfaces.

7. Coat the valve tips, rocker arm and fulcrum contact areas with Lubriplate® or equivalent.

8. For each valve, rotate the engine until the lifter is on the base circle of the cam (valve closed).

9. Install the rocker arm and components and tighten the rocker arm bolts in 2 steps, the first to 6–8 ft. lbs. (8–12 Nm) and the second torque to 20–26 ft. lbs. (28–35 Nm). Be sure the lifter is on the base circle of the cam for each rocker arm as it is installed. For the final tightening, the camshaft may be in any position. Check the valve lash.

10. Install a new rocker arm cover gasket, using suitable sealer, unless the cover is equipped with a moulded-in gasket, in which case no sealer should be used.

NOTE: If the moulded-in gasket is damaged by cuts and/or nicks less than ⅛ in. long in a maximum of 2 places, the damaged area can be filled in with RTV sealant. If the nicks or cuts are longer than ⅛ in. or there are more than 3 of

any size, the entire rocker arm cover should be replaced.

11. Install the rocker arm cover and tighten the bolts to 6–8 ft. lbs. (8–12 Nm).

12. Install the throttle cable(s), PCV hose and oil filler cap. Connect the negative battery cable.

3.0L Engine

1. Disconnect the negative battery cable. Disconnect and tag the spark plug wires.

2. Remove the ignition wire/separator assembly from the rocker arm attaching bolt studs. If the left rocker arm cover is being removed, remove the oil fill cap, disconnect the air cleaner closure system hose and remove the fuel injector harness from the inboard rocker arm cover studs.

3. If the right rocker arm cover is being removed, remove the PCV valve, loosen the lower EGR tube, if equipped, retaining nut and rotate the tube aside, remove the throttle body and move the fuel injection harness aside.

4. Remove the rocker arm cover attaching screws and the covers and gaskets from the vehicle.

5. Remove the rocker arm bolts, fulcrums, rocker arms and fulcrum washers. Keep all parts in order so they can be reinstalled to their original position.

To install:

6. Coat the valve tips, rocker arm and fulcrum contact areas with Lubriplate® or equivalent. Lightly oil all the bolt and stud threads before installation.

7. Rotate the engine until the lifter is on the base circle of the cam (valve closed).

8. Install the rocker arm and components and torque the rocker arm fulcrum bolts in 2 steps: the first to 8 ft. lbs. (11 Nm) and the final to 24 ft. lbs. (32 Nm). Be sure the lifter is on the base circle of the cam for each rocker arm as it is installed.

9. Clean the cylinder head and rocker arm cover sealing surfaces of all dirt and old sealer. If not equipped with integral gaskets, make sure all old gasket material is removed.

10. Apply a bead of silicone sealant at the cylinder head to intake manifold rail step. If not equipped with integral gaskets, install a new rocker arm cover gasket.

11. Install the rocker arm cover and the bolts and studs. Tighten to 9 ft. lbs. (12 Nm). On 1991–92 vehicles, tighten the cover in the proper sequence.

12. Install the remaining components in the reverse order of their removal.

3.8L Engine

1. Disconnect the negative battery cable.

2. Tag and disconnect the spark plug wires from the spark plugs.

3. If the left cover is being removed, remove the oil fill cap.

4. If the right cover is being removed, position the air cleaner assembly aside and remove the PCV valve.

5. Remove the rocker arm cover mounting bolts and remove the rocker arm cover.

6. Remove the rocker arm bolt, fulcrum and rocker arm. Keep all parts in order so they can be reinstalled in their original positions.

To install:

7. Coat the valve tips, rocker arm and fulcrum contact areas with Lubriplate® or equivalent. Install the rocker arm, fulcrum and rocker arm bolt.

8. Rotate the crankshaft until the lifter rests on the base circle of the camshaft lobe, then tighten the rocker arm bolt. Tighten in 2 steps, first to 62–132 inch lbs. (7–15 Nm) and finally to 19–25 ft. lbs. (25–35 Nm).

9. Clean the rocker arm cover and cylinder head mating surfaces of old gasket material and dirt.

10. Position a new gasket onto the cylinder head. install the rocker arm cover and the mounting bolts. Note the location of the spark plug wire routing clip stud bolts. Tighten the bolts to 80–106 inch lbs. (9–12 Nm).

11. Install the remaining components in the reverse order of their removal.

Intake Manifold

REMOVAL & INSTALLATION

2.5L Engine

1. Open and secure the hood.

2. Disconnect the negative battery cable. Properly relieve the fuel system pressure.

3. Drain the cooling system.

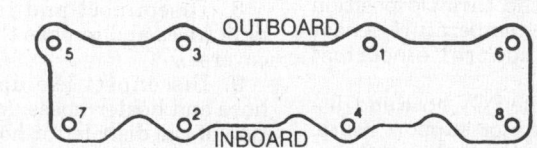

Rocker arm cover bolt torque sequence—1991–92 3.0L engine

4. Remove accelerator cable and the cruise control cable, if equipped.

5. Remove air cleaner assembly and heat stove tube at heat shield.

6. Remove required vacuum lines and electrical connections.

7. On 1988–90 vehicles, disconnect the thermactor check valve hose at the tube assembly and remove the bracket to EGR valve attaching nuts.

8. Disconnect the fuel supply and return lines.

9. On 1988–90 vehicles, disconnect the water inlet tube at the intake manifold. On 1991–92 vehicles, remove the exhaust manifold heat shroud assembly.

10. Disconnect EGR tube at EGR valve.

11. Remove the intake manifold retaining bolts. Remove the intake manifold. Remove the gasket and clean the gasket contact surfaces.

To install:

12. Install intake manifold with gasket and retaining bolts. Tighten the retaining bolts to 15–22 ft. lbs. (20–30 Nm) in the proper sequence.

13. On 1988–90 vehicles, connect water inlet tube at intake manifold, connect thermactor check valve hose at tube assembly and install bracket to EGR valve attaching nuts.

14. Connect EGR tube to EGR valve.

15. Connect the fuel supply and return lines.

16. Install vacuum lines and connect electrical connectors.

17. On 1991–92 vehicles, install the heat shroud.

18. Install air cleaner assembly and heat stove tube.

19. Install accelerator cable and cruise control cable, if equipped.

20. Connect negative battery cable and fill the cooling system.

21. Start engine and check for leaks.

3.0L Except SHO Engine

1. Disconnect the negative battery cable and drain the engine cooling system. Relieve the fuel system pressure.

2. Loosen the hose clamp attaching the flex hose to the throttle body. Remove the air cleaner flex hose.

3. Identify, tag and disconnect and all vacuum connections to the throttle body.

4. Loosen the lower EGR tube nut and rotate the tube away from the valve. Disconnect the throttle and TV cable from the throttle linkage.

5. Disconnect the throttle position sensor, air charge temperature sensor and idle speed control electrical connectors.

6. Disconnect the PCV hose and disconnect the alternator support brace. Remove the throttle body retaining bolts and the throttle body.

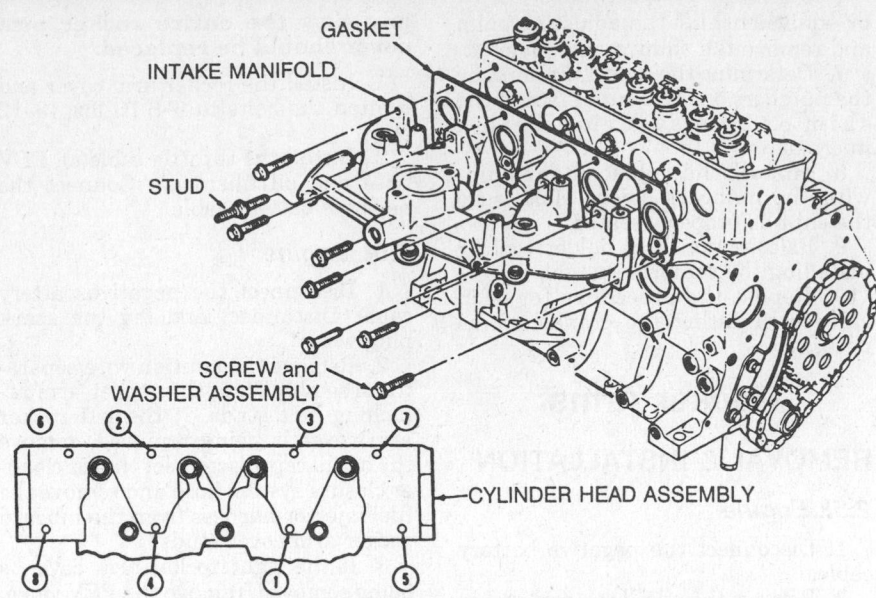

Intake manifold bolt torque sequence—2.5L engine

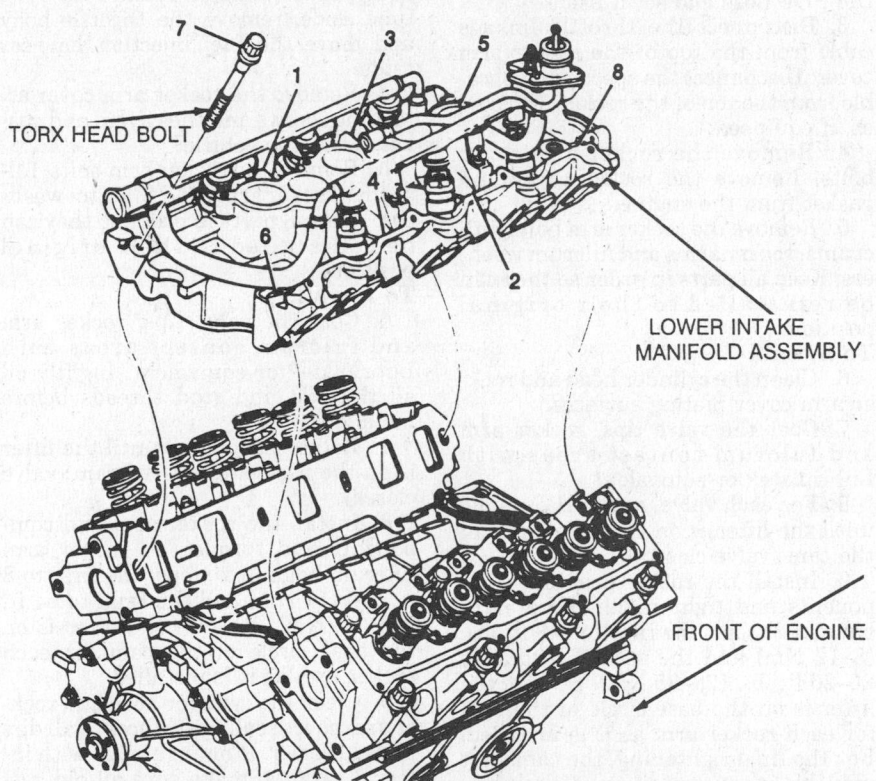

Intake manifold bolt torque sequence—3.0L engine

7. Disconnect the fuel lines. Remove the fuel injection wiring harness from the engine. Remove the fuel supply manifold and injectors.

8. Disconnect and tag the spark plug wires and remove the rocker arm covers.

9. Disconnect the upper radiator hose and heater hoses. Mark the position of the distributor housing and rotor and remove the distributor assembly.

10. Disconnect the engine coolant temperature sensor and temperature sending unit connector. Loosen the intake valve retaining bolt from cylinder No. 3 and rotate the rocker arm from the retainer and remove the pushrod.

11. Remove the intake manifold attaching bolts. Use a suitable prybar to loosen the intake manifold. Remove the manifold and old gaskets and seals.

To install:

NOTE: Lightly oil all the attaching bolts and stud threads before installation. When using a silcone rubber sealer, assembly must occur within 15 minutes after the sealer has been applied. After this time, the sealer may start to set-up and its sealing quality may be reduced. In high temperature/humidty conditions, the sealant will start to set up in approximately 5 minutes.

12. The intake manifold, cylinder head and cylinder block mating surfaces should be clean and free of old silicone rubber sealer. Use a suitable solvent to clean these surfaces.

13. Apply a suitable silicone rubber sealer to the intersection of the cylinder block end rails and cylinder heads.

14. Install the front and rear intake manifold end seals in place and secure. Install the intake manifold gaskets.

15. Carefully lower the intake manifold into position on the cylinder block and cylinder heads to prevent smearing the silicone sealer and causing gasket voids.

16. Install the retaining bolts and tighten the bolts, in sequence, to 11 ft. lbs. (15 Nm), then retorque to 21 ft. lbs. (28 Nm).

17. Install the fuel supply manifold and injectors. Apply lubricant to the injector holes in the intake manifold and fuel supply manifold prior to injector installation. Install the fuel supply manifold retaining bolts and tighten to 7 ft. lbs. (10 Nm).

18. Install the thermostat housing and a new gasket, if removed. Tighten the retaining bolts to 9 ft. lbs. (12 Nm).

19. Install the No. 3 cylinder intake valve pushrod. Apply Lubriplate® or equivalent to the pushrod and valve stem prior to installation. Position the lifter on the base circle of the camshaft and tighten the rocker arm bolt in 2 steps, first to 8 ft. lbs. (11 Nm) and then to 24 ft. lbs. (32 Nm).

20. Install the rocker arm covers. Install the fuel injector harness and attach to the injectors.

21. Install the throttle body with new gaskets.

22. Connect the PCV line at the PCV valve. Connect all necessary electrical connections and vacuum lines.

23. Connect the EGR tube and the fuel lines.

24. Install the distributor assembly, aligning the marks that were made during the removal procedure.

25. Install the coil and bracket. Install the upper radiator and heater hose.

26. Install and connect the air cleaner assembly and outlet tube. Fill the cooling system.

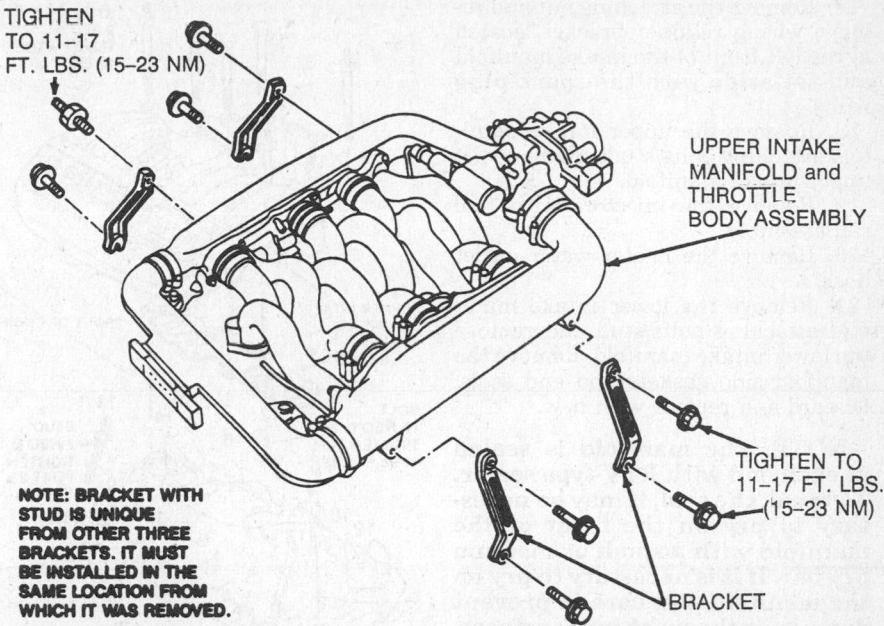

TIGHTEN TO 11–17 FT. LBS. (15–23 NM)

UPPER INTAKE MANIFOLD and THROTTLE BODY ASSEMBLY

TIGHTEN TO 11–17 FT. LBS. (15–23 NM)

NOTE: BRACKET WITH STUD IS UNIQUE FROM OTHER THREE BRACKETS. IT MUST BE INSTALLED IN THE SAME LOCATION FROM WHICH IT WAS REMOVED.

BRACKET

Intake manifold removal and Installation—3.0L SHO engine

27. Reconnect the negative battery cable, start the engine and check for coolant, fuel and oil leaks.

28. Check and if necessary, adjust the engine idle speed, transaxle throttle linkage and speed control.

3.0L SHO Engine

1. Disconnect the negative battery cable. Properly relieve the fuel system pressure.

2. Partially drain the engine cooling system.

3. Disconnect all electrical connectors and vacuum lines from the intake assembly.

4. Remove the air cleaner tube.

5. Disconnect the coolant lines and cables from the throttle body.

6. Remove the bolts retaining the upper intake brackets.

7. Loosen the lower bolts and remove the brackets.

8. Remove the bolts retaining the intake to the cylinder heads.

9. Remove the intake assembly and the gaskets.

10. Installation is the reverse of the removal procedure.

11. Lightly oil the attaching bolts and stud threads before installation.

NOTE: The intake gasket is reuseable.

12. Install the retaining bolts and tighten to 11–17 ft. lbs. (15–23 Nm).

3.8L Engine

1. Disconnect the negative battery cable. Drain the cooling system.

2. Properly relieve the fuel system pressure. Remove the air cleaner assembly including air intake duct and heat tube.

3. Disconnect the accelerator cable at throttle body assembly. Disconnect speed control cable, if equipped.

4. Disconnect the transaxle linkage at the upper intake manifold.

5. Remove the attaching bolts from accelerator cable mounting bracket and position cables aside.

6. Disconnect the thermactor air supply hose at the check valve.

7. Disconnect the flexible fuel lines from steel lines over rocker arm cover.

8. Disconnect the fuel lines at injector fuel rail assembly.

9. Disconnect the radiator hose at thermostat housing connection.

10. Disconnect the coolant bypass hose at manifold connection.

11. Disconnect the heater tube at the intake manifold. Remove the heater tube support bracket attaching nut. Remove the heater hose at rear of heater tube. Loosen hose clamp at heater elbow and remove heater tube with hose attached. Remove heater tube with fuel lines attached and set the assembly aside.

12. Disconnect vacuum lines at fuel rail assembly and intake manifold.

13. Identify, tag and disconnect all necessary electrical connectors.

14. If equipped with air conditioning, remove the air compressor support bracket.

15. Disconnect the PCV lines. One is located on upper intake manifold. The second is located at the left rocker cover and the lower intake stud.

16. Remove the throttle body assembly and remove the EGR valve assembly from the upper manifold.

17. Remove the attaching nut and remove wiring retainer bracket located at the left front of the intake manifold and set aside with the spark plug wires.

18. Remove the upper intake manifold attaching bolts/studs. Remove the upper intake manifold.

19. Remove the injectors with fuel rail assembly.

20. Remove the heater water outlet hose.

21. Remove the lower intake manifold attaching bolts/stud and remove the lower intake manifold. Remove the manifold side gaskets and end seals. Discard and replace with new.

NOTE: The manifold is sealed at each end with RTV-type sealer. To break the seal, it may be necessary to pry on the front of the manifold with a small or medium pry bar. If it is necessary to pry on the manifold, use care to prevent damage to the machined surfaces.

To install:

22. Lightly oil all attaching bolt and stud threads before installation.

NOTE: When using silicone rubber sealer, assembly must occur within 15 minutes after sealer application. After this time, the sealer may start to set-up and its sealing effectiveness may be reduced. The lower intake manifold, cylinder head and cylinder block mating surfaces should be clean and free of oil gasketing material. Use a suitable solvent to clean these surfaces.

23. Apply a bead of contact adhesive to each cylinder head mating surface. Press the new intake manifold gaskets into place, using locating pins as necessary to aid in assembly alignment.

24. Apply a ⅛ in. bead of silicone sealer at each corner where the cylinder head joins the cylinder block.

25. Install the front and rear intake manifold end seals.

26. Carefully lower the intake manifold into position on cylinder block and cylinder heads. Use locating pins as necessary to guide the manifold.

27. Install the retaining bolts and stud bolts in their original locations. Torque the retaining bolts in numerical sequence to the following specifications in 3 steps.

   a. Step 1 — 8 ft. lbs. (10 Nm)
   b. Step 2 — 15 ft. lbs. (20 Nm)
   c. Step 3 — 24 ft. lbs. (32 Nm)

28. Connect the rear PCV line to upper intake tube and install the front PCV tube so the mounting bracket sits over the lower intake stud.

29. Install the injectors and fuel rail assembly. Tighten the screws to 6–8 ft. lbs. (8–11 Nm).

Lower intake manifold bolt torque sequence—3.8L engine

30. Position the upper intake gasket and manifold on top of the lower intake. Use locating pins to secure position of gasket between manifolds.

31. Install bolts and studs in their original locations. Tighten the 4 center bolts, then tighten the end bolts. Repeat Step 27.

32. Install the EGR valve assembly on the manifold. Tighten the attaching bolt to 15–22 ft. lbs. (20–30 Nm).

33. Install the throttle body. Cross-tighten the retaining nuts to 15–22 ft. lbs. (20–30 Nm).

34. Connect the rear PCV line at PCV valve and upper intake manifold connections. If equipped with air conditioning, install the compressor support bracket. Tighten attaching fasteners to 15–22 ft. lbs. (20–30 Nm).

35. Connect all electrical connectors and vacuum hoses.

36. Connect the heater tube hose to the heater elbow. Position the heater tube support bracket and tighten attaching nut to 15–22 ft. lbs. (20–30 Nm). Connect the heater hose to the rear of the heater tube and tighten hose clamp.

37. Connect coolant bypass and upper radiator hoses and secure with hose clamps.

38. Connect the fuel line(s) at injector fuel rail assembly and connect the flexible fuel lines to steel lines.

39. Position the accelerator cable mounting bracket and install and tighten attaching bolts to 15–22 ft. lbs. (20–30 Nm).

40. Connect the speed control cable,

if equipped. Connect the transaxle linkage at upper intake manifold.

41. Fill the cooling system to the proper level.

42. Start the engine and check for coolant or fuel leaks.

43. Check and, if necessary, adjust engine idle speed, transaxle throttle linkage and speed control.

44. Install the air cleaner assembly and air intake duct.

Exhaust Manifold

REMOVAL & INSTALLATION

2.5L Engine

1. Open and secure the hood.
2. Disconnect the negative battery cable.
3. Drain the cooling system.
4. Remove the accelerator cable and the cruise control cable, if equipped.
5. Remove air cleaner assembly and heat stove tube at heat shield.
6. Identify, tag and disconnect all necessary vacuum lines and electrical connections.
7. Disconnect the exhaust pipe-to-exhaust manifold retaining nuts.
8. Remove exhaust manifold heat shroud. Disconnect the oxygen sensor wire at the connector.
9. Disconnect the fuel supply and return lines.
10. On 1988–90 vehicles, disconnect the thermactor check valve hose at tube assembly, remove bracket-to-EGR valve attaching nuts and disconnect water inlet tube at intake manifold.
11. Disconnect EGR tube from the EGR valve.
12. Remove the intake manifold.
13. Remove the exhaust manifold retaining nuts. Remove the exhaust manifold from the vehicle.

To install:

14. Position exhaust manifold to the cylinder head using guide bolts in holes 2 and 3.
15. Install the remaining attaching bolts.
16. Tighten the attaching bolts until snug, then remove guide bolts and install attaching bolts in holes 2 and 3.
17. Tighten all exhaust manifold bolts to specification using the following tightening procedure: torque retaining bolts, in sequence, to 5–7 ft. lbs. (7–10 Nm), then retorque, in sequence, to 20–30 ft. lbs. (27–41 Nm).
18. Install the intake manifold gasket and bolts. Tighten the intake manifold retaining bolts to 15–23 ft. lbs. (20–30 Nm).
19. On 1988–90 vehicles, connect the water inlet tube at intake manifold, connect thermactor check valve hose

at tube assembly and install bracket to EGR valve attaching nuts.

20. Connect the oxygen sensor wire.
21. Connect the EGR tube to EGR valve.
22. Install exhaust manifold studs.
23. Connect exhaust pipe to exhaust manifold.
24. Install vacuum lines and electrical connectors.
25. Install air cleaner assembly and heat stove tube.
26. Install accelerator cable and cruise control cable, if equipped.
27. Connect the negative battery cable.
28. Fill the cooling system.
29. Start engine and check for leaks.

3.0L Engine

LEFT SIDE

1. Disconnect the negative battery cable. Remove the oil level indicator support bracket.
2. On 1988–89 vehicles, remove the power steering pump pressure and return hoses.
3. Raise and safely support the vehicle. Remove the manifold-to-exhaust pipe retaining nuts.
4. Lower the vehicle. Remove the exhaust manifold attaching bolts and the manifold.
5. Installation is the reverse of the removal procedure. Clean all mating surfaces and lightly oil all bolt and stud threads prior to installation. Tighten the exhaust manifold retaining bolts to 19 ft. lbs. (25 Nm) and tighten the exhaust pipe attaching nuts to 30 ft. lbs. (41 Nm).

RIGHT SIDE

1. Disconnect the negative battery cable. Remove the heater hose support bracket.
2. Disconnect and plug the heater hoses. Remove the EGR tube from the exhaust manifold. Use a back-up wrench on the lower adapter.
3. Raise the vehicle and support it safely. Remove the manifold-to-exhaust pipe attaching nuts and remove the pipe from the manifold.
4. Lower the vehicle. Remove the exhaust manifold attaching bolts and remove the exhaust manifold from the vehicle.
5. Installation is the reverse of the removal procedure. Clean all mating surfaces and lightly oil all bolt and stud threads prior to installation. Tighten the exhaust manifold retaining bolts to 19 ft. lbs. (25 Nm) and tighten the exhaust pipe attaching nuts to 30 ft. lbs. (41 Nm). Tighten the EGR tube to the exhaust manifold to 31 ft. lbs. (42 Nm).

3.0L SHO Engine

LEFT SIDE

1. Disconnect the negative battery cable.
2. Remove the oil level indicator tube support bracket.
3. Remove the power steering pump pressure and return hoses.
4. Remove the manifold to exhaust pipe attaching nuts.
5. Remove the heat shield retaining bolts.
6. Remove the exhaust manifold retaining nuts and manifold.
7. Installation is the reverse of the removal procedure. Clean all mating surfaces and lightly oil all bolt and stud threads before installation. Tighten the manifold retaining nuts to 26–38 ft. lbs. (35–52 Nm), the heat shield retaining bolts to 11–17 ft. lbs. (15–23 Nm) and the exhaust pipe to manifold nuts to 16–24 ft. lbs. (21–32 Nm).

RIGHT SIDE

1. Disconnect the negative battery cable.
2. Remove the right cylinder head.
3. Remove the heat shield retaining bolts.
4. Remove the exhaust manifold retaining nuts and manifold.
5. Installation is the reverse of the removal procedure. Clean all mating surfaces and lightly oil all bolt and stud threads prior to installation. Tighten the manifold retaining nuts to 26–38 ft. lbs. (35–52 Nm). Tighten the heat shield retaining bolts to 11–17 ft. lbs. (15–23 Nm).

3.8L Engine

LEFT SIDE

1. Disconnect the negative battery cable. Remove the oil level dipstick tube support bracket.
2. Tag and disconnect the spark plug wires.
3. Raise the vehicle and support safely.
4. Remove the manifold-to-exhuast pipe attaching nuts.
5. Lower the vehicle.
6. Remove the exhaust manifold retaining bolts and remove the manifold from vehicle.

To install:

7. Lightly oil all bolt and stud threads before installation. Clean the mating surfaces on the exhaust manifold, cylinder head and exhaust pipe.
8. Position the exhaust manifold on the cylinder head. Install the lower front bolt hole on No. 5 cylinder as a pilot bolt.
9. Install the remaining manifold retaining bolts. Tighten the bolts 15–22 ft. lbs. (20–30 Nm).

NOTE: A slight warpage in the exhaust manifold may cause a misalignment between the bolt holes in the head and the manifold. Elongate the holes in the exhaust manifold as necessary to correct the misalignment, if apparent. Do not elongate the pilot hole, the lower front bolt on No. 5 cylinder.

10. Raise the vehicle and support safely.
11. Connect the exhaust pipe to the manifold. Tighten the attaching nuts to 16–24 ft. lbs. (21–32 Nm).
12. Lower the vehicle.
13. Connect the spark plug wires. Install dipstick tube support bracket attaching nut. Tighten to 15–22 ft. lbs. (20–30 Nm).
14. Start the engine and check for exhaust leaks.

RIGHT SIDE

1. Disconnect the negative battery cable. Remove the air cleaner outlet tube assembly. On Taurus and Sable, disconnect the thermactor hose from the downstream air tube check valve.
2. Tag and disconnect the coil secondary wire from coil and the wires from spark plugs. Remove the spark plugs.
3. Disconnect the EGR tube.
4. Raise the vehicle and support safely.
5. Remove the transaxle dipstick tube. On Taurus and Sable, remove the thermactor air tube by cutting the tube clamp at the underbody catalyst fitting with a suitable cutting tool.
6. Remove the manifold-to-exhaust pipe attaching nuts.
7. Lower the vehicle.
8. Remove the exhaust manifold retaining bolts and the exhaust manifold.

To install:

9. Lightly oil all bolt and stud threads before installation. Clean the mating surfaces on exhaust manifold cylinder head and exhaust pipe.
10. Position the inner half of the heat shroud, if equipped, and exhaust manifold on cylinder head. Start 2 attaching bolts to align the manifold with the cylinder head. Install the remaining retaining bolts and tighten to 15–22 ft. lbs. (20–30 Nm).
11. Raise the vehicle and support safely.
12. Connect the exhaust pipe to manifold. Tighten the attaching nuts to 16–24 ft. lbs. (21–32 Nm). On Taurus and Sable, position the thermactor hose to the downstream air tube and clamp tube to the underbody catalyst fitting.
13. Install the transaxle dipstick tube and lower vehicle.

14. Install the outer heat shroud and tighten the retaining screws to 50–70 inch lbs. (5–8 Nm).

15. Install the spark plugs. Connect the wires to their respective spark plugs and connect coil secondary wire to coil.

16. Connect the EGR tube. On Taurus and Sable, connect the thermactor hose to the downstream air tube and secure with clamp. Install the air cleaner outlet tube assembly.

17. Start the engine and check for exhaust leaks.

Timing Chain Front Cover

REMOVAL & INSTALLATION

2.5L Engine

1. Disconnect the negative battery cable.

2. Remove the engine and transaxle assembly from the vehicle and position in a suitable holding fixture. Remove the dipstick.

3. Remove accessory drive pulley, if equipped. Remove the crankshaft pulley attaching bolt and washer and remove pulley.

4. Remove front cover attaching bolts from front cover. Pry the top of the front cover away from the block.

5. Remove the oil pan.

6. Clean all dirt and old gasket material from all mating surfaces.

To install:

7. Clean and inspect all parts before installation. Clean the oil pan, cylinder block and front cover of gasket material and dirt.

8. Apply oil resistant sealer to a new front cover gasket and position gasket into front cover.

9. Remove the front cover oil seal and position the front cover on the engine.

10. Position front cover alignment tool T84P–6019–C or equivalent, onto the end of the crankshaft, ensuring the crank key is aligned with the keyway in the tool. Bolt the front cover to the engine and tighten the bolts to 6–8 ft. lbs. (10–12 Nm). Remove the front cover alignment tool.

11. Install a new front cover oil seal using a suitable seal installer. Lubricate the hub of the crankshaft pulley with polyethylene grease to prevent damage to the seal during installation and initial engine start. Install crankshaft pulley.

12. Install the oil pan.

13. Install the accessory drive pulley, if equipped.

14. Install crankshaft pulley attaching bolt and washer. Tighten to 140–170 ft. lbs. (190–230 Nm).

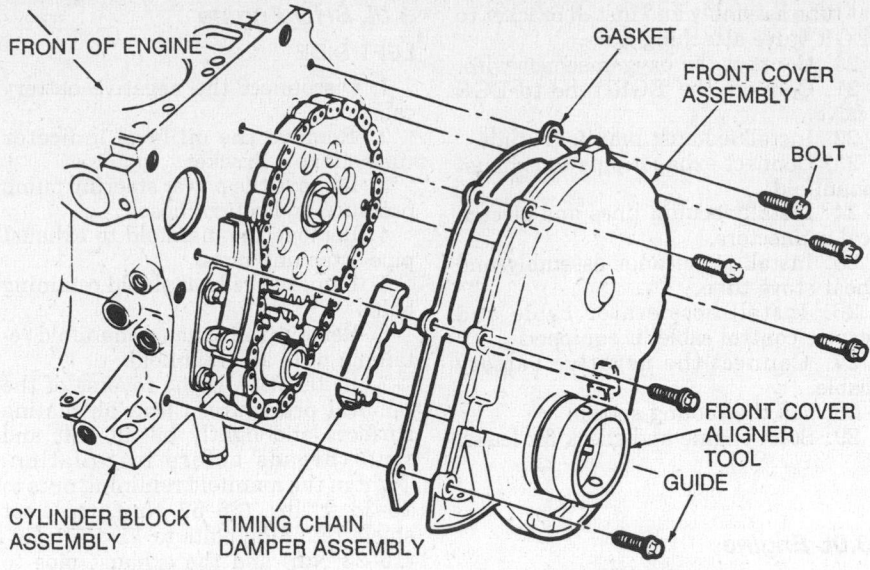

Timing chain front cover removal and installation—2.5L engine

15. Install the engine and transaxle assembly in the vehicle. Connect the negative battery cable.

3.0L Engine

1. Disconnect the negative battery cable.

2. Loosen the 4 water pump pulley bolts while the water pump drive belt is in place.

3. Loosen the alternator belt-adjuster jackscrew to provide enough slack in the alternator drive belt for removal.

4. Using a ½ in. drive breaker bar, rotate the automatic belt tensioner down and to the left to remove the water pump drive belt.

5. Drain the cooling system.

6. Remove the lower radiator hose and the heater hose from the water pump.

7. Remove the crankshaft pulley and damper.

8. Drain and remove the oil pan.

9. Remove the retaining bolts from the timing cover to the block and remove the timing cover.

NOTE: The timing cover and water pump may be removed as an assembly by not removing bolts 11–15.

To install:

10. Lightly oil all bolt and stud threads except those specifying special sealant.

11. Clean all old gasket material and sealer from the timing cover, oil pan and cylinder block.

12. Inspect the timing cover seal for wear or damage and replace if necessary.

13. Align a new timing cover gasket over the cylinder block dowels.

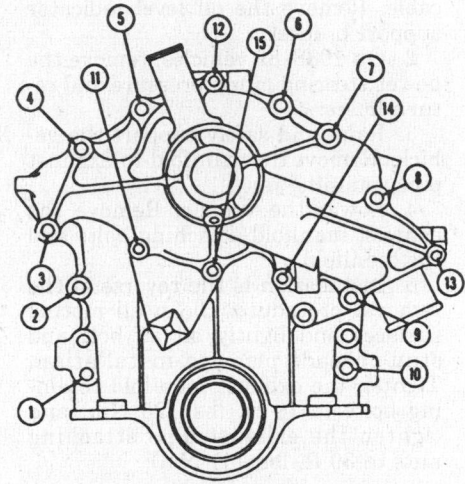

Water pump and front cover bolt identification—3.0L engine

14. Install the timing cover/water pump assembly onto the cylinder block with the water pump pulley loosely attached to the water pump hub.

15. Apply pipe sealant to bolt numbers 1, 2 and 3 and hand start them along with the rest of the cover retaining bolts. Tighten bolts 1–10 to 19 ft. lbs. (25 Nm) and 11–15 to 7 ft. lbs. (10 Nm).

16. Install the oil pan and tighten the retaining bolts to 9 ft. lbs. (12 Nm).

17. Hand tighten the water pump pulley retaining bolts.

18. Install the crankshaft damper and pulley. Torque the damper bolt to 107 ft. lbs. (145 Nm) and the 4 pulley bolts to 26 ft. lbs. (35 Nm).

19. Install the automatic belt tensioner. Tighten the 2 retaining nuts and bolt to 35 ft. lbs. (48 Nm).

20. Install the water pump and accessory drive belts. Torque the water

pump pulley retaining bolts to 16 ft. lbs. (21 Nm).

21. Install the lower radiator hose and the heater hose and tighten the clamps.

22. Fill the crankcase with the correct amount and type of engine oil. Connect the negative battery cable. Fill and bleed the cooling system.

23. Start the engine and check for coolant and oil leaks.

3.8L Engine

1. Disconnect the negative battery cable. Drain the cooling system and crankcase.

2. Remove the air cleaner assembly and air intake duct.

3. Loosen the accessory drive belt idler. Remove the drive belt and water pump pulley.

4. Remove the power steering pump mounting bracket attaching bolts. Leaving the hoses connected, place the pump/bracket assembly in a position that will prevent the loss of power steering fluid.

5. If equipped with air conditioning, remove the compressor front support bracket. Leave the compressor in place.

6. Disconnect coolant bypass and heater hoses at the water pump. Disconnect radiator upper hose at thermostat housing.

7. Disconnect the coil wire from distributor cap and remove cap with secondary wires attached. Remove the distributor retaining clamp and lift distributor out of the front cover.

8. Raise the vehicle and support safely.

9. Remove the crankshaft damper and pulley.

NOTE: If the crankshaft pulley and vibration damper have to be separated, mark the damper and pulley so they may be reassembled in the same relative position. This is important as the damper and pulley are initially balanced as a unit. If the crankshaft damper is being replaced, check if the original damper has balance pins installed. If so, new balance pins (E0SZ-6A328-A or equivalent) must be installed on the new damper in the same position as the original damper. The crankshaft pulley must also be installed in the original installation position.

10. Remove the oil filter, disconnect the radiator lower hose at the water pump and remove the oil pan.

11. Lower the vehicle.

12. Remove the front cover attaching bolts.

NOTE: Do not overlook the cover attaching bolt located behind the oil filter adapter. The front cover will break if pried upon if all attaching bolts are not removed.

13. Remove the ignition timing indicator.

14. Remove the front cover and water pump as an assembly. Remove the cover gasket and discard.

NOTE: The front cover houses the oil pump. If a new front cover is to be installed, remove the water pump and oil pump from the old front cover.

To install:

15. Lightly oil all bolt and stud threads before installation. Clean all gasket surfaces on the front cover, cylinder block and fuel pump. If reusing the front cover, replace crankshaft front oil seal.

16. If a new front cover is to be installed, complete the following:

    a. Install the oil pump gears.

    b. Clean the water pump gasket surface. Position a new water pump gasket on the front cover and install the water pump. Install the pump attaching bolts and tighten to 15–22 ft. lbs.

17. Install the distributor drive gear.

18. Lubricate the crankshaft front oil seal with clean engine oil.

19. Position a new cover gasket on the cylinder block and install the front cover/water pump assembly using dowels for proper alignment. A suitable contact adhesive is recommended to hold the gasket in position while the front cover is installed.

20. Position the ignition timing indicator.

21. Install the front cover attaching bolts. Apply Loctite® or equivalent, to the threads of the bolt installed below the oil filter housing prior to installation. This bolt is to be installed and tightened last. Tighten all bolts to 15–22 ft. lbs. (20–30 Nm).

22. Raise the vehicle and support safely.

23. Install the oil pan. Connect the radiator lower hose. Install a new oil filter.

24. Coat the crankshaft damper sealing surface with clean engine oil.

25. Position the crankshaft pulley key in the crankshaft keyway.

26. Install the damper with damper washer and attaching bolt. Tighten the bolt to 104–132 ft. lbs. (140–180 Nm).

27. Install the crankshaft pulley and tighten the attaching bolts 19–28 ft. lbs. (26–28 Nm).

28. Lower the vehicle.

29. Connect the coolant bypass hose.

30. Install the distributor with rotor pointing at No. 1 distributor cap tower. Install the distributor cap and coil wire.

31. Connect the radiator upper hose at thermostat housing.

32. Connect the heater hose.

33. If equipped with air conditioning, install compressor and mounting brackets.

34. Install the power steering pump and mounting brackets.

35. Position the accessory drive belt over the pulleys.

36. Install the water pump pulley. Position the accessory drive belt over water pump pulley and tighten the belt.

37. Connect battery ground cable. Fill the crankcase and cooling system to the proper level.

38. Start the engine and check for leaks.

39. Check the ignition timing and curb idle speed, adjust as required.

40. Install the air cleaner assembly and air intake duct.

Timing Chain and Sprockets

REMOVAL & INSTALLATION

2.5L Engine

1. Remove the engine and transaxle from the vehicle as an assembly and position in a suitable holding fixture. Remove the dipstick.

2. Remove accessory drive pulley, if equipped, Remove the crankshaft pulley attaching bolt and washer and remove pulley.

3. Remove front cover attaching bolts from front cover. Pry the top of the front cover away from the block.

4. Clean any gasket material from the surfaces.

5. Check timing chain and sprockets for excessive wear. If the timing chain and sprockets are worn, replace with new.

6. Check timing chain tensioner blade for wear depth. If the wear depth exceeds specification, replace tensioner.

7. Turn engine over until the timing marks are aligned. Remove camshaft sprocket attaching bolt and washer. Slide both sprockets and timing chain forward and remove as an assembly.

8. If equipped, check timing chain vibration damper for excessive wear. Replace if necessary; the damper is located inside the front cover.

9. Remove the oil pan.

To install:

10. Clean and inspect all parts before installation. Clean the oil pan, cylinder

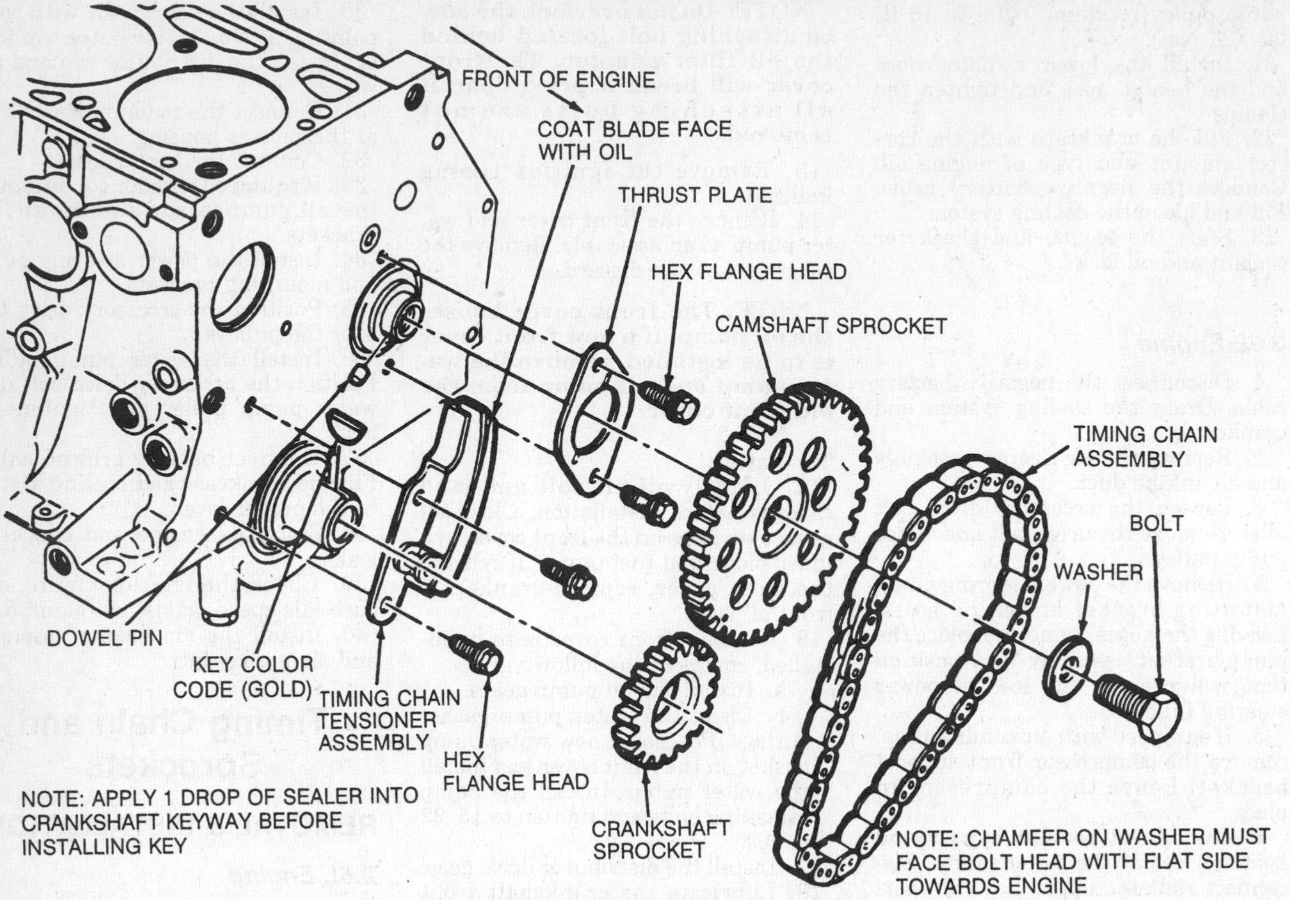

FRONT OF ENGINE

COAT BLADE FACE WITH OIL

THRUST PLATE

HEX FLANGE HEAD

CAMSHAFT SPROCKET

TIMING CHAIN ASSEMBLY

BOLT

WASHER

DOWEL PIN

KEY COLOR CODE (GOLD)

TIMING CHAIN TENSIONER ASSEMBLY

HEX FLANGE HEAD

CRANKSHAFT SPROCKET

NOTE: APPLY 1 DROP OF SEALER INTO CRANKSHAFT KEYWAY BEFORE INSTALLING KEY

NOTE: CHAMFER ON WASHER MUST FACE BOLT HEAD WITH FLAT SIDE TOWARDS ENGINE

Timing chain and sprockets installation—2.5L engine

block and front cover of gasket material and dirt.

11. Slide both sprockets and timing chain onto the camshaft and crankshaft with timing marks aligned. Install camshaft bolt and washer and tighten to 41–56 ft. lbs. (55–75 Nm). Oil timing chain, sprockets and tensioner after installation with clean engine oil.

12. Apply oil resistant sealer to a new front cover gasket and position gasket into front cover.

13. Remove the front cover oil seal and position the front cover on the engine.

14. Position front cover alignment tool T84P–6019–C or equivalent, onto the end of the crankshaft, ensuring the crank key is aligned with the keyway in the tool. Bolt the front cover to the engine and tighten the bolts to 6–8 ft. lbs. (8–12 Nm). Remove the front cover alignment tool.

15. Install a new front cover oil seal using a suitable seal installer. Lubricate the hub of the crankshaft pulley with polyethylene grease to prevent damage to the seal during installation and initial engine start. Install crankshaft pulley.

16. Install the oil pan.

17. Install the accessory drive pulley, if equipped.

18. Install crankshaft pulley attaching bolt and washer. Tighten to 140–170 ft. lbs. (190–230 Nm).

19. Remove engine from work stand and install in vehicle.

3.0L Engine

1. Disconnect the negative battery cable. Drain the cooling system and crankcase. Remove the crankshaft pulley and front cover assemblies.

2. Rotate the crankshaft until the No. 1 piston is at the TDC on its compression stroke and the timing marks are aligned.

3. Remove the camshaft sprocket attaching bolt and washer. Slide both sprockets and timing chain forward and remove as an assembly.

4. Check the timing chain and sprockets for excessive wear. Replace if necessary.

To install:

NOTE: Before installation, clean and inspect all parts. Clean the gasket material and dirt from the oil pan, cylinder block and front cover.

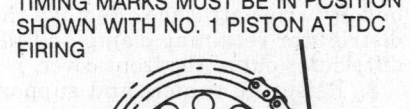
TIMING MARKS MUST BE IN POSITION SHOWN WITH NO. 1 PISTON AT TDC FIRING

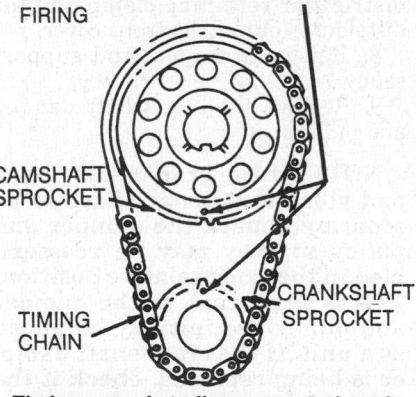
CAMSHAFT SPROCKET

TIMING CHAIN

CRANKSHAFT SPROCKET

Timing sprocket alignment—3.0L and 3.8L engines

5. Slide both sprockets and timing chain onto the camshaft and crankshaft with the timing marks aligned. Install the camshaft bolt and washer and torque to 46 ft. lbs. (63 Nm). Apply clean engine oil to the timing chain and sprockets after installation.

NOTE: The camshaft bolt has a drilled oil passage in it for timing chain lubrication. If the bolt is damaged, do not replace it with a standard bolt.

6. Install the timing cover and the crankshaft pulley and damper. Tighten the crankshaft danmper bolt to 107 ft. lbs. (145 Nm) and the pulley bolts to 26 ft. lbs. (35 Nm).

7. Fill the crankcase with the proper type and quantity of oil and the cooling system with coolant. Connect the negative battery cable.

3.8L Engine

1. Disconnect the negative battery cable. Drain the cooling system and crankcase.

2. Remove the air cleaner assembly and air intake duct.

3. Loosen the accessory drive belt idler. Remove the drive belt and water pump pulley.

4. Remove the power steering pump mounting bracket attaching bolts. Leaving the hoses connected, place the pump/bracket assembly in a position that will prevent the loss of power steering fluid.

5. If equipped with air conditioning, remove the compressor front support bracket. Leave the compressor in place.

6. Disconnect coolant bypass and heater hoses at the water pump. Disconnect radiator upper hose at thermostat housing.

7. Disconnect the coil wire from distributor cap and remove cap with secondary wires attached. Remove the distributor retaining clamp and lift distributor out of the front cover.

8. Raise the vehicle and support safely.

9. Remove the crankshaft damper and pulley.

NOTE: If the crankshaft pulley and vibration damper have to be separated, mark the damper and pulley so they may be reassembled in the same relative position. This is important as the damper and pulley are initially balanced as a unit. If the crankshaft damper is being replaced, check if the original damper has balance pins installed. If so, new balance pins (E0SZ–6A328–A or equivalent) must be installed on the new damper in the same position as the original damper. The crankshaft pulley must also be installed in original installation position.

10. Remove the oil filter, disconnect the radiator lower hose at the water pump and remove the oil pan.

11. Lower the vehicle.

12. Remove the front cover attaching bolts.

NOTE: Do not overlook the cover attaching bolt located behind the oil filter adapter. The front cover will break if pried upon if all attaching bolts are not removed.

13. Remove the ignition timing indicator.

14. Remove the front cover and water pump as an assembly. Remove the cover gasket and discard.

15. Remove the camshaft bolt and washer from end of the camshaft. Remove the distributor drive gear.

16. Remove the camshaft sprocket, crankshaft sprocket and timing chain.

17. Remove the chain tensioner assembly from the front of the cylinder block. This is accomplished by pulling back on the ratcheting mechanism and installing a pin through the hole in the bracket to relieve tension.

NOTE: The front cover houses the oil pump. If a new front cover is to be installed, remove the water pump and oil pump from the old front cover.

To install:

18. Lightly oil all bolt and stud threads before installation. Clean all gasket surfaces on the front cover, cylinder block and fuel pump. If reusing the front cover, replace crankshaft front oil seal.

19. If a new front cover is to be installed, complete the following:

a. Install the oil pump gears.

b. Clean the water pump gasket surface. Position a new water pump gasket on the front cover and install water pump. Install the pump attaching bolts and tighten to 15–22 ft. lbs.

20. Rotate the crankshaft as necessary to position piston No. 1 at TDC and the crankshaft keyway at the 12 o' clock position.

21. Install the tensioner assembly. Make sure the ratcheting mechanism is in the retracted position with the pin pointing outward from the hole in the bracket assembly. Tighten the retaining bolts to 6–10 ft. lbs. (8–14 Nm).

22. Lubricate timing chain with clean engine oil. Install the camshaft sprocket, crankshaft sprocket and timing chain.

23. Remove the pin from the tensioner assembly. Make certain the timing marks are positioned across from each other.

24. Install the distributor drive gear.

25. Install the washer and bolt at end of camshaft and tighten to 30–37 ft. lbs. (40–50 Nm).

26. Lubricate the crankshaft front oil seal with clean engine oil.

27. Position a new cover gasket on the cylinder block and install the front cover/water pump assembly using dowels for proper alignment. A suitable contact adhesive is recommended

to hold the gasket in position while the front cover is installed.

28. Position the ignition timing indicator.

29. Install the front cover attaching bolts. Apply Loctite® or equivalent, to the threads of the bolt installed below the oil filter housing prior to installation. This bolt is to be installed and tightened last. Tighten all bolts to 15–22 ft. lbs. (20–30 Nm).

30. Raise the vehicle and support safely.

31. Install the oil pan. Connect the radiator lower hose. Install a new oil filter.

32. Coat the crankshaft damper sealing surface with clean engine oil.

33. Position the crankshaft pulley key in the crankshaft keyway.

34. Install the damper with damper washer and attaching bolt. Tighten the bolt to 104–132 ft. lbs. (140–180 Nm).

35. Install the crankshaft pulley and tighten the attaching bolts 19–28 ft. lbs. (26–28 Nm).

36. Lower the vehicle.

37. Connect the coolant bypass hose.

38. Install the distributor with rotor pointing at No. 1 distributor cap tower. Install the distributor cap and coil wire.

39. Connect the radiator upper hose at thermostat housing.

40. Connect the heater hose.

41. If equipped with air conditioning, install compressor and mounting brackets.

42. Install the power steering pump and mounting brackets.

43. Position the accessory drive belt over the pulleys.

44. Install the water pump pulley. Position the accessory drive belt over water pump pulley and tighten the belt.

45. Connect battery ground cable. Fill the crankcase and cooling system to the proper level.

46. Start the engine and check for leaks.

47. Check the ignition timing and curb idle speed, adjust as required.

48. Install the air cleaner assembly and air intake duct.

Timing Belt Front Cover

REMOVAL & INSTALLATION

3.0L SHO Engine

NOTE: The front cover on the 3.0L SHO engine is made up of 3 sections.

1. Disconnect the battery cables and remove the battery. Remove the right engine roll damper.

2. Disconnect the wiring to the ignition module. Remove the intake manifold crossover tube bolts, loosen the crossover tube clamps and remove the crossover tube.

3. Loosen the alternator/air conditioner belt tensioner pulley and relieve the tension on the belt by backing out the adjustment screw. Remove the belt.

4. Loosen the water pump/power steering belt tensioner pulley and relieve the tension on the belt by backing out the adjustment screw. Remove the belt.

5. Remove the alternator/air conditioner belt tensioner pulley and bracket assembly. Remove the water pump/power steering belt tensioner pulley only.

6. Remove the upper timing belt cover.

7. Disconnect the crankshaft sensor connectors.

8. Raise and safely support the vehicle. Remove the right front wheel and tire assembly.

9. Loosen the fender splash shield and move aside. Remove the crankshaft damper using a suitable puller.

10. Remove the center and lower timing belt covers.

11. Installation is the reverse of the removal procedure. Tighten the timing belt cover retaining bolts to 60–90 inch lbs. (7–11 Nm) and the crankshaft damper bolt to 113–126 ft. lbs. (152–172 Nm).

OIL SEAL REPLACEMENT

3.0L SHO Engine

1. Loosen the accessory drive belts.
2. Raise the vehicle and support it safely.
3. Remove the right front wheel.
4. Remove the damper attaching bolt and the accessory drive belts from the crankshaft damper.
5. Using a suitable puller, remove the crankshaft damper from the crankshaft.
6. Remove the timing belt.
7. Remove the crankshaft timing gear using a suitable puller.

NOTE: Be careful not to damage the crankshaft sensor or shutter.

8. Remove the crankshaft front oil seal using a suitable puller.
To install:
9. Inspect the front cover and shaft seal surface of the crankshaft damper for damage, nicks, burrs or other roughness which may cause the new seal to fail. Repair or replace as necessary.
10. Using suitable tools, install a new

crankshaft front oil seal and the crankshaft timing gear.

11. Install the timing belt.

12. Install the crankshaft damper using a suitable tool. Tighten the damper attaching bolt to 113–126 ft. lbs. (152–172 Nm).

13. Install the accessory drive belts.

14. Lower the vehicle.

15. Start the engine and check for oil leaks.

Timing Belt and Tensioner

REMOVAL & INSTALLATION

3.0L SHO Engine

1. Disconnect the battery cables.
2. Remove the battery.
3. Remove the ight engine roll damper.
4. Disconnect the wiring to the ignition module.
5. Remove the intake manifold crossover tube bolts. Loosen the intake manifold tube hose clamps. Remove the intake manifold crossover tube.
6. Loosen the alternator/air conditioning belt tensioner pulley and relieve the tension on the belt by backing out the adjustment screw. Remove the alternator/air conditioning belt.
7. Loosen the water pump/power steering belt tensioner pulley and relieve the tension on the belt by backing out the adjustment screw. Remove the water pump/power steering belt.
8. Remove the alternator/air conditioning belt tensioner pulley and bracket assembly.
9. Remove the water pump/power steering belt tensioner pulley only.
10. Remove the upper timing belt cover.
11. Disconnect the crankshaft sensor connectors.
12. Place the gear selector in **N**.
13. Set the engine to the TDC on No. 1 cylinder position. Make sure the white mark on the crankshaft damper aligns with the **0** degree index mark on the lower timing belt cover and that the marks on the intake camshaft pulleys align with the index marks on the metal timing belt cover.
14. Raise the vehicle and support safely.
15. Remove the right front wheel and tire assembly.
16. Loosen the fender splash shield and place it aside.
17. Using a suitable puller, remove the crankshaft damper.
18. Remove the lower timing belt cover.
19. Remove the center timing belt cover and disconnect the crankshaft

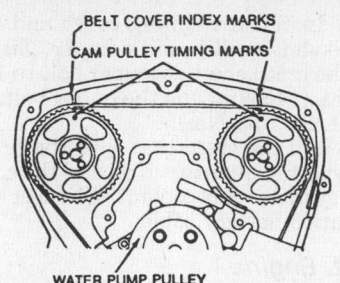

Camshaft pulley to belt cover index marks—3.0L SHO engine

Crankshaft damper to lower timing cover index mark alignment—3.0L SHO engine

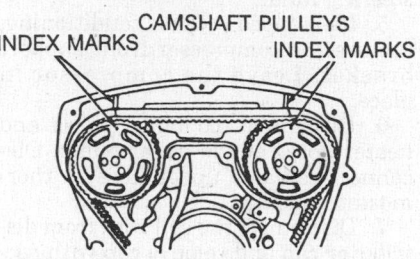

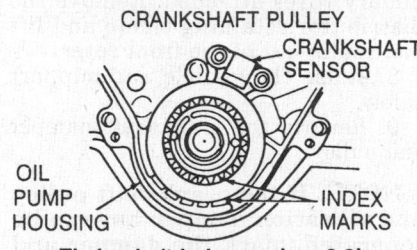

Timing belt index marks—3.0L SHO engine

sensor wire and grommet from the slot in the cover and the stud on the water pump.

20. Loosen the timing belt tensioner, rotate the pulley 180 degrees clockwise and tighten the tensioner nut to hold the pulley in the unload position.

21. Lower the vehicle and remove the timing belt.
To install:

NOTE: Before installing the timing belt, inspect it for cracks, wear or other damage and replace, if necessary. Do not allow the timing belt to come into contact with gasoline, oil, water, coolant or steam. Do not twist or turn the belt inside out.

22. Make sure the engine is at TDC on the No. 1 cylinder. Check that the camshaft pulley marks line up with the index marks on the upper steel belt

cover and that the crankshaft pulley aligns with the idex mark on the oil pump housing.

NOTE: The timing belt has 3 yellow lines. Each line aligns with the index marks.

23. Install the timing belt over the crankshaft and camshaft pulleys. The lettering on the belt **KOA** should be readable from the rear of the engine; top of the lettering to the front of the engine. Make sure the yellow lines are aligned with the index marks on the pulleys.

24. Release the tensioner locknut and leave the nut loose.

25. Raise the vehicle and support safely.

26. Install the center timing belt cover. Make sure the crankshaft sensor wiring and grommet are installed and routed properly. Tighten the mounting bolts to 60–90 inch lbs. (7–11 Nm).

27. Install the lower timing belt cover. Tighten the bolts to 60–90 inch lbs. (7–11 Nm).

28. Using a suitable tool, install the crankshaft damper. Tighten the damper attaching bolt to 113–126 ft. lbs. (152–172 Nm).

29. Rotate the crankshaft 2 revolutions in the clockwise direction until the yellow mark on the damper aligns with the 0 degree mark on the lower timing belt cover.

30. Remove the plastic door in the lower timing belt cover. Tighten the tensioner locknut to 25–37 ft. lbs. (33–51 Nm) and install the plastic door.

31. Rotate the crankshaft 60 degrees more in the clockwise direction until the white mark on the damper aligns with the 0 degree mark on the lower timing belt cover.

32. Lower the vehicle.

33. Make sure the index marks on the camshaft pulleys align with the marks on the rear metal timing belt cover.

34. Route the crankshaft sensor wiring and connect with the engine wiring harness.

35. Install the upper timing belt cover. Tighten the bolts to 60–90 inch lbs. (7–11 Nm).

36. Install the water pump/power steering tensioner pulley. Tighten the nut to 11–17 ft. lbs. (15–23 Nm).

37. Install the alternator/air conditioning tensioner pulley and bracket assembly. Tighten the bolts to 11–17 ft. lbs. (15–23 Nm).

38. Install the water pump/power steering and alternator/air conditioning belts and set the tension. Tighten the idler pulley nut to 25–36 ft. lbs. (34–50 Nm).

39. Install the intake manifold crossover tube. Tighten the bolts to 11–17 ft. lbs. (15–23 Nm).

40. Install the engine roll damper and the battery.

41. Connect the battery cables.

42. Raise the vehicle and support safely.

43. Install the splash shield and the right front wheel and tire assembly.

44. Lower the vehicle.

Timing Sprockets

REMOVAL & INSTALLATION

3.0L SHO Engine

1. Disconnect the negative battery cable.

2. Remove the timing belt.

3. Remove the camshaft and crankshaft timing belt sprockets.

4. Install in the reverse order of removal. Tighten the camshaft timing belt sprocket bolts to 15–18 ft. lbs. (21–25 Nm) and the crankshaft pulley bolt to 113–126 ft. lbs. (152–172 Nm).

Camshaft

REMOVAL & INSTALLATION

2.5L Engine

1. Drain the cooling system and the crankcase. Relieve the fuel system pressure.

2. Remove the engine from the vehicle and position in a suitable holding fixture. Remove the engine oil dipstick.

3. Remove necessary drive belts and pulleys.

4. Remove cylinder head.

5. Using a magnet, remove the hydraulic lifters and label them so they can be installed in their original positions. If the tappets are stuck in the bores by excessive varnish, etc., use a suitable claw-type puller to remove the tappets.

6. Loosen and remove the drive belt, fan and pulley and crankshaft pulley.

7. Remove the oil pan.

8. Remove the cylinder front cover and gasket.

9. Check the camshaft endplay as follows:

   a. Push the camshaft toward the rear of the engine and install a dial indicator tool, so the indicator point is on the camshaft sprocket attaching screw.

   b. Zero the dial indicator. Position a small prybar or equivalent, between the camshaft sprocket or gear and block.

   c. Pull the camshaft forward and release it. Compare the dial indicator reading with the camshaft endplay specification of 0.009 in.

   d. If the camshaft endplay is over

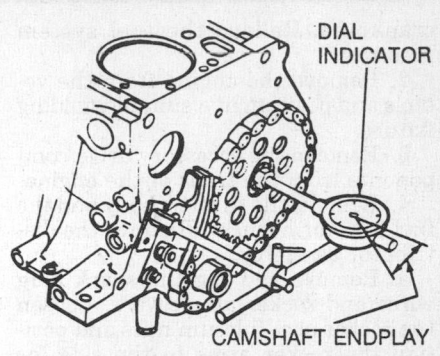

DIAL INDICATOR

CAMSHAFT ENDPLAY

Checking camshaft endplay

the amount specified, replace the thrust plate.

10. Remove the timing chain, sprockets and timing chain tensioner.

11. Remove camshaft thrust plate. Carefully remove the camshaft by pulling it toward the front of the engine. Use caution to avoid damaging bearings, journals and lobes.

To install:

12. Clean and inspect all parts before installation.

13. Lubricate camshaft lobes and journals with heavy engine oil. Carefully slide the camshaft through the bearings in the cylinder block.

14. Install the thrust plate. Tighten attaching bolts to 6–9 ft. lbs. (8–12 Nm).

15. Install the timing chain, sprockets and timing chain tensioner.

16. Install the cylinder front cover and crankshaft pulley.

17. Clean the oil pump inlet tube screen, oil pan and cylinder block gasket surfaces. Prime oil pump by filling the inlet opening with oil and rotate the pump shaft until oil emerges from the outlet tube. Install oil pump, oil pump inlet tube screen and oil pan.

18. Install the accessory drive belts and pulleys.

19. Lubricate the lifters and lifter bores with heavy engine oil. Install tappets into their original bores.

20. Install cylinder head.

21. Install the engine assembly.

22. Position No. 1 piston at TDC after the compression stroke. Position distributor in the block with the rotor at the No. 1 firing position. Install distributor retaining clamp.

23. Connect engine temperature sending unit wire. Connect coil primary wire. Install distributor cap. Connect spark plug wires and the coil high tension lead.

24. Fill the cooling system and crankcase to the proper levels. Connect the negative battery cable.

25. Start the engine. Check and adjust ignition timing. Check for leaks.

3.0L Except SHO Engine

1. Drain the cooling system and

crankcase. Relieve the fuel system pressure.

2. Remove the engine from the vehicle and position in a suitable holding fixture.

3. Remove the accessory drive components from the front of the engine.

4. Remove the throttle body and the fuel injector harness. Remove the distributor assembly.

5. Remove and tag the spark plug wires and rocker arm covers. Loosen the rocker arm fulcrum nuts and position the rocker arms to the side for easy access to the pushrods. Remove the pushrods and label so they may be installed in their original positions.

6. Remove the intake manifold.

7. Using a suitable magnet or lifter removal tool, remove the hydraulic lifters and keep them in order so they can be installed in their original positions. If the lifters are stuck in the bores by excessive varnish use a hydraulic lifter puller to remove the lifters.

8. Remove the crankshaft pulley and damper using a suitable removal tool. Remove the oil pan assembly.

9. Remove the front cover assembly. Align the timing marks on the camshaft and crankshaft gears. Check the camshaft endplay as follows:

a. Push the camshaft toward the rear of the engine and install a dial indicator tool, so the indicator point is on the camshaft sprocket attaching screw.

b. Zero the dial indicator. Position a small prybar or equivalent, between the camshaft sprocket or gear and block.

c. Pull the camshaft forward and release it. Compare the dial indicator reading with the camshaft endplay service limit specification of 0.005 in.

d. If the camshaft endplay is over the amount specified, replace the thrust plate.

10. Remove the timing chain and sprockets.

11. Remove the camshaft thrust plate. Carefully remove the camshaft by pulling it toward the front of the engine. Remove it slowly to avoid damaging the bearings, journals and lobes.

To install:

12. Clean and inspect all parts before installation.

13. Lubricate camshaft lobes and journals with heavy engine oil. Carefully insert the camshaft through the bearings in the cylinder block.

14. Install the thrust plate. Tighten the retaining bolts to 7 ft. lbs. (10 Nm).

15. Install the timing chain and sprockets. Check the camshaft sprocket bolt for blockage of drilled oil passages prior to installation and clean, if necessary.

16. Install the front timing cover and crankshaft damper and pulley.

17. Lubricate the lifters and lifter bores with a heavy engine oil. Install the lifters into their original bores.

18. Install the intake manifold assembly.

19. Lubricate the pushrods and rocker arms with heavy engine oil. Install the pushrods and rocker arms into their original positions. Rotate the crankshaft to set each lifter on its base circle, then tighten the rocker arm bolt. Tighten the rocker arm bolts to 24 ft. lbs. (32 Nm).

20. Install the oil pan and the rocker covers.

21. Install the fuel injector harness and the throttle body. Install the distributor and connect the spark plug wires to the spark plugs.

22. Install the accessory drive components and install the engine assembly.

23. Connect the negative battery cable. Start the engine and check for leaks. Check and adjust the ignition timing.

3.0L SHO Engine

1. Disconnect the negative battery cable. Properly relieve the fuel system pressure.

2. Set the engine on TDC on No. 1 cylinder.

3. Remove the intake manifold assembly.

4. Remove the timing cover and belt.

5. Remove the cylinder head covers.

6. Remove the camshaft pulleys, noting the location of the dowel pins.

7. Remove the upper rear timing belt cover.

8. Uniformly loosen the camshaft bearing caps.

NOTE: If the camshaft bearing caps are not uniformly loosened, camshaft damage may result.

9. Remove the bearing caps and note their positions for installation.

10. Remove the camshaft chain tensioner mounting bolts.

11. Remove the camshafts together with the chain and tensioner.

12. Remove and discard the camshaft oil seal.

13. Remove the chain sprocket from the camshaft.

To install:

14. Align the timing marks on the chain sprockets with the camshaft and install the sprockets. Tighten the bolts to 10–13 ft. lbs. (14–18 Nm).

15. Install the chain over the camshaft sprockets. Align the white painted link with the timing mark on the sprocket.

16. Rotate the camshafts 60 degrees

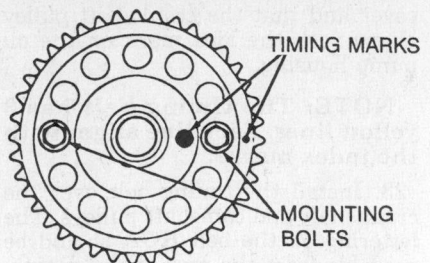

Timing chain sprocket and camshaft alignment—3.0L SHO engine

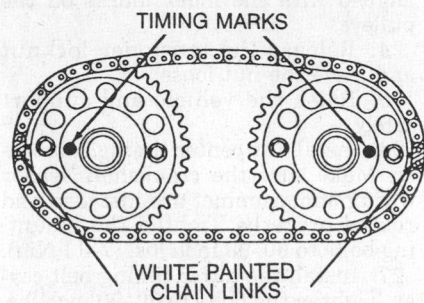

Aligning the timing chain with the timing marks—3.0L SHO engine

counterclockwise. Set the chain tensioner between the sprockets and install the camshafts on the cylinder head.

NOTE: The left and right chain tensioners are not interchangeable.

17. Apply a thin coat of engine oil to the camshaft journals and install bearing caps No. 2 through No. 5 and loosely install the bolts.

NOTE: The arrows on the bearing caps point to the front of the engine when installed.

18. Apply silicone sealer to outer diameter of the new camshaft seal and the seal seating area on the cylinder head. Install the camshaft seal.

19. Apply silicone sealer to the No. 1 bearing cap and install the bearing cap.

20. Tighten the bearing caps in sequence using a 2 step method. Tighten to 12–16 ft. lbs. (16–22 Nm).

NOTE: For left camshaft installation, apply pressure to the chain tensioner to avoid damage to the bearing caps.

21. Install the chain tensioner and tighten the bolts to 11–14 ft. lbs. (15–19 Nm). Rotate the camshafts 60 degrees clockwise and check for proper alignment of the timing marks. Marks on the camshaft sprockets should align with the cylinder head cover mating surface.

22. Install the camshaft positioning tool T89P-6256-C or equivalent, on

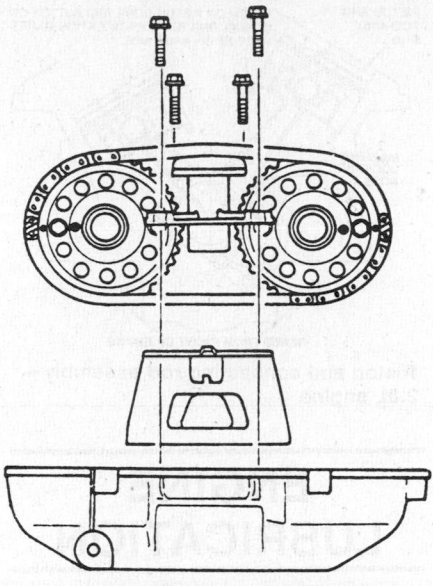

LEFT SIDE CHAIN TENSIONER

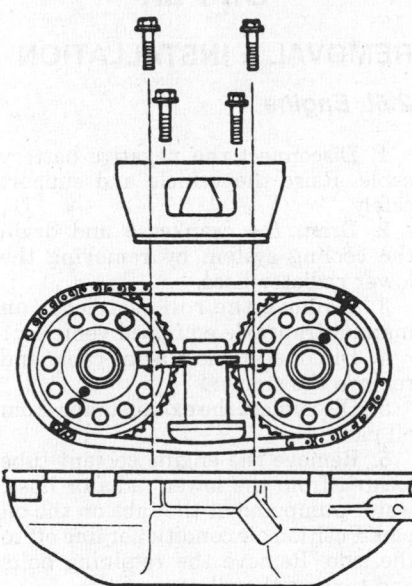

RIGHT SIDE CHAIN TENSIONER

Chain tensioner installation—3.0L SHO engine

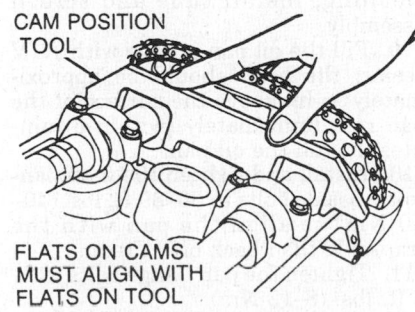

Camshaft positioning tool—3.0L SHO engine

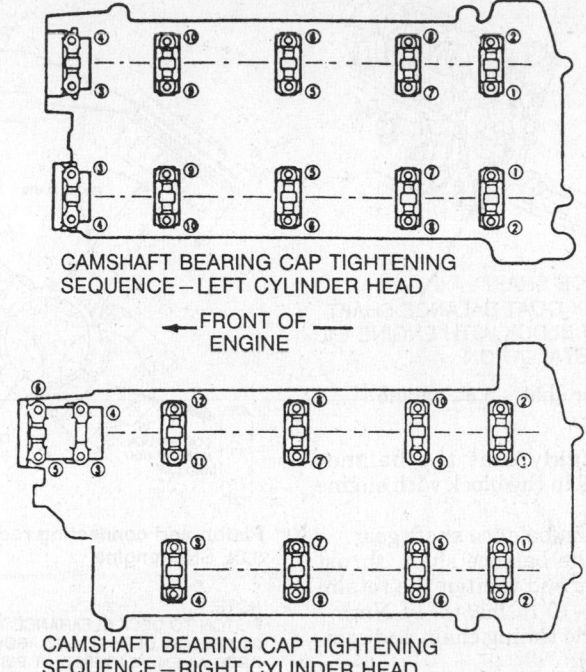

CAMSHAFT BEARING CAP TIGHTENING
SEQUENCE—LEFT CYLINDER HEAD

←FRONT OF
ENGINE

CAMSHAFT BEARING CAP TIGHTENING
SEQUENCE—RIGHT CYLINDER HEAD

Camshaft bearing cap tightening sequence—3.0L SHO engine

the camshafts to check for correct positioning. The flats on the tool should align with the flats on the camshaft. If the tool does not fit and/or timing marks will not line up, repeat the procedure from Step 14.

23. Install the timing belt rear cover and tighten the bolts to 70 inch lbs. (8.8 Nm).

24. Install the camshaft pulleys and tighten the bolts to 15–18 ft. lbs. (21–25 Nm).

25. Install the timing belt and cover.

26. Install the cylinder head covers and tighten the bolts to 8–11 ft. lbs. (10–16 Nm).

27. Install the intake manifold assembly.

3.8L Engine

1. Disconnect the negative battery cable.

2. Properly relieve the fuel system pressure.

3. Drain the cooling system and crankcase.

4. Remove the engine from the vehicle and position in a suitable holding fixture.

5. Remove the intake manifold.

6. Remove the rocker arm covers, rocker arms, pushrods and lifters.

7. Remove the oil pan.

8. Remove the front cover and timing chain.

9. Remove the thrust plate. Remove the camshaft through the front of the engine, being careful not to damage bearing surfaces.

To install:

10. Lightly oil all attaching bolts and stud threads before installation. Lubricate the cam lobes, thrust plate and bearing surfaces with a suitable heavy engine oil.

11. Install the camshaft being careful not to damage bearing surfaces while sliding into position. Install the thrust plate and tighten the bolts to 6–10 ft. lbs. (8–14 Nm).

12. Install the front cover and timing chain.

13. Install the oil pan.

14. Install the lifters.

15. Install the upper and lower intake manifolds.

16. Install the engine assembly.

17. Fill the cooling system and crankcase to the proper level and connect the negative battery cable.

18. Start the engine. Check and adjust the ignition timing and engine idle speed as necessary. Check for leaks.

Balance Shaft

REMOVAL & INSTALLATION

3.8L Engine

1. Remove the engine from the vehicle.

2. Remove the intake manifolds.

3. Remove the oil pan.

4. Remove the front cover and timing chain and camshaft sprocket.

5. Remove the balance shaft drive gear and spacer.

6. Remove the balance shaft gear, thrust plate and shaft assembly.

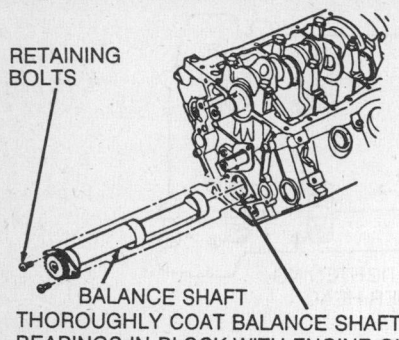

RETAINING BOLTS

BALANCE SHAFT

THOROUGHLY COAT BALANCE SHAFT BEARINGS IN BLOCK WITH ENGINE OIL PRIOR TO INSTALLATION

Balancer shaft—3.8L engine

To install:

7. Thoroughly coat the balance shaft bearings in the block with engine oil.

8. Install the balance shaft gear.

9. Install the balance shaft, thrust plate and gear and tighten the retaining bolts to 6–10 ft. lbs. (8–14 Nm).

10. Install the timing chain and camshaft sprocket.

11. Install the oil pan.

12. Install the timing cover.

13. Install the intake manifolds.

14. Install the engine in the vehicle.

Piston and Connecting Rod
POSITIONING

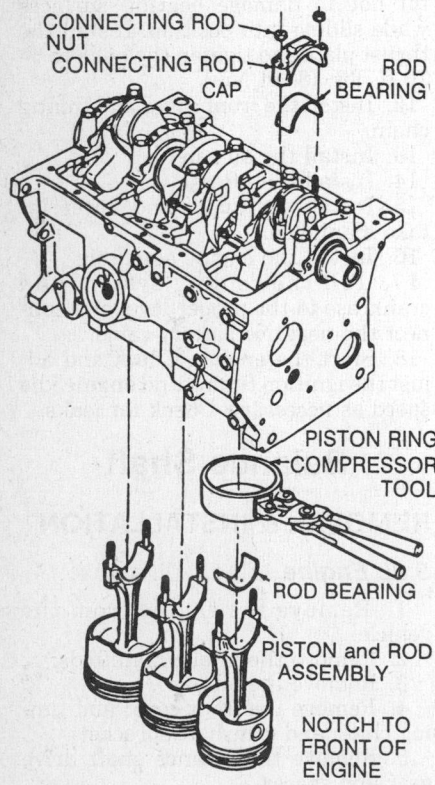

CONNECTING ROD NUT
CONNECTING ROD CAP
ROD BEARING

PISTON RING COMPRESSOR TOOL

ROD BEARING

PISTON and ROD ASSEMBLY

NOTCH TO FRONT OF ENGINE

Piston and connecting rod assembly—2.5L engine

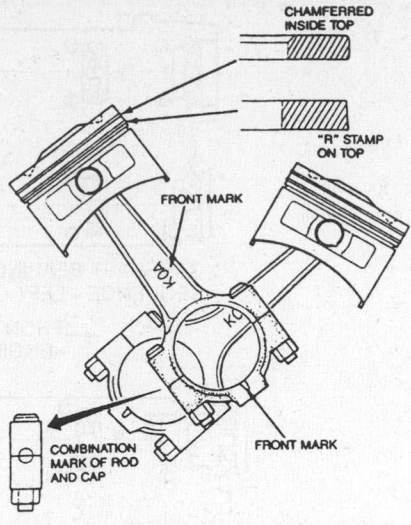

CHAMFERRED INSIDE TOP

"R" STAMP ON TOP

FRONT MARK

COMBINATION MARK OF ROD AND CAP

FRONT MARK

Piston and connecting rod assembly—3.0L SHO engine

NOTE:
PISTON TO DECK CLEARANCE TO BE 0.27 BELOW DECK TO 0.25 ABOVE DECK WHEN MEASURED AT PISTON T.D.C. PARALLEL TO CRANKSHAFT ON TRUE CENTERLINE OF PISTON. (AVERAGE OF TWO READINGS)

NOTE:
DOME AND BUTTON IDENTIFICATION MUST BE ON SAME SIDE AND TOWARDS FRONT OF ENGINE (AS SHOWN)

PISTON AND ROD ASSY 6100

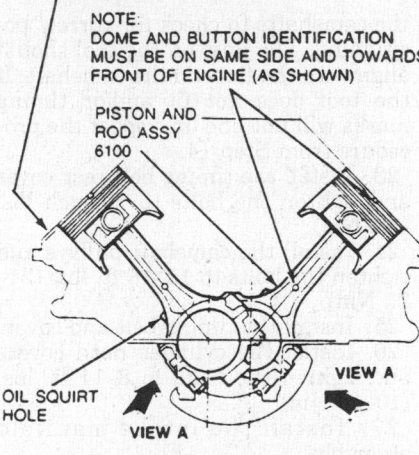

OIL SQUIRT HOLE

VIEW A

VIEW A

NOTE:
STAMP CORRESPONDING BORE NUMBERS ON CAP AND ROD IN THESE AREAS FOR NUMBER SIZE REFER TO 6100 PISTON AND ROD ASSY

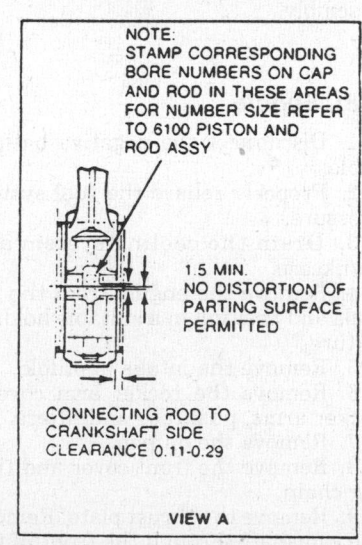

1.5 MIN.
NO DISTORTION OF PARTING SURFACE PERMITTED

CONNECTING ROD TO CRANKSHAFT SIDE CLEARANCE 0.11-0.29

VIEW A

Piston and connecting rod assembly—3.8L engine

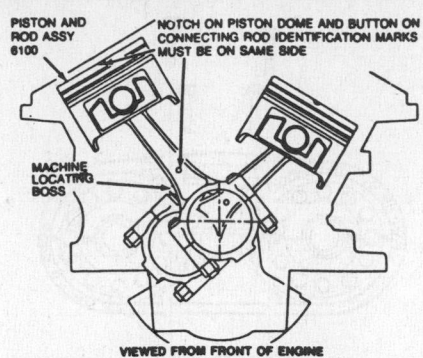

PISTON AND ROD ASSY 6100

NOTCH ON PISTON DOME AND BUTTON ON CONNECTING ROD IDENTIFICATION MARKS MUST BE ON SAME SIDE

MACHINE LOCATING BOSS

VIEWED FROM FRONT OF ENGINE

Piston and connecting rod assembly—3.0L engine

ENGINE LUBRICATION

Oil Pan

REMOVAL & INSTALLATION

2.5L Engine

1. Disconnect the negative battery cable. Raise the vehicle and support safely.

2. Drain the crankcase and drain the cooling system by removing the lower radiator hose.

3. Remove the roll restrictor on manual transaxle equipped vehicles.

4. Disconnect the starter cable and remove the starter.

5. Disconnect the exhaust pipe from oil pan.

6. Remove the engine coolant tube located from the lower radiator hose, water pump and at the tabs on the oil pan. Position air conditioner line off to the side. Remove the retaining bolts and remove the oil pan.

To install:

7. Clean both mating surfaces of oil pan and cylinder block making certain all traces of RTV sealant are removed.

8. Remove and clean oil pump pickup tube and screen assembly. After cleaning, install tube and screen assembly.

9. Fill the oil pan groove with RTV sealer; the bead should be approximately ⅛ in. above the surface of the pan rail. Immediately (within 5 minutes) install the oil pan.

10. Install and tighten the 2 oil pan-to-transaxle bolts to 30–39 ft. lbs. (40–50 Nm) to align the pan with the transaxle then back off ½ turn.

11. Tighten the pan flange bolts to 6–8 ft. lbs. (8–12 Nm).

12. Tighten the 2 oil pan-to-transaxle bolts to 30–39 ft. lbs. (40–50 Nm).

13. Install the remaining compo-

nents in the reverse order of their removal.

14. Fill the crankcase and cooling system to the proper level.

15. Start the engine and inspect for leaks.

3.0L Except SHO Engine

1. Disconnect the negative battery cable and remove the oil level dipstick.

2. Raise the vehicle and support safely. If equipped with a low level sensor, remove the retainer clip at the sensor. Remove the electrical connector from the sensor.

3. Drain the crankcase. Remove the starter motor and disconnect the electrical connector from the oxygen sensor.

4. Remove the catalyst and pipe assembly. Remove the lower engine/flywheel dust cover from the torque converter housing.

5. Remove the oil pan attaching bolts and slowly remove the oil pan from the engine block. Remove the oil pan gasket.

To install:

6. Clean the gasket surfaces on the cylinder block and oil pan. Apply a $^3/_{16}$ in. bead of silicone sealer to the junction of the rear main bearing cap and cylinder block junction of the front cover assembly and cylinder block.

NOTE: When using a silicone sealer, the assembly process should occur within 15 minutes after the sealer has been applied. After this time, the sealer may start to set-up and its sealing effectiveness may be affected.

7. Position the oil pan gasket over the oil pan and secure the gasket with a suitable sealer contact adhesive.

8. Position the oil pan on the engine block and install the oil pan attaching bolts. Torque the bolts to 8 ft. lbs. (10 Nm).

9. Install the lower engine/flywheel dust cover to the torque converter housing. Install the catalyst and pipe assembly. Connect the oxygen sensor connector.

10. Install the starter motor. Install the low oil level sensor connector to the sensor and install the retainer clip. Lower the vehicle and replace the oil level dipstick.

11. Connect the negative battery cable. Fill the crankcase. Start the engine and check for oil and exhaust leaks.

3.0L SHO Engine

1. Disconnect the negative battery cable.

2. Remove the oil level dipstick.

3. Remove the accessory drive belts.

4. Remove the timing belt.

5. Raise the vehicle and support it safely.

6. If equipped with a low oil level sensor, remove the retainer clip and the electrical connector from the sensor.

7. Drain the engine oil.

8. Remove the starter motor.

9. Disconnect the oxygen sensors.

10. Remove the catalyst and pipe assembly.

11. Remove the lower flywheel dust cover from the converter housing.

12. Remove the oil pan attaching bolts and the oil pan.

To install:

13. Clean the gasket surfaces of the cylinder block and the oil pan.

14. Position the oil pan gasket on the oil pan and secure with silicone sealer.

15. Position the oil pan and tighten the retaining bolts to 11–17 ft. lbs. (15–23 Nm).

16. Install the lower flywheel dust cover to the converter housing.

17. Install the catalyst and pipe assembly and connect the oxygen sensors.

18. Install the starter and connect the low oil level sensor connector to the sensor. Install the retainer clip.

19. Lower the vehicle and install the accessory drive belts.

20. Replace the oil level dipstick and connect the negative battery cable.

21. Fill the crankcase with the proper type and quantity of oil. Start the vehicle and check for leaks.

3.8L Engine

1. Disconnect the negative battery cable.

2. Raise the vehicle and support safely.

3. Drain the crankcase and remove the oil filter element.

4. Remove the converter assembly, starter motor and converter housing cover.

5. Remove the retaining bolts and remove the oil pan.

To install:

6. Clean the gasket surfaces on cylinder block, oil pan and oil pickup tube.

7. Trial fit oil pan to cylinder block. Ensure enough clearance has been provided to allow oil pan to be installed without sealant being scraped off when pan is positioned under engine.

8. Apply a bead of silicone sealer to the oil pan flange. Also apply a bead of sealer to the front cover/cylinder block joint and fill the grooves on both sides of the rear main seal cap.

NOTE: When using silicone rubber sealer, assembly must occur within 15 minutes after sealer application. After this time, the sealer may start to harden and its

sealing effectiveness may be reduced.

9. Install the oil pan and secure to the block with the attaching screws. Tighten the screws to 7–9 ft. lbs. (9–12 Nm).

10. Install a new oil filter element. Install the converter housing cover and starter motor.

11. Install the converter assembly and lower the vehicle.

12. Fill the crankcase and connect the negative battery cable.

13. Start the engine and check for leaks.

Oil Pump

REMOVAL & INSTALLATION

2.5L Engine

1. Remove the oil pan.

2. Remove oil pump attaching bolts and remove oil pump and intermediate driveshaft.

To install:

3. Prime oil pump by filling inlet port with engine oil. Rotate pump shaft until oil flows from outlet port.

4. If screen and cover assembly have been removed, replace gasket. Clean screen and reinstall screen and cover assembly and tighten attaching bolts.

5. Position intermediate driveshaft into distributor socket.

6. Insert intermediate driveshaft into oil pump. Install pump and shaft as an assembly.

NOTE: Do not attempt to force the pump into position if it will not seat. The shaft hex may be mis-aligned with the distributor shaft. To align, remove the oil pump and rotate the intermediate driveshaft into a new position.

7. Tighten the oil pump attaching bolts to 15–22 ft. lbs. (20–30 Nm).

8. Install the oil pan.

9. Fill the crankcase. Start engine and check for leaks.

3.0L Except SHO Engine

1. Remove the oil pan.

2. Remove the oil pump attaching bolts. Lift the oil pump off the engine and withdraw the oil pump driveshaft.

To install:

3. Prime the oil pump by filling either the inlet or the outlet port with engine oil. Rotate the pump shaft to distribute the oil within the oil pump body cavity.

4. Insert the oil pump intermediate shaft assembly into the hex drive hole in the oil pump assembly until the retainer "clicks" into place. Place the oil pump in the proper position with a

new gasket and install the retaining bolt.

5. Torque the oil pump retaining bolt to 35 ft. lbs. (48 Nm).

6. Install the oil pan with new gasket.

7. Fill the crankcase. Start engine and check for leaks.

3.0L SHO Engine

1. Remove the oil pan.

2. Remove the crankshaft timing belt pulley.

3. Remove the sump to oil pump bolts.

4. Remove the oil pump to block bolts and remove the pump.

To install:

5. Align the oil pump on the crankshaft and install the oil pump retaining bolts. Tighten the bolts to 11–17 ft. lbs. (15–23 Nm).

6. Install the oil sump to oil pump retaining bolts and tighten to 6–8 ft. lbs. (7–11 Nm).

7. Install the crankshaft timing belt pulley.

8. Install the oil pan with a new gasket.

9. Fill the crankcase with the proper type and quantity of oil.

10. Start the engine and check for leaks.

3.8L Engine

NOTE: The oil pump, oil pressure relief valve and drive intermediate shaft are contained in the front cover assembly.

1. Disconnect the negative battery cable. Drain the cooling system and crankcase.

2. Remove the air cleaner assembly and air intake duct.

3. Loosen the accessory drive belt idler. Remove the belt and water pump pulley.

4. Remove the power steering pump mounting bracket attaching bolts. Leaving the hoses connected, place the pump/bracket assembly in a position that will prevent the loss of power steering fluid.

5. If equipped with air conditioning, remove the compressor front support bracket. Leave the compressor in place.

6. Disconnect coolant bypass and heater hoses at the water pump. Disconnect radiator upper hose at thermostat housing.

7. Disconnect the coil wire from distributor cap and remove cap with secondary wires attached. Remove the distributor hold-down clamp and lift distributor out of the front cover.

8. Raise the vehicle and support safely.

9. Remove the crankshaft damper and pulley.

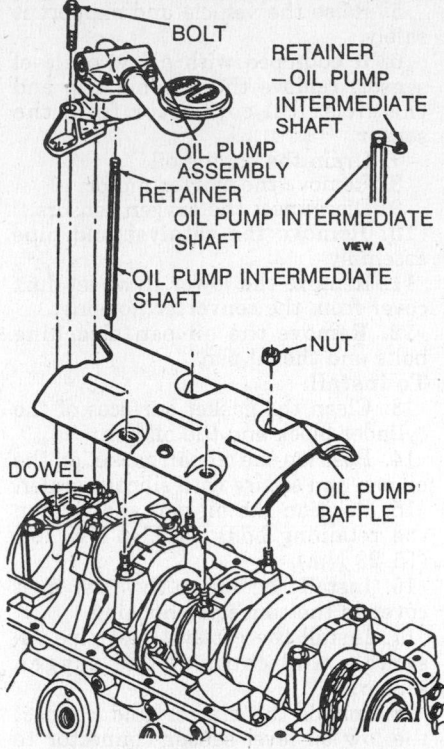

Oil pump Installation—3.0L engine

NOTE: If the crankshaft pulley and vibration damper have to be separated, mark the damper and pulley so they may be reassembled in the same relative position. This is important as the damper and pulley are initially balanced as a unit. If the crankshaft damper is being replaced, check if the original damper has balance pins installed. If so, new balance pins (E0SZ–6A328–A or equivalent) must be installed on the new damper in the same position as the original damper. The crankshaft pulley must also be installed in original installation position.

10. Remove the oil filter, disconnect the radiator lower hose at the water pump and remove the oil pan.

11. Lower the vehicle.

12. Remove the front cover.

NOTE: Do not overlook the cover attaching bolt located behind the oil filter adapter. The front cover will break if pried upon if all attaching bolts are not removed.

13. Remove the oil pump cover attaching bolts and remove the cover. Lift the pump gears off the front cover pocket. Remove the cover gasket and replace with new.

To install:

14. Clean the front cover oil pump gasket contact surface. Place a straight edge across the oil pump cover mounting surface and check for wear or warpage using a feeler gauge. If the surface is out of flat by more than 0.0016 in. (0.04mm), replace the cover.

15. Lightly pack the gear pocket with petroleum jelly or coat all pump gear surfaces with oil conditioner.

16. Install the gears in the pocket. Make certain the petroleum jelly fills the gap between the gears and the pocket.

17. Position the oil pump cover gasket and install the oil pump cover. Tighten the oil pump cover retaining bolts to 18–22 ft. lbs. (25–30 Nm).

18. Clean the gasket surfaces of the front cover and cylinder block.

19. Position a new gasket and the front cover on the cylinder block.

20. Install the front cover attaching bolts. Apply Loctite® or equivalent, to the threads of the bolt installed below the oil filter housing prior to installation. This bolt is to be installed and tightened last. Tighten all bolts to 15–22 ft. lbs. (20–30 Nm).

21. Raise the vehicle and support safely.

22. Install the oil pan. Connect the radiator lower hose. Install a new oil filter.

23. Coat the crankshaft damper sealing surface with clean engine oil.

24. Position the crankshaft pulley key in the crankshaft keyway.

25. Install the damper with damper washer and attaching bolt. Tighten the bolt to 104–132 ft. lbs. (140–180 Nm).

26. Install the crankshaft pulley and tighten the attaching bolts 19–28 ft. lbs. (26–28 Nm).

27. Lower the vehicle.

28. Connect the coolant bypass hose.

29. Install the distributor with rotor pointing at No. 1 distributor cap tower. Install the distributor cap and coil wire.

30. Connect the radiator upper hose at thermostat housing.

31. Connect the heater hose.

32. If equipped with air conditioning, install compressor and mounting brackets.

33. Install the power steering pump and mounting brackets.

34. Position the accessory drive belt over the pulleys.

35. Install the water pump pulley. Position the accessory drive belt over water pump pulley and tighten the belt.

36. Connect battery ground cable. Fill the crankcase and cooling system to the proper level.

37. Start the engine and check for leaks.

38. Check the ignition timing and curb idle speed, adjust as required.

39. Install the air cleaner assembly and air intake duct.

Rear Main Bearing Oil Seal

REMOVAL & INSTALLATION

1. Disconnect the negative battery cable.
2. Raise the vehicle and support it safely. Remove the transaxle.
3. Remove flywheel. Remove the cover plate, if necessary.
4. With a suitable tool, remove the oil seal.

NOTE: Use caution to avoid damaging the oil seal surface.

To install:

5. Inspect the crankshaft seal area for any damage which may cause the seal to leak. If damage is evident, service or replace the crankshaft as necessary.
6. Coat the crankshaft seal area and the seal lip with engine oil.
7. Using a seal installer tool, install the seal. Tighten the 2 bolts of the seal installer tool evenly so the seal is straight and seats without misalignment.
8. Install the flywheel. Tighten attaching bolts to 54–64 ft. lbs. (73–87 Nm) on all except the 3.0L SHO engine. On the 3.0L SHO engine, tighten the bolts to 51–58 ft. lbs. (69–78 Nm).
9. Install rear cover plate, if necessary.
10. Install the transaxle and connect the negative battery cable.

ENGINE COOLING

Radiator

REMOVAL & INSTALLATION

1. Disconnect the negative battery cable.
2. Drain the cooling system by removing the radiator cap and opening the draincock located at the lower rear corner of the radiator inlet tank.
3. Remove the rubber overflow tube from the coolant recovery bottle and detach it from the radiator. On Taurus SHO, disconnect the tube from the radiator and remove the recovery bottle.
4. Remove 2 upper shroud retaining screws and lift the shroud out of the lower retaining clips.
5. Disconnect the electric cooling

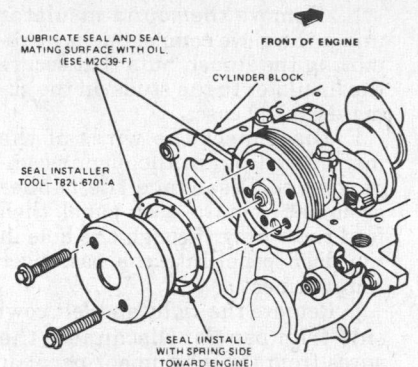

LUBRICATE SEAL AND SEAL MATING SURFACE WITH OIL (ESE-M2C39-F)

FRONT OF ENGINE

CYLINDER BLOCK

SEAL INSTALLER TOOL–T82L-6701-A

SEAL (INSTALL WITH SPRING SIDE TOWARD ENGINE)

NOTE: REAR FACE OF SEAL MUST BE WITHIN 0.127mm (0.005 INCH) OF THE REAR FACE OF THE BLOCK

Rear main oil seal installation – 3.8L engine

fan motor wires and remove the fan and shroud assembly.
6. Loosen the upper and lower hose clamps at the radiator and remove the hoses from the radiator connectors.
7. If equipped with an automatic transaxle, disconnect the transmission oil cooling lines from the radiator fittings using disconnect tool T82L-9500-AH or equivalent.
8. If equipped with 3.0L and SHO engines, remove 2 radiator upper retaining screws. If equipped with the 3.8L engine, remove 2 hex nuts from the right radiator support bracket and 2 screws from the left radiator support bracket and remove the brackets.
9. Tilt the radiator rearward approximately 1 in. and lift it directly upward, clear of the radiator support.
10. Remove the radiator lower support rubber pads, if pad replacement is necessary.

To install:

11. Position the radiator lower support rubber pads to the lower support, if removed.
12. Position the radiator into the engine compartment and to the radiator support. Insert the moulded pins at the bottom of each tank through the slotted holes in the lower support rubber pads.
13. Make sure the plastic pads on the bottom of the radiator tanks are resting on the rubber pads. Install 2 upper retaining bolts to attach the radiator to the radiator support. Tighten the bolts to 46–60 inch lbs. (5–7 Nm). If equipped with the 3.8L engine, tighten the bolts to 13–20 ft. lbs. (17–27 Nm).
14. If equipped with the 3.8L engine, fasten the left radiator support bracket to the radiator support with 2 screws. Tighten the screws to 8.7–17.7 ft. lbs. (11.8–24 Nm). Attach the right support bracket to the radiator support with 2 hex nuts. Tighten the nuts to 8.7–17.7 ft. lbs. (11.8–24 Nm).
15. Attach the radiator upper and lower hoses to the radiator. Position the hose on the radiator connector so

the index arrow on the hose is in line with the mark on the connector. Tighten the clamps to 20–30 inch lbs. (2.3–3.4 Nm) if equipped with the 2.5L engine. If equipped with the 3.8L and 3.0L SHO engines, install constant tension hose clamps between the alignment marks on the hoses.
16. If equipped with automatic transaxles, connect the transmission cooler lines using oil resistant pipe sealer.
17. Install the fan and shroud assembly by connecting the fan motor wiring and positioning the assembly on the lower retainer clips. Attach the top of the shroud to the radiator with 2 screw, nut and washer assemblies. Tighten to 35 inch lbs. (4 Nm).
18. Attach the rubber overflow tube to the radiator filler neck overflow nipple and coolant recovery bottle. On Taurus SHO, install the coolant recovery bottle and connect the overflow hose.
19. Refill the cooling system. If the coolant is being replaced, refill with a 50/50 mixture of water and antifreeze. Connect the negative battery cable. Operate the engine for 15 minutes and check for leaks. Check the coolant level and add, as required.

Heater Core

REMOVAL & INSTALLATION

Without Air Conditioning

TAURUS AND SABLE

1. Disconnect the negative battery cable.
2. Remove the instrument panel on 1988–89 vehicles as follows:
   a. Remove the 4 screws retaining the steering column opening cover and remove the cover.
   b. Remove the sound insulator under the glove compartment by removing the 2 push nuts securing the insulator to the studs on the climate control case.
   c. Remove the steering column trim shrouds and disconnect all electrical connections from the steering column switches.
   d. Remove the 4 screws at the steering column bracket to remove the steering column.
   e. Remove the screws retaining the lower left and radio finish panels and remove the panels by snapping out.
   f. Remove the cluster opening finish panel retaining screws. On Taurus remove 1 jam nut behind the headlight switch and 1 screw behind the clock or clock cover. Remove the finish panel by rocking the upper edge toward the driver.

g. Disconnect the speedometer cable by reaching up under the instrument panel and pressing on the flat surface of the plastic connector. The panel can be removed with the cluster installed.

h. Release the glove compartment assembly by depressing the side of the glove compartment bin and swinging the door/bin down.

i. Using the steering column, cluster and glove compartment openings and by reaching under the instrument panel, tag and disconnect all electrical connections, vacuum hoses, heater control cables and the radio antenna cable.

j. Disconnect all underhood electrical connectors of the main wire loom. Disengage the rubber grommet from the dash panel and push the wire and connectors into the instrument panel area.

k. Remove the right and left speaker opening covers by snapping out.

l. Remove the 2 lower instrument panel-to-cowl side retaining screws from the right and left side. Remove the 1 instrument panel brace retaining screw from under the radio area. On Sable, remove the defroster grille by snapping out.

m. Remove the 3 instrument panel upper retaining screws and remove the instrument panel.

3. Remove the instrument panel on 1990–92 vehicles as follows:

a. Position the front wheels in the straight-ahead position.

b. Remove the ignition lock cylinder and, if equipped, remove the tilt lever.

c. Remove the steering column trim shrouds. Disconnect all electrical connections from the steering column switches.

d. Remove the 4 bolts and opening cover and the 2 bolts and reinforcement from under the steering column.

e. Disengage the insulator retainer and remove the insulator. Remove the 4 nuts and reinforcement from under the steering column.

NOTE: Do not rotate the steering column shaft.

f. Remove the 4 nuts retaining the steering column to the instrument panel, disconnect the shift indicator cable and lower the column on the front seat. Install the lock cylinder to make sure the steering column shaft does not turn.

g. Remove 1 bolt at the steering column opening attaching the instrument panel to the brace. Remove 1 instrument panel brace retaining bolt from under the radio area.

h. Remove the sound insulator under the glove compartment by removing the 2 push nuts that secure the insulator to the studs on the climate control case.

i. Disconnect the wires of the main wire loom in the engine compartment. Disengage the rubber grommet from the dash panel, then feed the wiring through the hole in the dash panel into the passenger compartment.

j. Remove the right and left cowl side trim panels. Disconnect the wires from the instrument panel at the right and left cowl sides.

k. Remove 1 screw each from the left and right side retaining the instrument panel. Pull up to unsnap the right and left speaker opening covers and remove.

l. Release the glove compartment assembly by depressing the side of the glove compartment bin and swinging the door/bin down.

m. Using the steering column and glove compartment openings and by reaching under the instrument panel, tag and disconnect all electrical connections, vacuum hoses, heater control cables, speedometer cable and radio antenna cable.

n. Close the glove compartment door, support the panel and remove the 3 screws attaching the top of the instrument panel to the cowl top and disconnect any remaining wires. Remove the panel from the vehicle.

4. Drain the coolant from the radiator.

5. Disconnect and plug the heater hoses at the heater core. Plug the heater core tubes.

6. Disconnect the vacuum supply hose from the in-line vacuum check valve in the engine compartment. Remove the screw holding the instrument panel shake brace to the heater case and remove the shake brace.

7. Remove the floor register and rear floor ducts from the bottom of the heater case. Remove the 3 nuts attaching the heater case to the dash panel in the engine compartment.

8. Remove the 2 screws attaching the brackets to the cowl top panel. Pull the heater case assembly away from the dash panel and remove from the vehicle.

9. Remove the vacuum source line from the heater core tube seal and remove the seal from the heater core tubes.

10. Remove the 4 heater core access cover attaching screws and remove the access cover from the heater case. Lift the heater core and seals from the heater case.

To install:

11. Transfer the 3 foam core seals to the new heater core. Install the heater core and seals into the heater case.

12. Position the heater case access cover on the case and install the 4 screws.

13. Install the seal on the heater core tubes and install the vacuum source line through the seal.

14. Position the heater case assembly to the dash panel and cowl top panel at the air inlet opening. Install the 2 screws to attach the support brackets to the cowl top panel.

15. Install the 3 nuts in the engine compartment to attach the heater case to the dash panel. Install the floor register and rear floor ducts on the bottom of the heater case.

16. Install the instrument panel shake brace and screw to the heater case. Install the instrument panel by reversing the removal procedure.

17. Connect the heater hoses to the heater core. Connect the black vacuum supply hose to the vacuum check valve in the engine compartment.

18. Fill the radiator and bleed the cooling system.

19. Connect the negative battery cable and check the system for proper operation.

With Air Conditioning

NOTE: It is necessary to remove the evaporator case in order to remove the heater core. Whenever an evaporator case is removed, it will be necessary to replace the suction accumulator/drier.

TAURUS AND SABLE

1. Disconnect the negative battery cable.

2. Remove the instrument panel on 1988–89 vehicles as follows:

a. Remove the 4 screws retaining the steering column opening cover and remove the cover.

b. Remove the sound insulator under the glove compartment by removing the 2 push nuts securing the insulator to the studs on the climate control case.

c. Remove the steering column trim shrouds and disconnect all electrical connections from the steering column switches.

d. Remove the 4 screws at the steering column bracket to remove the steering column.

e. Remove the screws retaining the lower left and radio finish panels and remove the panels by snapping out.

f. Remove the cluster opening finish panel retaining screws. On Taurus remove 1 jam nut behind the headlight switch and 1 screw behind the clock or clock cover. Remove the finish panel by rocking the upper edge toward the driver.

g. Disconnect the speedometer cable by reaching up under the instrument panel and pressing on the flat surface of the plastic connector. The panel can be removed with the cluster installed.

h. Release the glove compartment assembly by depressing the side of the glove compartment bin and swinging the door/bin down.

i. Using the steering column, cluster and glove compartment openings and by reaching under the instrument panel, tag and disconnect all electrical connections, vacuum hoses, heater/air conditioner control cables and the radio antenna cable.

j. Disconnect all underhood electrical connectors of the main wire loom. Disengage the rubber grommet from the dash panel and push the wire and connectors into the instrument panel area.

k. Remove the right and left speaker opening covers by snapping out.

l. Remove the 2 lower instrument panel-to-cowl side retaining screws from the right and left side. Remove the 1 instrument panel brace retaining screw from under the radio area. On Sable, remove the defroster grille by snapping out.

m. Remove the 3 instrument panel upper retaining screws and remove the instrument panel.

3. Remove the instrument panel on 1990–92 vehicles as follows:

a. Position the front wheels in the straight-ahead position.

b. Remove the ignition lock cylinder and, if equipped, remove the tilt lever.

c. Remove the steering column trim shrouds. Disconnect all electrical connections from the steering column switches.

d. Remove the 4 bolts and opening cover and the 2 bolts and reinforcement from under the steering column.

e. Disengage the insulator retainer and remove the insulator. Remove the 4 nuts and reinforcement from under the steering column.

NOTE: Do not rotate the steering column shaft.

f. Remove the 4 nuts retaining the steering column to the instrument panel, disconnect the shift indicator cable and lower the column on the front seat. Install the lock cylinder to make sure the steering column shaft does not turn.

g. Remove 1 bolt at the steering column opening attaching the instrument panel to the brace. Remove 1 instrument panel brace retaining bolt from under the radio area.

h. Remove the sound insulator under the glove compartment by removing the 2 push nuts that secure the insulator to the studs on the climate control case.

i. Disconnect the wires of the main wire loom in the engine compartment. Disengage the rubber grommet from the dash panel, then feed the wiring through the hole in the dash panel into the passenger compartment.

j. Remove the right and left cowl side trim panels. Disconnect the wires from the instrument panel at the right and left cowl sides.

k. Remove 1 screw each from the left and right side retaining the instrument panel. Pull up to unsnap the right and left speaker opening covers and remove.

l. Release the glove compartment assembly by depressing the side of the glove compartment bin and swinging the door/bin down.

m. Using the steering column and glove compartment openings and by reaching under the instrument panel, tag and disconnect all electrical connections, vacuum hoses, heater/air conditioner control cables, speedometer cable and radio antenna cable.

n. Close the glove compartment door, support the panel and remove the 3 screws attaching the top of the instrument panel to the cowl top and disconnect any remaining wires. Remove the panel from the vehicle.

4. Drain the coolant from the radiator. Properly discharge the air conditioning system.

5. Disconnect and plug the heater hoses at the heater core. Plug the heater core tubes.

6. Disconnect the vacuum supply hose from the in-line vacuum check valve in the engine compartment.

7. Disconnect the air conditioning lines from the evaporator core at the dash panel. Cap the lines and the core to prevent entrance of dirt and moisture.

8. Remove the screw holding the instrument panel shake brace to the evaporator case and remove the shake brace.

9. Remove the 2 screws attaching the floor register and rear seat duct to the bottom of the evaporator case. Remove the 3 nuts attaching the evaporator case to the dash panel in the engine compartment.

10. Remove the 2 screws attaching the support brackets to the cowl top panel. Carefully pull the evaporator assembly away from the dash panel and remove the evaporator case from the vehicle.

11. Remove the vacuum source line from the heater core tube seal and remove the seal from the heater core tubes.

12. If equipped with automatic temperature control, remove the 3 screws attaching the blend door actuator to the evaporator case and remove the actuator.

13. Remove the 4 heater core access cover attaching screws and remove the access cover and seal from the evaporator case. Lift the heater core and seals from the evaporator case.

To install:

14. Transfer the seal to the new heater core. Install the heater core into the evaporator case.

15. Position the heater core access cover on the evaporator case and install the 4 attaching screws. If equipped with automatic temperature control, position the blend door actuator to the blend door shaft and install the 3 attaching screws.

16. Install the seal on the heater core tubes and install the vacuum source line through the seal.

17. Position the evaporator case assembly to the dash panel and cowl top panel at the air inlet opening. Install the 2 screws attaching the support brackets to the cowl top panel.

18. Install the 3 nuts in the engine compartment attaching the evaporator case to the dash panel. Install the floor register and rear seat duct to the evaporator case and tighten the 2 attaching screws.

19. Install the instrument panel shake brace and screw to the evaporator case. Install the instrument panel in the reverse order of removal.

20. Connect the air conditioning lines to the evaporator core and the heater hoses to the heater core.

21. Connect the black vacuum supply hose to the vacuum check valve in the engine compartment.

22. Fill and bleed the cooling system. Connect the negative battery cable.

23. Leak test, evacuate and charge the air conditioning system. Observe all safety precautions.

24. Check the system for proper operation.

CONTINENTAL

1. Disconnect the negative battery cable.

2. Remove the instrument panel on 1988 vehicles as follows:

a. Remove the 4 screws retaining the lower instrument panel steering column cover and remove the cover.

b. Remove the 3 screws retaining the upper steering column shroud and remove. Remove the 1 screw retaining the tilt wheel lever and remove the lever.

c. Remove the lock cylinder by pushing a small Allen wrench into the groove located under the lock cylinder. Place the key into the ignition and gently wiggle to work the cylinder free.

d. Remove the lower steering column shroud by pulling out.

e. Remove the 2 screws retaining the steering wheel pad and disconnect the horn electrical connection from the pad. Remove the steering wheel.

f. Remove the bolt retaining the shift indicator cable to the steering column. Disconnect all electrical connectors.

g. Disconnect the hood and brake release cables. Remove the 3 plastic retainers from each side holding the left and right lower close-out panels in place and remove the panels.

h. Disconnect the ignition switch wiring.

i. Remove the 2 lower shaft universal joint nuts and pull the lower shaft away from the steering column. Remove the 4 nuts retaining the steering column and lower the column.

j. Remove the 1 screw and clip retaining the transaxle shift cable. Disconnect the 2 vacuum hoses and all electrical connectors. Remove the steering column from the vehicle.

k. Remove the center finish panel moulding by snapping out to release from the 5 clips. Remove the left finish panel retaining screw and remove the panel. Disconnect 2 light sockets and all electrical connectors.

l. Remove the 5 screws retaining the cluster opening finish panel and snap out to release. Disconnect the light socket and electrical connectors. Remove the 2 screws at the right finish panel and remove the panel. Disconnect all electrical connectors.

m. Remove the 4 screws retaining the radio and storage bin. Disconnect the electrical connectors. Remove the 4 screws retaining the air conditioning control. Disconnect the electrical connectors and 1 vacuum connector.

n. Remove the 2 screws at the cluster reinforcement brace and remove the brace. Remove the 4 screws retaining the cluster. Disconnect the electrical connectors.

o. Remove the 3 screws and 2 push-in plastic clips to remove the glove compartment assembly. Remove both speaker grilles by snapping out to release. Disconnect the electrical connector at the right grille.

p. Remove the 2 screws seated in

plastic push clips and remove the center defrost grille.

q. Working under the hood, disconnect all electrical connectors of the main wire loom. Disengage the rubber grommet from the dash panel and feed the wiring and connectors through the hole into the instrument panel area.

r. Remove the 3 screws (2 located on sill plate) at both right and left cowl trim panels and remove the panels. Remove the lower 3 screws at the instrument panel, 1 at each end and middle.

s. Remove the 3 upper instrument panel retaining screws and carefully lower the instrument panel. Disconnect all remaining electrical and vacuum connectors. Remove the instrument panel from the vehicle.

3. Remove the instrument panel on 1989–92 vehicles as follows:

a. Open the glove compartment door and depress the sides inward. Lower the glove compartment assembly toward the floor. On 1990–92 vehicles, disconnect the air bag backup power supply located to the right of the glove compartment opening.

NOTE: The backup power supply allows air bag deployment if the battery or battery cables are damaged in an accident before the crash sensors close. The power supply is a capacitor that will leak down in approximately 15 minutes after the battery is disconnected or in 1 minute if the battery positive cable is grounded. The backup power supply must be disconnected before any air bag related service is performed.

b. Remove the 4 nut and washer assemblies retaining the driver air bag module to the steering wheel. Disconnect the driver air bag module connector and attach a jumper wire to the air bag terminals on the clockspring. Remove the air bag from the vehicle.

c. Disconnect the passenger air bag connector, located behind the glove compartment. Attach a jumper wire to the air bag terminals on the wiring harness side of the passenger air bag module connector. Remove the 4 screws attaching the passenger air bag module to the instrument panel and remove the air bag from the vehicle.

CAUTION

When carrying a live air bag module, make sure the bag and trim cover are pointed away from the body. In the unlikely event of an accidental deployment, the bag will then

deploy with minimal chance of injury. In addition, when placing a live air bag module on a bench or other surface, always face the bag and trim cover up, away from the surface. This will reduce the motion of the module if it is accidentally deployed.

d. Remove the right finish moulding by pulling upward to unsnap the 6 clips. Disconnect the wiring. Remove the left finish moulding by pulling upward to unsnap the 2 clips.

e. Remove the right and left lower insulators. Remove the screws retaining the lower instrument panel steering column cover and remove the cover. Remove the 4 screws retaining the lower instrument panel steering column reinforcement and remove the reinforcement.

f. If equipped, remove the 4 retaining nuts and the absorber assembly. Remove the 3 screws retaining the upper steering column shroud and remove. Remove the 1 screw retaining the tilt wheel lever and remove the lever.

g. Remove the lock cylinder by pushing a small Allen wrench into the groove located under the lock cylinder. Place the key into the ignition and gently wiggle to work the cylinder free.

h. Remove the lower steering column shroud by pulling out. Remove the bolt retaining the shift indicator cable to the steering column. Remove the steering wheel. Disconnect all electrical connectors.

i. Disconnect the hood and brake release cables. Remove the 4 nuts retaining the steering column and lower the column. Remove the screw(s) at the steering column opening retaining the instrument panel to the brake pedal support.

j. Remove the 2 screws under the ash tray that hold the instrument panel to the air conditioning plenum case. Remove the headlight switch knob.

k. Remove the 5 screws from the cluster opening finish panel and remove the panel. Remove the 4 screws retaining the air conditioning control. Disconnect the electrical connectors and 1 vacuum connector.

l. Remove the 4 screws retaining the cluster and disconnect the electrical connectors. Remove the 3 screws to remove the glove compartment assembly.

m. Remove the 3 screws located above the left side of the glove compartment. Remove both speaker grilles by snapping out to release. Remove the 2 screws seated in the plastic push clips and remove the center defrost grille.

n. Working under the hood, disconnect all electrical connectors of the main wire loom. Disengage the rubber grommet from the dash panel and feed the wiring and connectors through the hole into the instrument panel area.

o. Remove the 3 screws (2 located on the sill plate) at both right and left cowl trim panels and remove the panels. Disconnect the wiring at the right and left cowl side panels.

p. Remove the lower 2 screws from the instrument panel, 1 at each end. Remove the 3 upper instrument panel retaining screws and carefully lower the instrument panel.

q. Disconnect the remaining electrical and vacuum connectors and remove the instrument panel.

4. Drain the coolant from the radiator and properly discharge the refrigerant from the air conditioning system.

5. Disconnect and plug the heater hoses at the heater core. Plug the heater core tubes. Disconnect the vacuum supply hose from the in-line vacuum check valve in the engine compartment.

6. Disconnect the air conditioner lines from the evaporator core at the dash panel. Cap the lines and the core to prevent the entrance of dirt and moisture.

7. Remove the screw holding the instrument panel shake brace to the evaporator case. Remove the shake brace. Remove the 2 screws attaching the floor register to the evaporator case.

8. Disconnect the vacuum line, electrical connections and aspirator hose from the evaporator case.

9. Remove the 3 nuts retaining the evaporator case to the dash panel in the engine compartment. Remove the 2 screws retaining the support brackets to the cowl top panel.

10. Carefully pull the evaporator case assembly away from the dash panel and remove from the vehicle.

11. Remove the vacuum source line from the heater core tube seal. Remove the seal from the heater core tubes.

12. Remove the 3 screws attaching the blend door actuator to the evaporator case and remove the actuator. On 1988 vehicles, remove the cold engine lock out switch.

13. Remove the 4 heater core access cover retaining screws and remove the access cover and seal from the evaporator case. Lift the heater core and seals from the evaporator case.

To install:

14. Transfer 3 foam core seals to the new heater core. Install the heater core in the evaporator case.

15. Position the heater core access cover on the evaporator case and install the 4 retaining screws.

16. Position the blend door actuator to the blend door shaft. Install the 3 retaining screws. On 1988 vehicles, attach the cold engine lock out switch by snapping the spring clip in place on the outermost heater core tube.

17. Install the seal on the heater core tubes. Install the vacuum source line through the seal.

18. Position the evaporator case assembly against the dash panel and cowl top panel at the air inlet opening. Install the 2 screws retaining the support brackets to the cowl top panel.

19. Install the 3 nuts in the engine compartment retaining the evaporator case to the dash panel. Install the floor register to the evaporator case and tighten the 2 retaining screws.

20. Connect the vacuum line, electrical connections and aspirator hose at the evaporator case. Install the instrument panel shake brace.

21. Install the instrument panel in the reverse order of removal.

22. Connect the air conditioner lines at the evaporator core and the heater hoses at the heater core. Connect the black vacuum supply hose to the vacuum checl valve in the engine compartment.

23. Fill and bleed the cooling system. Connect the negative battery cable.

24. Leak test, evacuate and charge the air conditioning system. Observe all safety precautions.

25. Check the system for proper operation.

Water Pump

REMOVAL & INSTALLATION

2.5L Engine

1. Disconnect the negative battery cable.

2. Remove the radiator cap and position a drain pan under the bottom radiator hose.

3. Raise and support the vehicle safely. Remove the lower radiator hose from the radiator and drain the coolant into the drain pan.

4. Remove the water pump inlet tube. Loosen the belt tensioner by inserting a ½ in. flex handle in the square hole of the tensioner and rotate the tensioner counterclockwise and remove the belt from the pulleys.

5. Disconnect the heater hose from the water pump. Remove the water pump retaining bolts and remove the pump from the engine.

6. Installation is the reverse of the removal procedure. Torque the water pump-to-engine block retaining bolts to 15–23 ft. lbs. (20–30 Nm).

7. Refill the cooling system to the proper level. Start the engine and allow to reach normal operating temperature and check for leaks.

3.0L Engine Except SHO

1. Disconnect the negative battery cable and place a drain pan under the radiator drain cock.

2. Remove the radiator cap, open

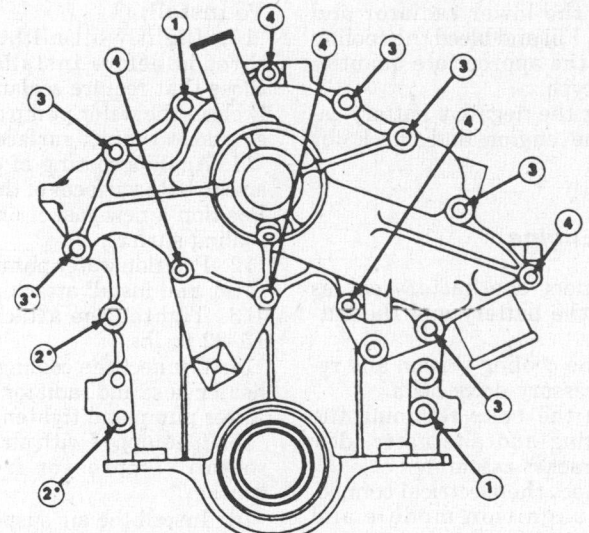

NUMBER	PART NUMBER	SIZE	QTY	N·m	LB-FT
1	N605909-SB	M8 x 1.25 x 42.0	2	25	19
2	N804113-S8	M8 x 1.25 x 43.5 (LARGE HEX)	2	25	19
3	N804811-S8	M8 x 1.25 x 70.0	6	25	19
4	N804168-S8	M6 x 1.0 x 25.0	5	10	7

* APPLY PIPE SEALANT D6AZ-19558-A TO THE THREADS OF THESE BOLTS.

Water pump bolt torque specifications—3.0L engine

the drain cock on the radiator and drain the cooling system.

3. Loosen the 4 water pump pulley retaining bolts while the accessory drive belts are still tight.

4. Loosen the alternator belt adjuster jack screw to provide enough clearance for removal of the alternator belt.

5. Using a ½ in. breaker bar, rotate the automatic tensioner down and to the left. Remove the power steering/air conditioner belt.

6. Remove the 2 nuts and 1 bolt retaining the automatic tensioner to the engine.

7. Disconnect and remove the lower radiator and heater hose from the water pump.

8. Remove the water pump to engine retaining bolts and lift the water pump and pulley up and out of the vehicle.

To install:

9. Clean the gasket surfaces on the water pump and front cover.

10. Install the water pump with the pulley loosely positioned on the hub, using a new gasket.

11. Install and tighten the retaining bolts as indicated. Apply a suitable pipe sealant prior to installation.

12. Hand tighten the water pump pulley retaining bolts.

13. Install the automatic belt tensioner assembly. Tighten the 2 retaining nuts and bolt to 35 ft. lbs. (48 Nm).

14. Install the alternator and power steering belts. Final tighten the water pump pulley retaining bolts to 16 ft. lbs. (21 Nm).

15. Install the lower radiator and heater hoses. Fill and bleed the cooling system with the appropriate quantity and coolant type.

16. Connect the negative battery cable. Start the engine and check for leaks.

3.0L SHO Engine

1. Disconnect the battery cables and remove the battery and the battery tray.

2. Drain the cooling system and remove the accessory drive belts.

3. Remove the bolts retaining the air conditioning and alternator idler pulley and bracket assembly.

4. Disconnect the electrical connector from the ignition module and ground strap.

5. Loosen the clamps on the upper intake connector tube, remove the retaining bolts and remove the connector tube.

6. Raise and safely support the vehicle. Remove the right wheel and tire assembly.

7. Remove the splash shield.

8. Remove the upper timing belt cover, crankshaft pulley and lower timing belt cover.

9. Remove the bolts from the center timing belt cover and position it aside.

10. Remove the water pump attaching bolts and remove the water pump.

11. To install, reverse the removal procedure. Tighten the water pump bolts to 12–16 ft. lbs. (15–23 Nm). Tighten the crankshaft pulley bolt to 113–126 ft. lbs. (152–172 Nm).

3.8L Engine

1. Disconnect the negative battery cable. Drain the cooling system.

2. Remove the lower nut on both right engine mounts. Raise and safely support the engine.

3. Loosen the accessory drive belt idler. Remove the drive belt and water pump pulley.

4. Remove the air suspension pump.

5. Remove the power steering pump mounting bracket attaching bolts. Leaving hoses connected, place pump/bracket assembly aside in a position to prevent fluid from leaking out.

6. If equipped with air conditioning, remove the compressor front support bracket.

7. Leave the compressor in place, if removed.

8. Disconnect coolant bypass and heater hoses at the water pump.

9. Remove the water pump-to-engine block attaching bolts and remove the pump from the vehicle. Discard the gasket and replace with new.

To install:

10. Lightly oil all bolt and stud threads before installation except those that require sealant. Thoroughly clean the water pump and front cover gasket contact surfaces.

11. Apply a coating of contact adhesive to both surfaces of the new gasket. Position a new gasket on water pump sealing surface.

12. Position water pump on the front cover and install attaching bolts.

13. Tighten the attaching bolts to 15–22 ft. lbs.

14. Connect the cooling bypass hose, heater hose and radiator lower hose to water pump and tighten the clamps.

15. If equipped with air conditioning, install compressor front support bracket.

16. Install the air suspension pump.

17. Position the accessory drive belt over the pulleys.

18. Install the water pump pulley, fan/clutch assembly and fan shroud. Cross-tighten fan/clutch assembly attaching bolts to 12–18 ft. lbs.

19. Position accessory drive belt over pump pulley and adjust drive belt tension.

20. Lower the tengine.

21. Install and tighten the lower right engine mount nuts.

22. Fill cooling system to the proper level.

23. Start engine and check for coolant leaks.

Thermostat

REMOVAL & INSTALLATION

2.5L Engine

1. Disconnect the negative battery cable.

2. Position a suitable drain pan below the radiator. Remove the radiator cap and open the draincock. Drain the radiator to a corresponding level below the water outlet connection. Close the draincock.

3. Remove the vent plug from the water outlet connection.

4. Loosen the top hose clamp at the radiator, remove the water outlet connection retaining bolts, lift clear of the engine and remove the thermostat by pulling it out of the water outlet connection.

NOTE: Do not pry the housing off.

To install:

5. Make sure the water outlet connection and cylinder head mating surfaces are clean and free from gasket material. Make sure the water outlet connection pocket and air vent passage are clean and free from rust. Clean the vent plug and gasket.

6. Place the thermostat in position, fully inserted to compress the gasket and pressed into the water outlet connection to secure. Install the water outlet connection to the cylinder head using a new gasket. Tighten the bolts to 12–18 ft. lbs. (16–24 Nm). Position the top hose to the radiator and tighten the clamps.

7. Refill the cooling system. Connect the negative battery cable. Start the engine and check for leaks. Check the coolant level and add as required.

3.0L Engine

1. Disconnect the negative battery cable.

2. Place a suitable drain pan under the radiator.

3. Remove the radiator cap and open the draincock. Drain the cooling system.

4. Remove the upper radiator hose from the thermostat housing.

5. Remove the 3 retaining bolts from the thermostat housing.

6. Remove the housing and the thermostat as an assembly.

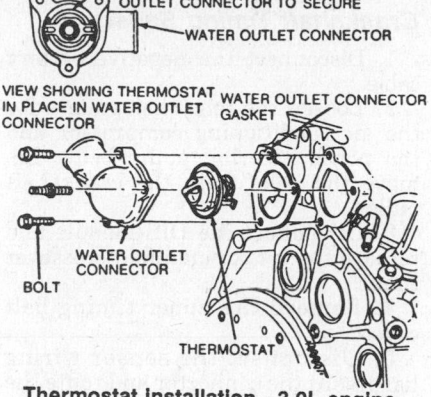

ROTATE THERMOSTAT CLOCKWISE INTO WASHER OUTLET CONNECTOR TO SECURE

WATER OUTLET CONNECTOR

VIEW SHOWING THERMOSTAT IN PLACE IN WATER OUTLET CONNECTOR

WATER OUTLET CONNECTOR GASKET

WATER OUTLET CONNECTOR

BOLT

THERMOSTAT

Thermostat Installation — 3.0L engine

To install:

7. Make sure all sealing surfaces are free of old gasket material.

8. Install the thermostat into the housing.

9. Position a new gasket onto the housing using the bolts as a holding device. Install the thermostat assembly and tighten the bolts to 9 ft. lbs. (12 Nm).

10. Install the upper radiator hose and tighten the clamp.

11. Fill and bleed the cooling system. Connect the negative battery cable, start the engine and check for coolant leaks. Check the coolant level and add as required.

3.0L SHO Engine

1. Disconnect the negative battery cable.

2. Place a suitable drain pan below the radiator. Remove the radiator cap and open the draincock. Partially drain the cooling system and then close the draincock.

3. Remove the air cleaner tube.

4. Disconnect the hose from the water outlet tube.

5. Remove the 2 retaining nuts and remove the water outlet tube.

6. Remove the thermostat and seal from the water outlet housing.

To install:

7. Install the seal around the outer rim of the thermostat and install the thermostat into the water outlet housing. Align the jiggle valve of the thermostat with the upper bolt on the water outlet housing.

8. Install the water outlet tube. Tighten the 2 retaining nuts to 5–8 ft. lbs. (7–11 Nm).

9. Install the air cleaner tube.

10. Refill the cooling system. Connect the negative battery cable. Start the engine and check for leaks. Check the coolant level and add as necessary.

3.8L Engine

1. Disconnect the negative battery cable.

2. Place a suitable drain pan below the radiator.

3. Remove the radiator cap and open the draincock. Drain the radiator to a level below the water outlet connection and then close the draincock.

4. Loosen the top hose clamp at the radiator, remove the water outlet connection retaining bolts and lift the water outlet clear of the engine. Remove the thermostat by rotating it counterclockwise in the water outlet connection until the thermostat becomes free to remove.

NOTE: Do not pry the housing off.

To install:

5. Make sure the water outlet connection pocket and all mating surfaces are clean.

6. Install the thermostat into the water outlet connection by rotating it clockwise until the engaging ramps on the thermostat are secure. Install water outlet connection on the intake manifold with a new gasket and tighten the mounting bolts to 15–22 ft. lbs. (20–30 Nm). Position the top hose to the radiator and tighten the clamps.

7. Refill the cooling system. Connect the negative battery cable. Start the engine and check for leaks. Check the coolant level and add as required.

Cooling System Bleeding

When the entire cooling system is drained, the following procedure should be used to ensure a complete fill.

1. Install the block drain plug, if removed and close the draincock. With the engine off, add anti-freeze to the radiator to a level of 50 percent of the total cooling system capacity. Then add water until it reaches the radiator filler neck seat.

NOTE: On 2.5L engine, remove the vent plug on the water connection outlet. The vent plug must be removed before the radiator is filled or the engine may not fill completely. Do not turn the plastic cap under the vent plug or the gasket may be damaged. Do not try to add coolant through the vent plug hole. Install the vent plug after filling the radiator and before starting the engine.

2. Install the radiator cap to the first notch to keep spillage to a minimum.

3. Start the engine and let it idle until the upper radiator hose is warm. This indicates that the thermostat is open and coolant is flowing through the entire system.

4. Carefully remove the radiator cap and top off the radiator with water. Install the cap on the radiator securely.

5. Fill the coolant recovery reservoir to the FULL COLD mark with anti-freeze, then add water to the FULL HOT mark. This will ensure that a proper mixture is in the coolant recovery bottle.

6. Check for leaks at the draincock, block plug and at the vent plug on 2.5L engine.

ENGINE ELECTRICAL

NOTE: Disconnecting the negative battery cable on some vehicles may interfere with the functions of the on board computer systems and may require the computer to undergo a relearning process, once the negative battery cable is reconnected.

Distributor

REMOVAL

1. Disconnect the negative battery cable.

2. Disconnect the wiring connector from the distributor.

3. Remove distributor cap and position it and the attached wires aside, so as not to interfere with removing the distributor.

4. Mark the position of the rotor in relation to the distributor housing and mark the position of the distributor housing on the engine.

5. Remove the rotor.

6. Remove the distributor hold-down bolt and clamp and remove the distributor.

INSTALLATION

NOTE: Before installation, inspect the distributor O-ring and drive gear for wear and/or damage. Rotate the distributor shaft to make sure it moves freely, without binding.

Timing Not Disturbed

1. Install the distributor, aligning the distributor housing and rotor with the marks that were made during the removal procedure.

2. Install the distributor hold-down bolt and clamp. Only snug the bolt at this time.

3. Connect the distributor to the wiring harness.

4. Install the rotor and the distributor cap. Make sure the ignition wires are securely connected to the distributor cap and spark plugs. Tighten the distributor cap screws to 18–23 inch lbs. (2.0–2.6 Nm).

5. Connect a suitable timing light and set the initial timing.

6. Tighten the distributor hold-down bolt to 17–25 ft. lbs. (23–34 Nm) on the 2.5L engine, 14–21 ft. lbs. (19–28 Nm) on the 3.0L engine or 20–29 ft. lbs. (27–40 Nm) on the 3.8L engine.

7. Recheck the initial timing and adjust if necessary.

Timing Disturbed

1. Disconnect the spark plug wire from the No. 1 cylinder spark plug and remove the spark plug.

2. Place a finger over the spark plug hole. Rotate the engine clockwise until compression is felt at the spark plug hole.

3. Align the timing pointer with the TDC mark on the crankshaft damper.

4. Install the rotor on the distributor shaft. Rotate the distributor shaft so the rotor tip is pointing to the distributor cap No. 1 spark plug tower position.

5. While installing the distributor, continue rotating the rotor slightly so the leading edge of the vane is centered in the vane switch stator assembly.

6. Rotate the distributor in the block to align the leading edge of the vane and vane switch stator assembly. Make sure the rotor is pointing to the distributor cap No. 1 spark plug tower position.

NOTE: If the vane and vane switch stator cannot be aligned by rotating the distributor in the block, remove the distributor just enough to disengage the distributor gear from the camshaft gear. Rotate the rotor enough to engage the distributor gear on another tooth of the camshaft gear. Repeat Steps 1 and 2, if necessary.

7. Install the distributor hold-down bolt and clamp. Only snug the bolt at this time.

8. Connect the distributor to the wiring harness and install the distributor cap. Tighten the distributor cap hold-down screws to 18–23 inch lbs. (2.0–2.6 Nm).

9. Install the No. 1 cylinder spark plug and connect the spark plug wire.

10. Connect a suitable timing light and set the initial timing.

11. Tighten the distributor hold-down bolt to 17–25 ft. lbs. (23–34 Nm) on the 2.5L engine, 14–21 ft. lbs. (19–

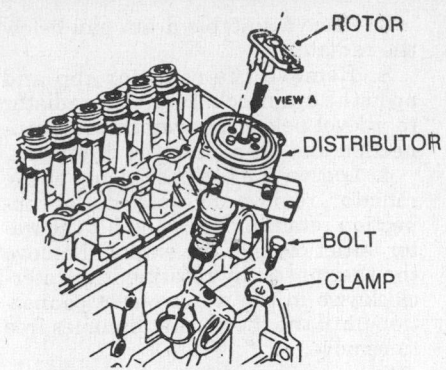

Distributor installation—2.5L engine

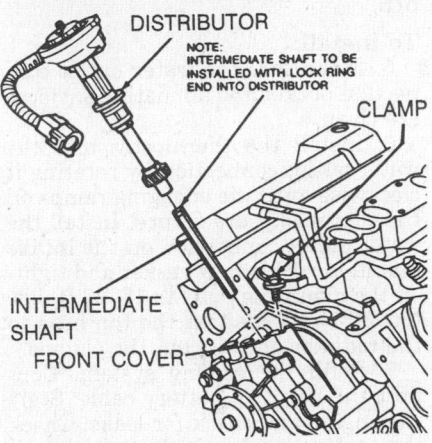

Distributor installation—3.8L engine

28 Nm) on the 3.0L engine or 20–29 ft. lbs. (27–40 Nm) on the 3.8L engine.

12. Recheck the initial timing and adjust if necessary.

Distributorless Ignition

The 3.0L SHO engine is equipped with a Distributorless Ignition System (DIS) which consists of the following components:
  Crankshaft timing sensor
  Camshaft sensor
  DIS ignition module
  Ignition coil pack
  The spark angle portion of the EEC-IV module

REMOVAL & INSTALLATION

Crankshaft Timing Sensor

1. Disconnect the negative battery cable.

2. Loosen the tensioner pulleys for the air conditioning compressor and the power steering pump belts. Remove the belts from the crankshaft pulley.

3. Disconnect the DIS module and remove the intake manifold crossover tube.

4. Remove the upper timing belt cover.

5. Disconnect the sensor wiring harness at the connector and route the wiring harness through the belt cover.

6. Raise the vehicle and support it safely.

7. Remove the right front wheel and tire assembly.

8. Remove the crankshaft pulley using universal puller T67L–3600–A or equivalent.

9. Remove the center and lower timing belt covers.

10. Rotate the crankshaft by hand, to position the metal vane of the shutter outside of the sensor air gap.

11. Remove the crankshaft sensor mounting screws and remove the sensor.

To install:

12. Route the sensor wiring harness through the belt cover. Install the sensor assembly on the mounting pad and install but do not tighten, the retaining screws.

13. Use a 0.03 in. (0.8mm) feeler gauge to set the clearance between the crankshaft sensor assembly and 1 vane on the crankshaft timing pulley and vane assembly. Tighten the screws to 22–31 inch lbs. (2.5–3.5 Nm).

NOTE: This is a critical torque. Overtightening can cause damage to the timing sensor.

14. Install the lower timing belt cover. Install the crankshaft pulley using a suitable tool. Tighten the pulley bolt to 112–127 ft. lbs. (152–172 Nm).

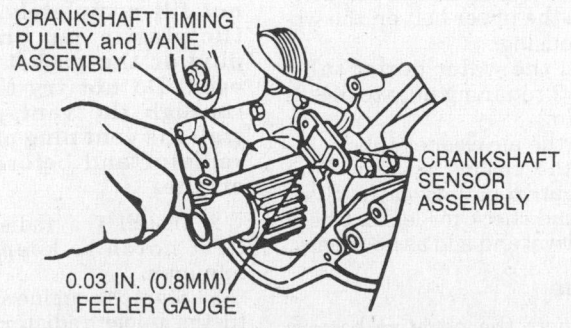

Adjusting crankshaft sensor-to-vane clearance

15. Install the center timing belt cover.

16. Install the right front wheel and tire assembly. Lower the vehicle.

17. Route and connect the sensor wiring harness.

18. Install the upper timing belt cover.

19. Install the intake manifold crossover tube and connect the DIS module.

20. Install the air conditioning and power steering belts and adjust them to the proper tension.

21. Connect the negative battery cable.

Camshaft Sensor Assembly

1. Disconnect the negative battery cable.

2. Remove the engine torque strut.

3. Remove the power steering belt and the pump pulley.

4. Disconnect the camshaft sensor wiring connector.

5. Remove the mounting bolts and remove the sensor.

6. To install, reverse the removal procedure. Tighten the mounting bolts to 22–31 inch lbs. (2.5–3.5 Nm).

DIS Ignition Module

1. Disconnect the negative battery cable.

2. Disconnect the wiring connectors at the module.

3. Remove the module mounting bolts and remove the module.

4. To install, reverse the removal procedure. Apply a uniform coating of heat sink grease to the mounting surface of the DIS module before it is installed. Tighten the mounting bolts to 22–31 inch lbs. (2.5–3.5 Nm).

Ignition Coil Pack

1. Disconnect the negative battery cable.

2. Remove the cover from the coil pack and disconnect the electrical connector.

3. Remove the spark plug wires by squeezing the locking tabs to release the coil boot retainers.

4. Remove the coil pack mounting screws and remove the coil pack.

5. To install, reverse the removal procedure. Tighten the mounting screws to 40–62 inch lbs. (4.5–7 Nm).

Ignition Timing

ADJUSTMENT

Except 3.0L SHO Engine

The timing marks on the 2.5L engine are visible through a hole loin the top of the transaxle case. The 3.0L and 3.8L engines have the timing marks on the crankshaft pulley and a timing pointer near the pulley.

1. Place the transaxle in the **P** or **N** position. Firmly apply the parking brake and block the wheels. The air conditioner and heater must be in the **OFF** position.

2. Open the hood, locate the timing marks and clean with a stiff brush or solvent. On vehicles with manual transaxle, it will be necessary to remove the cover plate which allows access to to the timing marks.

3. Using white chalk or paint, mark the specified timing mark and pointer.

4. Remove the in-line spout connector or remove the shorting bar from the double wire spout connector. The spout connector is the center wire between the electronic control assembly (ECA) connector and the Thick Film Ignition (TFI) module.

5. Connect a suitable inductive type timing light to the No. 1 spark plug wire. Do not, puncture and ignition wire with any type of probing device.

NOTE: The high ignition coil charging currents generated in the EEC–IV ignition system may falsely trigger timing lights with capacitive or direct connect pick-ups. It is necessary that an inductive type timing light be used in this procedure.

6. Connect a suitable tachometer to the engine. The ignition coil connector allows a test lead with an alligator clip to be connected to the Distributor Electronic Control (DEC) terminal without removing the connector.

7. Start the engine and let it run until it reaches normal operating temperature.

NOTE: Only use the ignition key to start the vehicle. Do not use a remote starter, as disconnecting the start wire at the starter relay will cause the TFI module to revert to start mode timing, after the vehicle is started. Reconnecting the start wire after the vehicle is running will not correct the timing.

8. Check the engine idle rpm, if it is not within specifications, adjust as necessary. After the rpm has been adjusted or checked, aim the timing light at the timing marks. If they are not aligned, loosen the distributor clamp bolt slightly and rotate the distributor body until the marks are aligned under timing light illumination.

9. Tighten the distributor clamp bolt and recheck the ignition timing. Readjust the idle speed. Shut the engine off, remove all test equipment, reconnect the in-line spout connector to the distributor and, if necessary, reinstall the cover plate on the manual transaxle vehicles.

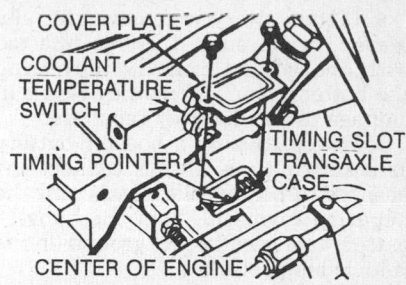

Timing marks location—2.5L engine with manual transaxle

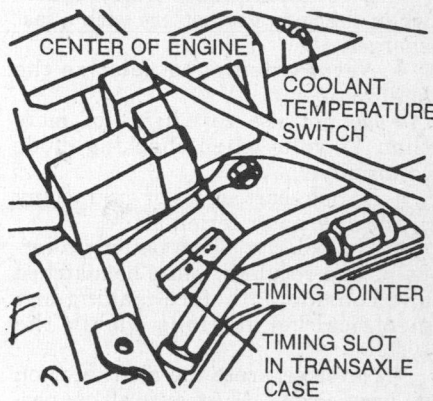

Timing marks location—2.5L engine with automatic transaxle

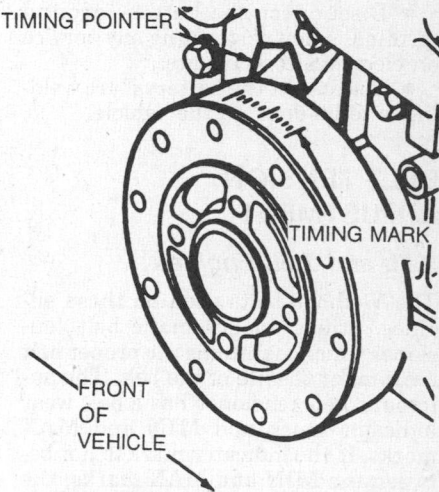

Timing marks location—3.0L engine

3.0L SHO Engine

The base ignition timing is set at 10 degrees BTDC and is not adjustable.

Alternator

PRECAUTIONS

Several precautions must be observed with alternator equipped vehicles to avoid damage to the unit.

- If the battery is removed for any reason, make sure it is reconnected with the correct polarity. Reversing the battery connections may result in damage to the one-way rectifiers.
- When utilizing a booster battery as a starting aid, always connect the positive to positive terminals and the negative terminal from the booster battery to a good engine ground on the vehicle being started.
- Never use a fast charger as a booster to start vehicles.
- Disconnect the battery cables when charging the battery with a fast charger.
- Never attempt to polarize the alternator.
- Do not use test lights of more than 12 volts when checking diode continuity.
- Do not short across or ground any of the alternator terminals.
- The polarity of the battery, alternator and regulator must be matched and considered before making any electrical connections within the system.
- Never separate the alternator on an open circuit. Make sure all connections within the circuit are clean and tight.
- Disconnect the battery ground terminal when performing any service on electrical components.
- Disconnect the battery if arc welding is to be done on the vehicle.

BELT TENSION ADJUSTMENT

2.5L and 3.8L Engines

The V-ribbed belts used on these engines, utilize an automatic belt tensioner which maintains the proper belt tension for the life of the belt. The automatic belt tensioner has a belt wear indicator mark and **MIN** and **MAX** marks. If the indicator mark is not between the **MIN** and **MAX** marks, the belt is worn or an incorrect belt is installed.

3.0L Engine

1. Disconnect the negative battery cable.
2. Loosen the alternator adjustment and pivot bolts.
3. Apply tension to the belt using the adjusting screw.
4. Using a belt tension gauge, set the belt to the proper tension. The tension should be 150 lbs. for a new belt or 120 lbs. for a used belt.
5. When the belt is properly tensioned, tighten the alternator adjustment bolt to 27 ft. lbs. (37 Nm).
6. Remove the tension gauge and run the engine for 5 minutes.

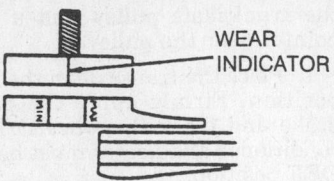

Automatic tensioner drive belt wear indicator

7. With the engine **OFF** and the belt tension gauge in place, check that the adjusting screw is in contact with the bracket before loosening the alternator adjustment bolt. Rotate the adjustment screw until the belt is tensioned to 120 lbs.
8. Tighten the alternator adjustment bolt to 27 ft. lbs. (37 Nm) and the pivot bolt to 43 ft. lbs. (58 Nm).

3.0L SHO Engine

1. Disconnect the negative battery cable.
2. Loosen the idler pulley nut.
3. Turn the adjusting bolt until the belt is adjusted properly.

NOTE: Turning the wrench to the right tightens the belt adjustment and turning the wrench to the left loosens the belt tension.

4. Tighten the idler pulley nut to 25–37 ft. lbs. (34–50 Nm) and check the belt tension.

REMOVAL & INSTALLATION

Except 3.0L SHO Engine

1. Disconnect the negative battery cable.
2. Tag and disconnect the wire harness from the alternator.
3. If equipped with an automatic belt tensioner, rotate the tensioner counterclockwise and remove the drive belt from the pulley.
4. If not equipped with an automatic tensioner, loosen the alternator pivot bolt and remove the adjustment arm bolt from the alternator. Remove the alternator belt from the pulley.
5. Remove the alternator mounting bolts or the pivot bolt, as required, and remove the alternator.
6. Installation is the reverse of the removal procedure. Adjust the belt tension, if not equipped with an automatic belt tensioner.

3.0L SHO Engine

1. Disconnect the battery cables and remove the battery and battery tray.
2. Tag and disconnect the wire harness from the alternator.
3. Loosen the belt tensioner and remove the alternator belt from the pulley.

4. Remove the mounting bolts and the alternator.
5. Installation is the reverse of the removal procedure. Tighten the front mounting bolt to 36–53 ft. lbs. (48–72 Nm) and the rear mounting bolts to 25–37 ft. lbs. (34–50 Nm). Adjust the belt tension.

Voltage Regulator

ADJUSTMENT

The electronic voltage regulator is calibrated and preset by the manufacturer. No adjustment is required or possible.

REMOVAL & INSTALLATION

1. Disconnect the negative battery cable.
2. Disconnect the electrical connectors from the wiring harness.
3. Remove the regulator mounting screws and the regulator.
4. Installation is the reverse of the removal procedure.
5. Connect the negative battery cable. Test the system for proper voltage regulation.

Starter

REMOVAL & INSTALLATION

1. Disconnect the negative battery cable and the cable connection at the starter.
2. Raise and support the vehicle safely.
3. Remove the cable support and ground cable connection from the upper starter stud bolt, if necessary.
4. If equipped, remove the starter brace from the cylinder block and the starter.
5. Remove the starter-to-bell housing bolts and remove the starter.
6. Installation is the reverse of the removal procedure.

EMISSION CONTROLS

Please refer to "Emission Controls" in the Unit Repair section for system maintenance procedures. Due to the complex nature of modern electronic engine control systems, comprehensive diagnosis and testing procedures fall

outside the confines of this repair manual. For complete information on diagnosis, testing and repair procedures concerning all modern engine and emission control systems, please refer to "Chilton's Guide to Fuel Injection and Electronic Engine Controls".

Emission Warning Lamps

These vehicles have a "Check Engine" light that will light when there is a fault in the engine control system. This light cannot be reset without diagnosing the fault in the system. When the system has been diagnosed and the problem corrected, the light will go out.

FUEL SYSTEM

Fuel System Service Precautions

Safety is the most important factor when performing not only fuel system maintenance but any type of maintenance. Failure to conduct maintenance and repairs in a safe manner may result in serious personal injury or death. Maintenance and testing of the vehicle's fuel system components can be accomplished safely and effectively by adhering to the following rules and guidelines.

• To avoid the possibility of fire and personal injury, always disconnect the negative battery cable unless the repair or test procedure requires that battery voltage be applied.

• Always relieve the fuel system pressure prior to disconnecting any fuel system component (injector, fuel rail, pressure regulator, etc.), fitting or fuel line connection. Exercise extreme caution whenever relieving fuel system pressure to avoid exposing skin, face and eyes to fuel spray. Please be advised that fuel under pressure may penetrate the skin or any part of the body that it contacts.

• Always place a shop towel or cloth around the fitting or connection prior to loosening to absorb any excess fuel due to spillage. Ensure that all fuel spillage (should it occur) is quickly removed from engine surfaces. Ensure that all fuel soaked cloths or towels are deposited into a suitable waste container.

• Always keep a dry chemical (Class B) fire extinguisher near the work area.

• Do not allow fuel spray or fuel vapors to come into contact with a spark or open flame.

• Always use a backup wrench when loosening and tightening fuel line connection fittings. This will prevent unnecessary stress and torsion to fuel line piping. Always follow the proper torque specifications.

• Always replace worn fuel fitting O-rings with new. Do not substitute fuel hose or equivalent where fuel pipe is installed.

RELIEVING FUEL SYSTEM PRESSURE

1. The pressure in the fuel system must be released before attempting to disconnect any fuel lines.

2. A special valve is incorporated in the fuel rail assembly for the purpose of relieving the pressure in the fuel system.

3. Attach pressure gauge tool T80L–9974–A or equivalent, to the fuel pressure valve on the fuel rail assembly and release the pressure from the system into a suitable container.

Fuel Tank

REMOVAL & INSTALLATION

1. Disconnect the negative battery cable.

2. Relieve the fuel system pressure.

3. Siphon or pump the fuel from the fuel tank, through the filler neck, into a suitable container.

NOTE: There are reservoirs inside the fuel tank to maintain fuel near the fuel pickup during cornering and under low fuel operating conditions. These reservoirs could block siphon tubes or hoses from reaching the bottom of the tank. A few repeated attempts using different hose orientations can overcome this situation.

4. Raise and safely support the vehicle.

5. Loosen the filler pipe and vent hose clamps at the tank and remove the hoses from the tank.

6. Place a safety support under the fuel tank and remove the bolts from the rear of the fuel tank straps. The straps are hinged at the front and will swing aside.

7. Partially remove the tank. Remove the hairpin clips from the push connect fitting and disconnect the fuel lines. Disconnect the electrical connector from the fuel sender/pump assembly.

8. Remove the fuel tank.

To install:

9. Raise the fuel tank into position. Connect the fuel lines and the electrical connector.

10. Bring the fuel tank straps around the tank and start the retaining bolt. Align the tank as far forward in the vehicle as possible while securing the retaining bolts.

NOTE: If equipped with a heat shield, make sure it is installed with the straps and positioned correctly on the tank.

11. Check the hoses and wiring mounted on the tank top, to make sure they are correctly routed and will not be pinched between the tank and the body.

12. Tighten the fuel tank strap retaining bolts to 21–29 ft. lbs. (28–40 Nm).

13. Install the fuel filler hoses and tighten the clamps. Refill the fuel tank.

14. Check all connections for leaks. Connect the negative battery cable.

Fuel Filter

REMOVAL & INSTALLATION

1. Disconnect the negative battery cable. Relieve the fuel system pressure.

2. Remove the push connect fittings at both ends of the fuel filter. This is accomplished by removing the hairpin clips from the fittings. Remove the hairpin clips by first bending and then breaking the shipping tabs on the clips. Then spread the 2 clip legs approximately 1/8 in. to disengage the body and push the legs into the fitting. Pull on the triangular end of the clip and work it clear of the fitting.

3. Remove the filter from the mounting bracket by loosening the worm gear mounting clamp enough to allow the filter to pass through.

To install:

4. Install the filter in the mounting bracket, ensuring that the flow direction arrow is pointing forward. Locate the fuel filter against the tab at the lower end of the bracket.

5. Insert a new hairpin clip into any 2 adjacent openings on each push con-

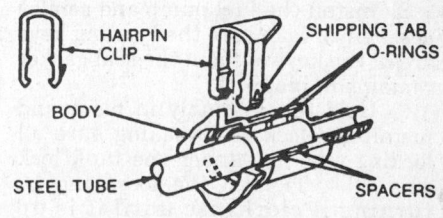

Push connect fittings

nect fitting, with the triangular portion of the clip pointing away from the fitting opening. Install the clip to fully engage the body of the fitting. This is indicated by the legs of the hairpin clip being locked on the outside of the fitting body. Apply a light coat of engine oil to the ends of the fuel filter and then push the fittings onto the ends of the fuel filter. When the fittings are engaged, a definite click will be heard. Pull on the fittings to ensure that they are fully engaged.

6. Tighten the worm gear mounting clamp to 15–25 inch lbs. (1.7–2.8 Nm).

7. Start the engine and check for leaks.

Electric Fuel Pump

PRESSURE TESTING

1. Ground the fuel pump lead of the self-test connector through a jumper wire at the FP lead.

2. Connect a suitable fuel pressure tester to the fuel pump outlet.

3. Turn the ignition key to the **RUN** position to operate the fuel pump.

4. The fuel pressure should be 35–45 psi for all engines escept the 2.5L engine. On the 1988–90 2.5L CFI engine, the fuel pressure should be 13–17 psi. On the 1991–92 2.5L SEFI engine, the fuel pressure should be 45–60 psi.

REMOVAL & INSTALLATION

1. Disconnect the negative battery cable.

2. Relieve the fuel system pressure.

3. Remove the fuel tank from the vehicle and place it on a work bench. Remove any dirt around the fuel pump attaching flange.

4. Turn the fuel pump locking ring counterclockwise and remove the lock ring.

5. Remove the fuel pump from the fuel tank and discard the flange gasket.

To install:

6. Clean the fuel pump mounting flange and fuel tank mounting surface and seal ring groove.

7. Put a light coating of grease on the new seal gasket to hold it in place during assembly and install it in the fuel ring groove.

8. Install the fuel pump and sender assembly. Make sure the locating keys are in the keyways and the seal gasket remains in place.

9. Hold the assembly in place and install the lock ring making sure all locking tabs are under the tank lock ring tabs. Tighten the lock ring by turning it clockwise until it is up against the stops.

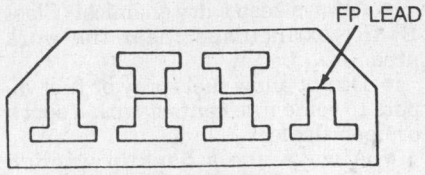

Self test connector

Electric fuel pump assembly and terminal locations

10. Install the fuel tank.

11. Fill the tank with a minimum of 10 gallons of fuel and check for leaks.

12. Connect a suitable fuel pressure gauge. Turn the ignition switch to the **ON** position 5–10 times, leaving it on for 3 seconds at a time, until the pressure gauge reads at least 30 psi. Check for leaks at the fittings.

13. Remove the pressure gauge, start the engine and recheck for leaks.

Fuel Injection

IDLE MIXTURE ADJUSTMENT

Idle mixture is controlled by the electronic control unit. No adjustment is possible.

Fuel Injector

REMOVAL & INSTALLATION

1988–90 2.5L CFI Engine

1. Disconnect the negative battery cable.

2. Relieve the fuel system pressure.

3. Remove the fuel injector retaining screw and retainer.

4. Remove the injector and lower O-ring. Discard the O-ring.

To install:

5. Lubricate a new lower O-ring and the injector seat area with clean engine oil; do not use transmission fluid. Install the lower O-ring on the injector.

6. Lubricate the upper O-ring and install the injector by centering and applying a steady downward pressure with a slight rotational force.

7. Install the injector retainer and retaining screw. Tighten the screw to 18–22 inch lbs. (2.0–2.5 Nm).

1991–92 2.5L SEFI Engine

1. Disconnect the negative battery cable.

2. Relieve the fuel pressure in the system.

3. Disconnect the engine air cleaner outlet tube from the air intake throttle body and the TP sensor from the wiring harness.

4. Disconnect the vacuum lines from the upper manifold and disconnect the EGR tube at the manifold connection.

5. Disconnect the air bypass valve connector, remove the accelerator and, if equipped, speed control cables and remove the manifold upper support bracket top bolt.

6. Remove the fuel supply manifold shield and the 4 upper manifold retaining bolts and 1 retaining shoulder stud.

7. Remove the upper manifold assembly and gasket and set it aside.

8. Disconnect the fuel supply and return lines and the vacuum line at the pressure regulator.

9. Disconnect the fuel injector wiring harness and disconnect the connectors from the injectors.

10. Remove the fuel supply manifold retaining bolts and remove the fuel supply manifold.

11. Grasping the injector body, pull up while gently rocking the injector from side-to-side.

12. Inspect the injector O-rings, the injector plastic hat and washer for signs of deterioration. Replace as necessary. If the hat is missing, look for it in the intake manifold.

13. Installation is the reverse of the removal procedure. Lubricate new O-rings with light engine oil and install on the injectors prior to installation. Tighten the fuel supply manifold retaining bolts and the upper intake manifold retaining bolts to 15–22 ft. lbs. (20–30 Nm).

3.0L EFI Engine

1. Disconnect the negative battery cable.

2. Relieve the fuel system pressure.

3. Remove the air intake throttle body.

4. Disconnect the fuel supply and fuel return lines.

5. Disconnect the wiring harness from the injectors.

6. Disconnect the vacuum line from the fuel pressure regulator valve.

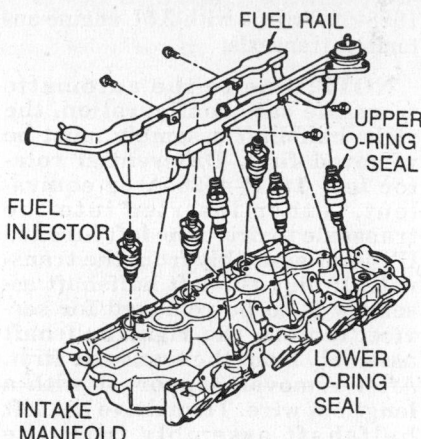

FUEL RAIL

UPPER O-RING SEAL

FUEL INJECTOR

LOWER O-RING SEAL

INTAKE MANIFOLD

Injector removal and installation— 3.0L engine

7. Remove the 4 fuel injector manifold retaining bolts.

8. Carefully disengage the fuel rail assembly from the fuel injectors by lifting and gently rocking the rail.

9. Remove the injectors by lifting while gently rocking from side to side.

To install:

10. Lubricate new O-rings with engine oil and install 2 on each injector.

11. Make sure the injector cups are clean and undamaged.

12. Install the injectors in the fuel rail using a light twisting-pushing motion.

13. Carefully install the rail assembly and injectors into the lower intake manifold, 1 side at a time. Make sure the O-rings are seated by pushing down on the fuel rail.

14. While holding the fuel rail assembly in place, install the 2 retaining bolts and tighten to 7 ft. lbs. (10 Nm).

15. Connect the fuel supply and fuel return lines.

16. Before connecting the fuel injector harness, turn the ignition switch to the **ON** position. This will pressurize the fuel system.

17. Using a clean paper towel. check for leaks where the injector connects to the fuel rail.

18. Install the air intake throttle body and connect the vacuum line to the fuel pressure regulator valve.

19. Connect the fuel injector harness, start the engine and let it idle for 2 minutes.

20. Using a clean paper towel, check for leaks where the injector is installed into the intake manifold.

3.0L SHO Engine

1. Disconnect the negative battery cable.

2. Relieve the fuel system pressure.

3. Remove the intake manifold as follows:

a. Drain the cooling system.

b. Remove the intake air tube from the throttle body and MAF sensor. Disconnect the throttle cables.

c. Disconnect the electrical connectors at the TP sensor, air bypass valve, vacuum switching valve and DIS module.

d. Disconnect the coolant bypass hoses and vacuum lines.

e. Disconnect the EGR pipe from the EGR valve.

f. Remove the 8 bolts at the intake manifold support brackets and remove the brackets.

g. Remove the bolt retaining the coolant hose bracket and disconnect the PCV hoses.

h. Remove the 12 manifold retaining bolts and remove the intake manifold and throttle body assembly.

4. Disconnect the electrical connectors at the fuel injectors.

5. Remove the fuel rail retaining bolts.

6. Raise and slightly rotate the fuel rail assembly and remove the injectors.

To install:

7. Lubricate new O-rings with engine oil and install them on the fuel injectors.

8. Install the injectors in the fuel rail by lightly twisting and pushing the injectors into position.

9. Install the fuel rail, making sure the injectors seat properly in the cylinder head.

10. Install the fuel rail retaining bolts and tighten to 11–17 ft. lbs. (15–23 Nm).

11. Connect the electrical connectors at the injectors. Install the intake manifold by reversing the removal procedure.

12. Run the engine and check for leaks.

3.8L Engine

1. Disconnect the negative battery cable.

2. Remove the fuel cap at the tank and release the pressure.

3. Relieve the pressure from the fuel system.

4. Remove the upper intake manifold and the fuel supply manifold as follows:

a. Disconnect the electrical connectors at the air bypass valve, TP sensor and EGR position sensor.

b. Disconnect the throttle linkage at the throttle ball and the transmission linkage from the throttle body. Remove the 2 bolts securing the bracket to the intake manifold and position the bracket with the cables aside.

c. Disconnect the upper intake

manifold vacuum fitting connections by disconnecting all vacuum lines to the vacuum tree, EGR valve and pressure regulator.

d. Disconnect the PCV hose and remove the nut retaining the EGR transducer to the upper intake manifold.

e. Loosen the EGR tube at the exhaust manifold and disconnect at the EGR valve.

f. Remove 2 bolts retaining the EGR valve to the upper intake manifold and remove the EGR valve and EGR transducer as an assembly.

g. Remove the 2 canister purge lines from the fittings on the throttle body and remove the 6 upper intake manifold retaining bolts.

h. Remove 2 retaining bolts on the front and rear edges of the upper intake manifold where the manifold support brackets are located.

i. Remove the nut retaining the alternator bracket to the upper intake manifold and the 2 bolts retaining the alternator bracket to the water pump and alternator.

j. Remove the upper intake manifold and throttle body as an assembly.

k. Disconnect the fuel supply and return lines from the fuel rail assembly.

l. Remove the fuel rail assembly retaining bolts, carefully disengage the fuel rail from the fuel injectors and remove the fuel rail.

5. Remove the injector retaining clips.

6. Remove the electrical connectors from the fuel injectors.

7. To remove the injector, pull it up while gently rocking it from side-to-side.

8. Inspect the injector pintle protection cap (plastic hat) and washer for deterioration and replace, as required.

To install:

9. Lubricate new engine O-rings with engine oil and install 2 on each injector.

10. Install the injectors, using a light, twisting, pushing motion to install them.

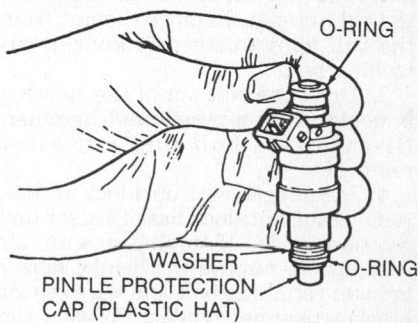

O-RING

WASHER

PINTLE PROTECTION CAP (PLASTIC HAT)

O-RING

Fuel injector—3.8L engine

11. Reconnect the injector retaining clips.

12. Install the fuel rail assembly.

13. Install the electrical harness connectors to the injectors.

14. Install the upper intake manifold by reversing the removal procedure.

15. Install the fuel cap at the tank.

16. Connect the negative battery cable.

17. Turn the ignition switch from **ON** to **OFF** position several times without starting the engine to check for fuel leaks.

DRIVE AXLE

Halfshaft

When removing both the left and right halfshafts, install suitable shipping plugs to prevent dislocation of the differential side gears. Should the gears become misaligned, the differential will have to be removed from the transaxle to re-align the side gears.

NOTE: Due to the automatic transaxle case configuration, the right halfshaft assembly must be removed first. Differential Rotator T81P–4026–A or equivalent, is then inserted into the transaxle to drive the left inboard CV-joint assembly from the transaxle. If only the left halfshaft assembly is to be removed for service, remove only the right halfshaft assembly from the transaxle. After removal, support it with a length of wire. Then drive the left halfshaft assembly from the transaxle.

REMOVAL & INSTALLATION

1. Disconnect the negative battery cable. Remove the wheel cover/hub cover from the wheel and tire assembly and loosen the wheel nuts.

2. Raise the vehicle and support safely. Remove the wheel assembly, remove the hub nut and washer. Discard the old hub nut. Remove the nut from the ball joint to steering knuckle attaching bolts.

3. Drive the bolt out of the steering knuckle using a punch and hammer. Discard this bolt and nut after removal.

4. If equipped with anti-lock brakes, remove the anti-lock brake sensor and position aside. If equipped with air suspension, remove the height sensor bracket retaining bolt and wire sensor bracket to inner fender. Position the sensor link aside.

5. Separate the ball joint from the steering knuckle using a suitable prybar. Position the end of the prybar outside of the bushing pocket to avoid damage to the bushing. Use care to prevent damage to the ball joint boot. Remove the stabilizer bar link at the stabilizer bar.

6. The following removal procedure applies to the right side halfshaft/link shaft for the 1988–90 2.5L engine Taurus. For all other automatic transaxles, proceed to Step 7:

a. Remove the bolts attaching the bearing support to the bracket. Slide the link shaft out of the transaxle. Support the end of the shaft by suspending it from a convenient underbody component with a piece of wire. Do not allow the shaft to hang unsupported, damage to the outboard CV-joint may occur.

b. Separate the outboard CV-joint from the hub using front hub remover tool T81P–1104–C or equivalent and metric adapter tools T83–P–1104–BH, T86P–1104–Al and T81P–1104–A or equivalent.

NOTE: Never use a hammer to separate the outboard CV-joint stub shaft from the hub. Damage to the CV-joint threads and internal components may result. The right side link shaft and halfshaft assembly is removed as a complete unit.

7. The following removal procedure applies to the right and left side halfshafts for the automatic transaxle, except 1988–90 2.5L engine Taurus and the left side halfshaft on the manual transaxle:

a. Install the CV-joint puller tool T86P–3514–A1 or equivalent, between CV-joint and transaxle case. Turn the steering hub and/or wire strut assembly aside.

b. Screw extension tool T86P–3514–A2 or equivalent, into the CV-joint puller and hand tighten. Screw an impact slide hammer onto the extension and remove the CV-joint.

c. Support the end of the shaft by suspending it from a convenient underbody component with a piece of wire. Do not allow the shaft to hang un-supported, damage to the outboard CV-joint may occur.

d. Separate the outboard CV-joint from the hub using front hub remover tool T81P–1104–C or equivalent and metric adapter tools T83–P–1104–BH, T86P–1104–Al and T81P–1104–A or equivalent.

e. Remove the halfshaft assembly from the vehicle.

8. The following removal procedure applies to the left side halfshaft for the 1988–90 Taurus with 2.5L engine automatic transaxle:

NOTE: Due to the automatic transaxle case configuration, the right halfshaft assembly must be removed first. Differential rotator tool T81P–4026–A or equivalent, is then inserted into the transaxle to drive the left inboard CV-joint assembly from the transaxle. If only the left halfshaft assembly is to be removed for service, remove the right halfshaft assembly from the transaxle first. After removal, support it with a length of wire. Then drive the left halfshaft assembly from the transaxle.

a. Support the end of the shaft by suspending it from a convenient underbody component with a piece of wire. Do not allow the shaft to hang unsupported as damage to the outboard CV-joint may occur.

b. Separate the outboard CV-joint from the hub front hub remover tool T81P–1104–C or equivalent and metric adapter tools T83–P–1104–BH, T86P–1104–Al and T81P–1104–A or equivalent.

c. Remove the halfshaft assembly from the vehicle.

To install:

9. Install a new circlip on the inboard CV-joint stub shaft and/or link shaft. The outboard CV-joint does not have a circlip. When installing the circlip, start one end in the groove and work the circlip over the stub shaft end into the groove. This will avoid over expanding the circlip.

NOTE: The circlip must not be re-used. A new circlip must be installed each time the inboard CV-joint is installed into the transaxle differential.

10. Carefully align the splines of the inboard CV-joint stub shaft with the splines in the differential. Exerting some force, push the CV-joint into the differential until the circlip is felt to seat in the differential side gear. Use care to prevent damage to the differential oil seal. If equipped, torque the link shaft bearing to 16–23 ft. lbs.

NOTE: A non-metallic mallet may be used to aid in seating the circlip into the differential side gear groove. If a mallet is necessary, tap only on the outboard CV-joint stub shaft.

11. Carefully align the splines of the outboard CV-joint stub shaft with the splines in the hub and push the shaft into the hub as far as possible.

12. Temporarily fasten the rotor to the hub with washers and 2 wheel lug

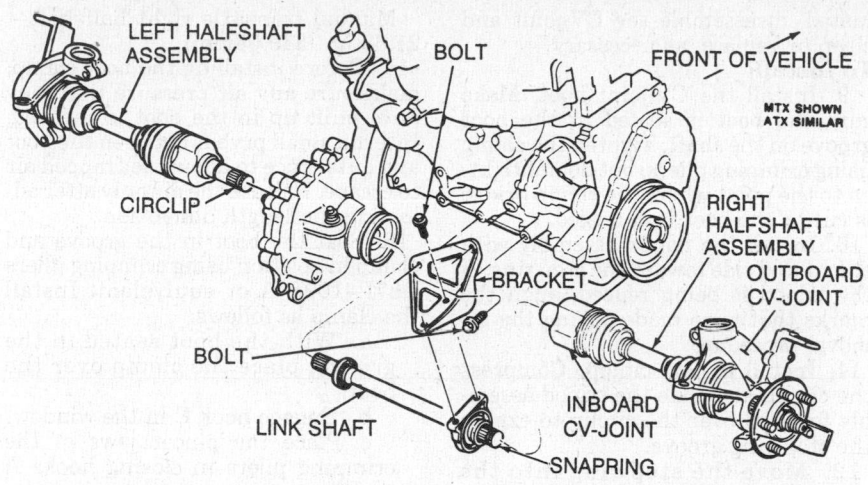

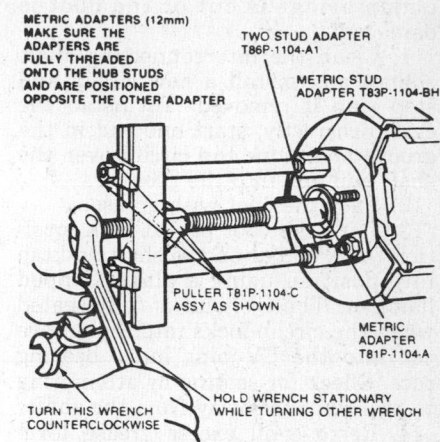

Halfshaft assembly removal and installation—Taurus/Sable with manual transaxle

Removing stub shaft from hub assembly

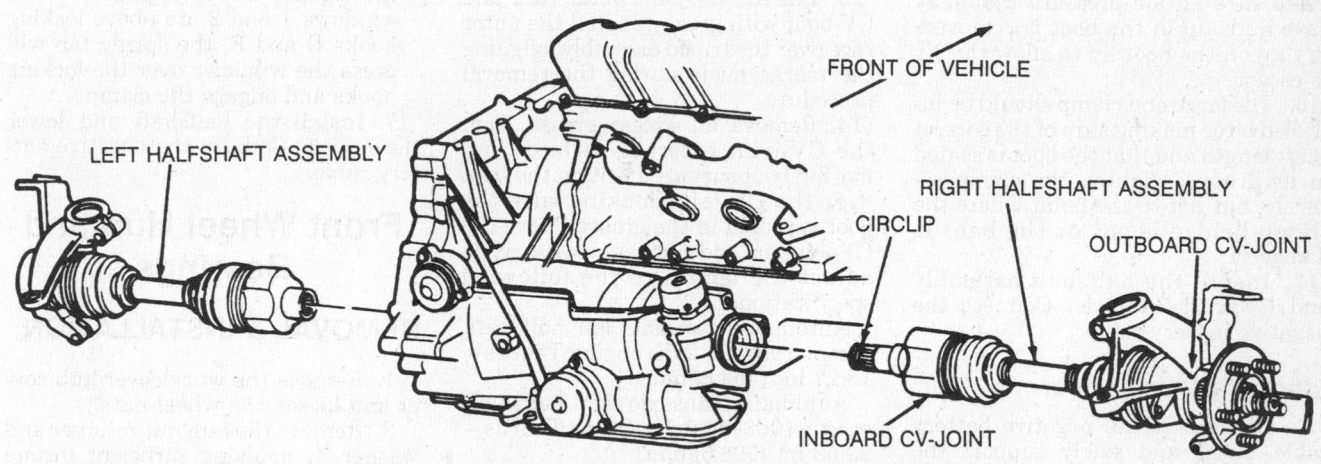

Halfshaft assembly removal and installation—Continental

nuts. Insert a steel rod into the rotor and rotate clockwise to contact the knuckle to prevent the rotor from turning during the CV-joint installation.

13. Install the hub nut washer and a new hub nut. Manually thread the retainer onto the CV-joint as far as possible.

14. Connect the control arm to the steering knuckle and install a new nut and bolt. Tighten the nut to 40–55 ft. lbs. (54–74 Nm). A new bolt must be installed also.

15. Install the anti-lockbrake sensor and/or the ride height sensor bracket, if equipped.

16. Connect the stabilizer link to the stabilizer bar. Tighten to 35–48 ft. lbs. (47–65 Nm).

17. Tighten the hub retainer nut to 180–200 ft. lbs. (245–270 Nm). Remove the steel rod.

18. Install the wheel and tire assembly and lower the vehicle. Tighten the wheel nuts to 80–105 ft. lbs. Fill the transaxle to the proper level with the specified fluid.

CV-Boot

REMOVAL & INSTALLATION

Outboard CV-Joint Boot

1. Disconnect the negative battery cable. Raise and safely support the vehicle.

2. Remove the halfshaft assembly from the vehicle.

3. Clamp the halfshaft in a vise that is equipped with soft jaw covers. Do not allow the vise jaws to contact the boot or boot clamp.

4. Cut the large boot clamp with a pair of side cutters and peel the clamp away from the boot. Roll the boot back over the shaft after the clamp has been removed.

5. Clamp the interconnecting shaft in a soft jawed vise with the CV-joint pointing downward so the inner bearing race is exposed.

6. Use a brass drift and hammer, give a sharp tap to the inner bearing race to dislodge the internal snaping and separate the CV-joint from the in-

terconnecting shaft. Take care to secure the CV-joint so it does not drop after separation. Remove the clamp and boot from the shaft.

7. Remove and discard the circlip at the end of the interconnecting shaft. The stop ring, located just below the circlip should be removed and replaced only if damaged or worn.

To install:

8. Clean the joint and repack with fresh grease. Do not reuse the old grease Install a new boot or reinstall the old boot with a new clamp.

9. The left and right interconnecting shafts are different, depending on year and vehicle application. The outboard end of the shaft is shorter from the end of the shaft to the end of the boot groove than the inboard end. Take a measurement to insure correct installation.

10. Install the new boot. Make sure the boot is seated in the mounting groove and secure it in position with a new clamp. Tighten the clamp securely, but not to the point where the

clamp bridge is cut or the boot is damaged.

11. Clean the interconnecting shaft splines and install a new circlip and stop ring if removed. To install the circlip correctly, start one end in the groove and work the circlip over the shaft end and into the groove.

12. Pack the boot with grease.

13. With the boot peeled back, position the CV-joint on the shaft and tap into position using a plastic tipped hammer. The CV-joint is fully seated when the circlip locks into the groove cut into the CV-joint inner bearing race. Check for seating by attempting to pull the joint away from the shaft.

14. Remove all excess grease form the CV-joint external surface and position the boot over the joint.

15. Before installing the boot clamp, make sure all air pressure that may have built up in the boot is removed. Pry up on the boot lip to allow the air to escape.

16. The large end clamp should be installed after making sure of the correct shaft length and that the boot is seated in its groove. Tighten the clamp securely, but not to the point where the clamp bridge is cut or the boot is damaged.

17. Install the halfshaft assembly and lower the vehicle. Connect the negative battery cable.

Inboard CV-Joint Boot

1. Disconnect the negative battery cable. Raise and safely support the vehicle.

2. Remove the halfshaft assembly from the vehicle.

3. Clamp the halfshaft in a vise that is equipped with soft jaw covers. Do not allow the vise jaws to contact the boot or boot clamp.

4. Cut and remove both boot clamps and slide the boot back on the shaft. Remove the clamp by engaging the pincer jaws of boot clamp pliers D87P-1090-A or equivalent, in the closing hooks on the clamp and draw together. Disengage the windows and locking hooks and remove the clamp.

5. Mark the position of the outer race in relation to the shaft and remove the outer race.

6. Move the stop ring back on the shaft using snapring pliers. Move the tripod assembly back on the shaft to allow access to the circlip.

7. Remove the circlip from the shaft. Mark the position of the tripod on the shaft and remove the tripod assembly. Remove the boot.

8. Check the CV-joint grease for contamination. If the CV-joints are operating properly and the grease is not contaminated, add grease and replace the boot. If the grease appears contam-

inated, disassemble the CV-joint and clean or replace, as necessary.

To install:

9. Install the CV-joint boot. Make sure the boot is seated in the boot groove on the shaft. Tighten the clamp using crimping pliers, but do not tighten to the point where the clamp bridge is cut or the boot is damaged.

10. Install the tripod assembly with chamfered side toward the stop ring. If the tripod is being reused, align the marks that were made during the removal procedure.

11. Install a new circlip. Compress the circlip and slide the tripod assembly forward over the circlip to expose the stop ring groove.

12. Move the stop ring into the groove using snapring pliers, making sure it is fully seated in the groove.

13. Fill the CV-joint outer race and CV-boot with grease. Install the outer race over the tripod assembly, aligning the marks made during the removal procedure.

14. Remove all excess grease from the CV-joint external surfaces and mating boot surface. Position the boot over the CV-joint making sure the boot is seated in the groove. Move the CV-joint in and out, as necessary, to adjust the length to the following specifications:

Automatic transaxle left halfshaft, except 1988–90 2.5L engine Taurus— 18.27 in. (463.65mm)

Automatic transaxle right halfshaft, except 1988–90 2.5L engine Taurus— 23.58 in. (598.55mm)

Automatic transaxle left halfshaft, 1988–90 2.5L engine Taurus—22.80 in. (578.75mm)

Automatic transaxle right halfshaft, 1988–90 2.5L engine Taurus—20.09 in. (510.05mm)

Manual transaxle left halfshaft— 21.24 in. (539.05mm)

Manual transaxle right halfshaft— 21.63 in. (549.05mm)

15. Before installing the boot clamp, make sure any air pressure that may have built up in the boot is relieved. Insert a small prybar between the boot and outer race to allow the trapped air to escape. Release the air only after adjusting the length dimension.

16. Seat the boot in the groove and clamp in position using crimping pliers D87P-1098-A or equivalent. Install the clamp as follows:

a. With the boot seated in the groove, place the clamp over the boot.

b. Engage hook C in the window.

c. Place the pincer jaws of the crimping pliers in closing hooks A and B.

d. Secure the clamp by drawing the closing hooks together. When windows 1 and 2 are above locking hooks D and E, the spring tab will press the windows over the locking hooks and engage the clamp.

17. Install the halfshaft and lower the vehicle. Connect the negative battery cable.

Front Wheel Hub and Bearings

REMOVAL & INSTALLATION

1. Remove the wheelcover/hub cover and loosen the wheel nuts.

2. Remove the hub nut retainer and washer by applying sufficient torque to the nut to overcome the prevailing torque feature of the crimp in the nut collar. Do not use an impact-type tool to remove the hub nut retainer. The hub nut retainer is not reuSable and must be discarded after removal.

3. Raise the vehicle and support it safely. Remove the wheel.

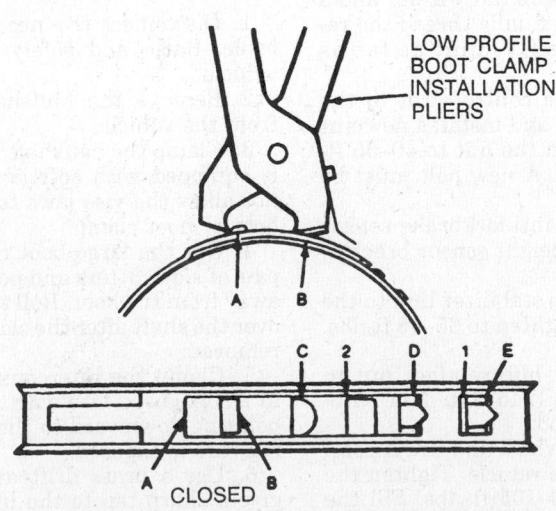

LOW PROFILE BOOT CLAMP INSTALLATION PLIERS

Inboard CV-joint boot clamp tightening procedure

4. Remove the brake caliper by loosening the caliper locating pins and rotating the caliper off of the rotor, starting from the lower end of the caliper and lifting upwards. Do not remove the caliper pins from the caliper assembly. Once the caliper is free of the rotor, support it with a length of wire. Do not allow the caliper to hang from the brake hose.

5. Remove the rotor from the hub by pulling it off of the hub bolts. If the rotor is difficult to remove, strike it sharply between the studs with a rubber or plastic hammer. If the rotor will not pull off, apply a suitable rust penetrator to the inboard and outboard rotor hub mating surfaces. Install a suitable 3-jaw puller and remove the rotor by pulling on the rotor outside diameter and pushing on the hub center. If excessive force is required to remove the rotor, check it for lateral runout prior to installation. Lateral runout must be checked with the nuts clamping the stamped hat section of the rotor.

6. Remove the rotor splash shield.

7. Disconnect the lower control arm and tie rod from the knuckle but leave the strut attached. Loosen the 2 strut top mount-to-apron nuts.

8. Install hub remover/installer adapter T81P–1104–A with front hub remover/installer T81P–1104–C and wheel bolt adapters T83P–1104–BH and 2 stud adapter T86P–1104–A1 or equivalent, and remove the hub, bearing and knuckle assembly by pushing out the CV-joint outer shaft until it is free of the assembly.

9. Support the knuckle with a length of wire, remove the strut bolt and slide the hub/bearing/knuckle assembly off of the strut. Remove the support wire and carry the hub/bearing/knuckle assembly to a bench.

10. Install front hub puller D80L–1002–L and shaft protector D80L–625–1 or equivalent, with the jaws of the puller on the knuckle bosses. Make sure the shaft protector is centered, clears the bearing inside diameter and rests on the end face of the hub journal. Remove the hub.

11. Remove the snapring that retains the bearing in the knuckle assembly and discard.

12. Using a suitable hydraulic press, place front bearing spacer T86P–1104–A2 or equivalent, on the press plate with the step side facing up and position the knuckle with the outboard side up on the spacer. Install front bearing remover T83P–1104–AH2 or equivalent, centered on the bearing inner race and press the bearing out of the knuckle and discard.

To install:

13. Remove all foreign material from the knuckle bearing bore and hub bearing journal to ensure correct seating of the new bearing.

NOTE: If the hub bearing journal is scored or damaged it must be replaced. The front wheel bearings are pregreased and sealed and require no scheduled maintenance. The bearings are preset and cannot be adjusted. If a bearing is disassembled for any reason, it must be replaced as a unit, as individual service seals, rollers and races are not available.

14. Place front bearing spacer T86P–1104–A2 or equivalent, with the step side down on the hydraulic press plate and position the knuckle with the outboard side down on the spacer. Position a new bearing in the inboard side of the knuckle. Install bearing installer T86P–1104–A3 or equivalent, with the undercut side facing the bearing, on the bearing outer race and press the bearing into the knuckle. Make sure the bearing seats completely against the shoulder of the knuckle bore.

NOTE: Bearing installer T86P–1104–A3 or equivalent, must be positioned as indicated above to prevent bearing damage during installation.

15. Install a new snapring (part of the bearing kit) in the knuckle groove.

16. Place front bearing spacer T86P–1104–A2 or equivalent, on the press plate and position the hub on the tool with the lugs facing downward. Position the knuckle assembly with the outboard side down on the hub barrel. Place bearing remover T83P–1104–AH2 or equivalent, flat side down, centered on the inner race of the bearing and press down on the tool until the bearing is fully seated onto the hub. Make sure the hub rotates freely in the knuckle after installation.

17. Prior to hub/bearing/knuckle installation, replace the bearing dust seal on the outboard CV-joint with a new seal from the bearing kit. Make sure the seal flange faces outboard toward the bearing. Use drive tube T83T–3132–A1 and front bearing dust seal installer T86P–1104–A4 or equivalent.

18. Suspend the hub/bearing/knuckle assembly on the vehicle with wire and attach the strut loosely to the knuckle. Lubricate the CV-joint stub shaft with SAE 30 weight motor oil and insert the shaft into the hub splines as far as possible using hand pressure only. Make sure the splines are properly engaged.

19. Temporarily fasten the rotor to the hub with washers and 2 wheel lug nuts. Insert a suitable tool into the rotor diameter and rotate clockwise to contact the knuckle.

20. Install the hub nut washer and a new hub nut retainer. Rotate the nut clockwise to seat the CV-joint. Tighten the nut to 180–200 ft. lbs. (245–270 Nm). Remove the washers and lug nuts.

NOTE: Do not use power or impact-type tools to tighten the hub nut.

21. Install the remainder of the front suspension components and the rotor splash shield.

22. Install the disc brake rotor and caliper. Make sure the outer brake pad spring hook is seated under the upper arm of the knuckle.

23. Install the wheel and tighten the wheel nuts finger tight.

24. Lower the vehicle. Tighten the wheel nuts to 85–105 ft. lbs. (115–142 Nm). Install the wheelcover/hub cover.

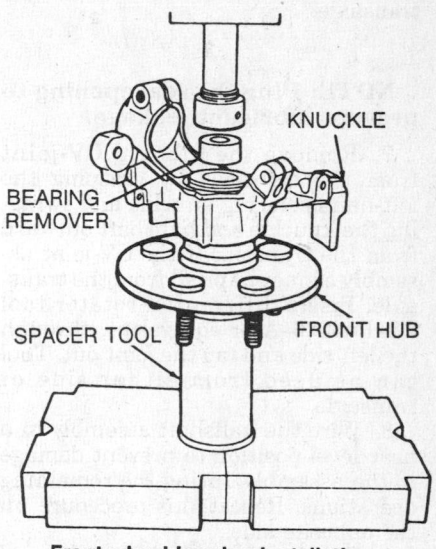

Front wheel bearing Installation

MANUAL TRANSAXLE

For further information on transmissions/transaxles, please refer to "Chilton's Guide to Transmission Repair".

Transaxle Assembly

REMOVAL & INSTALLATION

1988

1. Disconnect the negative battery cable. Wedge a wood block approximately 7 in. long under the clutch ped-

al to hold the pedal up slightly beyond its normal position. Grasp the clutch cable and pull forward, disconnecting it from the clutch release shaft assembly. Remove the clutch casing from the rib on the top surface of the transaxle case.

2. Remove the 2 top transaxle-to-engine mounting bolts.

3. Raise the vehicle and support safely.

4. Remove the nut and bolt that secures the lower control arm ball joint to the steering knuckle assembly. Discard the nut and bolt. Repeat this procedure on the opposite side.

5. Using a large prybar, pry the lower control arm away from the knuckle.

NOTE: Exercise care not to damage or cut the ball joint boot. Prybar must not contact the lower arm.

6. Using a large prybar, pry the left inboard CV-joint assembly from the transaxle.

NOTE: Plug the seal opening to prevent lubricant leakage.

7. Remove the inboard CV-joint from the transaxle by grasping the left-hand steering knuckle and swinging the knuckle and halfshaft outward from the transaxle.If the CV-joint assembly cannot be pried from the transaxle, insert differential rotator tool T81P–4026–A or equivalent, through the left side and tap the joint out. Tool can be used from either side of transaxle.

8. Wire the halfshaft assembly in a near level position to prevent damage to the assembly during the remaining operations. Repeat this procedure on the opposite side.

9. Disengage the locking tabs and remove the backup light switch connector from the transaxle backup light switch.

10. Remove the starter stud bolts.

11. Remove the shift mechanism to shift shaft attaching nut and bolt and control selector indicator switch arm. Remove from the shift shaft.

12. Remove the bolts attaching the shift cable and bracket assembly to the transaxle.

13. Remove the speedometer cable from the transaxle.

14. Remove the stiffener brace attaching bolts from the lower position of the clutch housing.

15. Remove the sub-frame.

16. Position a suitable jack under the transaxle.

17. Lower the transaxle support jack.

18. Remove the lower engine to transaxle attaching bolts.

19. Remove the transaxle from the rear face of the engine and lower it from the vehicle.

To install:

20. Raise the transaxle into position with the support jack. Engage the input shaft spline into the clutch disc and work the transaxle onto the dowel sleeves. Make sure the transaxle assembly is flush with the rear face of the engine prior to installation of the attaching bolts.

21. Install the lower engine to transaxle attaching bolts and tighten them to 28–31 ft. lbs. (38–42 Nm).

22. Install the speedometer cable.

23. Install the 10M and 12M bolts attaching the the shift cable and bracket to the transaxle. Tighten the 10M bolt to 16–22 ft. lbs. (22–30 Nm) and the 12M bolt to 22–35 ft. lbs. (31–48 Nm).

24. Install the bolt attaching the shift mechanism-to-shift shaft and tighten to 7–10 ft. lbs. (9–13 Nm).

25. Install the 2 bolts that attach the stiffener brace to the lower portion of the clutch housing and tighten to 15–21 ft. lbs. (21–28 Nm).

26. Install the starter stud bolts and tighten to 30–40 ft. lbs. (41–54 Nm).

27. Install the backup light switch connector to the transaxle switch.

28. Remove the seal plugs and install the inner CV-joints into the transaxle.

29. Install the center bearing to the bracket on the right side halfahaft.

NOTE: New circlips are required on both inner CV-joints prior to installation. Make sure both CV-joints are seated in the transaxle.

30. Attach the sub-frame and the lower ball joint to the steering knuckle. Insert a new service pinch bolt and a new nut. Tighten the nut to 37–44 ft. lbs. (50–60 Nm) but do not tighten the bolt.

31. Fill the transaxle with the proper type and quantity of transmission fluid.

32. Install the top transaxle to engine mounting bolts and tighten to 28–31 ft. lbs. (38–42 Nm).

33. Connect the clutch cable to the clutch release shaft assembly.

34. Remove the wood block from under the clutch pedal. Prior to starting the engine, set the hand brake and pump the clutch pedal a minimum of 2 times to ensure proper clutch adjustment.

1989–92

1. Disconnect the negative battery cable.

2. Wedge a 7 in. block of wood under the clutch pedal to hold the pedal up beyond it's normal position.

3. Remove the air cleaner hose.

4. Grasp the taulch cable and pull it forward, disconnecting it from the clutch release shaft assembly.

5. Disconnect the clutch cable casing from the rib on top of the transaxle case.

6. Install engine lifting eyes.

7. Tie up the wiring harness and power steering cooler hoses.

8. Disconnect the speedometer cable and speed sensor wire.

9. Support the engine using a suitable engine support fixture.

10. Raise the vehicle and support it safely. Remove the wheel and tire assemblies.

11. Remove the nut and bolt retaining the lower control arm ball joint to the steering knuckle assembly. Discard the removed nut and bolt. Repeat the procedure on the opposite side.

12. Using a suitable halfshaft remover, pry the lower control arm away from the knuckle.

NOTE: Be careful not to damage or cut the ball joint boot.

13. Remove the upper nut from the stabilizer bar and separate the stabilizer bar from the knuckle.

14. Remove the tie rod nut and separate the tie rod end from the knuckle.

15. Disconnect the oxygen sensor.

16. Remove the exhaust catalyst assembly.

17. Disconnect the power steering cooler from the subframe and place it aside.

18. Disconnect the battery cable bracket from the subframe.

19. Using a suitable prybar, pry the left inboard CV-joint assembly from the transaxle. Install a plug into the seal to prevent fluid leakage. Remove the CV-joint from the transaxle by grasping the left steering knuckle and swinging the knuckle and halfshaft outward from the transaxle. Repeat the procedure on the right side.

NOTE: If the CV-joint assembly cannot be pried from the transaxle, insert a suitable tool through the left side and tap the joint out. The tool can be used from either side of the transaxle.

20. Support the halfshaft assembly with wire in a near level position to prevent damage to the assembly during the remaining operations. Repeat the procedure on the opposite side.

21. Remove the retaining bolts from the center support bearing and remove the right halfshaft from the transaxle.

22. Remove the 2 steering gear retaining nuts from the sub-frame. Support the steering gear by wiring up the tie rod ends to the coil springs.

23. Remove the transaxle to engine retaining bolts.

24. Disconnect the 2 shift cables from the transaxle.

25. Remove the engine mount bolts.

26. Position jacks under the body mount positions and remove the 4 bolts, lower the sub-frame and position it aside.

27. Remove the starter motor assembly.

28. Remove the left engine vibration dampener lower bracket.

29. Remove the backup light switch connector from the transaxle backup light switch, located on top of the transaxle and remove the backup light switch.

30. Position a suitable support jack under the transaxle.

31. Lower the transaxle, remove it from the engine and lower it from the vehicle.

To install:

32. Raise the transaxle into position. Engage the input shaft spline into the clutch disc and work the transaxle onto the dowel sleeves. Make sure the transaxle assembly is flush with the rear face of the engine before installation of the retaining bolts.

33. Install the engine to transaxle retaining bolts. Tighten to 28–31 ft. lbs. (38–42 Nm).

34. Install the backup light switch and connect the electrical connector.

35. Install the starter motor. Tighten the retaining bolts to 30–40 ft. lbs. (41–54 Nm).

36. Using jacks, position the sub-frame and raise it into position. Install the 4 bolts and tighten to 65–85 ft. lbs. (90–115 Nm).

37. Install the left vibration dampener lower bracket.

38. Install the engine mount bolts and tighten to 40–55 ft. lbs. (54–75 Nm).

39. Connect the shift cables to the transaxle.

40. Install the engine to transaxle bolts and tighten to 28–31 ft. lbs. (38–42 Nm).

41. Install the steering gear retaining nuts and tighten to 85–100 ft. lbs. (115–135 Nm).

42. Install the center support bearing retaining bolts and tighten to 85–100 ft. lbs. (115–135 Nm).

43. Install the right halfshaft into the transaxle.

44. Install the left inboard CV-joint assembly into the transaxle.

45. Connect the battery cable bracket to the sub-frame.

46. Connect the power steering cooler to the subframe.

47. Install the exhaust catalyst retaining bolts and tighten to 25–34 ft. lbs. (34–47 Nm).

48. Connect the oxygen sensor.

49. Install the tie rod in the knuckle and the tie rod retaining nut. Tighten to 35–47 ft. lbs. (47–64 Nm).

50. Position the stabilizer bar to the knuckle and install the nut.

51. Install the lower control arm ball joint to steering knuckle assembly. Install and tighten a new retaining nut and bolt to 37–44 ft. lbs. (50–60 Nm).

52. Install the wheel and tire assemblies.

53. Check the transaxle fluid level.

54. Lower the vehicle.

55. Remove the engine support tool.

56. Install the speedometer cable. Connect the speedometer cable and speed sensor wire.

57. Remove the engine lifting eyes.

58. Connect the clutch cable to the transaxle.

59. Install the air cleaner hose and remove the wood block from the clutch pedal.

60. Connect the negative battery cable and check the transaxle for fluid leaks.

LINKAGE ADJUSTMENT

The manual shift mechanism and cables incorporate no adjustable features, therefore adjustments are neither possible or necessary.

CLUTCH

Clutch Assembly

REMOVAL & INSTALLATION

1. Disconnect the negative battery cable. Raise the vehicle and support it safely. Remove the transaxle.

2. Mark the pressure plate assembly and the flywheel so they can be assembled in the same position.

3. Loosen the attaching bolts 1 turn at a time, in sequence, until spring tension is relieved to prevent pressure plate cover distortion.

4. Support the pressure plate and remove the bolts. Remove the pressure plate and clutch disc from the flywheel.

5. Inspect the flywheel, clutch disc, pressure plate, throwout bearing and the clutch fork for wear. Replace parts as required. If the flywheel shows any signs of overheating (blue discoloration) or if it is badly grooved or scored, it should be refaced or replaced.

To install:

6. Install the flywheel, if removed. Tighten attaching bolts to 54–64 ft. lbs. (73–87 Nm) on all except the 3.0L SHO engine. On the 3.0L SHO engine,

tighten the bolts to 51–58 ft. lbs. (69–78 Nm).

7. Clean the pressure plate and flywheel surfaces thoroughly. Place the clutch disc and pressure plate into the installed position. Align the marks made during the removal procedure if components are being reused. Support the clutch disc and pressure plate with a suitable dummy shaft or clutch aligning tool.

8. Install the pressure plate-to-flywheel bolts. Tighten them gradually in a cross pattern to 12–24 ft. lbs. (17–32 Nm). Remove the alignment tool.

9. Lubricate the release bearing and install it in the fork.

10. Install the transaxle and connect the negative battery cable.

PEDAL HEIGHT/FREE-PLAY ADJUSTMENT

The free-play in the clutch is adjusted by a built in mechanism that allows the clutch controls to be self-adjusted during normal operation. The self-adjusting feature should be checked every 5000 miles. This is accomplished by insuring that the clutch pedal travels to the top of its upward position. Grasp the clutch pedal by hand or put a foot under the clutch pedal; pull up on the pedal until it stops. Very little effort is required (about 10 lbs.). During the application of upward pressure, a click may be heard which means an adjustment was necessary and has been accomplished.

Clutch Cable

REMOVAL & INSTALLATION

NOTE: Whenever the clutch cable is disconnected for any reason, such as transaxle removal or clutch, clutch pedal components or clutch cable replacement, the proper method for installing the clutch cable must be followed.

1. Disconnect the negative battery cable.

2. Prop up the clutch pedal to lift the pawl free of the quadrant which is part of the self-adjuster mechanism.

3. Remove the air cleaner assembly to gain access to the clutch cable.

4. Grasp the end of the clutch cable using a suitable tool and unhook the clutch cable from the clutch bearing release lever.

NOTE: Do not grasp the wire strand portion of the inner cable since this might cut the wires and result in cable failure.

5. Disconnect the cable from the in-

sulator that is located on the rib of the transaxle.

6. Position the clutch shield away from the mounting plate bracket by removing the rear retaining screw. Loosen the front retaining screw located near the toe board and rotate the shield aside.

7. With the clutch pedal lifted up to release the pawl, rotate the gear quadrant forward. Unhook the clutch cable from the gear quadrant. Let the quadrant swing rearward but do not let it snap back.

8. Remove the cable by withdrawing it through the engine compartment.

To install:

NOTE: The clutch pedal must be lifted to disengage the adjusting mechanism during cable installation. Failure to do so will result in damage to the self adjuster mechanism. A prying instrument must never be used to install the cable into the quadrant.

9. Insert the clutch cable assembly from the engine or passenger compartment through the dash panel and dash panel grommet. Make sure the cable is routed inboard of the brake lines and not trapped at the spring tower by the brake lines.

10. Push the clutch cable through the insulator on the stop bracket and through the recess between the pedal and the gear quadrant.

11. With the clutch pedal lifted up to release the pawl, rotate the gear quadrant forward. Hook the cable into the gear quadrant.

12. Secure the clutch shield on the clutch mounting plate.

13. Using a suitable device, secure the pedal in the upper most position.

14. Install the clutch cable in the insulator on the rib of the transaxle.

15. Hook the cable into the clutch release lever in the engine compartment.

16. Remove the device that was used to temporarily secure the pedal against it's stop.

17. Adjust the clutch by depressing the clutch pedal several times.

18. Install the air cleaner and connect the negative battery cable.

AUTOMATIC TRANSAXLE

For further information on transmissions/transaxles, please refer to "Chilton's Guide to Transmission Repair".

Transaxle Assembly

REMOVAL & INSTALLATION

1988–90

EXCEPT TAURUS WITH 2.5L ENGINE

1. Disconnect the negative battery cable. Raise and support the vehicle safely. Remove the air cleaner assembly.

2. Remove the bolt retaining the shift cable and bracket assembly to the transaxle.

NOTE: Hold the bracket with a prybar in the slot to prevent the bracket from moving.

3. Remove the shift cable bracket bolts and bracket from the transaxle. Disconnect the electrical connector from the neutral safety switch.

4. Disconnect the electrical bulkhead connector from the rear of the transaxle. Remove the dipstick. If with 3.8L engines, remove the throttle valve cable cover. Unsnap the throttle valve cable from the throttle body lever. Remove the throttle valve cable from the transaxle case.

5. Carefully pull up on the throttle valve cable and disconnect the throttle valve cable from the TV link.

NOTE: Pulling to hard on the throttle valve may bend the internal TV bracket.

6. Install engine lifting brackets.

7. Disconnect the power steering pump pressure and return line bracket.

8. Remove the converter housing bolts from the top of the transaxle.

9. Install a suitable engine support fixture.

10. Raise the vehicle and support it safely. Remove both front wheels. Remove the left outer tie rod end.

11. On Continental, remove the suspension height sensor and disconnect the brake line support brackets.

12. Remove the lower ball joint attaching nuts and bolts. Remove the lower ball joints and remove the lower control arms from each spindle. Remove stabilizer bar bolts.

13. Remove the nuts securing the steering rack to the sub-frame.

14. If equipped with 3.8L engine, disconnect the oxygen sensor electrical connection and remove the exhaust pipe, converter assembly and mounting bracket.

15. Remove the two 15mm bolts from the transaxle mount. Remove the four 15mm bolts from the left engine support and remove the bracket.

16. Position the sub-frame removal tool.

17. Remove the steering gear from the sub-frame and secure to the rear of the engine compartment. Remove the sub-frame.

18. Remove the dust cover and the starter assembly.

19. Rotate the engine by the crankshaft pulley bolt to align the torque converter bolts with the starter drive hole. Remove the torque converter-to-flywheel retaining nuts.

20. Remove the transaxle cooler line retaining clips. Disconnect the transaxle cooler lines.

21. Remove the engine to transaxle retaining bolts.

22. Remove the speedometer sensor heat shield.

23. Remove the vehicle speed sensor from the transaxle.

NOTE: Vehicles with electronic instrument clusters do not use a speedometer cable.

24. Position the transmission jack. Remove the halfshafts.

25. Remove the last 2 torque converter housing bolts.

26. Seperate the transaxle from the engine and carefully lower the transaxle from the vehicle.

To install:

27. Installation is the reverse of the removal procedure. During installation be sure to observe the following:

   a. Clean the transaxle oil cooler lines.

   b. Install new circlips on the CV-joint seals.

   c. Carefully install the halfshafts in the transaxle by aligning the splines of the CV-joint with the splines of the differential.

   d. Attach the lower ball joint to

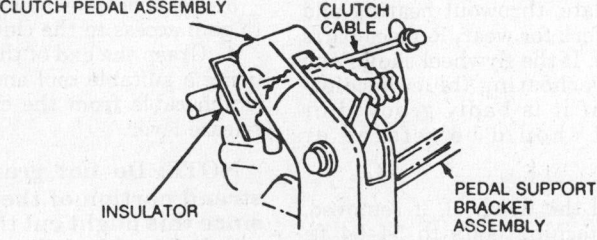

Clutch cable installation

the steering knuckle with a new nut and bolt. Tighten the nut to 37–44 ft. lbs.

e. When installing the transaxle to the engine, verify that the converter-to-transaxle engagement is maintained. Prevent the converter from moving forward and disengaging during installation.

f. Adjust the TV and manual linkages. Check the transaxle fluid level.

g. Tighten the following bolts to the torque specifications listed:

Transaxle-to-engine bolts: 41–50 ft. lbs. (55–68 Nm)

Control arm-to-knuckle bolts: 36–44 ft. lbs. (50–60 Nm)

Stabilizer U-clamp-to-bracket bolts: 60–70 ft. lbs. (81–95 Nm)

Tie rod-to-knuckle nut: 23–35 ft. lbs. (31–47 Nm)

Starter-to-transaxle bolts: 30–40 ft. lbs. (41–54 Nm)

Converter-to-flywheel bolts: 23–39 ft. lbs. (31–53 Nm)

Insulator-to-bracket bolts: 55–70 ft. lbs. (75–90 Nm)

TAURUS WITH 2.5L ENGINE

1. Disconnect the negative battery cable and remove the air cleaner assembly.

2. Position the engine control wiring harness away from the transaxle converter housing.

3. Disconnect the TV linkage and manual lever cable at the respective levers. Failure to disconnect the linkage during transaxle removal and allowing the transaxle to hang will fracture the throttle valve cam shaft joint, which is located under the transaxle cover.

4. Remove the power steering hose brackets.

5. Remove the upper transaxle-to-engine attaching bolts.

6. Install suitable engine lifting brackets to the right and left areas of the cylinder head and attach with bolts. Install 2 suitable engine support bars.

NOTE: An engine support bar may be fabricated from a length of 4 × 4 wood cut to 57 in.

7. Place 1 of the engine support bars across the vehicle in front of each engine shock tower. Place another support bar across the vehicle approximately between the alternator and valve cover. Attach chains to the lifting brackets. Raise the vehicle and support safely. Remove the wheel and tire assemblies.

8. Remove the catalytic converter inlet pipe and disconnect the exhaust air hose assembly.

9. Remove each tie rod end from it's spindle. Separate the lower ball joints from the struts and remove the lower control arm from each spindle.

10. Disconnect the stabilizer bar by removing the retaining nuts.

11. Disconnect and remove the rack and pinion and auxiliary cooler from the sub-frame. Position the rack and pinion away from the sub-frame and secure with wire.

12. Remove the right front axle support and bearing assembly retaining bolts.

13. Remove the halfshaft and link shaft assembly out of the right side of the transaxle.

14. Disengage the left halfshaft from the differential side gear. Pull the halfshaft from the transaxle.

NOTE: Support and secure the halfshaft from an underbody component with a length of wire. Do not allow the halfshafts to hang unsupported.

15. Plug the seal holes.

16. Remove the front support insulator and position the left front splash shield aside.

17. Properly support the sub-frame and lower the vehicle onto the sub-frame support. Remove the sub-frame and disconnect the neutral start switch wire assembly.

18. Raise the vehicle after the sub-frame is removed. Disconnect the speedometer cable.

19. Disconnect and remove the shift cable from the transaxle.

20. Disconnect the oil cooler lines and remove the starter.

21. Remove the dust cover from the torque converter housing and remove the torque converter-to-flywheel housing nuts.

22. Position a suitable transaxle jack under the transaxle.

23. Remove the remaining transaxle-to-engine attaching bolts.

NOTE: Before the transaxle can be lowered from the vehicle, the torque converter studs must be clear of the flywheel. Insert a suitable tool between the flywheel and converter and carefully guide the transaxle and converter away from the engine.

24. Lower the transaxle from the engine.

To install:

25. Installation is the reverse of the removal procedure. During installation be sure to observe the following:

a. Clean the transaxle oil cooler lines.

b. Install new circlips on the CV-joint seals.

c. Carefully install the halfshafts in the transaxle by aligning the splines of the CV-joint with the splines of the differential.

d. Attach the lower ball joint to

the steering knuckle with a new nut and bolt. Tighten the nut to 37–44 ft. lbs. Torquing of the bolt is not required.

e. When installing the transaxle to the engine, verify that the converter-to-transaxle engagement is maintained. Prevent the converter from moving forward and disengaging during installation.

f. Adjust the TV and manual linkages. Check the transaxle fluid level.

g. Tighten the following bolts to the torque specifications listed:

Transaxle-to-engine bolts: 25–33 ft. lbs. (34–45 Nm)

Control arm-to-knuckle bolts: 36–44 ft. lbs. (50–60 Nm)

Stabilizer U-clamp-to-bracket bolts: 60–70 ft. lbs. (81–95 Nm)

Tie rod-to-knuckle nut: 23–35 ft. lbs. (31–47 Nm)

Starter-to-transaxle bolts: 30–40 ft. lbs. (41–54 Nm)

Converter-to-flywheel bolts: 23–39 ft. lbs. (31–53 Nm)

Insulator-to-bracket bolts: 55–70 ft. lbs. (75–90 Nm)

1991–92

1. Disconnect the battery cables and remove the battery and battery tray.

2. Remove the air cleaner assembly, hoses and tubes.

3. Disconnect the electrical connectors from the engine and remove the bolt retaining the main wiring harness bracket.

4. Remove the shift lever. Remove the EGR bracket and throttle body bracket retaining bolts and install engine lifting eyes.

5. Secure the wiring harness aside and remove the radiator sight shield. Position a suitable engine support fixture.

6. If equipped with air suspension, turn the air suspension switch located in the luggage compartment to the **OFF** position.

7. Remove the dipstick and disconnect the power steering line bracket. Remove the 4 torque converter housing bolts from the top of the transaxle.

8. Raise and safely support the vehicle. Remove the front wheel and tire assemblies.

9. Disconnect the left outer tie rod end. Remove the suspension height sensor, if equipped. Disconnect the brake line support brackets.

10. Remove the retaining bolts from the front stabilizer bar assembly. Disconnect the right and left lower arm assemblies.

11. Remove the steering gear retaining nuts from the sub-frame. Remove the front oxygen sensor, exhaust pipe, converter assembly and mounting bracket.

12. Remove 2 bolts from the transaxle mount and the 4 bolts from the left engine support. Remove the engine support.

13. Position a suitable sub-frame removal tool. Remove the steering gear from the sub-frame and secure to the rear of the engine compartment. Remove the sub-frame-to-body bolts and lower the sub-frame.

14. Remove the starter and the dust cover.

15. Rotate the engine at the crankshaft pulley to align the torque converter bolts with the starter drive hole. Remove the 4 torque converter-to-flywheel retaining nuts.

16. Disconnect the transaxle cooler lines. Remove the engine-to-transaxle retaining bolts.

17. Remove the speedometer sensor heat shield. Remove the vehicle speed sensor from the transaxle.

NOTE: Vehicles with electronic instrument clusters do not use a speedometer cable.

18. Position a suitable transaxle jack. Remove the halfshafts.

19. Remove the last 2 torque converter housing bolts, carefully separate the transaxle from the engine and lower out of the vehicle.

20. Installation is the reverse of the removal procedure. During installation be sure to observe the following:

   a. Clean the transaxle oil cooler lines.

   b. Install new circlips on the CV-joint seals.

   c. Carefully install the halfshafts in the transaxle by aligning the splines of the CV-joint with the splines of the differential.

   d. Attach the lower ball joint to the steering knuckle with a new nut and bolt. Tighten the nut to 37–44 ft. lbs.

   e. When installing the transaxle to the engine, verify that the converter-to-transaxle engagement is maintained. Prevent the converter from moving forward and disengaging during installation.

   f. Adjust the TV and manual linkages. Check the transaxle fluid level.

   g. Tighten the following bolts to the torque specifications listed:

   Transaxle-to-engine bolts: 41–50 ft. lbs. (55–68 Nm)

   Control arm-to-knuckle bolts: 36–44 ft. lbs. (50–60 Nm)

   Stabilizer U-clamp-to-bracket bolts: 60–70 ft. lbs. (81–95 Nm)

   Tie rod-to-knuckle nut: 23–35 ft. lbs. (31–47 Nm)

   Starter-to-transaxle bolts: 30–40 ft. lbs. (41–54 Nm)

   Converter-to-flywheel bolts: 23–39 ft. lbs. (31–53 Nm)

Insulator-to-bracket bolts: 55–70 ft. lbs. (75–90 Nm)

SHIFT CABLE ADJUSTMENT

Except 1988–90 Taurus with 2.5L Engine

1. Position the selector lever in the **OD** position against the rearward stop. The shift lever must be held in the rearward position using a constant force of 3 lbs. (1.4 Kg) while the linkage is being adjusted.

2. Loosen the manual lever-to-control cable retaining nut.

3. Move the transaxle manual lever to the **OD** position, second detent from the most rearward position.

4. Tighten the retaining nut to 11–19 ft. lbs. (14–27 Nm).

5. Check the operation of the transaxle in each selector lever position. Make sure the park and neutral start switch are functioning properly.

1988–90 Taurus with 2.5L Engine

1. Position the selector lever in the **D** position against the drive stop. The shift lever must be held in the **D** position while the linkage is being adjusted.

2. Loosen the transaxle manual lever-to-control cable adjustment trunnion bolt.

3. Move the transaxle manual lever to the **D** position, second detent from the most rearward position.

4. Tighten the adjustment trunnion bolt to 12–20 ft. lbs. (16–27 Nm).

5. Check the operation of the transaxle in each selector lever position. Make sure the neutral start switch functions properly in **P** and **N** and the back-up lights are on in **R**.

THROTTLE CABLE ADJUSTMENT

NOTE: Transaxle downshift control is controlled through the throttle position switch on 1991–92 vehicles equipped with the electronic automatic overdive transaxle.

1988–90 3.0L and 3.8L Engines

The Throttle Valve (TV) cable normally does not need adjustment. The cable should be adjusted only if one of the following components is removed for service or replacement:

Main control assembly
Throttle valve cable
Throttle valve cable engine mounting bracket
Throttle control lever link or lever assembly

Engine throttle body
Transaxle assembly

1. Connect the TV cable eye to the transaxle throttle control lever link and attach the cable boot to the chain cover.

2. If equipped with the 3.0L engine, with the TV cable mounted in the engine bracket, make sure the threaded shank is fully retracted. To retract the shank, pull up on the spring rest with the index fingers and wiggle the top of the thread shank while pressing the shank through the spring with the thumbs.

3. If equipped with the 3.8L engine, the TV cable must be unclipped from the right intake manifold clip. To retract the shank, span the crack between the two 180 degree segments of the adjuster spring rest with a suitable tool. Compress the spring by pushing the rod toward the throttle body with the right hand. While the spring is compressed, push the threaded shank toward the spring with the index and middle fingers of the left hand. Do not pull on the cable sheath.

4. Attach the end of the TV cable to the throttle body.

5. If equipped with the 3.8L engine, rotate the throttle body primary lever by hand, the lever to which the TV-driving nailhead is attached, to the wide-open-throttle position. The white adjuster shank must be seen to advance. If not, look for cable sheath/foam hang-up on engine/body components. Attach the TV cable into the top position of the right intake manifold clip.

NOTE: The threaded shank must show movement or "ratchet" out of the grip jaws. If there is no movement, inspect the TV cable system for broken or disconnected components and repeat the procedure.

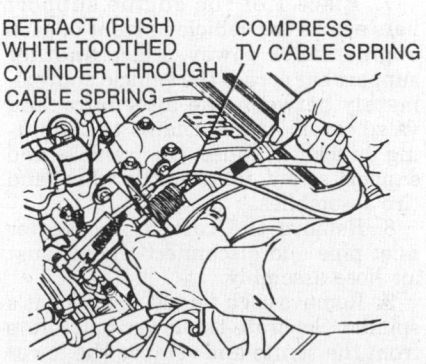

RETRACT (PUSH)   COMPRESS
WHITE TOOTHED   TV CABLE SPRING
CYLINDER THROUGH
CABLE SPRING

Throttle valve cable adjustment

FRONT SUSPENSION

MacPherson Strut

REMOVAL & INSTALLATION

Taurus and Sable

1. Place the ignition switch in the **OFF** position and the steering column in the **UNLOCKED** position.
2. Remove the hub nut. Loosen the 3 top mount-to-shock tower nuts; do not remove the nuts at this time.
3. Raise and support the vehicle safely.

NOTE: When raising the vehicle, do not lift by using the lower control arms.

4. Remove the tire and wheel assembly. Remove the brake caliper, supporting it on a wire. Remove the rotor.
5. At the tie rod end, remove the cotter pin and the castle nut. Discard the cotter pin and nut and replace with new.
6. Using tie rod end remover tool 3290-D and the tie rod remover adapter tool T81P-3504-W or equivalents, separate the tie rod from the steering knuckle.
7. Remove the stabilizer bar link nut and the link from the strut.
8. Remove the lower arm-to-steering knuckle pinch bolt and nut; it may be necessary to use a drift punch to remove the bolt. Using a suitable tool, spread the knuckle-to-lower arm pinch joint and remove the lower arm from the steering knuckle. Discard the pinch nut/bolt and replace with new.
9. Remove the halfshaft from the hub and support it on a wire.

NOTE: When removing the halfshaft, do not allow it to move outward as the internal parts of the tripod CV-joint could separate, causing failure of the joint.

10. Remove the strut-to-steering knuckle pinch bolt. Using a small prybar, spread the pinch bolt joint and separate the strut from the steering knuckle. Remove the steering knuckle/hub assembly from the strut.
11. Remove the 3 top mount-to-shock tower nuts and the strut assembly from the vehicle.
12. Compress the coil spring using a suitable spring compressor. Use a 10mm box end wrench to hold the top of the strut shaft while removing the nut with a 21mm 6-point crow foot wrench and ratchet.

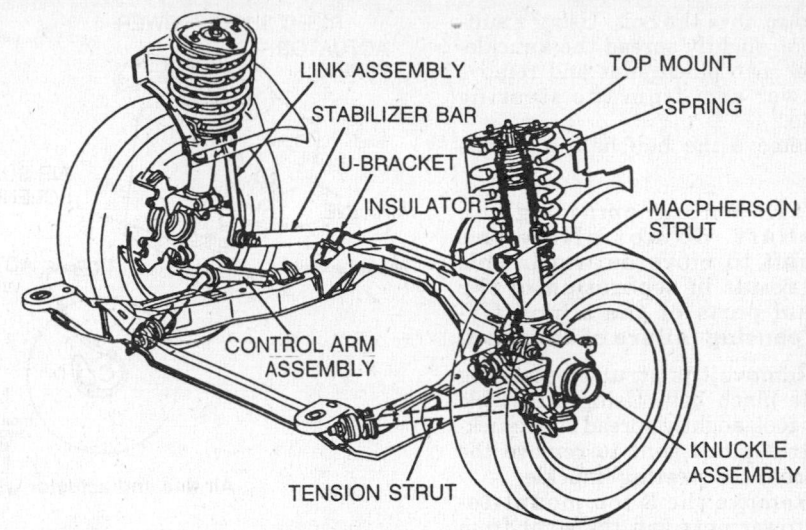

Front suspension — Taurus and Sable

13. Loosen the spring compressor, then remove the top mount bracket assembly, bearing plate assembly and spring.

To install:
14. Install the spring compressor. Install the spring, bearing plate assembly, lower washer and top mount bracket assembly.
15. Compress the spring. Install the upper washer and nut on the shock strut shaft. Tighten the nut with the 21mm 6-point crow foot wrench and ratchet while holding the shaft with the 10mm box end wrench.
16. Install the strut assembly and the 3 top mount-to-shock tower nuts.
17. Install the steering knuckle and hub assembly to the strut.
18. Install a new strut-to-steering knuckle pinch bolt. Tighten the bolt to 70–95 ft. lbs. (95–129 Nm).
19. Install the halfshaft into the hub.
20. Install the lower arm to the steering knuckle and install a new pinch bolt and nut. Tighten to 40–55 ft. lbs. (54–74 Nm).
21. Install the stabilizer link to the strut and install a new stabilizer bar link nut. Tighten to 55–75 ft. lbs. (75–101 Nm).
22. Install the tie rod end onto the knuckle using a new castle nut. Tighten the castle nut to 23–35 ft. lbs. (31–47 Nm). Retain the castle nut with a new cotter pin.
23. Install the disc brake rotor, caliper and tire and wheel assembly.
24. Tighten the 3 top mount-to-shock tower nuts to 20–30 ft. lbs. (27–40 Nm).
25. Lower the vehicle and tighten the hub nut to 180–200 ft. lbs. (244–271 Nm).
26. Check the front end alignment.

Continental

1. Turn OFF the air suspension switch, located in the left side of the luggage compartment.
2. Place the ignition switch in the **OFF** position and the steering column in the **UNLOCKED** position.
3. Remove the plastic cover from the shock tower to gain access to the upper mounting nuts and dual damping actuator.
4. Remove the actuator retaining screws. Remove the actuator and place it aside.
5. Remove the hub nut.
6. Loosen the 3 top mount-to-shock tower nuts but do not remove them at this time.
7. Raise the vehicle and support it safely.

NOTE: Do not raise the vehicle by the lower control arms.

8. Remove the tire and wheel assembly.
9. Remove the brake line bracket from the strut assembly.
10. Disconnect the height sensor link from the ball stud pin at the lower control arm.
11. Disconnect the air line from the solenoid valve.
12. Disconnect the electrical connector at the solenoid valve.
13. Remove the brake caliper and the disc brake rotor. Support the caliper with wire.
14. Remove the cotter pin and castle nut from the tie rod end. Discard the cotter pin and castle nut.
15. Using tie rod end remover TOOL-3290-D and tie rod end remover adapter T81P-3504-W or equivalent, remove the tie rod from the knuckle.
16. Remove the stabilizer bar link nut and the link from the strut.
17. Remove and discard the lower arm-to-steering knuckle pinch bolt and nut. A suitable drift punch may be

used to remove the bolt. Using a suitable tool, slightly spread the knuckle-to-lower arm pinch joint and remove the lower arm from the steering knuckle.

18. Remove the halfshaft from the hub.

NOTE: When removing the halfshaft, do not allow the halfshaft to move outward. This could result in seperation of the internal parts of the tripod CV-joint, causing failure of the joint.

19. Remove the strut-to-steering knuckle pinch bolt. Using a suitable prying tool, slightly spread the knuckle-to-strut pinch joint to remove the strut from the steering knuckle.
20. Remove the 3 top mount-to-shock tower nuts and the strut from the vehicle.

To install:

21. Install the strut with the 3 top mount-to-shock tower nuts and leave the nuts loose.
22. Install the steering knuckle and hub assembly to the strut. Install a new strut-to-steering knuckle pinch bolt. Tighten the bolt to 70–95 ft. lbs. (95–129 Nm).
23. Install the halfshaft into the hub.
24. Install the lower arm to the steering knuckle and install a new pinch bolt and nut. Tighten to 40–55 ft. lbs. (54–74 Nm).
25. Install the stabilizer bar link to the strut and install a new stabilizer bar link nut. Tighten to 55–75 ft. lbs. (75–101 Nm).
26. Install the tie rod end onto the knuckle using a new castle nut. Before tightening the nut, make sure the steering wheel and wheels are in the straight-ahead position. Tighten the castle nut to 23–35 ft. lbs. (31–47 Nm). Install a new cotter pin in the castle nut.
27. Install the brake caliper and rotor.
28. Connect the electrical connector and the air line to the solenoid valve and position them properly.
29. Install the height sensor link on the ball stud pin on the control arm.
30. Install the brake line bracket to the strut assembly.
31. Install the wheel and tire assembly.
32. Tighten the 3 top mount-to-shock tower nuts to 20–30 ft. lbs. (27–40 Nm).
33. Install the dual damping actuator and the plastic shock tower cover. Correctly position the actuator wiring.
34. Refill the air spring prior to fully lowering the vehicle. The refill procedure is as follows:

a. Place the air suspension service switch in the **ON** position.

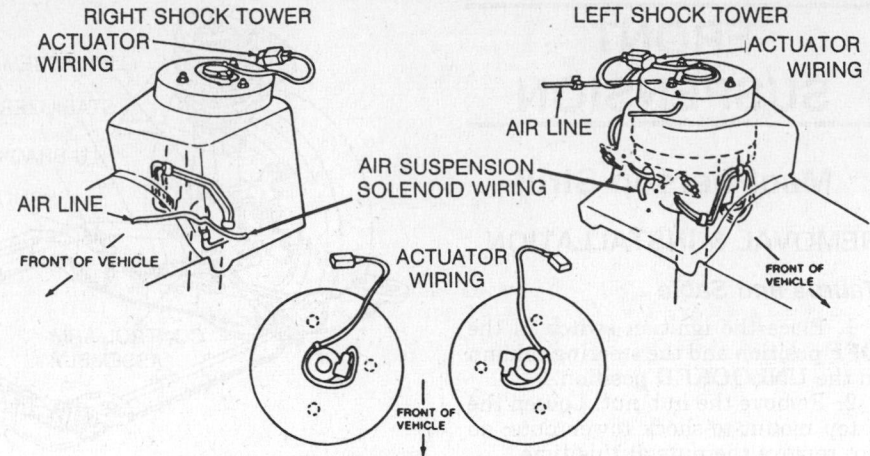

Air line and actuator wiring positioning—Continental

Code	Description
21	Vent R.F.
22	Vent L.F.
23	Vent R.R.
24	Inflate R.F.
25	Inflate L.F.
26	Inflate R.R.
27	Vent L.R.
28	Inflate L.R.

Air suspension spring fill codes—Continental

b. Turn the ignition switch **OFF**.
c. Connect a battery charger to reduce battery drain.
d. Open the access door in the left-hand luggage compartment trim panel to plug the super star II tester or equivalent, into the air suspension diagnostics wiring harness connector.
e. The tester button should be in the **HOLD** (up) position.
f. With the brake pedal depressed hard, turn the ignition switch to the **RUN** position.
g. Move the tester button to the **TEST** (down) position.
h. The air suspension control module will now start sending out the spring fill selection codes to be displayed on the tester. These codes will be displayed in a scrolling manner.
i. Select the desired spring fill operation by releasing the tester button when the desired code is displayed. Select either Code 24 or Code 25 to inflate either the right front or left front air spring. As long as the tester button is released the inflation will continue. To stop inflation, move the tester button back down to the **TEST** position. The spring fill codes will again be displayed.

NOTE: Do not apply a load to the suspension until after the air spring has been inflated at least 60 seconds.

j. To exit the spring fill mode, turn the ignition switch to the **OFF** position and unplug the tester.
35. Lower the vehicle and tighten the hub nut to 180–200 lbs. (244–271 Nm).
36. Turn on the air suspension.
37. Check the front end alignment.

Lower Ball Joints

INSPECTION

1. Turn **OFF** the air suspension switch, located in the left side of the luggage compartment on Continental.
2. Raise the vehicle and safely support it so the wheels fall to the full-down position.
3. Have an assistant grasp the lower edge of the tire and move the wheel and tire assembly in and out.
4. Observe the lower end of the knuckle and the lower control arm as the wheel is being moved in and out. Any movement indicates abnormal ball joint wear.
5. If there is any movement, install a new lower control arm assembly.

6. Lower the vehicle. On Continental, turn **ON** the air suspension.

REMOVAL & INSTALLATION

Ball joints are integral parts of the lower control arms. If an inspection reveals an unsatisfactory ball joint, the entire lower control arm assembly must be replaced.

Lower Control Arms

REMOVAL & INSTALLATION

1. Turn **OFF** the air suspension switch, located in the left side of the luggage compartment on Continental.
2. Raise and support the front of the vehicle safely. Remove the wheel and tire assembly. Position the steering column in the unlocked position.
3. Disconnect the height sensor link from the ball stud pin on Continental.
4. Remove the tension strut-to-control arm nut and the dished washer.
5. Remove and discard the control arm-to-steering knuckle pinch bolt. Using a small prybar, spread the pinch joint and separate the control arm from the steering knuckle.

NOTE: When separating the control arm from the steering knuckle, do not use a hammer. Be careful not to damage the ball joint boot seal.

6. Remove the control arm-to-frame nut/bolt, then the control arm from the frame and the tension strut.

NOTE: Do not allow the halfshaft to move outward or the tripod CV-joint internal parts could separate, causing failure of the joint.

7. To install, use a new pinch nut/bolt and reverse the removal procedures. Tighten the bolts to the following torque specifications:
Control arm-to-frame 70–95 ft. lbs. (95–129 Nm)
Control arm-to-steering knuckle 40–55 ft. lbs. (54–74 Nm)
Tension strut-to-control arm 70–95 ft. lbs. (95–129 Nm)
Wheel lug nuts 80–105 ft. lbs. (109–142 Nm)
8. Check the front end alignment.

Stabilizer Bar

REMOVAL & INSTALLATION

1. Raise and support the front of the vehicle on jackstands behind the subframe.
NOTE: Do not raise or support the vehicle on the front control arms.

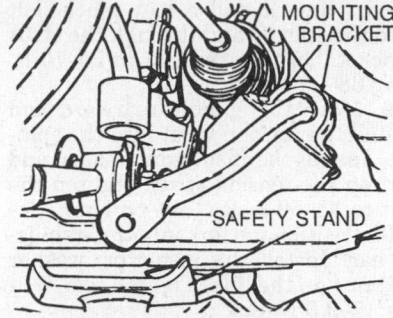

Stabilizer bar removal procedure

2. Remove and discard the stabilizer bar link-to-stabilizer bar nut, the stabilizer bar link-to-strut nut and the link from the vehicle.
3. Remove the steering gear-to-subframe nuts and move the gear from the sub-frame.
4. Position another set of jackstands under the subframe and remove the rear subframe-to-frame bolts. Lower the subframe rear to gain access to the stabilizer bar brackets.
5. Remove the stabilizer bar U-bracket bolts and the stabilizer bar from the vehicle.

NOTE: When removing the stabilizer bar, replace the insulators and the U-bracket bolts with new ones.

To install:

6. To install, reverse the removal procedure. Tighten the bolts to the folowing torque specifications:
U-bracket-to-subframe 21–32 ft. lbs. (28–43 Nm)
Subframe-to-steering gear 85–100 ft. lbs. (115–135 Nm)
Stabilizer bar-to-stabilizer bar link 35–48 ft. lbs. (47–65 Nm)
Stabilizer bar-to-strut 55–75 ft. lbs. (75–101 Nm)
7. Prior to assembly, coat the inside diameter of the new insulators with No. E25Y-19553-A or equivalent lubricant. Do not use any mineral or petroleum base lubricants as they will cause deterioration of the rubber insulators.

REAR SUSPENSION

Shock Absorbers

REMOVAL & INSTALLATION

Taurus and Sable Wagon

1. Raise and support the vehicle safely.

2. Remove the wheel and tire assembly.
3. Position a jack stand under the lower suspension arm. Remove the 2 nuts retaining the shock absorber to the lower suspension arm.
4. From inside the vehicle, remove the rear compartment access panels.
5. Remove and discard the top shock absorber attaching nut using a crow's foot wrench and ratchet while holding the shock absorber shaft stationary with an open end wrench.

NOTE: If the shock absorber is to be reused, do not grip the shaft with pliers or vise grips. Gripping the shaft in this manner will damage the shaft surface finish and will result in severe oil leakage.

6. Remove the rubber insulator from the shock and the shock from the vehicle.

NOTE: The shocks are gas filled. It will require an effort to remove the shock from the lower arm.

To install:

7. Install a new washer and insulator on the upper shock absorber rod.
8. Maneuver the upper part of the shock absorber into the shock tower opening in the body. Push slowly on the lower part of the shock absorber until the mounting studs are aligned with the mounting holes in the lower suspension arm.
9. Install new lower attaching nuts but do not tighten at this time.
10. Install a new insulator, washer and nut on top of the shock absorber. Torque the nut to 19–27 ft. lbs. (26–37 Nm.).
11. Install the rear compartment access panel.
12. Torque the 2 lower attaching nuts to 13–20 ft. lbs. (17–27 Nm).
13. Install the wheel and tire assembly. Remove the safety stand supporting the lower suspension arm and lower the vehicle.

MacPherson Strut

REMOVAL & INSTALLATION

Taurus and Sable Sedan

1. Raise and support the rear of the vehicle safely. Remove the wheel and tire.

NOTE: Do not raise or support the vehicle using the tension struts.

2. Raise the luggage compartment lid and loosen but do not remove, the upper strut-to-body nuts.
3. Remove the brake differential

control valve-to-control arm bolt. Using a wire, secure the control arm to the body to ensure proper support leaving at least 6 in. clearance to aid in the strut removal.

4. Remove the brake hose-to-strut bracket clip and move the hose aside.

5. If equipped, remove the stabilizer bar U-bracket from the vehicle.

6. If equipped, remove the stabilizer bar-to-stabilizer link nut, washer and insulator, then separate the stabilizer bar from the link.

NOTE: When removing the strut, be sure the rear brake flex hose is not stretched or the steel brake tube is not bent.

7. Remove the tension strut-to-spindle nut, washer and insulator. Move the spindle rearward to separate it from the tension strut.

8. Remove the shock strut-to-spindle pinch bolt. If necessary, use a medium prybar, spread the strut-to-spindle pinch joint to remove the strut. Discard the bolt and replace it.

9. Lower the jackstand and separate the shock strut from the spindle.

10. Support the shock strut, then loosen the top strut-to-body nuts completely and remove the strut from the vehicle.

11. Remove the nut, washer and insulator attaching link to strut and remove link. Mark the location of the insulator to top mount, then compress the spring, using a suitable spring compressor.

12. Use a 10mm box end wrench to hold the top of the strut shaft while removing the nut with a 21mm 6-point crow foot wrench and ratchet. Loosen the spring compressor, then remove the top mount bracket assembly, spring insulator and spring.

To install:

13. Using the spring compressor, install the spring, spring insulator, bottom washer, if equipped, top mount, upper washer and nut on the strut shaft. Make sure the spring is properly located in the upper and lower spring seats and the mount washers are positioned correctly.

14. Tighten the rod nut to 35–50 ft. lbs. (48–68 Nm). Use a 21mm crow foot wrench to turn the nut and a 10mm box end wrench to hold the shaft. Do not use pliers or vise grips.

15. Position the stabilizer bar link in the strut bracket. Install the insulator, washer and nut and tighten to 5–7 ft. lbs. (7–9.5 Nm).

16. Insert the 3 upper mount studs into the strut tower in the apron and hand start 3 new nuts. Do not tighten the nuts at this time.

17. Partially raise the vehicle.

18. Install the strut into the spindle

pinch joint. Install a new pinch bolt into the spindle and through the strut bracket. Tighten the bolt to 50–70 ft. lbs. (68–95 Nm).

19. Move the spindle rearward and install the tension strut into the spindle. Install the insulator, washer and nut on the tension strut. Tighten the nut to 35–50 ft. lbs. (48–68 Nm).

20. Position the link into the stabilizer bar. Install the insulator, washer and nut on the link. Tighten to 5–7 ft. lbs. (7–9.5 Nm).

21. Position the stabilizer bar U-bracket on the body. Install the bolt and tighten to 25–37 ft. lbs. (34–50 Nm).

22. Install the brake hose to the strut bracket.

23. Install the brake control differential valve on the control arm and remove the retaining wire.

24. Install the top mount-to-body nuts and tighten to 19–26 ft. lbs. (26–35 Nm).

25. Install the wheel and tire assembly and lower the vehicle.

Continental

1. Turn off the air suspension switch located in the luggage compartment.

2. From inside the luggage compartment, disconnect the electrical connector from the dual dampening actuator.

3. Loosen but do not remove the 3 nuts retaining the strut to the upper body.

4. Raise and support the vehicle safely. Remove the wheel and tire assembly.

NOTE: Do not raise the vehicle by the tension strut.

5. Disconect the air line and electrical connector from the solenoid valve.

6. Remove the brake hose retainer at the strut bracket.

7. Disconnect the parking brake cable from the brake caliper. Remove all the wire retainers and parking brake cable retainers from the lower suspension arm.

8. Disconnect the height sensor link from the ball stud pin on the lower arm.

9. Remove the caliper assembly from the spindle and position it off to the side with a piece of wire. Do not kink or place a load on the brake hose.

10. Bleed the air spring by performing the following:

   a. Remove the solenoid clip.

   b. Rotate the solenoid counterclockwise to the first stop.

   c. Slowly pull the solenoid straight out to the second stop and bleed the air from the system.

── **CAUTION** ──
Do not fully release the solenoid until the air is fully bled from the spring or personal injury may result.

   d. After the air is fully bled from the system, rotate the solenoid to the third stop and remove the solenoid from the housing.

11. Mark the position of the notch on the toe adjustment cam.

12. Remove the torsion spring clamp from the spindle-to-strut bolt.

13. Remove the nut from the inboard bushing on the suspension arm.

14. Install torsion spring remover tool T88P–5310–A or equivalent, on the suspension arm. Pry up on the tool and arm using a ¾ in. drive ratchet to relieve the pressure on the pivot bolt. An assistant may be required to pull outboard on the spindle simultaneously to fully relieve the tension on the bolt. Remove the bolt and lower arm. Repeat this procedure for the opposite arm.

15. Remove the torsion spring from the arms.

16. Remove the stabilizer U-bracket from the body.

17. Remove the nut, washer and insulator attaching the stabilizer bar to the link. Separate the stabilizer bar from the link.

18. Remove the nut washer and insulator retaining the tension strut to the spindle. Move the spindle rearward enough to separate it from the tension strut.

19. Remove and discard the strut-to-spindle pinch bolt. With a suitable prying tool, spread the strut-to-spindle pinch joint as required to assist in removing the bolt.

20. Separate the spindle from the strut. Remove the spindle as an assembly with the arms attached.

21. From inside the luggage compartment area, support the shock strut by hand and remove and discard the 3 upper mount-to-body nuts. Care should be taken not to drop the strut when removing the upper nuts. Guide the electric actuator wire through the opening to prevent snagging and damage while removing the strut assembly.

To install:

22. Install the solenoid valve on the air spring.

23. Guide the electric actuator wire through the opening and install the strut assembly. Install 3 new upper mount nuts.

24. Install the spindle and arms to the strut. Install a new strut-to-spindle pinch bolt. Do not tighten the bolt until the control arms are attached to the body and the cams are centered.

25. Position the tension strut to the

spindle. Install the insulator, washer and nut retaining the tension strut to the spindle. Tighten the nut to 35–50 ft. lbs. (48–68 Nm).

26. Install the stabilizer bar to the link. Install the insulator, washer and retaining nut. Tighten the nut to 5–7 ft. lbs. (7–9.5 Nm).

27. Install the stabilizer U-bracket to the body. Tighten the bolt to 25–37 ft. lbs. (34–50 Nm).

28. Install the torsion spring to the arms.

29. Position the inboard bushing using torsion spring remover T88P–5310–A or equivalent, and install the bolt. An assistant may be required to pull outboard on the spindle to align the bushing so the bolt can be inserted. Repeat the procedure for the opposite lower arm.

30. Install the nut to the inboard bushing on the suspension arm but do not tighten at this time.

31. Tighten the spindle-to-strut bolt to 51–70 ft. lbs. (68–95 Nm).

32. Set the toe adjustment cam to the alignment mark.

33. Remove the wire from the caliper and install the caliper to the spindle.

34. Connect the height sensor link to the ball stud pin on the lower arm.

35. Install the torsion spring clamp and secure.

36. Install all wire retainers and parking brake cable retainers to the lower suspension arm.

37. Connect the parking brake cable to the brake caliper and install the brake hose retainer at the strut bracket.

38. Connect the air line and the electrical connector to the solenoid valve.

39. Install the wheel and tire assembly and partially lower the vehicle.

40. Tighten the 3 nuts retaining the strut to the upper body to 19–26 ft. lbs. (26–35 Nm).

41. From inside the luggage compartment, connect the electrical connector for the dual dampening actuator.

42. Turn on the air suspension switch and fill the air spring as follows:

   a. Place the air suspension service switch in the **ON** position.

   b. Turn the ignition switch **OFF**.

   c. Connect a battery charger to reduce battery drain.

   d. Open the access door in the left-hand luggage compartment trim panel to plug the super star II tester or equivalent, into the air suspension diagnostics wiring harness connector.

   e. The tester button should be in the **HOLD** (up) position.

   f. With the brake pedal depressed hard, turn the ignition switch to the **RUN** position.

   g. Move the tester button to the **TEST** (down) position.

   h. The air suspension control module will now start sending out the spring fill selection codes to be displayed on the tester. These codes will be displayed in a scrolling manner.

   i. Select the desired spring fill operation by releasing the tester button when the desired code is displayed. Select either Code 26 or Code 28 to inflate either the right rear or left rear air spring. As long as the tester button is released the inflation will continue. To stop infla-

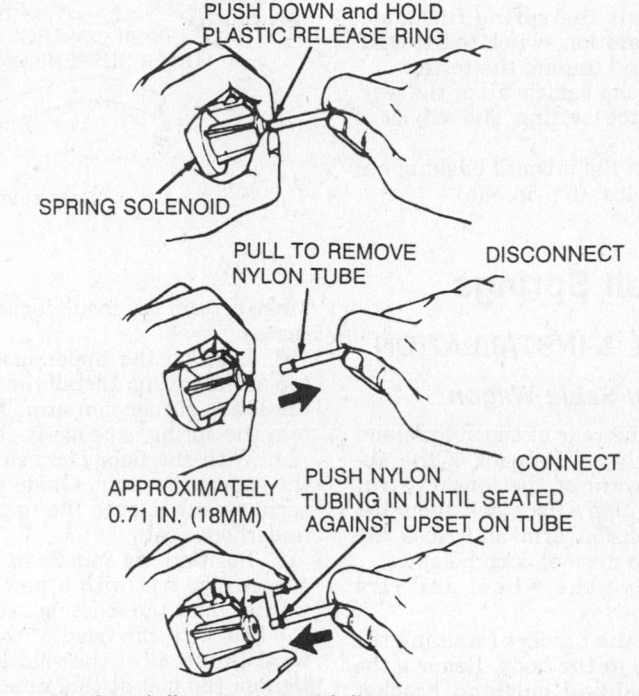

Air suspension air line connect and disconnect procedure—Continental

Code	Description
21	Vent R.F.
22	Vent L.F.
23	Vent R.R.
24	Inflate R.F.
25	Inflate L.F.
26	Inflate R.R.
27	Vent L.R.
28	Inflate L.R.

Air suspension spring fill codes

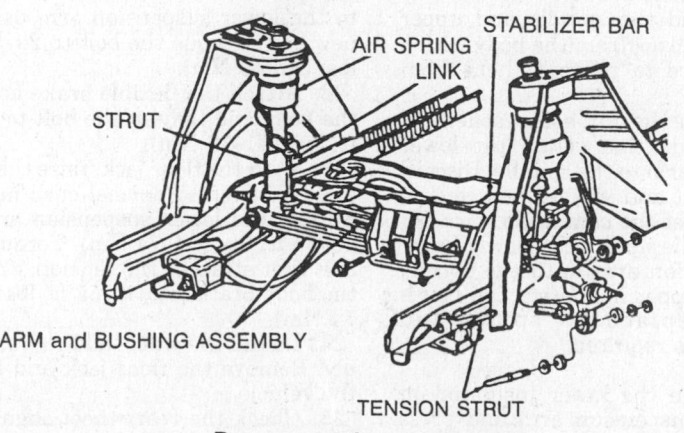

Rear suspension—Continental

tion, move the tester button back down to the **TEST** position. The spring fill codes will again be displayed.

NOTE: Do not apply a load to the suspension until after the air spring has been inflated at least 60 seconds.

j. To exit the spring fill mode, turn the ignition switch to the **OFF** position and unplug the tester.
43. Lower the vehicle all of the way. Check the toe setting and adjust if necessary.
44. Tighten the inboard bushing nut to 45–65 ft. lbs. (61–88 Nm).

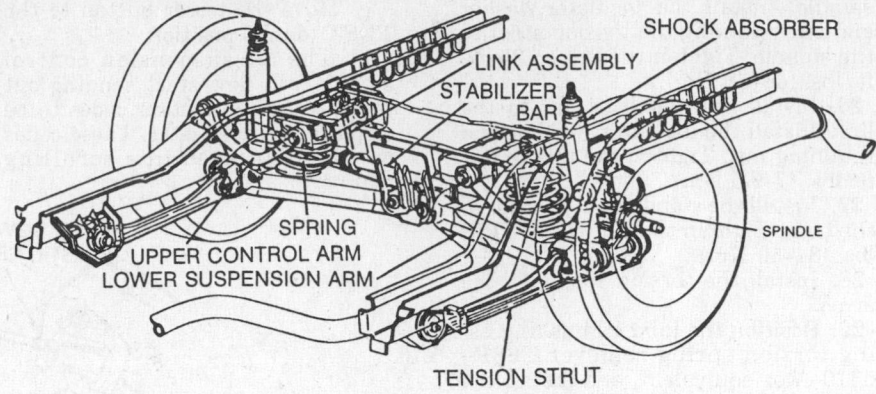

Rear suspension—Taurus and Sable wagon

Coil Springs

REMOVAL & INSTALLATION

Taurus and Sable Wagon

1. Raise the rear of the vehicle and support safely on the pads of the underbody forward of the tension strut bracket. Position a floor jack under the lower suspension arm and raise the lower arm to normal curb height.
2. Remove the wheel and tire assembly.
3. Locate the bracket retaining the flexible hose to the body. Remove the bracket retaining bolt and bracket from the body.
4. Remove the stabilizer bar U-bracket from the lower suspension arm.
5. Remove and discard the nuts attaching the shock absorber to the lower suspension arm.
6. Disconnect and remove the parking brake cable and clip from the lower suspension arm.
7. If equipped with rear disc brakes, remove the ABS cable from the clips on the lower suspension arm.
8. Remove and discard the bolt and nut attaching the tension strut to the lower suspension arm.
9. Suspend the spindle and upper suspension arms from the body with a piece of wire to prevent them from dropping.
10. Remove the nut, bolt, washer and adjusting cam that retain the lower suspension arm to the spindle. Discard the nut, bolt and washer and replace with new. Set the cam aside.
11. With the floor jack, slowly lower the suspension arm until the spring, lower and upper insulators can be removed. Replace the spring and insulators as required.
To install:
12. Position the lower insulator on the lower suspension arm and press the insulator downward into place.

Make certain the insulator is properly seated.
13. Position the upper insulator on top of the spring. Install the spring on the lower suspension arm. Make certain the spring is properly seated.
14. With the floor jack, slowly raise the suspension arm. Guide the upper spring insultor onto the upper spring underbody seat.
15. Position the spindle in the lower suspension arm with a new bolt, nut washer, and the existing cam. Install the bolt with the head of the bolt toward the front of the vehicle. Do not tighten the bolt at this time.
16. Remove the wire supporting the spindle and suspension arms.
17. Install the tension strut in the lower suspension arm using a new nut and bolt; do not tighten at this time.
18. Attach the parking brake cable and clip to the lower suspension arm.

19. If equipped with rear disc brakes, install the ABS cable into the clips on the lower suspension arm.
20. Position the shock absorber on the lower suspension arm and install 2 new nuts. Torque the nuts to 13–20 ft. lbs. (17–27 Nm).
21. Attach the stabilizer U-bracket to the lower suspension arm using a new bolt. Torque the bolt to 20–30 ft. lbs. (27–40 Nm).
22. Attach the flexible brake hose to the body and tighten the bolt to 8–12 ft. lbs. (11–16 Nm).
23. With the floor jack, raise the lower suspension to normal curb height. Torque the lower suspension arm to 40–55 ft. lbs. (54–74 Nm). Torque the bolt that attaches the tension strut to the body bracket to 40–55 ft. lbs. (54–74 Nm).
24. Install the wheel and tire assembly. Remove the floor jack and lower the vehicle.
25. Check the rear wheel alignment and adjust if necessary.

Rear Control Arms

REMOVAL & INSTALLATION

Taurus and Sable Sedan

1. Raise the vehicle and support it safely. Do not raise the vehicle by the tension strut.
2. Disconnect the brake proportioning valve from the left side front arm.
3. Disconnect the parking brake cable from the front arms.
4. Remove and discard the arm-to-spindle bolt, washer and nut.
5. Remove and discard the arm-to-body bolt and nut.
6. Remove the arm from the vehicle.
To install:

NOTE: When installing new control arms, the offset on all arms must face up. The arms are stamped "bottom" on the lower edge. The flange edge of the right side rear arm stamping must face the front of the vehicle. The other 3 must face the rear of the vehicle. The rear control arms have 2 adjustment cams that fit inside the bushings at the arm-to-body attachment. The cam is installed from the rear on the left arm and from the front on the right arm.

7. Position the arm and cam where required, at the center of the vehicle. Insert a new bolt and nut but do not tighten at this time.
8. Move the arm end up to the spindle and insert a new bolt, washer and nut. Tighten the nut to 42–57 ft. lbs. (57–77 Nm).
9. Tighten the arm-to-body nut to 45–65 ft. lbs. (61–88 Nm).
10. Attach the parking brake cable to the front arms and the brake proportioning valve to the left side front arm.
11. Lower the vehicle and check the alignment.

Continental

1. Turn **OFF** the air suspension switch located in the luggage compartment.

2. Raise and support the vehicle safely.

3. Remove all wire retainers and parking brake cable retainers from the lower suspension arm.

4. Disconnect the height sensor link from the ball stud pin on the lower arm.

5. Mark the position of the notch on the toe adjustment cam.

6. Remove the torsion spring retaining clamp at the spindle.

7. Remove the nut from the inboard bushing on the suspension arm.

8. Install torsion spring remover T88P-5310-A or equivalent, on the arm. Using a ¾ in. ratchet, pry up on the tool and arm to relieve the pressure on the pivot bolt. An assistant may also be required to pull outward on the spindle at the same time to fully relieve the tension on the bolt. Remove the bolt and lower the arm.

9. Remove the nut retaining the torsion spring to the arm and seperate the spring from the arm.

10. Remove the outboard attaching bolt at the spindle.

11. Repeat the removal procedure for the other arm.

To install:

NOTE: **When installing new control arms, the offset must face up. The arms are stamped "bottom" on the lower edge. The rear control arms have adjustment cams that fit inside the bushings at the arm-to-body attachment. The cams are installed from the front of both arms.**

12. Loosely attach the arm(s) at the spindle. Attach the torsion spring(s) to the arm(s).

13. Position the inboard bushing using torsion spring remover T88P-5310-A or equivalent, and install the bolt. It may be required to have an assistant pull outward on the spindle to align the bushing so the bolt can be inserted. Repeat this step for the opposite side.

14. Set the toe adjustment cam to the alignment mark for rear arm only.

15. Connect the height sensor link to the ball stud pin on the lower arm for right front only.

16. Install all wire retainers and parking brake cable retainers to the lower suspension arm.

17. Lower the vehicle and then turn **ON** the air suspension switch.

18. With the vehicle suspension at curb height, tighten the control arm-to-spindle bolt to 42–57 ft. lbs. (57–77 Nm) and the control arm-to-body bolt to 45–65 ft. lbs. (61–88 Nm).

19. Check the rear toe setting.

Taurus and Sable Wagon

UPPER ARMS

1. Raise the vehicle and support it with wood blocks on jackstands so the suspension is at normal curb height.

2. Remove the wheel and tire assembly.

3. Remove the brake line flexible hose bracket from the body.

4. Loosen, but do not remove the nuts attaching the spindle to the upper and lower suspension arms.

5. Remove and discard the nuts and bolts attaching the front and rear upper suspension arms to the body brackets. Make sure the spindle does not fall outward.

6. Tilt the top of the spindle outward, letting it pivot on the lower suspension arm attaching bolt until the ends of the upper suspension arms are clear of the body bracket. Support the spindle with wire in this position.

7. Remove and discard the nut attaching the upper suspension arms to the spindle and remove the arms from the vehicle.

To install:

8. Install the upper suspension arms on the spindle and install a new nut but do not tighten the nut at this time.

9. Position the upper suspension arm ends to the body bracket and install new nuts and bolts. Tighten to 70–95 ft. lbs. (95–129 Nm). Remove the wire from the spindle.

10. Tighten the nut attaching the upper suspension arms to the spindle to 150–190 ft. lbs. (204–257 Nm). Tighten the nut attaching the lower suspension arm to the spindle to 40–55 ft. lbs. (54–74 Nm).

11. Install the brake line bracket to the body.

12. Install the wheel and tire assembly, remove the jackstand and wood block and lower the vehicle.

13. Check the rear wheel alignment.

Lower Arm

1. Raise and support the vehicle safely on the lifting pads on the underbody forward of the tension strut body bracket.

2. Remove the wheel and tire assembly.

3. Place a floor jack under the lower suspension arm.

4. Remove the bracket retaining the flexible brake hose to the body.

5. Remove the stabilizer bar U-bracket from the lower suspension arm.

6. Remove and discard the nuts attaching the shock absorber to the lower suspension arm.

7. Remove the parking brake cable and clip from the lower suspension arm.

8. Remove and discard the bolt and nut attaching the tension strut to the lower suspension arm.

9. Support the spindle and upper suspension arms by wiring them to the body, to prevent them from dropping down.

10. Remove the nut, bolt, washer and adjusting cam retaining the lower suspension arm to the spindle. Discard the nut, bolt and washer.

11. Lower the suspension arm with the floor jack until the spring can be removed.

12. Remove and discard the bolt and nut attaching the lower suspension arm to the center body bracket and remove the arm.

To install:

13. Position the lower suspension arm-to-center body bracket and install but do not tighten a new bolt and nut with the bolt head toward the front of the vehicle.

14. Position the lower insulator on the lower suspension arm and press the insulator downward into place. Make sure the insulator is properly seated.

15. Position the upper insulator on top of the spring. Install the spring on the lower suspension arm, making sure the spring is properly seated.

16. Raise the suspension arm with the floor jack and guide the upper spring insulator onto the upper spring seat on the underbody.

17. Position the spindle in the lower suspension arm and install, but do not tighten, a new bolt, nut, washer and the existing cam, with the bolt head toward the front of the vehicle.

18. Remove the wire from the spindle and suspension arms.

19. Install the tension strut in the lower suspension arm using a new bolt and nut but do not tighten at this time.

20. Install the parking brake cable and clip to the lower suspension arm.

21. Position the shock absorber on the lower suspension arm and install 2 new nuts. Tighten the nuts to 13–20 ft. lbs. (17–27 Nm).

22. Install the stabilizer bar and U-bracket to the lower suspension arm using a new bolt. Tighten the bolt to 20–30 ft. lbs. (27–40 Nm).

23. Install the flexible brake hose bracket to the body. Tighten the bolt to 8–12 ft. lbs. (11–16 Nm).

24. Using the floor jack, raise the lower suspension arm to normal curb

height. Tighten the following to 40–55 ft. lbs. (54–74 Nm):

    Lower suspension arm-to-body bracket nut

    Lower suspension arm-to-spindle nut

    Tension strut-to-body bracket bolt

25. Install the wheel and tire assembly and lower the vehicle.
26. Check the rear wheel alignment.

Rear Wheel Bearings

REMOVAL & INSTALLATION

Drum Brakes

1988–89

1. Raise the vehicle and support it safely. Remove the wheel from the hub and drum.
2. Remove the grease cap from the hub. Remove the cotter pin, nut retainer, adjusting nut and keyed flat washer from the spindle. Discard the cotter pin.
3. Pull the hub and drum assembly off of the spindle. Remove the outer bearing assembly.
4. Using seal remover tool 1175–AC or equivalent, remove and discard the grease seal. Remove the inner bearing assembly from the hub.
5. Wipe all lubricant from the spindle and inside of the hub. Cover the spindle with a clean cloth and vacuum all loose dust and dirt from the brake assembly. Carefully remove the cloth to prevent dirt from falling on the spindle.
6. Clean both bearing assemblies and cups using a suitable solvent. Inspect the bearing assemblies and cups for excessive wear, scratches, pits or other damage and replace as necessary.
7. If the cups are to be replaced, remove them with impact slide hammer T50T–100–A and bearing cup puller T77F–1102–A or equivalent.

To install:

8. If the inner and outer bearing cups were removed, install the replacement cups using driver handle T80T–4000–W and bearing cup replacers T73T–1217–A and T77F–1217–A or equivalent. Support the drum hub on a block of wood to prevent damage. Make sure the cups are properly seated in the hub.

NOTE: Do not use the cone and roller assembly to install the cups. This will result in damage to the bearing cup and the cone and roller assembly.

9. Make sure all of the spindle and bearing surfaces are clean.
10. Using a bearing packer, pack the bearing assemblies with a suitable wheel bearing grease. If a packer is not available, work in as much grease as possible between the rollers and cages. Grease the cup surfaces.

NOTE: Allow all of the cleaning solvent to dry before repacking the bearings. Do not spin-dry the bearings with air pressure.

11. Install the inner bearing cone and roller assembly in the inner cup. Apply a light film of grease to the lips of a new grease seal and install the seal with rear hub seal replacer T56T–4676–B or equivalent. Make sure the retainer flange is seated all around.
12. Apply a light film of grease on the spindle shaft bearing surfaces. Install the hub and drum assembly on the spindle. Keep the hub centered on the spindle to prevent damage to the grease seal and spindle threads.
13. Install the outer bearing assembly and the keyed flatwasher on the spindle. Install the adjusting nut and adjust the wheel bearings. Install a new cotter pin. Install the grease cap.
14. Install the wheel and tire assembly and lower the vehicle.

1990–92

1. Raise the vehicle and support it safely.
2. Remove the wheel.
3. Remove the 2 pushnuts retaining the drum to the hub and remove the drum.
4. Remove the grease cap from the bearing and hub assembly and discard it.

5. Remove the hub retaining nut and remove the bearing and hub assembly from the spindle.
6. Install in the reverse order of removal. Use coil remover T89P–19623–FH or equivalent, to install the new grease cap. Tap on the tool to make sure the grease cap is fully seated. Tighten the hub retaining nut to 188–254 ft. lbs. (255–345 Nm).

Disc Brakes

1988–89

1. Raise the vehicle and support it safely. Remove the tire and wheel assembly from the hub.
2. Remove the brake caliper by removing the 2 bolts that attach the caliper support to the cast iron brake adapter. Do not remove the caliper pins from the caliper assembly. Lift the caliper off of the rotor and support it with a length of wire. Do not allow the caliper assembly to hang from the brake hose.
3. Remove the rotor from the hub by pulling it off the hub bolts. If the rotor is difficult to remove, strike the rotor sharply between the studs with a rubber or plastic hammer.
4. Remove the grease cap from the hub. Remove the cotter pin, nut retainer, adjusting nut and keyed flat washer from the spindle. Discard the cotter pin.
5. Pull the hub assembly off of the spindle. Remove the outer bearing assembly.
6. Using seal remover tool 1175–AC or equivalent, remove and discard the

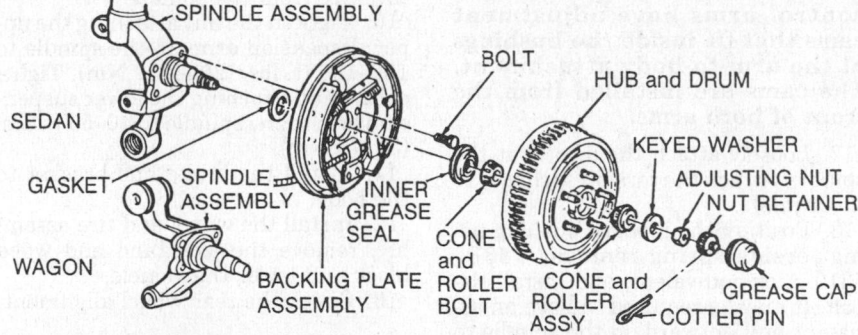

Rear wheel hub and bearing assembly—1988–89 vehicles with drum brakes

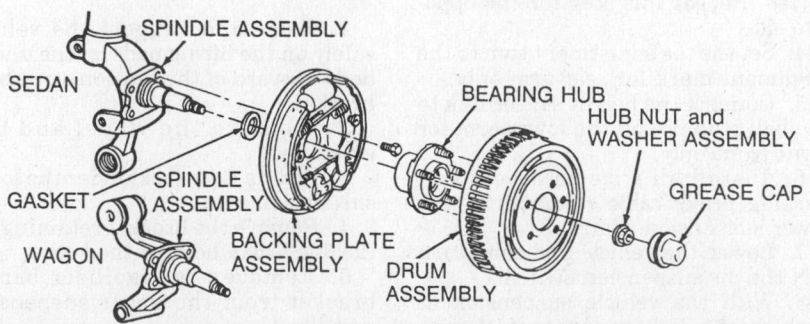

Rear wheel hub and bearing assembly—1990–92 vehicles with drum brakes

grease seal. Remove the inner bearing assembly from the hub.

7. Wipe all of the lubricant from the spindle and inside of the hub. Cover the spindle with a clean cloth and vacuum all of the loose dust and dirt from the brake assembly. Carefully remove the cloth to prevent dirt from falling on the spindle.

8. Clean both bearing assemblies and cups using a suitable solvent. Inspect the bearing assemblies and cups for excessive wear, scratches, pits or other damage and replace as necessary.

9. If the cups are being replaced, remove them with impact slide hammer tool T50T-100-A and bearing cup puller tool T77F-1102-A or equivalent.

To install:

10. If the inner and outer bearing cups were removed, install the replacement cups using driver handle tool T80T-4000-W and bearing cup replacer tools T73F-1217-A and T77F-1217-B or equivalent. Support the hub on a block of wood to prevent damage. Make sure the cups are properly seated in the hub.

NOTE: Do not use the cone and roller assembly to install the cups. This will result in damage to the bearing cup and the cone and roller assembly.

11. Make sure all of the spindle and bearing surfaces are clean.
12. Pack the bearing assemblies with a suitable wheel bearing grease using a bearing packer. If a packer is not available, work in as much grease as possible between the rollers and the cages. Grease the cup surfaces.

NOTE: Allow all of the cleaning solvent to dry before repacking the bearings. Do not spin-dry the bearings with air pressure.

13. Place the inner bearing cone and roller assembly in the inner cup. Apply a light film of grease to the lips of a new grease seal and install the seal with rear hub seal replacer tool T56T-4676-B or equivalent. Make sure the retainer flange is seated all around.
14. Apply a light film of grease on the spindle shaft bearing surfaces. Install the hub assembly on the spindle. Keep the hub centered on the spindle to prevent damage to the grease seal and spindle threads.
15. Install the outer bearing assembly and keyed flat washer on the spindle. Install the adjusting nut and adjust the wheel bearings. Install a new cotter pin. Install the grease cap.
16. Install the disc brake rotor to the hub assembly. Install the disc brake caliper over the rotor.

17. Install the wheel and tire assembly and lower the vehicle.

1990–91

1. Raise the vehicle and support it safely.
2. Remove the wheel and tire assembly.
3. Remove the caliper assembly from the brake adapter. Support the caliper assembly with a length of wire.
4. Remove the push on nuts that retain the rotor to the hub and remove the rotor.
5. Remove the grease cap from the bearing and hub assembly and discard the grease cap.
6. Remove the bearing and hub assembly retaining nut and remove the bearing and hub assembly from the spindle.
7. Install in the reverse order of removal. Install a new grease cap using coil remover tool T89P-19623-FH or equivalent. Tap on the tool until the grease cap is fully seated. Tighten the hub retaining nut to 188–254 ft. lbs. (255–345 Nm).

ADJUSTMENT

The following procedure applies only to 1988–89 vehicles. Adjustment is not possible on 1990–91 vehicles. This procedure should be performed whenever the wheel is excessively loose on the spindle or it does not rotate freely.

NOTE: The rear wheel uses a tapered roller bearing which may feel loose when properly adjusted; this condition should be considered normal.

1. Raise and support the rear of the vehicle until tires clear the floor.
2. Remove the wheel cover or the ornament and nut covers. Remove the hub grease cap.

NOTE: If the vehicle is equipped with styled steel or aluminum wheels, the wheel/tire assembly must be removed to remove the dust cover.

3. Remove the cotter pin and the nut retainer.
4. Back off the hub nut 1 full turn.
5. While rotating the hub/drum assembly, tighten the adjusting nut to 17–25 ft. lbs. (23–24 Nm). Back off the adjusting nut ½ turn, then retighten it to 24–28 inch lbs. (2.7–3.2 Nm).
6. Position the nut retainer over the adjusting nut so the slots are in line with cotter pin hole, without rotating the adjusting nut.
7. Install the cotter pin and bend the ends around the retainer flange.
8. Check the hub rotation. If the hub rotates freely, install the grease

cap. If not, check the bearings for damage and replace, as necessary.

9. Install the wheel and tire assembly and the wheel cover, if necessary. Lower the vehicle.

STEERING

Steering Wheel
CAUTION

If equipped with an air bag, the negative battery cable must be disconnected before working on the system. On 1990–92 vehicles, the backup power supply must also be disconnected. Failure to do so may result in deployment of the air bag and possible personal injury. Always wear safety glasses when servicing an air bag vehicle and when handling an air bag.

REMOVAL & INSTALLATION

1988–89 Taurus and Sable and 1988 Continental

1. Disconnect the negative battery cable.
2. Remove the steering wheel horn pad cover by removing 2 screws from the back of the steering wheel. If equipped with cruise control, disconnect the connector from the slip ring terminal.
3. Remove and discard the steering wheel retaining bolt.
4. Remove the steering wheel from the upper shaft by grasping the rim of the steering wheel and pulling off. A steering wheel puller is not required.
To install:
5. Position the steering wheel on the end of the shaft. Align the mark on the steering wheel with the mark on the shaft to ensure the straight-ahead steering wheel position corresponds to the straight-ahead position of the front wheels.

NOTE: The combination switch lever must be in the middle position before installing the steering wheel or damage to the switch cam may result.

6. Install a new steering wheel retaining bolt and tighten to 23–33 ft. lbs. (31–45 Nm).
7. If equipped with cruise control, connect the connector to the slip ring terminal. Install the steering wheel horn pad cover with the 2 screws. Tighten to 5–10 inch lbs. (0.5–1.1 Nm).
8. Connect the negative battery cable.

1989–90 Continental

1. Center the front wheels to the straight-ahead position. Disconnect the negative battery cable. On 1990 vehicles, lower the glove compartment past it's stops and disconnect the air bag backup power supply (blue box, 1 connector).

2. Remove the lower instrument panel cover.

3. Remove the steering column lock cylinder and the tilt release lever.

4. Remove the lower steering column shroud.

5. Disconnect the contact assembly at the body wire harness and remove the contact assembly ground wire screw located at the lock cylinder housing.

6. Remove the 4 air bag module retaining nuts and remove the air bag module from the steering wheel. Disconnect the contact as an assembly at the module.

— CAUTION —

When carrying a live air bag, make sure the bag and trim cover are pointed away from the body. In the unlikely event of an accidental deployment, the bag will then deploy with minimal chance of injury. In addition, when placing a live ai bag on a bench or other surface, always face the bag and trim cover up, away from the surface. This will reduce the motion of the module if it is accidentally deployed.

7. Remove and discard the steering wheel attaching bolt.

8. Remove the steering wheel and contact assembly. Make sure the contact assembly is locked in the straight-ahead position. Do not allow the contact assembly to rotate out of position.

To install:

9. Install the steering wheel and contact assembly onto the steering column. Make sure the drive pin on the speed control/horn brush assembly engages in the drive socket on the contact assembly housing.

10. Install a new steering wheel bolt and tighten to 23–33 ft. lbs. (31–45 Nm).

11. Install the ground wire and retaining screw.

12. Connect the contact assembly wire harness and connect the contact assembly at the module.

13. Install the module to the steering wheel and tighten the retaining nuts.

14. Install the lower steering column shroud, lock cylinder, tilt release lever and the lower instrument panel cover.

15. On 1990 vehicles, connect the air bag backup power supply. Connect the negative battery cable and check the steering column for proper operation.

1990–92 Taurus and Sable and 1991–92 Continental

1. Center the front wheels in the straight-ahead position.

2. Disconnect the negative battery cable. Lower the glove compartment past it's stops and disconnect the air bag backup power supply.

3. Remove the 4 air bag module retaining nuts and lift the module from the wheel. Disconnect the air bag wire harness from the air bag module and remove the module from the wheel.

— CAUTION —

When carrying a live air bag, make sure the bag and trim cover are pointed away from the body. In the unlikely event of an accidental deployment, the bag will then deploy with minimal chance of injury. In addition, when placing a live air bag on a bench or other surface, always face the bag and trim cover up, away from the surface. This will reduce the motion of the module if it is accidentally deployed.

4. Disconnect the cruise control wire harness from the steering wheel. Remove and discard the steering wheel retaining bolt.

5. Install a suitable steering wheel puller and remove the steering wheel. Route the contact assembly wire harness through the steering wheel as the wheel is lifted off the shaft.

To install:

6. Make sure the vehicle's front wheels are in the straight-ahead position.

7. Route the contact assembly wire harness through the steering wheel opening at the 3 o'clock position and install the steering wheel on the shaft. The steering wheel and shaft alignment marks should be aligned. Make sure the air bag contact wire is not pinched.

8. Install a new steering wheel retaining bolt and tighten to 23–33 ft. lbs. (31–48 Nm).

9. Connect the cruise control wire harness to the wheel and snap the connector assembly into the steering wheel clip. Make sure the wiring does not get trapped between the steering wheel and contact assembly.

10. Connect the air bag wire harness to the air bag module and install the module to the steering wheel. Tighten the module retaining nuts to 3–4 ft. lbs. (4–6 Nm).

11. Connect the air bag backup power supply and the negative battery cable. Verify the air bag warning indicator.

Steering Column

REMOVAL & INSTALLATION

1988–89 Taurus and Sable and 1988–90 Continental

1. Disconnect the negative battery cable. On 1990 Continental, lower the glove compartment past it's stops and disconnect the air bag backup power supply.

2. Remove the 4 self-tapping screws and remove the steering column cover from the lower portion of the instrument panel.

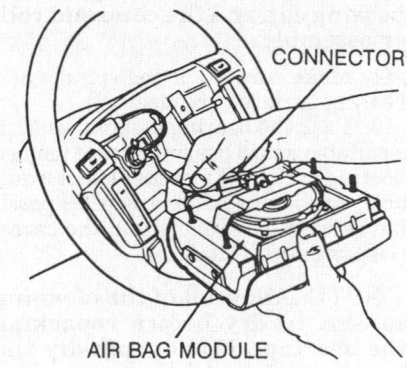

Air bag module removal

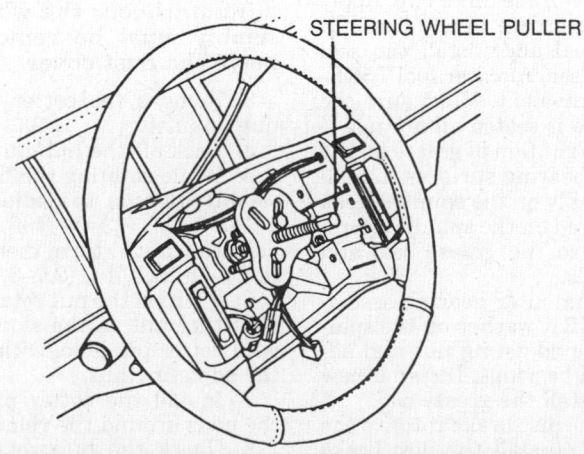

Steering wheel removal—1990–92 Taurus and Sable and 1991–92 Continental

3. Remove the retaining screw and the tilt release lever. Remove the ignition lock cylinder.

4. Remove the 3 self-tapping screws from the bottom of the lower shroud and remove the steering column trim shrouds. Remove the steering wheel.

5. If equipped with column shift, perform the following:

 a. Disconnect the shift position indicator cable from the lock cylinder housing by removing the retaining screw.

 b. On 1988 Continental, remove the hood release cable from the handle.

 c. Disconnect the shift position indicator cable from the shift socket.

 d. Remove the shift position indicator cable from the retaining hook on the bottom of the lock cylinder housing.

6. Using a punch, remove the shift lever-to-shift socket retaining pin and remove the shift lever.

7. Disconnect the cruise control/ horn brush wiring connector from the main wiring harness.

8. Remove the combination switch wiring harness retainer from the lock cylinder housing by squeezing the end of the retainer and pushing out. Disconnect the combination switch connector and remove the 2 self-tapping retaining screws. Remove the combination switch.

9. Disconnect the key warning buzzer switch wiring connector from the main wiring harness and disconnect the wiring connector from the ignition switch. On Continental, disconnect the steering column angular speed sensor wire connector at the cable bracket.

10. Disconnect the steering shaft from the intermediate shaft by removing 2 nuts and 1 U-clamp. If equipped with an air bag, wire the lower end of the steering shaft to the column housing to prevent rotation of the steering shaft. Rotating the steering shaft could damage the air bag contact clockspring if the steering wheel is attached to the column.

11. If equipped with column shift, perform the following:

 a. Remove the shift cable plastic terminal from the column selector lever pivot ball using a small prybar and prying between the plastic terminal and the selector lever. Be careful not to damage the cable during or after assembly.

 b. Remove the shift cable bracket, with shift cable still attached, from the lock cylinder housing by removing the 2 retaining screws.

12. If equipped with an automatic parking brake release mechanism, remove the vacuum hoses from the parking brake release switch. On 1988

Continental, remove the hood release cable grommet from the column bracket.

13. Remove the 2 nuts retaining the rear column assembly. Loosen the 2 nuts retaining the front column assembly to the end of the studs, but do not remove at this time.

14. Use a downward force to disengage the column assembly push-on clips from the rear attachments. Remove the remaining 2 nuts.

NOTE: When forcing downward, care should be taken to avoid damaging the safety slip-clips on the steering column.

15. Carefully lower the steering column assembly and remove from the vehicle.

To install:

16. Raise the steering column assembly into position and align the 4 mounting holes over the 4 support bracket studs. Hand start the 4 retaining nuts.

17. Center the column assembly in the instrument panel opening and tighten the 4 nuts to 15–25 ft. lbs. (21–33 Nm).

18. If equipped with an automatic parking brake release mechanism, install the vacuum hoses on the parking brake release switch.

19. If equipped with column shift, perform the following:

 a. Attach the cable shift bracket, with the shift cable attached, to the lock cylinder housing and tighten the retaining screws to 5–7 ft. lbs. (7–9 Nm).

 b. Snap the transaxle shift cable terminal to the selector lever pivot ball on the steering column.

20. Apply a generous amount of grease to the V-shaped steering shaft yoke. Connect the steering shaft to the intermediate shaft with 1 U-clamp and 2 hex nuts. When installing the steering column to the intermediate shaft, connect the intermediate shaft to the steering column with the retainer assembly and 2 nuts.

NOTE: Make sure the V-angle of the intermediate shaft fits correctly into the V-angle of the mating steering column yoke. If the V-angle is mis-aligned and the retainer is tightened, the retainer plate will be bent and then must be replaced.

21. After correctly installing the steering column to the intermediate shaft, tighten the nuts to 15–25 ft. lbs. (21–33 Nm).

NOTE: Tilt columns must be in the middle tilt position before the nuts are tightened.

22. Connect the main wiring harness connector to the ignition switch and connect the key warning buzzer switch wiring connector to the main harness. Install the steering sensor wire connector to the sensor lead connector.

23. Install the combination switch and tighten the 2 self-tapping screws to 18–26 inch lbs. (2.0–2.9 Nm). Install the combination switch wiring harness retainer over the shroud mounting boss and snap it into the slot in the lock cylinder housing.

24. Connect the cruise control/horn brush wiring connector to the main wiring harness.

25. If equipped with column shift, install the shift position indicator cable into the retaining hook on the lock cylinder housing, connect the cable to the shift socket and loosely install the cable onto the lock cylinder housing with 1 screw. On 1988 Continental, install the hood release cable grommet to the column bracket and secure the cable to the hood release handle. Adjust the shift position indicator cable as follows:

 a. Place the shift lever in **D** on Taurus equipped with the 2.5L engine. On all others, place the shift lever in **OD**. A weight of 8 lbs. should be hung on the shift selector lever to make sure the lever is firmly against the **D** or **OD** drive detent.

 b. Adjust the cable until the indicator pointer completely covers the **D** or **OD**, then tighten the screw to 18–30 inch lbs. (2.0–3.4 Nm).

 c. Cycle the shift lever through all positions and check that the shift position idicator completely covers the proper letter or number in each position.

26. Install the shift lever into the shift lever socket and install a new shift lever retaining pin. Use care to avoid damaging the shift position indicator post on the shift socket.

27. Place the combination switch in the middle position and install the steering wheel.

28. Install the shrouds with the retaining screws. Tighten to 6–10 inch lbs. (0.7–1.1 Nm). If equipped with tilt column, install the tilt release lever and tighten the screw to 6.5–9.0 ft. lbs. (8.8–12 Nm).

29. Install the ignition lock cylinder. Install the steering column cover on lower portion of instrument panel with the 4 self-tapping screws.

30. Connect the negative battery cable. Check the column function as follows:

 a. With the column shift lever in **P** position or the floor shift key release button depressed, and with the ignition switch in the **LOCK** position; make sure the steering column locks.

b. With the column shift lever in **D** or with the floor shift key release button extended, and with the ignition switch in the **RUN** position; rotate the ignition switch toward the **LOCK** position until it stops. In this position, make sure engine electrical off has been achieved and that the steering shaft does not lock.

c. On tilt columns, check column tilt travel through it's entire range to make sure there is no interference between the column and instrument panel.

d. Cycle the combination switch through all of it's functions.

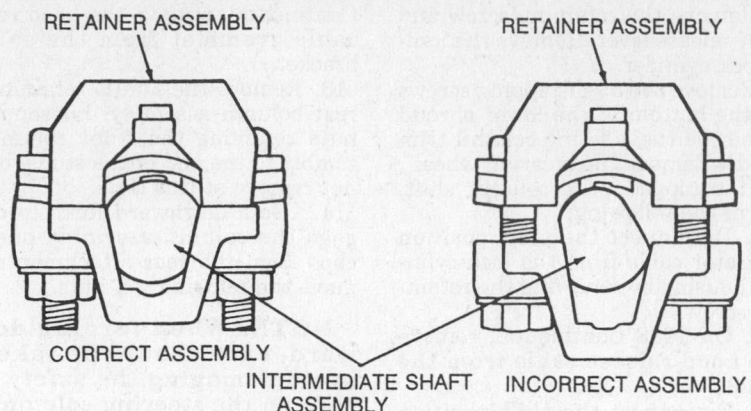

Steering column to intermediate shaft installation

1990–92 Taurus and Sable and 1991–92 Continental

1. Disconnect the negative battery cable. Lower the glove compartment past it's stops and disconnect the air bag backup power supply.
2. Remove the steering wheel.
3. Remove the left and right lower mouldings from the instrument panel by pulling up and snapping out of the retainers.
4. Remove the instrument panel lower trim cover and the lower steering column shroud.
5. Disconnect the air bag clockspring contact assembly wire harness. Apply 2 strips of tape across the contact assembly stator and rotor to prevent accidental rotation. Remove the 3 contact assembly retaining screws and pull the contact assembly off the steering column shaft.
6. Remove the tilt lever by unscrewing it from the columnn.
7. Rotate the ignition lock cylinder to the **RUN** position. Using an ⅛ in. drift, depress the lock cylinder retaining pin through the access hole and remove the lock cylinder.
8. Remove the 4 retaining screws from the lower shroud and remove the steering column shrouds.
9. Remove the 2 instrument panel reinforcement brace retaining bolts and remove the reinforcement.
10. If equipped with column shift, disconnect the shift position indicator cable from the actuator housing by removing 1 screw and disconnect the cable loop from the shift tube hook. If equipped with console shift, remove the interlock cable retaining screws and remove the cable.
11. Remove the 2 combination switch retaining screws and set the switch aside.
12. Remove the pinch bolt from the steering shaft flex coupling.
13. Disconnect the shift cable from the selector lever pivot. Remove the shift cable and bracket from the lower column mounting. Remove the column skid plate.

14. While supporting the column assembly, remove the 4 column assembly retaining nuts. Lower the column and disconnect the vacuum hoses at the parking brake release switch or remove the vacuum release assembly.
15. Remove the column from the vehicle.

To install:

16. Align the column lower universal joint to the lower shaft. Install 1 bolt and tighten to 29–41 ft. lbs. (40–56 Nm). Connect the parking brake release vacuum hoses.
17. Support the column assembly to the column support bracket. Install the 4 retaining nuts and tighten to 10–14 ft. lbs. (13–19 Nm).
18. Position the shift cable bracket, with the shift cable attached, to the lower 2 screws of the column. Tighten to 5–8 ft. lbs. (7–11 Nm). Snap the shift cable onto the shift selector pivot ball.
19. If equipped with automatic console shift, position the interlock cable and install the 2 retaining screws.
20. Position the combination switch and install the 2 retaining screws. Tighten to 18–26 inch lbs. (2–3 Nm). Connect all electrical connectors.
21. If equipped with column shift, attach the shift position indicator cable loop on the shift selector hook and install the cable bracket to the actuator housing. Install the retaining screw and tighten to 5–8 ft. lbs. (7–11 Nm).
22. Install the steering column skid plate and tighten the retaining nuts to 15–25 ft. lbs. (20–34 Nm).
23. Install the upper and lower column shrouds and the instrument panel reinforcement brace.
24. Install the lower instrument panel cover. Snap the right and left lower instrument panel mouldings into place.
25. Install the lock cylinder assembly and the tilt lever.
26. Install the air bag clockspring contact assembly screws and tighten

to 18–26 inch lbs. (2–3 Nm). Route the contact assembly down the column and connect to the wire harness.

NOTE: If a new contact assembly is being installed, remove the plastic lock mechanism after the contact assembly is secured to the column.

27. Install the steering wheel with a new bolt. Tighten to 23–33 ft. lbs. (31–48 Nm). Position the air bag module to the steering wheel. Install the 4 retaining nuts and tighten to 3–4 ft. lbs. (4–6 Nm).
28. Connect the air bag backup power supply and the negative battery cable. Verify the air bag warning indicator.

Power Rack and Pinion

ADJUSTMENT

Integral Power Rack and Pinion—Taurus and Sable, Except 1990–92 Taurus LX and Sable with 3.8L Engine

RACK YOKE PLUG CLEARANCE

NOTE: The rack yoke clearance adjustment is not a normal service adjustment. It is only required when the input shaft and valve assembly is removed.

1. Remove the steering gear from the vehicle. Clean the exterior of the steering gear thoroughly.
2. Install the steering gear in a suitable holding fixture. Do not remove the external transfer tubes unless they are leaking or damaged. If these lines are removed, they must be replaced with new ones.
3. Drain the power steering fluid by rotating the input shaft lock-to-lock twice, using a suitable tool. Cover the ports on the valve housing with a shop

cloth while draining the gear to avoid possible oil spray.

4. Insert an inch pound torque wrench with a maximum capacity of 60 inch lbs. (6.77 Nm) into pinion shaft torque adapter T74P–3504–R or equivalent. Position the adapter and wrench on the input shaft splines.

5. Loosen the yoke plug locknut and then the yoke plug.

6. With the rack at the center of travel, tighten the yoke plug to 45–50 inch lbs. (5–5.6 Nm). Clean the threads of the yoke plug prior to tightening to prevent a false reading.

7. Back off the yoke plug approximately ⅛ turn, 44 degrees minimum to 54 degrees maximum, until the torque required to initiate and sustain rotation of the input shaft is 7–18 inch lbs. (0.78–2.03 Nm).

8. Place a suitable wrench on the yoke plug locknut. While holding the yoke plug, tighten the locknut to 44–66 ft. lbs. (60–89 Nm). Do not allow the yoke plug to move while tightening or preload will be affected. Check the input shaft torque as in Step 7 after tightening the locknut.

9. Install the steering gear.

REMOVAL & INSTALLATION

Integral Power Rack and Pinion—Taurus and Sable, Except 1990–92 Taurus LX and Sable with 3.8L Engine

1. Disconnect the negative battery cable. Working from inside the vehicle, remove the nuts retaining the steering shaft weather boot to the dash panel.

2. Remove the bolts retaining the intermediate shaft to the steering column shaft. Set the weather boot aside.

3. Remove the pinch bolt at the steering gear input shaft and remove the intermediate shaft. Raise the vehicle and support safely.

4. Remove the left front wheel and tire assembly. Remove the heat shield. Cut the bundling strap retaining the lines to the gear.

5. Remove the tie rod ends from the spindles. Place a drain pan under the vehicle and remove the hydraulic pressure and return lines from the steering gear.

NOTE: The pressure and return lines are on the front of the housing. Do not confuse them with the transfer lines on the side of the valve.

6. Remove the nuts from the gear mounting bolts. The bolts are pressed into the gear housing and should not be removed during gear removal.

7. Push the weather boot end into the vehicle and lift the gear out of the

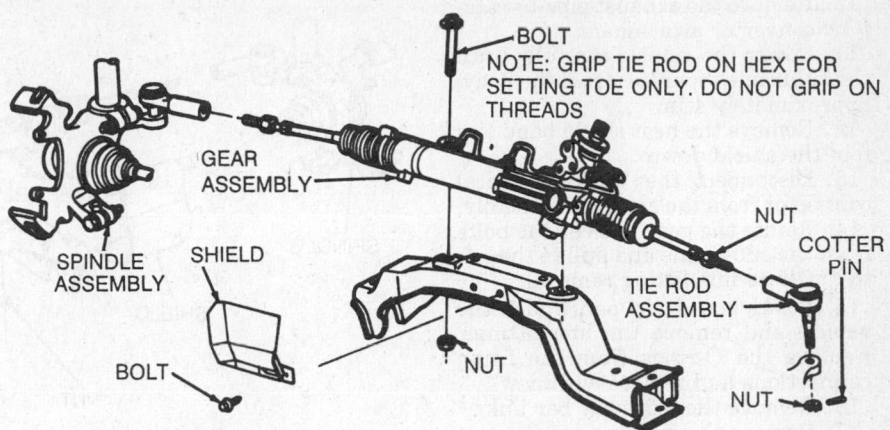

Integral power rack and pinion—Taurus and Sable, except Taurus LX and Sable with 3.8L engine

mounting holes. Rotate the gear so the input shaft will pass between the brake booster and the floor pan. Carefully start working the steering gear out through the left fender apron opening.

8. Rotate the input shaft so it clears the left fender apron opening and complete the removal of the steering gear. If the steering gear seems to be stuck, check the right tie rod to ensure the stud is not caught on anything.

To install:

9. Install new plastic seals on the hydraulic line fittings.

10. Insert the steering gear through the left fender apron. Rotate the input shaft forward to completely clear the fender apron opening.

11. To allow the gear to pass between the brake booster and the floorpan, rotate the input shaft rearward. Align the steering gear bolts to the bolt holes. Install the mounting nuts and torque them to 85–100 ft. lbs. (115–135 Nm). Lower the vehicle.

12. From inside the engine compartment, install the hydraulic pressure and return lines. Tighten the pressure line to 20–25 ft. lbs. (28–33Nm) and the return line to 15–20 ft. lbs. (20–28 Nm). Swivel movement of the lines is normal when the fittings are properly tightened.

13. Raise the vehicle and support safely. Secure the pressure and return lines to the transfer tube with the bundle strap. Install the heat shield.

14. Install the tie rod ends to spindles. Torque the castle nuts to 35 ft. lbs. (48 Nm) and if necessary, torque the nuts a little bit more to align the slot in the nut for the cotter pin. Install the cotter pin.

15. Install the left front wheel and tire assembly and lower the vehicle. Working from inside the vehicle, pull the weather boot end out of the vehicle and install it over the valve housing.

Install the intermediate shaft to the steering gear input shaft. Install the the inner weather boot to the floor pan.

16. Install the intermediate shaft to the steering column shaft. Fill the power steering system.

17. Check the system for leaks and proper operation. Adjust the toe setting as necessary.

Variable Assist Power Steering (VAPS) System—Continental and 1990–92 Taurus LX and Sable with 3.8L Engine

The Variable Assist Power Steering (VAPS) system used on these vehicles consists of a micro-processor based module, a power rack and pinion steering gear, an actuator valve assembly, hose assemblies and a high efficiency power steering pump.

1. Disconnect the negative battery cable. Remove the primary steering column boot attachments.

2. Remove the intermediate shaft retaining bolts and remove the intermediate shaft.

3. From inside the passenger compartment, remove the secondary steering column boot.

4. Raise the vehicle and support safely. Remove the front wheels. Support the vehicle under the rear edge of the sub-frame.

5. Remove the tie rod cotter pins and nuts. Remove the tie rod ends from the spindle.

6. Remove the tie rod ends from the shaft. Mark the position of the jam nut to maintain the alignment.

7. Remove the nuts from the gear-to-sub-frame attaching bolts.

8. Remove both height sensor attachments on Continental.

9. Remove the rear sub-frame-to-body attaching bolts.

10. Remove the exhaust pipe-to-catalytic converter attachment.

11. Lower the vehicle carefully until the subframe separates from the body; approximately 4 in.

12. Remove the heat shield band and fold the shield down.

13. Disconnect the VAPS electrical connector from the actuator assembly.

14. Rotate the gear to clear the bolts from the sub-frame and pull to the left to facilitate line fitting removal.

15. Position a drain pan under the vehicle and remove the line fittings. Remove the O-rings from the fiting connections and replace with new.

16. Remove the left sway bar link.

17. Remove the steering gear assembly through the left wheel well.

To install:

18. Install new O-rings into the line fittings.

19. Place the gear attachment bolts in the gear housing.

20. Install the steering gear assembly through the left wheel well.

21. Connect and tighten the line fittings to the steering gear assembly.

22. Connect the VAPS electrical connector.

23. Position the steering gear into the sub-frame.

24. Install the tie rod ends onto the shaft.

25. Install the heat shield band.

26. Attach the tie rod ends onto the spindle. Install the nuts and secure with new cotter pins.

27. Attach the sway bar link.

28. Raise the vehicle until the subframe contacts the body. Install the sub-frame attaching bolts.

29. Install the gear-to-sub-frame nuts and torque to 85–100 ft. lbs. (115–135 Nm).

30. Attach the exhaust pipe to the catalytic converter.

31. Attach the height sensors on Continental, install the wheels and lower the vehicle.

32. Fill the power steering system.

33. Install the secondary steering column boot and attach the intermediate shaft to the steering gear. Tighten the bolt to 30–38 ft. lbs. (41–51 Nm).

34. Install the primary steering column boot and attach the intermediate shaft to the steering column.

35. Bleed the system and align the front end.

Power Steering Pump

REMOVAL & INSTALLATION

2.5L and 3.8L Engines

1. Disconnect the negative battery cable. Loosen the tensioner pulley attaching bolts and using the ½ in. drive hole provided in the tensioner pulley,

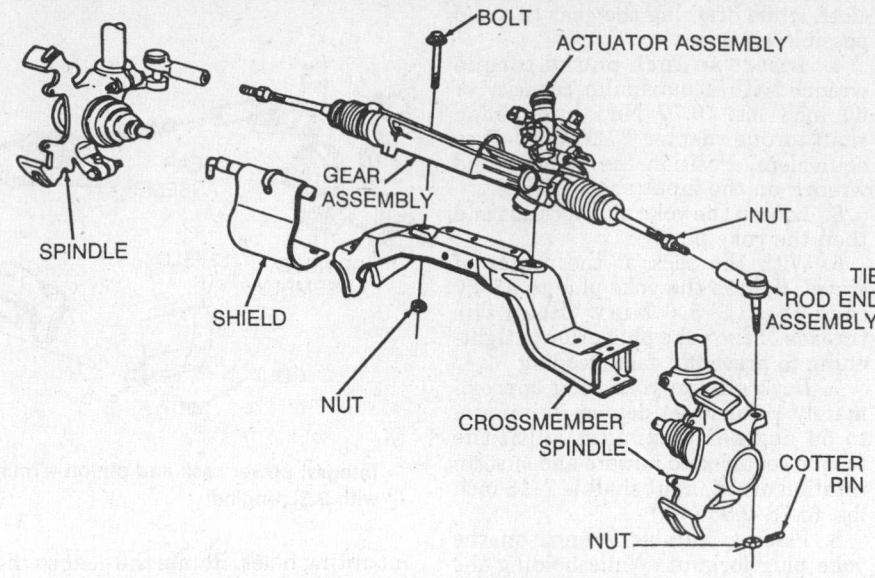

VAPS power rack and pinion

rotate the tensioner pulley clockwise and remove the belt from the alternator and power steering pulley.

2. Position a drain pan under the power steering pump from underneath the vehicle. Disconnect the hydraulic pressure and return lines.

3. Remove the pulley from the pump shaft using hub puller T69L–10300–B or equivalent. Remove the bolts retaining pump to bracket and remove the power steering pump.

4. Installation is the reverse of the removal procedure. Fill the pump with fluid and check the system for proper operation.

NOTE: To install the power steering pump pulley, use steering pump pulley replacer T65P–3A733–C or equivalent. When using this tool, the small diameter threads must be fully engaged in the pump shaft before pressing on the pulley. Hold the head screw and turn the nut to install the pulley. Install the pulley face flush with the pump shaft within ± 0.100 in. (0.25mm).

Except 3.0L SHO Engine

1. Disconnect the negative battery cable. Loosen the idler pulley and remove the power steering belt.

2. Remove the radiator overflow bottle in order to gain access to the 3 screws attaching the pulleys to the pulley hub.

3. Matchmark both pulley to hub positions.

4. Remove the pulleys from the pulley hub.

5. Remove the return line from the pump. Be prepared to catch any spilled fluid in a suitable container.

6. Back off the pressure line attaching nut completely. The line will separate from the pump connection when the pump is removed.

7. Remove the pump mounting bolts and remove the pump.

8. Installation is the reverse of the removal procedure. Fill the pump with fluid and check for proper operation.

3.0L SHO Engine

1. Disconnect the negative battery cable.

2. Remove the engine damper strut.

3. Remove the power steering belt.

4. Raise and support the vehicle safely.

5. Remove the right front wheel and tire assembly.

6. Position a jack under the engine and remove the right rear engine mount.

7. Remove the power steering pump pulley.

8. Place a drain pan under the pump and remove the pressure and return lines from the pump.

9. Remove the 4 pump retaining bolts and remove the pump.

10. Installation is the reverse of the removal procedure. Tighten the pump retaining bolts to 15–24 ft. lbs. (20–33 Nm).

BELT ADJUSTMENT

Except 3.0L SHO Engine

Belt tension is maintained by an automatic tensioner and does not require adjustment.

3.0L SHO Engine

1. Loosen the idler pulley nut and

turn the adjusting screw until the belt is adjusted. Measure the belt tension using a suitable offset belt tension gauge. New belts should measure 154–198 lbs. or used belts 112–157 lbs. The allowable minimum belt tension is 80 lbs.

2. Tighten the idler pulley nut to 25–37 ft. lbs. (34–50 Nm).

SYSTEM BLEEDING

If air bubbles are present in the power steering fluid, bleed the system by performing the following:

1. Fill the reservoir to the proper level.

2. Operate the engine until the fluid reaches normal operating temperature of 165–175°F.

3. Turn the steering wheel all the way to the left then all the way to the right several times. Do not hold the steering wheel in the far left or far right position stops.

4. Check the fluid level and recheck the fluid for the presence of trapped air. If apparent that air is still in the system, fabricate or obtain a vacuum tester and purge the system as follows:

a. Remove the pump dipstick cap assembly.

b. Check and fill the pump reservoir with fluid to the **COLD FULL** mark on the dipstick.

c. Disconnect the ignition wire and raise the front of the vehicle and support safely.

d. Crank the engine with the starter and check the fluid level. Do not turn the steering wheel at this time.

e. Fill the pump reservoir to the **COLD FULL** mark on the dipstick. Crank the engine with the starter while cycling the steering wheel lock-to-lock. Check the fluid level.

f. Tightly insert a suitable size rubber stopper and air evacuator pump into the reservoir fill neck. Connect the ignition coil wire.

g. With the engine idling, apply a 15 in. Hg vacuum to the reservoir for 3 minutes. As air is purged from the system, the vacuum will drop off. Maintain the vacuum on the system as required throughout the 3 minutes.

h. Remove the vacuum source. Fill the reservoir to the **COLD FULL** mark on the dipstick.

i. With the engine idling, re-apply 15 in. Hg vacuum source to the reservoir. Slowly cycle the steering wheel to lock-to-lock stops for approximately 5 minutes. Do not hold the steering wheel on the stops during cycling. Maintain the vacuum as required.

j. Release the vacuum and disconnect the vacuum source. Add fluid as required.

k. Start the engine and cycle the wheel slowly and check for leaks at all connections.

l. Lower the front wheels.

5. In cases of severe aeration, repeat the procedure.

Tie Rod Ends

REMOVAL & INSTALLATION

1. Remove and discard the cotter pin and nut from the worn tie rod end ball stud.

2. Disconnect the tie rod end from the steering spindle, using tie rod remover tool 3290–D or equivalent.

3. Hold the tie rod end with a wrench and loosen the tie rod jam nut.

4. Note the depth to which the tie rod is located, then grip the tie rod with a pair of suitable pliers and remove the tie rod end assembly from the tie rod.

To install:

5. Clean the tie rod threads. Thread the new tie rod end into the tie rod to the same depth as the removed tie rod end.

6. Place the tie rod end stud into the steering spindle. Make sure the front wheels are pointed straight-ahead before connecting the stud to the spindle.

7. Install a new nut on the tie rod end stud. Tighten the nut to 35 ft. lbs. (48 Nm) and continue tightening until the next castellation on the nut is aligned with the cotter pin hole in the stud. Install a new cotter pin.

8. Set the toe to specification. Tighten the jam nut to 35–50 ft. lbs. (47–68 Nm).

BRAKES

For all brake system repair and service procedures not detailed below, please refer to "Brakes" in the Unit Repair section.

Master Cylinder

REMOVAL & INSTALLATION

NOTE: The master cylinder on 1988–89 Continental is part of the anti-lock brake hydraulic actuation unit and cannot be removed separately.

1. Disconnect the negative battery cable. If equipped with anti-lock brakes, depress the brake pedal several times to exhaust all vacuum in the system.

2. Disconnect the brake lines from the primary and secondary outlet ports of the master cylinder and the pressure control valve.

3. Remove the nuts attaching the master cylinder to the brake booster assembly. Disconnect the brake warning light wire. If equipped with anti-lock brakes, disconnect the Hydraulic Control Unit (HCU) supply hose at the master cylinder and secure in a position to prevent loss of brake fluid.

4. Slide the master cylinder forward and upward from the vehicle.

To install:

5. Before installation, bench bleed the new master cylinder as follows:

a. Mount the new master cylinder in a holding fixture. Be careful not to damage the housing.

b. Fill the master cylinder reservoir with brake fluid.

c. Using a suitable tool inserted into the booster pushrod cavity, push the master cylinder piston in slowly. Place a suitable container under the master cylinder to catch the fluid being expelled from the outlet ports.

d. Place a finger tightly over each outlet port and allow the master cylinder piston to return.

e. Repeat the procedure until clear fluid only is expelled from the master cylinder. Plug the outlet ports and remove the master cylinder from the holding fixture.

6. Mount the master cylinder on the booster. Install a new seal in the groove in the master cylinder mounting face on vehicles equipped with anti-lock brakes. Attach the brake fluid lines to the master cylinder. If equipped with anti-lock brakes, install the HCU supply hose to the master cylinder.

7. Install the brake warning light wire.

8. Bleed the system. Operate the brakes several times, then check for external hydraulic leaks.

Proportioning Valve

REMOVAL & INSTALLATION

Taurus and Sable

The valve for the sedan is mounted to the floorpan near the left rear wheel. The valves for the station Wagon are screwed into the master cylinder.

Sedan

1. Raise the vehicle and support it safely.

2. Disconnect the brake lines from

the valve assembly and note their position.

3. Remove the screw retaining the valve bracket to the lower suspension arm. Remove the 2 screws retaining the valve bracket to the underbody and remove the assembly.

NOTE: The service replacement valve will have a red plastic gauge clip on the valve and must not be removed until it is installed on the vehicle.

To install:

4. Make sure the rear suspension is in the full rebound position.

5. Make sure the red plastic gauge clip is in position on the valve and that the operating rod lower adjustment screw is loose.

6. Position the valve lower mounting bracket to the lower suspension arm. Install 1 retaining screw. Make sure the valve adjuster is resting on the lower bracket and tighten the set screw.

7. Connect the brake lines in the same position as removed. Bleed the rear brakes.

8. Remove the red plastic gauge clip and lower the vehicle.

Wagon

1. Disconnect the primary or secondary brake line from the master cylinder, as necessary.

2. Loosen and remove the valve from the master cylinder housing.

3. Installation is the reverse of the removal procedure. Fill and bleed the brake system.

Continental

The proportioning valve is contained in the pressure control valve assembly along with a pressure switch. The control valve is located on a bracket below the hydraulic actuation unit on 1988–89 vehicles or the master cylinder on 1990–92 vehicles.

1. Disconnect the negative battery cable.

2. Disconnect the electrical connector from the pressure switch.

3. Disconnect the brake lines from the valve assembly.

4. Remove the retaining screw and remove the valve assembly.

5. Installation is the reverse of the removal procedure. Tighten the mounting screw to 8–11 ft. lbs. (11–16 Nm).

Power Brake Booster

REMOVAL & INSTALLATION

Without Anti-Lock Brakes

1. Disconnect the battery ground

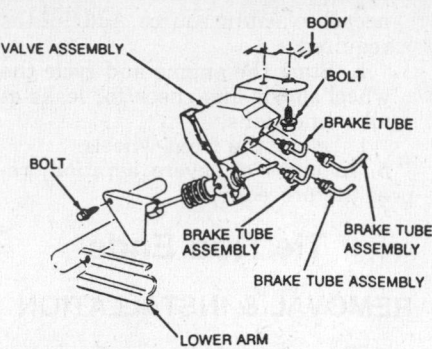

Brake pressure control valve—Taurus and Sable sedan

cable and remove the brake lines from the master cylinder.

2. Disconnect the manifold vacuum hose and the warning indicator. Remove the retaining nuts and the master cylinder.

3. From under the instrument panel, remove the stoplight switch wiring connector from the switch. Remove the pushrod retainer and outer nylon washer from the brake pin, slide the stoplight switch along the brake pedal pin, far enough for the outer hole to clear the pin.

4. Remove the switch by sliding it upward. Remove the booster to dash panel retaining nuts. Slide the booster pushrod and pushrod bushing off the brake pedal pin.

5. Remove the screws and position the vacuum fitting at the dash panel aside. Position the wire harness aside. Remove the transaxle shift cable and bracket.

6. Move the booster forward until the booster studs clear the dash panel and remove the booster.

7. Installation is the reverse of the removal procedure. Bleed the brake system.

NOTE: If equipped with speed control, the vacuum dump valve must be adjusted if the brake booster has been removed.

With Anti-Lock Brakes—Except 1988–89 Continental

1. Disconnect the negative battery cable. Pump the brake pedal until all vacuum is removed from the booster. This will prevent the O-ring from being sucked into the booster during disassembly.

2. Disconnect the manifold vacuum hose from the booster check valve and the electrical connector from the master cylinder reservoir cap.

3. Remove the brake lines from the primary and secondary outlet ports of the master cylinder and remove the Hydraulic Control Unit (HCU) supply hose. Plug the ports and reservoir feed

to prevent brake fluid from leaking onto paint and wiring.

4. Under the instrument panel, remove the stoplight switch wiring connector from the switch. Disengage the pedal position switch from the stud. Remove the hairpin retainer and outer nylon washer from the pedal pin. Slide the stoplight switch off the brake pedal just far enough for the outer arm to clear the pin. Remove the switch.

5. Remove the booster to dash panel attaching nuts. Slide the bushing and booster pushrod off the brake pedal pin.

6. Move the booster forward until the booster studs clear the dash panel. Remove the booster and master cylinder assembly.

7. Place the booster and master cylinder assembly on a bench. Remove the 2 nuts attaching the master cylinder to the booster and remove the master cylinder.

To install:

8. Slide the master cylinder onto the booster studs. Make sure the O-ring is in place in the groove on the master cylinder and install the 2 attaching nuts. Tighten the nuts to 13–25 ft. lbs. (18–34 Nm).

9. Under the instrument panel, install the booster pushrod and bushing on the brake pedal pin. Fasten the booster to the dash panel with self-locking nuts. Tighten the nuts to 13–25 ft. lbs. (18–34 Nm).

10. Position the stoplight switch so it straddles the booster pushrod with the switch slot towards the pedal blade and hole just clearing the pin. Slide the switch completely onto the pin.

11. Install the outer nylon washer on the pin and secure all parts to the pin with the hairpin retainer. Make sure the retainer is fully installed and locked over the pedal pin. Install the stoplight switch wiring connector.

12. Install the pedal travel switch. To adjust the switch, push the switch plunger fully into the switch housing. This zeros out the switch adjustment so it can be automatically reset to the correct dimension during the following steps:

a. Slowly pull the arm back out of the switch housing past the detent point. At this point it should be impossible to reattach the arm to the pin unless the brake pedal is forced down.

b. Depress the brake pedal until the switch hook can be snapped onto the pin. Snap the hook onto the pin and pull the brake pedal back up to it's normal at rest position. This automatically sets the switch to the proper adjustment.

13. Connect the brake lines to the master cylinder and tighten to 10–18

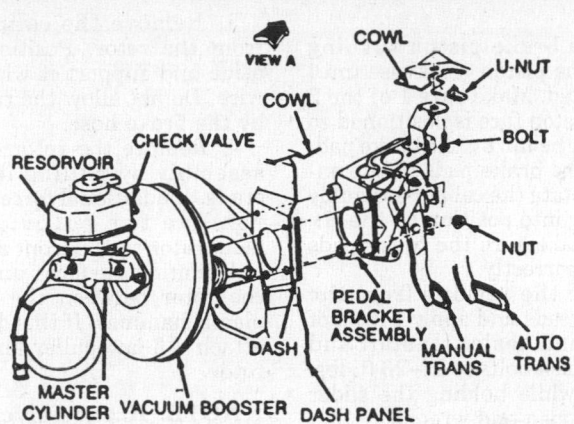

Brake booster assembly with anti-lock brakes—except 1988–89 Continental

ft. lbs. (14–24 Nm). Connect the HCU supply hose to the resorvoir.

14. Connect the manifold vacuum hose to the booster check valve and the electrical connector to the master cylinder reservoir cap.

15. Connect the negative battery cable and bleed the brake system.

Brake Caliper

REMOVAL & INSTALLATION

Front

1. Raise and support the vehicle safely.

2. Remove the wheel and tire assembly. Mark the caliper to ensure that it is reinstalled on the correct knuckle.

3. Disconnect the flexible brake hose from the caliper. Remove the hollow retaining bolt that connects the hose fitting to the caliper. Remove the hose assembly from the caliper and plug the hose.

4. Remove the caliper locating pins.

5. Lift the caliper off of the rotor, integral knuckle and anchor plate using a rotating motion.

NOTE: Do not pry directly against the plastic piston or damage to the piston will result.

To install:

6. Retract the piston fully in the piston bore. Position the caliper assembly above the rotor with the anti-rattle spring under the upper arm of the knuckle. Install the caliper over the rotor with a rotating motion. Make sure the inner and outer shoes are properly positioned and the outer anti-rattle spring is properly positioned. Make sure the correct caliper assembly, as marked during removal, is installed on the correct knuckle. The caliper bleed screw should be positioned on top of the caliper when assembled on the vehicle.

7. Lubricate the locating pins and the inside of the insulators with silicone grease. Install the locating pins through the caliper insulators and hand start the threads into the knuckle attaching holes. Tighten the locating pins to 18–25 ft. lbs. (24–34 Nm).

8. Remove the plug and install the brake hose on the caliper with a new copper washer on each side of the fitting outlet. Insert the attaching bolt through the washers and fittings and tighten to 30–45 ft. lbs. (40–60 Nm).

9. Bleed the brake system, filling the master cylinder as required.

10. Install the wheel and lower the vehicle. Pump the brake pedal prior to moving the vehicle to position the brake linings.

Rear

1. Raise and support the vehicle safely.

2. Remove the wheel and tire assembly.

3. Remove the brake flex hose from the caliper assembly.

4. Remove the retaining clip from the parking brake at the caliper. Disengage the parking brake cable end from the lever arm.

5. Hold the slider pin hex-heads with an open-end wrench and remove the pinch bolts. Lift the caliper assembly away from the anchor plate. Remove the slider pins and boots from the anchor plate.

To install:

6. Apply silicone dielectric compound to the inside of the slider pin boots and to the slider pins.

7. Position the slider pins and boots in the anchor plate. Position the caliper assembly on the anchor plate. Make sure the brake pads are installed correctly.

8. Remove the residue from the pich bolt threads and apply 1 drop of threadlock and sealer. Install the pinch bolts and tighten to 23–26 ft. lbs. (31–35 Nm) while holding the slider pins with an open-end wrench.

9. Attach the cable end to the parking brake lever. Install the cable retaining clip on the caliper assembly.

10. Using new washers, connect the brake flex hose to the caliper. Tighten the retaining bolt to 8–11 ft. lbs. (11–16 Nm).

11. Bleed the brake system, filling the master cylinder as required.

12. Install the wheel and lower the vehicle. Pump the brake pedal prior to moving the vehicle to position the brake pads.

Disc Brake Pads

REMOVAL & INSTALLATION

Front

1. Remove the master cylinder cap and check the fluid level in the reservoir. Remove the brake fluid until the reservoir is half full. Discard the removed fluid.

2. Raise the vehicle and support it

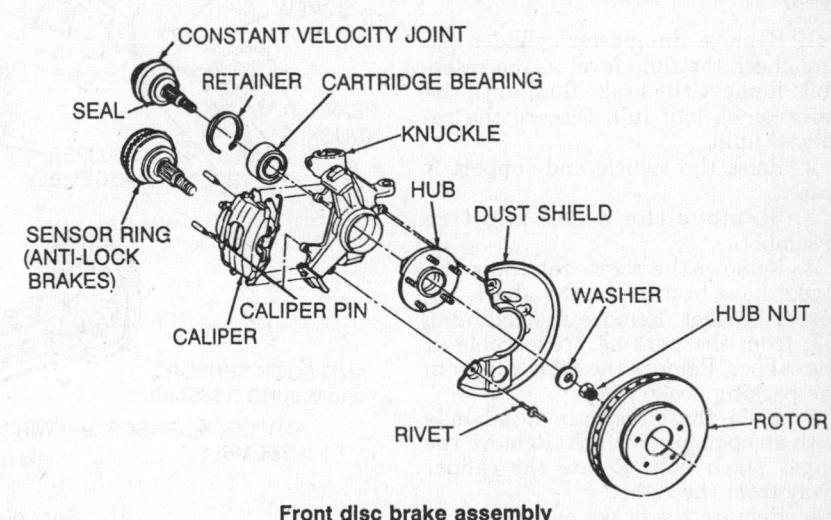

Front disc brake assembly

safely. Remove the wheel and tire assembly.

3. Remove the caliper locating pins. Lift the caliper assembly from the integral knuckle and anchor plate and rotor using a rotating motion. Suspend the caliper inside the fender housing with wire. Do not allow the caliper to hang from the brake hose.

NOTE: Do not pry directly against the plastic piston or damage will result.

4. Remove the inner and outer brake pads. Inspect the rotor braking surfaces for scoring and machine as necessary. Refer to the minimum rotor thickness specification when machining. If machining is not necessary, hand sand the glaze from the braking surfaces with medium grit sand paper.
To install:
5. Use a 4 in. C-clamp and a wood block 2¾ in. × 1 in. × ¾ in. thick to seat the caliper piston in it's bore. This must be done to provide clearance for the caliper assembly with the new brake pads to fit over the rotor during installation. Care must be taken during this procedure to prevent damage to the plastic piston. Do not allow metal or sharp objects to come into direct contact with the piston surface or damage will result.

6. Remove all rust buildup from the inside of the caliper legs. Install the inner pad in the caliper piston. Do not bend the pad clips during installation in the piston or distortion and rattles can occur. Install the outer pad. Make sure the clips are properly seated.

7. Install the caliper over the rotor and install the wheel. Lower the vehicle.

8. Pump the brake pedal prior to moving the vehicle to position the brake linings. Refill the master cylinder.

Rear

1. Remove the master cylinder cap and check the fluid level in the reservoir. Remove the brake fluid until the reservoir is half full. Discard the removed fluid.

2. Raise the vehicle and support it safely.

3. Remove the wheel and tire assembly.

4. Remove the screw retaining the brake hose bracket to the shock absorber bracket. Remove the retaining clip from the parking brake cable at the caliper. Remove the cable end from the parking brake lever.

5. Hold the slider pin hex-heads with an open-end wrench. Remove the upper pinch bolt. Rotate the caliper away from the rotor.

6. Remove the brake pads.

To install:

7. Using a brake piston turning tool, rotate the piston clockwise until it is fully seated. Make sure 1 of the 2 slots in the piston face is positioned so it will engage the nib on the brake pad.

8. Install the brake pads in the anchor plate. Rotate the caliper assembly over the rotor into position on the anchor plate. Make sure the brake pads are installed correctly.

9. Remove the residue from the pinch bolt threads and apply 1 drop of threadlock and sealer. Install and tighten the pinch bolts to 23–26 ft. lbs. (31–35 Nm) while holding the slider pins with an open-end wrench.

10. Attach the cable end to the parking brake lever. Install the cable retaining clip on the caliper assembly. Position the brake flex hose and bracket assembly to the shock absorber bracket and install the retaining screw. Tighten the screw to 8–11 ft. lbs. (11–16 Nm).

11. Install the wheel and tire assembly and lower the vehicle. Pump the brake pedal prior to moving the vehicle to position the brake linings. Refill the master cylinder.

Brake Rotor

REMOVAL & INSTALLATION

Front

1. Raise the vehicle and support it safely.

2. Remove the wheel and tire assembly.

3. Remove the caliper assembly from the rotor. Position the caliper aside and support it with a length of wire. Do not allow the caliper to hang by the brake hose.

4. Remove the rotor from the hub assembly by pulling it off the hub studs. If additional force is required to remove the rotor, apply rust penetrator on the front and rear rotor/hub mating surfaces and then strike the rotor between the studs with a plastic hammer. If this does not work, attach a 3-jaw puller and remove the rotor.

NOTE: If excessive force must be used to remove the rotor, it should be checked for lateral runout before installation.

5. Check the rotor for scoring and/or other wear. Machine or replace, as necessary. If machining, observe the minimum thickness specification.

6. Install the rotor in the reverse order of removal.

Rear

1. Raise the vehicle and support it safely.

2. Remove the wheel and tire assembly.

3. Remove the caliper assembly from the rotor and support it with a length of wire. Do not let the caliper hang from the brake line.

4. Remove the 2 rotor retaining nuts and remove the rotor from the hub.

5. Check the rotor for scoring and/

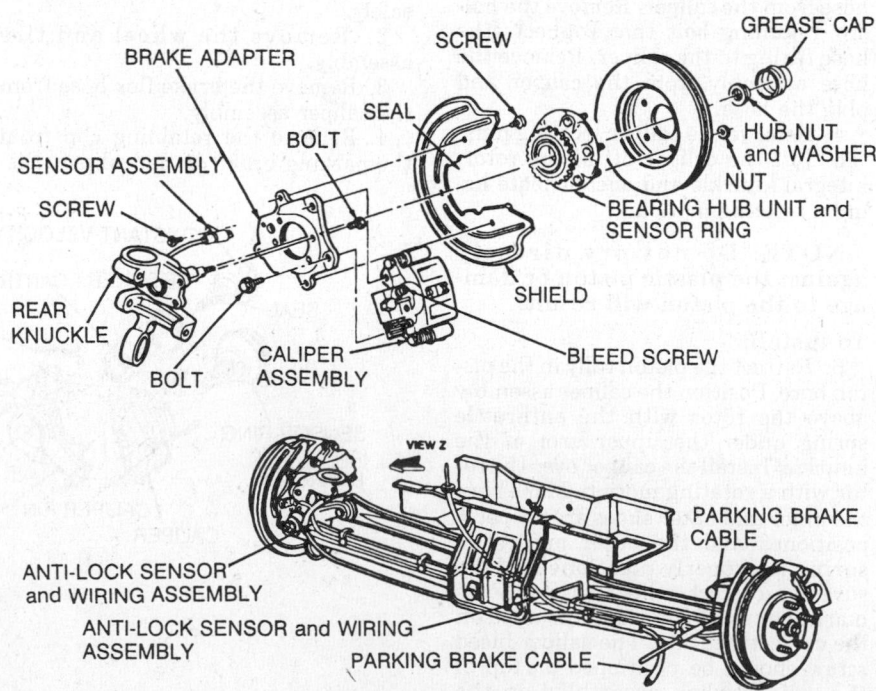

Rear disc brake assembly

or other wear. Machine or replace, as necessary. If machining, observe the minimum thickness specification.

6. Install the rotor in the reverse order of removal.

Brake Drums

REMOVAL & INSTALLATION

1988–89

1. Raise and safely support the vehicle.
2. Remove the wheelcover and nut covers, as required. Remove the wheel and tire assembly.
3. Remove the grease cap from the hub. Remove the cotter pin, nut lock, adjusting nut and keyed flatwasher from the spindle. Remove the outer bearing and discard the cotter pin.
4. Remove the hub/drum assembly as a unit. Be careful not to damage the grease seal and inner bearing during removal.
5. Inspect the drum for scoring and/or other wear. Machine or replace, as necessary. If machining, observe the maximum permissible drum diameter specification.

To install:

6. Inspect and lubricate the bearings, as necessary. Replace the grease seal if any damage is visible.
7. Clean the spindle stem and apply a thin coat of wheel bearing grease.
8. Install the hub and drum assembly on the spindle. Install the outer bearing, keyed flat washer and adjusting nut. Tighten the nut finger-tight.
9. Adjust the wheel bearings. Install the nut retainer and a new cotter pin. Install the grease cap.
10. Install the wheel and tire assembly. Install the wheel cover and nut covers, as required. Lower the vehicle.

1990–92

1. Raise the vehicle and support it safely.
2. Remove the wheel cover.
3. Remove the lugnuts and the wheel and tire assembly.
4. Remove the 2 drum retaining nuts and the drum.

NOTE: If the drum will not come off, pry the rubber plug from the backing plate inspection hole. Remove the brake line-to-axle retention bracket. This will allow sufficient room to insert suitable brake tools through the inspection hole to disengage the adjusting lever and back off the adjusting screw.

5. Inspect the drum for scoring and/or other wear. Machine or replace, as necessary. If machining, observe the

Brake shoe assembly

maximum permissible drum diameter specification.

6. Installation is the reverse of the removal procedure.

Brake Shoes

REMOVAL & INSTALLATION

1. Raise the vehicle and support it safely.
2. Remove the wheel and tire assembly and the brake drum.
3. Remove the 2 shoe hold-down springs and pins.
4. Lift the brake shoes, springs and adjuster assembly off the backing plate and wheel cylinder assembly. When removing the assembly, be careful not to bend the adjusting lever.
5. Remove the parking brake cable from the parking brake lever.
6. Remove the retracting springs from the lower brake attachments and upper shoe-to-adjusting lever attachment points. This will seperate the brake shoes and disengage the adjuster mechanism.
7. Remove the horse shoe retaining clip and spring washer and slide the lever off the parking brake lever pin on the trailing shoe.

To install:

8. Apply a light coating of disc brake caliper slide grease at the points where the brake shoes contact the backing plate.
9. Apply a thin coat of lubricant to the adjuster screw threads and socket end of the adjusting screw. Install the stainless steel washer over the socket end of the adjusting screw and install the socket. Turn the adjusting screw into the adjusting pivot nut to the limit of the threads and then back off ½ turn.
10. Assemble the parking brake lever to the trailing shoe by installing the spring washer and a new horse shoe retaining clip. Crimp the clip until it retains the lever to the shoe securely.
11. Attach the parking brake cable to the parking brake lever.

12. Attach the lower shoe retracting spring to the leading and trailing shoe and install to the backing plate. It will be necessary to stretch the retracting spring as the shoes are installed downward over the anchor plate to the inside of the shoe retaining plate.
13. Install the adjuster screw assembly between the leading shoe slot and the slot in the trailing shoe and parking brake lever. The adjuster socket end slot must fit into the trailing shoe and parking brake lever.
14. Assemble the adjuster lever in the groove located in the parking brake lever pin and into the slot of the adjuster socket that fits into the trailing shoe web.
15. Attach the upper retracting spring to the leading shoe slot. Using a suitable spring tool, stretch the other end of the spring into the notch on the adjuster lever. If the adjuster lever does not contact the star wheel after installing the spring, it is possible that the adjuster socket is installed incorrectly.

NOTE: The adjuster socket blade is marked R for the right-hand or L for the left-hand brake assemblies. The R or L adjuster blade must be installed with the letter R or L in the upright position, facing the wheel cylinder, on the correct side to ensure that the deeper of the 2 slots in the adjuster sockets fits into the parking brake lever.

16. Adjust the brake shoes.
17. Install the brake drum and wheel and tire assembly. Lower the vehicle.

Wheel Cylinder

REMOVAL & INSTALLATION

1. Raise and support the vehicle safely.
2. Remove the wheel and tire assembly.
3. Remove the brake drum.

4. Remove the brake shoes, retainers and springs from the backing plate.

5. Disconnect and plug the brake line at the rear-side of the wheel cylinder.

6. Remove the wheel cylinder-to-backing plate bolts and remove the wheel cylinder.

7. To install, reverse the order of removal. Tighten the wheel cylinder-to-backing plate bolts to 8–10 ft. lbs. (10–14 Nm). Bleed the rear brake system.

Parking Brake Cable

ADJUSTMENT

Except Taurus SHO and Continental

1. Make sure the parking brake is fully released. Place the transaxle in the **N** position.

2. Raise the vehicle and support it safely. Working in front of the left rear wheel, tighten the adjusting nut against the cable equalizer causing a rear wheel brake drag. Then loosen the adjusting nut until the rear brakes are fully released. There should be no brake drag.

3. If the brake cables were replaced, stroke the parking brake several times, then release control and repeat Step 2.

4. Check for operation of the parking brake with the vehicle supported and the parking brake fully released. If there is any slack in the cables or if the rear brakes drag when the wheels are turned, adjust as required.

5. Lower the vehicle.

Taurus SHO and Continental

1. Make sure the parking brake is fully released.

2. Raise and safely support the vehicle.

3. Tighten the adjusting nut against the cable adjuster bracket until there is a slight, less than $1/16$ in. (1.59mm), movement of either rear parking brake lever at the caliper.

4. If the brake cables were replaced, stroke the parking brake several times, then release the control and repeat Step 3.

5. Lower the vehicle.

Brake System Bleeding

Without Anti-Lock Brakes

1. Clean all dirt from the master cylinder filler cap.

2. If the master cylinder is known or suspected to have air in the bore, it must be bled before any of the wheel cylinders or calipers. To bleed the master cylinder, loosen the upper secondary left front outlet fitting approximately ¾ turn. Have an assistant depress the brake pedal slowly through it's full travel. Close the outlet fitting and let the pedal return slowly to the fully released position. Wait 5 seconds and then repeat the operation until all air bubbles disappear.

3. Repeat Step 2 with the right-hand front outlet fitting.

4. Continue to bleed the brake system by removing the rubber dust cap from the wheel cylinder bleeder fitting or caliper fitting at the right-hand rear of the vehicle. Place a suitable box wrench on the bleeder fitting and attach a rubber drain tube to the fitting. The end of the tube should fit snugly around the bleeder fitting. Submerge the other end of the tube in a container partially filled with clean brake fluid and loosen the fitting ¾ turn.

5. Have an assistant push the brake pedal down slowly through it's full travel. Close the bleeder fitting and allow the pedal to slowly return to it's full release position. Wait 5 seconds and repeat the procedure until no bubbles appear at the submerged end of the bleeder tube. Secure the bleeder fitting and remove the bleeder tube. Install the rubber dust cap on the bleeder fitting.

6. Repeat the procedure in Steps 4 and 5 in the following sequence: left front, left rear and right front. Refill the master cylinder reservoir after each wheel cylinder or caliper has been bled and install the master cylinder cover and gasket. When brake bleeding is completed, the fluid level should be filled to the maximum level indicated on the reservoir.

7. Always make sure the disc brake pistons are returned to their normal positions by depressing the brake pedal several times until normal pedal travel is established. If the pedal feels spongy, repeat the bleeding procedure.

With Anti-Lock Brakes—Except 1988–89 Continental

The anti-lock brake system must be bled in 2 steps.

1. The master cylinder and hydraulic control unit must be bled using the Rotunda Anti-Lock Brake Breakout Box/Bleeding Adapter tool No. T90P–50–ALA or equivalent. If this procedure is not followed, air will be trapped in the hydraulic control unit which will eventually lead to a spongy brake pedal. To bleed the master cylinder and the hydraulic control unit, disconnect the 55-pin plug from the electronic control unit and install the Anti-Lock Brake Breakout Box/Bleeding Adapter to the wire harness 55-pin plug.

   a. Place the Bleed/Harness switch in the **BLEED** position.

   b. Turn the ignition to the **ON** position. At this point the red off light should come ON.

   c. Push the motor button on the adapter down to start the pump motor. The red OFF light will turn OFF and the green ON light will turn ON. The pump motor will run for 60 seconds after the motor button is pushed. If the pump motor is to be turned off for any reason before the 60 seconds has elapsed, push the abort button to turn the pump motor off.

   d. After 20 seconds of pump motor operation, push and hold the valve button down. Hold the valve button down for 20 seconds and then release it.

   e. The pump motor will continue to run for an additional 20 seconds after the valve button is released.

2. The brake lines can now be bled in the normal fashion. Bleed the brake system by removing the rubber dust cap from the caliper fitting at the right-hand rear of the vehicle. Place a suitable box wrench on the bleeder fitting and attach a rubber drain tube to the fitting. The end of the tube should fit snugly around the bleeder fitting. Submerge the other end of the tube in a container partially filled with clean brake fluid and loosen the fitting ¾ turn.

3. Have an assistant push the brake pedal down slowly through it's full travel. Close the bleeder fitting and allow the pedal to slowly return to it's full release position. Wait 5 seconds and repeat the procedure until no bubbles appear at the submerged end of the bleeder tube. Secure the bleeder fitting and remove the bleeder tube. Install the rubber dust cap on the bleeder fitting.

4. Repeat the bleeding procedure at the left front, left rear and right front in that order. Refill the master cylinder reservoir after each caliper has been bled and install the master cylinder and gasket. When brake bleeding is completed, the fluid level should be filled to the maximum level indicated on the reservoir.

5. Always make sure the disc brake pistons are returned to their normal positions by depressing the brake pedal several times until normal pedal travel is established. If the pedal feels spongy, repeat the bleeding procedure.

With Anti-Lock Brakes—1988–89 Continental

The front brakes can be bled in the same manner as a vehicle without anti-lock brakes or they can be bled with a pressure bleeder. The rear brakes must be bled with a pressure bleeder or with a fully charged accumulator.

Anti-Lock Brake System Service

PRECAUTION

Failure to observe the following precautions may result in system damage.
- Before servicing any high pressure component, be sure to discharge the hydraulic pressure from the system.
- Do not allow the brake fluid to contact any of the electrical connectors.
- Use care when opening the bleeder screws due to the high pressures available from the accumulator.

RELIEVING ANTI-LOCK BRAKE SYSTEM PRESSURE

Before servicing any components which contain high pressure, it is mandatory that the hydraulic pressure in the system be discharged. To discharge the system, turn the ignition **OFF** and pump the brake pedal a minimum of 20 times until an increase in pedal force is clearly felt.

Hydraulic Control Unit (HCU)

REMOVAL & INSTALLATION

Except 1988–89 Continental

1. On all vehicles, except Taurus SHO, disconnect the battery cables and remove the battery from the vehicle. Remove the battery tray. Remove the 3 plastic push pins holding the acid shield to the HCU mounting bracket and remove the acid shield. On Taurus SHO, it is only necessary to disconnect the negative battery cable and remove the electronic control unit and it's mounting bracket from the top of the HCU mounting bracket.
4. Disconnect the 19-pin connector from the HCU to the wiring harness and disconnect the 4-pin connector from the HCU to the pump motor relay.
5. Remove the 2 lines from the inlet ports and the 4 lines from the outlet ports of the HCU. Plug each port to prevent brake fluid from spilling onto the paint and wiring.
6. Remove the 3 nuts retaining the HCU assembly to the mounting bracket and remove the assembly from the

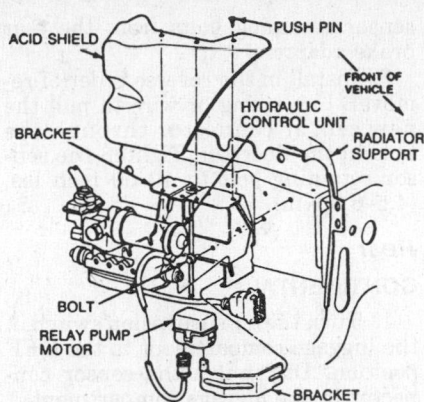

Anti-lock brake hydraulic control unit—except 1988–89 Continental

vehicle. The nut on the front of the HCU also retains the relay mounting bracket.
7. Install in the reverse order of removal. Tighten the 3 retaining nuts to 12–18 ft. lbs. (16–24 Nm) and the brake lines to 10–18 ft. lbs. (14–24 Nm). Bleed the brake system and check for fluid leaks.

1988–89 Continental

The hydraulic actuation assembly contains all the anti-lock brake hydraulic components: master cylinder and fluid reservoir, hydraulic pump motor and accumulator and solenoid valve block assembly.
1. Discharge the hydraulic pressure in the system.
2. Disconnect the negative battery cable.

3. Remove the air cleaner housing and duct assembly.
4. Tag and disconnect the electrical connectors from the fluid level indicator, main valve, solenoid valve block, pressure warning switch, hydraulic pump motor and ground connector from the master cylinder portion of the assembly.
5. Disconnect the 3 brake line fittings. Immediately plug each port to prevent fluid loss and contamination.

NOTE: Do not allow brake fluid to come in contact with any electrical connectors.

6. Remove the trim panel under the steering column. Disconnect the actuation assembly pushrod from the brake pedal by removing the hairpin connector next to the stoplight switch. Slide the switch, pushrod and plastic bushings off the pedal pin.
7. Remove the 4 retaining nuts that hold the actuation assembly to the brake pedal support bracket.
8. Remove the actuation assembly.

To install:
9. Mount the actuation assembly with the rubber boot and foam gasket to the engine side of the dash panel with the 4 mounting studs and pushrod inserted in the proper holes.
10. Working in the passenger compartment, loosely start 4 retaining locknuts attaching the actuation assembly to the pedal support bracket.
11. Connect the pushrod to the brake pedal pin by sliding the flanged plastic bushing, pushrod and washer onto the

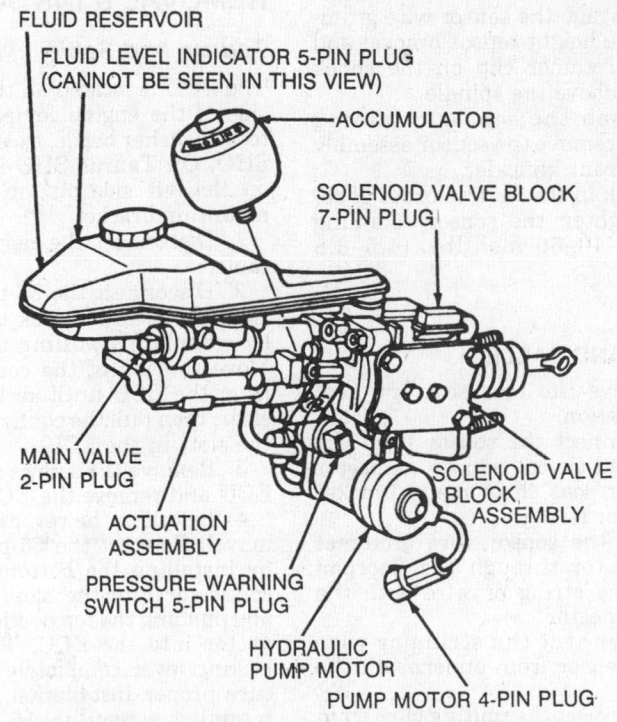

Hydraulic actuation assembly—1988–89 Continental

brake pedal pin. Position the stoplight switch so the slot on the switch bracket straddles the pushrod on the brake pedal pin, with the hole on the opposite leg of the switch bracket just clearing the pin. Slide the switch onto the pedal pin until it bottoms. Install the outer nylon bushing and secure the assembly with the hairpin retainer.

12. Tighten the 4 locknuts to 13–25 ft. lbs. (18–34 Nm).

13. From the engine compartment, connect the brake tubes and tighten the locknuts to 13–25 ft. lbs. (18–34 Nm). Connect the electrical connectors and install the air cleaner and duct assembly.

14. Connect the negative battery cable and bleed the brake system.

Wheel Sensors

REMOVAL & INSTALLATION

Front

1. Disconnect the sensor connector located in the engine compartment.

2. For the right front sensor, remove the 2 plastic push studs to loosen the front section of the splash shield in the wheel well. For the left front sensor, remove the 2 plastic push studs to loosen the rear section of the splash shield.

3. Thread the sensor wires through the holes in the fender apron. For the right front sensor, remove the 2 retaining clips behind the splash shield.

4. Raise and support the vehicle safely. Remove the wheel.

5. Disengage the sensor wire grommets at the height sensor bracket and from the retainer clip on the shock strut just above the spindle.

6. Loosen the sensor retaining screw and remove the sensor assembly from the front knuckle.

7. Install in the reverse order of removal. Tighten the sensor retaining screws to 40–60 inch lbs. (4.5–6.8 Nm).

Rear

TAURUS AND SABLE

1. Remove the rear seat and seat back insulation.

2. Disconnect the sensor from the harness and tie the sensor connector to the rear seat sheet metal bracket with wire or string.

3. Push the sensor wire grommet and connector through the floorpan drawing the string or wire with the sensor connector.

4. Disconnect the string or wire from the sensor from underneath the vehicle.

5. Disconnect the routing clips from the suspension arms and remove the

sensor retaining bolts from the rear brake adapters.

6. Install in the reverse order of removal. Use string or wire to pull the new sensor connector through the hole in the floorpan. Tighten the sensor retaining bolt to 40–60 inch lbs. (4.6–6.8 Nm).

Rear

CONTINENTAL

1. Turn the air suspension switch in the luggage compartment to the **OFF** position. Disconnect the sensor connector in the luggage compartment.

2. Push the rubber grommet through the sheet metal floorpan.

3. Raise and safely support the vehicle. Remove the retainer clips for the sensor wire and remove the wire from it's routing position.

4. Loosen the sensor retaining screw at the caliper anchor plate and remove the sensor.

5. On 1988–90 vehicles, install torsion spring replacement tool T88P-5310–A or equivalent, on the front suspension arm. Using a suitable breaker bar, lower the arm to provide clearance for the sensor wire connector to pass through.

6. Installation is the reverse of the removal procedure. Tighten the sensor retaining screw to 40–60 inch lbs. (4.5–6.8 Nm).

Electronic Control Unit (ECU)

REMOVAL & INSTALLATION

Taurus and Sable

The ECU is located on the front right side of the engine compartment next to the washer bottle, except on Taurus SHO. On Taurus SHO it is mounted on the left side on top of the HCU mounting bracket.

1. Disconnect the negative battery cable.

2. Disconnect the 55-pin connector from the ECU. Unlock the connector by completely pulling up the lever. Move the top of the connector away from the ECU until all terminals are clear, then pull the connector up out of the slots in the ECU.

3. Remove the screws attaching the ECU and remove the ECU.

4. Install in the reverse order of removal. Connect the 55-pin connector by installing the bottom part of the connector into the slots in the ECU and pushing the top portion of the connector into the ECU. Then pull the locking lever completely down to ensure proper installation. Tighten the retaining screws to 15–20 inch lbs. (1.7–2.3 Nm).

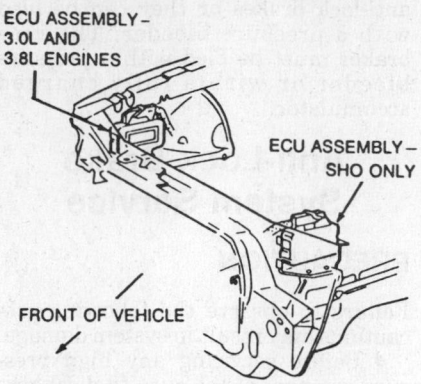

Anti-lock brake system electronic control unit location – Taurus and Sable

Continental

1. Disconnect the negative battery cable.

2. Remove the trim panel in the luggage compartment (behind the back seat) to gain access to the ECU.

3. Disconnect the connector from the ECU.

4. Remove the screws attaching the ECU to the panel and remove the ECU.

5. Installation is the reverse of the removal procedure. Tighten the retaining screws to 15–20 inch lbs. (1.7–2.3 Nm).

CHASSIS ELECTRICAL

Air Bag

DISARMING

1. Disconnect the negative battery cable.

2. On 1990–92 vehicles, open the glove compartment and lower it past its stops. Disconnect the backup power supply located to the right of the glove compartment opening.

NOTE: The backup power supply allows air bag deployment if the battery or battery cables are damaged in an accident before the crash sensors close. The power supply is a capacitor that will leak down in approximately 15 minutes after the battery is disconnected or in 1 minute if the battery positive cable is grounded. The backup power supply must be disconnected before any air bag related service is performed.

3. Remove the 4 nut and washer assemblies retaining the driver air bag module to the steering wheel.

4. Disconnect the driver air bag module connector and attach a jumper wire to the air bag terminals on the clockspring.

5. If equipped with a passenger air bag, disconnect the passenger air bag connector, located behind the glove compartment. Attach a jumper wire to the air bag terminals on the wiring harness side of the passenger air bag module connector.

Heater Blower Motor

REMOVAL & INSTALLATION

1. Disconnect the negative battery cable.

2. Open the glove compartment door, release the door retainers and lower the door.

3. Remove the screw attaching the recirculation duct support bracket to the instrument panel cowl.

4. If equipped with automatic temperature control, remove the nut holding the electrical connector bracket to the recirculation duct. Release the 3 connectors from the bracket and remove the bracket.

5. Remove the vacuum connection to the recirculation door vacuum motor. If equipped with automatic temperature control, disconnect the 2 aspirator hoses from the muffler.

6. Remove the screws attaching the recirculation duct to the heater assembly.

7. Remove the recirculation duct from the heater assembly, lowering the duct from between the instrument panel and the heater case.

8. Disconnect the blower motor electrical lead. Remove the blower motor wheel clip and remove the blower motor wheel.

9. Remove the blower motor mounting plate screws and remove the blower motor from the evaporator case.

10. Installation is the reverse of removal procedure.

Windshield Wiper Motor

REMOVAL & INSTALLATION

Front

1. Disconnect the negative battery cable.

2. Disconnect the power lead from the motor.

3. Remove the left wiper arm.

4. On Continental and 1991–92 Taurus and Sable, lift the water shield cover from the cowl on the passenger side. Remove the left cowl screen on 1988–90 Taurus and Sable.

5. Remove the linkage retaining clip from the operating arm on the motor.

6. Remove the attaching screws from the motor and bracket assembly and remove.

7. Installation is the reverse of removal procedure.

Rear – Station Wagon

1. Disconnect the negative battery cable.

2. Remove the wiper arm and blade.

3. Remove the pivot shaft retaining nut and spacers.

4. Disconnect the electrical connector to the wiper motor.

5. Remove the nut retaining the motor to the handle and remove the motor.

6. Installation is the reverse of the removal procedure.

Windshield Wiper Switch

REMOVAL & INSTALLATION

Front

The front wiper switch is a function of the combination switch.

Rear – Station Wagon

1988–89

1. Disconnect the negative battery cable.

2. Remove the 4 cluster opening finish panel retaining screws. Remove the finish panel by rocking the upper edge toward the driver.

3. Disconnect the wiring connector from the rear wiper switch.

4. Remove the wiper switch from the instrument panel. On Sable, the switch is retained with 2 screws.

5. Installation is the reverse of the removal procedure.

1990–92

1. Disconnect the negative battery cable.

2. Remove the cluster opening finish panel as follows:

   a. Engage the parking brake.

   b. Remove the ignition lock cylinder.

   c. If equipped with a tilt column, tilt the column to the full down position and remove the tilt lever.

   d. Remove the 4 bolts and the opening cover from under the steering column.

   e. Remove the steering column trim shrouds. Disconnect all electrical connections from the combination switch.

   f. Remove the 2 screws retaining the combination switch and remove the switch.

   g. Pull the gear shift lever to the full down position.

   h. Remove the 4 cluster opening finish panel retaining screws and the light switch knob and retaining nut.

   i. Remove the finish panel by pulling it toward the driver to unsnap the snap-in retainers and disconnect the wiring from the switches, clock and warning lights.

3. Remove the washer switch from the cluster opening finish panel.

To install:

4. Push the rear washer switch into the cluster finish panel until it snaps into place.

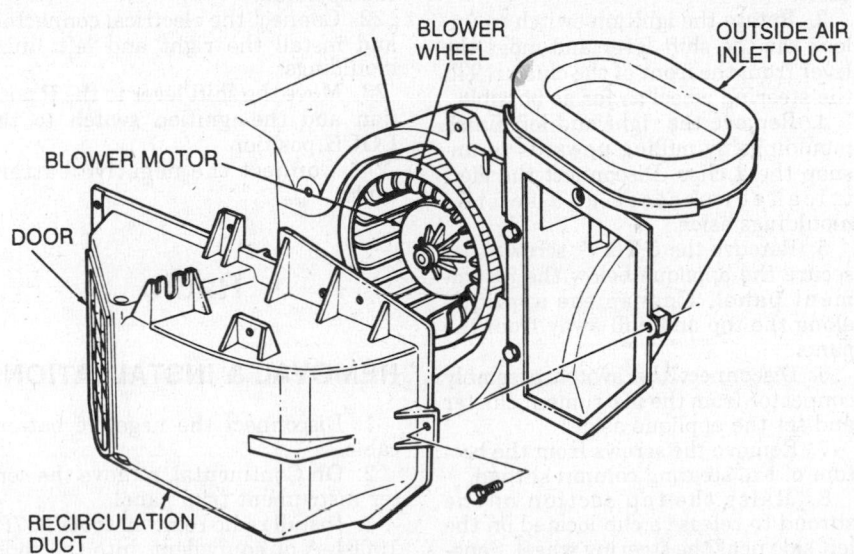

Exploded view of the heater motor and duct assembly

BLOWER WHEEL

OUTSIDE AIR INLET DUCT

BLOWER MOTOR

DOOR

RECIRCULATION DUCT

5. Install the cluster opening finish panel in the reverse order of removal.

6. Connect the negative battery cable.

Instrument Cluster

REMOVAL & INSTALLATION

Conventional

1. Disconnect the negative battery cable.

2. Remove the ignition lock cylinder to allow removal of the steering column shrouds.

3. Remove the steering column trim shrouds.

4. Remove the lower left and radio finish panel screws and snap the panels out.

5. On Taurus, remove the clock assembly (or clock cover) to gain access to the finish panel screw behind the clock.

6. Remove the cluster opening finish panel retaining screws and jam nut behind the headlight switch. Remove the finish panel by rocking the edge upward and outward.

7. On column shift vehicles, disconnect the transaxle selector indicator from the column by removing the retaining screw and cable loop.

8. Disconnect the upper speedometer cable from the lower speedometer cable in the engine compartment.

9. Remove the 4 cluster-to-instrument panel retaining screws and pull the cluster assembly forward.

10. Disconnect the cluster electrical connector and speedometer cable. Press the cable latch to disengage the cable from the speedometer head while pulling the cable away from the cluster. Remove the cluster.

To install:

11. Position the cluster in front of the cluster opening.

12. Connect the speedometer cable and electrical connectors.

13. Install the cluster and the 4 cluster-to-instrument panel retaining screws.

14. Connect the upper speedometer cable to the lower speedometer cable in the engine compartment.

15. On column shift vehicles, connect the transaxle selector indicator.

16. Install the cluster opening finish panel.

17. On Taurus vehicles, install the clock assembly or clock cover.

18. Install the lower left and radio finish panels.

19. Install the steering column trim shrouds.

20. Install the ignition lock cylinder and connect the negative battery cable.

Electronic

TAURUS AND SABLE

1. Disconnect the negative battery cable.

2. Remove the lower trim covers.

3. Remove the steering column cover and disconnect the shift indicator cable from the cluster by removing the retaining screws.

4. Disconnect the switch module and remove the cluster trim panel.

5. Remove the cluster mounting screws and pull the bottom of the cluster toward the steering wheel.

6. Reach behind and under the cluster, disconnect the 3 electrical connectors.

7. Swing the bottom of the cluster out to clear the top of the cluster from the crash pad and remove.

To install:

8. Insert the top of the cluster under the crash pad, leaving the bottom out.

9. Connect the 3 connectors.

10. Properly seat the cluster and install the retaining screws.

11. Connect the battery ground cable and check the cluster for proper operation.

12. Connect the shift indicator assembly to the cluster and secure with the retaining screw. Install the steering column cover.

13. Connect the switch module to the cluster and install the cluster trim panel.

14. Install the lower trim covers.

CONTINENTAL

1. Position the vehicle on a flat surface to prevent movement when the gear shift selector is out of position. Apply the parking brake and block the wheels.

2. Disconnect the negative battery cable.

3. Rotate the ignition switch to unlock the the shift lever and move the lever from the front of the cluster. Tilt the steering wheel as far as possible.

4. Remove the right and left finish mouldings by pulling upwards to unsnap the 2 clips. Disconnect the electrical connectors and set the mouldings aside.

5. Remove the 5 Torx® screws that secure the applique below the instrument panel. Unsnap the applique along the top and pull away from the panel.

6. Disconnect the switch assembly connector from the instrument cluster and set the applique aside.

7. Remove the screws from the bottom of the steering column shroud.

8. Raise the top section of the shroud to release a clip located on the left side near the steering wheel. Separate the upper section of the shroud from the side section near the ignition switch. Remove the upper section from the shift lever.

9. Remove the 4 Torx® screws attaching the instrument cluster to the substructure.

10. Place a clean, soft cloth on the steering column to prevent scratching the surface of the steering column as the instrument cluster is removed.

11. Tilt the top of the instrument cluster slightly toward the rear of the vehicle. Disconnect the shift indicator assembly from the cluster by undoing the 2 snaps located under the cluster.

12. Reach around the back of the instrument cluster to disconnect the 4 connectors. The connectors have locking tabs that must be pressed in to release.

13. Loosen the 2 clips retaining the shift indicator assembly to the cluster. Pull the shift indicator down and to the right to position it aside.

14. Push the bottom of the instrument cluster into the instrument panel cavity. Tilt the top of the instrument cluster toward the rear of the vehicle. Push the cluster up and out of the cavity.

To install:

15. Position the instrument cluster in front of the instrument panel cavity.

16. Connect the 4 electrical connectors.

17. Install the cluster into the instrument panel cavity.

18. Connect the shift indicator assembly to the instrument cluster.

19. Install the 4 Torx® screws.

20. Install the upper section of the steering column shroud and install the screws in the bottom section.

21. Connect the switch assembly connector and install the applique. Install the 5 Torx® screws that secure the applique.

22. Connect the electrical connectors and install the right and left finish mouldings.

23. Move the shift lever to the **P** position and the ignition switch to the **LOCK** position.

24. Connect the negative battery cable.

Radio

REMOVAL & INSTALLATION

1. Disconnect the negative battery cable.

2. On Continental, remove the center instrument trim panel.

3. Install radio removal tools T87P–19061–A or equivalent, into the radio face plate. Push the tools in approxi-

mately 1 in. (25.4mm) to release the retaining clips.

NOTE: Do not use excessive force when installing the radio removal tools as this will damage the retaining clips, making the radio difficult to remove.

4. Apply a light spreading force on the tools and pull the radio out of the dash. On Continental, raise the back end of the radio slightly to allow the rear support bracket to clear the track in the instrument panel.

5. Disconnect the wiring connectors and the antenna cable. Remove the rear support bracket on Continental.

6. Installation is the reverse of the removal procedure.

Headlight Switch

REMOVAL & INSTALLATION

1988–89 Taurus

1. Disconnect the negative battery cable.
2. Remove the bezel retaining nut and remove the bezel.
3. Remove the instrument cluster finish panel.
4. Remove the 2 screws retaining the headlight switch, pull the switch out of the instrument panel and disconnect the electrical connector.
5. Installation is the reverse of the removal procedure.

1988–89 Sable

1. Disconnect the negative battery cable.
2. Remove the lower left finish panel.
3. Remove the 2 screws retaining the headlight switch to the finish panel, disconnect the electrical connector and remove the switch.
4. Installation is the reverse of the removal procedure.

1990–92 Taurus and Sable

1. Disconnect the negative battery cable.
2. Pull off the headlight switch knob and remove the retaining nut.
3. Remove the instrument cluster finish panel as follows:
   a. Apply the parking brake.
   b. Remove the ignition lock cylinder.
   c. If equipped with a tilt column, tilt the column to the most downward position and remove the tilt lever.
   d. Remove the 4 bolts and opening cover from under the steering column.
   e. Remove the steering column trim shrouds. Disconnect all electri-

cal connections from the steering column combination switch.
   f. Remove the 2 screws retaining the combination switch and remove the switch.
   g. Pull the gear shift lever to the full down position.
   h. Remove the 4 cluster opening finish panel retaining screws. Remove the finish panel by pulling it toward the driver to unsnap the snap-in retainers and disconnect the wiring from the switches, clock and warning lights.
4. Remove the 2 screws retaining the headlight switch, pull the switch out of the instrument panel and disconnect the electrical connector.
5. Installation is the reverse of the removal procedure.

Continental

1. Disconnect the negative battery cable.
2. Gently pull off the headlight switch knob.
3. On 1988 vehicles, remove the finish panel and the moulding above the finish panel.
4. On 1989–92 vehicles, snap out the right and left mouldings, remove the 5 cluster opening finish panel retaining screws and the panel.
5. Remove the 2 screws retaining the headlight switch to the finish panel, disconnect the electrical connector and remove the switch.
6. Installation is the reverse of the removal procedure.

Combination Switch

The combination switch incorporates the turn signal, headlight dimmer, headlight flash-to-pass, hazard warning, cornering lights and windshield washer/wiper functions.

REMOVAL & INSTALLATION

1. Disconnect the negative battery cable. If equipped with a tilt steering column, set the tilt column to its lowest position and remove the tilt lever by removing the Allen head retaining screw.
2. Remove the ignition lock cylinder. Remove the steering column shroud screws and remove the upper and lower shrouds.
3. Remove the wiring harness retainer and disconnect the 3 electrical connectors.
4. Remove the self tapping screws attaching the switch to the steering column and disengage the switch from the steering column casting.
To install:
5. Align the turn signal switch mounting holes with the correspond-

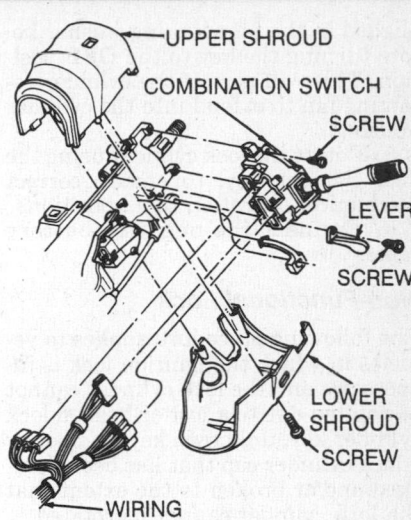

Combination switch removal and Installation—1988–89 Taurus and Sable

ing holes in the steering column and install self-tapping screws. Torque the screws to 17–26 inch lbs. (2–3 Nm).

6. Install the electrical connectors and install the wiring harness retainer.

7. Install the upper and lower steering column shroud and shroud retaining screws, torque the screws to 6–10 inch lbs. (0.7–1.1 Nm).

8. Install the ignition lock cylinder. Attach the tilt lever, if removed and torque the tilt lever Allen head retaining screw to 6–9 inch lbs. (0.7–1.0 Nm).

9. Connect the negative battery cable. Check the switch and the steering column for proper operation.

Ignition Lock Cylinder

REMOVAL & INSTALLATION

Functional Lock

The following procedure applies to vehicles that have functional lock cylinders. Lock cylinder keys are available for these vehicles or the lock cylinder key numbers are known and the proper key can be made.

1. Disconnect the negative battery cable.
2. Turn the lock cylinder key to the **RUN** position.
3. Using an ⅛ in. diameter wire pin or a small drift, depress the lock cylinder retaining pin through the access hole, while pulling out on the lock cylinder to remove it from the column.
To install:
4. Install the lock cylinder by turning it to the **RUN** position and depressing the retaining pin. Insert the lock cylinder into it's housing. Make sure the cylinder is fully seated and

aligned in the interlocking washer before turning the key to the **OFF** position. This will permit the cylinder retaining pin to extend into the cylinder housing.

5. Rotate the lock cylinder using the lock cylinder key, to ensure correct mechanical operation in all positions.

6. Connect the negative battery cable.

Non-Functional Lock

The following procedure applies to vehicles in which the ignition lock is inoperative and the lock cylinder cannot be rotated due to a lost or broken lock cylinder key, unknown key number or a lock cylinder cap that has been damaged and/or broken to the extent that the lock cylinder cannot be rotated.

1988–90

1. Disconnect the negative battery cable.

2. Remove the steering wheel.

3. Remove the 2 trim shroud halves by removing the 3 attaching screws.

4. Remove the electrical connector from the key warning switch.

5. Using an ⅛ in. diameter drill, drill out the retaining pin, being careful not to drill deeper than ½ in.

6. Place a suitable chisel at the base of the ignition lock cylinder cap and using a suitable hammer, strike the chisel with sharp blows to break the cap away from the lock cylinder.

7. Using a ⅜ in. diameter drill, drill down the middle of the ignition key slot approximately 1¾ in. until the lock cylinder breaks loose from the breakaway base of the lock cylinder. Remove the lock cylinder and drill shavings from the lock cylinder housing.

8. Remove the retainer, washer, ignition switch and actuator. Thoroughly clean all the drill shavings from the casting.

9. Inspect the lock cylinder housing for damage from the removal operation.

To install:

10. Replace the lock cylinder housing if it was damaged.

11. Install the actuator and ignition switch.

12. Install the trim and electrical parts.

13. Install a new ignition lock cylinder.

14. Install the steering wheel.

15. Connect the negative battery cable.

16. Check the lock cylinder operation.

1991–92

1. Disconnect the negative battery cable.

2. Remove the steering wheel.

3. Using channel lock or vise grip pliers, twist the lock cylinder cap until it separates from the lock cylinder.

4. Using a ⅜ in. diameter drill bit, drill down the middle of the ignition lock key slot approximately 1¾ in. (44mm) until the lock cylinder breaks loose from the breakaway base of the lock cylinder. Remove the lock cylinder and drill shavings from the lock cylinder housing.

5. Remove the retainer, washer, ignition switch and actuator. Thoroughly clean all drill shavings and other foreign materials from the casting.

6. Inspect the lock cylinder housing for damage from the removal operation. If the housing is damaged, it must be replaced.

To install:

7. Replace the lock cylinder housing, if damaged.

8. Install the actuator and ignition switch.

9. Install the trim and electrical parts.

10. Install the ignition lock cylinder.

11. Install the steering wheel.

12. Check the lock cylinder operation.

Ignition Switch

REMOVAL & INSTALLATION

1988–89 Taurus and Sable and 1988 Continental

1. Disconnect the negative battery cable.

2. Turn the ignition lock cylinder to the **RUN** position and depress the lock cylinder retaining pin through the access hole in the shroud with a ⅛ diameter punch.

3. Remove the lock cylinder. If equipped with tilt columns, remove the tilt release lever.

4. Remove the instrument panel lower cover and the steering column shroud.

5. Remove the 4 nuts attaching the steering column to the support bracket and lower the column.

6. Disconnect the ignition switch electrical connector.

7. Remove the lock actuator cover plate. The lock actuator assembly will slide freely out of the lock cylinder housing when the ignition switch is removed.

8. Remove the ignition switch and cover.

To install:

9. Make sure the ignition switch is in the **RUN** position by rotating the driveshaft fully clockwise to the **START** position and releasing.

10. Install the lock actuator assembly to a depth of 0.46–0.52 in. (11.75–13.25mm) from the bottom of the ac-

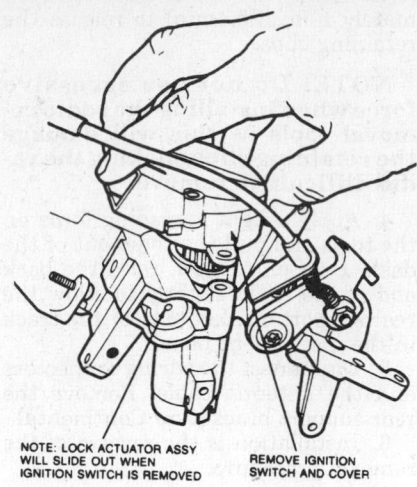

NOTE: LOCK ACTUATOR ASSY WILL SLIDE OUT WHEN IGNITION SWITCH IS REMOVED

REMOVE IGNITION SWITCH AND COVER

Ignition switch and cover removal– 1988–89 Taurus and Sable

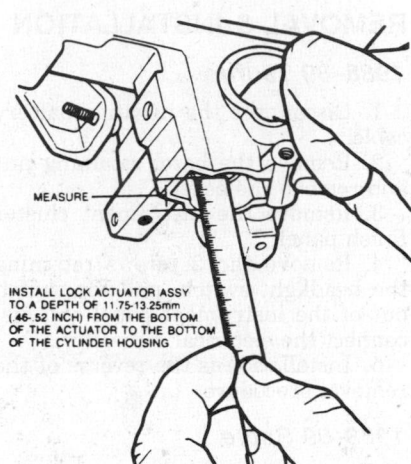

MEASURE

INSTALL LOCK ACTUATOR ASSY TO A DEPTH OF 11.75-13.25mm (.46-.52 INCH) FROM THE BOTTOM OF THE ACTUATOR TO THE BOTTOM OF THE CYLINDER HOUSING

Measuring lock cylinder depth–1988–89 Taurus and Sable

tuator assembly to the bottom of the lock cylinder housing.

11. While holding the actuator assembly at the proper depth, install the ignition switch. Install the ignition switch cover and tighten the retaining bolts to 30–48 inch lbs. (3.4–5.4 Nm).

12. Install the lock cylinder. Rotate the ignition lock cylinder to the **LOCK** position and measure the depth of the actuator assembly as in Step 10. The actuator assembly must be 0.92–1.00 in. (23.5–25.5mm) inside the lock cylinder housing. If the depth measured does not meet specification, the actuator assembly must be removed and installed again.

13. Install the lock actuator cover plate and tighten the bolts to 30–48 inch lbs. (3.4–5.4 Nm).

14. Install the ignition switch electrical connector.

15. Connect the negative battery cable. Check the ignition switch for proper function in all positions, including **START** and **ACC**.

16. Check the column function as follows:

a. With the column shift lever in the **P** position or with the floor shift key release button depressed and with the ignition lock cylinder in the **LOCK** position, make certain the steering column locks.

b. Position the column shift lever in the **D** position or the floor shift key release button fully extended and rotate the cylinder lock to the **RUN** position. Continue to rotate the cylinder toward the **LOCK** position until it stops. In this position, make certain the engine and all electrical accessories are **OFF** and that the steering shaft does not lock.

c. Turn the radio power button **ON**. Rotate the cylinder counterclockwise to the **ACC** position to verify that the radio is energized.

d. Place the shift lever in **P** and rotate the cylinder clockwise to the **START** position to verify that the starter energizes.

17. Remove the ignition lock cylinder.

18. Align the steering column mounting holes with the support bracket, center the steering column in the instrument panel opening and install the 4 nuts. Tighten the nuts to 15–25 ft. lbs. (20–34 Nm).

19. Install the column trim shrouds and the instrument panel lower cover. Install the tilt release lever, if equipped.

20. Install the ignition lock cylinder.

1989–90 Continental

1. Disconnect the negative battery cable.

2. Rotate the ignition lock cylinder to the **RUN** position and depress the lock cylinder retaining pin through the access hole in the shroud with a ⅛ in. drift punch or wire pin. Push on the pin and pull out on the lock cylinder.

3. Remove the lock cylinder.

4. If equipped with tilt steering columns, remove the tilt release lever by removing the Allen head cap screw that holds the tilt lever to the steering column.

5. Remove the lower steering column/instrument panel cover by removing the 4 Torx® head sheet metal screws.

6. Remove the steering column shroud.

7. Remove the bolts and nuts that attach the steering column to the support bracket and lower the column.

8. Remove the 3 screws from the diverter plate and remove it from the column.

9. Disconnect the ignition switch electrical connector.

10. Remove the ignition switch and

cover by removing the 2 tamper-resistant Torx® head bolts.

To install:

11. Make sure the ignition switch is in the **RUN** position by rotating the steering column shaft fully clockwise to the **START** position and releasing it.

12. Install the ignition switch and cover. Torque the cover retaining screws to 30–48 inch lbs. (3.4–5.4 Nm).

13. Install the ignition switch electrical connector.

14. Position the diverter plate on the column and secure it with 3 screws. Tighten to 30–48 inch lbs. (3.4–5.4 Nm).

15. Align the steering column mounting holes with the support bracket, center the steering column in the instrument panel opening and install the 4 nuts. Tighten to 15–25 ft. lbs. (20–34 Nm).

16. Install the 3 self-tapping screws and install the column trim shrouds. Tighten to 6–10 inch lbs. (0.7–1.1 Nm).

17. Install the instrument panel lower cover.

18. On tilt columns, install the tilt release lever. Tighten the retaining screw to 6.5–8.5 ft. lbs. (9–11 Nm). Check the column tilt travel through it's entire range to make sure there is no interference between the column and the instrument panel.

19. Connect the negative battery cable.

20. Check the column function as follows:

a. With the column shift lever in the **P** position or with the floor shift key release button depressed and with the ignition lock cylinder in the **LOCK** position, make certain the steering column locks.

b. Position the column shift lever in the **D** position or the floor shift key release button fully extended and rotate the cylinder lock to the **RUN** position. Continue to rotate the cylinder toward the **LOCK** position until it stops. In this position, make certain the engine and all electrical accessories are **OFF** and that the steering shaft does not lock.

c. Turn the radio power button **ON**. Rotate the cylinder counterclockwise to the **ACC** position to verify that the radio is energized.

d. Place the shift lever in **P** and rotate the cylinder clockwise to the **START** position to verify that the starter energizes.

1990–92 Taurus and Sable and 1991–92 Continental

1. Disconnect the negative battery cable.

2. Remove the steering column shroud by removing the self-tapping screws. Remove the tilt lever, if equipped.

3. Remove the instrument panel lower steering column cover.

4. Disconnect the ignition switch electrical connector.

5. Turn the ignition key lock cylinder to the **RUN** position.

6. Remove the 2 screws attaching the ignition switch and disengage the switch from the actuator pin.

To install:

7. Adjust the ignition switch by sliding the carrier to the switch **RUN** position. A new replacement switch assembly will already be set in the **RUN** position.

8. Make sure the ignition key lock cylinder is in the **RUN** position. The **RUN** position is achieved by rotating the key lock cylinder approximately 90 degrees from the lock position.

9. Install the ignition switch into the actuator pin. It may be necessary to move the switch slightly back and forth to align the switch mounting holes with the column lock housing threaded holes.

10. Install the attaching screws and tighten to 50–69 inch lbs. (5.6–7.9 Nm).

11. Connect the electrical connector to the ignition switch.

12. Connect the negative battery cable.

13. Check the ignition switch for proper function, including **START** and **ACC** positions. Make sure the column is locked with the switch in the **LOCK** position.

14. Install the instrument panel lower steering column cover, the steering column trim shrouds and the tilt lever, if equipped.

Stoplight Switch

The mechanical stoplight switch assembly is installed on the pin of the brake pedal arm, so it straddles the master cylinder pushrod.

REMOVAL & INSTALLATION

1. Disconnect the negative battery cable.

2. Disconnect the wire harness at the connector from the switch.

NOTE: The locking tab must be lifted before the connector can be removed.

3. Remove the hairpin retainer and white nylon washer. Slide the stoplight switch and the pushrod away from the pedal. Remove the switch by sliding the switch up/down.

NOTE: Since the switch side plate nearest the brake pedal is slotted, it is not necessary to remove the brake master cylinder pushrod black bushing and 1 white spacer washer nearest the pedal arm from the brake pedal pin.

To install:

4. Position the switch so the U-shaped side is nearest the pedal and directly over/under the pin. The black bushing must be in position in the push rod eyelet with the washer face on the side away from the brake pedal arm.

5. Slide the switch up/down, trapping the master cylinder pushrod and black bushing between the switch side plates. Push the switch and pushrod assembly firmly towards the brake pedal arm. Assemble the outside white plastic washer to pin and install the hairpin retainer to trap the whole assembly.

NOTE: Do not substitute other types of pin retainer. Replace only with production hairpin retainer.

6. Connect the wire harness connector to the switch.

7. Check the stoplight switch for proper operation. Stoplights should illuminate with less than 6 lbs. applied to the brake pedal at the pad.

NOTE: The stoplight switch wire harness must have sufficient length to travel with the switch during full stroke at the pedal.

Starter/Clutch Interlock Switch

ADJUSTMENT

1. Remove the panel above clutch pedal.

2. Disengage the wiring connector by flexing the retaining tab on the switch and withdrawing the connector.

3. Using a test light, check to see that the switch is open with the clutch pedal up (engaged) and closed at approximately 1 in. (25.4mm) from the clutch pedal full down position (disengaged).

4. If the switch does not operate as outlined in Step 3, check to see if the self-adjusting clip is out of position on the rod. It should be near the end of the rod.

5. If the self-adjusting clip is out of position, remove and reposition the clip approximately 1 in. (25.4mm) from the end of the rod.

6. Reset the switch by pressing the

clutch pedal to the floor. Repeat Step 3. If the switch is damaged or the clips do not remain in place replace the switch.

REMOVAL & INSTALLATION

1. Disconnect the negative battery cable.

2. Remove the panel above the clutch pedal.

3. Disconnect the switch wiring connector.

4. Remove clutch interlock attaching screw and hairpin clip and then remove the switch.

To install:

NOTE: Always install the switch with the self-adjusting clip about 1 in. (25.4mm) from the end of the rod. The clutch pedal must be fully up (clutch engaged). Otherwise, the switch may be misadjusted.

5. Insert the eyelet end of the rod over the clutch pedal pin and secure with the hairpin clip.

6. Align the mounting boss with the corresponding hole in the bracket and attach with a screw.

7. Reset clutch interlock switch by pressing the clutch pedal to the floor.

8. Connect the wiring connector.

9. Install the panel above the clutch.

Neutral Safety Switch

REMOVAL & INSTALLATION

1. Disconnect the negative battery cable and set the parking brake.

2. Disconnect the wire connector from the neutral safety switch.

3. Remove the nut and washer holding the Throttle Valve (TV) lever. Hold the lever stationary while loosening to prevent internal damage. Remove the lever from the TV shaft.

4. Remove the 2 neutral safety switch attaching bolts and remove the neutral safety switch.

To install:

5. Place the manual lever in N.

6. Install the neutral safety switch on the manual shaft.

7. Loosely install the 2 neutral safety switch attaching bolts, lockwashers and flatwashers.

8. Insert a No. 43 (0.089) drill bit through the hole provided in the switch. Tighten the attaching bolts to 7–9 ft. lbs. (9–12 Nm) and remove the drill bit.

9. Connect the neutral safety switch connector.

10. Install the TV lever, lockwasher and nut. Hold the lever stationary while tightening to prevent internal

damage. Tighten to 7.5–9.5 ft. lbs. (10–13 Nm).

11. Connect the negative battery cable.

12. Check the ignition switch for proper starting in P or N. Also make certain the start circuit cannot be actuated in the D or R position and that the column is locked in the LOCK position.

Fuses, Circuit Breakers and Relays

LOCATION

Fuses

All vehicles have a fuse panel located under the left side of the instrument panel. In addition, Continental is equipped with a high-current fuse panel located in the engine compartment on the left fender apron.

Fusible Links

Fusible links are used to prevent major wire harness damage in the event of a short circuit or an overload condition in the wiring circuits that are normally not fused, due to carrying high amperage loads or because of their locations within the wiring harness. Each fusible link is of a fixed value for a specific electrical load and should a fusible link fail, the cause of the failure must be determined and repaired prior to installing a new fusible link of the same value. Please be advised that the color coding of replacement fusible links may vary from the production color coding that is outlined in the text that follows.

TAURUS AND SABLE

Gray 12 Gauge Wire—located in left side of engine compartment at starter relay; used to protect battery to alternator circuit on all except 3.0L SHO engine.

Green 14 Gauge Wire—located in left side of engine compartment at starter relay; used to protect battery to alternator circuit if with 3.0L SHO engine.

Green 14 Gauge Wire—located in left side of engine compartment at starter relay; used to protect anti-lock brake system power relay circuit.

Black 16 Gauge Wire—located on the left shock tower; used to protect the battery feed to headlight switch and fuse panel circuits.

Black 16 Gauge Wire—located on the left shock tower; used to protect the battery feed to ignition switch and fuse panel circuits.

Black 16 Gauge Wire—located in left side of engine compartment at starter relay; used to protect rear window defrost circuit on 1988–90 vehicles and 1991–92 2.5L engine vehicles.

Brown 18 Gauge Wire—located in left side of engine compartment at starter relay; used to protect rear window defrost circuit on 1991–92 vehicles, except 2.5L engine.

Brown 18 Gauge Wire—located in right front of engine compartment at alternator output control relay; used to protect the alternator output control relay to heated windshield circuit.

Blue 20 Gauge Wire—located on the left shock tower; used to protect the ignition coil, ignition module and cooling fan controller circuits.

Blue 20 Gauge Wire—located in left rear of engine compartment; used to protect ignition switch to anti-lock brake system circuit.

CONTINENTAL

Except for the main alternator output, fusible links are not used. The fusible links have been replaced by the high current fuse panel.

NOTE: Always disconnect the negative battery cable before servicing the high current fuses or serious personal injury may result.

Gray 12 Gauge Wire—located at the front of the left fender apron, at the starter relay; used to protect the alternator output circuit.

Blue 20 Gauge Wire—located at the front of the left fender apron, at the starter relay; used to protect the alternator output circuit.

Circuit Breakers

Circuit breakers protect electrical circuits by interupting the current flow. A circuit breaker conducts current through an arm made of 2 types of metal bonded together. If the arm starts to carry too much current, it heats up. As 1 metal expands faster than the other the arm bends, opening the contacts and interupting the current flow.

TAURUS AND SABLE

Station Wagon Rear Window/Washer—One 4.5 amp circuit breaker located on the instrument panel brace, on the left side of the steering column on Taurus or on the left instrument panel end panel on Sable.

Windshield Wipers and Washer Pump—One 6 amp circuit breaker located on the fuse panel, on 1988 vehicles.

Windshield Wipers and Washer Pump—One 8.25 amp circuit breaker located on the fuse panel, on 1989–92 vehicles.

Cigar Lighters, Horn Relay and Horns—One 20 amp circuit breaker located on the fuse panel.

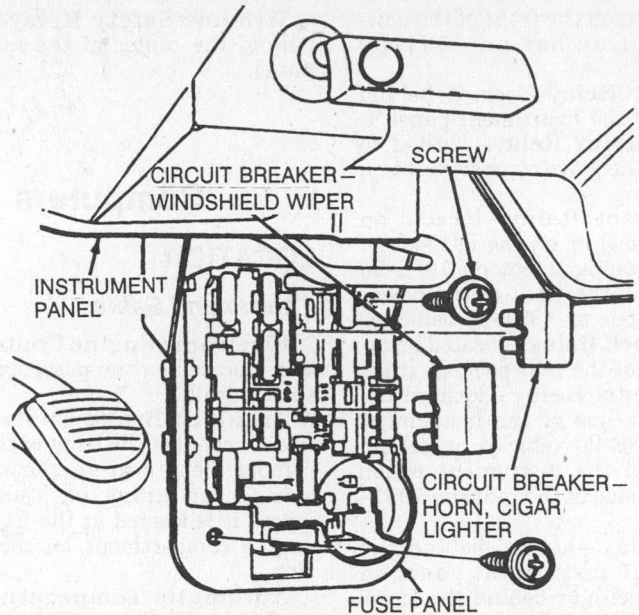

Fuse panel location—Taurus and Sable

Power Windows, Power Locks and Power Seats—One 20 amp circuit breaker located near the starter relay, on 1988–89 vehicles.

Power Windows, Power Locks and Power Seats—One 20 amp circuit breaker located on the fuse panel, on 1990–92 vehicles.

Headlights—One 22 amp circuit breaker incorporated in the headlight switch.

CONTINENTAL

Windshield Wipers and Washer Pump—One 6 amp circuit breaker located on the fuse panel, on 1988 vehicles.

Windshield Wipers and Washer Pump—One 8.25 amp circuit breaker located on the fuse panel, on 1989–92 vehicles.

Relays

TAURUS AND SABLE

Alternator Output Control Relay—located between the right front inner fender and fender splash shield, if with 3.0L or 3.8L engines and heated windshield.

Anti-lock Motor Relay—located in lower left front of engine compartment, if with anti-lock brakes.

Anti-lock Power Relay—located in left rear corner of engine compartment, if with anti-lock brakes.

Autolight Dual Coil Relay—located behind the center of the instrument panel on the instrument panel brace, if with automatic headlights.

Fog Light Relay—located behind the center of the instrument panel on the instrument panel brace.

Horn Relay—located behind the center of the instrument panel on the instrument panel brace.

LCD Dimming Relay—located behind the center of the instrument panel on the instrument panel brace, if with automatic headlights.

Low Oil Level Relay—located behind the center of the instrument panel on the instrument panel brace.

Moonroof Relay—located behind the right side of the instrument panel.

Police Accessory Relay—located behind the center of the instrument panel.

Starter Relay—located on the left fender apron, in front of the strut tower.

Window Safety Relay—located behind the right side of the instrument panel.

CONTINENTAL

Alternator Output Control Relay—if equipped with heated windshield, located between right front fender and apron, on the frame rail, on 1990 vehicles or in front of right fender on 1991–92 vehicles.

Anti-lock Motor Relay—if equipped with anti-lock brakes, located on the engine cowl on 1988–89 vehicles, on the right side of the engine compartment on the radiator support on 1990 vehicles or on the lower left front of the engine compartment, on the bracket behind the radiator on 1991–92 vehicles.

Anti-lock Power Relay—if equipped with anti-lock brakes, located on the engine cowl on 1988–90 vehicles or on the left side of the engine compartment, on the front of the power distribution box on 1991–92 vehicles.

Autolight Relay—located behind the center of the instrument panel.

Blower Motor Relay—located to the right of the glove compartment on 1990 vehicles.

Compressor Relay—located on the engine cowl or on the left side of the engine compartment on 1988–90 vehicles or on the front of the power distribution box on 1991–92 vehicles.

Hard Shock Relay—located below the left side of the rear package tray.

Hi-Lo Beam Relay—located behind the left side of the instrument panel on 1988–90 vehicles or behind the center of the instrument panel, near the left side of the radio on 1991–92 vehicles.

Horn Relay—located behind the center of the instrument panel on 1988–90 vehicles or behind the lower center of the instrument panel, near the left side of the warning chime module, on 1991–92 vehicles.

Interior Light Relay—located behind the right side of the instrument panel on 1988–90 vehicles.

LCD Dimming Relay—located behind the center of the instrument panel.

Soft Shock Relay—located below the left side of rear package tray.

Starter Relay—located on the left fender apron.

Window Safety Relay—located behind the center of the instrument panel.

Computers

LOCATION

Taurus and Sable

Electronic Engine Control Module—located on the passenger side of the firewall.

Anti-lock Brake Control Module—located at the front of the engine compartment next to the passenger side fender, except on Taurus SHO where it is located at the front of the engine compartment on the driver's side.

Automatic Temperature Control Module—located behind the center of the instrument panel.

Heated Windshield Control Module—located behind the left side of the instrument panel, to the right of the steering column.

Integrated Control Module—located at the front of the engine compartment, on the upper radiator support.

Air Bag Diagnostic Module—located behind the right side of the instrument panel, above the glove box.

Continental

Electronic Engine Control Module—located on the passenger side of the firewall.

Anti-lock Brake Control Module—located in the trunk on the passenger side under the package tray.

Air Bag Diagnostic Module—located behind the left side of the instrument panel, above the fuse panel.

Automatic Temperature Control Module—located behind the center of the instrument panel.

Heated Windshield Control Module—located behind the left side of the instrument panel.

Integrated Control Module—located at the front of the engine compartment, on the upper radiator support.

Air Suspension Control Module—located in left side of trunk.

Flashers

LOCATION

An electronic combination turn signal and emergency warning flasher is attached to the lower left instrument panel reinforcement above the fuse panel.

Ford Motor Co.
Front Wheel Drive
Mercury—Tracer

SPECIFICATIONS

VEHICLE IDENTIFICATION CHART

It is important for servicing and ordering parts to be certain of the vehicle and engine identification. The VIN (vehicle identification number) is a 17 digit number visible through the windshield on the driver's side of the dash and contains the vehicle and engine identification codes. The tenth digit indicates model year, and the eighth digit indicates engine code. It can be interpreted as follows:

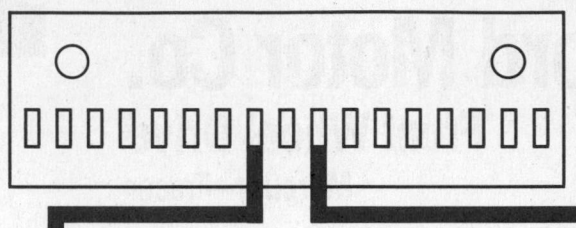

Engine Code						Model Year	
Code	Cu. In.	Liters	Cyl.	Fuel Sys.	Eng. Mfg.	Code	Year
5	98	1.6	4	EFI	Ford ①	J	1988
8	112	1.8	4	EFI	Mazda	K	1989
J	116	1.9	4	EFI	Ford	L	1990
① Mexico						M	1991
EFI—Electronic Fuel Injection						N	1992

ENGINE IDENTIFICATION

Year	Model	Engine Displacement cu. in. (liter)	Engine Series Identification (VIN)	No. of Cylinders	Engine Type
1988	Tracer	98 (1.6)	5	4	OHC
1989	Tracer	98 (1.6)	5	4	OHC
1990	Tracer	98 (1.6)	5	4	OHC
1991–92	Tracer	112 (1.8)	8	4	DOHC
	Tracer	116 (1.9)	J	4	OHC

OHC—Overhead Camshaft
DOHC—Double Overhead Camshaft

GENERAL ENGINE SPECIFICATIONS

Year	VIN	No. Cylinder Displacement cu. in. (liter)	Fuel System Type	Net Horsepower @ rpm	Net Torque @ rpm (ft. lbs.)	Bore × Stroke (in.)	Com- pression Ratio	Oil Pressure @ rpm
1988	5	4-98 (1.6)	EFI	82 @ 5000	92 @ 2500	3.07 × 3.29	9:3.1	50–64 ①
1989	5	4-98 (1.6)	EFI	82 @ 5000	92 @ 2500	3.07 × 3.29	9:3.1	50–64 ①
1990	5	4-98 (1.6)	EFI	82 @ 5000	92 @ 2500	3.07 × 3.29	9:3.1	50–64 ①
1991–92	8	4-112 (1.8)	EFI	127 @ 6500	114 @ 4500	3.27 × 3.35	9:0.1	43–57 @ 3000
	J	4-116 (1.9)	EFI	88 @ 4400	108 @ 3800	3.23 × 3.46	9:0.1	35–65 @ 2000

① 3000 rpm—hot
EFI—Electronic Fuel Injection

GASOLINE ENGINE TUNE-UP SPECIFICATIONS

Year	VIN	No. Cylinder Displacement cu. in. (liter)	Spark Plugs Type	Gap (in.)	Ignition Timing (deg.) MT	Ignition Timing (deg.) AT	Compression Pressure (psi)	Fuel Pump (psi)	Idle Speed (rpm) MT	Idle Speed (rpm) AT	Valve Clearance (in.) In.	Valve Clearance (in.) Ex.
1988	5	4-98 (1.6)	AGS32C	0.044	7B①	7B①	②	64–85	800–900	800–900	0.012H	0.012H
1989	5	4-98 (1.6)	AGS32C	0.044	7B①	7B①	②	64–85	800–900	800–900	0.012H	0.012H
1990	5	4-98 (1.6)	AGS32C	0.044	7B①	7B①	②	64–85	800–900	800–900	0.012H	0.012H
1991	8	4-112 (1.8)	AGSP-32C	0.043	9–11B	9–11B	②	64–85	700–800	700–800	Hyd.	Hyd.
	J	4-116 (1.9)	AGSF-34C6	0.054	10B	10B	②	35–45	③	③	Hyd.	Hyd.
1992		SEE UNDERHOOD SPECIFICATIONS STICKER										

H—Hot
Hyd.—Hydraulic
① Vacuum hose connected
② Lowest cylinder must be within 75% of the highest cylinder
③ Refer to underhood specifications sticker

FIRING ORDERS

NOTE: To avoid confusion, always replace spark plug wires one at a time.

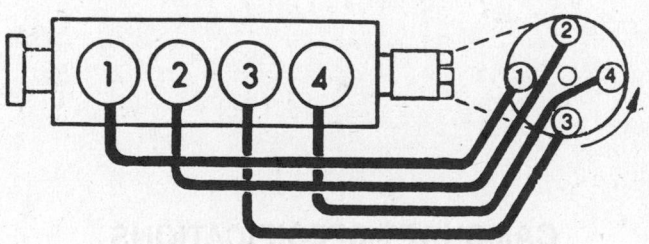

1.6L Engine
Engine Firing Order: 1–3–4–2
Distributor Rotation: Counterclockwise

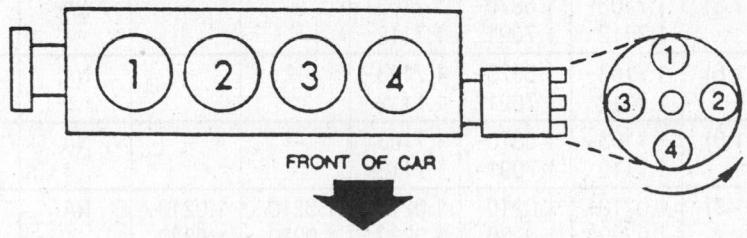

FRONT OF CAR

1.8L Engine
Engine Firing Order: 1–3–4–2
Distributor Rotation: Counterclockwise

1.9L Engine
Engine Firing Order: 1–3–4–2
Distributorless Ignition System

CAPACITIES

Year	Model	VIN	No. Cylinder Displacement cu. in. (liter)	Engine Crankcase (qts.) with Filter	Engine Crankcase (qts.) without Filter	Transmission (pts.) 4-Spd	Transmission (pts.) 5-Spd	Transmission (pts.) Auto.	Drive Axle (pts.)	Fuel Tank (gals.)	Cooling System (qts.)
1988	Tracer	5	4-98 (1.6)	3.5	3.2	—	6.8	12	NA	11.9	①
1989	Tracer	5	4-98 (1.6)	3.5	3.2	—	6.8	12	NA	11.9	①
1990	Tracer	5	4-98 (1.6)	3.5	3.2	—	6.8	12	NA	11.9	①
1991–92	Tracer	8	4-112 (1.8)	4.1	3.6	—	7.2	13.4	NA	13.2	①
	Tracer	J	4-116 (1.9)	4.0	3.5	—	5.6	13.4	NA	11.9	①

NA—Not available
① Manual transaxle—5.3 qts.
  Automatic transaxle—6.3 qts.

CAMSHAFT SPECIFICATIONS
All measurements given in inches.

Year	VIN	No. Cylinder Displacement cu. in. (liter)	Journal Diameter 1	Journal Diameter 2	Journal Diameter 3	Journal Diameter 4	Journal Diameter 5	Lobe Lift In.	Lobe Lift Ex.	Bearing Clearance	Camshaft End Play
1988	5	4-98 (1.6)	1.7103–1.7112	1.6870–1.7091	1.7103–1.7112	—	—	NA	NA	0.006	0.002–0.007
1989	5	4-98 (1.6)	1.7103–1.7112	1.6870–1.7091	1.7103–1.7112	—	—	NA	NA	0.006	0.002–0.007
1990	5	4-98 (1.6)	1.7103–1.7112	1.6870–1.7091	1.7103–1.7112	—	—	NA	NA	0.006	0.002–0.007
1991–92	8	4-112 (1.8)	1.0210–1.0220	1.0210–1.0220	1.0210–1.0220	1.0210–1.0220	1.0210–1.0220	NA	NA	0.001–0.003	0.003–0.007
	J	4-116 (1.9)	1.8007–1.8017	1.8007–1.8017	1.8007–1.8017	1.8007–1.8017	1.8007–1.8017	0.240	0.240	0.001–0.003	0.002–0.006

NA—Not available

CRANKSHAFT AND CONNECTING ROD SPECIFICATIONS

All measurements are given in inches.

| Year | VIN | No. Cylinder Displacement cu. in. (liter) | Crankshaft | | | | Connecting Rod | | |
			Main Brg. Journal Dia.	Main Brg. Oil Clearance	Shaft End-play	Thrust on No.	Journal Diameter	Oil Clearance	Side Clearance
1988	5	4-98 (1.6)	1.9661–1.9668	0.0011–0.0027	0.0031–0.0111	4	1.7693–1.7699	0.0009–0.0017	0.012
1989	5	4-98 (1.6)	1.9661–1.9668	0.0011–0.0027	0.0031–0.0111	4	1.7693–1.7699	0.0009–0.0017	0.012
1990	5	4-98 (1.6)	1.9661–1.9668	0.0011–0.0027	0.0031–0.0111	4	1.7693–1.7699	0.0009–0.0017	0.012
1991–92	8	4-112 (1.8)	1.9661–1.9668	0.0007–0.0014	0.003–0.011	4	1.7692–1.7699	0.0011–0.0027	NA
	J	4-116 (1.9)	2.2827–2.2835	①	0.004–0.008	3	1.7279–1.7287	0.0008–0.0015	0.004–0.011

NA—Not available
① Without cylinder head: 0.0018–0.0026 in.
 With cylinder head: 0.0011–0.0019 in.

VALVE SPECIFICATIONS

| Year | VIN | No. Cylinder Displacement cu. in. (liter) | Seat Angle (deg.) | Face Angle (deg.) | Spring Test Pressure (lbs.) | Spring Installed Height (in.) | Stem-to-Guide Clearance (in.) | | Stem Diameter (in.) | |
							Intake	Exhaust	Intake	Exhaust
1988	5	4-98 (1.6)	45	45	NA	NA	0.008	0.008	0.2744 0.2750	0.2742 0.2748
1989	5	4-98 (1.6)	45	45	NA	NA	0.008	0.008	0.2744 0.2750	0.2742 0.2748
1990	5	4-98 (1.6)	45	45	NA	NA	0.008	0.008	0.2744 0.2750	0.2742 0.2748
1991–92	8	4-112 (1.8)	45	45	NA	NA	0.0010–0.0024	0.0012–0.0026	0.2350–0.2356	0.2348–0.2354
	J	4-116 (1.9)	45	45.6	200 @ 1.09	1.44–1.48	0.0008–0.0027	0.0018–0.0037	0.3159–0.3167	0.3149–0.3156

NA—Not available

PISTON AND RING SPECIFICATIONS

All measurements are given in inches.

| Year | VIN | No. Cylinder Displacement cu. in. (liter) | Piston Clearance | Ring Gap | | | Ring Side Clearance | | |
				Top Compression	Bottom Compression	Oil Control	Top Compression	Bottom Compression	Oil Control
1988	5	4-98 (1.6)	0.006	0.006–0.012	0.006–0.012	0.008–0.028	0.001–0.003	0.001–0.003	Snug
1989	5	4-98 (1.6)	0.006	0.006–0.012	0.006–0.012	0.008–0.028	0.001–0.003	0.001–0.003	Snug
1990	5	4-98 (1.6)	0.006	0.006–0.012	0.006–0.012	0.008–0.028	0.001–0.003	0.001–0.003	Snug
1991–92	8	4-112 (1.8)	0.0015–0.0020	0.006–0.012	0.006–0.012	0.008–0.028	0.0012–0.0028	0.0012–0.0028	Snug
	J	4-116 (1.9)	0.0016–0.0024	0.010–0.020	0.010–0.020	0.016–0.055	0.0015–0.0032	0.0015–0.0035	Snug

TORQUE SPECIFICATIONS

All readings in ft. lbs.

Year	VIN	No. Cylinder Displacement cu. in. (liter)	Cylinder Head Bolts	Main Bearing Bolts	Rod Bearing Bolts	Crankshaft Pulley Bolts	Flywheel Bolts	Manifold Intake	Manifold Exhaust	Spark Plugs
1988	5	4-98 (1.6)	56–60 ①	40–43	37–41	②	71–76	14–19	12–17	11–17
1989	5	4-98 (1.6)	56–60 ①	40–43	37–41	②	71–76	14–19	12–17	11–17
1990	5	4-98 (1.6)	56–60 ①	40–43	37–41	②	71–76	14–19	12–17	11–17
1991–92	8	4-112 (1.8)	56–60	40–43	35–37	80–87	71–76	14–19	28–34	11–17
	J	4-116 (1.9)	③	67–80	26–30	81–96	54–67	12–15	16–19	8–15

① Using 2 steps
② Pulley bolts—36–45
   Sprocket bolt—80–94
③ Tighten in sequence to 44 ft. lbs.
   Loosen 2 turns
   Retighten in sequence to 44 ft. lbs.
   Turn all bolts 90 degrees
   Turn all bolts an additional 90 degrees

BRAKE SPECIFICATIONS

All measurements in inches unless noted.

Year	Model	Lug Nut Torque (ft. lbs.)	Master Cylinder Bore	Brake Disc Minimum Thickness	Brake Disc Maximum Runout	Standard Brake Drum Diameter	Minimum Lining Thickness Front	Minimum Lining Thickness Rear
1988	Tracer	65–87	0.875	①	0.003	7.870	0.120	0.040
1989	Tracer	65–87	0.875	①	0.003	7.870	0.120	0.040
1990	Tracer	65–87	0.875	①	0.003	7.870	0.120	0.040
1991–92	Tracer	65–87	0.870	②	0.004	9.00	0.080	0.040

① Front—0.630
   Rear—0.350
② Front—0.790
   Rear—0.280

WHEEL ALIGNMENT

Year	Model		Caster Range (deg.)	Caster Preferred Setting (deg.)	Camber Range (deg.)	Camber Preferred Setting (deg.)	Toe-in (in.)	Steering Axis Inclination (deg.)
1988	Tracer	Front	5/6P–2 2/3P	1 7/12P	1/20P–1 11/20P	12/15P	0.04N–0.20P	—
		Rear	—	—	3/4N–3/4P ①	0	0–0.16	—
1989	Tracer	Front	5/6P–2 2/3P	1 7/12P	1/20P–1 11/20P	12/15P	0.04N–0.20P	—
		Rear	—	—	3/4N–3/4P ①	0	0–0.16	—
1990	Tracer	Front	5/6P–2 2/3P	1 7/12P	1/20P–1 11/20P	12/15P	0.04N–0.20P	—
		Rear	—	—	3/4N–3/4P ①	0	0–0.16	—
1991–92	Tracer	Front	1P–2 7/8P	1 15/16P	3/4N–1/2P	1/16N	3/64N–13/64P	NA
		Rear	—	—	1N–1/2P	1/4N	3/64N–13/64P	—

N—Negative
P—Positive
① Not adjustable

ENGINE MECHANICAL

NOTE: Disconnecting the negative battery cable on some vehicles may interfere with the functions of the on board computer systems and may require the computer to undergo a relearning process, once the negative battery cable is reconnected.

Engine Assembly

REMOVAL & INSTALLATION

1.6L Engine

1. Using a scratch awl, matchmark the hood hinges to the hood. Remove the hood-to-hinge bolts and the hood.
2. Properly relieve the fuel system pressure.
3. Disconnect the cables from the battery; negative cable first. Remove the battery-to-vehicle bolts and the tray.
4. Using a clean drain pan, place it under the radiator. Remove the cooling system expansion tank cap, open the drain cock and drain the cooling system.
5. Drain the engine crankcase and the transaxle; discard the fluids.
6. Remove the air cleaner assembly and the dipstick.
7. Disconnect the electrical connector from the fan. Remove the fan shroud-to-radiator bolts, the fan and the shroud.
8. Disconnect the accelerator cable, the speedometer cable and the speed control cable, if equipped.
9. Disconnect and plug the fuel lines.
10. Disconnect the heater hoses and the radiator hoses from the engine.
11. From the power brake booster, disconnect the vacuum hose.
12. Disconnect the carbon canister hoses.
13. Disconnect the engine ground wire and the electrical harness connectors which will interfere with the engine removal.
14. Remove the exhaust pipe-to-exhaust manifold bolts and separate the pipe from the manifold.
15. If equipped with air conditioning, remove the compressor from the engine bracket and move it aside; do not disconnect the hoses.
16. If equipped with power steering, remove the pump from the engine bracket and move it aside; do not disconnect pressure hoses.
17. If equipped with a manual transaxle, disconnect the clutch control cable. Disconnect the shift control cable (automatic) or rod (manual).
18. Raise and support the vehicle safely.
19. Remove the engine splash shield-to-vehicle bolts and the shield. Remove the inner fender panel.
20. Remove the halfshafts from the vehicle.
21. Using a vertical lifting device, attach it to the engine and support its weight.
22. Remove the engine mount bolts and lift the engine/transaxle assembly from the vehicle. After removal, separate the engine from the transaxle.

To install:

23. Lower the engine/transaxle assembly into the vehicle.
24. Install the engine mount bolts.
25. Connect the halfshafts to the transaxle. Install the lower engine splash shield and the inner fender panel.
26. Connect the shift control cables. On manual transaxle equipped vehicles, connect the clutch cable.
27. Connect the exhaust system components to the manifold. Lower the vehicle.
28. Install the air conditioning compressor and/or power steering pump, if equipped.
29. Reconnect all hoses and wires. Install the fan and fan shroud.
30. Install the remaining components in the reverse order of removal.
31. Refill the cooling system, the crankcase and the transaxle. Connect the battery. Install the hood.
32. Start the engine, allow it to reach normal operating temperature and check for leaks.

1.8L Engine

WITH AUTOMATIC TRANSAXLE

The 1.8L engine equipped with an automatic transaxle can be removed without removing the transaxle from the vehicle. The engine can be split from the transaxle and lifted out of the engine compartment.

1. Disconnect the negative battery cable.
2. Mark the position of the hood hinges and remove the hood.
3. If equipped with air conditioning, properly discharge the system.
4. Drain the cooling system.
5. Remove the air duct connecting the throttle body and resonance chamber.
6. Disconnect the power brake vacuum supply hose from the power booster.
7. If equipped with speed control, disconnect the necessary vacuum hoses from the intake plenum.
8. Disconnect the electrical connectors from the power steering pump, water thermoswitch, temperature sending unit, oil pressure switch, fuel injector wiring harness, exhaust gas oxygen sensor, throttle position sensor and distributor.

NOTE: Mark the position of the connectors prior to removal to ease reinstallation.

9. Disconnect all engine ground straps.
10. Disconnect the ignition coil high-tension lead from the distributor.
11. Disconnect the accelerator and kickdown cables from the throttle cam.
12. Remove the accelerator and kickdown cable bracket from the intake plenum and set the assembly aside.
13. Disconnect the heater core inlet and outlet hoses at the bulkhead.
14. Relieve the fuel system pressure.
15. Remove the necessary fuel line clips and disconnect the fuel pressure and return lines.
16. Remove the upper radiator hose.
17. Disconnect the electrical connectors from the cooling fan and the radiator thermoswitch.
18. Remove the starter motor.
19. Raise and safely support the vehicle.
20. Remove the right upper and both left and right lower splash shields.
21. Remove the radiator lower hose.
22. Disconnect the 2 transaxle cooling lines from the radiator and plug the lines.
23. Remove the air conditioner line routing bracket from the radiator and position the line aside.
24. Remove the halfshaft bearing support.
25. Remove the inspection plate from the oil pan, place a wrench on the crankshaft pulley, and rotate the crankshaft to gain access to the torque converter nuts. Remove the nuts.
26. Remove the power steering and air conditioner drive belt.
27. Remove the crankshaft pulley.
28. Remove the exhaust flex-pipe and mounting flange assembly from the exhaust manifold.
29. If equipped with air conditioning, remove the compressor.
30. Remove the power steering pump and bracket assembly with the hoses still connected. Suspend the pump with wire, aside of the work area.
31. Remove all accessible transaxle-to-engine bolts from the engine block.
32. Lower the vehicle.
33. Remove the radiator mounting brackets and the resonance duct.
34. Remove the radiator, fan and shroud assembly from the vehicle.
35. Remove the vacuum chamber canister located next to the intake plenum.

36. Remove the pressure regulator and bracket assembly and set it aside.

37. Remove the shutter valve actuator and bracket assembly and set it aside.

38. Remove the alternator and water pump drive belt and remove the alternator.

39. Install a engine removal sling onto the engine lifting brackets. Place an engine hoist into position and support the engine.

40. Remove the oil pan-to-transaxle attaching bolts and the remaining transaxle-to-engine bolts from the engine block.

41. Remove the engine vibration dampener.

42. Remove the engine mount.

43. Carefully separate the engine from the transaxle, then remove the engine from the vehicle.

44. Install the engine onto an engine stand.

To install:

45. Install an engine removal sling onto the engine lifting brackets.

46. Place an engine hoist into position and install the engine sling. Remove the engine from the engine stand and lower it into the engine compartment.

47. Install the transaxle-to-engine upper right bolt and tighten to 41–59 ft. lbs. (55–80 Nm).

NOTE: Make sure the torque converter studs are properly seated in the flexplate mounting holes.

48. Install the engine mount. Tighten the bolt and nuts to 49–69 ft. lbs. (67–93 Nm).

50. Install the engine vibration dampener. Tighten the bolt and nuts to 41–50 ft. lbs. (55–80 Nm).

51. Remove the engine sling from the lifting brackets and remove the engine hoist.

52. Install the remaining transaxle-to-engine bolts and tighten to 41–59 ft. lbs. (55–80 Nm).

53. Install the alternator and the alternator and water pump drive belt.

54. Install the shutter valve actuator and bracket assembly.

55. Install the pressure regulator and bracket assembly.

56. Install the vacuum chamber canister located next to the intake plenum.

57. Place the power steering pump and bracket assembly into its mounting position.

58. Place the radiator, fan and shroud assembly into its mounting position.

59. Install the radiator mounting brackets along with the resonance duct. Tighten the mounting bolts to 69–95 inch lbs. (7.8–11.0 Nm).

60. Connect the cooling fan and radiator thermoswitch electrical connectors.

61. Raise and safely support the vehicle.

62. Install the oil pan-to-transaxle attaching bolts and tighten to 27–38 ft. lbs. (37–52 Nm).

63. Install the power steering pump and bracket assembly. Tighten the bolts to 27–38 ft. lbs. (37–52 Nm).

64. Install the lower radiator hose and clamps.

65. Connect the 2 transaxle cooling lines to the radiator.

66. If equipped, install the air conditioning compressor.

67. Install the air conditioning hose routing bracket to the radiator, if equipped. Tighten the bracket attaching nuts to 56–82 inch lbs. (6.4–9.3 Nm).

68. Install the crankshaft pulley and tighten the bolts to 109–152 inch lbs. (12–17 Nm).

69. Place a wrench on the crankshaft pulley and rotate the crankshaft to gain access to the torque converter studs. Install the torque converter nuts and tighten to 25–36 ft. lbs. (34–49 Nm). Install the transaxle inspection plate.

70. Install the power steering and air conditioning, if equipped, drive belt.

71. Install the halfshaft bearing support and tighten the bolts to 31–46 ft. lbs. (42–62 Nm).

72. Install the starter motor.

73. Connect the heater core inlet and outlet hoses at the bulkhead.

74. Install the exhaust flex-pipe, with a new gasket, to the exhaust manifold. Tighten the pipe-to-converter attaching nuts to 23–34 ft. lbs. (31–46 Nm).

75. Install the right and left lower splash shields and the right upper splash shield. Tighten the bolts to 69–95 inch lbs. (7.8–11.0 Nm).

76. Lower the vehicle.

77. Install the upper radiator hose and clamps.

78. Unplug the fuel pressure and return lines and connect them to the fuel rail. Install the necessary fuel line clips.

79. Install the accelerator and kickdown cable bracket onto the intake plenum. Tighten the bolts to 69–95 inch lbs. (7.8–11.0 Nm). Install the accelerator and kickdown cables onto the throttle cam.

80. Connect the power brake vacuum supply hose to the vacuum booster.

81. If equipped, connect the vehicle speed control vacuum hoses to the intake plenum.

82. Connect all engine ground straps.

83. Connect all remaining electrical connectors to their original locations, as marked during the removal procedure.

84. Connect the ignition coil high-tension lead into the distributor.

85. Install the air duct and resonance chamber assembly.

86. Fill the cooling system.

87. If equipped, recharge the air conditioning system according to the proper procedure.

88. Install the hood, aligning the marks that were made during the removal procedure.

89. Connect the negative battery cable.

90. Start the engine and check for leaks. Stop the engine and check the fluid levels.

WITH MANUAL TRANSAXLE

The 1.8L engine equipped with manual transaxle requires the engine and transaxle to be removed as an assembly. Lift the assembly out of the engine compartment.

1. Disconnect the negative battery cable.

2. Mark the position of the hood hinges and remove the hood.

3. If equipped with air conditioning, properly discharge the system.

4. Drain the cooling system.

5. Remove the resonance duct and the air cleaner assembly.

6. Remove the battery and the battery tray.

7. Disconnect the accelerator cable from the throttle cam and remove the accelerator cable bracket from the intake plenum.

8. Remove the upper radiator hose and disconnect the radiator overflow hose from the radiator filler neck.

9. Disconnect the radiator thermoswitch and cooling fan electrical connectors.

10. Remove the attaching nuts to the radiator mounting brackets and remove the brackets.

11. Disconnect the alternator, oil pressure switch, throttle position sensor, idle speed control, manual lever position switch, fuel injector wiring harness, back-up light switch, water thermoswitch, oxygen sensor, power steering pump and distributor electrical connectors.

NOTE: Mark the position of the connectors prior to removal to ease reinstallation.

12. Disconnect all engine ground straps.

13. Disconnect the ignition coil high-tension lead from the distributor.

14. Properly relieve the fuel system pressure.

15. Disconnect the fuel pressure and return lines.

16. Disconnect the heater core inlet and outlet, power brake vacuum sup-

ply, purge control vacuum and, if equipped, speed control vacuum hoses.

NOTE: Mark the position of the hoses prior to removal to ease reinstallation.

17. Raise and safely support the vehicle.

18. Remove the right upper and lower splash shields.

19. Remove the clutch slave cylinder pipe bracket from the transaxle with the hose still connected. Position the slave cylinder aside.

NOTE: Be careful not to damage the pipe or the hose.

20. Disconnect the shift control rod and the extension bar from the transaxle.

21. Remove the battery duct.

22. Remove the radiator lower hose.

23. Remove the power steering and, if equipped, air conditioning compressor drive belt.

24. Remove the power steering pump and bracket assembly with the hoses still connected. Suspend the pump with wire aside of the work area.

25. Remove the air conditioning hose routing bracket, if equipped, from the transaxle crossmember and position the air conditioning hose aside.

26. If equipped, remove the air conditioning compressor with the hoses still connected. Suspend the compressor with wire aside of the work area.

27. Disconnect the speedometer cable from the transaxle.

28. Remove the exhaust pipe front mounting flange and support bracket from the exhaust manifold.

29. Mark the location and disconnect the wires from the starter motor.

30. Remove the stabilizer bar.

31. Remove the tie rod ends from the steering knuckles.

32. Remove the halfshafts from the transaxle.

33. Remove the transaxle front and rear mount attaching nuts from the crossmember.

34. Lower the vehicle.

35. Remove the radiator, fan and shroud assembly from the vehicle.

36. Install an engine removal sling onto the engine lifting brackets.

37. Place an engine hoist into position and support the engine.

38. Remove the engine vibration dampener.

39. Remove the engine mount, transaxle upper mount and the transaxle support bracket.

40. Remove the engine and transaxle assembly.

41. Remove the intake plenum support bracket.

42. Remove the starter motor.

43. Remove the transaxle front mount.

44. Remove all oil pan-to-transaxle bolts and transaxle-to-engine attaching bolts from the engine block and separate the transaxle from the engine.

45. Remove the clutch assembly from the engine.

46. Install the engine onto an engine stand.

To install:

47. Install an engine removal sling onto the engine lifting brackets. Place an engine hoist into position and install the engine sling.

48. Remove the engine from the engine stand and lower the engine with the hoist still supporting it.

49. Install the clutch assembly.

50. Install the transaxle onto the engine.

51. Install the transaxle-to-engine bolts and tighten to 47–66 ft. lbs. (64–89 Nm).

52. Install the oil pan-to-transaxle attaching bolts and tighten to 27–38 ft. lbs. (37–52 Nm).

53. Position the transaxle front mount onto the transaxle and install the attaching bolts. Tighten the bolts to 27–38 ft. lbs. (37–52 Nm).

54. Position the starter motor into the transaxle housing and install the mounting bolts. Tighten the bolts to 27–38 ft. lbs. (37–52 Nm).

55. Install the intake plenum support bracket. Tighten the bolts to 27–38 ft. lbs. (37–52 Nm) and the nut to 14–19 ft. lbs. (19–25 Nm).

56. Using the engine hoist, position the engine and transaxle assembly into the engine compartment and align the engine mounting points with the engine mount and the mounting holes in the transaxle crossmember.

57. Install the attaching nuts to the transaxle front and rear mounts and the transaxle crossmember.

58. Position the engine mount into the vehicle.

59. Install the engine mount through-bolt and nut. Tighten them to 49–69 ft. lbs. (67–93 Nm).

60. Install the engine mount-to-engine attaching nuts. Tighten the nuts to 54–76 ft. lbs. (74–103 Nm).

61. Install the engine mount vibration dampener and attaching bolt and nut. Tighten the bolt and nut to 41–59 ft. lbs. (55–80 Nm).

62. Place the clutch slave cylinder and pipe assembly into its proper mounting position.

63. Install the engine support bracket and attaching bolts. Tighten the bolts to 41–59 ft. lbs. (55–80 Nm).

64. Install the transaxle upper mount and install the attaching bolts. Tighten the bolts to 32–45 ft. lbs. (43–61 Nm).

65. Install the transaxle upper

mount attaching nuts. Tighten the nuts to 49–69 ft. lbs. (67–93 Nm).

66. Place the radiator, fan and shroud assembly into its mounting position.

67. Install the radiator mounting brackets and tighten the nuts to 69–95 inch lbs. (7.8–11.0 Nm).

68. Install the upper radiator hose and connect the expansion reservoir overflow tube to the radiator filler neck.

69. Connect the cooling fan and radiator thermoswitch electrical connectors.

70. Raise and safely support the vehicle.

71. Install the lower radiator hose.

72. Install the halfshafts.

73. Install the tie rod ends into the steering knuckle.

74. Install the stabilizer bar.

75. Connect the wires to the starter motor according to their positions as marked during the removal procedure.

76. Install the exhaust front mounting flange to the exhaust manifold while making sure to install a new gasket. Tighten the flange-to-manifold attaching nuts to 23–34 ft. lbs. (31–46 Nm).

77. Install the exhaust pipe support bracket. Tighten the bracket attaching bolts to 27–38 ft. lbs. (37–52 Nm).

78. Install the speedometer cable into the transaxle.

79. If equipped, install the air conditioning compressor. Tighten the mounting bolts to 15–22 ft. lbs. (20–30 Nm).

80. Install the air conditioning routing bracket, if equipped, to the transaxle crossmember. Tighten the bolt to 56–82 inch lbs. (6.4–9.3 Nm).

81. Install the power steering pump and bracket assembly. Tighten the pump mounting bolts to 27–38 ft. lbs. (37–52 Nm).

82. Install the power steering and air conditioning, if equipped, drive belt.

83. Install the battery duct and tighten the attaching bolts to 69–95 inch lbs. (7.8–11.0 Nm).

84. Install the extension bar to the transaxle and tighten the attaching nut to 23–34 ft. lbs. (31–46 Nm).

85. Connect the shift control rod to the transaxle and tighten the attaching nut to 12–17 ft. lbs. (16–23 Nm).

86. Install the clutch slave cylinder attaching bolts and tighten to 12–17 ft. lbs. (16–23 Nm).

87. Position the slave cylinder pipe and install the routing bracket and attaching bolt. Tighten the bolt to 12–17 ft. lbs. (16–23 Nm).

88. Install the right upper and lower splash shields. Tighten the bolts to 69–95 inch lbs. (7.8–11.0 Nm).

89. Lower the vehicle.

90. Connect the heater core and vac-

uum hoses according to their original positions as marked during the removal procedure.

91. Connect the fuel pressure and return lines.

92. Connect the ignition coil high tension lead into the distributor.

93. Connect all engine ground straps.

94. Connect all remaining electrical connectors according to the locations marked during the removal procedure.

95. Install the accelerator cable bracket to the intake plenum and connect the accelerator cable to the throttle cam.

96. Install the battery tray and the battery.

97. Install the air cleaner assembly and the resonance duct.

98. Fill the cooling system.

99. If equipped, recharge the air conditioning system.

100. Install the hood, aligning the marks that were made during the removal procedure.

101. Connect the negative battery cable.

102. Start the engine and check for leaks. Stop the engine and check the fluid levels.

1.9L Engine

WITH AUTOMATIC TRANSAXLE

On automatic transaxle vehicles, the 1.9L engine assembly is removed without the transaxle attached. The engine is lifted from the engine compartment with the transaxle assembly remaining in the vehicle, attached to the mounts.

1. Mark the position of the hood hinges and remove the hood.

2. Disconnect the negative battery cable.

3. Drain the cooling system.

4. Remove the air intake duct.

5. Remove the crankcase ventilation hose from the valve cover and the vacuum hose from the bottom side of the throttle body.

6. Disconnect the power brake booster supply hose.

7. Disconnect the following electrical connectors:

   a. Fuel charging harness, located at the right shock tower.

   b. Alternator harness, from the back side of the alternator.

   c. Oxygen sensor.

   d. Ignition coil.

   e. Radio suppressor, mounted on the coil bracket.

   f. Coolant temperature sensor, cooling fan sensor and temperature gauge sending unit, mounted on a common water tube near the thermostat housing.

   g. Radiator cooling fan.

NOTE: Mark the position of the electrical connectors to aid reinstallation.

8. Remove the idle air bypass valve.

9. Remove the ground strap from the stud on the left side of the cylinder head near the ignition coil.

10. Disconnect the accelerator cable and the transaxle kickdown cable from the throttle lever. Remove the cable bracket from the intake manifold and position aside.

11. Disconnect both heater hoses at the engine compartment bulkhead.

12. Properly relieve the fuel system pressure and disconnect the fuel supply and return hoses at the fuel supply manifold.

13. Remove the upper radiator hose.

14. Raise and safely support the vehicle.

15. Remove the right side and the right and left front splash shields.

16. Remove the lower radiator hose from the radiator.

17. Position a drain pan under the radiator and remove the lower transaxle oil cooler line.

18. Remove the 2 oil cooler line retaining bracket bolts from the bottom of the radiator.

19. Remove the radiator fan shroud lower mounting bolts.

20. Lower the vehicle.

21. Remove the radiator fan shroud upper mounting bolts and remove the fan and shroud assembly from the vehicle.

22. Remove the upper transaxle oil cooler line from the radiator and remove the radiator from the vehicle.

23. If equipped with air conditioning, properly discharge the system.

24. Disconnect the air conditioning suction line at the suction accumulator/drier. Plug or cap the openings to prevent the entrance of dirt and moisture.

25. Remove the accessory drive belt.

26. Remove the power steering return hose from the pump reservoir and the high-pressure hose from the power steering pump.

27. Remove the power steering and air conditioner line retainer bracket bolts from the alternator bracket. Position the hoses aside.

28. Remove the accessory drive belt automatic tensioner assembly.

29. Raise and safely support the vehicle.

30. Remove the drive belt idler pulley.

31. If equipped, remove the 4 air conditioning compressor mounting bolts. Remove the compressor assembly with the lines attached and position aside. Safety wire the compressor to the vehicle sub-frame.

32. Remove the catalytic converter inlet pipe.

33. Remove the transaxle kickdown cable support bracket from the back side of the engine block. Position the cable and the bracket aside.

34. Disconnect the oil pressure switch.

35. Disconnect the relay wire and the positive battery cable from the starter.

36. Remove the flywheel inspection shield.

37. Remove the 4 torque converter attaching nuts.

38. Remove the crankshaft dampener.

39. Remove the 5 engine-to-transaxle bolts.

40. Lower the vehicle.

41. Remove the 3 starter motor mounting bolts and remove the starter out of the top of the engine compartment.

42. Remove the 2 transaxle-to-engine mounting bolts.

43. Connect an engine removal sling to engine lifting brackets. Position an engine hoist and support the engine.

44. Remove the right engine mount dampener and mount assembly.

45. With the engine assembly supported by the engine hoist, carefully separate the assembly from the transaxle.

46. Lift the engine assembly out of the vehicle.

47. Install the engine onto an engine stand.

To install:

48. Attach the engine removal sling to the engine lifting brackets and remove the engine from the stand with the engine hoist.

49. Carefully lower the engine into the vehicle and join the engine to the transaxle. Make sure the torque converter studs correctly engage the flywheel and the alignment dowels engage the transaxle housing.

50. Install the 2 transaxle-to-engine bolts but do not fully tighten them at this time.

51. Install the right engine mount insulator and dampener.

52. Position the engine hoist aside and remove the sling from the engine lifting brackets.

53. Raise and safely support the vehicle.

54. Install the 5 engine-to-transaxle bolts but do not fully tighten them at this time.

55. Install the crankshaft dampener and tighten the attaching bolt to 81–96 ft. lbs. (110–130 Nm).

56. Install the 4 torque converter attaching nuts and tighten to 25–36 ft. lbs. (34–49 Nm). Install the flywheel inspection plate.

57. Connect the oil pressure switch.

58. Install the kickdown cable support bracket.

59. If equipped, position the air con-

ditioning compressor on the bracket and install the 4 mounting bolts. Tighten the bolts to 15–22 ft. lbs. (20–30 Nm).

60. Install the catalytic converter inlet pipe.

61. Lower the vehicle.

62. From above, position the starter motor and install the 3 mounting bolts. Connect the positive battery cable and the relay wire to the starter.

63. Tighten the 2 transaxle-to-engine bolts to 40–59 ft. lbs. (55–80 Nm).

64. Install the power steering high-pressure hose on the pump.

65. Install the accessory drive belt idler pulley and automatic tensioner.

66. Install the power steering return hose on the pump reservoir.

67. Install the power steering hose retainer bracket on the alternator bracket.

68. Install the accessory drive belt.

69. If equipped, connect the air conditioner suction line to the accumulator.

70. Install the radiator assembly.

71. Connect the upper transaxle oil cooler line at the radiator.

72. Position the cooling fan and shroud assembly and install the upper mounting bolts.

73. Raise and safely support the vehicle.

74. Install the lower shroud bolts and connect the lower transaxle oil cooler line.

75. Install the oil cooler line retaining bracket bolts.

76. Install the lower radiator hose.

77. Tighten the 5 engine-to-transaxle bolts to 27–38 ft. lbs. (37–52 Nm).

78. Install the left and right front splash shields and the right side splash shield.

79. Lower the vehicle.

80. Install the upper radiator hose.

81. Connect both heater hoses at the engine compartment bulkhead.

82. Install the accelerator cable bracket and attach the accelerator and kickdown cables to the throttle lever.

83. Install the idle air bypass valve.

84. Install the ground strap on the stud at the front left side of the cylinder head, near the ignition coil.

85. Connect the remaining electrical connectors according to the positions marked during the removal procedure.

86. Connect the fuel supply and return lines. Be sure to install the fuel line safety clips.

87. Connect the power brake supply hose, the vacuum hose on the bottom side of the throttle body, and the crankcase ventilation hose to the valve cover.

88. Install the air intake duct.

89. Connect the negative battery cable.

90. Fill the cooling system.

91. Install the hood, aligning the marks that were made during the removal procedure.

92. Start the engine and check for leaks. Stop the engine and check the fluid levels.

93. If equipped, evacuate and recharge the air conditioning system according to the proper procedure.

WITH MANUAL TRANSAXLE

On manual transaxle vehicles, the 1.9L engine assembly is removed with the transaxle attached. The engine is lifted out of the engine compartment.

1. Mark the position of the hood hinges and remove the hood.

2. Disconnect the battery cables and remove the battery and the battery tray.

3. Drain the cooling system.

4. Remove the air cleaner.

5. Disconnect the crankcase ventilation hose from the valve cover and the vacuum hose from the bottom side of the throttle body.

6. Remove the power brake supply hose.

7. Disconnect the following electrical connectors:

    a. Fuel charging harness, located at the right shock tower.

    b. Alternator harness, from the back side of the alternator.

    c. Oxygen sensor.

    d. Ignition coil.

    e. Radio suppressor, mounted on the coil bracket.

    f. Coolant temperature sensor, cooling fan sensor and temperature gauge sending unit, mounted on a common water tube near the thermostat housing.

    g. Radiator cooling fan.

NOTE: Mark the position of the electrical connectors to aid reinstallation.

8. Remove the idle air bypass valve.

9. Remove the ground strap from the stud on the left side of the cylinder head near the ignition coil.

10. Disconnect the accelerator cable from the throttle lever. Remove the cable bracket from the intake manifold and position aside.

11. Disconnect both heater hoses at the engine compartment bulkhead.

12. Properly relieve the fuel system pressure and disconnect the fuel supply and return hoses at the fuel supply manifold.

13. Remove the upper radiator hose.

14. If equipped, properly discharge the air conditioning system and disconnect the suction line at the accumulator.

15. Remove the accessory drive belt and the automatic tensioner and idler pulley.

16. Disconnect the power steering re-

turn hose from the pump reservoir and the high pressure hose from the pump.

17. Remove the power steering hose and air conditioning line retainer brackets from the alternator bracket.

18. Raise and safely support the vehicle.

19. Remove the right and left side and front splash shields.

20. Disconnect the lower radiator hose from the radiator and remove the radiator fan shroud lower mounting bolts.

21. If equipped, remove the 4 air conditioning compressor mounting bolts. Remove the compressor assembly with the lines attached and position aside. Safety wire the compressor to the vehicle sub-frame.

22. Remove the catalytic converter inlet pipe.

23. Disconnect the oil pressure switch.

24. Disconnect the relay wire and the positive battery cable at the starter.

25. Remove the transaxle extension bar and shift control rod.

26. Remove the crankshaft dampener.

27. Remove the front wheel and tire assemblies.

28. Remove both halfshaft assemblies.

29. Install transaxle plugs into the differential side gears.

NOTE: Failure to install the transaxle plugs may allow the differential side gears to move out of position.

30. Disconnect the speedometer cable and the neutral switch on the transaxle.

31. Remove the clutch slave cylinder and line as an assembly from the transaxle and set it aside.

32. Remove the transaxle front and rear mount bolts.

33. Lower the vehicle.

34. Remove the radiator fan shroud upper mounting bolts and remove the fan shroud assembly from the vehicle.

35. Connect an engine removal sling to the engine lifting brackets. Connect the sling to an engine hoist, position the hoist and support the engine.

36. Remove the right engine mount dampener and mount assembly.

37. Remove the transaxle upper mount.

38. Lift the engine and transaxle assembly out of the vehicle and set it down on the floor.

To install:

39. Carefully lower the engine and transaxle assembly into the vehicle with the engine hoist.

40. Position the transaxle on its mounts and install the transaxle upper mount.

41. Install the right engine mount and mount dampener.

42. Remove the engine removal sling and the hoist.

43. Position the fan shroud assembly and install the upper mounting bolts.

44. Raise and safely support the vehicle.

45. Install the front and rear transaxle mount bolts.

46. Install the clutch slave cylinder and line assembly.

47. Connect the neutral switch and the speedometer cable.

48. Remove the transaxle plugs and install the halfshaft assemblies.

49. Install the crankshaft dampener and tighten the bolt to 81–96 ft. lbs. (110–130 Nm).

50. Install the transaxle extension bar and shift control rod.

51. Connect the relay wire and the positive battery cable to the starter.

52. Connect the oil pressure switch.

53. Install the catalytic converter inlet pipe.

54. If equipped, position the air conditioning compressor on its bracket and install the 4 mounting bolts.

55. Install the radiator fan shroud lower mounting bolts and install the lower radiator hose.

56. Install the left and right side and front splash shields.

57. Lower the vehicle.

58. Install the power steering hoses and install the power steering hose and air conditioner line retainer brackets.

59. Install the accessory drive belt idler pulley and automatic tensioner and install the accessory drive belt.

60. If equipped, connect the air conditioner suction line.

61. Install the upper radiator hose.

62. Connect the fuel supply and return hoses to the fuel supply manifold.

63. Connect both heater hoses.

64. Install the accelerator cable bracket on the intake manifold and connect the cable to the throttle lever.

65. Install the ground strap on the stud at the front left side of the cylinder head.

66. Install the idle air bypass valve.

67. Connect the remaining electrical connectors according to the positions marked during the removal procedure.

68. Connect the power brake supply hose, the crankcase ventilation hose and the vacuum line at the bottom of the throttle body.

69. Install the air cleaner assembly.

70. Install the battery tray and the battery. Connect the battery cables.

71. Fill the cooling system.

72. Install the hood, aligning the marks that were made during the removal procedure.

73. Start the engine and check for leaks. Stop the engine and check the fluid levels.

74. If equipped, evacuate and recharge the air conditioning system according to the proper procedure.

Engine Mounts

REMOVAL & INSTALLATION

1. Disconnect the negative battery cable.

2. Raise and support the vehicle safely. Drain the cooling system.

3. Disconnect the upper and lower radiator hoses.

4. Position a jack with a block of wood, under the engine.

5. Remove the engine-to-mount bolts and the mount-to-frame bolts.

6. Relieve the pressure from the mount by jacking the engine until the mount can be removed. Remove the mount.

7. Installation is the reverse order of the removal procedure.

8. Fill the cooling system to correct level. Connect the negative battery cable.

9. Road test the vehicle and check mounts for looseness.

Cylinder Head

REMOVAL & INSTALLATION

1.6L Engine

1. Disconnect the negative battery cable.

2. Remove the timing belt and rocker arm assembly.

3. Remove the exhaust manifold.

4. Using a clean drain pan, place it under the radiator. Open the drain cock and drain the cooling system.

5. Remove the spark plug wires and the spark plugs.

6. Remove the distributor-to-cylinder head bolts and the distributor from the engine.

7. From the front/rear of the engine, remove the engine lifting eyes. Disconnect the ground wire from the engine.

8. Disconnect the electrical harness connectors which may interfere with the cylinder head removal.

9. Remove the upper radiator hose, the water bypass hose and bracket.

10. Remove the cylinder head-to-engine bolts and the cylinder head.

11. Clean the gasket mating surfaces. Check and/or replace the damaged or worn parts.

To install:

12. Position the cylinder head gasket on the engine block and install the cylinder head.

13. Coat the cylinder head bolts with oil and install them into the head. Tighten the cylinder head bolts in sequence to 50–60 ft. lbs. (75–81 Nm) in 2 steps.

14. Install the water bypass hose and bracket and upper radiator hose.

15. Reconnect the wire harness connectors.

16. Install the front and rear engine lifting eyes to the cylinder head and the engine ground wire.

17. Install the distributor, spark plugs and spark plug wires.

18. Install the intake and exhaust manifolds. Torque the intake manifold-to-cylinder head bolts to 14–19 ft. lbs. (19–25 Nm) and the exhaust manifold-to-cylinder head bolts to 12–17 ft. lbs. (16–23 Nm).

19. Install the rocker arm cover, timing belt and timing belt cover.

20. Install the water pump pulley and drive belts.

21. Fill the cooling system to the correct level.

22. The remainder of the installation is the reverse order of the removal procedure.

23. Connect the negative battery cable.

24. Start the engine, allow it to reach normal operating temperature and check for leaks.

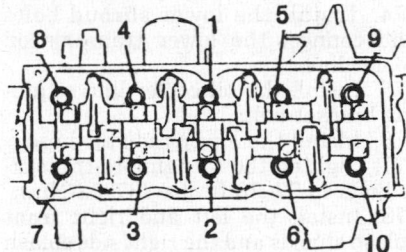

Cylinder head bolt torque sequence— 1.6L engine

1.8L Engine

1. Properly relieve the fuel system pressure.

2. Disconnect the negative battery cable.

3. Drain the cooling system.

4. Remove the bolts from the timing belt upper and middle covers. Remove the covers and gaskets.

5. Rotate the crankshaft by hand in the direction of normal engine rotation and align the timing marks located on the camshaft pulleys and seal plate.

6. Loosen the timing belt tensioner lock bolt and temporarily secure the tensioner spring in the fully extended position.

7. Remove the timing belt from the camshaft pulleys and secure it aside to prevent damage during the removal and installation of the cylinder head.

NOTE: Do not allow the timing belt to become contaminated by oil or grease.

8. Tag and disconnect the vacuum hoses from the cylinder head cover.

9. Tag and disconnect the spark plug wires from the spark plugs and position the wires aside.

10. Remove the cylinder head cover and gasket.

11. Remove the air duct from the resonance chamber and throttle body.

12. Disconnect the accelerator cable and, if equipped with automatic transaxle, the kickdown cable from the throttle cam. Remove the cable bracket from the intake plenum.

13. Tag and disconnect all vacuum lines from the intake plenum.

14. Tag and disconnect all necessary electrical connectors from the cylinder head, exhaust manifold, intake plenum, and throttle body. Disconnect the ground straps.

15. Remove the upper radiator hose.

16. Remove the transaxle-to-engine block upper-right bolt.

17. Disconnect the fuel pressure and return lines and plug the lines.

18. Disconnect the ignition coil high-tension lead from the distributor.

19. Tag and disconnect the necessary hoses connected to the cylinder head and intake plenum.

20. Remove the 2 bolts from the transaxle vent tube routing brackets.

21. Raise and safely support the vehicle.

22. Remove the bolt from the water pump-to-cylinder head hose bracket.

23. Remove the exhaust front mounting flange and exhaust pipe support bracket from the exhaust manifold.

24. Remove the intake plenum support bracket.

25. Lower the vehicle.

26. Remove the cylinder head bolts in the proper sequence.

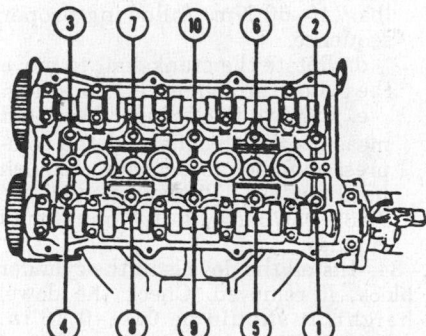

Cylinder head bolt removal sequence— 1.8L engine

27. Remove the cylinder head assembly, with the intake plenum and ex-

haust manifold attached, from the vehicle.

28. Remove the intake plenum and exhaust manifold.

29. Inspect the cylinder head for damage, cracks, and leakage of water and oil. Measure the cylinder head for warpage in 6 directions. The maximum distortion allowable is 0.004 in. (0.10mm).

30. If the cylinder head distortion exceeds specification, machine the cylinder head surface. The cylinder head must be replaced if the cylinder head height is not within 5.268–5.276 in. (133.8–134.0mm).

31. Inspect the manifold contact surface distortion in 4 directions. The maximum distortion allowable is 0.006 in. (0.15mm). If the distortion exceeds specification, machine the manifold contact surface or replace the cylinder head, as necessary.

To install:

32. Remove all dirt, oil and old gasket material from all gasket contact surfaces.

33. Install the intake plenum and exhaust manifold.

34. Install a new head gasket onto the top of the engine block, using the dowel pins for reference.

35. Place the cylinder head into its mounting position on top of the engine block.

36. Lubricate the cylinder head bolts with engine oil and install them finger-tight. Tighten the bolts in the proper sequence to 56–60 ft. lbs. (76–81 Nm).

37. Install the 2 bolts to the transaxle vent tube routing brackets.

38. Connect the heater hoses to the cylinder head and install the clamps.

39. Connect the ignition coil high-tension lead to the distributor.

40. Connect the fuel pressure and return lines to the fuel supply manifold and install the safety clips.

41. Install the transaxle-to-engine block upper-right bolt. If equipped with manual transaxle, tighten the bolt to 47–66 ft. lbs. (64–89 Nm). If equipped with automatic transaxle, tighten the bolt to 41–59 ft. lbs. (55–80 Nm).

42. Install the upper radiator hose and clamps.

43. Connect the ground straps and connect the electrical connectors that were disconnected at the cylinder head, exhaust manifold, intake plenum, and throttle body.

44. Connect the vacuum lines to the intake plenum.

45. Install the accelerator and kickdown cable bracket onto the intake plenum and tighten the bolts to 69–95 inch lbs. (7.8–11.0 Nm). Connect the cable(s) to the throttle cam.

46. Install the cylinder head cover and gasket, then connect the hose run-

ning from the plenum to the cylinder head cover. Tighten the cover bolts to 43–78 inch lbs. (4.9–8.8 Nm).

47. Install the air duct to the resonance chamber and throttle body and tighten the clamps. Connect the hose going from the air duct to the cylinder head cover.

48. Install and connect the spark plug wires.

49. Raise and safely support the vehicle.

50. Install the intake plenum support bracket. Tighten the bolts to 27–38 ft. lbs. (37–52 Nm) and the nut to 14–19 ft. lbs. (19–25 Nm).

51. Install the bolt to the water pump-to-cylinder head hose bracket.

52. Install the exhaust front mounting flange with a new gasket to the exhaust manifold. Tighten the flange-to-manifold attaching nuts to 23–34 ft. lbs. (31–46 Nm).

53. Install the exhaust pipe support bracket. Tighten the bracket attaching bolts to 27–38 ft. lbs. (37–52 Nm).

54. Make sure the yellow ignition timing mark on the crankshaft pulley is aligned with the TDC mark on the timing belt cover.

55. Lower the vehicle.

56. Make sure the timing marks on the camshaft pulleys and seal plate are aligned. Install the timing belt so there is no looseness at the idler pulley side or between the 2 camshaft pulleys.

NOTE: Do not turn the crankshaft counterclockwise.

57. Turn the crankshaft 2 turns clockwise by hand and verify that the yellow ignition timing mark on the crankshaft pulley is aligned with the timing mark on the timing belt cover. Verify that the timing marks on the camshaft pulley and seal plate are aligned.

NOTE: If the timing marks are not aligned, remove the timing belt and repeat the procedure beginning with Step 54.

58. Turn the crankshaft $1^5/_6$ turns clockwise by hand and align the 4th tooth to the right of the **I** and **E** timing marks on the camshaft pulleys with the seal plate alignment marks.

59. Loosen the timing belt tensioner lock bolt and apply tension to the timing belt. Tighten the tensioner lock bolt to 27–38 ft. lbs. (37–52 Nm).

60. Turn the crankshaft $2^1/_6$ turns clockwise and verify that the timing marks on the camshaft pulleys and the seal plate are aligned.

61. Install new gaskets onto the timing belt upper and middle covers and install the covers. Tighten the mounting bolts to 69–95 inch lbs. (8–11 Nm).

62. Fill the cooling system.

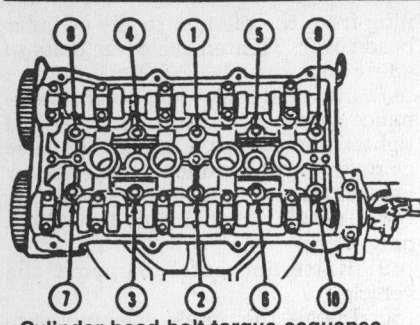

Cylinder head bolt torque sequence—
1.8L engine

63. Connect the negative battery cable.

64. Start the engine and check for leaks.

1.9L Engine

1. Properly relieve the fuel system pressure.

2. Disconnect the negative battery cable.

3. Drain the cooling system.

4. Remove the air intake duct.

5. Remove the crankcase breather hose from the rocker arm cover and the vacuum hose from the bottom of the throttle body.

6. Remove the power brake supply hose.

7. Disconnect the electrical connectors at the following:

   a. Fuel charging harness.

   b. Alternator harness.

   c. Crank angle sensor.

   d. Oxygen sensor.

   e. Ignition coil.

   f. Radio suppressor.

   g. Coolant temperature sensor, cooling fan sensor and temperature sending unit.

NOTE: Tag the connectors prior to removal to aid reinstallation.

8. Remove the ground strap from the stud on the left side of the cylinder head.

9. Disconnect the accelerator and the transaxle kickdown cables from the throttle lever and remove the cable bracket from the intake manifold.

10. Disconnect the heater hose containing the coolant temperature switches at the bulkhead.

11. Remove the upper radiator hose.

12. Disconnect the fuel supply and return lines.

13. Remove the oil level indicator tube mounting nut from the cylinder head stud.

14. Remove the power steering hose and the air conditioner line retainer bracket bolts from the alternator bracket.

15. Remove the accessory drive belt, alternator and the drive belt automatic tensioner.

16. Raise and safely support the vehicle.

17. Remove the right side splash shield and remove the crankshaft dampener.

18. Remove the catalytic converter inlet pipe.

19. Remove the starter wiring harness from the retaining clip below the intake manifold.

20. Set the engine No. 1 cylinder on TDC.

21. Lower the vehicle.

22. Support the engine with a floor jack.

23. Remove the right engine mount dampener and the right engine mount retaining bolts from the mount bracket on the engine. Loosen the right engine mount thru-bolt and roll the mount back aside.

24. Remove the timing belt cover.

25. Loosen the belt tensioner attaching bolt and pry the tensioner as far toward the rear of the engine as possible. Tighten the attaching bolt while in this position.

26. Remove the timing belt.

27. Roll the right engine mount back into position and install the mounting bolts. Lower the floor jack.

28. Remove the heater hose support bracket retaining bolt and the alternator bracket-to-cylinder head mounting bolt.

29. Remove the rocker arm cover.

30. Remove and discard the cylinder head bolts.

31. Remove the cylinder head with the exhaust and intake manifolds attached. Discard the cylinder head gasket.

NOTE: Do not lay the cylinder head flat. Damage to the spark plugs, valves or gasket surfaces may result.

To install:

32. Clean all gasket material from the mating surfaces on the cylinder head and block and clean out the head bolt holes in the block.

33. Before final installation of the cylinder head to the engine, check the piston squish height as follows:

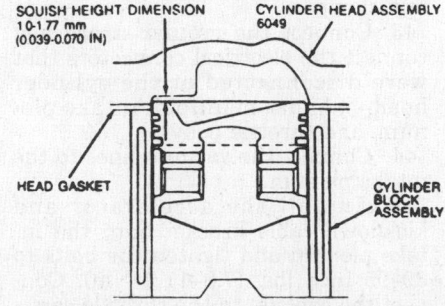

Piston squish height dimension location—
1.9L engine

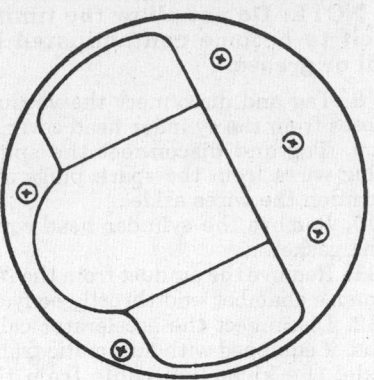

Lead solder placement positions

NOTE: Squish height is the clearance of the piston dome to the combustion chamber at piston TDC. No cylinder block deck machining or use of replacement crankshaft, piston or connecting rod causing the assembled squish height to be over or under tolerance specification, is permitted. If no parts other than the head gasket are replaced, the squish height should be within specification. If parts other than the head gasket are replaced, check the squish height. If the squish height is out of specification, replace the parts again and recheck the squish height.

   a. Place a small amount of soft lead solder or shot of an appropriate thickness on the piston spherical areas shown.

   b. Rotate the crankshaft to lower the piston in the bore and install the head gasket and cylinder head.

NOTE: A compressed (used) head gasket is preferred.

   c. Install used head bolts and tighten the head bolts to 30–44 ft. lbs. (40–60 Nm) following proper sequence.

   d. Rotate the crankshaft to move the piston through its TDC position.

   e. Remove the cylinder head and measure the thickness of the compressed solder to determine squish height at TDC. The solder should be 0.039–0.070 in. (1.0–1.77mm) for all engines.

34. Install the dowels in the cylinder block, if removed. Check the dowel height, it should be 0.41–0.46 in. (10.40–11.75mm) above the surface of the block. A dowel that is too long will not allow the cylinder head to sit properly.

35. Position the cylinder head gasket on the cylinder block.

36. Install the cylinder head and in-

stall new bolts and washers in the following order:

    a. Apply a light coat of engine oil to the threads of the new cylinder head bolts and install the new bolts into the head.

    b. Torque the cylinder head bolts in sequence to 44 ft. lbs. (60 Nm).

    c. Loosen the cylinder head bolts approximately 2 turns and then torque again to 44 ft. lbs. (60 Nm) using the same torque sequence.

    d. After setting the torque again, turn the head bolts 90 degrees in sequence and to complete the head bolt installation, turn the head bolts an additional 90 degrees in the same torque sequence.

NOTE: The cylinder head attaching bolts cannot be tightened to the specified torque more than once and must therefore be replaced when installing a cylinder head.

37. Install the rocker arm cover and the alternator bracket-to-cylinder head bolt.
38. Support the engine with a floor jack.
39. Remove the right engine mount-to-mount bracket bolts. Roll the mount aside.
40. Make sure cylinder No. 1 is at TDC.
41. Install the timing belt and the timing belt cover.
42. Roll the right engine mount into place and install the 2 mounting bolts and the mount dampener. Remove the floor jack.
43. Raise and safely support the vehicle.
44. Install the crankshaft dampener.
45. Install the starter wiring harness on the retaining clip below the intake manifold.
46. Install the catalytic converter inlet pipe and the right side splash shield.
47. Lower the vehicle.
48. Install the alternator and the accessory drive belt automatic tensioner. Install the accessory drive belt.
49. Install both the power steering hose and air conditioner line retainer bracket bolts. Install the oil level indicator tube retainer bolt.
50. Connect the fuel supply and return lines.
51. Install the upper radiator hose and connect the heater hose at the engine compartment bulkhead. Install the heater hose support bracket retaining bolt.
52. Install the accelerator cable bracket on the intake manifold and connect the accelerator and kickdown cables to the throttle lever.
53. Install the ground strap at the left side of the cylinder head.

54. Connect all remaining electrical connectors according to their positions marked during the removal procedure.
55. Connect the power brake supply hose, crankcase breather hose and the vacuum line at the bottom of the throttle body.
56. Install the air intake duct.
57. Connect the negative battery cable.
58. Fill and bleed the cooling system.
59. Start the engine and check for leaks. Stop the engine and check the coolant level.

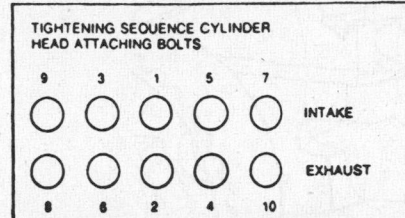

Cylinder head bolt torque sequence—1.9L engine

Valve Lifters

REMOVAL & INSTALLATION

1.8L Engine

NOTE: Hydraulic lash adjusters are used on the 1.8L engine between the valve stem and the camshaft to reduce noise and to provide maintenance-free valve clearance.

1. Disconnect the negative battery cable.
2. Remove the camshafts according to the proper procedure.
3. Mark the hydraulic lash adjusters and the cylinder head with alignment marks so the hydraulic lash adjusters can be installed in their original mounting positions.
4. Remove the hydraulic lash adjusters from the cylinder head.

To install:
5. Apply clean engine oil to the hydraulic lash adjuster friction surfaces.
6. If the hydraulic lash adjusters are being reused, install them in the positions from which they were removed.
7. Make sure the hydraulic lash adjusters move smoothly in their bores.
8. Install the camshafts.
9. Connect the negative battery cable.

1.9L Engine

1. Disconnect the negative battery cable.
2. Remove air cleaner assembly. Remove valve cover and gasket.

3. Remove rocker arms, lifter guides, lifter retainers and lifters.

NOTE: Always return lifters to the original bores unless they are being replaced.

To install:
4. Lubricate each lifter bore with engine oil.
5. If equipped with flat bottom lifters, install with oil hole in plunger upward. If equipped with roller lifters, install with plunger upward and position guide flats of lifters to be parallel with centerline of camshaft. Color orientation dots on lifters should be opposite the oil feed holes in cylinder head.
6. For roller lifters only, install lifter guide plates over tappet guide flats with notch toward exhaust side. For flat lifters, no guide plate is required.
7. Lubricate lifter plunger cap and valve tip with engine oil.
8. Install lifter guide plate retainers into rocker arm fulcrum slots, in both intake and exhaust side. Notch to be with exhaust valve lifter.
9. Install 4 rocker arms in lifter position No's 3, 6, 7 and 8.
10. Lubricate rocker arm surface that will contact fulcrum surface with engine oil.
11. Install 4 fulcrums. Fulcrums must be fully seated in slots of cylinder head.
12. Install 4 bolts. Tighten to 17–22 ft. lbs. (23–30 Nm).
13. Rotate the engine until the camshaft sprocket keyway is in the 6 o'clock position.
14. Repeat steps 9–12 in lifter position No's 1, 2, 4 and 5.
15. Install valve cover and gasket. Install air cleaner assembly.
16. Connect negative battery cable.

Valve Lash

ADJUSTMENT

Collapsed Lifter Clearance

1.9L ENGINE

1. Connect an auxiliary starter switch in the starting circuit. Crank the engine with the ignition switch **OFF** until the No. 1 piston is on TDC of the compression stroke.
2. With the crankshaft in position, place the hydraulic lifter compressor tool T81P–6500–A or equivalent, on the rocker arm. Slowly apply pressure to bleed down the lifter until it completely bottoms. Hold the lifter in this position and check the available clearance between the rocker arm and the valve stem tip with a feeler gauge. The feeler gauge width must not exceed ⅜ in., in order to fit between the rails on the rocker arm.

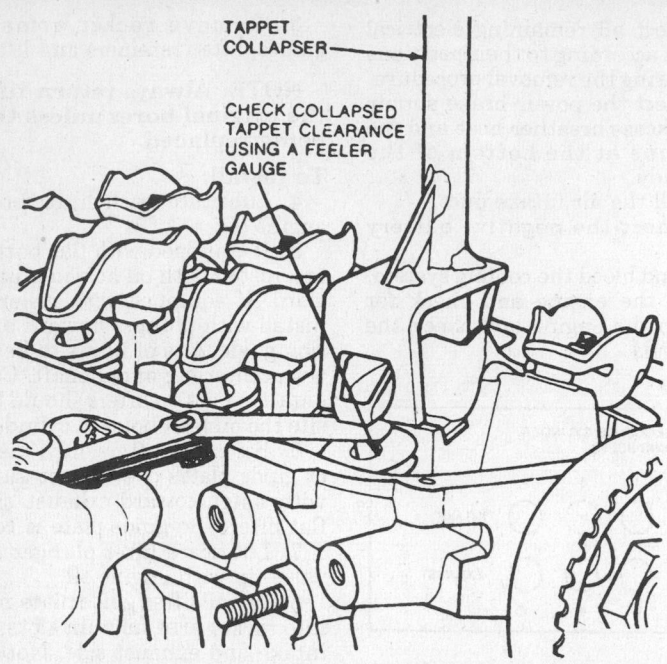

Collapsed lifter clearance checking—1.9L engine

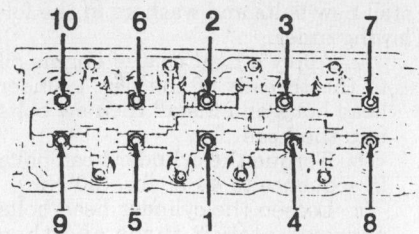

Rocker arm shaft torque sequence—1.6L engine

3. The clearance should be 0–4.5mm, 2.2mm nominal.

4. If the clearance is less than specifications, check the fulcrum, lifter, camshaft lobe and valve tip for wear.

5. With the No. 1 piston on TDC at the end of the compression stroke check the following valves: No. 1 intake, No. 1 exhaust, No. 2 intake.

6. Rotate the crankshaft 180 degrees and check the following valves: No. 3 intake, No. 3 exhaust.

7. Rotate the crankshaft another 180 degrees TDC and check the following valves: No. 4 intake, No. 4 exhaust, No. 2 exhaust.

Rocker Arms/Shafts

REMOVAL & INSTALLATION

1.6L Engine

1. Disconnect the negative battery cable. Remove the upper front cover.

2. Remove the air duct.

3. Remove the accelerator and cruise control cables, if equipped, from the rocker arm cover.

4. Disconnect the vent hose from the rocker arm cover and the spark plug wires from their clips.

5. Remove the rocker arm cover-to-cylinder head bolts, the cover and the gasket.

6. Remove the rocker arm shaft(s)-to-cylinder head bolts and the rocker arm shaft assemblies.

7. If necessary to separate the rocker arms from the rocker arm shafts, perform the following procedures:

a. Remove the bolts from the rocker arm(s).

b. Slide the rocker arm and springs from the shafts.

NOTE: Be sure to keep all the parts in order of disassembly for reinstallation purposes. The rocker arm shafts can only be installed in 1 position.

8. Clean the gasket mounting surfaces. Check and/or replace the parts if worn or damaged.

NOTE: To prevent damage to the O-ring on the hydraulic lash adjuster of the rocker arm, do not tamper with it unless replacement is necessary.

To install:

9. To install, use new gasket or sealant and reverse the removal proce-

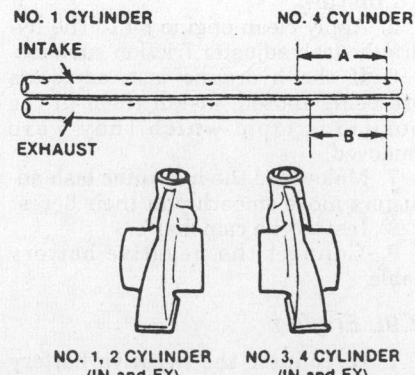

Rocker arms and shafts positioning—1.6L engine

dure. Torque the rocker arm shaft(s)-to-cylinder head (oil holes facing downward) bolts to 16–21 ft. lbs. (22–28 Nm) and the rocker arm cover-to-cylinder head bolts to 44–79 inch lbs. (5–9 Nm).

NOTE: When torquing the rocker arm shaft(s)-to-cylinder head bolts, start in the center and move outwards in both directions.

1.9L Engine

1. Disconnect the negative battery cable and remove the air cleaner assembly.

2. Remove and tag all necessary vacuum hoses from the rocker cover. Remove the spark plug wire retainers, if equipped. Remove the rocker cover from the cylinder head.

3. Remove the rocker cover and gasket from the engine.

4. Remove the rocker arm nuts, fulcrums, rocker arms and fulcrum washers. Keep all parts in order so they can be reinstalled to their original position.

To install:

5. Before installation, coat the valve tips, rocker arm and fulcrum contact areas with Lubriplate® or equivalent.

6. Rotate the engine until the lifter is on the base circle of the cam (valve closed).

NOTE: Be sure to turn the engine only in the normal rotation. Backward rotation will cause the camshaft belt to slip or lose teeth, altering the valve timing and causing serious engine damage.

7. Install the rocker arm and components and torque the rocker arm bolts to 17–22 ft. lbs. (23–30 Nm). Be sure the lifter is on the base circle of the cam for each rocker arm as it is installed.

8. Install a new gasket and the rocker arm cover. Install the 3 retaining bolts and tighten to 4–9 ft. lbs. (5–12 Nm).

NOTE: Do not use any type of sealer with the rocker arm cover silicone gasket.

9. Connect all vacuum hoses and install the spark plug wire retainers, if equipped. Connect the negative battery cable.

Intake Manifold

REMOVAL & INSTALLATION

1.6L Engine

1. Disconnect the negative battery cable.

2. Using a clean drain pan, place it under the radiator. Remove the cooling system expansion tank cap, open the drain cock and drain the cooling system.

3. Disconnect the accelerator cable and remove the air duct from the throttle body.

4. Label and disconnect all of the necessary wiring and hoses which may interfere with the intake manifold removal.

5. Remove the throttle body and the intake plenum.

6. Remove the intake manifold-to-cylinder head bolts, the intake manifold and the gasket.

7. Clean the gasket mounting surfaces. Clean and inspect the parts for damage and/or wear; replace if necessary.

8. To install, use new gaskets and reverse the removal procedures. Torque the intake manifold-to-cylinder head bolts to 14–19 ft. lbs. (19–25 Nm). Refill the cooling system. Start the engine, allow it to reach normal operating temperature and check for leaks.

1.8L Engine

1. Properly relieve the fuel system pressure.

2. Disconnect the negative battery cable.

3. Tag and disconnect the necessary vacuum hoses from the intake manifold and plenum.

4. Remove the vacuum chamber canister from the intake plenum.

5. Disconnect the idle speed control and bypass air hoses from the intake plenum.

6. Disconnect the accelerator cable and, if equipped with automatic transaxle, the kickdown cable from the throttle cam. Remove the cable bracket from the intake plenum.

7. Tag and disconnect the throttle body electrical connectors.

8. Disconnect the fuel pressure and return line spring lock couplings.

9. Disconnect the PCV hose from the intake plenum and cylinder head cover.

10. Disconnect the fuel pressure regulator vacuum hose and the fuel injec-tor wiring harness electrical connectors.

11. Remove the fuel rail mounting bolts and remove the fuel rail.

12. Remove the 2 bolts from the transaxle vent tube and remove the vent tube from the intake plenum.

13. Remove the intake manifold upper mounting nuts.

14. Raise and safely support the vehicle.

15. Remove the intake plenum support bracket and the intake manifold lower mounting nuts.

16. Lower the vehicle.

17. Remove the intake manifold, intake plenum and throttle body as an assembly from the vehicle.

18. Remove the intake manifold gasket.

19. If necessary, separate the intake plenum and throttle body from the intake manifold.

20. Clean all gasket mating surfaces.

To install:

21. If necessary, install the throttle body and intake plenum onto the intake manifold.

22. Install the intake manifold gasket.

23. Install the intake manifold, intake plenum and throttle body assembly onto the intake manifold mounting studs.

24. Install the mounting nuts and tighten to 14–19 ft. lbs. (19–25 Nm) in the proper sequence.

25. Raise and safely support the vehicle.

26. Install the intake plenum support bracket and tighten the bolts to specification.

27. Lower the vehicle.

28. Place the fuel rail into position and install the mounting bolts. Tighten the bolts to 14–19 ft. lbs. (19–25 Nm).

29. Connect the fuel injector wiring harness electrical connectors and connect the vacuum hose to the pressure regulator.

30. Connect the PCV hose to the intake plenum and cylinder head cover.

31. Connect the fuel pressure and return line spring lock couplings.

32. Install the transaxle vent tube that bolts onto the intake plenum.

33. Connect the electrical connectors to the throttle body and the necessary vacuum hoses to the intake plenum and throttle body.

34. Connect the idle speed control and bypass air hoses to the intake plenum.

35. Install the cable bracket onto the intake plenum and connect the accelerator and, if equipped, kickdown cables to the throttle cam.

36. Install the inlet air duct that connects to the throttle body and the resonance chamber.

37. Connect the negative battery cable.

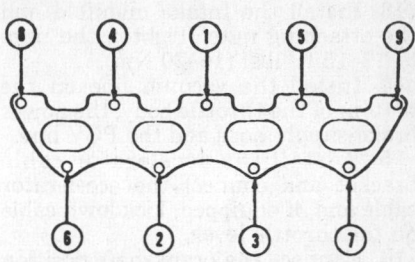

Intake manifold torque sequence—1.8L engine

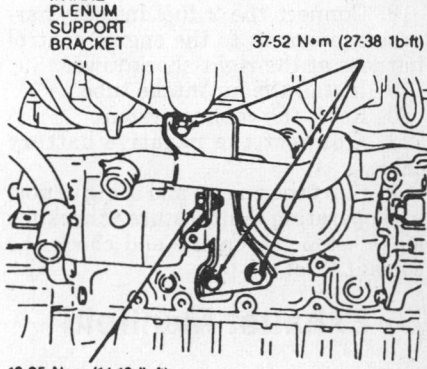

Intake plenum support bracket torque specifications—1.8L engine

1.9L Engine

1. Properly relieve the fuel system pressure.

2. Disconnect the negative battery cable.

3. Partially drain the cooling system.

4. Remove the air intake tube.

5. Disconnect the fuel injector harness from the engine control harness at the right shock tower.

6. Disconnect the crankshaft position sensor.

7. Disconnect the fuel supply and return lines.

8. Remove the accelerator cable and, if equipped with automatic transaxle, kickdown cable from the throttle lever. Remove the cable bracket from the intake manifold and position the cables aside.

9. Remove the power brake supply hose, PCV line and the vacuum line from the bottom of the throttle body.

10. Remove the 7 attaching nuts from the intake manifold studs, slide the manifold assembly off of the studs and remove it from the cylinder head. Remove and discard the intake manifold gasket.

To install:

11. Clean and inspect the mounting faces of the intake manifold and cylinder head. Both surfaces must be clean and flat.

12. Clean and oil the manifold studs and position a new gasket over them.

13. Install the intake manifold and the attaching nuts. Tighten the nuts to 12–15 ft. lbs. (16–20 Nm).

14. Install the vacuum line on the bottom of the throttle body, the power brake supply hose and the PCV line.

15. Install the accelerator cable bracket and connect the accelerator cable and, if equipped, kickdown cable on the throttle lever.

16. Connect the crankshaft position sensor electrical connector.

17. Connect the fuel supply and return lines. Install the fuel line retaining clips.

18. Connect the 2 fuel injector harness connectors to the engine control harness at the right shock tower.

19. Install the air intake tube.

20. Refill the cooling system.

21. Connect the negative battery cable.

22. Start the engine and bring to normal operating temperature. Check for leaks. Stop the engine and check the coolant level.

Exhaust Manifold

REMOVAL & INSTALLATION

1.6L Engine

1. Disconnect the negative battery cable.

2. Disconnect the electrical connector from the oxygen sensor.

3. Remove the exhaust insulators-to-exhaust manifold bolts and the insulators.

4. Remove the exhaust pipe-to-exhaust manifold nuts and separate the exhaust pipe from the manifold.

5. Remove the exhaust manifold-to-cylinder head bolts, the manifold and the gasket.

6. Clean the gasket mounting surfaces. Inspect the parts for damage and replace them, if necessary.

7. To install, use new gaskets and reverse the removal procedures. Torque the exhaust manifold-to-cylinder head bolts to 12–17 ft. lbs. (16–23 Nm) and the exhaust pipe-to-exhaust manifold bolts and nuts to 23–34 ft. lbs. (31–46 Nm).

8. Start the engine and check for exhaust leaks.

1.8L Engine

1. Disconnect the negative battery cable.

2. Remove the resonance duct.

3. Partially drain the cooling system and disconnect the upper radiator hose.

4. Remove the cooling fan.

5. Raise and safely support the vehicle.

6. Remove the exhaust pipe from the exhaust manifold and remove the gasket.

7. Remove the 2 bolts from the exhaust pipe support bracket.

8. Remove the left lower splash shield.

9. Lower the vehicle.

10. Disconnect the oxygen sensor electrical connector.

11. Remove the exhaust manifold heat shield mounting bolts and remove the shield.

12. Remove the exhaust manifold mounting nuts and remove the assembly.

13. Remove all gasket material from the cylinder head and exhaust manifold.

To install:

14. Install a new gasket onto the exhaust manifold mounting studs.

15. Place the exhaust manifold onto the mounting studs and install the manifold mounting nuts. Tighten the nuts to 28–34 ft. lbs. (38–46 Nm).

16. Place the heat shield into its mounting position and install the shield mounting bolts. Tighten the bolts to 69–95 inch lbs. (7.8–11.0 Nm).

17. Connect the oxygen sensor electrical connector.

18. Install the cooling fan.

19. Connect the upper radiator hose.

20. Install the resonance duct.

21. Raise and safely support the vehicle.

22. Install the exhaust pipe support bracket.

23. Install a new gasket and install the exhaust pipe to the exhaust manifold. Tighten the attaching nuts to 23–34 ft. lbs. (31–46 Nm).

24. Install the left lower splash shield and tighten the bolts to 69–95 inch lbs. (7.8–11.0 Nm).

25. Lower the vehicle.

26. Refill the cooling system.

27. Connect the negative battery cable.

1.9L Engine

1. Disconnect the negative battery cable.

2. Remove the accessory drive belt.

3. Remove the alternator.

4. Remove the radiator cooling fan and the shroud assembly.

5. Remove the exhaust manifold heat shield.

6. Raise and safely support the vehicle.

7. Remove the 2 catalytic converter inlet pipe-to-exhaust manifold attaching nuts.

8. Lower the vehicle.

9. Remove the 8 exhaust manifold attaching nuts and remove the exhaust manifold trand gasket.

To install:

10. Clean the cylinder head and exhaust manifold gasket surfaces.

11. Position the new gasket onto the manifold mounting studs.

12. Position the exhaust manifold on the cylinder head and install the attaching nuts. Tighten the nuts to 16–19 ft. lbs. (21–26 Nm).

13. Raise and safely support the vehicle.

14. Install the catalytic converter inlet pipe-to-exhaust manifold attaching nuts.

15. Lower the vehicle.

16. Install the exhaust manifold heat shield.

17. Install the radiator cooling fan and shroud assembly.

18. Install the alternator and the accessory drive belt.

19. Connect the negative battery cable.

Timing Belt Front Cover

REMOVAL & INSTALLATION

1.6L Engine

1. Disconnect the negative battery cable.

2. Remove the drive belt(s) from the front of the engine.

3. Remove the water pump pulley-to-water pump bolts and the pulley.

4. To remove the crankshaft pulley, perform the following procedures:

    a. Remove the right inner fender panel.

    b. Remove the crankshaft pulley-to-crankshaft bolts, outer spacer, outer pulley, inner spacer, inner pulley and baffle.

5. Remove the upper/lower front cover-to-engine bolts and the covers.

6. Clean the gasket mounting surfaces.

7. To install, use a new gasket and reverse the removal procedures. Torque the front cover-to-engine bolts to 69–95 inch lbs. (8–11 Nm), the crankshaft pulley-to-crankshaft bolts to 36–45 ft. lbs. (49–61 Nm) and the water pump pulley-to-water pump bolts to 36–45 ft. lbs. (49–61 Nm).

1.8L Engine

1. Disconnect the negative battery cable.

2. Remove the timing belt upper cover and gasket.

3. Loosen the water pump pulley bolts.

4. Remove the alternator and water pump accessory drive belt.

5. Remove the water pump pulley bolts and remove the pulley.

6. Raise and safely support the vehicle.

7. Remove the right wheel and tire assembly.

8. Remove the right upper and lower splash shields.

9. Remove the air conditioning, if equipped, and power steering accessory drive belt.

10. Remove the crankshaft pulley, crankshaft pulley guide plate and timing belt outer and inner guide plates.

11. Remove the timing belt middle and lower covers along with the gaskets.

To install:

12. Install the timing belt middle and lower covers along with the gaskets.

13. Install the timing belt inner and outer guide plates, the crankshaft pulley and the crankshaft pulley guide plate. Tighten the bolts to 109–152 inch lbs. (12–17 Nm).

14. Install the air conditioning, if equipped, and power steering accessory drive belt.

15. Install the splash shields and tighten the bolts to 69–95 inch lbs. (7.8–11.0 Nm).

16. Install the water pump pulley and tighten the bolts to 69–95 inch lbs. (7.8–11.0 Nm).

17. Install the alternator and water pump accessory drive belt.

18. Install the right wheel and tire assembly and lower the vehicle.

19. Install the timing belt upper cover and gasket. Tighten the bolts to 69–95 inch lbs. (7.8–11.0 Nm).

20. Connect the negative battery cable.

1.9L Engine

1. Disconnect the negative battery cable.

2. Remove the accessory drive belt.

3. Remove the drive belt automatic tensioner, if equipped.

4. Remove the timing cover retaining nuts.

5. Installation is the reverse of the removal procedure.

OIL SEAL REPLACEMENT

1.6L Engine

1. Disconnect the negative battery cable. Remove the timing belt.

2. Remove the crankshaft sprocket-to-crankshaft bolt, the sprocket and Woodruff® key.

3. Using a small prybar, pry the oil seal from the oil pump housing.

To install:

4. Using a clean shop cloth, clean the oil seal bore.

5. Lubricate the new oil seal with clean engine oil.

6. Using a front crankshaft seal installer tool, press the oil seal into the oil pump bore.

7. Install the crankshaft sprocket and the timing belt.

8. Connect the negative battery cable, start the engine and check for oil leaks.

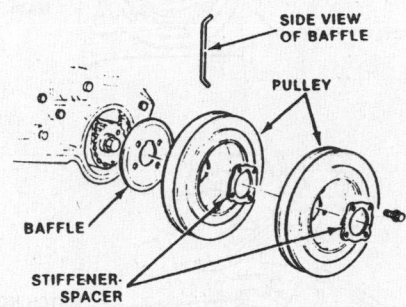

Crankshaft pulley assembly—1.6L engine

1.8L Engine

1. Disconnect the negative battery cable.

2. Remove the timing belt.

3. Remove the timing belt pulley locking bolt.

4. Remove the timing belt pulley. If necessary, use a suitable puller.

5. Remove the Woodruff® key.

6. If necessary, cut the lip of the front oil seal to ease removal.

7. Use a prybar to pry the oil seal from the engine.

To install:

8. Lubricate the lip of the new oil seal with clean engine oil.

9. Using a suitable installation tool, install the seal evenly until it is flush with the edge of the oil pump body.

10. Install the timing belt pulley onto the shaft while making sure to match the alignment grooves.

11. Install the Woodruff® key with the tapered end facing the oil pump.

12. Install the timing belt pulley locking bolt. Tighten the locking bolt to 80–87 ft. lbs. (108–118 Nm).

13. Install the timing belt.

14. Connect the negative battery cable.

1.9L Engine

1. Disconnect the negative battery cable.

2. Remove the accessory drive belt.

3. Raise and safely support the vehicle.

4. Remove the right side splash shield.

5. Remove the flywheel inspection shield.

6. Use a suitable tool to hold the flywheel in place.

7. Remove the crankshaft bolt and washer and remove the crankshaft dampener.

8. Remove the timing belt.

9. Remove the crankshaft sprocket and belt guide.

10. Using a seal remover tool, remove the crankshaft seal from the oil pump body.

To install:

11. Lubricate the lip of the new seal with clean engine oil.

12. Install the new seal using a suitable installation tool.

13. Install the belt guide and crankshaft sprocket.

14. Install the timing belt.

15. Position the crankshaft dampener on the crankshaft. Install the attaching bolt and washer and tighten to 81–96 ft. lbs. (110–130 Nm).

16. Remove the flywheel holding tool and install the inspection shield.

17. Install the right splash shield and lower the vehicle.

18. Install the accessory drive belt.

19. Connect the negative battery cable.

20. Start the engine and check for leaks.

Timing Belt and Tensioner

ADJUSTMENT

1.8L Engine

1. Disconnect the negative battery cable.

2. Remove the timing belt upper and middle covers and gaskets.

3. Place a wrench onto the crankshaft pulley and rotate the crankshaft clockwise so the timing marks located on the camshaft pulley and the seal plate are aligned.

4. Rotate the crankshaft clockwise 2 complete revolutions and align the timing marks on the camshaft pulleys and seal plate.

5. Make sure the yellow ignition timing mark on the crankshaft pulley is aligned with the TDC mark on the timing belt cover.

6. Measure the timing belt deflection by applying 22 lbs. of pressure on the belt, at a point between the camshaft pulleys. The timing belt deflection should be 0.35–0.45 in. (9.0–11.5mm).

7. If the deflection is not within specification, loosen the tensioner lock bolt. Use a prybar to move the tensioner, tighten or loosen the belt, as required, so the deflection will meet specification. Tighten the tensioner lock bolt to 27–38 ft. lbs. (37–52 Nm) and recheck the timing belt deflection beginning with Step 3.

8. If the timing belt will not meet specification, it must be replaced.

9. Install new gaskets onto the timing belt covers and install. Tighten the bolts to 69–95 inch lbs. (7.8–11.0 Nm).

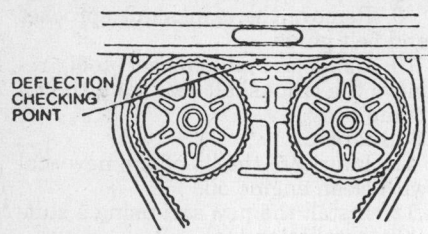

Timing belt deflection checking point—1.8L engine

1.6L and 1.9L Engines

Timing belt adjustment is not necessary on these engines. If there is a problem with timing belt tension, the tensioner assembly must be replaced.

REMOVAL & INSTALLATION

1.6L Engine

1. Disconnect the negative battery cable. Remove the timing belt covers.
2. Remove the No. 1 spark plug. Rotate the crankshaft to position the No. 1 cylinder on the TDC of its compression stroke.
3. Using a piece of chalk, mark the rotation direction on the timing belt.
4. Remove the timing belt tensioner spring, mounting bolt and tensioner.
5. Remove the timing belt.

To install:

6. Inspect the timing belt, tensioner and sprockets for signs of wear and replace, as necessary.
7. Check and/or align the camshaft and crankshaft sprockets with the cylinder head and oil pump alignment marks.

NOTE: If the No. 1 cylinder is not on the TDC of its compression stroke, rotate the crankshaft 1 complete revolution and realign the timing mark on the oil pump housing.

8. If reusing the timing belt, install it in the direction of the rotation mark.
9. Install the timing belt tensioner and spring; tighten the timing belt tensioner finger-tight.
10. Rotate the crankshaft 2 complete revolutions and realign the timing marks. Reaffirm that the timing marks are aligned, if not, repeat the alignment procedure.
11. Torque the tensioner bolt to 14–19 ft. lbs. (19–25 Nm) and check the timing belt deflection between the crankshaft and camshaft sprockets. The timing belt deflection should be 0.35–0.39 in. (9–10mm) at 22 lbs. pressure. If the deflection is not correct, repeat Steps 10 and 11.
12. Install the remaining components in the reverse order of their removal.

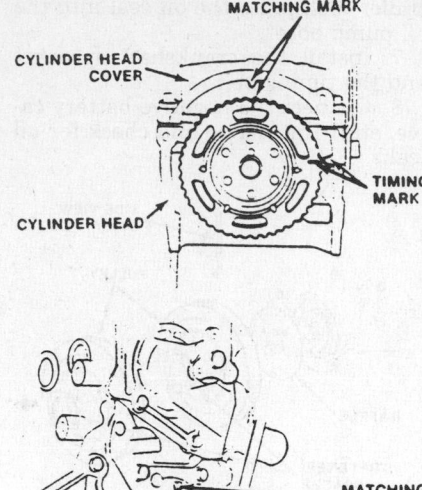

Timing sprocket alignment marks—1.6L engine

1.8L Engine

1. Disconnect the negative battery cable.
2. Remove the timing belt covers.
3. Rotate the crankshaft and align the timing marks located on the camshaft pulleys and the seal plate.
4. If the timing belt is to be reused, mark an arrow on the belt to indicate its rotational direction for installation reference.
5. Loosen the timing belt tensioner lock bolt and remove the timing belt.

To install:

6. Temporarily secure the timing belt tensioner in the far left position with the spring fully extended, then tighten the lock bolt.
7. Make sure the timing marks on the timing belt pulley and the engine block are aligned.
8. Make sure the timing marks on the camshaft pulleys and the seal plate are aligned.
9. Install the timing belt.
10. Loosen the tensioner lock bolt. Using a prybar, position the timing belt tensioner so the timing belt is taut, then tighten the tensioner lock bolt.
11. Turn the crankshaft 2 turns clockwise and align the timing belt pulley mark with the mark on the engine block.
12. Make sure the camshaft pulley marks are aligned with the seal plate marks.

NOTE: If the timing marks are not aligned, remove the belt and repeat the procedure.

13. Turn the crankshaft $1^5/_6$ turns clockwise and align the timing belt pulley mark with the tension set mark, at approximately the 10 o'clock position.
14. Apply tension to the timing belt tensioner and install the tensioner lock bolt. Tighten the bolt to 27–38 ft. lbs. (37–52 Nm).
15. Turn the crankshaft $2^1/_6$ turns clockwise and make sure the timing marks are aligned.
16. Measure the timing belt deflection by applying 22 lbs. of pressure on the belt between the camshaft pulleys. The timing belt deflection should be 0.35–0.45 in. (9.0–11.5mm). If necessary, adjust the timing belt deflection.
17. Turn the crankshaft 2 turns clockwise and make sure the timing marks are aligned.

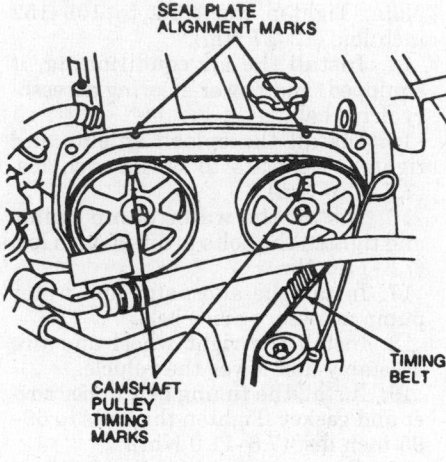

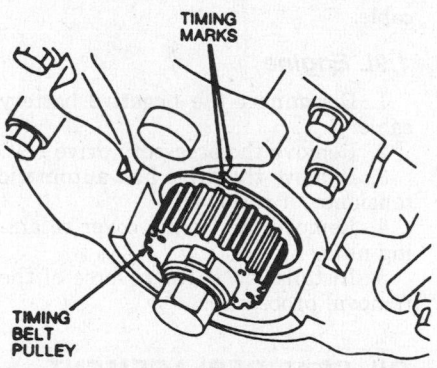

Timing sprocket alignment marks—1.8L engine

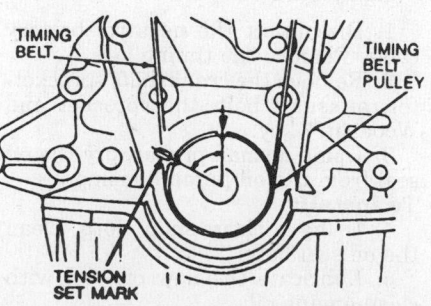

Tension set mark location—1.8L engine

NOTE: If the timing marks are not aligned, repeat the procedure beginning at Step 9.

18. Install the timing belt covers and the remaining components according to the proper procedure.

19. Connect the negative battery cable.

1.9L Engine

1. Disconnect the negative battery cable.

2. Remove the accessory drive belt automatic tensioner and the accessory drive belt.

3. Remove the timing belt cover.

4. Align the timing mark on the camshaft sprocket with the timing mark on the cylinder head.

5. Confirm that the timing mark on the crankshaft sprocket is aligned with the timing mark on the oil pump housing.

6. Loosen the belt tensioner attaching bolt, pry the tensioner away from the timing belt and retighten the bolt.

7. Remove the spark plugs.

8. Raise and safely support the vehicle.

9. Remove the right side splash shield.

10. Remove the flywheel inspection shield.

11. Use a suitable tool to hold the flywheel in place.

12. Remove the crankshaft dampener bolt and washer and remove the dampener.

13. Remove the timing belt.

NOTE: With the timing belt removed and the No. 1 piston at TDC, Do not rotate the camshaft. If the camshaft must be rotated, align the crankshaft dampener 90 degrees BTDC.

To install:

14. Install the timing belt over the sprockets in a counterclockwise direction starting at the crankshaft. Keep the belt span from the crankshaft to the camshaft tight while installing over the remaining sprocket.

15. Loosen the belt tensioner attaching bolt, allowing the tensioner to snap against the belt.

16. Rotate the crankshaft clockwise 2 complete revolutions. This will allow the tensioner spring to load the timing belt.

NOTE: Do not turn the engine counterclockwise to align the timing marks.

17. Tighten the tensioner attaching bolt to 17–22 ft. lbs. (23–30 Nm).

18. Install the crankshaft dampener and the bolt and washer. Tighten the bolt to 81–96 ft. lbs. (110–130 Nm).

19. Install the flywheel inspection shield.

20. Install the splash shield and lower the vehicle.

21. Install the spark plugs.

22. Install the timing belt cover.

23. Install the accessory drive belt automatic tensioner and the accessory drive belt.

24. Connect the negative battery cable.

Timing Sprockets

REMOVAL & INSTALLATION

1.6L Engine

1. Disconnect the negative battery cable.

2. Remove the timing belt.

3. Insert a suitable tool into a camshaft sprocket slot to keep the sprocket from turning. Remove the sprocket and dowel pin. If replacing the camshaft seal, pry the seal out with a prybar, being careful not to damage the camshaft or cylinder head.

4. If equipped with manual transaxle, place the shift lever in **4th** gear and apply the parking brake. If equipped with automatic transaxle, remove the flywheel dust cover and hold the flywheel, using a suitable tool.

5. Remove the crankshaft sprocket bolt and the sprocket and key.

To install:

6. Install the crankshaft sprocket and key. Install the sprocket bolt and tighten to 80–94 ft. lbs. (108–128 Nm). Remove the flywheel holding tool and replace the dust cover, if necessary.

7. If replacing the camshaft seal, lubricate the seal lip and outer surface with clean engine oil and install, using a seal installer tool.

8. Install the camshaft sprocket, dowel and bolt. Tighten the bolt to 36–45 ft. lbs. (49–61 Nm).

9. Install the timing belt and remaining components in the reverse order of removal.

1.8L Engine

1. Disconnect the negative battery cable.

2. Remove the timing belt according to the proper procedure.

3. Disconnect the vacuum hoses from the cylinder head cover.

4. Disconnect the spark plug wires from the spark plugs and position the wires aside.

5. Remove the cylinder head cover mounting bolts and remove the cover and gasket.

6. While holding the camshaft with a wrench, remove the camshaft pulley lock bolt. Remove the camshaft pulley.

7. Remove the timing belt crankshaft pulley locking bolt.

8. Remove the timing belt pulley. If necessary, use a suitable puller.

9. Remove the Woodruff® key from the crankshaft.

To install:

10. Install the timing belt pulley onto the shaft while making sure to match the alignment grooves.

11. Install the Woodruff® key with the tapered end facing the oil pump.

12. Install the timing belt pulley

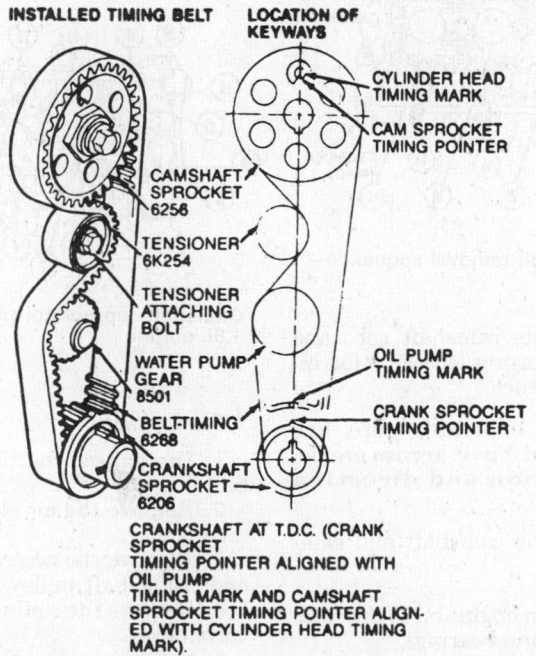

INSTALLED TIMING BELT   LOCATION OF KEYWAYS

CYLINDER HEAD TIMING MARK

CAM SPROCKET TIMING POINTER

CAMSHAFT SPROCKET 6256

TENSIONER 6K254

TENSIONER ATTACHING BOLT

WATER PUMP GEAR 8501

OIL PUMP TIMING MARK

BELT-TIMING 6268

CRANK SPROCKET TIMING POINTER

CRANKSHAFT SPROCKET 6206

CRANKSHAFT AT T.D.C. (CRANK SPROCKET TIMING POINTER ALIGNED WITH OIL PUMP TIMING MARK AND CAMSHAFT SPROCKET TIMING POINTER ALIGNED WITH CYLINDER HEAD TIMING MARK).

Timing belt and sprocket alignment position—1.9L engine

locking bolt and tighten to 80–87 ft. lbs. (108–118 Nm).

13. Install the camshaft pulley with the timing mark aligned with the timing mark on the seal plate.

14. While holding the camshaft with a wrench, install the camshaft pulley lock bolt. Tighten the bolt to 36–45 ft. lbs. (49–61 Nm).

15. Install a new cylinder head cover gasket onto the cylinder head.

16. Place the cylinder head cover into its mounting position and install the mounting bolts. Tighten the cylinder head cover bolts to 43–78 inch lbs. (4.9–8.8 Nm).

17. Connect the spark plug wires to the spark plugs and connect the vacuum hoses to the cylinder head cover.

18. Install the timing belt and timing belt covers according to the proper procedure.

1.9L Engine

1. Disconnect the negative battery cable.

2. Remove the timing belt cover and timing belt.

NOTE: With the timing belt removed and pistons at TDC, do not rotate the engine. If the camshaft must be rotated, align the crankshaft pulley to 90 degrees BTDC.

3. Remove the camshaft sprocket attaching bolt and washer and camshaft sprocket.

4. Remove the crankshaft sprocket.

5. Install the camshaft sprocket and attaching bolt and washer. Tighten to 37–46 ft. lbs. (50–62 Nm) on 1988 vehicles or 71–84 ft. lbs. (95–115 Nm) on 1989–92 vehicles.

6. Install the crankshaft sprocket.

7. Install the timing belt and cover.

8. Connect the negative battery cable.

Camshaft

REMOVAL & INSTALLATION

1.6L Engine

1. Disconnect the negative battery cable.

2. Remove the timing belt and rocker arm assembly.

3. Matchmark the distributor housing-to-cylinder head and rotor-to-distributor housing. Remove the distributor hold-down bolts and the distributor from the rear end of the camshaft.

4. Using a medium prybar (to prevent the camshaft from turning), remove the camshaft sprocket-to-camshaft bolt and the sprocket.

5. Using a small prybar, pry the camshaft oil seal from the cylinder head.

6. From the rear camshaft bearing

journal, remove the thrust plate-to-cylinder head bolt and the thrust plate.

7. Slide the camshaft from the cylinder head; be careful not to damage the journals and/or the lobes.

To install:

8. Clean the gasket mounting surfaces. Clean and inspect the parts for damage and/or wear and replace, if necessary.

9. To install, lubricate the parts with clean engine oil, use new gaskets or sealant and reverse the removal procedure. Torque the camshaft thrust plate bolt to 6–9 ft. lbs. (8–12 Nm), the camshaft sprocket-to-camshaft bolt to 36–45 ft. lbs. (49–61 Nm), the distributor hold-down bolt to 14–22 ft. lbs. (19–30 Nm).

10. Refill the cooling system. Check the ignition timing. Start the engine, allow it to reach normal operating temperature and check for leaks.

1.8L Engine

1. Disconnect the negative battery cable.

2. Remove the distributor assembly.

3. Remove the camshaft pulleys.

4. Remove the seal plate mounting bolts and remove the seal plate.

5. Loosen the camshaft cap bolts in the correct sequence.

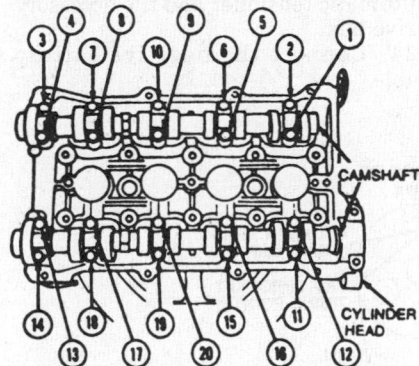

Camshaft cap bolt removal sequence—1.8L engine

6. Remove the camshaft caps and note their mounting locations for installation reference.

NOTE: The camshaft caps are numbered and have arrow marks for installation and direction reference.

7. Remove the camshaft and camshaft oil seal.

To install:

8. Apply clean engine oil to the camshaft journals and bearings.

9. Place the camshaft into its mounting position.

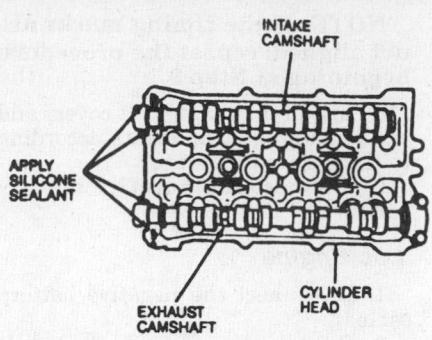

Sealant application locations—1.8L engine

NOTE: The exhaust camshaft has a groove which must be installed into the distributor drive gear.

10. Apply silicone sealant to the required areas.

11. Install the camshaft caps according to the cap numbers and arrow marks.

12. Install the camshaft cap bolts and tighten them in the proper sequence to 100–126 inch lbs. (11.3–14.2 Nm).

13. Apply a small amount of clean engine oil to the lip of a new camshaft oil seal. Using a installation tool, install the new seal.

14. Place the seal plate into its mounting position and install the mounting bolts. Tighten the bolts to 69–95 inch lbs. (7.8–11.0 Nm).

15. Install the camshaft pulleys and the distributor assembly.

16. Connect the negative battery cable.

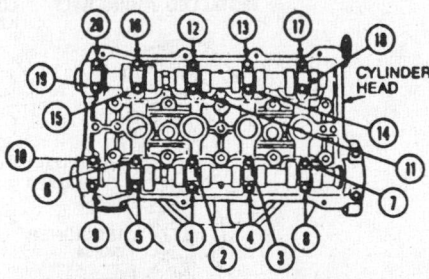

Camshaft cap bolt torque sequence—1.8L engine

1.9L Engine

1. Disconnect the negative battery cable.

2. Remove the air cleaner or air intake duct.

3. Remove the accessory drive belts and crankshaft pulley.

4. Remove the timing belt cover and valve cover.

5. Set the engine No. 1 cylinder at TDC prior to removing timing belt.

NOTE: Make sure the crankshaft is positioned at TDC and do not turn the crankshaft until the timing belt is installed.

6. Remove rocker arms and lifters as follows:
   a. Remove hex flange bolts.
   b. Remove fulcrums.
   c. Remove rocker arms.
   d. Remove lifter guide retainers.
   e. Remove lifters guides.
   f. Remove lifters.
7. Remove the ignition coil assembly.
8. Remove timing belt.
9. Remove the camshaft sprocket and key.
10. Remove the camshaft thrust plate.
11. Remove the cup plug from the back of the cylinder head.
12. Remove the camshaft through the back of the head toward the transaxle.
13. Replace camshaft seal.

To install:

14. Thoroughly coat the camshaft bearing journals, cam lobe surfaces and thrust plate groove with lubricant.
15. Install camshaft through the rear of the cylinder head. Rotate camshaft during installation.

NOTE: Before installing the camshaft, apply a thin film of lubricant to the lip of the camshaft seal.

16. Install the camshaft thrust plate. Tighten attaching bolts to 6–9 ft. lbs. (8–13 Nm).
17. Align and install the cam sprocket over the cam key. Install attaching washer and bolt. While holding camshaft stationary, tighten the bolt to 71–84 ft. lbs. (95–115 Nm).
18. Install the cup plug using sealer. Use the sealer sparingly, as excess sealer may clog the oil holes in the camshaft.
19. Install the timing belt.
20. Install the timing belt cover.
21. Install the rocker arm assembly as follows:

NOTE: Lubricate all the parts with a heavy engine oil before installation.

   a. Install the lifters.
   b. Install the lifter guides.
   c. Install the lifter retainers.
   d. Install the rocker arms.
   e. Install the fulcrums.
   f. Install the rocker arm bolts. Tighten to 17–22 ft. lbs. (23–30 Nm).
22. Install the ignition coil assembly.
23. Install new valve cover gasket, if required.

NOTE: Make sure the surfaces on the cylinder head and valve cover are clean and free of sealant material.

24. Install the valve cover and attaching bolts and tighten to 4–9 ft. lbs. (5–12 Nm).
25. Install the air intake duct or the air cleaner assembly.
26. Connect negative battery cable.

Piston and Connecting Rod

POSITIONING

NOTE: On 1.8L engine, the piston and rod assembly must be positioned in the engine block with the F mark facing the front of the engine.

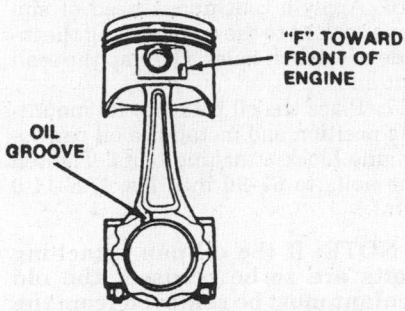

Piston and connecting rod positioning—1.6L engine

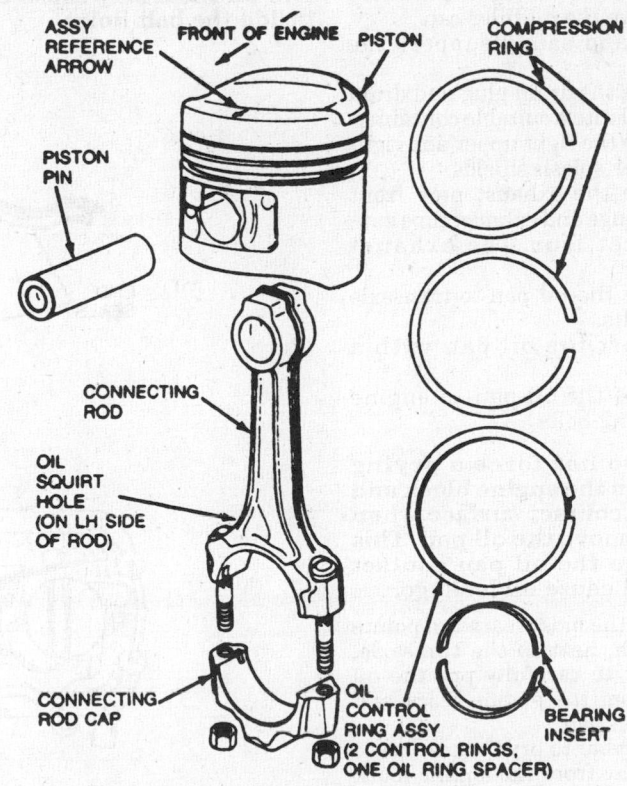

Piston and connecting rod positioning—1.9L engine

ENGINE LUBRICATION

Oil Pan

REMOVAL & INSTALLATION

1.6L Engine

1. Disconnect the negative battery cable.
2. Raise and support the vehicle safely.
3. Remove the under engine splash shields and the right front inner fender panel.
4. Using a drain pan, place it under the engine, remove the oil pan plug and drain the crankcase.
5. Remove the flywheel-to-engine support housing bracket and the dust cover from the flywheel housing.
6. Remove the oil pan-to-engine nuts, bolts and stiffeners. Remove the oil pan.

NOTE: If might be necessary to rotate the crankshaft to clear the oil pan.

To install:

7. Clean the gasket mounting surfaces.
8. Apply oil resistant sealer across the joint line of the cylinder block and

the front and rear engine covers. Install a new pan gasket, the oil pan and stiffeners and tighten the bolts and nuts to 69–78 inch lbs. (8–9 Nm).

9. Install the dust cover and tighten the bolts to 13–20 ft. lbs. (18–26 Nm).

10. Install the engine-to-flywheel housing support bracket. Tighten the bolts to 69–86 ft. lbs. (93–117 Nm).

11. Install the oil pan drain plug and tighten.

12. Install the splash shields and inner fender panel.

13. Lower the vehicle and add engine oil to the proper level.

14. Connect the negative battery cable, start the engine and check for leaks.

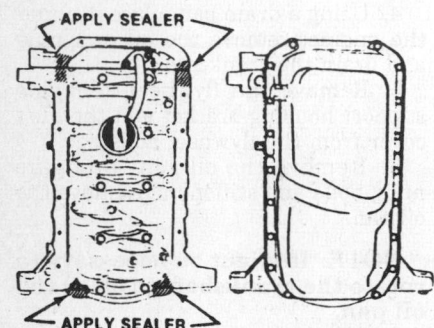

Sealant application locations – 1.6L engine

1.8L Engine

1. Disconnect the negative battery cable. Remove the oil filler cap.

2. Raise and safely support the vehicle.

3. Remove the drain plug and drain the engine oil into a suitable container.

4. Remove the right upper and right and left lower splash shields.

5. Remove the exhaust pipe front mounting flange and exhaust pipe support bracket from the exhaust manifold.

6. Remove the oil pan-to-transaxle attaching bolts.

7. Support the oil pan with a jackstand.

8. Remove the oil pan-to-engine block attaching bolts.

NOTE: Do not force a prying tool between the engine block and the oil pan contact surface when trying to remove the oil pan. This may damage the oil pan contact surface and cause oil leakage.

9. Only at the most rearward points of the oil pan, next to the transaxle, use a prybar to carefully pry the oil pan away from the engine block and remove the oil pan.

10. Use a prybar to pry the crankcase stiffeners away from the engine block and/or oil pan.

11. Remove the front and rear oil pan gaskets and end seals. Remove all sealant material from the engine block and oil pan.

NOTE: When removing the crankcase stiffeners and sealant material from the oil pan and engine block, be careful not to damage the oil pan and engine block contact surfaces.

To install:

12. Apply a bead of silicone sealant to the crankcase stiffeners along the inside of the bolt holes.

13. Install the crankcase stiffeners onto the oil pan.

14. Apply sealant to the proper areas of the end seals. Be sure to install the end seals with the projections in the notches.

15. Install the front and rear end seals onto the oil pan.

16. Apply a continuous bead of silicone sealant to the oil pan along the inside of the bolt holes. Overlap the sealant ends.

17. Place the oil pan into its mounting position and install the oil pan-to-engine block attaching bolts. Tighten the bolts to 69–95 inch lbs. (7.8–11.0 Nm).

NOTE: If the oil pan attaching bolts are to be reused, the old sealant must be removed from the bolt threads. Tightening the old attaching bolts with old sealant still on them may cause cracking inside the bolt holes.

18. Install the oil pan-to-transaxle attaching bolts and tighten to 27–38 ft. lbs. (37–52 Nm).

19. Install the oil drain plug and tighten to 22–30 ft. lbs. (29–41 Nm).

20. Install the exhaust front mounting flange to the exhaust manifold while making sure to install a new gasket. Tighten the mounting flange-to-exhaust manifold attaching nuts to 23–34 ft. lbs. (31–46 Nm).

21. Install the exhaust pipe support bracket and tighten the bolts to 27–38 ft. lbs. (37–52 Nm).

22. Install the splash shields. Tighten the bolts to 69–95 inch lbs. (7.8–11.0 Nm).

23. Lower the vehicle.

24. Fill the crankcase with the proper type and quantity of engine oil. Install the filler cap.

25. Connect the negative battery cable.

1.9L Engine

1. Disconnect negative cable at the battery.

2. Raise the vehicle and support safely.

3. Drain the crankcase.

4. Remove the 2 oil pan-to-transaxle bolts.

5. Disconnect the exhaust inlet pipe at the manifold and converter. Remove pipe.

6. Remove oil pan retaining bolts and oil pan.

7. Remove oil pan gasket and discard.

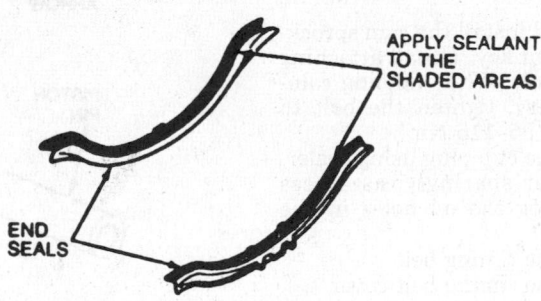

APPLY SEALANT TO THE SHADED AREAS

END SEALS

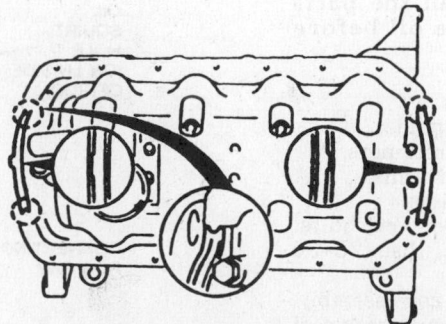

Sealant application areas – 1.8L engine

To install:

8. Clean the oil pan gasket surface and the mating surface on the cylinder block. Wipe the oil pan rail with a solvent-soaked cloth to remove oil traces.

9. Remove and clean the oil pump pick up tube and screen assembly. Install tube and screen assembly using a new gasket.

10. Apply a bead of silicone rubber sealer at the corner of the oil pan front and rear seals and at the seating point of the oil pump to the block retainer joint.

11. Install the gasket in oil pan ensuring press fit tabs are fully engaged in oil pan gasket channel.

12. Install the oil pan and the attaching bolts. Tighten the bolts lightly until the 2 oil pan-to-transmission bolts can be installed.

NOTE: **If the oil pan is installed on the engine outside of the vehicle, a transaxle case, or equivalent fixture must be bolted to the block to line the oil pan up, flush with the rear face of the block.**

13. Tighten the 2 pan-to-transaxle bolts to 30–40 ft. lbs. (40–54 Nm), then loosen ½ turn.

14. Tighten the oil pan flange-to-cylinder block bolts to 15–22 ft. lbs. (20–30 Nm) in the proper sequence. Retighten the 2 oil pan-to-transaxle bolts to 30–40 ft. lbs. (40–55 Nm).

15. Install the transaxle inspection plate.

16. Install the exhaust inlet pipe. Lower the vehicle and fill the crankcase.

17. Connect negative battery cable.

18. Start the engine and check for oil leaks.

2. Remove the timing belt and the oil pan.

3. Remove the crankshaft sprocket-to-crankshaft bolt and the sprocket.

4. Remove the oil pickup tube-to-oil pump bolts and the tube.

5. Remove the oil pump-to-engine bolts and the oil pump.

6. If necessary, remove the crankshaft seal with a prybar. Clean the seal bore.

To install:

7. Clean the gasket mounting surfaces. Clean and inspect the parts for wear and/or damage and replace, if necessary.

8. If replacing the crankshaft seal, lubricate the lip and outer surface of a new seal with clean engine oil and install, using a seal installer tool.

9. Apply a thin film of sealer to both sides of a new gasket and install the gasket and oil pump onto the cylinder block. Do not allow sealer to squeeze into the pump outlet hole in the pump or cylinder block.

10. Tighten the pump mounting bolts to 14–19 ft. lbs. (19–25 Nm).

11. Install the remaining components in the reverse order of removal.

12. Refill the crankcase with clean engine oil. Start the engine and check for leaks.

1.8L Engine

1. Disconnect the negative battery cable.

2. Remove the timing belt and crankshaft pulley.

3. Remove the oil pan.

4. Remove the oil strainer mounting bolts and remove the oil strainer and gasket.

5. If equipped, remove the air conditioning compressor mounting bolts and position the compressor so it is free from the work area.

6. Remove the air conditioning compressor mounting bracket.

7. Remove the mounting bolt from the engine oil dipstick tube bracket and remove the alternator lower mounting bolt.

8. Remove all oil pump mounting bolts and remove the oil pump. Remove all gasket material from the oil pump.

To install:

9. Install a new gasket onto the oil pump.

10. Place the oil pump into its mounting position and install the pump mounting bolts. Tighten the bolts to 14–19 ft. lbs. (19–25 Nm).

11. Place the dipstick tube bracket bolt into its mounting position and install the mounting bolt.

12. Install the alternator lower mounting bolt and tighten to 27–38 ft. lbs. (37–52 Nm).

13. Install a new gasket onto the oil strainer, place the strainer into its mounting position and install the mounting bolts. Tighten to 69–95 inch lbs. (7.8–11.0 Nm).

14. Install the oil pan.

15. If equipped, place the air conditioning compressor bracket into its mounting position and install the mounting bolts. Tighten the bolts to 15–22 ft. lbs. (20–30 Nm).

OIL PAN ASSEMBLY

BOLT TIGHTENING SEQUENCE

10 7 2 5 9

3 6 4 8 1

FRONT OF ENGINE

Oil pan bolt torque sequence—1.9L engine

Oil Pump

REMOVAL & INSTALLATION

1.6L Engine

1. Disconnect the negative battery cable.

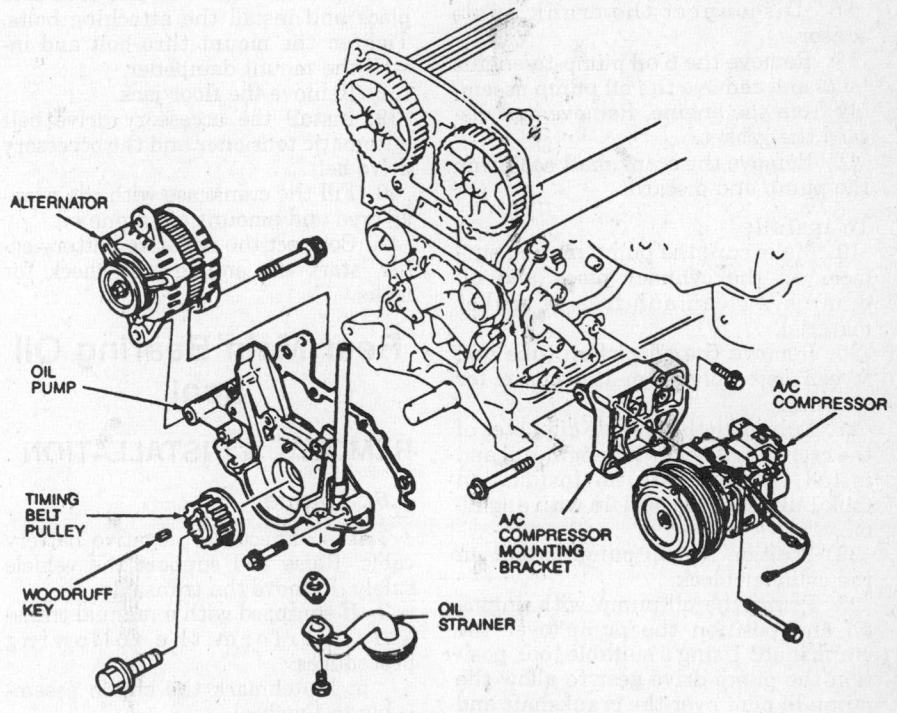

Oil pump removal and installation—1.8L engine

16. Install the crankshaft pulley and timing belt.

17. Connect the negative battery cable.

1.9L Engine

1. Disconnect the negative battery cable.

2. Remove the accessory drive belt and the automatic tensioner.

3. Support the engine with a floor jack.

4. Remove the right engine mount dampener and remove the right engine mount bolts from the mount bracket.

5. Loosen the mount thru-bolt and roll the mount aside.

6. Remove the timing belt cover.

7. Make sure the No. 1 cylinder is at TDC.

8. Roll the engine mount back into place and install the 2 mount bolts. Remove the floor jack.

9. Loosen the belt tensioner attaching bolt and pry the tensioner to the rear of the engine. Tighten the attaching bolt.

10. Raise and safely support the vehicle.

11. Remove the right side splash shield.

12. Remove the catalytic converter inlet pipe.

13. Drain and remove the oil pan. Remove the oil filter.

14. Remove the crankshaft dampener and the timing belt.

15. Remove the crankshaft sprocket and the timing belt guide from the crankshaft.

16. Disconnect the crank angle sensor.

17. Remove the 6 oil pump-to-engine bolts and remove the oil pump assembly from the engine. Remove and discard the gasket.

18. Remove the crankshaft seal from the pump and discard.

To install:

19. Make sure the pump mating surfaces on the cylinder block and oil pump are clean and free of gasket material.

20. Remove the oil pickup tube and screen assembly from the pump for cleaning.

21. Lubricate the outside diameter of the crankshaft seal with engine oil and install the seal with an installation tool. Lubricate the seal lip with engine oil.

22. Position the oil pump gasket on the cylinder block.

23. Prime the oil pump with engine oil and position the pump over the crankshaft. Using a suitable tool, position the pump drive gear to allow the pump to pilot over the crankshaft and seat firmly on the cylinder block.

NOTE: The pump drive gear can be accessed through the oil pickup hole in the body of the pump. Do not install the oil pump pickup tube and screen until the pump has been correctly installed on the cylinder block.

24. Install the 6 oil pump bolts and tighten to 8–12 ft. lbs. (11–16 Nm).

NOTE: When the oil pump bolts are tightened, the gasket must not be below the cylinder block sealing surface.

25. Install the pickup tube and screen assembly on the oil pump using a new gasket. Tighten the attaching screws to 7–9 ft. lbs. (10–13 Nm).

26. Install the timing belt guide over the end of the crankshaft and install the crankshaft sprocket.

27. Make sure the No. 1 cylinder is at TDC.

28. Position the timing belt over the sprockets.

29. Connect the crank angle sensor.

30. Install the oil pan and the crankshaft dampener.

31. Install the catalytic converter inlet pipe.

32. Install the splash shield and lower the vehicle.

33. Install the timing belt. Tighten the tensioner attaching bolt to 17–22 ft. lbs. (23–30 Nm).

34. Support the engine with a floor jack.

35. Remove the right engine mount bolts and roll the mount back.

36. Install the timing belt cover.

37. Roll the engine mount back into place and install the attaching bolts. Tighten the mount thru-bolt and install the mount dampener.

38. Remove the floor jack.

39. Install the accessory drive belt automatic tensioner and the accessory drive belt.

40. Fill the crankcase with the proper type and amount of engine oil.

41. Connect the negative battery cable, start the engine and check for leaks.

Rear Main Bearing Oil Seal

REMOVAL & INSTALLATION

1.6L Engine

1. Disconnect the negative battery cable. Raise and support the vehicle safely. Remove the transaxle.

2. If equipped with a manual transaxle, perform the following procedures:

   a. Matchmark the clutch assembly-to-flywheel.

   b. Remove the pressure plate-to-

flywheel bolts, evenly, a little at a time and the clutch assembly.

3. Remove the flywheel-to-crankshaft bolts, the flywheel and the spacer plates (automatic transaxle).

4. If necessary, remove the rear engine plate-to-engine bolts and the plate.

5. Remove the rear oil seal retainer-to-engine bolts, the oil pan-to-rear oil seal retainer bolts and the retainer.

6. Press the oil seal from the rear oil seal retainer.

To install:

7. Clean the gasket mounting surfaces. Using a clean shop cloth, clean the oil seal bore.

8. Using a new oil seal, lubricate it with clean engine oil and press it into the retainer until it seats.

9. Install the seal retainer with a new gasket and tighten the bolts to 69–95 inch lbs. (8–11 Nm). Trim excess gasket material from the retainer after installation.

10. Install the rear engine plate and tighten the bolts to 69–95 inch lbs. (8–11 Nm).

11. Install the remaining components in the reverse order of removal. Torque the flywheel-to-crankshaft bolts to 71–76 ft. lbs. (96–103 Nm).

1.8L Engine

1. Disconnect the negative battery cable.

2. Remove the transaxle assembly.

3. If equipped with manual transaxle, remove the clutch assembly and the flywheel. If equipped with automatic transaxle, remove the flexplate.

4. If necessary, remove the rear cover mounting bolts and remove the rear cover.

5. Using a prybar, remove the crankshaft rear oil seal.

To install:

6. If removed, install the rear cover and attaching bolts. Tighten the bolts to 69–95 inch lbs. (7.8–11.0 Nm).

7. Lubricate the lip of a new seal and install, using an installation tool.

8. Install the flywheel and clutch assembly or flexplate.

9. Install the transaxle.

10. Connect the negative battery cable.

1.9L Engine

1. Disconnect the negative battery cable.

2. Raise the vehicle and support it safely. Remove the transaxle.

3. Remove flywheel and the engine cover plate.

4. With a prybar, remove the oil seal.

NOTE: Use caution to avoid damaging the oil seal surface.

To install:

5. Inspect the crankshaft seal area for any damage which may cause the seal to leak. If damage is evident, service or replace the crankshaft, as necessary.

6. Coat the crankshaft seal area and the seal lip with engine oil.

7. Using a seal installer tool, install the seal.

8. Install the engine cover plate and the flywheel. Tighten the flywheel attaching bolts to 54–64 ft. lbs. (73–87 Nm).

9. Install the transaxle assembly.

10. Connect the negative battery cable, start the engine and check for leaks.

ENGINE COOLING

Radiator

REMOVAL & INSTALLATION

1988–90

1. Disconnect the negative battery cable.

2. Disconnect the coolant recovery hose from the filler neck.

3. Loosen the retaining clamp and disconnect the upper radiator hose from the radiator.

4. Disconnect the cooling fan wiring harness connector. Disengage the wiring harness from the routing clamp on the cooling fan shroud.

5. Remove the radiator cap from the filler neck. Raise the vehicle and support safely.

6. Position a fluid catch pan under the radiator. Open the drain valve and drain the cooling system.

7. Disconnect the radiator temperature switch wires. If equipped with automatic transaxle, disconnect and plug the oil cooler lines.

8. Loosen the retaining clamp and disconnect the lower radiator hose.

9. Lower the vehicle.

10. Support the radiator by hand and remove the 4 bolts attaching the radiator support brackets to the vehicle body. Raise the radiator/cooling fan/shroud assembly from the crossmember mounting insulator supports and remove from vehicle. Disconnect the cooling fan and shroud from the radiator as required.

To install:

11. Lower the radiator into the normal operating position making certain the mounting insulators engage with their supports. Attach the radiator to the support brackets with the 4 bolts.

12. Connect the upper radiator hose. Raise the vehicle and connect the lower radiator hose and temperature switch wires. If equipped, connect the transaxle oil cooler lines.

13. Close the drain valve and lower the vehicle. Connect the negative battery cable. Fill the cooling system and connect the cooling fan harness connector.

14. Start the engine and allow to reach normal operating temperature. Inspect for coolant leaks and correct as required.

1991–92

1. Disconnect the negative battery cable.

2. Raise and safely support the vehicle. Drain the cooling system.

3. Remove the right side and front splash shields and remove the lower radiator hose.

4. If equipped with automatic transaxle, remove the lower oil cooler line from the radiator. Remove the oil cooler line brackets from the bottom of the radiator.

5. Lower the vehicle.

6. If equipped with automatic transaxle and air conditioning, remove the seal located between the radiator and fan shroud.

7. If equipped with automatic transaxle, remove the upper oil cooler line from the radiator.

8. If equipped with 1.8L engine, remove the resonance duct from the radiator isomounts.

9. Disconnect the cooling fan motor electrical connector and the cooling fan thermoswitch electrical connector.

10. Remove the 3 fan shroud attaching bolts and remove the shroud assembly by pulling it straight up.

11. Remove the upper radiator hose and the 2 upper radiator isomounts. Remove the radiator by lifting it straight up.

To install:

12. Make sure the radiator lower isomounts are installed over the bolts on the radiator support.

13. Position the radiator to the radiator support, making sure the radiator lower brackets are positioned properly on the lower isomounts.

14. Install the radiator upper isomounts, making sure the radiator locating pegs are positioned correctly. Install the upper radiator hose.

15. Lower the cooling fan shroud assembly into place and install the 3 shroud attaching bolts.

16. Connect the cooling fan motor electrical connector and thermoswitch electrical connector.

17. If equipped with 1.8L engine, install the resonance duct on the radiator isomounts.

18. Install the upper oil cooler line on the radiator.

19. If equipped with automatic transaxle and air conditioning, install the seal between the radiator and fan shroud.

20. Raise and safely support the vehicle. Install the lower oil cooler line on the radiator.

21. Install the lower radiator hose and install the right side and front splash shields.

22. Lower the vehicle and fill the cooling system.

23. Connect the negative battery cable, start the engine and check for coolant leaks.

Heater Core

REMOVAL & INSTALLATION

1988–90

1. Disconnect the negative battery cable.

2. Remove the instrument panel as follows:

a. From under the instrument panel, remove both sound deadening panels and the lap duct register panel.

b. From the blower motor case and heater case, disconnect the 3 air door control cables.

c. From behind the instrument cluster, depress the speedometer lock tab and pull the speedometer cable from the cluster.

d. From behind the instrument cluster, depress the lock tab (located in the center of the connector) of the 3 electrical harness connectors and pull the connectors from the cluster.

e. From under the steering column, remove the lap duct brace-to-instrument panel screws, the brace, the lap duct and the driver's demister tube.

f. Remove the lower cover-to-steering column screws and the lower cover.

g. Remove the steering column-to-instrument panel bolts and lower the steering column.

h. Remove the glove box-to-instrument panel screws and the glove box.

i. Remove the hood release-to-instrument panel nut and move the release cable aside.

j. Remove the center floor console-to-chassis screws and the cover.

k. From below the radio, remove the lower trim panel-to-instrument panel screws and the lower panel.

l. Using a small prybar, pry the instrument panel-to-chassis bolt covers from the perimeter of the instrument panel. Remove the instrument panel-to-chassis nuts/bolts. Lift and pull the panel out slightly.

m. Disconnect the electrical connector from the blower motor assembly.

n. From the rear of the radio, disconnect the antenna cable.

o. From the left corner of the instrument panel, disconnect the 3 instrument panel harness connectors and remove the instrument panel.

3. Using a clean drain pan, place it under the radiator, open the radiator drain cock, remove the radiator cap and drain the cooling system to a level below the heater case.

4. Disconnect and plug the heater hoses at the heater case.

5. Remove the defroster tubes-to-heater case push pins and the defroster tubes from the heater case. Remove the main air duct-to-heater case push pins and the main air duct.

6. From under the heater case, remove the lower carpet panel push pins, screw and the panel.

7. From the heater case, disconnect the electrical harness braces and remove the lower brace screws and brace.

8. Remove the heater case-to-chassis nuts and bolts. Remove the lower duct-to-heater case push pins and lower duct. Remove the heater case by pulling it straight out; be careful not to damage the extension tubes.

9. Remove the heater core cover-to-heater case screws and the cover. Re-move the tube braces and pull the heater core straight out.

10. Remove the outlet extension tube clip and tube. Loosen the inlet extension tube clamp and the extension tube.

To install:

11. Install a new O-ring onto the outlet extension tube, position the tube into the heater core tube and install the clip.

12. Install the inlet extension tube to the heater core inlet and tighten the clamps.

13. Slide the heater core into the heater case and install the tube braces and the heater core cover.

14. Install the heater case onto the mounting studs, being careful not to damage the heater core extension tubes. Install the mounting nuts and bolts.

15. Install the lower duct to the heater case with the 2 push pins and install the lower brace with the 3 attaching screws.

16. Reroute and connect all wiring harness braces to the heater case. Install the lower carpet panel with the screw and 2 push pins.

17. Install the defroster tube and the main air duct with the push pins.

18. Make sure the rubber grommets for the extension tubes are still in place on the engine side of the bulkhead. Uncap the extension tubes and connect the heater hoses. Tighten the clamps.

19. Install the instrument panel in the reverse order of removal.

20. Fill the cooling system, connect the negative battery cable, start the engine and check for coolant leaks and proper heater system operation.

1991–92

1. Disconnect the negative battery cable and drain the cooling system.

2. Disconnect the heater hoses at the bulkhead.

3. Remove the instrument panel as follows:

a. Remove the 4 bolts securing the steering column to the instrument panel frame. Lower the steering column.

b. Remove the cap screws securing the instrument cluster bezel to the instrument panel and remove the instrument cluster bezel.

c. Disconnect the speedometer cable at the transaxle by pulling the cable out of the vehicle speed sensor.

d. Disconnect the screws and bolts securing the instrument cluster to the instrument panel. Pull the instrument cluster out slightly and disconnect the electrical connectors from the rear of the instrument cluster.

e. Disconnect the speedometer cable from the instrument cluster. Remove the instrument cluster from the instrument panel.

f. Detach the hood release cable from the left lower dash trim panel. Carefully pry out both dash side panels.

g. Remove the 4 retaining screws and the left lower dash trim panel. Disconnect all necessary electrical connectors.

h. Remove the 2 hinge-to-instrument panel retaining screws and remove the glove compartment.

i. Remove the climate control assembly and the ash tray.

j. Remove the 7 accessory console retaining screws. Disconnect the radio antenna, radio wire connectors and cigarette lighter connector.

k. Remove the retaining screws and the right lower dash trim panel. Disconnect the 3 amplifier wire connectors.

l. Remove the 4 bolts attaching the instrument panel frame to the floor pan. Remove the bolt from both of the lower instrument panel mounts.

m. Remove the 2 bolts from both of the upper instrument panel mounts. Remove the retaining screw and the defroster duct bezel.

n. Remove the 3 mounting bolts that attach the upper instrument panel to the cowl and remove the instrument panel from the vehicle.

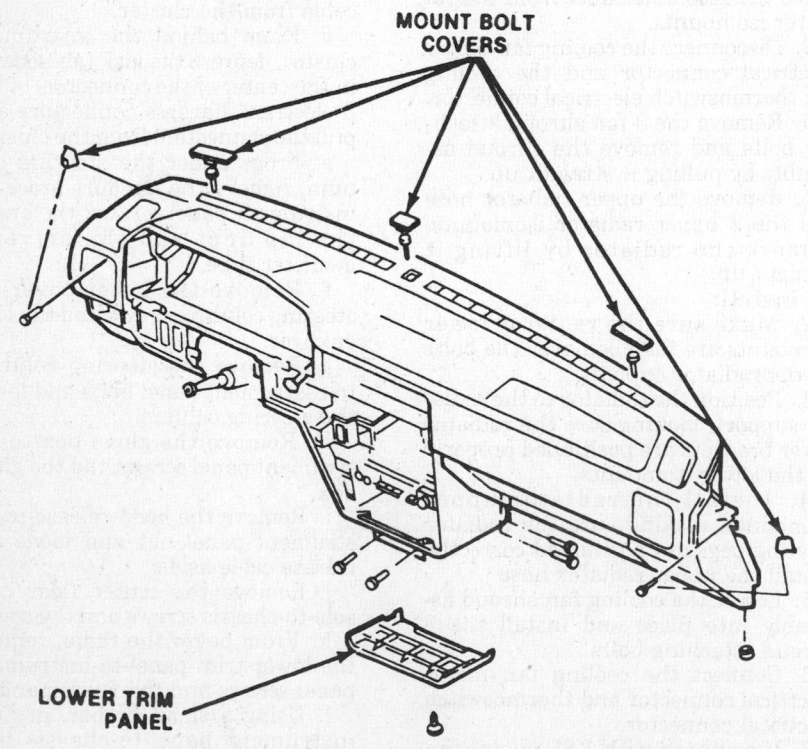

MOUNT BOLT COVERS

LOWER TRIM PANEL

Instrument panel mount bolt cover locations—1988-90 vehicles

NOTE: Use care to prevent any damage to the instrument panel or the surrounding interior trim.

4. Disconnect the mode selector and temperature control cables from the cams and retaining clips.

5. Remove the necessary defroster duct screws and loosen the capscrew that secures the heater-to-blower clamp.

6. Remove the 3 heater unit mounting nuts and disconnect the antenna lead from the retaining clip. Remove the heater unit.

7. Remove the insulator and the 4 brace capscrews. Remove the brace.

8. Remove the heater core from the heater unit.

To install:

9. Install the heater core into the heater unit and install the brace.

10. Install the brace capscrews and the insulator.

NOTE: If a new heater unit is being installed, save the keys that are found on the new unit for mode selector and temperature control cable adjustment.

11. Position the heater unit and attach the defroster and floor ducting. Install the heater unit mounting nuts.

12. Tighten the heater-to-blower clamp capscrew and install the defroster duct screws. Connect the antenna lead to the retaining clip.

13. Install the instrument panel by reversing the removal procedure.

14. Connect and adjust the mode selector and temperature control cables. Connect the heater hoses at the bulkhead.

15. Fill the cooling system and connect the negative battery cable. Start the engine and check for leaks. Check the coolant level and fill as necessary.

Water Pump

REMOVAL & INSTALLATION

1.6L Engine

1. Disconnect the negative battery cable. Remove the timing belt.

2. Place a clean drain pan under the radiator. Remove the radiator drain plug and the radiator cap; drain the cooling system to a level below the water pump.

3. Disconnect the lower radiator hose from the water pump inlet. Remove the coolant inlet pipe-to-water pump bolts and the inlet pipe.

4. Remove the water pump-to-engine bolts and the water pump.

5. Clean the gasket mounting surfaces. Inspect the parts for wear and/or damage and replace, if necessary.

To install:

6. To install, use new gaskets, seal-

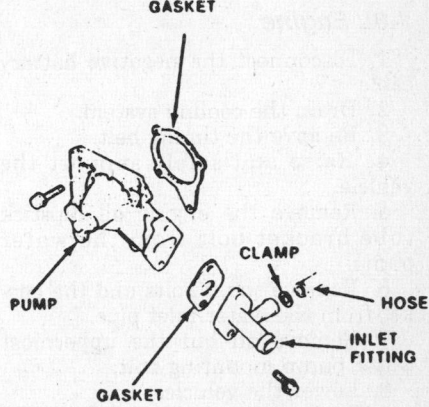

Water pump assembly—1.6L engine

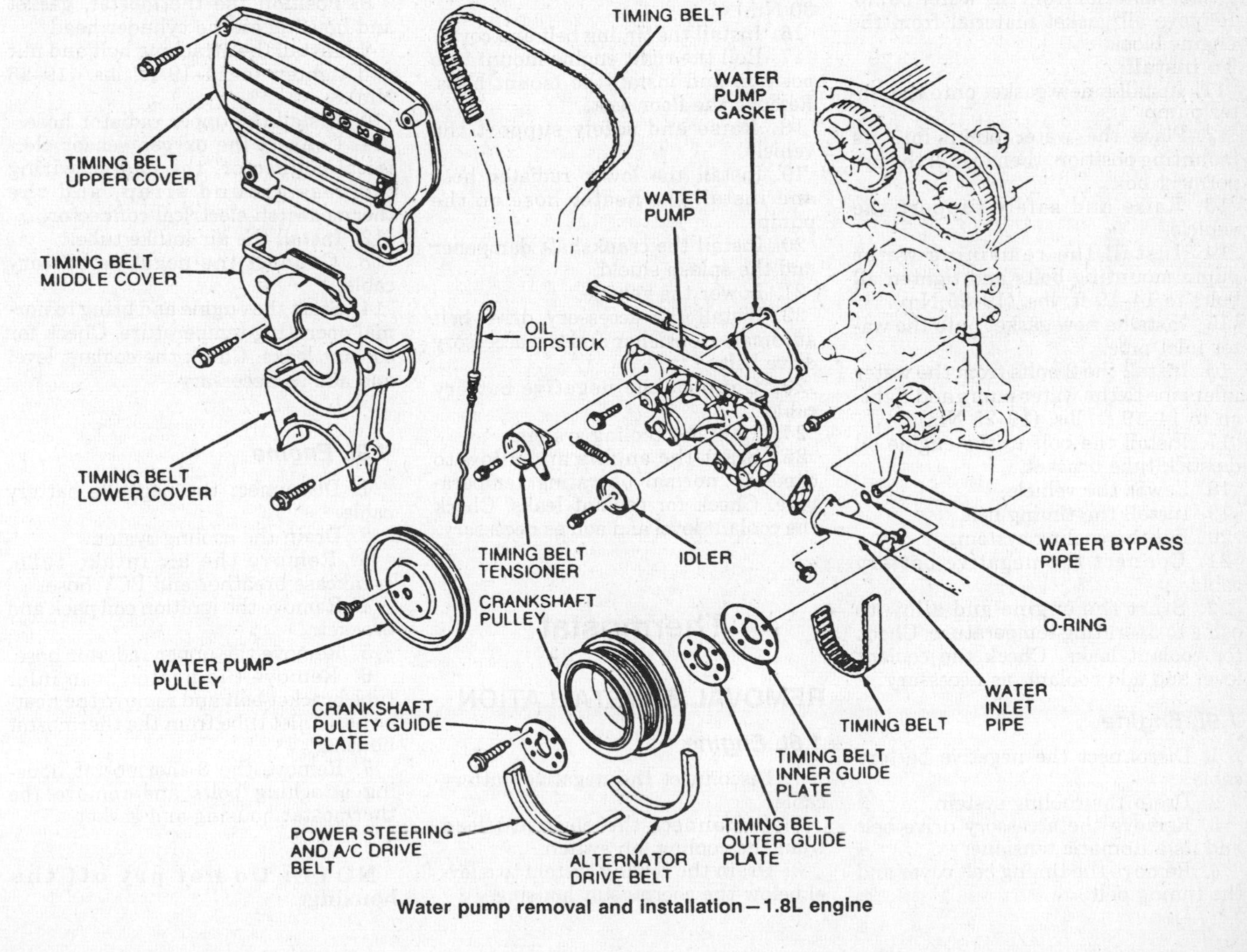

Water pump removal and installation—1.8L engine

ant and reverse the removal procedure.

7. Torque the water pump-to-engine bolts to 14–19 ft. lbs. (19–25 Nm), the water inlet pipe-to-water pump bolts to 14–19 ft. lbs. (19–25 Nm) and the water pump pulley-to-water pump bolts to 11–13 ft. lbs. (15–18 Nm).

8. Refill the cooling system. Start the engine, allow it to reach normal operating temperature and check for leaks.

1.8L Engine

1. Disconnect the negative battery cable.
2. Drain the cooling system.
3. Remove the timing belt.
4. Raise and safely support the vehicle.
5. Remove the engine oil dipstick tube bracket bolt from the water pump.
6. Remove the 2 bolts and the gasket from the water inlet pipe.
7. Remove all but the uppermost water pump mounting bolt.
8. Lower the vehicle.
9. Remove the remaining bolt and the water pump assembly.
10. If it is being reused, remove all gasket material from the water pump. Remove all gasket material from the engine block.

To install:

11. Install a new gasket onto the water pump.
12. Place the water pump into its mounting position, then install the uppermost bolt.
13. Raise and safely support the vehicle.
14. Install the remaining water pump mounting bolts and tighten all bolts to 14–19 ft. lbs. (19–25 Nm).
15. Install a new gasket onto the water inlet pipe.
16. Install the 2 bolts from the water inlet pipe to the water pump and tighten to 14–19 ft. lbs. (19–25 Nm).
17. Install the bolt to the engine oil dipstick tube bracket.
18. Lower the vehicle.
19. Install the timing belt.
20. Fill the cooling system.
21. Connect the negative battery cable.
22. Start the engine and allow to come to operating temperature. Check for coolant leaks. Check the coolant level and add coolant, as necessary.

1.9L Engine

1. Disconnect the negative battery cable.
2. Drain the cooling system.
3. Remove the accessory drive belt and its automatic tensioner.
4. Remove the timing belt cover and the timing belt.

5. Raise and safely support the vehicle.
6. Remove the lower radiator hose and remove the heater hose from the water pump.
7. Lower the vehicle.
8. Support the engine with a floor jack.
9. Remove the right engine mount attaching bolts and roll the engine mount aside.
10. Remove the water pump attaching bolts.
11. Using the floor jack, raise the engine enough to provide clearance for removing the water pump.
12. Remove the water pump and the gasket from the engine through the top of the engine compartment.

To install:

13. Make sure the mating surfaces of the cylinder block and water pump are clean and free of gasket material.
14. If the water pump is to be replaced, transfer the timing belt tensioner components to the new water pump.
15. With the engine supported and raised with a floor jack, place the water pump and the gasket on the cylinder block and install the 4 attaching bolts. Tighten the bolts to 15–22 ft. lbs. (20–30 Nm).
16. Install the timing belt and cover.
17. Roll the right engine mount into position and install the mount bolts. Remove the floor jack.
18. Raise and safely support the vehicle.
19. Install the lower radiator hose and install the heater hose on the pump.
20. Install the crankshaft dampener and the splash shield.
21. Lower the vehicle.
22. Install the accessory drive belt automatic tensioner and the accessory drive belt.
23. Connect the negative battery cable.
24. Refill the cooling system.
25. Start the engine and allow to come to normal operating temperature. Check for coolant leaks. Check the coolant level and add as necessary.

Thermostat

REMOVAL & INSTALLATION

1.6L Engine

1. Disconnect the negative battery cable.
2. Disconnect the electrical lead from the cooling fan switch.
3. Drain the cooling system to a level below the thermostat housing.

4. Disconnect the upper radiator hose from the housing.
5. Remove the thermostat housing bolts and remove the thermostat.
6. Clean the cylinder and housing mating surfaces.
7. Installation is the reverse of the removal procedure. Torque the housing bolts to 14–22 ft. lbs. (19–30 Nm).

1.8L Engine

1. Disconnect the negative battery cable.
2. Drain the cooling system.
3. Remove the air intake tube.
4. Disconnect the water thermoswitch connector, the engine wiring harness ground strap from the connector above the housing, and the exhaust gas oxygen sensor electrical connector.
5. Remove the upper radiator hose from the housing.
6. Remove the thermostat housing attaching bolt and nut and remove the housing. Remove the gasket and the thermostat.

To install:

7. Clean the thermostat housing and cylinder head gasket surfaces.
8. Position the thermostat, gasket and housing on the cylinder head.
9. Install the attaching bolt and nut and tighten to 14–19 ft. lbs. (19–26 Nm).
10. Install the upper radiator hose.
11. Connect the oxygen sensor electrical connector, the engine wiring harness ground strap, and the thermoswitch electrical connector.
12. Install the air intake tube.
13. Connect the negative battery cable.
14. Start the engine and bring to normal operating temperature. Check for coolant leaks. Check the coolant level and add as necessary.

1.9L Engine

1. Disconnect the negative battery cable.
2. Drain the cooling system.
3. Remove the air intake tube, crankcase breather and PCV hose.
4. Remove the ignition coil pack and bracket.
5. Remove the upper radiator hose.
6. Remove the heater hose inlet tube bracket bolt and remove the heater hose inlet tube from the thermostat housing.
7. Remove the 3 thermostat housing attaching bolts and remove the thermostat housing and gasket.

NOTE: Do not pry off the housing.

8. Remove the thermostat and the rubber seal from the housing.

To install:

9. Clean the thermostat housing pocket and cylinder head mating surfaces.

10. Place the thermostat into position, fully inserted to compress the rubber seal inside the housing.

NOTE: Make sure the thermostat tabs engage properly into the housing slots.

11. Position the thermostat housing and gasket on the cylinder head.

12. Install the 3 attaching bolts and tighten to 8.0–11.5 ft. lbs. (11–15 Nm).

13. Install the heater hose inlet pipe and the heater hose inlet pipe bracket bolt.

14. Install the upper radiator hose.

15. Install the ignition coil and bracket.

16. Install the crankcase breather, PCV hoses and the air intake tube.

17. Connect the negative battery cable.

18. Refill the cooling system.

19. Start the engine and bring to normal operating temperature. Check for coolant leaks. Check the coolant level and add as necessary.

Cooling System Bleeding

When the entire cooling system is drained, the following procedure should be used to ensure a complete fill.

1. Install the block drain plug, if removed, and close the draincock. With the engine off, add anti-freeze to the radiator to a level of 50 percent of the total cooling system capacity. Then add water until it reaches the radiator filler neck seat.

2. Install the radiator cap to the first notch to keep spillage to a minimum.

3. Start the engine and let it idle until the upper radiator hose is warm. This indicates that the thermostat is open and coolant is flowing through the entire system.

4. Carefully remove the radiator cap and top off the radiator with water. Install the cap on the radiator securely.

5. Fill the coolant recovery reservoir to the FULL COLD mark with anti-freeze, then add water to the FULL HOT mark. This will ensure that a proper mixture is in the coolant recovery bottle.

6. Check for leaks at the draincock and, if removed, the block drain plug.

ENGINE ELECTRICAL

NOTE: Disconnecting the negative battery cable on some vehicles may interfere with the functions of the on board computer systems and may require the computer to undergo a relearning process, once the negative battery cable is reconnected.

Distributor

REMOVAL

1. Disconnect the negative battery cable.

2. Disconnect the coil wire, remove the distributor cap screws and move the cap (wires attached) aside. Remove the gasket, if equipped.

3. Tag and disconnect the vacuum hoses, if equipped.

4. Disconnect the electrical connector(s) from the distributor.

5. Mark the position of the distributor on the cylinder head and the rotor to the distributor housing.

6. Remove the distributor mounting bolts and remove the distributor from the engine. Remove and discard the O-ring from the distributor.

INSTALLATION

Timing Not Disturbed

1. Lubricate a new O-ring with clean engine oil and install on the distributor.

2. Install the distributor, aligning the marks that were made during the removal procedure. Make sure the distributor drive tangs align with the camshaft slot.

3. Install the distributor mounting bolts and connect the wiring to the distributor.

4. Connect the vacuum hoses, if equipped.

5. Install the distributor cap gasket, if equipped. Install the distributor cap and tighten the screws.

6. Connect the coil wire to the distributor cap and connect the negative battery cable. Check the ignition timing.

Timing Disturbed

1. If the crankshaft was rotated while the distributor was removed, the engine must be brought to TDC on the compression stroke of the No. 1 cylinder.

2. Remove the No. 1 spark plug. Place a finger over the hole and rotate the crankshaft slowly in the direction

of normal rotation, until engine compression is felt.

NOTE: Turn the engine only in the direction of normal rotation. Backward rotation may cause the cam belt to slip or lose teeth, altering engine timing.

3. When engine compression is felt at the spark plug hole, indicating that the piston is approaching TDC, continue to turn the crankshaft until the timing mark on the pulley is aligned with the **0** mark on the engine front cover.

4. Turn the distributor shaft until the rotor aligns with the No. 1 spark plug wire on the distributor cap.

5. Place the distributor in the cylinder head, seating the distributor drive tangs in the camshaft slot.

6. Install the distributor mounting bolts and tighten them so the distributor can just barely be moved.

7. Install the rotor and distributor cap and connect all wiring.

8. Set initial timing.

9. After timing has been set, tighten the distributor mounting bolts.

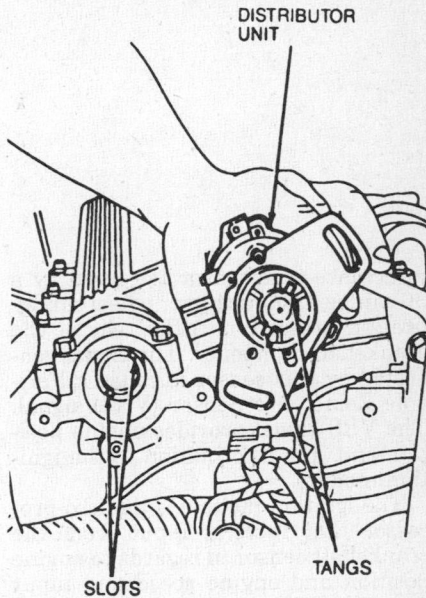

Distributor drive tang and camshaft slot locations

Distributorless Ignition System

The 1.9L engine is equipped with a Distributorless Ignition System (DIS). The DIS consists of the following components: crankshaft sensor, ignition module, ignition coil pack, the spark angle portion of the ECU and the related wiring.

The crankshaft sensor is a variable

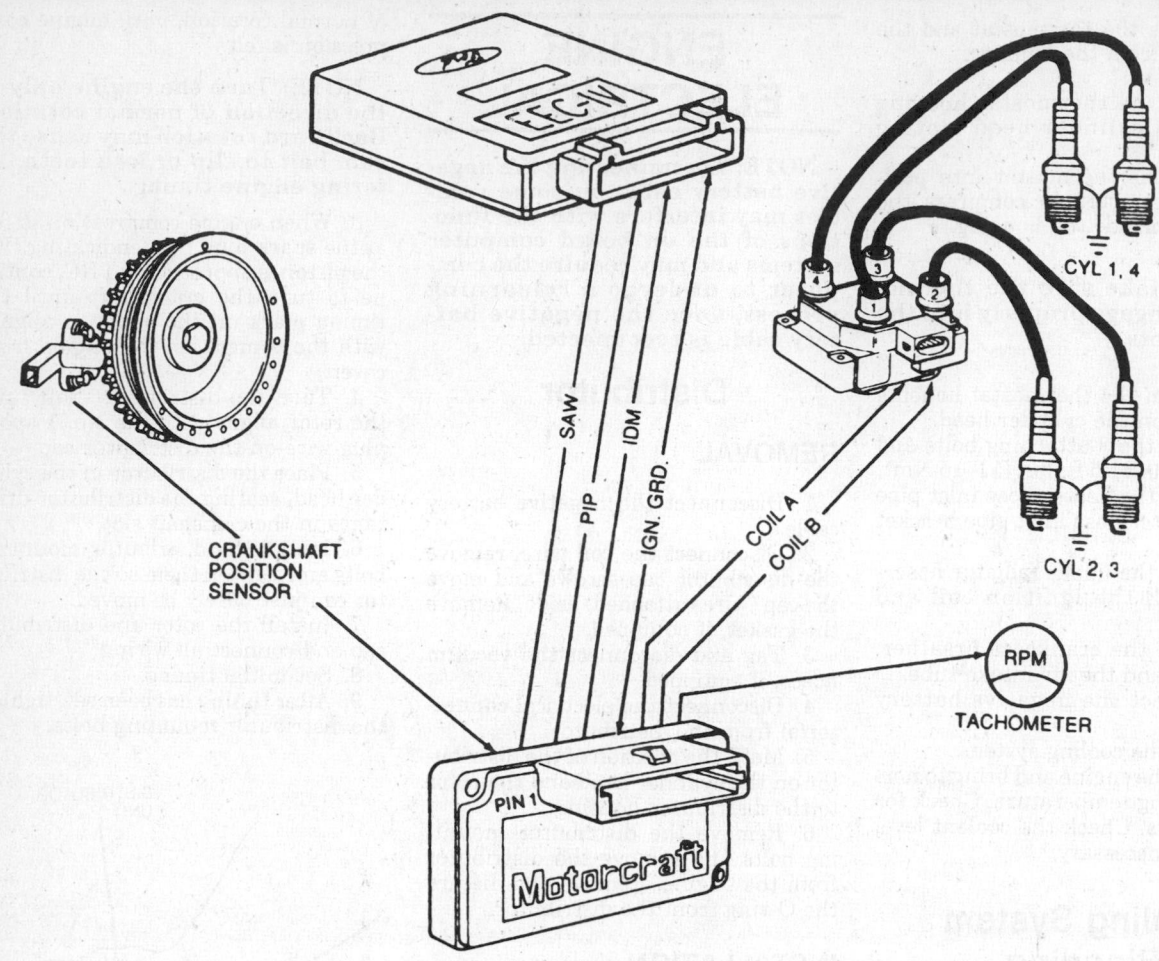

Distributorless ignition system — 1.9L engine

reluctance-type sensor triggered by a 36-minus-1 tooth trigger wheel configuration pressed onto the rear of the crankshaft dampener. The signal generated by this sensor is called a Variable Reluctance Sensor (VRS) signal. The VRS signal provides engine position and rpm information to the ignition module.

The ignition module is a micro-processor that receives input from the crankshaft sensor in regards to engine position and engine speed and input from the ECU pertaining to spark advance. The ignition module uses this information to direct which coil to fire and to calculate the turn on and turn off times of the coils required to achieve the correct dwell and spark advance.

The ignition coil pack contains 2 separate ignition coils. Each ignition coil fires 2 cylinders simultaneously. When 1 cylinder is firing on the compression stroke, the other is firing on the exhaust stroke. During the next engine revolution, the reverse occurs. The spark plug fired on the exhaust stroke uses very little of the ignition coil's stored energy; the majority of the energy is used by the spark plug on the compression stroke. Since these 2 spark plugs are connected in series, the firing voltage of 1 plug will be negative with respect to ground, while the voltage of the other will be positive with respect to ground.

REMOVAL & INSTALLATION

Crankshaft Sensor

1. Disconnect the negative battery cable.
2. Raise and safely support the vehicle.
3. Remove the right side splash shield.
4. Disconnect the sensor electrical connector from the wiring harness.
5. Remove the crankshaft sensor mounting screws and remove the sensor.
6. Installation is the reverse of the removal procedure. Tighten the sensor attaching screws to 40–61 inch lbs. (5–7 Nm).

Ignition Module

NOTE: The ignition module is

located on the left side of the engine compartment, in front of the left strut tower.

1. Disconnect the negative battery cable.
2. Remove the 3 module sub-bracket attaching nuts.
3. Gently pull the sub-bracket and module assembly straight up and disconnect the module electrical harness.
4. Remove the 2 module attaching screws from the sub-bracket. Remove the ignition module from the sub-bracket.
5. Installation is the reverse of the removal procedure. Tighten the module attaching screws to 24–35 inch lbs. (3–4 Nm). Tighten the sub-bracket attaching nuts to 62–88 inch lbs. (7–10 Nm).

Ignition Coil Pack

1. Disconnect the negative battery cable.
2. Disconnect the electrical connector from the coil pack.
3. Remove the spark plug wires by squeezing the locking tabs to release the coil boot retainers. Tag the wires

and mark their position on the coil pack prior to removal.

4. Remove the coil pack attaching bolts and remove the coil pack.

NOTE: Save the capacitor for installation with the new coil pack.

5. Installation is the reverse of the removal procedure. Tighten the attaching bolts to 40–62 inch lbs. (5–7 Nm).

Ignition Timing

ADJUSTMENT

1.6L Engine

1. Operate the engine until normal operating temperatures are reached.

2. Check and/or adjust the idle speed.

3. Turn off all of the accessories.

4. Disconnect and plug the vacuum lines at the distributor diaphragm. Mark the lines for correct installation.

5. Disconnect the black electrical connector at the distributor.

6. Connect a timing light to the engine according to the manufacturers instructions.

7. Aim the timing light at the crankshaft pulley/timing plate location and check that the timing marks line up. The timing should be 2 degrees ± 1 degree BTDC.

8. If necessary to adjust the ignition timing, perform the following procedures:

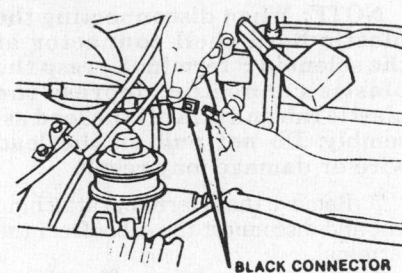

BLACK CONNECTOR

Black distributor connector location— 1.6L engine

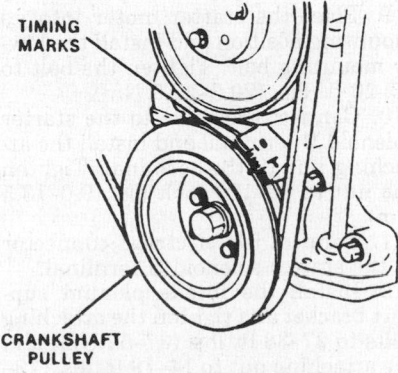

TIMING MARKS

CRANKSHAFT PULLEY

Ignition timing marks location

a. Loosen the distributor hold-down bolts, just enough so the distributor can be turned.

b. Rotate the distributor clockwise (to advance) or counterclockwise (to retard) the timing.

c. With the timing corrected, tighten the distributor hold-down bolts.

d. Recheck the timing.

9. Connect the vacuum lines and check the timing. The timing should be 7 degrees ± 1 degree with the vacuum hoses connected.

10. Connect the black electrical connector and remove the timing light.

1.8L Engine

1. Apply the parking brake and make sure the vehicle is in **N** or **P**.

2. Start the engine and warm it up to normal operating temperature.

3. Turn off all electrical loads and accessories.

4. Connect a timing light according to the manufacturers instructions.

5. Using a jumper wire, connect the **GROUND** terminal to the **TEN** terminal on the diagnosis connector.

6. Connect the positive lead of a tachometer to the **IG** terminal on the diagnosis connector and connect the negative lead to the negative battery post.

7. Using the timing light, inspect

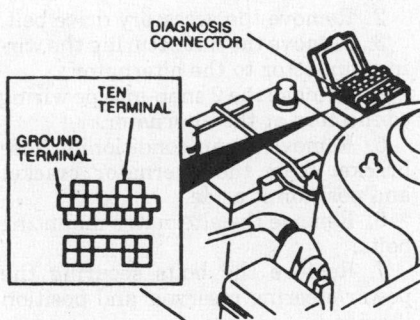

DIAGNOSIS CONNECTOR

TEN TERMINAL

GROUND TERMINAL

Diagnosis connector location—1.8L engine

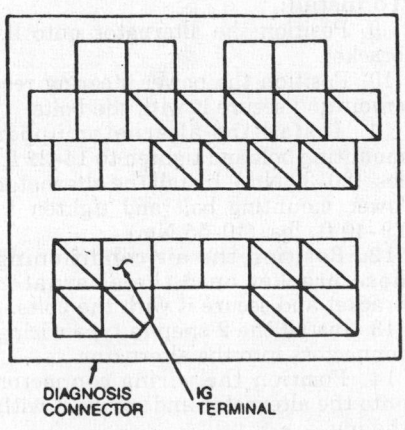

DIAGNOSIS CONNECTOR   **IG TERMINAL**

IG terminal location—1.8L engine

the ignition timing. The ignition timing should be 9–11 degrees at 700–800 rpm. The yellow mark on the crankshaft pulley should be aligned with the corresponding mark on the timing belt cover.

8. If the marks are not aligned, loosen the distributor mounting bolts and turn the distributor until the ignition timing is within specification.

9. Tighten the distributor mounting bolts to 14–19 ft. lbs. (19–25 Nm).

10. Remove the jumper wire from the diagnosis connector and remove the timing light and tachometer.

1.9L Engine

The initial timing on 1.9L engine with distributorless ignition system is fixed at 10 degrees ± 2 degrees and is not adjustable.

Alternator

PRECAUTIONS

Several precautions must be observed with alternator equipped vehicles to avoid damage to the unit.

• If the battery is removed for any reason, make sure it is reconnected with the correct polarity. Reversing the battery connections may result in damage to the one-way rectifiers.

• When utilizing a booster battery as a starting aid, always connect the positive to positive terminals and the negative terminal from the booster battery to a good engine ground on the vehicle being started.

• Never use a fast charger as a booster to start vehicles.

• Disconnect the battery cables when charging the battery with a fast charger.

• Never attempt to polarize the alternator.

• Do not use test lights of more than 12 volts when checking diode continuity.

• Do not short across or ground any of the alternator terminals.

• The polarity of the battery, alternator and regulator must be matched and considered before making any electrical connections within the system.

• Never separate the alternator on an open circuit. Make sure all connections within the circuit are clean and tight.

• Disconnect the battery ground terminal when performing any service on electrical components.

• Disconnect the battery if arc welding is to be done on the vehicle.

BELT TENSION ADJUSTMENT

1.6L and 1.8L Engines

1. Loosen the alternator adjusting bolt.
2. Raise and safely support the vehicle.
3. Loosen the alternator mounting bolt.

NOTE: Do not pry against the stator frame. Position the prybar against a stronger point, such as the area around a case bolt.

4. Position a belt tension gauge on the longest accessible span of belt and tighten the belt. Adjust the tension to 85.8–103.4 lbs. for a new belt or 68.2–85.8 lbs. for a used belt.
5. If a belt tension gauge is not available, adjust the tension to 0.31–0.35 in. (8–9mm) deflection, midway between the pulleys, for a new belt or 0.35–0.39 in. (9–10mm) deflection for a used belt.
6. Tighten the alternator adjusting bolt to 14–19 ft. lbs. (19–25 Nm).
7. Tighten the alternator mounting bolt to 27–38 ft. lbs. (37–52 Nm).
8. Lower the vehicle.

1.9L Engine

An automatic tensioner maintains correct belt tension during operation. No adjustment is necessary.

REMOVAL & INSTALLATION

1.6L Engine

1. Disconnect the negative battery cable.
2. Label and disconnect each alternator wiring connector.
3. Remove the alternator-to-adjusting bracket bolt. Loosen the alternator through bolt and allow it to pivot. Shift the alternator toward the block and remove the drive belt.
4. Remove the through bolt and the alternator.
5. To install, reverse the removal procedures. Torque the alternator through bolt to 27–38 ft. lbs. (32–52 Nm) and the adjusting bracket bolt to 14–19 ft. lbs. (19–26 Nm). Start the engine and check the operation.

1.8L Engine

1. Disconnect the negative battery cable.
2. Remove the nut securing the wiring connector to the alternator.
3. Remove the field terminal wiring connector.
4. Remove the upper mounting bolt securing the alternator to the alternator to the alternator bracket.
5. Loosen the lower alternator mounting bolt and pivot the alternator forward.
6. Remove the alternator belt from the pulley and position the belt aside.
7. Raise and safely support the vehicle.
8. Remove the lower splash shield located under the accessory belts.
9. Remove the alternator lower mounting bolt and remove the alternator from the vehicle.

To install:

10. Position the alternator into the vehicle and install the lower mounting bolt.
11. Install the lower splash shield.
12. Lower the vehicle.
13. Position the alternator belt onto the pulley and adjust the belt according to the proper procedure.
14. Tighten the upper mounting bolt to 14–19 ft. lbs. (19–25 Nm) and the lower mounting bolt to 27–38 ft. lbs. (37–52 Nm).
15. Install the field terminal wiring connector.
16. Position the wiring connector to the alternator and secure it with the nut.
17. Connect the negative battery cable.

1.9L Engine

1. Disconnect the negative battery cable.
2. Remove the accessory drive belt.
3. Remove the nut securing the wiring connector to the alternator.
4. Remove the 2 snap-in type wiring connectors at the alternator.
5. Remove the air conditioning hose bracket from the alternator bracket and position it aside.
6. Remove the alternator mounting bolts.
7. Remove the bolts securing the power steering reservoir and position it aside.
8. Remove the alternator from it's bracket.

To install:

9. Position the alternator onto its bracket.
10. Position the power steering reservoir and secure it with the bolts.
11. Install the alternator upper mounting bolt and tighten to 14–22 ft. lbs. (20–30 Nm). Install the alternator lower mounting bolt and tighten to 29–40 ft. lbs. (40–55 Nm).
12. Position the air conditioning hose bracket onto the alternator bracket and secure it with the bolts.
13. Install the 2 snap-in type wiring connectors into the alternator.
14. Position the wiring connector onto the alternator and secure it with the nut.
15. Install the accessory drive belt.

16. Connect the negative battery cable.

Starter

REMOVAL & INSTALLATION

1.6L Engine

1. Disconnect the negative terminal from the battery.
2. Disconnect the electrical connectors from the starter terminals.
3. Remove the starter-to-engine support bracket.
4. Remove the starter-to-transaxle bolts and remove the starter from the vehicle.
5. To install, reverse the removal procedure. Torque the starter-to-engine bolts to 23–30 ft. lbs. (31–41 Nm) and the support bracket nuts and thru bolt to 54–71 inch lbs. (6–8 Nm).

1.8L Engine

1. Disconnect the negative battery cable.
2. Remove the air duct that connects to the throttle body and resonance chamber.
3. Remove the starter motor upper mounting bolts.
4. Raise and safely support the vehicle.
5. Remove the intake plenum support bracket mounting bolts and remove the bracket.
6. Disconnect the **S** terminal connector from the starter solenoid.

NOTE: When disconnecting the plastic hard shell connector at the solenoid S terminal, grasp the plastic connector, depress the plastic tab and pull off the lead assembly. Do not pull on the lead wire or damage may result.

7. Remove the **B** terminal attaching nut and disconnect the cable from the terminal.
8. Remove the starter motor lower mounting bolt and remove the starter motor.

To install:

9. Place the starter motor into its mounting position and install the lower mounting bolt. Tighten the bolt to 15–20 ft. lbs. (20.3–27.0 Nm).
10. Connect the cable to the starter solenoid **B** terminal and install the attaching nut to the terminal. Tighten the nut to 80–120 inch lbs. (9.0–13.5 Nm).
11. Connect the electrical connector to the starter solenoid **S** terminal.
12. Install the intake plenum support bracket and tighten the attaching bolts to 27–38 ft. lbs. (37–52 Nm) and the attaching nut to 14–19 ft. lbs. (19–25 Nm).

13. Lower the vehicle.

14. Install the starter motor upper mounting bolts and tighten to 15–20 ft. lbs. (20.3–27.0 Nm).

15. Install the air duct that connects to the throttle body and resonance chamber.

16. Connect the negative battery cable.

1.9L Engine

1. Disconnect the negative battery cable.

2. If equipped with an automatic transaxle, remove the kickdown cable routing bracket from the engine block.

3. Disconnect the wire from the starter solenoid **S** terminal.

NOTE: When disconnecting the plastic hard shell connector at the solenoid S terminal, grasp the plastic connector, depress the plastic tab and pull off the lead assembly. Do not pull on the lead wire or damage may result.

4. Remove the attaching nut from the starter solenoid **B** terminal and disconnect the cable from the terminal.

5. Remove the starter motor mounting bolts and remove the starter motor.

To install:

6. Place the starter motor into its mounting position and install the mounting bolts. Tighten the bolts to 15–20 ft. lbs. (20.3–27.0 Nm).

7. Connect the cable to the starter solenoid **B** terminal and install the attaching nut. Tighten the nut to 80–120 inch lbs. (9–13.5 Nm).

8. Connect the wire to the starter solenoid **S** terminal.

9. If equipped with an automatic transaxle, install the kickdown cable routing bracket to the engine block.

10. Connect the negative battery cable.

EMISSION CONTROLS

Please refer to "Emission Controls" in the Unit Repair section for system maintenance procedures. Due to the complex nature of modern electronic engine control systems, comprehensive diagnosis and testing procedures fall outside the confines of this repair manual. For complete information on diagnosis, testing and repair procedures concerning all modern engine and emission control systems, please refer to "Chilton's Guide to Fuel Injection and Electronic Engine Controls".

FUEL SYSTEM

Fuel System Service Precautions

Safety is the most important factor when performing not only fuel system maintenance but any type of maintenance. Failure to conduct maintenance and repairs in a safe manner may result in serious personal injury or death. Maintenance and testing of the vehicle's fuel system components can be accomplished safely and effectively by adhering to the following rules and guidelines.

• To avoid the possibility of fire and personal injury, always disconnect the negative battery cable unless the repair or test procedure requires that battery voltage be applied.

• Always relieve the fuel system pressure prior to disconnecting any fuel system component (injector, fuel rail, pressure regulator, etc.), fitting or fuel line connection. Exercise extreme caution whenever relieving fuel system pressure to avoid exposing skin, face and eyes to fuel spray. Please be advised that fuel under pressure may penetrate the skin or any part of the body that it contacts.

• Always place a shop towel or cloth around the fitting or connection prior to loosening to absorb any excess fuel due to spillage. Ensure that all fuel spillage (should it occur) is quickly removed from engine surfaces. Ensure that all fuel soaked cloths or towels are deposited into a waste container.

• Always keep a dry chemical (Class B) fire extinguisher near the work area.

• Do not allow fuel spray or fuel vapors to come into contact with a spark or open flame.

• Always use a backup wrench when loosening and tightening fuel line connection fittings. This will prevent unnecessary stress and torsion to fuel line piping. Always follow the proper torque specifications.

• Always replace worn fuel fitting O-rings with new. Do not substitute fuel hose or equivalent where fuel pipe is installed.

RELIEVING FUEL SYSTEM PRESSURE

1. Start the engine.

2. Remove the rear seat cushion and disconnect the fuel pump electrical connectors.

3. Wait for the engine to stall, then turn **OFF** the ignition switch.

4. Connect the fuel pump electrical connectors and install the rear seat cushion.

Fuel Tank

REMOVAL & INSTALLATION

1988–90

1. Relieve the fuel system pressure and disconnect the negative battery cable.

2. Remove the rear seat cushion and disconnect the fuel pump connector and ground wire.

3. Remove the 4 mounting screws and the sending unit access cover.

4. Disconnect the fuel supply and return lines from the fuel tank.

5. Raise and safely support the vehicle.

6. Position a suitable container under the fuel tank drain plug. Remove the drain plug and drain the fuel tank.

7. Remove the clamps and hoses at the filler neck.

8. Support the fuel tank and remove the 4 mounting bolts. Lower the fuel tank enough to disconnect the vapor line.

9. Lower the fuel tank and remove it from the vehicle.

To install:

10. Raise the fuel tank into position and connect the vapor line. Install the tank with the 4 mounting bolts.

11. Install the filler neck hoses and tighten the clamps. Install the drain plug.

12. Lower the vehicle and connect the fuel supply and return lines. Tighten the clamps.

13. Install the access cover and connect the fuel pump connector and ground wire.

14. Install the rear seat cushion.

15. Fill the fuel tank and check for leaks.

16. Connect the negative battery cable.

1991–92

1. Relieve the fuel system pressure.

2. Disconnect the negative battery cable.

3. Completely drain the fuel tank by siphoning or pumping the fuel through the fuel filler hose.

4. Remove the rear seat cushion.

5. Remove the ground strap retaining screw and the 3 remaining fuel pump assembly cover screws.

6. Remove the fuel pump assembly cover.

7. Remove the clips from the fuel hoses.

8. Disconnect the fuel hoses from the fuel pump assembly.

9. Raise and safely support the vehicle.

10. Loosen the filler neck clamp and disconnect the filler neck hose from the filler neck.

11. Loosen the clamp and disconnect the filler neck overflow hose from the overflow tube.

12. Disconnect the vapor hoses from the vapor tubes.

13. Remove the exhaust middle pipe heat shield.

14. Support the fuel tank with a jack.

15. Remove the mounting bolts from the fuel tank straps, unclip the straps and remove them.

16. Remove the 3 fuel tank heat shield attaching bolts from the fuel tank and remove the fuel tank.

To install:

17. Place the fuel tank onto the fuel tank heat shield and install the heat shield attaching bolts into the fuel tank.

18. Clip the fuel tank straps into their mounting positions.

19. Install the fuel tank strap mounting bolts and tighten to 27–38 ft. lbs. (37–52 Nm).

20. Remove the support jack from under the fuel tank.

21. Install the exhaust middle pipe heat shield.

22. Connect the vapor hoses to the vapor tubes.

23. Connect the filler neck hose to the filler neck and install the attaching clamp.

24. Connect the filler neck overflow hose to the overflow tube and install the attaching clamp.

25. Lower the vehicle.

26. Connect the fuel hoses to the fuel pump assembly.

27. Install the clips onto the fuel hoses.

28. Position the fuel pump assembly cover and ground strap and install the retaining screws.

29. Connect the fuel pump assembly electrical connector.

30. Install the rear seat cushion.

31. Replace the fuel that was drained from the tank during the removal procedure. Check for leaks.

32. Connect the negative battery cable.

33. Turn the ignition switch **ON** to run the fuel pump and pressurize the system. Check for fuel leaks.

Fuel Filter
REMOVAL & INSTALLATION

1. Properly relieve the fuel system pressure.

2. Disconnect the negative battery cable.

3. Position a suitable container below the fuel filter to collect any excess fuel that may leak from the filter and lines.

4. Remove the retaining clip or loosen the clamp from the fuel filter upper hose.

5. Disconnect the upper hose from the fuel filter and drain any excess fuel into the container. Plug the hose.

6. Loosen the fuel filter mounting clamp.

7. Raise and safely support the vehicle, if necessary.

8. Remove the retaining clip or the bolt and washers from the fuel filter lower hose.

9. Disconnect the lower hose from the fuel filter and drain any excess fuel into the container. Plug the hose.

10. Lower the vehicle.

11. Remove the fuel filter.

To install:

12. Position the fuel filter and tighten the filter mounting clamp.

13. Connect the filter upper hose to the filter and install the upper hose retaining clip or tighten the clamp.

14. Raise and safely support the vehicle.

15. Connect the filter lower hose to the filter and install the lower hose retaining clip or a bolt and 2 new washers. If equipped, tighten the bolt to 18–25 ft. lbs. (25–34 Nm).

16. Lower the vehicle.

17. Connect the negative battery cable.

18. Start the engine and check for leaks.

Electric Fuel Pump

The electric fuel pump is located in the fuel tank, on the fuel sending unit.

PRESSURE TESTING

1.6L Engine

1. Relieve the fuel system pressure. Connect a fuel gauge in line with the fuel filter and the fuel rail.

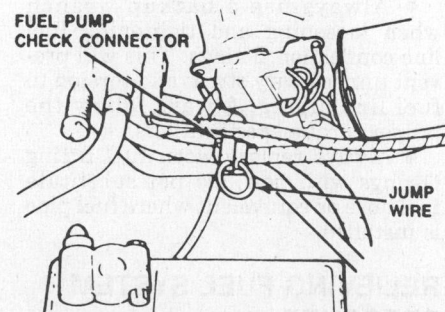

FUEL PUMP CHECK CONNECTOR

JUMP WIRE

Fuel pump check connector location— 1.6L engine

2. Jumper the fuel pump check connector, so the pump will run with the key in the **ON** position.

3. Observe the pressure reading. If the fuel pressure is below 60 psi, check the system for restrictions. Replace the pump as needed.

4. Relieve the fuel pressure. Remove the fuel pressure gauge.

1.8L Engine

1. Make sure the ignition key is in the **OFF** position.

2. Properly relieve the fuel system pressure.

3. Install a fuel pressure tester in the fuel line between the fuel filter and the fuel rail.

4. Jump the fuel pump test connector terminals together, terminal **LG** and terminal **BK** at the Self-Test connector.

5. Turn the ignition key to **RUN**, to operate the fuel pump, and observe the fuel pressure.

6. The fuel pressure should be 64–85 psi.

7. Turn the ignition key **OFF** and remove the jumper wire.

8. Properly relieve the fuel system pressure and remove the pressure tester.

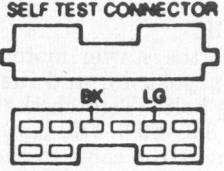

SELF TEST CONNECTOR

BK   LG

Self-test connector terminal locations— 1.8L engine

1.9L Engine

1. Make sure the ignition key is in the **OFF** position.

2. Properly relieve the fuel system pressure.

3. Disconnect the fuel pump output line and connect a suitable pressure tester.

4. Ground the fuel pump lead of the Self-Test connector through a jumper wire at the **FP** lead.

5. Turn the ignition key **ON**, to operate the fuel pump, and observe the fuel pressure.

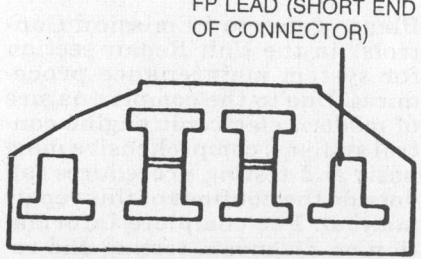

FP LEAD (SHORT END OF CONNECTOR)

Self-test connector—1.9L engine

6. The fuel pressure should be 35–45 psi.

7. Turn the ignition key **OFF** and remove the jumper wire.

8. Properly relieve the fuel system pressure and remove the pressure tester.

9. Reconnect the fuel line.

REMOVAL & INSTALLATION

1. Relieve the fuel system pressure and disconnect the negative battery cable.

2. Disconnect the ground wire and remove the fuel pump access cover.

3. Disconnect and plug the fuel supply and return lines.

4. On 1988–90 vehicles, remove the mounting bolts and remove the fuel pump/sending unit assembly and gasket from the fuel tank.

5. On 1991–92 vehicles, remove the spanner nut using a suitable tool and remove the fuel pump/sending unit assembly and gasket.

6. Separate the fuel pump from the sending unit.

To install:

7. Install the fuel pump onto the sending unit.

8. On 1988–90 vehicles, position a new gasket onto the fuel tank and install the fuel pump/sending unit assembly with the mounting bolts.

9. On 1991–92 vehicles, install a new gasket and the fuel pump/sending unit assembly. Install the spanner nut.

10. Unplug and connect the fuel supply and return lines.

11. Install the fuel pump access cover and connect the ground wire and fuel pump electrical connector.

12. Connect the negative battery cable, start the engine and check for fuel leaks.

13. Install the rear seat cushion.

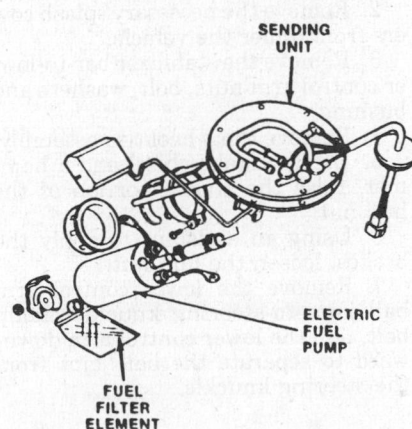

Electric fuel pump assembly—1988–90 vehicles.

Fuel Injection

IDLE SPEED ADJUSTMENT

1.6L Engine

NOTE: Before adjusting idle speed, check ignition timing and adjust, if necessary. Turn off all lights and unnecessary electrical loads. Adjust idle speed while cooling fan motor is not operating.

1. Operate the engine until normal operating temperature is reached.

2. Using a tachometer, connect it to Pin 1 (white) of the test connector and check the idle speed.

3. If necessary to adjust the idle speed, connect a jumper wire between Pin 1 (green) of the test connector and ground and turn the air adjustment screw to obtain the correct idle speed.

NOTE: Do not turn the adjustment screw located to the right of the idle adjustment screw, for it will affect the driveability and may damage the throttle body.

4. After adjustment, remove the jumper wire and the test equipment.

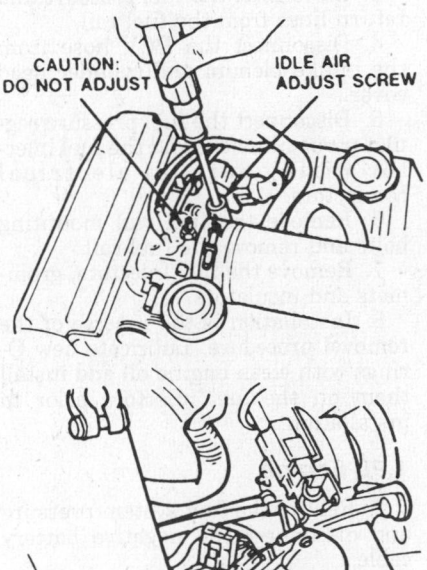

Idle speed adjustment procedure—1.6L engine

1.8L Engine

1. Apply the parking brake and make sure the vehicle is in **N** or **P**.

2. Start the engine and warm it up to normal operating temperature.

3. Turn off all electrical loads and accessories.

4. Using a jumper wire, connect the **GROUND** terminal to the **TEN** terminal on the diagnosis connector.

5. Connect the positive lead of a tachometer to the **IG** terminal on the diagnosis connector and the tachometer negative lead to the negative battery terminal.

6. Check the vehicle idle speed when the electric cooling fan is not operating. The idle speed should be 700–800 rpm.

NOTE: When the parking brake is not applied, the idle speed for automatic transaxle vehicles is approximately 800 rpm.

7. If the idle speed is not within specification, adjust the idle speed by turning the idle speed adjusting screw until the idle speed is within specification.

8. Remove the jumper wire from the diagnosis connector and remove the tachometer.

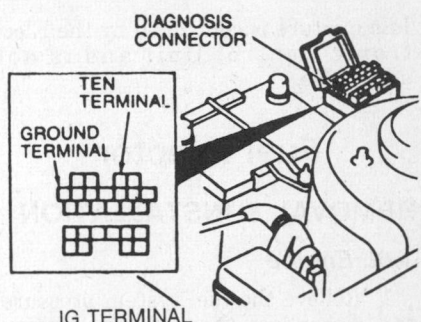

Diagnosis connector location—1.8L engine

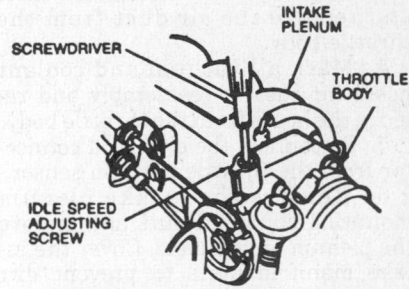

Idle speed adjusting screw location—1.8L engine

1.9L Engine

1. Place the automatic transaxle in **P** or the manual transaxle in neutral.

2. Disconnect the negative battery cable for 5 minutes, then reconnect it.

3. Start the engine and let it stabilize for 2 minutes, then goose the engine and let it return to idle, lightly depress and release the accelerator and let the engine idle. If the engine does not idle properly, proceed to Step 4.

NOTE: If the electric fan comes on, wait until it turns itself off to check the engine idle.

4. Stop the engine and disconnect the idle speed control air bypass solenoid.

5. Start the engine and run it at 2000 rpm for 60 seconds.

6. After the engine returns to idle, check the idle speed. It should be 600 ± 150 rpm.

7. If the idle rpm is incorrect, turn the plate stop screw until the idle speed is within specification.

8. Turn the engine **OFF** and reconnect the idle speed control-air bypass solenoid. Verify the throttle is not stuck in the bore and the linkage is not preventing the throttle from closing.

9. Start the engine and let it stabilize for 2 minutes, then goose the engine and let it return to idle; lightly depress and release the accelerator and let the engine idle. Check idle speed.

IDLE MIXTURE ADJUSTMENT

Idle mixture is controlled by the Electronic Control Unit and is not adjustable.

Fuel Injector

REMOVAL & INSTALLATION

1.6L Engine

1. Relieve the fuel system pressure and disconnect the negative battery cable.

2. Drain the cooling system.

3. Disconnect the accelerator cable and remove the air duct from the throttle body.

4. Mark all vacuum and coolant hoses for ease of reassembly and remove the hoses from the throttle body.

5. Disconnect the electrical connector from the throttle position sensor.

6. Remove the intake plenum mounting bolts and nuts and remove the plenum and gaskets. Cover the intake manifold ports to prevent dirt from entering.

7. Disconnect the fuel supply and return lines from the fuel rail.

8. Disconnect the electrical connectors from the fuel injectors and remove the pressure regulator.

9. Remove the mounting bolts and the fuel rail. Remove the fuel injectors and remove and discard the O-rings.

To install:

10. Lubricate new O-rings with gasoline and install on the injectors. Install the injectors into the cylinder head.

11. Install the fuel rail onto the injectors and install the mounting bolts.

Make sure the injectors seat in the manifold and fuel rail.

12. Install the pressure regulator and connect the fuel injector electrical connectors.

13. Connect the fuel supply and return lines to the fuel rail.

14. Clean the gasket mating surfaces of the intake plenum and intake manifold.

15. Install the intake plenum with the mounting bolts and nuts, using a new gasket.

16. Connect the throttle position sensor electrical connector. Connect the vacuum lines and coolant hoses to their original positions, as noted during the removal procedure.

17. Connect the accelerator cable and install the air duct to the throttle body.

18. Fill the cooling system.

19. Connect the negative battery cable and turn the ignition switch **ON** to activate the fuel pump. Check for fuel leaks.

1.8L Engine

1. Properly relieve the fuel system pressure.

2. Disconnect the negative battery cable.

3. Disconnect the fuel pressure and return lines from the fuel rail.

4. Disconnect the PCV hose from the intake plenum and cylinder head cover.

5. Disconnect the fuel pressure regulator vacuum hose and the fuel injector wiring harness electrical connectors.

6. Remove the fuel rail mounting bolts and remove the fuel rail.

7. Remove the fuel injectors, grommets and insulators.

8. Installation is the reverse of the removal procedure. Lubricate new O-rings with clean engine oil and install them on the fuel injectors prior to installation.

1.9L Engine

1. Relieve the fuel system pressure and disconnect the negative battery cable.

2. Remove spring-lock coupling retainer clips from fuel inlet and return fittings.

3. Disconnect fuel supply and return lines.

3. Remove vacuum line from fuel pressure regulator.

4. Disconnect the fuel injector wiring harness.

5. Remove the 2 bolts securing the injector manifold assembly and remove the assembly.

6. Carefully remove connectors from individual injectors(s) as required.

7. Grasping the injector's body, pull up while gently rocking the injector from side-to-side.

8. Inspect the injector O-rings (2 per injector) for signs of deterioration. Replace as required.

9. Inspect the injector "plastic hat" (covering the injector pintle) and washer for signs of deterioration. Replace as required. If hat is missing, look for it in intake manifold.

To install:

10. Use a light grade oil to lubricate new O-rings and install 2 on each injector.

11. Install the injector(s). Use a light, twisting, pushing motion to install the injector(s).

12. Carefully seat the fuel injector manifold assembly on the injectors and secure the manifold with the attaching bolts. Tighten to 15–22 ft. lbs. (20–30 Nm).

13. Connect the vacuum line to the fuel pressure regulator.

14. Connect fuel injector wiring harness.

15. Connect fuel supply and fuel return lines.

16. Reconnect spring-lock coupling retaining clips on fuel inlet and return fittings.

17. Check entire assembly for proper alignment and seating.

18. Connect negative battery cable.

DRIVE AXLE

Halfshaft

REMOVAL & INSTALLATION

1988–90

1. Raise and support the vehicle safely.

2. Remove the necessary splash covers from under the vehicle.

3. Remove the stabilizer bar-to-lower control arm nuts, bolt, washers and bushing.

4. Remove the wheel/tire assembly.

5. Using a stake chisel and a hammer, raise the staked portion of the hub nut.

6. Using an assistant to apply the brakes, loosen the hub nut.

7. Remove the lower control arm ball joint-to-steering knuckle clamp bolt, pull the lower control arm downward to separate the ball joint from the steering knuckle.

NOTE: When separating the ball joint, be careful not to damage the ball joint dust boot.

8. If equipped with a manual transaxle, use both hands, grasp the steering knuckle/hub assembly, apply even pressure (gradually increasing), pull both halfshafts from the transaxle. If equipped with an automatic transaxle, insert a medium prybar between the halfshaft and the transaxle (a notch is provided), pry both halfshafts from the transaxle.

NOTE: When removing the halfshafts, withdraw them completely from the transaxle (to prevent damage to the oil seal lips), do not move the CV-joints in excess of a 20 degree angle (damage to the boots and/or joint may occur) and use a wire to support the halfshaft in the horizontal position.

9. Remove the hub nut (discard it) and washer. Pull the halfshaft from the steering knuckle assembly.

NOTE: If the wheel hub binds on the halfshaft splines, use a puller tool, to press the halfshaft from the wheel hub. Never use a hammer to separate the halfshaft from the wheel hub, for damage to the CV-joint may occur.

10. Using differential plug tool(s), plug the transaxle bore(s) to prevent oil leakage.
11. Check the halfshaft for damage, wear and/or good working order; replace the halfshaft, if necessary.
To install:
12. To install the halfshaft into the transaxle, perform the following procedures:
   a. Install a new locking clip on the halfshaft spline; be sure the gap in the clip is at the top of the clip groove.
   b. Slide the halfshafts into the transaxle bore; be careful not to damage the oil seal lip.
   c. Make sure the locking clip can be felt as it snaps into the differential side gear groove.
   d. Position the halfshaft through the wheel hub and install a new attaching nut. Do not tighten the nut at this time.
   e. After installation, pull the front hub outward to confirm that the circlips are engaged.
13. Install the ball joint into the steering knuckle and install the clamp bolt and attaching nut. Tighten the nut to 32–40 ft. lbs. (43–54 Nm).
14. Install the bolt, nuts, washers and bushings connecting the stabilizer bar and lower control arm. Tighten the nut until $7/16$ in. (0.8mm) of bolt threads extend beyond the nut.
15. Install the splash covers.
16. Tighten the halfshaft nut to 116–174 ft. lbs. (157–235 Nm). Stake the

nut using a cold chisel with a rounded edge.

NOTE: If the nut splits or cracks after staking, it must be replaced with a new nut.

17. Install the wheel and tire assembly and lower the vehicle.

1991–92

LEFT SIDE – 1.8L AND 1.9L ENGINES
RIGHT SIDE – 1.9L ENGINE

1. Raise and safely support the vehicle.
2. Remove the wheel and tire assembly.
3. Remove the splash shield.
4. Carefully raise the staked portion of the halfshaft retaining nut using a small chisel. Remove and discard the retaining nut.
5. Remove the cotter pin and nut from the tie rod end and remove the tie rod end from the steering knuckle using a removal tool.
6. Remove the lower ball joint clamp bolt. Carefully pry down on the lower control arm to separate the ball joint from the steering knuckle.
7. Pull outward on the steering knuckle/brake assembly. Carefully pull the halfshaft from the steering knuckle and position it aside.
8. Removal of the left side halfshaft requires removal of the crossmember to allow access with a prybar. If the left side halfshaft is being removed, proceed as follows:
   a. Support the transaxle with a transmission jack.
   b. Remove the 4 transaxle mount-to-crossmember attaching nuts.
   c. Remove the 2 crossmember attaching nuts at the rear of the crossmember.
   d. While supporting the rear of the crossmember, remove the 2 front mounting bolts. Remove the crossmember.
9. Position a drain pan under the transaxle.
10. Insert a prybar between the halfshaft and the transaxle case. Gently pry outward to release the halfshaft from the differential side gears. Be careful no to damage the transaxle case, oil seal, CV-joint or CV-joint boot.
11. Remove the halfshaft.

NOTE: Install suitable plugs after removing the halfshafts to prevent the differential side gears from becoming mispositioned. Should the gears become misaligned, the differential will have to be removed from the transaxle to align the gears.

To install:
12. Position the circlip on the inner CV-joint spline so the circlip gap is at the top. Lubricate the splines lightly with suitable grease.
13. Remove the plugs that were installed in the differential side gears.
14. Position the halfshaft so the CV-joint splines are aligned with the differential side gear splines. Push the halfshaft into the differential.

NOTE: When seated properly, the circlip can be felt as it snaps into the differential side gear groove.

15. Pull outward on the steering knuckle/brake assembly and insert the halfshaft into the steering knuckle.
16. Pry downward on the control arm and position the lower ball joint in the steering knuckle.
17. For left side halfshaft installation, proceed as follows:
   a. Position the crossmember in place.
   b. Install the 2 mounting bolts and the 2 attaching nuts. Tighten the nuts and bolts to 47–66 ft. lbs. (64–89 Nm).
   c. Install the 4 transaxle mount-to-crossmember attaching nuts. Tighten the nuts to 27–38 ft. lbs. (37–52 Nm).
   d. Remove the transmission jack.
18. Install the lower ball joint clamp bolt and tighten to 32–43 ft. lbs. (43–59 Nm).
19. Install the tie rod end in the steering knuckle. Install the nut to the tie rod end and tighten to 31–42 ft. lbs. (42–57 Nm). Install a new cotter pin.
20. Install a new halfshaft retaining nut and tighten to 174–235 ft. lbs. (235–319 Nm). Stake the halfshaft retaining nut using a chisel with a rounded cutting edge.

NOTE: If the nut splits or cracks after staking, replace it with a new nut.

21. Install the splash shield.
22. Install the wheel and tire assembly and lower the vehicle.
23. Check and refill the transaxle with the proper type and quantity of fluid.

RIGHT SIDE – 1.8L ENGINE

NOTE: The right side halfshaft assembly is a 2 piece shaft with a bearing support bracket positioned between the 2 halves. The bearing support bracket is mounted on the cylinder block and must be unbolted if the entire halfshaft assembly is to be removed. If only the CV-joints/boots are to be serviced, the outboard shaft assembly may be removed, leaving the bearing support

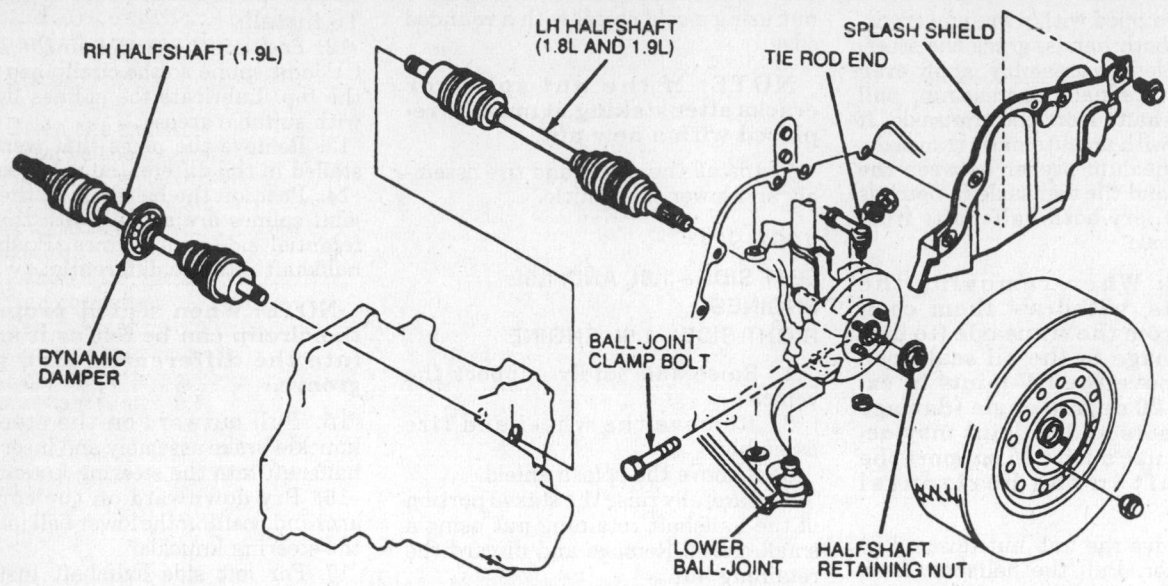

Halfshaft removal and installation—left side, 1.8L and 1.9L engines and right side, 1.9L engine

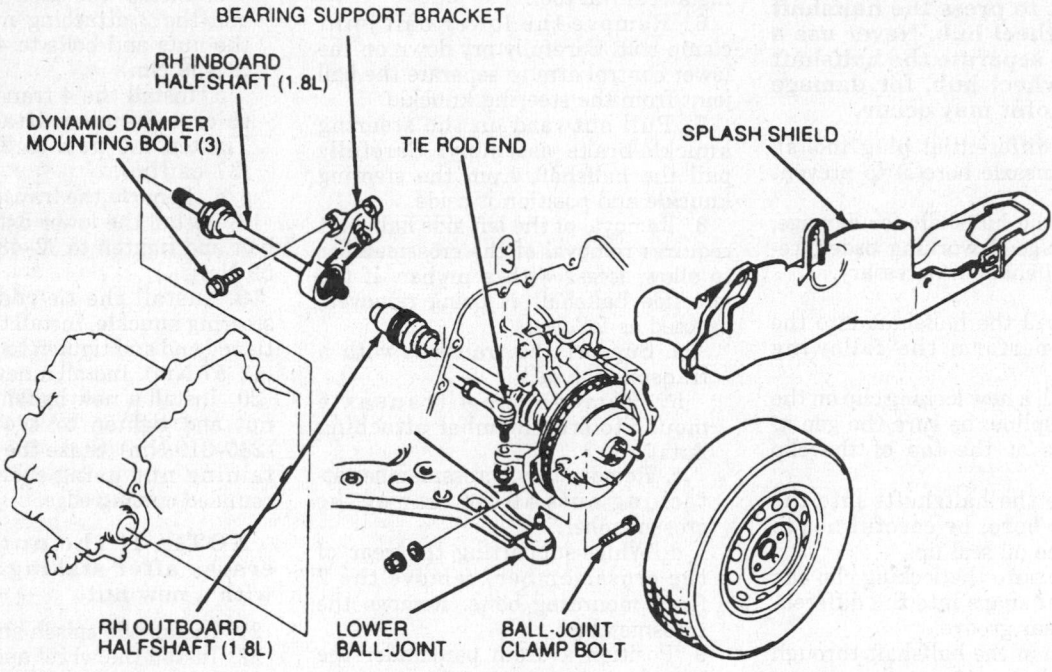

Halfshaft removal and installation—right side, 1.8L engine

bracket mounted on the engine cylinder block.

1. Raise and safely support the vehicle.
2. Remove the right front wheel and tire assembly.
3. Remove the splash shield.
4. Carefully raise the staked portion of the halfshaft retaining nut using a small chisel. Remove and discard the retaining nut.
5. Remove the cotter pin and nut from the tie rod end and remove the tie

rod end from the steering knuckle using a removal tool.
6. Remove the lower ball joint clamp bolt. Carefully pry down on the lower control arm to separate the ball joint from the steering knuckle.
7. Pull outward on the steering knuckle/brake assembly. Carefully pull the halfshaft from the steering knuckle and position it aside.
8. Position a drain pan under the transaxle.
9. Remove the 3 bearing support bracket mounting bolts.

10. Insert a prybar between the bearing support bracket and the starter bracket. Gently pry outward on the damper until the halfshaft disengages from the differential side gear.
11. Remove the halfshaft assembly. Install an appropriate differential plug in the differential side gear.
To install:
12. Position the circlip on the inner CV-joint spline so the circlip gap is at the top. Lubricate the splines lightly with a suitable grease.
13. Remove the differential plug

from the side gear. Position the halfshaft assembly so the shaft splines are aligned with the differential side gear splines. Push the halfshaft into the differential.

NOTE: When seated properly, the circlip can be felt as it snaps into the differential side gear groove.

14. Pull outward on the steering knuckle/brake assembly and insert the halfshaft into the steering knuckle.

15. Pry downward on the control arm and position the lower ball joint in the steering knuckle. Install the lower ball joint clamp bolt and tighten to 32–43 ft. lbs. (43–59 Nm).

16. Install the tie rod end in the steering knuckle. Install the nut to the tie rod end and tighten to 31–42 ft. lbs. (42–57 Nm). Install a new cotter pin.

17. Position the bearing support bracket and install the 3 mounting bolts. Tighten the bolts in the proper sequence to 31–46 ft. lbs. (42–62 Nm).

18. Install a new halfshaft retaining nut and tighten to 174–235 ft. lbs. (235–319 Nm). Stake the retaining nut using a suitable chisel with the cutting edge rounded off.

NOTE: If the nut splits or cracks after staking, it must be replaced with a new nut.

19. Install the splash shield.
20. Install the right front wheel and lower the vehicle.
21. Check and refill the transaxle with the proper type and quantity of fluid.

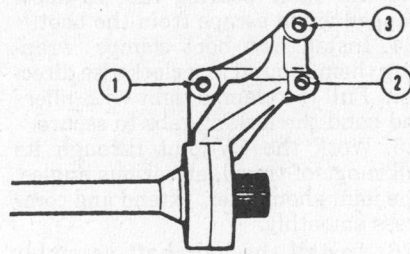

Bearing support bracket torque
sequence—1991–92 vehicles

CV-Boot

REMOVAL & INSTALLATION

NOTE: All vehicles are equipped with Birfield outer CV-joints. This type of joint cannot be disassembled. If a Birfield CV-joint boot needs replacement, the inner CV-joint must be removed in order to install a new outer CV-joint boot. All vehicles are equipped with Tripod inner CV-joints, except 1988–90 manual

transaxle vehicles which are equipped with Rzeppa inner CV-joints. The disassembly procedures vary accordingly.

Except 1988–90 Manual Transaxle Vehicles

1. Raise and safely support the vehicle.
2. Remove the halfshaft assembly from the vehicle.
3. Secure the halfshaft in a vise with protective jaw covers.
4. Using a suitable tool, remove the CV-joint boot clamps.
5. Slide the boot back to expose the tripod CV-joint. Mark the shaft and the CV-joint housing to ensure correct assembly.
6. Remove the retainer ring from the CV-joint housing and remove the CV-joint housing from the halfshaft.
7. Mark the tripod bearing and the shaft to ensure correct assembly. Using snapring pliers, remove the tripod snapring.
8. Using a soft-faced mallet, gently tap the tripod bearing from the shaft.
9. Wrap the shaft splines with tape to protect the CV-boot if the boot is to be reused.
10. Slide the inner CV-joint boot off the shaft. If the outer CV-joint boot is to be replaced, continue with the procedure.
11. On 1.9L engine right side halfshafts, pry up the rubber damper retaining band locking clip using a prybar. Remove the retaining band using pliers and remove the rubber damper from the shaft.
12. Using a suitable tool, remove the outer CV-boot clamps.
13. Slide the outer CV-boot off the shaft.

NOTE:When replacing a damaged boot, check the grease for contamination by rubbing it between 2 fingers. Any gritty feeling indicates a contaminated CV-joint. A contaminated inner CV-joint must be completely disassembled, cleaned and inspected. The outer CV-joint is not serviceable and should be replaced as an assembly, if necessary. If the grease is not contaminated and the CV-joint has been operating satisfactorily, replace only the boot and add the required lubricant.

To install:

14. Cover the halfshaft splines with tape and install the outer CV-joint boot.

NOTE: On 1991–92 vehicles, the outer and inner CV-joint boots are different. Failure to correctly install the boot on the proper end

	1.9L Engine		1.8L Engine	
	Right Side	Left Side	Right Side	Left Side
(A)	84.0 mm (3.31 in)	90.0 mm (3.54 in)	89.9 mm (3.54 in)	
(B)	89.0 mm (3.50 in)		85.2 mm (3.35 in)	

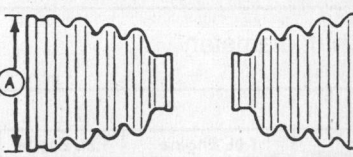

INNER CV BOOT   OUTER CV BOOT

CV-joint boots—1991–92 vehicles

of the halfshaft could lead to premature boot and/or CV-joint wear.

15. Fill the outer CV-joint housing with the proper type and amount of lubricant.

16. Position the CV-boot. Make sure the boot is fully seated in the shaft grooves and the CV-joint housing.

17. Insert a prybar between the boot and the CV-joint housing to allow trapped air to escape.

18. Position new clamps on the outer CV-joint boot.

NOTE: Always use new clamps. The clamps should be mounted in the direction opposite the forward revolving direction of the halfshaft.

19. Wrap the clamps around the boot in a clockwise direction, pull them tight with pliers and bend the locking tabs to secure the clamps in position.

20. Work the CV-joint through it's full range of travel at various angles. The CV-joint should flex, extend and compress smoothly.

21. On 1.9L engine right side halfshafts, position the rubber damper on the halfshaft. Position a new band on the damper. Pull the band tight with pliers and fold it back. Lock the end of the band by bending the locking clip.

22. Position the inner CV-joint boot on the halfshaft.

23. Align the marks on the tripod bearing and the halfshaft. Install the tripod bearing on the halfshaft. If necessary, using a soft-faced mallet, tap the bearing into place.

24. Install the snapring.

25. Fill the inner CV-joint housing with the proper type and amount of lubricant. Coat the tripod bearing with the same lubricant.

26. Position the inner CV-joint housing over the tripod bearing, making sure to align the alignment marks. Install the retainer ring in the CV-joint housing.

Item	Model	1.8L Engine	1.9L Engine
Halfshaft			
Length of joint (between center of joint)	Right side	631.2 mm (24.85 in)	918.7 mm (36.16 in)
	Left side	621.7 mm (24.48 in)	640.7 mm (25.22 in)
Shaft diameter	Right side	23.0 mm (0.91 in)	
	Left side	23.0 mm (0.91 in)	

Halfshaft Assemblies	1.9L Engine		1.8L Engine	
	Right Side	Left Side	Right Side	Left Side
Differential Side	220 g (7.77 oz) Lt. Yellow	140 g (4.94 oz) Yellow	145 g (5.12 oz) Yellow	
Wheel Side	140 g (4.94 oz) Black		90 g (3.18 oz) Black	

27. Slide the inner CV-boot in place. Make sure the boot is fully seated in the shaft grooves and in the housing. On 1988–90 vehicles, extend or compress the joint, as necessary, until the distance between the CV-joint boot clamp grooves measures 3.5 in. (90mm).

28. Insert a prybar between the boot and the CV-joint housing to allow trapped air to escape.

29. Position new clamps on the inner CV-joint boot.

NOTE: Always use new clamps. The clamps should be mounted in the direction opposite the forward revolving direction of the halfshaft.

30. Wrap the clamps around the boot in a clockwise direction, pull them tight with pliers and bend the locking tabs to secure the clamps in position.

31. Work the CV-joint through it's full range of travel at various angles. The CV-joint should flex, extend and compress smoothly.

32. On 1991–92 vehicles, measure the length of the assembled halfshaft. If the length is not as specified, check the CV-joints for freedom of movement to ensure that it was assembled correctly. Repair or replace any components as necessary.

33. Install the halfshaft assembly into the vehicle.

1988–90 Manual Transaxle Vehicles

1. Raise and safely support the vehicle.

2. Remove the halfshaft assembly.

3. Secure the halfshaft in a vise with protective jaw covers.

4. Using a side cutters, remove the large boot clamp from the inner CV-joint and roll the boot back over the shaft.

5. Mark the outer race and the shaft for assembly reference, remove the wire ring bearing retainer and remove the outer race.

6. Mark the inner race and the shaft for assembly reference and remove the snapring from the end of the halfshaft.

7. Remove the inner race, cage and ball bearings as an assembly.

8. Remove the small clamp and remove the inner CV-boot from the halfshaft.

9. If replacing the outer CV-joint boot, remove the clamps and slide the boot along the shaft, removing it from the inner CV-joint end.

NOTE: When replacing a damaged boot, check the grease for contamination by rubbing it between 2 fingers. Any gritty feeling indicates a contaminated CV-joint. A contaminated inner CV-joint must be completely disassembled, cleaned and inspected. The outer CV-joint is not serviceable and should be replaced as an assembly, if necessary. If the grease is not contaminated and the CV-joint has been operating satisfactorily, replace only the boot and add the required lubricant.

To install:

10. Cover the halfshaft splines with tape and install the outer CV-joint boot.

11. Fill the outer CV-joint housing with the proper type and amount of lubricant.

12. Position the CV-boot. Make sure the boot is fully seated in the shaft grooves and the CV-joint housing.

13. Insert a prybar between the boot and the CV-joint housing to allow trapped air to escape.

14. Position new clamps on the outer CV-joint boot.

NOTE: Always use new clamps. The clamps should be mounted in the direction opposite the forward revolving direction of the halfshaft.

15. Wrap the clamps around the boot in a clockwise direction, pull them tight with pliers and bend the locking tabs to secure the clamps in position.

16. Work the CV-joint through it's full range of travel at various angles. The CV-joint should flex, extend and compress smoothly.

17. Install the inner CV-boot.

18. Install the inner race, cage and balls on the halfshaft as an assembly. Make sure the chamfer on the bearing cage faces the snapring and the marks made during removal line up. Install the snapring.

19. Lubricate the outer race with 1.4–2.1 oz. of CV-joint grease. Install the race, aligning the marks that were made at removal.

20. Add another 0.7–1.0 oz. of grease to the outer race and install the wire ring bearing retainer.

21. Position the inner CV-boot, making sure it is fully seated in the grooves in the shaft and outer race.

22. Extend or compress the CV-joint, as necessary, until the distance between the boot clamp grooves is 3.5 in. (90mm). Do not allow this distance to change until the boot clamps are installed.

23. Insert a prybar between the boot and the outer bearing race to allow trapped air to escape from the boot.

24. Install new boot clamps, wrapping them around in a clockwise direction. Pull the clamps tight with pliers and bend the locking tabs to secure.

25. Work the CV-joint through its full range of travel, at various angles. The joint should flex, extend and compress smoothly.

26. Install the halfshaft assembly into the vehicle.

Front Wheel Hub, Knuckle and Bearings

REMOVAL & INSTALLATION

1988–90

1. Raise and support the vehicle safely. Remove the wheel and tire assembly.

2. Carefully raise the staked portion of the halfshaft retaining nut. Have an assistant apply the brakes to lock the hub and remove the nut and washer. Discard the nut.

3. Remove the stabilizer bar-to-control arm attaching bolt, nut, washers and bushings.

4. Remove the cotter pin and tie rod end attaching nut. Use a suitable tool to separate the tie rod end from the steering knuckle.

5. Disconnect the U-shaped clip from the center section of the caliper hose; do not disconnect the hose from the caliper. Remove the brake caliper-to-steering knuckle bolts and support the caliper on a length of wire; do not allow the caliper to hang by the brake hose.

6. Remove the lower ball joint clamp bolt and nut. Use a prybar to pry down on the lower control arm and separate the ball joint from the steering knuckle.

7. Remove the steering knuckle-to-strut mounting bolts. Slide the hub/steering knuckle assembly out of the strut and off the end of the halfshaft. Be careful not to damage the grease seals.

NOTE: If the hub is difficult to remove from the halfshaft splines, tap the end of the halfshaft with a plastic mallet. Never use a metal hammer. If the halfshaft splines are rusted to the hub, use a suitable hub puller to separate them.

8. Use a suitable tool to press the hub and rotor assembly from the steering knuckle.

9. Remove the bearing preload spacer from the hub and rotor assembly.

NOTE: The spacer located between the bearings determines bearing preload. It must not be discarded.

10. Mark the position of the rotor on the hub. Remove the mounting bolts and remove the rotor from the hub.

11. Using a suitable puller, remove the bearing from the wheel hub. Remove the outer grease seal from the hub and pry the inner grease seal from the steering knuckle.

12. Remove the bearing from the steering knuckle. Do not remove the dust shield as it is pressed onto the knuckle and is difficult to get off without damage.

13. If the wheel bearings are to be replaced, drive the bearing races from the steering knuckle using a brass drift.

14. Inspect the hub and steering knuckle for cracks, wear, scoring or other damage and replace as necessary.

To install:

15. If the wheel bearings are to be replaced, proceed as follows:

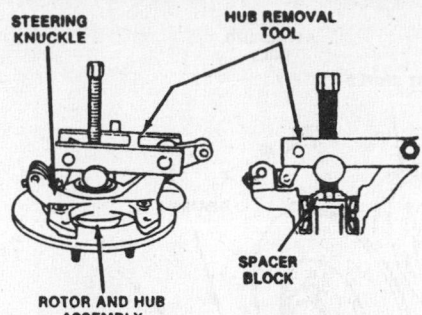

Removing the hub/rotor assembly from the steering knuckle—1988–90 vehicles

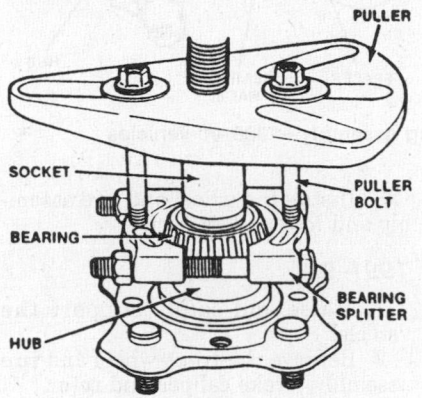

Removing the bearing from the hub—1988–90 vehicles

a. Install the new bearing races in the steering knuckle using a bearing race driver.

b. Install the bearing and preload spacer in the steering knuckle.

c. Install spacer selector tool T87C–1104–B or equivalent, in the steering knuckle and clamp the tool in a vise.

d. Tighten the center bolt in increments to 36, 72, 108 and 145 ft. lbs. (49, 98, 147 and 196 Nm). After tightening to each specified increment, seat the bearings by rotating the steering knuckle. Final torque the center bolt to 145 ft. lbs. (196 Nm).

e. Remove the tool/steering knuckle assembly from the vise and remount in the vise, clamping where the strut mounts.

f. Install a socket and inch pound torque wrench on the spacer selector tool and measure the amount of torque required to rotate the tool. The torque reading must be taken just as the tool starts to rotate.

g. If the torque wrench indicates 2.21–10.44 inch lbs. (0.25–1.8 Nm), the spacer is the correct thickness. If the indication is less than 2.21 inch lbs. (0.25 Nm), a thinner spacer must be installed. If the indication is more than 10.44 inch lbs. (1.8 Nm), a thicker spacer must be installed.

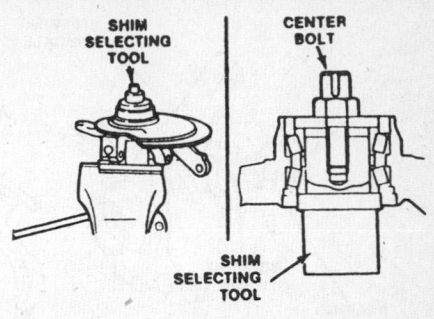

Spacer selector tool assembly—1988–90 vehicles

Stamped Mark	Thickness in. (mm)
1	0.2474 (6.285)
2	0.2490 (6.325)
3	0.2506 (6.365)
4	0.2522 (6.405)
5	0.2538 (6.445)
6	0.2554 (6.485)
7	0.2570 (6.525)
8	0.2586 (6.565)
9	0.2602 (6.605)
10	0.2618 (6.645)
11	0.2634 (6.685)
12	0.2650 (6.725)
13	0.2666 (6.765)
14	0.2682 (6.805)
15	0.2698 (6.845)
16	0.2714 (6.885)
17	0.2730 (6.925)
18	0.2746 (6.965)
19	0.2762 (7.005)
20	0.2778 (7.045)
21	0.2794 (7.085)

Bearing preload spacer selection chart

h. Each bearing spacer has a numerical code stamped on the outer diameter of the spacer that corresponds to a particular thickness. If the code number is not legible, determine the spacer thickness by measuring with a micrometer.

i. Changing the bearing thickness by 1 number, either higher or lower, will change the bearing preload by 1.7–3.5 inch lbs. (0.2–0.4 Nm). After selecting a spacer, verify the bearing preload using the spacer selector tool.

16. If removed, install the dust shield on the steering knuckle using tool T87C–1175–B or equivalent.

17. Pack the bearings and the hub area with a high temperature wheel bearing grease.

18. Place the inner bearing in the steering knuckle. Lubricate the lip of a new inner grease seal and install the seal in the knuckle, using a seal installer.

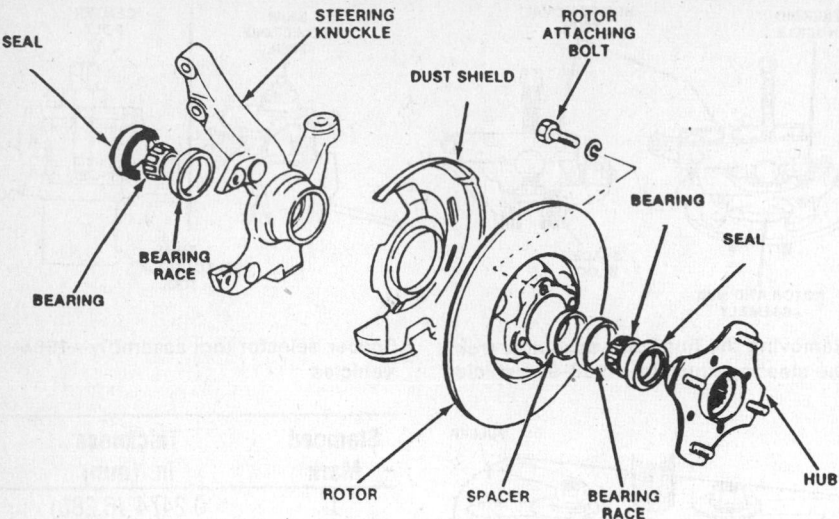

Steering knuckle/hub/wheel bearing assembly—1988–90 vehicles

19. Install the bearing preload spacer and the outer bearing in the steering knuckle. Lubricate the lip of a new outer grease seal and install the seal in the knuckle, using a seal installer.

20. Position the hub on the rotor, aligning the marks that were made during the removal procedure. Tighten the mounting bolts to 33–40 ft. lbs. (44–54 Nm).

21. Install the hub and rotor assembly in the steering knuckle using a hydraulic press and suitable fixtures.

22. Install the hub/steering knuckle assembly over the halfshaft and into the strut. Install the strut-to-steering knuckle mounting bolts and nuts and tighten to 69–86 ft. lbs. (93–97 Nm).

23. Install the lower control arm ball joint through the steering knuckle and install the clamp bolt and nut. Tighten the clamp bolt to 32–40 ft. lbs. (43–54 Nm).

24. Install the brake caliper with the mounting bolts. Tighten the bolts to 29–36 ft. lbs. (39–49 Nm). Install the U-clip on the caliper flex line.

25. Install a new halfshaft attaching nut and tighten to 116–174 ft. lbs. (157–235 Nm). Stake the nut using a chisel with a rounded cutting edge. If the nut splits or cracks after staking, it must be replaced with a new nut.

26. Connect the tie rod to the steering knuckle and install the mounting nut. Tighten the nut to 22–33 ft. lbs. (29–44 Nm) and install a new cotter pin. If the slots in the nut do not align with the hole in the ball joint stud, tighten the nut for proper alignment. Never loosen the nut.

27. Position the stabilizer bar and install the link assembly bolt, nut, washers, sleeve and rubber bushings. Tighten the nut until $7/16$ in. (10.8mm) of the bolt threads extend beyond the nut.

28. Install the wheel and tire assembly and lower the vehicle.

1991–92

1. Raise and safely support the vehicle.

2. Remove the front wheel and tire assembly, brake caliper and rotor.

3. Remove the nut securing the halfshaft to the hub.

4. Remove the outer tie rod end at the steering knuckle.

5. Remove the nuts and bolts and separate the shock/strut assembly from the steering knuckle.

6. Remove the nut and bolt and separate the lower ball joint from the steering knuckle.

7. Remove the front hub/steering knuckle assembly from the halfshaft.

8. Remove the oil seal from the rear of the hub/steering knuckle assembly.

9. Position the hub/steering knuckle assembly on a hydraulic press and press the front hub out of the steering knuckle using a removal tool.

NOTE: If the bearing inner race remains on the hub, use a grinder to grind a section of the bearing inner race until only 0.020 in. (0.5mm) remains. Remove the inner race with a chisel.

10. Remove the E-clip from the steering knuckle.

11. Position the steering knuckle onto a hydraulic press and, using an appropriate bearing remover, press the bearing out of the steering knuckle.

NOTE: If the dust cover is removed, it must be replaced.

12. Scribe a mark in the dust cover

and steering knuckle. Using a chisel, remove the dust cover.
To install:

13. Scribe a mark on the new dust cover in the same position as on the previous mark. Align the marks on the steering knuckle to the mark on the dust cover and, using a suitable tool, press the dust cover onto the steering knuckle.

14. Position the steering knuckle onto a press and press the bearing into the steering knuckle, using a bearing installer.

15. Install the E-clip.

16. Position the hub onto the knuckle and press the hub into the bearing and the steering knuckle, using a suitable installation tool.

17. Using an appropriate seal installer, install a new oil seal onto the inboard side of the steering knuckle. Make sure the oil seal mounts flush with the steering knuckle.

18. Install the hub/steering knuckle assembly onto the ball joint and install the nut and bolt. Tighten to 32–43 ft. lbs. (43–59 Nm).

19. Install the outer tie rod.

20. Install the steering knuckle to the shock/strut assembly. Tighten the nuts and bolts to 69–93 ft. lbs. (93–127 Nm).

21. Install a new locknut securing the halfshaft to the front hub. Tighten the locknut to 174–235 ft. lbs. (235–319 Nm). Stake the locknut to prevent it from loosening.

22. Install the brake rotor, caliper and wheel and tire assembly.

23. Lower the vehicle.

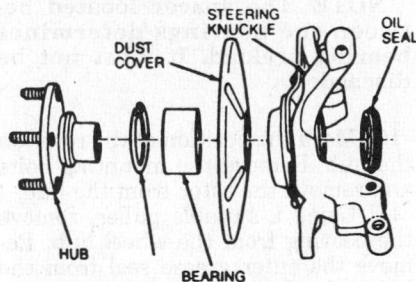

Hub/steering knuckle assembly—1991–92 vehicles

MANUAL TRANSAXLE

For further information on transmissions/transaxles, please refer to "Chilton's Guide to Transmission Repair".

Transaxle Assembly

REMOVAL & INSTALLATION

1988–90

1. Disconnect the negative battery cable.
2. Remove the air cleaner. Loosen the front wheel lug nuts.
3. From the transaxle, disconnect the speedometer cable.
4. From the clutch release lever, remove the adjusting nut, pin and the clutch cable. Remove the clutch cable bracket-to-transaxle bolts and the bracket. Remove the ground wire bolt and ground wire.
5. Remove the coolant pipe bracket bolt and the bracket.
6. Remove the secondary air pipe, the EGR pipe bracket and the electrical harness clip.
7. Disconnect the neutral switch/back-up light switch coupler and the body ground connector.
8. Remove the upper 2 transaxle-to-engine bolts.
9. Using a engine support bar tool, attach it to the rear engine lifting hook and support the engine's weight.
10. Raise and support the vehicle safely.
11. Place a drain pan under the transaxle, remove the drain plug and drain the transaxle.
12. Remove the front wheel lug nuts and the wheels. Remove the engine undercover and side covers.
13. Remove the front stabilizer bar. From both sides, remove the lower control arm ball joint-to-steering knuckle nut/bolt, pull the control arm downward and separate the lower control arm from the steering knuckle.

NOTE: When separating the ball joint, be careful not to damage the ball joint dust boot.

14. Using both hands, grasp the steering knuckle/hub assembly, apply even pressure (gradually increasing), pull both halfshafts from the transaxle.

NOTE: When removing the halfshafts, withdraw them completely from the transaxle (to prevent damage to the oil seal lips), do not move the CV-joints in excess of a 20 degree angle (damage to the boots and/or joint may occur) and use a wire to support the halfshaft in the horizontal position.

15. From under the vehicle, remove the crossmember-to-chassis bolts and the crossmember.
16. Remove the shift control rod-to-transaxle nut/bolt and slide the control rod aside. Remove the shift extension bar-to-bracket bolt and slide the extension bar off the bracket.
17. Remove the starter's positive cable-to-solenoid nut and the solenoid wire by pulling the wire from the connector.
18. Remove the starter-to-engine bolts and the starter. Remove the dust cover-to-clutch housing bolts and the cover.
19. Loosen the bracket bar on the engine support tool to lower the transaxle. Using a floor jack, support the transaxle.
20. Remove the No. 2 engine mount-to-transaxle nut/bolt, the transaxle-to-engine bolts and lower the transaxle from the vehicle.

To install:

21. Install the transaxle by performing the following procedure:
    a. Apply a small amount of clutch grease to the input shaft spline and reverse the removal procedure.
    b. Torque the transaxle-to-engine bolts to 47–66 ft. lbs. (63–89 Nm), the No. 2 engine mount-to-transaxle nut/bolt to 27–38 ft. lbs. (37–52 Nm), the starter to engine bolts to 23–34 ft. lbs. (31–46 Nm), the extension bar-to-transaxle bracket bolt to 23–34 ft. lbs. (31–46 Nm), the control rod-to-transaxle nut/bolt to 12–17 ft. lbs. (16–22 Nm), the cross-member-to-chassis bolts to 47–66 ft. lbs. (64–89 Nm) and the rear engine mount-to-crossmember nut to 20–34 ft. lbs. (28–46 Nm).
22. To install the halfshaft into the transaxle, perform the following procedure:
    a. Install a new locking clip on the halfshaft spline; be sure the gap in the clip is at the top of the clip groove.
    b. Slide the halfshafts into the transaxle bore; be careful not to damage the oil seal lip.
    c. Push firmly on the hub assembly, making sure the circlip snaps into place.
    d. After installation, pull the front hub outward to confirm that the circlips are engaged.
23. To complete the installation, reverse the removal procedure. Torque the lower control arm ball joint-to-steering knuckle nut/bolt to 32–40 ft. lbs. (43–54 Nm), the stabilizer bar-to-chassis nuts/bolts to 23–33 ft. lbs. (31–44 Nm), stabilizer bar-to-lower control arm nuts to 9–13 ft. lbs. (12–18 Nm).
24. Fill the transaxle with the proper fluid. Adjust the clutch pedal free-play. Test the vehicle performance.

1991–92

1. Disconnect the battery cables and remove the battery and the battery tray.
2. Remove the air hose and the resonance chamber.
3. Disconnect the speedometer cable at the transaxle.
4. Remove the retaining clip, then disconnect the slave cylinder line from the slave cylinder hose and plug the hose.
5. Disconnect the ground strap from the transaxle.
6. Remove the tie wrap and disconnect the 3 electrical connectors located above the transaxle. Remove the electrical connector support bracket.
7. Mount engine support bar D88L-6000-A or equivalent, and attach it to the engine hangers.
8. Remove the 3 nuts from the upper transaxle mount. Loosen the mount pivot nut and rotate the mount out of position. Remove the 3 bolts and the upper transaxle mount bracket.
9. Remove the 2 upper transaxle-to-engine bolts.
10. Raise and safely support the vehicle.
11. Remove the front wheel and tire assemblies.
12. Remove the inner fender splash shields.
13. Drain the transaxle fluid and install the drain plug.
14. Remove the halfshafts. Install 2 transaxle plugs between the differential side gears.

NOTE: Failure to install the transaxle plugs may cause the differential side gears to become improperly positioned.

15. Remove the plenum support bracket and remove the starter.
16. Remove the nut and the extension bar and the bolt and nut and shift control rod from the transaxle.
17. Remove both lower splash shields.
18. Remove the 2 transaxle mount-to-crossmember nuts and remove the lower crossmember and the front transaxle mount.
19. Position and secure a jack under the transaxle.
20. Remove the 5 lower engine-to-transaxle bolts and lower the transaxle out of the vehicle.

To install:

21. Apply a thin coating of grease to the spline of the input shaft.
22. Place the transaxle onto a jack. Make sure the transaxle is secure.
23. Raise the transaxle into position on the engine.
24. Install the 5 lower engine-to-transaxle bolts and tighten to 27–38 ft. lbs. (37–52 Nm).
25. Install the front transaxle mount and tubing bracket. Tighten the bolts to 12–17 ft. lbs. (16–23 Nm).
26. Install the lower crossmember.

Tighten the nuts and bolts to 47–66 ft. lbs. (64–89 Nm).

27. Install the 2 transaxle mount-to-crossmember nuts and tighten to 27–38 ft. lbs. (37–52 Nm).

28. Install both lower splash shields.

29. Install the shift control rod bolt and nut and tighten to 23–34 ft. lbs. (31–46 Nm).

30. Install the extension bar nut and tighten to 12–17 ft. lbs. (16–23 Nm).

31. Install the starter and the plenum support bracket.

32. Remove the transaxle plugs and install the halfshafts.

33. Install the inner fender splash shields.

34. Install the wheel and tire assemblies. Tighten the lug nuts to 65–87 ft. lbs. (88–118 Nm).

35. Lower the vehicle.

36. Install the 2 upper engine-to-transaxle bolts and tighten to 47–66 ft. lbs. (64–89 Nm).

37. Install the upper transaxle mount bracket and tighten the 3 bolts. Rotate the mount into position and tighten the pivot nut. Install and tighten the 3 upper mount nuts.

38. Remove the engine support bar.

39. Install the electrical connector support bracket. Connect the 3 electrical connectors and secure with the tie wrap.

40. Connect the ground strap to the transaxle.

41. Connect the slave cylinder line to the slave cylinder hose and install the retaining clip.

42. Add the proper type and amount of fluid to the transaxle.

43. Connect the speedometer cable.

44. Install the air hose and the resonance chamber.

45. Install the battery tray and the battery. Connect the battery cables.

46. Check for fluid leaks and proper operation.

CLUTCH

Clutch Assembly

REMOVAL & INSTALLATION

1. Disconnect the negative battery cable. Raise and safely support the vehicle. Remove the transaxle.

2. Matchmark the pressure plate assembly and the flywheel so they can be assembled in the same position.

3. Loosen the pressure plate-to-flywheel bolts 1 turn at a time, in sequence, until spring tension is relieved to prevent pressure plate cover distortion.

4. Support the pressure plate and remove the bolts. Remove the pressure plate and clutch disc from the flywheel.

5. Inspect the flywheel, clutch disc, pressure plate, throwout bearing, pilot bearing and the clutch fork for wear; replace parts, as required.

NOTE: If the flywheel shows any signs of overheating (blue discoloration) or if it is badly grooved or scored, it should be refaced or replaced.

To install:

6. If removed, install a new pilot bearing using an installation tool.

7. If removed, install the flywheel. Make sure the flywheel and crankshaft flange mating surfaces are clean. Tighten the flywheel bolts to 71–76 ft. lbs. (96–103 Nm).

8. Clean the pressure plate and flywheel surfaces thoroughly. Position the clutch disc and pressure plate into the installed position, aligning the matchmarks made previously; support them with a dummy shaft or clutch aligning tool.

9. Install the pressure plate-to-flywheel bolts. Tighten them gradually in a criss-cross pattern to 13–20 ft. lbs. (18–26 Nm). Remove the alignment tool.

10. Lubricate the release bearing and install it in the fork.

11. To complete the installation, reverse the removal procedures. Lower the vehicle and connect the negative battery cable.

PEDAL HEIGHT/FREE-PLAY ADJUSTMENT

Pedal Height

To determine if the pedal height requires adjustment, measure the distance from the bulkhead to the upper center of the pedal pad. The distance should be 8.4–8.6 in. (214–219mm) on 1988–90 vehicles or 7.72–8.03 in. (196–204mm) on 1991–92 vehicles. If adjustment is necessary, proceed as follows:

1988–90

1. Remove the necessary instrument panel components which block access to the clutch pedal.

2. Loosen the clutch pedal locknut.

3. Turn the stop bolt to obtain the correct pedal height of 8.4–8.6 in. (214–219mm) and tighten the locknut.

4. If components from the instrument panel were removed, reinstall them.

1991–92

1. Disconnect the clutch switch electrical connector.

2. Loosen the clutch switch locknut.

3. Turn the clutch switch until the correct height is achieved.

4. Tighten the locknut to 10–13 ft. lbs. (14–18 Nm).

5. Measure the pedal free-play.

6. Connect the electrical connector.

Pedal Free-Play

To determine if the pedal free-play requires adjustment, depress the clutch pedal by hand until clutch resistance is felt. Measure the distance between the upper pedal height and where the resistance is felt. Free-play should be 0.35–0.59 in. (9–15mm) on 1988–90 vehicles or 0.20–0.51 in. (5–13mm) on 1991–92 vehicles. If an adjustment is necessary, proceed as follows:

1988–90

1. Depress the clutch pedal and pull the pin away from the clutch lever (at the transaxle).

2. Turn the cable locknut to adjust the clutch pedal free-play of 0.35–0.59 in. (9–15mm).

3. Depress the pedal and check the disengagement height of 3.3 in. (85mm); the distance from the floor and the center of the clutch pedal pad.

1991–92

1. Loosen the pushrod locknut.

2. Turn the pushrod until the pedal free-play is within specification.

3. Check that the disengagement height is correct when the pedal is fully depressed. Minimum disengagement height is 1.6 in. (41mm).

4. Tighten the pushrod locknut to 9–12 ft. lbs. (12–17 Nm).

Clutch Cable

REMOVAL & INSTALLATION

1988–90

1. At the transaxle, remove the clutch cable adjusting nut and pin; separate the cable from the fork.

2. Remove the clutch cable bracket-to-cowl nuts and the bracket.

3. From under the instrument panel, separate the clutch cable from the top of the clutch pedal.

4. Pull the cable through the cowl and remove the cable assembly from the engine side.

5. Inspect the clutch cable housing for frayed wire, cracked or worn housing and the cable for smooth operation; replace the cable assembly, if necessary.

6. To install, lubricate the cable

with multi-purpose grease and reverse the removal procedures. Adjust the clutch pedal free-play.

Clutch Master Cylinder

REMOVAL & INSTALLATION

1991–92

1. Disconnect the battery cables and remove the battery and battery tray.
2. Disconnect the clutch pipe from the master cylinder using a line wrench.
3. Disengage the clamp and remove the master cylinder hose from the clutch master cylinder. Prevent excess fluid loss by plugging the hose.
4. Remove the external mounting nut.
5. Remove the internal mounting nut and remove the master cylinder.
To install:
6. Align the pushrod and install the clutch master cylinder.
7. Install the external and internal mounting nuts and tighten to 14–19 ft. lbs. (19–25 Nm).
8. Connect the clutch pipe and tighten the nut to 10–16 ft. lbs. (13–22 Nm).
9. Install the hose and the clamp to the master cylinder.
10. Install the battery and battery tray.
11. Bleed the air from the system.
12. Test the system and make sure there is no leakage.
13. Connect the negative battery cable.

Clutch Slave Cylinder

ADJUSTMENT

1991–92

The clutch slave cylinder is not adjustable. The only adjustments necessary on the clutch control system are pedal height and pedal free-play.

REMOVAL & INSTALLATION

1991–92

1. Disconnect the pressure line. Plug the line to prevent leaking.
2. Remove the attaching bolts and remove the slave cylinder.
To install:
3. Install the slave cylinder.
4. Install the attaching bolts and tighten to 12–17 ft. lbs. (16–23 Nm).
5. Connect the pressure line and tighten the nut to 10–16 ft. lbs. (13–22 Nm).
6. Bleed the air from the system.

7. Press on the clutch pedal and make sure there is no leakage.

Hydraulic Clutch System Bleeding

NOTE: The fluid level in the reservoir must be maintained at the ¾ level or higher during air bleeding.

1. Remove the bleeder cap from the slave cylinder and attach a vinyl hose to the bleeder screw.
2. Place the other end of the hose in a container.
3. Slowly pump the clutch pedal several times.
4. With the clutch pedal depressed, loosen the bleeder screw to release the fluid and air.
5. Tighten the bleeder screw.
6. Repeat the last 3 steps until no air bubbles appear in the fluid.

AUTOMATIC TRANSAXLE

For further information on transmissions/transaxles, please refer to "Chilton's Guide to Transmission Repair".

Transaxle Assembly

REMOVAL & INSTALLATION

1988–90

1. Disconnect the negative battery cable.
2. Remove the air cleaner. Loosen the front wheel lug nuts.
3. From the transaxle, disconnect the speedometer cable.
4. Disconnect the shift control cable-to-transaxle clip and 2 bracket bolts. Remove the ground wire from the cylinder head.
5. Remove the water pipe bracket bolt and the bracket.
6. Remove the secondary air pipe, the EGR pipe bracket and the electrical harness clip.
7. Disconnect the electrical connectors from the inhibitor switch, the neutral switch and the kickdown solenoid. Disconnect the body ground connector.
8. Remove the upper 2 transaxle-to-engine bolts.
9. Remove the vacuum hose from the vacuum diaphragm line. Disconnect and plug the oil cooler lines at the transaxle.

10. Using a engine support bar tool, attach it to the rear engine lifting hook and support the engine's weight.
11. Raise and support the vehicle safely.
12. Place a drain pan under the transaxle, remove the drain plug and drain the transaxle.
13. Remove the front wheel lug nuts and the wheels. Remove the engine undercover and side covers.
14. Remove the front stabilizer bar. From both sides, remove the lower control arm ball joint-to-steering knuckle nut/bolt, pull the control arm downward and separate the lower control arm from the steering knuckle.

NOTE: When separating the ball joint, be careful not to damage the ball joint dust boot.

15. Using a medium prybar, insert it between the halfshaft and the transaxle (a notch is provided), pry both halfshafts from the transaxle.

NOTE: When removing the halfshafts, withdraw them completely from the transaxle (to prevent damage to the oil seal lips), do not move the CV-joints in excess of a 20 degree angle (damage to the boots and/or joint may occur) and use a wire to support the halfshaft in the horizontal position.

16. From under the vehicle, remove the crossmember-to-chassis bolts and the crossmember.
17. Remove the starter's positive cable-to-solenoid nut and the solenoid wire by pulling the wire from the connector.
18. Remove the starter-to-engine bolts and the starter. Remove the dust cover bolts and the cover.
19. Matchmark the torque converter-to-flexplate location. Remove the torque converter-to-flexplate bolts and slide the torque converter back into the transaxle.
20. Loosen the bracket bar on the engine support tool to lower the transaxle. Using a floor jack, support the transaxle.
21. Remove the No. 2 engine mount-to-transaxle nut/bolt, the transaxle-to-engine bolts and lower the transaxle from the vehicle.
To install:
22. Hold the torque converter in an upright position and fill it with fluid. Install the torque converter on the transaxle input shaft. If the converter does not fit easily, remove the converter, realign the splines and refit the converter. Do not use force.
23. To make sure the converter is correctly installed, measure the clearance between the end of the converter

and the end of the converter housing. The clearance should be 0.79 in. (20mm).

24. Align the mounting studs on the engine with the holes in the transaxle case and install the transaxle. Install the bolts and tighten to 47–66 ft. lbs. (64–89 Nm).

25. Raise the transaxle to the proper position and install the bolt/nut to the No. 2 engine mount. Tighten to 27–38 ft. lbs. (37–52 Nm).

26. Align the matchmarks on the converter and flexplate and install the mounting bolts. Tighten to 25–36 ft. lbs. (34–49 Nm).

27. Install the starter and tighten the mounting bolts to 23–34 ft. lbs. (31–46 Nm). Connect the solenoid wire and positive battery cable to the starter.

28. Install the crossmember. Tighten the crossmember mounting bolts to 47–66 ft. lbs. (64–89 Nm) and the rear mount bolt to 21–34 ft. lbs. (28–46 Nm).

29. Install a new clip on both halfshaft ends, with the gap at the top of the clip groove. Slide the halfshafts into the transaxle, being careful not to damage the seals. Make sure the clips engage into the side gears.

30. Install the lower ball joints into the steering knuckles. Tighten the nuts to 32–40 ft. lbs. (43–54 Nm).

31. Install the under and side covers and the wheel and tire assemblies. Lower the vehicle.

32. Install the 2 remaining upper transaxle mounting bolts and tighten to 47–66 ft. lbs. (64–89 Nm).

33. Remove the engine support bar.

34. Connect the transaxle cooler lines, the vacuum line to the vacuum diaphragm, the neutral switch connector, body ground connector, inhibitor switch and kick-down solenoid wiring.

35. Install the wire harness clip, the secondary air pipe and EGR pipe bracket, and the engine ground wire.

36. Connect the change control cable, install the mounting bracket bolts and tighten. Connect the speedometer cable and install the hold-down bolt.

37. Install the air cleaner and all remaining components.

38. Connect the negative battery cable. Fill the transaxle with the proper type and quantity of fluid. Start the engine and check for leaks.

1991–92

1. Disconnect the battery cables and remove the battery and battery tray.

2. Disconnect the wiring harness retaining clip from the battery tray.

3. Remove the air cleaner assembly.

4. Disconnect the shift control cable from the manual lever.

5. Disconnect the speedometer cable from the transaxle by unsnapping the cable at the speedometer driven gear.

6. Disconnect the transaxle electronic control electrical connectors and separate the harness from the transaxle clips.

7. Remove the manual lever position switch wiring brackets and disconnect the ground cables from the top of the transaxle.

8. Remove the starter.

9. Disconnect the manual lever position switch wiring connectors.

10. Install engine support tool D88L–6000–A or equivalent, to support the engine.

11. Disconnect the kickdown cable at the throttle cam.

12. Place a drain pan under the transaxle and disconnect the transaxle cooler lines at the transaxle.

13. Remove the upper transaxle mount bolts, the mount and the upper transaxle housing bolts.

14. Disconnect the oxygen sensor electrical connector, the transaxle vent hose, and the electrical connector at the vehicle speed sensor.

15. Raise and safely support the vehicle.

16. Remove the front wheel and tire assemblies.

17. Using a hammer and a flat punch, straighten the detent in the halfshaft nut.

18. Remove the nuts securing the halfshafts to the steering knuckles and remove the nuts and bolts securing the lower ball joints to the steering knuckles. Separate the lower ball joints from the steering knuckles.

19. Disconnect the halfshaft midbearing bracket from the back of the engine.

20. Remove the halfshafts from both steering knuckles.

21. Remove the 3 engine/transaxle lower splash shields and the torque converter inspection plate. Remove the nuts securing the torque converter to the flexplate.

22. Disconnect the lower crossmember from the chassis and the transaxle mounts.

23. Remove the driver's side and then the passenger's side halfshafts. Install 2 transaxle plugs T88C–7025–AH or equivalent into the differential side gears.

NOTE: Failure to install the transaxle plugs may cause the differential side gears to become improperly positioned.

24. Position the drain pan and remove the drainplug from the transaxle. Drain the fluid from the differential cavity. Remove the transaxle pan and drain the transaxle fluid.

25. Position a transmission jack under the transaxle. Secure the transaxle to the jack.

26. Remove the lower bolts securing the transaxle to the engine and carefully lower the transaxle out of the vehicle.

To install:

NOTE: A pin is used for securing the throttle cam in a fixed position on new and rebuilt transaxles. This pin must be removed to allow proper transaxle operation. If the pin is not removed, the throttle lever will remain in a fixed position. After removing the pin, apply sealant to the bolt from the previous transaxle. Install the bolt and tighten to 69–95 inch lbs. (8–11 Nm).

27. Secure the transaxle on the transmission jack.

28. Raise the transaxle into position and install the lower transaxle-to-engine bolts. Tighten the bolts to 41–59 ft. lbs. (55–80 Nm).

29. Position the torque converter to the flexplate and install the nuts. Tighten the nuts to 25–36 ft. lbs. (34–49 Nm).

30. Install the halfshafts.

31. Connect the crossmember to the transaxle mounts and the chassis. Tighten the crossmember-to-transaxle mount nuts to 27–38 ft. lbs. (37–52 Nm). Tighten the crossmember-to-chassis nuts and bolts to 47–66 ft. lbs. (64–89 Nm).

32. Install the engine/transaxle splash shields and the starter.

33. Position the lower ball joints into the steering knuckles and secure with the nuts and bolts. Tighten the nuts and bolts to 32–43 ft. lbs. (43–59 Nm).

34. Position the tie rod ends into the steering knuckles and install the nuts. Tighten to 31–42 ft. lbs. (42–57 Nm).

35. Install the wheel and tire assemblies. Tighten the lugs to 65–88 ft. lbs. (88–118 Nm).

36. Lower the vehicle.

37. Install the transaxle-to-engine bolts and tighten to 41–59 ft. lbs. (55–80 Nm).

38. Install the upper transaxle mount and tighten the nuts to 49–69 ft. lbs. (67–93 Nm).

39. Connect the transaxle vent hose, the electrical connector at the speed sensor, the speedometer cable and the oxygen sensor connector.

40. Connect the transaxle cooler lines and connect the kickdown cable at the throttle body.

41. Remove the engine support.

42. Connect the ground wires to the transaxle and connect the manual le-

ver position switch bracket and wiring connectors.

43. Connect the shift control cable to the cable bracket and to the selector lever.

44. Install the battery tray and battery.

45. Install the air cleaner assembly.

46. Connect the battery cables.

47. Add the proper type and quantity of transaxle fluid.

48. Check the transaxle for proper operation.

SHIFT LINKAGE ADJUSTMENT

1988–90

1. Place the gear selector lever in the **N** position.

2. At the transaxle, remove the shift cable trunnion-to-transaxle shift lever spring clip and pin.

3. Rotate the transaxle shift lever fully counterclockwise to place it in the **P** position.

4. Move the transaxle shift lever clockwise 2 detents to place it in the **N** position.

NOTE: When moving the transaxle shift lever, be sure to position it between the ends of the shift cable trunnion.

5. If the trunnion holes align with the shift lever hole, the cable is adjusted; replace the pin and spring clip. If the holes are not aligned, proceed with the remaining adjustment procedures.

6. From inside the vehicle, remove the shift quadrant bezel-to-console screws. Lift the front of the bezel to disengage it from the console and rotate it to provide access to the cable adjusting nuts.

7. At the shift cable, loosen the adjusting nuts.

8. Position the gear selector lever in the **P** position and inspect the detent spring roller. If the spring is not centered, perform the following procedures:

a. Loosen the detent spring roller screws and move the spring to center it in the **P** position.

b. Position the shift quadrant and reinstall the screws.

9. Move the shift selector lever to the **N** position.

10. Move the shift cable adjuster nuts until the holes in the cable trunnion and transaxle shift lever are aligned. Torque the shift cable adjuster nuts to 69–95 inch lbs. (8–11 Nm).

11. Recheck the cable trunnion and transaxle shift lever holes for alignment. If aligned, install the pin and spring clip.

12. Using an assistant to watch the transaxle shift lever movement, start with the gear selector lever in the **N** position, push the shift interlock button and carefully move the shift lever forward until the transaxle shift lever begins to move; note the amount of shift selector movement.

13. With the gear selector lever in the **N** position, press in on the shift interlock button and carefully pull the lever rearward while an assistant watches the transaxle shift lever. When the transaxle lever begins to move, note the amount the shift lever has moved.

14. If the shift selector lever forward movement **a** does not equal the rearward movement **b**, turn the adjuster nuts until the movement is equal.

NOTE: Make sure the adjustment procedure does not affect the neutral safety switch operation. Apply the parking brakes and try to start the engine in the N and P positions. If the engine starts in any other gear selector lever positions, check and adjust the linkage adjustment and the neutral safety switch adjustment.

15. When adjustment is completed, tighten the adjuster nut to 69–95 inch lbs. (8–11 Nm). Position the shift quadrant bezel and install the attaching screws.

1991–92

1. Disconnect the negative battery cable. This will deactivate the shift-lock system.

2. Move the gear selector lever to **P**.

3. Remove the screw securing the gear selector knob to the gear selector lever. Remove the knob.

4. Remove the shift console as follows:

a. Remove the rear seat ash tray and position both front seats to the rear-most position.

b. Remove the 2 front retaining screws from the parking brake console and recline both front seats.

c. Remove the 2 rear retaining screws from the parking brake console.

d. With the parking brake engaged, remove the parking brake console.

e. Remove the 2 front retaining screws from the shift console and remove the console.

5. Remove the position indicator mounting screws and disconnect the illumination bulb from the position indicator.

6. Disconnect the shift-lock servo and park range switch electrical connectors.

7. Remove the position indicator.

NOTE: Make sure the detent spring roller is in the P detent.

8. Loosen the shift control cable bracket mounting bolts.

9. Push the gear selector lever against the **P** range and hold it.

10. Tighten the shift control cable bracket mounting bolts to 69–95 inch lbs. (8–11 Nm).

11. Lightly press the gear selector pushrod and make sure the guide plate and guide pin clearances are within specifications.

12. Check that the guide plate and guide pin clearances are within the appropriate specifications when the selector lever is shifted to **N** and **OD**. If the clearances are not as specified, readjust the shift control cable.

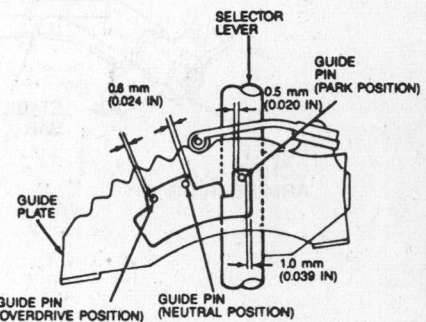

Shift control cable adjustment clearances—1991–92 Escort

13. Make sure the gear selector operates properly.

14. Connect the illumination bulb to the position indicator.

15. Connect the shift-lock servo and park range switch electrical connectors.

16. Install the position indicator and secure it with the mounting screws.

17. Install the shift console by reversing the removal procedure.

18. Position the gear selector knob onto the gear selector lever and secure the knob with the screw.

19. Connect the negative battery cable.

FRONT SUSPENSION

MacPherson Strut

REMOVAL & INSTALLATION

1. Raise and safely support the vehicle.

2. Remove the front wheel and tire assembly.

3. Remove the clip securing the

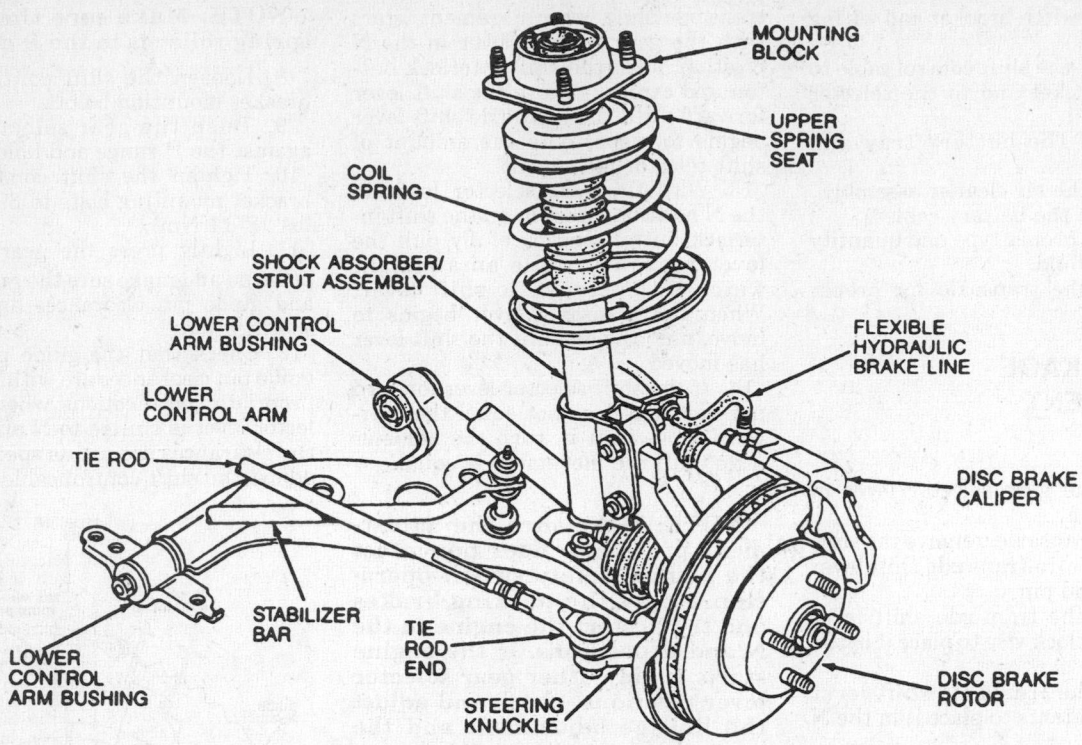

MOUNTING BLOCK

UPPER SPRING SEAT

COIL SPRING

SHOCK ABSORBER/ STRUT ASSEMBLY

LOWER CONTROL ARM BUSHING

LOWER CONTROL ARM

TIE ROD

FLEXIBLE HYDRAULIC BRAKE LINE

DISC BRAKE CALIPER

STABILIZER BAR

TIE ROD END

LOWER CONTROL ARM BUSHING

STEERING KNUCKLE

DISC BRAKE ROTOR

Front suspension assembly—1991–92 vehicles

flexible brake hose to the strut assembly.

4. On 1988–90 vehicles, paint a white aligning stripe on the inside of the strut mounting block.

5. Remove the 2 nuts and 2 bolts securing the strut assembly to the steering knuckle.

6. Remove the upper mounting block nuts and remove the strut assembly from the vehicle.

7. Remove the cap from the top of the strut assembly.

8. Secure the strut assembly mounting block in a vise. Turn the piston rod nut 1 revolution to loosen.

9. Install an appropriate spring compressor onto the strut spring and compress the spring.

10. Remove the nut, mounting block, thrust bearing, upper spring seat, rubber spring seat, coil spring and bound stopper.

To install:

11. Position the bound stopper onto the strut piston rod.

12. With the coil spring compressed, position the spring onto the strut assembly.

13. Install the rubber spring seat, upper spring seat, thrust bearing, mounting block and piston rod nut. Tighten the piston rod nut to 58–81 ft. lbs. (79–110 Nm).

14. With the nut tightened to specification, carefully remove the spring compressor from the spring while making sure the spring is properly

seated in the upper and lower spring seats.

15. Install the cap.

16. Position the strut assembly into the wheel housing. Make sure the direction indicator on the mounting block faces inboard.

17. Secure the upper mounting block to the strut tower with the nuts. Tighten the nuts to 22–30 ft. lbs. (29–40 Nm).

18. Attach the strut assembly to the steering knuckle and install the bolts and nuts. Tighten to 69–72 ft. lbs. (93–97 Nm).

19. Position the flexible brake hose to the strut assembly and secure it with the brake hose clip.

20. Install the front wheel and tire assembly.

21. Lower the vehicle and check the front wheel alignment.

Lower Ball Joints

INSPECTION

1. Raise and safely support the vehicle so wheels are in the full-down position.

2. Have an assistant grasp lower edge of the tire and move wheel and tire assembly in and out.

3. As wheel is being moved in and out, observe lower end of knuckle and lower control arm. Any movement indicates abnormal ball joint wear.

4. If any movement is observed, install new lower ball joint.

REMOVAL & INSTALLATION

1988–90

1. Raise the vehicle and support it safely.

2. Rmove the wheel and tire assembly.

3. Remove the brake caliper and support it aside with mechanics wire. Do not disconnect the brake line.

4. Separate the stabilizer bar from the control arm.

5. Remove the tie rod end cotter pin and nut. Separate the tie rod from the knuckle, using the proper removal tool.

6. Remove the ball joint-to-knuckle clamp bolt and pry the lower control arm from the knuckle.

7. Remove the ball joint-to-knuckle bolts and remove the ball joint.

8. Installation is the reverse of the removal procedure. Torque the ball joint clamp bolt to 32–40 ft. lbs. (43–54 Nm).

1991–92

1. Raise and safely support the vehicle.

2. Remove the wheel and tire assembly.

3. Remove the nut and bolt securing the ball joint to the steering knuckle.

4. Remove the nuts securing the lower ball joint to the lower control arm. Remove the lower ball joint.

5. Mount the lower ball joint in a vise.

6. Place a chisel between the ball joint and the dust boot. Lightly tap on the chisel to separate the dust boot from the ball joint.

To install:

7. Position the dust boot over the ball joint and, using a suitable tool, press down on the tool to secure the dust boot to the ball joint.

8. Install the ball joint into the lower control arm and install the mounting nuts. Tighten the nuts to 69–86 ft. lbs. (93–117 Nm).

9. Install the lower ball joint into the steering knuckle and secure it with the nut and bolt. Tighten the nut to 32–43 ft. lbs. (43–59 Nm).

10. Install the wheel and tire assembly and lower the vehicle.

Lower Control Arms

REMOVAL & INSTALLATION

1988–90

1. Raise and safely support the vehicle. Remove the wheel and tire assembly.

2. Remove the brake caliper and support it aside with mechanics wire. Do not disconnect the brake hose.

3. Remove the stabilizer link assembly.

4. Paint aligning stripes on the rear control arm bushing and mounting bracket and on the rear control arm bushing and control arm.

5. Remove the ball joint clamp bolt from the steering knuckle.

6. Loosen the lower control arm front bushing nut and rear bushing bolt.

7. Remove the lower control arm rear bushing bracket mounting bolts.

8. Remove the front bushing bracket and the rear bushing bolt.

9. Remove the lower control arm. Remove the front bushing nut and remove the bushing.

To install:

10. Install the front bushing on the lower control arm and install the nut, hand-tight.

11. Position the lower control arm on the vehicle and install the rear bushing bolt. Install the front bushing bracket.

12. Raise the lower control arm until the painted stripes align, then tighten the front bushing nut and rear bushing bolt.

13. Install the ball joint stud in the steering knuckle. Install the clamp bolt and tighten to 32–40 ft. lbs. (44–55 Nm).

14. Install the stabilizer link assembly. Tighten the nut until 0.43 in. (10.8mm) protrudes below the nut.

15. Install the brake caliper and the wheel and tire assembly. Lower the vehicle.

1991–92

1. Raise and safely support the vehicle.

2. Remove the front wheel and tire assembly.

3. Remove the front stabilizer nuts, washers, bushings, bolts and sleeves.

4. Remove the lower control arm front bushing bolt and washer.

5. Remove the bolts securing the lower control arm rear bushing retaining strap.

6. Remove the nut and bolt securing the lower ball joint to the steering knuckle. Separate the steering knuckle from the lower ball joint.

7. Remove the lower control arm.

8. Remove the nut and washers from the lower control arm rear pivot bolt.

9. Remove the lower control arm rear bushing.

To install:

10. Position the lower control arm rear bushing onto the rear pivot bolt.

11. Install the washers and nut onto the lower control arm pivot bolt. Tighten the nut to 69–86 ft. lbs. (93–117 Nm).

12. Install the ball joint into the steering knuckle. Install the ball joint retaining nut and bolt and tighten the nut to 32–43 ft. lbs. (43–59 Nm).

13. Install the lower control arm rear bushing retaining strap to the lower frame. Install the bolts and tighten to 69–86 ft. lbs. (93–117 Nm).

14. Install the lower control arm front pivot bolt and washer. Tighten the nut to 69–93 ft. lbs. (93–127 Nm).

15. Install the stabilizer bolts, washers, bushings, sleeves and nuts. Tighten the stabilizer nuts so 0.67–0.75 in. (17–19mm) of thread is exposed at the end of the bolt.

16. Install the wheel and tire assembly. Tighten the lug nuts to 65–87 ft. lbs. (88–118 Nm).

17. Lower the vehicle.

Stabilizer Bar

REMOVAL & INSTALLATION

1988–90

1. Raise and safely support the vehicle.

2. Disconnect the stabilizer links from the lower control arm.

3. Remove the stabilizer bushing bracket bolts and remove the brackets and bushings. Remove the stabilizer bar.

To install:

4. Install the stabilizer bar onto the vehicle with the bushings, brackets and bracket bolts. Install the bolts hand-tight at this time.

5. Install the stabilizer link assembly. Tighten the nut until 0.43 in. (10.8mm) protrudes below the nut.

6. Lower the vehicle. Now that the suspension is loaded, tighten the bushing bracket mounting bolts to 44–54 ft. lbs. (59–74 Nm).

1991–92

1. Support the engine with engine support D88L–6000–A or equivalent.

2. Raise and safely support the vehicle.

3. Remove the wheel and tire assemblies.

4. Remove the nuts securing the steering gear mounting brackets and position the steering gear slightly forward.

5. Remove the stabilizer bar nuts, washers, bushings, sleeves and bolts from the lower control arm.

6. Remove the rear crossmember nuts from the rear transaxle mount and the vehicle frame.

7. Loosen the front crossmember bolts and nuts from the front transaxle mount and the vehicle frame. Lower the rear end of the crossmember.

8. Remove the nuts and bolts securing the chassis frame to the vehicle frame. Lower the chassis frame.

NOTE: The engine and transaxle mounts will support the chassis frame when unbolting the chassis frame from the vehicle frame.

9. Unbolt the stabilizer bar from the chassis frame and remove the stabilizer bar from the vehicle.

To install:

10. Position the stabilizer bar into the vehicle.

11. Secure the stabilizer bar to the chassis frame with the bolts. Tighten the bolts to 32–43 ft. lbs. (43–59 Nm).

12. Install the chassis frame to the vehicle frame with the bolts and nuts. Tighten the bolts and nuts to 69–93 ft. lbs. (93–127 Nm).

13. Position the crossmember to the vehicle frame and the transaxle mounts. Tighten the bolts and nuts to the specified torque.

14. Install the stabilizer bar bolts, sleeves, bushings, washers and nuts. Tighten the stabilizer bolts so 0.67–0.75 in. (17–19mm) of thread is exposed at the end of the bolt.

15. Position the steering gear and secure it with the brackets and nuts. Tighten the nuts to 28–38 ft. lbs. (37–52 Nm).

16. Install the wheel and tire assem-

Tightening Torque:
A: 37-52 N·m (27-38 LB-FT)
B: 64-89 N·m (47-66 LB-FT)

Crossmember mounting bolts and nuts torque specifications

blies. Tighten the lug nuts to 65–87 ft. lbs. (88–118 Nm).

17. Lower the vehicle.

REAR SUSPENSION

MacPherson Strut

REMOVAL & INSTALLATION

1988–90

1. Raise and safely support the vehicle. Remove the wheel and tire assembly.

2. Remove the brake drum and backing plate or the disc brake caliper and rotor, as required.

3. Loosen the trailing arm bolt and the spindle-to-strut mounting bolts. Remove the trailing arm and spindle mounting bolts.

4. Paint a white index mark on the strut rubber mounting bracket. Remove the strut mounting nuts from inside the vehicle. Remove the strut assembly.

5. Install a coil spring compressor on the strut assembly. While the spring is compressed, remove the nut, rubber mounting bracket, spring upper seal and rubber spring seat.

6. Slowly release the coil spring and remove the spring compressor. Remove the coil spring, dust boot and rebound bumper from the strut.

To install:

7. Install the rebound bumpers and dust boot on the strut. Compress the coil spring with the spring compressor and install the spring on the strut.

8. Install the rubber seat, spring upper seat with rubber mounting bracket and the nut. Slowly release the spring compressor.

9. Install the strut assembly in the strut tower, aligning the index mark that was made during the removal procedure. Install and tighten the mounting nuts.

10. Install the spindle-to-strut mounting bolts and tighten the bolts hand-tight.

11. Install the rear brake assembly and the wheel and tire assembly. Lower the vehicle.

12. Now that the suspension is loaded, tighten the spindle-to-strut mounting bolts to 69–86 ft. lbs. (93–117 Nm).

1991–92

1. Raise and safely support the vehicle.

2. Remove the wheel and tire assembly.

3. Remove the clip securing the flexible brake hose to the rear strut assembly.

4. Remove the nuts and bolts securing the rear strut assembly to the rear wheel spindle assembly.

5. On hatchback and wagon, remove the quarter lower trim panel.

6. Remove the mounting block nuts and remove the rear strut assembly from the vehicle.

7. Position the strut assembly into a vise and secure the assembly at the mounting block.

8. Remove the cap and loosen the piston rod nut 1 turn. Do not remove the piston rod nut at this time.

9. Install an appropriate coil spring compressor onto the coil spring and compress the coil spring.

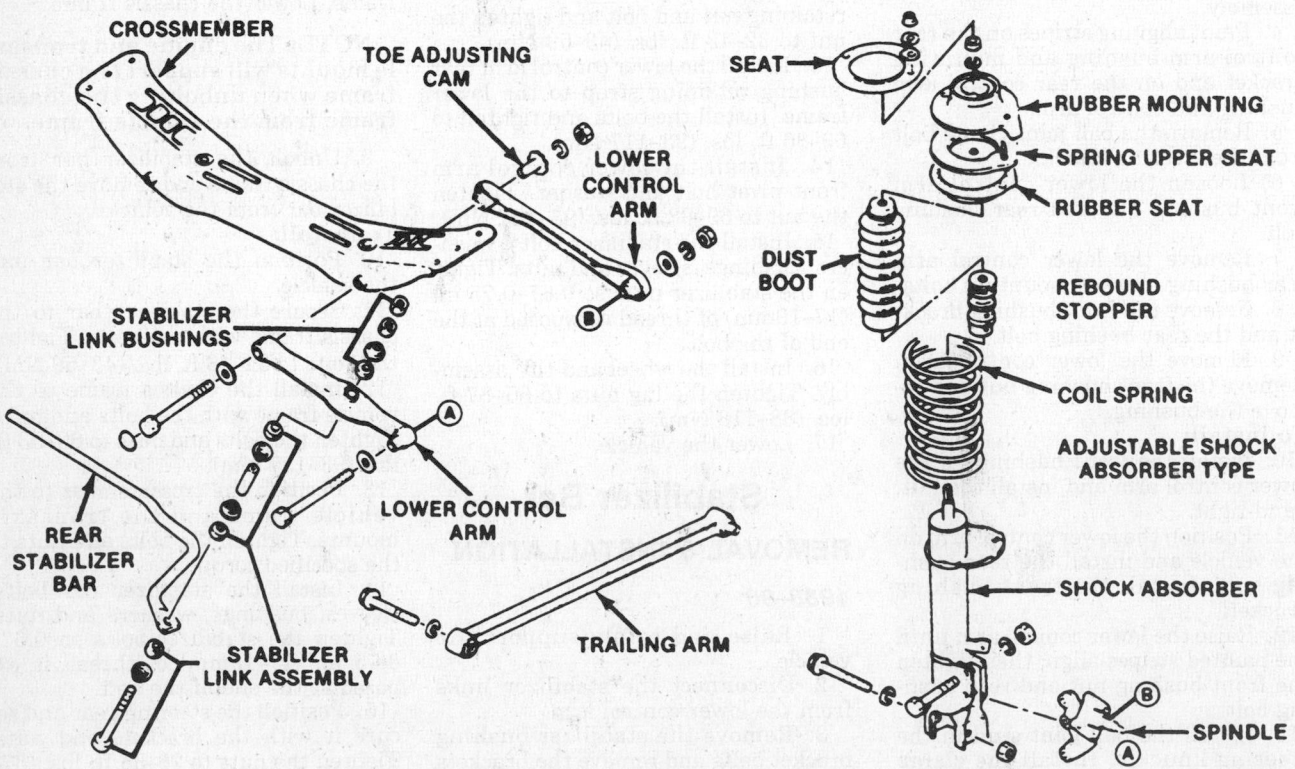

Rear suspension assembly—1988–90 vehicles, except station wagon

10. Remove the piston rod nut, washer, retainer and mounting block.

11. Remove the coil spring.

12. Remove the bound stopper seat and stopper from the strut piston.

To install:

13. Position the strut assembly into a vise and secure.

14. Install the bound stopper seat and stopper onto the strut piston rod.

15. Install the coil spring onto the strut assembly.

16. Install the mounting block, then align the mounting block studs and the lower bracket of the strut assembly.

17. Install the retainer, washer and piston rod nut. Tighten the nut to 41–50 ft. lbs. (55–68 Nm).

18. Make sure the spring is properly aligned and carefully release the spring into the seats of the strut.

19. Remove the spring compressor from the coil spring and install the cap.

20. Position the strut assembly into the vehicle wheel housing.

21. Install the mounting block nuts and tighten to 22–27 ft. lbs. (29–40 Nm).

22. On hatchback and wagon, install the quarter lower trim panel.

23. Install the nuts and bolts securing the strut assembly to the rear spindle assembly. Tighten the lower strut bolts to 69–93 ft. lbs. (93–127 Nm).

24. Install the wheel and tire assembly. Tighten the lug nuts to 65–87 ft. lbs. (88–118 Nm).

25. Check the rear alignment and lower the vehicle.

Rear Control Arms

REMOVAL & INSTALLATION

1988–90

1. Raise the vehicle and support it safely.

2. Remove the wheel and tire assembly. Remove the brake drum and backing plate, if equipped with drum brakes. If equipped with disc brakes, remove the caliper and rotor.

3. Matchmark the rear toe adjusting cam, the control arm bushings and the control arm to the crossmember.

5. Remove the stabilizer bar mounting bolts and remove the stabilizer bar.

6. Remove the trailing arm mounting bolts and remove the parking brake cable from the trailing arm.

7. Remove the control arm and the trailing arm from the vehicle.

To install:

8. Mount both control arms on the crossmember and hand-tighten the bolts.

9. Connect both control arms with

the outer control arm bolt but do not install the spindle yet. Raise the control arms to align the matchmarks and tighten the control arm bolts. Be sure to align the matchmarks on the control arm and alignment cam.

10. Install the spindle in the strut and tighten the bolts to 69–86 ft. lbs. (93–117 Nm). Install the control arm-to-spindle attaching bolt and tighten to 69–86 ft. lbs. (93–117 Nm).

11. Tighten the inner control arm bolt to 69–86 ft. lbs. (93–117 Nm).

12. Loosely install the stabilizer bar in the bushing, making sure the alignment mark on the bar aligns with the bushings. Do not fully tighten the bracket bolts yet.

13. Install the stabilizer link assembly. Tighten the stabilizer link bolt until 0.31 in. (18mm) of thread extends beyond the nut.

14. Install the brake assembly and the wheel and tire assembly. Lower the vehicle.

15. Now that the suspension is loaded, tighten the stabilizer bushing bracket bolts to 32–40 ft. lbs. (45–55 Nm).

1991–92

1. Raise and safely support the vehicle.

2. Remove the wheel and tire assembly.

3. Remove the stabilizer nuts, washers, bushings, sleeves and bolts.

4. Remove the bolts securing the stabilizer bar brackets and grommets to the rear suspension crossmember.

5. Remove the stabilizer bar.

6. Remove the cap covering the front and rear lateral link pivot bolts.

7. Position a floor jack stand under the rear suspension crossmember.

8. Remove the bolts securing the rear suspension crossmember to the vehicle frame.

9. Lower the floor jack stand to allow the rear suspension crossmember to be lowered from the vehicle frame.

10. Remove the front and rear lateral link pivot nut, washer and bolt from the rear suspension crossmember.

11. Remove the front and rear lateral links from the rear suspension crossmember.

12. Remove the bolt, washers and nut securing the front and rear lateral links to the rear wheel spindle and remove the lateral links.

13. Remove the nuts securing the parking brake cable and cable bracket to the trailing link.

14. Remove the rear trailing link bolts and washers from the vehicle frame and rear wheel spindle. Remove the rear trailing link.

To install:

15. Position the rear trailing link and install the bolts and washers. Tighten

the trailing link front bolt to 46–69 ft. lbs. (63–93 Nm) and the rear bolt to 69–93 ft. lbs. (93–127 Nm).

16. Position the parking brake cable and bracket to the trailing link and secure it with the nuts.

17. Position the front and rear lateral links to the rear wheel spindle and install the washers, bolt and nut. Tighten the front and rear lateral link nut at the rear wheel spindle to 63–86 ft. lbs. (85–117 Nm).

18. Position the front and rear lateral links to the rear suspension crossmember. Tighten the front and rear lateral link nut at the rear suspension crossmember to 50–70 ft. lbs. (68–95 Nm).

19. Install the cap.

20. Raise the floor jack stand to position the rear suspension crossmember to the vehicle frame. Install and tighten the bolts. Remove the floor jack stand from under the vehicle.

21. Position the grommets onto the stabilizer bar and align the grommets to the positions painted on the bar.

22. Position the stabilizer bar to the rear suspension crossmember and secure it in place with the straps and bolts. Tighten the bolts to 32–43 ft. lbs. (43–59 Nm).

23. Install the stabilizer bolts, washers, grommets, sleeves and nuts. Tighten the stabilizer nuts so 0.64–0.72 in. (16.2–17.0mm) of thread is exposed at the end of the bolt.

24. Install the wheel and tire assembly. Tighten the lug nuts to 65–87 ft. lbs. (88–118 Nm).

25. Check the wheel alignment and lower the vehicle.

Rear Wheel Bearings

REMOVAL & INSTALLATION

1988–90

1. Raise and support the rear of the vehicle safely.

2. Remove the wheel and tire assembly.

3. Remove the grease cup from the rear wheel hub.

4. Using a small cape chisel and a hammer, carefully raise the staked portion of the locknut. Discard the locknut.

NOTE: The locknuts are threaded left and right. The left hand threaded locknut is on the right side of the vehicle and is turned clockwise to loosen. The right hand threaded locknut is on the left side of the vehicle and is turned counterclockwise to loosen.

5. Remove the outer wheel bearing

from the hub and the brake drum/bearing hub assembly.

6. Using a small prybar, pry the grease seal from the rear of the drum. Remove the inner wheel bearing from the hub.

7. If the bearings are to be replaced, remove the inner and outer bearing races using a brass drift.

To install:

8. Install the new bearing races using a brass drift.

9. Pack the bearings and the hub with high temperature grease. Install the inner bearing in the hub.

10. Lubricate the lip of a new seal with grease and install the seal in the hub using a seal installer.

11. Install the drum/bearing hub assembly. Install the outer bearing and a new locknut. Properly adjust the bearing preload.

12. Install the grease cap and the wheel and tire assembly. Lower the vehicle.

1991–92

1. Raise and safely support the vehicle.

2. Remove the wheel and tire assembly.

3. Remove the brake drum or brake caliper and rotor, as necessary.

4. Remove the nut securing the rear wheel hub to the spindle and remove the hub and bearing assembly.

To install:

5. Install the rear wheel hub and bearing assembly onto the spindle.

6. Install the hub nut onto the spindle and tighten to 130–174 ft. lbs. (177–235 Nm).

7. Stake the hub nut and install the cap.

8. Install the brake drum or the brake caliper and rotor, as necessary.

9. Install the wheel and tire assembly. Tighten the lugnuts to 65–87 ft. lbs. (88–118 Nm).

10. Lower the vehicle.

ADJUSTMENT

1988–90

1. Raise and support the vehicle safely.

2. Remove the wheel and tire assembly.

3. Remove the grease cup from the rear wheel hub.

4. Rotate the brake drum to make sure there is no brake drag.

5. Using a small cape chisel and a hammer, carefully raise the staked portion of the locknut. Discard the locknut.

NOTE: The locknuts are threaded left and right. The left hand threaded locknut is on the right side of the vehicle and is turned clockwise to loosen. The right hand threaded locknut is on the left side of the vehicle and is turned counterclockwise to loosen.

6. Install the new locknut. Seat the bearings by tightening the locknut to 18–21 ft. lbs. (25–29 Nm) while rotating the brake drum or rotor. Loosen the locknut slightly until it can be turned by hand.

7. Before bearing preload can be set, the amount of seal drag must be measured and added to the required preload. Using an inch pound torque wrench, position it (12 o'clock position) on 1 of the lug nuts and measure the torque necessary to start the wheel hub to turn.

8. The bearing preload is the seal drag plus 1.3–4.3 inch lbs. (0.15–0.49 Nm). For example, if the seal drag measures 2.2 inch lbs. (0.25 Nm), this amount must be added to the required preload:

1.3 inch lbs. + 2.2 inch lbs. = 3.5 inch lbs. minimum.

0.15 Nm + 0.25 Nm = 0.40 Nm minimum.

4.3 inch lbs. + 2.2 inch lbs. = 6.5 inch lbs. maximum.

0.49 Nm + 0.25 Nm = 0.74 Nm maximum.

9. In this example, when seal drag is added, the required bearing preload becomes 3.5–6.5 inch lbs. (0.40–0.74 Nm).

10. Tighten the wheel bearing locknut a slight amount. Place the inch pound torque wrench onto a lug nut positioned at 12 o' clock and measure the amount of pull required to rotate the brake drum.

11. Continue tightening the attaching nut until the specified amount of preload is measured with the torque wrench.

1991–92

1. Raise and safely support the vehicle.

2. Remove the wheel and tire assembly.

3. Remove the brake drum or the brake caliper and rotor, as necessary.

4. Position a dial indicator to the wheel hub.

5. By hand, push and pull the wheel hub in the axial direction and measure the wheel bearing play.

6. If the wheel bearing play exceeds 0.002 in. (0.05mm), check and adjust the locknut torque or replace the wheel bearing, if necessary.

7. Install the brake drum or brake caliper and rotor, as necessary.

8. Install the wheel and tire assembly and lower the vehicle.

STEERING

Steering Wheel

REMOVAL & INSTALLATION

1. Disconnect the negative battery cable.

2. Remove the steering wheel cover retaining screws from the back side of the steering wheel and remove the cover.

NOTE: On 2-spoke steering wheels there are 2 retaining screws, and on 4-spoke steering wheels there are 4 retaining screws.

3. Disconnect the horn electrical connector and the speed control electrical connector, if equipped.

4. Remove the steering wheel mounting nut. On 1988–90 vehicles, remove the steering wheel cover pad mounting bracket.

5. Paint an aligning stripe on the steering wheel and steering shaft. Remove the steering wheel with a suitable puller. Do not attempt to remove the steering wheel by hitting the column shaft with a hammer; the column may collapse.

To install:

6. Position the steering wheel on the shaft, aligning the paint marks. On 1988–90 vehicles, install the cover pad mounting bracket.

7. Install the steering wheel attaching nut and tighten to 29–36 ft. lbs. (39–49 Nm).

8. Connect the horn electrical connector and the speed control electrical connector, if equipped.

9. Position the steering wheel cover and install the retaining screws. Connect the negative battery cable.

Steering Column

REMOVAL & INSTALLATION

1988–90

1. Disconnect the negative battery cable.

2. Remove the lap duct register panel screws, the lap duct brace screws, the brace and the lap duct.

3. Remove the combination switch lower cover screws and the cover.

4. Using paint, matchmark the lower universal joint-to-intermediate shaft.

5. Remove the lower steering column nuts, the lower steering column universal joint bolt and the upper steering column bolts.

6. Lower the steering column and

disconnect the electrical harness connectors from the lower steering column.

7. Remove the steering column from the vehicle.

8. To install, align the universal joint-to-intermediate shaft and reverse the removal procedures. Inspect the operation of the steering column.

1991–92

1. Disconnect the negative battery cable.

2. Remove the steering wheel.

3. Remove the combination switch and disconnect the ignition switch electrical connector.

4. Remove the shift-lock cable mounting bracket bolt and place the bracket and cable aside.

5. Remove the 4 steering column upper mounting bracket bolts and lower the column.

6. Remove the 5 set plate mounting nuts and remove the set plate.

7. Remove the intermediate shaft-to-pinion shaft bolt.

8. Remove the 2 steering column lower mounting bracket nuts and remove the column.

To install:

9. Position the steering column and install the 2 lower mounting bracket nuts.

10. Install the intermediate shaft-to-pinion shaft bolt and tighten to 30–36 ft. lbs. (40–50 Nm).

11. Position the set plate and install the 5 mounting nuts.

12. Install the 4 steering column upper mounting bracket bolts and tighten to 80–123 inch lbs. (9–14 Nm).

13. Position the shift-lock cable mounting bracket and install the bolt. Tighten the bolt to 37–55 inch lbs. (4–6 Nm).

14. Connect the ignition switch electrical connector and install the combination switch.

15. Install the steering wheel.

16. Connect the negative battery cable and inspect the shift-lock system.

Manual Rack and Pinion

ADJUSTMENT

1988–90

1. Remove the steering rack from the vehicle and place it in a vise.

2. Using an inch pound torque wrench and a pinion torque adapter tool place the assembly on the pinion and measure the pinion turning torque; the torque should be 8–11 inch lbs. (0.9–1.3 Nm).

3. If the pinion torque is not correct, adjust by tightening or loosening

the adjusting bolt. After the pinion torque is adjusted, tighten the adjusting bolt locknut to 7.2–10.8 ft. lbs. (10–15 Nm).

4. Install the steering rack and check for proper operation.

1991–92

1. Remove the rack and pinion assembly from the vehicle and mount it in a vise.

2. Loosen the locknut.

3. Tighten the adjusting bolt using yoke adjustment adapter T90P-3504–JH in the yoke plug to 8.7 inch lbs. (1 Nm), then loosen the adjusting bolt 10–40 degrees from that position.

4. Measure the pinion turning torque using pinion shaft adapting tool T86P-3504-K. The correct torque at the neutral position ± 90 degrees should be 9–12 inch lbs. (1.0–1.3 Nm). At any other position the torque should be 14.7 inch lbs. (1.6 Nm) or less.

5. If the pinion torque is not within specification, re-adjust the adjusting bolt to achieve the correct pinion torque. Tighten the adjusting bolt locknut.

REMOVAL & INSTALLATION

1988–90

1. Disconnect the terminals from the battery (negative cable first) and remove the battery from the vehicle.

2. Raise and support the vehicle safely. Remove the front wheel assemblies.

3. Remove the tie rod end-to-steering knuckle cotter pins and nuts. Using a tie rod separator tool, separate the tie rod end from the steering knuckle.

4. From the right side lower inner fender, remove the plastic dust shield.

5. Using a pair of diagonal cutters, cut the steering column dust boot-to-steering gear plastic wire tie clamp. Pull the dust boot back. Have an assistant turn the steering wheel until the steering column shaft bolt is accessible and lock the steering column.

6. Using white paint, matchmark the steering rack pinion shaft-to-intermediate lower shaft universal joint.

7. Remove the steering gear pinion shaft-to-intermediate lower shaft universal joint clamp bolt.

8. Remove the steering rack-to-chassis bolts, lower the steering gear to disengage it from the intermediate shaft universal joint. Carefully slide the steering gear out through the right side fender well.

To install:

9. Slide the steering rack into position through the right side lower inner fender well opening.

10. Guide the pinion shaft into the intermediate shaft lower universal joint, aligning the marks that were made during the removal procedure. Install the steering rack mounting bolts and tighten to 23–34 ft. lbs. (32–47 Nm).

11. Install the clamp bolt in the intermediate shaft universal joint.

12. Connect the tie rod ends to the steering knuckle arms and install the nuts. Tighten the nuts to 25–29 ft. lbs. (35–40 Nm) and install new cotter pins.

NOTE: If the slots in the nut do not align the with the hole in the ball joint stud, tighten the nut for alignment. Never loosen the nut.

13. Slide the steering column dust boot over the steering gear and install a new plastic tie strap.

14. Install the plastic dust shield and the wheel and tire assemblies. Lower the vehicle.

15. Install the battery and connect the negative battery cable.

1991–92

1. Working inside the vehicle, remove the nuts securing the set plate and remove the set plate.

2. Remove the intermediate shaft-to-pinion shaft bolt from inside the vehicle.

3. Raise and safely support the vehicle.

4. Remove the front wheel and tire assemblies.

5. Remove the cotter pins and nuts securing the tie rod ends to the steering knuckles. Separate the tie rod ends from the steering knuckles using a suitable tool.

6. If equipped with manual transaxle, disconnect the extension bar.

7. Remove the nuts securing the steering gear brackets to the bulkhead. Remove the brackets.

8. Remove the steering gear from the vehicle.

To install:

9. Position the steering gear into its mounting position and install the brackets and nuts. Tighten the nuts to 27–38 ft. lbs. (37–52 Nm).

10. If equipped with a manual transaxle, connect the extension bar. Tighten the nut to 23–34 ft. lbs. (31–46 Nm).

11. Attach the tie rod ends to the steering knuckles. Install the nuts and tighten to 31–42 ft. lbs. (42–57 Nm). Install new cotter pins.

12. Install the front wheel and tire assemblies.

13. Lower the vehicle.

14. Install the intermediate shaft-to-pinion shaft bolt and tighten to 13–20 ft. lbs. (18–27 Nm).

15. Position the set plate and secure it with the nuts.

Power Rack and Pinion

ADJUSTMENT

1988–90

1. Remove the power steering rack from the vehicle and place it in a vise.
2. Using an inch lb. torque wrench and a pinion torque adapter tool, place the assembly on the pinion and measure the pinion turning torque; the torque should be 0.52–1.3 inch lbs. (0.6–1.5 Nm).
3. If the pinion torque is not as specified, adjust by tightening or loosening the adjusting plug.
4. Install the steering rack in the vehicle.

REMOVAL & INSTALLATION

1988–90

1. Disconnect the terminals from the battery (negative cable first) and remove the battery from the vehicle.
2. Raise and support the vehicle safely. Remove the front wheel assemblies.
3. Remove the tie rod end-to-steering knuckle cotter pins and nuts. Using a tie rod separator tool, separate the tie rod end from the steering knuckle.
4. From the right side lower inner fender, remove the plastic dust shield.
5. Using a pair of diagonal cutters, cut the steering column dust boot-to-steering gear plastic wire tie clamp. Pull the dust boot back. Have an assis-tant turn the steering wheel until the steering column shaft bolt is accessible and lock the steering column.
6. Using white paint, matchmark the steering gear pinion shaft-to-intermediate lower shaft universal joint.
7. Remove the steering gear pinion shaft-to-intermediate lower shaft universal joint clamp bolt.
8. Using a 17mm crowsfoot tubing wrench, disconnect and plug the fluid return line from the power steering rack.
9. Using a 14mm socket, remove the banjo bolt from the pressure line at the power steering gear and discard the copper washers.

NOTE: Be sure to position the lines out of the way.

10. Remove the steering rack-to-chassis bolts, lower the steering gear to disengage it from the intermediate shaft universal joint. Carefully slide the steering rack out through the right side fender well.
To install:
11. Slide the steering rack into position through the right side lower inner fender well opening.
12. Guide the pinion shaft into the intermediate shaft lower universal joint, aligning the marks that were made during the removal procedure. Install the steering rack mounting bolts and tighten to 23–34 ft. lbs. (32–47 Nm).
13. Install the clamp bolt in the intermediate shaft universal joint.
14. Connect the tie rod ends to the steering knuckle arms and install the nuts. Tighten the nuts to 25–29 ft. lbs. (35–40 Nm) and install new cotter pins.

NOTE: If the slots in the nut do not align the with the hole in the ball joint stud, tighten the nut for alignment. Never loosen the nut.

15. Attach the hoses from the power steering pump to the steering gear. Install new copper washers at the return line banjo fitting. Tighten the pressure hose fitting with a 17mm crowsfoot tubing wrench.
16. Install the plastic dust shield and the wheel and tire assemblies. Lower the vehicle.
17. Install the battery and connect the negative battery cable. Fill the reservoir and bleed the hydraulic system.
18. Check for leaks.

1991–92

1. From inside the passenger compartment, remove the 5 set plate nuts and remove the set plate.
2. Remove the intermediate shaft-to-pinion shaft bolt.
3. Raise and safely support the vehicle.
4. Remove the front wheel and tire assemblies.
5. Remove the cotter pins and attaching nuts from the tie rod ends. Using a suitable tool, separate the tie rod ends from the steering knuckles.
6. If equipped with the 1.8L engine, remove the 2 screws from the power steering line retaining bracket and remove the bracket from the steering gear housing. If equipped with the 1.9L engine, remove the strap that holds the power steering lines to the steering gear housing and discard the strap.
7. Disconnect the high-pressure and return lines from the steering gear and plug the lines.
8. If equipped with manual transaxle, disconnect the extension bar and shift control rod from the transaxle.
9. Remove the nuts from the 2 steering gear mounting brackets.
10. Remove the splash shield from the left wheel well.
11. Remove the steering gear from the left side of the vehicle.
To install:
12. Position the steering gear in its mounting location and install the splash shield in the left wheel well.
13. Position the 2 steering gear mounting brackets and install the 2 nuts to each bracket. Tighten the nuts to 27–38 ft. lbs. (37–52 Nm).
14. If equipped with a manual transaxle, connect the extension bar and shift control rod. Tighten the extension bar nut to 23–34 ft. lbs. (31–46 Nm) and the shift control rod nut to 12–17 ft. lbs. (16–23 Nm).
15. Remove the plugs and connect the pressure and return lines to the

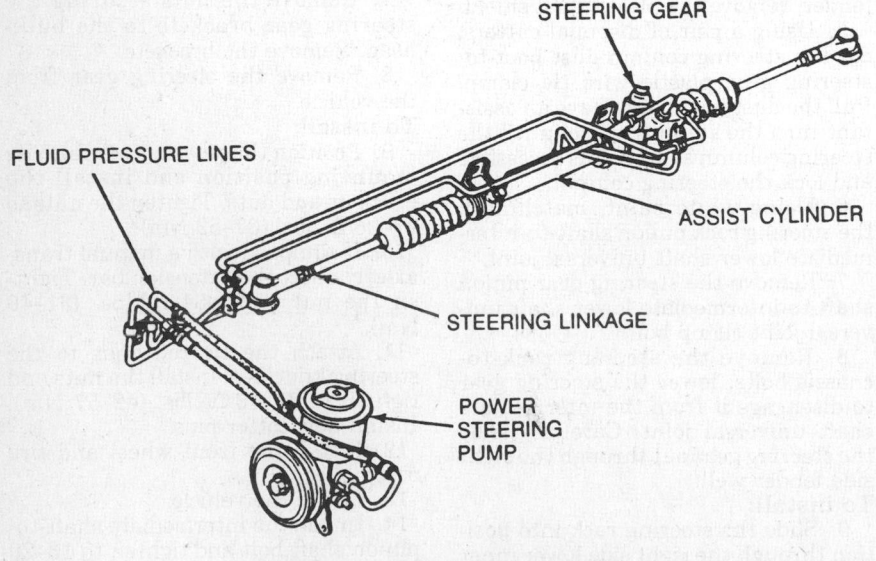

STEERING GEAR

FLUID PRESSURE LINES

ASSIST CYLINDER

STEERING LINKAGE

POWER STEERING PUMP

Power steering system—1988–90 vehicles

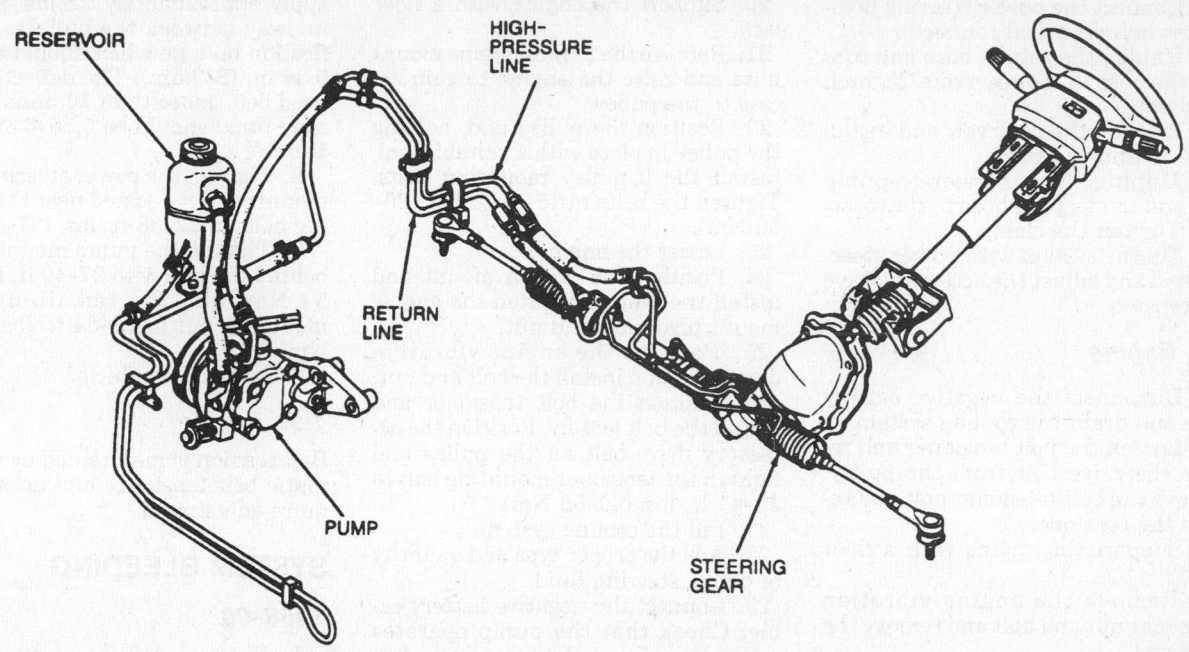

RESERVOIR

HIGH-PRESSURE LINE

RETURN LINE

PUMP

STEERING GEAR

Power steering system—1991–92 vehicles with 1.8L engine

steering gear. Tighten the flare nuts to 22–28 ft. lbs. (29–39 Nm).

16. If equipped with 1.8L engine, position the power steering line retaining bracket and install the 2 screws. If equipped with 1.9L engine, install a new strap to hold the power steering lines to the steering gear housing.

17. Position the tie rod ends in the steering knuckles and install the attaching nuts. Tighten the nuts to 31–42 ft. lbs. (42–57 Nm). Install new cotter pins.

18. Install the wheel and tire assemblies and lower the vehicle.

19. From inside the vehicle, install the intermediate shaft-to-pinion shaft bolt. Tighten the bolt to 13–20 ft. lbs. (18–27 Nm).

20. Position the set plate and install the 5 set plate nuts.

21. Fill the system with steering fluid.

Power Steering Pump

REMOVAL & INSTALLATION

1.6L Engine

1. Disconnect the negative battery cable.

2. At the power steering pump, loosen the locknut and adjuster bolt. Move the pump toward the engine and remove the drive belt.

3. From the engine lifting eye, remove the ground wire.

4. Disconnect and plug the hoses from the power steering pump. Dis-

connect the electrical connector from the pump's pressure switch.

5. Remove the adjusting screw, nut, block, pivot bolt and pump; if necessary, remove the pump pulley.

6. To install, reverse the removal procedures. Adjust the drive belt tension. Using Dexron® II automatic transmission fluid, fill the power steering pump reservoir. Bleed the power steering system.

1.8L Engine

1. Disconnect the negative battery cable.

2. Loosen the power steering fluid reservoir-to-pump hose clamp and pull the hose from the reservoir. Plug the hose.

3. Remove the 2 reservoir mounting bolts and lift the reservoir from its mounting position.

4. Loosen the return hose clamp and pull the return hose from the reservoir. Plug the hose and remove the reservoir.

5. Disconnect the electrical connector from the power steering pressure switch.

6. Loosen the high-pressure line flare nut and disconnect the line from the pump. Plug the line.

7. Raise and safely support the vehicle.

8. Remove the 5 right front undercover bolts and remove the undercover.

9. Remove the belt tensioner adjustment bolt and remove the accessory drive belt from the pulley.

10. Lower the vehicle.

11. Remove the 3 pump mounting bracket bolts and remove the pump and the bracket.

12. Remove the bolt that attaches the pump to the mounting bracket.

13. Remove the nut and bolt that attaches the tensioner to the pump mounting bracket and remove the nut and bolt that attaches the tensioner to the pump.

To install:

14. Position the tensioner to the pump and install the bolt and nut. Tighten the nut to 14–19 ft. lbs. (19–25 Nm).

15. Position the tensioner to the pump mounting bracket and install the bolt and nut. Tighten the nut to 23–34 ft. lbs. (31–46 Nm).

16. Install the bolt that attaches the pump to the mounting bracket and tighten to 27–40 ft. lbs. (36–54 Nm).

17. Position the pump and bracket and install the 3 pump mounting bracket bolts. Tighten the bolts to 27–38 ft. lbs. (37–54 Nm).

18. Raise and safely support the vehicle.

19. Position the accessory drive belt on the pulley and install the belt tensioner adjustment bolt.

20. Position the right front undercover and install the 5 bolts.

21. Lower the vehicle.

22. Unplug the high-pressure line and connect the line to the pump. Tighten the flare nut to 12–17 ft. lbs. (16–24 Nm).

23. Connect the power steering pressure switch electrical connector.

24. Unplug the return hose and connect the hose to the reservoir. Tighten the clamp.

25. Position the reservoir and install the 2 mounting bolts.

26. Unplug the reservoir-to-pump hose and connect the hose to the reservoir. Tighten the clamp.

27. Fill the system with power steering fluid and adjust the accessory drive belt tension.

1.9L Engine

1. Disconnect the negative battery cable and drain the cooling system.

2. Loosen the belt tensioner and remove the drive belt from the pulley. Remove the belt tensioner bolt and remove the tensioner.

3. Support the engine with a floor jack.

4. Remove the engine vibration dampener nut and bolt and remove the dampener.

5. Remove the 2 front engine mount nuts. Loosen the engine mount pivot bolt and nut and position the engine mount aside.

6. Raise the engine to gain access to the power steering pump pulley.

7. Hold the pulley in position with a suitable tool and remove the 3 pulley mounting bolts. Remove the pulley and lower the engine.

8. Position the engine mount and install the 2 nuts.

9. Loosen the clamp and disconnect the return line from the pump. Loosen the flare nut from the high-pressure line and disconnect the line from the pump.

10. Raise and safely support the vehicle.

11. Remove the 2 passenger side splash shields.

12. If equipped, remove the 4 compressor mounting bolts and position the air conditioning compressor aside.

13. Remove the lower radiator hose.

14. Remove the 3 power steering pump mounting bolts and remove the pump.

To install:

15. Position the power steering pump and install the 3 mounting bolts. Tighten the bolts to 30–45 ft. lbs. (40–62 Nm).

16. Install the lower radiator hose.

17. If equipped, position the air conditioning compressor and install the 4 mounting bolts. Tighten the bolts to 30–40 ft. lbs. (40–55 Nm).

18. Install the 2 passenger side splash shields and lower the vehicle.

19. Connect the high-pressure line to the power steering pump and tighten the nut. Connect the return line to the pump and position the clamp.

20. Support the engine with a floor jack.

21. Remove the 2 front engine mount nuts and raise the engine to gain access to the pulley.

22. Position the pulley and, holding the pulley in place with a suitable tool, install the 3 pulley mounting bolts. Tighten the bolts to 15–22 ft. lbs. (20–30 Nm).

23. Lower the engine.

24. Position the engine mount and install the 2 nuts. Tighten the engine mount pivot bolt and nut.

25. Position the engine vibration dampener and install the bolt and nut.

26. Position the belt tensioner and install the bolt loosely. Position the accessory drive belt on the pulley and tighten the tensioner mounting bolt to 30–41 ft. lbs. (40–55 Nm).

27. Fill the cooling system.

28. Add the proper type and quantity of power steering fluid.

29. Connect the negative battery cable. Check that the pump operates properly and that there are no leaks.

BELT ADJUSTMENT

1.6L Engine

1. Inspect the condition of the drive belt; replace it, if necessary.

2. At the power steering pump, loosen the locknut and adjuster bolt.

3. Using a drive belt tension gauge tool, position it between the power steering pump pulley and the crankshaft pulley. The drive belt deflection should be 0.31–0.35 in. (8–9mm) for a new belt or 0.35–0.39 in. (9–10mm) for a used belt at 22 lbs. pressure.

NOTE: A used belt is one that has at least 10 minutes run time.

4. If the power steering pump locknut was loosened, torque it to 32–45 ft. lbs. (43–61 Nm).

1.8L Engine

1. Raise and safely support the vehicle.

2. Loosen the power steering pump mounting bolt and nuts.

3. Adjust the belt tension by turning the pump adjusting bolt.

4. Tighten the power steering pump mounting nut near the pump adjusting bolt.

5. Check the belt tension using either a belt tension gauge or using the deflection method.

6. If using a belt tension gauge, position the gauge on the longest accessible span of belt. The tension for a new belt should be 110–132 lbs. The tension for a used belt (more than 10 minutes running time) should be 95–110 lbs.

7. If using the deflection method, apply approximately 22 lbs. pressure midway between the pulleys. The deflection on a new belt should be 0.31–0.35 in. (8–9mm). The deflection on a used belt (more than 10 minutes running time) should be 0.35–0.39 in. (9–10mm).

8. Tighten the power steering pump mounting nut, located near the adjusting bolt, to 27–38 ft. lbs. (37–52 Nm).

9. Tighten the pump mounting bolt behind the pulley to 27–40 ft. lbs. (36–54 Nm) and the remaining pump mounting nut to 23–34 ft. lbs. (31–46 Nm).

10. Lower the vehicle.

1.9L Engine

Belt tension is maintained by an automatic belt tensioner and does not require adjustment.

SYSTEM BLEEDING

1988–90

1. Raise and support the vehicle safely.

2. Using Dexron® II automatic transmission fluid, fill the power steering reservoir to the **L** mark on the dipstick.

3. Start the engine and allow it to reach normal operating temperature.

4. Slowly turn the steering wheel (back and forth) lock-to-lock about 10 times, until all of the air is bled from the system and the reservoir is maintaining a full level.

NOTE: When bleeding the system, be sure to check the reservoir level several times.

5. Position the wheels in the straight-ahead position and turn the engine off.

6. Refill the power steering reservoir until the fluid level is between the **L** and the **H** marks on the dipstick.

7. Lower the vehicle, start the engine, check for leaks and road test the vehicle.

1991–92

1. Fill the reservoir to the proper level.

2. Operate the engine until the fluid reaches normal operating temperature (165–175°F).

3. Turn the steering wheel all the way to the left then all the way to the right several times. Do not hold the steering wheel in the far left or far right position stops.

4. Check the fluid level and recheck the fluid for the presence of trapped air. If apparent that air is still in the system, fabricate or obtain a vacuum tester and purge the system as follows:

a. Remove the pump dipstick cap assembly.

b. Check and fill the pump reservoir with fluid to the **COLD FULL** mark on the dipstick.

c. Disconnect the ignition coil wire or the coil pack electrical connector if equipped with distributorless ignition, and raise the front of the vehicle and support safely.

d. Crank the engine with the starter and check the fluid level. Do not turn the steering wheel at this time.

e. Fill the pump reservoir to the **COLD FULL** mark on the dipstick. Crank the engine with the starter while cycling the steering wheel lock-to-lock. Check the fluid level.

f. Tightly insert a suitable size rubber stopper and air evacuator pump into the reservoir fill neck. Connect the ignition coil wire or coil pack electrical connector.

g. With the engine idling, apply a 15 in. Hg vacuum to the reservoir for 3 minutes. As air is purged from the system, the vacuum will drop off. Maintain the vacuum on the system as required throughout the 3 minutes.

h. Remove the vacuum source. Fill the reservoir to the **COLD FULL** mark on the dipstick.

i. With the engine idling, re-apply 15 in. Hg vacuum source to the reservoir. Slowly cycle the steering wheel to lock-to-lock stops for approximately 5 minutes. Do not hold the steering wheel on the stops during cycling. Maintain the vacuum as required.

j. Release the vacuum and disconnect the vacuum source. Add fluid, as required.

k. Start the engine and cycle the wheel slowly and check for leaks at all connections.

l. Lower the front wheels.

5. In cases of severe aeration, repeat the procedure.

Tie Rod Ends

REMOVAL & INSTALLATION

1. Raise the vehicle and support it safely. Remove the wheel and tire assembly.

2. Remove the tie rod-to-steering knuckle cotter and nut.

3. Using a tie rod separator tool, separate the tie rod from the knuckle.

4. Matchmark the tie rod to the tie rod end.

5. Loosen the tie rod jam nut and, counting the number of turns, remove the tie rod end.

To install:

6. Install the new tie rod end into the tie rod, the same number of turns as required for removal.

7. Install the tie rod to the steering knuckle and tighten the nut to 22–33 ft. lbs. (29–42 Nm) on 1988–90 vehicles or 31–42 ft. lbs. (42–57 Nm) on 1991–92 vehicles. Install a new cotter pin.

NOTE: If the slots in the nut do not align the with the hole in the ball joint stud, tighten the nut for alignment. Never loosen the nut.

8. Tighten the jam nut and install the wheel and tire assembly. Lower the vehicle.

BRAKES

For all brake system repair and service procedures not detailed below, please refer to "Brakes" in the Unit Repair section.

Master Cylinder

REMOVAL & INSTALLATION

1. On 1991–92 vehicles, disconnect the battery cables and remove the battery.

2. Disconnect the low fluid level sensor electrical connector.

3. Loosen the brake line fittings and disconnect the brake lines from the master cylinder.

4. On 1991–92 vehicles equipped with manual transaxle, remove the clamp and pull the clutch hose from the brake/clutch fluid reservoir.

5. Cap the lines and the master cylinder ports.

6. Remove the 2 mounting nuts and remove the master cylinder assembly.

To install:

7. Adjust the piston to pushrod clearance as follows:

a. Insert a pencil in the pushrod socket of the master cylinder. Mark the point on the pencil that is even with the end of the master cylinder with a hacksaw blade.

b. Measure the length of the pencil to the saw mark with a ruler.

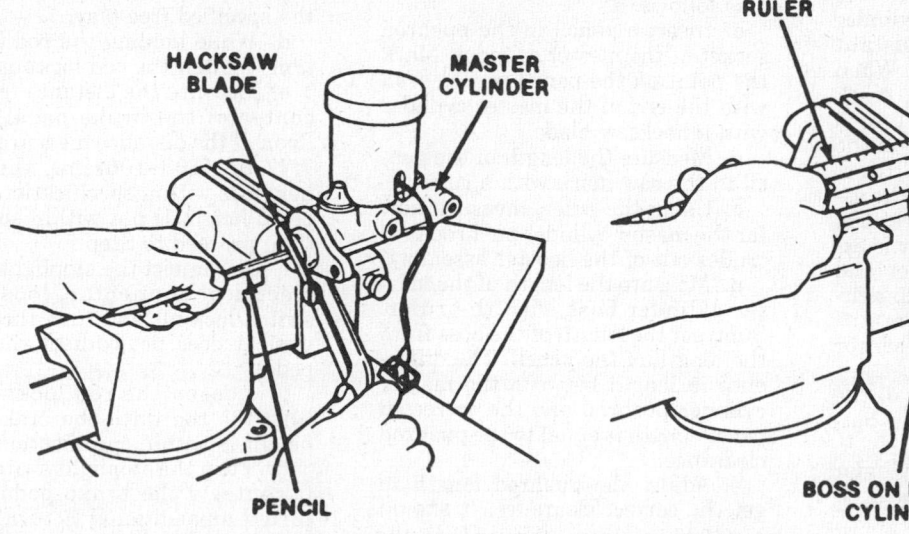

Master cylinder pushrod adjustment procedure

c. Using the ruler, measure how far the master cylinder pushrod protrudes out of the booster assembly.

d. Measure the length of the master cylinder boss with the ruler. Subtract the length of the boss from the length of the pencil. The difference in length between the master cylinder pushrod and the corrected pencil length is equal to the pushrod clearance.

e. Adjust the pushrod length to get the correct clearance. It should be 0.025 in. (1mm) shorter than the pushrod socket.

8. Before installation, bench bleed the new master cylinder as follows:

a. Mount the new master cylinder in a holding fixture. Be careful not to damage the housing.

b. Fill the master cylinder reservoir with brake fluid.

c. Using a suitable tool inserted into the booster pushrod cavity, push the master cylinder piston in slowly. Place a suitable container under the master cylinder to catch the fluid being expelled from the outlet ports.

d. Place a finger tightly over each outlet port and allow the master cylinder piston to return.

e. Repeat the procedure until clear fluid only is expelled from the master cylinder. Plug the outlet ports and remove the master cylinder from the holding fixture.

9. Position the master cylinder over the booster pushrod and booster mounting studs. Install the nuts and tighten to 8–12 ft. lbs. (10–16 Nm).

10. On 1991–92 vehicles equipped with manual transaxle, connect the clutch hose onto the brake/clutch fluid reservoir and install the clamp.

11. Remove the caps and connect the brake lines. Tighten the fittings.

12. Make sure the master cylinder reservoir is full. Have an assistant push down on the brake pedal. When the pedal is all the way down, crack open the brake line fittings, 1 at a time, to expel any remaining air in the master cylinder and brake lines. Tighten the fittings, then have the assistant allow the brake pedal to return.

13. Repeat Step 12 until all air is expelled from the master cylinder and brake lines. Tighten the brake line fittings to 10–16 ft. lbs. (13–22 Nm).

14. Connect the low fluid level sensor electrical connector.

15. On 1991–92 vehicles, install the battery and connect the negative battery cable.

16. Make sure the master cylinder reservoir is full. Bleed the brakes, if necessary.

17. Check for brake fluid leaks and for proper brake operation.

Proportioning Valve

REMOVAL & INSTALLATION

1. Disconnect and plug the brake lines leading to the valve.

2. Remove the valve mounting bolts from the firewall.

3. Remove the valve.

4. Installation is the reverse of the removal procedure.

5. Bleed the brake system.

Power Brake Booster

REMOVAL & INSTALLATION

1988–90

1. Remove the cables from the battery (negative cable first) and the battery from the vehicle.

2. Remove the master cylinder.

3. Remove the vacuum hose from the brake booster.

4. From under the instrument panel, remove the spring clip and the clevis pin from the brake pedal.

5. Remove the brake booster-to-cowl nuts and the brake booster.

To install:

6. Have an assistant place the booster on the firewall so the 4 retaining studs protrude into the passenger compartment. Working under the dash, install the 4 retaining nuts.

7. Apply lithium grease to the clevis pin and install it through the brake pedal pushrod and the brake pedal. Install the clevis pin spring clip.

8. Working under the hood, connect the vacuum hose to the booster. A vacuum check valve is located in the center of the hose, so the arrow on the hose must point toward the engine.

9. Adjust the master cylinder pushrod as follows:

a. Insert a pencil in the pushrod socket of the master cylinder. Mark the point on the pencil that is even with the end of the master cylinder with a hacksaw blade.

b. Measure the length of the pencil to the saw mark with a ruler.

c. Using the ruler, measure how far the master cylinder pushrod protrudes out of the booster assembly.

d. Measure the length of the master cylinder boss with the ruler. Subtract the length of the boss from the length of the pencil. The difference in length between the master cylinder pushrod and the corrected pencil length is equal to the pushrod clearance.

e. Adjust the pushrod length to get the correct clearance. It should be 0.025 in. (1mm) shorter than the pushrod socket.

10. Install the master cylinder and bleed the brakes.

11. Install the battery and connect the battery cables.

1991–92

1. Disconnect the negative battery cable.

2. Remove the master cylinder assembly.

3. Loosen the vacuum hose clamp and remove the hose from the power brake booster.

4. From inside the vehicle, remove the pin and discard.

5. Remove the clevis pin.

6. Remove the 4 booster mounting nuts and remove the booster. Remove and discard the gasket.

To install:

7. Install a new gasket over the studs and position the power brake booster.

8. From inside the vehicle, install the 4 mounting nuts and tighten to 14–19 ft. lbs. (19–25 Nm).

9. Lubricate the clevis pin with white lithium grease and install. Install a new pin.

10. Position the vacuum hose to the booster and install the clamp.

11. Install the master cylinder, making sure to check the master cylinder pushrod clearance.

12. Adjust the brake pedal as follows:

a. Press the brake pedal several times to eliminate the vacuum in the booster.

b. Carefully press the pedal and measure the amount of free-play until resistance is felt. If the free-play is 0.16–0.28 in. (4–7mm), the pedal free-play is within specification. If the free-play is not within specification, proceed to Step c.

c. Loosen the rod locknut and rotate the rod either in or out to obtain the specified free-play.

d. While holding the rod in position, tighten the rod locknut.

e. Measure the distance from the center of the brake pedal to the floor. If the distance measures 7.60–7.72 in. (193–196mm), the pedal height is within specification. If the pedal height is not within specification, proceed to Step f.

f. Disconnect the stoplight switch electrical connector, loosen the switch locknut and turn the switch until it does not contact the brake pedal.

g. Loosen the rod locknut and turn the rod until the brake pedal height is within specification.

h. Turn the stoplight switch until it contacts the brake pedal, then turn it an additional ½ turn. Tighten the stoplight locknut and the rod locknut.

i. Connect the stoplight switch electrical connector and check the operation of the stoplights and brake system.

Brake Caliper

REMOVAL & INSTALLATION

Front

1. Raise and safely support the vehicle. Remove the wheel and tire assembly.
2. Remove the brake pads.
3. Clamp the brake hose and remove the brake hose attaching bolt.
4. Disconnect the brake hose from the caliper and discard the 2 copper washers.
5. Remove the 2 caliper mounting bolts and remove the caliper. Remove the guide pin bushing dust boots and push out the caliper guide pin bushings.
To install:
6. Lubricate the guide pin bushings

with high temperature grease and install them in the caliper with the dust boots. Position the caliper and install the 2 caliper mounting bolts. Tighten the bolts to 29–36 ft. lbs. (39–49 Nm).
7. Install 2 new copper washers to the brake hose. Position the brake hose onto the caliper and install the attaching bolt. Tighten the bolt to 16–22 ft. lbs. (22–29 Nm).
8. Remove the clamp from the brake hose.
9. Install the brake pads.
10. Bleed the brake system.
11. Install the wheel and tire assembly and lower the vehicle.

Rear

1988–90

1. Raise and safely support the vehicle. Remove the wheel and tire assembly.
2. Remove the brake pads and remove the retaining clip from the brake flex hose.

3. Remove the banjo bolt attaching the brake hose to the caliper. Remove and discard the copper washers.
4. Remove the lower caliper mounting bolt.
5. Using a chisel, remove the upper caliper guide pin dust cap to gain access to the Allen head on the guide pin. Use an Allen head wrench to remove the upper caliper guide pin.
6. Remove the caliper from the rotor. Remove the caliper upper guide pin and the lower guide pin bushing and the dust boots.
To install:
7. Install the brake pads and shims.
8. Install the guide pin and guide pin bushing dust boots. Lubricate the upper guide pin and lower guide pin bushing with high temperature grease and install in the caliper.
9. Position the caliper over the rotor. It may be necessary to rotate the piston to provide clearance.
10. Tighten the upper guide pin with an Allen wrench and install the dust cap with a plastic hammer. Install the lower caliper mounting bolt and tighten to 29–36 ft. lbs. (39–49 Nm).
11. Install the brake hose with the banjo bolt, using new copper washers. Tighten the bolt to 16–22 ft. lbs. (22–29 Nm).
12. Bleed the brakes and install the wheel and tire assembly.

1991–92

1. Raise and safely support the vehicle. Remove the wheel and tire assembly.
2. Remove the brake pads.
3. Remove the parking brake cable bracket bolt and position the bracket aside.
4. Remove the parking brake cable from the operating lever.
5. Clamp the brake hose, remove the brake line attaching bolt and remove the 2 washers. Discard the washers.
6. Disconnect the brake line and slide the caliper off the mounting bracket.
To install:
7. Position the caliper on the mounting bracket.
8. Install 2 new washers to the brake line. Position the brake line to the caliper and install the attaching bolt. Tighten the bolt to 16–22 ft. lbs. (22–29 Nm).
9. Remove the clamp from the brake hose.
10. Attach the parking brake cable to the operating lever and position the bracket and install the bracket bolt.
11. Install the brake pads.
12. Bleed the brake system.
13. Install the wheel and tire assembly and lower the vehicle.

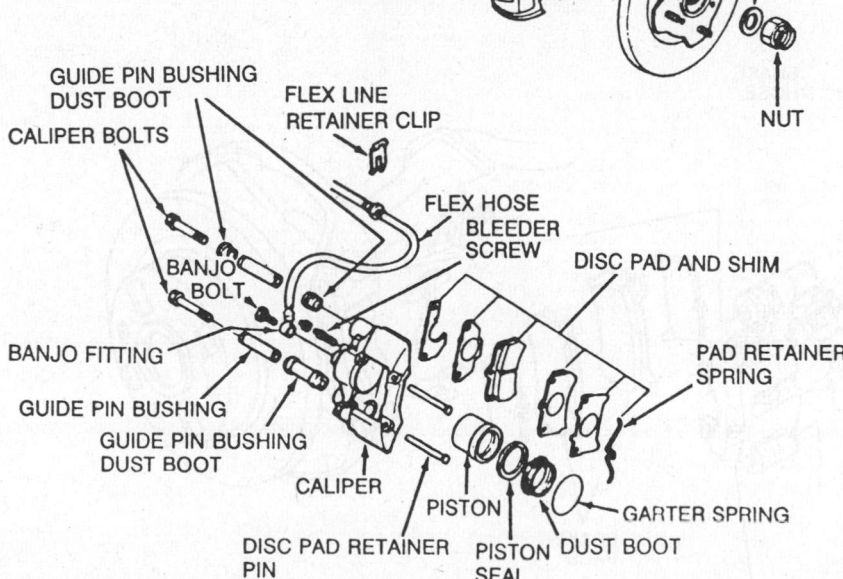

Front disc brake assembly – 1988–90 vehicles

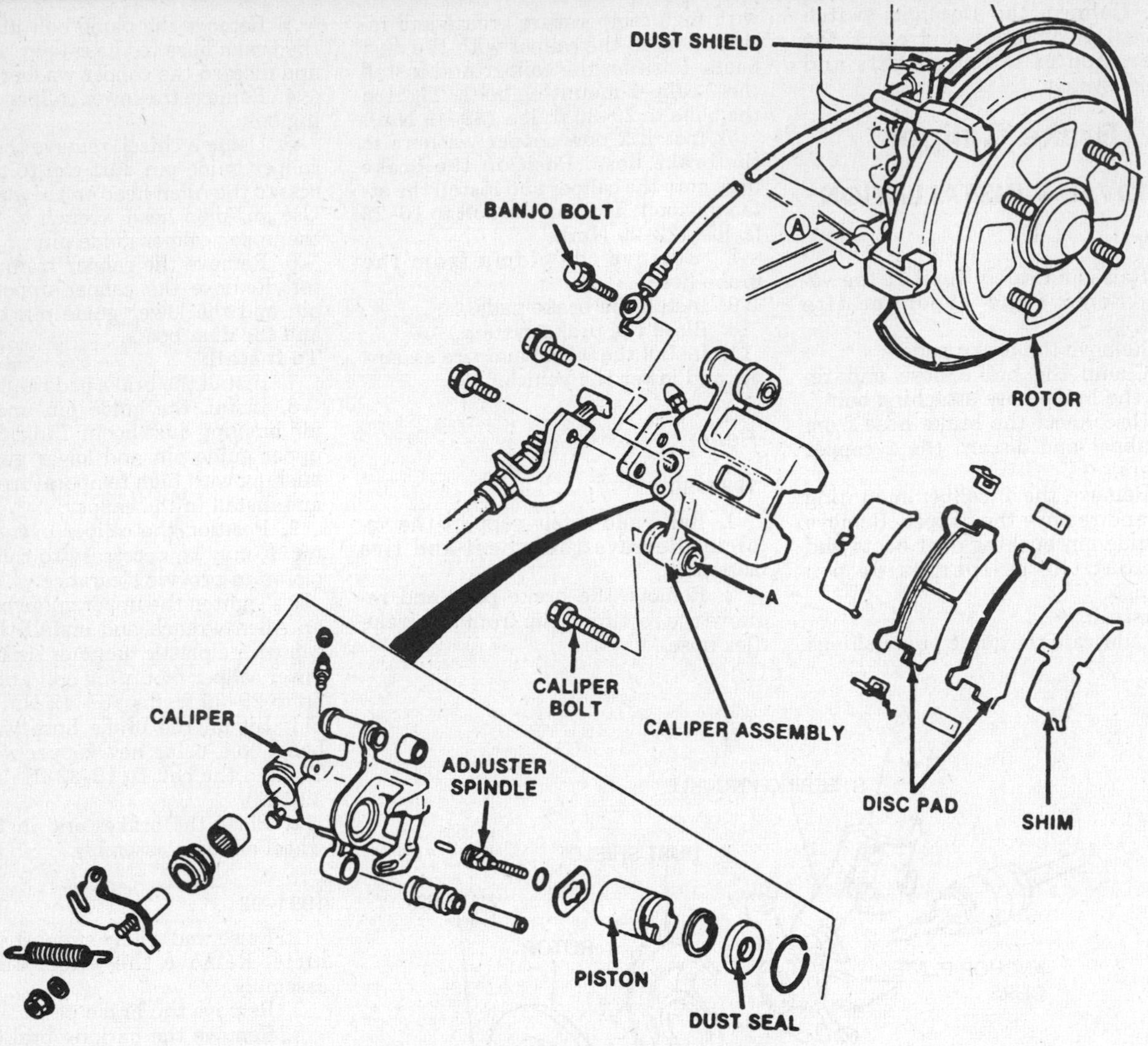

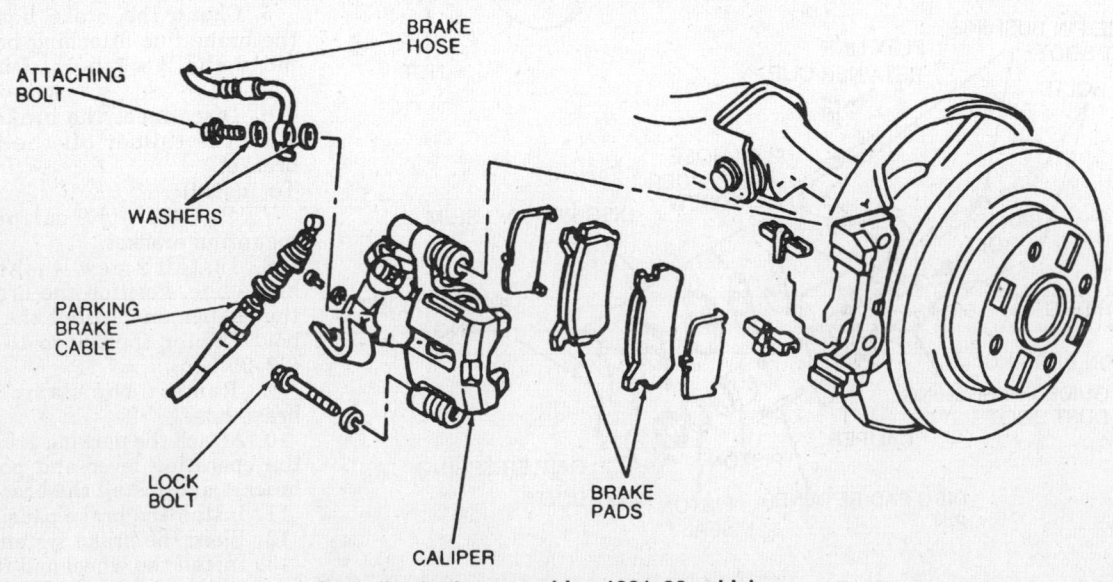

DUST SHIELD

BANJO BOLT

ROTOR

A

CALIPER BOLT

CALIPER ASSEMBLY

DISC PAD

SHIM

CALIPER

ADJUSTER SPINDLE

PISTON

DUST SEAL

Rear disc brake assembly—1988–90 vehicles

BRAKE HOSE

ATTACHING BOLT

WASHERS

PARKING BRAKE CABLE

LOCK BOLT

CALIPER

BRAKE PADS

Rear disc brake assembly—1991–92 vehicles

Disc Brake Pads

REMOVAL & INSTALLATION

Front

1988–90

1. Remove approximately ⅔ of the brake fluid from the master cylinder. Raise and support the vehicle safely.

2. Remove the tire and wheel assembly.

3. Using needle-nose pliers, remove the pad retainer spring that locks in the disc pad retainer pins.

4. Use a hammer and a pin punch to remove the retainer pins.

5. Using a small prybar, pry the caliper outwards and remove the outer brake pad and shim. Tag the shims so they can be reinstalled in their original positions.

6. Push the caliper inward by hand and remove the inner brake shoe and shims. Be careful not to damage the caliper piston dust boot.

7. Remove the anchor plate clips from the caliper anchor plate. Tag the plates to indicate the top and bottom positions.

To install:

8. Install the anchor plate clips in their original positions.

9. Push the caliper inward and install the inner brake pad and shims. Make sure the spring tabs on the back of the brake pad are properly aligned and fully seated in the caliper piston.

10. Pry the caliper outward and install the outer brake pad and shim.

11. Install the retaining pins and the spring.

12. Install the wheel and tire assembly and lower the vehicle.

13. Pump the brake pedal several times to seat the brake pads. Check the fluid level in the master cylinder reservoir.

1991–92

1. Remove master cylinder cap and check fluid level in reservoir. Remove brake fluid until reservoir is ½ full. Discard removed fluid.

2. Raise and safely support the vehicle.

3. Remove wheel and tire assembly.

4. Remove the 2 brake pad pins and remove the M-spring and the W-spring.

5. Remove the brake pads and shims from the caliper.

To install:

6. Use a suitable tool to push the piston into the caliper bore.

7. Apply grease between the shims and the brake pad guide plates and position the brake pads and shims into the caliper.

8. Install the W-spring and the M-spring. Install the 2 brake pad pins.

9. Install the wheel and tire assembly and lower the vehicle.

10. Pump brake pedal prior to moving vehicle to position brake linings. Check the fluid level in the master cylinder.

11. Road test vehicle.

Rear

1988–90

1. Remove approximately ⅔ of the brake fluid from the master cylinder. Raise and support the vehicle safely.

2. Remove the tire and wheel assembly.

3. Using needle-nose pliers, remove the parking brake return springs at the back of the caliper.

4. Loosen the parking brake cable housing adjusting nut and remove the cable housing from the bracket on the rear lower control arm.

5. Loosen the mounting bolt connecting the parking brake cable bracket to the caliper and remove the cable from the caliper.

6. Loosen the lower caliper bolt and pivot the caliper upward on the upper caliper guide pin.

7. Remove the disc pad retaining spring and remove the disc pads and shims.

8. Remove the anchor plate clips and label their positions, indicating top and bottom.

To install:

9. Install the anchor plate clips in their original locations. Lubricate the clips with lithium grease.

10. Install the shims on the backs of the pads and install the pads in the anchor plate. Install the disc pad retaining spring.

11. Pivot the caliper down over the pads. If necessary, rotate the disc brake piston to provide clearance. Install the lower caliper bolt.

12. Install the parking brake cable in the caliper parking brake lever. Position the cable bracket against the rear caliper and install the cable mounting bolt.

13. Install the wheel and tire assembly and lower the vehicle.

14. Pump the brake pedal several times to seat the brake pads. Check the fluid level in the master cylinder reservoir.

15. Raise and safely support the vehicle. Spin each wheel several times to be sure the calipers are not frozen and the parking brake is not adjusted too tight.

1991–92

1. Remove master cylinder cap and check fluid level in reservoir. Remove brake fluid until reservoir is ½ full. Discard removed fluid.

2. Raise and safely support the vehicle.

3. Remove wheel and tire assembly.

4. If necessary, remove the screw plug and turn the adjustment gear counterclockwise with a hex wrench to pull the piston fully inward.

5. Remove the caliper lock bolt.

6. Using a suitable tool, pivot the caliper on its mounting bracket to access the brake pads. If the upper lock bolt requires lubrication or service, remove it and suspend the caliper with mechanics wire.

7. Remove the brake pads, shims, spring and guides.

To install:

8. Apply an appropriate brake pad grease between the shims and the brake pads.

9. Using a suitable tool, pivot the caliper on its mounting bracket and position the brake pads, shims, spring and guides to the rotor.

10. Lubricate and install the lock bolt. Tighten the bolt to 33–43 ft. lbs. (45–59 Nm).

11. If necessary, turn the adjustment gear clockwise with an Allen wrench until the brake pads just touch the rotor, then loosen the gear ⅓ of a turn. Install the screw plug and tighten to 9–12 ft. lbs. (12–16 Nm).

12. Install the wheel and tire assembly and lower the vehicle.

13. Pump brake pedal prior to moving vehicle to position brake linings. Check the fluid level in the master cylinder.

14. Road test vehicle.

Brake Rotor

REMOVAL & INSTALLATION

Front

1988–90

1. Raise and safely support the vehicle.

2. Remove the wheel and tire assembly.

3. Remove the disc brake pads and the caliper. Do not disconnect the brake hose from the caliper. Support the caliper aside with mechanics wire.

4. Using a chisel, unstake and remove the halfshaft nut and washer. Discard the nut.

5. Remove the cotter pin and nut from the tie rod end. Using a separator tool, disconnect the tie rod end from the steering knuckle.

6. Remove the clinch bolt from the lower ball joint and push the lower control arm down away from the steering knuckle. Remove the strut-to-steering knuckle bolts and remove the steering knuckle/hub/rotor assembly from the vehicle.

7. Use a suitable tool to press the rotor and hub assembly from the steering knuckle.

8. If the rotor is to be reused, mark the position of the rotor on the hub. Unbolt the rotor from the hub. If the rotor is to be machined, observe the minimum thickness specification.

To install:

9. Install the rotor on the hub, tightening the bolts to 33–40 ft. lbs. (44–54 Nm). If the rotor is being reused, be sure to align the marks made during the removal procedure.

10. Press the hub and rotor assembly into the steering knuckle.

11. Slide the hub over the halfshaft and install the steering knuckle on the strut. Install the mounting bolts and tighten to 69–86 ft. lbs. (93–117 Nm).

12. Raise the lower control arm and install the ball joint stud in the steering knuckle. Install the cinch bolt and tighten to 32–40 ft. lbs. (43–54 Nm).

13. Install the caliper and brake pads.

14. Install the washer and a new halfshaft nut. Tighten the nut to 116–174 ft. lbs. (157–235 Nm). Stake the nut using a cold chisel with a rounded cutting edge.

NOTE: If the nut splits or cracks after staking, it must be replaced with a new nut.

15. Install the wheel and tire assembly and lower the vehicle.

1991–92

1. Raise and safely support the vehicle.

2. Remove the wheel and tire assembly.

3. Remove the 2 caliper mounting bolts.

4. Secure the caliper aside with mechanics wire.

5. Pull the rotor from the hub. Inspect the rotor and refinish or replace, as necessary. If refinishing, check the minimum thickness specification.

To install:

6. Position the rotor onto the hub.

7. Remove the mechanics wire and position the caliper.

8. Install the 2 caliper mounting bolts and tighten to 29–36 ft. lbs. (39–49 Nm).

9. Install the wheel and tire assembly and lower the vehicle.

Rear

1988–90

1. Disconnect the negative battery cable.

2. Remove the wheel and tire assembly.

3. Remove the dust cap and, using a chisel, unstake and remove the wheel bearing nut and washer.

4. Remove the disc brake pads and the caliper but do not disconnect the brake hose from the caliper. Support the caliper aside with mechanics wire.

5. Remove the rotor/hub/bearing assembly.

To install:

6. Install the rotor/hub/bearing assembly on the spindle. Install a new locknut and adjust the bearing preload. Install the dust cap.

7. Install the caliper and brake pads.

8. Install the wheel and tire assembly and lower the vehicle.

1991–92

1. Raise and safely support the vehicle.

2. Remove the wheel and tire assembly.

3. Remove the brake pads.

4. Remove the 2 rotor mounting screws.

5. Using a suitable tool, pivot the caliper on its mounting bracket and remove the rotor. Inspect the rotor and refinish or replace, as necessary. If refinishing, check the minimum thickness specification.

To install:

6. Using a suitable tool, pivot the caliper on its mounting bracket and position the rotor.

7. Install the 2 mounting screws.

8. Install the brake pads.

9. Install the wheel and tire assembly and lower the vehicle.

Brake Drums

REMOVAL & INSTALLATION

1988–90

1. Raise and safely support the vehicle.

2. Remove the tire and wheel assembly.

3. Remove the grease cap from the brake drum.

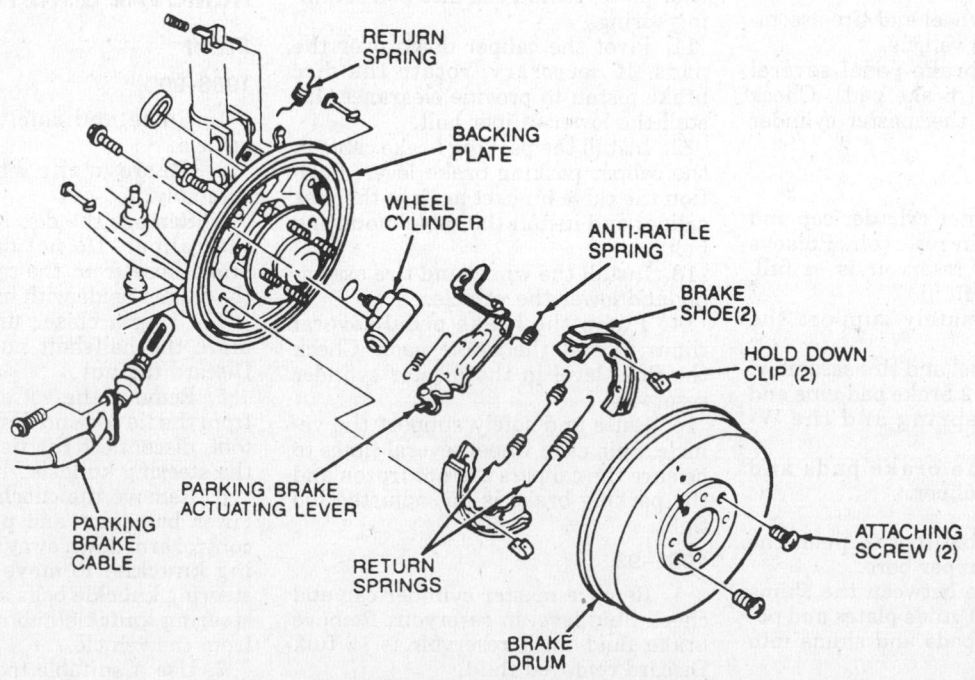

Drum brake assembly–1991–92 vehicles

4. Using a chisel, raise the staked portion of the locknut. Discard the locknut.

NOTE: The locknuts are threaded left and right. The left hand threaded locknut is on the right side of the vehicle and should be turned clockwise to loosen. The right hand threaded locknut is on the left side of the vehicle and is turned counter-clockwise to loosen.

5. Remove the brake drum and bearings as an assembly.
To install:
6. Position the brake drum/bearing/hub assembly on the spindle.
7. Install a new locknut and adjust the bearing preload. Install the grease cap.
8. Install the wheel and tire assembly and lower the vehicle.

1991–92

1. Raise and safely support the vehicle.
2. Remove wheel and tire assembly.
3. Remove the spring nut or attaching screws, if necessary.
4. Pull the brake drum from the hub. Inspect the drum and refinish or replace, as necessary. If refinishing, check the maximum inside diameter specification.
To install:
5. Position the brake drum on the hub.
6. Install the 2 drum attaching screws, if applicable.
7. Install the wheel and tire assembly and lower the vehicle.

Brake Shoes

REMOVAL & INSTALLATION

1. Raise and safely support the vehicle.
2. Remove the wheel and tire assembly and remove the brake drum.
3. Remove the 2 brake shoe return springs.
4. Remove the anti-rattle spring.
5. Push and turn the 2 brake shoe hold-down clips and remove the clips.
6. Remove the leading and trailing shoes from the backing plate.
To install:
7. Use a high temperature grease to lubricate the brake shoe contact points on the backing plate.
8. Position the trailing brake shoe on the backing plate and install 1 of the brake shoe hold-down clips.
9. Position the leading brake shoe on the backing plate and install the other brake shoe hold-down clip.
10. Install the anti-rattle spring.

11. Install the 2 brake shoe return springs.
12. On 1988–90 vehicles, insert a prybar between the knurled quadrant and the parking brake strut. Twist the prybar until the quadrant just touches the backing plate.
13. On 1991–92 vehicles, press the brake pedal to verify operation of the automatic brake adjuster.
14. Install the brake drum and the wheel and tire assembly. Lower the vehicle.
15. Firmly apply the brakes 2–3 times to adjust the rear brakes.

Wheel Cylinder

REMOVAL & INSTALLATION

1. Raise and safely support the vehicle.
2. Remove the wheel and tire assembly and remove the brake drum.
3. On 1988–90 vehicles, remove the brake shoes. On 1991–92 vehicles, remove the upper brake shoe return spring.
4. Clamp the wheel cylinder brake hose.
5. Using a flare nut wrench, loosen the wheel cylinder-to-brake line flare nut.
6. On 1991–92 vehicles, pull the clip from the brake hose retaining bracket and remove the brake hose from the retaining bracket.
7. Remove the brake line from the wheel cylinder.
8. Remove the 2 wheel cylinder mounting bolts and remove the wheel cylinder from the backing plate.
9. On 1991–92 vehicles, remove and discard the wheel cylinder gasket.
To install:
10. On 1991–92 vehicles, install a new wheel cylinder gasket onto the backing plate.
11. Position the wheel cylinder onto the backing plate and install the 2 mounting bolts. Tighten the bolts to 89–115 inch lbs. (10–13 Nm).
12. Position the brake line into the wheel cylinder fitting and tighten the wheel cylinder-to-brake line flare nut to 12–16 ft. lbs. (16–22 Nm).
13. On 1991–92 vehicles, position the brake hose into the retaining bracket and install the clip. Remove the clamp from the wheel cylinder brake hose.
14. On 1988–90 vehicles, install the brake shoes. On 1991–92 vehicles, install the brake shoe return spring.
15. Press the brake pedal to verify the operation of the automatic brake adjuster.
16. Install the brake drum and the wheel and tire assembly.
17. Bleed the brake system and lower the vehicle.

Parking Brake Cable

ADJUSTMENT

1988–90

1. Remove the attaching screws in the center console and remove the console.
2. Make sure the parking brake lever is fully released.
3. Tighten the adjusting nut on the left side of the lever to shorten the equalizer cable. Tighten the nut until it takes 10 notches to fully set the parking brake.
4. Install the center console.

1991–92

1. Start the engine and place the transaxle in **R**.
2. With the vehicle moving in reverse, depress the brake pedal several times.
3. Stop the vehicle and place the transaxle in **P**. Stop the engine.
4. Remove the parking brake console as follows:
   a. Remove the rear seat ash tray.
   b. Position both front seats to the rear-most position.
   c. Remove the 2 front retaining screws from the parking brake console.
   d. Recline both front seats.
   e. Remove the 2 rear retaining screws and with the parking brake engaged, remove the parking brake console.
   f. Release the parking brake lever.
5. Turn the adjusting nut until the parking brake lever stroke is 5–7 notches when pulled with a force of 22 lbs.
6. Install the parking brake console by reversing the removal procedure.

REMOVAL & INSTALLATION

1988–90

1. Raise and safely support the vehicle.
2. With a pair of needle-nose pliers, remove the parking brake return spring at each backing plate. Be careful not to over-extend the spring.
3. Remove the mounting bolts from the parking brake cable housing and pull it away from the backing plate.
4. Disconnect the parking brake cables from the backing plate parking brake levers.
5. Unbolt the cable housing clamps from the rear suspension trailing arm and the trailing arm support bracket.
6. Loosen the cable housing adjustment nut at the end of the cable housing and remove the cable housing from the bracket.

7. Using a small prybar, pry the plastic cable bushings out of their brackets.

8. Disconnect the parking brake return spring from the cable equalizer and remove each cable from the equalizer.

To install:

9. Install the 2 cable ends in the parking brake equalizer and connect the return spring.

10. Using a plastic hammer, install the 2 plastic cable guides in their bracket.

11. Install the threaded end of the cable housings in their brackets and install the adjuster nuts.

12. Install the cable housing support clamps at the trailing support bracket and on each trailing arm.

13. Connect the parking brake cables to the backing plate parking brake levers.

14. Position the parking brake cable against the backing plate and install the mounting bolts.

15. Lower the vehicle and adjust the parking brake cables.

1991–92

1. Remove the parking brake console as follows:
   a. Remove the rear seat ash tray.
   b. Position both front seats to the rear-most position.
   c. Remove the 2 front retaining screws from the parking brake console.
   d. Recline both front seats.
   e. Remove the 2 rear retaining screws and with the parking brake engaged, remove the parking brake console.
   f. Release the parking brake lever.

2. Remove the cable adjusting nut.

3. Raise and safely support the vehicle.

4. Remove the rear exhaust pipe and resonator heat shields.

5. Disconnect the equalizer return spring and remove the cables from the equalizer.

6. Remove the clip that attaches the cable to the retaining bracket located near the equalizer. Remove the cable from the bracket.

7. Remove the cable routing bracket bolt from the floorpan and remove the bracket.

8. Remove the 2 cable routing bracket nuts from the trailing link and remove the bracket.

9. Remove the 2 cable retaining bracket bolts from the backing plate and remove the bracket.

10. Remove the cable from the parking brake actuating lever.

To install:

11. Position the cable onto the parking brake actuating lever.

12. Position the parking brake cable bracket onto the backing plate and install the 2 bolts. Tighten the bolts to 14–19 ft. lbs. (19–25 Nm).

13. Position the cable routing bracket onto the trailing link and install the 2 nuts. Tighten the nuts to 12–17 ft. lbs. (16–23 Nm).

14. Position the cable routing bracket to the floor pan and install the mounting bolt. Tighten the bolt to 14–19 ft. lbs. (19–25 Nm).

15. Position the cable into the retaining bracket near the equalizer and install the clip.

16. Install the cables into the equalizer and install the cable return spring.

17. Install the rear exhaust pipe and resonator heat shields.

18. Lower the vehicle.

19. Install the adjusting nut.

20. Adjust the parking brake cable and install the parking brake console in the reverse order of removal.

Brake System Bleeding

1. Clean all the dirt from around the master cylinder filler cap.

2. Fill the reservoir with brake fluid. The reservoir must be at least ¾ full throughout the bleeding procedure.

3. If the master cylinder is known or suspected to have air in bore, it must be bled before any wheel cylinders or calipers.

4. To bleed the master cylinder, loosen 1 outlet fitting aproximately ¾ turn. Have an assistant push the brake pedal down slowly through full travel. Close the outlet fitting, then return the pedal slowly to the full released position. Wait 5 seconds, then repeat the operation until the air bubbles cease to appear.

5. Loosen the other outlet fitting approximately ¾ turn and repeat Step 4.

6. To continue to bleed the system, remove the rubber cap dust cap from the wheel cylinder bleeder fitting or caliper fitting. Check to make sure the bleeder fitting is positioned at the upper half on the front of the caliper, if not the caliper is located on the wrong side.

7. Attach a length of rubber hose to the fitting. Submerge the free end of the hose in a container partially filled with clean brake fluid and loosen the bleeder fitting approximately ¾ of a turn.

8. Have the assistant push brake pedal down slowly through full travel.

Close the bleeder fitting, then return the pedal to the full release position. Wait 5 seconds, then repeat this operation until the air bubbles cease to appear at the submerged end of the bleeder hose.

9. When the fluid is completely free of air bubbles, properly tighten the bleeder fitting and reinstall the rubber dust cap. Repeat this process on the opposite diagonal system. Refill the master cylinder reservoir after each wheel cylinder or caliper is bled and reinstall the master cylinder cap.

NOTE: If all wheels are to be bled, proceed as follows: right rear, left front, left rear and right front.

10. When the bleeding operation is completed, the fluid level should be filled to the maximum fill level indicated on the reservoir. Always ensure the disc brake pistons are returned to their normal positions by depressing the brake pedal several times until the normal pedal travel is established. Check the pedal feel. If the pedal feels spongy, repeat the bleeding procedure.

CHASSIS ELECTRICAL

Heater Blower Motor

REMOVAL & INSTALLATION

1988–90

1. Disconnect the negative battery cable.

2. From the passenger's side, remove the sound deadening panel.

3. Disconnect the electrical connector from the blower motor assembly.

4. Remove the blower motor-to-blower case screws, the cover, the cooling tube and the motor.

5. Remove the blower wheel-to-motor nut and pull the wheel straight off the motor. Remove the gasket from the motor.

6. To install, reverse the removal procedure. Check the blower motor operation.

1991–92

1. Disconnect the negative battery cable.

2. Remove the trim panel below the glove compartment.

3. Remove the wiring bracket and bolt.

4. Disconnect the blower motor electrical connector.

5. Remove the 3 blower motor mounting bolts and remove the blower motor.

6. Remove the blower wheel retaining clip and remove the blower wheel from the blower motor.

7. Installation is the reverse of the removal procedure.

Windshield Wiper Motor

REMOVAL & INSTALLATION

FRONT

1988–90

1. Disconnect the negative battery cable.

2. From the top left side of the cowl, remove the windshield wiper motor shield-to-chassis plastic retainers and the shield.

3. From the windshield wiper motor shaft, remove the drive link nut and split washer.

4. Disconnect the electrical connector from the windshield wiper motor.

5. Remove the windshield wiper motor-to-cowl bolts, the motor and rubber insulators.

6. To install, make sure the windshield wiper motor is in the PARK position and reverse the removal procedures. Inspect the operation of the windshield wiper system.

1991–92

1. Disconnect the negative battery cable.

2. Remove the windshield wiper arms.

3. With the hood closed, remove the 7 screw covers.

4. Remove the 7 cowl grille retaining screws and remove the cowl grille.

5. Pry up the 4 baffle retaining clips and remove the baffle trim piece.

NOTE: Make sure the motor is in the PARK position before disconnecting the linkage.

6. Remove the wiper linkage retaining clip and disconnect the wiper linkage from the motor.

7. Disconnect the 2 motor electrical connectors.

8. Remove the 3 motor mounting bolts until they are loose from the sheetmetal mounting surface. Remove the motor.

9. Installation is the reverse of the removal procedure. Tighten the 3 motor mounting bolts to 61–87 inch lbs. (7–9 Nm).

Rear

1988–90

1. Disconnect the negative battery cable.

2. From the liftgate, remove the wiper arm/blade assembly and pull the luggage compartment end trim from inside.

3. Remove the seal cap, nut, outer bushings, packings and inner bushings.

4. Disconnect the electrical connector and the ground wire from the windshield wiper motor.

5. Remove the windshield wiper motor-to-liftgate bolts, the motor and rubber insulators.

6. To install, make sure the windshield wiper motor is in the PARK position and reverse the removal procedures. Inspect the operation of the windshield wiper system.

1991–92

1. Disconnect the negative battery cable.

2. Remove the wiper arm by lifting the wiper arm attaching nut cover, removing the attaching nut and pulling the wiper arm from the pivot shaft.

3. Remove the shaft seal from the outer bushing attaching nut.

4. Remove the outer bushing attaching nut and remove the outer bushing.

5. Remove the liftgate trim panel.

6. Disconnect the wiper motor electrical connector.

7. Remove the 3 wiper motor mounting bolts and washers and remove the wiper motor.

8. Installation is the reverse of the removal procedure. Tighten the mounting bolts to 61–87 inch lbs. (7–9 Nm) and the outer bushing attaching nut to 35–52 inch lbs. (4–6 Nm).

Instrument Cluster

REMOVAL & INSTALLATION

1988–90

1. Disconnect the negative battery cable.

2. Remove the steering wheel.

3. Remove the instrument cluster bezel-to-instrument panel screws and bezel.

4. Remove the instrument cluster-to-instrument panel screws.

5. From under the dash, disconnect the speedometer cable from the cluster.

6. Pull the instrument cluster outward and disconnect the wiring connectors. Remove the cluster.

7. To install, reverse the removal procedure. Inspect the operation of the instruments.

1991–92

1. Disconnect the negative battery cable.

2. If equipped with a standard col-umn, remove the 4 bolts securing the steering column to the instrument panel frame.

3. Lower the steering column.

4. Remove the cap screws securing the instrument cluster bezel to the instrument panel and remove the instrument cluster bezel.

5. Disconnect the speedometer cable at the transaxle by pulling the cable out of the vehicle speed sensor.

6. Remove the screws and bolts securing the instrument cluster to the instrument panel.

7. Pull the instrument cluster out slightly and disconnect the electrical connectors from the rear of the instrument cluster.

8. Disconnect the speedometer cable from the instrument cluster and remove the cluster from the instrument panel.

9. Installation is the reverse of the removal procedure. Make sure the instrument cluster is held in its forward most position while attaching the 2 upper screws.

Speedometer

REMOVAL & INSTALLATION

1988–90

1. Disconnect the negative battery cable.

2. Remove the instrument cluster.

3. Remove the lens from the instrument cluster.

4. Remove the indicator light overlay from the instrument cluster.

5. Remove the speedometer-to-instrument cluster screws and the speedometer.

6. To install, reverse the removal procedure. Check the operation of the speedometer.

1991–92

1. Disconnect the negative battery cable.

2. Remove the instrument cluster.

3. Remove the instrument cluster lens and shroud.

4. Remove the speedometer from the instrument cluster.

5. Installation is the reverse of the removal procedure. Be careful when handling the speedometer so as not to disturb the factory calibration.

Radio

REMOVAL & INSTALLATION

1988–90

1. Disconnect the negative battery cable.

2. Remove the radio trim cover from the instrument panel.

3. Remove the radio retaining screws and pull the radio from the instrument panel.

4. Disconnect the electrical and antenna leads from the radio.

5. Installation is the reverse of the removal procedure.

1991–92

1. Disconnect the negative battery cable.

2. Using radio removal tool T87P-19061–A or equivalent, pull the radio out from its mounting position so the antenna and the electrical connectors are accessible.

3. Disconnect the antenna lead and the radio electrical connectors from the radio.

4. Remove the radio.

5. Installation is the reverse of the removal procedure.

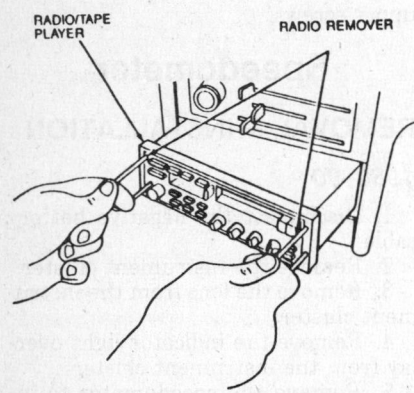

Radio removal and installation—1991–92 vehicles

Combination Switch

The combination switch assembly controls the turn signal, headlight, dimmer and windshield wiper switch functions.

REMOVAL & INSTALLATION

1988–90

1. Disconnect the negative battery cable and remove the steering wheel.

2. Remove the steering column covers-to-steering column screws and the covers.

3. Depress the small tang on the electrical harness clip and disconnect the clip; move the electrical harness aside.

4. Loosen the combination switch-to-steering column clamp, slide the switch slightly forward and disconnect the electrical connector from the rear of the combination switch.

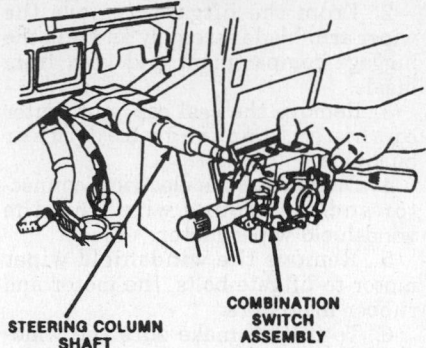

Combination switch removal and installation—1988–90 vehicles

5. Remove the combination switch from the steering column.

6. To install, reverse the removal procedure. Check the switch operations.

1991–92

1. Disconnect the negative battery cable.

2. Remove the steering wheel cover retaining screws from the back side of the steering wheel and remove the cover.

3. Disconnect the horn electrical connector and the speed control electrical connectors, if equipped.

4. Remove the steering wheel mounting nut.

NOTE: Do not attempt to remove the steering wheel by hitting the column shaft with a hammer. The column may collapse.

5. Remove the steering wheel using a suitable puller.

6. Remove the 4 retaining screws from the steering column lower cover and remove the cover. Remove the upper cover.

7. Disconnect the 3 combination switch electrical connectors.

8. Remove the combination switch retaining screw, pull the electrical connectors from the retaining brackets and remove the switch.

9. Installation is the reverse of the removal procedure. Tighten the steering wheel mounting nut to 29–36 ft. lbs. (39–49 Nm).

Ignition Lock

REMOVAL & INSTALLATION

1991–92

1. Disconnect the negative battery cable.

2. Remove the steering wheel cover retaining screws from the back side of the steering wheel and remove the cover.

3. Disconnect the horn electrical connector and the speed control electrical connectors, if equipped.

4. Remove the steering wheel mounting nut.

NOTE: Do not attempt to remove the steering wheel by hitting the column shaft with a hammer. The column may collapse.

5. Remove the steering wheel using a suitable puller.

6. Remove the combination switch.

7. Disconnect the ignition switch electrical connector.

8. Remove the shift-lock cable mounting bracket bolt and position the bracket and cable aside.

9. Remove the 4 steering column upper mounting bracket bolts and lower the column.

10. Using a hammer and chisel, make a groove in the head of each of the 2 column lock mounting bracket bolts.

11. Remove the bolts with a prybar and discard the bolts.

12. Remove the steering column lock and mounting bracket.

To install:

13. Position the steering column lock and mounting bracket and install 2 new bolts, tightening them only enough to hold the column lock in position.

14. With the key in the ignition, verify the operation of the column lock. If necessary, reposition the column lock until it operates properly.

15. Tighten the mounting bracket bolts until the bolt heads break off.

16. Position the steering column and install the 4 upper mounting bracket bolts. Tighten the bolts to 80–123 inch lbs. (9–14 Nm).

17. If equipped with a tilt steering wheel, remove the upper mounting bracket retaining pin.

18. Position the shift-lock cable mounting bracket and install the bolt. Tighten to 37–55 inch lbs. (4–6 Nm).

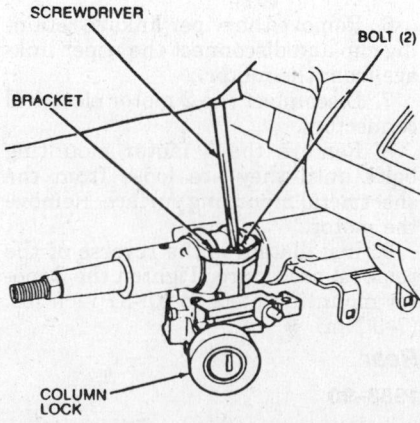

Ignition lock removal—1991–92 vehicles

19. Connect the ignition switch electrical connector.

20. Install the combination switch.

21. Install the steering wheel and the mounting nut. Tighten to 29–36 ft. lbs. (39–49 Nm).

22. Connect the horn electrical connector and the speed control electrical connectors, if equipped.

23. Position the steering wheel cover and install the retaining screws.

24. Connect the negative battery cable.

Ignition Switch

REMOVAL & INSTALLATION

1988–90

1. Disconnect the negative battery cable.

2. Grasp the black trim ring around the ignition lock switch and pull it straight out.

3. From the driver's side, remove the sound deadening panel and the lap duct cover.

4. If equipped with air conditioning, remove the air conditioning duct assembly-to-access panel support bracket center screw, the access panel support bracket screws and the bracket.

5. From under the steering column, grasp the side window defogger duct ends, pull it outward, while twisting it slightly. From the ignition switch, located under the steering column, disengage the plastic strap connector locking tang and remove the plastic strap.

6. Remove the steering column-to-instrument panel bolts and lower the column.

7. Lift the upper steering column shroud and remove it from the steering column.

8. Remove the ignition switch-to-ignition switch housing screw, grasp the ignition switch body and pull it straight outward.

9. To disengage the electrical connectors from the ignition switch, perform the following procedure:

    a. Disengage the electrical connector locking tang.

    b. Grasp an electrical connector in each hand and pull them straight apart.

NOTE: Be aware of the electrical connector cavity position for reassembly purposes.

10. Using a straightened paper clip, disengage the 2 in-key buzzer wires from the 4-terminal connector; the wire colors are red and red wire/orange tracer.

To install:

11. To install wires and connector, perform the following procedures:

    a. Align the wire end flat sides with the grooved portion of the connector and push the wire inward until the locking tang engages wire end.

    b. Push the 4-terminal connector into the housing connector until the locking tangs are in place.

    c. Using electrical tape, wrap it around the ignition switch wires.

12. Install the ignition switch-to-ignition switch housing screw and the plastic snap connector around the ignition switch wiring. Attach the connector peg to the steering column mounting bracket.

13. Using electrical tape, wrap it around the ignition switch wiring and steering column.

14. To complete the installation, reverse the removal procedure. Check the ignition switch operation.

1991–92

1. Disconnect the negative battery cable.

2. Remove the combination switch.

3. Disconnect the ignition switch electrical connector.

4. Remove the 3 ignition switch mounting screws and remove the ignition switch.

5. Installation is the reverse of the removal procedure. Check the switch for proper operation.

Stoplight Switch

ADJUSTMENT

1988–90

1. Measure the distance from the center of the brake pedal pad to the floor. The distance should be 8.62–8.82 in. (219–224mm). If not, proceed to Step 2.

2. Disconnect the stoplight switch electrical connector.

3. Loosen the switch locknut and rotate the switch until the correct pedal height is reached.

4. Tighten the locknut and connect the switch electrical connector.

1991–92

1. Measure the distance from the center of the brake pedal pad to the floor. The distance should be 7.60–7.72 in. (193–196mm). If not, proceed to Step 2.

2. Disconnect the stoplight switch electrical connector.

3. Loosen the stoplight locknut and turn the stoplight switch until it does not contact the brake pedal.

4. Loosen the rod locknut and turn the rod until the brake pedal height is within specification.

5. Turn the stoplight switch until it

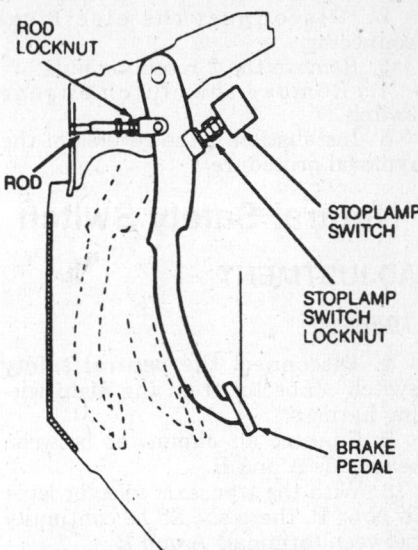

Stoplight switch adjustment—1991–92 vehicles

contacts the brake pedal, then turn it an additional ½ turn.

6. Tighten the stoplight locknut and the rod locknut.

7. Connect the stoplight switch electrical connector.

8. Check the operation of the stoplights.

REMOVAL & INSTALLATION

1. Disconnect the negative battery cable.

2. Disconnect the stoplight switch electrical connector.

3. Remove the stoplight switch retaining nuts and remove the stoplight switch.

4. Installation is the reverse of the removal procedure. Adjust the switch according to the proper procedure.

Clutch Switch

ADJUSTMENT

1. Disconnect the negative battery cable.

2. Disconnect the clutch engage electrical connector.

3. Using a ohmmeter, check the resistance between the connector terminals.

4. The ohmmeter should show continuity when the switch rod is pushed into the switch. The ohmmeter should show no continuity with the switch rod released.

5. Replace the switch if it does not perform as specified.

REMOVAL & INSTALLATION

1. Disconnect the negative battery cable.

2. Disconnect the electrical connector.

3. Remove the 2 retaining nuts.

4. Remove the clutch engage switch.

5. Installation is the reverse of the removal procedure.

Neutral Safety Switch

ADJUSTMENT

1988–90

1. Disconnect the neutral safety switch connector from the main wiring harness.

2. Connect an ohmmeter between terminals **A** and **B**.

3. With the transaxle selector lever in **N** or **P**, there should be continuity between terminals **A** and **B**.

4. Replace the switch if it does not perform as specified.

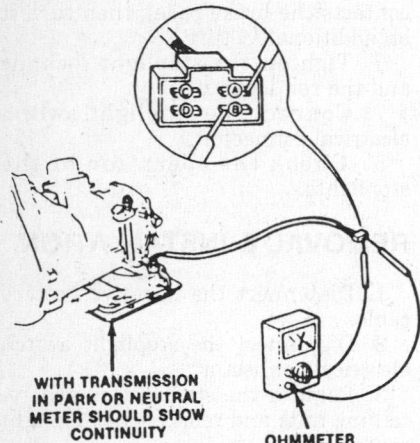

WITH TRANSMISSION IN PARK OR NEUTRAL METER SHOULD SHOW CONTINUITY

OHMMETER

Neutral safety switch testing—1988–90 vehicles

1991–92

The neutral safety switch function is performed by the Manual Lever Position Switch (MLPS). The MLPS is an adjustable switch that informs the automatic transaxle control unit of the position of the transaxle manual shaft. The MLPS will allow the vehicle to be started with the gear selector in the **P** or **N** positions when properly adjusted. The MLPS is located externally on the transaxle housing and is positioned on the manual shaft.

1. Remove the air cleaner assembly and air inlet tube.

2. Remove the nut securing the manual shaft lever to the transaxle manual shaft.

3. Remove the lever from the manual shaft.

4. Turn the manual shaft to the neutral position.

5. Loosen the MLPS mounting bolts.

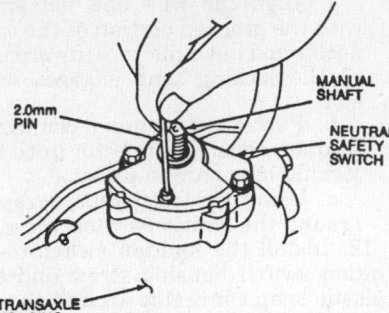

2.0mm PIN

MANUAL SHAFT

NEUTRAL SAFETY SWITCH

TRANSAXLE HOUSING

Neutral safety switch adjustment—1991–92 vehicles

6. Align the hole of the MLPS with the hole on the manual shaft lever by inserting a 0.079 in. (2.0mm) outside diameter pin.

7. Tighten the MLPS mounting bolts to 69–95 inch lbs. (8–11 Nm). Remove the pin.

8. Check the continuity of the switch as follows:

    a. Disconnect the switch connector.

    b. Using a ohmmeter, check the switch for continuity at the connector.

    c. There should be continuity between the **BK/BL** and **BK/R** terminals with the shift lever in the **P** or **N** position.

9. If the continuity is incorrect, replace the MLPS.

10. Position the manual shaft lever to the manual shaft and install the nut. Tighten to 33–47 ft. lbs. (44–64 Nm).

11. Install the air cleaner assembly and air inlet tube.

REMOVAL & INSTALLATION

1988–90

1. Disconnect the negative battery cable.

2. Raise and support the vehicle safely.

3. Disconnect the electrical connector from the neutral safety switch.

4. Using a wrench, remove the switch from the transaxle.

5. To install, apply sealant to the switch threads and reverse the removal procedures. Torque the neutral safety switch to 14–19 ft. lbs. (19–26 Nm).

1991–92

1. Disconnect the negative battery cable.

2. Remove the air cleaner assembly and air inlet tube.

3. Remove the nut securing the manual shaft lever to the transaxle manual shaft and remove the lever.

4. Disconnect the 3 electrical connectors located on the top of the top of the transaxle.

5. Disconnect the electrical connector located on the front side of the transaxle.

6. Remove the 2 bolts securing the MLPS and the bolts securing the electrical connector brackets to the top of the transaxle.

7. Remove the MLPS from the manual shaft.

To install:

8. Position the MLPS onto the manual shaft.

9. Install the bolts securing the electrical connectors to the transaxle housing.

10. Install the bolts securing the MLPS but do not tighten yet.

11. Connect the electrical connectors located on the top and on the side of the transaxle.

12. Adjust the MLPS according to the proper procedure.

13. Install the air cleaner assembly and the air inlet tube.

14. Connect the negative battery cable.

Fuses, Circuit Breakers and Relays

LOCATION

Fuses

All vehicles are equipped with a fuse panel mounted inside the vehicle, under the left side of the instrument panel as well as a fuse box mounted in the engine compartment.

Fuse Link Cartridges

Fuse link cartridges are used on 1991–92 vehicles. Fuse link cartridges have a colored plastic housing with a clear "window" at the top. To check a fuse cartridge, look at the fuse element through the clear "window". The fuse link cartridges are located in the engine compartment fuse box. The following fuse link cartridges are listed according to their labels in the fuse box.

Fuel Inj—Pink 30 amp: to protect the electronic engine control circuit.

Head—Pink 30 amp: to protect the headlight circuit and the daytime running lights circuit.

Main—Black 80 amp for 1.8L engine or Dark Blue for 1.9L engine: to protect all circuits, except starter and starter solenoid circuits.

BTN—Yellow 60 amp for 1.8L engine or Green 40 amp for 1.9L engine: to protect the courtesy lights, electronic automatic transaxle, electronic engine control, exterior lights, horn, interior lights, passive restraint, power door locks, radio, shift lock and warning chime circuits.

Cooling Fan — Pink 30 amp for 1.8L engine or Green 40 amp for 1.9L engine: to protect cooling fans circuit.

Circuit Breakers

A circuit breaker is mounted on the interior fuse panel. This breaker controls the blower motor circuit.

Various Relays

1988–90

Horn Relay — located in the engine compartment on the left inner fender.

A/C Cut-out Relay — located in the front of the left front shock tower in the engine compartment.

A/C Relay No. 1 — located on the left front shock tower in the engine compartment.

A/C Relay No. 2 — located on the left front shock tower in the engine compartment.

A/C Relay No. 3 — located on the left front shock tower in the engine compartment.

Cooling Fan Relay — located in the left front side of the engine compartment, next to the coolant recovery bottle.

Door Buzzer Relay — located in the electrical equipment panel, above the fuse block.

Fuel Pump Relay — mounted under the center of the instrument panel.

1991–92

Air Conditioning Relay — located on the rear of the right fender apron.

Cooling Fan Relay — located on the front of the left fender apron.

Door Lock Relay — located above the left cowl.

Daytime Running Lights Relay — located behind the right side of the instrument panel, near the blower motor.

Electronic Engine Control Power Relay — located behind the center of the instrument panel.

Fuel Pump Relay — located behind the center of the instrument panel.

Headlight Relay — located above the left cowl.

Horn Relay — located above the left cowl.

Parking Light Relay — located behind the left side of the instrument panel.

Vane Air Flow Meter Relay — located behind the center of the instrument panel.

Wide Open Throttle Cutout Relay — located on the rear of the right fender apron.

Computers

LOCATION

The Electronic Control Unit (ECU) is located behind the center of the instrument panel.

Flashers

LOCATION

On 1988–90 vehicles, the turn signal/hazard flasher is located on the interior fuse panel. On 1991–92 vehicles, the flasher unit is located near the combination switch.

Cruise Control

ADJUSTMENT

Actuator Cable

1988–90

1. With the engine off, remove the clip from the actuator cable and adjust the locknut while pressing down on the cable until free-play is 0.04–0.12 in. (1–3mm).
2. Check the system operation and adjust as needed.

1991–92

1. Remove the cable adjusting clip from the cable housing.
2. Pull tightly on the cable until all of the slack is taken out.
3. Install the cable adjusting clip.

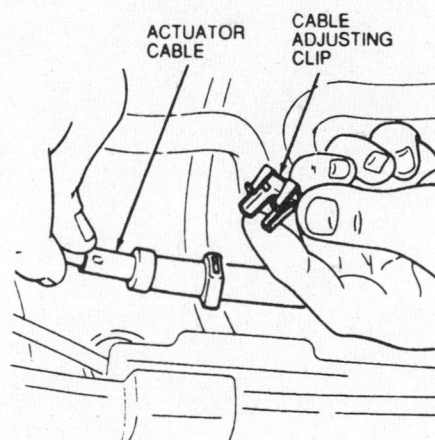

Actuator cable adjustment — 1991–92 vehicles

Clutch Pedal Height

1988–90

Pedal height is the distance from the cowl to the center of the clutch pedal pad.

1. Remove the necessary instrument panel components which block access to the clutch pedal.
2. Loosen the clutch pedal locknut.
3. Turn the stop bolt to obtain the correct pedal height of 8.4–8.6 in. Tighten the locknut.
4. If components from the instrument panel were removed, reinstall them.

Brake Pedal Height

1988–90

Measure the distance from the center of the brake pedal to lower dash panel. Pedal height must be 8.62–8.82 in. (214.5–219.5mm). If the brake pedal height is not within these specifications, adjustment is necessary.

1. Disconnect the negative battery cable.
2. Adjust the pedal height by adjusting the stoplight switch.
3. Disconnect the connector on the stoplight switch.
4. Loosen the stoplight switch locknut and rotate the switch until the pedal height is 8.62–8.82 in. (214.5–219.5mm).
5. Tighten the switch locknut.
6. Connect the stoplight switch connector.
7. Connect the negative battery cable and check stoplight operation.

Vacuum Dump Valve

1991–92

1. Firmly depress the brake pedal and hold in position.
2. Push in the dump valve until the valve collar bottoms against the retaining clip.
3. Place a 0.050–0.10 in. (1.27–2.54mm) shim between the white button of the valve and the pad on the brake pedal.
4. Firmly pull the brake pedal rearward to its normal position, allowing the dump valve to ratchet backwards in the retaining clip.

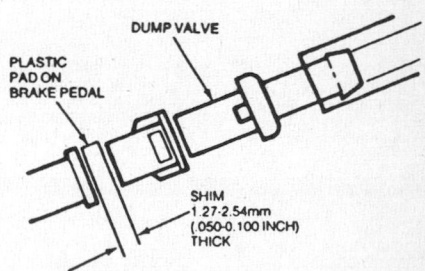

Vacuum dump valve adjustment — 1991–92 vehicles

Clutch Switch

1991–92

1. Measure the distance from the bulkhead to the upper center of the clutch pedal pad. The distance should be 7.72–8.03 in. (196–204mm). If not proceed to Step 2.

2. Disconnect the clutch switch electrical connector.

3. Loosen the switch locknut and turn the switch until the specified distance is achieved. Tighten the locknut to 10–13 ft. lbs. (14–18 Nm).

4. Push the clutch pedal down by hand until clutch resistance is felt.

5. Measure the distance between the upper pedal height and where resistance is felt. The free-play should be 0.20–0.51 in. (5–13mm). If not, proceed to Step 6.

6. Loosen the pushrod locknut and turn the pushrod until the specified free-play is achieved.

7. Check that disengagement height is correct when the pedal is fully depressed. Minimum disengagement height is 1.6 in. (41mm).

8. Tighten the pushrod locknut to 9–12 ft. lbs. (12–17mm) and connect the clutch switch electrical connector.

Buick/Chevrolet Oldsmobile/Pontiac

Rear Wheel Drive

Buick—Estate Wagon, Roadmaster, Roadmaster Wagon
Chevrolet—Caprice, Caprice Wagon, Monte Carlo
Oldsmobile—Cutlass Custom Cruiser   **Pontiac**—Safari

13-1

SPECIFICATIONS

VEHICLE IDENTIFICATION CHART

It is important for servicing and ordering parts to be certain of the vehicle and engine identification. The VIN (vehicle identification number) is a 17 digit number visible through the windshield on the driver's side of the dash and contains the vehicle and engine identification codes. The tenth digit indicates model year, and the eighth digit indicates engine code. It can be interpreted as follows:

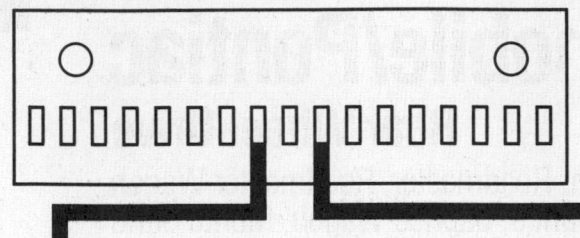

		Engine Code						Model Year	
Code	Cu. In.	Liters	Cyl.	Fuel Sys.	Eng. Mfg.		Code		Year
Z	262	4.3	6	TBI	C.P.C.		J		1988
E	305	5.0	8	TBI	C.P.C.		K		1989
H	305	5.0	8	4 bbl	C.P.C.		L		1990
Y	307	5.0	8	4 bbl	B.O.C.		M		1991
7	350	5.7	8	EFI	B.O.C.		N		1992

B.O.C.—Buick, Oldsmobile, Cadillac
C.P.C.—Chevrolet, Pontiac, Canada
TBI—Throttle Body Injection
EFI—Electronic Fuel Injection

ENGINE IDENTIFICATION

Year	Model	Engine Displacement cu. in. (liter)	Engine Series Identification (VIN)	No. of Cylinders	Engine Type
1988	Estate Wagon	307 (5.0)	Y	8	OHV
	Caprice Sedan	262 (4.3)	Z	6	OHV
	Caprice Sedan	305 (5.0)	H	8	OHV
	Caprice Wagon	307 (5.0)	Y	8	OHV
	Monte Carlo	262 (4.3)	Z	6	OHV
	Monte Carlo	305 (5.0)	H	8	OHV
	Monte Carlo	307 (5.0)	Y	8	OHV
	Cutlass	307 (5.0)	Y	8	OHV
	Custom Cruiser	307 (5.0)	Y	8	OHV
	Safari	307 (5.0)	Y	8	OHV
1989	Estate Wagon	307 (5.0)	Y	8	OHV
	Caprice Sedan ①	262 (4.3)	Z	6	OHV
	Caprice Sedan	305 (5.0)	E	8	OHV
	Caprice Wagon	307 (5.0)	Y	8	OHV
	Caprice Sedan	350 (5.7)	7	8	OHV
	Custom Cruiser	307 (5.0)	Y	8	OHV
	Safari	307 (5.0)	Y	8	OHV

ENGINE IDENTIFICATION

Year	Model	Engine Displacement cu. in. (liter)	Engine Series Identification (VIN)	No. of Cylinders	Engine Type
1990	Estate Wagon	307 (5.0)	Y	8	OHV
	Caprice Sedan ①	262 (4.3)	Z	6	OHV
	Caprice Sedan	305 (5.0)	E	8	OHV
	Caprice Wagon	307 (5.0)	Y	8	OHV
	Caprice Sedan ②	350 (5.7)	7	8	OHV
	Custom Cruiser	307 (5.0)	Y	8	OHV
1991–92	Roadmaster Sedan	350 (5.7)	7	8	OHV
	Roadmaster Estate Wagon	305 (5.0)	E	8	OHV
	Roadmaster Estate Wagon	350 (5.7)	7	8	OHV
	Caprice Sedan	305 (5.0)	E	8	OHV
	Caprice Wagon	305 (5.0)	E	8	OHV
	Caprice Wagon	350 (5.7)	7	8	OHV
	Caprice Sedan ②	350 (5.7)	7	8	OHV
	Custom Cruiser	305 (5.0)	E	8	OHV
	Custom Cruiser	350 (5.7)	7	8	OHV

OHV—Overhead Valves
① Fleet only
② Police

GENERAL ENGINE SPECIFICATIONS

Year	VIN	No. Cylinder Displacement cu. in. (liter)	Fuel System Type	Net Horsepower @ rpm	Net Torque @ rpm (ft. lbs.)	Bore × Stroke (in.)	Compression Ratio	Oil Pressure @ rpm
1988	Z	6-262 (4.3)	TBI	140 @ 4200	225 @ 2000	4.000 × 3.480	9.3:1	45 @ 2000
	H	8-305 (5.0)	4 bbl	165 @ 4200	245 @ 2400	3.736 × 3.480	8.6:1	45 @ 2000
	Y	8-307 (5.0)	4 bbl	148 @ 3800	250 @ 2400	3.800 × 3.385	8.0:1	40 @ 2000
1989	Z	6-262 (4.3)	TBI	140 @ 4000	225 @ 2000	4.000 × 3.480	9.3:1	18 @ 2000
	E	8-305 (5.0)	TBI	170 @ 4400	255 @ 2400	3.740 × 3.480	9.3:1	18 @ 2000
	Y	8-307 (5.0)	4 bbl	148 @ 3800	250 @ 2400	3.800 × 3.385	8.0:1	40 @ 2000
	7	8-350 (5.7)	TBI	195 @ 4200	295 @ 2400	4.000 × 3.500	9.8:1	18 @ 2000
1990	Z	6-262 (4.3)	TBI	140 @ 4000	225 @ 2000	4.000 × 3.480	9.3:1	18 @ 2000
	E	8-305 (5.0)	TBI	170 @ 4400	255 @ 2400	3.740 × 3.480	9.3:1	18 @ 2000
	Y	8-307 (5.0)	4 bbl	140 @ 3200	255 @ 2000	3.800 × 3.385	8.0:1	18 @ 2000
	7	8-350 (5.7)	TBI	195 @ 4200	295 @ 2400	4.000 × 3.500	9.8:1	18 @ 2000
1991–92	E	8-305 (5.0)	TBI	170 @ 4200	255 @ 2400	3.740 × 3.480	9.3:1	18 @ 2000
	7	8-350 (5.7)	TBI	195 @ 4200	295 @ 2400	4.000 × 3.500	9.8:1	18 @ 2000

TBI—Throttle Body Injection

GASOLINE ENGINE TUNE-UP SPECIFICATIONS

Year	VIN	No. Cylinder Displacement cu. in. (liter)	Spark Plugs Type	Spark Plugs Gap (in.)	Ignition Timing (deg.) MT	Ignition Timing (deg.) AT	Compression Pressure (psi)	Fuel Pump (psi)	Idle Speed (rpm) MT	Idle Speed (rpm) AT	Valve Clearance In.	Valve Clearance Ex.
1988	Z	6-262 (4.3)	R-45TS	0.035	—	①	②	9.0–13.0	—	400	Hyd.	Hyd.
	H	8-305 (5.0)	R-45TS	0.035	—	①	②	7.5–9.0	—	500 ①	Hyd.	Hyd.
	Y	8-307 (5.0)	FR3L56	0.060	—	①	②	6.0–7.5	—	450	Hyd.	Hyd.

GASOLINE ENGINE TUNE-UP SPECIFICATIONS

Year	VIN	No. Cylinder Displacement cu. in. (liter)	Spark Plugs Type	Spark Plugs Gap (in.)	Ignition Timing (deg.) MT	Ignition Timing (deg.) AT	Com-pression Pressure (psi)	Fuel Pump (psi)	Idle Speed (rpm) MT	Idle Speed (rpm) AT	Valve Clearance In.	Valve Clearance Ex.
1989	Z	6-262 (4.3)	R-45TS	0.035	—	①	②	9–13	—	①	Hyd.	Hyd.
	E	8-305 (5.0)	R-45TS	0.035	—	①	②	11	—	①	Hyd.	Hyd.
	Y	8-307 (5.0)	FR3L56	0.060	—	①	②	6.0–7.5	—	①	Hyd.	Hyd.
	7	8-350 (5.7)	R-45TS	0.035	—	①	②	11	—	①	Hyd.	Hyd.
1990	Z	6-262 (4.3)	R-45TS	0.035	—	①	②	9–13	—	①	Hyd.	Hyd.
	E	8-305 (5.0)	R-45TS	0.035	—	①	②	9–13	—	①	Hyd.	Hyd.
	Y	8-307 (5.0)	FR3LS6	0.060	—	①	②	6.0–7.5	—	①	Hyd.	Hyd.
	7	8-350 (5.7)	R-43TS	0.035	—	①	②	9–13	—	①	Hyd.	Hyd.
1991	E	8-305 (5.0)	R-45TS	0.035	—	①	②	9–13	—	①	Hyd.	Hyd.
	7	8-350 (5.7)	CR43TS ③	0.035	—	①	②	9–13	—	①	Hyd.	Hyd.
1992						See Underhood Specifications Sticker						

Hyd.—Hydraulic
① See the Emission Control Label
② The lowest cylinder compression reading should not be less than 70% of the highest reading, and no cylinder should be less than 100 PSI
③ Police vehicle

FIRING ORDERS

NOTE: To avoid confusion, always replace spark plug wires one at a time.

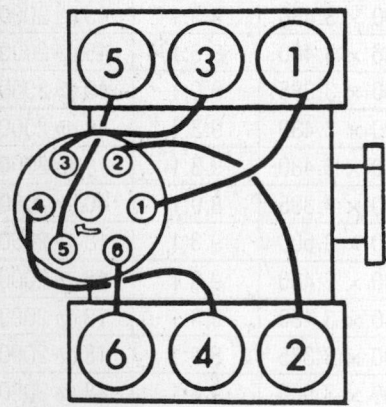

4.3L Engine
Engine Firing Order: 1–6–5–4–3–2
Distributor Rotation: Clockwise

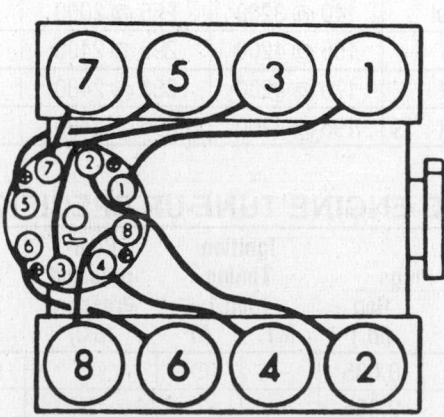

5.0L (VIN E, H) and 5.7L Engines
Engine Firing Order: 1–8–4–3–6–5–7–2
Distributor Rotation: Clockwise

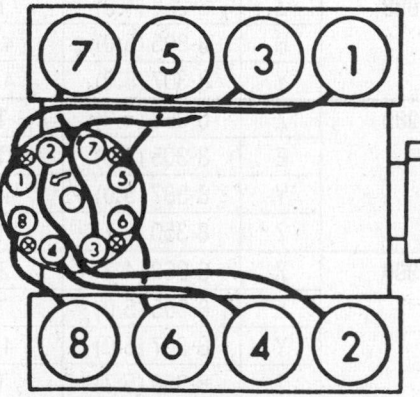

5.0L (VIN Y) Engine
Engine Firing Order: 1–8–4–3–6–5–7–2
Distributor Rotation: Counterclockwise

CAPACITIES

Year	Model	VIN	No. Cylinder Displacement cu. in. (liter)	Engine Crankcase with Filter	Engine Crankcase without Filter	Transmission (pts.) 4-Spd	Transmission (pts.) 5-Spd	Transmission (pts.) Auto. ①	Drive Axle (pts.)	Fuel Tank (gal.)	Cooling System (qts.)
1988	Estate Wagon	Y	8-307 (5.0)	5	4	—	—	10.1	4.25	22	15
	Caprice Sedan	Z	6-262 (4.3)	5	4	—	—	7④	②	25③	12.5
	Caprice Sedan	H	8-305 (5.0)	5	4	—	—	7④	②	25③	16.8
	Caprice Wagon	Y	8-307 (5.0)	5	4	—	—	7④	②	25③	17.1
	Monte Carlo	Z	6-262 (4.3)	5	4	—	—	7④	②	17.6	13.1
	Monte Carlo	H	8-305 (5.0)	5	4	—	—	7④	②	18.1	16.5
	Monte Carlo	Y	8-307 (5.0)	5	4	—	—	7④	②	18.1	16.7
	Cutlass	Y	8-307 (5.0)	5	4	—	—	7	②	18.1	15.5
	Custom Cruiser	Y	8-307 (5.0)	5	4	—	—	7	②	22	15.5
	Safari	Y	8-307 (5.0)	5	4	—	—	10.1	4.25	22	15
1989	Estate Wagon	Y	8-307 (5.0)	5	4	—	—	7④	②	22	17.1
	Caprice Sedan	Z	6-262 (4.3)	4⑧	4	—	—	7④	②	24.5	12.0
	Caprice Sedan	E	8-305 (5.0)	5	4	—	—	7④	②	25	16.7
	Caprice Wagon	Y	8-307 (5.0)	5	4	—	—	7④	②	22	17.1
	Caprice Sedan ⑦	7	8-350 (5.0)	5	4	—	—	7④	②	25	14.9
	Custom Cruiser	Y	8-307 (5.0)	5	4	—	—	7④	②	22	17.1
	Safari	Y	8-307 (5.0)	5	4	—	—	7④	②	22	17.1
1990	Estate Wagon	Y	8-307 (5.0)	5	4	—	—	7⑨	②	22	16.4
	Caprice Sedan ⑥	Z	6-262 (4.3)	4⑧	4	—	—	7⑨	②	24.5	12.0
	Caprice Sedan	E	8-305 (5.0)	5	4	—	—	7⑨	②	24.5	16.7
	Caprice Wagon	Y	8-307 (5.0)	5	4	—	—	7⑨	②	24.5	16.7
	Caprice Sedan ⑦	7	8-350 (5.7)	5	4	—	—	7⑨	②	22	14.8
	Custom Cruiser	Y	8-307 (5.0)	5	4	—	—	7⑨	②	22	16.4
1991-92	Roadmaster Sedan	7	8-350 (5.7)	5	4	—	—	10	②	23	14.6
	Roadmaster Estate Wagon	E	8-305 (5.0)	5	4	—	—	10	②	23	16.7⑤
	Caprice Sedan	E	8-305 (5.0)	5	4	—	—	10	②	23	16.7⑤
	Caprice Wagon	E	8-305 (5.0)	5	4	—	—	10	②	23	16.7⑤
	Caprice Sedan ⑦	7	8-350 (5.7)	5	4	—	—	10	②	23	14.6
	Custom Cruiser	E	8-305 (5.0)	5	4	—	—	10	②	23	16.7⑤

① Additional transmission fluid may be required to bring level to full mark if overhauled or torque converter drained
② 7½ in. ring gear—3.5 pts.
　8½ in. ring gear—4.25 pts.
　8¾ in. ring gear—5.4 pts.
③ Wagon—22 gals.
④ 700R4—10 pts.
⑤ With Heavy Duty Radiator add 0.6 qts.
⑥ Fleet only
⑦ Police
⑧ Add as necessary to bring to appropriate level.
⑨ Hydra-matic 4L60-10.0 pts.

CAMSHAFT SPECIFICATIONS

Year	VIN	No. Cylinder Displacement cu. in. (liter)	Journal Diameter 1	Journal Diameter 2	Journal Diameter 3	Journal Diameter 4	Journal Diameter 5	Lobe Lift In.	Lobe Lift Ex.	Bearing Clearance	Camshaft End Play
1988	Z	6-262 (4.3)	1.8682–1.8692	1.8682–1.8692	1.8682–1.8692	1.8682–1.8692	1.8682–1.8692	0.234	0.257	NA	0.004–0.012
	H	8-305 (5.0)	1.8682–1.8692	1.8682–1.8692	1.8682–1.8692	1.8682–1.8692	1.8682–1.8692	0.234	0.257	NA	0.004–0.012
	Y	8-307 (5.0)	2.0362	2.0360	1.9959	1.9759	1.9559	0.247	0.251	0.0038–	0.006–0.022

CAMSHAFT SPECIFICATIONS

Year	VIN	No. Cylinder Displacement cu. in. (liter)	Journal Diameter 1	2	3	4	5	Lobe Lift In.	Ex.	Bearing Clearance	Camshaft End Play
1989	Z	6-262 (4.3)	1.8682–1.8692	1.8682–1.8692	1.8682–1.8692	1.8682–1.8692	1.8682–1.8692	0.234	0.257	NA	0.004–0.012
	E	8-305 (5.0)	1.8682–1.8692	1.8682–1.8692	1.8682–1.8692	1.8682–1.8692	1.8682–1.8692	NA	NA	NA	0.004–0.012
	Y	8-307 (5.0)	2.0362	2.0360	1.9959	1.9759	1.9559	0.247	0.251	0.0038	0.006–0.022
	7	8-350 (5.7)	1.8682–1.8692	1.8682–1.8692	1.8682–1.8692	1.8682–1.8692	1.8682–1.8692	0.257	0.269	NA	0.004–0.012
1990	Z	6-262 (4.3)	1.8682–1.8692	1.8682–1.8692	1.8682–1.8692	1.8682–1.8692	1.8682–1.8692	0.234	0.257	NA	0.004–0.012
	E	8-305 (5.0)	1.8682–1.8692	1.8682–1.8692	1.8682–1.8692	1.8682–1.8692	1.8682–1.8692	0.234	0.257	NA	0.004–0.012
	Y	8-307 (5.0)	2.0362	2.0360	1.9959	1.9759	1.9559	0.247	0.251	0.0038	0.006–0.022
	7	8-350 (5.7)	1.8682–1.8692	1.8682–1.8692	1.8682–1.8692	1.8682–1.8692	1.8682–1.8692	0.257	0.269	NA	0.004–0.012
1991–92	E	8-305 (5.0)	1.8682–1.8692	1.8682–1.8692	1.8682–1.8692	1.8682–1.8692	1.8682–1.8692	0.234	0.257	NA	0.004–0.012
	7	8-350 (5.7)	1.8682–1.8692	1.8682–1.8692	1.8682–1.8692	1.8682–1.8692	1.8682–1.8692	0.257	0.269	NA	0.004–0.012

NA—Not available

CRANKSHAFT AND CONNECTING ROD SPECIFICATIONS

All measurements are given in inches.

Year	VIN	No. Cylinder Displacement cu. in. (liter)	Crankshaft Main Brg. Journal Dia.	Main Brg. Oil Clearance	Shaft End-play	Thrust on No.	Connecting Rod Journal Diameter	Oil Clearance	Side Clearance
1988	Z	6-262 (4.3)	2.4484–2.4493 ①	0.0008–0.0020 ③	0.002–0.006	4	2.2487–2.2498	0.0013–0.0035	0.006–0.014
	H	8-305 (5.0)	2.4484–2.4493 ①	0.0008–0.0020 ③	0.002–0.006	5	2.0986–2.0998	0.0013–0.0035	0.006–0.014
	Y	8-307 (5.0)	2.4985–2.4995 ①	0.0005–0.0021 ③	0.003–0.013	3	2.1238–2.1248	0.0004–0.0033	0.006–0.020
1989	Z	6-262 (4.3)	2.4484–2.4493 ①	0.0008–0.0020 ③	0.002–0.006	4	2.2487–2.2498	0.0013–0.0035	0.006–0.014
	E	8-305 (5.0)	2.4481–2.4490 ④	0.0011–0.0020 ②	0.001–0.007	5	NA	0.0013–0.0035	0.006–0.014
	Y	8-307 (5.0)	2.4985–2.4995 ①	0.0005–0.0021 ③	0.003–0.013	3	2.1238–2.1248	0.0004–0.0033	0.006–0.020
	7	8-350 (5.7)	2.4481–2.4990 ④	0.0011–0.0020 ②	0.001–0.007	5	NA	0.0013–0.0035	0.006–0.014

CRANKSHAFT AND CONNECTING ROD SPECIFICATIONS

All measurements are given in inches.

| | | No. Cylinder Displacement cu. in. (liter) | Crankshaft | | | | Connecting Rod | | |
| | | | Main Brg. Journal Dia. | Main Brg. Oil Clearance | Shaft End-play | Thrust on No. | Journal Diameter | Oil Clearance | Side Clearance |
Year	VIN								
1990	Z	6-262 (4.3)	2.4484–2.4493 ①	0.0008–0.0020 ③	0.002–0.006	4	2.2487–2.2498	0.0013–0.0035	0.006–0.014
	E	8-305 (5.0)	2.4481–2.4490 ④	0.0011–0.0020 ②	0.001–0.007	5	NA	0.0013–0.0035	0.006–0.014
	Y	8-307 (5.0)	2.4985–2.4995 ①	0.0005–0.0021 ③	0.003–0.013	3	2.1238–2.1248	0.0004–0.0033	0.006–0.020
	7	8-350 (5.7)	2.4481–2.4990 ④	0.0011–0.0020 ②	0.001–0.007	5	NA	0.0013–0.0035	0.006–0.014
1991–92	E	8-305 (5.0)	2.4481–2.4490 ④	0.0011–0.0020 ②	0.001–0.007	5	NA	0.0013–0.0035	0.006–0.014
	7	8-350 (5.7)	2.4481–2.4490 ④	0.0011–0.0020 ②	0.001–0.007	5	NA	0.0013–0.0035	0.006–0.014

① Intermediate—2.4481–2.4490
   Rear—2.4479–2.4488
② Rear: 0.0020–0.0032
③ Intermediate—0.0011–0.0034
   Rear—0.0015–0.0031
④ Front: 2.4488–2.4493
   Rear: 2.4481–2.4488

VALVE SPECIFICATIONS

| | | No. Cylinder Displacement cu. in. (liter) | Seat Angle (deg.) | Face Angle (deg.) | Spring Test Pressure (lbs.) | Spring Installed Height (in.) | Stem-to-Guide Clearance (in.) | | Stem Diameter (in.) | |
| | | | | | | | Intake | Exhaust | Intake | Exhaust |
Year	VIN									
1988	Z	6-262 (4.3)	46	45	200	1.70	0.0010–0.0027	0.0010–0.0027	0.3414	0.3414
	H	8-305 (5.0)	46	45	200	1.70	0.0010–0.0027	0.0010–0.0027	0.3414	0.3414
	Y	8-307 (5.0)	①	②	194	1.70	0.0010–0.0027	0.0015–0.0032	0.3425–0.3432	0.3420–0.3427
1989	Z	6-262 (4.3)	46	45	200	1.70	0.0010–0.0027	0.0010–0.0027	0.3414	0.3414
	E	8-305 (5.0)	46	45	200	1.72	0.0011–0.0027	0.0011–0.0027	NA	NA
	Y	8-307 (5.0)	①	②	194	1.72	0.0010–0.0027	0.0015–0.0032	0.3425–0.3432	0.3420–0.3427
	7	8-350 (5.7)	46	45	200	1.72	0.0011–0.0027	0.0011–0.0027	NA	NA
1990	Z	6-262 (4.3)	46	45	200	1.70	0.0010–0.0027	0.0010–0.0027	0.3414	0.3414
	E	8-305 (5.0)	46	45	200	1.70	0.0011–0.0027	0.0011–0.0027	NA	NA
	Y	8-307 (5.0)	①	②	194	1.70	0.0010–0.0027	0.0015–0.0032	0.3425–0.3432	0.3420–0.3427
	7	8-350 (5.7)	46	45	200	1.70	0.0011–0.0027	0.0011–0.0027	NA	NA

VALVE SPECIFICATIONS

Year	VIN	No. Cylinder Displacement cu. in. (liter)	Seat Angle (deg.)	Face Angle (deg.)	Spring Test Pressure (lbs.)	Spring Installed Height (in.)	Stem-to-Guide Clearance (in.) Intake	Stem-to-Guide Clearance (in.) Exhaust	Stem Diameter (in.) Intake	Stem Diameter (in.) Exhaust
1991–92	E	8-305 (5.0)	46	45	200	1.70	0.0011–0.0027	0.0011–0.0027	NA	NA
	7	8-350 (5.7)	46	45	200	1.70	0.0011–0.0027	0.0011–0.0027	NA	NA

NA—Not available
① Intake—45°, Exhaust—31°
② Intake—44°, Exhaust—30°

PISTON AND RING SPECIFICATIONS

Year	VIN	No. Cylinder Displacement cu. in. (liter)	Piston Clearance	Ring Gap Top Compression	Ring Gap Bottom Compression	Ring Gap Oil Control	Ring Side Clearance Top Compression	Ring Side Clearance Bottom Compression	Ring Side Clearance Oil Control
1988	Z	6-262 (4.3)	0.0027	0.010–0.020	0.010–0.025	0.015–0.055	0.0012–0.0032	0.0012–0.0032	0.0020–0.0070
	H	8-305 (5.0)	0.0027	0.010–0.020	0.010–0.025	0.015–0.055	0.0012–0.0032	0.0012–0.0032	0.0020–0.0070
	Y	8-307 (5.0)	0.0008–0.0018	0.009–0.019	0.009–0.019	0.015–0.055	0.0018–0.0038	0.0018–0.0038	0.0010–0.0050
1989	Z	6-262 (4.3)	0.0027	0.010–0.020	0.010–0.025	0.015–0.055	0.0012–0.0032	0.0012–0.0032	0.0020–0.0070
	E	8-305 (5.0)	0.0070–0.0170	0.010–0.020	0.010–0.025	0.015–0.055	0.0012–0.0032	0.0012–0.0032	0.0020–0.0070
	Y	8-307 (5.0)	0.0008–0.0018	0.009–0.019	0.009–0.019	0.015–0.055	0.0018–0.0038	0.0018–0.0038	0.0010–0.0050
	7	8-350 (5.7)	0.0070–0.0170	0.010–0.020	0.010–0.025	0.015–0.055	0.0012–0.0032	0.0012–0.0032	0.0020–0.0070
1990	Z	6-262 (4.3)	0.0012 0.0021	0.010–0.020	0.010–0.020	0.015–0.055	0.0012–0.0032	0.0012–0.0032	0.0020–0.0070
	E	8-305 (5.0)	0.0070–0.0021	0.010–0.020	0.010–0.025	0.015–0.055	0.0012–0.0032	0.0012–0.0032	0.0020–0.0070
	Y	8-307 (5.0)	0.0008–0.0018	0.009–0.019	0.009–0.019	0.015–0.055	0.0018–0.0038	0.0018–0.0038	0.0010–0.0050
	7	8-350 (5.7)	0.0070–0.0021	0.010–0.020	0.010–0.025	0.015–0.055	0.0012–0.0032	0.0012–0.0032	0.0020–0.0070
1991–92	E	8-305 (5.0)	0.0070–0.0021	0.010–0.020	0.010–0.025	0.015–0.055	0.0012–0.0032	0.0012–0.0032	0.0020–0.0070
	7	8-350 (5.7)	0.0070–0.0021	0.010–0.020	0.010–0.025	0.015–0.055	0.0012–0.0032	0.0012–0.0032	0.0020–0.0070

TORQUE SPECIFICATIONS

All readings in ft. lbs.

Year	VIN	No. Cylinder Displacement cu. in. (liter)	Cylinder Head Bolts	Main Bearing Bolts	Rod Bearing Bolts	Crankshaft Pulley Bolts	Flywheel Bolts	Manifold Intake	Manifold Exhaust	Spark Plugs
1988	Z	6-262 (4.3)	60–75	70–85	42–47	—	50–70	25–45	20	22
	H	8-305 (5.0)	60–75	70–85	42–47	—	70	25–45	14–26	22
	Y	8-307 (5.0)	130③	①	18⑤	200–310	60	40③	25	25

TORQUE SPECIFICATIONS
All readings in ft. lbs.

Year	VIN	No. Cylinder Displacement cu. in. (liter)	Cylinder Head Bolts	Main Bearing Bolts	Rod Bearing Bolts	Crankshaft Pulley Bolts	Flywheel Bolts	Manifold Intake	Manifold Exhaust	Spark Plugs
1989	Z	6-262 (4.3)	60–75	70–85	42–47	—	50–70	25–45	20	22
	E	8-305 (5.0)	68	77	44	70	74	35	26	22
	Y	8-307 (5.0)	40④	①	18⑤	200–310	60	40③	25	25
	7	8-350 (5.7)	68	77	44	70	74	35	26	22
1990	Z	6-262 (4.3)	65	65	44	70②	70	35	⑥	22
	E	8-305 (5.0)	68	77	44	70②	74	35	⑥	22
	Y	8-307 (5.0)	40④	①	18⑤	200–310	60	40③	25	25
	7	8-350 (5.7)	68	77	44	70②	74	35	⑥	22
1991-92	E	8-305 (5.0)	68	77	44	70②	74	35	⑥	22
	7	8-350 (5.7)	68	77	44	70②	74	35	⑥	22

① 80 ft. lbs. on Nos. 1–4; 120 ft. lbs. on No. 5
② Torque listed is for torsioner damper, crankshaft pulley is 43 ft. lbs.
③ Dip in clean engine oil before tightening
④ Rotate position 1, 7 & 9—120°
Rotate position 8 & 10—95°
⑤ Torque in 2 steps:
1st step—18 ft.lbs.
2nd step—additional 70 degrees turn further
⑥ Bolts—26 ft. lbs.
Studs—20 ft. lbs.

BRAKE SPECIFICATIONS
All measurements in inches unless noted

Year	Model	Lug Nut Torque (ft. lbs.)	Master Cylinder Bore	Brake Disc Minimum Thickness	Brake Disc Maximum Runout	Standard Brake Drum Diameter	Minimum Lining Thickness Front	Minimum Lining Thickness Rear
1988	Estate Wagon	①	1.125	0.965	0.004	11.00	0.125	0.125
	Monte Carlo, Caprice	80②	1.125④	0.965	0.004	11.00	0.030	0.030
	Cutlass	100	0.931	0.980	0.004	9.50	0.030	0.030③
	Custom Cruiser	100	1.125	0.980	0.004	11.00	0.030	0.030③
	Safari	100	1.125	0.980	0.004	9.50	0.030	—
1989	Estate Wagon	①	1.125	0.980	0.004	11.00	0.030	0.030③
	Caprice	80	1.125	0.980	0.004	11.00⑤	0.030	0.030③
	Custom Cruiser	100	1.125	0.980	0.004	11.00	0.030	0.030③
	Safari	100	1.125	0.980	0.004	11.00	0.030	0.030③
1990	Estate Wagon	①	1.125	0.980	0.004	11.00	0.030	0.030③
	Caprice	⑥	1.125	0.980	0.004	11.00	0.030	0.030③
	Custom Cruiser	100	1.125	0.980	0.004	11.00	0.030	0.030③
1991-92	Roadmaster	①	1.125	0.980	0.004	11.00	0.030	0.030③
	Caprice	100	1.125	0.980	0.004	11.00	0.030	0.030③
	Custom Cruiser	100	1.125	0.980	0.004	11.00	0.030	0.030③

① Wheel lug type: ½ × 20—100 ft. lbs.
7/16 × 20 Steel—80 ft. lbs.
7/16 × 20 Aluminum—90 ft. lbs.
② 88 with 7/16 in. stud; 100 with 11 in. brake drums
③ If Bonded use .062
④ Hydroboost—1 3/16 in.
⑤ Sedan—9.50
⑥ Sedan—81 ft. lbs.
Wagon and Police Sedan—103 ft. lbs.

WHEEL ALIGNMENT

Year	Model	Caster Range (deg.)	Caster Preferred Setting (deg.)	Camber Range (deg.)	Camber Preferred Setting (deg.)	Toe-in (in.)	Steering Axis Inclination (deg.)
1988	Monte Carlo	1 13/16P–3 13/16P	2 13/16P	5/16N–1 5/16P	1/2P	3/64	7 7/8
	Caprice	1 13/16P–3 13/16P	2 13/16P	5/16N–1 5/16P	13/16P	3/32	7 7/8
	Estate Wagon	1 13/16P–3 13/16P	2 13/16P	0–1 5/8P	13/16P	3/64	NA
	Cutlass	1 13/16P–3 13/16P	2 13/16P	5/16N–1 5/16P	1/2P	3/64	—
	Custom Cruiser	1 13/16P–3 13/16P	2 13/16P	0–1 5/8P	13/16P	3/64	—
	Safari	1 13/16P–3 13/16P	2 13/16P	0–1 5/8P	13/16P	1/16	7 9/16
1989	Estate Wagon	2P–4P	3P	0–1 5/8P	13/16P	1/32	NA
	Caprice	2P–4P	3P	0–1 5/8P	13/16P	1/32	7 7/8
	Custom Cruiser	2P–4P	3P	0–1 5/8P	13/16P	1/32	—
	Safari	2P–4P	3P	0–1 5/8P	13/16P	1/16	7 9/16
1990	Estate Wagon	2P–4P	3P	0–1 5/8P	13/16P	1/32	NA
	Custom Cruiser	2P–4P	3P	0–1 5/8P	13/16P	1/32	—
	Caprice	2P–4P	3P	0–1 5/8P	13/16P	1/32	7 7/8
1991–92	Estate Wagon	2P–4P	3P	0–1 5/8P	13/16P	1/32	NA
	Custom Cruiser	2P–4P	3P	0–1 5/8P	13/16P	1/32	—
	Roadmaster	2.5P–4.5P	3.5P	0–1.6P	0.8P	0.16	—
	Caprice	1 25/32P–3 25/32P	2 25/32P	0–1 7/32P	25/32P	1/16–3/8	NA

NA—Not Available
N—Negative
P—Positive

ENGINE MECHANICAL

NOTE: Disconnecting the negative battery cable on some vehicles may interfere with the functions of the on board computer systems and may require the computer to undergo a relearning process, once the negative battery cable is reconnected.

Engine Assembly

REMOVAL & INSTALLATION

4.3L Engine

1. Remove the hood.
2. Relieve the fuel system pressure and disconnect the negative battery cable.
3. Drain coolant into a suitable container.
4. Remove the air cleaner.
5. If equipped with air conditioning, disconnect compressor ground wire from the mounting bracket. Remove the electrical connector from the compressor clutch, remove the compressor to mounting bracket attaching bolts and position the compressor aside.
6. Remove fan blade, pulleys and belts.
7. Disconnect radiator and heater hoses from engine.
8. Remove fan shroud assembly.
9. Remove power steering pump to mounting bracket bolts and position pump assembly aside.
10. Disconnect fuel pump hoses and plug.
11. Disconnect battery ground cable from the engine.
12. Disconnect the vacuum supply hoses that supply all non-engine mounted components with engine vacuum. If equipped, the vacuum modulator, load leveler and power brake vacuum hoses should all be disconnected at the engine.
13. Disconnect accelerator cable from carburetor and intake manifold bracket.
14. Disconnect alternator, oil and coolant sending unit switch connections at the engine. Remove the alternator.
15. Disconnect engine to body ground strap(s) at engine.
16. Raise and safely support the vehicle, disconnect the cable shield from the engine, if equipped.
17. Disconnect exhaust pipes from exhaust manifolds.
18. Remove the lower flywheel or converter cover.
19. Remove the flywheel to converter attaching bolts. Scribe chalk mark on the flywheel and converter for reassembly alignment.
20. Remove transmission to engine attaching bolts.
21. Remove motor mount through bolts and cruise control bracket if equipped.
22. Lower the vehicle and support the automatic transmission.
23. Attach a safe lifting device to the engine and raise the engine enough so mounting through bolts can be removed. Ensure the wiring harness, vacuum hoses and other parts are free and clear before lifting engine out of the vehicle.
24. Raise engine far enough to clear engine mounts, raise transmission support accordingly and alternately until engine can be disengaged from the transmission and removed.

To install:
25. With the engine and transmission safely supported, lower into position and align with the transmission.
26. Install the motor mount through bolts and cruise control bracket if equipped.
27. Install transmission to engine attaching bolts.
28. Install the flywheel to converter attaching bolts, matching the scribe

chalk marks on the flywheel and converter.

29. Install the lower flywheel or converter cover.

30. Connect the exhaust pipes to the exhaust manifolds.

31. Lower the vehicle, connect the cable shield to the engine, if equipped.

32. Connect the engine to body ground strap(s), at engine.

33. Install the alternator. Connect alternator, oil and coolant sending unit switch connections at the engine.

34. Connect the accelerator cable and secure the cable housing to the intake bracket.

35. Connect the vacuum supply hoses.

36. Connect the battery ground cable to the engine.

37. Connect the fuel pump hoses.

38. Install the power steering pump to mounting bracket bolts.

39. Install the fan shroud assembly.

40. Connect the radiator and heater hoses to the engine.

41. Install fan blade, pulleys and belts.

42. If equipped with air conditioning, install the air conditioning compressor.

43. Install the air cleaner and fill the cooling system to the proper level.

44. Connect the battery cable and install the hood.

5.0L and 5.7L Engines

1. Drain the cooling system.

2. Remove air cleaner and hot air pipe.

3. Remove hood from hinges, mark hood for reassembly.

4. Disconnect the negative battery cable. Relieve the fuel system pressure.

5. Disconnect radiator hoses, automatic transmission cooler lines, heater hoses, vacuum hoses, power steering hose bracket from engine, air conditioning compressor with brackets and hoses attached, fuel hose from fuel line, engine wiring harness connectors.

6. Remove the radiator assembly, fan and shroud.

7. Disconnect the engine ground straps.

8. Set the engine on TDC and remove the distributor.

9. Disconnect the transmission, cruise control and throttle cable.

10. Remove the wiper motor assembly, if necessary.

11. Raise and support the vehicle safely.

12. Disconnect exhaust pipes at manifold.

13. Remove torque converter cover and the bolts holding converter to flywheel.

14. Remove engine mount bolts and nuts.

15. Remove the transmission to engine retaining bolts and remove the starter.

16. Disconnect the torque converter clutch wiring at the transmission, if required.

17. Lower the vehicle. Secure a suitable lifting device to the engine.

18. Support the transmission and remove the engine.

To install:

19. With the engine safely supported, lower into position with the lifting device.

20. Install the transmission to engine retaining bolts and install the starter.

21. Install the engine mount bolts or nuts.

22. Install the bolts holding converter to flywheel and the torque converter cover.

23. Connect the exhaust system.

24. Connect the torque converter clutch wiring at the transmission.

25. Lower the vehicle.

26. Install the wiper motor assembly.

27. Connect the engine ground straps.

28. Set the engine on TDC and install the distributor.

29. Install the radiator, fan and shroud.

30. Connect radiator hoses, automatic transmission cooler lines, heater hoses, vacuum hoses, power steering hose bracket from engine, air conditioning compressor with brackets and hoses attached, fuel hose from fuel line, wiring cruise control, transmission and throttle cables.

31. Connect the negative battery cable and ground straps.

32. Install the hood, aligning the marks made during removal.

33. Fill the cooling system to the proper level.

34. Install the air cleaner assembly.

35. Inspect the vehicle fluid levels, specifications and verify there are no fluid leaks.

Engine Mounts

REMOVAL & INSTALLATION

1. Disconnect the negative battery cable.

2. Raise and support the vehicle safely.

3. Properly support the weight of the engine at the forward edge of the oil pan.

4. Remove the mount to engine block bolts.

NOTE: Verify the clearance between the rear of the engine and the firewall is sufficient enough to avoid possible damage to the distributor.

5. Raise the engine slightly and remove the mount to mount bracket bolt and nut. Remove the engine mount.

6. Installation is the reverse of the removal procedure.

Cylinder Head

REMOVAL & INSTALLATION

4.3L Engine

1. Relieve the fuel system pressure and disconnect negative battery cable.

2. Disconnect the fuel lines, vacuum hoses, wiring connectors and remove the intake manifold.

3. Drain cooling system into a suitable container, loosen and remove belt(s).

4. When removing left cylinder head:

   a. Remove oil dipstick.

   b. Remove air and vacuum pumps with mounting bracket if present and move aside with hoses attached.

5. When removing right cylinder head:

   a. Remove the alternator.

   b. Disconnect power steering gear pump and brackets attaching to cylinder head.

6. Disconnect wires from spark plugs, and remove the spark plug wire clips from the rocker arm cover studs.

7. Remove exhaust manifold bolts from the head being removed.

8. With air hose and cloths, clean dirt off cylinder head and adjacent area to avoid getting dirt into engine.

9. Remove rocker arm cover and rocker arm and shaft assembly from cylinder head. Lift out pushrods.

10. Loosen all cylinder head bolts, remove bolts and lift off the cylinder head.

11. Installation is the reverse of the removal procedure. Torque the cylinder head bolts, in sequence, to 65 ft. lbs.

5.0L and 5.7L Engines

1. Relieve the fuel system pressure and disconnect the negative battery cable.

2. Drain the cooling system into a suitable container. Remove the distributor, intake and exhaust manifolds.

3. Remove the valve cover. Remove the ground strap from the left cylinder head.

4. Remove the power steering pump and alternator bracket.

5. Remove the air conditioner mounting bracket.

6. Remove rocker arm bolts, pivots, rocker arms and pushrods. Scribe pivots and keep rocker arms separated so they can be installed in their original locations.

7. Remove cylinder head bolts and cylinder head.

To install:

8. Installation is the reverse of the removal procedure.

9. Cylinder heads using a steel gasket should have both sides of the new gasket coated with a good sealer. Composite type gaskets need no sealer.

10. Torque the cylinder head bolts, in sequence, and in 3 passes, to 68 ft. lbs.

11. Inspect the engine for fluid leakage.

Valve Lifters

REMOVAL & INSTALLATION

1. Disconnect the negative battery cable.

2. Drain the coolant.

3. Remove rocker arm covers.

4. Remove the intake manifold assembly.

5. Remove the rocker arm assembly or the rocker and pivot. Remove the pushrods. Be sure to keep them in order as they must be installed in the same bores as they were removed.

6. Remove the valve lifter retainer and the restrictor, if equipped.

7. Remove the valve lifters, using the proper valve lifter removal tool.

8. Installation is the reverse of the removal procedure.

9. Soak the lifter assemblies with clean engine oil prior to installation. Coat the valve lifter rollers with Molykote® or equivalent. Use new gaskets as required.

Valve Lash

ADJUSTMENT

The Buick and Oldsmobile engines use hydraulic valve lifters, which are not adjustable. The rocker arm shaft assembly or the rockers, with pivot, are bolted to the cylinder head with a specific torque pressure, automatically positioning the lifter internal components for correct hydraulic operation. If there is excess play in the valve train, check for worn pushrods, rocker arms, valve springs and/or collapsed lifters.

Chevrolet engines do not require any routine valve lash adjustments. However, if the rocker arms are removed, the initial valve lash must be adjusted before the engine is started.

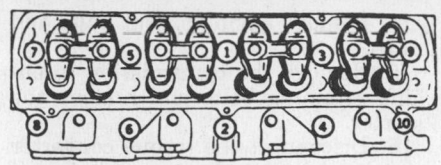

Cylinder head bolt torque sequence— 5.0L engine (VIN Y)

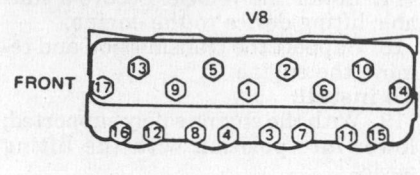

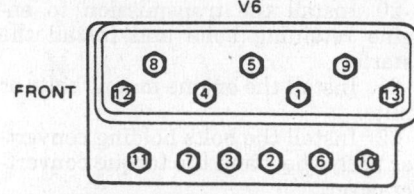

Cylinder head bolt torque sequence— 5.0L and 5.7L engines (VIN E, H)

Use the following procedure for Chevrolet engines.

1. Remove the valve covers.

2. Rotate the crankshaft, positioning each valve lifter on its base circle of the camshaft, remove the lash from each rocker arm and pushrod.

3. To adjust the valves, crank the engine until the mark on the vibration damper aligns with the top dead center or **0** mark on the timing tab of the front cover and the engine is on the No. 1 firing position or its opposite cylinder No. 6 on V8 engine and No. 4 on V6 engine, firing position.

NOTE: The firing cylinder may be determined by placing a finger on the No. 1 cylinder valve rocker arms as the mark on the damper comes near the 0 mark on the crankcase front cover. If the valve rocker arms moves as the mark comes up to the timing tab, the engine is on the opposite cylinder, No. 6 on V8 engine and No. 4 on V6 engine firing position and should be turned over a complete revolution to reach the No. 1 cylinder firing position.

4. With the engine in the No. 1 firing position, adjust the following valves:

   a. V8 engine—Exhaust—1, 3, 4, 8
   b. V8 engine—Intake—1, 2, 5, 7
   c. V6 engine—Exhaust—1, 5, 6
   d. V6 engine—Intake—1, 2, 3

5. Back out adjusting nut until lash is felt at the pushrod, then turn in adjusting nut until all lash is removed. This can be determined by rotating pushrod while turning adjusting nut. When play has been removed, turn adjusting nut a full additional turn clock-

wise, the lifter plunger will now be centered.

6. Crank the engine 1 revolution until the pointer **0** mark and the vibration damper mark are again in alignment. This is the No. 6 (V8 engine) or No. 4 (V6 engine) firing position.

7. With the engine in this position, adjust the following valves:

   a. V8 engine—Exhaust—2, 5, 6, 7
   b. V8 engine—Intake—3, 4, 6, 8
   c. V6 engine—Exhaust—2, 3, 4
   d. V6 engine—Intake—4, 5, 6

8. Install the rocker arm covers.

9. Start the engine and adjust the idle speed, as required.

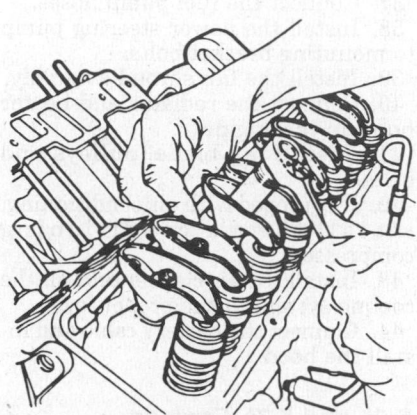

Typical valve adjustment—4.3L, 5.0L (VIN E, H) and 5.7L engines

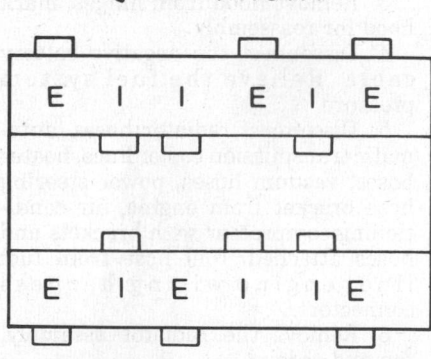

Intake and exhaust valve arrangement— 4.3L engine

Intake and exhaust valve arrangement— 5.0L (VIN E, H) and 5.7L engines

Rocker Arms/Shafts

REMOVAL & INSTALLATION

4.3L, 5.0L (VIN E, H) and 5.7L Engines

1. Disconnect the negative battery cable.
2. Remove the air cleaner.
3. Disconnect the Computer Command Control (CCC) harness from the intake manifold and the oxygen sensor.
4. Disconnect the power brake vacuum line at the carburetor and booster.
5. Disconnect the Air Injection Reaction (AIR) hose from the exhaust manifold.
6. Disconnect the alternator and choke wires.
7. Disconnect the wiring harness attached to the rocker arm cover and route away from the cover.
8. Disconnect the PCV valve and remove the EGR solenoid bracket.
9. Disconnect the oxygen sensor electrical lead.
10. Loosen the air conditioning compressor upper rear bracket mounting bolts with drive belt and position the unit to the side so the right side rocker arm cover retaining bolts are accessible and there is sufficient clearance to remove the right side rocker arm cover, only if required.
11. Remove the spark plug wires. Loosen the retaining bolts and remove the left and right rocker arm covers.
12. Remove the rocker arm nuts, rocker arm balls and rocker arm pushrods. Mark each rocker arm assembly to ensure installation in the original positions.

To install:

NOTE: If new rocker arms or rocker arm balls are being installed, coat the bearing surfaces with Molycoat® or equivalent.

13. Install the pushrods. Ensure the pushrods seat properly in the lifter socket.
14. Install the rocker arms, rocker arm balls and nuts. Tighten the rocker arm nuts until all the valve lash is eliminated.
15. Adjust the valves to there proper specification.
16. Install the rocker arm covers and all associated components in the reverse of the removal procedure.
17. Install the air conditioning bracket, compressor and adjust the belt to the proper tension.
18. Start the engine. Check the curb idle speed and adjust, if necessary.

5.0L (VIN Y) Engine

1. Disconnect the negative battery cable.
2. Remove the air cleaner from the crankcase inlet pipe and filter.
3. Disconnect the heater hoses and position them to the side.
4. Disconnect the PCV valve and hose.
5. Disconnect the Idle Load Compensator (ILC) anti-dieseling solenoid vacuum hoses.
6. Disconnect the following wires from their respective connectors:
   a. Alternator
   b. ILC anti-dieseling solenoid
   c. Air Injection Reaction (AIR) and oxygen sensor
   d. Rear Vacuum Break (RVB), Idle Load Compensator (ILC), Exhaust Gas Recirculation (EGR) solenoid assembly
   e. Oil pressure, temperature and coolant sensor
7. Remove the ILC anti-dieseling solenoid.
8. Remove the alternator drive belt and rear bracket.
9. Remove the canister purge hose.
10. Loosen the right side exhaust manifold upper shroud. Remove the EGR valve and the oil level indicator.
11. Disconnect and remove the following AIR system components:
    a. AIR hoses to the AIR switching valve and catalytic converter pipe
    b. AIR switching valve
    c. AIR/AC drive belt
    d. AIR pump pulley

12. Remove the air conditioning compressor rear bracket.
13. Loosen the retaining screws/nuts and remove the left and right valve covers.
14. Remove the rocker arm bolts, pivots and rocker arms. Indentify each rocker arm assembly to ensure installation in the original position.

To install:

15. Lubricate all wear points with 1050169 lubricant or equivalent.
16. Install the pivots, rocker arms and bolts.
17. Tighten the bolts evenly to 22 ft. lbs.
18. Install the valve covers and associated components in reverse of the removal procedure.

Intake Manifold

REMOVAL & INSTALLATION

1. Relieve the fuel system pressure and disconnect the negative battery cable.
2. Drain the engine coolant into a suitable container and remove the air cleaner.
3. Disconnect the upper radiator hose, bypass hose and heater hose from the manifold.
4. Disconnect all necessary electrical connections.
5. If equipped, disconnect the Computer Command Control (CCC) harness and lay it to the side aside.

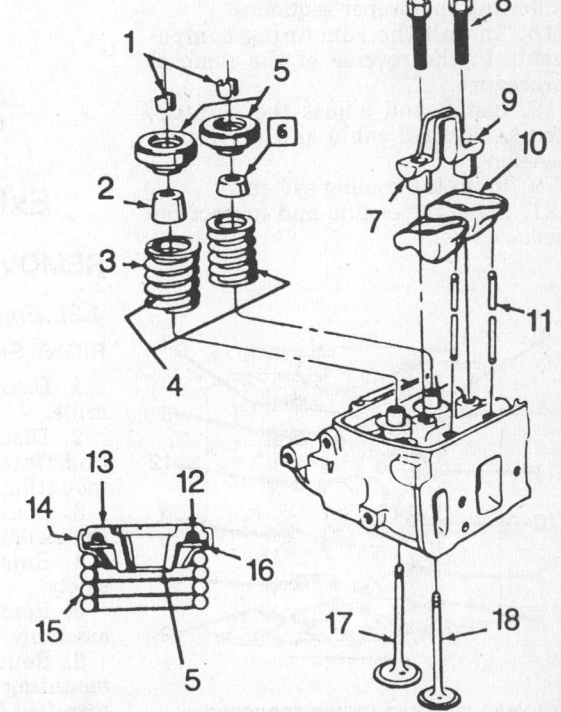

1. Valve keys
2. Intake valve seal
3. Spring
4. Dampener rotator
5. Valve rotator
6. Exhaust valve seal
7. Identification pad
8. 22 ft. lbs.
9. Rocker arm pivot
10. Rocker arms
11. Pushrods
12. Coil spring
13. Body
14. Collar
15. Valve spring
16. Flat washer
17. Intake valve
18. Exhaust valve

Disassembled view of the cylinder head—5.0L (VIN Y) engine

6. Remove the distributor cap and mark the position of the rotor, then remove the distributor. On some V6 engines, remove the coil.

7. Remove the accessory mounting brackets, as required.

8. Disconnect the throttle linkage and cruise control cable.

9. Disconnect the vacuum, fuel and the brake booster line from the carburetor.

10. Remove the mass air flow sensor on fuel injected engines.

NOTE: On fuel injected engines, the fuel system must be depressurized before disconnecting any fuel lines.

11. Remove the manifold bolts and remove the intake manifold. Remove and discard the intake manifold gaskets.

12. Thoroughly clean the intake manifold and cylinder block surfaces to remove any trace of gasket material or sealant.

To install:

13. Apply an even coat of sealer to both sides of the new intake manifold gasket. Coat the new front and rear seals with RTV sealant.

14. Position and install the intake manifold gasket onto the cylinder head surface. Install the front and rear seals.

15. Support the intake manifold and lower into position.

16. Coat the intake manifold retaining bolts with clean engine oil and tighten by hand.

17. Torque the bolts to specification, following the proper sequence.

18. Install the remaining components in the reverse of the removal procedure.

19. Install and adjust the throttle, cruise control cable and drive belt tension.

20. Refill the cooling system.

21. Start the engine and inspect for leaks.

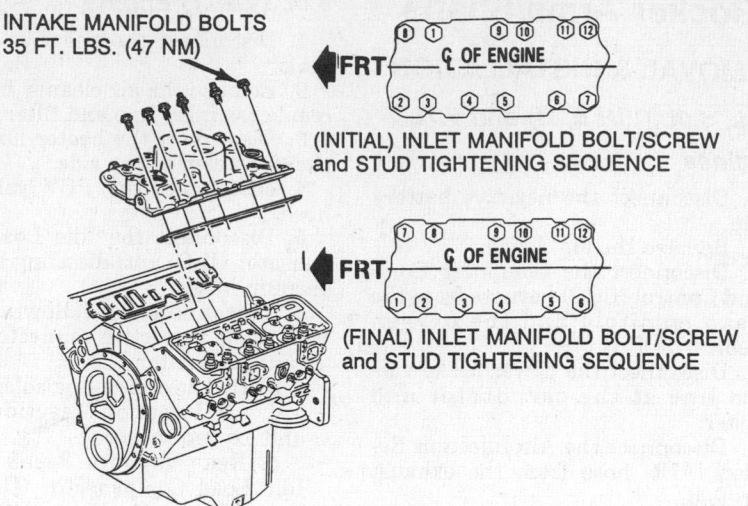

Intake manifold torque sequence—4.3L engine

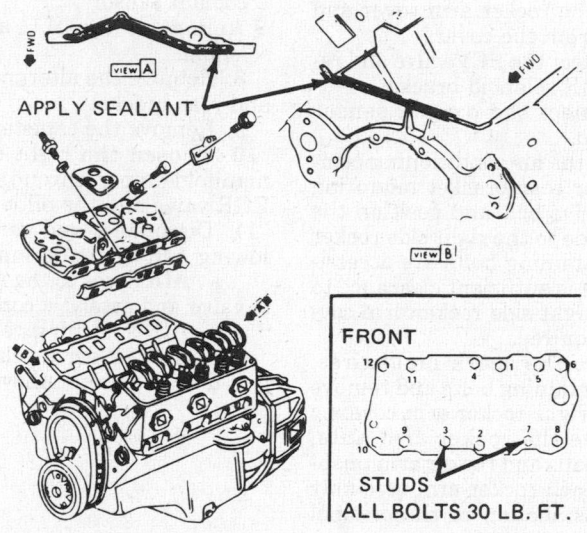

Intake manifold torque sequence—5.0L (VIN E, H) and 5.7L engines

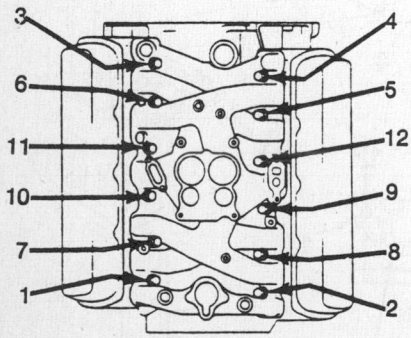

Intake manifold torque sequence—5.0L engine (VIN Y)

Exhaust Manifold

REMOVAL & INSTALLATION

4.3L Engine

RIGHT SIDE

1. Disconnect the negative battery cable.

2. Disconnect the oxygen sensor and flatten the exhaust manifold mounting bolt lock tabs.

3. Disconnect the exhaust pipe at the exhaust manifold flange.

4. Raise and support the vehicle safely.

5. Remove the right front wheel assembly.

6. Remove the exhaust manifold mounting bolts. Remove the exhaust manifold from the engine.

7. Installation is the reverse of the removal procedure. Bend the exhaust manifold bolt's lock tabs after installing the mounting bolts.

LEFT SIDE

1. Disconnect the negative battery cable.

2. On carbureted engines, remove the air cleaner.

3. Disconnect the EFE pipe and flatten the exhaust manifold lock tabs.

4. Disconnect the exhaust pipe from the exhaust manifold. On turbocharged engines, disconnect the exhaust manifold from the crossover pipe.

5. Remove the air conditioning compressor and rear adjusting bracket.

6. Remove the power steering pump and the lower rear adjusting bracket.

7. Disconnect the spark plug wires at the plugs.

8. Remove the exhaust manifold mounting bolts.

9. Remove the exhaust manifold from the engine.

10. Installation is the reverse of the removal procedure.

11. Start the engine and check for leaks.

5.0L and 5.7L Engines

RIGHT SIDE

1. Disconnect the negative battery cable.

2. Raise and support the vehicle safely. Remove the right front wheel, if necessary, and remove the exhaust crossover pipe(s) at the manifold.

3. Flatten the lock tabs on the manifold bolts and remove the bolts. Disconnect the oxygen sensor lead.

4. Remove the lower engine mounting bolt and raise the engine slightly, if necessary, for clearance.

5. Remove the spark plug wires and hoses as required.

6. Remove the AIR hoses.

7. Remove the exhaust manifold from either the above or below the engine.

8. To install, use new gaskets and reverse the removal procedure. Torque the manifold bolts to specification, bend bolt lock tabs back into position.

LEFT SIDE

1. Disconnect the negative battery cable.

2. Remove the air cleaner assembly.

3. Remove the hot air shroud and the hot air tube, if equipped.

4. Remove the lower alternator bracket, air conditioning drive belt and the air pump pulley and brackets, if required.

5. Remove all air pump hoses and the AIR switching valve.

6. Remove the spark plug wires.

7. Raise the vehicle and remove the crossover pipe.

8. Lower the vehicle and, flatten the lock tabs on the manifold bolts and remove.

9. To install, use a new gasket and reverse the removal procedures. Torque the exhaust manifold mounting bolts to specifications, bend bolt lock tabs back into position.

Timing Chain Front Cover

REMOVAL & INSTALLATION

4.3L, 5.0L (VIN E, H) and 5.7L Engines

1. Disconnect the negative battery cable.

2. Remove the vibration damper assembly.

3. Drain the cooling system into a suitable container and remove the water pump assembly.

4. Raise and safely support the vehicle. Remove the oil pan assembly.

5. Remove the crankcase front cover retaining bolts. Remove the front cover and discard the gasket.

6. Installation is the reverse of the removal procedure. Replace the timing cover gasket and oil seal.

5.0L (VIN Y) Engine

1. Disconnect the negative battery cable.

2. Drain the coolant. Disconnect the radiator hose and the bypass hose. Remove the fan, belts and pulley. Remove the air conditioner compressor, if equipped, and bracket.

3. Remove the vibration damper and crankshaft pulley.

4. Remove the front cover attaching bolts and remove the cover, timing indicator and water pump from the front of the engine.

5. Remove the dowel pins. If necessary grind a flat surface on the dowel pins to aid in removal. When installing the dowel pins, they must be inserted chamfered end first.

6. Install in the reverse order of removal using new gaskets. Apply RTV sealer around the coolant holes of the new cover gasket. Trim about 1/8 in. from each end of the new front pan seal and trim any excess material from the front edge of the oil pan gasket. Be sure all mating surfaces are clean.

Front Cover Oil Seal

REPLACEMENT

Except 5.0L (VIN Y) Engine

1. Disconnect the negative battery cable.

2. With the torsional damper removed, remove the old seal using a suitable prying tool.

To install:

3. Support the rear of the cover at the seal area. Drive in the new seal with the open end toward the inside of the cover using tool J–35468.

4. Reinstall torsion damper, check cover for oil leaks.

5.0L (VIN Y) Engine

1. Disconnect the negative battery cable.

2. Remove the crankshaft pulley and balancer.

3. Remove the oil seal using tool BT–6406 or J–23129 and J–1859–03 or their equivalents.

To install:

4. Coat the outside diameter of the new seal with sealer.

5. Install seal with lip facing the engine, using tool BT–6405 or J–25264–A.

6. Install crankshaft pulley and balancer.

7. Install the belts and adjust tension.

8. Reconnect the negative battery cable. Inspect cover for oil leaks.

Timing Chain and Sprockets

REMOVAL & INSTALLATION

5.0L (VIN Y) Engine

1. Disconnect the negative battery cable.

2. Remove the front cover and gasket.

3. Remove the crankshaft oil slinger.

4. Remove the camshaft thrust button and spring.

5. Remove the fuel pump, fuel pump gasket and fuel pump eccentric.

6. Remove the camshaft sprocket and timing chain.

7. Remove the spark plugs.

NOTE: The crankshaft key has a blind keyway. The key must be removed before removing the crankshaft sprocket.

8. Remove the crankshaft key.

9. Remove the crankshaft sprocket, using removal tool BT–6812, J–25287, J–21052 or equivalent.

To install:

10. Place the timing chain on a flat surface.

11. Insert the camshaft and crankshaft timing sprockets into the timing chain with the timing marks aligned. Ensure this alignment is maintained through out the remaining procedure.

12. Place the sprockets with the timing chain into position.

13. Rotate the camshaft sprocket as required until it engages with the camshaft. With the camshaft sprocket engaged, install the fuel pump eccentric with the flat side toward the engine.

14. Install the camshaft sprocket bolts until finger-tight.

15. Rotate the crankshaft until the crankshaft sprocket and keyway are in alignment. When the keyway is aligned, tap the crankshaft key into place with a brass hammer until the key bottoms.

16. Check the timing marks are still in alignment.

NOTE: When the timing marks are aligned, the No. 6 piston is at TDC. When the timing marks are on top, the No. 1 piston is in the firing position.

17. After the timing gear alignment is verified, torque the camshaft sprocket (fuel pump eccentric) bolt to 65 ft. lbs.

18. Install the remaining components in the reverse of the removal procedure using new gaskets where required.

4.3L, 5.0L (VIN E, H) and 5.7L Engines

1. Disconnect the negative battery cable.

2. Drain the coolant into a suitable container and remove the water pump.

3. Remove the torsion damper and remove the engine front cover.

4. Rotate the engine until the marks on the camshaft sprocket and crankshaft sprocket are aligned with the shaft centers.

5. Remove the camshaft sprocket retaining bolts. Remove the camshaft sprocket along with the timing chain.

6. If necessary for replacement, use a crankshaft sprocket removal tool and remove the crankshaft sprocket. Remove the crankshaft sprocket key, if required.

To install:

7. Installation is the reverse of the removal procedure.

8. Ensure that the timing marks on the crankshaft sprocket and the camshaft sprocket are aligned with the shaft centers.

9. Torque the camshaft sprocket bolts to 21 ft. lbs.

Camshaft

REMOVAL & INSTALLATION

4.3L Engine

NOTE: This procedure requires that the air conditioner system be discharged and recharged.

1. Disconnect the negative battery cable.

2. Drain the radiator.

3. Remove the intake manifold.

4. Remove the valve covers.

5. Remove the rocker arm assemblies, pushrods and valve lifters, noting location.

6. Remove the radiator and the air condition condenser, as required.

7. Remove timing chain cover, timing chain and sprocket.

8. Align timing marks of camshaft and crankshaft sprocket.

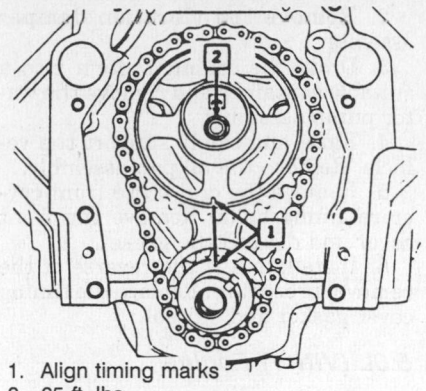

1. Align timing marks
2. 65 ft. lbs.

Timing gear alignment—5.0L (VIN Y)

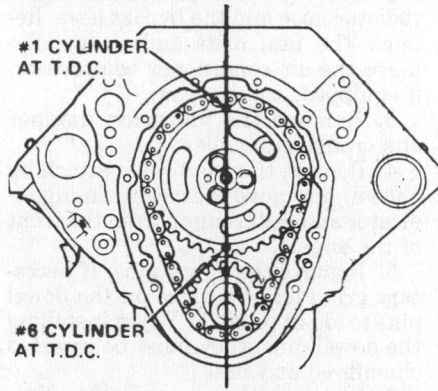

#1 CYLINDER AT T.D.C.

#6 CYLINDER AT T.D.C.

Timing gear alignment—4.3L, 5.0L (VIN E, H) and 5.7L engines

9. Slide camshaft forward out of bearing bores carefully to avoid marring the bearing surfaces.

To install:

10. Installation is the reverse of the removal procedure.

11. Before installing the camshaft and the lifters, be sure to coat them with clean engine oil.

12. Be sure to use new gaskets and seals as required.

5.0L and 5.7L Engines

1. Properly discharge the air conditioning system. Disconnect the negative battery cable.

2. Drain the cooling system into a suitable container.

3. Remove the upper radiator baffle.

4. Disconnect the upper radiator hose.

5. Remove the radiator assembly.

6. Properly disconnect the fuel line.

7. Remove the air cleaner assembly.

8. Disconnect the throttle and cruise control cables.

9. Remove the alternator belt. Remove the alternator bracket attaching bolts.

10. Remove the power steering pump bracket mounting bolts and remove the pump.

11. Remove the air conditioning compressor mounting bracket attaching bolts and lay compressor to side. The air conditioning lines at the compressor are flexible and should be left attached to the compressor.

12. Disconnect and plug the air conditioner condenser lines.

13. Remove the condenser assembly.

14. Disconnect the thermostat bypass hose at the water pump.

15. Disconnect the electrical and vacuum connections.

16. Remove distributor with cap and wiring intact.

17. Remove balancer pulley. Remove balancer.

18. Remove engine front cover.

19. Remove both valve covers.

20. Remove intake manifold and gaskets.

21. Remove rocker arms, pushrods and valve lifters.

NOTE: All parts for each assembly must be kept together and re-installed in the same location.

22. Remove bolt securing fuel pump eccentric, remove eccentric, camshaft gear, oil slinger and timing chain.

23. Remove camshaft by carefully sliding it out the front of the engine.

To install:

24. Install camshaft into journals, applying a light coat of oil to all lobes.

25. Install lifters, pushrods and rockers.

26. Install timing chain and gears.

27. Install timing cover and water pump.

28. Install the torsion damper and pulleys.

29. Adjust all valves to specification and install valve covers.

30. Install intake manifold, fuel lines and distributor.

31. Connect all vacuum hoses and wire connectors.

32. Install the power steering pump, bracket and belt.

33. Install the altenator, bracket and belt.

34. Install the air conditioner compressor, bracket and belt.

35. Install the condensor and radiator assembly.

37. Connect air conditioner lines, water hoses and transmission oil cooler lines.

38. Fill cooling system with antifreeze and connect the negative battery cable.

39. Recharge air conditioning system.

40. Set engine to specifications and verify there are no leaks.

Piston and Connecting Rod

POSITIONING

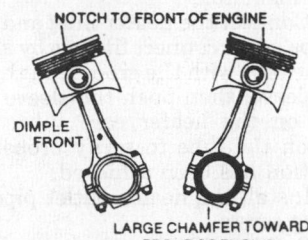

Piston assembly—5.0L (VIN E, H) and 5.7L engines

ENGINE LUBRICATION

Oil Pan

REMOVAL & INSTALLATION

1. Disconnect the negative battery cable and remove the air cleaner assembly.
2. Remove the distributor cap.
3. Remove the transmission and oil dipstick.
4. Remove the upper radiator support and the fan shroud attaching screws.
5. Raise and safely support the vehicle.
6. Remove the flywheel cover.
7. Remove the starter motor assembly.
8. Disconnect the exhaust pipes and the crossover pipe.
9. Disconnect the engine mounts and raise the front of the engine as far as possible.
10. Drain the engine oil.
11. Disconnect the oil level sensor wire connector and remove the sensor, if equipped.
12. Remove the oil pan attaching bolts.
13. Position the crankshaft timing mark to the 6 o'clock position and remove the pan.

To install:

14. Install the oil pan, using new gaskets and seals.
15. Lower the engine and connect the engine mounts.
16. Install the starter assembly.
17. Install the flywheel cover.
18. Install the exhaust pipes.
19. Install the oil sensor and connect the wire harness.
20. Lower the vehicle.
21. Install the fan shroud and radiator support.

1. Rear oil pan seal
2. Side gaskets
3. Fully seat bolt
4. Front oil pan seal
5. 17 ft. lbs.
6. 10 ft. lbs.
7. Apply sealer

Oil pan and gasket—5.0L (VIN Y) engine

22. Install the transmission and oil dipsticks.
23. Install the distributor cap.
24. Reconnect the battery cable and check engine for leaks.

Oil Pump

REMOVAL & INSTALLATION

1. Disconnect the negative battery cable.
2. Drain the engine oil, remove the oil pan and baffle, if equipped.
3. Remove the pump attaching bolts and remove the pump.
4. Reinstall in reverse order. Tighten the pump retaining bolts to 35 ft. lbs. for 5.0L VIN Y engine, 77 ft. lbs. for 5.7L and 5.0L VIN E, H engines or 65 ft. lbs. for the 4.3L engine. To insure immediate oil pressure on start up, the oil pump gear cavity should be packed with petroleum jelly.

Rear Main Bearing Oil Seal

REMOVAL & INSTALLATION

5.0L (VIN Y) Engines

1. Remove the oil pan. Remove the oil pump, if required. Remove the rear main bearing cap.
2. Pry the lower seal out of the bearing cap with a suitable tool, being careful not to gouge the cap surface.
3. Remove the upper seal by lightly tapping on 1 end with a brass pin punch until the other end can be grasped and pulled out.
4. Clean the bearing cap, cylinder block and crankshaft mating surfaces with solvent. Inspect all these surfaces for gouges, nicks and burrs.
5. Apply a sealer comparable to Loctite® 414 or Fel-Pro Mighty Grip® to the seal groove.
6. Insert the seal with tool J25285, BT-7923 or equivalent seal installer into the grooves.

7. Cut the excess seal material flush with the surface.
8. Apply sealer to the cylinder block only where the cap mates to the surface.
9. Install the rear cap and torque the bolts to specifications.

4.3L, 5.0L (VIN E, H) and 5.7L Engines

ONE PIECE SEAL

1. Remove the transmission from the vehicle.
2. Using the notches provided in the rear seal retainer, pry out the seal using a suitable tool.

NOTE: Care should be taken when removing the seal so as not to nick the crankshaft sealing surface.

3. Before installation, lubricate the inside and outside diameter of the new seal with clean engine oil.
4. Install the seal on tool J–3561 or equivalent. Thread the tool into the rear of the crankshaft. Tighten the screws snugly, this is to insure the seal will be installed squarely over the crankshaft. Tighten the tool wing nut until it bottoms.
5. Remove the tool from the crankshaft.
6. Install the transmission.

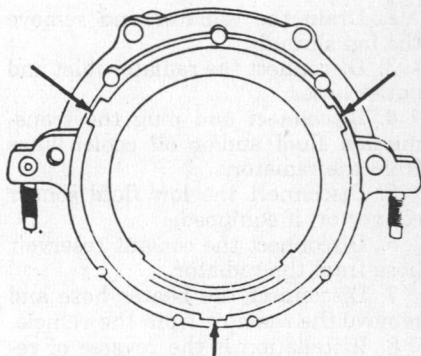

Rear main seal removal—4.3L, 5.0L (VIN E, H) and 5.7L engines

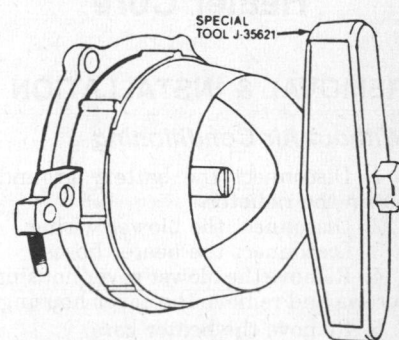

Installing rear main seal—4.3L, 5.0L (VIN E, H) and 5.7L engines

ONE PIECE SEAL RETAINER AND GASKET

1. Remove the transmission from the vehicle.
2. Remove the oil pan bolts. Lower the oil pan.
3. Remove the retainer and seal assembly.
4. Remove the gasket.

NOTE: Whenever the retainer is removed a new retainer gasket and rear main seal must be installed.

5. Installation is the reverse of the removal procedure. Once the oil pan has been installed the new rear main oil seal can be installed.

ENGINE COOLING

Radiator

REMOVAL & INSTALLATION

1. Disconnect the negative battery cable.
2. Drain the radiator and remove the fan shrouds.
3. Disconnect the radiator inlet and outlet hoses.
4. Disconnect and plug the transmission fluid and/or oil cooler lines from the radiator.
5. Disconnect the low fluid sensor connector, if equipped.
6. Disconnect the coolant reservoir hose from the radiator.
7. Disconnect the heater hose and remove the radiator from the vehicle.
8. Installation is the reverse of removal. Add coolant and check system for leaks.

Heater Core

REMOVAL & INSTALLATION

Without Air Conditioning

1. Disconnect the battery ground. Drain the radiator.
2. Disconnect the blower wiring.
3. Disconnect the heater hoses.
4. Remove the blower cover housing screws and remove the cover housing.
5. Remove the heater core.
6. Installation is the reverse of removal. Remove and replace sealer, as necessary.

With Air Conditioning

EXCEPT 1991–92 ROADMASTER, CAPRICE AND CUSTOM CRUISER

1. Disconnect the negative battery cable and drain the engine coolant into a suitable container.
2. Disconnect the blower motor wire connector.
3. Remove the thermostatic switch and diagnostic connector, if required.
4. Remove the hood cowl seal and the air inlet screen attaching screws.
5. Remove the right side windshield wiper arm.
6. Remove the air conditioning module ground strap from the dash panel.
7. Remove the flange mounting screws and lift the upper case off.
8. Remove the heater core hoses and remove the core.
9. Installation is the reverse of the removal procedure. Inspect system for proper operation and/or leaks.

1991–92 ROADMASTER, CAPRICE AND CUSTOM CRUISER

1. Disconnect the negative battery cable.
2. Drain the radiator coolant into a suitable container for use when done.
3. Remove the screw attaching the heater outlet pipe to the cowl panel.
4. Remove the heater inlet and outlet pipe quick connect fittings by squeezing both release tabs at the base of the heater core tube and pulling on the pipe to disengage the fitting.
5. Remove the right side panel insulator.
6. Remove the instrument panel lower reinforcement.
7. Remove the 2 vacuum harness connectors at the lower evaporator case.
8. Remove the right side trim panel.
9. Remove the 7 lower evaporator case attaching screws.
10. Remove the lower evaporator housing.
11. Remove the heater core mounting straps and screws.
12. Remove the heater core by pulling it rearward and working the heater core tubes out of the seal.

To install:
13. Install heater core into position, carefully guiding the heater core tubes through the seals at the cowl panel.

NOTE: If installing new heater core, transfer the quick connect tabs to the tubes of the replacement core.

14. Install the heater core mounting straps and screws.
15. Install the lower evaporator housing.
16. Install the right side trim panel.

17. Reconnect the 2 vacuum harness connectors at the lower evaporator case.
18. Install the instrument panel lower reinforcement and the right side panel insulator.
19. Connect the heater inlet and outlet pipe quick connect fittings by aligning the tabs with the grooves in the fitting sleeve, then push the sleeve into place on the heater core tube. Pull back on the tube to verify proper installation has been achieved.
20. Install the heater outlet pipe retaining screw.
21. Refill coolant to proper level.
22. Reconnect negative battery terminal.
23. Start the vehicle and inspect system for proper operation and/or leakage.

Water Pump

REMOVAL & INSTALLATION

1. Disconnect the negative battery cable.
2. Drain the cooling system.
3. Unfasten the heater bypass and lower radiator hoses from the pump.
4. Loosen and remove the drive belt or belts.
5. Remove the fan shroud, unbolt and remove the fan assembly and pulley.

NOTE: If equipped with a clutch fan, keep the fan in an upright position during repairs to prevent the silicone fluid from leaking out.

6. Remove the alternator, air conditioning compressor and power steering brackets, if required.
7. Remove the bolts securing the water pump and remove the pump.
8. Installation is the reverse of the removal procedure.

Thermostat

REMOVAL & INSTALLATION

1. Disconnect the negative battery cable.
2. Drain the coolant from the radiator into a suitable container until the level is below the thermostat housing.
3. Remove the air cleaner assembly and remove the radiator inlet hose from the thermostat housing.
4. Remove the thermostat housing bolts and remove the thermostat.
5. Installation is the reverse of removal. Clean the sealing surfaces, use a new gasket and tighten the housing retaining bolts to 13 ft. lbs. on the V6 engine or 19–21 ft. lbs. on the V8 engine.

Cooling System Bleeding

1. With the cooling system completely drained, begin adding a combination of ethylene glycol antifreeze and water until achieving a mixture of at least 50 percent and not exceeding 70 percent antifreeze.
2. Fill the radiator up to the lower portion of the filler neck.
3. Fill the coolant recovery reservoir to the **COLD FILL** mark and install the coolant recovery cap.
4. Start the vehicle and run the engine with the radiator cap **OFF** and the heating system temperature controls set to **HOT** until normal engine operating temperature is reached.
5. With the engine idling, add coolant to the radiator until the level reaches the bottom of the filler neck.
6. Reinstall the radiator cap, ensuring the arrow on the cap is aligned with coolant recovery hose.
7. Inspect the system for leaks.

ENGINE ELECTRICAL

NOTE: Disconnecting the negative battery cable on some vehicles may interfere with the functions of the on board computer systems, which may require the computer to undergo a relearning process when the negative battery cable is reconnected.

Distributor

All vehicle use a High Energy Ignition (HEI) distributor with Electronic Spark Timing (EST). The HEI system incorporates a distributor cap, rotor, ignition module, pole piece with internal teeth and pickup coil. All vehicles use an internally distributor mounted coil or externally mounted ignition coil. Spark timing changes are done electronically by the Engine Control Module (ECM), which monitors various engine sensors, computes the desired spark timing and signals the distributor to change the timing accordingly.

REMOVAL

1. Disconnect the negative battery cable.
2. Disconnect and tag the ignition wire (pink), tachometer wire, if equipped, and 3 terminal connector from distributor cap or externally mounted coil.

NOTE: Use care when releasing the connector locking tabs on the distributor cap.

3. Remove distributor cap with the spark plug wires attached and position it aside.
4. Disconnect the 4 terminal connector from the distributor.
5. Remove the distributor holddown bolt and clamp. Mark the position of the rotor in relation to the engine. Pull the distributor from the engine until the rotor just stops turning counterclockwise. Again mark the position of rotor.

INSTALLATION
Timing Not Disturbed

NOTE: To insure correct ignition timing if the engine has not been disturbed, the distributor must be installed with the rotor in the same position as when removed.

1. Align the rotor to the last mark made and install the distributor in the engine.
2. The rotor should turn and end up at the first mark made.
3. Check the timing when finished.

Timing Disturbed

1. Remove the No. 1 spark plug. Place a finger over the spark plug hole and rotate the engine in the normal direction of rotation slowly, until compression is felt.
2. Align the timing mark on the crankshaft pulley to the **0** on the engine timing indicator by rotating the engine in the same direction slowly.
3. Position the rotor between No. 1 and No. 8 spark plug towers on the V8 engine and the No. 1 and No. 6 spark plug towers on the V6 engine.
4. Install the distributor, distributor cap, spark plug, wiring and connectors.
5. Check the engine timing and adjust, as required.

Ignition Timing

ADJUSTMENT
4.3L Engine

1. Run the engine until normal operating temperature is reached. Be sure the air cleaner is installed and the air conditioner is OFF.
2. Connect a timing light with the pickup lead on the No. 1 plug wire.
3. Disconnect the 4 wire electrical connector at the distributor. The check engine light will come on.

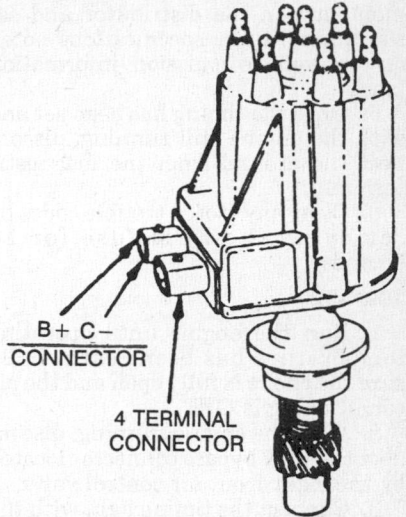

HEI distributor with externally mounted coil

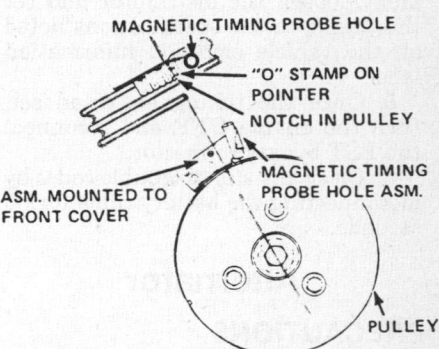

Timing marks and timing probe hole

4. If the timing requires adjustment, loosen the distributor and set the timing to the specifications as noted on the vehicle emission information label.
5. After the timing has been set, remove the ECM fuse from the fuse block for 15 seconds to cancel any stored trouble codes.

NOTE: Some engines will incorporate a magnetic timing probe hole for the use with electronic timing equipment. Be sure to consult the equipment manufactures instructions for the use of this equipment.

5.0L and 5.7L Engines
1988–90

1. Run the engine until operating temperature has been reached. Be sure the choke is fully open and the air conditioning is OFF.
2. With the engine running, ground the diagnostic terminal of the 12 terminal ALDL connector.
3. Connect the timing light with the pickup lead on the No. 1 plug wire.
4. If the timing requires adjust-

ment, loosen the distributor and set the timing to the specifications noted on the vehicle emission information label.

5. Once the timing has been set and with the engine still running, disconnect the ground from the diagnostic terminal.

6. Clear any stored trouble codes by removing the ECM fuse for 15 seconds.

1991–92

1. Run the engine until operating temperature has been reached. Be sure the choke is fully open and the air conditioning is OFF.

2. With the engine running, disconnect the EST bypass connector located by the right front air control valve.

3. Connect the timing light with the pickup lead on the No. 1 plug wire.

4. If the timing requires adjustment, loosen the distributor and set the timing to the specifications noted on the vehicle emission information label.

5. Once the timing has been set, turn the engine **OFF** and reconnect the EST bypass connector.

6. Clear any stored trouble codes by disconnecting the battery cable for 30 seconds.

Alternator

PRECAUTIONS

Several precautions must be observed with alternator equipped vehicles to avoid damage to the unit.

• If the battery is removed for any reason, make sure it is reconnected with the correct polarity. Reversing the battery connections may result in damage to the 1-way rectifiers.

• When utilizing a booster battery as a starting aid, always connect the positive to positive terminals and the negative terminal from the booster battery to a good engine ground on the vehicle being started.

• Never use a fast charger as a booster to start vehicles.

• Disconnect the battery cables when charging the battery with a fast charger.

• Never attempt to polarize the alternator.

• Do not use test lamps of more than 12 volts when checking diode continuity.

• Do not short across or ground any of the alternator terminals.

• The polarity of the battery, alternator and regulator must be matched and considered before making any electrical connections within the system.

• Never separate the alternator on an open circuit. Make sure all connec-

tions within the circuit are clean and tight.

• Disconnect the battery ground terminal when performing any service on electrical components.

• Disconnect the battery if arc welding is to be done on the vehicle.

BELT TENSION ADJUSTMENT

V-Belts are normally adjusted by loosening the bolts of the accessory being driven and moving that accessory on its pivot points until the proper tension is applied to the belt. The accessory is held in this position while the bolts are tightened. To determine proper belt tension, a belt tension gauge will be needed or simply use the deflection method. To determine deflection, press inward on the belt at the mid-point of the longest straight run. The belt should deflect (move inward) 3/8–1/2 in. Some long V-belts have idler pulleys which are used for adjusting purposes. With these systems, loosen the idler pulley and move it to take up tension on the belt.

Serpentine belts are automatically adjusted by the tensioner on the engine. If the belt is loose, check the condition of the belt and tensioner. The tensioner should place enough tension on the belt so it can only be twisted 90 degrees at it's longest run.

REMOVAL & INSTALLATION

1. Disconnect the negative battery cable.

2. Disconnect and tag the electrical connections.

3. With V-Belts, remove the bolt holding the slotted adjusting bracket to the alternator and remove the belt.

4. With serpentine belts, loosen and rotate the tensioner to release the drive belt.

5. Remove the thru-bolt to release the alternator from the engine.

To install:

6. When reinstalling, reverse the removal procedure.

7. Adjust the drive belt to allow 1/2 in. play on the longest run between pulleys.

8. On some vehicles, it may be necessary to loosen and rotate the fan shroud.

9. On models with air conditioning, it may be necessary to remove the compressor bracket. Do not discharge the air conditioning system.

Starter

REMOVAL & INSTALLATION

1. Disconnect the negative battery cable.

2. Raise and support the vehicle safely.

3. Remove upper support attaching bolts and the brace and wire guide tube bolt, if equipped.

4. Remove the flywheel housing cover, as required.

5. If necessary, remove the exhaust crossover pipe.

6. If necessary, disconnect the transmission oil cooler lines.

7. Remove the starter mounting bolts and lower the starter.

8. Disconnect the wiring and remove starter.

9. If equipped with dual exhaust, it may be necessary to remove the left exhaust pipe.

10. Install by reversing the removal procedure.

11. If shims were removed, they must be installed in their original location to assure proper drive pinion to flywheel engagement.

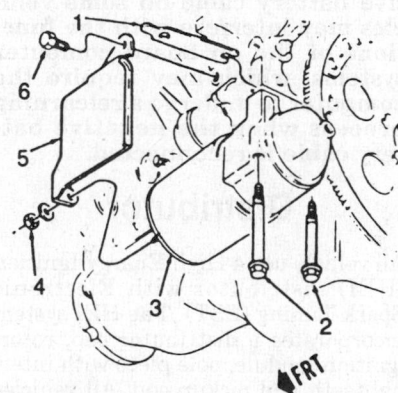

1. Shim
2. Bolts 35 ft. lbs.
3. Starter assy.
4. Nut — 13 ft. lbs.
5. Shield
6. Bolt — 20 ft. lbs.

Starter mounting — except 1991–92 Roadmaster, Caprice and Custom Cruiser

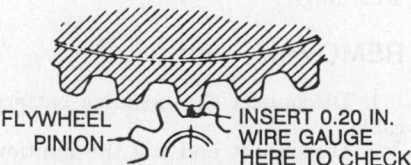

FLYWHEEL PINION   INSERT 0.20 IN. WIRE GAUGE HERE TO CHECK

Flywheel to pinion gear clearance

1. Nut
2. Washer
3. Bracket only on 5.7L engine
4. Bolt
5. Starter motor
6. Bolts
7. Double shim on 5.0L (VIN E, G, H) engine
8. Single shim on 5.0L (VIN E, G, H) and 5.7L engine
9. Double shim on 5.7L engine

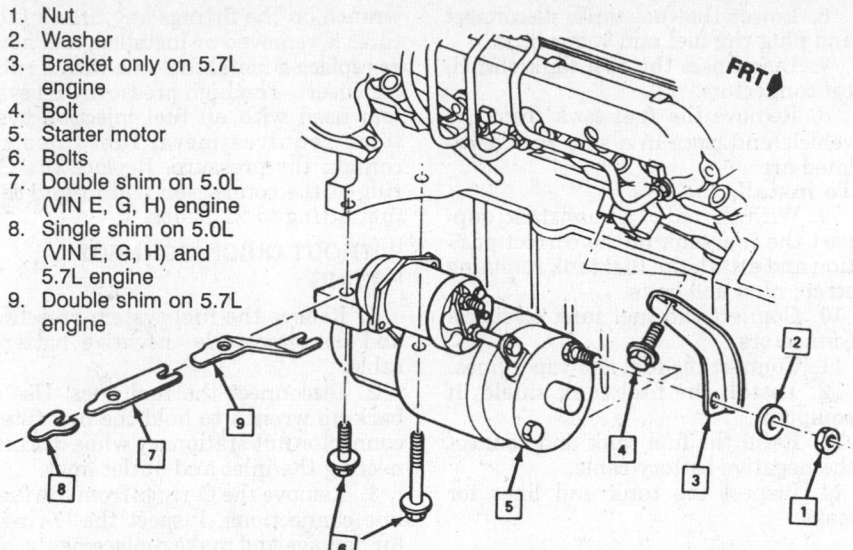

FRT

Starter mounting—1991–92 Roadmaster, Caprice and Custom Cruiser

EMISSION CONTROLS

Please refer to "Emission Controls" in the Unit Repair section for system maintenance procedures. Due to the complex nature of modern electronic engine control systems, comprehensive diagnosis and testing procedures fall outside the confines of this repair manual. For complete information on diagnosis, testing and repair procedures concerning all modern engine and emission control systems, please refer to "Chilton's Guide to Fuel Injection and Electronic Engine Controls".

Emission Warning Lamps

RESETTING

The emission light located on the instrument panel and has 2 functions:

1. The light indicates to the driver that a problem has occurred and the vehicle should be taken for service as soon as reasonably possible.

2. The light is used by technicians to monitor "Trouble Codes" when the system is in the diagnostic mode.

To verify the system is operating properly (bulb and wiring), the "Check Engine/Service Engine Soon" light will come on with the key **ON** and the engine not running. When the engine is started, the "Check Engine/Service Engine Soon" light will turn off if the system is operating properly.

If the "Check Engine/Service Engine Soon" light remains on, the self-diagnostic system has detected a problem. If the problem goes away, the light will go out in most cases after 10 seconds but a Trouble Code will remain in the ECM memory.

CLEARING TROUBLE CODES

When the ECM finds a problem, the "Check Engine/Service Engine Soon" light will come on and a trouble code will be recorded in the ECM memory. If the problem is intermittent, the "Check Engine/Service Engine Soon" light will go out after 10 seconds, when the fault goes away. However, the trouble code will stay in the ECM memory until the battery voltage to the ECM is removed. Removing battery voltage for 10 seconds will clear all stored trouble codes. Do this by disconnecting the ECM harness from the positive battery pigtail for 10 seconds with the ignition OFF or by disconnecting the ECM fuse, designated ECM or ECM/Bat., from the fuse holder.

NOTE: To prevent ECM damage, the key must be OFF when disconnecting or reconnecting power to ECM (for example battery cable, ECM pigtail, ECM fuse, jumper cables, etc.).

ECM LEARNING ABILITY

The ECM has a "learning" ability. If the battery is disconnected to clear diagnostic codes, or for repair, the "learning" process has to begin all over again. A change may be noted in the vehicle's performance to "teach" the vehicle, make sure the vehicle is at operating temperature and drive at part throttle, with moderate acceleration and idle conditions, until normal performance returns.

FUEL SYSTEM

Fuel System Service Precautions

Safety is the most important factor when performing not only fuel system maintenance but any type of maintenance. Failure to conduct maintenance and repairs in a safe manner may result in serious personal injury or death. Maintenance and testing of the vehicle's fuel system components can be accomplished safely and effectively by adhering to the following rules and guidelines.

• To avoid the possibility of fire and personal injury, always disconnect the negative battery cable unless the repair or test procedure requires that battery voltage be applied.

• Always relieve the fuel system pressure prior to disconnecting any fuel system component (injector, fuel rail, pressure regulator, etc.), fitting or fuel line connection. Exercise extreme caution whenever relieving fuel system pressure to avoid exposing skin, face and eyes to fuel spray. Please be advised that fuel under pressure may penetrate the skin or any part of the body it contacts.

• Always place a shop towel or cloth around the fitting or connection prior to loosening to absorb any excess fuel due to spillage. Ensure that all fuel spillage (should it occur) is quickly removed from engine surfaces. Ensure that all fuel soaked cloths or towels are deposited into a suitable waste container.

• Always keep a dry chemical (Class B) fire extinguisher near the work area.

• Do not allow fuel spray or fuel vapors to come into contact with a spark or open flame.

• Always use a backup wrench when loosening and tightening fuel line connection fittings. This will prevent unnecessary stress and torsion to fuel line piping. Always follow the proper torque specifications.

• Always replace worn fuel fitting O-rings with new. Do not substitute fuel hose or equivalent where fuel pipe is installed.

RELIEVING FUEL SYSTEM PRESSURE

1. Disconnect the negative battery cable.
2. Loosen the fuel filler cap to relieve tank vapor pressure.

NOTE: The internal constant bleed feature of the Model 220 TBI relieves fuel pump system pressure when the engine is turned OFF. Therefore, no further relief procedure is required.

3. Disconnect the negative battery cable.

Fuel Tank

REMOVAL & INSTALLATION

Except Monte Carlo and Cutlass

1. Disconnect the negative battery cable.
2. Drain the fuel tank and store the fuel in a safe location.
3. Properly relieve fuel system pressure.
4. Raise and support the vehicle safely.
5. Remove the fuel tank shield and fuel filler tube, if necessary.
6. Wipe the surrounding area of the fuel and vapor lines. Disconnect and remove the lines.
7. Disconnect the fuel tank electrical connectors.
8. With the aid of an assistant, support the fuel tank and remove the fuel tank retaining strap nuts and bolts.
9. Remove the fuel tank and place in a safe well ventilated area.
To install:
10. With the aid of an assistant, support the fuel tank in the correct position and attach the fuel tank retaining strap, nuts and bolts.
11. Connect the fuel tank electrical connectors.
12. Connect the fuel and vapor lines.
13. Install the fuel tank shield, if equipped.
14. Refill the fuel tank and connect the negative battery cable.
15. Inspect the tank and lines for leaks.

Monte Carlo and Cutlass

1. Disconnect the negative battery cable.
2. Drain the fuel tank and store the fuel in a safe location.
3. Properly relieve fuel system pressure, if fuel injected.
4. Raise and support the vehicle safely.
5. With the aid of an assistant, support the fuel tank and remove the fuel tank retaining strap nuts and bolts.

6. Lower the fuel tank, disconnect and plug the fuel and vapor lines.
7. Disconnect the fuel tank electrical connectors.
8. Remove the fuel tank from the vehicle and place in a safe well ventilated area.
To install:
9. With the aid of an assistant, support the fuel tank in the correct position and attach the fuel tank retaining strap, nuts and bolts.
10. Connect the fuel tank electrical connectors.
11. Connect the fuel and vapor lines.
12. Install the fuel tank shield, if equipped.
13. Refill the fuel tank and connect the negative battery cable.
14. Inspect the tank and lines for leaks.

Fuel Filter

REMOVAL & INSTALLATION

Carbureted Engine

1. Disconnect the negative battery cable.
2. Disconnect the fuel line connection at the fuel inlet filter nut on the carburetor.
3. Remove the fuel inlet filter nut and gasket from the carburetor.
4. Remove the filter, filter check valve and spring.
5. Remove the gasket from the fuel inlet nut. Discard the gasket, filter check valve and filter.
To install:
6. Install the fuel filter spring first and then the fuel filter with the check valve facing out, into the carburetor opening.
7. Ensure that the filter assembly is installed with the check valve end facing the fuel inlet line. Ribs on the closed end of the filter prevent the filter from being installed incorrectly.
8. Install a new gasket onto the fuel line nut and tighten the nut into the carburetor opening.
9. Reconnect and tighten the fuel inlet line to the fuel nut.
10. Start the engine and inspect for leaks. Repair all fuel leaks immediately.

Fuel Injected Engine

The fuel injection system uses an inline filter located in the fuel feed line under the hood, attached to the frame rail or on the rear crossmember of the vehicle. There are 2 different styles of fuel line used, the first style is the traditional metal line with fittings to secure the lines or filter and the other is a new style nylon lines with quick disconnect fittings. Always use a backup

wrench on the fittings any time a fuel filter is removed or installed, and never replace a metal fuel line with a rubber insert. The high pressure fuel system used with all fuel injection systems requires metal fuel lines to contain the pressure. Replace the O-ring at the connection and torque the fuel fitting to 22 ft. lbs.

WITHOUT QUICK-CONNECT FITTING

1. Relieve the fuel system pressure and disconnect the negative battery cable.
2. Disconnect the fuel lines. Use a back-up wrench to hold the fuel filter connector nut stationary while disconnecting the inlet and outlet lines.
3. Remove the O-rings from the fuel line connections. Inspect the O-rings for damage and make replacements, as required.
4. Remove the fuel filter from the retainer. Discard the filter.
To install:
5. Installation is the reverse of the removal procedure. The filter has an arrow (fuel flow direction) on the side of the case to ensure proper installation.
6. Install the filter in the retainer with the arrow facing away from the fuel tank, toward the front of the engine.
7. Start the engine and inspect for leaks. Correct fuel leaks immediately.

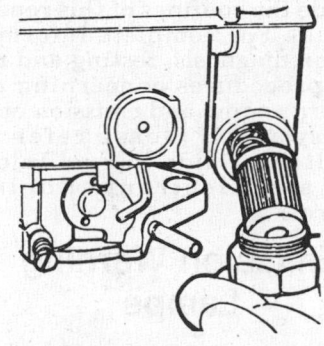

Fuel filter installation — carbureted vehicles

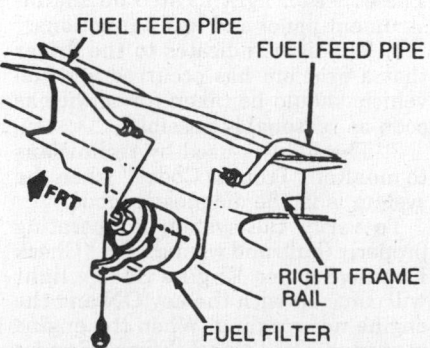

In-line fuel filter with standard fittings — fuel injected vehicles

WITH QUICK-CONNECT FITTINGS

1. Disconnect the negative battery cable.
2. Properly relieve fuel system pressure.
3. Raise and support the vehicle safely.
4. Remove the filter bracket attaching bolt.
5. While grasping the fuel filter and 1 of the fuel lines, twist the line approximately ¼ turn in each direction to loosen any dirt in the fitting and use compressed air to blow dirt out of fitting.
6. Squeeze the plastic tabs of the male connector on the fuel lines and pull connection apart.
7. Remove the fuel feed and return line body harness clips.
8. Remove the filter.

To install:

9. Position the fuel filter in original location with arrow pointing in correct direction.
10. Secure fuel filter to plastic retainers.
11. Apply a few drops of clean engine oil to both ends of fuel filter.
12. Push the fuel line connectors onto the fuel filter tubes until tabs snap into place.
13. Once installed, pull on both ends of the lines to verify they are secure.
14. Secure the filter and bracket to the frame with the attaching bolt.
15. Reconnect the battery negative cable and inspect fuel system for leaks.

Mechanical Fuel Pump

PRESSURE TESTING

1. Disconnect the negative battery cable. Disconnect the fuel line at the carburetor and install a rubber hose approximately 8–10 in. long over the line and attach a low reading pressure gauge.
2. Hold the gauge up so it is approximately 16 inches. above the fuel pump. Pinch the fuel return line, if equipped.
3. Start the engine and run at slow idle using the fuel in the carburetor.
4. Note the reading on the pressure gauge, if the pump is operating properly, the pressure should be 5½–6½ psi constant.

REMOVAL & INSTALLATION

1. Disconnect the fuel inlet hose from the pump.
2. Disconnect the vapor return hose, if equipped.
3. Disconnect the inlet hose.
4. Remove the 2 mounting bolts.
5. Remove the fuel pump.

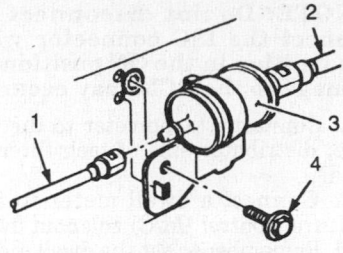

1. Fuel feed line
2. Fuel feed line
3. In-line fuel filter
4. Attaching bolt

In-line fuel filter with quick-connect fittings—fuel injected vehicles

To install:

6. Clean the mating surfaces and install a new gasket.
7. Install the fuel pump and tighten the bolts.
8. Reconnect the hoses, start the engine and check for leaks.

Electric Fuel Pump

PRESSURE TESTING

1988 TBI

1. Turn the ignition **OFF** to relieve the fuel pressure.
2. Remove the air cleaner and plug the THERMAC vacuum port on the TBI unit.
3. Uncouple the fuel supply flexible hose in the engine compartment. Install fuel pressure gauge J–29658A/BT8205 and adapter 29658A–85 between the steel line and the flexible hose.
4. Tighten the gauge in the line to ensure no leaks occur during testing.
5. Start the vehicle and observe the fuel pressure reading. It should be between 9–13 psi.
6. Relieve the fuel system pressure and remove the gauge.
7. Reinstall the fuel line, start the car and check for leaks.
8. Remove the plug from the vacuum port and install the air cleaner.

1989–92 TBI

When the ignition switch is turned **ON**, the ECM will turn **ON** the in-tank fuel pump. It will remain on as long as the engine is cranking or running and the ECM is receiving ignition reference pulses. If there are no reference pulses, the ECM will shut the fuel pump **OFF** within 2 seconds after the key is turned **ON**. The pump will deliver fuel to the TBI unit at a pressure controlled by the internal regulator to approximately 9–13 psi. Excess fuel is then returned to the fuel tank.

The fuel pump pressure test terminal is located in the passenger side cowl of the engine compartment.

NOTE: Fuel pressure should be noted while the fuel pump is running. Fuel pressure will drop immediately after the fuel pump stops running due to the controlled bleed in the fuel system.

1. Turn the ignition **OFF** and relieve fuel system pressure by removing fuel filler cap.
2. Uncouple the fuel supply flexible hose in the engine compartment. Install fuel pressure gauge between the steel line and the flexible hose and/or quick conneect fitting.
3. Tighten the gauge in the line to ensure no leaks occur during testing.
4. Apply battery voltage to the fuel pump test connector.
5. The fuel pressure should be 9–13 psi.

REMOVAL & INSTALLATION

1. Relieve the fuel system pressure.
2. Disconnect the negative battery cable.
3. Raise and support the vehicle safely.
4. Remove the fuel tank.
5. Remove the fuel tank sending unit and pump assembly by turning the cam lock ring counterclockwise. Lift the assembly from the fuel tank and remove the fuel pump from the sending unit.
6. Pull the fuel pump up into the attaching hose while pulling outward away from the bottom support. Take care to prevent damage to the rubber insulator and strainer during removal. After the pump assembly is clear of the bottom support, pull the pump assembly out of the rubber connector for removal.
7. Inspect the pump attaching hose for any signs of deterioration and replace, if necessary. Check the rubber sound insulator at the bottom of the pump and replace, as required.

To install:

8. Push the fuel pump into the attaching hose.
9. Install the tank sending unit and pump assembly into the fuel tank. Use a new O-ring during assembly.
10. Install the cam lock over the assembly and lock into place by turning clockwise.
11. Install the fuel tank.
12. Connect all wire connectors, fuel hoses and lines.
13. Lower the vehicle and reconnect the battery cable.
14. Turn the ignition **ON**, inspect the system for leaks.

Carburetor

REMOVAL & INSTALLATION

1. Disconnect the negative battery cable. Remove air cleaner.
2. Disconnect accelerator linkage.
3. Disconnect transmission detent cable.
4. Disconnect cruise control, if equipped.
5. Disconnect all necessary electrical connectors.
6. Disconnect and tag all necessary vacuum lines.
7. Disconnect fuel line at carburetor inlet.
8. Remove the attaching bolts and remove carburetor.
9. Installation is the reverse of the removal procedure, inspect the carburetor for leaks and adjust as required.

IDLE SPEED ADJUSTMENT

Idle Speed Control (ISC)

The Idle Speed Control (ISC) is controlled by the Electronic Control Module (ECM), which has the desired idle speed programmed in its memory. The ECM compares the actual idle speed to the desired idle speed and the plunger is moved in or out. This automatically adjusts the throttle to hold an idle rpm independent of the engine loads.

An integral part of the ISC is the throttle contact switch. The position of the switch determines whether or not the ISC should control idle speed. When the throttle lever is resting against the ISC plunger, the switch contacts are closed, at which time the ECM moves the ISC to the programmed idle speed. When the throttle lever is not contacting the ISC plunger, the switch contacts are open; the ECM stops sending idle speed commands and the drive controls engine speed.

NOTE: Before starting engine, place transmission selector lever in P or N position, set the parking brake and block the drive wheels.

When a new ISC assembly is installed, a base (minimum authority) and high (maximum authority) rpm speed check must be performed and adjustments made as required. These adjustments limit the low and high rpm speeds to the ECM. When making a low and high speed adjustment, the low speed adjustment is always made first. Do not use the ISC plunger to adjust curb idle speed as the idle speed is controlled by the ECM.

NOTE: Do not disconnect or connect the ISC connector with the ignition in the ON position or damage to the ECM may occur.

1. Connect a tachometer to the engine; distributor side of tach filter, if used.
2. Connect a dwell meter to the Mixture Control (M/C) solenoid dwell lead. Remember to set the dwell meter on the 6 cylinder scale, regardless of the engine being tested.
3. Turn the air conditioning OFF.
4. Start and run the engine until it is stabilized by entering closed loop. The dwell meter needle will start to vary.
5. Turn the ignition OFF.
6. Unplug the connector from ISC motor.
7. Fully retract the ISC plunger by applying 12 volts DC (battery voltage) to terminal C of the ISC motor connection and ground lead to terminal D of the ISC motor connection. It may be necessary to install jumper leads from the ISC motor in order to make proper connections.

NOTE: Do not apply battery voltage to the motor longer than necessary to retract the ISC plunger. Prolonged contact will damage the motor. Also, never connect a voltage source across terminals A and B or damage to the internal throttle contact switch will result.

8. Start the engine and wait until the dwell meter needle starts to vary, indicating "closed loop" operation.
9. With the parking brake applied and the drive wheels blocked, place the transmission in D.
10. With the ISC plunger fully retracted, adjust carburetor base (slow) idle stop screw to the minimum idle specified rpm. The ISC plunger should not be left in the fully retracted position.
11. Place the transmission in the P or N position and fully extend the ISC plunger by applying 12 volts DC to terminal D of the ISC motor connection and ground lead to terminal C of the ISC motor connection.

NOTE: Never connect voltage source across terminals A and B as damage to the internal throttle contact switch will result.

12. With the transmission in P, using tool J-29607, BT-8022 or equivalent, preset ISC plunger to obtain 1500 rpm.
13. With parking brake set and drive wheels blocked, place transmission in D position. Using tool J-29607, BT-8022 or equivalent, turn ISC plunger to obtain ISC adjustment rpm (maximum authority).
14. Recheck ISC adjustment rpm with voltage applied to motor. Motor will ratchet at full extension with power applied.
15. Fully retract ISC plunger. Place transmission in P or N position and turn ignition key to the OFF position. Disconnect 12 volt power source, ground lead, tachometer and dwell meter. With ignition in OFF position, reconnect 4 terminal harness connector to ISC motor. To prevent internal damage to ISC, apply finger pressure to ISC plunger while retracting.
16. Remove block from drive wheels.

Idle Air Bleed Valve

1. Engage the parking brake and block the drive wheels. Disconnect and plug the hoses as directed on the vehicle emission control label.
2. Check and adjust ignition timing. Connect a dwell meter to the carburetor mixture solenoid and a tachometer to the engine's distributor electrical system.
3. Start engine and with transmission in P, run engine at idle until fully warm and a varying dwell is noted on the dwell meter. The engine now in closed loop operation. It is essential that the engine is operated for a sufficient length of time to ensure that the engine coolant sensor and the oxygen sensor in the exhaust, are at full operational temperature.
4. Check engine idle speed and compare to specifications on the underhood emission label. If necessary, adjust curb idle speed. On models with Idle Speed Control (ISC) or Idle Load Compensator (ILC), idle speeds are controlled by signals from the computer.
5. With engine idling in D, observe dwell reading on the 6 cylinder scale. If varying within the 10–50 degree range, adjustment is correct. If not, perform the following:

    a. Remove the idle air bleed valve cover. If the cover is staked in place, pry it off using a suitable tool.

    b. If the cover is riveted, cover the internal bowl vents to the bleed valve with masking tape.

    c. Cover carburetor air intakes with masking tape to prevent metal chips from entering carburetor and engine.

    d. Carefully align a No. 35 (0.110 in.) drill bit on a steel rivet head holding the idle air bleed valve cover in place. Drill only enough to remove rivet head. Drill the remaining rivet head located on the other side of the tower. Use a drift and small hammer to drive the remainder of the rivets out of the idle air bleed valve tower in the air horn casting. Use care in drilling to prevent damage to the air horn casting.

e. Lift out cover over the idle air bleed valve and remove the rivet pieces from inside the idle air bleed valve tower.

f. Using shop air, carefully blow out any remaining chips from inside the tower. Discard cover after removal. A missing cover indicates the idle air bleed valve setting has been changed from its original factory setting.

g. With cover removed, look for presence or absence, of a letter identification on top of idle air bleed valve.

h. If an identifying letter appears on top of the valve proceed to the procedure outlined under Type 2. If an identifying letter does not appear on the top of the valve proceed to the procedure outlined under Type 1.

TYPE 1

1. Presetting the idle air bleed valve to a gauge dimension if the idle air bleed valve was serviced prior to on-vehicle adjustment.

a. Install idle air bleed valve gauging tool J–33815–2, BT–8253–B or equivalent, in throttle side D-shaped vent hole in the air horn casting. The upper end of the tool should be positioned over the open cavity next to the idle air bleed valve.

b. While holding the gauging tool down lightly, so the solenoid plunger is against the solenoid stop, adjust the idle air bleed valve so the gauging tool will pivot over and just contact the top of the valve. The valve is now preset for on-vehicle adjustment.

c. Remove the gauging tool.

2. Adjusting the idle air bleed valve on the vehicle to obtain correct dwell reading.

a. Start engine and allow it to reach normal operating temperature.

b. While idling in D, use a suitable tool to slowly turn valve counterclockwise or clockwise, until the dwell reading varies within the 25–35 degree range, attempting to be as close to 30 degrees as possible. Perform this Step carefully. The air bleed valve is very sensitive and should be turned in ⅛ turn increments only.

c. If, after performing Steps a and b above, the dwell reading does not vary and is not within the 25–35 degree range, it will be necessary to remove the plugs and to adjust the idle mixture needles.

3. Idle mixture needle plug removal, only if necessary.

a. Disconnect the negative battery cable and remove the carburetor from the engine, following normal service procedures, to gain access to the plugs covering the idle mixture needles.

b. Invert carburetor and drain fuel into a suitable container.

c. Place carburetor on a suitable holding fixture, with manifold side up. Use care to avoid damaging linkage, tubes and parts protruding from air horn.

d. Make 2 parallel cuts in the throttle body, on each side of the locator points beneath the idle mixture needle plug on the manifold side, with a hacksaw.

e. The cuts should reach down to the steel plug but should not extend more than ⅛ in. beyond the locator points. The distance between the saw cuts depends on the size of the punch to be used.

f. Place a flat punch near the ends of the saw marks in the throttle body. Hold the punch at a 45 degrees angle and drive it into the throttle body until the casting breaks away, exposing the steel plug.

g. The hardened plug will break, rather than remaining intact. It is not necessary to remove the plug as a whole but remove the loose pieces.

h. Repeat this procedure with the other mixture needle.

4. Proceed with setting of the idle mixture needles, if necessary, where correct dwell reading could not be obtained with the idle air bleed valve adjustment.

a. Using tool J–29030, BT–7610B or equivalent, turn both idle mixture needles clockwise until they are lightly seated, then turn each mixture needle counterclockwise the number of turns specified.

b. Reinstall carburetor on engine using a new flange mounting gasket but do not install air cleaner and air horn gasket at this time.

5. Readjust the idle air bleed valve to finalize correct dwell reading. The following is necessary if idle mixture needles required setting in Step 4, above.

a. Start engine and run until fully warm, and repeat Step 2, above.

b. If unable to set dwell to 25–35 degrees and the dwell is below 25 degrees, turn both mixture needles counterclockwise an additional turn. If dwell is above 35 degrees, turn both mixture needles clockwise an additional turn. Readjust idle air bleed valve to obtain dwell limits.

c. After adjustments are complete, seal the idle mixture needle openings in the throttle body, using silicone sealant, RTV rubber or equivalent. The sealer is required to

discourage unnecessary adjustment of the setting and to prevent fuel vapor loss in that area.

d. On vehicles without a carburetor-mounted idle speed control or idle load compensator, adjust curb idle speed, if necessary.

e. Check and only if necessary adjust, fast idle speed as described on emission control information label.

TYPE 2

1. Setting the idle air bleed valve to a gauge dimension:

a. Install air bleed valve, gauging tool J–33815–2, BT–8253–B or equivalent, in throttle side D shaped vent hole in the air horn casting. The upper end of the tool should be positioned over the open cavity next to the idle air bleed valve.

b. While holding the gauging tool down lightly, so the solenoid plunger is against the solenoid stop, adjust the idle air bleed valve so the gauging tool will pivot over and just contact the top of the valve.

c. The valve is now set properly. No further adjustment of the valve is necessary.

d. Remove gauging tool.

2. Adjusting the idle mixture needles on the vehicle to obtain correct dwell readings.

a. Remove idle mixture needle plugs, following instructions in the information given for type 1.

b. Using tool J–29030–B, BT–7610–B or equivalent, turn each idle mixture needle clockwise until lightly seated, then turn each mixture needle counterclockwise 3 turns.

c. Reinstall carburetor on engine, using a new flange mounting gasket, but do not install air cleaner or gasket.

d. Start engine and allow it to reach normal operating temperature.

e. While idling in D, adjust both mixture needles equally, in ⅛ turn increments, until dwell reading varies within the 25–35 degree range, attempting to be as close to 30 degrees as possible. If reading is too low, turn mixture needles counterclockwise. If reading is too high, turn mixture needles clockwise. Allow time for dwell reading to stabilize after each adjustment.

f. After adjustments are complete, seal the idle mixture needle openings in the throttle body, using silicone sealant, RTV rubber or equivalent. The sealer is required to discourage unnecessary readjustment of the setting, and to prevent fuel vapor loss in that area.

g. On vehicles without a carburetor-mounted idle speed control or idle load compensator, adjust curb idle speed, if necessary.

h. Check and, if necessary, adjust fast idle speed, as described on the emission control information label.

Idle Load Compensator (ILC)

1. Prepare vehicle for adjustments—see vehicle emission information label.

2. Connect tachometer to distributor side of tach filter, if used.

3. Remove air cleaner and plug vacuum hose to Thermal Vacuum Valve (TVV).

4. Disconnect and plug vacuum hose to EGR.

5. Disconnect and plug vacuum hose to canister purge port.

6. Disconnect and plug vacuum hose to ILC.

7. Back out idle stop screw on carburetor 3 turns.

8. Turn the air conditioning to **OFF** position.

NOTE: Before starting engine, place transmission in P, set parking brake and block drive wheels.

9. With engine running, engine warm, choke off, transmission in **D** and ILC plunger fully extended (no vacuum applied), using tool J–29607, BT–8022 or equivalent, adjust plunger to obtain 725 rpm on E4MC carburetor. Jam nut on plunger must be held with wrench to prevent damage to guide tabs when tightening.

10. Remove plug from vacuum hose, reconnect hose to ILC and observe idle speed. Idle speed should be 500 rpm in **D**.

11. If rpm in Step 10 is correct, proceed to Step 13. No further adjustment of the ILC is necessary.

12. If rpm in Step 10 is not correct:

a. Stop engine and remove the ILC. Plug vacuum hose to ILC.

b. With the ILC removed, remove the rubber cap from the center outlet tube and then remove the metal plug if used from this same tube.

c. Install ILC on carburetor and re-attach throttle return spring and any other related parts removed during disassembly. Remove plug from vacuum hose and reconnect hose to ILC.

d. Using a spare rubber cap with hole punched to accept a 0.090 in. (³/₃₂ in.) hex key wrench, install cap on center outlet tube, to seal against vacuum loss and insert wrench through cap to engage adjusting screw inside tube. Start engine and turn adjusting screw with wrench to obtain 550 rpm in **D**. Turning the adjusting screw will change the idle speed approximately 75–100 rpm for each complete turn. Turning the

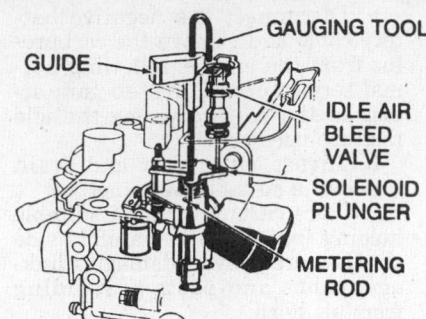

GUIDE — GAUGING TOOL — IDLE AIR BLEED VALVE — SOLENOID PLUNGER — METERING ROD

Idle air bleed adjusting

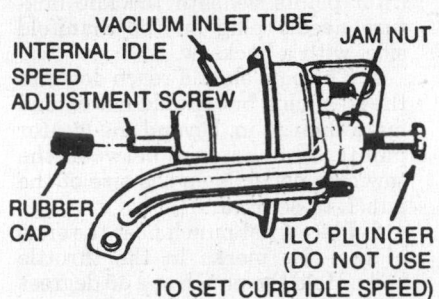

VACUUM INLET TUBE — INTERNAL IDLE SPEED ADJUSTMENT SCREW — JAM NUT — RUBBER CAP — ILC PLUNGER (DO NOT USE TO SET CURB IDLE SPEED)

Idle load compensator—5.0L (VIN Y) engine

screw counterclockwise will increase the engine speed.

e. Remove wrench and cap, with hole, from center outlet tube and install new rubber cap.

f. Engine running, transmission in **D**, observe idle speed. If a final adjustment is required, it will be necessary to repeat Steps 12a through 12e.

13. After adjustment of the ILC plunger, measure distance from the jam nut to tip of the plunger, dimension must not exceed 1.000 in. (25mm).

14. Disconnect and plug vacuum hose to ILC. Apply vacuum source, such as hand vacuum pump J–23768, BT–7517 or equivalent, to ILC vacuum inlet tube to fully retract the plunger.

15. Adjust the idle stop on the carburetor float bowl to obtain 500 rpm in **D**.

16. Place transmission in **P** and stop engine.

17. Remove plug from vacuum hose and install hose on ILC vacuum inlet tube.

18. Remove plugs and reconnect all vacuum hoses.

19. Install air cleaner and gasket.

20. Remove block from drive wheels.

Throttle Position Sensor (TPS)

NOTE: The plug covering the TPS adjustment screw is used to provide a tamper resistant design and retain the factory setting during vehicle operation. Do not

remove the plug unless diagnosis indicates the TPS sensor is not adjusted correctly or it is necessary to replace the air horn assembly, float bowl, TPS sensor to TPS adjustment screw. This is a critical adjustment that must be performed accurately and carefully to ensure proper vehicle performance and control of exhaust emissions.**

1. If necessary to adjust the TPS sensor, perform the following:

a. Using a ⁵/₆₄ in. (2mm) drill, drill hole in aluminum plug covering TPS adjustment screw, drilling only enough to start self-tapping screw.

b. Use care in drilling to prevent damage to adjustment screw head.

c. Start a No. 8, ½ in. long self-tapping screw in drilled hole in plug, turning screw in only enough to ensure good thread engagement in hole.

d. Using a suitable tool placed between the screw head and air horn casting, pry against screw head to remove plug. Discard plug.

e. Using tool J–28696, BT–7967A or equivalent, remove screw.

f. Connect digital voltmeter, such as J–29125 or equivalent from TPS connector center terminal **B** to bottom terminal **C**. Jumpers for access can be made using terminals 12014836 and 12014837.

g. With ignition **ON**, engine stopped, reinstall TPS adjustment screw and with tool J–28696, BT7967A or equivalent turn screw to obtain specified voltage at specified throttle position with air conditioning **OFF**.

h. After adjustment, install new plug (supplied in service kits) in air horn, driving plug in place until flush with raised pump lever boss on casting.

2. Remove ECM fuse from fuse block to clear any stored trouble codes.

NOTE: After TPS screw is adjusted, a new plug should be installed. If a new plug is not available, a locking type of sealer should be placed on the screw threads to prevent movement of the screw after installation.

Fuel Injection

On the 4.3L, 5.0L and 5.7L engines, the EFI system centrally locates a single Model 220 Throttle Body Injection (TBI) unit on the intake manifold where air and fuel are distributed through a single bore in the unit. The air used for combustion is controlled by a single throttle valve which is connected to the accelerator pedal linkage through a throttle shaft and lever assembly. A special plate is located under the throttle valve to aid in uniform mixture distribution. Fuel for combustion is supplied by a single fuel injector mounted on the TBI unit. The metering tip of the fuel injector is positioned directly above the throttle valve. The injector metering tip is "pulsed" or "timed" open or closed by an electronic output signal received from the ECM. The ECM receives inputs from the the various engine sensors concerning engine operating conditions, coolant temperature, exhaust gas oxygen content, etc. The ECM uses this information to calculate the engines fuel requirements by controlling the injector pulse openings to provide an ideal fuel/air mixture ratio.

IDLE SPEED ADJUSTMENT

Except 1988 Vehicles

NOTE: If installing a new IAC valve measure and adjust the valve accordingly. If reinstalling a used IAC valve, do not push or pull on the pintle to adjust pintle length or damage to the IAC worm gear might occur. The valve is preset at the factory and will self adjust when the following procedure is performed.

1. On new IAC valve only, measure the distance between the tip of the pintle and the valve mounting surface.
2. If greater than 1.10 in. (28mm), use light finger pressure to slowly retract the pintle. The force required to retract a new valve will not damage the valve.
3. Install and connect the valve wire connector.
4. Reset the IAC valve pintle position by depressing the accelerator pedal slightly, start the engine and run for 5 seconds, turn the key OFF for 10 seconds, then restart the vehicle and check for proper idle operation.

1988 Vehicles

1. Plug any vacuum ports as required.
2. Remove the idle speed stop screw cover, if installed.
3. Connect a tachometer to the engine.
4. Leaving the IAC valve connected, ground the ALDL diagnostic terminal.
5. Turn the ignition switch to the ON position but do not start the engine. Allow the ignition switch to remain in the ON position (engine OFF) for a period of 45 seconds. This allows the IAC valve pintle to extend and seat in the valve body.

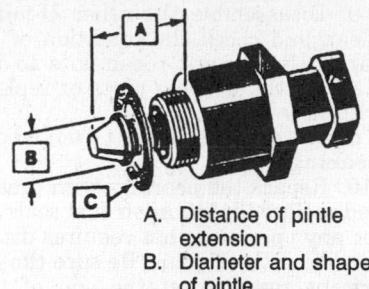

A. Distance of pintle extension
B. Diameter and shape of pintle
C. IAC valve gasket

New IAC valve pintle adjustment— Except 1988 vehicles

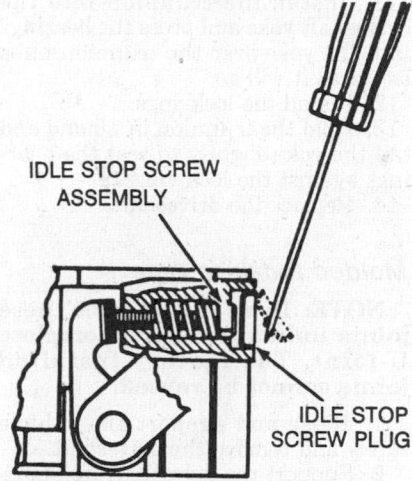

IDLE STOP SCREW ASSEMBLY

IDLE STOP SCREW PLUG

Removal of idle speed plug—1988 vehicles

6. With the ignition switch ON (engine stopped) and the diagnostic terminal grounded, disconnect the IAC valve electrical connector.
7. Remove the diagnostic terminal ground and start the engine. Place the transmission in P and allow the engine rpm to stabilize.
8. The tachometer should read between 400–450 rpm. Adjust the idle stop screw as required until the idle speed is within the specified range.
9. Turn the ignition switch OFF and reconnect the IAC valve connector.
10. Apply a bead of RTV sealant to cover the idle stop screw hole. Unplug and reconnect the vacuum hoses.

Fuel Injector

REMOVAL & INSTALLATION

NOTE: Exercise care when removing the fuel injectors to prevent damage to the electrical connector terminals, the injector filter and the fuel nozzle. Also, since the injectors are electrical components, they should not be immersed in any type of liquid solvent of cleaner as damage may occur.

1. Relieve the fuel system pressure. Disconnect the electrical connectors from the fuel injectors by squeezing the plastic tabs and pulling straight up.
2. Remove the fuel meter cover assembly in the same manner as the electrical connectors.
3. With fuel meter cover gasket in place, to prevent damage to casting, carefully lift out each injector and set aside.
4. Remove the lower (small) O-rings from the injector nozzles. Discard the O-rings and replace with new.
5. Remove the fuel meter cover gasket and discard.
6. Remove the upper (large) O-rings and steel backup washers from top of each fuel injector cavity. Discard the O-rings and replace with new.

To install:
7. Inspect the fuel injector filter for evidence of dirt and contamination. If present, check for presence of dirt in fuel lines and fuel tank.

NOTE: If replacements are required, ensure that the injector is replaced with an identical part. The model 220 TBI is capable of accepting other types of injectors but other injectors are calibrated for different flow rates.

8. Lubricate new lower (small) O-ring with automatic transmission fluid and push on nozzle end of injector until it seats against injector fuel filter.
9. Install the steel injector backup washer in counterbore of fuel meter body.
10. Lubricate new upper (large) O-ring with automatic transmission fluid and install directly over the backup washer. Ensure the O-ring is seated properly and is flush with top of fuel meter body surface.

NOTE: Backup washers and O-rings must be installed before the injectors or improper seating of large O-ring could cause fuel to leak.

11. Align the raised lug on each injector base with notch in fuel meter body cavity and install the injector. Push down with moderate pressure on injector until it is fully seated in fuel meter body. The electrical terminals of injector should be parallel with throttle shaft.
12. Install the fuel meter cover gasket.
13. Install the fuel meter cover.
14. Coat the threads of the fuel meter attaching screw with a suitable thread locking compound. Install and tighten the screws.
15. Reconnect the electrical connectors to their respective fuel injectors.

16. Turn the ignition switch to the **ON** position, engine not running and check for fuel leaks.

DRIVE AXLE

Driveshaft

REMOVAL & INSTALLATION

1. Raise the vehicle and support it safely.
2. Mark the relationship of the driveshaft to the differential flange.
3. Unbolt the straps or flange. Tape the bearing caps in place to prevent losing the bearing rollers. Support the driveshaft to prevent excessive strain on the universal joint.
4. Position a suitable drain pan under the transmission end to catch any fluid that may leak out when the driveshaft is removed. Pull the shaft back and remove it. Be careful not to damage the splines at the transmission end.
5. If the transmission splined slip yoke does not have a vent hole at the center, it should be lubricated for installation with engine oil. If it does have a vent hole, it should be lubricated with grease. Slide the slip yoke into place.
6. Align the driveshaft marks and tighten the bolts. Tighten the U-bolts to 16 ft. lbs.

Universal Joints

REMOVAL & INSTALLATION

Snapring Type

1. Raise and support the vehicle safely. Mark and remove the driveshaft.
2. Remove the lock rings from the yoke and remove the lubrication fitting.
3. Support the yoke in a bench vise. Never clamp the driveshaft tube.
4. Use a socket to press against 1 trunnion bearing to press the opposite bearing from the yoke.
5. Grasp the cap and work it out.
6. Support the other side of the yoke and press the bearing cap from the yoke and as in previous steps.
7. Remove the trunnion from the driveshaft yoke.
8. If equipped with a sliding sleeve, remove the trunnion bearings from the sleeve yoke. Remove the seal retainer from the end of the sleeve and pull the seal and washer from the retainer.

9. Disassemble the other U-joint. Clean and check the condition of all parts. Use U-joint repair kits to replace all the wearing parts or replace with new U-joint.

To assemble the trunnion bearings:

10. Repack the bearings with grease and replace the trunnion dust seals after any operation that requires disassembly of the U-joint. Be sure the lubricant reservoir at the end of the trunnion is full of lubricant. Fill the reservoirs with lubricant from the bottom.
11. Install the trunnion into the driveshaft yoke and press the bearings into the yoke over the trunnion hubs as far as it will go.
12. Install the lockrings.
13. Hold the trunnion in 1 hand and tap the yoke slightly to seat the bearings against the lock rings.
14. Replace the driveshaft.

Molded Retainer Type

NOTE: Don't disassemble these joints unless replacing complete U-joint. The factory installed joints cannot be reused.

1. Raise and support the vehicle safely and remove the driveshaft.
2. Support the dirveshaft in a horizontal position. Place the U-joint so the lower ear of the shaft yoke is supported by a 1⅛ in. socket. Press the lower bearing cup of the yoke ear. This will shear the plastic retaining the lower bearing cup.

NOTE: Never clamp the driveshaft tubing in a vise.

3. If the bearing cup is not completely removed, lift the cross, insert a spacer and press the cup completely out.
4. Rotate the driveshaft, shear the opposite plastic retainer, and press the other bearing cup out in the same manner.
5. Remove the cross from the yoke.

NOTE: Production U-joints cannot be reassembled. There are no bearing retainer grooves in the cups. Discard all parts that were removed and substitute those in the overhaul kit.

6. Remove the sheared plastic bearing retainer from the yoke. Drive a small pin or punch through the injection holes to aid in removal.
7. If the other U-joint is to be serviced, remove the bearing cups from the slip yoke.
8. Be sure the seals are installed on the service bearing cups to hold the needle bearings in place for handling. Grease the bearings, if not pregreased.

9. Install 1 bearing cup part way into 1 side of the yoke and turn this ear to the bottom.
10. Insert the cross into the yoke so the trunnion seats freely in the bearing cup.
11. Install the opposite bearing cup part way. Be sure both trunnions are started straight into the bearing cups.
12. Press against opposite bearing cups, working the cross constantly to be sure it is free in the cups. If binding occurs, check the needle rollers to be sure 1 needle has not become lodged under and end of the trunnion.
13. As soon as 1 bearing retainer groove is exposed, stop pressing and install the bearing retainer snapring.
14. Continue to press until the opposite bearing retainer can be installed. If difficulty installing the snaprings is encountered, rap the yoke with a hammer to spring the yoke ears slightly.
15. Assemble the other half of the U-joint in the same manner.
16. Check that the cross is free in the cups. If it is too tight, rap the yoke ears again to help seat the bearing retainers.
17. Reinstall the driveshaft.

Rear Axle Shaft, Bearing and Seal

REMOVAL & INSTALLATION

1. Raise vehicle and support it safely. Remove the tire and wheel assembly. Remove the brake drum.
2. Drain the fluid. Remove the rear carrier cover. Discard the gasket.
3. Remove the rear axle pinion shaft lock screw and the rear axle pinion shaft.
4. Push flanged end of axle shaft toward center of the vehicle and remove C-lock from button end of shaft.
5. Remove axle shaft from housing, being careful not to damage oil seal.
6. Remove seal from housing with a pry bar behind steel case of seal, being careful not to damage housing.
7. Insert tool J–23689 or equivalent, into bore and position it behind bearing so tangs on tool engage bearing outer race. Remove bearing, using slide hammer.

To install:

8. Lubricate the new bearing with gear lubricant and install bearing so tool bottoms against shoulder in housing, using tool J–23690 or equivalent.
9. Lubricate seal lips with gear lubricant. Position seal on tool J–21128 or equivalent, and position seal into housing bore. Tap seal into place so it is flush with axle tube.
10. Insert the axle into the place while engaging the splines on the end of the shaft with the splines of the rear

1. Cover bolt
2. Cover gasket
3. Differential bearing cap bolt
4. Differential bearing cap
5. Drive pinion
6. Shim
7. Rear pinion bearing
8. Inner race
9. Spacer
10. Rear axle housing
11. Outer race
12. Front pinion bearing
13. Pinion yoke oil seal
14. Pinion yoke
15. Washer
16. Pinion yoke nut
17. Axle shaft
18. Bearing assy.
19. Oil seal
20. Backing plate
21. Shim
22. Side bearing
23. Bolt
24. Pinion gear shaft
25. Differential case
26. Lock bolt
27. Ring gear
28. Thrust washer
29. Pinion gear
30. Side gear
31. ABS sensor ring

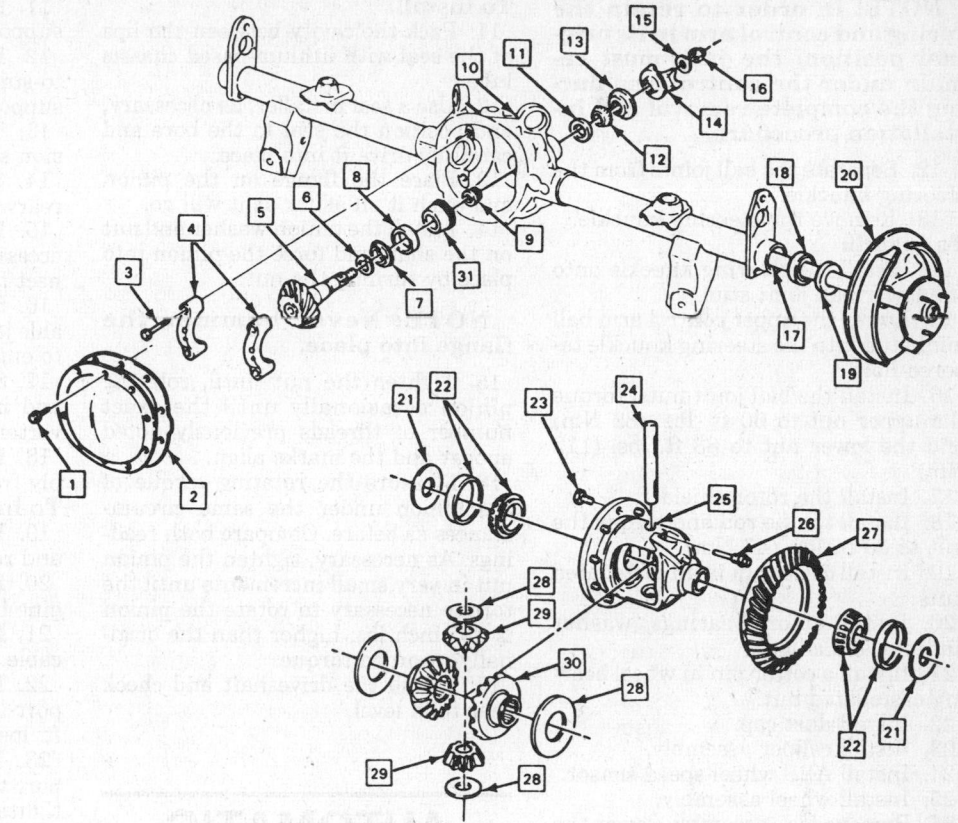

Exploded view of standard rear axle

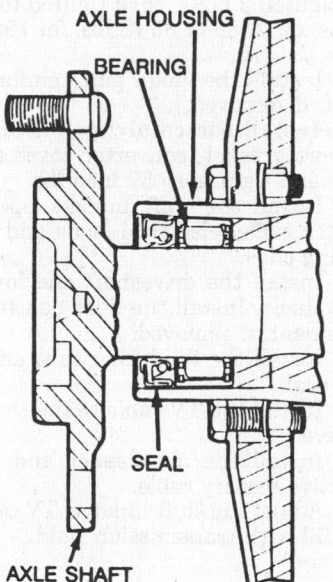

Rear axle shaft, bearing and seal— cut away view

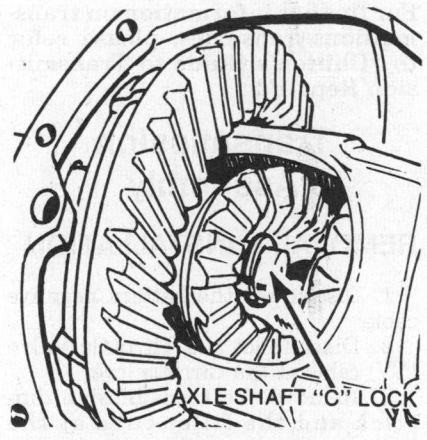

Removing and Installing C-locks

axle side gear. Be careful not to damage the seal.

11. Install the C-lock on the bottom of the axle shaft and push the shaft outward so the lock seats in the counterbore of the rear axle side gear.

12. Install the rear axle pinion gear shaft through the differential case, thrust washers and pinions, align the hole in the shaft with the lock bolt hole. Install the lock bolt and tighten to 24 ft. lbs. for 7½ in. ring gears or 20 ft. lbs. for 8½ in. ring gears.

13. Install the carrier cover and bolts using a new gasket.

14. Fill the rear assembly with the proper grade and type gear oil.

15. Install the brake drum and wheel and lower the vehicle.

Front Wheel Hub, Knuckle/Spindle and Bearings

REMOVAL & INSTALLATION

1. Disconnect the battery negative cable.

2. Raise and support the vehicle safely.

3. Disconnect the ABS wheel speed sensor, if equipped.

4. Remove the tire and wheel assembly.

5. Remove the caliper assembly.

6. Remove the dust cap from the hub assembly.

7. Remove the cotter pin, nut and washer from the spindle.

8. Remove the hub from the spindle assembly.

9. Remove the rotor shield from the spindle.

10. Separate the tie rod from the steering knuckle.

11. Position a floor jack under the control arm near the spring seat and raise the jack until it just supports the lower control arm.

NOTE: In order to retain the spring and control arm in its original position, the jack must remain under the control arm during the complete removal and installation procedure.

12. Seperate the ball joints from the steering knuckle.
13. Remove the steering knuckle.

To install:
14. Install the steering knuckle onto the lower ball joint stud.
15. Lower the upper control arm ball joint stud into the steering knuckle tapered hole.
16. Install the ball joint nuts, torque the upper nut to 60 ft. lbs. (82 Nm) and the lower nut to 83 ft. lbs. (112 Nm).
17. Install the rotor shield.
18. Install the tie rod and torque the nut to 35 ft. lbs. (47 Nm).
19. Install cotter pin in all castellated nuts.
20. Install rotor, bearings, washer and nut and adjust.
21. Install a cotter pin in wheel bearing castellated nut.
22. Install dust cap.
23. Install caliper assembly.
24. Install ABS wheel speed sensor.
25. Install wheel assembly.
26. Remove jack assembly, lower the vehicle.
27. Reconnect the negative battery cable.
28. Road test vehicle.

Pinion Seal

REMOVAL & INSTALLATION

1. Raise and support the vehicle safely. It would help to have the front end slightly higher than the rear to avoid fluid loss.
2. Mark and remove the driveshaft.
3. Release the parking brake.
4. Remove the rear wheels. Rotate the rear wheels by hand to make sure there is absolutely no brake drag. If there is brake drag, remove the drums.
5. Using an inch lb. torque wrench on the pinion nut, record the force needed to rotate the pinion.
6. Mark the pinion shaft, nut and flange. Count the number of exposed threads on the pinion shaft.
7. Install a holding tool on the pinion. A very large adjustable wrench will do or if 1 is not available, put the drums back on and set the parking brake as tightly as possible.
8. Remove the pinion nut.
9. Slide the flange off of the pinion. A puller may be necessary.
10. Center punch the oil seal to distort it and pry the seal out, being careful to avoid scratching the bore.

To install:
11. Pack the cavity between the lips of the seal with lithium-based chassis lube.
12. Use a seal installer, as necessary, and position the seal in the bore and carefully drive it into place.
13. Place the flange on the pinion and push it on as far as it will go.
14. Install the pinion washer and nut on the shaft and force the pinion into place by turning the nut.

NOTE: Never hammer the flange into place.

15. Tighten the nut until, rotating pinion occasionally until the exact number of threads previously noted appear and the marks align.
16. Measure the rotating torque of the pinion under the same circumstances as before. Compare both readings. As necessary, tighten the pinion nut in very small increments until the torque necessary to rotate the pinion is 3–5 inch lbs. higher than the originally recorded torque.
17. Install the driveshaft and check rear fluid level.

AUTOMATIC TRANSMISSION

For further information on transmissions/transaxles, please refer to "Chilton's Guide to Transmission Repair".

Transmission Assembly

REMOVAL & INSTALLATION

1. Disconnect the battery negative cable.
2. Disconnect the Throttle Valve (TV) cable at the throttle lever.
3. Remove the transmission dipstick and the filler tube at the transmission.
4. Raise and support the vehicle safely.
5. Remove the driveshaft and the floor pan reinforcement, if equipped.
6. Disconnect the speedometer cable and the shift linkage at the transmission.
7. Disconnect all electrical leads at the transmission and any clips that retain the leads to the transmission.
8. Remove the flywheel cover.
9. Mark flywheel and converter for installation reference.
10. Remove the torque converter to flywheel bolts.

11. Remove the catalytic converter support bracket.
12. Remove the transmission mount-to-support bolt and the transmission support-to-frame bolts and insulators.
13. Support and raise the transmission slightly.
14. Slide the transmission support rearward.
15. Lower the transmission to gain access to the oil cooler lines. Disconnect the lines and cap all openings.
16. Support the engine with a suitable jack and remove the transmission to engine bolts.
17. Slide the transmission rearward and install tool J–21366 to the converter to hold it in place.
18. Remove the transmission assembly from the vehicle.

To install:
19. Raise the transmission into place and remove tool J–21366.
20. Install the transmission to engine bolts and tighten to 35 ft. lbs.
21. Install the oil cooler pipe and TV cable.
22. Install the the transmission support to frame bolts and tighten to 41 ft. lbs.
23. Install the the transmission support to transmission mount bolts and tighten to 35 ft. lbs.
24. Remove the transmission jack and install the converter to flywheel in the original position marked and finger-tighten 3 bolts, then tighten to 46 ft. lbs. for 1988 or 35 ft. lbs. for 1989–92.
25. Install the floor pan reinforcement, if removed.
26. Install the catalytic converter support bracket, converter cover and bolts and tighten to 89 inch lbs.
27. Install the shift linkage, speedometer cable, electrical leads and retaining clips.
28. Install the driveshaft and lower the vehicle. Install the floor pan reinforcement, if removed.
29. Install the fluid filler tube and a new seal.
30. Install the TV cable at the throttle lever.
31. Install the air cleaner and the negative battery cable.
32. Adjust the shift linkage, TV cable and fill with transmission fluid.

THROTTLE VALVE (TV) CABLE ADJUSTMENT

Setting of the TV cable must be done by rotating the throttle lever at the carburetor or throttle body. Do not use the accelerator pedal to rotate the throttle lever.

1. With the engine OFF, depress and hold the reset tab at the engine end of the TV cable.

2. Move the slider until it stops against the fitting.

3. Release the rest tab.

4. Rotate the throttle lever to its full travel.

5. The slider must move (ratchet) toward the lever when the lever is rotated to its full travel position.

6. Recheck after the engine is hot and road test the vehicle.

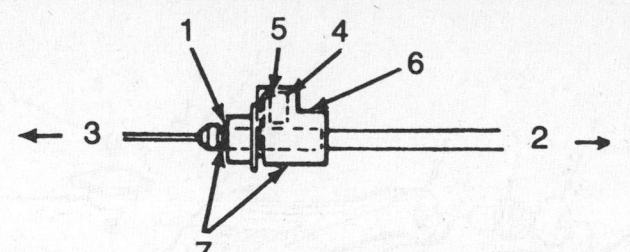

1.	Slider against fitting (zero or reset position)
2.	Direction of cable actuating lever
3.	Reset direction
4.	Reset tab
5.	Fitting
6.	Cable
7.	Slider

Throttle valve cable adjustment

FRONT SUSPENSION

Shock Absorbers

REMOVAL & INSTALLATION

1. Raise and support the vehicle safely.

2. Hold the shock absorber upper stem from turning and remove the upper nut, retainer and grommet.

3. Remove the 2 bolts and lock washers securing the shock to the lower control arm.

4. To install, reverse the removal procedure.

NOTE: Purge new shock absorbers of air by repeatedly extending in their normal position and compressing while inverted. This is not necessary if installing gas pressurized shock absorbers.

Coil Spring

REMOVAL & INSTALLATION

1988 Vehicles

1. Raise and support the vehicle safely.

2. Disconnect the shock absorber from the lower control arm and push the shock up into the spring.

3. Install a safety chain around the spring and control arm.

4. While the vehicle is supported so the control arms hang free, place the spring removal/installer tool into position cradling the inner bushings. The tool should be bolted to a suitable floor jack.

5. Remove the stabilizer to the lower control arm attachment.

6. Raise the jack to remove the tension on the lower control arm pivot bolts. Remove the nuts and bolts. Remove the rear bolt first.

7. Lower the control arm by slowly lowering the jack.

8. When the tension is released from the spring, remove the chain and spring.

To install:

9. Properly position the spring onto the lower control arm, using installer tool.

10. Position the control arm into the frame and install the pivot bolts. Front bolt installed from front to rear first.

11. Torque the lower control arm nuts to 65 ft. lbs. for the Monte Carlo or 92 ft. lbs. for all other vehicles.

12. Lower the jack and install the stabilizer bar link and tighten the bolt to 15 ft. lbs.

13. Install the shock absorber and tighten the lower bolts to 20 ft. lbs.

14. Lower the vehicle.

1989–92 Vehicles

1. Raise and support the vehicle safely.

2. Disconnect the ABS wheel speed sensor, if equipped, and secure aside.

3. Remove the wheel and shock absorber.

4. Remove the stabilizer linkage from the lower control arm.

5. Remove the steering knuckle from the tie rod end, using a suitable puller tool.

6. Install a universal spring compressor and compress the spring.

7. Support the lower control arm and remove the lower control arm to frame bolts.

8. Pivot the lower control arm rearward and remove the compressor and spring.

To install:

9. Properly position the spring onto the lower control arm, using spring compressor tool.

10. Position the control arm into the frame and install the pivot bolts. Front bolt installed from front to rear first. Tighten to 92 ft. lbs.

11. Remove the spring compressor tool, install the steering knuckle to tie rod end and tighten the nut to 35 ft. lbs. Install the cotter pin.

12. Remove the support from the lower control arm and install the stabilizer linkage and tighten the bolt/nut to 13 ft. lbs.

13. Install the ABS wheel speed sensor, if equipped.

14. Install the shock absorber and tighten the lower attaching bolts to 20 ft. lbs. and the upper nut to 97 inch lbs.

15. Install the wheel and lower the vehicle. Tighten the wheel lug nuts to 100 ft. lbs.

Torsion Bars

REMOVAL & INSTALLATION

1. Raise and support the vehicle safely.

2. Disconnect each side of the torsion bar by removing the nut from the link bolt. Pull the bolt from the linkage and remove retainers, grommets and spacer.

3. Remove bracket to frame or body bolts and remove torsion bar, rubber bushings and brackets.

4. Installation is the reverse of the removal procedure. Install the torsion bar with the identification forming on the right side of the vehicle and the slit in the rubber bushings facing the front of the vehicle.

5. Tighten torsion bar link nut to 13 ft. lbs. (18 Nm) and the bracket bolts to 24 ft. lbs. (33 Nm).

Upper Ball Joints

INSPECTION

1. Raise the vehicle and position floor stands under the left and right lower control arm as near as possible to each lower ball joint. There should be sufficient space between the upper control arm bumper and frame.

2. Position a dial indicator against the wheel rim.

3. Grasp the front wheel and push in on bottom of the tire while pulling out at the top. Read the gauge, then reverse the push-pull procedure. Horizontal deflection on the dial indicator should not exceed 1.125 in. (3.18mm).

4. If the indicator exceeds 1.125 in. (3.18mm) or if the ball stud, when disconnected from the knuckle assembly, can be twisted in its socket by hand, replace the ball joint.

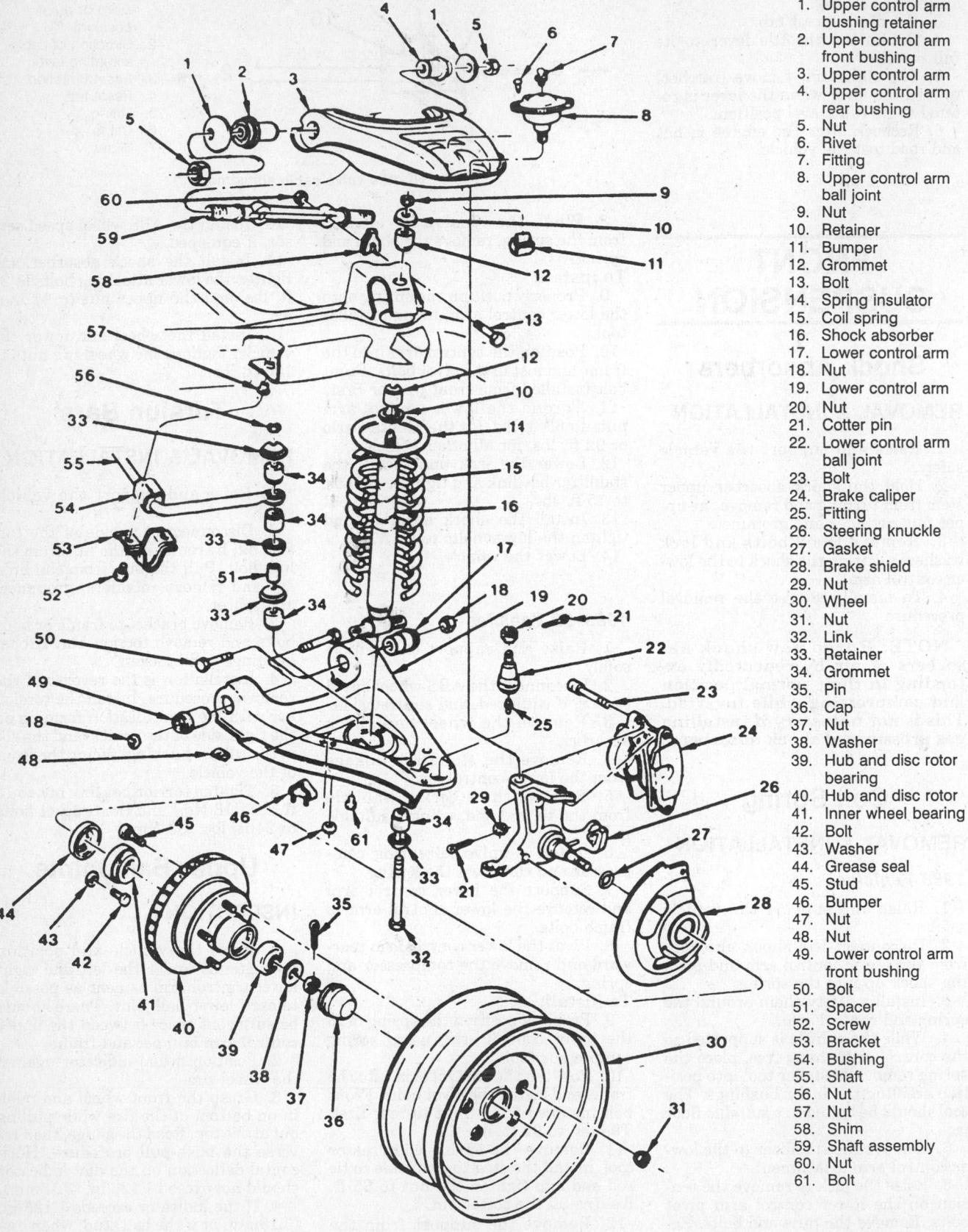

1. Upper control arm bushing retainer
2. Upper control arm front bushing
3. Upper control arm
4. Upper control arm rear bushing
5. Nut
6. Rivet
7. Fitting
8. Upper control arm ball joint
9. Nut
10. Retainer
11. Bumper
12. Grommet
13. Bolt
14. Spring insulator
15. Coil spring
16. Shock absorber
17. Lower control arm
18. Nut
19. Lower control arm
20. Nut
21. Cotter pin
22. Lower control arm ball joint
23. Bolt
24. Brake caliper
25. Fitting
26. Steering knuckle
27. Gasket
28. Brake shield
29. Nut
30. Wheel
31. Nut
32. Link
33. Retainer
34. Grommet
35. Pin
36. Cap
37. Nut
38. Washer
39. Hub and disc rotor bearing
40. Hub and disc rotor
41. Inner wheel bearing
42. Bolt
43. Washer
44. Grease seal
45. Stud
46. Bumper
47. Nut
48. Nut
49. Lower control arm front bushing
50. Bolt
51. Spacer
52. Screw
53. Bracket
54. Bushing
55. Shaft
56. Nut
57. Nut
58. Shim
59. Shaft assembly
60. Nut
61. Bolt

Front suspension components

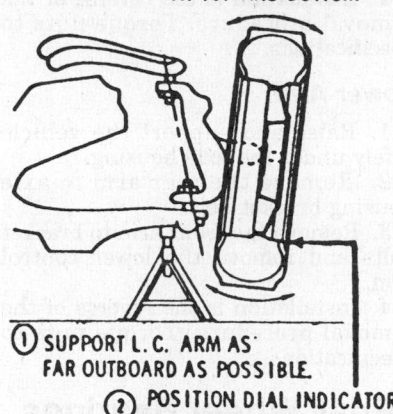

⑶ ROCK WHEEL IN AND OUT AT TOP AND BOTTOM

① SUPPORT L.C. ARM AS FAR OUTBOARD AS POSSIBLE.

② POSITION DIAL INDICATOR TO CHECK MOVEMENT AT THIS POINT

Upper and lower ball joint inspection

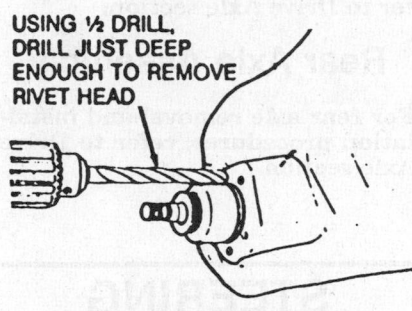

USING ½ DRILL, DRILL JUST DEEP ENOUGH TO REMOVE RIVET HEAD

Removing the upper ball joint

REMOVAL & INSTALLATION

1. Raise and safely support the vehicle; place floor stands under the lower control arm between the spring seats and the ball joints.

NOTE: Leave the jack under the spring seat during removal and installation, in order to keep the spring and control arm positioned.

2. Remove the wheel.
3. Remove the cotter pin from the upper ball joint stud and loosen the upper ball joint nut.
4. Using a ball joint splitter tool, break the stud loose and remove the nut and pull the stud out of the knuckle. Support the steering knuckle to prevent damage to the brake line.
5. Using a ⅛ in. diameter drill bit, drill into each of the 4 rivet heads to a depth of ½ in.
6. Drill off the rivet heads with a ½ in. diameter bit.
7. Punch out the rivets and remove the ball joint.
To install:
8. Place the new ball joint in the upper control arm and secure it with 4 bolts and nuts in place of rivets. Torque the nuts to specifications.

9. Connect the ball joint to steering knuckle. Torque the nut to specifications.

NOTE: When replacing the ball joints, use only high-quality replacement parts; bolts and nuts specified to be strong enough to endure the stress. Always turn the ball stud nut to align the cotter pin hole.

10. Install the grease fitting and lubricate until grease appears at the seal.
11. Install the wheel and road test the vehicle.

Lower Ball Joints

INSPECTION

These lower ball joints contain a visual wear indicator. The lower ball joint grease plug screws into the wear indicator which protrudes from the bottom of the ball joint housing. As long as the wear indicator extends out of the ball joint housing, the ball joint is not worn. If the tip of the wear indicator is parallel with or recessed into the ball joint housing, the ball joint is defective.

REMOVAL & INSTALLATION

1. Raise the vehicle and support the frame safely.
2. Remove the tire and wheel.
3. Place a floor jack under the control arm spring seat.

NOTE: Leave the jack under the spring seat during removal and installation, in order to keep the spring and control arm positioned.

4. Remove the cotter pin from the ball joint stud. Using a ball joint split-

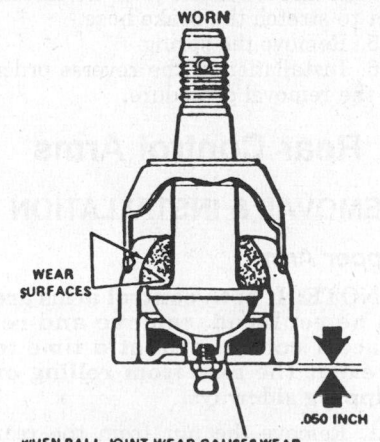

WORN

WEAR SURFACES

.050 INCH

WHEN BALL JOINT WEAR CAUSES WEAR INDICATOR SHOULDER TO RECEDE WITHIN THE SOCKET HOUSING REPLACEMENT IS REQUIRED

Lower ball joint wear indicator

ter tool, separate the ball joint from the steering knuckle.
5. When the stud comes loose, remove the stud nut.
6. Guide the lower control arm through the opening in the splash shield using a suitable tool.
7. Block the steering knuckle aside by using a block of wood between the frame and the upper control arm.
8. Pry the retainer off the ball joint seal with a drift pin and remove the seal.
9. Using a ball joint remover, remove the lower ball joint from the control arm.
To install:
10. Press in a new ball joint until it bottoms on the lower control arm.

NOTE: On disc brake vehicles, make sure the grease purge on the seal faces away from the brakes.

11. Assemble the suspension and torque the nut to specifications. Install the cotter pin and bend it to the side, not over the top of the nut. The cotter pin on the Cutlass must be installed parallel to the center line of the vehicle.
12. Install the ball joint fitting and lube until grease appears at the seal.
13. Install the tire and wheel assembly.
14. Road test the vehicle.

Upper Control Arms

REMOVAL & INSTALLATION

1. Raise and support the vehicle safely.
2. Remove the tire and wheel assembly.
3. Place a floor jack under the lower control arm spring seat.

NOTE: Leave the floor jack under the spring seat during removal and installation, in order to keep the spring and control arm positioned.

4. Disconnect the wheel speed sensor and remove the ball joint from the steering knuckle. Support the hub assembly to prevent damage to the brake line.
5. Loosen the pivot shaft to frame nuts and remove alignment shims. Remove the bolts to allow clearance and remove the control arm assembly from the vehicle.
6. Installation is the reverse of the removal procedure. Install alignment shims in the same position from which they were removed and tighten all bolts to specifications. Check the wheel alignment.

Lower Control Arm

REMOVAL & INSTALLATION

1. Raise and support the vehicle safely. Remove the wheel.
2. Remove the front coil spring and the lower ball joint.
3. Remove the lower control arm attaching bolts and remove the assembly.
4. Installation is the reverse of removal. Torque the lower control arm frame bolts to to 65 ft. lbs for the Monte Carlo or 92 ft. lbs. for all other vehicles. Torque at curb height. Torque the lower ball joint steering knuckle attaching nut to 83 ft. lbs.

Front Wheel Bearings

ADJUSTMENT

1. Raise the vehicle so the wheel can spin freely. Remove the dust cap.
2. Tighten the adjusting nut to 12 ft. lbs. (16 Nm) while turning the wheel, this will seat the bearings and remove any burrs on the threads.
3. Back off the nut until it is just loose.
4. Finger-tighten the nut and install the cotter pin thru the retaining ring or castle nut.

NOTE: If the cotter pin cannot be installed, back off the nut until the slot aligns with the serrations on the nut. Do not back off the nut more than ¼ of a turn.

5. Once adjusted, the front wheel bearings should have 0.001–0.005 in. (0.03–0.13mm) endplay.

REMOVAL & INSTALLATION

1. Remove the caliper and hub assembly.
2. Remove the dust cap, spindle nut and outer roller bearing assembly.
3. Remove the rotor assembly.
4. Pry the inner bearing seal from the hub, then remove the inner roller bearing assembly.
5. If necessary, remove the inner bearing outer race using tool J–29117–A or a suitable brass punch.
6. To remove outer bearing outer race, insert tool into the hub, indexing end of drift with notches in hub and tap with a hammer.
7. Using clean solvent, clean all old grease from hub, spindle and bearings.
8. If outer races were removed, press the races into the hub using tool J–8092 with J–8850 or J–8457.
9. Pack the bearings with a high temperature wheel bearing grease and reassemble the hub. Install the hub on

the steering knuckle and adjust the wheel bearings.

REAR SUSPENSION

Shock Absorbers

REMOVAL & INSTALLATION

NOTE: Purge new shock absorbers of air by repeatedly extending in their normal position and compressing while inverted. If replacing with a gas pressurized shock absorber, purging of the shocks is not necessary.

1. Raise and safely support the vehicle. Also support the rear axle housing.
2. Remove the lower shock mounting bolt from the shock absorber eye.
3. Unfasten the upper mounting bracket bolts and remove the shock.
4. Installation is in the reverse order of removal, except the upper attaching bolts should remain loose until the lower eye is tightened.

Coil Springs

REMOVAL & INSTALLATION

1. Raise the rear of the vehicle on the axle housing and place jack stands under the frame.
2. Disconnect the brake line at the axle housing and at the differential housing.
3. Disconnect the upper control arms at the differential.
4. Remove the shock absorber lower mount and lower the jack. Be careful not to stretch the brake hose.
5. Remove the spring.
6. Installation is the reverse order of the removal procedure.

Rear Control Arms

REMOVAL & INSTALLATION

Upper Arm

NOTE: If both control arms are to be replaced, remove and replace 1 control arm at a time to prevent the axle from rolling or slipping sideways.

1. Remove the nut from the rear arm to rear axle housing bolt and rock the axle to remove the bolt.
2. Remove front and rear arm attaching nuts and bolts.

3. Remove the suspension arm and inspect the bushing for damage.
4. Installation is the reverse of the removal procedure. Torque nuts to specifications.

Lower Arm

1. Raise and support the vehicle safely under the axle housing.
2. Remove the rear arm to axle housing bracket bolt.
3. Remove the front arm to bracket bolts and remove the lower control arm.
4. Installation is the reverse of the removal procedure. Torque nuts to specifications.

Rear Wheel Bearings

For all rear wheel bearing removal and installation procedures, refer to Drive Axle section.

Rear Axle Assembly

For rear axle removal and installation procedures, refer to Drive Axle section.

STEERING

— CAUTION —
If equipped with the Supplemental Inflatable Restraint System (SIR), the system must be disabled, before working on any part of the system. Failure to do so may result in deployment of the air bag and possible personal injury.

Steering Wheel

REMOVAL & INSTALLATION

Except 1991–92 Roadmaster, Caprice and Custom Cruiser

NOTE: Do not pound on the steering wheel or the steering shaft. The collapsible column could be damaged enough to require replacement.

1. Disconnect the negative battery cable.
2. On the stock wheel, remove the screws attaching the horn pad assembly to the wheel. Disconnect the horn contact from the pad assembly.
3. On the deluxe wheel, remove the pad attaching screws, lift up the pad and disconnect the horn wire by pushing on the insulator and turning counterclockwise.
4. On the sport steering wheel, pull up on the emblem to remove it. Re-

move the contact assembly attaching screws and the contact assembly.

5. On all columns, remove the steering wheel nut retainer.

6. Remove the retaining nut and the steering wheel, using a puller.

7. Installation is the reverse of removal. Align the marks on the wheel hub and the steering shaft. Torque the attaching bolt to 30 ft. lbs.

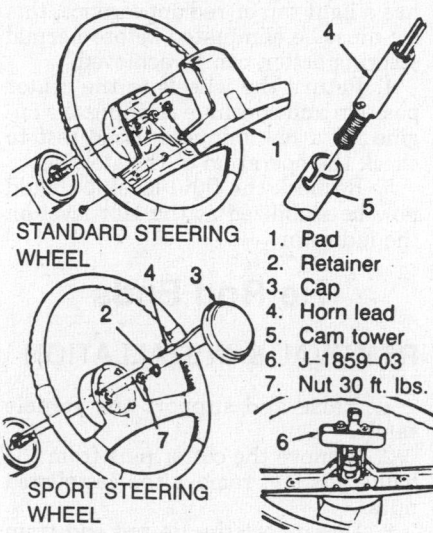

STANDARD STEERING WHEEL

1. Pad
2. Retainer
3. Cap
4. Horn lead
5. Cam tower
6. J-1859-03
7. Nut 30 ft. lbs.

SPORT STEERING WHEEL

REMOVE STEERING WHEEL

Steering wheel assembly—except 1991–92 Roadmaster, Caprice and Custom Cruiser

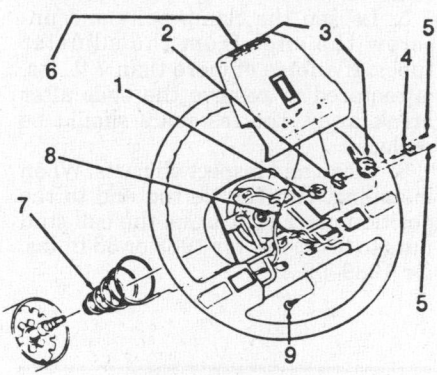

1. Steering wheel
2. Nut retainer
3. Telescoping adjuster lever
4. Steering shaft lock knob bolt
5. Steering shaft lock knob bolt positioning screw
6. Steering wheel pad
7. Horn contact spring
8. Horn lead
9. Horn pad attaching screw

Tilt and telescoping wheel assembly— except 1991–92 Roadmaster, Caprice and Custom Cruiser

1991–92 Roadmaster, Caprice and Custom Cruiser

1. Properly disable the SIR system.
2. Remove the Torx® screws from the back of the steering wheel, disconnect the connector and remove the the inflator module.

CAUTION

To avoid personal injury when carrying a live inflator module, make sure the bag and trim cover are pointed away. When placing a live inflator module on a bench or other surface, always face the bag and trim cover up and away from the surface. Never carry the inflator module by the wires or connector on the underside of the module, otherwise personal injury may result if bag is deployed.

3. Disconnect the negative battery cable.
4. Remove the hexagon locking nut.
5. Use a suitable wheel puller and remove the steering wheel and horn contact.

NOTE: When attaching the wheel puller to the wheel, use care to prevent threading the side screws into the coil assembly and damaging the coil assembly.

To install:

6. Route the coil assembly connector thru the steering wheel.
7. Install the horn contact.
8. Install the steering wheel by aligning the block tooth on the steering wheel with the block tooth on the steering shaft with 1 female serration and install the locking nut.
9. Connect the coil assembly, install the inflator module to the wheel and tighten the bolts to 25 inch lbs.
10. Properly enable the SIR system.

1. Terminal from inflatable restraint module
2. Terminal from coil assembly
3. Connector position assurance
4. Inflator module
5. Steering wheel
6. Screw

Steering Column

REMOVAL & INSTALLATION

Except 1991–92 Roadmaster, Caprice and Custom Cruiser

NOTE: Handle the steering column very carefully. Rapping on the end of it or leaning on it could shear off the inserts which allow the column to collapse in a crash.

On 1990–91 vehicles the wheels must be in the straight-ahead position and the key must be in the LOCK position when removing or installing the steering column

1. Disconnect negative battery cable.
2. Disconnect flexible coupling.
3. Remove cover and toe-pan attaching screws.
4. If necessary, remove instrument panel lower trim.
5. Disconnect shift linkages, wiring and vacuum hoses.
6. Remove upper and lower column mounts and pull the column up and out of the vehicle.
7. When installing, verify the flexible coupling aligned correctly.

NOTE: When installing, use only the specified hardware. Over length bolts could prevent the column from properly collapsing in a crash.

1991–92 Roadmaster, Caprice and Custom Cruiser

NOTE: The wheels of the vehicle must be in the straight-ahead position and the key must be in the LOCK position when remov-

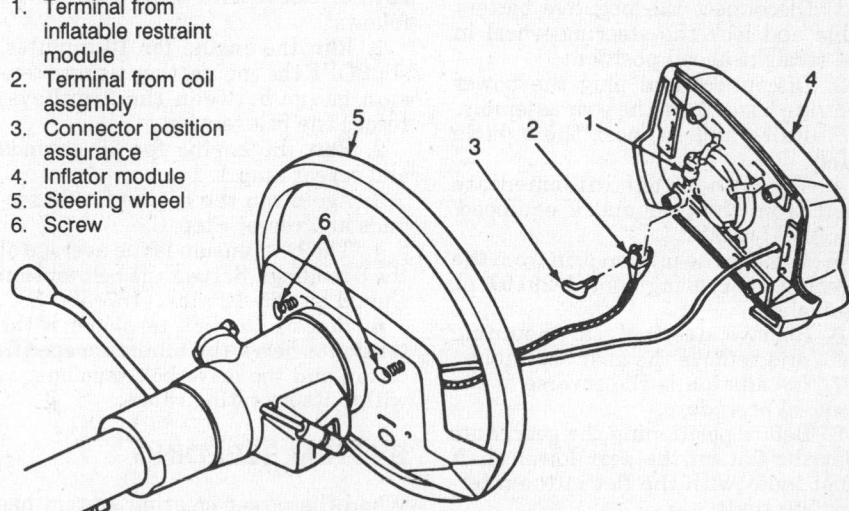

Steering wheel assembly with inflator module—1991–92 Roadmaster, Caprice and Custom Cruiser

ing or installing the steering column. Failure to do so will cause the coil assembly in the steering column to become off center and possibly damage the coil or deploy the SIR module.

1. Properly disable the SIR system.
2. Remove the stoplight switch.

NOTE: Failure to remove the stoplight switch may cause damage to the switch or the switch to be thrown out of adjustment.

3. Remove the bolt and nut from the joint coupler attaching the intermediate shaft to the steering column.
4. Disconnect the shift linkage from the steering column.
5. If necessary, remove the steering wheel.
6. Remove the steering column opening filler and the driver knee bolster and deflector.
7. Remove the bolts attaching the toe plate to the cowl and remove the shift indicator cable from the steering column.
8. Disconnect all of the steering column electrical connections.
9. Remove the capsule nuts attaching the column support bracket to the instrument panel and remove the steering column from the vehicle.
10. Installation is the reverse of removal. Tighten the capsule nuts attaching the column support bracket to the instrument panel to 20 ft. lbs. Tighten the bolt and nut at the joint coupling attaching the upper intermediate shaft to the steering column to 40 ft. lbs. Enable the SIR system.

Power Steering Gear

REMOVAL & INSTALLATION

1. Disconnect the negative battery cable and lock the steering wheel in the straight-ahead position.
2. Disconnect and plug the power steering hoses from the gear assembly.
3. Raise and support the vehicle safely.
4. Disconnect the intermediate shaft from the gear and if equipped, the ABS bracket.
5. Remove the pitman arm from the steering gear using tool J–29107 or equivalent.
6. Remove steering gear mounting bolts and remove the gear assembly.
7. Installation is the reverse of the removal procedure.
8. Before positioning the gear, note that the flat on the gear lower shaft must index with the flat in the intermediate shaft.
9. Make certain there is a minimum of 0.040 in. (1.02mm) clearance between intermediate shaft coupling and

steering gear upper seal. Tighten mounting bolts to 80 ft. lbs. for 1988 or 70 ft. lbs. for 1989–92 vehicles.

Power Steering Pump

REMOVAL & INSTALLATION

1. Disconnect the negative battery cable.
2. Loosen and remove power steering pump belt or serpentine belt.
3. Remove the power steering pump pulley, if required.
4. Remove bolts and nut from adjuster bracket.
4. Disconnect both lines at the pump and cap the lines.
5. Remove the pump assembly.
6. Installation is the reverse of the removal procedure. Connect the power steering pressure and return lines.
7. Fill with fluid, bleed the system and adjust belt to proper tension.

BELT ADJUSTMENT

V-BELT

EXCEPT 5.0L (VIN Y) ENGINE

Use a belt tension gauge and if the tension is below 50 lbs. (222 Nm), adjust to 70–80 ft. lbs. (312–356 Nm) on used belts or 120–130 ft. lbs. (534–578 Nm) on new belts.

5.0L (VIN Y) ENGINE

Use a belt tension gauge and if the tension is below 67.5 ft. lbs. (300 Nm), adjust to 90 ft. lbs. (400 Nm) on used belts or 135 ft. lbs. (600 Nm) on new belts.

Serpentine Belt

If belt slippage occurs and the drive belt tensioner is within its operating range, check the belt tension as follows:

1. Run the engine for 10 minutes, shut OFF the engine, then using a tension gauge between the 2 pulleys, record the belt tension.
2. Run the engine for 30 seconds and repeat Step 1.
3. Again run the engine for 30 seconds and repeat Step 1.
4. The belt tension is the average of the 3 readings. Serpentine belt tension should be 105–125 lbs. (467–556 N).
5. Replace the belt tensioner if the tension is below the minimum specification and the drive belt tensioner is within its operating range.

SYSTEM BLEEDING

When the power steering system has been serviced, air must be bled from the system by using the following procedure:

1. Turn wheels all the way to the left, add power steering fluid to the cold mark on the level indicator.
2. Start engine and run at fast idle momentarily, shut engine OFF and recheck the fluid level. If necessary add fluid to the cold mark.
3. Start engine and bleed system by turning wheels from side to side without hitting stops. Keep the fluid level at the cold mark. Fluid with air in it has a light tan or red appearance, this air must be eliminated before normal steering action can be achieved.
4. Return the wheels to the center position and continue running the engine for a few minutes. Road test to check the operation of the steering.
5. Recheck the fluid level it should now be stabilized at the Hot level on the indicator.

Tie Rod Ends

REMOVAL & INSTALLATION

1. Raise and support the vehicle safely.
2. Remove the cotter pins from the ball studs and remove the castellated nuts.
3. Disconnect the tie rod end from the steering arm or knuckle with a tie rod joint separator.
4. Remove the inner ball stud from the intermediate rod with a puller. Mark the tie rod end position before removal.
5. Loosen the clamp bolts and unscrew the ends from the adjuster tubes. If a force of more than 7 ft. lbs. is required to remove the ends after break away, the fasteners should be replaced.
6. Clean and inspect all parts. When installing, run the tie rod end to the position marked. Torque the ball stud nuts to 30 ft. lbs. for 1988 or 35 ft. lbs. for 1989–92.

BRAKES

For all brake system repair and service procedures not detailed below, please refer to "Brakes" in the Unit Repair section.

Master Cylinder

REMOVAL & INSTALLATION

NOTE: Be sure the area where the master cylinder is mounted is clean, before beginning removal.

1. Disconnect and cap or plug hy-

draulic lines. Disconnect the electrical lead, if equipped.

2. On non-power brakes, disconnect the pushrod at the brake pedal.

3. Remove the attaching bolts and remove the master cylinder.

4. Install in the reverse order of removal. Fill with approved brake fluid and bleed system.

Combination Valve

REMOVAL & INSTALLATION

1. Remove the electrical wire connector from the pressure differential switch.

2. Disconnect the hydraulic lines at the combination valve.

3. Plug the lines to prevent loss of fluid and entrance of dirt.

4. Remove the combination valve.

5. Installation is the reverse of the removal procedure. Bleed the entire brake system.

Power Brake Booster

REMOVAL & INSTALLATION

1. Remove the 2 nuts holding the master cylinder to the power unit. Carefully position the master cylinder aside, being careful not to kink any of the hydraulic lines. It is not necessary to disconnect the brake lines.

2. Disconnect the vacuum hose from the vacuum check valve on the front housing. Plug the hose. Plug the lines immediately.

3. Loosen the 4 nuts that hold the power unit mounted on the firewall.

4. Disconnect the pushrod from the brake pedal. Do not force the pushrod to the side when disconnecting.

5. Remove the 4 mounting nuts and lift the power unit off the studs.

6. Installation is in the reverse order of removal. Torque the master cylinder-to-power brake unit mounting studs to specifications.

Brake Caliper

REMOVAL & INSTALLATION

1. Remove ⅔ of the brake fluid from the master cylinder.

2. Raise the vehicle and support it safely.

3. Remove the tire and wheel assembly.

4. Position a C-clamp over the outboard shoe and lining and the caliper housing and bottom the piston into the caliper bore.

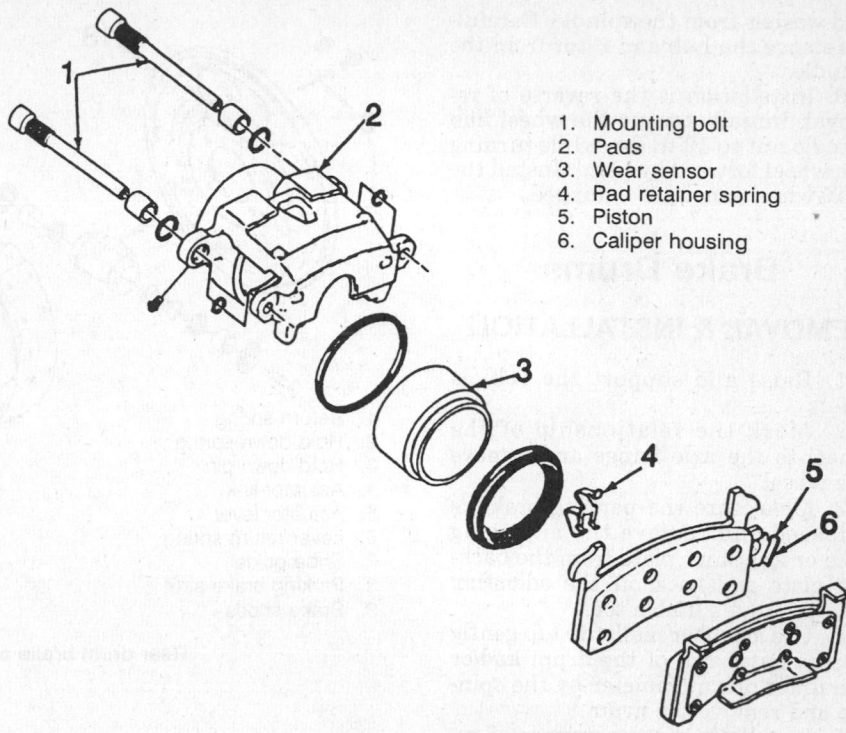

1. Mounting bolt
2. Pads
3. Wear sensor
4. Pad retainer spring
5. Piston
6. Caliper housing

Front brake caliper assembly

NOTE: If removing the caliper assembly only to access other brake parts go to Step 5. If removing the caliper from the vehicle, the brake system will have to be bled.

5. Remove the bolt, copper washers and inlet fitting from the caliper housing. Plug the line to prevent fluid loss and contamination.

6. Remove the mounting bolts and the sleeves and remove the caliper from the rotor.

7. Installation is the reverse of removal. Tighten the mounting bolts to 38 ft. lbs.

Disc Brake Pads

REMOVAL & INSTALLATION

1. Raise the vehicle and support it safely.

2. Remove the tire and wheel assembly.

3. Remove the caliper assembly and suspend it from the front suspension using a fabricated wire hanger.

4. Remove the inner and outer brake pads from the caliper.

5. Remove the bushings from the grooves in the caliper housing.

6. Remove the shoe retainer spring from the inboard pad.

To install

7. Clean and then lubricate bushings and sleeves with silicone grease.

8. Install the bushings in the grooves in the caliper housing.

9. Install the shoe retainer spring on the inboard pad.

10. Install the inboard pad. Be sure to seat the shoe retainer spring in the piston.

NOTE: The wear sensor should be at the leading edge of the inboard pad during forward wheel rotation.

11. Install the outboard pad.

12. Install the caliper assembly.

13. Apply the brake pedal 3 times to seat the pads.

14. With the aid of an assistant holding pressure on the brake pedal, use a pair of channel lock pliers to clinch down the outboard shoe ears.

15. Install the front wheels on the vehicle.

Brake Rotor

REMOVAL & INSTALLATION

1. Disconnect the negative battery cable.

2. Raise and support the vehicle safely.

3. Disconnect the ABS wheel speed sensor from the steering knuckle and secure, if equipped.

4. Remove the brake caliper and position aside.

5. Remove the dust cap from the hub then remove the cotter pin, nut

and washer from the spindle. Carefully remove the hub and rotor from the spindle.

6. Installation is the reverse of removal. Initially torque the wheel hub spindle nut to 12 ft. lbs. while turning the wheel forward by hand. Install the ABS wheel sensor, if equipped.

Brake Drums

REMOVAL & INSTALLATION

1. Raise and support the vehicle safely.
2. Mark the relationship of the wheel to the axle flange and remove the wheel.
3. Make sure the parking brake is released and remove the adjusting hole or knockout plate from the backing plate and back off the adjusting screw with a suitable tool.
4. Use a rubber mallet to tap gently on the outer rim of the drum and/or the inner drum diameter by the spindle and remove the drum.
5. Installation is the reverse of removal. Adjust the brakes as necessary.

Brake Shoes

REMOVAL & INSTALLATION

1. Raise the vehicle and support it safely.
2. Remove the tire and wheel assembly.
3. Remove the brake drum. If the brake drum cannot be removed, try the following:
  a. Make sure the parking brake is released.
  b. Back off the parking brake cable adjustment.
  c. Remove the adjusting hole knockout plate from the backing plate and back off the adjusting screw.
  d. Use a rubber mallet to tap on the outer rim of the drum and around the inner drum diameter by the spindle.
4. Remove the return springs.
5. Remove the hold-down springs and hold-down pins.
6. Remove the actuator lever pivot.
7. Lift up on the actuator lever to remove the actuator link. Remove the lever and the return spring.
8. Remove the shoe guide and the parking brake strut and spring.
9. Remove the brake shoes from the backing plate and the parking brake cable.
10. Remove the adjusting screw assembly and spring from the brake shoes.

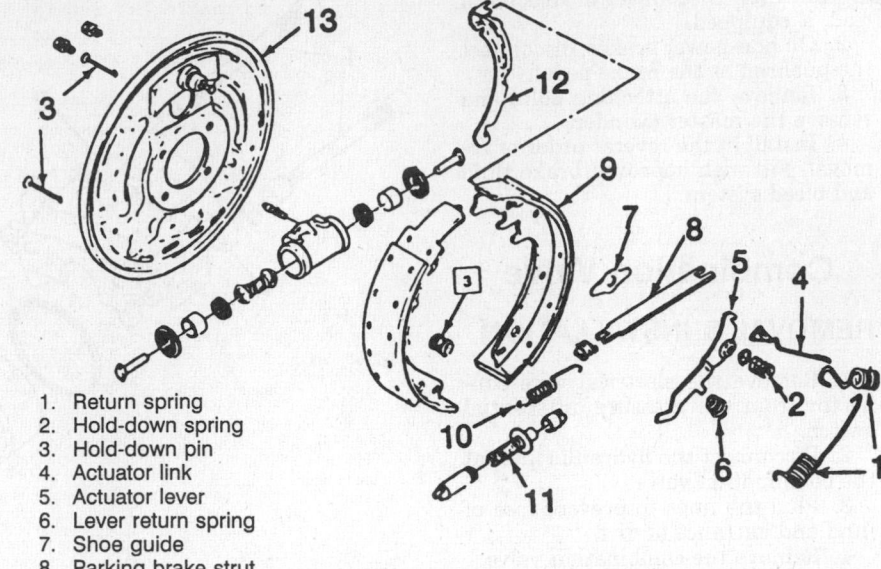

1. Return spring
2. Hold-down spring
3. Hold-down pin
4. Actuator link
5. Actuator lever
6. Lever return spring
7. Shoe guide
8. Parking brake strut
9. Brake shoes

Rear drum brake assembly—standard duty

11. Remove the parking brake lever by unhooking the lever tab from the slot in the brake shoe.
12. Clean the backing plate and lubricate the shoe pads.
13. Clean the adjusting screw and lubricate the screw threads for smooth rotation over the full length.
14. Install in the reverse order of removal.
15. Adjust the brake shoes. The outside diameter of both shoe and linings should be 0.050 in. (1.27mm) less than the inside diameter of the brake drum on each wheel.

Wheel Cylinder

REMOVAL & INSTALLATION

1. Raise the vehicle and support it safely.
2. Remove the wheel and brake drum.
3. Clean dirt and foreign material from around wheel cylinder assembly.
4. Disconnect the inlet tube nut and line from the cylinder.
5. Plug the opening in the line to prevent fluid loss or contamination.
6. Remove the cylinder to shoe links.
7. Remove the 2 attaching bolts and the cylinder assembly.
8. Installation is the reverse of the removal procedure. Bleed the brake system.

Parking Brake Cable

ADJUSTMENT

NOTE: Before attempting to adjust the parking brake, verify the rear brakes are correctly adjusted. If rear brakes are adjusted properly, the parking brake original adjustment should not have to be changed.

1. Apply the parking brake exactly 2–6 clicks, then raise and support the vehicle safely.
2. Loosen the locknut at the rear of the equalizer adjusting nut. Tighten the adjusting nut until the rear wheels can barely be turned backward, using 2 hands, but lock up when moved forward. Rear disc brakes will not lock up but will have a drag. Tighten the nut against the adjusting nut.
3. With the parking brake disengaged the rear wheel should turn freely in either direction with no brake drag.

REMOVAL & INSTALLATION

Front

1. Raise the vehicle and support it safely.
2. Loosen adjuster nut and disconnect the cable from the connector. Loosen the retainer at the frame.
3. Lower the vehicle.
4. Remove the lower rear bolt from the wheel house panel and pull the panel out to gain access to the front cable.

5. Disconnect the cable from the parking brake pedal assembly and remove the cable.

6. Installation is the reverse of the removal procedure. Adjust the parking brake.

Rear

1. Raise the vehicle and support it safely.

2. Remove the adjuster nut, compress the retainer fingers and loosen the cable from all retainers. On the left side remove the cable from the equalizer.

3. Mark the relationship of the wheel to the axle flange and remove the tire and wheel assembly.

4. Remove the brake drum, the primary shoe return spring and the parking brake strut. On the right side also remove the secondary shoe hold-down spring.

5. Compress the retainer fingers and loosen the cable from the backing plate. Disconnect the cable from the parking brake lever and remove the cable.

6. Installation is the reverse of the removal procedure. Adjust the parking brake.

Brake System Bleeding

The brake system must be bled when any brake line is disconnected or there is air in the system.

NOTE: Never bleed a wheel cylinder when a drum is removed.

1. Clean the master cylinder of excess dirt and remove the cylinder cover and the diaphragm.

2. Fill the master cylinder to the proper level. Check the fluid level periodically during the bleeding process and replenish it, as necessary. Do not allow the master cylinder fall below ½ full.

3. If the master cylinder is suspected or known to have air in the bore, bleed it before any wheel cylinder or caliper as follows:

   a. Disconnect the forward brake line connection at the master cylinder.

   b. Allow brake fluid to fill the master cylinder bore until it begins to flow from the forward line connector port.

   c. Connect the forward brake line to the master cylinder and tighten.

   d. Have an assistant depress the brake pedal slowly, 1 at a time and hold while loosening the forward brake line connection at the master cylinder to purge the air from the bore. Tighten the connection and

then have an assistant release the brake pedal slowly. Wait 15 seconds and repeat the sequence. Repeat the sequence, including the 15 second wait until all air is removed from the bore.

   e. After all air is removed at the forward connection, repeat the above procedure for the rear connection at the master cylinder.

4. Bleed the individual wheel cylinders or calipers only after all air is removed from the master cylinder.

   a. Attach the proper size box end wrench over the bleeder valve.

   b. Attach a length of vinyl hose to the bleeder screw of the brake to be bled. Insert the other end of the hose into a clear jar half full of clean brake fluid, so the end of the hose is beneath the level of fluid. The correct sequence for bleeding is to work from the brake farthest from the master cylinder to the 1 closest; right rear, left rear, right front, left front.

5. Have an assistant depress and release the brake pedal 1 time and hold. Loosen the bleeder valve to purge the air from the cylinder. Tighten the bleeder screw and slowly release the pedal and wait 15 seconds. Repeat the sequence including the 15 second wait until all air is removed.

NOTE: Make sure an assistant presses the brake pedal to the floor slowly. Rapid pumping of the brake pedal pushes the master cylinder secondary piston down the bore in a way that makes it difficult to bleed the rear side of the system.

6. Repeat this procedure at each of the wheels. Remember to check the master cylinder level occasionally. Use only fresh fluid to refill the master cylinder, not the fluid bled from the system.

7. When the bleeding process is complete, refill the master cylinder, install its cover and diaphragm and discard the fluid bled from the brake system.

Anti-Lock Brake System Service

PRECAUTION

Failure to observe the following precautions may result in system damage.

● Before performing electric arc welding on the vehicle, disconnect the Electronic Brake Control Module (EBCM) and the hydraulic modulator connectors.

● When performing painting work on the vehicle, do not expose the Elec-

tronic Brake Control Module (EBCM) to temperatures in excess of 185°F (85°C) for longer than 2 hours. The system may be exposed to temperatures up to 200°F (95°C) for less than 15 minutes.

● Never disconnect or connect the Electronic Brake Control Module (EBCM) or hydraulic modulator connectors with the ignition switch ON.

● Never disassemble any component of the Anti-Lock Brake System (ABS) which is designated non-serviceable; the component must be replaced as an assembly.

● When filling the master cylinder, always use Delco Supreme 11 brake fluid or equivalent, which meets DOT-3 specifications; petroleum base fluid will destroy the rubber parts.

Modulator Valve

REMOVAL & INSTALLATION

——— **CAUTION** ———

The modulator is not repairable and no screws on the modulator may be loosened. If the screws are loosened, it will not be possible to to get the brake circuits leak-tight and personal injury injury may result.

1. Disconnect the negative battery cable.

2. Remove the air intake duct and resonator and move the upper coolant hose aside.

3. Disconnect the canister purge line at the canister and move aside.

4. Remove the retaining screw and remove the modulator valve cover.

5. Unlock the tab and disconnect the modulator valve electrical connector.

6. Remove the nut and disconnect the ground wire from the modulator.

7. Note the hydraulic brake pipe locations then disconnect and plug the lines from the modulator.

8. Remove the 3 nuts retaining the modulator to the bracket.

9. Remove the insulators from the modulator valve.

To install:

10. Install the insulators to the modulator valve.

11. Install the modulator valve to the bracket and tighten the 3 nuts to 89 inch lbs.

12. Connect the hydraulic brake pipes to their original locations in the modulator and tighten to 11 ft. lbs.

——— **CAUTION** ———

If brake pipes are switched (inlet vs. outlet) wheel lockup will occur and personal injury may result.

13. Install the ground wire and nut to the modulator.

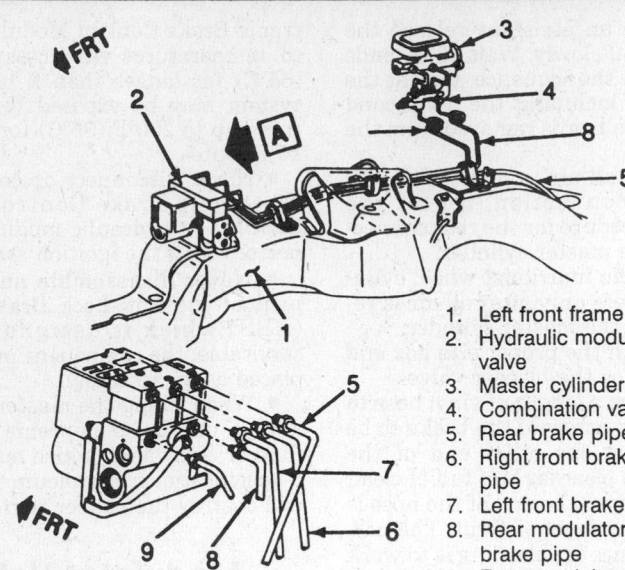

1. Left front frame rail
2. Hydraulic modulator valve
3. Master cylinder
4. Combination valve
5. Rear brake pipe
6. Right front brake pipe
7. Left front brake pipe
8. Rear modulator valve brake pipe
9. Front modulator valve brake pipe

ABS brake line routing

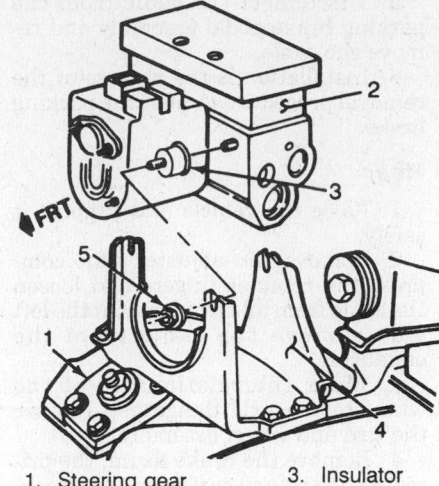

1. Steering gear
2. Hydraulic modulator valve
3. Insulator
4. Bracket
5. Nut

ABS Hydraulic modulator valve removal

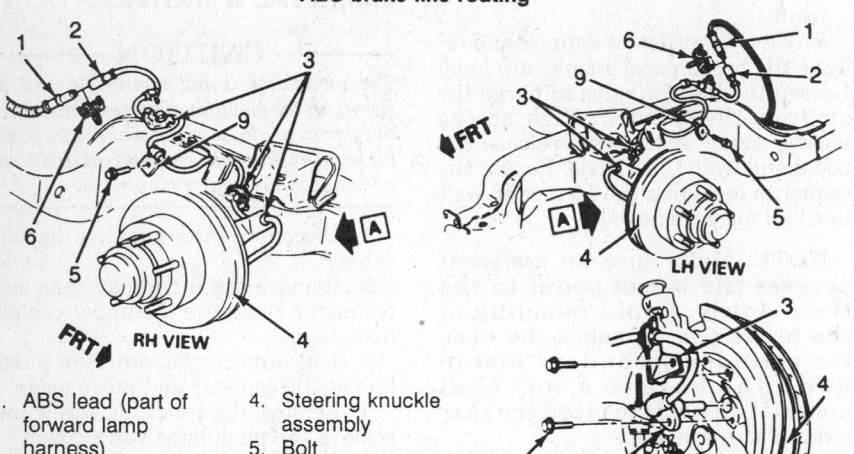

1. ABS lead (part of forward lamp harness)
2. Sensor assembly connector
3. Bracket
4. Steering knuckle assembly
5. Bolt
6. Clip
7. Wheel speed sensor
8. Bolt
9. Bracket

ABS front wheel speed sensor

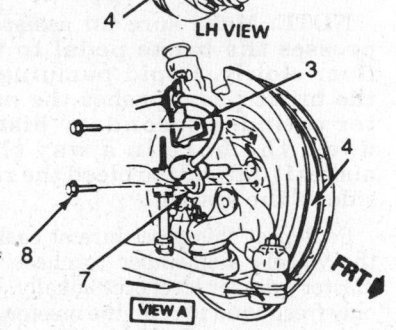

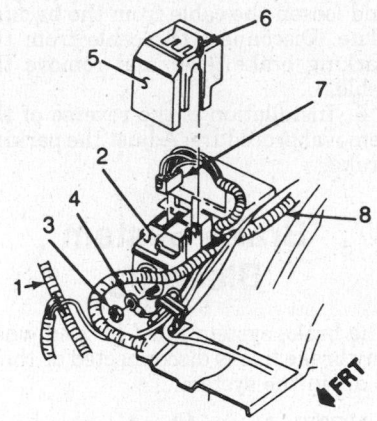

1. Forward lamp harness
2. Hydraulic modulator valve
3. Nut
4. Ground wire
5. Modulator valve cover
6. Screw
7. Modulator valve electrical connector
8. ABS wiring harness

ABS Hydraulic modulator valve electrical connections

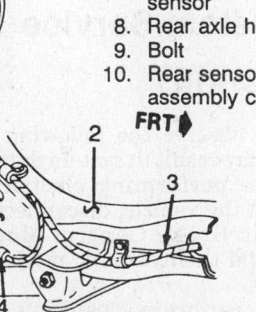

1. Frame cross bar
2. Left frame rail
3. ABS wiring harness
4. Differential sensor connector
5. Bracket
6. Clip
7. Rear axle speed sensor
8. Rear axle housing
9. Bolt
10. Rear sensor assembly connector

Rear wheel speed sensor mounting

14. Install the modulator valve electrical connector.
15. Install the the modulator valve cover with the retaining screw.
16. Connect the canister purge line to the canister.
17. Install the air intake duct and resonator and move the upper coolant hose into position.
18. Connect the negative battery cable.
19. Use only DOT 3 hydraulic brake fluid, fill and bleed the brake system.
20. Road test the vehicle.

Electronic Brake Control Module (EBCM)

The EBCM is located near the left rear quarter trim panel or the rear compartment.

Front Wheel Speed Sensor

REMOVAL & INSTALLATION

1. Disconnect the negative battery cable.
2. For the right side speed sensor, unclip the connectors from the clip and separate.
3. For the left side speed sensor, raise and support the vehicle safely, then unclip the connectors from the clip and separate.
4. Remove the sensor wiring harness mounting bolt from the frame rail.
5. Remove the sensor retaining bolt and remove the sensor from the knuckle assembly.
6. Installation is the reverse of the removal procedure. Sensor retaining bolt torque is 71 inch lbs. Be sure to coat the sensor body with anti corrosion compound GM part 1052856 or equivalent, at the knuckle contact point.

NOTE: Proper installation of the wheel speed sensor cables is critical to proper operation of the ABS system. Make sure the cables are installed in the retainers. Failure to do this may result in contact with moving parts and the over extension of the cables, resulting in an open circuit.

Rear Axle Speed Sensor

REMOVAL & INSTALLATION

1. Disconnect the negative battery cable. Raise and support the vehicle safely.
2. Disconnect the rear sensor assembly from the differential sensor connector.
3. Remove the sensor wiring harness from the retainer brackets.
4. Remove the sensor retaining bolt and remove the speed sensor from the rear axle housing.
5. Installation is the reverse of the removal procedure. Sensor retaining bolt torque is 71 inch lbs.

NOTE: Proper installation of the wheel speed sensor cables is critical to proper operation of the ABS system. Make sure the cables are installed in the retainers. Failure to do this may result in contact with moving parts and the over extension of the cables, resulting in an open circuit.

CHASSIS ELECTRICAL

Air Bag

DISARMING

1. Align the steering wheel so the vehicle wheels are pointing in the straight-ahead position.
2. Turn the ignition switch to the **LOCK** position.
3. Remove the SIR fuse from the fuse block.
4. Disconnect the yellow 2-way SIR harness wire connector at the base of the steering column.
To enable system:
5. Turn the ignition switch to the **LOCK** position.
6. Reconnect the yellow 2-way connector at the base of the steering column.
7. Reinstall the SIR fuse.
8. Turn the ignition switch to the **RUN** position.
9. Verify the SIR indicator light flashes 7–9 times, if not as specified, inspect system for malfunction.

Air Bag Coil Assembly

NOTE: The coil assembly must remain centered in order to avoid accidental deployment of the air bag after any repair procedures to the internals of the steering column. There are 2 different styles of coil assemblies, 1 rotates clockwise and the other rotates counterclockwise.

ADJUSTMENT

1. With the system properly disarmed, hold the coil assembly with the clear bottom up to see the coil ribbon.
2. While holding the coil assembly, depress the lock spring and rotate the hub in the direction of the arrow until it stops. The coil should now be wound up snug against the center hub.
3. Rotate the coil assembly in the opposite direction approximately 2½ turns and release the lock spring be-

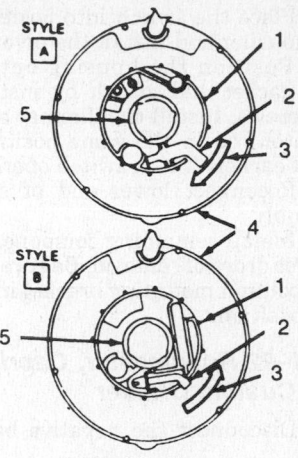

1. Locking tab
2. Spring
3. Hub direction
4. Coil housing
5. Coil hub

Centering the coil assembly

tween the locking tabs in front of the arrow.
4. Install the coil assembly onto the steering shaft.

Heater Blower Motor

REMOVAL & INSTALLATION

Without Air Conditioning

1. Disconnect the negative battery cable.
2. Disconnect the blower motor wiring harness.
3. Remove the blower motor retaining screws and pull the blower motor and fan straight forward out of the heater module.
4. Installation is the reverse of removal. Clean and replace sealer as necessary.

With Air Conditioning

EXCEPT 1991–92 ROADMASTER, CAPRICE AND CUSTOM CRUISER

1. Disconnect the negative battery cable.
2. Disconnect the blower motor wiring harness.
3. Remove the blower motor cooling tube.
4. Remove the blower motor retaining screws and lift the blower motor and fan straight up and out of the upper case of the air conditioning module.
5. Installation is the reverse of removal. Clean and replace sealer as necessary.

1991–92 ROADMASTER, CAPRICE AND CUSTOM CRUISER

1. Disconnect the negative battery cable.

6. Place the switch into position in the housing and install the pivot pin.

7. Position the housing onto the mast jacket and attach by installing the screws. Install the dimmer switch actuator rod in the same position as noted earlier. Check switch operation.

8. Reconnect lower end of switch assembly.

9. Install remaining components in reverse order of removal. Be sure to attach column mounting bracket in original position.

1991–92 Roadmaster, Caprice and Custom Cruiser

1. Disconnect the negative battery cable.

NOTE: This vehicle is equipped with an air bag system, make certain to follow the recommended disarming procedure before proceeding.

2. Remove the steering column cover behind the windshield wiper switch.

3. Disconnect the switch wire connector.

4. Verify all switches are in the **OFF** position.

5. Remove the switch lever by pulling it straight out of the switch.

6. Installation is the reverse of the removal procedure.

Instrument Cluster

REMOVAL & INSTALLATION

1988–90 Estate Wagon

1. Disconnect the negative battery cable. Remove the defroster grille.

2. Remove the 10 screws retaining the instrument panel top cover to the instrument panel.

3. If equipped with a twilight sentinel, pop up the photocell retainer, turn the photocell counterclockwise in the retainer and pull it down and out.

4. Slide the instrument panel top cover out far enough to disconnect the aspirator hose, electrical connector to the in-vehicle sensor and the electrical connector to the electro-luminescent inverter.

5. Remove the instrument panel top cover from the instrument panel. If equipped with quartz electronic speedometer clusters, remove the steering column trim cover, so the shift indicator can be removed.

6. Remove the 5 screws from the instrument cluster to the instrument panel carrier.

7. Disconnect the speedometer cable and pull the cluster housing assembly straight out, this will also separate the electrical connectors to the cluster.

NOTE: If equipped with tilt steering wheel, it may be helpful to tilt the wheel all the way down and pull the gear select lever to low, when removing the cluster.

8. Installation is the reverse of the removal procedure.

1988–90 Caprice and 1988–89 Safari

1. Disconnect the negative battery cable.

2. Remove the steering column lower cover screws and the cover.

3. If equipped with automatic transmission, disconnect the shift indicator cable from the steering column.

4. Remove the steering column to instrument panel screws. Lower the steering column.

NOTE: Use extreme care when lowering the steering column in order to prevent damage to column assembly.

5. Remove the screws and the snap in fasteners from the perimeter of the instrument cluster lens.

6. Remove the screws from the lower corner of the cluster.

7. Remove the stud nuts from the lower corner of the cluster.

8. Disconnect the speedometer cable and pull cluster from the instrument panel.

9. Disconnect the electrical connectors from the cluster and remove the assembly from the vehicle.

10. Installation is the reverse of the removal procedure.

Monte Carlo

1. Disconnect the negative battery cable. Remove the clock set stem and radio knobs.

2. Remove the instrument bezel retaining screws.

3. Lightly pull the bezel rearward. Disconnect the rear defogger switch. Remove the remote control mirror control knob, if equipped.

4. Remove the dash panel bezel. Remove the speedometer assembly retaining screws. Pull the assembly from the cluster, disconnect the speedometer cable from the assembly and remove the speedometer from the vehicle.

5. Remove the fuel gauge or the tachometer retaining screws, disconnect the electrical connectors and remove the components.

6. Remove the clock or voltmeter retaining screws, disconnect the electrical connectors and remove the components. Mark the electrical connectors.

7. Disconnect the transmission shift indicator cable from the steering column.

8. Disconnect all wiring connectors and remove the cluster case. Ensure that all electrical connectors are indentified to ensure proper reinstallation.

9. Installation is the reverse of the removal procedure.

1988–90 Cutlass and Custom Cruiser

1. Disconnect the negative battery cable. Disconnect speedometer cable at transducer, if equipped with cruise control.

2. Remove the right and left side trim covers by pulling outward. The covers are retained by clips.

3. Remove screws attaching cluster pad to panel adapter.

4. Pull pad assembly away from panel adapter.

5. Remove pad assembly.

6. Remove steering column trim cover.

7. Disconnect shift indicator clip from steering column shift bowl.

8. Remove screws holding instrument cluster to panel adapter.

9. Pull instrument cluster assembly rearward far enough to reach behind cluster and disconnect speedometer cable.

10. Disconnect speed sensor, if equipped.

11. Remove instrument cluster.

12. Installation is the reverse of removal.

1991–92 Roadmaster, Caprice and Custom Cruiser

1. Disconnect the negative battery cable.

NOTE: This vehicle is equipped with an air bag system, make certain to follow the recommended disarming and coil centering procedure before and after repairs.

2. Remove the left side trim plate:

   a. Remove the steering column opening filler.

   b. Open the instrument panel compartment door and unsnap the right side molding from the carrier.

   c. Loosen the capsule nuts attaching the steering column support bracket to the carrier, to the end of the threads but do not remove from the bolts. Gently lower the steering column.

   d. Remove the 6 screws attaching the trim plate to the carrier and carefully unsnap and pull away.

3. Remove the 5 screws attaching the cluster to the carrier.

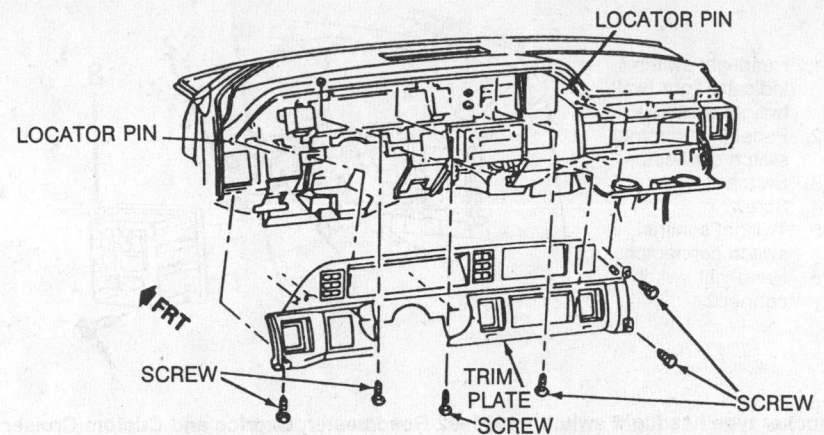

Left side trim plate assembly—1991–92 Roadmaster, Caprice and Custom Cruiser

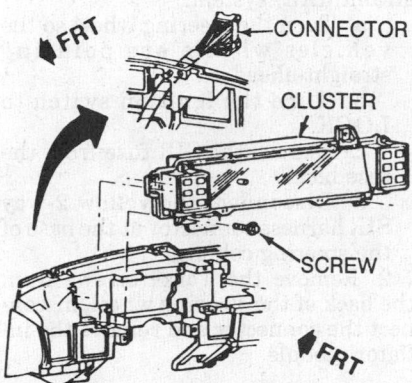

Instrument cluster assembly—1991–92 Roadmaster, Caprice and Custom Cruiser

4. Disconnect the shift indicator cable from the steering column.

5. Remove the cluster from the vehicle and disconnect the electrical connector.

6. Installation is the reverse of removal. Adjust the shift indicator as follows:

a. Remove the steering column opening filler.

b. The shift lever should be in the **N** gate notch.

c. Position the guide clip on the edge of the gearshift lever bowl to centrally position the pointer on **N**. Push the guide clip onto the gearshift lever bowl.

Speedometer

REMOVAL & INSTALLATION

NOTE: The 1991–92 Roadmaster, Caprice and Custom Cruiser speedometer assemblies must either be replaced as a unit or sent to an authorized repair facility. No individual parts are available for replacement.

Except 1991–92 Roadmaster, Caprice and Custom Cruiser

1. Disconnect the negative battery cable.

2. Remove the instrument cluster.

3. Remove the speedometer retaining screws. Pull the assembly forward in order to disconnect the speedometer cable. To gain slack, it may be necessary to disconnect the cable at the cruise control transducer or the transmission.

4. Remove the speedometer assembly from the vehicle.

5. Installation is the same as the removal procedure.

Radio

REMOVAL & INSTALLATION

Except Caprice

1. Disconnect the negative battery cable.

2. Remove the knobs from the radio and pull out the cigarette lighter.

3. Remove the 2 trim cover attaching screws and remove the cover.

4. Remove the radio bracket attaching screw from the lower tie bar.

5. Remove the 4 mounting plate screws and pull the radio out to obtain access to the electrical connections. Detach the wiring harness and the antenna lead.

6. Remove the mounting plate nuts and remove the radio. Installation is in the reverse order of removal.

1988–90 Caprice

1. Disconnect the negative battery cable.

2. Remove the ash tray.

3. Remove the instrument panel compartment.

4. Disconnect the air conditioning and heater cables.

5. Remove the air conditioning and radio trim plate.

6. Remove the radio and air conditioning heater control assembly.

7. Remove the screws holding the bracket to the trim plate and remove the bracket.

8. Remove the screws holding the radio to the trim plate and remove the radio.

9. Installation is the reverse of removal.

1991–92 Roadmaster, Caprice and Custom Cruiser

1. Disconnect the negative battery cable.

NOTE: This vehicle is equipped with an air bag system, make certain to follow the recommended disarming and coil centering procedure before and after repairs.

2. Remove the left side trim plate:

a. Remove the steering column opening filler.

b. Open the instrument panel compartment door and unsnap the right side molding from the carrier.

c. Loosen the capsule nuts attaching the steering column support bracket to the carrier, to the end of the threads but do not remove from the bolts. Gently lower the steering column.

d. Remove the 6 screws attaching the trim plate to the carrier and carefully unsnap and pull away.

3. Remove the 3 screws attaching the bracket to the carrier and remove the bracket and the attached radio from the carrier.

4. Disconnect the body harness connector and the antenna lead from the radio.

5. Remove the 3 nuts attaching the bracket to the radio and remove the 3 bolts from the radio, if necessary.

6. Installation is the reverse of removal.

Headlight Switch

REMOVAL & INSTALLATION

Knob Type Switch

1. Disconnect the negative battery cable.

2. Pull the headlight switch to the **ON** position.

3. Depending upon the switch mechanism, pull the trim knob from the switch by either reaching under the dash and depressing the switch shaft release button while pulling the knob and shaft from the light switch or by using a suitable tool and pushing

the tang under the trim knob while pulling the knob from the shaft.

4. Remove the ferrule nut retaining the switch to the dash panel. Disconnect the electrical connector and remove the switch.

5. Installation is the reverse of the removal procedure.

Rocker Type Switch

EXCEPT 1991–92 ROADMASTER, CAPRICE AND CUSTOM CRUISER

1. Disconnect the negative battery cable.

2. Remove the left side sound insulator and trim plate or cover, as necessary.

3. Remove the left side switch trim cover.

4. Remove the screws. The center screw shares the top of the switch and the bottom of the interior light rheostat.

5. Pull the switch and rheostat straight out, disconnect the wiring and remove the switch.

6. Installation is the reverse of the removal procedure.

1991–92 ROADMASTER, CAPRICE AND CUSTOM CRUISER

1. Disconnect the negative battery cable.

NOTE: This vehicle is equipped with an air bag system, make certain to follow the recommended disarming and coil centering procedure before and after repairs.

2. Remove the left side trim plate:

a. Remove the steering column opening filler.

b. Open the instrument panel compartment door and unsnap the right side moulding from the carrier.

c. Loosen the capsule nuts attaching the steering column support bracket to the carrier, to the end of the threads but do not remove from the bolts. Gently lower the steering column.

d. Remove the 6 screws attaching the trim plate to the carrier and carefully unsnap and pull away.

3. Remove the 3 screws attaching the switch to the instrument carrier and remove the switch.

4. Installation is the reverse of removal.

Dimmer Switch

REMOVAL & INSTALLATION

1. Disconnect the negative battery cable.

1. Headlight switch indicator light (with twilight sentinel)
2. Panel light dimmer switch connector
3. Switch
4. Screw
5. Twilight sentinel switch connector
6. Headlight switch connector

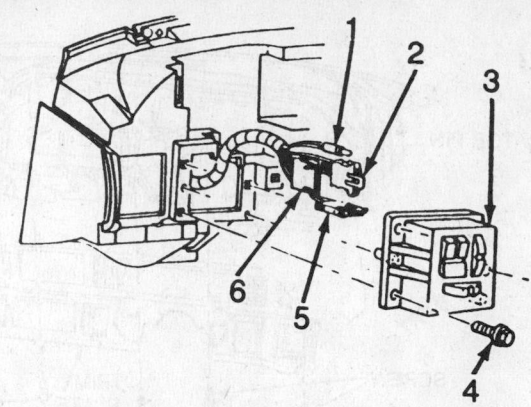

Rocker type headlight switch—1991-92 Roadmaster, Caprice and Custom Cruiser

NOTE: If vehicle is equipped with an air bag system, make certain to follow the recommended disarming and coil centering procedure before and after repairs.

2. The dimmer switch is attached to the lower steering column jacket. Disconnect all electrical connections from the switch.

3. Remove the nut and screw that attach the switch to the steering column jacket and remove switch.

4. Install the dimmer switch and depress it slightly to insert a $\frac{3}{32}$ in. drill. Force switch up to remove lash, then tighten screw and nut to 4.0 ft. lbs.

Turn Signal Switch

REMOVAL & INSTALLATION

Except 1991–92 Roadmaster, Caprice and Custom Cruiser

1. Disconnect the negative battery cable.

2. Disconnect turn signal switch wire connector at lower end of steering column.

3. Remove the steering wheel and shaft lock cover.

4. Remove retaining ring and shaft lock using tool J-23653 or equivalent.

5. Remove canceling cam and spring assembly.

6. Remove screws attaching the switch to the housing. Pull switch out. If equipped with cruise control, the wiring harness will have to be pulled up through the steering column.

7. Installation is the reverse of the removal procedure.

1991–92 Roadmaster, Caprice and Custom Cruiser

NOTE: This vehicle is equipped with an air bag system, make certain to follow the recommended disarming and coil centering procedure before and after repairs.

1. Disable the Supplemental Air Restraint (SIR) system.

a. Turn the steering wheel so the vehicles wheels are pointing straight-ahead.

b. Turn the ignition switch to **LOCK**.

c. Remove the SIR fuse from the fuse block.

d. Disconnect the yellow 2-way SIR harness connector at the base of the steering column.

2. Remove the Torx® screws from the back of the steering wheel, disconnect the connector and remove the inflator module.

— **CAUTION** —

To avoid personal injury when carrying a live inflator module, make sure the bag and trim cover are pointed away. Always face the air bag assembly up, also never carry the inflator module by the wires or connector, otherwise personal injury may result if the module should deploy.

3. Disconnect the negative battery cable.

4. Remove the steering wheel.

5. Remove the coil assembly retaining ring and allow it to hang.

— **CAUTION** —

Use a ½ in. wrench to hold the shaft of tool J–23653–C stationary when releasing the nut. Failure to do so may cause tool J–23653–C to fly off and cause personal injury.

6. Remove the wave washer.

7. Remove the shaft lock retaining ring using tool J–23653–C or equivalent.

8. Remove the shaft lock, turn signal canceling cam, upper bearing spring, inner race seat and inner race.

9. Remove the multifunction turn signal lever.

10. Remove the signal switch arm retaining screw.

11. Remove the hazard warning knob retaining screw.

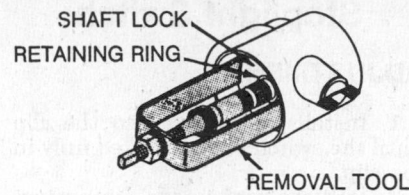

Removal of coil assembly from steering shaft—1991–92 Roadmaster, Caprice and Custom Cruiser

1. Shaft lock
2. Turn signal cancelling cam
3. Upper bearing spring
4. Upper bearing inner race seat
5. Inner race

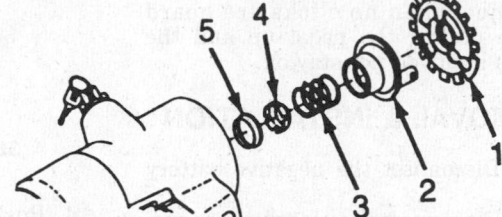

Removal of upper shaft bearing components—1991–92 Roadmaster, Caprice and Custom Cruiser

1. Multi-function lever
2. Screw
3. Hazard warning button
4. Spring
5. Hazard warning knob
6. Screw
7. Screw
8. Signal switch arm
9. Turn signal and hazard warning switch

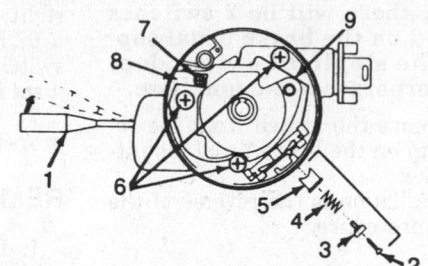

Turn signal and hazard switch removal—1991–92 Roadmaster, Caprice and Custom Cruiser

12. Remove the wiring protector from the steering column then disconnect the switch connector.

13. Remove the screws retaining the turn signal and hazard warning switch to the steering column, using care not to drop the screws in the column.

14. Installation is the reverse of removal. Enable the SIR system as follows:

   a. Reconnect the yellow 2-way SIR harness connector located at the base of the steering column.

   b. Install the SIR fuse to the fuse block.

   c. Turn the ignition switch to **RUN** and verify that the inflatable restraint indicator flashes 7–9 times and then turns off.

Ignition Switch

REMOVAL & INSTALLATION

NOTE: If vehicle is equipped with an air bag system, make certain to follow the recommended disarming and coil centering procedure before and after repairs.

1. Disconnect the negative battery cable and disable the air bag system, if equipped, as follows:

   a. Turn the steering wheel so the vehicles wheels are pointing straight-ahead.

   b. Turn the ignition switch to **LOCK**.

   c. Remove the SIR fuse from the fuse block.

   d. Disconnect the yellow 2-way SIR harness connector located at the base of the steering column.

2. Remove the column to instrument panel trim plates and attaching nuts.

3. Lower the steering column and disconnect the shift indicator cable.

4. Disconnect the ignition and dimmer switch wire connectors.

NOTE: The steering column must be supported at all times to prevent damage.

5. Remove the switch attaching screws and remove the ignition switch.

To install:

6. Move the key lock to the **LOCK** position.

7. Move the actuator rod hole in the switch to the **LOCK** position.

8. Install the switch with the rod in the hole. Adjust the ignition switch as follows:

   a. 1988—Move the switch slider to the extreme left **ACC** position, then move the slider 2 detents to the right to the **OFF-UNLOCK** position.

   b. 1989–90—Place a $^{3}/_{32}$ in. drill bit in the hole on the switch to lock the switch into position. Move the switch slider to the extreme left position then move the slider 1 detent to the right **OFF LOCK** position. Remove the drill bit.

   c. 1991–92—Install the ignition switch in the **LOCK** position. Move the switch slider to the extreme right position and move the slider 1 detent to the left **LOCK** position.

9. Position and reassemble the steering column in reverse of the disassembly procedure following the proper coil centering procedure. Enable the air bag system, if equipped, as follows:

   a. Connect the yellow 2-way SIR harness connector located at the base of the steering column.

   b. Install the SIR fuse to the fuse block.

   c. Turn the ignition switch to **RUN** and verify the inflatable restraint indicator flashes 7–9 times and then turns off.

Ignition Lock Cylinder

REMOVAL & INSTALLATION

Except 1991–92 Roadmaster, Caprice and Custom Cruiser

1. Disconnect the negative battery cable.

2. Position the ignition lock cylinder in the **RUN** position.

3. Remove the steering wheel. Remove the lock plate, turn signal switch and the buzzer switch.

4. Remove the lock cylinder retaining screw. Remove the lock cylinder.

5. To install, rotate the lock cylinder clockwise to align the cylinder key with the keyway in the lock housing.

6. Push the lock all the way in. Install the screw.

7. Continue the installation in the reverse order of the removal procedure.

1991–92 Roadmaster, Caprice and Custom Cruiser

NOTE: This vehicle is equipped with an air bag system, make certain to follow the recommended disarming and coil centering procedure before and after repairs.

1. Disconnect the negative battery cable.

2. Disable the SIR system as follows:

   a. Turn the steering wheel so the wheels point straight-ahead.

   b. Turn the ignition switch to the **LOCK** position.

   c. Remove the SIR fuse from the fuse block.

   d. Disconnect the yellow 2-way SIR harness wire connector located at the base of the steering column.

3. Remove the steering wheel cover.

4. Remove the steering wheel and knee bolster.

5. Remove the coil assembly retaining ring.

6. Remove the coil assembly, allowing the coil to hang freely.

7. Remove the wave washer.

8. Remove the steering shaft retaining ring using tool J23653-C or equivalent to depress the shaft lock.

9. Remove the shaft lock plate.

10. Remove the turn signal assembly.

11. Remove the key from the lock cylinder and the buzzer switch retaining clip.

12. Reinsert the key into the lock cylinder and place the key in the **LOCK** position.

13. Remove the lock cylinder retaining screw.

14. Remove the lock cylinder from the steering column.

To install:

15. Reinstall the lock cylinder with retaining screw into the steering column.

16. Insert the ignition key and turn it to the **RUN** position, and install the buzzer switch retaining clip.

17. Install the turn signal assembly and the shaft lock plate with retaining ring.

18. Install the wave washer.

NOTE: To avoid damaging the coil assembly, set the steering wheel so the block teeth on the shaft are at the 12 o'clock and 6 o'clock positions and the vehicles wheels are straight-ahead. Ensure the ignition switch is in the locked position.

19. Install the coil assembly and retaining ring, while following the proper centering procedure.

20. Install the steering wheel assembly.

21. Enable the SIR system by reconnecting the yellow 2-way connector located at the base of the steering wheel.

22. Reconnect the battery negative cable.

23. Verify the SIR system is functioning properly.

Stoplight Switch

ADJUSTMENT

1. Install the switch into the clip until the switch body is seated fully in the clip.

2. Pull the brake pedal up against internal pedal stop.

3. The switch will move in the retainer clip until the pedal assembly is against the stop.

4. Proper adjustment has been reached when no clicks are heard while pulling the pedal up and the brake lights do not stay on.

REMOVAL & INSTALLATION

1. Disconnect the negative battery cable.

2. Disconnect the electrical connection from the switch.

NOTE: If equipped with cruise control, there will be 2 switches mounted on the brake pedal support. The stoplight switch does not incorporate a vacuum hose.

3. Remove the switch from the retainer clip on the brake pedal mounting bracket.

4. Installation is the reverse of the removal procedure.

Neutral Safety Switch

All steering columns use a mechanical and electrical neutral start system. The mechanical system relies on a block which prevents starting the engine in positions other than **P** or **N**. The mechanical block is achieved by a wedge shaped finger added to the ignition switch actuator rod. The finger will only pass through the bowl plate notches when the shifter is in the **P** and **N** quadrant positions, which then allows the lock cylinder to rotate to the **START** position.

The electrical switch permits voltage to pass to the starter when the vehicles rear selector is in **P** or **N**. The switch also permits voltage to the backup lights when the gear selector is in the **R** postion.

ADJUSTMENT

1. Block the drive wheels and place the transmission in **N**.

2. Align the actuator on the switch assembly with the hole on the shaft tube.

3. Position the connctor side of the switch assembly to fit into the cutout on the steering column shaft.

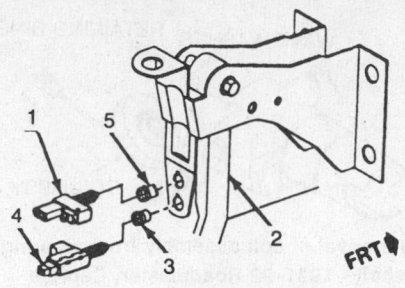

1. Release switch—cruise control
2. Brake pedal
3. Retainer
4. Stoplight switch
5. Retainer—cruise control switch

Stoplight switch assembly

4. Push down on the switch assembly to lock the tangs into place in the steering shaft.

5. Move the switch assembly to the right, **LOW** gear position.

6. Place the transmission in **P**, the switch assembly will ratchet as it adjusts itself.

7. Verify the switch is adjusted properly.

REMOVAL & INSTALLATION

1. Disconnect the negative battery cable.

2. Disconnect the wiring at the switch assembly.

3. Remove the switch assembly.

4. Installation is the reverse of the removal procedure. properly adjust the switch.

Fuses, Circuit Breakers and Relays

LOCATION

Fusible Links

Fusible links are used to prevent major wire harness damage in the event of a short circuit or an overload condition in the wiring circuits which are normally not fused. Each fusible link is of a fixed value for a specific electrical load and should a link fail, the cause of the failure must be determined and repaired prior to installing a new fusible link of the same value.

Circuit Breakers

Various circuit breakers are located under the instrument panel. In order to gain access to these components it may be necessary to first remove the under dash padding.

Fuse Panel

The fuse panel is located on the left side of the vehicle. It is under the instrument panel assembly. In order to gain access to the fuse panel, it may be necessary to first remove the under dash padding.

Various Relays

A/C Blower—located in the right side front of the dash, behind the accumulator.

A/C Compressor Control—located on the right side front of the dash, near the blower motor.

Antenna—located behind the right side side of the instrument panel compartment.

Choke Heater—located on the left side front of the dash, beside the brake booster.

Power Door Locks—located on the lower right side shroud, lower access hole.

Early Fuel Evaporation (EFE)—located in the right side side of the engine compartment, top of the wheel house.

Electronic Level Control—located on the left side front fender behind the battery.

Rear Defogger Timer Relay—located on top of the fuse block.

Wiper Motor Relay—located inside the connector, on the wiper/washer assembly.

Horn—is attached to the left side side of the fuse block.

A/C Programmer Unit—located behind the right side of the instrument panel.

A/C Temperature Cut-Out Relay—located behind the right front fender apron or at the center of the firewall.

Fast Idle Relay—located on the same bracket as the electronic spark control module, which is located at the top of the right front fender well.

Fast Idle Relay (4.3L Engine)—located in the center of the firewall in the engine compartment.

Fuel Pump Relay—located on a bracket in the right side of the engine compartment.

Headlight Relay—located at the front side of the engine compartment, near the headlight.

Power Master Brake Relay—located on top of the electro-hydraulic pump motor below the master cylinder.

Starter Interrupt Relay—located under the left side of the dash panel, above the steering column.

Wide Open Throttle Relay—located in the engine compartment at the center of the firewall.

Theft Deterrent Relay—located behind the instrument panel to the left of the steering column.

Power Seat Relay—located on under the right or left seat.

Rear Hatch Release Relay—located at the rear hatch release latch.

ABS Overvoltage Protection—Sedan–located inside the luggage compartment, attached to the support on the left wheelhouse; Station Wagon–located near center of left side "D" pillar on ABS harness.

Electronic Control Module (ECM)

LOCATION

The electronic control module is located in the passenger compartment. It is positioned in front of the right side kick panel. In order to gain access to the assembly, first remove the trim panel.

Flasher

LOCATION

Turn Signal

The turn signal flasher is located inside the convenience center. In order to gain access to the turn signal flasher it may be necessary to first remove the under dash padding.

Hazard

The hazard flasher is located in the convenience center and is positioned on the lower right side corner of the fuse block assembly.

Cruise Control

ADJUSTMENT

Servo

5.0L (VIN Y) ENGINE

Adjust the rod length to minimum slack with the carburetor lever on the slow idle screw and the engine not running. The idle load control must be fully retracted when the retainer is installed.

Electric Brake Release Switch and Vacuum Release Valve

The switch and valve can only be adjusted when the brake booster pushrod is assembled to the brake pedal assembly. To adjust use the following procedure.

1. Depress brake pedal and insert switch and valve into their proper retaining clips until fully seated.

2. Slowly pull pedal back to its fully retracted position, the switch and valve will move within the retainers to their adjusted position.

3. Verify the switches are adjusted properly.

  a. Cruise control switch contacts must open at $\frac{1}{8}$–$\frac{1}{2}$ in. (3.5–12.5mm) brake pedal travel, measured at the centerline of the brake pedal pad.

  b. Vacuum release valve assembly, if equipped, must open at $1\frac{1}{16}$–$1\frac{5}{16}$ in. (27.0–33.0mm) brake pedal travel, measured at the centerline of the brake pedal pad.

NOTE: Nominal actuation of stoplight contacts is about $\frac{3}{16}$ in. (4.5mm) after cruise control contacts open.

Cruise Control Cable

1991–92 ROADMASTER, CAPRICE AND CUSTOM CRUISER

1. Disconnect the negative battery cable.

2. Remove the air cleaner and resonator.

3. Lift the cable lock up to the unlocked position.

4. Pull the cable from the throttle body side until the throttle valve is beginning to close.

5. When the cable is at the desired position, depress the cable lock into the locked position.

6. Reinstall the air cleaner and resonator.

7. Reconnect the negative battery cable.

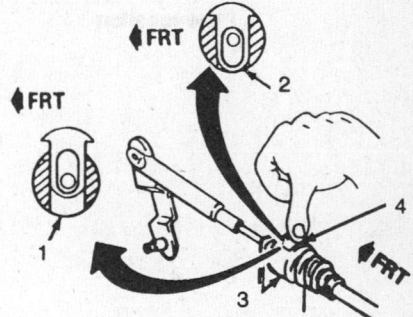

1. Unlocked position
2. Locked position
3. Conduit
4. Cable lock

Cruise control cable adjustment

FUSE	DESCRIPTION
1	Lighted Mirror/Courtesy/Reading Lamps, Underhood Lamp, Interior Courtesy Lighting, Power Antenna Relay, Power Door Lock Switch, Power Mirror Switch, Vanity Mirrors Lamps, Rear Glass Release Relay, Tailgate Lock Switch
2	Audio Alarm Module, Cigar Lighter, I/P Compartment Lamp, Rear Wiper Motor, Radio, Instrument Cluster
3	Overvoltage Protection Relay
4	Delco-Bose® Amplifier
5	Interior Dimmer Lighting
6	Generator, Daytime Running Lights (DRL) Module, Heater and A/C Selector Switch
7	Not Used
8	Rear Window Wiper/Washer Switch, Horn Relay
9	Radio
10	Windshield Wiper/Washer Switch
11	Fuel Injectors
12	Arming Sensor, Diagnostic/Energy Reserve Module (DERM)
13	Electronic Control Module (ECM), Electronic Spark Control Module, Emission Control Solenoids, Instrument Cluster, Vehicle Speed Sensor Module
14	Panel Dimmer, Park/Marker/Tail Lights
15	Rear Defogger
16	Heated Mirrors
17	Audio Alarm Module, Defogger Relay/Timer, Diagnostic/Energy Reserve Module (DERM), Headlight Switch Illumination, Electronic Level Control (ELC) Height Sensor, Instrument Cluster, Overvoltage Protection Relay, TCC/Cruise Release Switch, Daytime Running Lights (DRL) Module
18	Backup Lights, Turn Flasher, Turn Lights, Cruise Control
19	Brake Lights, Hazard Flasher
20	Diagnostic/Energy Reserve Module (DERM), Electronic Control Module (ECM)

Fuse location

Cadillac
Rear Wheel Drive
Brougham

SPECIFICATIONS

VEHICLE IDENTIFICATION CHART

It is important for servicing and ordering parts to be certain of the vehicle and engine identification. The VIN (vehicle identification number) is a 17 digit number visible through the windshield on the driver's side of the dash and contains the vehicle and engine identification codes. The tenth digit indicates model year and the eighth digit indicates engine code. It can be interpreted as follows:

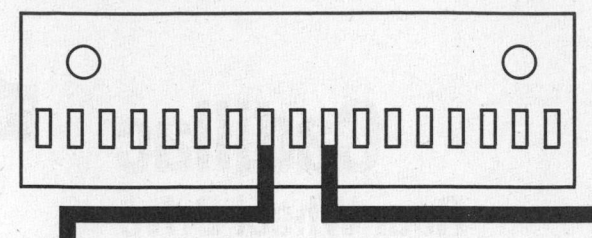

Engine Code							Model Year	
Code	Cu. In.	Liters	Cyl.	Fuel Sys.	Eng. Mfg.		Code	Year
Y	307	5.0	8	Carburetor	B.O.C.		J	1988
E	305	5.0	8	TBI	C.P.C.		K	1989
7	350	5.7	8	TBI	C.P.C.		L	1990
							M	1991
							N	1992

TBI—Throttle Body Injection
B.O.C.—Buick, Oldsmobile, Cadillac
C.P.C.—Chevrolet, Pontiac, Canada

ENGINE IDENTIFICATION

Year	Model	Engine Displacement cu. in. (liter)	Engine Series Identification (VIN)	No. of Cylinders	Engine Type
1988	Brougham	8-307 (5.0)	Y	8	OHV
1989	Brougham	8-307 (5.0)	Y	8	OHV
1990	Brougham	8-307 (5.0)	Y	8	OHV
	Brougham	8-350 (5.7)	7	8	OHV
1991–92	Brougham	8-305 (5.0)	E	8	OHV
	Brougham	8-350 (5.7)	7	8	OHV

OHV—Overhead valves

GENERAL ENGINE SPECIFICATIONS

Year	VIN	No. Cylinder Displacement cu. in. (liter)	Fuel System Type	Net Horsepower @ rpm	Net Torque @ rpm (ft. lbs.)	Bore × Stroke (in.)	Compression Ratio	Oil Pressure @ rpm
1988	Y	8-307 (5.0)	4 bbl	140 @ 3200	255 @ 2000	3.800 × 3.385	8.0:1	30–45 [1]
1989	Y	8-307 (5.0)	4 bbl	140 @ 3200	255 @ 2000	3.800 × 3.385	8.0:1	30–45 [1]
1990	Y	8-307 (5.0)	4 bbl	140 @ 3200	255 @ 2000	3.800 × 3.385	8.0:1	30–45 [1]
	7	8-350 (5.7)	TBI	185 @ 3800	300 @ 2400	4.000 × 3.480	9.3:1	18 [2]
1991–92	E	8-305 (5.0)	TBI	170 @ 4200	255 @ 2400	3.740 × 3.480	9.3:1	18 [2]
	7	8-350 (5.7)	TBI	185 @ 3800	300 @ 2400	4.000 × 3.480	9.3:1	18 [2]

[1] @ 1500 rpm
[2] @ 2000 rpm
TBI—Throttle Body Injection

GASOLINE ENGINE TUNE-UP SPECIFICATIONS

Year	VIN	No. Cylinder Displacement cu. in. (liter)	Spark Plugs Type	Gap (in.)	Ignition Timing (deg.) MT	AT	Compression Pressure (psi)	Fuel Pump (psi)	Idle Speed (rpm) MT	AT	Valve Clearance In.	Ex.
1988	Y	8-307 (5.0)	FR3LS6	0.060	—	20B	NA	6.0–7.5	—	450	Hyd.	Hyd.
1989	Y	8-307 (5.0)	FR3LS6	0.060	—	20B	NA	6.0–7.5	—	450	Hyd.	Hyd.
1990	Y	8-307 (5.0)	FR3L56	0.060	—	①	NA	6.0–7.5	—	①	Hyd.	Hyd.
	7	8-350 (5.7)	CR43TS	0.035	—	①	NA	9.0–13.0	—	①	Hyd.	Hyd.
1991	E	8-305 (5.0)	R45TS	0.035	—	①	NA	9.0–13.0	—	①	Hyd.	Hyd.
	7	8-350 (5.7)	CR43TS	0.035	—	①	NA	9.0–13.0	—	①	Hyd.	Hyd.
1992	SEE UNDERHOOD SPECIFICATIONS STICKER											

Hyd.—Hydraulic
① Use data on underhood vehicle specification sticker

FIRING ORDERS

NOTE: To avoid confusion, always replace spark plugs and wires one at a time.

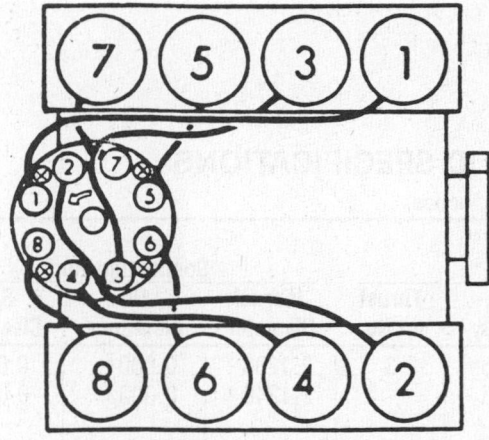

5.0L (VIN Y) Engine
Engine Firing Order: 1–8–4–3–6–5–7–2
Distributor Rotation: Counterclockwise

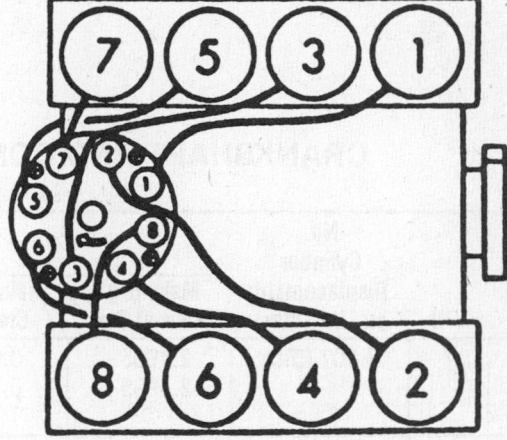

5.0L (VIN E) and 5.7L Engines
Engine Firing Order: 1–8–4–3–6–5–7–2
Distributor Rotation: Clockwise

CAPACITIES

Year	Model	VIN	No. Cylinder Displacement cu. in. (liter)	Engine Crankcase qts. with Filter	without Filter	Transmission (pts.) 4-Spd	5-Spd	Auto.	Drive Axle (pts.)	Fuel Tank (gal.)	Cooling System (qts.)
1988	Brougham	Y	8-307 (5.0)	5.0	4.0	—	—	⑤	4.25	20.7	15.2①
1989	Brougham	Y	8-307 (5.0)	5.0	4.0	—	—	⑤	3.50②	20.7	15.2①
1990	Brougham	Y	8-307 (5.0)	5.0	4.5	—	—	③	⑥	25.0	15.2①
	Brougham	7	8-350 (5.7)	5.0	4.0	—	—	④	⑥	25.0	16.5
1991–92	Brougham	E	8-305 (5.0)	5.0	4.0	—	—	⑦	⑥	25.0	16.7
	Brougham	7	8-350 (5.7)	5.0	4.0	—	—	④	⑥	25.0	16.5

① Heavy duty—15.6
② Optional 3.23—4.25
③ Pan 6.0, overhaul 20.0
④ Pan 9.0, overhaul 21.0
⑤ Pan 10.6, overhaul 22.0
⑥ Fill to Flush or ¼″ (6mm) of Filler Hole
⑦ Pan 10.0, overhaul 22.4

CAMSHAFT SPECIFICATIONS

All measurements given in inches.

Year	VIN	No. Cylinder Displacement cu. in. (liter)	Journal Diameter					Lobe Lift		Bearing Clearance	Camshaft End Play
			1	2	3	4	5	In.	Ex.		
1988	Y	8-307 (5.0)	2.0352–2.0365	2.0152–2.0166	1.9952–1.9965	1.9752–1.9765	1.9552–1.9565	0.247	0.251	0.0020–0.0058	0.006–0.022
1989	Y	8-307 (5.0)	2.0352–2.0365	2.0152–2.0166	1.9952–1.9965	1.9752–1.9765	1.9552–1.9565	0.247	0.251	0.0020–0.0058	0.006–0.022
1990	Y	8-307 (5.0)	2.0352–2.0365	2.0152–2.0166	1.9952–1.9965	1.9752–1.9765	1.9552–1.9565	0.247	0.251	0.0020–0.0058	0.006–0.022
	7	8-350 (5.7)	1.8682–1.8692	1.8682–1.8692	1.8682–1.8692	1.8682–1.8692	1.8682–1.8692	NA	NA	NA	0.004–0.012
1991–92	E	8-305 (5.0)	1.8680–1.8690	1.8682–1.8692	1.8682–1.8692	1.8682–1.8692	1.8682–1.8692	0.234	0.251	NA	0.004–0.012
	7	8-350 (5.7)	1.8682–1.8692	1.8682–1.8692	1.8682–1.8692	1.8682–1.8692	1.8682–1.8692	NA	NA	NA	0.004–0.012

NA—Not Available

CRANKSHAFT AND CONNECTING ROD SPECIFICATIONS

All measurements are given in inches.

Year	VIN	No. Cylinder Displacement cu. in. (liter)	Crankshaft				Connecting Rod		
			Main Brg. Journal Dia.	Main Brg. Oil Clearance	Shaft Endplay	Thrust on No.	Journal Diameter	Oil Clearance	Side Clearance
1988	Y	8-307 (5.0)	2.4985–2.4995 ②	0.0005–0.0021 ①	0.0035–0.0135	3	2.1238–2.1248	0.0004–0.0033	0.006–0.020
1989	Y	8-307 (5.0)	2.4985–2.4995 ②	0.0005–0.0021 ①	0.0035–0.0135	3	2.1238–2.1248	0.0004–0.0033	0.006–0.020
1990	Y	8-305 (5.0)	2.4985–2.4995 ②	0.0005–0.0021 ①	0.0035–0.0135	3	2.1238–2.1248	0.0004–0.0033	0.006–0.020
	7	8-350 (5.7)	④	③	0.001–0.007	5	0.0013–0.0035	NA	0.006–0.014
1991–92	E	8-305 (5.0)	④	0.0011–0.0020 ⑤	0.001–0.007	5	NA	0.0013–0.0035	0.006–0.014
	7	8-350 (5.7)	④	③	0.001–0.007	5	0.0013–0.0035	NA	0.006–0.014

NA—Not Available
① No. 5—0.0015–0.0031
② No. 1—2.4988–2.4998
③ No. 1—0.0008–0.0020
   No.'s 2, 3, 4—0.0011–0.0020
   No. 5—0.0017–0.0032
④ No. 1—2.4488–2.4493
   No.'s 2, 3, 4—2.4481–2.4490
   No. 5—2.4481–2.4488
⑤ Rear—0.0020–0.0032

VALVE SPECIFICATIONS

Year	VIN	No. Cylinder Displacement cu. in. (liter)	Seat Angle (deg.)	Face Angle (deg.)	Spring Test Pressure (lbs.)	Spring Installed Height (in.)	Stem-to-Guide Clearance (in.)		Stem Diameter (in.)	
							Intake	Exhaust	Intake	Exhaust
1988	Y	8-307 (5.0)	45 ①	44 ①	180–194 @ 1.27	1⁴³⁄₆₄	0.0010–0.0027	0.0015–0.0032	0.3425–0.3432	0.3420–0.3427
1989	Y	8-307 (5.0)	45 ①	44 ①	180–194 @ 1.27	1⁴³⁄₆₄	0.0010–0.0027	0.0015–0.0032	0.3425–0.3432	0.3420–0.3427
1990	Y	8-307 (5.0)	45 ①	44 ①	180–194 @ 1.27	1⁴³⁄₆₄	0.0010–0.0027	0.0015–0.0032	0.3425–0.3432	0.3420–0.3427
	7	8-350 (5.7)	46	45	194–206 @ 1.25	1²³⁄₃₂	0.0011–0.0027	0.0011–0.0027	NA	NA
1991–92	E	8-305 (5.0)	46	45	194–206 @ 1.25	1²³⁄₃₂	0.0011–0.0027	0.0011–0.0027	NA	NA
	7	8-350 (5.7)	46	45	194–206 @ 1.25	1²³⁄₃₂	0.0011–0.0027	0.0011–0.0027	NA	NA

NA—Not Available
① Exhaust Valve—31° Seat, 30° Face

PISTON AND RING SPECIFICATIONS

All measurements are given in inches.

Year	VIN	No. Cylinder Displacement cu. in. (liter)	Piston Clearance	Ring Gap			Ring Side Clearance		
				Top Compression	Bottom Compression	Oil Control	Top Compression	Bottom Compression	Oil Control
1988	Y	8-307 (5.0)	0.00075–0.00175 ①	0.009–0.019	0.009–0.019	0.015–0.055	0.0018–0.0038	0.0018–0.0038	0.001–0.005
1989	Y	8-307 (5.0)	0.00075–0.00175 ①	0.009–0.019	0.009–0.019	0.015–0.055	0.0018–0.0038	0.0018–0.0038	0.001–0.005
1990	Y	8-307 (5.0)	0.00075–0.00175 ①	0.009–0.019	0.009–0.019	0.015–0.055	0.0018–0.0038	0.0018–0.0038	0.001–0.005
	7	8-350 (5.7)	0.0007–0.0021	0.010–0.020	0.010–0.025	0.015–0.055	0.0012–0.0032	0.0012–0.0032	0.002–0.007
1991–92	E	8-305 (5.0)	0.0007–0.0021	0.010–0.020	0.010–0.025	0.015–0.055	0.0012–0.0032	0.0012–0.0032	0.002–0.007
	7	8-350 (5.7)	0.0007–0.0021	0.010–0.020	0.010–0.025	0.015–0.055	0.0012–0.0032	0.0012–0.0032	0.002–0.007

① Clearance to bore (selective)

TORQUE SPECIFICATIONS

All readings in ft. lbs.

Year	VIN	No. Cylinder Displacement cu. in. (liter)	Cylinder Head Bolts	Main Bearing Bolts	Rod Bearing Bolts	Crankshaft Pulley Bolts	Flywheel Bolts	Manifold		Spark Plugs
								Intake	Exhaust	
1988	Y	8-307 (5.0)	130 ①	80 ④	18 ⑦	255	60	40 ①	25	25
1989	Y	8-307 (5.0)	40 ① ③	70 ⑥	18 ⑦	255	60	40 ①	25	25
1990	Y	8-307 (5.0)	40 ① ③	70 ⑥	18 ⑦	255	60	40 ①	25	25
	7	8-350 (5.7)	70	77	44	70 ⑤	75	35	26 ②	22

TORQUE SPECIFICATIONS
All readings in ft. lbs.

Year	VIN	No. Cylinder Displacement cu. in. (liter)	Cylinder Head Bolts	Main Bearing Bolts	Rod Bearing Bolts	Crankshaft Pulley Bolts	Flywheel Bolts	Manifold Intake	Manifold Exhaust	Spark Plugs
1991–92	E	8-305 (5.0)	68	77	44	70 ⑤	75	35	26 ②	22
	7	8-350 (5.7)	70	77	44	70 ⑤	75	35	26 ②	22

① Dip bolt in oil before installation
② Stud 20
③ Rotate bolts 1–7 and 9—120°
   Roate bolts 8 and 10—95°
④ Rear main bearing torque 120
⑤ Torque listed is for torsional damper; crankshaft pulley is torqued to 43 ft. lbs.
⑥ Rear main bearing torque 105
⑦ Rotate 70°

BRAKE SPECIFICATIONS
All measurements in inches unless noted

Year	Model	Lug Nut Torque (ft. lbs.)	Master Cylinder Bore	Brake Disc Minimum Thickness	Brake Disc Maximum Runout	Standard Brake Drum Diameter	Minimum Lining Thickness Front	Minimum Lining Thickness Rear
1988	Brougham	100	NA	0.972	0.0005	11.000	0.030	0.030 ①
1989	Brougham	100	NA	0.972	0.0005	11.000	0.030	0.030 ①
1990	Brougham	100	NA	0.972	0.0005	11.000	0.030	0.030 ①
1991–92	Brougham	100	NA	0.972	0.0005	11.000	0.030	0.030 ①

① Bonded lining 0.062
NA—Not Available

WHEEL ALIGNMENT

Year	Model	Caster Range (deg.)	Caster Preferred Setting (deg.)	Camber Range (deg.)	Camber Preferred Setting (deg.)	Toe-in (in.)	Steering Axis Inclination (deg.)
1988	Brougham	2P–4P	3P	3/16P–13/16P	5/16P	3/64	NA
1989	Brougham	2P–4P	3P	3/16P–13/16P	5/16P	3/64	NA
1990	Brougham	2P–4P	3P	3/16P–13/16P	5/16P	3/64	NA
1991–92	Brougham	2P–4P	3P	3/16P–13/16P	5/16P	3/64	NA

NA—Not Available

ENGINE MECHANICAL

NOTE: Disconnecting the negative battery cable on some vehicles may interfere with the functions of the on board computer systems and may require the computer to undergo a relearning process.

Engine Assembly

REMOVAL & INSTALLATION

5.0L (VIN Y) Engine

1. Disconnect the negative battery cable. Properly relieve the fuel system pressure. Remove the air cleaner assembly.

2. Remove the hood, first marking the hinge-to-hood for proper realignment.

3. Remove the air cleaner and heat shroud. Disconnect and plug the transmission oil cooler lines.

4. Drain the cooling system into a suitable container. Disconnect the heater hoses from the engine. Unfasten the fender struts from the radiator shroud. Remove the radiator hose bracket, radiator shroud and fan. Remove the radiator hoses. Remove the radiator.

5. Disconnect the throttle and cruise control linkage at the carburetor.

6. Disconnect the brake vacuum hose from the vacuum pipe. Remove the cruise control power unit, if equipped.

7. Disconnect the power steering pump bracket and position the pump aside with the hoses still connected.

8. Remove the air conditioning compressor bracket bolts and position the compressor aside with the hoses still connected. Do not discharge the system.

9. Disconnect all electrical wires and vacuum lines that will interfere with the removal of the engine.

10. Disconnect the automatic level control line, if equipped. Remove the alternator assembly. Remove the air pump, if equipped.

11. Raise and support the vehicle safely. Remove the engine to transmission bolts. Remove each engine mount through bolt.

12. Remove the starter assembly. Disconnect the exhaust pipes from the exhaust manifolds. Drain the engine oil.

13. Remove the bolts attaching the flywheel inspection cover to the transmission. Remove the cover. Remove the bolts attaching the flywheel to the converter.

16. Disconnect and plug the fuel line and the vapor return line at the fuel pump, as required.

17. Lower the vehicle. Install the lifting equipment to the engine. Support the transmission properly. Raise the engine slightly and pull it forward to disengage it from the transmission. Remove the engine from the vehicle.

18. Installation is the reverse of the removal procedure.

To install:

19. Lower the engine assembly into the engine compartment; align the transmission bellhousing dowels and motor mounts.

20. Loosely install 2 transmission-to-engine bolts.

21. Remove the chain and hoist assembly.

22. Raise and safely support the vehicle, install the engine mount through bolts and torque to 55 ft. lbs. (73 Nm).

23. Route the wiring harness into its original location and reconnect all sensors and connectors.

24. Reconnect the AIR pipe to the catalytic converter.

25. Connect the ground straps to the engine.

26. Install and torque all transmission-to-engine bolts to 35 ft. lbs. (47 Nm).

27. Install the starter assembly and reconnect the wiring.

28. Install the flywheel-to-torque converter bolts and torque the bolts to

60 ft. lbs. (81 Nm). Install the flywheel cover.

29. Reconnect the exhaust and exhaust hangers.

30. Lower the vehicle.

31. Install heater hoses, fuel lines and power steering lines.

32. Reconnect the throttle cable brackets and cables.

33. Install the distributor cap.

34. Connect the negative battery cable at the cowl.

35. Install the fan and fan shroud assembly.

36. Install the radiator assembly. Connect the transmission and oil cooler lines.

37. Connect the radiator hoses. Install the radiator cover and secure the fan shroud.

38. Install the radiator tie struts. Install the air cleaner assembly.

39. Install the air conditioning compressor and belts.

40. Fill the cooling system and connect the battery cables.

41. Check all fluid levels, start the engine and inspect for leaks.

42. Reinstall the hood assembly and align with previous marks.

5.0L (VIN E) and 5.7L Engines

1. Disconnect the negative battery cable. Properly relieve the fuel system pressure. Remove the air cleaner assembly.

2. Remove the hood, after scribing hood hinge outline for proper alignment.

3. Drain the cooling system. Disconnect the radiator hoses. Disconnect the heater hose from the radiator. Disconnect and plug the transmission cooler lines.

4. Remove the radiator cover and tie struts. Disconnect the fan shroud from the radiator assembly and position it aside. Remove the radiator from the vehicle.

5. Remove the cooling fan assembly. Remove the fan shroud assembly from the vehicle. Disconnect the heater hose at the rear of the intake manifold.

6. As required, disconnect and plug the power steering hoses at the power steering gear.

7. Remove the serpentine drive belt. Remove the air conditioning compressor and position it aside. Remove the alternator assembly.

8. Disconnect the accelerator, cruise control and throttle valve cables from their mountings. Remove the retaining brackets from the intake manifold and position them aside.

9. Disconnect the fuel line clips at the thermostat housing and air pump. Position the fuel lines aside. As required, remove the air pump assembly.

10. Disconnect and plug all required electrical connectors and vacuum hoses. Remove the distributor cap. Remove the negative battery cable at the cylinder head.

11. Raise and support the vehicle safely. Drain the engine oil. Disconnect the exhaust pipes at the crossover pipe.

12. Disconnect the starter electrical connectors. Remove the starter retaining bolts. Remove the starter from the vehicle. As required, disconnect and plug the fuel line.

13. Remove the flywheel cover. Remove the torque converter to flywheel retaining bolts. Remove the motor mount bolts.

14. Disconnect the transmission oil cooler lines at the clip on the oil pan. Disconnect all necessary electrical connectors. Remove the oil cooler hose shield.

15. Drain the engine oil into a suitable container.

16. Remove the transmission to engine retaining bolts. Lower the vehicle.

17. Install the lifting equipment to the engine. Support the transmission properly.

18. Raise the engine slightly and pull it forward to disengage it from the transmission. Remove the engine from the vehicle.

To install:

19. Lower the engine assembly into the engine compartment; align the transmission bellhousing dowels and motor mounts.

20. Loosely install 2 transmission-to-engine bolts.

21. Remove the chain and hoist assembly.

22. Raise and safely support the vehicle, install the engine mount through bolts.

23. Route the wiring harness into its original location and reconnect all sensors and connectors.

24. Reinstall the oil cooler line bracket and heat shield.

25. Connect the ground straps to the back of the cylinder heads.

26. Install and torque all transmission-to-engine bolts to 55 ft. lbs. (75 Nm).

27. Install the starter assembly and reconnect the wiring.

28. Install the flywheel-to-torque converter bolts and torque the bolts to 45 ft. lbs. (62 Nm). Install the flywheel cover.

29. Reconnect the exhaust and exhaust hangers.

30. Lower the vehicle.

31. Install heater hoses, fuel lines and power steering lines.

32. Reconnect the throttle cable brackets and cables.

33. Install the distributor cap.

34. Install the alternator bracket and

Cylinder head bolt torque sequence— 5.0L (VIN Y) engine

connect the negative battery cable at the cylinder head.

35. Install the fan and fan shroud assembly.

36. Install the radiator assembly. Connect the transmission and oil cooler lines.

37. Connect the radiator hoses. Install the radiator cover and secure the fan shroud.

38. Install the radiator tie struts. Install the air cleaner assembly.

39. Install the air conditioning compressor and serpentine belt.

40. Fill the cooling system and connect the battery cables.

41. Check all fluid levels, start engine and inspect for leaks.

42. Reinstall the hood assembly and align with previous marks.

Engine Mounts

REMOVAL & INSTALLATION

1. Disconnect the negative battery cable. Raise and support the vehicle safely.

2. Remove the engine mount through bolt and nut.

3. Using a suitable lifting device, raise the front of the engine far enough to remove the engine mount retaining bolts and the engine mount.

4. Installation is the reverse of the removal procedure.

Cylinder Head

REMOVAL & INSTALLATION

5.0L (VIN Y) Engine

1. Disconnect the negative battery cable. Properly relieve the fuel system pressure. Remove the air cleaner assembly.

2. Drain the radiator. Remove the intake manifold. Remove the exhaust manifolds. Remove the alternator,

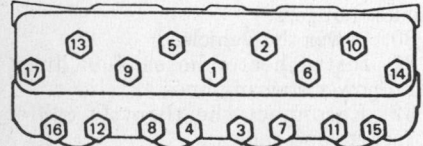

Cylinder head bolt torque sequence— 5.0L (VIN E) and 5.7L engines

power steering pump, air pump and air conditioning compressor, as required.

3. Remove the valve covers, rocker assemblies and pushrods. Note the location of the valve train components so they can be reassembled in the proper location.

4. Remove the cylinder head retaining bolts. Remove the cylinder head and discard the old gasket.

5. Installation is the reverse of the removal procedure. Before installing the cylinder head bolts, coat the bolts with clean engine oil. Torque the cylinder head bolts in the proper sequence to the torque specification.

5.0L (VIN E) and 5.7L Engines

1. Disconnect the negative battery cable. Properly relieve the fuel system pressure. Remove the air cleaner assembly.

2. Drain the radiator. Remove the intake manifold. Remove the exhaust manifolds. Remove the alternator, power steering pump, air pump and air conditioning compressor, as required. Remove the diverter valve.

3. Remove the valve covers, rocker assemblies and pushrods. Note the location of each valve train component so they can be reassembled in the proper location.

4. Remove the cylinder head retaining bolts. Remove the cylinder head and discard the old gasket.

5. Installation is the reverse of the removal procedure. Before installing the cylinder head bolts, coat the threads with sealing compound, GM part 1052080 or equivalent. Torque the cylinder head bolts in proper sequence to the torque specification.

Valve Lifters

REMOVAL & INSTALLATION

1. Disconnect the negative battery cable. Properly relieve the fuel system pressure. Remove the air cleaner assembly.

2. Remove the valve covers. Remove the rocker arms and pivot assemblies. Remove the pushrods. Mark all valve train components, so they may be installed into there original positions.

3. Remove the intake manifold.

4. Remove the lifter guide retainer bolts and remove the lifter guide. Using the proper valve lifter removal tool, remove the valve lifters.

5. Installation is the reverse of the removal procedure. Be sure to coat the lifters in clean engine oil before installing them.

6. Adjust the valves, as required.

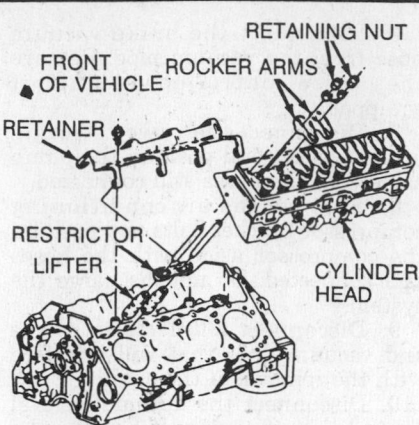

Rocker arm assembly and related components—5.0L (VIN E) and 5.7L engines

Valve Lash

ADJUSTMENT

5.0L (VIN Y) Engine

The rocker arm assembly on this engine is equipped with rocker arm pivots. The hydraulic lifters are properly position in there bores once the rocker arm pivots are torqued to specification, thereby eliminating the need for valve adjustment.

5.0L (VIN E) and 5.7L ENGINES

1. Disconnect the negative battery cable. Remove the valve covers.

2. Tighten the rocker arm nuts until rocker arm lash is eliminated, if necessary.

3. Adjust the valves when the lifter is on the base circle of the camshaft lobe. To do this, crank the engine until the mark on the vibration damper lines up with the center or 0 mark on the timing tab, which is fastened to the crankcase front cover. Ensure the engine is in the No.1 firing position.

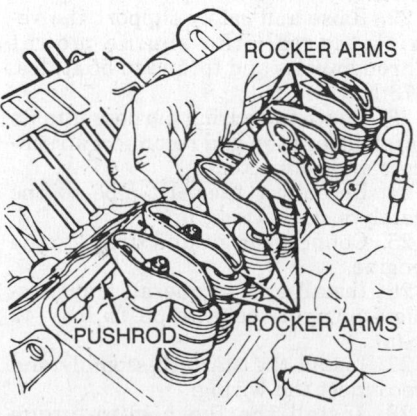

Valve adjustment procedure

NOTE: This may be determined by placing a finger on the No. 1 valve as the mark on the damper comes near the 0 mark on the crankcase front cover. If the valves move as the mark comes up to the timing tab, the engine is in the No. 6 firing position and should be turned 1 full turn to reach the No. 1 firing position.

4. With the engine in the No. 1 firing position, adjust the following valves. Exhaust—1, 3, 4, 8; Intake—1, 2, 5, 7.

5. Back out adjusting nut until lash is felt at the pushrod, then turn in adjusting nut until all lash is removed. This can be determined by rotating pushrod while turning the adjusting nut. When the pushrod stops turning, turn the adjusting nut in an additional turn.

6. Crank the engine one revolution until the pointer, **0** mark and the vibration damper mark are again in alignment. This is the No. 6 firing position.

7. With the engine in this position, repeat the above procedure, adjusting the following valves: Exhaust—2, 5, 6, 7; Intake—3, 4, 6, 8.

8. Install the valve covers.

9. Start the engine and adjust the idle speed, as required.

Rocker Arms

REMOVAL & INSTALLATION

5.0L (VIN Y) Engine

1. Disconnect the negative battery cable. Remove the air cleaner assembly.

2. Disconnect and tag all electrical leads or hoses preventing access to the valve cover retaining bolts. Remove the spark plug wires.

3. Remove the accessory drive belts and brackets, as required. On the left side remove the EGR valve and the exhaust manifold upper shroud.

4. Remove the AIR system components and drive belts, as required.

5. Remove the valve cover retaining bolts. Install tool BT-8315 or J-34144 midway between the ends of the valve cover on the upper side. Tighten the

tool screw and apply pressure on the valve cover.

6. Using a rubber mallet, hit the side of the valve cover above where the tool is installed. Be sure to use a shop towel to absorb the blow from the mallet, otherwise damage to the valve cover may result.

7. Remove the valve cover from the engine. Remove the rocker arm bolts, pivots and the rocker arms.

8. Installation is the reverse of the removal procedure. Torque the rocker arms to 22 ft. lbs. (28 Nm.) Apply a ¼ in. (6mm) bead of RTV sealant or equivalent to the valve cover and torque to specification.

5.0L (VIN E) and 5.7L Engines

1. Disconnect the negative battery cable. Remove the air cleaner assembly.

2. Disconnect the computer command control harness from the intake manifold and oxygen sensor. Remove the required electrical and vacuum connections. Remove the spark plug wires.

3. Disconnect the harness from the right valve cover. Remove the crankcase air inlet hose and connector.

4. Remove the EGR valve solenoid bracket. Disconnect the power brake vacuum pipe. Remove the AIR hose at the manifold check valve. Remove the PCV valve and hose from the valve cover.

5. Disconnect the fuel lines, as required. Remove the alternator rear support bracket and wire harness.

6. Remove the valve cover retaining bolts. Remove the valve covers from the engine. Discard the gaskets.

7. Remove the rocker arm nuts, rocker arm balls and rocker arms, marking each component to ensure they are installed in their original location.

8. Installation is the reverse of the removal procedure. Be sure to use new gaskets or RTV sealant, as required.

Intake Manifold

REMOVAL & INSTALLATION

5.0L (VIN Y) Engine

1. Disconnect the negative battery cable. Properly relieve the fuel system pressure. Remove the air cleaner assembly.

2. Drain the radiator. Remove the upper radiator hose, thermostat, bypass hose, and heater hose at the rear of the manifold. Remove and tag all vacuum lines from the intake manifold.

3. Disconnect and plug the fuel line. Remove the throttle cable, detent cable and cruise control rod.

4. Remove the drive belts, alternator rear brace, air conditioning compressor rear brace and all necessary electrical leads.

5. Remove the computer command control solenoid assembly and the idle load compensator and bracket assembly. Remove the EGR valve, as necessary.

6. As required, remove the carburetor assembly from the intake manifold.

7. Remove the intake manifold retaining bolts. Remove the intake manifold from the vehicle and discard the gasket.

8. Installation is the reverse order of the removal procedure. Apply a suitable RTV sealant to both sides of the manifold gasket and to the front and rear corners of the manifold seals. Tighten the intake manifold attaching bolts in sequence to the proper specification.

5.0L (VIN E) and 5.7L Engines

1. Disconnect the negative battery cable. Properly relieve the fuel system pressure. Remove the air cleaner assembly.

2. Drain the radiator fluid into a suitable container. Remove the radiator hose at the thermostat housing. Remove the heater hose from the intake manifold.

3. Remove the thermostat housing and gasket, as required. Remove the throttle body assembly.

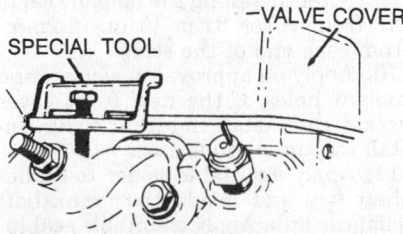

Valve cover removal tool positioning— 5.0L (VIN Y) engine

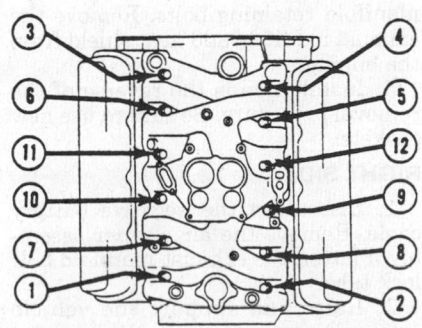

Intake manifold torque sequence— 5.0L (VIN Y) engine

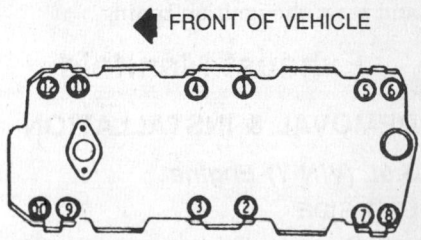

Intake manifold torque sequence— 5.0L (VIN E) and 5.7L engines

4. Disconnect the computer command control harness and position it aside. Disconnect the power brake vacuum pipe. Disconnect the accelerator cable and throttle valve cable retaining bracket.

5. Disconnect the fuel line clips, as required. Remove the spark plug wires at the distributor cap.

6. Remove the distributor cap. Matchmark and remove the distributor assembly from the engine. Remove the coil.

7. Remove the coolant temperature sensor. Disconnect the air conditioning compressor brace and the alternator brace, as required.

8. Remove the intake manifold retaining bolts. Remove the intake manifold from the engine. Discard the gaskets.

To install:

9. Position new gaskets on the cylinder heads. Apply a $\frac{3}{16}$ in. (5mm) bead of RTV sealant, part number 1052289 or equivalent on the front and rear of the cylinder block.

10. Extend the bead of RTV sealant $\frac{1}{2}$ in. (13mm) up each cylinder head to seal and retain the gaskets in position.

11. Install the intake manifold on the engine. Install the retaining bolts. Torque the bolts to specification and in the proper sequence.

12. Install the coolant temperature sensor. Connect the air conditioning compressor brace and the alternator brace.

13. Install the distributor assembly. Install the distributor cap. Install the coil.

14. Connect the fuel line clips, as required. Install the spark plug wires at the distributor cap.

15. Connect the computer command control harness. Connect the power brake vacuum pipe. Connect the accelerator cable and throttle valve cable retaining bracket.

16. Install the thermostat housing and gasket, as required. Install the throttle body assembly.

17. Install the radiator hose at the thermostat housing. Install the heater hose at the intake manifold. Fill the radiator, using the proper quantity and type coolant.

18. Connect the negative battery cable. Install the air cleaner assembly and reset the ignition timing.

Exhaust Manifold

REMOVAL & INSTALLATION

5.0L (VIN Y) Engine

LEFT SIDE

1. Disconnect the negative battery cable. Remove the air cleaner assembly.

2. Flatten the exhaust manifold bolt lock tabs. Remove the heat shroud shield retaining bolts. Remove the heat shroud shield.

3. Raise and support the vehicle safely. Remove the exhaust pipe from the exhaust manifold. Lower the vehicle.

4. Remove the spark plug wires. Loosen the alternator bracket bolts and move the bracket aside, as required.

5. Remove the exhaust manifold retaining bolts and remove the exhaust manifold from the engine.

6. Installation is the reverse order of the removal procedure. Be sure to install new gaskets, as required.

RIGHT SIDE

1. Disconnect the negative battery cable. Remove the air cleaner assembly. Remove the spark plug wires.

2. Remove the oxygen sensor lead wire. Raise and support the vehicle safely.

3. Remove the crossover pipe. Disconnect the exhaust pipe from the exhaust manifold. Remove the oil filter.

4. Remove the front wheel to gain access to the exhaust manifold bolts, if necessary. Flatten the exhaust manifold bolt lock tabs.

5. Remove the exhaust manifold retaining bolts and remove the exhaust manifold from the engine.

6. Installation is the reverse order of the removal procedure. Be sure to install new gaskets.

5.0L (VIN E) and 5.7L Engines

LEFT SIDE

1. Disconnect the negative battery cable. Remove the air cleaner assembly.

2. Raise and support the vehicle safely. Remove the crossover pipe. Lower the vehicle.

3. Remove the spark plug wires and wire clips. Disconnect the oxygen sensor connector. Disconnect the AIR hose at the check valve.

4. Flatten the exhaust manifold bolt lock tabs. Remove the exhaust manifold retaining bolts. Remove the exhaust manifold and heat shield from the engine.

5. Installation is the reverse of the removal procedure. Be sure to use new gaskets.

RIGHT SIDE

1. Disconnect the negative battery cable. Remove the air cleaner assembly. Flatten the exhaust manifold bolt lock tabs.

2. Raise and support the vehicle safely. Disconnect the crossover pipe at both exhaust manifolds.

3. Remove the exhaust mount at the rear of the catalytic converter, as required. Remove the back two exhaust manifold retaining bolts. Lower the vehicle.

4. Disconnect the spark plug wires at the spark plugs. Remove the diverter valve.

5. Remove the remaining exhaust manifold bolts. Remove the exhaust manifold and heat shield from the engine.

6. Installation is the reverse of the removal procedure. Be sure to use new gaskets.

Timing Chain Front Cover

REMOVAL & INSTALLATION

5.0L (VIN Y) Engine

1. Disconnect the negative battery cable. Properly relieve the fuel system pressure.

2. Drain the cooling system. Remove the drive belts. Remove the fan shroud. Remove the fan assembly and fan pulley. Use hub balancer puller J-8614 or equivalent and remove the hub balancer.

3. Remove the power steering pump and position it to the side with the hoses attached.

4. Remove the AIR pump pulley. Remove the air conditioning compressor front bracket and position the compressor to the side.

5. For convenience, the water pump assembly may be removed from the front cover assembly.

6. Remove the front cover retaining bolts, timing indicator and front cover. Discard the gasket.

NOTE: It may be necessary to grind a flat surface on the dowel pins to aid in the removal of the front cover.

7. Clean the front cover and engine mating surfaces. Remove the front cover oil seal using an appropriate oil seal removal tool.

To install:

8. Coat the outside diameter of the new seal with an approved sealer and install the seal using an appropriate tool.

9. After installing the oil pan seal to the front cover, trim $\frac{1}{8}$ in. (3.2mm) from each end of the seal.

10. Apply an approved sealer around coolant holes of the new front cover gasket and install the front cover. Install the timing indicator.

11. Apply a suitable sealer to crankshaft key and inside the crankshaft balancer hub. Apply a suitable seal lubricant to the seal contact area of the balancer hub.

12. Install the crankshaft balancer on the crankshaft. Check the clearance between the front of the engine and balancer hub while installing the hub. The proper balancer to engine clearance is 0.0007–0.001 in.

13. Torque the crankshaft hub bolt to specification and torque the crankshaft pulley bolts to 28 ft. lbs.

14. Complete installation by reversing the removal procedure.

5.0L (VIN E) and 5.7L Engines

1. Disconnect the negative battery cable. Properly relieve the fuel system pressure.

2. Remove the serpentine drive belt. Raise and support the vehicle safely. Remove the vibration damper retaining bolt.

3. Remove the crankshaft pulley bolts and the crankshaft pulley. Using tool J–23523 or equivalent remove the vibration damper.

4. Drain the engine oil. Remove the oil pan retaining bolts. Remove the oil pan.

5. Lower the vehicle. Remove the water pump assembly.

6. Remove the front cover retaining bolts. Remove the front cover. Discard the gasket.

7. Clean the front cover and engine mating surfaces. Remove the front cover oil seal using an appropriate oil seal removal tool.

To install:

8. Coat the outside diameter of the new seal with an approved sealer and install the seal using an appropriate tool.

9. Position the cover and the gasket over the crankshaft end.

10. Loosely install the cover to block upper retaining bolts.

NOTE: Do not force the cover over the dowels to the point where the cover flange or the dowels become distorted.

11. Tighten the bolts in a alternate pattern and evenly while pressing downward on the cover so the dowels in the block are aligned with the corresponding holes in the cover. Position the engine front cover so the dowels enter the holes in the cover without binding.

12. Continue the installation in the reverse of the removal procedure.

Front Cover Oil Seal

REPLACEMENT

5.0L (VIN Y) Engine
COVER INSTALLED

1. Disconnect the negative battery cable. Properly relieve the fuel system pressure.

2. Remove the crankshaft pulley retaining bolts. Remove the crankshaft pulley.

3. Remove the harmonic balancer, using the proper tools.

4. Using seal removal tool J–23129 and J–185903 remove the oil seal from the front cover.

5. Installation is the reverse of the removal procedure. Coat the outside diameter of the new seal with sealing compound.

COVER REMOVED

1. Disconnect the negative battery cable. Properly relieve the fuel system pressure.

2. Remove the front cover from the engine.

3. Using the proper tool remove the old seal and discard it.

4. Coat the new seal with clean engine oil. Using the proper installation tool install the new seal in the cover.

5. Install the front cover to the engine.

5.0L (VIN E) and 5.7L Engines
COVER INSTALLED

1. Disconnect the negative battery cable. Properly relieve the fuel system pressure.

2. Remove the serpentine drive belt. Raise and support the vehicle safely. Remove the vibration damper retaining bolt.

3. Remove the crankshaft pulley bolts and the crankshaft pulley. Using tool J–23523 or equivalent remove the vibration damper.

4. Pry the old seal from the cover, using the proper tool. Care should be used as not to damage the cover.

5. Installation is the reverse of the removal procedure. Be sure to use tool J–35468 or equivalent to properly align and install the new oil seal.

COVER REMOVED

1. Disconnect the negative battery cable. Properly relieve the fuel system pressure.

2. Remove the front cover from the engine.

3. Using the proper tool remove the old seal and discard it.

4. Coat the new seal with clean engine oil. Using the proper installation tool install the new seal in the cover.

5. Install the front cover to the engine.

6. Reconnect the negative battery cable.

Timing Chain and Sprockets

REMOVAL & INSTALLATION

5.0L (VIN Y) Engine

1. Disconnect the negative battery cable. Properly relieve the fuel system pressure. Remove the engine front cover.

2. Remove the fuel pump.

3. Remove the crankshaft oil slinger, camshaft thrust button and spring.

4. Remove the camshaft sprocket retaining bolt and the fuel pump eccentric. Remove the camshaft sprocket and timing chain assembly.

5. Remove the crankshaft key before attempting to remove the crankshaft sprocket. Using an appropriate puller tool, remove the crankshaft sprocket.

To install:

6. Insert the camshaft sprocket and crankshaft sprocket into the timing chain, with the timing marks aligned. Lube the thrust surface with Molykote or equivalent.

7. Grasp both sprockets and the timing chain together and put them into their prospective places. Rotate the camshaft sprocket and engage it on the camshaft, while maintaining timing mark alignment.

8. Install the fuel pump eccentric, flat side toward the engine. Install the camshaft sprocket bolt finger tight. Rotate the crankshaft until the keyways are aligned. Install the crankshaft sprocket key, tap it in with a brass hammer until the key bottoms.

9. When the timing marks are in alignment, the No. 6 cylinder should be at TDC. When both timing marks are on the top, the No. 1 cylinder is at TDC of the compression stroke.

10. Slowly and evenly draw the camshaft sprocket onto the camshaft, us-

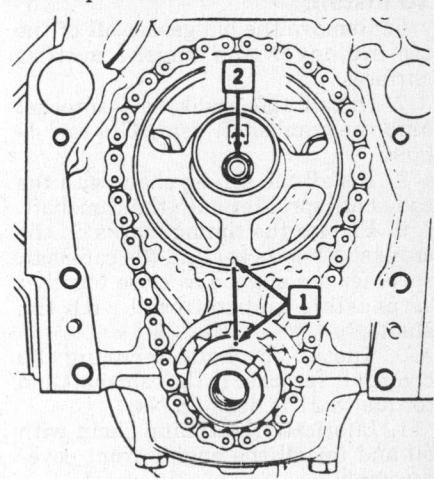

Timing mark alignment—5.0L (VIN Y) engine

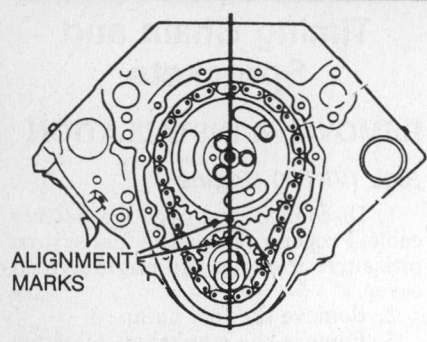

Timing mark alignment—5.0L (VIN E) and 5.7L engines

ALIGNMENT MARKS

ing the mounting bolt and torque the bolt to 65 ft. lbs. (88 Nm).

11. Lubricate the timing chain and finish the installation by reversing the order of the removal procedure.

12. When installing the front cover, apply a suitable RTV sealant around the coolant holes of the new front cover. Be sure to trim the ends of the oil pan seal and install the seal onto the timing chain cover.

5.0L (VIN E) and 5.7L Engines

1. Disconnect the negative battery cable. Properly relieve the fuel system pressure.

2. Remove the engine front cover and water pump assembly.

3. Rotate the engine until the marks on both the camshaft sprocket and crankshaft sprocket align with the shaft centers.

4. Remove the camshaft sprocket retaining bolts. Remove the camshaft sprocket along with the timing chain.

5. Using a crankshaft sprocket removal tool, remove the crankshaft sprocket. Remove the crankshaft sprocket key, if required.

To install:

6. Remove the old gasket off of the timing cover and engine mating surfaces.

7. Install the crankshaft sprocket onto the crankshaft using GM tool J–5590.

8. Install the timing chain with the camshaft sprocket onto the camshaft.

9. Ensure the timing marks on the crankshaft sprocket and the camshaft sprocket are aligned as close together as possible and centered with the shafts.

10. Install the bolts securing the camshaft sprocket to the camshaft and torque to 21 ft. lbs. (28 Nm).

11. Lubricate the timing chain with oil and install the engine front cover assembly.

12. Reconnect the negative battery cable and set the ignition timing.

Camshaft

REMOVAL & INSTALLATION

5.0L (VIN Y) Engine

1. Disconnect the negative battery cable. Properly relieve the fuel system pressure. Drain the cooling system.

2. Remove the intake manifold. Remove the valve covers, rocker arm assemblies, pushrods and lifters. Be sure to note the location of each component for proper installation.

3. Remove the radiator shroud assembly. Remove the radiator. Remove the front grille, if necessary. Remove the cooling fan and water pump pulley.

4. Remove the power steering pump and position it to the side. Remove the alternator assembly. Remove the air pump. Remove the crankshaft pulley and vibration damper.

5. Properly discharge the air conditioning system. Remove the compressor mount bolts, brackets and compressor assembly. Remove the condenser assembly and seal all openings.

6. Remove the water pump. Remove the fuel pump. Remove the front engine cover. Remove the camshaft thrust button and spring. Rotate the crankshaft and align the timing marks.

7. Remove the camshaft retaining bolt, gear and chain. Remove the camshaft retaining plate and camshaft flange adapter and carefully remove the camshaft. The camshaft sprocket is a tight fit. If the sprocket does not come off easily, a light blow on the lower edge of the sprocket with a soft face mallet should dislodge the sprocket.

8. Installation is the reverse of the removal procedure. Lubricate camshaft journals with a suitable engine oil supplement, before installing the camshaft.

5.0L (VIN E) and 5.7L Engines

1. Disconnect the negative battery cable. Properly relieve the fuel system pressure. Drain the cooling system.

2. Remove the intake manifold. Remove the valve covers, rocker arm assemblies, pushrods and lifters Be sure to note the location of each component for proper installation.

3. Remove the radiator tie struts at the radiator cradle. Remove the radiator. Remove the serpentine drive belt. Remove the clutch fan assembly.

4. Remove the engine front cover. Remove the timing chain. Remove the camshaft sprocket.

5. As required, properly discharge the air conditioning system and remove the condenser assembly. As required, remove the grille assembly.

6. Install three $5/16$–18×4 in. bolts in the camshaft timing gear bolt holes. Carefully pull the camshaft partially out of the engine using the bolts as a handle. Remove the bolts from the camshaft. Remove the camshaft from the engine.

7. Installation is the reverse of the removal procedure. Lubricate the camshaft journals with a suitable engine oil supplement, before installing the camshaft.

Piston and Connecting Rod

POSITIONING

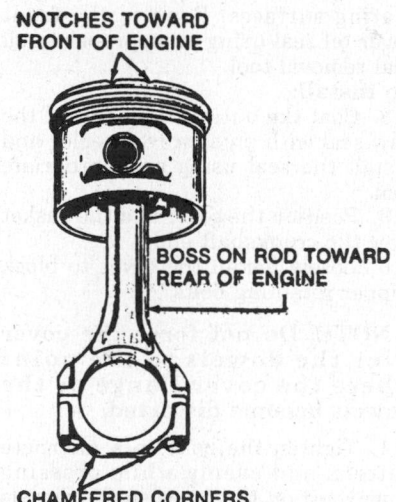

NOTCHES TOWARD FRONT OF ENGINE

BOSS ON ROD TOWARD REAR OF ENGINE

CHAMFERED CORNERS TOWARD FRONT OF ENGINE

Piston positioning and identification— 5.0L (VIN Y) engine

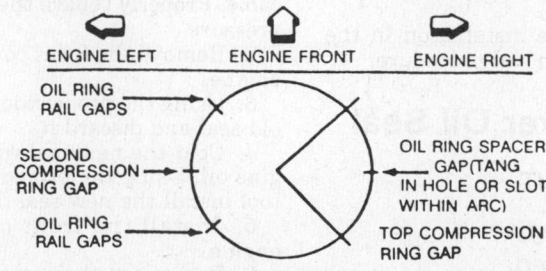

ENGINE LEFT   ENGINE FRONT   ENGINE RIGHT

OIL RING RAIL GAPS

SECOND COMPRESSION RING GAP

OIL RING RAIL GAPS

OIL RING SPACER GAP(TANG IN HOLE OR SLOT WITHIN ARC)

TOP COMPRESSION RING GAP

Piston ring gap locations—5.0L (VIN E) and 5.7L engines

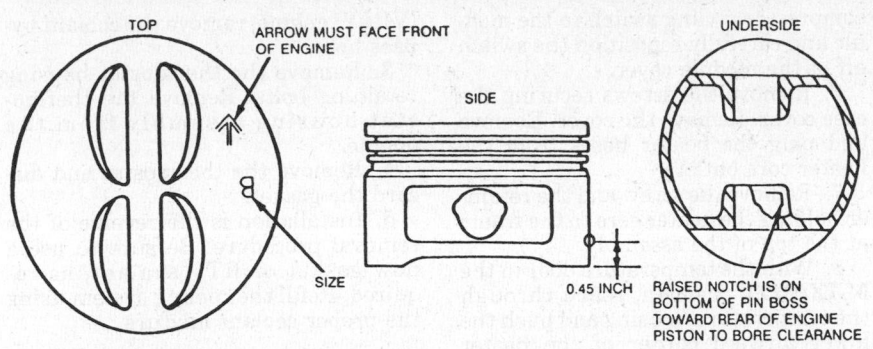

Piston positioning and identification—5.0L (VIN E) and 5.7L engines

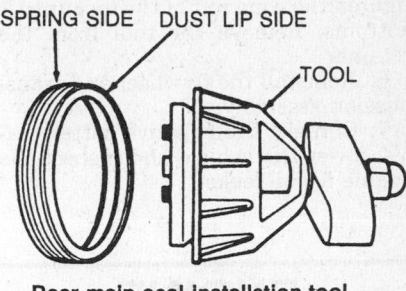

Rear main seal installation tool—
5.0L (VIN E) and 5.7L engines

ENGINE LUBRICATION

Oil Pan

REMOVAL & INSTALLATION

1. Disconnect the negative battery cable. Remove the air cleaner assembly.
2. If required, remove the AIR pipe diverter valve outlet hose.
3. If required, remove the oil level dipstick.
4. Remove the fan shroud attaching screws. Position the fan shroud backward and out of the way. Raise and support the vehicle safely.
5. Drain the engine oil. Remove the flywheel cover. Remove the exhaust crossover pipe.
6. Remove the starter assembly. Using the proper jack, with a block of wood on top, place it under the crankshaft hub to support the engine. Remove both engine mount retaining bolts.
7. Carefully raise the front of the engine. Remove the oil pan retaining bolts and remove the oil pan from the engine.
8. Clean all the gasket material from the pan and the block mating surfaces. Use a new gasket kit and sealer. Ensure the seals are firmly positioned on the flange surfaces with each seal properly located in the cut out notches of the pan gasket.
9. Installation is the reverse of the removal procedure.

Oil Pump

REMOVAL & INSTALLATION

1. Disconnect the negative battery cable. Raise and support the vehicle

safely. Drain the engine oil. Remove the oil pan.
2. Remove the oil pump retaining bolts. Remove the oil pump with the pump driveshaft from the engine.
3. Before installing the oil pump to the engine, remove the pump cover and fill the cavity with petroleum jelly.
4. Installation is the reverse of the removal procedure. Ensure the oil pump driveshaft extension is fully engaged.
5. After completing installation, remove the oil pressure sending unit and install an oil pressure gauge. Start the engine and ensure the oil pressure is within specification.

Rear Main Bearing Oil Seal

REMOVAL & INSTALLATION

5.0L (VIN Y) Engine

NOTE: If upper seal replacement is necessary, crankshaft removal will be required. The following procedure is intended for engine-in-vehicle upper oil seal repair.

1. Disconnect the negative battery cable. Raise and support the vehicle safely. Drain the engine oil. Remove the oil pan. Remove the rear main bearing cap.
2. Using packing tool BT-6433 or J-25282-2 or equivalent, drive both sides of the old seal gently into the groove until it is packed tight.
3. Measure the amount of the seal that was driven up on one side and add $\frac{1}{16}$ in. Cut this length from the old seal that was removed from the main bearing cap.
4. Measure the amount of the seal that was driven up on the other side. Add a $\frac{1}{16}$ in. Cut another length from the old seal. Use the main bearing cap as a holding fixture when cutting the seal.

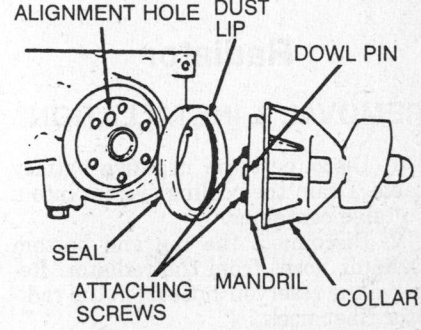

Installing one piece rear main seal—
5.0L (VIN E) and 5.7L engines

5. Work these 2 pieces of the seal into the cylinder block (one piece on each side). Pack the short pieces up into the block using tool BT-6436 or equivalent.
6. Place a piece of shim stock between the seal and the crankshaft to protect the bearing surface before trimming the seal.
7. Form a new rope seal in the rear main bearing cap. Place a drop of a suitable sealer on each end of the seal and cap. Install the main bearing cap. Do not use the attaching bolts to pull down the bearing cap. Tap the cap gently into place with a suitable tool.
8. Continue the installation in the reverse order of the removal procedure.

5.0L (VIN E) and 5.7L Engines

1. Disconnect the negative battery cable. Raise and support the vehicle safely. Remove the transmission and the flywheel assembly.
2. Using the notches in the rear crankshaft seal retainer, pry the old seal out.
To install:
3. Coat the new seal with engine oil and install the seal onto the rear main seal installer tool J-35621.
4. Install tool J-35621 onto the rear of the crankshaft and tighten the screws snugly to ensure the seal will be installed squarely over the crankshaft.
5. Install the seal onto the crankshaft and into the rear seal retainer by

tighten the wing nut of the tool until it bottoms. Remove the tool from the retainer.

6. Reinstall the flywheel and transmission assembly.

7. Connect the negative battery cable, lower the vehicle and inspect the engine for oil leaks.

ENGINE COOLING

Radiator

REMOVAL & INSTALLATION

1. Disconnect the negative battery cable. Drain the cooling system into a suitable container.

2. Disconnect the top and bottom radiator hoses from the radiator. Remove the reservoir hose from the radiator filler neck.

3. Disconnect and plug the transmission fluid cooler lines. Disconnect and plug the oil cooler lines, if equipped.

4. Remove the bolts retaining the engine compartment support rod to the radiator core support. Loosen each anchor bolt and position the support rods aside.

5. Remove the fan shroud retaining bolts. Position the fan shroud assembly aside.

6. Remove the radiator core support cover retaining bolts. Remove the radiator core support cover.

7. Carefully lift the radiator assembly upward and out of the vehicle.

8. Installation is the reverse of the removal procedure. Refill the cooling system with the proper type and quantity of coolant mixture.

Heater Core

REMOVAL & INSTALLATION

1. Disconnect the negative battery cable. Disconnect and tag all electrical wiring from the heater core housing, as required.

2. Remove the right windshield washer nozzle.

3. Remove the right air inlet screen from the plenum. Partially remove the rubber molding above the plenum (1 screw on the right side). Drain the radiator.

4. Remove the remaining screws and remove the primary inlet screen. Remove the blower motor.

5. Remove the 2 screws holding the compressor cycling switch to the module and carefully reposition the switch off of the module cover.

6. Remove the screws securing the case cover. Remove the cover. Remove and plug the heater hoses from the heater core outlets.

7. Remove the screw and the retainer holding the heater core to the frame at the top of the assembly.

8. With the temperature door in the **MAX/HOT** position, reach through the temperature housing and push the lower forward corner of the heater core away from the housing.

9. Rotate the core parallel to the housing. This will cause the core to snap out of the lower clamp. The core can now be removed in a vertical direction due to the configuration of the component.

10. Installation is the reverse order of the removal procedure. Be sure to install a new cover seal as required.

Water Pump

REMOVAL & INSTALLATION

1. Disconnect the negative battery cable.

2. Drain the radiator. Disconnect the lower radiator hose at the water pump. On the 5.0L (VIN Y) engine, remove the bypass hose and the heater hose from the water pump.

3. Remove the drive belts on the 5.0L (VIN Y) engine. Remove the serpentine drive belt on the 5.0L (VIN E) and 5.7L engines.

4. Remove the radiator fan and water pump pulley.

5. Remove the front air conditioning compressor bracket. Remove the front alternator bracket. Remove the power steering pump adjusting bracket. Remove the air pump mounting bracket.

6. Remove the water pump retaining bolts. Remove the water pump from the engine.

7. Installation is the reverse of the removal procedure. Use a new gasket or RTV sealant, as required.

8. Install the drive belts and tighten to the proper tension.

9. Refill the cooling system with the correct mixture of antifreeze and water. Start the engine and check for leaks.

Thermostat

REMOVAL & INSTALLATION

1. Disconnect the negative battery cable. Drain the cooling system.

2. Remove the radiator hose from the thermostat housing. On the 5.0L (VIN Y) engine, remove the coolant bypass hose.

3. Remove the thermostat housing retaining bolts. Remove the thermostat housing assembly from the engine.

4. Remove the thermostat and discard the gasket.

5. Installation is the reverse of the removal procedure. Be sure to use a new gasket or RTV sealant, as required. Refill the cooling system using the proper coolant mixture.

Cooling System Bleeding

1. With the engine turned **OFF**, fill the cooling system to just below the filler neck. Use a mixture of at least 50/50 ethylene glycol antifreeze and water; not exceeding 70 percent antifreeze.

2. Fill the coolant recovery reservoir to the **COLD FILL** mark.

3. Run the engine with the radiator cap removed until normal operating temperature is reached (radiator inlet hose becomes hot).

4. With the engine idling, add coolant to the radiator until it reaches the bottom of the filler neck.

5. Install the radiator cap, with the arrows on the cap aligned with the coolant reservoir hose.

ENGINE ELECTRICAL

NOTE: Disconnecting the nega-

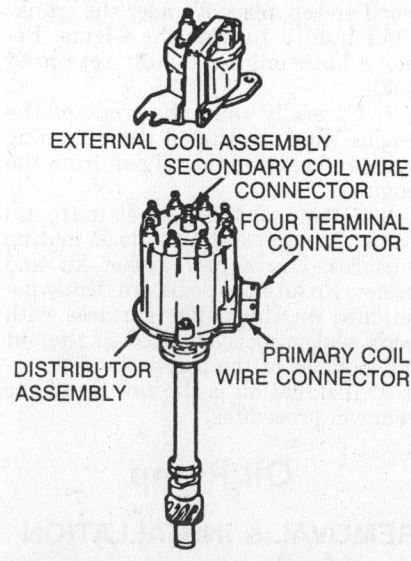

Distributor assembly—5.0L (VIN E) and 5.7L engines

tive battery cable on some vehicles may interfere with the functions of the on board computer systems and may require the computer to undergo a relearning process.

Distributor

REMOVAL

1. Disconnect the negative battery cable. Remove the air cleaner assembly, as required.
2. On 5.0L (VIN Y) engine, disconnect the coil locking tab connectors. On the 5.0L (VIN E) and 5.7L engines, disconnect the wiring harness connectors at the side of the distributor cap.
3. Remove the distributor cap retaining screws. Remove the distributor cap with spark plug wires attached and position it to the side.
4. On the 5.0L (VIN Y) engine, disconnect the 4 terminal ECM harness from the distributor.
5. Remove the distributor assembly retaining bolt.
6. Note and mark the position of the rotor. Pull the distributor upward until the rotor stops turning and again note and mark the position of the rotor. Remove the distributor assembly from the vehicle.

INSTALLATION

Timing Not Disturbed

1. To install the distributor, position the rotor in the last position as marked and lower the assembly into the distributor bore of the engine. When the distributor rotor stops turning and the unit is seated, the rotor should be pointing to the first position marked.
2. Tighten the distributor retaining bolt.
3. Install the distributor cap. Connect all required electrical connections.
4. As required, install the air cleaner assembly. Connect the negative battery cable.
5. Check the ignition timing.

Timing Disturbed

1. If the engine has been accidently cranked with the distributor out, remove the No. 1 spark plug. Place a finger over the No. 1 spark plug hole and crank the engine slowly until a compression build up can be felt in that cylinder.
2. Carefully align the timing mark on the crankshaft pulley to the **O** mark on the timing indicator of the engine. Turn the distributor rotor to

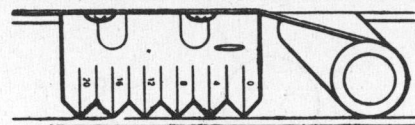
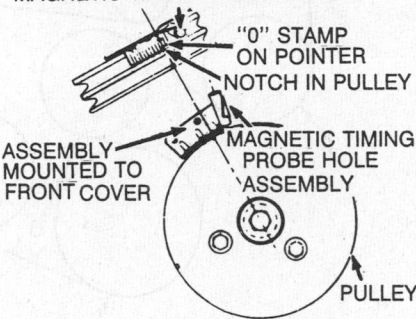

MAGNETIC TIMING PROBE HOLE
"0" STAMP ON POINTER
NOTCH IN PULLEY
ASSEMBLY MOUNTED TO FRONT COVER
MAGNETIC TIMING PROBE HOLE ASSEMBLY
PULLEY

Typical timing mark location

point between the No. 1 and No. 8 spark plug towers on the distributor cap.
3. Lower the assembly into the distributor bore of the engine. When the distributor rotor stops turning and the unit is seated, the rotor should be pointing to No. 1 cylinder segment of the distributor cap.
4. Tighten the distributor retaining bolt.
5. Install the distributor cap. Install the spark plug wires. Connect all required electrical connections.
6. As required, install the air cleaner assembly. Connect the negative battery cable.
7. Check the ignition timing.

Ignition Timing

ADJUSTMENT

NOTE: Always refer back to the timing procedures listed on the Emission Control Information Label.

1988–90

1. Connect a tachometer and a tim-

ing light to the engine. Ensure to connect the timing light to the No. 1 spark plug wire.
2. Start the engine and operate until normal operating temperature is reached.
3. Turn all accessories **OFF**. Ground the diagnostic terminal of the ALDL terminal using a jumper.
4. Check the ignition timing at the specified rpm. If the ignition timing is not within specification, loosen the distributor clamp bolt and rotate the distributor until the specified timing is obtained.
5. Tighten the distributor clamp bolt making sure the distributor does not change position. Recheck the ignition timing.
6. With the engine still running, remove the jumper from the ALDL terminal.
7. Adjust the carburetor idle speed, as required.
8. Turn the engine **OFF**. Remove the tachometer and timing light.

1991–92

1. Run the engine until operating temperature has been reached. Be sure the choke is fully open and the air conditioning is OFF.
2. With the engine running, disconnect the EST bypass connector located by the right front air control valve.
3. Connect the timing light with the pickup lead on the No. 1 plug wire.
4. If the timing requires adjustment, loosen the distributor and set the timing to the specifications noted on the vehicle emission information label.
5. Once the timing has been set, turn the engine **OFF** and reconnect the EST bypass connector.
6. Clear any stored trouble codes by disconnecting the negative battery cable for 30 seconds.

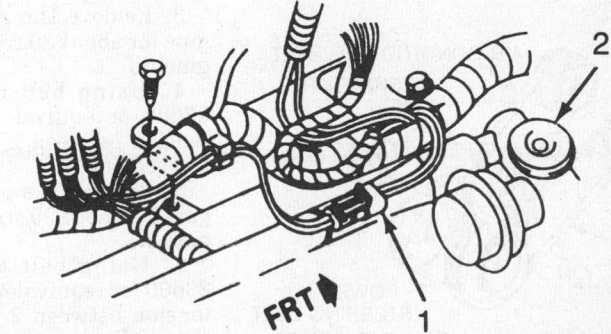

1. EST bypass connector
2. Air control valve

FRT

Electronic Spark Timing (EST) bypass connector—5.0L (VIN E) and 5.7L engines

Alternator

PRECAUTIONS

Several precautions must be observed with alternator equipped vehicles to avoid damage to the unit.

- If the battery is removed for any reason, ensure it is reconnected with the correct polarity. Reversing the battery connections may result in damage to the one-way rectifiers.
- When utilizing a booster battery as a starting aid, always connect the positive to positive terminals and the negative terminal from the booster battery to a good engine ground on the vehicle being started.
- Never use a fast charger as a booster to start vehicles.
- Disconnect the battery cables when charging the battery with a fast charger.
- Never attempt to polarize the alternator.
- Do not use test lights of more than 12 volts when checking diode continuity.
- Do not short across or ground any of the alternator terminals.
- The polarity of the battery, alternator and regulator must be matched and considered before making any electrical connections within the system.
- Never separate the alternator on an open circuit. Make sure all connections within the circuit are clean and tight.
- Disconnect the battery ground terminal when performing any service on electrical components.
- Disconnect the battery if arc welding is to be done on the vehicle.

BELT TENSION ADJUSTMENT

5.0L (VIN Y) Engine

1. Using belt tension gauge J–23600 or equivalent, adjust the alternator belt as follows:
   a. If the belt is used, the correct

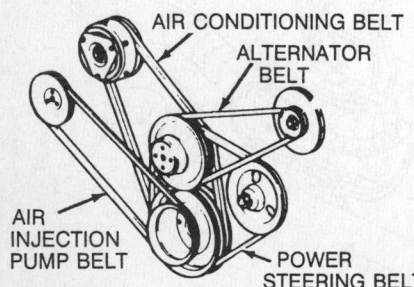

Accessory drive belts—5.0L (VIN Y) engine

AIR CONDITIONING BELT
ALTERNATOR BELT
AIR INJECTION PUMP BELT
POWER STEERING BELT

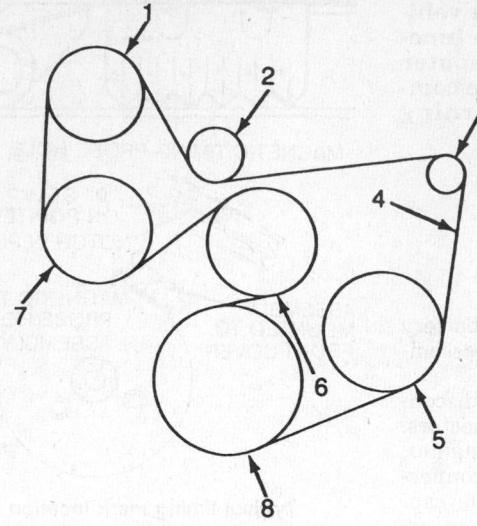

1. Air conditioning compressor belt
2. Drive belt tensioner
3. Alternator pulley
4. Serpentine drive belt
5. Power steering pump pulley
6. Water pump pulley
7. Air injection pump pulley
8. Crankshaft pulley

Accessory drive belt—5.0L (VIN E) and 5.7L engines

belt tension is 112 lbs. (500 N), as indicated on the gauge.
   b. If the belt is new, the correct tension is 157 lbs. (700 N), as indicated on the gauge.
2. Adjust the air conditioning compressor and air pump belt tension as follows:
   a. If the belt is used, the correct belt tension is 112 lbs. (500 N), as indicated on the gauge.
   b. If the belt is new, the correct tension is 167 lbs. (750 N), as indicated on the gauge.
3. Adjust the power steering belt tension as follows:
   a. If the belt is used, the correct belt tension is 90 lbs. (400 N), as indicated on the gauge.
   b. If the belt is new, the correct tension is 146 lbs. (650 N), as indicated on the gauge.

5.0L (VIN E) and 5.7L Engines

1. Run the engine for about 10 minutes. Shut the engine off.
2. Using belt tension gauge J–23600 or equivalent, check the belt tension between 2 pulleys and record the reading.
3. Remove the gauge. Run the engine for about 30 seconds. Shut the engine off.
4. Using belt tension gauge J–23600 or equivalent, check the belt tension between 2 pulleys and record the reading.
5. Remove the gauge. Run the engine for about 30 seconds. Shut the engine off.
6. Using belt tension gauge J–23600 or equivalent, check the belt tension between 2 pulleys and record the reading.
7. Take the average of the 3 record-

ed readings. The belt tension should be 105–125 lbs. (467–556 N).
8. Replace the belt tensioner if the belt tension is below specification.

REMOVAL & INSTALLATION

NOTE: Serpentine belts are automatically adjusted by the tensioner on the engine. If the belt is loose, check the condition of the belt and tensioner. The tensioner should place enough tension on the belt so it can only be twisted 90 degrees at it's longest run.

1. Disconnect the negative battery cable.
2. Disconnect and tag the electrical connections.
3. With V-Belts, remove the bolt holding the slotted adjusting bracket to the alternator and remove the belt.
4. With serpentine belts, loosen and rotate the tensioner to release the drive belt.
5. Remove the thru-bolt to release the alternator from the engine.

To install:

6. When reinstalling, reverse the removal procedure.
7. Adjust the drive belt to specification.
8. On some vehicles, it may be necessary to loosen and rotate the fan shroud.
9. On models with air conditioning, it may be necessary to remove the compressor bracket. Do not discharge the air conditioning system.

Starter

REMOVAL & INSTALLATION

5.0L (VIN Y) Engine

1. Disconnect the negative battery cable.
2. Raise and support the vehicle safely.
3. Remove the starter braces, shields, flywheel housing cover and if necessary, remove the exhaust crossover pipe.
4. Remove the starter motor retaining bolts and lower the starter. Remove any starter shims, if equipped.
5. Disconnect the solenoid wires and the battery cable while supporting the starter. Be sure to note the position of the wires for reinstallation. Remove the starter from the vehicle.
6. Installation is the reverse of the removal procedure. Ensure to install the starter shims, if equipped.

5.0L (VIN E) and 5.7L Engines

1. Disconnect the negative battery cable.
2. Raise and support the vehicle safely.
3. Remove upper support attaching bolts and the brace and wire guide tube bolt, if equipped.
4. Remove the flywheel housing cover, as required.
5. If necessary, remove the exhaust crossover pipe.
6. If necessary, disconnect the transmission oil cooler lines.
7. Remove the starter mounting bolts and lower the starter.
8. Disconnect the wiring and remove starter.
9. If equipped with dual exhaust, it may be necessary to remove the left exhaust pipe.
10. Install by reversing the removal procedure.
11. If shims were removed, they must be installed in their original location to assure proper drive pinion to flywheel engagement.

EMISSION CONTROLS

Please refer to "Emission Controls" in the Unit Repair section for system maintenance procedures. Due to the complex nature of modern electronic engine control systems, comprehensive diagnosis and testing procedures fall outside the confines of this repair manual. For complete information on diagnosis, testing and repair procedures concerning all modern engine and emission control systems, please refer to "Chilton's Guide to Fuel Injection and Electronic Engine Controls".

Emission Warning Lamps

RESETTING

A service engine soon light, located on the instrument panel, alerts the driver that the vehicle should be taken for service as soon as possible. If the light remains on, the self-diagnostic system has detected a problem. After the system has been repaired, all trouble codes must be cleared from the ECM memory. To clear the trouble codes, remove the 3 amp ECM fuse for 10 seconds with the ignition switch turned **OFF**.

FUEL SYSTEM

Fuel System Service Precautions

Safety is the most important factor when performing not only fuel system maintenance but any type of maintenance. Failure to conduct maintenance and repairs in a safe manner may result in serious personal injury or death. Maintenance and testing of the vehicle's fuel system components can be accomplished safely and effectively by adhering to the following rules and guidelines.

- To avoid the possibility of fire and personal injury, always disconnect the negative battery cable unless the repair or test procedure requires that battery voltage be applied.
- Always relieve the fuel system pressure prior to disconnecting any fuel system component (injector, fuel rail, pressure regulator, etc.), fitting or fuel line connection. Exercise extreme caution whenever relieving fuel system pressure to avoid exposing skin, face and eyes to fuel spray. Please be advised that fuel under pressure may penetrate the skin or any part of the body that it contacts.
- Always place a shop towel or cloth around the fitting or connection prior to loosening to absorb any excess fuel due to spillage. Ensure that all fuel spillage, should it occur, is quickly removed from engine surfaces. Ensure that all fuel soaked cloths or towels are deposited into a suitable waste container.
- Always keep a dry chemical (Class B) fire extinguisher near the work area.
- Do not allow fuel spray or fuel vapors to come into contact with a spark or open flame.
- Always use a backup wrench when loosening and tightening fuel line connection fittings. This will prevent unnecessary stress and torsion to fuel line piping. Always follow the proper torque specifications.
- Always replace worn fuel fitting O-rings with new. Do not substitute fuel hose or equivalent where fuel pipe is installed.

RELIEVING FUEL SYSTEM PRESSURE

5.0L (VIN Y) Engine

1. Release the fuel vapor pressure in the fuel tank by removing the fuel tank cap.
2. Ensure the engine is cold. Disconnect the negative battery cable.
3. Cover the fuel line with an absorbent shop cloth and loosen the connection slowly, using the proper tool, to release the fuel pressure gradually.

5.0L (VIN E) and 5.7L Engines

1. Release the fuel vapor pressure in the fuel tank by removing the fuel tank cap.
2. Ensure the engine is cold. Disconnect the negative battery cable.
3. The internal constant bleed feature to throttle body injection relieves the fuel pump system pressure when the engine is not running. Therefore, no further action is required.

Fuel Filter

REMOVAL & INSTALLATION

5.0L (VIN Y) Engine

1. Ensure the engine is cold. Disconnect the negative battery cable.
2. Properly relieve the fuel pump pressure. Remove the air cleaner assembly.
3. Using the proper tools disconnect the fuel line from the fuel filter at the base of the carburetor.
4. Using the proper tools remove the fuel filter housing from the carburetor. Remove the fuel filter, gasket and spring.
5. Installation is the reverse of the removal procedure.

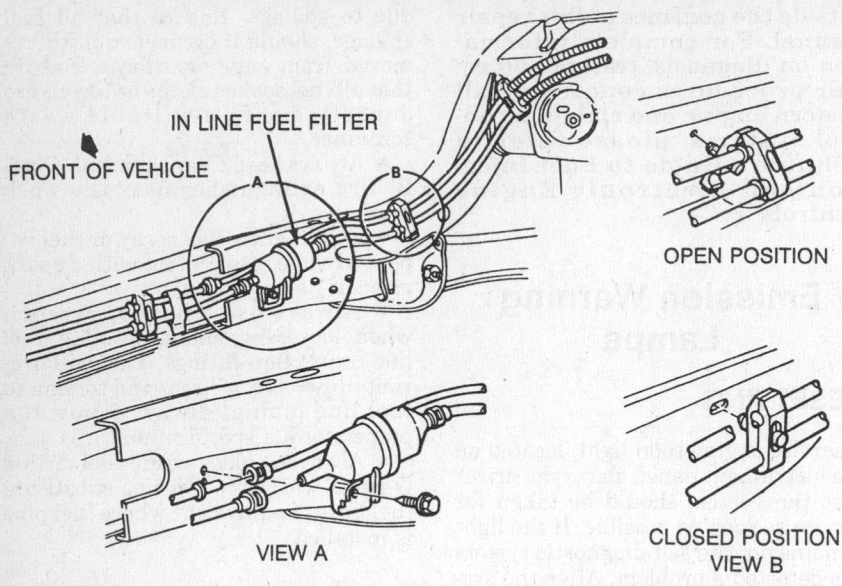

Fuel filter location and related fuel lines—5.0L (VIN E) and 5.7L engines

5.0L (VIN E) and 5.7L Engines

1. Ensure the engine is cold. Disconnect the negative battery cable.
2. Properly relieve the fuel system pressure. Raise and support the vehicle safely.
3. Grasp the fuel filter and one of the fuel line fittings. Twist the quick connect assembly about a ¼ turn in each direction to loosen any dirt within the fitting.
4. Grasp the fuel filter and the other fuel line fitting. Twist the quick connect assembly about a ¼ turn in each direction to loosen any dirt within the fitting.
5. Squeeze the plastic tabs of the connector and pull the connection apart.
6. Remove the fuel filter retaining bolts. Remove the fuel filter from its mounting.
7. Installation is the reverse of the removal procedure. Prior to installation, apply a few drops of clean engine oil to both tube ends of the filter.

NOTE: The application of clean engine oil will ensure proper reconnection and prevent a possible fuel leak. During normal operation the O-ring that is located in the connector will swell and may prevent proper reconnection if not lubricated. If the new filter is nicked, scratched or damaged during installation it must be replaced.

Mechanical Fuel Pump

The mechanical fuel pump was used on the 5.0L (VIN Y) carbureted engine until 1990 when the engine was replaced by the 5.0L (VIN E) TBI engine.

PRESSURE TESTING

1. Ensure the engine is cold. Disconnect the negative battery cable.
2. Properly relieve the fuel pump pressure. Remove the air cleaner assembly.
3. Using the proper tools disconnect the fuel line from the fuel filter at the base of the carburetor.
4. Install a rubber hose about 8–10 in. in length, over the fuel line and connect a low reading fuel pressure gauge.
5. Position the pressure gauge upward about 16 in. above the fuel pump. If equipped with a fuel return line, pinch it.
6. Start the engine and run at slow idle using the fuel in the carburetor.
7. If the fuel pump is performing properly the pressure should be 5.5–6.5 psi. If not within specification, replace the fuel pump.

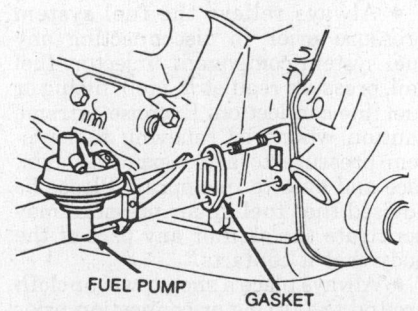

Fuel pump location—5.0L (VIN Y) engine

REMOVAL & INSTALLATION

1. Ensure the engine is cold. Disconnect the negative battery cable. Properly relieve the fuel pump pressure.
2. Remove the air conditioning compressor drive belt.
3. If equipped with an air pump, loosen the air pump pulley bolts and remove the air pump hoses and electrical leads to the air pump. Remove the air pump pulley and the air pump from the engine.
4. Remove the compressor front bracket. Remove the fuel inlet hose from the fuel pump. Disconnect the vapor return hose, if equipped.
5. Remove the fuel outlet pipe. Remove the nuts securing the fuel pump to the engine. Remove the fuel pump from the engine. Discard the gasket.
6. Installation is the reverse order of the removal procedure. Be sure to use a new fuel pump gasket. Start the engine and check for leaks.

Electric Fuel Pump

PRESSURE TESTING

NOTE: The fuel pressure should be recorded while the fuel pump is operating. Fuel pump pressure will drop immediately after the fuel pump stops running due to a controlled bleed within the fuel system. The fuel pump test location is on the right side of the engine compartment.

REMOVAL & INSTALLATION

1. Ensure the engine is cold. Disconnect the negative battery cable.
2. Properly relieve the fuel system pressure. Drain the fuel tank. Raise and support the vehicle safely.

NOTE: If the nylon fuel feed or return lines become kinked and cannot be straightened, they must be replaced. Do not attempt to repair sections of nylon fuel lines.

3. Disconnect and plug the fuel line fittings at the fuel meter assembly. Disconnect all electrical connectors at the tank.
4. Properly support the fuel tank. Remove the fuel tank retaining strap bolts. Remove the retaining straps. Carefully remove the empty fuel tank from the vehicle.
5. Using tool J–24187 or equivalent, remove the fuel level meter assembly retaining cam. Remove the fuel

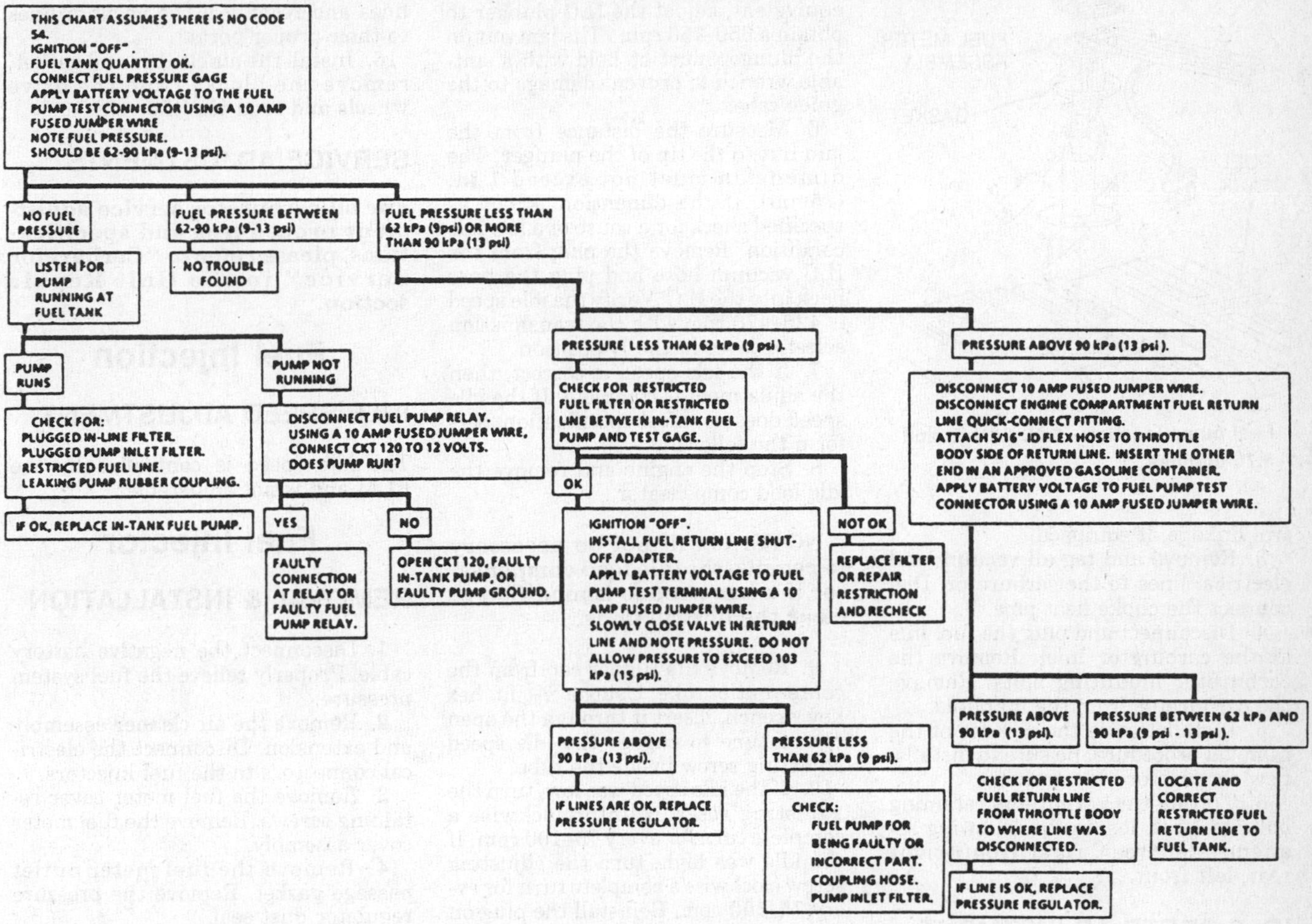

THIS CHART ASSUMES THERE IS NO CODE 54.
IGNITION "OFF".
FUEL TANK QUANTITY OK.
CONNECT FUEL PRESSURE GAGE
APPLY BATTERY VOLTAGE TO THE FUEL PUMP TEST CONNECTOR USING A 10 AMP FUSED JUMPER WIRE
NOTE FUEL PRESSURE.
SHOULD BE 62-90 kPa (9-13 psi).
NO FUEL PRESSURE
FUEL PRESSURE BETWEEN 62-90 kPa (9-13 psi)
FUEL PRESSURE LESS THAN 62 kPa (9psi) OR MORE THAN 90 kPa (13 psi)
LISTEN FOR PUMP RUNNING AT FUEL TANK
NO TROUBLE FOUND
PUMP RUNS
PUMP NOT RUNNING
CHECK FOR: PLUGGED IN-LINE FILTER. PLUGGED PUMP INLET FILTER. RESTRICTED FUEL LINE. LEAKING PUMP RUBBER COUPLING.
DISCONNECT FUEL PUMP RELAY. USING A 10 AMP FUSED JUMPER WIRE, CONNECT CKT 120 TO 12 VOLTS. DOES PUMP RUN?
IF OK, REPLACE IN-TANK FUEL PUMP.
YES
NO
FAULTY CONNECTION AT RELAY OR FAULTY FUEL PUMP RELAY.
OPEN CKT 120, FAULTY IN-TANK PUMP, OR FAULTY PUMP GROUND.
PRESSURE LESS THAN 62 kPa (9 psi).
CHECK FOR RESTRICTED FUEL FILTER OR RESTRICTED LINE BETWEEN IN-TANK FUEL PUMP AND TEST GAGE.
OK
NOT OK
IGNITION "OFF". INSTALL FUEL RETURN LINE SHUT-OFF ADAPTER. APPLY BATTERY VOLTAGE TO FUEL PUMP TEST TERMINAL USING A 10 AMP FUSED JUMPER WIRE. SLOWLY CLOSE VALVE IN RETURN LINE AND NOTE PRESSURE. DO NOT ALLOW PRESSURE TO EXCEED 103 kPa (15 psi).
REPLACE FILTER OR REPAIR RESTRICTION AND RECHECK
PRESSURE ABOVE 90 kPa (13 psi).
PRESSURE LESS THAN 62 kPa (9 psi).
IF LINES ARE OK, REPLACE PRESSURE REGULATOR.
CHECK: FUEL PUMP FOR BEING FAULTY OR INCORRECT PART. COUPLING HOSE. PUMP INLET FILTER.
PRESSURE ABOVE 90 kPa (13 psi).
DISCONNECT 10 AMP FUSED JUMPER WIRE. DISCONNECT ENGINE COMPARTMENT FUEL RETURN LINE QUICK-CONNECT FITTING. ATTACH 5/16" ID FLEX HOSE TO THROTTLE BODY SIDE OF RETURN LINE. INSERT THE OTHER END IN AN APPROVED GASOLINE CONTAINER. APPLY BATTERY VOLTAGE TO FUEL PUMP TEST CONNECTOR USING A 10 AMP FUSED JUMPER WIRE.
PRESSURE ABOVE 90 kPa (13 psi).
PRESSURE BETWEEN 62 kPa AND 90 kPa (9 psi - 13 psi).
CHECK FOR RESTRICTED FUEL RETURN LINE FROM THROTTLE BODY TO WHERE LINE WAS DISCONNECTED.
LOCATE AND CORRECT RESTRICTED FUEL RETURN LINE TO FUEL TANK.
IF LINE IS OK, REPLACE PRESSURE REGULATOR.

Fuel system diagnosis—5.0L (VIN E) and 5.7L engines

CONSTANT BLEED
J-37287-2
STEEL
NYLON
RETURN LINE
PRESSURE LINE
STAINLESS STEEL
FUEL FILTER
IN-TANK PUMP
FUEL PRESSURE CONNECTION
FUEL TANK

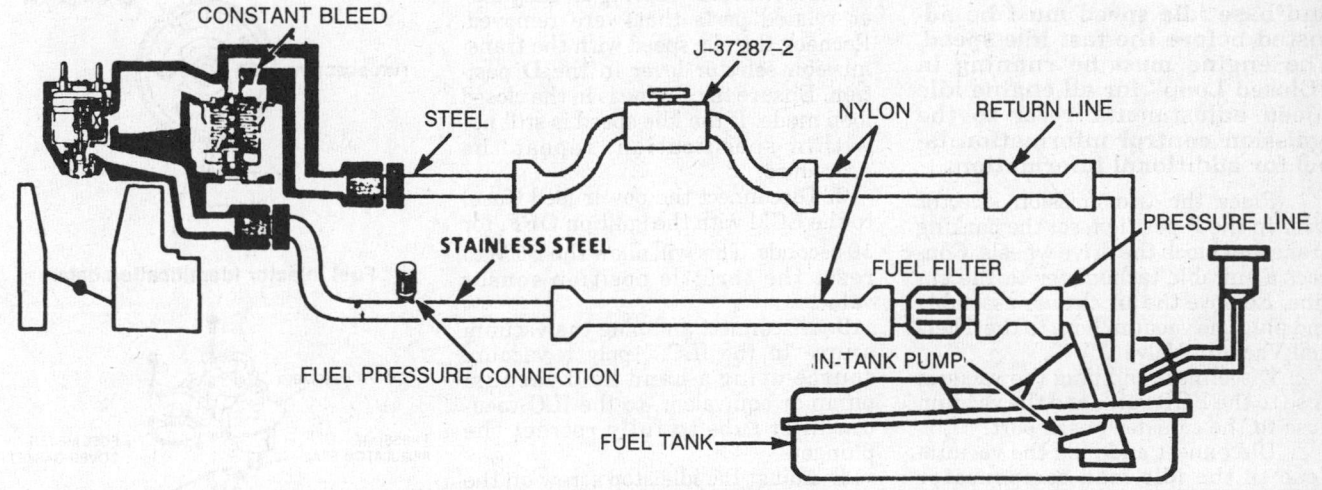

Fuel line pressure connector location—5.0L (VIN E) and 5.7L engines

level meter assembly from the tank and discard the O-ring.

6. Installation is the reverse of the removal procedure. Be sure to use a new O-ring.

Carburetor

REMOVAL & INSTALLATION

1. Disconnect the negative battery cable. Remove the air cleaner assembly. Disconnect the accelerator linkage.

2. Disconnect the transmission detent cable. Disconnect the cruise con-

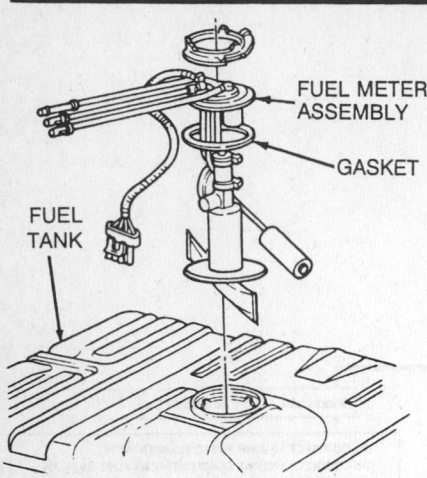

Fuel pump location—5.0L (VIN E) and 5.7L engines

trol linkage, if equipped.

3. Remove and tag all vacuum and electrical lines to the carburetor. Disconnect the choke heat pipe.

4. Disconnect and plug the fuel line at the carburetor inlet. Remove the carburetor mounting bolts. Remove the carburetor from the manifold.

5. Installation is the reverse of the removal procedure. Be sure to install a new carburetor base gasket.

6. Torque the carburetor retaining bolts to 12 ft. lbs. in the following sequence, left rear, right front, right rear, left front.

IDLE SPEED ADJUSTMENT

NOTE: The Idle Load Compensator (ILC) minimum/maximum and base idle speed must be adjusted before the fast idle speed. The engine must be running in "Closed Loop" for all engine idle speed adjustments, refer to the emission control information label for additional information.

1. Place the transmission selector lever in the **P** position, set the parking brake and block the drive wheels. Connect a suitable tachometer to the engine. Remove the air cleaner assembly and plug the vacuum hose to the Thermal Vacuum Valve (TVV).

2. Disconnect and plug the vacuum hose to the EGR valve and the vacuum hose to the canister purge port.

3. Disconnect and plug the vacuum hose to the idle load compensator (ILC). Back out the idle stop screw on the carburetor 3 turns.

4. Turn the air conditioning control switch to the **OFF** position. With the engine running and at normal operating temperature, place the transmission selector lever in the **D** position. Fully extend the idle load compensator plunger (no vacuum applied).

5. Using tool J–29607, BT–8022 or equivalent, adjust the ILC plunger to obtain a 650–750 rpm . The jam nut on the plunger must be held with a suitable wrench to prevent damage to the guide tabs.

6. Measure the distance from the jam nut to the tip of the plunger. The dimension must not exceed 1 in. (25mm). If the dimension is not as specified check for a cause of a low idle condition. Remove the plug from the ILC vacuum hose and plug the hose back into the ILC. Verify the idle speed is 425–475 rpm with the transmission selector lever in the **D** position.

7. If the idle speed is correct, then the adjustment is complete. If the idle speed does not meet specification, perform the following:

8. Stop the engine and remove the idle load compensator.

NOTE: It will not be necessary to remove the idle load compensator if a hex wrench is modified to clear the obstructions.

9. Remove the rubber cap from the center outlet tube. Using a $^3/_{32}$ in. hex key wrench, insert it through the open center tube to engage the idle speed adjusting screw inside the tube.

10. If the idle speed was low, turn the adjusting screw counterclockwise a complete turn for every 75–100 rpm. If the idle was high, turn the adjusting screw clockwise a complete turn for every 75–100 rpm. Reinstall the plug on the center of the outlet tube.

11. Reinstall the idle load compensator on the carburetor and attach the throttle return spring and any other related parts that were removed. Recheck the idle speed with the transmission selector lever in the **D** position. Ensure the engine is in the closed loop mode. If the idle speed is still not within specification, repeat the procedure.

12. Disconnect the power feed (fuse) to the ECM with the ignition **OFF**, for 10 seconds. This will allow the ECM to reset the throttle position sensor value.

13. Disconnect and plug the vacuum source to the ILC. Apply a vacuum source using a hand held vacuum pump or equivalent, to the ILC vacuum inlet tube to fully retract the plunger.

14. Adjust the idle stop screw on the carburetor float bowl to obtain 450 rpm, with the transmission selector lever in the **D** position. When the base idle is set, place the transmission selector lever in the **P** and turn **OFF** the engine.

15. Remove the plug from the vacuum hose and install the hose on the ILC vacuum inlet tube. Remove all the

plugs from the disconnected vacuum lines and reconnect the vacuum lines to their proper ports.

16. Install the air cleaner and gasket, remove the blocks from the drive wheels and road test the vehicle.

SERVICE ADJUSTMENTS

For all carburetor service adjustment procedures and specifications, please refer to "Carburetor Service" in the Unit Repair section.

Fuel Injection

IDLE SPEED ADJUSTMENT

The idle speed is controlled by the ECM and is not adjustable.

Fuel Injector

REMOVAL & INSTALLATION

1. Disconnect the negative battery cable. Properly relieve the fuel system pressure.

2. Remove the air cleaner assembly and extension. Disconnect the electrical connectors to the fuel injectors.

3. Remove the fuel meter cover retaining screws. Remove the fuel meter cover assembly.

4. Remove the fuel meter outlet passage gasket. Remove the pressure regulator dust seal.

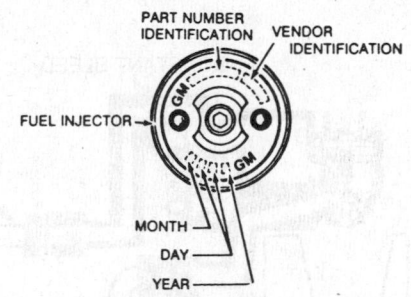

Fuel Injector identification data

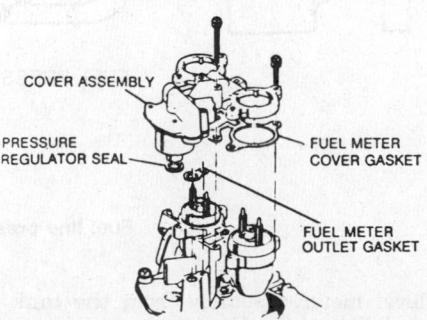

Fuel meter cover assembly and related components—5.0L (VIN E) and 5.7L engines

NOTE: If the fuel meter cover gasket is stuck to the fuel meter body, leave it in place. If it is stuck to the fuel meter cover, remove it and place it on the fuel meter body.

5. Once the fuel meter cover gasket is on the fuel meter body, use the proper tool and fulcrum to carefully pry out the fuel injector.

6. Remove and discard the small O-ring from the nozzle end of the fuel injector.

7. Remove and discard the fuel meter cover gasket. Remove and discard the large O-ring and fuel injector washer from the top of the injector cavity.

8. Installation is the reverse of the removal procedure.

NOTE: When installing the injectors, install the fuel injector washer and large O-ring before the injector, to be sure the O-ring is properly seated. Reversing these procedures could result in a fuel leak and possible engine fire.

DRIVE AXLE

Driveshaft and U-Joints

REMOVAL & INSTALLATION

1. Raise and support the vehicle safely. Position a drain pan under the transmission.

2. Matchmark the driveshaft to the axle pinion flange so it can be reinstalled in its original position. Remove the rear driveshaft flange capscrews.

NOTE: Never let the full weight of the driveshaft be supported only by the front universal joint.

3. Push the driveshaft forward to clear the pinion flange, then pull the driveshaft rearward to disengage the slip yoke from the transmission. Plug the transmission to prevent oil leakage or the entry of dirt.

4. Installation is the reverse of the removal procedure.

Rear Axle Shaft, Bearing and Seal

REMOVAL & INSTALLATION

1. Raise and support the vehicle safely. Remove the wheel and tire assembly. Remove the brake drum.

2. Clean any dirt from the differential cover. Loosen the cover attaching bolts. Drain the lubricant.

3. Remove the pinion shaft lockbolt and remove the pinion shaft.

4. Push in on the flanged end of the axle shaft and remove the C-lock from the splined end of the axle shaft.

5. Remove the axle shaft from the housing, being cautious not to damage the oil seal.

6. Use a prybar to remove the oil seal from the bore. Use an axle shaft bearing puller or a slide hammer to remove the axle bearing from the bearing bore.

To install:

7. Lubricate the new bearing with gear lubricant. Use bearing installer tool J–23690 or equivalent, and install the bearing so the tool bottoms out against the shoulder in the housing. Lubricate the lips of the seal with gear lubricant. Position the new seal on seal installer tool J–23771 or equivalent, and position the seal into the housing bore. Tap the seal into place so it is flush with the axle tube.

8. Slide the axle shaft into the housing until the splines on the end of the shaft engage the splines of the differential side gear. Handle the shaft gently when trying to engage to splines.

9. Install the axle shaft C-lock on the splined end of the axle shaft in the differential. Push the shaft outward so the shaft lock seats in the counterbore of the differential side gear.

10. Install the pinion shaft through the differential case and pinion gears. Align the lockbolt hole and install the lock screw, tightening it to 25 ft. lbs.

11. Clean the differential housing and cover mating surfaces and install the cover with a new gasket.

12. Fill the differential with lubricant. Install the brake drum. Install the tire and wheel assembly. Lower the vehicle.

Pinion Seal

REMOVAL & INSTALLATION

1. Raise and support the vehicle safely. Matchmark and remove the driveshaft.

3. Matchmark the position of the pinion yoke, pinion shaft and nut so the proper pinion bearing preload can be maintained.

4. Position a drain pan under the assembly to catch any fluid that may drain from the rear assembly. Remove the pinion yoke nut and washer. Remove the pinion yoke.

5. Remove the oil seal by driving it out of the carrier using a blunt chisel.

6. Installation is the reverse of the removal procedure. Coat the outside diameter of the yoke and the sealing lip of the new seal with seal lubricant, part number 1050169 or equivalent.

7. Tighten the yoke nut to the position marked previously. While holding the pinion yoke tighten the nut an additional $\frac{1}{16}$ in. beyond the alignment marks.

Axle Housing

REMOVAL & INSTALLATION

1. Raise and support the vehicle safely. Remove the tire and wheel assemblies. Remove the brake drums. Properly support the rear axle.

2. Disconnect the shock absorbers from axle. Matchmark the driveshaft and disconnect it from the rear axle pinion flange.

3. Remove the brake line junction block bolt at axle housing. Disconnect and plug the brake lines at the junction block. If equipped with ABS, disconnect all required electrical connectors.

4. Disconnect the upper control arms from axle housing. Lower the rear axle assembly slightly and remove the springs.

5. Continue lowering the rear axle assembly and remove it from the vehicle.

6. Installation is the reverse of the removal procedure.

AUTOMATIC TRANSMISSION

For further information on transmissions/transaxles, please refer to "Chilton's Guide to Transmission Repair".

Transmission Assembly

REMOVAL & INSTALLATION

1. Disconnect the negative battery cable. Position the selector lever in the **N** detent position.

2. Remove the air cleaner assembly. Disconnect the accelerator cable and detent cable, as required.

3. Remove the transmission dipstick retaining bolt and the dipstick tube.

4. Raise and support the vehicle safely.

5. Matchmark the driveshaft for reinstallation in its original position and remove the driveshaft.

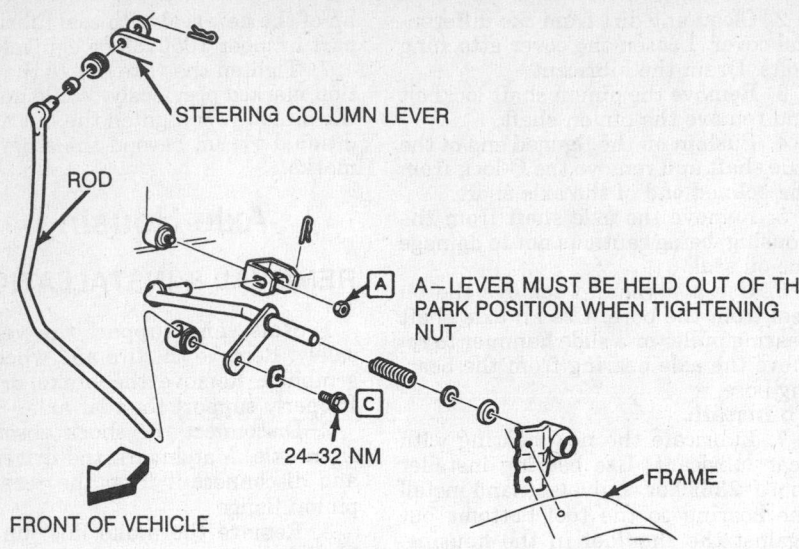

STEERING COLUMN LEVER

ROD

A — LEVER MUST BE HELD OUT OF THE PARK POSITION WHEN TIGHTENING NUT

24–32 NM

FRAME

FRONT OF VEHICLE

Shift linkage adjustment

6. Disconnect the shift linkage, speedometer cable and all electrical connections at the transmission. Remove the starter.

7. Remove the flexplate cover and mark the flexplate and converter so they can be realigned in their original location. Remove the flexplate to converter bolts.

8. Position a transmission jack under the transmission and remove the transmission mount.

9. Remove the crossmember attaching bolts and remove the crossmember. If necessary, remove the floor pan reinforcement.

10. Remove the transmission to engine bolts. Support the engine with a suitable tool.

11. Lower the transmission slightly. Disconnect the transmission lines and required cables. Plug all openings.

12. Install a torque converter holding tool and remove the transmission assembly from the vehicle.

13. Installation is the reverse of the removal procedure.

SHIFT LINKAGE ADJUSTMENT

1. With the transmission selector

lever in the **N** detent, tighten the linkage rod retaining bolt to the proper torque.

2. The linkage is correctly adjusted if at final vehicle inspection with the selector lever raised and centered in the **N** detent, the column lever can be lowered and will engage in the column neutral notch.

3. Adjustment is unacceptable if any rotation of the selector lever is required to engage the column neutral notch.

DETENT CABLE ADJUSTMENT

1. Remove the air cleaner assembly.

2. Depress and hold-down the metal readjustment tab at the engine end of the throttle valve cable.

3. Move the slider until it stops against the fitting. Release the readjustment tab.

4. Rotate the throttle lever to its full travel position.

5. The slider must ratchet toward the lever when the lever is rotated to its full travel position.

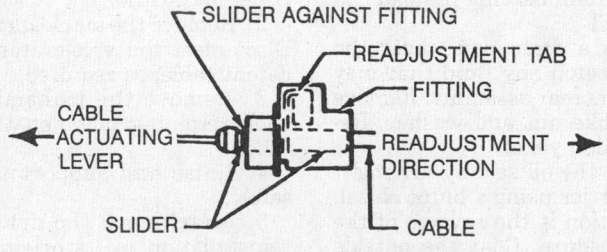

SLIDER AGAINST FITTING

READJUSTMENT TAB

FITTING

CABLE ACTUATING LEVER

READJUSTMENT DIRECTION

SLIDER

CABLE

Throttle valve cable adjustment

FRONT SUSPENSION

Shock Absorbers

REMOVAL & INSTALLATION

1. Disconnect the negative battery cable.

2. Remove the top shock absorber retaining nut.

3. Raise and support the vehicle safely.

4. Remove the bottom shock absorber retaining bolts. Remove the shock absorber from the vehicle.

5. Installation is the reverse of the removal procedure.

Coil Springs

REMOVAL & INSTALLATION

1. Raise and support the vehicle safely. Be sure the vehicle is supported so the lower control arms hang free.

2. Remove the lower shock absorber retaining bolts and push the shock absorber up through the control arm and into the spring.

3. Secure tool J–23028–01 to a suitable jack. Position the tool to cradle the lower control arm inner bushings.

4. Remove the stabilizer to lower control arm attachment.

5. Raise the jack to relieve tension on the lower control arm pivot bolts. Install a safety chain around the spring and through the lower control arm as a safety precaution.

6. Remove the lower control arm rear pivot bolt first, then remove the lower control arm front pivot bolt.

7. Lower the control arm assembly from its mounting. Remove the spring and the safety chain.

8. Installation is the reverse of the removal procedure. Properly position the spring in the frame.

NOTE: The lower end of the coil must cover all or part of one inspection hole in the lower control arm. The second hole must be partly or completely uncovered.

Upper Ball Joints

INSPECTION

1. Raise and support the vehicle safely.

2. Position jack stands under the lower control arms as near as possible to each lower ball joint.

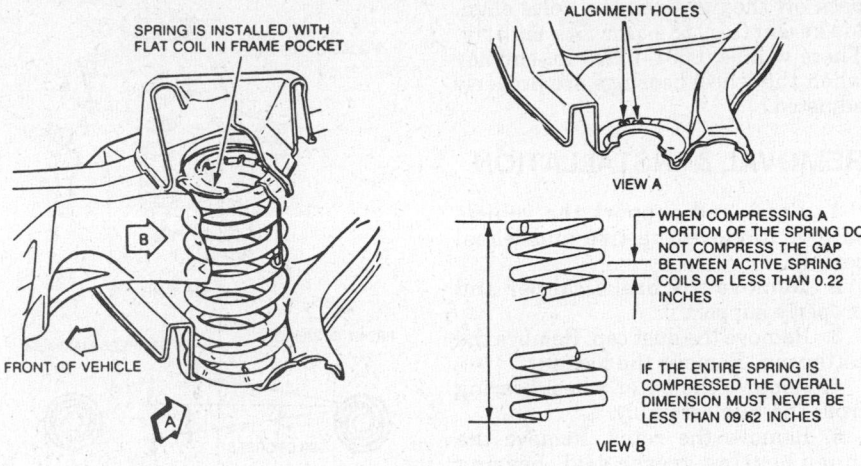

SPRING IS INSTALLED WITH
FLAT COIL IN FRAME POCKET

FRONT OF VEHICLE

ALIGNMENT HOLES

VIEW A

WHEN COMPRESSING A
PORTION OF THE SPRING DO
NOT COMPRESS THE GAP
BETWEEN ACTIVE SPRING
COILS OF LESS THAN 0.22
INCHES

IF THE ENTIRE SPRING IS
COMPRESSED THE OVERALL
DIMENSION MUST NEVER BE
LESS THAN 09.62 INCHES

VIEW B

Front coil spring positioning

3. The upper control arm bumpers must not contact the frame. Position a dial indicator gauge against the wheel rim.

4. Grasp the front wheel and push in on the bottom of the tire while pulling out at the top.

5. Read and record the gauge reading. Reverse the push pull procedure. Read and record the gauge reading. Horizontal deflection should not exceed 0.125 in.

6. If the horizontal deflection exceeds 0.125 in., if the ball joint has been disconnected from the knuckle assembly and looseness is detected or the joint can be twisted in the socket, ball joint replacement is necessary.

REMOVAL & INSTALLATION

1. Raise and support the vehicle safely. Remove the tire and wheel assembly.

2. Remove the caliper and properly support it so the brake hose is not damaged.

3. Remove the cotter pin from the ball joint. Loosen but do not remove the locknut. Properly support the lower control arm.

4. Using the proper ball joint separating tool, separate the ball joint stud

from the steering knuckle. Remove the locknut.

5. Lift the upper control arm upward and position a block of wood between the frame and the upper arm to act as a support.

6. Grind off the heads of the rivets retaining the ball joint in place. Remove the ball joint from its mounting.

7. Installation is the reverse of the removal procedure. Be sure to use new bolts to hold the ball joint in place. Check and adjust the front end alignment, as required.

Lower Ball Joints

INSPECTION

The vehicle must be supported by the wheel so the weight of the vehicle will properly load the ball joints. The lower ball joint is checked for wear by visual inspection. Wear is indicated by protrusion of the ½ in. diameter nipple into which the grease fitting is threaded. The round nipple projects 0.050 in. beyond the surface of the ball joint cover on a new ball joint. Normal wear will result in the surface of this nipple retreating slowly inward.

REMOVAL & INSTALLATION

1. Raise and support the vehicle safely. Remove the tire and wheel assembly.

2. Remove the cotter pin from the lower ball joint. Loosen, but do not remove the locknut. Properly support the lower control arm.

3. Using the proper ball joint separator tool, separate the ball joint from the steering knuckle. Remove the locknut.

4. Lift the upper control arm with the knuckle and hub assembly attached and position a block of wood between the frame and the upper arm to act as a support.

NOTE: Do not pull on the brake hose when lifting the knuckle and hub assembly as damage may occur.

5. As required, remove the tie rod end from the steering knuckle. Using the proper ball joint removal tool press the ball joint from the lower control arm.

6. Installation is the reverse of the removal procedure. Be sure to adjust the front end alignment, as required.

Upper Control Arms

REMOVAL & INSTALLATION

1. Raise and support the vehicle safely. Remove the tire and wheel assembly.

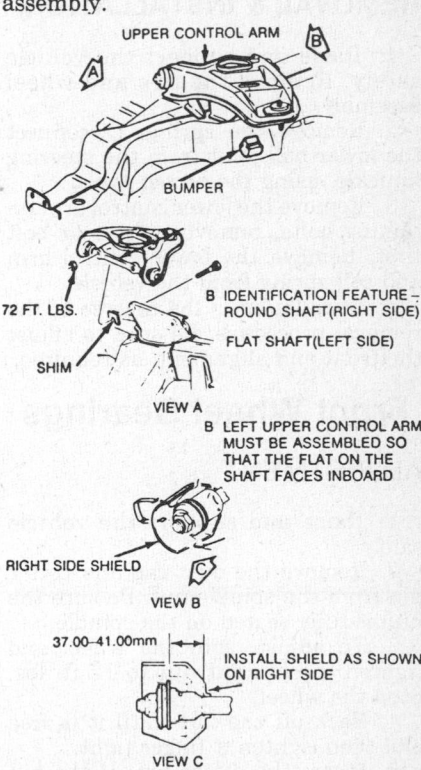

UPPER CONTROL ARM

BUMPER

72 FT. LBS.

SHIM

B IDENTIFICATION FEATURE —
ROUND SHAFT(RIGHT SIDE)

FLAT SHAFT(LEFT SIDE)

VIEW A

B LEFT UPPER CONTROL ARM
MUST BE ASSEMBLED SO
THAT THE FLAT ON THE
SHAFT FACES INBOARD

RIGHT SIDE SHIELD

VIEW B

37.00–41.00mm

INSTALL SHIELD AS SHOWN
ON RIGHT SIDE

VIEW C

Upper control arm adjustment data

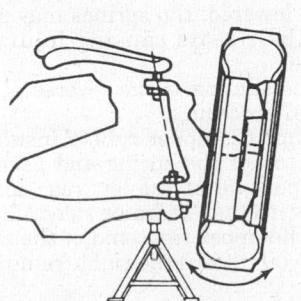

Checking upper ball joints

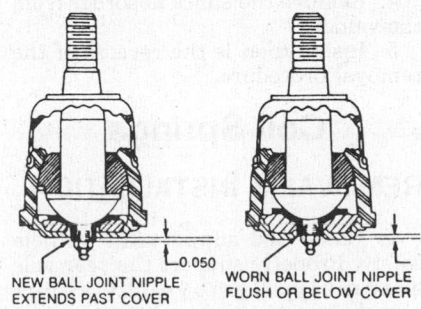

0.050

NEW BALL JOINT NIPPLE
EXTENDS PAST COVER

WORN BALL JOINT NIPPLE
FLUSH OR BELOW COVER

Checking lower ball joints

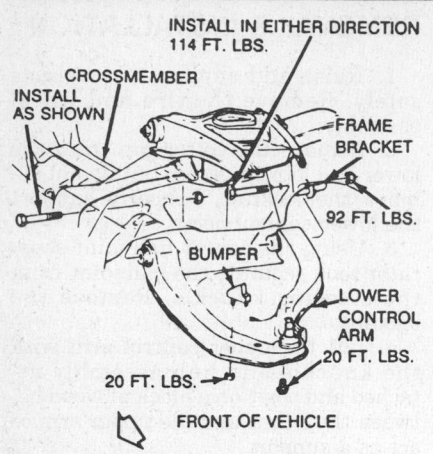

INSTALL IN EITHER DIRECTION 114 FT. LBS.

CROSSMEMBER

INSTALL AS SHOWN

FRAME BRACKET

92 FT. LBS.

BUMPER

LOWER CONTROL ARM

20 FT. LBS. **20 FT. LBS.**

FRONT OF VEHICLE

Lower control arm and related components

2. Properly support the lower control arm assembly.

3. Separate the upper ball joint from the steering knuckle, using the proper tools.

4. Remove the upper control arm shaft to frame bracket nuts. Be sure to replace the shims exactly in the same position they were removed.

5. Remove the upper control arm assembly from the vehicle.

6. Installation is the reverse of the removal procedure. Be sure to adjust the front end alignment, as required.

Lower Control Arms

REMOVAL & INSTALLATION

1. Raise and support the vehicle safely. Remove the tire and wheel assembly.

2. Remove the spring. Disconnect the lower ball joint from the steering knuckle, using the proper tools.

3. Remove the lower control arm retaining bolts, removing the rear bolt first. Remove the lower control arm and coil spring from the vehicle.

4. Installation is the reverse of the removal procedure. Be sure to adjust the front end alignment, as required.

Front Wheel Bearings

ADJUSTMENT

1. Raise and support the vehicle safely.

2. Remove the dust cap and cotter pin from the spindle nut. Be sure the hub is fully seated on the spindle.

3. To adjust, spin the wheel and tighten the locknut nut to 12 ft. lbs. Stop the wheel.

4. Back off the nut until it is free and then tighten it finger tight.

5. Insert the cotter pin. If the pin cannot be installed in this position,

back off the nut until the holes align. Make certain the pin fits tightly. There will be 0.001–0.005 in. endplay when the wheel bearings are properly adjusted.

REMOVAL & INSTALLATION

1. Raise and support the vehicle safely. Remove the tire and wheel sembly.

2. Remove the brake caliper and properly support it.

3. Remove the dust cap. Remove the cotter pin. Remove the locknut.

4. Remove the outer wheel bearing from the hub assembly.

5. Remove the rotor. Remove the inner bearing grease seal, bearing assembly.

6. Using a suitable brass punch, insert the punch into the hub recessed areas behind the inner or outer race, knock the inner and outer bearing races out.

7. When installing the outer and inner bearing cups, use an appropriate bearing cup installation tool.

8. Repack the bearings and lubricate all parts with an approved wheel bearing grease. Complete installation by reversing the removal procedure. Adjust the wheel bearings.

REAR SUSPENSION

Shock Absorbers

REMOVAL & INSTALLATION

1. Raise and support the vehicle safely. Properly support the rear axle assembly.

2. Disconnect the air lines, as required. Remove the upper shock absorber retaining bolts.

3. Remove the lower shock absorber retaining bolts.

4. Remove the shock absorber from the vehicle.

5. Installation is the reverse of the removal procedure.

Coil Springs

REMOVAL & INSTALLATION

1. Raise and support the vehicle safely. Properly support the rear axle assembly. Remove the shock absorbers.

2. Remove the stabilizer shaft.

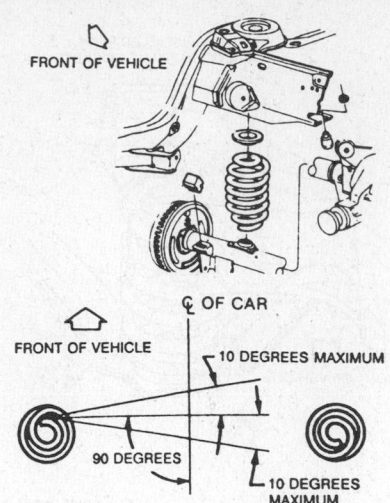

FRONT OF VEHICLE

C OF CAR

FRONT OF VEHICLE

10 DEGREES MAXIMUM

90 DEGREES

10 DEGREES MAXIMUM

Rear coil spring positioning

NOTE: Do not lower the axle assembly to the point at which the brake hoses become taut, as damage may result.

3. Remove the bolt that secures the brake hose junction block to the rear axle assembly. Remove the link on the height sensor arm, if equipped.

4. Remove the driveshaft retaining bolts. Remove the driveshaft and position it to the side.

5. Remove the upper control arm pivot bolts at the rear axle housing.

6. Disconnect the left side parking brake cable at the equalizer and disconnect the cable at the frame by removing the clip. Slide the cable through the hole.

7. Disconnect the cable from the clip at the center of the rear crossmember and disconnect the cable at the C connector which is located at the left of the frame between No. 2 and No. 3 body bolts.

8. Support both frame rails at the rear of the frame. Lower the axle to the point where the springs can be pried out. Be careful not to stretch the brake line or cable.

9. Remove the springs from the vehicle. If the axle is allowed to wind up as it is lowered, the springs may snap from their seats causing injury or damage.

10. Installation is the reverse of the removal procedure.

11. Tape the upper rubber insulator to the top of the spring and position the upper end of the left rear spring coil toward the left frame side rail. Position the upper right end of the right spring coil toward the right frame side rail.

12. Adjust the parking brakes, as required

Upper Control Arms

REMOVAL & INSTALLATION

NOTE: If both control arms are being replaced, only replace them one at a time to prevent the rear axle from slipping sideways

1. Raise and support the vehicle safely.
2. If equipped with electronic level control, remove the height sensor link retaining nut.
3. Properly support the rear axle assembly.
4. Remove the upper control arm retaining bolts. Remove the upper control arm from the vehicle.
5. Installation is the reverse of the removal procedure.

Lower Control Arms

REMOVAL & INSTALLATION

NOTE: If both control arms are being replaced, only replace them one at a time to prevent the rear axle from slipping sideways

1. Raise and support the vehicle safely. Properly support the rear axle assembly.
2. Disconnect the stabilizer arm bracket at the lower control arm. Remove the front and rear lower control arm retaining bolts.
3. Remove the lower control arm assembly from the vehicle.
4. Installation is the reverse of the removal procedure.

STEERING

Steering Wheel

REMOVAL & INSTALLATION

1. Disconnect the negative battery cable.
2. Remove the horn pad retaining screws. Remove the horn pad assembly.
3. Remove the horn contact wire from the plastic tower by pushing in on the wire and turning counterclockwise. The wire will spring out of the tower. It may be necessary to turn the ignition to the **ON** position in order to facilitate removal.
4. Remove the screws that secure the telescope locking lever assembly to the adjuster. Unscrew and remove the adjuster from the steering shaft.
5. Remove the locking lever assembly. Scribe an alignment mark on the

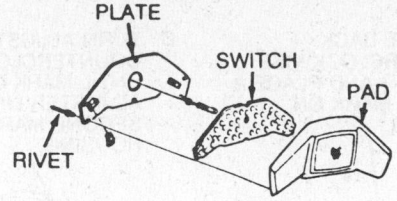

PAD AND HORN SWITCH

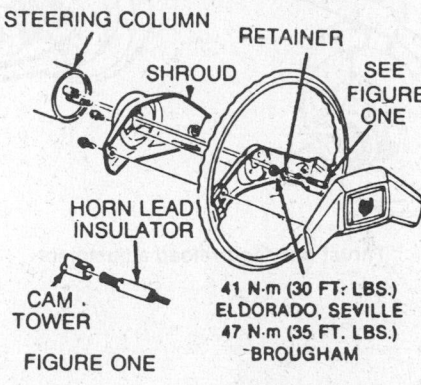

FIGURE ONE

STANDARD COLUMN

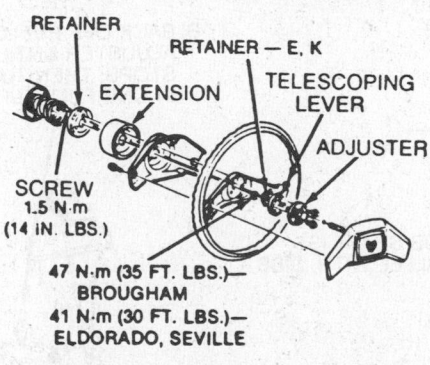

TILT AND TELESCOPING COLUMN

Steering wheel and related components

steering wheel hub in line with the slash mark on the steering shaft.
6. Loosen the locknut on the steering shaft and position it flush with the end of the shaft. Using the proper steering wheel removal tool, remove the wheel from its mounting on the steering shaft.
7. Remove the steering wheel removal tool from the steering wheel. Remove the locknut from the steering shaft. Remove the steering wheel from the vehicle.
8. Installation is the reverse of the removal procedure. When installing the steering wheel, it should not be driven on the steering shaft as damage to the steering column and components is possible.

Steering Column

REMOVAL & INSTALLATION

1. Disconnect the negative battery cable.
2. Center the steering wheel and remove the upper coupling pinch bolt and nut.
3. Disconnect the transmission shift linkage at the lower shift lever.
4. Remove the steering column lower cover from the instrument panel, exposing the upper support bolts.
5. Disconnect the turn signal wiring connector. If equipped with cruise control, disconnect the harness.
6. As required, remove the lower instrument panel trim cover. Remove the screw securing the shift cable to the shift bowl.
7. Loosen bolts at the steering column upper support. Do not completely remove the upper support nuts or bolts as the steering column could bend under its own weight.
8. Move the rubber carpet seal up the steering column as far as possible and position the carpet to gain access to the toe plate.
9. Remove the screws retaining the toe plate to the floor pan.
10. Remove the bolts at the upper column bracket, disconnect the remaining electrical connectors and vacuum connectors while supporting the column.
11. Carefully pull the steering column up and out of the vehicle. If the shaft hangs up in the upper coupling, secure the upper mounting bracket and free the coupling from the steering shaft. Remove the column assembly.
12. Installation is the reverse of the removal procedure. A clearance of $^5/_{16}$ in. should exist between the shaft and the upper coupling when the installation is complete or lower steering column bearing damage could result.

Power Steering Gear

ADJUSTMENT

NOTE: Adjustment of the steering gear, in the vehicle, is not recommended because of the difficulty encountered in adjusting the worm bearing preload and the effects of hydraulic fluid in the gear.

1. With the steering gear removed from the vehicle, proceed with the thrust bearing preload adjustment as follows:

    a. Using spanner wrench J–7624 or equivalent, tighten the adjuster

A. USING SPANNER WRENCH J-7624. TIGHTEN ADJUSTER PLUG UNTIL THRUST BEARING IS FIRMLY BOTTOMED, 27 N·m (20 FT. LBS.)

MARK HOUSING AND FACE OF ADJUSTER PLUG.

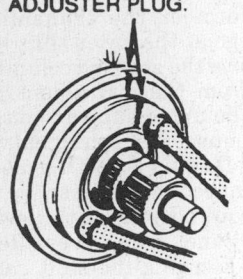

B. MEASURE BACK COUNTERCLOCKWISE 13mm (½") AND PLACE A SECOND MARK ON HOUSING.

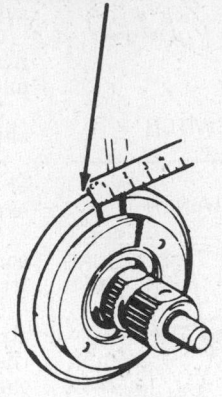

C. TURN ADJUSTER COUNTERCLOCKWISE UNTIL MARK ON FACE OF ADJUSTER LINES UP WITH SECOND MARK ON HOUSING.

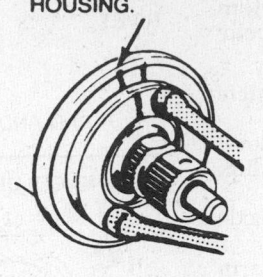

D. USING PUNCH IN NOTCH TIGHTEN LOCK NUT SECURELY. HOLD ADJUSTER PLUG TO MAINTAIN ALIGNMENT OF THE MARKS.

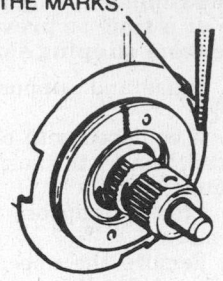

Thrust bearing preload adjustment

A.

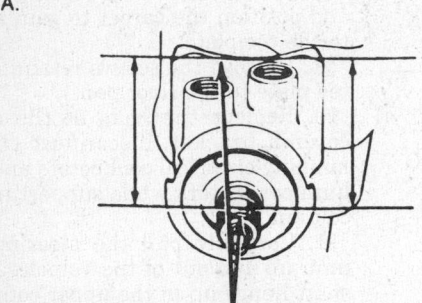

WHEN GEAR IS ON CENTER FLAT ON STUB SHAFT IS NORMALLY ON SAME SIDE AS, AND PARALLEL WITH, SIDE COVER.

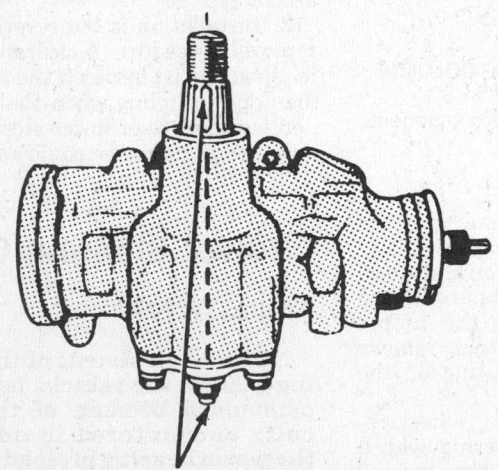

THE BLOCK TOOTH ON THE PITMAN SHAFT IS IN LINE WITH THE OVER-CENTER PRELOAD ADJUSTER.

B. BACK OFF PRELOAD ADJUSTER UNTIL IT STOPS, THEN TURN IT IN ONE FULL TURN.

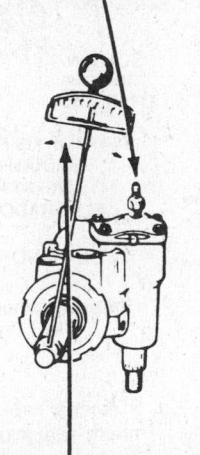

WITH GEAR AT CENTER OF TRAVEL, CHECK TORQUE TO TURN STUB SHAFT (READING #1).

C. TURN ADJUSTER IN UNTIL TORQUE TO TURN STUB SHAFT IS 0.6 TO 1.2 N·m (6 TO 11 IN. LBS.) MORE THAN READING #1.

TORQUE ADJUSTER LOCK NUT TO 27 N·m (20 FT. LBS.)

PREVENT ADJUSTER SCREW FROM TURNING WHILE TORQUING LOCK NUT.

Pitman shaft "over-center" sector adjustment

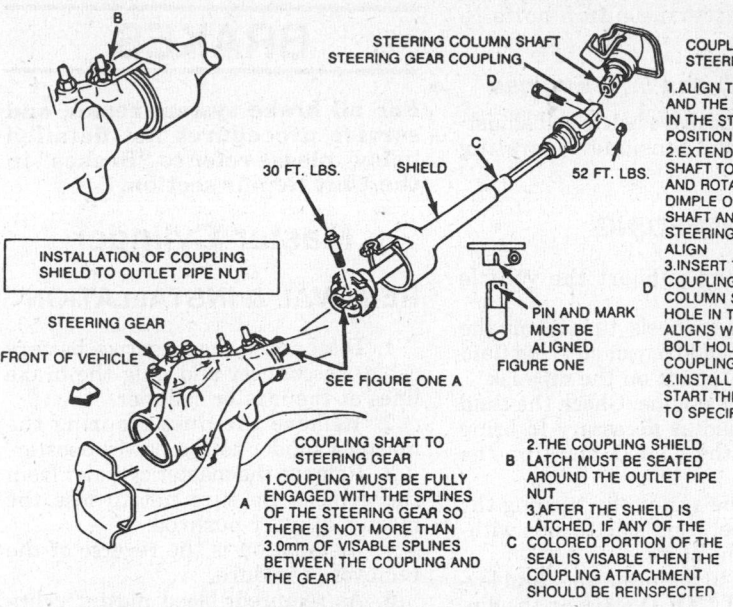

Steering gear and alignment data

Figure labels within diagram:

STEERING COLUMN SHAFT STEERING GEAR COUPLING

COUPLING SHAFT TO STEERING COLUMN SHAFT

1. ALIGN THE FRONT WHEELS AND THE STEERING COLUMN IN THE STRAIGHT AHEAD POSITION
2. EXTEND THE COUPLING SHAFT TO STEERING COLUMN AND ROTATE IT UNTIL THE DIMPLE ON THE COUPLING SHAFT AND FLAT ON THE STEERING COLUMN SHAFT ALIGN
3. INSERT THE STEERING GEAR COUPLING ONTO THE COLUMN SHAFT UNTIL THE HOLE IN THE COLUMN SHAFT ALIGNS WITH THE CROSS BOLT HOLES IN THE SHAFT COUPLING
4. INSTALL THE BOLT AND START THE NUT AND TORQUE TO SPECIFICATION

30 FT. LBS. — SHIELD — 52 FT. LBS.

INSTALLATION OF COUPLING SHIELD TO OUTLET PIPE NUT

STEERING GEAR

FRONT OF VEHICLE

PIN AND MARK MUST BE ALIGNED FIGURE ONE

SEE FIGURE ONE A

COUPLING SHAFT TO STEERING GEAR

1. COUPLING MUST BE FULLY ENGAGED WITH THE SPLINES OF THE STEERING GEAR SO THERE IS NOT MORE THAN 3.0mm OF VISABLE SPLINES BETWEEN THE COUPLING AND THE GEAR

2. THE COUPLING SHIELD LATCH MUST BE SEATED AROUND THE OUTLET PIPE NUT
3. AFTER THE SHIELD IS LATCHED, IF ANY OF THE COLORED PORTION OF THE SEAL IS VISABLE THEN THE COUPLING ATTACHMENT SHOULD BE REINSPECTED

plug until the thrust bearing is firmly bottomed, 20 ft. lbs. (27 Nm).

b. Mark the housing and face of the adjuster plug.

c. Measure back counterclockwise ½ in. (13mm) and place a second mark on the housing.

d. Turn the adjuster counterclockwise until the mark on the face of the adjuster lines up with the second mark on the housing.

e. Place a punch in the notch on the lockplate and tighten securely while preventing the adjuster plug from turning.

2. With the steering gear removed from the vehicle, proceed with the pitman shaft "over-center" sector adjustment as follows:

a. Place the steering gear in its centered position with the center flat on the stub shaft parallel with the preload adjuster cover and the block tooth on the pitman shaft in line with the preload adjuster cover.

b. Back off the preload adjuster until it stops, then turn the screw in 1 full turn.

c. With the gear at the center point of travel; using a suitable inch lb. torque wrench, check and record the torque required to turn the stub shaft.

d. Turn the adjuster in until the torque required to turn the stub shaft is 6–11 inch lbs. (0.6–1.2 Nm) more than the previously recorded reading.

e. Tighten the adjuster locknut while preventing the adjuster screw from turning.

REMOVAL & INSTALLATION

1. Disconnect the negative battery cable. Position a drain pan under the steering gear. Disconnect the pressure and return lines from the steering gear assembly. Plug the opening to prevent the entrance of dirt.

2. If equipped, disconnect the stone shield from the return pipe.

3. Remove the pinch bolt from the flex coupling and disconnect the coupling from the gear.

NOTE: Failure to disconnect the flexible coupling from the steering gear stub shaft can result in damage to the steering gear and or the intermediate shaft. This damage can cause the loss of steering control which could result in a vehicle crash and bodily injuries.

4. Raise the vehicle and support it safely.

5. Remove the pitman arm nut and washer. Remove the pitman arm from the sector shaft using a pitman arm puller tool.

6. Remove the retaining bolts and washers holding the steering gear to the side rail. Lower the gear assembly from the vehicle.

7. The installation is the reverse of the removal procedure. Tighten the pitman arm nut to 185 ft. lbs., the mounting bolts to 70 ft. lbs. and the flex coupling pinch bolt to 30 ft. lbs.

Power Steering Pump

REMOVAL & INSTALLATION

5.0L (VIN Y) Engine

1. Disconnect the negative battery cable. Disconnect and relocate the air cleaner inlet tube and the upper radiator hose to gain access to the pump.

2. Remove the alternator belt. Loosen the alternator mounting bolts, except for the long bolt. Rotate the unit upward to gain access by pivoting the long bolt.

3. Remove and plug the pressure and return hoses from the pump. Remove the front pump bracket mounting bolts and spacer. Remove the rear pump mounting nut.

4. Remove the power steering pump drive belt. Remove the pump and bracket from the engine as an assembly.

5. Installation is the reverse order of the removal procedure. Be sure to bleed the air from the system.

5.0L (VIN E) and 5.7L Engines

1. Disconnect the negative battery cable. Remove the serpentine drive belt.

2. Remove the power steering pump pulley. Raise and support the vehicle safely.

20 FT. LBS.

30 FT. LBS.

SPACER

19 FT. LBS.

PUMP SUPPORT

70 FT. LBS.

22 FT. LBS.

SPACER

* LOOSEN BOLTS FOR PUMP ADJUSTMENT

35 FT. LBS.

30 FT. LBS.

PUMP ADJUSTMENT BRACKET

Power steering pump mounting — 5.0L (VIN Y) engine

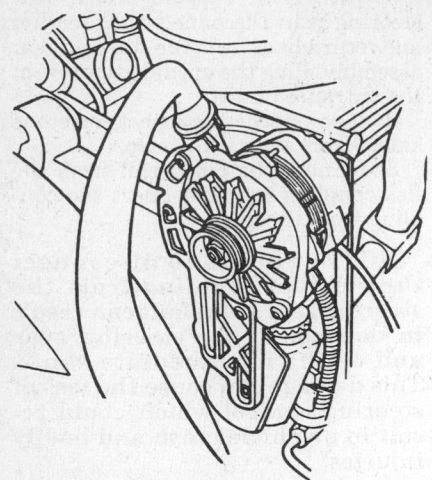

Power steering pump mounting—5.0L (VIN E) and 5.7L engines

3. Remove and plug the pressure and return hoses from the pump. Remove the 3 Torx head screws an separate the pump from the bracket.

4. Remove the power steering pump from the vehicle.

5. Installation is the reverse of the removal procedure. Bleed the system, as required.

BELT ADJUSTMENT

5.0L (VIN Y) Engine

1. Disconnect the negative battery cable.

2. Loosen the power steering pump mounting bolts.

3. Install the belt tension gauge. Correct tension is 90 lbs. minimum for a used belt and 170 lbs. maximum for a new belt.

4. Adjust the belt by prying the power steering pump away from the engine.

NOTE: When adjusting the power steering pump belt be sure not to pry against the pump reservoir. Only the power steering pump bracket should be pryed against when adjusting the belt.

5. Once the belt is adjusted, tighten

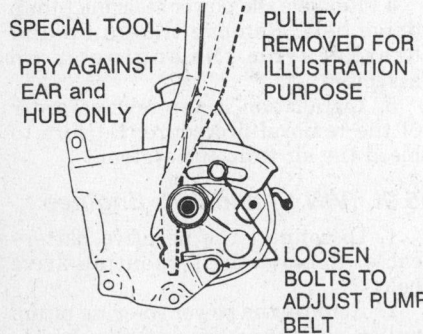

SPECIAL TOOL→

PRY AGAINST EAR and HUB ONLY

PULLEY REMOVED FOR ILLUSTRATION PURPOSE

LOOSEN BOLTS TO ADJUST PUMP BELT

Power steering pump adjustment—5.0L (VIN Y) engine

the power steering pump bolts to specification.

5.0L (VIN E) and 5.7L Engines

The serpentine drive belt is self adjusting within the tensioner operating limits.

SYSTEM BLEEDING

1. Raise and support the vehicle safely.

2. With the wheels turned all the way to the left add power steering fluid to the **COLD** mark on the dipstick.

3. Start the engine. Check the fluid level. Add fluid as necessary to bring the level to the **COLD** mark on the dipstick.

4. Bleed the system by turning the steering wheel from side to side without hitting the stops.

5. Be sure to maintain the fluid level at the **HOT/COLD** mark on the dipstick. Fluid with air in it will have a light tan appearance. This air must be expelled from the system before normal steering action can be obtained.

6. Return the wheels to the center position. Allow the engine to run for several minutes and then turn the engine **OFF**.

7. Road test the vehicle and make sure the steering performs properly and there is no noise from the power steering pump.

8. Recheck the power steering level. Be sure the fluid level is at the **HOT** mark on the dipstick after the system has stabilized at its normal operating temperature.

Tie Rod Ends

REMOVAL & INSTALLATION

1. Raise and support the vehicle safely.

2. Remove the cotter pin and castellated nut from the outer tie rod end.

3. Using the proper tool disconnect the tie rod end from the steering knuckle.

4. Using the proper tool remove the inner ball stud from the intermediate rod.

NOTE: When disconnecting a linkage joint no attempt should be made to disengage the joint by driving a wedge between the joint and the retained part as seal damage may result.

5. Remove the tie rod from the adjuster tube by loosening the clamp bolts and unscrewing the end assemblies.

6. Installation is the reverse of the removal procedure. Check and adjust front end alignment, as required.

BRAKES

For all brake system repair and service procedures not detailed below, please refer to "Brakes" in the Unit Repair section.

Master Cylinder

REMOVAL & INSTALLATION

1. Disconnect the negative battery cable. Disconnect and plug the brake lines at the master cylinder.

2. Remove the nuts securing the master cylinder to the power booster.

3. Remove the master cylinder from the vehicle. Be sure not to lose the master cylinder pushrod.

4. Installation is the reverse of the removal procedure.

5. As required, bleed master cylinder before installing on the vehicle. Bleed the system, as required.

Combination Valve

REMOVAL & INSTALLATION

1. Disconnect the negative battery cable. Disconnect the electrical connector from the valve assembly.

2. Disconnect and plug the brake lines at the valve.

3. Remove the valve retaining bolt. Remove the valve from its mounting.

4. Installation is the reverse of the removal procedure. Bleed the brake system.

Power Brake Booster

REMOVAL & INSTALLATION

1. Disconnect the negative battery cable. Remove the master cylinder retaining nuts and position the assembly aside.

2. Be sure not to lose the master cylinder pushrod. Disconnect vacuum line from vacuum check valve on unit.

3. Remove the steering column lower cover inside the vehicle.

4. Remove cotter pin, washer and spring spacer that secures power unit pushrod to brake pedal arm.

5. Remove the nuts that secure the power unit to the firewall. Remove the power brake unit.

6. Installation is the reverse of the removal procedure. As required, bleed the system.

Brake Caliper

REMOVAL & INSTALLATION

1. Disconnect the negative battery

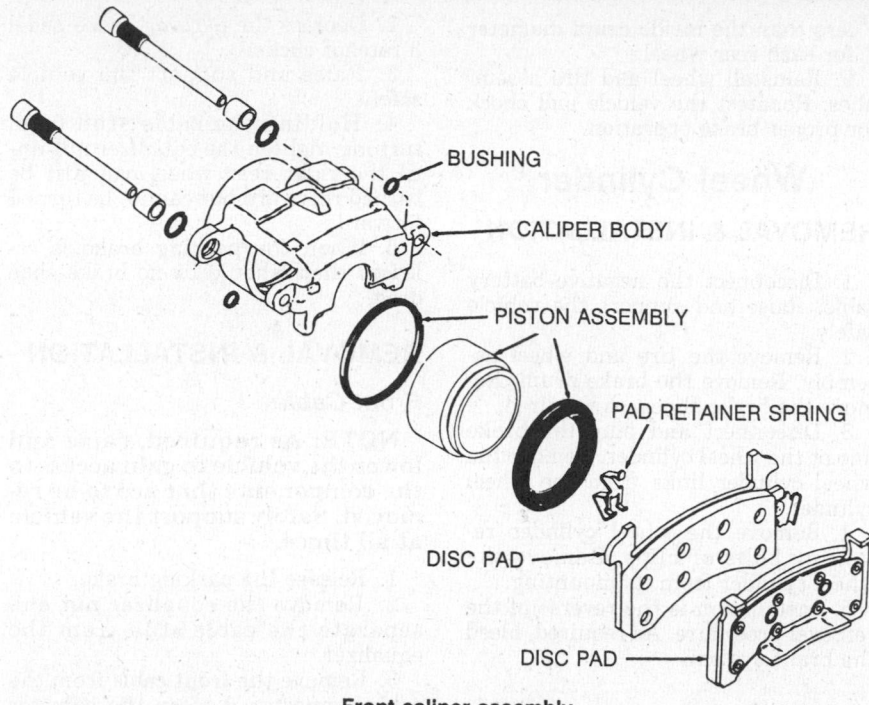

Front caliper assembly

BUSHING

CALIPER BODY

PISTON ASSEMBLY

PAD RETAINER SPRING

DISC PAD

DISC PAD

cable. Partially drain the master cylinder assembly.

2. Raise and support the vehicle safely. Remove the tire and wheel assembly.

3. Position a C-clamp over the inboard brake shoe tab and and the inboard caliper housing, carefully compress the piston assembly back into the caliper housing.

4. Disconnect and plug the brake line hose at the caliper. If equipped, disconnect any required electrical connectors at the caliper assembly.

5. Remove the caliper retaining bolts. Remove the caliper assembly from its mounting.

6. Installation is the reverse of the removal procedure. As required, bleed the brake system.

Disc Brake Pads

REMOVAL & INSTALLATION

1. Disconnect the negative battery cable. Partially drain the master cylinder assembly.

2. Raise and support the vehicle safely. Remove the tire and wheel assembly.

3. Position a C-clamp over the inboard brake shoe tab and and the inboard caliper housing, carefully compress the piston assembly back into the caliper housing.

4. Remove the caliper retaining bolts. Remove the caliper assembly from its mounting. Do not allow the caliper to hang by the brake line hose.

5. Remove the outboard shoe and lining. Remove the inboard shoe and lining.

To install:

6. Installation is the reverse of the removal procedure. As required, bleed the brake system. Use a new inboard shoe retainer springs and ensure caliper sliding points are free of rust and dirt; apply grease to the slide points.

7. Apply about 175 lbs. of force to the brake pedal 3 times to seat the linings.

8. Position a pair of channel lock pliers over the brake shoe ears and bottom edge of the caliper. While applying about 50 lbs. of force on the brake pedal, clinch the outboard shoe ears to the caliper.

Brake Rotor

REMOVAL & INSTALLATION

1. Disconnect the negative battery cable. Partially drain the master cylinder assembly.

2. Raise and support the vehicle

1. Return spring
2. Return spring
3. Hold down spring
4. Lever pivot
5. Hold down pin
6. Actuator link
7. Actuator lever
8. Pawl
9. Lever return spring
10. Shoe guide
11. Parking brake
12. Strut spring
13. Primary shoe
14. Secondary shoe
15. Adjusting screw spring
16. Parking brake lever
17. Backing plate
18. Adjusting screw assembly
19. Anchor pin

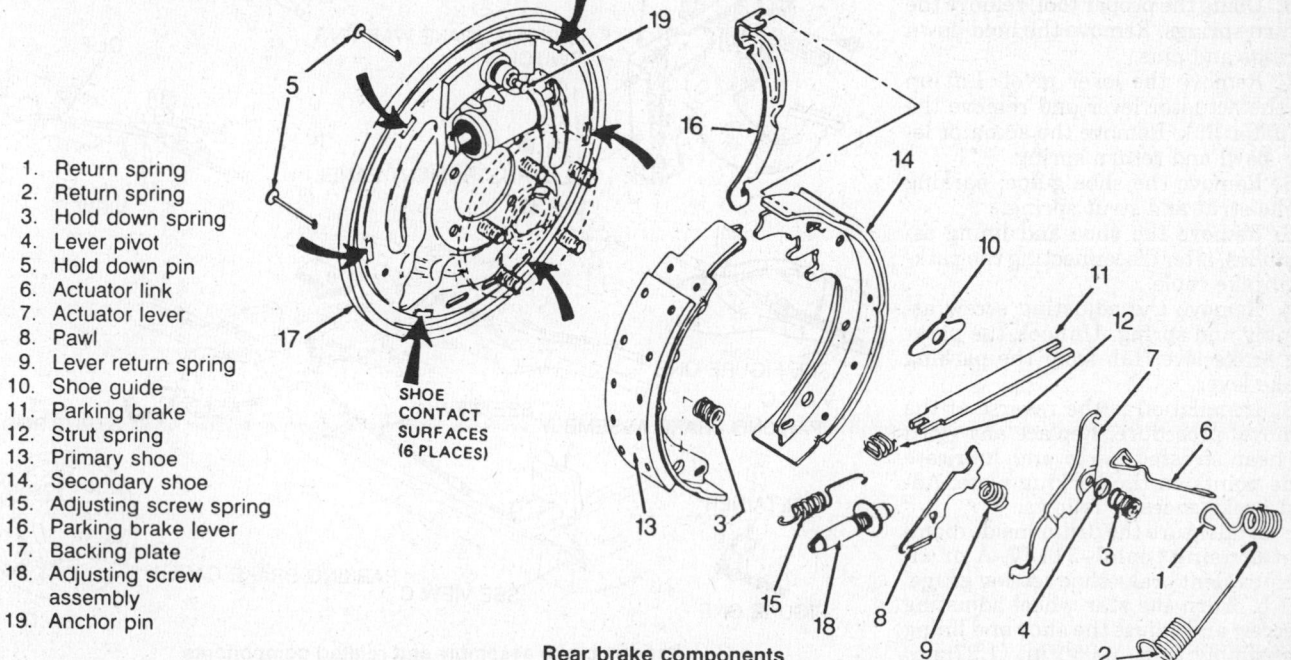

SHOE CONTACT SURFACES (6 PLACES)

Rear brake components

safely. Remove the tire and wheel assembly.

3. Remove the caliper retaining bolts. Remove the caliper assembly from its mounting and position it aside.

4. Remove the wheel bearing dust cap. Remove the cotter pin, locknut, washer and outer bearing assembly.

5. Remove the rotor assembly from the spindle.

6. Installation is the reverse of the removal procedure. Adjust the wheel bearings, as required.

Brake Drums

REMOVAL & INSTALLATION

1. Disconnect the negative battery cable.

2. Raise and support the vehicle safely. Remove the tire and wheel assembly.

3. Remove the drum locking tabs, if equipped. Remove the rear brake drum from the vehicle.

4. Ensure parking brake is released and if required, back off automatic brake adjuster to aid in drum removal.

5. Installation is the reverse of the removal procedure.

Brake Shoes

REMOVAL & INSTALLATION

1. Disconnect the negative battery cable. Raise and support the vehicle safely.

2. Remove the tire and wheel assembly. Remove the brake drum.

3. Using the proper tool, remove the return springs. Remove the hold-down springs and pins.

4. Remove the lever pivot. Lift up on the actuator lever and remove the actuator link. Remove the actuator lever, pawl and return spring.

5. Remove the shoe guide, parking brake strut and strut spring.

6. Remove the shoe and lining assemblies, after disconnecting the parking brake cable.

7. Remove the adjusting screw assembly and spring. Unhook the parking brake lever tab from the parking brake lever.

8. Installation is the reverse of the removal procedure. Replace any worn or heat stressed parts and lubricate slide points on the backing plate. Adjust brake shoes as follows:

   a. Measure the drum inside diameter using tool J–21177–A or an equivalent brake shoe setting gauge.

   b. Turn the star wheel adjusting screw and adjust the shoe and lining assembly to be 0.050 in. (1.27mm)

less than the inside drum diameter for each rear wheel.

9. Reinstall wheel and tire assemblies. Roadtest the vehicle and check for proper brake operation.

Wheel Cylinder

REMOVAL & INSTALLATION

1. Disconnect the negative battery cable. Raise and support the vehicle safely.

2. Remove the tire and wheel assembly. Remove the brake drum. Remove the brake shoes, as required.

3. Disconnect and plug the brake line at the wheel cylinder. Remove the wheel cylinder links from the wheel cylinder.

4. Remove the wheel cylinder retaining bolts or clips. Remove the wheel cylinder from its mounting.

5. Installation is the reverse of the removal procedure. As required, bleed the brake system.

Parking Brake Cable

ADJUSTMENT

1. Be sure the rear brakes are properly adjusted before adjusting the parking brake. Check the parking brake linkage for the free movement of all the cables. Lubricate or replace, as necessary.

2. Depress the parking brake pedal 3 ratchet clicks.

3. Raise and support the vehicle safely.

4. Holding the cable stud from turning, tighten the equalizer nut until the right rear wheel can just be turned rearward but cannot be turned forward.

5. When the parking brake is released there should be no brake shoe drag.

REMOVAL & INSTALLATION

Front Cable

NOTE: As required, raise and lower the vehicle to gain access to the components that are to be removed. Safely support the vehicle at all times.

1. Release the parking brake.

2. Remove the equalizer nut and separate the cable stud from the equalizer.

3. Remove the front cable from the cable connector. Loosen the adjuster nut and disconnect the front cable from the connector.

4. Compress the cable retainer fingers and loosen the assembly at the frame.

5. Remove the cable at the pedal assembly. Remove the cable end from the parking brake assembly clevis.

6. Pull the cable through the hole in the frame and remove it from the vehicle.

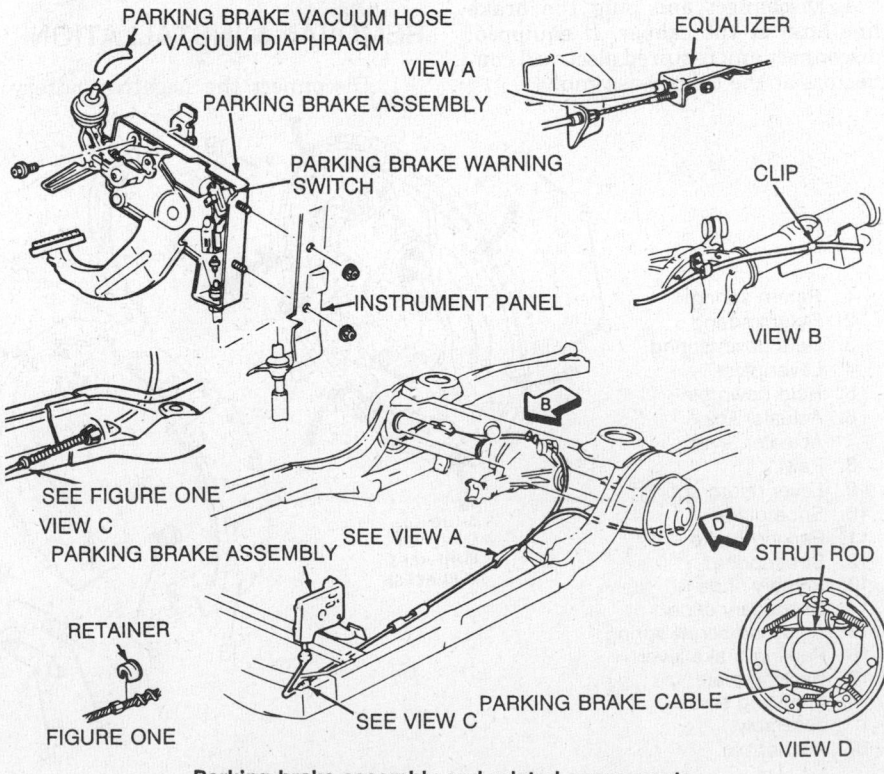

Parking brake assembly and related components

7. Installation is the reverse of the removal procedure. Adjust the parking brakes.

Rear Cable

1. Release the parking brake. Raise and support the vehicle safely.
2. Remove the tire and wheel assembly. Remove the brake drum.
3. Remove the equalizer nut and the retainer. Separate the equalizer from the right rear cable stud.
4. Remove the end of the left rear cable from the cable connector and equalizer.
5. Remove the clip retaining the right rear cable to the control arm bracket. Pull the cable rearward and remove it from the bracket.
6. Remove the rear brake shoe assembly, as required. Remove the cable from the brake backing plate. Remove the pawl spring and pawl lever from the actuating lever.
7. Remove the cable end from the operating lever. Remove the cable from the backing plate.
8. Installation is the reverse of the removal procedure. Adjust the parking brakes.

Brake System Bleeding

1. Fill the master cylinder to within ¼ in. of the reservoir rim.
2. Raise and support the vehicle safely.
3. Bleed the system in the following sequence: right rear, left rear, right front and left front.
4. Bleed one wheel at a time.
5. Install a transparent tube on the bleeder screw of the caliper or wheel cylinder to be bled and place the opposite end of the hose in a clear container partially filled with brake fluid.
6. Open the bleeder screw ¾ turn. Depress the brake pedal to the floor, then tighten the bleeder screw. Slowly release the brake pedal.
7. Repeat the bleeding operation until clear brake fluid flows without air bubbles.

NOTE: Check the master cylinder fluid level frequently during the bleeding procedure and refill, if necessary.

8. After bleeding operation is completed, discard the fluid in the container. Fill the master cylinder to ¼ in. from the reservoir rim and check the brake system operation.

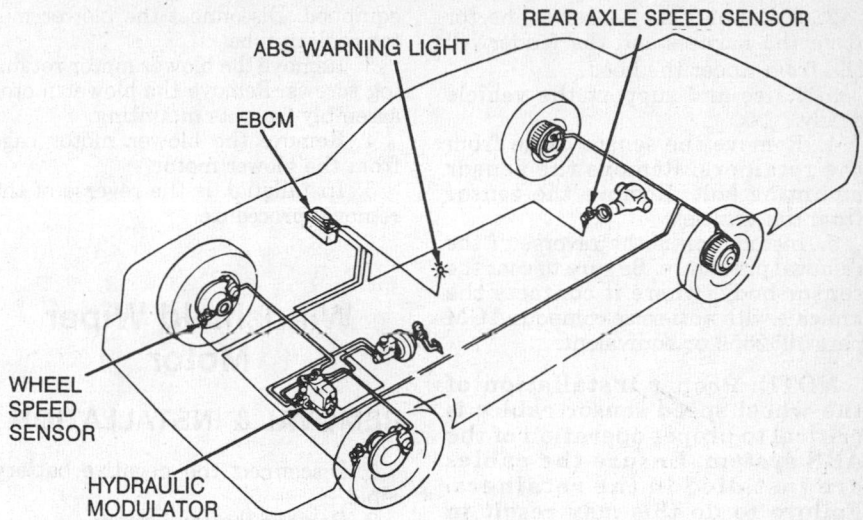

ABS braking system component layout

Anti-Lock Brake System Service

PRECAUTIONS

Failure to observe the following precautions may result in system damage.
● Before performing electric arc welding on the vehicle, disconnect the Electronic Brake Control Module (EBCM) and the hydraulic modulator connectors.
● When performing painting work on the vehicle, do not expose the Electronic Brake Control Module (EBCM) to temperatures in excess of 185°F (85°C) for longer than 2 hours. The system may be exposed to temperatures up to 200°F (95°C) for less than 15 minutes.
● Never disconnect or connect the Electronic Brake Control Module (EBCM) or hydraulic modulator connectors with the ignition switch ON.
● Never disassemble any component of the Anti-Lock Brake System (ABS) which is designated non-servicable; the component must be replaced as an assembly.
● When filling the master cylinder, always use Delco Supreme 11 brake fluid or equivalent, which meets DOT-3 specifications; petroleum base fluid will destroy the rubber parts.

RELEIVING ANTI-LOCK BRAKE SYSTEM PRESSURE

When servicing and bleeding ABS components, follow normal manual or pressure bleeding procedures. Special service procedures for bleeding the brake system with a hydraulic modulator are not required.

Hydraulic Modulator

REMOVAL & INSTALLATION

1. Disconnect the negative battery cable. Remove the left front radiator brace.
2. Remove the air cleaner intake hose. Remove the ABS modulator relay cover.
3. Disconnect the modulator 12-pin connector and ground strap.
4. Disconnect and plug the brake line connections at the modulator assembly.
5. Remove the modulator retaining bolts. Remove the modulator assembly from its mounting.
6. Installation is the reverse of the removal procedure. Bleed the brake system.

Electronic Brake Control Module

REMOVAL & INSTALLATION

1. Disconnect the negative battery cable.
2. Remove the passenger side close-out panel. Remove the glove box liner.
3. Disconnect the EBCM wire harness connector and remove the EBCM from its retaining bracket.
4. Installation is the reverse of the removal procedure.

Front Wheel Speed Sensor

REMOVAL & INSTALLATION

1. Disconnect the negative battery cable.

2. Disconnect the sensor connector from the harness and the fenderwell clip from under the hood.

3. Raise and support the vehicle safely.

4. Remove the sensor cable from the retainers. Remove the sensor mounting bolt. Remove the sensor from the vehicle.

5. Installation is the reverse of the removal procedure. Be sure to coat the sensor body, where it contacts the knuckle with anti-seize compound GM part 1052856 or equivalent.

NOTE: Proper installation of the wheel speed sensor cables is critical to proper operation of the ABS system. Ensure the cables are installed in the retainers. Failure to do this may result in contact with moving parts and the over extension of the cables, resulting in an open circuit.

Rear Axle Speed Sensor

REMOVAL & INSTALLATION

1. Disconnect the negative battery cable. Raise and support the vehicle safely.

2. Disconnect the sensor connector. Remove the sensor cable from the retainer brackets.

3. Remove the sensor mounting bolt. Remove the sensor, plastic spacer and the O-ring from the vehicle.

4. Installation is the reverse of the removal procedure.

NOTE: Proper installation of the wheel speed sensor cable is critical to proper operation of the ABS system. Ensure the cables are installed in the retainers. Failure to do this may result in contact with moving parts and the over extension of the cable, resulting in an open circuit.

CHASSIS ELECTRICAL

Heater Blower Motor

REMOVAL & INSTALLATION

1. Disconnect the negative battery cable.

2. Disconnect the electrical connector from the blower motor assembly. Disconnect the ground wire, if

equipped. Disconnect the blower motor cooling tube.

3. Remove the blower motor retaining screws. Remove the blower motor assembly from its mounting.

4. Remove the blower motor cage from the blower motor.

5. Installation is the reverse of the removal procedure.

Windshield Wiper Motor

REMOVAL & INSTALLATION

1. Disconnect the negative battery cable.

2. Remove the cowl screen.

3. Reach through the opening and disengage the transmission drive link from the wiper crank arm by loosening the nuts.

4. Disconnect the electrical wiring and washer hoses.

5. Remove the bolts that secure the wiper/washer unit to the firewall.

6. Remove the entire assembly.

7. Installation is the reverse of the removal procedure. Be sure the wiper crank arm is in the park position.

Windshield Wiper Switch

REMOVAL & INSTALLATION

1. Disconnect the negative battery cable.

2. Loosen the set screw in the left climate control outlet door knob and remove the knob.

3. Remove the left climate control air outlet grille.

4. Remove the left trim plate attaching screws.

5. Remove the lower steering column cover.

6. Disconnect the steering column seal from lower surface and remove the trim plate.

7. Remove the wiper switch mounting screws. Remove the switch and separate the electrical connector.

8. Installation is the reverse of the removal procedure.

Instrument Cluster

REMOVAL & INSTALLATION

1. Disconnect the negative battery cable.

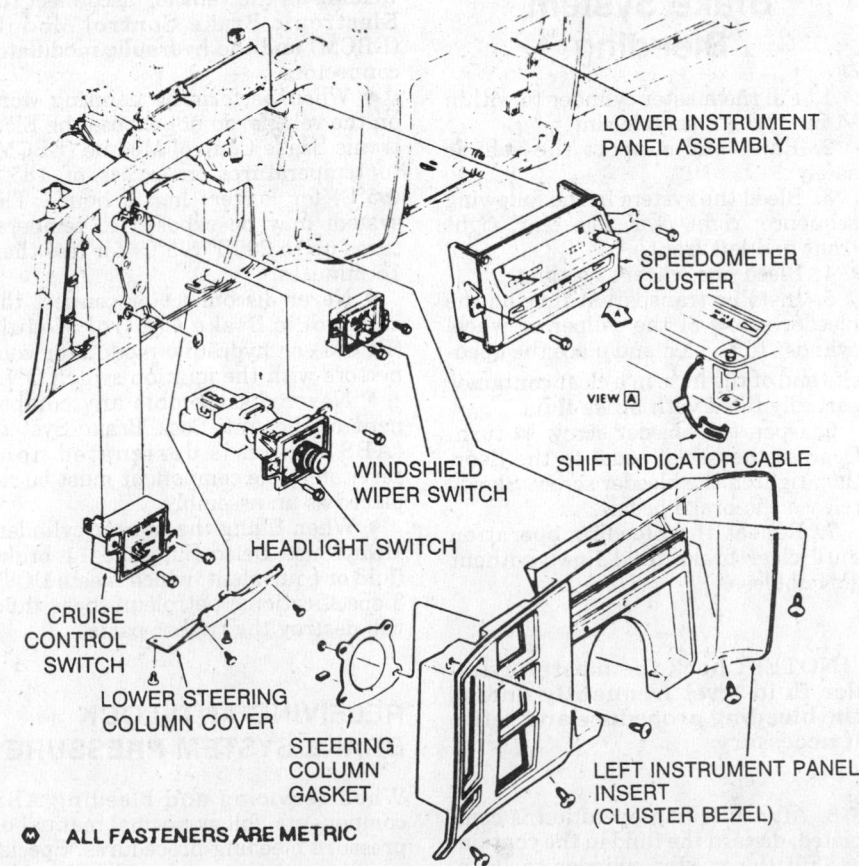

LOWER INSTRUMENT PANEL ASSEMBLY

SPEEDOMETER CLUSTER

VIEW Ⓐ

SHIFT INDICATOR CABLE

WINDSHIELD WIPER SWITCH

HEADLIGHT SWITCH

CRUISE CONTROL SWITCH

LOWER STEERING COLUMN COVER

STEERING COLUMN GASKET

LEFT INSTRUMENT PANEL INSERT (CLUSTER BEZEL)

Ⓜ **ALL FASTENERS ARE METRIC**

Instrument cluster and related components

2. Remove the instrument panel insert and left side trim panel, as required.

3. With the shift lever in the **P** position, remove the shift indicator cable and clip retaining screw from the steering column.

4. Remove the upper and lower cluster assembly retaining screws. Remove the screw directly above the steering column which retains the cluster to the speedometer mounting plate.

5. Pull the cluster outward and disengage the speedometer cable, if equipped. Disconnect all electrical connections.

6. If equipped, disconnect the speed control sensor from the cluster assembly.

7. Place the shift lever in the **L** and the tilt wheel in the lower position. Remove the cluster assembly from the dash.

8. Installation is the reverse of the removal procedure. Set the shift, indicator cable in the **N** position and adjust the cable accordingly.

Radio

REMOVAL & INSTALLATION

1. Disconnect the negative battery cable.

2. Remove the center air conditioning outlets, using tool J–24612 or equivalent.

3. Remove the center panel retaining screws, once the outlets are removed.

4. Remove the radio assembly retaining knobs, as required.

5. Remove the remaining trim plate screws and remove the trim plate from the vehicle.

6. Remove the radio retaining screws and pull the radio assembly forward. Disconnect the electrical connections. Disconnect the radio antenna. Remove the radio assembly from the vehicle.

7. Installation is the reverse of the removal procedure.

Headlight Switch

REMOVAL & INSTALLATION

1. Disconnect the negative battery cable.

2. Remove the instrument panel insert.

3. Remove the screws securing the switch to the instrument panel.

4. If equipped with cruise control and twilight sentinel, remove the screws securing the cruise control switch to the instrument panel.

5. Slide the cruise control switch forward to remove the light switch. If equipped, disconnect the 2 piece connector from the headlight switch. Disconnect the guidematic and twilight sentinel electrical connectors from under the instrument panel.

6. Remove the headlight switch retaining screws. Remove the switch assembly from the vehicle.

7. Installation is the reverse of the removal procedure.

Dimmer Switch

REMOVAL & INSTALLATION

1. Disconnect negative battery cable.

2. Remove left sound insulator.

3. Remove 2 nuts securing steering column to upper mounting bracket.

4. Lower steering column and remove 2 screws securing ignition switch and dimmer switch.

5. Disconnect electrical connections and remove switch.

6. Installation is the reverse of the removal procedure. Adjust the dimmer switch as necessary.

ADJUSTMENT

1. Insert the proper alignment tool (drill bit) through the locating hole.

2. Loosen both screws attaching the dimmer switch mounting bracket.

3. Slide the dimmer switch firmly against the actuator arm and tighten both adjusting screws.

Turn Signal Switch

REMOVAL & INSTALLATION

Except Tilt and Telescopic Column

1. Disconnect the negative battery cable.

2. Remove the steering wheel.

3. Using a suitable tool, remove the lock plate cover assembly.

4. Install a suitable lock plate tool onto the steering shaft. Tighten the

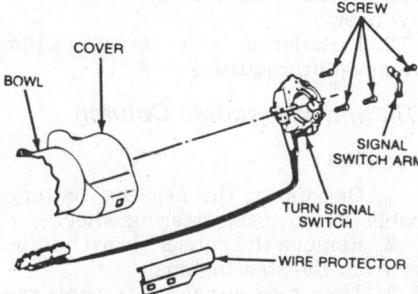

Turn signal switch assembly—standard column

tool to compress the lock plate and the spring. Remove the snapring from the groove in the shaft.

5. Remove the lock plate, the turn signal cam, the upper bearing preload spring and the thrust washer from the upper steering shaft.

6. Remove the steering column lower cover.

7. Remove the turn signal lever from the column.

8. If equipped with cruise control, disconnect the cruise control wire from the harness near the bottom of the column. Remove the harness protector from the cruise control wire. Do not remove the wire from the column.

9. Remove the vertical bolts at the steering column upper support. Remove the shim packs. Keep the shims in order for reinstallation.

10. Remove the screws securing the column upper mounting bracket to the column. Remove the bracket.

11. Disconnect the turn signal wiring and remove the wires from the plastic protector.

12. Remove the turn signal switch mounting screws.

13. Slide the switch connector out of the bracket on the steering column.

14. If the switch is known to be bad, cut the wires and discard the switch. Tape the connector of the new switch to the old wires, and pull the new harness down through the steering column while removing the old wires.

15. If the original switch is to be reused, wrap tape around the connector and pull the harness up through the column. It may be helpful to attach a length of string to the harness connector before pulling it up through the column to help with installation.

16. After freeing the switch wiring protector from its mounting, pull the turn signal switch straight up and remove the switch, switch harness and the connector from the column.

17. Installation is the reverse of the removal procedure.

Tilt and Telescopic Column

1. Disconnect the negative battery cable. Remove the steering wheel.

2. Remove the rubber sleeve bumper from the steering shaft.

3. Remove the plastic retainer and disengage the tabs on the retainer from the C-ring.

4. Compress the upper steering shaft preload spring with a spring compressor and remove the C-ring. When installing the spring compressor, pull the upper shaft up about 1 in. and turn the ignition to the **LOCK** position to hold the shaft in place.

5. Remove the spring compressor and remove the upper steering shaft lock plate, horn contact carrier and the preload spring.

6. Remove the steering column lower cover. Unscrew and remove the turn signal lever.

7. If equipped with cruise control, disconnect the cruise control wire from the harness near the bottom of the steering column. Slide the protector off the cruise control wire. Remove the lever attaching screw and carefully pull the lever out enough to allow the removal of the turn signal switch.

8. Remove the nuts and shim packs from the upper column support. Keep the shims together as a unit for reinstallation.

9. Remove the bracket from the steering column by removing the 2 attaching screws from each side.

10. Disconnect the turn signal wiring harness and remove the wires from the plastic protector.

11. Remove the turn signal switch retaining screws and pull the switch up and out of the steering column.

12. If the switch is to be replaced, cut the wires from the switch and tape the new switch connector to the old wires. Carefully pull the new harness down through the column as the old wires are removed.

13. If the old switch is to be reused, tape the connector to the wires and carefully pull the harness up out of the column.

14. Feed the wiring harness down through the steering column to replace the old switch.

15. Secure the switch in the steering column.

16. Install the upper shaft preload spring.

17. Install the lock plate and carrier assembly. Make sure the flat on the lower end of the steering shaft is pointing up and the small plastic tab on the carrier is up or nearest the top of the column. The flat surface of the lock plate must be installed facing down against the turn signal switch.

18. Install the spring compressor, compress the preload spring and lock plate and install the C-ring with the wide side toward the keyway.

19. Remove the spring compressor and install the plastic retainer on the C-ring.

20. Install the rubber sleeve bumper over the steering shaft and install the steering wheel.

21. Install the turn signal lever. If equipped with cruise control, secure the lever to the switch with the retaining screw and install the wiring harness.

22. Remove the tape from the end of the harness and connect the switch and cruise control, if equipped, to the wire harness.

23. Cover both harnesses with the plastic protector and position it to the column. The turn signal connector slides on the tabs of the column.

24. Position the steering column upper bracket over the turn signal switch harness plastic protector.

25. Install the mounting bracket nuts and shims in their original positions.

26. Install the steering column lower cover.

Ignition Lock

REMOVAL & INSTALLATION

Except Tilt and Telescopic Column

1. Disconnect the negative battery cable. Remove the steering wheel.

2. Remove the lock plate cover assembly.

3. After compressing the lock plate spring, remove the snapring from the groove in the shaft.

4. Remove the lock plate and slide the turn signal cam and the upper bearing preload spring off the upper steering shaft.

5. Remove the thrust washer from the shaft.

6. Remove the hazard warning switch knob from the column along with the turn signal lever.

7. If equipped with cruise control, attach a piece of wire to the connector on the cruise control switch harness. Gently pull the harness up and out of the column.

8. Remove the turn signal switch mounting screws.

9. Slide the switch connector out of the bracket on the steering column.

10. As required, free the turn signal switch wiring protector from its mounting after disconnecting the turn signal switch electrical connectors, then pull the turn signal switch straight up and out of the steering column along with the switch harness and the connector from the steering column.

11. Turn the ignition switch to the ON or RUN position. Insert a small drift pin into the slot next to the switch mounting screw boss. Push the lock cylinder tab and remove the lock cylinder.

12. Installation is the reverse of the removal procedure.

Tilt and Telescopic Column

1988-89

1. Disconnect the negative battery cable. Remove the steering wheel.

2. Remove the rubber sleeve bumper from the steering shaft.

3. Using an appropriate tool, remove the plastic retainer.

4. Using a spring compressor, compress the upper steering shaft spring and remove the C-ring. Remove the steering shaft lock plate, the horn contact carrier and the upper steering shaft preload spring.

5. Remove the 4 screws which hold the upper mounting bracket and then remove the bracket.

6. Slide the harness connector out of the bracket on the steering column. Tape the upper part of the harness and connector.

7. Disconnect the hazard button and position the shift bowl in P. Remove the turn signal lever from the column.

8. If equipped with cruise control, remove the harness protector from the harness. Attach a piece of wire to the switch harness connector. Before removing the turn signal lever, loop a piece of wire and insert it into the turn signal lever opening. Use the wire to pull the cruise control harness out through the opening. Pull the rest of the harness up through and out of the column. Remove the guide wire from the connector and secure the wire to the column. Remove the turn signal lever.

9. Pull the turn signal switch up until the end connector is within the shift bowl. Remove the hazard flasher lever. Allow the switch to hang.

10. Place the ignition key in the RUN position.

11. Depress the center of the lock cylinder retaining tab with a suitable tool and then remove the lock cylinder.

12. Installation is the reverse of the removal procedure.

1990-92

1. Disconnect the negative battery cable.

2. Remove the steering wheel assembly.

3. Insert the key into the ignition switch and turn the ignition switch ON.

4. Remove the key warning buzzer switch and retaining clip.

5. Remove the ignition cylinder retaining screw located inside the lock housing cover.

6. Remove the lock cylinder from the column.

7. Installation is the reverse of the removal procedure.

Ignition Switch

REMOVAL & INSTALLATION

1. Disconnect the negative battery cable.

2. Position lock cylinder in LOCK position.

3. Remove steering column lower cover.

4. Loosen the nuts on the upper steering column. Allow the column to drop and be supported by the seat.

5. Disconnect the ignition switch connector at switch.

6. Remove the screws securing the dimmer switch and ignition switch to the steering column. Position the dimmer switch aside and remove the screw securing the ignition switch to the steering column. Remove the ignition switch from the vehicle.

To install:

7. Assemble the ignition switch on the actuator rod and adjust it to the **LOCK** position.

8. If equipped with a standard column, hold the switch actuating rod stationary with while moving the switch toward the bottom of the column until it reaches the end of its travel, which is the **ACC** position. Back off 2 detents to the right, which is the **OFF/UNLOCK** position, then with the key also in the **OFF/UNLOCK** position, tighten the switch mounting screws to 35 inch lbs.

9. If equipped with a tilt wheel, hold the switch actuating rod stationary with one hand while moving the switch toward the upper end of column until it reaches the end of its travel, which is the **ACC** position. Back off 1 detent and with the key in **LOCK** position, tighten the switch mounting screws to 35 inch lbs.

10. Continue the installation in the reverse order of the removal procedure. Test the starting system to start in **P** and **N** only.

Stoplight Switch

ADJUSTMENT

1. Place the stoplight switch clip in the bore on the pedal assembly bracket.

2. With the brake pedal depressed, insert the switch into the clip and depress the switch body. Clicks can be heard as the threaded portion of the switch is pushed through the clip towards the brake pedal.

3. Pull the brake pedal fully rearward against the pedal stop until the clicking sounds cannot be heard. The switch can be moved in the clip to correct the adjustment.

4. Release the brake pedal and repeat Step 5 to ensure the switch is correctly adjusted.

5. Install the harness connector and verify the stoplights operate correctly.

REMOVAL & INSTALLATION

NOTE: The cruise control re-

lease switch and the stoplight switch are adjusted or replaced in the same manner.

1. Disconnect the negative battery cable. Remove the underneath trim panel, as required. Disconnect the wire harness connector from the switch.

2. Remove the switch from the clip and then remove the clip from the bracket.

To install:

3. Place the clip in its bore on the bracket.

4. With the brake pedal depressed, insert the switch into the clip and depress the switch body. Clicks can be heard as the threaded portion of the switch is pushed through the clip towards the brake pedal.

5. Pull the brake pedal fully rearward against the pedal stop until the clicking sounds cannot be heard. The switch can be moved in the clip to correct the adjustment.

6. Release the brake pedal and repeat Step 5 to assure that no clicking sounds remain. The switch is now correctly adjusted.

7. Install the harness connector and verify the stoplights operate correctly.

Neutral Safety Switch

These vehicles incorporate a mechanical neutral start system. This system relies on a mechanical block, rather than the starter safety switch to prevent starting the engine in other than **P** or **N** positions.

The mechanical block is achieved by a cast in finger added to the switch actuator rack, which interferes with the bowl plate in all shift positions except **N** or **P**. This interference prevents rotation of the lock cylinder into the **START** position.

In either **P** or **N**, this finger passes through the bowl plate slots allowing the lock cylinder full rotational travel into the **START** position.

Fuses, Circuit Breakers and Relays

LOCATION

Fusible Links

Fusible links are used to prevent major wire harness damage in the event of a short circuit or an overload condition in the wiring circuits which are normally not fused, due to carrying high amperage loads or because of their locations within the wiring harness. Each fusible link is of a fixed value for a specific electrical load and should a link fail, the cause of the failure must be determined and repaired

prior to installing a new fusible link of the same value.

Circuit Breakers

Various circuit breakers are located under the instrument panel. In order to gain access to these components, it may be necessary to first remove the under dash padding.

Fuse Panel

The fuse panel is located on the left side of the vehicle. It is under the instrument panel assembly. In order to gain access to the fuse panel, it may be necessary to first remove the under dash padding.

Relays

All vehicles use a combination of the following electrical relays in order to function properly.

Overvoltage Protection Relay— located near the right instrument panel under the glove box.

Defogger Relay—located on the relay panel under the instrument panel to the left of the fuse block.

Door Lock Relay—attached to the lower right shroud panel behind the kick panel.

Power Antenna Relay—located on the relay panel under the instrument panel to the left of the fuse block.

Fuel Pump Relay—located on the relay panel under the instrument panel to the left of the fuse block.

Horn Relay—located on the relay panel under the instrument panel to the left of the fuse block.

Starter Interrupt Relay—located on a bracket under the left side of the dash panel, to the left of the steering column.

Power Seat Relay—located under the seat.

Stop/Turn Light Relay—located at the left rear quarter panel.

Guidematic Power Relay—located under the dash panel, near the fuse block.

Theft Deterrent Relay—located behind a bracket under the left side of the instrument panel.

Air Condition Compressor Control Relay—located on the right side of the firewall in the engine compartment.

Electronic Level Control Relay—located to the left of the level control compressor.

Memory Disable Relay—located behind the right side of the dash near the connector.

Low Brake Vacuum Relay—located behind the right side of the dash near the connector.

Illuminated Entry Timer—located behind the right side of the dash near the connector.

Horn Relay — located under the left side of the dash panel, to the left of the steering column.

Headlight Washer Relay — located on the fluid reservoir on the front of the right front shock tower.

High Mount Stop Light Relays — are located on the left rear wheelwell inside the trunk.

Computers
LOCATION
ECM

The electronic control module is locat-

ed on the right side of the vehicle. It is positioned in front of the right kick panel. In order to gain access to the assembly, the trim panel must first be removed.

BCM

The body computer module is located under the right side of the dash above the relay center. In order to gain access to the assembly, the trim panel must first be removed.

Flashers
LOCATION
Turn Signal

The turn signal flasher is located be-

hind the instrument panel bracket to the right of the steering column. In order to gain access to the turn signal flasher, it may be necessary to first remove the under dash padding.

Hazard Flasher

The hazard flasher is located in the fuse block. It is positioned on the lower right corner of the fuse block assembly. In order to gain access to the turn signal flasher, it may be necessary to first remove the under dash padding.

Chevrolet-GM
Front Wheel Drive
Beretta, Corsica

SPECIFICATIONS

VEHICLE IDENTIFICATION CHART

It is important for servicing and ordering parts to be certain of the vehicle and engine identification. The VIN (vehicle identification number) is a 17 digit number visible through the windshield on the driver's side of the dash and contains the vehicle and engine identification codes. The tenth digit indicates model year and the eighth digit indicates engine code. It can be interpreted as follows:

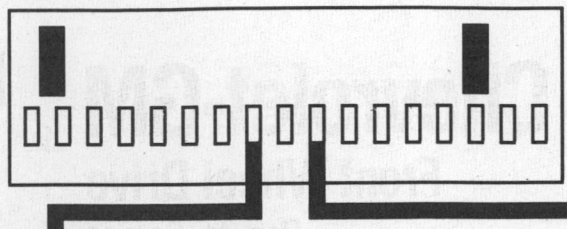

Engine Code						Model Year	
Code	Cu. In.	Liters	Cyl.	Fuel Sys.	Eng. Mfg.	Code	Year
1	121	2.0	4	TBI	Chevrolet	J	1988
G	133	2.2	4	TBI	Chevrolet	K	1989
A	138	2.3	4	PFI	Chevrolet	L	1990
W	173	2.8	V6	PFI	Chevrolet	M	1991
T	191	3.1	V6	PFI	Chevrolet	N	1992

TBI—Throttle Body Injection
PFI—Port Fuel Injection

ENGINE IDENTIFICATION

Year	Model	Engine Displacement cu. in. (liter)	Engine Series Identification (VIN)	No. of Cylinders	Engine Type
1988	Beretta	121 (2.0)	1	4	OHV
	Beretta	173 (2.8)	W	6	OHV
	Corsica	121 (2.0)	1	4	OHV
	Corsica	173 (2.8)	W	6	OHV
1989	Beretta	121 (2.0)	1	4	OHV
	Beretta	173 (2.8)	W	6	OHV
	Corsica	121 (2.0)	1	4	OHV
	Corsica	173 (2.8)	W	6	OHV
1990	Beretta	133 (2.2)	G	4	OHV
	Beretta	138 (2.3)	A	4	OHC
	Beretta	191 (3.1)	T	6	OHV
	Corsica	133 (2.2)	G	4	OHV
	Corsica	191 (3.1)	T	6	OHV
1991-92	Beretta	133 (2.2)	G	4	OHV
	Beretta	138 (2.3)	A	4	OHC
	Beretta	191 (3.1)	T	6	OHV
	Corsica	133 (2.2)	G	4	OHV
	Corsica	191 (3.1)	T	6	OHV

OHV Overhead Valves
OHC Overhead Cam

GENERAL ENGINE SPECIFICATIONS

Year	VIN	No. Cylinder Displacement cu. in. (liter)	Fuel System Type	Net Horsepower @ rpm	Net Torque @ rpm (ft. lbs.)	Bore × Stroke (in.)	Compression Ratio	Oil Pressure @ rpm
1988	1	4-121 (2.0)	PFI	90 @ 5600	108 @ 3200	3.500 × 3.150	9.0:1	63-77 @ 1200
	W	6-173 (2.8)	PFI	125 @ 4500	160 @ 3600	3.500 × 2.990	8.9:1	50-65 @ 1200
1989	1	4-121 (2.0)	TBI	90 @ 5600	108 @ 3200	3.500 × 3.150	9.0:1	63-77 @ 1200
	W	6-173 (2.8)	PFI	125 @ 4500	160 @ 3600	3.500 × 2.990	8.9:1	50-65 @ 1200
1990	G	4-133 (2.2)	TBI	95 @ 5200	120 @ 3200	3.500 × 3.460	9.0:1	63-77 @ 1200
	A	4-138 (2.3)	PFI	180 @ 6200	160 @ 5200	3.622 × 3.460	10.0:1	30 @ 2000
	T	6-191 (3.1)	PFI	135 @ 4200	180 @ 3600	3.500 × 3.310	8.8:1	50-65 @ 2400
1991–92	G	4-133 (2.2)	PFI	110 @ 5200	130 @ 3200	3.500 × 3.460	9.0:1	63-77 @ 1200
	A	4-138 (2.3)	PFI	180 @ 6200	160 @ 5200	3.622 × 3.460	10.0:1	30 @ 2000
	T	6-191 (3.1)	PFI	140 @ 4200	185 @ 3200	3.500 × 3.310	8.8:1	50-65 @ 2400

PFI Port Fuel Injection
TBI Throttle Body Injection

GASOLINE ENGINE TUNE-UP SPECIFICATIONS

Year	VIN	No. Cylinder Displacement cu. in. (liter)	Spark Plugs Type	Gap (in.)	Ignition Timing (deg.) MT	AT	Compression Pressure (psi)	Fuel Pump (psi)	Idle Speed (rpm) MT	AT	Valve Clearance In.	Ex.
1988	1	4-121 (2.0)	R44LTSM	0.035	①	①	②	9–13	①	①	Hyd.	Hyd.
	W	6-173 (2.8)	R43LTSE	0.045	①	①	②	9–13	①	①	Hyd.	Hyd.
1989	1	4-121 (2.0)	R44LTSM	0.035	①	①	②	9–13	①	①	Hyd.	Hyd.
	W	6-173 (2.8)	R43LTSM	0.045	①	①	②	9–13	①	①	Hyd.	Hyd.
1990	G	4-133 (2.2)	R44LTSM	0.035	①	①	②	9–13	①	①	Hyd.	Hyd.
	A	4-138 (2.3)	FR3LS	0.035	①	①	②	③	①	①	Hyd.	Hyd.
	T	6-191 (3.1)	R43LTSE	0.045	①	①	②	③	①	①	Hyd.	Hyd.
1991	G	4-133 (2.2)	R44LTSM	0.035	①	①	②	9–13	①	①	Hyd.	Hyd.
	A	4-138 (2.3)	FR3LS	0.035	①	①	②	③	①	①	Hyd.	Hyd.
	T	6-191 (3.1)	R44LTSM	0.045	①	①	②	③	①	①	Hyd.	Hyd.
1992			SEE UNDERHOOD SPECIFICATIONS STICKER									

Hyd.—Hydraulic

① Ignition timing and idle speed is controlled by the electronic control module. No adjustments are possible

② The lowest cylinder compression reading should not be less than 70% of the highest reading and no cylinder should be less than 100 psi.

③ 1—Connect fuel pressure gauge, engine at normal operating temperature.
2—Turn ignition switch on.
3—After approx. 2 seconds; pressure should read 41-47 psi and hold steady.
4—Start engine and idle; pressure should drop 3-10 psi from static pressure.

FIRING ORDERS

NOTE: To avoid confusion, always replace spark plug wires 1 at a time.

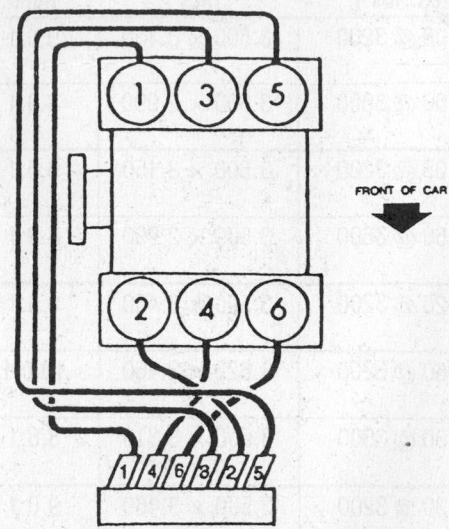

2.8L and 3.1L Engines
Engine Firing Order: 1–2–3–4–5–6
Distributorless Ignition System

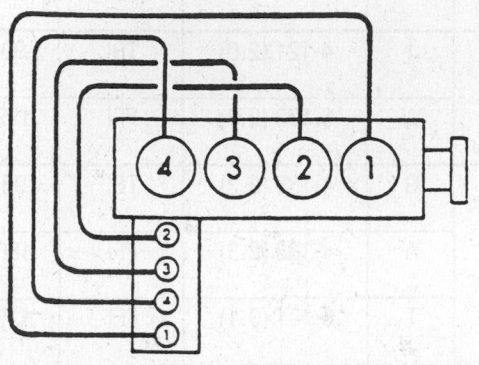

2.0L (VIN 1) and 2.2L (VIN G) Engines
Engine Firing Order: 1–3–4–2
Distributorless Ignition System

2.3L Engine
Engine Firing Order: 1–3–4–2
Distributorless Ignition System

CAPACITIES

Year	Model	VIN	No. Cylinder Displacement cu. in. (liter)	Engine Crankcase (qts.) with Filter	without Filter	Transmission (pts.) 4-Spd	5-Spd	Auto.	Drive Axle (pts.)	Fuel Tank (gal.)	Cooling System (qts.)
1988	Beretta	1	4-121 (2.0)	4.0⑪	4.0	NA	5.4①	8.0②	NA	14	9.6③
		W	6-173 (2.8)	4.0⑪	4.0	NA	5.4①	8.0②	NA	14	11④
	Corsica	1	4-121 (2.0)	4.0⑪	4.0	NA	5.4①	8.0②	NA	14	9.6③
		W	6-173 (2.8)	4.0⑪	4.0	NA	5.4①	8.0②	NA	14	11④
1989	Beretta	1	4-121 (2.0)	4.0⑪	4.0	NA	4.0	8.0⑦	NA	14	14.1⑤
		W	6-173 (2.8)	4.0⑪	4.0	NA	4.0	8.0⑦	NA	14	⑥
	Corsica	1	4-121 (2.0)	4.0⑪	4.0	NA	4.0	8.0⑦	NA	14	14.1⑤
		W	6-173 (2.8)	4.0⑪	4.0	NA	4.0	8.0⑦	NA	14	⑥
1990	Beretta	G	4-133 (2.2)	4.0⑪	4.0	NA	4.0	14.0⑫	NA	15.6	⑧ ⑨
		T	6-191 (3.1)	4.0⑪	4.0	NA	4.0	14.0⑫	NA	15.6	⑩
	Corsica	G	4-133 (2.2)	4.0⑪	4.0	NA	4.0	14.0⑫	NA	15.6	⑧ ⑨
		T	6-191 (3.1)	4.0⑪	4.0	NA	4.0	14.0⑫	NA	15.6	⑩

CAPACITIES

Year	Model	VIN	No. Cylinder Displacement cu. in. (liter)	Engine Crankcase (qts.) with Filter	without Filter	Transmission (pts.) 4-Spd	5-Spd	Auto.	Drive Axle (pts.)	Fuel Tank (gal.)	Cooling System (qts.)
1991–92	Beretta	G	4-133 (2.2)	4.0⑪	4.0	NA	4.2	14.0⑫	NA	15.6	⑧ ⑨
		A	4-138 (2.3)	4.0⑪	4.0	NA	4.2	14.0⑫	NA	15.6	⑨
		T	6-191 (3.1)	4.0⑪	4.0	NA	4.2	14.0⑫	NA	15.6	⑩
	Corsica	G	4-133 (2.2)	4.0⑪	4.0	NA	4.2	14.0⑫	NA	15.6	⑧ ⑨
		T	6-191 (3.1)	4.0⑪	4.0	NA	4.2	14.0⑫	NA	15.6	⑩

NA Not applicable
① 5 speed (Getrag)—4 pts.
② This figure is for drain and refill. After a complete overhaul, use 12.0 pts. If the torque converter is replaced, use 18.0 pts.
③ With air conditioning—9.8 qts.
④ With air conditioning—11.1 qts.
⑤ Without air conditioning—13.1 qts.

⑥ Automatic transaxle
   With air conditioning—16.6 qts.
   Without air conditioning—16.7 qts.
   Manual transaxle
   With air conditioning—16.1 qts.
   Without air conditioning—16.2 qts.
⑦ This figure is for drain and refill. After a complete overhaul, use 12.0 pts.

⑧ Automatic transaxle
   With air conditioning—9.5 qts.
   Without air conditioning—9.6 qts.
⑨ Manual transaxle
   With or without air conditioning—9.5 qts.
⑩ Automatic transaxle—12.4 qts.
   Manual transaxle—11.8 qts.
⑪ Additional oil will be required if filter is changed.
⑫ This figure is for drain and refill. If equipped with HM-3T40, after complete overhaul, use 18.0 pts.

CRANKSHAFT AND CONNECTING ROD SPECIFICATIONS

All measurements are given in inches.

Year	VIN	No. Cylinder Displacement cu. in. (liter)	Crankshaft Main Brg. Journal Dia.	Main Brg. Oil Clearance	Shaft End-play	Thrust on. No.	Connecting Rod Journal Diameter	Oil Clearance	Side Clearance
1988	1	4-121 (2.0)	2.4945–2.4954	0.0006–0.0019	0.002–0.008	1	1.9983–1.9994	0.0010–0.0030	0.0040–0.0150
	W	6-173 (2.8)	2.6473–2.6483	0.0016–0.0033	0.002–0.008	4	1.9983–1.9993	0.0010–0.0030	0.0060–0.0170
1989	1	4-121 (2.0)	2.4945–2.4954	0.0006–0.0019	0.002–0.008	1	1.9983–1.9994	0.0010–0.0030	0.0040–0.0150
	W	6-173 (2.8)	2.6473–2.6483	0.0016–0.0033	0.002–0.008	4	1.9983–1.9993	0.0010–0.0030	0.0060–0.0170
1990	G	4-133 (2.2)	2.4945–2.4954	0.0006–0.0019	0.002–0.007	4	1.9983–1.9994	0.0010–0.0030	0.0040–0.0150
	A	4-138 (2.3)	2.0470–2.0480	0.0005–0.0023	0.003–0.009	3	1.8887–1.8897	0.0005–0.0020	0.0059–0.0177
	T	6-191 (3.1)	2.6473–2.6483	0.0012–0.0030	0.002–0.008	3	1.9983–1.9994	0.0010–0.0040	0.0140–0.0270
1991–92	G	4-133 (2.2)	2.4945–2.4954	0.0006–0.0019	0.002–0.007	4	1.9983–1.9994	0.0010–0.0030	0.0040–0.0150
	A	4-138 (2.3)	2.0470–2.0480	0.0005–0.0023	0.003–0.009	3	1.8887–1.8897	0.0005–0.0020	0.0059–0.0177
	T	6-191 (3.1)	2.6473–2.6483	0.0012–0.0030	0.002–0.008	3	1.9983–1.9994	0.0010–0.0040	0.0140–0.0270

PISTON AND RING SPECIFICATIONS

All measurements are given in inches.

Year	VIN	No. Cylinder Displacement cu. in. (liter)	Piston Clearance	Ring Gap			Ring Side Clearance		
				Top Compression	Bottom Compression	Oil Control	Top Compression	Bottom Compression	Oil Control
1988	1	4-121 (2.0)	0.0010–0.0022	0.010–0.020	0.010–0.020	0.010–0.050	0.001–0.003	0.001–0.003	0.0080
	W	6-173 (2.8)	0.0020–0.0029	0.010–0.020	0.010–0.020	0.020–0.055	0.001–0.003	0.001–0.003	0.0080
1989	1	4-121 (2.0)	0.0010–0.0022	0.010–0.020	0.010–0.020	0.010–0.050	0.001–0.003	0.001–0.003	0.0080
	W	6-173 (2.8)	0.0009–0.0022	0.010–0.020	0.010–0.020	0.020–0.055	0.001–0.003	0.001–0.003	0.0080
1990	G	4-133 (2.2)	0.0007–0.0017	0.010–0.020	0.010–0.020	0.010–0.050	0.002–0.003	0.002–0.003	0.0020–0.0082
	A	4-138 (2.3)	0.0007–0.0020	0.014–0.024	0.016–0.026	0.016–0.055	0.003–0.005	0.002–0.003	—
	T	6-191 (3.1)	0.0009–0.0022	0.010–0.020	0.020–0.028	0.010–0.030	0.002–0.003	0.002–0.003	0.0080
1991–92	G	4-133 (2.2)	0.0007–0.0017	0.010–0.020	0.010–0.020	0.010–0.050	0.002–0.003	0.002–0.003	0.0020–0.0082
	A	4-138 (2.3)	0.0007–0.0020	0.014–0.024	0.016–0.026	0.016–0.055	0.003–0.005	0.002–0.003	—
	T	6-191 (3.1)	0.0009–0.0022	0.010–0.020	0.020–0.028	0.010–0.030	0.002–0.003	0.002–0.003	0.0080

VALVE SPECIFICATIONS

Year	VIN	No. Cylinder Displacement cu. in. (liter)	Seat Angle (deg.)	Face Angle (deg.)	Spring Test Pressure (lbs.)	Spring Installed Height (in.)	Stem-to-Guide Clearance (in.)		Stem Diameter (in.)	
							Intake	Exhaust	Intake	Exhaust
1988	1	4-121 (2.0)	46	45	176–188 ① @ 1.33	1.60 ②	0.0011–0.0026	0.0014–0.0030	NA	NA
	W	6-173 (2.8)	46	45	215 ① @ 1.29 in.	1.70 ②	0.0010–0.0027	0.0010–0.0027	NA	NA
1989	1	4-121 (2.0)	46	45	208–222 ① @ 1.22 in.	1.61 ②	0.0011–0.0026	0.0014–0.0030	NA	NA
	W	6-173 (2.8)	46	45	215 ① @ 1.29 in.	1.70 ②	0.0010–0.0027	0.0010–0.0027	NA	NA
1990	G	4-133 (2.2)	46	45	208–222 ① @ 1.22 in.	1.61 ②	0.0011–0.0026	0.0014–0.0030	NA	NA
	A	4-138 (2.3)	45	44	193–207 ① @ 1.04	1.44 ②	0.0010–0.0027	0.0015–0.0032	0.274–0.275	0.274–0.275
	T	6-191 (3.1)	46	45	90 ①	1.60 ②	0.0010–0.0027	0.0010–0.0027	NA	NA

VALVE SPECIFICATIONS

Year	VIN	No. Cylinder Displacement cu. in. (liter)	Seat Angle (deg.)	Face Angle (deg.)	Spring Test Pressure (lbs.)	Spring Installed Height (in.)	Stem-to-Guide Clearance (in.)		Stem Diameter (in.)	
							Intake	Exhaust	Intake	Exhaust
1991–92	G	4-133 (2.2)	46	45	208–222① @ 1.22 in.	1.61②	0.0011–0.0026	0.0014–0.0031	NA	NA
	A	4-138 (2.3)	46	45	193–207① @ 1.04	1.44②	0.0010–0.0027	0.0015–0.0032	0.274–0.275	0.274–0.275
	T	6-191 (3.1)	46	45	215① @ 1.29 in.	1.57②	0.0010–0.0027	0.0010–0.0027	NA	NA

NA Not available
① With valve open
② With valve closed

TORQUE SPECIFICATIONS
All readings in ft. lbs.

Year	VIN	No. Cylinder Displacement cu. in. (liter)	Cylinder Head Bolts	Main Bearing Bolts	Rod Bearing Bolts	Crankshaft Pulley Bolts	Flywheel Bolts	Manifold		Spark Plugs
								Intake	Exhaust	
1988	1	4-121 (2.0)	62–70①	63–77	34–43	66–89	45–59②	15–22	6–13	20
	W	6-173 (2.8)	③	63–83	34–44	67–85	45–59②	18	15–23	20
1989	1	4-121 (2.0)	62–70①	63–77	34–43	66–89	45–59②	15–22	6–13	20
	W	6-173 (2.8)	③	63–83	34–44	67–85	45–59②	18	15–23	20
1990	G	4-133 (2.2)	⑫	70	38	85④	52⑤	18	⑥	11
	A	4-138 (2.3)	26⑦	15⑧	18⑨	74⑩	22⑪	18⑭	27⑮	17
	T	6-191 (3.1)	③	73	39	66–85	45–59②	⑯	18	20
1991–92	G	4-133 (2.2)	⑬	70	38	77④	52⑤	18	⑥	11
	A	4-138 (2.3)	26⑦	15⑧	18⑨	74⑩	22⑪	18⑭	31⑮	17
	T	6-191 (3.1)	③	73	39	76	52	⑯	18	18

① Specification is for the shorter bolts. Torque the longer bolts to 73–83 ft. lbs.
② Specification is for automatic transaxle. Torque the manual transaxle bolts to 47–63 ft. lbs.
③ Cylinder head bolts should first be torqued to 33 ft. lbs. Then tighten the bolts by rotating the torque wrench an additional 90 degrees.
④ Specification is for the crankshaft center bolt. Torque the pulley to hub bolts to 37 ft. lbs.
⑤ Specification is for automatic transaxle. Torque the manual transaxle bolts to 55 ft. lbs.
⑥ Nuts to 115 inch lbs.
Studs to 89 inch lbs.
⑦ Cylinder head bolts should first be torqued in sequence to 26 ft. lbs. Then tighten the bolts by rotating the torque wrench an additional:
100 degrees for short bolts
110 degrees for long bolts.

⑧ Main bearing bolts should first be torqued to 15 ft. lbs. Then tighten the bolts by rotating the torque wrench an additional 90 degrees.
⑨ Connecting rod bolts should first be torqued to 18 ft. lbs. Then tighten the bolts by rotating the torque wrench an additional 80 degrees.
⑩ Crankshaft balancer to crankshaft bolt should first be torqued to 74 ft. lbs. Then tighten an additional 90 degrees.
⑪ Flywheel bolts should first be torqued to 22 ft. lbs. Then tighten an additional 45 degrees.
⑫ Tighten all bolts in sequence to 41 ft. lbs. Then tighten all bolts 45 degrees in sequence. Then tighten all bolts an additional 45 degrees in sequence.
Then tighten the long bolts (8, 4, 1, 5 & 9) an addition 20 degrees and the short belts (7, 3, 2, 6 & 10) an additional 10 degrees.

⑬ Tighten all bolts in sequence to: long bolts (8, 4, 1, 5 & 9) to 46 ft. lbs. and short bolts (7, 3, 2, 6 & 10) to 43 ft. lbs. Then tighten all bolts an additional 90 degrees in sequence.
⑭ Cylinder head studs—96 inch lbs.
⑮ Cylinder head studs—106 inch lbs.
⑯ Tighten all bolts to 15 ft. lbs. Then tighten all bolts to 24 ft. lbs.

BRAKE SPECIFICATIONS

All measurements in inches unless noted.

Year	Model	Lug Nut Torque (ft. lbs.)	Master Cylinder Bore	Brake Disc Minimum Thickness	Brake Disc Maximum Runout	Standard Brake Drum Diameter	Minimum Lining Thickness Front	Minimum Lining Thickness Rear
1988	Beretta	100	0.945	0.830	0.004	7.879	3/32	3/32
	Corsica	100	0.945	0.830	0.004	7.879	3/32	3/32
1989	Beretta	100	0.945	0.830	0.004	7.879	3/32	3/32
	Corsica	100	0.945	0.830	0.004	7.879	3/32	3/32
1990	Beretta	100	0.875	0.830	0.004	7.879	3/32	3/32
	Corsica	100	0.875	0.830	0.004	7.879	3/32	3/32
1991–92	Beretta	100	0.875	0.830	0.004	7.879	3/32	3/32
	Corsica	100	0.875	0.830	0.004	7.879	3/32	3/32

WHEEL ALIGNMENT

Year	Model		Caster Range (deg.)	Caster Preferred Setting (deg.)	Camber Range (deg.)	Camber Preferred Setting (deg.)	Toe-in (in.)	Steering Axis Inclination (deg.)
1988	Beretta	Front	7/10P–2 7/10P	1 7/10P ①	3/10P–1 3/10P	8/10P	1/10N–1/10P	—
		Rear	—	—	—	—	1/8P	—
	Corsica	Front	7/10P–2 7/10P	1 7/10P ①	3/10P–1 3/10P	4/5P	1/16N–1/16P	—
		Rear	—	—	3/10P–1 2/10P	1/4P	1/8P	—
1989	Beretta	Front	2/5P–1 9/10P	1 2/10P	0–1 2/10P	6/10P	0	—
		Rear	—	—	9/10N–4/10P	1/4P	5/16P	—
	Corsica	Front	2/5P–1 9/10P	1 2/10P	0–1 2/10P	6/10P	0	—
		Rear	—	—	8/10N–3/10P	1/4P	1/4P	—
1990	Beretta	Front	2/5P–1 9/10P	1 2/10P	0–1 2/10P	6/10P	0	—
		Rear	—	—	9/10N–4/10P	1/4P	5/16P	—
	Corsica	Front	2/5P–1 9/10P	1 2/10P	0–1 2/10P	6/10P	0	—
		Rear	—	—	8/10N–3/10P	1/4P	1/4P	—
1991–92	Beretta	Front	13/32P–1 29/32	1 5/32P	②	③	①	14
		Rear	—	—	13/16N–5/16P	1/4N	1/16	—
	Corsica	Front	13/32P–1 29/32	1 5/32P	②	③	①	14
		Rear	—	—	7/8N–3/8P	1/4N	3/16	—

N—Negative
P—Positive
① Not adjustable

② Except Beretta GTZ and Corsica w/Sport Susp.: 9/16P–13/16P
Beretta GTZ and Corsica w/Sport Susp.: 13/16N–7/16P

③ Except Beretta GTZ and Corsica w/Sport Susp.: 1/8P
Beretta GTZ and Corsica w/Sport Susp.: 13/16N

ENGINE MECHANICAL

NOTE: Disconnecting the negative battery cable on some vehicles may interfere with the functions of the on board computer systems and may require the computer to undergo a relearning process, once the negative battery cable is reconnected.

Engine Assembly
REMOVAL & INSTALLATION

2.0L, 2.2L and 2.3L Engines

1. Relieve the fuel system pressure. Disconnect the battery cables (negative cable first). Remove the battery from the vehicle.
2. Position a clean drain pan under the radiator, open the drain cock and drain the cooling system. Remove the air intake hose.
3. Disconnect the TV and accelerator cables from the throttle body. Disconnect the ECM electrical harness connector from the engine.
4. Remove all vacuum hoses, not a part of the engine assembly, the upper/lower radiator hoses and the heater hoses from the engine.
5. Remove the heat shield from the exhaust manifold. Disconnect and label the engine wiring harness from the firewall.
6. Disconnect the windshield washer hoses and the bottle. Rotate the tensioner pulley, to reduce the belt tension and remove the serpentine drive belt.
7. Disconnect and plug the fuel hoses. Raise and safely support the vehicle.
8. Remove the right side inner fender splash shield.
9. Remove the air conditioning compressor-to-bracket bolts and move it aside, so it will not interfere with the engine removal; do not disconnect the refrigerant lines.
10. Remove the flywheel splash shield. Label and disconnect electrical wires from the starter.
11. Remove the front starter brace, the starter-to-engine bolts and the starter.
12. If equipped with an automatic transaxle, remove the torque converter-to-flywheel bolts and push the converter back into the transaxle.
13. Remove the crankshaft pulley-to-crankshaft bolt. Using a crankshaft pulley hub remover tool, press the pulley from the crankshaft.
14. Remove the oil filter. Remove the engine-to-transaxle support bracket.
15. Disconnect the right rear engine mount.
16. Remove the exhaust pipe-to-exhaust manifold bolts, the exhaust pipe from the center hanger and loosen the muffler hanger.
17. Remove the TV and shift cable bracket. Remove both lower engine-to-transaxle bolts.
18. Lower the vehicle. From the intake manifold, remove the TV and accelerator cable bracket.
19. Remove the right front engine mount nuts. Disconnect the electrical connectors. Remove the alternator-to-bracket bolts and the alternator.
20. Remove the master cylinder-to-booster nuts, move the master cylinder and support it aside; do not disconnect the brake lines.
21. Using a vertical lifting device, install to the engine and lift it slightly.
22. Remove the right front engine mount bracket. Remove the remaining engine-to-transaxle bolts.
23. Remove the power steering pump-to-engine bolts and move it aside; do not disconnect the high pressure hoses.
24. Carefully lift and remove the engine from the vehicle.

To install:
25. Secure the engine on a engine suitable lifting device.
26. Support the transaxle with floor jack.
27. Carefully lower the engine into the vehicle, aligning it to the transaxle.
28. Install the engine-to-transaxle bolts. Install the right front engine mount bracket and attaching nuts.
29. Install the right rear engine mount and attaching bolts.
30. Install the engine-to-transaxle support bracket and attaching bolts.
31. Lower the transaxle jack and remove it from the vehicle.
32. Install the power steering pump and pump-to-engine attaching bolts.
33. Install the master cylinder and the master cylinder-to-booster attaching nuts.
34. Install the alternator, bracket and attaching bolts. Connect the electrical connectors to the alternator.
35. Install the TV and accelerator cable bracket.
36. Install the TV and shift cable bracket.
37. Raise and safely support the vehicle.
38. Install the exhaust pipe to the exhaust manifold and install the center hanger. Install the exhaust pipe-to-exhaust manifold attaching bolts.
39. Install the oil filter.
40. Install the crankshaft pulley on the crankshaft and install the pulley-to-crankshaft attaching bolt.
41. If equipped with an automatic transaxle, install the torque converter-to-flywheel attaching bolts.
42. Install the starter, front starter brace and the starter-to-engine attaching bolts.
43. Connect the electrical wires to the starter.
44. Install the flywheel splash shield.
45. Lower the vehicle.
46. Install the air conditioning compressor, with the refrigerant lines attached. Install the air conditioning compressor-to-bracket bolts.
47. Install the right side inner fender splash shield.
48. Connect the fuel hoses.
49. Install the windshield washer bottle and connect the washer hoses.
50. Rotate the tensioner pulley and install the serpentine drive belt.
51. Install the heat shield to the exhaust manifold. Connect the engine wiring harness to the firewall.
52. Install all vacuum hoses, the upper and lower radiator hoses and heater hoses to the engine.
53. Connect the TV and accelerator cables to the throttle body.
54. Connect the ECM electrical harness connector to the engine.
55. Install the air intake hose.
56. Close the radiator pet cock and refill the cooling system.
57. Install the battery and secure it in place. Connect the battery cables (the negative cable first).
58. Start the engine, allow it to reach normal operating temperatures and check for leaks.

2.8L and 3.1L Engines

1. Relieve the fuel pressure. Disconnect the battery cables (negative cable first). Remove the battery from the vehicle.
2. Remove the air cleaner, the air inlet hose and the mass air flow sensor.
3. Position a clean drain pan under the radiator, open the drain cock and drain the cooling system. Remove the exhaust manifold crossover assembly bolts and separate the assembly from the exhaust manifolds.
4. Remove the serpentine belt tensioner and the drive belt. Remove the power steering pump-to-bracket bolts and support the pump aside.
5. Disconnect the radiator hose from the engine.
6. Disconnect the TV and accelerator cables from the throttle valve bracket on the plenum.
7. Disconnect the electrical connectors. Remove the alternator-to-bracket bolts and the alternator. Label and disconnect the electrical wiring harness from the engine.
8. Disconnect and plug the fuel hoses. Remove the coolant overflow and bypass hoses from the engine.

9. From the charcoal canister, disconnect the purge hose. Label and disconnect all the necessary vacuum hoses.

10. Using a engine holding fixture tool, support the engine.

11. Raise and safely support the vehicle.

12. Remove the right inner fender splash shield. Remove the crankshaft pulley-to-crankshaft bolt. Using a wheel puller, press the crankshaft pulley from the crankshaft.

13. Remove the flywheel cover. Label and disconnect the starter wires. Remove the starter-to-engine bolts and the starter.

14. Disconnect the wires from the oil pressure sending unit.

15. Remove the air conditioning compressor-to-bracket bolts and the bracket-to-engine bolts. Support the compressor so it will not interfere with the engine; do not disconnect the refrigerant lines.

16. Disconnect the exhaust pipe from the rear of the exhaust manifold.

17. If equipped with an automatic transaxle, remove the torque converter-to-flywheel bolts and push the converter into the transaxle.

18. Remove the front and rear engine mount bolts along with the mount brackets.

19. Remove the intermediate shaft bracket from the engine.

20. Disconnect the shifter cable from the transaxle.

21. Remove the lower engine-to-transaxle bolts and lower the vehicle.

22. Disconnect the heater hoses from the engine.

23. Using an vertical engine lift, install it to the engine and lift it slightly. Remove the engine holding fixture. Using a floor jack, support the transaxle.

24. Remove the upper engine-to-transaxle bolts. Remove the front engine mount bolts and transaxle mounting bracket.

25. Remove the engine from the vehicle.

To install:

26. Secure the engine on a engine suitable lifting device.

27. Carefully lower the engine into the vehicle, aligning it to the transaxle.

28. Install the upper engine-to-transaxle bolts. Tighten bolts to 55 ft. lbs. (75 Nm).

29. Install the transaxle mount bracket and front engine mount attaching bolts. Tighten the bolts to 65 ft. lbs. (88 Nm).

30. Using a floor jack, support the transaxle and remove the engine lifting device from the engine.

31. Install the lower engine-to-transaxle.

32. Connect the heater hoses to the engine.

33. Connect the shifter cable to the transaxle.

34. Install the intermediate shaft bracket to the engine.

35. Install the front and rear engine mount bolts along with the mount brackets.

36. Lower the jack and remove it from the transaxle.

37. Raise the vehicle and support it safely.

38. If equipped with an automatic transaxle, install the torque converter-to-flywheel bolts.

39. Install the flywheel cover and attaching bolts.

40. Connect the exhaust pipe to the the exhaust manifold and install the attaching bolts.

41. Lower the vehicle.

42. Position the air conditioning compressor, with the lines attached, in place and install the compressor-to-bracket bolts.

43. Install the compressor bracket-to-engine bolts.

44. Connect the wires to the oil pressure sending unit.

45. Connect the starter wires. Position the starter in place and install the starter-to-engine bolts.

46. Install the crankshaft pulley and install the pulley-to-crankshaft bolt. Install the right inner fender splash shield.

47. Connect the purge hose to the charcoal canister. Connect all the necessary vacuum hoses.

48. Connect the coolant overflow and bypass hoses to the engine.

49. Connect the fuel delivery hoses to the engine.

50. Position the alternator in place and install the alternator-to-bracket bolts. Connect the electrical connectors to the alternator.

51. Connect all electrical wiring harnesses to the engine.

52. Connect the TV and accelerator cables to the throttle valve bracket on the plenum.

53. Connect the radiator hoses to the engine.

54. Install the serpentine belt tensioner and the drive belt.

55. Position the power steering pump in place and install the power steering pump-to-bracket bolts.

56. Connect the crossover pipe to the exhaust manifold and install the attaching bolts.

57. Install the air cleaner, air inlet hose and the mass air flow sensor.

58. Close the radiator cock and refill the cooling system.

59. Install the battery and secure it in place. Connect the battery cables (the negative cable last).

60. Start the engine, allow it to reach

normal operating temperatures and check for leaks.

Cylinder Head

REMOVAL & INSTALLATION

2.0L and 2.2L Engines

1. Relieve the fuel pressure. Disconnect the negative battery cable.

2. Drain the cooling system. Remove the TBI cover.

3. Raise and safely support the vehicle.

4. Disconnect the exhaust pipe-to-exhaust manifold bolts and separate the exhaust pipe from the manifold.

5. Lower the vehicle. Disconnect the heater hose from the intake manifold.

6. Disconnect the TV and accelerator cable bracket.

7. Label and disconnect the vacuum hoses from the intake manifold and thermostat.

8. Disconnect the accelerator linkage from the TBI unit.

9. Label and disconnect the electrical wiring from the engine.

10. Disconnect the upper radiator hose from the thermostat. Remove the serpentine belt.

11. Remove the power steering pump-to-bracket bolts and support the pump aside; do not disconnect the high pressure hoses from the pump.

12. Disconnect and plug the fuel lines. Remove the alternator-to-bracket bolts and the alternator. Position it aside, with electrical connectors attached.

13. Remove the alternator rear brace.

14. Remove the rocker arm cover-to-cylinder head bolts and the cover. Remove the rocker arm bolts, the rocker arms and pushrods; be sure to keep valve train components in the order that they were removed.

15. Starting with the outer bolts, remove the cylinder head-to-engine bolts.

To install:

16. Clean and inspect the gasket mounting surfaces. Make sure the threads on the cylinder head bolts and in the block are clean.

17. Position the cylinder head gasket in place on the engine block dowel pins.

18. Install the cylinder head and tighten the head bolts hand tight.

19. Following the torquing sequence, tighten the head bolts in 3 steps to 73–83 ft. lbs. (99–113 Nm) on the intake side and 62–70 ft. lbs. (84–95 Nm) on the exhaust side.

20. Install the pushrods and rocker arms in the same order they were re-

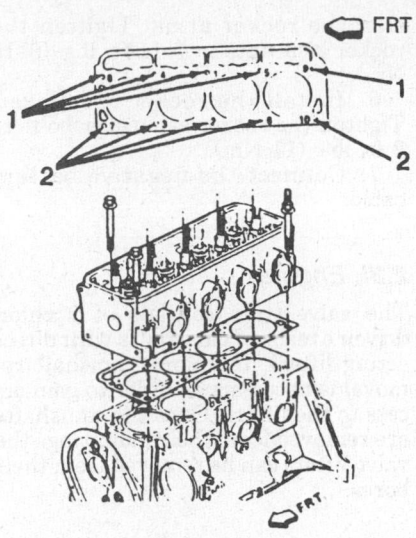

1. 73–83 ft. lbs.
2. 62–70 ft. lbs.

Cylinder head bolt torque sequence – 2.0L and 2.2L engines

moved. Tighten the rocker arm nuts to 7–11 ft. lbs. (9–15 Nm).

21. Install the alternator rear bracket.

22. Install the alternator and alternator-to-bracket bolts. Connect the fuel lines.

23. Install the power steering pump and the pump-to-bracket bolts.

24. Connect the upper radiator hose to the thermostat. Install the serpentine belt.

25. Connect the electrical wiring to the engine.

26. Connect the accelerator linkage to the TBI unit.

27. Connect the vacuum hoses to the intake manifold and thermostat.

28. Connect the TV and accelerator cable bracket.

29. Connect the heater hose to the intake manifold.

30. Raise and safely support the vehicle.

31. Connect the exhaust pipe and install the pipe-to-exhaust manifold bolts.

32. Lower the vehicle.

33. Connect the negative battery cable.

34. Fill cooling system and check for leaks. Start the engine and allow to come to normal operating temperature. Recheck for leaks. Top-up coolant.

2.3L Engine

1. Relieve the fuel system pressure. Disconnect the negative battery cable. Drain the engine coolant into a clean container for reuse.

2. Disconnect heater inlet and throttle body heater hoses from water

outlet. Disconnect upper radiator hose from water outlet.

3. Remove exhaust manifold.

4. Remove intake and exhaust camshaft housings.

5. Remove oil cap and dipstick. Pull oil fill tube upward to unseat from block.

6. Disconnect and tag injector harness electrical connector.

7. Disconnect throttle body to air cleaner duct. Remove throttle cable and bracket and position aside.

8. Remove throttle body from intake manifold with electrical harness, hoses, cable attached and position aside.

9. Disconnect and tag MAP sensor vacuum hose from intake manifold.

10. Remove intake manifold bracket to block bolt.

11. Disconnect and tag 2 coolant sensor connections.

12. Remove cylinder head to block bolts.

NOTE: When removing cylinder head to block bolts follow reverse of tighten sequence.

13. Remove cylinder head and gasket.

NOTE: Clean all gasket surfaces with plastic or wood scraper. Do not use any sealing material.

To install:

14. Install the cylinder head gasket to the cylinder block and carefully position the cylinder head in place.

15. Coat the head bolt threads with clean engine oil and allow the oil to drain off before installing.

16. Tighten the cylinder head bolts in sequence in 2 steps as follows:

Step 1: in sequence, tighten the long and short cylinder head to block bolts – 26 ft. lbs. (35 Nm).

Step 2: in sequence, tighten the short bolts – 80 degree turn and the long bolts – 90 degree turn.

17. Install the intake manifold-to-block bracket bolt and bracket.

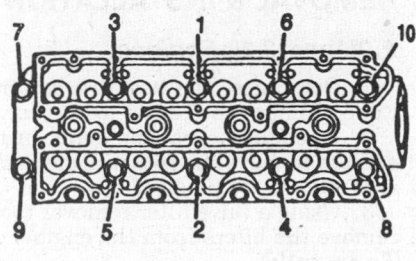

◄ FRONT OF ENGINE

Cylinder head bolt torque sequence – 2.3L engine

18. Connect the MAP sensor vacuum hose to the intake manifold.

19. Install the throttle body on the intake manifold with electrical harness, hoses and cable attached.

20. Connect the throttle body-to-air cleaner duct. Install the throttle cable and bracket.

21. Connect the injector harness electrical connector.

22. Connect the 2 coolant sensor connections.

23. Install the oil cap and dipstick. Install the oil fill tube into the block.

24. Install the exhaust and intake camshaft housings.

25. Install the exhaust manifold.

26. Connect the heater inlet and throttle body heater hoses to the water outlet. Connect the upper radiator hose to the water outlet.

27. Fill the cooling system and connect the negative battery cable.

28. Start the engine, allow it to reach operating temperature and check for leaks.

2.8L and 3.1L Engines

LEFT SIDE

1. Relieve the fuel pressure.

2. Disconnect the negative battery cable.

3. Place a drain pan under the radiator and drain the cooling system.

4. Remove the rocker cover attaching bolts and remove rocker cover.

5. Remove the intake manifold-to-cylinder head bolts and the remove the intake manifold.

6. Remove the fuel plenum and fuel rail assembles.

7. Disconnect the exhaust crossover from the right exhaust manifold.

8. Disconnect the oil level indicator tube bracket.

9. Loosen the rocker arms nuts, turn the rocker arms and remove the pushrods.

NOTE: Be sure to keep the parts in order for installation purposes.

10. Remove the cylinder head-to-engine bolts; start with the outer bolts and work toward the center. Remove the cylinder head with the exhaust manifold.

To install:

11. Clean the gasket mounting surfaces. Inspect the surfaces of the cylinder head, block and intake manifold damage and/or warpage. Clean the threaded holes in the block and the cylinder head bolt threads.

12. Using new gaskets, align the new cylinder head gasket over the dowels on the block with the note **This Side Up** facing the cylinder head.

13. Install the cylinder head and ex-

HEAD TORQUE SEQUENCE

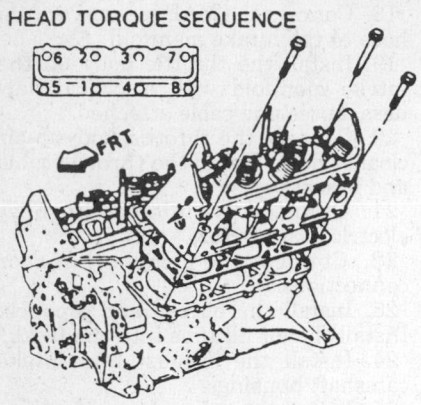

Cylinder head bolt torque sequence—
2.8L and 3.1L engines

haust manifold crossover assembly on the engine.

14. Coat the cylinder head bolt threads with engine oil and install the hand tight.

15. Using the proper torque sequence, tighten the bolts to 33 ft. lbs. (45 Nm). After all bolts are torqued to 33 ft. lbs. (45 Nm), rotate the torque wrench another 90 degrees or ¼ turn. This will apply the correct torque to the bolts.

16. Install the pushrods in the same order that they were removed. Tighten the rocker arm nuts to 14–20 ft. lbs. (19–27 Nm).

17. Install the intake manifold using a new gasket and following the correct sequence, tighten the bolts to 24 ft. lbs. (33 Nm) and nuts to 18 ft. lbs. (24 Nm).

18. Install the fuel plenum and fuel rail. Tighten the plenum bolts to 16 ft. lbs. (22 Nm).

19. Connect the exhaust crossover to the right exhaust manifold.

20. Connect the oil level indicator tube bracket.

21. Refill the cooling system. Connect the negative battery cable.

22. Operate the engine until normal operating temperatures are reached and check for leaks.

RIGHT SIDE

1. Relieve the fuel pressure. Disconnect the negative battery cable. Drain the cooling system.

2. Raise and safely support the vehicle. Remove the exhaust manifold-to-exhaust pipe bolts and separate the pipe from the manifold.

3. Lower the vehicle. Remove the exhaust manifold-to-cylinder head bolts and exhaust manifold.

4. Remove the rocker arm cover. Remove the intake manifold-to-cylinder head bolts and the intake manifold.

5. Loosen the rocker arms nuts,

turn the rocker arms and remove the pushrods.

NOTE: Be sure to keep the components in order for reassembly purposes.

6. Remove the cylinder head-to-engine bolts, starting with the outer bolts, working towards the center of the head.

7. Lift the cylinder head from the engine.

To install:

8. Clean the gasket mounting surfaces. Inspect the parts for damage and/or warpage; if necessary, machine or replace the parts.

9. Clean the engine block's threaded holes and the cylinder head bolt threads.

10. Using new gaskets, reverse the removal procedures. Using sealant, coat the cylinder head bolts and install the bolts hand tight.

11. Using the torquing sequence, tighten the bolts to 33 ft. lbs. (45 Nm). After all bolts are torqued to 33 ft. lbs. (45 Nm), rotate the torque wrench another 90 degrees or ¼ turn; this will apply the correct torque to the bolts.

12. Install the pushrods in the same order as they were removed. Tighten the rocker arm nuts to 14–20 ft. lbs. (19–27 Nm).

13. Follow the torquing sequence, use a new gasket and install the intake manifold.

14. Install the exhaust manifold and exhaust manifold-to-cylinder head bolts.

15. Raise the vehicle and support it safely.

16. Connect the exhaust pipe to the exhaust manifold and install the exhaust manifold-to-exhaust pipe bolts.

17. Lower the vehicle. Refill the cooling system.

18. Connect the negative battery cable. Start the engine, allow it to reach normal operating temperatures and check for leaks.

Valve Lifters

REMOVAL & INSTALLATION

2.0L and 2.2L Engines

1. Disconnect the negative battery cable. Remove the rocker arm cover.

2. Loosen the rocker arms nuts enough to move the rocker arms aside and remove the pushrods.

3. Using a valve lifter remover tool, remove the lifters from the engine.

To install:

4. Using Molykote® or equivalent, coat the base of the new lifters. Using a valve lifter remover tool, install the lifters into the engine.

5. Install the pushrods and reposi-

tion the rocker arms. Tighten the rocker arm nuts to 7–11 ft. lbs. (9–15 Nm).

6. Install the rocker arm cover. Tighten the rocker arm cover bolts to 8 ft. lbs. (11 Nm).

7. Connect the negative battery cable.

2.3L Engine

The valve train consists of 2 chain driven overhead camshafts with direct acting lifters, therefore, camshaft removal is necessary in order to gain access to the lifters. Once the camshafts are removed from their mountings the valve lifters can be removed from their bores.

2.8L and 3.1L Engines

1. Disconnect the negative battery cable.

2. Drain the cooling system.

3. Remove the rocker arm covers and intake manifold.

4. Loosen the rocker arms nuts enough to move the rocker arms aside and remove the pushrods.

5. Remove the lifters from the engine.

To install:

6. Using Molykote® or equivalent, coat the base of the new lifters and install them into the engine.

7. Install the pushrods and the reposition the rocker arms. Tighten the rocker arm nuts to 18 ft. lbs. (24 Nm).

8. Install the rocker arm covers.

9. Install the intake manifold. Tighten the intake manifold-to-cylinder head bolts to 20 ft. lbs. (27 Nm).

10. Connect the negative battery cable.

Valve Lash

ADJUSTMENT

Hydraulic valve lifters are used in the 2.0L, 2.2L, 2.8L and 3.1L engines and are not adjustable. If valve system noise is present, check the torque on the rocker arm nuts. The correct torque is 7–11 ft. lbs. (9–15 Nm) for 2.0L and 2.2L engines or 14–20 ft. lbs. (19–27 Nm) for 2.8L and 3.1L engines. If noise is still present, check the condition of the camshaft, lifters, rocker arms, pushrods and valves.

On the 2.3L engine, direct acting hydraulic valve lifters are used. The valve lifter body includes a harden iron contact foot bonded to a steel shell. These lifters are not serviceable or adjustable.

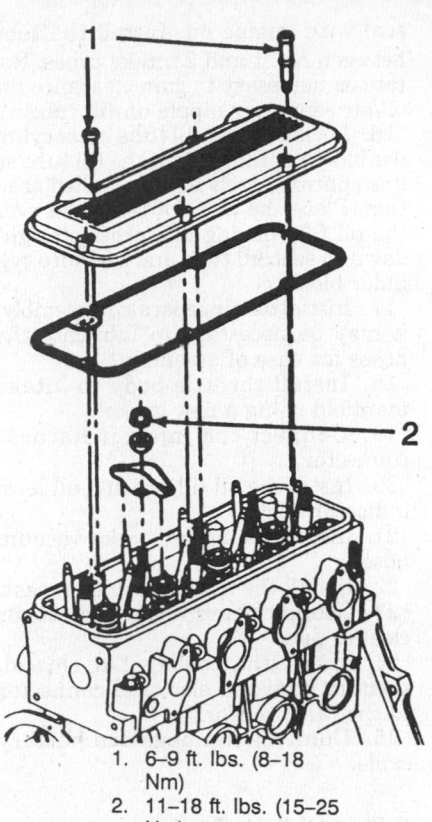

1. 6–9 ft. lbs. (8–18 Nm)
2. 11–18 ft. lbs. (15–25 Nm)

Rocker arm and cover installation — 2.0L and 2.2L engines

Rocker Arms

REMOVAL & INSTALLATION

2.0L and 2.2L Engines

1. Disconnect the negative battery cable. Remove the air hose from the TBI unit and the air cleaner.
2. Remove the intake manifold-to-rocker cover hose.
3. Remove the rocker arm cover bolts and the cover.
4. Remove the rocker arm nuts and the rocker arms.

NOTE: Be sure to keep the components in order for installation purposes.

To install:

5. Install the rocker arms and rocker arm nuts. Tighten to 7–11 ft. lbs. (9–15 Nm).
6. Install the rocker arm cover.
7. Connect the intake manifold-to-rocker cover.
8. Connect the hose to the TBI unit and air cleaner.
9. Connect the negative battery cable.

2.3L Engine

The valve train consists of 2 chain driven overhead camshafts with direct acting lifters.

2.8L and 3.1L Engines

LEFT SIDE

1. Disconnect the negative battery cable. Disconnect the bracket tube from the rocker cover.
2. Remove the spark plug wire cover. Drain the cooling system and remove the heater hose from the filler neck.
3. Remove the rocker arm cover-to-cylinder head bolts and the rocker cover.

NOTE: If the rocker arm cover will not lift off the cylinder head easily, strike the end with the palm of the hand or a rubber mallet.

4. Remove the rocker arm nuts and the rocker arms; be sure to keep the components in order for installation purposes.

To install:

5. Clean the gasket mounting surfaces.
6. Install the rocker arms and rocker arm nuts. Tighten to 14–20 ft. lbs. (19–27 Nm).
7. Install the rocker cover.
8. Install the spark plug wire cover.
9. Connect the negative battery cable.
10. Fill cooling system and check for leaks. Start the engine and allow to come to normal operating temperature. Recheck for leaks. Top-up coolant.

RIGHT SIDE

1. Disconnect the negative battery cable. Disconnect the brake booster vacuum line from the bracket.
2. Disconnect the cable bracket from the plenum.
3. Disconnect the vacuum line bracket from the cable bracket.
4. Disconnect the lines from the alternator brace stud.
5. Remove the rear alternator brace and the serpentine drive belt.
6. Remove the alternator and support it aside.
7. Remove the PCV valve.
8. Loosen the alternator bracket.
9. Disconnect the spark plug wires from the spark plugs. Remove the rocker cover-to-cylinder head bolts and the rocker cover.

NOTE: If the rocker arm cover will not lift off the cylinder head easily, strike the end with the palm of the hand or a rubber mallet.

10. Remove the rocker arm nuts and the rocker arms; be sure to keep the components in order for installation purposes.

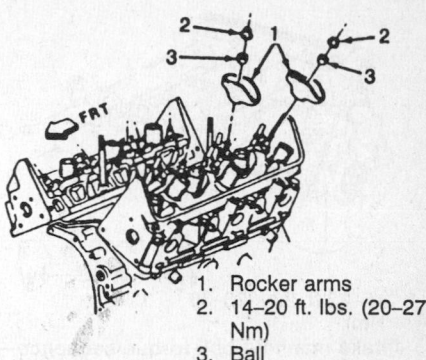

1. Rocker arms
2. 14–20 ft. lbs. (20–27 Nm)
3. Ball

Rocker arm installation — 2.8L and 3.1L engines

To install:

11. Clean the gasket mounting surfaces.
12. Install the rocker arm and rocker arm nuts. Tighten the rocker arm nuts to 14–20 ft. lbs. (19–27 Nm).
13. Install the rocker cover and the rocker cover-to-cylinder head bolts. Connect the spark plug wires to the spark plugs.
14. Tighten the alternator bracket.
15. Install the PCV valve.
16. Install the alternator.
17. Install the rear alternator brace and the serpentine drive belt.
18. Connect the lines to the alternator brace stud.
19. Connect the vacuum line bracket to the cable bracket.
20. Connect the cable bracket to the plenum.
21. Connect the brake booster vacuum line to the bracket.
22. Connect the negative battery cable.

Intake Manifold

REMOVAL & INSTALLATION

2.0L and 2.2L Engines

1. Disconnect the negative battery cable. Relieve the fuel pressure. Remove the TBI cover.
2. Drain the cooling system. Label and disconnect the vacuum lines and electrical connectors from the intake manifold.
3. Disconnect and plug the fuel line.
4. Disconnect the TBI linkage. Remove the throttle body-to-intake manifold bolts and the throttle body.
5. Remove the serpentine drive belt. Remove the power steering pump-to-bracket bolts and support the pump aside; do not disconnect the pressure hoses.
6. Raise and safely support the vehicle.
7. Disconnect the TV cable, accelerator cable and brackets.

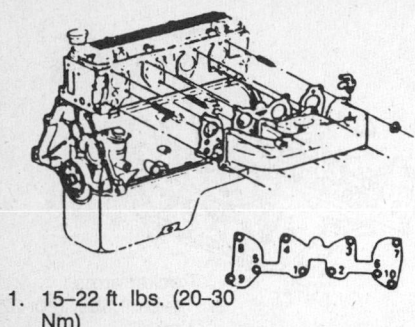

1. 15–22 ft. lbs. (20–30 Nm)

Intake manifold bolt torque sequence—2.0L and 2.2L engines

8. Disconnect the heater hose from the bottom of the intake manifold. Lower the vehicle.

9. Remove the intake manifold-to-cylinder head nuts/bolts and the manifold.

To install:

10. Clean the gasket mounting surfaces. Install new intake manifold gaskets.

11. Install the intake manifold and the intake manifold-to-cylinder head nuts/bolts. Tighten the intake manifold-to-cylinder heads bolts, in the proper sequence to 15–22 ft. lbs. (20–30 Nm).

12. Raise and safely support the vehicle.

13. Connect the heater hose to the bottom of the intake manifold.

14. Connect the TV cable, accelerator cable and brackets.

15. Lower the vehicle.

16. Install the serpentine drive belt. Install the power steering pump-to-bracket bolts and support the pump aside; do not disconnect the pressure hoses.

17. Install the throttle body-to-intake manifold bolts and the throttle body. Connect the TBI linkage.

18. Connect the fuel line.

19. Connect the vacuum lines and electrical connectors to the intake manifold.

20. Install the TBI cover.

21. Connect the negative battery cable.

22. Fill cooling system and check for leaks. Start the engine and allow to come to normal operating temperature. Recheck for leaks. Top-up coolant.

2.3L Engine

1. Disconnect the negative battery cable.

2. Remove the coolant fan shroud, vacuum hose and electrical connector from the MAP sensor.

3. Disconnect the throttle body to air cleaner duct.

4. Remove the throttle cable bracket.

5. Remove the power brake vacuum hose, including the retaining bracket to power steering bracket and position it to the side.

6. Remove the throttle body from the intake manifold with electrical harness, coolant hoses, vacuum hoses and throttle cable attached. Position these components aside.

7. Remove the oil/air separator bolts and hoses. Leave the hoses attached to the separator, disconnect from the oil fill, chain housing and the intake manifold. Remove as an assembly.

8. Remove the oil fill cap and oil level indicator stick.

9. Pull the oil tube fill upward to unseat from block and remove.

10. Disconnect the injector harness connector.

11. Remove the fill tube out top, rotating as necessary to gain clearance for the oil/air separator nipple between the intake tubes and fuel rail electrical harness.

12. Remove the intake manifold support bracket bolts and nut. Remove the intake manifold attaching nuts and bolts.

13. Remove the intake manifold.

NOTE: Intake manifold mounting hole closest to chain housing is slotted for additional clearance.

To install:

14. Install the intake manifold and gasket. Tightening the intake manifold bolts/nuts in sequence and to 18 ft. lbs. (25 Nm). Tighten intake manifold brace and retainers hand tight. Tighten to specifications in the following sequence:

    a. Nut to stud bolt—18 ft. lbs. (25 Nm).

    b. Bolt to intake manifold—40 ft. lbs. (55 Nm).

    c. Bolt to cylinder block—40 ft. lbs. (55 Nm).

15. Lubricate a new oil fill tube ring seal with engine oil. Install the tube between No. 1 and 2 intake tubes. Rotate as necessary to gain clearance for oil/air separator nipple on fill tube.

16. Locate the oil fill tube in its cylinder block opening. Align the fill tube so it is approximately in its installed position. Place the palm of the hand over the oil fill opening and press straight down to seat fill tube and seal into cylinder block.

17. Install oil/air separator assembly, it may be necessary to lubricate the hoses for ease of assembly.

18. Install throttle body to intake manifold using a new gasket.

19. Connect the injector harness connector.

20. Instal the oil fill cap and oil level indicator stick.

21. Install the power brake vacuum hose.

22. Install the throttle cable bracket.

23. Connect the throttle body to air cleaner duct.

24. Install the cooling fan shroud, vacuum hose and electrical connector to the MAP sensor.

25. Connect the negative battery cable.

2.8L and 3.1L Engines

1. Disconnect the negative battery cable. Relieve the fuel pressure. Drain the cooling system.

2. Disconnect the TV and accelerator cables from the plenum.

3. Remove the throttle body-to-plenum bolts and the throttle body. Remove the EGR valve.

4. Remove the plenum-to-intake manifold bolts and the plenum. Disconnect and plug the fuel lines and return pipes at the fuel rail.

5. Remove the serpentine drive belt. Remove the power steering pump-to-bracket bolts and support the pump aside; do not disconnect the pressure hoses.

6. Remove the alternator-to-brack-

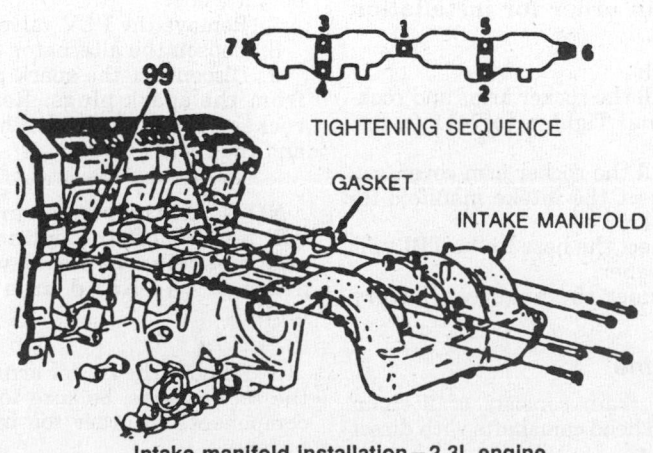

TIGHTENING SEQUENCE

GASKET

INTAKE MANIFOLD

Intake manifold Installation—2.3L engine

et bolts and support the alternator aside.

7. Loosen the alternator bracket. From the throttle body, disconnect the idle air vacuum hose.

8. Label and disconnect the electrical connectors from the fuel injectors. Remove the fuel rail.

9. Remove the breather tube. Disconnect the runners.

10. Remove both rocker arm cover-to-cylinder head bolts and the covers. Remove the radiator hose from the thermostat housing.

11. Label and disconnect the electrical connectors from the coolant temperature sensor and oil pressure sending unit. Remove the coolant sensor.

12. Remove the bypass hose from the filler neck and cylinder head.

13. Remove the intake manifold-to-cylinder head bolts and the manifold.

14. Loosen the rocker arm nuts, turn them 90 degrees and remove the pushrods; be sure to keep the components in order for installation purposes.

To install:

15. Clean all gasket mounting surfaces.

16. Place a ³⁄₁₆ in. bead of RTV sealant on the ridges where the manifold contacts the block.

17. Install a new intake manifold gasket.

18. Install the pushrod. Ensure proper seat in the lifter. Tighten the rocker arm nuts to 18 ft. lbs. (25 Nm).

19. Install the intake manifold and intake manifold bolts. Tighten the intake manifold-to-cylinder head bolts, following the torquing sequence, to 15 ft. lbs. (20 Nm) and retighten to 24 ft. lbs. (33 Nm).

20. Connect the bypass hose to the filler neck and cylinder head.

21. Install the coolant sensor. Connect the electrical connectors to the coolant temperature sensor and oil pressure sending unit.

22. Install both rocker arm covers and rocker arm cover-to-cylinder head bolts. Install the radiator hose to the thermostat housing.

23. Connect the runners. Install the breather tube.

24. Install the fuel rail. Connect the electrical connectors to the fuel injectors.

25. Install the alternator and the alternator-to-bracket bolts.

26. Connect the idle air vacuum hose to the throttle body.

27. Install the power steering pump-to-bracket bolts. Install the serpentine drive belt.

28. Install the plenum and the plenum-to-intake manifold bolts. Connect the fuel lines and return pipes at the fuel rail.

29. Install the throttle body-to-plenum bolts and the throttle body. Install the EGR valve.

30. Connect the TV and accelerator cables to the plenum.

31. Connect the negative battery cable.

32. Fill cooling system and check for leaks. Start the engine and allow to come to normal operating temperature. Recheck for leaks. Top-up coolant.

Exhaust Manifold

REMOVAL & INSTALLATION

2.0L and 2.2L Engines

1. Disconnect the negative battery cable.

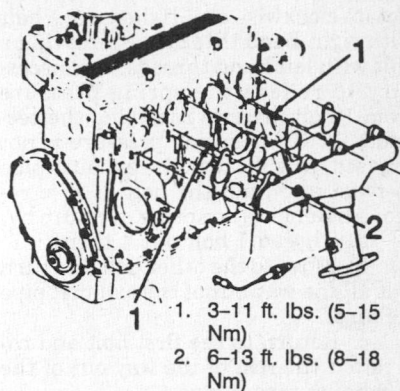

1. 3–11 ft. lbs. (5–15 Nm)
2. 6–13 ft. lbs. (8–18 Nm)

Exhaust manifold installation—2.0L and 2.2L engines

2. Disconnect the oxygen sensor wire.

3. Remove the serpentine belt.

4. Remove the alternator-to-bracket bolts and position the alternator aside with the wires attached.

5. Raise and safely support the vehicle.

6. Disconnect the exhaust pipe-to-exhaust manifold bolts and lower the vehicle.

7. Remove the exhaust manifold-to-cylinder head bolts.

8. Remove the exhaust manifold from the exhaust pipe flange and the manifold from the vehicle.

To install:

9. Clean the gasket mounting surfaces.

10. Using new gaskets, install the exhaust manifold and connect to the exhaust pipe flange. Tighten the exhaust manifold-to-cylinder head nuts to 3–11 ft. lbs. (4–15) Nm and bolts to 6–13 ft. lbs. (8–18 Nm).

11. Raise and safely support the vehicle.

12. Install the exhaust pipe-to-exhaust manifold bolts and lower the vehicle.

13. Install the alternator and the alternator-to-bracket bolts.

14. Install the serpentine belt.

15. Connect the oxygen sensor wire.

16. Connect the negative battery cable.

2.3L Engine

1. Disconnect the negative battery cable and oxygen sensor connector.

2. Remove upper and lower exhaust manifold heat shields.

3. Remove exhaust manifold brace to manifold bolt.

4. Break loose the manifold to exhaust pipe spring loaded bolts using a 13mm box wrench.

5. Raise and support vehicle safely.

6. Remove the manifold-to-exhaust pipe bolts from the exhaust pipe flange, using a ⁷⁄₃₂ in. (5.5mm) socket.

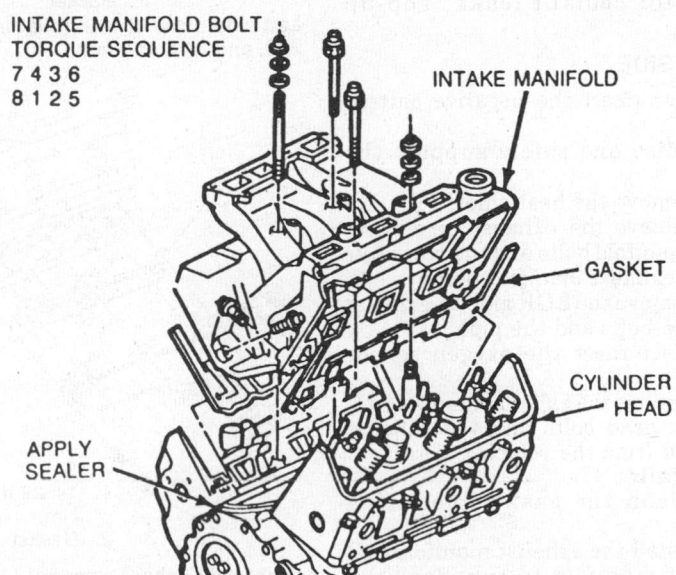

INTAKE MANIFOLD BOLT TORQUE SEQUENCE
7 4 3 6
8 1 2 5

INTAKE MANIFOLD

GASKET

CYLINDER HEAD

APPLY SEALER

Intake manifold installation—2.8L and 3.1L engines

Rotate clockwise as if tightening a bolt with right hand threads or removing a bolt with left hand threads. It is necessary to relieve the spring pressure from 1 bolt prior to removing the second bolt. If the spring pressure is not relieved, it will cause the exhaust pipe to twist and bind the bolt as it is removed. Relieve the spring pressure by:

   a. Thread 1 bolt out 4 turns.

   b. Move to the other bolt and turn it all the way out of the exhaust pipe flange.

   c. Return to the first bolt and rotate it the rest of the way out of the exhaust pipe flange.

7. Pull down and back on the exhaust pipe to disengage it from the exhaust manifold bolts.

8. Lower vehicle.

9. Remove exhaust manifold to cylinder head attaching nuts and remove exhaust manifold.

To install:

10. Clean all sealing surfaces. Install a new exhaust manifold gasket, the exhaust manifold and the exhaust manifold-to-cylinder head attaching nuts. Tighten, in sequence, to 27 ft. lbs. (37 Nm).

11. Raise and safely support the vehicle.

12. Connect the exhaust pipe to the exhaust manifold flange.

13. Install the manifold to exhaust pipe bolts to the exhaust pipe flange. Tighten to 22 ft. lbs. (30 Nm). Turn the nuts evenly to prevent binding.

14. Lower the vehicle.

15. Install the exhaust manifold brace-to-manifold bolt. Tighten to 19 ft. lbs. (26 Nm).

16. Install upper and lower exhaust manifold heat shields.

17. Connect the oxygen sensor connector.

18. Connect the negative battery cable.

2.8L and 3.1L Engines

LEFT SIDE

1. Disconnect the negative battery cable. Drain the cooling system.

2. Remove the air cleaner, air inlet hose and the mass air flow sensor.

3. Remove the coolant bypass pipe. Remove the manifold heat shield.

4. Disconnect the exhaust manifold crossover assembly at the right manifold.

5. Remove the exhaust manifold-to-cylinder head attaching bolts.

6. Remove the exhaust manifold with the crossover assembly.

To install:

7. Clean the gasket mounting surfaces.

8. Install the exhaust manifold with the crossover assembly. Tighten the exhaust manifold-to-cylinder head

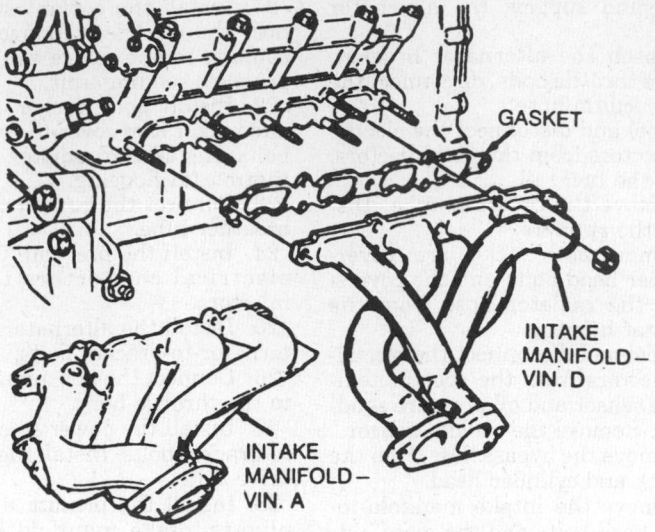

GASKET

INTAKE MANIFOLD— VIN. D

INTAKE MANIFOLD— VIN. A

TIGHTENING SEQUENCE

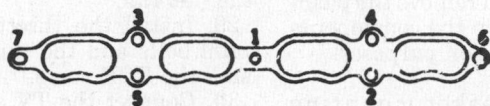

Exhaust manifold Installation—2.3L engine

bolts to 19 ft. lbs. (26 Nm) and the crossover bolts to 25 ft. lbs. (34 Nm).

9. Connect the exhaust manifold crossover assembly to the right manifold.

10. Install the coolant bypass pipe. Install the manifold heat shield.

11. Install the mass air flow sensor, air inlet hose and air cleaner.

12. Connect the negative battery cable.

13. Fill cooling system and check for leaks. Start the engine and allow to come to normal operating temperature. Check for exhaust leaks and recheck for coolant leaks. Top-up coolant.

RIGHT SIDE

1. Disconnect the negative battery cable.

2. Raise and safely support the vehicle.

3. Remove the heat shield.

4. Remove the exhaust pipe-to-exhaust manifold bolts and the crossover pipe-to-exhaust manifold bolts.

5. Remove the EGR pipe-to-exhaust manifold bolts and the pipe.

6. Disconnect the oxygen sensor wire.

7. Remove the exhaust manifold-to-cylinder head bolts and the exhaust manifold from the vehicle.

To install:

8. Clean the gasket mounting surfaces.

9. Install the exhaust manifold and exhaust manifold-to-cylinder head bolts. Tighten the exhaust mani-

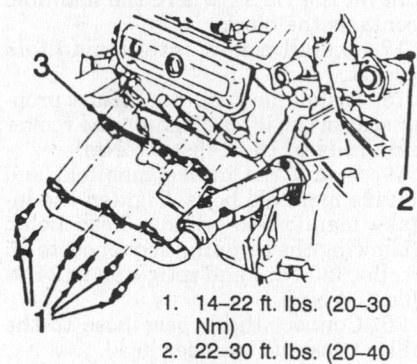

1. 14–22 ft. lbs. (20–30 Nm)
2. 22–30 ft. lbs. (20–40 Nm)
3. Gasket

Left side exhaust manifold installation—2.8L and 3.1L engines

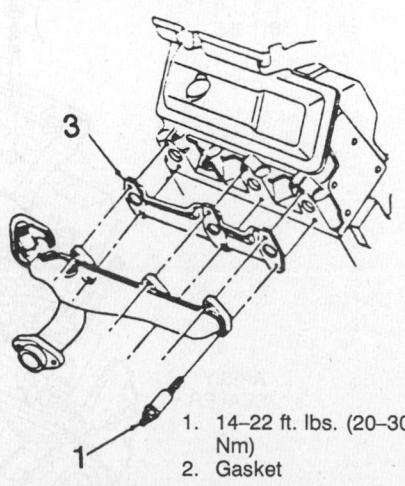

1. 14–22 ft. lbs. (20–30 Nm)
2. Gasket

Right side exhaust manifold installation—2.8L and 3.1L engines

to-cylinder head bolts to 19 ft. lbs. (26 Nm) and the crossover pipe bolts to 25 ft. lbs. (34 Nm).

10. Connect the oxygen sensor wire.

11. Install the EGR pipe and the EGR pipe-to-exhaust manifold bolts.

12. Install the exhaust pipe-to-exhaust manifold bolts and the crossover pipe-to-exhaust manifold bolts.

13. Install the heat shield.

14. Lower the vehicle.

15. Connect the negative battery cable. Start the engine and check for leaks.

Timing Chain Front Cover

REMOVAL & INSTALLATION

2.0L and 2.2L Engines

1. Disconnect the negative battery cable.

2. Raise and safely support the vehicle.

3. Drain the engine oil and remove the oil pan.

4. Lower the vehicle.

5. Remove the serpentine belt and the belt tensioner.

6. Remove the crankshaft pulley attaching bolt. Using a crankshaft pulley puller tool, remove the crankshaft pulley.

7. Remove the timing case cover bolts. Tap the cover with a rubber mallet and remove the cover.

To install:

8. Clean gasket mounting surfaces.

9. Using new gaskets, install the timing case cover over the dowels on the block and reverse the removal procedures. Tighten the timing case cover-to-engine bolts to 6–9 ft. lbs. (8–2 Nm).

10. Using a crankshaft pulley installer tool, press the pulley onto the crankshaft. Tighten the crankshaft pulley bolt to 66–88 ft. lbs. (89–119 Nm).

11. Install the belt tensioner and serpentine belt.

12. Raise and safely support the vehicle.

13. Install the oil pan.

14. Lower the vehicle.

15. Fill the crankcase with oil to specification.

16. Connect the negative battery cable.

17. Start the engine and check for leaks.

18. Stop the engine, allow to stand for several minutes and check oil level.

2.3L Engine

1. Disconnect the negative battery cable from the battery. Remove coolant recovery reservoir.

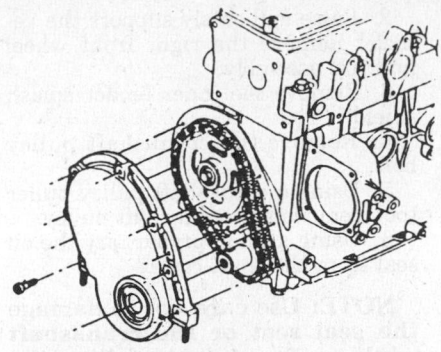

Front cover installation—2.0L and 2.2L engines

2. Remove the serpentine drive belt.

NOTE: To avoid personal injury when rotating the serpentine belt tensioner, use a 13mm wrench that is at least 24 inch long.

3. Remove upper cover fasteners.

4. Raise and safely support the vehicle.

5. Remove right front wheel assembly.

6. Remove right lower splash shield.

7. Remove crankshaft balancer assembly.

8. Remove lower cover fasteners and lower vehicle.

9. Remove the front cover.

To install:

10. Install the front cover using new gaskets. Tighten to 106 inch lbs. (12 Nm).

11. Raise and safely support the vehicle. Install the remaining front cover bolts. Tighten to 106 inch lbs. (12 Nm).

12. Install crankshaft balancer assembly. Tighten the attaching bolt and washer for balancer assembly to 74 ft. lbs. (100 Nm).

NOTE: The automatic transaxle crankshaft balancer must not be installed on a manual transaxle engine.

13. Install right lower splash shield.

14. Install right front wheel assembly.

15. Lower the vehicle.

16. Install upper cover fasteners.

17. Install the serpentine drive belt.

NOTE: To avoid personal injury when rotating the serpentine belt tensioner, use a 13mm wrench that is at least 24 inch long.

18. Install coolant recovery reservoir.

19. Connect the negative battery cable.

2.8L and 3.1L Engines

1. Disconnect the negative battery cable. Drain the cooling system.

2. Remove the serpentine belt and the belt tensioner.

3. Remove the alternator-to-bracket bolts and with the wires attached to the alternator, position it aside.

4. Remove the power steering pump-to-bracket bolts and support it aside; do not disconnect the pressure hoses.

5. Raise and safely support the vehicle.

6. Remove the right side inner fender splash shield and the flywheel dust cover.

7. Using a crankshaft pulley puller tool, remove the crankshaft damper.

8. Label and disconnect the starter wires and remove the starter.

9. Loosen the front 5 oil pan bolts, on both sides, enough to lower the oil pan ½ in.

10. Lower the vehicle. Disconnect the radiator hose from the water pump.

11. Disconnect the heater coolant hose from the cooling system filler pipe.

12. Remove the bypass and overflow hoses.

13. Remove the water pump pulley. Disconnect the canister purge hose.

14. Remove the spark plug wire shield from the water pump.

15. Remove the upper timing case cover-to-engine bolts and the timing case cover.

To install:

16. Clean gasket mounting surfaces.

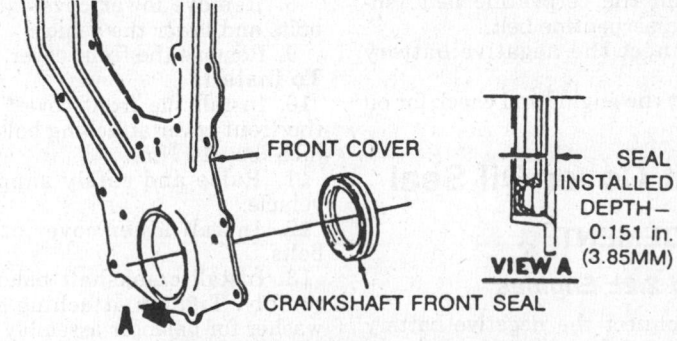

FRONT COVER

SEAL INSTALLED DEPTH— 0.151 in. (3.85MM)

VIEW A

CRANKSHAFT FRONT SEAL

Front cover installation—2.3L engine

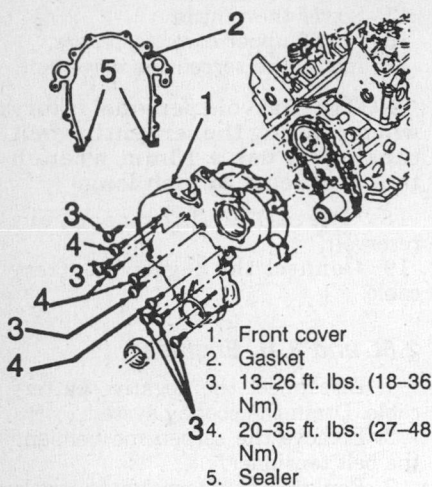

1. Front cover
2. Gasket
3. 13–26 ft. lbs. (18–36 Nm)
4. 20–35 ft. lbs. (27–48 Nm)
5. Sealer

Front cover installation—2.8L and 3.1L engines

17. Using silicone sealant and a new gasket, apply a thin bead to the front cover mating surface, install the timing case cover on the engine. Apply silicone sealant to the sections of the oil pan rails that were lowered and install the mounting bolts.
18. Using a crankshaft pulley installer tool, press the damper pulley onto the crankshaft.
19. Install the spark plug wire shield to the water pump.
20. Install the water pump pulley. Connect the canister purge hose.
21. Install the bypass and overflow hoses.
22. Connect the heater coolant hose to the cooling system filler pipe.
23. Raise and safely support the vehicle. Connect the radiator hose to the water pump.
24. Raise the oil pan into position and tighten the front 5 oil pan bolts.
25. Install the starter and connect the starter wires.
26. Install the right side inner fender splash shield and the flywheel dust cover.
27. Lower the vehicle.
28. Install the power steering pump-to-bracket bolts.
29. Install the alternator-to-bracket bolts.
30. Install the serpentine belt tensioner and serpentine belt.
31. Connect the negative battery cable.
32. Start the engine and check for oil leaks.

Front Cover Oil Seal

REPLACEMENT

2.0L and 2.2L Engines

1. Disconnect the negative battery cable. Remove the serpentine belt.

2. Raise and safely support the vehicle. Remove the right front wheel and tire assembly.
3. Remove the inner fender splash shield.
4. Remove the crankshaft pulley bolt.
5. Using a crankshaft pulley puller tool, remove the crankshaft pulley.
6. Using a small prybar, pry the oil seal from the front cover.

NOTE: Use care not to damage the seal seat or the crankshaft while removing or installing the seal. Inspect the sealing surface of the crankshaft for grooves or other wear.

To install:
7. Using an oil seal centering tool, drive the new seal into the cover with the lip facing towards the engine.
8. Install a crankshaft pulley installer tool, onto the crankshaft pulley and press the pulley onto the crankshaft. Install the pulley bolt and tighten to 66–88 ft. lbs. (89–119 Nm).
9. Remove the inner fender splash shield.
10. Remove the right front wheel and tire assembly.
11. Lower the vehicle.
12. Install the serpentine belt.
13. Connect the negative battery cable.

2.3L Engine

1. Disconnect the negative battery cable from the battery. Remove coolant recovery reservoir.
2. Remove the serpentine drive belt.

NOTE: To avoid personal injury when rotating the serpentine belt tensioner, use a 13mm wrench that is at least 24 in. long.

3. Remove upper cover attaching bolts.
4. Raise vehicle and support it safely.
5. Remove right front wheel assembly.
6. Remove right lower splash shield.
7. Remove crankshaft balancer assembly.
8. Remove lower cover attaching bolts and lower the vehicle.
9. Remove the front cover.
To install:
10. Install the front cover. Tighten the front cover attaching bolts to 106 inch lbs. (12 Nm).
11. Raise and safely support the vehicle.
12. Install lower cover attaching bolts.
13. Install crankshaft balancer assembly. Tighten attaching bolt and washer for balancer assembly to 74 ft. lbs. (100 Nm).

NOTE: The automatic transaxle crankshaft balancer must not be installed on a manual transaxle engine.

14. Install right lower splash shield.
15. Install right front wheel assembly.
16. Lower the vehicle.
17. Install upper cover attaching bolts.
18. Install the serpentine drive belt.

NOTE: To avoid personal injury when rotating the serpentine belt tensioner, use a 13mm wrench that is at least 24 in. long.

19. Install coolant recovery reservoir.
20. Connect the negative battery cable.

2.8L and 3.1L Engines

1. Disconnect the negative battery cable. Remove the serpentine belt.
2. Raise and safely support the vehicle. Remove the right side inner fender splash shield.
3. Remove the damper attaching bolt.
4. Using a crankshaft pulley puller tool, press the damper pulley from the crankshaft.
5. Using a small prybar, pry out the seal in the front cover.

NOTE: Use care not to damage the seal seat or the crankshaft while removing or installing the seal. Inspect the crankshaft seal surface for signs of grooves or wear.

To install:
6. Using a seal installer tool, drive the new seal in the cover with the lip facing towards the engine.
7. Using a crankshaft pulley installer tool, press the crankshaft pulley onto the crankshaft. Tighten the damper bolt to 67–85 ft. lbs. (90–115 Nm).
8. Install the right side inner fender splash shield. Lower the vehicle.
9. Install the serpentine belt.
10. Connect the negative battery cable.

Timing Chain and Sprockets

REMOVAL & INSTALLATION

2.0L and 2.2L Engines

1. Disconnect the negative battery cable. Remove the timing case cover.
2. Rotate the crankshaft to until the marks on the crankshaft and camshaft sprockets are aligned.

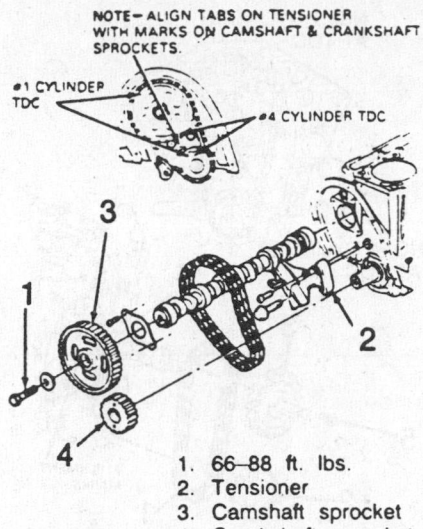

NOTE—ALIGN TABS ON TENSIONER WITH MARKS ON CAMSHAFT & CRANKSHAFT SPROCKETS.

#1 CYLINDER TDC

#4 CYLINDER TDC

1. 66–88 ft. lbs.
2. Tensioner
3. Camshaft sprocket
4. Crankshaft sprocket

Timing chain and sprockets installation —2.0L and 2.2L engines

3. Remove the timing chain tensioner upper bolt.

4. Loosen the timing chain tensioner nut as far as possible but do not remove the nut.

5. Remove the timing chain and camshaft sprocket.

6. Using a gear puller, remove the crankshaft sprocket.

To install:

7. Before installing the camshaft sprocket, lubricate the thrust side with Molykote® or equivalent. Using a sprocket installer tool, install the crankshaft sprocket.

8. Align the camshaft sprocket mark with the crankshaft sprocket marks. Install the timing chain and camshaft sprocket.

9. Press the camshaft sprocket onto the camshaft using the camshaft sprocket bolt. Tighten the camshaft sprocket bolt to 66–88 ft. lbs. (89–119 Nm).

10. Align the tabs on the tensioner with the marks on the camshaft and crankshaft sprockets and tighten the tensioner.

11. Install the timing case cover.

12. Connect the negative battery cable.

2.3L Engine

NOTE: Prior to removing the timing chain, review the entire procedure.

1. Disconnect the negative battery cable.

2. Remove front engine cover and crankshaft oil slinger.

3. Rotate the crankshaft clockwise, as viewed from front of engine/normal rotation until the camshaft sprockets' timing dowel pin holes line up with the holes in the timing chain housing. The mark on the crankshaft sprocket should line up with the mark on the cylinder block. The crankshaft sprocket keyway should point upwards and line up with the centerline of the cylinder bores. This is the timed position.

4. Remove 3 timing chain guides.

5. Raise vehicle and support in safely.

6. Gently pry off timing chain tensioner spring retainer and remove spring.

NOTE: Two styles of tensioner are used. One with a spring post, early production and 1 without a spring post, late production. Both styles are identical in operation and are interchangeable.

7. Remove timing chain tensioner shoe retainer.

8. Make sure all the slack in the timing chain is above the tensioner assembly; remove the chain tensioner shoe. The timing chain must be disengaged from the wear grooves in the tensioner shoe in order to remove the shoe. Slide a prybar under the timing chain while pulling shoe outward.

9. If difficulty is encountered removing chain tensioner shoe, proceed as follows:

a. Lower the vehicle.

b. Hold the intake camshaft sprocket with a holding tool and remove the sprocket bolt and washer.

c. Remove the washer from the bolt and re-thread the bolt back into the camshaft by hand, the bolt provides a surface to push against.

d. Remove intake camshaft sprocket using a 3-jaw puller in the 3 relief holes in the sprocket. Do not attempt to pry the sprocket off the camshaft or damage to the sprocket or chain housing could occur.

10. Remove tensioner assembly attaching bolts and tensioner.

— CAUTION —
Tensioner piston is spring loaded and could fly out causing personal injury.

11. Remove chain housing to block stud, timing chain tensioner shoe pivot.

12. Remove timing chain.

NOTE: Failure to follow this procedure could result in severe engine damage.

To install:

13. Tighten intake camshaft sprocket attaching bolt and washer, to specification while holding sprocket in place.

14. Install a special tool through holes in camshaft sprockets into holes in timing chain housing, this positions the camshafts for correct timing.

15. If the camshafts are out of position and must be rotated more than 1/8 turn in order to install the alignment dowel pins, perform the following:

a. The crankshaft must be rotated 90 degrees clockwise off of TDC in order to give the valves adequate clearance to open.

b. Once the camshafts are in position and the dowels installed, rotate the crankshaft counterclockwise back to top dead center. Do not rotate the crankshaft clockwise to TDC, valve or piston damage could occur.

16. Install timing chain over exhaust camshaft sprocket, around idler sprocket and around crankshaft sprocket.

17. Remove the alignment dowel pin from the intake camshaft. Using a dowel pin remover tool rotate the intake camshaft sprocket counterclockwise enough to slide the timing chain over the intake camshaft sprocket. Release the camshaft sprocket wrench. The length of chain between the 2 camshaft sprockets will tighten. If properly timed, the intake camshaft alignment dowel pin should slide in easily. If the dowel pin does not fully index, the camshafts are not timed correctly and the procedure must be repeated.

18. Leave the alignment dowel pins installed.

19. With slack removed from chain between intake camshaft sprocket and crankshaft sprocket, the timing marks on the crankshaft and the cylinder block should be aligned. If marks are not aligned, move the chain 1 tooth forward or rearward, remove slack and recheck marks.

20. Tighten chain housing to block stud, timing chain tensioner shoe pivot. Stud is installed under the timing chain. Tighten to 19 ft. lbs. (26 Nm).

21. Reload timing chain tensioner assembly to its zero position as follows:

a. Assemble restraint cylinder, spring and nylon plug into plunger. Index slot in restraint cylinder with peg in plunger. While rotating the restraint cylinder clockwise, push the restraint cylinder into the plunger until it bottoms. Keep rotating the restraint cylinder clockwise but allow the spring to push it out of the plunger. The pin in the plunger will lock the restraint in the loaded position.

b. Install a special plunger installer tool into plunger assembly.

c. Install plunger assembly into tensioner body with the long end toward the crankshaft when installed.

22. Install tensioner assembly to chain housing. Recheck plunger assembly installation. It is correctly in-

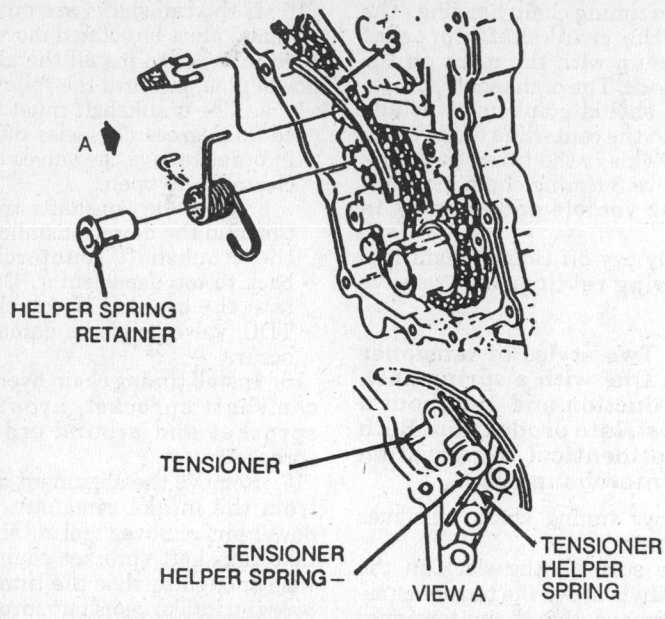

HELPER SPRING RETAINER

TENSIONER

TENSIONER HELPER SPRING

VIEW A

TENSIONER HELPER SPRING

Installing the timing chain tensioner—2.3L engine

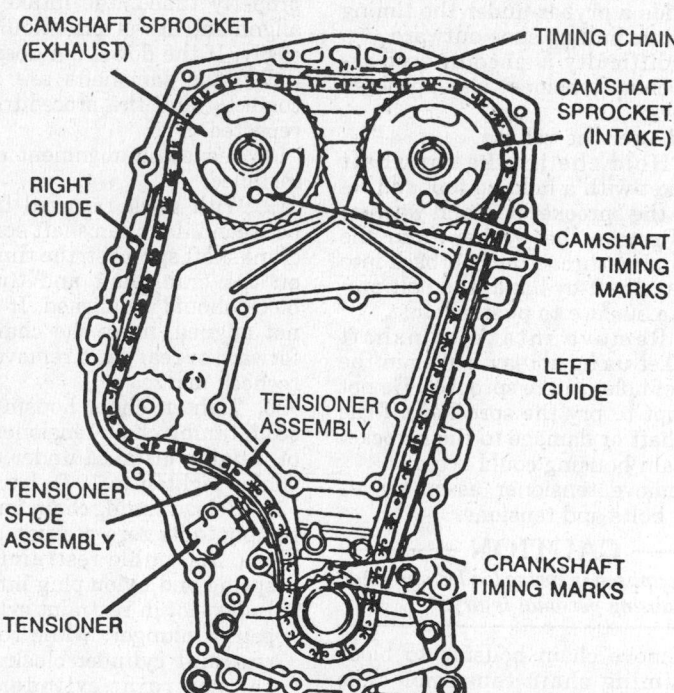

CAMSHAFT SPROCKET (EXHAUST)

TIMING CHAIN

CAMSHAFT SPROCKET (INTAKE)

RIGHT GUIDE

CAMSHAFT TIMING MARKS

LEFT GUIDE

TENSIONER ASSEMBLY

TENSIONER SHOE ASSEMBLY

TENSIONER

CRANKSHAFT TIMING MARKS

Timing chain installation—2.3L engine

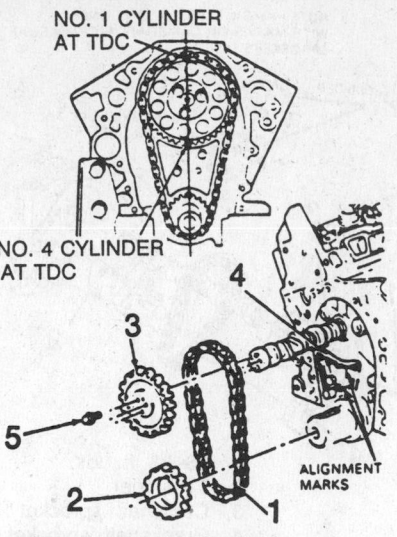

NO. 1 CYLINDER AT TDC

NO. 4 CYLINDER AT TDC

ALIGNMENT MARKS

NOTE—ALIGN TIMING MARKS ON CAM & CRANK SPROCKETS USING ALIGNMENT MARKS ON DAMPER STAMPING OR CAST ALIGNMENT MARKS ON CYL & CASE.

1. Timing chain
2. Crankshaft sprocket
3. Camshaft sprocket
4. Damper
5. 15–20 ft. lbs.

Timing chain and sprockets installation —2.8L and 3.1L engines

stalled when the long end is toward the crankshaft.

23. Install and tighten timing chain tensioner bolts and tighten to 10 ft. lbs. (14 Nm).

24. Install tensioner shoe and tensioner shoe retainer.

25. Remove the special tool from the plunger and squeeze plunger assembly into tensioner body to unload the plunger assembly.

26. Lower vehicle enough to reach and remove the alignment dowel pins. Rotate crankshaft clockwise 2 full rotations. Align crankshaft timing mark with mark on cylinder block and reinstall alignment dowel pins. Alignment dowel pins will slide in easily if engine is timed correctly.

NOTE: If the engine is not correctly timed, severe engine damage could occur.

27. Install 3 timing chain guides and crankshaft oil slinger.

28. Install engine front cover.

29. Connect the negative battery cable. Start engine and check for oil leaks.

2.8L and 3.1L Engines

1. Disconnect the negative battery cable. Remove the front cover.

2. Rotate the crankshaft to position the No. 1 piston at TDC with the crankshaft and camshaft sprockets aligned.

NOTE: When the camshaft and crankshaft marks are aligned, the No. 4 piston is on the TDC of its compression stroke.

3. Remove the camshaft sprocket bolts, the sprocket and the timing chain.

4. Remove the crankshaft sprocket.

To install:

5. Before installing the sprockets, apply Molykote® or equivalent, to the thrust face of the sprocket(s).

6. Install the sprocket on the crankshaft.

7. Hold the camshaft sprocket with the chain hanging down. Align the marks on the camshaft and crankshaft sprockets.

8. Align the dowel in the camshaft with the sprocket. Install the sprocket and timing chain using a camshaft bolt to pull the sprocket into position.

9. Tighten the camshaft bolts to 15–20 ft. lbs. (20–27 Nm).

10. Lubricate the new timing chain with clean engine oil.

11. Install the front cover.

12. Connect the negative battery cable. Start the engine and check for leaks.

Camshaft

REMOVAL & INSTALLATION

2.0L and 2.2L Engines

1. Relieve the fuel pressure. Disconnect the negative battery cable. Remove the engine and attach it to an engine stand.

2. Remove the timing chain and sprocket from the engine.

3. Drain the engine oil and remove the oil filter.

4. Remove the rocker cover. Loosen the rocker arms and turn the rocker arms 90 degrees. Remove the pushrods and lifters; note the position of the valve train components for reassembly purposes.

5. Remove the oil pump drive.

6. Remove the camshaft thrust plate-to-engine bolts and carefully pull the camshaft from the engine.

NOTE: Use care when removing and installing the camshaft; do not damage the camshaft bearings or the bearing surfaces on the camshaft.

To install:

7. Clean gasket mounting surfaces.

8. Lubricate the lobes of the new camshaft and insert the camshaft into the engine.

NOTE: If a new camshaft is being installed, replace the lifters. Reused lifters must be reinstalled on the same camshaft and lobe location in which they were originally installed.

9. Align the marks on the camshaft and crankshaft sprockets. Install the timing chain and sprocket.

10. Install the oil pump drive.

11. Install the lifters, pushrods and reposition the rocker arms. Tighten the rocker arm nuts to 11–18 ft. lbs. (15–24 Nm).

12. Install the rocker covers.

13. Install the timing chain and sprockets to the engine.

14. Install the engine.

15. Install the oil filter and add engine oil to specification.

16. Connect the negative battery cable.

2.3L Engine

INTAKE CAMSHAFT

NOTE: Any time the camshaft housing to cylinder head bolts are loosened or removed, the camshaft housing to cylinder head gasket must be replaced.

1. Relieve the fuel system pressure. Disconnect the negative battery cable.

2. Remove ignition coil and module assembly electrical connections mark or tag, if necessary.

3. Remove 4 ignition coil and module assembly to camshaft housing bolts and remove assembly by pulling straight up. Use a special spark plug boot wire remover tool to remove connector assemblies if stuck to the spark plugs.

4. Remove the idle speed power steering pressure switch connector.

5. Loosen 3 power steering pump pivot bolts and remove drive belt.

6. Disconnect the 2 rear power steering pump bracket to transaxle bolts.

7. Remove the front power steering pump bracket to cylinder block bolt.

8. Disconnect the power steering pump assembly and position aside.

9. Using special tools remove power steering pump drive pulley from intake camshaft.

10. Remove oil/air separator bolts and hoses. Leave the hoses attached to the separator, disconnect from the oil fill, chain housing and intake manifold. Remove as an assembly.

11. Remove vacuum line from fuel pressure regulator and fuel injector harness connector.

12. Disconnect fuel line retaining clamp from bracket on top of intake camshaft housing.

13. Remove fuel rail to camshaft housing attaching bolts.

14. Remove fuel rail from cylinder head. Cover injector openings in cylinder head and cover injector nozzles. Leave fuel lines attached and position fuel rail aside.

15. Disconnect timing chain and housing but do not remove from the engine.

16. Remove intake camshaft housing cover to camshaft housing attaching bolts.

17. Remove intake camshaft housing to cylinder head attaching bolts. Use the reverse of the tightening procedure when loosening camshaft housing to cylinder head attaching bolts. Leave 2 bolts loosely in place to hold the camshaft housing while separating camshaft cover from housing.

18. Push the cover off the housing by threading 4 of the housing to head attaching bolts into the tapped holes in the cam housing cover. Tighten the bolts in evenly so the cover does not bind on the dowel pins.

19. Remove the 2 loosely installed camshaft housing to head bolts and remove cover, discard gaskets.

20. Note the position of the chain sprocket dowel pin for reassembly. Remove camshaft being careful not to damage the camshaft oil seal from camshaft or journals.

21. Remove intake camshaft oil seal from camshaft and discard seal. This seal must be replaced any time the housing and cover are separated.

To install:

NOTE: If the camshaft is being replaced, the lifters must also be replaced. Lube camshaft lobes, journals and lifters with camshaft and lifter prelube. The camshaft lobes and journals must be adequately lubricated or engine damage could occur upon start up.

22. Install camshaft in same position as when removed. The timing chain sprocket dowel pin should be straight up and line up with the centerline of the lifter bores.

23. Install new camshaft housing to camshaft housing cover seals into cover. Do not use sealer.

NOTE: Cam housing to cover seals are all different.

24. Apply locking type sealer to camshaft housing and cover attaching bolt threads.

25. Install bolts and tighten to 11 ft. lbs. (15 Nm). Rotate the bolts an additional 75 degrees in sequence.

NOTE: Tighten the 2 rear bolts that hold fuel pipe to camshaft housing to 11 ft. lbs. (15 Nm), then rotate the bolts an additional 25 degrees.

26. Install timing chain housing and timing chain.

27. Uncover fuel injectors and install new fuel injector ring seals lubed with engine oil.

28. Install fuel rail to cylinder head.

29. Install fuel rail to camshaft housing attaching bolts.

30. Connect fuel line retaining clamp to bracket on top of intake camshaft housing.

31. Install vacuum line to fuel pressure regulator and fuel injector harness connector.

32. Install oil/air separator bolts and hoses.

33. Install power steering pump drive pulley to intake camshaft.

34. Install the power steering pump assembly.

35. Install the front power steering pump bracket to cylinder block bolt.

36. Connect the 2 rear power steering pump bracket to transaxle bolts.

37. Tighten the 3 power steering pump pivot bolts and install serpentine belt.

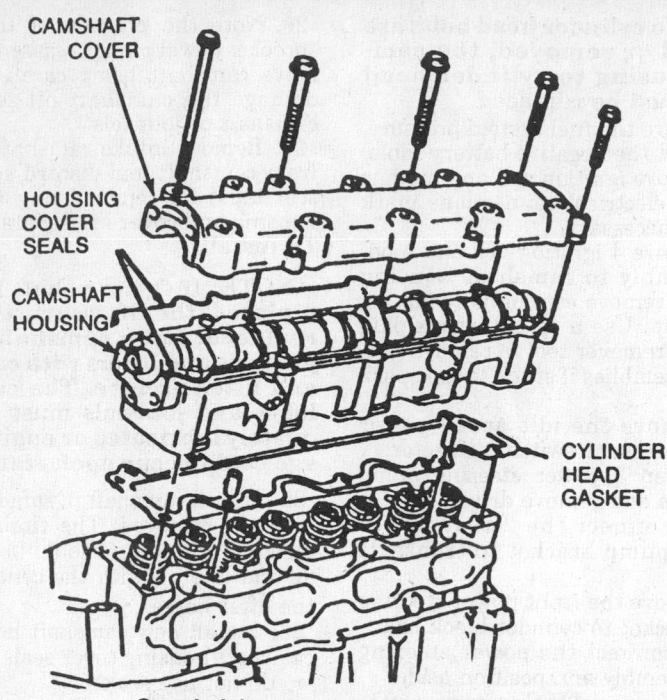

Camshaft housing assembly—2.3L engine

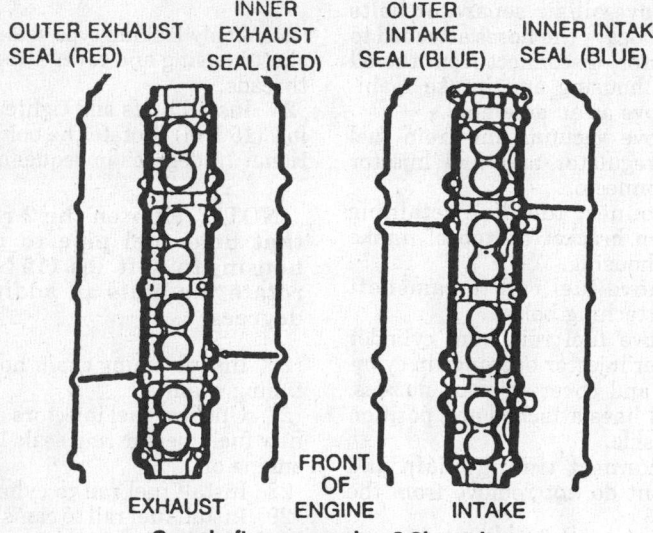

OUTER EXHAUST SEAL (RED) INNER EXHAUST SEAL (RED) OUTER INTAKE SEAL (BLUE) INNER INTAKE SEAL (BLUE)

EXHAUST FRONT OF ENGINE INTAKE

Camshaft cover seals—2.3L engine

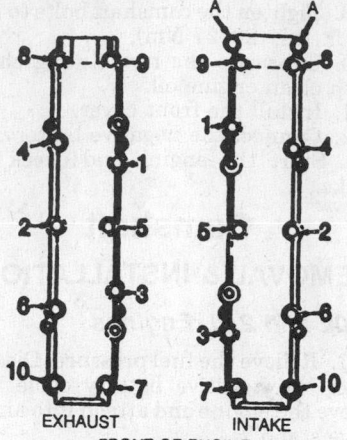

EXHAUST INTAKE
FRONT OF ENGINE

A. 16 ft. lbs. (15 Nm), rotate and additional 25 degrees

Camshaft housing bolt torque sequence —2.3L engine

38. Connect the idle speed power steering pressure switch connector.

39. Install ignition module assembly and the 4 ignition coil and module assembly to camshaft housing bolts.

NOTE: Clean any loose lubricant that is present on the ignition coil and module assembly to camshaft housing bolts. Apply Loctite® 592 or equivalent onto the ignition coil and module assembly to camshaft housing bolts. Install the bolts and tighten to 13 ft. lbs. (18 Nm).

40. Connect ignition coil and module assembly electrical connectors.

41. Connect the negative battery cable.

EXHAUST CAMSHAFT

NOTE: Any time the camshaft housing to cylinder head bolts are loosened or removed the camshaft housing to cylinder head gasket must be replaced.

1. Relieve the fuel system pressure. Disconnect the negative battery cable.

2. Remove electrical connection from ignition coil and module assembly.

3. Remove 4 ignition coil and module assembly to camshaft housing bolts and remove assembly by pulling

straight up. Use a special tool to remove connector assembly if stuck to the spark plugs.

4. Remove electrical connection from oil pressure switch.

5. Remove transaxle fluid level indicator tube assembly from exhaust camshaft cover and position aside.

6. Remove exhaust camshaft cover and gasket.

7. Disconnect timing chain and housing but do not remove from the engine.

8. Remove exhaust camshaft housing to cylinder head bolts. Use the reverse of the tightening procedure when loosening camshaft housing while separating camshaft cover from housing.

9. Push the cover off the housing by threading 4 of the housing to head attaching bolts into the tapped holes in the camshaft cover. Tighten the bolts in evenly so the cover does not bind on the dowel pins.

10. Remove the 2 loosely installed camshaft housing to cylinder head bolts and remove cover, discard gaskets.

11. Loosely reinstall 1 camshaft housing to cylinder head bolt to hold the camshaft housing in place during camshaft and lifter removal.

12. Note the position of the chain sprocket dowel pin for reassembly. Remove camshaft being careful not to damage the camshaft or journals.

13. If removing the camshaft housing, remove the valve lifters. Keep the lifters in order so they can be reinstalled in the same location.

14. Remove the camshaft housing and gasket.

To install:

15. Install the camshaft housing and gasket.

16. Loosely install one camshaft

housing-to-cylinder head bolt to hold the housing in place.

NOTE: Used lifters must be returned to their original position in the camshaft. If the camshaft is being replaced, the lifters must also be replaced. Lube camshaft lobe, journals and lifters with camshaft and lifter prelube. The camshaft lobes and journals must be adequately lubricated or engine damage could occur upon start up.

17. Install the lifters into the lifter bores.

18. Install camshaft in same position as when removed. The timing chain sprocket dowel pin should be straight up and line up with the centerline of the lifter bores.

19. Install new camshaft housing-to-camshaft housing cover seals into cover, no sealer is needed.

NOTE. Cam housing to cover seals are all different.

20. Remove the bolt holding the housing in place. Apply locking type sealer to camshaft housing and cover attaching bolt threads.

21. Install camshaft housing cover to camshaft housing.

22. Install bolts and tighten in sequence to 11 ft. lbs. (15 Nm), then rotate an additional 75 degrees in sequence.

23. Install timing chain housing and timing chain.

24. Install exhaust camshaft housing cover and new gasket and tighten to 10 ft. lbs. (14 Nm).

25. Connect the oil pressure switch electrical connector.

26. Reinstall any spark plug boot connector that was stuck to a spark plug back onto the ignition coil assembly.

27. Locate the ignition coil and module assembly over the spark plugs and push straight down.

NOTE: Clean any loose lubricant that is present on the ignition coil and module assembly to camshaft housing bolts.

28. Apply Loctite® 592 or equivalent to the ignition coil and module assembly to camshaft housing bolts. Install and hand start the ignition coil and module assembly bolts. Tighten to 15 ft. lbs. (20 Nm).

29. Connect the ignition coil and module assembly electrical connectors.

30. Connect the negative battery cable.

2.8L and 3.1L Engines

1. Relieve the fuel pressure. Disconnect the negative battery cable. Remove the engine and attach it to an engine stand.

2. Remove the intake manifold, the timing chain and sprockets.

NOTE: Be sure to keep the valve train components in order for reassembly purposes.

3. Remove the valve lifters.
4. Carefully pull the camshaft from the front of the engine.

NOTE: The camshaft journals are all the same size. Use extreme care when removing or installing the camshaft not to damage the camshaft bearings or the bearing journals of the camshaft.

To install:

5. Clean gasket mounting surfaces.
6. If installing a new camshaft, lubricate the camshaft lobes and insert the camshaft in the engine.

NOTE: If a new camshaft is being used, replace all of the lifters. Used lifters can only be used on the camshaft that they were originally installed with; provided they are installed in the exact same position they were removed.

7. Align the camshaft and crankshaft sprocket marks. Install the timing chain and sprocket.

8. Install the front cover and valve train components. Tighten the rocker arm nuts to 14–20 ft. lbs. (19–27 Nm).

9. Install the intake manifold.
10. Install the engine in the vehicle.
11. Connect the negative battery cable.

12. Fill cooling system and check for leaks. Start the engine and allow to come to normal operating temperature. Recheck for leaks. Top-up coolant.

Piston and Connecting Rod

POSITIONING

NOTCH TOWARD FRONT OF ENGINE

Piston and c position–typical

ENGINE LUBRICATION

Oil Pan

REMOVAL & INSTALLATION

2.0L and 2.2L Engines

1. Disconnect the negative battery cable. Remove the exhaust pipe shield.
2. Raise and safely support the vehicle. Drain the engine oil.
3. Disconnect the air conditioning brace from the starter and the air conditioning bracket.
4. Disconnect the starter brace from the block. Label and disconnect the starter wires. Remove the starter.
5. Remove the flywheel dust cover.
6. Remove the right support bolts and lower the support for clearance to remove the oil pan. If equipped with an automatic transaxle, remove the oil filter and extension.
7. Remove the oil pan-to-engine bolts and the oil pan.

To install:

8. Clean gasket mounting surfaces.
9. Install a new gasket. Apply a small bead of RTV sealant to the oil pan-to-engine block sealing surface. Apply a thin layer of RTV sealant on the ends of the oil pan rear seal.
10. Install the oil pan and attaching bolts. Tighten the oil pan-to-engine bolts to 6 ft. lbs. (8 Nm).
11. If removed, install the oil filter and extension.
12. Install the starter. Connect the electrical connectors. Install the starter brace to the block.
13. Install the air conditioning bracket and install the brace to the starter.
14. Lower the vehicle.
15. Connect the negative battery cable.
16. Refill the engine with the clean engine oil. Start the engine and check for leaks.

2.3L Engine

1. Disconnect the negative battery cable.
2. Raise and support the vehicle safely.
3. Remove the flywheel inspection cover.
4. Remove the splash shield-to-suspension support bolt.
5. Remove the exhaust manifold brace.
6. Remove the radiator outlet pipe-to-oil pan bolt.
7. Remove the transaxle-to-oil pan nut and stud using a 7mm socket.

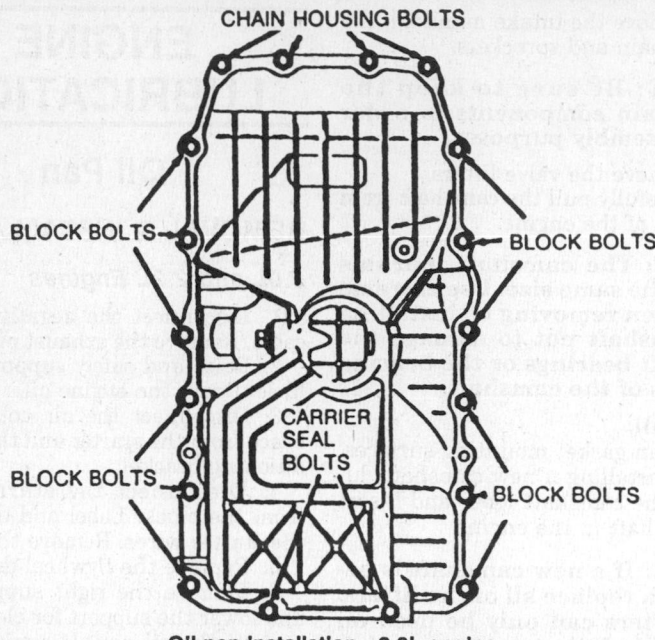

Oil pan installation—2.3L engine

8. Gently pry the spacer out from between oil pan and transaxle.

9. Remove the oil pan bolts. Remove the oil pan from the engine.

To install:

10. Install the oil pan to the engine. Install the oil pan bolts. Tighten the chain housing and carrier seal bolts to 106 inch lbs. (12 Nm). Tighten the oil pan-to-block bolts to 17 ft. lbs. (23 Nm).

11. Install the spacer and install the stud.

12. Install the oil pan-to-transaxle nut and bolt.

13. Install the splash shield-to-suspension support bolt.

14. Install the radiator outlet pipe-to-oil pan bolt.

15. Install the exhaust manifold brace.

16. Install the flywheel inspection cover.

17. Lower the vehicle.

18. Fill the crankcase with oil to specification.

19. Connect the negative battery cable. Start the engine and check for leaks.

20. Turn the engine off and allow to stand. Check oil level, add as necessary.

2.8L and 3.1L Engines

1. Disconnect the negative battery cable.

2. Raise and safely support the vehicle. Drain the engine oil.

3. Remove the flywheel dust cover.

4. Label and disconnect the starter wires. Remove the starter.

5. Remove the oil pan-to-engine nuts/bolts and the oil pan.

To install:

6. Clean gasket mounting surfaces.

7. Install a new gasket. Install the oil pan and attaching bolts. Tighten the oil pan nuts to 6-9 ft. lbs. (8-12 Nm) or bolts to 15-22 ft. lbs. (20-30 Nm).

8. Install the starter. Connect the electrical connectors.

9. Install the flywheel dust cover.

10. Install a new oil filter.

11. Lower the vehicle.

12. Fill the crankcase with oil to specification.

13. Connect the negative battery cable. Start the engine and check for leaks.

14. Turn the engine off and allow to stand. Check oil level, add as necessary.

Oil Pump

REMOVAL & INSTALLATION

Except 2.3L Engine

1. Disconnect the negative battery cable. Raise and safely support the vehicle. Drain the engine oil.

2. Remove the oil pan-to-engine bolts and the oil pan.

3. Remove the oil pump-to-rear main bearing cap bolt, the oil pump and extension shaft.

To install:

4. Install the extension shaft, oil pump and pump-to-rear main cap bolt. Tighten the oil pump-to-bearing cap bolt to 25-38 ft. lbs. (34-52 Nm) and the upper oil pump drive bolt to 14-22

ft. lbs. (19-30 Nm), on the 2.0L and 2.2L engines or to 25-38 ft. lbs. (34-52 Nm) on the 2.8L and 3.1L engines.

5. Install the oil pan and attaching bolts.

6. Lower the vehicle.

7. Fill the crankcase with oil to specification.

8. Connect the negative battery cable. Start the engine and check oil pressure and check for leaks.

9. Turn the engine off and allow to stand. Check oil level, add as necessary.

2.3L Engine

1. Disconnect the negative battery cable.

2. Raise and support the vehicle safely.

3. Remove the attaching bolts and the oil pan.

4. Remove the oil pump assembly retainers, bolts and nut.

5. Remove the oil pump assembly and shims if equipped.

NOTE: Oil pump drive gear backlash must be checked when any of the following components are replaced: oil pump assembly, oil pump drive gear, crankshaft and cylinder block.

To install:

6. Check and adjust oil pump drive gear backlash as follows:

a. With oil pump assembly off engine, remove 3 attaching bolts and separate the driven gear cover and screen assembly from the oil pump.

b. Install the oil pump on the block using the original shims. Tighten the bolts to 33 ft. lbs. (45 Nm).

c. Install the dial indicator assembly to measure backlash between oil pump to drive gear.

d. Record oil pump drive to driven gear backlash correct backlash clearance is 0.0091-0.0201 in. (0.23-0.51mm). When taking measurement crankshaft cannot move.

e. Remove oil pump from block reinstall driven gear cover and screen assembly to pump and tighten to 106 inch lbs. (12 Nm).

f. Reinstall the pump assembly on block. Tighten oil pump-to-block bolts 33 ft. lbs. (45 Nm).

7. Install the oil pump assembly, including shims if removed.

8. Tighten oil pump to block bolts to 33 ft. lbs. (45 Nm).

9. Install the oil pan and attaching bolts.

10. Lower the vehicle.

11. Fill the crankcase with oil to specification.

12. Connect the negative battery ca-

ble. Start the engine and check oil pressure and check for leaks.

13. Turn the engine off and allow to stand. Check oil level, add as necessary.

Rear Main Bearing Oil Seal

REMOVAL & INSTALLATION

NOTE: This procedure should only be performed if rear crankshaft seal installer tool J–34686 for 2.0L (VIN 1), 2.2L, 2.8L or 3.1L engines, or J–36005 for the 2.3L engine or equivalent, is available.

1. Disconnect the negative battery cable. Remove the transaxle.

2. If equipped with a manual transaxle, matchmark and remove the clutch/flywheel assembly. If equipped with an automatic transaxle, remove the flywheel.

3. Using a small prybar, pry the rear main seal from the engine.

NOTE: Use care when removing or installing the seal to avoid damage to the crankshaft sealing surface. If equipped with a manual transaxle, inspect the condition of the clutch to insure that the clutch was not damaged by oil loss from the rear main seal.

To install:

4. To install the rear main oil seal, perform the following procedures:

a. Lubricate the seal bore and seal surface with engine oil.

b. Using a seal installation tool, press the new rear oil seal into the engine. The seal must fit squarely against the back of the tool.

c. Align the dowel pin of the tool with the dowel pin in the crankshaft and tighten the attaching screws on the tool to 2–5 ft. lbs. (3–7 Nm).

d. Tighten the T-handle of the tool to push the seal into the seal bore.

e. Loosen the T-handle. Remove the attaching screws and tool.

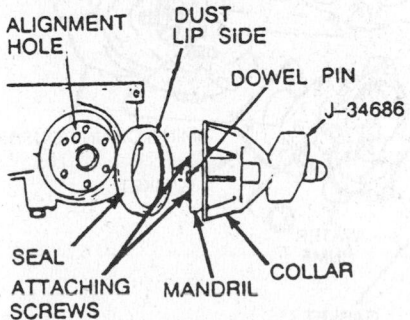

ALIGNMENT HOLE
DUST LIP SIDE
DOWEL PIN
J–34686
SEAL ATTACHING SCREWS
MANDRIL
COLLAR

Rear main bearing oil seal installation— Typical

f. Check the seal to make sure it is seated squarely in the bore.

5. Install the flywheel. Tighten the flywheel-to-crankshaft bolts to 45–59 ft. lbs. (61–80 Nm) for automatic transaxles or to 47–63 ft. lbs. (64–85 Nm) for manual transaxles.

6. Lower the vehicle and connect the negative battery cable.

7. Start the engine and check for leaks.

ENGINE COOLING

Radiator

REMOVAL & INSTALLATION

─────── **CAUTION** ───────

Before attempting any work on the cooling system, allow the engine to first cool sufficiently. To avoid personal injury, do not remove the radiator cap while the engine is at or above normal operation temperature.

1. Disconnect the negative battery cable.

2. Remove the air cleaner assembly.

3. Drain the engine coolant into a clean container for reuse.

4. Disconnect the electrical connection from the electric fan.

5. Remove the fan-to-chassis mounting bolts and remove the fan assembly.

6. Disconnect the radiator upper and lower hoses at the radiator end.

7. If equipped with an automatic transaxle, disconnect the transaxle cooler lines and plug.

8. Remove the upper radiator mounting bolts.

9. Remove the condenser-to-radiator mounting bolts.

10. Carefully lift the radiator out.

To install:

11. Install the radiator in the vehicle.

12. Install the condenser-to-radiator mounting bolts.

13. Install the upper radiator mounting bolts.

14. Tighten the radiator mounting bolts to 89 inch lbs. (10 Nm).

15. If equipped with an automatic transaxle, connect the transaxle cooler lines.

16. Connect the radiator upper and lower hoses at the radiator end.

17. Install the fan assembly and the fan-to-chassis mounting bolts.

18. Connect the electrical connection to the electric fan.

19. Connect the negative battery cable.

20. Fill cooling system and check for

leaks. Start the engine and allow to come to normal operating temperature. Recheck for leaks. Top-up coolant.

21. Install the air cleaner assembly.

Heater Core

REMOVAL & INSTALLATION

Without Air Conditioning

1. Disconnect the negative battery cable. Drain the engine coolant into a clean container for reuse.

2. Disconnect the heater hoses from the heater core.

3. Remove the right and left sound insulators and the steering column trim cover.

4. Remove the heater air outlet deflector.

5. Remove the heater core cover, the heater core and retaining straps.

To install:

6. Install the heater core cover, the heater core and retaining straps.

7. Install the heater air outlet deflector.

8. Install the right and left sound insulators and the steering column trim cover.

9. Connect the heater hoses to the heater core.

10. Connect the negative battery cable.

11. Fill cooling system and check for leaks. Start the engine and allow to come to normal operating temperature. Recheck for leaks. Top-up coolant.

With Air Conditioning

1. Disconnect the negative battery cable. Drain the engine coolant into a clean container for reuse.

2. Raise and safely support the vehicle.

3. Remove the drain tube from the heater case and the heater hoses from the heater core. Lower the vehicle.

4. Remove the right and left side sound insulators and the steering column trim cover.

5. Remove the heater air outlet deflector and the glove box.

6. Remove the heater core cover, the heater core and retaining straps.

To install:

7. Install the heater core cover, the heater core and retaining straps.

8. Install the heater air outlet deflector and the glove box.

9. Install the right and left side sound insulators and the steering column trim cover.

10. Raise and safely support the vehicle. Connect the heater hoses to the heater core and connect the drain tube to the heater case.

11. Lower the vehicle.

12. Connect the negative battery cable.

13. Fill cooling system and check for leaks. Start the engine and allow to come to normal operating temperature. Recheck for leaks. Top-up coolant.

Water Pump

REMOVAL & INSTALLATION

Except 2.3L Engine

1. Disconnect the negative battery cable.

2. Drain the engine coolant into a clean container for reuse.

3. Remove the serpentine drive belt.

4. If equipped with the 2.0L engine, remove the alternator and bracket with wires attached and position it aside.

5. If equipped with the 3.1L engine, remove the radiator and heater hoses.

6. Remove the water pump pulley bolts and the pulley.

7. Remove the water pump-to-engine bolts and the pump.

To install:

8. Clean the gasket mounting surfaces.

9. Install the water pump and the water pump attaching bolts. Tighten the water pump-to-engine bolts to 14–22 ft. lbs. (19–30 Nm) on the 2.0L and 2.2L engines or to 6–9 ft. lbs. (8–12 Nm) on the 2.8L and 3.1L engines.

10. Install the water pump pulley and attaching bolts.

11. If equipped with the 3.1L engine, install the radiator and heater hoses.

12. If equipped with the 2.0L engine, install the alternator and bracket.

13. Install the serpentine drive belt.

14. Connect the negative battery cable.

15. Fill cooling system and check for leaks. Start the engine and allow to come to normal operating temperature. Recheck for leaks. Top-up coolant.

2.3L Engine

1. Disconnect the negative battery cable.

2. Drain the engine coolant into a clean container for reuse.

NOTE: Remove the heater hose from the thermostat housing for additional draining.

3. Remove the oxygen sensor connector.

4. Remove the upper and lower exhaust manifold heat shield attaching bolts and remove the shields.

5. Remove the exhaust manifold brace-to-manifold attaching bolt.

6. Using a 13mm box wrench, loosen the exhaust pipe-to-manifold spring bolts from the engine compartment.

7. Raise and safely support the vehicle.

8. Remove the bolts from the exhaust flange using a 7/32 in. (5.5mm) socket an 1 bolt rotate clockwise first.

NOTE: Rotating the bolt clockwise is necessary to relieve the spring pressure from 1st bolt prior to removing the 2nd bolt otherwise the exhaust pipe will twist and bind the bolt as it is removed.

9. Thread the bolt with least pressure on it out 4 turns.

10. Move the other bolt and turn it all the way out of the exhaust pipe flange.

11. Return to the 1st bolt and rotate it the rest of the way out.

12. Pull the exhaust pipe back from the exhaust manifold.

13. Remove the radiator outlet pipe from the oil pan and transaxle.

14. Remove the exhaust manifold brace.

15. Pull down on the radiator outlet pipe to disengage it from the water pump.

16. Lower the vehicle.

17. Remove the exhaust manifold-to-cylinder head attaching nuts.

18. Remove the exhaust manifold, seals and gaskets.

19. Remove the water pump cover-to-engine attaching bolts.

20. Remove the water pump-to-timing chain housing attaching nuts.

21. Remove the water pump and cover assembly from the engine.

22. Remove the water pump cover-to-radiator pump assembly.

To install:

NOTE: Before installing the water pump it is important to first read over the entire procedure. Pay special attention to the

tightening sequence, to avoid part damage and to insure proper sealing.

23. Clean all mating surfaces throughly and use new gaskets.

24. Position the water pump cover to the radiator pump assembly and install the attaching bolts. Do not tighten.

25. Lubricate the splines of the radiator pump drive with the an approved chassis grease and install the pump and cover assembly.

26. Install the pump cover-to-engine attaching bolts. Do not tighten.

27. Install the timing chain housing nuts. Do not tighten.

28. Lubricate the O-ring on the radiator outlet pipe with a solution of antifreeze and slide the pipe into the radiator pump cover. Install the attaching bolts. Do not tighten.

29. Tighten the bolts and nuts in following order:

   a. Pump assembly-to-timing chain housing nuts—19 ft. lbs. (26 Nm).

   b. Water pump-to-pump cover assembly—106 inch lbs. (12 Nm).

   c. Water Pump cover-to-engine (tighten the bottom bolt first)—19 ft. lbs. (26 Nm).

   d. Radiator outlet pipe assembly-to-pump cover—125 ft. lbs. (14 Nm).

30. Install the exhaust manifold with new gaskets.

31. Install the exhaust manifold-to-cylinder head attaching nuts. Tighten the attaching nuts in sequence to 22 ft. lbs. (30 Nm).

32. Raise and safely support the vehicle.

33. Seat the exhaust manifold bolts into the exhaust pipe flange.

34. Using a 7/32 in. (5.5mm) socket start both bolts. Rotate the bolts counterclockwise.

35. Turn both bolts in evenly to avoid cocking the exhaust pipe and binding

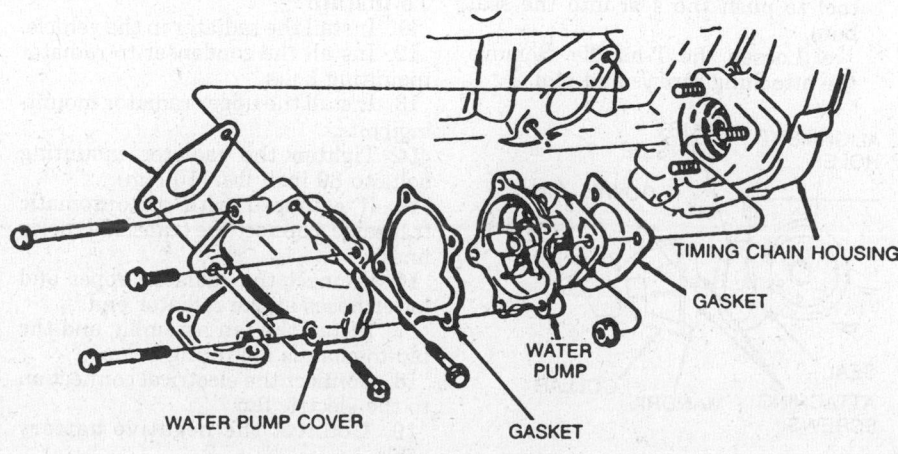

Water pump removal and installation—2.3L engine

the bolts. Turn the bolts in until fully seated.

36. Install the radiator outlet pipe to the transaxle and to the oil pan and install the exhaust manifold brace.

37. Lower the vehicle.

38. Install the exhaust manifold brace-to-manifold attaching bolt.

39. Using a 13mm wrench, tighten the exhaust pipe-to-manifold nuts to 22 ft. lbs. (30 Nm).

40. Install the lower heat shields.

41. Connect the oxygen connector to the oxygen sensor.

42. Connect the negative battery cable.

43. Fill cooling system and check for leaks. Start the engine and allow to come to normal operating temperature. Recheck for leaks. Top-up coolant.

Thermostat
REMOVAL & INSTALLATION

1. Disconnect the negative battery cable.

2. Remove the air cleaner assembly.

3. Drain the engine coolant level below the thermostat housing.

4. Remove the upper radiator hose from the thermostat water outlet and position it to the side.

5. On the 2.3L engine, remove the heater and throttle body coolant hoses from the thermostat housing and disconnect the electrical connector from the coolant temperature sensor.

6. Remove the thermostat attaching bolts.

7. Remove the thermostat housing gasket and thermostat.

To install:

8. Throughly clean the mating surfaces of the engine and thermostat.

9. Install the new thermostat, gasket and housing, being careful not to allow the thermostat to slip out of position.

10. Install the attaching bolts and tighten to 6–9 ft. lbs. (8–12 Nm) for 2.0L and 2.2L engines, 15–22 ft. lbs. (20–30 Nm) for the 2.8L and 3.1L engines or 19 ft. lbs. (26 Nm) for the 2.3L engine.

11. On 2.3L engine, connect the heater and throttle body coolant hoses the thermostat housing and connect the coolant temperature sensor connector.

12. Connect the upper radiator hose to the thermostat housing water outlet.

13. Refill and bleed the cooling system. Start the engine, allow it to reach normal operating temperature and check for leaks.

14. Allow time for the thermostat to open, recheck the coolant level and top up, as required.

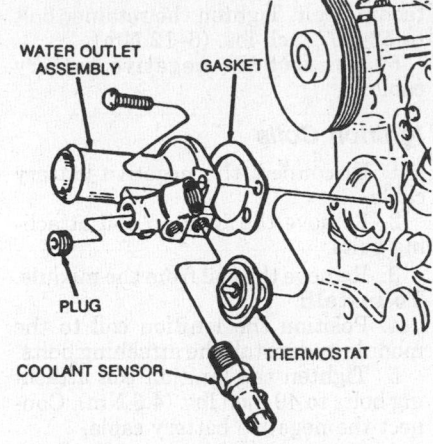

Thermostat removal and installation— 2.3L engine

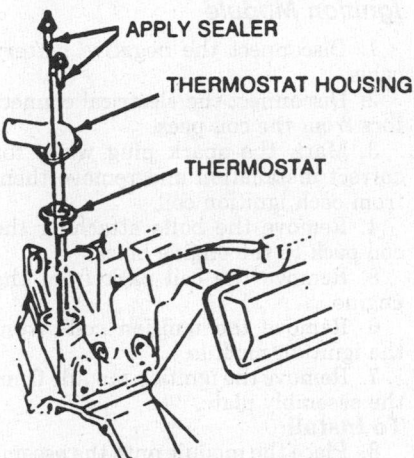

Thermostat removal and installation— 2.8L and 3.1L engines

Cooling System Bleeding

After working on the cooling system, even to replace the thermostat, the system must be bled. Air trapped in the system will prevent proper coolant circulation and leave the system coolant level low, causing a risk of overheating.

1. To bleed the system, start with the system cool, the radiator cap off and the radiator filled to about an inch below the filler neck.

2. Start the engine and run it at slightly above normal idle speed. This will insure adequate circulation. If air bubbles appear and the coolant level drops, fill the system with a mixture of anti-freeze and water to bring the level back to the proper level.

3. Run the engine this way until the thermostat opens. When this happens, the coolant will move abruptly across the top of the radiator and the temperature of the upper radiator tank and upper radiator hose will rise suddenly.

4. At this point, air is often expelled and the level may drop quite a bit. Keep refilling the system until the level is near the top of the radiator and remains constant.

5. If the vehicle has an overflow tank, fill the radiator up to the top of the filler neck and check the coolant the level in the overflow tank.

ENGINE ELECTRICAL

NOTE: Disconnecting the negative battery cable on some vehicles may interfere with the functions of the on board computer systems and may require the computer to undergo a relearning process, once the negative battery cable is reconnected.

Distributorless Ignition System

Two types of distributorless ignition systems will be covered in this section. The first is the Direct Ignition System (DIS) and the second is the Integrated Direct Ignition System (IDIS).

The distributorless ignition systems use of a waste spark method of spark distribution. Each cylinder is paired with its companion cylinder in the firing order. This places 1 cylinder on the compression stroke with the companion cylinder on the exhaust stroke. The cylinder that is on the exhaust stroke uses very little spark allowing most of the spark to go to the cylinder on the compression stroke. This process reverses when the cylinder roles reverse. There are 2 coils for the 4 cylinder engines and 3 coils for the 6 cylinder engines. The Direct Ignition System (DIS) is used on the 2.0L, 2.2L, 2.8L and 3.1L engines. The Integrated Direct Ignition System (IDIS) is used on the 2.3L engine.

Since no distributor is used, the timing references are gathered from the engine sensors.

REMOVAL & INSTALLATION
Coil Pack

1. Disconnect the negative battery cable.

2. Disconnect the electrical connectors from the coil pack.

3. Mark the spark plug wires for correct installation and remove them from each coil.

4. Remove the coil pack-to-engine

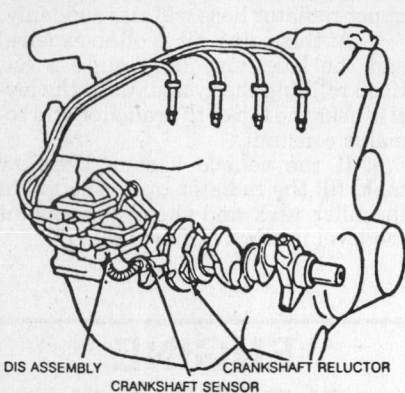

View of the Direct Ignition System (DIS) 2.0L, 2.2L, 2.8L and 3.1L engines

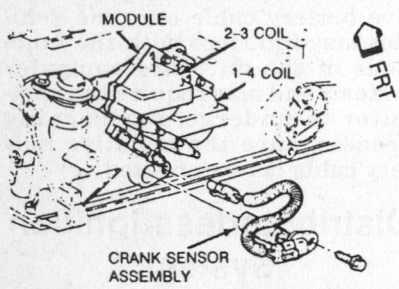

Removing the coil pack from the engine

bolts and remove the coil pack from the engine.

To install:

5. Position the coil pack to the engine block and install the attaching bolts. Tighten the attaching bolts to 15–21 ft. lbs. (20–30 Nm).

6. Connect the spark plug wires to their respective positions, marked during removal, on the ignition coils.

7. Connect the electrical connectors to the coil pack.

8. Connect the negative battery cable.

9. Start the engine and test the engine performance.

Crankshaft Sensor

1. Disconnect the negative battery cable.

2. Disconnect the sensor harness connector at the ignition module.

3. Remove the crankshaft sensor attaching bolt and remove the sensor from the engine.

NOTE: Prior to installing the crankshaft sensor, inspect the O-ring for wear, cracks or signs of leakage. Replace it, if necessary. If it is necessary to replace the seal, lubricate it with engine oil prior to installation.

To install:

4. Position the crankshaft sensor to the engine block.

5. Install the crankshaft sensor at-

taching bolt. Tighten the retainer bolt to 53–107 inch lbs. (6–12 Nm).

6. Connect the negative battery cable.

Ignition Coils

1. Disconnect the negative battery cable.

2. Remove the ignition coil attaching bolts.

3. Remove the coil from the module.

To install:

4. Position the ignition coil to the module and install the attaching bolts.

5. Tighten the ignition coil attaching bolts to 40 inch lbs. (4.5 Nm). Connect the negative battery cable.

6. Start the engine and check performance.

Ignition Module

1. Disconnect the negative battery cable.

2. Disconnect the electrical connectors from the coil pack.

3. Mark the spark plug wires for correct installation and remove them from each ignition coil.

4. Remove the bolts attaching the coil pack to the engine block.

5. Remove the coil pack from the engine.

6. Remove the ignition coils from the ignition module.

7. Remove the ignition module from the assembly plate.

To install:

8. Place the module onto the assembly plate and assemble the ignition coils to the module with the attaching bolts.

9. Tighten the attaching bolts to 40 inch lbs. (4.5 Nm).

10. Position the coil pack to the engine block and install the attaching bolts. Tighten the attaching bolts to 15–21 ft. lbs. (20–30 Nm).

11. Connect the spark plug wires to the ignition coils (marked during removal).

12. Connect the electrical connectors to the coil pack.

13. Connect the spark plug wires and electrical connectors to their respective places.

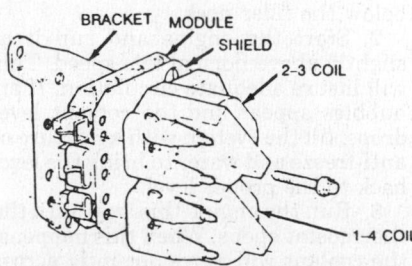

Removing the ignition coils and module from the assembly plate

14. Connect the negative battery cable.

15. Start the engine and check performance.

INTEGRATED DIRECT IGNITION SYSTEM

The Integrated Direct Ignition System (IDIS) consist of 2 separate ignition coils an ignition module and a secondary conductor housing mounted to an aluminum cover plate. A crankshaft sensor, related connecting wires and an Electronic Spark Timing (EST) portion of the Electronic Control Module (ECM) make up the remainder of the system.

This system, being a distributorless ignition system, uses a waste spark method of spark distribution. Each cylinder is paired with its companion cylinder in the firing order. This places 1 cylinder on the compression stroke with the companion cylinder on the exhaust stroke. The cylinder that is on the exhaust stroke uses very little spark allowing most of the spark to go to the cylinder on the compression stroke. This process reverses when the cylinder roles reverse.

Because of the direction of current flow in the primary winding and thus, into the secondary winding, 1 plug will fire from the center electrode to the side electrode while the other fires from the side electrode.

The magnetic pick-up sensor is mounted on the side of engine block, in proximity to the crankshaft reluctor ring. Notches in the crankshaft reluctor ring trigger the magnetic pick-up sensor to provide timing information to the Electronic Control Module. The magnetic pick-up sensor provides a cam signal to identify correct

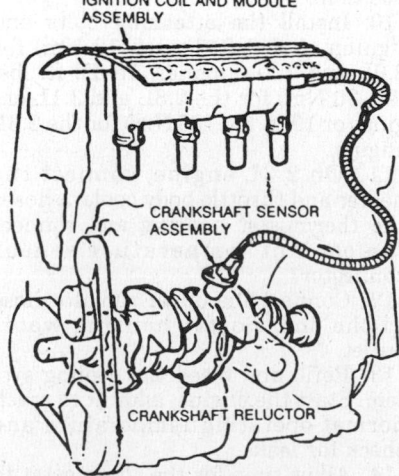

View of the Integrated Direct Ignition System (IDIS) 2.3L engine

firing sequence and crank signals to trigger each coil at the proper time.

The Electronic Control Module (ECM) sends a signal from the Electronic Spark Control (EST) to control spark timing. Under 700 rpm, the ignition module controls spark timing when the system is in the module or bypass timing mode. When over 700 rpm, the Electronic Control Module controls spark timing, when the system is in the EST mode.

Since no distributor is used, the timing references are gathered from the engine sensors.

REMOVAL & INSTALLATION

Ignition Assembly

1. Disconnect the negative battery cable.
2. Disconnect the harness connector from the coil and module assembly.
3. Remove the ignition assembly-to-camshaft housing attaching bolts.
4. Carefully remove the ignition assembly from the the engine.

NOTE: If the spark plug boots present a problem coming off, it may be necessary to use a special removal tool, first twisting and pulling upward on the retainers.

To install:

5. Install the spark plug boots and retainers on the ignition assembly housing secondary terminals.

NOTE: If the boots and retainers are not in place on the housing secondary terminals prior to installing the ignition assembly, damage to the ignition system may result.

6. Position the ignition assembly to the engine while carefully aligning the boots to the spark plug terminals.

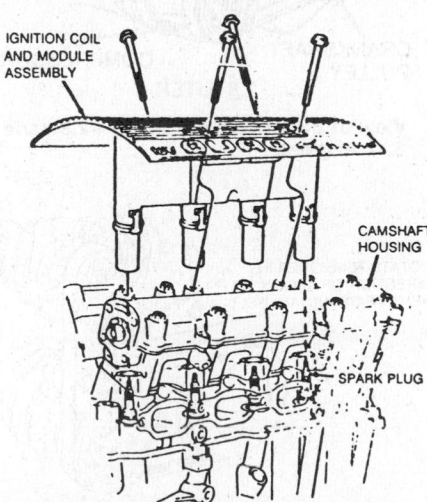

Removing the IDIS Ignition assembly from the engine

7. Coat the ignition assembly-to-camshaft housing attaching bolts with an approved lubricant and install them into the housing.
8. Tighten the attaching bolts to 19 ft. lbs. (26 Nm).
9. Connect the harness connector to the ignition coil module assembly.
10. Connect the negative battery cable.
11. Start the engine and test the engine performance.

Crankshaft Sensor

1. Disconnect the negative battery cable.
2. Disconnect the harness connector at the crankshaft sensor.
3. Remove the sensor attaching bolt.
4. Remove the crankshaft sensor from the engine.
5. Inspect the sensor O-ring for wear, cracks or signs of leakage. Replace it if necessary.

To install:

6. Lubricate the O-ring with engine oil and install it on the sensor.
7. Position the sensor to the engine block and install the attaching bolt. Tighten the attaching bolt to 88 inch lbs. (10 Nm).
8. Connect the sensor harness connector.
9. Connect the negative battery cable.
10. Start the engine and test engine performance.

Ignition Coil

1. Disconnect the negative battery cable.
2. Disconnect the harness connector from the coil and module assembly.
3. Remove the ignition assembly-to-camshaft housing attaching bolts.
4. Carefully remove the ignition assembly from the engine.

NOTE: If the spark plug boots present a problem coming off, it may be necessary to use a special

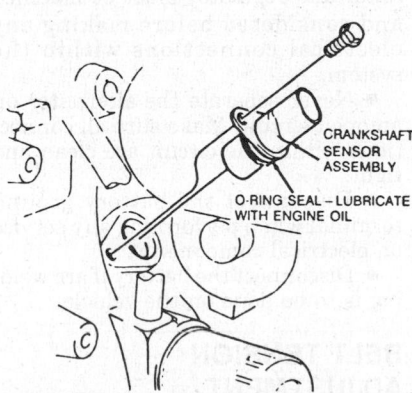

Removing crankshaft sensor from the engine

removal tool, first twisting and pulling upward on the retainers.

5. Remove the ignition coil housing-to-cover bolts.
6. Remove the cover from the coil housing.
7. Disconnect the ignition coil harness connectors from the coil pack assembly.
8. Carefully lift the coil pack out and remove the contacts and seals from the housing.

To install:

9. Install new coil seals into the coil housing.
10. Install the coil contacts to the coil housing and retain with petroleum jelly.
11. Place the coil pack into the housing and connect the harness connectors.
12. Assemble the cover to the coil housing and install the attaching bolts. Tighten the attaching bolts to 35 inch lbs. (4 Nm).
13. Install the spark plug boots and retainers on the ignition assembly housing secondary terminals.

NOTE: If the boots and retainers are not in place on the housing secondary terminals prior to installing the ignition assembly, damage to the ignition system may result.

14. Position the ignition assembly to the engine while carefully aligning the boots to the spark plug terminals.
15. Coat the ignition assembly-to-camshaft housing attaching bolts with an approved lubricant and install them into the housing.
16. Tighten the attaching bolts to 19 ft. lbs. (26 Nm).
17. Connect the harness connector to the ignition coil module assembly.
18. Connect the negative battery cable.
19. Start the engine and test the engine performance.

Ignition Module

1. Disconnect the negative battery cable.
2. Disconnect the harness connector from the coil and module assembly.
3. Remove the ignition assembly-to-camshaft housing attaching bolts.
4. Carefully remove the ignition assembly from the engine.

NOTE: If the spark plug boots present a problem coming off, it may be necessary to use a special removal tool, first twisting and pulling upward on the retainers.

5. Remove the ignition coil housing-to-cover bolts.
6. Remove the cover from the coil housing.

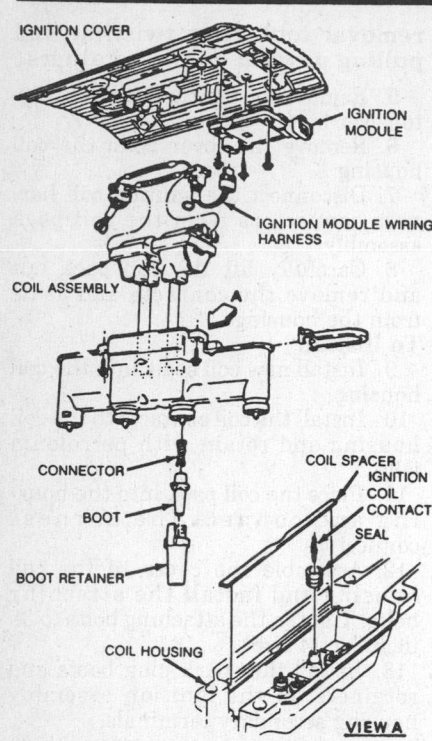

Exploded view of the IDIS ignition assembly

7. Disconnect the coil harness connector from the module.

8. Remove the screws attaching the module to the ignition assembly cover.

NOTE: If the same module is going to be replaced, take care not to remove the grease from the module or coil. If a new module is to be installed, a package of silicone grease will be included with it. This grease aids in preventing the module from overheating.

To install:

9. Place the module on the ignition cover and install the attaching bolts. Tighten the attaching bolts to 35 inch lbs. (4 Nm).

10. Connect the coil harness connector to the module.

11. Assemble the module cover to the coil housing and install the attaching bolts. Tighten the attaching bolts to 35 inch lbs. (4 Nm).

12. Install the spark plug boots and retainers on the ignition assembly housing secondary terminals.

NOTE: If the boots and retainers are not in place on the housing secondary terminals prior to installing the ignition assembly, damage to the ignition system may result.

13. Position the ignition assembly to the engine while carefully aligning the boots to the spark plug terminals.

14. Coat the ignition assembly-to-camshaft housing attaching bolts with an approved lubricant and install them into the housing.

15. Tighten the attaching bolts to 19 ft. lbs. (26 Nm).

16. Connect the harness connector to the ignition coil module assembly.

17. Connect the negative battery cable.

18. Start the engine and test the engine performance.

Ignition Timing

ADJUSTMENT

Ignition timing is controlled by the Electronic Control Module (ECM). No adjustments are possible.

Alternator

Precautions

Several precautions must be observed with alternator equipped vehicles to avoid damage to the unit.

• If the battery is removed for any reason, make sure it is reconnected with the correct polarity. Reversing the battery connections may result in damage to the one-way rectifiers.

• When utilizing a booster battery as a starting aid, always connect the positive to positive terminals and the negative terminal from the booster battery to a good engine ground on the vehicle being started.

• Never use a fast charger as a booster to start vehicles.

• Disconnect the battery cables when charging the battery with a fast charger.

• Never attempt to polarize the alternator.

• When checking diode continuity, ensure the tester does not exceed 12 volts.

• Do not short across or ground any of the alternator terminals.

• The polarity of the battery, alternator and regulator must be matched and considered before making any electrical connections within the system.

• Never separate the alternator on an open circuit. Make sure all connections within the circuit are clean and tight.

• Disconnect the battery ground terminal when performing any service on electrical components.

• Disconnect the battery if arc welding is to be done on the vehicle.

BELT TENSION ADJUSTMENT

A single (serpentine) belt is used to drive all engine mounted components. Drive belt tension is maintained by a spring loaded tensioner.

The serpentine drive belt may be removed or installed by rotating the tensioner using a 15mm socket for the 2.0L, 2.2L and 3.1L engines. Use a 13mm open end wrench for the 2.3L engine or a ¾ in. open end wrench for the 2.8L engine. This will eliminate

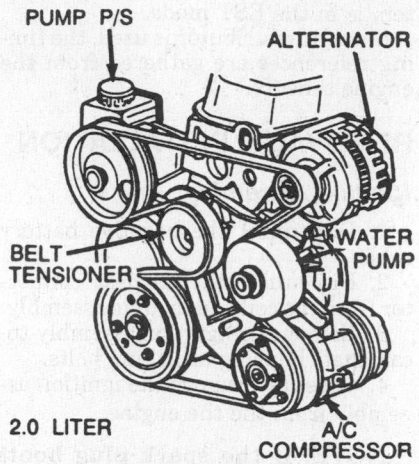

View of the drive belt routing—2.0L and 2.2L engines

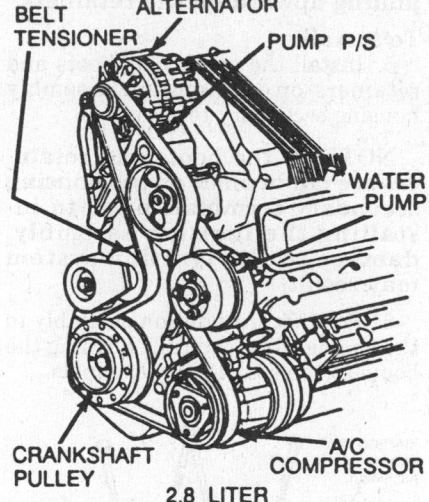

View of the drive belt routing—2.8L and 3.1L engines

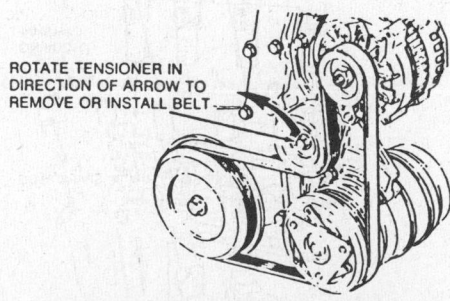

View of the drive belt routing—2.3L engine

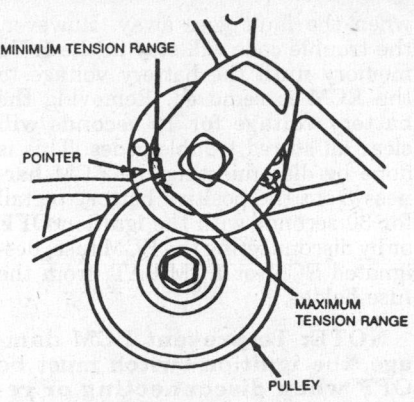

Tensioner operating range

the belt tension and will allow the belt to be removed or installed.

CAUTION

To avoid personal injury when rotating the serpentine belt tensioner on the 2.3L engine, be sure to use a tight fitting 13mm wrench at least 24 inch long.

NOTE: The drive belt tensioner can control belt tension over a wide range of belt lengths; however, there are limits to the tensioner's ability to compensate for various belt lengths. Installing the wrong size belt and using the tensioner outside its operating range can result in poor tension control and/or damage to the tensioner, belt and driven components.

REMOVAL & INSTALLATION

Except 2.3L Engine

1. Disconnect the negative battery cable.
2. Remove the serpentine drive belt.
3. Label and disconnect the electrical connectors from the back of the alternator.
4. Remove the alternator mounting bolts.
5. Remove the alternator-to-bracket bolts and the alternator.
To install:
6. Position the alternator to the the mounting bracket and install the attaching bolts.
7. Connect the alternator electrical connectors to the rear of the alternator.
8. Install the serpentine drive belt.
9. Connect the negative battery cable.
10. Start the engine and perform a charging system test.

2.3L Engine

1. Disconnect the negative battery cable.
2. Remove the serpentine drive belt.

CAUTION

To avoid personal injury when rotating the serpentine belt tensioner, be sure to use a tight fitting 13mm wrench at least 24 inch long.

3. Remove the coolant and washer reservoir attaching screws.
4. Disconnect the washer pump electrical connector and position the reservoir to the side.
5. Remove the air conditioner line rail clip.
6. Disconnect the 2 vacuum lines at the front of the engine and remove vacuum harness attaching bracket, as required.
7. Disconnect and tag electrical connections from injector harness and alternator.
8. Remove the rear alternator mounting bolts.
9. Remove the front alternator mounting bolt and engine harness clip.
10. Carefully remove the alternator from between the mounting bracket and the air conditioning and condenser hose.

NOTE: Extreme care must be taken when removing or installing the alternator as not to damage the air conditioner compressor and condenser hoses.

To install:
11. Place the alternator between the air conditioner compressor and condenser hoses and install it on the bracket.
12. Install the rear mounting bolt. Tighten the mounting bolt to 19 ft. lbs. (26 Nm).
13. Install the front mounting bolts. Tighten the upper mounting bolt to 37 ft. lbs. (50 Nm) and the lower mounting bolt to 19 ft. lbs. (26 Nm).
14. Install the serpentine drive belt.

CAUTION

To avoid personal injury when rotating the serpentine belt tensioner, be sure to use a tight fitting 13mm wrench at least 24 in. long.

15. Install the air conditioner rail clip.
16. Connect the washer pump electrical connector.
17. Install the coolant and washer pump reservoir.
18. Connect the electrical connections for the alternator and injector harness.
19. If removed, install the vacuum harness attaching bracket and connect the vacuum lines at the front of the engine.
20. Connect the negative battery cable.
21. Start the engine and perform a charging system test.

Starter

REMOVAL & INSTALLATION

Except 2.3L Engine

1. Disconnect the negative battery cable.
2. Remove the air cleaner assembly, as required.
3. Raise and safely support the vehicle.

NOTE: If equipped with an oil cooler, remove the oil filter and position the hose next to the starter to the side.

4. Remove the air conditioning compressor brace attaching nuts and remove the brace from the engine, as required.
5. Remove flywheel inspection cover bolts and remove the inspection cover, as required.
6. Remove the starter attaching bolts.
7. Carefully lower the starter and remove the shims. Note the number and position of any shims.
8. Disconnect the electrical wiring connections at the starter.
To install:
9. Connect the electrical connections to the starter.
10. Position the shims in place and install the starter and mounting bolts. Tighten the bolts to 32 ft. lbs. (43 Nm).

NOTE: If equipped with an oil cooler, position the cooler hose next to the starter motor and install the oil filter.

11. If removed, install the flywheel inspection cover and attaching bolts.
12. If removed, install the air conditioning compressor brace and attaching nuts.
13. Lower vehicle and connect the negative battery cable.
14. If removed, install the air cleaner assembly.
15. Crank the engine and check the starter operation.

2.3L Engine

1990

1. Disconnect the negative battery cable.
2. Disconnect the electrical connector from the cooling fan.
3. Remove the cooling fan mounting bolts and remove the fan assembly.
4. Remove the intake manifold-to-engine brace bolts and remove the the brace from the engine.
5. Remove the starter mounting bolts.
6. Carefully lift the starter away

from the engine with the solenoid harness attached to it.

7. When the starter is clear, disconnect the solenoid harness connections and lift the starter up and out toward the front of the vehicle.

To install:

8. Connect the solenoid harness connections to the starter, while supporting the starter toward the mounting position.

9. Rotate the starter so the solenoid faces the engine at a slight angle to clear the bottom of the intake manifold.

10. Install the starter to the engine and install the mounting bolts. Tighten the bolts to 32 ft. lbs. (43 Nm).

11. Install the intake manifold-to-engine brace and the attaching bolts.

12. Install the cooling fan and attaching bolts. Tighten bolts to 89 inch lbs. (10 Nm).

13. Connect the negative battery cable.

14. Crank the engine and check starter operation.

1991

1. Disconnect the negative battery cable.

2. Remove the serpentine drive belt.

3. Remove the coolant reservoir.

4. Remove the air conditioner rail clip.

5. Remove the alternator.

6. Remove the dipstick, bolt and oil filler tube.

7. Remove the alternator bracket.

8. Remove the air cleaner assembly.

9. Remove the upper transaxle-to-starter mounting bolt.

10. Remove the oil filter.

11. Remove the lower starter mounting bolt.

12. Position the starter for access to the solenoid wiring.

13. Disconnect the electrical wiring.

14. Remove the starter from the vehicle, routing through the front of the engine between the intake manifold and engine block.

To install:

15. Install the starter by lowering between the intake manifold and engine block.

16. Connect the starter electrical connectors.

17. Position the starter to the engine.

18. Install the starter mounting bolts. Tighten the lower mounting bolt to 46 ft. lbs. (63 Nm). Tighten the upper transaxle-to-starter mounting bolt to 71 ft. lbs. (96 Nm).

19. Install the oil filter.

20. Install the air cleaner assembly.

21. Install the alternator bracket.

22. Install the dipstick, bolt and oil fill tube.

23. Install the alternator.

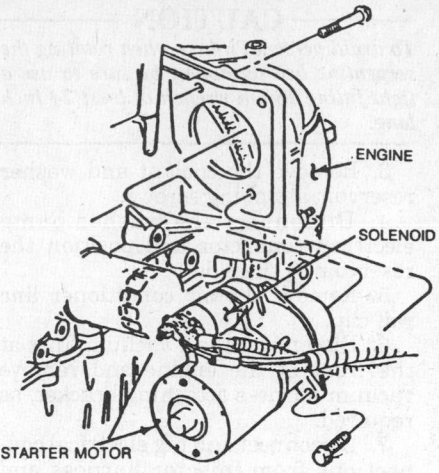

Starter removal and installation—2.8L and 3.1L engines

24. Install the air conditioner rail clip.

25. Install the coolant reservoir.

26. Install the serpentine belt.

27. Connect the negative battery cable.

28. Refill the engine oil as necessary and check for leaks.

EMISSION CONTROLS

Please refer to "Emission Controls" in the Unit Repair section for system maintenance procedures. Due to the complex nature of modern electronic engine control systems, comprehensive diagnosis and testing procedures fall outside the confines of this repair manual. For complete information on diagnosis, testing and repair procedures concerning all modern engine and emission control systems, please refer to "Chilton's Guide to Fuel Injection and Electronic Engine Controls".

Emission Warning Lamps

RESETTING

When the ECM finds a problem, the Service Engine Soon light will turn ON and a trouble code will be recorded in the ECM memory. If the problem is intermittent, the Service Engine Soon light will light go out after 10 seconds, when the fault goes away. However, the trouble code will stay in the ECM memory until the battery voltage to the ECM is removed. Removing the battery voltage for 10 seconds will clear all stored trouble codes. This is done by disconnecting the ECM harness from the positive battery pigtail for 30 seconds with the ignition **OFF** or by disconnecting the ECM fuse, designated ECM or ECM/BAT, from the fuse holder.

NOTE: To prevent ECM damage, the ignition switch must be OFF when disconnecting or reconnecting power to ECM (for example battery cable, ECM pigtail, ECM fuse, jumper cables, etc.).

FUEL SYSTEM

Fuel System Service Precautions

Safety is the most important factor when performing not only fuel system maintenance but any type of maintenance. Failure to conduct maintenance and repairs in a safe manner may result in serious personal injury or death. Maintenance and testing of the vehicle's fuel system components can be accomplished safely and effectively by adhering to the following rules and guidelines.

• To avoid the possibility of fire and personal injury, always disconnect the negative battery cable unless the repair or test procedure requires that battery voltage be applied.

• Always relieve the fuel system pressure prior to disconnecting any fuel system component (injector, fuel rail, pressure regulator, etc.), fitting or fuel line connection. Exercise extreme caution whenever relieving fuel system pressure to avoid exposing skin, face and eyes to fuel spray. Please be advised that fuel under pressure may penetrate the skin or any part of the body that it contacts.

• Always place a shop towel or cloth around the fitting or connection prior to loosening to absorb any excess fuel due to spillage. Ensure that all fuel spillage (should it occur) is quickly removed from engine surfaces. Ensure that all fuel soaked cloths or towels are deposited into a suitable waste container.

• Always keep a dry chemical (Class B) fire extinguisher near the work area.

• Do not allow fuel spray or fuel vapors to come into contact with a spark or open flame.

- Always use a backup wrench when loosening and tightening fuel line connection fittings. This will prevent unnecessary torsional stress to fuel line piping. Always follow the proper torque specifications.
- Always replace worn fuel fitting O-rings with new. Do not substitute fuel hose or equivalent where fuel pipe is installed.

RELIEVING FUEL SYSTEM PRESSURE

Throttle Body Injection

1. Disconnect the negative battery cable.
2. Remove the fuel filler cap to relieve tank vapor pressure.
3. Wrap a shop towel around the fuel line fitting.
4. Open the fuel line and absorb any excess fuel remaining in the line.
5. When the line fitting is reconnected, use a new O-ring.

Port Fuel Injection
EXCEPT 2.3L ENGINE

1. Disconnect the negative battery cable.
2. Remove the fuel filler cap to relieve tank vapor pressure.
3. Connect a fuel gauge to the fuel pressure test fitting.

NOTE: **Be sure to wrap a shop cloth around the fuel line fitting when connecting the fuel gauge tool to the fuel pressure connector.**

4. Place the bleeder hose and shop cloth in an approved fuel container. Open the pressure valve to bleed the fuel pressure from the system.
5. After the fuel pressure is bled, retighten the fuel pressure valve.

2.3L ENGINE

1. Loosen the fuel filler cap to relieve the tank pressure.
2. Raise and safely support the vehicle.
3. Disconnect the fuel pump electrical connector.
4. Lower the vehicle.
5. Start the engine and run until the fuel supply remaining in the fuel lines is consumed. Engage the starter for 3.0 seconds to assure relief of any remaining pressure.
6. Raise and safely support the vehicle.
7. Connect the fuel pump electrical connector.
8. Lower the vehicle.
9. Disconnect the negative battery cable to avoid possible fuel discharge if an accidental attempt is made to start the engine.

Fuel Filter

REMOVAL & INSTALLATION

An inline fuel filter is used on all engines. It is located on a frame crossmember near the rear of the vehicle.

Threaded Fuel Line Fitting

1. Relieve the fuel system pressure.
2. Raise and safely support the vehicle.
3. Using a backup wrench, remove the fuel line fittings from the fuel filter.
4. Remove the fuel filter-to-crossmember screws and the filter from the vehicle.

To install:
5. Install the fuel filter and the attaching screws.
6. Replace the fuel filter O-rings.
7. Connect the fuel lines. Using a backup wrench, tighten the fuel line fittings to 22 ft. lbs. (30 Nm).
8. Lower the vehicle.
9. Connect the negative battery cable.

Quick-Connect Fitting

1. Disconnect the negative battery cable.
2. Relieve the fuel system pressure.
3. Raise and safely support the vehicle.
4. Remove the fuel filter attaching screw.

NOTE: **If the nylon fuel feed or return connecting lines become kinked and cannot be straightened, they must be replaced.**

5. Grasp the filter and 1 nylon fuel connecting line fitting. Twist the quick-connect fitting ¼ turn in each direction to loosen any dirt within the fitting. Repeat for the other nylon fuel connecting line fitting.
6. Using compressed air, blow out any accumulated dirt from the quick-connect fittings at both ends of the fuel filter.
7. Squeeze the plastic tabs of the male end connector and pull connection apart. Repeat for the other fitting.
8. Remove the fuel filter.
To install:
9. Apply a few drops of clean engine oil to the male tube ends of the filter.
10. Remove the protective caps from the new filter.
11. Install new plastic connector retainers on the filter inlet and outlet tubes.
12. Push the connectors together to cause the retaining tabs/fingers to snap into place.
13. Once installed, pull on both ends of each connection to ensure they are secure.

14. Place the fuel filter into position and install the attaching screw.
15. Tighten the fuel filler cap.
16. Connect the negative battery cable.

Electric Fuel Pump

REMOVAL & INSTALLATION

The fuel pump is located in the fuel tank. Removal and installation procedures require the fuel tank to be removed from the vehicle.

——————— CAUTION ———————
The fuel system pressure must be relieved before attempting any service procedures. Use caution to avoid the risk of fire by disposing of any fuel and fuel soaked rags properly.
————————————————————————

1. Relieve the fuel pressure.
2. Disconnect the negative battery cable.
3. Using a siphon hose and pump, drain the fuel from the fuel tank.
4. Raise and safely support the vehicle.
5. Support the fuel tank and disconnect the retaining straps.
6. Lower the tank enough to disconnect the sending unit wire, the hoses and the ground strap. Remove the fuel tank from the vehicle.
7. Using a locking cam tool, remove the sending unit retaining cam from the fuel tank.
8. Remove the fuel pump and sending unit assembly from the tank. Remove and discard the O-ring gasket.
To install:
9. Install a new O-ring and gasket. Carefully install the fuel pump and sending unit assembly into the fuel tank.
10. Install the retaining cam and lock and secure the sending unit in place to the fuel tank.
11. Raise the tank in position to connect the sending unit wire, the hoses and the ground strap. Install the tank retaining straps and secure the tank in place.
12. Lower the vehicle and refill the tank with fuel.

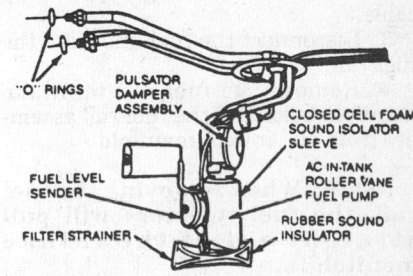

Fuel pump and sending unit assemblies

13. Connect the negative battery cable. Turn the ignition switch to the **ON** position, to restore system pressure.

14. Start the engine and check for fuel leaks.

Fuel Injection

Idle Speed and Idle Mixture Adjustment

Idle speed and mixture are controlled by the Electronic Control Module (ECM). No adjustments are possible.

Fuel Injector

REMOVAL & INSTALLATION

2.0 and 2.2L Engines

1. Relieve the fuel system pressure.
2. Remove the air cleaner. Disconnect the negative battery cable from the battery.
3. Disconnect the electrical connector from the fuel injector.
4. Remove the injector retainer-to-throttle body screw and the retainer.
5. Using a small prybar and a fulcrum, carefully lift the injector until it is free from the fuel meter body.
6. Remove the O-rings form the nozzle end of the injector.
7. Inspect the fuel injector filter for dirt and/or contamination.

To install:

8. Lubricate the O-rings with automatic transmission fluid and place them on the fuel injector.
9. Push the fuel injector straight into the fuel meter body, apply thread locking compound on the fuel injector retainer screw and install.
10. Connect the electrical connector to the fuel injector.
11. Install the air cleaner. Connect the negative battery cable.
12. Turn the ignition switch to the **ON** position, to restore system pressure. Start the engine and check for fuel leaks.

2.3L, 2.8L and 3.1L Engines

1. Relieve the fuel system pressure.
2. Disconnect the negative battery cable.
3. Disconnect the fuel line from the fuel rail.
4. Remove the fuel rail-to-intake manifold bolts and the fuel rail assembly from the intake manifold.

NOTE: When removing the fuel rail, the fuel injectors will pull straight out of the intake manifold.

5. Remove the fuel injector-to-fuel rail retaining clips and the injectors from the fuel rail.

To install:

6. Replace the O-rings on the fuel injectors.
7. Install the injectors to the fuel rail and fuel injector-to-fuel rail retaining clips.
8. Install the fuel rail assembly to the intake manifold and the fuel rail-to-intake manifold bolts.
9. Connect the fuel line to the fuel rail.
10. Connect the negative battery cable.

DRIVE AXLE

Halfshaft

REMOVAL & INSTALLATION

If equipped with an automatic transaxle, the inner joint on the right side halfshaft uses a male spline that locks into the transaxle gears. The left side halfshaft uses a female spline that is installed over the stub shaft on the transaxle.

An intermediate shaft is installed between the transaxle and the right halfshaft.

Except Intermediate Shaft

1. With the weight of the vehicle on the tires, loosen the hub nut.
2. Raise and safely support the vehicle.
3. Remove the hub nut.
4. Install boot protectors on the boots.
5. Remove the brake caliper with the line attached and support it (on a wire) aside; do not allow the caliper to hang from the line.
6. Remove the brake rotor and caliper mounting bracket.
7. Remove the strut to steering knuckle bolts. Pull the steering knuckle out of the strut bracket.
8. Using a halfshaft removal tool and an extension, remove the halfshafts from the transaxle and support them safely.
9. Using a spindle remover tool, remove the halfshaft from the hub and bearing.

To install:

10. Loosely place the halfshaft on the transaxle and in the hub and bearing.
11. Properly position the steering knuckle to the strut bracket and install the bolt. Tighten the bolts to 133 ft. lbs. (181 Nm).
12. Install the brake rotor, caliper bracket and caliper. Place a holding

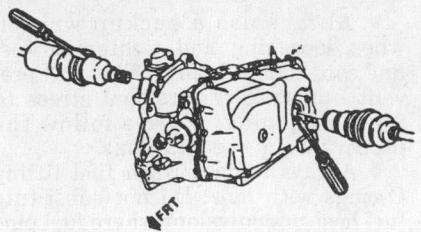

Removing the halfshafts from transaxle

device in the rotor to prevent it from turning.

13. Install the hub nut and washer. Tighten the nut to 71 ft. lbs. (96 Nm).
14. Seat the halfshafts into the transaxle using a prybar on the groove on the inner retainer.
15. Verify that the shafts are seated by grasping the CV-joint and pulling outwards; do not grasp the shaft. If the snapring is seated, the halfshaft will remain in place.
16. Remove the boot protectors.
17. Lower the vehicle.
18. When the vehicle is lowered with the weight on the wheels, tighten the hub nut to 191 ft. lbs. (259 Nm).

Intermediate Shaft

1. Raise and safely support the vehicle. Remove the front right wheel and tire assembly.
2. Drain the transaxle.
3. Using a modified boot protector, place it over the outer boot.
4. Remove the stabilizer bar from the right control arm.
5. Remove the right ball joint-to-steering knuckle cotter pin and nut. Using a ball joint remover tool, separate the ball joint from the steering knuckle.
6. Pull the steering knuckle outward and separate the halfshaft from the intermediate shaft.
7. Remove the intermediate shaft housing-to-bracket bolts and the lower bracket-to-engine bolt. Loosen the upper bracket-to-engine bolt and swing the bracket aside.
8. Remove the intermediate shaft housing-to-transaxle bolts, disengage the housing from the transaxle and remove the intermediate shaft assembly.

To install:

9. Lubricate the intermediate shaft splines with grease and install the intermediate shaft. Tighten the intermediate shaft housing-to-transaxle bolts to 18 ft. lbs. (25 Nm), the intermediate shaft housing-to-bracket bolts to 37 ft. lbs. (50 Nm) and the bracket-to-engine bolts to 37 ft. lbs. (50 Nm).
10. Install the intermediate shaft housing-to-bracket bolts and the lower bracket-to-engine bolt.
11. Connect the halfshaft to the intermediate shaft.

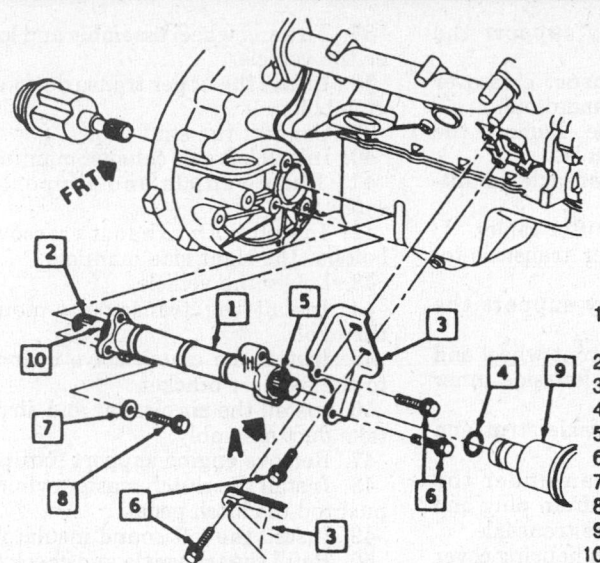

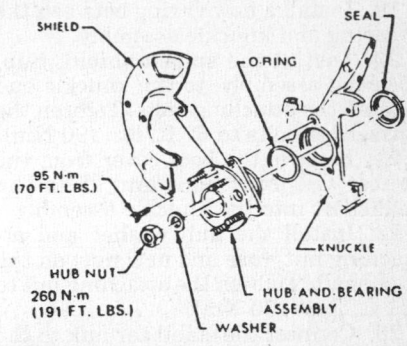

VIEW A

1. Intermediate shaft assembly
2. Intermediate axle shaft
3. Bracket
4. Axle shaft retaining ring
5. Lip seal
6. Bolt — 37 ft. lbs. (50 Nm)
7. Washer
8. Bolt — 18 ft. lbs.
9. Right drive axle
10. O-ring seal

Exploded view of the intermediate shaft assembly

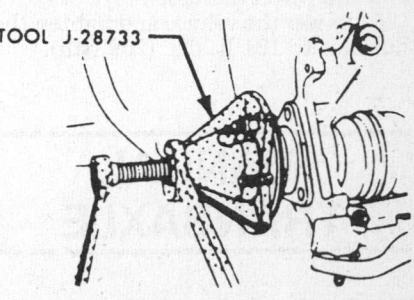

Exploded view of the hub, knuckle and bearing

TOOL J-28733

Removing halfshaft from steering knuckle bearing

12. Connect the ball joint to the steering knuckle. Install the ball joint-to-steering knuckle nut and cotter pin.
13. Install the stabilizer bar to the right control arm.
14. Remove the boot protector.
15. Install the right front tire and wheel assembly.
16. Lower the vehicle.
17. Refill the transaxle.
18. Start the engine and allow to come to normal operating temperature. Check the automatic transaxle fluid level. Top-up as necessary.

CV-Boot

REMOVAL & INSTALLATION

Inner

1. Remove the halfshaft.
2. Cut the seal retaining clamps.
3. Using a pair of snapring pliers, remove the retaining ring from the shaft and remove the spider assembly.
4. Remove the old boot from the shaft.
To install:
5. Using solvent, clean the splines of the shaft and repack the joint.
6. Install the inner boot clamp first and the new boot second.
7. Push the CV-joint assembly onto the shaft until the retaining ring is seated on the shaft.
8. Slide the boot onto the joint. Install both the inner and outer clamps.
9. Install the halfshaft.

Outer

1. Remove the halfshaft from the vehicle.
2. Cut off the boot retaining clamps

and discard them. Remove the old boot.
3. If equipped with a deflector ring, use a brass drift and carefully tap it off.
4. Using a pair of snapring pliers, spread the retaining ring inside the outer CV-joint and tap the joint off the halfshaft.
To install:
5. Using solvent, clean the splines of the halfshaft and the CV-joint and repack the joint. Install a new retaining ring inside the joint.
6. Install the inner boot clamp first, the new boot second.
7. Push the joint assembly onto the halfshaft until the ring is seated on the shaft.
8. Slide the boot onto the joint and install the clamps on both the inner and outer part of the boot.
9. Install the halfshaft.

Front Wheel Hub, Knuckle and Bearings

REMOVAL & INSTALLATION

The hub and bearing are replaced as an assembly only.
1. With the vehicle weight on the tires, loosen the hub nut.
2. Raise and safely support the vehicle. Remove the wheel and tire assembly.
3. Install a boot cover over the outer CV-joint boot.
4. Remove the hub nut. Remove the brake caliper and support it aside (on a wire); do not allow the caliper to hang on the brake line.
5. Remove the hub and bearing mounting bolts.

6. Remove the brake rotor splash shield.
7. Using a hub puller tool, press the hub and bearing from the halfshaft.
8. Disconnect the stabilizer link from the lower control arm.
9. Remove the cotter pin and the ball joint-to-knuckle attaching nut.
10. Disconnect the ball joint from the steering knuckle, using a ball joint separator tool.
11. Remove the halfshaft from the knuckle and support it aside.
12. Matchmark the strut in relationship to the knuckle, for alignment purposes and remove the strut-to-knuckle attaching nuts.
13. Remove the knuckle from the strut.
14. Using a brass drift, remove the inner knuckle seal.
To install:
15. Clean and inspect the steering knuckle bore and the bearing mating surfaces.
16. Using a seal driver tool, install it into the steering knuckle; be sure to lubricate the new seal and the bearing with a high temperature wheel bearing grease.
17. Connect the ball joint to the knuckle and install the ball joint-to-knuckle attaching nut, hand tight.
18. Position the knuckle to the strut and install the attaching bolts. Align the matchmarks and tighten the attaching bolts to 129 ft. lbs. (175 Nm). Tighten the ball joint-to-knuckle attaching nut to 55 ft. lbs (75 Nm).

19. Install a new O-ring between the bearing and knuckle assembly.

20. Install the splash shield, hub/bearing assembly, to the knuckle and install the attaching bolts. Tighten the attaching bolts to 67 ft. lbs. (90 Nm).

21. Remove the boot cover from the outer CV-joint boot and slide the halfshaft into the knuckle assembly.

22. Install the hub washer and attaching nut, (use and new nut) on the halfshaft. Tighten the attaching nut to 71 ft. lbs. (100 Nm).

23. Connect the stabilizer link to the lower control arm.

24. Install the brake rotor, caliper and the wheel/tire assembly.

25. Lower the vehicle and tighten the hub nut to 191 ft. lbs. (259 Nm).

MANUAL TRANSAXLE

For further information on transmissions/transaxles, please refer to "Chilton's Guide to Transmission Repair".

Transaxle Assembly

REMOVAL & INSTALLATION

NOTE: Before performing any maintenance that requires the removal of the slave cylinder, transaxle or clutch housing, the clutch master cylinder pushrod must first be disconnected from the clutch pedal. Failure to disconnect the pushrod will result in permanent damage to the slave cylinder if the clutch pedal is depressed with the slave cylinder disconnected.

Muncie

1. Disconnect the negative battery cable.

2. Using an engine support fixture tool and an adapter, install them on the engine and raise the engine enough to take the engine weight off of the engine mounts.

3. Remove the left side sound insulator.

4. Disconnect the clutch master cylinder pushrod from the clutch pedal.

5. Remove the air cleaner and duct assembly.

6. Disconnect the clutch slave cylinder-to-transaxle support bolts and position the cylinder aside.

7. Remove the transaxle-to-mount through bolt.

8. Raise and safely support the vehicle.

9. Remove both exhaust crossover bolts at the right side manifold.

10. Lower the vehicle. Remove the left side exhaust manifold.

11. Disconnect the transaxle mounting bracket.

12. Disconnect the shifter cables.

13. Remove the upper transaxle-to-engine bolts.

14. Raise and safely support the vehicle.

15. Remove the left front wheel and tire assembly and the left side inner splash shield.

16. Remove the transaxle strut and bracket.

17. Place a drain pan under the transaxle, remove the drain plug and drain the fluid from the transaxle.

18. Remove the clutch housing cover bolts.

19. Disconnect the speedometer wire.

20. From the left suspension support and control arm, disconnect the stabilizer shaft.

21. Remove the left suspension support mounting bolts and move the support aside.

22. Disconnect both halfshafts from the transaxle and remove the left halfshaft from the vehicle.

23. Using a transmission jack, attach it to and support the transaxle.

24. Remove the remaining transaxle-to-engine bolts.

25. Slide the transaxle away from the engine, lower it and remove the right side halfshaft.

To install:

26. Raise the vehicle and support it safely.

27. Support the transaxle assembly on a transaxle jack.

28. Raise the transaxle in position and guide the right halfshaft into the bore of the transaxle.

NOTE: The right halfshaft cannot be readily installed after the transaxle is connected to engine.

29. Install the transaxle to engine and install the mounting bolts. Tighten the bolts to 60 ft. lbs. (81 Nm).

30. Install the left halfshaft into its bore and seat both halfshafts to the transaxle securely.

31. Install the suspension support-to-body bolts.

32. Install the stabilizer shaft-to-suspension support and install the control arm.

33. Install the speedometer wire connector.

34. Install the clutch housing cover bolts.

35. Install the strut bracket to transaxle and install the strut.

36. Install the inner splash shield.

37. Tire and wheel assembly and lower the vehicle.

38. Install the upper transaxle-to-engine bolts.

39. Connect the shift cables.

40. Install left side exhaust manifold.

41. Raise vehicle and support it safely.

42. Install both exhaust crossover bolts at the right side manifold.

43. Lower the vehicle.

44. Install the transaxle-to-mount thru bolt.

45. Install the clutch slave cylinder to the support bracket.

46. Install the air cleaner and air intake duct assembly.

47. Remove engine support fixture.

48. Install the clutch master cylinder pushrod to clutch pedal.

49. Install the left sound insulator.

50. Refill the transaxle and check for leaks.

51. Connect the negative battery cable.

Isuzu

1. Disconnect the negative battery cable.

2. Using an engine support fixture tool and an adapter, install them on the engine and raise the engine enough to take the engine weight off the engine mounts.

3. Remove the left side sound insulator.

4. Disconnect the clutch master cylinder pushrod from the clutch pedal.

5. Disconnect the clutch slave cylinder-to-transaxle support bolts and position the cylinder aside.

6. Remove the wiring harness from the transaxle mount bracket and the shift wire electrical connector.

7. Remove the transaxle-to-mount bolts and the transaxle mount bracket-to-chassis nuts/bolts.

8. Disconnect the shift cables and remove the retaining clamp from the transaxle. Remove the ground cables from the transaxle mounting studs.

9. Raise and safely support the vehicle.

10. Remove the left front wheel and tire assembly and the left side inner splash shield.

11. Remove the transaxle front strut and bracket.

12. Remove the clutch housing cover bolts. Disconnect the speedometer wire connector.

13. From the left suspension support and control arm, disconnect the stabilizer shaft.

14. Remove the left suspension support mounting bolts and move the support aside.

15. Disconnect both halfshafts from the transaxle and remove the left halfshaft from the vehicle.

16. Place a drain pan under the transaxle, remove the drain plug and drain the fluid from the transaxle.

17. Using a transmission jack, attach it to and support the transaxle.

18. Remove the transaxle-to-engine bolts.

19. Slide the transaxle away from the engine, lower it and remove the right side halfshaft.

To install:

20. Raise the vehicle and support it safely.

21. Support the transaxle assembly on a transaxle jack.

22. Raise the transaxle in position and guide the right halfshaft into the bore of the transaxle.

NOTE: The right halfshaft cannot be readily installed after the transaxle is connected to engine.

23. Install the transaxle to engine and install the mounting bolts. Tighten the bolts to 60 ft. lbs. (81 Nm).

24. Install the left halfshaft into its bore and seat both halfshafts to the transaxle securely.

25. Install the suspension support-to-body bolts.

26. Install the stabilizer shaft-to-suspension support and install the control arm.

27. Install the speedometer wire connector.

28. Install the clutch housing cover bolts.

29. Install the front strut bracket to transaxle and install the front strut.

30. Install the inner splash shield.

31. Install the tire and wheel assembly and lower the vehicle.

32. Install the ground cables at the mounting studs.

33. Install the electrical connections for the shift light.

34. Install the slave cylinder to the transaxle bracket aligning the pushrod into the pocket of the clutch release lever. Install the attaching nuts and tighten evenly to prevent damage to the cylinder.

35. Install the transaxle mount bracket.

36. Install the transaxle mount to the side frame and install the attaching bolts.

37. Connect the wire harness at the mount bracket.

38. Remove the engine support.

39. Install the shift cables.

40. Refill the transaxle and check for leaks.

41. Connect the negative battery cable.

LINKAGE ADJUSTMENT

No adjustments are possible on the manual transaxle shifting cables or linkage. If the transaxle is not engaging completely, check for stretched cables or broken shifter components or a faulty transaxle.

CLUTCH

REMOVAL & INSTALLATION

1. Raise and safely support the vehicle. Disconnect the negative battery cable.

2. Remove the left side sound insulator panel.

3. Disconnect the clutch master cylinder pushrod from the clutch pedal.

4. Remove the transaxle.

5. Using paint or chalk, matchmark the pressure plate and flywheel assembly to insure proper balance during reassembly.

6. Loosen the pressure plate-to-flywheel bolts, 1 turn at a time, until the spring pressure is removed.

7. Support the pressure plate and remove the bolts.

8. Remove the pressure plate and disc assembly; be sure to note the flywheel side of the clutch disc.

9. Clean and inspect the clutch assembly, flywheel, release bearing, clutch fork and pivot shaft for signs of wear. Replace any necessary parts.

To install:

10. Position the clutch disc and pressure plate in the appropriate position, support the assembly with an alignment tool.

NOTE: Ensure the clutch disc is facing the same direction it was when removed. The driven plate is installed with the damper springs offset toward the transaxle. If the same pressure plate is being reused, align the marks made during the removal, install the pressure plate attaching bolts. Tighten them gradually and evenly.

11. Remove the alignment tool and tighten the pressure plate-to-flywheel bolts to 15 ft. lbs. (20 Nm). Lightly lubricate the clutch fork ends. Fill the recess ends of the release bearing with grease. Lubricate the spline input shaft with a light coat of grease.

NOTE: On 5-speed Isuzu transaxles, ensure the bearing pads are located on the fork ends and both spring ends are in the fork holes with the spring completely seated in the bearing groove.

12. Install the transaxle in the vehicle.

NOTE: The clutch lever must not be moved towards the flywheel until the transaxle is bolted to the engine. Damage to the transaxle, release bearing and clutch fork could occur if this is not followed.

13. Connect the clutch master cylinder pushrod to the clutch pedal.

14. Install the left side sound insulator.

15. Connect the negative battery cable.

PEDAL HEIGHT/FREE-PLAY ADJUSTMENT

Push the clutch pedal all the way to the floor; the distance of travel should be 6.4 ± 0.5 in. (163 ± 13mm). If the measurement is not correct check the following areas:

Clutch pedal assembly distorted

Incorrect clutch master cylinder pushrod length

Dash mat under the neutral start switch

Mislocated neutral start switch

Clutch Master and Slave Cylinder

REMOVAL & INSTALLATION

The clutch master and slave cylinders are removed from the vehicle as an assembly. After installation the clutch hydraulic system must be bled.

1. Disconnect the negative battery cable.

2. From inside the vehicle, remove the left side sound insulator.

NOTE: If equipped with a 2.8L or 3.1L engine, remove the air cleaner, the mass air flow sensor and the air intake duct as an assembly.

3. Disconnect the clutch master cylinder pushrod from the clutch master cylinder.

4. From the front of the dash, remove the trim cover.

5. Remove the clutch master cylinder-to-clutch pedal bracket nuts and the remote reservoir-to-chassis screws.

6. Remove the slave cylinder-to-transaxle nuts and the slave cylinder.

7. Remove the hydraulic system (as a unit) from the vehicle.

To install:

8. Install the slave cylinder-to-transaxle support, align the pushrod to the clutch fork outer lever pocket. Tighten the slave cylinder-to-transaxle support nuts to 14–20 ft. lbs. (19–27 Nm).

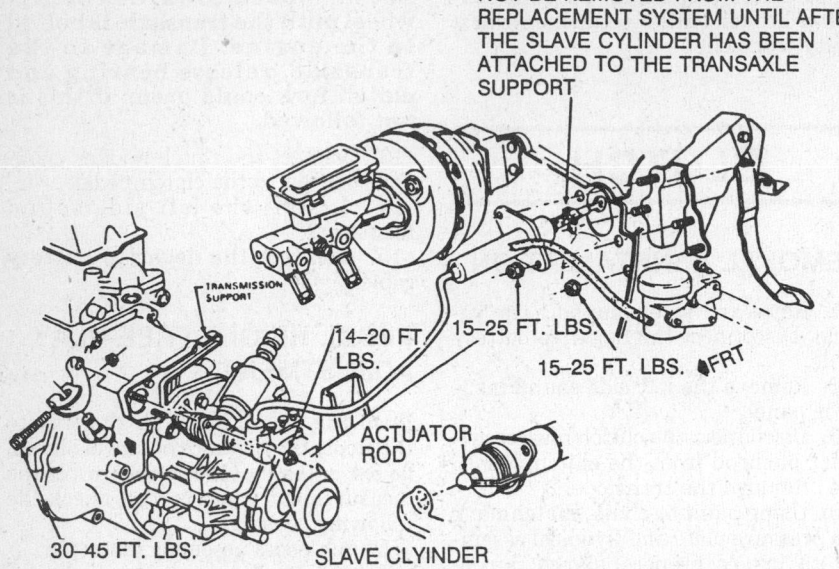

NOTICE: PEDAL RESTRICTOR SHOULD NOT BE REMOVED FROM THE REPLACEMENT SYSTEM UNTIL AFTER THE SLAVE CYLINDER HAS BEEN ATTACHED TO THE TRANSAXLE SUPPORT

TRANSMISSION SUPPORT

14–20 FT. LBS. 15–25 FT. LBS.

15–25 FT. LBS. ◀FRT

ACTUATOR ROD

30–45 FT. LBS. SLAVE CLYINDER

Clutch master and slave clylinder

NOTE: If installing a new clutch hydraulic system, do not break the pushrod plastic retainer; the straps will break on the first pedal application.

9. Install the master cylinder-to-clutch pedal bracket. Tighten the nuts evenly, to prevent damaging the master cylinder, to 15–20 ft. lbs. (20–27 Nm). Remove the pedal restrictor from the pushrod. Lubricate the pushrod bushing on the clutch pedal; if the bushing is cracked or worn, replace it.
10. If equipped with cruise control, check the switch adjustment at the clutch pedal bracket.

NOTE: When adjusting the cruise control switch, do not exert more than 20 lbs. of upward force on the clutch pedal pad for damage to the master cylinder pushrod retaining rod can result.

11. Depress the clutch pedal several times to break the plastic retaining straps; do not remove the plastic button from the end of the pushrod.

NOTE: If equipped with a 2.8L or 3.1L engine, install the air cleaner, the mass air flow sensor and the air intake duct as an assembly.

12. Install the left side sound insulator.
13. Connect the negative battery cable.
14. If necessary, bleed the clutch hydraulic system.

Hydraulic Clutch System Bleeding

PROCEDURE

1988–89

1. Remove any dirt or grease around the reservoir cap so dirt cannot enter the system.
2. Fill the reservoir with an approved DOT 3 brake fluid.
3. Loosen but do not remove, the bleeder screw on the slave cylinder.
4. Fluid will now flow from the master cylinder to the slave cylinder.

NOTE: It is important that the reservoir remain filled throughout the procedure.

5. Air bubbles should now appear at the bleeder screw.
6. Continue this procedure until a steady stream of fluid without any air bubbles is present.
7. Tighten the bleeder screw. Check the fluid level in the reservoir and refill to the proper mark.
8. The system is now fully bled. Check the clutch operation by starting the engine, pushing the clutch pedal to the floor and placing the transmission in reverse.
9. If any grinding of the gears is noted, repeat the procedure.

NOTE: Never under any circumstances reuse fluid that has been in the system. The fluid may be contaminated with dirt and moisture.

1990–92

1. Disconnect the slave cylinder from the transaxle.
2. Loosen the master cylinder mounting attaching nuts. Do not remove the master cylinder.
3. Remove any dirt or grease around the reservoir cap so dirt cannot enter the system. Fill the reservoir with an approved DOT 3 brake fluid.
4. Depress the hydraulic actuator cylinder pushrod approximately 0.787 in. (20.0mm) into the slave cylinder bore and hold.
5. install the diaphragm and cap on the reservoir while holding the slave cylinder pushrod.
6. Release the slave cylinder pushrod.
7. Hold the slave cylinder vertically with the pushrod end facing the ground.

NOTE: The slave cylinder should be lower than the master cylinder.

8. Press the pushrod into the slave cylinder bore with short 0.390 in. (10.0mm) strokes.
9. Observe the reservoir for air bubbles. Continue until air bubbles no longer enter the reservoir.
10. Connect the slave cylinder to the transaxle.
11. Tighten the master cylinder attaching nuts.
12. Top-up the clutch master cylinder reservoir.
13. To test the system, start the engine and push the clutch pedal to the floor. Wait 10 seconds and select reverse gear. There should be no gear clash. If clash is present, air may still be present in the system. Repeat bleeding procedure.

AUTOMATIC TRANSAXLE

For further information on transmissions/transaxles, please refer to "Chilton's Guide to Transmission Repair".

Transaxle Assembly

REMOVAL & INSTALLATION

2.0L, 2.2L and 2.3L Engines

1. Disconnect the negative battery cable. Remove the air cleaner and air intake assembly.
2. Disconnect the TV cable from the throttle lever and the transaxle.

3. Remove the fluid level indicator and the filler tube.

4. Using an engine support fixture tool and an adapter, install them onto the engine.

5. Remove the wiring harness-to-transaxle nut.

6. Label and disconnect the electrical connectors for the speed sensor, TCC connector and the neutral safety/backup light switch.

7. Disconnect the shift linkage from the transaxle.

8. Remove the upper transaxle-to-engine bolts, the transaxle mount and bracket assembly.

9. Disconnect the rubber hose that runs from the transaxle to the vent pipe.

10. Raise and safely support the vehicle.

11. Remove the front wheels and tire assemblies.

12. Disconnect the shift linkage and bracket from the transaxle.

13. Remove the left side splash shield.

14. Using a modified halfshaft boot protector tool, install 1 on each halfshaft to protect the boot from damage and the joint from possible failure.

15. Using care not to damage the halfshaft boots, disconnect the halfshafts from the transaxle.

16. Remove the transaxle strut. Remove the left side stabilizer link pin bolt and bushing clamp nuts from the support.

17. Remove the left frame support bolts and move it aside.

18. Disconnect the speedometer wire from the transaxle.

19. Remove the transaxle converter cover and matchmark the torque converter-to-flywheel for reassembly.

20. Disconnect and plug the transaxle cooler pipes.

21. Remove the transaxle-to-engine support.

22. Using a transmission jack, position and secure the jack to the transaxle. Remove the remaining transaxle-to-engine bolts.

23. Making sure the torque converter does not fall out, remove the transaxle from the vehicle.

NOTE: The transaxle cooler and lines should be flushed any time the transaxle is removed for overhaul or replacing the pump, case or converter.

To install:

24. Put a small amount of grease on the pilot hub of the converter and make sure the converter is properly engaged with the pump.

25. Raise the transaxle to the engine while guiding the right side halfshaft into the transaxle.

26. Install the lower transaxle mounting bolts and remove the jack.

27. Align the converter with the matchmarks on the flywheel and install the bolts hand tight.

28. Tighten the converter bolts to 46 ft. lbs. (62 Nm); retighten the first bolt after the others.

29. Connect the transaxle cooler pipes.

30. Connect the speedometer wire to the transaxle.

31. Install the left frame support bolts.

32. Install the left side stabilizer link pin bolt and bushing clamp nuts to the support. Install the transaxle strut.

33. Connect the halfshafts to the transaxle.

34. Remove the halfshaft boot protector.

35. Install the left side splash shield.

36. Connect the shift linkage and bracket to the transaxle.

37. Install the front wheels and tire assemblies.

38. Lower the vehicle.

39. Connect the rubber hose that runs from the transaxle to the vent pipe.

40. Install the upper transaxle-to-engine bolts, the transaxle mount and bracket assembly.

41. Connect the shift linkage to the transaxle.

42. Connect the electrical connectors for the speed sensor, TCC connector and the neutral safety/backup light switch.

43. Install the wiring harness-to-transaxle nut.

44. Remove the engine support fixture tool and an adapter.

45. Install the fluid level indicator and the filler tube.

46. Connect the TV cable to the throttle lever and the transaxle.

47. Install the air cleaner and air intake assembly.

48. Connect the negative battery cable. Check the fluid level when finished.

2.8L and 3.1L Engines

1. Disconnect the negative battery cable. Remove the air cleaner, bracket, Mass Air Flow (MAF) sensor and air tube as an assembly.

2. Disconnect the exhaust crossover from the right side manifold and remove the left side exhaust manifold. Raise and support the manifold/crossover assembly.

3. Disconnect the TV cable from the throttle lever and the transaxle.

4. Remove the vent hose and the shift cable from the transaxle.

5. Remove the fluid level indicator and the filler tube.

6. Using an engine support fixture tool and an adapter, install them on the engine.

7. Remove the wiring harness-to-transaxle nut.

8. Label and disconnect the wires for the speed sensor, TCC connector and the neutral safety/backup light switch.

9. Remove the upper transaxle-to-engine bolts.

10. Remove the transaxle-to-mount through bolt, the transaxle mount bracket and the mount.

11. Raise and safely support the vehicle.

12. Remove the front wheel and tire assemblies.

13. Disconnect the shift cable bracket from the transaxle.

14. Remove the left side splash shield.

15. Using a modified halfshaft boot protector tool, install 1 on each halfshaft to protect the boot from damage and the joint from possible failure.

16. Using care not to damage the halfshaft boots, disconnect the halfshafts from the transaxle.

17. Remove the torsional and lateral strut from the transaxle. Remove the left side stabilizer link pin bolt.

18. Remove the left frame support bolts and move it aside.

19. Disconnect the speedometer wire from the transaxle.

20. Remove the transaxle converter cover and matchmark the converter-to-flywheel for assembly.

21. Disconnect and plug the transaxle cooler pipes.

22. Remove the transaxle-to-engine support.

23. Using a transmission jack, position and secure it to the transaxle. Remove the remaining transaxle-to-engine bolts.

24. Make sure the torque converter does not fall out and remove the transaxle from the vehicle.

NOTE: The transaxle cooler and lines should be flushed any time the transaxle is removed for overhaul, to replace the pump, case or converter.

To install:

25. Put a small amount of grease on the pilot hub of the converter and make sure the converter is properly engaged with the pump.

26. Raise the transaxle to the engine while guiding the right side halfshaft into the transaxle.

27. Install the lower transaxle mounting bolts and Install the jack.

28. Align the converter with the matchmarks on the flywheel and install the bolts hand tight.

29. Tighten the converter bolts to 46

ft. lbs. (62 Nm); retighten the first bolt after the others.

30. Connect the transaxle cooler pipes.

31. Connect the speedometer wire to the transaxle.

32. Install the left frame support bolts.

33. Install the left side stabilizer link pin bolt. Install the torsional and lateral strut to the transaxle.

34. Connect the halfshafts to the transaxle.

35. Remove the boot protectors.

36. Install the left side splash shield.

37. Connect the shift cable bracket to the transaxle.

38. Install the front wheel and tire assemblies.

39. Lower the vehicle.

40. Install the transaxle-to-mount through bolt, the transaxle mount bracket and the mount.

41. Install the upper transaxle-to-engine bolts.

42. Connect the wires for the speed sensor, TCC connector and the neutral safety/backup light switch.

43. Install the wiring harness-to-transaxle nut.

44. Remove the engine support fixture tool.

45. Install the fluid level indicator and the filler tube.

46. Install the vent hose and the shift cable to the transaxle.

47. Connect the TV cable to the throttle lever and the transaxle.

48. Install the left side exhaust manifold. Connect the exhaust crossover to the right side manifold.

49. Install the air cleaner, bracket, Mass Air Flow (MAF) sensor and air tube as an assembly.

50. Connect the negative battery cable. Check the fluid level when finished.

SHIFT CONTROL LINKAGE ADJUSTMENT

1. Loosen the cable-to-transaxle shift lever nut so the cable is free.

2. Position the gear shift selector and the transaxle shift lever into the **N** position.

3. While holding transaxle's shift lever out of the **P** position, tighten the shift cable-to-shift lever nut to 11 ft. lbs. (15 Nm) for floor shift or 15 ft. lbs. (20 Nm) for column shift.

THROTTLE VALVE ADJUSTMENT

1. Turn the engine **OFF**.

2. At the end of the TV cable (engine side), depress and hold down the cable's metal readjustment tab.

3. Move the slider until it stops

against the fitting and release the readjustment tab.

4. Rotate the throttle lever (by hand) to it's full travel position; the TV slider should move (ratchet) toward the lever when the lever. Release the TV.

5. After adjustment, make sure the cable moves freely and road test the vehicle.

NOTE: Even if the cable appears to function properly when the engine is cold or stopped, recheck it after the engine is hot.

Park/Neutral and Backup Light Switch

The switch assembly is located on top of the transaxle.

1. Place the transaxle's shift control lever in the **N** notch in the detent plate.

2. Loosen the switch-to-transaxle screws.

3. Rotate the switch on the shifter assembly to align the service adjustment hole with the carrier tang hole.

4. Using a $\frac{3}{32}$ in. drill bit or gauge pin, insert it into the service adjustment hole to a depth of $\frac{3}{8}$ in. (9mm).

5. Tighten the switch-to-transaxle screws and remove the drill bit or gauge pin.

FRONT SUSPENSION

MacPherson Strut

REMOVAL & INSTALLATION

1. Disconnect the upper strut-to-body attaching bolts.

2. Raise and safely support the vehicle. Allow the suspension to hang free. Remove the wheel assembly. Install a halfshaft boot protector.

3. Remove the cotter pin and tie rod attaching nut. Using a tie rod separator tool, separate the tie rod from the strut.

4. Support the steering knuckle to prevent tension from being applied to the brake line.

5. Matchmark the strut in relationship to the knuckle and remove both strut-to-knuckle attaching bolts. Remove the strut assembly from the vehicle.

To install:

6. Installation is the reverse of the removal procedures. When installing the mounting bolts be sure to place the

flats of the bolts in the horizontal position.

7. Tighten the strut-to-knuckle bolts to 129 ft. lbs. (175 Nm) and the upper strut-to-body attaching bolts to 18 ft. lbs. (25 Nm).

8. Lower the vehicle. Check and adjust the alignment as required.

Lower Ball Joints

INSPECTION

1. Raise and safely support the vehicle; be sure the weight of the vehicle does not rest on the lower control arm assemblies.

2. With the ball joint installed to the steering knuckle, grasp the top and bottom of the wheel, then move the wheel using an in and out shaking motion. Observe any movement between the steering knuckle and the control arm. If movement exists, replace the ball joint.

REMOVAL & INSTALLATION

1. Raise and safely support the vehicle and remove the wheel assembly.

2. If no countersink is found on the lower side of the rivets, carefully locate the center of the rivet body and mark it using a punch.

3. Properly drill out the rivets of the ball joint assembly. Using a ball joint separator tool, separate the ball joint from the steering knuckle.

4. Disconnect the stabilizer bar from the lower control arm. Remove the ball joint from the vehicle.

To install:

5. Installation is the reverse of the removal procedures.

6. Attach the ball joint the lower control arm with the attaching bolts and nuts. Tighten the attaching bolts and nuts to 50 ft. lbs. (68 Nm).

7. Lower the vehicle. Check and align the front end as required.

Lower Control Arms

REMOVAL & INSTALLATION

1. Raise and safely support the vehicle and remove the wheel assembly.

2. Disconnect the stabilizer bar from the lower control arm assembly. Using a ball joint separator tool, separate the ball joint from the steering knuckle.

3. Remove the lower control arm attaching bolts and remove the lower control arm from the vehicle.

To install:

4. Installation is the reverse of the removal procedures. Tighten the lower control arm bolts to 63 ft. lbs. (85 Nm).

Tighten the ball joint-to-knuckle attaching nut to 55 ft. lbs. (75 Nm). Check and align the front end as required.

Sway Bar

REMOVAL & INSTALLATION

1. Open the hood and install an engine support tool. Raise and safely support the vehicle; allow the suspension to hang free. Remove the left front wheel assembly.

2. Disconnect the stabilizer link bolts and nuts from the control arms. Disconnect the stabilizer shaft from the support assemblies.

3. Loosen the front bolts and remove the bolts from the rear and center of the support assemblies, allowing the supports to be lowered enough to remove the stabilizer bar assembly. Remove the assembly from the vehicle.

To install:

4. Installation is the reverse of the removal procedures. Loosely assemble all components while insuring that the stabilizer bar is centered, side-to-side. Tighten the stabilizer bar support assemblies to 14 ft. lbs. (19 Nm). Tighten the stabilizer link bolts and nuts 14 ft. lbs. (19 Nm).

5. Lower the vehicle.

REAR SUSPENSION

Shock Absorbers

REMOVAL & INSTALLATION

1. Open the trunk and remove the shock absorber trim cover, if equipped. Remove the upper shock absorber attaching bolt. Remove each shock absorber separately when both assemblies are being replaced.

2. Raise and safely support the vehicle and the rear axle assembly.

3. Remove the lower shock attaching bolts. Remove the shock absorber from the vehicle.

To install:

4. Installation is the reverse of the removal procedures. Tighten the lower shock attaching bolt to 35 ft. lbs. (48 Nm) and the upper shock attaching bolt to 22 ft. lbs. (30 Nm).

Coil Springs

REMOVAL & INSTALLATION

1. Raise and safely support the vehicle under the rear control arms. Support the rear axle assembly with and jack.

2. Remove the wheel assembly. Remove the right and left brake line bracket attaching screws from the body and allow the brake line to hang free.

3. Remove the shock absorber lower attaching bolts. Lower the rear axle assembly to remove the coil springs. Do not allow the axle assembly to hang in this position.

To install:

4. Installation is the reverse of the removal procedures. Before installing the coil springs it is necessary to install the insulators to the body using adhesive.

5. Position the spring and insulator in the spring seat and raise the axle. The upper ends of the coil must be positioned properly in the seat of the body.

6. Tighten the shock absorber lower attaching bolts to 21 ft. lbs. (28 Nm).

7. Install the wheel assemblies and lower the vehicle.

Stabilizer Bar

REMOVAL & INSTALLATION

1. Raise and safely support the vehicle.

2. Remove the nuts and bolts at both axle and control arm attachments.

3. Remove the bracket. Remove the insulator and the stabilizer bar assembly assembly.

To install:

4. Installation is the reverse of the removal procedures. Tighten the bracket-to-axle bolts to 13 ft. lbs. (18 Nm) and the bracket-to-control arm bolts to 16 ft. lbs. (22 Nm).

Rear Wheel Bearings

REMOVAL & INSTALLATION

The rear wheel hub and bearing are replaced as an assembly only.

1. Raise and safely support the vehicle.

2. Remove the wheel and tire assembly and the brake drum.

3. Remove the hub/bearing assembly-to-rear axle nuts/bolts.

NOTE: The top mounting bolt will not clear the brake shoe when removing the hub and bearing. The hub and bearing must be

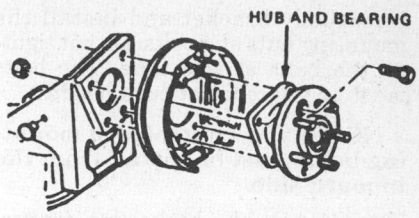

Rear wheel hub and bearing mounting

partially removed while the top bolt is being turned out.

To install:

4. To install, insert and turn the top bolt in while installing the hub and bearing. Install the attaching bolts.

5. Tighten the hub/bearing assembly-to-rear axle nuts/bolts to 38 ft. lbs. (52 Nm).

6. Install the brake drum and wheel and tire assembly.

Rear Axle Assembly

REMOVAL & INSTALLATION

1. Raise and safely support the vehicle under the rear control arms. Support the rear axle assembly with and jack.

2. Remove the wheel assembly. Remove the right and left brake line bracket attaching screws from the body and allow the brake line to hang free.

3. Remove the stabilizer bar brackets. Remove the insulator and the stabilizer bar assembly.

4. Remove the shock absorber lower attaching bolts. Lower the rear axle assembly to remove the coil springs. Do not allow the axle assembly to hang in this position.

5. Remove the control arm attaching bolts from the underbody bracket and lower the axle.

6. Remove the hub attaching bolts an remove the hub, bearing and backing plate assembly.

NOTE: Be careful not to drop the hub/bearing assembly, damage to the bearing could result.

To install:

7. Install the backing plate and hub/bearing assembly to the rear axle assembly. Install the attaching bolts and nuts and tighten to 38 ft. lbs. (52 Nm).

8. Install the stabilizer bar to the rear axle assembly and install the attaching nuts and bolts. Tighten the bracket-to-axle bolts to 13 ft. lbs. (18 Nm) and the bracket-to-control arm bolts to 16 ft. lbs. (22 Nm).

9. Secure the axle assembly on a transmission jack and raise it into position.

10. Install the control arms to the

underbody bracket and install the mounting nuts and bolts. Do not tighten the bolts at this time. The bolts must be tightened at curb height.

NOTE: The control arm mounting bolts must be install from the inboard side.

11. Connect the brake line connections and install the brake cable to the rear axle assembly.

12. Position the springs and insulators in the spring seat and raise the axle. The upper ends of the coil must be positioned properly in the seat of the body.

13. Connect the shock absorber at the lower end and install attaching bolt. Tighten the attaching bolt to 35 ft. lbs. (48 Nm).

14. Connect the parking brake to the guide hook. Adjust the cable as required.

15. Bleed the brake system and refill the reservoir. Adjust the brakes as required.

16. Lower the axle to curb height and tighten the axle-to-body mounting bolts. Tighten the bolts to 66 ft. lbs. (90 Nm).

17. Install the wheel assemblies and lower the vehicle. Tighten the lug nuts to 100 ft. lbs. (140 Nm).

STEERING

Steering Wheel

REMOVAL & INSTALLATION

1. Disconnect the negative battery cable.

2. If equipped with SIR, disable the system and remove the inflator module.

3. Without SIR, remove the horn cover-to-steering wheel screws.

4. Disconnect the horn electrical connector from the steering wheel and remove the horn contact.

5. Remove the steering wheel-to-column retainer, nut, washer.

6. Mark the steering wheel alignment with the steering shaft for installation purposes.

7. Using a steering wheel puller, press the steering wheel from the steering column.

NOTE: Under no circumstances should the steering wheel or shaft be hammered on. Sharp blows to the steering column could loosen the plastic injections which maintain column rigidity.

To install:

8. If equipped with SIR, feed the coil assembly connector through the steering wheel.

9. Align the matchmarks made during removal and install the steering wheel.

10. Align the steering wheel with the turn signal cancelling cam assembly.

11. Install the hexagon locking nut. Tighten the steering wheel nut to 31 ft. lbs. (42 Nm).

12. If equipped with SIR, install the inflator module and enable the system.

13. Connect the negative battery cable.

Steering Column

REMOVAL & INSTALLATION

1. Disconnect the negative battery cable.

2. Remove the left side sound insulator.

3. If equipped, disable the SIR system.

4. If equipped, disconnect the cruise switch terminal connection.

5. Remove the steering column support bracket bolts.

6. Remove the flange and coupling pinch bolt.

7. Remove the upper and lower steering column support bolts.

8. Disconnect the dimmer switch and turn signal switch electrical connectors.

9. If equipped with park lock, remove the park lock cable from the ignition switch.

NOTE: If equipped with park lock, the park lock cable must be disconnected by pressing the locking tab at the ignition switch inhibitor before removing the column from the vehicle.

10. Remove the steering column assembly from the vehicle.

To install:

11. Position the steering column in the vehicle.

12. Connect the park lock cable to the ignition switch.

13. Connect the ignition and dimmer switch electrical connectors.

14. Position the steering column into the flange and coupling assembly.

15. Install the lower steering column support bolts. Tighten to 21 ft. lbs. (28 Nm).

16. Install the steering column support bracket-to-steering column bolts. Tighten to 22 ft. lbs. (30 Nm).

17. Install the upper steering column support bolts. Tighten to 21 ft. lbs. (28 Nm).

18. Install the flange coupling pinch bolt. Tighten to 30 ft. lbs. (41 Nm).

19. Connect the hazard/turn signal switch electrical connectors.

20. If equipped, connect the cruise control electrical connector.

21. If equipped, enable the SIR system.

22. Install the left side sound insulator.

23. Connect the negative battery cable.

Power Steering Rack

ADJUSTMENT

1. Disconnect the negative battery cable. Raise and safely support the vehicle.

2. With the front tires off the ground, loosen the locknut on the bottom of the steering rack.

3. Turn the adjuster plug clockwise until it bottoms out in the housing.

4. Turn the adjuster plug in the opposite direction 50–70 degrees.

5. While holding the adjuster plug, tighten the locknut to 50 ft. lbs. (68 Nm).

NOTE: If the adjuster plug is not held, damage to the pinion teeth on the steering rack may occur.

6. Check to make sure the steering wheel returns to center.

REMOVAL & INSTALLATION

1. Disconnect the negative battery cable. From inside the vehicle, remove the left side lower sound insulator.

2. Remove the upper steering shaft-to-steering rack coupling pinch bolt.

3. Place a drain pan under the steering gear and disconnect the pressure lines from the steering gear.

4. Raise and safely support the vehicle.

5. Remove both front wheel and tire assemblies.

6. Using a ball joint remover, disconnect the tie rod ends from the steering knuckles.

7. Lower the vehicle.

8. Remove both steering gear-to-chassis clamps.

9. Slide the steering gear forward and remove the lower steering shaft-to-steering rack coupling pinch bolt.

10. From the firewall, disconnect the coupling and seal from the steering gear.

11. Raise and safely support the vehicle.

12. Through the left wheel opening, remove the steering gear with the tie rods.

To install:

13. Installation is the reverse of the removal procedures. Lower the vehicle.

14. Tighten the steering gear-to-chassis clamp bolts to 28 ft. lbs. (38 Nm), the tie rod nut to 44 ft. lbs. (60 Nm) and the fluid lines to 18 ft. lbs. (24 Nm).

15. Refill power steering pump reservoir and bleed the power steering system. Connect the negative battery cable.

16. Check and adjust the front end alignment as required.

Power Steering Pump
REMOVAL & INSTALLATION

1. Disconnect the negative battery cable.

2. Remove the pressure and return hoses from the pump and drain the system into a suitable container.

3. Cap the fittings at the pump.

4. Remove the serpentine belt.

5. Locate the pump attaching bolts through the pulley and remove the bolts.

6. Remove the pump assembly.

To install:

7. Installation is the reverse of the removal procedures.

8. Tighten the power steering pump bolts to 20 ft. lbs. (27 Nm).

9. Refill power steering pump reservoir and bleed the system.

10. Connect the negative battery cable.

Belt Adjustment

1. Install a belt tension gauge on the power steering belt.

2. Loosen pump adjustment bolts.

3. Tighten the front bracket-to-engine bolt A to 9 inch lbs. (1 Nm).

4. Set the belt tension by turning adjustment stud.

NOTE: The adjustment bolts are all tighten to different torque specifications. Tighten each bolt as follows:

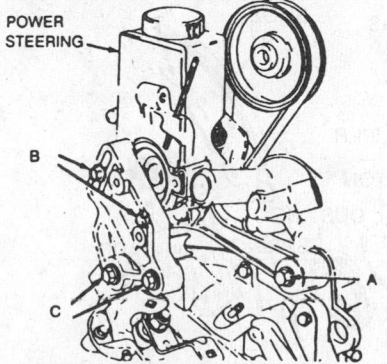

Adjusting the power steering belt tension — 2.3L engine

5. Tighten adjustment bolts A to 67 ft. lbs. (91 Nm), bolts B to 19 ft. lbs. (26 Nm) and bolts C to 40 ft. lbs. (54 Nm).

6. Start engine and run it for a minimum of 2 minutes. Re-adjust the belt tension.

SYSTEM BLEEDING

NOTE: Automatic transmission fluid is not compatible with the seals and hoses of the power steering system. Under no circumstances should automatic transmission be used in place of power steering fluid in this system.

1. With the engine turned **OFF**, turn the wheels all the way to the left.

2. Fill the reservoir with power steering fluid until the level is at the **COLD** mark on the reservoir.

3. Start and operate the engine at fast idle for 15 seconds. Turn the engine **OFF**.

4. Recheck the fluid level and fill it to the **COLD** mark.

5. Start the engine and bleed the system by turning the wheels in both directions slowly to the stops.

6. Stop the engine and check the fluid. Fluid that still has air in it will be a light tan color.

7. Repeat this procedure until all air is removed from the system.

BRAKES

For all brake system repair and service procedures not detailed below, please refer to "Brakes" in the Unit Repair section.

Master Cylinder
REMOVAL & INSTALLATION

1. Disconnect the negative battery cable and the electrical connector from the fluid level sensor.

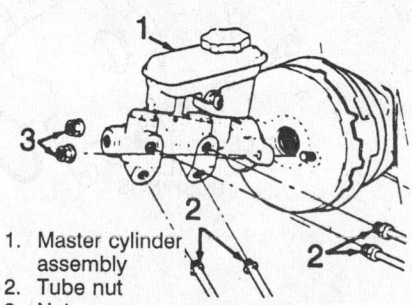

1. Master cylinder assembly
2. Tube nut
3. Nut

Master cylinder Installation

2. Disconnect and plug the 4 brake lines on the master cylinder.

3. Remove the master cylinder-to-power booster nuts and the master cylinder with the reservoir attached.

To install:

4. Bench bleed the new master cylinder and reverse the removal procedures. Tighten the master cylinder-to-power booster nuts to 20 ft. lbs. (27 Nm) and the brake lines-to-master cylinder to 13 ft. lbs. (18 Nm).

5. Connect the fluid level electrical sensor wires. Refill the reservoir with an approved DOT 3 brake fluid and bleed the brake system. Connect the negative battery cable.

Bench Bleeding

This procedure is used to bench bleed the master cylinder.

1. Refill the master cylinder reservoir.

2. Push the plunger several times to force fluid into the piston.

3. Continue pumping the plunger until the fluid is free of the air bubbles.

4. Plug the outlet ports and install the master cylinder.

Proportioning Valve

REMOVAL & INSTALLATION

NOTE: It may be necessary to remove the reservoir in order to remove the proportioning valve. If the reservoir is removed, bleed the brake system when finished.

1. Disconnect the negative battery cable.

2. Remove the proportioning valve cap on the master cylinder.

3. Remove and discard the O-rings.

4. Remove the springs, the proportioning valve pistons and the seals from the valves.

5. Inspect the valves for corrosion or abnormal wear, replace as required.

6. Clean all parts in denatured alcohol or an equivalent. Dry all parts with air before reassembling.

To install:

7. Assemble the springs, proportioning valve pistons and the seals on the valves.

8. Install new O-ring seals.

9. Install the proportioning valve cap on the master cylinder.

10. Tighten the caps to 20 ft. lbs. (27 Nm). Refill the reservoir and bleed the brake system.

11. Connect the negative battery cable.

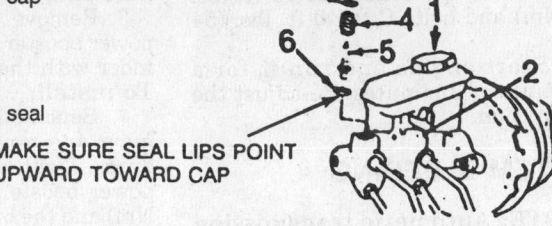

1 Master cylinder
2 Proportioning valve cap
3 O-ring
4 Spring
5 Proportioning valve piston
6 Proportioning valve seal

MAKE SURE SEAL LIPS POINT
UPWARD TOWARD CAP

Proportioning valve installation

Power Brake Booster

REMOVAL & INSTALLATION

1. Disconnect the negative battery cable. Remove the master cylinder.

NOTE: Place the master cylinder in an upright position to prevent fluid loss.

2. Remove the lower-left trim panel inside the vehicle and disconnect the brake pedal-to-booster pushrod from the brake pedal.
3. Disconnect the vacuum line from the booster.
4. Remove the brake booster mounting nuts and the booster.

To install:

5. Installation is the reverse of the removal procedures. Tighten the master cylinder-to-power booster to 20 ft. lbs. (27 Nm) and the power booster mounting nuts to 20 ft. lbs. (27 Nm).
6. Bleed the brake system. Connect the negative battery cable.

Brake Caliper

REMOVAL & INSTALLATION

1. Disconnect the negative battery cable.
2. Remove half of the brake fluid from the master cylinder.
3. Raise and support the vehicle safely and remove the wheel assembly.

NOTE: Remove the brake hose attaching bolt, only if the caliper is going to be overhauled or replaced.

4. Position a large C-clamp over the caliper with the screw end against the outboard brake pad. Tighten the clamp until the caliper piston is pushed out enough to bottom the piston.
5. Remove the C-clamp. Remove the caliper guide pins and lift the caliper off of the rotor.
6. Support the caliper so there is no strain on the brake hose.
7. Press the inboard pad outward and remove it from the caliper.

8. Remove and discard the O-ring bushings and steel sleeves, new parts are to be installed.
9. Check the condition of the rotor. If rotor measurements exceed manufacturer's specifications or has mild scoring, machine the rotor.

To install:

10. Lubricate and install the O-ring bushings. Install the sleeves by pressing them through the O-rings until the sleeve end on the pad side is flush with caliper ear.
11. Position the inboard pads so the pad contacts the piston and the support spring ends. The inboard and outboard pads are similar but not interchangeable.
12. Press down on the ears at the top of the inboard pad until the pad lies flat and the spring ends are just inside the lower edge of the pad.
13. Position the outboard pad with the ears toward the positioning pin holes and the tab on the inner edge of the pad resting in the notch in the edge of the caliper. Bend the ears to provide a slight interference fit in the caliper.
14. Press the outboard pad tightly into position and clinch the ears of the outboard pad over the outboard caliper half.
15. Position the caliper over the rotor.
16. Install the caliper over the rotor.

17. Install the caliper mounting bolts and tighten to 38 ft. lbs. (51 Nm).
18. If the brake hose attaching nut was disconnected, reconnect it and tighten to 33 ft. lbs. (45 Nm).
19. Install the wheel assembly and lower the vehicle.
20. Fill the master cylinder with brake fluid and bleed the system.
21. Connect the negative battery cable.

Disc Brake Pads

REMOVAL & INSTALLATION

1. Raise and safely support the front of the vehicle. Remove the wheel assembly; reinstall 2 lug nuts to retain the rotor to the axle hub.
2. Using a siphon, remove $\frac{2}{3}$ of the brake fluid from the master cylinder.
3. Using a pair of large adjustable pliers, position the jaws over the inboard pad tab and the inboard caliper housing. Squeeze the pliers to bottom the piston in the caliper housing.
4. Remove the caliper-to-caliper bracket boots, bolt coverings, the bolts and sleeve assemblies.
5. Remove the caliper from the rotor. Using a wire, suspend the caliper from the strut.
6. Remove the outboard pad, the inboard pad and the bushing from the mounting bolt hole groves.

To install:

7. Using silicone grease, lubricate the new mounting bolt bushings and install them in the holes.
8. Install the retainer spring onto the inboard pad and the pad into the caliper by snapring the retaining spring into the piston; the inboard pad must lay flat against the piston.

NOTE: On some models, the retaining spring is already staked to the inboard pad.

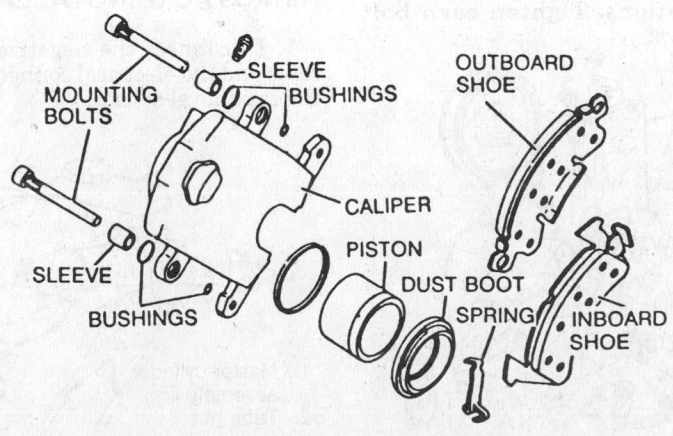

Exploded view of the brake caliper assembly

9. Install the outboard pad with the wear sensor at the leading edge of the forward wheel rotating; the pad must lay flat against the caliper.

10. Position the caliper assembly over the rotor in the mounting bracket. Tighten the caliper-to-bracket bolts to 38 ft. lbs. (51 Nm).

11. Using a small prybar, position it between the outboard pad and the rotor hub to hold the pad in position. Have an assistant, apply approximately 50 lbs. pressure on the brake pedal.

12. While the assistant is applying pressure, position a ball peen hammer on the outboard pad tab and tap it with another hammer to drive the tab downward to a 45 degree angle to lock the pad into position.

13. Remove the 2 rotor-to-wheel hub nuts and install the wheel.

14. Lower the vehicle. Refill the master cylinder reservoir and road test the vehicle.

Brake Rotor

REMOVAL & INSTALLATION

1. Raise and safely support the vehicle. Remove wheel assembly.

2. Remove caliper attaching bolts and remove the caliper. Using a wire suspend the caliper from the strut.

3. Remove the rotor by sliding it off of the hub assembly.

To install:

4. Slide the caliper on the hub assembly and install 2 lug nuts to hold it in place.

5. Install the brake pads into the caliper and place the caliper assembly over the rotor.

6. Install the caliper mounting bolts and tighten to 38 ft. lbs. (51 Nm).

7. Remove the lug nuts from the caliper and install the wheel assembly.

8. Lower the vehicle.

9. Fill the master cylinder with brake fluid.

10. Depress the brake pedal 3-4 times to seat the brake linings and to restore pressure in the system.

Brake Drums

REMOVAL & INSTALLATION

1. Raise and safely support the vehicle.

2. Remove the wheel assembly.

3. Remove the brake drum from the spindle.

To install:

4. Using a brake adjusting tool, adjust the brake shoe to 0.50 in. (1.27mm) less than the brake drum diameter. Install the brake drum to the axle.

5. Install the wheel assembly and lower the vehicle.

Brake Shoes

REMOVAL & INSTALLATION

1. Raise and safely support the vehicle. Remove the wheel assembly.

2. If the brake drum is difficult to remove, perform the following procedures:

   a. Make sure the parking brake is released.

   b. Back off the parking brake cable adjustment.

   c. Remove the adjusting hole knockout plate and back off the adjusting screw.

NOTE: On some drum designs, the knockout plate must be drilled out using a $7/16$ in. (11mm) drill bit. A rubber adjusting hole cover is available for installation purposes.

   d. Using a rubber mallet, tap the drum from the spindle.

3. Remove the return springs, the hold-down springs and the lever pivot. While lifting up on the actuator lever, remove the actuator link.

4. Remove the actuator lever, the lever return spring, the parking brake strut and the strut spring.

5. Disconnect the parking brake cable and remove the primary brake shoe.

6. Remove the adjusting screw, the spring, the retaining ring, the pin, the parking brake lever and the secondary shoe.

7. If any parts are of doubtful strength or quality, due to discoloration from heat, stress or wear, replace them.

To install:

8. Clean all of the parts in denatured alcohol. Lubricate the necessary parts.

9. To install, reverse the removal procedures and install all of the parts, except the brake drum.

10. Using a brake adjusting tool, adjust the brake shoe to 0.50 in. (1.27mm) less than the brake drum diameter. Install the brake drum.

11. To complete the installation, reverse the removal procedures. Road test the vehicle.

Wheel Cylinder

REMOVAL & INSTALLATION

1. Raise and safely support the vehicle. Remove the wheel assembly and brake drum Remove the brake shoes and attaching hardware.

2. Clean any dirt from around the wheel cylinder.

3. Disconnect and plug the brake line from the wheel cylinder.

4. Remove the wheel cylinder-to-backing plate bolt and lockwasher.

5. Remove the wheel cylinder.

To install:

6. Apply a liquid gasket to the shoulder of the wheel cylinder that faces the backing plate and reverse the removal procedures. Tighten the wheel cylinder-to-backing plate bolt to 106 inch lbs. (12 Nm) and the brake line-to-wheel cylinder to 13 ft. lbs. (18 Nm).

7. Bleed the brake system. Lower the vehicle and check the brake operation.

Parking Brake Cable

ADJUSTMENT

1. Apply and release the parking brake lever (10 clicks) at least 6 times. Apply the parking brake lever 4 clicks.

2. Raise and safely support the vehicle.

3. Locate the access hole in the backing plate and adjust the parking brake cable until a $1/8$ in. drill bit can be inserted between the the brake shoe webbing and the parking brake lever.

4. Check to make sure a $1/4$ in. drill bit will not fit in the same position.

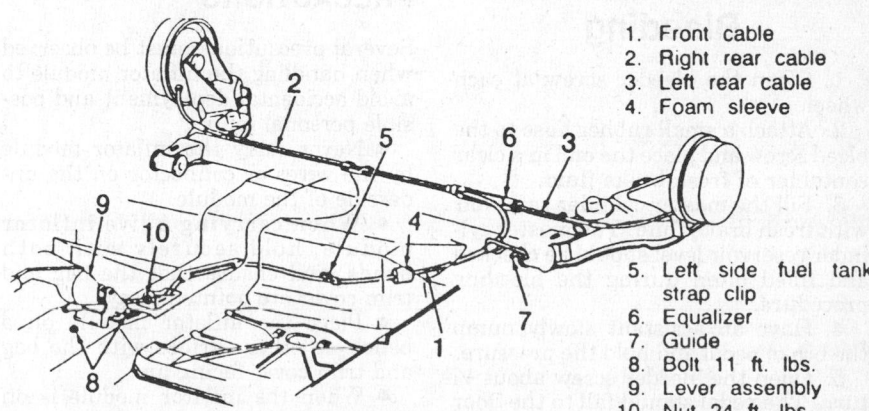

1. Front cable
2. Right rear cable
3. Left rear cable
4. Foam sleeve

5. Left side fuel tank strap clip
6. Equalizer
7. Guide
8. Bolt 11 ft. lbs.
9. Lever assembly
10. Nut 24 ft. lbs.

Parking brake cables — Beretta shown, Corsica similar

5. Release the parking brake and check to see if both wheels turn freely by hand.

6. Lower the vehicle.

REMOVAL & INSTALLATION

Front Cable

1. Raise and safely support the vehicle.

2. Loosen but do not remove the equalizer nut to remove the cable.

3. Disconnect the cable from the equalizer and right side cable.

4. Remove the hand grip from the parking brake lever inside the vehicle.

5. Remove the console.

6. Disconnect the cable from the parking brake lever.

7. Remove the nut holding the cable to the floor.

8. Remove the exhaust hanger bracket mounting nuts.

9. Remove the catalytic converter shield.

10. Remove the cable.

11. To install, Lubricate the cable and reverse the removal procedures. Adjust the parking brake.

Rear Cable

1. Raise and safely support the vehicle.

2. Loosen the equalizer nut until the cable tension is released.

3. Disconnect the right side cable button from the connector.

4. Disconnect the conduit end of the cable from the bracket on the axle.

5. Remove the wheel/tire assembly and the brake drum.

6. Disconnect the cable from the parking brake lever attached to the brake shoes.

7. Remove the conduit end from the brake shoe backing plate.

8. To install, lubricate the cable and reverse the removal procedures. Adjust the parking brake.

Brake System Bleeding

1. Clean the bleeder screw at each wheel.

2. Attach a small rubber hose to the bleed screw and place the end in a clear container of fresh brake fluid.

3. Fill the master cylinder reservoir with fresh brake fluid. The master cylinder reservoir level should be checked and filled often during the bleeding procedure.

4. Have an assistant slowly pump the brake pedal and hold the pressure.

5. Open the bleeder screw about ¼ turn. The pedal should fall to the floor as air and fluid are pushed out. Close the bleeder screw while the assistant holds the pedal to the floor. Slowly release the pedal and wait 15 seconds. Repeat the process until no more air bubbles are forced from the system when the brake pedal is applied. It may be necessary to repeat this 10 or more times to get all of the air from the system.

6. Repeat this procedure on the remaining wheel cylinders and calipers. Make sure the master cylinder does not run out of brake fluid.

NOTE: Remember to wait 15 seconds between each bleeding and do not pump the pedal rapidly. Rapid pumping of the brake pedal pushes the master cylinder secondary piston down the bore in a manner that makes it difficult to bleed the system.

7. Check the brake pedal for sponginess and the brake warning light for an indication of unbalanced pressure. Repeat the entire bleeding procedure to correct either of these conditions. Check the fluid level when finished.

CHASSIS ELECTRICAL

Air Bag
CAUTION

Some vehicles are equipped with the Supplemental Inflatable Restraint (SIR) or air bag system. The SIR system must be disabled before performing service on or around SIR system components, steering column, instrument panel components, wiring and sensors. Failure to follow safety and disabling procedures could result in accidental air bag deployment, possible personal injury and unnecessary SIR system repairs.

PRECAUTIONS

Several precautions must be observed when handling the inflator module to avoid accidental deployment and possible personal injury.

• Never carry the inflator module by the wires or connector on the underside of the module.

• When carrying a live inflator module, hold securely with both hands, and ensure that the bag and trim cover are pointed away.

• Place the inflator module on a bench or other surface with the bag and trim cover facing up.

• When the inflator module is on the bench, never place anything on or close to the module which may be thrown in the event of an accidental deployment.

DISARMING

1. Disconnect the negative battery cable.

2. Remove the SIR fuse from the fuse panel.

3. Remove the left side sound insulator.

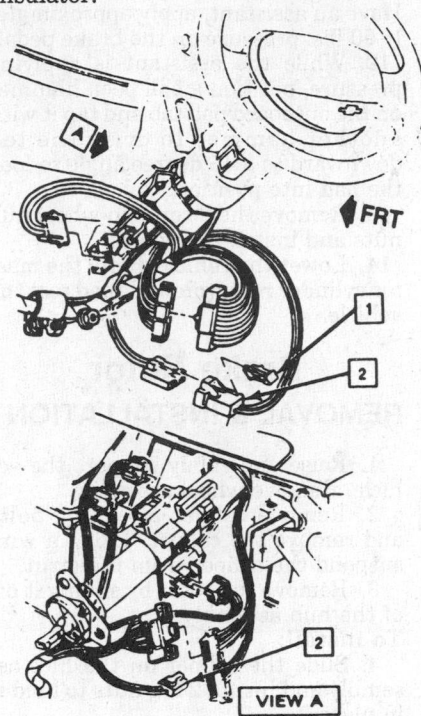

1. Connector Positive Assurance (CPA)
2. Yellow 2-way SIR harness connector

Location of yellow 2-way SIR harness connector.

4. Remove the Connector Positive Assurance (CPA) from the yellow 2-way SIR harness connector at the base of the steering column and separate the connector.

ARMING

1. Connect the yellow 2-way SIR connector at the base of the steering column and insert the Connect Positive Assurance (CPA).

2. Install the left side sound insulator.

3. Install the SIR fuse in the fuse panel.

4. Connect the negative battery cable.

REMOVAL & INSTALLATION

Inflator Module

1. Disconnect the negative battery cable.

2. Disarm the SIR system.

NOTE: Rotate the steering wheel so the access holes on the back of the steering wheel are at the 12 and 6 o'clock positions. This will allow tool access and reduce the possibility of marring the steering column cover.

3. Remove the 4 screws from the back of the inflator module.

4. Remove the inflator module from the steering wheel.

5. Remove the Connector Positive Assurance (CPA) from the inflator module electrical connector and disconnect the connector.

To install:

6. Connect the coil assembly connector. Install the CPA into the connector.

NOTE: Ensure that no wires at the back of the inflator module are pinched when aligning the inflator module to the steering wheel.

7. Install the inflator module and the 4 attaching bolts.

8. Arm the SIR system.

9. Connect the negative battery cable.

CENTERING THE COIL ASSEMBLY

In the event the coil becomes uncentered, perform the following:

1. With the steering wheel removed, remove the coil assembly from the steering column.

2. Hold the coil assembly with the clear bottom up in order to see the coil ribbon.

3. There are 2 styles of coils: one rotates clockwise and the other rotates counterclockwise.

4. While holding the coil assembly, depress the spring lock to rotate the hub in the direction of the arrow until it stops. The coil ribbon should be wound up snug against the center hub.

6. Rotate the coil hub in the opposite direction approximately 2½ turns.

7. Release the spring lock between the locking tabs adjacent to the arrow.

Heater Blower Motor

REMOVAL & INSTALLATION

1. Disconnect the negative battery cable.

2. Disconnect the electrical connections from the blower motor and resistor.

3. Remove the plastic water shield from the right side of the cowl, if equipped.

4. If equipped with the 3.1L engine,

it may be necessary to remove the alternator.

5. Remove the blower motor cooling hose.

6. Remove the blower attaching screws and pull the blower motor from the cowl.

7. Remove the fan attaching nut and the fan from the motor.

To install:

8. Install the fan on the new blower motor with the opening facing away from the motor and install the attaching nut.

9. Position the blower motor assembly to the cowl and install the attaching screws.

10. Install the blower motor cooling hose.

11. If equipped with the 3.1L engine, install the alternator, if removed.

12. Install the plastic water shield on the right side of the cowl, if equipped.

13. Connect the electrical connections to the blower motor and resistor.

14. Connect the negative battery cable.

Removal and Installation of the blower motor assembly

Windshield Wiper Motor

REMOVAL & INSTALLATION

1. Disconnect the negative battery cable.

2. Remove the left and right side wiper arms.

3. Disconnect the wiper motor drive link from the crank arm.

4. Disconnect the electrical connectors and washer hoses.

5. Remove the wiper motor-to-chassis bolts and the wiper motor by guiding the crank arm through the hole.

6. Remove the crank arm from the motor.

To install:

7. Install the crank arm on the new wiper motor shaft and install the attaching nut.

8. Install the wiper motor while guiding the crank arm through cowl opening.

9. Install the wiper motor to the chassis and install the attaching bolts.

10. Connect the blower motor electrical connectors to the wiper harness connectors and connect the washer hoses.

11. Connect the wiper arm drive link to the crank arm.

12. Install the top vent screen shroud in place to the cowl area.

13. Install the left and right wiper arms.

14. Connect the negative battery cable.

Windshield Wiper Switch

REMOVAL & INSTALLATION

1988–90

1. Disconnect the negative battery cable.

2. Remove the switch by gently prying behind the switch.

3. Disconnect the electrical connectors.

4. Remove the switch assembly.

To install:

5. Connect the electrical connectors to the new switch and press it into the instrument panel to the same depth as the old switch.

6. Reconnect the negative battery cable and check the wiper operation.

1991–92

1. Disconnect the negative battery cable.

2. Remove the instrument cluster bezel.

3. Squeeze the small knob at the side and pull straight out.

4. Insert a small flat blade into the slots adjacent to the center of the inner knob to disengage the knob from the switch.

5. Remove the screws attaching the switch to the bezel.

6. Remove the switch.

To install:

7. Install the switch to the bezel. Install the attaching screws.

8. Position the inner knob on the switch. Ensure the tabs are lined up with the slots and press to secure the knob.

9. Position the outer knob on the switch and align the D-shaped hole in the knob to the shaft on the switch and press to secure the knob.

10. Install the instrument cluster bezel.

11. Connect the negative battery cable.

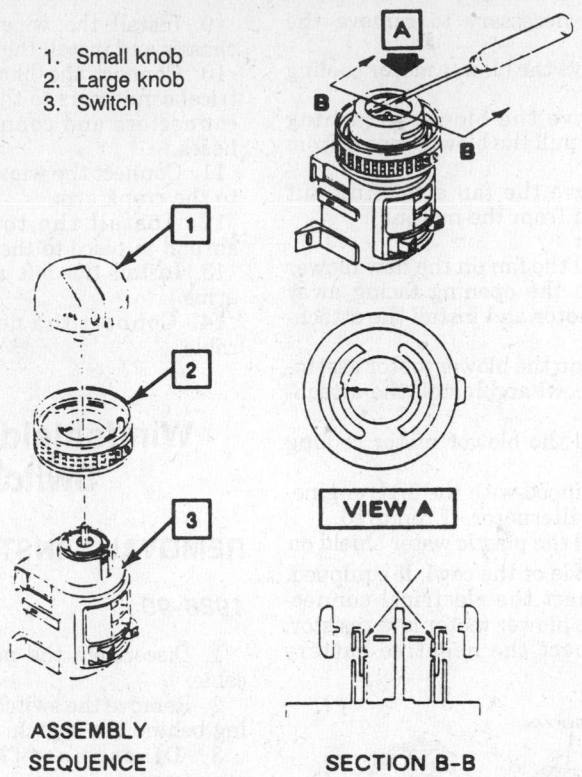

1. Small knob
2. Large knob
3. Switch

VIEW A

ASSEMBLY SEQUENCE

SECTION B-B

Headlight and windshield wiper switch assembly sequence—1991–92

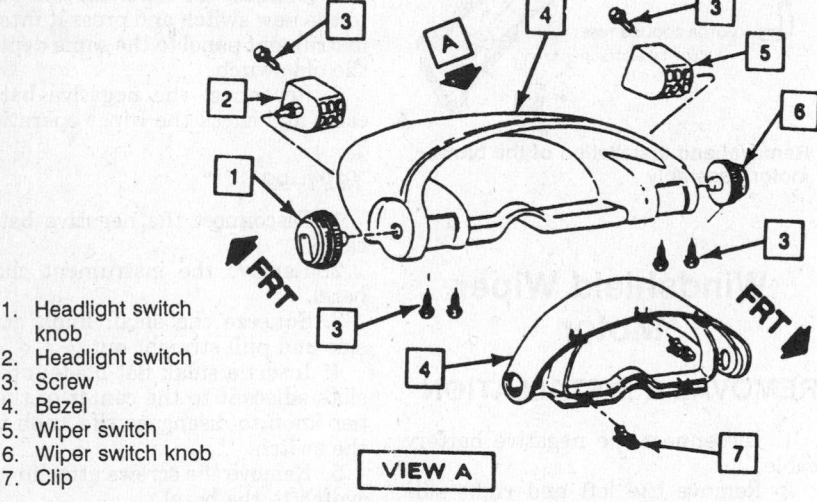

1. Headlight switch knob
2. Headlight switch
3. Screw
4. Bezel
5. Wiper switch
6. Wiper switch knob
7. Clip

VIEW A

Instrument cluster bezel—1991–92

Instrument Cluster Bezel

REMOVAL & INSTALLATION

1991–92

1. Disconnect the negative battery cable.
2. Remove the bezel-to-instrument panel screws.
3. Pull the bezel to the rear to disengage the retaining clips.

4. Disconnect the headlight and windshield wiper switch electrical connectors.
5. If removing the switches, remove the screws attaching the switches to the bezel.
6. Remove the clips, as required.
To install:
7. If removed, install the clips to the bezel.
8. If removed, install the headlight and windshield wiper switches to the bezel and install the attaching screws.
9. Install the switch knobs.

10. Connect the electrical connectors.
11. Position the bezel and press in to engage the retaining clips.
12. Install the instrument panel-to-bezel screws.
13. Connect the negative battery cable.

Instrument Cluster

REMOVAL & INSTALLATION

The speedometer and gauge cluster are replaced as an assembly.

NOTE: Whenever working on any electronic equipment, make sure to have a clean, static free environment in which to work. Always cover the work surface with a mat that is grounded and static free. Static electricity from walking across the floor or sliding across a car seat is enough to damage any equipment.

1988–90

1. Disconnect the negative battery cable.
2. Remove the left side sound insulator attaching screws and remove the insulator from the lower dash and cowl.
3. Remove the 2 screws from the top of the trim cover. Remove the trim cover.
4. Remove the 2 bolts from the upper part of the column and bolt(s) from the lower part of the column. Lower and support the column to prevent tension on the flex joint.
5. Remove the instrument cluster trim panel attaching hardware. Pull the trim panel rearward.
6. Disconnect the electrical connectors.
7. Remove the trim panel.
8. Remove the 4 screws attaching instrument cluster.
9. Pull the cluster rearward to disconnect the electrical connectors.
10. Remove the instrument cluster.
To install:
11. Position the instrument cluster close to the wiring harness and connect to the harness. Ensure the cluster connectors plug in securely to the connectors in the cluster carrier.
12. Slide the instrument cluster into position and install the mounting screws.
13. Position the instrument cluster trim panel close to the wiring harness and connect the electrical connectors. Install the cluster trim panel and attaching hardware.
14. Raise the steering wheel into position and install the bolt(s) at the bottom part of the column and 2 bolts at the top of the column.

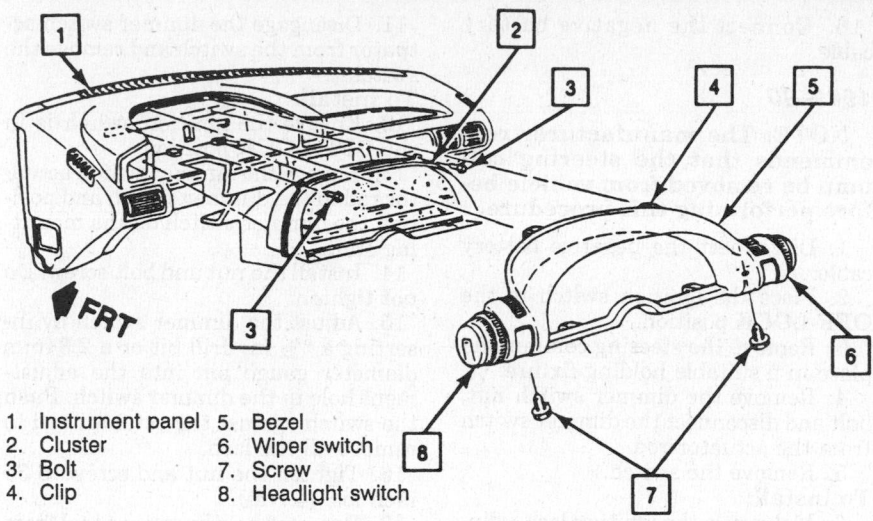

1. Instrument panel
2. Cluster
3. Bolt
4. Clip
5. Bezel
6. Wiper switch
7. Screw
8. Headlight switch

Instrument cluster installation — 1991–92

15. Install the trim cover and 2 attaching screws.
16. Install the left side sound insulator.
17. Connect the negative battery cable.

1991–92

1. Disconnect the negative battery cable.
2. Remove the instrument cluster bezel.
3. Remove the instrument cluster-to-instrument panel attaching screws.
4. Remove the instrument cluster. The electrical connector will release as the cluster is removed.
To install:
5. Carefully, install the instrument cluster. The electrical connector will align and engage as the cluster is pushed into position.
6. Install the instrument cluster bezel.
7. Connect the negative battery cable.

Radio

REMOVAL & INSTALLATION

1988–90

The radio is part of the accessory center, which also includes the heater and air conditioning controls.

1. Disconnect the negative battery cable.
2. Remove left side sound insulator attaching screws and remove the insulator from the lower dash.
3. On the Beretta, remove the lower trim panel screws and pull the trim panel out to release the taps at the top and remove the trim panel.

4. On Corsica, the trim panel has no attaching screws. Carefully pull the trim panel out and release it from the retaining taps and remove the trim panel.
5. Remove the accessory center attaching screws from the top and from the bottom.
6. Pull the accessory center away from the carrier.
7. On vehicles with air conditioning, remove the electrical and vacuum harness from the back of the heater and air conditioning control assembly.
8. On vehicles without air conditioning, remove the cables from the control module.
9. Disconnect the antenna connection and unplug a label the attaching electrical connections.
10. Pull the accessory center assembly from the dash.
11. Place the assembly on a clean working area.
12. Remove all controls knobs by pulling off.
13. Remove the screws attaching the trim plate to the radio. Separate the radio from the trim plate.
To install:
14. Assemble the radio to the trim plate and accessory center. Install the attaching screws.
15. Install the knobs by pushing in place.
16. Position the accessory center assembly to the dash and connect the electrical harness connections.
17. On vehicles with air conditioning, connect the electrical and vacuum harnesses at the rear of the heater and air conditioning control assembly.
18. On vehicles without air conditioning, connect the control cables to the control module.
19. Connect the antenna lead to the radio.

20. Slide the accessory center into place in the dash. Install the attaching screws at top and bottom.
21. Install the trim panel in place.
22. Install the left side sound insulator to lower dash and cowl and install the attaching screws.
23. Connect the negative battery cable.

1991–92

1. Disconnect the negative battery cable.
2. Remove the accessory center bezel by inserting a flat blade tool to separate the bezel from the panel and disengage the clips. Remove the bezel.

NOTE: If equipped with air conditioning, the outlets need not be removed prior to removing the bezel.

3. Remove the bracket-to-panel screws.
4. Remove the radio receiver.
5. Disconnect the antenna cable.
6. Remove the bracket, clip-retained bolts and clip-retained rear guide.
7. Remove the radio receiver.
To install:
8. Install the clip-retained bolts, rear guide and bracket to the radio receiver.
9. Position the receiver and connect the antenna lead.
10. Press the guide receiver into the opening.

NOTE: Ensure the rear guide is engaged into the slot in the instrument panel.

11. Install the attaching screws.
12. Place the trim bezel into position, align the clips to the holes in the panel and press in to secure.
13. Connect the negative battery cable.

Headlight Switch

REMOVAL & INSTALLATION

1988–90

1. Disconnect the negative battery cable.
2. Remove the switch by gently prying behind the switch.
3. Disconnect and label the wiring.
To install:
4. Connect the wires to the new switch and press it into the instrument panel.
5. Connect the negative battery cable and test the switch operation.

1991–92

1. Disconnect the negative battery cable.

2. Remove the instrument cluster bezel.

3. Squeeze the small knob at the side and pull straight out.

4. Insert a small flat blade into the slots adjacent to the center of the inner knob to disengage the knob from the switch.

5. Remove the screws attaching the switch to the bezel.

6. Remove the switch.

To install:

7. Install the switch to the bezel. Install the attaching screws.

8. Position the inner knob on the switch. Ensure the tabs are lined up with the slots and press to secure the knob.

9. Position the outer knob on the switch and align the D-shaped hole in the knob to the shaft on the switch and press to secure the knob.

10. Install the instrument cluster bezel.

11. Connect the negative battery cable.

Dimmer Switch

REMOVAL & INSTALLATION

1988

1. Disconnect the negative battery cable.

2. If equipped with a column shift, place the transaxle selector in **P**.

3. Place the lock cylinder in the **ACC** position.

4. Remove the left side sound insulator panel.

5. Remove the steering column-to-support bolts. Gently lower the steering column onto the driver's seat.

6. Disconnect the dimmer and ignition switch connectors.

7. Remove the dimmer switch nut, bolt and disconnect the dimmer switch from the actuator rod.

8. Remove the switch.

To install:

9. Make sure the ignition lock cylinder is in the **ACC** position.

10. Insert the dimmer switch actuator rod into the hole in the dimmer switch.

11. Position the dimmer switch in place on the stud and install the retainer nut and screw. Do not tighten.

12. Position the dimmer switch so a click is heard when the turn signal lever is pulled. Tighten the nut and screw to 35 inch lbs. (4 Nm).

13. Raise the steering column into place and connect the dimmer switch and turn signal switch connectors.

14. Install the steering column-to-support screws and secure the steering column in place.

15. Install the left side lower trim panel.

16. Connect the negative battery cable.

1989–90

NOTE: The manufacturer recommends that the steering column be removed from vehicle before performing this procedure.

1. Disconnect the negative battery cable.

2. Place the ignition switch in the **OFF-LOCK** position.

3. Remove the steering column and place in a suitable holding fixture.

4. Remove the dimmer switch nut, bolt and disconnect the dimmer switch from the actuator rod.

5. Remove the switch.

To install:

6. Make sure the ignition lock cylinder is in the **OFF-LOCK** position.

7. Insert the dimmer switch actuator rod into the hole in the dimmer switch.

NOTE: Should the actuator rod become disengaged from the rod cap, upon installation the tab on the rod must engage the wide slot in the rod cap and snap into place.

8. Position the dimmer switch in place on the stud and install the retainer nut and screw. Do not tighten.

9. Adjust the dimmer switch by inserting a $\frac{3}{32}$ in. drill bit or a 2.34mm diameter gauge pin into the adjustment hole in the dimmer switch. Push the switch against the actuator rod to remove all the lash.

10. Tighten the nut and screw to 35 inch lbs. (4 Nm).

11. Remove the adjustment tool from the dimmer switch.

12. Install the steering column in the vehicle.

13. Connect the negative battery cable.

1991–92

1. Disconnect the negative battery cable.

2. Disable the Supplemental Inflatable Restraint (SIR) system.

3. Place the ignition switch in the **OFF-LOCK** position.

4. Remove the left side sound insulator panel.

5. Remove the bolts from the lower steering column support.

6. Remove the flange and coupling pinch bolt.

7. Remove the upper and lower bolts from the upper steering column support.

8. Disconnect the dimmer and ignition switch electrical connectors.

9. Lower the steering column.

10. Remove the hexagonal nut and bolt/screw attaching the dimmer switch.

11. Disengage the dimmer switch actuator from the switch and remove the switch.

To install:

12. Ensure the ignition switch is in the **OFF-LOCK** position.

13. Engage the dimmer switch actuator rod in the dimmer switch and position the dimmer switch on the mounting stud.

14. Install the nut and bolt/screw. Do not tighten.

15. Adjust the dimmer switch by inserting a $\frac{3}{32}$ in. drill bit or a 2.34mm diameter gauge pin into the adjustment hole in the dimmer switch. Push the switch against the actuator rod to remove all the lash.

16. Tighten the nut and screw to 35 inch lbs. (4 Nm).

17. Remove the adjustment tool from the dimmer switch.

18. Support the steering column and install the column into the flange and coupling assembly.

19. Connect the dimmer and ignition switch electrical connectors.

20. Raise the column into position and loosely install the lower bolts to the upper steering column support bracket.

21. Install the lower steering column support bracket bolts. Tighten to 22 ft. lbs. (30 Nm).

22. Install the upper bolts to the upper steering column support bracket. Tighten the upper and lower bolts to 21 ft. lbs. (28 Nm).

23. Install the flange and coupling assembly pinch bolt. Tighten to 30 ft. lbs. (41 Nm).

24. Install the right side sound insulator panel.

25. Enable the SIR system.

26. Connect the negative battery cable.

Turn Signal Switch

REMOVAL & INSTALLATION

1988–90

NOTE: A special terminal remover tool is required to remove the terminals from the connector on the turn signal switch.

1. Disconnect the negative battery cable. Remove the steering wheel.

2. Remove the turn signal canceling cam assembly from the steering shaft.

3. Remove the hazard warning knob-to-steering column screw and the knob.

4. Disconnect the dimmer switch actuator rod.

NOTE: Before removing the turn signal assembly, position the turn signal lever so the turn signal assembly-to-steering column screws can all be removed.

5. Remove the column housing cover-to-column housing bowl screw and the cover.

NOTE: If equipped with cruise control, disconnect the cruise control electrical connector.

6. Remove the turn signal lever attaching screw and the lever.

7. Using a terminal remover tool, disconnect and label the wires F and G on the connector at the buzzer switch assembly from the turn signal switch electrical harness connector.

8. Remove the turn signal switch-to-steering column screws and the switch.

To install:

9. Position the turn signal switch into the steering column and install the switch-to-steering column screws. Tighten the screws to 35 inch. lbs. (4 Nm).

10. Connect wires F and G of the buzzer switch assembly, to the turn signal switch harness connector.

11. Install the dimmer actuator assembly.

12. Install the turn signal lever and attaching screw to the pivot assembly. Tighten the attaching screw to 18 inch lbs. (2 Nm).

13. Install the actuator pivot assembly and screw. Tighten to 20 inch lbs. (2.3 Nm).

NOTE: If equipped with cruise control, connect the cruise control electrical connector.

14. Install the column housing cover and attaching screw. Tighten the attaching screw to 35 inch lbs. (4 Nm).

15. Install the hazard warning knob and attaching screw. Tighten to 7 inch lbs. (0.8 Nm).

16. Install the turn signal canceling cam assembly over the steering shaft. Install the steering wheel and attaching nut. Tighten the nut to 30 ft. lbs. (41 Nm).

17. Connect the negative battery cable. Check the switch operation.

1991–92

1. Disconnect the negative battery cable.

2. Disable the SIR system.

3. Remove the steering wheel.

4. Remove the coil assembly retaining ring.

5. Lift the coil assembly from the end of the steering shaft and allow coil to hang freely.

6. Remove the wave washer.

7. If equipped with a standard column, remove the spacer shaft lock.

8. Remove the shaft lock retaining ring using tool J–23653–C or equivalent, to compress the shaft lock.

9. Pry off the retaining ring.

10. Remove the shaft lock.

11. Remove the turn signal cancelling cam assembly.

12. Remove the upper bearing spring.

13. Position the turn signal lever to the right turn position.

14. Remove the multi-function lever by performing the following:

   a. Ensure the lever is in the center or **OFF** position.

   b. If equipped with cruise control, disconnect the cruise control connector from the steering column assembly.

   c. Pull the lever straight out of the turn signal switch.

15. Remove the hazard knob assembly.

16. Remove the screw and signal switch arm. If equipped with tilt column and cruise control, allow the switch arm to hang freely.

17. Remove the turn signal switch screws. Allow the switch to hang freely.

18. Disconnect the turn signal/hazard switch assembly terminal from the instrument panel harness.

19. If equipped with tilt column, disconnect the buzzer switch assembly terminals from the turn signal/hazard assembly connector. Remove the tan/black wire lead from cavity E and the light green wire from the cavity F.

20. Remove the upper steering column bolts.

21. Remove the wiring protector.

22. Connect a length of wire to the turn signal/hazard assembly terminal connector to aid in reassembly.

23. Gently pull the wire harness through the steering column housing shroud, steering column housing and lock assembly cover.

24. Disconnect the wire from the connector.

To install:

25. Connect the wire to the turn signal/hazard switch assembly connector.

26. Gently pull the connector through the steering column housing shroud, steering column housing and lock assembly cover.

27. Remove the wire.

28. Install the wiring protector.

29. If disconnected, connect the buzzer switch terminals to the turn signal/hazard switch assembly connector. Insert the tan/black wire lead into cavity E and the light green wire into cavity F.

30. Connect the turn signal/hazard switch assembly connector to the instrument panel harness.

31. Install the steering column support bracket bolts to the steering column. Tighten to 22 ft. lbs. (30 Nm).

32. Install the steering column upper support bolts. Tighten to 20 ft. lbs. (28 Nm).

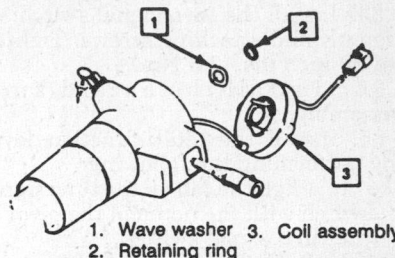

1. Wave washer 3. Coil assembly
2. Retaining ring

SIR coil assembly installation

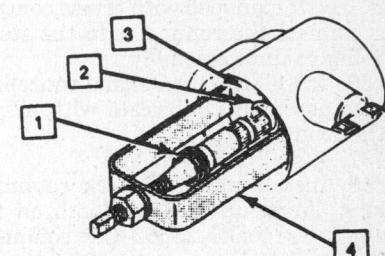

1. Extra rings
2. Retaining ring
3. Shaft lock
4. Shaft lock retaining ring compressor

Shaft lock retaining ring removal—SIR steering column

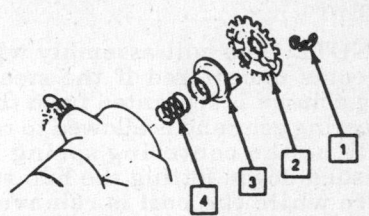

1. Shaft lock spacer
2. Shaft lock
3. Turn signal cancelling cam
4. Upper bearing spring

Lock plate steering column

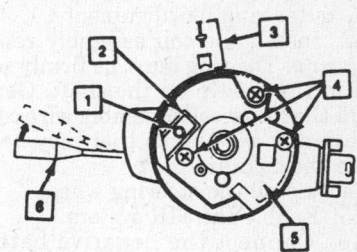

1. Screw
2. Signal switch arm
3. Hazard knob assembly
4. Screw
5. Turn signal/hazard switch assembly
6. Multi-function turn signal lever

Upper SIR steering column component locations—tilt column shown, standard similar

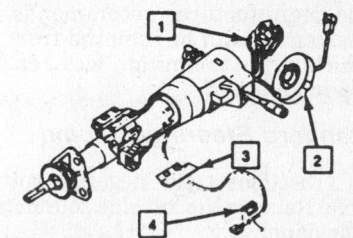

1. Turn signal/hazard switch assembly
2. Coil assembly
3. Wiring protector
4. Connector shroud

SIR steering column assembly

33. Install the turn signal switch assembly and attaching screws. Tighten to 20 inch lbs. (2.3 Nm).

34. Install the hazard knob assembly.

35. Install the multi-function lever by performing the following:

a. Align the tab on the turn signal switch with the notch in the pivot of the turn signal switch.

b. Push the lever into the turn signal switch.

c. If equipped with cruise control, connect the connector to the steering column assembly.

36. Install the turn signal cancelling cam assembly. Lubricate with a synthetic grease.

37. Install the shaft lock.

38. Install the shaft lock retaining ring, lining up to block tooth on the shaft. Use tool J–23653–C to compress the shaft lock.

39. If equipped with a standard column, install the spacer shaft lock.

40. Install the wave washer.

41. Ensure the coil assembly is centered.

NOTE: The coil assembly will become uncentered if the steering column is separated from the steering gear and is allowed to rotate or the centering spring is pushed down, letting the hub rotate while the coil is removed from the steering column.

42. Install the coil assembly using the horn tower on the cancelling cam assembly inner ring and projections on the outer ring for alignment.

43. Install the coil assembly retaining ring. The ring must be firmly seated in the groove on the shaft. Gently pull the lower coil assembly wire to remove any wire kinks that may be inside the column.

44. Install the steering wheel.

45. Enable the SIR system.

46. Connect the negative battery cable.

Ignition Lock/Switch

REMOVAL & INSTALLATION

1988–90

The manufacturer recommends the steering column be removed from the vehicle prior to ignition lock removal and installation.

Standard Steering Column

1. Disconnect the negative battery cable. Remove the left side sound insulator panel.

2. Remove the steering column-to-support screws and lower the steering column.

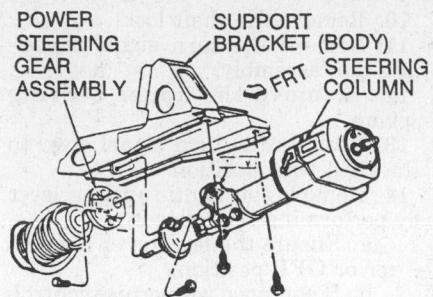

POWER STEERING GEAR ASSEMBLY — SUPPORT BRACKET (BODY) — FRT — STEERING COLUMN

Steering column mounting

3. Disconnect the dimmer and turn signal switch connectors.

4. Remove the wiring harness-to-firewall nuts and steering column.

5. If equipped with a park lock cable, insert a small prybar into the ignition switch inhibitor switch access hole, depress the locking tab and disconnect the park lock cable from the inhibitor switch.

6. Remove the steering column-to-steering gear bolt and the steering column from the vehicle.

7. Remove the combination switch.

8. Place the lock cylinder in the **RUN** position.

9. Remove the steering shaft assembly and turn signal switch housing as an assembly.

10. Using a terminal remover tool, disconnect and label the wires F and G on the connector at the buzzer switch assembly from the turn signal switch electrical harness connector.

11. Place the lock cylinder in the **RUN** position and remove the buzzer switch.

12. Place the lock cylinder in the **ACC** position. Remove the lock cylinder attaching screw and the lock cylinder.

13. Remove the dimmer switch nut/bolt, the dimmer switch and actuator rod.

14. Remove the dimmer switch mounting stud, the mounting nut was mounted to it.

15. Remove the ignition switch-to-steering column screws and the ignition switch.

16. Remove the lock bolt screws and the lock bolt.

17. Remove the switch actuator rack and ignition switch.

18. Remove the steering shaft lock and spring.

To install:

19. To install the lock bolt, lubricate it with lithium grease and install the lock bolt, spring and retaining plate.

20. Lubricate the teeth on the switch actuator rack. Install the rack and the ignition switch through the opening in the steering bolt until it rests on the retaining plate.

21. Install the steering column lock cylinder set by holding the barrel of the lock cylinder, insert the key and turn it to the **ACC** position.

22. Install the lock set in the steering column while holding the rack against the lock plate.

23. Install the lock attaching screw and tighten the screw to 27 inch lbs. (3 Nm). Insert the key in the lock cylinder and turn the lock cylinder to the **START** position and the rack will extend.

24. Center the slotted holes on the ignition switch mounting plate and install the ignition switch mounting screw and nut.

25. Install the dimmer switch and actuator rod into the center slot on the switch mounting plate. Tighten the dimmer switch stud to 35 inch lbs. (4 Nm).

26. Install the buzzer switch and turn the lock cylinder to the **RUN** position. Push the switch in until it is bottomed out with the plastic tab that covers the lock attaching screw.

27. Install the steering shaft and turn signal housing as an assembly.

28. Install the turn signal switch. Tighten the turn signal switch housing screws to 88 inch lbs. (10 Nm), the turn signal switch screws to 35 inch lbs. (4 Nm) and the steering wheel locknut to 30 ft.lbs. (41 Nm).

29. To complete the installation, reverse the removal procedures.

Tilt Steering Column

1. Disconnect the negative battery cable. Tilt the column up as far as it will go and remove the left side lower trim panel.

2. Remove the steering column-to-support screws and lower the steering column.

3. Disconnect the dimmer switch and turn signal switch connectors.

4. Remove the wiring harness-to-firewall nuts and steering column.

5. If equipped with a park lock cable, insert a small prybar into the ignition switch inhibitor switch access hole, depress the locking tab and disconnect the park lock cable from the inhibitor switch.

6. Remove the steering column-to-steering gear bolt and the steering column from the vehicle.

7. Remove the combination switch.

8. Using a flat type pry blade, position it in the square opening of the spring retainer, push downward to the left, to release the spring retainer. Remove the wheel tilt spring.

9. Remove the spring retainer, the tilt spring and the tilt spring guide.

10. Remove the shoe pin retaining cap. Using a pivot pin removal tool, remove both pivot pins.

11. Place the lock cylinder in the **RUN** position.

12. Pull the shoe release lever and release the steering column housing.

13. Remove the column housing, the steering shaft assembly and turn signal switch housing as an assembly.

14. Using a terminal remover tool, disconnect and label the wires F and G on the connector at the buzzer switch assembly from the turn signal switch electrical harness connector.

15. Place the lock cylinder in the **RUN** position and remove the buzzer switch.

16. Place the lock cylinder in the **ACC** position. Remove the lock cylinder attaching screw and the lock cylinder.

17. Remove the dimmer switch nut/bolt, the dimmer switch and actuator rod.

18. Remove the dimmer switch mounting stud, the mounting nut was mounted to it.

19. Remove the ignition switch-to-steering column screws and the ignition switch.

20. Remove the lock bolt screws and the lock bolt.

21. Remove the switch actuator rack and ignition switch.

22. Remove the steering shaft lock and spring.

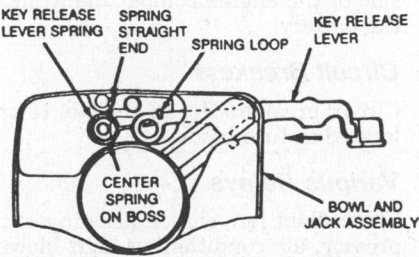

Key release lever spring installation position

To install:

23. To install the lock bolt, lubricate it with lithium grease and install the lock bolt, spring and retaining plate.

24. Lubricate the teeth on the switch actuator rack. Install the rack and the ignition switch through the opening in the steering bolt until it rests on the retaining plate.

25. Install the steering column lock cylinder set by holding the barrel of the lock cylinder, insert the key and turn the key to the **ACC** position.

26. Install the lock set in the steering column while holding the rack against the lock plate.

27. Install the lock attaching screw and tighten the screw to 27 inch lbs. (3 Nm). Insert the key in the lock cylinder. Turn the lock cylinder to the **START** position and the rack will extend.

28. Center the slotted holes on the ignition switch mounting plate. Install the ignition switch mounting screw and nut.

29. Install the dimmer switch and actuator rod into the center slot on the switch mounting plate. Tighten the dimmer switch stud to 35 inch lbs. (4 Nm).

30. Install the buzzer switch and turn the lock cylinder to the **RUN** position. Push the switch in until it is bottomed out with the plastic tab that covers the lock attaching screw.

31. Install the steering shaft and turn signal housing as an assembly.

32. Install the turn signal switch. Tighten the turn signal switch housing screws to 88 inch lbs. (10 Nm), the turn signal switch screws to 35 inch lbs. (4 Nm) and the steering wheel locknut to 30 ft. lbs. (41 Nm).

33. Connect the negative battery cable.

1991–92

IGNITION SWITCH

1. Disconnect the negative battery cable.

2. Disable the SIR system.

3. Place the ignition switch in the **OFF-LOCK** position.

4. Remove the left side sound insulator panel.

5. Remove the bolts from the lower steering column support.

6. Remove the flange and coupling pinch bolt.

7. Remove the upper and lower bolts from the upper steering column support.

8. Disconnect the dimmer and ignition switch electrical connectors.

9. Lower the steering column.

10. Remove the hexagonal nut and bolt/screw attaching the dimmer switch.

11. Disengage the dimmer switch actuator from the switch and remove the switch.

12. Remove the ignition switch stud.

13. Disconnect the ignition switch actuator rod.

14. Disconnect the park lock cable from the ignition switch.

15. Remove the ignition switch.

To install:

16. Ensure the ignition switch is in the **OFF-LOCK** position.

17. Adjust the ignition switch by performing the following:

    a. Place the ignition switch slider in the far left position and move back 1 detent to the right of the **OFF-LOCK** position.

    b. Insert a $^{3}/_{32}$ in. drill bit or a 2.34mm gauge pin into the adjustment hole in the ignition switch to hold the switch slider in the proper position during installation.

18. Connect the park lock switch to the ignition switch.

19. Connect the ignition switch actuator rod.

20. Install the ignition switch mounting stud. Tighten to 35 inch lbs. (4 Nm).

21. Remove the adjustment tool from the ignition switch.

22. Engage the dimmer switch actuator rod in the dimmer switch and position the dimmer switch on the mounting stud.

23. Install the nut and bolt/screw. Do not tighten.

24. Adjust the dimmer switch by inserting a $^{3}/_{32}$ in. drill bit or a 2.34mm diameter gauge pin into the adjustment hole in the dimmer switch. Push the switch against the actuator rod to remove all the lash.

25. Tighten the nut and screw to 35 inch lbs. (4 Nm).

26. Remove the adjustment tool from the dimmer switch.

27. Support the steering column and install the column into the flange and coupling assembly.

28. Connect the dimmer and ignition switch electrical connectors.

29. Raise the column into position and loosely install the lower bolts to the upper steering column support bracket.

30. Install the lower steering column support bracket bolts. Tighten to 22 ft. lbs. (30 Nm).

31. Install the upper bolts to the upper steering column support bracket. Tighten the upper and lower bolts to 21 ft. lbs. (28 Nm).

32. Install the flange and coupling assembly pinch bolt. Tighten to 30 ft. lbs. (41 Nm).

33. Install the right side sound insulator panel.

34. Enable the SIR system.

35. Connect the negative battery cable.

LOCK CYLINDER

1. Disconnect the negative battery cable.

2. Disable the SIR system.

3. Remove the steering wheel.

4. Remove the coil assembly retaining ring.

5. Lift the coil assembly from the end of the steering shaft and allow coil to hang freely.

6. Remove the wave washer.

7. If equipped with a standard column, remove the spacer shaft lock.

8. Remove the shaft lock retaining ring using tool J–23653–C or equivalent, to compress the shaft lock.

9. Pry off the retaining ring.

10. Remove the shaft lock.

11. Remove the turn signal cancelling cam assembly.

12. Remove the upper bearing spring.

13. Position the turn signal lever to the right turn position.

14. Remove the multi-function lever by performing the following:

a. Ensure the lever is in the center or **OFF** position.

b. If equipped with cruise control, disconnect the cruise control connector from the steering column assembly.

c. Pull the lever straight out of the turn signal switch.

15. Remove the hazard knob assembly.

16. Remove the screw and signal switch arm. If equipped with tilt column and cruise control, allow the switch arm to hang freely.

17. Remove the turn signal switch screws. Allow the switch to hang freely.

18. Disconnect the turn signal/hazard switch assembly terminal from the instrument panel harness.

19. If equipped with tilt column, disconnect the buzzer switch assembly terminals from the turn signal/hazard assembly connector. Remove the tan/black wire lead from cavity E and the light green wire from the cavity F.

20. Remove the upper steering column bolts.

21. Remove the wiring protector.

22. Connect a length of wire to the turn signal/hazard assembly terminal connector to aid in reassembly.

23. Gently pull the wire harness through the steering column housing shroud, steering column housing and lock assembly cover.

24. Disconnect the wire from the connector.

25. Ensure the lock cylinder is in the **LOCK** position. Remove the lock cylinder attaching screw.

26. Remove the lock cylinder.

To install:

27. Install the lock cylinder and attaching screw. Tighten to 40 inch lbs. (4 Nm).

28. Turn the ignition key to the **RUN** position.

29. Install the buzzer switch.

30. Connect the wire to the turn signal/hazard switch assembly connector.

31. Gently pull the connector through the steering column housing shroud, steering column housing and lock assembly cover.

32. Remove the wire.

33. Install the wiring protector.

34. If disconnected, connect the buzzer switch terminals to the turn signal/hazard switch assembly connector. Insert the tan/black wire lead into cavity E and the light green wire into cavity F.

35. Connect the turn signal/hazard switch assembly connector to the instrument panel harness.

36. Install the steering column support bracket bolts to the steering column. Tighten to 22 ft. lbs. (30 Nm).

37. Install the steering column upper support bolts. Tighten to 20 ft. lbs. (28 Nm).

38. Install the turn signal switch assembly and attaching screws. Tighten to 20 inch lbs. (2 Nm).

39. Install the hazard knob assembly.

40. Install the multi-function lever by performing the following:

a. Align the tab on the turn signal switch with the notch in the pivot of the turn signal switch.

b. Push the lever into the turn signal switch.

c. If equipped with cruise control, connect the connector to the steering column assembly.

41. Install the turn signal cancelling cam assembly. Lubricate with a synthetic grease.

42. Install the shaft lock.

43. Install the shaft lock retaining ring, lining up to block tooth on the shaft. Use tool J–23653–C to compress the shaft lock.

44. If equipped with a standard column, install the spacer shaft lock.

45. Install the wave washer.

46. Ensure the coil assembly is centered.

NOTE: The coil assembly will become uncentered if the steering column is separated from the steering gear and is allowed to rotate or the centering spring is pushed down, letting the hub rotate while the coil is removed from the steering column.

47. Install the coil assembly using the horn tower on the cancelling cam assembly inner ring and projections on the outer ring for alignment.

48. Install the coil assembly retaining ring. The ring must be firmly seated in the groove on the shaft. Gently pull the lower coil assembly wire to remove any wire kinks that may be inside the column.

49. Install the steering wheel.

50. Enable the SIR system.

51. Connect the negative battery cable.

Stoplight Switch

ADJUSTMENT

1. Disconnect the negative battery cable.

2. Remove the lower, left trim panel and locate the stoplight switch on the brake pedal support.

3. Disconnect the electrical connector from the switch and remove the switch by twisting it out of the tubular retaining clip.

4. Pull back on the brake pedal and push the switch through the retaining clip noting the clicks; repeat this procedure until no more clicks can be heard.

5. Connect the electrical connector to the switch.

6. Connect the negative battery cable and check the switch operation.

Fuses, Circuit Breakers and Relays

LOCATION

Fuse Panel

The fuse panel is located on the left side of the instrument panel assembly. In order to gain access to the fuse panel, it is necessary to first remove the lower trim panel.

Fusible Links

Fusible links—A and E are located rear of the engine compartment, at the battery junction box.

Fusible links—B, C and D are located at the front section of the engine at the starter solenoid.

Fusible link—F is located on the left side of the engine compartment, near the battery.

Circuit Breakers

Circuit breakers No. 12 and No. 15 are located in fuse block.

Various Relays

The coolant fan, air conditioning compressor, air conditioning high blower speed and fuel pump relays are all located in the engine compartment mounted to the center of the firewall on the relay bracket.

Computers

LOCATION

The electronic control module is located on the right side of the vehicle. It is positioned up behind the glove box. In order to gain access to the electronic control module, remove the right side trim panel and/or glove box assembly.

Flashers

LOCATION

Turn Signal Flasher

The turn signal flasher is located behind the lower left side of the instrument panel on the steering column.

Chevrolet
Rear Wheel Drive
Corvette

16

SPECIFICATIONS

VEHICLE IDENTIFICATION CHART

It is important for servicing and ordering parts to be certain of the vehicle and engine identification. The VIN (vehicle identification number) is a 17 digit number visible through the windshield on the driver's side of the dash and contains the vehicle and engine identification codes. The tenth digit indicates model year, and the eighth digit indicates engine code. It can be interpreted as follows:

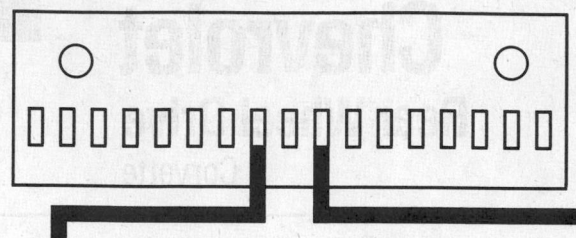

Engine Code

Code	Cu. In.	Liters	Cyl.	Fuel Sys.	Eng. Mfg.
8	350	5.7	8	PFI	Chevrolet
J	350	5.7	8	PFI	Chevrolet

PFI—Port Fuel Injection

Model Year

Code	Year
J	1988
K	1989
L	1990
M	1991
N	1992

ENGINE IDENTIFICATION

Year	Model	Engine Displacement cu. in. (liter)	Engine Series Identification (VIN)	No. of Cylinders	Engine Type
1988	Corvette	350 (5.7)	8	8	OHV
1989	Corvette	350 (5.7)	8	8	OHV
1990	Corvette	350 (5.7)	8	8	OHV
	Corvette ZR-1	350 (5.7)	J	8	DOHV
1991-92	Corvette	350 (5.7)	8	8	OHV
	Corvette ZR-1	350 (5.7)	J	8	DOHV

OHV—Overhead Valves
DOHV—Dual Overhead Valves

GENERAL ENGINE SPECIFICATIONS

Year	VIN	No. Cylinder Displacement cu. in. (liter)	Fuel System Type	Net Horsepower @ rpm	Net Torque @ rpm (ft. lbs.)	Bore × Stroke (in.)	Compression Ratio	Oil Pressure @ rpm
1988	8	8-350 (5.7)	PFI	245 @ 4300	340 @ 3200	4.000 × 3.480	9.5:1	50–65 @ 2000
1989	8	8-350 (5.7)	PFI	245 @ 4300	340 @ 3200	4.000 × 3.480	9.5:1	50–65 @ 2000
1990	8	8-350 (5.7)	PFI	245 @ 4000	345 @ 3200	4.000 × 3.480	10.25:1	50–65 @ 2000
	J	8-350 (5.7)	PFI	375 @ 5800	370 @ 5600	3.897 × 3.661	11:1	50–60 @ 2000
1991-92	8	8-350 (5.7)	PFI	245 @ 4000	345 @ 3200	4.000 × 3.480	10.25:1	50–65 @ 2000
	J	8-350 (5.7)	PFI	375 @ 5800	370 @ 5600	3.897 × 3.661	11:1	50–60 @ 2000

PFI—Port Fuel Injection

GASOLINE ENGINE TUNE-UP SPECIFICATIONS

Year	VIN	No. Cylinder Displacement cu. in. (liter)	Spark Plugs Type	Gap (in.)	Ignition Timing (deg.) MT	AT	Compression Pressure (psi)	Fuel Pump (psi)	Idle Speed (rpm) MT	AT	Valve Clearance In.	Ex.
1988	8	8-350 (5.7)	FR3CLS	0.035	②	②	①	34③	②	②	Hyd.	Hyd.
1989	8	8-350 (5.7)	FR3CLS	0.035	②	②	①	34③	②	②	Hyd.	Hyd.
1990	8	8-350 (5.7)	FR5LS	0.035	②	②	①	34③	②	②	Hyd.	Hyd.
	J	8-350 (5.7)	FR2LS	0.035	②	②	①	43③	②	②	Hyd.	Hyd.
1991	8	8-350 (5.7)	FR5LS	0.035	②	②	①	34③	②	②	Hyd.	Hyd.
	J	8-350 (5.7)	FR2LS	0.035	②	②	①	43③	②	②	Hyd.	Hyd.
1992			SEE UNDERHOOD SPECIFICATIONS STICKER									

Hyd—Hydraulic
① When checking cylinder compression pressures, the throttle should be open, all spark plugs should be removed and the battery should be near or at full charge. The lowest reading cylinder should not be less than 70% of the highest cylinder. No individual cylinder reading should be less than 100 lbs.

② Refer to Vehicle Emission Control Information label for ignition timing and idle specifications. If no specifications are shown, no adjustment is required.
③ Ignition key on, engine idling, pressure regulator vacuum hose connected, reading will vary 5 psi depending on barometric pressure.

FIRING ORDERS

NOTE: To avoid confusion, always replace spark plug wires one at a time.

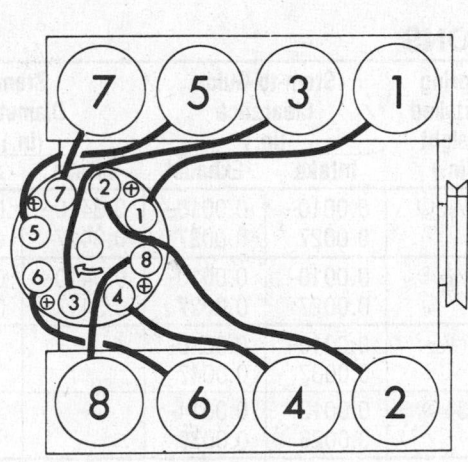

5.7L Engine (VIN 8)
Engine Firing Order: 1–8–4–3–6–5–7–2
Distributor Rotation: Clockwise

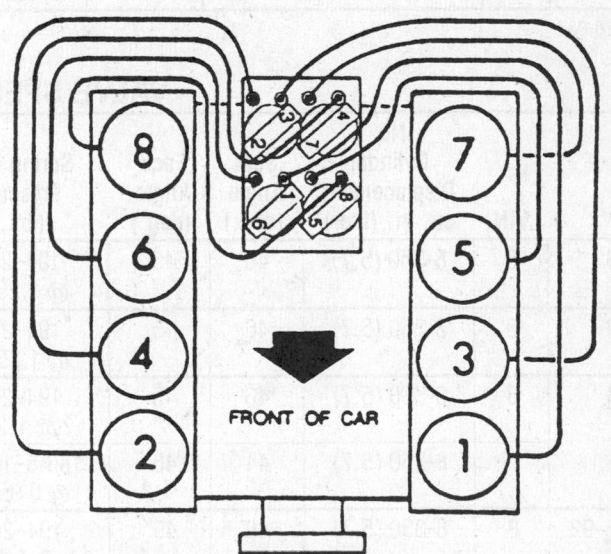

5.7L Engine (VIN J)
Engine Firing Order: 1–8–4–3–6–5–7–2
Distributorless Ignition System

CAPACITIES

Year	Model	VIN	No. Cylinder Displacement cu. in. (liter)	Engine Crankcase (qts.) with Filter	Engine Crankcase (qts.) without Filter	Transmission (pts.) 4-Spd	Transmission (pts.) 5-Spd	Transmission (pts.) Auto.	Drive Axle (pts.)	Fuel Tank (gal.)	Cooling System (qts.)
1988	Corvette	8	8-350 (5.7)	5.0	4.0	3.5①	—	10	3.75③	20	14.0
1989	Corvette	8	8-350 (5.7)	5.0	4.0	—	②	10	3.75③	20	14.0
1990	Corvette	8	8-350 (5.7)	5.0	4.0	—	②	10	3.75③	20	14.7
	Corvette ZR-1	J	8-350 (5.7)	8.6	7.6	—	②	—	3.75③	20	17.8
1991–92	Corvette	8	8-350 (5.7)	5.0	4.0	—	②	10	③	20	14.7
	Corvette ZR-1	J	8-350 (5.7)	8.6	7.6	—	②	—	③	20	17.8

① 4 speed overdrive uses Dexron®II in the overdrive section and 80WGL5 in the transmission section.
② ZF 6 speed transmission—4.4 pts.
③ Fluid level should be no lower than a ¼" (6mm) below filler plug opening.

CAMSHAFT SPECIFICATIONS

All measurements given in inches.

Year	VIN	No. Cylinder Displacement cu. in. (liter)	Journal Diameter 1	Journal Diameter 2	Journal Diameter 3	Journal Diameter 4	Journal Diameter 5	Lobe Lift In.	Lobe Lift Ex.	Bearing Clearance	Camshaft End Play
1988	8	8-350 (5.7)	1.8680–1.8690	1.8680–1.8690	1.8680–1.8690	1.8680–1.8690	1.8680–1.8690	0.2733①	0.2820①	—	0.004–0.012
1989	8	8-350 (5.7)	1.8680–1.8690	1.8680–1.8690	1.8680–1.8690	1.8680–1.8690	1.8680–1.8690	0.2733①	0.2820①	—	0.004–0.012
1990	8	8-350 (5.7)	1.8680–1.8690	1.8680–1.8690	1.8680–1.8690	1.8680–1.8690	1.8680–1.8690	0.2750①	0.2856①	—	0.004–0.012
	J	8-350 (5.7)	1.1400–1.1410	1.1400–1.1410	1.1400–1.1410	1.1400–1.1410	1.1400–1.1410	0.3898①	0.3898①	—	0.006–0.014
1991–92	8	8-350 (5.7)	1.8682–1.8692	1.8682–1.8692	1.8682–1.8692	1.8682–1.8692	1.8682–1.8692	0.2750①	0.2856①	—	0.004–0.012
	J	8-350 (5.7)	1.1400–1.1410	1.1400–1.1410	1.1400–1.1410	1.1400–1.1410	1.1400–1.1410	0.3898①	0.3898①	—	0.006–0.014

① ±0.002

VALVE SPECIFICATIONS

Year	VIN	No. Cylinder Displacement cu. in. (liter)	Seat Angle (deg.)	Face Angle (deg.)	Spring Test Pressure (lbs.)	Spring Installed Height (in.)	Stem-to-Guide Clearance (in.) Intake	Stem-to-Guide Clearance (in.) Exhaust	Stem Diameter (in.) Intake	Stem Diameter (in.) Exhaust
1988	8	8-350 (5.7)	46	45	194–206 @ 1.25①	1 23/32②	0.0010–0.0027	0.0010–0.0027	0.3410–0.3417	0.3410–0.3417
1989	8	8-350 (5.7)	46	45	194–206 @ 1.25①	1 23/32②	0.0010–0.0027	0.0010–0.0027	0.3410–0.3417	0.3410–0.3417
1990	8	8-350 (5.7)	46	45	194–206 @ 1.25	1 23/32	0.0010–0.0037	0.0010–0.0047	—	—
	J	8-350 (5.7)	44	45	146.8–166.4 @ 0.95④	1.34③	0.0012–0.0026	0.0014–0.0030	—	—
1991–92	8	8-350 (5.7)	46	45	194–206 @ 1.25	1 23/32	0.0010–0.0037	0.0010–0.0047	—	—
	J	8-350 (5.7)	44	45	146.8–166.4 @ 0.95④	1.34③	0.0012–0.0026	0.0014–0.0030	—	—

① Exhaust valve—1.16
② Exhaust—1 19/32
③ Inner spring—1.18 in.
④ Inner spring—75.5–81.8 lb. @ 0.79 in.

CRANKSHAFT AND CONNECTING ROD SPECIFICATIONS

All measurements are given in inches.

Year	VIN	No. Cylinder Displacement cu. in. (liter)	Crankshaft				Connecting Rod		
			Main Brg. Journal Dia.	Main Brg. Oil Clearance	Shaft End-play	Thrust on. No.	Journal Diameter	Oil Clearance	Side Clearance
1988	8	8-350 (5.7)	2.4484– 2.4493 ①	0.0008– 0.0020 ②	0.0020– 0.0060	5	2.0988– 2.0998	0.0013– 0.0035	0.006– 0.014
1989	8	8-350 (5.7)	2.4484– 2.4493 ①	0.0008– 0.0020 ②	0.0020– 0.0060	5	2.0988– 2.0998	0.0013– 0.0035	0.006– 0.014
1990	8	8-350 (5.7)	2.4484– 2.4493 ①	0.0008– 0.0030 ④	0.0020– 0.0060	5	2.0988– 2.0998	0.0035 ③	0.006– 0.014
	J	8-350 (5.7)	2.7550– 2.7560	0.0007– 0.0023	0.0006– 0.0010	3	2.0993– 2.1000	0.0007– 0.0027	0.008 0.028
1991–92	8	8-350 (5.7)	2.4484– 2.4493 ①	0.0008– 0.0030 ④	0.0020– 0.0060	5	2.0988– 2.0998	0.0035 ③	0.006– 0.014
	J	8-350 (5.7)	2.7550– 2.7560	0.0007– 0.0023	0.0006– 0.0010	3	2.0993– 2.1000	0.0007– 0.0027	0.008 0.028

① Specification applies to the No. 1 bearing.
Nos. 2, 3, 4—2.4481–2.4490
No. 5—2.4479–2.4488
② Specification applies to the No. 1 bearing.
Nos 2, 3, 4—0.0011–0.0023
No. 5—0.0017–0.0032
Specifications shown apply to new components
③ Maximum clearance
④ Specification applies to the No. 1 bearing.
Nos 2, 3, 4—0.0011–0.0033
No. 5—0.0017–0.0042

PISTON AND RING SPECIFICATIONS

All measurements are given in inches.

Year	VIN	No. Cylinder Displacement cu. in. (liter)	Piston Clearance	Ring Gap			Ring Side Clearance		
				Top Compression	Bottom Compression	Oil Control	Top Compression	Bottom Compression	Oil Control
1988	8	8-350 (5.7)	0.0007– 0.0017 ①	0.010– 0.020	0.013– 0.025	0.015– 0.055	0.0012– 0.0029	0.0012– 0.0029	0.002– 0.008
1989	8	8-350 (5.7)	0.0007– 0.0017 ①	0.010– 0.020	0.013– 0.017	0.010– 0.030	0.0012– 0.0029	0.0012– 0.0029	0.0012– 0.0029
1990	8	8-350 (5.7)	②	0.010– 0.030	0.013– 0.027	0.010– 0.040	0.0012– 0.0039	0.0012– 0.0039	0.0012– 0.0039
	J	8-350 (5.7)	—	0.016– 0.026	0.031– 0.039	0.012– 0.024	0.0020– 0.0030	0.0020– 0.0030	0.0010– 0.0020
1991–92	8	8-350 (5.7)	②	0.010– 0.030	0.013– 0.027	0.010– 0.040	0.0012– 0.0039	0.0012– 0.0039	0.0012– 0.0039
	J	8-350 (5.7)	—	0.016– 0.026	0.031– 0.039	0.012– 0.024	0.0020– 0.0030	0.0020– 0.0030	0.0010– 0.0020

① 0.0025 maximum
② 0.0027 maximum

TORQUE SPECIFICATIONS
All readings in ft. lbs.

Year	VIN	No. Cylinder Displacement cu. in. (liter)	Cylinder Head Bolts	Main Bearing Bolts	Rod Bearing Bolts	Crankshaft Pulley Bolts	Flywheel Bolts	Manifold Intake	Manifold Exhaust	Spark Plugs
1988	8	8-350 (5.7)	65	80	45	59–81	60	35	20	22
1989	8	8-350 (5.7)	67	80	45	70	60	35	19	22
1990	8	8-350 (5.7)	67	80	45	70	74	35②	19	20
	J	8-350 (5.7)	③	⑤	22⑥	148	74	④	22①	15
1991–92	8	8-350 (5.7)	67	80	45	70	74	35②	19	20
	J	8-350 (5.7)	③	⑤	22⑥	148	74	④	22①	15

① Manifold studs only, all others; 11 ft. lbs.
② All except Nos. 1 and 4; 1 and 4, 45 ft. lbs.
③ Torque bolts in 3 steps: 1st at 45 ft. lbs.; 2nd at 74 ft. lbs.; and final at 118 ft. lbs.
④ Injector Housing Bolts & Fuel Rail Bolts; 20 ft. lbs.
⑤ Torque bolts on No. 1, 3 and 5 to 30 ft. lbs. (40Nm) plus 45–50° turn
Torque bolts on No. 2 and 4 to 15 ft. lbs. (20Nm), plus 77.5–82.5° turn
⑥ Plus 80–85° turn

BRAKE SPECIFICATIONS
All measurements in inches unless noted

Year	Model	Lug Nut Torque (ft. lbs.)	Master Cylinder Bore	Brake Disc Minimum Thickness	Brake Disc Maximum Runout	Standard Brake Drum Diameter	Minimum Lining Thickness Front	Minimum Lining Thickness Rear
1988	Corvette	100	—	0.724	0.006	—	0.062	0.062
1989	Corvette	100	—	0.724②	0.006	—	0.062	0.062
1990	Corvette	100①	—	0.724②	0.006	—	0.062	0.062
1991–92	Corvette	100①	—	0.724②	0.006	—	0.062	0.062

① Compact spare 80 ft. lbs.
② Heavy duty—1.039

WHEEL ALIGNMENT

Year	Model		Caster Range (deg.)	Caster Preferred Setting (deg.)	Camber Range (deg.)	Camber Preferred Setting (deg.)	Toe-in (in.)	Axis Inclination (deg.)
1988	Corvette	Front	4¹¹/₁₆P–6⁵/₁₆P	5½P	⁵/₁₆P–1⁵/₁₆P	¹³/₁₆P	³/₃₂	8¾
		Rear	—	—	¹/₁₆N–²⁹/₃₂P	¹³/₃₂P	³/₃₂	—
1989	Corvette	Front	5⁵/₁₆P–6⁵/₁₆P	5¹³/₁₆P	0–1	½P	0±0.10°①	
		Rear	—	—	⁵/₁₆N–¹¹/₁₆P	³/₁₆P		—
1990	Corvette	Front	5½P–6½P	6P	0–1P	½P	0±0.10°①	
		Rear	—	—	½N–½P	0	0±0.10°①	—
1991–92	Corvette	Front	5½P–6½P	6P	0–1P	½P	0±0.10°①	
		Rear	—	—	½N–½P	0	0±0.10°①	—

N Negative
P Positive
① In degrees

ENGINE MECHANICAL

NOTE: Disconnecting the negative battery cable on some vehicles may interfere with the functions of the on board computer systems and may require the computer to undergo a relearning process.

Engine Assembly

REMOVAL & INSTALLATION

5.7L (VIN 8) Engine

1. Mark the relationship between each hood hinge and the hood. Remove the hood.
2. Disconnect the negative battery cable.
3. Drain the coolant into a suitable container.
4. Disconnect the throttle, transmission and cruise control cables at the engine.
5. Remove the plenum extension. Disconnect the spark plug wires from the plugs. Remove the wires and distributor cap as an assembly.
6. Remove the distributor from the engine. Remove the EGR Pipe from the intake and exhaust manifold, if required.
7. Remove the cowl screen and the nut from the wiper motor arm, if required.
8. Disconnect the wiper motor wires and remove the wiper motor cover. Remove the wiper motor, if required.
9. Disconnect the oil pressure switches, if required. Remove the air intake duct.
10. Disconnect the brake booster vacuum hose.
11. Disconnect the canister hose at the PCV pipe.
12. Disconnect all necessary wiring and vacuum hoses from the engine.
13. Disconnect the fuel injection harness at the intake manifold.
14. Disconnect the heater hoses at the pipe.
15. Disconnect the upper radiator hose at the thermostat housing.
16. Remove the colant pump damper and remove the serpentine belt and coolant pump pulley.
17. Remove the AIR control valve with the bracket attached at the air conditioning compressor.
18. Relieve the fuel system pressure.
19. Disconnect and plug the fuel lines at the rail.
20. Disconnect the catalytic converter AIR pipe. Remove the right side

wheelhouse lower center panel, if required.
21. Remove the air conditioning compressor braces.
22. Remove the accumulator at the fan shroud and brace.
23. Disconnect the fuel lines at the block.
24. Disconnect the lower radiator hose and the heater hose from the water pump. Disconnect heater hose from the oil cooler pipe, if equipped.
25. Remove the alternator.
26. Remove the AIR pump with the bracket.
27. Remove the power steering pump and reservoir and wire it aside.
28. Remove the water pump pulley, the water pump and the crankshaft pulley.
29. Raise and safely support the vehicle.
30. Disconnect the wires harness connection at the oxygen, Electronic Spark Control (ESC) system harness and temperature sensors.
31. Remove the temperature sensor wire retainer at the block.
32. Disconnect the ground wires at the engine.
33. Disconnect the transmission oil cooler lines at the transmission, if equipped.
34. Remove the starter and flywheel cover.
35. Disconnect the exhaust system.
36. Drain the engine oil into a suitable container.
37. Remove the oil filter.
38. Remove the oil cooler adapter and lines at the block.
39. Disconnect the exhaust system at the converter hanger.
40. Disconnect the clutch system, if equipped with a manual transmission.
41. Remove the engine mount through bolts and nuts.
42. Remove the transmission to engine bolts. Remove the torque converter to flywheel bolts, if equipped with an automatic transmission.
43. Lower the vehicle.
44. Support the transmission with a transmission jack.
45. Install a suitable lifting device and remove the engine from the vehicle.

To install:
46. Using a suitable lifting device, install the engine.
47. Remove the lifting device, then raise and support vehicle.
48. Support the transmission with a suitable jack.
49. Install the engine mount through bolts.
50. Install the engine to transmission attaching bolts.
51. If equipped with automatic transmission, install the torque converter to flywheel bolts.

52. Install the exhaust system, if removed.
53. Connect the oil cooler line at the oil pan.
54. Install the starter, then the transmission cooler lines and the flywheel cover.
55. Connect the catalytic converter AIR pipe at the manifold.
56. Connect the ground wire to the engine block.
57. Connect the temperature sensor, oxygen sensor and ESC electrical connectors.
58. Lower vehicle, then install the crankshaft pulley and water pump and pulley.
59. Install the power steering pump, then the power steering pump reservoir at the fan shroud.
60. Install the alternator, then connect the heater hoses to the water pump.
61. Connect fuel lines to engine, then install the accumulator.
62. Install the air conditioning compressor and brackets.
63. Install the catalytic converter AIR pipe, then the AIR control valve.
64. Connect fuel lines at the fuel rail.
65. Install the serpentine belt, then the radiator upper hose.
66. Connect the heater hoses and the injector wire harness.
67. Connect all vacuum hoses and wires connectors previously removed.
68. Connect the canister hose to the PCV, then the brake booster hose.
69. Install the wiper motor and cowl screen, if removed.
70. Install the distributor, the cap and wires.
71. Connect the throttle, transmission and cruise control cables.
72. Fill crankcase with oil, install a new oil filter. Fill the cooling system with the proper type and quantity of antifreeze.
73. Connect the battery negative cable. Start engine and check for leaks.

5.7L (VIN J) Engine

1. Mark the relationship between each hood hinge and the hood. Remove the hood.
2. Disconnect the battery negative cable, then relieve fuel system pressure.
3. Raise and support the vehicle safely.
4. Drain engine coolant into a suitable container.
5. Drain the engine oil.
6. Remove the exhaust system, then the driveshaft.
7. Position a suitable transmission support stand under transmission and remove the transmission support beam.
8. Remove transmission from vehicle.

9. Remove the clutch actuator cylinder, left side converter shield, clutch housing cover, then the clutch cover and disc.

10. Install a suitable engine lift hook to rear of engine.

11. Remove the AIR tube center section from the AIR hose and oil pan.

12. Disconnect oxygen sensor electrical connectors.

13. Remove the power steering lower hose from the oil cooler.

14. Remove the battery negative cable from the cylinder case.

15. Remove the nuts attaching the engine mounts to the driveline and the frame.

16. Lower the vehicle.

17. Remove the air cleaner assembly and air duct.

18. Disconnect the engine oil cooler lines from the oil filter housing.

19. Raise the rear of the engine.

20. Loosen fuel tank filler cap, then release fuel system pressure.

22. Disconnect the fuel lines from the fuel rail.

23. Remove the evaporator housing panel and the resistor.

24. Remove the bolts attaching the right bulkhead connector. Remove the engine right side wiring harness.

25. Remove the instrument panel right lower sound insulator panel.

26. Disconnect the bulkhead wiring harness connectors from under the dash.

27. Remove the air bleed hose from the plenum.

28. Remove the radiator upper and lower hoses, then disconnect the vacuum lines attached to power steering pump.

29. Properly discharge the air conditioning system.

30. Remove the air conditioning suction and discharge lines flange from the compressor.

31. Remove the air conditioning compressor to accumulator line from the accumulator.

32. Remove the air conditioning accumulator and position aside.

33. Remove the air conditioning accumulator bracket from the vehicle.

34. Disconnect and plug the power steering pressure line at the power steering gear.

35. Disconnect the throttle body linkage shield, then remove the throttle body cable to plenum retainers.

36. Disconnect the accelerator and cruise control cables from the throttle body.

37. Install a suitable engine lift hook to front of the engine.

38. Remove the ECM from the ECM bracket, then disconnect ECM harness connectors.

39. Remove the left front fender attaching bolts, shims and seal. Remove the left fender.

40. Remove the positive cable from the battery, battery hold-down clamp, then the battery from the vehicle.

41. Disconnect the engine left side bulkhead block electrical connector.

42. Disconnect the engine wiring harness fusible links at the junction block.

43. Disconnect the engine harness connectors from the following:
    Secondary injector modules
    Battery positive cable at junction block
    Differential pressure switch vacuum and electrical connectors
    Air conditioning cutout relay
    Air conditioning high blower relay
    Transmission shift solenoid relay
    Fuel pump fuse
    Forward light link connector
    Battery positive lead
    Air conditioning blower resistor
    Air conditioning pressure sensor
    Air conditioning cooling fan switch
    Windshield washer pump
    Low coolant sensor
    Blower motor
    ESC knock sensor
    ESC knock sensor relay

44. Disconnect hoses from the secondary port throttle vacuum pump.

45. Disconnect the front and rear vacuum connections. Reposition engine harness aside.

46. Remove the braided ground strap from the left side frame rail. Reposition the positive battery cable aside.

47. Remove the left side plenum panel screen.

48. Disconnect the brake booster vacuum hose.

49. Remove the windshield wiper motor from the vehicle.

50. Disconnect the MAP electrical connector, then remove the MAP sensor bracket from the plenum.

51. Disconnect the AIR hose from the left exhaust manifold.

52. Install a suitable lifting device and remove the engine from the vehicle.

To install:
53. Install the engine mounts.

54. Using a suitable lifting device, install the engine into vehicle.

55. Install the engine mount/bracket bolts, then remove the the lifting device and the lifting brackets.

56. Connect the AIR hose to the left exhaust manifold.

57. Install the MAP sensor and MAP sensor bracket.

58. Install the wiper motor.

59. Install the left side plenum panel screen.

60. Route the left side wiring harness into position, then install the braided ground strap to the frame rail.

61. Connect the left side bulkhead block connector.

62. Connect the engine harness fusible links and relays.

63. Install the battery and hold-down clamps.

64. Connect the battery positive cable, then install the left front fender.

65. Install the ECM to the ECM bracket, then connect the ECM electrical connector.

66. Connect power brake booster vacuum hose to the plenum.

67. Connect the cruise control and throttle cables to the throttle body.

68. Install cable shield.

69. Install cable retainers to the plenum.

70. Install the power steering pressure line to the power steering gear.

71. Connect the engine oil cooler lines to the engine.

72. Install the accumulator to the accumulator bracket.

73. Connect the air conditioning lines.

74. Attach the vacuum lines to the power steering pump.

75. Connect the radiator upper and lower hoses.

76. Connect the air bleed hose to the plenum.

77. Connect the bulkhead wire connector to the bulkhead.

78. Connect the evaporator housing panel resistor electrical connector.

79. Install secondary port throttle hose onto the vacuum pump, then the front and rear vacuum connections.

80. Connect the engine harness connectors to the following:

    Air conditioning blower resistor
    Air conditioning pressure sensor
    Air conditioning cooling fan
    Windshield washer pump
    Low coolant sensor
    Blower motor
    ESC knock sensor
    ESC knock sensor relay
    Differential pressure switch

81. Connect the fuel lines to the fuel rail.

82. Install the engine right side wiring harness under the dash.

83. Install the instrument panel right sound insulator panel.

84. Raise and safely support the vehicle, then install the engine/bracket bolts and nuts. Torque to 40 ft. lbs. (54 Nm).

85. Lower the vehicle and install the power steering hose to power steering oil cooler.

86. Install the oxygen sensor wire connectors.

87. Connect the AIR tube center section to the AIR hose and oil pan.

88. Connect the negative battery cable, then remove the engine rear lift hook.

89. Raise and safely support the vehicle, then install the clutch assembly and housing.

90. Install transmission and support beam, then the driveshaft.
91. Install the exhaust system.
92. Lower the vehicle and fill the cooling system with the proper type and quantity of antifreeze.
93. Add the proper quantity and type of engine oil to the crankcase.
94. Recharge the air conditioning system, then reset the change oil monitor if the oil was drained during removal.
95. Start the vehicle and inspect for leaks.

Engine Mounts

REMOVAL & INSTALLATION

1988

1. Disconnect the negative battery cable, then raise and support vehicle.
2. Remove engine mount through bolt.
3. Disconnect the AIR pipe at the exhaust manifold, converter and exhaust pipe.
4. Raise the engine enough to gain sufficient clearance.
5. Remove 1 exhaust manifold shroud screw, if replacing the left mount.
6. Remove engine mount to block bolts.
7. Remove the engine mounts.

To install:

8. Replace mount to engine and lower engine into place.
9. Install and tighten the retaining bolts.
10. When installing the left engine mount, install the exhaust manifold shroud screw. Connect the battery negative cable.

1989–92

5.7L (VIN 8) ENGINE

1. Disconnect battery negative cable.
2. Remove the air intake duct from the air cleaner assembly.
3. Raise and support the vehicle safely.
4. When removing the right engine mount, remove the Electronic Spark Control (ESC) sensor shield.
5. Remove the engine mount through bolts and nuts.
6. Raise engine slightly for sufficient clearance, then remove engine mount bolts and engine mount from vehicle.
7. Installation is the reverse of the removal procedure. Torque engine mount bolts and through bolts to 40 ft. lbs. (54 Nm) Connect the battery negative cable.

5.7L (VIN J) ENGINE

1. Disconnect the battery negative cable, then remove the right and left exhaust manifold.
2. Remove the nut attaching the engine mount to drivetrain and frame.

3. Raise the engine slightly and safely support it to allow for sufficient clearance to the engine mounts.
4. Remove the engine mount bracket nut and bolt from bracket.
5. Remove the engine mount and heat shield from the vehicle.

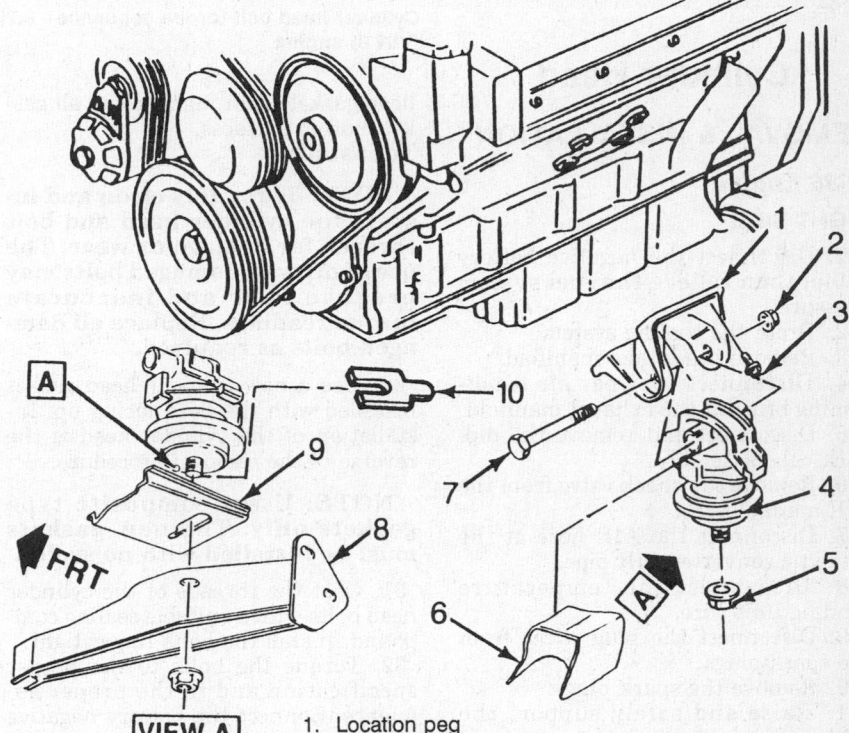

1. Location peg	6. Engine mount heat shield
2. Engine mount bracket	7. Engine mount through bolt
3. Engine mount bracket bolt	8. Front side member
4. Engine hydraulic mount	9. Frame
5. Engine mount nut	10. Engine mount spacer

Engine mount removal—5.7L (VIN J) engine

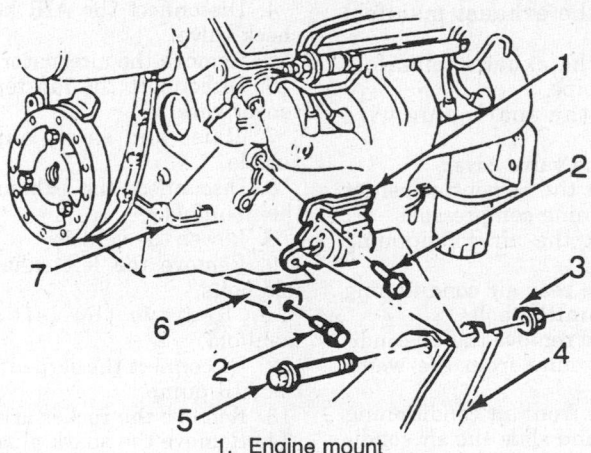

1. Engine mount
2. Bolt
3. Engine mount nut
4. Frame
5. Engine mount bolt
6. Support brace
7. Engine block

Engine mount removal—5.7L (VIN 8) engine

6. Remove the bolts attaching the bracket to the cylinder case, then the bracket from the vehicle.

7. Reverse procedure to install. Torque engine mount bracket bolts to 38 ft. lbs. (52 Nm). Torque engine mount nuts and bolts to 40 ft. lbs. (54 Nm). Connect the battery negative cable.

Cylinder Head

REMOVAL & INSTALLATION

1988 Engine

RIGHT SIDE

1. Disconnect the negative battery cable, then relieve the fuel system pressure.
2. Drain the cooling system.
3. Remove the intake manifold.
4. Disconnect the rear air conditioning brace at the exhaust manifold.
5. Disconnect and remove the dipstick tube assembly.
6. Remove the check valve from the AIR manifold.
7. Disconnect the AIR hose at the catalytic converter AIR pipe.
8. Disconnect the temperature sending unit wire.
9. Disconnect the plug wires from the spark plugs.
10. Remove the spark plugs.
11. Raise and safely support the vehicle.
12. Disconnect the converter AIR pipe clamp at the manifold.
13. Disconnect the exhaust pipe at both manifolds.
14. Remove the front catalytic converter hanger bolts.
15. Remove the converter AIR pipe.
16. Lower the vehicle.
17. Remove the exhaust manifold bolts.
18. Remove the exhaust manifold with the EGR pipe.
19. Remove the spark plug wire retainers.
20. Remove the valve cover.
21. Disconnect the serpentine belt at the air conditioning compressor.
22. Disconnect the air conditioning wire connectors.
23. Loosen the rear air conditioning compressor mounting bolts.
24. Loosen and remove the air conditioning bracket nuts from the water pump studs.
25. Loosen the front air conditioning mounting bolt and slide the air conditioning compressor unit with bracket forward.
26. Remove the pushrods.
27. Loosen and remove the cylinder head bolts.
28. Remove the right cylinder head.
29. Remove and discard the cylinder

Cylinder head bolt torque sequence – 5.7L (VIN 8) engine

head gasket. Thoroughly clean all gasket mating surfaces.
To install:

NOTE: Throughly clean and inspect the cylinder head and bolt threads for damage or wear. The use of dirty or damaged bolts may produce false and inaccurate torque readings. Replace all damaged bolts as required.

30. Use a new cylinder head gasket installed with the bead facing up. Installation of the cylinder head is the reverse of the removal procedure.

NOTE: Use a composite type gaskets only. The new gaskets must be installed with no sealer.

31. Coat the threads of the cylinder head bolts with a suitable sealing compound. Install the bolts finger-tight.
32. Torque the bolts to the proper specification and in the proper sequence. Connect the battery negative cable.

LEFT SIDE

1. Disconnect the negative battery cable, then relieve the fuel system pressure.
2. Drain the cooling system into a suitable container.
3. Remove the intake manifold.
4. Disconnect the AIR hose at the check valve.
5. Remove the alternator brace.
6. Disconnect the fan temperature sensor wire.
7. Raise and safely support the vehicle.
8. Disconnect the exhaust pipe at the manifold.
9. Lower the vehicle.
10. Remove the left exhaust manifold bolts.
11. Remove the left exhaust manifold.
12. Disconnect the serpentine belt at the AIR pump.
13. Remove the rocker arm cover.
14. Remove the spark plugs.
15. Disconnect the power steering/alternator mounting bracket at the cylinder head.
16. Remove the pushrods.
17. Loosen and remove the left cylinder head bolts. Remove the left cylinder head from the engine block.

18. Remove the cylinder head gasket and discard. Thoroughly clean all gasket mating surfaces.
To install:

NOTE: Throughly clean and inspect the cylinder head and bolt threads for damage or wear. The use of dirty or damaged bolts may produce false and inaccurate torque readings. Replace all damaged bolts as required.

19. Use a new cylinder head gasket installed with the bead facing up. Installation of the cylinder head is the reverse of the removal procedure.

NOTE: Use a composite type head gasket only. The new composite gasket must be installed with no sealer.

20. Coat the threads of the cylinder head bolts with a suitable sealing compound. Install the bolts finger-tight.
21. Torque bolts in the specified pattern to 65 ft. lbs. (88 Nm).

1989–92 5.7L (VIN 8) Engine

RIGHT SIDE

1. Disconnect battery negative cable, then relieve the fuel system pressure.
2. Drain the cooling system into a suitable container. Remove the intake manifold.
3. Remove the right exhaust manifold.
4. Remove the air conditioning compressor mounting bracket attaching bolts, then the mounting bracket.
5. Remove the valve cover, rocker arms and pushrods.
6. Remove the cylinder head bolts, washers and the cylinder head.
7. Installation is the reverse of the removal procedure. Torque cylinder head bolts 67 ft. lbs. (91 Nm). Connect battery negative cable and inspect the engine for leaks.

LEFT SIDE

1. Disconnect battery negative cable and relieve fuel system pressure.
2. Remove the left exhaust manifold.
3. Remove the alternator.
4. Remove the 2 bolts and 1 nut attaching the AIR pump bracket to the head. Position pump and bracket assembly aside.
5. Remove the bolt attaching the power steering pump bracket to the cylinder head.
6. Remove the valve cover, rocker arms and pushrods.
7. Remove the cylinder head bolts, washers and the cylinder head.
8. Installation is the reverse of the removal procedure. Torque cylinder head bolts 67 ft. lbs. (97 Nm). Connect

the battery negative cable and check for leaks.

1990–92 5.7L (VIN J) Engine

RIGHT SIDE

1. Disconnect the battery negative cable.
2. Drain engine coolant into a suitable container.
3. Relieve fuel system pressure.
4. Remove plenum assembly.
5. Disconnect the fuel lines from the right fuel rail.
6. Disconnect electrical connectors from the fuel injectors.
7. Remove the bolts attaching the fuel rail assembly to the injector housing.
8. Remove the injectors from the housing, then the fuel rail assembly from the vehicle.
9. Remove and clamp the hose from the right coolant outlet pipe.
10. Remove the oil pressure sensor from the oil filter housing.
11. Remove the bolt attaching the outlet pipe bracket to the alternator bracket.
12. Remove the screws attaching the outlet pipe to the injector housing, then remove the outlet pipe and gasket.
13. Remove PCV grommet from the injector housing. Plug the ventilation hose from the injector housing.
14. Remove the bolt attaching the alternator rear support bracket to the alternator.
15. Remove the bolt attaching the alternator rear support bracket and right side ventilation pipe to the injector housing.
16. Remove the ventilation pipe and bracket from the vehicle.
17. Remove the bolts attaching the injector housing to the cylinder head.
18. Remove the injector housing and gasket from the vehicle.
19. Remove the right bank camshafts and valve lifters.
20. Remove the alternator assembly.
21. Remove the right exhaust manifold, if necessary.
22. Remove the fuel filter heat shield.
23. Remove the vacuum hose from secondary port throttle valve actuator.
24. Remove the access plug from the right cylinder head.
25. Remove the bolt attaching the right secondary timing chain guide.
26. Remove cylinder head bolts and remove the cylinder head and gasket from the vehicle.

To install:

27. Install the cylinder head locating dowels into block, if loosened or removed.
28. Install the cylinder head gasket, head washers and bolts. Coat bolt threads with engine oil.

NOTE: **Cylinder head gaskets are not interchangeable between cylinder banks.**

29. Torque cylinder head bolts in sequence as follows:
    1st pass–45 ft. lbs. (60 Nm)
    2nd pass–74 ft. lbs. (100 Nm)
    3rd pass–118 ft. lbs. (160 Nm)
30. Install the fixed guide bolt. Apply Loctite® to bolt threads and torque to 20 ft. lbs. (26 Nm).
31. Install the access plug into the cylinder head and torque to 15 ft. lbs. (20 Nm).
32. Connect the vacuum hose to the actuator.
33. Raise and support vehicle, drain engine oil and lower the vehicle.
34. Install the fuel filter heat shield, then the exhaust manifold.
35. Install the alternator.
36. Install valve lifters and camshafts, then the injector housing.
37. Install the intake manifold assembly.
38. Fill crankcase with engine oil and connect the battery negative cable.

LEFT SIDE

1. Disconnect battery negative cable.
2. Drain engine coolant into a suitable container.
3. Relieve fuel system pressure.
4. Remove plenum assembly.
5. Disconnect fuel lines from the left fuel rail.
6. Disconnect electrical connectors from the injectors.
7. Remove the bolts attaching the fuel rail assembly to the injector housing.
8. Remove the injectors from the housing, then the fuel rail assembly from the vehicle.

9. Remove and plug hose from left coolant outlet pipe.
10. Remove the bolt attaching the outlet pipe bracket to the power steering pump bracket.
11. Remove bolts attaching the outlet pipe to the injector housing, then remove the outlet pipe and gasket from vehicle.
12. Remove PCV grommet from injector housing.
13. Disconnect and plug ventilation hose from injector housing.
14. Disconnect the coolant temperature sensor and cooling fan switch electrical connectors.
15. Remove bolts attaching the injector housing to the cylinder head, then remove the injector housing and gasket from vehicle.
16. Remove the vacuum hose from the secondary port throttle valve actuator.
17. Remove the power brake booster assembly.
18. Remove the left bank valve lifters and camshafts.
19. Remove the AIR control valve hoses, then disconnect the electrical connectors.
20. Disconnect the camshaft sensor electrical connector.
21. Remove the left exhaust manifold, if necessary.
22. Remove the access plug from the left cylinder head.
23. Remove the bolt attaching the left secondary timing chain guide.
24. Remove the cylinder head bolts. Remove the cylinder head and gasket from the vehicle.

To install:

25. Install cylinder head locating dowels into cylinder head, if loosened or removed during removal.
26. Install cylinder head gasket, head, washers and bolts. Coat bolt threads with engine oil.

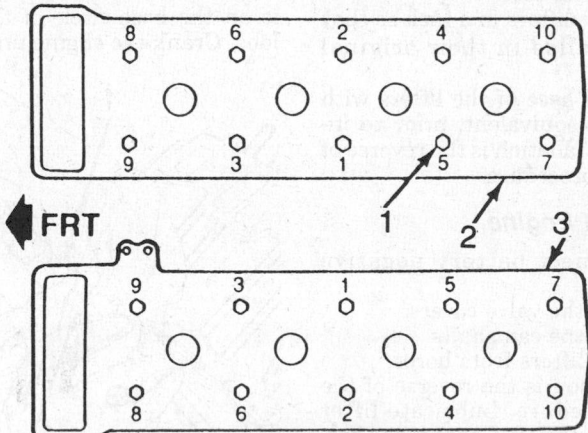

1. Cylinder head bolt
2. Right cylinder head
3. Left cylinder head

Cylinder head bolt torque sequence—5.7L (VIN J) engine

27. Torque cylinder head bolts in sequence as follows:
    1st pass—45 ft. lbs. (60 Nm)
    2nd pass—74 ft. lbs. (100 Nm)
    3rd pass—118 ft. lbs. (160 Nm)
28. Install fixed guide bolt. Coat bolt threads with Loctite®. Torque bolt to 20 ft. lbs. (26 Nm).
29. Install access plug into cylinder head. Torque plug to 15 ft. lbs. (20 Nm).
30. Connect vacuum hose to actuator.
31. Raise and support vehicle, then drain engine oil. Lower vehicle.
32. Install exhaust manifold, then connect the camshaft position sensor electrical connector.
33. Connect the AIR control valve hoses and electrical connectors.
34. Install the valve lifters and camshaft, then the injector housing.
35. Fill engine crankcase with oil. Reset the change oil monitor if oil was drained during removal. Connect battery and check for leaks.

Valve Lifters

REMOVAL & INSTALLATION

Except 5.7L (VIN J) Engine

1. Disconnect the negative battery cable.
2. Drain the cooling system.
3. Remove the intake manifold assembly.
4. Remove the rocker arm covers.
5. Remove the rocker arms and pushrods. Be sure to keep them in order as they must be installed in the same bores as they were removed.
6. As required, remove the valve lifter guide retaining bolts. Remove the valve lifter guide, if equipped with roller lifters.
7. Remove the valve lifters using a suitable removal tool.
8. Place the lifters in a rack so they may be installed in their original locations.
9. Coat the base of the lifters with Molycoat® or equivalent, prior to installation. Installation is the reverse of the removal procedure.

5.7L (VIN J) Engine

1. Disconnect battery negative cable.
2. Remove the valve covers.
3. Remove the camshafts.
4. Remove lifters from bores.
5. Installation is the reverse of the removal procedure. Lubricate lifter bores with engine oil. Connect the battery negative cable

NOTE: Ensure that the lifters, that are to be reused, are retained in proper order so each lifter can

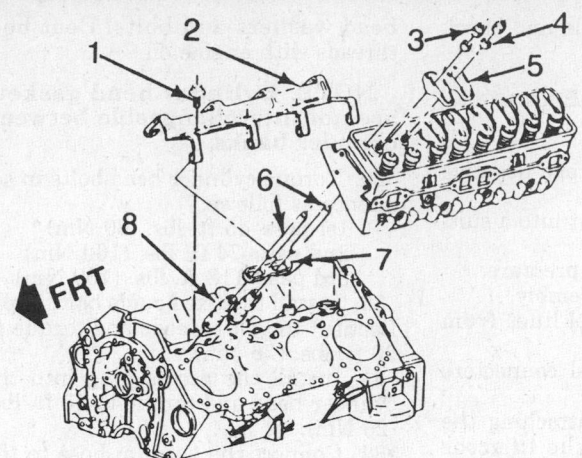

Rocker arms and pushrod assembly—5.7L (VIN 8) engine

1. Retainer bolt
2. Valve lifter restrictor retainer
3. Valve rocker arm adjustment nut
4. Rocker arm ball
5. Valve rocker arm
6. Pushrod
7. Valve lifter guide
8. Lifter

be returned to its original bore. Pre-oil any new lifter being installed.

Valve Lash

ADJUSTMENT

Except 5.7L (VIN J) Engine

NOTE: The 5.7L (VIN 8) engine utilizes hydraulic lifters that normally require very little maintenance or adjustment. These components are simple in design and are best maintained through regular, scheduled engine oil changes. If the engine is running well and no audible clicking sounds are heard from the valve train, do not attempt to remove or disassemble the valve lifters.

1. Disconnect the negative battery cable.
2. Remove the valve covers.
3. Tighten the rocker arm nuts until all lash is eliminated.
4. Adjust the valves when the lifter is on the base circle of the camshaft lobe. Crank the engine until the mark

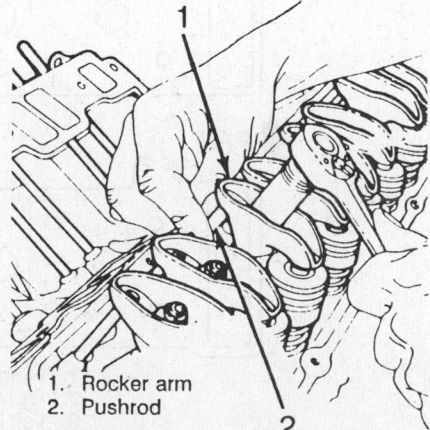

1. Rocker arm
2. Pushrod

Valve adjusting—5.7L (VIN 8) engine

on the vibration damper aligns with the **0** mark on the timing chain cover and the engine is in the No. 1 firing position.

NOTE: This may be determined by placing a finger over the No. 1 spark plug hole as the mark on the damper comes near the 0 mark on the crankcase front cover. If both the intake and exhaust valves are closed as the mark comes up to the timing tab, the engine is in the No. 1 firing position.

If either valve is open, the engine is in No. 6 firing position and should be turned over one full revolution in order to reach the No. 1 firing position.

5. With the engine in the No. 1 firing position, adjust the following valves:
    a. Exhaust—1, 3, 4, 8
    b. Intake—1, 2, 5, 7
6. Back out adjusting nut until lash is felt at the pushrod then turn in adjusting nut until all lash is removed. This can be determined by rotating pushrod while turning adjusting nut. When play has been removed, the pushrod will not turn. Then, turn the adjusting nut a full additional turn.
7. Crank the engine one revolution until the **0** mark and the vibration damper mark are again in alignment. This is the No. 6 firing position.
8. With the engine in this position, adjust the following valves:
    a. Exhaust—2, 5, 6, 7
    b. Intake—3, 4, 6, 8
9. Install the rocker arm covers.
10. Connect the battery negative cable, then start the engine and adjust the idle speed as required.

5.7L (VIN J) Engine

This engine is equipped with hydraulic lifters which maintain zero lash between the camshaft lobes and the valve stem. The lifter is non-adjustable and upon failure, the lifter can only be replaced.

Rocker Arms

REMOVAL & INSTALLATION

5.7L (VIN 8) Engine

1. Disconnect the negative battery cable. Remove the air cleaner.
2. Remove the right rocker arm cover as follows:
   a. Remove the EGR pipe assembly.
   b. Remove the crankcase vent pipe.
   c. Loosen the spark plug wire retainer on the right cylinder head and remove the remaining wire retainers.
   d. Remove the injector harness retaining nuts and position the harness off to the side.
   e. On 1988 vehicles, disconnect the heater control valve harness vacuum hose.
   f. On 1989–92 vehicles, remove the engine cooling air pipe assembly. Disconnect the heater core to plenum coolant hose and AIR hoses from the control valve.
   g. On 1989–92 vehicles, remove the bolts attaching the air conditioning compressor and position aside.
   h. Remove the rocker arm retaining bolts. Remove the rocker arm cover and gasket. Replace the gasket as required.
3. Remove the left rocker arm cover as follows:
   a. Disconnect the PCV valve and hose.
   b. Disconnect the injector harness and position to the side.
   c. On 1989–92 vehicles, remove spark plug wire retainers, then the spark plug wires. Remove the serpentine belt from the AIR pump.
   d. On 1988 vehicles, disconnect the canister hose at the purge pipe.
   e. On 1988 vehicles, disconnect the power brake booster vacuum line. Disconnect the brake booster pipe at the plenum.
   f. Loosen the AIR pump pulley bolts and disconnect the serpentine belt from the AIR pump.
   g. Remove the AIR pump pulley and loosen the AIR pump lower mounting bolt.
   h. Remove the rocker cover retaining bolts and remove the cover and cover gasket.
4. Remove the rocker arm nuts, rocker arm balls and the rocker arms. Identify the rocker arm nuts, rocker arm balls, rocker arms and pushrods so they may be installed in their original locations. If new rocker arms and/

or rocker arm balls are being installed, place a coat of Molycoat® or equivalent, onto the bearing surfaces prior to installation.

To install:
5. Install the pushrods making certain they seat in the lifter sockets.
6. Install the rocker arms, rocker arm balls and rocker arm nuts in their original positions.
7. Tighten the rocker arm nuts until all lash is eliminated.
8. Adjust the valves and the air conditioning belt tension.
9. Reinstall the rocker arm covers and connect the battery negative cable.
10. Start the engine and inspect for leaks. Check and adjust the curb idle speed.

5.7L (VIN J) Engine

This engine utilizes an overhead cam design, thus eliminating the need for any rocker arm assembly. This design also improves and smoothens engine operation.

Intake Manifold

REMOVAL & INSTALLATION

5.7L (VIN 8) Engine

1988
1. Disconnect the negative battery cable.
2. Relieve the fuel system pressure.
3. Drain the cooling system.
4. Remove the fuel injection subassembly: mass air flow sensor, plenum, runners and the fuel rail assembly.
5. Disconnect and mark all necessary vacuum and electrical connections.
6. Remove the distributor cap, mark the position of the rotor and the distributor. Remove the distributor.
7. Disconnect the upper radiator hose at the thermostat outlet opening.
8. Remove the air pump brace.
9. Disconnect the EGR pipe at the inlet opening.
10. Disconnect the heater control vacuum line at the intake.
11. Remove the thermostat outlet.
12. Remove the intake manifold retaining bolts.
13. Lift the intake manifold upward and away from the intake surface.
14. Remove the gaskets from the cylinder head surfaces.
To install:
15. Thoroughly clean the cylinder block, intake manifold and cylinder head surfaces and remove any traces of gasket material and RTV sealant. Any material left on these surfaces will cause installation interference and improper sealing.

16. Install the cylinder head gaskets so the blocked openings are positioned toward the rear of the engine. Locate the gasket tabs and bend the tabs so they are flush with the rear surface of the cylinder head. After the tabs are bent into place, apply a $^{3}/_{16}$ bead of RTV onto the front and rear cylinder case ridges.
17. Apply Loctite to the threads of the intake manifold bolts.
18. Install the intake manifold in the reverse of the removal procedure. Torque the intake manifold retaining bolts in sequence to 35 ft. lbs. (47 Nm). Connect battery negative cable.

1989–92

1. Disconnect battery negative cable.
2. Relieve system fuel pressure.
3. Drain engine coolant into a suitable container.
4. Disconnect the throttle and cruise control cables from the throttle body and cable bracket.
5. Disconnect the transmission cable from the throttle body and bracket.
6. Remove the bracket from the plenum and position aside.
7. Disconnect the following electrical connectors:
   Throttle position sensor
   Idle air control valve
   Mass air flow sensor (1989)
   Coolant temperature sensor (1990–92)
   Manifold absolute pressure sensor (1990–92)
   Manifold air temperature sensor (1990–92)
8. Remove the air intake duct from the throttle body.
9. Disconnect vacuum hoses from the throttle body and the plenum.
10. Disconnect the heater hoses from the throttle body.

INITIAL TIGHTENING SEQUENCE

FINAL TIGHTENING SEQUENCE
25–45 FT. LBS.

Intake manifold bolt torque sequence—1988 vehicles

11. Remove the power brake vacuum booster fitting from the plenum.

12. Remove the runner to plenum bolts, then remove the plenum from the vehicle.

13. On 1990–92 vehicles, remove the vacuum harness from the EGR solenoid.

14. On 1990–92 vehicles, remove the EGR solenoid assembly from the coolant outlet, then the EGR valve from the intake manifold.

15. Remove the injector harness attaching bolts, then disconnect the injector harness connectors.

16. Remove the runner to manifold attaching bolts, then the runner.

17. Disconnect the fuel lines.

18. Remove the fuel rail and injector assembly.

19. Remove the distributor.

20. Remove the radiator upper hose, then the AIR pump brace.

21. Remove the EGR valve pipe and position aside.

22. Remove the PCV valve hose from the manifold.

23. Remove the crankshaft vent tube from the manifold.

24. Remove the intake manifold attaching bolts, then the intake manifold and gaskets.

To install:

25. Thoroughly clean the cylinder block, intake manifold and cylinder head surfaces with the proper cleaning compound to remove any traces of gasket material and RTV sealant. Any material left on these surfaces will cause installation interference and improper sealing.

26. Install the manifold gaskets on the cylinder head so the blocked openings are positioned at the front of the engine. Locate the gasket tabs and bend the tabs so they are flush with the rear surface of the cylinder head. After the tabs are bent into place, apply a $^3/_{16}$ bead of RTV onto the front and rear cylinder case ridges.

27. Apply Loctite to the threads of the intake manifold retaining bolts.

28. Install the intake manifold in the reverse order of the removal procedure. Torque the intake manifold retaining bolts in sequence to 35 ft. lbs. (47 Nm), except positions 1 and 4. Torque positions 1 and 4 to 45 ft. lbs. (61 Nm). Connect the battery negative cable.

5.7L (VIN J) Engine

1. Disconnect the battery negative cable.

2. Drain the engine coolant into a suitable container.

3. Relieve the fuel system pressure.

4. Remove the air intake duct and plenum assembly as follows:

   a. Remove the throttle, cruise control and transmission cables.

   b. Remove the fresh air hose from the left and right side of the throttle body extension.

   c. Remove all electrical connector at plenum assembly.

   d. Remove the coolant air bleed hose from the plenum.

   e. Disconnect all vacuum hose at plenum assembly.

   f. Disconnect the fuel lines from the fuel rail assembly.

   g. Remove the bolts attaching the plenum assembly to the injector housing.

   h. Raise the plenum assembly and disconnect the electrical connectors.

   i. Disconnect the purge solenoid and PCV valve hose fitting from the plenum.

   j. Remove the plenum assembly.

5. Disconnect the electrical connectors from the fuel injectors.

6. Remove the bolts attaching the fuel rail assembly to the injector housing.

7. Remove the injectors from the housing, then the fuel rail assembly from the vehicle.

8. Disconnect and plug the hose from the coolant outlet pipe.

9. Remove the oil pressure sensor from the oil filter housing.

10. Remove the bolt attaching the outlet pipe to the injector housing. Remove the outlet pipe and gasket from the vehicle.

11. Remove the PCV grommet from the injector housing.

12. Remove the clamp and ventilation hose from the injector housing.

13. Remove the bolt attaching the alternator rear support bracket to the alternator.

14. Remove the bolt attaching the alternator rear support bracket and right side ventilation pipe to the injector housing.

15. Remove the ventilation pipe and bracket from the vehicle. Disconnect the temperature sensor and fan switch wire connector.

16. Remove the injector housings attaching bolts. Remove the injector housings and gasket from the vehicle.

To install:

17. Install injector housings, gasket, alternator bracket, ventilation pipe and bolts. Torque bolts to 20 ft. lbs. (26 Nm).

18. Install ventilation hose.

19. Install PCV grommet into injector housing.

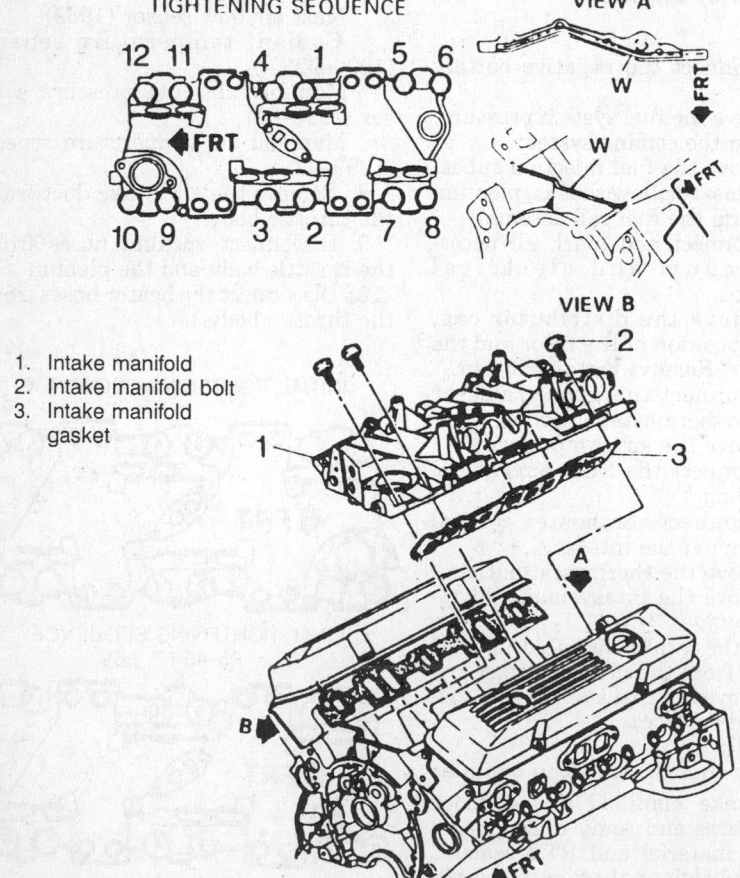

TIGHTENING SEQUENCE

VIEW A

VIEW B

1. Intake manifold
2. Intake manifold bolt
3. Intake manifold gasket

Intake manifold bolt torque sequence—1989–92 5.7L (VIN 8) engines

20. Install coolant outlet and screws. Torque screws to 89 inch lbs. (10 Nm).

21. Install bolt attaching the outlet pipe bracket to alternator bracket. Torque bolt to 38 ft. lbs. (52 Nm).

22. Install the oil pressure sensor. Apply Loctite to sensor threads.

23. Install the hose and clamp onto the right coolant outlet pipe.

24. Install the fuel rail assembly to the injector housing. Torque bolts to 20 ft. lbs. (26 Nm).

25. Connect injector electrical connectors.

26. Connect the fuel lines to the fuel rail.

27. Install the air intake duct and plenum assembly as follows:

    a. Connect the temperature sensor and fan switch wire connector.

    b. Position the plenum assembly with the MAP sensor over the fuel pressure regulator.

    c. Connect the vacuum and electrical wire connector to the MAP sensor.

    d. Install plenum bolts and new gasket. Torque bolts to 20 ft. lbs. (26 Nm).

    e. Connect all vacuum hoses onto the plenum assembly.

    f. Install the fresh air hose onto the left and right side of the throttle.

    g. Connect the purge solenoid and PCV valve hose fitting onto the plenum.

    h. Connect all electrical connectors.

    i. Connect the purge canister connection to the rear right side.

    j. Connect the throttle, cruise control, transmission cables and securing brackets to the plenum.

    k. Install the cable shield and coolant air bled hose.

    l. Install the air intake duct.

28. Fill cooling system, add oil, if drained.

29. Connect the battery negative cable. Inspect the engine for leaks and bleed the cooling system.

Exhaust Manifold

REMOVAL & INSTALLATION

5.7L (VIN 8) Engine

RIGHT SIDE

1. Disconnect the negative battery cable.

2. Remove the plenum extension.

3. Disconnect the EGR sensor wire.

4. Remove the EGR pipe bolts at the intake manifold.

5. Remove the rear air conditioning compressor brace and allow it to hang from the compressor.

6. Disconnect the dipstick tube at the manifold and remove the dipstick/tube as an assembly.

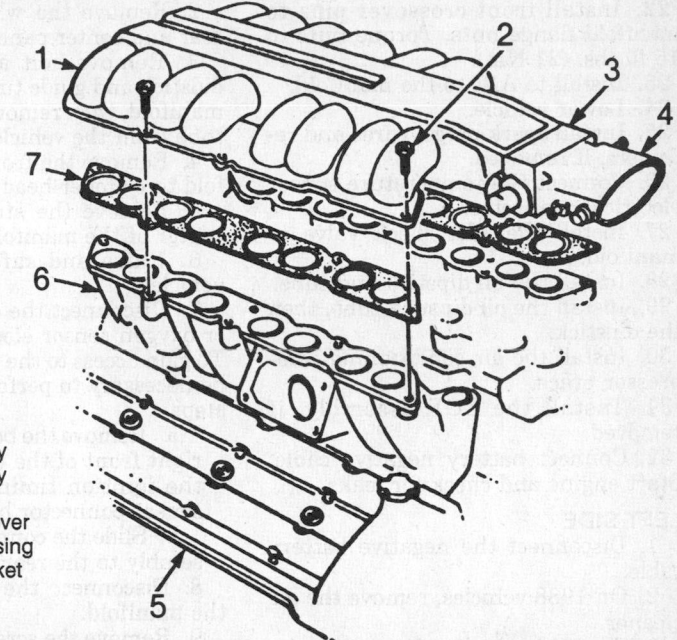

1. Plenum
2. Plenum bolt
3. Throttle body
4. Throttle body extension
5. Camshaft cover
6. Injector housing
7. Plenum gasket

Injector housing and plenum assembly—1990–92 5.7L (VIN J) engine

7. Remove the AIR check valve at the manifold.

8. Disconnect the AIR hose at the catalytic air pipe opening.

9. Disconnect the temperature sending unit wire.

10. Disconnect the spark plug wires from the plugs, cylinder head and the valve covers.

11. Remove the spark plugs.

12. Raise and safely support the vehicle.

13. Remove the catalytic converter AIR pipe at the manifold.

14. Disconnect the exhaust crossover pipe at the manifold.

15. Remove the bolts from the catalytic converter front support hanger.

16. Remove the catalytic converter AIR pipe.

17. Lower the vehicle.

18. Support the exhaust manifold and remove the retaining bolts.

19. Remove the exhaust manifold and EGR assembly from the vehicle. If the manifold is being replaced, remove the EGR pipe clamp and EGR pipe.

To install:

20. Install exhaust manifold, gasket and bolts. Torque bolts to 19 ft. lbs. (26 Nm).

21. Raise and support vehicle.

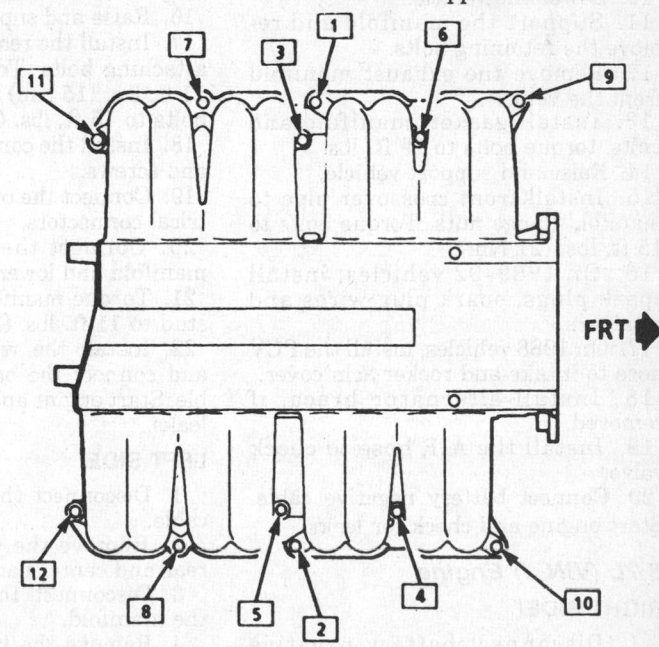

Plenum assembly torque sequence—1990–92 (VIN J) engine

22. Install front crossover pipe to manifold flange nuts. Torque nuts to 15 ft. lbs. (21 Nm).

23. Install to AIR to the manifold.

24. Lower vehicle.

25. Install spark plugs, wires and retainers, if removed.

26. Connect the temperature sensor electrical connector.

27. Install the AIR check valve to manifold.

28. Install the oil dipstick and tube.

29. Install the oil dipstick tube, then the dipstick.

30. Install the air conditioning compressor brace.

31. Install the EGR assembly, if removed.

32. Connect battery negative cable. Start engine and check for leaks.

LEFT SIDE

1. Disconnect the negative battery cable.

2. On 1988 vehicles, remove the air cleaner.

3. On 1988 vehicles, disconnect the PCV hose from the intake and rocker arm cover.

4. Disconnect the AIR hose at the exhaust check valve.

5. Disconnect the rear alternator brace and allow to hang from the alternator.

6. On 1989–92 vehicles, remove the spark plug wires from the plugs, then from the wire retainers and position aside.

7. On 1989–92 vehicles, remove the spark plugs.

8. Raise and properly support the vehicle.

9. Disconnect the exhaust pipe at the manifold.

10. Lower the vehicle.

11. Support the manifold and remove the retaining bolts.

12. Remove the exhaust manifold from the vehicle.

13. Install gasket, manifold and bolts. torque bolts to 19 ft. lbs.

14. Raise and support vehicle.

15. Install front crossover pipe to manifold flange nuts. Torque bolts to 15 ft. lbs. (21 Nm).

16. On 1989–92 vehicles, install spark plugs, spark plug wires and retainers.

17. On 1988 vehicles, install the PCV hose to intake and rocker arm cover.

18. Install alternator brace, if removed.

19. Install the AIR hose to check valve.

20. Connect battery negative cable. Start engine and check for leaks.

5.7L (VIN J) Engine

RIGHT SIDE

1. Disconnect battery negative cable.

2. Remove the wheel house lower rear and center panels.

3. Remove bolt attaching the oil dipstick and guide tube to the exhaust manifold, then remove oil dipstick and tube from the vehicle.

4. Remove the front exhaust manifold to cylinder head attaching bolts.

5. Remove the stud nut from the center of the manifold.

6. Raise and safely support the vehicle.

7. Disconnect the catalytic converter oxygen sensor electrical connector. To gain access to the connector, it may be necessary to perform the following steps:

    a. Remove the bolts located at the right front of the oil pan, attaching the ignition timing sensor/oxygen sensor connector bracket.

    b. Slide the connector bracket assembly to the rear.

8. Disconnect the AIR hose from the manifold.

9. Remove the screws attaching the manifold/converter flanges to the exhaust pipe flanges.

10. Remove the rear manifold to head attaching bolts. Remove the manifold and gasket from the vehicle.

11. Install manifold/converter assembly to engine.

12. Install bolts attaching manifold/converter flanges to the exhaust pipe flanges. Finger-tighten bolts.

13. Lower vehicle.

14. Install manifold gasket to head, then install the front manifold to head attaching bolts.

15. Install stud nut at center manifold, if removed. Coat threads with Loctite and torque to 11 ft. lbs. (15 Nm).

16. Raise and support vehicle.

17. Install the rear manifold to head attaching bolts. Torque rear bolts to 11 ft. lbs. (15 Nm) and exhaust flange bolts to 15 ft. lbs. (20 Nm).

18. Install the converter heat shields and screws.

19. Connect the oxygen sensors electrical connectors.

20. Connect the AIR hose to the manifold and lower vehicle.

21. Torque manifold front bolts and stud to 11 ft. lbs. (15 Nm).

22. Install the wheel house panels and connect the battery negative cable. Start engine and check for exhaust leaks.

LEFT SIDE

1. Disconnect the battery negative cable.

2. Remove the wheelhouse lower rear and center panels.

3. Disconnect the AIR hose from the manifold.

4. Remove the bolts attaching the manifold to the head.

5. Remove the stud nut from the center of the manifold.

6. Remove the gasket from the manifold.

7. Raise and safely support the vehicle.

8. Disconnect the converter oxygen sensor electrical connector.

9. Remove the screws attaching the converter heat shields, then remove the heat shield from the vehicle.

10. Remove the bolts attaching the manifold/converter flange to the front exhaust pipe flange, then remove manifold/converter from the vehicle.

To install:

11. Install manifold/converter to the engine.

12. Install bolts attaching manifold/converter flange to the exhaust pipe flange. Finger-tight only.

13. Lower vehicle.

14. Install the manifold gasket to head.

15. Install bolts attaching manifold to head, finger-tight.

16. Install stud nut, if removed. Torque manifold bolts and stud nut to 11 ft. lbs. (15 Nm).

17. Raise and support vehicle.

18. Torque exhaust pipe flange bolts to 15 ft. lbs. (20 Nm).

19. Install the converter heat shields and screws.

20. Connect the oxygen sensors electrical connectors.

21. Lower vehicle, then connect the AIR hose to the exhaust manifold.

22. Install wheel house panels and connect the battery negative cable. Start engine and check for leaks.

Timing Chain Front Cover

REMOVAL & INSTALLATION

5.7L (VIN 8) Engine

1988

1. Disconnect the negative battery cable.

2. Remove the drive belt and the crankshaft pulley.

3. Remove the coolant pump damper.

4. Remove the crankshaft pulley.

5. Install tool J–23523 onto the vibration damper assembly.

6. Remove the vibration damper from the face of the crankcase front cover.

NOTE: The use of pullers, such as the universal claw type, that pull on the outside of the hub may damage the torsional damper. The outside ring of the damper is bonded to the hub with rubber. The use of the improper type puller may destroy this bond.

7. Raise and safely support the vehicle.

8. Drain the engine oil and remove the oil pan.

9. Lower the vehicle. Remove the AIR control valve, pipe and silencer as an assembly.

10. Remove the AIR pump pulley and air pump retaining bolts. Remove the air pump.

11. Relieve the fuel system pressure and disconnect the fuel inlet and return pipes.

12. Disconnect the air conditioner compressor mounting bracket nuts at the water pump. Slide the mounting bracket forward and remove the compressor mounting bolt. Disconnect the electrical wires and position the unit to the side.

13. Disconnect the AIR hose from the right exhaust manifold.

14. Remove the air conditioning compressor mounting bracket.

15. Remove the upper AIR pump bracket with the power steering reservoir.

16. Remove the lower AIR pump bracket.

17. Drain the radiator and disconnect the radiator and heater hoses at the water pump. Remove the water pump.

18. Remove the front cover retaining screws. Remove the front cover and discard the gasket.

19. Thoroughly clean the gasket mating surfaces on the cylinder block and front cover. Inspect the front cover for damage and distortion. Replace the front cover, if necessary. Replace the oil seal.

20. With a suitable cutting tool, remove any excess gasket material that may be protruding at the oil pan to engine block surface.

To install:

21. Coat the new cover gasket with a suitable sealing compound and apply the gasket onto the front cover sealing surface.

22. Position the front cover and gasket onto the cylinder block surface and hold in place. Install the cover retaining screws and make them finger-tight.

23. Tighten the retaining screws evenly in an alternate pattern. While tightening the retaining screws, readjust the position of the front cover, as required, to ensure the cylinder block locating dowels are evenly aligned with the holes in the cover. Do not force the cover over the locating dowels.

24. When the front cover is properly in place, torque the retaining screws to 90 inch lbs. (9 Nm).

25. Prior to installing the oil pan, apply an even coating of GM sealant 1052080 or equivalent, onto the front

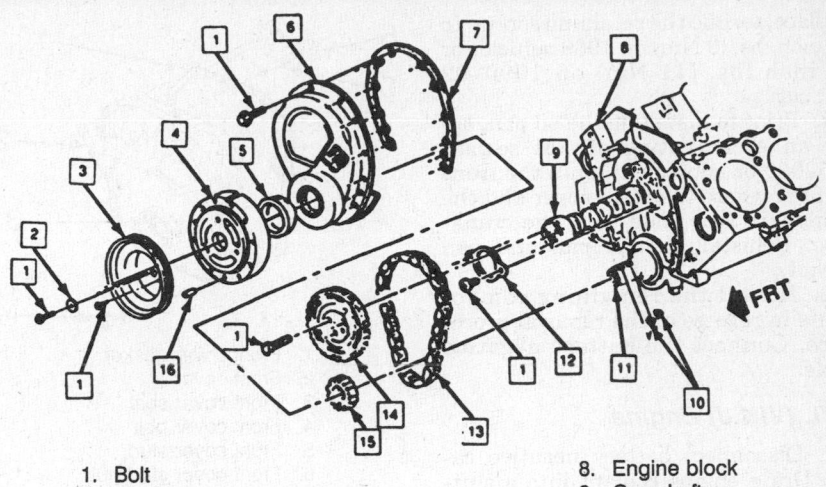

1.	Bolt		8.	Engine block
2.	Washer		9.	Camshaft
3.	Pulley		10.	Keys
4.	Torsional damper		11.	Crankshaft
5.	Front cover seal		12.	Camshaft retainer
6.	Timing chain cover		13.	Tiing chain
7.	Timing chain cover		14.	Camshaft sprocket
	gasket		15.	Crankshaft sprocket
			16.	Pin

Timing chain, camshaft and front cover assembly—5.7L (VIN 8) engine

corners where the front cover and the rear main seal mate with the crankcase. Reinstall the oil pan and pan gasket.

26. Install the remaining components in reverse of the removal procedure. Connect the battery negative cable.

1989–92

1. Disconnect the negative battery cable.

2. Remove the drive belt and the crankshaft pulley.

3. Remove the coolant pump damper.

4. Remove the crankshaft pulley.

5. Remove the power steering gear line, if required.

6. Install tool J–23523 onto the vibration damper assembly. Remove the vibration damper from the face of the crankcase front cover.

NOTE: The use of pullers, such as the universal claw type, that pull on the outside of the hub may damage the torsional damper. The outside ring of the damper is bonded to the hub with rubber. The use of the improper type puller may destroy this bond.

7. Raise and safely support the vehicle.

8. Drain the engine oil and remove the oil pan.

9. Lower the vehicle. Remove the AIR control valve, pipe and silencer as an assembly.

10. On 1988 vehicles, remove the AIR pump pulley and air pump retaining bolts. Remove the air pump.

11. Relieve the fuel system pressure and disconnect the fuel inlet and return pipes.

12. Remove the air conditioner compressor mounting bracket.

13. Drain the radiator and disconnect the radiator and heater hoses at the water pump. Remove the water pump.

14. Remove the front timing cover retaining screws. Remove the front cover and discard the gasket.

15. Thoroughly clean the gasket mating surfaces on the cylinder block and front cover. Inspect the front cover for damage and distortion. Replace the front cover, if necessary. Replace the oil seal.

16. With a suitable cutting tool, remove any excess gasket material that may be protruding at the oil pan to engine block surface.

To install:

17. Coat the new cover gasket with a suitable sealing compound and apply the gasket onto the front cover sealing surface.

18. Position the front cover and gasket onto the cylinder block surface and hold in place. Install the cover retaining screws and make them finger-tight.

19. Tighten the retaining screws evenly in an alternate pattern. While tightening the retaining screws, readjust the position of the front cover, as required, to ensure the cylinder block locating dowels are evenly aligned with the holes in the cover. Do not force the cover over the locating dowels.

20. When the front cover is properly

in place, torque the retaining screws to 90 inch lbs. (9 Nm) on 1989 vehicles or 98 inch lbs. (11 Nm) on 1990–92 vehicles.

21. Prior to installing the oil pan, apply an even coating of GM sealant 1052080 or equivalent, onto the front corners where the front cover and the rear main seal mate with the crankcase. Reinstall the oil pan and pan gasket.

22. Install the remaining components in reverse of the removal procedure. Connect the battery negative cable.

5.7L (VIN J) Engine

1. Disconnect battery negative cable. Drain engine coolant into a suitable container.

2. Remove water pump assembly.

3. Discharge the air conditioning system and remove the air conditioning compressor.

4. Remove the power steering pump assembly, then the serpentine belt.

5. Remove the bolt and washer attaching the torsional damper to the crankshaft.

6. Using tool J–24420–C or equivalent, remove the torsional damper and drift key from the crankshaft.

7. Remove the front cover attaching nuts and bolts, then remove the front cover and gasket from the vehicle.

8. Installation is the reverse of the removal procedure. Apply Loctite to bolt threads.

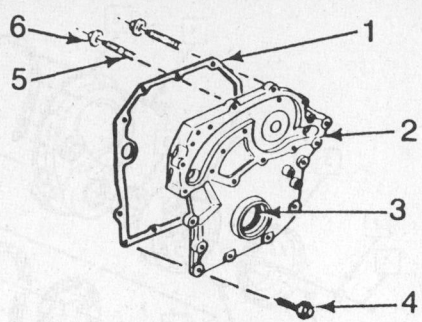

1. Front cover gasket
2. Front cover
3. Front cover seal
4. Front cover bolt
5. Front cover stud
6. Front cover stud nut

Front timing cover assembly—5.7L (VIN J) engine

9. Torque crankshaft damper bolt to 148 ft. lbs. (200 Nm).

10. Torque front cover attaching nuts and bolts to 20 inch lbs. (26 Nm). Connect the battery negative cable.

Front Cover Oil Seal

REPLACEMENT

5.7L (VIN 8) Engine

FRONT COVER REMOVED

1. Remove the old seal from the front cover, using the appropriate tool.

2. Discard the oil seal. Exercise cau-

tion when removing the seal to prevent damaging the front cover.

3. With a clean rag, ensure the front cover sealing surfaces are free from dirt and grease.

4. Support the rear of the front cover and position the new seal so the open end of the seal is toward the the inside of the front cover.

5. With tool J–35468 or equivalent, drive the new seal into the front cover. Visually inspect the seal to ensure it is seated evenly in the front cover.

FRONT COVER INSTALLED

1. Disconnect the negative battery cable.

2. Loosen and remove the drive belt from the crankshaft pulley.

3. Remove the crankshaft pulley.

4. Install tool J–23523 onto the vibration damper assembly. Remove the vibration damper from the crankcase front cover.

NOTE: The use of pullers, such as the universal claw type, that pull on the outside of the hub may damage the torsional damper. The outside ring of the damper is bonded to the hub with rubber. The use of the wrong type puller may destroy this bond.

5. Using the appropriate tool, remove the old seal from the front cover. Discard the old oil seal. Exercise caution when removing the seal to prevent damaging the front cover and crankshaft surfaces.

6. With a clean rag, ensure the front cover sealing surfaces are free from dirt and grease.

7. With tool J–35468 or equivalent, drive the new seal into the front cover. Visually inspect the seal to ensure it is seated evenly in the front cover.

8. Reinstall the vibration damper, using the appropriate tool.

9. Reinstall the crankshaft pulley and the drive belt. Adjust the drive belt tension. Connect the battery negative cable.

5.7L (VIN J) Engine

1. Disconnect the battery negative cable.

2. Remove the front timing cover assembly.

3. Using tool J–29077–A or equivalent, remove the seal from the front cover.

4. Install the new seal coated with engine oil using tool J–37309 or equivalent.

5. Installation is the reverse of the removal procedure.

NOTE: Do not remove seal installing tool J–37309, until the front cover has been installed and the bolts torqued.

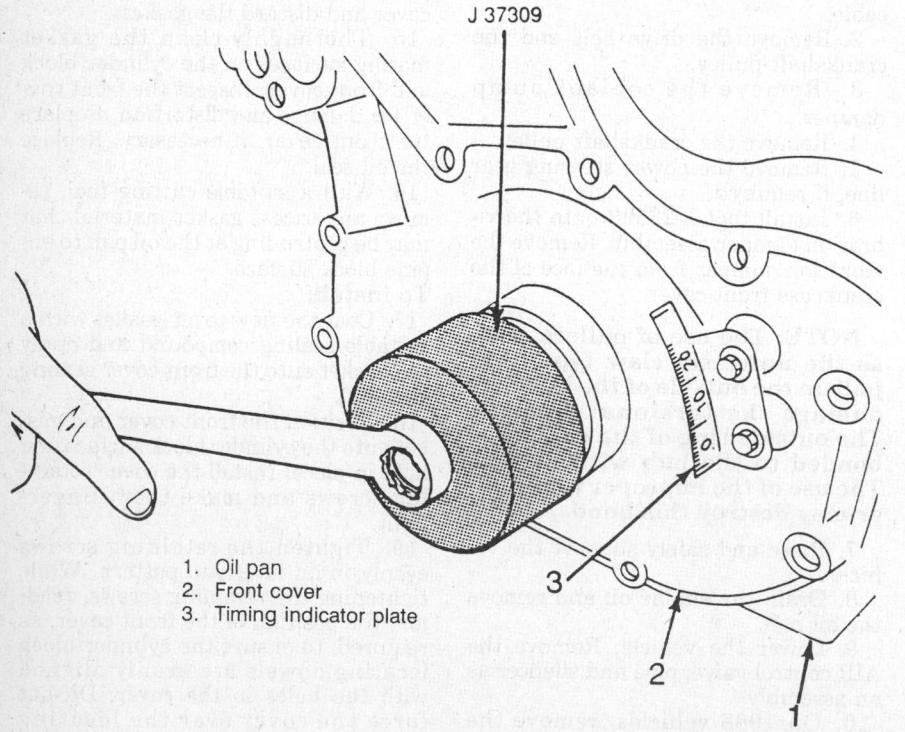

J 37309

1. Oil pan
2. Front cover
3. Timing indicator plate

Front timing cover installation tool—5.7L (VIN J) engine

Timing Chain and Sprockets

REMOVAL & INSTALLATION

5.7L (VIN 8) Engine

1. Disconnect the negative battery cable. Remove the front timing cover.
2. Rotate the crankshaft and align the timing marks.
3. Remove the camshaft gear mounting bolts. Remove the camshaft gear.
4. Remove the timing chain.
5. Using a suitable puller, carefully remove the crankshaft gear sprocket. Remove the key from the end of the crankshaft keyway. Inspect the keyway surface for excessive wear or rounding. Replace if necessary.
6. Visually inspect the crankshaft and camshaft gear teeth for chipped, missing and cracked teeth. Replace all damaged parts.
7. Installation is the reverse of the removal procedure. Connect the battery negative cable.

5.7L (VIN J) Engine

PRIMARY TIMING CHAIN AND SPROCKET

1. Disconnect battery negative cable. Remove the front timing cover assembly.
2. Remove the left and right intake camshafts as follows:
   a. Remove the oil filter housing.
   b. Remove the air conditioning compressor.
   c. Remove the left and right valve covers.
   d. Raise and support the vehicle safely.
   e. Remove the ignition timing sensor from the cylinder case.
   f. Insert tool J–38098 into the ignition timing sensor opening. Verify tool is fully seated with indicating pin inserted in deep notch of crankshaft timing disc.
   g. Lower the vehicle.
   h. Remove the bolts attaching the secondary timing chain tensioner housing to the cylinder head.
   i. Remove the housing, O-ring and tensioner from the cylinder case.
   j. Remove the bolts and washers attaching the camshaft sprockets to the camshaft.

NOTE: To avoid exerting excess pressure on tool J–38098, use a wrench on the rear of the camshaft when removing the sprocket bolts.

   k. Remove the intake camshaft timing plates and pins.

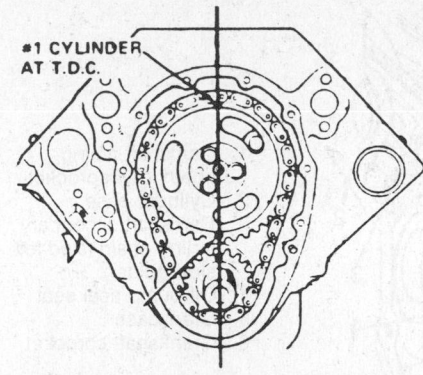

Timing mark alignment—5.7L (VIN 8) engines

   l. Remove the intake camshaft retainers and thrust washers.
   m. Remove the intake camshaft and sprockets. Insert tool J–38099 or equivelant to hold the secondary chain in position.
3. Remove the bolts attaching the primary chain guide to the oil pump, then remove the guide from the vehicle.
4 Remove the idler sprocket assembly attaching bolts, then the primary chain from the idler sprocket and crankshaft sprocket.
5. Using tool J–38211 or equivalent, remove the crankshaft sprocket.
6. Remove the key and oil pump seal from crankshaft.

To install:

7. Inspect the primary chain guide for excessive wear. Wear groove should not exceed a depth of 0.040 in. (1.0mm). If necessary, replace wear strip. Apply Loctite® to chain guide bolt threads. Torque bolts to 89 inch lbs. (10 Nm).
8. Inspect primary chain and sprocket for wear or damage. If abnormal wear or damage is present on either the idler sprocket, chain or crankshaft sprocket, all 3 must be replaced as an assembly.

NOTE: When installing guide, do not use any leverage tools, finger pressure is sufficient.

9. Install crankshaft sprocket using tool J–38132 or equivalent.
10. Installation is the reverse of the removal procedure.

NOTE: Ensure that sprocket is installed with long shoulder facing the front.

SECONDARY TIMING CHAINS AND SPROCKET

1. Disconnect battery negative cable.
2. Remove the camshafts.
3. Remove the primary timing chain and crankshaft sprocket.

4. Remove the left and right secondary timing chain tensioners.
5. Remove the left and right secondary chains from the idler sprocket. Remove idler sprocket attaching bolts and the idler sprocket.
6. Remove the left and right secondary chains from the vehicle.

To install:

7. Always install new timing chain tensioners. Coat tensioner with engine oil.
8. Ensure the oil hole in the tensioner piston is installed in a vertical position.
9. Check that the fork on the end of the tensioner is properly engaged onto the chain guide.
10. Torque chain tensioner bolts to 89 inch lbs. (10 Nm). Apply Loctite to bolt threads.
11. Inspect chains and sprockets for abnormal wear or damage. If abnormal wear or damage is present on either the secondary timing chain, cam sprockets or idler sprockets, the entire assembly must be replaced.
12. Inspect the idler sprocket shaft bearings for wear or damage. If necessary, replace idler sprocket shaft bearings as follows:
   a. Using tool J–37328 or equivalent, remove bearings from idler sprocket.
   b. When installing bearings, ensure the manufacture's name and part No. are visible from either end of the sprocket assembly.
   c. Using a suitable press, press in bearings until bearings are flush with idler sprocket. Apply minimum pressure to obtain a fit 0.0–1.3mm below the surface.
13. Installation is the reverse of the removal procedure.

Camshaft

REMOVAL & INSTALLATION

5.7L (VIN 8) Engine

1. Disconnect the negative battery cable.
2. Properly relieve the fuel and air conditioning system pressure.
3. Drain the cooling system.
4. Remove the radiator, fan and condenser assembly.
5. Align the timing marks on the crankshaft pully and timing chain cover to the **TDC** position or 0 mark. Mark the distributor and remove the assembly.
6. Remove the intake manifold assembly and valve covers. Remove the valve rockers, pushrods and lifters.
7. Remove the sepentine belt.
8. Remove the air conditioning compressor and brackets.

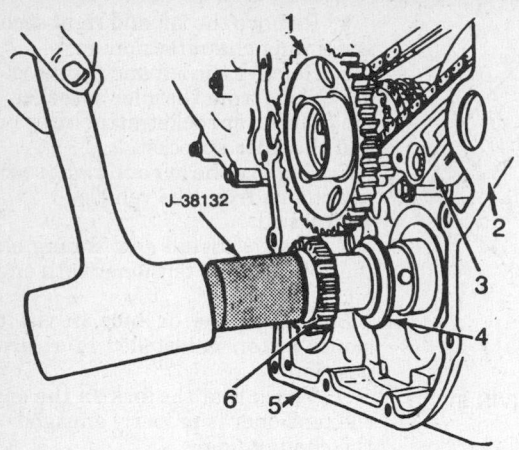

1. Camshaft timing chain idler sprocket
2. Cylinder case
3. Camshaft secondary timing chain fixed left side guide
4. Oil pump seal seat
5. Crankcase
6. Crankshaft sprocket

J–38132

Crankshaft sprocket installation—5.7L (VIN J) engine

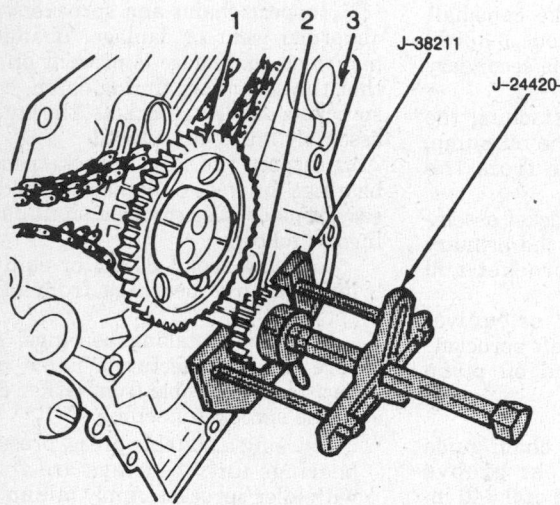

J–38211
J–24420–C

1. Camshaft timing chain idler sprocket
2. Crankshaft sprocket
3. Cylinder case

Crankshaft sprocket removal—5.7L (VIN J) engine

9. Rmove the Power steering line, if required.

10. Disconnect and remove the AIR pump assembly, if required

11. Remove the alternator assembly and position unit aside.

12. Remove the water pump assembly.

13. Remove the crankshaft pulley and torsional damper.

14. Remove the timing cover.

15. Remove the timing chain and camshaft gear.

16. Reinstall the camshaft gear and rotate the camshaft while pulling the assembly out of the cylinder block.

To install:

NOTE: When installing a new camshaft, coat camshaft lobes and distributor gear with Molykote® or equivalent.

17. Using a micrometer, check the camshaft bearing journals for an out-of-round condition. If journals exceed 0.001 in. out-of-round, the camshaft should be replaced.

18. Inspect the camshaft bearings for wear or damage. Replace as necessary.

19. Install the camshaft.

20. Install the timing chain on the camshaft sprocket. Position the sprocket vertically with the chain hanging down and align the marks on the camshaft with the marks on the camshaft sprockets.

21. Install the timing sprocket on the camshaft. Torque bolts to 20 ft. lbs. (27 Nm).

22. Install the timing chain front cover, torsional damper and crankshaft pulley.

23. Install the valve lifters. When in-

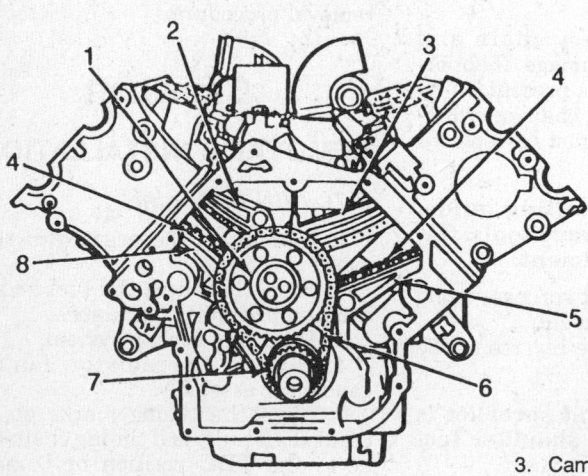

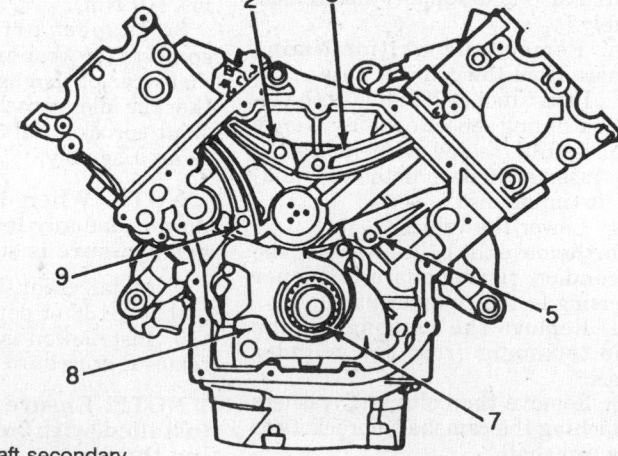

1. Camshaft timing chain idler sprocket assembly
2. Camshaft secondary timing chain fixed right side guide
3. Camshaft secondary timing chain fixed right side guide
4. Camshaft secondary timing chain
5. Camshaft secondary timing chain fixed left side guide
6. Camshaft primary timing chain
7. Crankshaft sprocket
8. Oil pump
9. Camshaft timing chain pivot right side guide

Primary and secondary timing assembly—5.7L (VIN J) engine

stalling a new camshaft, install new lifters. Install the pushrods, rockers and valve covers.

24. Install the intake manifold and distributor assembly.

25. Install the water pump assembly.

26. Install the AIR pump, air conditioning compressor, power steering pump and brackts.

27. Install the serpentine belt.

28. Install the condenser, radiator and hoses.

29. Install the fan and shroud assembly.

30. Change the engine oil and filter.

31. Fill the cooling system with the proper type and quantity of antifreeze.

32. Connect the negative battery cable. Start the engine, set the rpm and timing to specification. Inspect the engine for leaks.

5.7L (VIN J) Engine

1. Disconnect battery negative cable.

2. Remove the oil filter housing and right valve cover to gain access to the right camshaft as follows:

   a. Drain the engine coolant into a suitable container.

   b. Remove the air intake duct.

   c. Remove the hoses and clamps from the coolant outlets, radiator inlet and inlet pipe.

   d. Remove the hoses and inlet pipe assembly from the vehicle.

   e. Remove the water pump pulley.

   f. Release the belt tensioner and remove the serpentine belt.

   g. Remove the belt tensioner from the engine.

   h. Remove the oil filter.

   i. Disconnect the electrical connectors from the oil filter housing.

   j. Remove the oil pressure sensor from the housing.

   k. Remove the alternator bracket from the oil filter housing.

   l. Disconnect and plug the oil cooler lines.

   m. Remove the oil filter housing mounting bolts and remove the assembly.

   n. Disconnect the electrical connector from the blower motor resistor block.

   o. Remove spark plug wires from plugs.

   p. Remove the screws attaching the evaporator housing quarter panel, then remove the panel.

   q. Remove the bolts attaching the coolant outlet pipe to injector housing, then position aside.

   r. Remove the bolt attaching the fresh air pipe bracket to the injector housing.

   s. Remove the camshaft cover at-

1. Camshaft idler sprocket assembly screw
2. Camshaft timing chain idler sprocket washer
3. Camshaft timing chain idler sprocket
4. Camshaft timing chain idler sprocket shaft
5. Camshaft idler sprocket bolt
6. Camshaft idler sprocket bolt

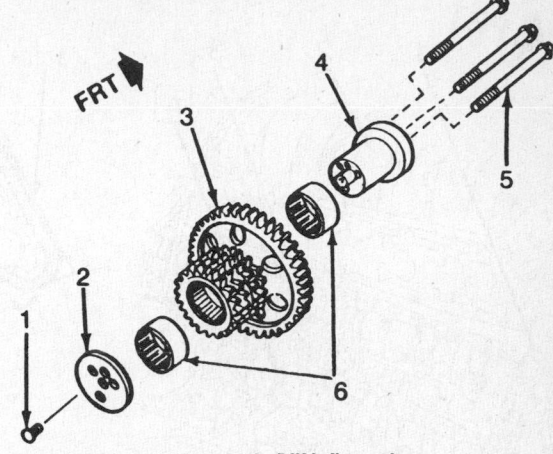

Timing chain idler sprocket assembly—5.7L (VIN J) engine

taching bolts and the camshaft cover.

3. Remove the air conditioning compressor and left valve cover to gain access to the left camshafts as follows:

   a. Drain the cooling system into a suitable container.

   b. Properly discharge the air conditioning system.

   c. Remove the throttle body assembly.

   d. Remove the alternator assembly.

   e. Disconnect and plug the refrigerant hose from the rear of the compressor.

   f. Remove the compressor mounting bolts.

   g. Disconnect the compressor electrical connectors.

   h. Remove the compressor assembly from the engine.

   i. Remove the power steering pump and reservoir unit from the engine.

   j. Remove the spark plug wires from the plugs.

   k. Remove the ventilation breather pipe from the valve cover.

   l. Remove the throttle cable and cruise control hold-down clamps from the plenum.

   m. Remove the vacuum hose from the power brake booster.

   n. Remove the left valve cover attaching bolts and remove the cover.

4. Remove the right and left camshafts as follows:

   a. Raise and support vehicle.

   b. Disconnect the electrical connector from the crankshaft ignition timing sensor.

   c. Install crankshaft timing slot locator tool J–38098 into the ignition timing sensor opening. Ensure the tool head is fully seated with the indicating pin is inserted in deep notch of the crankshaft timing disc.

   d. Lower vehicle.

   e. Remove the bolts attaching the

secondary timing chain tensioner housing to the cylinder head, then remove the O-ring and tensioner from the cylinder case.

   f. Remove the bolts and washers attaching the camshaft to the sprockets.

NOTE: Install a wrench on the rear camshaft hex when removing the sprocket bolts, to prevent the camshafts from exerting force on the crankshaft timing slot locator tool J–38098.

   g. Remove the camshaft timing plates and pins.

   h. Remove the camshaft retainers and thrust washers.

   i. Remove the camshafts and sprockets from the vehicle. Install timing chain retainers J–38099 to retain secondary chain loops.

   j. Remove lifters from bores. Make sure any lifters, to be reused, are retained in proper order so each one can be returned to its original bore.

To install:

5. Inspect the camshaft bearing journals for wear or damage.

6. Inspect the camshaft bearing surfaces in the cylinder head and camshaft cover for wear or damage.

NOTE: The camshaft cover and cylinder head must be replaced as a set if excessive wear or damage to the bearing surfaces is found.

7. Lubricate lifters and bores with engine oil, then install lifters into bores.

8. Install the camshaft sprocket onto the secondary timing chain, while removing the timing chain retainers J–38099.

9. Slide the camshaft into the sprocket, noting the position of the alignment hole for timing pin tool installation. Position the camshaft in the neutral position, no valves opened.

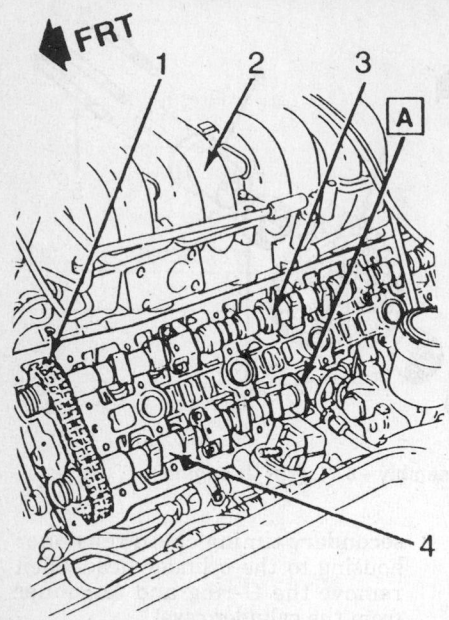

A. Camshaft sensor
   reluctor disc
1. Camshaft secondary
   timing chain
2. Plenum
3. Intake left side
   camshaft
4. Exhaust left side
   camshaft

**Camshaft assembly–left cylinder
pictured–5.7L (VIN J) engine**

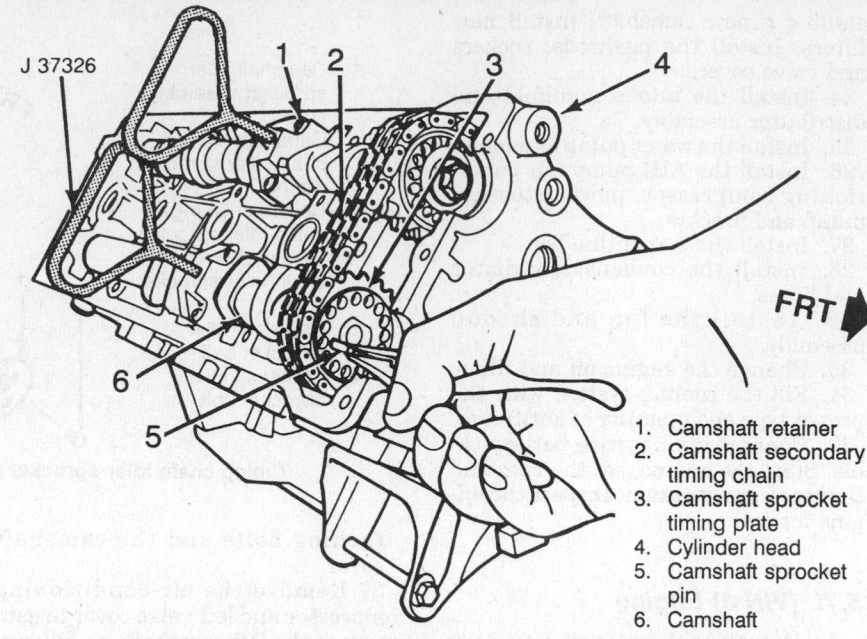

1. Camshaft retainer
2. Camshaft secondary
   timing chain
3. Camshaft sprocket
   timing plate
4. Cylinder head
5. Camshaft sprocket
   pin
6. Camshaft

Installing camshaft sprocket pin–5.7L (VIN J) engine

10. Lubricate camshaft journals, lobes, thrust washers and retainers with clean engine oil.

11. Install the camshaft thrust washers, retainers and bolts. Torque bolts to 89 inch lbs. (10 Nm).

12. Repeat the steps for the remaining camshafts.

13. Install timing pins J–37326 into camshaft retainers and indexing holes in camshafts.

14. Install camshaft secondary chain pre-tensioner J–37305. Tighten until all the slack has been removed from the timing chain.

15. Install timing plate pin. If no

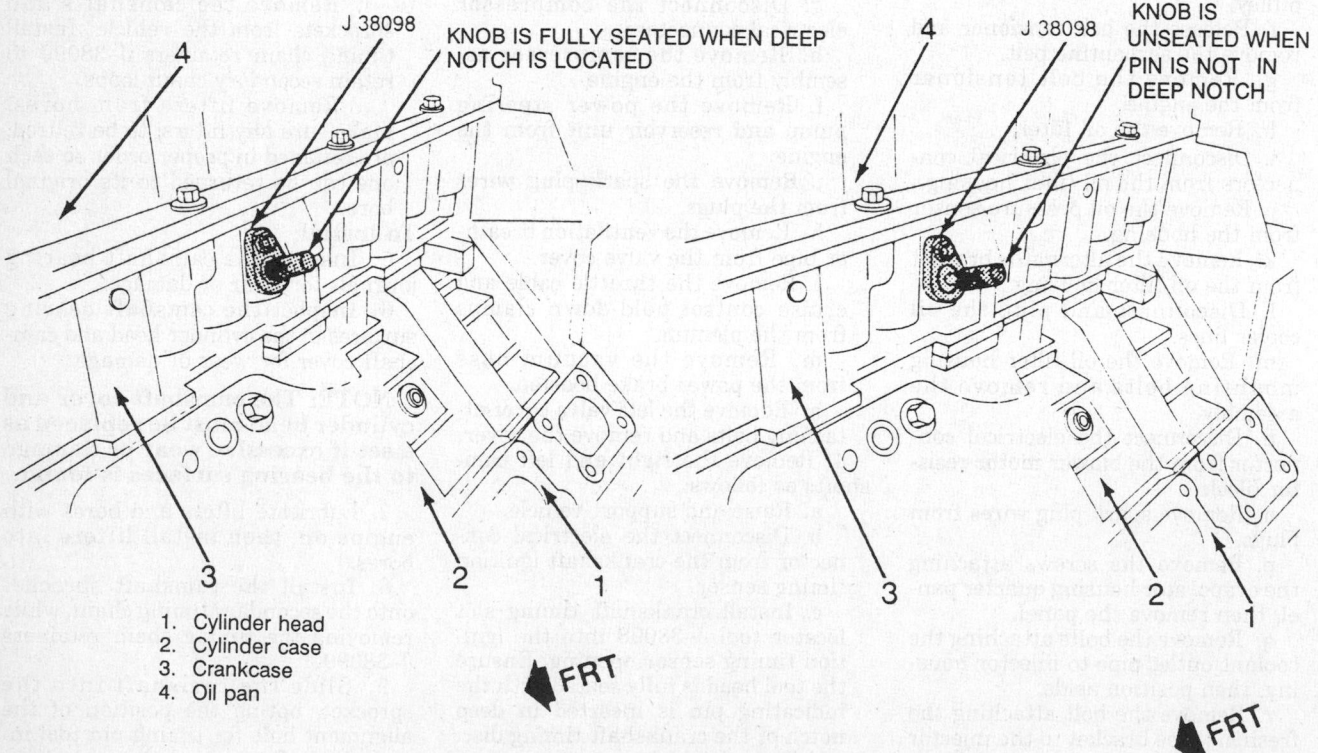

1. Cylinder head
2. Cylinder case
3. Crankcase
4. Oil pan

Crankshaft timing slot locator tool–5.7L (VIN J) engine

holes line up on the timing plate, reverse plate.

16. Install bolts finger-tight. New camshaft bolts should be used.

17. Apply Loctite on the camshaft sprocket bolts and torque bolts to 19 ft. lbs. (25 Nm) and turn 80–85 degrees using torque angle meter J–36660. Install a backup wrench on the camshaft hex on the rear of the camshaft.

18. Remove timing pins J–37326.

19. Remove the secondary timing chain pre-tensioner tool J–37305.

20. Install the secondary timing chain tensioner, housing, new O-ring and bolt. Always install a new timing chain tensioner. Lubricate tensioner with engine oil.

21. Ensure that oil hole in tensioner piston be installed in a vertical position. Check that fork on end of tensioner must be properly engaged onto the chain guide.

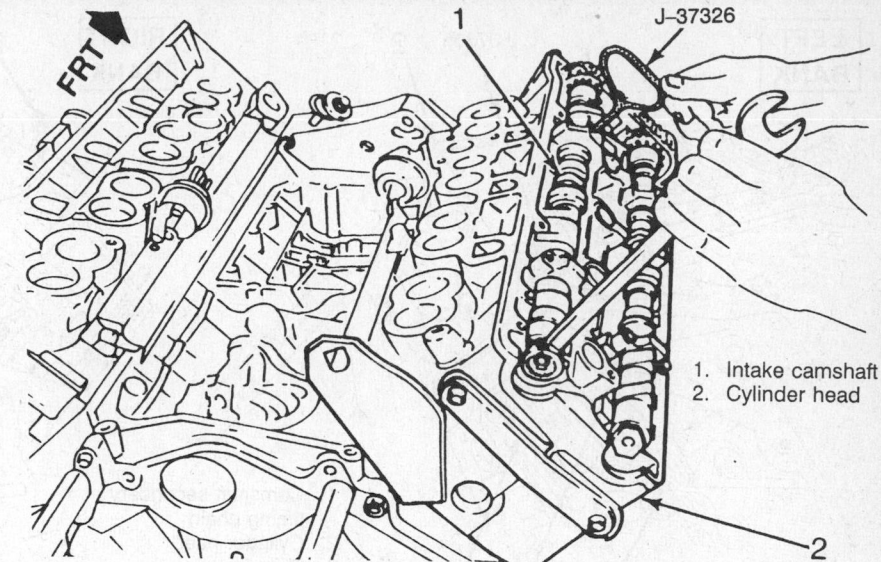

1. Intake camshaft
2. Cylinder head

Installing camshaft timing pins—5.7L (VIN J) engine

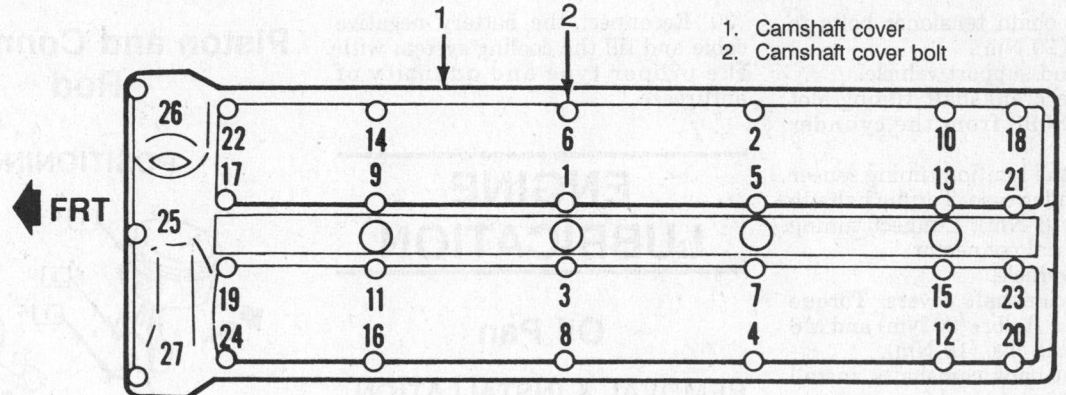

1. Camshaft cover
2. Camshaft cover bolt

Valve cover torque sequence—5.7L (VIN J) engine

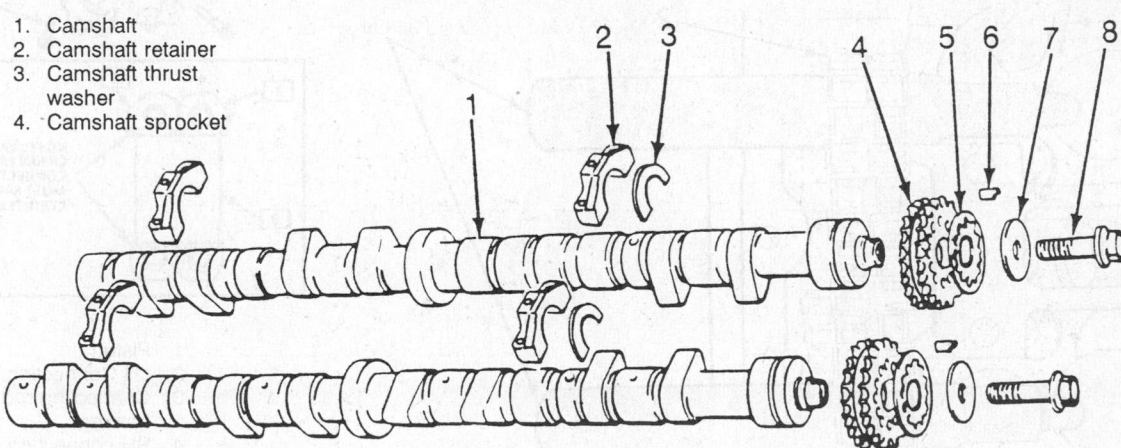

1. Camshaft
2. Camshaft retainer
3. Camshaft thrust washer
4. Camshaft sprocket

5. Camshaft sprocket timing plate
6. Camshaft sprocket pin
7. Camshaft sprocket washer
8. Camshaft sprocket washer

Cylinder head camshaft assembly—5.7L (VIN J) engine

16–23

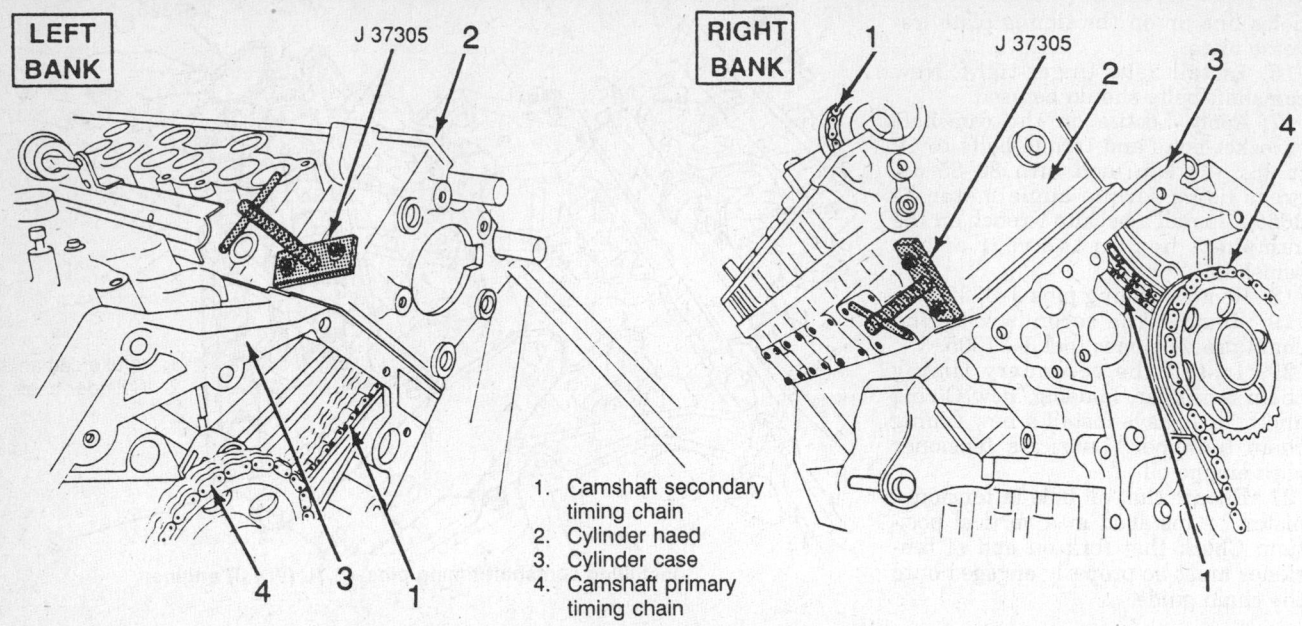

LEFT BANK

RIGHT BANK

J 37305

1. Camshaft secondary timing chain
2. Cylinder haed
3. Cylinder case
4. Camshaft primary timing chain

Camshaft chain pretensioner tool—5.7L (VIN J) engine

22. Torque chain tensioner bolts to 89 inch lbs. (10 Nm).

23. Raise and support vehicle.

24. Remove crankshaft timing slot locator J-38098 from the cylinder case.

25. Install the ignition timing sensor into the cylinder case. Torque bolts to 71 inch lbs. (8 Nm). Connect timing sensor electrical connector.

26. Lower vehicle.

27. Install camshaft covers. Torque M8 bolts to 15 ft. lbs. (20 Nm) and M6 bolts to 89 inch lbs. (10 Nm).

28. On right bank camshafts, install oil filter housing assembly.

29. On left bank camshafts, install air conditioning compressor and power steering pump assembly.

30. Reconnect the battery negative cable and fill the cooling system with the proper type and quantity of antifreeze.

ENGINE LUBRICATION

Oil Pan

REMOVAL & INSTALLATION

5.7L (VIN 8) Engine

1. Disconnect the negative battery

Piston and Connecting Rod

POSITIONING

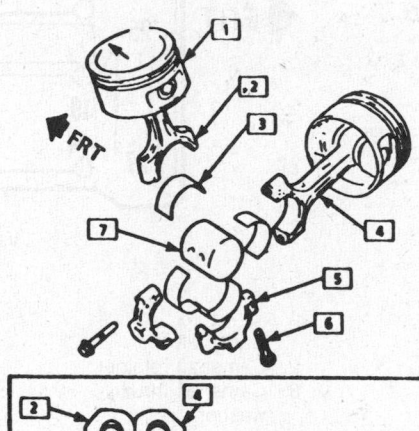

FRT

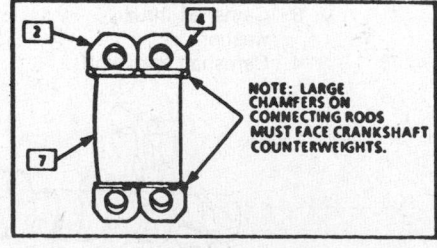

NOTE: LARGE CHAMFERS ON CONNECTING RODS MUST FACE CRANKSHAFT COUNTERWEIGHTS.

1. Piston
2. LH Connecting rod
3. Connecting rod bearing
4. RH connecting rod
5. Connecting rod bearing cap
6. Connecting rod bearing cap bolt
7. Crankshaft

Piston assembly—1990–92 vehicles

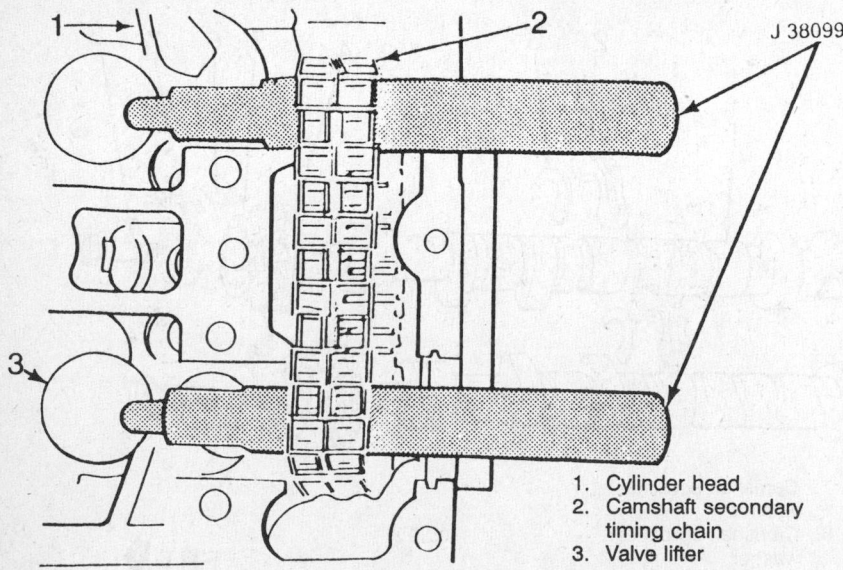

J 38099

1. Cylinder head
2. Camshaft secondary timing chain
3. Valve lifter

Secondary timing chain retainers—5.7L (VIN J) engine

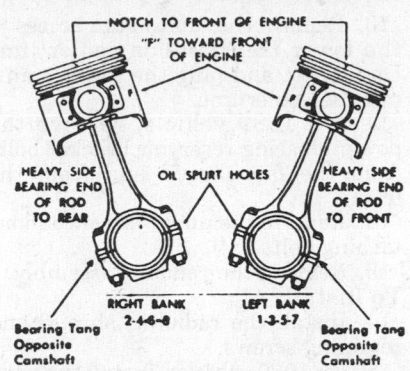

Piston assembly—1988–89 5.7L (VIN 8) engines

cable. Raise and safely support the vehicle.

2. Drain the engine oil and remove the oil filter.

3. Remove the cooler adapter at the block and set aside.

4. Disconnect the oil level sensor, if equipped.

5. If equipped with an automatic transmission, remove the torque converter cover.

6. If equipped with a manual transmission, remove the starter and the clutch housing cover.

7. Disconnect the oil cooler pipe at the oil pan.

8. Remove the Electronic Spark Control (ESC) knock sensor shield.

9. Remove the front crossmember braces.

10. Remove the oil pan bolts and the pan.

11. Installation is the reverse of the removal procedure. Tighten the bolts to 16 ft. lbs. (22 Nm).

5.7L (VIN J) Engine

1. Disconnect battery negative cable, then disconnect the oil lever indicator from the guide tube.

2. Raise and safely support the vehicle. Drain the engine oil.

3. Disconnect and remove the low oil sensor from the oil pan.

4. Remove the clutch housing cover attaching bolts, then remove the cover from the vehicle.

5. Remove the bolts attaching the AIR pipe bracket to the oil pan, then remove the right and left converter heat shields.

6. Remove the nuts attaching the engine mounts at the front crossmember rear brace on the left and right sides. Remove the bolts attaching the front crossmember to the rear brace.

7. Remove the bolts attaching the left front crossmember rear brace to the left front side member and brace from the vehicle.

8. Remove the bolts attaching the right front crossmember rear brace to the right front side member and brace from the vehicle.

9. Remove the bolts attaching the oil pan and crankcase. Remove the oil pan and gasket from the vehicle.

To install:

10. Apply Loctite to bolt threads. Torque oil pan front screws to 89 inch lbs. (10 Nm). Torque oil pan bolts to 20 ft. lbs. (26 Nm).

11. Torque left and right front crossmember rear brace to front crossmember bolts to 59 ft. lbs. (80 Nm).

12. Torque the left and right front crossmember rear brace to front side member bolts to 46 ft. lbs. (62 Nm).

13. Torque engine mounts to front crossmember bolts to 40 ft. lbs. (54 Nm).

14. Installation is the reverse of the removal procedure.

Oil Pump

REMOVAL & INSTALLATION

5.7L (VIN 8) Engine

1. Disconnect the negative battery cable.

2. Raise the vehicle and support it safely.

3. Drain the engine oil and remove the oil pan.

4. Remove the oil pan baffle nuts.

5. Support the oil pump by hand and remove the main bearing cap bolt.

6. Carefully remove the baffle and oil pump.

To install:

NOTE: The oil pump pickup should be submerged in oil and the pump primed prior to installation. Failure to prime the pump may result in oil pump failure or internal engine damage.

7. Support the oil pump with extension shaft and assemble to the rear main bearing cap.

8. Ensure the slot on the top of the extension shaft is aligned with the drive tang on the lower end of the distributor driveshaft.

9. Torque the main bearing cap bolt to specification. Connect the battery negative cable and refill the crankcase with oil.

5.7L (VIN J) Engine

1. Disconnect battery negative cable.

2. Remove the primary timing chain and crankshaft sprocket.

3. Remove bolts attaching the oil pump to cylinder case, then remove the oil pump from the vehicle.

4. Remove O-ring from crankshaft.

To install:

5. Install new O-ring onto crankshaft.

6. Install oil pump and bolts, finger-tight only.

NOTE: Apply Loctite to bolt threads. Ensure the 2 flats of the pump drive gear are aligned with the 2 flats on the crankshaft. Do not force pump onto crankshaft.

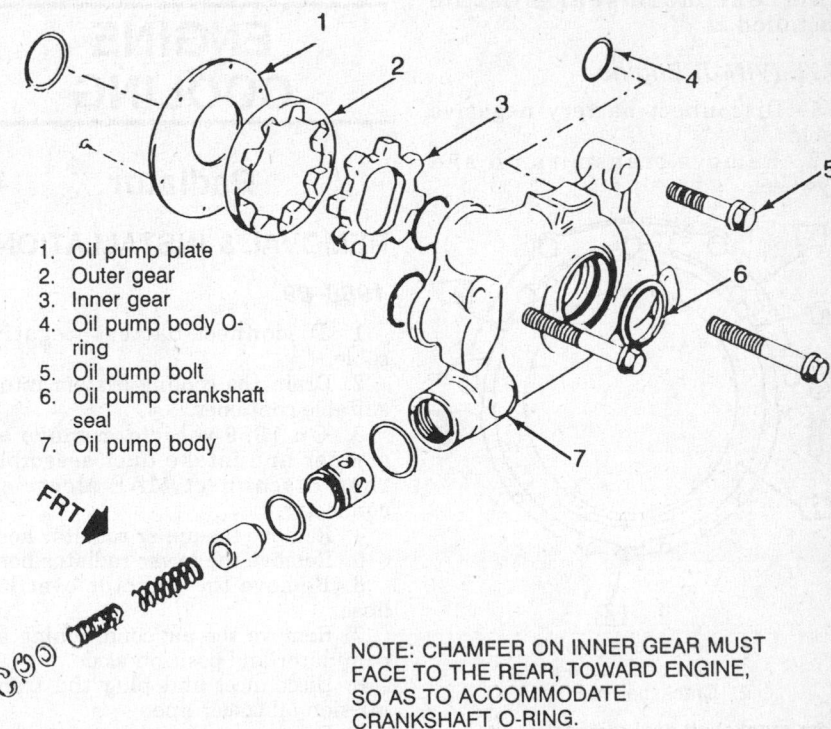

1. Oil pump plate
2. Outer gear
3. Inner gear
4. Oil pump body O-ring
5. Oil pump bolt
6. Oil pump crankshaft seal
7. Oil pump body

NOTE: CHAMFER ON INNER GEAR MUST FACE TO THE REAR, TOWARD ENGINE, SO AS TO ACCOMMODATE CRANKSHAFT O-RING.

Oil pump assembly—5.7L (VIN J) engine

7. Using oil pump aligning tool J–38383, align oil pump on the crankshaft. Torque oil pump bolts to 20 ft. lbs. (26 Nm).

8. Install a new oil pump shaft seal using tools J–38135 and J–38463.

NOTE: Install a new oil pump shaft seal whenever the pump is removed from the vehicle.

Rear Main Bearing Oil Seal

REMOVAL & INSTALLATION

5.7L (VIN 8) Engine

1. Disconnect the negative battery cable.

2. Raise the vehicle and support it safely.

3. Remove the transmission assembly.

4. Using the notches provided in the seal retainer, pry the old seal out.

To install:

5. Lubricate the inside and outside of a new seal with engine oil.

6. Install the seal on tool J–35621.

7. Thread the screw of the tool into the rear of the crankshaft.

8. Tighten the tools wingnut until it bottoms and then remove the tool.

9. Reinstall the transmission assembly.

NOTE: Whenever the seal retainer is removed, a new gasket and rear main seal must be installed.

5.7L (VIN J) Engine

1. Disconnect battery negative cable.

2. Remove transmission and flywheel.

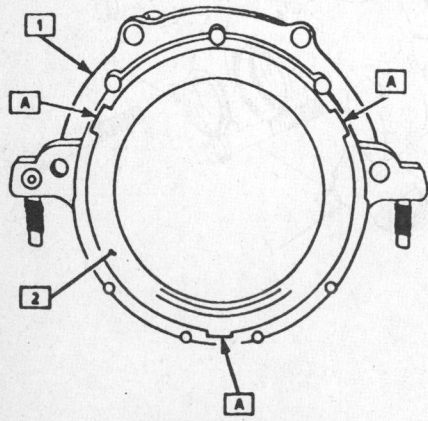

A. Seal retainer notch
1. Rear seal retainer
2. Crankshaft rear seal

Rear crankshaft seal removal points— 5.7L (VIN 8) engine

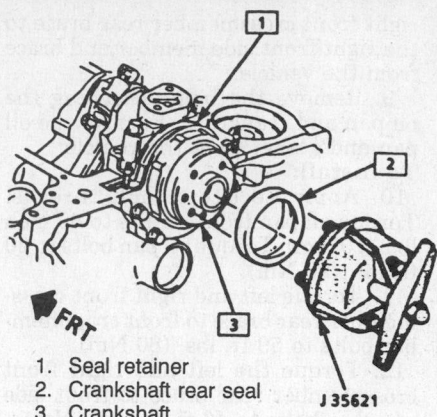

1. Seal retainer
2. Crankshaft rear seal
3. Crankshaft

Rear crankshaft seal installation—5.7L (VIN 8) engine

3. Remove the screws attaching the crankshaft rear main oil seal/housing assembly to the cylinder case.

4. Remove the seal/housing assembly from the engine.

5. Remove the seal from the housing.

To install:

6. Lubricate seal lip with engine oil.

7. Install seal into housing using crankshaft rear seal tool J–37312.

8. Seal should be installed 1.0–1.5mm below housing surface.

9. Torque seal housing bolts to 89 inch lbs. (10 Nm).

10. Install the flywheel and transmission assembly.

ENGINE COOLING

Radiator

REMOVAL & INSTALLATION

1988–89

1. Disconnect battery negative cable.

2. Drain the cooling system into a suitable container.

3. On 1989 vehicle, remove air cleaner and intake duct assembly, then disconnect MAF electrical connector.

4. Remove the upper radiator hose.

5. Remove the lower radiator hose.

6. Remove the radiator overflow hose.

7. Remove the air conditioning accumulator and position aside.

8. Disconnect and plug the transmission oil cooler line.

9. Disconnect the cooling fan wires from the fan and fan shroud.

10. Remove the fan to gain access to the lower transmission cooler line. Disconnect and plug the lower transmission cooler line.

11. On 1989 vehicle, remove the power steering reservoir bracket bolts from the fan shroud and from the frame.

12. Remove the upper fan shroud attaching bolts.

13. Remove the radiator assembly.

To install:

14. Install the radiator, shroud and retaining screws.

15. On 1989 vehicles, install the power steering reservoir bracket bolts at the shroud and frame.

16. Install the wiring harness and relay onto the shroud.

17. Connect the automatic transmission oil cooler line.

18. Install the air conditioning accumulator bracket.

19. Install cooling fan, if removed.

20. Install the radiator overflow hose, then the upper and lower radiator hoses.

21. Install the air cleaner and intake duct assembly.

22. Connect the MAF electrical connector.

23. Reconnect the negative battery cable.

24. Fill cooling system with the proper type and quantity of antifreeze. Start engine and check the cooling system for leaks.

1990–92

5.7L (VIN 8) ENGINE

1. Disconnect battery negative cable.

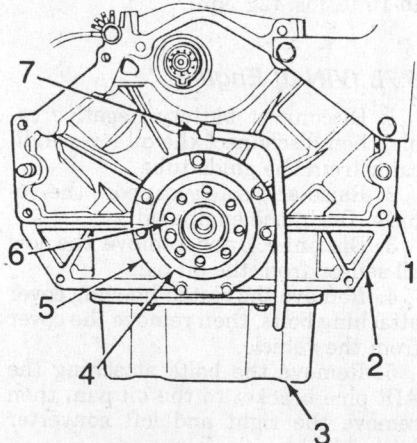

1. Cylinder case
2. Crankcase
3. Oil pan
4. Crankcase rear seal
5. Rear oil seal housing
6. Crankshaft
7. Cylinder case drain pipe

Crankshaft rear seal and housing—5.7L (VIN J) engine

2. Remove radiator pressure cap.

3. Drain coolant into a suitable container.

4. Remove the air cleaner assembly.

5. Disconnect the electrical connectors from the cooling fan relays.

6. Remove the bolts attaching the accumulator bracket to the radiator upper support.

7. Remove the fan shroud to upper support attaching bolts.

8. Remove the rubber access plug from the top of the radiator.

9. Remove the radiator air bleed hose.

10. Remove the upper bolts attaching the upper support to the front side member.

11. Remove the screws attaching the upper support to the lower support. Remove upper support.

12. Remove the radiator upper and lower hose clamps, then the hoses.

13. Disconnect and plug the transmission oil cooler lines from the radiator, if equipped.

14. Remove the radiator from the vehicle.

To install:

15. Install the radiator.

16. Connect transmission cooler lines to the radiator, if equipped.

17. Install the upper and lower radiator hoses.

18. Install upper support. Torque the upper support to front side member nuts and bolts to 18 ft. lbs. (25 Nm).

19. Connect cooling fan electrical connectors.

20. Install fan shroud to upper support attaching bolts. Torque bolts to 80 inch lbs. (9 Nm).

21. Install the accumulator bracket to upper support attaching bolts. Torque bolts to 80 inch lbs. (9 Nm).

22. Connect the radiator air bleed hose.

23. Install access plug and the air cleaner assembly.

24. Fill cooling system with the proper type and quantity of antifreeze.

25. Connect battery negative cable, then start engine and check for leaks.

5.7L (VIN J) ENGINE

1. Disconnect battery negative cable.

2. Remove radiator pressure cap.

3. Drain coolant into a suitable container.

4. Remove the air cleaner assembly and remove the radiator upper air deflector.

5. Disconnect the electrical connectors from the cooling fan relays.

6. Remove the bolts attaching the accumulator bracket to the radiator upper support.

7. Remove the fan shroud to upper support attaching bolts.

8. Remove the rubber access plug from the top of the radiator.

9. Remove the radiator air bleed hose.

10. Remove the upper bolts attaching the upper support to the front side member.

11. Disconnect the oil cooler lines at the oil cooler.

12. Remove the seal retainers and seal from the oil cooler and air conditioning line.

13. Remove the air pump assembly.

14. Remove the air pump brackets and air pump intake duct.

15. Remove the screws attaching the upper support to the lower support. Remove upper support.

16. Remove the radiator upper and lower hose clamps, then the hoses.

17. Remove the radiator from the vehicle.

To install:

18. Install the radiator.

19. Install the upper and lower radiator hoses.

20. Install upper support. Torque the upper support to front side member nuts and bolts to 18 ft. lbs. (25 Nm).

21. Install the air pump assembly.

22. Install the seal retainers and seal onto the oil cooler and air conditioning line.

23. Connect the oil cooler lines at the oil cooler.

24. Connect cooling fan electrical connectors.

25. Install fan shroud to upper support attaching bolts. Torque bolts to 80 inch lbs. (9 Nm).

26. Install the accumulator bracket to upper support attaching bolts. Torque bolts to 80 inch lbs. (9 Nm).

27. Connect the radiator air bleed hose.

28. Install the rubber access plug on the radiator support, the upper air deflector and the air cleaner assembly.

29. Fill cooling system with the proper type and quantity of antifreeze.

30. Connect battery negative cable, then start engine and check for leaks.

Electric Cooling Fans

REMOVAL & INSTALLATION

1988–89

1. Disconnect battery negative cable.

2. Remove the air cleaner and the intake air duct assembly.

3. Disconnect the MAF sensor and cooling fan electrical connector.

4. Remove fan assembly upper screws. Raise and support vehicle.

5. Remove cooling fan lower mounting screws, then remove engine cooling fan.

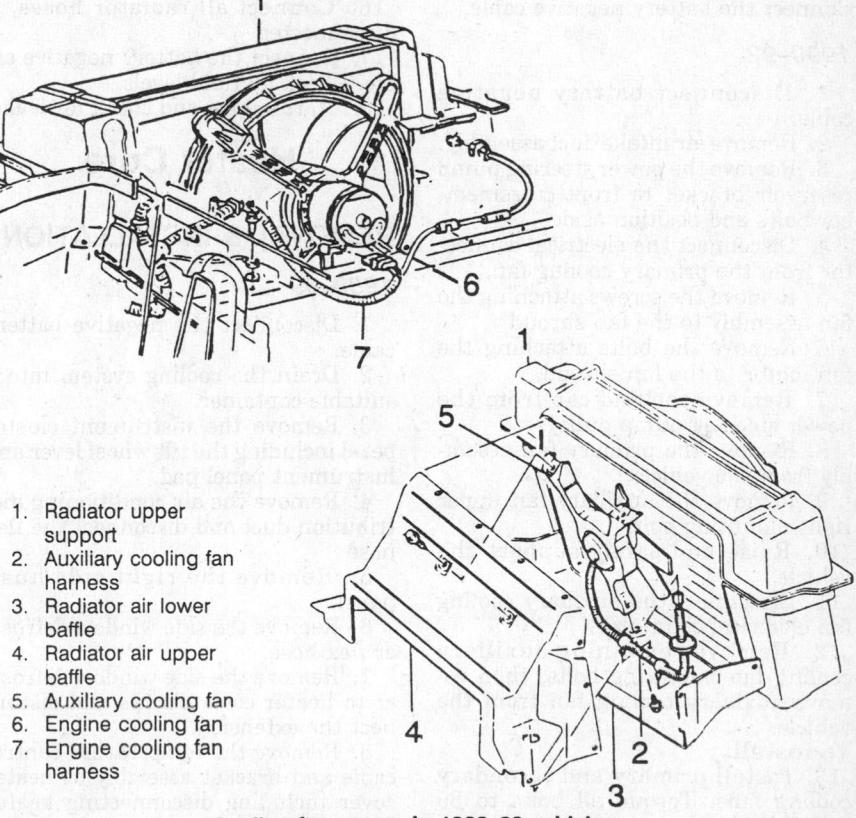

1. Radiator upper support
2. Auxiliary cooling fan harness
3. Radiator air lower baffle
4. Radiator air upper baffle
5. Auxiliary cooling fan
6. Engine cooling fan
7. Engine cooling fan harness

Cooling fan removal–1988–89 vehicles

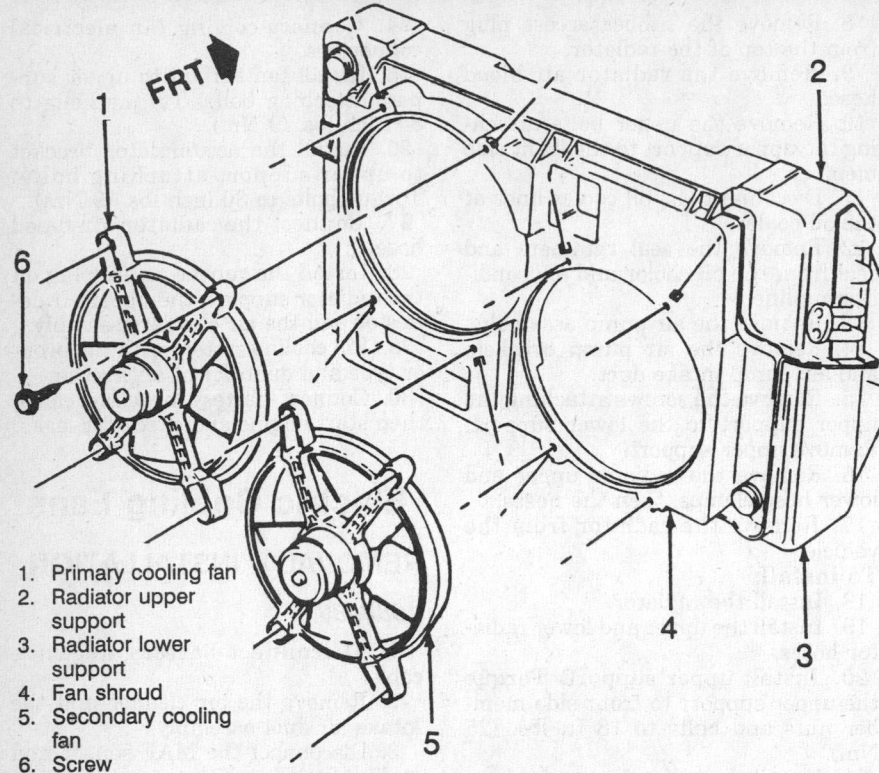

1. Primary cooling fan
2. Radiator upper support
3. Radiator lower support
4. Fan shroud
5. Secondary cooling fan
6. Screw

Cooling fan removal–1990–92 vehicles

6. Remove auxiliary cooling fan upper and lower mounting bolts.

7. Disconnect electrical connector and remove auxiliary fan from vehicle.

8. Reverse procedure to install. Connect the battery negative cable.

1990–92

1. Disconnect battery negative cable.

2. Remove air intake duct assembly.

3. Remove the power steering pump reservoir bracket to front crossmember bolts and position aside.

4. Disconnect the electrical connector from the primary cooling fan.

5. Remove the screws attaching the fan assembly to the fan shroud.

6. Remove the bolts attaching the fan motor to the fan shroud.

7. Remove the end cap from the power steering pump pulley.

8. Remove the primary fan assembly from the vehicle.

9. Remove the auxiliary fan upper right mounting bolt.

10. Raise and safely support the vehicle.

11. Disconnect the auxiliary cooling fan electrical connector.

12. Remove remaining auxiliary cooling fan mounting bolts, then remove auxiliary cooling fan from the vehicle.

To install:

13. Install primary and secondary cooling fans. Torque all bolts to 89 inch lbs. (10 Nm).

14. Connect cooling fan electrical connectors.

15. Install air conditioning discharge line bolt and the end cap on the power steering pump pulley.

16. Connect all radiator hoses, if disconnected.

17. Connect the battery negative cable. Check all fluid levels.

18. Start engine and check for leaks.

Heater Core

REMOVAL & INSTALLATION

1988–89

1. Disconnect the negative battery cable.

2. Drain the cooling system into a suitable container.

3. Remove the instrument cluster bezel including the tilt wheel lever and instrument panel pad.

4. Remove the air conditioning distribution duct and disconnect the flex hose.

5. Remove the right side hush panel.

6. Remove the side window defroster flex hose.

7. Remove the side window defroster to heater cover screws and disconnect the extension.

8. Remove the temperature control cable and bracket assembly at heater cover including disconnecting heater door control shaft.

9. Remove the Electronic Control Module (ECM) and disconnect the electrical connectors. Be sure the ignition switch is **OFF** when disconnecting the ECM.

10. Remove the tubular support brace from the door pillar to the instrument panel reinforcement brace.

11. Remove heater core cover attaching screws.

12. Remove heater pipe and heater water control bracket attaching screws.

13. Remove heater hose at the heater core pipes.

14. Remove the heater core from the case.

To install:

15. Install heater core, then retainers.

16. Install heater core case, then connect heater hoses.

17. Connect the heater control cable.

18. Install tubular brace, then the side window defroster duct.

19. Install ECM, then the instrument panel pad.

20. Install the floor heat deflector, right side hush panel and fuse panel bezel.

21. Install the instrument panel outlet duct, panel brace and upper pad.

22. Connect battery negative cable. Fill cooling system. Start engine and check for leaks.

1990–92

1. Properly disable the SIR air bag system and then disconnect the battery negative cable.

2. Remove the instrument panel upper trim, instrument panel cluster and console.

3. Drain engine coolant into a suitable container.

4. Disconnect the in-vehicle temperature senor aspirator hose.

5. Disconnect the in-vehicle temperature sensor electrical connector.

6. Remove the floor heat deflector attaching screws, right side knee bolster brace, then the floor heat deflector.

7. Disconnect the relays from the multi-use relay bracket.

8. Loosen the nuts attaching the wiring harness retainer to the radio, then remove the wiring harness retainer from the radio.

9. Remove the harnesses from the wiring harness retainer, then remove the wiring harness retainer.

10. Remove the carrier nuts from the right side pillar.

11. Remove the multi-use relay bracket.

12. Remove the passenger knee bolster brace attachments.

13. Unclip the side window defroster duct clip, then remove the duct hose from the knee bolster brace.

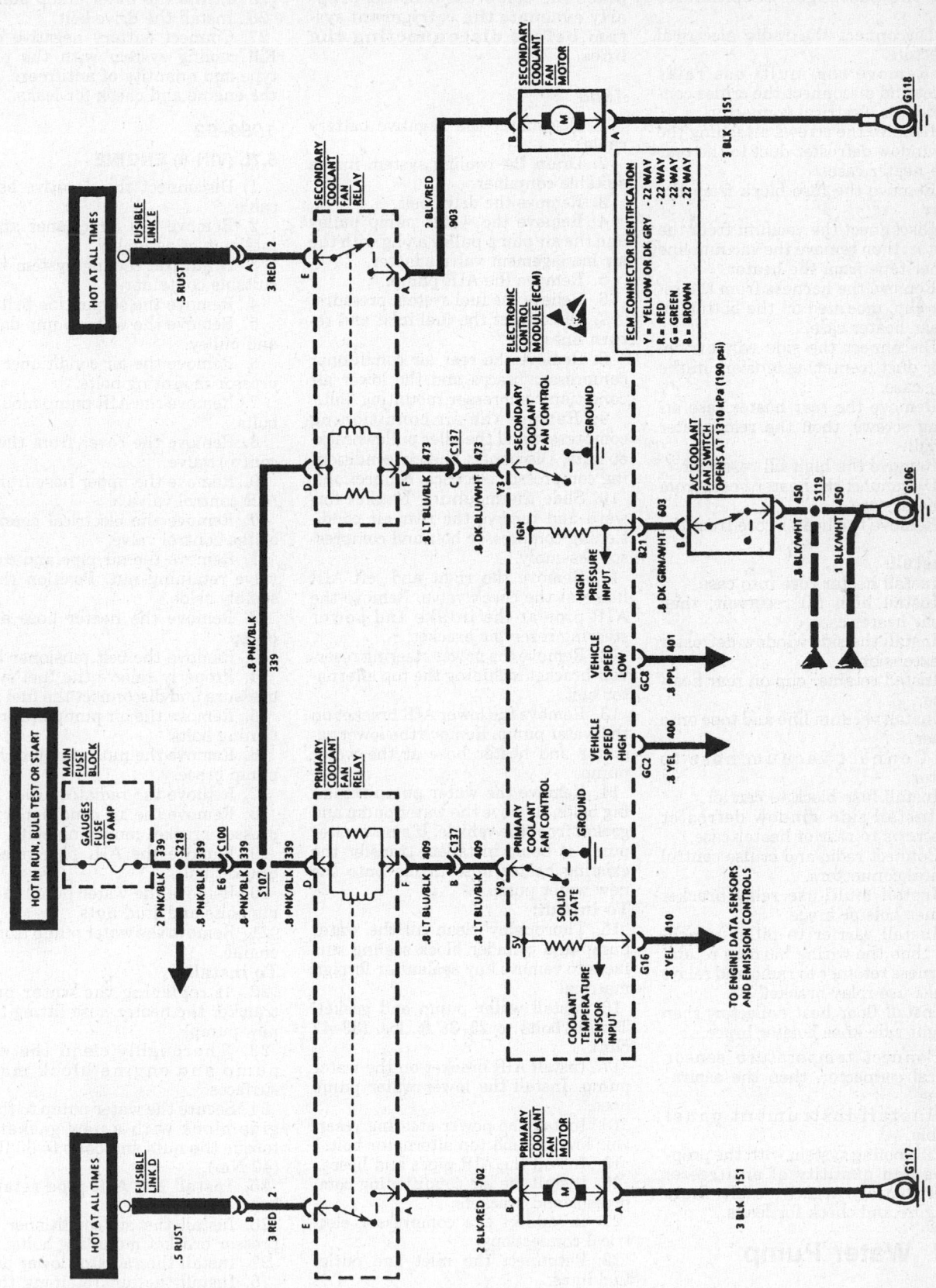

Cooling fan electrical schematic – 1990-92 vehicles

14. Pull the carrier back, then remove the passenger knee bolster brace.

15. Disconnect the radio electrical connectors.

16. Remove the multi-use relay bracket and disconnect the cruise control module electrical connector.

17. Remove the screws attaching the side window defroster duct to the rear of the heater case.

18. Remove the fuse block from the carrier.

19. Disconnect the vacuum from the actuator, then remove the vacuum line retainer tape from the heater.

20. Remove the harness from the retainer clip, mounted on the bottom of the rear heater case.

21. Disconnect the side window defroster duct (center) extension, in the heater case.

22. Remove the rear heater case attaching screws, then the rear heater case half.

23. Remove the high fill reservoir.

24. Disconnect the heater hoses from the heater core.

25. Remove the heater core from the case.

To install:

26. Install heater core into case.

27. Install high fill reservoir, then the rear heater case.

28. Install the side window defroster duct extension.

29. Install retainer clip on rear heater case.

30. Install vacuum line and tape onto retainer.

31. Connect vacuum hose to actuator.

32. Install fuse block to carrier.

33. Install side window defroster duct screws to rear of heater case.

34. Connect radio and cruise control electrical connectors.

35. Install multi-use relay bracket and knee bolster brace.

36. Install carrier to pillar attachment, then the wiring harness retainer, harness retainer to radio and relays to multi-use relay bracket.

37. Install floor heat deflector, then the right side knee bolster brace.

38. Connect temperature sensor electrical connector, then the aspirator hose.

39. Install instrument panel assembly.

40. Fill cooling system with the proper type and quantity of antifreeze. Connect battery negative cable. Start the engine and check for leaks.

Water Pump

REMOVAL & INSTALLATION

NOTE: If the air conditioner

compressor lines are to short to place the compressor aside, properly evacuate the refrigerant system before disconnecting the lines.

1988

1. Disconnect the negative battery cable.

2. Drain the cooling system into a suitable container.

3. Remove the drive belt.

4. Remove the water pump pulley and the air pump pulley along with the air management valve adapter.

5. Remove the AIR pump.

6. Relieve the fuel system pressure.

7. Disconnect the fuel inlet and return lines.

8. Remove the rear air conditioner compressor braces and the lower air condition compressor mounting bolt.

9. Remove the air conditioning compressor and the idler pulley bracket nuts. Disconnect the air conditioning compressor electrical connection.

10. Slide the mounting bracket forward and remove the rear air conditioning compressor bolt and compressor assembly.

11. Remove the right and left AIR hoses at the check valve. Remove the AIR pipe at the intake and power steering reservoir bracket.

12. Remove the power steering reservoir bracket including the top alternator bolt.

13. Remove the lower AIR bracket on the water pump. Remove the lower radiator and heater hose at the water pump.

14. Remove the water pump retaining bolts. Remove the water pump and gasket from the vehicle. If a new water pump is being installed, transfer the existing heater hose fitting onto the new water pump.

To install:

15. Thoroughly clean all the water pump and cylinder block sealing surfaces to remove any sealant or foreign material.

16. Install water pump and gasket. Torque bolts to 25–35 ft. lbs. (33–47 Nm).

17. Install AIR bracket on the water pump. Install the lower water pump hose.

18. Install the power steering reservoir bracket and top alternator bolt.

19. Install the AIR pipes and hoses.

20. Install the air conditioning compressor and brackets.

21. Reconnect the compressor electrical connections.

22. Reconnect the inlet and outlet fuel lines.

23. Install the air pump assembly and connect the air management valve.

24. Install radiator lower hose.

25. Install the water pump pulley.

26. Install the drive belt.

27. Connect battery negative cable. Fill cooling system with the proper type and quantity of antifreeze. Start the engine and check for leaks.

1989–92

5.7L (VIN 8) ENGINE

1. Disconnect the negative battery cable.

2. Remove the air cleaner and air intake duct assembly.

3. Drain the cooling system into a suitable container.

4. Remove the serpentine belt.

5. Remove the water pump damper and pulley.

6. Remove the air conditioner compressor mounting bolts.

7. Remove the AIR pump mounting bolts.

8. Remove the cover from the AIR control valve.

9. Remove the upper hose from the AIR control valve.

10. Remove the electrical connector at the control valve.

11. Remove the air pipe and control valve retaining nut. Position the assembly aside.

12. Remove the heater hose at the pump.

13. Remove the belt tensioner bolt.

14. Properly relieve the fuel system pressure and disconnect the fuel lines.

15. Remove the air pump bracket retaining bolts.

16. Remove the nut retaining the air pump brace.

17. Remove the radiator lower hose.

18. Remove the air conditioner compressor bracket mounting bolts.

19. Remove the AIR pipe retaining bracket bolt.

20. Remove the water pump mounting bolts and stud nuts.

21. Remove the water pump from the engine.

To install:

22. If replacing the water pump, transfer the heater hose fitting to the new pump.

23. Thoroughly clean the water pump and engine block mating surfaces.

24. Secure the water pump to the engine block with a new gasket and torque the nuts and bolts to 30 ft. lbs. (40 Nm).

25. Install the AIR pipe retaining bracket bolt.

26. Install the air conditioner compressor bracket mounting bolts.

27. Install the radiator lower hose.

28. Install the nut retaining the air pump brace.

29. Install the air pump bracket retaining bolts.

30. Connect the fuel lines.
31. Install the belt tensioner bolt.
32. Install the heater hose at the pump.
33. Install the air pipe and control valve retaining nut. Connect the electrical connector to the control valve.
34. Install the upper hose and the cover onto the AIR control valve.
35. Install the AIR pump mounting bolts.
36. Install the air conditioner compressor mounting bolts.
37. Install the water pump damper and pulley and install the serpentine belt.
38. Install the air cleaner and air intake duct assembly.
39. Refill the cooling system with the proper type and quantity of antifreeze.
40. Reconnect the negative battery cable. Start the engine and inspect system for leaks.

5.7L (VIN J) ENGINES

1. Disconnect the battery negative cable.
2. Drain engine coolant into a suitable container.
3. Disconnect the air intake duct.
4. Remove the screws attaching the throttle body extension to the throttle body, then remove the throttle body extension and gasket.
5. Remove clamps and hoses from the coolant outlets, radiator inlet and inlet pipe.
6. Remove the inlet pipe assembly and hose from the vehicle.
7. Loosen the coolant pump pulley attaching bolts, then rotate the belt tensioner.
8. Remove the bolts from the pulley, then the pulley from the vehicle.
9. Release the belt tensioner, then remove the belt from the vehicle. Re-

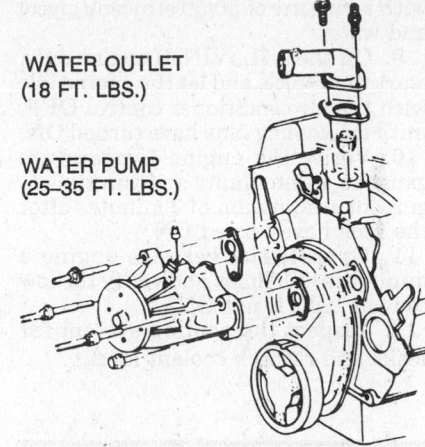

WATER OUTLET (18 FT. LBS.)

WATER PUMP (25–35 FT. LBS.)

Water pump assembly—5.7L (VIN 8) engine

move the belt tensioner bolt and remove the tensioner.
10. Remove the water pump hose clamp, then the hose from the water pump.
11. Remove the alternator lower bracket mounting bolts, then remove the bracket from the vehicle.
12. Remove the water pump attaching bolts and the bolt attaching the air conditioning compressor to the water pump, then remove the water pump from the vehicle.

To install:
13. Clean pump and front cover sealing surfaces.
14. Install water pump, gasket and bolts, finger-tight only.
15. Install bolt attaching air conditioning compressor to pump.
16. Torque air conditioning compressor bolt and water pump attaching bolts to 20 ft. lbs. (26 Nm).
17. Install engine hoses and clamps.
18. Install alternator bolts. Apply

Loctite to bolt threads. Torque alternator mounting bolts to 39 ft. lbs. (52 Nm) and bracket bolts to 20 ft. lbs. (26 Nm).
19. Install the belt tensioner. Torque belt tensioner bolt to 45 ft. lbs. (60 Nm).
20. Install serpentine belt. Torque water pump pulley bolts to 89 inch lbs. (10 Nm).
21. Install water pump hose and inlet pipe assembly.
22. Install throttle body extension and gasket. Torque bolts to 53 inch lbs. (6 Nm).
23. Install air intake duct. Refill the cooling system with the proper type and quantity of antifreeze.
24. Connect battery negative cable. Start the engine and inspect the system for leaks.

Thermostat

REMOVAL & INSTALLATION

5.7L (VIN 8) Engine

1. Disconnect the battery negative cable.
2. Drain the engine coolant into a suitable container.
3. Remove the air cleaner and intake duct assembly, then disconnect the radiator upper hose from the outlet.
4. Remove the coolant hose from the throttle body.
5. Disconnect EGR electrical connector and vacuum harness from the EGR solenoid.
6. Remove the thermostat housing attaching bolts. Remove the thermostat and housing.
7. Installation is the reverse of the

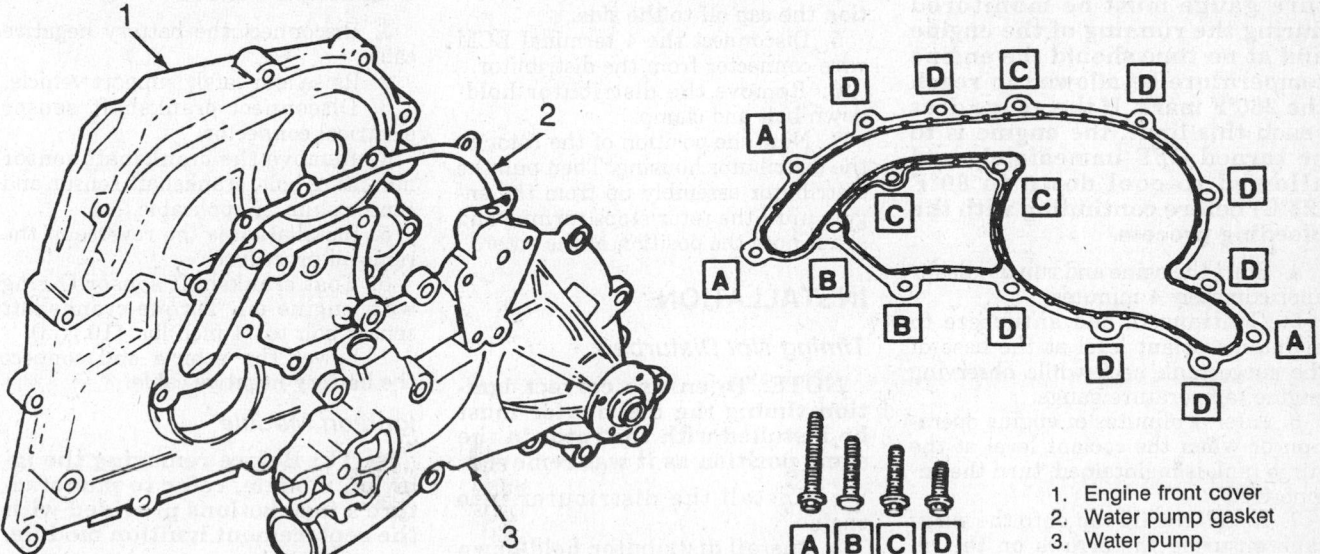

1. Engine front cover
2. Water pump gasket
3. Water pump

Water pump assembly—5.7L (VIN J) engine

removal procedure. Torque the thermostat housing mounting bolts to 25 ft. lbs. (34 Nm).

5.7L (VIN J) Engine

1. Disconnect the battery negative cable.
2. Drain the engine coolant into a suitable container.
3. Raise and safely support the vehicle.
4. Remove thermostat housing attaching bolts.
5. Loosen the bolts attaching the housing assembly bracket to the front side member.
6. Remove the thermostat and gasket from the housing.
7. Installation is the reverse of the removal procedure. Torque the thermostat housing and bracket bolts to 18 ft. lbs. (25 Nm).

Cooling System Bleeding

1. With the cooling system completely drained, the engine **OFF** and radiator drainplug closed, begin adding antifreeze. Use a combination of 50 percent ethylene glycol antifreeze and 50 percent water for system refills (no flush) and, if the system was flushed, use a 100 percent ethylene glycol and then add water to bring system to the correct level.
2. Slowly fill the cooling system through the opening in the high fill surge tank.
3. Wait 2 minutes and recheck the level. Add water to bring the level up to the base of the surge tank neck.

NOTE: The coolant temperature gauge must be monitored during the running of the engine and at no time should the engine temperature be allowed to reach the 260°F mark. If the gauge does reach this limit, the engine is to be turned OFF immediately and allowed to cool down to 80°F (27°C) before continuing with the bleeding process.

4. Start the egine and run at idle for approximately 4 minutes.
5. Continue to add antifreeze to maintain coolant level at the base of the surge tank neck while observing engine temperature gauge.
6. After 4 minutes of engine operation or when the coolant level at the surge tank is maintained, turn the engine **OFF**.
7. Install the fill cap onto the surge tank ensuring the arrows on the fill cap line up with the overflow tube.
8. Fill the coolant reservoir tank

with a mixture of 50/50 ethylene glycol and water.
9. On the 5.7L (VIN J) engine only, start the vehicle and let the engine idle with the air conditioner control **OFF**, until the cooling fans have turned **ON**.
10. Verify the engine temperature gauge is within limits and run the engine for a minimum of 3 minutes after the fans have turned **ON**.
11. Operate the vehicles engine a minimum of 3 times and verify the low coolant light is not **ON**.
12. Inspect the cooling system for leaks and recheck coolant level.

ENGINE ELECTRICAL

NOTE: Disconnecting the negative battery cable on some vehicles may interfere with the functions of the on board computer system and may require the computer to undergo a relearning process when the negative battery cable is reconnected.

Distributor

REMOVAL

1. Disconnect the negative battery cable.
2. Remove the intake manifold plenum extension.
3. Disconnect the battery feed wire and tachometer wire from the distributor cap.
4. Turn the distributor cap retaining screws counterclockwise and position the cap off to the side.
5. Disconnect the 4 terminal ECM wire connector from the distributor.
6. Remove the distributor hold-down bolt and clamp.
7. Note the position of the rotor to the distributor housing. Then pull the distributor assembly up from the engine until the rotor stops turning and again note the position of the rotor.

INSTALLATION

Timing Not Disturbed

NOTE: To ensure correct ignition timing the distributor must be installed with the rotor in the same position as it was removed.

1. Install the distributor into engine.
2. Install distributor hold-down clamp and bolt.
3. Install the distributor cap, then

connect the ECM wire connector to distributor.
4. Connect the battery feed wire and tachometer wire to the distributor cap.
5. Install spark plug wires, if removed.
6. Install the intake manifold plenum extension.
7. Connect battery negative cable and check the ignition timing.

Timing Disturbed

1. If the engine has been cranked with the distributor out, remove the No. 1 spark plug.
2. Verify the transmission is **N** or **P**. Place a finger over the spark plug hole and crank the engine slowly until compression is felt.
3. Align the timing mark on the pulley to **0** on the engine timing indicator. Position the rotor between the No. 1 and No. 8 spark plug towers.
4. The distributor can now be correctly installed in the engine.
5. Install distributor hold-down clamp and bolt.
6 Install the distributor cap, then connect the ECM wire connector to distributor.
7. Connect the battery feed wire and tachometer wire to the distributor cap.
8. Install spark plug wires, if removed.
9. Install the intake manifold plenum extension.
10. Connect battery negative cable and check the ignition timing.

Distributorless Ignition

REMOVAL & INSTALLATION

Crankshaft Sensor

1. Disconnect the battery negative cable.
2. Raise and safely support vehicle.
3. Disconnect crankshaft sensor electrical connector.
4. Remove the crankshaft sensor mounting bolt, crankshaft sensor and sensor shim, if applicable.
5. Installation is the reverse of the removal procedure.
6. Coat crankshaft sensor O-ring with engine oil. Torque crankshaft sensor bolt to 71 inch lbs. (10 Nm).
7. Lower the vehicle and connect the battery negative cable.

Ignition Module

NOTE: Before removing the ignition module, refer to manufacture's instructions provided with the replacement ignition module.

1. Disconnect the battery negative cable.

2. Remove the intake manifold plenum. The ignition module is mounted on the bottom of the plenum.

3. Disconnect the electrical connectors from the ignition module.

4. Remove the mounting bolts, then the ignition module.

5. Installation is the reverse of the removal procedure.

6. Apply a suitable dielectric grease to the back of the ignition module. Torque the 4 mounting bolts to 89 inch lbs. (10 Nm).

Ignition Coil

1. Disconnect the battery negative cable.

2. Remove the intake manifold plenum.

3. Disconnect and tag the spark plug wires from the ignition coil pack.

4. Remove the 2 mounting bolts and the ignition coil pack.

5. Installation is the reverse of the removal procedure.

6. Torque mounting bolts to 40 inch lbs. (4.5 Nm).

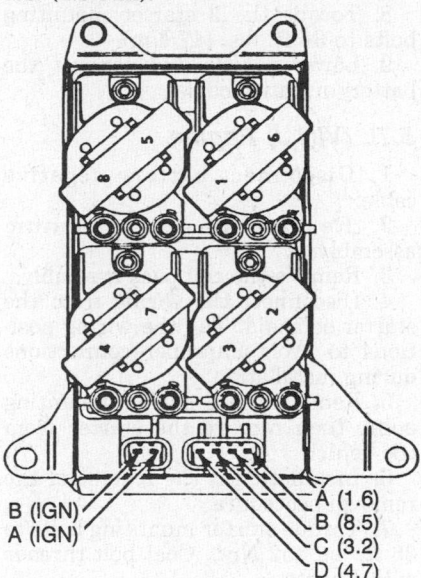

B (IGN)
A (IGN)
A (1.6)
B (8.5)
C (3.2)
D (4.7)

Ingition coil pack

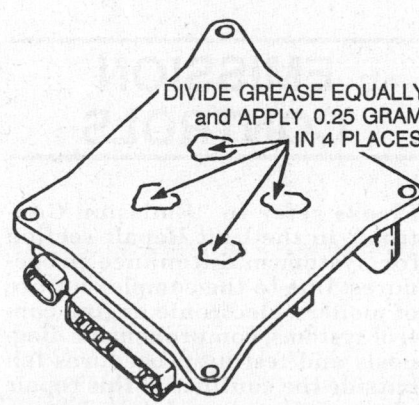

DIVIDE GREASE EQUALLY and APPLY 0.25 GRAM IN 4 PLACES

Ignition module dielectric grease application

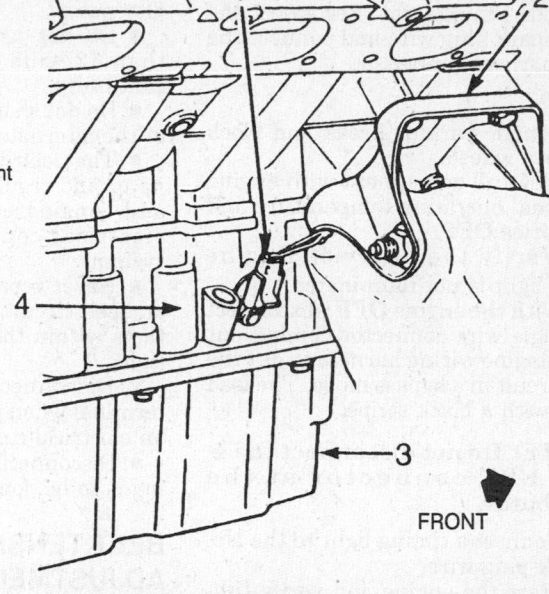

1. Crankshaft sensor
2. Right engine mount bracket
3. Oil pan
4. Engine block

FRONT

Crankshaft sensor location

Ignition Housing

1. Disconnect battery negative cable.

2. Remove intake manifold plenum.

3. Disconnect and tag electrical connectors and spark plug wires.

4. Remove the 4 ignition housing bracket mounting bolts. Remove the ignition coil mounting bolts, Noting the position of each coil.

5. Remove the coils from the ignition housing, then the ignition housing from the bracket.

6. Installation is the reverse of the removal procedure.

7. Torque coil retaining bolts to 40 inch lbs. (4.5 Nm). Torque ignition housing mounting bolts as follows: M6–16 bolts—89 inch lbs.; M8–20 bolts—19 ft. lbs.

Ignition Timing

Electronic Spark Control (ESC) retards the spark advance when engine detonation occurs. If the controller fails, the result could be no ignition, no retard or full retard. Some engines will also have a magnetic timing probe hole for use with electronic timing equipment. The use of an inductive pick-up timing light is recommended.

ADJUSTMENT

5.7L (VIN 8) Engine

1988–89

NOTE: It is not necessary to adjust the idle speed prior to the timing adjustment, though the engine must be at normal operating temperature.

1. With the ignition **OFF**, connect the pick-up lead of the timing light to the No. 1 spark plug. Use a jumper lead between the wire and the plug or an inductive type pick-up.

2. Connect the timing light power leads according to the manufacturer's instructions.

3. Disconnect the ECM 4-wire harness connector at the distributor.

4. Start the engine and run it at idle speed. Aim the timing light at the timing mark. The line on the balancer or pulley will align with the timing mark.

5. If required, adjust the timing by loosening the securing clamp hold-down bolt and rotating the distributor until the desired specification is achieved, then tighten the bolt.

6. To advance the timing, rotate the distributor opposite to the normal direction of rotor rotation. Retard the timing by rotating the distributor in the normal direction of rotor rotation.

7. Tighten the hold-down bolt to 25 ft. lbs. (34 Nm).

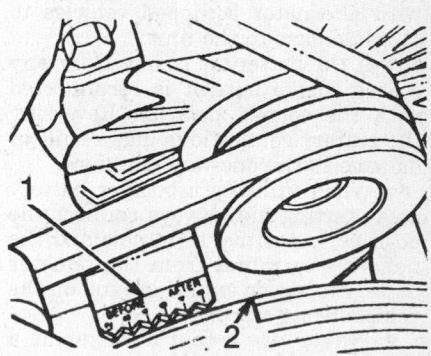

1. Timing indicator
2. Timing mark

Ignition timing mark

8. Turn the ignition **OFF** and remove the timing light. Reconnect the No. 1 spark plug wire and connect the ECM harness connector.

1990–92

1. Set the parking break and block the drive wheels.
2. Make all adjustment with engine at normal operating temperature and accessories **OFF**.
3. Verify the "Service Engine Soon" light is not illuminated.
4. With the engine **OFF**, disconnect the single wire connector coming out of the engine wiring harness to put the EST circuit in a bypass mode. The lead is tan with a black stripe.

NOTE: **Do not disconnect the 4-wire EST connector at the distributor.**

5. Connect a timing light to the No. 1 spark plug wire.
6. Start the engine and verify timing is at 6° BTDC at idle.
7. If adjustment is necessary, loosen the distributor clamp bolt and turn distributor to obtain the specified setting.
8. Retighten the hold-down bolt to 25 ft. lbs. (34 Nm.) and recheck the timing.
9. Turn the engine **OFF** and reconnect EST bypass connector.
10. Clear ECM codes and confirm there is no "Service Engine Soon" light illuminated.

5.7L (VIN J) Engine

Base timing is preset at the factory, no adjustment is possible. Timing advance and retard are accomplished by the ECM with EST and Electronic Spark Control.

Alternator

PRECAUTIONS

Several precautions must be observed with alternator equipped vehicles to avoid damage to the unit.
- If the battery is removed for any reason, make sure it is reconnected with the correct polarity. Reversing the battery connections may result in damage to the one-way rectifiers.
- When utilizing a booster battery as a starting aid, always connect the positive to positive terminals and the negative terminal from the booster battery to a good engine ground on the vehicle being started.
- Never use a fast charger as a booster to start vehicles.
- Disconnect the battery cables when charging the battery with a fast charger.

- Never attempt to polarize the alternator.
- Do not use test lights of more than 12 volts when checking diode continuity.
- Do not short across or ground any of the alternator terminals.
- The polarity of the battery, alternator and regulator must be matched and considered before making any electrical connections within the system.
- Never separate the alternator on an open circuit. Make sure all connections within the circuit are clean and tight.
- Disconnect the battery ground terminal when performing any service on electrical components.
- Disconnect the battery if arc welding is to be done on the vehicle.

BELT TENSION ADJUSTMENT

A single serpentine belt is used to drive all accessories. Belt tension is maintained by a spring loaded tensioner which has the ability to maintain belt tension over a broad range of belt lengths. There is an indicator to make sure the tensioner is adjusted to within its operating range.

To check the belt tension, install belt tension tool J–23600 between the alternator and the air pump. The correct belt tension should be 120–140 ft. lbs. (534–623 Nm.) on 1988–89 vehicles or 60–90 ft. lbs. (267–400 Nm.) on 1990–92 vehicles.

REMOVAL & INSTALLATION

1. Disconnect the battery negative cable.
2. Remove the air intake duct hose.
3. Retract the tensioner using a ½ inch breaker bar, then remove the belt.
4. Installation is the reverse of the removal procedure.

Starter

REMOVAL & INSTALLATION

Except 5.7L (VIN J) Engine

1. Disconnect the negative battery cable.
2. Raise and support the vehicle safely.
3. Remove the flywheel cover.
4. Disconnect the wiring from the starter solenoid. Tag the wiring positions to avoid improper connections during installation.
5. Loosen the 2 starter mounting

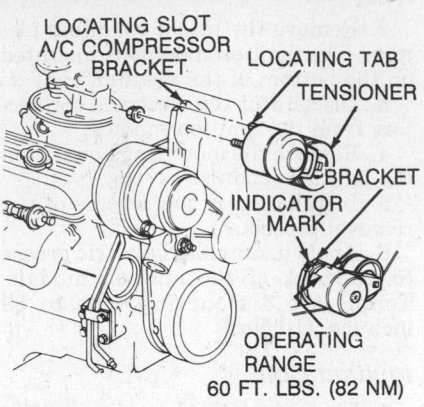

Drive belt tensioner location and adjustments

bolts, support the starter and remove the bolts. Lower the starter from the vehicle and remove shims, if equipped.
6. Installation is the reverse of the removal procedure.
7. Check the flywheel to pinion clearance. The clearance should be 0.020 in. (0.5mm).
8. Torque the 2 starter mounting bolts to 34 ft. lbs. (47 Nm).
9. Lower vehicle and connect the battery negative cable.

5.7L (VIN J) Engine

1. Disconnect battery negative cable.
2. Remove the intake plenum assembly.
3. Remove the coil pack assembly.
4. Disconnect the wiring from the starter solenoid. Tag the wiring positions to avoid improper connections during installation.
5. Remove the 2 starter mounting bolts, then remove the starter from the vehicle.
6. Installation is the reverse of the removal procedure.
7. Torque starter mounting bolts to 38 ft. lbs. (52 Nm). Coat bolt threads with Loctite.

EMISSION CONTROLS

Please refer to "Emission Controls" in the Unit Repair section for system maintenance procedures. Due to the complex nature of modern electronic engine control systems, comprehensive diagnosis and testing procedures fall outside the confines of this repair manual. For complete information on diagnosis, testing and repair procedures concerning all modern engine and emission control systems, please refer to "Chilton's Guide to Fuel Injection and Electronic Engine Controls".

Emission Warning Lamps

RESETTING

When the Service Engine Soon light located on the instrument panel illuminates, it can be used to indicates any of the following malfunctions:

The light informs the driver that a problem has occurred in the fuel, ignition or emissions and the vehicle should be serviced as soon as possible.

It displays trouble codes stored by the ECM.

It indicates Open Loop or Closed Loop operation.

The light will come ON when the key is turned ON and the engine has not yet been started.

If the light remains ON, the self-diagnostic system has detected a problem. If the problem is intermittent, the light will go out in most cases after 10 seconds but a trouble code will remain stored in the ECM.

To determine if a malfunction will occur again or because repairs have been completed, the codes must be cleared from the memory of the ECM.

To clear the codes the ECM power feed must be disconnected for at least 30 seconds.

Depending on how the vehicle is equipped, the ECM power feed can be disconnected at the positive battery terminal pigtail. The inline fuse holder that originates at the positive cable or the ECM fuse in the fuse block.

If the negative battery terminal is disconnected, other on-board memory data, such as pre-set radio tuning will have to be reset.

FUEL SYSTEM

Fuel System Service Precautions

Safety is the most important factor when performing not only fuel system maintenance but any type of maintenance. Failure to conduct maintenance and repairs in a safe manner may result in personal injury or death.

Maintenance and testing of the vehicle's fuel system components can be accomplished safely and effectively by adhering to the following rules and guidelines.

• To avoid the possibility of fire and personal injury, always disconnect the negative battery cable unless the repair or test procedure requires that battery voltage remain connected.

• Always relieve the fuel system pressure prior to disconnecting any fuel system component (injector, fuel rail, pressure regulator, etc.), fitting or fuel line connection. Exercise extreme caution whenever relieving fuel system pressure to avoid exposing skin, face and eyes to fuel spray. Please be advised that fuel under pressure may penetrate the skin or any part of the body that it contacts.

• Always place a shop towel or cloth around the fitting or connection prior to loosening, to absorb any excess fuel due to spillage.

Ensure that all fuel spillage (should it occur) is quickly removed from engine surfaces. Ensure that all fuel soaked cloths or towels are deposited into a suitable waste container.

• Always keep a dry chemical (Class B) fire extinguisher near the work area.

• Do not allow fuel spray or fuel vapors to come into contact with a spark or open flame.

• Always use a backup wrench when loosing and tightening fuel line connection fittings. This will prevent unnecessary stress and torsion to fuel line piping. Always follow the proper torque specifications.

• Always replace worn fuel fitting O-rings with new. Do not substitute fuel hose or equivalent, where fuel pipe is installed.

RELIEVING FUEL SYSTEM PRESSURE

1. Disconnect the negative battery cable.
2. Loosen the fuel filler cap to relieve the tank pressure.
3. Remove the MAP sensor and bracket to gain access to the fuel pressure valve, if necessary.
4. Connect a fuel gauge J–34730–1 or equivalent, to the fuel pressure tap.
5. Wrap a shop towel around the fitting while connecting the gauge to catch any fuel spray.
6. Install a bleed hose into a suitable container, then open the valve to bleed the fuel system pressure.

Fuel Tank

REMOVAL & INSTALLATION

NOTE: Ensure an approved dry chemical (Class B) fire extinguisher is near the work area.

1. Disconnect the negative battery cable.

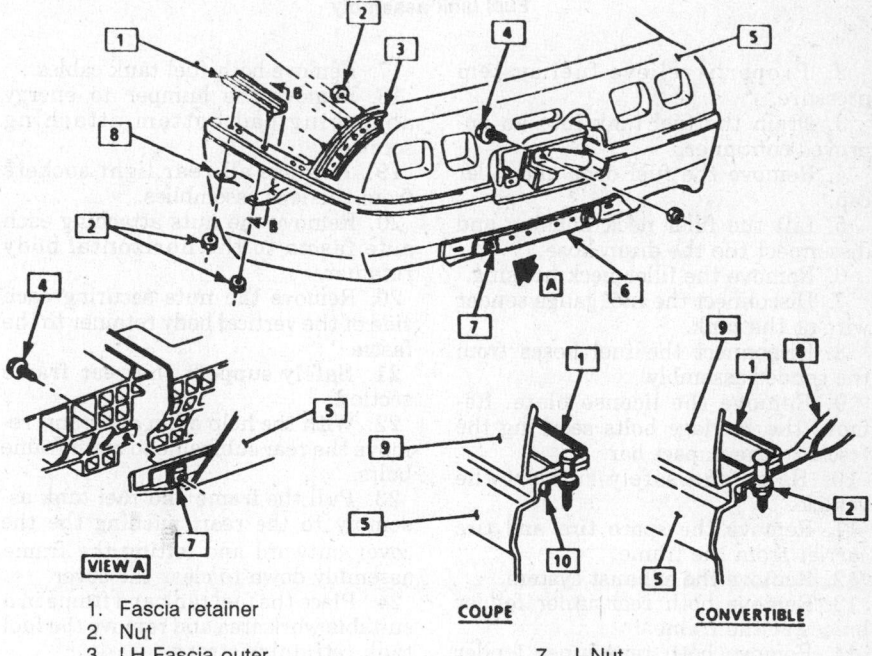

VIEW A

1. Fascia retainer
2. Nut
3. LH Fascia outer retainer
4. Bolt
5. Rear bumper fascia
6. Fascia lower reinforcement

COUPE   **CONVERTIBLE**

7. J–Nut
8. LH Fascia support
9. Upper rear body panel
10. LH Fascia upper retainer

Rear fascia assembly

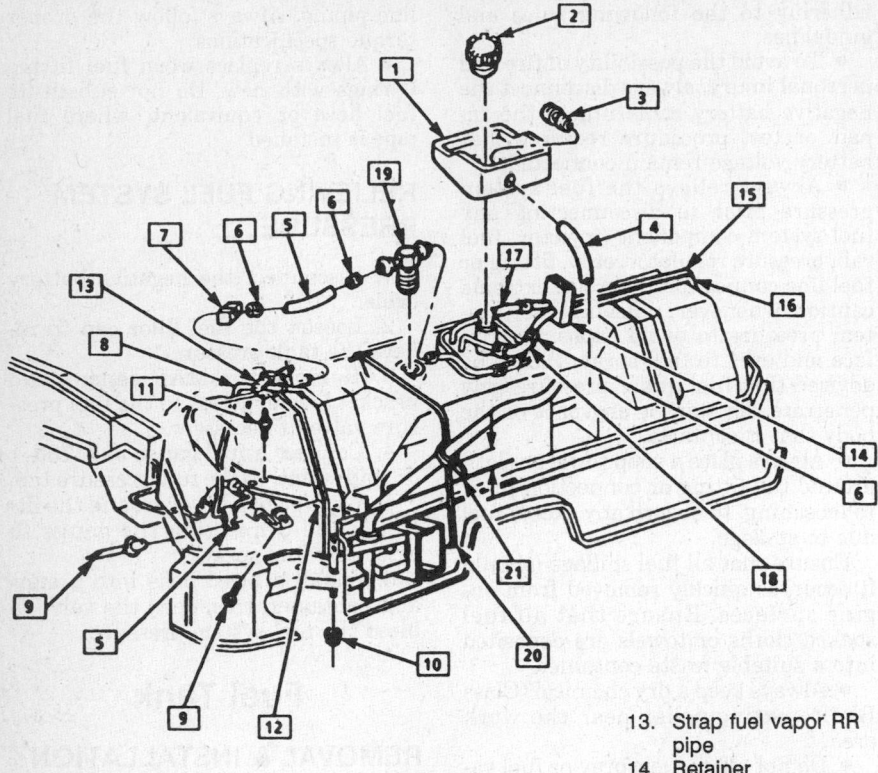

1. Filler neck housing
2. Fuel cap
3. Nipple
4. Drain hose
5. Fuel vapor pipe clamp
6. Clamp
7. Retainer
8. Vapor pipe
9. Rivet
10. Nut
11. Rivet
12. Support
13. Strap fuel vapor RR pipe
14. Retainer
15. Fuel feed pipe
16. Fuel return pipe
17. Fuel feed hose
18. Fuel return hose
19. Fuel vapor connector
20. Strap
21. Strap

Fuel tank assembly

2. Properly relieve fuel system pressure.

3. Drain the fuel tank into an approved container.

4. Remove the fuel door and filler cap.

5. Lift the filler neck housing and disconnect the the drain hose.

6. Remove the filler neck housing.

7. Disconnect the fuel gauge sender wire at the tank.

8. Disconnect the fuel hoses from the sender assembly.

9. Remove the license plate. Remove the carriage bolts securing the fascia to the impact bar.

10. Raise and safely support the vehicle.

11. Remove the spare tire and tire carrier from the frame.

12. Remove the exhaust system.

13. Remove both rear inner fender braces at the frame.

14. Remove both rear inner fender panels.

15. Remove the antenna ground strap and clip at the antenna base and frame.

16. Disconnect the fuel vapor hose at the bottom of the tank.

17. Remove both fuel tank cables.

18. Remove the bumper to energy absorbing pad bottom attaching screws.

19. Remove all rear light sockets from the lens assemblies.

20. Remove the nuts attaching each side fascia to the horizontal body retainer.

20. Remove the nuts securing each side of the vertical body retainer to the fascia.

21. Safely support the rear frame section.

22. With the help of an assistant, remove the rear subframe to main frame bolts.

23. Pull the frame and fuel tank assembly to the rear, pushing the the cover outward and letting the frame assembly down to clear the cover.

24. Place the fuel tank and frame in a suitable work area and remove the fuel tank retraining straps.

25. Remove the fuel tank.

To install:

26. Install the fuel tank into the rear subframe assembly and secure with retaining straps. Torque the straps bolts to 11 ft. lbs. (15 Nm).

27. Install the rear subframe and fuel tank assembly to the main frame with the mounting bolts.

28. Install rear and side fascia to body retainer with attaching nuts.

29. Install all rear light sockets into lens assemblies.

30. Attach screws from bottom of bumper to energy absorber pad.

31. Reattach fuel tank cables.

32. Attach ground straps from antenna to frame.

33. Connect fuel vapor hose to bottom of tank vapor pipe.

34. Install both inner fender braces and panels.

35. Install the exhaust system.

36. Install the spare tire and carrier assembly.

37. Lower the vehicle.

38. Install the carriage bolts securing the fascia and install the license plate.

39. Connect the fuel lines and hoses.

40. Connect the fuel sender wire connector.

41. Install the filler neck housing.

42. Connect the filler housing drain hose.

43. Refill the tank and connect the negative battery cable.

44. Inspect system for leaks.

Fuel Filter

REMOVAL & INSTALLATION

1. Properly relieve the fuel system pressure.

2. Disconnect the battery negative cable.

3. Raise and safely support the vehicle.

4. Remove fuel filter shield attaching screws, then the fuel filter shield, if equipped.

5. Disconnect fuel lines from the fuel filter.

6. Remove the fuel filter attaching screws, then the fuel filter.

Electric Fuel Pump

PRESSURE TESTING

1. Properly relieve fuel system pressure.

2. With ignition **OFF**, install fuel pressure gauge J–34730–1 or equivalent, to the fuel rail pressure connection.

3. Remove the secondary fuel pump fuse, if equipped with 5.7L (VIN J) engine.

4. Turn the ignition switch **ON**. The fuel pump will operate for 2 seconds and then turn **OFF**.

5. Fuel pressure should be 41–47 psi, if equipped with the 5.7L (VIN 8) engine or 48–55 psi, if equipped with the 5.7L (VIN J) engine.

6. The pressure should hold steady when the pump stops, with little or no pressure drop.

7. Start the engine and the pressure should drop approximately 3–10 psi.

REMOVAL & INSTALLATION

1988–89

NOTE: **The fuel pump(s) are mounted to the fuel gauge sender inside the fuel tank.**

1. Disconnect the negative battery cable.
2. Relieve the fuel system pressure.
3. Remove the fuel tank filler door and drain tube.
4. Disconnect the feed, return and vapor hoses from the sending unit.
5. Remove the license plate to gain access for removal of the 2 bolts securing the fascia to impact bar.
6. Raise and safely support the vehicle.
7. Remove the spare tire and carrier from the frame.
8. Disconnect the intermediate exhaust pipe at the converter. Remove the intermediate pipe and mufflers as an assembly from the vehicle.
9. Remove both rear inner fender braces at the frame.
10. Remove both rear inner fender panels.
11. Remove the antenna ground strap and clip.
12. Disconnect the fuel vapor pipe from the left side fuel tank strap.
13. Disconnect the fuel tank cables from the rear stabilizer shaft brackets.
14. Remove the screws securing the bottom edge of the fascia to the energy absorber pad.
15. Remove all rear lights.
16. Disconnect each side of the fascia from the horizontal body retainer.
17. Disconnect the right and left vertical retainers securing the fascia to the body.
18. Remove the 6 frame bolts and loosen the 2 front frame bolts.
19. Remove the front frame bolts. Pull the tank and frame assembly to the rear pushing the cover outward and letting the rear of the frame assembly down to clear the cover.
20. Remove the vapor hose from the vapor connector and remove the tank and frame assembly.
21. Remove the fuel sending unit and pump assembly by turning the cam lock ring counterclockwise. Lift the assembly from the fuel tank.
22. Disconnect the fuel pump from the fuel level sending unit.

To install:

23. Installation is the reverse of the removal procedure, noting the following:

a. Always use a new gasket when

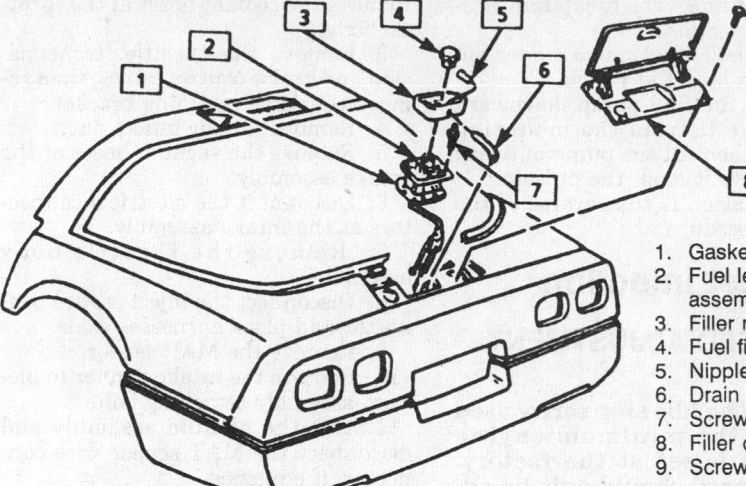

1. Gasket
2. Fuel level meter assembly
3. Filler neck housing
4. Fuel filler cap
5. Nipple
6. Drain hose
7. Screw and O-ring
8. Filler door bezel
9. Screw

Fuel sender and pump assembly removal—1990–92 vehicles

the sending unit is removed from the vehicle.

b. Do not fold or twist the strainer when installing the sender, as this could cause fuel restriction.

c. Ensure to install insulator pads on the top of the fuel tank to reduce rattle or any noises.

d. Peel off release paper on rear insulators and apply to the tank or underbody.

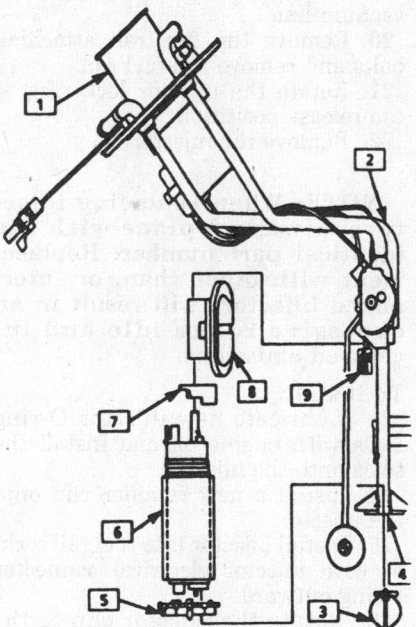

1. Fuel level meter
2. Inlet tube
3. Fuel pump filter
4. Mounting bracket
5. Rubber insulator
6. Fuel pump
7. Rubber bumper
8. Pulsator
9. Electrical connector

Fuel pump and sender assembly— 1990–92 5.7L (VIN 8) engines

e. Torque fuel tank retaining strap bolts to 26 ft. lbs. (35 Nm).

f. Torque rear strap bolts to 8 ft. lbs. (11 Nm).

NOTE: **Install the left side tank strap first, to prevent the tank flange from grounding out against the tank strap bracket.**

g. Connect the battery negative cable and check system for leaks.

1990–92

NOTE: **Vehicles equipped with the 5.7L (VIN J) engine use 2 fuel pumps which are not serviced separately, therefore if one fuel pump is not operational, the fuel sender assembly with both pumps must be replaced as a unit.**

1. Disconnect the negative battery cable.
2. Properly relieve the fuel system pressure.
3. Remove the filler door bezel attaching screws, then the filler door bezel.
4. Remove gas cap, then lift the fuel tank filler neck housing and disconnect the drain hose.
5. Remove filler neck housing, then disconnect and plug the fuel pipes and fuel vapor pipe.
6. Disconnect the sending unit electrical connector, remove the attaching bolts, then the sending unit assembly from the vehicle.
7. If equipped with the 5.7L (VIN J) engine, replace fuel sender assembly.
8. If equipped with the 5.7L (VIN 8) engine, proceed as follows:

a. Note the position of the fuel filter on the pump.

b. Support the pump with one hand and grasp the filter with the other. Turn the filter in one direction and pull the filter off the pump and discard it.

c. Diconnnect the fuel pump electrical connection.

d. Place the fuel gauge sender upside down on a flat bench.

e. Pull the fuel pump downward to remove it from the mounting bracket, then tilt the pump outward and remove it from the pulsator.

9. Installation is the reverse of the removal procedure.

Fuel Injection

IDLE SPEED ADJUSTMENT

NOTE: The idle stop screw used to adjust the minimum engine idle speed is set at the factory. The idle speed should only be adjusted if the manufacturer lists a specification and it is absolutely necessary.

Prior to adjusting the idle speed, ensure the ignition timing is correct and the throttle body is clean around the throttle plates. The 1988 Corvette is the only vehicle with an adjustment procedure.

1. Using a suitable tool, pierce the idle stop plug and remove it.

2. With the idle air control motor connected, ground the diagnostic lead. Turn the ignition to the **ON** position but do not start the engine.

3. Wait 30 seconds, then disconnect the idle air control connector.

4. Disconnect the distributor setting-timing connector.

5. Start the engine and allow to go into closed loop.

6. Remove the ground from the diagnostic terminal.

7. Adjust the idle stop screw to 450 rpm in **N**.

8. Turn the ignition **OFF** and reconnect the idle air control connector.

9. Adjust the throttle position sensor as follows:

a. With the ignition switch in the **ON** position, connect a scan tool or 3 jumper wires to the TPS.

b. Adjust the TPS to obtain a reading from 0.46–0.62 volts.

c. Tighten the screws and recheck the TPS reading.

10. Start the engine and check for proper idle operation.

Fuel Injector

REMOVAL & INSTALLATION

5.7L (VIN 8) Engine

1. Disconnect the negative battery cable.

2. Partially drain the cooling sytem.

Remove the coolant hoses at the throttle body.

3. Remove the throttle, transmission and cruise control cables, then remove the cable retaining bracket.

4. Remove the air intake duct.

5. Remove the vacuum hoses at the intake assembly.

6. Disconnect the electrical connectors at the intake assembly.

7. Remove the throttle body assembly.

8. Disconnect the injector wire connector and place harnesses aside.

9. Remove the MAP sensor.

10. Remove the intake runner to plenum assembly attaching bolts.

11. Lift the plenum assembly and disconnect the MAT sensor wire connector, if equipped.

12. Remove the plenum assembly.

13. Remove the left and right runner assembly attaching bolts.

14. Disconnect the PCV hose and EGR solenoid.

15. Remove the left and right side runners.

16. Disconnect the fuel feed and return lines.

17. Remove the fuel tube bracket bolt.

18. Remove the cold start valve, if equipped.

19. Remove the pressure regulator vacuum line.

20. Remove the fuel rail attaching bolts and remove the fuel rail.

21. Rotate the injector lock ring to the release positition.

22. Remove the injector.

NOTE: When replacing injectors, always replace with the identical part number. Replacement with other than, or intermixed injectors will result in an excessive rough idle and increased emissions.

To install:

23. Lubricate new injector O-ring seals with engine oil and install the seals onto the injector.

24. Install a new retainer clip onto the injector.

25. Install injector into fuel rail socket with injector electrical connector facing outward.

26. Rotate the injector clip to the **LOCK** position.

27. Install the fuel rail assembly.

28. Temporarily connect the battery cable and turn the ignition switch **ON** for 2 seconds, then turn the ignition switch to the **OFF** position for 10 seconds. Repeat the procedure once again, then disconnect the battery negative cable and inspect the fuel system for leaks.

29. Install the intake plenum and runner assemblies.

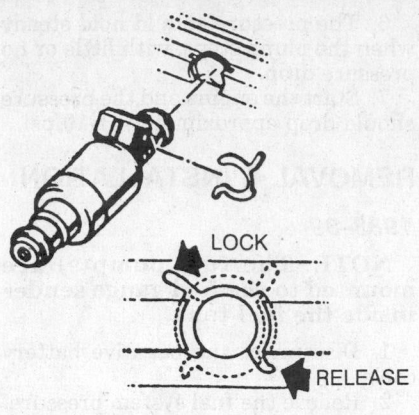

Fuel injector removal—5.7L (VIN 8) engine

30. Connect the negative battery cable.

5.7L (VIN J) Engine

1. Disconnect the negative battery cable.

2. Drain the cooling system into a suitable container.

3. Remove the air intake duct.

4. Remove the throttle cable cover and attaching hardware.

5. Remove the throttle and cruise control cables.

6. Disconnect the electrical connectors from the IAC, TPS and MAT sensor.

7. Remove the coolant air bleed hose from the plenum.

8. Remove the power brake booster and fuel pressure regulator vacuum hose from the plenum.

9. Remove the left and right vacuum hoses at the mid-plenum.

10. Remove the MAP sensor.

11. Remove the plenum assembly attaching bolts.

12. Remove the PCV dual hose fitting from the plenum.

13. Remove the canister hose from the plenum.

14. Lift the front of the plenum assembly and disconnect the following:

a. Left and right fresh air hoses from the throttle body extension.

b. Canister vacuum signal hose from the throttle body extension.

c. Electrical connector from the ignition module.

15. Remove the plenum assembly.

16. Properly relieve the fuel system pressure.

17. Remove the bolts securing the fuel feed and return lines.

18. Disconnect the fuel lines and discard the O-rings.

19. Disconnect the fuel injector wire connectors.

20. Remove the bolts securing the fuel rail to the injector housing.

21. Remove the fuel rail and spacers, if equipped.

Fuel Injector removal–5.7L (VIN J) engine

22. Remove the injector retaining clip, the injector and discard the O-ring seals.

To install:

23. Lubricate new injector O-rings with engine oil and install injector with retaining clip onto fuel rail.

24. Verify the injector wire connection is facing outward.

25. Install the fuel rail into the injector housing and connect the fuel lines and injector wire connectors.

26. Temporarily connect the negative battery cable and turn the ignition switch **ON** for 2 seconds, then turn the ignition switch to the **OFF** position for 10 seconds. Repeat the procedure once again, then disconnect the battery negative cable and inspect the fuel system for leaks.

27. Install the intake plenum assembly.

28. Connect all electrical connectors to the modules and sensors.

29. Connect all vacuum lines.

30. Refill the cooling system with the proper type and quantity of antifreeze.

31. Reconnect the negative battery cable.

DRIVE AXLE

Driveshaft and U-Joints

REMOVAL & INSTALLATION

1. Raise and support the vehicle safely, then remove the complete exhaust system.

2. Remove the bolts attaching the support beam at the axle and transmission. Remove support beam from the vehicle.

3. Mark relationship of shaft to companion flange and disconnect the rear universal joint by removing trunnion bearing straps. Tape bearing cups

to trunnion to prevent dropping and loss of bearing rollers.

4. Slide slip yoke from the transmission and remove shaft. Place a suitable drain pan under the transmission for oil leakage.

5. Remove the universal joint as follows:

   a. Remove the snapring, then place the drive shaft horizontally into suitable holding fixture.

   b. Support the lower ear of the universal joint with a 1⅛ inch socket.

   c. Press the lower bearing cap out from the yoke by pushing on the upper bearing cap.

   d. Rotate the driveshaft, then remove the the opposite bearing cap.

   e. Remove the universal joint from the yoke.

To install:

6. Install one bearing cap partially into 1 side of the yoke, then turn this side to the bottom.

7. Install the joint into yoke so that the trunnion seats freely in the bearing cap.

8. Install the opposite bearing cap

partially into the yoke, verifying the trunnion is straight and true in the bearing cap.

9. Press against the opposite bearing cap, while verifying the joint is not binding and turns freely.

10. If the joint begins to bind, there is probably a needle bearing out of place.

11. Install a snapring into the yoke as soon as the bearing cap clears the groove.

12. Continue to press the opposite side until a snapring can be inserted.

13. Assemble the other half of the joint in the same manner.

14. Install and align driveline support beam as follows:

   a. To ensure proper alignment of the driveline, a clearance of 1.52–2.02 in. (39–51mm) must be maintained between the top of the beam to the underbody and a clearance of 0.85–1.35 in. (22–34mm) from the passenger side of the beam to the side wall.

   b. Take the measurements directly above and to the right of the driveshaft yoke.

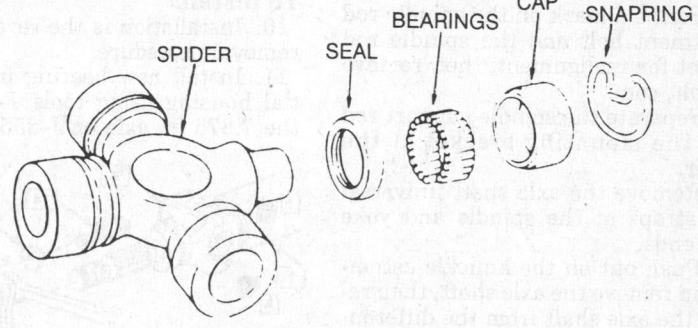

Universal joint assembly

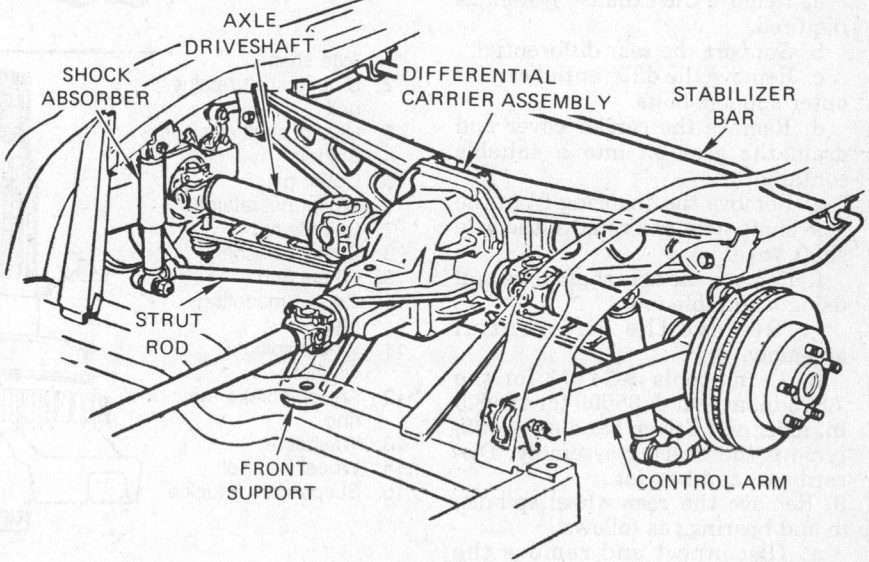

Driveshaft alignment

15. Reinstall the driveshaft in the reverse of the removal procedure.

Rear Axle Shaft, Bearing and Seal

REMOVAL & INSTALLATION

1. Raise and support the vehicle safely.

NOTE: Do not support the vehicle by means of the differential or the transverse leaf springs.

2. Remove the rear leaf spring from the knuckle as follows:

a. Remove the rear wheel assembly.

b. Install tool J–33432 onto the rear transverse spring and compress the spring.

c. Remove the cotter pin, nut, rubber grommet and bolt attaching spring to knuckle.

d. Release and remove spring compressor.

3. Remove the outer tie rod from the knuckle.

4. Scribe a mark on the spindle rod adjustment bolt and the spindle rod bracket for realignment, then remove the bolt, cam.

5. Separate the spindle support rod from the mounting bracket at the carrier.

6. Remove the axle shaft universal joint straps at the spindle and yoke shaft ends.

7. Push out on the knuckle assembly and remove the axle shaft, then remove the axle shaft from the differential case.

8. Remove rear axle yoke, oil seal and bearing as follows:

a. Remove the exhaust system as required.

b. Support the rear differential.

c. Remove the differential carrier outer support bolts.

d. Remove the carrier cover and drain the gear oil into a suitable container.

e. Remove the snapring from the axle shaft yoke and remove the axle shaft yoke.

f. Remove the axle shaft yoke seal using a suitable tool.

g. Remove the differential assembly.

h. Using tools J–34171 for the 7.875 in. axle or J–35509 for the 8.5 in. axle, and driver handle J–8592, remove the bearing assembly. Discard seal and bearing.

9. Remove the rear wheel spindle, hub and bearings as follows:

a. Disconnect and remove the rear wheel speed sensor.

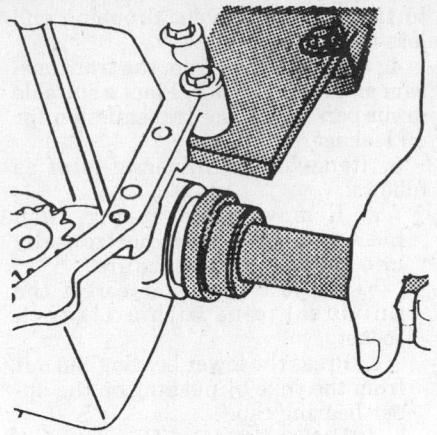

Installing axle shaft bearing into differential housing

b. Remove the spindle nut, washer, nut retainer and cotter pin.

c. Remove the wheel spindle from the hub and knuckle assembly.

d. Remove the brake caliper and parking brake assembly.

e. Remove the brake rotor.

f. Remove the wheel hub mounting bolts and remove the hub and bearing assembly.

To install:

10. Installation is the reverse of the removal procedure.

11. Install new bearing in differential housing using tools J–34172 for the 7.875 in. axle or J–35510 for the

8.5 in. axle and driver J–8592. Lubricate bearings with a suitable hypoid lubricate.

12. Apply a light coat of hypoid lubricant on the lip of the axle shaft seal.

13. Install axle shaft seal using tools J–26938 for 7.875 in. axle or J–35511 for 8.5 in. axle and driver J–8592.

14. Torque the universal joint strap retaining bolts to 26 ft. lbs. (35 Nm). Tighten the spindle bolt to 59 ft. lbs. (80 Nm).

15. Torque the spindle nut to 164 ft. lbs. (223 Nm).

16. Torque the hub mounting bolts to 66 ft. lbs. (90 Nm).

Front Wheel Hub, Knuckle/Spindle and Bearings

REMOVAL & INSTALLATION

1. Raise and safely support the vehicle, then remove the tire and wheel assembly.

2. Remove the brake caliper and rotor.

3. Remove the wheel speed sensor and cable bracket.

4. Remove the wheel hub attaching bolts, then the wheel hub.

5. Using a suitable jack, safely support the lower control arm assembly.

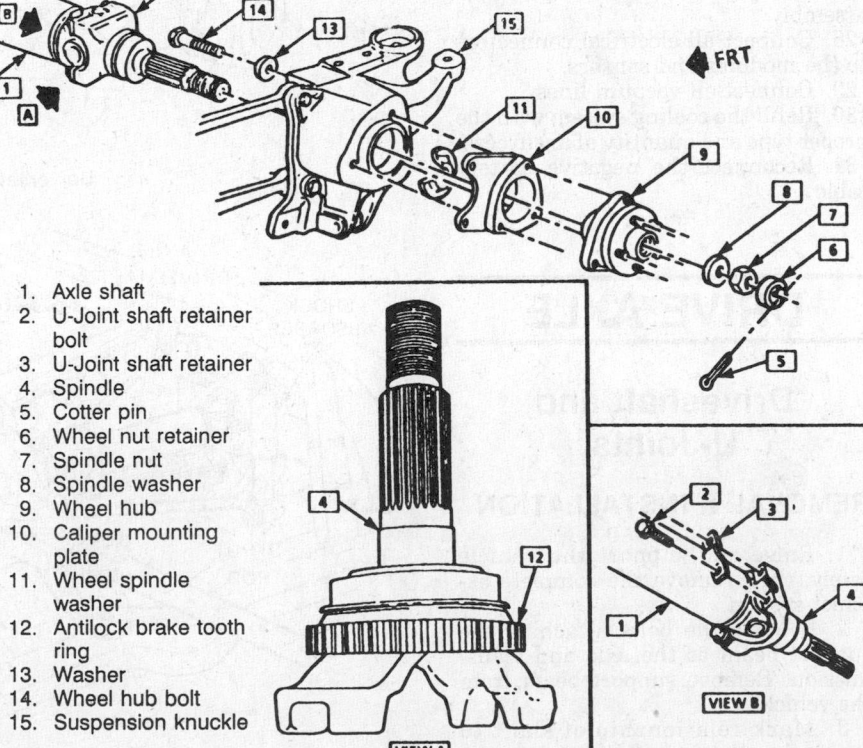

1. Axle shaft
2. U-Joint shaft retainer bolt
3. U-Joint shaft retainer
4. Spindle
5. Cotter pin
6. Wheel nut retainer
7. Spindle nut
8. Spindle washer
9. Wheel hub
10. Caliper mounting plate
11. Wheel spindle washer
12. Antilock brake tooth ring
13. Washer
14. Wheel hub bolt
15. Suspension knuckle

Rear wheel hub, bearing and spindle assembly

6. Remove the ball stud nut, washer and cotter pin from the upper and lower control arms using tool J–33436.

7. Using a suitable tie rod splitter, disconnect tie rod from the steering knuckle.

8. Remove the knuckle from the vehicle.

To install:

9. Installation is the reverse of the removal procedure.

10. Torque wheel hub nuts to 46 ft. lbs. (62 Nm).

11. Torque tie rod ball stud to 33 ft. lbs. (45 Nm).

12. Torque upper ball stud nut to 33 ft. lbs. (45 Nm) and lower ball stud nut to 50 ft. lbs. (68 Nm).

Pinion Seal

REMOVAL & INSTALLATION

1. Raise and safely support the vehicle.

2. Remove the exhaust system, as required.

3. Remove the drive shaft and driveline support beam.

4. Remove the pinion nut and yoke.

5. Inspect the yoke seal area for wear, replace the yoke if necessary.

6. Remove the pinion yoke seal from the differential housing.

7. Installation is the reverse of the removal procedure.

8. Apply a light coat of engine oil to the lip of the new pinion seal.

9. Using a torque wrench, torque pinion nut to 200 ft. lbs. (271 Nm) if equipped with an automatic transmission or 250 ft. lbs. (339 Nm) if equipped with a manual transmission.

Differential Carrier

REMOVAL & INSTALLATION

1. Disconnect the negative battery cable.

2. Raise and support the vehicle safely.

3. Remove the rear axle assembly.

4. Remove the differential cover and drain the gear oil into a suitable container.

5. Remove the snapring from the axle shaft yoke in the differential carrier. Mark each snapring to indicate which side it was removed from. The snaprings come in several different sizes.

6. Remove the axle shaft yokes.

7. Remove the differential bearing caps and note the matched letters stamped on the caps and carrier.

8. Mount the carrier and housing spreader tools J–24385-01 and J–24385-20 to the carrier.

9. Measure the carrier spread using

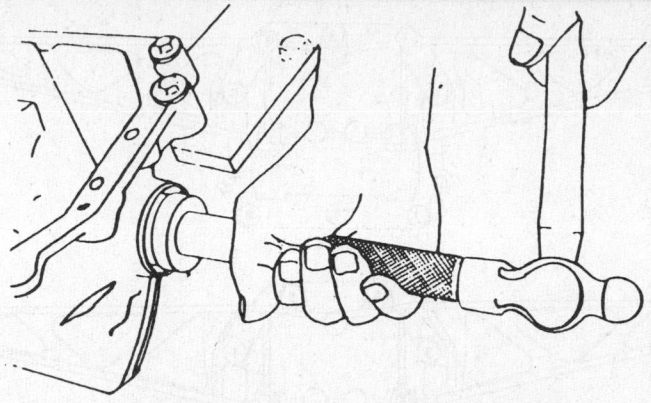

Installing axle seal shaft into differential housing

the dial indicator mounted to the assembly.

10. Spread the case, not exceeding 0.010 in. (0.25mm) of spread.

11. Using a suitable tool, remove the carrier assembly from the vehicle.

12. Remove the spreader after the assembly has been removed.

To install:

13. With carrier and housing spreader mounted to carrier, spread the carrier, not exceeding 0.010 in. (0.25mm) of spread.

14. Lubricate and assemble the bearing cups to the differential bearing.

15. Install the differential assembly into the carrier.

16. Install the bearing caps and bolts, ensure the letters stamped on the caps and carrier assembly coincide in both the direction and letter.

17. Torque the caps to the proper specification.

18. Measure the ring gear backlash at 3 equally spaced points:

  a. Backlash tolerance is 0.006–0.009 in. (0.15–0.23mm) and cannot vary more than 0.001–0.0015 in. (0.03–0.04mm).

  b. High backlash is corrected by moving some shims opposite side of the case to the ring gear side, thus

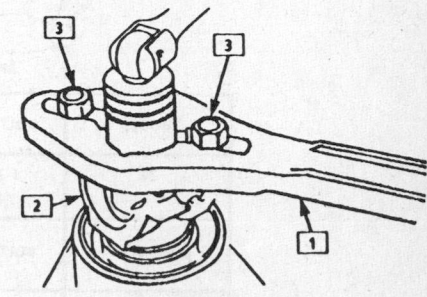

1. Pinion flange remover, tool J–8614–01
2. Propeller shaft yoke
3. Pinion flange shoulder bolts

Removal and installation of differential pinion yoke

moving the ring gear closer to the pinion.

  c. Low backlash is corrected by moving shims from the ring gear side of the case to the opposite side, thus moving the ring gear away from the pinion.

19. Install axle shaft yokes and snapring, ensuring snaprings are installed on the side they were removed from.

20. Install carrier cover with gasket

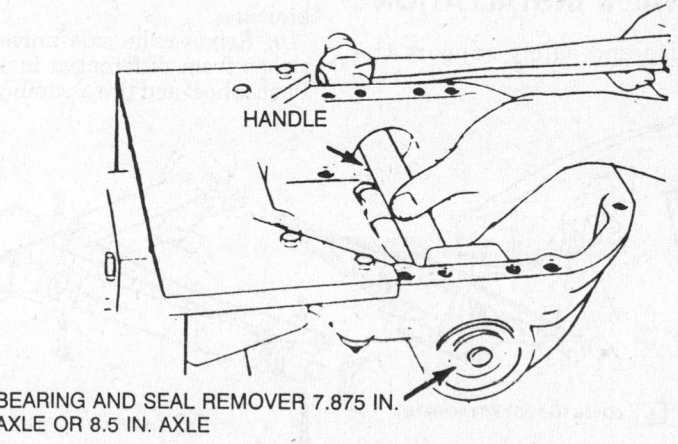

BEARING AND SEAL REMOVER 7.875 IN. AXLE OR 8.5 IN. AXLE

Removing axle shaft seal from differential housing

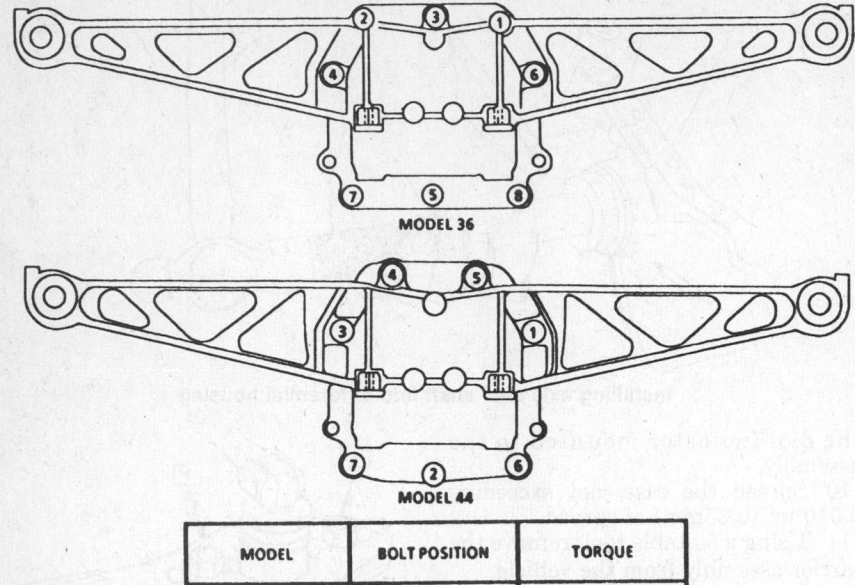

MODEL 36

MODEL 44

MODEL	BOLT POSITION	TORQUE
36	1-2-3-4-5-6	25 N·m (19 lb. ft.)
	7-8	40 N·m (30 lb. ft.)
MODEL	BOLT POSITION	TORQUE
44	1-2-3-4-5	40 N·m (30 lb. ft.)
	6-7	47 N·m (35 lb. ft.)

Differential carrier cover torques and sequence

and torque in sequence to specification.

21. Install the carrier assembly into vehicle.

22. Lower the vehicle and connect the negative battery cable.

Axle Housing

There are 2 differential assemblies used, a Dana model 36 with a 7.875 inch ring gear used on vehicles equipped with an automatic transmission and a Dana model 44 with a 8.5 inch ring gear used on vehicles equipped with a manual transmission.

REMOVAL & INSTALLATION

1. Raise and safely support the vehicle.

2. Remove the spare tire. Remove tire cover by removing support hooks.

3. Remove the upper and lower underbody braces, if equipped.

4. Remove the complete exhaust system, as required.

5. Remove the transverse leaf springs from the knuckles and the differential cover. Remove the leaf spring from vehicle.

6. Remove exhaust hangers, as required.

7. Scribe alignment marks on spindle bolts and brackets for ease of installation.

8. Remove the spindle, bolts and mounting bracket from the carrier.

9. Remove both tie rod ends from knuckles.

10. Remove the axle universal joint straps from differential inside yokes. Push wheel and tire assembly outward

to disengage joints from differential yokes.

11. Scribe alignment marks on driveshaft and pinion yoke for ease of installation.

12. Remove the driveshafts universal joint straps from the pinion flange. Push driveshaft forward into transmission and tie shaft to driveline support beam.

13. Support the differential carrier safely with a suitable jack.

14. Remove the differential carrier cover attaching bolts at the frame brackets.

15. Remove the drive line support beam attaching bolts at the differential housing.

16. Remove the differential carrier assembly from the vehicle.

To install:

17. Install differential carrier cover assembly onto vehicle.

18. Apply suitable sealant to driveline support and differential carrier.

19. Install the driveline support attaching bolts at the differential carrier cover. Torque cover attaching bolts to specification.

20. Align marks on driveshaft and yoke, and install driveshaft. Torque universal joint straps to 18 ft. lbs. (24 Nm).

21. Install wheel axle shaft joints into yoke.

22. Install rear wheel driveshaft universal joint retainers onto yoke shaft. Torque retainers to 26 ft. lbs. (35 Nm).

23. Install tie rod ends into knuckle. Torque tie rod nut to 33 ft. lbs. (45 Nm) to align slot in nut with hole in stud.

24. Install mounting bracket onto carrier. Torque bolts to 60 ft. lbs. (80 Nm).

25. Install transverse leaf spring, then the exhaust system hangers and nuts.

26. Install the exhaust system.

27. If equipped with convertible top, install upper and lower underbody braces.

28. Install spare tire cover and spare tire.

29. Fill rear axle with suitable lubricant, then lower the vehicle.

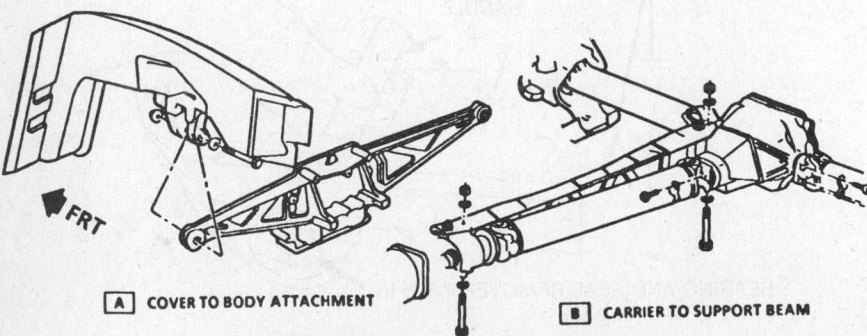

A COVER TO BODY ATTACHMENT

B CARRIER TO SUPPORT BEAM

Differential carrier cover attachments

MANUAL TRANSMISSION

For further information on transmissions/transaxles, please refer to "Chilton's Guide to Transmission Repair".

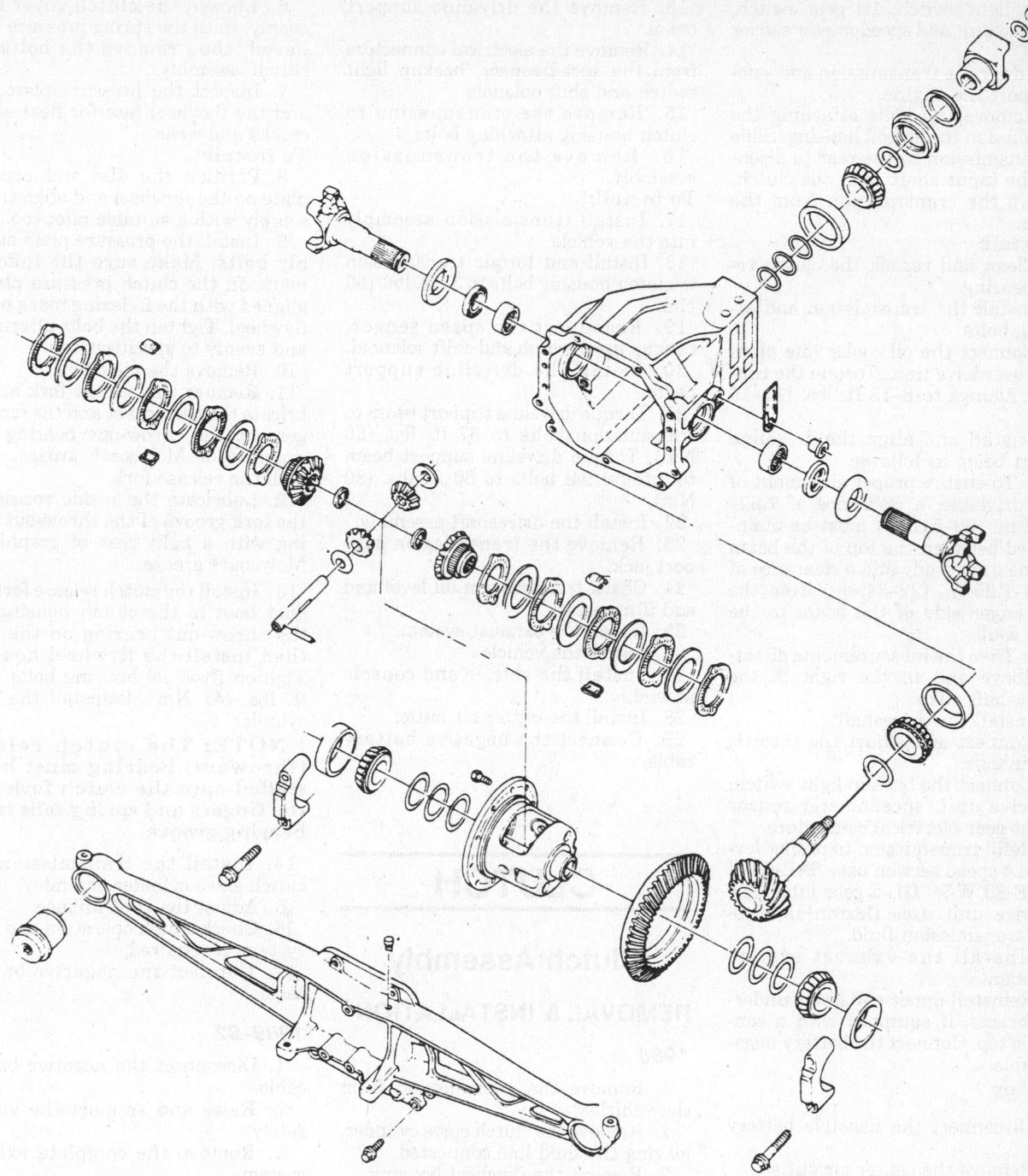

Exploded view of rear axle assembly

Transmission Assembly

REMOVAL & INSTALLATION

1988

1. Disconnect the negative battery cable.

2. Remove the distributor cap and lay aside.

3. Raise the vehicle and support it safely.

4. If equipped with a convertible top, remove the upper and lower underbody braces.

5. Remove the complete exhaust system as an assembly.

6. Remove the exhaust hanger at the transmission.

7. Support the transmission safely using a suitable jack.

8. Remove the bolts attaching the driveline support beam at the axle and the transmission. Remove the driveline support beam from the vehicle.

9. Mark the relationship of the driveshaft to the axle flange. Remove the universal joint bearing straps and disengage the rear universal joint from the axle. Slide the driveshaft yoke out of the overdrive unit and remove the driveshaft from the vehicle.

10. Disconnect the transmission cooler lines at the overdrive unit.

11. Disconnect the shift linkage at the side cover.

12. Disconnect the electrical connectors at the side cover. Disconnect the

backup light switch, 1st gear switch, overdrive unit and speedometer sensor switch.

13. Lower the transmission and safely support the engine.

14. Remove the bolts attaching the transmission to the bell housing. Slide the transmission to the rear to disengage the input shaft from the clutch. Remove the transmission from the vehicle.

To install:

15. Clean and repack the clutch release bearing.

16. Install the transmission and attaching bolts.

17. Connect the oil cooler line pipes to the overdrive unit. Torque the connector fittings to 8–12 ft. lbs. (11–16 Nm).

18. Install and align the driveline support beam as follows:

a. To ensure proper alignment of the driveline, a clearance of 1.52–2.02 in. (39–51mm) must be maintained between the top of the beam to the underbody and a clearance of 0.85–1.35 in. (22–34mm) from the passenger side of the beam to the side wall.

b. Take the measurements directly above and to the right of the driveshaft yoke.

19. Install the driveshaft.

20. Connect and adjust the throttle shift linkage.

21. Connect the backup light switch, overdrive unit, speedometer sensor and 1st gear electrical connectors.

22. Refill transmission to proper level. The 4 speed section uses SAE-80 W or SAE-80 W-90 GL-5 gear lube. The overdrive unit uses Dexron®II automatic transmission fluid.

23. Install the exhaust system components.

24. Reinstall upper and lower underbody braces, if equipped with a convertible top. Connect the battery negative cable.

1989–92

1. Disconnect the negative battery cable.

2. Remove the center air outlet.

3. Remove the console trim and accessory plate.

4. Remove the control lever button.

5. Remove the shift lever knob assembly.

6. Remove the center console trim plate.

7. Remove the shift lever snapring.

8. Remove the shifter retainer nuts.

9. Raise and safely support the vehicle.

10. Remove the complete exhaust assembly.

11. Remove the drive shaft.

12. Support the transmission safely with a jack.

13. Remove the driveline support beam.

14. Remove the electrical connectors from the speed sensor, backup light switch and shift solenoid.

15. Remove the transmission to clutch housing attaching bolts.

16. Remove the transmission assembly.

To install:

17. Install transmission assembly into the vehicle.

18. Install and torque transmission to clutch housing bolts to 37 ft. lbs. (50 Nm).

19. Reconnect the speed sensor, backup light switch and shift solenoid.

20. Install the driveline support beam.

21. Torque driveline support beam to transmission bolts to 37 ft. lbs. (50 Nm). Torque driveline support beam to differential bolts to 60 ft. lbs. (80 Nm).

22. Install the driveshaft assembly.

23. Remove the transmission support jack.

24. Check transmission oil level and add if necessary.

25. Install the exhaust system.

26. Lower the vehicle.

27. Install the shifter and console assembly.

28. Install the center air outlet.

29. Connect the negative battery cable.

CLUTCH

Clutch Assembly

REMOVAL & INSTALLATION

1988

1. Remove the transmission from the vehicle.

2. Remove the clutch slave cylinder, leaving the fluid line connected.

3. Remove the flywheel housing.

4. Slide the clutch fork from the ball stud and remove the fork from the dust boot. The ball stud is threaded into the clutch housing and is easily replaced, if necessary.

5. Install a clutch pilot tool to support the clutch assembly during removal.

NOTE: Look for an "X" mark on the flywheel and the clutch assembly. If there are no indexing marks or a letter is not evident, mark the flywheel and clutch cover for easy indexing during reassembly.

6. Loosen the clutch cover bolts evenly, until the spring pressure is relieved, then remove the bolts and clutch assembly.

7. Inspect the pressure plate, disc and the flywheel face for heat stress, cracks and wear.

To install:

8. Position the disc and pressure plate on the flywheel and align the assembly with a suitable pilot tool.

9. Install the pressure plate assembly bolts. Make sure the indexing mark on the clutch pressure plate is aligned with the indexing mark on the flywheel. Tighten the bolts alternately and evenly to specifications.

10. Remove the pilot tool.

11. Remove the release fork and lubricate the ball socket and the fork fingers at the throw-out bearing with graphite or Molycoat® grease. Reinstall the release fork.

12. Lubricate the inside recess and the fork groove of the throw-out bearing with a light coat of graphite or Molycoat® grease.

13. Install the clutch release fork and dust boot in the clutch housing and the throw-out bearing on the fork, then install the flywheel housing. Tighten flywheel housing bolts to 30 ft. lbs. (41 Nm). Reinstall the slave cylinder.

NOTE: The clutch release (throwout) bearing must be installed onto the clutch fork with the fingers and spring tabs in the bearing groove.

14. Install the transmission and clutch slave cylinder assembly.

15. Adjust the shift linkage.

16. Check clutch operation and bleed system, if required.

17. Connect the negative battery cable.

1989–92

1. Disconnect the negative battery cable.

2. Raise and support the vehicle safely.

3. Remove the complete exhaust system.

4. Remove the transmission assembly.

5. Remove the ground wire connected to the clutch housing left side, if equipped.

6. Remove the clutch slave cylinder at the housing and support the unit off to the side.

7. Remove the starter assembly, 5.7L (VIN 8) engine only.

8. Remove the left side converter shield, if required.

9. Remove the clutch housing cover.

NOTE: Excessive clutch wear may require removal of the ball

stud locking screw and loosening of the ball stud to disengage the fork and housing.

10. Mark the alignment of the clutch cover and flywheel for installation.

11. Loosen the clutch cover bolts evenly, 1 turn at a time until spring pressure is released.

12. Remove the clutch plate and disc assembly.

To install:

13. Inspect flywheel, clutch plate and disc for heat stress, cracks or worn parts.

14. Install the clutch assembly using a suitable alignment tool.

15. Verify the reference marks on the flywheel and pressure plate are alined to retain proper engine balance.

16. Install the clutch assembly retaining bolts and tighten in the specified sequence.

17. Tighten bolts 1 turn at a time until spring pressure is built up.

18. Torque the bolts to 30 ft. lbs. (41 Nm).

19. Position the clutch housing to the engine block, engaging the fork onto the release bearing.

20. Torque the clutch housing bolts to 37 ft. lbs. (50 Nm).

21. Reconnect the ground harness connection to the housing cover.

22. Install the heat shields, if removed.

23. Install the clutch slave cylinder.

24. Install the transmission assembly and check the fluid level.

25. Install the starter assembly, if removed.

26. Install the exhaust system and lower the vehicle.

27. Connect the battery negative cable. Check clutch operation.

Clutch Master Cylinder

REMOVAL & INSTALLATION

1. Disconnect the negative battery cable.

2. Remove the hush panel from under the dash.

3. Disconnect the pushrod from the clutch pedal.

4. Disconnect and plug the hydraulic line at the clutch master cylinder.

5. Remove the clutch master cylinder retaining bolts. Remove the clutch master cylinder from the vehicle.

6. Installation is the reverse of the removal procedure. Bleed the system, as required. Connect battery negative cable.

Clutch Slave/Actuator Cylinder

REMOVAL & INSTALLATION

1. Disconnect the negative battery cable.

2. Raise and support the vehicle safely.

3. Disconnect the hydraulic line at the slave/actuator cylinder on the clutch housing.

4. Remove the slave/actuator cylinder mounting bolts from the clutch housing.

5. Remove the pushrod and the slave/actuator cylinder from the vehicle.

6. Installation is the reverse of the removal procedure. Bleed the system, as required. Connect battery negative cable.

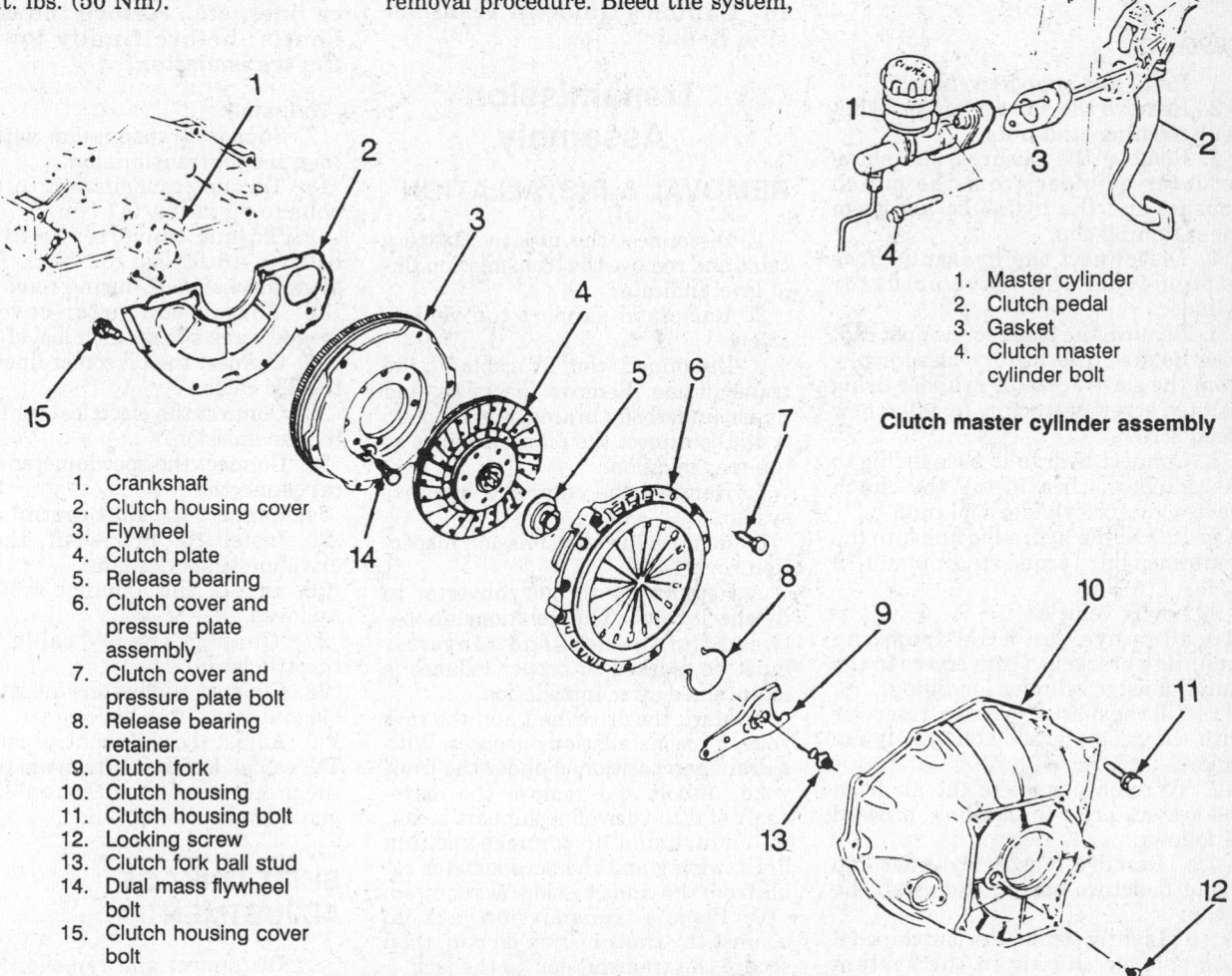

1. Master cylinder
2. Clutch pedal
3. Gasket
4. Clutch master cylinder bolt

Clutch master cylinder assembly

1. Crankshaft
2. Clutch housing cover
3. Flywheel
4. Clutch plate
5. Release bearing
6. Clutch cover and pressure plate assembly
7. Clutch cover and pressure plate bolt
8. Release bearing retainer
9. Clutch fork
10. Clutch housing
11. Clutch housing bolt
12. Locking screw
13. Clutch fork ball stud
14. Dual mass flywheel bolt
15. Clutch housing cover bolt

Exploded view of clutch assembly—1989–92 vehicles

Hydraulic Clutch System Bleeding

1988–89

1. Fill the master cylinder reservoir with the proper grade and type brake fluid.
2. Raise the vehicle and support it safely.
3. Remove the clutch slave cylinder and proceed as follows:

    a. On 1988 vehicles, hold the slave cylinder at a 45 degree angle with the bleeder valve at the highest point.

    b. On 1989 vehicles, remove the factory bleeder valve on the slave/actuator cylinder using a suitable fluted screw extractor. Install a new bleeder valve and then reinstall the clutch slave/actuator cylinder.

4. Fully depress the clutch pedal and open the bleeder valve. Close the bleeder valve and release the clutch pedal.
5. Repeat the last step until all air is expelled from the system.
6. Check the fluid reservoir and re-fill as required.

1990–92

1. Raise and support vehicle.
2. Remove the slave/actuator cylinder attaching stud nuts.
3. Remove the pushrod and slave/actuator cylinder from the clutch housing and the hydraulic line from the retaining clip.
4. Disconnect the hydraulic hose fitting from the slave/actuator cylinder.
5. Remove the bleed screw dust cap.
6. Remove the factory bleed screw from the slave/actuator cylinder using a fluted screw extractor. Install a new bleed screw.
7. Connect hydraulic hose fitting to the actuator, then install the clutch slave/actuator cylinder and nuts.
8. Install the hydraulic line into the retaining clip. Torque stud nuts to 19 ft. lbs. (25 Nm).
9. Lower vehicle.
10. Remove the ECM from the mounting bracket to gain access to the clutch master cylinder for filling.
11. Fill the master cylinder reservoir with the proper grade and type brake fluid.
12. To remove some of the air from the system prior to bleeding, proceed as follows:

    a. Remove master cylinder cap and moisture barrier. Reinstall the cap.

    b. Lightly stroke the clutch pedal to release the air in the system through the master cylinder.

    c. Remove the master cylinder cap and reinstall moisture barrier.

    d. Reinstall the cap.

13. Raise and support vehicle.
14. Fully depress the clutch pedal and open the bleeder screw. Close the bleed screw and release the clutch pedal.
15. Repeat Step 13 until all the air is expelled from the system.
16. Check the fluid reservoir and re-plenish, as required.
17. Torque bleeder screw until screw breaks, requires for body clearance. Screw should break at approximately 10–14 ft. lbs. (14–19 Nm).
18. Install dust cap on the bleeder screw.
19. Lower the vehicle and install the ECM.

AUTOMATIC TRANSMISSION

For further information on transmissions/transaxles, please refer to "Chilton's Guide to Transmission Repair".

Transmission Assembly

REMOVAL & INSTALLATION

1. Disconnect the negative battery cable and remove the transmission fluid level indicator.
2. Raise and support the vehicle safely.
3. Disconnect the TV cable at the transmission. Remove the lower and upper underbody braces, if equipped.
4. Disconnect the oil cooler lines at the transmission.
5. Remove the complete exhaust system.
6. Remove the transmission inspection cover.
7. Remove the torque convertor to flywheel bolts. The relationship between the flywheel and convertor must be marked so proper balance is maintained after installation.
8. Mark the driveshaft and the rear yoke, for reinstallation purposes. With a drain pan positioned under the front yoke, unbolt and remove the driveshaft and the driveline support beam.
9. Mark and disconnect vacuum lines, wiring and the speedometer cable from the transmission, as required.
10. Place a transmission jack up against the transmission oil pan, then secure the transmission to the jack.
11. Remove the transmission mount-

ing pad bolt(s), then carefully raise the transmission just enough to take the weight, of the transmission, off the supporting crossmember. Remove the transmission mounting pad.

NOTE: Exercise extreme care to avoid damage to underhood components while raising or lowering the transmission.

12. Remove the transmission filler tube.
13. Disconnect the floor shift cable. Disconnect the oil cooler lines from the transmission.
14. Support the engine using a jack placed beneath the engine oil pan. Be sure to put a block of wood between the jack and the oil pan, to prevent damage to the pan.
15. With the proper gauge wire, fasten the torque convertor to the transmission case.
16. Remove the transmission to engine mounting bolts, then carefully move the transmission rearward, downward and out from under the vehicle.

NOTE: If interference is encountered with the cable(s), cooler lines, etc., remove the component(s) before finally lowering the transmission.

To install:

17. Support transmission with jack, then install transmission.
18. Torque transmission to engine bolts to 35 ft. lbs. (47 Nm).
19. Torque converter to flywheel bolts to 46 ft. lbs. (62 Nm). Ensure align marks made during removal.
20. Install converter cover and torque screws to 89 inch lbs. (10 Nm).
21. Connect the oil cooler lines, then the TV cable.
22. Connect the electrical connectors to transmission.
23. Connect the speedometer electrical connector.
24. Connect the shift control cable.
25. Install the driveshaft, then the driveline support beam.
26. Install the exhaust system, if removed.
27. Connect the TV cable to the throttle lever.
28. Connect the battery negative cable and install the air cleaner.
29. Adjust the shift control cable and TV cable. Refill the transmission to the proper level with Dexron®II automatic transmission fluid.

SHIFT LINKAGE ADJUSTMENT

1. Disconnect and remove the battery negative cable.

2. Place the control lever in the **N** position.

3. Raise and support vehicle.

4. Loosen the cable attachment at the shift lever.

5. Rotate the shift lever clockwise to **P** detent and then back to **N**.

6. Tighten the cable attachment to 15 ft. lbs. (20 Nm).

NOTE: The lever must be be held out of the P position when tightening the nut.

7. Lower the vehicle.

8. Check the cable adjustment by rotating the control lever through the detents.

9. Connect the battery negative cable.

THROTTLE LINKAGE ADJUSTMENT

1. Check that the cable slider is in the 0 or fully reset position.

2. Rotate the throttle idler lever to the wide open throttle stop position.

3. Slider must move toward the lever, when the lever is rotated to the wide open throttle stop position.

4. Release the lever.

5. Check TV cable operation. If readjustment is necessary, proceed as follows:

    a. Depress and hold the metal reset tab.

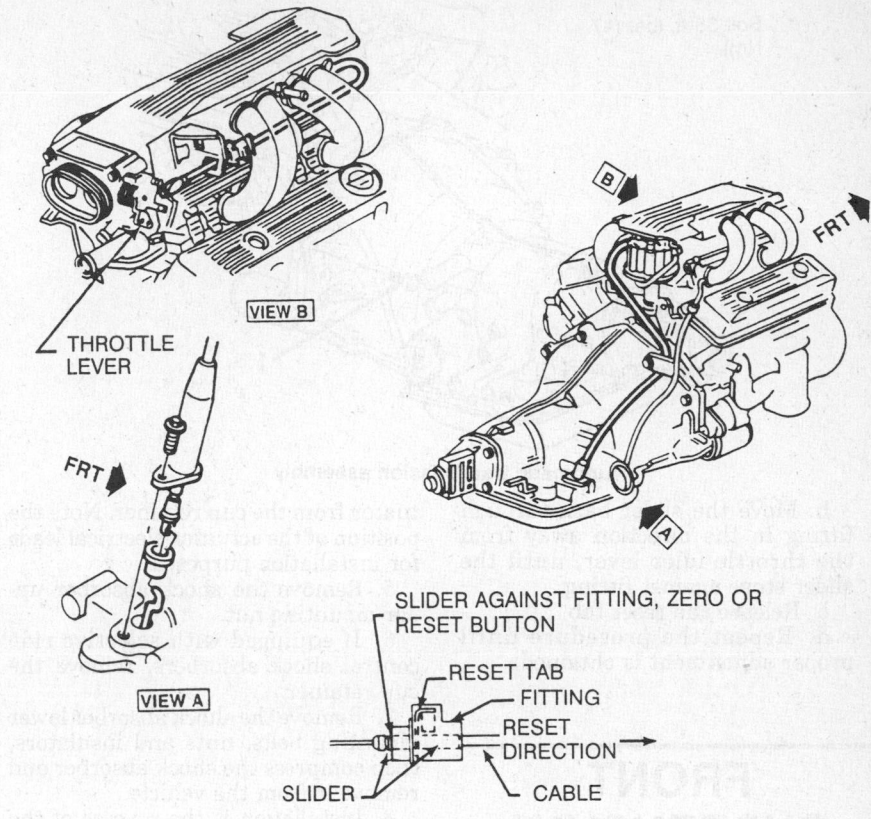

Automatic transmission throttle valve cable

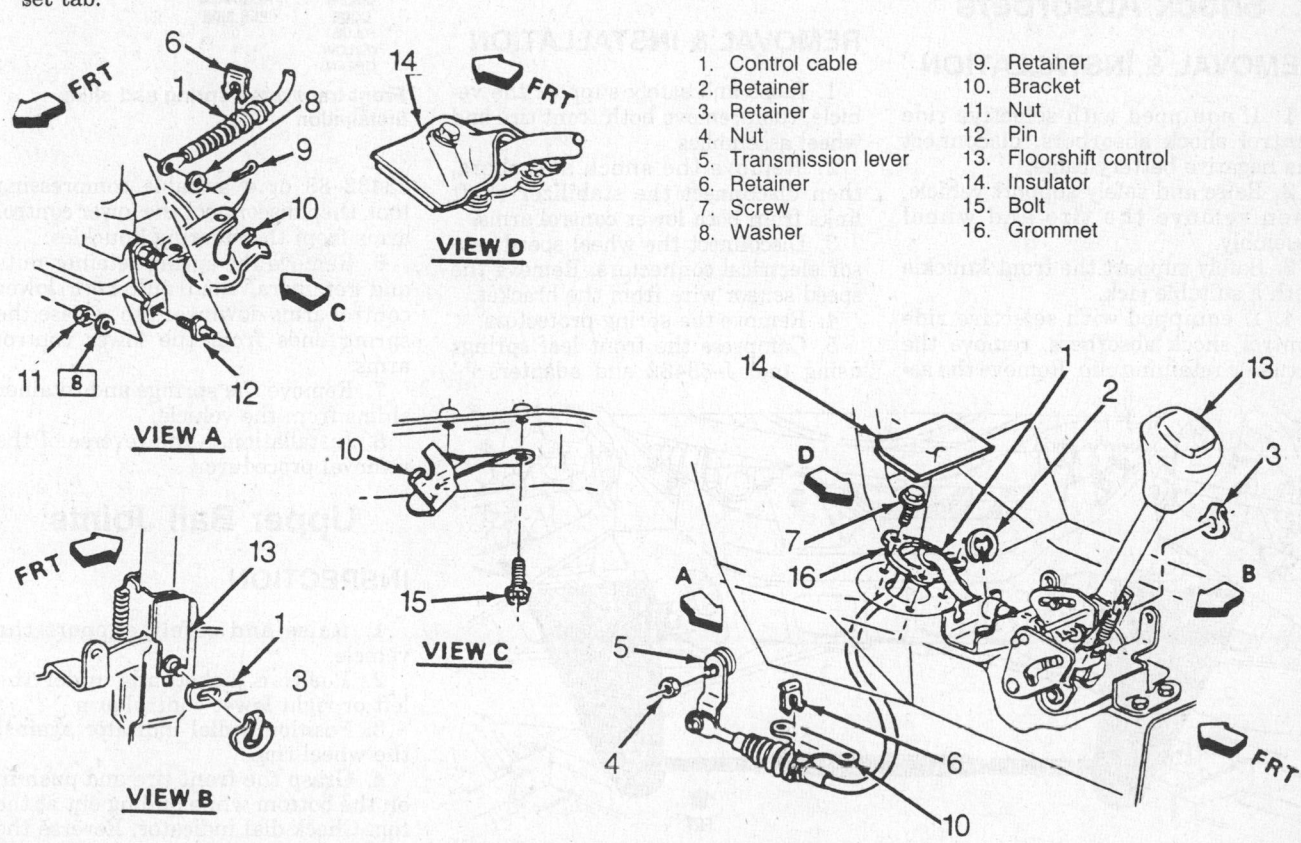

1. Control cable	9. Retainer
2. Retainer	10. Bracket
3. Retainer	11. Nut
4. Nut	12. Pin
5. Transmission lever	13. Floorshift control
6. Retainer	14. Insulator
7. Screw	15. Bolt
8. Washer	16. Grommet

Automatic transmission shifter cable

1. Bolt 35 ft. lbs. (47 Nm)

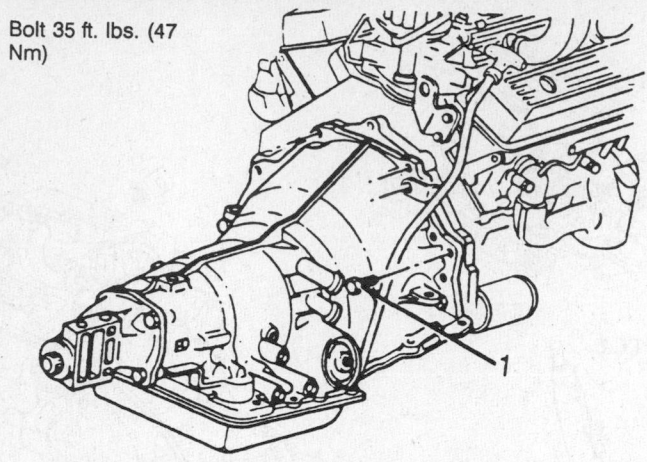

Automatic transmission assembly

b. Move the slider back through fitting in the direction away from the throttle idler lever, until the slider stops against fitting.

c. Release the reset tab.

d. Repeat the procedure until proper adjustment is obtained.

FRONT SUSPENSION

Shock Absorbers

REMOVAL & INSTALLATION

1. If equipped with selective ride control shock absorbers, disconnect the negative battery cable.

2. Raise and safely support vehicle, then remove the tire and wheel assembly.

3. Safely support the front knuckle with a suitable jack.

4. If equipped with selective ride control shock absorbers, remove the actuator retaining clip. Remove the ac-

tuator from the cup retainer. Note the position of the actuator electrical leads for installation purposes.

5. Remove the shock absorber upper mounting nut.

6. If equipped with selective ride control shock absorbers, remove the cup retainer.

7. Remove the shock absorber lower mounting bolts, nuts and insulators, then compress the shock absorber and remove it from the vehicle.

8. Installation is the reverse of the removal procedure.

Transverse Springs

REMOVAL & INSTALLATION

1. Raise and safely support the vehicle, then remove both front tire and wheel assemblies.

2. Remove the shock absorbers, then disconnect the stabilizer shaft links from both lower control arms.

3. Disconnect the wheel speed sensor electrical connectors. Remove the speed sensor wire from the bracket.

4. Remove the spring protectors.

5. Compress the front leaf springs using tool J–33432 and adapters J–

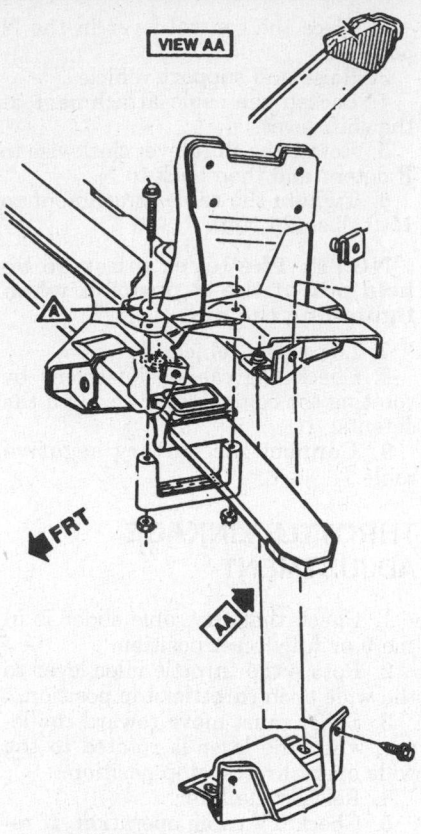

VIEW AA

SPRING SHIM REQUIREMENTS

SPRING COLOR CODE	NUMBER OF SHIMS REQUIRED PER SIDE
BLUE	0
YELLOW	1
GREEN	2

Front transverse spring and shim installation

33432–88 or a suitable compressing tool, then disconnect the lower control arms from the steering knuckles.

6. Remove the spring retainer nuts and retainers, then pull both lower control arms downward to release the spring ends from the lower control arms.

7. Remove the springs and retainer shims from the vehicle.

8. Installation is the reverse of the removal procedure.

Upper Ball Joints

INSPECTION

1. Raise and safely support the vehicle.

2. Position jackstands under the left or right lower control arm.

3. Position a dial indicator against the wheel rim.

4. Grasp the front tire and push in on the bottom while pulling out at the top. Check dial indicator. Reverse the push-pull procedure.

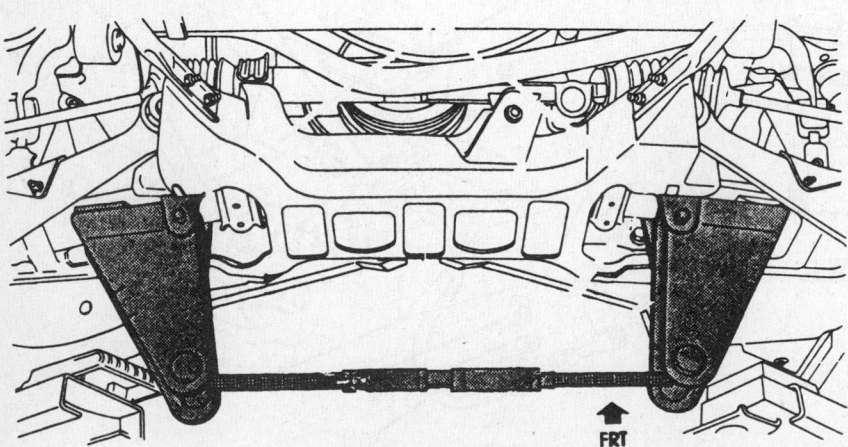

Front transverse spring removal

5. Horizontal deflection on the dial indicator must not exceed 0.125 in. (3.18mm).

6. If specifications are not as indicated, replace the ball joint.

REMOVAL & INSTALLATION

1. Raise and safely support vehicle.
2. Safely support the lower control arm with jackstands.
3. Remove the tire and wheel assembly.
4. Using tool J–33436, remove the ball joint from the knuckle.
5. Grind or chisel off the ball joint to control arm attaching rivets.
6. Install a new ball joint.
7. Reverse the procedure to install.
8. Torque the 4 new upper ball joint mounting nuts to 13 ft. lbs. (18 Nm).
9. Torque upper ball joint nut to 33 ft. lbs. (45 Nm).

NOTE: Do not exceed 63 ft. lbs. (85 Nm) to align the cotter pin holes.

Lower Ball Joints

INSPECTION

1. Check the wear indicator on the lower ball joint.

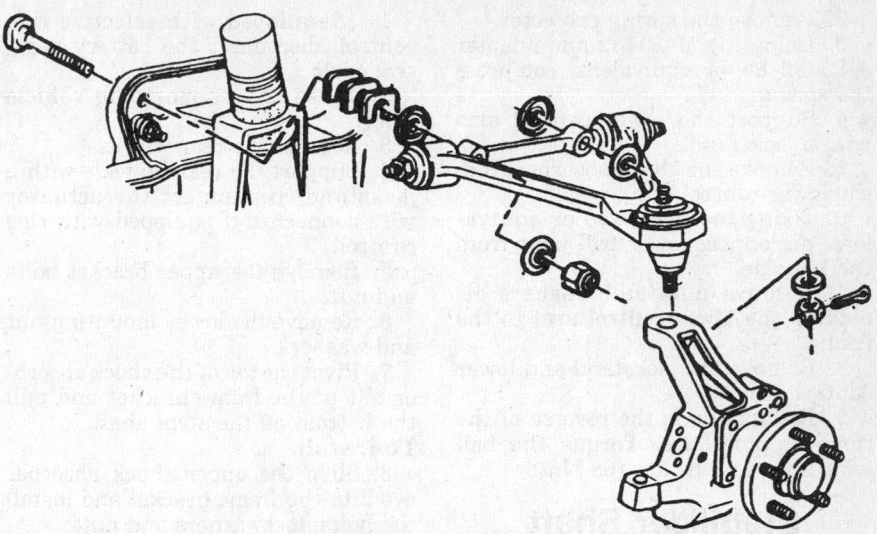

Upper control assembly

2. The wear indicator should protrude 0.050 in. (1.27mm) when new.
3. When the wear indicator shoulder retreats below the surface, the ball joint must be replaced.

REMOVAL & INSTALLATION

1. Raise and support vehicle safely.
2. Remove the wheel and tire assembly.

3. Support the lower control arm using jackstands.
4. Using tool J–33436, remove the ball joint from the knuckle.
5. Using tool J–9519–E, remove the ball joint from the control arm.
6. Reverse the procedure to install.
7. Torque the lower control arm ball stud nut to 50 ft. lbs. (68 Nm).

NOTE: Do not exceed 88 ft. lbs. (120 Nm) to align the cotter pin holes.

Upper Control Arms

REMOVAL & INSTALLATION

1. Raise and support the vehicle safely. Remove the tire and wheel assembly.
2. Remove the front wheelhouse panel seal and lower center panel.
3. Remove the shock absorber actuator wire connector, if equipped.
4. Support the lower control arm with a jackstand.
5. Disconnect the wheel speed sensor wire connector. Remove the wire from the bracket.
6. Use tool J–33436 or equivalent, disconnect the upper ball joint from the knuckle.
7. Remove the upper control arm attaching bolts, shims and nuts. Remove the control arm.
8. Installation is the reverse of the removal procedure. Torque the upper control arm bolts and the ball joint nut to specification.

Lower Control Arms
REMOVAL & INSTALLATION

1. Raise and support the vehicle safely. Remove the tire and wheel assembly.

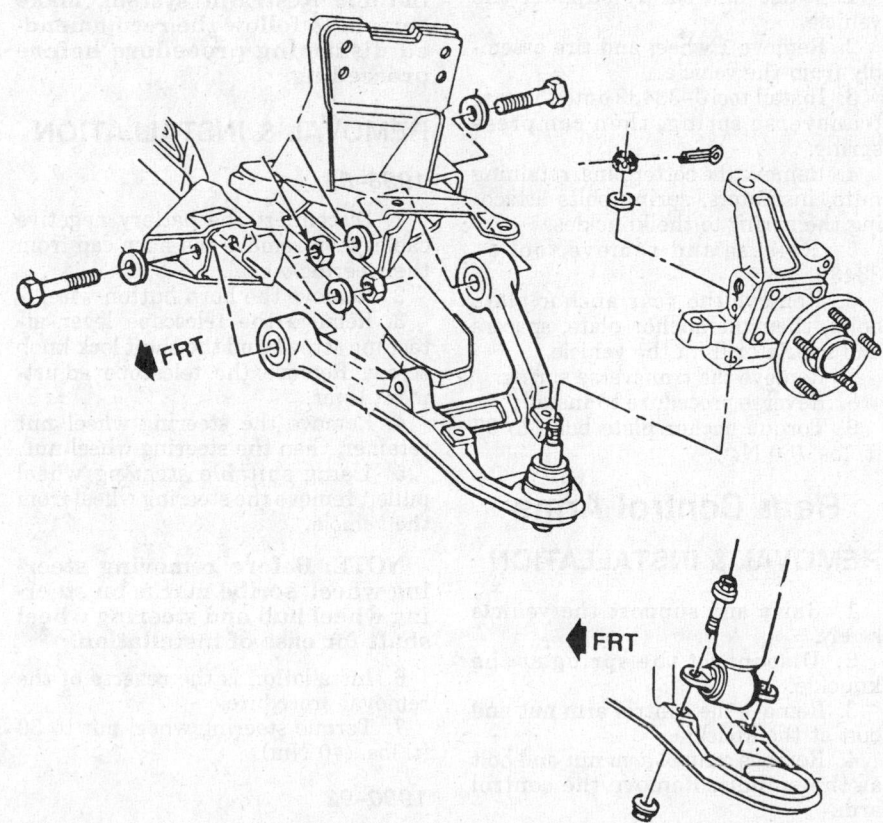

Lower control assembly

2. Remove the spring protector.

3. Using tool J–33432 and adapter J–33432–88 or equivalent, compress the spring.

4. Support the lower control arm with a jackstand.

5. Remove the shock absorber from the lower control arm.

6. Using tool J–33436 or equivalent, disconnect lower ball joint from the knuckle.

7. Remove nuts and washers attaching the lower control arm to the frame.

8. Remove the jackstand and lower control arm.

9. Installation is the reverse of the removal procedure. Torque the ball joint nut to 50 ft. lbs. (68 Nm).

Stabilizer Shaft

REMOVAL & INSTALLATION

1. Raise and support vehicle safely.

2. Remove the tire and wheel assembly.

3. Support the lower control arms using jackstands.

4. Remove the stabilizer shaft insulator clamp bolts and brackets from the frame.

5. Remove the stabilizer shaft to links attaching bolts.

6. Remove the stabilizer shaft from the vehicle.

7. Reverse the procedure to install.

Front Wheel Bearings

REMOVAL & INSTALLATION

1. Raise and support the vehicle safely.

2. Remove the tire and wheel assembly.

3. Remove the caliper assembly and position it aside.

4. Remove the hub and bearing assembly.

5. Installation is the reverse of the removal procedure. The bearings do not require adjustment.

REAR SUSPENSION

Shock Absorbers
REMOVAL & INSTALLATION

NOTE: Purge new shocks of air by repeatedly extending them in their normal position and compressing them while inverted.

---

1. If equipped with selective ride control, disconnect the battery negative cable.

2. Raise and support the vehicle safely.

3. Remove the rear wheels.

4. Support the rear knuckle with a jackstand. Disconnect the actuator wire connector, if equipped with ride control.

5. Remove the upper bracket bolts and nuts.

6. Remove the lower mounting nut and washers.

7. Pivot the top of the shock absorber out of the frame bracket and pull the bottom off the strut shaft.

To install:

8. Slide the upper shock absorber eye into the frame bracket and install the bolts, lockwashers and nuts.

9. Install the rubber grommets on the lower shock eye and place the shock over the strut shaft. Install the washers and nut.

10. Connect the actuator wire connector, if equipped with ride control.

11. Lower the vehicle.

12. Connect the battery negative cable.

Transverse Spring

REMOVAL & INSTALLATION

1. Raise and safely support the vehicle.

2. Remove 1 wheel and tire assembly from the vehicle.

3. Install tool J–33432 onto the rear transverse spring, then compress spring.

4. Remove the cotter pins, retaining nuts, insulators, spring bolts attaching the spring to the knuckles.

5. Release and remove tool J–33432.

6. Remove the rear anchor plate bolts, then the anchor plate, spacers and insulator from the vehicle.

7. Remove the transverse spring.

8. Reverse procedure to install.

9. Torque anchor plate bolts to 37 ft. lbs. (50 Nm).

Rear Control Arms

REMOVAL & INSTALLATION

1. Raise and support the vehicle safely.

2. Disconnect the spring at the knuckle.

3. Remove the control arm nut and bolt at the knuckle.

4. Remove control arm nut and bolt at the spindle. Remove the control arm.

5. Installation is the reverse of the removal procedure.

Spindle/Support Rod

REMOVAL & INSTALLATION

1. Raise and support vehicle.

2. Scribe alignment marks on the wheel spindle/support rod adjustment bolt and the spindle/support rod bracket so they can be installed in the same position.

3. Remove the adjustment bolt, cam and nut, then separate the spindle/support rod from the bracket.

4. Remove the spindle/support bolt, washer and nut at the spindle, then remove the spindle/support rod from the vehicle.

5. Installation is the reverse of the removal procedure.

6. Torque spindle/support rod to knuckle nut to 107 ft. lbs. (145 Nm). Torque spindle/support rod adjustment nut to 186 ft. lbs. (253 Nm).

STEERING

Steering Wheel

NOTE: 1990-92 vehicles are equipped with a Supplemental Inflatable Restraint system, make certain to follow the recommended disarming procedure before proceeding.

REMOVAL & INSTALLATION

1988–89

1. Disconnect the battery negative cable, then remove the horn cap from the steering wheel.

2. Remove the horn button wire.

3. Remove the telescope lever attaching screws and the shaft lock knob screw. Remove the telescope adjustment lever.

4. Remove the steering wheel nut retainer, then the steering wheel nut.

5. Using suitable steering wheel puller, remove the steering wheel from the vehicle.

NOTE: Before removing steering wheel, scribe marks on steering wheel hub and steering wheel shaft for ease of installation.

6. Installation is the reverse of the removal procedure.

7. Torque steering wheel nut to 30 ft. lbs. (40 Nm).

1990–92

1. Ensure the ignition switch is in the **OFF** position, disconnect battery

negative cable and remove the left sound insulator.

2. Remove the air bag fuse and disconnect the Supplemental Inflatable Restraint (SIR) harness yellow 2–way connector at base of the steering column.

3. Remove screws attaching the back of the inflator module to the steering wheel, then remove the inflator module from the steering wheel.

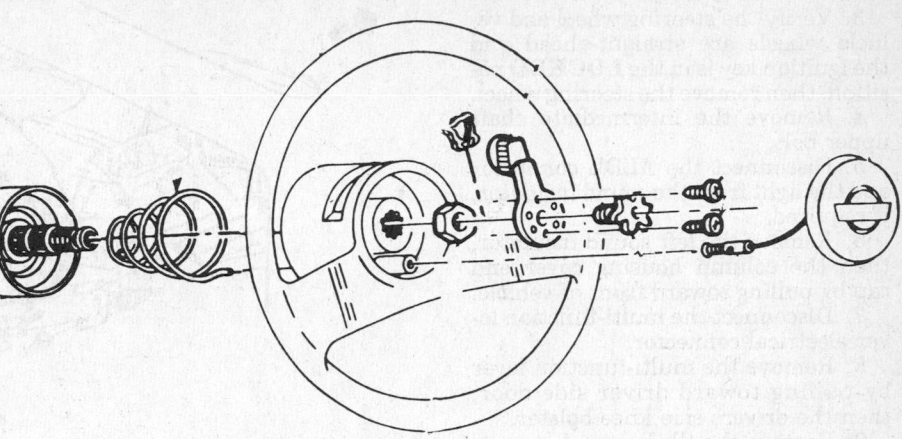

Steering wheel assembly–1988–89 vehicles

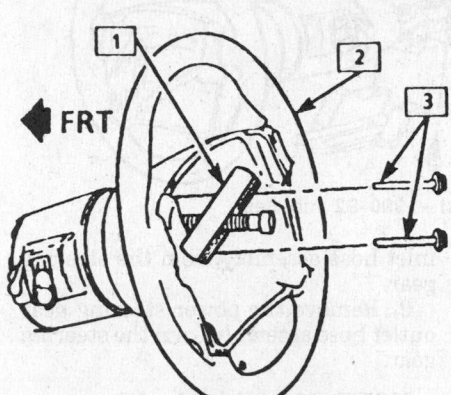

1. Steering wheel puller
2. Steering wheel
3. Puller side screws

Steering wheel removal–1990–92 vehicles

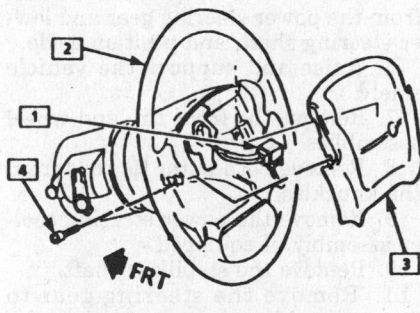

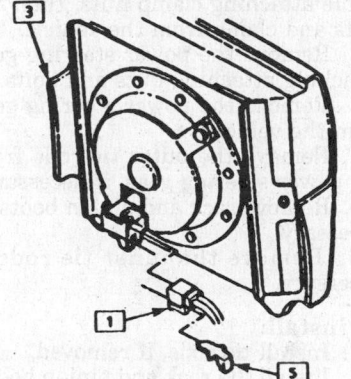

1. Connector
2. Steering wheel
3. Inflator module
4. Torx screw
5. Connector Pin Assurance (CPA) clip

Inflator module removal–1990–92 vehicles

4. Disconnect the SIR electrical connector at inflator module, then remove the steering wheel attaching nut.

NOTE: To avoid damaging the SIR coil, do not use any steering wheel puller other than those recommended.

5. Using steering wheel puller tool J–1859–03 and puller screws J–38720, remove the steering wheel. Disconnect the horn electrical connector.

6. Reverse procedure to install.

7. Torque steering wheel nut to 30 ft. lbs. (41 Nm).

Steering Column

REMOVAL & INSTALLATION

1988–89

NOTE: Handle the steering column very carefully. Hammering or leaning on the end of the column could shear off the plastic type inserts which allow the column to collapse in a collision.

1. Disconnect the negative battery cable, then remove the steering wheel.

2. Remove the nut and bolt from the upper intermediate shaft coupling.

3. Separate the coupling from the lower end of steering column.

4. Remove the left side instrument panel sound insulator and lower trim pad, then disconnect all electrical connectors from the column assembly.

5. Remove the nuts securing the lower plate to the floor.

6. Remove the nuts securing the bracket to the instrument panel.

7. Remove the steering column assembly.

8. Installation is the reverse of the removal procedure.

NOTE: These vehicles are equipped with an air bag system, make certain that the SIR system is disabled.

Also ensure the steering wheel and the front wheels of the vehicle are locked in the straight-ahead position before disconnecting the intermediate shaft.

Failure to follow these procedures will cause improper alignment of some internal components and also result in the SIR coil assembly being damaged.

1990–92

1. Properly disable the SIR sytem.

2. Disconnect battery negative cable.

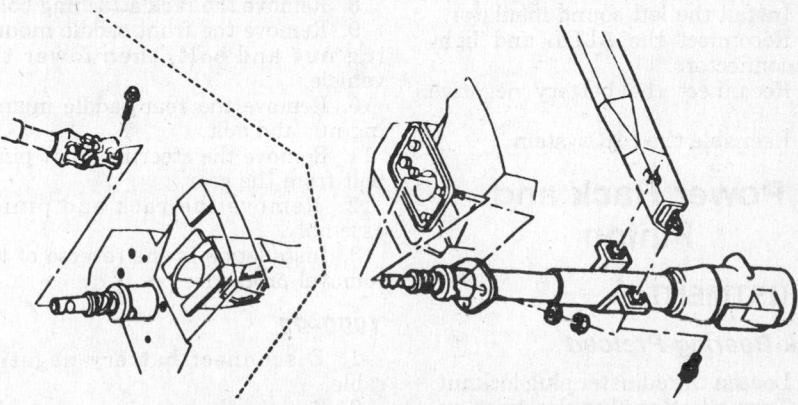

Steering column removal–1988–89 vehicles

3. Verify the steering wheel and vehicle wheels are straight-ahead and the ignition key is in the **LOCKED** position, then remove the steering wheel.

4. Remove the intermediate shaft upper bolt.

5. Disconnect the ALDL connector and the light from the sound insulator, if required.

6. Remove the left sound insulator, then the column housing cover end cap by pulling toward front of vehicle.

7. Disconnect the multi-function lever electrical connector.

8. Remove the multi-function lever by pulling toward driver side door, then the drivers side knee bolster.

9. Remove the tilt lever.

10. Remove the nuts from the lower support plate and capsule bolts from the reinforcement assembly, then disconnect all electrical connectors from the steering column.

11. Remove the sound insulator to steering column lower support bracket and the accelerator pedal bracket nuts.

12. Remove the steering column assembly from the vehicle.

To install:

13. Position steering column assembly into the vehicle and insert the lower steering shaft assembly into the joint of the intermediate shaft.

14. Install the intermediate shaft upper bolt and torque to 26 ft. lbs. (34Nm).

15. Attach the steering column and upper support plate to the instrument panel reinforcement assembly.

16. Attach the steering column nuts to the support plate studs.

17. Connect all electrical connectors, except SIR connector.

18. Install drivers side knee bolster.

19. Install the multifunction lever, harness and grommet.

20. Install the column housing cover end cap.

21. Install the tilt lever.

22. Install the sound insulator and accelerator pedal bracket nuts.

23. Install the steering wheel assembly.

24. Install the inflator module.

25. Install the left sound insulator.

26. Reconnect the ALDL and light wire connectors.

27. Reconnect the battery negative cable.

28. Reenable the SIR system.

Power Rack and Pinion

ADJUSTMENT

Rack Bearing Preload

1. Loosen the adjuster plug locknut.

2. Turn adjuster plug clockwise un-

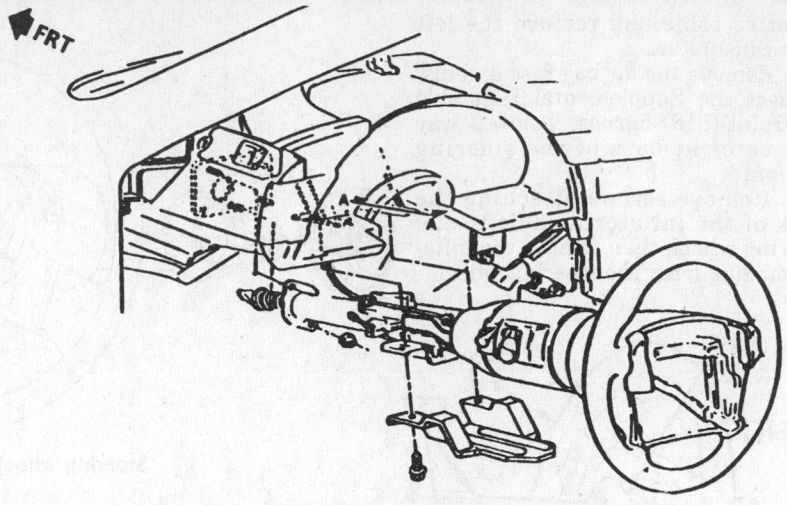

Steering column removal—1990–92 vehicles

til it bottoms, then back off 50–70 degrees.

3. Make adjustment with vehicle front wheels raised and steering wheel centered.

4. Inspect the steering wheel returnability to center after adjustment.

5. Tighten the locknut to 50 ft. lbs. (70 Nm).

REMOVAL & INSTALLATION

1988–89

1. Disconnect the battery negative cable.

2. Disconnect the inlet and outlet hoses, and the pipe from the gear valve. Place a suitable drain pan under the disconnected lines.

3. Raise and support vehicle safely, then remove both front wheel assemblies.

4. Using tool J-24319-01, disconnect both tie rod ends.

5. Remove the stabilizer bar and brackets assembly.

6. Remove the return pipe and brackets, mounted on the gear.

7. Disconnect the left cylinder feed pipe from the cylinder.

8. Remove the rack attaching bolts.

9. Remove the front saddle mounting nut and bolt, then lower the vehicle.

10. Remove the rear saddle mounting nut and bolt.

11. Remove the steering shaft pinch bolt from the gear.

12. Remove the rack and pinion assembly.

13. Installation is the reverse of the removal procedure.

1990–92

1. Disconnect battery negative cable.

2. Remove the power steering gear

inlet hose assembly from the steering gear.

3. Remove the power steering gear outlet hose assembly from the steering gear.

NOTE: If equipped with a power steering fluid cooling pipe, disconnect fluid cooling pipe outlet hose from the fluid cooling pipe.

4. Remove the steering gear coupling shield.

5. Remove the intermediate shaft from the power steering gear and lower steering shaft, and position aside.

6. Raise and support the vehicle safely.

7. Remove the front tire and wheel assemblies.

8. Remove both outer tie rods from the knuckles.

9. Remove the power steering cooler assembly, if equipped.

10. Remove the stabilizer shaft.

11. Remove the steering gear to frame attaching clamp nuts, then the bolts and clamp from the vehicle.

12. Remove the power steering gear attaching attaching nuts and bolts.

13. Remove the power steering gear from the vehicle.

14. Remove the outer tie rods from the power steering gear, if necessary.

15. Remove rack and pinion boots, if necessary.

16. Remove the inner tie rods if necessary.

To install:

17. Install tie rods, if removed.

18. Install the rack and pinion boots, if removed.

19. Connect the power steering cooling pipe, if removed.

20. Install the power steering gear, nuts and bolts. Torque nuts to 30 ft. lbs. (40 Nm). Torque the steering gear clamp nuts to 18 ft. lbs. (25 Nm).

21. Install the stabilizer shaft and

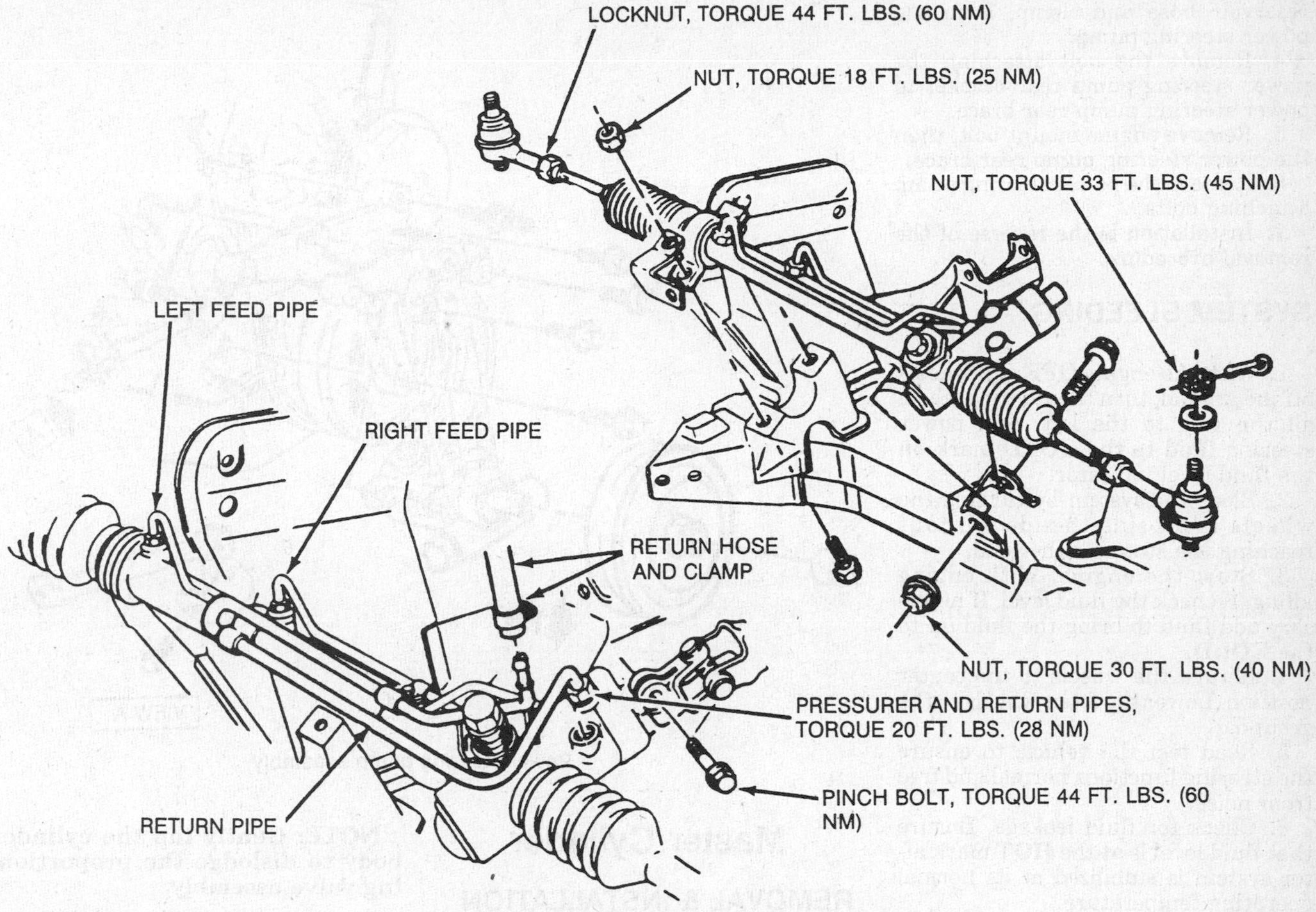

LOCKNUT, TORQUE 44 FT. LBS. (60 NM)

NUT, TORQUE 18 FT. LBS. (25 NM)

NUT, TORQUE 33 FT. LBS. (45 NM)

LEFT FEED PIPE

RIGHT FEED PIPE

RETURN HOSE
AND CLAMP

NUT, TORQUE 30 FT. LBS. (40 NM)

PRESSURER AND RETURN PIPES,
TORQUE 20 FT. LBS. (28 NM)

RETURN PIPE

PINCH BOLT, TORQUE 44 FT. LBS. (60
NM)

Power rack and pinion assembly–1988–89 vehicles

the power steering cooler assembly, if removed.

22. Install both outer tie rods to the steering knuckle.

23. Install tire and wheel assemblies and lower the vehicle.

24. Install the intermediate shaft and the steering gear coupling shield.

25. Install the power steering gear outlet hose assembly to the power steering gear. Torque fitting to 21 ft. lbs. (28 Nm).

26. Install the power steering gear inlet hose assembly to the power steering gear. Torque fitting to 21 ft. lbs. (28 Nm).

27. Refill power steering reservoir and bleed system. Connect battery negative cable and check operation.

Power Steering Pump

REMOVAL & INSTALLATION

1. Disconnect battery negative cable, then remove serpentine belt.

2. Remove the power steering pump pulley, then disconnect the power steering gear inlet hose assembly from the power steering pump.

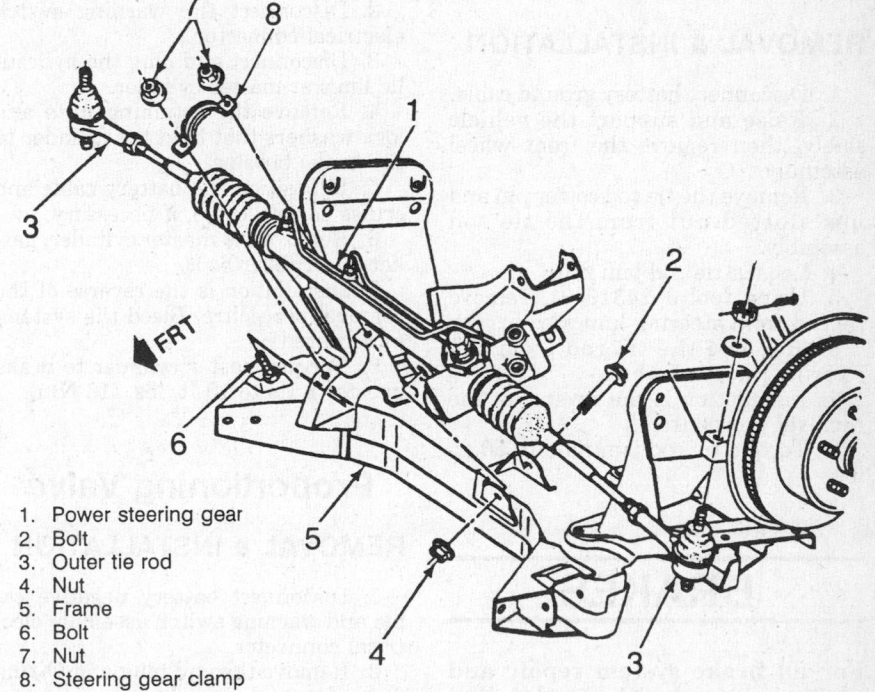

FRT

1. Power steering gear
2. Bolt
3. Outer tie rod
4. Nut
5. Frame
6. Bolt
7. Nut
8. Steering gear clamp

Power rack and pinion assembly–1990–92 vehicles

3. Disconnect the power steering reservoir hose and clamp, from the power steering pump.

4. Remove the bolt attaching the power steering pump rear bracket to power steering pump rear brace.

5. Remove engine mount bolt, then the power steering pump rear brace.

6. Remove the power steering pump attaching bolts.

7. Installation is the reverse of the removal procedure.

SYSTEM BLEEDING

1. With the engine **OFF** and wheels off the ground, turn the steering wheel all the way to the left. Add power steering fluid to the **COLD** mark on the fluid level indicator.

2. Bleed the system by turning the wheels from side-to-side without reaching the stop at either end.

3. Start the engine. With engine idling, recheck the fluid level. If necessary add fluid to bring the fluid up to the **COLD**.

4. Return the wheels to the center position. Lower the front wheels to the ground.

5. Road test the vehicle to ensure the steering functions normal and free from noise.

6. Check for fluid leakage. Ensure that fluid level is at the **HOT** mark after system is stabilized at its normal operating temperature.

Tie Rod Ends

REMOVAL & INSTALLATION

1. Disconnect battery ground cable.

2. Raise and support the vehicle safely, then remove the front wheel assembly.

3. Remove the tie rod cotter pin and hex slotted nut from the tie rod assembly.

4. Loosen tie rod jam nut.

5. Using tool J–24319–01, remove tie rod from steering knuckle.

6. Remove the tie rod from the steering rack assembly.

7. Installation is the reverse of removal procedure.

8. Torque tie rod jam nuts to 50 ft. lbs.

BRAKES

For all brake system repair and service procedures not detailed below, please refer to "Brakes" in the Unit Repair section.

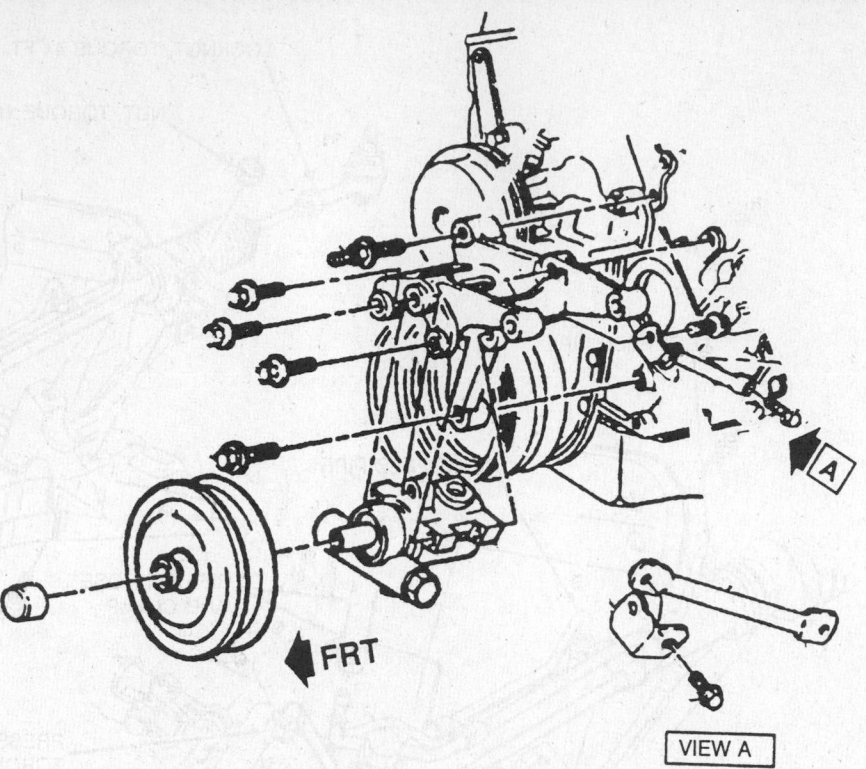

Power steering pump assembly

Master Cylinder

REMOVAL & INSTALLATION

1. Disconnect the negative battery cable.

2. Disconnect the warning switch electrical connector.

3. Disconnect and plug the hydraulic lines at master cylinder.

4. Remove the retaining nuts and lock washers that hold the cylinder to the brake booster.

5. Reposition the battery cable and cruise control cable, if necessary.

6. Remove the master cylinder, gasket and rubber boot.

7. Installation is the reverse of the removal procedure. Bleed the system, as required.

8. Torque master cylinder to brake booster nuts to 13 ft. lbs. (18 Nm).

Proportioning Valve

REMOVAL & INSTALLATION

1. Disconnect battery negative cable and warning switch assembly electrical connector.

2. Remove the end plug and O-ring from the master cylinder.

3. Remove the proportioning valve with the ground spring attached.

NOTE: Gently tap the cylinder body to dislodge the proportioning valve assembly.

4. Reverse the procedure to install. Connect the battery negative cable and bleed system.

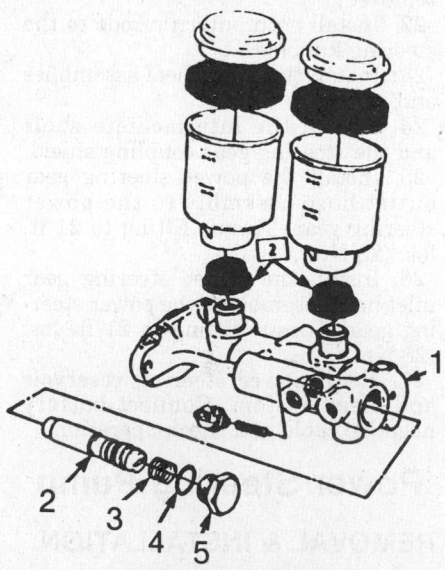

1. Master cylinder body
2. Proportioning valve
3. Ground spring
4. O-Ring seal
5. End plug

Master cylinder and proportioning valve

Power Brake Booster

REMOVAL & INSTALLATION

1988–89

1. Disconnect battery negative cable.
2. Remove the master cylinder from the vehicle.
3. Disconnect the vacuum hose from the vacuum check valve.
4. Remove the pushrod end of the valve assembly from the brake pedal by removing the retaining clip.
5. Remove the brake booster attaching nuts, then remove the power booster assembly from the vehicle.
6. Installation is the reverse of the removal procedure.
7. Torque power booster attaching bolts to 15 ft. lbs. (21 Nm). Connect battery negative cable and bleed system.

1990–92

1. Disconnect battery negative cable.
2. Remove the ECM, then the ECM housing bracket attaching bolt.
3. Remove the cruise control cable from the cruise control servo, then the servo mounting bracket.
4. Disconnect the pressure differential sensor electrical connector and the vacuum hose.
5. Disconnect the master cylinder warning switch electrical connector, then remove the nuts attaching the master cylinder to the power booster assembly. Position master cylinder, the cruise control cable and battery cable aside.
6. Remove the power booster vacuum check valve from the power booster assembly.
7. Remove the instrument panel left sound insulator, then remove the

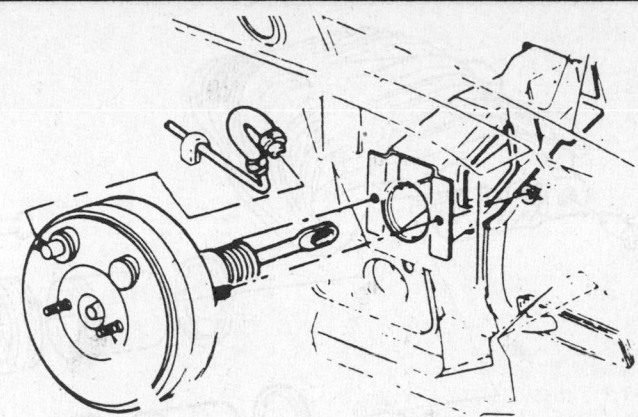

Power brake booster assembly—1988–89 vehicles

input pushrod assembly retaining ring and washer from the brake pedal.
8. Remove the power booster assembly attaching nuts and washers, then remove the power booster with seals and the ECM bracket attached.
9. Installation is the reverse of the removal procedure.
10. Torque power booster attaching nuts to 15 ft. lbs. (21 Nm). Connect battery negative cable and bleed system.

Brake Caliper

REMOVAL & INSTALLATION

Front

1. Remove ⅔ of the brake fluid from the master cylinder reservoir.
2. Raise and support the vehicle safely.
3. Mark the relationship between the wheel and axle flange.
4. Remove the tire and wheel assembly. Install 2 wheel nuts to retain the brake rotor.
5. Depress the caliper pistons into

the caliper bores to provide clearance between the pads and the rotor.
6. Disconnect and plug the brake line fitting at the caliper by removing the bolt, 2 gaskets then brake hose inlet fitting.
7. Remove the circlip and the retainer pin, then the caliper housing from the rotor and the caliper mounting bracket.
8. Installation is the reverse of the removal procedure.

Rear

1. Disengage the parking brake automatic adjuster as follows:
   a. Remove the drivers seat cushion.
   b. Remove the parking brake lever cover and screws.
   c. Using a suitable tool, disengage and hold the drive pawl from the drive sector.
   d. Insert a nail or drift through the hole in the anchor plate to retain the drive pawl in the disengaged position.
   e. Move the parking brake lever until it aligns with the lock pawl.

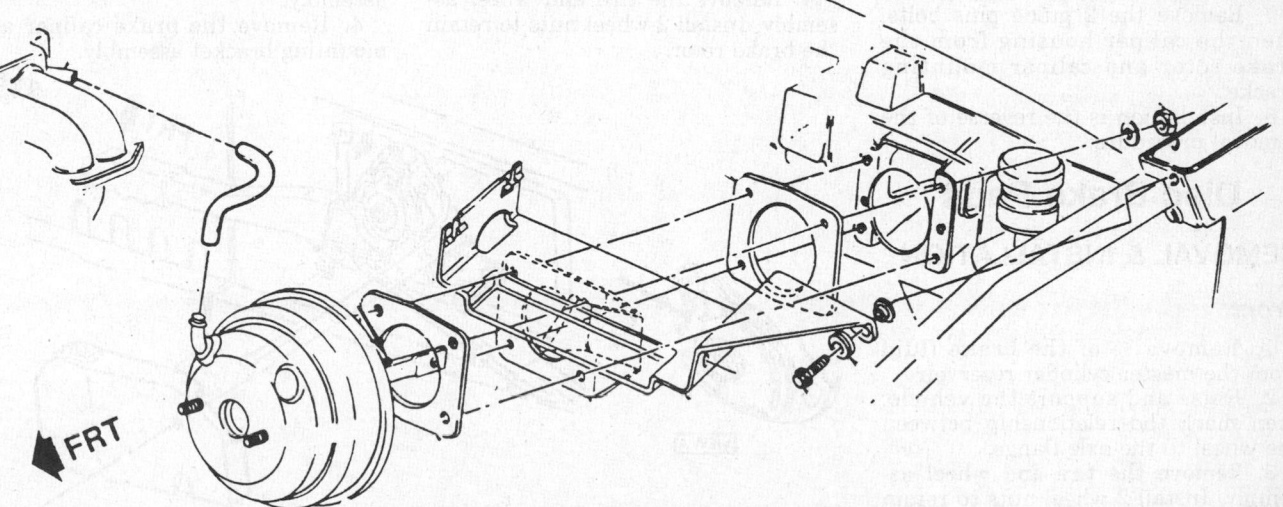

Power brake booster assembly—1990–92 vehicles

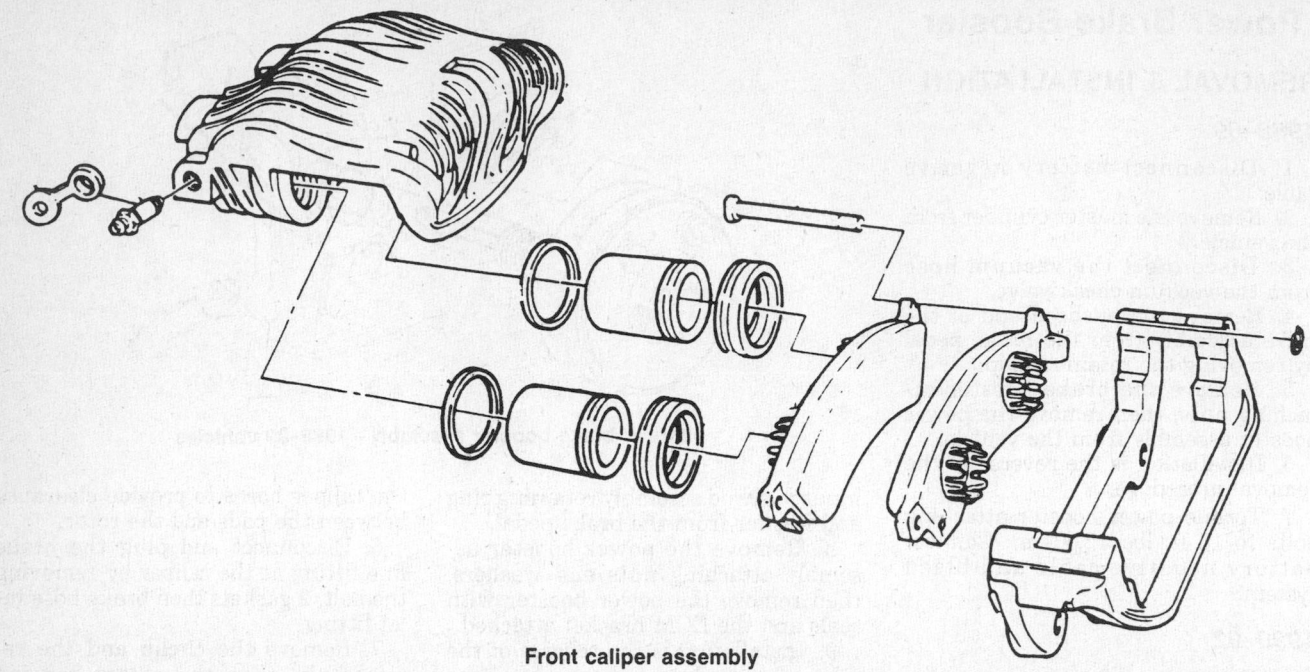

Front caliper assembly

f. Depress the button on the lever and move the lever to the down position.

g. Verify the anchor plate is against the stud on the parking brake lever, if not as specified, repeat the procedure.

2. Remove ⅔ of the brake fluid from the master cylinder reservoir.

3. Raise and support the vehicle.

4. Mark the relationship between the wheel and axle flange, then remove the tire and wheel assembly. Install 2 wheel nuts to retain the brake rotor.

5. Remove the brake line inlet fitting bolt, inlet fitting and 2 gaskets. Plug all lines.

6. Remove the lever return spring, then the brake cable from the lever and bracket, only after disengaging the parking brake automatic adjuster.

7. Remove the 2 guide pins bolts, then the caliper housing from the brake rotor and caliper mounting bracket.

8. Installation is the reverse of the removal procedure.

Disc Brake Pads

REMOVAL & INSTALLATION

Front

1. Remove ⅔ of the brake fluid from the master cylinder reservoir.

2. Raise and support the vehicle, then mark the relationship between the wheel to the axle flange.

3. Remove the tire and wheel assembly. Install 2 wheel nuts to retain the brake rotor.

4. Depress the caliper pistons into the caliper bores to provide clearance between the pads and the rotor, then remove the circlip and the retainer clip.

5. Lift the caliper from the rotor and the mounting bracket.

6. Remove the pads from the caliper.

7. Suspend the caliper from the upper control arm with wire to avoid damage to the brake hose.

8. Reverse procedure to install.

Rear

1. Remove ⅔ of the brake fluid from the master cylinder reservoirs.

2. Raise and support the vehicle.

3. Mark the relationship between the wheel to the axle flange.

4. Remove the tire and wheel assembly. Install 2 wheel nuts to retain the brake rotor.

5. Depress the caliper pistons into the caliper bores to provide clearance between the pads and the rotor.

6. Remove the caliper upper guide pin bolt, then rotate the caliper at the lower guide pin bolt and guide pin within the caliper mounting bracket.

7. Remove the pads from the caliper.

8. Reverse the procedure to install.

Brake Rotor

REMOVAL & INSTALLATION

Front and Rear

1. Disconnect battery negative cable.

2. Raise and support vehicle.

3. Remove the tire and wheel assembly.

4. Remove the brake caliper and mounting bracket assembly.

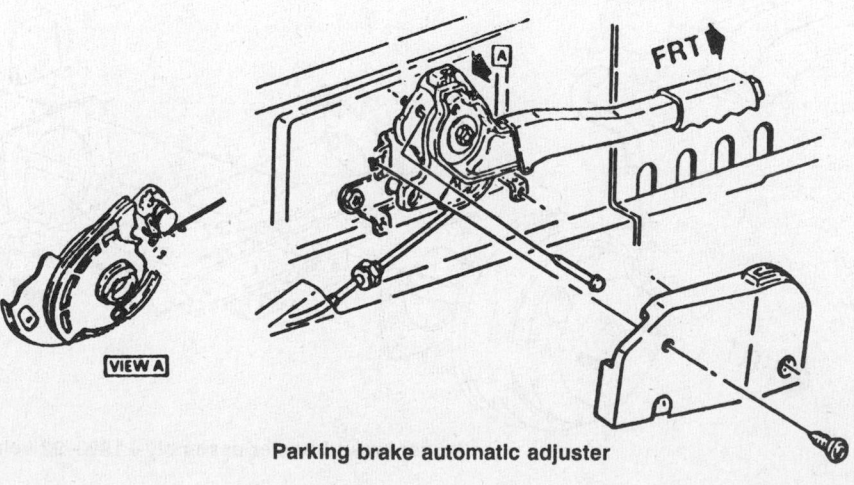

Parking brake automatic adjuster

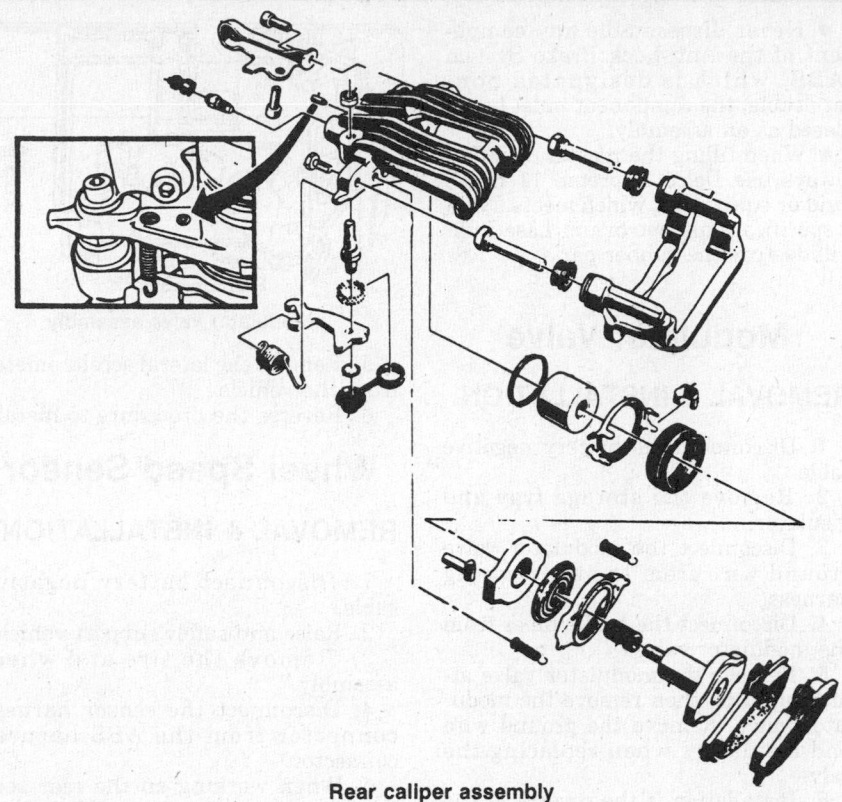

Rear caliper assembly

5. Remove the rotor from the vehicle.

6. Reverse the procedure to install.

Parking Brake Cable
ADJUSTMENT

The parking brake lever/cable adjustment is automatic and only needs to be disabled and enabled during rear brake pad, caliper or parking brake cable service.

1. To disable the automatic parking brake adjuster, proceed as follows:

a. Remove the drivers seat cushion.

b. Remove the parking brake lever cover and screws.

c. Using a suitable tool, disengage and hold the drive pawl from the drive sector.

d. Insert a nail or drift through the hole in the anchor plate to retain the drive pawl in the disengaged position.

e. Move the parking brake lever until it aligns with the lock pawl.

f. Depress the button on the lever and move the lever to the down position.

g. Verify the anchor plate is against the stud on the parking brake lever, if not as specified, repeat the procedure.

2. To enable the automatic parking brake adjuster, proceed as follows:

a. Remove the nail or drift pin from the anchor plate.

b. Apply and release the parking brake 3 times.

c. Release the parking brake.

d. There should be no rear brake drag and no gap between the caliper housings and caliper parking brake levers.

e. Install the wheel and tire assembly.

f. Install the parking brake lever cover and screws.

NOTE: Applying the brake lever 3 times should result in parking brake lever movement of 3–5 ratchet clicks when a 61 lb. (270 N) force is applied.

REMOVAL & INSTALLATION

Front Cable

1. Remove the drivers seat cushion and disable the parking brake automatic adjuster.

2. Raise the vehicle and support it safely.

3. Disconnect the front cable from the front cable connector.

4. Disconnect the front cable from the front cable return spring.

5. Disconnect the left rear cable from the parking brake front cable assembly connector.

6. Remove the front cable attaching clip bolt and clip.

7. Lower the vehicle and remove the front cable from the automatic adjuster.

8. Remove the front cable attaching

nut and washer. Remove the front cable from the vehicle.

9. Installation is the reverse of the removal procedure.

Intermediate Cable

1. Disable the parking brake automatic adjuster.

2. Raise the vehicle and support it safely.

3. Disconnect the parking brake intermediate cable from the cable connectors and front cable guide.

4. Installation is the reverse of the removal procedure.

Rear Cable

LEFT

1. Disable the automatic parking brake adjuster.

2. Raise and support the vehicle safely. Remove the tire and wheel assembly.

3. Remove the front cable from the front cable return spring.

4. Disconnect the front and rear cable assembly connector.

5. Disconnect the rear cable from the left rear cable bracket.

6. Disconnect the left rear cable from the caliper mounting bracket and lever.

7. Installation is the reverse of the removal procedure.

RIGHT

1. Disable the parking brake automatic adjuster.

2. Raise the vehicle and support it safely. Remove the tire and wheel assembly.

3. Disconnect the right rear cable from the intermediate cable.

4. Disconnect the right rear cable from the right rear cable bracket.

5. Disconnect the right rear cable from the caliper mounting bracket and lever.

6. Installation is the reverse of the removal procedure.

Brake System Bleeding

1. Clean the bleed screw at the caliper.

2. Start the wheel furthest from the master cylinder, the right-rear.

3. Attach a small rubber hose to the bleed screw and place the end in a clear glass container of brake fluid.

4. Fill the master cylinder with brake fluid, check the fluid during bleeding. Have an assistant slowly pump up the brake pedal and hold the pressure.

5. Open the bleed screw approximately ¼ turn, press the brake pedal to the floor, close the bleed screw and

slowly release the pedal. Continue until no more air bubbles are forced from the cylinder or caliper when depressing the brake pedal.

6. Repeat procedure on remaining wheel cylinders or calipers, still working from cylinder/caliper furthest from the master cylinder.

NOTE: The bleeder valve at the wheel cylinder must be closed at the end of each stroke and before the brake pedal is released, to insure that no air can enter the system. It is also important that the brake pedal be returned to the full up position so the piston in the master cylinder moves back enough to clear the bypass outlets.

Anti-Lock Brake System Service

PRECAUTIONS

Failure to observe the following precautions may result in system damage.

- Before performing electric arc welding on the vehicle, disconnect the Electronic Brake Control Module (EBCM) and the hydraulic modulator connectors.
- When performing painting work on the vehicle, do not expose the Electronic Brake Control Module (EBCM) to temperatures in excess of 185°F (85°C) for longer than 2 hrs. The system may be exposed to temperatures up to 200°F (95°C) for less than 15 min.
- Never disconnect or connect the Electronic Brake Control Module (EBCM) or hydraulic modulator connectors with the ignition switch ON.

- Never disassemble any component of the Anti-Lock Brake System (ABS) which is designated non-servicable; the component must be replaced as an assembly.
- When filling the master cylinder, always use Delco Supreme 11 brake fluid or equivalent, which meets DOT-3 specifications; petroleum base fluid will destroy the rubber parts.

Modulator Valve

REMOVAL & INSTALLATION

1. Disconnect the battery negative cable.
2. Remove the storage tray and insulator.
3. Disconnect the modulator valve ground wire from the body wiring harness.
4. Disconnect the brake pipes from the modulator valve.
5. Remove the modulator valve attaching nuts, then remove the modulator valve. Remove the ground wire and insulators when replacing the valve.
6. Installation is the reverse of the removal procedure.

Lateral Accelerometer

REMOVAL & INSTALLATION

1. Disconnect battery negative cable.
2. Remove the console trim plate.
3. Remove the radio assembly.
4. Remove the lateral accelerometer-to-instrument panel carrier assembly attaching screws. Disconnect electrical connector.

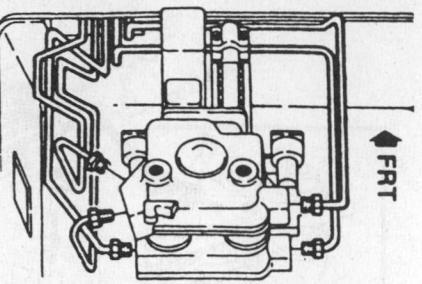

Modulator valve assembly

5. Remove the lateral accelerometer from the vehicle.
6. Reverse the procedure to install.

Wheel Speed Sensor

REMOVAL & INSTALLATION

1. Disconnect battery negative cable.
2. Raise and safely support vehicle.
3. Remove the tire and wheel assembly.
4. Disconnect the sensor harness connector from the ABS harness connector.
5. When working on the rear sensors, remove the bracket and bolt from the knuckle.
6. Remove sensor wiring harness with the grommets from the bracket.
7. Remove the wheel speed sensor attaching bolt, then the wheel speed sensors from the knuckle.
8. Reverse the procedure to install.

Control Module

REMOVAL & INSTALLATION

1. Disconnect the negative battery cable.
2. Remove the storage tray and insulator.
3. Disconnect the electrical connector from the control module.
4. Remove the module relay from the brackets.
5. Remove the control module attaching bolts, then remove the control module from the vehicle.
6. Reverse the procedure to install.

Pump Motor Relay and Solenoid Relay

REMOVAL & INSTALLATION

1. Disconnect battery negative cable.
2. Remove the storage tray and insulator.
3. Remove the retainer screw and retainer, then the appropriate relay.
4. Reverse the procedure to install.

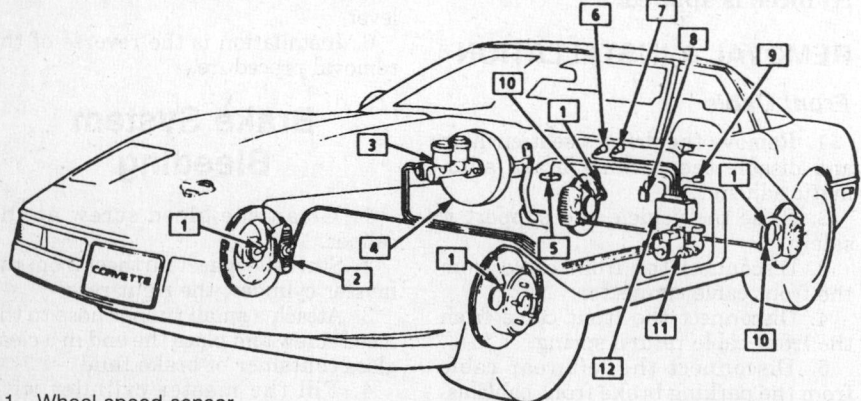

1. Wheel speed sensor
2. Caliper
3. Master cylinder
4. Power booster
5. Lateral accelerometer
6. ABS service light
7. ABS active light
8. Electronic brake control relay
9. ABS active indicator relay
10. Toothed ring
11. Electronic brake control module
12. Modulator valve

Antilock brake system components

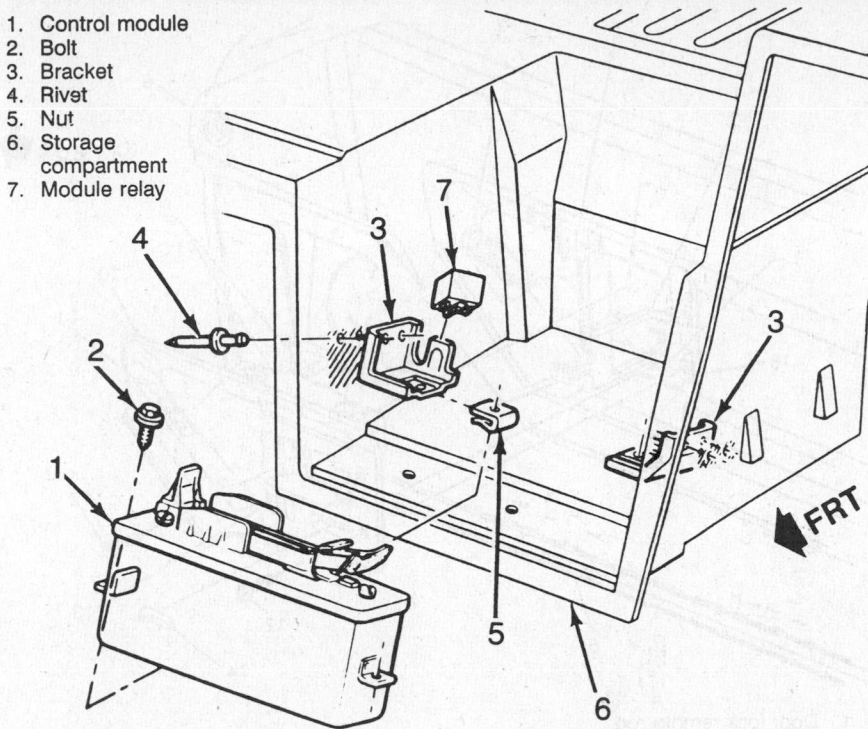

1. Control module
2. Bolt
3. Bracket
4. Rivet
5. Nut
6. Storage compartment
7. Module relay

Antilock electronic brake control module

Module Relay and Active indicator Relay

The electronic brake control relay and active indicator relay are located on the wiring harness in the left rear storage compartment.

CHASSIS ELECTRICAL

Air Bag

DISARMING

1. Turn the steering wheel to align the wheels in the straight-ahead position.
2. Turn the ignition switch to the **LOCK** position.
3. Remove the SIR air bag fuse from the fuse block.
4. Remove the left side trim panel and disconnect the yellow 2-way SIR harness wire connector at the base of the steering column.
To enable system:
5. Turn the ignition switch to the **LOCK** position.
6. Reconnect the yellow 2-way connector at the base of the steering column.

7. Reinstall the SIR fuse and the left side trim panel.
8. Turn the ignition switch to the **RUN** position.
9. Verify the SIR indicator light flashes 7–9 times, if not as specified, inspect system for malfunction.

SUPPLEMENTAL INFLATABLE RESTRAINT (SIR) COIL ASSEMBLY

NOTE: After performing repairs on the internals of the steering column the coil assembly must be centered in order to avoid damaging the coil or accidental deployment of the air bag.

ADJUSTMENT

1. With the system properly disarmed, hold the coil assembly with the

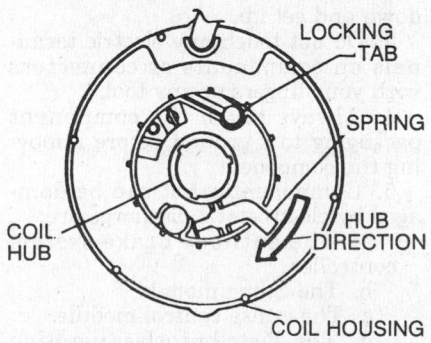

Centering the SIR coil assembly

clear bottom up to see the coil ribbon.
2. While holding the coil assembly, depress the lock spring and rotate the hub in the direction of the arrow until it stops. The coil should now be wound up snug against the center hub.
3. Rotate the coil assembly in the opposite direction approximately 2½ turns and release the lock spring between the locking tabs in front of the arrow.
4. Install the coil assembly onto the steering shaft.

Heater Blower Motor

REMOVAL & INSTALLATION

1. Disconnect the negative battery cable.
2. Remove the front wheel house rear panel. Move the wheel house seal aside.
3. Remove the blower motor cooling tube.
4. Disconnect the motor electrical connectors and remove the blower motor relay, if required.
5. Remove the motor retaining screws and remove the motor assembly from the vehicle.
6. Installation is the reverse of the removal procedure.

Windshield Wiper Motor

REMOVAL & INSTALLATION

1988–90 5.7L (VIN 8) Engine

1. Raise the hood.
2. Remove the wiper arms.
3. Remove the left plenum screen.
4. Turn ignition **ON** and activate wiper motor with the switch. Allow motor crank arm to rotate to the 4–5 o'clock position as viewed from passenger compartment. Stop the crank arm in this position by turning the ignition switch **OFF**.
5. Disconnect the battery negative cable.
6. Disconnect upper motor electrical connectors.
7. Remove the nuts securing the motor crank arm to the transmission link sockets and the motor mounting bolts.
8. With crank arm in position described in Step 4 above, motor may now be removed from vehicle. Disconnect the lower electrical connectors from the motor.
9. Installation is the reverse of the removal procedure. Reconnect the battery negative cable and check motor operation.

1991–92 5.7L (VIN 8) and 1990–92 5.7L (VIN J) Engines

1. Raise the hood and disconnect the negative battery cable.
2. Remove the left side plenum screen.
3. Remove the wiper transmission nuts and sockets.
4. Remove the vacuum booster supply hose at the plenum, if equipped.
5. Disconnect the wiper mounting bolts.
6. Disconnect motor electrical connectors while removing wiper motor assembly.
7. Installation is the reverse of the removal procedure. Reconnect the battery negative cable and check motor operation.

Windshield Wiper Switch

REMOVAL & INSTALLATION

1988–89

1. Disconnect the negative battery cable.
2. Remove the drivers door armrest and trim plate.
3. Disconnect the electrical connections from the switch.
4. Remove the retaining screws and the switch from the panel.
5. Installation is the reverse of the removal procedure.

1990–92

NOTE: This vehicle is equipped with an Supplemental Inflatable Restraint system, make certain to follow the recommended disarming procedure before proceeding.

1. Disconnect the negative battery cable.
2. Remove the column housing end cover by pulling toward front of the vehicle.
3. Disconnect the electrical connector and grommet.
4. Pull the multi-function lever toward the drivers door to remove.
5. Installation is the reverse of the removal procedure.

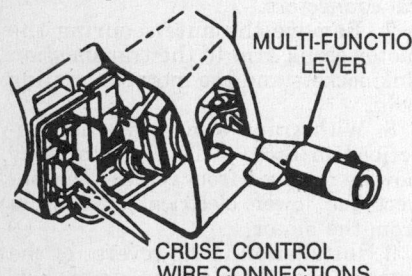

Windshield wiper switch removal— 1990–92 vehicles

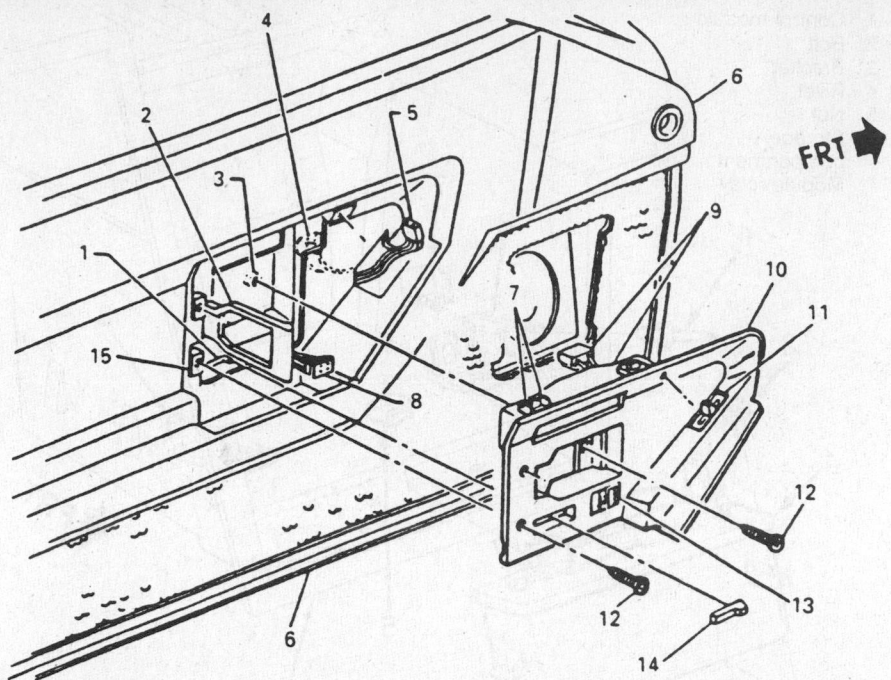

1. Door lock remote rod
2. Door inside handle
3. Accessory trim plate nut
4. Courtesy light electrical harness
5. Windshield wiper/ washer electrical harness
6. Door trim panel
7. Accessory trim plate retainer tabs
8. Door lock electrical harness
9. Courtesy light
10. Accessory trim plate
11. Windshield/wiper washer switch
12. Accessory trim plate screw
13. Door lock switch
14. Door lock rod knob
15. Accessory trim plate U-nut

Windshield wiper switch removal—1988–89 vehicles

Instrument Cluster

PRELIMINARY PROCEDURES

Many electrical components are sensitive to static electricity discharge; in order to avoid damaging any components certain precautions should be taken:

1. To discharge personal static electricity, touch a ground point on the vehicle.
2. Personal static discharge should be performed any time you walk across the shop, slide across the seat or sit down and get up.
3. Do not touch any electric terminals on components or connectors with your fingers or any tool.
4. Always touch the component packaging to a ground before removing the component.
5. Components that can be damaged by electrostatic discharge are:
   a. The antilock brake system controller.
   b. The chime module.
   c. The cruise control module.
   d. The distributorless ignition system module.
   e. The electronic digital instrument clusters.
   f. The electronic control module and attributing parts.
   g. The low tire pressure warning system module.
   h. The radio assembly.
   i. The theft deterrent modules.
   j. The electronic automatic air conditioning assembly.

REMOVAL & INSTALLATION

1988–89

1. Disconnect the battery ground cable.
2. Remove light switch knob (spring loaded).
3. Remove steering column trim cover.
4. Remove the steering column attaching bolts and lower steering column for access.
5. Remove cluster bezel front and left side attaching screws.
6. Remove cluster bezel from instrument panel.
7. Remove the cluster to instrument panel attaching screws.
8. Pull cluster rearward for access to disconnect cluster electrical connec-

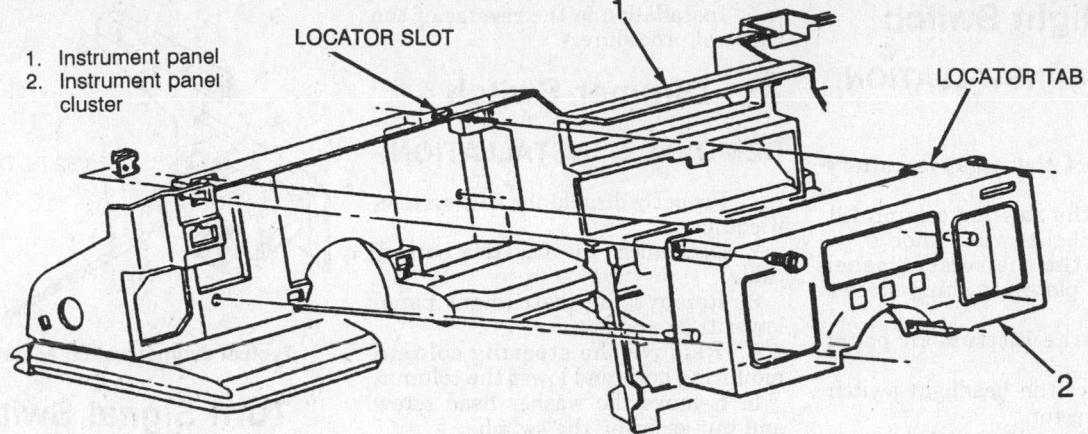

1. Instrument panel
2. Instrument panel cluster

LOCATOR SLOT

LOCATOR TAB

Instrument panel cluster—1988–89 vehicles

tors. Metal retaining clips are located at back side of connectors.

9. Remove cluster from instrument panel.

10. Installation is the reverse of the removal procedure.

1990–92

1. Disconnect battery negative cable.

2. Remove shifter button, snapring and shift knob.

3. Remove center console trim plate attaching screws, disconnect electrical connectors from trim plate, then remove trim plate.

4. Remove the center air outlet.

5. Remove the accessory trim plate attaching screws and clips, then lower trim panel carefully and disconnect the instrument panel electrical connectors. Remove the ALDL connector assembly screw.

6. Remove the steering column attaching bolts, then lower the steering column for access.

7. Remove the cluster bezel attaching screws, cluster screws and disconnect the instrument panel cluster electrical connector. Remove the instrument panel cluster.

8. Reverse procedure to install. Torque cluster bezel screws to 29 inch lbs. (3.3 Nm).

Radio

REMOVAL & INSTALLATION

1. Disconnect battery negative cable.

2. Remove the instrument panel trim plates as necessary.

3. Remove radio attaching screws and pull radio carefully outwards.

4. Disconnect electrical connectors antenna lead from radio.

5. Remove the radio from the vehicle.

6. Installation is the reverse of the removal procedure.

Concealed Headlights

MANUAL OPERATION

The vehicle is equipped with concealed headlights, the headlight doors can be opened automatically by turning the headlights switch to the **ON** position, then turn the switch back one click to the parking lights **ON** position and the headlight doors will stay open. To open the headlight doors manually, raise the hood and turn the headlight manual control knob, located on the headlamp door, in the direction of the arrow until the door is fully opened.

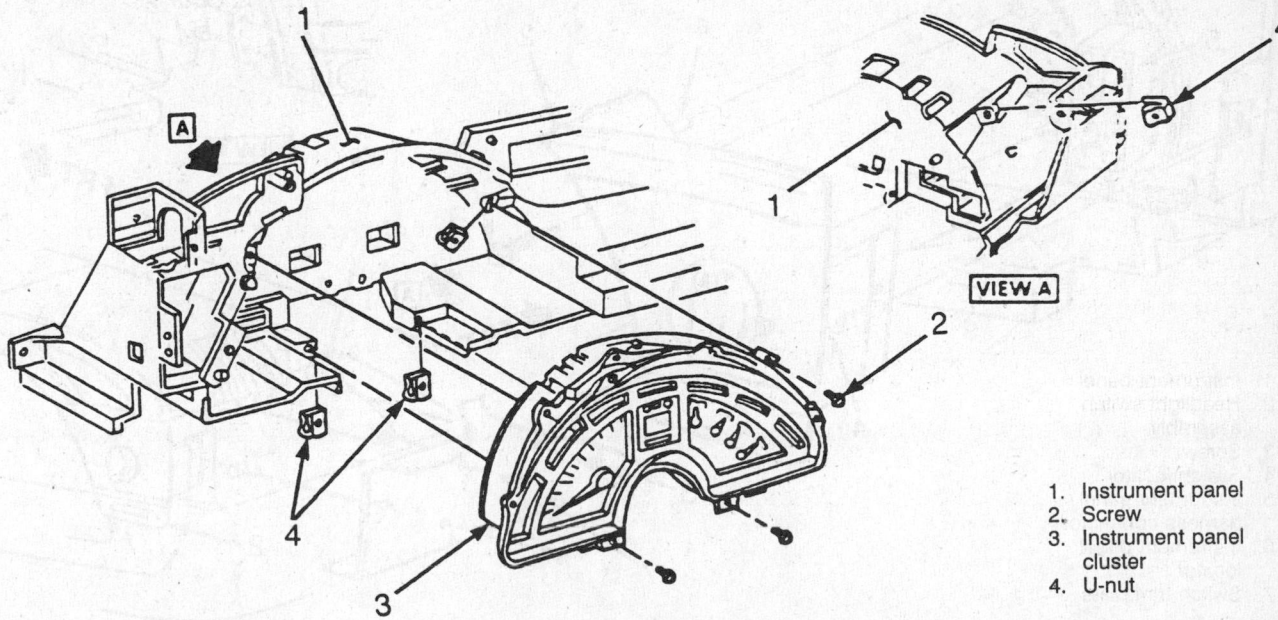

VIEW A

1. Instrument panel
2. Screw
3. Instrument panel cluster
4. U-nut

Instrument panel cluster—1990–92 vehicles

Headlight Switch

REMOVAL & INSTALLATION

1988–89

1. Disconnect the negative battery terminal.
2. Remove the steering column tilt lever and headlight switch knob.
3. Remove the instrument panel courtesy trim plate attaching screws and trim plate.
4. Remove the instrument panel upper pad.
5. Disconnect the headlight switch electrical connector.
6. Remove the switch retaining nut.
7. Remove the switch assembly.
8. Installation is the reverse of the removal procedure. Connect the battery negative cable and check operation of switch.

1990–92

1. Disconnect battery negative cable.
2. Remove the instrument cluster trim plate screws and reposition the trim plate to gain access to the right side of the headlight switch.
3. Remove headlight switch attaching screws and disconnect electrical connectors.
4. Remove the headlight switch.

5. Installation is the reverse of the removal procedure.

Dimmer Switch

REMOVAL & INSTALLATION

1. Properly disable the SIR system, if equipped.
2. Disconnect the negative battery cable.
3. Remove the instrument panel lower trim and lower.
4. Remove the steering column mounting bolts and lower the column.
5. Remove the washer head screw and nut securing the switch.
6. Remove the horn ground strap attached to the switch, if equipped.
7. Remove the switch assembly and disconnect the wire connector.

To install:

8. Install switch with attaching nuts finger-tight.
9. Connect the horn ground strap and switch wire connector.
10. Insert a $\frac{3}{32}$ in. drill bit in the switch hole to limit travel.
11. Push against dimmer switch to remove all free play.
12. Tighten the switch nut and screw.
13. Remove the drill bit.
14. Reenable the SIR system and connect the negative battery cable.

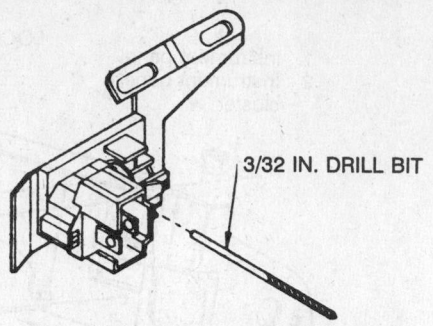

3/32 IN. DRILL BIT

Typical dimmer switch adjustment

Turn Signal Switch

REMOVAL & INSTALLATION

1988–89

1. Disconnect the negative battery cable.
2. Remove the steering wheel.
3. Remove the steering column left side lower trim cover. Disconnect the turn signal harness at the base of the steering column.
4. Remove the spacers and snap ring retainer.
5. Install the lock plate compressing tool over the steering shaft and remove the lock plate retainer.
6. Remove the tool and lift out the lock plate, horn contact carrier and the upper bearing preload spring.

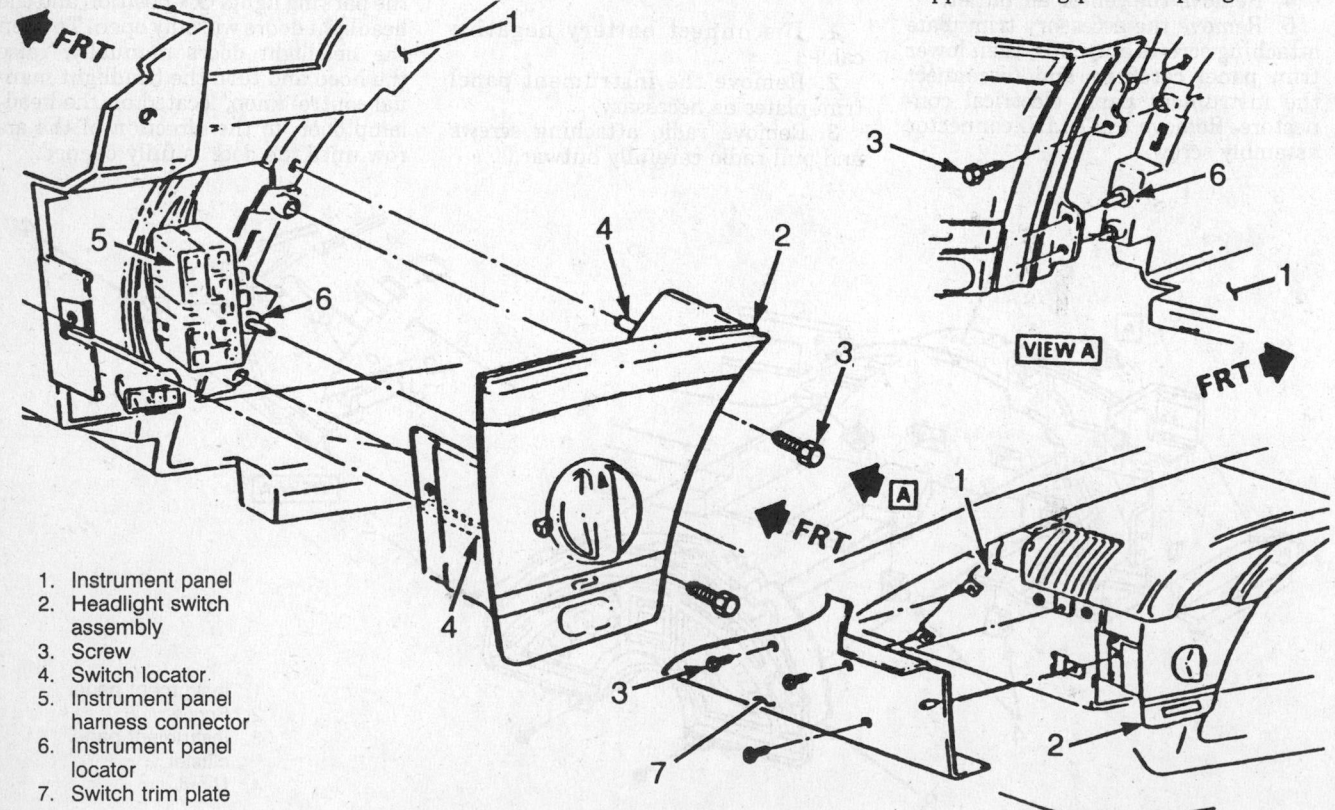

1. Instrument panel
2. Headlight switch assembly
3. Screw
4. Switch locator
5. Instrument panel harness connector
6. Instrument panel locator
7. Switch trim plate

Headlight switch assembly—1990–92 vehicles

7. Position the turn signal switch to the right and remove the switch actuator pivot and switch screws.

8. Remove the switch wiring from the protective jacket.

9. Remove the turn signal lever and hazard knob.

10. Remove the switch by pulling it straight up while guiding the wiring harness out of the housing.

To install:

11. Install the replacement switch by working the harness connector down through the housing and under the mounting bracket.

12. Install the harness cover and clip the connector to the column.

13. Install the switch mounting screws, signal lever and the flasher knob.

14. With the turn signal lever in neutral and the flasher knob out, install the upper bearing preload spring, horn contact carrier and lock plate onto the shaft.

15. Position the tool as in Step 4 and compress the plate far enough to allow the C-ring to be installed.

16. Remove the tool. Install the plastic C-ring retainer.

17. Install the lower dash trim cover and the steering wheel.

18. Connect the negative battery cable.

1990–92

NOTE: The vehicle is equipped with a Supplemental Inflatable Restraint system, make certain to follow the recommended disarming procedure before proceeding.

1. Properly disable the SIR system.
2. Place the ignition switch to the lock position to prevent uncentering of the coil assembly.
3. Disconnect the negative battery cable.
4. Remove the steering wheel and place the wheel upright on a bench.
5. Remove the coil assembly and let the coil hang freely.

NOTE: The coil assembly will become uncentered if the steering column is separated from the steering gear and allowed to rotate or the center spring of the coil assembly is pushed down, letting the hub rotate while the coil is removed from the steering column. In the event this should occur, follow the recommended procedure for recentering of the coil in order to avoid accidental deployment of the air bag or damage to the internal components of the steering column.

6. Remove the wave washer.
7. Remove the shaft lock retaining

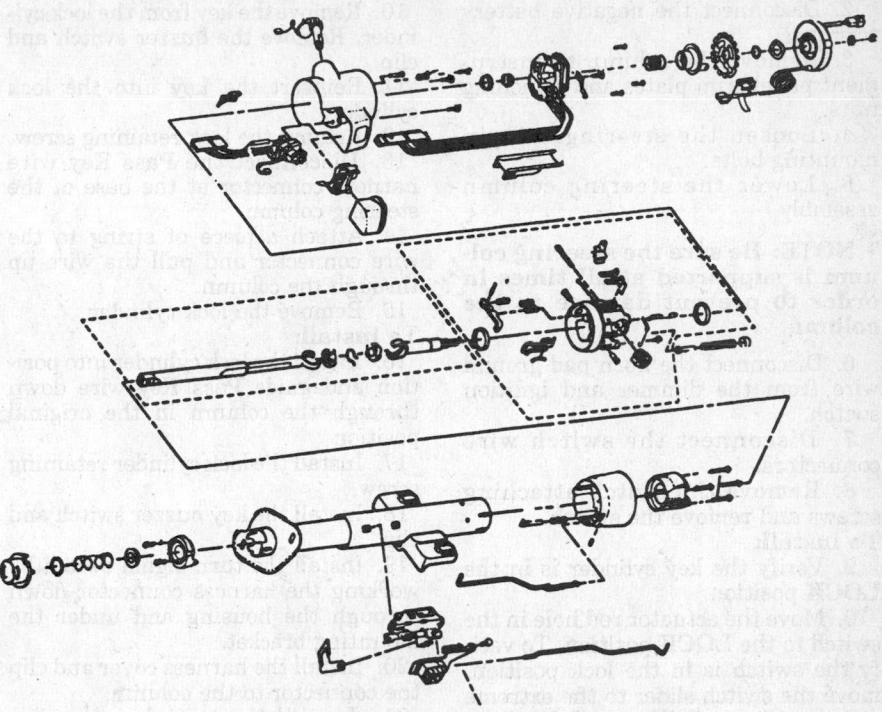

Turn signal assembly with SIR—1990–92 vehicles

ring using tool J23653-C or equivalent.

8. Remove the shaft lock, turn signal cancelling cam and upper bearing assembly.

9. Remove the turn signal lever and the hazard knob.

10. Remove the turn signal switch actuator and screws.

11. Remove the column lower trim and disconnect the switch harness connector.

12. Remove the wiring protector and gently pull the assembly up through the housing.

To install:

13. Route the switch harness through the column and connect to the harness.

14. Reconnect the harness wiring protector and lower trim panel.

15. Install the switch mounting screws and actuator arm.

16. Install the turn signal lever and the hazard knob.

17. Install the shaft lock, turn signal cancelling cam and upper bearing assembly.

18. Install the shaft lock retaining ring using tool J23653-C or equivalent.

19. Install the wave washer.

20. Install the coil assembly, so the steering shaft block teeth are set at the 6 and 12 o'clock positions and wheels straight-ahead, then set the ignition switch to the **LOCK** position. Verify the coil is centered properly (revolutions, not side-to-side positioning).

21. Install the steering wheel.

22. Reconnect the negative battery cable.

Ignition Switch

REMOVAL & INSTALLATION

1. Properly disable the SIR system, if equipped.

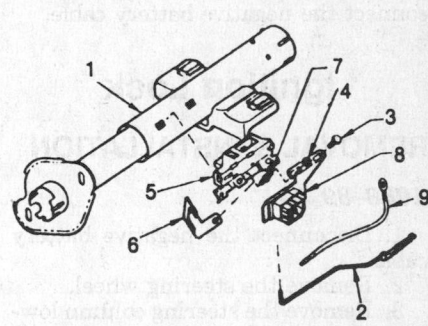

1. Steering column housing
2. Actuator rod
3. Screw
4. Nut
5. Ignition switch
6. Cable bracket
7. Stud
8. Dimmer switch
9. Horn pad ground

Ignition and dimmer switch assembly

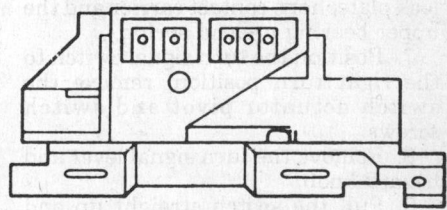

Ignition switch adjustment

2. Disconnect the negative battery terminal.

3. Remove the column to instrument panel trim plates and attaching nuts.

4. Loosen the steering column mounting bolts.

5. Lower the steering column assembly.

NOTE: Be sure the steering column is supported at all times in order to prevent damage to the column.

6. Disconnect the horn pad ground wire from the dimmer and ignition switch.

7. Disconnect the switch wire connectors.

8. Remove the switch attaching screws and remove the switch.

To install:

9. Verify the key cylinder is in the **LOCK** position.

10. Move the actuator rod hole in the switch to the **LOCK** position. To verify the switch is in the lock position, move the switch slider to the extreme right position and then move the slider 1 detent to the left.

11. Install the switch with the rod in the hole and secure with the screw and nut.

12. Connect the switch wire harness.

13. Reconnect the horn ground strap.

14. Raise and secure the steering column.

15. Install the instrument panel trim.

16. Re-enable the SIR system and connect the negative battery cable.

Ignition Lock

REMOVAL & INSTALLATION

1988–89

1. Disconnect the negative battery cable.

2. Remove the steering wheel.

3. Remove the steering column lower trim cover.

4. Remove the spacers and snap ring retainer from the steering shaft.

5. Install the lock plate compressing tool over the steering shaft and remove the lock plate retainer.

6. Remove the tool and lift out the lock plate, horn contact carrier and the upper bearing preload spring.

7. Position the turn signal switch to the right turn position, remove the switch actuator pivot and switch screws.

8. Remove the turn signal lever and hazard knob.

9. Pull the switch straight up and position the switch off to the side.

10. Remove the key from the lock cylinder. Remove the buzzer switch and clip.

11. Reinsert the key into the lock cylinder.

12. Remove the lock retaining screw.

13. Disconnect the Pass Key wire harness connector at the base of the steering column.

14. Attach a piece of string to the wire connector and pull the wire up through the column.

15. Remove the lock cylinder.

To install:

16. Install the lock cylinder into position and guide Pass Key wire down through the column in the original position.

17. Install the lock cylinder retaining screw.

18. Install the key buzzer switch and clip.

19. Install the turn signal switch by working the harness connector down through the housing and under the mounting bracket.

20. Install the harness cover and clip the connector to the column.

21. Install the switch mounting screws, signal lever and the flasher knob.

22. With the turn signal lever in neutral and the flasher knob out, install the upper bearing preload spring, horn contact carrier and lock plate onto the shaft.

23. Position the tool as in Step 4 and compress the plate far enough to allow the C-ring to be installed.

24. Remove the tool. Install the plastic C-ring retainer.

25. Install the lower dash trim cover and the steering wheel.

26. Connect the negative battery cable.

1990–92

NOTE: The vehicle is equipped with a Supplemental Inflatable Restraint system, make certain to follow the recommended disarming procedure before proceeding.

1. Properly disable the SIR system.

2. Place the ignition switch to the lock position to prevent uncentering of the coil assembly.

3. Disconnect the negative battery cable.

4. Remove the steering wheel and place the wheel upright on a bench.

5. Remove the coil assembly and let the coil hang freely.

NOTE: The coil assembly will become uncentered if the steering column is separated from the steering gear and allowed to rotate or the center spring of the coil assembly is pushed down, letting the hub rotate while the coil

is removed from the steering column. In the event this should occur, follow the recommended procedure for recentering of the coil in order to avoid accidental deployment of the air bag or damage to the internal components of the steering column.

6. Remove the wave washer.

7. Remove the shaft lock retaining ring using tool J–23653–C or equivelent.

8. Remove the shaft lock, turn signal cancelling cam and upper bearing assembly.

9. Remove the turn signal lever and the hazard knob.

10. Remove the turn signal switch actuator and screws.

11. Remove the column lower trim and disconnect the switch harness connector.

12. Remove the wiring protector and gently pull the turn signal switch up and position the assembly to the side.

13. Remove the key from the lock cylinder. Remove the buzzer switch and clip.

14. Reinsert the key into the lock cylinder.

15. Remove the lock retaining screw.

16. Disconnect the Pass Key wire harness connector at the base of the steering column.

17. Attach a piece of string to the wire connector and pull the wire up through the column.

18. Remove the lock cylinder.

To install:

19. Install the lock cylinder into position and guide Pass Key wire down through the column in the original position using the string as a guide.

20. Install the lock cylinder retaining screw.

21. Install the key buzzer switch and clip.

22. Install the turn signal switch by working the harness connector down through the housing and under the mounting bracket.

23. Reconnect the harness wiring, the protector and lower trim panel.

24. Install the switch mounting screws and actuator arm.

25. Install the turn signal lever and the hazard knob.

26. Install the shaft lock, turn signal cancelling cam and upper bearing assembly.

27. Install the shaft lock retaining ring using tool J–23653–C or equivelent.

28. Install the wave washer.

29. Install the coil assembly, with the steering shaft block teeth set at the 6 and 12 o'clock positions and wheels straight-ahead, then set the ignition switch to the **LOCK** position. Verify the coil is centered properly (according

to revolutions, not side-to-side positioning).

30. Install the steering wheel.
31. Reconnect the negative battery cable.

Stoplight Switch

ADJUSTMENT

While depressing the brake pedal, insert the switch into the retainer until seated. Pull the brake pedal fully rearward against the pedal stop until the clicking sounds are not heard. The switch will move into the retainer providing proper adjustment. Release the brake pedal and repeat, to ensure that no clicking sounds are heard.

REMOVAL & INSTALLATION

1. Disconnect the battery negative cable. Remove the lower trim panel.
2. Disconnect the electrical connector from the switch.
3. Remove the retainer and the switch from the vehicle.
4. Installation is the reverse of the reverse of the removal procedure.

Clutch Switch

REMOVAL & INSTALLATION

1. Disconnect the negative battery cable.
2. Remove the sound insulator panel from under the dash.
3. Remove the bolt and switch from the clutch bracket. Rotate the switch and remove the clip retainer from the actuating rod.
4. Disconnect the wire connector form the switch.
5. Remove the switch assembly.

To install:

6. Place a new switch in position with the actuating rod in line with the hole in the clutch pedal and reinstall the bolt.
7. Insert the rod into the bracket and secure with the retainer.
8. Connect the electrical connector to the switch and depress the clutch pedal fully to the floor to check adjustment of the switch.
9. Connect the battery negative cable.

Neutral Safety Switch

ADJUSTMENT

1. Disconnect battery negative cable.
2. Remove shifter knob assembly.
3. Remove console.

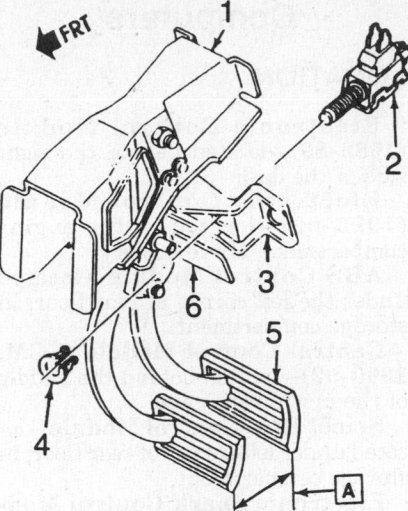

1. Normal travel
2. Brake pedal bracket
3. Stoplight switch
4. Retainer
5. Brake pedal
6. Actuator

Stoplight switch assembly

4. Remove neutral switch mounting bolts, then the switch.
5. Position the shift control lever in N.
6. Align tang on switch with the tang slot on the shift control.
7. Insert gauge pin $\frac{3}{32}$ in. (2.34mm) in service adjustment hole and rotate switch until pin drops to a depth of $\frac{19}{32}$ inch (15mm).
8. Install switch mounting bolts and torque to 26 inch lbs. (3 Nm). Remove the gage pin.
9. Install console and shifter knob assembly. Connect battery negative cable and verify the engine starts in P or N only.

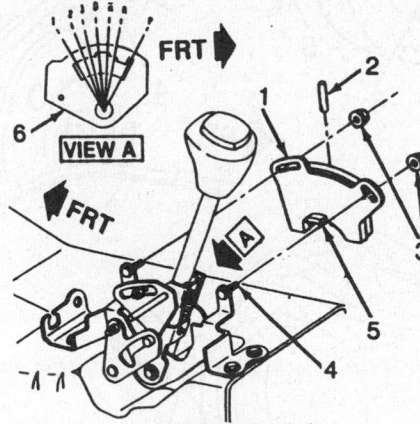

1. Neutral start switch
2. Gauge pin
3. Nut
4. Mounting stud
5. Carrier tang
6. Detent plate

Neutral safety switch

REMOVAL & INSTALLATION

1. Disconnect battery ground cable.
2. Remove the shifter knob assembly.
3. Remove the console assembly.
4. Remove the neutral switch mounting bolts.
5. Remove the switch.

To install:

6. Position shifter lever in the N position.
7. Insert carrier tang on the switch in slot on shifter.
8. If not installing a new switch, adjust switch as specified.
9. Torque switch mounting bolts to 26 inch lbs. (3 Nm).

NOTE: If holes do not align with shifter control, check that the shifter control lever is in N. If installing a new switch, do not rotate the switch. The switch is pinned in the N position. If new switch was rotated and the pin broken, switch adjustment must be performed.

10. If installing a new switch, move the shift control lever out of the N position and the factory installed set pin will shear.
11. Check that engine starts only in the P or N positions. If engine starts in any other position, readjust neutral switch.
12. Install console and shifter knob assembly.
13. Connect battery negative cable.

Fuses, Circuit Breakers and Relays

LOCATION

Fusible Links

Fusible links are located at the jump start junction block. The fusible links which are normally not fused, are used to prevent wire harness damage in the event of a short circuit or an overload condition. Each fusible link is of a fixed value for a specific electrical load. Should a link fail, the cause of the failure must be determined and repaired prior to installing a new fusible link of the same value.

Circuit Breakers

There are 2 different style circuit breakers used. A standard heat activated circuit breaker and the other style is a solid state design called a Positive Temperature Coefficient (PTC) circuit breaker. Various circuit breakers are located in the fuse block.

Fuse Block

The fuse block assembly is located behind the right side of the instrument panel.

Relays

Listed below are relays that are mounted on the instrument panel harness.

Power Antenna Relay (1988–89 Coupe)—located in the left side rear of cargo compartment, on the end panel.

Power Antenna Relay (Convertible and 1990–92 Coupe)—located on the left side of the cargo compartment above the rear of the wheel house.

Anti-Lock Brake Module Relay—located under the left side of the cargo compartment, behind the driver's seat.

Amplifier Relay—located behind the left side of the instrument panel or to the right of the radio.

Primary and Secondary Engine Cooling Fan Relay—located in the left rear corner of the engine compartment or on the left side of the radiator shroud.

Fuel Pump Relay (1988–89)—located to the right of the master cylinder.

Fuel Pump Relays (1990–92)—located below right side of the instrument panel. The 5.7L (VIN J) engine uses a second relay located below the left side of the instrument panel.

ABS Control Module Relay (1988–90)—located under the corner of the rear floor, in storage compartment.

Computers

LOCATION

Electronic Control Module (1988–89)—located behind the right side of the dash.

Electronic Control Module (1990–92)—located in the engine compartment, above battery.

ABS Control Module—located under the left corner of rear floor, in storage compartment.

Central Control Module (CCM, 1990–92)—located behind the middle of the instrument panel.

Select Ride Control Module—located under left corner of rear floor, in storage compartment.

Electronic Spark Control Module (1988–90)—located on the right side of air conditioning heater blower housing.

Flashers

LOCATION

Turn Signal (1988–89)—located behind the right side of the dash panel, near the fuse panel.

Turn Signal (1990–92)—located below left side of instrument panel.

Hazard Flasher—located near the radio on the right side of the instrument panel.

Cruise Control

ADJUSTMENT

Servo Linkage

1988–89

1. With the cable attached to the cable bracket and throttle lever, install the cable to the clip and servo bracket using the first ball on the servo chain.
2. Connect the servo chain to the cable assembly connector leaving a space of 4 ball links.
3. Verify the ignition is **OFF** and the throttle closed completely.
4. Adjust the cable jam nuts until the cable sleeve at the throttle lever is tight but not holding the throttle open.
5. Tighten the jam nuts.
6. Pull the servo boot over the washer on the cable.

1990–92

1. With the cruise control cable installed in the cable clips and throttle lever, insert cable into servo bracket.
2. Pull servo assembly end of cable toward servo without moving the throttle lever.
3. If 1 out of the 5 holes in the servo assembly tab aligns with the cable pin, push pin through hole and connect pin to tab with retainer.
4. If the tab holes does not align with the pin, move the cable away from the servo assembly until the next closest tab hole aligns and connect the pin to the tab with the retainer.

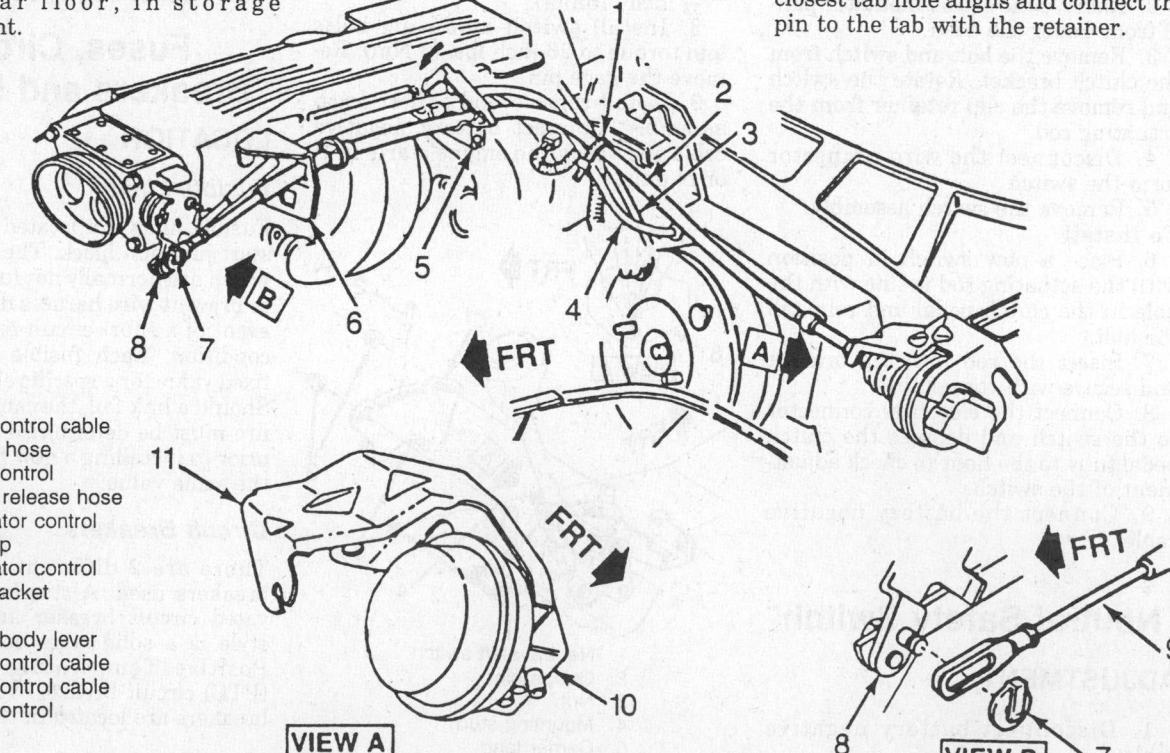

1. Strap
2. Cruise control cable
3. Vacuum hose
4. Cruise control vacuum release hose
5. Accelerator control cable clip
6. Accelerator control cable bracket
7. Retainer
8. Throttle body lever
9. Cruise control cable
10. Cruise control cable
11. Cruise control bracket

Cruise control cable routing and adjustment–1990–92 5.7L (VIN 8) engine

GM—Pontiac
Rear Wheel Drive
Fiero

SPECIFICATIONS

VEHICLE IDENTIFICATION CHART

It is important for servicing and ordering parts to be certain of the vehicle and engine identification. The VIN (vehicle identification number) is a 17 digit number visible through the windshield on the driver's side of the dash and contains the vehicle and engine identification codes. The tenth digit indicates model year, and the eighth digit indicates engine code. It can be interpreted as follows:

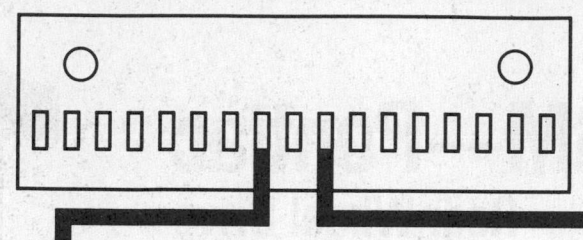

Engine Code						Model Year	
Code	Cu. In.	Liters	Cyl.	Fuel Sys.	Eng. Mfg.	Code	Year
R	151	2.5	4	TBI	Pontiac	J	1988
9	173	2.8	6	MFI	Chevrolet		

TBI—Throttle Body Injection
MFI—Multiport Fuel Injection

ENGINE IDENTIFICATION

Year	Model	Engine Displacement cu. in. (liter)	Engine Series Identification (VIN)	No. of Cylinders	Engine Type
1988	Fiero	151 (2.5)	R	4	OHV
	Fiero	173 (2.8)	9	4	OHV

OHV—Overhead Valve

GENERAL ENGINE SPECIFICATIONS

Year	VIN	No. Cylinder Displacement cu. in. (liter)	Fuel System Type	Net Horsepower @ rpm	Net Torque @ rpm (ft. lbs.)	Bore × Stroke (in.)	Compression Ratio	Oil Pressure @ rpm
1988	R	4-151 (2.5)	TBI	98 @ 4500	134 @ 2800	4.000 × 3.000	8.3:1	36-41 @ 2000
	9	6-173 (2.8)	MFI	135 @ 4500	170 @ 3600	3.500 × 3.000	8.5:1	30-45 @ 2000

TBI—Throttle Body Fuel Injection
MFI—Multiport Fuel Injection

GASOLINE ENGINE TUNE-UP SPECIFICATIONS

Year	VIN	No. Cylinder Displacement cu. in. (liter)	Spark Plugs Type	Spark Plugs Gap (in.)	Ignition Timing (deg.) MT	Ignition Timing (deg.) AT	Compression Pressure (psi)	Fuel Pump (psi)	Idle Speed (rpm) MT	Idle Speed (rpm) AT	Valve Clearance In.	Valve Clearance Ex.
1988	R	4-151 (2.5)	R42 CTS	.060	①	①	NA	9-13	①	①	Hyd.	Hyd.
	9	6-173 (2.8)	R42 CTS	.045	①	①	NA	41-47	①	①	Hyd.	Hyd.

NOTE: The underhood specifications sticker often reflects tune-up specification changes made in production. Sticker figures must be used if they disagree with those in this chart.
NA—Not available
Hyd.—Hydraulic
① See underhood sticker

FIRING ORDERS

NOTE: To avoid confusion, always replace spark plug wires one at a time.

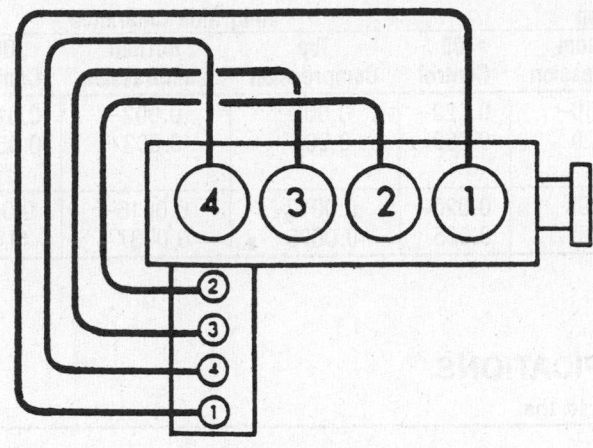

2.5L Engine
Engine Firing Order: 1-3-4-2
Distributorless Ignition

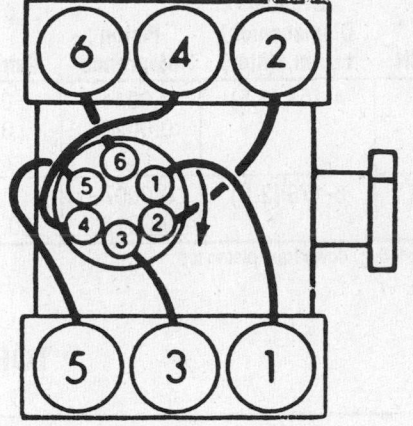

2.8L Engine
Engine Firing Order: 1-2-3-4-5-6
Distributor Rotation: Clockwise

CAPACITIES

Year	Model	VIN	No. Cylinder Displacement cu. in. (liter)	Engine Crankcase with Filter	Engine Crankcase without Filter	Transmission (pts.) 4-Spd	Transmission (pts.) 5-Spd	Transmission (pts.) Auto.	Drive Axle (pts.)	Fuel Tank (gals.)	Cooling System (qts.)
1988	Fiero	R	4-151 (2.5)	4	4	—	4.1②	8.0①	—	11.9	13.8
	Fiero	9	6-173 (2.8)	4	4	—	5.3②	8.0①	—	11.9	13.8

① Overhaul—10.0
② Isuzu—5.3

CRANKSHAFT AND CONNECTING ROD SPECIFICATIONS

All measurements are given in inches.

Year	VIN	No. Cylinder Displacement cu. in. (liter)	Crankshaft Main Brg. Journal Dia.	Crankshaft Main Brg. Oil Clearance	Crankshaft Shaft End-play	Crankshaft Thrust on No.	Connecting Rod Journal Diameter	Connecting Rod Oil Clearance	Connecting Rod Side Clearance
1988	R	4-151 (2.5)	2.3000–2.3005	0.0005–0.0022	0.0035–0.0085	5	2.0000–2.0005	0.0005–0.0026	0.006–0.022
	9	6-173 (2.8)	2.6473–2.6482	0.0016–0.0031	0.0023–0.0082	3	1.9984–1.9994	0.0014–0.0037	0.006–0.017

VALVE SPECIFICATIONS

Year	VIN	No. Cylinder Displacement cu. in. (liter)	Seat Angle (deg.)	Face Angle (deg.)	Spring Test Pressure (lbs.)	Spring Installed Height (in.)	Stem-to-Guide Clearance (in.) Intake	Stem-to-Guide Clearance (in.) Exhaust	Stem Diameter (in.) Intake	Stem Diameter (in.) Exhaust
1988	R	4-151 (2.5)	46	45	178	1.44	0.0010–0.0027	0.0010–0.0027	0.3130–0.3140	0.3120–0.3130
	9	6-173 (2.8)	46	45	195	1.57	0.0010–0.0027	0.0010–0.0027	0.3410–0.3425	0.3410–0.3426

PISTON AND RING SPECIFICATIONS

All measurements are given in inches.

Year	VIN	No. Cylinder Displacement cu. in. (liter)	Piston Clearance	Ring Gap			Ring Side Clearance		
				Top Compression	Bottom Compression	Oil Control	Top Compression	Bottom Compression	Oil Control
1988	R	4-151 (2.5)	0.0014– 0.0022 ①	0.010– 0.020	0.010– 0.020	0.020– 0.060	0.002– 0.003	0.002– 0.003	0.015– 0.055
	9	6-173 (2.8)	0.0007– 0.0017	0.0098– 0.0197	0.0098– 0.0197	0.020– 0.055	0.0012– 0.0028	0.0016– 0.0037	0.008 Max

① Measured 1.8 in. down from piston top

TORQUE SPECIFICATIONS

All readings in ft. lbs.

Year	VIN	No. Cylinder Displacement cu. in. (liter)	Cylinder Head Bolts	Main Bearing Bolts	Rod Bearing Bolts	Crankshaft Pulley Bolts	Flywheel Bolts	Manifold		Spark Plugs
								Intake	Exhaust	
1988	R	4-151 (2.5)	①	70	32	162	②	25	35	7-15
	9	6-173 (2.8)	65-90	63-74	34-40	66-84	45-55	③	18	7-15

① Step 1: 18 ft. lbs.—Except position #9
   Step 2: 26 ft. lbs.—Except position #9
   Step 3: 18 ft. lbs.—Only position #9
   Step 4: + 90 degrees turn—All positions
② Automatic Transaxle: 55 ft. lbs.
   Manual Transaxle: 69 ft. lbs.
③ Lower: 19 ft. lbs.
   Middle: 15 ft. lbs.
   Upper: 18 ft. lbs.

BRAKE SPECIFICATIONS

All measurements in inches unless noted.

Year	Model	Lug Nut Torque (ft. lbs.)	Master Cylinder Bore	Brake Rotor		Standard Brake Drum Diameter	Minimum Lining Thickness	
				Minimum Thickness	Maximum Runout		Front	Rear
1988	Fiero	①	1.00	③	0.003	②	0.062	0.062

① Steel 80; aluminum 100
② Cars equipped with 4 wheel disc brakes
③ Minimum refinish thickness (front)—0.445
   Discard thickness (front)—0.390
   Minimum refinish thickness (rear)—0.0500
   Discard thickness (rear)—0.450

WHEEL ALIGNMENT

Year	Model	Caster		Camber		Toe-in (in.)	Steering Axis Inclination (deg.)
		Range (deg.)	Preferred Setting (deg.)	Range (deg.)	Preferred Setting (deg.)		
1988	Fiero	3N-7P	5P	5/16N–1 5/16P	½P	1/16 ± 1/32	NA

N—Negative
P—Positive

ENGINE MECHANICAL

NOTE: Disconnecting the negative battery cable on some vehicles may interfere with the functions of the on board computer systems and may require the computer to undergo a relearning process.

Engine Assembly

REMOVAL & INSTALLATION

NOTE: The engine assembly is removed from underneath the vehicle.

1. Disconnect the negative battery cable.
2. Drain the engine coolant.
3. Remove the rear compartment lid and also the side panels. Do not remove the torsion rod retaining bolts.
4. Remove the air cleaner assembly.
5. Disconnect the throttle and shift cables.
6. Disconnect the heater hose at the intake manifold.
7. Disconnect the vacuum hoses from all non-engine components.
8. Properly relieve the fuel system pressure. Disconnect the fuel lines and filter.
9. Disconnect the fuel pump relay and the oxygen sensor.
10. If equipped with automatic transaxle, disconnect the transaxle cooler lines.
11. Disconnect the slave cylinder from the manual transaxle equipped vehicles.
12. Disconnect the engine to chassis ground strap.
13. If equipped with air conditioning, properly discharge the system. Disconnect and plug the refrigerant lines at the compressor and seal the end.
14. Remove the rear console.
15. Remove the ECM harness through the bulkhead panel.
16. Install an engine support fixture.
17. Remove the engine strut bracket and mark the bolt and bracket for reassembly.
18. Raise and support the vehicle safely.
19. Remove the rear wheels.
20. If equipped with an automatic transaxle, remove the torque converter bolts.
21. Remove the parking brake cable and calipers. Do not disconnect the brake hoses. Support the caliper out of the way.
22. Remove the strut bolts and mark the struts for realignment.
23. Disconnect the air conditioning wiring, if equipped.
24. Loosen the 4 engine cradle bolts.
25. On the 2.5L engine, release the parking brake cables at the cradle using tool J–34065 or equivalent.

NOTE: Support the engine/transaxle and cradle assembly on the proper equipment. Be sure to support the outboard ends of the lower control arms. Disconnect the engine support fixture.

26. Lower the vehicle and attach the engine/transaxle assembly to a dolly. Remove the cradle bolts. Raise the vehicle and roll the dolly from under the vehicle.
27. Separate the engine and transaxle.
28. Installation is the reverse of removal.

Engine Mounts

REMOVAL & INSTALLATION

1. Disconnect the negative battery cable. Safely support the engine.
2. Remove the bolt for the forward torque reaction rod.
3. Raise the vehicle and support it safely.
4. Remove the engine mount to chassis nuts.
5. Remove the upper engine mount to support bracket nuts.
6. Remove the mount.
7. Installation is the reverse order of the removal procedure.
8. Torque engine mount to specification.
9. Torque engine mount to support bracket to specification.

Cylinder Head

REMOVAL & INSTALLATION

2.5L Engine

1. Relieve the fuel system pressure. Disconnect the negative battery cable. Drain the cooling system.
2. Raise the vehicle and support it safely.
3. Remove the exhaust pipe.
4. Lower the vehicle.
5. Remove the oil level indicator tube.
6. Remove the air cleaner assembly.
7. Disconnect the EFI electrical connections and vacuum hoses.
8. Remove the EGR base plate.
9. Remove the heater hose from the intake manifold.
10. Remove the ignition coil lower mounting bolt and wiring connections.
11. Remove all wiring connections from the intake manifold and cylinder head.
12. Remove the engine strut bolt from the upper support. Remove the power steering pump and position it to the side, as required.
13. Remove the alternator belt. Remove the air conditioning compressor and position to the side, as required.
14. Remove the throttle cables from the intake manifold. Remove the intake manifold.
15. Remove the valve cover, rocker arms and pushrods.
16. Remove the cylinder head bolts and remove the cylinder head.
17. Before installing, clean the gasket surfaces of the head and block.
18. Make sure the retaining bolt threads and the cylinder block threads are clean since dirt could affect bolt torque.

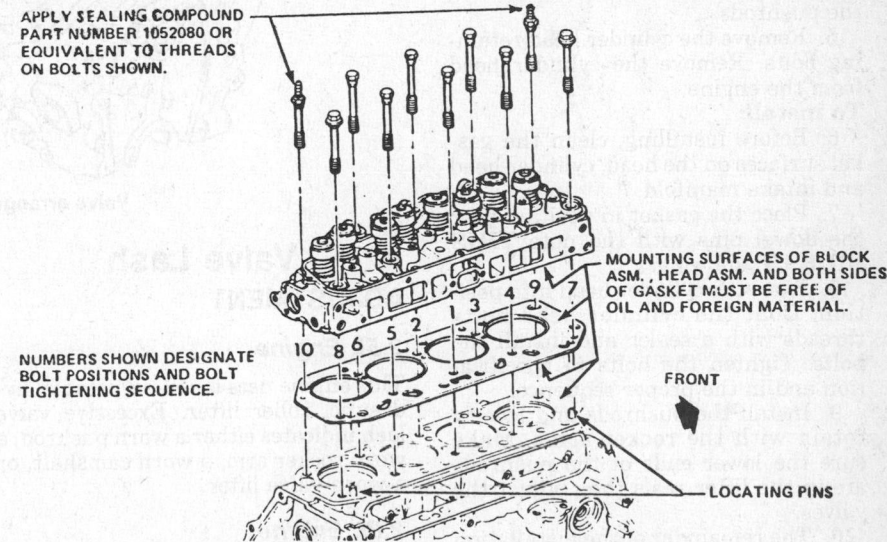

APPLY SEALING COMPOUND PART NUMBER 1052080 OR EQUIVALENT TO THREADS ON BOLTS SHOWN.

MOUNTING SURFACES OF BLOCK ASM., HEAD ASM. AND BOTH SIDES OF GASKET MUST BE FREE OF OIL AND FOREIGN MATERIAL.

FRONT

LOCATING PINS

NUMBERS SHOWN DESIGNATE BOLT POSITIONS AND BOLT TIGHTENING SEQUENCE.

Cylinder head torque sequence—2.5L engine

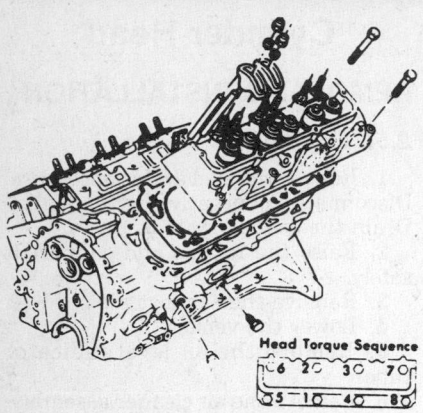

Cylinder head torque sequence—2.8L engine

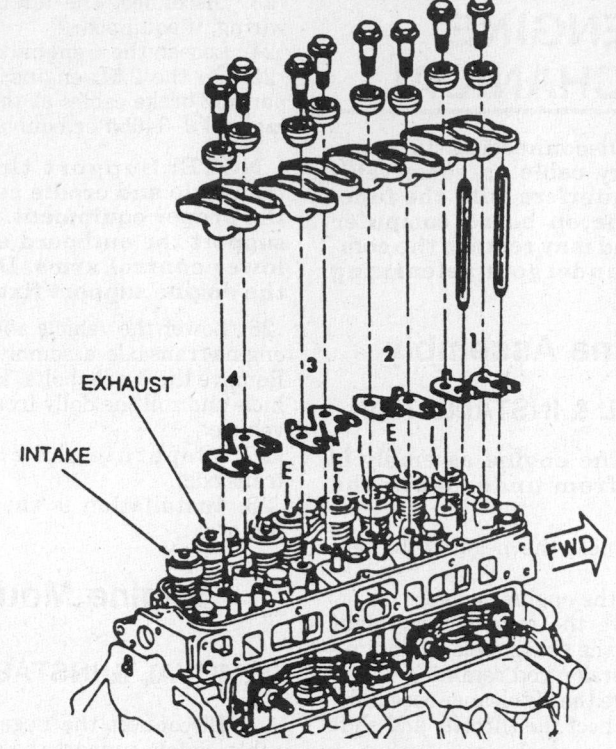

Valve arrangement—2.5L engine

19. Install a new gasket over the dowel pins in the cylinder block. Install the cylinder head into place over the dowel pins.

20. Tighten the cylinder head bolts to specification gradually and in the proper sequence.

21. Installation is the reverse order of the removal procedure.

2.8L Engine

1. Relieve the fuel system pressure. Disconnect the negative battery cable. Drain the radiator.

2. Remove the intake manifold. Remove the exhaust manifolds, as necessary.

3. If removing the left cylinder head, disconnect the alternator bracket and the oil level indicator tube. If removing the right cylinder head, disconnect the cruise control servo bracket.

4. Remove the valve covers. Remove the pushrods.

5. Remove the cylinder head retaining bolts. Remove the cylinder head from the engine.

To install:

6. Before installing, clean the gasket surfaces on the head, cylinder head and intake manifold.

7. Place the gasket in position over the dowel pins with the note **THIS SIDE UP** showing.

8. Place the cylinder head into position. Coat the cylinder head bolts threads with a sealer and install the bolts. Tighten the bolts to specification and in the proper sequence.

9. Install the pushrods and loosely retain with the rocker arms. Make sure the lower ends of the pushrods are in the lifter seals then adjust the valves.

10. The remainder of the installation is the reverse of the removal.

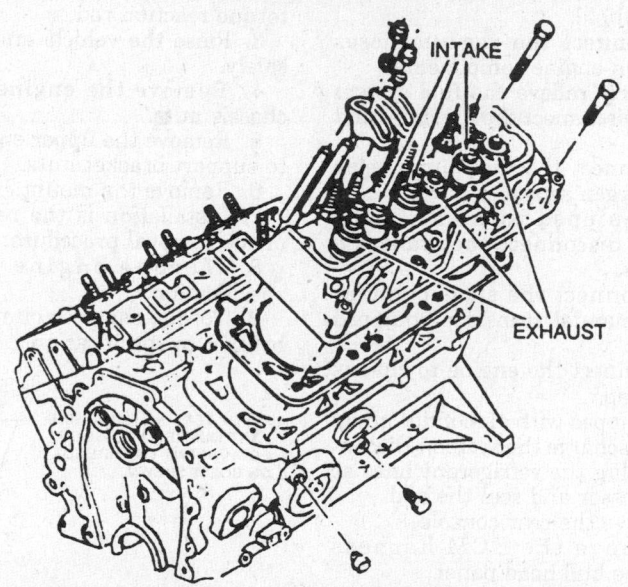

Valve arrangement—2.8L engine

Valve Lash

ADJUSTMENT

2.5L Engine

This engine uses a non-adjustable, hydraulic, roller lifter. Excessive valve lash indicates either a worn pushrod, a worn rocker arm, a worn camshaft, or a worn valve lifter.

2.8L Engine

1. Rotate engine until mark on tor-

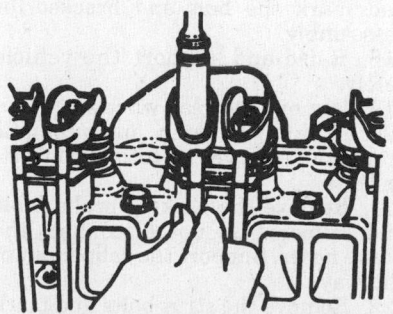

Valve adjustment—2.8L engine

sional damper lines up with **0** on the timing tab, with the engine in the No. 1 firing position. This can be determined by placing fingers on the No. 1 rocker arms as the mark on the damper comes near the **0** mark. If the valves are not moving, the engine is in the No. 1 firing position.

2. With the engine in the No. 1 firing position the following valves may be adjusted: Exhaust–1, 2, 3; Intake–1, 5, 6.

3. Back out the adjusting nut until lash is felt at the pushrod.

4. Turn in adjusting nut until all lash is removed.

5. When all lash has been removed, turn in adjusting nut 1½ additional turns.

6. Crank the engine 1 revolution until the timing tab **0** mark and torsional damper mark are again in alignment. This is the No. 4 firing position.

7. With the engine in the No. 4 firing position the following valves may be adjusted: exhaust–4, 5, 6; intake–2, 3, 4.

8. Install rocker arm covers.

9. Start engine, check timing and idle speed, check for oil leaks.

Rocker Arms/ Pushrods

REMOVAL & INSTALLATION

2.5L Engine

1. Disconnect the negative battery cable. Remove the air cleaner.

2. Remove the PCV valve and hose.

3. Disconnect the wires from the spark plugs and clips.

4. Remove the valve cover retaining bolts.

5. Remove the valve cover by tapping lightly with a rubber hammer. Prying on the cover could cause damage to the sealing surfaces.

6. Remove the rocker arm bolt.

7. If replacing the pushrod only, loosen the rocker arm bolt and swing the arm clear of the pushrod.

8. Remove the rocker arm and pushrod.

9. Installation is the reverse order of the removal procedure.

10. Torque the rocker arm bolt to specification. Apply a continuous $^3/_{16}$ in. diameter bead of RTV sealant or equivalent around the cylinder head sealant surfaces inboard at the bolt holes.

2.8L Engine

1. Disconnect the negative battery cable.

2. Remove the engine compartment lid and both side covers. Do not remove the torsion rod retaining bolts.

3. Disconnect the vacuum boost line and tube.

4. Disconnect the throttle and downshift cables and bracket.

5. Disconnect the cruise control cable, if applicable.

6. Disconnect the ground cable.

7. Remove the PCV from the cover.

8. Remove the oil dipstick tube.

9. Disconnect the plug wires and bracket.

10. Remove the engine lift hook.

11. Remove the rocker arm cover bolts and carefully remove the cover by taping with a rubber mallet. If prying is necessary do not distort the sealing flange.

12. Remove the rocker arm nuts. Keep all components in order so they may be reinstalled in the same location.

13. Remove the rocker arm pivot balls, arms and pushrods.

14. Before installation, coat the bearing surfaces of the rocker arms and pivot balls with Molykote® or equivalent.

15. Insert the pushrods, rocker arms and pivot balls. Make sure the pushrods are seated in the valve lifters.

16. Turn the rocker arm nuts until lash is eliminated.

17. Install the rocker arm covers. Clean the surfaces on the cylinder head and rocker arm cover. Place a ⅛ in. of RTV sealant or an equivalent, at the intake manifold and cylinder head split line. Install the rocker arm cover gasket, using care to line up the holes in the gasket with the bolt holes in the cylinder head.

18. Install the rocker arm cover bolts and torque to specification.

19. The remainder of the installation is the reverse order of the removal procedure

Pushrod Cover

REMOVAL & INSTALLATION

2.5L Engine

1. Disconnect the negative battery cable.

2. Remove the intake manifold assembly.

3. Remove the pushrod cover retaining bolts. Remove the pushrod cover.

4. Installation is the reverse of the removal procedure. Be sure to use new gaskets or RTV sealant, as required.

Intake Manifold

REMOVAL & INSTALLATION

2.5L Engine

1. Relieve the fuel pump pressure.

Disconnect the negative battery cable. Remove the air cleaner assembly.

2. Remove the PCV valve and hose.

3. Drain the cooling system.

4. Disconnect the fuel lines.

5. Disconnect the vacuum hoses.

6. Disconnect the wiring and the throttle linkage from the throttle body assembly.

7. Disconnect the cruise control and linkage, if equipped.

8. Disconnect the throttle linkage and bell crank and place to the side.

9. Disconnect the heater hose.

10. Remove the generator upper bracket.

11. Remove the ignition coil.

12. Remove the retaining bolts and remove the manifold.

13. Installation is the reverse of removal. Be sure to use new gaskets, as required. Torque all bolts in the proper sequence.

2.8L Engine

1. Position the engine at TDC on the compression stroke. Properly relieve the fuel system pressure.

2. Disconnect the negative battery cable. Remove the valve covers.

3. Drain the engine coolant.

4. Disconnect the throttle body to elbow intake hose.

5. Remove the distributor. Disconnect the vacuum booster pipe and bracket.

6. Disconnect the shift and throttle linkage.

7. Remove the throttle body to upper plenum.

8. Disconnect the heater and radiator hoses.

9. Disconnect all wiring harness and vacuum hoses while noting their locations for reassembly.

10. Disconnect the EGR pipe.

11. Unbolt and remove the upper manifold plenum and gaskets.

12. Unbolt and remove the intermediate intake manifold and gasket.

13. Unbolt and remove the lower intake manifold and gaskets.

To install:

14. Clean all gasket surfaces on the intake manifolds and cylinder head.

15. Install the lower intake manifold and gasket and torque to specification in the proper sequence.

16. Install the intermediate intake manifold and gaskets and torque in sequence to specification.

17. Install the upper manifold plenum and gaskets and torque in sequence.

18. The remainder of the installation is the reverse of the removal. Check engine timing, coolant level and for leaks.

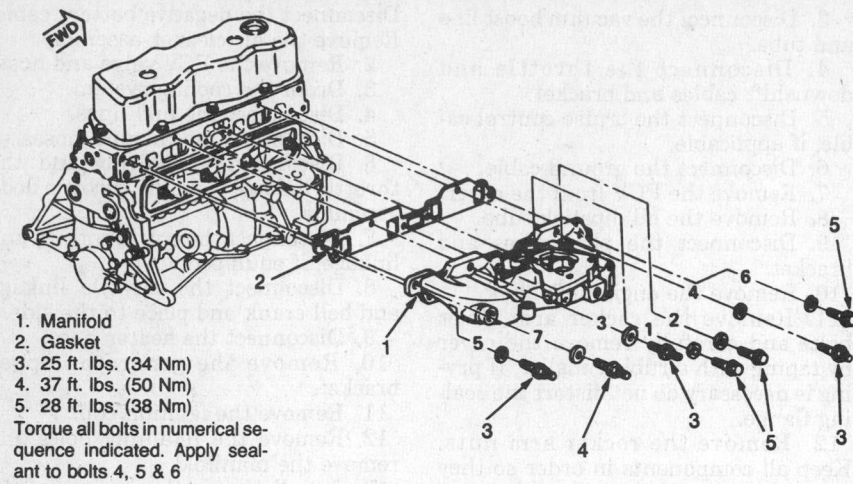

1. Manifold
2. Gasket
3. 25 ft. lbs. (34 Nm)
4. 37 ft. lbs. (50 Nm)
5. 28 ft. lbs. (38 Nm)

Torque all bolts in numerical sequence indicated. Apply sealant to bolts 4, 5 & 6

Intake manifold torque sequence—2.5L engine

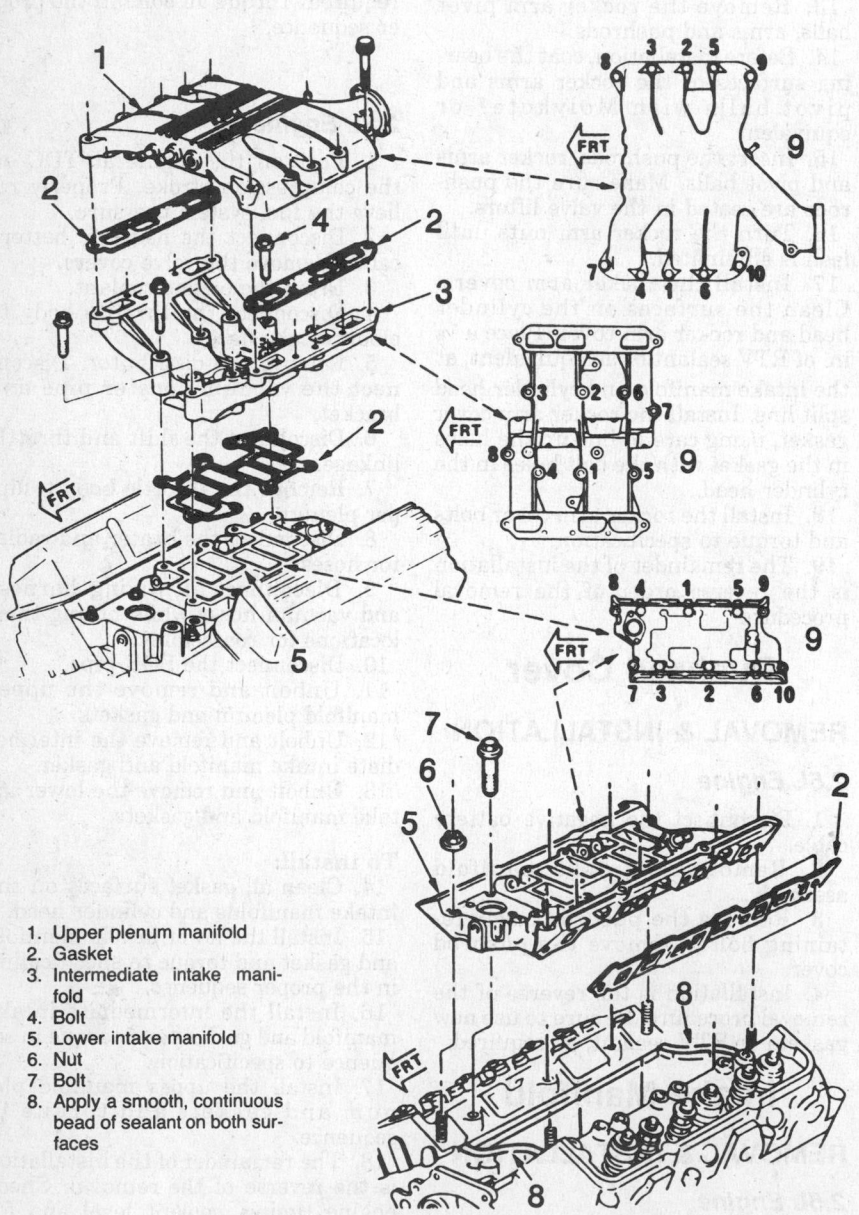

1. Upper plenum manifold
2. Gasket
3. Intermediate intake manifold
4. Bolt
5. Lower intake manifold
6. Nut
7. Bolt
8. Apply a smooth, continuous bead of sealant on both surfaces

Intake manifold torque sequence—2.8L engine

Exhaust Manifold

REMOVAL & INSTALLATION

2.5L Engine

1. Disconnect the negative battery cable. Remove the air cleaner.
2. Raise and support the vehicle safely.
3. Remove the exhaust pipe and lower the vehicle. As required, remove the battery side cover.
4. Disconnect the oxygen sensor connector and remove the dipstick tube. Remove the exhaust manifold retaining bolts. Remove the exhaust manifold and gasket from the engine.
5. Installation is the reverse of removal. Clean the sealing surfaces and use a new gasket. Torque the retaining bolts in sequence.

2.8L Engine

FRONT

1. Disconnect the negative battery cable.
2. Remove the rear compartment lid. Do not remove the torsion rod retaining bolts.
3. Remove the brake vacuum hose.
4. Remove the manifold heat shield.
5. Remove the front crossover bolts.
6. Raise and support the vehicle safely. Remove the front converter heat shield and the lower manifold bolts.
7. Lower the vehicle and remove the upper manifold bolts then remove the manifold.
8. Installation is the reverse of the removal procedure. Be sure to use a new gasket. Torque the manifold to specification.

REAR

1. Disconnect the negative battery cable. Disconnect the manifold to crossover bolts.
2. Remove the manifold retaining bolts. remove the manifold.
3. Installation is the reverse order of the removal procedure.
4. Torque the manifold bolts to specification.

Timing Chain Front Cover and Oil Seal

REMOVAL & INSTALLATION

2.5L Engine

1. Disconnect the negative battery cable. Remove the engine compartment lid and side panels. Remove the trim at the sail panel below the battery side panel.
2. Remove the drive belt. Raise and support the vehicle safely.

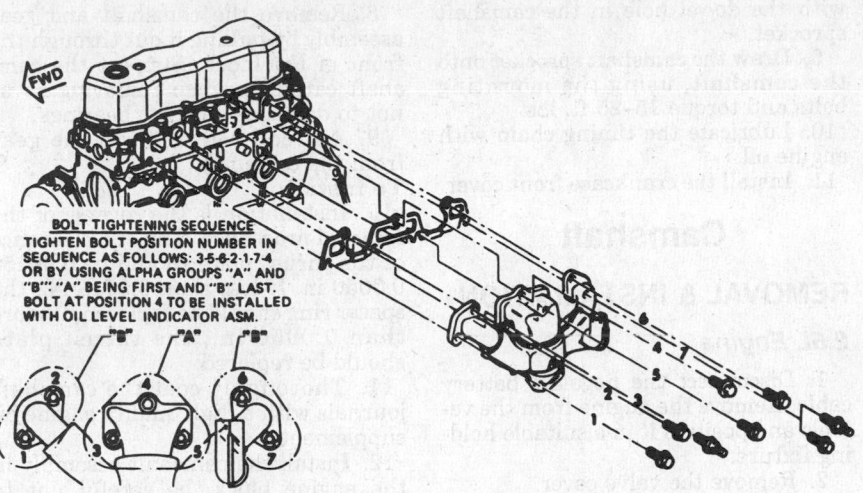

BOLT TIGHTENING SEQUENCE
TIGHTEN BOLT POSITION NUMBER IN
SEQUENCE AS FOLLOWS: 3-5-6-2-1-7-4
OR BY USING ALPHA GROUPS "A" AND
"B." "A" BEING FIRST AND "B" LAST.
BOLT AT POSITION 4 TO BE INSTALLED
WITH OIL LEVEL INDICATOR ASM.

Exhaust manifold torque sequence—2.5L engine

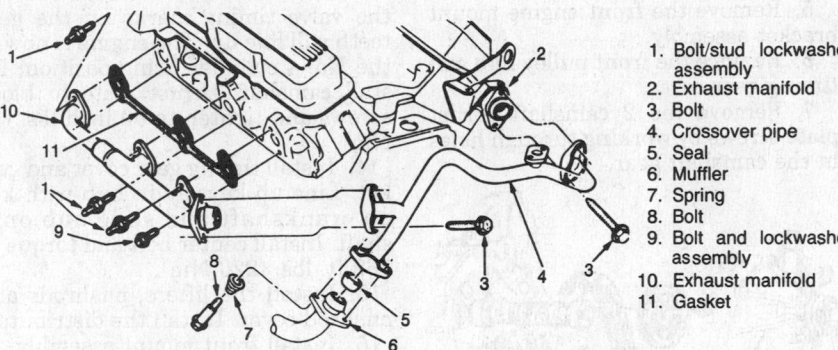

1. Bolt/stud lockwasher assembly
2. Exhaust manifold
3. Bolt
4. Crossover pipe
5. Seal
6. Muffler
7. Spring
8. Bolt
9. Bolt and lockwasher assembly
10. Exhaust manifold
11. Gasket

Exhaust manifold torque sequence—2.8L engine

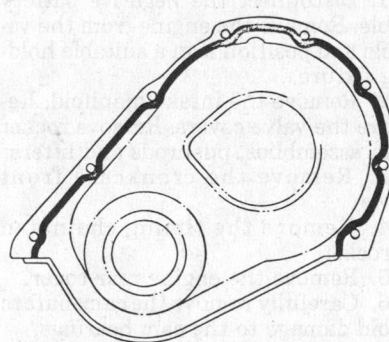

Timing cover seal application—2.5L engine

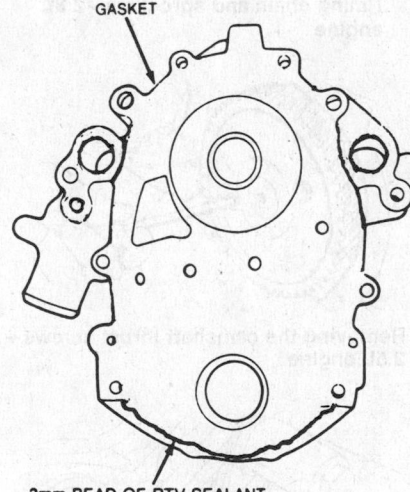

GASKET

3mm BEAD OF RTV SEALANT
#1052366 OR EQUIVALENT

Timing cover seal application—2.8L engine

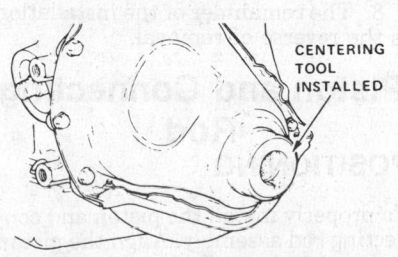

CENTERING TOOL INSTALLED

Front cover centering tool installation—2.5L engine

5. Remove the crankshaft pulley and hub. Lower the vehicle.

6. Properly support the engine using the required equipment. Remove the engine torque strut.

7. Raise and support the vehicle safely. Remove the engine mounts.

8. Remove the timing gear cover bolts. Remove the timing gear cover.

9. Installation is the reverse of the removal procedure. Be sure to use new gaskets or RTV sealant, as required.

2.8L Engine

1. Disconnect the negative battery cable.

2. Remove the air conditioning compressor and bracket, without disconnecting the refrigerant lines and position out of the way.

3. Remove the water pump.

4. Raise the vehicle and support it safely. Remove the torsional damper.

5. If replacing the seal, pry the old seal out using a suitable tool. When installing a new seal, lubricate the seal with clean engine oil. Insert the seal in the front cover with the lip facing the engine. Using the proper tool, drive the seal into place.

6. Remove the oil pan to cover bolts.

7. Lower the vehicle and remove the front cover.

8. Before installing, clean the sealing surfaces on the front cover and cylinder block. Install a new gasket and apply a ⅛ in. bead of RTV sealer to the oil pan sealing surface of the front cover.

9. Place the front cover on the engine and install the stud bolt and bolts.

10. The remainder of the installation is the reverse order of the removal procedure.

Timing Chain and Sprockets

REMOVAL & INSTALLATION

2.5L Engine

CAMSHAFT SPROCKET

1. Disconnect the negative battery cable. Remove the engine from the vehicle.

2. Position the engine assembly in a suitable holding fixture.

3. Position the engine at TDC on the compression stroke. Remove the front cover. Remove the camshaft.

4. Using the proper equipment, press the camshaft sprocket from the camshaft.

5. Installation is the reverse of the removal procedure.

6. The end clearance of the thrust plate should be 0.0015–0.0050 in. If less than 0.0015 in., the spacer ring should be replaced. If more than

3. Remove the right rear tire and wheel assembly. Remove the inner splash shield.

4. Remove the starter assembly. Remove the flywheel cover.

0.0050 in., the thrust plate should be replaced.

CRANKSHAFT SPROCKET

1. Disconnect the negative battery cable. Position the engine at TDC on the compression stroke.
2. Remove the engine front cover.
3. Remove the crankshaft gear from its mounting.
4. Installation is the reverse of the removal procedure.

2.8L Engine

1. Disconnect the negative battery cable. Remove the crankcase front cover.
2. Align the No. 1 piston at TDC, with the marks on the camshaft and crankshaft sprockets aligned.
3. Remove the camshaft sprocket and chain. It may be necessary to use a plastic mallet on the lower edge of the sprocket to dislodge it.
4. Remove the crankshaft sprocket, using the proper tool.

To install:

5. Install the crankshaft sprockets.
6. Apply Molykote® or equivalent to the sprocket thrust surface.
7. Hold the camshaft sprocket with the chain hanging down and align the marks on the camshaft and crankshaft sprockets.
8. Align the dowel in the camshaft

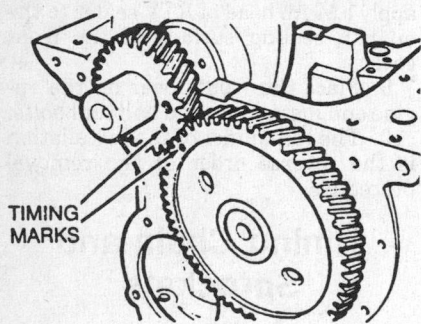

Timing gear alignment—2.5L engine

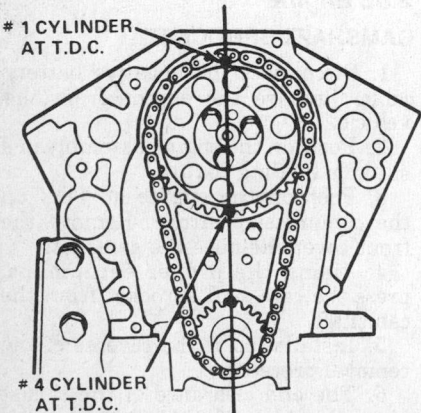

1 CYLINDER AT T.D.C.

4 CYLINDER AT T.D.C.

Timing chain and gear alignment—2.8L engine

with the dowel hole in the camshaft sprocket.

9. Draw the camshaft sprocket onto the camshaft, using the mounting bolts and torque 15–25 ft. lbs.
10. Lubricate the timing chain with engine oil.
11. Install the crankcase front cover.

Camshaft

REMOVAL & INSTALLATION

2.5L Engine

1. Disconnect the negative battery cable. Remove the engine from the vehicle and position it in a suitable holding fixture.
2. Remove the valve cover.
3. Remove the pushrods.
4. Remove the pushrod cover, and valve lifters.
5. Remove the front engine mount bracket assembly.
6. Remove the front pulley hub and timing gear cover.
7. Remove the 2 camshaft thrust plate screws by working through holes in the camshaft gear.

Timing chain and sprockets—2.8L engine

Removing the camshaft thrust screws—2.5L engine

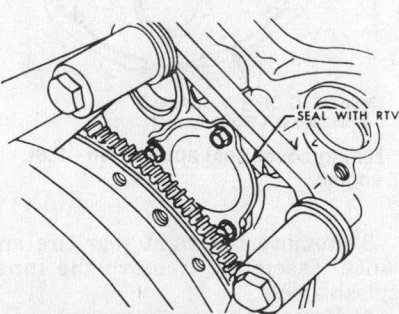

SEAL WITH RTV

Camshaft rear cover—2.8L engine

8. Remove the camshaft and gear assembly by pulling it out through the front of the block. Support the camshaft carefully when removing so as not to damage camshaft bearings.
9. As required, remove the gear from the camshaft.

To install:

10. Installation is the reverse of the removal procedure. The end clearance of the thrust plate should be 0.0015–0.0050 in. If less than 0.0015 in., the spacer ring should be replaced. If more than 0.0050 in., the thrust plate should be replaced.
11. Thoroughly coat the camshaft journals with a high quality engine oil supplement.
12. Install the camshaft assembly in the engine block, be careful not to damage the cam bearings or the camshaft.
13. Turn crankshaft and camshaft so the valve timing marks on the gear teeth will line up. The engine is now in the No. 4 cylinder firing position. Install camshaft thrust plate to block screws and tighten to 90 inch lbs. (10 Nm).
14. Install timing gear cover and gasket. Line up keyway in hub with key on crankshaft and slide hub onto shaft. Install center bolt and torque to 162 ft. lbs. (220 Nm).
15. Install the lifters, pushrods and pushrod cover. Install the distributor.
16. Install front mount assembly.
17. Install the engine in the vehicle.

2.8L Engine

1. Disconnect the negative battery cable. Remove the engine from the vehicle and position it in a suitable holding fixture.
2. Remove the intake manifold. Remove the valve covers. Remove rocker arm assemblies, pushrods and lifters.
3. Remove the crankcase front cover.
4. Remove the timing chain and sprocket.
5. Remove the engine rear cover.
6. Carefully remove the camshaft to avoid damage to the cam bearings.
7. Before installation, lubricate the camshaft journals with engine oil. If a new camshaft is to be installed, coat the lobes with clean engine oil.
8. The remainder of the installation is the reverse of removal.

Piston and Connecting Rod

POSITIONING

To properly install the piston and connecting rod assembly. Align the piston and connecting rod assembly with the piston mark (notch) toward the front of the engine.

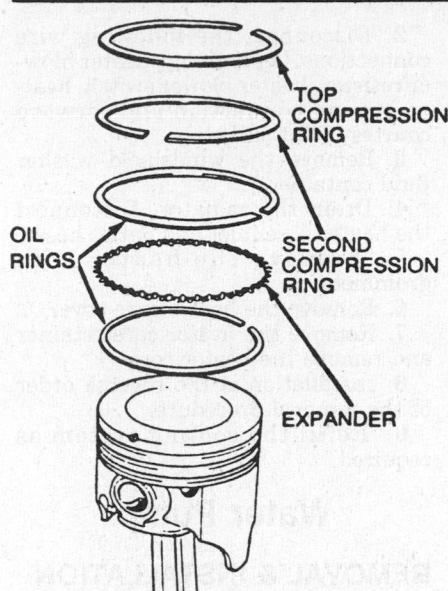

Typical piston and rod assembly

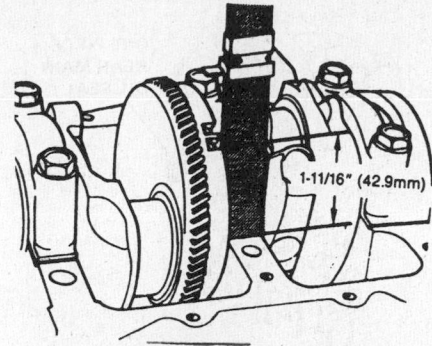

Crankshaft positioning during force balancer installation—2.5L engine

ENGINE LUBRICATION

Force Balancer Assembly

REMOVAL & INSTALLATION

2.5L Engine

NOTE: It is not necessary to remove the balancer assembly for oil pump service.

1. Disconnect the negative battery cable. Raise and support the vehicle safely.
2. Drain the engine oil. Remove the oil pan.
3. Remove the balancer assembly.
4. Installation is the reverse of the removal procedure. Torque the short bolts to 9 ft. lbs. plus a 75 degree turn. Torque the long bolts 9 ft. lbs. plus a 90 degree turn.
5. Rotate the engine to TDC on the No. 1 and No. 4 cylinders. Measure from the engine block to the first cut of the double notch on the reluctor ring of the crankshaft.
6. The measurement should be $1\frac{11}{16}$ in. (42.8mm). Mount the balancer with the counterweights parallel and pointing away from the crankshaft. Be sure not to move the crankshaft.
7. Be sure to use new gaskets or RTV sealant, as required.

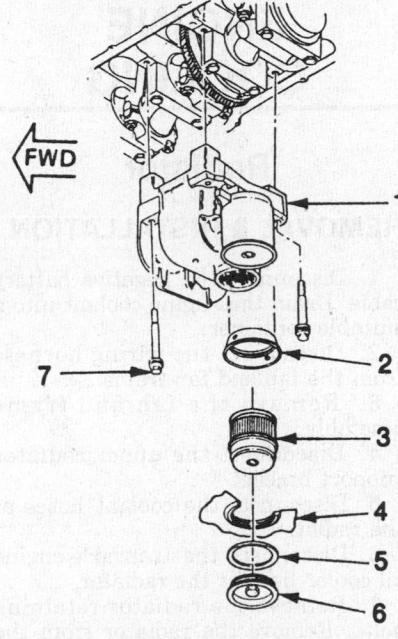

Force balancer assembly—2.5L engine

Force balancer correct counterweight installation—2.5L engine

Oil Pan

REMOVAL & INSTALLATION

2.5L Engine

1. Disconnect the negative battery cable. Remove the engine compartment lid and side panels.

2. Remove the sail panel below the battery side panel trim. Remove the battery side shield.
3. Raise and support the vehicle safely. Drain the engine oil. Remove the oil filter. Remove the serpentine drive belt.
4. Remove the engine mount to cradle nuts. Remove the flywheel cover. Remove the starter.
5. As required, remove the right rear tire and wheel assembly. Remove the splash shield. Loosen the alternator bracket. As required, remove the alternator.
6. Remove the heat shield at the air conditioning compressor. Remove the air conditioning compressor mounting bolts. Position the compressor to the side.
7. Lower the vehicle. Remove the engine strut. Properly support the engine using tool J–28467 or equivalent.
8. Raise and support the vehicle safely. Remove the engine front support bracket and mount.
9. Remove the oil pan retaining bolts. Remove the oil pan from the vehicle.
10. Installation is the reverse of the removal procedure. Be sure to use new gaskets or RTV sealant, as required.

2.8L Engine

1. Disconnect the negative battery cable.
2. Raise the vehicle and support it safely.
3. Drain the oil.
4. Remove the flywheel shield or clutch housing cover.
5. Remove the starter.
6. Remove the oil pan retaining bolts. Remove the oil pan from the engine.

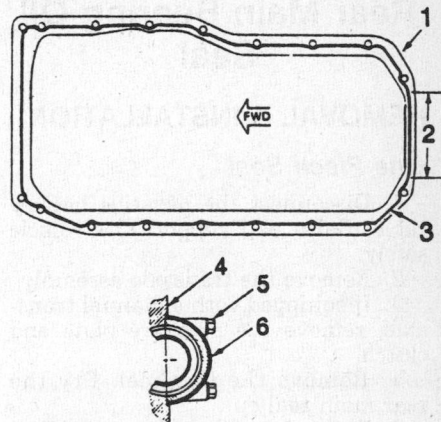

1. Oil pan
2. Apply a ⅜" thick bead of RTV sealer in area indicated
3. Apply a ³⁄₁₆" wide by ⅛" thick bead of RTV sealer in area indicated
4. Engine block assy.
5. Rear bearing
6. Groove in main bearing cap must be filled flush to ⅛" above surface with RTV

Oil pan sealer application—2.5L engine

7. Before installation, clean all mating surfaces.

8. Place a ⅛ in. bead of RTV sealant or an equivalent, on the oil pan sealing flange.

9. Install the oil pan and torque the 1 in. bolts to 6–9 ft. lbs. and the 1.5 in. bolts to 14–22 ft. lbs.

10. The remainder of the installation is the reverse of removal.

Oil Pump

REMOVAL & INSTALLATION

2.5L Engine

1. Disconnect the negative battery cable. Raise and support the vehicle safely.

2. Remove the engine and remove the oil pan.

NOTE: It is not necessary to remove the balancer assembly in order to service the oil pump.

3. Remove the restrictor and the filter.

4. Remove the oil pump cover assembly and remove the gears.

5. Installation is the reverse of removal procedure.

2.8L Engine

1. Disconnect the negative battery cable.

2. Raise and support the vehicle safely.

3. Remove the oil pan.

4. Remove the oil pump retaining bolts.

5. Remove the oil pump and driveshaft extension.

6. Installation is the reverse of the removal procedure.

Rear Main Bearing Oil Seal

REMOVAL & INSTALLATION

One Piece Seal

1. Disconnect the negative battery cable. Raise and support the vehicle safely.

2. Remove the transaxle assembly.

3. If equipped with a manual transaxle, remove the pressure plate and clutch.

4. Remove the flywheel. Pry the rear main seal out.

5. Before installing, clean the block and crankshaft to seal mating surfaces.

6. Lubricate the outside of the seal for ease of installation and press into the block with fingers.

7. Installation is the reverse of the removal procedure.

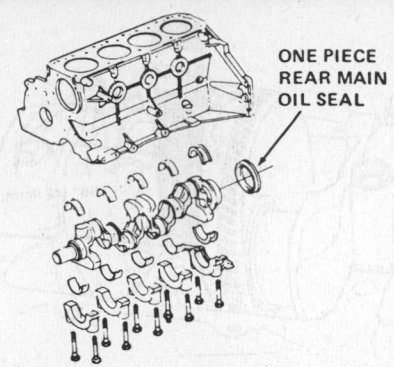

ONE PIECE REAR MAIN OIL SEAL

Lower block assembly—2.5L engine

ENGINE COOLING

Radiator

REMOVAL & INSTALLATION

1. Disconnect the negative battery cable. Drain the engine coolant into a suitable container.

2. Disconnect the wiring harness from the fan and fan frame.

3. Remove the fan and frame assembly.

4. Disconnect the upper radiator support bracket.

5. Disconnect the coolant hoses at the radiator.

6. Disconnect the transaxle/engine oil cooler lines at the radiator.

7. Remove the radiator retaining bolts. Remove the radiator from the vehicle.

8. Installation is the reverse of removal. After installation run the engine and check for leaks.

Heater Core

REMOVAL & INSTALLATION

With Air Conditioning

1. Disconnect the negative battery cable. Drain the cooling system. Disconnect and plug the heater hoses at the heater.

2. Remove the speaker grille and the speaker.

3. Remove the heater core cover retainers and the heater core.

4. Installation is the reverse order of the removal procedure.

5. Refill the cooling system as required.

Without Air Conditioning

1. Disconnect the negative battery cable.

2. Disconnect the following wire connections, heater relay, heater blower resistor, heater blower switch, heater ground connection and forward courtesy light socket.

3. Remove the windshield washer fluid container.

4. Drain the radiator. Disconnect the heater core inlet and outlet hoses.

5. Remove the heater core grommets.

6. Remove the heater case cover.

7. Remove the heater core retainer and remove the heater core.

8. Installation is the reverse order of the removal procedure.

9. Refill the cooling system as required.

Water Pump

REMOVAL & INSTALLATION

2.5L Engine

1. Disconnect battery negative cable. Drain the engine coolant.

2. Remove accessory drive belts. Remove all components in order to gain access to the water pump retaining bolts.

3. Remove the water pump attaching bolts and remove the pump.

4. If installing a new water pump, transfer the pulley from the old unit. With sealing surfaces cleaned, place a ⅛ in. bead of RTV gasket sealant or an equivalent, on the water pump sealing

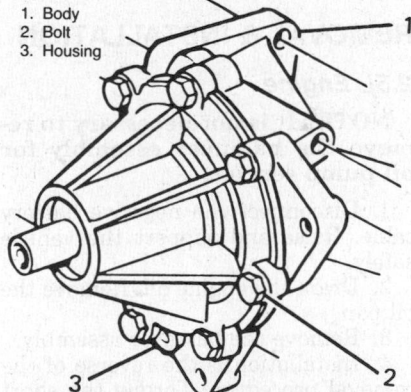

1. Body
2. Bolt
3. Housing

Water pump assembly—2.5L engine

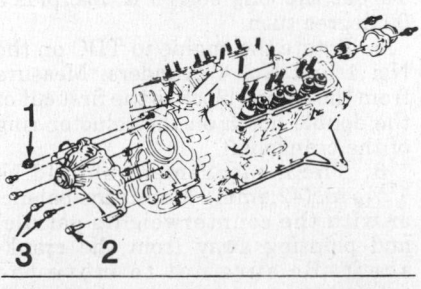

1— 30 N·m (22 FT-LBS)	2— 30 N·m (22 FT-LBS)	3— 10 N·m (7 FT-LBS)

Water pump assembly—2.8L engine

surface. While sealer is still wet, install pump and torque bolts to 25 ft. lbs. (34 Nm). Ensure to coat the bolts with RTV sealant in order to prevent coolant leakage.

5. Install accessory drive belts.

6. Connect battery negative cable.

2.8L Engine

1. Disconnect the negative battery cable. Drain the engine coolant.

2. Remove the fan shroud and the drive belts. Remove all the necessary components in order to gain access to the water pump retaining bolts.

3. Remove the radiator hoses and the heater hose running to the water pump.

4. Remove the bolts attaching the water pump to the engine block and remove the water pump and gasket.

5. Installation is the reverse order of the removal procedure, be sure to apply a thin bead of RTV sealant or an equivalent to the water pump mounting surface and the water pump bolts.

6. Torque the bolts to the following specifications:

    a. M6 × 1.0 bolts – 6–9 ft. lbs. (8–12 Nm)

    b. M8 × 1.25 bolts – 13–18 ft. lbs. (18–24 Nm)

    c. M10 × 1.5 bolts – 20–30 ft. lbs. (27–41 Nm)

7. Do not over torque the water pump bolts, the pump is aluminum and will crack very easily.

Thermostat

REMOVAL & INSTALLATION

1. Disconnect the negative battery cable. As required, drain the coolant. Remove the thermostat cap.

2. Grasp the thermostat handle and gently pull up.

3. Before installing, clean the thermostat housing and O-ring. Apply a suitable lubricant to the O-ring for easier installation.

4. Push the thermostat down into the housing until it is properly seated and install the cap.

Cooling System Bleeding

1. With engine completely cool and cooling system drained, remove the thermostat and reinstall the cap.

2. Fill the radiator to the base of the filler neck with the proper type and quantity of antifreeze. Add sufficient coolant to the reservoir tank to bring the fluid to the **FULL** mark. Reinstall the reservoir cap.

3. Start the engine and run until normal operating temperature is achieved with the radiator cap or thermostat cap removed.

4. With the engine idling, fill the radiator until the coolant level reaches the bottom of the filler neck. Reinstall the cap, making certain the arrows on the cap align with the reservoir hose.

5. Ensure antifreeze protection is approximately −34°F (−37°C).

6. Allow engine to sufficiently cool and reinstall the thermostat.

ENGINE ELECTRICAL

NOTE: Disconnecting the negative battery cable on some vehicles may interfere with the functions of the on board computer systems and may require the computer to undergo a relearning process.

Distributor

REMOVAL & INSTALLATION

Timing Not Disturbed

1. Position the engine at TDC on the compression stroke. Disconnect the negative battery cable.

2. If required, remove the air cleaner assembly. Tag and disconnect all electrical wires.

3. Remove the spark plug wires. Remove the distributor cap.

4. Matchmark the distributor assembly in relation to where the rotor is pointing, to aid in reassembly.

5. Remove the distributor retaining bolt. Carefully remove the distributor from the engine.

6. Installation is the reverse of the removal procedure. Check and adjust the engine timing as required.

Timing Disturbed

If the engine was cranked while the distributor was removed, place the engine on TDC of the compression stroke of cylinder number 1, to obtain proper ignition timing.

1. Remove the No. 1 spark plug.

2. Place a thumb or finger over the spark plug hole. Crank the engine slowly until compression is felt.

3. Align the timing mark on the crankshaft pulley with the 0 degree mark on the timing scale attached to the front of the engine. This places the engine at TDC of the compression stroke.

4. Turn the distributor shaft until the rotor points between the No. 1 and No. 3 spark plug towers on the cap for

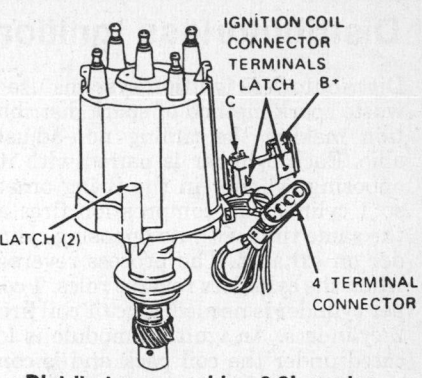

Distributor assembly—2.8L engine

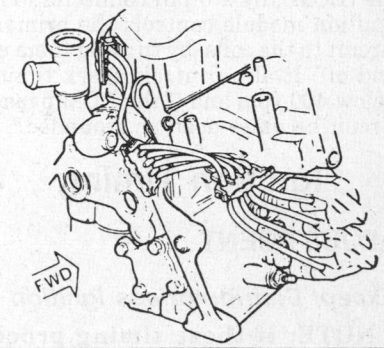

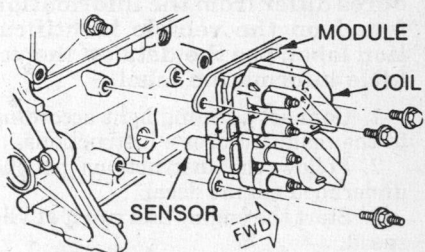

Distributorless ignition assembly— 2.5L engine

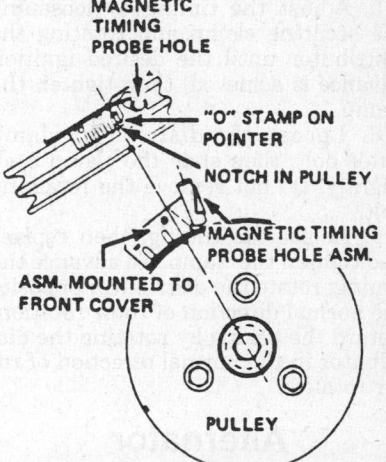

Typical timing mark location

the 2.5L engine. If equipped with the 2.8L engine, turn the distributer shaft until the rotor points between the No. 1 and No. 6 spark plug towers on the cap.

5. Continue the installation in the reverse order of the removal.

Distributorless Ignition

Distributorless ignition systems use a waste spark method of spark distribution making the timing non-adjustable. Each cylinder is paired with its opposing cylinder in the firing order, so 1 cylinder on compression fires at the same time with its opposing cylinder on exhaust. The process reverses when the cylinders reverse roles. 1 coil per cylinder is needed, since 1 coil fires 2 cylinders. An ignition module is located under the coil pack and is connected to the Electronic Control Module (ECM), by a 6 pin connector. The ignition module controls the primary circuit to the coils, by turning them on and off. It also controls spark timing below 400 rpm and if the ECM bypass circuit becomes open or grounded.

Ignition Timing

ADJUSTMENT

Except Distributorless Ignition

NOTE: If these timing procedures differ from the information found on the vehicle identification label, use the data on the vehicle information label.

1. Connect a timing light according to the manufacturer's instructions.
2. Follow the instructions on the underhood engine decal.
3. Start the engine and run it at idle speed.
4. Aim the timing light at the degree scale just over the harmonic balancer.
5. Adjust the timing by loosening the securing clamp and rotating the distributor until the desired ignition advance is achieved, then tighten the clamp.
6. Loosen the distributor clamp outer bolt, then slide the clamp back slightly. Do not remove the retaining bolt.
7. Adjust the timing, then replace and tighten the clamp. To advance the timing, rotate the distributor opposite the normal direction of rotor rotation. Retard the timing by rotating the distributor in the normal direction of rotor rotation.

Alternator

PRECAUTIONS

Several precautions must be observed with alternator equipped vehicles to avoid damage to the unit.

• If the battery is removed for any reason, make sure it is reconnected with the correct polarity. Reversing the battery connections may result in damage to the one-way rectifiers.

• When utilizing a booster battery as a starting aid, always connect the positive to positive terminals and the negative terminal from the booster battery to a good engine ground on the vehicle being started.

• Never use a fast charger as a booster to start vehicles.

• Disconnect the battery cables when charging the battery with a fast charger.

• Never attempt to polarize the alternator.

• Do not use test lights of more than 12 volts when checking diode continuity.

• Do not short across or ground any of the alternator terminals.

• The polarity of the battery, alternator and regulator must be matched and considered before making any electrical connections within the system.

• Never separate the alternator on an open circuit. Make sure all connections within the circuit are clean and tight.

• Disconnect the battery ground terminal when performing any service on electrical components.

• Disconnect the battery if arc welding is to be done on the vehicle.

BELT TENSION ADJUSTMENT

2.5L Engine

A single (serpentine) belt is used to drive all engine mounted components. Drive belt tension is maintained by a spring loaded tensioner. The drive belt tensioner can control belt tension over a wide range of belt lengths; however, there are limits to the tensioner's ability to compensate. Using the tensioner outside its operating range can result in poor tension control and/or damage to the tensioner.

2.8L Engine

The drive belt tension is 145 lbs. for a new belt and 70 lbs. for a used belt.

REMOVAL & INSTALLATION

2.5L Engine

1. Disconnect the negative battery cable.
2. Remove the rear brace bolt.
3. Remove the bolt at the tensioner.
4. Remove the alternator adjusting bolts and remove the drive belt.
5. Disconnect the wiring from the back of the alternator.
6. Remove the alternator from the vehicle.

7. Installation is the reverse of the removal procedure. Adjust the drive belt, as required.

2.8L Engine

1. Disconnect the negative battery cable.
2. Loosen the top alternator bracket retaining bolts.
3. Raise and support the vehicle safely. Remove the right rear tire and wheel assembly.
4. Remove the splash guards. Remove the toe link rod outer end and swing it up and to the left.
5. Remove the lower alternator bracket. Remove the alternator adjusting bolt.
6. Remove the drive belt. Disconnect the upper alternator bracket bolt.
7. Disconnect the electrical connections from the alternator.
8. Rotate the alternator bracket lower end toward the engine. Remove the alternator and shield from the vehicle.
9. Installation is the reverse of the removal procedure. Adjust the drive belt, as required.

Starter

REMOVAL & INSTALLATION

1. Disconnect the negative battery cable.
2. Raise and support the vehicle safely.
3. Disconnect all wires at solenoid terminals. Note color coding of wires for reinstallation.
4. Remove the starter support bracket mount bolts, as required. Loosen the front bracket bolt or nut and rotate the bracket clear.
5. Remove the starter retaining bolts. Remove the starter. Note the location of any shims so they may be replaced in the same positions upon installation.
6. Installation is the reverse of the removal procedure.

EMISSION CONTROLS

Please refer to "Emission Controls" in the Unit Repair section for system maintenance procedures. Due to the complex nature of modern electronic engine control systems, comprehensive diagnosis and testing procedures fall outside the confines of this repair manual. For complete information on diagnosis, testing and repair procedures concerning all modern engine and emission control systems, please refer to "Chilton's Guide to Fuel Injection and Electronic Engine Controls".

FUEL SYSTEM

Fuel System Service Precautions

Safety is the most important factor when performing not only fuel system maintenance but any type of maintenance. Failure to conduct maintenance and repairs in a safe manner may result in serious personal injury or death. Maintenance and testing of the vehicle's fuel system components can be accomplished safely and effectively by adhering to the following rules and guidelines.

● To avoid the possibility of fire and personal injury, always disconnect the negative battery cable unless the repair or test procedure requires that battery voltage be applied.

● Always relieve the fuel system pressure prior to disconnecting any fuel system component (injector, fuel rail, pressure regulator, etc.), fitting or fuel line connection. Exercise extreme caution whenever relieving fuel system pressure to avoid exposing skin, face and eyes to fuel spray. Please be advised that fuel under pressure may penetrate the skin or any part of the body that it contacts.

● Always place a shop towel or cloth around the fitting or connection prior to loosening to absorb any excess fuel due to spillage. Ensure that all fuel spillage (should it occur) is quickly removed from engine surfaces. Ensure that all fuel soaked cloths or towels are deposited into a suitable waste container.

● Always keep a dry chemical (Class B) fire extinguisher near the work area.

● To not allow fuel spray or fuel vapors to come into contact with a spark or open flame.

● Always use a backup wrench when loosening and tightening fuel line connection fittings. This will prevent unnecessary stress and torsion to fuel line piping. Always follow the proper torque specifications.

● Always replace worn fuel fitting O-rings with new. Do not substitute fuel hose or equivalent where fuel pipe is installed.

RELIEVING FUEL SYSTEM PRESSURE

2.5L Engine

1. Be sure the engine is cold. Remove the fuel pump fuse from the fuse panel.
2. Start the engine and let it run until all fuel in the line is used.
3. Crank the starter an additional 3 seconds to relieve any residual pressure.
4. With the ignition **OFF**, replace the fuse.
5. Disconnect the negative battery cable. Disable the fuel pump by disconnecting the electrical connector at the pump.

2.8L Engine

1. Be sure the engine is cold. Connect the fuel gauge J–34730–1 or equivalent to the fuel pressure valve, located on the fuel rail.
2. Wrap a shop towel around the fitting while connecting the gauge to avoid any spillage.
3. Install the bleed hose into a suitable container and open the valve to bleed off the fuel pressure.
4. Disconnect the negative battery cable. Disable the fuel pump by disconnecting the electrical connectors at the pump.

Fuel Filter

REMOVAL & INSTALLATION

1. Properly relieve the fuel system pressure. Disconnect the negative battery cable.
2. Disconnect the inlet and outlet hoses from the fuel filter.
3. Remove the filter retaining clamps, as required.
4. Remove the filter from the engine.
5. Installation is the reverse of the removal procedure.

Electric Fuel Pump

REMOVAL & INSTALLATION

1. Relieve the fuel system pressure. Disconnect the negative battery cable.
2. Drain the fuel tank into a suitable container.
3. Raise and support the vehicle safely.
4. Remove the ground wire retaining screw from under the body.
5. Disconnect the filler neck hoses from the tank.
6. Properly support the tank and remove the retaining strap nuts.
7. Remove the fuel tank from the

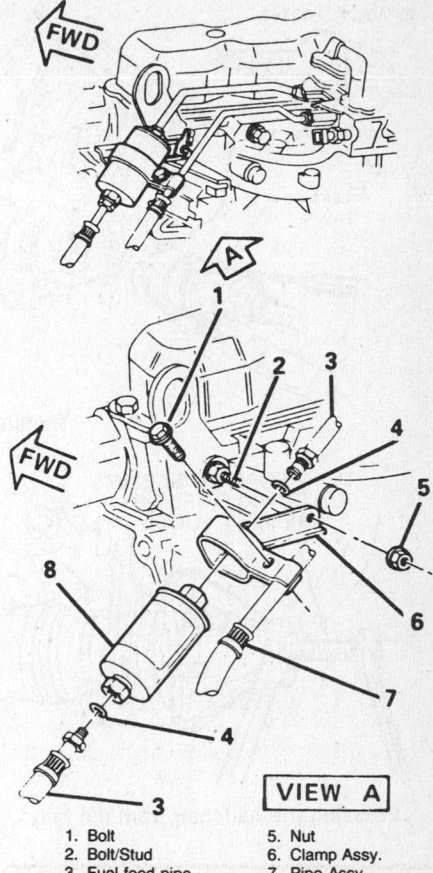

1. Bolt
2. Bolt/Stud
3. Fuel feed pipe
4. "O-ring
5. Nut
6. Clamp Assy.
7. Pipe Assy.
8. Filter Assy.

Fuel filter and related components

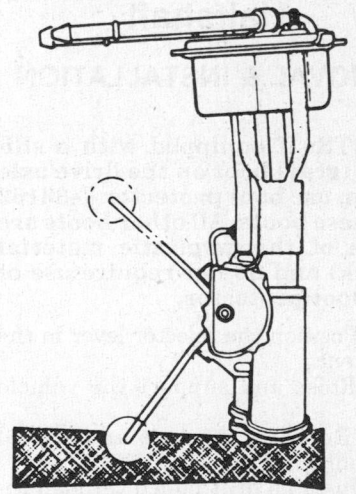

Electric fuel pump and sending unit

vehicle. Disconnect the fuel lines and the sending unit wires from the tank.

8. Remove the fuel gauge/pump retaining ring using spanner wrench tool J–24187 or equivalent.
9. Remove the gauge unit and the pump.
10. Installation is the reverse of removal. Always replace the O-ring under the gauge/pump retaining ring.

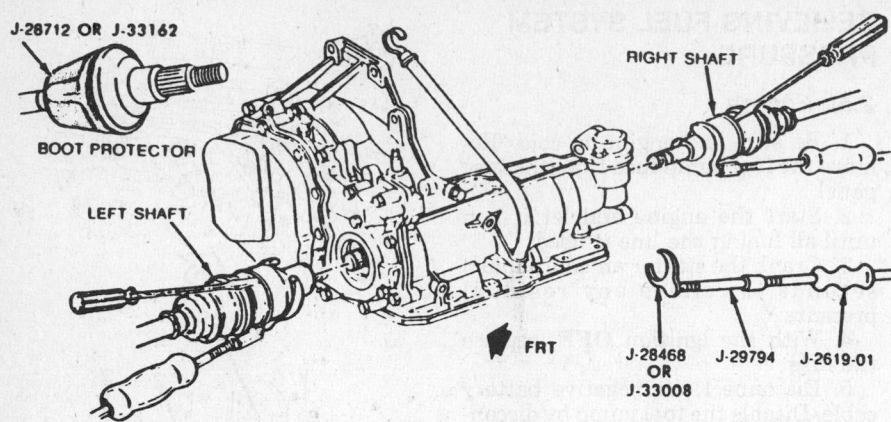

Halfshaft removal

TURN FORCING SCREW UNTIL AXLE SPLINES ARE JUST LOOSE

Pressing the halfshaft from the hub

DRIVE AXLE

Halfshaft

REMOVAL & INSTALLATION

NOTE: If equipped with a silicone (gray) boot on the drive axle joints, use boot protector J–33162 on these boots. All other boots are made of thermoplastic material (black) and do not require use of the boot protector.

1. Position the selector lever in the N detent.
2. Raise and support the vehicle safely.
3. Remove the tire and wheel assembly.
4. Install a drift punch through rotor and remove hub nut and washer (discard nut).
5. Remove the caliper and rotor.
6. Disconnect the trailing arm at the knuckle.
7. Remove the fixed adjusting link, lateral control arm through bolt.
8. Scribe a matchmark on the strut and knuckle assembly.
9. Remove the strut mounting bolts.
10. Press the hub from the halfshaft.

NOTE: If equipped with Tri-Pot joints, care must be exercised not to allow Tri-Pot joints to become overextended. When either end or both ends of the shaft are disconnected, overextending the joint could result in separation of internal compounds. This could cause failure of the joint. Therefore, it is important to handle the drive axle in a manner that prevents overextending.

11. Install special tools J–28468 or J–33008 with J–29794 and J–2619–01 or equivalent slide and remove halfshaft from the transaxle.
12. Installation is the reverse of removal. Install the hub and washer and replace with a new nut. Torque the nut to 183–208 ft. lbs. (250–285 Nm).

Rear Axle Shaft, Bearing and Seal

REMOVAL & INSTALLATION

1. Remove the hub cap and loosen the hub nut.
2. Raise and support the vehicle safely and remove the tire and wheel assembly.
3. Install the drive axle boot protectors. Remove and discard the hub nut.
4. Remove the caliper and rotor and remove the hub and bearing attaching bolts. If the bearing assembly is being reused, mark the attaching bolts and corresponding holes for installation.
5. Install tool J–28733 or equivalent and remove the hub and bearing assembly.
6. If installing a new bearing, be sure to replace the knuckle seal. Clean and inspect the bearing mating surfaces and knuckle bore for dirt, nicks and burrs.
7. If installing a knuckle seal, use tool J–28671 or equivalent and apply grease to the seal and knuckle bore.
8. Place the hub and bearing on the

axle shaft and install all other components at this time.
9. Apply a torque of 74 ft. lbs. (100 Nm) to the new hub nut, until the hub and bearing assembly is seated properly.
10. Install the rotor and caliper and torque the hub nut to specification.
11. Install the tire and wheel assembly and lower the vehicle.

MANUAL TRANSAXLE

For further information on transmissions/transaxles, please refer to "Chilton's Guide to Transmission Repair".

Transaxle Assembly

REMOVAL & INSTALLATION

Isuzu 76mm

1. Disconnect the negative battery cable.
2. Remove the deck lid and louvered panels.
3. Remove the upper rear engine support bolt.
4. Install engine support J–28467–A or equivalent.
5. Raise and safely support the vehicle.
6. Remove the slave cylinder from the clutch, but do not disconnect the lines.
7. Disconnect the shift cables from the transaxle.
8. Disconnect the EGR valve output pipe from the exhaust manifold.
9. Remove the rear wheel and tire assemblies.
10. Remove the parking brake cables from the calipers and the body.
11. Remove the lateral control arm/fixed adjusting link through bolt.
12. Remove the trailing arms.
13. Remove the axle shafts from the transaxle.
14. Remove the rubber skirts from the splash shield cradle retainers.
15. Remove the rear transaxle bracket mount bolts.
16. Remove the motor mount nuts from the cradle and front engine mount shock.
17. Disconnect the crossover pipe at the catalytic converter.
18. Remove the cradle bolts and the cradle from the engine and support the cradle on an adjustable stand.
19. Disconnect the oxygen sensor wire connector.

20. Remove the exhaust crossover pipe and heat shields.
21. Remove the upper transaxle-to-engine bolts.
22. Remove the clutch inspection cover.
23. Remove the remaining transaxle to engine retaining bolts.
24. Lower the transaxle from the vehicle.

To install:

25. Hoist transaxle into position.
26. Install clutch inspection cover.
27. Install transaxle-to-engine bolts and torque to specification.
28. Remove transaxle support.
29. Reconnect all wire connectors.
30. Install the axle shafts.
31. Install the engine cradle and torque the 4 front bolts to 67 ft. lbs. (90 Nm) and the rear bolts to 76 ft. lbs. (103 Nm).
32. Install the rear transaxle bracket bolts and motor mount nuts.
33. Install the engine shock.
34. Install the lateral control arm/fixed adjusting link and the trailing arm.
35. Install and adjust the parking brake cables.
36. Install the splash shield retainers.
37. Install the wheels and tires.
38. Remove the engine support fixture.
39. Install the upper rear engine support bolt.
40. Install the crossover pipe, oxygen sensor wire and heat shields.
41. Install the EGR valve pipe.
42. Install the clutch slave cylinder-to-transaxle bolts.
43. Install the transaxle shift cables.
44. Install the intake duct to the throttle body.
45. Check and fill the transaxle with the proper type and quantity of fluid.
46. Install the deck lid and louvered panels.
47. Reconnect the negative battery cable.

Muncie 282

1. Disconnect the negative battery cable.
2. Drain the transaxle fluid.
3. Remove the select and shift cables from the transaxle brackets.
4. Disconnect the backup light switch wire and remove the switch.
5. Remove the shift cables from the transaxle.
6. Remove the clutch slave cylinder bracket.
7. Remove the exhaust crossover pipe.
8. Remove the upper bolts attaching the transaxle to the engine.
9. Install engine support fixture J–28467–A and J–35563 and raise en-

gine enough to take pressure off the engine mounts.
10. Remove the front and rear transaxle mounts.
11. Raise and support the vehicle safely.
12. Remove the clutch inspection cover.
13. Lower the frame and tilt away.
14. Remove the axle shafts.
15. Remove the remaining transaxle-to-engine mounting bolts.
16. Tilt engine down to gain clearance for transaxle removal.
17. Remove the transaxle assembly.

To install:

18. Install the transaxle to the engine and secure with the 2 lower studs.
19. Raise the frame and install the axle shafts.
20. Install the clutch inspection cover. Install the front and rear transaxle mounts.
21. Lower the vehicle.
22. Lower the engine and remove the support fixture.
23. Install the remaining transaxle-to-engine bolts and torque to specification.
24. Install the exhaust pipe.
25. Install the clutch slave cylinder bracket.
26. Install the select and shift cables.
27. Install the backup light switch.
28. Fill the transaxle with the proper type and quantity of fluid.

LINKAGE ADJUSTMENT
Isuzu 76mm

1. Disconnect the negative battery cable.
2. Place the transaxle in 1st gear.
3. Loosen the shift cable attaching nuts at the transaxle levers.
4. Remove the console and trim plates as required for access to shifter.
5. With the shifter lever in the 1st gear position, insert the proper alignment pins.
6. Remove the lash from transaxle by first compressing the selector cable and then tightening the nut. The levers should be kept from moving during this process. Similarly, the shift cable is first compressed and then the nut is tightened. Again the levers are to remain stationary. The nut on these levers is tightened to 20 ft. lbs. (27 Nm).
7. Ensure that the reverse inhibit cam is against roller and align if necessary.
8. Remove the alignment pins at shifter assembly.

NOTE: While cycling from 1st to 2nd and 2nd to 1st, the select cable should not move. Difficulty in shifting the transaxle to reverse may be corrected by moving select lever inboard toward the 1st-3rd-Reverse position during the shift cable adjustment.

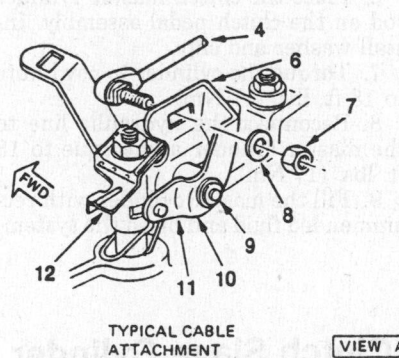

TYPICAL CABLE ATTACHMENT | VIEW A

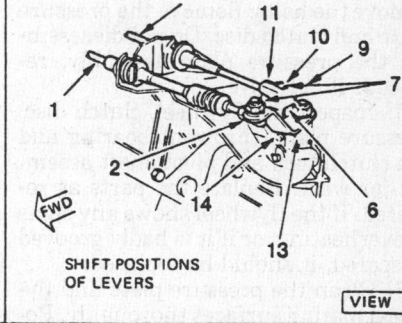

SHIFT POSITIONS OF LEVERS | VIEW B

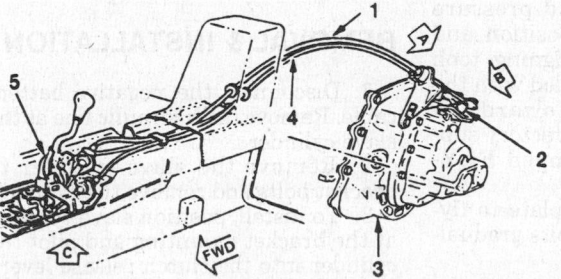

VIEW C

1. Cable A	5. Trans. control assy.	9. R	13. R/3rd/1st
2. Lever F	6. Lever D	10. 1st/2nd	14. 2nd/4th
3. Tansaxle assy.	7. Nut E	11. 4th/3rd	15. Alignment pin F
4. Cable B	8. Washer P	12. Retainer clip J	16. Alignment pin G

Shift linkage cable adjustment—Isuzu 76mm transaxle

Muncie 282

NOTE: Only the shift cable is adjustable and it is adjusted at the transaxle. Do not adjust the select cable.

1. Loosen the nut on the transaxle shift lever ball stud on the shift cable only.

2. Place the transaxle in 3rd gear.

3. Remove the shifter knob, front trimplate and shifter trimplate.

4. Pin the floor shift mechanism in 3rd gear.

5. Tighten the nut on the shift cable ball stud to 18 ft. lbs. (25 Nm).

6. Check adjustment and reinstall trimplates and shifter knob.

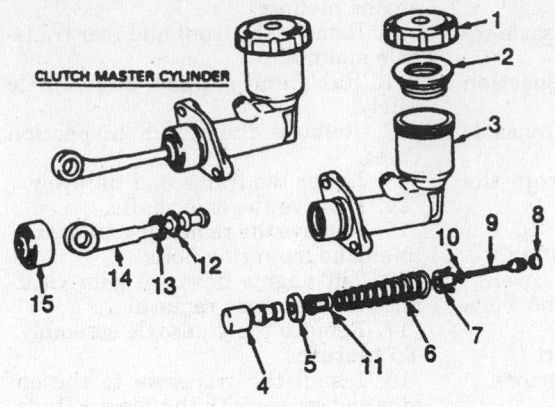

1. Reserve Cap
2. Baffle
3. Cylinder Body and Reservoir Assembly
4. Plunger
5. Seal
6. Spring
7. Valve Spacer
8. Center Valve Seal
9. Valve Stem
10. Spring
11. Spring Retainer
12. Retaining Washer
13. Circlip
14. Push Rod
15. Dust Cover

Clutch master cylinder

CLUTCH

Clutch Assembly

REMOVAL & INSTALLATION

1. Disconnect the negative battery cable. Raise and support the vehicle safely. Remove the transaxle.

2. Mark the pressure plate assembly and the flywheel so they can be assembled in the same position..

3. Loosen the attaching bolts 1 turn at a time until spring tension is relieved.

4. Support the pressure plate and remove the bolts. Remove the pressure plate and clutch disc. Do not disassemble the pressure plate assembly, replace it if defective.

5. Inspect the flywheel, clutch disc, pressure plate, throwout bearing and the clutch fork and pivot shaft assembly for wear. Replace the parts as required. If the flywheel shows any signs of overheating, or if it is badly grooved or scored, it should be replaced.

6. Clean the pressure plate and flywheel mating surfaces thoroughly. Position the clutch disc and pressure plate into the installed position and support with a clutch aligning tool. The clutch plate is assembled with the damper springs offset toward the transaxle. One side of the factory supplied clutch disc is stamped **FLYWHEEL SIDE**.

7. Install the pressure plate to flywheel bolts. Torque the bolts gradually in a crisscross pattern.

8. Lubricate the outside groove and the inside recess of the release bearing with high temperature grease. Wipe off any excess. Install the release bearing.

9. Install the transaxle assembly.

Clutch Master Cylinder

REMOVAL & INSTALLATION

1. Disconnect the negative battery cable. Disconnect clutch pushrod at clutch pedal assembly.

2. Remove the hydraulic line at the clutch master cylinder.

3. Remove the 2 nuts attaching the cylinder to cowl wall.

4. Remove the clutch cylinder.

5. To install, position clutch rod through the the cowl opening and install the cylinder to cowl nuts loosely.

6. Place the clutch master cylinder rod on the clutch pedal assembly. Install washer and clip.

7. Torque the cylinder to cowl nuts to 13 ft. lbs. (17 Nm).

8. Reconnect the hydraulic line to the master cylinder and torque to 13 ft. lbs. (17 Nm).

9. Fill the master cylinder with recommended fluid and bleed the system.

Clutch Slave Cylinder

REMOVAL & INSTALLATION

1. Disconnect the negative battery cable. Remove the hydraulic line at the slave cylinder.

2. Remove the slave cylinder to bracket bolts and remove the cylinder.

3. To install, position slave cylinder at the bracket mounting and pilot the cylinder into the clutch release lever.

4. Install slave cylinder to bracket nuts and torque to 16 ft. lbs.

5. Install hydraulic line to slave cylinder, torque to 13 ft. lbs.

6. Fill master cylinder with recommended fluid and bleed system.

1. Bleedscrew Dust Cover
2. Bleedscrew
3. Cylinder Bolt
4. Spring
5. Seal
6. Plunger
7. Retaining Ring
8. Dust Cover
9. Retaining Band
10. Push Rod

Clutch slave cylinder

Hydraulic Clutch System Bleeding

1. Fill the master cylinder reservoir with brake fluid.

2. Have an assistant pump the clutch pedal 2–3 times and hold to the floor.

3. Open the bleeder screw at the slave cylinder ½ turn and allow all air in the system to escape, close the bleeder screw as soon as brake fluid begins to flow.

4. Repeat this procedure until all air is completely out of the system.

AUTOMATIC TRANSAXLE

For further information on transmissions/transaxles, please refer to "Chilton's Guide to Transmission Repair".

Transaxle Assembly

REMOVAL & INSTALLATION

1. Disconnect the negative battery cable.
2. Remove the air cleaner assembly.
3. Remove the right engine vent cover.
4. Remove the left engine vent cover.
5. Remove the throttle valve cable.
6. Remove the shift cable at the transaxle bracket.
7. Disconnect the neutral start switch electrical connection.
8. Disconnect the transaxle converter clutch electrical connection.
9. Disconnect the speedometer pick-up electrical connection.
10. Remove the wire harness at the transaxle to engine retaining bolts.
11. Remove the transaxle cooler line support bracket.
12. Remove the transaxle to engine retaining bolts.
13. Remove the shift cable bracket to remove the neutral start switch harness.
14. Install the engine fixture tool J–28467–A or equivalent.
15. Raise the vehicle and support it safely.
16. Remove the rear wheels.
17. Install rear axle boot protectors.
18. Remove the fixed adjusting link/lateral control arm through bolts.
19. Disconnect the trailing arms at knuckles.

NOTE: If equipped with Tri-Pot joints, care must be exercised not to allow the Tri-Pot joints to become overextended. When either end or both ends of the shaft are disconnected, overextending the joint could result in separation of internal components. This could cause failure of the joint. Therefore, it is important to handle the drive axle in a manner that prevents overextending.

20. Remove rear axle shafts from transaxle.
21. Support the rear axle shafts.
22. Remove the splash shields.
23. Disconnect the brake cables at the calipers.
24. Disconnect the brake control cable at the frame.
25. Disconnect the exhaust pipe at the exhaust manifold.
26. Remove the engine mounts to cradle nuts.
27. Remove the transaxle mounts to cradle nuts.
28. Remove the front cradle retaining bolts.
29. Remove the rear cradle retaining bolts.
30. Remove the cradle from the vehicle.
31. Remove the flywheel shield. Remove the starter.
32. Remove the flexplate-to-converter bolts.
33. Disconnect and plug the cooler lines.
34. Install the transaxle support jack.
35. Remove the transaxle support bracket at the right rear.
36. Remove the remaining transaxle to engine retaining bolts including the ground wire.
37. Lower the transaxle from the vehicle.
38. Installation is the reverse of the removal procedure.

FRONT SUSPENSION

Shock Absorbers

REMOVAL & INSTALLATION

1. Raise the vehicle and support safely.
2. Remove the wheel and tire assembly.
3. Remove the lower retaining bolts.
4. Remove the nut and bolt from the top of the shock absorber.
5. Remove the shock from the vehicle through the lower control arm.
6. Installation is the reverse of the removal procedure.

Coil Springs

REMOVAL & INSTALLATION

1. Raise the vehicle and support it safely. Properly support the lower control arm, using a suitable jack.
2. Remove wheel and tire assembly.
3. Remove the shock absorber. Disconnect the stabilizer bar from the lower control arm.
4. Install a safety chain through the spring, as a safety precaution. Remove the lower control arm pivot bolts.
5. Slowly lower the jack under the lower control arm and remove the spring from the vehicle.
6. Installation is the reverse of the removal procedure.

Upper Ball Joints

INSPECTION

1. Raise the front of the vehicle with a lift placed under the engine cradle. The front wheels should be clear of the ground.
2. Grasp the wheel at the top and bottom and shake the wheel in and out.
3. If any movement is seen of the steering knuckle relative to the control arm, the ball joints are defective and must be replaced. Note, movement elsewhere may be due to loose wheel bearings or other troubles. Watch the knuckle-to-control arm connection.
4. If the ball stud is disconnected from the steering knuckle and any looseness is noted or the ball joint stud can be twisted in its socket with your fingers, replace the ball joints.

REMOVAL & INSTALLATION

1. Raise the vehicle and support it safely. Properly support the lower control arm, using a suitable jack.
2. Remove wheel and tire assembly.
3. Remove the bolt that retains the brake line clip to the upper control arm.
4. Disconnect the tie rod end from the steering knuckle, swing the knuckle outboard.
5. Using a ball joint removal tool, remove the upper ball joint from the steering knuckle.
6. Drill out the ball joint retaining rivets. Remove the ball joint from the control arm.
7. Install a new ball joint, using the retaining bolts provided with the ball joint.
8. Installation is the reverse of the removal procedure. Be sure to check the front end alignment.

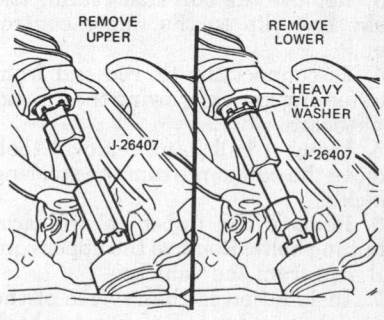

Ball joint removal

Lower Ball Joints

INSPECTION

1. Raise the front of the vehicle with a lift placed under the engine cradle. The front wheels should be clear of the ground.
2. Grasp the wheel at the top and bottom and shake the wheel in and out.
3. If any movement is seen of the steering knuckle relative to the control arm, the ball joints are defective and must be replaced. Note, movement elsewhere may be due to loose wheel bearings or other troubles. Watch the knuckle-to-control arm connection.
4. If the ball stud is disconnected from the steering knuckle and any looseness is noted or the ball joint stud can be twisted in its socket with your fingers, replace the ball joints.

REMOVAL & INSTALLATION

1. Raise the vehicle and support it safely. Properly support the lower control arm, using a suitable jack.
2. Remove wheel and tire assembly.
3. Disconnect the tie rod end from the steering knuckle, using the proper tool.
4. Using a ball joint removal tool, remove the lower ball joint from the steering knuckle.
5. Position the knuckle with the caliper and rotor to the side.
6. Using a suitable ball joint press, remove the ball joint from the control arm.
7. Installation is the reverse of the removal procedure. Be sure to check the front end alignment, as required.

Upper Control Arms

REMOVAL & INSTALLATION

1. Raise the vehicle and support it safely. Properly support the lower control arm, using a suitable jack.
2. Remove wheel and tire assembly.
3. Remove the bolt that retains the brake line clip to the upper control arm.
4. Disconnect the tie rod end from the steering knuckle, swing the knuckle outboard.
5. Using a ball joint removal tool, remove the ball joint from the steering knuckle.
6. Remove the upper control arm retaining bolts. Remove the upper control arm from the vehicle.
7. Installation is the reverse of the removal procedure. Be sure to check the front end alignment.

Lower Control Arms

REMOVAL & INSTALLATION

1. Raise the vehicle and support it safely. Properly support the lower control arm, using a suitable jack.
2. Remove wheel and tire assembly. Remove the control arm retaining bolts and remove the coil spring assembly.
3. Remove the lower control arm ball joint from the steering knuckle. Remove the lower control arm from the vehicle.
4. Installation is the reverse of the removal procedure. Be sure to check the front end alignment, as required.

Front Wheel Bearings

ADJUSTMENT

Sealed Wheel Bearings

1. Raise and support the vehicle safely. Remove the tire and wheel assemblies.
2. Remove the disc brake linings and position the calipers out of the way. Complete removal of the caliper, may be required.
3. Reinstall the disc, use 2 wheel nuts to secure the disc to the bearing.
4. Mount a dial indicator gauge, tool J-8001 to the disc and hub assembly.
5. Grasp the disc and use a push pull movement to check the specification.
6. Specification exceeds 0.005 in. (0.1270mm), replace the hub and bearing assembly.

REMOVAL & INSTALLATION

1. Raise the vehicle and support it safely.
2. Remove the wheel and tire assembly.
3. Remove the bolt attaching the brake line clip to the upper control arm.
4. Remove the caliper and suspend with a wire.
5. Remove the rotor.
6. Remove the 3 bolts attaching the hub and bearing assembly to the steering knuckle.
7. Remove the bearing and hub assembly.
To install:
8. Install the hub and bearing assembly. Torque the hub and bearing to steering knuckle bolt to 220 ft. lbs. (260 Nm).

NOTE: When ever the brake rotor has been separated from the wheel bearing, remove any rust or other foreign material from the mating surfaces of the wheel bearing flange and rotor. Failure to do so may result in lateral runout of the rotor, causing brake pedal pulsation.

9. Install the rotor and caliper.
10. Install the bolt attaching the brake line clip to the upper control arm.
11. Install the wheel and tire assembly.
12. Lower the vehicle to the ground.

REAR SUSPENSION

MacPherson Strut

REMOVAL & INSTALLATION

1. Remove the engine compartment cover.
2. Remove the 3 upper strut nuts and washers.
3. Loosen the wheel lug nuts.
4. Raise the vehicle and support it safely. Support the rear control arm with a floor jack.
5. Remove the wheel and tire. Remove the brake line clip.
6. Scribe the strut and knuckle. Remove the 2 strut mounting nuts and bolts. Remove the strut assembly and spacer plate from the vehicle.
7. Installation is the reverse of removal. Align the scribe marks on the strut and knuckle and replace the bolts in the same order in which they were removed. Tighten the strut mounting nuts to 140 ft. lbs. and the upper strut nuts to 18 ft. lbs.

Rear Wheel Bearings

REMOVAL & INSTALLATION

For all rear wheel bearing removal and installation procedures, please refer to Rear Axle Shaft, Bearing and Seal in the Drive Axle section.

STEERING

Steering Wheel

REMOVAL & INSTALLATION

1. Disconnect the negative battery cable. Remove the horn pad and retaining screws.

2. Remove the steering wheel retaining nut. Remove the wheel using a steering wheel puller.

3. When installing, align the index mark on the steering wheel with the index mark on the steering shaft. Torque the retaining nut to 35 ft. lbs.

4. The cancelling cam tower must be centered in the slot of the lock plate cover before assembling the wheel.

Manual Steering Rack

REMOVAL & INSTALLATION

1. Disconnect the negative battery cable.

2. Raise the vehicle and support it safely.

3. Disconnect the flexible coupling pinch bolt to the shaft.

4. Remove the outer tie rod cotter pins and nuts on the left and right sides.

5. Disconnect the tie rods from the steering knuckle.

6. Remove the 4 bolts retaining the steering assembly to the crossmember and remove the steering assembly.

7. Installation is the reverse of removal. Tighten the flexible coupling bolt to 46 ft. lbs. (62 Nm), the 4 new steering assembly bolts to 21 ft. lbs., the 4 crossmember brace bolts to 20 ft. lbs. (27 Nm) and the tie rod nut at each knuckle to 29 ft. lbs. (39 Nm), turn nut to align the cotter pin.

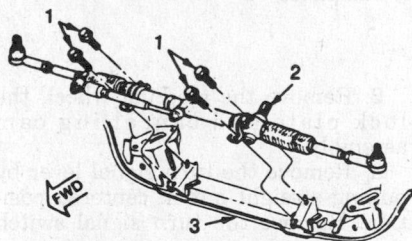

1. Bolt (21 ft. lbs.)
2. Steering assy.
3. Cross member
4. Nut (32 ft. lbs.)
5. Washer
6. Stud assy (36 ft. lbs.)
7. Steering link damper

Rack and pinion assembly

Tie Rod Ends

REMOVAL & INSTALLATION

1. Raise and support the vehicle safely. Loosen the jam nut and remove the tie rod from the steering knuckle using tool J–24319–01 or BT7101.

2. Count the number of threads showing on the tie rod, inboard of the jam nut. This number will be a reference for installing the new tie rod end. Remove the outer tie rod.

3. Install the outer tie-rod in the reverse or removal. Do not tighten the jam nut.

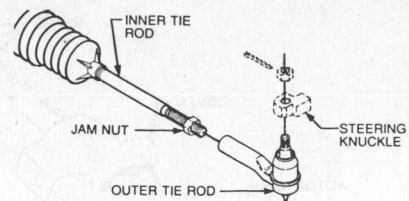

Outer tie rod end removal

4. Adjust the toe-in by turning the inner tie-rod the required number of turns.

5. Make sure the boot is not twisted then torque the jam nut to 50 ft. lbs.

BRAKES

For all brake system repair and service procedures not detailed below, please refer to "Brakes" in the Unit Repair section.

Master Cylinder

REMOVAL & INSTALLATION

1. Disconnect the negative battery cable. Disconnect and plug the hydraulic lines at the master cylinder.

2. Remove the 2 nuts attaching the master cylinder to its mounting. Remove the master cylinder from the vehicle.

3. Installation is the reverse of the removal procedure. Be sure to bleed the master cylinder, prior to installation.

Combination Valve

REMOVAL & INSTALLATION

NOTE: The combination valve is not repairable and must be replaced if found defective.

1. Disconnect the negative battery cable. Disconnect and plug the hydraulic lines at the combination valve.

2. Disconnect the warning switch wiring harness from the valve switch terminal.

3. Remove the bolt attaching the valve to the bracket.

4. Remove the combination valve.

5. Installation is the reverse of removal.

6. Bleed the system.

Power Brake Booster

REMOVAL & INSTALLATION

1. Disconnect the negative battery cable. Disconnect the master cylinder

from the booster and position it to the side.

2. Remove the booster attaching nuts.

3. Remove the booster pushrod from the brake pedal assembly.

4. Remove the booster from the vehicle.

5. Installation is the reverse of removal.

NOTE: If any hydraulic component is removed or brake line disconnected, bleed the brake system after installation, as necessary.

Disc Brake Pads

REMOVAL & INSTALLATION

Front Pads

1. Partially drain the master cylinder.

2. Raise and support the vehicle safely.

3. Remove the tire and wheel assembly.

4. Bottom the piston in the caliper bore to provide clearance between the linings and rotor.

5. Remove the spring pins by connecting spring pin removal tool J–36620 to a slide hammer tool.

6. Remove the threaded tip from the rod on the spring pin tool. Insert the rod completely through the pin and install the threaded tip as far as it will go.

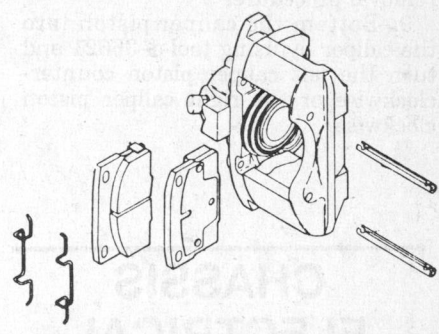

Front caliper and brake pad assembly

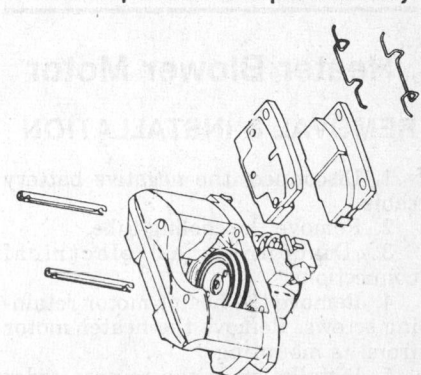

Rear caliper and brake pad assembly

7. Use the slide hammer to drive out the pin.

NOTE: Be prepared to catch the springs when removing the spring pins, as the springs may fly out.

8. Remove the springs from the shoe flanges. Remove the disc brake linings from the caliper.

9. Installation is the reverse of the removal procedure.

Rear Pads

1. Partially drain the master cylinder.

2. Release the parking brake. Raise and support the vehicle safely.

3. Remove the tire and wheel assembly.

4. Remove the spring pins by connecting the spring pin removal tool J-36620 to a slide hammer tool.

5. Remove the threaded tip from the rod on the spring pin tool. Insert the rod completely through the pin and install the threaded tip as far as it will go.

6. Use the slide hammer to drive out the pin.

NOTE: Be prepared to catch the springs when removing the spring pins, as the springs may fly out.

7. Remove the springs from the shoe flanges. Remove the disc brake linings from the caliper.

8. Installation is the reverse of the removal procedure.

9. Bottom the caliper piston into the caliper by using tool J-36621 and turn the left caliper piston counterclockwise or the right caliper piston clockwise.

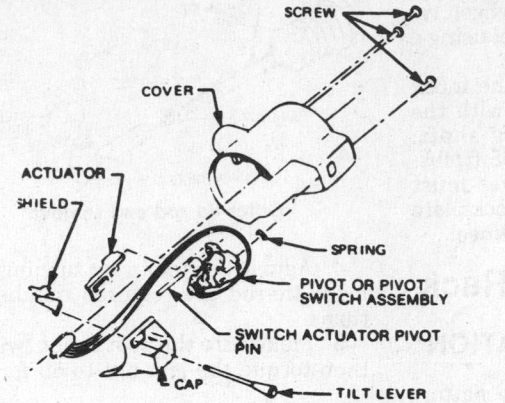

Wiper switch removal—standard steering column

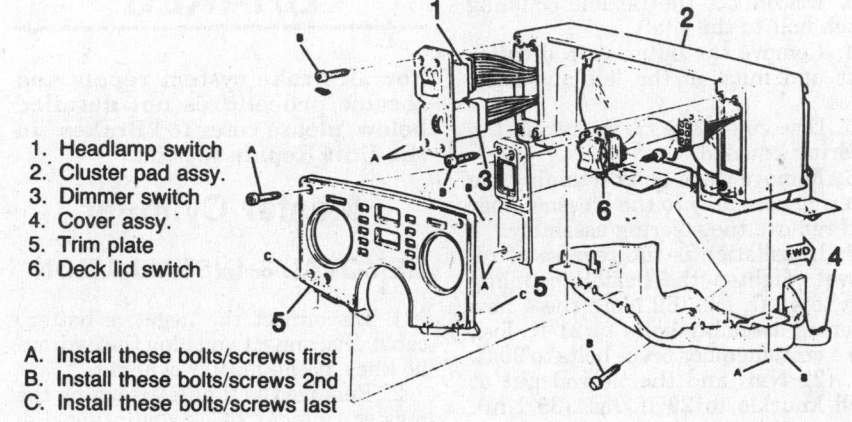

1. Headlamp switch
2. Cluster pad assy.
3. Dimmer switch
4. Cover assy.
5. Trim plate
6. Deck lid switch

A. Install these bolts/screws first
B. Install these bolts/screws 2nd
C. Install these bolts/screws last

Instrument cluster trim plates

CHASSIS ELECTRICAL

Heater Blower Motor

REMOVAL & INSTALLATION

1. Disconnect the negative battery cable.
2. Remove the cooling tube.
3. Disconnect all electrical connections.
4. Remove the heater motor retaining screws. Remove the heater motor from its mounting.
5. Installation is the reverse order of the removal procedure.

Windshield Wiper Motor

REMOVAL & INSTALLATION

1. Disconnect the negative battery cable.
2. Loosen, but do not remove the transmission drive link to motor crank arm retaining nuts.
3. Detach the drive link from the motor crank arm.
4. Disconnect the electrical leads.
5. Remove the attaching screws and remove the wiper motor.
6. Installation is the reverse of removal. Be sure the wiper motor is in the park position before installing the wiper arms and the shroud top screen.

Windshield Wiper Switch

REMOVAL & INSTALLATION

1. Disonnect the negative battery cable.

2. Remove the steering wheel, the lock plate and cancelling cam assembly.
3. Remove the turn signal lever by pulling straight out in centered position. Remove the turn signal switch assembly.
4. Remove the ignition lock and key warning buzzer.
5. Remove the steering column housing attaching screws and remove the housing.
6. Remove the pivot and switch pin.
7. Remove the pivot and switch assembly.
8. Installation is the reverse of the removal procedure.

Instrument Cluster

REMOVAL & INSTALLATION

1. Disconnect the negative battery cable.
2. Remove the rear cluster cover.
3. Remove the front trim plate.
4. Remove the steering column cover.
5. Remove the cluster attaching

screws. Disconnect the wiring harness and remove the cluster assembly.

6. The speedometer, tachometer and gauges may be serviced by removing the front cluster lens.

7. Installation is the reverse of the removal procedure.

Radio

REMOVAL & INSTALLATION

1. Disconnect the negative battery cable. Remove the console trim plate assembly.

2. Disconnect the side retaining nuts and the rear retaining bolt.

3. Disconnect the electrical and the antenna connections.

4. Remove the radio from its mounting.

5. Installation is the reverse of the removal procedure.

Concealed Headlights

Manual Operation

In the event of system failure, the headlights may be raised or lowered manually.

1. Locate the headlight door motors in each headlight assembly under the front compartment.

2. Rotate the knob on each headlight door motor in the direction of the arrow until each headlight door fully open.

3. Reverse the procedure to lower the headlights doors.

Headlight Switch

REMOVAL & INSTALLATION

1. Disconnect the negative battery cable.

2. Remove the headlight/dimmer switch trim plate screws.

3. Disconnect the electrical connector and remove the switch assembly.

4. Installation is the reverse order of the removal procedure.

Combination Switch

REMOVAL & INSTALLATION

1. Disconnect the negative battery cable. Remove the steering wheel.

2. Remove the trim cover from the steering column.

3. Position a lockplate remover on the end of the steering shaft and compress the lock plate by turning the shaft nut clockwise. Pry the wire snapring out of the shaft groove.

4. Remove the tool and lift the lockplate off the shaft.

5. Slip the cancelling cam, upper bearing preload spring and thrust washer off the shaft.

6. Remove the turn signal lever. Remove the hazard flasher button retaining screw and remove the button, spring and knob.

7. Pull the switch connector out of the mast jacket and tape the upper part of the connector to aid in switch removal. Attach a long piece of wire to the turn signal switch connector. On vehicles equipped with tilt wheel, place the turn signal and shifter housing in low position and remove the harness cover.

8. Remove the 3 switch mounting screws. Remove the switch by pulling it straight up while guiding the wiring harness cover through the column.

To install:

9. Install the replacement switch by connecting the long piece of wire from the previously removed switch onto the new switch. Use the wire to aid in guiding the connector and cover down through the housing and under the bracket. If equipped with tilt wheel, the connector is worked down through the housing, under the bracket and then the cover in installed on the harness.

10. Install the switch mounting screws and the connector on the mast jacket bracket. Install the column to dash trim plate.

11. Install the flasher knob and the turn signal lever.

12. With the turn signal lever in neutral and the flasher knob out, slide the thrust washer, upper bearing preload spring and cancelling cam onto the shaft.

13. Position the lock plate on the shaft and press it down until a new snapring can be inserted in the shaft groove. Always use a new snapring when assembling.

14. Install the cover and the steering wheel.

Stoplight Switch

NOTE: Both the stoplight switch and the cruise control switch are mounted on the brake pedal support bracket, the stoplight switch is located directly above the cruise control switch.

ADJUSTMENT

1. With the brake pedal depressed, insert the switch into the retainer until the switch body seats on the retainer.

2. Note that audible clicks can be heard as the threaded portion of the switch is pushed through the retainer toward the brake pedal.

3. Pull the brake pedal fully rearward against the pedal stop until the audible click sound can no longer be heard. The switch will be moved in the retainer providing the correct adjustment.

4. Release the the brake pedal and repeat Step 3 to assure that no audible click sounds remain. Check the brake light operation.

REMOVAL & INSTALLATION

1. Disconnect the negative battery cable. Disconnect the wiring harness from the switch which is located under the instrument panel at the brake pedal support.

2. Remove the retaining nut from the switch and remove the switch from the bracket.

3. Installation is the reverse of the removal procedure.

Clutch Switch

REMOVAL & INSTALLATION

1. Disconnect the negative battery cable. Disconnect the electrical connector at the clutch switch, which is located at the top of the clutch pedal.

2. Remove the bolt attaching the switch to the clutch bracket.

3. Rotate the clutch switch slightly to disconnect the shaft from the clutch pedal hole.

4. Installation is the reverse of the removal procedure.

Neutral Safety Switch

This vehicle is equipped with a mechanical neutral start system. This system relies on a mechanical block, rather than the starter safety switch to prevent starting the engine in any gear except **P** or **N**. This unit is mounted in the steering column and has a manual actuated rod preventing the ignition switch from moving to the start position unless the shifter is in the **P** or **N** position.

Fuses, Circuit Breakers and Relays

Fusible Link

Added protection is provided in all battery feed circuits and other selected circuits by a fusible link. This link is a short piece of copper wire approximately 4 in. long inserted in series with the circuit and acts as a fuse. The link is 2 or more gauges smaller in size than the circuit wire it is protecting and will burn out without damage to the circuit in case of current overload.

LOCATION

1. Fuse block—Behind the left side of instrument panel.
2. Fusible link B—right front of engine compartment at battery junction block.
3. Fusible link C—In front lights harness, right of master cylinder.
4. Fusible link D—In front lights harness, right of master cylinder.

Fuse Panel

The fuse panel is a swing down unit located in the underside of the instrument panel, left of the steering column. The fuse panel uses miniaturized fuses, designed for increased circuit protection and greater reliability. Various convenience connectors, which snap into the fuse panel, add to the serviceability of this unit.

Computer

The electronic control module (ECM) is located between the seats and is mounted to the rear bulkhead. Access to the computer can be gained by removing the console. The computer is not a serviceable part and can only be replaced if diagnosed to be faulty.

Turn Signal/Hazard Flasher

The turn signal and hazard flasher is located on the left side of the steering column bracket

Convenience Center

The convenience center is a stationary unit. It is located on the right side of the heater or air conditioning module in the vehicle under the instrument panel. This location provides easy access to the audio alarm, hazard warnings, the horn relay and the seatbelt key and headlight warning alarm. All units are serviced by plug-in replacements.

GM—Pontiac
Front Wheel Drive
Pontiac—LeMans

SPECIFICATIONS

VEHICLE IDENTIFICATION CHART

It is important for servicing and ordering parts to be certain of the vehicle and engine identification. The VIN (vehicle identification number) is a 17 digit number visible through the windshield on the driver's side of the dash and contains the vehicle and engine identification codes. The tenth digit indicates model year and the eighth digit indicates engine code. It can be interpreted as follows:

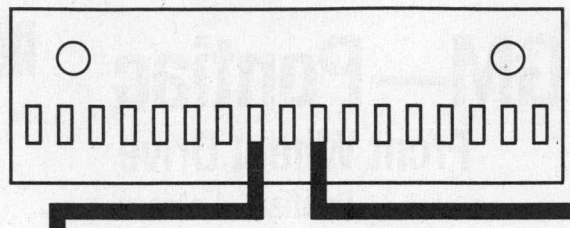

Engine Code							Model Year	
Code	Cu. In.	Liters	Cyl.	Fuel Sys.	Eng. Mfg.		Code	Year
6	98	1.6	4	TBI	GM		J	1988
K	121	2.0	4	TBI	GM		K	1989
							L	1990
							M	1991
							N	1992

TBI—Throttle body injection

ENGINE IDENTIFICATION

Year	Model	Engine Displacement cu. in. (liter)	Engine Series Identification (VIN)	No. of Cylinders	Engine Type
1988	Lemans	98 (1.6)	6	4	OHC
1989	Lemans	98 (1.6)	6	4	OHC
	Lemans	121 (2.0)	K	4	OHC
1990	Lemans	98 (1.6)	6	4	OHC
	Lemans	121 (2.0)	K	4	OHC
1991–92	Lemans	98 (1.6)	6	4	OHC

OHC—Overhead cam

GENERAL ENGINE SPECIFICATIONS

Year	VIN	No. Cylinder Displacement cu. in. (liter)	Fuel System Type	Net Horsepower @ rpm	Net Torque @ rpm (ft. lbs.)	Bore × Stroke (in.)	Compression Ratio	Oil Pressure @ rpm
1988	6	4-98 (1.6)	TBI	74 @ 5200	88 @ 3400	3.11 × 3.20	8.5:1	55 @ 2000
1989	6	4-98 (1.6)	TBI	74 @ 5600	88 @ 3400	3.11 × 3.20	8.5:1	55 @ 2000
	K	4-121 (2.0)	TBI	96 @ 4800	118 @ 3600	3.39 × 3.39	8.8:1	55 @ 2000
1990	6	4-98 (1.6)	TBI	74 @ 5600	90 @ 2800	3.11 × 3.20	8.5:1	55 @ 2000
	K	4-121 (2.0)	TBI	96 @ 4800	118 @ 3600	3.39 × 3.39	8.8:1	55 @ 2000
1991–92	6	4-98 (1.6)	TBI	74 @ 5600	90 @ 2800	3.11 × 3.20	8.5:1	55 @ 2000

TBI—Throttle body injection

ENGINE TUNE-UP SPECIFICATIONS

Year	VIN	No. Cylinder Displacement cu. in. (liter)	Spark Plugs Type	Gap (in.)	Ignition Timing (deg.) MT	AT	Compression Pressure (psi)	Fuel Pump (psi)	Idle Speed (rpm) MT	AT	Valve Clearance In.	Ex.
1988	6	4-98 (1.6)	R44XLS6	0.060	①	①	②	9-13	①	①	HYD.	HYD.
1989	6	4-98 (1.6)	R44XLS6	0.060	①	①	②	9-13	①	①	HYD.	HYD.
	K	4-121 (2.0)	R44XLS6	0.060	①	①	②	9-13	①	①	HYD.	HYD.
1990	6	4-98 (1.6)	R45XLS	0.045	①	①	②	9-13	①	①	HYD.	HYD.
	K	4-121 (2.0)	R45XLS	0.045	①	①	②	9-13	①	①	HYD.	HYD.
1991	6	4-98 (1.6)	R45XLS	0.045	①	①	②	9-13	①	①	HYD.	HYD.
1992			SEE UNDERHOOD SPECIFICATIONS STICKER									

HYD.—Hydraulic
① See underhood specifications sticker.
② Lowest reading not less than 70% of highest.
No reading less than 100 psi.

FIRING ORDERS

NOTE: To avoid confusion, always replace spark plug wires one at a time.

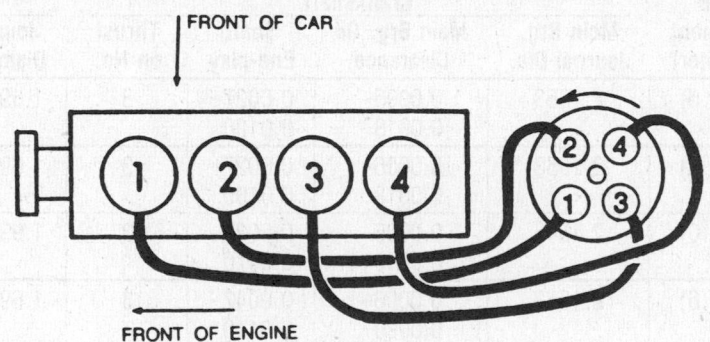

FRONT OF CAR

FRONT OF ENGINE

1.6L and 2.0L Engines
Engine Firing Order: 1–3–4–2
Distributor Rotation: Counterclockwise

CAPACITIES

Year	Model	VIN	No. Cylinder Displacement cu. in. (liter)	Engine Crankcase with Filter	without Filter	Transmission (pts.) 4-Spd.	5-Spd.	Auto.	Drive Axle (pts.)	Fuel Tank (gal.)	Cooling System (qts.)
1988	Lemans	6	4-98 (1.6)	4.0	3.5	3.5	3.5	8①	—	13.2	8.1
1989	Lemans	6	4-98 (1.6)	4.0	3.5	3.5	3.5	8①	—	13.0	8.1
	Lemans	K	4-121 (2.0)	4.0	3.5	3.5	4.5	8①	—	13.0	8.1
1990	Lemans	6	4-98 (1.6)	4.0	3.5	3.5	3.5	8①	—	13.0	8.1
	Lemans	K	4-121 (2.0)	4.0	3.5	3.5	4.5	8①	—	13.0	8.1
1991-92	Lemans	6	4-98 (1.6)	4.0	3.5	3.5	3.5	8①	—	13.0	8.1

① Overhaul—12 pts.

CAMSHAFT SPECIFICATIONS

All measurements given in inches.

Year	VIN	No. Cylinder Displacement cu. in. (liter)	Journal Diameter 1	2	3	4	5	Lobe Lift In.	Ex.	Bearing Clearance	Camshaft End Play
1988	6	4-98 (1.6)	1.578–1.577	1.588–1.590	1.598–1.597	1.608–1.607	1.618–1.617	0.220	0.241	0.0018–0.0020	0.0016 0.0064
1989	6	4-98 (1.6)	1.578–1.577	1.588–1.590	1.598–1.597	1.608–1.607	1.618–1.617	0.220	0.241	0.0018–0.0020	0.0016 0.0064
	K	4-121 (2.0)	1.867–1.869	1.867–1.869	1.867–1.869	1.867–1.869	1.867–1.869	0.259	0.259	0.0010–0.0039	0.002 0.006
1990	6	4-98 (1.6)	1.578–1.577	1.588–1.587	1.600–1.597	1.608–1.607	1.618–1.617	0.220	0.241	0.0020–0.0035	0.0016 0.0064
	K	4-121 (2.0)	1.670–1.671	1.671–1.672	1.691–1.692	1.701–1.702	1.710–1.711	0.237	0.251	0.0011–0.0035	0.002 0.006
1991–92	6	4-98 (1.6)	N/A	N/A	N/A	N/A	N/A	0.220	0.241	N/A	0.0083

N/A—Not available

CRANKSHAFT AND CONNECTING ROD SPECIFICATIONS

All measurements are given in inches.

Year	VIN	No. Cylinder Displacement cu. in. (liter)	Crankshaft Main Brg. Journal Dia.	Main Brg. Oil Clearance	Shaft End-play	Thrust on No.	Connecting Rod Journal Diameter	Oil Clearance	Side Clearance
1988	6	4-98 (1.6)	2.1653	0.0005–0.0018	0.0027–0.0100	3	1.6929	0.0014–0.0031	0.0027–0.0095
1989	6	4-98 (1.6)	2.1653	0.0006–0.0018	0.0027–0.0100	3	1.6929	0.0014–0.0031	0.0027–0.0095
	K	4-121 (2.0)	2.4945	0.0006–0.0019	0.0020–0.0070	3	1.9983	0.0010–0.0031	0.0039–0.0149
1990	6	4-98 (1.6)	2.1653	0.0006–0.0020	0.0047–0.0140	3	1.6929	0.0007–0.0025	0.0028–0.0095
	K	4-121 (2.0)	2.2828	0.0006–0.0016	0.0028–0.0118	3	1.9279	0.0007–0.0025	0.0028–0.0095
1991–92	6	4-98 (1.6)	2.1700	0.0006–0.0020	0.0047–0.0138	3	1.6918–1.6920	0.0017–0.0035	0.0028–0.0095

VALVE SPECIFICATIONS

Year	VIN	No. Cylinder Displacement cu. in. (liter)	Seat Angle (deg.)	Face Angle (deg.)	Spring Test Pressure (lbs.)	Spring Installed Height (in.)	Stem-to-Guide Clearance (in.) Intake	Exhaust	Stem Diameter (in.) Intake	Exhaust
1988	6	4-98 (1.6)	45	46	140 @ 0.85	1.26	0.0006–0.0017	0.0014–0.0025	0.275	0.275
1989	6	4-98 (1.6)	45	46	140 @ 0.85	1.26	0.0006–0.0017	0.0014–0.0025	0.275	0.275
	K	4-121 (2.0)	45	46	165–197 @ 1.043	1.48	0.0011–0.0026	0.0014–0.0030	0.275	0.276
1990	6	4-98 (1.6)	46	46	140 @ 0.85	1.24	0.0008–0.0020	0.0016–0.0028	0.276	0.275
	K	4-121 (2.0)	45	46	165–197 @ 1.043	1.24	0.0008–0.0020	0.0016–0.0028	0.276	0.275

VALVE SPECIFICATIONS

Year	VIN	No. Cylinder Displacement cu. in. (liter)	Seat Angle (deg.)	Face Angle (deg.)	Spring Test Pressure (lbs.)	Spring Installed Height (in.)	Stem-to-Guide Clearance (in.)		Stem Diameter (in.)	
							Intake	Exhaust	Intake	Exhaust
1991–92	6	4-98 (1.6)	46	46	140 @ 0.85	1.24	0.0008–0.0020	0.0016–0.0028	0.276	0.275

PISTON AND RING SPECIFICATIONS

All measurements are given in inches.

Year	VIN	No. Cylinder Displacement cu. in. (liter)	Piston Clearance	Ring Gap			Ring Side Clearance		
				Top Compression	Bottom Compression	Oil Control	Top Compression	Bottom Compression	Oil Control
1988	6	4-98 (1.6)	0.0008–0.0016	0.012–0.020	0.012–0.020	0.016–0.055	0.0012–0.0027	0.0012–0.0032	0.0000–0.0050
1989	6	4-98 (1.6)	0.0008	0.012–0.020	0.012–0.020	0.016–0.055	0.0012–0.0027	0.0012–0.0032	0.0000–0.0050
	K	4-121 (2.0)	0.0098–0.0022	0.010–0.020	0.010–0.020	0.010–0.050	0.0019–0.0027	0.0019–0.0027	0.0019–0.0032
1990	6	4-98 (1.6)	0.0008	0.012–0.020	0.012–0.020	0.016–0.055	0.0019–0.0027	0.0012–0.0032	0.0000–0.0050
	K	4-121 (2.0)	0.0004–0.0012	0.010–0.018	0.012–0.020	0.010–0.050	0.0019–0.0027	0.0019–0.0027	0.0019–0.0032
1991–92	6	4-98 (1.6)	0.0008	0.012–0.020	0.012–0.020	N/A	0.0024–0.0036	0.0019–0.0032	N/A

N/A—Not available

TORQUE SPECIFICATIONS

All readings in ft. lbs.

Year	VIN	No. Cylinder Displacement cu. in. (liter)	Cylinder Head Bolts	Main Bearing Bolts	Rod Bearing Bolts	Crankshaft Pulley Bolts	Flywheel Bolts	Manifold		Spark Plugs
								Intake	Exhaust	
1988	6	4-98 (1.6)	18①	36②	18③	40	25④	16	16	18
1989	6	4-98 (1.6)	18①	36②	18③	40	25④	16	16	18
	K	4-121 (2.0)	18①	70	38	20	⑤	16	16	18
1990	6	4-98 (1.6)	18①	36②	18③	41	25④	16	16	15
	K	4-121 (2.0)	18①	70	38	20	⑤	16	16	15
1991–92	6	4-98 (1.6)	18①	37②	18③	41	26④	16	16	15

① Cold—plus 2 turns of 60 degrees each and 1 turn of 30 degrees
Warm—plus 30–50 degree turn after warm up (thermostat open)
② Plus a 45–60 degree turn
③ Plus a 30 degree turn
④ Plus a 30–45 degree turn
⑤ Automatic—52 ft. lbs.
Manual —55 ft. lbs.

BRAKE SPECIFICATIONS

All measurements in inches unless noted.

Year	Model	Lug Nut Torque (ft. lbs.)	Master Cylinder Bore	Brake Disc Minimum Thickness	Brake Disc Maximum Runout	Standard Brake Drum Diameter	Minimum Lining Thickness Front	Minimum Lining Thickness Rear
1988	Lemans	65	0.813	0.460	0.004	7.900	0.28②	0.02①
1989	Lemans	66	0.813	0.460	0.004	7.900	0.28②	0.02①
1990	Lemans	66	0.874	0.420	0.004	7.900	0.28②	0.02①
1991–92	Lemans	65	0.874	0.420	0.004	7.900	0.28②	0.02①

① Above rivet head
② Shoe and lining together

WHEEL ALIGNMENT

Year	Model		Caster Range (deg.)	Caster Preferred Setting (deg.)	Camber Range (deg.)	Camber Preferred Setting (deg.)	Toe-in (in.)	Steering Axis Inclination (deg.)
1988	Lemans	Front	¾P–2¾P	NA	1¼N–¼P	NA	0	—
		Rear	—	—	1N	NA	⅓	—
1989	Lemans	Front	¾P–2¾P	NA	1¼N–¼P	NA	0	—
		Rear	—	—	1N	NA	⅓	—
1990	Lemans	Front	¾P–2¾P	NA	1¼N–¼P	NA	0	—
		Rear	—	—	1N	NA	⅓	—
1991–92	Lemans	Front	¾P–2¾P	NA	1¼N–¼P	NA	0	—
		Rear	—	—	1N	NA	⅓	—

NA Not adjustable
P Positive
N Negative

ENGINE MECHANICAL

NOTE: Disconnecting the negative battery cable on some vehicles may interfere with the functions of the on board computer systems and may require the computer to undergo a relearning process.

Engine Assembly

REMOVAL & INSTALLATION
1.6L Engine

1. Relieve the fuel system pressure.
2. Disconnect the terminals from the battery and chassis ground wire.
3. Position a clean drain pan under the radiator, remove the lower radiator hose and drain the cooling system. Remove the upper radiator hose and the heater hoses.
4. Remove the air cleaner. Detach the cable from the throttle body.
5. Remove the vacuum hoses from the power brake booster, vacuum sensor, intake manifold and TBI unit.
6. If equipped with automatic transaxle, disconnect the throttle valve to transaxle cable.
7. Remove the fuel lines from the throttle body. Disconnect and plug the transaxle cooler lines, if equipped.
8. Disconnect the electrical connectors from the distributor, oxygen sensor, oil pressure switch, intake manifold temperature sensor, speed sensor, injector nozzle, IAC, throttle valve. Remove the ground wires from the camshaft housing and the intake manifold, remove the the wiring harness retaining strap.
9. Disconnect the ignition coil plugs and cable, the instrument panel wiring harness multi-connector, the TCC connector, vehicles with automatic transaxle and the neutral safety switch.
10. Raise the vehicle and support it safely.
11. Remove the exhaust pipe to manifold bolts and disconnect the rear exhaust pipe from the catalytic converter. Remove the exhaust pipe/catalytic converter assembly from the vehicle.
12. Remove the closure plug and the halfshaft from the transaxle.
13. Remove the clutch cover plate and the clutch housing to lower engine block bolts.
14. Lower the vehicle.
15. Using an engine sling, attach it to

the engine hooks and support the engine weight.

16. Using the proper equipment, support the transaxle.

17. Remove the engine mounts and the clutch housing to upper engine bolts. Separate the engine from the clutch housing and lift the engine from the vehicle.

To install:

18. Lower the engine into the engine compartment and align it with the clutch housing.

19. Install the upper clutch housing bolts and the front engine mounts.

20. Remove transaxle support and engine sling. Torque the engine mount bolts to 29 ft. lbs. (39 Nm).

21. Raise the vehicle and support it safely.

22. Install the lower clutch cover bolts and the clutch cover plate.

23. Install the clutch driveshaft.

24. Install the interference suppression capacitor cable to the transmission.

25. Install the exhaust pipe with the catalytic converter.

26. Install the lower radiator hose.

27. Install the neutral safety switch.

28. If equipped with automatic transaxle, connect the TCC connector.

29. Install the instrument panel harness multiple connector.

30. Lower the vehicle.

31. Connect the fuel lines to the throttle body.

32. Reconnect all the wires and hoses. Refill the cooling system and the engine with the proper type and quantity of fluid, if drained.

33. Start the engine and check operation.

2.0L Engine

1. Relieve the fuel system pressure.

2. Disconnect the terminals from the battery and chassis ground wire.

3. Position a clean drain pan under the radiator, remove the lower radiator hose and drain the cooling system. Remove the upper radiator hose and the heater hoses.

4. Remove the air cleaner.

5. Remove or disconnect all wires and connectors at the following components:

    Engine harness bulk head
    Master cylinder
    Air conditioning relay cluster switches
    Wiper motor
    Cooling fan, relay and ground connection
    ECM
    Temperature switch at the thermostat housing.

6. Remove the vacuum hoses from the power brake booster, vacuum sensor, intake manifold to vapor canister

and throttle valve body to vapor canister.

7. Disconnect the throttle cable and shift cable.

8. Disconnect the hoses at the power steering cut off switch and return line to the pump.

9. Raise and support the vehicle safely.

10. Disconnect the exhaust pipe at the exhaust manifold and hangers.

11. Disconnect the heater hoses and fuel lines. If equipped with automatic transaxle, disconnect the cooler lines.

12. Remove the front wheels.

13. Remove the brake calipers and suspend by a wire or hanger.

14. Disconnect wire connections at the air conditioning compressor and discharge the system.

15. Remove the suspension supports by removing the 2 center bolts on each side, removing 1 bolt at each end and loosen the remaining bolt.

NOTE: Ensure that the vehicle, engine and transaxle weight are supported properly during the the following procedure.

16. If equipped with automatic transaxle, remove the rear transaxle lateral strut.

17. Remove the front transaxle strut.

18. Support the vehicle under the radiator support frame section.

19. Reposition the jack to the rear of the cowl with a 4 × 4 × 6 ft. board spanning the width of the vehicle.

20. Raise the vehicle enough to position a dolly under the transaxle with the 4 × 4 × 6 block as support.

21. Lower the vehicle weight onto the dolly.

22. Remove the remaining bolt at each end of the right and left front suspension supports.

23. Remove the transaxle mount and the rear engine mount.

24. Remove the strut to knuckle bolts.

NOTE: Carefully scribe the position of the strut on the hub to maintain camber adjustments for reassembly purposes.

25. Raise the vehicle leaving the engine, transaxle and suspension on the dolly.

26. Separate the engine from the transaxle.

To install:

27. Raise and support the vehicle safely.

28. Assemble the engine to the transaxle.

29. Place the engine and transaxle assembly on a dolly and position it in the chassis.

30. Install the transaxle mount, rear engine mount and front engine mounts and related bolts.

31. Loosely install each end of the right and left suspension supports.

32. Install the steering knuckle-to-strut bolts and position the strut in the previously scribed location.

33. Raise the vehicle and remove the dolly.

34. Install the remaining bolts in the right and left suspension supports. Install the front and rear transaxle struts.

35. Install the rear transaxle lateral strut.

36. Install the suspension supports and install the 2 center bolts on each side, install the 1 bolt at each end and tighten the remaining bolt.

37. Connect the wire connections at the air conditioning compressor.

38. Install the brake calipers and bleed the brake system as required.

39. Install the front wheel assemblies.

40. Connect the heater hoses and fuel lines. If equipped with automatic transaxle, connect the cooler lines to the transaxle.

41. Connect the exhaust pipe to the exhaust manifold with the bracket hangers.

42. Lower the vehicle.

43. Connect the power steering hoses to the pump.

44. Connect the throttle cable and shift cable. Adjust as required.

45. Install the vacuum hoses to the power brake booster, vacuum sensor, intake manifold to vapor canister and throttle valve body to vapor canister.

46. Install or connect all wires and connectors at the following components:

    Engine harness bulk head
    Master cylinder
    Air conditioning relay cluster switches
    Wiper motor
    Cooling fan, relay and ground connection
    ECM
    Temperature switch at the thermostat housing.

47. Install the air cleaner.

48. Install the radiator and heater hoses. Fill the cooling system with coolant.

49. Connect the fuel lines to the throttle body.

50. Connect the chassis ground wire. Reconnect all remaining wires.

51. Refill the engine with oil. Connect the battery cables.

52. Start the engine and check proper operation.

Engine Mounts

REMOVAL & INSTALLATION

1. Disconnect the negative battery cable.

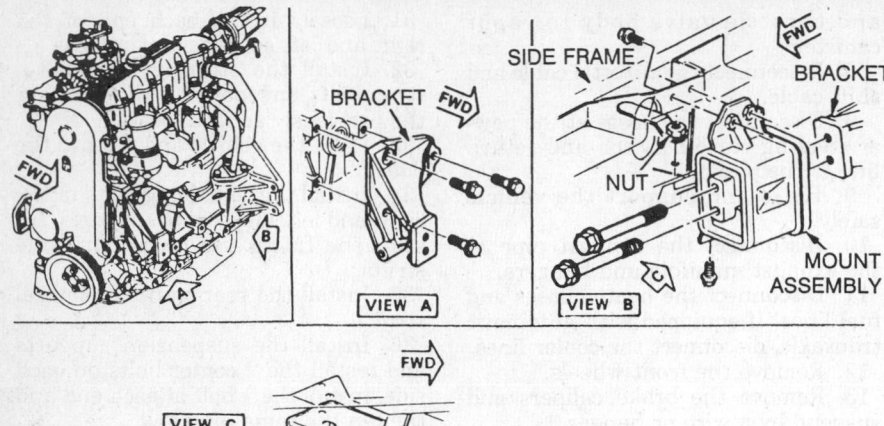

View A, View B, View C

Front engine mount—2.0L engine

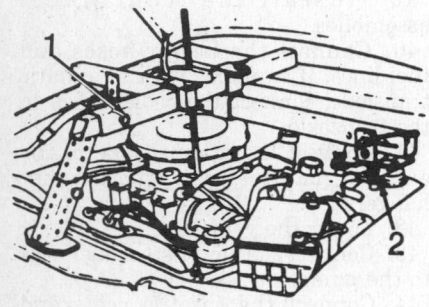

1. Engine support fixture
2. Adapters

Supporting the engine

2. Using an engine support fixture tool, center it on the cowl and attach it to the engine. Raise the engine slightly to take the weight off of the engine mounts.

3. Remove the engine mount bolts and the mount.

4. Inspect the engine mount for deterioration and replace it, if necessary.

5. To install, support the engine using a engine support fixture tool.

6. Install the engine mounts and the retaining bolts to the engine.

7. Torque the engine mount to bracket bolts to 29 ft. lbs. (39 Nm) and the engine mount to engine bolts to 29 ft. lbs. (39 Nm).

8. Connect the negative battery cable.

Cylinder Head

REMOVAL & INSTALLATION

NOTE: Cylinder head gasket replacement is necessary if cylinder head/camshaft carrier bolts are loosened. These bolts should only be loosened when the engine is cold. New cylinder head bolts must be used, because the head bolts are of the stretch bolt design.

1. Relieve the fuel system pressure.
2. Disconnect the negative battery cable.
3. Remove the lower radiator hose and drain the cooling system. Remove the upper radiator hose and the heater hoses.
4. Remove the air cleaner. Detach the throttle cable and remove it from the intake manifold. Remove the downshift cable.
5. Disconnect the electrical wiring connectors from the throttle body, the intake manifold and the oxygen sensor. Disconnect the engine wiring harness at the thermostat housing.
6. Disconnect the exhaust pipe from the exhaust manifold. Remove the alternator bracket and lay the alternator aside.
7. Remove the accessory drive belts, the front covers and the timing belt.
8. Remove the camshaft carrier and cylinder head bolts in sequence.
9. Remove the camshaft carrier, the rocker arms and the valve lash compensators.
10. Remove the cylinder head with the intake and exhaust manifolds attached.
11. Clean the cylinder head gasket mounting surfaces.

To install:

12. Apply a 3mm bead of anerobic sealant to the camshaft carrier sealing surface.
13. Assemble the intake and exhaust manifolds to the cylinder head. Use a new head gasket and install the assembly in place on the engine.
14. Install the valve lash compensators and the rocker arms.
15. Install the camshaft carrier on the cylinder head. Use new camshaft carrier and cylinder head bolts.

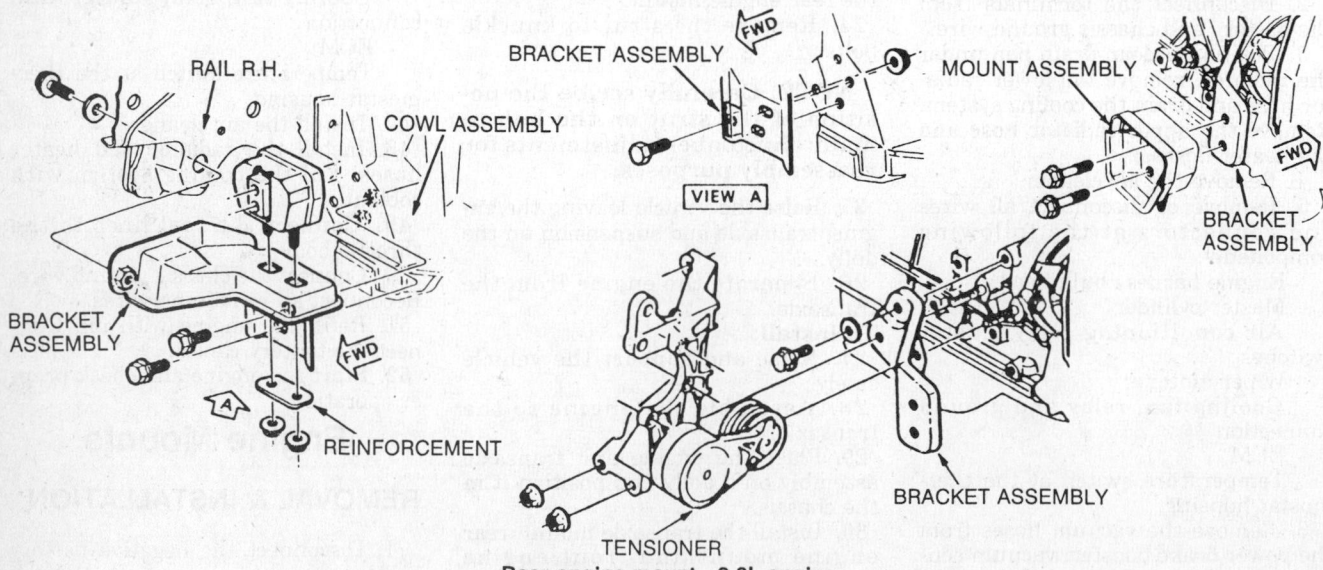

Rear engine mount—2.0L engine

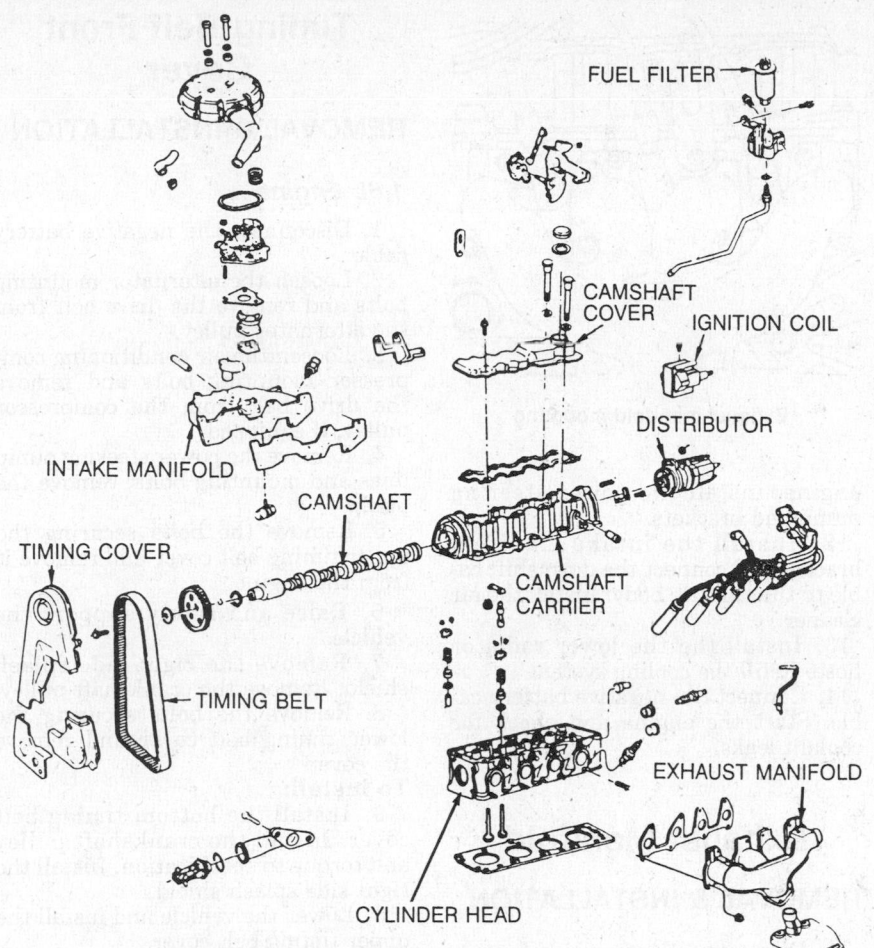

INTAKE MANIFOLD

FUEL FILTER

CAMSHAFT COVER

IGNITION COIL

DISTRIBUTOR

CAMSHAFT

TIMING COVER

CAMSHAFT CARRIER

TIMING BELT

EXHAUST MANIFOLD

CYLINDER HEAD

Exploded view of the upper engine assembly—1.6L engine

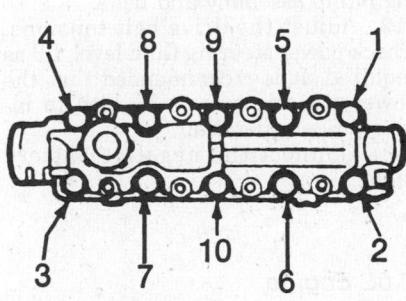

Removal sequence for camshaft carrier/cylinder head bolts

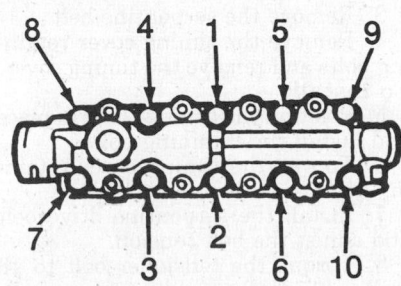

Camshaft carrier/cylinder head bolt torqueing sequence

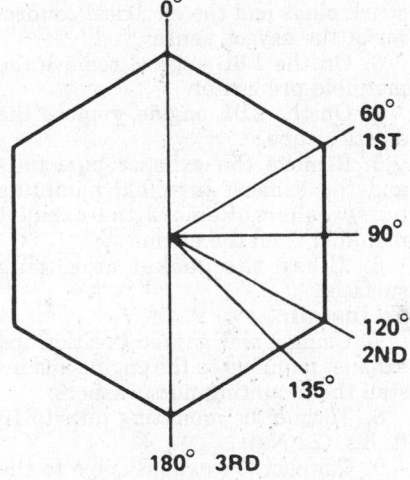

Camshaft carrier/cylinder head bolt torque degree chart

NOTE: New cylinder head bolts must be used, because the head bolts are of the stretch bolt design.

16. Torque the camshaft carrier and cylinder head bolts to 18 ft. lbs. (25 Nm) in sequence, turn an additional 60 degrees, turn another 60 degrees (to 120 degrees) and finally to a total of 150 degrees for the 1.6L engine or 180 degrees for the 2.0L engine.

17. Install the rear cover, timing belt, the front cover and the accessory drive belts.

18. Connect the exhaust pipe to the exhaust manifold. Install the alternator bracket and the thermostat housing.

19. Connect the electrical wiring connectors to the throttle body, the intake manifold and the oxygen sensor. Connect the engine wiring harness to the thermostat housing.

20. Install the air cleaner. Connect the throttle cable to the throttle body. Connect the downshift cable.

21. Install the lower radiator hose, the upper radiator hose and the heater hoses.

22. Refill the cooling system. Connect the negative battery cable.

23. Start the engine and allow it to reach normal operating temperature. After operating temperature is reached, torque the camshaft carrier and cylinder head bolts an additional 30–50 degrees. Check for coolant and oil leaks.

24. Check the ignition timing and adjust, if necessary.

Valve Lifters/Rocker Arms

REMOVAL & INSTALLATION

1. Disconnect the negative battery cable.

2. Remove the camshaft carrier cover bolts and the cover.

3. Remove the spark plugs.

4. Using a valve spring compressor tool, compress the valve springs.

5. Remove the rocker arms and the valve lifters; it is important that all of the valve train parts are kept in the order that they were removed.

6. Inspect and/or replace the worn parts.

7. Clean the gasket mounting surfaces of the camshaft carrier and cover.

To install:

8. Using a special valve spring compressing tool, compress the valve springs.

9. Coat the lifters with an engine oil prior to installation.

10. If using the same lifters, install each lifter in the same bore which it was removed from.

11. If using the same rocker arms, install each rocker arm in its original position.

12. Release the valve spring compressor tool and remove it from the carrier.

13. Install a new camshaft carrier cover gasket, install the carrier cover and retaining bolts. Torque the camshaft carrier cover bolts to 72 inch lbs. (8 Nm).

14. Install the spark plugs and wires. Connect the negative battery cable.

Valve Lash

ADJUSTMENT

The valve train uses hydraulic valve compensators, located in the cylinder head, which eliminate the need for valve lash adjustment.

Intake Manifold

REMOVAL & INSTALLATION

1. Relieve the fuel system pressure.
2. Disconnect the negative battery cable.
3. Position a clean drain pan under the radiator, remove the lower radiator hose and drain the cooling system.
4. Remove the air cleaner. Detach the cable from the throttle body, the intake manifold bracket and the downshift cable.
5. Loosen the alternator and swing it aside. On 2.0L engine, remove the power steering pump and bracket.
6. Disconnect the electrical wiring connectors from the throttle body, the intake manifold, the engine wiring harness and the thermostat housing.
7. Remove the intake manifold mounting nuts and washers, remove the manifold from the engine.
8. Clean the gasket mounting surfaces.

To install:

9. Install a new manifold gasket. Install the intake manifold on the engine and install the mounting nuts and washers. Torque the manifold mounting nuts to 16 ft. lbs. (22 Nm).
10. Connect the electrical wiring connectors to the throttle body, the intake manifold, the engine wiring harness and the thermostat housing.
11. Position the alternator in place and install the retaining bolts. On 2.0L

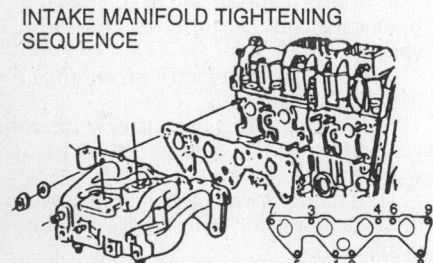

INTAKE MANIFOLD TIGHTENING SEQUENCE

Installation of the intake manifold— 2.0L engine

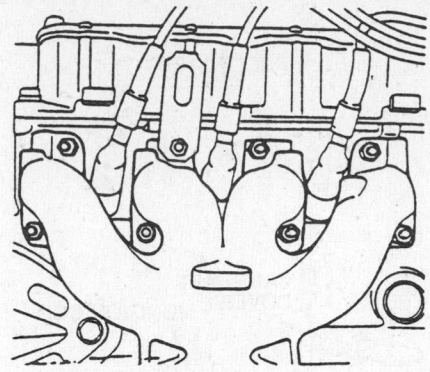

Exhaust manifold mounting

engine, install the power steering pump and brackets.

12. Install the intake manifold bracket and connect the downshift cable to the throttle body. Install the air cleaner.
13. Install the the lower radiator hose. Refill the cooling system.
14. Connect the negative battery cable. Start the engine and check for coolant leaks.

Exhaust Manifold

REMOVAL & INSTALLATION

1. Disconnect the negative battery cable.
2. Remove the air cleaner. Disconnect the spark plug wires from the spark plugs and the electrical connector at the oxygen sensor.
3. On the 1.6L engine, remove the manifold pre-heater.
4. On the 2.0L engine, remove the dipstick tube.
5. Remove the exhaust pipe nuts and the exhaust manifold mounting nuts/washers. Remove the exhaust manifold from the engine.
6. Clean the gasket mounting surfaces.

To install:

7. Using a new gasket, position the exhaust manifold to the engine and install the mounting nuts/washers.
8. Torque the mounting nuts to 16 ft. lbs. (22 Nm).
9. Connect the exhaust pipe to the exhaust manifold and install the retaining nuts. Torque the retaining nuts to 9.5 ft. lbs. (13 Nm).
10. Connect the spark plug wires to the spark plugs. Connect the electrical connector at the oxygen sensor.
11. Install the manifold pre-heater, dipstick tube and air cleaner.
12. Connect the negative battery cable. Start the engine and check for exhaust leaks.

Timing Belt Front Cover

REMOVAL & INSTALLATION

1.6L Engine

1. Disconnect the negative battery cable.
2. Loosen the alternator mounting bolts and remove the drive belt from the alternator pulley.
3. Loosen the air conditioning compressor mounting bolts and remove the drive belt from the compressor pulley, if equipped.
4. Remove the power steering pump lines and mounting bolts, remove the pump.
5. Remove the bolts securing the upper timing belt cover and remove it from the engine.
6. Raise and safely support the vehicle.
7. Remove the right side splash shield. Remove the crankshaft pulley.
8. Remove the bolts securing the lower timing belt cover and remove the cover.

To install:

9. Install the bottom timing belt cover. Install the crankshaft pulley and torque to specification. Install the right side splash shield.
10. Lower the vehicle and install the upper timing belt cover.
11. Install the alternator, air conditioning compressor and power steering pump assembly and belts.
12. Adjust the drive belt tensions. Check power steering fluid level, fill as required. It is recommended that the power steering system be bled to insure proper operation.
13. Connect the negative battery cable.

2.0L Engine

1. Disconnect the negative battery cable.
2. Loosen the serpentine belt tensioner bolt and allow the tensioner to swing downward.
3. Remove the serpentine belt.
4. Remove the timing cover retaining bolts and remove the timing cover.

To install:

5. Position the timing cover in place and install the retaining bolts.
6. Torque the timing cover retaining bolts to 90 inch lbs. (10 Nm).
7. Install the serpentine drive belt and adjust the belt tension.
8. Torque the tensioner bolt to 18 ft. lbs. (25 Nm).
9. Connect the negative battery cable.

Timing Belt Rear Cover

REMOVAL & INSTALLATION

1. Disconnect the negative battery cable.
2. Remove front timing belt cover.
3. Loosen timing belt tensioner and remove timing belt.
4. Remove camshaft sprocket.
5. Remove rear timing cover bolts and remove cover by slipping it around the water pump.
6. To install, slip cover around water pump and install the retaining bolts. Install the camshaft sprocket, tighten to 34 ft. lbs. (46 Nm). Slide the timing belt over sprocket and adjust tension to specification. Install front cover.
7. Connect the negative battery cable.

OIL SEAL REPLACEMENT

1. Disconnect the negative battery cable.
2. Remove the timing belt front cover.
3. Mark the relationship of the timing belt to the crankshaft and camshaft sprockets. Loosen the timing belt tensioner and remove the timing belt.
4. Remove the crankshaft pulley retaining bolt. Remove the crankshaft

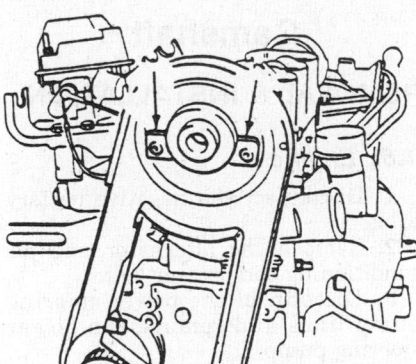

Timing belt cover attachment

Installation of the front cover seal

sprocket and the rear thrust washer. Remove the rear timing cover mounting screws and remove the cover.
5. Using a suitable tool, pry the front oil seal from the oil pump housing.
6. Using the protective sleeve of the seal installation tool, install the sleeve onto the crankshaft.

To install:

7. Using a new front oil seal, lubricate the seal lips with engine oil and install it onto the protective sleeve.
8. Using the seal installation tool, install the new oil seal into the oil pump until it seats.
9. Install the rear timing cover and mounting screws. Install the crankshaft sprocket, thrust washer retaining bolt. Torque the crankshaft sprocket bolt to 114 ft. lbs. (155 Nm).
10. Install the crankshaft pulley and retaining bolt. Torque the retaining bolt to 40 ft. lbs. (55 Nm).
11. Align the timing marks made prior to disassembly and install the timing belt. Adjust the belt tension.
12. Install the front timing cover and adjust all accessory drive belts to specification.
13. Connect the negative battery cable. Check ignition timing and engine for leaks.

Timing Belt and Tensioner

ADJUSTMENT

1.6L Engine

1. Disconnect the negative battery cable.
2. Remove the drive belts from the alternator and air conditioning compressor pulleys.
3. Remove the power steering pump lines and mounting bolts, remove the pump.
4. Remove the front timing covers from the engine.
5. Loosen the water pump retaining bolts and release the timing belt tension.
6. Ensure the camshaft and crankshaft timing marks are aligned properly.
7. Using a special timing belt tension adjustment tool, turn the water pump eccentric clockwise until the tensioner contacts the high torque stop. Tighten the water pump retaining bolts slightly.
8. Rotate the crankshaft 2 rotations counterclockwise to insure the belt is fully seated into the gear.
9. Rotate the water pump eccentric counterclockwise until the hole in the tensioner arm is aligned with the hole in the base.

11. Tighten the water pump retaining bolts to 18 ft. lbs. (25 Nm). Make sure the tensioner holes remain in alignment. If not readjust.
12. Position the front covers into place, bottom first. Install the power steering pump and lines.
13. Install the alternator and air conditioning compressor drive belts.
14. Install the power steering pump. Connect the power steering lines and

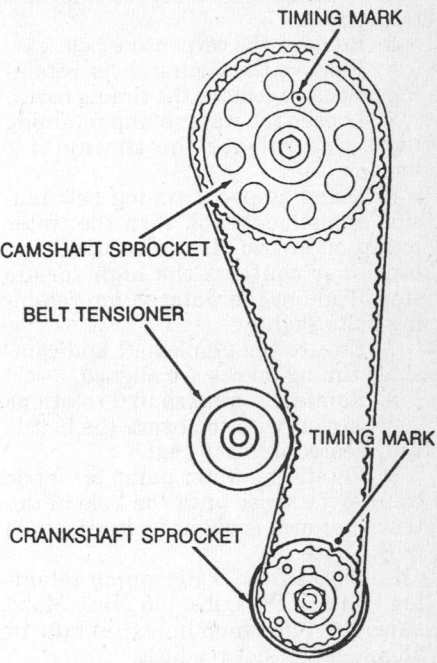

Timing mark positions—1.6L engine

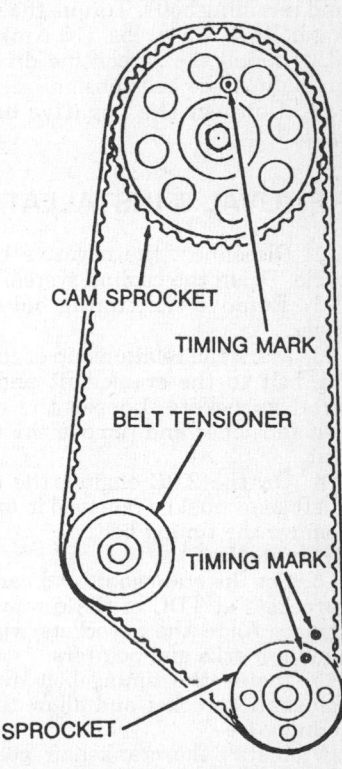

Timing mark positions—2.0L engine

fill the fluid level as required. It is recommended that the power steering system be bled to insure proper operation.

15. Connect the negative battery cable.

2.0L Engine

1. Disconnect the negative battery cable. Drain the cooling system.

2. Remove the power steering pump, bracket and serpentine belt tensioner.

3. Remove the serpentine belt.

4. Remove the timing cover retaining bolts and remove the timing cover.

5. Loosen the water pump retaining bolts and release the timing belt tension.

6. Using a special timing belt tension adjustment tool, turn the water pump eccentric clockwise until the tensioner contacts the high torque stop. Tighten the water pump retaining bolts slightly.

7. Ensure the crankshaft and camshaft timing marks are aligned.

8. Rotate the crankshaft 2 rotations counterclockwise to ensure the belt is fully seated into the gear.

9. Rotate the water pump eccentric counterclockwise until the hole in the tensioner arm is aligned with the hole in the base.

10. Tighten the water pump retaining bolts to 18 ft. lbs. (25 Nm). Make sure the tensioner holes remain in alignment. If not readjust.

11. Install the timing belt front cover and retaining bolts. Torque the retaining bolts to 7.5 ft. lbs. (10 Nm).

12. Install the serpentine drive belt and adjust the belt tension.

13. Connect the negative battery cable.

REMOVAL & INSTALLATION

1. Disconnect the negative battery cable. Drain the cooling system.

2. Remove the timing belt front cover.

3. Mark the relationship of the timing belt to the crankshaft and camshaft sprockets. Loosen the timing belt tensioner and remove the timing belt.

4. On the 2.0L engine, the crankshaft gear must be removed in order to remove the timing belt.

To install:

5. Set the crankshaft and camshaft sprockets at TDC of the compression stroke. Align the sprockets with the timing marks and pointers.

6. Install the timing belt over the camshaft sprocket and allow the belt to hang free.

7. Install the crankshaft gear and position the timing belt over it. Adjust

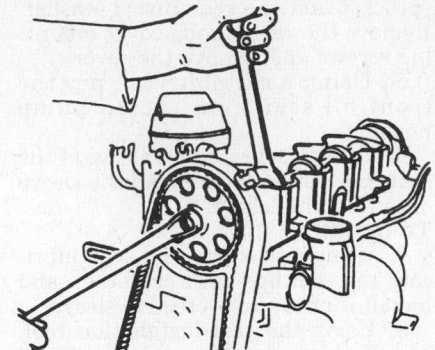

Removing the camshaft sprocket

the timing belt tension.

8. Install the front timing cover. Install the power steering pump, drive belts and hoses.

9. Connect the negative battery cable. Fill the cooling system.

10. Check and adjust the ignition timing as required.

Timing Sprockets

REMOVAL & INSTALLATION

Camshaft Sprocket

1. Disconnect the negative battery cable.

2. Remove the timing belt front cover.

3. Mark the relationship of the timing belt to the crankshaft and camshaft sprockets. Loosen the timing belt tensioner and remove the timing belt.

4. Remove the camshaft carrier cover to camshaft carrier bolts and remove the cover from the engine.

5. Hold the camshaft in place with an open end wrench and remove the sprocket bolt, washer and sprocket.

To install:

6. Install the camshaft sprocket, washer and retaining bolt. Hold the camshaft in place with an open end wrench and tighten the sprocket retaining bolt. Torque the camshaft sprocket bolt to 34 ft. lbs. (45 Nm).

7. Install the timing belt, aligning the marks made prior to removal.

8. Install the front timing cover.

9. Install the power steering pump, drive belts and hoses.

10. Connect the negative battery cable. Fill the cooling system with the proper type and quantity of antifreeze.

11. Install the camshaft carrier cover and retaining bolts. Torque the bolts to 72 inch lbs. (8 Nm). Check and adjust the ignition timing as needed.

12. Connect the negative battery cable.

Crankshaft Sprocket

1. Disconnect the negative battery cable.

2. Remove the timing belt front cover.

3. Mark the relationship of the timing belt to the crankshaft and camshaft sprockets. Loosen the timing belt tensioner and remove the timing belt.

4. Remove the crankshaft pulley bolt, pulley and thrust washer. Remove the crankshaft sprocket and the key.

NOTE: It is recommended to replace the front oil seal when the crankshaft pulley is removed.

To install:

5. Insert the key in the crankshaft and install the sprocket over it.

6. Install the crankshaft sprocket thrust washer and retaining bolt. Torque the retaining bolt to 33 ft. lbs. (45 Nm) on the 1.6L engine or 114 ft. lbs. (155 Nm) on the 2.0L engine.

7. Install the timing belt and adjust the belt tension.

8. Install the front timing cover. Install the crankshaft pulley and retaining bolts. Torque the pulley bolts to 41 ft. lbs. (55 Nm) for the 1.6L engine or 17 ft. lbs. (13.5 Nm) for the 2.0L engine.

9. Install the power steering pump, drive belts and hoses.

10. Connect the negative battery cable. Fill the cooling system and check ignition timing.

Camshaft

REMOVAL & INSTALLATION

1.6L Engine

1. Disconnect the negative battery cable.

2. Remove the alternator and air conditioning compressor belt.

3. Disconnect the power steering pump lines and remove the power steering pump.

4. Remove the front cover, upper half first and remove it from the engine.

5. Remove the camshaft carrier cover bolts and the cover.

6. Remove the spark plugs.

NOTE: When working on a particular cylinder, be sure to rotate the crankshaft until the piston of that cylinder is located on the TDC of its compression stroke.

7. Using a air line adapter tool, install it into the spark plug hole of the cylinder being serviced and apply compressed air.

CAUTION

Keep hands clear of the belts and pulleys when applying air pressure to the cylinder. When air pressure is applied to the cylinder, it may cause the crankshaft to move, causing belts and pulleys to rotate unexpectedly.

8. Using a valve spring compressor tool, compress the valve springs.

9. Remove the rocker arms and the valve lifters; it is important that all of the valve train parts are kept in the order that they were removed.

10. Hold the camshaft firmly with the proper tool and loosen the camshaft retaining bolt.

11. Remove the washer and camshaft sprocket.

12. Mark the distributor position in relationship to the engine and remove the distributor.

13. Remove the camshaft thrust plate from the rear carrier.

14. Slide the camshaft out from the rear.

15. Clean the gasket mounting surfaces of the carrier and the cover. Inspect and replace any worn or damaged parts.

To install:

16. Using a seal installer tool, install a new camshaft carrier front seal.

17. Coat camshaft bearings and camshaft with an approved camshaft lubricant. Carefully slide the camshaft into the carrier from the rear. Be careful not to damage the front carrier seal.

18. Install the camshaft thrust plate at the rear of the carrier. Torque the trust plate retaining bolts to 71 inch lbs. (8 Nm).

19. Using a feeler gauge, check the camshaft endplay; it should be 0.016–0.064 in. (0.04–0.16mm), if not, replace the rear thrust washer.

20. Install the camshaft sprocket and retaining bolt. Hold the camshaft firmly with the proper tool and tighten the camshaft retaining bolt. Torque the bolt to 34 ft. lbs. (46 Nm).

21. Align the marks on the crankshaft gear and camshaft sprocket with the timing marks on the rear timing cover.

22. Install the timing belt and cover.

23. Install the distributor in the engine, aligning the marks previously made during disassembly.

24. Using a air line adapter tool, install it into the spark plug hole of the cylinder being serviced and apply compressed air.

NOTE: When working on a particular cylinder, be sure to rotate the crankshaft until the piston of that cylinder is located on the TDC of its compression stroke.

25. Using a valve spring compressor tool, compress the valve springs.

26. Install the rocker arms and the valve lifters; it is important that all of the valve train parts installed in the same order that they were removed.

27. Disconnect the air line and adapter from the engine. Remove the valve spring compressor tool.

28. Install the spark plugs.

29. Using a new gasket install the camshaft carrier cover and retaining bolts. Torque the bolts to 72 inch lbs. (9 Nm).

30. install the front cover in place on the engine, lower half first and then the upper cover.

31. Install the power steering pump and connect the power steering pump lines.

32. Install the alternator and air conditioning compressor belts. Check and adjust the ignition timing as needed.

33. Connect the negative battery cable.

2.0L Engine

1. Disconnect the negative battery cable.

2. Remove the air cleaner and breather hoses.

3. Remove the camshaft carrier cover bolts and the cover.

4. Remove the serpentine belt.

5. Remove the alternator and bracket from the carrier.

6. Using a valve spring compressor tool, compress the valve springs.

7. Remove the rocker arms and the valve lifters; it is important that all of the valve train parts are kept in the order that they were removed.

8. Hold the camshaft firmly with the proper tool and loosen the camshaft retaining bolt.

9. Using a sprocket remover tool, remove the washer and camshaft sprocket.

10. Mark the distributor position in relationship to the engine and remove it.

11. Remove the camshaft thrust plate from the rear carrier.

12. Slide the camshaft out from the rear.

13. Clean the gasket mounting surfaces of the carrier and the cover. Inspect and replace any worn or damaged parts.

To install:

14. Oil the camshaft carrier seal with oil prior to installing the camshaft.

15. Coat camshaft bearings and camshaft with an approved camshaft lubricant. Carefully slide the camshaft into the carrier from the rear. Be careful not to damage the front carrier seal.

16. Install the camshaft thrust plate at the rear of the carrier. Torque the

trust plate retaining bolts to 70 inch lbs. (8 Nm).

17. Using a feeler gauge, verify the camshaft endplay is 0.016–0.064 in. (0.04–0.16mm); if not as specified, replace the rear thrust washer.

18. Install the camshaft sprocket and retaining bolt. Hold the camshaft firmly with the proper tool and tighten the camshaft retaining bolt. Torque the bolt to 34 ft. lbs. (46 Nm).

19. Align the marks on the crankshaft gear and camshaft sprocket with the timing marks on the rear timing cover.

20. Install the timing belt, front cover and retaining bolts. Torque the front cover bolts to 7.5 ft. lbs. (10 Nm).

21. Install the distributor in the engine, aligning the marks previously made during disassembly.

22. Using a valve spring compressor tool, compress the valve springs.

23. Install the rocker arms and the valve lifters; it is important that all of the valve train parts installed in the same order that they were removed.

24. Using a new gasket install the camshaft carrier cover and retaining bolts. Torque the bolts to 6 ft. lbs. (9 Nm).

25. Install the alternator and bracket to the carrier.

26. Install the power steering pump and serpentine belt tensioner as an assembly. Install the serpentine belt.

27. Install the alternator and air conditioning compressor belts.

28. Install the air cleaner and breather hoses.

29. Connect the negative battery cable. Check and adjust the ignition timing as needed.

Piston and Connecting Rod

POSITIONING

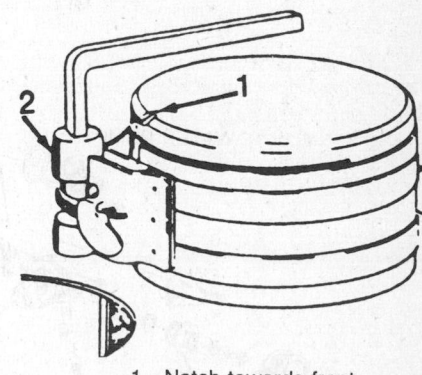

1. Notch towards front of engine
2. Tool

Installing the piston assembly into the engine

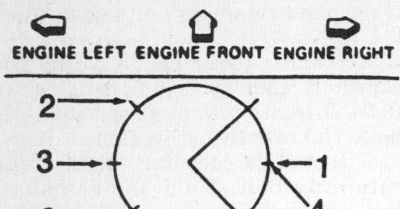

1. Oil ring spacer gap (tang in hole or slot with arc)
2. Oil ring rail gaps
3. 2nd compression ring
4. Top compression ring gap

Location of piston ring gaps

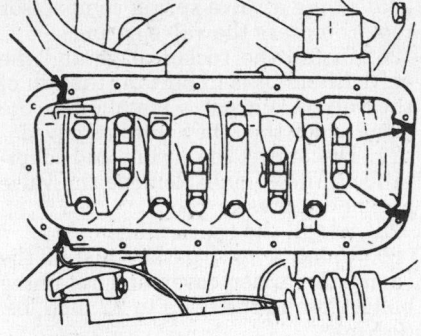

Installing sealer at oil pan rail seams

ENGINE LUBRICATION

Oil Pan

REMOVAL & INSTALLATION

1. Disconnect the negative battery cable.
2. Raise and safely support the vehicle.
3. Place a drain pan under the engine, remove the drain plug from the oil pan and drain the oil from the crankcase.
4. Remove the exhaust pipe from the exhaust manifold. Remove the clutch cover, if equipped with a manual transmission.

5. Remove the oil pan to engine bolts and remove the oil pan from the engine.
6. Clean the gasket mounting surfaces.
7. To install, use a new gasket, RTV sealant and reverse the removal procedures. Be sure to coat the oil pan rail seams with sealant.
8. Torque the oil pan bolts to 4 ft. lbs. (5 Nm) and the oil pan plug to 34 ft. lbs. (46 Nm).
9. Install the exhaust pipe to manifold and tighten the nuts to 19 ft. lbs. (26 Nm).
10. Refill the crankcase with new oil.

Oil Pump

REMOVAL & INSTALLATION

1. Disconnect the negative battery cable.
2. Remove the accessory drive belts on the 1.6L engine or the serpentine belt on the 2.0L engine. Remove the crankshaft pulley assembly.

3. Remove the front timing belt cover and the timing belt.
4. Remove the rear timing belt cover bolts and the cover from the engine.
5. Disconnect the electrical connector from the oil pressure switch.
6. Raise and support the vehicle safely.
7. Drain the engine oil.
8. Remove the oil pan bolts and the oil pan from the engine.
9. Remove the oil filter and the pick up tube. Remove the oil pump bolts and the oil pump from the engine.
To install:

NOTE: When the oil pump has been removed from the engine, it is recommended to replace the front oil seal.

10. Clean the gasket mounting surfaces and the oil pump, if it is going to be re-used.
11. To install, use new oil pan and pump gaskets. Use an approved sealant on the gasket surfaces. Torque the oil pump to engine bolts to 62 inch lbs. (7 Nm) and the oil pan bolts to 72 inch lbs. (8 Nm).
12. Install the timing belt cover, the crankshaft sprocket and the timing belt. Install and adjust the accessory drive belts. Use a new oil filter and fill the crankcase with clean engine oil.
13. Run the engine and check for proper oil pressure, leaks and check ignition timing. Connect the negative battery cable.

CHECKING

1. If foreign matter is present, determine it's source.
2. Check the pump cover and hous-

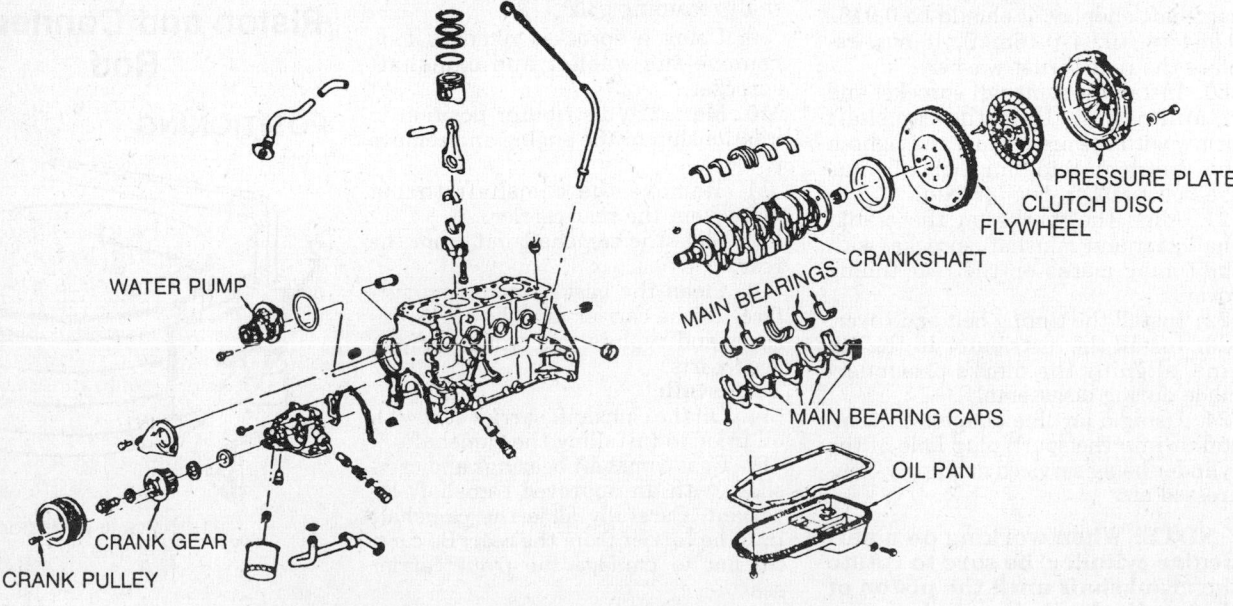

Exploded view of the lower engine assembly—1.6L engine

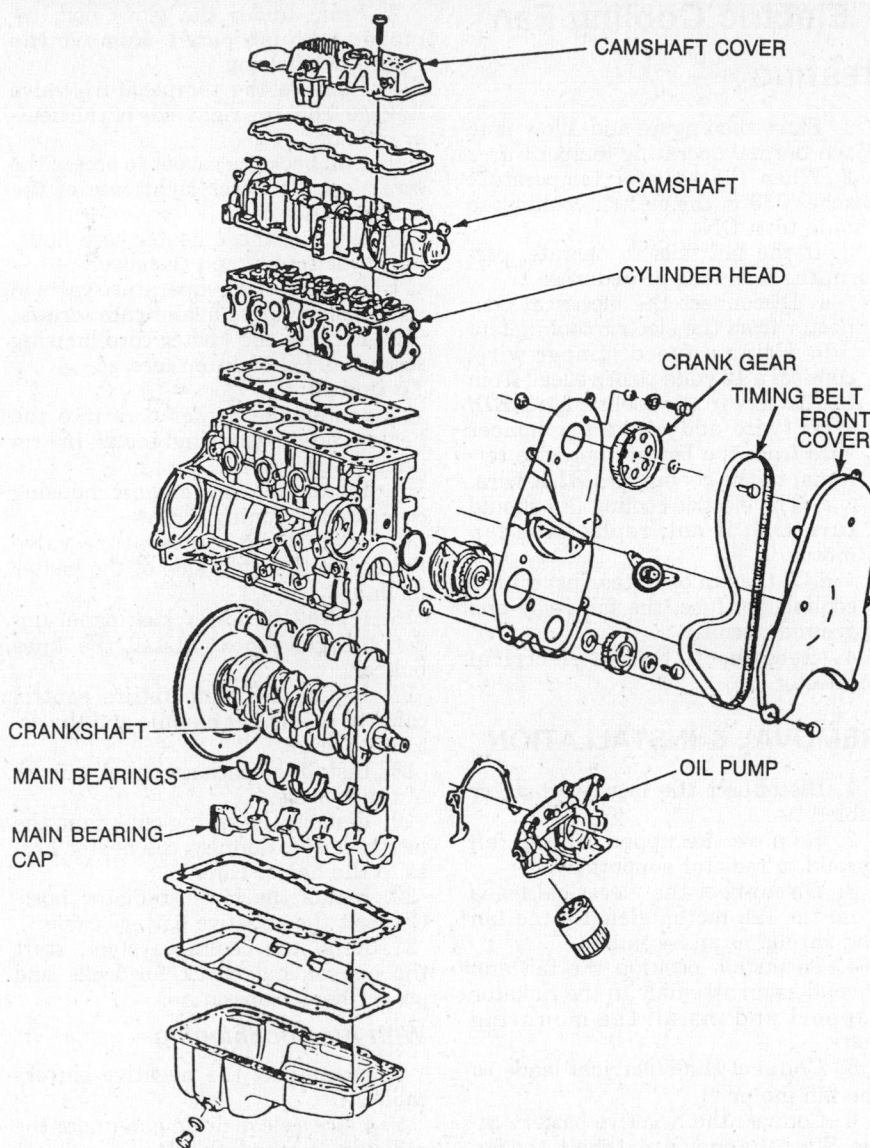

CAMSHAFT COVER

CAMSHAFT

CYLINDER HEAD

CRANK GEAR

TIMING BELT
FRONT
COVER

CRANKSHAFT

MAIN BEARINGS

MAIN BEARING
CAP

OIL PUMP

Exploded view of the engine assembly—2.0L engine

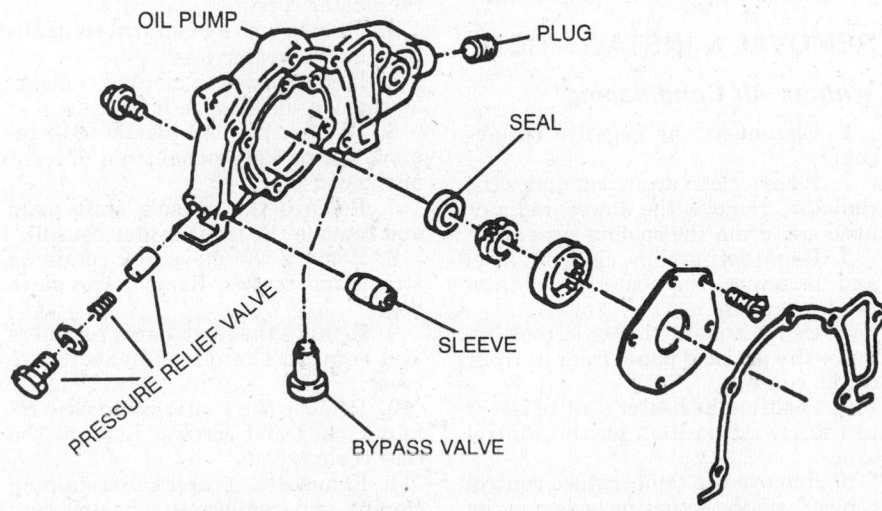

OIL PUMP

PLUG

SEAL

SLEEVE

PRESSURE RELIEF VALVE

BYPASS VALVE

Exploded view of the oil pump—1.6L and 2.0L engines

ing for cracks, scoring and/or damage; if necessary, replace the housings.

3. Inspect the idler gear shaft for looseness in the housing; if necessary, replace the pump.

4. Inspect the pressure regulator valve for scoring or sticking; if burrs are present, remove them with an oil stone.

5. Inspect the pressure regulator valve spring for loss of tension or distortion; if necessary, replace it.

6. Inspect the suction pipe for looseness, if pressed into the housing, check the screen for broken wire mesh; if necessary, replace them.

7. Inspect the gears for chipping, galling and/or wear; if necessary, replace them.

Rear Main Bearing Oil Seal

REMOVAL & INSTALLATION

The rear seal used in the 1.6L or 2.0L engine is of the 1 piece lip seal design.

1. Disconnect the negative battery cable.

2. Raise and support the vehicle safely. Remove the transaxle.

3. If equipped with an automatic transaxle, remove flexplate retaining bolts and remove the flexplate from the engine.

4. If equipped with a manual transaxle, remove the pressure plate and clutch disc assembly. Remove the flywheel bolts and remove the flywheel from the engine.

5. Using a small prybar, pry the rear main oil seal from its retainer. Clean all gasket mating surfaces thoroughly.
To install:

6. Lubricate the new oil seal lips with engine oil. Using the seal installation tool, drive the new rear oil seal into the block until seated.

NOTE: When installing the flywheel, new retaining bolts must be used.

7. If equipped with a manual transaxle, install the flywheel and retaining bolts. Torque the flywheel bolts to 25 ft. lbs. (35 Nm) plus an additional 30–45 degree turn for 1.6L engine or 55 ft. lbs. (75 Nm) for 2.0L engine. Install the clutch disc/pressure plate assembly.

8. If equipped with automatic transaxle torque the flexplate bolts to 25 ft. lbs. (35 Nm) for 1.6L engine or 52 ft. lbs. (70 Nm) for 2.0L engine.

9. Install the transaxle.

10. Connect the negative battery cable. Fill the engine with oil and check for leaks.

ENGINE COOLING

Radiator

REMOVAL & INSTALLATION

— CAUTION —

Before attempting any work on the cooling system, allow the engine to first cool sufficiently. To avoid personal injury, do not remove the radiator cap while the engine is at normal operation temperature. The cooling system is under pressure and will release scalding fluid and steam if the cap is removed.

1. Disconnect the negative battery cable.
2. Place a drain pan under the radiator. Remove the lower radiator hose and drain the cooling system.
3. Remove the coolant recovery hose and the upper and lower radiator hose.
4. Disconnect the electrical connections from the electric fan motor, the oxygen sensor and the temperature sensor.
5. If equipped with automatic transaxle, disconnect the transaxle cooler lines and plug.
6. Remove the radiator support mounting bolts. Carefully lift the radiator assembly from the vehicle.
7. Remove the upper fan mounting bolts and remove the fan and shroud as an assembly.

To install:

8. Install the fan and shroud as an assembly and install the upper fan mounting bolts.
9. Carefully lower the radiator into the engine compartment. Install the support mounting bolts.
10. If equipped with automatic transaxle, connect the transaxle cooler lines.
11. Connect the electrical connections to the electric fan motor, the oxygen sensor and the temperature sensor.
12. Install the coolant recovery hose and the upper radiator hose.
13. Connect the lower radiator hoses. Refill the cooling system with an approved coolant mixture.
14. Connect the negative battery cable.
15. Start the engine, allow it to reach operating temperature and check for leaks.

Electric Cooling Fan

TESTING

1. Start the engine and allow it to reach normal operating temperature.
2. When the radiator temperature reaches 230°F, the electric cooling fan should turn **ON**.
3. If the fan fails to operate, perform the following procedures:
   a. Disconnect the electrical connector from the electric cooling fan.
   b. Using a fused jumper wire, connect a 12 volts positive lead from the battery to the cooling fan **RED/WHT** wire and connect a jumper wire from the battery negative terminal to the cooling fan **BLK** wire.
   c. The electric cooling fan should turn **ON**; if not, replace the fan motor.
   d. If the fan operates, inspect the cooling fan fuse, the fan relay and ground circuit.
4. Reconnect the fan electrical connector.

REMOVAL & INSTALLATION

1. Disconnect the negative battery cable.
2. Remove the upper cooling fan shroud to radiator support bolts.
3. Disconnect the electrical leads from the fan motor. Remove the fan and shroud as an assembly.
4. To install, position the fan and shroud as an assembly to the radiator support and install the mounting bolts.
5. Connect the electrical leads to the fan motor.
6. Connect the negative battery cable. Start the engine and check the fan operation.

Heater Core

REMOVAL & INSTALLATION

Without Air Conditioning

1. Disconnect the negative battery cable.
2. Place a clean drain pan under the radiator, remove the lower radiator hose and drain the cooling system.
3. Using spring clips, close off, label and disconnect the heater hoses from the heater core.
4. Using a pointed plastic tool, remove the package panel from in front of the console.
5. Position the heater control levers to the lowest position on the control unit.
6. Remove the temperature control cables from the actuating lever and the heater module.

7. From under the glove box, remove the hush panel. Remove the right knee bolster.
8. Remove the temperature valve linkage from the right side of the heater module.
9. Pull back the carpet to access the screw on the lower right side of the heater outlet case.
10. Remove the 4 heater case housing cover screws and the cover.
11. Position the temperature valve to access the 2 upper heater core screws.
12. Remove the heater core housing screws and the heater core.

To install:

13. Install the heater core into the heater core housing and install the retaining screws.
14. Install the heater core housing cover and retaining screws.
15. Install the temperature valve linkage to the right side of the heater module.
16. Install the lower kick panel under the glove box. Install the knee bolster.
17. Install the temperature control cable to the heater module and the actuating lever.
18. Install the package panel in front of the console.
19. Remove the spring clips from the heater hoses. Connect the heater hoses to the heater core.
20. Install the lower radiator hose. Connect the negative battery cable.
21. Refill the cooling system, start the engine and check for leaks and proper heater operation.

With Air Conditioning

1. Disconnect the negative battery cable.
2. Place a clean drain pan under the radiator, remove the lower radiator hose and drain the cooling system.
3. Using spring clips, close off, label and disconnect the heater hoses from the heater core.
4. Remove the evaporator drain hose at the heater case.
5. If equipped with manual transaxle, remove the gear shift boot.
6. Using a pointed plastic tool, remove the package panel from in front of the console.
7. Remove the console shift plate and remove the front center console.
8. Remove the glove box retaining straps and screws. Remove the glove box.
9. Remove the hush panel retainers and remove the outer heater case cover.
10. Remove the heater cover case retainer clips and screws. Remove the case cover.
11. Remove the heater core retaining clamps and remove the heater core from the case.

To install:

12. Position the heater core into the case housing and secure in place with retaining clamps.

13. Install the heater case cover and secure in place with the retaining clips and screws. Install the hush panel and retainers.

14. Install the glove box, retaining straps and screws.

15. Install the front center console and install the console shift plate.

16. Install the package panel in front of the console.

17. If equipped with manual transaxle, install the gear shift boot.

18. Install the evaporator drain hose to the heater case.

19. Remove the spring clips and connect the heater hoses to the heater core.

20. Install the lower radiator hose. Connect the negative battery cable.

21. Refill the cooling system, start the engine, allow it to reach normal operating temperature and check for leaks and proper heater operation.

Water Pump

REMOVAL & INSTALLATION

1. Disconnect the negative battery cable. Drain the cooling system.

2. Remove the front cover timing belt covers, the timing belt and the rear cover.

3. Remove the water pump mounting bolts and the water pump from the engine.

4. Clean the seal mounting surfaces.

To install:

5. Use a new water pump seal. Coat the sealing surface and the seal ring with grease.

6. Torque the water pump bolts to 71 inch lbs. (8 Nm). Refill the cooling system.

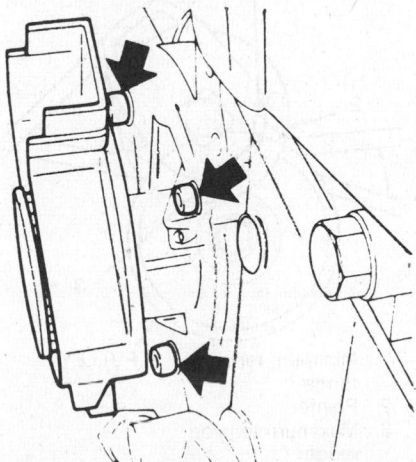

Water pump attaching bolt location

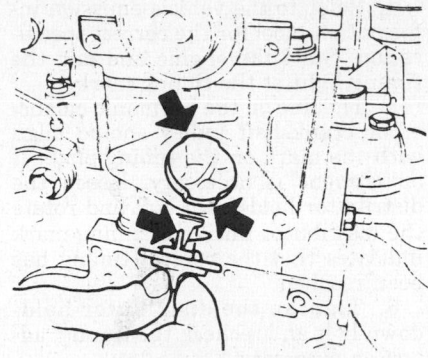

Thermostat positioning

7. Start the engine, allow it to reach normal operating temperature and check for leaks, bleed cooling system.

8. Check and adjust the ignition timing if needed.

Thermostat

REMOVAL & INSTALLATION

1.6L Engine

1. Disconnect the negative battery cable.

2. Using a clean drain pan, place it under the radiator, remove the lower radiator hose and drain the cooling system.

3. Remove the front cover, the timing belt and the timing belt rear cover; slip the rear cover over the water pump rear cover piece.

4. Remove the water inlet bolts, the water inlet housing and the thermostat.

5. Clean the gasket mounting surfaces.

6. To install, use a new seal. Torque the thermostat housing bolts to 90 inch lbs. (10 Nm). Refill the cooling system and bleed the air from the system. Connect the negative battery cable.

2.0L Engine

1. Disconnect the negative battery cable.

2. Using a clean drain pan, place it under the radiator, remove the lower radiator hose and drain the cooling system.

3. Remove the upper radiator hose.

4. Remove the thermostat housing bolts and remove the thermostat.

To install:

5. Clean the sealing surfaces of the cylinder head and thermostat housing.

6. Install the thermostat to the cylinder head.

7. Install the thermostat housing using a new O-ring seal. Torque the housing bolts to 90 inch lbs. (10 Nm).

8. Install the upper and lower radiator hoses.

9. Refill the cooling system and connect the battery cable. Start the engine and check thermostat operation and for leaks.

Cooling System Bleeding

After working on the cooling system, even when replacing the thermostat, the system must bled. Air trapped in the system will prevent proper filling and leave the radiator coolant level low, causing a risk of overheating.

1. To bleed the system, start with the system cool, the radiator cap OFF and fill the radiator with the upper hose disconnected and held in position until coolant reaches upper hose position.

2. Reconnect the upper hose and start the engine. Run the engine at slightly above normal idle speed. This will insure adequate circulation. If air bubbles appear and the coolant level drops, fill the system with a mixture of anti-freeze and water to bring the level back to the proper level.

3. Run the engine this way until the thermostat opens. When the thermostat opens, the coolant will move abruptly across the top of the radiator and the temperature of the radiator will increase.

4. At this point, air is often expelled and the coolant level may drop. Keep refilling the system until the level is near the top of the radiator and remains constant.

5. If the vehicle has an overflow tank, fill the radiator to the top of the filler neck.

ENGINE ELECTRICAL

NOTE: Disconnecting the negative battery cable on some vehicles may interfere with the functions of the on board computer systems and may require the computer to undergo a relearning process.

Distributor

REMOVAL

1. Disconnect the negative battery cable.

2. Remove the distributor cap.

3. Mark and remove all electrical leads connected to the distributor assembly.

4. Mark the relationship of the ro-

tor to the distributor housing and the distributor housing to the engine.

5. Remove the hold-down bolt, clamp and distributor.

INSTALLATION

Timing Not Disturbed

1. Turn the distributor shaft until the rotor points to the No. 1 spark plug tower on the cap.

2. Align the marks on the distributor housing and the engine.

3. Install the distributor into the engine.

4. Tighten the distributor hold-down bolt and reconnect the electrical connections.

5. Check and/or adjust the ignition timing.

Timing Disturbed

If the engine was cranked with the distributor removed, it will be necessary to position the No. 1 cylinder at TDC on the compression stroke. Follow the procedure listed here. This will enable the proper setting of the ignition timing.

1. Remove the No. 1 spark plug.

2. Place a finger over the spark plug hole. Crank the engine slowly until compression is felt.

3. Align the timing mark on the crankshaft pulley with the **0** degree mark on the timing scale attached to the front of the engine. This places the No. 1 cylinder at the TDC of the compression stroke.

4. Turn the distributor shaft until the rotor points to the No. 1 spark plug tower on the cap.

5. Install the distributor into the engine.

6. Tighten the distributor hold-down bolt and reconnect the electrical connections. Check the timing and adjust, as necessary.

Ignition Timing

ADJUSTMENT

1.6L Engine

1. Make sure the ignition switch is turned **OFF** when connecting electrical equipment to the engine.

2. Using an induction type timing light, connect the pickup lead of the light to the No. 1 spark plug wire.

3. Connect the timing light positive and negative leads to the battery positive and negative terminals.

NOTE: When connecting a non-inductive timing light to the No. 1 spark plug wire, be sure to use a jumper wire between the spark plug and boot; do not pierce or cut the high tension wire.

4. Refer to the vehicle emission information label for the correct specification. Start the engine and aim the timing light at the timing mark.

5. The line on the harmonic balancer or crankshaft pulley should align with the mark on the timing plate. If adjustment is necessary, loosen the distributor hold-down bolt and rotate the distributor until the timing mark indicates that the correct timing has been reached.

6. Tighten the distributor hold-down bolt and recheck the timing, adjust as necessary.

2.0L Engine

Timing procedure and specifications are provided on the vehicle emission information label located in the engine compartment.

Alternator

PRECAUTIONS

Several precautions must be observed with alternator equipped vehicles to avoid damage to the unit.

• If the battery is removed for any reason, make sure it is reconnected with the correct polarity. Reversing the battery connections may result in damage to the one-way rectifiers.

• When utilizing a booster battery as a starting aid, always connect the positive to positive terminals and the negative terminal from the booster battery to a good engine ground on the vehicle being started.

• Never use a fast charger as a booster to start vehicles.

• Disconnect the battery cables when charging the battery with a fast charger.

• Never attempt to polarize the alternator.

• Do not use test lights of more than 12 volts when checking diode continuity.

• Do not short across or ground any of the alternator terminals.

• The polarity of the battery, alternator and regulator must be matched and considered before making any electrical connections within the system.

• Never separate the alternator on an open circuit. Make sure all connections within the circuit are clean and tight.

• Disconnect the battery ground terminal when performing any service on electrical components.

• Disconnect the battery if arc welding is to be done on the vehicle.

BELT TENSION ADJUSTMENT

NOTE: The following proce-

dure requires the use of a belt tension gauge.

1. If the belt is cold, operate the engine at idle speed, until it reaches normal operating temperature; the belt will seat itself in the pulleys allowing the belt fibers to relax or stretch. If the belt is hot, allow it to cool, until it is warm to the touch.

NOTE: A used belt is one that has been rotated at least 1 complete revolution on the pulleys. This begins the belt seating process and it must never be tightened to the new belt specifications.

2. Loosen the alternator mounting bolts.

3. Using a belt tension gauge tool, place the tension gauge at the center of the belt between the pulleys on its longest section.

4. While applying pressure on the component, adjust the drive belt tension to the following:

New belt – 155 lbs.
Used belt – 80 lbs.

5. While holding tension on the component, tighten the component mounting bolts and remove the tension gauge.

REMOVAL & INSTALLATION

1. Disconnect the negative battery cable.

2. Disconnect and label the electrical terminal plug and the battery lead from the rear of the alternator.

3. Loosen the mounting bolts. Push the alternator inwards and slip the drive belt off the pulley.

4. Remove the mounting bolts and the alternator.

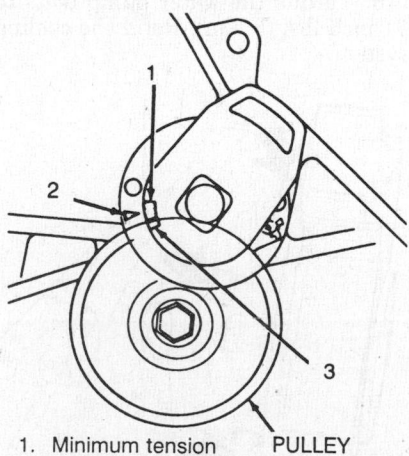

1. Minimum tension range
2. Pointer
3. Maximum tension range

Adjusting the serpentine belt tension

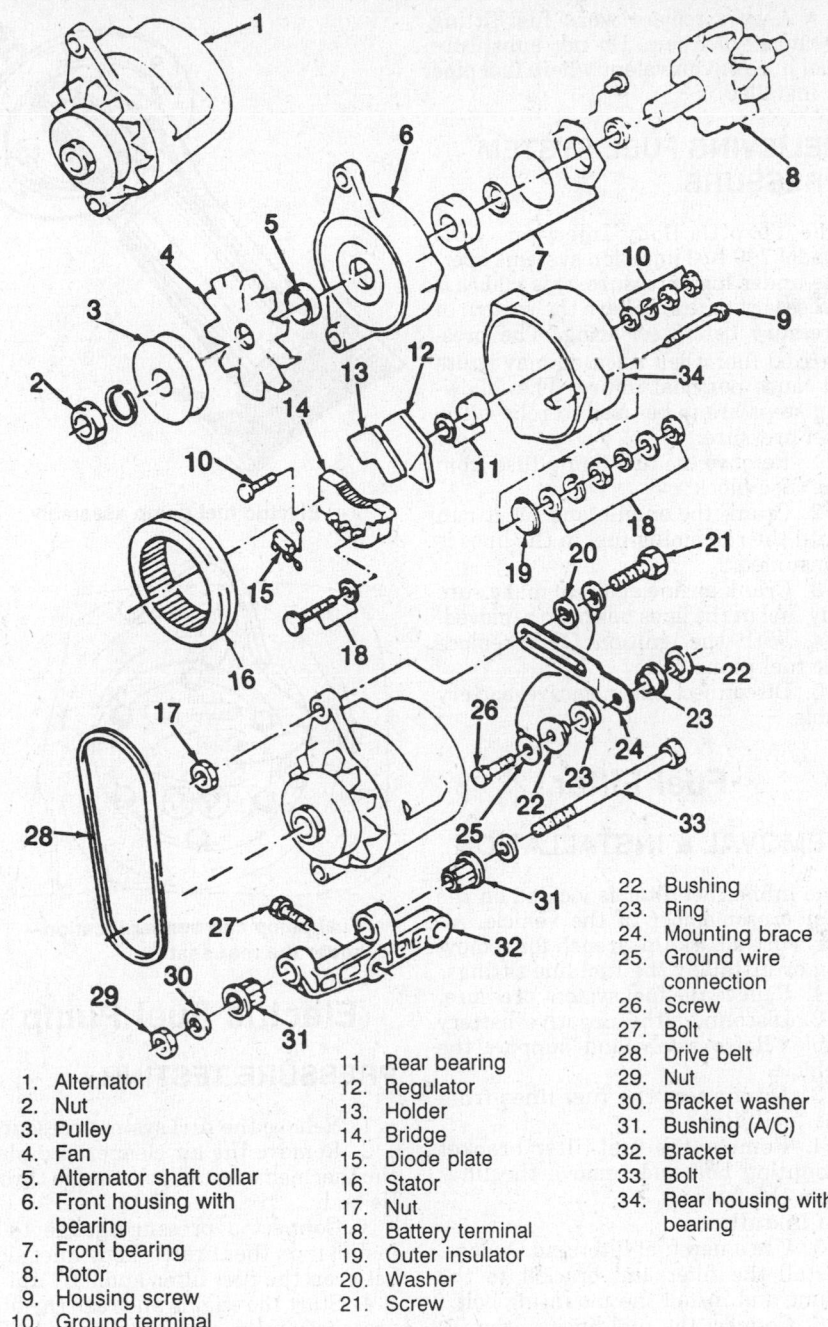

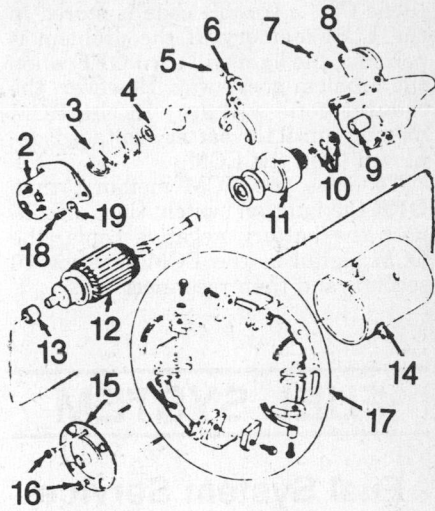

1. Starter motor   5. Pin
   assembly   6. Lever
2. Starter solenoid   7. Shaft
   switch   8. Housing
3. Spring   9. Starter drive end
4. Plunger   bushing

10. Starter drive pinion   15. Frame end
    bushing   16. Bolt
11. Starter drive pinion   17. Brush and
12. Armature   holder
13. Starter commutator   package
    end bushing   18. Bolt
14. Frame and field   19. washer

Exploded view of the starter assembly

1. Alternator	11. Rear bearing
2. Nut	12. Regulator
3. Pulley	13. Holder
4. Fan	14. Bridge
5. Alternator shaft collar	15. Diode plate
6. Front housing with	16. Stator
bearing	17. Nut
7. Front bearing	18. Battery terminal
8. Rotor	19. Outer insulator
9. Housing screw	20. Washer
10. Ground terminal	21. Screw

22. Bushing	
23. Ring	
24. Mounting brace	
25. Ground wire	
connection	
26. Screw	
27. Bolt	
28. Drive belt	
29. Nut	
30. Bracket washer	
31. Bushing (A/C)	
32. Bracket	
33. Bolt	
34. Rear housing with	
bearing	

Exploded view of the alternator assembly

5. Install in the reverse order of removal. Install the electrical leads and the negative battery cable. Adjust the belt tension.

Starter

REMOVAL & INSTALLATION

1. Disconnect the negative battery cable.
2. Disconnect the ignition switch lead wire and the battery cable from the starter motor terminal.
3. Remove the starter mounting bolts and the starter.
4. To install, hold the starter in place and install the mounting bolts. Torque the starter mounting bolts to 33 ft. lbs. (45 Nm).
5. Connect the electrical wiring to the starter.
6. Connect the negative battery cable. Start the engine and test starter amperage draw.

EMISSION CONTROLS

Please refer to "Emission Controls" in the Unit Repair section for system maintenance procedures. Due to the complex nature of modern electronic engine control systems, comprehensive diagnosis and testing procedures fall outside the confines of this repair manual. For complete information on diagnosis, testing and repair procedures concerning all modern engine and emission control systems, please refer to "Chilton's Guide to Fuel Injection and Electronic Engine Controls".

Emission Warning Lamps

RESETTING

When the "Service Engine Soon" light

turns **ON**, a trouble code is stored in the ECM memory. If the problem is periodic, the light will turn **OFF** when the problem goes away. However, the trouble code will stay in the ECM memory until the battery voltage is removed from the ECM.

To erase the ECM memory, turn **OFF** the ignition switch, then disconnect the battery negative cable, the ECM pigtail or the ECM fuse for 10 seconds and then reconnect.

FUEL SYSTEM

Fuel System Service Precautions

Safety is the most important factor when performing not only fuel system maintenance but any type of maintenance. Failure to conduct maintenance and repairs in a safe manner may result in serious personal injury or death. Maintenance and testing of the vehicle's fuel system components can be accomplished safely and effectively by adhering to the following rules and guidelines.

• To avoid the possibility of fire and personal injury, always disconnect the negative battery cable unless the repair or test procedure requires elsewise.

• Always relieve the fuel system pressure prior to disconnecting any fuel system component (injector, fuel rail, pressure regulator, etc.), fitting or fuel line connection. Exercise extreme caution whenever relieving fuel system pressure to avoid exposing skin, face and eyes to fuel spray. Please be advised that fuel under pressure may penetrate the skin or any part of the body that it contacts.

• Always place a shop towel or cloth around the fitting or connection prior to loosening to absorb any excess fuel due to spillage. Ensure that all fuel spillage (should it occur) is quickly removed from engine surfaces. Ensure that all fuel soaked cloths or towels are deposited into a suitable waste container.

• Always keep a dry chemical (Class B) fire extinguisher near the work area.

• Do not allow fuel spray or fuel vapors to come into contact with a spark or open flame.

• Always use a backup wrench when loosening and tightening fuel line connection fittings. This will prevent unnecessary stress and torsion to fuel line piping. Always follow the proper torque specifications.

• Always replace worn fuel fitting O-rings with new. Do not substitute fuel hose or equivalent where fuel pipe is installed.

RELIEVING FUEL SYSTEM PRESSURE

The Throttle Body Injection (TBI) model 700 fuel injection systems operate under high pressure, this makes it necessary to first relieve the system of pressure before servicing. The pressurized fuel when released may ignite or cause personal injury. The following steps are to be used to relieve the fuel pressure.

1. Remove the fuel pump fuse from the fuse block.
2. Crank the engine and let it run until the remaining fuel in the lines is consumed.
3. Crank engine again to make sure any fuel in the lines has been removed.
4. With the ignition **OFF** replace the fuel pump fuse.
5. Disconnect the negative battery cable.

Fuel Filter

REMOVAL & INSTALLATION

The inline fuel filter is located on the rear crossmember of the vehicle. Always use a back-up wrench for removing or installing the fuel line fittings.

1. Relieve the fuel system pressure.
2. Disconnect the negative battery cable. Raise safely and support the vehicle.
3. Disconnect the fuel lines from the fuel filter.
4. Remove the fuel filter bracket mounting bolt and remove the filter from the frame.

To install:
5. Use a new fuel filter and O-rings. Install the filter and bracket to the frame and install the mounting bolt.
6. Connect the fuel lines to the filter. Torque the fuel line-to-filter connectors to 22 ft. lbs.
6. Lower the vehicle. Connect the negative battery cable.
7. Start the engine and check for leaks.

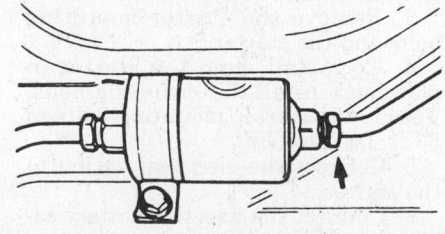

In-line fuel filter assembly

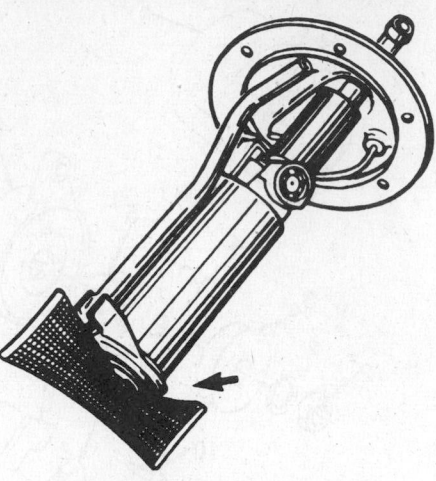

Electric fuel pump assembly

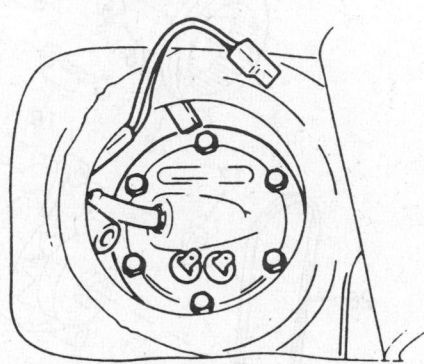

Fuel pump and sender location — under the rear seat

Electric Fuel Pump

PRESSURE TESTING

1. Relieve the fuel system pressure.
2. Remove the air cleaner and plug the thermac vacuum port on the throttle body.
3. Connect a pressure gauge tool, install it on the throttle body inlet side between the fuel filter and TBI unit.
4. Start the engine and read the fuel pressure on the gauge, it should be 9–13 psi.
5. Turn the ignition **OFF**, relieve the fuel system pressure and remove the fuel pressure gauge. Reconnect all fuel and vacuum lines. Install the air cleaner assembly.

REMOVAL & INSTALLATION

The electric fuel pump is attached to the sending unit and is located in the fuel tank.

1. Relieve the fuel system pressure.
2. Disconnect the negative battery cable.
3. Raise the rear seat and remove the floor pan cover.
4. Disconnect the electrical connec-

1. Throttle stop screw
2. Throttle stop screw plug

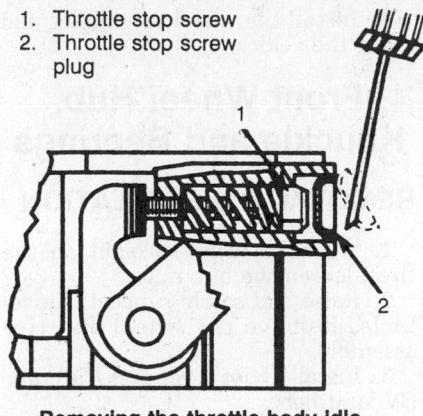

Removing the throttle body idle stop screw

1. Air filter gasket
10. Flange gasket
20. Fuel meter assembly
21. Fuel meter body screw/washer assembly
25. Fuel meter-to-throttle body gasket
35. Injector retainer screw
36. Injector retainer
40. Fuel injector
42. Upper fuel injector O-ring
43. Lower fuel injector O-ring
60. Pressure regulator cover assembly

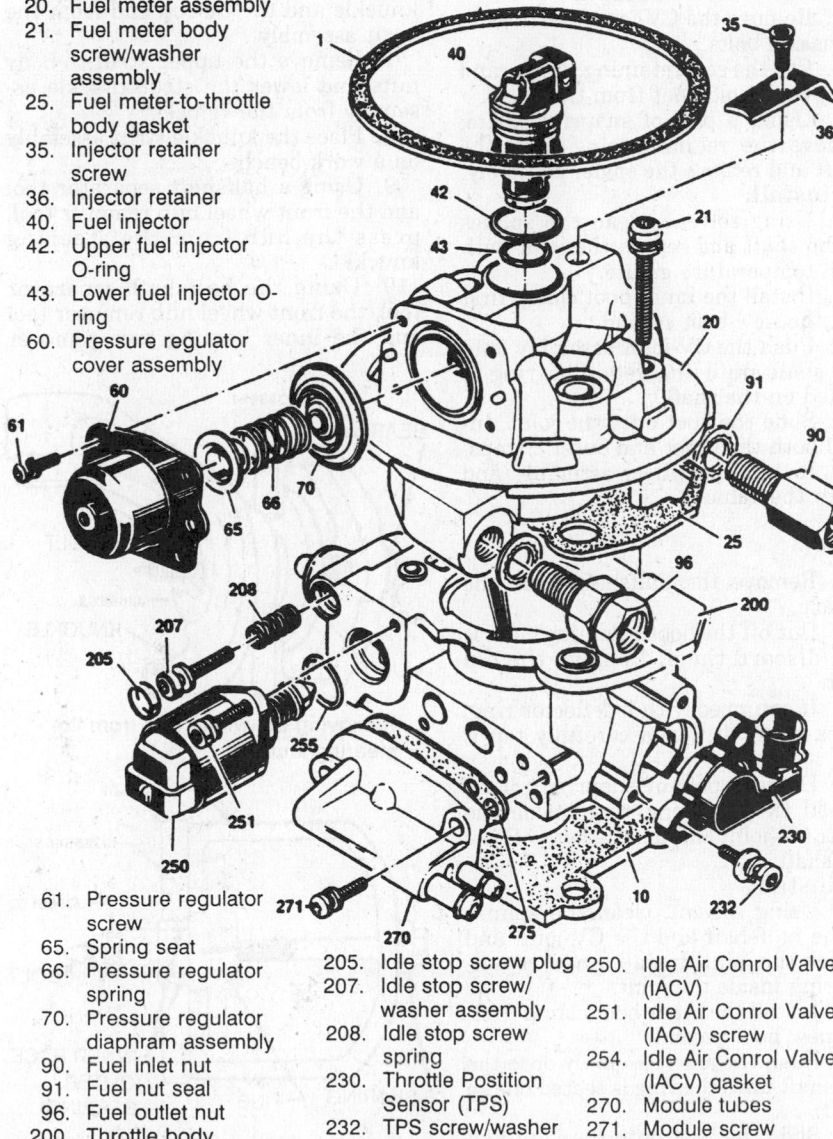

61. Pressure regulator screw
65. Spring seat
66. Pressure regulator spring
70. Pressure regulator diaphram assembly
90. Fuel inlet nut
91. Fuel nut seal
96. Fuel outlet nut
200. Throttle body assembly

205. Idle stop screw plug
207. Idle stop screw/washer assembly
208. Idle stop screw spring
230. Throttle Postition Sensor (TPS)
232. TPS screw/washer assembly

250. Idle Air Conrol Valve (IACV)
251. Idle Air Conrol Valve (IACV) screw
254. Idle Air Conrol Valve (IACV) gasket
270. Module tubes
271. Module screw
275. Module tube gasket

Exploded view of the throttle body—Model 700

tors. Disconnect and plug the fuel lines from the pump.

5. Remove the fuel pump retaining bolts and lift the fuel pump and sending unit assembly from the fuel tank.

6. Remove the fuel pre-filter and the fuel pump with rubber mounting.
To install:

7. Install new pre-filter, rubber mounting and gasket.

8. Place the fuel pump in position to the fuel tank and install the retaining screws.

9. Connect the fuel line and electrical connectors to the pump. Connect the negative battery cable.

10. Turn the ignition switch to the **ON** position to pressurize the fuel system. Start the engine and check for leaks and proper operation.

11. Install the floor pan cover and seat.

Fuel Injection

IDLE SPEED ADJUSTMENT

The throttle body is adjusted and sealed at the factory, no adjustment should be performed. All fuel control functions are controlled by the Electronic Control Module (ECM). However, if it is necessary to adjust the minimum idle speed, perform the following procedures.

1. Remove the air cleaner.

2. Using a scratch awl, pierce the idle stop screw plug, apply leverage and remove the plug.

3. Using a tachometer, follow the manufacturer's recommendations and connect the tachometer to the engine.

4. Position the transaxle in **P**, for automatic transaxle or **N**, for manual transaxle. Start the engine and allow the rpm to stabilize.

5. Using the special tool BT–8528A or equivalent, position it fully into the idle air passage so no air leak exists.

6. Using a No. 20 Torx® Bit tool, turn the idle stop screw until the engine speed is 525–575 rpm, for automatic transaxle or 575–625 rpm, for manual transaxle.

7. After adjustment, stop the engine, remove the tool BT–8528A or equivalent. Using silicone sealant, cover the idle stop screw.

IDLE MIXTURE ADJUSTMENT

The idle mixture is electronically controlled by the Electronic Control Module (ECM), no adjustment is possible.

Fuel Injector

REMOVAL & INSTALLATION

NOTE: When removing the injector, be careful not to damage the electrical connector pins, on top of the injector, the injector fuel filter and the nozzle. The fuel injector is serviced as a complete assembly only. It is an electrical component and should not be immersed in any kind of cleaner.

1. Relieve the fuel system pressure.

2. Remove the air cleaner. Disconnect the negative battery cable.

3. Disconnect the electrical connector from the fuel injector.

4. Remove the injector retainer to throttle body screw and the retainer.

5. Using a suitable tool under the

ledge opposite the connector, carefully lift the injector until it is free from the fuel meter body.

6. Remove the O-rings from the nozzle end of the injector.

7. Inspect the fuel injector filter for dirt and/or contamination.

To install:

8. Lubricate the O-rings with automatic transmission fluid and place them on the fuel injector.

9. Push the fuel injector straight into the fuel meter cavity, place thread locking compound on the fuel injector retainer bolt and install. Torque the retainer bolt to 27 inch lbs. (3 Nm).

10. Connect the electrical connector to the fuel injector.

NOTE: It is important that the replacement injector is identical. Although different injectors may fit in the 700 TBI unit, the calibrated flow rates may be different.

11. Connect the negative battery cable.

12. Turn the ignition switch to the **ON** position to pressurize the fuel system.

13. Start the engine and check for leaks. Install the air cleaner assembly.

DRIVE AXLE

Halfshaft

REMOVAL & INSTALLATION

1. Raise the hood. Loosen the upper strut to body nuts.

2. Remove the hub cap, then loosen the wheel lug nuts. Remove the cotter pin, the wheel hub nut and thrust washer.

3. Raise and safely support the vehicle. Remove the wheel/tire assembly.

4. From the lower ball joint, remove retaining clip and the nut.

5. Using the ball joint separator tool, separate the lower ball joint from the steering knuckle.

6. Remove the tie rod end-to-ball joint nut.

7. Using a special tie rod end separator tool, separate the tie rod end from the steering knuckle.

8. Using a halfshaft to hub separator tool, drive the halfshaft from the hub assembly.

9. Position a drain pan under the transaxle. Using an axle shaft to transaxle separator tool and the slide hammer tool, separate and remove the halfshaft from the transaxle.

To install:

10. Use new retaining clips, cotter pins. Torque the lower ball joint-to-steering knuckle nut to 50 ft. lbs. (70 Nm), the tie rod end-to-steering knuckle nut to 45 ft. lbs. (60 Nm).

11. Lower the vehicle to the floor. Torque the hub-to-halfshaft nut to 74 ft. lbs. (100 Nm), then back nut off and retorque to 15 ft. lbs. (20 Nm) plus an additional 90 degree turn. Torque the wheel lug nuts to 65 ft. lbs. (90 Nm) and the upper strut-to-body nuts to 22 ft. lbs. (30 Nm). Check and/or refill the transaxle.

CV-Boot

REMOVAL & INSTALLATION

Inner

1. Remove the halfshaft.

2. Remove the CV-joint housing-to-transaxle bolts.

3. Cut the seal retaining clamps and remove the old boot from the shaft.

4. Using a pair of snapring pliers, remove the retaining ring from the shaft and remove the spider assembly.

To install:

5. Using solvent, clean the splines of the shaft and repack the joint with high temperature grease.

6. Install the inner boot clamp first and the new boot second.

7. Push the CV-joint assembly onto the shaft until the retaining ring is seated on the shaft.

8. Slide the boot onto the joint. Install both the inner and outer clamps.

9. Install the wheel assembly and lower the vehicle.

Outer

1. Remove the halfshaft from the vehicle.

2. Cut off the boot retaining clamps and discard them. Remove the old boot.

3. If equipped with a deflector ring, use a brass drift and carefully tap it off.

4. Using a pair of snapring pliers, spread the retaining ring inside the outer CV-joint and tap the joint off the halfshaft.

To install:

5. Using solvent, clean the splines of the halfshaft and the CV-joint and repack the joint. Install a new retaining ring inside the joint.

6. Install the inner boot clamp first, the new boot second.

7. Push the joint assembly onto the halfshaft until the ring is seated on the shaft.

8. Slide the boot onto the joint and install the clamps on both the inner and outer part of the boot.

9. Install the wheel assemblies and lower the vehicle.

Front Wheel Hub, Knuckle and Bearings

REMOVAL & INSTALLATION

1. With the vehicle weight on the tires, loosen the hub nut.

2. Raise and safely support the vehicle. Remove the wheel and tire assembly.

3. Install a boot cover over the outer CV-joint boot.

4. Remove the hub nut. Remove the brake caliper and support it aside (on a wire); do not allow the caliper to hang on the brake line.

5. Using a special tool, separate the halfshaft from the hub assembly.

6. Disconnect the ball joint from the knuckle and the tie rod end from the strut assembly.

7. Remove the upper strut-to-body nuts and lower the strut/knuckle assembly from the vehicle.

8. Place the knuckle/strut assembly on a work bench.

9. Using a halfshaft separator tool and the front wheel hub remover tool, press the hub from the steering knuckle.

10. Using the halfshaft separator tool, the front wheel hub remover tool and the inner bearing race remover

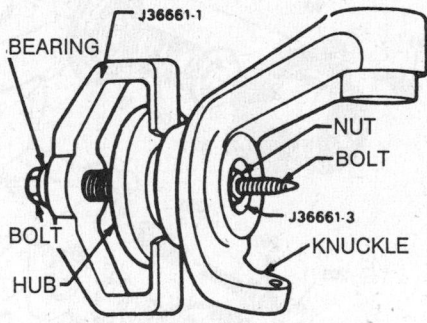

Removing the wheel hub from the steering knuckle

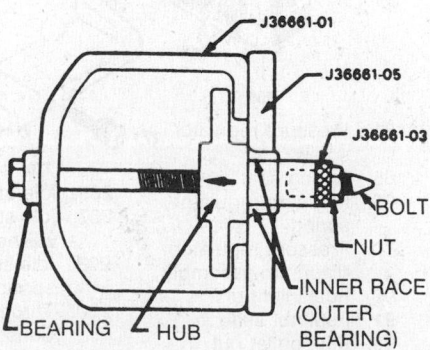

Removing the inner bearing race from the hub

tool, remove the inner bearing race from the hub.

11. From inside the steering knuckle, remove the internal snaprings.

12. Using the halfshaft separator tool and the bearing remover/installer tool, press the bearing from the steering knuckle.

To install:

NOTE: Whenever the wheel bearing is removed from the steering knuckle, it must be discarded and replaced with a new one.

13. Using solvent, clean all of the parts and blow dry with compressed air.

14. Before assembling the parts, be sure to coat them with a layer of wheel bearing grease.

15. Using snapring pliers, install the outer internal snapring into the steering knuckle.

16. Using the halfshaft separator tool and the bearing remover/installer tool, press the new wheel bearing into the steering knuckle until it butts against the snapring.

17. Using snapring pliers, install the inner internal snapring into the steering knuckle.

18. Install the strut onto the body. Remove the seal protector from the halfshaft. Install the halfshaft into the steering knuckle/strut assembly. Install the new washer and new halfshaft nut onto the halfshaft.

19. Use new cotter pins and reverse

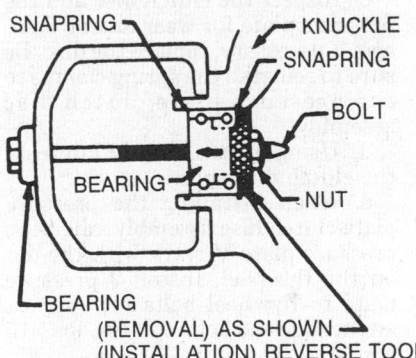

Removing the wheel bearing from the steering knuckle

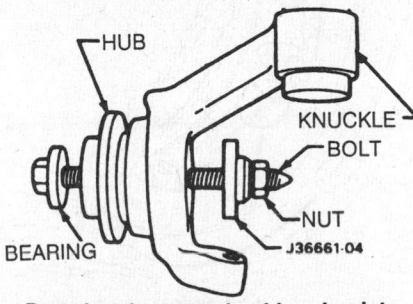

Pressing the new wheel bearing into the steering knuckle

the removal procedures. Torque the steering knuckle/strut assembly to body nuts to 22 ft. lbs. (30 Nm).

20. Tighten the lower ball joint-to-steering knuckle nut 50 ft. lbs. (68 Nm), the tie rod end-to-steering knuckle nut to 45 ft. lbs. (61 Nm), the disc rotor-to-hub screw to 3 ft. lbs. (4 Nm), the caliper-to-steering knuckle bolts to 70 ft. lbs. (95 Nm) and the wheel lug nuts to 65 ft. lbs. (88 Nm).

21. Torque the halfshaft-to-hub nut to 74 ft. lbs. (100 Nm), back it off, re-tighten to 15 ft. lbs. (20 Nm), then, tighten another 90 degrees. Check and/or adjust the front end alignment.

NOTE: When tightening the halfshaft nut, be sure to have the vehicle resting on its wheels. If the castellated nut does not align with a shaft hole, back off the nut until it does; do not tighten the nut to locate another shaft hole.

MANUAL TRANSAXLE

For further information on transmissions/transaxles, please refer to "Chilton's Guide to Transmission Repair".

Transaxle Assembly

REMOVAL & INSTALLATION

NOTE: It is not necessary to remove the transaxle from the vehicle to replace the clutch assembly.

1. Disconnect the negative battery cable.

2. Remove the clutch cable from the release lever.

3. From the shifter universal joint, remove the retaining clip and bolt.

4. From the transaxle, remove the speedometer cable, the speed sensor and the backup light connector.

5. Remove the upper transaxle to engine bolts. Install the engine support fixture tool and support the engine.

6. Raise the vehicle and support it safely. Remove the both front wheels.

7. From the transaxle cover, remove the plug and drain the fluid into suitable container. Remove the ground wire from the transaxle housing.

8. Remove the flywheel cover bolts and the cover.

9. To remove the lower ball joint, perform the following procedures:

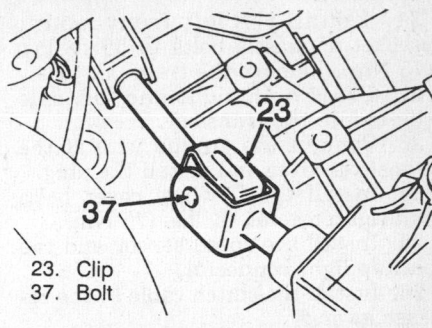

23. Clip
37. Bolt

Removing the universal joint clip and bolt

a. Remove the retainer clip from the ball joint.

b. Remove the ball joint to steering knuckle nut.

c. Using a ball joint separator tool, separate the ball joint from the steering knuckle.

10. To remove the tie rod end, perform the following procedures:

a. Remove the tie rod end ball joint to steering knuckle nut.

b. Using a steering linkage puller tool, separate the tie rod end ball joint from the steering knuckle.

11. Using a halfshaft separator tool, separate both halfshafts from the transaxle.

NOTE: When removing the halfshafts from the transaxle, be sure to swing the left strut assembly outward.

12. Using a transaxle jack, position it under the transaxle and support it.

13. Remove the remaining transaxle-to-engine bolts.

14. Remove the left front and rear bracket to transaxle bolts.

15. Remove the lower transaxle to engine bolts. Move the transaxle away from the engine and downward; guide the right halfshaft from the transaxle, then support it on a wire.

To install:

16. Secure the transaxle on a transaxle jack and raise it towards the engine.

17. Guide the right halfshaft into the transaxle, while installing the transaxle to the engine.

NOTE: The right halfshaft cannot be readily installed after the transaxle is connected to the engine.

18. Install the lower transaxle-to-engine bolts and tighten to 55 ft. lbs. (75 Nm).

19. Install the the left rear mount bracket-to-transaxle bolts and tighten to 55 ft. lbs. (75 Nm).

20. Install the left front mount bracket-to-transaxle bolts and tighten to 47 ft. lbs. (65 Nm).

21. Install the left front mount bracket-to-chassis bolts to 55 ft. lbs. (75 Nm). Remove the transaxle jack.

22. Press both right and left axle shafts into the transaxle.

23. Connect the ground wire at the transaxle cover and install the plug.

24. Install the flywheel cover bolts and tighten to 55 ft. lbs. (75 Nm).

25. Install the speed sensor and the backup light connector.

26. Install the clutch cable to the release lever.

27. Install the shifter universal and install the retaining clip and bolt.

28. Connect and tighten the ball joint-to-steering knuckle nut to 50 ft. lbs. (68 Nm).

29. Connect and tighten the tie rod end-to-steering knuckle nut to 45 ft. lbs. (61 Nm).

30. Lower the vehicle and install the remaining transaxle-to-engine bolts.

31. Remove the engine support fixture tool and the adaptor tool from the engine.

32. Check and replenish the transaxle fluid level. Connect the negative battery cable.

LINKAGE ADJUSTMENT

1. Disconnect the negative battery cable.

2. Position the gear shift lever in the **N** position.

3. Loosen the shift rod clamp bolt.

4. From the shift lever cover, remove the adjustment hole plug. Turn

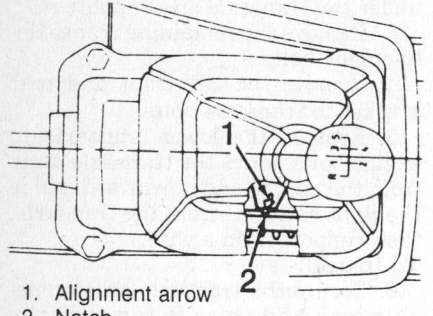

1. Alignment arrow
2. Notch

Aligning the shift lever

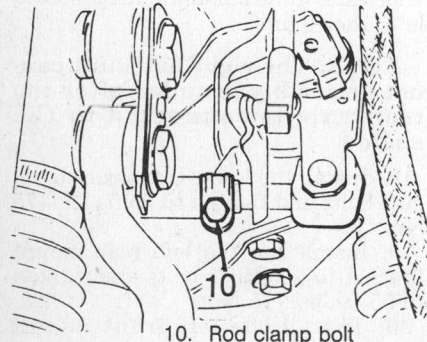

10. Rod clamp bolt

Removing the rod clamp

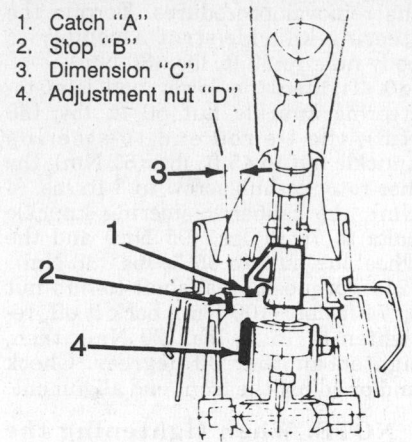

1. Catch "A"
2. Stop "B"
3. Dimension "C"
4. Adjustment nut "D"

Adjusting the shift lever

the shift rod left until a $\frac{3}{16}$ in. gauge pin can be inserted into the adjustment hole into the intermediate shift lever.

5. Remove the boot from the console, then, pull it upward to expose the shift control lever mechanism.

6. Position the shift lever to the **1ST/2ND** gear position, while the transaxle is still **N**. With the lever against the stop and the arrow aligned with the notch, torque the shift rod clamp bolt to 10 ft. lbs. (14 Nm), turn the bolt another 90–180 degrees.

7. Using a 0.120 in. (3mm) diameter gauge pin, check the clearance between the **A** catch and the **B** stop.

8. Remove the gauge pin and install the plug.

9. To further adjust the shift lever, bend back the adjusting nut locking tabs and turn the adjusting nut **D** until the **C** dimension is 0.449–0.465 in. (11.4–11.8mm). Bend up the locking tabs to secure the adjusting nut.

10. Install the boot to the console and then install the center console.

11. Connect the negative battery cable.

CLUTCH

Clutch Assembly

REMOVAL & INSTALLATION

NOTE: It is not necessary to remove the transaxle from this vehicle to replace the clutch assembly.

1. Disconnect the negative battery cable. Remove the clutch cable from the release lever. Raise and safely support the vehicle.

2. From the transaxle cover, re-

move the plug and the ground wire. Using a pair of internal snapring pliers, remove the snapring from the end of the input shaft; mark the position of the input shaft in relation to the cluster gear.

3. Using an input shaft retaining screw tool, remove the screw from the end of the input shaft.

4. Using the input shaft removal and installation tool, with the slide hammer tool, screw the assembly into the end of the input shaft. Using the slide hammer assembly, disengage the input shaft from the cluster gear.

5. Remove the clutch cover from the bottom of the transaxle and push back the clutch release bearing. Using a 3 pressure plate spring clamps, rotate the flywheel and install a clamp every 120 degrees.

NOTE: The pressure plate and clutch disc cannot be removed without installing the 3 spring clamps.

6. Remove the pressure plate to flywheel bolts, the pressure plate/clutch disc assembly. Be sure to support the assembly when removing the last bolt.
To install:

7. To replace the clutch disc assembly, perform the following procedures:

   a. Using a hydraulic press, apply pressure to the pressure plate spring fingers, remove the spring clamps.

   b. Reduce the pressure of the press and separate the clutch disc from the pressure plate.

   c. Inspect the clutch disc and the pressure plate for wear and/or damage; if necessary, replace the disc. Be sure to reinstall the spring clamps to the pressure plate/clutch disc assembly.

   d. Using grease, lightly lubricate the clutch disc spline.

   e. When installing the pressure plate/clutch disc assembly, align the pressure plate **V** mark with the dot on the flywheel. Install 2 pressure plate to flywheel bolts for support and torque the bolts to 11 ft. lbs. (15 Nm).

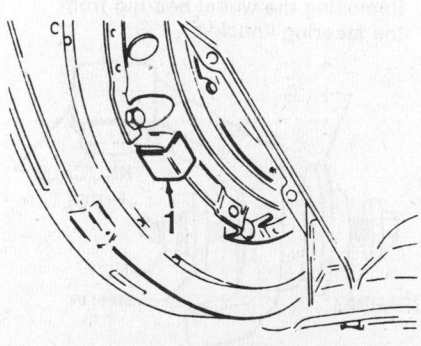

1. J-36554

Pressure plate spring clip installation

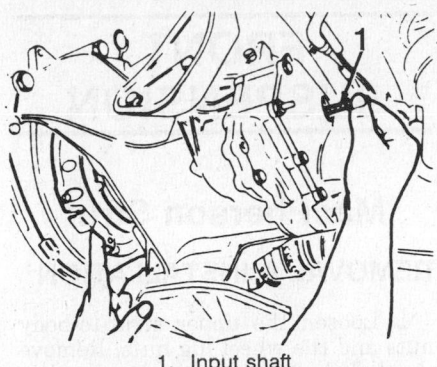

1. Input shaft

Installing the input shaft into the clutch assembly

f. Install the input shaft by aligning the input shaft with the mark on the cluster gear. Using an input shaft removal/installation tool and a slide hammer tool, seat the input shaft with the cluster gear.

g. Install the screw into the end of the input shaft; torque it to 11 ft. lbs. (15 Nm).

h. Install the snapring on the end of the input shaft; the sharp edges must face the cover.

i. Install the remainder of the pressure plate retaining bolts and torque to 11 ft. lbs. (15 Nm).

NOTE: When installing a new clutch disc, make sure the long part of the clutch disc hub faces the transaxle.

8. Using Teflon® pipe thread sealer, coat the threads of the input shaft cover plug and tighten to 36 ft. lbs. (50 Nm).

9. Remove the spring clamps from the pressure plate/clutch disc assembly.

10. Install the clutch cover and retaining bolts. Torque the retaining bolts to 62 inch lbs. (7 Nm). Install the left tire and wheel assembly, lower the vehicle. Install the clutch cable to the release lever. Check clutch engagement and operation.

11. Connect the negative battery cable.

Clutch Cable

ADJUSTMENT

1. Measure the distance from the center of the clutch pedal to the bottom edge of the steering wheel.

2. Depress the clutch pedal (fully) and measure the distance again. Subtract the 1st measurement from the 2nd to determine the pedal travel.

3. If the pedal travel is not 5.43–5.71 in. (138–146mm), remove the clutch cable clip and adjust the nut to bring the measurement within specifications.

NOTE: With the correct adjustment, the clutch pedal will be higher than the brake pedal and there will be no free-play operation. As the clutch disc wears, the clutch pedal will move further away from the brake pedal.

REMOVAL & INSTALLATION

1. Disconnect the negative battery cable.

2. Measure the threaded end of the clutch cable at the release lever and record the measurement for pre-adjustment procedures.

3. Remove the clutch cable clip and loosen the clutch cable adjusting nut. Disconnect the cable from the release arm and cable bracket.

4. Disconnect the clutch start safety switch.

5. At the clutch pedal, remove the cable return spring at the brace. Remove the clutch pedal retaining nut, spring and shaft. Remove the pedal and pull the cable through the firewall.

6. Disconnect the spring and the cable from the pedal.

7. Remove the clutch cable by pulling it from the engine compartment.

To install:

8. From the engine compartment, install the clutch cable through the front dash. Place the grommet and washer the clutch cable.

9. From under the dash, connect the cable and return spring to the clutch pedal.

10. Position the clutch pedal assembly to the support brace and install the

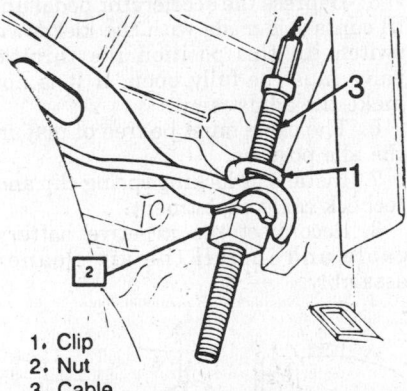

1. Clip
2. Nut
3. Cable

View of the clutch cable adjuster

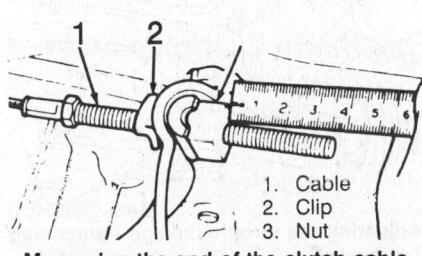

1. Cable
2. Clip
3. Nut

Measuring the end of the clutch cable

pedal support shaft through the brace and pedal. Coat the shaft prior to installing it.

11. Install the washer, shaft retaining nut, shaft spring and return spring to the brace.

12. Install the clutch start switch.

13. Install the clutch cable to the release lever and install the adjustment nut.

14. Adjust the clutch cable to the measurement taken earlier. Adjust the clutch pedal travel.

15. Connect the negative battery cable. Check the clutch operation.

AUTOMATIC TRANSAXLE

For further information on transmissions/transaxles, please refer to "Chilton's Guide to Transmission Repair".

Transaxle Assembly

REMOVAL & INSTALLATION

1. Disconnect the negative battery cable.

2. Remove the air cleaner.

3. Remove the TV cable from the transaxle and the throttle body. Remove the cable.

4. Disconnect the shift selector cable from the transaxle lever and the cable bracket; leave the cable attached to the bracket.

5. Disconnect electrical connectors from the speed sensor, the TCC and the park/neutral/backup light switch.

6. Disconnect the speedometer drive cable from the transaxle.

7. Remove the top transaxle to engine bolts. Remove the starter assembly. Remove the transaxle vent hose.

8. Using the engine support fixture tool and the adaptor tool, attach it to the engine and support the weight.

9. Raise and safely support the vehicle.

10. Remove both front wheels. Remove the both lower ball joint-to-knuckle bolts. Using the ball joint separator tool, separate the ball joints (both sides) from the steering knuckles.

11. Remove the left and right axle shafts from the transaxle. Disconnect and plug transaxle cooler lines.

12. Remove the converter cover. Matchmark and remove the torque converter to flywheel bolts.

13. Remove the transaxle to engine

mount bolts and the remaining transaxle-to-engine bolts.

14. Lower the transaxle from the vehicle.

To install:

15. Apply a light coating of grease to the torque converter pilot hub. Place the torque converter onto the hub and seat it securely into the pump.

16. Secure the transaxle on a transaxle jack and position it to the engine.

17. Install lower transaxle to engine bolts and torque to engine bolts to 54 ft. lbs. (73 Nm).

18. Install the right transaxle to engine mount bolts and torque to 30 ft. lbs. (41 Nm).

19. Install the left and rear transaxle to engine mount bolts and torque to 16 ft. lbs. (22 Nm).

20. Install the torque converter to flywheel bolts and torque to 44 ft. lbs. (60 Nm). Torque the remaining transaxle to engine bolts to 55 ft. lbs. (75 Nm).

21. Check and/or adjust the selector cable adjustment. Check and/or adjust the TV cable adjustment. Check and refill the transaxle with fluid.

22. Install the right transaxle to engine mount bolts and the remaining transaxle to engine bolts.

23. Align the marks on the torque converter and flywheel, made during disassembly and install the torque converter to flywheel bolts. Torque the bolts to 44 ft. lbs. (60 Nm).

24. Install the left transaxle to engine mount bolts. Install the torque converter cover.

25. Connect and the oil cooler lines to the transaxle.

26. Install the left and right axleshafts into the transaxle and seat securely.

27. Install front wheel assemblies. Connect the right and left upper ball joints to the steering knuckles.

28. Lower the vehicle.

29. Install the top transaxle to engine bolts.

30. Remove the engine support fixture tool and the adaptor tool from the engine. Install the starter assembly.

31. Connect the speedometer drive cable to the transaxle.

32. Connect the electrical connectors to the speed sensor, the TCC and the park/neutral/backup light switch.

33. Connect the shift selector cable to the transaxle lever and the cable bracket assembly.

34. Install the TV cable to the transaxle and the throttle body. Connect the cable.

35. Check and replenish the transaxle fluid level.

36. Install the air cleaner. Connect the negative battery cable.

SHIFT LINKAGE ADJUSTMENT

1. Disconnect the negative battery cable.

2. Remove the center console.

3. Loosen the clamp at the selector lever and cable.

4. Adjust the cable so the selector lever will catch in each position; **P, R, N, D, 2** and **1**.

5. Adjust the cable at the selector lever. Check the shift in all positions.

6. Install the console and connect the negative battery cable.

THROTTLE LINKAGE ADJUSTMENT

1. Disconnect the negative battery cable.

2. Remove the air cleaner.

3. Remove the locking spring clip adjuster.

4. Pull the adjuster pin out, to relieve tension on the cable.

5. Depress the accelerator pedal until contact is made with the kickdown switch. In this position the throttle valve must be fully open. If it is not make the adjustment.

6. The cable must be free of play in the idle position.

7. Install the locking spring clip and recheck cable adjustment.

8. Reconnect the negative battery cable and install the air cleaner assembly.

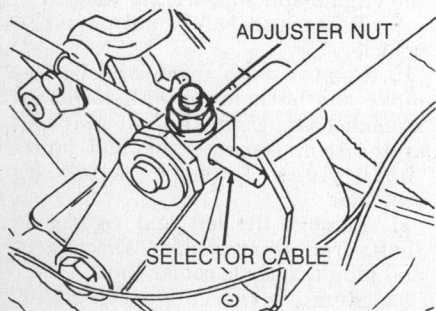

Adjusting the shift linkage—automatic transaxle

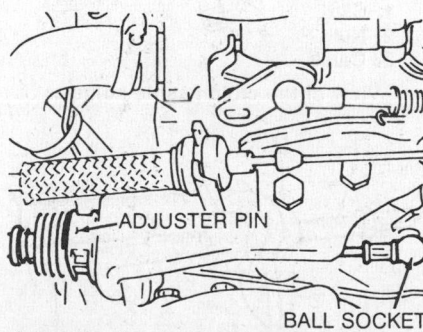

Adjusting the throttle linkage—automatic transaxle

FRONT SUSPENSION

MacPherson Strut

REMOVAL & INSTALLATION

1. Loosen the upper strut-to-body nuts and the wheel lug nuts. Remove the halfshaft-to-hub cotter pin, nut and washer.

2. Raise and safely support the vehicle; allow the wheels to hang free. Remove the wheel assembly. Using a drive axle boot seal protector tool, place them on the outer CV-joints.

3. Remove the caliper-to-steering knuckle bolts. Remove the caliper and position it aside. Remove the rotor-to-wheel hub screw and the rotor.

4. Remove the outer tie rod end-to-steering knuckle nut. Using tie rod end remover tool, separate the outer tie rod end-to-steering knuckle arm.

5. Remove the lower ball joint-to-steering knuckle retaining clip and nut. To remove the clip, lift up on the rear of the clip, while pulling outward on the loops.

6. Using ball joint separator tool, separate the lower ball joint from the steering knuckle arm.

7. Using a front wheel hub remover tool, separate the halfshaft from the steering knuckle hub. Properly support the halfshaft.

8. Remove the upper strut-to-body nuts and washers. Remove the strut assembly from the vehicle.

To install:

9. Install the strut assembly to the strut tower and install the upper strut-to-body nuts and washers.

10. Install the halfshaft to the steering knuckle hub.

11. Connect the lower ball joint to the steering knuckle arm and install the retaining nut.

12. Connect the outer tie rod end to the steering knuckle and install the retaining nut.

13. Install the rotor and hub nut. Install brake caliper and the caliper-to-steering knuckle bolts.

14. Remove the drive axle boot seal protector tool from the outer CV-joints.

15. Torque the following:
Steering knuckle/strut assembly nuts—18 ft. lbs. (25 Nm)
Lower ball nut—50 ft. lbs. (70 Nm)
Tie rod end nut—45 ft. lbs. (60 Nm)
Disc rotor retaining bolts—36 inch lbs. (4 Nm)

Caliper retaining bolts — 7 ft. lbs. (9.5 Nm)

Halfshaft-to-hub nut — 74 ft. lbs. (100 Nm), back off the nut and retorque to 15 ft. lbs. (20 Nm), tighten another 90 degrees. Install the cotter pin.

16. Install the wheel assembly and lower the vehicle. Torque the wheel lug nuts 65 ft. lbs. (88 Nm). Check the front end alignment and adjust, as required.

Lower Ball Joints

INSPECTION

1. Raise and safely support the vehicle.
2. With the ball joint installed to the steering knuckle, grasp the top and bottom of the wheel; then, move the wheel using an in and out shaking motion.
3. Observe any movement between the steering knuckle and the control arm. If movement exists, replace the ball joint.

REMOVAL & INSTALLATION

1. Raise and safely support the vehicle. Lower the vehicle slightly so the weight does not rest on the control arm. Remove the wheel assembly.
2. If a silicone (gray) boot is used on the inboard axle joint, install a boot seal protector tool. If a thermoplastic (black) boot is used, no protector is necessary.
3. Remove the retaining clip from the ball joint castle nut.
4. Remove the castle nut. Using ball joint separator tool, disconnect the ball joint from the steering knuckle arm.
5. Using a 0.47 in. drill, drill out the ball joint-to-control arm rivets. Be careful not to damage the halfshaft boot when drilling out the ball joint rivets.
6. Loosen the stabilizer shaft bushing assembly nut. Remove the ball joint from the control arm.
To install:
7. Position the new ball joint to the lower control arm. Install new retaining bolts and nuts.
8. Torque the lower ball joint retaining nuts to 51 ft. lbs. (68 Nm). Install and torque the stabilizer bolts to 13 ft. lbs. (18 Nm).
9. Install and torque the ball joint-to-steering knuckle nut to 50 ft. lbs. (68 Nm).

Lower Control Arms

REMOVAL & INSTALLATION

1. Raise and safely support the vehicle. Lower the vehicle slightly so the weight does not rest on the control arm. Remove the wheel assembly.
2. Disconnect the stabilizer shaft from the control arm and the support assembly.
3. Remove the ball joint-to-steering knuckle cotter pin and nut. Using ball joint separator tool, separate the ball joint from the steering knuckle.
4. Remove the control arm-to-support arm bolts. Remove the control arm from the vehicle.
To install:
5. Install the control arm and mounting bolts, torque the front control arm-to-support arm bolt to 100 ft. lbs. (140 Nm) and the rear control arm-to-support arm bolt to 50 ft. lbs. (70 Nm).
6. Connect the ball joint to the steering knuckle and install the retaining nut. Torque the nut to 51 ft. lbs. (70 Nm).
7. Connect the stabilizer shaft to the control arm and the support assembly. Install the retaining nuts and torque to 13 ft. lbs. (17 Nm).
8. Install the wheel assembly and lower the vehicle.

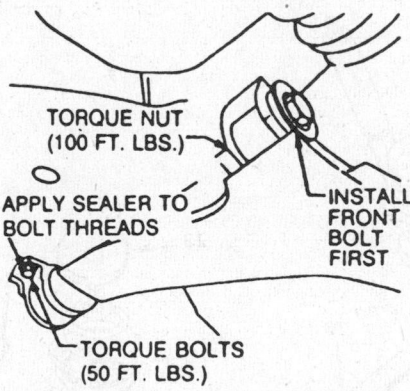

TORQUE NUT (100 FT. LBS.)

APPLY SEALER TO BOLT THREADS

INSTALL FRONT BOLT FIRST

TORQUE BOLTS (50 FT. LBS.)

Lower control arm mounting bolt location

Sway Bar

REMOVAL & INSTALLATION

1. Loosen the left front wheel lug nuts. Raise and safely support the vehicle; allow the suspension to hang free. Remove the wheel assembly.
2. Remove the stabilizer shaft link assemblies from the control arms.
3. Remove the stabilizer shaft bushings and brackets from the body.
4. Remove the stabilizer shaft and bushings.
To install:
5. Install the stabilizer bar bushings in place on the bar, position the stabilizer to the frame and install the mounting brackets a bolts. Do not tighten.
6. Install the stabilizer end links, bushings and retaining bolts. Center the stabilizer bar in the proper installation position and tighten the link retaining nuts.
7. Torque the stabilizer bar bracket retaining bolts to 29 ft. lbs. (40 Nm).
8. Install the wheel assemblies and lower the vehicle.

REAR SUSPENSION

Shock Absorbers

REMOVAL & INSTALLATION

NOTE: When replacing both rear shocks, remove only 1 shock at a time. Otherwise damage to the brake lines and hoses will occur when the axle is fully extended.

1. Open the trunk. If equipped, remove the trim cover. Remove the upper shock absorber-to-body nut.
2. Raise and safely support the vehicle and the rear axle assembly before unbolting the shock absorbers.

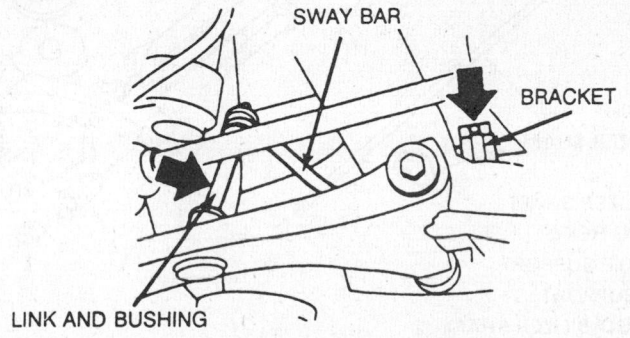

SWAY BAR

BRACKET

LINK AND BUSHING

Installation of the sway bar

3. Remove the shock absorber-to-rear axle assembly bolt. Remove the shock absorber from the vehicle.

To install:

4. Install lower mounting bolt through the lower end of the shock absorber and attach the lower end of the shock absorber to the shock support.

5. Install the retaining nut loosely.

6. Install a washer and rubber grommet on the shock stud and guide the stud through the opening in the body.

7. Lower the vehicle and install the a rubber grommet, washer and retaining nut over the stud.

8. Torque the lower mounting nut to 51 ft. lbs. (70 Nm) and the upper stud nut until 0.36 in. (9mm) of thread is visible. Lower the vehicle.

Coil Springs

REMOVAL & INSTALLATION

1. Raise and safely support the ve-hicle under the rear control arms. Support the rear axle assembly with a jack.

2. Remove the wheel assembly. Remove the right and left brake line bracket retaining screws from the body and allow the brake line to hang free.

3. Remove the shock absorber lower retaining bolts. Lower the rear axle assembly to remove the coil springs. Do not allow the axle assembly to hang in this position.

To install:

4. Installation is the reverse of the removal procedures. Before installing the coil springs, it is necessary to install the insulators to the body using adhesive.

5. Position the spring and insulator in the spring seat and raise the axle. The upper ends of the coil must be positioned properly in the seat of the body.

6. Torque the shock absorber lower retaining bolts to 51 ft. lbs. (70 Nm).

7. Install the wheel assemblies and lower the vehicle.

Rear Wheel Bearings

REMOVAL & INSTALLATION

1. Raise and safely support the vehicle.

2. Remove the wheel assembly.

3. Remove the brake drum detent screw and the drum.

4. Remove the hub/bearing assembly to axle spindle grease cap, cotter pin, hub nut, thrust washer and the outer bearing from the axle spindle.

5. Using a small prybar, remove the grease seal from the inside of the hub. Remove the inner and outer bearing from the hub.

6. If replacing the wheel bearings, perform the following procedures:

   a. Using a hammer and a drift punch, drive both outer bearing races (in opposite directions) from the wheel hub.

   b. Using a cleaning solvent (not gasoline), clean the bearings, races

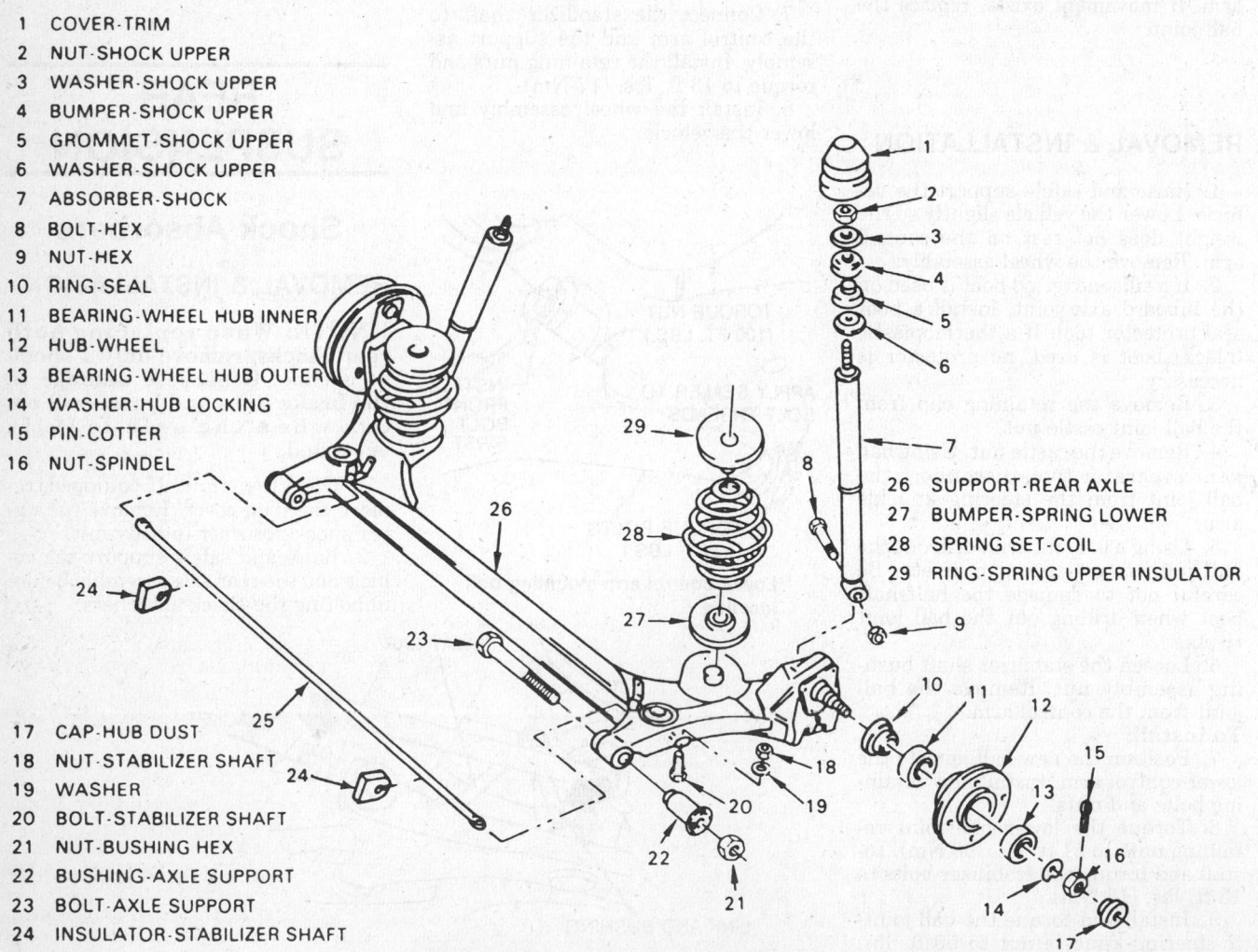

1	COVER-TRIM
2	NUT-SHOCK UPPER
3	WASHER-SHOCK UPPER
4	BUMPER-SHOCK UPPER
5	GROMMET-SHOCK UPPER
6	WASHER-SHOCK UPPER
7	ABSORBER-SHOCK
8	BOLT-HEX
9	NUT-HEX
10	RING-SEAL
11	BEARING-WHEEL HUB INNER
12	HUB-WHEEL
13	BEARING-WHEEL HUB OUTER
14	WASHER-HUB LOCKING
15	PIN-COTTER
16	NUT-SPINDEL
17	CAP-HUB DUST
18	NUT-STABILIZER SHAFT
19	WASHER
20	BOLT-STABILIZER SHAFT
21	NUT-BUSHING HEX
22	BUSHING-AXLE SUPPORT
23	BOLT-AXLE SUPPORT
24	INSULATOR-STABILIZER SHAFT
25	SHAFT-STABILIZER

26	SUPPORT-REAR AXLE
27	BUMPER-SPRING LOWER
28	SPRING SET-COIL
29	RING-SPRING UPPER INSULATOR

Exploded view of the rear axle assembly

and hub. Using compressed air, blow dry the parts.

c. Inspect the parts for damage and/or wear. If necessary, replace any defective parts.

d. Press, the rear hub inner and outer bearing race into wheel hub until it seats.

NOTE: Before installing the wheel bearings, be sure to pack the wheel bearing with grease.

e. Lubricate the lips of the new grease seal. Using the rear hub seal installation, press the new seal into the hub.

f. Install the wheel bearing hub onto the axle spindle, followed by the outer, inner bearing, thrust washer and hub nut.

g. Adjust the wheel bearing play. Adjust the parking brake.

7. Install the wheel assemblies. Torque the wheel lug nuts to 66 ft. lbs. (89 Nm).

ADJUSTMENT

1. Raise and support the vehicle.

2. Remove the grease cap from the rear wheel hub.

3. Remove the cotter pin from the spindle and the spindle nut.

4. While turning the wheel, by hand, in the forward direction, tighten the spindle nut to 18 ft. lbs. (25 Nm).

NOTE: The tightening procedure will remove any grease or burrs which could cause excessive wheel bearing play.

5. Back off the nut to the "JUST LOOSE" position.

6. Hand tighten the spindle nut and loosen it until a spindle hole aligns with a slot in the nut.

7. Install a new cotter pin and bend the ends around the nut.

8. Measure the endplay. If it is within 0.001–0.005 in. (0.03–0.13mm), it is properly adjusted. Install the dust cap.

9. Install the wheel assembly and lower the vehicle. Torque the wheel lug nuts to 66 ft. lbs. (89 Nm).

Rear Axle Assembly

REMOVAL & INSTALLATION

1. Raise and safely support the vehicle under the rear control arms. Support the rear axle assembly with a jack.

2. Remove the wheel assembly. Remove the right and left brake line bracket retaining screws from the body and allow the brake line to hang free.

3. Remove the stabilizer bar brackets. Remove the insulator and the stabilizer bar assembly.

4. Remove the shock absorber lower retaining bolts. Lower the rear axle as-

sembly to remove the coil springs. Do not allow the axle assembly to hang in this position.

5. Remove the control arm retaining bolts from the underbody bracket and lower the axle.

6. Remove the hub retaining bolts and remove the hub, bearing and backing plate assembly.

NOTE: Be careful not to drop the hub/bearing assembly, damage to the bearing could result.

To install:

7. Install the backing plate and hub/bearing assembly to the rear axle assembly. Install the retaining bolts and nuts and torque to 21 ft. lbs. (28 Nm).

8. Install the stabilizer bar to the rear axle assembly and install the attaching nuts and bolts. Torque the nuts to 59 ft. lbs. (80 Nm).

9. Secure the axle assembly on a transmission jack and raise it into position.

10. Install the control arms to the underbody bracket and install the mounting nuts and bolts. Do not torque the bolts at this time. The bolts must be torque and curb height.

NOTE: The control arm mounting bolts must be install from the inboard side.

11. Connect the brake line connections and install the brake cable to the rear axle assembly.

12. Position the springs and insulators in the spring seat and raise the axle. The upper ends of the coil must be positioned properly in the seat of the body.

13. Connect the shock absorber at the lower end and install retaining bolt. Torque the retaining bolt to 51 ft. lbs. (69 Nm).

14. Connect the parking brake to the guide hook. Adjust the cable, as required.

15. Bleed the brake system and refill the reservoir. Adjust the brakes, as required.

16. Lower the axle to curb height and torque the axle-to-body mounting bolts. Torque the bolts to 70 ft. lbs. (95 Nm).

17. Install the wheel assemblies and lower the vehicle. Torque the lug nuts to 100 ft. lbs. (140 Nm).

STEERING

Steering Wheel

REMOVAL & INSTALLATION

1. Disconnect the negative battery cable.

2. Remove the horn pad and wires from the steering wheel.

3. Using a scratch awl, mark the alignment of the steering wheel with the steering column.

4. Remove the steering wheel retaining nut and washer.

5. Using a steering wheel puller, remove the steering wheel from the steering column.

NOTE: It may be necessary to disconnect the horn contact ring from the steering wheel. When installing the steering wheel, be sure the turn signal return segment is positioned on the upper left side, facing the steering column.

To install:

6. Align the steering wheel on the column with marks made during removal and install the steering wheel.

7. Torque the steering wheel column nut to 18 ft. lbs. (24 Nm) and bend the retaining tabs.

8. Install the horn pad and wires to the steering wheel.

9. Connect the negative battery cable.

Steering Column

REMOVAL & INSTALLATION

Standard Column

1. Disconnect the negative battery cable.

2. Remove the upper/lower steering column switch cover panel screws and the panels.

NOTE: When removing the steering column cover panels, it will be necessary to turn the steering 90 degrees in both directions so that the screws become visible.

3. From the instrument panel, remove the lower steering column trim.

4. Disconnect the electrical harness connectors from the steering column and ignition lock switches.

5. If equipped with an automatic transmission, disconnect the park lock actuation cable.

6. Unclip the turn signal switch (left side) and the wiper switch (right side).

7. Position the steering wheel in the straight-ahead position.

8. Remove the steering column shaft to steering shaft flange pinch bolt and the steering column to dash bolts.

9. To remove the steering column to instrument panel shear bolts, perform the following procedures:

a. Using a center punch, center punch the left side shear bolt.

b. Using an ⅛ in. drill bit, drill a hole through the shear bolt.

c. Using a bolt extractor tool, drive the extractor tool into the shear bolt and unscrew it.

10. Guide the steering column out of the vehicle.

NOTE: The plastic washer seated loosely on the steering shaft serves to center the steering shaft. It is placed in the steering shaft prior to mounting the assembly and must be removed after the assembly has been completed.

To install:

11. Center the steering wheel, use a new shear bolt. Torque the steering column-to-instrument panel bolt to 16 ft. lbs. (22 Nm), the steering column shaft-to-coupling pinch bolt to 16 ft. lbs. (22 Nm), the steering pinion-to-steering coupling pinch bolt to 16 ft. lbs. (22 Nm) and the steering wheel-to-steering column nut to 18 ft. lbs. (24 Nm).

Tilt Column

1. Disconnect the negative battery terminal from the battery.

2. From the steering wheel, remove the cover cap with the horn button.

3. Position the steering wheel on the straight-ahead position. Remove the steering wheel nut and washer.

4. Using a steering wheel puller tool, pull the steering wheel from the steering column. Do not pound on the steering wheel for damage may occur to the steering column.

NOTE: When removing the steering wheel, make sure the puller hook claws face outwards. If necessary, unclip the contact ring from the steering wheel and replace it with a new one. When installing, make sure the turn signal return segment points to the left. Lubricate the contact finger to contact plate surface.

5. Remove the steering wheel height adjustment lever, the turn signal switch cover screws and both covers.

6. Place the ignition switch (with key) in **2ND** position, press detent spring downward and remove the ignition lock cylinder.

7. Disconnect the electrical harness connector from the ignition contact switch, then, remove the headless set screw and the ignition switch. Remove the switch housing screws and push the switch towards the steering wheel.

8. Compress the locking tabs, re-move the turn signal switch (left side) and the wiper switch (right side) from the switch housing.

9. Remove the steering column shaft-to-steering shaft flange pinch bolt and the steering column-to-dash bolts.

10. To remove the steering column-to-instrument panel shear bolts, perform the following procedures:

a. Using a center punch, center punch the left side shear bolt.

b. Using an ⅛ in. drill bit, drill a hole through the shear bolt.

c. Using the bolt extractor tool, drive the extractor tool into the shear bolt and unscrew it.

11. Guide the steering column out of the vehicle.

NOTE: The plastic washer seated loosely on the steering shaft serves to center the steering shaft. It is placed in the steering shaft prior to mounting the assembly and must be removed after the assembly has been completed.

To install:

12. Center the steering wheel, use a new shear bolt. Torque the steering column-to-instrument panel bolt to 16 ft. lbs. (22 Nm), the steering column shaft-to-coupling pinch bolt to 16 ft. lbs. (22 Nm), the steering pinion-to-steering coupling pinch bolt to 16 ft. lbs. (Nm) and the steering wheel-to-steering column nut to 18 ft. lbs. (24 Nm).

Manual Steering Rack

ADJUSTMENT

1. Raise and support the vehicle safely.

2. Center the steering wheel.

3. Loosen the adjuster plug locknut. Turn the adjuster plug clockwise until it bottoms, then, back it out 50–70 degrees.

4. Inspect the steering pinion torque, it should be 8–20 inch lbs.

5. After adjusting the pinion torque, hold the adjuster stationary and torque the adjuster plug locknut to 50 ft. lbs. (68 Nm).

REMOVAL & INSTALLATION

1. Disconnect the negative battery cable.

2. Position the steering wheel in the straight-ahead position. Remove the steering column-to-coupling pinch bolt and the pinion shaft-to-coupling pinch bolt.

3. Remove the air cleaner.

4. Raise and support the vehicle safely.

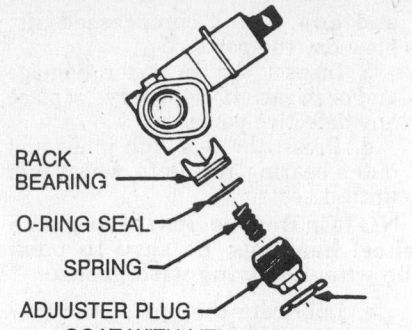

RACK BEARING
O-RING SEAL
SPRING
ADJUSTER PLUG
COAT WITH LITHIUM BASE GREASE BEFORE ASSEMBLY
ADJUSTER PLUG LOCKNUT

Exploded view of the manual steering gear adjustment assembly

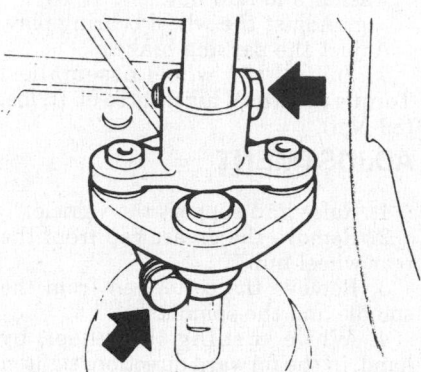

View of the steering column pinch bolts

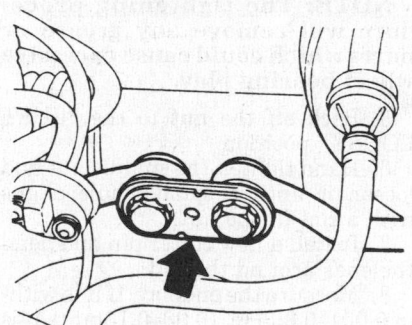

Manual steering rack and pinion lock plate

5. At the center of the manual steering assembly, cut the plate lock in half before attempting to remove the lock plate bolts; do not attempt to re-use the lock plate. Remove both tie rod end-to-steering rack/pinion center bolts.

6. If equipped, remove both steering damper brackets.

7. Remove the steering assembly-to-chassis bolts and the dash seal from the rack/pinion. Remove the steering assembly through the right wheel opening.

To install:

NOTE: If the studs were removed with the mounting studs, reinstall the studs and retorque.

If removed for the 2nd time, reuse the stud with thread locking compound.

8. Use new self locking nuts, a new lock plate. Pay attention to the direction of the notches in the lock plate. Torque the steering rack/pinion-to-chassis nuts to 28 ft. lbs. (38 Nm) and the tie rod end-to-steering rack/pinion bolts to 65 ft. lbs. (88 Nm). Adjust the manual steering rack/pinion play.

NOTE: When installing the coupling onto the steering pinion, push it downward and torque the pinion-to-coupling pinch bolt to 18 ft. lbs. (25 Nm). When installing the steering spindle, pull it upward until it stops on the spindle ball bearing and torque the upper pinch bolt to 34 ft. lbs. (46 Nm).

Power Steering Rack

ADJUSTMENT

1. Raise and support the vehicle safely.
2. Center the steering wheel.
3. Loosen the adjuster plug locknut. Turn the adjuster plug clockwise until it bottoms, then, back it out 50–70 degrees.
4. Inspect the steering pinion torque is 16 inch lbs.
5. After adjusting the pinion torque, hold the adjuster stationary and torque the adjuster plug locknut to 50 ft. lbs. (68 Nm).

REMOVAL & INSTALLATION

1. Disconnect the negative battery cable.
2. Position the steering wheel in the straight-ahead position. Remove the steering column-to-coupling pinch bolt and the pinion shaft-to-coupling pinch bolt.
3. Remove the air cleaner.
4. At the center of the steering rack, cut the plate lock in half before attempting to remove the lock plate bolts; do not attempt to reuse the lock plate. Remove both tie rod end-to-steering rack center bolts.
5. Remove the high pressure hoses from the steering rack.
6. Remove the rack-to-chassis bolts and remove the dash seal from the steering rack. Remove the steering rack through the rightwheel opening.
To install:

NOTE: If the studs were removed with the mounting studs, reinstall the studs and retorque. If removed for the 2nd time, reuse the stud with thread locking compound.

7. Use new self locking nuts, a new lock plate. Pay attention to the direction of the notches in the lock plate. Torque the steering rack-to-chassis nuts to 28 ft. lbs. (38 Nm) and the tie rod end-to-steering rack bolts to 65 ft. lbs. (88 Nm). Adjust the steering rack play.

NOTE: When installing the coupling onto the steering pinion, push it downward and torque the pinion-to-coupling pinch bolt to 37 ft. lbs. (51 Nm). When installing the steering spindle, pull it upwards until it stops on the spindle ball bearing and torque the upper pinch bolt to 34 ft. lbs. (46 Nm).

8. Refill the power steering system reservoir and bleed the system.

Power Steering Pump

REMOVAL & INSTALLATION

1.6L Engine

1. Disconnect the negative battery cable.
2. Remove the power steering belt.
3. Remove the pressure and return lines from the power steering pump and plug.
4. Remove the upper timing belt cover.
5. Remove the pump retaining bolts and remove the pump.
To install:
6. Install the power steering pump and retaining bolts. Torque the pump retaining bolts to 20 ft. lbs. (27 Nm).
7. Connect the pressure and return lines. Fill the pump with new fluid and bleed the system.
8. Install the drive belt, adjust the belt tension and connect the negative battery cable.

2.0L Engine

1. Disconnect the negative battery cable.
2. Remove the serpentine belt.
3. Remove the pressure and return lines from the power steering pump and plug.
4. Remove the pump retaining bolts which are accessible through the holes of the pulley and remove the pump from the bracket.
To install:
5. Install the power steering pump and retaining bolts. Torque the pump retaining bolts to 18 ft. lbs. (25 Nm).
6. Connect the pressure and return lines. Fill the pump with new fluid and bleed the system.
7. Install the serpentine belt, adjust the belt tension and connect the negative battery cable. Tighten the pump retaining bolts to 18 ft. lbs. (25 Nm).

8. Fill the pump with new fluid and bleed the system.

BELT ADJUSTMENT

The drive belt should be inspected and/or replaced every 30,000 miles or 24 mouths.
1. Disconnect the negative battery cable.
2. Using a belt tension gauge tool, position it on the center of the longest belt span.
3. If the belt tension is not correct, loosen the alternator adjuster bolt, move the alternator until the correct tension is attained.
4. Tighten the alternator bolts. Connect the negative battery cable.

SYSTEM BLEEDING

NOTE: If the power steering system has been serviced, an accurate fluid level reading cannot be obtained until the air is bled from the system.

1. Position the wheels the the extreme left and add an approved power steering fluid to the **COLD** mark on the fluid level indicator.
2. Start the engine and run it at fast idle, recheck the fluid level. If necessary bring the level up to the **COLD** mark.
3. Bleed the system by turning the wheels from left to right without reaching the stop at either end. Maintain the fluid level at the **COLD** mark or just above the pump casting.

NOTE: Power steering fluid with air in it will appear to be light tan or red. This air must be eliminated from the fluid before normal steering action can be obtained.

4. Return the the wheels to the center position. Continue to run the engine for 2 or 3 minutes.
5. Road test the vehicle to be sure the steering functions normal and is free of noise.
6. Recheck the fluid level. When the engine has reached normal operating temperature, the level should be at the **HOT** mark. Add fluid if necessary.
7. Check the system for leaks.

Tie Rod Ends

REMOVAL & INSTALLATION

Outer

1. Raise and safely support the vehicle.
2. Remove the tie rod end-to-steering knuckle nut.

3. Loosen the tie rod end pinch bolt.

4. Using a steering linkage puller tool, separate the tie rod end from the steering knuckle.

5. Remove the tie rod end from the tie rod adjuster sleeve. When unscrewing the tie rod end, record the number of revolutions to remove it, this will aid in installation.

To install:

6. Turn the new tie rod end in the same amount of turns to remove it. Torque the tie rod end-to-steering knuckle nut to 43–50 ft. lbs. (58–68 Nm).

7. Check and adjust the front end alignment.

Inner

1. Raise and safely support the vehicle.

2. Remove the outer tie rod end-to-steering knuckle nut.

3. Loosen the tie rod end pinch bolt.

4. Using a steering linkage puller tool, separate the outer tie rod end from the steering knuckle.

5. Remove the outer tie rod end from the tie rod adjuster sleeve. When unscrewing the tie rod end, record the number of revolutions to remove it, this will aid in installation.

6. Cut the center bolt lock plate and discard.

7. Remove the inner tie rod end retaining bolt and slide the tie rod end from behind the bolt support. Loosen the pinch bolt at the outer tie rod end and remove the inner tie rod end.

NOTE: If both tie rod ends are to be removed, the rack and pinion and other parts must be kept aligned. This can be accomplished by installing a retaining bolt in the tie rod end bolt hole of the rack after removing the first the tie rod end.

To install:

8. Install the inner tie rod end to the outer tie rod end.

9. Torque the pinch bolt to 16 ft. lbs. (22 Nm). Slip the inside end of the

tie rod end behind the bolt support and install the lock plate and bolt.

10. Torque the mounting bolt to 65 ft. lbs. (88 Nm). Lower the vehicle.

BRAKES

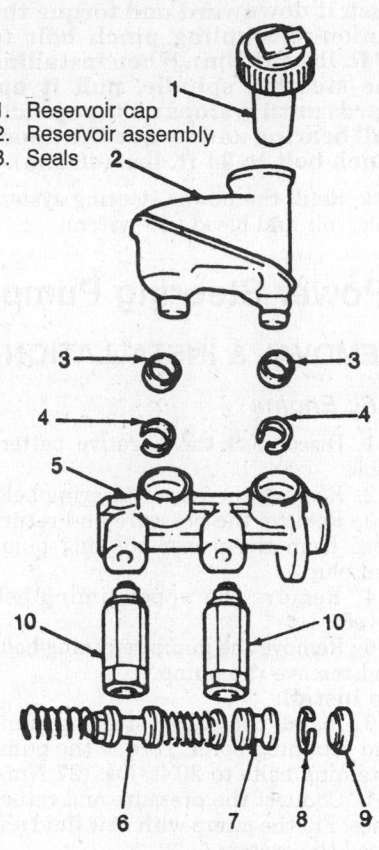

1. Reservoir cap
2. Reservoir assembly
3. Seals
4. Retaining clamps
5. Cylinder body
6. Secondary piston assembly
7. Primary piston
8. Retainer
9. Seal ring
10. Proportioning valves

Master cylinder assembly

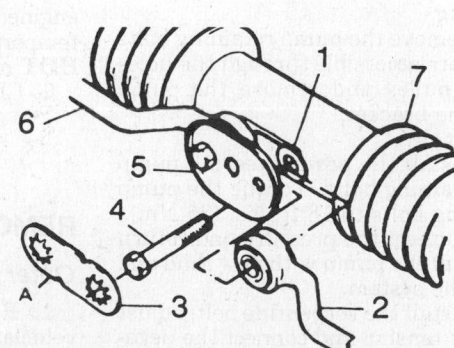

1. Washer
2. Inner tie rod end
3. Locking plate
4. Bolt
5. Inner tie rod end plate
6. Inner tie rod end
7. Rack and pinion boot

Installation of the inner tie rod end assembly

For all brake system repair and service procedures not detailed below, please refer to "Brakes" in the Unit Repair section.

Master Cylinder

REMOVAL & INSTALLATION

1. Disconnect the negative battery cable.

2. Disconnect the electrical connector from the reservoir cap.

3. Disconnect and plug the brake lines from the master cylinder.

4. Remove the master cylinder to power brake booster nuts.

5. Remove the master cylinder from the vehicle.

To install:

6. Use new self locking nuts. Torque the master cylinder-to-brake booster nuts to 13 ft. lbs. (18 Nm).

7. Connect master cylinder brake lines. Refill the master cylinder reservoir with clean brake fluid. Bleed the brake system.

8. Connect the electrical connection to the reservoir cap. Connect the negative battery cable and test the brake operation.

Proportioning Valve

Since the valves are factory set in pairs, they must be replaced in pairs.

REMOVAL & INSTALLATION

1. Disconnect the negative battery cable.

2. Disconnect and plug the brake lines from the proportioning valves.

3. Remove the proportioning valves from the master cylinder.

NOTE: Be sure the valves are stamped with identical part numbers.

4. To install, Torque the proportioning valve-to-master cylinder bolts to 30 ft. lbs. (40 Nm). Refill the master cylinder reservoir with clean brake fluid. Bleed the brake system.

Power Brake Booster

REMOVAL & INSTALLATION

1. Disconnect the negative battery cable.

2. Remove the master cylinder from the booster; do not disconnect the brake lines.

3. Remove the vacuum hose union nut from the intake manifold, on 1.6L engine only. Disconnect the vacuum hose from the brake booster assembly.

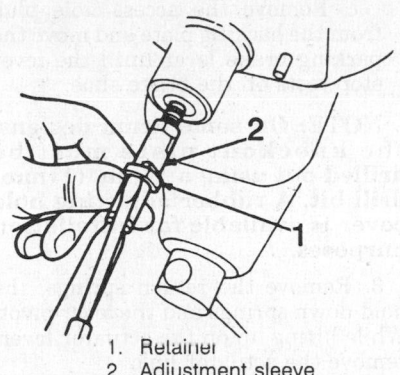

1. Retainer
2. Adjustment sleeve

Removing the retaining ring from the pushrod

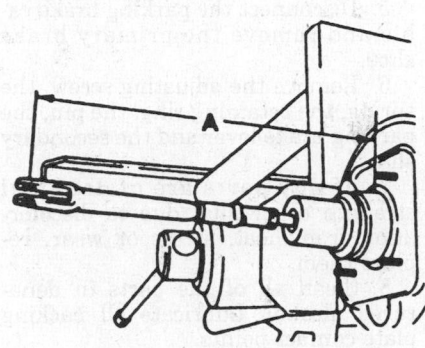

Measuring pushrod length

4. Remove the windshield washer reservoir.

5. From under the dash, remove the brake light switch and the brake pedal spring.

6. Remove the pushrod to pedal pin retainer and the pin.

7. If not equipped with power steering, remove the brake pedal bracket-to-dash nuts. If equipped with power steering, remove the lower mounting screw from behind the fluid lines using a flat head socket wrench.

8. To remove the power brake booster and bracket, tilt the brake servo slightly and remove it upwards.

9. From the power brake booster, remove the 2-part support bracket and the rubber boot.

10. Remove the pushrod retainer and pushrod, then, unscrew and remove the adjuster sleeve from the piston rod.

11. Unscrew the hex nut.

To install:

12. Install the adjuster sleeve to the pushrod and verify the distance from the booster flange to the end of the adjustment sleeve is 10.96 in. (278.5mm). Adjust as required.

13. Position the booster and mounting bracket to the firewall and install mounting bolts. Tighten the mounting bolts to 16 ft. lbs. (22 Nm).

14. Install the master cylinder to the booster and the mounting nuts.

Torque the master cylinder-to-power booster nuts to 13 ft. lbs. (18 Nm).

15. Connect the negative battery cable. Start the engine and check the brake operation.

Brake Caliper

REMOVAL & INSTALLATION

1. Disconnect the negative battery cable.

2. Remove half of the brake fluid from the master cylinder.

3. Raise and support the vehicle safely and remove the wheel assembly.

NOTE: Remove the brake hose retaining bolt, only if the caliper is going to be overhauled or replaced.

4. Position a large C-clamp over the caliper with the screw end against the outboard brake pad. Tighten the clamp until the caliper piston is pushed in enough to bottom the piston.

5. Remove the C-clamp. Remove the caliper guide pins and lift the caliper off of the rotor.

6. Support the caliper so there is no strain on the brake hose.

7. Press the inboard pad outward and remove it from the caliper.

8. Remove and discard the O-ring bushings and steel sleeves, new parts are to be installed.

9. Check the condition of the rotor. If rotor measurements exceed manufacture's specifications or has mild scoring, machine the rotor.

To install:

10. Lubricate and install the O-ring bushings. Install the sleeves by pressing them through the O-rings until the sleeve end on the pad side is flush with caliper ear.

11. Position the inboard pads so the pad contacts the piston and the support spring ends. The inboard and outboard pads are similar but not interchangeable.

12. Press down on the ears at the top of the inboard pad until the pad lies flat and the spring ends are just inside the lower edge of the pad.

13. Position the outboard pad with the ears toward the positioning pin holes and the tab on the inner edge of the pad resting in the notch in the edge of the caliper. Bend the ears to provide a slight interference fit in the caliper.

14. Press the outboard pad tightly into position and clinch the ears of the outboard pad over the outboard caliper half.

15. Position the caliper over the rotor.

16. Install the caliper over the rotor.

17. Install the caliper mounting bolts and tighten to 70 ft. lbs. (95 Nm).

18. If the brake hose retaining nut was disconnected, reconnect it and torque to 18 ft. lbs. (25 Nm).

19. Install the wheel assembly and lower the vehicle.

20. Fill the master cylinder with brake fluid and bleed the system.

21. Connect the negative battery cable.

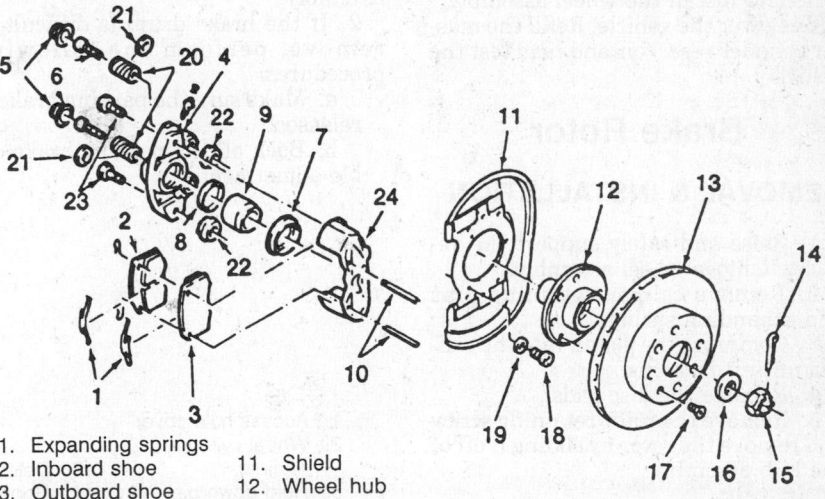

1. Expanding springs
2. Inboard shoe
3. Outboard shoe
4. Caliper housing
5. Bleeder valve cap
6. Bleeder valve
7. Piston boot
8. Piston seal
9. Piston caliper
10. Retaining pins
11. Shield
12. Wheel hub
13. Rotor
14. Cotter pin
15. Wheel hub nut
16. Wheel hub washer
17. Detent screw
18. Shield screw
19. Shield washer
20. Bleeder valve plug
21. Bleeder valve washer
22. Caliper housing spacer
23. Caliper housing bolt
24. Retaining frame

Exploded view of the front disc brake and related components

Disc Brake Pads

REMOVAL & INSTALLATION

NOTE: Caliper removal is not necessary for removal disc pads.

1. Raise and safely support the vehicle. Remove the wheel assembly; reinstall 2 lug nuts to retain the rotor to the axle hub.
2. Using a siphon, remove ⅔ of the brake fluid from the master cylinder.
3. Using a C-clamp, position the screw end against the outboard brake pad. Tighten the clamp until the caliper piston is pushed in enough to bottom the piston.
4. Using a punch, drive out the retaining pins from the inside-out.
5. Remove the expanding springs from under retainer pins.

NOTE: The expanding springs are under tension, use caution when removing retainer pins.

6. Remove the outboard pad and the inboard pad from the caliper assembly.

To install:

7. Install the pads into the caliper assembly with the wear sensor on the top of the inboard pad.
8. Install the upper retainer pin from the outside-in.
9. Install the expanding springs with the long end under the retaining pins.
10. Install the lower retaining pin, making sure the long side of the expanding spring is under the pin.
11. Remove the 2 rotor-to-wheel hub nuts and install the wheel assembly.
12. Lower the vehicle. Refill the master cylinder reservoir and road test the vehicle.

Brake Rotor

REMOVAL & INSTALLATION

1. Raise and safely support the vehicle. Remove wheel assembly.
2. Remove caliper retaining pins and expanding springs.
3. Compress the piston into the caliper until it bottoms.
4. Remove the disc pads.
5. Remove the rotor retaining screw and remove the rotor by sliding it off of the hub assembly.

To install:

6. Slide the rotor on the hub assembly and install 2 lug nuts to hold it in place.
7. Install the brake pads into the caliper.
8. Remove the lug nuts from the rotor and install the wheel assembly.
9. Lower the vehicle.

10. Fill the master cylinder with brake fluid.
11. Depress the brake pedal 3–4 times to seat the brake linings and to restore pressure in the system.

Brake Drums

REMOVAL & INSTALLATION

1. Raise and safely support the rear of the vehicle.
2. Remove the wheel assembly.
3. Remove the brake drum from the axle.
4. If difficulty is encountered in removing the drum, perform the following:
   a. Ensure the parking brake is released.
   b. Loosen the parking brake cable.
   c. Remove the access hole plug from the backing plate and move the parking brake lever until the lever stop rests on the brake shoe.

To install:

5. Using a brake adjusting tool, adjust the brake shoe to 0.50 in. (1.27mm) less than the brake drum diameter. Install the brake drum to the axle.
6. Install the wheel assembly and lower the vehicle.

Brake Shoes

REMOVAL & INSTALLATION

1. Raise and safely support the rear of the vehicle. Remove the wheel assembly.
2. If the brake drum is difficult to remove, perform the following procedures:
   a. Make sure the parking brake is released.
   b. Back off the parking brake cable adjustment.

c. Remove the access hole plug from the backing plate and move the parking brake lever until the lever stop rests on the brake shoe.

NOTE: On some drum designs, the knockout plate must be drilled out using a ⁷⁄₁₆ in. (11mm) drill bit. A rubber adjusting hole cover is available for installation purposes.

3. Remove the return springs, the hold-down springs and the lever pivot. While lifting up on the actuator lever, remove the actuator link.
4. Remove the actuator lever, the lever return spring, the parking brake strut and the strut spring.
5. Disconnect the parking brake cable and remove the primary brake shoe.
6. Remove the adjusting screw, the spring, the retaining ring, the pin, the parking brake lever and the secondary shoe.
7. If any parts are of doubtful strength or quality, due to discoloration from heat, stress or wear, replace them.
8. Clean all of the parts in denatured alcohol. Lubricate all backing plate contact points.

To install:

9. Install the primary brake shoe and connect the parking brake cable.
10. Install the adjusting screw, the spring, the retaining ring, the pin, the parking brake lever and the secondary shoe.
11. Install the actuator lever, the lever return spring, the parking brake strut and the strut spring.
12. Install the return springs, the hold-down springs and the lever pivot. While lifting up on the actuator lever, install the actuator link.
13. Using a brake adjusting tool, adjust the brake shoe to 0.50 in. (1.27mm) less than the brake drum diameter. Install the brake drum.

1. Access hole cover
2. Wheel cylinder retainer
3. Hold-down pin
4. Backing plate
5. Wheel cylinder assembly
6. Shoe and lining
7. Hold-down spring
8. Hold-down washer
9. Brake drum
10. Detent screw
11. Upper return spring
12. Lower return spring
13. Connecting spring link
14. Adjuster assembly
15. Adjuster spring
16. Actuator adjuster
17. Shoe adjuster

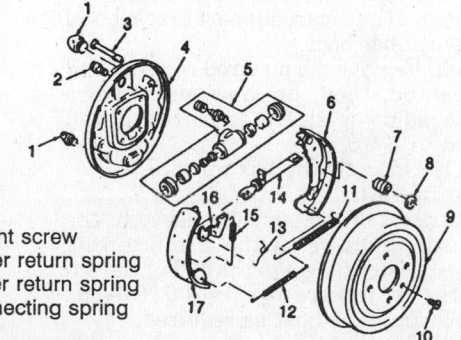

Exploded view of the rear drum and related components

14. Lower the vehicle. Road Test the vehicle.

Wheel Cylinder

REMOVAL & INSTALLATION

1. Raise and safely support the vehicle.

2. Using a piece of chalk, mark the relative position of the wheels to the wheel hub. Remove the wheel/tire assemblies.

3. Remove the brake drum-to-wheel hub detent screw and the brake drum.

4. Remove the upper return spring and push the brake shoes slightly outward.

NOTE: Note the position of the adjuster assembly and the adjuster actuator-to-actuator spring.

5. Remove and plug the brake line from the wheel cylinder.

6. Remove the wheel cylinder-to-backing plate bolt and the wheel cylinder.

To install:

7. Position the wheel cylinder-to-the backing plate and install the retaining bolts and brake line.

8. Torque the wheel cylinder-to-backing plate bolt to 7 ft. lbs. (9.5 Nm).

9. Adjust the rear wheel brakes and the parking brake.

10. Bleed the brake system. Road test the vehicle.

Parking Brake Cable

ADJUSTMENT

1. Raise and safely support the vehicle.

2. Release the parking brake.

3. Inspect the parking brake cable for free movement and ensure rear brakes are adjusted properly.

4. At the equalizer, adjust the self-

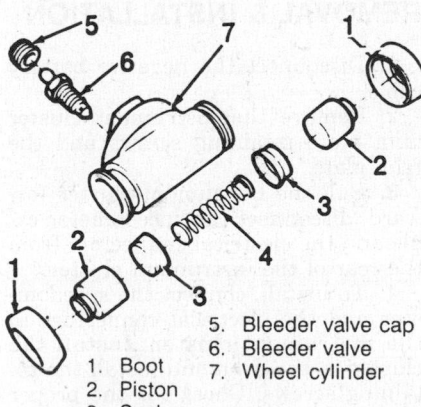

5. Bleeder valve cap
6. Bleeder valve
7. Wheel cylinder

1. Boot
2. Piston
3. Seal
4. Spring assembly

Exploded view of the rear wheel cylinder

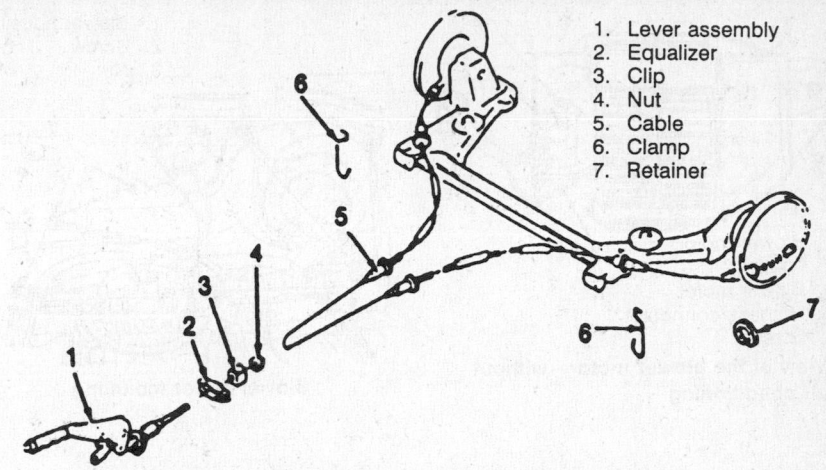

1. Lever assembly
2. Equalizer
3. Clip
4. Nut
5. Cable
6. Clamp
7. Retainer

Exploded view of the parking brake cable system

locking nut until the wheels are difficult to turn.

5. Back-off the self-locking nut until the rear wheels are just free to turn.

6. Check the parking brake for proper operation.

REMOVAL & INSTALLATION

1. Release parking brake lever.

2. Raise and safely support the vehicle. Remove the rear wheel assembly.

3. Remove the brake drum mounting screw and the brake drum.

4. At the transaxle tunnel, remove the parking brake cable from the guides.

5. Remove the plastic guides from the fuel tank bracket. Remove the cable from the parking brake equalizer.

6. Remove the parking brake cable from the rear axle assembly guides.

7. Using a pointed tool, remove the retaining ring from the plastic sleeve in the backing plate.

8. Remove the parking brake cable from the parking brake shoe lever and the brake anchor plate.

To install:

9. Install the parking brake cable to the parking brake shoe lever and the brake anchor plate.

10. Install the retaining ring into the plastic sleeve and into the backing plate.

11. Install the parking brake cable to the rear axle assembly guides.

12. Install the plastic guides in the fuel tank bracket.

13. At the transaxle tunnel, secure the parking brake cable to the guides.

14. Install the brake drum mounting screw and the brake drum.

15. Install the wheel assembly and lower the vehicle.

16. Check the adjustment and operation of the parking brake.

Brake System Bleeding

On diagonally split brake systems, start the bleeding procedure with the right rear, then the left front, the left rear and the right front.

1. Clean the bleeder screw at each wheel.

2. Attach a small rubber hose to the bleeder screw and place the other end in a clear container filled partially with new brake fluid.

3. Refill the master cylinder with new brake fluid. The master cylinder reservoir should be checked and topped off frequently during the bleeding procedure.

4. Have an assistant slowly pump the brake pedal and hold the pressure.

5. Open the bleeder screw about ¼ turn. The pedal should fall to the floor as air and fluid are pushed out. Close the bleeder screw while the assistant holds the pedal to the floor, then, slowly release the pedal and wait 15 seconds. Repeat the process until no air bubbles are forced from the system when the brake pedal is applied. It may be necessary to repeat this numerous times to get all of the air from the system.

6. Repeat this procedure on the remaining wheel cylinders and calipers.

NOTE: Remember to wait 15 seconds between each bleeding and do not pump the pedal rapidly. Rapid pumping of the brake pedal pushes the master cylinder secondary piston down the bore in a manner that makes it difficult to bleed the system.

7. Check the brake pedal for sponginess and the brake warning light for an indication of unbalanced pressure.

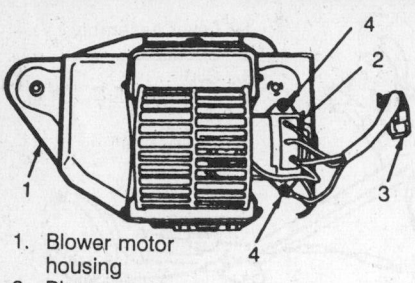

1. Blower motor housing
2. Blower motor
3. Harness connector
4. Screw

View of the blower motor—without air conditioning

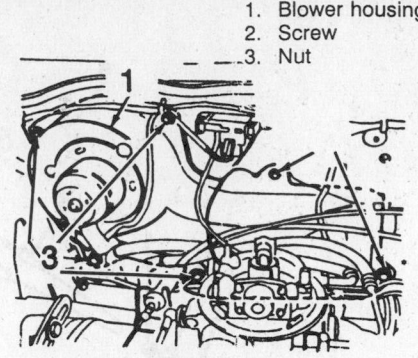

1. Blower housing
2. Screw
3. Nut

Blower motor mounting

CHASSIS ELECTRICAL

Heater Blower Motor

The heater blower motor is located in the engine compartment, attached to the cowl.

REMOVAL & INSTALLATION

Without Air Conditioning

1. Disconnect the negative battery cable.
2. Remove the wiper arms.
3. Remove the wind deflector screws and the deflector halves.
4. Remove the windshield washer nozzles from the water deflectors.
5. Remove the dash panel seal and clip.
6. Remove the right wiper bearing nut.
7. Remove the water deflector.
8. Disconnect the electrical connectors from the blower motor. Remove the heater blower motor retaining screws.
9. Remove the housing, the motor and the housing cover.
To install:
10. Align the motor by first inserting the lower screw, then the upper.
11. Connect the electrical connectors to the blower motor. Install remainder of the heater blower motor retaining screws.
12. Install the water deflector.
13. Install the right wiper bearing nut.
14. Install the dash panel seal and clip.
15. Connect the windshield washer nozzles to the water deflectors.
16. Install the wind deflector screws and the deflector halves.
17. Install the wiper arms.
18. Connect the negative battery cable. Check blower operation.

With Air Conditioning

1. Disconnect the negative battery cable.
2. Disconnect the electrical connections and cooling hose from the blower motor.
3. Remove the blower motor attaching screws and pull the blower motor from the cowl.
4. Remove the fan retaining nut and the fan from the motor.
To install:
5. Install the fan on the new blower motor with the opening facing away from the motor and install the retaining nut.
6. Position the blower motor assembly to the cowl and install the attaching screws.
7. Install the blower motor cooling hose.
8. Connect the electrical connections to the blower motor.
9. Connect the negative battery cable.

Windshield Wiper Motor

REMOVAL & INSTALLATION

1. Disconnect the negative battery cable. At the wiper arm shaft, remove the plastic pivot cap by moving it upwards.
2. Remove the wiper arm shaft nut and the wiper arm from the shaft.
3. Remove the cowl vent grille screws and the grille.
4. Disconnect the electrical connectors from the wiper motor.
5. Remove the crank arm to wiper motor nut and disconnect the crank arm from the wiper motor.
6. Remove the wiper motor to cowl bolts and the motor from the vehicle.

To install:
7. Position the wiper motor to cowl and install the mounting bolts.
8. Connect the crank arm to wiper motor and install the retaining nut.
9. Connect the electrical connectors to the wiper motor.
10. Install the cowl vent grille and retaining screws.
11. Install the wiper arm to the wiper shaft and install the retaining nut.
12. At the wiper arm shaft, install the plastic pivot cap. Connect the negative battery cable.

Windshield Wiper Switch

REMOVAL & INSTALLATION

1. Disconnect the negative battery cable.
2. Remove the lower instrument panel trim.
3. Remove the upper steering column panel screws from both sides; turn the steering wheel 90 degrees for right and left access.
4. Remove the screws from the lower cover panel and remove the panel.
5. Pull the handle from the lock release lever and unscrew the tilt lever, if equipped.
6. Disconnect the electrical connector from the switch housing; push inward on either side of the switch to release it from the retaining clips.
7. To install, push the switch into the retaining clips and connect the electrical lead. Install the lock release lever and tilt lever, if equipped. Install the column cover panels and check switch operation.
8. Connect the negative battery cable.

Instrument Cluster

REMOVAL & INSTALLATION

1. Disconnect the negative battery cable.
2. Remove the instrument cluster trim plate retaining screws and the trim plate.
3. Pull the instrument cluster forward, disconnect the speedometer cable and the electrical connectors from the rear of the instrument cluster.
4. To install, connect the speedometer and the electrical connectors to the instrument cluster. Install the cluster in the dash and install the retaining screws. Check for the proper operation of the speedometer and the gauges.
5. Connect the negative battery cable.

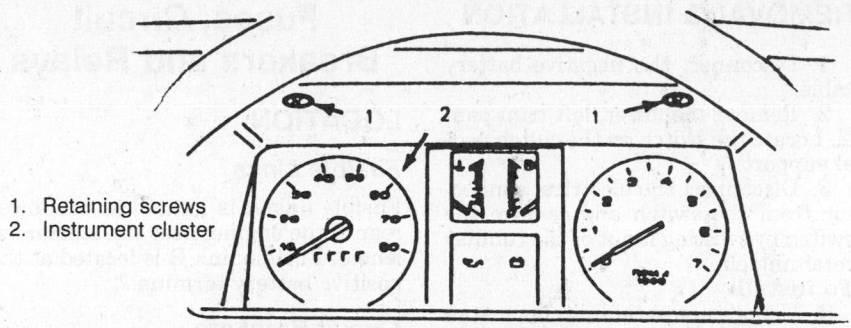

1. Retaining screws
2. Instrument cluster

Instrument cluster mounting

Speedometer

REMOVAL & INSTALLATION

1. Disconnect the negative battery cable.
2. Remove the instrument panel cluster lens, the face plate and the instrument cluster from the instrument panel.
3. Press the speedometer cable retainer and separate the speedometer cable from the speedometer.
4. Remove the speedometer retaining screws. Remove the speedometer from the cluster.

To install:
5. Install the speedometer into the cluster and install the retaining screws.
6. Position the cluster assembly to the dash, while inserting the speedometer cable into the speedometer, push in securely.
7. Install the instrument cluster face plate and lens to the panel.
8. Connect the negative battery cable.

Radio

REMOVAL & INSTALLATION

1. Disconnect the negative battery cable.
2. Turn the key to the **ON** position. Move the shift lever into **1ST** gear.
3. Remove the storage pocket panel and the front center console.
4. Remove the retaining screws at the top of the console and slide the center console back.
5. Using a center punch on each side of the radio bracket, release the radio by inserting the center punch into the each hole in the control panel and push to the right.
6. Reach behind the radio and push rearward toward the shift lever until the radio is free. Disconnect the antenna and harness connections from the radio.

To install:
7. Connect the antenna and harness connections. Position the radio to the dash and slide it in until it seats into the bracket.
8. Slide the console forward to the dash and install the retaining screws.
9. Move the shift lever in to **P** or **N** position.
10. Install the front center console and storage pocket panel.
11. Connect the negative battery cable. Check radio operation.

Headlight Switch

REMOVAL & INSTALLATION

1. Disconnect the negative battery cable.
2. Using an offset tool, depress the headlight switch retaining clips and pull the switch from the dash.
3. Disconnect the electrical connector from the rear of the switch. Remove the switch from the vehicle.
4. To install, connect the electrical lead and push switch into position in the dash. Connect the negative battery cable.

Combination Switch

The combination switch consists of the turn signal switch and the dimmer

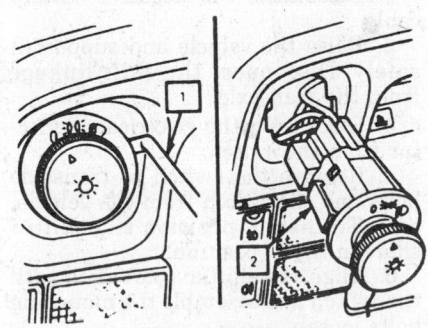

1. Offset screwdriver
2. Retainer

Replacing the headlight switch

switch which is connected to the steering column.

REMOVAL & INSTALLATION

1. Disconnect the negative battery cable.
2. Remove the lower instrument panel trim.
3. Remove the upper steering column panel screws from both sides; turn the steering wheel 90 degrees for right and left access.
4. Remove the screws from the lower cover panel and remove the panel.
5. Pull the handle from the lock release lever and unscrew the tilt lever, if equipped.
6. Disconnect the electrical connector from the switch housing; push inward on either side of the switch to release it from the retaining clips.
7. To install, push the switch into the retaining clips and connect the electrical lead. Install the lock release lever and tilt lever, if equipped. Install the column cover panels and check switch operation.
8. Connect the negative battery cable.

Ignition Lock

REMOVAL & INSTALLATION

1. Disconnect the negative battery cable.
2. Remove the lower instrument cluster trim.
3. Remove the turn signal/wiper switch cover panels.
4. With the key in the ignition switch, turn the key to the **2ND** position.
5. Using a small Allen wrench, press downward on the detent spring and remove the ignition lock cylinder.
6. To install, place the key in the cylinder and install it into the ignition switch. Install the trim panels. Connect the negative battery cable.

Ignition Switch

REMOVAL & INSTALLATION

1. Disconnect the negative battery cable.
2. Remove the lower instrument cluster trim.
3. Remove the turn signal/wiper switch cover panels.
4. Disconnect the electrical connector, remove the set screw and remove the ignition switch.
5. Install the ignition switch into place and attach the wiring.
6. Install the set screw. Install the trim panels and connect the negative battery cable.

Stoplight Switch

ADJUSTMENT

1. Disconnect the negative battery cable.
2. Remove the lower, left trim panel and locate the stoplight switch on the brake pedal support.
3. Disconnect the electrical connector from the switch and remove the switch by twisting it out of the tubular retaining clip.
4. Pull back on the brake pedal and push the switch through the retaining clip noting the clicks; repeat this procedure until no more clicks can be heard.
5. Connect the electrical connector to the switch.
6. Connect the negative battery cable and check the switch operation.

REMOVAL & INSTALLATION

1. Disconnect the negative battery cable.
2. Disconnect the electrical connector from the brake light switch, located above the brake pedal.
3. Remove the switch from the tubular clip on the brake pedal mounting bracket.
To install:
4. Insert the switch into the clip until the switch body seats on the clip.
5. Pull the brake pedal rearward against the internal pedal stop. The switch will be moved in the tubular clip providing the proper adjustment.
6. Connect the electrical connector to the switch.
7. Connect the negative battery cable and check the switch operation.

Clutch Switch

ADJUSTMENT

1. Disconnect the negative battery cable.
2. Remove the lower, left trim panel and locate the switch on the clutch pedal support.
3. Disconnect the electrical connector from the switch and remove the switch by twisting it out of the tubular retaining clip.
4. Pull back on the clutch pedal and push the switch through the retaining clip noting the clicks; repeat this procedure until no more clicks can be heard.
5. Connect the electrical connector to the switch.
6. Disconnect the negative battery cable and check the switch operation.

REMOVAL & INSTALLATION

1. Disconnect the negative battery cable.
2. Remove the lower, left trim panel. Locate the switch on the clutch pedal support.
3. Disconnect the electrical connector from the switch and remove the switch by twisting it out of the tubular retaining clip.
To install:
4. Using a new retaining clip, install the switch and connect the electrical connector.
5. To adjust the switch, pull back on the clutch pedal, push the switch through the retaining clip noting the clicks; repeat this procedure until no more clicks can be heard.
6. Connect the negative battery cable and check the switch operation.

Neutral Safety Switch

ADJUSTMENT

1. Disconnect the negative battery cable.
2. Raise the vehicle and support it safely.
3. Loosen the switch to transaxle bolts. Insert $3/32$ in. (2.34mm) drill in the service adjustment hole and rotate the switch until the drill drops into the a depth of $9/64$ in. (9mm).
4. Torque the mounting bolts to 22 ft. lbs. (30 Nm). Remove the drill gauge.
5. Lower the vehicle and connect the negative battery cable.
6. Start the engine and verify the engine will start in **P** and **N** positions only.

NOTE: If the engine will start in any other position other than P or N, readjust the switch.

REMOVAL & INSTALLATION

1. Disconnect the negative battery cable.
2. Raise the vehicle and support it safely. Disconnect the shift linkage from the transaxle.
3. Disconnect the electrical connector from the switch.
4. Remove the switch to transaxle bolts and the switch from the vehicle.
5. To install, position the shifter shaft in the **N** position.
6. Align the shifter shaft flats with the switch and assemble the mounting bolts loosely.
7. Adjust the switch. Connect the negative battery cable and test operation.

Fuses, Circuit Breakers and Relays

LOCATION

Fusible Links

Fusible link **A** is located at the lower rear of the engine, near the starter solenoid. Fusible link **B** is located at the positive battery terminal.

Circuit Breakers

Circuit breakers **12** and **15** are located in the fuse panel.

Fuse Panel

The fuse panel is located at the left side of the instrument panel and is reached by pulling the release handle and swinging the panel downward. Always return the fuse panel to its full upward, latched position before driving the vehicle.

Various Relays

All relays for this vehicle are located on the fuse block, under the left side of the instrument panel.

Computers

LOCATION

The Electronic Control Module (ECM) is located behind the kick panel at the passenger-side door jam (under the dash).

Memory Calibration Unit

The memory calibration unit is located inside the ECM. It contains programmed information tailored to the vehicle's weight, engine, transmission, axle ratio and etc. Even though a single ECM unit can be used for many vehicles, a specific memory calibration unit must be used for each vehicle.

Flashers

LOCATION

Turn Signal Flasher

The turn signal flasher is located on the fuse block, under the left side of the instrument panel.

Hazard Warning Flasher

The hazard warning flasher is located on the fuse block, under the left side of the instrument panel.

GM "A" Body
Front Wheel Drive
Buick—Century   **Chevrolet**—Celebrity
Oldsmobile—Cutlass Ciera, Cutlass Cruiser
Pontiac—600

SPECIFICATIONS

VEHICLE IDENTIFICATION CHART

It is important for servicing and ordering parts to be certain of the vehicle and engine identification. The VIN (vehicle identification number) is a 17 digit number visible through the windshield on the driver's side of the dash and contains the vehicle and engine identification codes. The tenth digit indicates model year, and the eighth digit indicates engine code. It can be interpreted as follows:

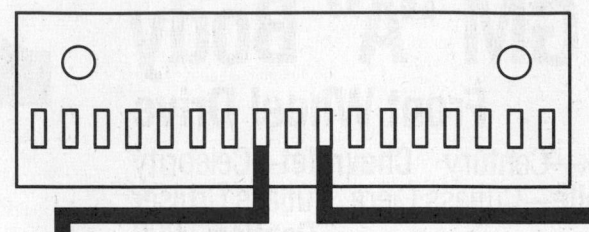

Engine Code						Model Year	
Mode	Cu. In.	Liters	Cyl.	Fuel Sys.	Eng. Mfg.	Code	Year
R	151	2.5	4	TBI	CPC	J	1988
W	173	2.8	6	PFI	CPC	K	1989
T	192	3.1	6	PFI	CPC	L	1990
N	204	3.3	6	PFI	BOC	M	1991
3	231	3.8	6	SFI	BOC	N	1992

TBI—Throttle Body Injection   CPC—Chevrolet, Pontiac, Canada
PFI—Port Fuel Injection   BOC—Buick, Oldsmobile, Cadillac
SFI—Sequential Fuel Injection

ENGINE IDENTIFICATION

Year	Model	Engine Displacement Cu. In. (liter)	Engine Series Identification (VIN)	No. of Cylinders	Engine Type
1988	Celebrity	151 (2.5)	R	4	OHV
	Celebrity	173 (2.8)	W	6	OHV
	Century	151 (2.5)	R	4	OHV
	Century	173 (2.8)	W	6	OHV
	Century	231 (3.8)	3	6	OHV
	Cutlass ①	151 (2.5)	R	4	OHV
	Cutlass ①	173 (2.8)	W	6	OHV
	Cutlass ①	231 (3.8)	3	6	OHV
	6000	151 (2.5)	R	4	OHV
	6000	173 (2.8)	W	6	OHV
1989	Celebrity	151 (2.5)	R	4	OHV
	Celebrity	173 (2.8)	W	6	OHV
	Century	151 (2.5)	R	4	OHV
	Century	173 (2.8)	W	6	OHV
	Century	204 (3.3)	N	6	OHV
	Cutlass ①	151 (2.5)	R	4	OHV
	Cutlass ①	173 (2.8)	W	6	OHV
	Cutlass ①	204 (3.3)	N	6	OHV
	6000	151 (2.5)	R	4	OHV
	6000	173 (2.8)	W	6	OHV
	6000	192 (3.1)	T	6	OHV

ENGINE IDENTIFICATION

Year	Model	Engine Displacement Cu. In. (liter)	Engine Series Identification (VIN)	No. of Cylinders	Engine Type
1990	Celebrity	151 (2.5)	R	4	OHV
	Celebrity	192 (3.1)	T	6	OHV
	Century	151 (2.5)	R	4	OHV
	Century	204 (3.3)	N	6	OHV
	Cutlass ①	151 (2.5)	R	4	OHV
	Cutlass ①	204 (3.3)	N	6	OHV
	6000	151 (2.5)	R	4	OHV
	6000	192 (3.1)	T	6	OHV
1991–92	Century	151 (2.5)	R	4	OHV
	Century	204 (3.3)	N	6	OHV
	Cutlass ①	151 (2.5)	R	4	OHV
	Cutlass ①	204 (3.3)	N	6	OHV
	6000	151 (2.5)	R	4	OHV
	6000	192 (3.1)	T	6	OHV

① Ciera & Cruiser
TBI—Throttle Body Injection
MFI—Multiport Fuel Injection
SFI—Sequential Fuel Injection
HO—High output
OHV—Overhead Valves

GENERAL ENGINE SPECIFICATIONS

Year	VIN	No. Cylinder Displacement cu. in. (liter)	Fuel System Type	Net Horsepower @ rpm	Net Torque @ rpm (ft. lbs.)	Bore × Stroke (in.)	Compression Ratio	Oil Pressure @ rpm
1988	R	4-151 (2.5)	TBI	92 @ 4000	134 @ 2800	4.000 × 3.000	8.3:1	37.5 @ 2000
	W	6-173 (2.8)	PFI	130 @ 4800	155 @ 3600	3.503 × 2.992	8.9:1	50–65 @ 1200
	3	6-231 (3.8)	SFI	150 @ 4400	200 @ 2000	3.800 × 3.400	8.0:1	37 @ 2400
1989	R	4-151 (2.5)	TBI	92 @ 4000	134 @ 2800	4.000 × 3.000	8.3:1	37.5 @ 2000
	W	6-173 (2.8)	PFI	130 @ 4800	155 @ 3600	3.503 × 2.992	8.9:1	50–65 @ 1200
	T	6-192 (3.1)	PFI	120 @ 4200	175 @ 2200	3.503 × 3.312	8.8:1	50–65 @ 2400
	N	6-204 (3.3)	PFI	160 @ 5200	185 @ 2000	3.700 × 3.160	9.0:1	45 @ 2000
1990	R	4-151 (2.5)	TBI	92 @ 4400	134 @ 2800	4.000 × 3.000	8.3:1	37.5 @ 2000
	T	6-192 (3.1)	PFI	120 @ 4200	175 @ 2200	3.503 × 3.312	8.8:1	50–65 @ 2400
	N	6-204 (3.3)	PFI	160 @ 5200	185 @ 2000	3.700 × 3.160	9.0:1	45 @ 2000
1991–92	R	4-151 (2.5)	TBI	110 @ 5200	135 @ 3200	4.000 × 3.000	8.3:1	26 @ 800
	T	6-192 (3.1)	PFI	140 @ 4400	185 @ 3200	3.503 × 3.312	8.8:1	15 @ 1100
	N	6-204 (3.3)	PFI	160 @ 5200	185 @ 2000	3.700 × 3.160	9.0:1	60 @ 1850

TBI—Throttle Body Injection
PFI—Port Fuel Injection
SFI—Sequential Multi-port Fuel Injection

GASOLINE ENGINE TUNE-UP SPECIFICATIONS

Year	VIN	No. Cylinder Displacement cu. in. (liter)	Spark Plugs Type	Gap (in.)	Ignition Timing (deg.) MT	AT	Compression Pressure (psi)	Fuel Pump (psi)	Idle Speed (rpm) MT	AT	Valve Clearance In.	Ex.
1988	R	4-151 (2.5)	R-43TS6	0.060	①	①	NA	6.0–7.0	①	①	Hyd.	Hyd.
	W	6-173 (2.8)	R-43LTSE	0.045	①	①	NA	40.0–46.0	①	①	Hyd.	Hyd.
	3	6-231 (3.8)	R-44LTS	0.080	①	①	NA	34.0–40.0	①	①	Hyd.	Hyd.
1989	R	4-151 (2.5)	R-43TS6	0.060	①	①	NA	6.0–7.0	①	①	Hyd.	Hyd.
	W	6-173 (2.8)	R-43LTSE	0.045	①	①	NA	40.0–46.0	①	①	Hyd.	Hyd.
	T	6-192 (3.1)	R-43LTSE	0.045	①	①	NA	34.0–47.0	①	①	Hyd.	Hyd.
	N	6-204 (3.3)	R-44LTS6	0.060	①	①	NA	37.0–43.0	①	①	Hyd.	Hyd.
1990	R	4-151 (2.5)	R-43TS6	0.060	①	①	NA	6.0–7.0	①	①	Hyd.	Hyd.
	T	6-192 (3.1)	R-43LTSE	0.045	①	①	NA	34.0–47.0	①	①	Hyd.	Hyd.
	N	6-204 (3.3)	R-44LTS6	0.060	①	①	NA	37.0–43.0	①	①	Hyd.	Hyd.
1991	R	4-151 (2.5)	R-43TS6	0.060	①	①	NA	9.0–13.0	①	①	Hyd.	Hyd.
	T	6-192 (3.1)	R-43LTSE	0.045	①	①	NA	40.5–47.0	①	①	Hyd.	Hyd.
	N	6-204 (3.3)	R-45LTS6	0.060	①	①	NA	41.0–47.0	①	①	Hyd.	Hyd.
1992		SEE UNDERHOOD SPECIFICATIONS STICKER										

Hyd.—Hydraulic
① Refer to underhood specifications sticker

FIRING ORDERS

NOTE: To avoid confusion, always replace spark plug wires one at a time.

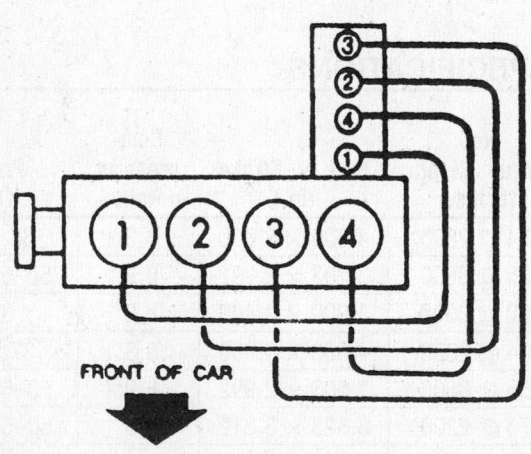

2.5L Engine
Engine Firing Order: 1–3–4–2
Distributorless Ignition System

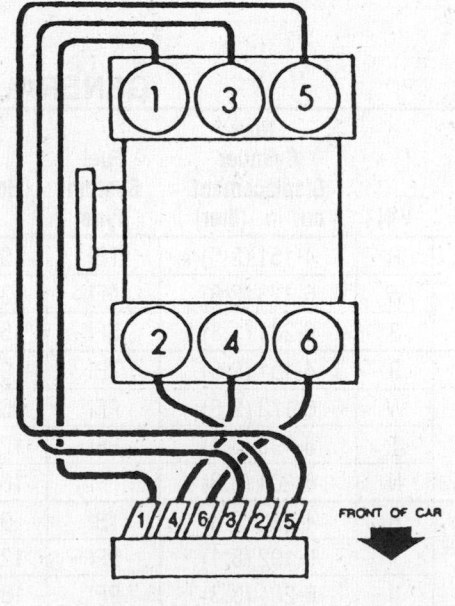

2.8L and 3.1L Engine
Engine Firing Order: 1–2–3–4–5–6
Distributorless Ignition System

FIRING ORDERS

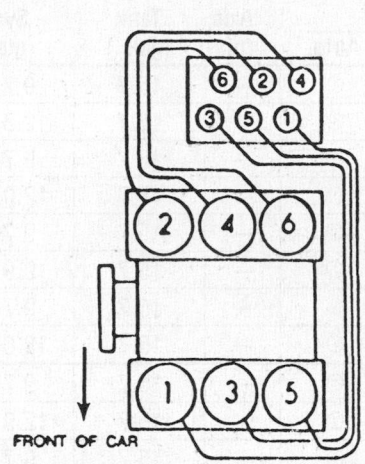

3.3L Engine
Engine Firing Order: 1–6–5–4–3–2
Distributorless Ignition System

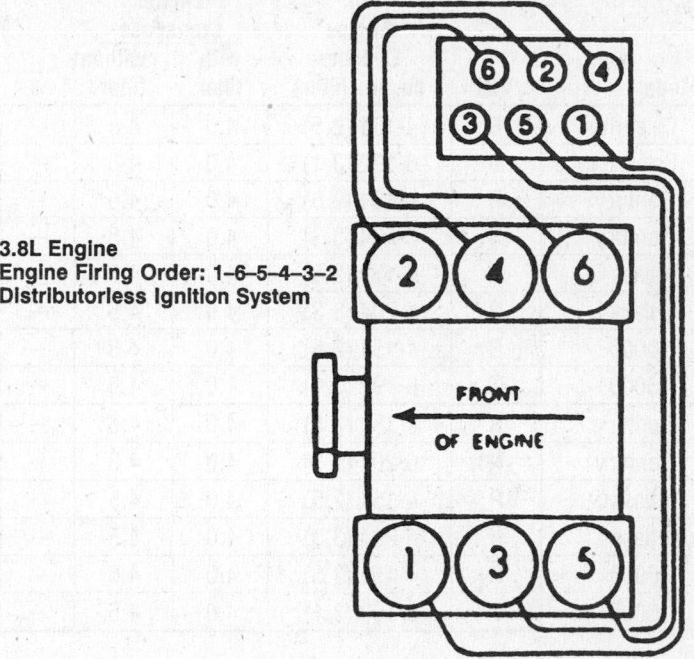

3.8L Engine
Engine Firing Order: 1–6–5–4–3–2
Distributorless Ignition System

CAPACITIES

Year	Model	VIN	No. Engine Cylinder cu. in. (liter)	Engine Crankcase with filter	without filter	Transmission (pts.) 4	5	Auto.	Drive Axle (pts.)	Fuel Tank (gals.)	Cool Sys. (qts.)
1988	Celebrity	R	4-151 (2.5)	3.0	3.5	—	4	①	—	15.7	9.7②
	Celebrity	W	6-173 (2.8)	4.0	4.0	—	4	①	—	15.7	13.5
	Century	R	4-151 (2.5)	3.0	3.5	—	—	①	—	15.7	9.7②
	Century	W	6-173 (2.8)	4.0	4.5	—	—	①	—	15.7	13.5
	Century	3	6-231 (3.8)	4.0	4.5	—	—	①	—	15.7	12.7
	Cutlass⑤	R	4-151 (2.5)	3.0	3.5	6	6	①	—	15.7	12.0
	Cutlass⑤	W	6-173 (2.8)	4.0	4.5	6	6	①	—	15.7	13.5
	Cutlass⑤	3	6-231 (3.8)	4.0	4.5	6	6	①	—	15.7	12.7
	6000	R	4-151 (2.5)	4.0	4.5	—	—	①	—	15.7	9.7②
	6000	W	6-173 (2.8)	4.0	4.5	—	—	①	—	15.7	13.5
1989	Celebrity	R	4-151 (2.5)	4.0	4.5	—	—	①	—	15.7	9.7
	Celebrity	W	6-173 (2.8)	4.0	4.5	—	—	①	—	15.7	13.2
	Century	R	4-151 (2.5)	4.0	4.5	—	—	①	—	15.7	9.7
	Century	W	6-173 (2.8)	4.0	4.5	—	—	①	—	15.7	13.2
	Century	N	6-204 (3.3)	4.0	4.5	—	—	①	—	15.7	12.0③
	Cutlass⑤	R	4-151 (2.5)	4.0	4.5	—	—	①	—	15.7	9.7
	Cutlass⑤	W	6-173 (2.8)	4.0	4.5	—	—	①	—	15.7	13.2
	Cutlass⑤	N	6-204 (3.3)	4.0	4.5	—	—	①	—	15.7	12.0③
	6000	R	4-151 (2.5)	4.0	4.5	—	—	①	—	15.7	9.7
	6000	W	6-173 (2.8)	4.0	4.5	—	—	①	—	15.7	13.2
	6000	T	6-192 (3.1)	4.0	4.5	—	—	①	—	15.7	12.6

CAPACITIES

Year	Model	VIN	No. Engine Cylinder cu. in. (liter)	Engine Crankcase with filter	Engine Crankcase without filter	Transmission (pts.) 4	5	Auto.	Drive Axle (pts.)	Fuel Tank (gals.)	Cool Sys. (qts.)
1990	Celebrity	R	4-151 (2.5)	4.0	4.5	—	—	①	—	15.7	9.7
	Celebrity	T	6-192 (3.1)	4.0	4.5	—	—	①	—	15.7	12.8
	Century	R	4-151 (2.5)	4.0	4.5	—	—	①	—	15.7	9.7
	Century	N	6-204 (3.3)	4.0	4.5	—	—	①	—	15.7	12.0③
	Cutlass⑤	R	4-151 (2.5)	4.0	4.5	—	—	①	—	15.7	9.7
	Cutlass⑤	N	6-204 (3.3)	4.0	4.5	—	—	①	—	15.7	12.9④
	6000	R	4-151 (2.5)	4.0	4.5	—	—	①	—	15.7	9.7
	6000	T	6-192 (3.1)	4.0	4.5	—	—	①	—	15.7	12.6
1991-92	Century	R	4-151 (2.5)	4.0	4.5	—	—	①	—	15.7	9.7
	Century	N	6-204 (3.3)	4.0	4.5	—	—	①	—	15.7	12.9
	Cutlass⑤	R	4-151 (2.5)	4.0	4.5	—	—	①	—	15.7	9.7
	Cutlass⑤	N	6-204 (3.3)	4.0	4.5	—	—	①	—	15.7	12.9
	6000	R	4-151 (2.5)	4.0	4.5	—	—	①	—	15.7	9.7
	6000	T	6-192 (3.1)	4.0	4.5	—	—	①	—	15.7	12.6

① 125C—8 pts.
  Overhaul—12 pts.
  440—T4—13 pts.
  Overhaul—20 pts.
② Heavy Duty—12 qts.
③ Air Cond.—12.7
④ Air Cond. & H.D. Rad—13.2
⑤ Ciera and Cruiser

CAMSHAFT SPECIFICATIONS

All measurements given in inches.

Year	VIN	No. Cylinder Displacement cu. in. (liter)	Journal Diameter 1	2	3	4	5	Lobe Lift In.	Ex.	Bearing Clearance	Camshaft End Play
1988	R	4-151 (2.5)	1.8690	1.8690	1.8690	—	—	0.232	0.232	0.0007–0.0027	0.0015–0.0050
	W	6-173 (2.8)	1.8678–1.8815	1.8678–1.8815	1.8678–1.8815	1.8678–1.8815	—	0.262	0.273	0.0010–0.0040	NA
	3	6-231 (3.8)	1.7850–1.7860	1.7850–1.7860	1.7850–1.7860	1.7850–1.7860	—	0.245	0.245	0.0005–0.0035	NA
1989	R	4-151 (2.5)	1.8690	1.8690	1.8690	—	—	0.232	0.232	0.0007–0.0027	0.0015–0.0050
	W	6-173 (2.8)	1.8678–1.8815	1.8678–1.8815	1.8678–1.8815	1.8678–1.8815	—	0.262	0.273	0.0010–0.0040	NA
	T	6-192 (3.1)	1.8678–1.8815	1.8678–1.8815	1.8678–1.8815	1.8678–1.8815	—	0.263	0.273	0.0010–0.0040	NA
	N	6-204 (3.3)	1.7850–1.7860	1.7850–1.7860	1.7850–1.7860	1.7850–1.7860	—	0.250	0.255	0.0005–0.0035	NA
1990	R	4-151 (2.5)	1.8690	1.8690	1.8690	—	—	0.248	0.248	0.0007–0.0027	0.0015–0.0050
	T	6-192 (3.1)	1.8678–1.8815	1.8678–1.8815	1.8678–1.8815	1.8678–1.8815	—	0.263	0.273	0.0010–0.0040	NA
	N	6-204 (3.3)	1.7850–1.7860	1.7850–1.7860	1.7850–1.7860	1.7850–1.7860	—	0.250	0.255	0.0005–0.0035	NA

CAMSHAFT SPECIFICATIONS

All measurements given in inches.

Year	VIN	No. Cylinder Displacement cu. in. (liter)	Journal Diameter					Lobe Lift		Bearing Clearance	Camshaft End Play
			1	2	3	4	5	In.	Ex.		
1991–92	R	4-151 (2.5)	1.8690	1.8690	1.8690	—	—	0.248	0.248	0.0007–0.0027	0.0015–0.0050
	T	6-192 (3.1)	1.8677–1.8815	1.8677–1.8815	1.8677–1.8815	1.8677–1.8815	—	0.263	0.273	0.0010–0.0040	NA
	N	6-204 (3.3)	1.7850–1.7860	1.7850–1.7860	1.7850–1.7860	1.7850–1.7860	—	0.250	0.255	0.0005–0.0035	NA

NA—Not available

CRANKSHAFT AND CONNECTING ROD SPECIFICATIONS

All measurements are given in inches.

Year	VIN	No. Cylinder Displacement cu. in. (liter)	Crankshaft				Connecting Rod		
			Main Brg. Journal Dia.	Main Brg. Oil Clearance	Shaft End-play	Thrust on No.	Journal Diameter	Oil Clearance	Side Clearance
1988	R	4-151 (2.5)	2.3000	0.0005–0.0022	0.003–0.008	5	1.9995–2.0005	0.0005–0.0026	0.006–0.022
	W	6-173 (2.8)	2.6473–2.6483	0.0016–0.0033	0.002–0.008	3	1.9983–1.9993	0.0013–0.0026	0.006–0.017
	3	6-231 (3.8)	2.4988–2.4998	0.0003–0.0018	0.003–0.011	2	2.2487–2.2495	0.0005–0.0026	0.006–0.023
1989	R	4-151 (2.5)	2.3000	0.0005–0.0022	0.003–0.008	5	1.9995–2.0005	0.0005–0.0026	0.006–0.022
	W	6-173 (2.8)	2.6473–2.6483	0.0016–0.0033	0.002–0.008	3	1.9983–1.9993	0.0013–0.0026	0.006–0.017
	T	6-192 (3.1)	2.6473–2.6483	0.0012–0.0027	0.002–0.008	3	1.9983–1.9994	0.0013–0.0031	0.014–0.027
	N	6-204 (3.3)	2.4988–2.4998	0.0003–0.0018	0.003–0.011	3	2.2487–2.2499	0.0003–0.0026	0.003–0.015
1990	R	4-151 (2.5)	2.3000	0.0005–0.0022	0.003–0.008	5	1.9995–2.0005	0.0005–0.0026	0.006–0.022
	T	6-192 (3.1)	2.6473–2.6483	0.0012–0.0027	0.002–0.008	3	1.9983–1.9994	0.0013–0.0031	0.014–0.027
	N	6-204 (3.3)	2.4988–2.4998	0.0003–0.0018	0.003–0.011	3	2.2487–2.2499	0.0003–0.0026	0.003–0.015
1991–92	R	4-151 (2.5)	2.3000	0.0005–0.0022	0.005–0.001	5	2.0000–0.0030	0.0005–0.0030	0.006–0.024
	T	6-192 (3.1)	2.6473–2.6483	0.0012–0.0030	0.002–0.008	3	1.9983–1.9994	0.0011–0.0034	0.014–0.027
	N	6-204 (3.3)	2.4988–2.4998	0.0003–0.0018	0.003–0.011	3	2.2487–2.2499	0.0003–0.0026	0.003–0.015

VALVE SPECIFICATIONS

Year	VIN	No. Cylinder Displacement cu. in. (liter)	Seat Angle (deg.)	Face Angle (deg.)	Spring Test Pressure (lbs.)	Spring Installed Height (in.)	Stem-to-Guide Clearance (in.)		Stem Diameter (in.)	
							Intake	Exhaust	Intake	Exhaust
1988	R	4-151 (2.5)	46	46	176 @ 1.254	1.440	0.0010–0.0028	0.0013–0.0041	0.3130–0.3140	0.3120–0.3130
	W	6-173 (2.8)	46	45	215 @ 1.291	1.727	0.0010–0.0027	0.0010–0.0027	0.3412–0.3416	0.3412–0.3416
	3	6-231 (3.8)	45	45	195 @ 1.340	1.727	0.0015–0.0035	0.0015–0.0032	0.3405–0.3412	0.3405–0.3412
1989	R	4-151 (2.5)	46	46	176 @ 1.254	1.440	0.0010–0.0028	0.0013–0.0041	0.3130–0.3140	0.3120–0.3130
	W	6-173 (2.8)	46	45	215 @ 1.291	1.727	0.0010–0.0027	0.0010–0.0027	0.3412–0.3416	0.3412–0.3416
	T	6-192 (3.1)	46	45	215 @ 1.291	1.575	0.0010–0.0027	0.0010–0.0027	NA	NA
	N	6-204 (3.3)	46	45	215 @ 1.291	1.701	0.0010–0.0027	0.0010–0.0027	NA	NA
1990	R	4-151 (2.5)	46	46	176 @ 1.254	1.440	0.0010–0.0028	0.0013–0.0041	NA	NA
	T	6-192 (3.1)	46	45	215 @ 1.291	1.575	0.0010–0.0027	0.0010–0.0027	NA	NA
	N	6-204 (3.3)	46	45	215 @ 1.291	1.701	0.0010–0.0027	0.0010–0.0027	NA	NA
1991-92	R	4-151 (2.5)	46	45	173 @ 1.240	1.680	0.0010–0.0028	0.0013–0.0041	NA	NA
	T	6-192 (3.1)	46	45	215 @ 1.291	1.5758	0.0010–0.0027	0.0010–0.0027	NA	NA
	N	6-204 (3.3)	45	45	210 @ 1.315	1.690–1.720	0.0015–0.0035	0.0015–0.0032	NA	NA

NA—Not Available

PISTON AND RING SPECIFICATIONS

All measurements are given in inches.

Year	VIN	No. Cylinder Displacement cu. in. (liter)	Piston Clearance	Ring Gap			Ring Side Clearance		
				Top Compression	Bottom Compression	Oil Control	Top Compression	Bottom Compression	Oil Control
1988	R	4-151 (2.5)	0.0014–0.0022 ①	0.010–0.020	0.010–0.020	0.020–0.060	0.002–0.003	0.001–0.003	0.015–0.055
	W	6-173 (2.8)	0.0020–0.0028	0.010–0.020	0.010–0.020	0.020–0.055	0.001–0.003	0.001–0.003	0.005–0.008
	3	6-231 (3.8)	0.0010–0.0020	0.013–0.023	0.013–0.023	0.015–0.035	0.003–0.005	0.003–0.005	0.003
1989	R	4-151 (2.5)	0.0014–0.0022 ①	0.010–0.020	0.010–0.020	0.020–0.060	0.002–0.003	0.001–0.003	0.015–0.055
	W	6-173 (2.8)	0.0020–0.0028	0.010–0.020	0.010–0.020	0.020–0.055	0.001–0.003	0.001–0.003	0.005–0.008
	T	6-192 (3.1)	0.0022–0.0028	0.010–0.020	0.010–0.020	0.010–0.050	0.002–0.004	0.002–0.004	0.008 ②
	N	6-204 (3.3)	0.0004–0.0022 ③	0.010–0.025	0.010–0.025	0.010–0.040	0.001–0.003	0.001–0.003	0.001 0.008

PISTON AND RING SPECIFICATIONS

All measurements are given in inches.

Year	VIN	No. Cylinder Displacement cu. in. (liter)	Piston Clearance	Ring Gap			Ring Side Clearance		
				Top Compression	Bottom Compression	Oil Control	Top Compression	Bottom Compression	Oil Control
1990	R	4-151 (2.5)	0.0014–0.0022 ①	0.010–0.020	0.010–0.020	0.020–0.060	0.002–0.003	0.001–0.003	0.015–0.055
	T	6-192 (3.1)	0.0022–0.0028	0.010–0.020	0.010–0.020	0.010–0.050	0.002–0.004	0.002–0.004	0.008 ②
	N	6-204 (3.3)	0.0004–0.0022 ③	0.010–0.025	0.010–0.025	0.010–0.040	0.001–0.003	0.001–0.003	0.001–0.008
1991–92	R	4-151 (2.5)	0.0014–0.0022	0.010–0.020	0.010–0.020	0.020–0.060	0.002–0.003	0.001–0.003	0.015–0.055
	T	6-192 (3.1)	0.0009–0.0022	0.010–0.020	0.010–0.028	0.010–0.030	0.0020–0.0035	0.0020–0.0035	0.008 ②
	N	6-204 (3.3)	0.0004–0.0022 ③	0.010–0.025	0.010–0.025	0.010–0.040	0.0013–0.0031	0.0013–0.0031	0.0011–0.0081

① Measured 1/8 in. down from piston top
② Maximum clearance
③ 44 mm from top of piston

TORQUE SPECIFICATIONS

All readings in ft. lbs.

Year	VIN	No. Cylinder Displacement cu. in. (liter)	Cylinder Head Bolts	Main Bearing Bolts	Rod Bearing Bolts	Crankshaft Pulley Bolts	Flywheel Bolts	Manifold		Spark Plugs
								Intake	Exhaust	
1988	R	4-151 (2.5)	①	70	32	162	55	25	③	15
	W	6-173 (2.8)	②	73	39	76	52	⑨	15–23	10–25
	3	6-231 (3.8)	⑤	100	45	219	60	32	37	20
1989	R	4-151 (2.5)	①	70	32	162	55	25	③	15
	W	6-173 (2.8)	②	73	39	76	52	⑨	15–23	10–25
	T	6-192 (3.1)	②	73	34–40	76	52	⑨	19	10–25
	N	6-204 (3.3)	⑥	⑦	⑧	219	④	88 ⑩	41	20
1990	R	4-151 (2.5)	①	65	29	162	55	25	③	15
	T	6-192 (3.1)	②	73	39	76	52	⑨	19	10–25
	N	6-204 (3.3)	⑥	⑦	⑧	219	④	88 ⑩	41	20
1991–92	R	4-151 (2.5)	①	65	29	162	55	25	③	20
	T	6-192 (3.1)	②	73	39	76	52	⑨	19	18
	N	6-204 (3.3)	⑥	⑦	⑧	219	④	89 ⑩	41	20

① Step 1: All bolts to 18 ft. lbs.
  Step 2: Except position "I or 9" to 26 ft. lbs.
  Step 3: Retorque position "i or 9" to 18 ft. lbs.
  Step 4: All bolts + 90° turn
② Step 1: 33 ft. lbs.
  Step 2: + 90° turn
③ Inner bolts: 37 ft. lbs.
  Outer bolts: 28 ft. lbs.
④ Step 1: 89 inch lbs.
  Step 2: + 90° turn

⑤ Step 1: 25 ft. lbs.
  Step 2: + 90° turn
  Step 3: + 90° turn
  NOTE: If at any time 60 ft. lbs. is reached during the sequence, STOP!; Do not continue turning the bolt
⑥ Step 1: 35 ft. lbs.
  Step 2: + 130° turn
  Step 3: + 30° turn on 4 center bolts only

⑦ Step 1: 26 ft. lbs.
  Step 2: + 45° turn
⑧ Step 1: 20 ft. lbs.
  Step 2: + 50° turn
⑨ Manifold-to-cylinder head: 24 ft. lbs.
  Manifold-to-plenum: 16 ft. lbs.
⑩ Inch lbs.

BRAKE SPECIFICATIONS

All measurements in inches unless noted.

Year	Model	Lug Nut Torque (ft. lbs.)	Master Cylinder Bore	Brake Disc Minimum Thickness	Brake Disc Maximum Runout	Standard Brake Drum Diameter	Minimum Lining Thickness Front	Minimum Lining Thickness Rear
1988	Celebrity	100	0.874	0.830	0.002	8.858	0.030	①
	Century	100	0.874	0.830	0.002	8.858	0.030	①
	Cutlass ⑥	100	0.874	0.830	0.002	8.858	0.030	①
	6000	100	0.874	0.830	0.002	8.858	0.030	①
1989	Celebrity	100	0.874 ②	0.830	0.004	8.863	0.030	①
	Century	100	0.874 ②	0.830	0.004	8.863	0.030	①
	Cutlass ⑥	100	0.874 ②	0.830	0.004	8.863	0.030	①
	6000	100	0.874 ②	0.830	0.004	8.863	0.030	①
1990	Celebrity	100	0.874 ②	0.830 ③ ④	0.004 ⑤	8.863	0.030	0.030
	Century	100	0.874 ②	0.830 ③	0.004	8.863	0.030	0.030
	Cutlass ⑥	100	0.874 ②	0.830 ③	0.004	8.863	0.030	0.030
	6000	100	0.874 ②	0.830 ③ ④	0.004 ⑤	8.863	0.030	0.030
1991–92	Century	100	0.874 ②	0.830 ③ ④	0.004	8.863	0.030	0.030
	Cutlass ⑥	100	0.874 ②	0.830 ③	0.004	8.863	0.030	0.030
	6000	100	0.874 ②	0.830 ③	0.004 ⑤	8.863	0.030	0.030

① 0.030 in. over rivet head; if bonded, 0.062 in. over shoe
② Medium and heavy duty—0.944
③ Medium & heavy duty—0.972
④ Rear disc—0.756
⑤ Rear disc—0.003
⑥ Ciera and Cutlass

WHEEL ALIGNMENT

Year	Model	Caster Range (deg.)	Caster Preferred Setting (deg.)	Camber Range (deg.)	Camber Preferred Setting (deg.)	Toe-in (in.)	Steering Axis Inclination (deg.)
1988	Celebrity	23/32P–2 23/32	1 23/32P	1/2N–1/2P	0	3/32N–3/32P	NA
	Century	3/4P–2 3/4P	1 3/4P	1/2N–1/2P	0	3/32N–3/32P	NA
	Cutlass ①	1 1/2P–2 1/2P	2P	3/16P–1 3/16P	11/16P	3/32N–3/32P	NA
	6000	23/32P–2 23/32P	1 23/32P	1/2N–1/2P	0	3/32N–3/32P	NA
1989	Celebrity	23/32P–2 23/32	1 23/32P	1/2N–1/2P	0	3/32N–3/32P	NA
	Century	3/4P–2 3/4P	1 3/4P	1/2N–1/2P	0	3/32N–3/32P	NA
	Cutlass ①	1 5/16P–2 5/16P	1 13/16P	3/16P–1 3/16P	11/16P	3/32N–3/32P	NA
	6000	11/16P–2 11/16P	1 11/16P	1/2N–1/2P	0	3/32N–3/32P	NA
1990	Celebrity	11/16P–2 11/16	1 11/16P	1/2N–1/2P	0	3/32N–3/32P	NA
	Century	3/4P–2 3/4P	1 3/4P	1/2N–1/2P	0	3/32N–3/32P	NA
	Cutlass ①	1 1/2P–2 1/2P	2P	3/16P–1 3/16P	11/16P	3/32N–3/32P	NA
	6000	11/16P–2 11/16P	1 11/16P	1/2N–1/2P	0	3/32N–3/32P	NA
1991–92	Century	3/4P–2 3/4P	1 3/4P	1/2N–1/2P	0	3/32N–3/32P	NA
	Cutlass ①	1 1/2P–2 1/2P	2P	3/16P–1 3/16P	11/16P	3/32N–3/32P	NA
	6000	11/16P–2 11/16P	1 11/16P	1/2N–1/2P	0	3/32N–3/32P	NA

NA—Not available
N—Negative
P—Positive
① Ciera and Cruiser

ENGINE MECHANICAL

NOTE: Disconnecting the negative battery cable on some vehicles may interfere with the functions of the on board computer systems and may require the computer to undergo a relearning process, once the negative battery cable is reconnected.

Engine Assembly

REMOVAL & INSTALLATION

2.5L ENGINE

WITH MANUAL TRANSAXLE

1. Relieve the fuel system pressure. Disconnect the negative battery cable.
2. Scribe reference marks at the hood supports and remove the hood. Install covers on both fenders.
3. Raise and support the vehicle safely. Remove front mount-to-cradle nuts.
4. Remove forward exhaust pipe.
5. Remove starter assembly.
6. Remove flywheel inspection cover.
7. Lower the vehicle.
8. Remove the air cleaner assembly.
9. Remove all bellhousing retaining bolts.
10. Remove forward torque reaction rod from engine and core support.
11. If equipped with air conditioning, remove the serpentine belt and compressor and position to the side.
12. Remove emission hoses at canister.
13. Remove power steering hose, if equipped.
14. Remove vacuum hoses and electrical connectors at solenoid.
15. Remove heater blower motor.
16. Disconnect throttle cable.
17. Drain cooling system.
18. Disconnect heater hose.
19. Disconnect engine harness at bulkhead connector.
20. Install an engine lift support tool and partially hoist the engine. Remove heater hose at intake manifold and disconnect fuel line.
21. Remove the engine from the vehicle.
To install:
22. Install move the engine in the vehicle.
23. Connect the heater hose at intake manifold and connect fuel line. Lower the engine and remove the engine lift tool.
24. Connect engine harness at bulkhead connector.
25. Connect heater hose.

26. Connect throttle cable.
27. Install heater blower motor.
28. Install vacuum hoses and electrical connectors at solenoid.
29. Install power steering hose, if equipped.
30. Install emission hoses at canister.
31. If equipped with air conditioning, install the compressor and serpentine belt.
32. Install the forward torque reaction rod to the engine and core support.
33. Install all bellhousing retaining bolts.
34. Install the air cleaner assembly.
35. Raise and safely support the vehicle.
36. Install flywheel inspection cover.
37. Install starter assembly.
38. Install forward exhaust pipe.
39. Lower the vehicle. Install front mount-to-cradle nuts.
40. Install the hood observing the scribe reference marks made upon removal.
41. Connect the negative battery cable.
42. Fill cooling system and check for leaks. Start the engine and allow to come to normal operating temperature. Recheck for leaks. Top-up coolant.

WITH AUTOMATIC TRANSAXLE

1. Relieve the fuel system pressure. Disconnect the negative battery cable.
2. Scribe reference marks at the hood supports and remove the hood. Install covers on both fenders.
3. Drain the cooling system. Remove the air cleaner assembly and preheat tube.
4. Disconnect engine harness connector.
5. Disconnect all external vacuum hose connections.
6. Remove throttle and transaxle linkage at TBI assembly and intake manifold.
7. Remove upper radiator hose.
8. If equipped with air conditioning, remove the air conditioning compressor from mounting brackets and set aside. Do not discharge the air conditioning system.
9. Remove front engine strut assembly.
10. Disconnect heater hoses.
11. Raise the vehicle and support it safely. Remove transaxle-to-engine bolts leaving the upper 2 bolts in place.
12. Remove front mount-to-cradle nuts.
13. Remove forward exhaust pipe.
14. Remove flywheel inspection cover and remove starter motor.
15. Remove torque converter-to-flywheel bolts.
16. Remove power steering pump

and bracket with hoses attached and set aside.
17. Remove lower radiator hose.
18. Remove the 2 rear transaxle support bracket bolts.
19. Remove fuel supply line at fuel filter.
20. Using a floor jack and a block of wood placed under the transaxle, raise engine and transaxle until engine front mount studs clear cradle.
21. Connect engine lift equipment and put tension on engine.
22. Remove the 2 remaining transaxle bolts.
23. Slide engine forward and remove from the vehicle.
To install:
24. Position the engine in the engine compartment, aligning the engine with the transaxle bellhousing.
25. With the engine supported by the lifting tool, install the 2 upper bell housing bolts. Do not lower the engine while the jack is supporting the transaxle.
26. Remove the transaxle support jack and lower the engine onto the engine mounts. Remove the engine lift tool.
27. Raise and safely support the vehicle. Install the front mount-to-cradle nuts.
28. Install fuel supply line at fuel filter.
29. Install the 2 rear transaxle support bracket bolts.
30. Install lower radiator hose.
31. Install power steering pump and bracket.
32. Install torque converter-to-flywheel bolts.
33. Install starter motor and flywheel inspection cover.
34. Install forward exhaust pipe.
35. Lower the vehicle.
36. Connect heater hoses.
37. Install front engine strut assembly.
38. If equipped with air conditioning, install the air conditioning mounting brackets and compressor.
39. Connect upper radiator hose.
40. Install throttle and transaxle linkage at TBI assembly and intake manifold.
41. Connect all external vacuum hose connections.
42. Connect engine harness connector.
43. Drain the cooling system. Remove the air cleaner assembly and preheat tube.
44. Install the hood.
45. Connect the negative battery cable.
46. Fill cooling system and check for leaks. Start the engine and allow to come to normal operating temperature. Recheck for leaks. Top-up coolant.

2.8L and 3.1L Engines

1. Relieve the fuel system pressure.
2. Disconnect the negative battery cable. Scribe reference marks at the hood supports and remove the hood. Install covers on both fenders.
3. Remove the airflow tube at the air cleaner and throttle valve.
4. Drain the cooling system.
5. Disconnect vacuum hoses from all non-engine mounted components.
6. Disconnect the accelerator linkage and TV cable. Disconnect the cruise control cable, if equipped.
7. Disconnect the engine harness connector from the ECM and pull the connector through the front of dash. Disconnect the engine harness from the junction block at the dash panel.
8. Remove the engine strut bracket from the radiator support and position aside, as required.
9. Disconnect the radiator hoses from radiator and heater hoses from engine. Disconnect and plug the transaxle cooler lines.
10. Remove the serpentine belt cover and belt.
11. On vehicles with the air conditioning compressor mounted on the upper portion of the engine, remove the AIR pump and bracket. Then, remove the air conditioning compressor from the mounting bracket and position aside.
12. If equipped, remove power steering pump from engine and set it aside.
13. Disconnect and plug the fuel lines.
14. Disconnect the EGR at the exhaust, as required.
15. Raise and safely support the vehicle.
16. On vehicles with the air conditioning compressor mounted on the lower portion of the engine, remove the air conditioning compressor from the engine. Do not discharge the air conditioning system.
17. Remove the engine front mount-to-cradle and mount-to-engine bracket retaining nuts, as required.
18. Disconnect and tag all electrical wiring at the starter. Remove the starter retaining bolts and remove the starter.
19. If equipped with automatic transaxle, remove the transaxle inspection cover and disconnect the torque converter from the flexplate.
20. Disconnect the exhaust pipe.
21. Remove the 1 transaxle-to-engine bolt from the back side of the engine.
22. Disconnect the power steering cut-off switch, if equipped.
23. Lower the vehicle.
24. Remove the exhaust crossover pipe.

25. Remove the remaining transaxle-to-engine bolts.
26. Support the transaxle by positioning a floor jack and a block of wood under the transaxle. Install an engine lift tool and remove the engine from the vehicle.

To install:

27. Position the engine in the vehicle while aligning the transaxle. Install the transaxle-to-engine bolts.
28. Position the front engine mount studs in the cradle and engine bracket.
29. Remove the engine lift tool. Raise and support the vehicle safely.
30. Install the engine mount retaining nuts.
31. If equipped, connect the power steering cut-off switch.
32. Install the 1 transaxle-to-engine bolt from the back side of the engine.
33. Connect the exhaust pipe.
34. If equipped with automatic transaxle, connect the torque converter to the flexplate and install the transaxle inspection cover.
35. Install the starter and retaining bolts. Connect the starter electrical connectors.
36. Install the engine front mount-to-cradle and mount-to-engine bracket retaining nuts.
37. On vehicles with the air conditioning compressor mounted on the lower portion of the engine, install the air conditioning compressor.
38. Lower the vehicle.
39. Connect the EGR at the exhaust, if removed.
40. Connect the fuel lines.
41. If equipped, install power steering pump.
42. On vehicles with the air conditioning compressor mounted on the upper portion of the engine, install the air conditioner compressor. Install the AIR pump and bracket.
43. Install the serpentine belt cover and belt.
44. Connect the radiator and heater hoses. Connect the transaxle cooler lines.
45. Install the engine strut bracket to the radiator support.
46. Connect the engine harness connector to the ECM.
47. Connect the accelerator linkage and TV cable. Connect the cruise control cable, if equipped.
48. Connect vacuum hoses to all non-engine mounted components.
49. Install the airflow tube at the air cleaner and throttle valve.
50. Install the hood using the reference marks made upon removal.
51. Connect the negative battery cable.
52. Fill cooling system and check for leaks. Start the engine and allow to

come to normal operating temperature. Recheck for leaks. Top-up coolant.

3.3L Engine

1. Disconnect the negative battery cable. Scribe reference marks at the hood supports and remove the hood. Install covers on both fenders.
2. Relieve the fuel system pressure.
3. Disconnect the negative battery cable.
4. Drain the cooling system. Disconnect the radiator and heater hoses. Disconnect and plug the transaxle cooler lines.
5. Remove the upper engine strut and engine cooling fan.
6. Remove the intake duct from the throttle body. Disconnect vacuum hoses from all non-engine mounted components. Disconnect all electrical connections.
7. Remove the cable bracket and cables from the throttle body.
8. Remove the serpentine belt. If equipped, remove the power steering pump and locate to the side.
9. Remove the upper transaxle-to-engine retaining bolts.
10. Raise and support the vehicle safely.
11. Remove the air conditioning compressor and locate to the side.
12. Remove the engine mount-to-frame nuts, flywheel dust cover and flywheel-to-converter bolts.
13. Remove the lower engine-to-transaxle bolts; 1 bolt is located behind the transaxle case and engine block.
14. Lower the vehicle. Install an engine lift tool and remove the engine from the vehicle.

To install:

15. Install the engine in the engine compartment. Install the upper engine-to-transaxle bolts. Remove the engine lift tool.
16. Raise and safely support the vehicle.
17. Install the lower engine-to-transaxle bolts; 1 bolt is located behind the transaxle case and engine block.
18. Install the flywheel-to-converter bolts, flywheel dust cover and engine mount-to-frame nuts.
19. Install the air conditioning compressor.
20. Lower the vehicle.
21. If equipped, install the power steering pump. Install the serpentine belt.
22. Install the cable bracket and cables to the throttle body.
23. Install the intake duct to the throttle body. Connect vacuum hoses to all non-engine mounted components. Connect all electrical connections.

24. Install the upper engine strut and engine cooling fan.

25. Connect the radiator and heater hoses. Connect the transaxle cooler lines.

26. Install the hood.

27. Connect the negative battery cable.

28. Fill cooling system and check for leaks. Start the engine and allow to come to normal operating temperature. Recheck for leaks. Top-up coolant.

3.8L Engine

1. Disconnect the negative battery cable. Scribe reference marks at the hood supports and remove the hood. Install covers on both fenders.

2. Remove the air cleaner assembly and drain the cooling system.

3. Disconnect vacuum hoses to all non-engine mounted components.

4. Disconnect the detent cable and accelerator linkage.

5. Disconnect the engine electrical harness and ground strap.

6. Disconnect the heater hoses from the engine and radiator hoses from the radiator. Disconnect the transaxle cooler lines, if equipped.

7. If equipped, remove the power steering pump and bracket assembly.

8. Raise and safely support the vehicle.

9. Disconnect the exhaust pipe from the manifold.

10. Disconnect the fuel lines.

11. Remove the engine front mount-to-cradle retaining nuts.

12. Disconnect and tag all electrical wiring at the starter. Remove the starter retaining bolts and remove the starter.

13. If equipped with automatic transaxle, remove the flexplate cover and disconnect the flexplate from the torque converter.

14. Remove the retaining bolts from the transaxle rear support bracket.

15. Lower the vehicle and place a support under the transaxle rear extension.

16. Remove the engine strut bracket from the radiator support and position aside.

17. Remove the transaxle-to-engine retaining bolts.

18. If equipped with air conditioning, remove the air conditioning compressor from the mounting bracket and lay aside.

19. Install an engine lift tool and remove the engine from the vehicle.

To install:

20. Position the engine in the vehicle and align the engine front mount studs. Align the transaxle and install the transaxle-to-engine retaining bolts.

21. Remove the engine lift tool.

22. If equipped, install the air conditioning compressor to the mounting bracket.

23. Raise and safely support the vehicle. Install the retaining bolts to the transaxle rear support bracket.

24. If equipped with automatic transaxle, connect the flexplate to the torque converter and install the flexplate cover.

25. Install the starter and retaining bolts. Connect the starter electrical connectors.

26. Install the engine front mount-to-cradle retaining nuts.

27. Connect the fuel lines.

28. Connect the exhaust pipe to the manifold.

29. Lower the vehicle.

30. If equipped, install the power steering bracket assembly and pump.

31. Connect the heater and radiator hoses. Connect the transaxle cooler lines, if equipped.

32. Connect the engine electrical harness and ground strap.

33. Connect the detent cable and accelerator linkage.

34. Connect vacuum hoses to all non-engine mounted components.

35. Install the air cleaner assembly.

36. Install the hood.

37. Connect the negative battery cable.

38. Fill cooling system and check for leaks. Start the engine and allow to come to normal operating temperature. Recheck for leaks. Top-up coolant.

Engine Mounts

REMOVAL & INSTALLATION

1. Disconnect the negative battery cable.

2. Raise and support the vehicle safely.

3. Using a suitable tool, support the engine and remove the engine mounting bracket nuts.

4. Raise the engine slightly until the engine mount is free from the vehicle chassis.

5. Remove the nuts holding the engine mount to the frame.

6. Remove the engine mounts and discard.

To install:

7. Install the engine mounts.

8. Install the nuts holding the engine mount to the frame.

9. Lower the engine onto the mount and install the engine mounting bracket nuts.

10. Remove the engine lift tool.

11. Connect the negative battery cable.

Cylinder Head

REMOVAL & INSTALLATION

2.5L Engine

1. Relieve the pressure in the fuel system before disconnecting any fuel line connections.

2. Disconnect the negative battery cable.

3. Drain the cooling system. Remove the air cleaner and the oil level indicator tube.

4. Disconnect the throttle linkage and fuel lines.

5. Disconnect the oxygen sensor connector. Remove the intake and exhaust manifolds.

6. Remove the alternator bracket-to-cylinder head bolts, as required.

7. If equipped with air conditioning, remove the compressor bracket bolts and position the compressor aside. Do not disconnect any of the refrigerant lines.

8. Disconnect and tag all vacuum and electrical connections from the cylinder head.

9. Disconnect the radiator hoses and engine strut rod bolt from the upper support.

10. Remove the power steering pump bracket, if top mounted.

11. Remove the rocker arm cover, rocker arms and pushrods.

12. Remove the cylinder head bolts and remove the cylinder head from the engine.

To install:

13. Clean the cylinder head and block from any foreign matter, nicks or heavy scratches. Clean the cylinder head bolt threads and threads in the cylinder block.

14. Position the new cylinder head gasket over the dowel pins.

15. Carefully guide the cylinder head into place. Coat the cylinder head bolts with sealing compound and install finger-tight.

16. Torque the cylinder head bolts as follows:

  a. Torque the cylinder head bolts gradually to 25 ft. lbs. in the proper sequence.

  b. Torque all bolts except No. 9 in sequence again to 22 ft. lbs. Torque No. 9 to 29 ft. lbs.

  c. Repeat sequence. Turn all bolts, except No. 9, 120 degrees (2 flats). Turn No. 9 a ¼ turn (90 degrees).

17. Install the pushrods, rocker arms and rocker arm cover.

18. If removed, install the power steering pump bracket and pump.

19. Connect the radiator hoses and engine strut rod bolt to the upper support.

20. Connect all vacuum and electrical connections to the cylinder head.

21. If equipped with air conditioning, install the compressor bracket bolts and install the compressor.

22. If removed, install the alternator bracket-to-cylinder head bolts.

23. Connect the oxygen sensor connector. Install the intake and exhaust manifolds.

24. Connect the throttle linkage and fuel lines.

25. Install the air cleaner and the oil level indicator tube.

26. Connect the negative battery cable.

27. Connect the negative battery cable.

28. Fill cooling system and check for leaks. Start the engine and allow to come to normal operating temperature. Recheck for leaks. Top-up coolant.

2.8L and 3.1L Engines

LEFT SIDE

1. Relieve the pressure in the fuel system before disconnecting any fuel line connections. Disconnect the fuel lines.

2. Disconnect the negative battery cable. Raise and safely support the vehicle.

3. Drain the cylinder block and lower the vehicle.

4. Remove the oil level indicator tube, rocker arm cover, intake manifold and plenum, as required.

5. Remove the exhaust crossover, alternator bracket, AIR pump and brackets.

6. Disconnect and tag all electrical wiring and vacuum hoses that may interfere with the removal of the left cylinder head.

7. Loosen the rocker arm until the pushrods can be removed. Remove the pushrods. Keep the pushrods in the same order as removed.

8. Remove the cylinder head bolts. Remove the cylinder head. Do not pry on the head to loosen it.

To install:

9. Clean the cylinder head and block from any foreign matter, nicks or heavy scratches. Clean the cylinder head bolt threads and threads in the cylinder block.

10. Position the new cylinder head gasket over the dowel pins with the words "This Side Up" facing upwards. Carefully guide the cylinder head into place.

11. Install the cylinder head bolts and tighten in sequence to 33 ft. lbs. (45 Nm). Turn an additional 90 degrees in sequence.

12. Install the pushrods. Make sure the lower ends of the pushrods are in the lifter seats. Install the rocker arm

nuts and torque the nuts to 14–20 ft. lbs. (20–27 Nm).

13. Install the intake manifold.

14. Connect all electrical wiring and vacuum hoses.

15. Install the exhaust crossover, alternator and AIR pump brackets, alternator and AIR pump.

16. If removed, install the oil level indicator tube, rocker arm cover, intake manifold and plenum.

17. Connect the fuel lines.

18. Connect the negative battery cable.

19. Adjust the valve lash, as required.

RIGHT SIDE

1. Relieve the pressure in the fuel system before disconnecting any fuel line connections.

2. Disconnect the negative battery cable. Raise the vehicle and support it safely.

3. Drain the cylinder block and lower the vehicle.

4. If equipped, remove the cruise control servo bracket, the air management valve and hose and the intake manifold.

5. Remove the exhaust pipe at crossover, crossover and heat shield, as required.

6. Disconnect and tag all electrical wiring and vacuum hoses that may interfere with the removal of the right cylinder head.

7. Remove the rocker cover. Loosen the rocker arm nuts and remove the pushrods. Keep the pushrods in the order in which they were removed.

8. Remove the cylinder head bolts. Remove the cylinder head. Do not pry on the head to loosen it.

To install:

9. Clean the cylinder head and block from any foreign matter, nicks or heavy scratches. Clean the cylinder head bolt threads and threads in the cylinder block.

10. Position the new cylinder head gasket over the dowel pins with the words "This Side Up" facing upwards. Carefully guide the cylinder head into place. Install the pushrods and loosely retain with the rocker arms.

11. Install the cylinder head bolts and tighten in sequence to 33 ft. lbs. (45 Nm). Turn an additional 90 degrees in sequence.

12. Install the pushrods. Make sure the lower ends of the pushrods are in the lifter seats. Install the rocker arm nuts and torque the nuts to 14–20 ft. lbs. (20–27 Nm).

13. Install the intake manifold.

14. Install the rocker cover.

15. Connect all electrical wiring and vacuum hoses.

16. If removed, install the crossover exhaust pipe and heat shield.

17. If equipped, install the cruise control servo bracket, the air management valve and hose.

18. Connect the negative battery cable.

19. Fill cooling system and check for leaks. Start the engine and allow to come to normal operating temperature. Recheck for leaks. Top-up coolant.

20. Adjust the valve lash, as required.

3.3L Engine

1. Relieve the pressure in the fuel system before disconnecting any fuel line connections.

2. Disconnect the negative battery cable. Raise the vehicle and support it safely.

3. Drain the cylinder block and lower the vehicle.

4. Remove the intake manifold and exhaust manifold.

5. Remove the valve cover.

6. Remove the ignition module and coils as a unit.

7. Disconnect and tag all electrical wiring and vacuum hoses, as necessary.

8. If equipped with air conditioning, remove the air conditioning compressor and position to the side.

9. Remove the alternator and power steering pump and position to the side. Remove the belt tensioner assembly.

10. Remove the rocker arm assembly, guide plate and pushrods.

11. Remove the cylinder head bolts and remove the cylinder head.

To install:

12. Clean the cylinder head and block of any foreign matter, nicks or heavy scratches. Clean the cylinder head bolt threads and threads in the cylinder block.

13. Position the new cylinder head gasket on the block.

14. Carefully guide the cylinder head into place.

15. Coat the cylinder head bolts with sealing compound and install into the head. Tighten the cylinder head bolts according to the following procedure:

    a. Tighten in sequence to 35 ft. lbs. (47 Nm).

    b. Using an appropriate torque angle gauge, rotate each bolt in sequence an additional 130 degrees.

    c. Rotate the center 4 bolts an additional 30 degrees in sequence.

16. Install the pushrods, guide plate and rocker arm assembly. Tighten the rocker arm pivot bolts to 28 ft. lbs. (38 Nm).

17. Install the intake manifold and exhaust manifold.

18. Install the valve cover.

19. Remove the ignition module and coils as a unit, as required.

1. Apply sealing compound No. 102080 or equivalent to bolts shown

2. Mounting surfaces of block assy., head assy. and both sides of gasket must be free of oil.

3. Locating pins

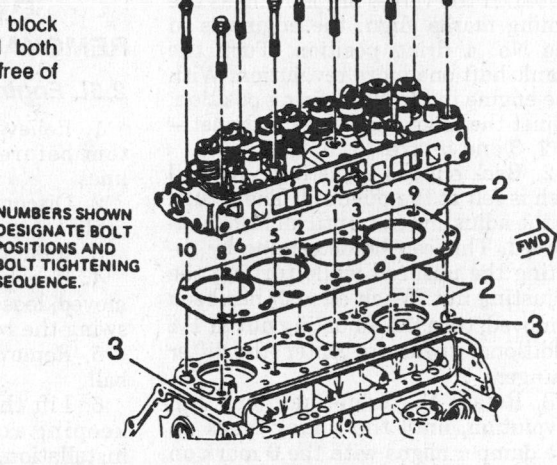

NUMBERS SHOWN DESIGNATE BOLT POSITIONS AND BOLT TIGHTENING SEQUENCE.

Cylinder head torque sequence—2.5L engine

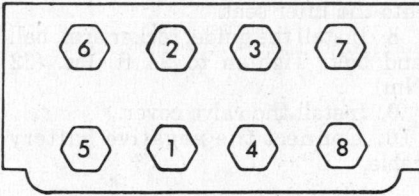

Cylinder head torque sequence—2.8L and 3.1L engines

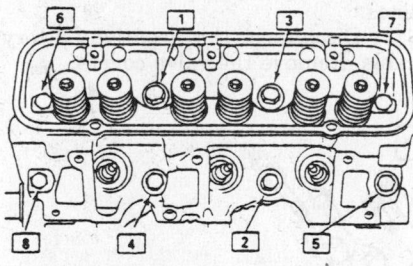

Cylinder head torque sequence—3.3L and 3.8L engines

20. Connect all electrical wiring and vacuum hoses.

21. If equipped with air conditioning, install the air conditioning compressor.

22. Install the alternator and power steering pump. Remove the belt tensioner assembly.

23. Connect the negative battery cable.

24. Fill cooling system and check for leaks. Start the engine and allow to come to normal operating temperature. Recheck for leaks. Top-up coolant.

3.8L Engine

1. Relieve the pressure in the fuel system before disconnecting any fuel line connections.

2. Disconnect the negative battery cable. Raise the vehicle and support it safely.

3. Drain the cylinder block and lower the vehicle.

4. Remove the serpentine belt.

5. Remove the alternator, AIR pump, oil indicator and power steering pump, as required. Position to the side.

6. Remove the throttle cable. Remove the cruise control cable, if equipped.

7. Disconnect the fuel lines and fuel rail, as required.

8. Remove the heater hoses and radiator hoses.

9. Disconnect and tag all vacuum and electrical wiring.

10. Remove the radiator and cooling fan, if necessary.

11. Remove the intake manifold and valve cover.

12. Remove the exhaust manifold(s).

13. Remove the rocker arm assembly and pushrods.

14. Remove the cylinder head bolts and remove the cylinder head.

To install:

15. Clean the cylinder head and block from any foreign matter, nicks or heavy scratches. Clean the cylinder head bolt threads and threads in the cylinder block.

16. Position the new cylinder head gasket on the block.

17. Carefully guide the cylinder head into place. Coat the cylinder head bolts with sealing compound and install.

18. Tighten the cylinder head bolts according to the following procedure:

    a. Tighten the cylinder head bolts in sequence to 25 ft. lbs. (34 Nm).

    b. Do not exceed 60 ft. lbs. (81 Nm) at any point during the next 2 steps.

    c. Using a torque angle gauge, tighten each bolt an additional 90 degrees in sequence.

    d. Tighten each bolt an additional 90 degrees in sequence.

19. Install the exhaust manifold.

20. Install the intake manifold, pushrods and rocker arm assembly.

21. Install the valve cover.

22. Install the radiator and cooling fan, as required.

23. Connect all vacuum and electrical wiring.

24. Install the heater hoses and radiator hoses.

25. Connect the fuel lines and fuel rail.

26. Install the throttle cable. Install the cruise control cable, if equipped.

27. Install the alternator, AIR pump, oil indicator and power steering pump.

28. Install the serpentine belt.

29. Connect the negative battery cable.

30. Fill cooling system and check for leaks. Start the engine and allow to come to normal operating temperature. Recheck for leaks. Top-up coolant.

Valve Lifters

REMOVAL & INSTALLATION

2.5L Engine

1. Disconnect the negative battery cable.

2. Remove the intake manifold and valve cover.

3. Loosen the rocker arms and rotate to clear the pushrods.

4. Remove the pushrods, retainer and guide.

5. Remove the lifters. Keep all components separated so they may be reinstalled in the same location.

To install:

6. Lubricate the lifters with engine oil and install the lifters in their bore.

7. Install the guides, retainers and pushrods.

8. With the lifter on the base circle of the camshaft, tighten the rocker arm bolts to 24 ft. lbs. (32 Nm).

9. Install the intake manifold and valve cover.

10. Connect battery negative cable.

Except 2.5L Engine

1. Disconnect the negative battery cable.

2. Drain the cooling system.

3. Remove the valve cover and the intake manifold.

4. If the engine is equipped with individual rocker arms, loosen the rocker arm adjusting nut and rotate the arm so as to clear the pushrod.

5. If the engine is equipped with a

rocker shaft assembly, remove the rocker shaft retaining bolts/nuts and remove the shaft assembly.

NOTE: Be sure to keep all valve train parts in order so they may be reinstalled in their original locations and with the same mating surfaces as when removed.

6. Remove the pushrods and valve lifters using tool J–3049 or equivalent.
To install:
7. Lubricate the bearing surfaces with Molykote® or equivalent.
8. Install the lifters in their original locations.
9. With the lifter on the base circle of the camshaft, tighten the rocker arm bolts to 14–20 ft. lbs. (20–27 Nm).
10. Connect the negative battery cable.
11. Adjust the valves, as required.

Valve Lash

ADJUSTMENT

Except 2.8L Engine

Hydraulic valve lifter keep all parts of the valve train in constant contact and adjust automatically to maintain zero lash under all conditions.

2.8L Engine

Anytime the valve train has been disturbed, the valve lash must be readjusted.
1. Crank the engine until the timing mark on the damper aligns with the **0** mark on the timing scale. Both valves in the No. 1 cylinder should be closed. If the valves are moving as the timing marks align, the engine is in the No. 4 firing position. Turn the crankshaft one more revolution. With the engine in the No. 1 firing position, adjust the following valves: exhaust—1, 2, 3 and intake—1, 5, 6.
2. Back out the adjusting nut until lash is felt at the pushrod. Then, turn in the adjusting nut until all lash is removed. This can be determine by rotating the pushrod while turning the adjusting nut. When all lash has been removed, turn the adjusting nut in 1½ additional turns to center the lifter plunger.
3. Rotate the crankshaft one full revolution, until the timing mark on the damper aligns with the **0** mark on the timing scale once again. This is the No. 4 firing position. Adjust the fol-

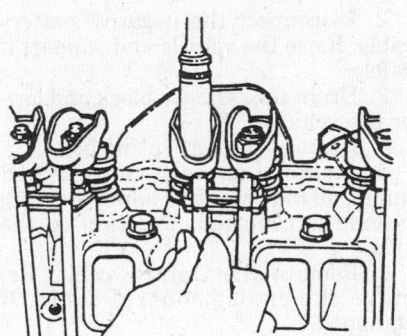

Adjusting valve lash—2.8L engine

lowing valves: exhaust—4, 5, 6 and intake—2, 3, 4.

Rocker Arms

REMOVAL & INSTALLATION

2.5L Engine

1. Relieve pressure in the fuel system before disconnecting any fuel lines.
2. Disconnect the negative battery cable.
3. Remove the valve cover.
4. If only the pushrod is being removed, loosen the rocker arm bolt and swing the rocker arm aside.
5. Remove the rocker arm nut and ball.
6. Lift the rocker arm off the stud, keeping rocker arms in order for installation.
To install:
7. If the pushrod was removed, install through the cylinder head and into the lifter seat.
8. Install the guide, rocker arm, ball and bolt. Tighten to 24 ft. lbs. (32 Nm).
9. Install the valve cover.
10. Connect the negative battery cable.

2.8L and 3.1L Engines

1. Relieve pressure in the fuel system before disconnecting any fuel lines.
2. Disconnect the negative battery cable. Remove the valve covers.

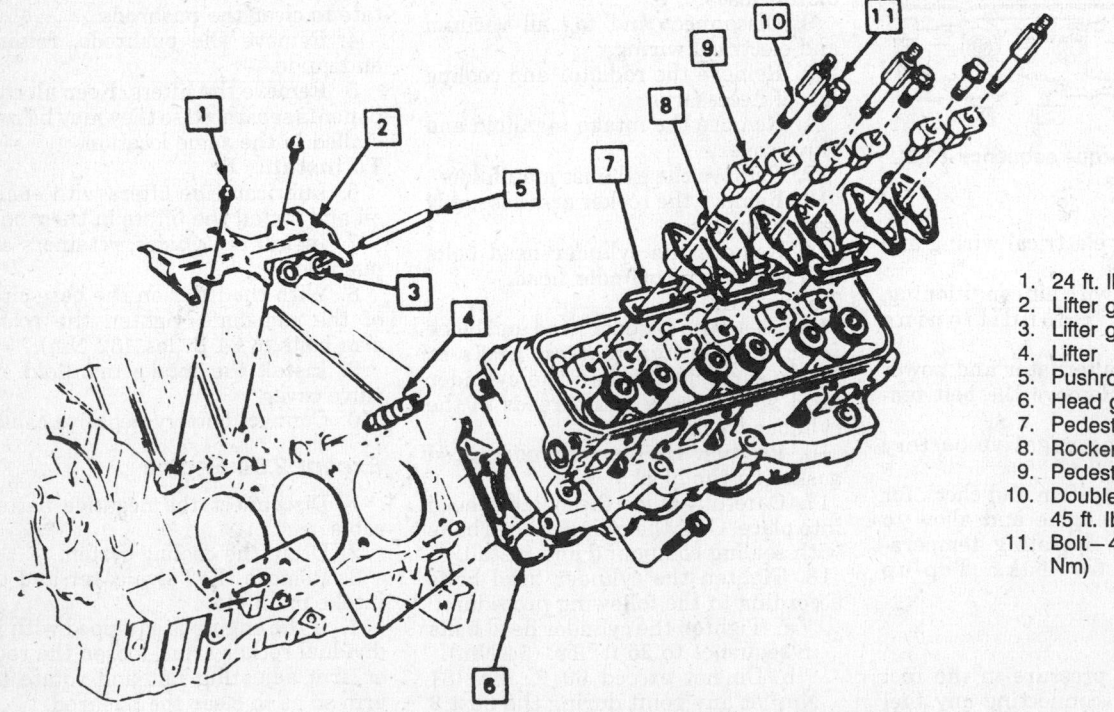

1. 24 ft. lbs. (33 Nm)
2. Lifter guide retainer
3. Lifter guide
4. Lifter
5. Pushrod
6. Head gasket
7. Pedestal retainer
8. Rocker arm
9. Pedestal
10. Double ended bolt—45 ft. lbs. (60 Nm)
11. Bolt—45 ft. lbs. (60 Nm)

Valve train and cylinder head installation—3.8L engine

3. Remove the rocker arm nuts, pivot balls, rocker arms and pushrods. Keep all components separated so they may be reinstalled in the same location.

NOTE: The intake and exhaust pushrods are of different lengths.

To install:

4. Install the pushrods in their original location. Be sure they are seated in the lifter.

5. Coat the bearing surfaces of the rocker arms and pivots balls with Molykote or equivalent.

6. If equipped with adjustable lifters, install the rocker arms and pivot balls. Loosely retain with the rocker arms nuts until the valve lash is eliminated.

7. If equipped with non-adjustable lifters, install the pushrods. Make sure the lower ends of the pushrods are in the lifter seats. Install the rocker arm nuts and torque the nuts to 14–20 ft. lbs. (20–27 Nm).

8. Install the valve cover.

9. Connect the negative battery cable.

10. Adjust valve lash, as required.

3.3L and 3.8L Engines

1. Relieve pressure in the fuel system before disconnecting any fuel lines. Disconnect the negative battery cable.

2. Remove the valve covers.

3. Remove the rocker arm bolts, pivots, and rocker arms assembly. Keep all components separated so they may be reinstalled in the same location.

To install:

4. Install the rocker arms, pivots and bolts. On 3.3L engine, tighten bolts to 37 ft. lbs. (51 Nm). On 3.8L engine, tighten bolts to 45 ft. lbs. (60 Nm).

5. Install the valve covers.

6. Connect the negative battery cable.

Intake Manifold

REMOVAL & INSTALLATION

2.5L Engine

1. Relieve the pressure in the fuel system before disconnecting any fuel line connections.

2. Disconnect the negative battery cable.

3. Remove the air cleaner and the PCV valve.

4. Drain the cooling system into a clean container.

5. Disconnect the fuel and vacuum lines and the electrical connections.

6. Disconnect the throttle linkage

at the EFI unit. Disconnect the transaxle downshift linkage and cruise control linkage.

7. Remove the bell crank and the throttle linkage. Position to the side for clearance.

8. Remove the heater hose at the intake manifold.

9. Remove the pulse air check valve bracket from the manifold, as required.

10. Remove the manifold attaching bolts and remove the manifold.

To install:

11. Clean the cylinder head and intake manifold surfaces from any foreign matter, nicks or heavy scratches.

12. Install the intake manifold with a new gasket and tighten the retaining bolts in sequence to the specified torque value.

13. If removed, install the pulse air check valve bracket to the manifold.

14. Install the heater hose to the intake manifold.

15. Install the bell crank and the throttle linkage.

16. Connect the throttle linkage to the EFI unit. Connect the transaxle downshift linkage and cruise control linkage.

17. Connect the fuel and vacuum lines and the electrical connections.

18. Install the air cleaner and the PCV valve.

19. Connect the negative battery cable.

20. Fill cooling system and check for leaks. Start the engine and allow to come to normal operating temperature. Recheck for leaks. Top-up coolant.

2.8L and 3.1L Engines

1. Relieve the pressure in the fuel system before disconnecting any fuel line connections.

2. Disconnect the negative battery cable.

3. Disconnect the accelerator and TV cable bracket at the plenum.

4. Disconnect the throttle body at the plenum.

5. Disconnect the EGR valve at the plenum.

6. Remove the plenum.

7. Disconnect the fuel inlet and return pipes at the fuel rail.

8. Remove the serpentine belt.

9. Remove the power steering pump and lay it aside.

10. Disconnect the alternator and lay it aside.

11. Loosen the alternator bracket.

12. Disconnect the idle air vacuum hose at the throttle body.

13. Disconnect the wires at the injectors.

14. Disconnect the fuel rail.

15. Remove the breather tube.

16. Remove both rocker covers.

17. Drain the cooling system.

18. Disconnect the radiator hose at the thermostat housing.

19. Disconnect the wires at the coolant sensor and the oil sending switch.

20. Remove the coolant sensor.

21. Disconnect the bypass hose at the fill neck and head.

22. Loosen the rocker arms and remove the pushrods.

23. Remove the intake manifold bolts and remove the intake manifold.

To install:

24. Place a ³⁄₁₆ in. (5mm) diameter bead GM sealer 1052917 or equivalent, on each ridge.

25. Position a new intake manifold gasket.

26. Install the pushrods and tighten the rocker arm nuts to 14–20 ft. lbs. (19–27 Nm).

27. Install the intake manifold and torque the bolts to specifications.

28. Connect the bypass hose to the filler neck and head.

29. Install the coolant sensor.

30. Connect the wires to the coolant sensor and the oil sending switch.

31. Connect the radiator hose to the thermostat housing.

32. Install both rocker covers.

33. Install the breather tube.

34. Connect the fuel rail.

35. Connect the wires to the injectors.

36. Connect the idle air vacuum hose to the throttle body.

37. Tighten the alternator bracket.

38. Connect the alternator electrical connectors.

39. Install the power steering pump.

40. Install the serpentine belt.

41. Connect the fuel inlet and return pipes to the fuel rail.

42. Install the plenum.

43. Connect the EGR valve to the plenum.

44. Connect the throttle body to the plenum.

45. Connect the accelerator and TV cable bracket to the plenum.

46. Connect the negative battery cable.

47. Fill cooling system and check for leaks. Start the engine and allow to come to normal operating temperature. Recheck for leaks. Top-up coolant.

48. Adjust the valve, as required.

3.3L Engine

1. Relieve the pressure in the fuel system before disconnecting any fuel line connections.

2. Disconnect the negative battery cable.

3. Drain the cooling system.

4. Remove the serpentine belt, alternator and braces and power steering pump braces.

5. Remove the coolant bypass hose, heater pipe and upper radiator hose.

6. Remove the air inlet duct, throttle cable bracket and cables.

7. Disconnect and tag all vacuum hoses and electrical connectors, as necessary.

8. Remove the fuel rail, vapor canister purge line and heater hose from the throttle body.

9. Remove the intake manifold retaining bolts and intake manifold.

To install:

10. Clean the cylinder head and intake manifold surfaces from any foreign matter, nicks or heavy scratches.

11. Apply sealer 12345336 or equivalent, to the ends of the manifold seals. Clean the intake manifold bolts and bolt holes. Apply thread lock compound 1052624 or equivalent, to the intake manifold bolt threads before assembly.

12. Install the new gasket and intake manifold. Tighten the intake manifold bolts twice to 88 inch lbs. (10 Nm) in the proper sequence.

13. Install the fuel rail, vapor canister purge line and heater hose from the throttle body.

14. Connect all vacuum hoses and electrical connectors.

15. Install the air inlet duct, throttle cable bracket and cables.

16. Install the coolant bypass hose, heater pipe and upper radiator hose.

17. Install the serpentine belt, alternator and braces and power steering pump braces.

18. Connect the negative battery cable.

19. Fill cooling system and check for leaks. Start the engine and allow to come to normal operating temperature. Recheck for leaks. Top-up coolant.

3.8L Engine

1. Relieve the pressure in the fuel system before disconnecting any fuel line connections.

2. Disconnect the negative battery cable.

3. Remove the mass air flow sensor and air intake duct.

4. Remove the serpentine accessory drive belt, alternator and bracket.

5. Remove the ignition coil module, TV cable, throttle cable and cruise control cable.

6. Disconnect and tag all vacuum hoses and electrical wiring, as necessary.

7. Drain the cooling system. Remove the heater hoses from the throttle body and upper radiator hose.

8. Disconnect the fuel lines from the fuel rail and injectors.

9. Remove the intake manifold retaining bolts and remove the intake manifold and gasket.

To install:

10. Clean the cylinder head and intake manifold surfaces from any foreign matter, nicks or heavy scratches.

11. Install the intake manifold gasket and rubber seals. Apply sealer 1050026 or equivalent, on the gasket. Apply sealer/lubricant 1052080 or equivalent, to all pipe thread fitting.

12. Carefully install the intake manifold to cylinder block. Install the intake manifold bolts and torque in sequence to the specified value.

13. Connect the fuel lines to the fuel rail and injectors.

14. Install the heater hoses to the throttle body and upper radiator hose.

15. Connect all vacuum hoses and electrical wiring.

16. Install the ignition coil module, TV cable, throttle cable and cruise control cable.

17. Install the bracket, alternator and serpentine belt.

18. Install the mass air flow sensor and air intake duct.

19. Connect the negative battery cable.

20. Fill cooling system and check for leaks. Start the engine and allow to come to normal operating temperature. Recheck for leaks. Top-up coolant.

Exhaust Manifold

REMOVAL & INSTALLATION

2.5L Engine

1. Disconnect the negative battery cable. Remove the air cleaner and the TBI preheat tube.

2. Remove the manifold strut bolts from the radiator support panel and the cylinder head.

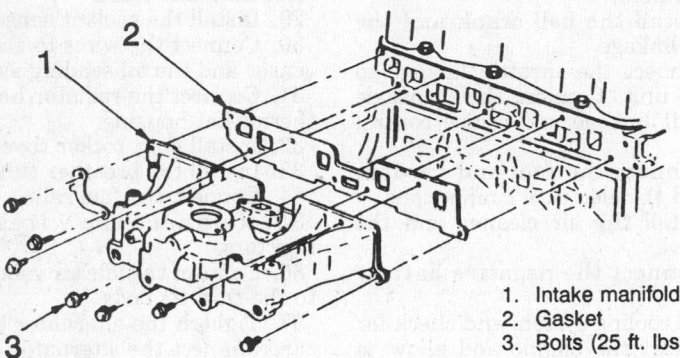

1. Intake manifold
2. Gasket
3. Bolts (25 ft. lbs.)

Intake manifold assembly—2.5L engine

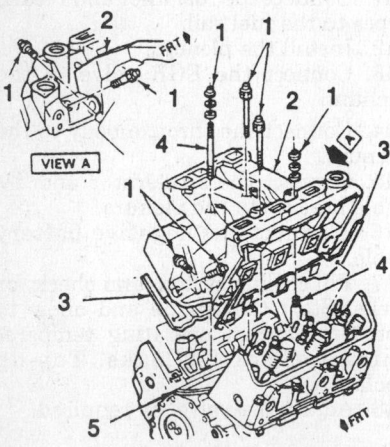

1. 16 ft. lbs. (22 Nm)
   Then 23 ft. lbs. (32 Nm)
   Retorque 23 ft. lbs. (32 Nm) in sequence
2. Intake manifold
3. Gasket
4. 24 ft. lbs. (33 Nm)
5. Sealer

Intake manifold assembly—2.8L and 3.1L engines

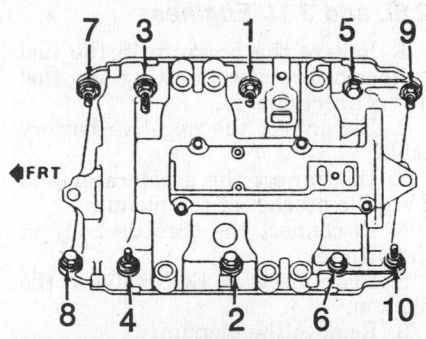

Intake manifold assembly—3.3L engine

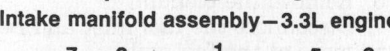

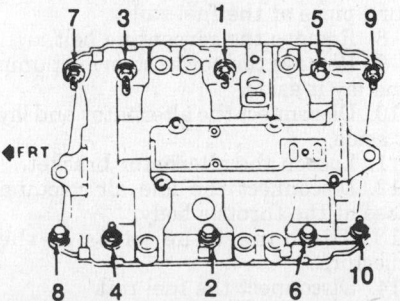

Intake manifold bolt tightening sequence—3.8L engine

3. Remove the air conditioning compressor bracket to one side. Do not disconnect any of the refrigerant lines.

4. If necessary, remove the dipstick tube attaching bolt and the engine mount bracket from the cylinder head.

5. Raise the vehicle and support safely. Disconnect the exhaust pipe from the manifold.

6. Remove the manifold attaching bolts and remove the manifold.

To install:

7. Install the exhaust manifold and gasket to the cylinder head. Torque all bolts in sequence to the specified torque value.

8. Connect the exhaust pipe to the manifold. Lower the vehicle.

9. Install the dipstick tube attaching bolt and the engine mount bracket to the cylinder head.

10. Install the air conditioning compressor bracket.

11. Install the manifold strut bolts to the radiator support panel and the cylinder head.

12. Install the air cleaner and the TBI preheat tube.

13. Connect the negative battery cable.

2.8L and 3.1L Engines

LEFT SIDE

1. Disconnect the negative battery cable.

2. Remove the air supply plumbing from the exhaust manifold, as required.

3. Remove the coolant recovery bottle, if necessary.

4. Remove the serpentine belt cover and belt, as required.

5. Remove the air conditioning compressor and lay aside, if necessary.

6. Remove the right side torque strut, air conditioning and torque strut mounting bracket, as required.

7. Remove the heat shield, if equipped.

8. Remove the exhaust crossover pipe at the manifold.

9. Remove the exhaust manifold retaining bolts and manifold.

To install:

10. Install the exhaust manifold and retaining bolts. Tighten to 15–22 ft. lbs. (20–30).

11. Install the exhaust crossover pipe at the manifold.

12. Install the heat shield, if equipped.

13. If removed, install the right side torque strut, air conditioning and torque strut mounting bracket.

14. If removed, install the air conditioning compressor.

15. If removed, install the serpentine belt and cover.

16. If removed, install the coolant recovery bottle.

17. Install the air supply plumbing to the exhaust manifold.

18. Connect the negative battery cable.

RIGHT SIDE

1. Disconnect the negative battery cable.

2. Raise and safely support the vehicle.

3. Disconnect the exhaust pipe and lower the vehicle.

4. Remove the air cleaner assembly, breather, mass air flow sensor and heat shield.

5. Remove the crossover at the manifold.

6. Remove the accelerator and TV cables and brackets, as required.

7. Remove the exhaust manifold retaining bolts and remove the manifold.

To install:

8. Install the exhaust manifold and retaining bolts. Tighten to 15–22 ft. lbs. (20–30 Nm).

9. If removed, install the accelerator, TV cables and brackets.

10. Install the crossover at the manifold.

11. Install the air cleaner assembly, breather, mass air flow sensor and heat shield.

12. Raise and safely support the vehicle.

13. Connect the exhaust pipe and lower the vehicle.

14. Connect the negative battery cable.

3.3L ENGINE

LEFT SIDE

1. Disconnect the negative battery cable.

2. Remove the air cleaner inlet ducting. Install the spark plug wires.

3. Remove the 2 bolts attaching the exhaust crossover pipe to the manifold.

4. Remove the engine lift hook, manifold heat shield and oil level indicator.

5. Remove the exhaust manifold retaining bolts and remove the manifold.

To install:

6. Install the exhaust manifold and retaining bolts. Tighten to 30 ft. lbs. (41 Nm).

7. Install the engine lift hook, manifold heat shield and oil level indicator.

8. Install the 2 bolts attaching the exhaust crossover pipe to the manifold.

9. Install the air cleaner inlet ducting. Install the spark plug wires.

10. Connect the negative battery cable.

RIGHT SIDE

1. Disconnect the negative battery cable.

2. Remove the spark plug wires, oxygen sensor connector, throttle cable bracket and cables.

3. Remove the brake booster hose from the manifold.

4. Remove the 2 bolts attaching the exhaust crossover pipe to the manifold.

5. Remove the exhaust pipe-to-manifold bolts, engine lift hook and transaxle oil level indicator tube.

6. Remove the manifold heat shield. Remove the exhaust manifold retaining bolts and remove the manifold.

To install:

7. Install the exhaust manifold and retaining bolts. Tighten to 30 ft. lbs. (41 Nm). Install the manifold heat shield.

8. Install the exhaust pipe-to-manifold bolts, engine lift hook and transaxle oil level indicator tube.

9. Install the 2 bolts attaching the exhaust crossover pipe to the manifold.

10. Install the brake booster hose to the manifold.

11. Install the spark plug wires, oxy-

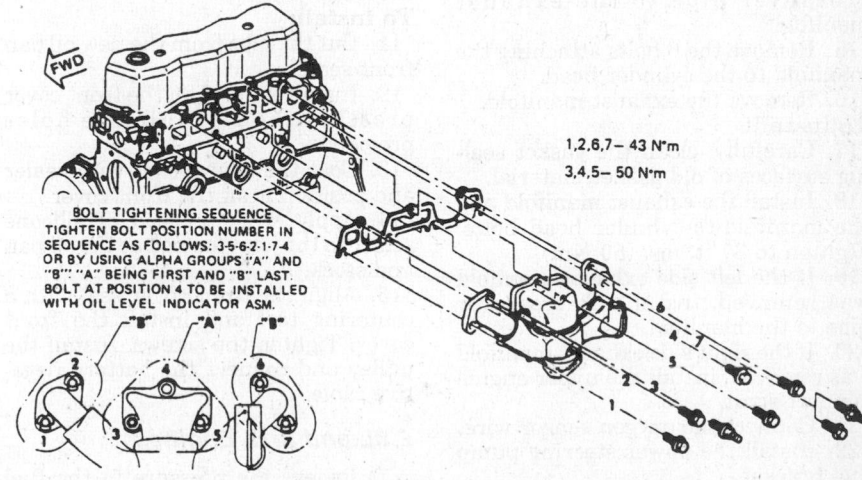

BOLT TIGHTENING SEQUENCE
TIGHTEN BOLT POSITION NUMBER IN SEQUENCE AS FOLLOWS: 3-5-6-2-1-7-4 OR BY USING ALPHA GROUPS "A" AND "B". "A" BEING FIRST AND "B" LAST. BOLT AT POSITION 4 TO BE INSTALLED WITH OIL LEVEL INDICATOR ASM.

1,2,6,7 – 43 N·m
3,4,5 – 50 N·m

Exhaust manifold assembly—2.5L engine

gen sensor connector, throttle cable bracket and cables.

12. Connect the negative battery cable.

3.8L Engine

1. Disconnect the negative battery cable.

NOTE: Failure to disconnect the intermediate shaft from the rack and pinion stub shaft may result in damage to the steering gear and/or intermediate shaft.

2. Remove the pinch bolt from the intermediate shaft and separate the intermediate shaft from the stub shaft.

3. Raise and safely support the vehicle.

4. Remove the 2 bolts attaching the exhaust pipe to the manifold.

5. Lower the vehicle.

6. Remove the upper engine support strut.

7. Place a jack under the front crossmember of the cradle and raise the jack until it starts to raise the vehicle.

8. Remove the 2 front body mount bolts.

9. With the cushions removed, thread the body mount bolts and retainers a minimum of 3 turns into the cage nuts.

10. Release the jack slowly.

NOTE: To avoid damage, do not lower the cradle without it being restrained.

11. Remove the power steering pump and bracket from the cylinder head and exhaust manifold.

12. Disconnect the oxygen sensor.

13. If removing the left side exhaust manifold, remove the upper engine support strut.

14. Remove the 2 nuts retaining the crossover pipe to the exhaust manifold.

15. Remove the 6 bolts attaching the manifold to the cylinder head.

16. Remove the exhaust manifold.

To install:

17. Carefully, clean the gasket sealing surfaces of old gasket material.

18. Install the exhaust manifold and the manifold-to-cylinder head bolts. Tighten to 37 ft. lbs. (50 Nm).

19. If the left side exhaust manifold was removed, install the crossover pipe to the manifold.

20. If the right side exhaust manifold was removed, install the upper engine support strut.

21. Connect the oxygen sensor wire.

22. Install the power steering pump and bracket.

23. Support the cradle with the jack.

Remove the 2 body mount bolts and install the cushions.

24. Raise the cradle into position and install the 2 body mount bolts.

25. Remove the jack.

26. Connect the intermediate shaft to the stub shaft and install the pinch bolt.

27. Raise and safely support the vehicle.

28. Install the exhaust pipe-to-manifold bolts.

29. Lower the vehicle.

30. Connect the negative battery cable.

Timing Chain Front Cover and Seal

REMOVAL & INSTALLATION

2.5L Engine

1. Relieve the pressure in the fuel system before disconnecting any fuel line connections.

2. Disconnect the negative battery cable.

3. Remove the inner fender splash shield. Remove the crankshaft pulley.

4. Remove the alternator lower bracket and the front engine mounts.

5. Using a floor jack, raise the engine.

6. Remove the engine mount mounting bracket-to-cylinder block bolts. Remove the bracket and mount as an assembly.

7. Remove the oil pan-to-front cover screws and front cover-to-block screws.

8. Pull the cover slightly forward, just enough to allow cutting of the oil pan front seal flush with the block on both sides.

9. Remove the front cover and attached portion of the pan seal.

10. Clean the gasket surfaces thoroughly.

To install:

11. Cut the tabs from the new oil pan front seal.

12. Install the seal on the front cover pressing the tips into the holes provided.

13. Coat the new gasket with sealer and position it on the front cover.

14. Apply a ⅛ in. bead of silicone sealer to the joint formed at the oil pan and stock.

15. Align the front cover seal with a centering tool and install the front cover. Tighten the screws. Install the pulley and connect the battery negative cable.

2.8L and 3.1L Engines

1. Relieve the pressure in the fuel system before disconnecting any fuel

line connections. Disconnect the negative battery cable.

2. Drain the cooling system.

3. Remove the serpentine belt and tensioner.

4. Remove the alternator and power steering pump. Locate and support these accessories to the side.

5. Raise and support the vehicle safely.

6. Remove the inner splash shield. Remove the torsion damper using tool J–24420–B or equivalent.

7. Remove the flywheel cover at the transaxle and starter.

8. Remove the serpentine belt idler pulley.

9. Drain the engine oil. Remove the oil pan and lower front cover bolts.

10. Lower the vehicle.

11. Remove the radiator hose at the water pump. Remove the heater hose at fill pipe.

12. Remove the bypass hose and overflow hoses. Remove the canister purge hose.

13. Remove the upper front cover retaining bolts and remove the front cover.

14. After removing the timing cover, pry oil seal from front of cover. Lubricate the seal lip and install new lip seal with lip, open side of seal, facing toward the cylinder block. Carefully drive or press seal into place.

To install:

15. Clean the mating surfaces of the front cover and cylinder block.

16. Install a new gasket. Make sure not to damage the sealing surfaces. Apply sealer 1052080 or equivalent, to the sealing surface of the front cover.

17. Position the front cover on the engine block and install the upper cover bolt.

18. Raise and safely support the vehicle. Install the oil pan and lower cover bolts.

19. Install the serpentine belt idler pulley.

20. Install the flywheel cover to the transaxle. Install the starter.

21. Install the torsion damper. Install the inner splash shield.

22. Lower the vehicle.

23. Install the bypass hose and overflow hoses. Install the canister purge hose.

24. Connect the radiator hose to the water pump. Connect the heater hose to fill pipe.

25. Install the alternator and power steering pump.

26. Install the tensioner serpentine belt.

27. Connect the negative battery cable.

3.3L Engine

1. Relieve the pressure in the fuel

system before disconnecting any fuel line connections. Disconnect the negative battery cable.

2. Drain the cooling system.
3. Remove the serpentine belt.
4. Remove the heater pipes. Remove the coolant bypass hose and lower radiator hose from cover.
5. Raise and support the vehicle safely.
6. Remove the inner splash shield.
7. Remove the crankshaft balancer.
8. Disconnect all electrical connectors at the camshaft sensor, crankshaft sensor and oil pressure sender.
9. Remove the oil pan-to-front cover retaining bolts, front cover retaining bolts and remove the front cover.
10. After removing the timing cover, pry oil seal from front of cover. Lubricate the seal lip and install new lip seal with lip, open side of seal, facing toward the cylinder block. Carefully drive or press seal into place.

To install:
11. Clean the mating surfaces of the front cover and cylinder block.
12. Install a new gasket on the cylinder block. Install the front cover. Apply sealer to the threads of the cover retaining bolts and secure the cover. Tighten the bolts to 22 ft. lbs. (30 Nm).
13. Install the oil pan-to-front cover bolts. Tighten the bolts to 88 inch lbs. (10 Nm).
14. Reconnect the camshaft sensor, crankshaft sensor and oil pressure sender electrical connectors. Adjust the crankshaft sensor using tool J–37087 or equivalent.
15. Install the crankshaft balancer.
16. Install the inner splash shield.
17. Lower the vehicle.
18. Install the heater pipes. Install the coolant bypass hose and lower radiator hose from cover.
19. Install the serpentine belt.
20. Connect the negative battery cable.
21. Fill cooling system and check for leaks. Start the engine and allow to come to normal operating temperature. Recheck for leaks. Top-up coolant.

3.8L Engine

1. Relieve the pressure in the fuel system before disconnecting any fuel line connections. Disconnect the negative battery cable.
2. Drain the cooling system.
3. Disconnect the lower radiator hose and the heater hose from the water pump.
4. Remove the 2 nuts from the front engine mount from the cradle and raise the engine using a suitable lifting device.
5. Remove the water pump pulley and the serpentine belt.

6. Remove the alternator and brackets.
7. Remove the balancer bolt and washer. Using a puller, remove the balancer.
8. Remove the cover-to-block bolts. Remove the 2 oil pan-to-cover bolts.
9. Remove the cover and gasket.
10. After removing the timing cover, pry oil seal from front of cover. Lubricate the seal lip and install new lip seal with lip, open side of seal, facing toward the cylinder block. Carefully drive or press seal into place.

To install:
11. Clean the mating surfaces of the front cover and cylinder block.

NOTE: Remove the oil pump cover and pack the space around the oil pump gears completely with petroleum jelly. There must be no air space left inside the pump. If the pump is not packed, it may not begin to pump oil as soon as the engine is started and engine damage may result.

12. Install a new gasket to the oil pan and cylinder block. Install the front cover. Apply sealer to the threads of the cover retaining bolts and secure the cover.
13. Install the cover-to-block bolts. Install the 2 oil pan-to-cover bolts.
14. Install the balancer, washer and balancer bolt.
15. Install the alternator brackets and install the alternator.
16. Install the water pump pulley and the serpentine belt.
17. Lower the engine into position and install the 2 nuts to the front engine mount at the cradle.
18. Connect the lower radiator hose and the heater hose to the water pump.
19. Connect the negative battery cable.
20. Fill cooling system and check for leaks. Start the engine and allow to come to normal operating temperature. Recheck for leaks. Top-up coolant.

Timing Chain and Sprockets

REMOVAL & INSTALLATION

2.5L Engine

NOTE: The camshaft gear is press fitted on the camshaft. If replacement of the camshaft gear is necessary, the engine must be removed from the vehicle and the camshaft and gear removed from the engine.

1. Relieve the pressure in the fuel

system before disconnecting any fuel line connections.
2. Disconnect the negative battery cable.
3. Remove the engine from the vehicle.
4. Remove the camshaft and gear assembly from the engine block.
5. Using an arbor press and adapter, remove the gear from the camshaft. Position the thrust plate to avoid damage by interference with the Woodruff® key as the gear is removed.

To install:
6. Support the camshaft at the back of the front journal in the arbor press using press plate adapters.
7. Position the spacer ring thrust plate over the end of the shaft and Woodruff® key in keyway.
8. Press the gear on the shaft with the bottom against the spacer ring. Measure the end clearance at the thrust plate. Clearance should be within 0.0015–0.0050 in. (0.0381–1.270mm).
9. If the clearance is less than 0.0015 in. (0.0381mm), replace the spacer ring.
10. If more than 0.0050 in. (1.270mm), make certain the gear is seated properly against the spacer. If the clearance is still excessive, replace the thrust plate.
11. Measure the backlash at position outside the 2 retainer plate access holes and at 2 other areas 90 degrees from these holes. If the backlash is not within specifications, replace the camshaft and crankshaft gears.
12. Lubricate the camshaft journals with a high quality engine oil supplement. Install the camshaft and gear into the engine block.
13. Rotate the camshaft and crankshaft so the timing marks on the gear teeth align. The engine is now in No. 4 cylinder firing position.
14. Install the camshaft thrust plate-to-block screws and tighten to 90 inch lbs. (10 Nm).
15. Install the engine in the vehicle.
16. Connect the negative battery cable.

2.8L and 3.1L Engines

1. Relieve the pressure in the fuel system before disconnecting any fuel line connections. Disconnect the negative battery cable.
2. Remove the crankcase front cover.
3. Place the No. 1 piston at TDC with the marks on the camshaft and crankshaft sprockets aligned.
4. Remove the camshaft sprocket and chain.

NOTE: If the sprocket does not come off easily, a light blow with a plastic mallet on the lower edge of

the sprocket should dislodge the sprocket.

5. Remove the crankshaft sprocket.

To install:

6. Install the crankshaft sprocket. Apply Molykote® or equivalent, to the sprocket thrust surface.

7. Hold the sprocket with the chain hanging down and align the marks on the camshaft and crankshaft sprockets.

8. Align the dowel in the camshaft with the dowel hole in the camshaft sprocket.

9. Draw the camshaft sprocket onto the camshaft using the mounting bolts. Tighten the camshaft sprocket mounting bolts to 18 ft. lbs. (25 Nm).

10. Lubricate the timing chain with engine oil. Install the crankcase front cover. Connect battery negative cable.

3.3L Engine

1. Relieve the pressure in the fuel system before disconnecting any fuel line connections. Disconnect the negative battery cable.

2. Remove the crankcase front cover and camshaft thrust bearing.

3. Turn the crankshaft so the timing marks are aligned.

4. Remove the timing chain damper and camshaft sprocket bolts.

5. Remove the camshaft sprocket and chain. Remove the crankshaft sprocket.

To install:

6. Make sure the crankshaft is positioned so No. 1 piston is at TDC on compression stroke.

7. Rotate the camshaft with the sprocket temporarily installed, so the timing mark is straight down.

8. Assembly the timing chain on the sprockets with the timing marks aligned. Install the timing chain and sprocket.

9. Install the camshaft sprocket bolts. Torque the bolts to 27 ft. lbs. (37 Nm).

10. Install the timing chain damper and engine front cover. Connect battery negative cable.

3.8L Engine

1. Relieve the pressure in the fuel system before disconnecting any fuel line connections. Disconnect the negative battery cable.

2. Remove the crankcase front cover.

3. Turn the crankshaft so the timing marks are aligned.

4. Remove the crankshaft oil slinger, as required.

5. Remove the camshaft sprocket bolts.

6. Remove the cam sensor magnet assembly.

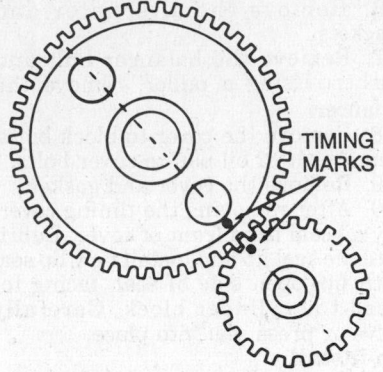

Timing gear alignment—2.5L engine

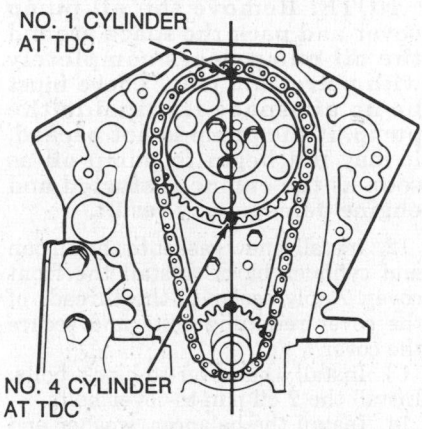

Timing gear alignment—2.8L and 3.1L engines

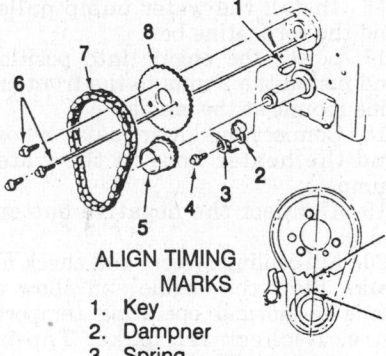

ALIGN TIMING MARKS
1. Key
2. Dampner
3. Spring
4. Bolt 16 ft. lbs. (22 Nm)
5. Crankshaft sprocket
6. Bolt 26 ft. lbs. (35 Nm)
7. Timing chain
8. Camshaft sprocket

Timing chain and sprockets—3.8L engine

7. Use 2 prybars to alternately pry the camshaft and crankshaft sprocket free along with the chain.

To install:

8. Make sure the crankshaft is positioned so No. 1 piston is at TDC.

9. Rotate the camshaft with the sprocket temporarily installed, so the timing mark is straight down.

10. Assembly the timing chain on the sprockets with the timing marks aligned. Install the timing chain and sprocket.

11. Install the cam sensor magnet assembly.

12. Install the oil slinger with the large part of the cone toward the front of the engine, as required.

13. Install the camshaft sprocket bolt, thrust button and spring.

14. Install the timing chain damper and engine front cover.

Camshaft

REMOVAL & INSTALLATION

2.5L Engine

1. Relieve the pressure in the fuel system before disconnecting any fuel line connections.

2. Disconnect the negative battery cable.

3. Remove the engine from the vehicle and support on a suitable engine stand.

4. Remove the rocker cover, rocker arms and pushrods.

5. Remove the spark plugs and fuel pump.

6. Remove the pushrod cover and gasket. Remove the lifters.

7. Remove the alternator, the alternator lower bracket and the front engine mount bracket assembly.

8. Remove the oil pump driveshaft and gear assembly.

9. Remove the crankshaft hub and timing gear cover.

10. Remove the 2 camshaft thrust plate screws by working through the holes in the gear.

11. Remove the camshaft and gear assembly by pulling it through the front of the block. Take care not to damage the bearings.

12. If replacement of the camshaft gear is necessary, use the following procedure:

a. Remove the camshaft gear using an arbor press and adapter.

b. Position the thrust plate to avoid damage by interference with the Woodruff® key as the gear is removed.

c. When assembling the gear onto the camshaft, support the camshaft at the back of the front journal in the arbor press using press plate adapters.

d. Press the gear on the shaft until it bottoms against the spacer ring.

e. Measure the end clearance of the thrust plate. End clearance should be 0.0015–0.0050 in.

f. If clearance is less than 0.0015 in., replace the spacer ring.

g. If clearance is more than 0.0050 in., replace the thrust plate.

To install:

13. Lubricate the camshaft journals with a high quality engine oil supplement and carefully install the camshaft and gear into the cylinder block.

14. Rotate the camshaft and crankshaft so the timing marks on the gear teeth align. The engine is now in No. 4 cylinder firing position.

15. Install the camshaft thrust plate-to-block screw. Torque the screw to 90 inch lbs. (10 Nm).

16. Install the crankshaft hub and timing gear cover.

17. Install the oil pump driveshaft and gear assembly.

18. Install the lower alternator bracket, alternator and the front engine mount bracket assembly.

19. Install the spark plugs and fuel pump.

20. Install the lifters. Install the pushrod cover and gasket.

21. Install the pushrods, rocker arms and rocker cover.

22. Install the engine in the vehicle.

23. Connect the negative battery cable.

2.8L, 3.1L and 3.3L Engines

1. Relieve the pressure in the fuel system before disconnecting any fuel line connections.

2. Disconnect the negative battery cable.

3. Remove the engine from the vehicle and support on a suitable engine stand.

4. Remove the intake manifold, valve cover, rocker arms, pushrods and valve lifters.

5. Remove the crankshaft balancer and front cover.

6. Remove the timing chain and sprockets.

7. Carefully remove the camshaft. Avoid marring the camshaft bearing surfaces.

To install:

8. Coat the camshaft with lubricant 1052365 or equivalent, and install the camshaft.

9. Install the timing chain and sprocket.

10. Install the camshaft thrust button and front cover.

11. Install the crankshaft balancer.

12. Install the intake manifold, valve cover, rocker arms, pushrods and valve lifters.

13. Install the engine in the vehicle.

14. Connect the negative battery cable.

15. Adjust the valves, as required.

3.8L Engine

1. Relieve the pressure in the fuel

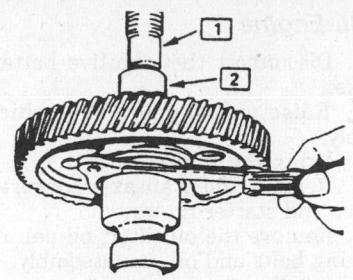

1. Arbor Press
2. Tool J–21474–13 or J–21795–1

Camshaft timing gear/thrust plate end clearance—2.5L engine

system before disconnecting any fuel line connections.

2. Disconnect the negative battery cable.

3. Remove the engine from the vehicle and support on a suitable engine stand.

4. Remove the intake manifold.

5. Remove the rocker arm covers.

6. Remove the rocker arm assemblies, pushrods and lifters.

7. Remove the timing chain cover.

8. Remove the timing chain, camshaft sensor magnet assembly and sprockets.

To install:

9. Coat the camshaft with lubricant 1052365 or equivalent, and install the camshaft.

10. Install the timing chain, camshaft sensor magnet assembly and sprockets.

11. Install the camshaft thrust button and front cover.

12. Complete installation by reversing the removal procedure. Connect battery negative cable.

Piston and Connecting Rod

POSITIONING

FRONT OF ENGINE

NOTCH

Piston Identification—2.8L engine

ENGINE LUBRICATION

Oil Pan

REMOVAL & INSTALLATION

2.5L Engine

1. Disconnect the negative battery cable. Raise and support the vehicle safely. Drain the oil.

2. Remove cradle-to-front engine mount nuts.

3. Disconnect exhaust pipe at manifold and at rear transaxle mount.

4. Disconnect starter and remove flywheel housing inspection cover.

5. Remove upper alternator bracket. Remove the splash shield, if equipped, in order to gain working clearance.

6. Install suitable engine support equipment and raise engine.

7. Remove lower alternator bracket and engine support bracket.

8. Remove oil pan retaining bolts and remove oil pan.

To install:

9. Thoroughly clean all gasket sealing surfaces.

10. Install rear oil pan gasket in rear main bearing cap and apply a small quantity of sealer in depressions where pan gasket engages into block.

11. Install front oil pan gasket on timing gear cover pressing tips into holes provided in cover.

12. Install side gaskets on oil pan using grease as a retainer.

13. Apply a ⅛ inch by ¼ inch long bead of sealer at split lines of front and side gaskets.

14. Install oil pan. Bolts into timing gear cover should be installed last. They are installed at an angle and holes align after rest of pan bolts are snugged up.

15. Install lower alternator bracket and engine support bracket.

16. Lower the engine and remove engine support equipment.

17. Install upper alternator bracket. If removed, install the splash shield.

18. Install flywheel housing inspection cover. Connect starter.

19. Connect exhaust pipe at manifold and at rear transaxle mount.

20. Install cradle-to-front engine mount nuts.

21. Lower the vehicle.

22. Fill the crankcase with oil.

23. Connect the negative battery cable.

24. Start the engine and check for leaks.

2.8L and 3.1L Engines

1. Disconnect the battery ground.
2. Remove the serpentine belt cover, belt and tensioner.
3. Support the engine with tool J–28467–A or equivalent, using an extra support leg.
4. Raise and safely support the vehicle.
5. Drain the oil.
6. Remove the right tire and wheel assembly. Remove the splash shield.
7. Remove the steering gear pinch bolt, as required.
8. Remove the transaxle mount retaining nuts and engine-to-frame mount retaining nuts, as required.
9. Remove the front engine horse collar bracket from the block, as required.
10. Remove the bellhousing cover and remove the starter.
11. Position a jackstand under the frame front center crossmember.
12. Loosen but do not remove the rear frame bolts.
13. Remove the front frame bolts and lower the front frame.
14. Remove the oil pan retaining bolts and remove the oil pan.

To install:

NOTE: The oil pan on some vehicles may not require a gasket. If a gasket is not required, the oil pan is installed using RTV gasket material. Make sure the sealing surfaces are free of old RTV material. Use a ⅛ inch bead of RTV material on the pan sealing flange. Torque the pan bolts to 8–10 ft. lbs.

15. Install the oil pan using a new gasket or RTV gasket material.
16. Raise the front frame and install the the front frame bolts.
17. Tighten the rear frame bolts.
18. Remove the jackstand from the front center crossmember.
19. Install the starter and bellhousing cover.
20. If removed, install the front engine horse collar bracket from the block.
21. Install the transaxle mount retaining nuts and engine to frame mount retaining nuts.
22. If removed, install the steering gear pinch bolt.
23. Install the splash shield. Install the right tire and wheel assembly.
24. Lower the vehicle.
25. Install the tensioner, serpentine belt and cover.
26. Fill the crankcase with oil.
27. Connect the negative battery cable.

3.3L Engine

1. Disconnect the negative battery cable.
2. Raise and support the vehicle safely.
3. Drain the engine oil.
4. Remove the transaxle converter cover and starter motor.
5. Remove the oil filter, oil pan retaining bolts and oil pan assembly.

To install:

6. Clean the oil pan and cylinder block mating surfaces.
7. Install a new oil pan gasket to the oil pan flange.
8. Install the oil pan and torque the retaining bolts 8–10 ft. lbs.
9. Lower the vehicle.
10. Fill the crankcase with oil.
11. Connect the negative battery cable.

3.8L Engine

1. Disconnect the battery ground cable.
2. Raise and safely support the vehicle.
3. Drain the oil.
4. Remove the bellhousing cover.
5. Unbolt and remove the oil pan.

To install:

6. RTV gasket material is used in place of a gasket. Make sure the sealing surfaces are free of all old RTV material. Use a ⅛ inch bead of RTV material on the oil pan sealing flange. Torque the pan bolts to 10–14 ft. lbs.
7. Install the bellhousing cover.
8. Lower the vehicle.
9. Fill the crankcase with oil.
10. Connect the negative battery cable.

Oil Pump

REMOVAL & INSTALLATION

2.5L Engine

1. Disconnect the negative battery cable.
2. Raise and support the vehicle safely.
3. Drain the engine oil and remove the oil pan.
4. Remove the 2 flange mounting bolts and nut from the main bearing cap bolt.
5. Remove the pump and screen as an assembly.

To install:

6. Remove the 4 cover attaching screws and cover from the oil pump assembly.
7. Pack the space around the oil pump gears completely full of petroleum jelly. There must be no air space left inside the pump. If the pump is not packed, it may not begin to pump oil as

soon as the engine is started and engine damage may result.
8. Align the oil pump shaft to match with the oil pump drive shaft tang, then install the oil pump to the block positioning the flange over the oil pump driveshaft lower bushing. Do not use any gasket. Torque the bolts to 20 ft. lbs. (30 Nm).
9. Install the oil pan using a new gasket and seals.
10. Install the 2 flange mounting bolts and nut to the main bearing cap bolt.
11. Lower the vehicle.
12. Fill the crankcase with oil.
13. Connect the negative battery cable.

2.8L and 3.1L Engines

1. Disconnect the negative battery cable.
2. Raise and support the vehicle safely.
3. Drain the engine oil and remove the oil pan.
4. Remove the pump-to-rear main bearing cap bolt and remove the pump and extension shaft.

To install:

5. Remove the 4 cover attaching screws and cover from the oil pump assembly.
6. Pack the space around the oil pump gears completely full of petroleum jelly. There must be no air space left inside the pump. If the pump is not packed, it may not begin to pump oil as soon as the engine is started and engine damage may result.
7. Assemble the pump and extension shaft with retainer to rear main bearing cap, aligning the top end of the extension shaft with the lower end of the drive gear.
8. Install the pump-to-the rear bearing cap bolt. Tighten to 30 ft. lbs. (40 Nm).
9. Install the oil pan.
10. Lower the vehicle.
11. Fill the crankcase with oil.
12. Connect the negative battery cable.

3.3L and 3.8L Engines

1. Disconnect the negative battery cable.
2. Drain the engine oil.
3. Remove the oil filter adapter, pressure regulator valve and spring.
4. Remove the oil pump cover attaching screws and cover.
5. Remove the gears.

To install:

6. Lubricate the gears with petroleum jelly.
7. Assemble the gears in the housing.
8. Pack the gear cavity with petroleum jelly.

9. Install the oil pump cover and screws. Tighten to 97 inch lbs. (11 Nm).

10. Install the pressure regulator and spring valve.

11. Install the oil filter adapter with a new gasket. Tighten the oil filter adapter bolts to 24 ft. lbs. (33 Nm).

12. Install the front cover on the engine.

13. Fill the crankcase with oil.

14. Connect the negative battery cable.

Rear Main Bearing Oil Seal

REMOVAL & INSTALLATION

2.5L Engine

1. Disconnect the negative battery cable.

2. Support the engine. Remove the transaxle and flywheel.

3. Being careful not to scratch the crankshaft, pry out the old seal with an suitable pry tool.

To install:

4. Coat the new seal with clean engine oil and install it by hand or use seal installer tool J–34924 onto the crankshaft. The seal backing must be flush with the block opening.

5. Install the flywheel.

6. Install the transaxle.

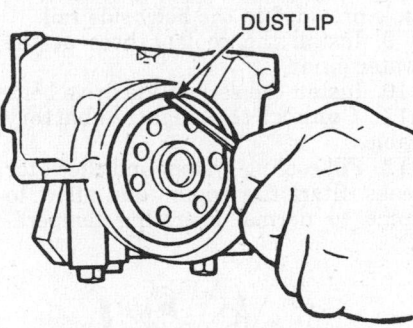

Remove rear seal—2.8L and 3.1L engines

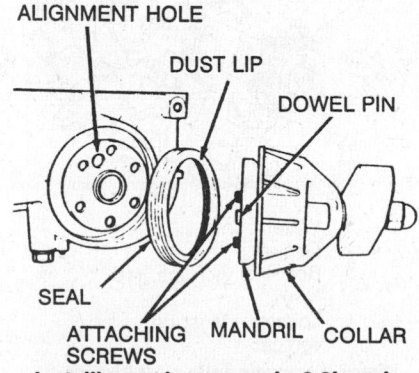

Installing main rear seal—2.8L and 3.1L engines

SPRING SIDE   DUST LIP SIDE

J–34686

SEAL BORE TO SEAL SURFACE TO BE LUBRICATED WITH ENGINE OIL BEFORE ASSEMBLY

Rear main seal and tool—2.8L and 3.1L engines

7. Connect the negative battery cable.

2.8L and 3.1L Engines

1. Disconnect the negative battery cable.

2. Support the engine with tool J–28467–A or equivalent.

3. Remove the transaxle and flywheel.

4. Carefully remove the old seal by inserting a prying tool through the dust lip at an angle. Pry out the old seal with an suitable pry tool.

To install:

5. Coat the new seal with clean engine oil, and install it using seal installer tool J–34686 or equivalent.

6. Install the flywheel.

7. Install the transaxle.

8. Remove the engine support tool.

9. Connect the negative battery cable.

3.3L and 3.8L Engines

1. Disconnect the negative battery cable.

2. Raise and support the vehicle safely.

3. Drain the engine oil and remove the oil pan.

4. Remove the rear main bearing cap. Remove the oil seal from the bearing cap.

To install:

5. Insert a packing tool J–21526–2 or equivalent against one end of the seal in the cylinder block. Pack the old seal in until it is tight. Pack the other end of seal in the same manner.

6. Measure the amount the seal was driven up into the block on one side and add approximately $\frac{1}{16}$ in. (2mm). With a single edge razor blade, cut this amount off of the old lower seal. The bearing cap can be used as a holding fixture.

7. Install the packing guide tool J–21526–1 or equivalent, onto the cylinder block.

8. Using the packing tool, work the short pieces of the seal into the guide tool and pack into the cylinder block until the tool hits the built in stop. Repeat this step on the other side. A small amount of oil on the pieces of seal may be helpful when packing into the cylinder block.

9. Remove the guide tool.

10. Install a new rope seal in the bearing cap and install the cap. Torque the retaining bolts to specifications.

11. Install the oil pan.

12. Fill the crankcase with oil.

13. Connect the negative battery cable.

ENGINE COOLING

Radiator

REMOVAL & INSTALLATION

1. Disconnect the battery ground cable, then drain the cooling system.

2. Remove the engine forward strut bracket to the radiator. Loosen the bolt to prevent damage to the bushing, then swing the strut rearward.

3. Disconnect the headlight harness connectors from the fan frame. Disconnect the electrical connector from the fan.

4. Remove the cooling fan attaching bolts, then the cooling fan.

5. Scribe the hood latch location on the radiator support, then remove the latch.

6. Disconnect the coolant hoses from the radiator. If equipped with automatic transaxle, disconnect the oil cooler lines.

7. Remove the radiator attaching bolts, then the radiator. If equipped with air conditioning, it may be necessary to raise the left side of the radiator so the radiator neck will clear the compressor.

To install:

8. Install the radiator and attaching bolts.

9. Connect the coolant hoses to the radiator. If equipped with automatic transaxle, connect the oil cooler lines.

10. Install the hood latch observing the scribe marks made upon removal.

11. Remove the cooling fan and attaching bolts.

12. Connect the headlight harness connectors to the fan frame. Connect the electrical connector to the fan.

13. Install the engine forward strut bracket to the radiator.

14. Connect the negative battery cable.

15. Fill cooling system and check for leaks. Start the engine and allow to come to normal operating temperature. Recheck for leaks. Top-up coolant.

Heater Core

REMOVAL & INSTALLATION

Without Air Conditioning

1. Disconnect the negative battery cable. Drain the cooling system.
2. Remove the heater inlet and outlet hoses.
3. Remove the radio noise suppression strap.
4. Remove the heater core cover retaining screws. Remove the cover.
5. Remove the heater core.

To install:

6. Install the heater core.
7. Install the heater core cover and retaining screws.
8. Install the radio noise suppression strap.
9. Install the heater inlet and outlet hoses.
10. Fill cooling system and check for leaks. Start the engine and allow to come to normal operating temperature. Recheck for leaks. Top-up coolant.
11. Connect the negative battery cable.

With Air Conditioning

1. Disconnect the negative battery cable. Drain the cooling system.
2. Disconnect the heater hoses at the heater core.
3. Remove the heater duct and the lower side covers.
4. Remove the lower heater outlet.
5. Remove the housing cover-to-air valve housing clips.
6. Remove the housing cover bolts. Remove the housing cover.
7. Remove the heater core retaining straps. Remove the heater core tubing retainers. Lift out the heater core.

To install:

8. Install the heater core, tubbing retainers and retaining straps.
9. Install the housing cover and retaining bolts.
10. Install the housing cover-to-air valve housing clips.
11. Install the lower heater outlet.
12. Install the heater duct and the lower side covers.
13. Connect the heater hoses to the heater core.
14. Fill cooling system and check for leaks. Start the engine and allow to come to normal operating tempera-

ture. Recheck for leaks. Top-up coolant.

15. Connect the negative battery cable.

Water Pump

REMOVAL & INSTALLATION

2.5L Engine

1. Disconnect the negative battery cable.
2. Drain the cooling system.
3. Remove serpentine drive belt.
4. Remove water pump attaching bolts and remove pump.

To install:

5. If installing a new water pump, transfer pulley from old unit. With sealing surfaces cleaned, place a ⅛ in. (3mm) bead of RTV sealant or equivalent, on the water pump sealing surface. While sealer is still wet, install pump and torque bolts to 6 ft. lbs.
6. Install serpentine drive belt.
7. Connect the negative battery cable.
8. Fill cooling system and check for leaks. Start the engine and allow to come to normal operating temperature. Recheck for leaks. Top-up coolant.

2.8L Engine

1. Disconnect the negative battery cable.
2. Drain cooling system and remove heater hose.
3. Remove serpentine belt.
4. Remove water pump attaching bolts and nut and remove pump.

To install:

5. Clean the sealing surfaces and place a ³⁄₃₂ in. (2mm) bead of RTV sealant or equivalent on the water pump sealing surface.
6. Coat bolt threads with pipe sealant 1052080 or equivalent.
7. Install pump and torque bolts to 10 ft. lbs.
8. Connect the negative battery cable.
9. Fill cooling system and check for leaks. Start the engine and allow to come to normal operating temperature. Recheck for leaks. Top-up coolant.

3.1L Engine

1. Disconnect the negative battery cable.
2. Drain cooling system.
3. Remove the serpentine belt.
4. Remove the heater hose and radiator hose.
5. Remove the water pump cover attaching bolts and remove the cover.

6. Remove the water pump attaching bolts and remove the water pump.

To install:

7. Position the water pump on the engine and install the attaching bolts. Torque bolts to 89 inch lbs. (10 Nm).
8. Install the water pump cover and attaching bolts.
9. Install the heater hose and radiator hose.
10. Install the serpentine belt.
11. Connect the negative battery cable.
12. Fill cooling system and check for leaks. Start the engine and allow to come to normal operating temperature. Recheck for leaks. Top-up coolant.

3.3L Engine

1. Disconnect the negative battery cable.
2. Drain cooling system.
3. Remove the serpentine drive belt.
4. Remove the coolant hose at the water pump.
5. Remove the water pump pulley bolts. The long bolt should be removed through the access hole provided in the body side rail. Remove the pulley.
6. Remove the water pump attaching bolts and remove the water pump.

To install:

7. Install the water pump attaching bolts and install the water pump.
8. Install the pulley. Install the water pump pulley bolts. The long bolt should be installed through the access hole provided in the body side rail.
9. Install the coolant hose at the water pump.
10. Install the serpentine drive belt.
11. Connect the negative battery cable.
12. Fill cooling system and check for leaks. Start the engine and allow to come to normal operating tempera-

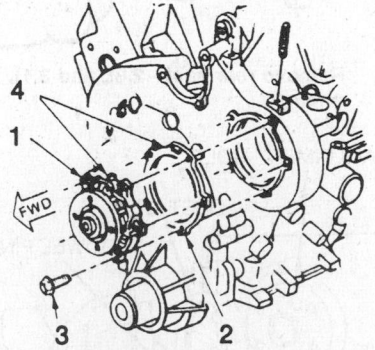

1. Water pump
2. Gasket
3. Bolt - 89 inch lbs. (10 Nm)
4. Locator - Must be vertical

Water pump mounting—2.8L and 3.1L engines

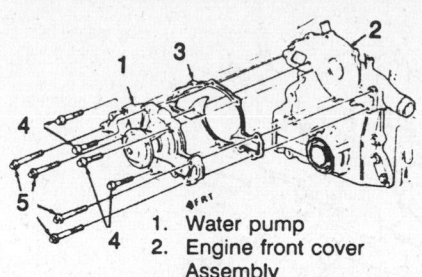

1. Water pump
2. Engine front cover Assembly
3. Gasket
4. Bolts - 97 inch lbs. (11 Nm)
5. Bolts - 29 ft. lbs. (39 Nm)

Water pump mounting—3.3L engine

ture. Recheck for leaks. Top-up coolant.

3.8L Engine

1. Disconnect the negative battery cable.
2. Drain cooling system.
3. Remove the serpentine drive belt.
4. Disconnect the radiator and heater hoses at the water pump.
5. Remove the water pump pulley bolts, long bolt removed through access hole provided in the body side rail. Remove the pulley.
6. Remove the water pump attaching bolts. Remove the water pump.

To install:
7. Clean all gasket mating surfaces.
8. Using a new gasket, install the water pump to the engine.
9. Install the water pump pulley.
10. Connect the radiator and heater hoses to the water pump.
11. Install the serpentine drive belt.
12. Connect the negative battery cable.
13. Fill cooling system and check for leaks. Start the engine and allow to come to normal operating temperature. Recheck for leaks. Top-up coolant.

Thermostat

REMOVAL & INSTALLATION

1. Disconnect the battery negative cable. Drain the cooling system.
2. If equipped with cruise control, remove the vacuum modulator from the thermostat housing.
3. On all vehicles, except 2.5L engine, unbolt the water outlet from the intake manifold. Remove the outlet and lift the thermostat from the the intake manifold.
4. If equipped with a 2.5L engine, unbolt the water outlet from the thermostat housing. Remove the outlet and lift the thermostat from the vehicle.

To install:
5. Clean mating surfaces throughly. Apply a ⅛ inch bead of suitable RTV sealant in the groove of the water outlet.
6. Install the thermostat with the spring toward the engine. Install the water outlet. Torque bolts to 21 ft. lbs.
7 If equipped with cruise control, install the vacuum modulator to the thermostat housing.
8. Connect the negative battery cable.
9. Fill cooling system and check for leaks. Start the engine and allow to come to normal operating temperature. Recheck for leaks. Top-up coolant.

ENGINE ELECTRICAL

NOTE: Disconnecting the negative battery cable on some vehicles may interfere with the functions of the on board computer systems and may require the computer to undergo a relearning process, once the negative battery cable is reconnected.

Direct Ignition System (DIS)

REMOVAL & INSTALLATION

DIS Assembly

2.5L, 2.8L AND 3.1L ENGINES

1. Disconnect the negative battery cable.
2. Disconnect the DIS electrical connectors.
3. Tag and disconnect the spark plug wires.
4. Remove the DIS assembly attaching bolts.
5. Remove the DIS assembly from the engine.

To install:
6. Install the DIS assembly and attaching bolts.
7. Connect the spark plug wires.
8. Connect the DIS electrical connectors.
9. Connect the negative battery cable.

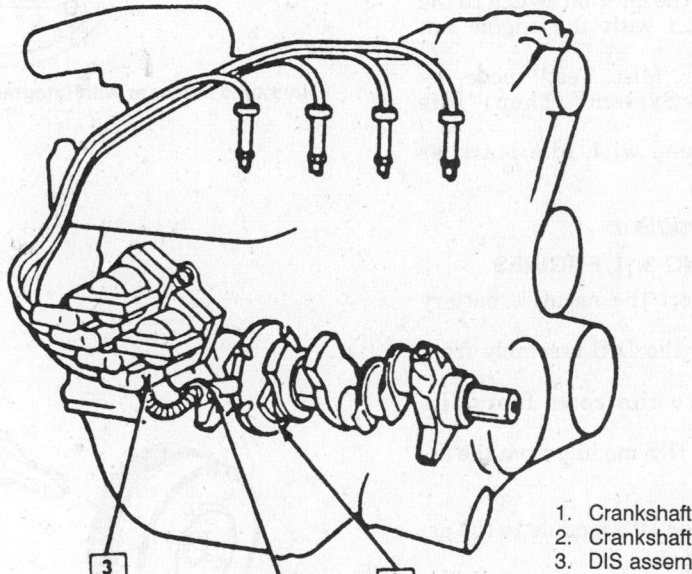

1. Crankshaft reluctor
2. Crankshaft sensor
3. DIS assembly

Direct Ignition System (DIS) components

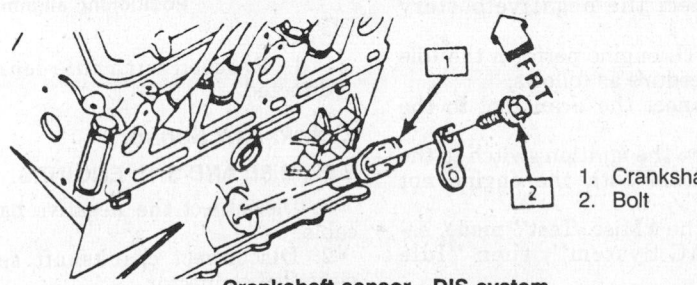

1. Crankshaft sensor
2. Bolt

Crankshaft sensor—DIS system

10. On 3.1L engine, perform the Idle Learn Procedure as follows:

    a. Connect the Scan tool to the ALDL.

    b. Turn the ignition switch to the **ON** position with the engine not running.

    c. In the "Misc. Test" mode, select "IAC System", then "Idle Learn".

    d. Proceed with idle learn as directed.

Ignition Coil

2.5L, 2.8L AND 3.1L ENGINES

1. Disconnect the negative battery cable.

2. Disconnect and tag spark plug wires.

3. Remove ignition coil(s) attaching bolts, then the ignition coil from the module.

To install:

4. Install the coil(s) and attaching bolts.

5. Connect the spark plug wires.

6. Connect the negative battery cable.

7. On 3.1L engine perform the Idle Learn Procedure as follows:

    a. Connect the Scan tool to the ALDL.

    b. Turn the ignition switch to the **ON** position with the engine not running.

    c. In the "Misc. Test" mode, select "IAC System", then "Idle Learn".

    d. Proceed with idle learn as directed.

Ignition Module

2.5L, 2.8L AND 3.1L ENGINES

1. Disconnect the negative battery cable.

2. Remove the DIS assembly from the engine.

3. Remove the coils from the assembly.

4. Remove DIS module from the assembly plate.

To install:

5. Install the DIS module to the assembly plate.

6. Install the coils to the assembly.

7. Install the DIS assembly to the engine.

8. Connect the negative battery cable.

9. On 3.1L engine perform the Idle Learn Procedure as follows:

    a. Connect the Scan tool to the ALDL.

    b. Turn the ignition switch to the **ON** position with the engine not running.

    c. In the "Misc. Test" mode, select "IAC System", then "Idle Learn".

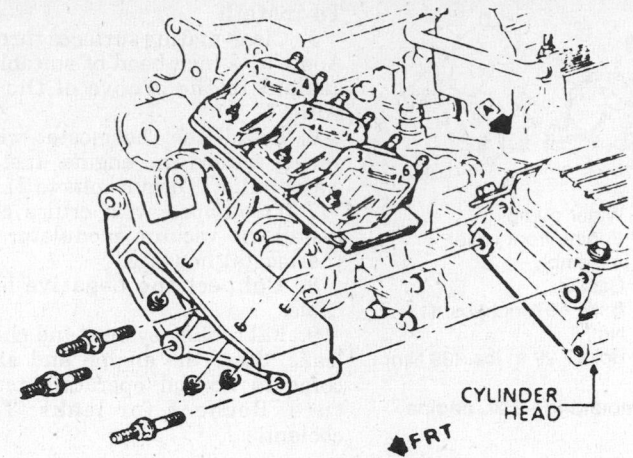

Ignition system coils and module assemblies — C³I system

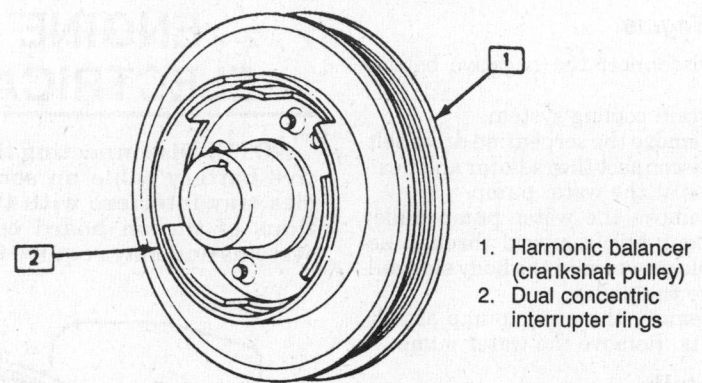

1. Harmonic balancer (crankshaft pulley)
2. Dual concentric interrupter rings

Harmonic balancer with integral concentric interrupter rings — C³I system

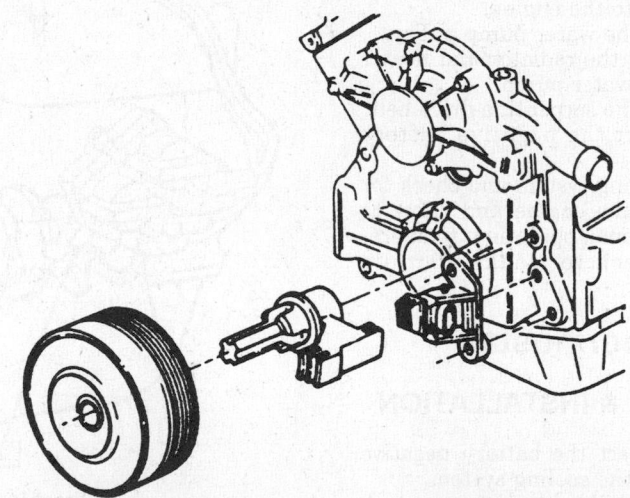

Positioning alignment tool on crankshaft — C³I system

    d. Proceed with idle learn as directed.

Crankshaft Sensor

2.5L, 2.8L AND 3.1L ENGINES

1. Disconnect the negative battery cable.

2. Disconnect crankshaft sensor electrical connector.

3. Remove the crankshaft sensor attaching bolt, then remove the crankshaft sensor from the vehicle.

To install:

4. Inspect the sensor O-ring for wear, cracks or leakage. Replace as necessary. Lubricate the new O-ring with engine oil prior to installation.

5. Install the sensor into the hole in

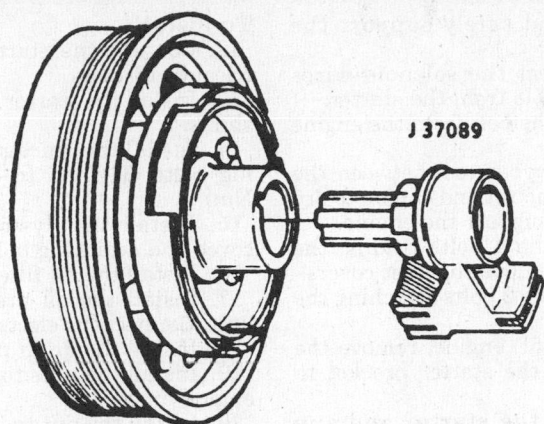

J 37089

Checking harmonic balancer vanes—C³I system

the engine block and install retaining bolt.

6. Connect the electrical connector.
7. Connect the negative battery cable.
8. On 3.1L engine perform the Idle Learn Procedure as follows:

a. Connect the Scan tool to the ALDL.

b. Turn the ignition switch to the **ON** position with the engine not running.

c. In the "Misc. Test" mode, select "IAC System", then "Idle Learn".

d. Proceed with idle learn as directed.

Computer Controlled Coil Ignition (C³I) System

REMOVAL & INSTALLATION

C³I Module

3.3L AND 3.8L ENGINES

1. Disconnect the negative battery cable.
2. Disconnect the 14-way connector at the ignition module.
3. Tag and disconnect the spark plug wires at the coil assembly.
4. Remove the nuts and washers securing the C³I module assembly to the bracket.
5. Remove the 6 bolts attaching the coil assemblies to the ignition module.
To install:
6. Install the coil assemblies to the ignition module and install the 6 attaching bolts.
7. Install the nuts and washers attaching the assembly to the bracket.
8. Connect the spark plug wires.
9. Connect the 14-way connector to the module.
10. Connect the negative battery cable.

Ignition Coil(s)

3.3L AND 3.8L ENGINES

1. Disconnect the negative battery cable.
2. Tag and disconnect spark plug wires.
3. Remove ignition coil(s) attaching bolts, then the ignition coil from the module.
To install:
4. Install the coil(s) and attaching bolts.
5. Connect the spark plug wires.
6. Connect the negative battery cable.

Dual Crankshaft Sensor

3.3L AND 3.8L ENGINES

1. Disconnect battery negative cable.
2. Disconnect serpentine belt from crankshaft pulley.
3. Raise and safely support the vehicle.
4. Remove right front tire and wheel assembly, then the inner access cover.
5. Remove crankshaft harmonic balancer retaining bolt and crankshaft harmonic balancer.
6. Disconnect electrical connector from sensor and remove the crankshaft sensor from the vehicle.
To install:
7. Loosely install the crankshaft sensor on the pedestal.
8. Position the sensor with the pedestal attached on special tool J–37089.
9. Position the tool on the crankshaft.
10. Install the bolts to hold the pedestal to the block face. Tighten to 14–28 ft. lbs. (20–40 Nm).
11. Tighten the pedestal pinch bolt to 30–35 inch lbs. (3–4 Nm).
12. Remove special tool J–37089.
13. Place special tool J–37089 on the harmonic balancer and turn. If any vane of the harmonic balancer touches

the tool, replace the balancer assembly.
14. Install the balancer on the crankshaft and install the crankshaft balancer bolt. Tighten to 200–239 ft. lbs. (270–325 Nm).
15. Install the inner fender shield.
16. Install the tire and wheel assembly. Tighten to 100 ft. lbs. (140 Nm).
17. Lower the vehicle.
18. Install the serpentine belt.
19. Connect the negative battery cable.

Ignition Timing

All vehicles are equipped with either the Direct Ignition System (DIS) or the Computer Controlled Coil Ignition (C³I) system. The systems consist of a coil pack, ignition module, crankshaft reluctor or interrupter ring(s), magnetic sensor and an Electronic Control Module (ECM). Timing advance and retard are accomplished through the ECM with the Electronic Spark Timing (EST) and Electronic Spark Control (ESC) circuitry. No ignition timing adjustment is required or possible.

Alternator

PRECAUTIONS

Several precautions must be observed with alternator equipped vehicles to avoid damage to the unit.

• If the battery is removed for any reason, make sure it is reconnected with the correct polarity. Reversing the battery connections may result in damage to the one-way rectifiers.

• When utilizing a booster battery as a starting aid, always connect the positive to positive terminals and the negative terminal from the booster battery to a good engine ground on the vehicle being started.

• Never use a fast charger as a booster to start vehicles.

• Disconnect the battery cables when charging the battery with a fast charger.

• Never attempt to polarize the alternator.

• Do not use test lamps of more than 12 volts when checking diode continuity.

• Do not short across or ground any of the alternator terminals.

• The polarity of the battery, alternator and regulator must be matched and considered before making any electrical connections within the system.

• Never separate the alternator on an open circuit. Make sure all connections within the circuit are clean and tight.

- Disconnect the battery ground terminal when performing any service on electrical components.
- Disconnect the battery if arc welding is to be done on the vehicle.

BELT TENSION ADJUSTMENT

Serpentine Belt

A single serpentine belt is used to drive all engine accessories. The belt tension is maintained by a spring loaded tensioner. The belt tensioner has the ability to control the belt tension over a broad range of belt lengths. However, there are limits to which the tensioner can compensate for varying lengths. If the belt tension is below the minimum specifications, replace the belt tensioner.

Check the serpentine belt tension with tool J–23600B or equivalent belt tension gauge in the following manner:

1. Start the engine and run until operating temperature is reached.
2. Shut the engine **OFF** and place the tension gauge midway between the pulleys. Install the gauge on the longest belt span possible. If the belt is notched on the inner surface, place the middle finger of the tensioner gauge into 1 of the notches. Correct belt tension readings should be approximately:

40 lbs. (178 N) – 2.5L engine
70 lbs. (311 N) – 2.8L engine
50–70 lbs. (225–315 N) – 3.1L engine
67 lbs. (298 N) – 3.3L and 3.8L engines.

REMOVAL & INSTALLATION

1. Disconnect the negative battery cable.
2. Remove the alternator electrical connectors.
3. Remove the serpentine belt.
4. Remove the alternator mounting bolt(s) and pivot bolt.
5. Remove the alternator.
To install:
6. Position the alternator in the bracket.
7. Install the alternator pivot bolt and mounting bolt(s).
8. Install the serpentine belt.
9. Connect the alternator electrical connectors.
10. Connect the negative battery cable.

Starter
REMOVAL & INSTALLATION

2.5L and 2.8L Engines
1. Disconnect the negative battery cable.

2. Raise and safely support the vehicle.
3. Disconnect the solenoid wires and battery cable from the starter.
4. Remove the bolt from the engine cross brace.
5. Place a prybar tool between the upper engine mount and engine to pry rearward and support the engine.
6. Remove the 4 bolts holding the dust covers. Remove the dust covers.
7. Remove the 2 bolts attaching the starter.
8. On the 2.5L engine, remove the bolt attaching the starter bracket to the engine.
9. Remove the starter and any shims.

To install:

NOTE: If replacing the starter, transfer the starter bracket to the new starter.

10. Install the starter and any shims.
11. Install the 2 starter attaching bolts.
12. On the 2.5L engine, install the bolt attaching the starter bracket to the engine.
13. Install the dust cover and 4 attaching bolts.
14. Connect the battery cable and solenoid wires.
15. Lower the vehicle.
16. Roll the engine forward and replace the engine brace bolts.
17. Connect the negative battery cable.

3.1L Engine

1. Disconnect the negative battery cable.
2. Raise and safely support the vehicle.
3. If equipped, remove the nut from the brace at the air conditioning compressor.
4. If equipped, remove the nuts from the starter-to-engine brace.
5. Remove the drain pan under the engine oil pan.
6. Disconnect the oil pressure sending unit electrical connector. Remove the oil pressure sending unit.
7. Remove the oil filter.
8. Remove the bolts from the flywheel inspection cover. Remove the inspection cover.
9. Remove the bolts from the starter motor.
10. Remove the starter motor and any shims.
11. Disconnect the starter motor electrical connectors.

To install:
12. Connect the starter motor electrical connectors.
13. Install the starter motor and any shims.
14. Install the starter motor attaching bolts. Tighten to 32 ft. lbs. (43 Nm).
15. Install the flywheel inspection cover and attaching bolts.
16. Install the oil filter.
17. Install the oil pressure sending unit. Connect the electrical connector.
18. Install the drain pan.
19. Install the nuts to the starter-to-engine brace.
20. Install the nut to brace at the air conditioner compressor.
21. Lower the vehicle.
22. Connect the negative battery cable.
23. Check the engine oil level, add as required.

3.3L and 3.8L Engines

1. Disconnect the negative battery cable.
2. Properly discharge the air conditioning system.
3. On 3.3L engine, remove the cooling fan assembly.
4. Remove the front exhaust manifold.
5. Raise and support the vehicle safely.
6. Remove the bolts from the flywheel inspection cover and remove the cover.
7. Disconnect the air conditioner condenser hose from the compressor and position aside.
8. Disconnect the starter motor electrical connectors.
9. Remove the 2 bolts attaching the starter.
10. Remove the starter and any shims.

To install:
11. Install the starter motor and any shims.
12. Install the 2 bolts attaching the starter. Tighten to 30 ft. lbs. (40 Nm).
13. Connect the starter motor electrical connectors.
14. Replace the condenser O-ring. Lubricate with refrigerant oil. Connect the air conditioner condenser hose to the compressor.
15. Install the flywheel inspection cover and attaching bolts.
16. Lower the vehicle.
17. Install the front exhaust manifold.
18. On 3.3L engine, install the cooling fan assembly.
19. Evacuate, recharge and leak test the air conditioning system.
20. Connect the negative battery cable.

EMISSION CONTROLS

Please refer to "Emission Controls" in the Unit Repair section for system maintenance procedures. Due to the complex nature of modern electronic engine control systems, comprehensive diagnosis and testing procedures fall outside the confines of this repair manual. For complete information on diagnosis, testing and repair procedures concerning all modern engine and emission control systems, please refer to "Chilton's Guide to Fuel Injection and Electronic Engine Controls".

Emission Warning Lamps

RESETTING

1989–92 Ciera

Vehicles equipped with an engine oil life index display as a part of the Driver Information System (DIS), have a display that will show when to change the engine oil.

The oil change interval is determined by the driver information system and will usually fall at or between the 2 recommended alternative intervals of 3000 miles and 7500 miles but it could be shorter than 3000 miles under some severe driving conditions. The driver information system will also signal the need for an oil change at 7500 miles or one year passed since the last oil change. If the drive information system does not indicate the need for an oil change after 7500 miles or one year if the engine oil life index display fails to appear, the oil should be changed and the driver information system serviced.

When the engine oil life index reaches 10 percent or less, the change oil light display will function as a reserve trip odometer, indicating the distance to an oil change. Until the engine oil lift index reset is performed, the driver information system will display the distance to the oil change and sound a beep when the ignition switch is turned to the **ACCESSORY** or **RUN** position the first time each day.

When the distance to the next oil change reaches 0, the driver information system will display the change oil now light. Until an engine oil life index reset is performed the the driver information system will display the change

oil now light and sound a beep when the ignition switch is turned to the **ACCESSORY** or **RUN** position the first time each day.

The driver information system will not detect dusty conditions or engine malfunctions which may affect the engine oil. If driving in severe conditions exists, change the engine oil every 3000 miles or 3 months which ever comes first, unless instructed otherwise by the driver information system. The driver information center does not measure the engine oil level, it remains the owner's responsibility to check the engine oil level. After the oil has been changed, the engine oil life index light must be reset. Reset the can be accomplished as follows:

a. The engine oil life index can be reset by pressing the **RESET** and **OIL** buttons simultaneously for at least 5 seconds while on the engine oil life index display. The driver information system will reset the engine oil life index to 100 percent and display an engine oil life index of 100 percent.

b. Oil life index 100 message appears.

NOTE: The Engine Oil Life Index is stored on a non-volatile memory chip and will not reset by disconnecting the battery and or fuse.

6000

The Service Reminder section of the Driver Information Center (DIC) display shows how many miles remain until service is needed. When the reset button is pressed twice, the type of service and the number of miles remaining until the service is needed will be displayed. Each time the Reset button is pressed, another type of service and the miles remaining until service will be displayed.

With the ignition switch in the Run, Bulb Test or Start positions, voltage is applied from the ECM fuse through the Pink/Black wire to the ECM. As the vehicle moves, the speed sensor sends electrical pulses (signals) to the ECM. The ECM then sends a signal to the speed signal input of the DIC module. The DIC module converts these pulses (signals) into miles. The module subtracts the miles travelled from the distance remaining for each item of the service reminder.

When the miles remaining for a service approaches 0, that service will be displayed on the DIC display. All 4 types of service can be shown at the same time.

SERVICE REMINDER RESET

To reset the service light, it will be nec-

essary to subtract the mileage from the service interval light that is illuminated. The miles remaining for a certain type of service can be decreased by holding the Reset button. The miles remaining will be decreased in steps of 500 miles every 5 seconds. In the first step, the miles will decrease to a multiple of 500. For example, 2880 miles will decrease to 2500 miles. If the Reset button is held in and the miles remaining reach 0, the DIS display will show the service interval for the service selected. The service intervals are as follows:

1. Change oil – 7500 miles
2. Oil filter change – 7500 miles
3. Next filter change – 15,000 miles
4. Rotate tires – 7500 miles
5. Next tire rotation – 15,000 miles
6. Tune Up – 30,000 miles

If the Reset button is still held down, the miles will decrease in steps of 500 miles from the service interval. When the Reset button is released, the mile display shown will be the new distance until the service should be performed.

When a service distance reaches 0, the service reminder item will be displayed. If the service interval is reset within 10 miles, the display will go out immediately. If more than 10 miles passes before the service interval is reset, the item will remain displayed for another 10 miles after being reset before going out.

NOTE: On some models it may be necessary to depress the system Recall button, in order to display the service interval light on the driver information center in order to decrease the mileage from it, so as to reset the interval light.

FUEL SYSTEM

Fuel System Service Precautions

Safety is the most important factor when performing not only fuel system maintenance but any type of maintenance. Failure to conduct maintenance and repairs in a safe manner may result in serious personal injury or death. Maintenance and testing of the vehicle's fuel system components can be accomplished safely and effectively by adhering to the following rules and guidelines.

• To avoid the possibility of fire and personal injury, always disconnect the negative battery cable unless the repair or test procedure requires that battery voltage be applied.

• Always relieve the fuel system pressure prior to disconnecting any fuel system component (injector, fuel rail, pressure regulator, etc.), fitting or fuel line connection. Exercise extreme caution whenever relieving fuel system pressure to avoid exposing skin, face and eyes to fuel spray. Please be advised that fuel under pressure may penetrate the skin or any part of the body that it contacts.

• Always place a shop towel or cloth around the fitting or connection prior to loosening to absorb any excess fuel due to spillage. Ensure that all fuel spillage (should it occur) is quickly removed from engine surfaces. Ensure that all fuel soaked cloths or towels are deposited into a suitable waste container.

• Always keep a dry chemical (Class B) fire extinguisher near the work area.

• Do not allow fuel spray or fuel vapors to come into contact with a spark or open flame.

• Always use a backup wrench when loosening and tightening fuel line connection fittings. This will prevent unnecessary stress and torsion to fuel line piping. Always follow the proper torque specifications.

• Always replace worn fuel fitting O-rings with new. Do not substitute fuel hose or equivalent where fuel pipe is installed.

RELIEVING FUEL SYSTEM PRESSURE

Throttle Body Injection (TBI)

1. On a cold engine, remove the fuse marked "Fuel Pump" from the fuse block in the passenger compartment.
2. Start the engine and run until the fuel supply remaining in the fuel lines is exhausted. When the engine stops, engage the starter again for 3.0 seconds to assure dissipation of any remaining pressure.
3. With the ignition **OFF**, replace the fuel pump fuse.
4. Disconnect the negative battery cable.

Port Fuel Injection (PFI)

1. Disconnect the negative battery cable to avoid possible fuel discharge if an accidental attempt is made to start the engine.
2. Loosen the fuel filler cap to relieve the tank pressure.
3. Connect a suitable fuel pressure gauge to the fuel pressure test fitting. Wrap a shop towel around the fitting while connecting gauge to avoid spillage.

4. Place the bleed hose in an approved container and open the valve on the pressure gauge to relieve system pressure.
5. Dispose of the discharged liquid fuel promptly.

Fuel Tank

REMOVAL & INSTALLATION

1. Disconnect the negative battery cable.
2. Relieve fuel system pressure.
3. Drain the fuel tank into an approved container.
4. Raise and safely support the vehicle.
5. Remove the filler tube and clamp.
6. Remove the fuel tank vent tube and clamp at the fuel tank.
7. Disconnect the electrical connectors.
8. Disconnect the vapor hose connector and clamp from the fuel tank.
9. Disconnect the fuel line hoses from the tank meter assembly.
10. If equipped with quick-connect fuel line fittings, perform the following:
   a. Grasp the fuel level meter feed tube and fuel feed line quick-connect fitting. Twist the quick-connect fitting ¼ turn in each direction to loosen any dirt. Repeat for the fuel return line quick-connect fitting.
   b. Squeeze the plastic tabs of the male end of the connector and pull the connection apart. Repeat for the other fitting.
11. With the aid of an assistant, support the fuel tank and remove the 2 front fuel tank retaining strap attaching bolts, 2 rear fuel tank strap attaching nuts and bolts, bolt fuel tank retaining straps and remove the tank.
To install:
12. With the aid of an assistant, position and support the fuel tank. Install the 2 fuel tank retaining straps, front attaching bolts and rear attaching bolts and nuts.
13. Connect the fuel lines to the tank meter assembly.
14. If equipped with quick-connect fuel line connectors, perform the following:
   a. Apply a few drops of clean engine oil to the male connector tube ends.
   b. Push the connectors together to cause the retaining tabs/fingers to snap into place.
   c. Once installed, pull on both ends of each connection to make sure the connection is secure.
   d. Repeat for the other fittings.
15 Connect the vapor hose and clamp.

16. Connect the fuel level meter electrical connector.
17. Connect the fuel tank vent tube and clamp.
18. Connect the filler tube and clamp.
19. Lower the vehicle.
20. Add fuel to the tank and install the fuel filler cap.
21. Connect the negative battery cable.
22. Turn the ignition switch to the **ON** position for 2 seconds, then turn to the **OFF** position for 10 seconds. Turn the ignition switch back to the **ON** position and check for fuel leaks.

Fuel Filter

REMOVAL & INSTALLATION

Threaded Fitting

The filter is an inline unit located just ahead of the TBI unit or to the left of the fuel tank.

1. Ensure the engine is cold, then unclamp and remove the fuel hose.
2. Unscrew the filter from the fuel line.
To install:
3. Place the new filter into position and connect the fuel lines.
4. Tighten the retaining clamp.
5. Start the engine and check for fuel leaks.

Quick-Connect Fitting

1. Disconnect the negative battery cable.
2. Relieve the fuel system pressure.
3. Raise and safely support the vehicle.
4. Remove the filter bracket attaching screw and filter bracket.
5. Grasp the filter and 1 fuel line fitting. Twist the quick-connect fitting ¼ turn in each direction to loosen any dirt within the fitting. Repeat for the other fuel line fitting.
6. Use compressed air, blow out dirt from the quick-connect fittings at both ends of the fuel filter.
7. To disconnect the fuel line fittings, squeeze the plastic tabs of the male end of the connector and pull the connector apart. Repeat for the other fitting.
8. Remove the fuel filter.
To install:
9. Remove the protective caps from the new filter.
10. Install new plastic connector retainers on the filter inlet and outlet tubes. Observe the positions on the old filter and duplicate with new filter.
11. Connect the quick-connect fittings by performing the following:
   a. Apply a few drops of clean engine oil to the male tube ends of the

fuel filter and fuel level meter assembly.

b. Push the connectors together until the retaining tabs/fingers snap into place.

c. Once installed, pull on both ends of each connector to ensure a tight connection.

12. Align the fuel filter bracket on the frame with the brake line mounting bracket and install the filter bracket attaching screw.

13. Lower the vehicle.

14. Tighten the fuel filler cap.

15. Connect the negative battery cable.

Electric Fuel Pump

PRESSURE TESTING

1. Disconnect the fuel line from the EFI unit.

2. Install a suitable pressure gauge to the fuel line.

3. Connect a jumper wire from the positive terminal on the battery to the G terminal of the ALDL.

4. Fuel pressure gauge should be 9–13 psi if equipped with TBI or 34–46 psi if equipped with PFI or SFI.

NOTE: If fuel pressure does not meet specifications, check the fuel line for restrictions or the fuel pump for malfunctions.

REMOVAL & INSTALLATION

The fuel pump is attached to the fuel sending unit located inside the fuel tank.

1. Relieve the fuel system pressure, then disconnect the negative battery cable.

2. Raise and support the vehicle safely. Drain the fuel tank.

3. Disconnect wiring from the tank, then remove the ground wire retaining screw from under the body.

4. Disconnect all hoses from the tank.

5. Support the tank on a jack and remove the retaining strap nuts.

6. Lower the tank and remove it from the vehicle.

7. Remove the fuel gauge/pump retaining ring using a suitable spanner wrench.

8. Remove the gauge unit and the pump.

To install:

9. Install the gauge unit and the pump.

10. Install the fuel gauge/pump retaining ring using a suitable spanner.

11. Raise the tank and and install it to the vehicle.

12. Support the tank on a jack stand and install the retaining strap nuts.

13. Connect the hoses to the tank.

14. Connect the electrical connectors and the ground wire, if equipped.

15. Lower the vehicle.

16. Fill the fuel tank.

17. Turn the ignition switch to the **ON** position for 2 seconds, then turn to the **OFF** position for 10 seconds. Turn the ignition switch back to the **ON** position and check for fuel leaks.

Fuel Injection

IDLE SPEED ADJUSTMENT

Throttle Body Injection (TBI)

NOTE: This procedure should be performed only only after throttle body parts have been replaced.

1. Block the drive wheels and apply the parking brake.

2. Connect a Scan tool to the ALDL connector.

3. Turn the ignition switch to the **ON** position.

4. Select the "Field Service Mode" on the Scan tool. This will cause the IAC valve pintle to seat in the throttle body. Wait at least 45 seconds, disconnect the IAC valve connector and exit the "Field Service Mode."

5. Place the transmission in **P**, if equipped with an automatic transmission or **N**, if equipped with a manual transmission. Start the engine and allow to come to normal operating temperature.

6. Confirm the following prior to checking idle speed:

a. Engine at normal operating temperature and in Closed Loop.

b. All accessories and cooling fan off.

c. Ensure that throttle and cruise control cables do not hold the throttle open.

7. Select "Engine rpm" on the Scan tool. Observe the engine speed and adjust as necessary to 600 ± 50 rpm.

8. Turn the ignition switch to the **OFF** position.

9. Connect the IAC valve electrical connector.

10. Reset the IAC valve pintle position by performing the following:

a. Select "Engine rpm" on the Scan tool.

b. Start the engine and hold speed above 2000 rpm. Select "Field Service Mode" for 10 seconds.

c. Exit "Field Service Mode" and allow the engine to return to idle.

d. Turn the ignition switch to the **OFF** position. Restart the engine and check for proper idle operation.

11. Disconnect the Scan tool.

12. Remove the block from the drive wheels.

Port Fuel Injection (PFI) and Sequential Fuel Injection (SFI)

1. Using an suitable tool, pierce the idle stop screw plug, located on the side of the throttle body, and remove it by prying it from the housing.

2. Using a jumper wire, ground the diagnostic lead of the IAC motor.

3. Turn the ignition **ON**. Do not start the engine. After 30 seconds, disconnect the IAC electrical connector. Remove the diagnostic lead ground lead and start the engine. Allow the system to go to closed loop.

4. Adjust the idle set screw to 550 rpm on automatic transaxle in **D** or 650 rpm on manual transaxle.

5. Turn the ignition **OFF** and reconnect the IAC motor lead.

6. Using a voltmeter, adjust the TPS to 0.55 ± 0.1 volt and secure the TPS.

7. Recheck the setting, then start the engine and check for proper idle operation.

8. Seal the idle stop screw with silicone sealer.

Fuel Injector

All fuel injectors are serviced as a complete assembly only. Since it is an electrical component, it should not be immersed in any type of cleaner.

REMOVAL & INSTALLATION

Throttle Body Injection

1. Relieve fuel system pressure. Disconnect the negative battery cable.

2. Remove the air cleaner assembly.

3. Squeeze the 2 tabs on the injector electrical connector together and pull straight upward.

4. Remove the fuel meter cover retaining screws. The 2 front retaining screws are shorter than the 3 rear retaining screws. Remove the fuel meter cover.

5. With the fuel meter cover gasket in place, use a prying tool and carefully lift the injector until it is free from the fuel meter body.

6. Remove the small O-ring from the injector nozzle end. Carefully rotate the injector fuel filter back and forth and remove the filter from the base of the injector.

7. Remove and discard the fuel meter cover gasket. Remove the large O-ring and steel back-up washer from the top counterbore of the fuel meter body injector cavity.

To install:

8. Install the fuel injector nozzle filter on the nozzle end of the fuel injector, with the larger end of the filter

facing the injector, so the filter covers raised rib at the base of the injector.

9. Lubricate the new small O-ring with automatic transmission fluid and push the O-ring on the nozzle end of the injector until it presses against the injector fuel filter.

10. Install the steel backup washer in the top counterbore of the fuel meter body injector cavity.

11. Lubricate the new large O-ring with automatic transmission fluid and install it directly over the backup washer. Be sure the O-ring is seated properly in the cavity and is flush with the top of the fuel meter body casting surface.

12. Install the injector into the cavity, aligning the raised lug on the injector base with cast-in notch in the fuel meter body cavity. Push down on the injector until it is fully seated in the cavity. The electrical terminals of the injector will be approximately parallel to the throttle shaft.

13. Install a new dust seal into the recess on the fuel meter body.

14. Install a new fuel outlet passage gasket on the fuel meter cover and a new cover gasket on the fuel meter body.

15. Install the fuel meter cover, making sure the pressure regulator dust seal and cover gaskets are in place; then, apply a thread locking compound to the threads on the fuel meter cover attaching screws. Install the fuel meter cover attaching screws and lock washers and torque to 28 inch lbs. (3 Nm). The 2 short screws go to the front of the injector. Connect battery negative cable.

Port Fuel Injection (PFI) and Sequential Fuel Injection (SFI)

NOTE: Always support the fuel rail to avoid damaging other components while removing the injectors.

1. Relieve fuel system pressure. Disconnect the negative battery cable.
2. Remove the intake manifold plenum.
3. Remove the fuel rail.
4. Remove the injector retaining clips and remove the injectors.
5. Remove the injector O-ring seals from both ends of the injector and discard.

To install:

6. Lubricate the new injector seals with clean engine oil and install on the injectors.
7. Install new injector retaining clips on the injectors. Position the open end of the clip facing the injector electrical connector.
8. Install the injectors into the fuel rail assembly. Push in far enough to

engage the retainer clip with the machined slots on the injector socket.

9. Install the fuel rail assembly and intake manifold plenum.
10. Complete installation by reversing the removal procedure. Connect battery negative cable.

DRIVE AXLE

Halfshaft

REMOVAL & INSTALLATION

Except Rear Axle—6000 STE AWD

1. Remove the hub nut and discard. A new hub nut must be used for reassembly.
2. Raise and safely support the vehicle. Remove the wheel and tire assembly.
3. Install an halfshaft boot seal protector onto the seal.
4. Disconnect the brake hose clip from the MacPherson strut but do not disconnect the hose from the caliper. Remove the brake caliper from the spindle and support the caliper with a length of wire. Do not allow the caliper to hang by the brake hose unsupported.
5. Mark the camber alignment cam bolt for reassembly. Remove the cam bolt and the upper attaching bolt from the strut and spindle.
6. Pull the steering knuckle assembly from the strut bracket.
7. Remove the halfshaft from the transaxle.
8. Using spindle remover tool J-28733 or equivalent, remove the halfshaft from the hub and bearing assembly. Do not allow the halfshaft to hang unsupported. If necessary, support using a length of wire in order to prevent component damage.

To install:

9. If a new halfshaft is to be installed, a new knuckle seal should be installed first along with a boot seal protector when necessary.
10. Loosely install the halfshaft into the transaxle and steering knuckle.
11. Loosely attach the steering knuckle to the suspension strut.
12. The halfshaft is an interference fit in the steering knuckle. Press the axle into place, then install the hub nut. When the shaft begins to turn with the hub, insert a drift through the caliper into one of the cooling slots in the rotor to keep it from turning.

NOTE: On some vehicles, the hub flange has a notch in it which

can be used to prevent the hub and the shaft from turning, when one of the hub bearing retainer bolts is removed, by placing a longer bolt put in its place through the notch

13. Tighten the hub nut to 70 ft. lbs. (95 Nm) to completely seat the shaft.
14. Install the brake caliper. Tighten the caliper mounting bolts to 30 ft. lbs. (41 Nm).
15. Load the hub assembly by lowering it onto a jackstand. Align the camber cam bolt marks made during removal, install the bolt and tighten to 140 ft. lbs. (190 Nm). Tighten the upper nut to the same value.
16. Install the halfshaft all the way into the transaxle using a suitable tool inserted into the groove provided on the inner retainer. Tap the tool until the shaft seats in the transaxle. Remove the boot seal protector.
17. Connect the brake hose clip the the strut. Install the tire and wheel, lower the vehicle and tighten the hub nut to 192 ft. lbs. (261 Nm).

Rear Axle—6000 STE AWD

1. Raise and safely support the vehicle.
2. Remove the tire and wheel assembly.
3. Disconnect the parking brake cable end from the bracket.
4. Insert a suitable tool through the caliper into the rotor to prevent the rotor from turning.
5. Remove the shaft nut and washer using special tool J-34826. Discard the shaft nut.
6. Remove the anti-lock brake sensor bolt and move the sensor aside.
7. Remove the 2 brake caliper bolts and remove the caliper. Support the caliper using a length of wire.

NOTE: Do not allow the caliper to hang by the brake hose unsupported.

8. Remove the rotor from the hub and bearing assembly.
9. Install leaf spring compression tool J-33432 or equivalent.
10. Remove the 3 bolts mounting the hub and bearing to the knuckle.
11. Remove the hub and bearing assembly from the knuckle using special tool J-28733-A or equivalent.
12. Remove the bolts and nut plate attaching the lower strut mount to the knuckle. Scribe the position of the upper bolt prior removing.
13. Install a suitable CV-boot protector to prevent damage to the boot.
14. Swing the knuckle downward and away from the driveshaft.
15. Remove the drive axle from the differential using a suitable slide hammer.

To install:

16. Install the drive axle to the differential. Ensure positive engagement by pulling outward on the inner axle end. Grasp the housing only. Do not grasp and pull on the axle shaft.

17. Swing the knuckle up to the lower strut mount.

18. Position the nut plate and install the lower strut mount bolts to the knuckle. Align the top bolt with scribe marks before tightening the bolts. Tighten the bolts to 148 ft. lbs. (200 Nm).

19. Remove the CV boot protector.

20. Install the hub and bearing assembly to the knuckle and axle spline.

21. Install the hub and bearing attaching bolts. Tighten to 61 ft. lbs. (84 Nm).

22. Remove the leaf spring compression tool.

23. Install the rotor to hub and bearing assembly.

24. Install the brake caliper to the rotor and install the retaining bolts. Tighten the bolts to 38 ft. lbs. (51 Nm).

25. Install the anti-lock brake sensor to the knuckle and install the retaining bolt. Using a non-ferrous feeler gauge, adjust the sensor gap to 0.028 in. (0.7mm). Tighten the adjustment screw to 19 inch lbs. (2.2 Nm).

26. Connect the parking brake cable end into the bracket.

27. Install the shaft washer and new torque prevailing nut. Hold the rotor with a suitable to prevent the axle from turning while tightening. Tighten to 185 ft. lbs. (260 Nm).

28. Install the tire and wheel assembly.

29. Check the rear wheel camber. Adjust as necessary.

NOTE: If the lower strut to knuckle bolts are properly aligned with the scribe marks, no camber adjustment should be necessary.

30. Lower the vehicle.

CV-Boot

REMOVAL & INSTALLATION

Outer Boot

1. Raise and support the vehicle safely.

2. Remove the front tire and wheel assembly.

3. Remove the caliper bolts. Remove the caliper and support using a length of wire.

4. Remove the hub nut, washer and wheel bearing.

5. Using a brass drift, lightly tap around the seal retainer to loosen it. Remove the seal retainer.

6. Remove the seal retaining clamp or ring and discard.

7. Using snapring pliers, remove the race retaining ring from the halfshaft.

8. Pull the outer joint assembly and the outboard seal away from the halfshaft.

9. Flush the grease from the joint and repack with half of the grease provided. Put the remainder of the grease in the seal.

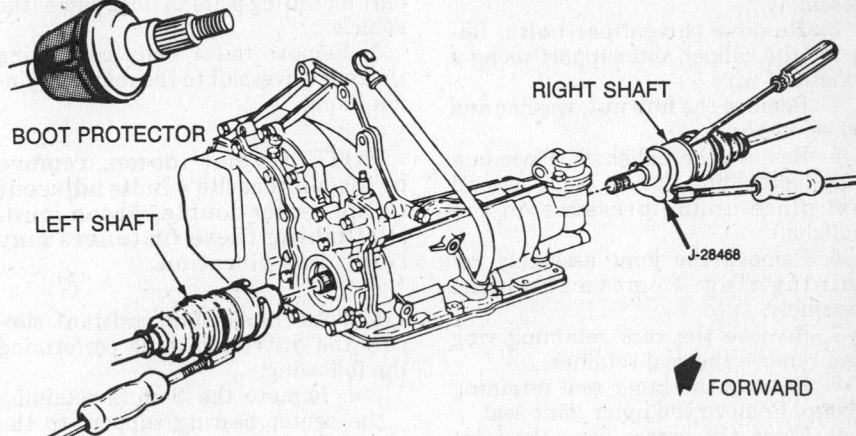

Halfshaft removal using special tools attached to slide hammers

To install:

10. Assemble the inner seal retainer, outboard seal and outer seal retainer to the halfshaft. Push the joint assembly onto the shaft until the retaining ring is seated in the groove.

11. Slide the outboard seal onto the joint assembly and secure using the outer seal retainer. Using seal clamp tool J–35910 or equivalent, torque the outer clamp to 130 ft. lbs. (176 Nm) and the inner clamp to 100 ft. lbs. (136 Nm).

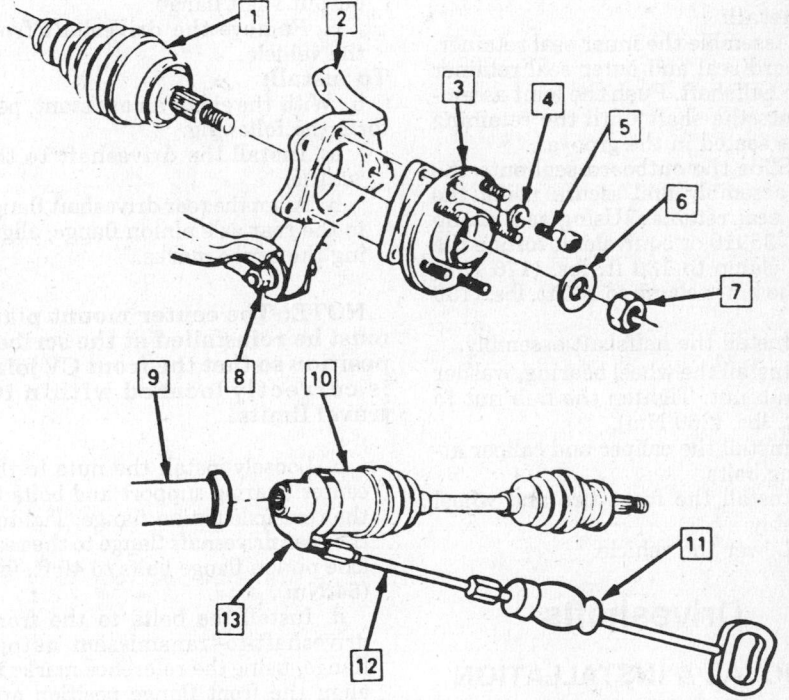

1. Outer joint
2. Knuckle
3. Hub and bearing assembly
4. Washer
5. 61 ft. lbs. (84 Nm)
6. Washer
7. Shaft nut 185 ft. lbs. (260 Nm)
8. 125 ft. lbs. (170 Nm)
9. Differential
10. Inner joint
11. Slide hammer
12. Adapter
13. Adapter

Rear drive axle removal—6000 STE AWD

12. Install the wheel bearing, washer and hub nut. Tighten the hub nut to 192 ft. lbs. (260 Nm).

13. Install the caliper and caliper attaching bolts.

14. Install the front tire and wheel assembly.

15. Lower the vehicle.

Inner Boot

1. Raise and safely support the vehicle.

2. Remove the front tire and wheel assembly.

3. Remove the caliper bolts. Remove the caliper and support using a length of wire.

4. Remove the hub nut, washer and wheel bearing.

5. Remove the halfshaft. Place in a suitable holding fixture being careful not place undue pressure on the halfshaft.

6. Remove the joint assembly retaining ring. Remove the joint assembly.

7. Remove the race retaining ring and remove the seal retainer.

8. Remove the inner seal retaining clamp. Remove the inner joint seal.

9. Flush the grease from the joint and repack with half of the grease provided. Put the remainder of the grease in the seal.

To install:

10. Assemble the inner seal retainer, outboard seal and outer seal retainer to the halfshaft. Push the joint assembly onto the shaft until the retaining ring is seated in the groove.

11. Slide the outboard seal onto the joint assembly and secure using the outer seal retainer. Using seal clamp tool J–35910 or equivalent, torque the outer clamp to 130 ft. lbs. (176 Nm) and the inner clamp to 100 ft. lbs. (136 Nm).

12. Install the halfshaft assembly.

13. Install the wheel bearing, washer and hub nut. Tighten the hub nut to 192 ft. lbs. (260 Nm).

14. Install the caliper and caliper attaching bolts.

15. Install the front tire and wheel assembly.

16. Lower the vehicle.

Driveshafts

REMOVAL & INSTALLATION

6000 SE AWD

ORIGINAL DRIVESHAFT

1. Raise and safely support the vehicle.

NOTE: The relationship between the center bearing support and the floor pan must be main-

tained. Established at the assembly plant, the relationship has an influence on the front joint stoke capacity and remains constant for the vehicle, regardless of the driveshaft installed.

2. Scribe the transmission output shaft flange opposite the "painted" marking on the front driveshaft. The rear pinion flange must be scribed opposite the "painted" marking on the rear driveshaft. Scribe the center support mounting plate to floor pan of the vehicle.

3. Remove the 4 bolts connecting the rear driveshaft to the rear axle pinion flange.

NOTE: Do not loosen, remove or disconnect the 4 bolts adjacent to the center double cardan joint. Disturbing these fasteners may result in a vibration.

4. With the aid of an assistant, support the driveshaft while performing the following:

a. Remove the 3 nuts retaining the center bearing support to the underbody of the vehicle.

b. Remove the 4 bolts connecting the front driveshaft to the transaxle output shaft flange.

c. Remove the driveshaft from the vehicle.

To install:

5. With the aid of an assistant, perform the following:

a. Install the driveshaft to the vehicle.

b. Align the rear driveshaft flange to the rear axle pinion flange, aligning the scribe marks.

NOTE: The center mount plate must be reinstalled at the scribed position so that the front CV joint is correctly located within its travel limits.

c. Loosely install the nuts to the center bearing support and bolts to the rear axle pinion flange. Tighten the rear driveshaft flange to the rear axle pinion flange bolts to 40 ft. lbs. (54 Nm).

d. Install the bolts to the front driveshaft-to-transmission output flange, using the reference marks to align the front flange position and center bracket location. Tighten the center bearing support bolts using the scribed reference marks to 25 ft. lbs. (34 Nm); nuts to 20 ft. lbs. (27 Nm). Tighten the front propeller shaft-to-transmission output flange bolts to 40 ft. lbs. (54 Nm).

6. Remove the support and lower the vehicle.

NEW DRIVESHAFT

1. Remove the original driveshaft from the vehicle.

To install:

2. Using a dial indicator, measure and mark the bolt hole corresponding to the high point of radial runout on both the transmission output flange and rear axle pinion flange.

3. Transfer the center bearing bracket from the old driveshaft to the new driveshaft assembly.

4. With the aid of an assistant, perform the following:

a. Install the driveshaft to the vehicle.

b. Align the point marks supplied on the flanges of the propeller shaft to the scribe marks made during removal.

c. Loosely install the nuts to the center bearing support and bolts to the rear axle pinion flange. Tighten the rear driveshaft flange to the rear axle pinion flange bolts to 40 ft. lbs. (54 Nm).

d. Push the propeller shaft forward until a click is heard. Temporarily install a $^{15}/_{16}$ in. (23mm) thick spacer between the output flange and the front driveshaft flange. Clamp in this position.

5. Install the remaining center bracket nuts. Tighten to 20 ft. lbs. (27 Nm).

6. Remove the spacer, extend the front CV joint to meet the output flange and tighten the bolts. Tighten the front propeller shaft-to-transmission output flange bolts to 40 ft. lbs. (54 Nm).

7. Verify correct location of the front CV joint plunge.

8. Remove the supports. Lower the vehicle.

Front Wheel Hub, Knuckle and Bearings

REMOVAL & INSTALLATION

1. Disconnect the negative battery cable.

2. Remove the wheel cover, loosen the hub nut. Raise and safely support the vehicle.

3. Remove the tire and wheel assembly.

4. Install boot cover tool J–28712, for a double off-set joint or tool J–33162 for a tri-pot joint.

5. Remove and discard the hub nut. A new hub nut must be used during assembly.

6. Remove the brake caliper and rotor.

7. Remove the 3 hub and bearing attaching bolts. Remove the hub.

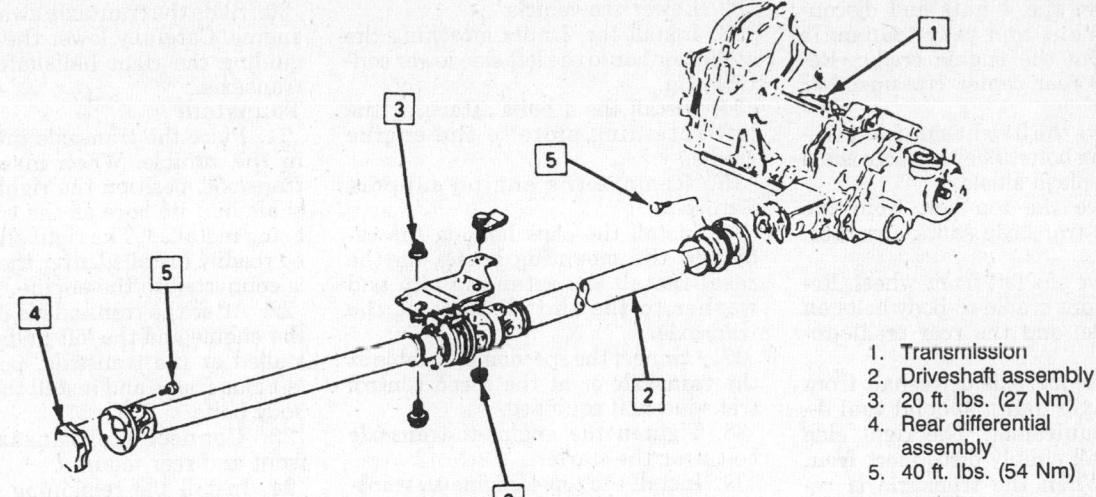

1. Transmission
2. Driveshaft assembly
3. 20 ft. lbs. (27 Nm)
4. Rear differential assembly
5. 40 ft. lbs. (54 Nm)

Driveshafts—6000 STE AWD

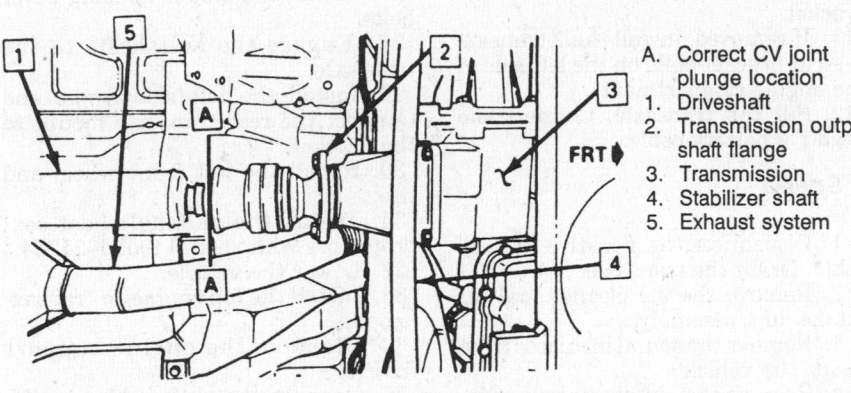

A. Correct CV joint plunge location
1. Driveshaft
2. Transmission output shaft flange
3. Transmission
4. Stabilizer shaft
5. Exhaust system

Verifying correct CV joint plunge location—6000 STE AWD

NOTE: If the old bearing is to be reused, make matchmarks on the bolts and holes for installation purposes.

8. Attach bearing puller J–28733 or equivalent, then remove the bearing.
To install:
9. Clean the mating surfaces of all dirt and corrosion. Check the knuckle bore and seal for damage. If a new bearing is to be installed, remove the old knuckle seal and install a new one. Grease the lips of new seal.
10. Push the bearing onto the halfshaft. Install a new washer and hub nut.
11. Tighten the hub nut on the halfshaft until the new bearing is seated. If the rotor and hub start to rotate as the hub nut is tightened, insert a drift through the caliper and into the rotor cooling fins to prevent rotation.

NOTE: Do not apply full torque to the hub nut at this time.

12. Install the brake shield, if re-moved, and the bearing retaining bolts. Torque bolts to 63 ft. lbs. (85 Nm).
13. Install the caliper and rotor. En-sure the caliper hose is not twisted. Install caliper bolts.
14. Install the wheel assembly. Torque the hub nut to 192 ft. lbs. (260 Nm).
15. Connect battery negative cable.

MANUAL TRANSAXLE

For further information on trans-missions/transaxles, please refer to "Chilton's Guide to Transmission Repair".

Transaxle Assembly

REMOVAL & INSTALLATION

4 Speed

1. Disconnect the negative battery cable. Drain the transaxle.
2. Remove the 2 transaxle strut bracket bolts on the left side of the en-gine compartment, if equipped.
3. If equipped with a V6 engine, dis-connect the fuel lines and fuel line clamps at the clutch cable bracket, as required.
4. Remove the top 4 engine-to-transaxle bolts and the one at the rear near the firewall. The one at the rear is installed from the engine side.
5. Loosen the engine-to-transaxle bolt near the starter but do not remove.
6. Disconnect the speedometer ca-ble at the transaxle or at the speed control transducer, if equipped.
7. Remove the retaining clip and washer from the shift linkage at the transaxle. Remove the clips holding the cables to the mounting bosses on the case.
8. Support the engine with a lifting chain.
9. Unlock the steering column. Raise and safely support the vehicle. Drain the transaxle. Remove the 2 nuts attaching the stabilizer bar to the left lower control arm. Remove the 4 bolts which attach the left retaining plate to the engine cradle. The retain-ing plate covers holds the stabilizer bar.
10. Loosen the 4 bolts holding the right stabilizer bracket.
11. Disconnect and remove the ex-haust pipe and crossover, if necessary.
12. Pull the stabilizer bar down on the left side.

13. Remove the 4 nuts and disconnect the front and rear transaxle mounts from the engine cradle. Remove the 2 rear center crossmember bolts.

14. Remove the 3 right side front cradle attaching bolts. They are accessible under the splash shield.

15. Remove the top bolt from the lower front transaxle shock absorber, if equipped.

16. Remove the left front wheel. Remove the front cradle-to-body bolts on the left side, and the rear cradle-to-body bolts.

17. Pull the left side driveshaft from the transaxle using special tool J–28468 or equivalent. The right side halfshaft will simply disconnect from the case. When the transaxle is removed, the right shaft can be swung aside. A boot protector should be used when disconnecting the driveshafts.

18. Swing the cradle to the left side. Secure aside, outboard of the fender well.

19. Remove the flywheel and starter shield bolts and remove the shields.

20. Remove the 2 transaxle extension bolts from the engine-to-transaxle bracket, if equipped.

21. Place a jack under the transaxle case. Remove the last engine-to-transaxle bolt. Pull the transaxle to the left, away from the engine, then down and out from under the vehicle.

To install:

22. Place the transaxle under the vehicle and raise into position. Position the right halfshaft into its bore as the transaxle is being installed. When the transaxle is bolted to the engine, swing the cradle into position and install the cradle-to-body bolts immediately. Be sure to guide the left halfshaft into place as the cradle is moved back into position.

23. If removed, install the 2 transaxle extension bolts to the engine-to-transaxle bracket.

24. Install flywheel, starter shield and attaching bolts.

25. Install the front cradle-to-body bolts on the left side and the rear cradle-to-body bolts. Install the left front wheel.

26. If removed, install the top bolt to the lower front transaxle shock absorber.

27. Install the 3 right side front cradle attaching bolts.

28. Connect the front and rear transaxle mounts to the engine cradle and install the 4 nuts. Install the 2 rear center crossmember bolts.

29. Place the stabilizer bar in its mounting position.

30. Install and connect the exhaust pipe and crossover.

31. Tighten the 4 bolts holding the right stabilizer bracket.

32. Lower the vehicle.

33. Install the 2 nuts attaching the stabilizer bar to the left side lower control arm.

34. Install the 4 bolts attaching the left retaining plate to the engine cradle.

35. Remove the engine support fixture.

36. Install the clips holding the cables to the mounting bosses on the case. Install the retaining clip and washer to the shift linkage at the transaxle.

37. Connect the speedometer cable at the transaxle or at the speed control transducer, if equipped.

38. Tighten the engine-to-transaxle bolt near the starter.

39. Install the top 4 engine-to-transaxle bolts and the one at the rear near the firewall.

40. If removed, connect the fuel lines and fuel line clamps to the clutch cable bracket.

41. If removed, install the 2 transaxle strut bracket bolts on the left side of the engine compartment.

42. Fill the transaxle. Connect the negative battery cable.

5 Speed

1. Disconnect the negative battery cable. Drain the transaxle.

2. Remove the air cleaner and air intake duct assembly.

3. Remove the sound insulator from inside the vehicle.

4. Remove the clutch master cylinder pushrod from the clutch pedal.

5. Remove the clutch slave cylinder from the transaxle.

6. Disconnect the exhaust crossover pipe.

7. Disconnect the shift cables at the transaxle.

8. Install the engine support fixture J–28467.

9. Remove the top engine-to-transaxle bolts.

10. Raise and safely support the vehicle.

11. Install the halfshaft boot seal protectors with special tool J–34754.

12. Remove the left front wheel and tire.

13. Remove the left side frame and disconnect the rear transaxle mount from the bracket.

14. Drain the transaxle.

15. Disengage the halfshafts from the transaxle.

16. Remove the clutch housing cover bolts.

17. Disconnect the speedometer cable.

18. Attach a jack to the transaxle case.

19. Remove the remaining transaxle-to-engine bolts.

20. Slide the transaxle away from the engine. Carefully lower the jack while guiding the right halfshaft from the transaxle.

To install:

21. Place the transaxle into position in the vehicle. When installing the transaxle, position the right halfshaft shaft into its bore as the transaxle is being installed. The right shaft cannot be readily installed after the transaxle is connected to the engine.

22. After the transaxle is fastened to the engine and the left halfshaft is installed at the transaxle, position the left side frame and install the frame to body bolts.

23. Connect the transaxle to the front and rear mounts.

24. Install the remaining transaxle-to-engine bolts.

25. Remove the jack.

26. Connect the speedometer cable.

27. Install the clutch housing cover bolts.

28. Engage the halfshafts to the transaxle.

29. Install the left side frame and Connect the rear transaxle mount to the bracket.

30. Install the left front wheel and tire.

31. Install the halfshaft boot seal protectors with special tool J–34754.

32. Lower the vehicle.

33. Install the top engine-to-transaxle bolts.

34. Remove the engine support fixture.

35. Connect the shift cables to the transaxle.

36. Connect the exhaust crossover pipe.

37. Install the clutch slave cylinder to the transaxle.

38. Install the clutch master cylinder pushrod to the clutch pedal.

39. Install the sound insulator from inside the vehicle.

40. Install the air cleaner and air intake duct assembly.

41. Connect the negative battery cable.

LINKAGE ADJUSTMENT

4 Speed

1. Remove the shifter boot and retainer from inside the vehicle. Ensure that shifter is in 1st gear.

2. Install two No. 22 drill bits or two $5/32$ inch rods into the 2 alignment holes in the shifter assembly to hold it in 1st gear.

3. Place the transaxle into 1st gear by pushing the rail selector down, just to the point of feeling the resistance of the inhibitor spring. Rotate the shifter lever all the way counterclockwise.

4. Install the stud, with the cable

attached, into the slotted area of the select lever, while gently pulling on the lever to remove all the lash.

5. Remove the 2 drill bits or pins from the shifter.

6. Check the shifter for proper operation. It may be necessary to readjust after road test.

CLUTCH

Clutch Assembly

REMOVAL & INSTALLATION

1. Disconnect the negative battery cable.

2. Disconnect the clutch cable from the transaxle.

3. On 1988–89 vehicles, remove the hush panel from inside the vehicle and disconnect the clutch master cylinder pushrod from the clutch pedal.

4. Remove the transaxle assembly.

5. Mark the relationship between the pressure plate and flywheel.

6. Evenly and carefully loosen the pressure plate bolts until the spring pressure is relieved.

7. Support the pressure plate and remove the pressure plate retaining bolts. Remove the pressure plate and disc.

NOTE: Do not disassemble the pressure plate and disc assembly. If the unit is defective, replace as an assembly.

To install:

8. Clean and lubricate all parts as required.

9. Install the pressure plate and disc. Center the disc using tool J–35822 or equivalent.

NOTE: The disc is installed with the damper springs offset towards the transaxle. Most discs are marked FLYWHEEL SIDE. If the old pressure plate is to be reused, align the marks made previously.

10. Evenly and carefully tighten the pressure plate bolts. Tighten to 15 ft. lbs. (20 Nm).

11. Install the transaxle.

12. On 1988–89 vehicles, connect the clutch master cylinder pushrod to the clutch pedal and install the hush panel.

13. Connect the clutch cable from the transaxle.

14. Connect the negative battery cable.

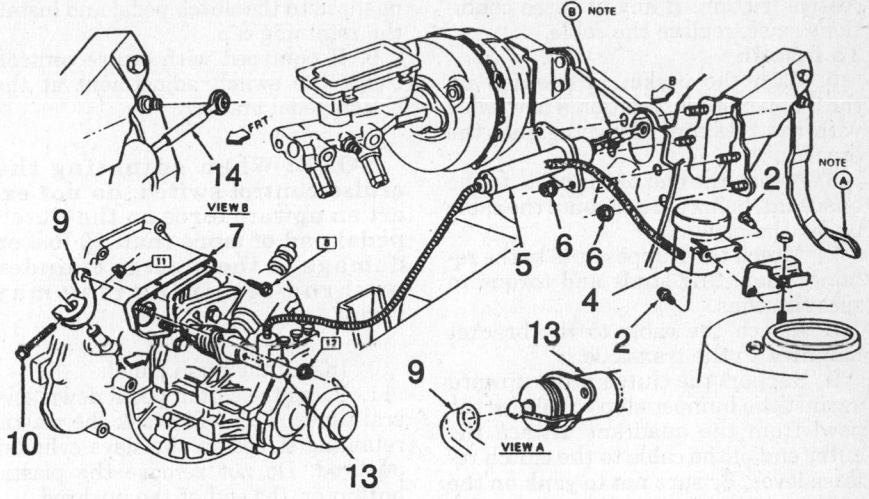

1. Clutch pedal assembly	6. 20 ft. lbs.	11. Nut
2. Bolt	7. Slave cylinder bracket	12. Nut 40 ft. lbs.
3. Bracket	8. Bolt 40 ft. lbs.	13. Slave cylinder
4. Reservoir	9. Lever	14. Master cylinder push rod.
5. Master cylinder	10. Bolt 37 ft. lbs.	

Clutch hydraulic system

PEDAL HEIGHT/FREE-PLAY ADJUSTMENT

All vehicles with a cable-actuated clutch use a self-adjusting clutch mechanism which may be checked as follows. As the clutch friction material wears, the cable must be lengthened. This is accomplished by simply pulling the clutch pedal up to its rubber bumper. This action forces the pawl against its stop and rotates it from mesh with the quadrant teeth, allowing the cable to play out until the quadrant spring load is balanced against the load applied by the release bearing. This adjustment procedure is required every 5000 miles or less.

1. With engine running and brake on, hold the clutch pedal approximately ½ inch from floor mat and move shift lever between first and reverse several times. If this can be done smoothly without clashing into reverse, the clutch is fully releasing. If the shift is not smooth, the clutch is not fully releasing and the linkage should be inspected and corrected as necessary.

2. Check clutch pedal bushings for sticking or excessive wear.

3. Have an assistant fully apply the clutch pedal to the floor. Observe the clutch fork level travel at the transaxle. The end of the clutch fork lever should have a total travel of approximately 1.5–1.7 in.

4. If fork lever is not correct, check the adjusting mechanism by depressing the clutch pedal and looking for pawl to firmly engage with the teeth in the quadrant.

5. To check the self-adjusting mechanism for proper operation, proceed as follows:

a. Depress the clutch pedal and look for the pawl to firmly engage with the teeth in the quadrant.

b. Release the clutch pedal and look for the pawl to be lifted off the quadrant teeth by the bracket stop.

Clutch Cable

REMOVAL & INSTALLATION

1. Support the clutch pedal upward against the bumper stop to release the pawl from the quadrant. Disconnect the end of the cable from the clutch release lever at the transaxle. Be careful to prevent the cable from snapping rapidly toward the rear of the vehicle. The quadrant in the adjusting mechanism can be damaged by allowing the cable to snap back.

2. Disconnect the clutch cable from the quadrant. Lift the locking pawl away from the quadrant, then slide the cable out on the right side of the quadrant.

3. From the engine side of the cowl disconnect the 2 upper nuts holding the cable retainer to the upper studs. Disconnect the cable from the bracket mounted to the transaxle and remove the cable.

4. Inspect the clutch cable for frayed wires, kinks, worn ends and ex-

cessive friction. If any of these conditions exist, replace the cable.

To install:

5. With the gasket in position on the 2 upper studs, position a new cable with the retaining flange against the bracket.

6. Attach the end of the cable to the quadrant, being sure to route the cable under the pawl.

7. Attach the 2 upper nuts to the retainer mounting studs and torque to specifications.

8. Attach the cable to the bracket mounted to the transaxle.

9. Support the clutch pedal upward against the bumper stop to release the pawl from the quadrant. Attach the outer end of the cable to the clutch release lever. Be sure not to yank on the cable, since overloading the cable could damage the quadrant.

10. Check clutch operation and adjust by lifting the clutch pedal up to allow the mechanism to adjust the cable length. Depress the pedal slowly several times to set the pawl into mesh with the quadrant teeth.

Clutch Master and Slave Cylinder

NOTE: The clutch hydraulic system is serviced as a complete unit, it has been bled of air and filled with fluid.

REMOVAL & INSTALLATION

1. Disconnect the negative battery cable.

2. Remove the hush panel from inside the vehicle.

3. Remove the clutch master cylinder retaining nuts at the front of the dash.

4. Remove the slave cylinder retaining nuts at the transaxle.

5. Remove the hydraulic system as a unit from the vehicle.

To install:

6. Install the slave cylinder to the transaxle support bracket aligning the pushrod into the pocket on the clutch fork outer lever. Tighten the retaining nuts evenly to prevent damage to the slave cylinder. Torque the nuts to 40 ft. lbs. (54 Nm).

NOTE: Do not remove the plastic pushrod retainer from the slave cylinder. The straps will break on the first clutch pedal application.

7. Position the clutch master cylinder to the front of the dash. Torque the nuts evenly to 20 ft. lbs.

8. Remove the pedal restrictor from the pushrod. Lube the pushrod bushing on the clutch pedal. Connect the

pushrod to the clutch pedal and install the retaining clip.

9. If equipped with cruise control, check the switch adjustment at the clutch pedal bracket.

NOTE: When adjusting the cruise control switch, do not exert an upward force on the clutch pedal pad of more than 20 lbs. or damage to the master cylinder pushrod retaining ring may result.

10. Install the hush panel.

11. Press the clutch pedal down several times. This will break the plastic retaining straps on the slave cylinder pushrod. Do not remove the plastic button on the end of the pushrod.

12. Connect the negative battery cable.

Hydraulic Clutch System Bleeding

1. Clean the dirt and grease from the cap to ensure no foreign substances enter the system.

2. Remove the cap and diaphragm and fill the reservoir to the top with hydraulic clutch fluid or an equivalent fluid that meets DOT 3 specifications.

3. Fully loosen the bleed screw which is in the slave cylinder body next to the inlet connection.

4. The fluid will now begin to move from the master cylinder down the tube to the slave cylinder. It is important, when using the gravity bleed method, that the master cylinder reservoir be kept full at all times.

5. Air bubbles will be noticeable at the bleed screw outlet. This means that the air is being expelled from the line. When the slave clyinder is full and all air is out of the line, the flow of fluid will be a continuous stream with no bubbles.

6. Tighten the bleed screw.

7. Top-up the fluid level and install the diaphragm and cap to the reservoir.

AUTOMATIC TRANSAXLE

For further information on transmissions/transaxles, please refer to "Chilton's Guide to Transmission Repair".

Transaxle Assembly

REMOVAL & INSTALLATION

THM 125C

NOTE: By September 1, 1991, Hydra-matic will have changed the name designation of the THM 125C automatic transaxle. The new name designation for this transaxle will be Hydra-matic 3T40. Transaxles built between 1989 and 1990 will serve as transitional years in which a dual system, made up of the old designation and the new designation will be in effect.

1. Disconnect the negative battery cable.

2. Remove the air cleaner.

3. Unbolt the detent cable attaching bracket at the transaxle.

4. Pull up on the detent cable cover at the transaxle until the cable is exposed. Disconnect the cable from the rod.

5. Remove the 2 transaxle strut bracket bolts at the transaxle, if equipped.

6. Remove all the engine-to-transaxle bolts except the one near the starter. The one nearest the firewall is installed from the engine side of the vehicle.

7. Loosen but do not remove the engine-to-transaxle bolt near the starter.

8. Disconnect the speedometer cable at the upper and lower coupling. If equipped with cruise control, remove the speedometer cable at the transducer.

9. Remove the retaining clip and washer from the shift linkage at the transaxle. Remove the 2 shift linkages at the transaxle. Remove the 2 shift linkage bracket bolts.

10. Disconnect and plug the cooler lines at the transaxle.

11. Install an engine holding fixture. Raise the engine enough to take its weight off the mounts.

12. Unlock the steering column. Raise and safely support the vehicle.

13. Remove the 2 nuts holding the anti-sway bar to the left lower control arm (driver's side).

14. Remove the 4 bolts attaching the covering plate over the stabilizer bar to the engine cradle on the left side of vehicle.

15. Loosen but do not remove the 4 bolts holding the stabilizer bar bracket to the right side of the engine cradle. Pull the bar downward.

16. Disconnect the front and rear transaxle mounts at the engine cradle.

17. Remove the 2 rear center crossmember bolts.

18. Remove the 3 right (passenger)

side front engine cradle attaching bolts. The nuts are accessible under the splash shield next to the frame rail.

19. If equipped with V6 engine, remove the top bolt from the lower front transaxle shock absorber, as required.

20. Remove the left side front and rear cradle-to-body bolts.

21. Remove the left front wheel. Attach an halfshaft removing tool J–28468 or equivalent, to a slide hammer. Place the tool behind the halfshaft cones and pull the cones out away from the transaxle. Remove the right shaft in the same manner. Set the shafts aside. Plug the openings in the transaxle to prevent fluid leakage and the entry of dirt.

22. Swing the partial engine cradle to the left (driver) side and wire it aside outboard of the fender well.

23. Remove the 4 torque converter and starter shield bolts. Remove the 2 transaxle extension bolts from the engine-to-transaxle bracket.

24. Attach a transaxle jack to the case.

25. Use a felt pen to matchmark the torque converter and flywheel. Remove the 3 torque converter-to-flywheel bolts.

26. Remove the transaxle-to-engine bolt near the starter. Remove the transaxle by sliding it to the left, away from the engine.

To install:

27. As the transaxle is installed, slide the right halfshaft into the case. Install the cradle-to-body bolts before the stabilizer bar is installed. To aid in stabilizer bar installation, a pry hole has been provided in the engine cradle.

28. Install the 3 torque converter-to-flywheel bolts. Tighten to 46 ft. lbs. (62 Nm).

29. Install the 4 torque converter and starter shield bolts. Install the 2 transaxle extension bolts to the engine-to-transaxle bracket.

30. Swing the partial engine cradle into position and install attaching bolts.

31. Install the halfshafts. Install the left front wheel.

32. Install the left side front and rear cradle-to-body bolts.

33. If equipped with V6 engine, Install the top bolt to the lower front transaxle shock absorber, as required.

34. Install the 3 right (passenger) side front engine cradle attaching bolts.

35. Install the 2 rear center crossmember bolts.

36. Disconnect the front and rear transaxle mounts at the engine cradle.

37. Tighten the 4 bolts holding the stabilizer bar bracket to the right side of the engine cradle.

38. Install the 4 bolts attaching the covering plate over the stabilizer bar to the engine cradle on the left side of vehicle.

39. Install the 2 nuts holding the anti-sway bar to the left lower control arm (driver's side).

40. Lower the vehicle.

41. Remove the engine holding fixture.

42. Connect the cooler lines to the transaxle.

43. Install the 2 shift linkage bracket bolts. Install the 2 shift linkages at the transaxle. Install the retaining clip and washer to the shift linkage on the transaxle.

44. Connect the speedometer cable at the upper and lower coupling. If equipped with cruise control, connect the speedometer cable to the transducer.

45. Tighten the engine-to-transaxle bolt near the starter.

46. Install all the engine-to-transaxle bolts except the one near the starter. The one nearest the firewall is installed from the engine side of the vehicle.

47. Install the 2 transaxle strut bracket bolts at the transaxle, if equipped.

48. Connect the detent cable.

49. Install the air cleaner.

50. Connect the negative battery cable.

THM 440–T4

NOTE: By September 1, 1991, Hydra-matic will have changed the name designation of the THM 440–T4 automatic transaxle. The new name designation for this transaxle will be Hydra-matic 4T60. Transaxles built between 1989 and 1990 will serve as transitional years in which a dual system, made up of the old designation and the new designation will be in effect.

1. Disconnect the negative battery cable.

2. Remove the air cleaner and disconnect the TV cable at the throttle body.

3. Disconnect the shift linkage at the transaxle.

4. Remove the engine support fixture tool J–28467 or equivalent.

5. Disconnect all electrical connectors.

6. Remove the 3 bolts from the transaxle to the engine.

7. Disconnect the vacuum line at the modulator.

8. Raise and safely support the vehicle.

9. Remove the left front wheel and tire assembly.

10. Remove the left side ball joint from the steering knuckle.

11. Disconnect the brake line bracket at the strut.

NOTE: A halfshaft seal protector tool J–34754 should be modified and installed on any halfshaft prior to service procedures on or near the halfshaft. Failure to do so could result in seal damage or joint failure.

12. Remove the halfshafts from the transaxle.

13. Disconnect the pinch bolt at the intermediate steering shaft. Failure to do so could cause damage to the steering gear.

14. Remove the frame to stabilizer bolts.

15. Remove the stabilizer bolts at the control arm.

16. Remove the left front frame assembly.

17. Disconnect the speedometer cable or wire connector from the transaxle.

18. Remove the extension housing to engine block support bracket.

19. Disconnect the cooler pipes.

20. Remove the converter cover and converter-to-flywheel bolts.

21. Remove all of the remaining transaxle-to-engine bolts except one.

22. Position a jack under the transaxle.

23. Remove the remaining transaxle-to-engine bolt and remove the transaxle.

To install:

24. Install the transaxle in the vehicle. Install the engine-to-transaxle bolt accessible from under the vehicle. Tighten to 55 ft. lbs. (75 Nm).

25. Install all of the remaining transaxle-to-engine bolts. Tighten to 55 ft. lbs. (75 Nm).

26. Remove the jack.

27. Install the converter-to-flywheel bolts and the converter cover.

28. Connect the cooler pipes.

29. Install the extension housing to engine block support bracket.

30. Connect the speedometer cable or wire connector to the transaxle.

31. Install the left front frame assembly.

32. Install the stabilizer bolts at the control arm.

33. Install the frame-to-stabilizer bolts.

34. Connect the pinch bolt at the intermediate steering shaft.

35. Install the halfshafts to the transaxle.

36. Connect the brake line bracket at the strut.

37. Install the left side ball joint to the steering knuckle.

38. Install the left front wheel and tire assembly.

39. Lower the vehicle.

40. Connect the vacuum line at the modulator.
41. Install the 3 bolts from the transaxle to the engine.
42. Connect all electrical connectors.
43. Remove the engine support tool.
44. Connect the shift linkage to the transaxle.
45. Connect the TV cable at the throttle body and adjust as necessary. Install the air cleaner.
46. Connect the negative battery cable.

SHIFT CONTROL CABLE ADJUSTMENT

1. Place the shift lever in **N**. To determine the **N** position, rotate the selector shaft clockwise from **P** through **R** to **N**.
2. Place the shift control assembly in **N**.
3. Push the tab on the cable adjuster to adjust the cable in cable mounting bracket.

PARK/LOCK CONTROL CABLE ADJUSTMENT

The shifter lever must not be able to move to any other positions with the shift lever in **P** and the key in the **LOCK** position. Also, with the key in the **RUN** position and the shift lever in **N**, ensure that the key cannot be turned to the **LOCK** position. If these conditions cannot be met, adjustment is necessary.
1. If the key cannot be removed in the **P** position, snap the connector lock button to the **UP** position.
2. Move the cable connector nose rearward until the key can be removed from the ignition.
3. Push the snap lock button down.

TV DETENT CABLE ADJUSTMENT

1. With the engine **OFF**, depress and hold-down the readjust tab at the TV cable adjuster.
2. Move the cable conduit until it stops against the fitting. Release the readjustment tab.
3. Rotate the throttle lever by hand to its full travel position. The slider must ratchet toward the lever when the lever is rotated to its full travel.

NOTE: Check that the cable moves freely. The cable may appear to function properly with the engine OFF and COLD. Recheck after the engine is HOT.

FRONT SUSPENSION

MacPherson Strut

REMOVAL & INSTALLATION

1. Loosen the wheel nuts, raise and support vehicle, then remove the wheel and tire assembly.
2. Remove the brake hose clip-to-strut bolt, if equipped. Do not disconnect the hose from the caliper. Install a halfshaft cover to protect the axle boot.
3. Mark the camber cam eccentric adjuster for assembly.
4. Remove the 2 lower strut-to-steering knuckle bolts and the 3 upper strut-to-body nuts. Remove the strut assembly.
To install:
5. Install the strut assembly. Install the 2 lower strut-to-steering knuckle bolts and 3 upper strut-to-body nuts. Realign the camber marks made upon removal.
6. Install the brake hose clip-to-strut bolt.
7. Remove the axle boot protector.
8. Install the tire and wheel assembly.
9. Lower the vehicle.

Lower Ball Joints

INSPECTION

1. Raise and support vehicle safely.

2. Grasp the wheel at the top and bottom and shake the wheel in and out.
3. If any movement is seen of the steering knuckle relative to the control arm, the ball joints are defective and must be replaced. Note that movement elsewhere may be due to loose wheel bearings or other problems; watch the knuckle-to-control arm connection.
4. If the ball stud is disconnected from the steering knuckle and any looseness is noted, often the ball joint stud can be twisted in its socket with your fingers, replace the ball joints.

REMOVAL & INSTALLATION

1. Loosen the wheel nuts, raise and support vehicle safely, then remove the tire and wheel assembly.
2. Using an ⅛ inch drill bit, drill a hole approximately ¼ inch deep in the center of each of the 3 ball joint rivets.
3. Using a ½ inch drill bit, drill off the rivet heads. Drill only enough to remove the rivet head.
4. Using a hammer and punch, remove the rivets, driving them out from the bottom.
5. Loosen the ball joint pinch bolt in the steering knuckle, then remove the ball joint.
To install:
6. Install a new ball joint in the control arm. Torque new bolts to 13 ft. lbs.
7. Install the ball stud into the steering knuckle pinch bolt fitting. It should go in easily; if not, check the stud alignment. Install the pinch bolt from the rear to the front. Torque to 45 ft. lbs.

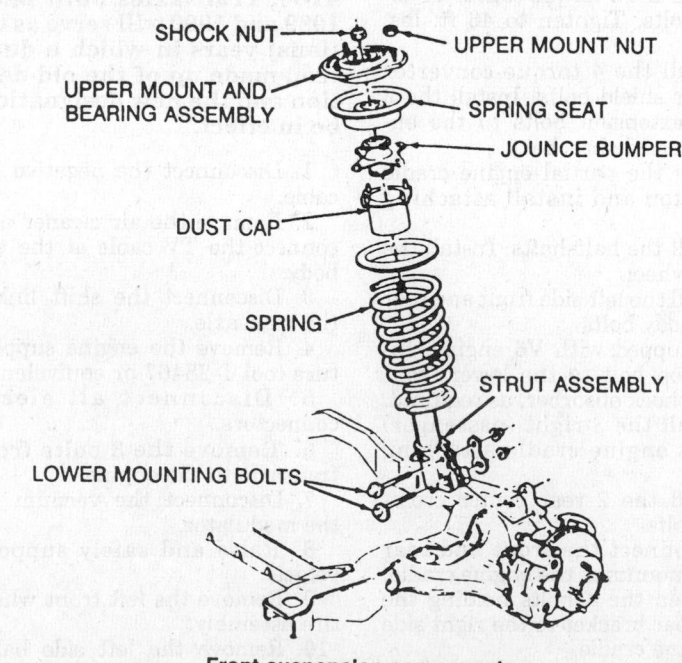

SHOCK NUT
UPPER MOUNT NUT
UPPER MOUNT AND BEARING ASSEMBLY
SPRING SEAT
JOUNCE BUMPER
DUST CAP
SPRING
STRUT ASSEMBLY
LOWER MOUNTING BOLTS

Front suspension components

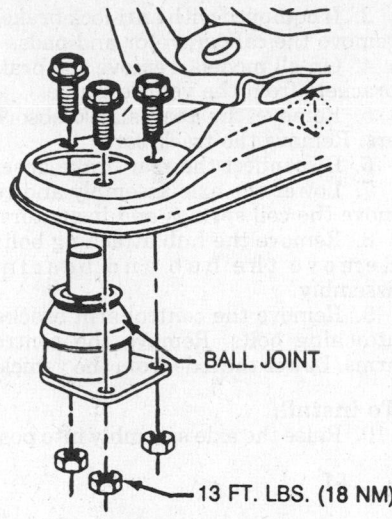

BALL JOINT

13 FT. LBS. (18 NM)

Ball joint installation

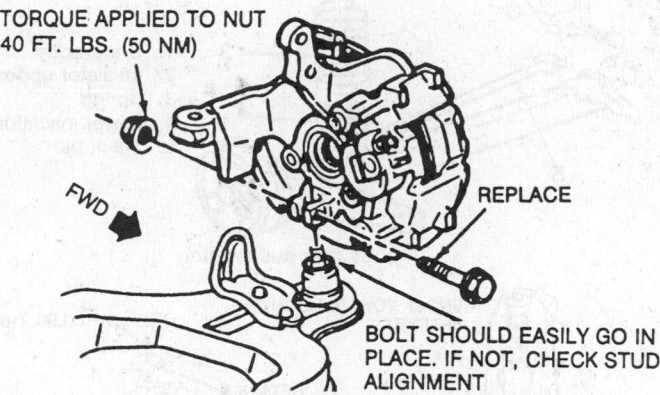

TORQUE APPLIED TO NUT 40 FT. LBS. (50 NM)

FWD

REPLACE

BOLT SHOULD EASILY GO IN PLACE. IF NOT, CHECK STUD ALIGNMENT

Ball joint stud should go in easily

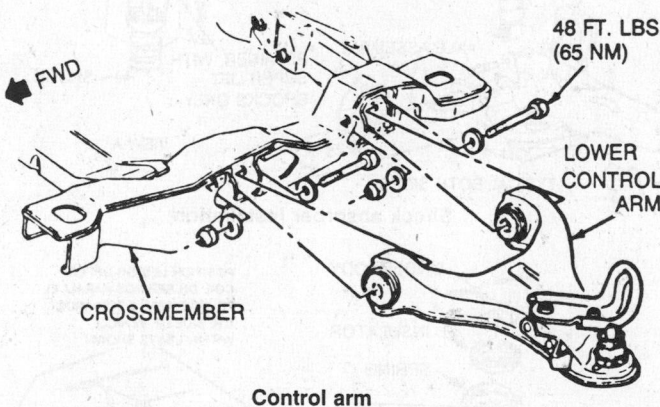

FWD

48 FT. LBS. (65 NM)

LOWER CONTROL ARM

CROSSMEMBER

Control arm

8. Install the wheel and lower the vehicle.

Lower Control Arms

REMOVAL & INSTALLATION

1. Loosen the wheel nuts, raise and support vehicle safely, then remove the tire and wheel assembly.

2. Remove the stabilizer bar from the control arm.

3. Remove the ball joint from the steering knuckle.

4. Remove the control arm pivot bolts, then the control arm from the vehicle.

To install:

5. Install the control into the fittings. Install the pivot bolts from the rear to the front. Torque bolts to 50 ft. lbs. (68 Nm).

6. Install the ball stud into the pinch bolt fitting. It should go in easily; if not, check the ball joint stud alignment.

7. Install the pinch bolt from the rear to the front. Torque bolts 40 ft. lbs. (54 Nm).

8. Install the stabilizer bar attachment. Torque bolts to 35 ft. lbs. (47 Nm).

9. Install the tire and wheel assembly and lower the vehicle.

Stabilizer Shaft

REMOVAL & INSTALLATION

1. Disconnect battery negative cable.

2. Raise and safely support the vehicle.

3. Remove the stabilizer shaft insulator clamp and insulator at the control arms. Do not remove the studs from the control arm.

4. Remove the plate from the frame at each side, then the stabilizer shaft and insulator bushings from the vehicle.

To install:

5. Install the stabilizer insulator bushings, stabilizer shaft and plate to the frame at each side. Tighten plate to frame bolts to 40 ft. lbs. (55 Nm)

6. Install the insulators at the control arms. Install the stabilizer shaft insulator clamp. Tighten insulator clamp nuts to 33 ft. lbs. (45 Nm).

7. Lower the vehicle.

8. Connect the negative battery cable.

REAR SUSPENSION

Shock Absorbers

REMOVAL & INSTALLATION

1. Disconnect the negative battery cable.

2. Open the deck or trunk lid, then remove the trim cover and the upper shock nut. Remove and replace 1 shock at a time when replacing both shocks.

3. Raise and support the vehicle safely.

4. Remove the shock lower attaching bolt, then remove the shock. If equipped with air shocks, disconnect the air lines.

To install:

NOTE: Purge new shocks of air by repeatedly compressing them while inverted and extending them in their normal installed position.

5. Install the shock absorber and attaching bolts. Tighten to 43 ft. lbs. (58 Nm).

6. Lower the vehicle.

7. Install the upper shock absorber nut. Tighten the upper nut to 13 ft. lbs. (18 Nm).

8. Install the trim cover.

9. Connect the negative battery cable.

Coil Springs

REMOVAL & INSTALLATION

1. Disconnect the battery negative cable.
2. Raise and safely support the vehicle using jacks that can be raised and lowered.
3. Remove the brake hose attaching brackets (right and left), allowing the hoses to hang freely. Do not disconnect the hoses.
4. Remove the track bar attaching bolts from the rear axle.
5. Lower the axle, then remove the coil spring and insulator.

NOTE: Do not suspend the rear axle by the brake hose.

To install:
6. Position the spring and insulator on the axle. The leg on the upper coil of the spring must be parallel to the axle, facing the left side of the vehicle.
7. Install the shock absorber bolts. Torque bolts to 43 ft. lbs. Install track bar, if equipped, and torque to 33 ft. lbs. Install the brake line brackets and torque to 8 ft. lbs.

Rear Wheel Bearings

REMOVAL & INSTALLATION

1. Disconnect battery negative cable, then raise and support vehicle.
2. Remove the wheel and brake drum. Do not hammer on the brake drum as damage to the bearing may result.
3. If equipped with anti-lock brakes, remove caliper, rotor and pads.
3. On all vehicles, remove the hub and bearing assembly from the rear axle attaching bolts and remove the rear axle.

NOTE: The bolts attaching the hub and bearing assembly also support the brake assembly. When removing these bolts, support the brake assembly with a length of wire. Do not let the brake assembly hang by the brake line unsupported.

To install:
4. On all vehicles, install the rear axle and install the hub and bearing assembly attaching bolts.
5. If equipped with anti-lock brakes, install the rotor, caliper and pads.
6. Install the brake drum. Install the tire and wheel assembly.
7. Lower the vehicle.
8. Connect the negative battery cable.

Rear Axle Assembly

REMOVAL & INSTALLATION

1. Raise and safely support the vehicle.

NOTE: If removing the rear axle on a twin post lift, the axle assembly must be supported securely to prevent the possibility of the axle assembly slipping from the lift when certain fasteners are removed.

2. Remove the rear wheels. Remove the rear brake drums. Disconnect the parking brake from the rear axle.
3. If equipped with anti-lock brakes, remove the caliper, rotor and pads.
4. On all models, remove the brake brackets from the vehicle frame.
5. Remove the rear shock absorbers. Remove the track bar.
6. Disconnect the rear brake hoses.
7. Lower the axle assembly and remove the coil springs and insulators.
8. Remove the hub attaching bolts. Remove the hub and bearing assembly.
9. Remove the control arm bracket attaching bolts. Remove the control arms. Lower the axle from the vehicle.

To install:
10. Raise the axle assembly into posi-

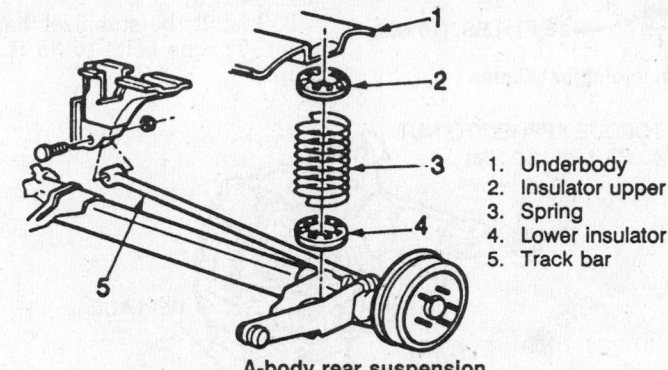

1. Underbody
2. Insulator upper
3. Spring
4. Lower insulator
5. Track bar

A-body rear suspension

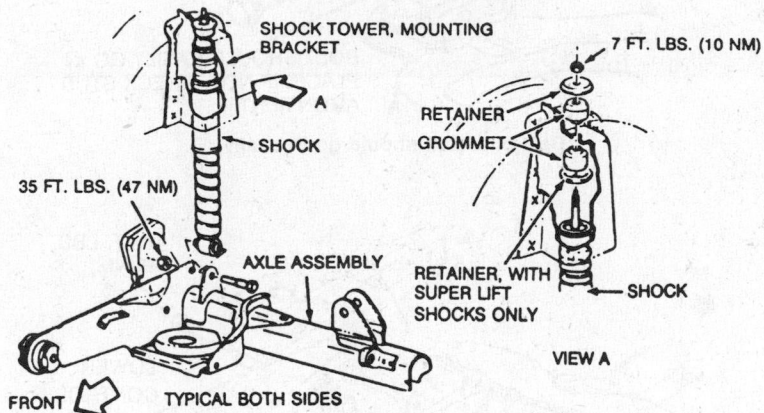

SHOCK TOWER, MOUNTING BRACKET

A

SHOCK

35 FT. LBS. (47 NM)

AXLE ASSEMBLY

FRONT   TYPICAL BOTH SIDES

7 FT. LBS. (10 NM)

RETAINER

GROMMET

RETAINER, WITH SUPER LIFT SHOCKS ONLY   SHOCK

VIEW A

Shock absorber installation

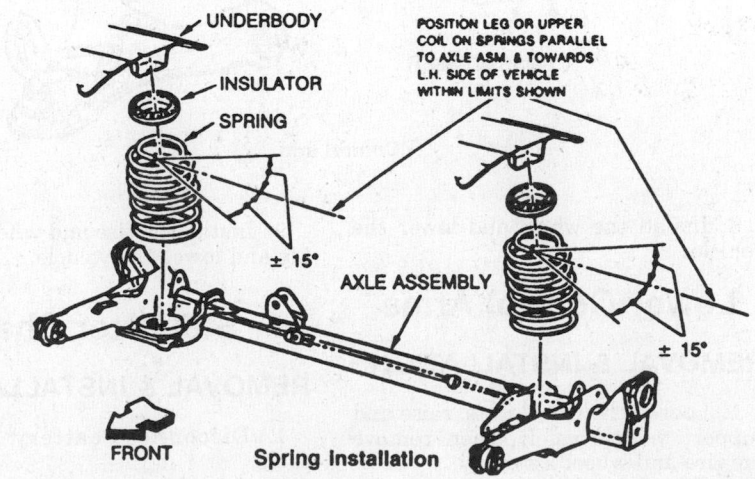

UNDERBODY

INSULATOR

SPRING

POSITION LEG OR UPPER COIL ON SPRINGS PARALLEL TO AXLE ASM. & TOWARDS L.H. SIDE OF VEHICLE WITHIN LIMITS SHOWN

± 15°

AXLE ASSEMBLY

± 15°

FRONT   **Spring Installation**

tion. Install the control arms and control arm bracket attaching bolts.

11. Install the hub and bearing assembly. Install the hub attaching bolts.

12. Install the coil springs and insulators. Raise the axle assembly.

13. Install the shock absorbers. Install the track bar.

14. Connect the brake hoses.

15. Install the brake brackets to the vehicle frame.

16. If equipped with anti-lock brakes, install the rotor, caliper and pads.

17. Connect the parking brake to the rear axle. Install the rear brake drums. Install the rear tire and wheel assemblies.

18. Bleed the brake system and adjust the parking brake, as required.

19. Lower the vehicle.

STEERING

Steering Wheel

REMOVAL & INSTALLATION

NOTE: When installing the steering wheel, always make sure the turn signal lever is in the neutral position.

1. Disconnect the negative battery cable. Remove the trim retaining screws from behind the wheel. On steering wheels with a center cap, pull off the cap.

2. Lift the trim off and pull the horn wires from the turn signal cancelling cam.

3. Remove the retainer and the steering wheel nut.

4. Mark the wheel-to-shaft relationship and then remove the wheel with a puller.

To install:

5. Install the wheel on the shaft, aligning the previously made marks. Tighten the nut to 30 ft. lbs. (41 Nm).

6. Insert the horn wires into the cancelling cam.

7. Install the center trim and reconnect the battery cable.

Steering Column

NOTE: Once the steering column is removed from the vehicle, the column is extremely susceptible to damage. Dropping the column assembly on its end could collapse the steering shaft or loosen the plastic injections which maintain column rigidity. Leaning on the column assembly could

cause the jacket to bend or deform. Any of the above damage could impair the column's collapsible design. If it is necessary to remove the steering wheel, use a standard wheel puller. Under no condition should the end of the shaft be hammered upon, as hammering could loosen the plastic injection which maintains column rigidity.

REMOVAL & INSTALLATION

1. Disconnect the negative battery cable.

2. If column repairs are to be made, remove the steering wheel.

3. Remove the nuts and bolts attaching the flexible coupling to the bottom of the steering column. Remove the safety strap and bolt if equipped.

4. Remove the steering column trim shrouds and column covers.

5. Disconnect all wiring harness connectors. Remove the dust boot mounting screws and column mounting bracket bolts.

6. Lower the column to clear the mounting bracket and carefully remove from the vehicle.

To install:

7. Install the column in the vehicle.

8. Install the column mounting bracket bolts. Install the dust boot mounting screws. Connect all wiring harness connectors.

9. Install the steering column trim shrouds and column covers.

10. Install the safety strap and bolt if equipped. Install the nuts and bolts attaching the flexible coupling to the bottom of the steering column.

11. If removed, install the steering wheel.

12. Connect the negative battery cable.

Power Steering Rack

BEARING PRELOAD ADJUSTMENT

1. Raise and safely support vehicle.

2. When adjusting, ensure front wheels are raised and the steering wheel centered.

3. Loosen the adjuster plug locknut, turn adjuster plug clockwise until it bottoms in the housing, then back adjuster plug approximately 50–70 degrees (approximately 1 flat).

4. After adjustment, check the returnabilty of the steering wheel.

REMOVAL & INSTALLATION

1. Disconnect the negative battery cable. Raise and safely support vehicle.

Allow the front suspension to hang freely. Disconnect the power steering hoses from the gear, where equipped.

2. Move the intermediate shaft seal upward and remove the intermediate shaft-to-stub shaft pinch bolt.

3. Remove both front wheels.

4. Remove the cotter pins and nut from both tie rod ends. Disconnect the tie rod ends from the steering knuckles.

5. Remove the air management system pipe bracket bolt from the crossmember.

6. Support the engine cradle with a floor jack. Remove the 2 rear cradle mount bolts and, using a jack, lower the rear of the engine cradle about 4–5 in.

NOTE: Do not lower the engine cradle too far or damage to surrounding components will result.

7. Remove the rack and pinion heat shield, then the 2 rack and pinion mount bolts.

8. Remove the rack and pinion assembly through the left wheel opening.

To install:

9. Install the rack and pinion assembly through the left wheel opening.

10. Install the rack and pinion heat shield, then the 2 rack and pinion mount bolts. Torque the mount bolts to 70 ft. lbs. (95 Nm).

11. Raise the engine cradle into position and install the 2 rear cradle mount bolts.

12. Install the air management system pipe bracket bolt to the crossmember.

13. Connect the tie rod ends to the steering knuckles. Install the cotter pin and nut to both tie rod ends. Tighten the tie rod end nuts to 30 ft. lbs. (41 Nm).

14. Install both front wheels.

15. Install the intermediate shaft-to-stub shaft pinch bolt. Tighten the pinch bolt to 45 ft. lbs. (61 Nm).

16. If equipped, connect the power steering hoses to the steering gear. Refill the system.

17. Lower the vehicle. Connect the negative battery cable.

Power Steering Pump

REMOVAL & INSTALLATION

2.5L Engine

1. Disconnect the negative battery cable. Raise and safely support the vehicle.

2. Remove the serpentine belt and siphon the fluid from the pump reservoir.

3. Disconnect the hydraulic lines from the pump.

4. Remove the radiator hose clamp bolt.

5. Remove the upper and lower bolts and nuts from the front pump bracket.

6. Remove the pump and bracket from the engine.

To install:

7. Install the pump and bracket to the engine.

8. Install the upper and lower bolts and nuts to the front pump bracket.

9. Install the radiator hose clamp bolt.

10. Connect the hydraulic lines to the pump.

11. Install serpentine belt. Refill the system.

12. Lower the vehicle.

13. Connect the negative battery cable.

14. Bleed the system.

Except 2.5L Engine

1. Disconnect the negative battery cable at the battery. Remove air cleaner, if necessary.

2. Disconnect the blower motor wiring and remove the blower motor.

3. Remove the coolant hose from the water pump.

4. Siphon the fluid from the pump reservoir, then disconnect the lines from the pump.

5. Remove the pump drive belt.

6. Remove the 1 nut which attaches the rear pump bracket to the engine bracket.

7. Remove the 2 front pump bracket-to-engine bolts, then remove the pump and bracket assembly.

To install:

8. Install the pump and bracket assembly. Install the 2 front pump bracket-to-engine bolts.

9. Install the 1 nut which attaches the rear pump bracket to the engine bracket.

10. Install the serpentine belt.

11. Connect the lines to the pump. Refill the system.

12. Install the coolant hose to the water pump.

13. Install the blower motor. Connect the blower motor wiring.

14. If removed, install the air cleaner.

15. Connect the negative battery cable.

16. Bleed the system.

BELT ADJUSTMENT

The accessories are driven by a single serpentine belt. Belt tension is controlled automatically by the spring-loaded tensioner. No adjustment is necessary.

SYSTEM BLEEDING

1. Fill the fluid reservoir.

2. Let fluid stand undisturbed for 2 minutes, then crank engine for about 2 seconds. Refill reservoir if necessary.

3. Repeat Steps 1 and 2 until fluid level remains constant after cranking the engine.

4. Raise the front of the vehicle until both wheels are off the ground, then start the engine. Increase engine speed to 1500 rpm.

5. Turn the wheels lightly against the stop to the left and right, checking the fluid level and refilling, as necessary.

Tie Rod Ends

REMOVAL & INSTALLATION

1. Loosen the jam nut on the steering rack inner tie rod.

2. Remove the tie rod end nut. Separate the tie rod end from the steering knuckle using a suitable puller.

3. Unscrew the tie rod end, counting the number of turns.

To install:

4. Screw the tie rod end onto the steering rack inner tie rod the same number of turns as counted for removal. This will give approximately correct toe.

5. Install the tie rod end into the knuckle. Install nut and torque to 40 ft. lbs. (54 Nm).

6. If the toe must be adjusted, use pliers to expand the boot clamp. Turn the inner tie rod to adjust. Replace clamp.

7. Tighten the jam nut to 59 ft. lbs. (80 Nm).

BRAKES

For all brake system repair and service procedures not detailed below, please refer to "Brakes" in the Unit Repair section.

Master Cylinder

REMOVAL & INSTALLATION

1. Disconnect the negative battery cable. Disconnect and plug the hydraulic lines at master cylinder.

2. Remove the master cylinder retaining nuts and lock washers.

3. Remove the master cylinder from the vehicle.

To install:

4. Install the cylinder on the booster. Install nuts and lock washers.

Tighten the attaching nuts to 25 ft. lbs. (34 Nm).

5. Install hydraulic lines.

6. Bleed the brakes system.

Power Brake Booster

REMOVAL & INSTALLATION

1. Disconnect the negative battery cable. Disconnect vacuum hose from vacuum check valve.

2. Unbolt the master cylinder and carefully move it aside without disconnecting the hydraulic lines.

3. Working inside the vehicle, disconnect the pushrod at the brake pedal assembly. Remove the nuts and lock washers that secure the booster to firewall.

4. Remove the booster from the engine compartment.

To install:

5. Install the booster from the engine compartment.

6. Working inside the vehicle, install the nuts and lock washers that secure the booster to the firewall. Tighten the mounting nuts to 25 ft. lbs. (34 Nm). Connect the pushrod at the brake pedal.

7. Position the master cylinder on the booster and install the attaching nuts.

8. Connect vacuum hose to the vacuum check valve.

9. Connect the negative battery cable. Check operation of stop lights. Allow engine vacuum to build before applying brakes.

Brake Caliper

REMOVAL & INSTALLATION

1. Disconnect the negative battery cable.

2. Raise and safely support the vehicle. Remove the tire and wheel assembly.

3. Remove ⅔ of the brake fluid from the master cylinder.

4. Position a 12 inch adjustable pliers over the inboard brake shoe tab and the inboard caliper housing. Squeeze the pliers to compress the piston back into the caliper bore and to provide clearance between the lining and rotor.

5. If equipped with rear disc brakes, disconnect the parking brake cable and return spring from the parking brake lever, then the parking brake cable from the bracket.

6. On all models, remove the caliper mounting bolts, then lift caliper from bracket and remove the inner and outer pads with the anti-rattle springs.

7. Disconnect the hydraulic hose. Remove the caliper from the vehicle.

To install:

8. Connect the hydraulic hose.

9. On all models, install the caliper on the mounting brackets and install the mounting bolts. Install the inner and outer pads with the anti-rattle springs.

10. If equipped with rear disc brakes, connect the parking brake cable to the bracket and the cable and return spring to the parking brake lever.

11. Install the tire and wheel assembly.

12. Lower the vehicle.

13. Fill the master cylinder reservoir.

14. Connect the negative battery cable.

15. Bleed the brake system.

Disc Brake Pads

REMOVAL & INSTALLATION

1. Disconnect the negative battery cable.

2. Raise and safely support the vehicle. Remove the tire and wheel assembly.

3. Remove ⅔ of the brake fluid from the master cylinder.

4. Position a 12 inch adjustable pliers over the inboard brake shoe tab and the inboard caliper housing. Squeeze the pliers to compress the piston back into the caliper bore and to provide clearance between the lining and rotor.

5. If equipped with rear disc brakes, disconnect the parking brake cable and return spring from the parking brake lever, then the parking brake cable from the bracket.

6. On all models, remove the caliper mounting bolts, then lift caliper from bracket and remove the inner and outer pads with the anti-rattle springs.

To install:

7. On all models, install the caliper and mounting bolts. Install the inner and outer pads with the anti-rattle springs.

8. If equipped with rear disc brakes, connect the parking brake cable to the bracket and the cable and return spring to the parking brake lever.

9. Install the tire and wheel assembly.

10. Lower the vehicle.

11. Fill the master cylinder reservoir with fresh brake fluid.

12. Apply the brake pedal until the pedal is firm and steady.

13. If the pedal remains spongy and/or sinks to the floor, bleed the brake system.

14. Top-up fluid level in the master cylinder, as necessary.

Brake Rotor

REMOVAL & INSTALLATION

1. Disconnect the negative battery cable.

2. Raise and safely support the vehicle.

3. Remove the tire and wheel assembly.

4. Remove the caliper. Support the caliper using a length of wire.

5. Remove the rotor.

To install:

6. Install the rotor.

7. Install the caliper.

8. Install the tire and wheel assembly.

9. Lower the vehicle.

10. Apply the brake pedal until the pedal is firm and steady.

11. If the pedal remains spongy and/or sinks to the floor, bleed the brake system.

12. Top-up fluid level in the master cylinder, as necessary.

13. Connect the negative battery cable.

Brake Drums

REMOVAL & INSTALLATION

1. Disconnect battery negative cable.

2. Raise and safely support the vehicle. Remove the tire and wheel assembly.

3. Remove the brake drum. If the drum is difficult to remove, remove the access plug from the backing plate and back off the adjusting screw.

To install:

4. Install the brake drum.

5. Adjust the brakes.

6. Install the tire and wheel assembly.

7. Lower the vehicle.

8. Connect the negative battery cable.

Brake Shoes

REMOVAL & INSTALLATION

1. Disconnect the negative battery cable.

2. Raise and safely support vehicle. Remove the tire and wheel assembly.

3. Remove the brake drum. If the drum is difficult to remove, remove the access plug from the backing plate and back off the adjusting screw.

4. Remove the return springs from the anchor using appropriate brake spring pliers.

5. Remove the hold-down springs and retaining pins. Remove the lever pivot, actuator link, actuator lever, ac-

tuator pivot and lever return spring, parking brake strut and strut spring.

6. Remove the brake shoes, then disconnect the parking brake cable.

7. Remove the adjusting screw assembly and spring. Note position of adjusting spring.

NOTE: Do not interchange the adjusting screws or adjusting screw springs from right to left brake assembly.

8. Remove the retaining ring, pin and parking brake lever from the secondary shoe.

To install:

9. Lubricate the shoe contact surfaces on the backing plate and adjusting screw assembly.

10. Install the parking brake lever on the secondary shoe with the pin and retaining ring.

11. Install the adjusting screw assembly and spring. The coil of the spring must not be over the star wheel.

12. Install the shoe and lining assemblies after attaching the parking brake cable.

13. Install the parking brake strut and spring by spreading the shoes apart. Ensure the strut is properly positioned. The end with the spring engages the primary shoe and the end without the spring engages the parking brake lever.

14. Install the actuator pivot, actuator lever and return spring.

15. Install the actuator link in the shoe retainer.

16. Install the link into the lever while holding up on the lever.

17. Install the hold-down pins, lever pivot and hold-down springs.

18. Install the shoe return springs.

19. Install the brake drum, wheel and tire assembly, then lower vehicle. Apply the brake pedal several times to seat the brake shoes. Check and adjust the parking brake, as required.

20. Check the master cylinder reservoir. Connect the negative battery cable.

Wheel Cylinder

REMOVAL & INSTALLATION

1. Disconnect the negative battery cable.

2. Loosen the wheel lug nuts, raise and safely support the vehicle. Remove the tire and wheel assembly. Remove the drum and brake shoes. Leave the hub and wheel bearing assembly in place.

3. Remove any dirt from around the brake line fitting, then disconnect the brake line.

4. Remove the wheel cylinder retainer by using 2 awls or punches into

the slots between the wheel cylinder pilot and retainer locking tabs. Bend both tabs away simultaneously. Remove the wheel cylinder from the backing plate.

To install:

5. Position the wheel cylinder against the backing plate and hold it in place with a wooden block between the wheel cylinder and the hub and bearing assembly.

6. Install a new retainer over the wheel cylinder abutment on the rear of the backing plate by pressing it into place with an 1⅛ inch 12-point socket and an extension.

7. Install a new bleeder screw into the wheel cylinder. Install brake line and torque to 10–15 ft. lbs. (14–20 Nm).

8. Install the brake shoes and drum. Adjust the brakes.

9. Install the tire and wheel assembly.

10. Lower the vehicle.

11. Bleed the brake system.

12. Connect the negative battery cable.

Parking Brake Cable

ADJUSTMENT

1. Raise the rear of the vehicle and support it safely using jackstands, with both rear wheels off the ground.

2. Apply the parking brake 3 ratchet clicks from the fully released position.

3. Loosen the equalizer locknut, the tighten the adjusting nut until a light to moderate drag is felt when the rear wheels are rotate. Tighten the locknut.

4. Fully release parking brake and rotate rear wheels; no drag should be felt.

REMOVAL & INSTALLATION

Front Cable

1. Raise and safely support the vehicle.

2. Loosen the equalizer nut.

3. Disconnect the front cable from the connector and equalizer.

4. Remove the clip at the frame.

5. Remove the cable from the hanger.

6. Lower the vehicle.

7. Remove the 3 screws and 1 nut and lower the driver's side sound insulator panel.

8. Remove the carpet finish molding. Lift the carpet.

9. Remove the cable retaining clip at the lever assembly.

10. Depress the retaining tangs and remove the cable and casing from the lever assembly.

11. Remove the cable from the retaining clips.

12. Remove the grommet retainer from the floor pan.

13. Unseat the grommet and pull the cable through the floor pan.

To install:

14. Insert the cable through the floor pan and grommet.

15. Seat the grommet. Install the grommet retainer to the floor pan.

16. Fasten the cable in the retaining clips.

17. Connect the cable and casing to the lever assembly. Seat the retaining tangs.

18. Install the cable retaining clip at the lever assembly.

19. Place the carpet into position. Install the carpet finish molding.

20. Install the driver's side sound insulator panel and attaching screws and nuts.

21. Raise and safely support the vehicle.

22. Fasten the cable to the hanger.

23. Install the clip to the frame.

24. Connect the front cable to the equalizer and connector.

25. Adjust the parking brake cable.

26. Lower the vehicle.

Rear Cables

1. Raise and safely support the vehicle.

2. Loosen the equalizer nut.

3. Disconnect the cable at the equalizer and connector.

4. Remove the tire, wheel and brake drum.

5. Disconnect the cable from the parking brake lever.

6. Depress the retaining tangs on the cable. Remove the cable and casing from the backing plate.

To install:

7. Install the cable through the rear of the backing plate. Seat the retaining tangs in the backing plate.

8. Connect the cable to the parking brake lever.

9. Install the brake drum, tire and wheel.

10. Connect the cable at the equalizer and connector.

11. Adjust the parking brake cable.

12. Lower the vehicle.

Anti-Lock Brake System Service

PRECAUTION

Failure to observe the following precautions may result in system damage.

• Before performing electric arc welding on the vehicle, disconnect the Electronic Brake Control Module (EBCM) and the hydraulic modulator connectors.

• When performing painting work on the vehicle, do not expose the Electronic Brake Control Module (EBCM) to temperatures in excess of 185°F (85°C) for longer than 2 hrs. The system may be exposed to temperatures up to 200°F (95°C) for less than 15 min.

• Never disconnect or connect the Electronic Brake Control Module (EBCM) or hydraulic modulator connectors with the ignition switch ON.

• Never disassemble any component of the Anti-Lock Brake System (ABS) which is designated non-serviceable; the component must be replaced as an assembly.

• When filling the master cylinder, always use Delco Supreme 11 brake fluid or equivalent, which meets DOT-3 specifications; petroleum base fluid will destroy the rubber parts.

RELIEVING ANTI-LOCK BRAKE SYSTEM PRESSURE

——— CAUTION ———
Failure to fully depressurize the accumulator before performing any repairs could result in injury, and/or damage to the system.

With the ignition switch in the **OFF** position, apply and release the brake pedal a minimum of 20 times using approximately 50 lbs. (222 N) of force on the pedal. A change in the pedal feel will occur when the accumulator is completely discharged.

Hydraulic Unit

REMOVAL & INSTALLATION

1. Depressurize the system, then disconnect the negative battery cable.

2. Disconnect the electrical connectors from the unit, then remove the fluid from the reservoir.

3. Remove the wire clip from the return hose fitting, then the return hose from the pump.

4. Remove the pressure hose attaching bolt, then the pressure hose and O-ring from the pump.

5. Remove the pump mounting bolt, then the energy unit from the hydraulic unit.

6. Disconnect the 4 brake lines from the valve block and hydraulic unit.

7. Disconnect the pushrod from the brake pedal, then push the dust boot forward off the rear half of the pushrod and unthread the 2 halfs of the pushrod.

8. Remove the hydraulic unit attaching bolts from the pushrod bracket, then the hydraulic unit from the vehicle.

To install:

9. Install the hydraulic unit to the pushrod bracket. Torque bolts to 37 ft. lbs. (50 Nm).

10. Thread the 2 halves of pushrod together, reposition the dust boot, then install pushrod to the brake pedal. Torque bolts to 27 ft. lbs. (37 Nm).

11. Connect the 4 brake lines to the valve block and hydraulic unit. Torque brake lines to 11 ft. lbs. (15 Nm).

12. Install energy unit to hydraulic unit, then the pump mounting bolt.

13. Install the pressure hose and O-ring to the pump, then the pressure hose bolt. Torque bolt to 15 ft. lbs.

14. Install return hose to pump, then the wire clip to the return hose fitting.

15. Connect electrical connectors to hydraulic unit, then the battery negative cable.

Valve Block

REMOVAL & INSTALLATION

1. Depressurize the system.
2. Remove the hydraulic unit.
3. Remove the valve block attaching nuts and bolts, then the valve block and O-rings from the vehicle.

To install:

4. Replace the O-ring. Install the valve block and attaching nuts and bolts. Tighten the valve block bolts to 18 ft. lbs. (25 Nm).
5. Install the hydraulic unit.
6. Bleed the brake system.

Pump Motor

REMOVAL & INSTALLATION

1. Depressurize system, then disconnect the battery negative cable.
2. Remove the brake fluid from the reservoir, then disconnect the electrical connectors from the pressure switch and the pump motor.
3. Remove the hydraulic accumulator and O-ring.
4. Remove the pressure hose attaching bolt, then the pressure hose and O-ring from pump.
5. Remove the wire clip and return hose fitting, then the return hose from the pump.
6. Remove the pump attaching bolts and grommets, then the pump from the hydraulic unit.

To install:

7. Install the pump to the hydraulic unit. Install the attaching bolts and grommets. Tighten the pump mounting bolts to 71 inch lbs. (8 Nm).
8. Connect the return hose to the pump. Install the wire clip.
9. Replace the pressure hose O-ring and install the pressure hose attaching

bolt. Tighten the pressure hose bolt to 15 ft. lbs. (20 Nm).

10. Replace the hydraulic accumulator O-ring and install the hydraulic accumulator. Tighten the accumulator to 17 ft. lbs. (23 Nm).

11. Connect the pump motor and pressure switch electrical connectors.

12. Fill the reservoir with brake fluid.

13. Bleed the brake system.

14. Connect the negative battery cable.

Pressure Switch

REMOVAL & INSTALLATION

1. Depressurize system, then disconnect the battery negative cable.
2. Disconnect the electrical connector from the pressure switch.
3. Using tool J–35804–A or equivalent, remove the pressure switch and O-ring.

To install:

4. Replace the pressure switch O-ring. Install the pressure switch. Tighten to 17 ft. lbs. (23 Nm).
5. Connect the pressure switch electrical connector.
6. Bleed the brake system.
7. Connect the negative battery cable.

Hydraulic Accumulator

REMOVAL & INSTALLATION

1. Depressurize system, then disconnect the battery negative cable.
2. Remove the hydraulic accumulator bolts, the hydraulic accumulator and O-ring from the vehicle.

To install:

3. Replace the accumulator O-ring.
4. Install the accumulator and at-

taching bolts. Tighten the accumulator bolts to 17 ft. lbs. (23 Nm).

5. Bleed the brake system.
6. Connect the negative battery cable.

Wheel Speed Sensor

REMOVAL & INSTALLATION

1. Raise and support vehicle, then remove the tire and wheel assembly.
2. Disconnect the speed sensor electrical connector.
3. Remove the speed sensor attaching bolt, disconnect the sensor and cable from brackets, then the sensor and cable from the vehicle.

To install:

4. Install the sensor and cable into the bracket. Install the attaching bolt. Tighten to 53 inch lbs. (6 Nm).
5. Connect the speed sensor electrical connector.
6. Install the tire and wheel assembly.
7. Lower the vehicle.

CHASSIS ELECTRICAL

Heater Blower Motor

REMOVAL & INSTALLATION

Without Air Conditioning

1. Disconnect the negative battery cable.

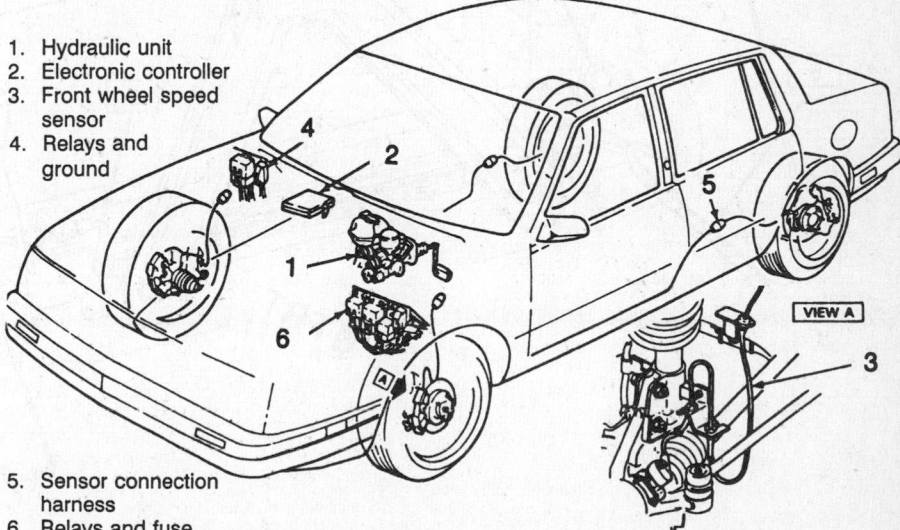

1. Hydraulic unit
2. Electronic controller
3. Front wheel speed sensor
4. Relays and ground
5. Sensor connection harness
6. Relays and fuse

Anti-Lock brake system components – 1990 Celebrity and 1990–92 6000

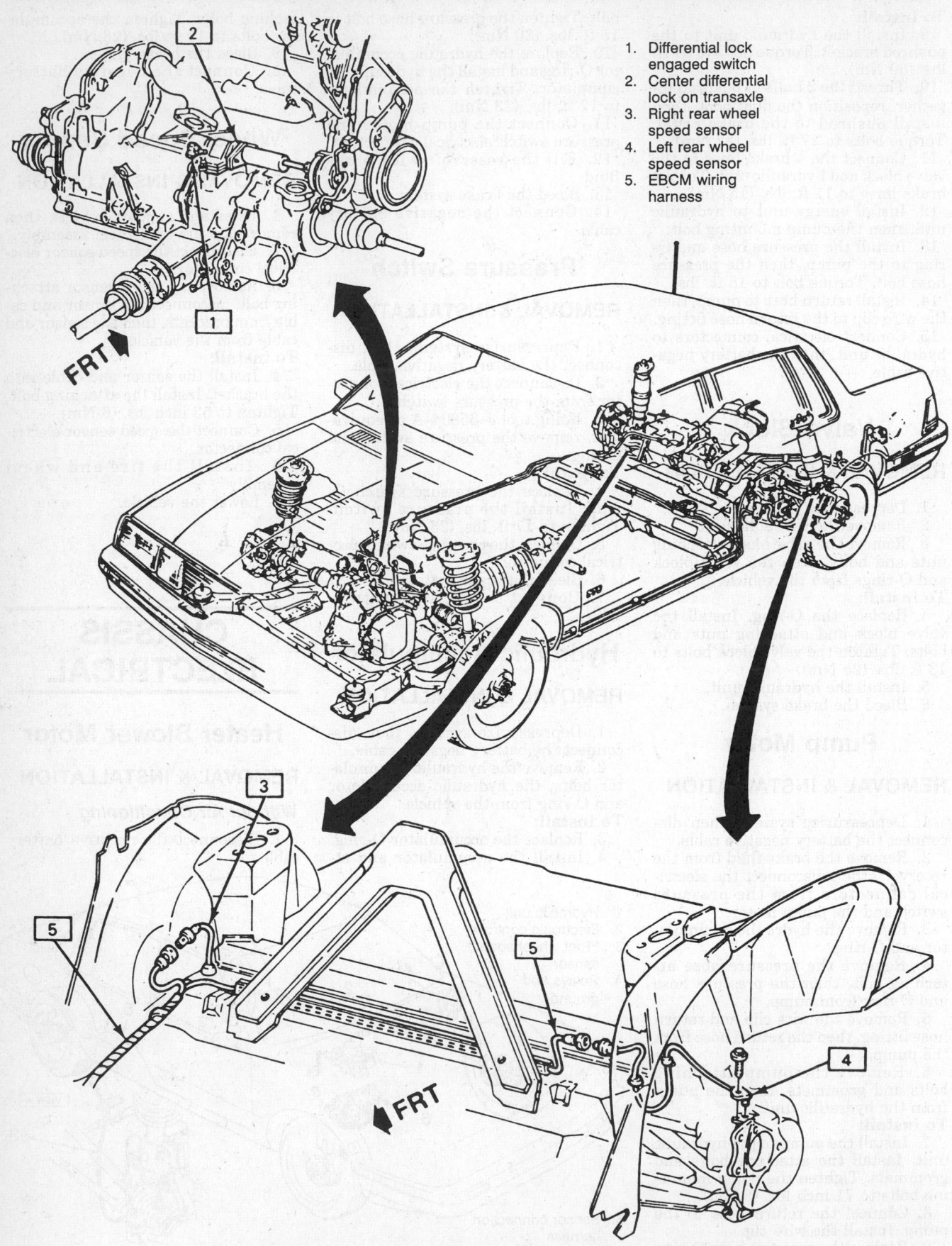

1. Differential lock engaged switch
2. Center differential lock on transaxle
3. Right rear wheel speed sensor
4. Left rear wheel speed sensor
5. EBCM wiring harness

Anti-lock brake system components—1990 Celebrity and 1990–92 6000

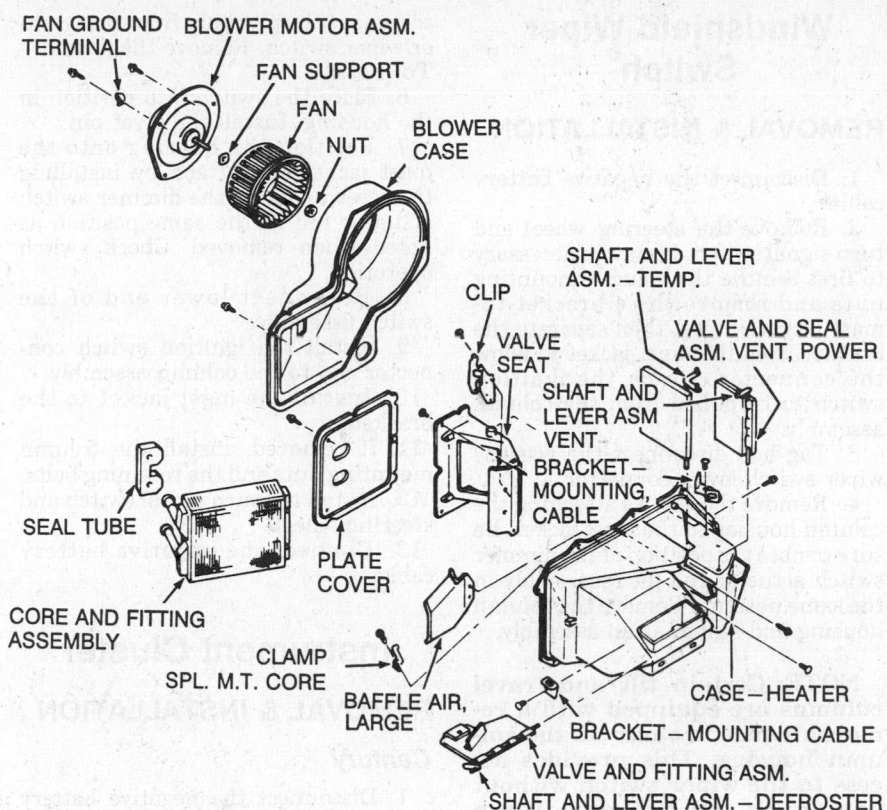

Heater core, blower motor and related components—without air conditioning

reaching through the plenum opening to hold the fan and insert the blower motor shaft into the fan.

9. Install the fan retaining nut to the blower motor shaft.

10. Install the blower motor retaining screws.

11. Connect the blower motor electrical connector and vent tube.

12. Install the cowl panel.

13. Install the wiper arms.

14. Connect the negative battery cable.

Windshield Wiper Motor

REMOVAL & INSTALLATION

1. Disconnect the negative battery cable.

2. Remove the wiper arms.

3. Remove the cowl cover.

4. Disconnect the wiper arm drive link from the crank arm.

5. Disconnect the wiper motor electrical connector.

6. Remove the wiper motor attaching bolts.

7. Remove the wiper motor, guiding the crank arm through the hole.

To install:

8. Insert the wiper motor, guiding the crank arm through the hole.

9. Install the wiper motor attaching bolts.

2. Tag and disconnect the blower motor electrical leads.

3. Remove the motor retaining bolts and remove the blower motor.

4. If the blower motor is to be replaced, separate the fan from the blower motor by removing the retaining nut and sliding the fan from the shaft.

To install:

5. If the blower motor is to be replaced, install the fan to the new blower motor and install the retaining nut.

6. Install the fan in the heater module and install the retaining bolts.

7. Connect the electrical connector.

8. Connect the negative battery cable.

With Air Conditioning

1. Disconnect the negative battery cable.

2. Remove the wiper arms.

3. Remove the cowl panel.

4. Disconnect the blower motor electrical connector and vent tube.

5. Remove the blower motor retaining screws.

6. Remove the fan retaining nut from the blower motor shaft by reaching through the plenum opening.

7. While reaching through the plenum opening, hold the fan to separate the blower motor from the fan and remove the blower motor from the air conditioner/heater module.

To install:

8. Install the blower motor to the air conditioner/heater module by

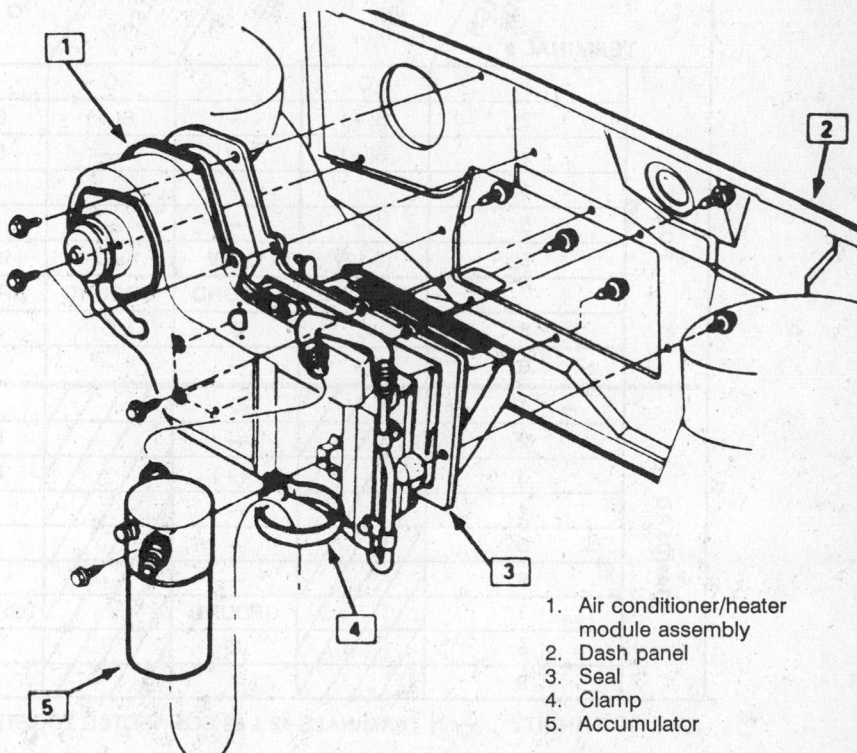

1. Air conditioner/heater module assembly
2. Dash panel
3. Seal
4. Clamp
5. Accumulator

Module assembly and accumulator—with air conditioning

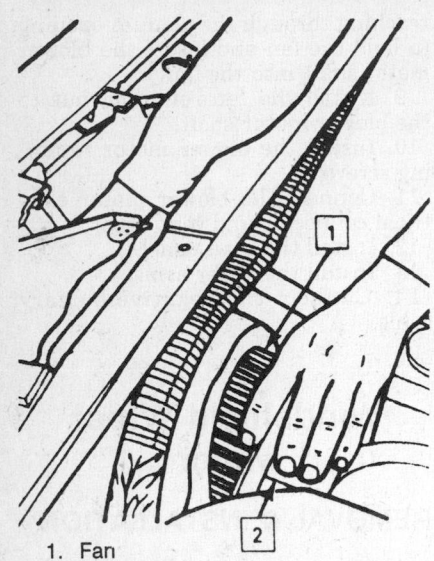

1. Fan
2. Blower motor

Blower motor removal—with air conditioning

10. Connect the electrical connectors.
11. Connect the wiper arm drive link to the crank arm.
12. Install the cowl cover.
13. Install the wiper arms.
14. Connect the negative battery cable.

Windshield Wiper Switch

REMOVAL & INSTALLATION

1. Disconnect the negative battery cable.
2. Remove the steering wheel and turn signal switch. It may be necessary to first remove the column mounting nuts and remove the 4 bracket-to-mast jacket screws, then separate the bracket from the mast jacket to allow the connector clip on the ignition switch to be pulled from the column assembly.
3. Tag and disconnect the washer/wiper switch lower connector.
4. Remove the screws attaching the column housing to the mast jacket. Be sure to note the position of the dimmer switch actuator rod for reassembly in the same position. Remove the column housing and switch as an assembly.

NOTE: Certain tilt and travel columns are equipped with a removable plastic cover on the column housing. This provides access to the wiper switch without removing the entire column housing.

5. Turn upside down and use a drift to remove the pivot pin from the washer/wiper switch. Remove the switch.
To install:
6. Place the switch into position in the housing. Install the pivot pin.
7. Position the housing onto the mast jacket and attach by installing the screws. Install the dimmer switch actuator rod in the same position as noted when removed. Check switch operation.
8. Reconnect lower end of the switch assembly.
9. Install the ignition switch connector clip to the column assembly.
10. Install the mast jacket to the bracket.
11. If removed, install the column mounting nuts and the retaining bolts.
12. Install the turn signal switch and steering wheel.
13. Connect the negative battery cable.

Instrument Cluster

REMOVAL & INSTALLATION

Century

1. Disconnect the negative battery cable.
2. Disconnect the speedometer cable and pull it through the firewall.

SWITCH MODE	TERMINAL #	MIST	OFF	PULSE	LO	HI †	WASH
PULSE	1	C	C	C	C	C	C
	2	B(+)	—	B(+)	B(+)	—	•B(+)
	3	B(+)	B(+)	—	B(+)	—	•B(+)
	4	—	—	—	—	—	—
	5	—	—	—	—	—	—
	6	10-12V	10-12V	10-12V	10-12V	10-12V	B(+)
	7	GROUND	GROUND	GROUND	GROUND	GROUND	GROUND
	8	C	C	C	C	C	C
	9	—	—	—	—	B(+)	—
STANDARD	1		C		C	C	C
	2		—		B(+)	—	•B(+)
	3		B(+)		B(+)	—	•B(+)
	4		—		—	—	—
	5		—		—	—	—
	6		—		—	—	B(+)
	7		GROUND		GROUND	GROUND	GROUND
	8		C		C	C	C
	9		—		—	B(+)	—

C = CONTINUITY   † TERMINALS #2 & #3 CONNECTED TOGETHER.   •EXCEPT ON HI.

Wiper-washer switch check chart

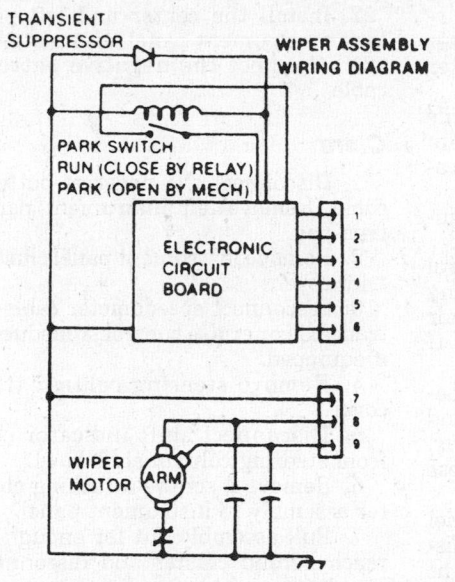

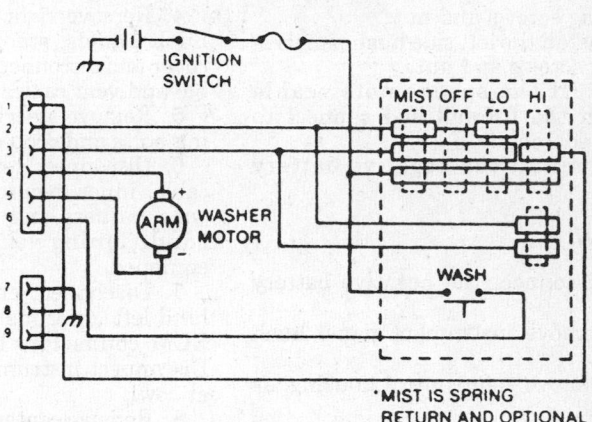

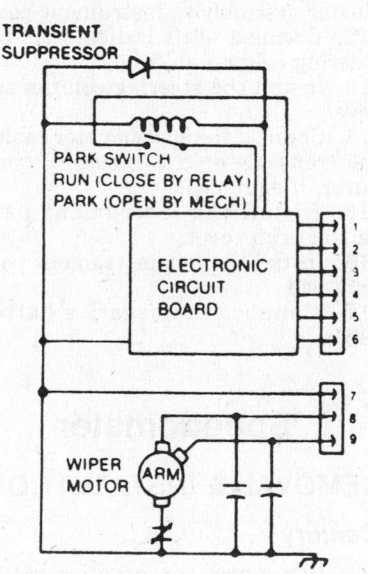

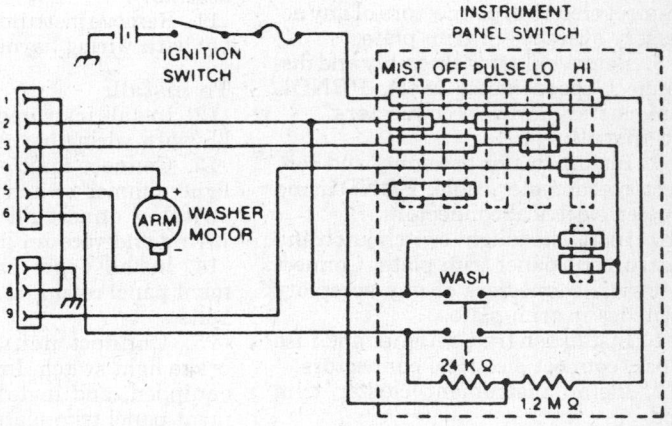

Wiper-washer wiring diagrams

3. Remove the left side hush panel retaining screws and nut.

4. Remove the right side hush panel retaining screws and nut.

5. Remove the shift indicator cable clip.

6. Remove the steering column trim plate.

7. Put the gear selector in **L**. Remove the retaining screws and gently pull out the instrument panel trim plate.

8. Disconnect the parking brake cable at the lever by pushing it forward and sliding it from its slot.

9. Unbolt and lower the steering column.

10. Remove the gauge cluster retaining screws. Pull the cluster out far enough to disconnect any wires. Remove the instrument cluster.

To install:

11. Install the gauge cluster, connect the electrical connectors and install the retaining screws.

12. Position the steering column and install the retaining bolts.

13. Connect the parking brake cable at the lever.

14. Put the gear selector in **L**. Install

the instrument panel trim plate.

15. Install the steering column trim plate.

16. Install the shift indicator cable clip.

17. Install the right side hush panel retaining screws and nut.

18. Install the left side hush panel retaining screws and nut.

19. Pull the speedometer cable through the firewall and connect to the speedometer.

20. Connect the negative battery cable.

Celebrity

1. Disconnect the negative battery cable.

2. Remove instrument panel hush panel.

3. Remove vent control housing, as required.

4. On non-air conditioning vehicles, remove steering column trim cover screws and lower cover with vent cables attached. On air conditioning vehicles, remove trim cover attaching screws and remove cover.

5. Remove instrument cluster trim pad.

6. Remove ash tray, retainer and fuse block, disconnect wires as necessary.

7. Remove headlight switch knob and instrument panel trim plate. Disconnect electrical connectors of any accessory switches in trim plate.

8. Remove cluster assembly and disconnect speedometer cable, **PRNDL** and cluster electrical connectors.

To install:

9. Install cluster assembly and connect speedometer cable, **PRNDL** and cluster electrical connectors.

10. Install headlight switch knob and instrument panel trim plate. Connect electrical connectors of any accessory switches in trim plate.

11. Install ash tray, retainer and fuse block, connect electrical connectors.

12. Install instrument cluster trim pad.

13. On non-air conditioned vehicles, raise the cover with vent cables attached and install steering column trim cover screws. On air conditioned vehicles, install trim cover and attaching screws.

14. If removed, install vent control housing.

15. Install instrument panel hush panel.

16. Connect the negative battery cable.

6000

1. Disconnect the negative battery cable, and remove the center and left side lower instrument panel trim plate.

2. Remove the screws holding the instrument cluster to the instrument panel carrier.

3. Remove the instrument cluster lens to gain access to the speedometer head and gauges.

4. Remove right side and left side hush panels, steering column trim cover and disconnect parking brake cable and vent cables, if equipped.

5. Remove steering column retaining bolts and drop steering column.

6. Disconnect temperature control cable, inner-to-outer air conditioning wire harness and inner-to-outer air conditioning vacuum harness, if equipped.

7. Disconnect chassis harness behind left lower instrument panel and ECM connectors behind glove box. Disconnect instrument panel harness at cowl.

8. Remove center instrument panel trim plate, radio, if equipped, and disconnect neutral switch and brake light switch.

9. Remove upper and lower instrument panel retaining screws, nuts and bolts.

10. Pull instrument panel assembly out far enough to disconnect ignition switch, headlight dimmer switch and turn signal switch. Disconnect all other accessory wiring and vacuum lines necessary to remove instrument panel assembly.

11. Remove instrument panel assembly with wiring harness.

To install:

12. Install instrument panel assembly with wiring harness.

13. Connect ignition switch, headlight dimmer switch and turn signal switch. Connect all other accessory wiring and vacuum lines.

14. Install upper and lower instrument panel retaining screws, nuts and bolts.

15. Connect neutral switch and brake light switch. Install the radio, if equipped, and install center instrument panel trim plate.

16. Connect chassis harness behind left lower instrument panel and ECM connectors behind glove box. Connect instrument panel harness at cowl.

17. Connect temperature control cable, inner-to-outer air conditioning wire harness and inner-to-outer air conditioning vacuum harness, if equipped.

18. Raise the steering column and install retaining bolts.

19. Install right side and left side hush panels, steering column trim cover and connect parking brake cable and vent cables, if equipped.

20. Install the instrument cluster lens.

21. Install the screws holding the instrument cluster to the instrument panel carrier.

22. Install the center and left side lower instrument panel trim plate.

23. Connect the negative battery cable.

Ciera

1. Disconnect the negative battery cable. Remove left instrument panel trim pad.

2. Remove instrument panel cluster trim cover.

3. Disconnect speedometer cable at transaxle or cruise control transducer, if equipped.

4. Remove steering column trim cover.

5. Disconnect shift indicator clip from steering column shift bowl.

6. Remove 4 screws attaching cluster assembly to instrument panel.

7. Pull assembly out far enough to reach behind cluster and disconnect speedometer cable.

8. Remove cluster assembly.

To install:

9. Install the cluster assembly.

10. Connect the speedometer cable.

11. Install the 4 screws attaching cluster assembly to instrument panel.

12. Connect shift indicator clip to steering column shift bowl.

13. Install the steering column trim cover.

14. Connect the speedometer cable at the transaxle or cruise control transducer, if equipped.

15. Install the instrument panel cluster trim cover.

16. Install the left instrument panel trim pad.

17. Connect the negative battery cable.

Speedometer

REMOVAL & INSTALLATION

Century

1. Disconnect the negative battery cable.

2. Remove the left side trim plate.

3. Remove the instrument cluster housing screws. Remove the instrument cluster. If the vehicle is equipped with tilt-wheel steering, working room can be gained by removing the tilt-wheel cover.

4. Remove the speedometer lens screws and remove the speedometer lens.

5. Disconnect the speedometer cable by pushing in on the retaining clip and pulling back on the cable.

6. Remove the screws holding the speedometer to the instrument and remove the speedometer assembly.

To install:

7. Install the speedometer assembly and the screws holding the speedometer assembly to the instrument panel.

8. Connect the speedometer cable.

9. Install the speedometer lens and retaining screws.

10. Install the instrument cluster housing and retaining screws. If equipped with tilt-wheel steering and the tilt-wheel cover was removed, replace the cover.

11. Install the left side trim plate.

12. Connect the negative battery cable.

Celebrity

1. Disconnect the negative battery cable.

2. Remove the cluster trim panel.

3. Remove the cluster lens screws. Remove the cluster lens.

4. Remove the speedometer-to-cluster attaching screws. Remove the speedometer from the instrument cluster.

5. Disconnect the speedometer cable and remove the speedometer assembly.

To install:

6. Position the speedometer assembly and connect the speedometer cable.

7. Install the speedometer-to-instrument cluster and install the attaching screws.

8. Install the cluster lens and attaching screws.

9. Install the cluster trim panel.

10. Connect the negative battery cable.

Ciera

1. Disconnect the negative battery cable.

2. Remove the instrument cluster assembly.

3. Remove the vehicle speed sensor screw from the rear of the speedometer. Remove the vehicle speed sensor, if equipped.

4. Remove the speedometer lens screws and remove the speedometer lens. Remove the bezel.

5. Remove the screw that holds the speedometer to the instrument cluster.

6. Remove the speedometer head by pulling forward. Disconnect the speedometer cable by prying gently on the retainer and pulling the speedometer cable from the speedometer head.

To install:

7. Position the speedometer head and connect the speedometer cable.

8. Install the screw that holds the speedometer to the instrument cluster.

9. Install the speedometer lens and attaching screws. Install the bezel.

10. Install the vehicle speed sensor and the attaching screw at the rear of the speedometer.

11. Install the instrument cluster assembly.

12. Connect the negative battery cable.

6000

1. Disconnect the negative battery cable.

2. Remove the center and left lower trim plates.

3. Remove the screws holding the instrument cluster assembly to the dash assembly. Remove the instrument cluster.

4. Remove the instrument cluster lens screws. Remove the instrument cluster lens.

5. Remove the screws holding the speedometer to the instrument cluster. Remove the speedometer.

6. Disconnect the speedometer cable from the rear of the speedometer.

To install:

7. Connect the speedometer cable at the rear of the speedometer.

8. Install speedometer and the screws holding the speedometer to the instrument cluster.

9. Install the instrument cluster lens and attaching screws.

10. Install the instrument cluster and the screws holding the instrument cluster assembly to the dash assembly.

11. Install the center and left lower trim plates.

12. Connect the negative battery cable.

Radio

REMOVAL & INSTALLATION

1. Disconnect battery negative cable.

2. Remove the accessory trim plate.

3. Remove the instrument cluster trim plate.

4. Remove the radio attaching bolts, then disconnect electrical connectors.

5. Remove radio from vehicle.

To install:

6. Install the radio in the vehicle.

7. Connect the electrical connectors. Install the radio attaching bolts.

8. Install the instrument cluster trim plate.

9. Install the accessory trim plate.

10. Connect the negative battery cable.

Headlight Switch

REMOVAL & INSTALLATION

Century

1. Disconnect the negative battery cable.

2. Remove the instrument panel trim plate.

3. Remove the left side instrument panel switch trim panel by removing the 3 screws and gently rocking the panel out.

4. Remove the three screws and pull the switch straight out.

To install:

5. Install the switch and the 3 attaching screws.

6. Install the left side instrument panel switch trim panel and 3 attaching screws.

7. Install the instrument panel trim plate.

8. Connect the negative battery cable.

Celebrity

1. Disconnect the negative battery cable.

2. Remove the headlight switch knob.

3. Remove the instrument panel trim pad.

4. Unbolt the switch mounting plate from the instrument panel carrier.

5. Disconnect the wiring from the switch.

6. Remove the switch.

To install:

7. Install the switch.

8. Connect the wiring to the switch.

9. Install the bolts attaching the switch mounting plate to the instrument panel carrier.

10. Install the instrument panel trim pad.

11. Install the headlight switch knob.

12. Connect the negative battery cable.

Ciera

1. Disconnect the negative battery cable.

2. Remove the left side instrument panel trim pad.

3. Unbolt the switch from the instrument panel.

4. Pull the switch rearward and remove it.

To install:

5. Install the switch and connect the electrical connectors.

6. Install the bolts attaching the switch to the instrument panel.

7. Install the left side instrument panel trim pad.

8. Connect the negative battery cable.

6000

1. Disconnect the negative battery cable.

2. Remove the steering column trim cover and headlight rod and knob by

reaching behind the instrument panel and depressing the lock tab.

3. Remove the left instrument panel trim plate.

4. Unbolt and remove the switch and bracket assembly from the instrument panel.

5. Loosen the bezel and remove the switch from the bracket.

To install:

6. Install the switch to the bracket and install the bezel.

7. Install the switch and bracket assembly to the instrument panel and install the attaching bolts.

8. Install the left instrument panel trim plate.

9. Install the headlight rod and knob. Install the steering column trim cover.

10. Connect the negative battery cable.

Dimmer Switch

REMOVAL & INSTALLATION

1. Disconnect the negative battery cable.

2. Remove the steering wheel. Remove the trim cover.

3. Remove the turn signal switch assembly.

4. Remove the ignition switch stud and screw. Remove the ignition switch.

5. Remove the dimmer switch actuator rod by sliding it from the switch assembly.

6. Remove the dimmer switch bolts and remove the dimmer switch.

To install:

7. Install the dimmer switch and attaching bolts.

8. Install the dimmer switch actuator rod by sliding it into the switch assembly.

9. Adjust the dimmer switch by depressing the switch slightly and inserting a $^3/_{32}$ in. drill bit into the adjusting hole. Push the switch up to remove any play and tighten the dimmer switch adjusting screw.

10. Install the ignition switch, stud and screw.

11. Install the turn signal switch assembly.

12. Install the trim cover. Install the steering wheel.

13. Connect the battery negative cable.

Combination Switch

REMOVAL & INSTALLATION

1. Disconnect the negative battery cable. Remove the steering wheel and trim cover.

2. Loosen the cover screws. Pry the cover upward and remove it from the shaft.

3. Position U-shaped lock plate compressing tool J–23653–C on the end of the steering shaft and compress the lockplate by turning the shaft nut clockwise. Pry the wire snapring from the shaft groove.

4. Remove the tool and lift the lock plate off the shaft.

5. Slip the cancelling cam, upper bearing preload spring and thrust washer off the shaft.

6. Remove the turn signal lever. Push the flasher knob in and unscrew it. Remove the button retaining screw and remove the button, spring and knob.

7. Pull the switch connector out the mast jacket and tape the upper part to facilitate switch removal. Attach a long piece of wire to the turn signal switch connector. When installing the turn signal switch, feed this wire through the column first, and then use this wire to pull the switch connector into position. If equipped with tilt-wheel, place the turn signal and shifter housing in the lowest position and remove the harness cover.

8. Remove the 3 switch mounting screws. Remove the switch by pulling it straight up while guiding the wiring harness cover through the column.

To install:

9. Install the replacement switch by working the connector and cover down through the housing and under the bracket. If equipped with tilt-wheel, work the connector down through the housing, under the bracket and install the harness cover.

10. Install the switch mounting screws and the connector on the mast jacket bracket. Install the column-to-dash trim plate.

11. Install the flasher knob and the turn signal lever.

12. With the turn signal lever in the middle position and the flasher knob out, slide the thrust washer, upper bearing preload spring and cancelling cam onto the shaft.

13. Position the lock plate on the shaft and press it down until a new snapring can be inserted in the shaft groove. Always use a new snapring when assembling.

14. Install the cover and the steering wheel. Connect the battery negative cable.

Ignition Lock

REMOVAL & INSTALLATION

1. Disconnect the negative battery cable. Place the lock in the **RUN** position. Remove the steering wheel.

2. Remove the lock plate, turn signal switch and buzzer switch.

3. Remove the screw and lock cylinder.

NOTE: Be careful not to drop the screw which could fall into the column assembly requiring complete disassembly of the column to retrieve the screw.

To install:

4. Rotate the cylinder clockwise to align cylinder key with the keyway in the housing.

5. Push the lock all the way in.

6. Install the screw. Tighten the screw to 14 inch lbs. for adjustable columns or 25 inch lbs. for standard columns. Connect battery negative cable.

Ignition Switch

REMOVAL & INSTALLATION

The switch is located inside the channel section of the brake pedal support and is accessible only by lowering the steering column. The switch is actuated by a rod and rack assembly. A gear on the end of the lock cylinder engages the toothed upper end of the rod.

1. Disconnect the negative battery cable. Lower and properly support the steering column.

2. Put the ignition switch in the **OFF–UNLOCKED** position. With the cylinder removed, the rod is in the **LOCK** position when it is in the next to the uppermost detent. **OFF–UNLOCKED** position is 2 detents from the top.

3. Remove the 2 switch screws and remove the switch assembly.

To install:

4. Before installing, place the new switch in **OFF–UNLOCKED** position

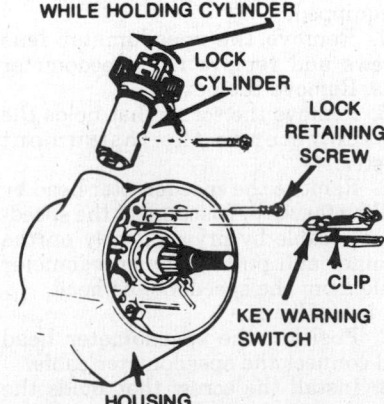

TO ASSEMBLE, ROTATE TO STOP WHILE HOLDING CYLINDER

LOCK CYLINDER

LOCK RETAINING SCREW

CLIP

KEY WARNING SWITCH

HOUSING

Mounting of the ignition lock cylinder assembly, removal of the key warning switch, is shown in the inset

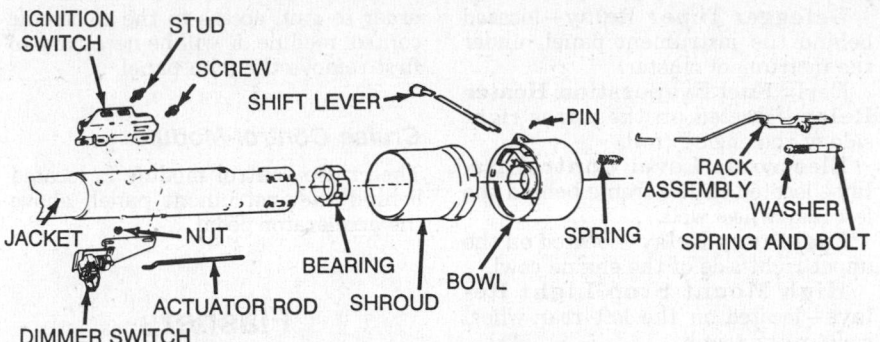

Installation of the ignition switch

and make sure the lock cylinder and actuating rod are in **OFF-UN-LOCKED**, third detent from the top.

5. Install the activating rod into the switch and assemble the switch on the column. Tighten the mounting screws. Use only the specified screws since over-length screws could impair the collapsibility of the column.

6. Reinstall the steering column. Connect battery negative cable.

Stoplight Switch

ADJUSTMENT

1. The switch is mounted on the brake pedal bracket.

2. To adjust, depress the pedal and push the switch through the circular retaining clip until it contacts the brake pedal, then pull the pedal up against the internal pedal stop. This places the switch in the correct position within the clip.

REMOVAL & INSTALLATION

1. Disconnect the negative battery cable. Disconnect the electrical connector to the switch.

2. Remove the switch from the brake pedal bracket.

To install:

3. Install the new switch into the bracket.

4. Connect the electrical connector.

5. Adjust the switch. Connect battery negative cable.

Clutch Switch

ADJUSTMENT

1. Lift the clutch pedal to its uppermost position. Check the operation of the pawl and the quadrant. Make sure the pawl disengages from the quadrant when the pedal is pulled to this position.

2. Check the quadrant for free rotation in both directions.

3. Depress the clutch pedal slowly several times to set the pawl into mesh with the quadrant teeth.

REMOVAL & INSTALLATION

1. Disconnect the negative battery cable. Support the clutch pedal against the bumper stop in order to release the pawl from the quadrant. Disconnect the clutch cable from the release lever at the transaxle assembly.

—————— CAUTION ——————

Be careful to prevent the cable from snapping towards the rear of the vehicle possibly causing bodily injury. The quadrant in the adjusting mechanism can also be damaged by allowing the cable to snap rearward.

———————————————

2. From inside the vehicle, disconnect the clutch cable from the quadrant. Lift the locking pawl away from the quadrant. Slide the cable away from the pedal along the right side of the quadrant.

3. Remove the clutch switch from the pedal. Remove the pedal pivot nut, bolt and clutch pedal from the mounting bracket.

4. Note the position of the adjusting mechanism, the pawl and quadrant springs. Remove the E-ring.

5. Inspect the components for tooth damage and replace any components found to be defective.

To install:

6. Install the clutch switch to the pedal. Install the E-ring, pedal pivot nut, bolt and clutch pedal to the mounting bracket.

7. Connect the clutch cable to the quadrant.

8. Connect the clutch cable to the release lever at the transaxle assembly.

9. Connect the negative battery cable.

10. Check the clutch operation and adjust by lifting the clutch pedal up to

allow the mechanism to adjust the cable length. Depress the pedal slowly several times to set the pawl into mesh with the quadrant teeth.

Neutral Safety Switch

ADJUSTMENT

1. After the switch is installed, move the housing towards the **L** gear position.

2. Shift the gear selector into the **P** position.

3. The main housing and the housing back should ratchet. This will provide proper switch adjustment.

4. Repeat if necessary.

REMOVAL & INSTALLATION

Vehicles With Console Shift

1. New switches include a small plastic alignment pin. Leave this pin in place. Position the shifter assembly in **N**.

2. Disconnect the negative battery cable. Remove the old switch and install the replacement, align the pin on the shifter with the slot in the switch and fasten with the 2 screws.

3. Move the shifter from the **N** position. This shears the plastic alignment pin and frees the switch.

4. If the switch is to be adjusted, insert a $\frac{3}{32}$ in. drill bit or similar size pin and align the hole switch. Position switch, adjust as necessary. Remove the pin before shifting from **N**. Connect battery negative cable.

Vehicles With Column Shift

1. Disconnect the negative battery cable. Disconnect the electrical connectors from the combination back-up and neutral safety switch.

2. Remove the 2 screws attaching the switch to the steering column.

3. Remove the switch.

To install:

4. Install the new switch and 2 attaching screws.

5. Adjust the switch by performing the following:

   a. Position the shift lever in **N**.

   b. Loosen the attaching screws. Install a 0.090 in. gauge pin into the outer hole in the switch cover.

   c. Rotate the switch until the pin goes into the alignment hole in the inner plastic slide.

   d. Tighten the switch-to-column attaching screws and remove the gauge pin. Torque the screws to 20 inch lbs. maximum.

6. Connect battery negative cable.

7. Make sure the engine starts only in the **P** and **N** positions.

Fuses, Circuit Breakers and Relays

LOCATION

Fusible Links

There are several locations where fusible links can be found. They are located ahead of the left side front shock tower, near the positive battery connection or at the starter solenoid near the front of the engine.

Circuit Breakers

Circuit breakers are used along with the fusible links to protect the various components of the electrical system, such as headlights, the windshield wipers and electric windows. The circuit breakers are located either in the switch or mounted on or near the lower lip of the instrument panel, to the right or left of the steering column.

Fuse Panel

The fuse panel is located on the left side of the vehicle. It is under the instrument panel assembly. In order to gain access to the fuse panel, it may be necessary to first remove the under dash padding.

Relays

CELEBRITY, CIERA AND 6000

Air Conditioner Compressor Relay—located on the upper right corner of the engine cowl.

Air Conditioner Delay Relay—located in the upper right corner of the engine cowl.

Air Conditioner/Heater Blower Relay—located on the plenum, on the right side of the firewall.

Altitude Advance Relay—located on the left inner fender, in front of the shock tower.

Charging System Relay—located behind the instrument panel, near the fuse block.

Constant Run Relay—located on the left inner fender wheel well.

Coolant Fan Low-Speed Relay—located on the left inner fender wheel well, on a bracket on the 2.5L engine and on the fender panel in front of the left front shock tower on all except 2.5L engine.

Coolant Fan Relay—located on the left front wheel well on the bracket on the 2.5L engine and on the fender panel ahead of the left front shock tower on all except 2.5L engine.

Defogger Timer Relay—located behind the instrument panel, under the instrument cluster.

Early Fuel Evaporation Heater Relay—located on the upper right side of the engine cowl.

Electronic Level Control Relay—located on the frame behind the left rear wheel well.

Fuel Pump Relay—located on the upper right side of the engine cowl.

High Mount Stop Light Relays—located on the left rear wheel well, in the trunk.

Horn Relay—located on the convenience center.

Low Brake Vacuum Relay—taped to the instrument panel above the fuse block.

Rear Wiper Relay—located in the top center of the tailgate.

Starter Interrupt Relay—located above the ashtray, taped to the instrument panel harness.

CENTURY

Air Conditioner Coolant Fan Relay (2.5L engine)—located on the right side of the firewall.

Blower Relay—located on the right side of the firewall.

Coolant Fan Delay Relay (SFI)—located in front of the left front shock tower, on a bracket.

Coolant Fan Relay—located in front of the left front shock tower.

Fuel Pump Relay (2.5L engine)—located in the relay bracket on the right side of the firewall.

High Speed Coolant Fan Relay—located on the left front side of the engine.

Horn Relay—located under the instrument panel, in the convenience center.

Low Speed Coolant Fan Relay—located near the battery, on the left side of the radiator shroud.

Rear Wiper Relay—located in the top center of the tailgate.

Starter Interrupt Relay—taped to the instrument panel harness, above the right side ashtray.

Computers

LOCATION

Electronic Control Module (ECM)

The electronic module is located on the right side of the vehicle. It is positioned under the instrument panel. In order to gain access to the electronic control module, it will be necessary to first remove the trim panel.

Cruise Control Module

The cruise control module is located behind the instrument panel, above the accelerator pedal.

Flashers

LOCATION

Hazard

The hazard flasher is located in the convenience center. The convenience center is a swing down type, located under the instrument panel near the fuse block.

Turn Signal

The turn signal flasher is located behind the instrument panel, to the right of the steering column.

Cruise Control

NOTE: To keep the vehicle under control and to prevent possible vehicle damage, it is not advisable to use the cruise control on slippery roads. Disengage the cruise control in conditions such as varying or heavy traffic or when traveling down a steep graded hill.

ADJUSTMENTS

1. Adjust the throttle lever to the idle position with the engine **OFF**. If equipped with the idle control solenoid, the solenoid must be de-energized.

2. Pull the servo assembly end of the cable towards the servo blade.

3. Align the holes in the servo blade with the cable pin. Install the cable pin.

4. If equipped with the 2.8L engine, it will be necessary to position the ball of the chain assembly into the chain retainer. This will allow a slight slack to occur not to exceed one ball diameter. Remove the excess chain outside of the chain retainer.

GM "C" & "H" Body

Front Wheel Drive

Buick—Electra, LeSabre, Park Avenue
Cadillac—DeVille, Fleetwood
Oldsmobile—Delta 88, Ninety-Eight
Pontiac—Bonneville

SPECIFICATIONS

VEHICLE IDENTIFICATION CHART

It is important for servicing and ordering parts to be certain of the vehicle and engine identification. The VIN (vehicle identification number) is a 17 digit number visible through the windshield on the driver's side of the dash and contains the vehicle and engine identification codes. The tenth digit indicates model year and the eighth digit indicates engine code. It can be interpreted as follows:

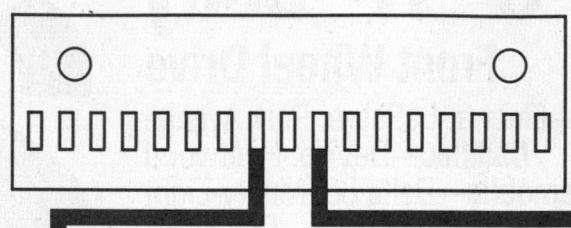

Engine Code						Model Year	
Code	Cu. In.	Liters	Cyl.	Fuel Sys.	Eng. Mfg.	Code	Year
3 ('88)	231	3.8	6	SFI	Buick	J	1988
C	231	3.8	6	SFI	Buick	K	1989
L	231	3.8	6	SFI	Buick	L	1990
5	273	4.5	8	TBI	Cadillac	M	1991
3 ('90)	273	4.5	8	PFI	Cadillac	N	1992
B	300	4.9	8	PFI	Cadillac		

PFI—Port Fuel Injection
SFI—Sequential Fuel Injection
TBI—Throttle Body Injection

GENERAL ENGINE SPECIFICATIONS

Year	VIN	No. Cylinder Displacement cu. in. (liter)	Fuel System Type	Net Horsepower @ rpm	Net Torque @ rpm (ft. lbs.)	Bore × Stroke (in.)	Compression Ratio	Oil Pressure @ rpm
1988	3	6-231 (3.8)	SFI	150 @ 4400	200 @ 2000	3.800 × 3.400	8.5:1	37 @ 2000
	C	6-231 (3.8)	SFI	165 @ 5200	210 @ 2000	3.800 × 3.400	8.5:1	37 @ 2400
	5	8-273 (4.5)	TBI	155 @ 4000	240 @ 2800	3.622 × 3.307	9.0:1	37 @ 1500
1989	C	6-231 (3.8)	SFI	165 @ 5200	210 @ 2000	3.800 × 3.400	8.5:1	37 @ 2400
	5	8-273 (4.5)	TBI	155 @ 4000	240 @ 2800	3.622 × 3.307	9.0:1	37 @ 1500
1990	C	6-231 (3.8)	SFI	165 @ 5200	210 @ 2000	3.800 × 3.400	8.5:1	40 @ 1850
	3	8-273 (4.5)	PFI	180 @ 4300	245 @ 3000	3.622 × 3.307	9.5:1	37 @ 1500
1991-92	C	6-231 (3.8)	SFI	165 @ 4800	210 @ 2000	3.800 × 3.400	8.5:1	60 @ 1850
	L	6-231 (3.8)	SFI	170 @ 4800	220 @ 3200	3.800 × 3.400	8.5:1	60 @ 1850
	B	8-300 (4.9)	PFI	200 @ 4100	275 @ 3000	3.623 × 3.623	9.5:1	53 @ 2000

PFI—Port Fuel Injection
SFI—Sequential Fuel Injection
TBI—Throttle Body Injection

ENGINE IDENTIFICATION

Year	Model	Engine Displacement cu. in. (liter)	Engine Series Identification (VIN)	No. of Cylinders	Engine Type
1988	DeVille	273 (4.5)	5	8	OHV
	Fleetwood	273 (4.5)	5	8	OHV
	Electra	231 (3.8)	C	6	OHV
	Park Avenue	231 (3.8)	C	6	OHV
	LeSabre	231 (3.8)	3	6	OHV
	Ninety Eight	231 (3.8)	C	6	OHV
	Delta 88	231 (3.8)	3	6	OHV
	Bonneville	231 (3.8)	3	6	OHV
	Bonneville	231 (3.8)	C	6	OHV
1989	DeVille	273 (4.5)	5	8	OHV
	Fleetwood	273 (4.5)	5	8	OHV
	Electra	231 (3.8)	C	6	OHV
	Park Avenue	231 (3.8)	C	6	OHV
	LeSabre	231 (3.8)	C	6	OHV
	Ninety Eight	231 (3.8)	C	6	OHV
	Delta 88	231 (3.8)	C	6	OHV
	Bonneville	231 (3.8)	C	6	OHV
1990	DeVille	273 (4.5)	3	8	OHV
	Fleetwood	273 (4.5)	3	8	OHV
	Electra	231 (3.8)	C	6	OHV
	Park Avenue	231 (3.8)	C	6	OHV
	LeSabre	231 (3.8)	C	6	OHV
	Ninety Eight	231 (3.8)	C	6	OHV
	Delta 88	231 (3.8)	C	6	OHV
	Bonneville	231 (3.8)	C	6	OHV
1991-92	DeVille	300 (4.9)	B	8	OHV
	Fleetwood	300 (4.9)	B	8	OHV
	Park Avenue	231 (3.8)	L	6	OHV
	LeSabre	231 (3.8)	C	6	OHV
	Ninety Eight	231 (3.8)	L	6	OHV
	88 Royale	231 (3.8)	L	6	OHV
	Bonneville	231 (3.8)	C	6	OHV

OHV—Overhead Valve

GASOLINE ENGINE TUNE-UP SPECIFICATIONS

Year	VIN	No. Cylinder Displacement cu. in. (liter)	Spark Plugs Type	Spark Plugs Gap (in.)	Ignition Timing (deg.) MT	Ignition Timing (deg.) AT	Compression Pressure (psi)	Fuel Pump (psi)	Speed (rpm) MT	Speed (rpm) AT	Valve Clearance In.	Valve Clearance Ex.
1988	3	6-231 (3.8)	R44LTS	0.045	①	①	②	34–40	①	①	Hyd.	Hyd.
	C	6-231 (3.8)	R44LTS6	0.060	①	①	②	40–47	①	①	Hyd.	Hyd.
	5	8-273 (4.5)	R44LTS6	0.060	①	①	140–165	9–12	①	①	Hyd.	Hyd.
1989	C	6-231 (3.8)	R44LTS6	0.060	①	①	②	40–47	①	①	Hyd.	Hyd.
	5	8-273 (4.5)	R44LTS6	0.060	①	①	140–165	9–12	①	①	Hyd.	Hyd.

GASOLINE ENGINE TUNE-UP SPECIFICATIONS

Year	VIN	No. Cylinder Displacement cu. in. (liter)	Spark Plugs Type	Gap (in.)	Ignition Timing (deg.) MT	AT	Compression Pressure (psi)	Fuel Pump (psi)	Speed (rpm) MT	AT	Valve Clearance In.	Ex.
1990	C	6-231 (3.8)	R44LTS6	0.060	①	①	②	40–47	①	①	Hyd.	Hyd.
	3	8-273 (4.5)	R44LTS6	0.060	①	①	140–165	40–47	①	①	Hyd.	Hyd.
1991	C	6-231 (3.8)	R44LTS6	0.060	①	①	②	40–47	①	①	Hyd.	Hyd.
	L	6-231 (3.8)	R44LTS6	0.060	①	①	②	40–47	①	①	Hyd.	Hyd.
	B	8-300 (4.9)	R45LTS6K	0.060	①	①	140–165	40–50	①	①	Hyd.	Hyd.
1992		SEE UNDERHOOD STICKER SPECIFICATIONS										

① These vehicles are equipped with computerized emissions systems which have no distributor vacuum advance unit. The idle speed and ignition timing are controlled by the ECM/PCM.

② The lowest cylinder reading should be no less than 70% of the highest and no cylinder should be less than 100 P.S.I.
Hyd.—Hydraulic

FIRING ORDERS

NOTE: To avoid confusion, always replace spark plug wires one at a time.

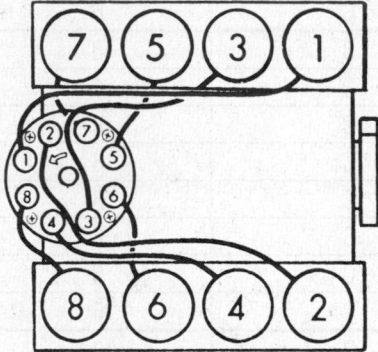

4.5L and 4.9L Engines
Engine Firing Order: 1–8–4–3–6–5–7–2
Distributor Rotation: Counterclockwise

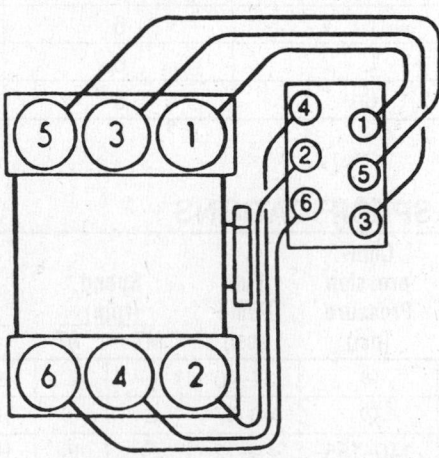

3.8L Engine VIN C and L
Engine Firing Order: 1–6–5–4–3–2
Distributorless Ignition System

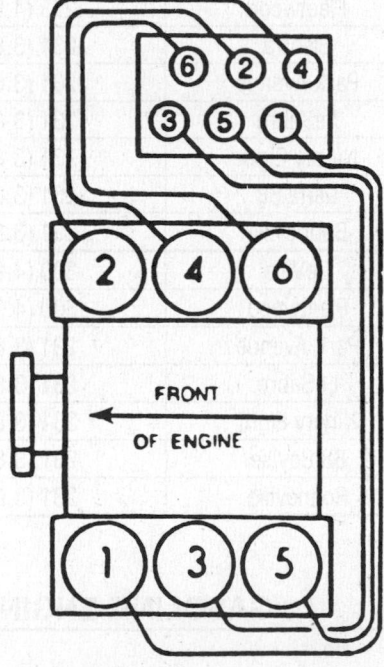

3.8L Engine VIN 3
Engine Firing Order: 1–6–5–4–3–2
Distributorless Ignition System

CAPACITIES

Year	Model	VIN	No. Cylinder Displacement cu. in. (liter)	Engine Crankcase (qts.) with Filter	Engine Crankcase (qts.) without Filter	Transmission (pts.) 4-Spd	Transmission (pts.) 5-Spd	Transmission (pts.) Auto.	Drive Axle (pts.)	Fuel Tank (gal.)	Cooling System (qts.)
1988	DeVille	5	8-273 (4.5)	5.5	5.0	—	—	22④	—	18	13.0③
	Fleetwood	5	8-273 (4.5)	5.5	5.0	—	—	22④	—	18	13.0③
	Electra	C	6-231 (3.8)	4.0①	4.0	—	—	22④	—	18	13.0
	Park Avenue	C	6-231 (3.8)	4.0①	4.0	—	—	22④	—	18	13.0
	LeSabre	3	6-231 (3.8)	4.0①	4.0	—	—	22④	—	18	13.0
	Ninety Eight	C	6-231 (3.8)	4.0①	4.0	—	—	22④	—	18	12.50
	Delta 88 Royale	3	6-231 (3.8)	4.0①	4.0	—	—	22④	—	18	13.25
	Bonneville	3	6-231 (3.8)	4.0①	4.0	—	—	22④	—	18	13.0
	Bonneville	C	6-231 (3.8)	4.0①	4.0	—	—	22④	—	18	13.0
1989	DeVille	5	8-273 (4.5)	5.5	5.0	—	—	22④	—	18	13.0③
	Fleetwood	5	8-273 (4.5)	5.5	5.0	—	—	22④	—	18	13.0③
	Electra	C	6-231 (3.8)	4.0①	4.0	—	—	22④	—	18	13.0
	Park Avenue	C	6-231 (3.8)	4.0①	4.0	—	—	22④	—	18	13.0
	LeSabre	C	6-231 (3.8)	4.0①	4.0	—	—	22④	—	18	13.0
	Ninety Eight	C	6-231 (3.8)	4.0①	4.0	—	—	22④	—	18	12.50
	Delta 88 Royale	C	6-231 (3.8)	4.0①	4.0	—	—	22④	—	18	13.25
	Bonneville	C	6-231 (3.8)	4.0①	4.0	—	—	22④	—	18	13.0
1990	DeVille	3	8-273 (4.5)	5.5①	5.0	—	—	22④	—	18	13.2③
	Fleetwood	3	8-273 (4.5)	5.5	5.0	—	—	22④	—	18	13.2③
	Electra	C	6-231 (3.8)	4.0①	4.0	—	—	22④	—	18	13.0
	Park Avenue	C	6-231 (3.8)	4.0①	4.0	—	—	22④	—	18	13.0
	LeSabre	C	6-231 (3.8)	4.0①	4.0	—	—	22④	—	18	13.0
	Ninety Eight	C	6-231 (3.8)	4.0①	4.0	—	—	22④	—	18	13.0
	Delta 88 Royale	C	6-231 (3.8)	4.0①	4.0	—	—	22④	—	18	13.0
	Bonneville	C	6-231 (3.8)	4.0②①	4.0	—	—	22④	—	18	13.0
1991-92	DeVille	B	8-300 (4.9)	5.5	5.0	—	—	22④	—	18	13.2③
	Fleetwood	B	8-300 (4.9)	5.5	5.0	—	—	22④	—	18	13.2③
	Park Avenue	L	6-231 (3.8)	4.0①	4.0	—	—	22④	—	18	13.0
	LeSabre	C	6-231 (3.8)	4.0①	4.0	—	—	22④	—	18	13.0
	Ninety Eight	L	6-231 (3.8)	4.0①	4.0	—	—	22④	—	18	13.0
	88 Royale	L	6-231 (3.8)	4.0①	4.0	—	—	22④	—	18	13.0
	Bonneville	C	6-231 (3.8)	4.0①	4.0	—	—	22④	—	18	13.0

① Additional oil may be necessary to bring the level to full
② SSE—5.5 qts.
③ Use coolant solution specifically designed for use in aluminum engines.
④ Specification for transaxle overhaul 12 pts. for fluid and filter change.

CAMSHAFT SPECIFICATIONS

All measurements given in inches.

Year	VIN	No. Cylinder Displacement cu. in. (liter)	Journal Diameter 1	2	3	4	5	Lobe Lift In.	Ex.	Bearing Clearance	Camshaft End Play
1988	3	6-231 (3.8)	1.7850–1.7860	1.7850–1.7860	1.7850–1.7860	1.7850–1.7860	—	0.245	0.245	0.0005–0.0035	NA
	C	6-231 (3.8)	1.7850–1.7860	1.7850–1.7860	1.7850–1.7860	1.7850–1.7860	—	0.272	0.272	0.0005–0.0035	NA
	5	8-273 (4.5)	NA	NA	NA	NA	NA	0.384①	0.396①	0.0018–0.0037 ②	NA
1989	C	6-231 (3.8)	1.7850–1.7860	1.7850–1.7860	1.7850–1.7860	1.7850–1.7860	—	0.250	0.255	0.0005–0.0035	NA
	5	8-273 (4.5)	NA	NA	NA	NA	NA	0.384①	0.396①	0.0018–0.0037 ②	NA
1990	C	6-231 (3.8)	1.7850–1.7860	1.7850–1.7860	1.7850–1.7860	1.7850–1.7860	—	0.250	0.255	0.0005–0.0035	NA
	3	8-273 (4.5)	NA	NA	NA	NA	NA	0.384①	0.396①	0.0018–0.0037 ②	NA
1991–92	C	6-231 (3.8)	1.7850–1.7860	1.7850–1.7860	1.7850–1.7860	1.7850–1.7860	—	0.250	0.255	0.0005–0.0035	NA
	L	6-231 (3.8)	1.7850–1.7860	1.7850–1.7860	1.7850–1.7860	1.7850–1.7860	—	0.250	0.255	0.0005–0.0035	NA
	B	8-300 (4.9)	NA	NA	NA	NA	NA	0.384①	0.396①	0.0018–0.0037 ②	NA

NA—Not available
① Specification is for valve lift—rocker arm ratio 1.6:1
② Max—0.004

CRANKSHAFT AND CONNECTING ROD SPECIFICATIONS

All measurements are given in inches.

Year	VIN	No. Cylinder Displacement cu. in. (liter)	Crankshaft Main Brg. Journal Dia.	Main Brg. Oil Clearance	Shaft End-play	Thrust on. No.	Connecting Rod Journal Diameter	Oil Clearance	Side Clearance
1988	3	6-231 (3.8)	2.4988–2.4988	0.0003–0.0018	0.0030–0.0110	2	2.2487–2.2499	0.0005–0.0026	0.003–0.015
	C	6-231 (3.8)	2.4998–2.4998	0.0003–0.0018	0.0030–0.0110	2	2.2487–2.2499	0.0005–0.0026	0.003–0.015
	5	8-273 (4.5)	2.6354–2.6364	①	0.0010–0.0070	3	2.0520–2.0540	0.0005–0.0028	0.008–0.020
1989	C	6-231 (3.8)	2.4988–2.4998	0.0003–0.0018	0.0030–0.0110	2	2.2487–2.2499	0.0003–0.0026	0.003–0.015
	5	8-273 (4.5)	2.6354–2.6364	①	0.0010–0.0070	3	2.0520–2.0540	0.0005–0.0028	0.008–0.020

CRANKSHAFT AND CONNECTING ROD SPECIFICATIONS

All measurements are given in inches.

Year	VIN	No. Cylinder Displacement cu. in. (liter)	Crankshaft				Connecting Rod		
			Main Brg. Journal Dia.	Main Brg. Oil Clearance	Shaft End-play	Thrust on. No.	Journal Diameter	Oil Clearance	Side Clearance
1990	C	6-231 (3.8)	2.4988–2.4998	0.0003–0.0018	0.0030–0.0110	2	2.2487–2.2499	0.0003–0.0026	0.003–0.015
	3	8-273 (4.5)	2.6354–2.6364	①	0.0010–0.0070	3	2.0520–2.0540	0.0005–0.0028	0.008–0.020
1991-92	C	6-231 (3.8)	2.4988–2.4998	0.0003–0.0018	0.0030–0.0110	2	2.2487–2.2499	0.0003–0.0026	0.003–0.015
	L	6-231 (3.8)	2.4988–2.4998	0.0003–0.0018	0.0030–0.0110	2	2.2487–2.2499	0.0003–0.0026	0.003–0.015
	B	8-300 (4.9)	2.6354–2.6364	①	0.0010–0.0080	3	2.0520–2.0530	0.0005–0.0028	0.008–0.020

① No. 1 bearing—0.0008–0.0031
Nos. 2–5 bearing—0.0016–0.0039

VALVE SPECIFICATIONS

Year	VIN	No. Cylinder Displacement cu. in. (liter)	Seat Angle (deg.)	Face Angle (deg.)	Spring Test Pressure (lbs.)	Spring Installed Height (in.)	Stem-to-Guide Clearance (in.)		Stem Diameter (in.)	
							Intake	Exhaust	Intake	Exhaust
1988	3	6-231 (3.8)	46	45	185 ± 10 @ 1.340 in.	1.720①	0.0015–0.0035	0.0015–0.0032	0.3401–0.3412	0.3405–0.3412
	C	6-231 (3.8)	45	45	225 ± 11 @ 1.255 in.	1.720①	0.0015–0.0035	0.0015–0.0032	0.3401–0.3412	0.3405–0.3412
	5	8-273 (4.5)	45	44	②	NA	0.0010–0.0030	0.0010–0.0030	0.3420–0.3413	0.3401–0.3408
1989	C	6-231 (3.8)	45	45	200–220 @ 1.315	1.690–1.750	0.0015–0.0035	0.0015–0.0032	0.3401–0.3412	0.3405–0.3412
	5	8-273 (4.5)	45	44	②	NA	0.0010–0.0030	0.0010–0.0030	0.3420–0.3413	0.3401–0.3408
1990	C	6-231 (3.8)	45	45	200–220 @ 1.315	1.690–1.750	0.0015–0.0035	0.0015–0.0032	0.3401–0.3412	0.3405–0.3412
	3	8-273 (4.5)	45	44	214–232 @ 1.35 in.	NA	0.0010–0.0030	0.0020–0.0040	0.3420–0.3413	0.3401–0.3408
1991–92	C	6-231 (3.8)	45	45	210 @ 1.315 in.	1.690–1.720	0.0015–0.0035	0.0015–0.0035	NA	NA
	L	6-231 (3.8)	45	45	210 @ 1.315 in.	1.690–1.720	0.0015–0.0035	0.0015–0.0035	NA	NA
	B	8-300 (4.9)	45	45	214–232 @ 1.35 in.	NA	0.0010–0.0030	0.0020–0.0040	0.3420–0.3413	0.3401–0.3408

NA—Not available
① ± 0.030
② I—209–216 @ 1.28 in.
   E—204–221 @ 1.28 in.

PISTON AND RING SPECIFICATIONS

All measurements are given in inches.

Year	VIN	No. Cylinder Displacement cu. in. (liter)	Piston Clearance	Ring Gap			Ring Side Clearance		
				Top Compression	Bottom Compression	Oil Control	Top Compression	Bottom Compression	Oil Control
1988	3	6-231 (3.8)	0.0004–0.0022	0.010–0.020	0.010–0.022	0.015–0.055	0.0010–0.0030	0.0010–0.0030	0.0005–0.0065
	C	6-231 (3.8)	①	0.010–0.025	0.010–0.055	0.015–0.055	0.0013–0.0031	0.0013–0.0031	0.0011–0.0081
	5	8-273 (4.5)	0.0010–0.0018	0.015–0.024	0.015–0.024	0.010–0.050	0.0016–0.0037	0.0016–0.0037	None (side sealing)
1989	C	6-231 (3.8)	①	0.010–0.025	0.010–0.025	0.015–0.055	0.0013–0.0031	0.0013–0.0031	0.0011–0.0081
	5	8-273 (4.5)	0.0010–0.0018	0.015–0.024	0.015–0.024	0.010–0.050	0.0016–0.0037	0.0016–0.0037	None (side sealing)
1990	C	6-231 (3.8)	0.0004–0.0022	0.010–0.025	0.010–0.025	0.015–0.055	0.0013–0.0031	0.0013–0.0031	0.0011–0.0081
	3	8-273 (4.5)	0.0010–0.0018	0.015–0.024	0.015–0.024	0.010–0.050	0.0016–0.0037	0.0016–0.0037	None (side sealing)
1991–92	C	6-231 (3.8)	0.0004–0.0022	0.010–0.025	0.010–0.025	0.015–0.055	0.0013–0.0031	0.0013–0.0031	0.0011–0.0081
	L	6-231 (3.8)	0.0004–0.0022	0.010–0.025	0.010–0.025	0.015–0.055	0.0013–0.0031	0.0013–0.0031	0.0011–0.0081
	B	8-300 (4.9)	0.0004–0.0020	0.012–0.022	0.012–0.022	0.004–0.020	0.0016–0.0037	0.0016–0.0037	None (side sealing)

① Skirt Top: 0.0007–0.0027
Skirt Bottom: 0.0010–0.0045

TORQUE SPECIFICATIONS

All readings in ft. lbs.

Year	VIN	No. Cylinder Displacement cu. in. (liter)	Cylinder Head Bolts	Main Bearing Bolts	Rod Bearing Bolts	Crankshaft Pulley Bolts	Flywheel Bolts	Manifold		Spark Plugs
								Intake	Exhaust	
1988	3	6-231 (3.8)	60⑤	100	40	219	60	80④	37	20
	C	6-231 (3.8)	60⑤	100	40	219	60	80④	37	20
	5	8-273 (4.5)	①	85	22	65	70	③	18	11
1989	C	6-231 (3.8)	60⑤	90	43	219	61	88	41	20
	5	8-273 (4.5)	①	85	24	65	70	②	18	11
1990	C	6-231 (3.8)	35	90	43	219	61	88④	41	20
	3	8-273 (4.5)	①	85	24	65	70	②	18	11

TORQUE SPECIFICATIONS
All readings in ft. lbs.

Year	VIN	No. Cylinder Displacement cu. in. (liter)	Cylinder Head Bolts	Main Bearing Bolts	Rod Bearing Bolts	Crankshaft Pulley Bolts	Flywheel Bolts	Manifold		Spark Plugs
								Intake	Exhaust	
1991–92	C	6-231 (3.8)	⑥	⑦	⑧	⑨	61	88④	41	20
	L	6-231 (3.8)	⑥	⑦	⑧	⑨	61	88④	41	20
	B	8-300 (4.9)	①	85	25	70	70	②	18	23

① Tighten in 3 Steps
  1. Tighten bolts in sequence to 38 ft. lbs.
  2. Tighten bolts in sequence to 68 ft. lbs.
  3. Tighten bolts, 1, 3 and 4 to 90 ft. lbs.
② Tighten in 3 Steps
  1. Tighten bolts 1, 2, 3, 4 in sequence to 8 ft. lbs.
  2. Tighten bolts 5 through 16 in sequence to 8 ft. lbs.
  3. Retighten all bolts in sequence to 12 ft. lbs.
③ Tighten in 3 Steps
  1. Tighten bolts 1, 2, 3, 4 in sequence to 15 ft. lbs.

  2. Tighten bolts 5 through 16 in sequence to 22 ft. lbs.
  3. Retighten all bolts in sequence to 22 ft. lbs.
④ Inch lbs.
⑤ 3 Step procedure: Should you reach 60 ft. lbs. at any time in step 2 or 3, stop tightening. Do not complete the balance of the 90 degree turn of this bolt.
  Step 1: 25 ft. lbs.
  Step 2: 90 degrees
  Step 3: 90 degrees

⑥ Tighten in 3 steps
  1. Tighten bolts in sequence to 35 ft. lbs.
  2. Rotate each bolt an additional 130 degrees in sequence
  3. Rotate the center 4 bolts an additional 30 degrees in sequence
⑦ 26 ± 3 ($+ 50° \pm 3$)
⑧ 20 ± 3 ($+ 50° \pm 3$)
⑨ 105 ± 7 ($+ 56° \pm 4$)

BRAKE SPECIFICATIONS
All measurements in inches unless noted

Year	Model	Lug Nut Torque (ft. lbs.)	Master Cylinder Bore	Brake Disc		Standard Brake Drum Diameter	Minimum Lining Thickness	
				Minimum Thickness	Maximum Runout		Front	Rear
1988	DeVille	100	0.937②	0.972	0.004	8.900	0.030	0.030
	Fleetwood	100	0.937②	0.972	0.004	8.900	0.030	0.030
	Electra	100	0.937②	0.972	0.004	8.860	0.030	0.030
	Park Ave	100	0.937②	0.972	0.004	8.860	0.030	0.030
	LeSabre	100①	0.937②	0.972	0.004	8.860	0.030	0.030
	Ninety Eight	100	0.937②	0.972	0.004	8.860	0.030	0.030
	Delta 88	100①	0.937②	0.972	0.004	8.860	0.030	0.030
	Bonneville	100①	0.937②	0.972	0.004	8.860	0.030	0.030
1989	DeVille	100	0.937	0.972	0.004	8.860	0.030	0.030
	Fleetwood	100	0.937	0.972	0.004	8.860	0.030	0.030
	Electra	100	0.937	0.972	0.004	8.860	0.030	0.030
	Park Ave	100	0.937	0.972	0.004	8.860	0.030	0.030
	LeSabre	100	0.937	0.972	0.004	8.860	0.030	0.030
	Ninety Eight	100	0.937	0.972	0.004	8.860	0.030	0.030
	Delta 88	100	0.937	0.972	0.004	8.860	0.030	0.030
	Bonneville	100	0.937	0.972	0.004	8.860	0.030	0.030
1990	DeVille	100	0.937	0.972	0.004	8.860	0.030	0.030
	Fleetwood	100	0.937	0.972	0.004	8.860	0.030	0.030
	Electra	100	0.937	0.972	0.004	8.860	0.030	0.030
	Park Ave	100	0.937	0.972	0.004	8.860	0.030	0.030
	LeSabre	100	0.937	0.972	0.004	8.860	0.030	0.030
	Ninety Eight	100	0.937	0.972	0.004	8.860	0.030	0.030
	Delta 88	100	0.937	0.972	0.004	8.860	0.030	0.030
	Bonneville	100	0.937	0.972	0.004	8.860	0.030	0.030

BRAKE SPECIFICATIONS

All measurements in inches unless noted

Year	Model	Lug Nut Torque (ft. lbs.)	Master Cylinder Bore	Brake Disc Minimum Thickness	Brake Disc Maximum Runout	Standard Brake Drum Diameter	Minimum Lining Thickness Front	Minimum Lining Thickness Rear
1991-92	DeVille	100	1.000	1.224	0.004	8.860	0.030	0.030
	Fleetwood	100	1.000	1.224	0.004	8.860	0.030	0.030
	Park Ave	100	1.000	1.204	0.004	8.860	0.030	0.030
	LeSabre	100	1.000	0.972	0.004	8.860	0.030	0.030
	Ninety Eight	100	1.000	1.204	0.004	8.860	0.030	0.030
	88 Royale	100	1.000	0.972	0.004	8.860	0.030	0.030
	Bonneville	100	1.000	0.972	0.004	8.860	0.030	0.030

① 7/16 in. stud—80 ft. lbs.
② Anti-lock brakes: standard brakes—0.945 in.

WHEEL ALIGNMENT

Year	Model	Caster Range (deg.)	Caster Preferred Setting (deg.)	Camber Range (deg.)	Camber Preferred Setting (deg.)	Toe-in (in.)	Steering Axis Inclination (deg.)
1988	DeVille	2P–3P	2½P	②	②	0	—
	Fleetwood	2P–3P	2½P	②	②	0	—
	Electra	2P–3P	2½P	5/16N–11/16P	3/16P	0	—
	Park Ave	2P–3P	2½P	5/16N–11/16P	3/16P	0	—
	LeSabre	2P–3P	2½P	5/16N–11/16P	3/16P	0	—
	Ninety Eight	2P–3P	2½P	5/16N–11/16P	3/16P	0	12 13/16P
	Delta 88	2P–3P	2½P	5/16N–11/16P	3/16P	0	12 13/16P
	Bonneville	2P–3P	2½P	5/16N–11/16P	3/16P	0	—
1989	DeVille	2½P–3½P	3P	②	②	0	—
	Fleetwood	2½P–3½P	3P	②	②	0	—
	Electra	2½P–3½P	3P	5/16N–11/16P	3/16P	0	—
	Park Ave	2½P–3½P	3P	5/16N–11/16P	3/16P	0	—
	LeSabre	2½P–3½P	3P	5/16N–11/16P	3/16P	0	—
	Ninety Eight	2½P–3½P	3P	5/16N–11/16P	3/16P	0	½P
	Delta 88	2½P–3½P	3P	5/16N–11/16P	3/16P	0	½P
	Bonneville	2½P–3½P	3P	5/16N–11/16P	3/16P	0	½P
1990	DeVille	2½P–3½P	3P	②	②	0	—
	Fleetwood	2½P–3½P	3P	②	②	0	—
	Electra	2½P–3½P	3P	5/16N–11/16P	3/16P	0	—
	Park Ave	2½P–3½P	3P	5/16N–11/16P	3/16P	0	—
	LeSabre	2½P–3½P	3P	5/16N–11/16P	3/16P	0	—
	Ninety Eight	2½P–3½P	3P	5/16N–11/16P	3/16P	0	½P
	Delta 88	2½P–3½P	3P	5/16N–11/16P	3/16P	0	½P
	Bonneville	2½P–3½P	3P	5/16N–11/16P	3/16P	0	½P

WHEEL ALIGNMENT

Year	Model	Caster Range (deg.)	Caster Preferred Setting (deg.)	Camber Range (deg.)	Camber Preferred Setting (deg.)	Toe-in (in.)	Steering Axis Inclination (deg.)
1991–92	DeVille	2½P–3½P	3P	②	②	0	—
	Fleetwood	2½P–3½P	3P	②	②	0	—
	Park Ave	2½P–3½P	3P	5/16N–11/16P	3/16P	0	—
	LeSabre	2½P–3½P	3P	5/16N–11/16P	3/16P	0	—
	Ninety Eight	2½P–3½P	3P	5/16N–11/16P	3/16P	0	½P
	88 Royale	2½P–3½P	3P	5/16N–11/16P	3/16P	0	½P
	Bonneville	2½P–3½P	3P	5/16N–11/16P	3/16P	0	½P

N Negative
P Positive
① In or out pref: 0
② Left wheel
   Min.—1N
   Pref.—½N
   Max.—0
  Right wheel
   Min.—0
   Pref.—½P
   Max.—1P

ENGINE MECHANICAL

NOTE: Disconnecting the negative battery cable on some vehicles may interfere with the functions of the on board computer systems and may require the computer to undergo a relearning process, once the negative battery cable is reconnected.

Engine Assembly

REMOVAL & INSTALLATION

3.8L Engine

1. Disconnect the negative battery cable. Using a scribing tool, matchmark the hood hinges and remove the hood.
2. Label and disconnect the air flow sensor wiring. Depressurize the fuel system.
3. Remove the air intake duct. Remove the throttle cable and bracket from the throttle body. Place a clean drain pan under the radiator, open the drain cock and drain the cooling system.
4. Raise and safely support the vehicle.
5. Remove the exhaust pipe-to-exhaust manifold bolts and separate the exhaust pipe.
6. Remove the engine mount bolts.
7. If equipped with a driveline vibration absorber, remove the bolts and disconnect the absorber.

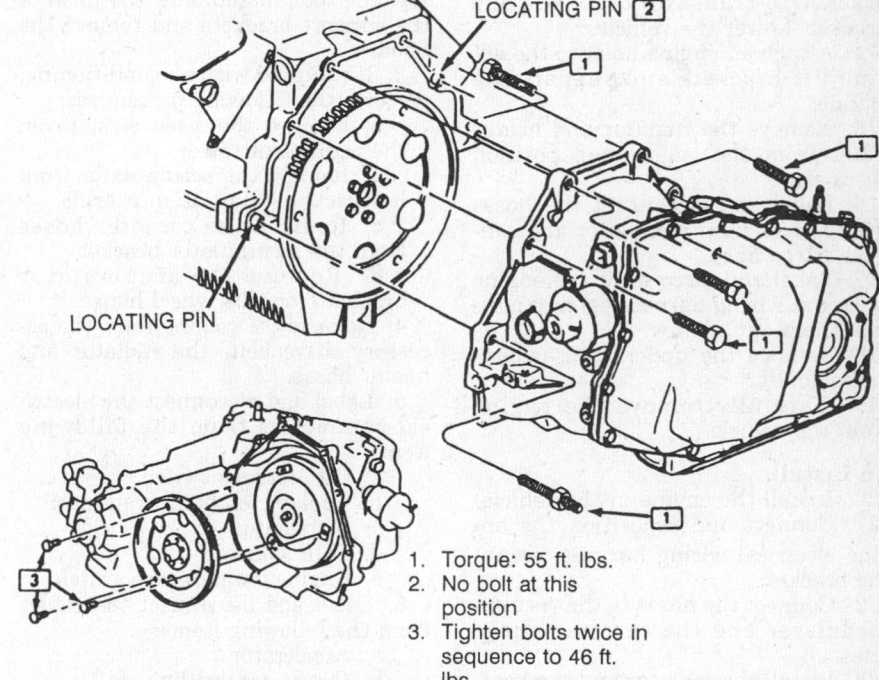

1. Torque: 55 ft. lbs.
2. No bolt at this position
3. Tighten bolts twice in sequence to 46 ft. lbs.

Engine-to-transaxle mounting location—3.8L engine

8. Label and disconnect the electrical connectors from the starter. Remove the starter-to-engine bolts and the starter.
9. If equipped with air conditioning, remove the compressor mounting bolts and position aside. Do not discharge the system or disconnect the refrigerant lines.
10. Place a pan under the power steering gear. Disconnect the hydraulic lines and drain the fluid. Use a length of wire to hold the hoses aside.

11. Remove the lower transaxle-to-engine bolts.

NOTE: One bolt is situated between the transaxle case and the engine block. It is installed in the opposite direction to the other bolts.

12. Remove the flywheel cover. Matchmark the flexplate-to-torque converter relationship to insure proper alignment upon installation. Re-

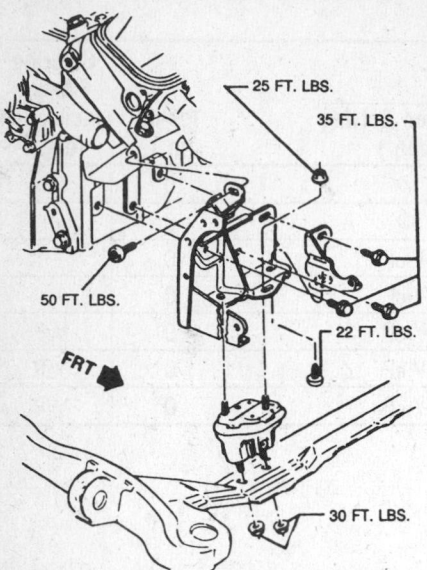

25 FT. LBS.

35 FT. LBS.

50 FT. LBS.

22 FT. LBS.

FRT

30 FT. LBS.

Engine mounting—3.8L engine

move the flexplate-to-torque converter bolts.

13. Remove the engine support bracket-to-transaxle bolts and the bracket. Lower the vehicle.

14. Attach an engine hoist to the engine lift brackets and support the engine.

15. Remove the radiator and heater hoses from the engine and position them aside.

16. Label and disconnect the hoses from the vacuum modulator and canister purge lines.

17. Label and disconnect the engine electrical wiring harness(es) and position them out of way.

18. Remove the upper transaxle-to-engine bolts.

19. Carefully remove the engine from the vehicle.

To install:

20. Install the engine in the vehicle.

21. Connect and reposition the engine electrical wiring harness, secure the bracket.

22. Connect the hoses to the vacuum modulator and the canister purge lines.

23. Install the radiator and the heater hoses.

24. Remove the engine hoist. Raise and safely support the vehicle.

25. Replace the engine support bracket-to-transaxle bolts.

26. Replace the upper transaxle-to-engine bolts. Install the engine mount bolts and tighten to 70 ft. lbs. (95 Nm).

27. Install the torque converter and tighten the bolts to 46 ft. lbs. (62 Nm). Replace the flywheel cover.

28. Install the lower transaxle-to-engine bolts, tighten to 55 ft. lbs. (75 Nm).

29. Connect the power steering hy-

draulic lines to the power steering gear.

30. Install the air conditioning compressor, if equipped.

31. Install the starter and tighten the mounting bolts to 35 ft. lbs. (47 Nm).

32. Connect the electrical connections at the starter motor.

33. Connect the driveline vibration absorber, if equipped.

34. Install the exhaust pipe and replace the exhaust pipe-to-manifold bolts.

35. Lower the vehicle.

36. Connect the air flow sensor wiring. Install the hood assembly.

37. Connect the negative battery cable. Refill the cooling system.

38. Start the engine, allow it to reach normal operating temperatures and check for leaks.

4.5L and 4.9L Engines

1. Disconnect the negative battery cable. Drain the coolant into a clean container for reuse.

2. Remove the air cleaner. Using a scribing tool, matchmark the hood to the support brackets and remove the hood.

3. If equipped with air conditioning, perform the following procedures:

a. Remove the hose strap from the right-strut tower.

b. Remove the accumulator from its bracket and position it aside.

c. Remove the canister hoses from the accumulator bracket.

d. Remove the accumulator bracket from the wheel house.

4. Remove the cooling fans, the accessory drive belt, the radiator and heater hoses.

5. Label and disconnect the electrical connectors from the following items:

a. Oil pressure switch

b. Coolant temperature sensor

c. Distributor

d. EGR solenoid

e. Engine temperature switch

6. Label and disconnect the cables from the following items:

a. Accelerator

b. Cruise control linkage

c. Transaxle Throttle Valve (TV) cable

7. If equipped with cruise control, remove the diaphragm with the bracket attached and move it aside.

8. Remove the vacuum supply hose and the exhaust crossover pipe.

9. Disconnect the oil cooler lines from the oil filter adapter, the oil line cooler bracket from the transaxle and position them aside.

10. Remove the air cleaner mounting bracket.

11. Properly relieve the fuel system pressure. Disconnect the fuel lines

from the throttle body. Remove the fuel line bracket from the transaxle and secure the fuel lines aside.

12. Remove the small vacuum line from the brake booster.

13. Label and disconnect the AIR solenoid electrical and hose connections. Remove the AIR valves with the bracket.

14. Label and disconnect the electrical connectors from the following:

a. Idle Speed Control (ISC) motor

b. Throttle Position Switch (TPS)

c. Fuel injectors

d. Manifold Air Temperature (MAT) sensor

e. Oxygen sensor

f. Electric Fuel Evaporation (EFE) grid

g. Alternator bracket

15. Remove the power steering pump hose strap from the stud-headed bolt in front of the right cylinder head and the stud-headed bolt.

16. Remove the AIR pipe clip located near the No. 2 spark plug, if equipped.

17. Remove the power steering pump and belt tensioner with bracket attached; wire them aside.

18. Raise and safely support the vehicle.

19. Label and disconnect the electrical connectors from the starter and the ground wire from the cylinder block.

20. Remove the 2 flywheel covers. Remove the starter-to-engine bolts and the starter. Matchmark the flywheel-to-torque converter location. Remove the 3 flywheel-to-torque converter bolts and slide the converter back into the bell housing.

21. If equipped with air conditioning, perform the following procedures:

a. Remove the compressor lower dust shield.

b. Remove the right front wheel/tire assembly and outer wheelhouse plastic shield.

c. Remove the compressor-to-bracket bolts and lower the compressor from the engine. Do not disconnect the refrigerant lines.

22. Remove the lower radiator hose.

23. From the lower right front of the engine and cradle, remove the driveline vibration damper with the brackets, if equipped, and the engine-to-transaxle bracket bolts. Pull the alternator wire with the plastic cover down and aside.

24. Remove the exhaust pipe-to-manifold bolts with the springs attached and the AIR pipe-to-converter bracket from the exhaust manifold stud.

NOTE: Be careful not to lose the springs when detaching the exhaust pipe.

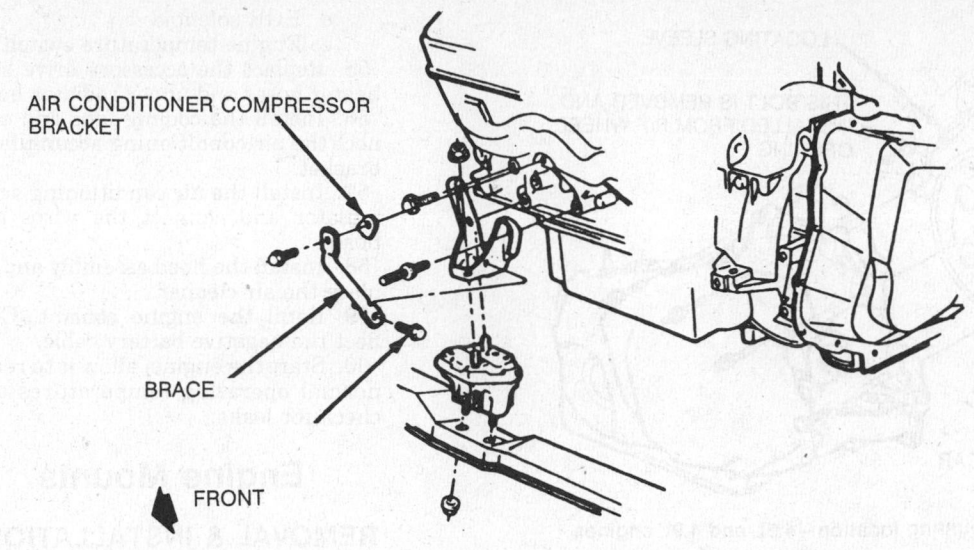

AIR CONDITIONER COMPRESSOR
BRACKET

BRACE

FRONT

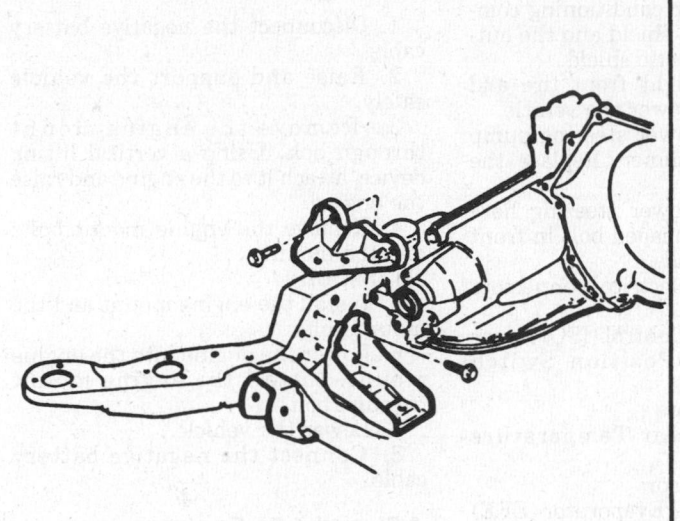

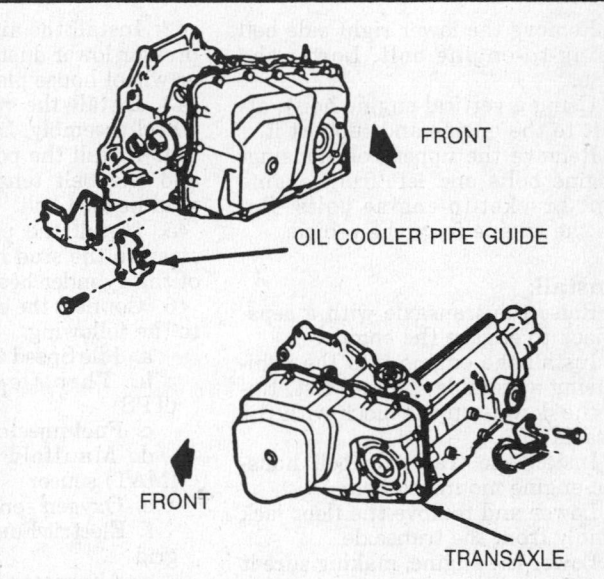

FRONT

OIL COOLER PIPE GUIDE

FRONT

TRANSAXLE

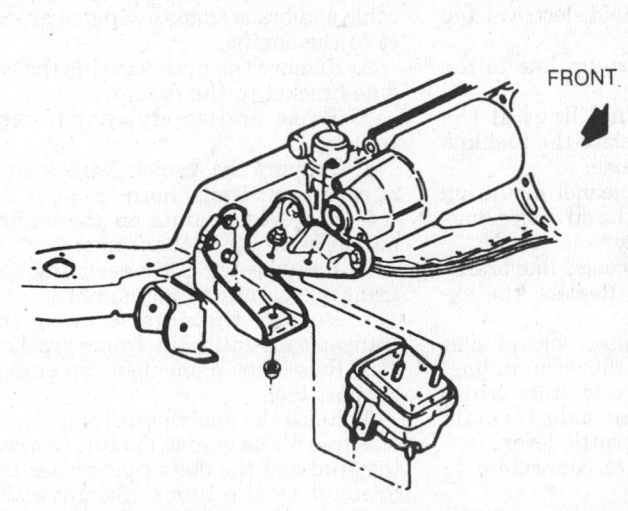

FRONT

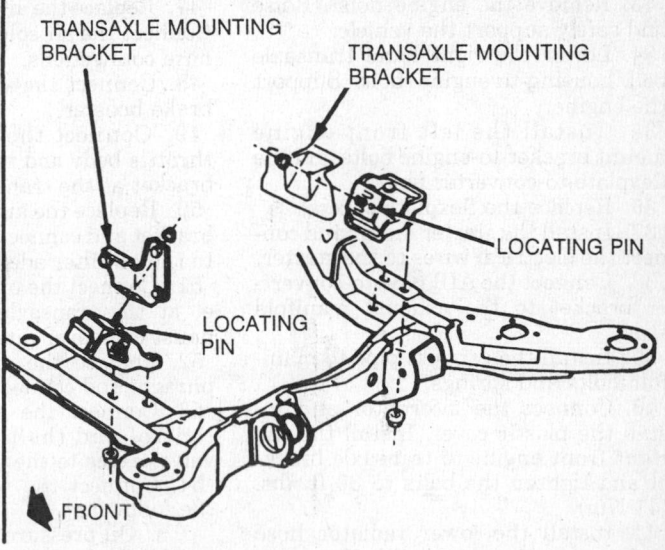

TRANSAXLE MOUNTING
BRACKET

TRANSAXLE MOUNTING
BRACKET

LOCATING PIN

LOCATING
PIN

LOCATING
PIN

FRONT

Engine and transmission mounts—4.5L and 4.9L engine

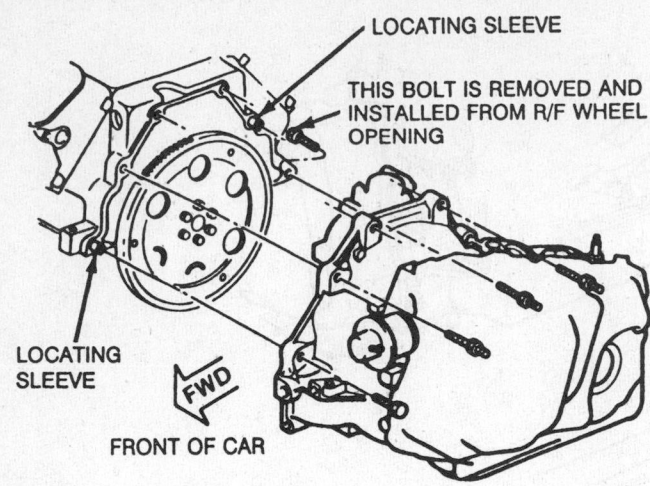

LOCATING SLEEVE

THIS BOLT IS REMOVED AND INSTALLED FROM R/F WHEEL OPENING

LOCATING SLEEVE

FWD

FRONT OF CAR

Engine-to-transaxle mounting location—4.5L and 4.9L engines

25. Remove the lower right side bell housing-to-engine bolt. Lower the vehicle.

26. Using a vertical engine hoist, attach it to the engine and support it.

27. Remove the upper bell housing-to-engine bolts and left front engine mount bracket-to-engine bolts. Remove the engine from the vehicle.

To install:

28. Raise the transaxle with a separate jack to engage the engine.

29. Install the engine into the vehicle, using a suitable engine hoist. Engage the dowels on the block with the transaxle case.

30. Install the transaxle bell housing-to-engine mounting bolts.

31. Lower and remove the floor jack assembly from the transaxle.

32. Lower the engine, making sure it is seated on the mount properly.

33. Remove the engine hoist. Raise and safely support the vehicle.

34. Lower the right hand transaxle bell housing-to-engine bolt. Support the engine.

35. Install the left front engine mount bracket-to-engine bolts and the flexplate-to-converter bolts.

36. Replace the flexplate covers.

37. Install the starter motor and connect the electrical wires to the starter.

38. Connect the AIR pipe-to-converter bracket to the exhaust manifold stud.

39. Install the exhaust pipe to manifold bolts and springs.

40. Connect the alternator and install the plastic cover. Install the the right front engine-to-transaxle bracket and tighten the bolts to 30 ft. lbs. (41 Nm).

41. Install the lower radiator hose and replace the air conditioning compressor mounting bolts.

42. Install the air conditioning compressor lower dust shield and the outer wheel house plastic shield.

43. Install the right front tire and wheel assembly. Lower the vehicle.

44. Install the power steering pump and the belt tensioner. Replace the stud headed bolt.

45. Install the power steering hose strap to the stud headed bolt in front of the cylinder head.

46. Connect the electrical connectors to the following:

   a. Idle Speed Control (ISC) motor

   b. Throttle Position Switch (TPS)

   c. Fuel injectors

   d. Manifold Air Temperature (MAT) sensor

   e. Oxygen sensor

   f. Electric Fuel Evaporator (EFE) grid

   g. Alternator bracket

47. Replace the air valve and bracket. Connect the air solenoid electrical and hose connections.

48. Connect the vacuum line to the brake booster.

49. Connect the fuel lines at the throttle body and replace the fuel line bracket at the transaxle.

50. Replace the air cleaner mounting bracket and connect the oil cooler lines to the oil filter adapter.

51. Connect the oil cooler line bracket at the transaxle. Replace the exhaust crossover pipe.

52. Replace the cruise control diaphragm and connect the vacuum line.

53. Connect the accelerator, cruise control and the transaxle throttle valve cables to the throttle lever.

54. Connect the wire connectors to the following:

   a. Oil pressure switch

   b. Coolant temperature sensor

   c. Distributor

   d. EGR solenoid

   e. Engine temperature switch

55. Replace the accessory drive belt, heater hoses and upper radiator hose.

56. Install the cooling fans and connect the air conditioning accumulator bracket.

57. Install the air conditioning accumulator and connect the wires and hoses.

58. Install the hood assembly and replace the air cleaner.

59. Refill the engine coolant. Connect the negative battery cable.

60. Start the engine, allow it to reach normal operating temperatures and check for leaks.

Engine Mounts

REMOVAL & INSTALLATION

3.8L Engine

1. Disconnect the negative battery cable.

2. Raise and support the vehicle safely.

3. Remove the engine mount through bolt. Using a vertical lifting device, attach it to the engine and raise the engine.

4. Remove the engine mount bolts and the mount.

To install:

5. Install the engine mount and the mount bolts.

6. Lower the engine into the engine mount. Install the engine mount through bolt.

7. Lower the vehicle.

8. Connect the negative battery cable.

4.5L and 4.9L Engines

RIGHT

1. Disconnect the negative battery cable and brace from the engine bracket to the engine.

2. Remove the nuts securing the engine bracket to the mount.

3. Raise and safely support the vehicle.

4. Support the vehicle with stands at each front frame horn.

5. Remove the nuts on the engine mount securing to the frame.

6. Remove the nuts securing the transaxle mount to the mount.

7. Remove the nuts securing the transaxle mount to the frame bracket.

8. Raise the engine using an engine support tool.

9. Raise the engine until the bracket is free of the engine mount. Remove the stud and the bolts that secure the bracket to the block. Remove the mount and bracket by pulling forward.

10. Remove the transaxle mounting bracket from the transaxle.

11. Remove the mount assembly.

To install:

12. Position the engine mount and bracket, in place between the transaxle and frame and secure the bracket to the transaxle with the 2 bolts and tighten to 34 ft. lbs. (46 Nm).

13. While lowering the engine, guide the motor mount into location and install the engine mount to frame and transaxle mount to frame bracket with the 2 nuts each and tighten to 22 ft. lbs. (30 Nm).

14. Install the nuts to the engine mount studs and the nuts to transaxle mount studs and tighten to 22 ft. lbs. (30 Nm).

15. Remove the brace from the engine bracket to engine.

16. Remove the stands and lower the hoist. Connect the negative battery cable.

LEFT

1. Raise the vehicle and support it safely. Disconnect the negative battery cable.

2. Support the vehicle with stands at each front frame horn.

3. Remove the nut securing the mount to the transaxle bracket and nuts securing the mount to the frame.

4. Lift the engine using engine support tool.

5. Remove the bolts securing the bracket to the transaxle.

6. Raise the engine assembly until the brackets are free.

7. Remove the mount and bracket by pulling it upward.

To install:

8. Position the engine mount and bracket in place between the transaxle and frame. Tighten the bracket to 41 ft. lbs. (56 Nm) and nuts to 22 ft lbs. (30 Nm).

9. Lower the transaxle onto the mount until it is seated.

10. Install the nut securing the mount to the bracket and tighten to 22 ft. lbs. (30 Nm).

11. Connect the negative battery cable.

Cylinder Head

REMOVAL & INSTALLATION

3.8L (VIN 3) Engine

1. Disconnect the negative battery cable. Drain the coolant into a clean container for reuse.

2. Properly relieve the fuel system pressure.

3. Disconnect the Mass Air Flow (MAF) sensor wiring and air intake duct.

4. Disconnect the TV and accelerator cables and, if equipped, the cruise control cable.

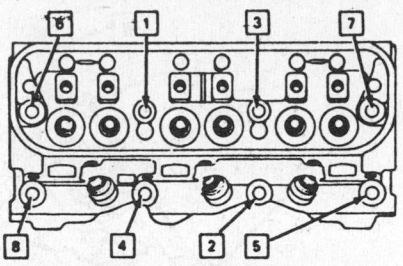

Cylinder head bolt tightening sequence— 3.8L engine

5. Remove crankcase ventilation pipe.

6. Remove all hoses, vacuum lines and wiring to gain access.

7. Remove the fuel rail.

8. Remove the intake manifold.

9. Disconnect the exhaust crossover pipe.

10. Remove the exhaust manifolds.

11. Remove the valve covers, rocker arms, guide plates and pushrods. Keep all parts in order so they may be reassembled in their original locations.

12. Loosen the cylinder head bolts in reverse of the torque sequence, then remove the bolts and lift off the cylinder head.

13. Clean all gasket mating surfaces and the cylinder head bolt holes in the block.

To install:

14. Install the cylinder head gasket and cylinder head.

15. Install the cylinder head bolts and tighten as follows:

   a. Tighten in the sequence to 25 ft. lbs. (34 Nm).

NOTE: Do not exceed 60 ft. lbs. (81 Nm) at any point during the completion of steps b and c.

   b. Rotate each bolt ¼ turn in sequence.

   c. Rotate each bolt an additional ¼ turn in sequence.

16. Install the rocker arms and pushrods. Tighten to 45 ft. lbs. (61 Nm).

17. Install the intake manifold. Tighten the bolts in sequence to 32 ft. lbs. (43 Nm).

18. Replace the valve covers and install the exhaust manifold. Tighten the bolts to 37 ft. lbs. (51 Nm).

19. Install the power steering pump and bracket. Connect the hydraulic lines.

20. Install the alternator and connect the electrical wires.

21. Replace the top radiator hose and the heater hoses. Install the fuel line and the wiring connector.

22. Install the exhaust crossover pipe and connect the vacuum lines and electrical connections.

23. Replace the crankcase ventilation pipe. Connect the TV and accelerator cables.

24. If equipped, connect the cruise control cable.

25. Replace the MAF sensor and air intake duct.

26. Connect the negative battery cable. Refill the engine coolant.

27. Start the engine and check for leaks.

3.8L (VIN C and L) Engine

1. Disconnect the negative battery cable.

2. Remove the intake and exhaust manifolds.

3. Remove the valve covers.

4. Label and disconnect the ignition module wires, spark plug wires and alternator bracket. Remove air conditioning compressor bracket bolt.

5. Remove the power steering pump, tensioner assembly and the fuel line heat shield.

6. Remove the rocker arm assemblies, guide plate and the pushrods.

7. Remove the cylinder head bolts and remove the cylinder head.

8. Clean all gasket mating surfaces and the cylinder head bolt holes in the block.

To install:

9. Install the cylinder head gasket and head onto the block.

10. Install the cylinder head bolts and tighten as follows:

   a. Tighten the cylinder head bolts, in sequence, to 35 ft. lbs. (47 Nm).

   b. Rotate each bolt 130 degrees, in sequence.

   c. Rotate the center 4 bolts an additional 30 degrees, in sequence.

11. Install the pushrods, guide plate and the rocker arm assemblies. Tighten the rocker arm pedestal bolts to 28 ft. lbs. (38 Nm).

12. Install the intake manifold, exhaust manifold and the valve covers.

13. Replace the air conditioning compressor bracket bolt and tighten to 52 ft. lbs. (71 Nm).

14. Install the alternator support bracket and replace the igniton module and spark plug wires.

15. Install the tensioner, power steering pump and the fuel line heat shield.

16. Connect the negative battery cable. Start the engine and check for leaks.

4.5L and 4.9L Engines

RIGHT

1. Disconnect the negative battery cable. Drain the coolant into a clean container for reuse. Properly relieve the fuel system pressure.

2. Remove the rocker arm covers and the intake manifold assembly.

3. Remove the right side exhaust manifold and disconnect the engine

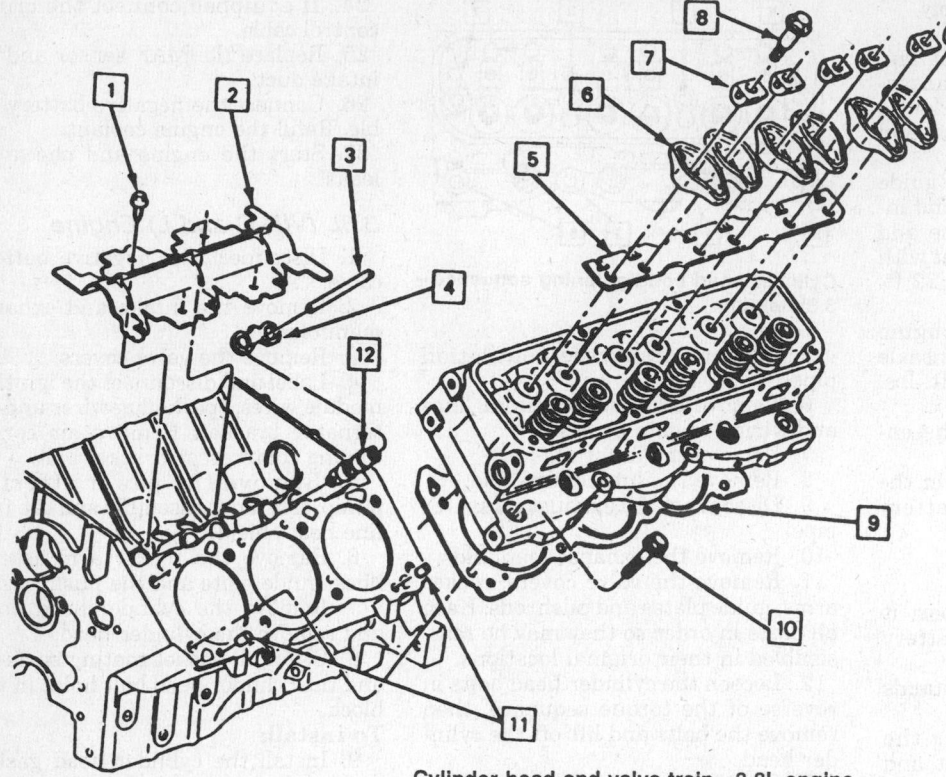

1. Bolt
2. Lifter guide retainer
3. Pushrod
4. Lifter guide
5. Pushrod guide
6. Rocker arm
7. Rocker arm pivot
8. Bolt
9. Head gasket
10. Head bolt
11. Dowel pin
12. Valve lifter

Cylinder head and valve train—3.8L engine

lift bracket and AIR pump bracket.

4. Remove the cylinder head bolts in the reverse order of the tightening sequence. Remove the cylinder head.

5. Clean all gasket mating surfaces and the cylinder head bolt holes in the block.

To install:

6. Install the cylinder head gasket and the cylinder head.

7. Tighten the cylinder head bolts as follows:

a. Tighten the cylinder head bolts, in sequence, to 38 ft. lbs. (50 Nm).

b. Tighten the cylinder head bolts, in sequence, to 68 ft. lbs. (90 Nm).

c. Tighten cylinder head bolts 1, 3 and 4 to 90 ft. lbs. (120 Nm).

8. Install the engine lift bracket and the air pump bracket.

9. Install the exhaust manifold, intake manifold and the rocker arm covers.

10. Refill the engine coolant. Connect the negative battery cable.

11. Start the engine and check for leaks.

LEFT

1. Disconnect the negative battery cable. Drain the engine coolant.

2. Remove the rocker arm covers and the intake manifold assembly.

3. Remove the left side exhaust manifold.

4. Remove the cooling fans and the dipstick tube.

5. Remove the cylinder head mounting bolts and remove the cylinder head.

6. Clean all gasket mating surfaces and the cylinder head bolt holes in the block.

To install:

7. Install a new head gasket over the dowels on the cylinder block.

8. Install the cylinder head and tighten the bolts as follows:

a. Tighten the cylinder head bolts, in sequence, to 38 ft. lbs. (50 Nm).

b. Tighten the cylinder head bolts, in sequence, to 68 ft. lbs. (90 Nm).

c. Tighten cylinder head bolts 1, 3 and 4 to 90 ft. lbs. (120 Nm).

9. Install the dipstick tube and replace the cooling fans.

10. Install the exhaust manifold, intake manifold and the rocker arm covers.

11. Refill the engine coolant. Connect the negative battery cable.

12. Start the engine and check for leaks.

Valve Lifters
REMOVAL & INSTALLATION
3.8L Engine

1. Disconnect the negative battery cable.

2. Remove the valve covers and the intake manifold.

3. Remove the rocker arm bolts, rocker arms and the pedestals.

4. Remove the pushrods, guide retainer bolts and the retainer.

5. Remove the lifter guides and lift out the lifters, using the proper tool.

To install:

6. Prior to installation dip the lifters in the proper prelube.

7. Install the lifters and lifter guides.

8. Install the pushrods, guide retainer bolts and the retainer.

9. Install the pedestals, rocker arms and rocker arms bolts.

10. Install the valve covers and the intake manifold.

11. Connect the negative battery cable.

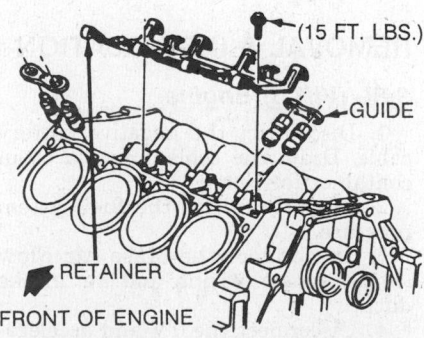

(15 FT. LBS.)
GUIDE
RETAINER
FRONT OF ENGINE

Lifter guides and retainer—4.5L and 4.9L engines

4.5L and 4.9L Engines

1. Disconnect the negative battery cable.
2. Remove the valve covers and the intake manifold.
3. Remove the rockers and the pushrods.
4. Disconnect the valve guide retainer.
5. Remove the valve lifter guides and pull out the lifters, using the proper tool.

To install:

6. Prior to installation dip the lifters in the proper prelube.
7. Install the valve lifters and lifter guides.
8. Connect the valve guide retainer. Tighten the retainer bolts to 15 ft. lbs.
9. Install the rockers and the pushrods.
10. Install the valve covers and the intake manifold.
11. Connect the negative battery cable.

Rocker Arms

REMOVAL & INSTALLATION

3.8L (VIN 3) Engine

RIGHT

1. Disconnect the negative battery cable.
2. Remove the spark plug cables, the wiring connector, the EGR solenoid wiring/hoses, the C³I module nuts and module.
3. Remove the serpentine drive belt. Disconnect the alternator wiring. Remove the alternator mounting bolt and rotate the alternator toward the front of the vehicle.
4. Remove the power steering pump and the belt tensioner.
5. Remove the engine lifting bracket and the rear alternator brace.
6. Remove the heater hoses from the throttle body.
7. Remove the rocker arm cover nuts, washers, seals, the cover and gasket. Discard the gasket.
8. Remove the rocker arm pedestal-to-cylinder head bolts, the pedestals, the rocker arms and the pedestal retainers.

NOTE: Be sure to keep the parts in order for reassembly purposes.

9. Clean the gasket mounting surfaces.

To install:

10. Install the rocker arms, pedestals and the pedestal retainers. Tighten the pedestal bolts to 45 ft. lbs. (61 Nm).
11. Install the rocker arm cover gasket and the cover. Replace the seals,

washers and the cover nuts. Tighten to 88 inch lbs. (10 Nm).
12. Connect the heater hoses to the throttle body.
13. Install the alternator brace and the engine lifting bracket.
14. Install the power steering pump and the belt tensioner.
15. Position the alternator in place and replace the alternator mounting bolt. Connect the alternator wiring.
16. Install the serpentine drive belt.
17. Connect the spark plug wires, EGR solenoid wiring/hoses and the connections at the ignition module.
18. Connect the negative battery cable.

LEFT

1. Disconnect the negative battery cable. Remove the PCV valve and pipe.
2. Remove the spark plug wiring harness cover and disconnect the spark plug wires from the spark plugs.
3. Remove the rocker arm cover nuts, washers, seals, the cover and gasket.
4. Remove the rocker arm pedestal-to-cylinder head bolts, the pedestals, the rocker arms and the pedestal retainers.

NOTE: Be sure to keep the parts in order for reassembly purposes.

To install:

5. Clean the gasket mounting surfaces.
6. Install the pedestal retainers, rocker arms, rocker arm pedestals and rocker arm pedestal-to-cylinder head bolts. Tighten to 45 ft. lbs. (60 Nm).
7. Install the rocker arm cover nuts, washers, seals, gasket and cover. Tighten to 88 ft. lbs. (10 Nm).
8. Connect the spark plug wires to the spark plugs. Install the spark plug wiring harness cover.
9. Install the PCV valve and pipe.
10. Disconnect the negative battery cable.

3.8L (VIN C and L) Engine

RIGHT

1. Disconnect the negative battery cable. Remove the accessory drive belt.
2. Loosen the power steering pump bolts and slide the pump forward. Disconnect the power steering bracket.
3. Disconnect the EGR pipe and remove EGR valve and adapter from the throttle body.
4. Disconnect the spark plug wires and remove the rocker arm cover bolts and cover.
5. Remove the rocker arm pedestal retaining bolts and lift out the pedestal and rocker arm assembly.

To install:

6. Install the pedestal, rocker arm

assembly and rocker arm pedestal retaining bolts. Tighten the pedestal bolts to 28 ft. lbs. (38 Nm).
7. Install the rocker arm cover and bolts. Tighten to 88 inch lbs. (10 Nm). Connect the spark plug wires.
8. Install the EGR valve and adapter to the throttle body. Connect the EGR pipe.
9. Connect the power steering bracket. Slide the power steering pump into position and install the bolts.
10. Install the serpentine drive belt. Connect the negative battery cable.

LEFT

1. Disconnect the negative battery cable. Remove the accessory drive belt.
2. Remove the alternator mounting bracket bolt and bracket.
3. Disconnect the spark plug wires. Remove the valve cover bolts and the valve cover.
4. Remove the rocker arm pedestal retaining bolts and lift out the pedestal and rocker arm assembly.

To install:

5. Install the rocker arm pedestal, rocker arm assembly and retaining bolts. Tighten to 28 ft. lbs. (38 Nm).
6. Install the valve cover and bolts. Tighten to 88 inch lbs. (10 Nm). Connect the spark plug wires.
7. Install the alternator mounting bracket and bolt.
8. Connect the negative battery cable.

4.5L and 4.9L Engines

RIGHT

1. Disconnect the negative battery cable. Remove the air cleaner and the AIR management valve with bracket, move the assembly aside.
2. From the throttle body, remove the Manifold Absolute Pressure (MAP) hose.
3. Remove the right side spark plug wires and conduit.
4. Remove the fuel vapor canister pipe bracket from the valve cover stud.
5. Drain the cooling system to a level below the thermostat housing. Remove the heater hose from the thermostat housing and move it aside.
6. Remove the brake booster vacuum hose from the intake manifold.
7. Remove the rocker arm cover-to-cylinder screws, the cover and the gasket/seals. Discard them.
8. Remove the rocker arm pivot-to-rocker arm support bolts, the pivots and the rocker arms.
9. If necessary, remove the rocker arm support-to cylinder head nuts/bolts and the support.
10. Clean the gasket mounting surfaces. Inspect the parts for wear and/or damage and replace the parts, if necessary.

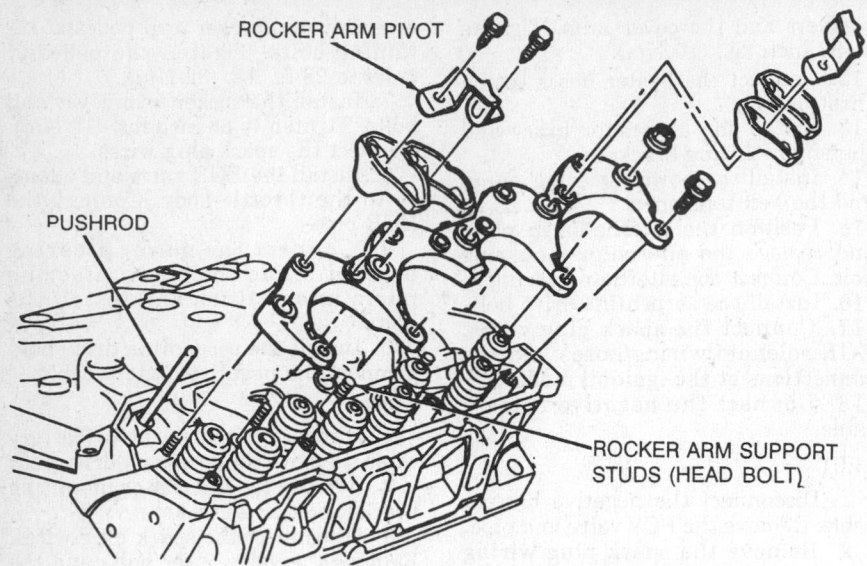

Rocker arm assembly—4.5L and 4.9L engines

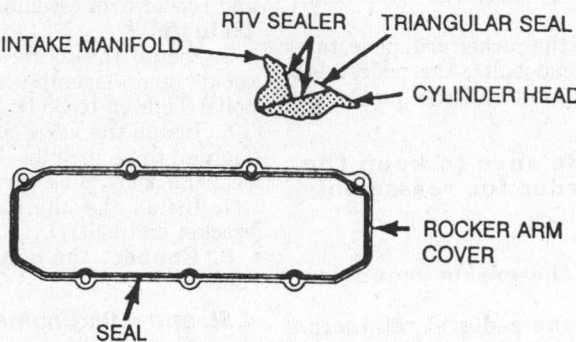

Rocker arm cover sealing—4.5L and 4.9L engines

To install:

11. Lubricate the parts with clean engine oil, use a new gasket and coat both sides with RTV sealant, install RTV sealant between the intake manifold-to-cylinder head mating surfaces.

12. Install the rocker arms and pivots to the rocker arm support. Tighten the pivot bolts to 22 ft. lbs. (30 Nm).

13. Install the rocker arm support and place each pushrod into the rocker arm seat.

14. Install the rocker arm support retaining nuts, tighten to 37 ft. lbs. (50 Nm).

15. Install the rocker arm support retaining bolts, tighten to 7 ft. lbs. (9 Nm).

16. Install the rocker arm cover seals and place the moulded seal into the groove in the rocker arm cover.

17. Install the rocker arm cover and tighten the mounting screws to 8 ft. lbs. (11 Nm).

18. Connect the brake booster vaccum hose and the EECS pipe bracket.

19. Install the spark plug wires and conduit. Connect the MAP hose to the throttle body.

20. Install the air management and bracket assembly.

21. Replace the heater hose and air cleaner assembly.

22. Connect the negative battery cable. Start the engine and check for leaks.

LEFT

1. Disconnect the negative battery cable. Remove the air cleaner, the PCV valve, the throttle return spring and the serpentine drive belt.

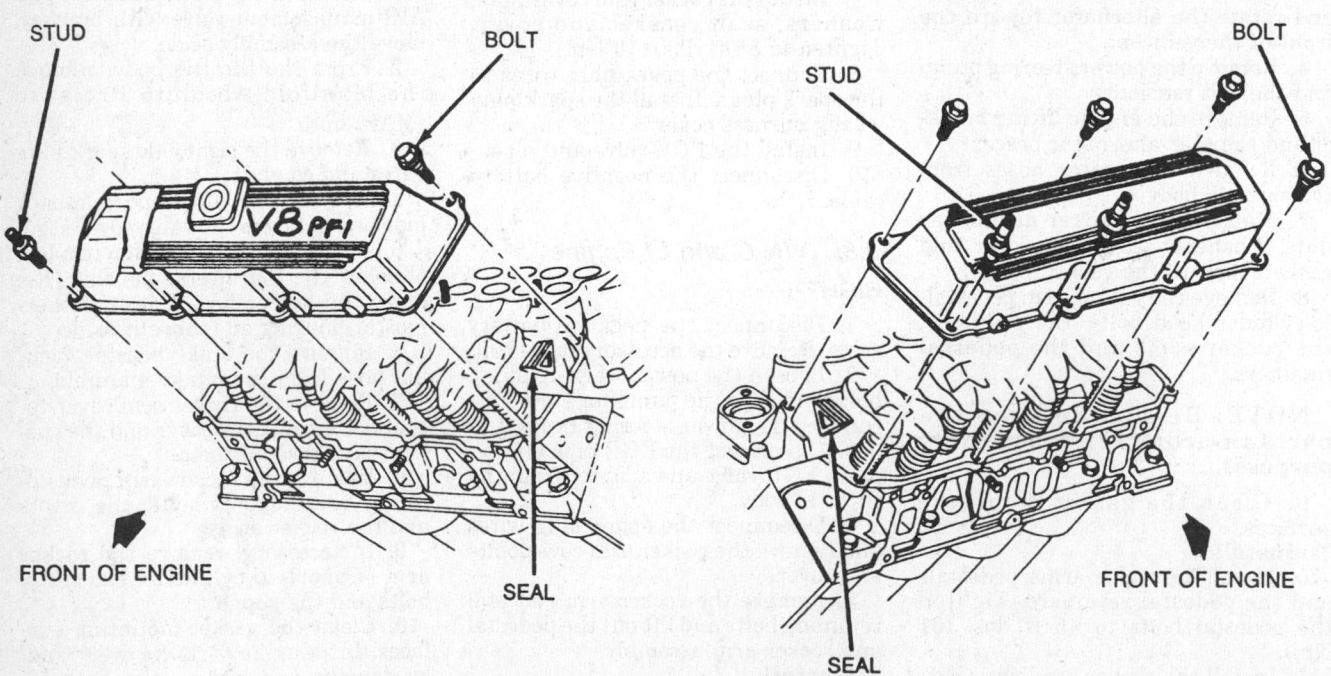

Rocker arm covers—4.5L and 4.9L engines

2. Loosen the lower power steering pump bracket nuts.

3. Remove the power steering pump, the belt tensioner, the bracket-to-engine bolts and the bracket. Move the power steering pump assembly toward the front of the vehicle; do not disconnect the pressure hoses.

4. Remove the left side spark plug wires and conduit.

5. Remove the rocker arm cover-to-cylinder screws, the cover and the gasket/seals. Discard them.

6. Remove the rocker arm pivot-to-rocker arm support bolts, the pivots and the rocker arms.

7. If necessary, remove the rocker arm support-to cylinder head nuts/bolts and the support.

8. Clean the gasket mounting surfaces. Inspect the parts for wear and/or damage and replace the parts, if necessary.

To install:

9. Lubricate the parts with clean engine oil, use a new gasket, coat both sides with RTV sealant, install RTV sealant between the intake manifold-to-cylinder head mating surfaces.

10. Install the rocker arms and pivots to the rocker arm support. Tighten the pivot bolts to 22 ft. lbs. (30 Nm).

11. Install the rocker arm support and place each pushrod into the rocker arm seat.

12. Install the rocker arm support retaining nuts, tighten to 37 ft. lbs. (50 Nm).

13. Install the rocker arm support retaining bolts, tighten to 7 ft. lbs. (9 Nm).

14. Install the rocker arm cover seals and place the moulded seal into the groove in the rocker arm cover.

15. Install the rocker arm cover and tighten the mounting screws to 8 ft. lbs. (11 Nm).

16. Install the spark plug wires and conduit.

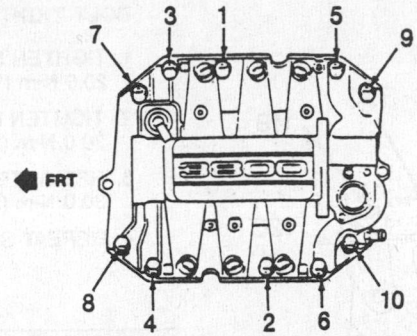

View of the intake manifold bolt torquing sequence—3.8L engine

17. Install the power steering pump, belt tensioner and bracket assembly. Replace the accessory drive belt.

18. Install the throttle return spring and the PCV valve.

19. Install the air cleaner and connect the negative battery cable.

20. Start the engine and check for leaks.

Intake Manifold

REMOVAL & INSTALLATION

3.8L Engine

1. Relieve the fuel system pressure.

2. Disconnect the negative battery cable. Place a clean drain pan under the radiator, open the drain cock and drain the cooling system.

3. Remove the serpentine drive belt, the alternator and the bracket.

4. Remove the power steering pump, the braces and move it aside; do not disconnect the pressure lines.

5. Remove the coolant bypass hose, the heater pipe and the upper radiator hose from the intake manifold.

6. Remove the vacuum hoses and disconnect the electrical connectors from the intake manifold.

7. Remove the EGR pipe, the EGR valve and the adapter from the throttle body.

8. Remove the throttle body coolant pipe, the throttle body and the throttle body adapter.

9. Disconnect the rear spark plug wires. Remove the intake manifold-to-engine bolts and the manifold.

10. Clean the gasket mounting surfaces.

To install:

11. Install new gaskets and the proper sealant on the ends of the manifold seals.

12. Install the intake manifold and tighten the mounting bolts in sequence twice to 88 inch lbs. (10 Nm).

13. Connect the rear spark plug wires.

14. Install the throttle body adapter, throttle body and the throttle body coolant pipe. Tighten the bolts to 20 ft. lbs. (27 Nm).

15. Install the EGR valve and adapter. Replace the EGR pipe.

16. Connect the vacuum hoses and the electrical connections to the intake manifold.

17. Install the coolant pipe, upper radiator hose and the upper bypass hose to the intake manifold.

18. Install the power steering pump and bracket assembly.

19. Install the alternator and bracket assembly. Replace the serpentine belt.

20. Refill the cooling system and connect the negative battery cable. Start the engine, allow it to reach normal operating temperatures and check for leaks.

4.5L and 4.9L Engines

1. Disconnect the negative terminal from the battery. Drain the cooling system to a level below the intake manifold. Disconnect the upper radiator hose from the thermostat housing.

1. Throttle body
2. Gasket
3. 20 ft. lbs.
4. Throttle body adapter
5. Gasket
6. Stud
7. Intake manifold

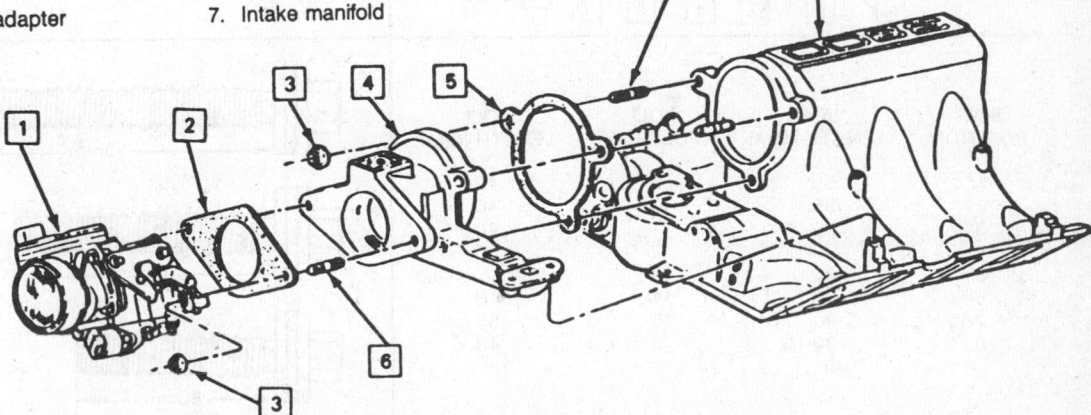

Throttle body and adapter to the intake manifold—3.8L engine

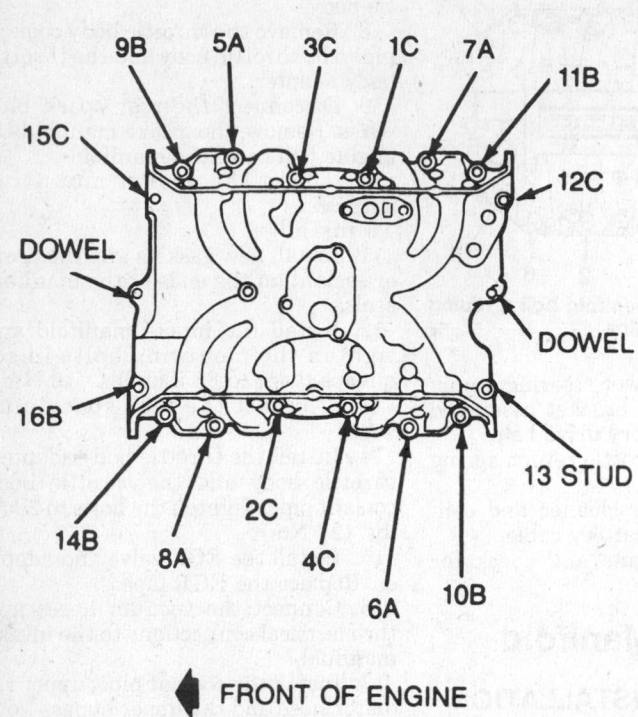

FRONT OF ENGINE

BOLT TIGHTENING SEQUENCE

1. TIGHTEN BOLTS 1, 2, 3, & 4 IN SEQUENCE TO 20.0 N·m (15 FT-LBS).
2. TIGHTEN BOLTS 5 THRU 16 IN SEQUENCE TO 30.0 N·m (22 FT-LBS).
3. RETIGHTEN ALL BOLTS IN SEQUENCE TO 30.0 N·m (22 FT-LBS).
4. REPEAT STEP 3.

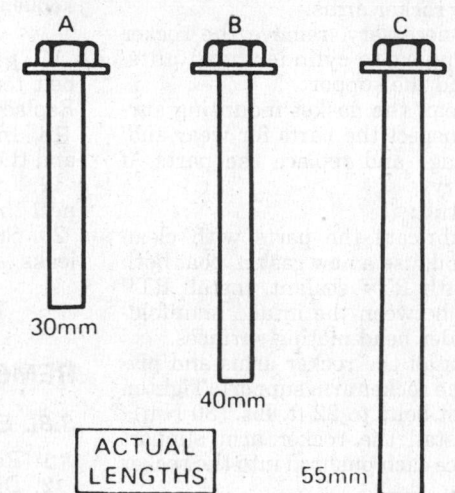

A   B   C

30mm   40mm   55mm

ACTUAL LENGTHS

Intake manifold bolt size and torque sequence—1988—4.5L engine

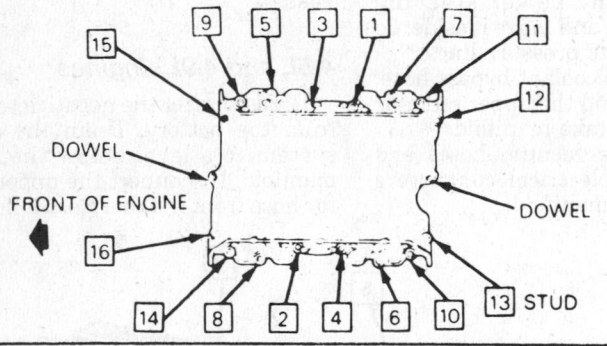

DOWEL

FRONT OF ENGINE

DOWEL

13 STUD

BOLT TIGHTENING SEQUENCE

1. TIGHTEN BOLTS 1, 2, 3, & 4 IN SEQUENCE TO 12.0 N·m (8 FT·LBS).
2. TIGHTEN BOLTS 5 THRU 16 IN SEQUENCE TO 12.0 N·m (8 FT·LBS).
3. RETIGHTEN ALL BOLTS IN SEQUENCE TO 16.0 N·m (12 FT·LBS).
4. REPEAT STEP 3 UNTIL TORQUE LEVEL IS MAINTAINED.

BOLT POSITION	BOLT LENGTH (MM)	BOLT POSITION	BOLT LENGTH (MM)
1	55	9	40
2	55	10	40
3	55	11	40
4	55	12	55
5	30	13	40 W Studhead
6	30	14	40
7	30	15	55
8	30	16	40

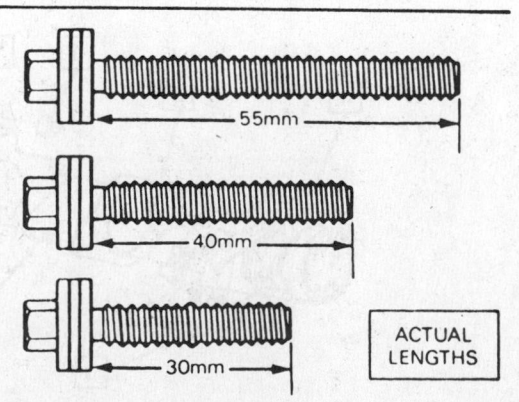

55mm

40mm

30mm

ACTUAL LENGTHS

Intake manifold bolt size and torque sequence—1989–90 4.5L and 4.9L engines

2. Remove the air cleaner and the serpentine drive belt. Label and disconnect the spark plug wires from the spark plugs.

3. Remove the upper power steering pump bracket-to-engine bolts and loosen the lower nuts.

4. Disconnect the following electrical connections and position the wiring harness aside: distributor, oil pressure switch, EGR solenoid, coolant sensor, mass airflow temperature sensor, throttle position sensor, 4-way connector at the distributor, electric fuel evaporator grid, idle speed control motor and fuel injectors.

5. From the throttle lever, disconnect the accelerator, cruise control, if equipped, and transaxle TV cables.

6. Using a shop rag at the fuel line Schraeder valve (test port), bleed off the fuel pressure. Disconnect the fuel inlet and return lines from the throttle body. From the transaxle, remove the fuel line brackets and move the lines aside; disconnect the modulator vacuum line.

7. Disconnect the heater hose from the nipple at the rear of the intake manifold.

8. From the intake manifold, remove the cruise control bracket, if equipped. Remove the vacuum line from the left rear engine lift bracket and the throttle body.

9. Disconnect the electrical connectors from the alternator and AIR management solenoid. Remove the alternator, the idler pulley, the AIR management valve/bracket and EGR solenoid/bracket. Disconnect the hose from the MAP hose.

10. From the right cylinder head, remove the power steering pipe and the AIR pipe. Raise and safely support the vehicle.

11. Drain the engine oil and remove the oil filter. Lower the vehicle.

12. Remove the distributor. Remove both rocker arm covers. Remove the rocker arm support with the rocker arms intact by first alternately and evenly removing the 4 bolts followed by the 5 nuts. Keep the pushrods in sequence so they may be reassembled in their original positions.

13. If equipped with air conditioning, partially remove the compressor; do not discharge the system. Remove the vacuum harness connections from the TVS at the rear of the intake manifold.

14. Remove the intake manifold bolts and remove the 2 bolts securing the lower thermostat housing to the front cover. Remove the engine lift brackets or bend them aside.

15. Remove the intake manifold and lower the thermostat housing as an assembly by lifting it straight up off of the dowels.

16. Clean the gasket mounting surfaces.

To install:

17. Install new gaskets and apply the proper RTV sealant to the 4 corners where the end seals meet.

18. Install the intake manifold, using new gaskets. For 1988, 4.5L engines, tighten the mounting bolts as follows:

    a. Torque the No. 1–4 bolts, in sequence, to 15 ft. lbs. (20 Nm).

    b. Torque the No. 5–16 bolts, in sequence, to 22 ft. lbs. (30 Nm).

    c. Retorque all bolts, in sequence, to 22 ft. lbs. (30 Nm).

    d. Recheck all bolts, in sequence, to 22 ft. lbs. (30 Nm).

19. For 1989–90, 4.5L and 1991–92, 4.9L engines, tighten the mounting bolts as follows:

    a. Torque the No. 1–4 bolts, in sequence, to 8 ft. lbs. (12 Nm).

    b. Torque the No. 5–16 bolts, in sequence, to 8 ft. lbs. (12 Nm).

    c. Retorque all bolts, in sequence, to 12 ft. lbs. (16 Nm).

    d. Repeat step c until torque level is maintained.

20. Install the right side engine lift brackets. Install the alternator and idler pulley mounting bracket and replace the brackets at the right cylinder head.

21. Install the pushrods and the rocker arm support assemblies.

22. Install the rocker arm covers, using new seals.

23. Replace the EGR valve and bracket assembly. Connect the MAP hose.

24. Connect the wire connectors at the ISC motor, TPS, the fuel injectors and the MAT sensor.

25. Connect the air management wires, valves and the bracket assembly.

26. Install the alternator and connect the electrical wires.

27. Install the belt tensioner, power steering pump and bracket assembly.

28. Connect the transaxle modulator vaccum line and the vacuum supply line at the throttle body.

29. Install the vacuum line bracket at the left rear engine lift bracket.

30. Install the cruise control servo bracket and connect the fuel lines at the throttle body. Connect the fuel line brackets at the transaxle.

31. Replace the upper radiator hose.

32. Connect the transmission TV, cruise control and accelerator cables at the throttle body.

33. Install the distributor cap, wires and conduit.

34. Connect the wire connectors at the distributor, oil pressure switch, coolant sensor and the EGR solenoid.

35. Replace the heater hose at the thermostat housing.

36. Raise and support the vehicle safely. Replace the oil filter and tighten the oil drain plug.

37. Install the upper left side power steering pump bracket bolts. Replace the accessory drive belt.

38. Install the air cleaner assembly and refill the cooling system.

39. Connect the negative battery cable. Start the engine and allow it to reach normal operating temperatures and check for leaks.

Exhaust Manifold

REMOVAL & INSTALLATION

3.8L Engine

RIGHT

1. Disconnect the negative battery cable.

2. If necessary, disconnect the Mass Air Flow (MAF) sensor, air intake duct, the crankcase ventilation pipe and the IAC connector from the throttle body.

3. Label and disconnect the wires from the spark plugs. Disconnect the oxygen sensor lead.

4. If equipped, disconnect the heater inlet pipe from the manifold stud. If equipped, remove the transaxle oil indicator tube.

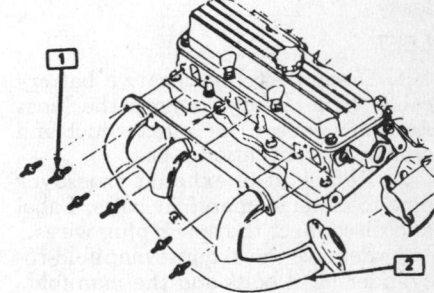

1. Stud—41 ft. lbs.
2. Left (front) exhaust manifold

Left side exhaust manifold—3.8L engine

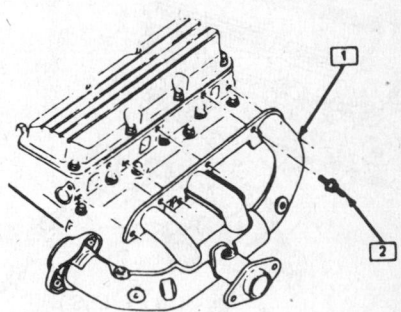

1. Right (rear) exhaust manifold
2. Stud—41 ft. lbs.

Right side exhaust manifold—3.8L engine

5. Remove the exhaust crossover pipe-to-exhaust manifold bolts and the pipe. Disconnect the alternator bracket, if necessary.

6. Raise and support the vehicle safely. Remove the exhaust pipe-to-manifold bolts, the exhaust manifold-to-cylinder head bolts and the manifold.

7. Remove the EGR pipe from the exhaust manifold.

8. Clean the gasket mounting surfaces.

To install:

9. Replace the EGR pipe to the exhaust manifold.

10. Install the exhaust manifold, using a new gasket. Tighten the mounting studs to 37–41 ft. lbs. (50–56 Nm).

11. Lower the vehicle. Connect the alternator bracket, if necessary.

12. Install the crossover pipe and replace the exhaust manifold-to-cylinder bolts.

13. If equipped, replace the transaxle oil indicator tube. Connect the heater inlet pipe to the manifold, if equipped.

14. Connect the oxygen sensor lead and the spark plug wires.

15. If removed, connect the Mass Air Flow sensor, air intake duct, the crankcase ventilation pipe and the IAC connector from the throttle body.

16. Connect the negative battery cable. Start the engine and check for leaks.

LEFT

1. Disconnect the negative battery cable. If necessary, remove the Mass Air Flow sensor, air intake duct and crankcase ventilation pipe.

2. Remove the exhaust crossover pipe-to-exhaust manifold bolts. Label and disconnect the spark plug wires.

3. Remove the exhaust manifold-to-cylinder head bolts and the manifold.

NOTE: It may be necessary to remove the oil dipstick tube to provide additional clearance.

To install:

4. Clean the gasket mounting surfaces and install a new gasket.

5. Install the exhaust manifold and tighten the manifold mounting studs to 37–41 ft. lbs. (50–56 Nm).

6. Connect the spark plug wires. Install the exhaust crossover pipe-to-exhaust manifold bolts.

7. If removed, install the Mass Air Flow sensor, air intake duct and crankcase ventilation pipe.

8. Connect the negative battery cable.

9. Start the engine and check for exhaust leaks.

4.5L and 4.9L Engines

RIGHT

1. Disconnect the negative battery cable. Remove the air cleaner.

2. Remove the exhaust crossover pipe. Disconnect the oxygen and coolant temperature sensors.

3. Remove the catalytic converter-to-AIR pipe clip bolt. Remove the upper manifold-to-cylinder head bolts. Raise and safely support the vehicle.

4. Disconnect the converter air pipe bracket from the stud and remove the converter-to-manifold exhaust pipe.

5. Support the engine cradle with screw jacks and remove the rear cradle bolts. Loosen the front cradle bolts and slightly lower the engine cradle.

6. Remove the remaining exhaust manifold-to-cylinder head bolts, the AIR pipe and the manifold.

7. Clean the gasket mounting surfaces.

To install:

8. Install the exhaust manifold and

replace the AIR pipe. Tighten the manifold mounting bolts to 16–18 ft. lbs.

9. Install the manifold-to-converter exhaust pipe and replace the converter air pipe bracket to the stud.

10. Raise the engine cradle and install the rear cradle bolts. Tighten to 75 ft. lbs (102 Nm).

11. Lower the vehicle. Replace the upper manifold-to-cylinder head bolts.

12. Replace the converter air pipe to AIR pipe clip bolt.

13. Connect the coolant temperature and oxygen sensor connectors. Replace the exhaust crossover pipe.

14. Replace the air cleaner and connect the negative battery cable.

15. Start the engine and check for leaks.

LEFT

1. Disconnect the negative battery cable. Remove both cooling fans and the exhaust crossover pipe.

2. Remove the serpentine drive belt and the AIR pump pivot bolt.

3. Remove the belt tensioner and the power steering pump brace.

4. Remove the exhaust manifold-to-cylinder head bolts, the AIR pipe and the manifold.

To install:

5. Clean the gasket mounting surfaces.

6. Install the manifold, AIR pipe and exhaust manifold-to-cylinder head bolts. Tighten to 16–18 ft. lbs. (22–24 Nm).

7. Install the belt tensioner and the power steering pump brace.

8. Install the AIR pump pivot bolt and the serpentine drive belt.

9. Install both cooling fans and the exhaust crossover pipe.

10. Connect the negative battery cable.

FRONT OF ENGINE

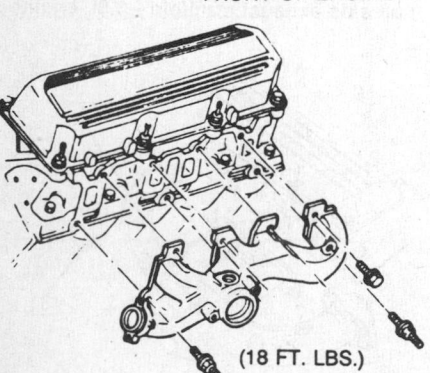

(18 FT. LBS.)

(8 FT. LBS.)

FRONT OF ENGINE

HEAT SHIELD

Exhaust manifolds—4.5L and 4.9L engines

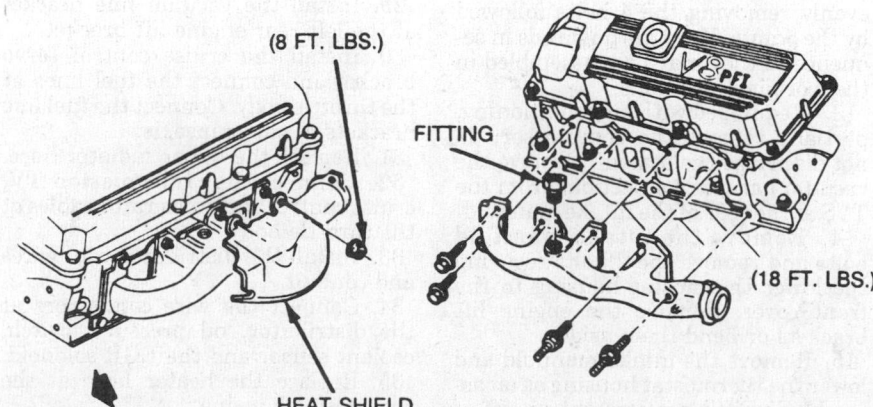

FRONT OF ENGINE

FITTING

(18 FT. LBS.)

Timing Chain Front Cover

REMOVAL & INSTALLATION

3.8L (VIN 3) Engine

1. Disconnect the negative battery cable. Drain the coolant into a clean container for reuse.

2. Remove the lower radiator hose and the coolant bypass hose from the front cover. Remove the heater pipes.

3. Remove the front engine cradle mount bolts. Using a vertical lifting device, secure it to the engine and raise it slightly. Remove the alternator and mounting bracket.

4. Remove the serpentine drive belt and the water pump pulley.

5. Label and disconnect the alternator wiring. Remove the alternator and the alternator bracket.

7. Remove the crankshaft balancer bolt/washer and the balancer.

8. Remove the front cover-to-engine bolts and the front cover.

9. Remove the front cover-to-oil pan bolts and the front chain cover.

To install:

10. Install the gasket to the cylinder block and replace the front cover. Tighten the front cover bolts to 22 ft. lbs. (30 Nm).

11. Install the balancer, washer and bolt. Tighten the bolt to 219 ft. lbs. (298 Nm).

12. Install the alternator and connect the alternator bracket.

13. Lower the engine and replace the front engine cradle bolts.

14. Install the serpentine belt and the water pump pulley.

15. Install the heater pipes, heater hoses and the radiator hoses.

16. Refill the cooling system and connect the negative battery cable.

17. Start the engine and check for leaks.

3.8L (VIN C and L) Engine

1. Disconnect the negative battery

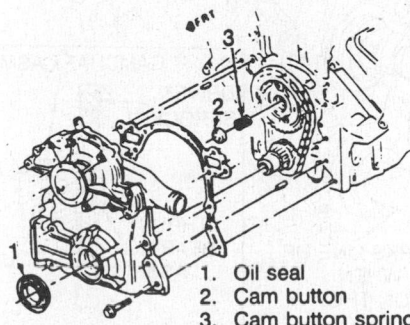

1. Oil seal
2. Cam button
3. Cam button spring

Exploded view of the front cover assembly—3.8L engine

cable. Drain the coolant into a clean container for reuse.

2. Remove the serpentine drive belt and the heater pipes.

3. Disconnect the lower radiator and bypass hoses from the cover.

4. Raise and safely support the vehicle. Remove the right front tire and wheel assembly and replace the inner splash shield.

5. Remove the crankshaft bolt and balancer.

6. Disconnect the electrical connections at the camshaft sensor, crankshaft sensor and the oil pressure switch.

7. Remove the oil pan-to-front cover bolts.

8. Remove the front cover mounting bolts and the cover assembly.

9. Clean the gasket surfaces at the cover and the cylinder block.

To install:

10. Install the gasket to the cylinder block.

11. Install the front cover and tighten the mounting bolts to 22 ft. lbs. (30 Nm).

12. Replace the oil pan-to-cover mounting bolts and tighten to 88 inch lbs. (10 Nm).

13. Connect the electrical connections and replace the crankshaft balancer and tighten the bolt to 219 ft. lbs. (298 Nm).

14. Install the inner fender splash shield and the right front tire and wheel assembly.

15. Lower the vehicle. Replace the coolant bypass hose and radiator hoses.

16. Connect the heater pipes and install the drive belt.

17. Refill the cooling system and connect the negative battery cable.

18. Start the engine and check for leaks.

4.5L and 4.9L Engines

1. Disconnect the negative battery cable. Remove the air cleaner.

2. Drain the coolant into a clean container for reuse.

3. Remove the serpentine belt.

4. Label and disconnect the alternator wiring. Remove the alternator and the alternator bracket.

5. Remove the air conditioner accumulator from the bracket and move it aside. Do not disconnect the fittings on the accumulator.

6. Remove the water pump pulley and pump. Remove the idler pulley, as required.

7. Raise and safely support the vehicle.

8. Remove the crankshaft pulley-to-crankshaft pulley bolt. Attach a puller to the crankshaft pulley; using the center bolt, press the crankshaft pulley from the crankshaft.

9. Remove the front cover-to-engine bolts, the oil pan-to-front cover bolts and the front cover.

10. Clean the gasket mounting surfaces.

To install:

11. Install the timing cover and tighten the mounting bolts to 15 ft. lbs. (20 Nm).

12. Install the crankshaft damper and tighten the bolt to 18 ft. lbs. (24 Nm).

13. Lower the vehicle. Replace the water pump and pulley.

14. If removed, install the idler pulley. Install the serpentine belt.

15. Connect the alternator wiring and install the alternator and bracket.

16. Replace the air conditioner accumulator and connect the bracket.

17. Replace the air cleaner and refill the cooling system.

18. Connect the negative battery cable. Start the engine and check for leaks.

Front Cover Oil Seal

REPLACEMENT

3.8L Engine

1. Disconnect the negative battery cable.

2. Remove the serpentine drive belt. Remove the crankshaft balancer-to-crankshaft bolt.

3. Using a small prybar, pry the oil seal from the front cover. Be careful not to damage the sealing surfaces.

To install:

4. Clean the oil seal mounting surface. Using the proper lubricant coat the outside of the seal and the crankshaft balancer.

5. Using the oil seal installation tool, drive the new seal into the front cover until it seats.

6. Install the crankshaft balancer-to-crankshaft bolts. Tighten to 219 ft. lbs. (298 Nm). Install the serpentine drive belt.

7. Connect the negative battery cable.

4.5L and 4.9L Engines

1. Disconnect the negative battery cable. Remove the serpentine belt.

2. Remove the crankshaft pulley-to-crankshaft pulley bolt.

3. Attach a puller tool to the crankshaft pulley. Using the center bolt, press the crankshaft pulley from the crankshaft.

4. Using the oil seal removal tools, press the oil seal from the front cover. Clean the oil seal mounting surface.

To install:

5. Lubricate the new seal with engine oil. Using a hammer and an oil

seal installation tool, drive the new oil seal into the front cover until it seats.

6. Install the crankshaft pulley to the crankshaft. Tighten the crankshaft pulley-to-crankshaft bolt to 18 ft. lbs. (24 Nm).

7. Install the serpentine belt.

8. Connect the negative battery cable.

Timing Chain and Sprockets

REMOVAL & INSTALLATION

3.8L Engine

1. Disconnect the negative battery cable. Remove the front cover.

2. Remove the button and spring from the center of the camshaft. Align the marks of the timing sprockets as they must be close together.

4. Remove the camshaft sprocket bolts, the sprocket and the timing chain.

5. Remove the crankshaft sprocket. Clean the gasket mounting surfaces.

To install:

6. Install the timing chain and sprockets by performing the following:

   a. Assemble the timing chain on the camshaft sprocket and crankshaft sprockets.

   b. Align the O-marks on the sprockets; they must face each other.

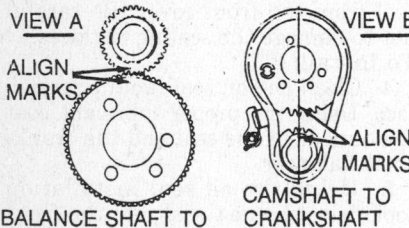

View of the timing chain, sprockets and balancer shaft alignment—3.8L engine

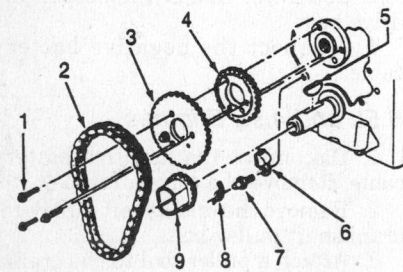

1. 27 ft. lbs.
2. Timing chain
3. Camshaft sprocket
4. Camshaft gear
5. Key
6. Damper
7. Special bolt (14 ft. lbs.)
8. Spring
9. Crankshaft sprocket

Exploded view of the timing chain, sprockets and balancer shaft sprocket—3.8L engine

c. Slide the assembly onto the camshaft and crankshaft. Install the camshaft sprocket-to-camshaft bolts. Tighten the camshaft sprocket-to-camshaft sprocket bolts to 27–28 ft. lbs. (37–38 Nm).

NOTE: If equipped with 3.8L (VIN C or L) engine, align the camshaft sprocket mark with the balancer shaft sprocket mark.

7. Install the camshaft button and spring.

8. Replace the front cover assembly.

9. Connect the negative battery cable.

10. Refill the cooling system. Start the engine, allow it to reach normal operating temperatures and check for leaks.

4.5L and 4.9L Engines

1. Disconnect the negative battery cable. Remove the front cover.

2. Remove the oil slinger from the crankshaft. Rotate the engine to align the sprocket timing marks; the No. 1 cylinder will be on the TDC of its compression stroke.

3. From the camshaft, remove the camshaft thrust button, replace it and the camshaft sprocket-to-camshaft screw. Slide the camshaft sprocket, the crankshaft sprocket and timing chain from the engine as an assembly.

To install:

4. Clean the gasket mounting surfaces. Inspect the parts for wear and/or damage; if necessary, replace the parts.

5. Install the timing chain and sprockets by performing the following:

   a. Assemble the timing chain on the camshaft sprocket and crankshaft sprockets.

   b. Align the timing marks on the sprockets; they must face each other.

   c. Align the dowel pin in the camshaft with the index hole in the sprocket.

   d. Slide the assembly onto the camshaft and crankshaft. Install the camshaft sprocket-to-camshaft bolts. Torque the camshaft sprocket-to-camshaft sprocket bolt to 37 ft. lbs. (50 Nm).

6. Install the oil slinger to the crankshaft.

7. Install the front cover. Connect the negative battery cable.

8. Refill the cooling system. Start the engine, allow it to reach normal operating temperatures and check for leaks.

Camshaft

REMOVAL & INSTALLATION

3.8L Engine

1. Disconnect the negative battery cable. Remove the engine assembly and position in a suitable holding fixture.

2. Remove the intake manifold, the front timing cover, timing chain and sprockets.

3. Remove the valve covers, the rocker arm shaft or rocker arm assemblies, the pushrods and the hydraulic lifters.

NOTE: Keep all valve components in order so they may be reinstalled in their original positions.

4. Carefully, slide the camshaft forward, out of the bearing bores; do not damage the bearing surfaces.

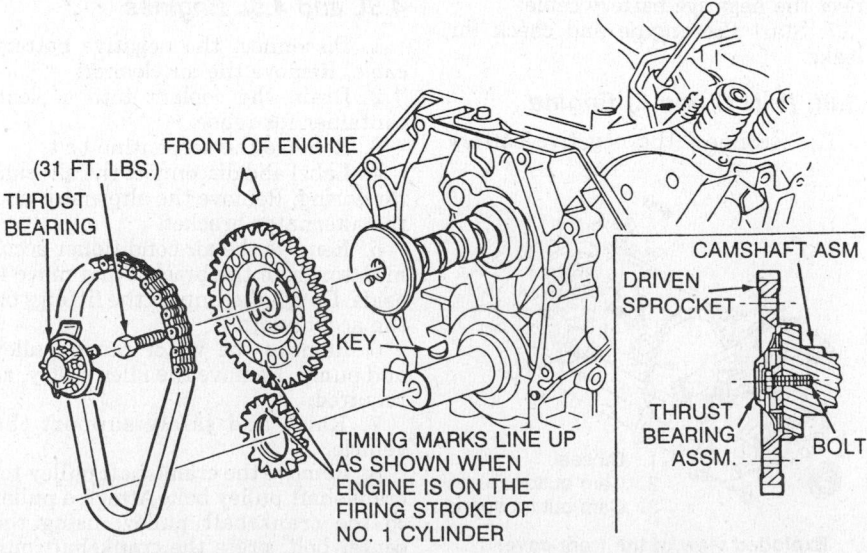

Camshaft and timing chain alignment—4.5L and 4.9L engines

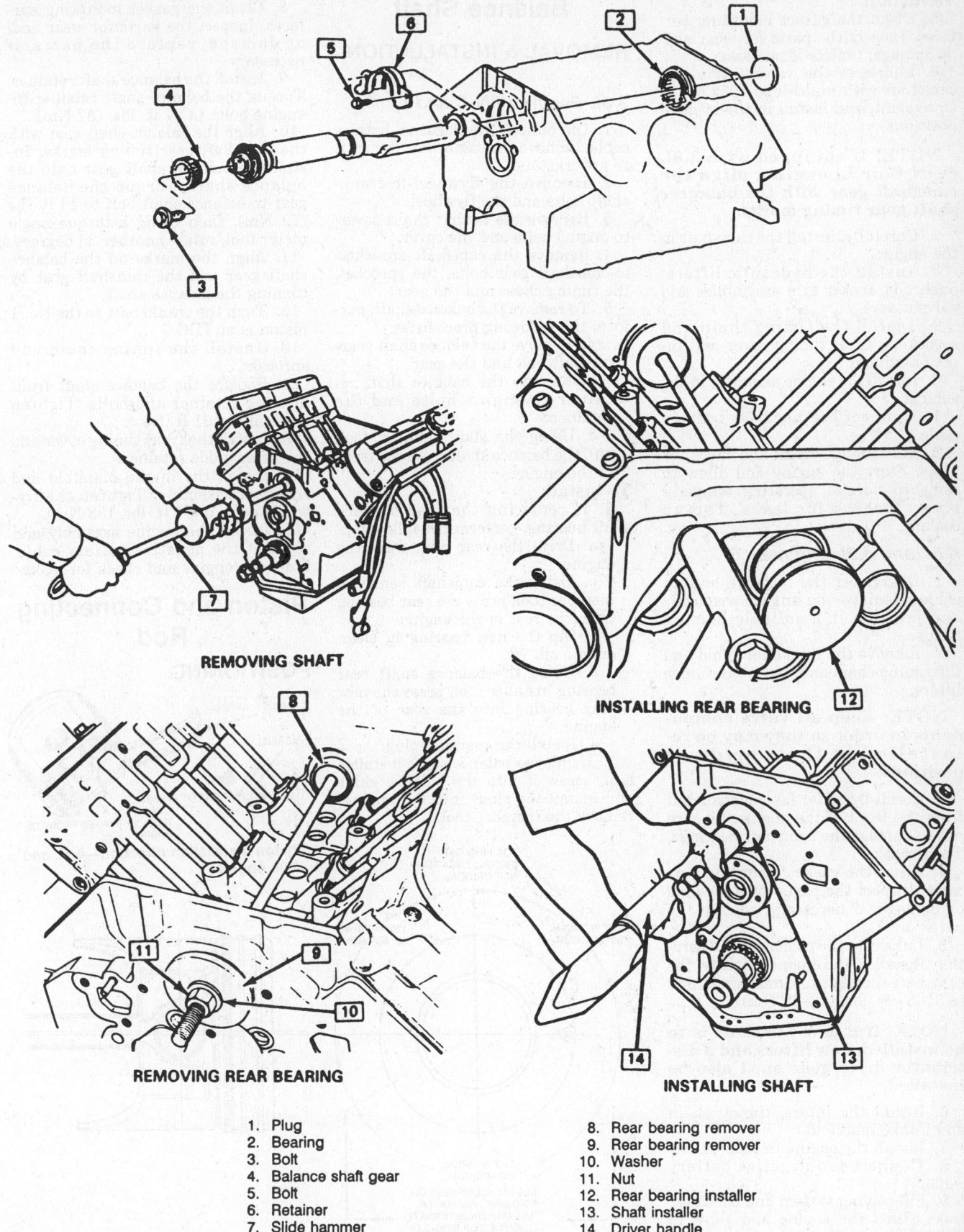

REMOVING SHAFT

INSTALLING REAR BEARING

REMOVING REAR BEARING

INSTALLING SHAFT

1. Plug
2. Bearing
3. Bolt
4. Balance shaft gear
5. Bolt
6. Retainer
7. Slide hammer

8. Rear bearing remover
9. Rear bearing remover
10. Washer
11. Nut
12. Rear bearing installer
13. Shaft installer
14. Driver handle

Balance shaft service—3.8L engine

To install:

5. Clean the gasket mounting surfaces. Inspect the parts for wear and/or damage, replace if necessary.

6. Lubricate the valve lifters and camshaft with multi-lube 1052365 or equivalent, and install in the original positions.

NOTE: If equipped with 3.8L (VIN C or L) engine, align the camshaft gear with the balancer shaft gear timing marks.

7. Carefully, install the camshaft in the engine.

8. Install the hydraulic lifters, pushrods, rocker arm assemblies and valve covers.

9. Install the timing chain and sprockets, front timing cover and intake manifold.

10. Install the engine assembly in the vehicle.

11. Connect the negative battery cable.

12. Fill cooling system and check for leaks. Start the engine and allow to come to normal operating temperature. Recheck for leaks. Top-up coolant.

4.5L and 4.9L Engines

1. Disconnect the negative battery cable. Remove the engine assembly and position in a suitable holding fixture.

2. Remove the intake manifold and the timing chain and remove the valve lifters.

NOTE: Keep all valve components in order so they may be reinstalled in their original positions.

3. Carefully slide the camshaft out from the front of the engine. Be sure not to damage the camshaft bearings.

To install:

4. Clean the gasket mounting surfaces. Inspect the parts for wear and/or damage; if necessary, replace the parts.

5. Lubricate the camshaft and carefully install in the engine. Tighten the camshaft sprocket-to-camshaft screws to 31–37 ft. lbs. (42–50 Nm).

NOTE: If a new camshaft is to be installed, new lifters and a distributor drive gear must also be installed.

6. Install the lifters, timing chain and intake manifold.

7. Install the engine in the vehicle.

8. Connect the negative battery cable.

9. Fill cooling system and check for leaks. Start the engine and allow to come to normal operating temperature. Recheck for leaks. Top-up coolant.

Balance Shaft

REMOVAL & INSTALLATION

3.8L Engine (VIN C and L)

1. Disconnect the negative battery cable. Remove the engine and secure it to a workstand.

2. Remove the flywheel-to-crankshaft bolts and the flywheel.

3. Remove the timing chain cover-to-engine bolts and the cover.

4. Remove the camshaft sprocket-to-camshaft gear bolts, the sprocket, the timing chain and the gear.

5. To remove the balance shaft, perform the following procedures:

   a. Remove the balance shaft gear-to-shaft bolt and the gear.

   b. Remove the balance shaft retainer-to-engine bolts and the retainer.

   c. Using the slide hammer tool, pull the balance shaft from the front of the engine.

To install:

6. If replacing the rear balance shaft bearing, perform the following:

   a. Drive the rear plug from the engine.

   b. Using the camshaft remover/installer tool, press the rear bearing from the rear of the engine.

   c. Dip the new bearing in clean engine oil.

   d. Using the balance shaft rear bearing installer tool, press the new rear bearing into the rear of the engine.

   e. Install the rear cup plug.

7. Using the balance shaft installer tool, screw it into the balance shaft and install the shaft into the engine; remove the installer tool.

8. Clean the gasket mounting surfaces. Inspect the parts for wear and/or damage; replace the parts, if necessary.

9. Install the balance shaft retainer. Torque the balance shaft retainer-to-engine bolts to 27 ft. lbs. (37 Nm).

10. Align the balance shaft gear with the camshaft gear timing marks. Install the balance shaft gear onto the balance shaft. Torque the balance gear-to-balance shaft bolt to 14 ft. lbs (19 Nm), then using a torque angle meter tool, rotate another 35 degrees.

11. Align the marks on the balance shaft gear and the camshaft gear by turning the balance shaft.

12. Turn the crankshaft so the No. 1 piston is at TDC.

13. Install the timing chain and sprocket.

14. Replace the balance shaft front bearing retainer and bolts. Tighten the bolts to 61 ft. lbs.

15. Install the front timing cover and the lifter guide retainer.

16. Install the intake manifold and flywheel assembly. Tighten the flywheel bolts to 61 ft. lbs. (83 Nm).

17. Install the engine assembly and connect the negative battery cable. Start the engine and check for leaks.

Piston and Connecting Rod

POSITIONING

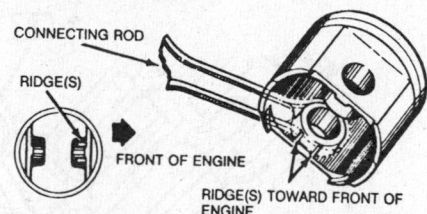

CONNECTING ROD
RIDGE(S)
FRONT OF ENGINE
RIDGE(S) TOWARD FRONT OF ENGINE

Piston installation direction—4.5L and 4.9L engines

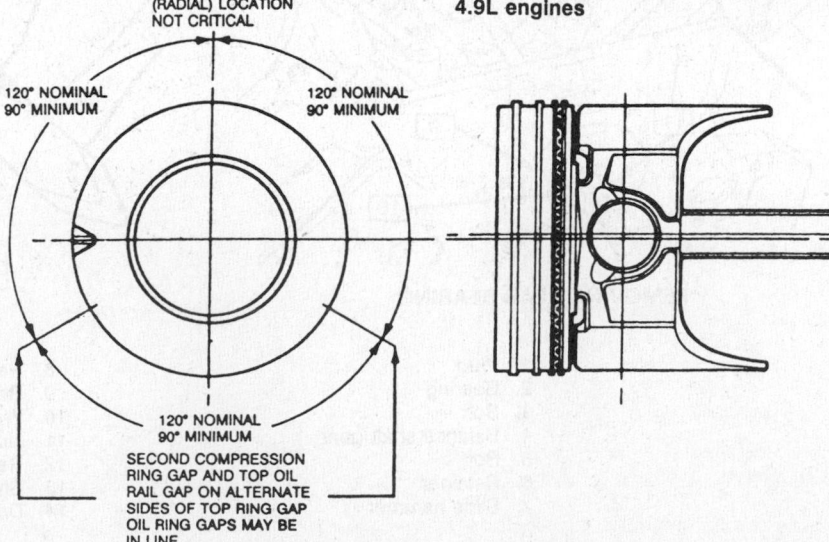

TOP RING GAP (RADIAL) LOCATION NOT CRITICAL

120° NOMINAL 90° MINIMUM

120° NOMINAL 90° MINIMUM

120° NOMINAL 90° MINIMUM

SECOND COMPRESSION RING GAP AND TOP OIL RAIL GAP ON ALTERNATE SIDES OF TOP RING GAP OIL RING GAPS MAY BE IN LINE

Piston ring orientation—4.5L and 4.9L engines

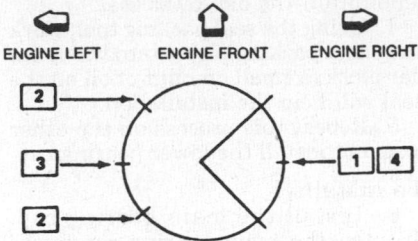

1. Oil ring spacer gap (tang in hole or slot with arc)
2. Oil ring rail gaps
3. 2nd compression ring gap
4. Top compression ring gap

Piston ring gap locations—3.8L engine

ENGINE LUBRICATION

Oil Pan

REMOVAL & INSTALLATION

3.8L Engine

1. Disconnect the negative battery cable. Raise and safely support the vehicle.
2. Drain the crankcase and remove the transaxle converter cover.
3. Remove the oil filter and the starter motor.
4. Remove the oil pan-to-engine bolts and the oil pan.

To install:

5. Clean the gasket mounting surfaces.
6. Install the oil pan and the oil pan-to-engine bolts. Tighten bolts according to the following:
   a. If equipped with VIN 3 engine, tighten to 88 inch lbs. (10 Nm).
   b. If equipped with VIN C or L engine, tighten to 124 inch lbs. (14 Nm).
7. Install a new oil filter and the starter motor.
8. Install the transaxle converter cover.
9. Lower the vehicle.
10. Fill the crankcase with oil.
11. Connect the negative battery cable.

4.1L and 4.5L Engines

1. Disconnect the negative battery cable. Raise and safely support the vehicle.
2. Drain the crankcase and remove the oil filter. Remove the flywheel inspection cover.

3. Remove the oil pan-to-engine bolts and the oil pan.

NOTE: If the pan is difficult to remove, lightly tap the edges with a plastic hammer.

To install:

4. Clean the gasket mounting surfaces.
5. Install a new oil pan gasket. Install the oil pan to the engine. Tighten the oil pan-to-engine bolts to 12 ft. lbs. (16 Nm).
6. Install the flywheel inspection cover.
7. Install a new oil filter.
8. Lower the vehicle.
9. Refill the crankcase with oil.
10. Connect the negative battery cable.
11. Start the engine and check for leaks.

Oil Pump

REMOVAL & INSTALLATION

3.8L Engine

1. Disconnect the negative battery cable. Remove the front cover from the engine.
2. Remove the oil filter adapter, pressure regulator valve and spring.
3. Remove the oil pump cover-to-front cover screws and the cover. Remove the inner and outer pump gears.

To install:

4. Using petroleum jelly, pack the pump and assemble the gears in the housing. Tighten the oil pump cover-to-front cover screws to 97 inch lbs. (11 Nm).
5. Install the pressure regulator spring and valve. Install the oil filter adapter.
6. Install the front cover to the engine.
7. Connect the negative battery cable.

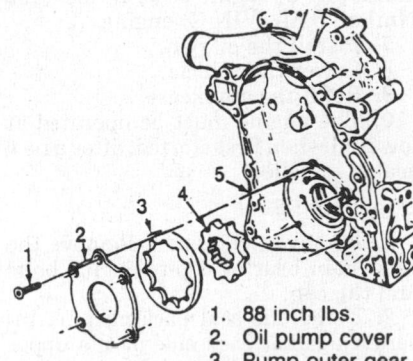

1. 88 inch lbs.
2. Oil pump cover
3. Pump outer gear
4. Pump inner gear
5. Front cover

Exploded view of the oil pump assembly—3.8L engine

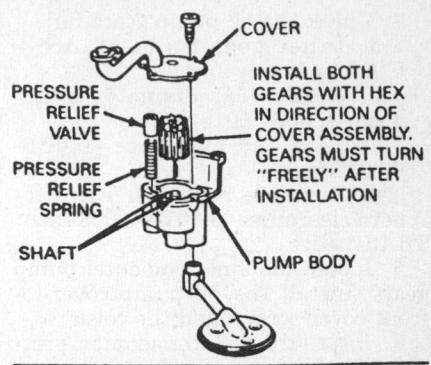

INSTALL BOTH GEARS WITH HEX IN DIRECTION OF COVER ASSEMBLY. GEARS MUST TURN "FREELY" AFTER INSTALLATION

COVER

PRESSURE RELIEF VALVE

PRESSURE RELIEF SPRING

SHAFT

PUMP BODY

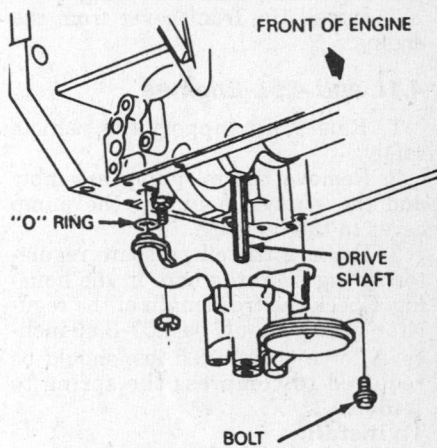

FRONT OF ENGINE

"O" RING

DRIVE SHAFT

BOLT

Oil pump assembly—4.5L and 4.9L engines

4.1L and 4.5L Engines

1. Disconnect the negative battery cable. Raise and safely support the vehicle.
2. Drain the crankcase. Remove the oil pan mounting bolts and remove the oil pan.
3. Remove the oil pump-to-engine screws/nut and the oil pump from the engine.

To install:

4. Clean the mounting surfaces. Install the pump assembly and tighten the mounting screws to 15 ft. lbs. (20 Nm) and the nut to 22 ft. lbs. (30 Nm).
5. Install a new oil pan gasket. Install the oil pan and bolts.
6. Lower the vehicle.
7. Connect the negative battery cable.
8. Refill the crankcase start the engine and check for leaks.

CHECKING

3.8L Engine

1. Remove the front cover from the engine.
2. Remove the oil filter adapter, pressure regulator valve and spring.
3. Remove the oil pump cover-to-front cover screws and the cover. Remove the inner and outer pump gears.

4. Check the oil pump gears for:

a. Inner gear tip clearance — 0.006 inch

b. Outer gear diameter clearance — 0.008–0.015 inch

c. Gear end clearance — 0.001–0.0035 inch

d. Pressure regulator valve-to-bore clearance — 0.0015–0.003 inch

To install:

5. Install the inner and outer pump gears. Install the oil pump cover-to-front cover screws and the cover.

6. Install the oil filter adapter, pressure regulator valve and spring.

7. Install the front cover from the engine.

4.1L and 4.5L Engines

1. Raise and support the vehicle safely.

2. Remove the oil pump assembly and the screws mounting the pump cover to the housing.

3. Remove the oil pressure regulator spring from the bore in the housing. Check the free length of the regulator spring, should be 2.57–2.69 inches. A force of 9.3–10.5 lbs. should be required to compress the spring to 1.46 inch.

To install:

4. Assemble the oil pump.

5. Replace the O-ring at the oil pump outlet pipe.

6. Position the oil pump to the engine block, engaging the drive rod to the distributor gear. Install the 2 screws and 1 nut. Tighten the nut to 22 ft. lbs. (30 Nm). and screws to 15 ft. lbs. (20 Nm).

7. Install the oil pan.

8. Lower the vehicle.

9. Refill the crankcase with oil.

Rear Main Bearing Oil Seal

REMOVAL & INSTALLATION

Rope Type

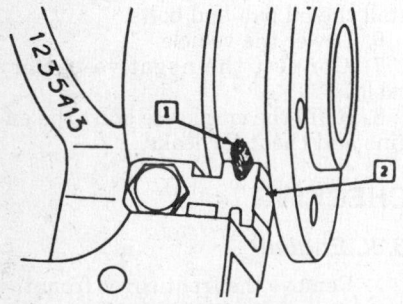

1. Short piece of rope seal
2. Guide tool installed

Installation of rope type seal — 3.8L engine

1988–90 3.8L (VIN 3 AND C) ENGINE

Lower Half-Seal

1. Remove the oil pan. Remove the rear main bearing cap-to-engine bolts and the cap.

2. Remove the old seal from the bearing cap.

3. To replace the oil seal, perform the following procedures:

a. Using a suitable sealant, apply it to the main bearing cap seal groove and wait for 1 minute.

b. Using a new rope seal and a wooden dowel or hammer handle, roll the new seal into the cap so both ends projecting above the parting surface of the cap; force the seal into the groove by rubbing it down, until the seal projects above the groove not more than $\frac{1}{16}$ in.

c. Using a sharp razor blade, cut the ends off flush with the surface of the cap.

d. Using chassis grease, apply a thin coat to the seals surface.

4. To install the neoprene sealing strips (side seals), perform the following procedures:

a. Using light oil or kerosene, soak the strips for 5 minutes.

NOTE: The neoprene composition seals will swell up once exposed to the oil and heat. It is normal for the seals to leak for a short time, until they become properly seated. The seals must not be cut to fit.

b. Place the sealing strips in the grooves on the sides of the bearing cap.

5. Using sealer the proper sealer, apply it to the main bearing cap mating surface; do not apply sealer to the bolt holes.

To install:

6. Install the main bearing cap. Torque the main bearing cap-to-engine bolts to 100 ft. lbs. (136 Nm) on 3.8L (VIN 3) engine or 90 ft. lbs. (122 Nm) on 3.8L (VIN C) engine.

7. Install the oil pan.

8. Lower the vehicle.

9. Refill the crankcase.

10. The engine must be operated at low rpm when first started, after a new seal is installed.

Upper Half-Seal

1. Remove the oil pan. Remove the rear main bearing cap-to-engine bolts and the cap.

2. Using the seal packing tool, insert it against each side of the upper seal and drive the seal in until it is tight.

3. Measure the amount the seal was driven into the engine and add about $\frac{1}{16}$ in. Using a razor blade, cut that amount off the old lower seal.

4. Using the seal packing tool, work the short packing pieces into the cylinder block; a small amount of oil on the seal will help the installation.

5. Repeat this process on the other side and install the lower bearing cap.

To install:

6. Install the main bearing cap. Tighten the main bearing cap-to-engine bolts to 100 ft. lbs. (136 Nm) on the 3.8L (VIN 3) engine or 90 ft. lbs. (122 Nm) on the 3.8L (VIN C) engine.

7. Install the oil pan.

8. Lower the vehicle.

9. Refill the crankcase.

10. The engine must be operated at low rpm when first started, after a new seal is installed.

One-Piece Lip Type

1991–92 3.8L (VIN C AND L) ENGINE

1. Disconnect the negative battery cable.

2. Raise and safely support the vehicle.

3. Remove the transaxle.

4. Remove the flywheel.

5. Insert a suitable prying tool through the dust lip and pry the seal out by moving the handle of the tool toward the end of the crankshaft pilot. Repeat the process, as required, around the seal until it is removed.

NOTE: Use care when prying out the seal to avoid damage to the OD and chamfer of the crankshaft.

To install:

6. Apply engine oil to the ID and OD of the new seal. Slide the new seal over the mandrel until the back of the seal bottoms squarely against the collar of the tool.

7. Align the dowel pin of the installation tool with the dowel pin in the crankshaft and attach the tool to the crankshaft by hand or by tightening the attaching screw to 60 inch lbs. (5 Nm).

8. Turn the T-handle of the tool so the collar pushes the seal into the bore. Continue turning until the collar is tight against the case. This will ensure that the seal is seated properly.

9. Loosen the T-handle of the tool until it comes to a stop. This will ensure that the collar will be in the proper position for install another new seal.

10. Remove the attaching screws.

11. Install the flywheel.

12. Install the transaxle.

13. Lower the vehicle.

14. Connect the negative battery cable.

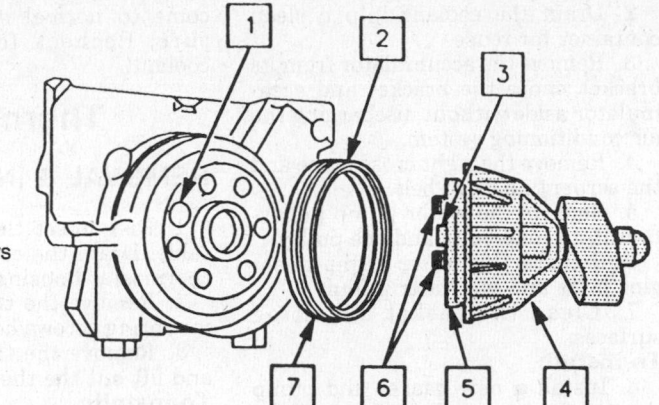

1. Alignment hole
2. Dust lip
3. Dowel pin
4. Collar
5. Mandril
6. Attaching screws
7. Seal

Installing rear main seal—1 piece lip type

4.1L and 4.5L Engines

1. Raise and safely support the vehicle.

2. Remove the transaxle assembly. Remove the flexplate from the crankshaft.

3. Using the proper tool, pry out the old seal from the rear of the engine.

To install:

4. Lubricate the new seal with wheel bearing grease and install on the crankshaft with the spring facing inside the engine.

5. Press the seal into position, using the proper tool.

NOTE: The seal should be flush with the block. It is necessary to use the proper tool because the seal must be installed square or an oil leak could result.

6. Install the flexplate to the crankshaft.

7. Install the transaxle assembly.

8. Lower the vehicle.

ENGINE COOLING

Radiator

REMOVAL & INSTALLATION

1. Disconnect the negative battery cable.

2. Drain the radiator coolant. Remove the upper radiator panel.

3. Disconnect and remove the cooling fans, as required.

4. Disconnect the coolant reservoir hoses and the radiator hoses.

5. If equipped, disconnect the engine coolant lines from the radiator.

6. Disconnect the transaxle cooler lines. Remove the radiator.

To install:

7. Install the radiator. Connect the transaxle cooler lines.

8. If removed, connect the engine coolant lines to the radiator.

9. Connect the coolant reservoir hoses and the radiator hoses.

10. If removed, install the cooling fans.

11. Install the upper radiator panel.

12. Connect the negative battery cable.

13. Fill cooling system and check for leaks. Start the engine and allow to come to normal operating temperature. Recheck for leaks. Top-up coolant.

Heater Core

REMOVAL & INSTALLATION

Except Cadillac

1. Disconnect the negative battery cable. Drain the coolant into a clean container for reuse.

2. Remove the right side sound insulator and disconnect the heater hoses at the heater core.

3. Remove the center and lower instrument panel trim plates.

4. If equipped with electronic climate control, perform the following procedures:

   a. Disconnect the wires and the hose from the programmer.

   b. Remove the programmer linkage cover and linkage.

   c. Remove the programmer mounting bolts and the programmer.

5. Remove the heater core cover and heater core assembly.

To install:

6. Install the heater core assembly and heater core cover.

7. If equipped with electronic climate control, perform the following:

   a. Install the programmer mounting bolts and the programmer.

   b. Install the programmer linkage and linkage cover.

   c. Connect the wires and the hose to the programmer.

8. Install the center and lower instrument panel trim plates.

9. Install the right side sound insulator and disconnect the heater hoses at the heater core.

10. Connect the negative battery cable.

11. Fill cooling system and check for leaks. Start the engine and allow to come to normal operating temperature. Recheck for leaks. Top-up coolant.

Cadillac

1. Disconnect the negative battery cable. Drain the coolant into a clean container for reuse.

2. Remove the right side sound insulator and disconnect the heater hoses at the heater core.

3. Remove the instrument panel and gauges.

4. Remove the glove box assembly and disconnect the programmer shield.

5. Disconnect the air mix valve link, the program vacuum and electrical connectors.

6. Remove the heater core cover with the programmer attached.

7. Remove the heater core retaining screws and the heater core assembly.

8. Clean the mounting surfaces.

To install:

9. Install the heater core assembly. Replace the heater core cover with the programmer attached.

10. Connect the vacuum and electrical connections.

11. Connect the air mix valve link and adjust the air mix.

12. Install the glove box assembly and connect programmer shield.

13. Install the instrument panel and the gauges.

14. Install the right side sound insulator and connect the heater hoses at the heater core.

15. Connect the negative battery cable.

16. Fill cooling system and check for leaks. Start the engine and allow to come to normal operating temperature. Recheck for leaks. Top-up coolant.

Water Pump

REMOVAL & INSTALLATION

3.8L Engine

1. Disconnect the negative battery cable. Drain the coolant into a clean container for reuse.

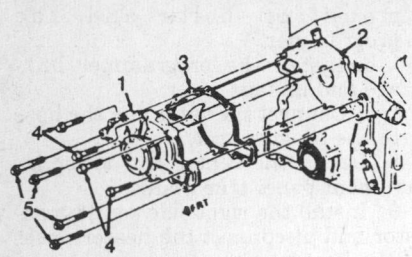

1. Water pump
2. Engine front cover assembly
3. Gasket
4. 97 inch lbs.
5. 29 ft. lbs.

Exploded view of the water pump—3.8L engine

2. Remove the serpentine drive belt and the coolant hoses from the water pump.

3. Remove the water pump pulley bolts and the pulley; the long bolt can be removed through the access hole in the body side rail.

4. Remove the water pump-to-engine bolts and the pump.

To install:

5. Clean the gasket mounting surfaces. Install a new gasket and pump assembly.

6. Install the water pump-to-engine mounting bolts and tighten to 29 ft. lbs. (26 Nm) for the long bolts and 97 inch lbs. (11 Nm) for the short bolts.

7. Connect the coolant hoses to the water pump and install the serpentine drive belt.

8. Connect the negative battery cable.

9. Fill cooling system and check for leaks. Start the engine and allow to come to normal operating temperature. Recheck for leaks. Top-up coolant.

4.5L and 4.9L Engines

1. Disconnect the negative battery cable.

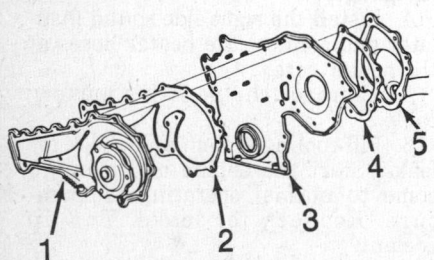

1. WATER PUMP ASSEMBLY
2. WATER PUMP GASKET
3. FRONT COVER
4. WATER PUMP INLET GASKET
5. WATER PUMP INLET

Exploded view of the water pump—4.5L and 4.9L engines

2. Drain the coolant into a clean container for reuse.

3. Remove the accumulator from its bracket, move the bracket and accumulator aside without discharging the air conditioning system.

4. Remove the right cross brace and the serpentine drive belt.

5. Remove the water pump pulley-to-water pump bolts and the pulley.

6. Remove the water pump-to-engine bolts and the water pump.

7. Clean the gasket mounting surfaces.

To install:

8. Install a new gasket and pump the water pump pulley, do not fully tighten the screws.

9. Install the right cross brace.

10. Install the air conditioning accumulator bracket and accumulator.

11. Install the serpentine drive belt.

12. Tighten the water pump pulley bolts fully.

13. Connect the negative battery cable.

14. Fill cooling system and check for leaks. Start the engine and allow to

come to normal operating temperature. Recheck for leaks. Top-up coolant.

Thermostat

REMOVAL & INSTALLATION

1. Disconnect the negative battery cable. Drain the coolant to below the thermostat housing.

2. Remove the thermostat housing mounting screws/bolts.

3. Remove the thermostat housing and lift out the thermostat.

To install:

4. Clean the mounting surfaces and install new gasket(s) or O-ring.

5. Install the thermostat and mounting screws/bolts.

6. Connect the negative battery cable.

7. Fill cooling system and check for leaks. Start the engine and allow to come to normal operating temperature. Recheck for leaks. Top-up coolant.

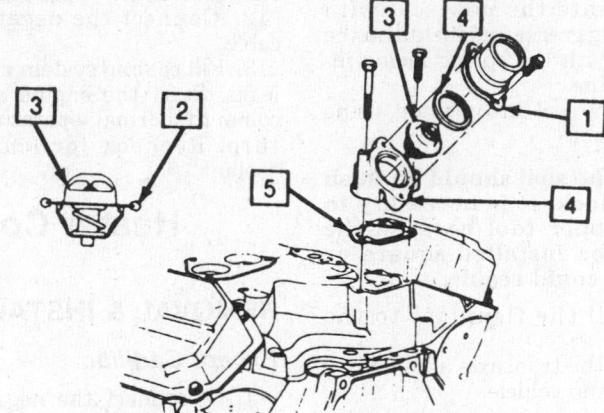

1. Upper housing
2. Gasket
3. Thermostat housing
4. Lower housing
5. Gasket

Location of thermostat—4.5L and 4.9L engines

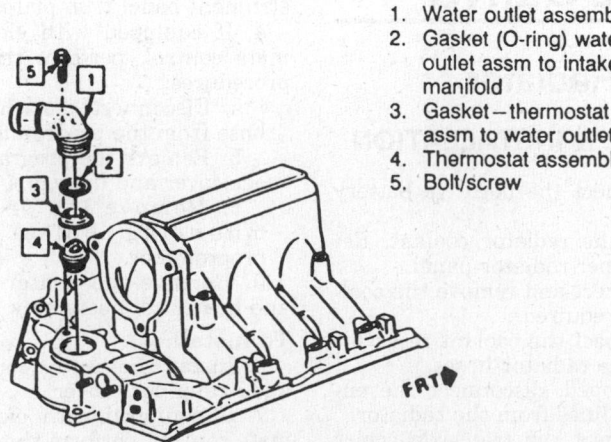

1. Water outlet assembly
2. Gasket (O-ring) water outlet assm to intake manifold
3. Gasket—thermostat assm to water outlet
4. Thermostat assembly
5. Bolt/screw

Location of thermostat—3.8L engine

ENGINE ELECTRICAL

NOTE: Disconnecting the negative battery cable on some vehicles may interfere with the functions of the on board computer systems and may require the computer to undergo a relearning process, once the negative battery cable is reconnected.

Distributor

The 4.5L and 4.9L engines are equipped with High Energy Ignition (HEI) system, utilizing Electronic Spark Timing (EST). The EST distributor uses no mechanical or vacuum advance and is easily identified by the absence of a vacuum advance.

All other engines are equipped with Computer Controlled Coil Ignition (C³I) system, which eliminates the distributor. The ECM provides sequential injection by processing signals received from the crankshaft and camshaft sensors.

The C³I system consists of the coil pack, ignition module, various hall effect sensors, interrupter rings and the Electronic Control Module (ECM). Since the ECM controls the ignition timing, no timing adjustments are necessary. These systems utilize the EST signal from the ECM to control spark timing.

REMOVAL

1. Disconnect the negative battery cable.
2. Label and disconnect all wires leading from the distributor cap.
3. Remove the distributor cap by turning the 4 latches counterclockwise. Lift off the distributor cap and carefully move it aside.
4. Disconnect the electrical connector harness from the distributor, if not already done.
5. Remove the distributor hold-down nut and clamp, using the proper tool.
6. Using a piece of chalk or paint, mark the rotor-to-distributor body and the distributor body-to-engine positions. Pull the distributor upward until the rotor just stops turning (counterclockwise); note the position of the rotor once again. Remove the distributor.

NOTE: Do not crank the engine with the distributor removed. On certain engines, a thrust washer is used between the distributor drive gear and the crankcase. This washer may stick to the bottom of the distributor when it is removed. Always make sure the washer is at the bottom of the distributor bore before installation. On Throttle Body Injection (TBI) systems, the malfunction trouble codes must be cleared after removal or adjustment of the distributor. This is accomplished by removing battery voltage to terminal R of the distributor for 10 seconds.

INSTALLATION

Timing Not Disturbed

1. To install the distributor, rotate the distributor shaft until the rotor aligns with the second mark, when the shaft stopped moving. Lubricate the drive gear with clean engine oil and install the distributor into the engine. As the distributor is installed, the rotor should rotate to the first alignment mark; this will ensure proper timing. If the marks do not align properly, remove the distributor and reset; be sure to install the thrust washer, if equipped.
2. Install the clamp and hold-down nut. Tighten the nut until the distributor can just be moved with a little effort.
3. Connect all wires and hoses. Install the distributor cap. Check and/or adjust the ignition timing.

Timing Disturbed

1. Remove the No. 1 spark plug.
2. Rotate the crankshaft until No. 1 piston is at the TDC of its compression stroke.

NOTE: The compression stroke can be determined by placing a thumb over the hole while slowly cranking the engine. Crank until compression is felt at the hole and continue cranking slowly until the timing mark on the crankshaft pulley aligns with the 0 degrees timing mark located on the timing chain cover.

3. Position the distributor in the block but do not, at this time, allow it to engage with the drive gear.
4. Rotate the distributor shaft until the rotor points between No. 1 and No. 8 spark plug towers and lower the distributor to engage the camshaft.

NOTE: It may be necessary to turn the rotor a small amount in either direction in order to achieve this engagement. The rotor will rotate slightly as the distributor gear engages. If installed correctly, the rotor should point toward the No. 1 spark plug terminal in the distributor cap.

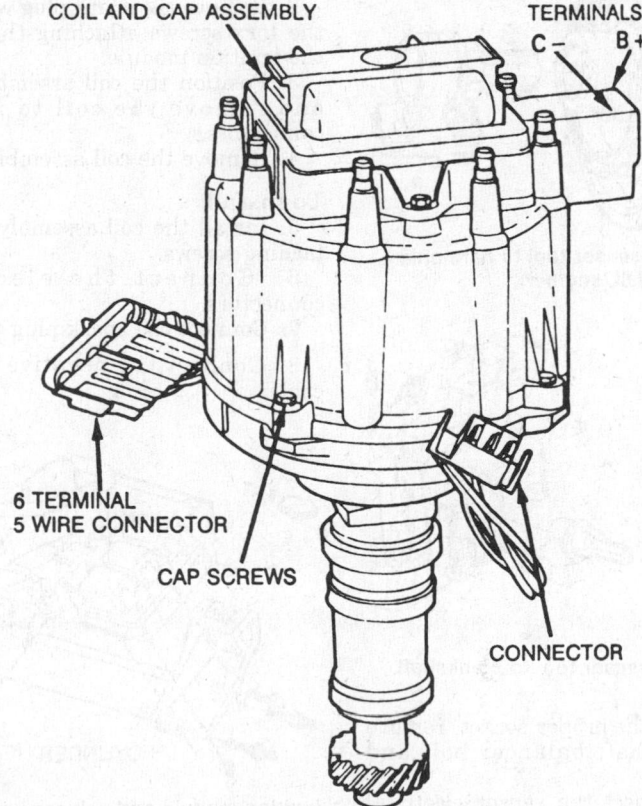

COIL AND CAP ASSEMBLY

TERMINALS

C − B+

6 TERMINAL
5 WIRE CONNECTOR

CAP SCREWS

CONNECTOR

Distributor Assembly—4.5L and 4.9L engines

5. Press down firmly on the distributor housing. This will ensure that the distributor shaft engages the oil pump shaft, thereby allowing the distributor to fully contact the engine block.

6. Install the hold-down clamp and tighten the nut until it is snug, do not tighten.

7. Install the distributor cap, making sure the rotor points to No. 1 terminal in the cap.

8. Attach all wires and hoses.

9. Start the engine. Check and/or adjust the ignition timing. Torque the distributor hold-down nut to 20 ft. lbs.

NOTE: Malfunction trouble codes must be cleared after removal or adjustment of the distributor. The ECM power feed must be disconnected for at least 30 seconds to clear the codes.

Distributorless Ignition

REMOVAL & INSTALLATION

Crankshaft Sensor

3.8L ENGINE (VIN 3)

1. Disconnect the negative battery cable.

2. Disconnect the crankshaft harness connector.

3. Rotate the harmonic balancer using the proper tool, until any window in the interrupter is aligned with the crank sensor.

4. Loosen the pinch bolt on the sensor pedestal until the sensor is free to slide in the pedestal.

5. Carefully remove the sensor and the pedestal as a unit.

To install:

6. Loosen the pinch bolt on the new sensor pedestal until the sensor is free to slide in the pedestal.

7. Verify that the window in the interrupter is still properly positioned and install the sensor and pedestal as a unit while making sure the interrupter ring is aligned with the proper slot.

8. Install the pedestal and torque the bolts to 22 ft. lbs.

9. Tighten the pinch bolt on the sensor pedestal.

10. Connect the crankshaft harness connector.

11. Connect the negative battery cable.

3.8L ENGINE (VIN C)

1. Disconnect the negative battery cable.

2. Remove the serpentine drive belt.

3. Raise the vehicle and support it safely.

4. Remove the right front tire and wheel assembly.

5. Remove the inner fender access panel.

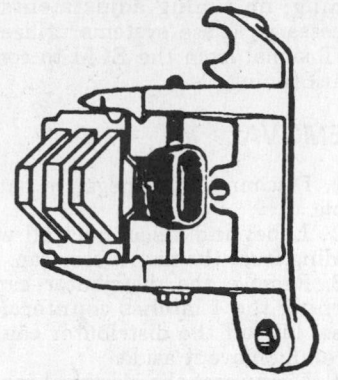

Crankshaft sensor—3.8L engine VIN C

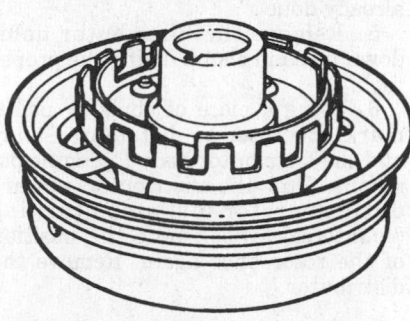

Crankshaft balancer with interrupter rings—3.8L engine

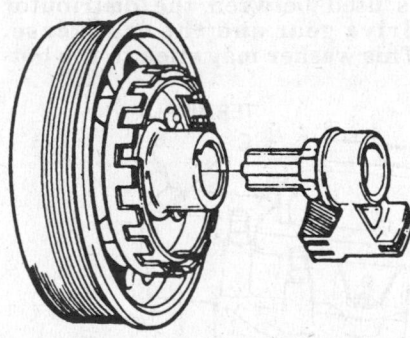

Crankshaft sensor tool to harmonic balancer—3.8L engine

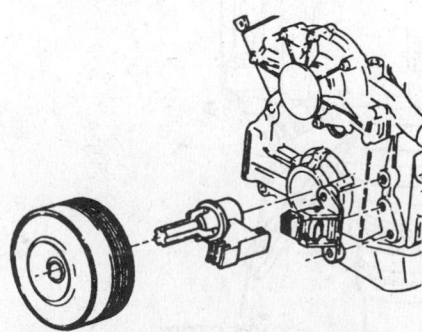

Crankshaft sensor tool to crankshaft

6. Using the proper socket, remove the crankshaft balancer bolt and balancer.

7. Disconnect the sensor electrical connector.

8. Remove the sensor and pedestal from the block face.

9. Remove the sensor from the pedestal.

To install:

10. Loosely install the crankshaft sensor on the pedestal.

11. Position the sensor with the pedestal attached, on the proper tool.

12. Position the special tool on the crankshaft.

13. Install the bolts to hold the pedestal to the block face and torque to 14–28 ft. lbs.

14. Torque the pedestal pinch bolt to 30–35 ft. lbs.

15. Remove the tool.

16. Place special tool on the harmonic balancer and turn. If any vane of the harmonic balancer touches the tool, replace the balancer assembly.

17. Install the balancer on the crankshaft.

18. Torque the crankshaft bolt to 200–239 ft. lbs.

19. Install the inner fender access panel.

20. Install the wheel and torque the lug nuts to 100 ft. lbs.

21. Lower the vehicle and install the serpentine belt.

22. Connect the battery cable.

Ignition Coil

1. Disconnect the negative battery cable.

2. Remove the spark plug wires and the torx screws attaching the coil to the ignition module.

3. Position the coil assembly aside and remove the coil to module connectors.

4. Remove the coil assembly.

To install:

5. Install the coil assembly and attaching screws.

6. Connect the electrical connectors.

7. Connect the spark plug wires.

8. Connect the negative battery cable.

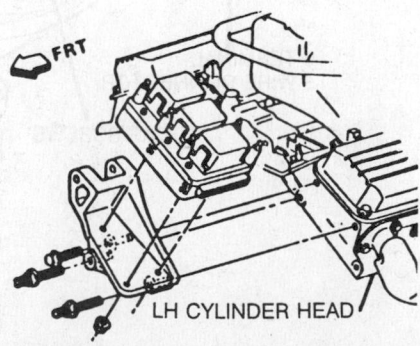

Ignition module and coil assembly—3.8L engine

Ignition Module

1. Disconnect the negative battery cable.
2. Remove the spark plug wires at the coil assembly.
3. Remove the ignition module bracket mounting nuts.
4. Remove the torx screws mounting the coil to the ignition module. Mark the position of the lead wires.
5. Disconnect the connecters between the coil and the ignition module.
6. Remove the ignition module.
To install:
7. Install the ignition module.
8. Connect the ignition module to coil electrical connectors.
9. Install the torx screws mounting the coil to the ignition module.
10. Install the ignition module bracket mounting nuts.
11. Install the spark plug wires at the coil assembly.
12. Connect the negative battery cable.

Ignition Timing

ADJUSTMENT

NOTE: The 4.5L and 4.9L engines incorporate a magnetic timing probe hole for use with special electronic timing equipment. Consult the manufacturer's in-structions before using this system. The following procedure is for use with the HEI-EST distributor.

1. Connect a timing light to the No. 1 spark plug wire according to the light manufacturer's instructions; do not pierce the spark plug wire to connect the timing light.
2. Follow the instructions on the Vehicle Emission Control Information label located in the engine compartment.
3. If equipped with an Electronic Spark Timing (EST) distributor, disconnect the 4-wire terminal plug from the distributor. Some models may require grounding the diagnostic connector located under the left side of the dash.
4. Start the engine and allow it to run at idle speed.
5. Aim the timing light at the degree scale just over the harmonic balancer.
6. Adjust the timing by loosening the hold-down clamp and rotate the distributor until the desired ignition advance is achieved. When the correct timing marks are aligned, tighten the clamp.
7. Adjust the timing, replace and tighten the hold-down clamp. To advance the timing, rotate the distributor opposite the normal direction of rotor rotation. Retard the timing by rotating the distributor in the normal direction of rotor rotation.

NOTE: If equipped with Throttle Body Injection (TBI), the malfunction trouble codes must be cleared after removal or adjustment of the distributor. This is accomplished by removing battery voltage to terminal R of the distributor for 10 seconds.

The 3.8L engine uses a Computer Controlled Coil Ignition (C³I) system. The C³I system components replace the conventional distributor and consist of a coil pack, ignition module, crankshaft sensor and camshaft sensor. No ignition timing adjustment is necessary or possible on the C³I system.

Alternator

PRECAUTIONS

Several precautions must be observed with alternator equipped vehicles to avoid damage to the unit.
- If the battery is removed for any reason, make sure it is reconnected with the correct polarity. Reversing the battery connections may result in damage to the one-way rectifiers.
- When utilizing a booster battery as a starting aid, always connect the positive to positive terminals and the negative terminal from the booster battery to a good engine ground on the vehicle being started.
- Never use a fast charger as a booster to start vehicles.
- Disconnect the battery cables when charging the battery with a fast charger.
- Never attempt to polarize the alternator.
- Do not use test lights of more than 12 volts when checking diode continuity.
- Do not short across or ground any of the alternator terminals.
- The polarity of the battery, alternator and regulator must be matched and considered before making any electrical connections within the system.
- Never separate the alternator on an open circuit. Make sure all connections within the circuit are clean and tight.
- Disconnect the battery ground terminal when performing any service on electrical components.
- Disconnect the battery if arc welding is to be done on the vehicle.

BELT TENSION ADJUSTMENT

A single serpentine belt is used to drive all engine mounted accessories. Drive

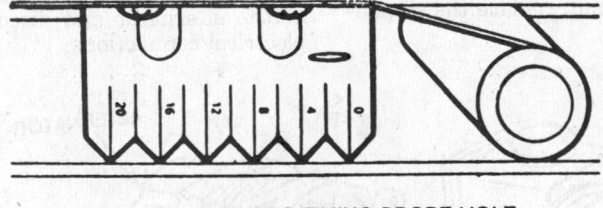
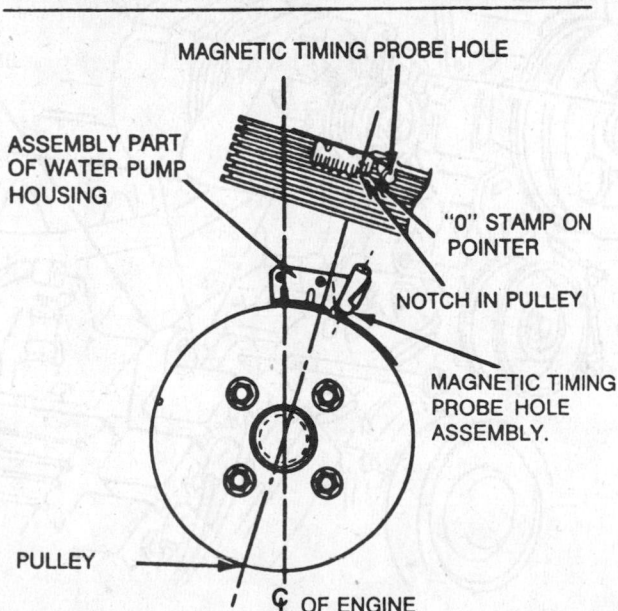

View of the magnetic timing probe hole—4.5L and 4.9L engines

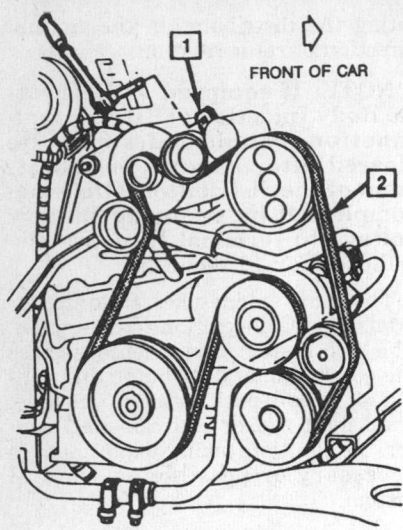

FRONT OF CAR

1. Drive belt tensioner
2. Serpentine drive belt

Drive belt—4.1L and 4.5L engines

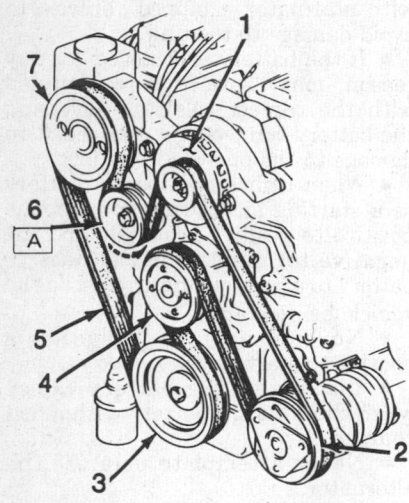

1. Generator pulley
2. A/C compressor
3. Crankshaft balancer
4. Water pump pulley
5. Serpentine belt
6. Belt tensioner
7. P/S pump pulley
A. Rotate the drive belt tensioner in direction of arrow in order to install or remove the drive belt

View of the serpentine drive belt routing—3.8L engine

belt tension is maintained by a spring loaded tensioner. A belt squeak when the engine is started or stopped is normal and has no effect on belt durability. The drive belt tensioner can control belt tension over a broad range of belt lengths; however, there are limits to the tensioner's ability to compensate.

1. Inspect tensioner markings to see if the belt is within operating lengths. Replace belt if the belt is ex-

cessively worn or is outside of the tensioner's operating range.

2. Run engine with the accessories **OFF** until the engine is warmed. Turn the engine **OFF** read belt tension with a proper belt tension gauge or equivalent placed halfway between the alternator and the air conditioning compressor. For non-air conditioning applications read tension between the power steering pump and crankshaft pulley. Remove tool.

3. Start the engine, with accessories **OFF**, and allow the system to stabilize for 15 seconds. Turn the engine **OFF**. Using the proper tool, apply clockwise force (tighten) to the tensioner pulley bolt. Release the force and immediately take a tension reading without disturbing belt tensioner position.

4. Apply a counterclockwise force to the tensioner pulley bolt and raise the pulley to the fully raised position. Slowly lower the pulley to engage the belt and take a tension reading without disturbing the belt tensioner position.

5. Average the 3 readings. If the average of the 3 readings is lower than the tension specified and the belt is within the tensioner's operating range, replace the belt tensioner. The drive belt tension should be 110 lbs. for 4.5L, 120 lbs. for 4.9L engines and never below 67 lbs. for 3.8L engine. If the belt tensioner is adjusted beyond it's movable limit, replace the serpentine drive belt.

REMOVAL & INSTALLATION

3.8L Engine

1. Disconnect the negative battery cable.
2. Label and disconnect the electrical connectors from the back of the alternator.
3. If equipped, remove the brace at the back of the alternator and the fuel rail cover.
4. Rotate the tensioner counterclockwise to remove the serpentine drive belt.
5. While supporting the alternator, remove the mounting bolts and the alternator.

To install:

6. Support the alternator in position and install the alternator.
7. Install the serpentine drive belt and rotate the tensioner into position.
8. If equipped, install the brace at the back of the alternator and the fuel rail cover.
9. Connect the electrical connectors at the back of the alternator.
10. Connect the negative battery cable.

4.5L and 4.9L Engines

1. Disconnect the negative battery cable.
2. Remove the air intake assembly at the throttle body.
3. Remove the serpentine belt from the tensioner pulley.
4. Remove the cover from the rear of the alternator and disconnect the electrical connections.

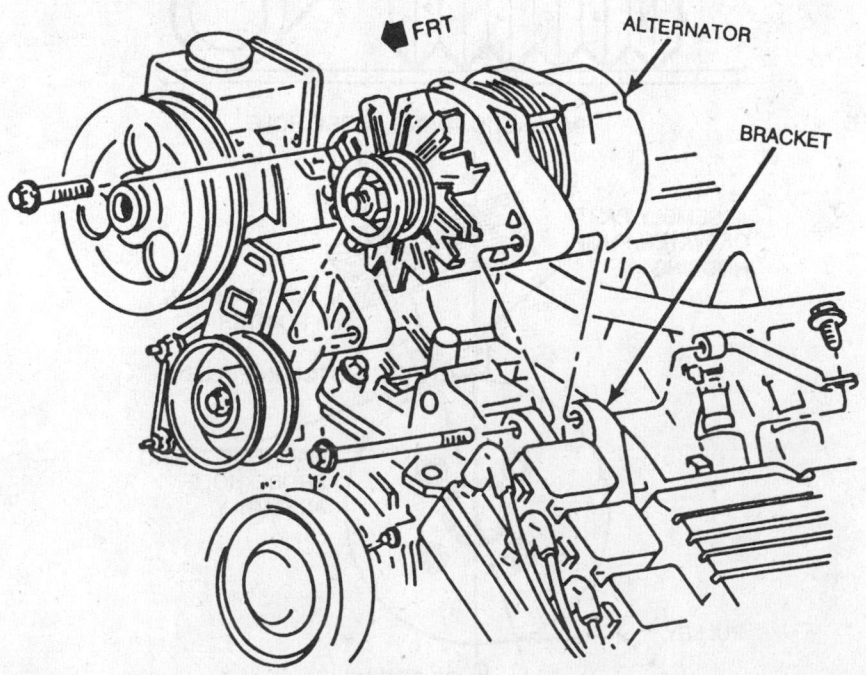

FRT   ALTERNATOR

BRACKET

Alternator mounting location—3.8L engine

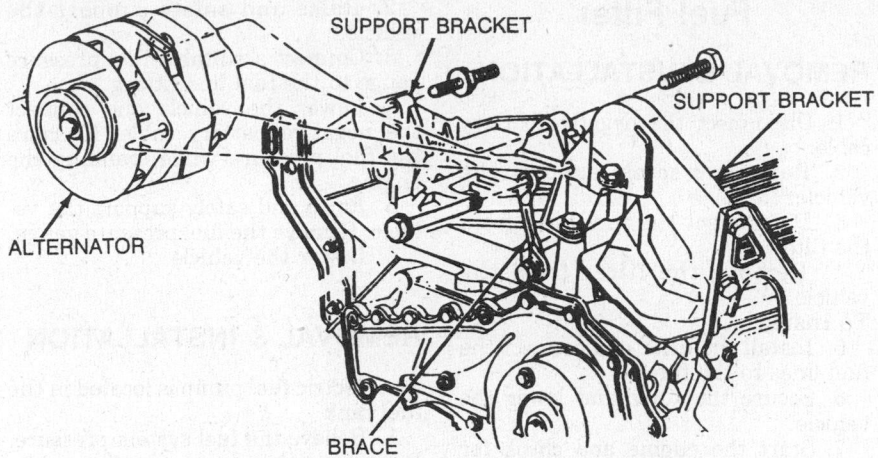

Alternator mounting location—4.1L and 4.5L engines

5. Disconnect the alternator mounting stud and the brace from the power steering pump.

6. Remove the rear alternator bolt and move the alternator upward and remove the connector.

7. Disconnect the heated windshield power module connection, if equipped.

8. Disconnect the front alternator bolt and remove the alternator.

To install:

9. Install the alternator and replace the front alternator bolt, tighten to 32 ft. lbs.

10. Connect the heated windshield power leads, if equipped.

11. Install the alternator connector and the rear mounting bolt. Tighten the bolt to 20 ft. lbs.

12. Install the power steering brace and replace the alternator mounting stud.

13. Connect the electrical connections and replace the cover.

14. Install the serpentine belt.

15. Replace the air intake assembly to the throttle body. Connect the negative battery cable.

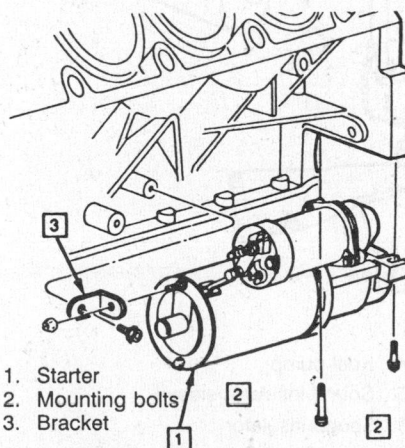

1. Starter
2. Mounting bolts
3. Bracket

Starter mounting location—4.1L and 4.5L engines

Starter

REMOVAL & INSTALLATION

1. Disconnect the negative battery cable.

2. Raise and support the vehicle safely.

3. If equipped, remove the flywheel shield.

4. Label and disconnect the electrical connectors from the starter.

5. Remove the starter-to-engine bolts and the starter.

NOTE: Note the location of any shims so they may be replaced in the same positions upon installation.

6. Install the starter and starter-to-engine bolts.

7. Connect the electrical connectors to the starter.

8. If removed, install the flywheel shield.

9. Lower the vehicle.

10. Connect the negative battery cable.

EMISSION CONTROLS

Please refer to "Emission Controls" in the Unit Repair section for system maintenance procedures. Due to the complex nature of modern electronic engine control systems, comprehensive diagnosis and testing procedures fall outside the confines of this repair manual. For complete information on diagnosis, testing and repair procedures concerning all modern engine and emission con-

trol systems, please refer to "Chilton's Guide to Fuel Injection and Electronic Engine Controls".

Emission Warning Lamps

The dash mounted "Service Soon" and "Service Now" lights are used to indicate a malfunction that the computer has detected in the vehicle's operation. The malfunctions can be related to the operating sensors or the Electronic Control Module (ECM). The service light will go out automatically if the trouble is cleared or intermittent.

The ECM, however will automatically store the trouble code until the diagnostic system is "Cleared".

CLEARING ECM TROUBLE CODES

Except Cadillac

With the ignition switch in the **OFF** position, disconnect battery voltage to the ECM for at least 30 seconds by performing 1 of the following:

1. Remove the ECM fuse from the fuse panel.

2. Disconnect the ECM pigtail.

3. Disconnect the negative battery cable.

NOTE: Disconnecting the negative battery cable should only be done as a last resort as it will also erase the memories for the digital radio, digital clock, trip odometer etc.

Cadillac

1. Turn the key to the **ON** position.

2. Simultaneously press the **OFF** and **HI** buttons on the climate control panel until E.O.O appears in the readout.

3. To clear the Body Computer Module (BCM) codes, depress the **OFF** and **LO** buttons simultaneously until F.O.O appears.

4. After E.O.O or F.O.O is displayed, .7.0 will appear. With the .7.0 displayed turn the ignition **OFF** for at least 10 seconds before re-entering the diagnostic mode.

FUEL SYSTEM

Fuel System Service Precautions

Safety is the most important factor

when performing fuel system maintenance. Failure to conduct maintenance and repairs in a safe manner may result in serious personal injury or death. Maintenance and testing of the vehicle's fuel system components can be accomplished safely and effectively by adhering to the following rules and guidelines.

• To avoid the possibility of fire and personal injury, always disconnect the negative battery cable unless the repair or test procedure requires that battery voltage be applied.

• Always relieve the fuel system pressure prior to disconnecting any fuel system component (injector, fuel rail, pressure regulator, etc.), fitting or fuel line connection. Exercise extreme caution whenever relieving fuel system pressure to avoid exposing skin, face and eyes to fuel spray. Please be advised that fuel under pressure may penetrate the skin or any part of the body that it contacts.

• Always place a shop towel or cloth around the fitting or connection prior to loosening to absorb any excess fuel due to spillage. Ensure that all fuel spillage (should it occur) is quickly removed from engine surfaces. Ensure that all fuel soaked cloths or towels are deposited in a suitable waste container.

• Always keep a dry chemical (Class B) fire extinguisher near the work area.

• Do not allow fuel spray or fuel vapors to come into contact with a spark or open flame.

• Always use a backup wrench when loosening and tightening fuel line connection fittings. This will prevent unnecessary stress and torsion to fuel line piping. Always follow the proper torque specifications.

• Always replace worn fuel fitting O-rings with new. Do not substitute fuel hose or equivalent where fuel pipe is installed.

RELIEVING FUEL SYSTEM PRESSURE

1. Disconnect the negative battery cable.
2. Loosen the fuel filler cap to relieve the tank vapor pressure.
3. Connect a suitable fuel pressure gauge to the fuel pressure connection. Wrap a shop towel around the gauge while connecting the gauge to avoid spillage.
4. Install a bleed hose into a container and open the valve to bleed the system pressure. The system is now safe for servicing.

Fuel Filter

REMOVAL & INSTALLATION

1. Disconnect the negative battery cable.
2. Raise and safely support the vehicle.
3. Disconnect the fuel lines from the filter.
4. Remove the filter from the vehicle.

To install:

5. Install the filter and connect the fuel lines to the filter.
6. Secure the filter and lower the vehicle.
7. Start the engine and check for leaks.

Electric Fuel Pump

PRESSURE TESTING

1. Disconnect the negative battery cable.

2. Raise and safely support the vehicle.
3. Connect a suitable fuel pressure gauge to the fuel line fitting.
4. Lower the vehicle and connect the negative battery cable. Measure the fuel pressure while cranking the engine.
5. Raise and safely support the vehicle. Remove the fuel pressure gauge.
6. Lower the vehicle.

REMOVAL & INSTALLATION

The electric fuel pump is located in the fuel tank.

1. Relieve the fuel system pressure. Disconnect the negative battery cable.
2. Drain the fuel from the tank. Raise and safely support the vehicle.
3. Support the tank and disconnect the tank retaining straps.
4. Loosen the exhaust heat shield and lower the exhaust at the rear hanger.
5. Lower the tank enough to disconnect the wires, hoses and ground

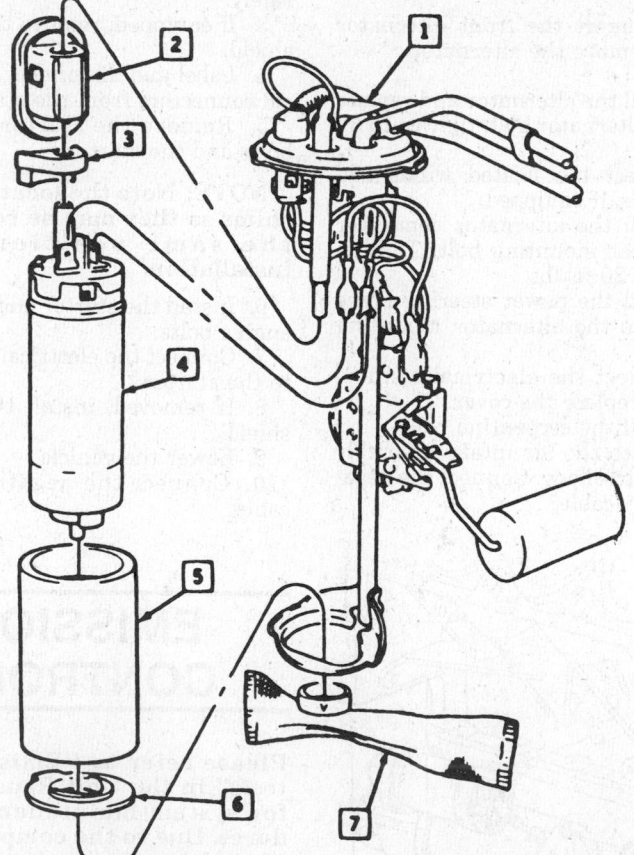

1. Fuel tank meter assembly
2. Pulsator (port injection only)
3. Bumper
4. Fuel pump
5. Sound insulator sleeve
6. Sound insulator
7. Filter

Exploded view of the fuel pump assembly

strap, if equipped. Remove the fuel tank.

6. Using a brass drift and a hammer, drive (turn) the cam lock ring-to-fuel tank counterclockwise and lift the assembly from the fuel tank.

7. Pull the fuel pump up into the attaching hose while pulling outward away from the bottom support. Take care to prevent damage to the rubber sound insulator and strainer during removal. Once the pump assembly is clear of the bottom support, pull it out of the rubber connector.

To install:

8. Install the pump into the fuel tank. Connect the fuel lines, wires and the ground strap, if equipped.

9. When installing the fuel tank, make sure all rubber sound isolators or anti-squeak spacers are replaced in their original locations.

10. Support the tank and install the tank retaining straps.

11. Lower the vehicle. Refill the fuel tank.

12. Connect the negative battery cable. Start the engine and check for fuel leaks.

Fuel Injection

ADJUSTMENTS

Idle speed and idle mixture are controlled automatically by the Electronic Control Module (ECM) and are not adjustable.

Fuel Injector

REMOVAL & INSTALLATION

3.8L Engine

1. Properly relieve the fuel system pressure. Remove the air cleaner assembly. Disconnect the negative battery cable.

2. Label and disconnect the fuel injector electrical connectors.

3. Remove the fuel rail retaining bolts. Disconnect the fuel injector electrical connectors and the fuel supply line.

4. Remove the fuel rail.

5. Separate the injector(s) from the fuel rail.

To install:

6. Replace the fuel injector O-rings.

7. Install the injector(s) into the fuel rail.

8. Install the fuel rail.

9. Install the fuel rail retaining bolts. Connect the the fuel injector electrical connectors and the fuel supply line.

10. Install the air cleaner assembly.

11. Connect the negative battery cable.

1988–89 4.5L Engines

NOTE: Care must be taken when removing injectors to prevent damage to the electrical connector pins on the injector and nozzle. The injectors are serviced as a complete assembly only. Injectors are an electrical component and should not be immersed in any type of cleaner.

1. Properly relieve the fuel system pressure.

2. Disconnect the negative battery cable. Remove the air cleaner assembly.

3. Disconnect the electrical connector from the fuel injector(s) by squeezing the 2 tabs together and pulling it straight up.

4. Remove the fuel meter cover-to-throttle body screws and the cover; be

sure to note the position of the 4 short screws. Allow the gasket to remain in place to prevent damage to the casting housing.

5. Using a small prybar and a ¼ in. rod, pry the fuel injector(s) from the throttle body; discard the O-rings.

To install:

6. Use new O-rings, lubricate with Dexron®II automatic transmission fluid, and install the injectors by pushing them into the sockets.

7. Install the fuel cover-to-throttle body screws, in the positions noted upon removal.

8. Connect the fuel injector electrical connectors.

9. Replace the air filter assembly and connect the negative battery cable.

10. Start the engine and check for leaks.

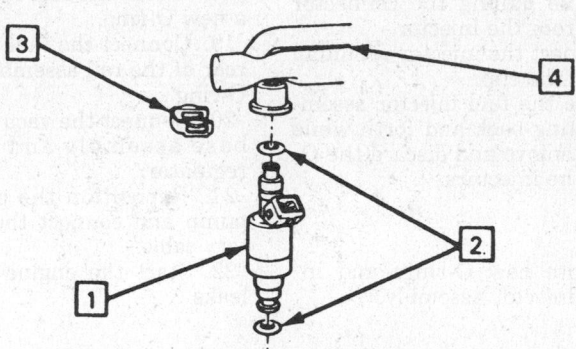

1. Injector assembly
2. Seal-O-ring injector
3. Clip-injector retainer
4. Fuel rail

Fuel Injector—1990 4.5L and 4.9L engines

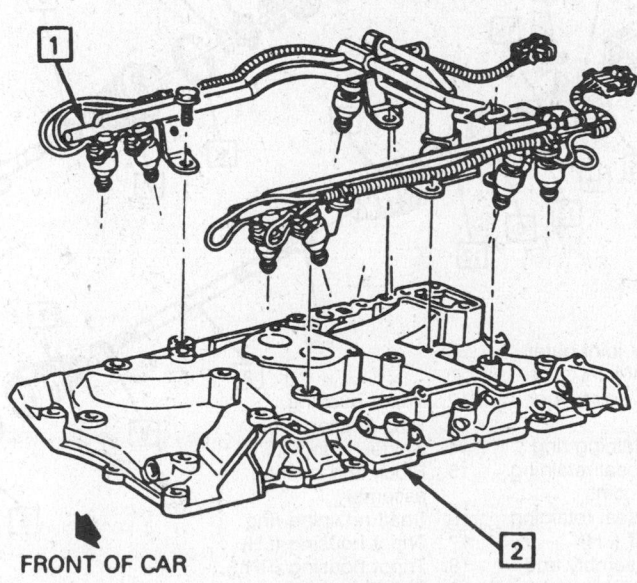

FRONT OF CAR

1. Fuel rail assembly
2. Intake manifold

Fuel rail assembly—1990 4.5L and 4.9L engines

1990 4.5L and 1991–92 4.9L Engines

1. Disconnect the negative battery cable.
2. Position the power steering pump aside.
3. Relieve the fuel system pressure.
4. Disconnect the vacuum lines from the pressure regulator and the base assembly.
5. Disconnect the fuel feed line from the rear of the rail assembly. Discard the O-ring.
6. Remove the fuel return line. Discard the O-ring.
7. Disconnect the electrical connectors at the front and the rear of the rail assembly.
8. Remove the rail support bracket mounting bolts and remove the rail assembly from the intake manifold.
9. Disconnect the electrical connector from the fuel injector by pushing in the clip while pulling the connector body away from the injector.
10. Disconnect the injector retaining clip. Discard the clip.
11. Remove the fuel injector assembly, by twisting back and forth while removing. Remove and discard the O-rings from the injectors.

To install:

12. Lubricate new O-rings and install on the injector assembly.

13. Install a new injector clip on the injector.
14. Install the fuel injector into the fuel rail socket. Push in to engage the retainer clip with the fuel rail cup.

NOTE: The electrical connectors should be facing the engine front for injectors 1–4. The connectors should be facing the rear of the engine for injectors 5–8.

15. Install the electrical connector to the injector assembly.
16. Install the fuel rail assembly and connect the support bracket mounting bolts.
17. Connect the electrical connectors at the front and rear of the rail assembly.
18. Install the fuel return line, using a new O-ring.
19. Connect the fuel feed line at the rear of the rail assembly, using a new O-ring.
20. Connect the vacuum lines to the base assembly and the pressure regulator.
21. Reposition the power steering pump and connect the negative battery cable.
22. Start the engine and check for leaks.

DRIVE AXLE

Halfshaft

REMOVAL & INSTALLATION

1988

NOTE: Use care when removing the halfshaft. Tri-pots can be damaged if the halfshaft is overextended.

1. Remove the hub nut.
2. Raise and safely support the vehicle. Remove the front tire and wheel assembly.
3. Install the halfshaft boot seal protector tool.
4. Disconnect the brake hose clip from the strut. Do not disconnect the hose from the caliper. Remove the brake caliper from the spindle and hang the caliper aside, using a length of wire. Do not allow the caliper to hang by the brake hose unsupported.
5. Mark the camber alignment cam bolt for reassembly. Remove the cam bolt and the upper strut-to-steering knuckle bolt.
6. Pull the steering knuckle assembly from the strut bracket.
7. Using spindle removal tool, remove the halfshaft from the hub/bearing assembly.

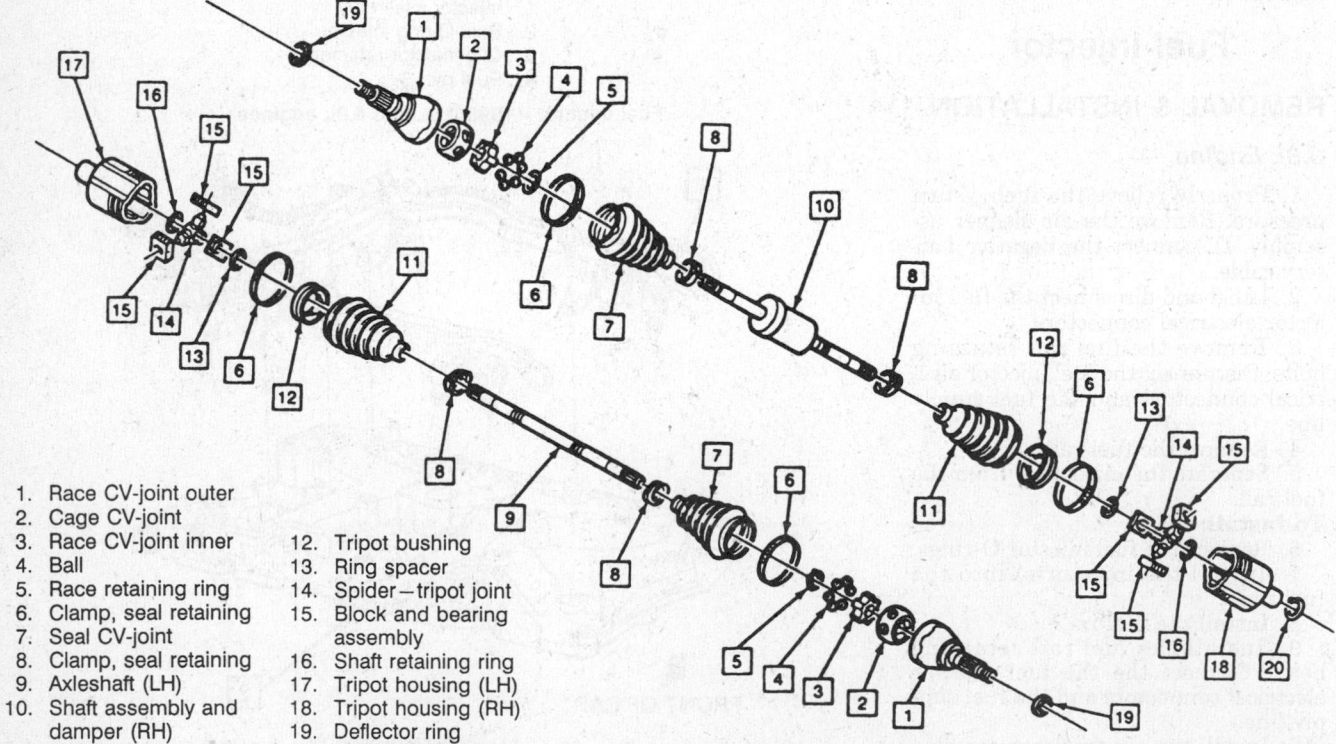

1. Race CV-joint outer
2. Cage CV-joint
3. Race CV-joint inner
4. Ball
5. Race retaining ring
6. Clamp, seal retaining
7. Seal CV-joint
8. Clamp, seal retaining
9. Axleshaft (LH)
10. Shaft assembly and damper (RH)
11. Tripot joint seal
12. Tripot bushing
13. Ring spacer
14. Spider – tripot joint
15. Block and bearing assembly
16. Shaft retaining ring
17. Tripot housing (LH)
18. Tripot housing (RH)
19. Deflector ring
20. Joint retaining ring

Exploded view of driveshaft assembly – 1990–92 Cadillac

To install:

8. If a new halfshaft is to be installed, a new steering knuckle seal should be installed first.

9. Loosely install the halfshaft into the transaxle and steering knuckle.

10. Loosely attach the steering knuckle-to-strut bolts.

11. The halfshaft is an interference fit in the steering knuckle. Press the axle into place and install the hub nut. When the shaft begins to turn with the hub, insert a drift through the caliper into one of the cooling slots in the rotor to keep it from turning. Insert a long bolt in the hub flange to prevent the shaft from turning. Torque the hub nut to 191 ft. lbs. (259 Nm) and tighten the brake caliper-to-steering knuckle bolts.

12. Load the hub assembly by lowering it onto a jackstand. Align the camber cam bolt marks made during removal, install the bolt and tighten to 140 ft. lbs. (190 Nm). Tighten the upper nut to the same value.

13. Install the halfshaft all the way into the transaxle, using a small prybar inserted into the groove provided on the inner retainer. Tap the prybar until the shaft seats in the transaxle. Remove the boot seal protector.

14. Connect the brake hose clip to the strut. Install the wheel assembly and lower the vehicle.

1989–92

NOTE: Use care when removing the halfshaft. Tri-pots can be damaged if the halfshaft is overextended.

1. Raise and safely support the vehicle. Remove the tire and wheel assembly.

2. Use an halfshaft boot seal protector tool and install it onto the seal.

3. Insert drift into rotor and caliper to prevent rotor from turning.

4. Remove hub nut and washer using a hub nut socket tool.

5. Remove the lower ball joint cotter pin and nut and loosen the joint using a ball joint separator tool. If removing the right halfshaft, turn the wheel to the left, if removing the left halfshaft turn the wheel to the right.

6. With a prybar between the suspension support and the lower control arm, separate the joint.

7. Pull out on the lower knuckle area and with a plastic or rubber mallet strike the end of the axle shaft to disengage the axle from the hub and bearing. The shaft nut can be partially installed to protect the threads.

8. Separate the hub and bearing assembly from the halfshaft and move the strut and knuckle assembly rearward. Remove the inner joint from the

transaxle using the proper tool from the intermediate shaft, if equipped.

NOTE: On vehicles equipped with the anti-lock brake system, care must be used to prevent damage to the toothed sensor ring on the halfshaft and the wheel speed sensor on the steering knuckle.

To install:

9. Seat the halfshaft into the transaxle by placing the proper tool into the groove on the joint housing and tapping until seated.

10. Verify the halfshaft is seated into the transaxle by grasping on the housing and pulling outboard. Do not pull on the halfshaft.

11. Install the halfshaft into the hub and bearing assembly.

12. Install the lower ball joint to the knuckle. Tighten the nut to 41 ft. lbs. (56 Nm) minimum and to 50 ft. lbs. (68 Nm) maximum to install the cotter pin.

13. Install the cotter pin.

14. Install the washer and new shaft nut.

15. Insert drift into rotor and caliper to prevent rotor from turning.

16. Torque the shaft nut to 185 ft. lbs. (251 Nm).

17. Remove the boot protector.

18. Install the tire and wheel assembly. Lower the vehicle.

CV-Boot

REMOVAL & INSTALLATION

Inner Boot (Inboard)

1. Raise and support the vehicle safely. Remove the halfshaft.

2. Remove the joint assembly retaining ring and the joint assembly.

3. Remove the bearing race retaining ring and the seal retainer.

4. Remove the inner seal retainer clamp and the inner joint seal.

To install:

5. Pack the joint with grease.

6. Install the inner seal retainer clamp.

7. Install the seal retainer and bearing race retaining ring.

8. Install the joint assembly and joint assembly retaining ring.

9. Install the halfshaft.

10. Lower the vehicle.

Outer Boot (Out Board)

1. Raise and support the vehicle safely. Remove the halfshaft.

2. Using a brass drift, lightly tap around the seal retainer to loosen it. Remove the seal retainer.

3. Remove the seal retainer clamp and discard.

4. Using snapring pliers, remove

the race retaining ring from the halfshaft.

5. Pull the outer joint assembly and the outboard seal away from the halfshaft.

To install:

6. Pack the joint with grease.

7. Install the outboard seal and outer joint assembly on the halfshaft.

8. Install the race retaining ring on the halfshaft.

9. Install a new seal retainer clamp.

10. Install the halfshaft.

11. Lower the vehicle.

Front Wheel Hub, Knuckle and Bearings

REMOVAL & INSTALLATION

1. Raise and support the vehicle safely. Place a suitable jacking device under the control arm and lower the vehicle slightly to rest the weight of the vehicle on the control arm.

2. Remove the tire and wheel assembly. Remove the caliper bolts, remove and support the caliper aside.

3. Remove the rotor and using the proper tool, separate the hub from the halfshaft.

4. Remove the hub and bearing retaining bolts, shield, hub and bearing assembly and the O-ring.

5. Disconnect the ball joint from the steering knuckle, using the proper tool.

6. Remove the halfshaft assembly and tap the seal from the steering knuckle. Remove the steering knuckle from the hub.

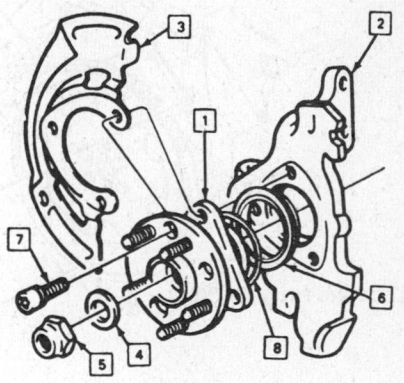

1. Hub and bearing assembly
2. Steering knuckle
3. Shield
4. Washer
5. Hub nut
6. Seal
7. Hub and bearing retaining bolt
8. O-ring

Front hub and bearing assembly

NOTE: The hub and bearing are replaced only as an assembly.

To install:

7. Install a new hub and bearing seal in the steering knuckle with the proper seal installer tool. Install the steering knuckle to the strut.

8. Lubricate the hub and bearing with grease and install the halfshaft.

9. Connect the ball joint to the steering knuckle and insert a new O-ring around the hub and bearing assembly.

10. Install the hub and bearing assembly into the steering knuckle. Tighten the bolts to 75 ft. lbs. (101 Nm).

11. Install the rotor and caliper assembly. Tighten the caliper bolts to 38 ft. lbs. (52 Nm).

12. Install the shaft washer and nut. Tighten the nut to 180 ft. lbs. (244 Nm).

13. Install the tire and wheel assembly.

14. Lower the vehicle.

AUTOMATIC TRANSAXLE

For further information on transmissions/transaxles, please refer to "Chilton's Guide to Transmission Repair".

Transaxle Assembly

REMOVAL & INSTALLATION

3.8L Engine

1. Disconnect the negative terminal from the battery. Disconnect the wire connector at the Mass Air Flow sensor, if equipped.

2. Remove the air intake duct and the Mass Air Flow sensor as an assembly.

3. Disconnect the cruise control assembly and the the shift control linkage.

4. Label and disconnect the following:
   a. Park/Neutral switch
   b. Torque converter clutch
   c. Vehicle speed sensor and fuel pipe retainers
   d. Vacuum modulator hose at the modulator

5. Remove the top transaxle-to-engine block bolts and install an engine support fixture.

6. Raise and safely support the vehicle. Remove both front tire and wheel assemblies and turn the steering wheel to the full left position.

7. Remove the right front ball joint nut and separate the control arm from the steering knuckle.

8. Remove the right halfshaft.

NOTE: Be careful not to allow the halfshaft splines to contact any portion of the lip seal.

9. Using a medium prybar, remove the left halfshaft. Be careful not to damage the pan. Install halfshaft boot seal protectors.

10. Remove the bolts at the transaxle and the nuts at the cradle member. Remove the left front transaxle mount.

11. Remove the right front mount-to-cradle nuts. Remove the left rear transaxle mount-to-transaxle bolts.

12. Remove the right rear transaxle mount. Remove the engine support bracket-to-transaxle case bolts.

13. Remove the flywheel cover, matchmark the flywheel-to-torque converter and remove the flywheel-to-converter bolts.

NOTE: Be sure to matchmark the flywheel-to-converter relationship for proper alignment upon reassembly.

14. Remove the rear cradle member-to-front cradle dog leg.

15. Remove the front left cradle-to-body bolt and the front cradle dog leg-to-right cradle member bolts.

16. Install a transaxle support fixture into position.

17. Remove the cradle assembly by swinging it aside and supporting it with jackstand.

18. Disconnect and plug the oil cooler lines at the transaxle.

NOTE: One bolt located between the transaxle and the engine block is installed in the opposite direction.

19. Remove the remaining lower transaxle-to-engine bolts and lower the transaxle from the vehicle.

To install:

20. Install the transaxle into the vehicle using the dowel pin as guide. Tighten the bolts to 55 ft. lbs. (75 Nm).

21. Connect the oil cooler lines and remove the support fixture.

22. Install the front left cradle-to-body bolts and replace the rear cradle-to-front cradle dog leg.

23. Install the flywheel and tighten the bolts to 46 ft. lbs. (62 Nm). Replace the flywheel cove and tighten the bolts to 136 inch lbs. (15 Nm).

24. Install the right rear transaxle mount. Replace the engine support bracket-to-transaxle case bolts and tighten to 40 ft. lbs. (54 Nm).

25. Install the right front mount-to-cradle nuts. Replace the left rear transaxle mount-to-transaxle bolts and tighten to 30 ft. lbs. (41 Nm).

26. Replace the bolts at the transaxle and the nuts at the cradle member. Replace the left front transaxle mount and tighten the bolts to 40 ft. lbs. (54 Nm).

27. Install both halfshafts.

28. Connect the control arm to the steering knuckle and tighten the right front ball joint nut.

29. Install the tire and wheel assemblies. Lower the vehicle.

30. Install the top transaxle-to-en-

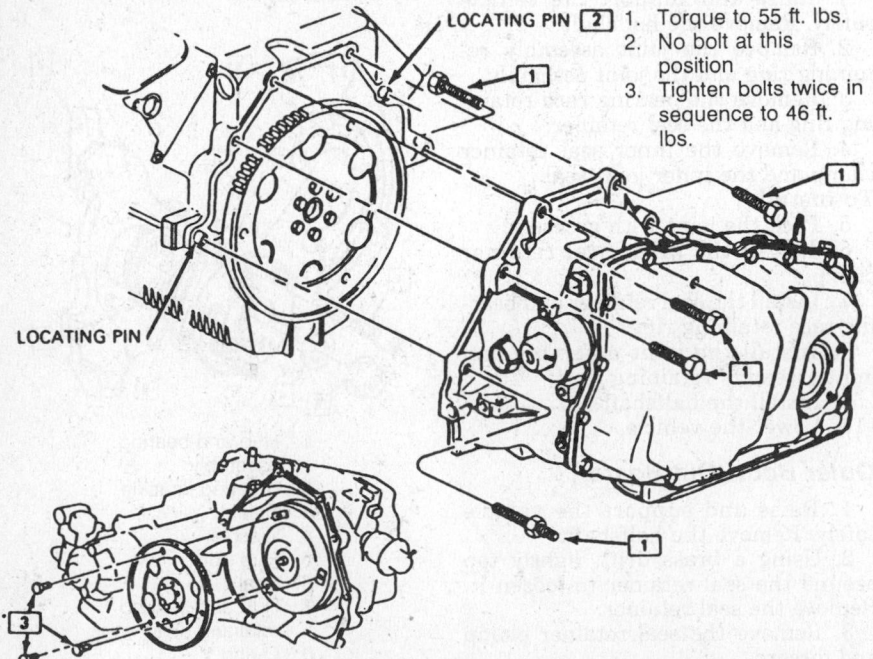

LOCATING PIN [2]
1. Torque to 55 ft. lbs.
2. No bolt at this position
3. Tighten bolts twice in sequence to 46 ft. lbs.

LOCATING PIN

Transaxle-to-engine mounting location—3.8L engine

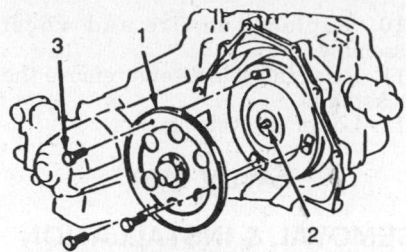

1. Flexplate
2. Torque converter
3. Bolt

Converter-to-flexplate attachments—4.5L and 4.9L engines

gine block bolts and remove the engine support fixture.

31. Connect the following:
    a. Park/Neutral switch
    b. Torque converter clutch
    c. Vehicle speed sensor and fuel pipe retainers
    d. Vacuum modulator hose at the modulator

32. Connect the cruise control assembly and the the shift control linkage.

33. Replace the air intake duct and the Mass Air Flow sensor as an assembly.

34. Connect the wire connector at the Mass Air Flow sensor, if equipped.

35. Connect the negative battery cable.

4.5L and 4.9L Engines

1. Disconnect the negative terminal from the battery. Remove the air cleaner and the TV cable.

2. Disconnect the shift linkage from the transaxle. Install an engine support fixture and support the engine.

3. Label and disconnect the electrical connectors from the following items:
    a. Converter clutch
    b. Vehicle speed sensor
    c. Vacuum line at the modulator

4. Remove the upper bell housing-to-engine bolts and studs.

5. Raise and support the vehicle safely. Remove both front wheels.

6. From the left side of the vehicle, disconnect the lower ball joint from steering knuckle. Remove both drive axles from the transaxle.

7. Remove the stabilizer bar-to-left control arm bolt.

8. Remove the left front cradle assembly.

9. Remove the extension housing-to-engine support bracket.

10. Disconnect and plug the oil cooler lines at the transaxle case.

11. Remove the right and left transaxle mount attachments.

12. Remove the flywheel splash shield. Matchmark the torque converter-to-flywheel and remove the converter-to-flywheel bolts.

13. Remove the lower bell housing bolts.

14. Using a floor jack, position it under the transaxle and remove the last bell housing bolt.

NOTE: To reach the last bell housing bolt, use a 3 in. socket wrench extension through the right wheel arch opening.

15. Remove the transaxle assembly.
To install:

16. Install the transaxle assembly and replace the lower bell housing bolts. Tighten the bolts to 55 ft. lbs. (75 Nm).

17. Replace the converter-to-flexplate bolts and tighten to 46 ft. lbs. (62 Nm). Install the flexplate splash shield.

18. Connect the oil cooler lines at the transaxle case.

19. Install the extension housing-to-engine support bracket. Replace the left front cradle assembly.

20. Replace the stabilizer bar-to-left control arm bolt.

21. Install both halfshafts. Fully seat the halfshafts by inserting the proper tool in the groove on the joint housing and tap until the joints are seated.

22. Connect the lower ball joint to the steering knuckle and replace the left and right front transaxle mount-to-cradle attachments. Tighten the nuts to 23 ft. lbs. (31 Nm).

23. Replace both front tire and wheel assemblies. Lower the vehicle.

24. Replace the upper bell housing-to-engine bolts/studs and tighten to 55 ft. lbs. (75 Nm).

25. Connect the electrical connectors to the converter clutch, vehicle speed sensor and the vacuum line to the modulator.

26. Connect the shift linkage to the transaxle. Remove the engine support fixture tool.

27. Replace the air cleaner and the TV cable.

28. Connect the negative battery cable. Check the fluid levels and start the engine and check for leaks.

SHIFT LINKAGE ADJUSTMENT

1. Position the shift lever in the **N** position.

2. Raise and safely support the vehicle.

3. Push the tab on the cable adjuster to adjust the cable in the cable mounting bracket on 3.8L engine.

4. Loosen and tighten the adjusting nut to 20 ft. lbs. (27 Nm) to adjust the cable on 4.1L and 4.5L engines.

5. Lower the vehicle.

THROTTLE LINKAGE ADJUSTMENT

1. Stop the engine. Raise and safely support the vehicle.

NOTE: Check the throttle body for full travel prior to any adjustments.

2. Depress and hold-down the metal readjust tab at the engine end of the TV cable.

3. Move the slider until it stops against the fitting.

4. Release the readjustment tab.

5. Rotate the throttle lever to the full travel position.

6. The slider must move toward the lever when the lever is rotated to it's full travel position.

FRONT SUSPENSION

MacPherson Strut

REMOVAL & INSTALLATION

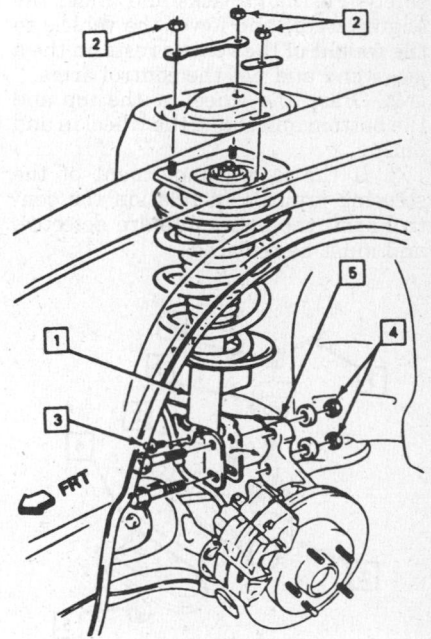

1. Strut assembly
2. Strut-to-body nuts
3. Brake line bracket bolt
4. Strut-to-steering knuckle nuts
5. Retain steering knuckle with wire once strut assembly is removed

Strut mount location

1. Disconnect the mounting nuts from the top of the strut assembly.

2. Raise and safely support the vehicle. Position a jackstand under the engine cradle and lower the vehicle so the weight of the vehicle rests on the a jackstand and not the control arms.

3. Remove the tire and wheel assemblies. If equipped with ABS, disconnect the front sensor.

4. Disconnect the brake line bracket from the strut assembly. Remove the strut-to-steering knuckle bolts.

5. Remove the strut assembly from the vehicle.

To install:

6. Install the strut assembly.

7. Install the strut-to-steering knuckle bolts. Connect the brake line bracket to the strut assembly.

8. If equipped with ABS, connect the front sensor. Install the tire and wheel assemblies.

9. Remove the jackstand. Lower the vehicle.

10. Connect the mounting nuts to the top of the strut assembly.

Lower Ball Joints

INSPECTION

1. Raise and support the vehicle safely. Position a jackstand under the engine cradle and lower the vehicle so the weight of the vehicle rests on the a jackstand and not the control arms.

2. Grasp the wheel at the top and the bottom and shake the wheel in and out.

3. If the is any movement of the steering knuckle in relation the control arm, the ball joints are defective and must be replaced.

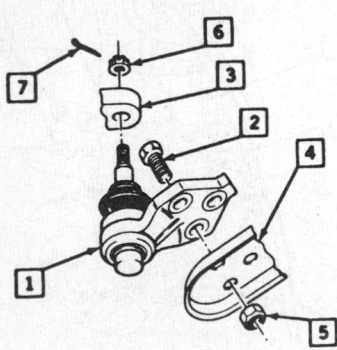

1. Ball joint
2. Ball joint mounting bolts must face down
3. Steering knuckle
4. Control arm
5. Ball joint mounting nuts
6. Ball joint-to-steering knuckle nut

Explode view of ball joint assembly

REMOVAL & INSTALLATION

1. Raise and safely support the vehicle. Position a jackstand under the engine cradle and lower the vehicle so the weight of the vehicle rests on the jackstand and not the control arms.

2. Remove the tire and wheel assembly.

3. Disconnect the ball joint from the steering knuckle, using the proper tool.

4. Drill out the rivets retaining the ball joint and loosen the stabilizer shaft bushing assembly nut.

5. Remove the ball joint from the steering knuckle and the control arm.

To install:

6. Install the ball joint to the steering knuckle and the control arm.

7. Install the 3 ball joint bolts facing down. Tighten nuts to 50 ft. lbs. (68 Nm). Tighten the stabilizer shaft bushing assembly nut to 13 ft. lbs. (17 Nm).

8. Connect the ball joint to the steering knuckle. Install a new cotter pin.

9. Install the tire and wheel assembly.

10. Lower the vehicle.

Lower Control Arms

REMOVAL & INSTALLATION

1. Raise and safely support the vehicle. Position a jackstand under the engine cradle and lower the vehicle so the weight of the vehicle rests on the jackstand and not the control arms.

2. Remove the tire and wheel assembly. Disconnect the stabilizer shaft-to-control arm bolt.

3. Remove the ball joint from the steering knuckle and the control arm.

4. Remove the control arm mounting bolts and remove the control arm from the engine cradle.

To install:

5. Install the control arm to the engine cradle. Do not tighten the control arm bolts at this time.

6. Install the stabilizer shaft bushings and connect the ball joint to the steering knuckle.

7. Raise the vehicle so the weight of the vehicle is supported by the control arm.

NOTE: The weight of the vehicle must be supported by the control arms when tightening the control arm mounting nuts.

8. Tighten the rear control arm mounting nut to 90 ft. lbs. (122 Nm) and the front mounting nut to 140 ft. lbs. (190 Nm).

9. Install the ball joint to the control arm and tighten the nut to 37 ft. lbs. (50 Nm).

10. Replace the tire and wheel assembly.

11. Raise the vehicle and remove the jackstand.

12. Lower the vehicle.

Sway Bar

REMOVAL & INSTALLATION

1. Raise and safely support the vehicle. Position a jackstand under the engine cradle and lower the vehicle so the weight of the vehicle rests on the jackstand and not the control arms.

2. Remove the tire and wheel assemblies.

3. Remove the bolts connecting the stabilizer bar bushings to the control arms.

4. Remove the stabilizer bar mounting bolts. Matchmark and disconnect the tie rod ends from the steering knuckles.

5. Disconnect the exhaust pipe from the exhaust manifold and turn the passenger side strut assembly completely to the right.

6. Slide the stabilizer bar over the steering knuckle and pull down until the stabilizer bar clears the frame.

7. Remove the stabilizer bar from the vehicle.

To install:

8. Install the stabilizer bar over the steering knuckle.

9. Raise the stabilizer bar over the frame and slide into position.

10. Loosely, install the stabilizer bar mount bushings, brackets and bolts.

11. Install the tie rod ends to the steering knuckles, tighten the nuts to 52 ft. lbs. (71 Nm). Tighten the stabilizer bar mounting bolts to 37 ft. lbs. (50 Nm).

12. Connect the exhaust pipe to the exhaust manifold and tighten the bolts to 15 ft. lbs. (20 Nm).

13. Replace the tire and wheel assemblies.

14. Raise the vehicle and remove the jackstands.

15. Lower the vehicle.

REAR SUSPENSION

MacPherson Strut

REMOVAL & INSTALLATION

1. Raise and safely support the vehicle.

2. Remove the trunk side cover. Remove the tire and wheel assemblies.

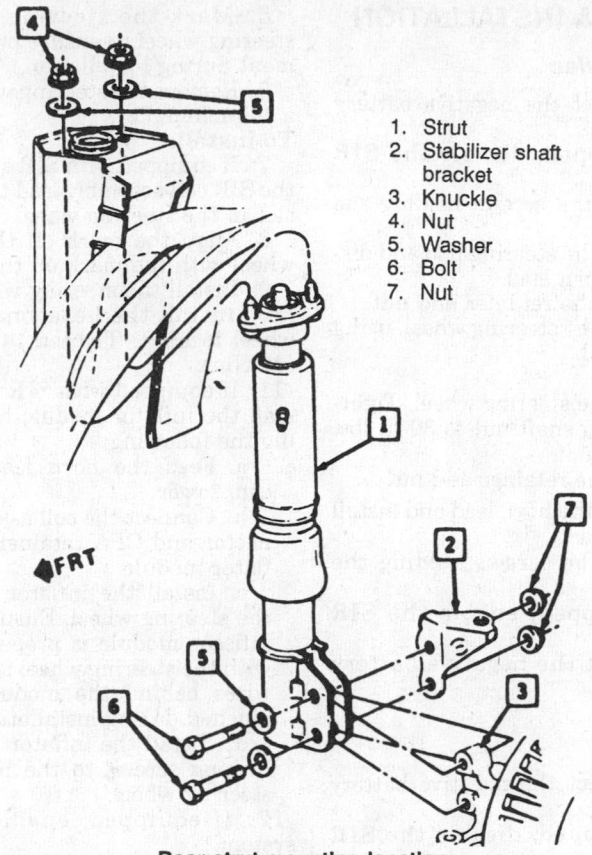

1. Strut
2. Stabilizer shaft bracket
3. Knuckle
4. Nut
5. Washer
6. Bolt
7. Nut

FRT

Rear strut mounting location

3. Support the control arm with a suitable jack.

4. Disconnect the Electronic Level Control (ELC) air tube and separate from the strut air tube.

5. Disconnect the strut tower mounting nuts. The nuts are located inside the trunk.

6. Remove the strut anchor bolts, washers and nuts from the steering knuckle and bracket.

7. Remove the strut assembly from the vehicle.

To install:

8. Install the strut assembly and connect the upper strut mounting nuts.

9. Replace the strut anchor bolts, washer, knuckle bracket and nuts.

10. Connect the ELC tube.

11. Tighten the upper mount nuts to 220 inch lbs. (25 Nm) and the strut-to-knuckle nuts to 125 ft. lbs. (170 Nm).

12. Replace the tire and wheel assemblies. Remove the jack from under the vehicle.

13. Replace the trunk side cover and lower the vehicle.

Coil Springs

REMOVAL & INSTALLATION

1. Raise and support the vehicle safely. Support the vehicle so the con-

trol arms hang free.

2. Remove both tire and wheel assemblies.

3. Disconnect the ELC height sensor on the right control arm and/or the parking brake cable retaining clip on the left control arm.

4. Place a proper tool and jack into position and remove the tension from the control arm pivot bolts.

NOTE: Place a chain around the spring and through the control arm as a safety measure.

5. Remove the pivot bolt and nut from the rear of the control arm.

6. Slowly, maneuver the jack to relieve in the front control arm pivot bolt.

7. Lower the jack to allow the control arm to pivot downward.

8. When all the compression is removed from the spring remove the safety chain, spring and the insulators.

NOTE: Do not apply force to the control arm and/or ball joint to remove the spring. Proper maneuvering of the spring will allow for easy removal.

To install:

9. Snap the upper insulator on the spring prior to installation.

10. Position the lower insulator and

spring in the vehicle. Install the coil springs so the upper end of the springs are positioned properly.

11. Raise the control arm into position, using the proper tool and jack.

12. Slowly, maneuver the jack to permit the installation of the pivot bolt and nut at the front of the control arm.

13. Install the pivot bolt and nut at the rear of the control arm.

14. Attach the rear stabilizer bar to the knuckle bracket. Connect the ELC height sensor link on the right control arm and/or the parking brake cable retaining clip on the left control arm.

15. Replace both tire and wheel assemblies.

16. Remove the jack from under the vehicle. Lower the vehicle.

17. Tighten the control arm pivot nuts to 85 ft. lbs. (115 Nm), the control arm pivot bolts to 125 ft. lbs. (170 Nm) and the stabilizer support bolt to 160 inch lbs. (18 Nm).

Rear Control Arms

REMOVAL & INSTALLATION

1. Raise and support the vehicle safely. Remove the tire and wheel assembly.

2. Disconnect the ELC height sensor on the right control arm and/or the parking brake cable retaining clip on the left control arm.

3. Disconnect the suspension adjustment link retaining nut and separate the link assembly from the control arm.

4. Remove the ball stud and the castellated nut. Turn over and install with the flat portion facing up. Do not tighten.

5. Separate the knuckle from the ball stud, using the proper tool. Remove the control arm.

To install:

6. Install the control arm. Connect the knuckle to the ball stud. Install the castellated nut.

7. Connect the link assembly to the control arm. Connect the suspension adjustment link retaining nut.

8. Connect the ELC height sensor on the right control arm and/or the parking brake cable retaining clip on the left control arm.

9. Install the tire and wheel assembly.

10. Lower the vehicle.

11. Tighten the control arm pivot nuts to 85 ft. lbs. (115 Nm) and the pivot bolts to 125 ft. lbs. (170 Nm). Tighten the pivot nuts and bolts with the vehicle unsupported and the wheels at normal height.

Rear Wheel Bearings

REMOVAL & INSTALLATION

1. Raise and support the vehicle safely. Remove the tire and wheel assembly.
2. Remove the brake drum from the vehicle.
3. Remove the hub and bearing assembly from the axle.

NOTE: The bolts that attach the hub and bearing assembly also support the brake assembly. Do not let the brake line support the brake assembly.

4. Remove the wheel bearings.
To install:
5. Install the wheel bearings.
6. Install the hub and bearing assembly to the axle. Tighten the hub and bearing bolts to 52 ft. lbs. (71 Nm).
7. Install the brake drum.
8. Install the tire and wheel assembly.
9. Lower the vehicle.

STEERING

Steering Wheel
—CAUTION—

Some vehicles are equipped with the Supplemental Inflatable Restraint or air bag system. The air bag system must be disabled before performing service on or around the air bag, instrument panel components, wiring and sensors. Failure to follow safety and disabling procedures could result in accidental air bag deployment, possible personal injury and unnecessary air bag system repairs.

REMOVAL & INSTALLATION

Except Cadillac

1. Disconnect the negative battery cable.
2. If equipped, disable the SIR system.
3. Remove the screws holding the steering pad.
4. Remove the steering pad and disconnect the horn lead.
5. Remove the retainer and nut.
6. Remove the steering wheel, using the proper tool.
To install:
7. Install the steering wheel. Tighten the steering shaft nut to 30 ft. lbs. (41 Nm).
8. Install the retainer and nut.
9. Connect the horn lead and install the steering pad.
10. Install the screws holding the steering pad.
11. If equipped, enable the SIR system.
12. Connect the negative battery cable.

Cadillac

1. Disconnect the negative battery cable.
2. If equipped, disable the SIR system.
3. If equipped with SIR, remove the inflator module by performing the following:
    a. Remove the inflator module attaching screws from the back of the steering wheel.
    b. Lift the inflator module from the steering wheel.
    c. Push down and twist the horn lead out of the cam tower.
    d. Remove the CPA retainer and coil assembly connector from the inflator module.
4. Remove the hexagonal steering wheel locknut.

5. Mark the steering shaft and steering wheel to ensure proper alignment during installation.
6. Remove the steering wheel, using a suitable puller.
To install:
7. If equipped with SIR system, feed the SIR coil assembly lead through the slot in the steering wheel.
8. Align the mark on the steering wheel with the mark on the shaft.
9. Install the steering wheel.
10. Install the hexagonal steering wheel locknut. Tighten to 30 ft. lbs. (41 Nm).
11. If equipped with SIR system, install the inflator module by performing the following:
    a. Feed the horn lead into the cam tower.
    b. Connect the coil assembly connector and CPA retainer to the inflator module.
    c. Install the inflator module to the steering wheel. Ensure that the inflator module is properly aligned with the steering wheel and that the wires behind the module are not pinched during installation.
    d. Install the inflator module attaching screws to the back of the steering wheel.
12. If equipped, enable the SIR system.
13. Connect the negative battery cable.

Steering Column
REMOVAL & INSTALLATION

Except Cadillac

1. Disconnect the negative battery cable.
2. Remove the lower instrument panel trim plates. Remove the left side sound insulator panel.
3. Remove the shift indicator cable from the shift bowl.

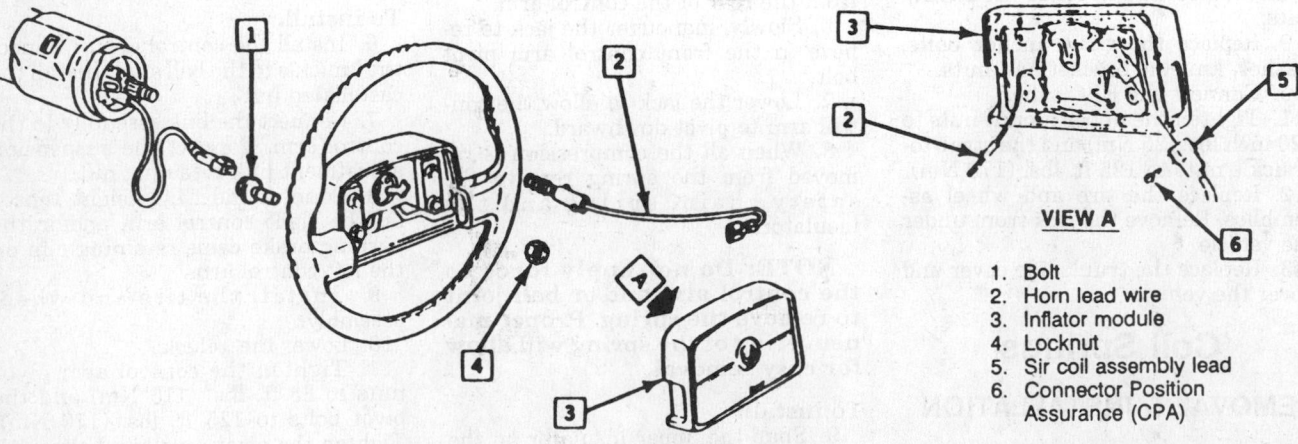

VIEW A

1. Bolt
2. Horn lead wire
3. Inflator module
4. Locknut
5. Sir coil assembly lead
6. Connector Position Assurance (CPA)

Steering wheel and inflator module – 1990–92 Cadillac

4. Label and disconnect the electrical connectors from the steering column. Remove the steering column-to-dash bolts.

5. Remove the steering shaft-to-intermediate shaft bolt and the steering column from the vehicle.

To install:

6. Install the steering column to the vehicle and steering shaft-to-intermediate shaft bolt.

7. Install the steering column-to-dash bolts. Connect the electrical connectors to the steering column.

8. Install the shift indicator cable to the shift bowl.

9. Install the lower instrument panel trim plates. Install the left side sound insulator panel.

10. Connect the negative battery cable.

Cadillac

1. Disconnect the negative battery cable.

2. If equipped, disable the SIR system and remove the inflator module.

3. Remove the steering column trim plate.

4. Remove the retaining filler, the column reinforcement plate and disconnect the electrical connections. Remove the shift control cable at the actuator.

5. Remove the bolts securing the seal assembly and the bolt from the upper knuckle of the intermediate steering shaft.

6. Disconnect the lower brace assembly and the lower support bracket.

7. Remove the bolts securing the column to the upper support and remove the column assembly.

To install:

8. Install the column assembly. Install the bolts securing the column to the upper support.

9. Connect the lower brace assembly and the lower support bracket.

10. Install the bolts securing the seal assembly and the bolt to the upper knuckle of the intermediate steering shaft.

11. Install the retaining filler, the column reinforcement plate and connect the electrical connections. Install the shift control cable at the actuator.

12. Install the steering column trim plate.

13. If equipped with SIR, install the inflator module and enable the SIR system.

14. Connect the negative battery cable.

Power Steering Rack

ADJUSTMENT

Rack Bearing Preload

1. Loosen the adjuster plug locknut and turn the adjuster plug clockwise until it bottoms in the housing. Then back off 50–70 degrees which is approximately one flat.

2. Raise and support the vehicle safely to make the proper adjustments. Be sure to check the returnability of the steering wheel to the center position after the adjustment.

3. Tighten the locknut to the adjuster plug to 50 ft. lbs.

REMOVAL & INSTALLATION

1. Raise and safely support the vehicle. Allow the front suspension to hang freely. Disconnect the pressure lines from the steering gear and drain the excess fluid into a container. Be sure to plug the openings.

2. Move the intermediate shaft cover upward and remove the intermediate shaft-to-stub shaft pinch bolt. Remove both front tire and wheel assemblies.

3. Disconnect the tie rod ends from the steering knuckles. Remove the line retainer, outlet and pressure hoses.

4. Remove the rack/pinion assembly-to-chassis bolts.

5. Loosen the front engine cradle mounting bolts and the lower the rear of the cradle about 3 inches, if necessary. Remove the rack and pinion assembly.

To install:

6. Install the rack and pinion assembly into the vehicle. Raise the front engine cradle into position and tighten the attaching bolts.

7. Install the rack/pinion assembly-to-chassis bolts. Tighten the rack mounting bolts to 50 ft. lbs. (68 Nm).

8. Connect the tie rod ends to the steering knuckles. Tighten the tie rod end nut to 35–52 ft. lbs. (47–71 Nm). Install the line retainer, outlet and pressure hoses.

9. Install the intermediate shaft-to-stub shaft pinch bolt. Move the intermediate shaft cover upward into position. Install both front tire and wheel assemblies.

10. Refill the power steering pump reservoir.

11. Bleed the power steering system and check for leaks.

12. Lower the vehicle.

13. Check and/or adjust the front wheel alignment.

Power Steering Pump

REMOVAL & INSTALLATION

3.8L Engine

1. Disconnect the negative battery cable.

2. Remove the serpentine drive belt and disconnect the pressure and return hoses.

3. Remove the power steering pump mounting bolts.

4. Remove the pump assembly. Transfer the pulley as necessary.

To install:

5. Install the pump assembly.

6. Remove the power steering pump mounting bolts.

7. Remove the drive belt and disconnect the pressure and return hoses.

8. Connect the negative battery cable.

9. Install the serpentine drive belt and bleed the power steering system.

4.5L and 4.9L Engines

1. Disconnect the negative battery cable.

2. Remove the serpentine drive belt and the power steering pump pulley, using the proper tool.

3. Disconnect and plug the high pressure and feed lines from the pump. Remove the belt tensioner, as required.

4. Remove the power steering pump-to-bracket bolts and the pump.

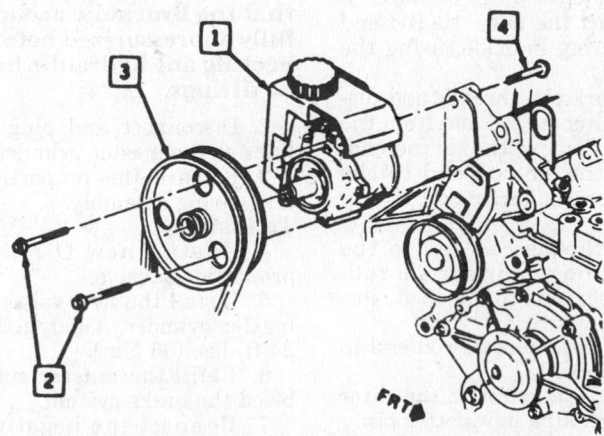

Power steering pump assembly—3.8L engine

To install:

5. Install the power steering pump-to-bracket bolts and the pump. Tighten the power steering pump mounting bolts to 18 ft. lbs. (24 Nm).

6. Connect the high pressure and feed lines to the pump. If removed, install the belt tensioner.

7. Install the power steering pump pulley and the serpentine drive belt.

8. Refill the power steering pump reservoir. Bleed the power steering system.

9. Connect the negative battery cable.

BELT ADJUSTMENT

The serpentine is self adjusting within the tensioner operating limits.

SYSTEM BLEEDING

1. Raise and support the vehicle safely. Fill the fluid reservoir.

2. Bleed the system by turning the wheels from side to side, without reaching the stop ay either end. Keep the fluid level at the FULL COLD mark. Continue this until the air is eliminated from the fluid.

3. Start the engine and run at fast idle. Recheck the fluid level.

4. Return the wheels to the center position and lower the vehicle.

5. Recheck the fluid level.

Tie Rod Ends

REMOVAL & INSTALLATION

1. Raise and safely support the vehicle.

2. Remove the cotter pin and loosen the jam nut from the outer tie rod end.

3. Disconnect the outer tie rod end from the steering knuckle, using the proper tool.

4. Matchmark the threads and disconnect the outer tie rod end from the inner tie rod. Remove the tie rod end.

5. Install the tie rod end to the matchmarks on the inner tie rod.

To install:

6. Install the tie rod end to the matchmarks on the inner tie rod. Tighten the hex nut to 35–45 ft. lbs. (47–61 Nm).

7. Connect the outer tie rod end to the steering knuckle.

8. Tighten the jam nut on the outer tie rod end. Install a new cotter pin.

9. Lower the vehicle.

BRAKES

For all brake system repair and service procedures not detailed below, please refer to "Brakes" in the Unit Repair section.

Master Cylinder

REMOVAL & INSTALLATION

1. Disconnect the negative battery cable and, if equipped, the electrical connector from the level sensor unit.

NOTE: If equipped with Antilock Brake System (ABS), ensure that the hydraulic accumulator is fully depressurized before disconnecting any hydraulic lines, hoses or fittings.

2. Disconnect and plug hydraulic lines from the master cylinder.

3. Remove the mounting bolts and the master cylinder assembly.

To install:

4. Install the master cylinder assembly and mounting bolts.

5. Connect the hydraulic lines to the master cylinder.

6. Connect the negative battery cable and, if equipped, the electrical connector to the level sensor unit.

7. Refill the master cylinder with clean brake fluid. Bleed the brake system.

Proportioning Valve

REMOVAL & INSTALLATION

1. Disconnect the negative battery cable.

NOTE: If equipped with Antilock Brake System (ABS), ensure that the hydraulic accumulator is fully depressurized before disconnecting any hydraulic lines, hoses or fittings.

2. Disconnect and plug the brake lines at the master cylinder.

3. Remove the proportioner valve and O-ring assembly.

To install:

4. Install new O-rings on the proportioner valve.

5. Install the new valve(s) into the master cylinder. Tighten the bolts to 24 ft. lbs. (33 Nm).

6. Refill the master cylinder and bleed the brake system.

7. Connect the negative battery cable.

Power Brake Booster

REMOVAL & INSTALLATION

1. Disconnect the negative battery cable.

2. From inside the vehicle, detach the brake pushrod from the brake pedal.

3. Remove the master cylinder-to-power brake booster bolts and move the master cylinder aside.

4. Disconnect the vacuum hose from the power brake booster.

5. Remove the power brake booster-to-cowl nuts and the booster.

To install:

6. Install the booster and the power brake booster-to-cowl nuts.

7. Connect the vacuum hose to the power brake booster.

8. Move the master cylinder into position and install the master cylinder-to-power brake booster bolts.

9. From inside the vehicle, attach the brake pushrod to the brake pedal.

10. Connect the negative battery cable.

Brake Caliper

REMOVAL & INSTALLATION

NOTE: If equipped with Antilock Brake System (ABS), ensure that the hydraulic accumulator is fully depressurized before disconnecting any hydraulic lines, hoses or fittings.

1. Raise and safely support the vehicle. Remove the tire and wheel assembly.

2. Push the piston into caliper, using the proper tool, to provide clearance between the pad and the rotor.

3. Disconnect and plug the brake line and remove the mounting bolts and sleeves.

4. Remove the caliper from the rotor and the mounting bracket.

To install:

5. Install the caliper onto the rotor and mounting bracket.

6. Connect the brake line to the caliper.

7. Install the caliper mounting nuts and sleeves. Tighten the caliper mounting bolts to 38 ft. lbs. (52 Nm).

8. With the caliper mounting bolts tight, ensure that the brake line fitting is tight.

9. Install the tire and wheel assembly.

10. Lower the vehicle.

11. Check the brake fluid level in the reservoir.

12. Before starting the engine, depress the brake pedal until the pedal is firm. Bleed the brake system, as required.

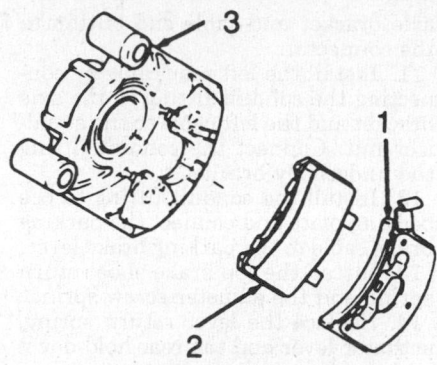

1. Outboard shoe and lining
2. Inboard shoe and lining
3. Caliper housing

Brake pad mounting

13. Recheck the master cylinder fluid level.

Disc Brake Pads

REMOVAL & INSTALLATION

1. If the brake pads are to be replaced, use a syringe or similar tool to remove the brake fluid from the master cylinder reservoir until it is approximately ⅓–½ full.
2. Raise and safely support the vehicle. Remove the tire and wheel assembly.
3. Remove the caliper from the rotor.
4. Remove the brake pads from the caliper.
5. Using a C-clamp, press the piston into caliper to provide additional clearance between the pad and rotor for the new pads.

To install:

6. Install new bushings into the grooves in the mounting bolt holes. Install the inboard pad by snapping the retainer spring into the piston.
7. Install the outboard pad with the back of the pad flat against the caliper.
8. Install the caliper to the rotor. Install the caliper attaching bolts.
9. Install the tire and wheel assembly.
10. Check the brake fluid level in the master cylinder reservoir. Top-up as necessary.
11. Before starting the engine, slowly depress the brake pedal until the pedal is firm. Bleed the brake system, as required.
12. Recheck the master cylinder fluid level.

Brake Rotor

REMOVAL & INSTALLATION

1. Raise and safely support the vehicle. Remove the tire and wheel assembly.
2. Remove the caliper from the rotor. Support the caliper aside using a length of wire. Do not allow the caliper to hang by the brake hose unsupported.
3. Remove the shaft nut and washer and remove the rotor assembly.

To install:

5. Install the rotor to the hub assembly. Tighten the shaft nut to 70 ft. lbs. (95 Nm).
6. Install the caliper.
7. Install the tire and wheel assembly.
8. Lower the vehicle.

Brake Drums

REMOVAL & INSTALLATION

1. Raise and safely support the vehicle.
2. Matchmark the wheel to the hub flange.
3. Remove the tire and wheel assembly. Matchmark the drum to the hub flange.
4. Remove the brake drum assembly. Make sure the parking brake is released.

To install:

5. Install the brake drum.
6. Install the tire and wheel assembly.
7. Lower the vehicle.

1. Actuator spring
2. Upper shoe return spring
3. Spring connecting link
4. Adjuster actuator
5. Spring washer
6. Lower return spring
7. Hold-down spring assembly
8. Hold-down pin
9. Adjuster shoe and lining
10. Shoe and lining
11. Adjuster socket
12. Spring clip
13. Adjuster nut
14. Adjuster screw
15. Retaining ring

16. Pin
17. Spring washer
18. Park brake lever
19. Screw and lockwasher
20. Boot
21. Piston
22. Seal
23. Spring assembly
24. Bleeder valve
25. Wheel cylinder
26. Bleeder valve cap
27. Backing plate assembly
28. Access hole plug

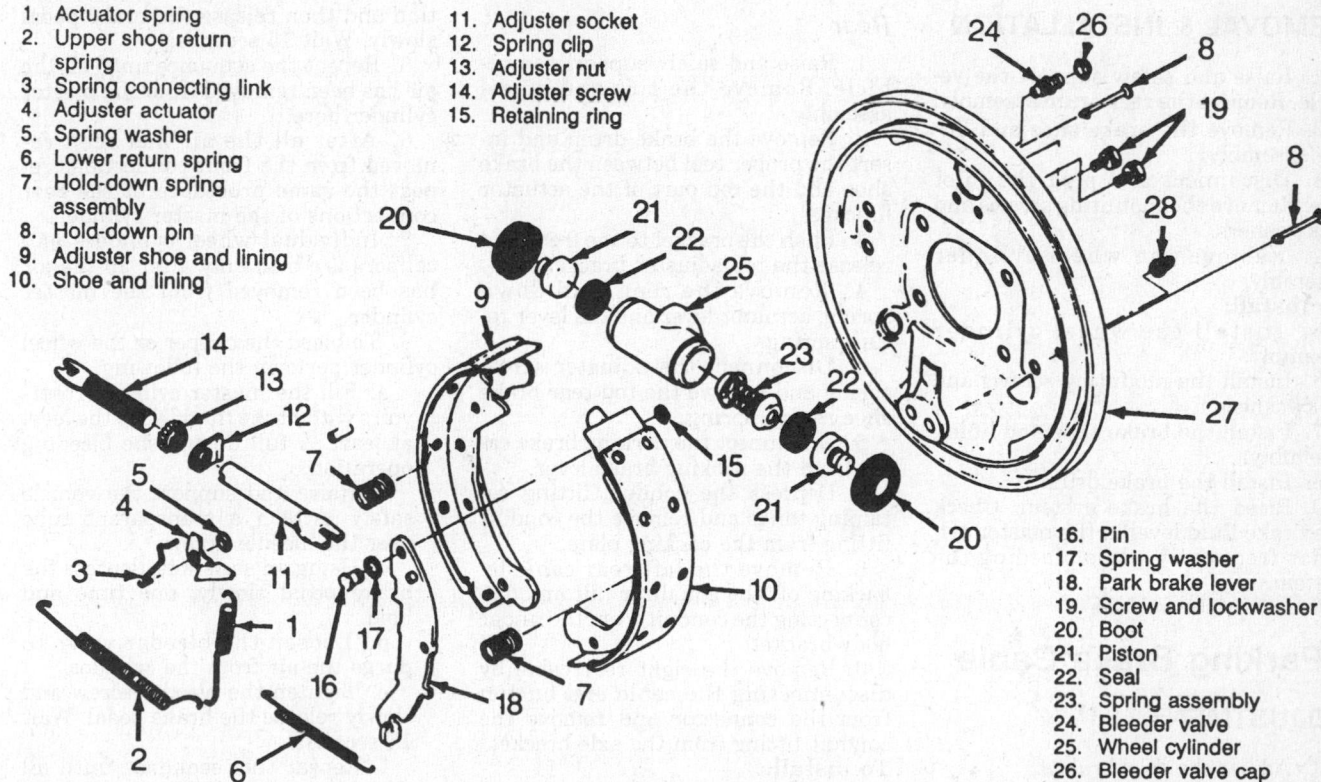

Exploded view of the rear brake assembly

Brake Shoes

REMOVAL & INSTALLATION

1. Raise and safely support the vehicle. Remove the brake drum assembly.
2. Remove the actuator and the upper return spring with the proper tools.
3. Disconnect the spring connecting link, adjuster actuator and the hold-down washer.
4. Remove the hold-down springs and the pins. Disconnect the brake shoes from the parking brake cable.
5. Remove the brake shoe and lining assemblies.

To install:

6. Install the brake shoe and lining assemblies.
7. Connect the brake shoes to the parking brake cable. Install the brake shoe hold-down pins and springs.
8. Connect the hold-down washer, adjuster actuator and spring connecting link.
9. Install the upper return springs and actuator.
10. Install the brake drum.
11. Install the tire and wheel assembly.
12. Adjust the brakes.
13. Lower the vehicle and road test.

Wheel Cylinder

REMOVAL & INSTALLATION

1. Raise and safely support the vehicle. Remove the rear drum assembly.
2. Remove the brake shoe and lining assembly.
3. Disconnect and plug the inlet line. Remove the mounting screws and lockwashers.
4. Remove the wheel cylinder assembly.

To install:

5. Install the wheel cylinder assembly.
6. Install the mounting screws and lockwashers.
7. Install the brake shoe and lining assembly.
8. Install the brake drum.
9. Bleed the brake system. Check the brake fluid level in the master cylinder frequently while bleeding the system.
10. Lower the vehicle.

Parking Brake Cable

ADJUSTMENT

1. Adjust the rear brakes.
2. Apply and release the parking brake 6 times to 10 clicks. Release the park brake pedal.
3. Raise and support the vehicle safely. Remove the access plug.
4. Adjust the park brake cable until a ⅛ drill can be inserted through the access hole into the space between the shoe web and the park brake lever.
5. Check for free wheel rotation. Replace the access plug.
6. Lower the vehicle.

REMOVAL & INSTALLATION

Front

1. Raise and safely support the vehicle.
2. Loosen the equalizer assembly at the front parking brake cable. Remove the front parking brake cable from the equalizer assembly.
3. Disconnect the cable casing retaining nut at the underbody. Remove the cable casing and cable from the control assembly.

To install:

4. Install the cable casing and cable to the control assembly. Connect the cable casing retaining nut at the underbody. Tighten the casing retaining nut to 22 ft. lbs. (30 Nm).
5. Install the front parking brake cable to the equalizer assembly. Tighten the equalizer assembly at the front parking brake cable.
6. Adjust the cable.
7. Lower the vehicle.

Rear

1. Raise and safely support the vehicle. Remove the tire and wheel assembly.
2. Remove the brake drum and insert the proper tool between the brake shoe and the top part of the actuator bracket.
3. Push the bracket to the front and release the top adjuster bracket rod.
4. Remove the rear hold-down spring, actuator lever and the lever return spring.
5. Disconnect the adjuster screw spring and remove the top rear brake shoe return spring.
6. Disconnect the parking brake cable from the parking brake lever.
7. Depress the conduit fitting retaining tangs and remove the conduit fitting from the backing plate.
8. Remove the left rear cable by backing off the equalizer nut and disconnecting the conduit from the under body bracket.
9. Remove the right rear cable by disconnecting the cable end button from the connector and remove the conduit fitting from the axle bracket.

To install:

10. Install the right rear cable by connecting the conduit fitting to the axle bracket and cable end button to the connector.
11. Install the left rear cable by connecting the conduit fitting to the axle bracket and the left cable to the equalizer nut. Connect the conduit fitting the underbody bracket.
12. Install the conduit fitting to the backing plate and connect the parking brake cable to the parking brake lever.
13. Install the top brake shoe return spring and the adjuster screw spring.
14. Replace the lever return spring, actuator lever and the rear hold-down spring.
15. Install the top adjuster bracket rod. Replace the brake drum assembly.
16. Install the tire and wheel assembly. Adjust the parking brake cable.
17. Lower the vehicle.

Brake System Bleeding

1. Fill the master cylinder reservoirs with brake fluid. Keep the level at least ½ full during the bleeding operation.
2. Disconnect and plug the brake lines. Fill the master cylinder until fluid begins to flow from the front pipe connector port.
3. Connect the brake lines to the master cylinder and tighten.
4. Depress the brake pedal slowly one time and hold, tighten the connection and then release the brake pedal slowly. Wait 15 seconds.
5. Repeat the sequence until all the air has been removed from the master cylinder bore.
6. After all the air has been removed from the front connections repeat the same procedure at the rear connections of the master cylinder.
7. Individual wheel cylinders and calipers are bled only after all the air has been removed from the master cylinder.
8. To bleed the caliper or the wheel cylinder perform the following:

a. Fill the master cylinder reservoirs with brake fluid. Keep the level at least ½ full during the bleeding operation.
b. Raise and support the vehicle safely. Attach a transparent tube over the bleeder screw.
c. Using an assistant, depress the brake pedal slowly, one time and hold.
d. Loosen the bleeder valve to purge the air from the cylinder.
e. Tighten the bleeder screw and slowly release the brake pedal. Wait 15 seconds.
f. Repeat this sequence until all the air is removed. The bleeding sequence is R/R, L/F, L/R and R/F.

g. Lower the vehicle and refill the master cylinder.

Anti-Lock Brake System Service

PRECAUTION

Failure to observe the following precautions may result in system damage.

• Before performing electric arc welding on the vehicle, disconnect the Electronic Brake Control Module (EBCM) and the hydraulic modulator connectors.

• When performing painting work on the vehicle, do not expose the Electronic Brake Control Module (EBCM) to temperatures in excess of 185°F (85°C) for longer than 2 hrs. The system may be exposed to temperatures up to 200°F (95°C) for less than 15 min.

• Never disconnect or connect the Electronic Brake Control Module (EBCM) or hydraulic modulator connectors with the ignition switch ON.

• Never disassemble any component of the Anti-Lock Brake System (ABS) which is designated non-serviceable; the component must be replaced as an assembly.

• When filling the master cylinder, always use Delco Supreme 11 brake fluid or equivalent, which meets DOT-3 specifications; petroleum base fluid will destroy the rubber parts.

RELIEVING ANTI-LOCK BRAKE SYSTEM PRESSURE

1. Disconnect the negative battery cable. Turn the ignition to the **OFF** position.

2. Pump the brake pedal a minimum of 25 times.

3. When a definite increase in pedal effort, stroke the pedal a few more times.

4. This should relieve all the hydraulic pressure from the system.

Hydraulic Unit

REMOVAL & INSTALLATION

1. Disconnect the negative battery cable.

2. Depressurize the hydraulic accumulator by applying and releasing the brake pedal a minimum of 20–25 times, using 50 lbs. of pedal force. A noticeable change in pedal feel will occur when the pressure is released.

3. Disconnect the electrical connectors at the hydraulic unit.

4. Remove the pump mounting bolts and move the energy unit to gain access to the brake lines.

5. Disconnect the brake lines at the valve block.

6. Remove the left and right sound insulators on Cadillac only.

7. Disconnect the pushrod from the brake pedal and push the dust boot forward past the hex on the pushrod.

8. Separate the pushrod halves by unthreading the 2 pieces.

9. From under the hood, remove the hydraulic unit-to-pushrod mounting bolts.

10. Remove the hydraulic unit from the vehicle, the front part of the pushrod will remain locked into the hydraulic unit.

To install:

11. Install the hydraulic unit to the pushrod bracket. Tighten the support bolts to 37 ft. lbs. (50 Nm).

12. Install the pushrod halves by threading the 2 pieces together.

13. Install the pushrod to the brake pedal and reposition the dust boot.

14. Install the left and the right sound insulators on Cadillac only.

15. Connect the brake lines to the valve block and tighten to 11 ft. lbs.

16. Reposition the energy unit and replace the pump mounting bolts. Tighten to 71 inch lbs.

17. Connect the electrical connections at the hydraulic unit.

18. Connect the negative battery cable and bleed the brake system.

Electronic Brake Control Module (EBCM)

REMOVAL & INSTALLATION

1. Disconnect the negative battery cable.

2. Lower the lower dash panel and disconnect the EBCM module from the bracket.

3. Disconnect the EBCM connector and remove the EBCM module.

To install:

4. Install the EBCM module and connect the EBCM electrical connector.

5. Connect the EBCM to the bracket and install the lower dash panel.

6. Connect the negative battery cable.

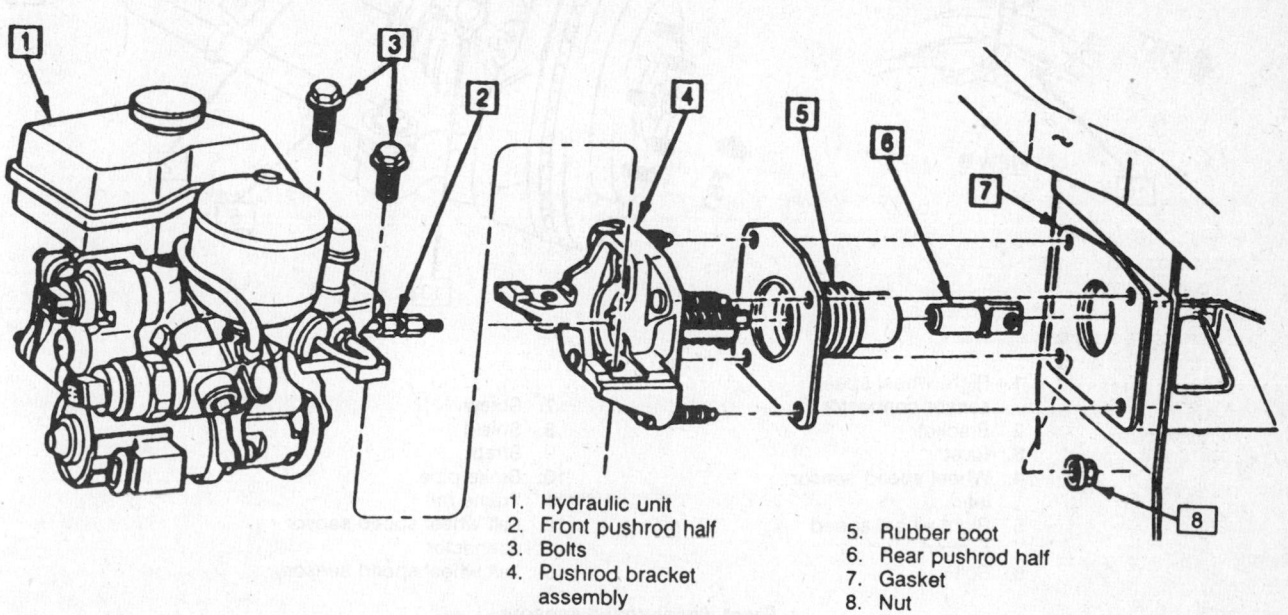

1. Hydraulic unit
2. Front pushrod half
3. Bolts
4. Pushrod bracket assembly
5. Rubber boot
6. Rear pushrod half
7. Gasket
8. Nut

Mounting of the hydraulic unit

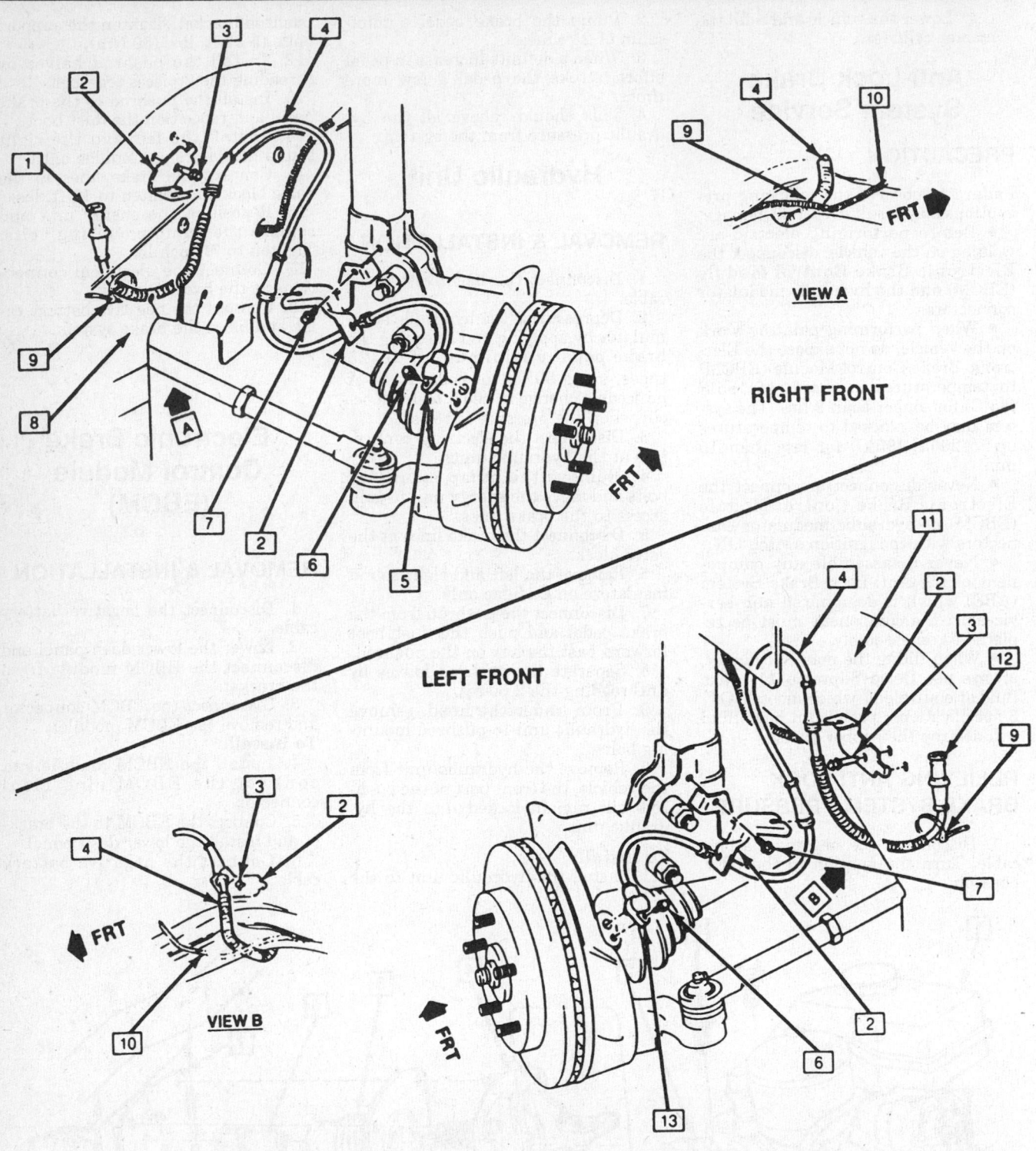

VIEW A

RIGHT FRONT

LEFT FRONT

VIEW B

1. Right wheel speed sensor connector
2. Bracket
3. Rivet
4. Wheel speed sensor lead
5. Right wheel speed sensor
6. Bolt
7. Screw
8. Shield
9. Strap
10. Brake pipe
11. Frame rail
12. Left wheel speed sensor connector
13. Left wheel speed sensor

Front wheel speed sensors

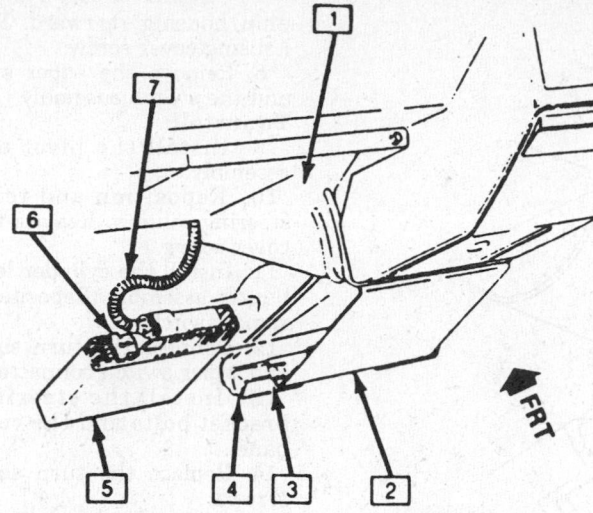

1. Right sound insulator
2. EBCM bracket
3. Locking tab
4. HVAC outlet
5. EBCM
6. Locking plate
7. EBCM harness

Location of the Electronic Brake Control Module (EBCM)—Cadillac equipped with ABS

Wheel Speed Sensor

REMOVAL & INSTALLATION

Front

1. Disconnect the negative battery cable.
2. Disconnect the sensor connector from the wiring harness.
3. Raise and safely support the vehicle. Remove the tire and wheel assembly.
4. Remove the sensor mounting screw and remove the sensor.

To install:

5. Install the sensor and sensor mounting screw.
6. Install the tire and wheel assembly. Lower the vehicle.
7. Connect the negative battery cable.

Rear

1. Disconnect the negative battery cable.
2. Disconnect the sensor connector located in the trunk.
3. Raise and safely support the vehicle. Remove the tire and wheel assembly.
4. Remove the grommet retaining screws.
5. Remove the sensor mounting bolts and remove the sensor.

To install:

6. Install the sensor and sensor mounting bolts.
7. Install the grommet retaining screws.
8. Install the tire and wheel assembly. Lower the vehicle.

9. Connect the sensor connector located in the trunk.
10. Connect the negative battery cable.

CHASSIS ELECTRICAL

Air Bag

CAUTION

Some vehicles are equipped with the Supplemental Inflatable Restraint (SIR) or air bag system. The SIR system must be disabled before performing service on or around SIR system components, steering column, instrument panel components, wiring and sensors. Failure to follow safety and disabling procedures could result in accidental air bag deployment, possible personal injury and unnecessary SIR system repairs.

PRECAUTIONS

Several precautions must be observed when handling the inflator module to avoid accidental deployment and possible personal injury.

- Never carry the inflator module by the wires or connector on the underside of the module.
- When carrying a live inflator module, hold securely with both hands, and ensure that the bag and trim cover are pointed away.
- Place the inflator module on a bench or other surface with the bag and trim cover facing up.

- With the inflator module on the bench, never place anything on or close to the module which may be thrown in the event of an accidental deployment.

DISABLING SIR SYSTEM

1. Disconnect the negative battery cable.
2. Remove the SIR fuse from the fuse panel.
3. Remove the left side sound insulator.
4. Remove the Connector Positive Assurance (CPA) from the yellow 2-way SIR harness connector at the base of the steering column and separate the connector.

ENABLING SIR SYSTEM

1. Connect the yellow 2-way SIR connector at the base of the steering column and insert theConnect Positive Assurance (CPA).
2. Install the left side sound insulator.
3. Install the SIR fuse in the fuse panel.
4. Connect the negative battery cable.

Heater Blower Motor

REMOVAL & INSTALLATION

1. Disconnect the negative battery cable.
2. Disconnect the electrical connections from the blower motor.
3. Disconnect the cooling hose from the blower motor.
4. Remove the mounting screws and the motor.
5. If necessary, remove the coil and spark plug wires.

To install:

6. Use a silicone sealer on the blower motor sealing surfaces.
7. If removed, connect the coil and spark plug wires.
8. Remove the motor and mounting screws.
9. Connect the cooling hose to the blower motor.
10. Connect the electrical connections to the blower motor.
11. Connect the negative battery cable.

Windshield Wiper Motor

REMOVAL & INSTALLATION

1. Disconnect the negative battery cable. Remove the wiper arms and the cowl cover.

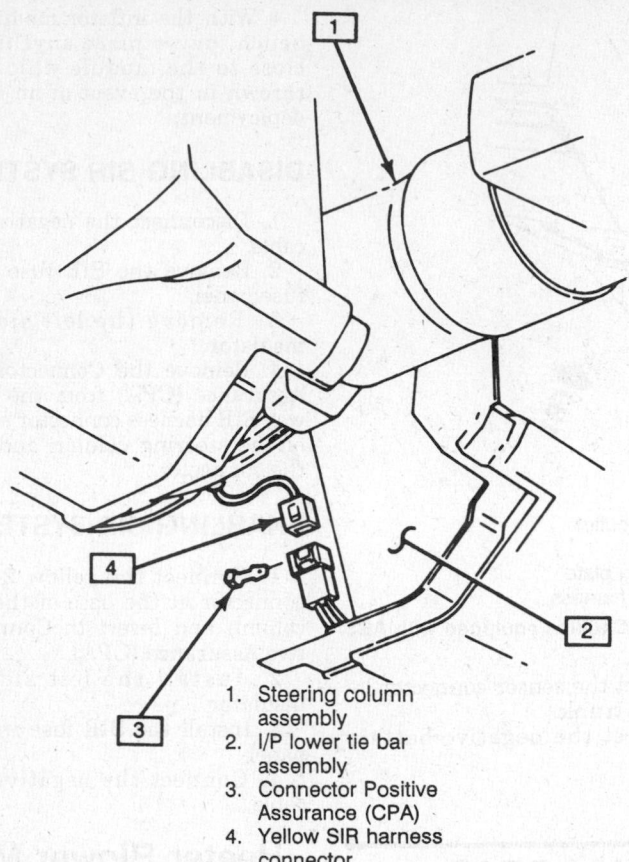

1. Steering column assembly
2. I/P lower tie bar assembly
3. Connector Positive Assurance (CPA)
4. Yellow SIR harness connector

Yellow 2-way SIR harness connector

2. Disconnect the wiper arm drive link from the crank arm.

3. Disconnect the electrical connectors and remove the wiper motor mounting bolts.

4. Guide the crank arm through the hole in the dash and remove the motor.

To install:

5. Guide the crank arm through the hole in the dash and place the motor into position.

6. Install the wiper motor mounting bolts and connect the electrical connectors.

7. Connect the wiper arm drive link to the crank arm. Install the wiper arms and the cowl cover.

8. Connect the negative battery cable.

Windshield Wiper Switch

REMOVAL & INSTALLATION

1. Disconnect the negative battery cable.

2. Remove the steering wheel, the cover and the lock plate assembly.

3. Remove the turn signal actuator arm, the lever and the hazard flasher button.

4. Remove the turn signal switch screws, the lower steering column trim panel and the steering column bracket bolts.

5. Disconnect the the turn signal switch and the wiper switch connectors.

6. Pull the turn signal switch rearward 6–8 inches, remove the key buzzer switch and cylinder lock assembly.

7. Remove and pull the steering col-

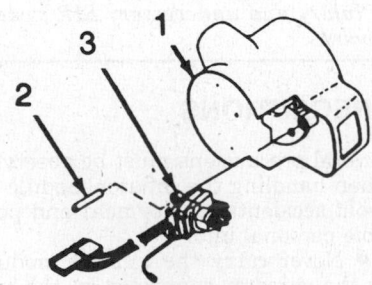

1. Cover assembly, lock housing
2. Pin, switch actuator pivot
3. Switch assembly, pivot and pulse

Windshield washer switch assembly— Cadillac

umn housing rearward. Remove the housing cover screw.

8. Remove the wiper switch pivot and the switch assembly.

To install:

9. Install the pivot and switch assembly.

10. Reposition and reinstall the steering column. Replace the housing cover screw.

11. Install the cylinder lock and key buzzer assembly. Reposition the turn signal switch.

12. Connect the turn signal switch and wiper switch connectors.

13. Install the steering column bracket bolts and the column trim panel.

14. Replace the turn signal switch screws.

15. Install the hazard flasher button, turn signal actuator arm and lever.

16. Install the lock plate assembly, cover and the steering wheel.

17. Connect the negative battery cable.

Instrument Cluster

REMOVAL & INSTALLATION

Except Cadillac

1. Disconnect the negative battery cable. Remove the defroster grille.

2. Remove the instrument panel top cover-to-instrument panel screws.

3. If equipped with a twilight sentinel, pop up the photocell retainer and turn the photocell counterclockwise in the retainer and pull it down-and-out.

4. Slide the instrument panel top cover out far enough to disconnect the aspirator hose and the electrical connector, if equipped.

5. Remove the instrument panel top cover from the instrument panel. If equipped with quartz electronic speedometer clusters, remove the steering column trim cover, so the shift indicator can be removed.

6. Remove the instrument cluster-to-instrument panel carrier screws. Pull the cluster housing assembly straight out; this will separate the electrical connectors from the cluster.

NOTE: It may be helpful to tilt the wheel all the way down and pull the gear select lever to low, when removing the cluster.

7. Disconnect the non-volatile memory chip, if equipped.

8. Remove the speedometer retaining screws and disconnect the speedometer cable or the electrical connection, if equipped.

9. Remove the speedometer assembly.

To install:

10. Install the speedometer assembly. Connect the speedometer cable or the electrical connection, if equipped.

11. Reconnect the non-volatile memory chip, if equipped.

12. Install the instrument cluster and connect the electrical connections.

13. Install the instrument panel top cover and the shift indicator, if equipped.

14. Connect the aspirator hose and the electrical connections.

15. Replace the photo cell and retainer, if equipped with a twilight sentinal.

16. Replace the defroster grille and connect the negative battery cable.

Cadillac

1. Disconnect the negative battery cable.

2. Remove the outlet screws and pry out each outlet, carefully.

3. Remove the glove box mounting screws and disconnect the electrical connections. Remove the glove box assembly.

4. Disconnect the in-vehicle temperature sensor electrical connector and aspirator tube. Remove the upper trim pad.

5. Disconnect the cluster mounting screws and the electrical connectors.

6. Remove the shift indicator cable clip.

7. Remove the instrument cluster.

NOTE: On a digital cluster, remove the memory chip for the season odometer before sending the unit to an authorized repair center. The printed circuit must be lifted to gain access to the memory chip.

8. Remove the lens mounting screws and the speedometer retaining screws.

9. Disconnect the speedometer cable or electrical connection, if equipped.

10. Remove the speedometer assembly.

To install:

11. Install the speedometer assembly. Connect the speedometer cable or electrical connection, if equipped.

12. Install the instrument cluster and connect the electrical connectors.

13. Install and adjust the shift indicator clip. Replace the cluster mounting screws.

14. Install the trim pad and connect the in-vehicle temperature sensor connector and the aspirator tube.

15. Install the glove box assembly and connect the electrical connectors.

16. Install the outlet and outlet screws.

17. Connect the negative battery cable.

Radio

REMOVAL & INSTALLATION

1. Disconnect the negative battery cable.

2. Remove the radio trim plate and the mounting screws from the mounting bracket.

3. Disconnect the electrical connectors and remove the antenna lead.

4. Remove the bracket mounting nuts and the bracket.

5. Remove the radio.

To install:

6. Install the radio.

7. Install the bracket and bracket mounting nuts.

8. Connect the electrical connectors and the antenna lead.

9. Install the mounting screws to the mounting bracket and install the radio trim plate.

10. Connect the negative battery cable.

Headlight Switch

REMOVAL & INSTALLATION

1. Disconnect the negative battery cable. Remove the steering column lower cover or the instrument panel trim plate covering the headlight switch, if equipped with a rocker-type headlight switch.

2. Disconnect the electrical harness retainer below headlight switch assembly. The switch connector is integral to the instrument panel. Pull the switch outward to disconnect it, Except on Cadillac.

3. On Cadillac, depress spring loaded release button on top of headlight switch and remove switch, knob and rod assembly with the switch in the **ON** position.

4. Remove screw with ground wire at bottom of switch housing and all other mounting screws.

5. Pull assembly down and rearward, disconnect wiring harness connectors, bulb(s) and remove assembly.

To install:

6. Connect wiring harness connectors, bulb(s) and install the assembly.

7. Install the screw with ground wire at the bottom of switch housing and all other mounting screws.

8. Connect the electrical harness retainer below headlight switch assembly. Push the switch inward to connect it, Except on Cadillac.

9. On Cadillac, install the switch, knob and rod assembly with the switch in the **ON** position.

10. Install the steering column lower cover or the instrument panel trim plate covering the headlight switch, if equipped with a rocker-type headlight switch.

11. Connect the negative battery cable.

Dimmer Switch

The dimmer switch is attached to the lower portion of the steering column and is controlled by an actuator rod connected to the turn signal lever.

1. Chart—pad assembly	4. Nut	
2. Screw	5. Grille	
3. Screw	6. Nut	
	7. Outlet	

Upper trim panel—Cadillac

REMOVAL & INSTALLATION

1. Disconnect the negative battery cable.
2. Remove the left side sound insulator.
3. Lower the steering column trim plate.
4. Remove the steering column-to-dash screws and lower the steering column.
5. Position the ignition switch in the **OFF-UNLOCKED** position. With the cylinder removed, the rod is in **LOCK** when it is in the next to the uppermost detent; **OFF-UNLOCKED** is 2 detents from the top.
6. Remove the mounting screws and disconnect the electrical connectors. Remove the ignition switch assembly along with the dimmer switch.
7. To adjust the dimmer switch, perform the following procedures:

   a. Install the dimmer switch-to-steering column screws loosely.

   b. Position the switch to firmly contact the actuator rod.

   c. Tighten the screws and test the actuator smoothness in all the tilt positions, if equipped with tilt wheel.

To install:

8. Install the dimmer switch and attach the mounting screws. Put the ignition switch in **OFF-UNLOCKED** position; make sure the lock cylinder and actuating rod are in **OFF-UNLOCKED** (third detent from the top) position.
9. Install the activating rod into the switch and assemble the switch on the column. Tighten the mounting screws.
10. Connect the electrical connections to the dimmer switch.
11. Position the steering column in place and install the column mounting screws.
12. Install the column trim plate and replace the sound insulator.
13. Connect the negative battery cable.

Turn Signal Switch

REMOVAL & INSTALLATION

1. Disconnect the negative battery cable and remove the steering wheel and the shroud.
2. Remove the inflation restraint (air bag module) coil assembly-to-steering shaft lock screw (home boss) and retaining ring. Remove the coil assembly from the shaft and allow it to hang freely.
3. Using the lock plate compression tool or equivalent, position it on the end of the steering shaft and compress the lock plate by turning the shaft nut

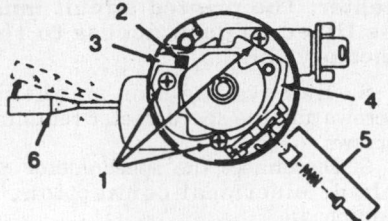

1. Screw, binding HD cross recess
2. Screw
3. Arm, signal switch
4. Switch assembly, turn signal
5. Multi-function lever
6. Hazard knob assembly

Turn signal switch

clockwise. Pry the wire snapring out of the shaft groove.
4. Remove the tool and lift the lock plate from the shaft.
5. Remove the cancelling cam, upper bearing preload spring, bearing seat and inner race from the shaft.
6. Position the turn signal switch in the right turn position. Remove the turn signal lever screw and the lever.
7. Remove the turn signal switch by performing the following:

   a. Remove the switch-to-steering column screws, pull the switch out and allow it to hang freely.

   b. From under the dash, remove the retainer spring and wiring protector.

   c. Remove the hazard knob.

   d. Disconnect the electrical connector from the lower steering column and gently pull the wiring connector through the gear shift lever bowl, the column housing and the lock housing cover. Remove the switch.

To install:

8. Install the turn signal switch harness through the steering column housing and connect the switch and screw.
9. Install the switch actuator arm and screw.
10. Install the inner race, bearing seat and the bearing preload spring. Replace the turn signal cancelling cam.
11. Install the lock plate, using a lock plate compression tool, compress the lock plate and install the shaft lock retaining ring.
12. Install the steering wheel and the shroud.
13. To install the inflation restraint coil, perform the following procedures:

   a. Install the home boss-to-steering column lock screw, allowing the hub to rotate.

   b. While holding the coil assembly (in one hand) with the steering wheel connector facing upwards, rotate the coil hub counterclockwise until it stops; the coil ribbon is now wound snug.

   c. Rotate the coil hub 2½ turns clockwise until the center lock hole is even with the notch in the coil housing.

   d. While holding the hub in position, install the lock screw into the center lock hole.

   e. Install the coil assembly using the horn tower on the inner ring cancelling cam and outer ring projections for alignment purposes.

14. Connect the negative battery cable.

Combination Switch

The combination switch is attached to the upper portion of the steering column and is part of the turn signal lever.

REMOVAL & INSTALLATION

1. Disconnect the negative terminal from the battery.
2. Remove the left side sound insulator.
3. Lower the steering column trim plate.
4. Remove the steering column-to-dash screws and lower the steering column.
5. Remove the inflation restraint (air bag module) and the combination switch assembly.
6. Position the ignition switch in the **OFF-UNLOCKED** position. With the cylinder removed, the rod is in **LOCK** when it is in the next to the uppermost detent; **OFF-UNLOCKED** is 2 detents from the top.
7. Remove the mounting screws and disconnect the electrical connectors. Remove the ignition switch assembly along with the dimmer switch.

To install:

8. Adjust the dimmer switch.
9. Install the dimmer switch and attach the mounting screws. Put the ignition switch in **OFF-UNLOCKED** position; make sure the lock cylinder and actuating rod are in **OFF-UNLOCKED** (third detent from the top) position.
10. Install the activating rod into the switch and assemble the switch on the column. Tighten the mounting screws.
11. Connect the electrical connections at the dimmer switch.
12. Install the combination switch and replace the air bag module.
13. Position the steering column in place and install the column mounting screws.
14. Install the column trim plate and replace the sound insulator.
15. Connect the negative battery cable.

Ignition Switch

REMOVAL & INSTALLATION

1. Disconnect the negative battery cable and lower the steering column; be sure to properly support it.

2. Position the switch in the **OFF-UNLOCKED** position. With the lock cylinder removed, the rod is in **LOCK** when it is in the next to the uppermost detent; **OFF-UNLOCKED** is 2 detents from the top.

3. Remove both switch screws and the switch assembly.

To install:

4. Place the new switch in the **OFF-UNLOCKED** position. Ensure the lock cylinder and actuating rod are in **OFF-UNLOCKED** position (3rd detent from the top).

5. Install the actuating rod into the switch and assemble the switch on the column. Tighten the mounting screws.

NOTE: Use only the specified screws since over-length screws could impair the collapsibility of the column.

6. Install the steering column.

7. Connect the negative battery cable.

Ignition Lock

REMOVAL & INSTALLATION

1. Disconnect the negative battery cable and remove the turn signal switch assembly.

2. Remove the key from the lock cylinder. Remove the buzzer switch and clip.

3. Reinsert the key into the lock cylinder and turn it to the **LOCK** position.

4. Remove the cylinder lock-to-steering column screw and the lock set.

To install:

5. Install the cylinder lock and tighten the lock-to-steering column screw to 22 inch lbs.

6. Position the key in the **RUN** position and reverse the removal procedures. Tighten the turn signal switch-to-steering column screws to 30 inch lbs. and the turn signal lever screw to 20 inch lbs.

7. Connect the negative battery cable.

Stoplight Switch

ADJUSTMENT

1. Install the switch into the tubular clip until the switch assembly seats itself on the tubular clip.

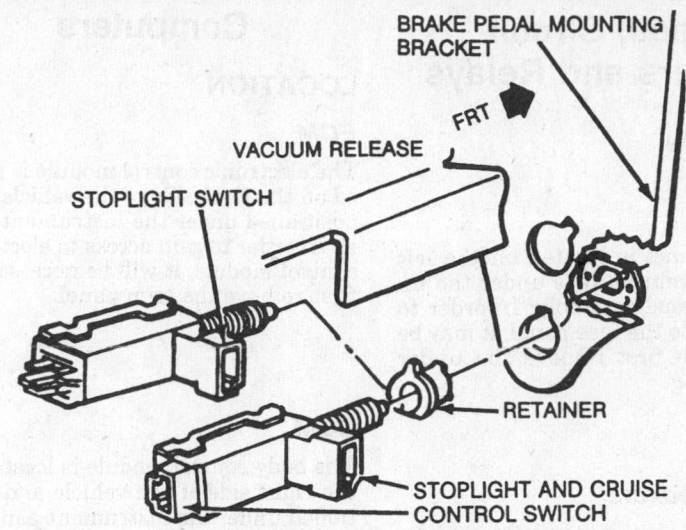

Stoplight switch location

2. Pull the brake pedal rearward against the pedal stop.

3. The switch will be moved in the tubular clip which will adjust itself properly.

4. The proper switch adjustment is achieved when no clicks are heard when the pedal is pulled upward and the brake lights stay off when the brake pedal is released.

REMOVAL & INSTALLATION

1. Disconnect the negative terminal from the battery. Remove the left side sound insulator.

2. Loosen the tubular clip from the stoplight switch assembly.

3. Disconnect the electrical connector from the rear of the switch assembly.

4. Remove the stoplight switch from the vehicle.

To install:

5. Install the stoplight switch to the vehicle.

6. Connect the electrical connector to the rear of the switch assembly.

7. Tighten the tubular clip to the stoplight switch assembly.

8. Install the left side sound insulator. Connect the negative battery cable.

Neutral Safety Switch

ADJUSTMENT

1. Disconnect the negative battery cable.

2. Place the transaxle shifter lever in the **N** position.

3. Loosen the switch mounting screws.

4. Rotate the switch on the shifter assembly to align the service adjustment holes.

5. Insert a gauge pin or equivalent, into the service slots. Tighten the mounting bolts.

6. Remove the gauge pin. Connect the negative battery cable.

REMOVAL & INSTALLATION

1. Disconnect the negative battery cable.

2. Disconnect the shift linkage and the electrical connectors at the switch.

3. Remove the mounting bolts and the switch assembly.

To install:

4. Place the transaxle shifter lever in the **N** position.

5. Rotate the switch on the shifter assembly to align the service adjustment holes.

6. Insert a gauge pin or the equivalent into the service slots. Tighten the mounting bolts.

7. Remove the gauge pin. Connect the negative battery cable.

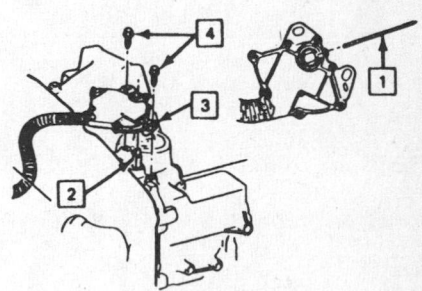

1. ³/₃₂ in. drill bit
2. Selector shaft
3. Neutral start and back up lamp switch
4. Bolts

Neutral safety switch adjustment

Fuses, Circuit Breakers and Relays

LOCATION

Fuses

The fuse panel is located on the left side of the vehicle. It is under the instrument panel assembly. In order to gain access to the fuse panel, it may be necessary to first remove the under dash padding.

Circuit Breakers

The convenience center is located on the underside of the instrument panel near the fuse panel. It provides a central location for various relays, hazard flasher units and warning buzzers/chimes. All units are replaced with plug-in modules.

Relays

The relay center is located on the right side of the instrument panel. The relay center is mounted behind the glove box assembly.

Computers

LOCATION

ECM

The electronic control module is located on the right side of the vehicle. It is positioned under the instrument panel. In order to gain access to electronic control module, it will be necessary to first remove the trim panel.

BCM

The body control module is located on the right side of the vehicle and positioned under the instrument panel. In order to gain access to body control module, it will be necessary to first remove the trim panel.

EBCM

The electronic brake control module is located on the right side of the vehicle and positioned under the right sound insulator panel. In order to gain access to electronic brake control module, it will be necessary to first remove the trim panel.

Flashers

LOCATION

The turn signal flasher unit is located behind the instrument panel near the steering column, along with the hazard flasher. It is secured in place with a plastic retainer. In order to gain access to components, it may first be necessary to remove certain under dash padding.

The hazard flasher is located on the fuse block. It is positioned on the lower right side corner of the fuse block assembly. In order to gain access to the turn signal flasher it may be necessary to first remove the under dash padding.

Cruise Control

ADJUSTMENT

1. Turn the ignition switch **OFF**.
2. Fully retract the idle speed control motor plunger.

NOTE: The throttle lever must not touch the idle speed control plunger.

3. Connect the cruise control cable to the hole in the servo blade that leaves the minimum slack.
4. Install the retainer at the servo.

GM "E," "K" & "V" Body

Front Wheel Drive

Buick—Reatta, Riviera
Cadillac—Allante, Eldorado, Seville
Oldsmobile—Toronado, Trofeo

SPECIFICATIONS

VEHICLE IDENTIFICATION CHART

It is important for servicing and ordering parts to be certain of the vehicle and engine identification. The VIN (vehicle identification number) is a 17 digit number visible through the windshield on the driver's side of the dash and contains the vehicle and engine identification codes. The tenth digit indicates model year, and the eighth digit indicates engine code. It can be interpreted as follows:

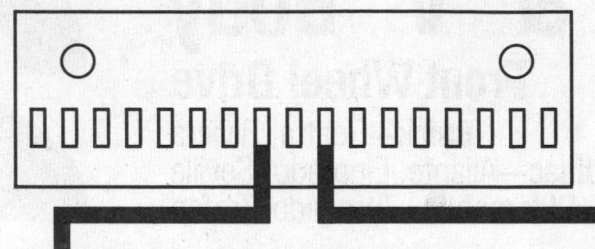

Engine Code

Code	Cu. In.	Liters	Cyl.	Fuel Sys.	Eng. Mfg.
C	231	3.8	6	SFI	Buick
L	231	3.8	6	SFI	Buick
7	250	4.1	8	SFI	Cadillac
5	273	4.5	8	TBI	Cadillac
8	273	4.5	8	SFI	Cadillac
3	273	4.5	8	SFI	Cadillac
B	300	4.9	8	SFI	Cadillac

Model Year

Code	Year
J	1988
K	1989
L	1990
M	1991
N	1992

SFI—Sequential Fuel Injection
TBI—Throttle Body Injection

ENGINE IDENTIFICATION

Year	Model	Engine Displacement cu. in. (liter)	Engine Series Identification (VIN)	No. of Cylinders	Engine Type
1988	Allante	250 (4.1)	7	8	OHV
	Eldorado	273 (4.5)	5	8	OHV
	Reatta	231 (3.8)	C	6	OHV
	Riviera	231 (3.8)	C	6	OHV
	Seville	273 (4.5)	5	8	OHV
	Toronado	231 (3.8)	C	6	OHV
1989	Allante	273 (4.5)	8	8	OHV
	Eldorado	273 (4.5)	5	8	OHV
	Reatta	231 (3.8)	C	6	OHV
	Riviera	231 (3.8)	C	6	OHV
	Seville	273 (4.5)	5	8	OHV
	Toronado	231 (3.8)	C	6	OHV
1990	Allante	273 (4.5)	8	8	OHV
	Eldorado	273 (4.5)	3	8	OHV
	Reatta	231 (3.8)	C	6	OHV
	Riviera	231 (3.8)	C	6	OHV
	Seville	273 (4.5)	3	8	OHV

ENGINE IDENTIFICATION

Year	Model	Engine Displacement cu. in. (liter)	Engine Series Identification (VIN)	No. of Cylinders	Engine Type
1990	Toronado	231 (3.8)	C	6	OHV
	Trofeo	231 (3.8)	C	6	OHV
1991–92	Allante	273 (4.5)	8	8	OHV
	Eldorado	300 (4.9)	B	8	OHV
	Reatta	231 (3.8)	L	6	OHV
	Riviera	231 (3.8)	L	6	OHV
	Seville	300 (4.9)	B	8	OHV
	Toronado	231 (3.8)	L	6	OHV
	Trofeo	231 (3.8)	L	6	OHV

OHV—Overhead Valve

GENERAL ENGINE SPECIFICATIONS

Year	VIN	No. Cylinder Displacement cu. in. (liter)	Fuel System Type	Net Horsepower @ rpm	Net Torque @ rpm (ft. lbs.)	Bore × Stroke (in.)	Compression Ratio	Oil Pressure @ rpm
1988	C	6-231 (3.8)	SFI	165 @ 5200	210 @ 2000	3.800 × 3.400	8.5:1	37 @ 2400
	7	8-250 (4.1)	SFI	170 @ 4300	235 @ 3200	3.460 × 3.310	8.5:1	①
	5	8-273 (4.5)	TBI	155 @ 4200	240 @ 2800	3.600 × 3.310	9.0:1	①
1989	C	6-231 (3.8)	SFI	165 @ 5200	210 @ 2000	3.800 × 3.400	8.5:1	37 @ 2400
	5	8-273 (4.5)	TBI	155 @ 4200	240 @ 2800	3.620 × 3.310	9.0:1	①
	8	8-273 (4.5)	SFI	200 @ 4400	270 @ 3200	3.620 × 3.310	9.0:1	①
1990	C	6-231 (3.8)	SFI	165 @ 5200	210 @ 2000	3.800 × 3.400	8.5:1	37 @ 2400
	8	8-273 (4.5)	SFI	200 @ 4400	230 @ 3200	3.620 × 3.310	9.0:1	①
	3	8-273 (4.5)	SFI	180 @ 4000	245 @ 3000	3.620 × 3.310	9.5:1	①
1991–92	C	6-231 (3.8)	SFI	165 @ 4800	210 @ 2000	3.800 × 3.400	8.5:1	40 @ 1850
	L	6-231 (3.8)	SFI	170 @ 4800	220 @ 3200	3.800 × 3.400	8.5:1	60 @ 1850
	8	8-273 (4.5)	SFI	200 @ 4400	270 @ 3200	3.620 × 3.310	9.0:1	①
	B	8-300 (4.9)	SFI	200 @ 4100	275 @ 3000	3.620 × 3.620	9.5:1	53 @ 2000

NOTE: Horsepower and torque are SAE net figures. They are measured with all accessories installed and operating. Since the figures vary when a given engine is installed in different models, some are representative rather than exact.
① 26–30 PSI at 30 MPH at normal operating
   temperature
SFI—Sequential Fuel Injection
TBI—Throttle Body Injection

GASOLINE ENGINE TUNE-UP SPECIFICATIONS

Year	VIN	No. Cylinder Displacement cu. in. (liter)	Spark Plugs Type	Spark Plugs Gap (in.)	Ignition Timing (deg.) MT	Ignition Timing (deg.) AT	Compression Pressure (psi)	Fuel Pump (psi)	Idle Speed (rpm) MT	Idle Speed (rpm) AT	Valve Clearance In.	Valve Clearance Ex.
1988	C	6-231 (3.8)	R44LTS6	0.060	—	①	④	31–42②	—	①	Hyd.	Hyd.
	7	8-250 (4.1)	R44LTS6K	0.060	—	①	140–165	40–50	—	①	Hyd.	Hyd.
	5	8-273 (4.5)	R44LTS6K	0.060	—	①	140–165	9–12	—	①	Hyd.	Hyd.

GASOLINE ENGINE TUNE-UP SPECIFICATIONS

Year	VIN	No. Cylinder Displacement cu. in. (liter)	Spark Plugs Type	Spark Plugs Gap (in.)	Ignition Timing (deg.) MT	Ignition Timing (deg.) AT	Compression Pressure (psi)	Fuel Pump (psi)	Idle Speed (rpm) MT	Idle Speed (rpm) AT	Valve Clearance In.	Valve Clearance Ex.
1989	C	6-231 (3.8)	R44LTS6	0.060	—	①	④	31–42 ②	—	①	Hyd.	Hyd.
	5	8-273 (4.5)	R44LTS6K	0.060	—	①	140–165	9–12	—	①	Hyd.	Hyd.
	8	8-273 (4.5)	R44LTS6K	0.060	—	①	140–165	40–50	—	①	Hyd.	Hyd.
1990	C	6-231 (3.8)	R44LTS6	0.060	—	①	④	③	—	①	Hyd.	Hyd.
	8	8-250 (4.1)	R44LTS6K	0.060	—	①	140–165	40–50	—	①	Hyd.	Hyd.
	3	8-273 (4.5)	R44LTS6K	0.060	—	①	140–165	40–50	—	①	Hyd.	Hyd.
1991	C	6-231 (3.8)	R44LTS6	0.060	—	①	④	③	—	①	Hyd.	Hyd.
	L	6-231 (3.8)	R44LTS6	0.060	—	①	④	③	—	①	Hyd.	Hyd.
	8	8-250 (4.1)	R44LTS6K	0.060	—	①	140–165	40–50	—	①	Hyd.	Hyd.
	B	8-300 (4.9)	R45LTS6K	0.060	—	①	140–165	40–50	—	①	Hyd.	Hyd.
1992					REFER TO UNDERHOOD STICKER							

Hyd.—Hydraulic
① Controlled by ECM
② Engine idling at normal operating temp.
③ 1—Connect fuel pressure gauge, engine at
    normal operating temperature
  2—Turn ignition switch on
  3—After approx. 2 seconds pressure should
    read 41–47 psi and hold steady
  4—Start engine and idle, pressure should drop
    3–10 psi from static pressure
④ The lowest cylinder compression reading
  should not be less than 70% of the highest
  reading and no cylinder should be less than
  100 psi

FIRING ORDERS

NOTE: To avoid confusion, always replace spark plug wires one at a time.

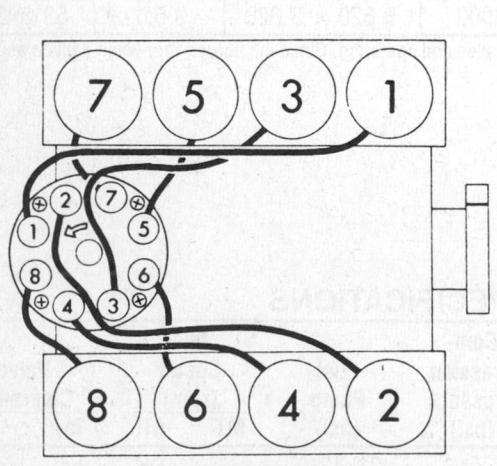

4.1L, 4.5L and 4.9L Engines
Engine Firing Order: 1-8-4-3-6-5-7-2
Distributor Rotation: Counterclockwise

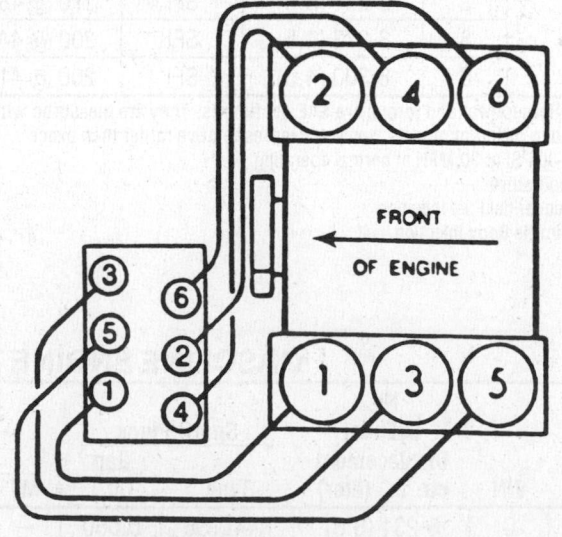

3.8L Engine VIN C
Engine Firing Order: 1-6-5-4-3-2
Distributorless Ignition System

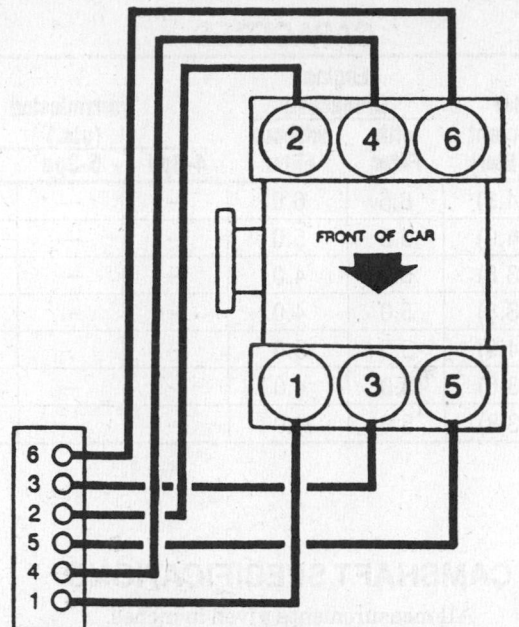

3.8L Engine VIN L
Engine Firing Order: 1–6–5–4–3–2
Distributorless Ignition System

CAPACITIES

Year	Model	VIN	No. Cylinder Displacement cu. in. (liter)	Engine Crankcase with Filter	Engine Crankcase without Filter	Transmission (pts.) 4-Spd	Transmission (pts.) 5-Spd	Transmission (pts.) Auto.	Drive Axle (pts.)	Fuel Tank (gals.)	Cooling System (qts.)
1988	Allante	7	8-250 (4.1)	5.5	5.0	—	—	13	—	22	12.1
	Eldorado	5	8-273 (4.5)	5.5	5.0	—	—	13	—	18.8	12.1
	Reatta	C	6-231 (3.8)	5.0	4.0	—	—	13	—	18	13
	Riviera	3	6-231 (3.8)	5.0	4.0	—	—	13	—	18	13
	Seville	5	8-273 (4.5)	5.5	5.0	—	—	13	—	18.8	12.1
	Toronado	C	6-231 (3.8)	5.0	4.0	—	—	13	—	18	13
1989	Allante	5	8-273 (4.5)	5.5	5.0	—	—	13	—	22	12.1
	Eldorado	5	8-273 (4.5)	5.5	5.0	—	—	13	—	18.8	12
	Reatta	C	6-231 (3.8)	5.0	4.0	—	—	12	—	18	13
	Riviera	C	6-231 (3.8)	5.0	4.0	—	—	12	—	18	13
	Seville	5	8-273 (4.5)	5.5	5.0	—	—	13	—	18.8	12.1
	Toronado	C	6-231 (3.8)	5.0	4.0	—	—	12	—	18	13
1990	Allante	8	8-273 (4.5)	5.5	6.0	—	—	13	—	22	12.1
	Eldorado	3	8-273 (4.5)	5.5	5.0	—	—	13	—	18.8	12.1
	Reatta	C	6-231 (3.8)	5.0	4.0	—	—	12	—	18	13
	Riviera	C	6-231 (3.8)	5.0	4.0	—	—	12	—	18	13
	Seville	3	8-273 (4.5)	5.5	5.0	—	—	13	—	18.8	12.1
	Toronado	C	6-231 (3.8)	5.0	4.0	—	—	12	—	18	13
	Trofeo	C	6-231 (3.8)	5.0	4.0	—	—	12	—	18	13

CAPACITIES

Year	Model	VIN	No. Cylinder Displacement cu. in. (liter)	Engine Crankcase with Filter	Engine Crankcase without Filter	Transmission (pts.) 4-Spd	Transmission (pts.) 5-Spd	Transmission (pts.) Auto.	Drive Axle (pts.)	Fuel Tank (gals.)	Cooling System (qts.)
1991-92	Allante	8	8-273 (4.5)	6.5	6.0	—	—	13	—	22	12.1
	Eldorado	B	8-300 (4.9)	5.5	5.0	—	—	13	—	18.8	12.1
	Reatta	L	6-231 (3.8)	5.0	4.0	—	—	12	—	18	13
	Riviera	L	6-231 (3.8)	5.0	4.0	—	—	12	—	18	13
	Seville	B	8-300 (4.9)	5.5	5.0	—	—	13	—	18.8	12.1
	Toronado	L	6-231 (3.8)	5.0	4.0	—	—	12	—	18	13
	Trofeo	L	6-231 (3.8)	5.0	4.0	—	—	12	—	18	13

CAMSHAFT SPECIFICATIONS

All measurements given in inches.

Year	VIN	No. Cylinder Displacement cu. in. (liter)	Journal Diameter 1	Journal Diameter 2	Journal Diameter 3	Journal Diameter 4	Journal Diameter 5	Lobe Lift In.	Lobe Lift Ex.	Bearing Clearance	Camshaft End Play
1988	C	6-231 (3.8)	1.785–1.786	1.785–1.786	1.785–1.786	1.785–1.786	—	0.250	0.250	0.0005–0.0035	NA
	7	8-250 (4.1)	NA	NA	NA	NA	NA	0.384	0.396	0.0018–0.0037	NA
	5	8-273 (4.5)	NA	NA	NA	NA	NA	0.384	0.396	0.0018–0.0037	NA
1989	C	6-231 (3.8)	1.785–1.786	1.785–1.786	1.785–1.786	1.785–1.786	—	0.250	0.255	0.0005–0.0035	NA
	5	8-273 (4.5)	NA	NA	NA	NA	NA	0.384	0.396	0.0018–0.0037	NA
	8	8-273 (4.5)	NA	NA	NA	NA	NA	0.384	0.396	0.0018–0.0037	NA
1990	C	6-231 (3.8)	1.785–1.786	1.785–1.786	1.785–1.786	1.785–1.786	—	0.250	0.255	0.0005–0.0035	NA
	8	8-273 (4.5)	NA	NA	NA	NA	NA	0.384	0.396	0.0018–0.0037	NA
	3	8-273 (4.5)	NA	NA	NA	NA	NA	0.384	0.396	0.0018–0.0037	NA
1991-92	C	6-231 (3.8)	1.785–1.786	1.785–1.786	1.785–1.786	1.785–1.786	—	0.250	0.255	0.0005–0.0035	NA
	L	6-231 (3.8)	1.785–1.786	1.785–1.786	1.785–1.786	1.785–1.786	—	0.250	0.255	0.0005–0.0035	NA
	8	8-273 (4.5)	NA	NA	NA	NA	NA	0.384	0.396	0.0018–0.0037	NA
	B	8-300 (4.9)	NA	NA	NA	NA	NA	0.384	0.396	0.0018–0.0037	NA

NA—Not available

CRANKSHAFT AND CONNECTING ROD SPECIFICATIONS

All measurements are given in inches.

Year	VIN	No. Cylinder Displacement cu. in. (liter)	Crankshaft				Connecting Rod		
			Main Brg. Journal Dia.	Main Brg. Oil Clearance	Shaft End-play	Thrust on No.	Journal Diameter	Oil Clearance	Side Clearance
1988	C	6-231 (3.8)	2.4988–2.4998	0.0018–0.0030	0.003–0.011	2	2.2487–2.2499	0.0003–0.0028	0.003–0.015
	7	8-250 (4.1)	2.6350–2.6360	0.0016–①	0.001–0.007	3	1.9290	0.0005–0.0028	0.008–0.020
	5	8-273 (4.5)	2.6350–2.6360	0.0016–①	0.001–0.007	3	1.9290	0.0005–0.0028	0.008–0.020
1989	C	6-231 (3.8)	2.4988–2.4998	0.0018–0.0030	0.003–0.011	2	2.2487–2.2499	0.0003–0.0028	0.003–0.015
	5	8-273 (4.5)	2.6350–2.6360	0.0016–①	0.001–0.007	3	1.9270	0.0005–0.0028	0.008–0.020
	8	8-273 (4.5)	2.6350–2.6360	0.0016–①	0.001–0.007	3	1.9270	0.0005–0.0028	0.008–0.020
1990	C	6-231 (3.8)	2.4988–2.4998	0.0018–0.0030	0.003–0.011	2	2.2487–2.2499	0.0003–0.0028	0.003–0.015
	8	8-273 (4.5)	2.6350–2.6360	0.0016–①	0.001–0.007	3	1.9270	0.0005–0.0028	0.008–0.020
	3	8-273 (4.5)	2.6350–2.6360	0.0016–①	0.001–0.007	3	1.9270	0.0005–0.0028	0.008–0.020
1991–92	C	6-231 (3.8)	2.4988–2.4998	0.0018–0.0030	0.003–0.011	2	2.2487–2.2499	0.0003–0.0028	0.003–0.015
	L	6-231 (3.8)	2.4988–2.4998	0.0018–0.0030	0.003–0.011	2	2.2487–2.2499	0.0003–0.0028	0.003–0.015
	8	8-273 (4.5)	2.6350–2.6360	0.0016–①	0.001–0.007	3	1.9270–1.9280	0.0005–0.0028	0.008–0.020
	B	8-300 (4.9)	2.6350–2.6360	0.0016–①	0.001–0.008	3	1.9270–1.9280	0.0005–0.0028	0.008–0.020

① No. 1—0.0008–0.0031

VALVE SPECIFICATIONS

Year	VIN	No. Cylinder Displacement cu. in. (liter)	Seat Angle (deg.)	Face Angle (deg.)	Spring Test Pressure (lbs.) ①	Spring Installed Height (in.)	Stem-to-Guide Clearance (in.)		Stem Diameter (in.)	
							Intake	Exhaust	Intake	Exhaust
1988	C	6-231 (3.8)	45	45	225 @ 1.255 ②	1.690–1.750	0.0015–0.0035	0.0015–0.0032	NA	NA
	7	8-250 (4.1)	45	44	② ③	④ ⑤	0.0010–0.0030	0.0010–0.0030	0.3413–0.3420	0.3401–0.3408
	5	8-273 (4.5)	45	44	② ③	④ ⑤	0.0010–0.0030	0.0010–0.0030	0.3413–0.3420	0.3401–0.3408
1989	C	6-231 (3.8)	45	45	200–220 @ 1.315 ②	1.690–1.750	0.0015–0.0035	0.0015–0.0032	NA	NA
	5	8-273 (4.5)	45	44	204–221 @ 1.28 ②	2.216 ④	0.0010–0.0030	0.0010–0.0030	0.3413–0.3420	0.3401–0.3408
	8	8-273 (4.5)	45	44	204–221 @ 1.28 ②	2.216 ④	0.0010–0.0030	0.0010–0.0030	0.3413–0.3420	0.3401–0.3408

VALVE SPECIFICATIONS

Year	VIN	No. Cylinder Displacement cu. in. (liter)	Seat Angle (deg.)	Face Angle (deg.)	Spring Test Pressure (lbs.) ①	Spring Installed Height (In.)	Stem-to-Guide Clearance (in.) Intake	Stem-to-Guide Clearance (in.) Exhaust	Stem Diameter (In.) Intake	Stem Diameter (In.) Exhaust
1990	C	6-231 (3.8)	45	45	200–220 @ 1.315②	1.690–1.750	0.0015–0.0035	0.0015–0.0032	NA	NA
	8	8-273 (4.5)	45	44	204–221 @ 1.28②	2.216④	0.0010–0.0030	0.0010–0.0030	0.3413–0.3420	0.3401–0.3408
	3	8-273 (4.5)	45	44	214–232 @ 1.35②	2.216④	0.0010–0.0030	0.0010–0.0030	0.3413–0.3420	0.3401–0.3408
1991–92	C	6-231 (3.8)	45	45	200–220 @ 1.315②	1.690–1.720	0.0015–0.0035	0.0015–0.0032	NA	NA
	L	6-231 (3.8)	45	45	210 @ 1.315②	1.690–1.720	0.0015–0.0035	0.0015–0.0032	NA	NA
	8	8-273 (4.5)	45	44	204–221 @ 1.28	2.216④	0.0010–0.0030	0.0020–0.0040	0.3413–0.3420	0.3401–0.3408
	B	8-300 (4.9)	45	45	214–232 @ 1.35	1.949④	0.0010–0.0030	0.002–0.004	0.3413–0.3420	0.3401–0.3408

① lbs. @ in.
② Load open
③ Intake—209–216 @ 1.28
   Exhaust—204–221 @ 1.28
④ Free length
⑤ Intake—2.311 in.
   Exhaust—2.126 in.
NA—Not available

PISTON AND RING SPECIFICATIONS

All measurements are given in inches.

Year	VIN	No. Cylinder Displacement cu. in. (liter)	Piston Clearance	Ring Gap Top Compression	Ring Gap Bottom Compression	Ring Gap Oil Control	Ring Side Clearance Top Compression	Ring Side Clearance Bottom Compression	Ring Side Clearance Oil Control
1988	C	6-231 (3.8)	①	0.010–0.025	0.010–0.025	0.015–0.055	0.0013–0.0031	0.0013–0.0031	0.0011–0.0081
	7	8-250 (4.1)	0.0010–0.0018	0.015–0.024	0.015–0.024	0.010–0.050	0.0016–0.0037	0.0016–0.0037	③
	5	8-273 (4.5)	0.0010–0.0018	0.015–0.024	0.015–0.024	0.010–0.050	0.0016–0.0037	0.0016–0.0037	③
1989	C	6-231 (3.8)	0.0004–0.0022 ②	0.010–0.025	0.010–0.025	0.015–0.055	0.0013–0.0031	0.0013–0.0031	0.0011–0.0081
	5	8-273 (4.5)	0.0010–0.0018	0.015–0.024	0.015–0.024	0.010–0.050	0.0016–0.0037	0.0016–0.0037	③
	8	8-273 (4.5)	0.0010–0.0018	0.015–0.024	0.015–0.024	0.010–0.050	0.0016–0.0037	0.0016–0.0037	③
1990	C	6-231 (3.8)	0.0004–0.0022 ②	0.010–0.025	0.010–0.025	0.015–0.055	0.0013–0.0031	0.0013–0.0031	0.0011–0.0081
	8	8-273 (4.5)	0.0010–0.0018	0.015–0.024	0.015–0.024	0.010–0.050	0.0016–0.0037	0.0016–0.0037	③
	3	8-273 (4.5)	0.0010–0.0018	0.015–0.024	0.015–0.024	0.010–0.050	0.0016–0.0037	0.0016–0.0037	③

PISTON AND RING SPECIFICATIONS

All measurements are given in inches.

Year	VIN	No. Cylinder Displacement cu. in. (liter)	Piston Clearance	Ring Gap			Ring Side Clearance		
				Top Compression	Bottom Compression	Oil Control	Top Compression	Bottom Compression	Oil Control
1991–92	C	6-231 (3.8)	0.0004–0.0022 ②	0.010–0.025	0.010–0.025	0.015–0.055	0.0013–0.0031	0.0013–0.0031	0.0011–0.0081
	L	6-231 (3.8)	0.0004–0.0022 ②	0.010–0.025	0.010–0.025	0.015–0.055	0.0013–0.0031	0.0013–0.0031	0.0011–0.0081
	8	8-273 (4.5)	0.0010–0.0018	0.015–0.024	0.015–0.024	0.010–0.050	0.0016–0.0037	0.0016–0.0037	③
	B	8-300 (4.9)	0.0004–0.0020	0.012–0.022	0.012–0.022	0.004–0.020	0.0016–0.0037	0.0016–0.0037	③

① Skirt top—0.0007–0.0027
  Skirt bottom—0.0010–0.0045
② 44mm from top of piston
③ None, side sealing

TORQUE SPECIFICATIONS

All readings in ft. lbs.

Year	VIN	No. Cylinder Displacement cu. in. (liter)	Cylinder Head Bolts	Main Bearing Bolts	Rod Bearing Bolts	Crankshaft Pulley Bolts	Flywheel Bolts	Manifold		Spark Plugs
								Intake	Exhaust	
1988	C	6-231 (3.8)	①	90	43	219 ⑤	61	88 ⑥	41	20
	7	8-250 (4.1)	②	85	22	18	37	⑦	18	11
	5	8-273 (4.5)	②	85	22	18	70	⑦	18	11
1989	C	6-231 (3.8)	③	90	43	219 ⑤	61	88 ⑥	41	20
	5	8-273 (4.5)	②	85	24 ④	18	70	⑧	18	11
	8	8-273 (4.5)	②	85	24 ④	18	70	⑧	18	11
1990	C	6-231 (3.8)	③	90	43	219 ⑤	61	88 ⑥	41	20
	8	8-273 (4.5)	②	85	24 ④	18	70	⑧	18	11
	3	8-273 (4.5)	②	85	24 ④	18	70	⑧	18	11
1991–92	C	6-231 (3.8)	③	90	43	219 ⑤	61	88 ⑥	41	20
	L	6-231 (3.8)	③	⑨	⑩	⑪	61	88 ⑥	41	20
	8	8-273 (4.5)	②	85	24 ④	18	70	⑧	18	11
	B	8-300 (4.9)	②	85	25	70 ⑤	70	⑧	16	23

① Torque in sequence to 25 ft. lbs.; then turn each bolt an add'l ¼ turn (90 degrees) in sequence; then turn each bolt an add'l ¼ turn (90 degrees) in sequence (if torque exceeds 60 ft. lbs. at any point during last 2 steps, do not complete 90 degrees turn).
② Torque in sequence to 38 ft. lbs.; then torque to 68 ft. lbs.; then torque No. 1, 3 and 4 bolts to 90 ft. lbs.
③ Torque in sequence to 35 ft. lbs.; then turn each bolt 130 degrees; then rotate each bolt an add'l 30 degrees.
④ Lubricate with engine oil
⑤ Crankshaft balancer assembly
⑥ Inch lbs.

⑦ Torque bolts 1, 2, 3 and 4 in sequence to 15 ft. lbs.; then tighten bolts 5 through 16 in sequence to 22 ft. lbs.; then retighten all bolts in sequence to 22 ft. lbs.; then retorque all bolts to sequence to 22 ft. lbs.
⑧ Torque bolts 1, 2, 3 and 4 in sequence to 8 ft. lbs.; then tighten bolts 5 through 16 in sequence to 8 ft. lbs; then retighten all bolts in sequence to 12 ft. lbs.; then retorque above step until torque level is maintained.
⑨ 26±3 ft. lbs. + 50°±3°
⑩ 20±3 ft. lbs. + 50°±3°
⑪ 105±7 ft. lbs. + 56°±4°

BRAKE SPECIFICATIONS
All measurements in inches unless noted.

Year	Model	Lug Nut Torque (ft. lbs.)	Master Cylinder Bore	Brake Disc		Standard Brake Drum Diameter	Minimum Lining Thickness	
				Minimum Thickness	Maximum Runout		Front	Rear
1988	Allante	100	①	0.971②	0.004③	—	0.030	0.030
	Eldorado	100	①	0.971②	0.004③	—	0.030	0.030
	Reatta	100	①	0.971②	0.004③	—	0.030	0.030
	Riviera	100	①	0.971②	0.004③	—	0.030	0.030
	Seville	100	①	0.971②	0.004③	—	0.030	0.030
	Toronado	100	①	0.971②	0.004③	—	0.030	0.030
1989	Allante	100	①	0.971②	0.004③	—	0.030	0.030
	Eldorado	100	①	0.971②	0.004③	—	0.030	0.030
	Reatta	100	①	0.971②	0.004③	—	0.030	0.030
	Riviera	100	①	0.971②	0.004③	—	0.030	0.030
	Seville	100	①	0.971②	0.004③	—	0.030	0.030
	Toronado	100	①	0.971②	0.004③	—	0.030	0.030
1990	Allante	100	①	0.971②	0.004③	—	0.030	0.030
	Eldorado	100	①	0.971②	0.004③	—	0.030	0.030
	Reatta	100	①	0.971②	0.004③	—	0.030	0.030
	Riviera	100	①	0.971②	0.004③	—	0.030	0.030
	Seville	100	①	0.971②	0.004③	—	0.030	0.030
	Toronado	100	①	0.971②	0.004③	—	0.030	0.030
	Trofeo	100	①	0.971②	0.004③	—	0.030	0.030
1991–92	Allante	100	1.000	0.971②	0.004③	—	0.030	0.030
	Eldorado	100	1.000	0.971②	0.004③	—	0.030	0.030
	Reatta	100	1.000	0.971②	0.004③	—	0.030	0.030
	Riviera	100	1.000	0.971②	0.004③	—	0.030	0.030
	Seville	100	1.000	0.971②	0.004③	—	0.030	0.030
	Toronado	100	1.000	0.971②	0.004③	—	0.030	0.030
	Trofeo	100	1.000	0.971②	0.004③	—	0.030	0.030

① Standard—1.126 in.
  Quick Take-up—1.574 in.
  Anti-Lock—1.000 in.
② Rear—0.444 in.
③ Rear—0.003 in.

WHEEL ALIGNMENT

Year	Model		Caster Range (deg.)	Caster Preferred Setting (deg.)	Camber Range (deg.)	Camber Preferred Setting (deg.)	Toe-in (in.)	Steering Axis Inclination (deg.)
1988	Allante	Front	$2^{5}/_{16}$P–$3^{1}/_{2}$P	$2^{13}/_{16}$P	$^{13}/_{16}$N–$^{13}/_{16}$P	0	0	$13^{5}/_{16}$
		Rear	—	—	$^{1}/_{8}$N–$^{1}/_{2}$P	$^{7}/_{32}$P	$^{3}/_{32}$	—
	Eldorado	Front	$1^{5}/_{16}$P–$3^{5}/_{16}$P	$2^{5}/_{16}$P	$^{13}/_{16}$N–$^{13}/_{16}$P	0	0	$13^{5}/_{16}$
		Rear	—	—	$^{13}/_{32}$N–$^{3}/_{16}$P	$^{3}/_{32}$N	$^{3}/_{32}$	—
	Reatta	Front	$1^{13}/_{16}$P–$3^{13}/_{16}$P	$2^{13}/_{16}$P	$^{13}/_{16}$N–$^{13}/_{16}$P	0	0	NA
		Rear	—	—	0–$1^{5}/_{16}$P	$^{5}/_{8}$P	$^{3}/_{32}$	—
	Riviera	Front	$1^{5}/_{16}$P–$3^{3}/_{16}$P	$2^{5}/_{16}$P	$^{13}/_{16}$N–$^{13}/_{16}$P	0	0	NA
		Rear	—	—	0–$1^{5}/_{16}$P	$^{5}/_{8}$P	$^{3}/_{32}$	—
	Seville	Front	$1^{5}/_{16}$P–$3^{5}/_{16}$P	$2^{5}/_{16}$P	$^{13}/_{16}$N–$^{13}/_{16}$P	0	0	$13^{5}/_{16}$
		Rear	—	—	$^{13}/_{32}$N–$^{3}/_{16}$P	$^{3}/_{32}$N	$^{3}/_{32}$	—
	Toronado	Front	$1^{5}/_{16}$P–$3^{5}/_{16}$P	$2^{5}/_{16}$P	$^{13}/_{16}$N–$^{13}/_{16}$P	0	0	NA
		Rear	—	—	$^{13}/_{32}$N–$^{7}/_{32}$P	$^{3}/_{32}$N	$^{7}/_{64}$	—
	Trofeo	Front	$1^{5}/_{16}$P–$3^{5}/_{16}$P	$2^{5}/_{16}$P	$^{13}/_{16}$N–$^{13}/_{16}$P	0	0	NA
		Rear	—	—	$^{13}/_{32}$N–$^{7}/_{32}$P	$^{3}/_{32}$N	$^{7}/_{64}$	—
1989	Allante	Front	$1^{13}/_{16}$P–$2^{13}/_{16}$P	$2^{5}/_{16}$P	$^{13}/_{16}$N–$^{13}/_{16}$P	0	$^{3}/_{32}$	$13^{5}/_{16}$
		Rear	—	—	$^{1}/_{2}$N–$^{1}/_{8}$P	$^{3}/_{16}$N	$^{3}/_{32}$	—
	Eldorado	Front	$1^{5}/_{16}$P–$3^{5}/_{16}$P	$2^{5}/_{16}$P	$^{13}/_{16}$N–$^{13}/_{16}$P	0	0	$13^{5}/_{16}$
		Rear	—	—	$^{13}/_{32}$N–$^{3}/_{16}$P	$^{3}/_{32}$N	$^{3}/_{32}$	—
	Reatta	Front	$1^{13}/_{16}$P–$3^{13}/_{16}$P	$2^{13}/_{16}$P	$^{13}/_{16}$N–$^{13}/_{16}$P	0	0	NA
		Rear	—	—	0–$1^{5}/_{16}$P	$^{5}/_{8}$P	$^{3}/_{32}$	—
	Riviera	Front	$1^{5}/_{16}$P–$3^{3}/_{16}$P	$2^{5}/_{16}$P	$^{13}/_{16}$N–$^{13}/_{16}$P	0	0	NA
		Rear	—	—	0–$1^{5}/_{16}$P	$^{5}/_{8}$P	$^{3}/_{32}$	—
	Seville	Front	$1^{5}/_{16}$P–$3^{5}/_{16}$P	$2^{5}/_{16}$P	$^{13}/_{16}$N–$^{13}/_{16}$P	0	0	$13^{5}/_{16}$
		Rear	—	—	$^{13}/_{32}$N–$^{3}/_{16}$P	$^{3}/_{32}$N	$^{3}/_{32}$	—
	Toronado	Front	$1^{5}/_{16}$P–$3^{5}/_{16}$P	$2^{5}/_{16}$P	$^{13}/_{16}$N–$^{13}/_{16}$P	0	0	NA
		Rear	—	—	$^{13}/_{32}$N–$^{7}/_{32}$P	$^{3}/_{32}$N	$^{7}/_{64}$	—
	Trofeo	Front	$1^{5}/_{16}$P–$3^{5}/_{16}$P	$2^{5}/_{16}$P	$^{13}/_{16}$N–$^{13}/_{16}$P	0	0	NA
		Rear	—	—	$^{13}/_{32}$N–$^{7}/_{32}$P	$^{3}/_{32}$N	$^{7}/_{64}$	—
1990	Allante	Front	$1^{13}/_{16}$P–$2^{13}/_{16}$P	$2^{5}/_{16}$P	$^{13}/_{16}$N–$^{13}/_{16}$P	0	$^{3}/_{32}$	$13^{5}/_{16}$
		Rear	—	—	$^{1}/_{2}$N–$^{1}/_{8}$P	$^{3}/_{16}$N	$^{3}/_{32}$	—
	Eldorado	Front	$1^{5}/_{16}$P–$3^{5}/_{16}$P	$2^{5}/_{16}$P	$^{13}/_{16}$N–$^{13}/_{16}$P	0	0	$13^{5}/_{16}$
		Rear	—	—	$^{13}/_{32}$N–$^{3}/_{16}$P	$^{3}/_{32}$N	$^{3}/_{32}$	—
	Reatta	Front	$1^{13}/_{16}$P–$3^{13}/_{16}$P	$2^{13}/_{16}$P	$^{13}/_{16}$N–$^{13}/_{16}$P	0	0	NA
		Rear	—	—	0–$1^{5}/_{16}$P	$^{5}/_{8}$P	$^{3}/_{32}$	—
	Riviera	Front	$1^{5}/_{16}$P–$3^{3}/_{16}$P	$2^{5}/_{16}$P	$^{13}/_{16}$N–$^{13}/_{16}$P	0	0	NA
		Rear	—	—	0–$1^{5}/_{16}$P	$^{5}/_{8}$P	$^{3}/_{32}$	—
	Seville	Front	$1^{5}/_{16}$P–$3^{5}/_{16}$P	$2^{5}/_{16}$P	$^{13}/_{16}$N–$^{13}/_{16}$P	0	0	$13^{5}/_{16}$
		Rear	—	—	$^{13}/_{32}$N–$^{3}/_{16}$P	$^{3}/_{32}$N	$^{3}/_{32}$	—
	Toronado	Front	$1^{5}/_{16}$P–$3^{5}/_{16}$P	$2^{5}/_{16}$P	$^{13}/_{16}$N–$^{13}/_{16}$P	0	0	NA
		Rear	—	—	$^{13}/_{32}$N–$^{7}/_{32}$P	$^{3}/_{32}$N	$^{7}/_{64}$	—
	Trofeo	Front	$1^{5}/_{16}$P–$3^{5}/_{16}$P	$2^{5}/_{16}$P	$^{13}/_{16}$N–$^{13}/_{16}$P	0	0	NA
		Rear	—	—	$^{13}/_{32}$N–$^{7}/_{32}$P	$^{3}/_{32}$N	$^{7}/_{64}$	—

WHEEL ALIGNMENT

Year	Model		Caster Range (deg.)	Caster Preferred Setting (deg.)	Camber Range (deg.)	Camber Preferred Setting (deg.)	Toe-in (in.)	Steering Axis Inclination (deg.)
1991–92	Allante	Front	1 13/16 P–2 13/16 P	2 5/16 P	13/16 N–13/16 P	0	0	13 5/16
		Rear	—	—	13/32 N–3/16 P	3/32 N	3/32	—
	Eldorado	Front	1 5/16 P–3 5/16 P	2 5/16 P	13/16 N–13/16 P	0	0	13 5/16
		Rear	—	—	13/32 N–3/16 P	3/32 N	3/32	—
	Reatta	Front	1 13/16 P–3 13/16 P	2 13/16 P	13/16 N–13/16 P	0	3/32	13 5/16
		Rear	—	—	5/8 N–3/8 P	1/8 N	3/32	—
	Riviera	Front	1 5/16 P–3 5/16 P	2 5/16 P	13/16 N–13/16 P	0	3/32	13 5/16
		Rear	—	—	1/2 N–1/2 P	0	3/32	—
	Seville	Front	1 5/16 P–3 5/16 P	2 5/16 P	13/16 N–13/16 P	0	0	13 5/16
		Rear	—	—	13/32 N–3/16 P	3/32 N	3/32	—
	Toronado	Front	1 5/16 P–3 5/16 P	2 5/16 P	13/16 N–13/16 P	0	3/32	13 15/16
		Rear	—	—	11/16 N–5/16 P	3/16 N	3/32	—
	Trofeo	Front	1 5/16 P–3 5/16 P	2 5/16 P	13/16 N–13/16 P	0	3/32	13 15/16
		Rear	—	—	11/16 N–5/16 P	3/16 N	3/32	—

P—Positive
N—Negative

ENGINE MECHANICAL

NOTE: Disconnecting the negative battery cable on some vehicles may interfere with the functions of the on board computer systems and may require the computer to undergo a relearning process, once the negative battery cable is reconnected.

Engine Assembly

REMOVAL & INSTALLATION

Riviera, Reatta, Toronado and Trofeo

1. Matchmark the hood hinge-to-hood and remove the hood.
2. Properly relieve the fuel pressure and disconnect the fuel lines from the fuel rail.
3. Disconnect the negative battery cable. Remove the air intake duct.
4. Remove the upper engine strut. From the throttle body, remove the throttle cable bracket and the cables.
5. Raise and safely support the vehicle.
6. Position a drain pan under the radiator, open the drain cock and drain the cooling system.

7. Remove the exhaust pipe from the rear exhaust manifold.
8. Using a vertical lifting device, secure it to the engine and support its weight. Remove the engine mounting bolts.
9. Disconnect the electrical connectors from the starter. Remove the starter-to-engine bolts and the starter.
10. Remove the serpentine drive belt. Remove the air conditioning compressor-to-bracket bolts and move the compressor aside; do not disconnect the pressure hoses.
11. Disconnect and plug the power steering hoses at the steering gear.
12. Remove the lower transaxle-to-engine bolts.

NOTE: One of the lower transaxle bolts is located between the transaxle case and the engine block and is installed in the opposite direction.

13. Remove the flywheel cover. Matchmark the torque converter-to-flywheel for alignment purposes. Remove the torque converter-to-flywheel bolts and slide the torque converter rearward.
14. Remove the engine support bracket-to-transaxle bolts and the bracket.
15. Lower the vehicle.
16. Disconnect the vacuum hoses from the vacuum modulator and the emission control canister. Disconnect and move aside any electrical harness connectors which may be in the way.
17. Remove the radiator and heater hoses from the engine.
18. Remove the remaining transaxle-to-engine bolts. Lift the engine assembly from the vehicle and attach it to a work stand.

To install:

19. Install engine assembly in vehicle. Install upper engine-to-transaxle bolts and tighten until snug. Do not tighten at this time.
20. Install radiator and connect heater hoses to engine.
21. Connect vacuum hoses to vacuum modulator and the emission control canister. Connect electrical harness connectors previously removed.
22. Raise and safely support the vehicle.
23. Install the engine support bracket-to-transaxle bolts and bracket.
24. Install the torque convertor-to-flywheel bolts. Tighten to 46 ft. lbs. (62 Nm). Install flywheel cover aligning marks made during removal.
25. Install lower transaxle bolts. Tighten to 55 ft. lbs. (75 Nm).

NOTE: One of the lower transaxle bolts is located between the transaxle case and the engine block; it is installed in the opposite direction.

26. Connect the power steering hoses at the steering gear.

27. Install the air conditioning compressor in the bracket and install the compressor-to-bracket bolts. Install the serpentine belt.

28. Install the starter on the engine and connect starter electrical connector.

29. Install engine mounting bolts. Tighten to 70 ft. lbs. (90 Nm). Remove lifting device.

30. Connect the exhaust pipe to the rear exhaust manifold.

31. Lower the vehicle.

32. Fill the cooling system.

33. Connect the throttle cable bracket and cables to the throttle body. Install the upper engine strut.

34. Install the air intake duct. Connect negative battery cable.

35. Connect the fuel lines to the fuel rail.

36. Install the hood at matchmarks made during removal.

37. Start the engine and check for fuel, coolant and transaxle leaks.

Allante

1. Disconnect the negative battery cable. Properly relieve the fuel system pressure. Position a drain pan under the radiator, open the drain cock and drain the cooling system.

2. Remove the air cleaner. Matchmark the hood hinge-to-hood position and remove the hood.

3. Remove the cooling fans and the accessory drive belt.

4. Remove the upper intake manifold. Remove the upper radiator hose and disconnect the heater hose from the thermostat housing.

5. Disconnect the following electrical connectors and position aside:
   a. Oil pressure sending unit
   b. Coolant temperature sensor
   c. Distributor
   d. EGR solenoid
   e. Engine temperature switch
   f. Idle speed control
   g. Throttle position sensor
   h. Injector electrical connections
   i. MAT sensor
   j. Oxygen sensor
   k. Throttle body base warmer
   l. Alternator
   m. Ground wires at the alternator mounting bracket

6. Disconnect the accelerator, the cruise control and the transaxle throttle valve cables from the throttle lever.

7. Disconnect the cruise control diaphragm/bracket and move them aside.

8. Disconnect the transaxle oil cooler lines from the radiator. Remove the radiator.

9. Disconnect and remove the oil cooler lines from the oil filter adapter.

10. Remove the oil cooler lines-to-transaxle bracket.

11. Remove the air cleaner bracket and the oil filter adapter.

12. Disconnect the air injection tubes from the diverter valve.

13. Remove the cross brace.

14. Remove the right front heater hose and the coolant reservoir.

15. Remove the Air Injection Reactor (AIR) filter and bracket.

16. Remove the power steering line brace from the right cylinder head. Remove the pump and belt tensioner as an assembly and position them forward of the engine.

17. Properly discharge and recover the refrigerant from the air conditioning system and remove the air conditioning lines from the accumulator and condenser.

18. Disconnect supply and return fuel lines from the fuel rail. Remove the fuel line bracket from the transaxle and move the fuel lines aside.

19. Raise and safely support the vehicle.

20. Label and disconnect the electrical connectors from the starter. Disconnect any ground wires still connected to the engine.

21. Disconnect the oxygen sensor wire and remove the oxygen sensors.

22. Disconnect and remove the exhaust Y-pipe. Remove the starter-to-engine bolts and the starter.

23. Remove the torque converter covers. Matchmark the torque converter-to-flywheel and remove the flywheel-to-torque converter bolts.

24. Remove the air conditioning compressor lower dust shield, the right front tire and the outer wheel house plastic shield.

25. Remove the right rear transaxle-to-engine mount bolt, the front engine mount nuts and the right rear transaxle mount bolts.

26. Remove the alternator. Remove the oxygen sensor wires. Remove the heater bypass bracket from the right side of the vehicle.

27. Remove the right side engine brace and lower the vehicle to the ground.

28. Remove the engine-to-transaxle bolts. The bolts are accessible from the top.

29. Run a chain from a lifting crane down to both lift points on top of the engine and ensure it is secure. Lift the engine out of the vehicle.

To install:

30. Situate a floor jack under the transaxle and raise it slightly so it will align with the engine. Lower the engine into the engine compartment and engage the dowels that are on the engine block with the corresponding holes in the transaxle.

31. Install the upper transaxle-to-engine bolts. Lower the engine, directing it squarely onto the mounts. Remove the lifting equipment.

NOTE: Ensure that converter is properly positioned to the flexplate and engaged in the front pump of the transaxle.

32. Install 5 upper transaxle bellhousing-to-engine bolts.

33. Lower floor jack and remove from transaxle.

34. Lower engine making sure it is properly seated on mounts.

35. Remove lifting equipment. Raise and safely support the vehicle.

36. Install right side engine brace.

37. Remove engine support.

38. Install alternator and oxygen sensor wires and heater bypass bracket to right side of vehicle.

39. Install front engine mount nuts and right rear transaxle mounting bolts.

40. Connect oil level sensor at oil pan and both oxygen sensors.

41. Install right rear transaxle-to-engine mounting bolt.

42. Install outer wheel house plastic shield.

43. Install right front tire and wheel assembly.

44. Install air conditioning compressor lower dust shield.

45. Install 3 flexplate-to-converter bolts. Install flexplate cover.

46. Install starter. Install exhaust "Y" pipe.

47. Install electrical connectors at starter and ground wires to block.

48. Lower vehicle.

49. Install fuel line bracket. Install fuel lines at fuel rail.

50. Install air conditioning lines to accumulator and condenser.

51. Install power steering pump and tensioner. Install power steering line brace on right cylinder head.

52. Install AIR system air filter and bracket.

53. Install coolant reservoir. Install right front heater hose.

54. Install front right and rear cross braces.

55. Install AIR tubes on diverter valve.

56. Install oil filter adapter.

57. Install air cleaner mounting bracket.

58. Install oil cooler line bracket at transaxle. Install oil cooler lines to oil filter adapter.

59. Install radiator. Install engine oil and transaxle oil cooler lines to radiator.

60. Connect cruise control diaphragm with bracket.

61. Install the following wiring connectors:
   a. Injectors
   b. Ground wires at alternator bracket
   c. Oil pressure switch
   d. Coolant temperature sensor
   e. Distributor

f. Engine temperature switch

62. Connect cables from throttle lever including: accelerator, cruise control and transaxle throttle valve.

63. Install accessory drive belt.

64. Install upper radiator hose and heater hose to thermostat housing.

65. Install cooling fan.

66. Install air conditioning accumulator hose brace.

67. Install vehicle hood.

68. Install air cleaner. Install engine coolant. Connect negative battery cable.

69. Evacuate, recharge and leak test the air conditioning system.

70. Start engine and check for oil, coolant and transaxle leaks.

Eldorado and Seville

1. Disconnect the negative battery cable. Properly relieve the fuel system pressure. Position a drain pan under the radiator, open the drain cock and drain the cooling system.

2. Remove the air cleaner. Matchmark the hood hinge-to-hood and remove the hood.

3. Remove the cooling fan and the accessory drive belt.

4. Remove the upper radiator hose and disconnect the heater hose from the thermostat housing.

5. Disconnect the following electrical connectors, if equipped and position the wires aside:
   a. Oil pressure sending unit
   b. Coolant temperature sensor
   c. Distributor
   d. EGR solenoid
   e. Engine temperature switch
   f. Idle speed control
   g. Throttle position sensor
   h. Injector electrical connections
   i. MAT sensor
   j. Oxygen sensor
   k. Throttle body base warmer
   l. Alternator
   m. Ground wires at the alternator mounting bracket

6. Disconnect the accelerator, the cruise control and the transaxle throttle valve cables from the throttle lever.

7. Disconnect the cruise control diaphragm/bracket and move them aside.

8. Disconnect the transaxle oil cooler lines from the radiator. Remove the radiator.

9. Disconnect and remove the oil cooler lines from the oil filter adapter.

10. Remove the oil cooler lines-to-transaxle bracket.

11. Remove the air cleaner bracket and the oil filter housing adapter.

12. If equipped, disconnect the air injection tubes from the diverter valve.

13. Remove the right front and right rear body braces.

14. Remove the right front heater hose and the coolant reservoir.

15. If equipped, remove the Air Injection Reactor (AIR) filter box and bracket. Remove the idler pulley for the accessory drive belt.

16. Remove the power steering line brace from the right cylinder head. Remove the pump and belt tensioner as an assembly and position them forward of the engine.

17. Properly discharge and recover the refrigerant from the air conditioning system and remove the air conditioning lines from the accumulator and condenser.

18. Disconnect supply and return fuel lines from the throttle body. Remove the fuel line bracket from the transaxle and move the fuel lines aside.

19. Remove the EGR lines and brackets. Remove the vacuum modulator line and the fuel filter; reposition them aside.

20. Raise and safely support the vehicle.

21. Remove the starter heat shield. Label and disconnect the electrical connectors from the starter. Disconnect any ground wires still connected to the engine.

22. Disconnect and remove the exhaust crossover pipe. Remove the starter-to-engine bolts and the starter.

23. Remove the torque converter covers. Matchmark the torque converter-to-flywheel and remove the flywheel-to-torque converter bolts.

24. Remove the air conditioning compressor lower dust shield, the right front tire and the outer wheel house plastic shield.

25. Remove the right rear transaxle-to-engine mount bolt and the lower engine mounting damper nut.

26. Remove the front engine mount nuts and the right rear transaxle mount nuts.

27. Remove the alternator. Remove the oxygen sensor wires. Remove the heater bypass bracket from the right side of the vehicle.

28. Remove the right side engine brace and lower the vehicle to the ground.

29. Remove the engine-to-transaxle bolts. The bolts are accessible from the top.

30. Run a chain from a lifting crane down to both lift points on top of the engine and ensure it is secure. Lift the engine out of the vehicle.

To install:

31. Situate a floor jack under the transaxle and raise it slightly so it will align with the engine. Lower the engine into the engine compartment and engage the dowels on the engine block with the corresponding holes in the transaxle.

NOTE: Ensure that converter is properly positioned to the flexplate and engaged in the front pump of transaxle.

32. Install upper 5 transaxle bellhousing-to-engine bolts.

33. Lower floor jack and remove from transaxle.

34. Lower engine making sure it is seated on the mount properly.

35. Remove lifting equipment.

36. Raise and safely support the vehicle.

37. Support the engine. Install right side engine brace. Remove engine support.

38. Install alternator and oxygen sensor wires and heater bypass bracket to right side of vehicle.

39. Install front engine mount nuts and right rear transaxle mount bolts.

40. Install lower engine damper nut.

41. Install right rear transaxle-to-engine mounting bolt.

42. Install outer wheel house plastic shield.

43. Install right front tire and wheel assembly.

44. Install air conditioner compressor lower dust shield.

45. Install 3 flexplate-to-converter bolts. Install 2 flexplate covers.

46. Install starter to engine and connect electrical connectors. Install engine ground connectors.

47. Install exhaust crossover pipe.

48. Install starter heat shield.

49. Lower vehicle.

50. Install vacuum modulator line and vacuum hose to power brake booster.

51. Install EGR lines and bracket.

52. Install fuel line bracket at transaxle. Install fuel lines at throttle body.

53. Install air conditioning lines to accumulator and condenser.

54. Install power steering pump and tensioner. Install power steering line brace on right cylinder head.

55. If equipped, install A.I.R. system air filter and bracket.

56. Install coolant reservoir. Install right front heater hose.

57. Install front right and rear cross braces.

58. If equipped, install A.I.R. tubes on diverter valve.

59. Install oil filter adapter.

60. Install air cleaner mounting bracket.

61. Install oil cooler line bracket at transaxle. Install oil cooler lines to oil filter adapter.

62. Install radiator. Install engine oil and transaxle oil cooler lines to radiator.

63. Connect cruise control diaphragm with bracket.

64. Install the following wiring connectors:
  a. ISC
  b. TPS
  c. Injectors
  d. MAT sensor
  e. Oxygen sensor
  f. Electric EFE grid
  g. Ground wires at alternator bracket
  h. Oil pressure switch
  i. Coolant temperature sensor
  j. Distributor
  k. EGR solenoid
  l. Engine temperature switch
65. Connect cables from throttle lever including: accelerator, cruise control and transaxle throttle valve.
66. Install accessory drive belt.
67. Install upper radiator hose and heater hose to thermostat housing.
68. Install cooling fan.
69. Install air conditioning accumulator hose brace.
70. Install vehicle hood.
71. Install air cleaner. Install engine coolant. Connect negative battery cable.
72. Evacuate, recharge and leak test the air conditioning system.
73. Start engine and check for oil, coolant and transaxle leaks.

Engine Mounts

REMOVAL & INSTALLATION

Reatta, Riviera, Toronado and Trofeo

1. Disconnect the negative battery cable.
2. Safely support the engine using a suitable engine holding fixture.
3. Raise and safely support the vehicle.
4. Remove the engine mount bracket nuts.
5. Raise the engine slightly.
6. Remove the engine mount retaining bolts. Remove the engine mount.
To install:
7. Install the engine mount and mount retaining bolts.
8. Lower the engine.
9. Install the engine mount bracket nuts.
10. Lower the vehicle.
11. Remove the engine holding fixture.
12. Connect the negative battery cable.

Allante

RIGHT SIDE ENGINE AND TRANSAXLE MOUNT

1. Disconnect the negative battery cable.
2. Raise and safely support the vehicle.

3. Remove 2 heat shield screws.
4. Remove screw from engine mount brace at engine mount bracket.
5. Loosen nut at top of brace to exhaust manifold and position brace aside.
6. Support the engine with a transaxle jack.
7. Remove 2 screws securing mount bracket to transaxle.
8. Remove 4 nuts at top and bottom of mount.
9. Raise engine with transaxle jack.
10. Remove engine mount.
To install:
11. Position transaxle mount and bracket in place between transaxle and frame. Secure bracket to transaxle with 2 bolts. Tighten to 50 ft. lbs. (70 Nm).

NOTE: Guide engine mount into location while lowering engine.

12. Lower engine.
13. Install mount to frame and transaxle bracket with 2 nuts each. Tighten to 30 ft. lbs. (40 Nm).
14. Install brace from bracket to engine. Tighten to 25 ft. lbs. (35 Nm).
15. Install heat shield.
16. Remove transaxle jack and lower hoist.

LEFT SIDE ENGINE MOUNT

1. Disconnect the negative battery cable.
2. Remove air cleaner assembly.
3. Remove serpentine belt.
4. Properly discharge and recover the refrigerant from the air conditioning system.
5. Lower center exhaust manifold nuts.
6. Raise and safely support the vehicle.
7. Remove right side engine compartment splash shield. Remove air conditioning splash shield.
8. Remove 2 air conditioning compressor brackets. Remove air conditioning compressor.
9. Remove engine mount bracket bolts from engine block and cradle.
10. Raise engine with transaxle jack and remove mount and bracket.
To install:
11. Place mount in vise and position mount bracket onto mount. Tighten 2 nuts to 30 ft. lbs. (40 Nm).
12. Install engine mount and bracket through right side wheel well.
13. Install engine mount bracket bolts to engine block. Tighten to 50 ft. lbs. (70 Nm).
14. Install engine mount to cradle nuts. Tighten to 30 ft. lbs. (40 Nm).
15. Install air conditioning compressor. Install 2 air conditioning compressor brackets.
16. Install air conditioning splash

shield. Install right side engine compartment splash shield.
17. Lower vehicle.
18. Install lower center exhaust manifold nut.
19. Install serpentine belt.
20. Connect negative battery cable.
21. Install air cleaner assembly.
22. Evacuate, recharge and leak test the air conditioning system.

Eldorado and Seville

RIGHT SIDE ENGINE AND TRANSAXLE MOUNT

1. Disconnect the negative battery cable.
2. Remove the brace from the engine bracket to engine.
3. Remove 2 engine bracket-to-mount nuts.
4. Raise and safely support the vehicle.
5. Remove 2 nuts securing the engine mount to the frame. Remove 2 nuts securing transaxle bracket to mount. Remove 2 nuts securing the transaxle mount to the frame bracket.
6. Using the engine support tool, raise the engine.
7. Raise the engine slowly until the bracket is free from the engine and transaxle mount. Remove the bracket-to-block stud and bolts. Remove the mount and bracket by pulling forward.
8. Remove the transaxle mounting bracket from the transaxle. Remove the mount assembly.
To install:
9. Position engine mount and bracket in place between cylinder block and frame. Secure bracket to block with 1 stud and 2 bolts. Tighten to 34 ft. lbs. (46 Nm).
10. Position transaxle mount and bracket in place between transaxle and frame. Secure bracket to transaxle with 2 bolts. Tighten to 34 ft. lbs. (46 Nm).

NOTE: Guide engine mount into location while lowering engine.

11. Lower engine.
12. Install engine mount to frame and transaxle mount to frame bracket with 2 nuts. Tighten to 22 ft. lbs. (31 Nm).
13. Install 2 nuts to engine mount studs and 2 nuts to transaxle mount studs. Tighten to 22 ft. lbs. (31 Nm).
14. Remove brace from engine bracket to engine.
15. Remove stands and lower vehicle.
16. Connect negative battery cable.

LEFT SIDE ENGINE MOUNT

1. Disconnect the negative battery cable. Remove the air cleaner assembly.
2. Remove the serpentine belt.

Properly discharge and recover the refrigerant from the air conditioning system.

3. Install the engine support tool.

4. Remove the lower center exhaust manifold nut and top nut of the engine damper.

5. Raise and safely support the vehicle.

6. Remove the right side engine compartment splash shield and air conditioning splash shield.

7. Remove the engine damper. Remove both air conditioning compressor brackets. Remove the air conditioning compressor.

8. Remove the water pipe bracket bolt.

9. Remove the engine mount bracket bolts from the engine block and cradle. Remove the engine mount and bracket through the right side wheel well.

To install:

10. Place mount in vice and position mount bracket onto mount. Tighten 2 nuts to 31 ft. lbs. (41 Nm).

11. Install engine mount and bracket through right wheel well.

12. Install engine mount bracket bolts to engine block. Tighten to 50 ft. lbs. (68 Nm).

13. Install engine mount to cradle nuts. Tighten bolts to 31 ft. lbs. (41 Nm).

14. Install water pipe bracket bolt.

15. Install air conditioning compressor bracket.

16. Install 2 air conditioning compressor brackets.

17. Install engine damper.

18. Install air conditioning compressor splash shield.

19. Install right side engine compartment splash shield.

20. Lower vehicle.

21. Install lower center exhaust manifold nut and top nut on engine damper.

22. Remove engine support tool.

23. Install serpentine belt.

24. Connect negative battery cable.

25. Install air cleaner assembly.

26. Evacuate, recharge and leak test the air conditioning system.

Cylinder Head

REMOVAL & INSTALLATION

3.8L Engine

1. Disconnect the negative battery cable.

2. Remove the intake and exhaust manifolds.

3. Remove the valve cover.

4. If removing the front (left) side cylinder head, perform the following:

a. Remove the C^3I and spark plug wires.

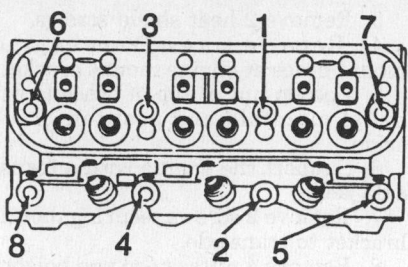

Cylinder head torque sequence— 3.8L engine

b. Remove the alternator bracket.

c. Remove 1 air conditioning compressor bracket bolt.

5. If removing the rear (right) side cylinder head, perform the following:

a. Remove the power steering pump.

b. Remove the belt tensioner assembly.

c. Remove fuel line heat shield.

6. Remove rocker arm assemblies, guide plate and pushrods.

7. Remove cylinder head bolts and cylinder head.

NOTE: Clean all gasket mating surfaces and cylinder head bolt holes in block.

To install:

8. Clean threads in block using an appropriate tap.

9. Install cylinder head gasket on block.

10. Apply an appropriate sealant to cylinder head bolt threads. Install cylinder head bolts.

11. On 1988 vehicles, tighten cylinder head bolts using the following steps:

a. Tighten each cylinder head bolt to 25 ft. lbs. (34 Nm) following the proper sequence.

NOTE: If 60 ft. lbs. (81 Nm) is reached at any time during steps b and c, STOP. Do not complete the balance of the 90 degree turn on this bolt.

b. Tighten each bolt ¼ turn (90 degree) in sequence.

c. Tighten each bolt an additional ¼ turn (90 degree) in sequence.

12. On 1989–92 vehicles, tighten cylinder head bolts using the following steps:

a. Tighten each cylinder head bolt to 35 ft. lbs. (47 Nm) following the proper sequence.

b. Rotate each bolt 130 degrees, in sequence, using an appropriate torque angle meter.

c. Rotate each bolt an additional 30 degrees, in sequence, using torque angle meter.

13. Install pushrods, guide plate and rocker arm assemblies. Apply an ap-

propriate high temperature, high strength thread sealant compound to the rocker arm pedestal bolts. Tighten to 28 ft. lbs. (38 Nm).

14. Install intake manifold.

15. Install valve cover.

16. Install exhaust manifold.

17. If the front (left) side cylinder head was removed, perform the following:

a. Install air conditioning compressor bracket bolt. Tighten to 52 ft. lbs. (80 Nm).

b. Install alternator support bracket to cylinder head.

c. Install alternator.

d. Install C^3I and spark plug wires.

18. If the rear (right) side cylinder head was removed, perform the following:

a. Install belt tensioner assembly.

b. Install power steering pump.

c. Install fuel line heat shield.

19. Connect negative battery cable.

20. Start engine and check for coolant, oil and fuel leaks. Allow engine to come to normal operating temperature and recheck for leaks.

4.1L, 4.5L and 4.9L Engines

RIGHT SIDE

1. Disconnect the negative battery cable. Drain the engine coolant.

2. Remove rocker arm covers.

3. Remove the lower intake and right side exhaust manifolds.

4. Remove engine lift bracket and oil dipstick tube.

5. Reposition AIR bracket.

6. Remove 10 cylinder head bolts.

7. Remove cylinder head.

To install:

NOTE: Clean sealing surfaces of cylinder head, block and liners. Clean cylinder head bolt holes with an appropriate tap. Ensure that bolt holes are free of shavings, oil and coolant.

8. Install new head gasket over dowels on cylinder block with either side facing up.

9. Install cylinder head.

10. Apply an appropriate lubricant to the threads of the head bolts. Install cylinder head bolts finger tight.

11. Tighten cylinder head bolts, in sequence, to 38 ft. lbs. (50 Nm).

12. Tighten cylinder head bolts, in sequence, to 68 ft. lbs. (90 Nm).

13. Tighten No. 1, 3 and 4 cylinder head bolts to 90 ft. lbs. (120 Nm).

14. Install engine lift bracket and AIR bracket.

15. Install lower intake and right side exhaust manifolds.

16. Install rocker arm covers.

17. Fill cooling system.

18. Connect negative battery cable.

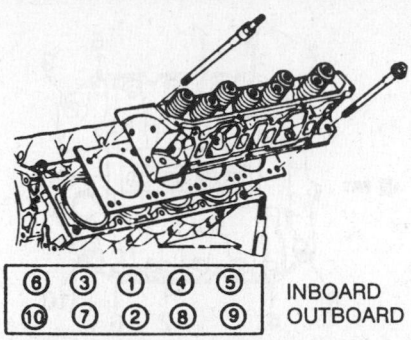

Cylinder head torque sequence—4.1L, 4.5L and 4.9L Engines

INBOARD
OUTBOARD

19. Start engine and check for coolant, oil and fuel leaks. Allow engine to come to normal operating temperature and recheck for leaks.

LEFT SIDE

1. Disconnect the negative battery cable.
2. Drain the cooling system.
3. Remove the rocker arm covers.
4. Remove the intake manifold-to-engine bolts and intake manifold.
5. Disconnect the exhaust manifold crossover pipe, the exhaust pipe-to-exhaust manifold bolts, the exhaust manifold-to-cylinder head bolts and the exhaust manifold.
6. Remove the engine lifting bracket and the dipstick tube.
7. Remove the AIR bracket-to-engine bolts and move the bracket aside.
8. Remove the cylinder head-to-engine bolts and the cylinder head.

To install:

9. Clean the gasket mounting surfaces.
10. Install new head gasket over dowels on cylinder block with either side facing up.
11. Install cylinder head.
12. Apply a suitable lubricant to the cylinder head bolt threads.
13. Install cylinder head bolts finger tight.
14. Tighten bolts, in sequence, to 38 ft. lbs. (50 Nm).
15. Tighten cylinder head bolts, in sequence, to 68 ft. lbs. (90 Nm).
16. Tighten No. 1, 3 and 4 cylinder head bolts to 90 ft. lbs. (120 Nm).
17. Install AIR bracket. Install dipstick tube and engine lift bracket.
18. Install exhaust manifold. Install lower intake manifold.
19. Install rocker arm covers.
20. Fill cooling system.
21. Connect negative battery cable.
22. Start engine and check for coolant, oil and fuel leaks. Allow engine to come to normal operating temperature and recheck for leaks.

Valve Lifters

REMOVAL & INSTALLATION

NOTE: When disassembling valve train components, ensure that all parts are kept in order so they can be reinstalled in their original locations and with the same mating surfaces.

1. Disconnect the negative battery cable. Remove the intake manifold.
2. Remove the rocker arm cover and discard the old gasket.
3. Remove the rocker arm assemblies. Remove the pushrods.
4. Remove the lifter guide retainer bolts and retainer.
5. Remove the lifter retainers.
6. Using the valve lifter removal tool, remove the valve lifters.

To install:

7. Clean the gasket mounting surfaces.
8. Lubricate the lifters with clean engine oil, use new gaskets and/or sealant.
9. Install the valve lifters.
10. Install the lifter guide, retainer and retainer bolts.
11. Install the pushrods. Install the rocker arm assemblies.
12. Install the rocker arm cover.
13. Install the intake manifold.
14. Connect the negative battery cable.

Valve Lash

All engines use hydraulic lifters which are non-adjustable. Hydraulic valve lifters keep all parts of the valve train in constant contact and adjust automatically to maintain **0** lash under all operating conditions.

Rocker Arms

REMOVAL & INSTALLATION

3.8L Engine

1. Disconnect the negative battery cable. Remove the rocker arm cover nuts, washers, seals, the cover and gasket, discard the gasket.
2. Remove the rocker arm pivot-to-cylinder head bolts, the pivots, the rocker arms and the pushrod guide.

NOTE: Be sure to keep the parts in order for reassembly purposes.

3. Clean the gasket mounting surfaces.
4. Install the pushrod guide, rocker arms, pivots and rocker arm pivot-to-cylinder head bolts. Tighten to 28 ft. lbs. (24 Nm).
5. Install the rocker arm cover using a new gasket.
6. Connect the negative battery cable.

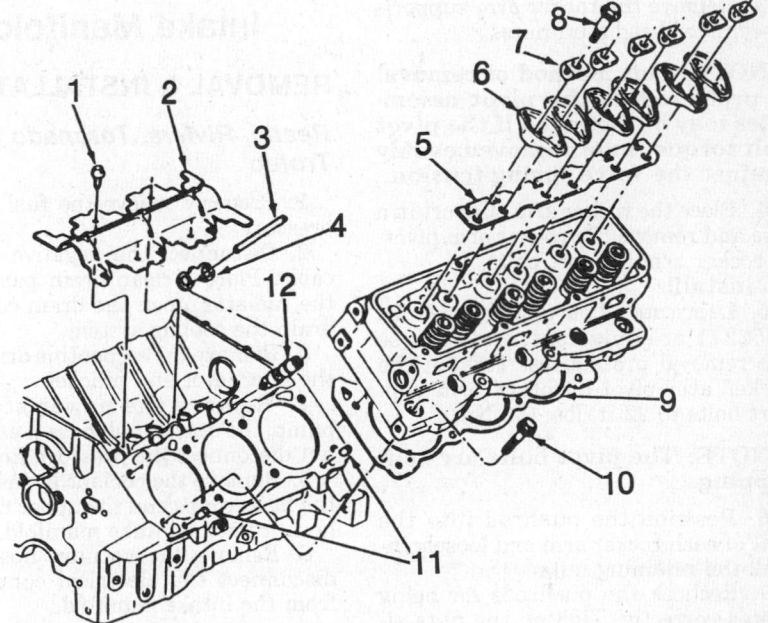

1.	Bolt (27 ft. lbs.)	5.	Pushrod guide	9.	Head gasket
2.	Lifter guide retainer	6.	Rocker arm	10.	Head bolt
3.	Pushrod	7.	Rocker arm pivot	11.	Dowel pin
4.	Lifter guide	8.	Bolt (28 ft. lbs.)	12.	Valve lifter

Exploded view of the rocker arm assembly—3.8L engine

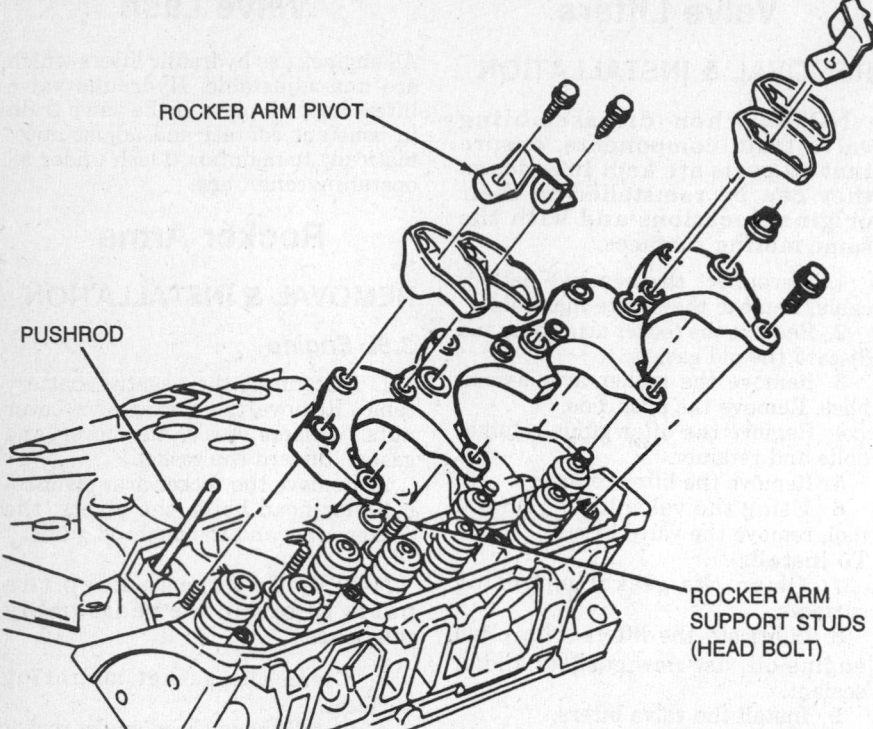

ROCKER ARM PIVOT

PUSHROD

ROCKER ARM SUPPORT STUDS (HEAD BOLT)

Exploded view of the rocker arm assembly—4.1L, 4.5L and 4.9L Engines

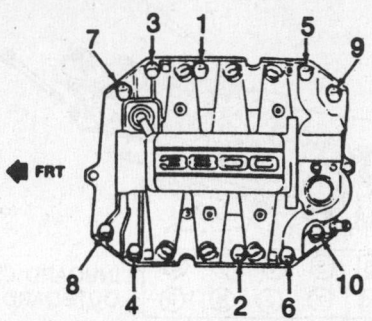

FRT

Intake manifold bolt torque sequence— 3.8L engine

4.1L, 4.5L and 4.9L Engines

1. Disconnect the negative battery cable. Remove the rocker arm cover.
2. Remove the rocker arm support-to-cylinder head bolts.
3. Remove the rocker arm support-to-cylinder head stud nuts.

NOTE: This method of removal is preferred as the pivot assemblies may be damaged if the pivot bolt torque is not removed evenly against the valve spring tension.

4. Place the rocker arm support in a vise and remove the rocker arm pivot-to-rocker arm support bolts.
To install:
5. Lubricate all parts with axle lube 1052271 or equivalent, and reverse the removal procedures. Tighten the rocker arm pivot-to-rocker arm support bolts to 22 ft. lbs. (30 Nm).

NOTE: The pivot bolts are self-tapping.

6. Position the pushrod into the seat of each rocker arm and loosely install the retaining nuts.
7. Recheck the pushrods for being seated correctly. Tighten the nuts alternately and evenly, checking the position of the pushrods while tightening.
8. When the nuts have been seated and the pushrods are correct, tighten the rocker arm support-to-cylinder head nuts to 37 ft. lbs. (50 Nm) and the bolts to 7 ft. lbs. (10 Nm).
9. Install the rocker arm cover.
10. Connect the negative battery cable.

Intake Manifold

REMOVAL & INSTALLATION

Reatta, Riviera, Toronado and Trofeo

1. Properly relieve the fuel system pressure.
2. Disconnect the negative battery cable. Place a clean drain pan under the radiator, open the drain cock and drain the cooling system.
3. Remove the serpentine drive belt, the alternator and bracket.
4. Remove the power steering pump, the braces and move it aside. Do not disconnect the pressure lines.
5. Remove the coolant bypass hose, the heater pipe and the upper radiator hose from the intake manifold.
6. Remove the vacuum hoses and disconnect the electrical connectors from the intake manifold.
7. Remove the EGR pipe, the EGR valve and adapter from the throttle body.
8. Remove the throttle body coolant pipe, the throttle body and the throttle body adapter.

9. Disconnect the rear spark plug wires. Remove the intake manifold-to-engine bolts and the manifold.
To install:
10. Clean the gasket mounting surfaces.
11. Using new gaskets and sealant 12345336 or equivalent, on the ends of the manifold seals, install the intake manifold. Tighten the intake manifold bolts, in sequence, to 88 inch lbs. (10 Nm).
12. Install the rear spark plug wires.
13. Install the throttle body adapter, throttle body and throttle body coolant pipe.
14. Install the EGR adapter to the throttle body, the EGR valve and the EGR pipe.
15. Install the vacuum hoses and connect the electrical connectors to the intake manifold.
16. Connect the upper radiator hose, heater pipe and coolant bypass hose to the intake manifold.
17. Install the power steering pump braces and power steering pump.
18. Install the alternator bracket and alternator. Install the serpentine drive belt.
19. Fill the cooling system.
20. Connect negative battery cable.
21. Start the engine and check for coolant, oil and fuel leaks. Allow the engine to come to normal operating temperature and recheck for leaks.

Eldorado and Seville

1. Disconnect the negative battery cable. Relieve fuel system pressure. Drain the cooling system to a level below the intake manifold. Remove the coolant reservoir. Disconnect the upper radiator hose from the thermostat housing.
2. Remove the air cleaner and the serpentine drive belt. Label and disconnect the spark plug wires from the spark plugs.
3. Remove the cross brace.
4. Remove power steering pump and tensioner bracket assembly and reposition toward the front of engine.
5. Remove alternator and bracket.

6. Remove cruise control servo with bracket and throttle valve cables and position aside.

7. Disconnect wire connections and reposition:
   a. Distributor
   b. Oil pressure switch
   c. Coolant temperature sensor
   d. EGR solenoid
   e. ISC motor
   f. Throttle position switch
   g. If equipped, electric EFE grid
   h. Injectors
   i. MAT sensor

8. If equipped, disconnect the MAP hoses. Remove upper radiator hose and heater hose. Remove air conditioning hose bracket.

9. Disconnect spark plug wire protectors and reposition cap.

10. Mark the distributor rotor position and remove distributor.

NOTE: Do not crank or in any other way rotate crankshaft with the distributor removed.

11. Disconnect fuel and vacuum lines from the throttle body. Disconnect the vacuum supply solenoid and lines.

12. Remove valve covers. Remove rocker arms and pushrods.

NOTE: Pushrods should be marked or retained in sequence so they may be reinstalled in their original positions.

13. Remove the right front and rear lift brackets. Remove intake manifold bolts and remove intake manifold, gaskets and seals. Discard gaskets and seals.

14. Clean sealing surfaces of intake manifold, cylinder head and cylinder block.

To install:

15. Install new end seals. Use RTV at 4 corners where end seals will meet side gaskets.

16. Install new intake to cylinder head gaskets. Use RTV at 4 corners of end seals.

17. On 1988 vehicles, tighten the intake manifold bolts by performing the following:
   a. Tighten bolts 1, 2, 3 and 4, in sequence, to 15 ft. lbs. (20 Nm).
   b. Tighten bolts 5 thru 16, in sequence, to 22 ft. lbs. (30 Nm).
   c. Retighten all bolts, in sequence, to 22 ft. lbs. (30 Nm).
   d. Repeat step c until torque level is maintained.

18. On 1989–92 vehicles, tighten the intake manifold bolts by performing the following:
   a. Tighten bolts 1, 2, 3 and 4, in sequence, to 8 ft. lbs. (12 Nm).
   b. Tighten bolts 5 thru 16, in sequence, to 8 ft. lbs. (12 Nm).

   c. Retighten all bolts, in sequence, to 12 ft. lbs. (16 Nm).
   d. Repeat step c until torque level is maintained.

19. Install pushrods and rocker arm assembly.

20. Install valve covers. Install vacuum supply solenoid and lines. Install fuel and vacuum lines to throttle body.

21. Install distributor in original position. Install distributor cap and wire protectors.

22. Install air conditioning hose bracket.

23. Install upper radiator hose and heater hose. If equipped, connect the MAP hoses.

24. Connnect following wire connectors:
   a. Distributor
   b. Oil pressure switch
   c. Coolant temperature sensor
   d. EGR solenoid
   e. ISC motor
   f. Throttle position switch
   g. If equipped, electric EFE grid
   h. Injectors
   i. MAT sensor

25. Install cruise control servo and throttle valve cables.

26. Install alternator bracket and alternator.

27. Install power steering pump and tensioner assembly. Install power steering line brace to right side cylinder head.

28. Install serpentine drive belt. Install coolant reservoir.

29. Install cross brace.

30. Fill cooling system.

31. Install air cleaner assembly.

32. Connect negative battery cable.

33. Start engine and check for coolant, oil and fuel leaks. Allow engine to come to normal operating temperature and recheck for leaks.

Allante

UPPER INTAKE MANIFOLD

1. Disconnect the negative battery cable.

2. Shock tower support bracket, as required.

3. Label and disconnect vacuum hoses.

4. Remove the transmission dipstick tube bolt.

5. Disconnect the MAT sensor electrical connector.

6. Remove the rear upper intake manifold support.

7. Remove the throttle body assembly from the upper intake manifold and discard the gasket.

8. Remove the throttle heater assembly and discard the gasket.

9. Remove the 4 upper intake manifold attaching nuts.

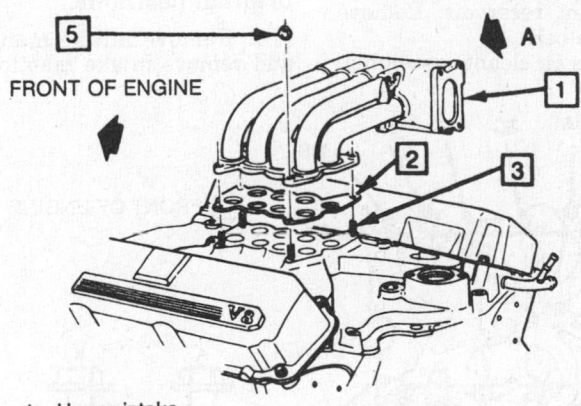

FRONT OF ENGINE

1. Upper intake manifold
2. Gasket
3. Lower intake manifold stud
4. Support
5. Nut

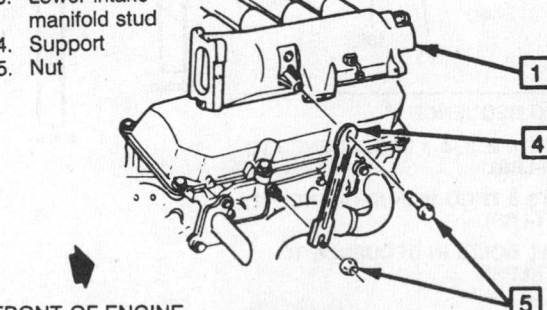

FRONT OF ENGINE

VIEW A

Upper intake manifold—Allante

10. Remove the upper intake manifold and discard the gasket.

To install:

11. Ensure that all gasket mating surfaces are free of old gasket material.

12. Install a new upper intake-to-lower intake gasket.

13. Install the upper intake manifold and attaching nuts. Tighten to 15 ft. lbs. (20 Nm).

14. Install a new throttle heater gasket. Install the throttle heater assembly.

15. Install a new throttle body gasket. Install the throttle body to the upper intake manifold. Tighten throttle body attaching bolts to 15 ft. lbs. (20 Nm).

16. Install the rear upper intake manifold support.

17. Connect the MAT sensor electrical connector.

18. Install the transmission dipstick tube bolt.

19. Connect all vacuum lines.

20. If removed, install the shock tower support bracket.

21. Connect the negative battery cable.

LOWER INTAKE MANIFOLD

1. Disconnect the negative battery cable. Relieve fuel system pressure. Drain the cooling system to a level below the lower intake manifold. Remove the coolant reservoir. Remove serpentine drive belt.

2. Remove the air cleaner assembly.

3. Label and disconnect appropriate vacuum lines.

4. Remove the upper intake manifold and fuel rails.

5. Remove the power steering line brace on the right side cylinder head.

6. Remove the power steering pump and tensioner bracket assembly and reposition toward the front of the engine.

7. Remove alternator with bracket and idler pulley.

8. Remove cruise control servo with bracket and cables. Reposition aside.

9. Disconnect wire connections as follows:

   a. Distributor
   b. Oil pressure switches
   c. Coolant temperature sensor
   d. Ground wires

10. Disconnect the upper radiator hose and 2 heater hose connections.

11. Disconnect spark plug wire protectors and reposition cap.

12. Mark distributor rotor position and remove distributor.

NOTE: Do not crank or in any other way rotate crankshaft with the distributor removed.

13. Remove valve covers. Remove rocker arms and pushrods.

NOTE: Pushrods should be marked or retained in sequence so they may be reinstalled in their original positions.

14. Remove intake manifold bolts and remove intake manifold, gaskets and seals. Discard gaskets and seals.

15. Clean sealing surfaces of intake manifold, cylinder head and cylinder block.

To install:

16. Install new end seals. Use RTV 1052915 or equivalent, at the 4 corners where end seals meet side gaskets.

17. Install new intake to cylinder head gaskets. Use RTV at 4 corners of end seals.

18. On 1988 vehicles, tighten the intake manifold bolts by performing the following:

   a. Tighten bolts 1, 2, 3 and 4, in sequence, to 15 ft. lbs. (20 Nm).
   b. Tighten bolts 5 thru 16, in sequence, to 22 ft. lbs. (30 Nm).
   c. Retighten all bolts, in sequence, to 22 ft. lbs. (30 Nm).
   d. Repeat step c until torque level is maintained.

19. On 1989–92 vehicles, tighten the intake manifold bolts by performing the following:

   a. Tighten bolts 1, 2, 3 and 4, in sequence, to 8 ft. lbs. (12 Nm).
   b. Tighten bolts 5 thru 16, in sequence, to 8 ft. lbs. (12 Nm).
   c. Retighten all bolts, in sequence, to 12 ft. lbs. (16 Nm).
   d. Repeat step c until torque level is maintained.

20. Install pushrods and rocker arm assembly.

21. Install valve covers.

22. Install distributor in original position. Install distributor cap and wire protectors.

23. Install air conditioning hose bracket.

24. Install upper radiator hose and heater hose.

25. Connect following wire connectors:

   a. Distributor
   b. Oil pressure switch
   c. Coolant temperature sensor
   d. Ground wires

26. Install cruise control servo and throttle valve cables.

27. Install alternator bracket and alternator.

28. Install power steering line brace to right side cylinder head. Install power steering pump and tensioner assembly.

29. Install serpentine drive belt. Install coolant reservoir.

30. Install fuel rail assembly and upper intake manifold to the lower intake manifold.

31. Connect vacuum lines.

32. Install heater assembly and gasket.

33. Install accelerator, cruise control and throttle valve cables.

34. Fill cooling system.

35. Install air cleaner assembly.

36. Connect negative battery cable.

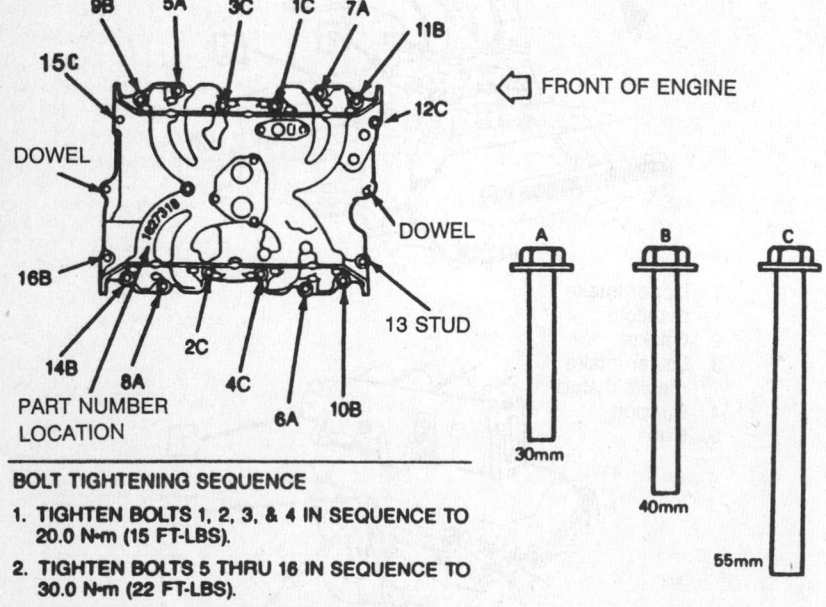

FRONT OF ENGINE

BOLT TIGHTENING SEQUENCE

1. TIGHTEN BOLTS 1, 2, 3, & 4 IN SEQUENCE TO 20.0 N·m (15 FT-LBS).

2. TIGHTEN BOLTS 5 THRU 16 IN SEQUENCE TO 30.0 N·m (22 FT-LBS).

3. RETIGHTEN ALL BOLTS IN SEQUENCE TO 30.0 N·m (22 FT-LBS).

4. REPEAT STEP 3.

Intake manifold bolt torque sequence—4.1L, 4.5L and 4.9L Engines

37. Start engine and check for coolant, oil and fuel leaks. Allow engine to come to normal operating temperature and recheck for leaks.

Exhaust Manifold

REMOVAL & INSTALLATION

3.8L Engine

LEFT SIDE

1. Disconnect the negative battery cable.
2. Remove the engine strut.
3. Remove the 2 bolts attaching the exhaust crossover pipe-to-exhaust manifold.
4. Remove the cooling fan assembly, as required.
5. Label and disconnect the spark plug wires.
6. Remove the oil dipstick tube to provide access to the manifold bolts, as required.
7. Remove the exhaust manifold-to-cylinder head bolts and the manifold.

To install:

8. Install the exhaust manifold gasket. Tighten the exhaust manifold-to-cylinder head bolts to 41 ft. lbs. (55 Nm).
9. If removed, install the oil dipstick tube.
10. Connect the spark plug wires.
11. If removed, install the cooling fan assembly.
12. Install the 2 exhaust crossover pipe-to-exhaust manifold attaching bolts. Tighten the exhaust crossover pipe-to-manifold bolts to 22 ft. lbs. (30 Nm).
13. Install the engine strut.
14. Connect the negative battery cable. Start the engine and check for exhaust leaks.

RIGHT SIDE

1. Disconnect the negative battery cable.
2. Label and disconnect the spark plug wires.
3. Remove the throttle cable bracket.
4. Remove the crossover pipe heat shield.
5. Remove the transaxle oil level indicator and indicator tube.
6. Disconnect the oxygen sensor lead.
7. Remove the 2 bolts attaching the exhaust crossover pipe to the manifold.
8. Remove the plastic vacuum tank mounted on the cowl, as required.
9. Remove the EGR pipe, as required.
10. Remove the 2 upper heat shield screws, as required.
11. Remove the upper exhaust manifold bolts.
12. Raise and safely support the vehicle.
13. Remove the 2 lower heat shield screws, as required.
14. Remove the lower exhaust manifold bolts.
15. Remove the front exhaust pipt-to-exhaust manifold attaching nuts.
16. Disconnect the front exhaust pipe from the exhaust manifold.
17. Lower the vehicle.
18. Remove the engine lift bracket.
19. Remove the exhaust manifold nuts and remove the manifold.

To install:

20. Install the exhaust manifold and manifold nuts. Tighten to 41 ft. lbs. (55 Nm).
21. Install the engine lift bracket.
22. Raise and safely support the vehicle.
23. Connect the front exhaust pipe to the exhaust manifold.
24. Install the front exhaust pipt-to-exhaust manifold attaching nuts.
25. Install the lower exhaust manifold bolts.
26. If removed, install the 2 lower heat shield screws.
27. Lower the vehicle.
28. Install the upper exhaust manifold bolts.
29. If removed, install the 2 upper heat shield screws.
30. If removed, install the EGR pipe.
31. If removed, install the plastic vacuum tank mounted on the cowl.
32. Install the 2 bolts attaching the exhaust crossover pipe to the manifold.
33. Connect the oxygen sensor lead.
34. Install the transaxle oil level indicator and indicator tube.
35. Install the crossover pipe heat shield.
36. Install the throttle cable bracket.
37. Connect the spark plug wires.
38. Connect the negative battery cable.

4.1L, 4.5L and 4.9L Engines

LEFT SIDE

1. Disconnect the negative battery cable.
2. Remove the air cleaner.
3. Remove the AIR pipe from the AIR pump and position aside.
4. Remove the starter shield.
5. Remove the serpentine belt.
6. Remove the power steering pump and tensioner bracket covering the manifold.
7. Remove both cooling fans.
8. Label and disconnect the spark plug wires.
9. Raise and safely support the vehicle.
10. Remove the exhaust Y-pipe and the air conditioning-to-manifold brace.
11. Remove the exhaust manifold-to-cylinder head bolts. Remove the exhaust manifold.

To install:

12. Clean the gasket mounting surfaces.
13. Apply graphite dry film lubricant to the exhaust manifold sealing surface.
14. Install the exhaust manifold to the cylinder head. Install the 7 attaching bolts and tighten to 16 ft. lbs. (20 Nm).
15. Install the exhaust Y-pipe and the air conditioning-to-manifold brace.
16. Lower the vehicle.
17. Connect the spark plug wires.
18. Install both cooling fans.
19. Install the power steering pump and tensioner bracket covering the manifold.
20. Install the serpentine belt.
21. Install the starter shield.
22. Install the AIR pipe to the AIR pump.
23. Install the air cleaner.
24. Connect the negative battery cable.

RIGHT SIDE

1. Disconnect the negative battery cable. Remove the air cleaner.
2. Remove the EGR pipe from the manifold, as required. Remove 2 heat shield screws.
3. Raise and safely support the vehicle.
4. Disconnect the Y-pipe from the manifold.
5. Remove the engine mount brace from the front of the manifold.
6. Disconnect the oxygen sensor wire. Remove heat shield.
7. Support engine cradle with screw jacks and remove rear cradle bolts on both sides. Loosen front cradle bolts. Slightly lower engine cradle.
8. Remove the exhaust manifold-to-cylinder head bolts and the manifold.

To install:

9. Clean the gasket mounting surfaces.
10. Apply graphite dry film lubricant to the exhaust manifold sealing surface.
11. Install the exhaust manifold to the cylinder head. Install the 7 attaching bolts and tighten to 16 ft. lbs. (20 Nm).
12. Install the heat shield.
13. Connect the oxygen sensor wire.
14. Install the engine brace on the right side of the manifold.
15. Raise the engine cradle and install the rear bolts. Tighten all mounting bolts to 75 ft. lbs. (100 Nm).
16. Install the exhaust crossover pipe.

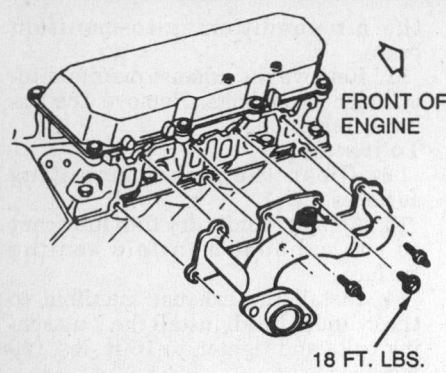

FRONT OF ENGINE

18 FT. LBS.

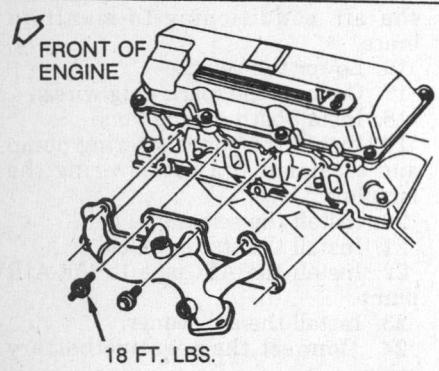

FRONT OF ENGINE

18 FT. LBS.

Exploded view of the exhaust manifolds—4.1L, 4.5L and 4.9L Engines

17. Lower the vehicle.
18. Install the 2 heat shield screws.
19. Install the air cleaner.
20. Connect the negative battery cable.

Timing Chain Front Cover

REMOVAL & INSTALLATION

3.8L Engine

1. Disconnect the negative battery cable.
2. Drain the cooling system. Remove the lower radiator hose and the coolant bypass hose from the timing case cover. Remove the heater pipes.
3. Remove the serpentine drive belt and the water pump pulley.
4. Raise and safely support the vehicle.
5. Remove right front tire and wheel assembly.
6. Remove the inner splash shield.
7. Remove the front engine-to-frame stabilizer and bracket, as required.
8. Remove the crankshaft balancer bolt/washer and the balancer.
9. Remove the sensor shield.
10. Disconnect the electrical connectors from the crankshaft sensor, the camshaft sensor and the oil pressure switch.
11. Remove the oil pan-to-timing case cover bolts, the timing case cover-to-engine bolts and the cover.
To install:
12. Clean the gasket mounting surfaces.
13. Install a new gasket and apply sealant 1052080 or equivalent. Install the front cover.
14. Install the oil pan-to-timing case cover bolts. Tighten to 124 inch lbs. (14 Nm). Install the timing case cover-to-engine bolts. Tighten to 22 ft. lbs. (30 Nm).
15. Connect the electrical connectors to the crankshaft sensor, the camshaft sensor and the oil pressure switch.
16. Install the sensor shield.
17. Install the crankshaft balancer bolt/washer and the balancer.
18. If removed, install the front engine-to-frame stabilizer and bracket.
19. Install the inner splash shield.
20. Install right front tire and wheel assembly.
21. Lower the vehicle.
22. Install the water pump pulley and serpentine drive belt.
23. Install the lower radiator hose and the coolant bypass hose to the timing case cover. Install the heater pipes.
24. Connect the negative battery cable.
25. Fill cooling system and check for leaks. Start the engine and allow to come to normal operating temperature. Recheck for leaks. Top-up coolant.

4.1L, 4.5L and 4.9L Engines

1. Disconnect the negative battery cable.
2. Drain the cooling system. Remove the air cleaner.
3. Remove the serpentine belt.
4. Remove the cross-car brace.
5. Remove the AIR air filter and bracket.
6. Remove the water pump pulley bolts and the pulley. If necessary, remove the idler pulley.
7. Raise and safely support the vehicle.
8. Remove the crankshaft damper by performing the following:
    a. Remove the crankshaft damper-to-crankshaft bolt.

NOTE: The use of shop air, applied to a cylinder on its compression stroke, may be required to prevent the crankshaft from turning while removing the crankshaft damper bolt. Remove a spark plug and rotate the crankshaft until that cylinder is on its compression stroke. Install the appropriate adapter finger-tight

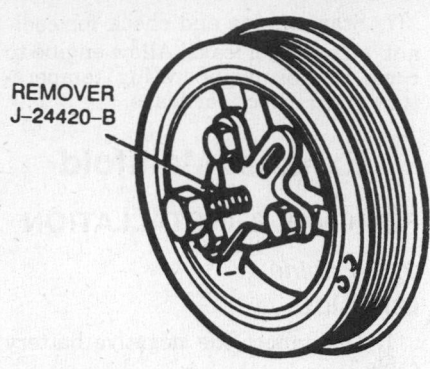

REMOVER J–24420–B

Using the wheel puller tool to remove the damper pulley—4.1L, 4.5L and 4.9L Engines

into the spark plug hole and apply shop air to the cylinder.
    b. Attach a wheel puller to the crankshaft damper.
    c. Using a pilot between the crankshaft and the center bolt, press the crankshaft damper from the crankshaft.
    d. Remove the Woodruff® key from the crankshaft.
9. Remove the timing case cover-to-engine bolts, the oil pan-to-timing case cover bolts and the cover.
To install:
10. Clean the gasket mounting surfaces.
11. To avoid oil leakage, apply RTV sealer according to the following:
    a. Apply a bead of RTV on the front cover lip on the oil pan sealing surface. Ensure that this bead is placed along the front cover lip behind the 2 oil pan-to-front cover bolts.
    b. Apply a ¼ in. bead of RTV on the oil pan where the oil pan, block and front cover join.
    c. Remove any excess RTV that is squeezed out of the sealing area.
12. Install the front cover.
13. Install the crankshaft damper by performing the following:
    a. Lubricate the bore of the hub and the inside diameter of the seal with EP lubricant.
    b. Install the Woodruff® key in the key slot in the crankshaft.
    c. Position the damper on the crankshaft, lining up the key slot with the key.
    d. Thread the installer into the end of the crankshaft. Position the thrust bearing with the inner race forward, washer next and installer nut last.
    e. Install the damper on the crankshaft by tightening the installer nut.

NOTE: The use of shop air, applied to a cylinder on its compres-

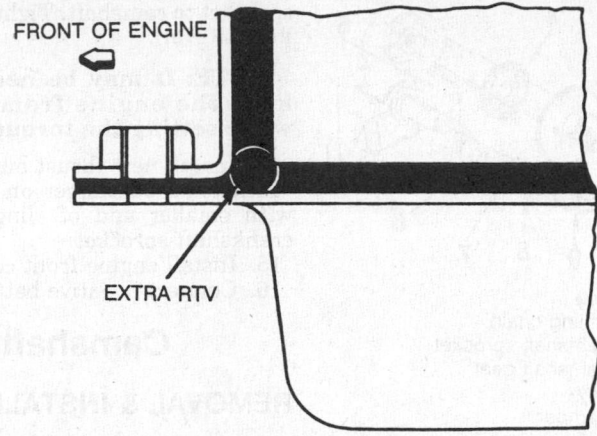

RTV application for front cover, oil pan installation—4.1L, 4.5L and 4.9L Engines

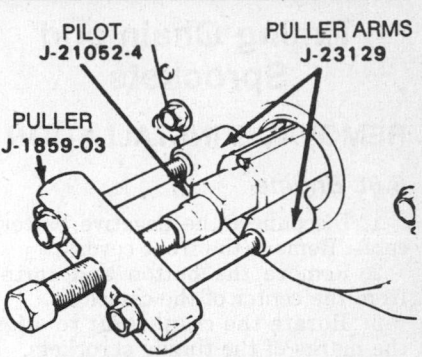

Removing the front oil seal—4.1L, 4.5L and 4.9L Engines

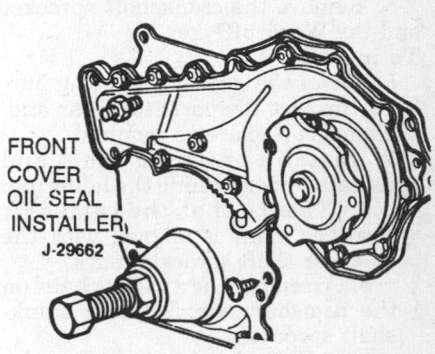

Installing the front oil seal—4.1L, 4.5L and 4.9L Engines

sion stroke, may be required to prevent the crankshaft from turning while installing the crankshaft damper bolt. Remove a spark plug and rotate the crankshaft until that cylinder is on its compression stroke. Install an adapter finger-tight into the spark plug hole and apply shop air to the cylinder.

   f. Tighten nut until the hub bottoms out on the crankshaft. Tighten the nut to 60–65 ft. lbs. (80–90 Nm) to fully seat the balancer and timing gear. Remove the installer and reinstall the bolt and washer into the crankshaft. Tighten to 60–65 ft. lbs. (80–90 Nm).

   g. Exhaust the shop air to the cylinder, remove the adapter and reinstall the spark plug.

14. Lower the vehicle.
15. Install the water pump.
16. Install the water pump pulley.
17. Install the serpentine belt.
18. Install the coolant reservoir and cross-car brace.
19. Connect the negative battery cable.
20. Fill cooling system and check for leaks. Start the engine and allow to come to normal operating temperature. Recheck for leaks. Top-up coolant.

Front Cover Oil Seal

REPLACEMENT

3.8L Engine

1. Disconnect the negative battery cable.
2. Remove the serpentine drive belt.
3. Remove the crankshaft balancer-to-crankshaft bolts.
4. Using a small prybar, pry the oil seal from the timing case cover; be careful not to damage the sealing surfaces.

To install:

5. Clean the oil seal mounting surface.
6. Using GM lubricant 1050169 or equivalent, coat the outside of the seal and the crankshaft balancer.
7. Using an appropriate oil seal installation tool, press the new seal into the timing case cover until it seats.
8. Install the crankshaft balancer-to-crankshaft bolt according to the following:

   a. On 3.8L (VIN C) engine, tighten the crankshaft balancer-to-crankshaft bolt to 219 ft. lbs. (297 Nm).

   b. On 3.8L (VIN L) engine, tighten the crankshaft balancer-to-crankshaft bolt to 105 ± 7 ft. lbs. plus 56 ± 4 degrees or 140 ± 10 Nm plus 56 ± 4 degrees.

9. Install the serpentine drive belt.
10. Connect the negative battery cable.

4.1L, 4.5L and 4.9L Engines

1. Disconnect the negative battery cable.
2. Remove the serpentine belt.
3. Raise and safely support the vehicle.
4. Remove right front tire. Remove right front air deflector.
5. Loosen and reposition the heater bypass line.
6. Remove the crankshaft pulley-to-crankshaft pulley bolt. Attach a wheel puller to the crankshaft pulley. Using a pilot between the crankshaft and the center bolt, press the crankshaft pulley from the crankshaft. Remove the Woodruff® key from the crankshaft.
7. Using a small prybar, pry the oil seal from the timing case cover, discard it.

To install:

8. Clean the oil seal mounting surface. Lubricate the new seal with engine oil.

9. Using a hammer and the oil seal installation tool, drive the new oil seal into the timing case cover until it seats.

10. Lubricate bore of hub and inside diameter of seal with EP lubricant to prevent seizure to crankshaft and provide lubrication of oil seal lip.

11. Position damper on crankshaft, lining up key slot in hub with key on crankshaft.

12. Position installer on end of crankshaft. Position thrust bearing with inner race forward, then washer and installer nut last. Install damper on crankshaft by tightening installer nut.

13. Hub will bottom out on crankshaft. Tighten installer nut to 65 ft. lbs. (90 Nm) to ensure balancer and timing gear are fully seated. Remove installer and reinstall bolt/washer in crankshaft. Tighten to 65 ft. lbs. (90 Nm).

14. Install heater bypass line.

15. Install right front air deflector. Install right front tire.

16. Install serpentine belt.

17. Connect negative battery cable.

Timing Chain and Sprockets

REMOVAL & INSTALLATION

3.8L Engine

1. Disconnect the negative battery cable. Remove the front cover.
2. Remove the button and spring from the center of the camshaft.
3. Rotate the crankshaft to align the marks of the timing sprockets.
4. Remove the camshaft sprocket bolts, the sprocket and the timing chain.
5. Remove the crankshaft sprocket and the Woodruff® key.

To install:

6. Clean the gasket mounting surfaces. Inspect the parts for wear and/or damage. Replace as required.
7. Install the timing chain and sprockets by performing the following:

    a. Ensure that the camshaft sprocket mark is aligned with the balancer shaft sprocket mark.

    b. Assemble the timing chain on the camshaft sprocket and crankshaft sprockets.

    c. Align the **0** marks on the sprockets; they must face each other.

    d. Slide the assembly onto the camshaft and crankshaft. Install the camshaft sprocket-to-camshaft bolts. Tighten the camshaft sprocket-to-camshaft sprocket bolts to 27 ft. lbs. (37 Nm).

8. Using petroleum jelly, pack the oil pump.
9. Install the button and spring to the center of the camshaft.
10. Install the front cover.
11. Connect the negative battery cable.
12. Fill cooling system and check for leaks. Start the engine and allow to come to normal operating temperature. Recheck for leaks. Top-up coolant.

4.1L, 4.5L and 4.9L Engines

1. Disconnect the negative battery cable. Drain the cooling system.
2. Remove engine front cover.
3. Remove oil slinger from crankshaft.
4. Rotate the engine until the crankshaft and camshaft timing marks are aligned.
5. Remove thrust button and screw securing camshaft sprocket to camshaft. Discard thrust button.
6. Remove camshaft and crankshaft sprockets with chain attached.

To install:

7. If timing was disturbed, rotate crankshaft until timing mark on crank sprocket is positioned straight up.

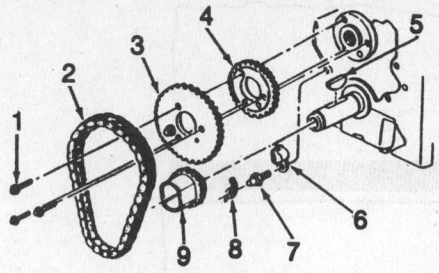

1. Bolt
2. Timing chain
3. Camshaft sprocket
4. Camshaft gear
5. Key
6. Damper
7. Bolt
8. Spring
9. Crankshaft sprocket

Exploded view of the timing chain assembly—3.8L engine

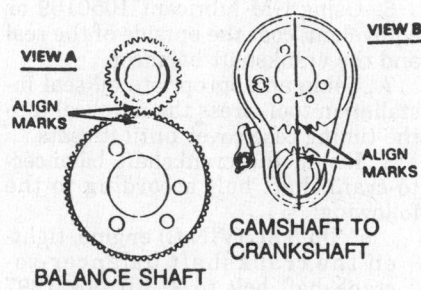

View of the timing sprocket alignment—3.8L engine

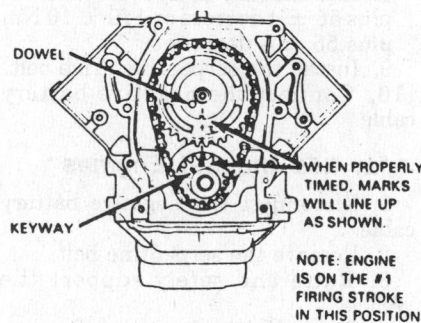

Aligning the timing marks—4.1L, 4.5L and 4.9L Engines

8. Install timing chain over camshaft sprocket.
9. Install cam sprocket, crank sprocket and timing chain over crankshaft, ensuring that timing marks are aligned.
10. Move camshaft until the dowel pin mates with the index hole in the sprocket.
11. Hold camshaft sprocket in position against end of camshaft and press sprocket onto camshaft by hand, being sure index pin in camshaft is aligned with index hole in sprocket.
12. Install screw securing camshaft

sprocket to camshaft. Tighten to 36 ft. lbs. (48 Nm).

NOTE: It may be necessary to keep the engine from rotating while setting the torque.

13. Install new thrust button.
14. Install oil slinger on crankshaft with smaller end of slinger against crankshaft sprocket.
15. Install engine front cover.
16. Connect negative battery cable.

Camshaft

REMOVAL & INSTALLATION

3.8L Engine

1. Disconnect the negative battery cable.
2. Remove the intake manifold.
3. Remove the rocker arm covers, rocker arms, pushrods and lifters.

NOTE: Keep all valve train components in order so they may be reinstalled in their original positions.

4. Remove the crankshaft pulley. Remove the crankshaft sensor cover.
5. Remove the front cover.
6. Remove the timing chain and sprockets.
7. Remove the camshaft thrust plate.
8. Carefully, remove the camshaft.

To install:

9. Coat the camshaft with prelube 10423565 or equivalent, prior to installation.
10. Carefully, install the thrust plate.
11. Install the timing chain and sprockets.
12. Install the front cover.
13. Install the crankshaft sensor cover.
14. Install the crankshaft pulley.
15. Install the lifters, pushrods and rocker arms in their original positions.
16. Install the rocker arm covers.
17. Install the intake manifold.
18. Connect the negative battery cable.

4.1L, 4.5L and 4.9L Engines

To perform this procedure, the engine must be removed from the vehicle and attached to an engine stand.

1. Disconnect the negative battery cable. Remove the intake manifold and the timing chain.
2. Remove the rocker arm covers, rocker arms, pushrods and valve lifters.

NOTE: Keep all valve train components in order so they may be reinstalled in their original positions.

3. Carefully slide the camshaft out from the front of the engine.

To install:

NOTE: If a new camshaft is to be installed, new lifters and a distributor drive gear must also be installed.

4. Lubricate the camshaft with camshaft prelube 1052365 or equivalent, on all camshaft lobes, distributor drive and driven gear teeth and bearing journals.

5. Carefully, install the camshaft into the engine.

6. Install the camshaft sprocket-to-camshaft bolt and tighten to 31 ft. lbs. (50 Nm).

7. Install the lifters, pushrods and rocker arms in their original positions. Install the rocker arm covers.

8. Install the timing chain and intake manifold.

9. Connect the negative battery cable.

Silent Shaft

REMOVAL & INSTALLATION

3.8L Engine

1. Disconnect the negative battery cable. Remove the engine and secure it to a workstand.

2. Remove the flywheel-to-crankshaft bolts and the flywheel.

3. Remove the intake manifold.

4. Remove the lifter guide retainer.

5. Remove the front cover-to-engine bolts and front cover.

6. Remove the camshaft sprocket-to-camshaft gear bolts, the sprocket, the timing chain and the gear.

7. To remove the balance shaft, perform the following:

    a. Remove the balance shaft gear-to-shaft bolt and the gear.

    b. Remove the balance shaft retainer-to-engine bolts and the retainer.

    c. Using the slide hammer tool, pull the balance shaft from the front of the engine.

8. If replacing the rear balance shaft bearing, perform the following procedures:

    a. Drive the rear plug from the engine.

    b. Using the camshaft remover/installer tool, press the rear bearing from the rear of the engine.

    c. Dip the new bearing in clean engine oil.

    d. Using the balance shaft rear bearing installer tool, press the new rear bearing into the rear of the engine.

    e. Install the rear cup plug.

To install:

9. Using the balance shaft installer tool, screw it into the balance shaft and install the shaft into the engine. Remove the installer.

10. Turn the camshaft so with the camshaft sprocket temporarily installed, the timing mark is straight down.

11. With the camshaft sprocket and the camshaft gear removed, turn the balance shaft so the timing mark on the gear points straight down.

12. Align the marks on the balance shaft gear and camshaft gear by turning the balance shaft. Install the camshaft gear.

13. Turn the crankshaft so No. 1 piston is on TDC.

14. Install the timing chain and camshaft sprocket.

15. Install the balance shaft front bearing retainer and bolts. Tighten to 22 ft. lbs. (30 Nm).

16. Install the front cover.

17. Install the lifter guide retainer.

18. Install the intake manifold.

19. Install the flywheel. Tighten the attaching bolts to 61 ft. lbs. (82 Nm).

20. Connect the negative battery cable.

Piston and Connecting Rod

POSITIONING

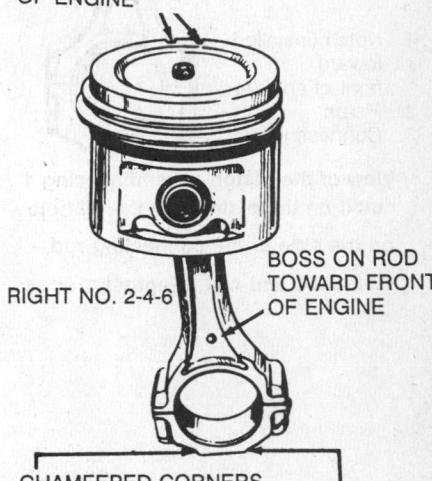

NOTCHES TOWARD FRONT OF ENGINE

RIGHT NO. 2-4-6

BOSS ON ROD TOWARD FRONT OF ENGINE

CHAMFERED CORNERS TOWARD REAR OF ENGINE

View of the right bank piston and rod positioning—3.8L engine

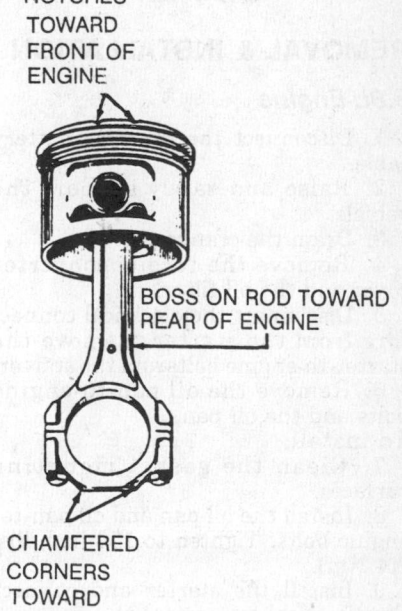

NOTCHES TOWARD FRONT OF ENGINE

BOSS ON ROD TOWARD REAR OF ENGINE

CHAMFERED CORNERS TOWARD FRONT OF ENGINE

LEFT NO. 1-3-5

View of the left bank piston and rod positioning—3.8L engine

1. Plug
2. Bearing
3. Bolt
4. Balance shaft gear
5. Bolt
6. Retainer

Exploded view of the balance shaft assembly—3.8L engine

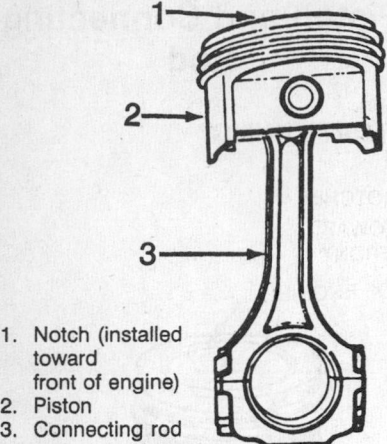

1. Notch (installed toward front of engine)
2. Piston
3. Connecting rod

View of the piston assembly using 1 notch on the piston and the oil hole on the side of the connecting rod—

4.1L, 4.5L and 4.9L Engines

ENGINE LUBRICATION

Oil Pan

REMOVAL & INSTALLATION

3.8L Engine

1. Disconnect the negative battery cable.
2. Raise and safely support the vehicle.
3. Drain the crankcase.
4. Remove the torque converter cover and the oil filter.
5. Disconnect the electrical connectors from the starter. Remove the starter-to-engine bolts and the starter.
6. Remove the oil pan-to-engine bolts and the oil pan.
To install:
7. Clean the gasket mounting surfaces.
8. Install the oil pan and oil pan-to-engine bolts. Tighten to 124 inch lbs. (14 Nm).
9. Install the starter and connect the electrical connectors.
10. Install the oil filter and torque converter cover.
11. Fill the crankcase.
12. Lower the vehicle.
13. Connect the negative battery cable.

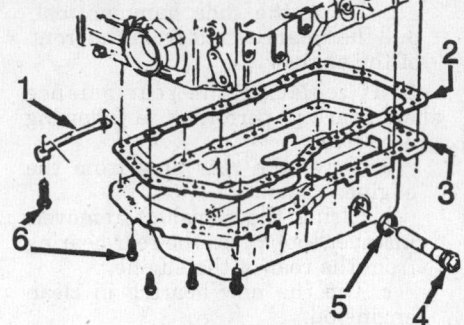

1. Spring tension
2. Oil pan gasket
3. Oil pan
4. Oil level indicator switch—40 ft. lbs.
5. Seal
6. Bolt

Exploded view of the oil pan assembly— 3.8L engine

4.1L, 4.5L and 4.9L Engines

ALLANTE

1. Disconnect the negative battery cable.
2. Raise and safely support the vehicle.
3. Drain the crankcase. Disconnect the oil level sensor, if equipped.
4. Remove the flywheel covers.
5. Remove the exhaust Y-pipe.
6. Remove the oil pan-to-engine bolts/nuts and the oil pan.
To install:
7. Clean the gasket mounting surfaces.

NOTE: Apply a ¼ in. bead of RTV at the rear main bearing cap and front cover to block joints.

8. Install the oil pan and oil pan-to-engine bolts/nuts. Tighten to 14 ft. lbs. (18 Nm).
9. Install the flywheel covers.
10. Install the exhaust Y-pipe.
11. If equipped, connect the oil level sensor.
12. Lower the vehicle.
13. Fill the crankcase.
14. Connect the negative battery cable.

ELDORADO AND SEVILLE

1. Disconnect the negative battery cable. Raise and safely support the vehicle. Drain the crankcase.
2. Remove the 2 torque converter covers from the lower side of the transaxle.
3. Remove the exhaust crossunder pipe and reposition.
4. Remove the oil pan-to-engine bolts and the oil pan.
To install:
5. Clean the gasket mounting surfaces.

NOTE: Apply a ¼ in. bead of RTV at the rear main bearing cap and front cover to block joints.

6. Install the oil pan and oil pan-to-engine bolts. Tighten to 12 ft. lbs. (16 Nm).
7. Install the exhaust crossunder pipe.
8. Install the 2 torque converter covers.
9. Lower the vehicle.
10. Fill the crankcase.
11. Connect the negative battery cable.

Oil Pump

REMOVAL & INSTALLATION

3.8L ENGINE

The oil pump is located in the bottom of the front cover. The oil pump is an integral part of the front cover with the crankshaft passing through the pump.

1. Disconnect the negative battery cable. Remove the front cover.
2. Clean the gasket mounting surfaces.
3. To inspect the pump gears, perform the following:
   a. Remove the oil pump cover-to-front cover screws and the cover.
   b. Remove the inner and outer pump gears.
   c. Using solvent, clean the gears.
   d. Inspect the gears for wear and/or damage; if necessary, replace the parts.
To install:
4. Using petroleum jelly, pack the pump and reinstall the parts. Tighten the oil pump cover-to-front cover screws to 88 inch lbs. (11 Nm).

NOTE: The oil pump must be primed this way or no pressure will be produced when the engine is started.

5. Install the front cover.
6. Connect the negative battery cable.

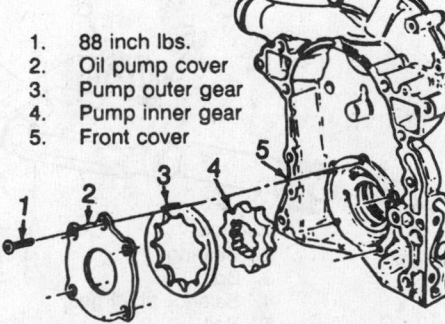

1. 88 inch lbs.
2. Oil pump cover
3. Pump outer gear
4. Pump inner gear
5. Front cover

Exploded view of the gerotor oil pump assembly—3.8L engine

7. Check and/or refill the oil level in the crankcase. Replace the oil filter. Start the engine and check for leaks.

4.1L, 4.5L and 4.9L Engines

1. Disconnect the negative battery cable. Remove the oil pan.

2. Remove the oil pump-to-engine screws/nut and the oil pump from the engine.

3. To disassemble, remove the oil pump cover-to-housing screws, slide the driveshaft, drive gear and driven gear from the pump housing.

4. Remove the oil pressure regulator valve and spring from the bore in the housing assembly.

5. Inspect the oil pressure regulator valve for nicks and burrs.

6. Measure the free length of the regulator valve spring. It should be 2.57–2.69 in. (65.28–68.32mm).

7. Inspect the drive gear and driven gear for nicks and burrs.

To install:

8. Assemble the pump drive gear over the driveshaft so the retaining ring is inside the gear. Position the drive gear over the pump housing shaft closest to the pressure regulator bore.

9. Slide the driven gear over the remaining shaft in the pump housing, meshing the driven gear with the drive gear.

10. Install the oil pressure regulator spring and valve in the bore of the pump housing assembly.

11. Install the pump cover-to-pump housing screws to 5 ft. lbs. (7 Nm), the oil pump-to-engine screws to 15 ft. lbs. (20 Nm) and nut to 22 ft. lbs. (30 Nm).

12. Install the oil pan.

13. Connect the negative battery cable.

Rear Main Bearing Oil Seal

REMOVAL & INSTALLATION

3.8L (VIN C) Engine

ROPE TYPE

If braided rope type seals are used, the upper seal half cannot be replaced without removing the crankshaft.

Lower Half-Seal

1. Disconnect the negative battery cable. Remove the oil pan.

2. Remove the rear main bearing cap-to-engine bolts and the cap.

3. Remove the old seal from the bearing cap.

To install:

4. To replace the oil seal, perform the following procedures:

   a. Using sealant GM 1052621,

Loctite® 414 or equivalent, apply it to the main bearing cap seal groove and wait for 1 minute.

   b. Using a new rope seal and a wooden dowel or hammer handle, roll the new seal into the cap so both ends projecting above the parting surface of the cap; force the seal into the groove by rubbing it down, until the seal projects above the groove not more than $\frac{1}{16}$ in.

   c. Using a sharp razor blade, cut the ends off flush with the surface of the cap.

   d. Using chassis grease, apply a thin coat to the seals surface.

5. To install the neoprene sealing strips (side seals), perform the following procedures:

   a. Using light oil or kerosene, soak the strips for 5 minutes.

NOTE: The neoprene composition seals will swell up once exposed to the oil and heat. It is normal for the seals to leak for a short time, until they become properly seated. The seals must not be cut to fit.

   b. Place the sealing strips in the grooves on the sides of the bearing cap.

6. Using sealer GM 1052621 or equivalent, apply it to the main bearing cap mating surface; do not apply sealer to the bolt holes.

7. To complete the installation, reverse the removal procedures. Tighten the main bearing cap-to-engine bolts to 90 ft. lbs. (122 Nm). Refill the crankcase. The engine must be operated at low rpm when first started, after a new seal is installed.

Upper Half-Seal

Engine and crankshaft removal are not necessary if the following time saver procedure is followed. While this procedure is effective for stopping leakage from the upper half of the seal, it is not a replacement procedure.

1. Disconnect the negative battery cable. Remove the oil pan.

2. Remove the rear main bearing cap-to-engine bolts and the cap.

To install:

3. Using the seal packing tool, insert it against each side of the upper seal and drive the seal until it is tight.

4. Measure the amount the seal was driven into the engine and add about $\frac{1}{16}$ in. Using a razor blade, cut that amount off the old lower seal.

5. Using the seal packing tool, work the short packing pieces into the cylinder block; a small amount of oil on the seal will help the installation.

6. Repeat this process on the other side.

7. Install the main bearing cap.

Tighten the main bearing cap-to-engine bolts to 90 ft. lbs. (122 Nm).

8. Install the oil pan.

9. Connect the negative battery cable.

10. Refill the crankcase. The engine must be operated at low rpm when first started, after a new seal is installed.

3.8L (VIN L), 4.1L, 4.5L and 4.9L Engines

LIP TYPE

NOTE: To perform this procedure, use a seal removal tool and a seal installer tool.

1. Disconnect the negative battery cable. Remove the transaxle.

2. Unbolt and remove the flexplate from the rear end of the crankshaft.

3. Using a seal removal tool, remove the old seal. Throughly clean the seal bore of any left over seal material with a clean rag.

To install:

4. Lubricate the lip of the new seal with wheel bearing grease. Position it over the crankshaft and into the seal bore with the spring facing inside the engine.

5. Using a seal installer tool, press the seal into place. The seal must be square and flush with the block to 1mm indented.

6. Install the flexplate and tighten according to the followng:

   a. On 1988 Allante tighten the flexplate-to-crankshaft bolts to 37 ft. lbs. (50 Nm).

   b. On 1989–92 Allante and Eldorado/Seville, tighten the flexplate-to-crankshaft bolts to 70 ft. lbs. (95 Nm).

   c. On the 3.8L (VIN L) engine, tighten the flexplate-to-crankshaft bolts to 61 ft. lbs. (82 Nm).

7. Install the transaxle.

8. Connect the negative battery cable.

ENGINE COOLING

Radiator

REMOVAL & INSTALLATION

Reatta and Riviera

1. Disconnect the negative battery cable.

2. Drain coolant from radiator.

3. Remove plastic radiator support cover.

4. Remove engine-to-radiator torque strut.

5. Remove rear cooling fan.

6. Remove coolant reservoir hose at filler neck.

7. Remove upper and lower radiator hoses from radiator.

8. Remove transaxle oil cooler lines at radiator.

9. Remove radiator top support, 3 remaining bolts with torque strut removed.

10. Remove radiator from vehicle; lift radiator straight up and out.

To install:

11. Install radiator in vehicle.

12. Install radiator top support, securing with 3 retaining bolts. Tighten to 18 ft. lbs. (25 Nm).

13. Connect oil cooler lines at radiator. Tighten to 20 ft. lbs. (27 Nm).

14. Install upper and lower radiator hoses to radiator, securing hose clamps.

15. Connect reservoir hose at filler neck, securing hose clamp.

16. Install rear cooling fan.

17. Install engine-to-radiator torque strut and 2 remaining strut/radiator support retaining bolts. Tighten radiator support retaining bolts to 18 ft. lbs. (25 Nm).

18. Install plastic radiator support cover.

19. Fill radiator with coolant.

20. Connect negative battery cable.

21. Start engine and check for leaks. Check transaxle fluid level and add, as necessary. Allow engine to come to normal operating temperature and check again for leaks.

Allante, Eldorado and Seville

1. Disconnect the negative battery cable.

2. Drain cooling system.

3. Remove right and left cooling fans. On Eldorado and Seville remove rear cooling fan.

4. Disconnect coolant reservoir hose at filler neck.

5. Remove upper and lower radiator hoses from radiator.

6. Remove engine oil cooler lines from left radiator end tank.

7. Remove transaxle oil cooler lines from right radiator end tank.

8. Remove radiator top support.

9. Remove radiator from car, lifting radiator straight up and out.

To install:

10. Install radiator in vehicle.

11. Install radiator top support. Tighten radiator support retaining bolts to 18 ft. lbs. (25 Nm).

12. Connect transaxle oil cooler lines at radiator. Tighten to 20 ft. lbs. (27 Nm).

13. Connect oil cooler lines at radiator. Tighten to 13 ft. lbs. (18 Nm).

14. Install upper and lower radiator hoses to radiator securing hose clamps.

15. Connect coolant reservoir hose at filler neck.

16. Install cooling fan(s).

17. Fill cooling system.

18. Connect negative battery cable.

19. Start engine and check for leaks. Check transaxle fluid level and add, as necessary. Allow engine to come to normal operating temperature and check again for leaks.

Toronado and Trofeo

1. Disconnect the negative battery cable.

2. Drain cooling system.

3. Remove plastic radiator support cover.

4. Remove engine-to-radiator torque strut.

5. Remove rear cooling fan.

6. Remove upper air cleaner duct and/or silencer, as necessary.

7. Remove coolant reservoir hose at filler neck.

8. Remove upper and lower radiator hoses from radiator.

9. Remove transaxle oil cooler lines.

10. Remove radiator top support, 3 remaining bolts with torque strut removed.

11. Remove radiator from vehicle, lifting straight up and out.

To install:

12. Install radiator in vehicle.

13. Install radiator top support, securing with 3 retaining bolts.

14. Connect transaxle oil cooler lines at radiator. Tighten to 20 ft. lbs. (27 Nm).

15. Connect upper and lower radiator hoses to radiator, securing with clamps.

16. Connect coolant reservoir hose at filler neck, securing with hose clamp.

17. Install rear cooling fan.

18. Install upper air cleaner duct and/or silencer, if removed.

19. Install engine-to-radiator torque strut and 2 remaining strut/radiator support retaining bolts. Tighten to 18 ft. lbs. (25 Nm).

20. Install plastic radiator support cover.

21. Fill cooling system.

22. Connect negative battery cable.

23. Start engine and check for leaks. Check transaxle fluid level and add, as necessary. Allow engine to come to normal operating temperature and check again for leaks.

Heater Core

REMOVAL & INSTALLATION

Reatta and Riviera

1988–89

1. Disconnect the negative battery cable.

2. Drain the cooling system.

3. Remove console and instrument panel, as required.

4. Disconnect the hoses from the heater core.

5. Remove the right side sound insulator and courtesy light.

6. Remove the glove box.

7. Disconnect the air conditioning programmer the electrical and vacuum connectors. Remove the air conditioning programmer screws and the programmer.

8. Disconnect the ECM electrical connectors. Remove the ECM and bracket.

9. Disconnect the BCM electrical connectors. Remove the BCM and bracket.

10. Remove the heater core cover screws, the cover, the retaining clip, the heater core screws and the heater core.

To install:

11. Install heater core cover screws, the cover, the retaining clip, heater core screws and heater core.

12. Install BCM bracket and connect BCM electrical connectors.

13. Install ECM bracket and connect ECM electrical connectors.

14. Install air conditioner programmer and connect air conditioning programmer electrical and vacuum connectors. Check adjustment of the programmer.

15. Install glove box.

16. Install right side sound insulator and courtesy light.

17. Connect the heater core hoses.

18. Connect negative battery cable.

19. Fill cooling system.

20. Start engine and check for coolant leaks. Allow engine to come to normal operating temperature. Recheck for coolant leaks.

1990–92

1. Disconnect the negative battery cable.

2. Drain the engine coolant into a clean container for reuse.

3. If equipped, disarm the SIR system.

4. Remove console and instrument panel.

5. Remove air conditioner programmer attaching screws. Disconnect the programmer electrical connectors. Remove the programmer.

6. Disconnect the BCM electrical

connectors. Remove BCM and mounting bracket.

7. Disconnect the ECM electrical connectors. Remove ECM and mounting bracket.

8. Remove heater core cover from housing.

9. Disconnect inlet and outlet heater hoses from heater core.

10. Remove 2 heater retaining screws.

11. Remove heater core from vehicle.

To install:

12. Install heater core to heater case, securing with 2 screws.

13. Connect inlet and outlet heater hoses to heater core.

14. Install heater core cover.

15. Install ECM mounting bracket and ECM. Connect the ECM electrical connectors. Install the ECM attaching screws.

16. Install BCM mounting bracket and BCM. Connect the BCM electrical connectors. Install the BCM attaching screws.

17. Install air conditioner programmer. Connect the programmer electrical connectors. Install the programmer attaching screws. Check adjustment of the programmer.

18. Install instrument panel and console.

19. If equipped, arm the SIR system.

20. Connect negative battery cable.

21. Fill cooling system and check for leaks. Start the engine and allow to come to normal operating temperature. Recheck for leaks. Top-up coolant.

Allante

1. Disconnect the negative battery cable.

2. Drain the cooling system to a level below the heater core.

3. Remove the glove box screws. Label and disconnect the electrical connectors from the glove box.

4. Remove the glove box assembly from the vehicle.

5. Remove the lower sound insulator to gain working clearance.

6. Remove the radio.

7. Remove the air conditioning programmer, the Electronic Control Module (ECM) screws and the ECM.

8. Remove the module assembly heater core cover. Disconnect the hoses from the heater core.

9. Remove the heater core screws and the heater core.

To install:

10. Install heater core in vehicle.

11. Install module assembly heater core cover. Connect hoses to the heater core.

12. Install ECM bracket, ECM and electrical connectors.

13. Install air conditioning program-

mer. Check adjustment of the programmer.

14. Install radio and lower sound insulator.

15. Install glove box assembly and glove box electrical connectors.

16. Fill cooling system.

17. Start engine and check for coolant leaks. Allow engine to come to normal operating temperature. Recheck for coolant leaks.

Eldorado and Seville

1. Disconnect the negative battery cable.

2. Drain the cooling system to a level below the heater core.

3. Remove the glove box screws. Label and disconnect the electrical connectors from the glove box.

4. Remove the glove box assembly from the vehicle.

5. Remove the lower sound insulator to gain working clearance.

6. Remove the air conditioner programmer, the Electronic Control Module (ECM) screws and the ECM.

7. Remove the module assembly heater core cover. Disconnect the hoses from the heater core.

8. Remove the heater core screws and the heater core.

To install:

9. Install the heater core in vehicle.

10. Connect hoses to heater core. Install module assembly heater core cover.

11. Install the air conditioner programmer and the ECM. Check adjustment of the programmer.

12. Install the lower sound insulator.

13. Install the glove box assembly to vehicle. Connect the electrical connectors to the glove box.

14. Fill cooling system.

15. Start engine and check for coolant leaks. Allow engine to come to normal operating temperature. Recheck for coolant leaks.

Toronado and Trofeo

1988–89

1. Disconnect the negative battery cable.

2. Drain the engine coolant into a clean container for reuse.

3. Remove the left side sound insulator. Disconnect the courtesy light.

4. Remove the right side sound insulator. Disconnect the courtesy light.

5. Remove the steering column opening filler panel screws and remove the filler panel.

6. Remove the steering column bolts. Lower the steering column and allow the steering wheel to rest on the seat.

7. Remove the windshield defroster nozzle grille. Remove the deflector housings.

8. Remove the screws attaching the top of the instrument panel that were under the deflector housings and windshield defroster nozzle grille. Remove the bolts attaching the bottom of the instrument panel.

9. Disconnect the bulkhead electrical connector.

10. Move the instrument panel rearward.

11. Remove the aspirator duct.

12. Disconnect the fuel filler door release electrical connector.

13. Disconnect the trunk lid release electrical connector.

14. Disconnect the antenna lead.

15. Remove the fuse panel attaching screws.

16. Remove the instrument panel from the car. Set the panel on a clean, protected surface.

17. Disconnect the heater hoses from the heater core.

18. Remove the air conditioner programmer attaching screws. Disconnect the programmer electrical and vacuum connectors. Remove the programmer.

19. Remove the power module attaching screws. Disconnect the power module electrical connectors. Remove the power module.

20. Remove the heater core cover attaching screws. Remove the heater core.

21. Remove the heater core retaining clip. Remove the heater core attaching screws. Remove the heater core.

To install:

22. Apply strip caulk to seal the heater core to the case.

23. Install the heater core and attaching screws. Install the retaining clip.

24. Install the heater core cover and attaching screws.

25. Install the power module. Connect the electrical connectors and the power module attaching screws.

26. Install the air conditioner programmer. Connect the vacuum and electrical connectors. Install the programmer attaching screws. Check adjustment of the programmer.

27. Install the instrument panel.

28. Install the fuse panel attaching screws.

29. Connect the antenna lead.

30. Connect the trunk lid release electrical connector.

31. Connect the fuel filler door release electrical connector.

32. Install the aspirator duct.

33. Move the instrument panel forward.

34. Connect the bulkhead electrical connector.

35. Install the bolts attaching the top and bottom of the instrument panel.

36. Install the windshield defroster

deflector housings. Install the nozzle grille.

37. Raise the steering column into position. Install the 4 steering column bolts.

38. Install the steering column opening filler panel. Install the filler panel attaching screws.

39. Install the right side sound insulator. Install the sound insulator attaching screws.

40. Install the left side sound insulator. Connect the courtesy light electrical connector. Install the sound insulator attaching screws.

41. Connect the negative battery cable.

42. Fill cooling system and check for leaks. Start the engine and allow to come to normal operating temperature. Recheck for leaks. Top-up coolant.

1990–92

1. Disconnect the negative battery cable.

2. Drain the engine coolant into a clean container for reuse.

3. If equipped, disarm the SIR system.

4. Remove console and instrument panel.

5. Remove air conditioner programmer attaching screws. Disconnect the programmer electrical connectors. Remove the programmer.

6. Disconnect the BCM electrical connectors. Remove BCM and mounting bracket.

7. Disconnect the ECM electrical connectors. Remove ECM and mounting bracket.

8. Remove heater core cover from housing.

9. Disconnect inlet and outlet heater hoses from heater core.

10. Remove 2 heater retaining screws.

11. Remove heater core from vehicle.

To install:

12. Install heater core to heater case, securing with 2 screws.

13. Connect inlet and outlet heater hoses to heater core.

14. Install heater core cover.

15. Install ECM mounting bracket and ECM. Connect the ECM electrical connectors. Install the ECM attaching screws.

16. Install BCM mounting bracket and BCM. Connect the BCM electrical connectors. Install the BCM attaching screws.

17. Install air conditioner programmer. Connect the programmer electrical connectors. Install the programmer attaching screws. Check adjustment of the programmer.

18. Install instrument panel and console.

19. If equipped, arm the SIR system.

20. Connect negative battery cable.

21. Fill cooling system and check for leaks. Start the engine and allow to come to normal operating temperature. Recheck for leaks. Top-up coolant.

Air Conditioner Programmer

ADJUSTMENT

1. Remove the right side sound insulator and glove box.

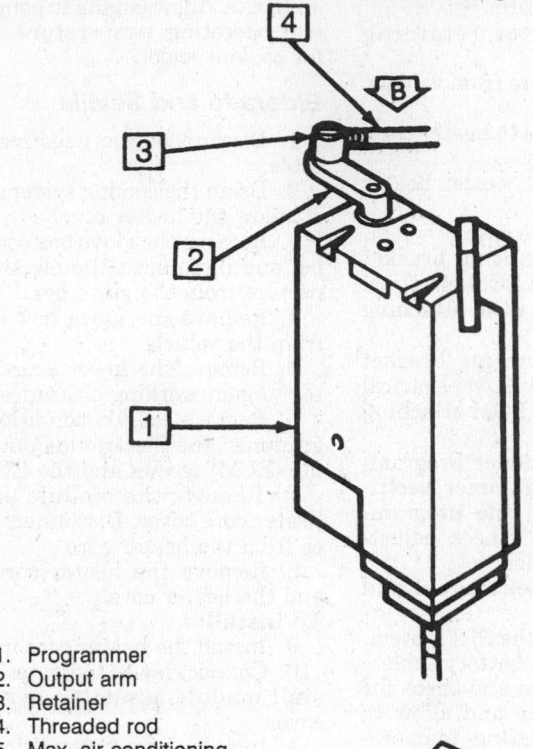

1. Programmer
2. Output arm
3. Retainer
4. Threaded rod
5. Max. air conditioning position
6. Max. heat position

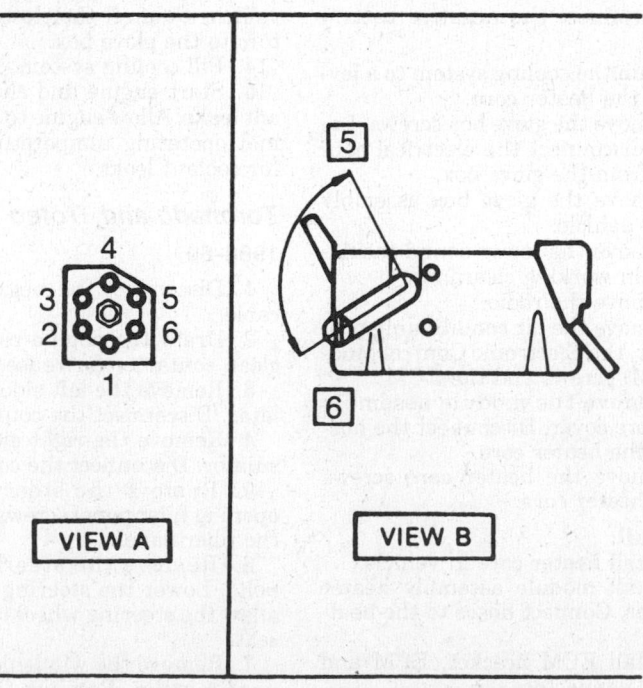

VIEW A

VIEW B

Electronic climate control programmer and linkage

2. On the temperature control panel, set the temperature for 90°F, allow 1–2 minutes for the programmer arm to travel to its maximum heat position.

3. Disconnect the threaded rod from the plastic retainer on the programmer output arm.

4. To check the air mixture valve for free travel, push the valve to the maximum air conditioning position and check for binding.

5. Place the pre-load air mixture valve in the maximum heat position; pull on the threaded rod to ensure the valve is seating. The programmer arm should be in the maximum heat position.

6. To avoid influencing the programmer arm or air mixture valve position, carefully snap the threaded rod into the plastic retainer.

7. Adjust the temperature setting to 60°F, then, check to verify the programmer arm and air mixture valve travel to the maximum air conditioning position.

Water Pump

REMOVAL & INSTALLATION

3.8L Engine

1. Disconnect the negative battery cable.

2. Position a drain pan under the radiator, open the drain cock and drain the cooling system.

3. Disconnect the hoses from the water pump.

4. Remove the serpentine drive belt.

5. Remove the water pump pulley bolts and the pulley.

NOTE: The long bolt is removed through the access hole provided in the body side rail.

6. Remove the water pump-to-engine bolts and the pump.

To install:

7. Clean the gasket mounting surfaces.

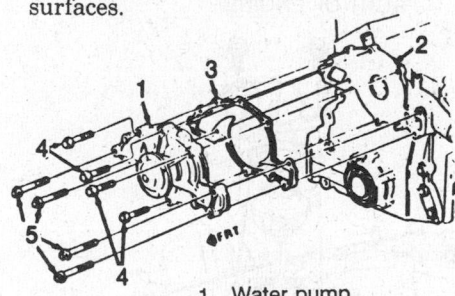

1. Water pump
2. Engine front cover assembly
3. Gasket
4. 97 inch lbs.
5. 29 ft. lbs.

Exploded view of the water pump—3.8L engine

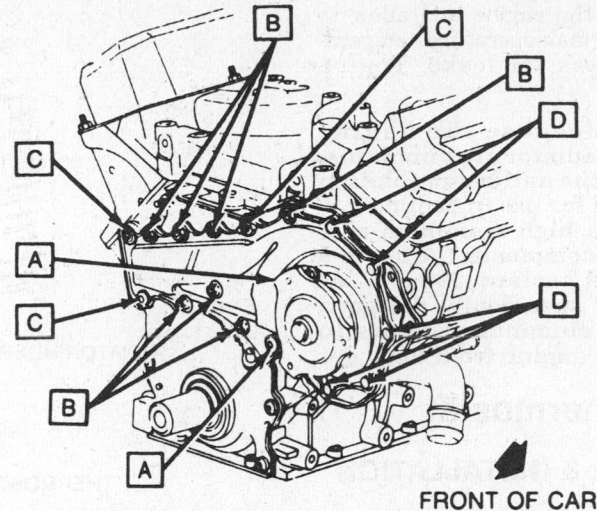

FRONT OF CAR

KEY	FASTENER TYPE	QTY.	TORQUE	
			N·m	FT. LBS.
A	TORX SCREW	2	40	30
B	NUT	7	7	5
C	HEX SCREW	3	40	30
D	HEX SCREW	5	7	5

Water pump fasteners—4.1L, 4.5L and 4.9L Engines

8. Install the water pump using a new gasket.

9. Tighten the water pump-to-engine long bolts to 29 ft. lbs. (39 Nm) and the short bolts to 97 inch lbs. (11 Nm).

10. Connect the hoses to the water pump.

11. Install the serpentine drive belt.

12. Connect the negative battery cable.

13. Fill cooling system and check for leaks. Start the engine and allow to come to normal operating temperature. Recheck for leaks. Top-up coolant.

NOTE: Because the radiator is made of aluminum and plastic, make sure the antifreeze solution is approved for use in cooling systems with a high aluminum content. GM recommends the use of a supplement/sealant 3634621 or equivalent, specifically designed for use in aluminum engines to protect the engine from damage.

4.1L, 4.5L and 4.9L Engines

1. Disconnect the negative battery cable.

2. Drain the engine coolant into a clean container for reuse.

3. Remove the air filter assembly. Disconnect and remove the coolant recovery tank.

4. Disconnect and remove the cross brace.

5. Remove the water pulley bolts.

6. Remove the serpentine drive belt and the water pump pulley.

7. Remove the water pump-to-engine bolts and the pump.

To install:

8. Clean the gasket mounting surfaces.

9. Place a new gasket over the water pump studs.

10. Install the water pump. Tighten the water pump bolts as follows:

Water pump-to-engine Torx® bolts to 30 ft. lbs. (40 Nm)

Water pump-to-engine stud nuts to 5 ft. lbs. (7 Nm)

Hex head bolts to 30 ft. lbs. (40 Nm)

Remaining hex head bolts to 5 ft. lbs. (7 Nm).

11. Install the water pump pulley. Install the water pump pulley bolts finger tight.

12. Install the serpentine drive belt.

13. Tighten the water pump pulley bolts to 22 ft. lbs. (30 Nm).

14. Install the cross brace.

15. Install the connect the coolant recovery tank. Install the air filter assembly.

16. Connect the negative battery cable.

17. Fill cooling system and check for

leaks. Start the engine and allow to come to normal operating temperature. Recheck for leaks. Top-up coolant.

NOTE: Because the engine block and radiator are aluminum, make sure the antifreeze solution is approved for use in cooling systems with a high aluminum content. GM recommends the use of a supplement/sealant 3634621 or equivalent, specifically designed for use in aluminum engines to protect the engine from damage.

Thermostat

REMOVAL & INSTALLATION

3.8L Engine

1. Drain the coolant until it is below the level of thermostat. Remove the thermostat housing. Observe the direction of the thermostat upon removal.

2. Replace the thermostat and O-ring ensuring the proper direction of new thermostat.

3. Fill cooling system using a 50/50 mixture of water and ethylene glycol antifreeze.

4. Start engine and check for coolant leaks. Allow engine to come to normal operating temperature. Recheck for coolant leaks.

4.1L, 4.5L and 4.9L Engines

1. Drain coolant to a level below the thermostat housing.

2. Remove 2 bolts securing thermostat housing to intake manifold.

3. Remove thermostat housing.

4. Remove thermostat and O-ring from intake manifold.

To install:

5. Install thermostat and a new O-ring to intake manifold.

6. Install thermostat housing to intake manifold. Tighten thermostat housing bolts to 18 ft. lbs. (25 Nm).

7. Refill cooling system using a 50/50 mixture of water and ethylene glycol antifreeze.

8. Start engine and check for coolant leaks. Allow engine to come to normal operating temperature. Recheck for coolant leaks.

Cooling System Bleeding

1. With the cooling system completely drained, fill the system with at least a 50/50 mixture of ethylene glycol antifreeze and water but no more than a 70/30 mixture of water to antifreeze.

2. Fill the radiator to just below the

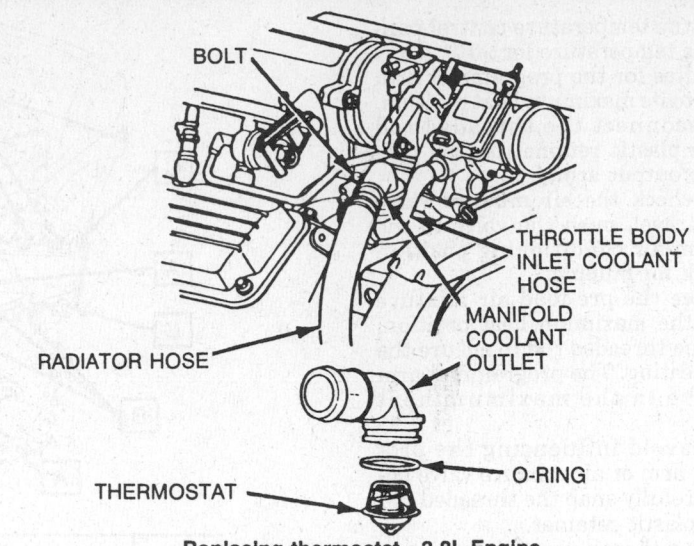

Replacing thermostat—3.8L Engine

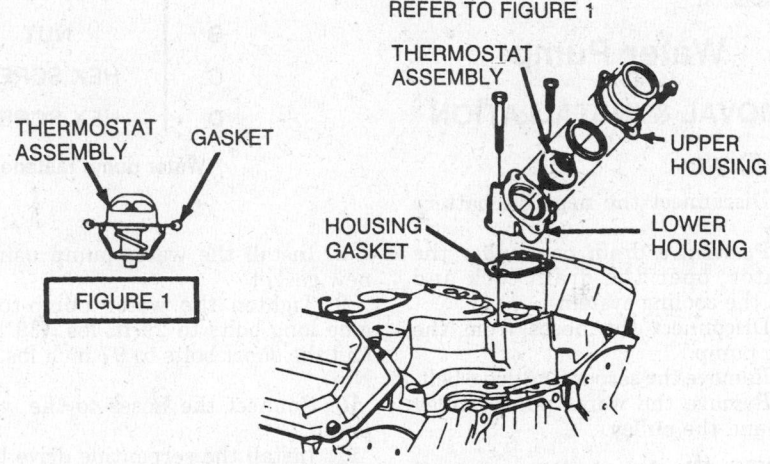

Replacing thermostat—Eldorado and Seville

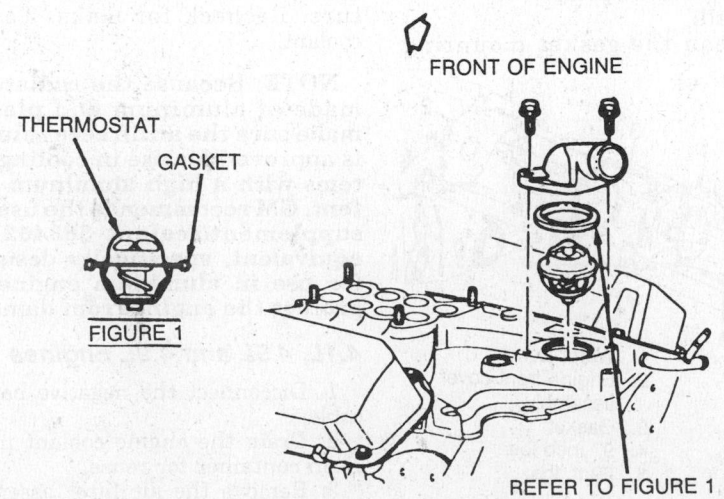

Replacing thermostat—Allante

filler neck. Fill the coolant recovery reservoir to the COLD FILL mark.

3. Run the engine with the radiator cap removed until normal operating temperature is reached, with the radiator inlet hose hot.

4. With the engine idling, add coolant to the radiator until it reaches the bottom of the filler neck.

5. Position the heating system controls on maximum; allowing coolant to circulate through the heater core.

6. Check the coolant level again and add, as necessary.

7. Install the radiator cap.

ENGINE ELECTRICAL

NOTE: Disconnecting the negative battery cable on some vehicles may interfere with the functions of the on board computer systems and may require the computer to undergo a relearning process, once the negative battery cable is reconnected.

Distributor

The High Energy Ignition (HEI) distributor with Electronic Spark Timing (EST) easily identified by the presence of a 6 terminal ECM connector.

REMOVAL

1. Disconnect the negative battery cable.
2. Set No. 1 cylinder to TDC of its compression stroke.
3. Remove distributor appearance cover and retainer, if equipped.
4. Remove ignition switch battery feed wire from distributor cap. Remove coil connectors from cap.

NOTE: Do not use a prybar to release locking tabs.

5. Remove 4 bolts from distributor cap and move cap off to the side. Note the location of the cap "doghouse" upon removal and reinstall in same position.
6. Remove 6 terminal ECM harness from distributor. Matchmark the rotor-to-housing and the housing-to-engine.
7. Remove distributor clamp nut and hold-down nut. Use special tool J-29791 or equivalent, to remove hold-down nut.
8. Note the position of rotor, then pull distributor up until rotor just

stops turning counterclockwise and again note position of rotor. Remove distributor.

Installation

Timing Not Disturbed

1. Insert the distributor into the engine, making sure the tip of the rotor is aligned with the alignment marks on the distributor housing and the engine.
2. Make sure the oil pump intermediate driveshaft is properly seated in the oil pump.
3. Install the distributor lock but do not tighten.
4. Reconnect the electrical harness connector(s) to the distributor, then, install distributor cap.
5. Start the engine and allow to come to normal operating temperature. Check and/or adjust the timing.

Timing Disturbed

1. Remove the No. 1 cylinder spark plug and place a finger over the hole. Using a wrench on the crankshaft pulley bolt, slowly turn the engine until compression is felt.
2. Align the timing marks so No. 1 cylinder is on TDC of the compression stroke.
3. Position the distributor in the engine with the rotor at No. 1 firing position. Make sure the oil pump intermediate driveshaft is properly seated in the oil pump.
4. Install the distributor retainer and lock bolt, tighten the lock bolt.
5. Reconnect the electrical harness connector(s) to the distributor and install distributor cap.

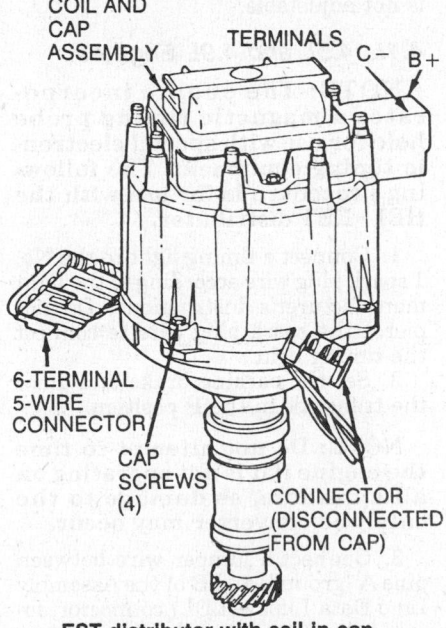

EST distributor with coil-in-cap

6. Start the engine and allow to come to normal operating temperature. Check and/or adjust the timing.

Distributorless Ignition

The Computer Controlled Coil Ignition (C^3I) system uses an ignition coil pack, ignition module, dual crankshaft sensor and associated wiring.

REMOVAL & INSTALLATION

IGNITION COIL

3.8L (VIN C) ENGINE

1. Disconnect the negative battery cable.
2. Label and remove spark plug wires.
3. Remove 6 Torx® screws securing the coil to ignition module.
4. Tilt coil assembly back.
5. Remove coil to module connectors.
6. Remove coil assembly.

NOTE: Ensure that the replacement coil pack is identical to the one being removed. The Type I coil pack, used on the 3.8L engine, will physically fit, however, the position of No. 1 coil, as noted on the coil pack, is in a different location, No. 1 and 4 on the 3.8L engine are closest to the module connector.

To install:
7. Install coil assembly and connectors.
8. Install 6 Torx® screws and tighten to 27 inch lbs. (3 Nm).
9. Install spark plug wires.

3.8L (VIN L) ENGINE

1. Disconnect the negative battery cable.
2. Label and disconnect the spark plug wires.
3. Remove the 2 screws securing the individual coil pack to the ignition module.
4. Remove the coil assembly.

To install:
5. Install the coil assembly.
6. Install the 2 screws and tighten to 40 inch lbs. (4–5 Nm).
7. Connect the spark plug wires.
8. Connect the negative battery cable.

Ignition Module

1. Disconnect the negative battery cable.
2. Remove 14-way connector at ignition module.
3. Label and disconnect the spark plug wires at coil assembly.
4. Remove nuts and washers (3) se-

curing ignition module assembly to bracket.

5. Remove 6 torx screws securing coil assembly to ignition module.

6. Note lead colors and mark for reassembly.

7. Disconnect connectors between coil and ignition module.

8. Remove ignition module.

To install:

9. Install coil and connectors to ignition module.

10. Install 6 torx screws and tighten to 27 inch lbs. (3 Nm).

11. Install nuts and washers securing assembly to bracket.

12. Install plug wires.

13. Connect 14-way connector to module.

14. Connect the negative battery cable.

Crankshaft Sensor

1. Disconnect the negative battery cable.

2. Remove nuts holding vibration damper support to ignition module bracket and vibration damper to engine bracket.

3. Remove support.

4. Remove bolts holding bracket to front of engine (2).

5. Remove nut from vibration damper to engine cradle.

6. Remove vibration damper and support assembly.

7. Remove serpentine belt from crankshaft pulley.

8. Raise and safely support the vehicle.

9. Remove right front tire and wheel assembly.

10. Remove right inner fender access cover.

11. Remove crankshaft harmonic balancer retaining bolt using 28mm socket.

12. Remove crankshaft harmonic balancer.

13. Disconnect sensor electrical connector.

14. Remove sensor and pedestal from block face.

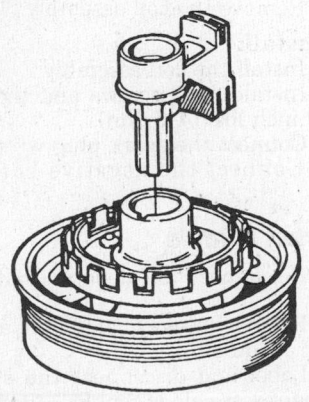

Checking vanes on harmonic balancer

15. Remove sensor from pedestal.

To install:

16. Loosely install crankshaft sensor on pedestal.

17. Position sensor with pedestal attached on special tool J-37089.

18. Position special tool on crankshaft.

19. Install bolts to hold pedestal to block face. Tighten to 14–28 ft. lbs. (20–40 Nm).

20. Tighten pedestal pinch bolt to 36–40 ft. lbs. (4–4.5 Nm).

21. Remove special tool J-37089.

22. Place special tool J-37089 on harmonic balancer and turn. If any vane of the harmonic balancer touches the tool, replace the balancer assembly.

23. Install balancer on crankshaft.

24. Tighten crankshaft bolt to 200–239 ft. lbs. (270–315 Nm).

25. Install inner fender shield.

26. Install tire and wheel assembly and tighten to 100 ft. lbs. (140 Nm).

27. Connect the negative battery cable.

Ignition Timing

ADJUSTMENT

NOTE: Always consult the Vehicle Emission Control Information label in the engine compartment before adjusting timing. If the underhood sticker differs from the following procedures, follow the sticker.

3.8L Engine

The 3.8L engines are equipped with a C³I ignition system which does not incorporate a distributor. Ignition timing is controlled by the ECM/PCM and is not adjustable.

4.1L, 4.5L and 4.9L Engines

NOTE: The engine incorporates a magnetic timing probe hole for use with special electronic timing equipment. The following procedure is for use with the HEI–EST distributor.

1. Connect a timing light to the No. 1 spark plug wire according to the light manufacturer's instructions. Do not pierce the spark plug wire to connect the timing light.

2. Set the parking brake and place the transaxle in the **P** position.

NOTE: Do not attempt to time the engine if it is not operating on all cylinders, as damage to the catalytic converter may occur.

3. Connect a jumper wire between pins **A** (ground) and **B** of the Assembly Line Data Link (ALDL) connector, located near the parking brake pedal un-

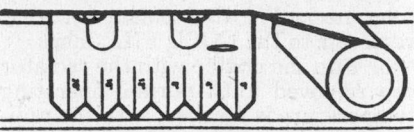

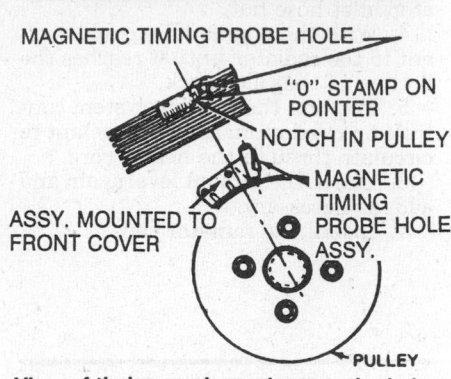

MAGNETIC TIMING PROBE HOLE

"0" STAMP ON POINTER

NOTCH IN PULLEY

MAGNETIC TIMING PROBE HOLE ASSY.

ASSY. MOUNTED TO FRONT COVER

PULLEY

View of timing marks and magnetic timing probe holder—4.1L, 4.5L and 4.9L Engines

der the dash. By jumping the Assembly Line Data Link (ALDL) connector, the ECM will command the BCM to display a SET TIMING message on the Climate Control Driver Information Panel (CCDIC). The engine will now operate at base timing. The timing can now be checked with a standard timing light at 10 degrees BTDC at 900 rpm or less. Varify proper timing setting with the Vehicle Emission Control Information label.

4. Start the engine and allow to come to normal operating temperature.

5. Aim the timing light at the degree scale just over the harmonic balancer; the line on the pulley should align with the mark on the timing plate.

6. If timing adjustment is necessary, use a distributor wrench to loosen the hold-down clamp. Rotate the distributor until the desired ignition advance is achieved. When the correct timing is set, tighten the hold-down clamp nut/bolt to 20 ft. lbs. (27 Nm).

NOTE: To advance the timing, rotate the distributor opposite the normal direction of rotor rotation. Retard the timing by rotating the distributor in the normal direction of rotor rotation.

Alternator

PRECAUTIONS

Several precautions must be observed with alternator equipped vehicles to avoid damage to the unit.

● If the battery is removed for any reason, make sure it is reconnected with the correct polarity. Reversing the battery connections may result in damage to the one-way rectifiers.

• When utilizing a booster battery as a starting aid, always connect the positive to positive terminals and the negative terminal from the booster battery to a good engine ground on the vehicle being started.

• Never use a fast charger as a booster to start vehicles.

• Disconnect the battery cables when charging the battery with a fast charger.

• Never attempt to polarize the alternator.

• Do not use test lights of more than 12 volts when checking diode continuity.

• Do not short across or ground any of the alternator terminals.

• The polarity of the battery, alternator and regulator must be matched and considered before making any electrical connections within the system.

• Never separate the alternator on an open circuit. Make sure all connections within the circuit are clean and tight.

• Disconnect the battery ground terminal when performing any service on electrical components.

• Disconnect the battery if arc welding is to be done on the vehicle.

BELT TENSION ADJUSTMENT

All accessories are driven by a single serpentine belt. The tension is maintained automatically by a spring-loaded tensioner. Periodic adjustment is not required.

Belt tension can be checked using a suitable belt tension gauge. The tensioner should maintain approximately 110 lbs. (490 N) of tension throughout its functional travel. If the tension is below specification and the tensioner is resting on the maximum travel stop, replace the serpentine belt.

REMOVAL & INSTALLATION

1. Disconnect the negative battery cable.
2. Label and disconnect the electrical connectors from the back of the alternator.
3. Release the tension from the drive belt and remove the belt from the alternator pulley. Do not remove the belt from any other pulleys.
4. Remove the alternator-to-bracket bolts and the alternator from the vehicle.
To install:
5. Install alternator on vehicle.
6. Reposition drive belt on alternator pulley.

7. Install electrical connectors.
8. Connect negative battery cable.

Starter

REMOVAL & INSTALLATION

1. Disconnect the negative battery cable.
2. Raise and safely support the vehicle.
3. If equipped, remove the starter motor shield.
4. Remove the 4 flywheel inspection cover bolts, as required.
5. Disconnect the solenoid wires and battery cables.
6. Remove starter motor mounting bolts and stud.

NOTE: If the starter is mounted using shims, note their position prior to removal and ensure that they are repositioned properly upon installation.

To install:

7. Install starter motor mounting bolts and stud. Tighten to 32 ft. lbs. (43 Nm).
8. If removed, install the 4 flywheel inspection cover bolts.
9. Connect solenoid wires and battery cable to the starter.
10. If equipped, install starter motor shield.
11. Lower the vehicle.
12. Connect the negative battery cable.

EMISSION CONTROLS

Please refer to "Emission Controls" in the Unit Repair section for system maintenance procedures. Due to the complex nature of modern electronic engine control systems, comprehensive diagnosis and testing procedures fall outside the confines of this repair manual. For complete information on diagnosis, testing and repair procedures concerning all modern engine and emission control systems, please refer to "Chilton's Guide to Fuel Injection and Electronic Engine Controls".

FUEL SYSTEM

Fuel System Service Precautions

Safety is the most important factor when performing not only fuel system maintenance but any type of maintenance. Failure to conduct maintenance and repairs in a safe manner may result in serious personal injury or death. Maintenance and testing of the vehicle's fuel system components can be accomplished safely and effectively by adhering to the following rules and guidelines.

• To avoid the possibility of fire and personal injury, always disconnect the negative battery cable unless the repair or test procedure requires that battery voltage be applied.

• Always relieve the fuel system pressure prior to disconnecting any fuel system component (injector, fuel rail, pressure regulator, etc.), fitting or fuel line connection. Exercise extreme caution whenever relieving fuel system pressure to avoid exposing skin, face and eyes to fuel spray. Please be advised that fuel under pressure may penetrate the skin or any part of the body that it contacts.

• Always place a shop towel or cloth around the fitting or connection prior to loosening to absorb any excess fuel due to spillage. Ensure that all fuel spillage (should it occur) is quickly removed from engine surfaces. Ensure that all fuel soaked cloths or towels are deposited into a suitable waste container.

• Always keep a dry chemical (Class B) fire extinguisher near the work area.

• Do not allow fuel spray or fuel vapors to come into contact with a spark or open flame.

• Always use a backup wrench when loosing and tightening fuel line connection fittings. This will prevent unnecessary stress and torsion to fuel line piping. Always follow the proper torque specifications.

• Always replace worn fuel fitting O-rings with new. Do not substitute fuel hose or equivalent where fuel pipe is installed.

RELIEVING FUEL SYSTEM PRESSURE

1. Disconnect the negative battery cable.
2. Loosen fuel filler cap to relieve tank vapor pressure. Do not tighten until service has been completed.
3. Connect a suitable fuel pressure gauge to fuel pressure connection on

fuel rail assembly. Wrap a shop towel around fitting while connecting gauge to avoid spillage.

4. Install bleed hose into an approved container and open valve to bleed system pressure. Fuel connections are now safe for servicing.

5. Drain any fuel into an approved container.

NOTE: When repairs to the fuel system have been completed, start the engine and check all connections that were loosened for possible leaks.

Fuel Filter

REMOVAL & INSTALLATION

1. Disconnect the negative battery cable.
2. Relieve fuel system pressure.
3. Raise and safely support the vehicle.
4. Remove bolt retaining fuel filter bracket or open fuel filter bracket release tabs, as required.
5. If equipped with quick-connect fuel fittings, perform the following procedures:

   a. Grasp filter and 1 fuel line fitting. Twist quick-connect fitting ¼ turn in each direction to loosen any dirt within fitting. Repeat for other fuel line fitting.

   b. Using compressed air, blow out dirt from quick-connect fittings at both ends of fuel filter.

   c. Remove quick-connect fittings by squeezing plastic tabs of male end connector and pull connection apart. Repeat for other fitting.

6. If equipped with threaded fuel fittings, perform the following:

   a. Using a backup wrench on fuel filter, loosen fuel line retaining nut. Repeat for other fuel line fitting.

   b. Using compressed air, blow out dirt from fuel line fittings at both ends of fuel filter.

   c. Back off nut completely so fuel line can be separated from filter at both ends.

7. Remove fuel filter.

To install:

NOTE: Before installing a new filter, always apply a few drops of clean engine oil to both ends of the filter. This will ensure proper reconnection and prevent a possible fuel leak.

8. Remove protective caps from new filter.
9. If equipped with quick-connect fuel fittings, perform the following:

   a. Install new plastic connector retainers on filter inlet and outlet tubes.

   b. Install filter in retainer noting direction of flow indicated on filter.

   c. Install quick-connect fittings by pushing connectors together to cause the retaining tabs/fingers to snap into place.

NOTE: Once installed, pull on both ends of each connection to make sure connection is secure.

10. If equipped with threaded fuel fittings, install new O-ring seals, install fuel lines into the filter.

NOTE: Use backup wrench when installing fuel lines into new filter to prevent filter O-ring or fuel line damage.

11. Install fuel filter into retainer and engage bracket tabs or install retainer bracket bolt, as required.
12. Lower vehicle.
13. Tighten fuel filler cap.
14. Connect negative battery cable.

NOTE: Before cranking the engine, turn ignition switch to the ON position for 2 seconds, then turn switch OFF for 5 seconds. Again turn ignition switch to ON position and check for fuel leaks.

Electric Fuel Pump

The fuel pump is mounted in the tank and is part of the fuel tank meter assembly. The tank must be removed from the vehicle in order to service the fuel pump and fuel tank meter assembly.

PRESSURE TESTING

1. Connect a suitable fuel pressure gauge to the fuel pressure test fitting on the fuel rail assembly. Wrap a shop towel around the fuel pressure tap to absorb any fuel leakage that may occur when installing the gauge.
2. Turn ignition switch to the **ON** position. Check to see that pressure is within specification.
3. Turn ignition switch **OFF**. Pressure should not leak down with fuel pump **OFF**.
4. Pressure at idle should be 3–10 psi (21–69 kPa) lower than static pressure.

REMOVAL & INSTALLATION

1. Disconnect the negative battery cable.
2. Relieve fuel system pressure.
3. Remove fuel filler cap to release fuel tank vapors. Leave cap off until repairs are completed.

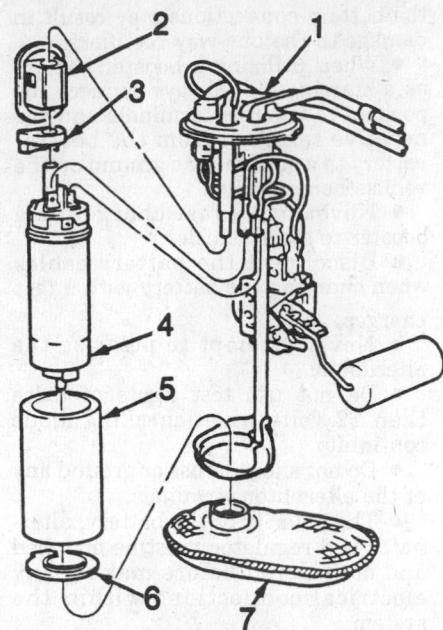

1. Fuel tank meter assembly
2. Pulsator
3. Bumper
4. Fuel pump
5. Sound isolator sleeve
6. Sound insulator
7. Filter strainer

Exploded view of the electric fuel pump/ sending unit assembly

---

— CAUTION —

Gasoline fuel vapors are extremely flammable. Ensure that fuel is stored in a container that can be properly sealed. Never store fuel in an open container. Store container in a safe place away from heat.

4. Remove fuel tank by performing the following:

   a. Drain fuel from the tank into an approved container for storage.

   b. Raise and safely support the vehicle.

   c. Remove rear stabilizer bar at links, pivot bar downward.

   d. Remove hoses and pipes from tank unit.

   e. Remove hoses at tank from filler and vent pipe.

   f. Disconnect tank unit harness from rear body harness.

   g. Support fuel tank and disconnect 2 fuel tank retaining straps.

   h. Remove tank from vehicle.

5. Remove sending unit, gasket and pump assembly by turning cam lock ring counterclockwise. Lift assembly from fuel tank and remove fuel pump from fuel tank sending unit.

6. Pull fuel pump up into attaching hose while pulling outward away from bottom support. Take care to prevent

damage to rubber insulator and strainer during removal. After pump assembly is clear of bottom support, pull pump assembly out of rubber connector for removal.

To install:

7. Push fuel pump assembly into attaching hose.

8. Install fuel tank sending unit and pump assembly into tank assembly. Use new O-ring seal during reassembly.

9. Install cam lock over assembly and lock by turning clockwise.

10. Support tank and position in vehicle. Install tank straps and secure with retaining bolts. Tighten to 25 ft. lbs. (33 Nm).

11. Connect tank unit harness to body harness.

12. Connect hoses to filler and vent pipes. Tighten clamps.

13. Connect hoses and pipes to tank unit.

14. Connect rear stabilizer bar to links. Tighten bolts to 42 ft. lbs. (58 Nm).

15. Lower vehicle.

16. Refill tank and install filler cap.

17. Connect negative battery cable.

18. Start engine and check for leaks.

Fuel Injection

IDLE SPEED ADJUSTMENT

Idle speed is automatically controlled by the ECM. Periodic adjustments are not required.

IDLE MIXTURE ADJUSTMENT

Idle mixture is automatically maintained by the ECM. Periodic adjustments are not required.

Fuel Rail

REMOVAL & INSTALLATION

Reatta, Riviera, Toronado and Trofeo

1. Disconnect the negative battery cable.

2. Properly relieve the fuel system pressure.

3. Using a shop towel to catch any fuel, disconnect the fuel supply and return lines from the inlet and outlet of the fuel rail. Use a backup wrench to avoid twisting the fittings on the fuel rail.

4. Disconnect the fuel injector electrical connectors.

5. Disconnect the vacuum line from the pressure regulator.

6. Remove the 4 bolts attaching the

fuel rail assembly to the intake manifold.

7. Carefully, remove the fuel rail assembly from the intake manifold.

NOTE: With the fuel rail removed, cover the injector openings to prevent the entry of dirt and other contaminants.

To install:

8. Install new the injector O-rings. Lubricate lightly with engine oil.

9. Carefully, install the fuel rail to the intake manifold. Seat each injector by hand.

10. Install the 4 bolts attaching the fuel rail to the intake manifold. Tighten to 7–14 ft. lbs. (10–20 Nm).

11. Connect the fuel supply and return lines to the fuel rail.

12. Connect the fuel injector electrical connectors.

13. Connect the vacuum line to the fuel pressure regulator.

14. Connect the negative battery cable.

15. Start the engine and check for fuel leaks.

Allante, Eldorado and Seville

1. Disconnect the negative battery cable.

2. Remove the air cleaner.

3. Properly relieve the fuel system pressure.

4. Remove the power steering pump.

5. Disconnect the vacuum line from the pressure regulator and base assembly.

6. Disconnect the accelerator cable, cruise control cable and bracket.

7. Disconnect the electrical connectors from the TPS, ISC, coolant and MAT sensors.

8. Disconnect the coolant hose to the thermostat housing.

NOTE: Wrap a shop cloth around the fuel lines to collect that leaks when disconnecting the fuel lines.

9. Disconnect the fuel feed line from the rear fuel rail assembly. Discard the O-ring.

10. Disconnect the fuel return line. Discard the O-ring.

11. Disconnect the EGR vacuum lines and remove the EGR valve.

12. Remove the 5 fuel rail support bracket attaching bolts.

13. Disconnect the front and rear fuel rail electrical connectors.

14. Carefully, remove the fuel rail from the intake manifold.

15. Remove the lower O-ring seal from the injectors and discard.

To install:

16. Install new injector O-rings. Lubricate lightly with engine oil.

17. Carefully, install the fuel rail assembly to the intake manifold.

18. Install the 5 fuel rail attaching bolts. Tighten to 18 ft. lbs. (24 Nm).

19. Connect the front and rear fuel rail electrical connectors.

20. Install the EGR valve and connect the vacuum lines.

21. Install new fuel feed line O-rings. Lubricate O-rings with petroleum based grease. Connect the fuel feed line. Using a backup wrench, tighten to 22 ft. lbs. (30 Nm).

22. Install new fuel return line O-rings. Lubricate O-rings with petroleum based grease. Connect the fuel return line. Using a backup wrench, tighten to 22 ft. lbs. (30 Nm).

23. Connect the coolant hose to the thermostat housing.

24. Connect the electrical connectors to the TPS, ISC, coolant and MAT sensors.

25. Connect the accelerator cable, and cruise control cable and bracket.

26. Connect the vacuum line to the pressure regulator and base assembly.

27. Install the power steering pump.

28. Install the air cleaner.

29. Connect the negative battery cable.

30. Start the engine and check for fuel leaks.

Fuel Injector

REMOVAL & INSTALLATION

NOTE: Care must be taken when removing injectors to prevent damage to the electrical connector pins on the injector and the nozzle. The injectors are serviced as a complete assembly only. Injectors are an electrical component and should not be immersed in any type of cleaner.

Sequential Fuel Injection (SFI)

1. Disconnect the negative battery cable.

2. Properly relieve the fuel pressure.

3. Remove the fuel rail.

4. On Allante, Eldorado and Seville, disconnect the injector electrical connector by pushing in the wire connector clip while pulling the connector body away from the injector.

5. Remove the injector retainer clip from the injector.

6. Separate the injector from the fuel rail.

7. Remove the injector O-rings and discard.

To install:

8. Lubricate new injector O-rings lightly and install on the injector.

9. If supplied, install new injector clip on the injector.

10. Install the fuel injector assembly into the fuel rail socket.

NOTE: On Allante, Eldorado and Seville, the electrical connectors should be facing the front of the engine for injectors 1–4 and the rear of the engine for injectors 5–8.

11. Connect the electrical connector to the injector assembly.

12. Install the fuel rail assembly.

13. Connect the negative battery cable.

14. Start the engine and check for fuel leaks.

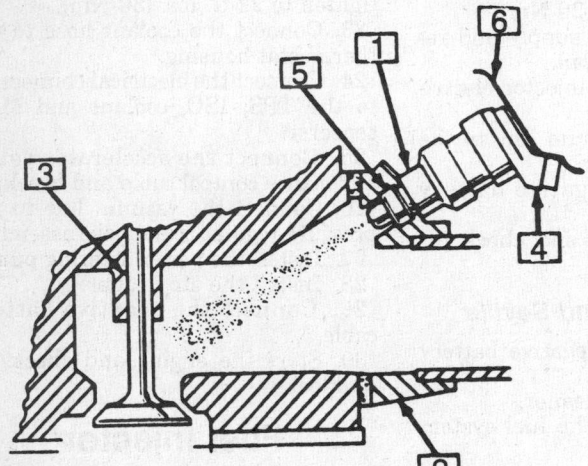

Typical port/sequential fuel injector installation

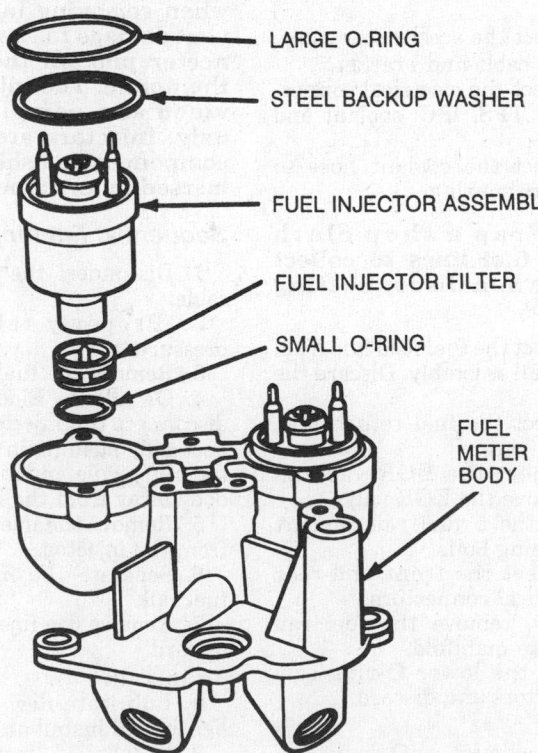

1. Fuel injector
2. Intake manifold
3. Intake valve
4. Electrical terminal
5. O-ring
6. Fuel rail

LARGE O-RING
STEEL BACKUP WASHER
FUEL INJECTOR ASSEMBLY
FUEL INJECTOR FILTER
SMALL O-RING
FUEL METER BODY

Throttle body fuel injector components

Throttle Body Injection (TBI)

1. Disconnect the negative battery cable. Remove the air cleaner.

2. Properly relieve the fuel system pressure.

3. Disconnect the electrical connector from the fuel injector(s) by squeezing both tabs together and pulling it straight up.

4. Remove the fuel meter cover-to-throttle body screws and the cover; be sure to note the position of the short screws. Allow the gasket to remain in place to prevent damage to the casting housing.

5. Using a small prybar and a ¼ in. rod, pry the fuel injector(s) from the throttle body; discard the O-rings.

To install:

6. Lubricate the new small O-ring with clean Dexron®II automatic transmission fluid. Push the new small O-ring onto the nozzle end of the injector pressing the ring up against the injector fuel filter.

7. Install the steel backup washer into the recess of the fuel meter body.

8. Lubricate the new large O-ring with clean Dexron®II automatic transmission fluid. Install the O-ring directly above the backup washer, pressing the O-ring down into the cavity recess. O-ring is located properly when it is flush with the fuel meter body casting surface.

NOTE: Do not attempt to reverse this procedure and install the backup washer and O-ring after the injector is located in the cavity. To do so will prevent the seating of the O-ring in the cavity recess which may result in a fuel leak.

9. Install the injector using a pushing/twisting motion to center the nozzle O-ring in the bottom of the injector cavity. Align the raised lug on the injector base with the notch cast into the fuel meter body. Push down on the injector making sure it is fully seated in the cavity. The injector installation is correct with the lug seated in the notch and the electrical terminals parallel to the throttle shaft in the throttle body.

10. Install the fuel meter cover.

11. Connect the injector electrical connector.

12. Install the air cleaner.

13. Connect the negative battery cable.

14. Start the engine and check for fuel leaks.

DRIVE AXLE

Halfshaft

REMOVAL & INSTALLATION

1. Remove the hub nut and washer.

2. Raise and safely support the vehicle. Remove the front wheel.

3. Remove the brake caliper and rotor.

4. Remove the stabilizer link from the control arm.

5. Remove the tie rod end-to-steering knuckle cotter pin and nut. Using a ball joint removal tool, separate the

tie rod end from the steering knuckle.

6. Remove the lower ball joint-to-steering knuckle cotter pin and nut. Using a ball joint removal tool, separate the lower ball joint from the steering knuckle.

7. Using a prybar and a wooden block, pry the halfshaft from the transaxle and suspend it on a wire.

NOTE: When removing the halfshaft, be careful not to allow the shaft to drop causing damage to the CV-joints. Do not allow the halfshaft to overextend because the Tri-Pot (S-plan) joint can disengage from the bearing blocks.

8. Using the halfshaft removal tool, press the halfshaft from the steering knuckle hub and remove it from the vehicle.

NOTE: If equipped with an anti-lock brake system, be careful not to damage the toothed sensor ring (on halfshaft) and the wheel speed sensor (on steering knuckle).

To install:

9. Install the drive axle into the transaxle. Verify that the drive axle snapring is properly seated by grasping the housing and pulling outboard.

NOTE: Do not pull on the drive axle.

10. Install the outer end of the drive axle into the hub and bearing assembly.

11. Install the lower ball joint stud to the steering knuckle.

12. Install the washer and new torque prevailing nut. Tighten to 183 ft. lbs. (245 Nm).

NOTE: To keep the halfshaft from turning, place a small drift pin in one of the rotor's slots.

13. Install the stabilizer link to the control arm.

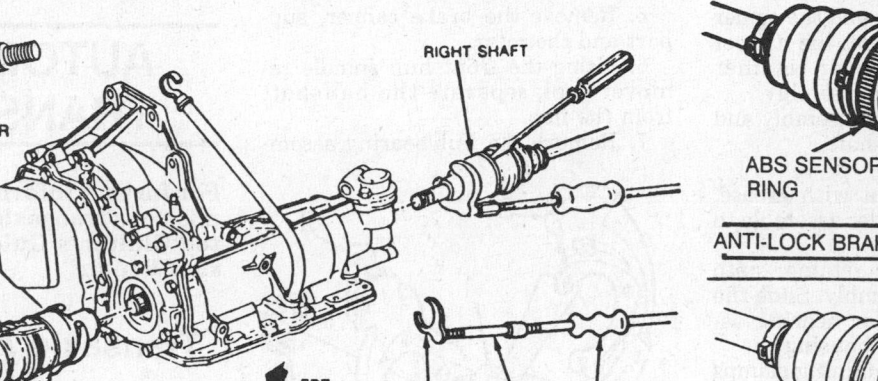

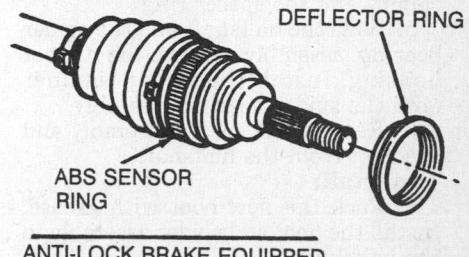

ABS SENSOR RING
DEFLECTOR RING
ANTI-LOCK BRAKE EQUIPPED

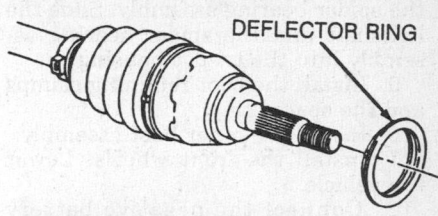

DEFLECTOR RING
STANDARD BRAKE EQUIPPED

Using a pry bar and special tools to pull the halfshafts from the transaxle. Make sure to support the axles at the center to avoid putting downward force on the outer joint

View of the 2 types of outer CV-joint assemblies and deflector rings

1. CV-joint outer race
2. CV-joint cage
3. CV-joint inner race
4. Balls
5. Race retaining ring
6. Seal retaining clamp
7. CV-joint seal
8. Seal retaining clamp
9. Left side halfshaft
10. Right side damper shaft assy.
11. Tri-Pot (S-plan) joint assy.

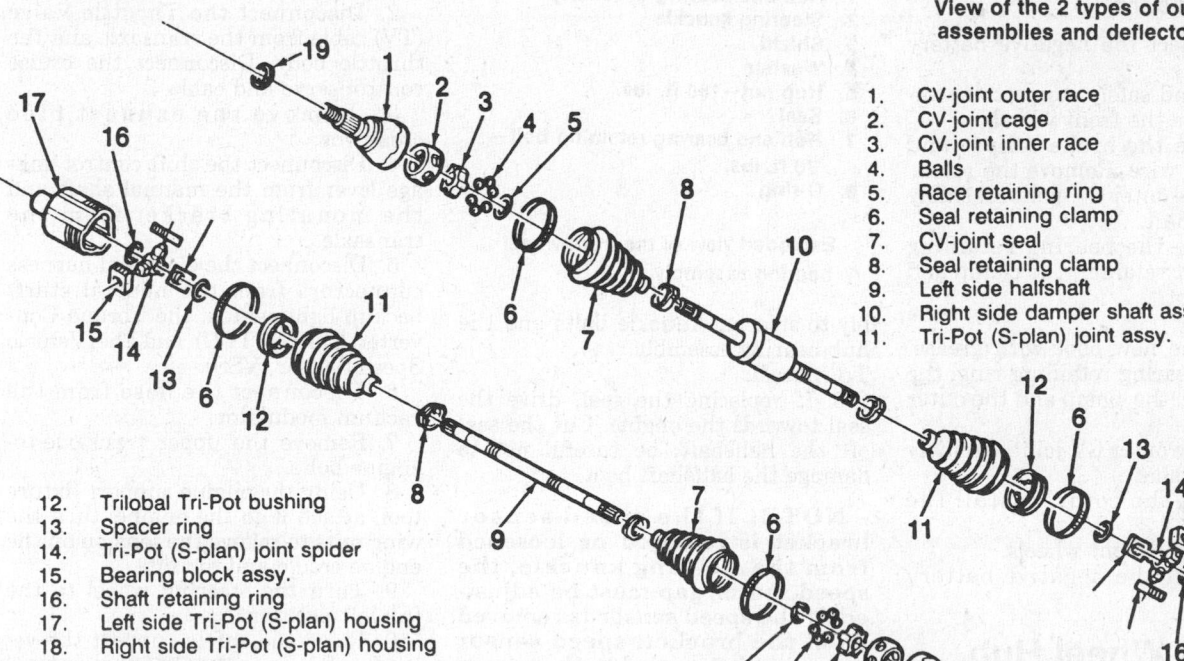

12. Trilobal Tri-Pot bushing
13. Spacer ring
14. Tri-Pot (S-plan) joint spider
15. Bearing block assy.
16. Shaft retaining ring
17. Left side Tri-Pot (S-plan) housing
18. Right side Tri-Pot (S-plan) housing
19. Deflector ring
20. Joint retaining ring

Exploded view of the halfshaft assemblies—Tri-Pot (S-plan)

14. Install the brake caliper and rotor.

15. Install the front wheel. Lower the vehicle

CV-Boot

REPLACEMENT

Inner (Inboard)

1. Disconnect the negative battery cable.

2. Raise and safely support the vehicle. Remove the front wheels.

3. Remove the outer boot assembly.

4. Remove the boot retaining clamps and the spacer ring.

5. Slide the halfshaft and the spider bearing assembly out of the tri-pot housing. Install the spider retainer onto the spider bearing assembly.

6. Remove the spider assembly and the boot from the halfshaft.

To install:

7. Pack the new boot with grease. Install the boot and spider assembly to the halfshaft.

8. Install the spider retainer onto the spider bearing assembly. Slide the halfshaft and the spider bearing assembly into the tri-pot housing.

9. Install the boot retaining clamps and the spacer ring.

10. Install the outer boot assembly.

11. Install the front wheels. Lower the vehicle.

12. Connect the negative battery cable.

Outer (Outboard)

1. Disconnect the negative battery cable.

2. Raise and safely support the vehicle. Remove the front wheels.

3. Remove the brake caliper and support on a wire. Remove the rotor.

4. Slide the outer CV-joint assembly off the halfshaft.

5. Remove the bearing retaining ring, the boot retainer, the clamp and the outer boot.

To install:

6. Pack the new boot with grease. Install the bearing retaining ring, the boot retainer, the clamp and the outer boot.

7. Slide the outer CV-joint assembly onto the halfshaft.

8. Install the rotor. Install the brake caliper.

9. Install the front wheels.

10. Connect the negative battery cable.

Front Wheel Hub, Spindle and Bearing

NOTE: The bearings are preadjusted and require no lubri-cation, maintenance or adjustment. There are darkened areas on the bearing assembly which are the result of a heat treating process.

REMOVAL & INSTALLATION

1. Raise and safely support the vehicle.

2. Place jackstands under the cradle and lower the vehicle slightly so the weight of the vehicle rests on the jackstands and not on the control arms.

3. Remove the wheel assembly.

4. Insert a drift punch into the rotor and remove the hub nut/washer.

5. Remove the brake caliper, support and the rotor.

6. Using the front hub spindle remover tool, separate the halfshaft from the hub.

7. Remove the hub/bearing assem-

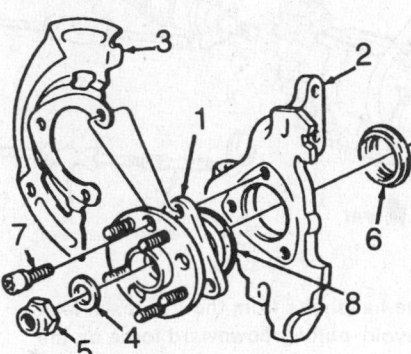

1. Hub and bearing assembly
2. Steering knuckle
3. Shield
4. Washer
5. Hub nut—180 ft. lbs.
6. Seal
7. Hub and bearing retaining bolt—70 ft. lbs.
8. O-ring

Exploded view of the front wheel bearing assembly

bly-to-steering knuckle bolts and the hub/bearing assembly.

To install:

8. If replacing the seal, drive the seal towards the engine. Cut the seal off the halfshaft; be careful not to damage the halfshaft boot.

NOTE: If the speed sensor bracket is removed or loosened from the steering knuckle, the speed sensor gap must be adjusted. If the speed sensor is removed from the bracket, speed sensor wax must be applied to the sensor before it is reinstalled in the bracket. Failure to apply the wax will permit corrosion and may result in sensor failure.

9. To install the new grease seal, lubricate the with wheel bearing grease and using the hub seal installer tool, install the seal.

10. Install the hub/bearing assembly-to-steering knuckle bolts and the hub/bearing assembly. Tighten the hub/bearing assembly-to-steering knuckle bolts to 70 ft. lbs. (95 Nm).

11. Install the halfshaft to the hub.

12. Install the rotor and caliper.

13. Install the hub nut and washer. Tighten to 183 ft. lbs. (245 Nm).

14. Install the wheel assembly.

15. Remove the jackstands and lower the vehicle.

AUTOMATIC TRANSAXLE

For further information on transmissions/transaxles, please refer to "Chilton's Guide to Transmission Repair".

Transaxle Assembly

REMOVAL & INSTALLATION

Reatta, Riviera, Toronado and Trofeo

1. Disconnect the negative battery cable. Remove the air intake duct.

2. Disconnect the Throttle Valve (TV) cable from the transaxle and the throttle body. Disconnect the cruise control servo and cable.

3. Remove the exhaust pipe crossover.

4. Disconnect the shift control linkage lever from the manual shaft and the mounting bracket from the transaxle.

5. Disconnect the electrical harness connectors from the neutral start/backup light switch, the Torque Converter Clutch (TCC) and the Vehicle Speed Sensor (VSS).

6. Disconnect the hose from the vacuum modulator.

7. Remove the upper transaxle-to-engine bolts.

8. Using the engine support fixture tool, attach it to the engine, turn the wing nuts to relieve the tension on the engine cradle and mounts.

9. Turn the steering wheel to the full left position.

10. Raise and safely support the vehicle. Remove both from wheel assemblies.

11. Using the halfshaft seal protector tool, install one on each halfshaft. Remove both front ball joint-to-steering

knuckle nuts and separate the control arms from the steering knuckles.

12. Using a medium prybar, pry the halfshaft from the transaxle and support it on a wire. Do not remove the halfshaft from the steering knuckle.

NOTE: When removing the halfshaft, be careful not to damage the seal lips.

13. Remove the right rear transaxle-to-frame nuts, the left rear transaxle mount-to-transaxle bolts and the right rear transaxle mount.
14. Remove the stabilizer shaft from the left control arm.
15. Remove the flywheel cover bolts and the cover.
16. Matchmark the torque converter-to-flywheel bolts for reinstallation purposes. Remove the torque converter-to-flywheel bolts and push the torque converter back into the transaxle.
17. Remove the partial frame-to-main frame bolts, the partial frame-to-body bolts and the partial frame.
18. Disconnect and plug the oil cooler tubes from the transaxle.
19. Remove the lower transaxle-to-engine bolts.

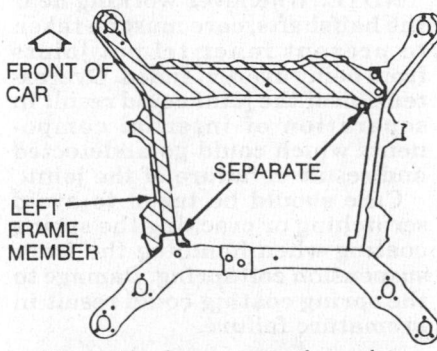

View of the frame separation points

NOTE: One bolt is located between the engine and the transaxle case and is positioned in the opposite direction.

20. Lower the transaxle from the vehicle. Be careful not to damage the hoses, lines and wiring.
To install:
21. Raise transaxle into position. Install the lower transaxle bolts.

NOTE: Ensure that the opposite-facing bolt is reinstalled in the proper direction.

22. Unplug and connect the oil cooler tubes to the transaxle.
23. Install the partial frame. Secure with the partial frame-to-body and the partial frame-to-main frame bolts.
24. Install the torque converter observing matchmarks made on disassembly and secure with torque converter-to-flywheel bolts. Tighten to 46

ft. lbs. (62 Nm). Install flywheel cover and secure with flywheel cover bolts.
25. Install left control arm and stabilizer shaft.
26. Install right rear transaxle mount, right rear transaxle-to-frame nuts and the left rear transaxle mount-to-frame nuts.
27. Install halfshaft into transaxle.

NOTE: When installing halfshafts, be sure not to damage seals.

28. Connect the control arms to the transaxle and secure with both front ball joint-to-steering knuckle nuts.
29. Install both wheel and tire assemblies.
30. Lower the vehicle.
31. Remove engine support fixture tool.
32. Install the upper transaxle-to-engine bolts.
33. Connect vacuum modulator hose.
34. Connect electrical harness connectors to neutral start/backup light switch, Torque Converter Clutch (TCC) and the Vehicle Speed Sensor (VSS).
35. Connect shift control linkage lever to manual shaft and mounting bracket to transaxle.
36. Install exhaust crossover pipe.
37. Connect Throttle Valve (TV) cable to the transaxle and throttle body. Connect cruise control servo.
38. Install air intake duct. Connect negative battery cable.
39. Start engine and check for transaxle leaks. Refill as necessary.

Eldorado, Seville and Allante

1. Disconnect the negative battery cable. Remove the air cleaner assembly. Disconnect the transaxle throttle valve cable.
2. Remove the cruise control servo and bracket assembly. Disconnect the electrical connectors going to the distributor, oil pressure sending unit and transaxle.
3. Remove the bracket for the engine oil cooler lines.
4. Remove the shift linkage bracket from the transaxle and the manual shift lever from the manual shift shaft; leave the cable attached to the lever and bracket.
5. Remove the fuel line bracket and disconnect the neutral safety switch connector.
6. Remove the vacuum modulator.
7. Remove the throttle valve cable support bracket and engine oil cooler line bracket. Remove the bellhousing bolts except the left and right side bolts; note the bolt lengths and positions.
8. Remove the air injection reactor

crossover pipe fitting and reposition the pipe. Remove the radiator hose bracket and transaxle mount-to-bracket nuts.
9. Install an engine support fixture, noting the positions of the hooks.
10. Raise and safely support the vehicle.
11. Remove both front wheels, the right and left stabilizer link bolts. Remove the ball joint cotter pins and nuts and press the ball joints from the steering knuckles.
12. Remove the air conditioner splash shield and the mount cover for the forward most cradle insulator.
13. Remove the hose connections from the ends of the air injection reactor pipes. Remove the vacuum hoses and the wire loom from the clips at the front of the cradle.
14. Remove the engine mount and dampener-to-cradle attachments. Remove the transaxle mount-to-cradle attachments. Remove the wire loom clip from the transaxle mount bracket and lower the vehicle.
15. Using both left side support hooks on the engine support fixture to raise the transaxle 2 in. from its normal position. Raise and safely support the vehicle.
16. Remove the right front and left rear transaxle-to-cradle bolts and the left stabilizer mount bolts. Remove the foremost cradle mount insulator bolt and the left cradle member, separate the right front corner first.
17. Remove the air injection reactor management valve/bracket assembly from the transaxle mount bracket and reposition the bracket to the transaxle stud bolts.
18. Lower the vehicle. Lower the transaxle to its normal position to gain access to the transaxle mounting bracket. Remove the mounting bracket.
19. Raise and safely support the vehicle. Remove the right rear transaxle mount-to-transaxle bracket. Remove the engine-to-transaxle brace bolts that pass into the transaxle VSS connector.
20. Mark the relationship between torque converter and flexplate for reassembly in the same position. Remove the flywheel covers, then, remove the torque converter bolts, rotating the crankshaft with a socket wrench as necessary to gain access. Position a jack under the transaxle to support it.
21. Remove the left and right bellhousing bolts; note the bolt lengths and positions.

NOTE: Access may be gained through the right wheelhouse opening to remove the bolt on the

right side; use a 3 foot long socket extension to reach it.

22. Disconnect the oil cooler lines at the transaxle, drain them and plug the openings. Then, install halfshaft boot seal protectors and disconnect the halfshafts at the transaxle. Suspend the halfshafts aside and remove the transaxle.

To install:

23. Install the transaxle. Remove the halfshaft boot seal protectors and connect the halfshafts to the transaxle. Connect the oil cooler lines at the transaxle.

24. Install the left and right bellhousing bolts in their original locations. Tighten the bellhousing bolts to 55 ft. lbs. (75 Nm).

25. Connect the torque converter to the flywheel, observing the matchmarks made upon removal. Install the converter-to-flexplate bolts and tighten to 46 ft. lbs. (63 Nm). Install the flywheel covers.

26. Install the engine-to-transaxle brace bolts that pass into the transaxle VSS connector. Install the right rear transaxle mount-to-transaxle bracket. Lower the vehicle.

27. Install the air injection reactor management valve/bracket assembly from the transaxle mount bracket.

28. Install the foremost cradle mount insulator bolt and the left cradle member. Install the right front and left rear transaxle-to-cradle bolts and the left stabilizer mount bolts.

29. Install the engine mount and dampener-to-cradle attachments. Install the transaxle mount-to-cradle attachments. Install the wire loom clip from the transaxle mount bracket and lower the vehicle.

30. Install the hose connections to the ends of the AIR pipes. Install the vacuum hoses and the wire loom to the clips at the front of the cradle.

31. Install the air conditioner splash shield and the mount cover for the forward most cradle insulator.

32. Install the ball joint cotter pins and nuts and press the ball joints to the steering knuckles. Tighten the ball joint nuts to 81 ft. lbs. (110 Nm). Install both front wheels, the right and left stabilizer link bolts.

33. Lower the vehicle.

34. Remove the engine support fixture.

35. Install the air injection reactor crossover pipe fitting and reposition the pipe. Install the radiator hose bracket and transaxle mount-to-bracket nuts.

36. Install the throttle valve cable support bracket and engine oil cooler line bracket. Install the bellhousing bolts. Tighten the bellhousing bolts to 55 ft. lbs. (75 Nm).

37. Install the vacuum modulator.

38. Install the fuel line bracket and connect the neutral safety switch connector.

39. Install the shift linkage bracket to the transaxle and the manual shift lever to the manual shift shaft.

40. Install the bracket for the engine oil cooler lines.

41. Install the cruise control servo and bracket assembly. Connect the electrical connectors going to the distributor, oil pressure sending unit and transaxle.

42. Install the air cleaner assembly. Disconnect the transaxle throttle valve cable. Connect the negative battery cable.

43. Adjust the transaxle valve cable and the shift linkage. Refill the transaxle to the proper level. Start engine and allow to come to normal operating temperature. Check transaxle fluid level and adjust as necessary.

SHIFT LINKAGE ADJUSTMENT

F7 Transaxle

1. Place the console shift lever in the **N** position.
2. Compress the lock tabs and pull the lock clip up.
3. Position the transaxle detent lever in **N**. Locate the neutral position by rotating the transaxle detent lever clockwise from the **P** position, through **R** to **N**. Verify the console shift lever is in **N**.
4. Push the lock clip down into the locked position.

Except F7 Transaxle

1. Move the shift lever to the **N** position. Neutral can be found by rotating the selector shaft clockwise from **P** through **R** to **N**.
2. Place the shift control assembly in **N**.
3. Push the cable adjuster tab to adjust the cable in the cable mounting bracket.

THROTTLE LINKAGE ADJUSTMENT

1. With the engine stopped, depress the accelerator pedal fully and have an assistant check the throttle body for wide open throttle.

NOTE: If the throttle body cannot achieve full throttle, repair the accelerator system.

2. At the engine end of the TV cable, depress and hold-down the metal readjust tab, move the slider until it stops against the fitting and release the readjustment tab.

3. Rotate the throttle lever, by hand, to it's full travel position.

4. The slider must move, ratchet, toward the lever when the lever is rotated to it's full travel position.

FRONT SUSPENSION

MacPherson Strut

REMOVAL & INSTALLATION

1. Disconnect the negative battery cable.
2. Remove nut attaching top of strut assembly to body.
3. If equipped, remove electrical connector from top of strut.
4. Raise and safely support the vehicle.
5. Remove tire and wheel assembly.

NOTE: Whenever working near the halfshafts, care must be taken to prevent inner tri-pot joints from being overextended. Overextension of the joint could result in separation of internal components which could go undetected and result in failure of the joint.

Care should be taken to avoid scratching or cracking the spring coating when handling the front suspension coil spring. Damage to the spring coating could result in premature failure.

6. In order to reassemble the knuckle and strut in the same relationship, make the following scribe marks:

   a. Using a sharp tool, scribe the inboard surface of the strut along the upper knuckle radius.
   b. Scribe the knuckle along the lower curve of the strut.
   c. Scribe mark across the strut and knuckle interface.

7. Remove brake line bracket from strut.
8. Remove stabilizer link from strut.
9. Remove strut-to-knuckle bolts and support knuckle with wire.
10. Remove strut from vehicle.

To install:

11. Install strut while aligning scribe marks.
12. Install strut-to-knuckle bolts.
13. Install stabilizer link to strut.
14. Install brake line bracket to strut.

15. Install nuts attaching top of strut to body. Tighten stabilizer link nuts to 48 ft. lbs. (65 Nm). Tighten strut assembly-to-body nuts to 18 ft. lbs. (24 Nm). Tighten steering knuckle-to-strut nuts to 136 ft. lbs. (184 Nm).

16. Install tire and wheel assembly.

17. Lower vehicle.

18. Tighten wheel mounting nuts to 100 ft. lbs. (140 Nm).

19. If equipped, connect electrical connector to top of strut.

20. Connect negative battery cable.

Lower Ball Joints

INSPECTION

1. Raise and safely support the vehicle. Install jackstands under both lower control arms as far outboard as possible.

2. Lower the vehicle onto the jackstands so the downward tension exerted by the stabilizer bar is relieved.

3. Install a dial indicator and clamp the assembly to the lower control arm.

4. Position the dial indicator plunger tip against the knuckle arm. Zero the dial indicator gauge.

5. Measure the axial travel of the knuckle arm with respect to the control arm, by raising and lowering the wheel using a prybar under the center of the tire.

6. During the measurement, if the axial travel of the control arm is 0.030 in. or more, relative to the knuckle arm, the ball joint should be replaced.

REMOVAL & INSTALLATION

1. Raise and safely support the vehicle.

2. Place jackstands under cradle and lower vehicle slightly so weight of the vehicle rests on the jackstands and not on the control arms.

3. Remove tire and wheel assembly.

4. Install a suitable outer CV-joint boot protector.

5. Remove stabilizer bar insulators, retainers, spacer and bolt.

6. Remove ball joint from knuckle.

NOTE: If equipped with anti-lock brakes, ensure that there is enough clearance between the ball joint stud and speed sensor ring. If not remove the halfshaft hub nut. Install special tool J-28733 or equivalent halfshaft remover. Tighten tool until halfshaft moves inboard enough to provide clearance for ball joint removal.

7. Drill out 3 rivets retaining ball joint starting with ¼ in. drill bit and finishing with ½ in. drill bit.

8. Remove ball joint.

To install:

9. Install new ball joint into control arm.

10. Install ball joint bolts.

11. Connect ball joint to knuckle. Tighten ball-joint bolts to 50 ft. lbs. (68 Nm). Tighten ball joint nut to 7 ft. lbs. (10 Nm). Tighten nut an additional ½ turn (3 flats.)

NOTE: When tightening nut, a minimum torque of 48 ft. lbs. (65 Nm) must be obtained. If 48 ft. lbs. (65 Nm) is not obtained, inspect for stripped threads. If threads are satisfactory, replace ball joint and knuckle. If required, turn the nut up to an additional ⅛ of a turn to allow for installation of the cotter pin. Bend both ends of the cotter pin.

12. If removed, tighten the hub nut to 183 ft. lbs. (245 Nm), to assure proper bearing clamp load.

13. Remove CV-joint boot protector.

14. Install tire and wheel assembly.

15. Raise vehicle enough to allow removal or jackstands.

16. Lower vehicle. Tighten wheel nuts to 100 ft. lbs. (140 Nm).

Lower Control Arms

REMOVAL & INSTALLATION

1. Raise and safely support the vehicle.

2. Place jackstands under cradle and lower vehicle slightly so weight of the vehicle rests on the jackstands and not on the control arms.

3. Remove the tire and wheel assembly.

NOTE: Care must be taken not to overextend Tri-Pot joints. Overextension of the joint could result in separation of internal components which could go undetected and result in failure of the joint.

4. Install a suitable CV-joint boot protector.

5. Remove stabilizer shaft insulator, retainers, spacer and bolt to control arm.

6. Lower ball joint from knuckle.

7. Remove control arm bushing bolt and front nut, retainer and insulator.

8. Remove control arm from frame.

To install:

9. Connect control arm to frame.

10. Install control arm bushing bolt and front nut, retainer and insulator. Do not tighten at this time.

11. Connect lower ball joint to knuckle.

12. Install stabilizer shaft insulator, retainers, spacer and bolt. Tighten stabilizer shaft nut and bolt to 13 ft. lbs. (17 Nm).

NOTE: Tighten ball joint nut to 7 ft. lbs. (10 Nm). Tighten nut an additional ½ turn (3 flats). When tightening nut a minimum torque of 48 ft. lbs. (65 Nm) must be obtained. If 48 ft. lbs. (65 Nm) is not obtained, inspect for stripped threads. If threads are satisfactory, replace ball joint and knuckle. If required, turn the nut up to an additional ⅛ of a turn to allow for installation of the cotter pin. Bend both ends of the cotter pin.

13. Remove outer CV-joint boot protector.

14. Install tire and wheel assembly.

15. Raise vehicle slightly so weight of vehicle is supported by the control arms. Tighten control arm bushing bolt to 100 ft. lbs. (140 Nm) or nut to 91 ft. lbs. (123 Nm). Tighten retainer to 52 ft. lbs. (70 Nm).

16. Remove jackstands and lower vehicle.

17. Tighten wheel nuts to 100 ft. lbs. (140 Nm).

Sway Bar

REMOVAL & INSTALLATION

1. Disconnect the negative battery cable.

2. Raise and safely support the vehicle.

3. Place jackstands under cradle and lower vehicle slightly so the weight of the vehicle rests on the jackstands and not on the control arms.

4. Remove right side wheel assembly.

5. Remove left and right insulators, retainers, spacers and bolts.

6. Remove left and right bracket bolts, brackets and insulators.

7. Remove exhaust pipe from rear manifold and move pipe up.

8. Remove stabilizer shaft.

To install:

9. Install stabilizer shaft.

10. Install exhaust pipe to rear manifold.

11. Install left and right insulators, brackets and loosely install bolts.

12. Install left and right insulators, retainers, spacers and bolts.

13. Center stabilizer on frame and check clearance. Tighten bracket to frame bolts to 33 ft. lbs. (45 Nm). Tighten nuts to 13 ft. lbs. (17 Nm).

14. Raise vehicle enough to allow for removal of jackstands.

15. Lower vehicle. Tighten wheel nuts to 100 ft. lbs. (140 Nm).

REAR SUSPENSION

MacPherson Strut

REMOVAL & INSTALLATION

1. Disconnect the negative battery cable.
2. Raise and safely support the vehicle.
3. Reinstall 2 wheel nuts to hold rotor on hub and bearing assembly.
4. Remove brake caliper and support with a length of wire.

NOTE: Do not allow caliper to hang by the brake hose unsupported.

5. Loosen knuckle pivot bolt on outboard end of control arm. Do not remove.
6. Remove upper strut rod cap, mounting nut, retainer and insulator.
7. Compress strut by hand and remove lower insulator.
8. Rotate strut and knuckle assembly outward by pivoting on knuckle pivot bolt.
9. Remove knuckle pinch bolt.
10. Remove strut from knuckle.

To install:
11. Position strut in knuckle. Strut must by fully seated in knuckle with tang on strut bottomed in knuckle slot.
12. Install knuckle pinch bolt. Tighten to 44 ft. lbs. (60 Nm).
13. Install lower insulator on strut and position strut rod in suspension support.
14. Install upper strut insulator, retainer and nut. Tighten upper strut nut to 65 ft. lbs. (88 Nm). Tighten knuckle pivot bolt to 59 ft. lbs. (80 Nm).
15. Install strut rod cap.
16. Install caliper and new caliper bracket mounting bolts.
17. Remove 2 wheel nuts previously installed to retain rotor.
18. Install wheel and tire assembly.
19. Lower vehicle. Tighten wheel nuts to 100 ft. lbs. (140 Nm).

Transverse-Mounted Leaf Spring

REMOVAL & INSTALLATION

NOTE: Removal and installa-tion of the transverse-mounted rear spring requires disassembly of either the left or right suspen-sion while leaving the other side intact. The spring may be re-moved from either side of the vehicle.

1. Disconnect the negative battery cable.
2. Raise and safely support the vehicle.
3. Remove tire and wheel assembly.
4. Disconnect height sensor link, if disassembling left control arm.
5. Remove stabilizer shaft mounting bolt at strut, if equipped with stabilizer.
6. Reinstall 2 wheel nuts to hold rotor on hub and bearing assembly.
7. Remove brake caliper and support with a length of wire.

NOTE: Do not allow caliper to hang by the brake hose unsupported.

8. Loosen knuckle pivot bolt on outboard end of control arm. Do not remove pivot bolt.
9. Support outboard end of control arm with a suitable lifting device to slightly compress spring.
10. Remove strut rod cap, mounting nut, retainer and upper insulator.
11. Slowly remove lifting device to relieve spring pressure.
12. Compress strut by hand and remove lower insulator.
13. Remove wheel speed sensor, if equipped with anti-lock brakes.
14. Remove inner control arm nuts.
15. While supporting the knuckle and control arm, remove inner control arm bolts and remove the control arm, knuckle, strut, hub and bearing and rotor from vehicle as an assembly.
16. Place a jackstand under the outboard end of spring.
17. Lower the vehicle so the weight loads the spring downward on jackstand.
18. Remove the 3 spring retainer bolts, retainer and lower insulator from retainer nearest the supported end of spring.
19. Slowly raise vehicle, allowing spring to deflect downward until spring no longer exerts force on the lifting device. Remove lifting device.
20. Remove spring retainer bolts, retainer and lower insulator from retainer on opposite side of vehicle.
21. Withdrawal spring from rear suspension support through disassembled side of vehicle suspension.
22. Remove upper spring insulators, as required.

NOTE: Inspect all spring insulators, insulator locating pads, retainers and control arm contact pads for cuts, cracks, tears or other damage. Replace worn or damaged parts.

To install:
23. Install spring insulators which were previously removed. Ensure that molded arrow on the insulator points toward the centerline of the vehicle when installing upper outboard insulators. Tighten center and upper outboard insulator nuts to 21 ft. lbs. (28 Nm).

NOTE: When positioning spring in suspension support, out-board and center insulator locat-ing bands must be centered on spring insulators. Failure to posi-tion spring correctly may result in reduced vehicle handling characteristics.

24. With spring properly located, install lower insulator and spring retainer on side of vehicle opposite the disassembled portion of suspension.
25. Place suitable lifting device under free end of spring.
26. Lower vehicle, allowing weight to load spring and deflect free end of spring into position in suspension support.
27. Install lower insulator and spring retainer on disassembled side of suspension support. Tighten spring retainer bolts to 21 ft. lbs. (28 Nm).
28. Raise the vehicle and remove spring lifting device.
29. Position the assembled control arm, knuckle, strut, hub and bearing and rotor assembly in suspension support and install inner control arm bolts and nuts. Do not tighten at this time.
30. Connect wheel sensor, if equipped with anti-lock brakes.
31. Install lower strut insulator and position strut rod in suspension support assembly.
32. Position suitable lifting under outboard end of lower control arm to slightly compress spring.
33. Install strut insulator, retainer and nut. Tighten upper strut nut to 65 ft. lbs. (88 Nm). Tighten knuckle pivot bolt to 59 ft. lbs. (80 Nm). Tighten inner control arm bolts to 66 ft. lbs. (90 Nm).
34. Remove lifting device.
35. Install strut rod cap.
36. Install stabilizer shaft mounting bolt, if equipped with stabilizer. Tighten stabilizer shaft mounting bolt to 43 ft. lbs. (58 Nm).
37. Remove 2 wheel nuts previously installed to retain motor.
38. Install caliper and new caliper mounting bracket bolts. Tighten caliper mounting bracket bolts to 83 ft. lbs. (113 Nm).
39. Connect height sensor link, if left

side of suspension was disassembled.

40. Install wheel and tire assembly.

41. Lower vehicle. Tighten wheel nuts to 100 ft. lbs. (140 Nm).

NOTE: Vehicle must have rear wheel alignment performed after removal and installation of rear spring.

Rear Control Arms

REMOVAL & INSTALLATION

1. Disconnect the negative battery cable.

2. Raise and safely support the vehicle.

3. Remove wheel and tire assembly.

4. If equipped with anti-lock brakes, remove speed sensor from knuckle.

5. Reinstall 2 wheel nuts to hold rotor on hub and bearing assembly.

6. Remove brake caliper and support with a length of wire.

NOTE: Do not allow caliper to hang by the brake hose unsupported.

7. Loosen knuckle pivot bolt on outboard end of control arm. Do not remove.

8. Remove upper strut rod cap, mounting nut, retainer and insulator.

9. Compress strut by hand and remove lower insulator.

10. While supporting the knuckle, remove knuckle pivot bolt and remove the knuckle, strut, hub, bearing and rotor from the vehicle as an assembly.

11. Remove both inner control arm bolts and remove control arm from vehicle.

To install:

12. Position control arm in vehicle and install both inner control arm bolts. Do not tighten bolts at this time.

13. Position the assembled knuckle, strut, hub and bearing and rotor assembly in control arm and install knuckle pivot bolt. Do not tighten bolt at this time.

14. Install lower strut insulator and position strut rod in suspension support.

15. Install upper strut insulator, retainer and nut. Tighten upper strut nut to 65 ft. lbs. (88 Nm). Tighten knuckle pivot bolt to 59 ft. lbs. (80 Nm). Tighten inner control arm bolts to 66 ft. lbs. (90 Nm).

16. Install strut rod cap.

17. Remove 2 wheel nuts previously installed to retain rotor.

18. Install caliper and new caliper bracket mounting bolts.

19. If equipped, install speed sensor to knuckle.

20. Install wheel and tire assembly.

21. Lower vehicle. Tighten wheel nuts to 100 ft. lbs. (140 Nm).

Rear Wheel Bearings

REMOVAL & INSTALLATION

1. Disconnect the negative battery cable.

2. Raise and safely support the vehicle.

3. Remove wheel and tire assembly.

4. If equipped with anti-lock brakes, remove speed sensor from knuckle.

5. Reinstall 2 wheel nuts to hold rotor on hub and bearing assembly.

6. Remove brake caliper and support with a length of wire.

NOTE: Do not allow caliper to hang by the brake hose unsupported.

7. Remove rotor.

8. Remove 4 hub mounting bolts.

9. Remove hub and bearing assembly.

To install:

10. Position hub and bearing assembly on knuckle.

11. Install 4 hub mounting bolts. Tighten to 52 ft. lbs. (70 Nm).

12. Install rotor.

13. Install caliper and new caliper bracket mounting bolts. Tighten to 83 ft. lbs. (113 Nm).

14. Install wheel and tire assembly.

15. Lower vehicle. Tighten wheel nuts to 100 ft. lbs. (140 Nm).

ADJUSTMENT

The hub and bearing are installed as an assembly. No periodic adjustment is required. If the bearing is found to have excessive play, the assembly must be replaced.

STEERING

Steering Wheel

——— CAUTION ———

Some vehicles are equipped with the Supplemental Inflatable Restraint (SIR) or air bag system. The SIR system must be disabled before performing service on or around SIR system components, steering column, instrument panel components, wiring and sensors. Failure to follow safety and disabling procedures could result in accidental air bag deployment, possible personal injury and unnecessary SIR system repairs.

REMOVAL & INSTALLATION

Reatta, Riviera, Toronado and Trofeo

1988–89

1. Disconnect the negative battery cable.

2. Remove the steering wheel-to-horn pad screws, located behind the steering wheel, and lift the pad from the steering wheel.

3. If the steering wheel is equipped with control buttons, disconnect the electrical connector(s).

4. Remove the steering wheel-to-shaft retainer, if equipped and nut.

5. Scribe an alignment mark on the steering wheel hub in line with the slash mark on the steering shaft.

6. Using the steering wheel puller, press the steering wheel from the steering shaft.

NOTE: If equipped with steering wheel controls, do not install steering wheel puller bolts beyond 5 turns as damage to electronic components behind the wheel may result.

To install:

7. Install the steering wheel to the shaft, observing the alignment mark made during removal.

8. Install the steering wheel-to-shaft retainer, if equipped and nut. Tighten the steering wheel-to-steering shaft nut to 35 ft. lbs. (47 Nm).

9. If the steering wheel is equipped with control buttons, connect the electrical connector(s).

10. Install the steering wheel-to-horn pad screws, located behind the steering wheel, and lift the pad from the steering wheel.

11. Connect the negative battery cable.

1990–92

1. Disconnect the negative battery cable. Ensure that ignition switch is in the **OFF** position.

2. Remove SIR fuse from fuse panel.

3. Remove left side sound insulator.

4. Remove left side courtesy light as required to ease removal of sound insulator.

5. Remove Connector Position Assurance (CPA) pin and yellow 2 way connector at the base of the steering column.

6. Loosen inflator module screws from back of steering wheel.

7. Remove horn contact by pushing slightly and twisting counterclockwise.

8. Remove Connector Position Assurance (CPA) pin and coil assembly connector from inflator module.

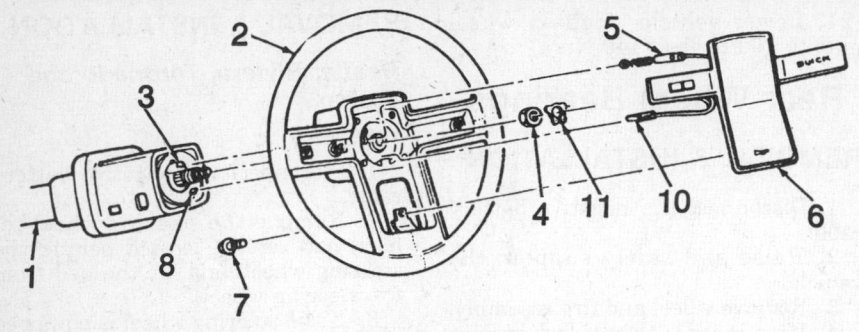

LEATHER WHEEL

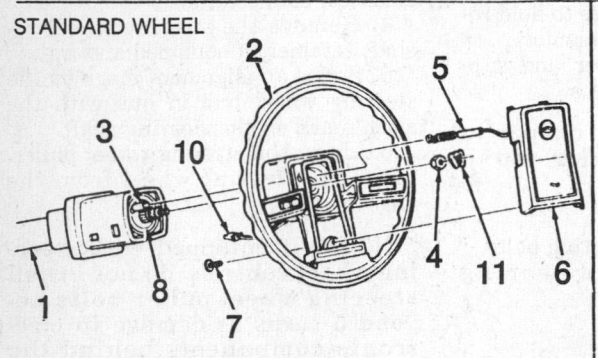

STANDARD WHEEL

REMOVING STEERING WHEEL

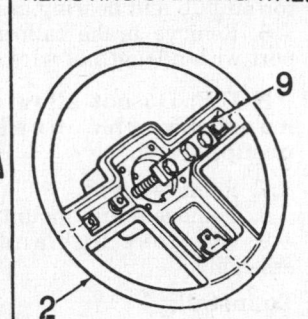

1. Steering column
2. Steering wheel
3. Cam tower
4. Nut—35 ft. lbs.
5. Horn lead
6. Horn pad
7. Horn pad mounting screws—13 inch lbs.
8. Cruise control connector (column)
9. Steering Wheel Puller tool
10. Cruise control connector
11. Retainer

Exploded view of the steering wheel assembly—Reatta, Riviera, Toronado and Trofeo

9. Remove steering column shaft nut.

10. Remove steering wheel using a suitable steering wheel puller.

To install:

11. Feed SIR coil assembly lead through slot in steering wheel.

12. Install steering wheel onto column shaft.

13. Install column shaft nut. Tighten to 30 ft. lbs. (41 Nm).

14. Install horn contact, coil assembly connector and CPA to inflator module.

15. Install inflator module onto steering wheel, securing with 4 screws behind steering wheel. Tighten to 27 inch lbs. (3 Nm).

16. Connect negative battery cable.

17. Connect yellow 2 way connector and CPA pin at the base of the steering column.

18. Install fuse in fuse panel.

19. Install left side sound insulator and connect courtesy light.

Allante, Eldorado and Seville

1988–89

1. Disconnect the negative battery cable.

2. For the Allante, pry the horn trim pad from the steering wheel. For the Eldorado and Seville, remove the steering wheel-to-horn pad screws, located behind the steering wheel, and the horn trim pad. Remove the horn contact wire, ground connector and cruise control wiring connector.

3. Remove the telescope locking lever assembly-to-adjuster screws. Unscrew and remove the telescoping adjuster from the steering shaft.

4. Remove the telescoping lever assembly. Scribe an alignment mark on the steering wheel hub-in-line with the slash mark on the steering shaft.

5. Remove the steering wheel-to-steering shaft locknut. Using the steering wheel puller, press the steering wheel from the steering shaft.

NOTE: When removing the steering wheel, be sure to remove the cruise control wire from it.

To install:

6. Feed the cruise control wire through the steering wheel, align the matchmark and install the steering wheel-to-steering shaft locknut.

Tighten the steering wheel-to-steering shaft to 35 ft. lbs. (47 Nm).

NOTE: For ease of installation, fully extend the steering shaft and install the lock plate compressor screw tool, hand-tight; this will keep the shaft extended when installing the steering wheel. Feed the cruise control wire through the wheel.

7. Remove the tool and place the telescoping lever in the 5 o'clock position.

8. Thread the telescope adjuster assembly finger tight onto the shaft. Install the screws into the telescoping adjuster lever.

9. Move the adjuster lever all the way to the right. The steering wheel should move freely in and out. Move the adjuster lever to the left. The steering wheel should be locked in place with the telescope lever approximately ¼ in. from the left side of the shroud opening. The lever must not contact the shroud in the full locked position. Loosen and adjust the lever as required.

10. For the Allante, install the horn trim pad to the steering wheel. For the Eldorado and Seville, install the steering wheel-to-horn pad screws, located behind the steering wheel, and the horn trim pad. Install the horn contact wire, ground connector and cruise control wiring connector.

11. Connect the negative battery cable.

1990–92

1. Disconnect the negative battery cable. Ensure that ignition switch is in the **OFF** position.

2. Remove SIR fuse from fuse panel.

3. Remove left side sound insulator.

4. Remove left side courtesy light, as required, to ease removal of sound insulator.

5. Remove Connector Position Assurance (CPA) pin and yellow 2 way connector at the base of the steering column.

6. Loosen inflator module screws from back of steering wheel.

7. Remove horn contact by pushing slightly and twisting counterclockwise.

8. Remove Connector Position Assurance (CPA) pin and coil assembly connector from inflator module.

9. Remove steering column shaft nut.

10. Remove steering wheel using a suitable steering wheel puller.

To install:

11. Feed SIR coil assembly lead through slot in steering wheel.

12. Install steering wheel onto column shaft.

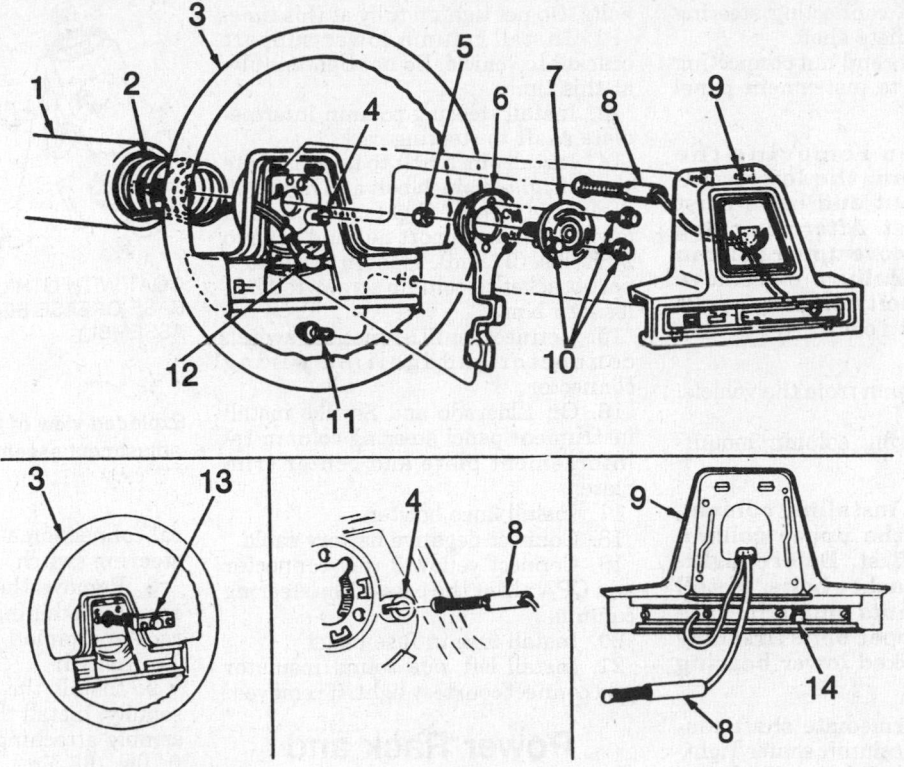

1.	Steering column	8.	Horn lead
2.	Telescoping spring	9.	Horn pad
3.	Steering wheel	10.	Telescope adjuster screws — 13 inch lbs.
4.	Cam tower	11.	Horn pad mounting screws — 13 inch lbs.
5.	Nut — 35 ft. lbs.	12.	Cruise control connector (column)
6.	Telescope lever	13.	Steering Wheel Puller tool No. J–23072
7.	Telescope adjuster	14.	Cruise control connector

Exploded view of the steering wheel assembly — Eldorado and Seville shown — Allante similar

13. Install column shaft nut. Tighten to 30 ft. lbs. (41 Nm).
14. Install horn contact, coil assembly, connector and CPA to inflator module.
15. Install inflator module onto steering wheel, securing with 4 screws behind steering wheel. Tighten to 27 inch lbs. (3 Nm).
16. Connect negative battery cable.
17. Connect yellow 2 way connector and CPA pin at the base of the steering column.
18. Install fuse in fuse panel.
19. Install left side sound insulator and connect courtesy light, if removed.

Steering Column

REMOVAL & INSTALLATION

1988–89

REATTA, RIVIERA, TORONADO AND TROFEO

1. Disconnect the negative battery cable.
2. Remove the left side sound insulator.

3. Remove the steering column trim cover.
4. Label and disconnect the electrical connectors from the steering column. Remove the wiring harness protector.
5. Remove the park lock cable from the ignition switch, if equipped.
6. Remove the lower column mounting bolts.

NOTE: On the Toronado, remove the pinch bolt.

7. If equipped with a column shifter, disconnect the shift linkage at the column.
8. Remove the upper steering column-to-instrument panel bolts and the column assembly from the vehicle.

To install:

9. Install the column assembly to the vehicle. Install the upper steering column-to-instrument panel bolts. Install the lower steering column-to-instrument panel bolts. Tighten the column bolts to 20 ft. lbs. (27 Nm).

NOTE: Failure to install the upper bolts first may result in a cracked lower bearing casting.

10. If equipped with a column shifter, connect the shift linkage to the column.
11 Install the lower column mounting bolts.

NOTE: On the Toronado, install the pinch bolt.

12. Install the park lock cable to the ignition switch, if equipped.
13. Connect the electrical connectors to the steering column. Install the wiring harness protector.
14. Install the steering column trim cover.
15. Install the left side sound insulator.
16. Connect the negative battery cable.

ALLANTE, ELDORADO AND SEVILLE

1. Disconnect the negative battery cable.
2. Remove left dash close-out panel.
3. Remove column wiring connector from left side hard shell grommet.
4. Remove park lock cable from ignition switch.

5. Remove bolt connecting steering shaft to intermediate shaft.

6. Remove bolts and nut connecting steering column to instrument panel bracket.

NOTE: When removing the steering column, the lower column bracket nut and bolts must be removed first. After removing lower bolts, remove upper column bracket bolts. Failure to remove lower nut and bolt first may result in a cracked lower bearing casting.

7. Remove column from the vehicle.

To install:

8. Install steering column mounting bolts and nut.

NOTE: When installing column, loosely install the upper column bracket bolts first. Before tightening upper bracket bolts, install lower column nuts and bolt. Failure to install upper bolts first may result in a cracked lower bearing casting.

9. Install intermediate shaft coupling to steering column shaft. Tighten intermediate shaft bolt to 35 ft. lbs. (47 Nm) and steering column bolts to 20 ft. lbs. (27 Nm).

10. Connect park lock cable to ignition switch.

11. Connect steering column wiring connector to left side hard shell grommet.

12. Install dash close-out panel.

13. Connect negative battery cable.

1990–92

1. Disconnect the negative battery cable. Ensure that ignition switch is in the **OFF** position.

2. Remove SIR fuse from fuse panel.

3. Remove left side sound insulator.

4. Remove left side courtesy light, as required, to ease removal of sound insulator.

5. Remove Connector Position Assurance (CPA) pin and yellow 2 way connector at the base of the steering column.

6. On Eldorado and Seville, remove center trim plate and instrument panel steering column reinforcing plate.

7. Remove knee bolster.

8. Disconnect ignition wiring connector and multi-function connector.

9. Remove pinch bolt from intermediate shaft.

10. Remove lower support bracket from vehicle. Remove upper column support from instrument panel and remove column from vehicle.

To install:

11. Install steering column into vehicle; support at upper bracket with 2

bolts. Do not tighten fully at this time.

12. Install column lower support bracket to vehicle. Do not tighten fully at this time.

13. Install steering column intermediate shaft to steering rack.

14. Install pinch bolt to intermediate shaft. Tighten pinch bolt and nut to 35 ft. lbs. (47 Nm). Tighten upper and lower column support nut and bolts to 20 ft. lbs. (27 Nm). Tighten lower support bracket-to-column screws to 12 ft. lbs. (16 Nm).

15. Connect multi-function switch connector and ignition wiring connector.

16. On Eldorado and Seville, install instrument panel steering column reinforcement plate and center trim plate.

17. Install knee bolster.

18. Connect negative battery cable.

19. Connect yellow 2 way connector and CPA pin at the base of the steering column.

20. Install fuse in fuse panel.

21. Install left side sound insulator and connect courtesy light, if removed.

Power Rack and Pinion

ADJUSTMENT

Rack Bearing Preload

NOTE: Make adjustment with front wheels raised and steering wheel centered. Be sure to check returnability of steering wheel to center after adjustment.

1. Disconnect the negative battery cable. Loosen the adjuster plug locknut.

2. Turn the adjuster plug clockwise until it bottoms and back it off 50–70 degrees.

3. While holding the adjuster plug, tighten the locknut to 50 ft. lbs. (70 Nm).

REMOVAL & INSTALLATION

1. Disconnect the negative battery cable.

2. Raise and safely support the vehicle.

3. Remove both front tire and wheel assemblies.

4. Remove the intermediate shaft lower pinch bolt.

5. Remove the tie rod ends from the steering knuckles.

6. Remove the line retainer. Disconnect and plug the return and pressure hose from the steering rack and pinion.

7. Label and disconnect the electri-

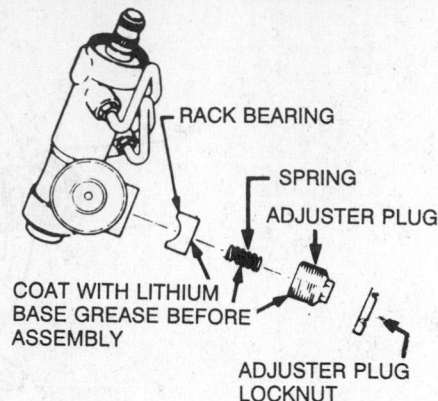

Exploded view of the power steering rack adjustment assembly

cal connection at the idle speed power steering switch.

8. Remove the rack and pinion assembly retaining bolts. Remove the rack and pinion assembly.

To install:

9. Install the rack and pinion assembly. Install the rack and pinion assembly attaching bolts. Tighten to 50 ft. lbs. (68 Nm).

10. Connect the electrical connection to the idle speed power steering switch.

11. Connect the return and pressure hose to the steering rack and pinion assembly. Install the line retainer.

12. Install the tie rod ends to the steering knuckles. Tighten nuts to 33 ft. lbs. (45 Nm).

13. Install the intermediate shaft lower pinch bolt. Tighten to 30 ft. lbs. (41 Nm).

14. Install both front tire and wheel assemblies.

15. Lower the vehicle.

16. Connect the negative battery cable.

17. Bleed the power steering system and check for leaks.

Power Steering Pump

REMOVAL & INSTALLATION

3.8L Engine

1. Disconnect the negative battery cable.

2. Remove the serpentine drive belt, the alternator bolts and the alternator.

3. Raise and safely support the vehicle.

4. Disconnect and plug the pressure and return lines from the pump.

5. Remove the rear pump adjustment bracket-to-pump nut.

6. Remove the alternator adjustment bracket and support brace.

7. Remove the rear pump adjust-

ment bracket and the pump assembly.

8. Remove the front pump adjustment bracket and the pulley.

To install:

9. Install the front pump adjustment bracket and the pulley.

10. Install the rear pump adjustment bracket and the pump assembly.

11. Install the alternator adjustment bracket and support brace.

12. Install the rear pump adjustment bracket-to-pump nut.

13. Connect the pressure and return lines to the pump.

14. Lower the vehicle.

15. Install the alternator, alternator bolts and the serpentine drive belt.

16. Connect the negative battery cable.

17. Refill the power steering pump reservoir. Bleed the power steering system.

4.1L, 4.5L and 4.9L Engines

1. Disconnect the negative battery cable.

2. Remove the serpentine drive belt, the power steering pump pulley.

3. Disconnect and plug the high pressure and feed lines from the pump.

4. Remove the power steering pump-to-bracket bolts and the pump.

To install:

5. Install the pump and the power steering pump-to-bracket bolts. Tighten to 30 ft. lbs. (41 Nm).

6. Connect the high pressure and feed lines to the pump.

7. Install the power steering pump pulley and the serpentine drive belt.

8. Connect the negative battery cable.

9. Refill the power steering pump reservoir. Bleed the power steering system.

BELT ADJUSTMENT

All accessories are driven by a single serpentine belt. The serpentine belt tension is maintained automatically by a spring tensioner. No adjustment is necessary or possible. If the belt tension is not within specification, replace the belt tensioner.

SYSTEM BLEEDING

1. Fill the fluid reservoir.

2. Let the fluid stand undisturbed for 2 minutes, crank the engine for about 2 seconds. Refill the reservoir, if necessary.

3. Repeat above steps until the fluid level remains constant after cranking the engine.

4. Raise and safely support the vehicle, until the wheels are off the ground. Start the engine and increase the engine speed to about 1500 rpm.

5. Turn the wheels lightly against the stops to the left and right, checking the fluid level and refilling, if necessary.

Outer Tie Rod Ends

REMOVAL & INSTALLATION

1. Disconnect the negative battery cable.

2. Raise and safely support the vehicle.

3. Remove cotter pin and hex slotted nut from outer tie rod assembly. Loosen jam nut.

4. Disconnect outer tie rod from steering knuckle using a suitable steering linkage separator tool.

5. Remove outer tie rod from inner tie rod.

To install:

6. Install outer tie rod assembly to inner tie rod. Do not tighten jam nut.

7. Connect outer tie rod to steering knuckle, hex slotted nut to outer tie rod stud. Tighten hex slotted nut to 35 ft. lbs. (50 Nm). Check for cotter pin slot alignment. Maximum torque is 45 ft. lbs. (60 Nm) to align slot. Do not back off for cotter pin insertion.

8. Install cotter pin into hole in tie rod stud.

9. Check toe and adjust by turning inner tie rod.

NOTE: Be sure rack and pinion boot is not twisted or puckered during toe adjustment.

10. Tighten jam nut against outer tie rod to 50 ft. lbs. (70 Nm).

BRAKES

For all brake system repair and service procedures not detailed below, please refer to "Brakes" in the Unit Repair section.

Master Cylinder

REMOVAL & INSTALLATION

1. Disconnect the negative battery cable.

2. If equipped with ABS, relieve the brake system pressure.

3. If equipped with a fluid level sensor, disconnect the electrical connector.

4. Disconnect and plug hydraulic lines. Drain the master cylinder.

5. Remove the master cylinder-to-power brake booster nuts and the master cylinder.

To install:

6. Install the master cylinder. Install the master cylinder-to-power brake booster nuts. Tighten the mounting nuts to 26 ft. lbs. (35 Nm).

7. Connect the hydraulic lines to the master cylinder.

8. If equipped with a fluid level sensor, connect the electrical connector.

9. Refill the master cylinder and bleed the system.

10. Connect the negative battery cable.

Proportioning Valve

REMOVAL & INSTALLATION

Diagonal Split System

NOTE: Individual proportioning valves are installed on the master cylinder outlets.

1. Disconnect the negative battery cable. Disconnect and plug the fluid lines from the proportioning valves.

2. Remove the proportioning valves and O-rings from the master cylinder.

To install:

3. Replace the O-rings and install the proportioning valves. Tighten the proportioning valve-to-master cylinder to 18–30 ft. lbs. (24–41 Nm). Refill the master cylinder reservoir with clean brake fluid. Bleed the brake system.

Teves Anti-Lock System

The Teves system uses a single proportioning valve located near the left rear wheel. The valve is not to be disassembled.

1. Disconnect the negative battery cable. Turn the ignition switch **OFF** throughout this procedure.

2. Using at least 50 lbs. (68 Nm) pressure on the brake pedal, depress the pedal at least 25 times; a noticeable change in pedal pressure will be noticed when the accumulator is discharged.

3. Disconnect the fluid lines from the proportioning valve and the valve from the vehicle.

To install:

4. Install the new proportioning valve and connect the brake fluid lines.

5. Bleed the brake system.

6. Connect the negative battery cable.

Bosch III System

The Bosch III system uses individual proportioning valves installed to the master cylinder. The valves are not to be disassembled.

1. Disconnect the negative battery

cable. Turn the ignition switch **OFF** throughout this procedure.

2. Using at least 50 lbs. pressure on the brake pedal, depress the pedal at least 25 times; a noticeable change in pedal pressure will be noticed when the accumulator is discharged.

3. Disconnect and plug the fluid line(s) from the proportioning valve(s).

4. Remove the proportioning valve(s) from the hydraulic unit.

To install:

5. Install the proportioning valve(s) to the hydraulic unit. Tighten the proportioning valve(s)-to-hydraulic unit to 11 ft. lbs. (15 Nm).

6. Connect the fluid line(s) to the proportioning valve(s).

7. Bleed the brake system.

8. Connect the negative battery cable.

Power Brake Booster

NOTE: This procedure is used only with the diagonal split system.

REMOVAL & INSTALLATION

1. Disconnect the negative battery cable. Remove the master cylinder-to-power booster nuts and move the master cylinder aside.

2. From inside the vehicle, detach the brake pushrod from the brake pedal.

3. Detach the vacuum hose at the vacuum cylinder.

4. Remove the nuts from the mounting studs which hold the unit to the dash panel. Remove the unit and clean it prior to installation.

To install:

5. Install the power brake booster and nuts to the mounting studs which hold the unit to the dash panel. Tighten the power booster-to-cowl nuts to 28 ft. lbs. (38 Nm).

6. Attach the vacuum hose to the vacuum cylinder.

7. From inside the vehicle, attach the brake pushrod to the brake pedal.

8. Install the master cylinder and master cylinder-to-power booster nuts. Tighten the master cylinder-to-

power booster nuts to 28 ft. lbs. (38 Nm).

9. Connect the negative battery cable.

10. Bleed the brake system.

Brake Caliper

REMOVAL & INSTALLATION

Front

1. Remove ⅔ of brake fluid from master cylinder assembly.

2. Raise and safely support the vehicle. Mark the relationship of the wheel to axle flange.

3. Remove wheel and tire assembly. Reinstall 2 wheel nuts to retain rotor.

4. Remove bolt attaching inlet fitting. Plug openings in caliper and pipe to prevent fluid loss and contamination.

5. Remove mounting bolts.

6. Remove caliper from rotor and mounting bracket.

To install:

7. Install caliper over rotor in

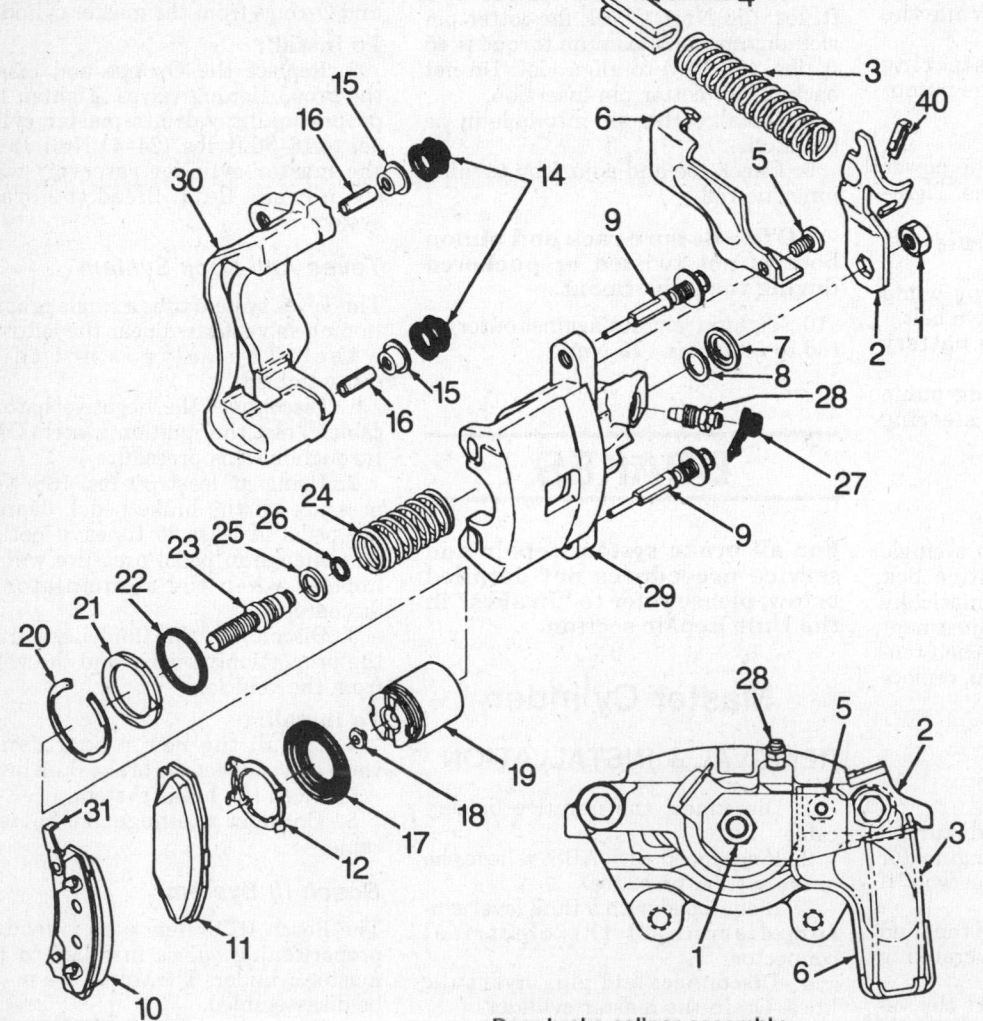

1.	Nut
2.	Park brake lever
3.	Return spring
4.	Damper
5.	Bolt
6.	Bracket
7.	Lever seal
8.	Anti-friction washer
9.	Mounting bolt
10.	Outboard shoe and lining
11.	Inboard shoe and lining
12.	Shoe retainer
14.	Bolt boot
15.	Support bushing
16.	Bushing
17.	Caliper piston boot
18.	Two-way check valve
19.	Piston assembly
20.	Retainer
21.	Piston locator
22.	Piston seal
23.	Actuator screw
24.	Balance spring and retainer
25.	Thrust washer
26.	Shaft seal
27.	Cap
28.	Bleeder valve
29.	Caliper housing
30.	Bracket
31.	Wear sensor
40.	Retaining clip

Rear brake caliper assembly

mounting bracket. Ensure that bolt boots are in place.

8. Lubricate entire shaft of mounting bolts with silicone grease. Tighten mounting bolts to 63 ft. lbs. (85 Nm).

9. Connect inlet fitting to 24 ft. lbs. (32 Nm).

10. Remove wheel nuts securing rotor to hub. Install wheels and tires, aligning previous marks.

11. Lower vehicle.

12. Tighten wheel nuts to 100 ft. lbs. (140 Nm).

13. Fill master cylinder to proper level with clean brake fluid.

14. Bleed caliper.

Rear

1. Remove ⅔ of brake fluid from master cylinder assembly.

2. Raise and safely support the vehicle.

3. Mark the relationship of wheel to axle flange. Remove wheel and tire assembly. Reinstall 2 wheel nuts to retain rotor.

4. Loosen tension on parking brake cable at equalizer.

5. Remove retaining clip from lever.

6. Remove cable, return spring and damper from return spring.

7. Remove locknut while holding lever.

8. Remove lever, lever seal and anti-friction washer.

9. Compress bottom piston into caliper bore to provide clearance between linings and rotor.

10. Reinstall anti-friction washer, lever seal (sealing bead against housing), lever and nut.

11. Remove bolt attaching inlet fitting. Plug openings in caliper and pipe to prevent fluid loss and contamination.

12. Remove mounting bolts.

13. Remove caliper from rotor and mounting bracket.

To install:

14. Install caliper over rotor in mounting bracket, making sure boots are in place.

15. Lubricate entire shaft of mounting bolts with silicone grease.

16. Install mounting bolts and tighten to 63 ft. lbs. (85 Nm).

17. Connect inlet fittings and tighten to 15 ft. lbs. (20 Nm).

NOTE: Ensure that parking brake components are clean and free of corrosion. Parts found to be corroded should be replaced. Do not try to polish corrosion away.

18. Install anti-friction washer.

19. Install lever seal with sealing bead against caliper housing. Lubricate seal prior to installation.

20. Install lever on actuator screw hex, with lever pointing down.

21. Install nut while holding rotated lever toward front of vehicle and tighten to 35 ft. lbs. (48 Nm). Rotate lever back against stop on caliper.

22. Install damper and return spring.

23. Connect parking brake cable and adjust.

24. Install retaining clip on lever so it retains parking brake cable from sliding out of the slot in lever.

25. Remove 2 wheel nuts securing rotor to hub. Install wheel and tire assembly aligning previous marks.

26. Lower vehicle.

27. Tighten wheel nuts to 100 ft. lbs. (140 Nm).

28. Fill master cylinder to proper level with clean brake fluid.

29. Bleed caliper.

Disc Brake Pads

REMOVAL & INSTALLATION

1. Remove disc brake caliper from mounting bracket and support with a length of wire. Do not allow caliper to hang by the brake line unsupported.

2. Remove outboard shoe and lining. Use a suitable tool to disengage shoe springs from holes in caliper housing.

3. Remove inboard shoe and lining, unsnapping shoe spring from piston.

4. If installing new shoe and linings, bottom piston in caliper bore using large pliers. Take care not to damage piston or piston boot.

5. Remove bushings from mounting bolt holes in bracket.

To install:

6. Install new bushings in mounting bolt holes in bracket. Lubricate bushings with silicone grease before installation.

7. Install inboard shoe and lining by snapping shoe retainer spring into piston. Shoe retainer spring is already staked to the inboard shoe. Shoe must lay flat against piston.

8. Install outboard shoe and lining by snapping shoe springs into holes in caliper housing. Wear sensor should be at the trailing edge of shoe during forward wheel rotation. Back of shoe must lay flat against caliper.

9. Install caliper.

10. Apply approximately 175 lbs. (778 N) of force 3 times to brake pedal to seat linings.

Brake Rotor

REMOVAL & INSTALLATION

1. Remove disc brake caliper from mounting bracket and support with a length of mechanics wire. Do not allow

caliper to hang by the brake line unsupported.

2. Remove 2 bolts retaining caliper mounting bracket, remove bracket and set aside.

3. Remove brake rotor taking care not to damage wheel nut threads.

To install:

4. Install the brake rotor.

5. Install the caliper mounting bracket and the 2 retaining bolts.

6. Install the caliper to the mounting bracket.

Parking Brake Cable

ADJUSTMENT

1. Lube the cables at the underbody rub points and at the equalizer hooks. Set and release the parking brake several times and check for free movement of all cables.

NOTE: With the ignition switch turned ON, the parking brake warning light should be OFF.

2. Set the parking brake pedal in the fully released position, raise and safely support the vehicle.

3. Hold the brake cable stud and tighten the equalizer nut until all cable slack is removed. Make sure the caliper levers are against the stops on the caliper housing; if not, loosen the cable until they are.

4. Operate the parking brake pedal several times to check the adjustment; the pedal should become firm after 3½ strokes.

5. Lower the vehicle and check that the caliper levers are still on their stops. If not, back off the parking brake adjuster until they are.

REMOVAL & INSTALLATION

The parking brake cable system consists of 4 separate cables: front, intermediate, left and right. The front and intermediate cables are joined at the adjuster screw. The left and right cables are joined to the intermediate cable through an equalizer.

1. Ensure that the parking brake is fully released.

2. Release the cable adjustment enough to allow removal of the desired cable(s).

NOTE: To prevent damage to threaded parking brake adjusting rod clean the exposed threads on each side of the nut and lubricate threads on the adjusting rod before turning the nut.

3. Remove old cable(s) and connect replacement cable(s).

4. Adjust new cable and check operation of parking brake.

Brake System Bleeding

Diagonal Split System

MASTER CYLINDER

1. Refill the master cylinder reservoir.
2. Push the plunger several times to force fluid into the piston.
3. Continue pumping the plunger until the fluid is free of the air bubbles.
4. Plug the outlet ports and install the master cylinder.

SYSTEM BLEEDING

1. Fill the master cylinder with fresh brake fluid. Check the level often during the procedure.
2. Starting with the right rear wheel, remove the protective cap from the bleeder, if equipped, and place where it will not be lost. Clean the bleed screw.

CAUTION

When bleeding the brakes, keep face away from the brake area. Spewing fluid may cause facial and/or visual damage. Do not allow brake fluid to spill on the car's finish; it will remove the paint.

---

3. If the system is empty, the most efficient way to get fluid down to the wheel is to loosen the bleeder about ½–¾ turn, place a finger firmly over the bleeder and have a helper pump the brakes slowly until fluid comes out the bleeder. Once fluid is at the bleeder, close it before the pedal is released inside the vehicle.

NOTE: If the pedal is pumped rapidly, the fluid will churn and create small air bubbles, which are difficult to remove from the system. These air bubbles will eventually congregate resulting in a spongy pedal.

4. Once fluid has been pumped to the caliper or wheel cylinder, open the bleed screw again, have the helper press the brake pedal to the floor, lock the bleeder and have the helper slowly release the pedal. Wait 15 seconds and repeat the procedure (including the 15 second wait) until no more air comes out of the bleeder upon application of the brake pedal. Remember to close the bleeder before the pedal is released inside the vehicle each time the bleeder is opened. If not, air will be induced into the system.

5. If a helper is not available, connect a small hose to the bleeder, place the end in a container of brake fluid and proceed to pump the pedal from inside the vehicle until no more air comes out the bleeder. The hose will prevent air from entering the system.

6. Repeat the procedure on remaining wheel cylinders in order:
   a. Left front
   b. Left rear
   c. Right front
7. Hydraulic brake systems must be totally flushed if the fluid becomes contaminated with water, dirt or other corrosive chemicals. To flush, bleed the entire system until all fluid has been replaced with the correct type of new fluid.
8. Install the bleeder cap(s) on the bleeder to keep dirt out. Always road test the vehicle after brake work of any kind is done.

Teves Anti-Lock Brake System

FRONT BRAKES

1. Turn the ignition switch **OFF** throughout this procedure.
2. Using at least 50 lbs. pressure on the brake pedal, depress the pedal at least 25 times; a noticeable change in pedal pressure will be noticed when the accumulator is discharged.
3. Remove the reservoir cap. Check and/or refill the master cylinder reservoir.
4. Using the bleeder adapter tool, install it onto the fluid reservoir.
5. Attach a diaphragm type pressure bleeder to the adapter and charge the bleeder to 20 psi.
6. Using a transparent vinyl tube, connect it to either front wheel caliper and insert the other end in a beaker ½ full of clean brake fluid.
7. Open the bleeder valve ½–¾ turn and purge the caliper until bubble free fluid flows from the hose.
8. Tighten the bleeder screw and remove the bleeder equipment.
9. Turn the ignition switch **ON** and allow the pump to charge the accumulator.
10. After bleeding, inspect the pedal for sponginess and the brake warning light for unbalanced pressure; if either of the conditions exist, repeat the bleeding procedure.

REAR BRAKES

1. Turn the ignition switch **OFF**.
2. Using at least 50 lbs. pressure on the brake pedal, depress the pedal at least 25 times; a noticeable change in pedal pressure will be noticed when the accumulator is discharged.
3. Check and/or refill the master cylinder reservoir.
4. Turn the ignition switch **ON** and allow the system to charge.

NOTE: The pump will turn OFF when the system is charged.

5. Using a transparent vinyl tube, connect it to a rear wheel bleeder valve and insert the other end in a beaker ½ full of clean brake fluid.

6. Open the bleeder valve ½–¾ turn and slightly depress the brake pedal for at least 10 seconds or until air is removed from the brake system. Close the bleeder valve.

NOTE: It is a good idea to check the fluid level several times during the bleeding operation. Remember, depressurize the system before checking the reservoir fluid.

7. Repeat the bleeding procedure for the other rear wheel.
8. After bleeding, inspect the pedal for sponginess and the brake warning light for unbalanced pressure; if either of the conditions exist, repeat the bleeding procedure.

Bosch III Anti-Lock Brake System

PUMP AND BOOSTER

1. Turn the ignition switch **OFF**.
2. Using at least 50 lbs. pressure on the brake pedal, depress the pedal at least 25 times; a noticeable change in pedal pressure will be noticed when the accumulator is discharged.
3. Check and/or refill the reservoir to the full mark.
4. Using a transparent vinyl hose, connect it to a pump bleeder screw and insert the other end in a beaker ½ full of clean brake fluid.
5. Loosen the bleeder screw ½–¾ turn. Turn the ignition switch **ON**; the pump should run forcing fluid from the hose. When the fluid becomes bubble-free, turn the ignition switch **OFF**, tighten the bleeder screw.
6. Move the transparent vinyl hose to the hydraulic unit bleeder screw. Loosen the bleeder screw ½–¾ turn. Turn the ignition switch **ON**; the pump should run forcing fluid from the hose. When the fluid becomes bubble-free, turn the ignition switch **OFF**, tighten the bleeder screw.
7. Disconnect the bleeder hose.
8. Turn the ignition switch **ON** and allow the hydraulic unit to charge; the pump should turn **OFF** after 30 seconds.

Anti-Lock Brake System Service

PRECAUTIONS

Failure to observe the following precautions may result in system damage.
- Before performing electric arc welding on the vehicle, disconnect the Electronic Brake Control Module (EBCM) and the hydraulic modulator connectors.
- When performing painting work

on the vehicle, do not expose the Electronic Brake Control Module (EBCM) to temperatures in excess of 185°F (85°C) for longer than 2 hrs. The system may be exposed to temperatures up to 200°F (95°C) for less than 15 min.

• Never disconnect or connect the Electronic Brake Control Module (EBCM) or hydraulic modulator connectors with the ignition switch ON.

• Never disassemble any component of the Anti-Lock Brake System (ABS) which is designated non-serviceable; the component must be replaced as an assembly.

• When filling the master cylinder, always use Delco Supreme 11 brake fluid or equivalent, which meets DOT-3 specifications; petroleum base fluid will destroy the rubber parts.

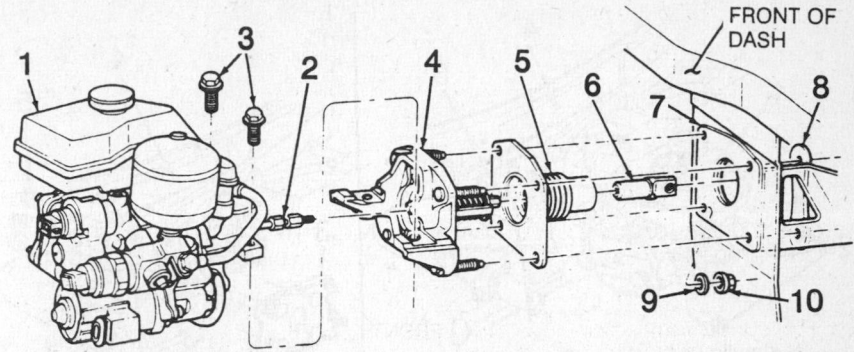

1. Hydraulic unit
2. Front pushrod half
3. Bolts – 37 ft. lbs.
4. Pushrod bracket assembly
5. Rubber boot
6. Rear pushrod half
7. Gasket
8. Reinforcement washer
9. Washer – used on lower right stud only
10. Nuts – 15 ft. lbs.

Exploded view of the anti-lock brake system hydraulic unit – Teves – except Allante

RELIEVING ANTI-LOCK BRAKE SYSTEM PRESSURE

NOTE: Unless otherwise specified, the hydraulic accumulator should be depressurized before disassembling any portion of the hydraulic system.

1. With the ignition switch in the OFF position, sensor block connector disconnected from the hydraulic unit or the negative battery cable disconnected, pump the brake pedal a minimum of 25 times using approximately 50 lbs. (222 N) of pedal force. When a noticeable change in pedal feel occurs, the accumulator is discharged.

2. When a definite increase in pedal effort is felt, stroke the pedal a few additional times.

Hydraulic Modulator

REMOVAL & INSTALLATION

Teves System

NOTE: The hydraulic accumulator is under pressure and must be depressurized before attempting to dismantle the system.

1. Disconnect the negative battery cable.
2. Firmly apply the parking brake.
3. Using at least 50 lbs. pressure on the brake pedal, depress the pedal at least 20 times; a noticeable change in pedal pressure will be noticed when the accumulator is discharged.
4. Disconnect the electrical connectors from the hydraulic brake unit.
5. Remove the pump-to-hydraulic unit bolt and move the unit aside to gain access to the hydraulic lines.
6. Using a back-up wrench, disconnect the hydraulic lines from the hydraulic unit.

7. From under the dash, disconnect the pushrod from the brake pedal.
8. Move the dust boot forward, past the pushrod hex and unscrew both pushrod halves.
9. Remove the hydraulic unit-to-pushrod bracket bolts and separate the hydraulic unit from the pushrod bracket; half of the pushrod will remain locked in the hydraulic unit.
10. Disassemble the master cylinder from the hydraulic unit.

To install:
11. Assemble the master cylinder to the hydraulic unit.
12. Install the hydraulic unit to the pushrod bracket and install the hydraulic unit-to-pushrod bracket bolts. Tighten the hydraulic unit-to-pushrod bracket bolts to 37 ft. lbs. (50 Nm).
13. Install the pushrod halves and move the dust boot into position.
14. From under the dash, connect the pushrod to the brake pedal.
15. Using a back-up wrench, connect the hydraulic lines to the hydraulic unit.
16. Install the hydraulic unit and pump-to-hydraulic unit bolt.
17. Connect the electrical connectors to the hydraulic brake unit. Bleed the brake system.
18. Release the parking brake.
19. Connect the negative battery cable.

Bosch III System

NOTE: The hydraulic accumulator is under pressure and must be depressurized before attempting to dismantle the system.

1. Disconnect the negative battery cable.
2. Firmly apply the parking brake.
3. Using at least 50 lbs. pressure on the brake pedal, depress the pedal at least 25 times; a noticeable change in

pedal pressure will be noticed when the accumulator is discharged.
4. On Allante, remove the air intake duct from the air cleaner and the throttle body, as required.
5. Remove the cross brace.
6. Disconnect the electrical connectors from the hydraulic brake unit and the pump motor. Using a siphon, remove as much fluid from the reservoir as possible.
7. Remove the pressure hose fitting (banjo bolt) from the hydraulic unit; be careful not to drop the fitting washers. Disconnect the return hose from the reservoir fitting.
8. Using a back-up wrench, disconnect the hydraulic lines from the hydraulic unit.
9. From under the dash, remove the driver's side sound insulator panel. From the pedal hub pin, remove the pushrod retainer and the foam washer.
10. From the engine compartment, remove the hydraulic unit-to-mounting adapter nuts.
11. Move the hydraulic unit to disengage the pushrod-to-pedal hub pin.
12. Remove the hydraulic unit from the vehicle.

To install:
13. Install the hydraulic unit to the vehicle.
14. Move the hydraulic unit to engage the pushrod-to-pedal hub pin.
15. Install the hydraulic unit-to-mounting adapter nuts. Tighten the hydraulic unit-to-mounting bracket nuts to 20 ft. lbs. (27 Nm).
16. From under the dash, install the pushrod retainer and the foam washer. Install the driver's side sound insulator panel.
17. From the engine compartment, connect the hydraulic lines to the hydraulic unit, using a back-up wrench.
18. Install the pressure hose fitting

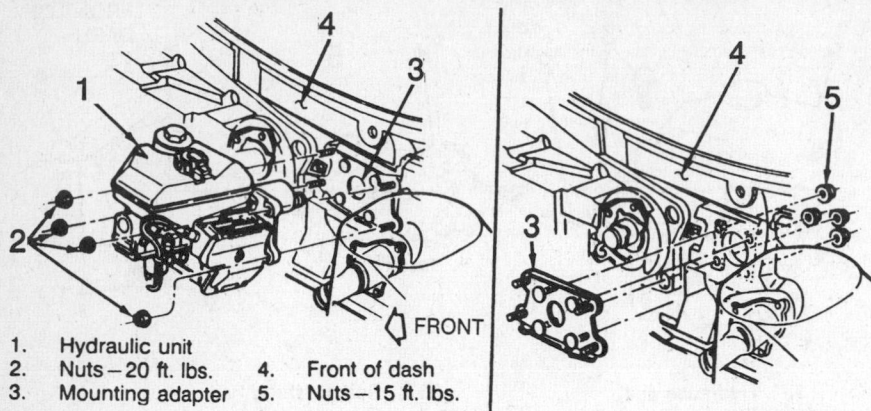

1. Hydraulic unit
2. Nuts — 20 ft. lbs.
3. Mounting adapter
4. Front of dash
5. Nuts — 15 ft. lbs.

⬇FRONT

View of the anti-lock brake system hydraulic unit and mounting bracket—Bosch III— Allante

(banjo bolt) to the hydraulic unit. Connect the return hose to the reservoir fitting.

19. Connect the electrical connectors to the hydraulic brake unit and the pump motor.
20. Refill the reservoir to the **FULL** mark.
21. Turn the ignition **ON** and allow the pump to charge the hydraulic accumulator. Bleed the brake system.
22. Install the cross brace.
23. On Allante, install the air intake duct to the air cleaner and the throttle body, if removed.
24. Release the parking brake.
25. Connect the negative battery cable.

Wheel Speed Sensor

REMOVAL & INSTALLATION

Front Sensor

1. Disconnect sensor connector from underhood area near strut tower.
2. Raise and safely support the vehicle.
3. Disengage sensor cable grommet from wheel house pass-through hole and remove sensor cable from retainers.
4. Remove sensor mounting bolt and remove sensor from vehicle.
To install:
5. Route sensor cable and install retainers. Install wheelhouse pass-through grommet.

NOTE: Proper installation of wheel speed sensor cables is critical to continued system operation. Be sure cables are installed in retainers. Failure to install cables in retainers properly may result in contact with moving parts and/or over-extension of cables, resulting in circuit damage.

6. Position sensor in knuckle and

install mounting bolt. Tighten mounting bolt to 9 ft. lbs. (12 Nm).

NOTE: If the wheel speed sensor is removed or replaced, the sensor body must be coated with a suitable anti-corrosion compound where the sensor comes in contact with the knuckle.

7. Lower vehicle.
8. Connect wheel speed sensor connector underhood.

Rear Sensor

1. Raise and safely support the vehicle.
2. Disconnect sensor connector and remove sensor cable from retainer brackets.
3. Remove sensor mounting bolt and remove sensor from vehicle.
To install:
4. Position sensor in knuckle and install mounting bolt. Tighten to 9 ft. lbs. (12 Nm).

NOTE: If the wheel speed sensor is removed or replaced, the sensor body must be coated with a suitable anti-corrosion compound where the sensor comes in contact with the knuckle.

5. Install wheel speed sensor cable in retainers.
6. Connect wheel speed sensor connector.
7. Lower vehicle.

Electronic Brake Control Module (EBCM)

REMOVAL & INSTALLATION

Except Allante

1. Disconnect the negative battery cable.

2. Open trunk lid. Remove left trunk carpet trim.
3. Remove velcro-attached cover concealing the EBCM.
4. Disconnect EBCM 35-pin connector.
5. Remove EBCM.
To install:
6. Install the EBCM.
7. Connect the 35-pin EBCM connector.
8. Install the velcro-attached cover concealing the EBCM.
9. Install the left trunk carpet trim. Close the trunk lid.
10. Connect the negative battery cable.

Allante

1. Disconnect the negative battery cable.
2. Remove driver's side insulator panel.
3. Remove EBCM connector by disengaging retainer and rotating connector toward the driver's seat.
4. Remove EBCM retaining bolts.
5. Disengage EBCM from mounting bracket and remove from vehicle.
To install:
6. Position EBCM in mounting bracket and install retaining bolts.
7. Install EBCM connector.
8. Install driver's side sound insulator panel.
9. Connect negative battery cable.

CHASSIS ELECTRICAL

Air Bag
— CAUTION —

Some vehicles are equipped with the Supplemental Inflatable Restraint (SIR) or air bag system. The SIR system must be disabled before performing service on or around SIR system components, steering column, instrument panel components, wiring and sensors. Failure to follow safety and disabling procedures could result in accidental air bag deployment, possible personal injury and unnecessary SIR system repairs.

PRECAUTIONS

Several precautions must be observed when handling the inflator module to avoid accidental deployment and possible personal injury.
● Never carry the inflator module by the wires or connector on the underside of the module.

- When carrying a live inflator module, hold securely with both hands, and ensure that the bag and trim cover are pointed away.
- Place the inflator module on a bench or other surface with the bag and trim cover facing up.
- With the inflator module on the bench, never place anything on or close to the module which may be thrown in the event of an accidental deployment.

DISARMING

1. Disconnect the negative battery cable.
2. Remove the SIR fuse from the fuse panel.
3. Remove the left side sound insulator.
4. Remove the Connector Positive Assurance (CPA) from the yellow 2-way SIR harness connector at the base of the steering column and separate the connector.

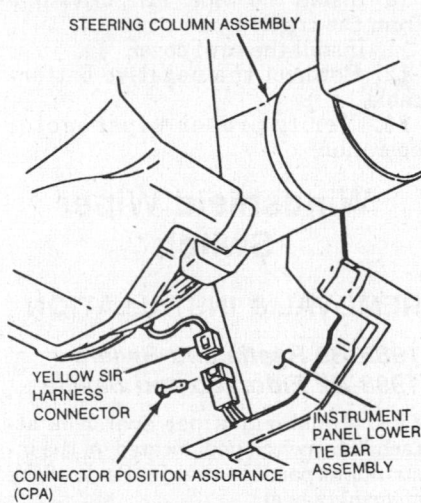

STEERING COLUMN ASSEMBLY

YELLOW SIR HARNESS CONNECTOR

INSTRUMENT PANEL LOWER TIE BAR ASSEMBLY

CONNECTOR POSITION ASSURANCE (CPA)

Yellow 2 way SIR harness connector

ARMING

1. Connect the yellow 2-way SIR connector at the base of the steering column and insert the Connect Positive Assurance (CPA).
2. Install the left side sound insulator.
3. Install the SIR fuse in the fuse panel.
4. Connect the negative battery cable.

REMOVAL & INSTALLATION

Inflator Module

1. Disconnect the negative battery cable.
2. Disarm the SIR system.

NOTE: Rotate the steering wheel so the access holes on the back of the steering wheel are at the 12 and 6 o'clock positions. This will allow tool access and reduce the possibility of marring the steering column cover.

3. Remove the 4 bolts from the back of the inflator module.
4. Remove the inflator module from the steering wheel.
5. Disconnect the horn contact by pushing slightly and twisting countercolockwise.
6. If equipped with steeri wheel controls, disconnect the steering wheel switch assembly connector from the steering column coil connector.
7. Disconnect the coil assembly from the inflator module.
To install:
8. If equipped, with the ignition switch in the **OFF** position, connect the steering wheel switch assembly connector to the coil connector.
9. Connect the horn contact.
10. Connect the coil assembly connector.

NOTE: Ensure that no wires at the back of the inflator module are pinched when aligning the inflator module to the steering wheel.

11. Install the inflator module and the 4 attaching bolts.
12. Arm the SIR system.
13. Connect the negative battery cable.

Heater Blower Motor

REMOVAL & INSTALLATION

Reatta and Riviera

1988–89

1. Disconnect the negative battery cable.
2. Remove the front of cowl shield(s).
3. Disconnect the electrical harness from the blower motor. Remove the harness from the retaining clips and move it aside.
4. Remove the cooling tube from the blower motor.
5. Remove the blower motor screws and the motor from the vehicle.
To install:
6. Install the blower motor to the housing and install the mounting screws.
7. Install the cooling tube to the blower motor.
8. Install the harness to the retaining clips. Connect the electrical harness to the blower motor.
9. Install the front of cowl shield(s).

10. Connect the negative battery cable.

1990–92

1. Disconnect the negative battery cable.
2. Remove cowl cross-tower brace; 2 nuts each side.
3. Remove both cowl relay center bracket nuts and position aside.
4. Remove blower motor electrical connector, cooling hose and mounting screws. Tilt blower motor in case and detach fan from motor.
5. Remove blower motor from case.
6. Remove fan from case.
To install:
7. Install fan to case.
8. Install blower motor to case.
9. Tilt blower motor in case and attach fan to motor. Install blower motor mounting screws, electrical connector and cooling hose.
10. Install both cowl relay center bracket nuts.
11. Install cowl cross-tower brace.
12. Connect the negative battery cable.

Allante

1. Disconnect the negative battery cable.
2. Remove the cross-tower brace.
3. Partially remove the upper intake manifold by performing the following procedures:
   a. Remove both right rear EGR pipe bolts.
   b. Remove the right rear transaxle dipstick bolt.
   c. Remove the right rear bracket bolt.
   d. Remove the right rear lower intake manifold nuts.
   e. Position the upper intake manifold aside.
4. Remove the electrical harness bracket and disconnect the electrical connector.
5. Remove the cooling hose, the mounting screws and the blower motor.
To install:
6. Install the blower motor, mounting screws and cooling hose.
7. Install the electrical harness bracket and connect the electrical connector.
8. Install the upper intake manifold by performing the following procedures:
   a. Place the upper intake manifold into position.
   b. Install the right rear lower intake manifold nuts.
   c. Install the right rear bracket bolt.
   d. Install the right rear transaxle dipstick bolt.
   e. Install both right rear EGR pipe bolts.

9. Install the cross-tower brace.
10. Connect the negative battery cable.

Eldorado and Seville

1. Disconnect the negative battery cable.
2. Remove the relay center bracket nuts and move the bracket aside.
3. Remove the air cleaner assembly and the cross-tower brace.
4. Disconnect the electrical harness support bracket.
5. On 1990–92 vehicles, remove the MAP sensor bracket.
6. Label and disconnect the electrical wiring connectors. Remove the cooling hose and mounting screws.
7. Tilt the blower motor in the case and remove the fan from the blower motor.

NOTE: Be careful not to bend the fan upon removal as a fan imbalance could result after reassembly.

8. Remove the blower motor and fan assembly from the vehicle.

To install:

9. Install the blower motor and fan assembly to the vehicle.
10. Tilt the blower motor in the case and install the fan to the blower motor.
11. Install the cooling hose and mounting screws. Connect the electrical wiring connectors.
12. On 1990–92 vehicles, install the MAP sensor bracket.
13. Connect the electrical harness support bracket.
14. Install the air cleaner assembly and the cross-tower brace.
15. Install the relay center bracket nuts.
16. Connect the negative battery cable.

Toronado and Trofeo

1988–89

1. Disconnect the negative battery cable.
2. Remove the front of the cowl shield.
3. Remove the bulkhead retaining screw and the bulkhead electrical connector.
4. Remove the Electronic Spark Control (ESC) module electrical connector.
5. Remove the ESC module and bracket assembly.
6. Remove the power steering pump bracket support.
7. Remove the coil bracket nuts. Label and disconnect the electrical connector from the coil.
8. Remove the plug wire guides. Remove the coil/bracket assembly and

move it aside. Remove the wiring harness conduit.
9. Remove the blower motor cooling tube.
10. Label and disconnect the electrical connectors from the blower motor. Remove the blower motor mounting screws.
11. Remove the blower motor mounting screws and the blower motor.

To install:

12. Install blower motor fan to blower motor.
13. Install blower motor using strip caulk type sealing material between the motor and heater and air conditioning module.
14. Install blower motor mounting screws (5).
15. Install blower motor cooling tube.
16. Install blower motor electrical connector and wiring harness conduits.
17. Install plug wire guides (2).
18. Install coil electrical connector.
19. Install plug wires to coil (3).
20. Install coil bracket nuts (3).
21. Install power steering pump bracket support and bracket support bolts.
22. Install ESC module and bracket assembly.
23. Install ESC module electrical connector.
24. Install bulkhead connector.
25. Install front of cowl shield.
26. Connect negative battery cable.

1990–92

1. Disconnect the negative battery cable.
2. Remove cowl cross-tower brace; 2 nuts each side.
3. Remove both cowl relay center bracket nuts and position aside.
4. Remove blower motor electrical connector, cooling hose and mounting screws. Tilt blower motor in case and detach fan from motor.
5. Remove blower motor from case.
6. Remove fan from case.

To install:

7. Install fan to case.
8. Install blower motor to case.
9. Tilt blower motor in case and attach fan to motor. Install blower motor mounting screws, electrical connector and cooling hose.
10. Install both cowl relay center bracket nuts.
11. Install cowl cross-tower brace.
12. Connect the negative battery cable.

Windshield Wiper Motor

REMOVAL & INSTALLATION

1. Disconnect the negative battery cable. Remove both wiper arms.
2. Remove the cowl cover.
3. Remove the wiper arm drive link from the crank arm.
4. Disconnect the electrical connectors.
5. Remove the air conditioning pipe shroud bracket, as required.
6. Remove the wiper motor-to-chassis bolts and the motor; guide the crank arm through the hole.

To install:

7. Guide the crank arm through the hole and install the wiper motor and the motor-to-chassis bolts.
8. If removed, install the air conditioning pipe shroud bracket.
9. Connect the electrical connectors.
10. Install the wiper arm drive link from the crank arm.
11. Install the cowl cover.
12. Connect the negative battery cable.
13. Verify proper wiper motor operation.

Windshield Wiper Switch

REMOVAL & INSTALLATION

1988–89 Reatta and Riviera
1988–92 Eldorado and Seville

The windshield wiper switch is attached to switch pod, located on the instrument panel to the right side of the steering wheel.

1. Disconnect the negative battery cable.
2. Remove the switch trim panel from the instrument panel.
3. Remove the switch-to-instrument panel screws.
4. Pull the switch outward and disconnect the electrical connectors from the rear of the switch.

To install:

5. Connect the electrical connectors to the rear of the switch and push the switch into position.
6. Install the switch-to-instrument panel screws.
7. Install the switch trim panel to the instrument panel
8. Connect the negative battery cable.

Allante

The windshield wiper switch is attached to switch pod, located on the instrument panel to the right side of the steering wheel.

1. Disconnect the negative battery cable.

2. Remove the bottom instrument panel trim plate.

3. Remove the switch pod-to-instrument panel screws, pull the pod outward and disconnect the electrical connectors. Remove the switch pod from the vehicle.

To install:

4. Connect the electrical connectors to the back of the pod and push the pod into position in the instrument panel.

5. Install the switch pod-to-instrument panel screws.

6. Install the bottom instrument panel trim plate.

7. Connect the negative battery cable.

1990–92 Reatta and Riviera 1988–92 Toronado and Trofeo

———— CAUTION ————

Replacing the windshield washer and wiper (pivot and pulse) switch necessitates removal of the steering wheel. If equipped with the Supplemental Inflatable Restraint (SIR) or air bag system, removing the steering wheel requires temporarily disabling the SIR system and removal of the inflator module. Failure to do so could result in accidental deployment of the air bag, possible personal injury and unnecessary SIR system repairs.

1. Disconnect the negative battery cable.

2. Place the ignition switch in the **LOCK** position to prevent uncentering of the coil assembly ring.

3. If equipped, disable the SIR system. Remove the inflator module.

4. Remove the steering wheel.

5. Remove coil assembly retaining ring.

6. Remove coil assembly from shaft end, allowing coil to hang freely.

NOTE: Coil assembly will become uncentered if the steering column is separated from steering gear and is allowed to rotate or if the centering spring is depressed, allowing hub to rotate while coil is removed from column.

7. Remove wave washer.

8. Remove shaft lock retaining ring using special tool J–23653–C to depress shaft lock.

9. Remove shaft lock.

10. Remove turn signal cancelling cam assembly.

11. Remove upper bearing spring, inner race seat and inner race.

12. Remove multi-function lever by performing the following:

a. Ensure that the switch is in the **OFF** position before removing the access cover from steering wheel.

b. Remove cruise control connector from lever. Note position of connector when installed in column.

c. Pull lever straight out of switch.

13. Remove screws and signal switch arm.

14. Remove turn signal switch screws.

15. Remove screw from end of hazard knob assembly. Remove button spring and knob from switch cavity.

16. Remove turn signal switch assembly and allow to hang freely.

17. Remove wiring protector at base of steering column.

18. Disconnect wiring harness at the base of the steering column.

19. Gently pull wire harness through instrument panel bracket and column housing.

20. Remove the coil assembly by performing the following:

a. Disconnect the yellow connector shroud from the black terminal connector.

b. Remove the wiring protector.

c. Attach a length of wire to the black terminal connector to aid in reassembly.

d. Gently pull the wire through the instrument panel bracket and column housing.

21. Remove the key from the pass key lock cylinder set.

22. Remove the buzzer switch assembly and buzzer switch retaining clip using a paper clip.

23. Reinsert the key in the pass key lock cylinder. Place the key in the **LOCK** position.

24. Remove the pass key lock cylinder by performing the following:

a. Disconnect the terminal connector.

b. Remove the wiring protector.

c. Attach a length of wire to the terminal connector to aid in reassembly.

d. Gently pull the wire through the instrument panel bracket and column housing.

25. Remove the lock housing cover screws. Remove the lock housing cover assembly.

26. Remove the tilt lever by gripping firmly and turning counterclockwise to remove from the steering column.

27. Remove the base plate and dimmer switch rod actuator.

28. Gently pull the pivot and pulse (wiper/washer) switch wire harness through the instrument panel bracket and column housing.

29. Remove the switch actuator pivot pin.

30. Remove the pivot and pulse switch assembly.

To install:

31. Install the pivot and pulse switch assembly to the cover.

32. Install the switch actuator pivot pin to the switch and cover.

33. Feed the pivot and pulse connector through the column housing and instrument panel bracket.

34. Connect the dimmer switch rod actuator to the base plate.

35. Install the base plate to the lock housing cover assembly.

NOTE: The bottom edge of the dimmer switch rod actuator should rest on the bend in the dimmer switch rod.

36. Install the lock housing cover assembly.

37. Install the multi-function lever by performing the following:

a. Plug the multi-function lever connector and cruise control wire together and mount on the base plate.

b. With the **WASH** paddle loose on the shaft, align the shaft with the switch notch and insert the shaft only.

c. Rotate the **WASH** paddle into position and push into the switch

d. Push on the knob to seat the lever into the switch.

38. Install the housing cover end cap. Install the screws and tighten the screw in the 12 o'clock position first, the screw in the 8 o'clock position second and the screw in the 3 o'clock position third. Tighten in the same sequence to 80 inch lbs. (9 Nm).

39. Install the pass key lock cylinder.

40. Install the lock retaining screw. Tighten to 22 inch lbs. (2.5 Nm).

41. Place the key in the **RUN** position.

42. Install the buzzer switch assembly and clip.

43. Route wiring assembly for the turn signal switch through column housing and instrument panel bracket.

44. Connect wiring assembly to connector at base of the steering column.

45. Connect coil assembly wire harness through column housing and instrument panel bracket. Allow coil to hang freely.

46. Install turn signal switch assembly and screws. Tighten to 30 inch lbs. (3.4 Nm).

47. Connect the yellow connector shroud to the black terminal connector.

48. Install wiring protector.

49. Install signal switch arm and screws. Tighten to 20.4 inch lbs. (2.3 Nm).

50. Install hazard knob, spring and button to hazard warning switch cavity. Install switch screw; drive in fully. Do not strip.

51. Install inner race, upper bearing

inner race seat and upper bearing spring.

52. Install turn signal cancelling cam assembly.

53. Install shaft lock.

NOTE: Inspect shaft lock retaining ring for damage or deformation. If damaged or deformed, replace with new retaining ring.

54. Install shaft lock retaining ring. Align to block tooth on shaft using special tool J–23653–C to depress shaft lock. Ring must be firmly seated in groove on shaft.

NOTE: Set steering shaft so block teeth on upper steering shaft are at the 12 o'clock and 6 o'clock positions. The alignment mark at the end of the shaft should be at the 12 o'clock position and vehicle wheels straight-ahead. Set the ignition switch to the LOCK position to ensure no damage occurs to the coil assembly.

55. Ensure coil assembly hub is centered by performing the following:

a. Hold coil assembly with clear bottom up to see coil ribbon.

b. There are 2 styles of coils. One rotates clockwise and the other rotates counterclockwise. While holding coil assembly, depress spring lock to rotate hub in direction of arrow until it stops.

c. The coil ribbon should be wound up snug against the center hub.

d. Rotate coil hub in opposite direction approximately 2½ turns. Release spring lock between locking tabs in front of arrow.

NOTE: If a new coil assembly is being installed, assemble the pre-centered coil assembly to column. Remove centering tab and dispose.

56. Install wave washer.

57. Install coil assembly using horn tower on cancelling cam assembly inner ring and projections on outer ring for alignment.

58. Install coil assembly retaining ring. Ring must be firmly seated in groove on shaft.

NOTE: Gently pull lower coil assembly wire to remove any wire kinks that may be inside column assembly.

59. Install steering wheel.

60. Install inflator module and enable SIR system.

61. Connect the negative battery cable.

Instrument Cluster

REMOVAL & INSTALLATION

Reatta and Riviera

1. Disconnect the negative battery cable.

2. Remove the center, left and right trim covers.

3. Remove the instrument cluster-to-dash screws, then, pull the cluster straight out of the housing.

To install:

4. Place the instrument cluster into position and install the instrument cluster-to-dash screws.

5. Install the center, left and right trim covers.

6. Connect the negative battery cable.

Allante

1. Disconnect the negative battery cable.

2. Remove the left and right switch pod trim plates.

3. Remove the cluster trim plate.

4. Remove the cluster assembly-to-dash screws, pull the cluster forward and disconnect the electrical connectors.

5. Remove the cluster assembly from the vehicle.

To install:

6. Install the cluster assembly to the vehicle.

7. Connect the electrical connectors. Install the instrument cluster assembly-to-dash screws.

8. Install the cluster trim plate.

9. Install the left and right switch pod trim plates.

10. Connect the negative battery cable.

Eldorado and Seville

1. Disconnect the negative battery cable. Remove the screws located along the top and remove the instrument panel trim plate.

2. Remove the mounting screws and the filter lens.

3. Remove the warning light lens screws and the lens. Remove the trip odometer reset button.

4. Remove the instrument panel cluster screws. Pull the cluster off the electrical connections and remove it. Using a pair of pliers, hold the retaining tabs at either end of the cluster board and remove the board.

To install:

5. Align the instrument cluster with the electrical connectors, push it into the instrument panel. Install the instrument cluster-to-dash panel screws.

6. Install the filter lens and mounting screws.

7. Install the instrument panel trim plate and the screws located along the top of the panel.

8. Connect the negative battery cable.

Toronado and Trofeo

1988

1. Disconnect the negative battery cable.

2. Remove the steering column trim cover. Lower the steering column.

3. Remove the instrument panel trim plate.

4. Remove the cluster-to-instrument panel screws.

5. Pull the cluster rearward and remove it.

To install:

6. Place the instrument panel into position. Install the cluster-to-instrument panel screws.

7. Install the instrument panel trim plate.

8. Raise the steering column into position. Install the steering column trim cover.

9. Connect the negative battery cable.

1989–92

1. Disconnect the negative battery cable.

2. Remove instrument panel cluster trim plate.

3. Remove screws retaining cluster to instrument panel.

4. Pull cluster out and disengage electrical connector.

To install:

5. Place the instrument cluster into position and engage the electrical connector.

6. Install the screws retaining the cluster to instrument panel.

7. Install the instrument cluster panel trim plate.

8. Connect the negative battery cable.

Radio

REMOVAL & INSTALLATION

Reatta and Riviera

The entertainment system on Reatta and Riviera vehicles consists of a remote radio receiver, an optional tape deck and an Electronic Control Center (ECC) monitor. The entertainment system is controlled by the Electronic Control Center (also known as CRT) through the use of "hard" and "soft" keys.

RADIO RECEIVER

1. Disconnect the negative battery cable.

2. Remove the gear selector handle, transaxle indicator assembly and storage compartment (lift lid) to reveal bolts retaining console assembly.

3. Remove 4 bolts retaining console assembly and remove assembly.

4. Disconnect antenna lead-in and radio harness connector from radio receiver.

5. Remove 2 bolts to radio support assembly top cover and remove radio receiver.

To install:

6. Install radio receiver and 2 bolts to radio support assembly top cover.

7. Connect radio harness connector and antenna lead-in to radio receiver.

8. Install console assembly and 4 retaining bolts.

9. Install storage compartment, transaxle indicator assembly and gear selector handle.

10. Connect negative battery cable.

Allante

The entertainment system on Allante vehicles consists of a remote radio receiver, a remote tape deck and a radio control head, below the Driver Information Center (DIC).

RADIO CONTROL HEAD AND COMBINATION PANEL

1. Disconnect the negative battery cable.

2. Remove pop-out air conditioner vent.

3. Remove 2 screws retaining the combination panel.

4. Remove left side sound insulation panel screws.

5. Remove left side sound insulation panel.

6. Remove 2 nuts and washers, at the back of tape player.

7. Remove combo panel.

8. Remove 3 electrical connectors, depress tabs, push in, then pull to release.

To install:

9. Connect electrical connectors.

10. Align combo panel.

11. Install 2 nuts and washers, at the back of the tape player.

12. Install left side sound insulation panel.

13. Install left side sound insulation panel screws.

14. Install 2 screws retaining combo panel.

15. Insert air conditioning vent.

TAPE PLAYER

1. Disconnect the negative battery cable.

2. Remove radio head/combo panel.

3. Remove 3 tape player retaining bolts, 1 on the side, 2 underneath.

4. Remove 3 bolts, open combo panel door.

5. Depress latch to open cassette door.

6. Release screw cover retaining tabs, access from inside cassette door.

7. Remove screw cover.

8. Remove 3 screws and washers.

9. Remove tape player door.

10. Remove 2 face plate retainer bolts.

11. Pull tape player out as far as it will go.

12. Disconnect 3 electrical connectors.

To install:

13. Install 3 electrical connectors.

14. Depress latch to open cassette door.

15. Slide tape player into place with door open.

16. Install 3 bolts retaining tape player in combo panel.

17. Close combo panel door and install 3 bolts.

18. Align face plate and install 2 bolts.

19. Align cassette door and install 3 screws.

20. Snap on screw cover.

21. Install radio head/combo panel.

22. Connect negative battery cable.

RADIO RECEIVER

1. Disconnect the negative battery cable.

2. Remove glove box assembly.

3. 2 screws and 1 nut and washer retaining the radio receiver.

4. Remove coaxial cable and 3 electrical connectors.

To install:

5. Install 3 electrical connectors and coaxial cable.

6. Install nut, washer and 3 screws retaining radio receiver.

7. Install glove box assembly.

8. Connect negative battery cable.

Eldorado and Seville

1. Disconnect the negative battery cable.

2. Remove radio trim plate.

3. Remove left side air conditioning vent.

4. Remove 7 screws attaching instrument panel trim plate.

5. Loosen lower 2 mounting nuts under radio; top 2 nuts do not have to be loosened.

6. Slide radio forward and disconnect electrical connectors.

7. Remove antenna lead-in.

To install:

8. Install antenna lead-in to radio.

9. Connect electrical connectors to radio.

10. Slide radio into instrument panel bracket and tighten lower mounting nuts.

11. Install instrument panel trim plate.

12. Install radio trim plate.

13. Install air conditioning vent.

14. Connect negative battery cable.

Toronado and Trofeo

RADIO HEAD

1. Disconnect the negative battery cable.

2. Remove driver's side lower hush panel.

3. Remove knee bolster.

4. Remove instrument panel trim panel.

5. Remove screws retaining radio/ECC bracket.

6. Remove nuts to remove radio bracket from radio.

7. Remove electrical connectors.

To install:

8. Install nuts attaching radio to mounting bracket.

9. Install electrical connections to radio and ECC.

10. Carefully reposition radio and mounting bracket to instrument panel.

NOTE: If the radio buttons operate unusually or intermittently, the condition may be due to poor alignment or uneven tightening of the radio or instrument panel trim panel screws. If this occurs, remove the trim panel, loosen the radio mounting bolts and realign the radio unit.

11. Install trim plate and knee bolster.

12. Install lower hush panel.

13. Connect negative battery cable.

REMOTE RADIO RECEIVER

1. Disconnect the negative battery cable.

2. Open console storage tray and remove CD, cassette holder or phone handset, if equipped.

3. Remove T-15 Torx® screws and remove storage tray liner.

4. Remove electrical connectors to console seat controls and handset connector, if equipped.

5. Open ashtray and take out cigar lighter and ashtray bucket.

6. Set emergency brake, place shift lever in **N**.

7. Pull console trim plate up, console trim plate has clip tabs in area to right and left of top edge of shifter plate.

8. Disconnect bulb and cigar lighter electrical connectors and remove console trim plate.

9. The remote chassis will be visible towards the front end of the console.

10. Remove 10mm nuts retaining chassis to CRTC bracket.

11. Remove electrical connectors to

remote chassis and coaxial lead-in connector.

12. Remove radio chassis.

To install:

13. Place radio chassis in top of CRTC bracket and install 10mm nuts. Make certain radio wiring is not trapped under radio receiver chassis.

14. Connect electrical connectors on left side of chassis.

15. Connect electrical connectors on right side of receiver. Best order is: lower white 4-pin, upper white 6-pin, lower black 6-pin and upper blue 4-pin.

16. Connect antenna coaxial lead-in.

17. Connect lower console trim plate over shift lever and reconnect electrical connectors to ashtray.

18. Carefully snap trim plate into place.

19. Return shift lever to park and release emergency brake.

20. Connect electrical connectors to remote chassis and coaxial lead-in connector.

21. Connect electrical connections to console mounted seat controls and connect phone handset connector, if equipped.

22. Connect lower console storage tray liner and fasten T-15 Torx® screws.

23. Reinsert CD/cassette bucket, handset, cigar lighter and ashtray.

24. Connect negative battery cable.

Concealed Headlights

MANUAL OPERATION

Reatta

1. Open the hood.
2. Turn the manual control knob in the direction of the arrow on the "Headlight Up" label. Turn the knob by hand until it stops.
3. Close the hood and check headlight operation.

Toronado and Trofeo

1. On 1988 vehicles, remove fuse No. 2 from engine compartment relay center. On 1989–92 vehicles, disconnect 3-way headlight door actuator connector.
2. Remove protective cover from the knob.
3. Rotate the knob clockwise until the headlight doors open.
4. To close the doors, rotate the knob counterclockwise until the headlight doors close.
5. Install protective cover over knob.
6. On 1988 vehicles, install fuse No. 2 in engine compartment relay center. On 1989–92 vehicles, connect 3-way headlight door actuator connectors.

Headlight Switch

REMOVAL & INSTALLATION

1988

The headlight switch is located on a switch pod, located on the left side of the instrument panel.

1. Disconnect the negative battery cable. Remove the instrument panel trim plate.
2. Remove the switch pod screws and pull the switch outward. Disconnect the electrical connectors and remove the switch from the vehicle.

To install:

3. Connect the electrical connectors and place the switch into position.
4. Install the switch pod screws.
5. Install the instrument panel trim plate.
6. Connect the negative battery cable.

1989–92

The headlight switch is located on the left side of the instrument panel.

1. Remove the left trim plate screws and the trim plate, if equipped.
2. If equipped, remove the left air vent.
3. Remove the headlight switch screws, pull the switch forward and disconnect the electrical connectors or the fiber optic lead, if equipped.
4. Remove the headlight switch.

To install:

5. Connect the electrical connectors or fiber optic lead, if equipped, to the headlight switch.
6. Push the switch into position and install the headlight attaching screws.
7. Install the left air vent, if equipped.
8. Install the left trim plate and attaching screws.

Dimmer Switch

The dimmer switch is attached to the lower steering column jacket. It is activated by a rod attached to the multi-function lever.

REMOVAL & INSTALLATION

1. Disconnect the negative battery cable. Remove the left side sound insulator panel.
2. If necessary, remove the lower steering column trim cover.
3. Disconnect the electrical connector from the dimmer switch.
4. Remove the dimmer switch-to-steering column screws and the dimmer switch.

To install:

5. Position the actuator rod into the

dimmer switch hole and install the dimmer switch-to-steering column screws.

6. Connect the electrical connector to the dimmer switch.
7. Adjust the dimmer switch by depressing the switch slightly and inserting a $\frac{3}{32}$ in. drill bit into the adjusting hole. Push the switch up to remove any play and tighten the dimmer switch adjusting screw.
8. If removed, install the lower steering column trim cover.
9. Install the left side sound insulator panel.
10. Connect the negative battery cable.

Turn Signal Switch

CAUTION

Replacing the turn signal switch necessitates removal of the steering wheel. If equipped with the Supplemental Inflatable Restraint (SIR) system, removing the steering wheel requires temporarily disabling the SIR system and removal of the inflator module. Failure to do so could result in accidental deployment of the air bag, possible personal injury and unnecessary SIR system repairs.

REMOVAL & INSTALLATION

1988–89

1. Disconnect the negative battery cable. Remove the steering wheel.
2. Remove the bumper and the carrier snapring retainer from the steering shaft.
3. Using the lock plate compressor screw tool, install in the upper steering shaft, tighten to 40 inch lbs., to keep the shaft from telescoping.
4. Using the lock plate compressor tool, install it on the upper steering shaft, tighten it to depress the shaft lock. Remove the shaft lock retainer, the compressor tool and the steering shaft lock.
5. Remove the turn signal cancelling cam assembly. Place the turn signal switch in the **N** position and remove the upper bearing spring.
6. Position the turn signal switch so the mounting screws can be removed through the holes in the switch and remove the turn signal lever.
7. Remove the turn signal switch-to-steering column screws and lift the turn signal switch. Remove the wire protector and disconnect the turn signal switch connector.
8. Using the terminal remover tool, disconnect the buzzer switch wires from the turn signal switch connector. Using needle-nose pliers, remove the buzzer switch assembly.
9. Place the lock cylinder in the **ACCESSORY** position, remove the lock

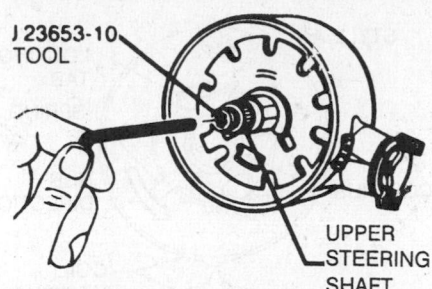

J 23653-10
TOOL

UPPER STEERING SHAFT

Installing the lock plate compressor screw

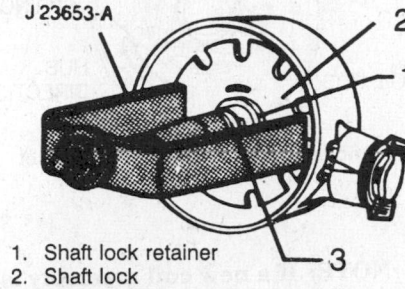

J 23653-A

1. Shaft lock retainer
2. Shaft lock
3. Upper steering shaft

Compressing the shaft lock

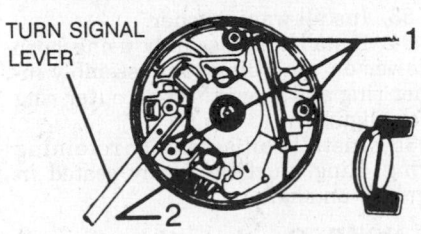

TURN SIGNAL LEVER

1. Screw
2. Turn signal switch assembly

Positioning the turn signal lever to remove the turn signal switch screws

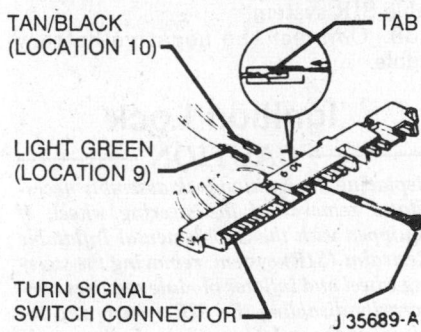

TAN/BLACK (LOCATION 10)

TAB

LIGHT GREEN (LOCATION 9)

TURN SIGNAL SWITCH CONNECTOR

J 35689-A

Separating the buzzer switch wires from the turn signal electrical connector

retaining screw and the lock cylinder set.

10. Lifting the turn signal switch assembly, gently pull the wires through the steering column shroud.

To install:

11. Install turn signal connector

through lock housing cover and steering column housing shroud.

12. Install steering column lock cylinder set while in **ACCESSORY** position.

13. Install lock retaining screw and tighten to 22 inch lbs. (2.5 Nm).

14. Install buzzer switch by pushing switch down into its retaining bore until bottomed with plastic tab covering lock retaining screw.

15. Install buzzer switch wires to turn signal switch connector: light green wire to location 9, tan/black wire to location 10.

NOTE: Wire terminal retainer must be removed and discarded from service buzzer switch wire.

16. Install wire connector retainer.

17. Install turn signal switch connector.

18. Install wire protector and turn signal switch. Install screws and tighten to 59 inch lbs. (6.8 Nm).

NOTE: Position turn signal switch so screws can be installed through openings in switch.

19. Install turn signal lever. Tighten screw to 53 inch lbs. (6 Nm).

20. Place turn signal switch in **OFF** position. Install upper bearing spring.

21. Install turn signal cancel cam assembly.

22. Install steering shaft lock.

23. Install shaft lock retainer to upper steering shaft using special tool J-23653-A to slightly depress shaft lock.

24. Install carrier snapring retainer. Install steering shaft bumper.

25. Extend shaft and lock in place. Install steering wheel and jam nut. Tighten to 30 ft. lbs. (41 Nm).

26. Remove lock plate compression screw J-23653-10.

1990–92

1. Disconnect the negative battery cable.

2. Place the ignition switch in the **LOCK** position to prevent uncentering of the coil assembly ring.

3. If equipped, disable the SIR system. Remove the inflator module.

4. Remove the steering wheel.

5. Remove coil assembly retaining ring.

6. Remove coil assembly from shaft end, allowing coil to hang freely.

NOTE: Coil assembly will become uncentered if the steering column is separated from steering gear and is allowed to rotate or if the centering spring is depressed, allowing hub to rotate while coil is removed from column.

7. Remove wave washer.

8. Remove shaft lock retaining ring using special tool J-23653-C to depress shaft lock.

9. Remove shaft lock.

10. Remove turn signal cancelling cam assembly.

11. Remove upper bearing spring, inner race seat and inner race.

12. Remove multi-function lever by performing the following:

a. Ensure that the switch is in the **OFF** position before access cover from steering wheel.

b. Remove cruise control connector from lever. Note position of connector when installed in column.

c. Pull lever straight out of switch.

13. Remove screws and signal switch arm.

14. Remove turn signal switch screws.

15. Remove screw from end of hazard knob assembly. Remove button spring and knob from switch cavity.

16. Remove turn signal switch assembly and allow to hang freely.

17. Remove wiring protector at base of steering column.

18. Disconnect wiring harness at the base of the steering column.

19. Gently pull wire harness through instrument panel bracket and column housing.

To install:

20. Route wiring assembly for new switch through column housing and instrument panel bracket.

21. Connect wiring assembly to connector at base of the steering column.

22. Connect coil assembly wire harness through column housing and instrument panel bracket. Allow coil to hang freely.

23. Install turn signal switch assembly and screws. Tighten to 30 inch lbs. (3.4 Nm).

24. Install wiring protector.

25. Install signal switch arm and screws. Tighten to 20.4 inch lbs. (2.3 Nm).

26. Install hazard knob, spring and button to hazard warning switch cavity. Install switch screw; drive in fully. Do not strip.

27. Install multi-function lever by performing the following:

NOTE: Ensure that the switch is in the OFF position before installation.

a. Install lever electrical connectors.

b. With "WASH" paddle loose on the metal shaft, align shaft with the switch notch and insert shaft only.

c. Rotate "WASH" paddle into position and push into switch.

d. Push on the knob to seat lever into switch.

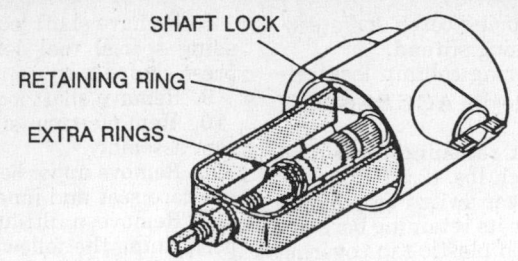

Removing shaft lock retaining ring

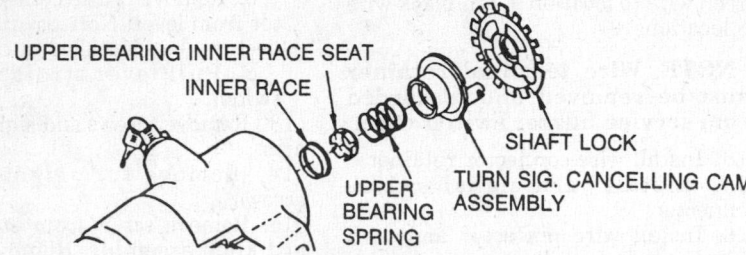

Removing upper shaft components

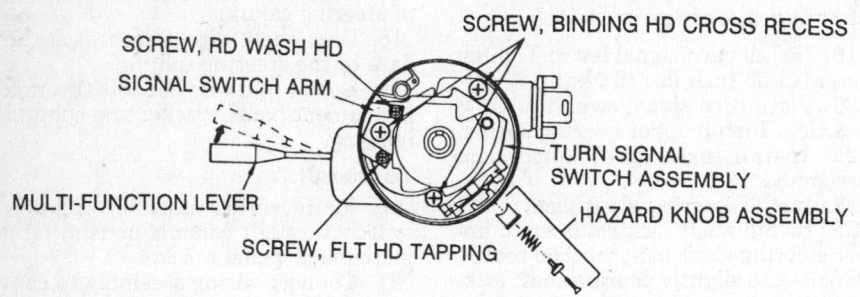

Turn signal switch installed

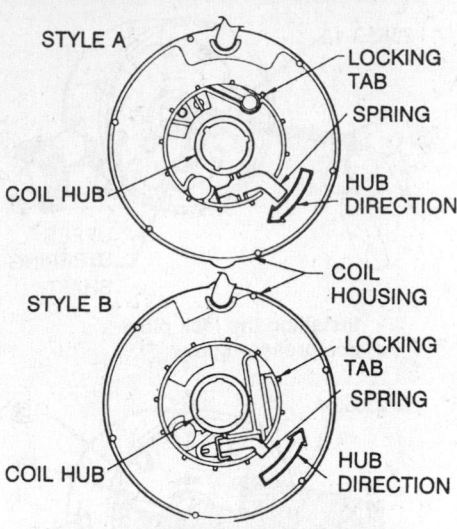

Removing pass key lock cylinder set

NOTE: If a new coil assembly is being installed, assemble the pre-centered coil assembly to column. Remove centering tab and dispose.

33. Install wave washer.

34. Install coil assembly using horn tower on cancelling cam assembly inner ring and projections on outer ring for alignment.

35. Install coil assembly retaining ring. Ring must be firmly seated in groove on shaft.

NOTE: Gently pull lower coil assembly wire to remove any wire kinks that may be inside column assembly.

36. Install steering wheel.
37. Install inflator module and enable SIR system.
38. Connect the negative battery cable.

Ignition Lock
— CAUTION —

Replacing the ignition lock assembly necessitates removal of the steering wheel. If equipped with the Supplemental Inflatable Restraint (SIR) system, removing the steering wheel and inflator module requires temporarily disabling the SIR system and removal of the inflator module. Failure to do so could result in accidental deployment of the air bag and possible personal injury.

REMOVAL & INSTALLATION
1988–89

1. Disconnect the negative battery cable. Remove the steering wheel.
2. Remove the bumper and the car-

e. Install cruise control connector.

f. Install access cover onto steering column.

28. Install inner race, upper bearing inner race seat and upper bearing spring.

29. Install turn signal cancelling cam assembly.

30. Install shaft lock.

NOTE: Inspect shaft lock retaining ring for damage or deformation. If damaged or deformed, replace with new retaining ring.

31. Install shaft lock retaining ring. Align to block tooth on shaft using special tool J–23653–C to depress shaft lock. Ring must be firmly seated in groove on shaft.

NOTE: Set steering shaft so block teeth on upper steering shaft are at the 12 o'clock and 6 o'clock positions. The alignment mark at the end of the shaft should be at the 12 o'clock position and vehicle wheels straight-ahead. Set the ignition switch to the LOCK position to ensure no damage occurs to the coil assembly.

32. Ensure coil assembly hub is centered by performing the following:

a. Hold coil assembly with clear bottom up to see coil ribbon.

b. There are 2 styles of coils. One rotates clockwise and the other rotates counterclockwise. While holding coil assembly, depress spring lock to rotate hub in direction of arrow until it stops.

c. The coil ribbon should be wound up snug against the center hub.

d. Rotate coil hub in opposite direction approximately 2½ turns. Release spring lock between locking tabs in front of arrow.

rier snapring retainer from the steering shaft.

3. Using the lock plate compressor screw tool, install it in the upper steering shaft, tighten to 40 inch lbs., to keep the shaft from telescoping.

4. Using the lock plate compressor tool, install it on the upper steering shaft, tighten it to depress the shaft lock. Remove the shaft lock retainer, the compressor tool and the steering shaft lock.

5. Remove the turn signal cancelling cam assembly. Place the turn signal switch in the **N** position and remove the upper bearing spring.

6. Position the turn signal switch so the mounting screws can be removed through the holes in the switch and remove the turn signal lever.

7. Remove the turn signal switch-to-steering column screws and lift the turn signal switch. Remove the wire protector and disconnect the turn signal switch connector.

8. Using the terminal remover tool, disconnect the buzzer switch wires from the turn signal switch connector. Using needle-nose pliers, remove the buzzer switch assembly.

9. Place the lock cylinder in the **ACC** position, remove the lock retaining screw and the lock cylinder set.

To install:

10. Reverse the removal procedures. Tighten the lock retaining screw to 22 inch lbs., the turn signal switch screws to 59 inch lbs. and the turn signal lever screw to 53 inch lbs.

11. Check the operation of the switches and the steering column.

1990–92

1. Disconnect the negative battery cable.

2. Place the ignition switch in the **LOCK** position to prevent uncentering of the coil assembly ring.

3. Disable the SIR system and remove the inflator module (air bag), if equipped.

4. Remove the steering wheel.

5. Remove the combination switch

LOCK HOUSING COVER ASSEMBLY
ROUTE WIRE FROM LOCK CYLINDER AS SHOWN (DO NOT TWIST WIRES)
LOCK CYLINDER SET
ROTATE PANEL CLIP UP TO 360 DEGREES TO LOOP WIRE AWAY FROM SECTOR GEAR AND SNAP CLIP INTO HOLE IN LOCK HOUSING

Routing pass key wire harness

assembly and allow to hang freely. Do not remove wiring harness and connector from steering column.

6. Remove key from pass key lock cylinder set.

7. Disconnect buzzer switch assembly.

8. Reinsert key in pass key lock cylinder. Turn key to **LOCK** position.

9. Remove lock retaining screw.

10. Disconnect pass key lock cylinder terminal connector.

11. Remove wiring protector.

12. Attach a length of mechanics wire to terminal connector to aid in reassembly.

13. Gently pull wire through instrument panel bracket and column housing.

14. Remove pass key lock cylinder.

To install:

NOTE: Route wire from lock cylinder through steering column using mechanics wire. Rotate panel clip 360 degrees and snap clip into hole in housing. Failure to do so may result in component damage or malfunction of pass key lock cylinder.

15. Install pass key lock cylinder.

16. Gently pull lower lock cylinder wire to remove any wire kinks that may be inside column assembly.

17. Install lock retaining screw. Tighten to 22 inch lbs. (2.5 Nm).

18. Turn key to **RUN** position.

19. Install buzzer switch assembly.

20. Install combination switch.

21. Install inflator module. Enable SIR system.

22. Install steering wheel.

23. Connect negative battery cable.

Ignition Switch

REMOVAL & INSTALLATION

1988

1. Disconnect the negative battery cable.

2. Place the ignition switch on the **OFF/UNLOCKED** or **ACC** if equipped with tilt-wheel.

3. Remove top pan cover, if equipped, and loosen the toe clamp bolts.

4. Remove lower instrument panel trim retaining screws. Remove the panel in order to gain working clearance.

5. Remove the automatic transaxle shift indicator needle.

6. Remove the steering column instrument panel bracket and allow the steering wheel to rest on the driver's seat.

7. Remove the dimmer switch retaining screws and the switch.

8. Remove the ignition switch attaching screws and lift the switch from the actuator rod.

9. Label and disconnect the electrical connector(s) from the ignition switch.

To install:

10. Before installation, place the slider on the new switch in one of the following positions, depending on the steering column and accessories:

a. Standard column with key release—extreme left detent.

b. Standard column with park lock—1 detent from extreme left **OFF/LOCK** position.

c. All other standard columns—2 detents from extreme left **OFF/UNLOCK** position.

d. Adjustable column with key release—extreme right detent.

e. Adjustable column with park lock—1 detent from extreme right **OFF/LOCK** position.

f. All other adjustable columns—2 detents from extreme right **OFF/UNLOCK** position.

11. Connect the electrical connector(s) to the switch.

12. Position the switch on actuator rod and install the ignition switch-to-steering column screws.

13. Connect the electrical connector(s) to the ignition switch.

14. Engage the actuator rod and install the ignition switch attaching screws.

15. Install the dimmer switch and retaining screws.

16. Raise the steering column into position and install the instrument panel bracket.

17. Install the automatic transaxle shift indicator needle.

18. Install lower instrument panel trim retaining screws.

19. Install top pan cover, if equipped, and tighten the toe clamp bolts.

20. Connect the negative battery cable.

1989–92

The ignition switch is hard-wired. The wiring harness with the column harness connector must be replaced with the ignition switch. Do not splice the new switch to the existing column wiring harness.

1. Disconnect the negative battery cable.

2. Remove the lower left sound insulator and the instrument panel steering column cover.

3. Remove the ignition switch wire protector and the switch-to-column screws.

4. Disconnect the ignition and turn signal switch column harness connectors from the dash connector.

5. Disconnect the turn signal har-

ness connector from the column harness connector.

6. Remove the steering column bolts and nuts and gently lower steering column to the seat.

7. Remove the ignition switch assembly with the switch, harness and connector.

To install:

8. Install the ignition switch assembly with the harness and connector.

9. Raise the steering column into position and install the bolts and nuts.

10. Connect the turn signal harness connector to the column harness connector.

11. Connect the ignition and turn signal switch column harness connectors to the dash connector.

12. Install the ignition switch wire protector and the switch-to-column screws.

13. Install the lower left sound insulator and the instrument panel steering column cover.

14. Connect the negative battery cable.

Stoplight Switch

ADJUSTMENT

NOTE: When the brake pedal is in the fully released position, the stoplight switch plunger should be fully depressed against the pedal arm. The switch is adjusted by moving it in or out.

1. Remove the stoplight switch from the brake pedal bracket.

2. Insert the switch into the retainer until the switch body seats on the tube clip.

3. Pull the brake pedal rearward against the internal pedal stop.

NOTE: The switch will be moved in the retainer resulting in proper adjustment.

4. When no further adjustment clicks are heard and the stoplights remain **OFF**, the stoplight switch will be properly seated.

REMOVAL & INSTALLATION

1. Disconnect the negative battery cable. Remove the underdash hush panel, if equipped.

2. Locate the stoplight switch on the brake pedal bracket.

3. Remove the tubular retaining clip.

4. Remove the stoplight switch electrical connectors.

5. Remove the switch assembly from the vehicle.

To install:

6. Install the switch assembly to the vehicle.

7. Connect the stoplight switch electrical connectors.

8. Install the tubular retaining clip.

9. Connect the negative battery cable.

10. Adjust stoplight switch.

Neutral Safety Switch

ADJUSTMENT

1. Place transaxle shift lever in **N**.

2. Loosen switch attaching screws.

3. Rotate switch on shifter assembly to align hole with carrier tang hole.

4. Insert $^3/_{32}$ in. (2.34mm) max. diameter gauge pin or drill bit to a depth of $^{15}/_{32}$ in. (12mm).

5. Tighten 2 attaching screws to 20 ft. lbs. (27 Nm).

6. Remove gauge pin.

7. Ensure that engine will start only in **P** and **N** positions. If engine will start in any other position, readjust switch.

REMOVAL & INSTALLATION

All neutral safety/back-up light switches come with a small plastic alignment pin installed. Leave this pin in place.

1. Place the shifter assembly in the **N** position.

2. Remove the shifter lever-to-switch nut and the lever.

3. Disconnect the electrical connector from the neutral safety/back-up light switch.

4. Remove the neutral safety/back-up light switch-to-transaxle bolts and the switch from the vehicle.

To install:

5. Position the shifter shaft in the **N** position.

NOTE: If using an old switch or the plastic pin (new switch) is broken, install a $^3/_{32}$ in. pin gauge (drill bit) into the neutral safety/back-up light switch; the switch is locked into its neutral position.

6. Align the flats of the shifter shaft and the neutral safety/back-up light, then, align the switch-to-tang on the transaxle. Tighten the switch-to-transaxle bolts to 22 ft. lbs. Remove the pin gauge.

7. To complete installation, reverse the removal procedures. Make sure the engine starts only in the **P** and **N** positions.

Fuses, Circuit Breakers and Relays

LOCATION

Fuse Panels

1988

Riviera and Reatta—behind center of instrument panel, front of console

Allante—center console, under ash tray

Eldorado and Seville—glove box

Toronado and Trofeo—right side of instrument panel

1989

Riviera and Reatta—Front right side of console

Allante—center console, under ash tray

Eldorado and Seville—glove box

Toronado and Trofeo—right side of instrument panel

1990

Riviera and Reatta—front left side of console

Allante—center console, under ash tray

Eldorado and Seville—glove box

Toronado and Trofeo—right side of instrument panel

1991–92

Riviera and Reatta—front left side of console

Allante—center console, under ash tray

Eldorado and Seville—glove box

Toronado and Trofeo—right side of instrument panel

Circuit Breakers

A circuit breaker is an electrical switch which breaks the circuit during an electrical overload. Some circuit breakers are designed to automatically reset after a specified period of time. Others must be manually reset after the electrical malfunction causing the overload has been corrected.

The majority of circuit breakers can be found in the fuse panel. Some, however, are installed in-line near the device they are intended to protect.

Relays

Relays are generally mounted in the vicinity of the device(s) they are intended to control. On the vehicles listed below, there is an Interior Relay Center (IRC).

Riviera and Reatta—below center of instrument panel, right front of console

Eldorado and Seville—behind right side of instrument panel, below glove box

Toronado and Trofeo—behind right side of instrument panel.

GM "F" Body

Rear Wheel Drive

Chevrolet—Camaro
Pontiac—Firebird

SPECIFICATIONS

VEHICLE IDENTIFICATION CHART

It is important for servicing and ordering parts to be certain of the vehicle and engine identification. The VIN (vehicle identification number) is a 17 digit number visible through the windshield on the driver's side of the dash and contains the vehicle and engine identification codes. The tenth digit indicates model year and the eighth digit indicates engine code. It can be interpreted as follows:

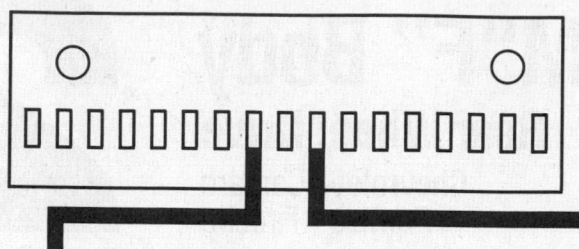

Engine Code

Code	Cu. In.	Liters	Cyl.	Fuel Sys.	Eng. Mfg.
S	173	2.8	6	MFI	Chevrolet
F	305	5.0	8	TPI	Chevrolet
E	305	5.0	8	TBI	Chevrolet
8	350	5.7	8	TPI	Chevrolet
T	191	3.1	6	MFI	CPC
7	231	3.8	6	SFI-Turbo	Buick

MFI—Multi Port Fuel Injection
SFI—Sequential Fuel Injection
TBI—Throttle Body Injection
TPI—Tuned Port Injection

Model Year

Code	Year
J	1988
K	1989
L	1990
M	1991
N	1992

ENGINE IDENTIFICATION

Year	Model	Engine Displacement cu. in. (liter)	Engine Series Identification (VIN)	No. of Cylinders	Engine Type
1988	Camaro	173 (2.8)	S	6	OHV
	Firebird	173 (2.8)	S	6	OHV
	Camaro	305 (5.0)	F	8	OHV
	Firebird	305 (5.0)	F	8	OHV
	Camaro	305 (5.0)	E	8	OHV
	Firebird	305 (5.0)	E	8	OHV
	Camaro	350 (5.7)	8	8	OHV
	Firebird	350 (5.7)	8	8	OHV
1989	Camaro	173 (2.8)	S	6	OHV
	Firebird	173 (2.8)	S	6	OHV
	Camaro	305 (5.0)	F	8	OHV
	Firebird	305 (5.0)	F	8	OHV
	Camaro	305 (5.0)	E	8	OHV
	Firebird	305 (5.0)	E	8	OHV
	Camaro	350 (5.7)	8	8	OHV
	Firebird	350 (5.7)	8	8	OHV
	Firebird	231 (3.8)	7	6	OHV

ENGINE IDENTIFICATION

Year	Model	Engine Displacement cu. in. (liter)	Engine Series Identification (VIN)	No. of Cylinders	Engine Type
1990	Camaro	191 (3.1)	T	6	OHV
	Firebird	191 (3.1)	T	6	OHV
	Camaro	305 (5.0)	F	8	OHV
	Firebird	305 (5.0)	F	8	OHV
	Camaro	305 (5.0)	E	8	OHV
	Firebird	305 (5.0)	E	8	OHV
	Camaro	350 (5.7)	8	8	OHV
	Firebird	350 (5.7)	8	8	OHV
1991-92	Camaro	191 (3.1)	T	6	OHV
	Firebird	191 (3.1)	T	6	OHV
	Camaro	191 (3.1)	T	6	OHV
	Firebird	305 (5.0)	F	8	OHV
	Camaro	305 (5.0)	E	8	OHV
	Firebird	305 (5.0)	E	8	OHV
	Camaro	350 (5.7)	8	8	OHV
	Firebird	350 (5.7)	8	8	OHV

OHV Overhead Valve

GENERAL ENGINE SPECIFICATIONS

Year	VIN	No. Cylinder Displacement cu. in. (liter)	Fuel System Type	Net Horsepower @ rpm	Net Torque @ rpm (ft. lbs.)	Bore × Stroke (in.)	Compression Ratio	Oil Pressure @ 2000 rpm
1988	S	6-173 (2.8)	MFI	135 @ 5100	165 @ 3600	3.500 × 3.000	8.9:1	55
	F	8-305 (5.0)	TPI	190 @ 4800	240 @ 3200	3.736 × 3.480	9:3:1	55
	E	8-305 (5.0)	TBI	150 @ 4000	240 @ 3200	3.736 × 3.480	9.3:1	55
	8	8-350 (5.7)	TPI	230 @ 4000	300 @ 3200	4.000 × 3.480	9.5:1	55
1989	S	6-173 (2.8)	MFI	135 @ 5100	165 @ 3600	3.500 × 3.000	8.9:1	55
	F	8-305 (5.0)	TPI	190 @ 4800	240 @ 3200	3.736 × 3.480	9.3:1	55
	E	8-305 (5.0)	TBI	150 @ 4000	240 @ 3200	3.736 × 3.480	9.3:1	55
	8	8-350 (5.7)	TPI	230 @ 4000	300 @ 3200	4.000 × 3.480	9.5:1	55
	7	6-231 (3.8)	SFI-Turbo	235 @ 4400	330 @ 2800	3.800 × 3.400	8.0:1	37
1990	T	6-191 (3.1)	MFI	140 @ 4400	180 @ 3600	3.500 × 3.310	8.9:1	55
	F	8-305 (5.0)	TPI	230 @ 4400	300 @ 3200	3.736 × 3.480	9.3:1	55
	E	8-305 (5.0)	TBI	170 @ 4000	255 @ 2400	3.736 × 3.480	9.3:1	55
	8	8-350 (5.7)	TPI	240 @ 4400	345 @ 3200	4.000 × 3.480	9.3:1	55
1991-92	T	6-191 (3.1)	MFI	140 @ 4400	180 @ 3600	3.500 × 3.310	8.9:1	55
	F	8-305 (5.0)	TPI	140 @ 4400	300 @ 3200	3.736 × 3.480	9.3:1	55
	E	8-305 (5.0)	TPI	170 @ 4000	255 @ 2400	3.736 × 3.480	9.3:1	55
	8	8-350 (5.7)	TPI	240 @ 4400	345 @ 3200	4.000 × 3.480	9.5:1	55

MFI—Multi-Port Fuel Injection
SFI—Sequential Fuel Injection
TBI—Throttle Body Injection
TPI—Tuned Port Injection

GASOLINE ENGINE TUNE-UP SPECIFICATIONS

Year	VIN	No. Cylinder Displacement cu. in. (liter)	Spark Plugs Type	Spark Plugs Gap (in.)	Ignition Timing (deg.) MT	Ignition Timing (deg.) AT	Compression Pressure (psi)	Fuel Pump (psi)	Speed (rpm) MT	Speed (rpm) AT	Valve Clearance In.	Valve Clearance Ex.
1988	S	6-173 (2.8)	R-42CTS	0.045	10	10	NA	40–47	450	400	Hyd.	Hyd.
	F	8-305 (5.0)	R-43TS	0.035	6	6	NA	40–47	500	500	Hyd.	Hyd.
	E	8-305 (5.0)	R-45TS	0.035	6	6	NA	9.0–13.0	450	400	Hyd.	Hyd.
	8	8-350 (5.7)	R-43TS	0.035	6	6	NA	40–47	450	400	Hyd.	Hyd.
1989	S	6-173 (2.8)	R-42CTS	0.045	10	10	NA	40–47	450	400	Hyd.	Hyd.
	F	8-305 (5.0)	R-43TS	0.035	6	6	NA	40–47	500	500	Hyd.	Hyd.
	E	8-305 (5.0)	R-45TS	0.035	6	6	NA	9.0–13.0	450	400	Hyd.	Hyd.
	8	8-350 (5.7)	R-43TS	0.035	6	6	NA	40–47	450	400	Hyd.	Hyd.
	7	6-231 (3.8)	R-43TS	0.035	①	①	NA	34–40	①	①	Hyd.	Hyd.
1990	T	6-191 (3.1)	R-43TSK	0.045	10	10	NA	40–47	①	①	Hyd.	Hyd.
	F	8-305 (5.0)	R-45TS	0.035	6	6	NA	40–47	①	①	Hyd.	Hyd.
	E	8-305 (5.0)	R-45TS	0.035	0	0	NA	9.0–13.0	①	①	Hyd.	Hyd.
	8	8-350 (5.7)	R-45TS	0.035	6	6	NA	40–47	①	①	Hyd.	Hyd.
1991	T	6-191 (3.1)	R-43TSK	0.045	10	10	NA	40–47	①	①	Hyd.	Hyd.
	F	8-305 (5.0)	R-45TS	0.035	6	6	NA	40–47	①	①	Hyd.	Hyd.
	E	8-305 (5.0)	R-45TS	0.035	0	0	NA	40–47	①	①	Hyd.	Hyd.
	8	8-350 (5.7)	R-45TS	0.035	6	6	NA	40–47	①	①	Hyd.	Hyd.
1992							REFER TO UNDERHOOD STICKER					

Hyd.—Hydraulic
① See Underhood Emission Decal
② Key on, Engine Off

FIRING ORDERS

NOTE: To avoid confusion, always replace spark plug wires one at a time.

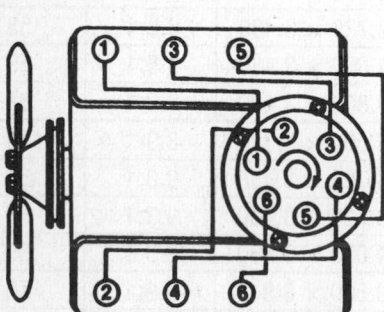

2.8L and 3.1L Engines
Engine Firing Order: 1–2–3–4–5–6
Distributor Rotation: Clockwise

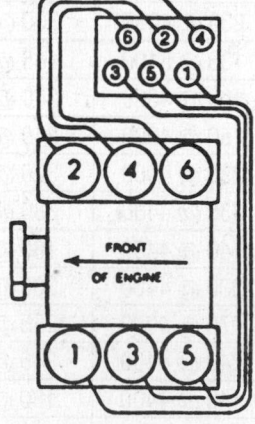

3.8L Engine
Engine Firing Order: 1–6–5–4–3–2
Distributorless Ignition System

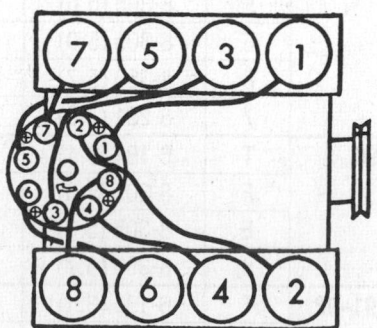

5.0L and 5.7L Engines
Engine Firing Order: 1–8–4–3–6–5–7–2
Distributor Rotation: Clockwise

CAPACITIES

Year	Model	VIN	No. Cylinder Displacement cu. in. (liter)	Engine Crankcase (qts.) with Filter	Engine Crankcase (qts.) without Filter	Transmission (pts.) 4-Spd	Transmission (pts.) 5-Spd	Transmission (pts.) Auto.	Drive Axle (pts.)	Fuel Tank (gal.)	Cooling System (qts.)
1988	Camaro	S	6-173 (2.8)	4.0	4.0	—	6.6	8.5①	3.5	16	13
	Camaro	F	8-305 (5.0)	5.0	4.0	—	6.6	8.5①	3.5	16	17
	Camaro	E	8-305 (5.0)	5.0	4.0	—	6.6	8.5①	3.5	16	15.5
	Camaro	8	8-350 (5.7)	5.0	4.0	—	6.6	8.5①	3.5	16	17
	Firebird	S	6-173 (2.8)	4.0	4.0	—	6.6	8.5①	3.5	16	13
	Firebird	F	8-305 (5.0)	5.0	4.0	—	6.6	8.5①	3.5	16	17
	Firebird	E	8-305 (5.0)	5.0	4.0	—	6.6	8.5①	3.5	16	15.5
	Firebird	8	8-350 (5.7)	5.0	4.0	—	6.6	8.5①	3.5	16	17
1989	Camaro	S	6-173 (2.8)	4.0	4.0	—	6.6	8.5①	3.5	16	13
	Camaro	F	8-305 (5.0)	5.0	4.0	—	6.6	8.5①	3.5	16	17
	Camaro	E	8-305 (5.0)	5.0	4.0	—	6.6	8.5①	3.5	16	15.5
	Camaro	8	8-350 (5.7)	5.0	4.0	—	6.6	8.5①	3.5	16	17
	Firebird	S	6-173 (2.8)	4.0	4.0	—	6.6	8.5①	3.5	16	13
	Firebird	F	8-305 (5.0)	5.0	4.0	—	6.6	8.5①	3.5	16	17
	Firebird	E	8-305 (5.0)	5.0	4.0	—	6.6	8.5①	3.5	16	15.5
	Firebird	8	8-350 (5.7)	5.0	4.0	—	6.6	8.5①	3.5	16	17
	Firebird	7	6-231 (3.8)	—	4.0	—	—	10	3.5	15.5	16.5
1990	Camaro	T	6-191 (3.1)	4.0	4.0	—	5.9	10	3.5	15.5	14.5
	Camaro	F	8-305 (5.0)	5.0	4.0	—	5.9	10	3.5	15.5	17.5
	Camaro	E	8-305 (5.0)	5.0	4.0	—	5.9	10	3.5	15.5	17.5
	Camaro	8	8-350 (5.7)	5.0	4.0	—	5.9	10	3.5	15.5	16.5
	Firebird	T	6-191 (3.1)	4.0	4.0	—	5.9	10	3.5	15.5	14.5
	Firebird	F	8-305 (5.0)	5.0	4.0	—	5.9	10	3.5	15.5	17.5
	Firebird	E	8-305 (5.0)	5.0	4.0	—	5.9	10	3.5	15.5	17.5
	Firebird	8	8-350 (5.7)	5.0	4.0	—	5.9	10	3.5	15.5	16.5
1991-92	Camaro	T	6-191 (3.1)	4.0	4.0	—	5.9	10	3.5	15.5	14.8
	Camaro	F	8-305 (5.0)	5.0	4.0	—	5.9	10	3.5	15.5	18.0
	Camaro	E	8-305 (5.0)	5.0	4.0	—	5.9	10	3.5	15.5	18.0
	Camaro	8	8-350 (5.7)	5.0	4.0	—	5.9	10	3.5	15.5	16.7
	Firebird	T	6-191 (3.1)	4.0	4.0	—	5.9	10	3.5	15.5	14.8
	Firebird	F	8-305 (5.0)	5.0	4.0	—	5.9	10	3.5	15.5	18.0
	Firebird	E	8-305 (5.0)	5.0	4.0	—	5.9	10	3.5	15.5	18.0
	Firebird	8	8-350 (5.7)	5.0	4.0	—	5.9	10	3.5	5.56	16.7

① 10.0 if equipped with overdrive transmission

CAMSHAFT SPECIFICATIONS

All measurements given in inches.

Year	VIN	No. Cylinder Displacement cu. in. (liter)	Journal Diameter 1	Journal Diameter 2	Journal Diameter 3	Journal Diameter 4	Journal Diameter 5	Lobe Lift In.	Lobe Lift Ex.	Bearing Clearance	Camshaft End Play
1988	S	6-173 (2.8)	1.8976–1.8996	1.8976–1.8996	1.8976–1.8996	1.8976–1.8996	—	0.2350	0.2660	NA	NA
	F	8-305 (5.0)	1.8682–1.8692	1.8682–1.8692	1.8682–1.8692	1.8682–1.8692	1.8682–1.8692	0.2690	0.2760	NA	0.004–0.012

CAMSHAFT SPECIFICATIONS

All measurements given in inches.

Year	VIN	No. Cylinder Displacement cu. in. (liter)	Journal Diameter 1	2	3	4	5	Lobe Lift In.	Ex.	Bearing Clearance	Camshaft End Play
1988	E	8-305 (5.0)	1.8682–1.8692	1.8682–1.8692	1.8682–1.8692	1.8682–1.8692	1.8682–1.8692	0.2340	0.2570	NA	0.004–0.012
	8	8-350 (5.7)	1.8682–1.8692	1.8682–1.8692	1.8682–1.8692	1.8682–1.8692	1.8682–1.8692	0.2730	0.2820	NA	0.004–0.012
1989	S	6-173 (2.8)	1.8976–1.8996	1.8976–1.8996	1.8976–1.8996	1.8976–1.8996	—	0.2350	0.2660	NA	NA
	F	8-305 (5.0)	1.8682–1.8692	1.8682–1.8692	1.8682–1.8692	1.8682–1.8692	1.8682–1.8692	0.2690	0.2760	NA	0.004–0.012
	E	8-305 (5.0)	1.8682–1.8692	1.8682–1.8692	1.8682–1.8692	1.8682–1.8692	1.8682–1.8692	0.2340	0.2570	NA	0.004–0.012
	8	8-350 (5.7)	1.8682–1.8692	1.8682–1.8692	1.8682–1.8692	1.8682–1.8692	1.8682–1.8692	0.2730	0.2820	NA	0.004–0.012
	7	6-231 (3.8)	1.7850–1.7860	1.7850–1.7860	1.7850–1.7860	1.7850–1.7860	—	NA	NA	①	NA
1990	T	6-191 (3.1)	1.8678–1.8697	1.8678–1.8697	1.8678–1.8697	1.8678–1.8697	—	0.2626	0.2732	0.0010–0.0040	NA
	F	8-305 (5.0)	1.8682–1.8692	1.8682–1.8692	1.8682–1.8692	1.8682–1.8692	1.8682–1.8692	0.2690	0.2760	NA	0.004–0.012
	E	8-305 (5.0)	1.8682–1.8692	1.8682–1.8692	1.8682–1.8692	1.8682–1.8692	1.8682–1.8692	0.2340	0.2570	NA	0.004–0.012
	8	8-350 (5.7)	1.8682–1.8692	1.8682–1.8692	1.8682–1.8692	1.8682–1.8692	1.8682–1.8692	0.2730	0.2820	NA	0.004–0.012
1991–92	T	6-191 (3.1)	1.8678–1.8697	1.8678–1.8697	1.8678–1.8697	1.8678–1.8697	—	0.2626	0.2732	0.0010–0.0040	NA
	F	8-305 (5.0)	1.8682–1.8692	1.8682–1.8692	1.8682–1.8692	1.8682–1.8692	1.8682–1.8692	0.2750	0.2850	NA	0.004–0.012
	E	8-305 (5.0)	1.8682–1.8692	1.8682–1.8692	1.8682–1.8692	1.8682–1.8692	1.8682–1.8692	0.2340	0.2570	NA	0.004–0.012
	8	8-350 (5.7)	1.8682–1.8692	1.8682–1.8692	1.8682–1.8692	1.8682–1.8692	1.8682–1.8692	0.2750	0.2850	NA	0.004–0.012

① No. 1—0.0005–0.0025   NA—Not available
No. 2, 3, 4—0.0005–0.0035

CRANKSHAFT AND CONNECTING ROD SPECIFICATIONS

All measurements are given in inches.

Year	VIN	No. Cylinder Displacement cu. in. (liter)	Crankshaft Main Brg. Journal Dia.	Main Brg. Oil Clearance	Shaft End-play	Thrust on No.	Connecting Rod Journal Diameter	Oil Clearance	Side Clearance
1988	S	6-173 (2.8)	2.6473–2.6483	0.0017–0.0029	0.0019–0.0066	3	1.998–1.999	0.0014–0.0035	0.0060–0.0170
	F	8-305 (5.0)	①	②	0.0020–0.0060	5	2.098–2.099	0.0018–0.0039	0.0080–0.0140
	E	8-305 (5.0)	①	②	0.0020–0.0060	5	2.098–2.099	0.0018–0.0039	0.0080–0.0140
	8	8-350 (5.7)	①	②	0.0020–0.0060	5	2.098–2.099	0.0013–0.0035	0.0060–0.0140

CRANKSHAFT AND CONNECTING ROD SPECIFICATIONS

All measurements are given in inches.

Year	VIN	No. Cylinder Displacement cu. in. (liter)	Crankshaft Main Brg. Journal Dia.	Crankshaft Main Brg. Oil Clearance	Crankshaft Shaft End-play	Crankshaft Thrust on No.	Connecting Rod Journal Diameter	Connecting Rod Oil Clearance	Connecting Rod Side Clearance
1989	S	6-173 (2.8)	2.6473– 2.6483	0.0017– 0.0029	0.0019– 0.0066	3	1.998– 1.999	0.0014– 0.0035	0.0060– 0.0170
	F	8-305 (5.0)	①	②	0.0020– 0.0060	5	2.098– 2.099	0.0018– 0.0039	0.0080– 0.0140
	E	8-305 (5.0)	①	②	0.0020– 0.0060	5	2.098– 2.099	0.0018– 0.0039	0.0080– 0.0140
	8	8-350 (5.7)	①	②	0.0020– 0.0060	5	2.098– 2.099	0.0013– 0.0035	0.0060– 0.0140
	7	6-231 (3.8)	2.4995	0.0003– 0.0018	0.003– 0.011	2	2.2487– 2.2495	0.0005– 0.0026	0.0030– 0.0150
1990	T	6-191 (3.1)	2.6473– 2.6483	③	0.0024– 0.0083	3	1.9983– 1.9994	0.0014– 0.0036	0.0140– 0.0290
	F	8-305 (5.0)	①	②	0.001– 0.007	5	2.089– 2.099	0.0013– 0.0035	0.0060– 0.0140
	E	8-305 (5.0)	①	②	0.001– 0.007	5	2.089– 2.099	0.0013– 0.0035	0.0060– 0.0140
	8	8-350 (5.7)	①	②	0.001– 0.007	5	2.089– 2.099	0.0013– 0.0035	0.0060– 0.0140
1991–92	T	6-191 (3.1)	2.6473– 2.6483	④	0.0024– 0.0083	3	1.9983– 1.9994	0.0011– 0.0033	0.0140– 0.0290
	F	8-305 (5.0)	①	②	0.001– 0.007	5	2.089– 2.099	0.0013– 0.0035	0.0060– 0.0140
	E	8-305 (5.0)	①	②	0.001– 0.007	5	2.089– 2.099	0.0013– 0.0035	0.0060– 0.0140
	8	8-350 (5.7)	①	②	0.001– 0.007	5	2.089– 2.099	0.0013– 0.0035	0.0060– 0.0140

① No. 1—2.4484-2.4493
   Nos. 2, 3, 4—2.4481-2.4490
   No. 5—2.4479-2.4488
② No. 1—0.0008-0020
   Nos. 2, 3, 4—0.0011-0.0020
   No. 5—.0-017-0.0032

③ Main Bearing Clearance—0.0012-0.0027
   Main Thrust Bearing Clearance—0.0016-0.0027
④ Main Bearing Clearance—0.0012-0.0030
   Main Thrust Bearing Clearance—0.0016-0.0030

VALVE SPECIFICATIONS

Year	VIN	No. Cylinder Displacement cu. in. (liter)	Seat Angle (deg.)	Face Angle (deg.)	Spring Test Pressure (lbs.)	Spring Installed Height (in.)	Stem-to-Guide Clearance (in.) Intake	Stem-to-Guide Clearance (in.) Exhaust	Stem Diameter (in.) Intake	Stem Diameter (in.) Exhaust
1988	S	6-173 (2.8)	46	45	194 @ 1.18	1.57	0.0010– 0.0027	0.3410– 0.0027	0.3410– 0.3420	0.3410– 0.3420
	E	8-305 (5.0)	46	45	194-206 @ 1.25	①	0.0010– 0.0027	0.0010– 0.0027	0.3410– 0.3420	0.3410– 0.3420
	F	8-305 (5.0)	46	45	194-206 @ 1.25	①	0.0010– 0.0027	0.0010– 0.0027	0.3410– 0.3420	0.3410– 0.3420
	8	8-350 (5.7)	46	45	194-206 @ 1.25	①	0.0010– 0.0027	0.0010– 0.0027	0.3410– 0.3420	0.3410– 0.3420

VALVE SPECIFICATIONS

Year	VIN	No. Cylinder Displacement cu. in. (liter)	Seat Angle (deg.)	Face Angle (deg.)	Spring Test Pressure (lbs.)	Spring Installed Height (in.)	Stem-to-Guide Clearance (in.)		Stem Diameter (in.)	
							Intake	Exhaust	Intake	Exhaust
1989	S	6-173 (2.8)	46	45	194 @ 1.18	1.57	0.0010–0.0027	0.3410–0.0027	0.3410–0.3420	0.3410–0.3420
	E	8-305 (5.0)	46	45	194-206 @ 1.25	①	0.0010–0.0027	0.0010–0.0027	0.3410–0.3420	0.3410–0.3420
	F	8-305 (5.0)	46	45	194-206 @ 1.25	①	0.0010–0.0027	0.0010–0.0027	0.3410–0.3420	0.3410–0.3420
	8	8-350 (5.7)	46	45	194-206 @ 1.25	①	0.0010–0.0027	0.0010–0.0027	0.3410–0.3420	0.3410–0.3420
	7	6-231 (3.8)	45	NA	185 @ 1.340	1.73	0.0015–0.0035	0.0015–0.0032	0.3420–0.3401	0.3412–0.3405
1990	T	6-191 (3.1)	46	45	190 @ 1.20	1.60	0.0014–0.0025	0.0016–0.0029	0.3410–0.3420	0.3410–0.3420
	F	8-305 (5.0)	46	45	194-206 @ 1.25	①	0.0011–0.0027	0.0011–0.0027	0.3410–0.3420	0.3410–0.3420
	E	8-305 (5.0)	46	45	194-206 @ 1.25	①	0.0011–0.0027	0.0011–0.0027	0.3410–0.3420	0.3410–0.3420
	8	8-350 (5.7)	46	45	194-206 @1.25	①	0.0011–0.0027	0.0011–0.0027	0.3410–0.3420	0.3410–0.3420
1991–92	T	6-191 (3.1)	46	45	190 @ 1.20	1.60	0.0014–0.0025	0.0016–0.0029	0.3410–0.3420	0.3410–0.3420
	F	8-305 (5.0)	46	45	194-206 @ 1.25	①	0.0011–0.0027	0.0011–0.0027	0.3410–0.3420	0.3410–0.3420
	E	8-305 (5.0)	46	45	194-206 @ 1.25	①	0.0011–0.0027	0.0011–0.0027	0.3410–0.3420	0.3410–0.3420
	8	8-350 (5.7)	46	45	194-206 @1.25	①	0.0011–0.0027	0.0011–0.0027	0.3410–0.3420	0.3410–0.3420

① Intake—1.72
Exhaust—1.59

PISTON AND RING SPECIFICATIONS
All measurements are given in inches.

Year	VIN	No. Cylinder Displacement cu. in. (liter)	Piston Clearance	Ring Gap			Ring Side Clearance		
				Top Compression	Bottom Compression	Oil Control	Top Compression	Bottom Compression	Oil Control
1988	S	6-173 (2.8)	0.017–0.043	0.0098–0.0196	0.0098–0.0196	0.0200–0.0550	0.0011–0.0027	0.0015–0.0037	0.0078 Max.
	F	8-305 (5.0)	NA	0.0100–0.0200	0.0100–0.0250	0.0150–0.0550	0.0012–0.0032	0.0012–0.0032	0.0020–0.0070
	E	8-305 (5.0)	0.0027	0.0100–0.0200	0.0100–0.0250	0.0150–0.0550	0.0012–0.0032	0.0012–0.0032	0.0020–0.0070
	8	8-350 (5.7)	0.0027	0.0100–0.0200	0.0100–0.0250	0.0150–0.0550	0.0012–0.0032	0.0012–0.0032	0.0020–0.0070
1989	S	6-173 (2.8)	0.017–0.043	0.0098–0.0196	0.0098–0.0196	0.0200–0.0550	0.0300–0.0070	0.0040–0.0950	0.0078 Max.
	F	8-305 (5.0)	NA	0.0100–0.0200	0.0100–0.0250	0.0150–0.0550	0.0012–0.0032	0.0012–0.0032	0.0020–0.0070
	E	8-305 (5.0)	0.0027	0.0100–0.0200	0.0100–0.0250	0.0150–0.0550	0.0012–0.0032	0.0012–0.0032	0.0020–0.0070

PISTON AND RING SPECIFICATIONS

All measurements are given in inches.

Year	VIN	No. Cylinder Displacement cu. in. (liter)	Piston Clearance	Ring Gap			Ring Side Clearance		
				Top Compression	Bottom Compression	Oil Control	Top Compression	Bottom Compression	Oil Control
1989	8	8-350 (5.7)	0.0027	0.0100–0.0200	0.0100–0.0250	0.0150–0.0550	0.0012–0.0032	0.0012–0.0032	0.0020–0.0070
	7	6-231 (3.8)	0.0013 0.0035	0.0100–0.0200	0.0100–0.0200	0.0150–0.0550	NA	NA	NA
1990	T	6-191 (3.1)	0.0012–0.0028	0.0100–0.0200	0.0100–0.0200	0.0100–0.0300	0.0020–0.0035	0.0020–0.0035	0.0075 Max.
	F	8-305 (5.0)	0.0007–0.0027	0.0100–0.0200	0.0100–0.0250	0.0650 Max.	0.0012–0.0032	0.0012–0.0032	0.0080 Max.
	E	8-305 (5.0)	0.0007–0.0027	0.0100–0.0200	0.0100–0.0250	0.0650 Max.	0.0012–0.0032	0.0012–0.0032	0.0080 Max.
	8	8-350 (5.7)	0.0007–0.0027	0.0100 0.0200	0.0180–0.0260	0.0650 Max.	0.0012–0.0032	0.0012–0.0032	0.0080 Max.
1991–92	T	6-191 (3.1)	0.0012–0.0028	0.0100–0.0200	0.0200–0.0280	0.0100–0.0300	0.0020–0.0035	0.0020–0.0035	0.0070 Max.
	F	8-305 (5.0)	0.0007–0.0027	0.0100–0.0200	0.0100–0.0250	0.0650 Max.	0.0012–0.0032	0.0012–0.0032	0.0080 Max.
	E	8-305 (5.0)	0.0007–0.0027	0.0100–0.0200	0.0100–0.0250	0.0650 Max.	0.0012–0.0032	0.0012–0.0032	0.0080 Max.
	8	8-350 (5.7)	0.0007–0.0027	0.0100 0.0200	0.0180–0.0260	0.0650 Max.	0.0012–0.0032	0.0012–0.0032	0.0080 Max.

NA—Not available

TORQUE SPECIFICATIONS

All readings in ft. lbs.

Year	VIN	No. Cylinder Displacement cu. in. (liter)	Cylinder Head Bolts	Main Bearing Bolts	Rod Bearing Bolts	Crankshaft Pulley Bolts	Flywheel Bolts	Manifold		Spark Plugs
								Intake	Exhaust	
1988	S	6-173 (2.8)	③	63–83	34–45	75	52	13–25	19–31	7–15
	F	8-305 (5.0)	60–75	63–85	42–47	70	74	25–45	②	15–20
	E	8-305 (5.0)	60–75	63–85	42–47	70	74	25–45	②	15–20
	8	8-350 (5.7)	60–75	63–85	42–47	70	74	25–45	②	15–20
1989	S	6-173 (2.8)	③	63–83	34–45	75	52	13–25	19–31	7–15
	F	8-305 (5.0)	60–75	63–85	42–47	70	74	25–45	②	15–20
	E	8-305 (5.0)	60–75	63–85	42–47	70	74	25–45	②	15–20
	8	8-350 (5.7)	60–75	63–85	42–47	70	74	25–45	②	15–20
	7	6-231 (3.8)	①	100	40	219	61	45	37	20
1990	T	6-191 (3.1)	③	73	39	70	52	④	25	25
	F	8-305 (5.0)	68	77	44	70	74	35	②	22
	E	8-305 (5.0)	68	77	44	70	74	35	②	22
	8	8-350 (5.7)	68	77	44	70	74	35	②	22

TORQUE SPECIFICATIONS

All readings in ft. lbs.

Year	VIN	No. Cylinder Displacement cu. in. (liter)	Cylinder Head Bolts	Main Bearing Bolts	Rod Bearing Bolts	Crankshaft Pulley Bolts	Flywheel Bolts	Manifold Intake	Manifold Exhaust	Spark Plugs
1991-92	T	6-191 (3.1)	③	73	39	70	52	④	25	25
	F	8-305 (5.0)	68	77	44	70	74	35	②	22
	E	8-305 (5.0)	68	77	44	70	74	35	②	22
	8	8-350 (5.7)	68	77	44	70	74	35	②	22

① Torque in 3 steps:
1st step: Tighten to 2.5 ft. lbs.
2nd step: Rotate wrench an additional 90 degrees.
3rd step: Rotate wrench an additional 90 degrees.
(Should 60 ft. lbs. be reached at any time in steps 2 & 3—STOP—do not turn any further)
② Outer bolts—14–26
Center bolts—20–32
③ Torque in 2 steps:
1st step: Tighten to 40 ft. lbs.
2nd step: Rotate wrench an additional 90 degrees.
④ Lower intake manifold—19 ft. lbs.
Center intake manifold—15 ft. lbs.

BRAKE SPECIFICATIONS

All measurements in inches unless noted.

Year	Model	Lug Nut Torque (ft. lbs.)	Master Cylinder Bore	Brake Disc Minimum Thickness	Brake Disc Maximum Runout	Standard Brake Drum Diameter	Minimum Lining Thickness Front	Minimum Lining Thickness Rear
1988	Camaro	80	NA	0.965	0.005	9.500	0.030	0.030 ① ③
	Firebird	80	NA	0.965	0.005	9.500	0.030	0.030 ① ③
1989	Camaro	80	NA	0.965	0.005	9.500	0.030	0.030 ① ③
	Firebird	80	NA	0.965	0.005	9.500	0.030	0.030 ① ③
1990	Camaro	81	④	②	0.005	9.500	0.030	0.030 ③
	Firebird	81	④	②	0.005	9.500	0.030	0.030 ③
1991-92	Camaro	81	④	②	0.005	9.500	0.030	0.030 ③
	Firebird	81	④	②	0.005	9.500	0.030	0.030 ③

① 0.062—bonded
② Front rotor—0.965
Rear rotor—0.724
③ Rear disc—0.030
④ Disc/Drum—0.945
Disc/Disc—1.00

WHEEL ALIGNMENT

Year	Model	Caster Range (deg.)	Caster Preferred Setting (deg.)	Camber Range (deg.)	Camber Preferred Setting (deg.)	Toe-in (in.)	Steering Axis Inclination (deg.)
1988	Camaro	4 1/2P–5 1/2P	5P	1/2N–1 1/2P	0	3/64P	NA
	Firebird	4 1/2P–5 1/2P	5P	1/2N–1 1/2P	0	3/64P	NA
1989	Camaro	4 3/16P–5 3/16P	4 11/16P	3/16N–13/16P	5/16P	0	NA
	Firebird	4 3/16P–5 3/16P	4 11/16P	3/16N–13/16P	5/16P	0	NA
1990	Camaro	4 1/2P–5 1/2P	5P	1/2P–1 1/2P	1P	3/64P	NA
	Firebird	4 1/2P–5 1/2P	5P	1/2P–1 1/2P	1P	3/64P	NA
1991-92	Camaro	4 5/16P–5 5/16P	4 3/4P	3/4N–1 1/2P	3/8P	0	NA
	Firebird	4 5/16P–5 5/16P	4 3/4P	3/4N–1 1/2P	3/8P	0	NA

NA—Not available
P—Positive
N—Negative

ENGINE MECHANICAL

NOTE: Disconnecting the negative battery cable on some vehicles may interfere with the functions of the on board computer systems and may require the computer to undergo a relearning process.

Engine Assembly

REMOVAL & INSTALLATION

2.8L and 3.1L Engines

1. Disconnect the negative battery cable.
2. Remove the air cleaner duct.
3. Mark the hood location on the hood supports and remove the hood.
4. Remove the water pump drive belt.
5. Drain the radiator and remove the radiator hoses. Disconnect the heater hoses and the transmission cooler lines.
6. Remove the fan shroud, fan and radiator.
7. Disconnect the throttle linkage, including the cruise control detent cable.
8. Disconnect the air conditioning compressor and lay aside. Remove the power steering pump and lay aside.
9. Remove the vacuum brake booster line.
10. Remove the distributor cap and spark plug wires.
11. Disconnect the necessary electical conctions and hoses.
12. Raise and safely support the vehicle.
13. Disconnect the exhaust pipes at the exhaust manifolds.
14. Remove the flywheel cover and remove the converter bolts.
15. Disconnect the starter wire connections.
16. Remove the bellhousing and the motor mount through bolts.
17. Lower the vehicle.
18. Relieve the fuel system pressure. Disconnect the fuel lines.
19. Support the transmission with a suitable jack. Attach an engine lifting device.
20. Remove the engine assembly.
To install:
21. Position the engine assembly in the vehicle.
22. Attach the motor mount to engine brackets and lower the engine in place. Remove the engine lifting device and the transmission jack.
23. Raise and support the vehicle safely.
24. Install the motor mount through

bolts and tighten the nuts to 50 ft. lbs. (68 Nm). Install the bellhousing bolts and tighten to 35 ft. lbs. (47 Nm).
25. On vehicles with automatic transmission, install the converter to flywheel attaching bolts to 46 ft. lbs. (63 Nm).
26. Install the flywheel spalsh shield and tighten to 89 inch lbs. (10 Nm).
27. Connect the starter wires and the fuel lines.
28. Install the exhaust pipe on the exhaust manifold.
29. Lower the vehicle.
30. Install the power steering pump and the air conditioning compressor.
31. Connect the necessary wires and hoses.
32. Install the radiator, fan and fan shroud. Connect the radiator and heater hoses and the transmission cooler lines.
33. Connect the vacuum brake booster line, the throttle linkage and cruise control cable. Install the distributor cap.
34. Fill the cooling system with the proper type and amount of coolant and the crankcase with the proper type of oil to the correct level.
35. Install the water pump drive belt, the air cleaner duct and the hood.
36. Connect the negative battery cable, start the engine and check for leaks.

3.8L Engine

1. Disconnect the negative battery cable.
2. Mark the location of the hood on the hood hinges and remove the hood.
3. Drain the engine coolant.
4. Remove the fan, pulleys and belts. Remove the radiator hoses and the radiator and fan shroud.
5. Disconnect the power steering pump and air conditioning compressor from their mounting brackets and position aside.
6. Relieve the fuel system pressure. Disconnect the fuel line and the battery ground cable from the engine.
7. Disconnect the necessary hoses and wiring.
8. Disconnect the throttle cable.
9. Remove the alternator assembly.
10. Disconnect the engine to body ground straps at the engine.
11. Raise and safely support the vehicle.
12. Disconnect the crossover pipe from the exhaust manifolds.
13. Remove the flywheel cover and use a scribe to mark the relationship of the torque converter to the flywheel. Remove the covnerter to flywheel bolts.
14. Disconnect the starter wiring.
15. Remove the transmission to engine attaching bolts and the motor

mount to frame bracket attaching bolts.
16. Lower the vehicle.
17. Support the transmission with a suitable jack and install and engine lifting device.
18. Remove the engine assembly.
To install:
19. Position the engine in the vehicle.
20. Install the transmission to engine attaching bolts and tighten to 35 ft. lbs. (48 Nm).
21. Raise and safely support the vehicle.
22. Install the motor mount to frame bracket attaching bolts and tighten to 48 ft. lbs. (65 Nm).
23. Install the converter to flywheel bolts making sure the scribed marks are aligned. Tighten the bolts to 46 ft. lbs. (63 Nm). Install the flywheel cover.
24. Connect the crossover pipe to the exhaust manifold.
25. Lower the vehicle.
26. Connect the engine to body ground straps and install the alternator.
27. Connect all necessary hoses and wiring.
28. Connect the throttle cable and the negative battery cable at the engine.
29. Connect the fuel lines.
30. Install the power steering pump and air conditioning compressor in their respective brackets.
31. Install the radiator, fan shroud, radiator and heater hoses and the fan, pulleys and belts.
32. Fill the cooling system with the proper type of quantity of coolant and the engine with the proper type of oil to the correct level.
33. Connect the negative battery cable, start the engine and check for leaks.

5.0L and 5.7L Engines

1. Disconnect the negative battery cable.
2. Mark the location of the hood on the hood hinges and remove the hood.
3. Remove the air cleaner.
4. Drain the cooling system.
5. Remove the radiator hoses.
6. Disconnect the transmission cooler lines, the electrical connectors and retaining clips at the fan and remove the fan and shroud.
7. Remove the radiator.
8. Remove the accessory drive belt.
9. Disconnect the throttle cable.
10. Remove the plenum extension screws and the plenum extension, if equipped.
11. Disconnect the spark plug wires at the distributor and remove the distributor.

12. Disconnect the necessary vacuum hoses and wiring.

13. Disconnect the power steering and air conditioning compressors from their respective brackets and lay them aside.

14. Relive the fuel system pressure. Disconnect the fuel lines.

15. Disconnect the negative battery cable at the engine block.

16. Raise and safely support the vehicle.

17. Remove the exhaust pipes at the exhaust manifolds.

18. Remove the flywheel cover and remove the converter to flywheel bolts.

19. Disconnect the starter wires.

20. Remove the bellhousing bolts and the motor mount through bolts.

21. Lower the vehicle.

22. Support the transmission with a suitable jack.

23. Remove the AIR/converter bracket and ground wires from the rear of the cylinder head.

24. Attach a suitable lifting device and remove the engine assembly.

To install:

25. Position the engine assembly in the vehicle.

26. Attach the motor mount to engine brackets and lower the engine into place.

27. Remove the engine lifting device and the transmission jack.

28. Raise and safely support the vehicle.

29. Install the motor mount through bolts and tighten to 50 ft. lbs. (68 Nm).

30. Install the bellhousing bolts and tighten to 35 ft. lbs. (47 Nm).

31. On vehicles with automatic transmission, install the converter to flywheel bolts. Tighten the bolts to 46 ft. lbs. (63 Nm). Install the flywheel cover.

32. Connect the starter wires and the fuel lines.

33. Connect the exhaust pipe at the exhaust manifold.

34. Lower the vehicle.

35. Connect the necessary wires and hoses.

36. Install the power steering pump and air conditioning compressor in their respective brackets.

37. Install the radiator, fan and fan shroud, radiator hoses and heater hoses.

38. Connect the transmission cooler lines and cooling fan electrical connectors.

39. Install the distributor.

40. Install the plenum extension, if equipped.

41. Fill the cooling system with the proper type and quantity of coolant and the crankcase with the proper type of oil to the correct level.

42. Install the air cleaner and the hood.

43. Connect the negative battery cable, start the engine, check for leaks and check timing.

Engine Mounts

REMOVAL & INSTALLATION

2.8L and 3.1L Engines

1. Disconnect the negative battery cable. Raise and support the vehicle safely.

2. Remove the engine mount through bolt. Using a suitable engine lift, safely raise the front of the engine and remove the engine mount bolts, nuts and washers from the crossmember.

3. Remove the engine mount.

NOTE: Raise the engine only enough for sufficient clearance. Check for interference between the rear of the engine and the cowl panel which could cause distributor damage.

4. Installation is the reverse of the removal procedure. Tighten the engine mount through bolt nut to 50 ft. lbs. (68 Nm) and the crossmember-to-mount bolts to 30 ft. lbs. (41 Nm).

3.8L Engine

1. Disconnect the negative battery cable.

2. Raise and safely support the vehicle.

3. Support the weight of the engine at the forward edge of the oil pan.

4. Remove the mount to cylinder block bolts.

5. Raise the engine slightly and remove the mount to mount bracket bolt and nut. Remove the engine mount.

6. Installation is the reverse of the removal procedure. Tighten the mount to cylinder block bolts to 59 ft. lbs. (80 Nm) and the bolt and nut to 48 ft. lbs. (65 Nm).

5.0L and 5.7L Engines

1. Disconnect the negative battery cable.

2. Raise and support the vehicle safely.

3. Support the engine with a suitable jack to unload the engine mount. Remove the engine mount retaining bolt from below the frame mounting bracket.

NOTE: Do not use a jack under the oil pan, crankshaft pulley or any sheet metal when supporting the engine. Due to the small clearance between the oil pan and oil pump screen, jacking against

the oil pan may cause it to be bent against the pump screen, resulting in a damaged oil pickup.

4. Using a suitable engine lift, raise the front of the engine and remove the engine mount and bracket bolts and nuts. Remove the engine mount.

NOTE: Raise the engine only engough for sufficient clearance. Check for interferance between the rear of the engine and the cowl panel which could cause distributor damage.

5. Installation is the reverse of the removal procedure. Tighten the mount and bracket bolts to 38 ft. lbs. (52 Nm) and the mount and bracket nuts to 30 ft. lbs. (41 Nm). Tighten the through bolt nut to 50 ft. lbs. (68 Nm).

Cylinder Head

REMOVAL & INSTALLATION

2.8L and 3.1L Engines

1. Disconnect the negative battery cable.

2. Relieve the fuel system pressure and drain the engine coolant from the radiator into a suitable continer.

3. Remove the intake manifold and the spark plugs.

4. Remove the dipstick tube and bracket. Raise and support the vehicle safely. Drain the oil and remove the oil filter. Lower the vehicle.

5. Remove the exhaust manifolds.

6. Remove the drive belt and remove the air conditioning compressor and lay aside.

7. Remove the power steering pump and bracket and lay aside.

8. Remove the ground cable from the rear of the cylinder head and remove the engine lift bracket.

9. Loosen the rocker arms until the pushrods can be removed.

10. Remove the belt tensioner, alternator and brackets.

11. Remove the AIR bracket, if equipped.

12. Remove the cylinder head bolts and remove the cylinder heads.

To install:

13. Clean the gasket mating surfaces of all components. Be careful not to nick or scratch any surfaces as this will allow leak paths. Clean the bolt threads in the cylinder block and on the head bolts. Dirt will affect bolt torque.

14. Place the head gaskets in position over the dowel pins, with the note "This Side Up" showing.

15. Install the cylinder heads.

16. Coat the cylinder head bolts threads with GM sealer 1052080 or equivalent, and install the bolts. Tighten the bolts in the proper se-

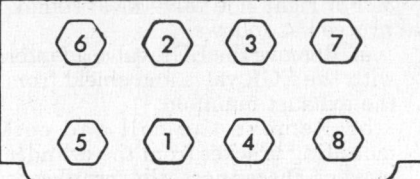

Cylinder head bolt torque sequence—2.8L and 3.1L engines

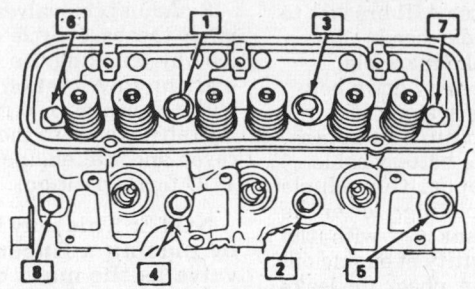

Cylinder head bolt torque sequence—3.8L engine

quence. Tighten the head bolts in 2 steps, first tighten to 40 ft. lbs. (55 Nm), then turn each bolt in sequence an additional ¼ turn (90 degrees).

17. Install the pushrods and loosely retain them with the rocker arms. Make sure the lower ends of the pushrods are in the lifter seats.

18. Install the power steering pump bracket and pump and the air conditioning compressor bracket and compressor.

19. Install the ground cable to the rear of the cylinder head.

20. Install the exhaust manifolds.

21. Install the dipstick tube and bracket.

22. Adjust the valve lash.

23. Install the intake manifold.

24. Install the AIR bracket and the belt tensioner.

25. Install the alternator bracket and alternator.

26. Install the accessory drive belt.

27. Install the spark plugs.

28. Fill the cooling system with the proper type and quantity of coolant. Install a new oil filter and fill the crankcase with the proper type and quantity of oil.

29. Connect the negative battery cable, start the vehicle and check for leaks.

3.8L Engine

1. Disconnect the negative battery cable. Drain the cooling system and relieve the fuel system pressure.

2. Raise and support the vehicle safely. Drain the engine oil and remove the oil filter. Lower the vehicle.

3. Remove the accessory drive belt.

4. If removing the right cylinder head, remove the air conditioning compressor with the hoses attached, the alternator and the alternator bracket.

5. If removing the left cylinder head, remove the oil level indicator, power steering pump with the hoses connected and the power steering pump mounting bracket.

6. Remove the spark plug wires and the exhaust manifolds.

7. Remove the intake manifold, valve covers, rocker arm shafts, pushrods, cylinder head bolts and the cylinder heads.

To install:

8. Clean the gasket mating surfaces

of all components. Be careful not to nick or scratch any surfaces at this will allow leak paths. Clean the bolt threads in the cylinder block and on the head bolts. Dirt will affect bolt torque.

9. Install the cylinder head gasket on the block and install the cylinder head.

10. Coat the head bolt threads with a suitable thread sealer and install. Tighten the head bolt in 3 steps as follows:

    a. Tighten in sequence to 25 ft. lbs. (34 Nm).

    b. Tighten in sequence an additional ¼ turn (90 degrees).

    c. Tighten in sequence an additional ¼ turn (90 degrees).

NOTE: If 60 ft. lbs. (81 Nm) is reached at any time in Steps b and c, stop at this point. Do not complete the balance of the 90 degree turn.

11. Install the exhaust manifolds, pushrods, rocker arm shaft, intake manifold, valve cover and spark plug wires.

12. When installing the right side cylinder head: install the alternator and it's mounting bracket and the air conditioning.

13. When installing the left side cylinder head: install the power steering pump and it's mounting bracket and the oil level indicator.

14. Install the accessory drive belt and fill the cooling system with the proper type and quantity of coolant.

15. Raise and safely support the vehicle. Install a new oil filter. Lower the vehicle and fill the crankcase with the proper type and quantity of engine oil.

16. Connect the negative battery cable, start the vehicle and check for leaks.

5.0L and 5.7L Engines

1. Disconnect the negative battery cable. Drain the cooling system and relieve the fuel system pressure.

2. Raise and support the vehicle safely. Drain the engine oil and remove the oil filter. Lower the vehicle.

3. Remove the accessory drive belt and remove the intake manifold.

4. Remove the power steering pump, alternator bracket or the air conditioning compressor mounting bracket, as necessary.

5. Remove the exhaust manifolds and the valve covers.

6. Remove the rocker arms and pushrods.

7. Disconnect the ground wires and the catalytic converter AIR pipe bracket at the rear of the cyliner heads.

8. Remove the cylinder head bolts and the cylinder head.

To install:

9. Clean the gasket mating surfaces of all components. Be careful not to nick or scratch any surfaces as this will allow leak paths. Clean the bolt threads in the cylinder block and on the head bolts. Dirt will affect bolt torque.

NOTE: When using a steel gasket, coat both sides of the new gasket with a thin even coat of sealer.

If using a composition gasket, do not use any sealer.

10. Position the head gasket over the dowel pins with the head up. Install the cylinder head over the dowel pins and gasket.

11. Coat the threads of the head bolts with GM 1052080 thread sealer or equivalent. Install the head bolts and tighten in sequence to 68 ft. lbs. (92 Nm).

12. Install the exhaust manifolds.

13. Install the pushrods and rocker arms and ajust the valve lash. Install the valve covers.

14. Install the power steering pump and alternator bracket or air conditioning compressor mounting bracket, as necessary.

15. Connect the ground wires and

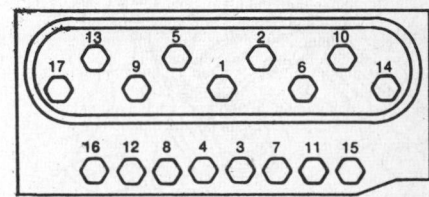

Cylinder head bolt torque sequence—5.0L and 5.7L engines

the catalytic converter AIR bracket to the rear of the cylinder head.

16. Install the intake manifold.

17. Install the accessory drive belt.

18. Fill the cooling system with the proper type and amount of colant. Connect the negative battery cable.

19. Raise and support the vehicle safely. Install a new oil filter, lower the vehicle. Fill the crankcase with the proper type and quantity of engine oil.

20. Start the engine, check for leaks and check the ignition timing.

Valve Lifters

REMOVAL & INSTALLATION

1. Disconnect the negative battery cable.

2. Drain the cooling system and relieve the fuel system pressure.

3. Remove the intake manifold assembly. Remove the valve covers.

4. Remove the pushrods and rocker arms or rocker shafts. Remove the valve lifter retainer assembly, if equipped with roller lifters. Using a suitable lifter removal tool, remove the valve lifters.

5. Installation is the reverse of the removal procedure. Install the lifters in the same lifter bore they were removed from. Coat the lifters in clean engine oil before installing them.

Valve Lash

ADJUSTMENT

3.8L Engine

Valve adjustment is not required, torque the rocker arms bolts to 28 ft. lbs. (38 Nm).

Except 3.8L Engine

1. Disconnect the negative battery cable. Remove the valve covers.

2. Tighten the rocker arm nuts until all lash is eliminated.

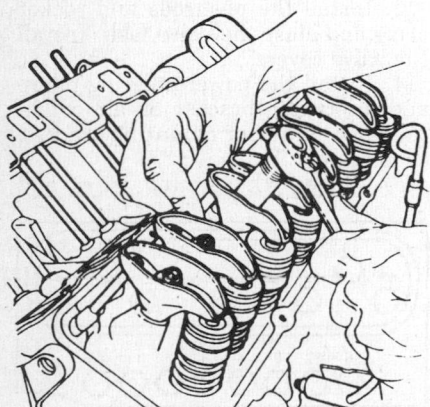

Valve adjustment procedure—5.0L and 5.7L engines

3. Adjust the valves when the lifter is on the base circle of the camshaft lobe by cranking the engine until the mark on the vibration damper lines up with the center or 0 mark on the timing tab fastened to the crankcase front cover and the engine is in the No. 1 TDC firing position.

NOTE: This may be determined by placing a finger on the No. 1 valve as the mark on the damper comes near the 0 mark on the crankcase front cover. If the valves move as the mark comes up to the timing tab, the engine is in the No. 6 position on the V8 engine or No. 4 position on the V6 engine and should be turned 1 full turn to reach to No. 1 firing position.

4. With the engine in the No. 1 firing position, adjust the following valves. V6 engine: intake–1, 5, 6 and exhaust–1, 2, 3; V8 engine: exhaust–1, 3, 4, 8 and intake–1, 2, 5, 7.

5. Back out adjusting nut until lash is felt at the pushrod, then turn in adjusting nut until all lash is removed. This can be determined by rotating pushrod while turning adjusting nut. When play has been removed, turn adjusting nut down 1 full additional turn.

6. Crank the engine 1 revolution until the pointer 0 mark and the vibration damper mark are again in alignment. This is the No. 6 position on the V8 engine or No. 4 position on the V6 engine.

7. With the engine in this position, adjust the following valves: V6 engine: intake–2, 3, 4 and exhaust–4, 5, 6; V8 engine: exhaust–2, 5, 6, 7 and intake–3, 4, 6, 8.

8. Install the valve covers and connect the negative battery cable.

9. Start the engine and adjust the idel speed as required.

Rocker Arms/Shafts

REMOVAL & INSTALLATION

2.8L and 3.1L Engines

1. Disconnect the negative battery cable.

2. For left side valve cover removal proceed as follows:

   a. Remove the accessory drive belt.

   b. Remove the transmission dipstick, if required.

   c. Remove the air management hose and air conditioning bracket, if equipped.

   d. Remove the intake plenum and throttle body assembly.

   e. Remove the valve cover reinforcments and nuts.

3. For right side valve cover removal proceed as follows:

   a. Remove the EGR valve adapter with the EGR valve and shield from the exhaust manifold.

   b. Remove the coil and coil mounting bracket from the cylinder head.c. Disconnect the crankcase vent pipe.

   d. Remove the intake plenum and throttle body assembly.

   e. Remove the valve cover reinforcments and nuts.

4. Remove the valve cover.

5. Remove the rocker arm nuts, rocker arm balls and rocker arms. Place the components in a rack so they can be reinstalled in the same location.

6. Installation is the reverse of the removal procedure. Ensure all gasket mating surfaces are clean before installing. Adjust the valve lash before installing the valve covers.

3.8L Engine

1. Disconnect the negative battery cable.

2. Remove the valve cover as follows:

   a. Remove the PCV pipe to the air cleaner, all necessary computer command control hoses wire connectors and the hot air tube.

   b. Remove the spark plug wires and the accessory mounting brackets, as required.

   c. Remove the valve cover attaching bolts and remove the valve cover.

3. Remove the rocker arm shaft retaining bolts and the rocker arm shaft.

4. If the rocker arms are being replaced, remove the nylon rocker arm retainers and discard. Install the replacement rocker arm using a new nylon retainer. Install the nylon retainers with a suitable drift of least ½ inch diameter.

NOTE: Service rocker arms are stamped (R) for right and (L) for left. Ensure the rocker arms are installed on the rocker shaft in the correct sequence.

5. Installation is the reverse of the removal procedure. Ensure all gasket mating surfaces are clean before installation. Tighten the rocker arm shaft retaining bolts to 25 ft. lbs. (35 Nm).

5.0L and 5.7L Engines

1. Disconnect the negative battery cable.

2. Remove the air cleaner, if necessary.

3. To remove the right side valve cover, perform the following:

   a. 1988 vehicles: Disconnect the wire EGR solenoid transfer tube from the plenum. Remove the coil and mounting bracket from the cylinder head. Remove the plenum, runners and throttle body assembly. Remove the valve cover retainers and nuts. Remove the valve cover.

b. 1989–92 vehicles: Remove the EGR pipe assembly, if necessary. Disconnect the electrical connections and wiring harnesses as necessary. Disconnect the spark plug wires from the distributor. Remove the crankcase vent hoses and valves. Remove the coil and disconnect the heater hose from the throttle body on 1991–92 vehicles. Remove the AIR control valve, check valve, pipes and hoses. Remove the valve cover bolts and remove the cover.

4. To remove the left side valve cover, perform the following:

a. 1988 vehicles: Remove the air management hose, if equipped. Remove the plenum and throttle body assembly. Remove the air conditioning bracket. Remove the valve cover reinforcements and nuts. Remove the valve cover.

b. 1989–92 vehicles: Disconnect the electrical connections and the wiring harnesses, as necessary. Remove the alternator and disconnect the crankcase hoses and the PCV valve. Remove the valve cover bolts and remove the valve cover.

5. Remove the rocker arm assemblies and place them in a rack so they may be reinstalled in the same location.

6. Install in the reverse order of removal. Make sure all gasket mating surfaces are clean before installation. Adjust the valve lash before installing the valve cover.

Intake Manifold

REMOVAL & INSTALLATION

2.8L and 3.1L Engines

1. Disconnect the negative battery cable.

2. Drain the cooling system and relieve the fuel system pressure.

3. Disconnect the air inlet duct at the throttle body and the crankcase vent pipe at the valve cover grommet.

4. Disconnect the vacuum harness connector from the throttle body.

5. Remove the throttle cable bracket bolt, the throttle body attaching bolts and remove the throttle body. Discard the throttle body gasket.

6. Remove the EGR transfer tube to plenum bolts and remove the EGR transfer tube. Discard the EGR transfer tube gasket.

7. Remove the air conditioning compressor to plenum bracket attaching hardware and the bracket.

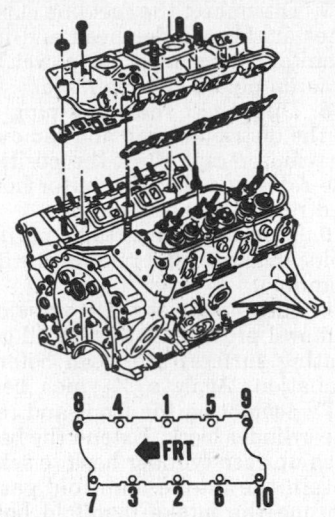

Intake manifold bolt tightening sequence—2.8L and 3.1L engines

8. Remove the plenum bolts/studs and the plenum. Discard the plenum gaskets.

9. Disconnect the fuel feed and return lines at the fuel rail. Discard the fuel line O-rings.

10. Disconnect the vacuum line at the pressure regulator and the injector electrical connectors.

11. Remove the fuel rail attaching bolts and the fuel rail assembly.

12. Remove the spark plug wires and the distributor cap. Mark the distributor position and remove the hold-down bracket and the distributor.

13. Remove the air management hose and bracket, if equipped.

14. Disconnect the emission canister hoses.

15. Remove the valve covers.

16. Remove the upper radiator hose at the manifold and disconnect the heater hose.

17. Disconnect the coolant switch sensors. Remove the transmission dipstick.

18. Remove the center intake manifold bolts and the center intake manifold.

To install:

19. Ensure all gasket mating surfaces are clean and free of oil or water prior to installation.

20. Place a $\frac{3}{16}$ inch diameter bead of RTV sealer on each ridge. Install new gaskets on the cylinder heads and hold in place by extending the ridge RTV bead up ¼ inch onto the gasket ends. The new gaskets will have to be cut, where indicated, to install behind the pushrods. Cut only those areas that are necessary.

21. Install the intake manifold on the engine. Make sure the areas between the case ridges and intake are completely sealed.

22. Install the intake manifold re-

taining bolts and nuts and torque to 13–25 ft. lbs. (18–34 Nm) for 1988 vehicles, 25–45 ft. lbs. (34–61 Nm) in the proper sequence on 1989 vehicles and 19 ft. lbs. (26 Nm) for 1990–92 vehicles.

23. Install the upper radiator hose and the valve covers and connect the heater hose and the coolant switch sensors.

24. Install the air management hose and bracket, if equipped.

25. Install the distributor, distributor cap and spark plug wires.

26. Install the fuel rail assembly in the intake manifold. Tighten the attaching bolts to 15 ft. lbs. (20 Nm).

27. Connect the injector electrical connectors and the vacuum line to the pressure regulator.

28. Install new O-rings on the fuel feed and return lines and connect the lines to the fuel rail. Tighten the fuel line nuts to 20 ft. lbs. (27 Nm).

29. Temporarily connect the negative battery cable. With the engine **OFF** and the ignition **ON**, check for fuel leaks. Disconnect the negative battery cable.

30. Install the plenum with a new gasket and install the bolts/studs. Tighten to 18 ft. lbs. (25 Nm).

31. Install the air conditioning compressor to plenum bracket and attaching hardware. Install the EGR transfer tube with a new gasket. Tighten the attaching bolts to 19 ft. lbs. (26 Nm).

32. Install the throttle body with a new gasket and tighten the retaining bolts to 20 ft. lbs. (27 Nm).

33. Install the throttle cable bracket bolts and connect the vacuum harness conneector to the throttle body.

34. Connect the air inlet duct to the throttle body and the crankcase vent pipe to the valve cover grommet.

35. Install the transmission dipstick and connect the necessary wires and hoses.

36. Connect the negative battery cable.

37. Fill the cooling system with the proper type and amount of coolant. Do not install the radiator cap.

38. Let the engine run until the upper radiator hose becomes hot (thermostat open). With the engine idling, add coolant to the radiator, if necessary, until the level reaches the bottom of the filler neck. Install the radiator cap, making sure the arrows on the cap line with the overflow tube.

3.8L Engine

1. Disconnect the negative battery cable.

2. Drain the cooling system and relieve the fuel system pressure.

3. Remove the air inlet tube.

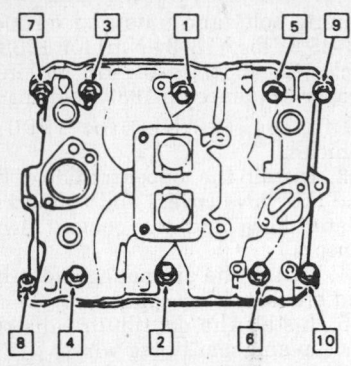

Intake manifold bolt tightening sequence—3.8L engine

4. Disconnect the fuel line at the fuel rail and at the pressure regulator.

5. Disconnect the injector wiring harness connectors located just behind the coil.

6. Disconnect the coolant temperature sensor wire connectors located at the front of the manifold.

7. Disconnect the heater, bypass and upper radiator hoses, the vacuum lines and hoses from the EGR, fuel pressure regulator and PCV valve and the throttle, cruise control and TV cables from the throttle body.

8. Remove the EGR vacuum control valve and the ignition wires from the spark plugs.

9. Remove the lower right side turbo mounting bracket to intake and bracket support to plenum. Remove the intake manifold bolts and remove the intake manifold.

10. Installation is the reverse of the removal procedure. Ensure all gasket mating surfaces are clean prior to installation. Torque the intake manifold bolts to 44 ft. lbs. (60 Nm) in the proper sequence.

5.0L TBI Engine

1. Disconnect the negative battery cable.

2. Drain the radiator, relieve the fuel system pressure and remove the air cleaner.

3. Disconnect the electrical connectors to the IAC valve, TPS and fuel injectors. Remove the injector wiring harness. Tag and disconnect the vacuum hoses. Disconnect all wires and hoses as necessary.

4. Disconnect the throttle, transmission control and cruise control cables. Disconnect the fuel feed and return lines and discard the O-rings.

5. Remove the TBI unit attaching bolts and remove the TBI unit.

6. Disconnect the ECM engine control harness and lay aside. Disconnect the upper radiator hose and heater hose at the manifold. Remove the EGR valve and solenoid and remove the thermostat housing and gasket.

7. Disconnect the fuel line clips and lines at the cylinder head and intake manifold. Disconnect the power brake vacuum pipe at the manifold.

8. Disconnect the spark plug wires at the distributor cap and remove the distributor cap. Mark the position of the rotor and the distributor housing and remove the distributor.

9. Remove the intake manifold bolts and studs and remove the intake manifold.

10. Installation is the reverse of the removal procedure. Ensure all gasket mating surfaces are clean before installation. Apply a $\frac{3}{16}$ inch bead of RTV sealant on the front and rear of the cylinder block. Extend the bead $\frac{1}{2}$ inch up each cylinder head to seal and retain the intake manifold gaskets. Torque the intake manifold bolts in the proper sequence to 25 ft. lbs. on 1988 vehicles or 35 ft. lbs. (47 Nm) on 1989–92.

5.0L and 5.7L TPI Engines

1. Disconnect the negative battery cable.

2. Drain the cooling system and relieve the fuel system pressure.

3. Disconnect the accelerator, TV and cruise control cables.

4. Remove the air intake duct.

5. Disconnect the heater hoses at the throttle body.

6. Disconnect the electrical connections at the throttle body and the intake manifold.

7. Disconnect the vacuum hoses and vent valve assembly.

8. Disconnect the fuel lines.

9. Disconnect the vapor pipe assembly.

10. Remove the plenum extension.

11. Remove the spark plug wires from the distributor cap and remove the distributor.

12. Remove the throttle body attaching bolts and remove the throttle body.

13. Disconnect the wiring harness from the fuel injectors and remove the harness from the manifold.

14. Disconnect the power brake vacuum hose at the plenum.

15. Remove the intake plenum bolts and disconnect the Manifold Absolute Pressure (MAP) sensor, if equipped.

16. Lift the plenum and disconnect the Manifold Air Temperature (MAT) sensor electrical connector, if equipped.

17. Remove the plenum and discard the plenum gaskets.

18. Remove the runner to manifold bolts, PCV valve and hose, EGR solenoid and the left and right side runners and gaskets. Discard the gaskets.

19. Remove the upper radiator hose.

20. Remove the fuel tube bracket bolt.

21. Disconnect the vacuum line at the pressure regulator.

22. Remove the fuel rail attaching bolts and the fuel rail assembly.

23. Rotate the injector retainer clip to the release position and remove the injector. Discard the O-rings and retainer clips.

24. Remove the ignition coil.

25. Remove the intake manifold bolts and studs and remove the intake manifold.

To install:

26. Ensure all gasket mating surfaces are clean and free of oil or water. Install the intake manifold gaskets. Apply a $\frac{1}{16}$ bead of RTV sealant to the front and rear ridges of the cylinder case. Extend the RTV bead $\frac{1}{2}$ inch up each cylinder head to seal and retain the intake manifold gaskets.

27. Install the intake manifold and tighten the bolts and studs to 35 ft. lbs. (47 Nm) in the proper sequence.

28. Connect the electrical wires and the upper radiator hose.

29. Install the EGR valve and pipe and the ignition coil and EGR solenoid.

30. Lubricate new injector O-ring seals with engine oil and install on the injector.

NOTE: **There are 2 injector part numbers used in production for the 5.0L engine and 2 different part numbers for the 5.7L engine. Do not intermix injectors with different part numbers, as this will result in engine roughness and excessive emissions.**

31. Install a new retainer clip onto the injector and install the injector into the fuel rail injector socket, with the electrical connector facing outward. Rotate the injector retainer clip to the locking position.

32. Install the fuel rail assembly in the intake manifold. Install the attaching bolts to 15 ft. lbs. (20 Nm). Install the fuel tube bracket bolt and tighten to 25 ft. lbs. (34 Nm).

33. Connect the vacuum line to the pressure regulator.

34. Install new O-rings on the fuel feed and return lines and connect the fuel lines to the fuel rail. Tighten the fuel line nuts to 20 ft. lbs. (27 Nm).

35. Temporarily connect the negative battery cable. Turn the ignition switch **ON** for 2 seconds and then **OFF**. Again turn to the **ON** position and check for fuel leaks and turn the ignition **OFF**. Disconnect the negative battery cable.

36. Install new gaskets on the runners and manifold.

37. Install the EGR solenoid.

38. Install the right and left runner to manifold bolts finger tight only.

39. Support the plenum above the

runners, connect the MAT sensor electrical connector, if equipped and lower the plenum into position. Start a few bolts to hold the plenum in position.

40. Connect the vacuum hoses and MAP sensor, if equipped.

41. Tighten all bolts to 25 ft. lbs. (34 Nm), starting in the center of the plenum/manifold and working outward.

42. Install the PCV valve and hose.

43. Connect the power brake vacuum hose to the fitting on the plenum, the left and right injector electrical harnesses, the attaching nuts and the electrical connectors to the injectors.

44. Install the throttle body with a new gasket and tighten the attaching bolts to 18 ft. lbs. (24 Nm).

45. Connect the electrical connectors to the TPS and IAC valve, coolant hoses, vacuum hoses, throttle cable bracket and the throttle, TV and cruise control cables.

46. Connect the upper radiator hose to the thermostat housing.

47. Install the distributor and the distributor cap. Connect the spark plug wires. Install the plenum extension.

48. Install the air intake duct and fill the cooling system with the proper type and quantity of coolant.

49. Connect the negative battery cable and start the engine and check for leaks. Check the ignition timing.

Exhaust Manifold

REMOVAL & INSTALLATION

2.8L and 3.1L Engines

1. Disconnect the negative battery cable.

2. Raise and safely support the vehicle.

3. Disconnect the exhaust crossover pipe and lower the vehicle.

4. To remove the right side manifold, perform the following procedure:

   a. 1988 vehicles: Remove the air management valve from the AIR pump, remove the alternator bracket and disconnect the air management hose.

   b. 1989-92 vehicles: Remove the throttle body air duct and the accessory drive belt. Disconnect the EGR transfer tube at the plenum and remove the EGR valve adapter with the EGR valve and shield from the exhaust manifold. Disconnect the vacuum line and the electrical connector from the diverter valve. Remove the AIR pump pulley and bracket and disconnect the AIR hose from the check valve. Remove the AIR pump bolt and AIR pump with diverter valve from the lower bracket. Remove the AIR pipe from the

exhaust manifold and remove the alternator brace.

5. To remove the left side manifold, perform the following:

   a. 1988 vehicles: Disconnect the air management hoses and wires and remove the power steering and fuel line bracket.

   b. 1989-92 vehicles: Remove the rear power steering pump bracket.

6. Remove the exhaust manifold bolts and nuts and remove the exhaust manifold.

7. Install in the reverse order of removal. Ensure all mating surfaces are clean before installation. Tighten the exhaust manifold bolts and nut to 25 ft. lbs. (34 Nm).

3.8L Engine

1. Disconnect the negative battery cable.

2. Remove the exhaust pipe from the turbocharger and disconnect the oxygen sensor wire on the right side exhaust manifold.

3. Raise and safely support the vehicle.

4. Remove the exhaust manifold to crossover pipe and lower the vehicle.

5. Remove the exhaust manifold to cylinder head bolts and remove the exhaust manifold.

6. Install in the reverse order of removal. Ensure all mating surfaces are clean before installation. Tighten the exhaust manifold to cylinder head bolds to 37 ft. lbs. (50 Nm).

5.0L and 5.7L Engines

1. Disconnect the negative battery cable.

2. Disconnect the spark plug wires, if necessary.

3. Disconnect the AIR pipes and remove the AIR valve.

4. Remove the air management valve, if equipped.

5. Raise and safely support the vehicle.

6. Remove the exhaust pipes from the exhaust manifolds and lower the vehicle.

7. Remove the exhaust manifold bolts and studs and remove the exhaust manifold.

8. Install in the reverse order or removal. Make sure all mating surfaces are clean before installation. Tighten the 4 outside exhaust manifold bolts and studs to 20 ft. lbs. (27 Nm) and the inside bolts to 26 ft. lbs. (35 Nm).

Turbocharger

REMOVAL & INSTALLATION

1. Disconnect the negative battery cable.

2. Remove the air inlet hose from the compressor section of the turbocharger.

3. Disconnect the compressor outlet pipe from the compressor.

4. Disconnect the oil breather and turbocharger head shields.

5. Remove the exhaust pipe from the turbine outlet.

6. Remove the oil breather vent from the valve cover. Disconnect and plug the oil pressure feed line at the turbocharger assembly.

7. Remove the turbocharger mounting bracket nuts. Disconnect the turbine inlet pipe from the exhaust manifold.

8. Disconnect the oil return line from turbocharger.

9. Disconnect the vacuum line from the turbocharger wastegate actuator.

10. Disconnect the intercooler outlet to throttle body pipe.

11. Remove the turbocharger assembly from the manifold adapter.

12. Install in the reverse order of removal. Always use new gaskets. Ensure all gasket mating surfaces are clean before installation. Tighten the turbocharger bracket to cylinder head bolts to 37 ft. lbs. (50 Nm). Tighten the turbocharger to bracket nuts to 20 ft. lbs. (27 Nm). Tighten the head shield retaining bolts to 20 ft. lbs. (27 Nm).

Timing Chain Front Cover

REMOVAL & INSTALLATION

2.8L and 3.1L Engines

1. Disconnect the negative battery cable.

2. Drain the cooling system into a suitable container and remove the accessory drive belt. Disconnect the lower radiator hose at the front cover and heater hose at the water pump.

3. Raise and safely support the vehicle.

4. Remove the oil pan.

5. Lower the vehicle.

6. On 1989-92 vehicles, remove the power steering pump and lay aside. Remove the power steering pump bracket.

7. Remove the water pump assembly.

8. Using a suitable puller, remove the vibration damper.

9. Remove the front cover bolts and remove the front cover.

10. Install in the reverse order of removal. Ensure all gasket mating surfaces are clean before installation. Tighten the front cover bolts to 15 ft. lbs. (21 Nm) and the vibration damper bolt to 76 ft. lbs. (103 Nm). Torque the water pump to specifications.

3.8L Engine

1. Disconnect the negative battery cable.
2. Drain the radiator.
3. Disconnect the radiator hoses and the heater return hose at the water pump.
4. Remove the fan assembly and pulleys.
5. Remove the crankshaft vibration damper.
6. Remove the alternator.
7. Remove the distributor, if equipped. If timing chain and sprockets are not going to be distrubed, note position of distributor rotor for reinstallation in same positon.
8. Loosen and slide front clamp on thermostat bypass hose rearward.
9. Remove bolts attaching timing chain cover to cylinder block.
10. Remove 2 oil pan to timing chain cover bolts.
11. Remove timing chain cover assembly and gasket.

To install:

12. Thoroughly clean the cover, taking care not to damage to the gasket surface.
13. Installation is the reverse of the removal procedure.
14. Remove oil pump cover and pack the space around the oil pump gears completely full of petroleum jelly.

There must be no air space left inside the pump. Reinstall cover using new gasket.

15. Tighten the front cover retaining bolts to 22 ft. lbs. (30 Nm) and the vibration damper bolt to 200 ft. lbs. (270 Nm).

5.0L and 5.7L Engines

1. Disconnect the negative battery cable.
2. Drain the cooling system and remove the accessory drive belt and pulleys.
3. Remove the water pump.
4. Remove the crankshaft pulley and vibration damper.
5. Raise and safely support the vehicle.
6. Remove the oil pan assembly.
7. Remove the front cover bolts and remove the timing cover.

To install:

8. Install in the reverse order or removal. Ensure all gasket mating surfaces are clean prior to installation.
9. Install a new gasket and seal. Align the front cover dowel pins.
10. Tighten the front cover and oil pan bolts to 100 inch lbs., (11 Nm). Tighten the vibration damper bolt to 70 ft. lbs. (95 Nm).

Front Cover Oil Seal

REPLACEMENT

Except 3.8L Engine
FRONT COVER REMOVED

1. Using a suitable tool, pry the seal out from the front of the cover.
2. Using a suitable tool, install the new seal with the open end of the seal toward the inside of the front cover. Support the rear of the cover at the seal area while installing.
3. Inspect the sealing area of the vibration damper and crankshaft for damage or grooving, repair as necessary. Coat the area which contacts the seal with oil prior to installing. Tighten the vibration damper bolt to 70 ft. lbs. (95 Nm).

FRONT COVER INSTALLED

1. Disconnect the negative battery cable.
2. Remove the accessory drive belt and pulleys.
3. Remove the vibration damper using a suitable puller.
4. Pry the seal out of the front cover with a suitable prying tool.
5. Using a suitable installation tool, install the new seal with the open end of the seal toward the inside of the front dover.
6. Inspect the sealing area of the vibration damper and crankshaft for damage or grooving, repair as necessary. Coat the area which contacts the seal with oil prior to installing. Tighten the vibration damper bolts to 70 ft. lbs. (95 Nm).
7. Install the accessory drive belt and pulleys and connect the negative battery cable. Inspect for leaks.

3.8L Engine

1. Disconnect the negative battery cable.
2. Remove the front cover.
3. Use a suitable drift to drive out the old seal and shedder from the front toward the rear of the cover.
4. Coil new packing around the front cover opening so the ends of the packing are at the top.
5. Drive in the shedder using a suitable punch and stake the shedder in place at 3 locations.
6. Size the packing by rotating a suitable tool around the pakcing until the vibration damper hub can be inserted through the opening.
7. Installation is the reverse of the removal procedure. Inspect the sealing area of the vibration damper for damage or grooving and replace, as necessary. Coat the area which contacts the seal prior to installing. Tighten the vibration damper bolt to 200 ft. lbs. (270 Nm).

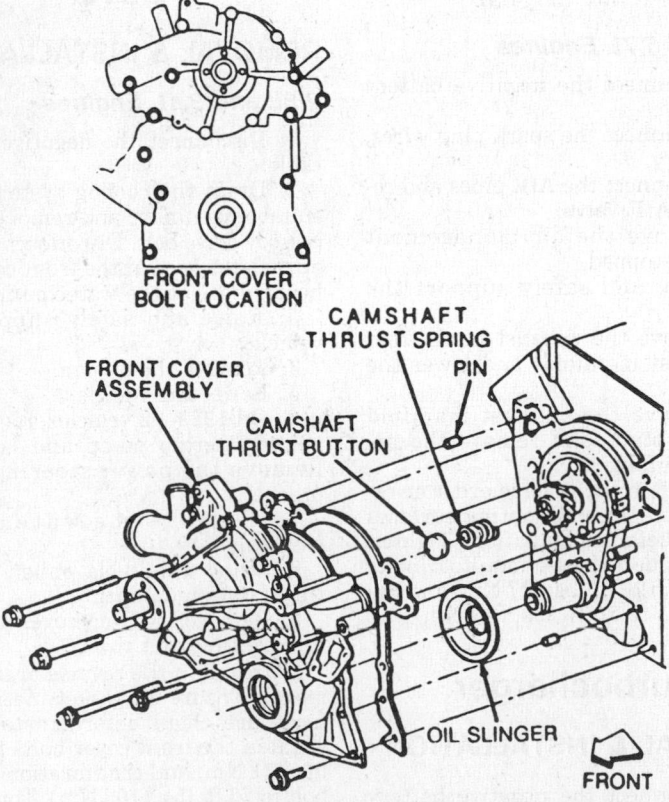

FRONT COVER BOLT LOCATION

CAMSHAFT THRUST SPRING
PIN
FRONT COVER ASSEMBLY
CAMSHAFT THRUST BUTTON
OIL SLINGER
FRONT

Timing chain front cover assembly—3.8L engine

Timing Chain and Sprockets

REMOVAL & INSTALLATION

1. Disconnect the negative battery cable. Remove the timing chain cover. Remove the crankshaft oil slinger, if equipped.

2. Turn the engine until the No. 1 piston is at TDC and the timing marks on the camshaft and crankshaft sprockets are aligned.

3. Remove the camshaft sprocket bolts and remove the camshaft sprocket and chain. Using a suitable puller, remove the crankshaft sprocket.

NOTE: The sprocket is a tight fit on the camshaft. If the sprocket does not come off easily, use a plastic mallet and strike the lower edge of the sprocket. This should dislodge the sprocket, allowing it to be removed from the shaft.

To install:

4. Install the crankshaft sprocket using a suitable installation tool. Install the timing chain on the camshaft sprocket and lube the thrust surface with Molykote® or equivalent.

5. Hold the sprocket vertically with

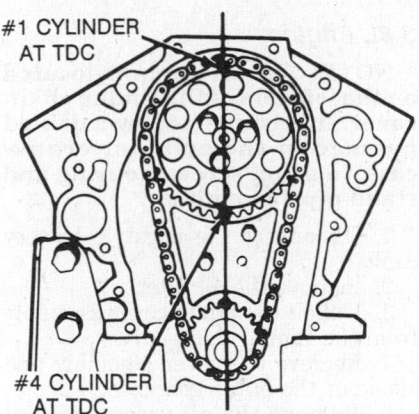

Timing mark alignment—2.8L and 3.1L engines

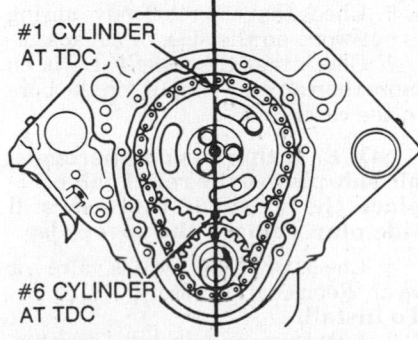

Timing mark alignment—5.0L and 5.7L engines

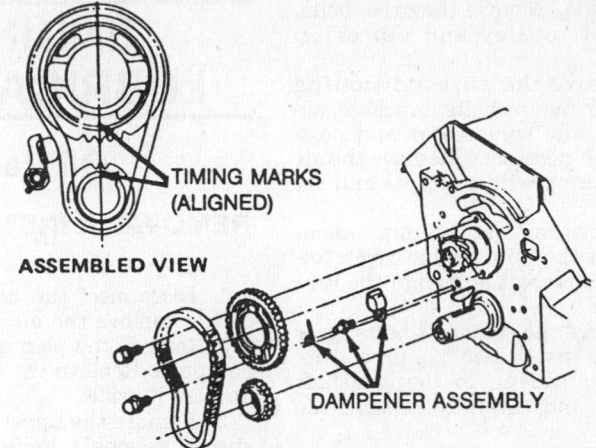

Timing mark alignment—3.8L engine

the chain hanging down and align the marks on the camshaft and crankshaft sprockets.

6. Align the dowel in the camshaft with the dowel hole in the camshaft sprocket and install the sprocket on the camshaft.

7. Slowly and evenly draw the camshaft sprocket onto the camshaft using the mounting bolts and torque the bolts to 21 ft. lbs. (28 Nm) except 3.8L engine which is torqued to 31 ft. lbs. (42 Nm).

NOTE: Do not drive the sprocket onto the camshaft, this could cause the rear camshaft core plug to be dislodged.

8. Lubricate the timing chain and install the timing chain cover.

9. Connect the negative battery cable, start the engine, set the timing and inspect for leaks.

Camshaft

REMOVAL & INSTALLATION

2.8L and 3.1L Engines

1. Disconnect the negative battery cable. Relieve fuel pressure, discharge the air conditioning system and drain the cooling system into a suitable container.

2. Remove the intake manifold and valve covers. Remove the rocker arm assemblies and pushrods.

3. Remove the valve lifters.

4. As required, remove the radiator, grille and air conditioning condenser.

5. Remove the front engine cover.

6. Remove the timing chain and sprockets. Carefully remove the camshaft.

7. Installation is the reverse of the removal procedure. Be sure to coat the camshaft lobes with Molykote® or equivalent and the camshaft journals

and lifters with clean engine oil before installation.

3.8L Engine

1. Disconnect the negative battery cable, drain the cooling system, relieve the fuel system pressure and properly discharge the air conditioning system.

3. Remove the intake manifold.

4. Remove the valve covers.

5. Remove the rocker arm assemblies, pushrods and valve lifters, noting location.

6. Remove the radiator and the air condition condenser, as required.

7. Remove the timing chain cover, timing chain and sprocket.

8. Align the timing marks of camshaft and crankshaft sprocket. This avoids burring of the camshaft journals by the crankshaft during removal.

9. Slide the camshaft forward out of the engine carefully to avoid marring the bearing surfaces.

10. Installation is the reerse of the removal procedure.

11. Before installing the camshaft and the lifters, be sure to coat the camshaft lobes with Molykote® or equivalent, and the camshaft journals and lifters with clean engine oil.

12. Use new gaskets and seals, as required.

5.0L and 5.7L Engines

1. Disconnect the negative battery cable. Relieve fuel pressure as necessary and drain the cooling system.

2. Remove the intake manifold. Remove the valve covers, rocker arm assemblies, pushrods and lifters, keep all parts in order.

3. Remove all necessary wires and hoses. Disconnect the upper and lower transmission cooler lines.

4. Remove the radiator shroud assembly and radiator. Remove the front fascia, if necessary. Remove the cooling fan.

5. Remove the power steering pump

and lay aside. Remove the drive belts, crankshaft pulley and vibration damper.

6. Remove the air conditioning compressor mount bolts, brackets, accumulator and compressor and position aside, if necessary. Remove the air injection pump with brackets and set it aside.

7. Remove the water pump assembly, remove the front engine cover. Rotate the crankshaft and align the timing marks.

8. Remove the camshaft bolts, gear and chain. Install two $^5/_{16}$ in. x 4 in. bolts or equivalent, in the camshaft bolt holes and carefully remove the camshaft.

9. Installation is the reverse of the removal procedure. Lubricate the camshaft lobes with Molykote® and the journals and lifters with a suitable engine oil supplement, before installing the camshaft.

Piston and Connecting Rod

POSITIONING

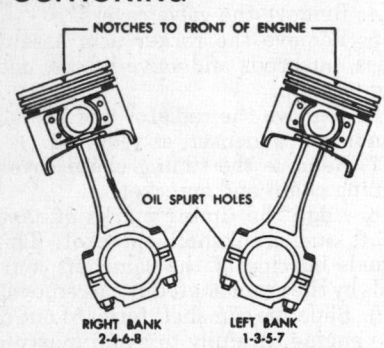

Piston and rod notches facing forward—5.0L and 5.7L engines

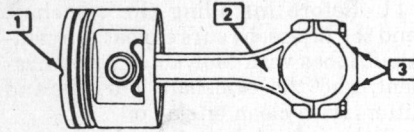

1. Notch on piston towards front of engine
2. Left bank: No. 1, 3 and 5—2 bosses on rod towards rear of engine
   Right bank: No. 2, 4 and 6—2 bosses on rod towards front of engine
3. Left bank: Chamfered corners on rod cap towards front of engine
   Right bank: Chamfered corners on rod cap towards rear of engine

Piston and rod positioning—3.8L engine

NOTE: The connecting rod bearing tang slots should be on the side opposite the camshaft.

ENGINE LUBRICATION

Oil Pan

REMOVAL & INSTALLATION

1. Disconnect the negative battery cable. Remove the air cleaner assembly. Remove the plenum extension, if equipped. Remove the distributor cap and lay it aside.

2. Remove the upper half of the fan shroud assembly. Remove the air conditioning compressor, if necessary, and lay aside.

3. Raise the vehicle and support it safely. Drain the engine oil.

4. Remove the air injector pipe at the catalytic convertor.

5. Remove the torque converter dust shield. If equipped with manual transmission, it may be necessary to remove the oil filter in order to remove the dust shield.

6. Remove the exhaust pipe at the manifolds.

7. Remove the starter bolts, loosen the starter brace, then lay the starter aside. On V8 engines, rmeove the front starter brace.

8. Disconnect the transmission oil cooler lines at the clips on the oil pan. Remove the engine mount through bolts.

9. Raise the engine enough to provide sufficient clearance for oil pan removal.

10. Remove the oil pan nuts and bolts. If the front of the crankshaft prohibits removal of the pan, turn the crankshaft timing mark to the 6 o'clock position.

11. Remove the oil pan from the vehicle.

12. Remove all old RTV from the oil pan and engine block.

To install:
13. Run a ⅛ in. bead of RTV around the oil pan sealing surface. Remember to keep the RTV on the inside of the bolt holes.

14. Install the oil pan. Tighten the 2 rear oil pan retaining bolts on the 2.8L and 3.1L engines to 18 ft. lbs. (25 Nm) and the rest of the retaining bolts and nuts to 89 inch lbs. (10 Nm). Tighten the oil pan retaining bolts on the 3.8L engine to 88 inch lbs. (10 Nm). Tighten the oil pan retaining bolts on 5.0L and 5.7L engines to 101 inch lbs. (11 Nm) and the retaining nuts to 17 ft. lbs. (23 Nm).

15. Lower the engine and install the remainder of the components.

16. Lower the vehicle and fill the en-

gine with the proper type of motor oil to the required level.

17. Connect the negative battery cable, start the engine and check for leaks.

Oil Pump

REMOVAL & INSTALLATION

Except 3.8L Engine

1. Raise the vehicle and support safely.
2. Drain the oil.
3. Remove the oil pan.
4. Remove the oil pump bolt. Remove the oil pump with the extension shaft.
5. Installation is the reverse of the removal procedure. Using a suitable tool, install a new pickup screen and pipe to the replacement pump.
6. Prime the pump by turning it upside down and pouring clean oil into the pickup screen while turning the pump extension clockwise.
7. Align the slot on the end of the shaft extension with the drive tang on the distributor shaft.
8. Tighten the oil pump bolt on 2.8L and 3.1L engines to 30 ft. lbs. (41 Nm) or to 65 ft. lbs. (88 Nm) on 5.0L and 5.7L engines.

3.8L Engine

NOTE: The oil pump is located on the left side of the timing chain cover. It is connected by a drilled passage in the cylinder crankcase, to an oil screen housing and stand pipe assembly.

1. Disconnect the negative battery cable.
2. Remove the oil filter.
3. Unbolt the pump cover assembly from the timing chain cover.
4. Remove the cover assembly and slide out the pump gears.
5. Remove the oil pressure relief valve cap, spring and valve. Do not remove the oil filter bypass valve and spring.
6. Check that the relief valve spring is not worn on the side, or collapsed.
7. Check that the relief valve is no more than an easy "slipfit" in the bore in the cover.

NOTE: If there is any perceptible side play in the relief valve, replace the valve. If there is still side play, replace the cover also.

8. Check the filter bypass valve for wear. Replace if necessary.
To install:
9. Lubricate and install the pressure relief valve and spring in the cover bore.

10. Install the gasket and cap, torquing the cap to 35 ft. lbs.

11. Install the gears and check that gear-to-cover end clearance is between 0.002–0.006 in. If the clearance is not as specified, check the timing cover gear pocket for wear. If the gear pocket is worn, the timing cover must be replaced.

12. Remove the gears and pack the gear pocket full of petroleum jelly. Don't use grease.

NOTE: Unless the pump is primed properly, the pump will not produce oil pressure when the engine is started.

13. Install the gears. Install a new gasket and the cover. Torque the bolts evenly to 10 ft. lbs. Replace the oil filter and check the oil level. Connect the negative battery cable.

Rear Main Bearing Oil Seal

REMOVAL & INSTALLATION

Except 3.8L Engine

1. Remove the transmission.
2. Remove the flywheel or flexplate.
3. Using a suitable prying tool, pry the oil seal from the engine.
4. Coat the new seal with clean oil. Using tool J–34686 or equivalent,

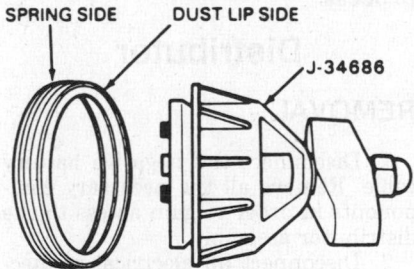

Rear main seal installation tool— except 3.8L engine

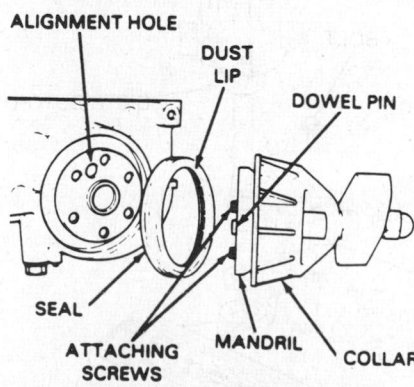

Installing rear main seal—except 3.8L engine

press the new seal onto the crankshaft.

5. Install the remainder of the components in the reverse order of their removal.

3.8L Engine

NOTE: The following procedure is only to be used as an oil seal repair while the engine is in the vehicle. Whenever possible the crankshaft should be removed and a complete seal installed.

1. Disconnect the negative battery cable.
2. Drain the engine oil and remove the oil pan.
3. Remove the rear main bearing cap.
4. Insert packing tool J–21526–2 or equivalent, against 1 end of the seal in the cylinder block. Drive the old seal gently into the groove until it is packed tight. This will vary from ¼ inch to ¾ inch depending on the amount of pack required.
5. Repeat the procedure on the other end of the seal.
6. Measure the amount the seal was driven up on 1 side and add ¹⁄₁₆ inch. Using a suitable cutting tool, cut that length from the old seal removed from the rear main bearing cap. Repeat the procedure for the other side. Use the rear main bearing cap as a holding fixture when cutting the seal.
7. Install guide tool J–21526–1 or equivalent, onto the cylinder block.
8. Using the packing tool, work the short pieces cut in Step 6 into the guide tool and then pack into the cylinder block. The guide tool and packing tool are machined to provide a built in stop. Use this procedure for both sides. It may help to use oil on the short pieces of the rope seal when packing them into the cylinder block.
9. Remove the guide tool.
10. Apply Loctite 414 or equivalent, to the seal groove in the rear main

bearing cap. Within 1 minute, insert a new seal into the groove and rool into place with a suitable tool until no more than ¹⁄₁₆ inch of the seal projects above the groove. Cut the excess seal material with a sharp cutting tool at the bearing cap parting line.

11. Apply a thin film of chassis grease to the rope seal. Apply a thin film of RTV sealant on the bearing cap mating surface around the seal groove. Use the sealer sparingly.

12. Coak the side sealing strips for 5 minutes in light oil or kerosene. Install the sealing strips into the grooves along the sides of the main bearing cap. Install the rear main bearing cap and tighten to 100 ft. lbs. (135 Nm).

13. Install all remaining components.

ENGINE COOLING

Radiator

REMOVAL & INSTALLATION

1. Disconnect the negative battery cable.
2. Drain the cooling system into a suitable container.
3. Remove the intake duct, air duct bracket and air cleaner top, if equipped.
4. Remove the Mass Air Flow (MAF) sensor, if equipped.
5. Remove the engine cooling fan. If equipped with a fan clutch, the clutch should be se aside in an upright position to prevent seal leakage.
6. Disconnect the radiator hoses and heater hoses from the radiator.
7. If equipped with an automatic transmission, disconnect and plug the

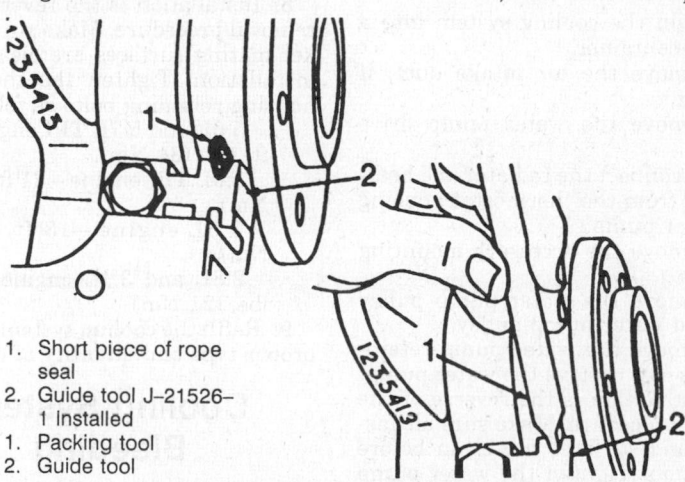

1. Short piece of rope seal
2. Guide tool J–21526–1 Installed
1. Packing tool
2. Guide tool

Rear main seal installation—3.8L engine

transmission cooler lines at the radiator.

8. If equipped, remove the fan shield assembly.

9. Remove the radiator and shroud assembly, then lift the radiator straight up and out of the vehicle.

10. Install radiator and shroud assembly by reversing removal procedures. Ensure the radiator is positioned in the lower cradle properly.

11. Reconnect all hoses, install the cooling fan and refill the cooling system with the proper type and quantity of coolant. Connect the negative battery cable.

Heater Core

REMOVAL & INSTALLATION

1. Disconnect the negative battery cable. Drain the cooling system and disconnect the heater hoses.

2. Remove the right lower dash panels. Remove the instrument panel lower trim pad and the console.

3. Disconnect the electronic control module retaining screws and position aside.

4. Remove the heater case retaining screws and remove the heater case.

5. Remove the core shroud screws and remove the core shroud, heater core and mounting strap as an assembly.

6. Remove the core mounting strap and remove the heater core from the core shroud.

7. Installation is the reverse of the removal procedure.

Water Pump

REMOVAL & INSTALLATION

1. Disconnect the negative battery cable.

2. Drain the cooling system into a suitable container.

3. Remove the air intake duct, if equipped.

4. Remove the water pump drive belt.

5. Disconnect the radiator and heater hoses from the thermostat housing and water pump.

6. Remove the accessory mounting brackets.

7. Remove the water pump pulley bolts and water pump pulley.

8. Remove the water pump retaining bolts and remove the water pump.

9. Installation is the reverse of the removal procedure. Make sure all gasket mating surfaces are clean before installation. Tighten the water pump retaining bolts as follows:

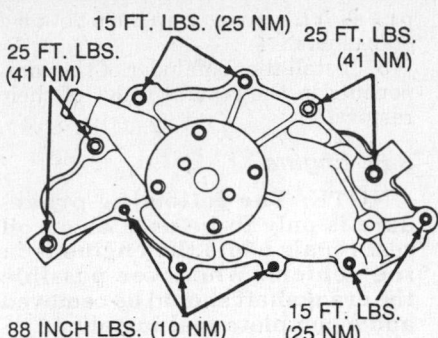

Water pump torque specification— 1989 2.8L and 1990–92 3.1L engines

15 FT. LBS. (25 NM)
25 FT. LBS. (41 NM)
25 FT. LBS. (41 NM)
88 INCH LBS. (10 NM)
15 FT. LBS. (25 NM)

5.0L and 5.7L engines—30 ft. lbs. (41 Nm)

3.8L engine—115 inch lbs. (13 Nm).

1988 2.8L engine: Small bolt—7 ft. lbs. (10 Nm), large bolt and nut—15 ft. lbs. (20 Nm).

1989 2.8L engine and 1990–92 3.1L engine—to specification.

10. Refill the cooling system with the proper type and quantity of coolant.

Thermostat

REMOVAL & INSTALLATION

1. Disconnect the negative battery cable.

2. Drain the cooling system to a level slightly below the thermostat.

3. Remove the air cleaner and intake duct, if required.

4. On some models it may be necessary to remove the throttle body and/or plenum, fuel lines and brackets.

5. Disconnect the radiator inlet hose.

6. Remove the thermostat housing retaining bolts and the thermostat housing.

7. Remove the thermostat.

8. Installation is the reverse of removal procedure. Make sure all gasket mating surfaces are clean before installation. Tighten the thermostat housing retaining bolts as follows:

5.0L and 5.7L TPI engines—25 ft. lbs. (34 Nm).

5.0L TBI engine—21 ft. lbs. (28 Nm).

3.8L engine—13 ft. lbs. (18 Nm).

2.8L and 3.1L engines—15 ft. lbs. (21 Nm).

9. Refill the cooling system with the proper type and quantity of coolant.

Cooling System Bleeding

1. Drain the cooling system.

2. Fill the cooling system with a 50/50 mix of ethylene glycol antifreeze and water to a level just below the filler neck.

3. Fill the coolant recovery reservoir to the **COLD** fill mark and install the reservoir cap.

4. Run the engine with the radiator cap removed until the normal operating temperature is reached.

— CAUTION —
Ethylene glycol in engine coolant can be flammable under some conditions. Do not spill coolant on the exhaust system or on hot engine parts.

5. With the engine idling, add coolant to the radiator until the level reaches the bottom of the filler neck.

6. Install the radiator cap. The arrows on the cap must line up with the coolant recovery reservoir hose.

ENGINE ELECTRICAL

NOTE: Disconnecting the negative battery cable on some vehicles may interfere with the functions of the on board computer systems and may require the computer to undergo a relearning process.

Distributor

REMOVAL

1. Disconnect the negative battery cable. Remove all the necessary components in order to gain access to the distributor assembly.

2. Disconnect all electrical connections from the distributor.

3. Remove the distributor cap re-

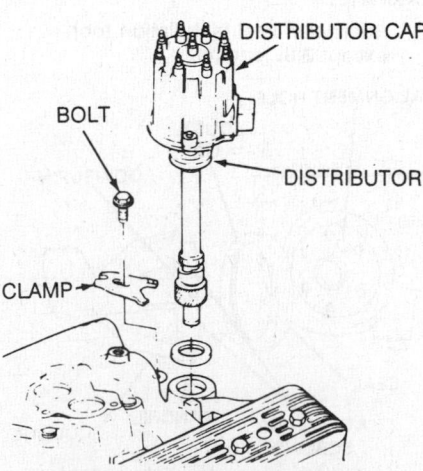

DISTRIBUTOR CAP
BOLT
DISTRIBUTOR
CLAMP

Typical HEI distributor assembly

taining screws and remove the cap. Mark the position of the distributor housing in relation to the engine block and the distributor rotor in relation to the distributor housing.

4. Remove the distributor hold-down clamp and bolt.

5. Pull the distributor assembly up from the engine, noting the position of the rotor as the distributor gear disengages from the camshaft.

Installation

Timing Not Disturbed

1. Install the distributor, aligning the marks that were made during the removal procedure.

2. Install the distributor hold-down clamp and bolt and temporarily tighten.

3. Install the distributor cap and connect the electrical connectors.

4. Install the remainder of the components that were removed to gain access to the distributor.

5. Connect the negative battery cable.

6. Start the engine and set the ignition timing. Tighten the hold-down clamp bolt to 27 ft. lbs. (36 Nm) and recheck the timing.

Timing Disturbed

1. Remove the No. 1 cylinder spark plug.

2. Place a finger over the spark plug hole and crank the engine slowly until compression is felt.

3. Align the timing mark on the crankshaft pulley with the 0 mark on the timing scale attached to the front cover of the engine. This places the engine at TDC of the compression stroke for No. 1 cylinder.

4. Rotate the distributor shaft until the rotor points to the No. 1 spark plug tower on the distributor cap.

5. Install the distributor in the engine.

6. Install the hold-down clamp and bolt and tighten temporarily.

7. Install the distributor cap and connect the electrical connectors.

8. Install the remainder of the components that were removed to gain access to the distributor.

9. Connect the negative battery cable.

10. Start the engine and set the ignition timing. Tighten the hold-down clamp bolt to 27 ft. lbs. (36 Nm) and recheck the timing.

Distributorless Ignition

The 20th anniversary Trans AM, offered during the 1989 model year, is equipped with a 3.8L SFI turbo-charged engine that features distributorless ignition. This system uses a "waste spark" method of spark distribution. Each cylinder is paired with it's opposite in the firing order, so a cylinder on the compression stroke fires simultaneously with it's opposing cylinder on the exhaust stroke. The cylinder on the exhaust stroke requires very little voltage to fire it's plug, so most of the available voltage is used to fire the cylinder that is on the compression stroke.

The distributorless ignition system consists of a coil pack, the ignition module, a dual hall effect sensor, interrupter rings and the Electronic Control Module (ECM). The coil pack contains 3 separate ignition coils (1 for each pair of cylinders) enclosed in 1 housing and serviced as a unit. The ignition module is located under the coil pack and is connected to the ECM. The ignition module controls the primary circuit to the ignition coils and the spark timing below 400 rpm, if the ECM bypass circuit becomes open or grounded. The dual hall effect sensor is a combination camshaft and crankshaft sensor, mounted on the front cover behind the crankshaft balancer. Interrupter rings, mounted on the crankshaft balancer, pass through slots in the dual hall effect sensor and provide timing information to the ECM.

The distributorless ignition system uses Electronic Spark Timing (EST). The ECM controls timing with inputs concerning crankshaft position, engine rpm, engine temperature and volume of intake air.

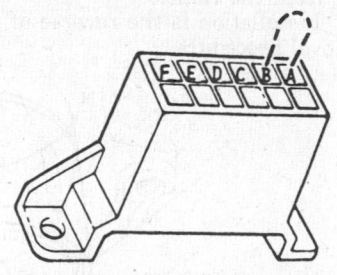

Grounding the ALDL terminals

Ignition Timing

ADJUSTMENT

NOTE: On all 1988 and 1991–92 vehicles, it will be necessary to put the Electronic Spark Timing (EST) in the bypass mode by disconnecting a single wire timing connector. This wire is tan with a black tracer and breaks out the wiring harness near the rear of the right side valve cover. Do not disconnect the 4 prong EST connector from the distributor assembly.

On 1989–90 vehicles, with the engine running, ground the diagnostic terminal (A and B) of the Assembly Line Diagnostic Link (ALDL) connector.

1. Refer to the Vehicle Emission Information label which is located on the radiator support panel, for the proper timing information.

2. With the ignition **OFF**, connect the pick-up lead of an inductive timing light to the No. 1 spark plug wire. Connect the timing light power leads according to the manufacturers instructions.

3. Run the engine to normal operating temperature. Disconnect the EST bypass mode connector or ground the diagnostic terminal of the ALDL connector.

4. Check the timing, by aiming the timing light at the timing mark and harmonic balancer. If the engine timing requires adjustment, loosen the distributor hold-down bolt and rotate the distributor slowly in either direction, to advance or retard the engine timing.

5. Tighten the hold-down bolt and recheck the engine timing.

6. With the engine still running, unground the diagnostic terminal or reconnect the EST bypass mode connector. An Electronic Control Moduel (ECM) code may be set when the EST bypass connector is disconnected or if

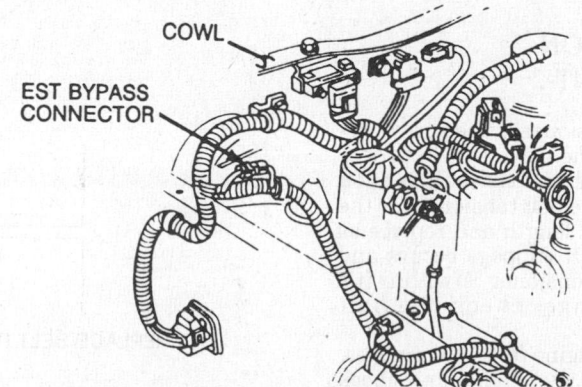

Electronic Spark Timing (EST) bypass connector location

the ALDL diagnostic terminal is un-grounded after the engine is shut OFF.

7. With the ignition **OFF**, clear the ECM code by disconnecting the negative battery cable for at least 30 seconds.

Alternator

PRECAUTIONS

Several precautions must be observed with alternator equipped vehicles to avoid damage to the unit.

- If the battery is removed for any reason, make sure it is reconnected with the correct polarity. Reversing the battery connections may result in damage to the one-way rectifiers.
- When utilizing a booster battery as a starting aid, always connect the positive to positive terminals and the negative terminal from the booster battery to a good engine ground on the vehicle being started.
- Never use a fast charger as a booster to start vehicles.
- Disconnect the battery cables when charging the battery with a fast charger.
- Never attempt to polarize the alternator.
- Do not use test lights or more than 12 volts when checking diode continuity.
- Do not short across or ground any of the alternator terminals.
- The polarity of the battery, alternator and regulator must be matched and considered before making any electrical connections within the system.
- Never separate the alternator on an open circuit. Make sure all connection with the circuit are clean and tight.
- Disconnect the battery ground terminal when performing any service on electrical components.
- Disconnect the battery if arc welding is to be done on the vehicle.

BELT TENSION ADJUSTMENT

A drive belt tensioner is used on all vehicles to keep the drive belt in proper adjustment. Check the belt length scale on the drive belt tensioner for the proper installed length and replace as necessary. If belt slippage occurs and the drive belt tensioner is within it's operating range, check the belt tension.

1. Run the engine for 5–10 minutes.
2. Shut OFF the engine and check the belt tension at the following loca-tions using J–23600–B V-belt tension gauge or equivalent.

   a. V6 engine (except 3.8L engine): On vehicles without air conditioning, check the belt tension between the tensioner and the power steering pump pulley. On vehicles with air conditioning, check the belt tension between the tensioner and the air conditioner compressor pulley.

   b. V8 engine: Check the belt tension between any 2 pulleys.
3. Run the engine for 30 seconds and recheck the belt tension.
4. Repeat Step No. 3. The belt tension is the average of the 3 readings.

   a. V6 engine (except 3.8L engine): Belt tension should be 116–142 lbs. (516–631 N) on vehicles without air conditioning. On vehicles with air conditioning the belt tension should be 103–127 lbs. (460–563 N).

   b. V8 engine: Belt tension should be 99–121 lbs. (440–538 N).
5. Replace the drive belt tensioner if the belt tension is below the minimum specified and if the tensioner is within it's operating range.

REMOVAL & INSTALLATION

1. Disconnect the negative battery cable.
2. Tag and disconnect the alternator wiring.
3. Remove the alternator brace bolt. If required, loosen the power steering pump brace and mount nuts. Remove the drive belt.
4. Support the alternator and remove the mount bolts. Remove the unit from the vehicle.
5. Installation is the reverse of the removal procedure.

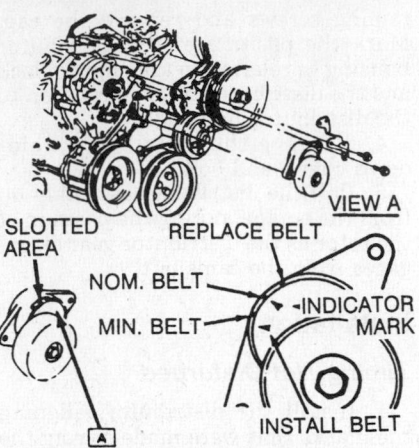

VIEW A

SLOTTED AREA   REPLACE BELT   NOM. BELT   MIN. BELT   INDICATOR MARK   INSTALL BELT

THE INDICATOR MARK ON THE MOVEABLE PORTION OF THE TENSIONER MUST BE WITHIN THE LIMITS OF THE SLOTTED AREA ON THE STATIOANRY PORTION OF THE TENSIONER. ANY READING OUTSIDE OF THESE LIMITS INDICATES EITHER A DEFECTIVE BELT OR TENSIONER.

Drive belt tensioner—2.8L and 3.1L engines

Starter
REMOVAL & INSTALLATION

1. Disconnect the negative battery cable.
2. Raise the vehicle and support it safely.
3. Disconnect all wiring from the starter.
4. Remove the flywheel housing cover. Remove the starter brace as required.
5. Remove the starter motor retaining bolts and any shims.

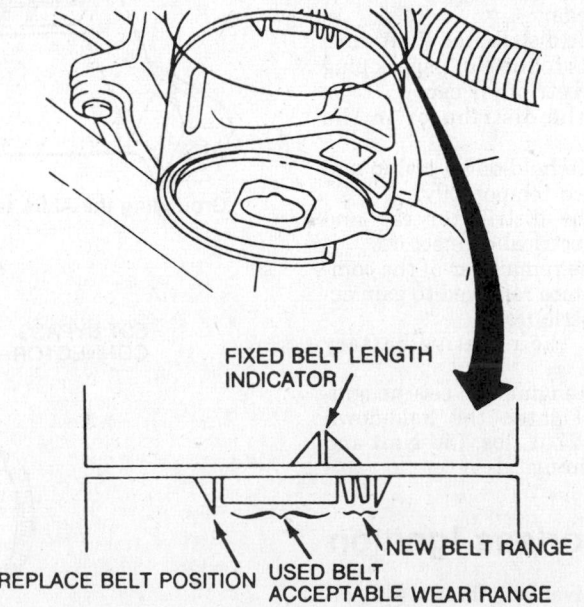

FIXED BELT LENGTH INDICATOR

NEW BELT RANGE

REPLACE BELT POSITION   USED BELT ACCEPTABLE WEAR RANGE

Drive belt tensioner—5.0L and 5.7L engines

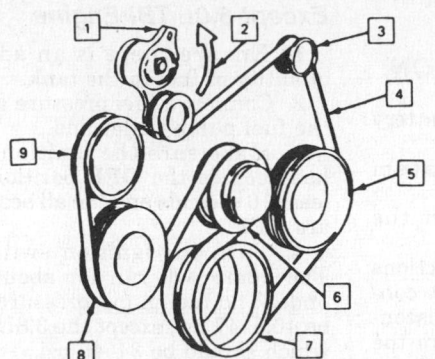

1. Tensioner assembly
2. Rotate tensioner in direction shown to install or remove belt
3. Alternator assembly
4. Accessory drive belt
5. Power steering belt
6. Water pump
7. Crankshaft
8. Air pump
9. Air conditioning compressor or belt idler

Drive belt and pulleys—5.0L and 5.7L engines

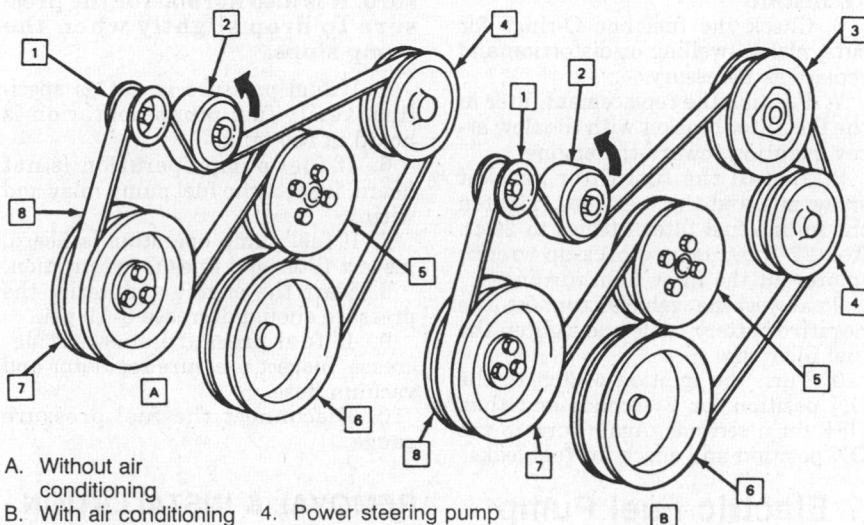

A. Without air conditioning
B. With air conditioning
1. Alternator
2. Tensioner
3. Air conditioning compressor
4. Power steering pump
5. Water pump
6. Crankshaft
7. AIR pump
8. Belt

Drive belt and pulleys—2.8L and 3.1L engines

6. Remove the starter from the vehicle.

7. Installation is the reverse of the removal procedure.

8. If shims were removed, they must be installed in their original locations.

EMISSION CONTROLS

Please refer to "Emission Controls" in the Unit Repair section for system maintenance procedures. Due to the complex nature of modern electronic engine control systems, comprehensive diagnosis and testing procedures fall outside the confines of this repair manual. For complete informa- tion on diagnosis, testing and repair procedures concerning all modern engine and emission control systems, please refer to "Chilton's Guide to Fuel Injection and Electronic Engine Controls".

Emission Warning Lamps

RESETTING

All vehicles feature a "Service Engine Soon" light on the instrument panel. This light has the following functions:

It informs the driver that a problem has occured and the vehicle should be taken in for service.

It displays diagnostic codes stored by the ECM.

It indicates "Open Loop" or "Closed Loop" operation.

When the ECM sets a trouble code, the "Service Engine Soon" light will come ON and a trouble code will be stored in the memory. The light will not go out until the malfunction is repaired.

If the problem is intermittent, the light will go out after 10 seconds, when the fault goes away. After the light goes out, there will still be a trouble code stored in the memory of the ECM.

To clear the code from memory, the ECM power feed must be disconnected for at least 30 seconds. The ECM power feed can be disconnected at the positive battery terminal pigtail, the inline fuse holder that originates at the positive connection at the battery or the ECM fuse in the fuse block. The negative battery cable may be disconnected but other on-board memory data, such as preset radio tuning, will also be deleted.

FUEL SYSTEM

Fuel System Service Precautions

Safety is the most important factor when performing not only fuel system maintenance but any type of maintenance. Failure to conduct maintenance and repairs in a safe manner may result in serious personal injury or death. Maintenance and testing of the vehicle's fuel system components can be accomplished safely and effectively by adhering to the following rules and guidelines.

• To avoid the possibility of fire and personal injury, always disconnect the negative battery cable unless the repair or test procedure requires that battery voltage be applied.

• Always releave the fuel system pressure prior to disconnecting any fuel system compoent (injector, fuel rail, pressure regulator, etc.),fitting or fuel line connection. Exercise extreme caution whenever relieving fuel system pressure to avoid exposing skin, face and eyes to fuel spray. Please be advised that fuel under pressure may penetrate the skin or any part of the body that it contacts.

• Always place a shop towel or cloth around the fitting or connection prior to loosening the absorb any excess fuel due to spillage. Ensure that all fuel spillage (should it occur) is quickly removed from engine surfaces. Ensure that all fuel soaked cloths or towels are deposited into a suitable waste container.

- Always keep a dry chemical (Class B) fire extinguisher near the work area.
- Do not allow fuel spary or fuel vapors to come into contact with a spark or open flame.
- Always use a backup wrench when loosening and tightening fuel line connection fittings. This will prevent unnecessary stress and torsion to fuel line piping. Always follow the proper torque specifications.
- Always replace worn fuel fitting O-rings with new. Do not substitute fuel hose or equivalent where fuel pipe is installed.

RELIEVING FUEL SYSTEM PRESSURE

Except TBI Engine

1. Disconnect the negative battery cable.
2. Loosen the fuel filler cap to relieve the tank pressure.
3. Connect J–34730–1 fuel pressure gauge or equivalent, to the fuel pressure valve. Wrap a shop cloth around the fitting while connecting the gauge to avoid spillage.
4. Place the end of the bleed hose into a suitable container and open the valve to relieve the fuel system pressure.

TBI Engine

1. Disconnect the negative battery cable.
2. Loosen the fuel filler cap to relieve the tank pressure.
3. Fuel system pressure is automatically relieved when the engine is turned OFF. No further action is necessary.

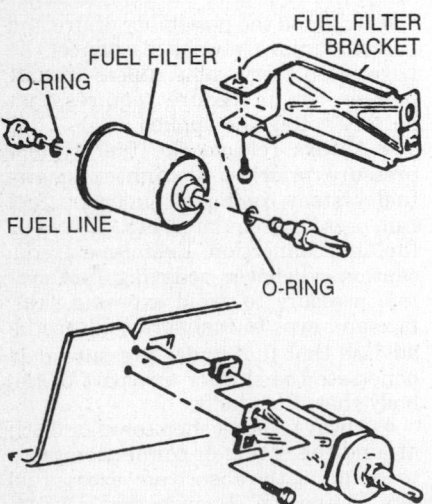

Fuel filter assembly

Fuel Filter
REMOVAL & INSTALLATION

1. Disconnect the negative battery cable.
2. Relieve the fuel suystem pressure.
3. Raise and safely support the vehicle.
4. Clean the fuel filter connections before disconnecting to prevent contamination of the fuel system. Disconnect and plug the fuel lines from the fuel filter.
5. Remove the filter bracket screw and remove the fuel filter from the fuel filter bracket.
To install:
6. Check the fuel line O-rings for cuts, nicks, swelling or distortion and replace as necessary.
7. Position the replacement filter in the fuel filter bracket with the flow arrow pointing toward the engine.
8. Install the fuel filter bracket screw and and the fuel lines. Tighten the in-line fuel filter fittings to 20 ft. lbs. (27 Nm), using a back-up wrench to prevent the filter from turning.
9. Lower the vehicle, connect the negative battery cable and tighten the fuel filler cap.
10. Turn the ignition switch to the ON position for 2 seconds and then OFF for 5 seconds. Again turn to the ON position and check for fuel leaks.

Electric Fuel Pump

PRESSURE TESTING

5.0L TBI Engine

1. Ensure there is an adequate quantity of fuel in the tank.
2. Ensure the ignition switch is in the OFF position.
3. Connect a fuel pressure gauge to the fuel pump outlet line.
4. Using a 10 amp fused jumper wire, apply battery voltage to the fuel pump test connector located on the passenger side of the engine compartment.
5. The fuel pressure should be 9–13 psi.
6. Fuel pump pressure will drop immediately after the fuel pump stops running due to an controlled bleed in the fuel system.
7. If there is no fuel pressure, listen for pump operation in tank.
8. If pump operation is heard, inspect lines and filter for restriction.
9. If there is no restriction, replace the fuel pump.
10. If pump is not heard running, inspect the fuel pump relay and wiring.
11. Disconnect the fuel pressure gauge.

Except 5.0L TBI Engine

1. Ensure there is an adequate quantity of fuel in the tank.
2. Connect a fuel pressure gauge to the fuel pump outlet line.
3. Make sure the ignition switch has been in the OFF position for at least 10 seconds and the all accessories are OFF.
4. Turn the ignition switch ON. The pump will run for about 2 seconds. The fuel pump pressure should be 40.5–47 psi, except the 3.8L engine, which should be 34–40 psi.

NOTE: The ignition switch may have to be cycled to ON more than once to obtain maximum pressure. It is also normal for the pressure to drop slightly when the pump stops.

5. If fuel pressure is not as specified, verify fuel pump operation is heard in the tank.
6. If fuel pump operation is not heard, inspect the fuel pump relay and wiring.
7. If fuel pump operation is heard, inspect filter and lines for restriction.
8. Start the engine and verify the pressure should decrease 3–10 psi.
9. If fuel pressure does not decrease, inspect pressure regulator and vacuum hose.
10. Disconnect the fuel pressure gauge.

REMOVAL & INSTALLATION

The electric fuel pump is located inside the fuel tank.

1. Release the fuel pressure and disconnect the negative battery cable. Drain the fuel from the tank into a safe container.
2. Raise and support the vehicle safely. Disconnect the exhaust pipe at the catalytic converter and the rear hanger. Allow the exhaust system to hang over the rear axle assembly.
3. Remove the tail pipe and muffler heat shields. Remove the fuel filler neck shield from behind the left rear tire.
4. Remove the rear track bar and brace.
5. Disconnect the fuel pump/sending unit electrical connector, at the body harness connector. Do not pry up on the cover connector, as the pump/sending unit wiring harness is an integral part of the sending unit.
6. Disconnect the fuel pipes. Remove the fuel pipe retaining bracket on the left side and the brake line clip from the retaiing bracket.
7. Position a jack under the rear axle assembly in order to support the rear axle.
8. Disconnect the lower ends of the

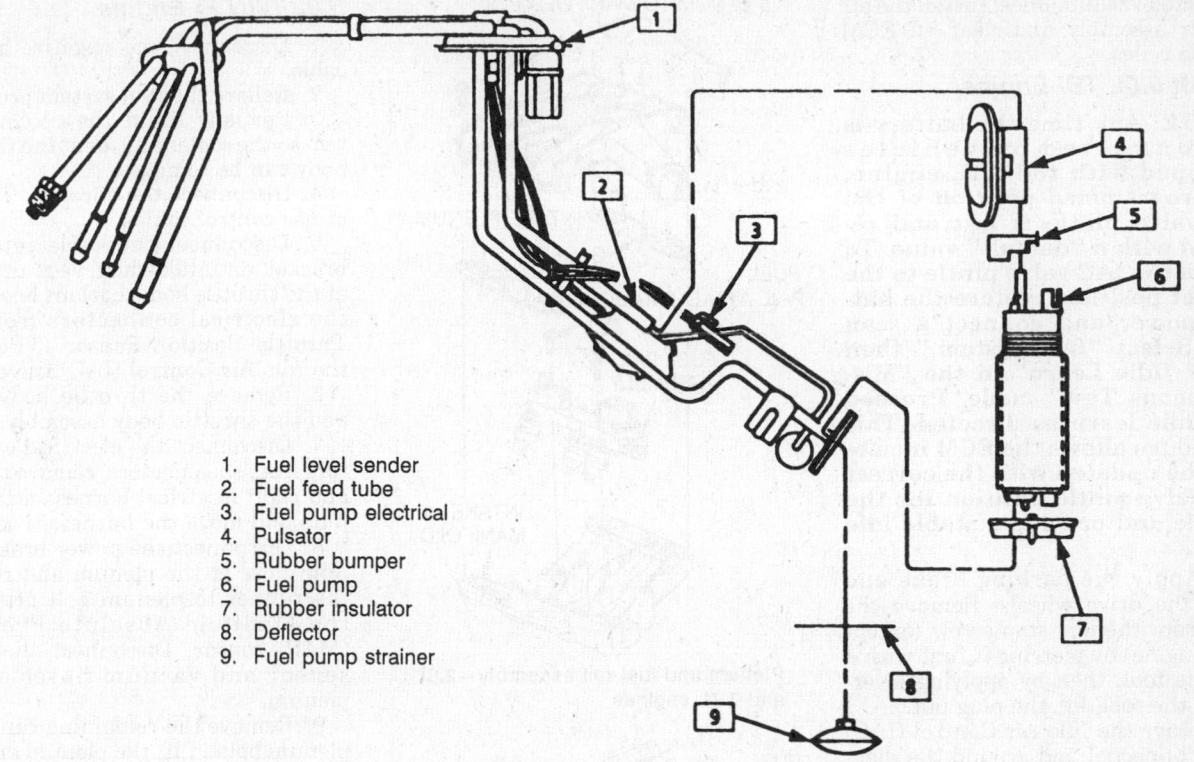

1. Fuel level sender
2. Fuel feed tube
3. Fuel pump electrical
4. Pulsator
5. Rubber bumper
6. Fuel pump
7. Rubber insulator
8. Deflector
9. Fuel pump strainer

Typical in-tank electric fuel pump assembly

shock absorbers, lower the axle assembly enough to release the tension on the coil springs. Remove the coil springs.

9. Lower the rear axle assembly as far as possible without causing damage to the brake lines and cables.

10. Remove the fuel tank strap bolts. Remove the tank by rotating the front of the tank downward and sliding it to the right side.

11. Remove the fuel pump/sending unit from the tank, by loosening the cam nut. When removing the cam nut, use brass tool or equivalent to tap the nut loose.

12. Remove the O-ring from under the unit. Replace the O-ring if defective.

13. Separate the fuel pump from the sending unit.

To install:

14. Install the fuel pump to the sending unit.

15. Install the O-ring in the groove around the tank opening and install the fuel pump/sending unit. Install the cam nut and tighten until is it against the stop.

16. Raise the fuel tank into position and install the tank strap bolts.

17. Install the coil springs and raise the rear axle into position. Connect the shcok absorbers.

18. Connect the fuel lines and the electrical connector.

19. Install the rear suspension track bar and the track bar brace.

20. Install the fuel filler neck shield and the tail pipe and muffler heat shields.

21. Connect the exhaust system and lower the vehicle.

22. Fill the fuel tank.

23. Connect the negative battery cable, start the engine and check for fuel leaks.

Fuel Injection

IDLE SPEED ADJUSTMENT

The idle speed and mixture are electronically controlled by the Electronic Control Module (ECM). All adjustments are preset at the factory. The only time the idle speed should need adjustment is when the throttle body assembly has been replaced. The throttle stop screw, used in regulating the minimum idle speed, is adjusted at the factory and should not require further adjustment. This adjustment should be performed only when the throttle body as been replaced.

5.0L TBI Engine

1. Block the drive wheels and apply the parking brake. Remove the air cleaner assembly and or air duct. Re-move and plug any vacuum hoses on the tube manifold assembly, if equipped. Disconnect the throttle cable.

2. Ground the diagnostic test terminal in the ALDL connector. Turn the ignition **ON** and leave the engine **OFF**. Wait at least 45 seconds, this will allow the Idle Air Control (IAC) pintle to seat in the throttle body.

3. With the ignition switch in the **ON** position, the engine **OFF** and the ALDL test terminal still grounded, disconnect the IAC valve electrical connector.

4. Connect a tachometer to the engine to monitor the engine speed.

5. Remove the ground from the ALDL connector test terminal.

6. Place the transmission in the **P** or **N** position. Start and run the engine until it reaches normal operating temperature. It may be necessary to depress the accelerator pedal in order to start the engine. Allow the engine idle speed to stabilize.

7. The idle speed should be 400–450 rpm. If not as specified, remove the idle speed stop screw plug and adjust as necessary.

8. Install the throttle cable, be sure the minimum idle speed is not affected by the throttle cable.

9. Turn the ignition **OFF** and reconnect the IAC valve electrical connector. Unplug and reconnect the dis-

connected vacuum lines. Install the air cleaner assembly and clear all ECM trouble codes.

Except 5.0L TBI Engine

NOTE: Any time the battery is disconnected on vehicles equipped with the 3.1L engine, the programmed position of the IAC valve pintle is lost and replaced with a "default" value. To return the IAC valve pintle to the correct position, restore the battery power and connect a scan tool. Select "IAC System," then select "Idle Learn" in the "Miscellaneous Test" mode. Proceed with idle learn as directed. This procedure allows the ECM memory to be updated with the correct IAC valve pintle position for the vehicle and provide a stable idle speed.

1. Apply the parking brake and block the drive wheels. Remove the plug from the idle stop screw (except 3.8L engine) by piercing it first with a suitable tool; then by applying leverage to the tool, lift the plug out.

2. Leave the Idle Air Control (IAC) valve connected and ground the diagnostic terminal (ALDL).

3. Turn the ignition switch to the **ON** position but do not start the engine. Wait for at least 30 seconds. This allows the IAC valve pintle to extend and seat in the throttle body.

4. With the ignition switch still in the **ON** position, disconnect IAC electrical connector.

5. Start the engine and allow the engine to reach normal operating temperature. Remove the ground from the diagnostic terminal. Disconnect the distributor set-timing connector on 5.0L and 5.7L engines.

6. With the engine in the D position, adjust the idle stop screw to obtain the correct specifications:

2.8L engine with automatic transmission: 550 rpm in D.

2.8L engine with manual transmission: 650 rpm in N.

3.8L engine with automatic transmission: 500 + 50 rpm in D.

5.0L engine, all transmissions: 400 rpm in N.

5.7L engine, all transmissins: 450 rpm in N.

7. Turn the ignition OFF and reconnect the connector at the IAC motor.

8. Adjust the TPS if necessary. Start the engine and inspect for proper idle operation.

Fuel Injector

REMOVAL & INSTALLATION

2.8L and 3.1L Engines

1. Disconnect the negative battery cable.

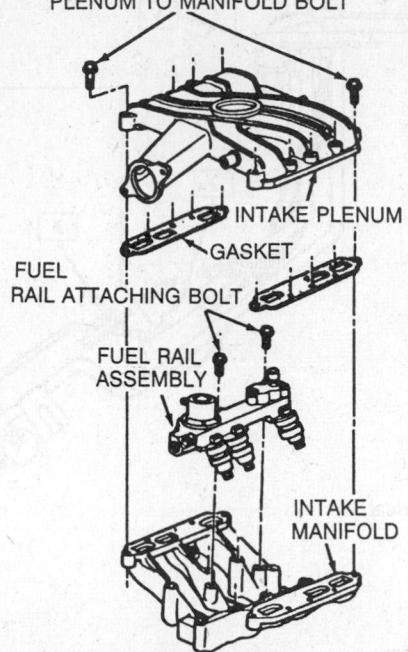

Plenum and fuel rail assembly—2.8L and 3.1L engines

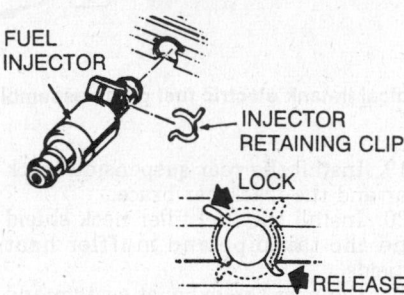

Fuel Injector and retaining clip assembly

2. Relieve the fuel system pressure.

3. Remove the intake manifold plenum.

4. Remove the fuel rail assembly.

5. Rotate the fuel injector retaining clip to the release position and remove the injector.

6. Installation is the reverse of the removal procedure.

3.8L Engine

1. Disconnect the negative battery cable.

2. Relieve the fuel system pressure.

3. Disconnect the electrical connectors from the fuel injectors.

4. Disconnect the vacuum hose from the pressure regulator.

5. Disconnect the fuel feed and return lines from the fuel rail.

6. Remove the fuel rail attaching bolts and remove the fuel rail and the fuel injectors.

7. Installation is the reverse of the removal procedure. Use new O-rings on the inejctors and coat the O-rings with engine oil.

5.0L (VIN F) Engine

1. Disconnect the negative battery cable.

2. Relieve the fuel system pressure.

3. Partially drain the cooling system so the coolant hoses at the throttle body can be removed.

4. Disconnect the throttle, TV and cruise control cables.

5. Disconnect the cable retaining bracket, air intake duct, vacuum hoses at the throttle body, coolant hoses and the electrical connectors from the Throttle Position Sensor (TPS) and the Idle Air Control (IAC) valve.

6. Remove the throttle body bolts and the throttle body assembly.

7. Disconnect the electrical connectors from the injectors, remove the left and right electrical harness attaching nuts and move the harnesses aside.

8. Disconnect the power brake vacuum hose at the plenum and remove the runner to plenum bolt attaching the Manifold Absolute Pressure (MAP) sensor. Disconnect the MAP sensor and vacuum hoses at the plenum.

9. Remove the remaining runner to plenum bolts. Lift the plenum and disconnect the Manifold Air Temperature sensor electrical connector. Remove the plenum and discard the plenum gaskets.

10. Remove the runner to manifold bolts, PCV valve and hose, EGR solenoid and the left and right side runners and gaskets. Discard the gaskets.

11. Disconnect the fuel feed and return lines. Discard the fuel line O-rings.

12. Remove the fuel tube bracket bolt.

13. Disconnect the vacuum line at the pressure regulator.

14. Remove the fuel rail attaching bolts and the fuel rail assembly.

15. Rotate the injector retainer clip to the release position and remove the injector. Discard the O-rings and retainer clips.

To install:

NOTE: There are 2 injector part numbers used in production for the 5.0L engine and 2 different part numbers for the 5.7L engine. Do not intermix injectors with different part numbers, as this will result in engine roughness and excessive emissions.

If the entire set of injectors are being replaced, either part number listed for that specific engine may be used.

16. Lubricate new injector O-ring seals with engine oil and install on the injector.

17. Install a new retainer clip onto the injector and install the injector

into the fuel rail injector socket, with the electrical connector facing outward. Rotate the injector retainer clip to the locking positon.

18. Install the fuel rail assembly in the intake manifold. Install the attaching bolts to 15 ft. lbs. (20 Nm). Install the fuel tube bracket bolt and tighten to 25 ft. lbs. (34 Nm).

19. Connect the vacuum line to the pressure regulator.

20. Install new O-rings on the fuel feed and return lines and connect the fuel lines to the fuel rail. Tighten the fuel line nuts to 20 ft. lbs. (27 Nm).

21. Temporarily connect the negative battery cable. Turn the ignition switch to **ON** for 2 seconds and then to **OFF**. Again turn the ignition switch to the **ON** position and check for fuel leaks. Disconnect the negative battery cable.

22. Clean all plenum and runner gasket mating surfaces.

23. Install new gaskets, the runners and manifold to runner bolts to the intake manifold. Tighten the bolts to 25 ft. lbs. (34 Nm).

24. Install the EGR solenoid.

25. Install the right and left side runner to manifold bolts finger tight only.

26. Support the plenum above the runners, connect the MAT sensor electrical connector and lower the plenum into position. Start a few bolts to hold the plenum in position.

27. Connect the vacuum hoses and MAP sensor.

28. Tighten all bolts to 25 ft. lbs. (34 Nm), starting in the center of the plenum/manifold and working outward.

29. Install the PCV valve and hose.

30. Connect the power brake vacuum hose to the fitting on the plenum, the left and right injector electrical harnesses, the attaching nuts and the electrical connectors to the injectors.

31. Install the throttle body with a new gasket and tighten the attaching bolts to 18 ft. lbs. (24 Nm).

32. Connect the electrical connectors to the TPS and IAC valve, coolant hoses, vacuum hoses, throttle cable bracket and the throttle, TV and cruise control cables.

33. Refill the cooling system, tighten the fuel filler cap and connect the negative battery cable.

5.0L TBI Engine

1. Disconnect the negative battery cable.

2. Relieve the fuel system pressure.

3. Remove the air cleaner assembly.

4. Remove the electrical connectors from the fuel injectors by squeezing the plastic tabs and pulling straight up.

5. Remove the fuel meter cover attaching screws and remove the fuel meter cover assembly.

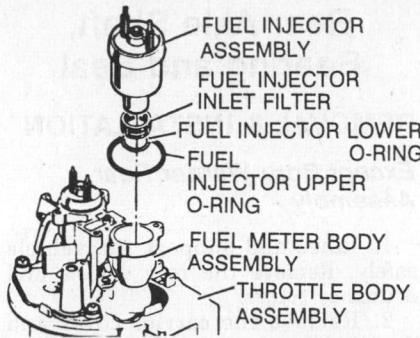

Fuel Injector and O-rings—TBI

6. Remove the fuel meter outlet passage gasket and pressure regulator dust seal. If the fuel meter cover gasket is stuck to the fuel meter body, leave it in place. If it is stuck to the fuel meter cover, remove it and place it on the fuel meter body.

7. With the fuel meter cover gasket in place to protect the fuel meter body, use a suitable prybar and fulcrum to carefully pry out the injector.

8. Discard both injector O-rings and the fuel meter cover gasket.

To install:

NOTE: **Be sure to replace the injector with an identical part. Injectors from other engines are calibrated for different flow rates. Service fuel injector packages may contain a fuel injector washer (spacer). The washer is not required for this application.**

9. Lubricate a new upper (large) O-ring with engine oil and install in the fuel meter body cavity. Make sure the O-ring is seated properly and is flush with the top of the fuel meter body surface.

10. Lubricate a new lower (small) O-ring with engine oil and install on the nozzle end of the injector. Push the O-ring on far enough to contact the filter.

11. Install the injector by aligning the raised lug on the injector base with the notch in the fuel meter body cavity. Push down on the injector until it is fully seated in the fuel meter body. The electrical terminals of the injector should be parallel with the throttle shaft.

12. Install a new pressure regulator dust seal, fuel meter outlet gasket and cover gasket.

13. Install the fuel meter cover assembly. Apply Loctite 262 or equivalent, to the retaining screws and tighten to 27 inch lbs. (3 Nm).

14. Connect the electrical connectors to the fuel injectors, tighten the fuel filler cap and connect the negative battery cable.

15. Turn the ignition switch to the **ON** position for 2 seconds, then turn it

to the **OFF** position for 5 seconds. Again turn the swith to **ON** and check for fuel leaks.

16. Install the air cleaner assembly and connect the negative battery cable.

DRIVE AXLE

Driveshaft and U-Joints

REMOVAL & INSTALLATION

1. Raise and support the vehicle safely. Matchmark the rear axle pinion flange and driveshaft for assembly.

2. Remove the driveshaft strap bolts and remove the retaining straps.

3. Drop the driveshaft down at the rear, then pull it backwards outs from the transmission extension housing. The transmission housing should be plugged to prevent leakage. If the bearing caps are loose, tape them together to prevent dropping and losing the bearing rollers.

4. To replace the U-joints proceed as follows:

   a. Support the driveshaft horizontally in line with the base plate of a press.

   b. Place the U-joint so the lower ear of the shaft yoke is supported on a 1⅛ in. socket.

   c. Remove the lower bearing cap out of the yoke ear by placing tool J–9522–3 or equivalent, on the open horizontal bearing caps and pressing the lower bearing cap out of the yoke ear.

NOTE: **This will shear the nylon injector ring, if the original U-joint is being removed. There are no bearing retaining grooves in the production bearing caps, therefore they cannot be reused. If a replacement U-joint is being removed, ensure to remove the retaining clips from the U-joint.**

   d. If the bearing cap is not completely removed, lift tool J–9522–3 and insert tool J–9522–5 or equivalent between the bearing cap and seal and continue pressing the U-joint out of the yoke.

   e. Repeat the procedure for the opposite side.

   f. Remove the spider from the yoke.

To install:

   g. Install 1 bearing cap part way into 1 side of the yoke. Turn this yoke ear to the bottom.

   h. Using tool J–9522–3 or equivalent, seat the trunnion into the bearing cap.

i. Install the opposite bearing cap partially onto the trunnion.

j. Ensure both trunnions are straight and true in the bearing caps.

k. Press the spider against the opposite bearing cap, while working the spider back and forth to ensure free movement of the trunnions in the bearing caps.

l. If trunnion is binding, the needle bearings have tipped over under the end of the cap.

m. Stop pressing when 1 bearing cap clears the retainer groove inside the yoke.

n. Install a retaining ring.

o. Repeat the procedure for the remaining bearing caps and u-joints.

5. Installation of the driveshaft is the reverse of the removal procedure. Tighten the strap bolts to 16 ft. lbs. (22 Nm).

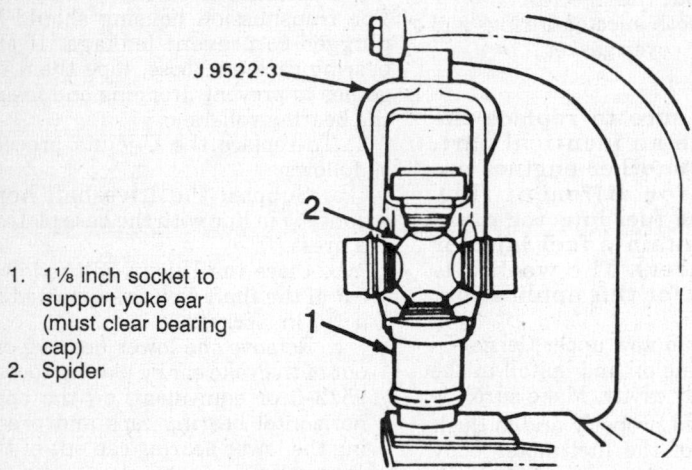

1. 1⅛ inch socket to support yoke ear (must clear bearing cap)
2. Spider

Universal joint removal and installation

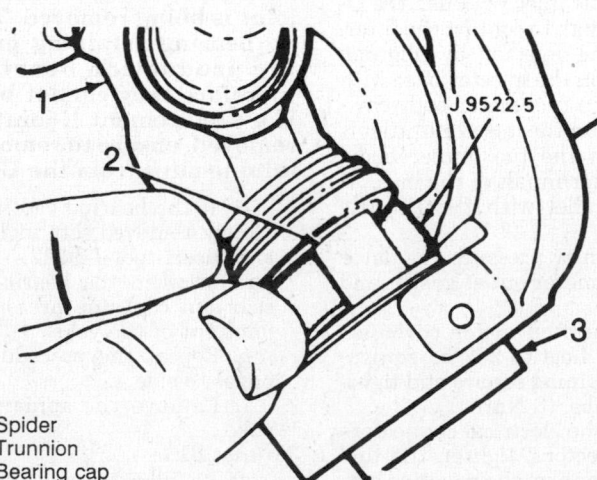

1. Spider
2. Trunnion
3. Bearing cap

Universal joint removal and installation

Rear Axle Shaft, Bearing and Seal

REMOVAL & INSTALLATION

Except Borg Warner Rear Assembly

1. Raise and support the vehicle safely. Remove the rear wheels and drums or rotors.

2. Remove the carrier cover and drain the gear oil into a suitable container.

3. Remove the rear axle pinion shaft lock screw. Remove the rear axle pinion shaft.

4. Push the flanged end of the axle shaft into the axle housing and remove the C-clip from the opposite end of the shaft.

5. Remove the axle shaft from the axle housing.

6. Using a suitable tool, remove the

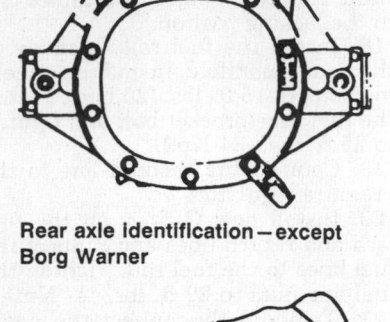

Rear axle identification—except Borg Warner

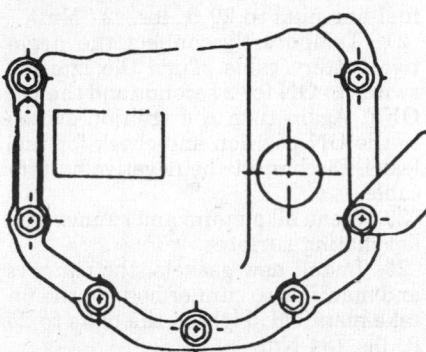

Rear axle identification—Borg Warner

oil seal from the axle housing. Be careful not to damage the housing.

7. Install tool J–22813–01 or equivalent, into the bore of the axle housing and position it behind the bearing, ensure the tangs of the tool engage the outer race. Remove the bearing using a slide hammer.

8. Installation is the reverse of the removal procedure. Lubricate the new bearing and sealing lips with gear lube before installing. Tighten the pinion gear shaft lock screw to 27 ft. lbs. (36 Nm). Tighten the carrier cover bolts to 22 ft. lbs. (30 Nm).

Borg Warner Rear Assembly

NOTE: The Borg Warner axle assembly can be quickly identified by checking the axle code. The Borg Warner axle numbers are 4EW, 4EU and 4ET on 1988 vehicles, BET, BEU and BEW on 1989 vehicles and 9EQ and 9ER on 1990 vehicles.

1. Raise the vehicle and support is safely.

2. Remove the rear wheels and drums or rotors. Remove the brake components as required.

3. Remove the 4 nuts attaching the brake anchor plate and outer bearing retainer to the axle housing.

4. Remove the axle shaft and wheel bearing assembly using axle shaft removal tool J–21595 and slide hammer J–2619 or equivalent.

5. To remove the inner bearing retainer and the bearing from the axle shaft, split the retainer with a chisel

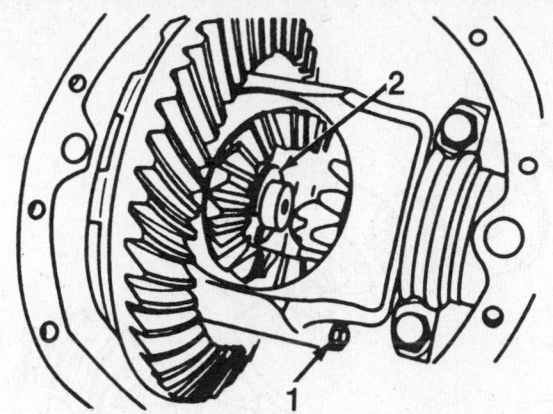

1. Lock screw
2. C-Lock

Pinion shaft lock screw and C-clip removal—except Borg Warner

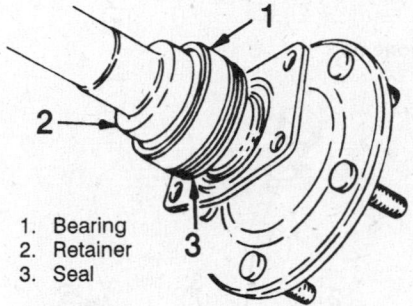

1. Bearing
2. Retainer
3. Seal

Axle shaft and bearing assembly— Borg Warner

and remove it from the shaft. Using tool J–22912–01, press the bearing off the shaft.

6. Installation is the reverse of the removal procedure. Ensure the axle seal is installed with the spring side facing the center of the axle. Tighten the backing plate bolts to 36 ft. lbs. (49 Nm).

NOTE: There are right (black banded) and left (gold banded) axle seals and they cannot be interchanged.

MANUAL TRANSMISSION

For further information on transmissions/transaxles, please refer to "Chilton's Guide to Transmission Repair".

Transmission Assembly

REMOVAL & INSTALLATION

1. Disconnect the negative battery cable.

2. Remove the shift control lever knob.

3. Raise and safely support the vehicle.

4. Drain the transmission oil.

5. Safely support the left side of the rear axle to avoid damaging the brake lines. Remove the torque arm and driveshaft assembly.

6. Remove the speed sensor connector or the speedometer cable.

7. Disconnect the transmission wire connectors.

8. Remove the catalytic converter hanger.

9. Safely support the engine. Remove the transmission support nuts and bolts. Remove the support.

10. Lower the transmission enough to remove the shift control bolts and lever. Remove the shift control lever from the extension housing.

11. Remove the transmission-to-bellhousing bolts.

12. Remove the transmission assembly.

To install:

13. Clean sealing surfaces of transmission-to-shifter location. Place a continuos ⅛ inch bead of RTV or equivalent to the shifter-to-transmission sealing surface.

14. Install the transmission and secure with attaching bolts to the bellhousing.

15. Torque the transmission-to-bellhousing bolts to 55 ft. lbs. (75 Nm).

16. Install the shifter onto the transmission and torque the bolts to 13 ft. lbs. (17 Nm).

17. Install the trnasmission support and torque the transmission mount nuts to 35 ft. lbs. (47 Nm) an the support bolts to 40 ft. lbs. (54 Nm).

18. Install the catalytic conerter hanger and torque the bolts to 35 ft. lbs. (40 Nm).

19. Reconnect all electrical connectors. Install speedometer cable, if equipped.

20. Install torque arm and driveshaft assembly.

21. Tighten the front torque arm bracket to 20 ft. lbs. (27 Nm) and the rear torque arm nuts to 98 ft. lbs. (133 Nm). Remove the engine support and left rear axle support stand.

22. Refill the transmission with Dexron® II or equivalent.

23. Lower the vehicle and connect the negative battery cable.

24. Start the vehicle and inspect transmission for leaks and proper operation.

LINKAGE ADJUSTMENT

The M39, MK6 and MB1 5-speed manual transmission are designed with an internal shift mechanism. Shifter control adjustments are not possible.

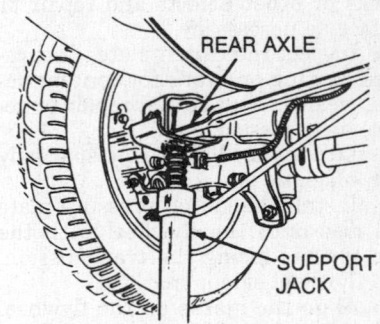

REAR AXLE

SUPPORT JACK

Supporting the rear axle during torque arm removal

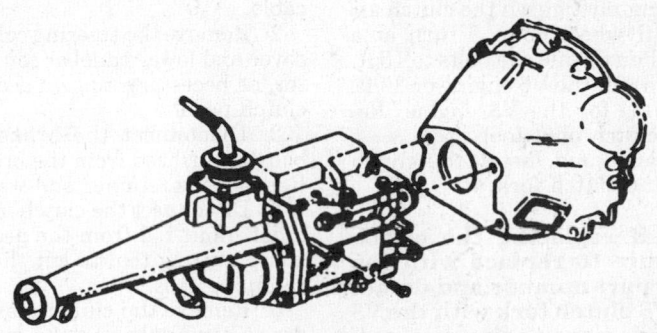

Transmission and bellhousing assembly

CLUTCH

Clutch Assembly

REMOVAL & INSTALLATION

1. Disconnect the negative battery cable.

2. Raise and safely support the vehicle.

3. Remove the transmission assembly.

4. Remove the clutch slave cylinder and heat shield assembly from the flywheel housing.

5. Remove the bolts attaching the flywheel housing to the engine and remove the housing.

6. Remove the clutch release bearing and clutch fork with spring.

7. Install tool J–33169 or equivalent to support the clutch assembly during removal. Locate marks or a white painted letter on the clutch housing and an **X** mark on the flywheel. If the marks are not visible, place a mark on the flywheel and clutch cover for alignment during installation.

8. Loosen the clutch-to-flywheel bolts evenly 1 turn at a time until all spring pressure is released.

9. Remove the clutch and pressure plate assembly.

To install:

10. Inspect flywheel for heat stress, cracks or other defects and repair or replace as necesssary.

11. Inspect the clutch plate, disc, release bearing and fork for contamination, wear or heat stress, repair or replace as necessary.

12. Lubricate pilot bearing sparingly with machine oil.

13. Install the clutch pressure plate and disc onto the flywheel with the disc springs facing the transmission. The flywheel side is marked.

14. Align the marks on the flywheel with the mark on the pressure plate.

15. Install the pressure plate-to-flywheel bolts loosly and install a suitable clutch pilot tool.

16. Alternately tighten the clutch assembly-to-flywheel bolts 1 turn at a time and then torque the bolts to 15 ft. lbs. (21 Nm) for the V6 or 30 ft. lbs. (40 Nm) for the V8 engine. Remove the clutch pilot tool.

17. Lubricate and install the clutch fork ball and clutch fork.

NOTE: If replacing the clutch fork, ensure to replace with the identical part number and do not mix the V6 clutch fork with the V8 clutch or damage to the slave cylinder may result.

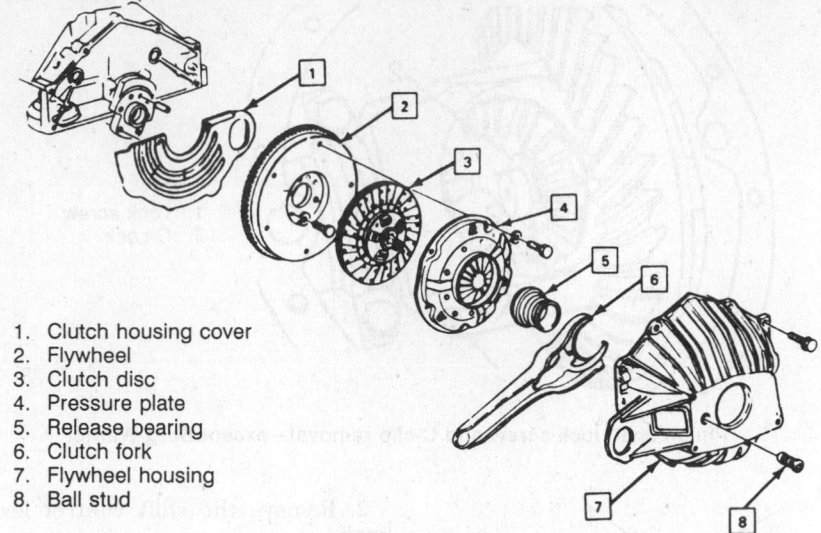

1. Clutch housing cover
2. Flywheel
3. Clutch disc
4. Pressure plate
5. Release bearing
6. Clutch fork
7. Flywheel housing
8. Ball stud

Clutch components

18. Install the dust boot, if removed.

19. Install the clutch release bearing onto the fork with the fork fingers and retaining spring tabs installed into the release bearing grooves.

20. Install the clutch housing and torque the bolts to 35 ft. lbs. (47 Nm) for the V6 engine or 70 ft. lbs. (95 Nm) for the V8 engine.

21. Install the transmission, clutch slave cylinder and heat shield.

22. Torque the slave cylinder bolts to 15 ft. lbs. (21 Nm).

23. Refill transmission with the proper type and quantity of oil. Lower the vehicle and check clutch operation.

PEDAL HEIGHT/FREE-PLAY ADJUSTMENT

The hydraulic clutch system locates the clutch pedal height and provides automatic clutch adjustment. No adjustment of clutch linkage or pedal position is required.

Clutch Master Cylinder

REMOVAL & INSTALLATION

1. Disconnect the negative battery cable.

2. Remove the steering column trim cover and lower panel or sound insulator, as necessary, to gain access to the clutch pedal.

3. Disconnect the brake vacuum booster pushrod from the brake pedal. Remove the retainer and washer.

4. Disconnect the clutch master cylinder input rod from the pedal. Use a sharp cutting tool to cut the bushing retaining tabs.

5. Remove the clutch master cylinder-to-cowl nuts and the brake vacuum booster-to-cowl nuts.

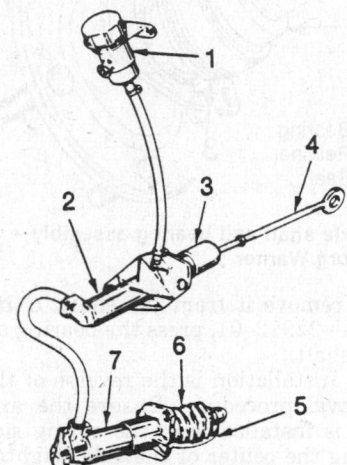

1. Fluid reservoir
2. Clutch master cylinder
3. Boot
4. Pushrod
5. Shipping strap
6. Boot
7. Clutch slave cylinder

Hydraulic clutch assembly

6. Remove the hose clamp and the clutch fluid reservoir hose. Place a suitable drain pan under hose to catch the fluid leaking from the reservoir.

7. Pull the brake vacuum booster forward to gain access to the clutch master cylinder. Remove the clutch master cylinder with U-bolt from the cowl. Lower the master cylinder down to the clutch housing area.

8. Raise and safely support the vehicle.

9. Disconnect the high pressure hose and remove the clutch master cylinder.

To install:

10. Connect the high pressure hose and place the master cylinder up near the brake vacuum booster. Lower the vehicle.

11. Install the clutch master cylinder to the cowl with the U-bolt.

12. Install a new bushing to the pedal. Install the flat end of the bushing toward the pedal.

13. Install the clutch master cylinder input rod to the pedal. Install the retainer and washer.

14. Attach the brake vacuum booster to the cowl, connect the clutch fluid reservoir hose with the hose clamp and install the clutch master cylinder-to-cowl nuts. Tighten the nuts to 115 inch lbs. (13 Nm). Tighten the brake vacuum booster-to-cowl nuts to 15 ft. lbs. (21 Nm).

15. Replace the sound insulator or steering column trim panel and lower panel.

16. Connect the brake vacuum pushrod to the brake pedal.

17. Fill and bleed the clutch hydraulic system. Connect the negative battery cable.

Clutch Slave Cylinder

REMOVAL & INSTALLATION

1. Disconnect the negative battery cable.

2. Remove the steering column trim cover and lower panel or sound insulator as necessary to gain access to the clutch pedal.

3. Disconnect the clutch master cylinder input rod from the clutch pedal. Using a sharp cutting tool, cut the bushing retaining tabs.

NOTE: Disconnect the clutch master cylinder input rod before removing the slave cylinder. If it is not disconnected, permanent damage to the slave cylinder will occur if the clutch pedal is depressed while the slave cylinder is disconnected.

4. Raise and safely support the vehicle.

5. Disconnect the pressure hose and catch the leaking hydraulic fluid in a suitable container.

6. Remove the slave cylinder bolts, then remove the heat shield and slave cylinder.

7. Installation is the reverse of the removal procedure. Tighten the slave cylinder bolts to 15 ft. lbs. (21 Nm). Install a new clutch pedal bushing with the flat side toward the clutch pedal. Bleed the hydraulic system.

Hydraulic Clutch System Bleeding

1. Clean all dirt and grease from the cap to make sure no foreign substances enter the system.

2. Remove the cap and diaphragm and fill the reservoir to the top with the approved DOT 3 brake fluid. Fully loosen the bleed screw which is in the slave cylinder body next to the inlet connection.

3. At this point, bubbles of air will appear at the bleed screw outlet. When the slave cylinder is full and a steady stream of fluid comes out of the slave cylinder bleeder, tighten the bleed screw to 18 inch lbs. (2 Nm).

4. Assemble the diaphragm and cap to the reservoir. Fluid in the reservoir should be level with the step. Exert a light load of about 20 lbs. to the slave cylinder piston by pushing the clutch fork towards the cylinder and loosening the bleed screw. Maintain a constant light load. Fluid and any air that is left will be expelled through the bleed port. Tighten the bleed screw when a steady flow of fluid and no air is being expelled.

5. Fill the reservoir fluid level back to normal capacity.

6. Exert a light load to the release lever but do not open the bleeder screw as the piston in the slave cylinder will move slowly down the bore. Repeat this operation 2–3 times. The fluid movement will force any air left in the system into the reservoir. The hydraulic system should now be fully bled.

7. Check the operation of the clutch hydraulic system and repeat this procedure if necessary. Check the pushrod travel at the slave cylinder to insure the minimum travel is 0.43 in. for 2.8L and 3.1L engines or 0.57 for 5.0L and 5.7L engines.

AUTOMATIC TRANSMISSION

For further information on transmissions/transaxles, please refer to "Chilton's Guide to Transmission Repair".

Transmission Assembly

REMOVAL & INSTALLATION

1. Disconnect the negative battery cable.

2. Remove the air cleaner assembly, if necessary.

3. Disconnect the Throttle Valve (TV) control cable at the throttle lever.

4. Remove the transmission oil dipstick. Unbolt and remove the dipstick tube.

5. Raise and support the vehicle safely.

6. Mark the relationship between the driveshaft and the rear pinion flange so the driveshaft may be reinstalled in its original position.

7. Remove the driveshaft from the vehicle.

8. Disconnect the catalytic converter support bracket at the transmission.

9. Disconnect the speedometer cable or speed sensor, the electrical connectors and the shift control cable from the transmission.

10. Remove the torque arm to transmission bolts.

NOTE: The rear spring force will cause the torque arm to move toward the floor pan. When disconnecting the arm from the transmission, carefully place a piece of wood between the floor pan and the torque arm. This will prevent possible personal injury and/or floor pan damage.

11. Remove the flywheel cover, then mark the relationship between the torque converter and the flywheel.

12. Remove the torque converter to flywheel attaching bolts.

13. Support the transmission with a jack, then remove the transmission mount bolt.

14. Unbolt and remove the transmission crossmember.

15. Lower the transmission slightly. Disconnect the TV cable and oil cooler lines from the transmission.

16. Support the engine. Remove the transmission to engine mounting bolts.

17. Remove the transmission from the vehicle. Keep the rear of the transmission lower then the front to avoid the possibility of the torque converter disengaging from the transmission.

To install:

18. Install the transmission in the vehicle.

19. Install and tighten the transmission to engine bolts to 35 ft. lbs. (47 Nm).

20. Connect the TV cable and oil cooler lines.

21. Install the crossmember.

22. Install and tighten the transmission crossmember to frame bolts to 40 ft. lbs. (54 Nm) and the crossmember to transmission mount nut to 35 ft. lbs. (47 Nm).

23. Install the torque converter to the flywheel, aligning the marks that were made prior to removal. Tighten the converter to flywheel bolts to 46 ft lbs. (63 Nm).

24. Install the flywheel cover.

25. Connect the torque arm to the transmission.

26. Connect the speedometer cable or speed sensor, the electrical connectors and the shift control cable.

27. Connect the catalytic converter support bracket.

28. Install the driveshaft, aligning the marks that were made on the driveshaft and pinion flange prior to removal.

29. Lower the vehicle.

30. Install the transmission dipstick and dipstick tube.

31. Connect the TV control cable at the throttle lever.

32. Install the air cleaner assembly, if necessary, and connect the negative battery cable.

33. Start the engine and check the transmission fluid level.

SHIFT CONTROL CABLE ADJUSTMENT

1. Raise and safely support the vehicle.

2. Loosen the shift control cable attachment at the shift lever.

3. Rotate the shift lever clockwise to the **P** detent and then back to the **N** detent.

4. Tighten the cable attaching nut to 11 ft. lbs. (15 Nm). The lever must be held out of **P** when tightening the nut.

5. Check cable adjustment by rotating the control lever through the detents.

THROTTLE VALVE CABLE ADJUSTMENT

NOTE: Setting of the TV cable must be done by rotating the throttle lever at the throttle body. Do not use the accelerator pedal to rotate the throttle body lever.

1. Ensure the engine is **OFF**.

2. Depress and hold-down the metal reset tab at the engine end of the TV cable.

3. Move the slider until it stops against the fitting.

4. Release the reset tab.

5. Rotate the throttle lever to it's full travel position.

6. The slider must move (ratchet)

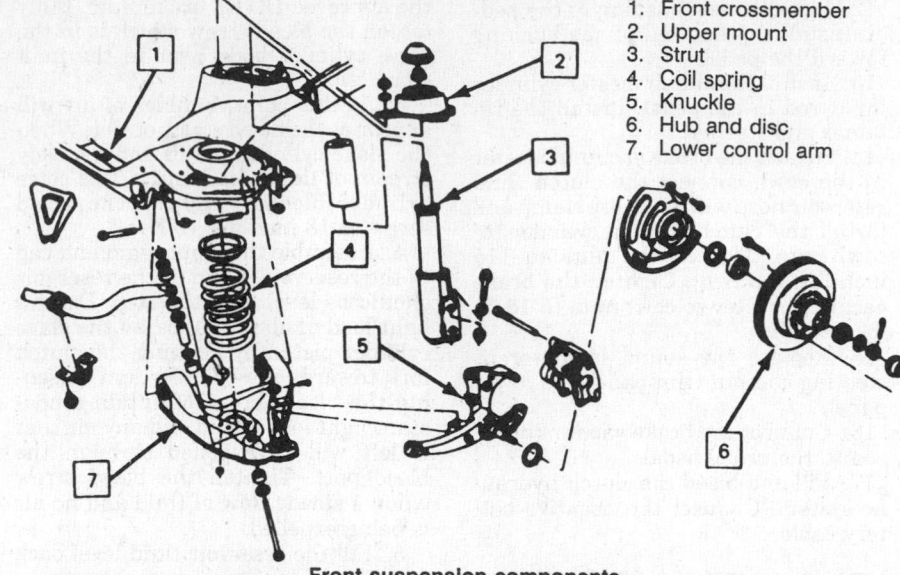

1. Front crossmember
2. Upper mount
3. Strut
4. Coil spring
5. Knuckle
6. Hub and disc
7. Lower control arm

Front suspension components

toward the lever when the lever is rotated to it's full travel position.

7. Ensure the cable moves freely. The cable may appear to function properly with the engine stopped and cold. Recheck after the engine is hot.

FRONT SUSPENSION

MacPherson Strut

REMOVAL & INSTALLATION

1. Raise and safely support the vehicle. Use a suitable device to support the lower control arm.

2. Remove the wheel and tire assembly.

3. Remove the brake hose bracket.

4. Remove and discard the strut-to-knuckle bolts, washers and nuts.

5. Remove the cover from the upper mount assembly.

6. Remove the nut from the upper end of the strut.

7. Remove the strut and shield.

8. Install in the reverse order of removal. Tighten the upper strut nut to 46 ft. lbs. (63 Nm). Use new strut-to-knuckle bolts, washers and nuts. Tighten the strut-to-knuckle nuts to 125 ft. lbs. (170 Nm) followed by a 120 degree turn. Final torque must exceed 148 ft. lbs. (200 Nm). Check the front end alignment.

Coil Springs

REMOVAL & INSTALLATION

1. Raise and safely support the vehicle.

2. Remove the wheel and tire assembly.

3. Remove the stabilizer link and bushings at the lower control arm.

4. Remove the cotter pin and nut from the tie rod end. Using a suitable tool, remove the tie rod ball joint stud from the steering knuckle.

5. Install a suitable spring compressor and compress the spring.

6. Remove the lower control arm pivot bolt and pivot the lower control arm rearward.

7. Remove the spring compressor and remove the spring.

To install:

8. Properly position the spring on the control arm, making sure the spring insulator is in place. The bottom of the spring is coiled helical and the top is coiled flat, with a gripper notch near the end of the wire. After assembly, the end of the spring coil must cover all or part of 1 inspection drain hole on the lower control arm. The other hole must be completely uncovered.

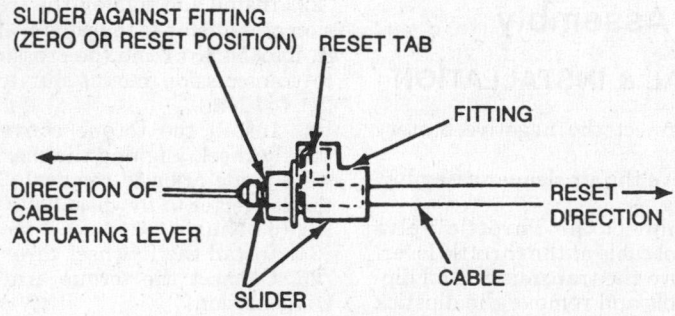

Throttle Valve (TV) cable adjustment

9. Install the spring compressor and compress the spring.

10. Pivot the lower control arm forward into position in the frame.

11. Install the bolts. The front bolt installs from front to rear first. Tighten the lower control arm pivot bolt/nuts to 66 ft. lbs. (90 Nm).

12. Remove the spring compressor.

13. Install the stabilizer linkage. Tighten the stabilizer link nut to 16 ft. lbs. (22 Nm).

14. Install the steering knuckle to the tie rod ball joint stud. Tighten the nut to 83 ft. lbs. (113 Nm).

15. Install a new cotter pin. If the hole in the stud does not line up with the slot in the nut, tighten the nut until it does. Do not back off the nut to install the cotter pin.

16. Install the wheel and tire assembly and lower the vehicle.

Lower Ball Joints

INSPECTION

1. Raise and safely support the vehicle. Use a suitable device to support the lower control arm in it's normal ride height position.

2. Grasp the wheel at the top and bottom. Alternately push and pull on the top and the bottom of the wheel in an attempt to move it toward and away from the vehicle. Check for any horizontal movement of the steering knuckle relative to the lower control arm. If there is any movement the ball joint must be replaced.

3. Check the ball joint when it is disconnected from the steering knuckle. If there is any looseness in the ball stud, if the ball stud can be twisted in it's socket using finger pressure or if there are any cuts or tears in the ball joint seal, replace the ball joint.

REMOVAL & INSTALLATION

1. Disconnect the negative battery cable. Raise and safely support the vehicle. Remove the wheel and tire assembly.

2. Support the lower control arm under the spring seat with a suitable jack.

3. Remove the cotter pin and loosen the lower ball joint nut. Remove the grease fitting.

4. Using a suitable tool, break the ball stud loose from the steering knuckle. Separate the lower control arm from the steering knuckle.

5. Using ball joint tools J–9519–23, J–9519–18 and adapter tool J–9519–9 or equivalent, press the ball joint from the lower control arm.

To install:

6. Install the new ball joint to the lower control arm. Using ball joint tools J–9519–23, J–9519–18 and adapter tool J–9519–9 or equivalents, press the ball joint into the lower control arm until it bottoms on the arm.

NOTE: When installing a new ball joint, position the purge vent on the rubber boot facing inward.

7. Connect the ball joint to the steering knuckle. Install the ball stud nut and tighten to 83 ft. lbs. (113 Nm). Then tighten enough to align the nut slot with the stud hold and install the cotter pin. Do not back off the nut to align the slot and hole.

8. Install and lubricate the ball joint fitting until grease appears at the seal.

9. Install the wheel and tire assembly.

10. Lower the vehicle and align the front end.

Lower Control Arms

REMOVAL & INSTALLATION

1. Raise and safely support the vehicle.

2. Remove the wheel and tire assembly.

3. Remove the stabilizer link and bushings at the lower control arm.

4. Remove the cotter pin and nut from the tie rod end. Using a suitable tool, remove the steering knuckle from the tie rod ball joint stud.

5. Install a suitable spring compressor and compress the coil spring.

6. Remove the lower control arm pivot bolts.

7. Remove the control arm and spring.

To install:

8. Properly position the spring on the control arm, making sure the spring insulator is in place. The bottom of the spring is coiled helical and the top is coiled flat, with a gripper notch near the end of the wire. After assembly, the end of the spring coil must cover all or part of 1 inspection drain hole. The other hole must be completely uncovered.

9. Install the spring compressor and compress the spring.

10. Pivot the lower control arm forward into position in the frame.

11. Install the bolts. The front bolt installs from front to rear first. Tighten the lower control arm pivot bolt/nuts to 66 ft. lbs. (90 Nm).

12. Remove the spring compressor.

13. Install the stabilizer linkage. Tighten the stabilizer link nut to 16 ft. lbs. (22 Nm).

14. Install the steering knuckle to the tie rod ball joint stud. Tighten the nut to 83 ft. lbs. (113 Nm).

15. Install a new cotter pin. If the hole in the stud does not line up with a slot in the nut, tighten the nut until it does. Do not back off the nut to install the cotter pin.

16. Install the wheel and tire assembly and lower the vehicle.

Sway Bar

REMOVAL & INSTALLATION

1. Raise and safely support the vehicle.

2. Remove each side of the sway bar linkage by removing the nut from the link bolt, pulling the bolt from the linkage and remove the retainers, grommets and spacer.

3. Remove the bracket-to-frame or body bolts and remove the sway bar, rubber bushings and brackets. Remove the lower structure brace if equipped.

4. Installation is the reverse of the removal procedure. Install the sway bar with the identification tag on the right side of the vehicle. The rubber bushings should be positioned squarely in the bracket with the slit in the bushings facing the front of the vehicle. Tighten the sway bar link nut/bolt to 16 ft. lbs. (22 Nm) and the bracket bolts to 39 ft. lbs. (53 Nm).

Spindle Assembly

REMOVAL & INSTALLATION

1. Raise and safely support the vehicle.

2. Remove the wheel and tire assembly.

3. Remove the caliper mounting bolts and remove the caliper. Support the caliper with mechanics wire. Do not let the caliper hang by the brake hose.

4. Remove the dust cap from the hub and remove the cotter pin, nut and washer from the spindle. Remove the hub and rotor assembly from the spindle. Remove the splash shield.

5. Remove the outer wheel bearing assembly from the hub. Using a suitable tool, pry out the inner bearing lip seal and remove the inner wheel bearing assembly.

6. Using a suitable tool, disconnect the tie rod from the spindle.

7. Support the lower control arm. Using a suitable tool, disconnect the ball joint from the spindle.

8. Remove and discard 2 bolts, washers and nuts attaching the strut to the spindle and remove the spindle.

To install:

9. Position the spindle to the strut. Install new strut-to-spindle bolts, washers and nuts to the strut. Tighten the 2 nuts to 125 ft. lbs. (170 Nm) followed by an additional 120 degree

turn. The final torque must exceed 148 ft. lbs. (200 Nm).

10. Connect the ball joint stud and nut to the spindle. Tighten the castle nut to 83 ft. lbs. (113 Nm) and install a new cotter pin. If the hole in the stud does not line up with the slot in the castle nut, continue to tighten the nut just enough to allow insertion of the cotter pin. Do not back off the nut to insert the cotter pin.

11. Connect the tie rod. Tighten the castle nut to 35 ft. lbs. (48 Nm). If the hole in the stud does not line up with alsot in the castle nut, continue to tighten the nut just enough to allow insertion of the cotter pin. Do not back off the nut to insert the cotter.

12. Install the splash shield.

13. Clean the hub, spindle and bearings of old grease and air dry. Inspect the bearings for cracked cages and worn or pitted rollers. Inspect the races for cracks or scores. Replace the bearings and races as assemblies, if necessary. If the bearing are to be replaced, remove the old bearing races with a suitable brass drift inserted behind the races.

14. If the races were removed, drive or press new races into the hub.

15. Apply a thin film of high temperature wheel bearing grease to the bearing and seal seat areas of the spindle. Put a small quantity of grease inboard of each bearing cup in the hub. Pack the wheel bearings with grease using a bearing packer. If a bearing packer is not available, the bearings may be greased by hand. If hand packing is used, it is extremely important to work the grease thoroughly into the bearings between the rollers, cone and cage.

16. Place the inner bearing into the hub. Put an additional quantity of grease outboard of the bearing. Install a new grease seal using a suitable tool Lubricate the seal lip with a thin layer of grease.

17. Install the hub and rotor assembly. Place the outer bearing in the outer bearing cup and install the washer and nut.

18. Tighten the nut to 12 ft. lbs. (16 Nm) while turning the hub and rotor assembly forward by hand. This will seat the bearing. Put an additional quantity of grease outboard of the bearing. Back off the nut to the just loose position. Hand tighten the spindle nut. Loosen the spindle nut just enough that either hole in the spindle lines up with a slot in the nut (not more than ½ a flat). Install a new cotter pin and the dust cap.

19. Using a dial indicator, check the hub assembly. There should be from 0.001–0.005 inch (0.03–0.13mm) endplay when properly adjusted.

20. Install the caliper to the steering knuckle. Tighten the caliper mounting bolts to 38 ft. lbs. (51 Nm).

21. Install the wheel and tire assembly and lower the vehicle. Check the brake fluid level.

Front Wheel Bearings

ADJUSTMENT

1. Raise and safely support the vehicle.

2. Remove the wheel cover or center cap.

3. Remove the dust cap from the hub.

4. Remove the cotter pin from the spindle nut.

5. Tighten the spindle nut to 12 ft. lbs. (16 Nm) while turning the wheel forward by hand to fully seat the bearings. This will remove any grease or burrs which could cause excessive wheel bearing play later.

6. Back off the nut to the just loose position.

7. Hand tighten the spindle nut. Loosen the spindle nut just enough that either hole in the spindle lines up with a slot in the nut, not more than ½ a flat.

8. Install a new cotter pin. Bend the ends of the cotter pin against the nut and cut off the extra length to ensure the ends will not interfere with the dust cap.

9. Using a dial indicator, check the hub assembly. There should be 0.001–0.005 inch (0.03–0.13mm) endplay when properly adjusted.

10. Install the dust cap on the hub. Install the wheel cover or center cap.

11. Lower the vehicle.

REMOVAL & INSTALLATION

1. Raise and support the vehicle safely.

2. Remove the wheel and tire assembly.

3. Remove the caliper assembly.

4. Remove the dust cap from the hub. Remove the cotter pin, nut and washer from the spindle.

5. Remove the hub and rotor assembly.

6. Remove the outer bearing assembly from the hub. The inner bearing assembly will remain in the hub and may be removed by prying out the inner seal. Discard the seal after removal.

7. Remove the old bearing races from the hub with a suitable brass drift inserted behind the races in a recessed slot.

8. Clean all parts in clean solvent and air dry. Do not spin the bearing with compressed air while drying or the bearing may be damaged.

To install:

9. Inspect the bearings for cracked cages and worn or pitted rollers. Check the bearing races for cracks, scores or pitting condition. Replace as necessary.

10. If the races were removed, drive or press the races into the hub.

11. Apply a thin film of high temperature wheel bearing grease to the spindle at the outer bearing seat and at the inner bearing seat, shoulder and seal seat.

12. Put a small quantity of grease inboard of each bearing cup in the hub.

13. Pack the wheel bearings using a suitable bearing packer. If a bearing packer is not available, the bearings can be packed by hand. If hand packing is used, it is extremely important to work the grease thoroughly into the bearings between the rollers, cone and cage.

14. Place the inner bearing cone and roller assembly in the hub. Put an additional quantity of grease outboard of the bearing.

15. Install a new grease seal using a suitable tool until the seal is flush with the hub. Lubricate the seal lip with a thin layer of grease.

16. Carefully install the hub and rotor assembly. Place the outer bearing assembly in the outer bearing race. Install the washer and nut and tighten to 12 ft. lbs. (16 Nm).

17. Install the remainder of the components in the reverse of the removal procedure. Adjust the wheel bearings.

REAR SUSPENSION

Shock Absorbers

REMOVAL & INSTALLATION

1. Raise and support the vehicle safely. Support the rear axle with a jackstand.

2. Pull back the carpet in the rear hatch and remove the upper shock attaching nut.

3. Remove the lower shock mounting nut. Remove the shock absorber from the vehicle.

4. Installation is the reverse of the removal procedure. Tighten the upper shock attaching nut to 13 ft. lbs. (17 Nm) and the lower attaching nut to 70 ft. lbs. (95 Nm).

Coil Springs

REMOVAL & INSTALLATION

1. Raise and safely support the ve-

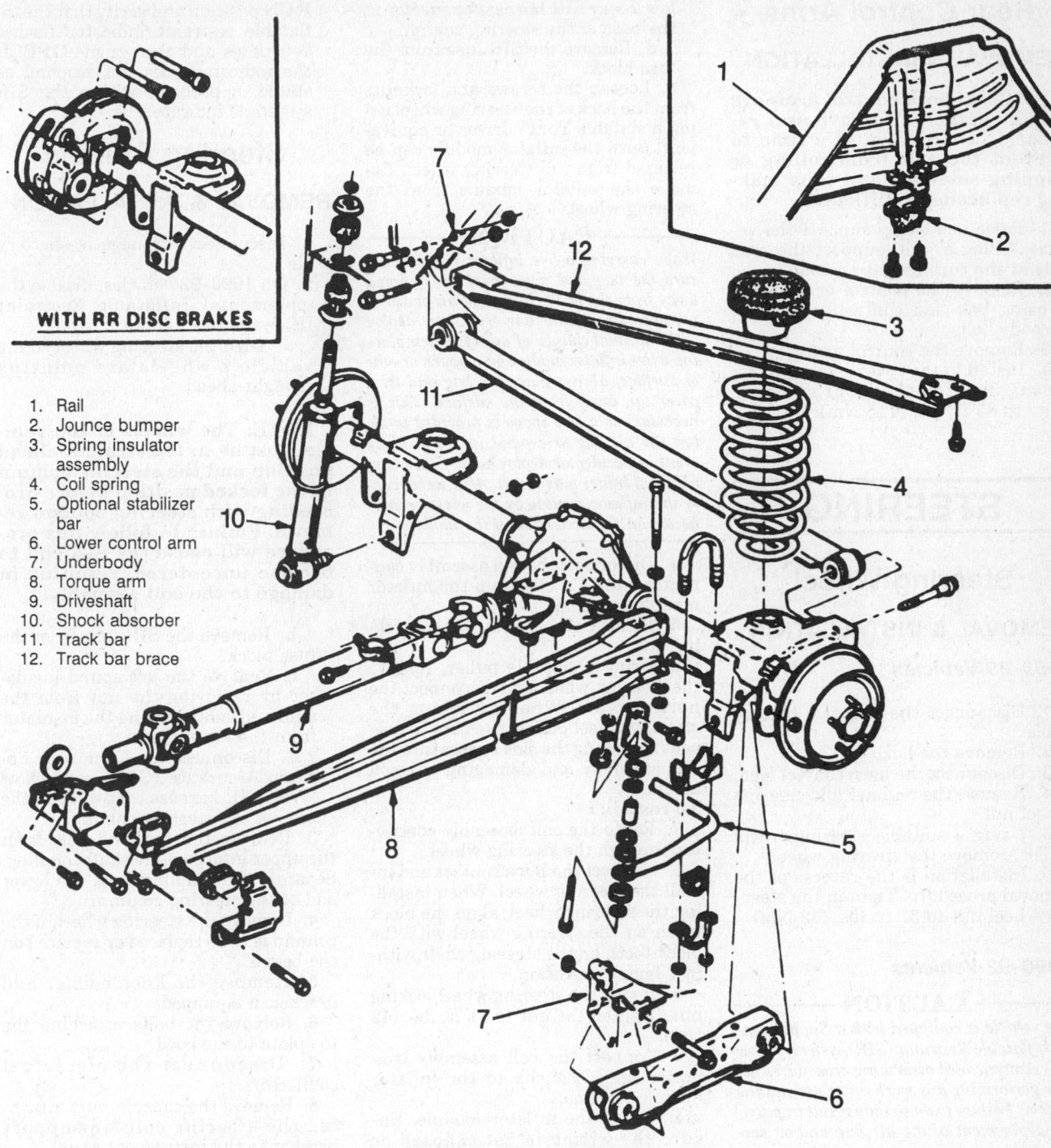

WITH RR DISC BRAKES

1. Rail
2. Jounce bumper
3. Spring insulator assembly
4. Coil spring
5. Optional stabilizer bar
6. Lower control arm
7. Underbody
8. Torque arm
9. Driveshaft
10. Shock absorber
11. Track bar
12. Track bar brace

Rear suspension components

hicle so the rear axle can be independently raised and lowered.

2. Support the rear axle with a suitable jack.

3. If equipped with brake hose attachment brackets, disconnect the brackets allowing the hoses to hang free. Do not disconnect the brackets allowing the hoses to hang free. Do not disconnect the hoses. Perform this step only if the hoses will be stretched

when the axle is lowered during spring removal.

4. Remove the track bar from the axle assembly and loosen the bar at the body brace.

5. Remove the lower shock absorber nuts and lower the axle. Ensure the axle is supported securely on the floor jack and that there is no chance of the axle slipping after the shock absorbers are disconnected.

6. Lower the axle and remove the coil spring. Do not lower the axle past the limits of the brake lines or the lines will be damaged.

7. To install, reverse the removal procedure. Ensure the spring is seated in the same position as before removal. Tighten the track bar mounting nut to 61 ft. lbs. (83 Nm) and bracket nut to 80 ft. lbs. (108 Nm). Tighten the shock mounting nuts to 70 ft. lbs. (95 Nm).

Rear Control Arms

REMOVAL & INSTALLATION

NOTE: If both control arms are being replaced, remove and replace 1 control arm at a time to prevent the axle from rolling or slipping sideways and thus making replacement difficult.

1. Raise and safely support the vehicle. Using a jack, support the rear axle at the curb height position.
2. Remove the control arm-to-axle housing bolt and control arm-to-underbody bolt.
3. Remove the control arm.
4. Install in the reverse order of removal. Tighten the front and rear bolts to 85 ft. lbs. (115 Nm).

STEERING

Steering Wheel

REMOVAL & INSTALLATION

1988–89 Vehicles

1. Disconnect the negative battery cable.
2. Remove the horn pad.
3. Disconnect the horn contact lead.
4. Remove the retainer and steering wheel nut.
5. Using a suitable steering wheel puller, remove the steering wheel.
6. Installation is the reverse of the removal procedure. Tighten the steering wheel nut to 31 ft. lbs. (42 Nm).

1990–92 Vehicles

— CAUTION —

The vehicle is equipped with a Supplemental Inflatable Restraint (SIR) system, follow the recommended disarming procedures before performing any work on or around the system. Failure to do so may result in possible deployment of the air bag and/or personal injury.

1. Disconnect the negative battery cable.
2. Disable the Supplemental Inflatable Restraint (SIR) system as follows:
   a. Turn the steering wheel so the vehicle's wheels are pointing straight-ahead.
   b. Remove the left sound insulator by removing the nut from the stud and gently prying the insulator from the knee bolster.
   c. Disconnect the Connector Position Assurance (CPA) clip and yellow 2-way SIR harness connector at the base of the steering column.
   d. Remove the SIR fuse from the fuse block.
3. Loosen the screws and locknuts from the back of the steering wheel using a suitable Torx® driver or equivalent, until the inflator module can be released from the steering wheel. Remove the inflator module from the steering wheel.

— CAUTION —

When carrying a live inflator module, ensure the bag and trim cover are pointed away from the body. In case of an accidental deployment, the bag will then deploy with minimal chance of injury. When placing a live inflator module on a bench or other surface, always place the bag and trim cover up, away from the surface. This is necessary so a free space is provided to allow the air bag to expand in the unlikely event of accidental deployment. Otherwise, personal injury may result. Also, never carry the inflator module by the wires or connector on the underside of the module.

4. Disconnect the coil assembly connector and CPA clip from the inflator module terminal.
5. Remove the steering wheel locking nut.
6. Using a suitable puller, remove the steering wheel and disconnect the horn contact. When attaching the steering wheel puller, use care to prevent threading the side screws into the coil assembly and damaging the coil assembly.

To install:

7. Route the coil assembly connector through the steering wheel.
8. Connect the horn contact and install the steering wheel. When installing the steering wheel, align the block tooth on the steering wheel with the block tooth on the steering shaft within 1 female serration.
9. Install the steering wheel locking nut. Tighten the nut to 31 ft. lbs. (42 Nm).
10. Connect the coil assembly connector and CPA clip to the inflator module terminal.
11. Install the inflator module. Ensure the wiring is not exposed or trapped between the inflator module and the steering wheel. Tighten the inflator module screws to 25 inch lbs. (2.8 Nm).
12. Connect the negative battery cable.
13. Enable the SIR system as follows:
   a. Connect the yellow 2-way SIR harness connector to base of the steering column and CPA.
   b. Install the left sound insulator.
   c. Install the SIR fuse in the fuse block.
   d. Turn the ignition switch to the RUN position and verify that the inflatable restraint indicator flashes 7–9 times and then turns OFF. If the indicator does not respond as stated, a problem within the SIR system is indicated.

Steering Column

REMOVAL & INSTALLATION

1. Disconnect the negative battery cable.
2. On 1990–92 vehicles, disable the Supplemental Inflatable Restraint (SIR) system as follows:
   a. Turn the steering wheel so the vehicle's wheels are pointing straight-ahead.

NOTE: The wheels of the vehicle must be in the straight-ahead position and the steering column in the locked position before proceeding with steering column removal. Failure to follow this procedure will cause the SIR coil to become uncentered, resulting in damage to the coil assembly.

   b. Remove the SIR fuse from the fuse block.
   c. Remove the left sound insulator by removing the nut from the stud and gently prying the insulator from the knee bolster.
   d. Disconnect the Connector Position Assurance (CPA) and yellow 2-way SIR harness connector at the base of the steering column.
3. Remove the nut and bolt from the upper intermediate shaft coupling. Separate the coupling from the lower end of the steering column.
4. Remove the steering wheel, if the column is to be replaced or repaired on the bench.
5. Remove the knee bolster and bracket, if equipped.
6. Remove the bolts attaching the toe plate to the cowl.
7. Disconnect the electrical connectors.
8. Remove the capsule nuts attaching the steering column support bracket to the instrument panel.
9. Disconnect the park lock cable from the ignition switch inhibitor, if equipped with automatic transmission.
10. Remove the steering column from the vehicle.

To install:

NOTE: If a replacement steering column is being installed, do not remove the anit-rotation pin until after the steering column has been connected to the steering gear. Removing the anit-rotation pin before the steering col-

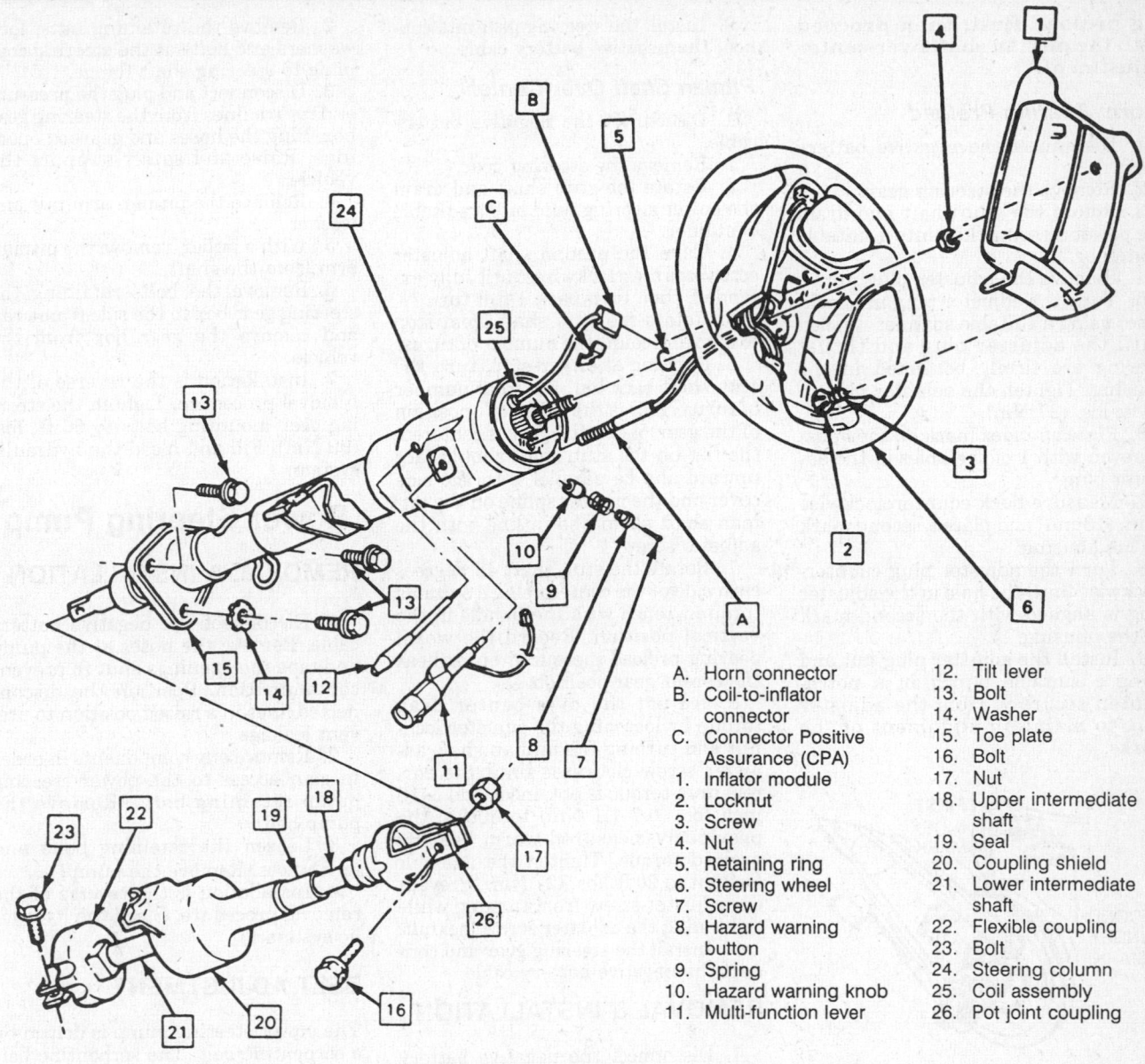

A. Horn connector
B. Coil-to-inflator connector
C. Connector Positive Assurance (CPA)
1. Inflator module
2. Locknut
3. Screw
4. Nut
5. Retaining ring
6. Steering wheel
7. Screw
8. Hazard warning button
9. Spring
10. Hazard warning knob
11. Multi-function lever
12. Tilt lever
13. Bolt
14. Washer
15. Toe plate
16. Bolt
17. Nut
18. Upper intermediate shaft
19. Seal
20. Coupling shield
21. Lower intermediate shaft
22. Flexible coupling
23. Bolt
24. Steering column
25. Coil assembly
26. Pot joint coupling

Steering column assembly with air bag

umn is connected to the steering gear may damage the SIR coil assembly.

11. Positon the steering column in the vehicle.

12. Connect the park lock cable to the ignition switch inhibitor on vehicles with automatic transmission.

13. Install the capsule nuts attaching the steering column support bracket to the instrument panel and tighten to 20 ft. lbs. (27 Nm).

14. Install the nut and bolt to the upper intermediate shaft coupling attaching the upper intermediate shaft to the steering column. Tighten the nut to 44 ft. lbs. (60 Nm).

15. Install the bolts attaching the toe plate to the cowl and tighten to 58 inch lbs. (6.5 Nm).

16. Connect the electrical connectors.

17. Remove the anti-rotation pin if a service replacement steering column is being installed.

18. Install the knee bolster and bracket, if equipped.

19. Install the sound insulator panel.

NOTE: If SIR coil has become uncentered by turning of the steering wheel without the column connected to the steering gear, follow the proper adjustment procedure for the SIR coil assembly before proceeding.

20. Install the steering wheel.

21. Connect the negative battery cable.

22. Enable the SIR system as follows:

a. Connect the yellow 2-way SIR harness connector to the base of the steering column and CPA clip and install the SIR fuse.

b. Install the left sound insulator.

c. Turn the ignition switch to the **RUN** position and verify that the inflatable restraint indicator flashes 7–9 times and then turns **OFF**. If the indicator does not respond as stated, a problem within the SIR system is indicated.

Power Steering Gear

ADJUSTMENT

NOTE: Adjust the worm bear-

ing preload first, then proceed with the pitman shaft over-center adjustment.

Worm Bearing Preload

1. Disconnect the negative battery cable.
2. Remove the steering gear.
3. Rotate the stub shaft and drain the power steering fluid into a suitable container.
4. Remove the adjuster plug nut.
5. Turn the adjuster plug in (clockwise) using a suitable spanner wrench until the adjuster plug and thrust bearing are firmly bottomed in the housing. Tighten the adjuster plug to 20 ft. lbs. (27 Nm).
6. Place an index mark on the housing even with 1 of the holes in the adjuster plug.
7. Measure back counterclockwise ½ in. (13mm) and place a second mark on the housing.
8. Turn the adjuster plug counterclockwise until the hole in the adjuster plug is aligned with the second mark on the housing.
9. Install the adjuster plug nut and using a suitable punch in a notch, tighten securely. Hold the adjuster plug to maintain alignment of the marks.

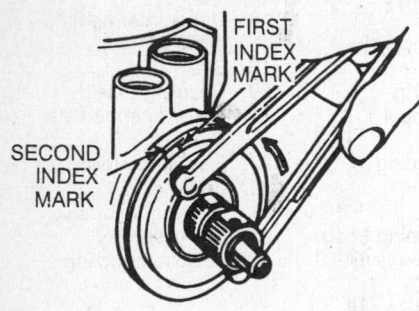

Adjusting the worm bearing preload

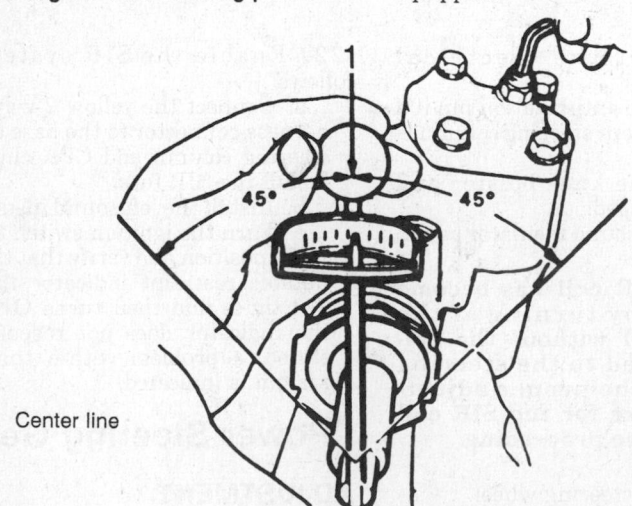

1. Center line

Pitman shaft over-center adjustment

10. Install the steering gear and connect the negative battery cable.

Pitman Shaft Over-Center

1. Disconnect the negative battery cable.
2. Remove the steering gear.
3. Rotate the stub shaft and drain the power steering fluid into a suitable container.
4. Turn the pitman shaft adjuster screw counterclockwise until fully extended, then turn back 1 full turn.
5. Rotate the stub shaft from stop to stop and count the number of turns.
6. Starting at either stop, turn the stub shaft back half the total number of turns. This is the "Center" position of the gear. When the gear is centered, the flat on the stub shaft should face upward and be parallel with the side cover and the master spline on the pitman shaft should be in line with the adjuster screw.
7. Rotate the stub shaft 45 degrees each side of the center using a suitable torque wrench with the handle in the vertical position. Record the worm bearing preload measured on or near the center gear position.
8. Adjust the over-center drag torque by loosening the adjuster locknut and turning the pitman shaft adjuster screw clockwise until the correct drag torque is obtained: Add 6–10 inch lbs. (0.7–1.1 Nm) torque to the previously measured worm bearing preload torque. Tighten the adjuster locknut to 20 ft. lbs. (27 Nm). Prevent the adjuster screw from turning while tightening the adjuster screw locknut.
9. Install the steering gear and connect the negative battery cable.

REMOVAL & INSTALLATION

1. Disconnect the negative battery cable. Remove coupling shield, if equipped.

2. Remove the retaining nuts, lock washers and bolts at the steering coupling to steering shaft flange.
3. Disconnect and plug the pressure and return lines from the steering gear box. Plug the hoses and gearbox openings. Raise and safely support the vehicle.
4. Remove the pitman arm nut and washer.
5. With a puller, remove the pitman arm from the shaft.
6. Remove the bolts retaining the steering gear box to the side frame rail and remove the gear box from the vehicle.
7. Installation is the reverse of the removal procedure. Tighten the steering gear mounting bolts to 66 ft. lbs. (90 Nm). Fill and bleed the hydraulic system.

Power Steering Pump

REMOVAL & INSTALLATION

1. Disconnect the negative battery cable. Remove the hoses at the pump and tape the openings shut to prevent contamination. Position the disconnected lines in a raised position to prevent leakage.
2. Remove any components in order to gain access to the power steering pump retaining bolts. Remove the pump belt.
3. Loosen the retaining bolts and any braces. Remove the pump.
4. Installation is the reverse of the removal procedure. Bleed the hydraulic system.

BELT ADJUSTMENT

The power steering pump is driven by a serpentine belt. The serpentine belt is kept in adjustment by an automatic tensioner. No adjustment of the belt is required.

SYSTEM BLEEDING

1. With the engine **OFF**, raise and safely support the vehicle high enough to get the wheels off the ground.
2. Turn the wheels all the way to the left, add power steering fluid to the **COLD** mark on the fluid level indicator.
3. Start the engine and run at fast idle momentarily, shut the engine **OFF** and recheck the fluid level. If necessary, add fluid to bring level to the **COLD** mark.
4. Start the engine and bleed the system by turning the wheels from side to side without hitting the stops.
5. Return the wheels to the center position and keep the engine running for a few minutes.

6. Road test the vehicle and recheck the fluid level ensuring the level is up to the **HOT** mark.

Tie Rod Ends

REMOVAL & INSTALLATION

1. Raise and safely support the vehicle.

2. Remove the cotter pin and the castle nut from the tie rod end.

3. Using a suitable tool, remove the tie rod end from the steering arm.

4. Loosen the clamp bolt on the adjuster tube. Unscrew the tie rod end from the adjuster tube, counting the number turns required to remove the tie rod end from the adjuster tube.
To install:

5. Inspect the threads on the ball stud and castle nut for damage and the ball stud taper for nicks. Check the seal for damage. Replace components as necessary.

6. Lubricate the tie rod adjuster tube threads with antiseize. Install the tie rod end into the tie rod adjuster tube, using the same number of turns counted during removal.

7. Install the tie rod end into the steering arm. Install the castle nut and tighten to 35 ft. lbs. (48 Nm). Tighten the nut again just enough to align the slot in the nut with the hole in the stud. Do not back off the nut to align the slot and hole.

8. Install a new cotter pin.

9. Position the clamp on the adjuster tube on the bolt to the bottom of the tube and the nut to the front of the vehicle. Tighten the adjuster clamp nut to 14 ft. lbs. (19 Nm).

10. Lower the vehicle and adjust the toe setting.

BRAKES

For all brake system repair and service procedures not detailed below, please refer to "Brakes" in the Unit Repair section.

Master Cylinder

REMOVAL & INSTALLATION

1. Disconnect the negative battery cable.

2. Disconnect and plug the brake lines at the master cylinder.

3. Remove the 2 master cylinder attaching nuts.

4. If equipped with manual transmission, move the clutch master cylinder reservoir and bracket aside.

5. Move the combination valve aside.

6. Remove the master cylinder.

7. Installation is the reverse of the removal procedure. Bench bleed the master cylinder prior to installation. Tighten the 2 attaching nuts to 18 ft. lbs. (25 Nm). Fill the master cylinder to the proper level and bleed the brake system.

Combination Valve

REMOVAL & INSTALLATION

1. Disconnect the negative battery cable.

2. Disconnect and plug the brake lines at the combination valve.

3. Disconnect the electrical connector from the combination valve switch terminal.

4. Remove the 2 nuts attaching the combination valve to the booster.

5. On vehicles with manual transmission, move the clutch master cylinder and bracket aside.

6. Remove the combination valve.

7. Installation is the reverse of the removal procedure. Tighten the 2 combination valve attaching nuts to 18 ft. lbs. (25 Nm). Bleed the brake system.

Power Brake Booster

REMOVAL & INSTALLATION

1. Disconnect the negative battery cable.

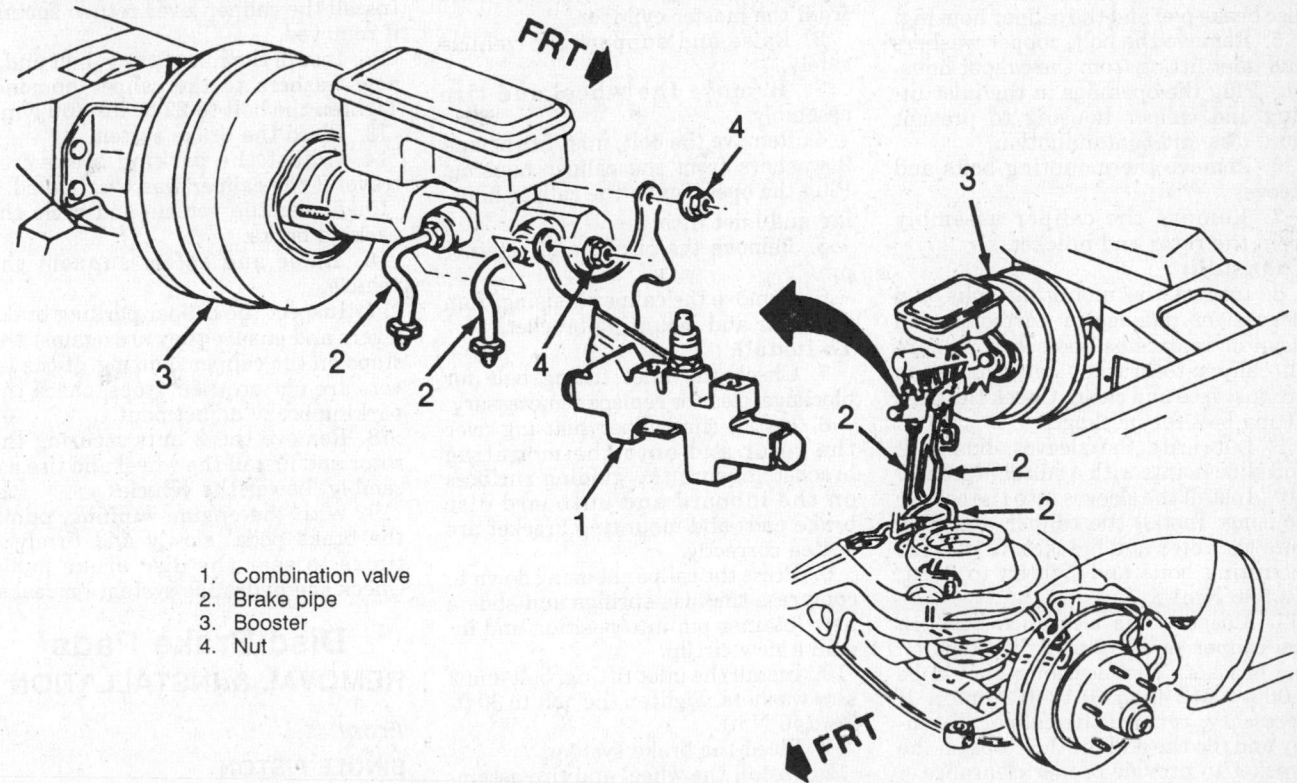

1. Combination valve
2. Brake pipe
3. Booster
4. Nut

Master cylinder and combination valve assembly

2. Disconnect the vacuum hose from the vacuum check valve.

3. On vehicles with manual transmission, remove the clutch master cylinder bracket with the clutch master cylinder reservoir and position aside.

4. Remove the combination valve with the attached combination valve bracket and position aside.

5. Remove the master cylinder from the booster.

6. Working inside the vehicle, remove the retainer and disconnect the booster pushrod from the brake pedal. Remove the booster attaching nuts.

7. Remove the power brake booster.

8. Installation is the reverse of the removal procedure. Tighten the booster attaching nuts to 15 ft. lbs. (21 Nm). Tighten the master cylinder attaching nuts to 18 ft. lbs. (25 Nm).

Brake Caliper

REMOVAL & INSTALLATION

Front

SINGLE PISTON

1. Remove 2/3 of the brake fluid from the master cylinder.

2. Raise and safely support the vehicle.

3. Remove the wheel and tire assembly.

4. Bottom the piston into the caliper bore using a suitable C-clamp. Position the C-clamp over the outboard disc brake pad and the caliper housing.

5. Remove the bolt, copper washers and inlet fitting from the caliper housing. Plug the openings in the inlet fitting and caliper housing to prevent fluid loss and contamination.

6. Remove the mounting bolts and sleeves.

7. Remove the caliper assembly from the rotor and bracket.

To install:

8. Inspect the mounting bolts and sleeves for damage or corrosion and clean or replace as necessary. Ensure all caliper-to-bracket contact points are rust free and clean. Check the inlet fitting bolt for blockage.

9. Lubricate the sleeves, bushings and slide points with a suitable grease.

10. Install the sleeves into the caliper housing. Install the caliper assembly onto the rotor and bracket. Install the mounting bolts and tighten to 37 ft. lbs. (50 Nm).

11. Measure the clearance between the caliper housing and the stops on the bracket. The clearance should be 0.005–0.012 inch (0.13–0.30mm). If necessary, remove the caliper assembly and file the ends of the stops on the bracket to provide proper clearance.

12. Install the bolt, new copper washers and the inlet fitting to the cal-

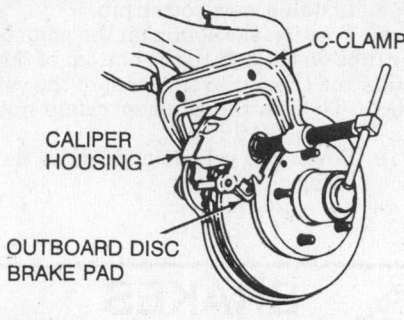

Compressing the front caliper housing

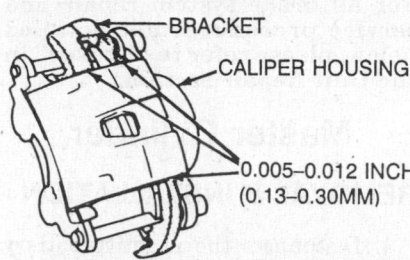

Front caliper housing to bracket clearance

iper housing. Tighten the bolt to 32 ft. lbs. (44 Nm).

13. Install the wheel and tire assembly and lower the vehicle.

14. Fill the master cylinder and bleed the brake system.

DUAL PISTON

1. Remove 2/3 of the brake fluid from the master cylinder.

2. Raise and support the vehicle safely.

3. Remove the wheel and tire assembly.

4. Remove the bolt, inlet fitting and 2 washers from the caliper housing. Plug the openings in the caliper housing and inlet fittings.

5. Remove the circlip and retainer pin.

6. Remove the caliper housing from the rotor and mounting bracket.

To install:

7. Check the inlet fitting bolt for blockage, clear or replace as necessary.

8. Install the caliper housing over the rotor and onto the mounting bracket. Ensure the guiding surfaces on the inboard and outboard disc brake pads and mounting bracket are seated correctly.

9. Press the caliper housing down to compress the bias springs and slide a new retainer pin into position and install a new circlip.

10. Install the inlet fitting, bolt and 2 new washers. Tighten the bolt to 30 ft. lbs. (40 Nm).

11. Bleed the brake system.

12. Install the wheel and tire assembly and lower the vehicle.

13. With the engine running, pump

the brake pedal slowly and firmly 3 times to seat the brake pads.

Rear

1. Raise and safely support the vehicle.

2. Loosen the parking brake cable at the equalizer.

3. Remove the wheel and tire assembly. Install 2 wheel nuts to retain the rotor.

4. Remove the bolt, inlet fitting and washers from the caliper housing. Plug the holes in the caliper housing and inlet fitting.

5. Remove the caliper lever return spring only if it is defective. Discard the spring if the coils are opened.

6. Disconnect the parking brake cable from the caliper lever and caliper bracket.

7. Remove the 2 caliper guide pin holes.

8. Remove the caliper housing from the rotor and mounting bracket.

To install:

9. Inspect the guide pins and boots and replace if corroded, worn or damaged. Check the inlet fitting bolt for blockage.

10. Install the caliper housing over the rotor and into the mounting bracket. Install the 2 caliper guide pin bolts. Tighten the upper caliper guide pin bolt to 26 ft. lbs. (35 Nm) and the lower guide pin bolt to 16 ft. lbs. (22 Nm).

11. Connect the parking brake cable to the caliper bracket and caliper lever. Install the caliper lever return spring, if removed.

12. Install the inlet fitting, bolt and 2 new washers to the caliper housing. Tighten the bolt to 22 ft. lbs. (30 Nm).

13. Bleed the brake system.

14. Adjust the parking brake free travel if the caliper was overhauled.

15. Lower the vehicle and cycle the parking brake.

16. Raise and safely support the vehicle.

17. Inspect the caliper parking brake levers and ensure they are against the stops on the caliper housing. If the levers are not on their stops, check the parking brake adjustment.

18. Remove the 2 nuts securing the rotor and install the wheel and tire assembly. Lower the vehicle.

19. With the engine running, pump the brake pedal slowly and firmly 3 times to seat the disc brake pads. Check the hydraulic system for leaks.

Disc Brake Pads
REMOVAL & INSTALLATION

Front

SINGLE PISTON

1. Raise and safely support the vehicle.

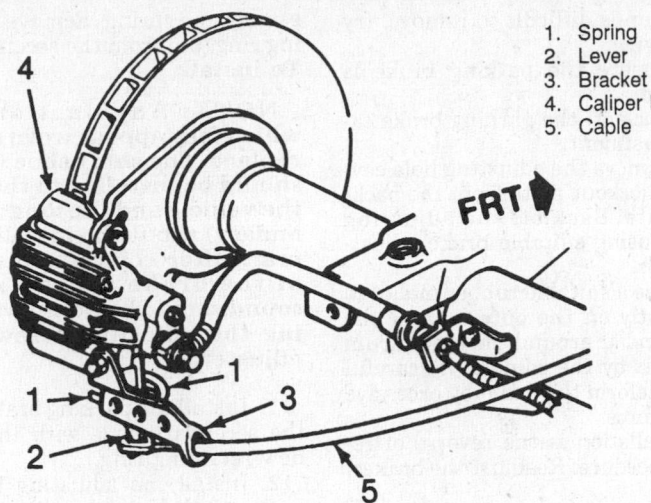

1. Spring
2. Lever
3. Bracket
4. Caliper housing
5. Cable

Rear caliper assembly

2. Remove the wheel and tire assemblies.

3. Remove the caliper mounting bolts.

4. Position a C-clamp over the outboard disc brake pad and the caliper housing. Tighten it until the caliper piston bottoms in its bore. Remove the C-clamp.

5. Pivot the caliper off of the rotor. Do not allow the caliper to hang by the brake hose, suspend it with a length of wire.

6. Remove the disc brake pads.

7. Remove the bushings from the mounting bolt holes.

To install:

8. Lubricate the bushings with silicone grease and install the bushings into the mounting bolt holes. Lubricate all caliper-to-bracket slide points with grease.

9. Install the retaining spring on the inboard pad in the correct position. Install the inboard pad into the caliper with the wear sensor at the leading edge of the pad during forward wheel rotation. Snap the retaining spring into position. The pad must lay flat against the piston.

10. Install the outboard pad. The pad must lay flat against the caliper housing.

11. Install the caliper onto the rotor. Tighten the mounting bolts to 37 ft. lbs. (50 Nm).

12. Apply the brake 3 times to seat the linings.

13. Clinch the outboard shoe retaining tabs to the caliper housing using a suitable tool, while a helper applies moderate pressure to the brake pedal. The outboard shoe should be locked in a fixed position.

14. Check the brake fluid. Install the tire and wheel assemblies. Lower the vehicle and road test to ensure correct brake operation.

DUAL PISTON

1. Remove the caliper assembly.

2. Position a C-clamp over the caliper housing and the center of the inboard disc brake pad. Compress the C-clamp until the pistons are bottomed.

3. Remove the disc brake pads.

To install:

4. Clean all residue from the mounting brackets and caliper pad contact surfaces.

5. Install the disc brake pads. The outboard disc brake pad with insulator is installed in the caliper housing. The inboard brake pad with wear sensor is pressed into the caliper pistons. Push the pads in firmly until they are flush and fully seated in the caliper housing.

6. Install the caliper and bleed the brake system.

Rear

1. Remove ²/₃ of the brake fluid from the master cylinder reservoir.

2. Raise and safely support the vehicle.

3. Remove the wheel and tire assembly. Install 2 wheel nuts to retain the rotor.

4. Position a C-clamp and tighten until the piston bottoms in the base of the caliper housing. Make sure 1 end of the C-clamp rests on the inlet fitting bolt and the other against the outboard disc brake pad.

NOTE: It is not necessary to remove the parking brake caliper lever return spring to replace the disc brake pads.

5. Remove the upper caliper guide pin bolt and discard.

6. Rotate the caliper housing. Be careful not to strain the hose or cable conduit.

7. Remove the disc brake pads.

To install:

8. Clean all residue from the pad guide surfaces on the mounting bracket and caliper housing. Inspect the guide pins for free movement in the mounting bracket. Replace the guide pins or boots, if they are corroded or damaged.

9. Install the disc brake pads. The outboard pad with insulator is installed toward the caliper housing. The inboard pad with the wear sensor is installed nearest the caliper piston. The wear sensor must be in the trailing position with forward wheel rotation.

10. Rotate the caliper housing into it's operating position. The springs on the outboard brake pad must not stick through the inspection hole in the caliper housing. If the springs are sticking through the inspection hole in the caliper housing, lift the caliper housing and make the necessary corrections to the outboard brake pad positions.

11. Install a new upper caliper guide pin bolt and tighten to 26 ft. lbs. (35 Nm). Tighten the lower caliper guide pin bolt to 16 ft. lbs. (22 Nm).

12. With the engine running, pump the brake pedal slowly and firmly to seat the brake pads.

13. Check the caliper parking brake levers to make sure they are against the stops on the caliper housing. If the levers are not on their stops, check parking brake adjustment.

14. Remove the 2 wheel nuts from the rotor and install the wheel and tire assembly.

15. Lower the vehicle, check the master cylinder fluid level and roadtest the vehicle.

Brake Rotor

REMOVAL & INSTALLATION

Front

1. Raise and support the vehicle safely.

2. Remove the wheel and tire assembly.

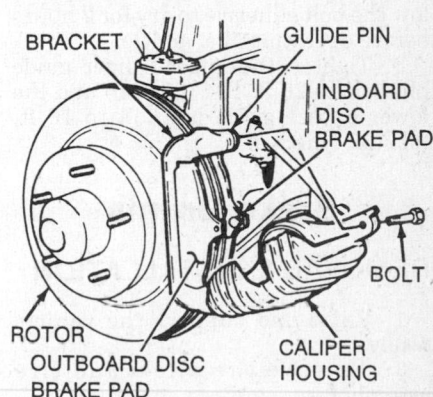

BRACKET — GUIDE PIN — INBOARD DISC BRAKE PAD — BOLT — CALIPER HOUSING — OUTBOARD DISC BRAKE PAD — ROTOR

Rear disc pad installation

3. Position the caliper assembly to the side.

4. Remove the dust cap from the hub. Remove the cotter pin, nut and washer from the spindle.

5. Remove the hub and rotor assembly.

To install:

6. Check the rotor for scoring or damage. Machine or replace the rotor assembly, as necessary. If machining is required, measure the rotor and check the minimum thickness specification.

7. Reinstall the hub and rotor assembly. Adjust the wheel bearings as follows:

a. Tighten the spindle nut to 12 ft. lbs. (16 Nm) while turning the rotor forward by hand to fully seat the bearings.

b. Back off the nut to the just loose position.

c. Hand tighten the spindle nut. Loosen the spindle nut just enough that either hole in the spindle lines up with a slot in the nut, not more then ½ a flat on the nut.

d. Install a new cotter pin.

e. Using a dial indicator, check the hub and rotor assembly. There should be 0.001–0.005 inch (0.03–0.13mm) endplay when properly adjusted.

8. Install the caliper assembly.

9. Install the wheel and tire assembly. Lower the vehicle.

Rear

1. Raise and support the vehicle safely.

2. Remove the wheel and tire assembly.

3. Remove the caliper assembly.

4. Remove the caliper mounting assembly.

5. Remove the brake rotor.

6. Install in the reverse order of removal.

7. Apply bolt adhesive to 2 new caliper mounting bracket bolts and tighten them to 70 ft. lbs. Recheck the torque on both bolts immediately. Allow the bolt adhesive to dry for 2 hours before operating the vehicle.

8. Tighten the upper caliper guide pin bolt to 26 ft. lbs. (35 Nm) and the lower caliper guide pin bolt to 16 ft. lbs. (22 Nm).

Brake Drums

REMOVAL & INSTALLATION

1. Raise and support the vehicle safely.

2. Remove the wheel and tire assembly.

3. Remove the brake drum. If the brake drum is difficult to remove, try the following:

a. Ensure the parking brake is released.

b. Back off the parking brake cable adjustment.

c. Remove the adjusting hole cover or knockout plate from the backing plate. Back off the adjusting screw, using suitable brake adjusting tools.

d. Use a suitable rubber mallet to tap gently on the outer rim of the drum and/or around the inner drum diameter by the spindle. Be careful not to deform the drum by excessive use of force.

4. Installation is the reverse of removal procedure. Readjust the brakes.

Brake Shoes

REMOVAL & INSTALLATION

1. Raise and safely support the vehicle.

2. Remove the wheel and tire assemblies.

3. Remove the brake drum.

4. Remove the return springs.

5. Remove the hold-down springs and pins. Remove the lever pivot.

6. Remove the actuator link while lifting up on the actuator lever.

7. Remove the actuator lever and lever return spring.

8. Remove the shoe guide, parking brake strut and strut spring.

9. Remove the brake shoes and disconnect the parking brake lever from the shoe.

10. Remove the adjusting screw assembly and spring. Remove the retaining ring, pin from the secondary shoe.

To install:

NOTE: Any part or spring which may appear worn should be replace. The short shoe (primary) should be installed to the front of the vehicle and the long shoe (secondary) should be installed to the rear. After complete installation of the brake shoes a clicking sound should be heard when turning the adjusting screw or self-adjuster.

11. Install the parking brake lever on the secondary shoe with the pin and new retaining ring.

12. Install the adjusting screw and spring. Lubricate the adjusting screw with brake (white) grease.

13. Clean and lubricate the contact points of the backing plate. Install the brake shoe assemblies after installing the parking brake cable on the shoe.

14. Install the parking brake strut and strut spring by spreading the shoes apart.

15. Install the shoe guide, actuator lever and lever return spring.

16. Install the hold-down pins, lever pivot and springs. Install the actuator link on the anchor pin.

17. Install the actuator link into the actuator lever while holding up on the lever.

18. Install the shoe return springs. Install the brake drum. Install the wheel and tire assemblies.

19. Adjust the brake and lower the vehicle. Check emergency brake for proper adjustment.

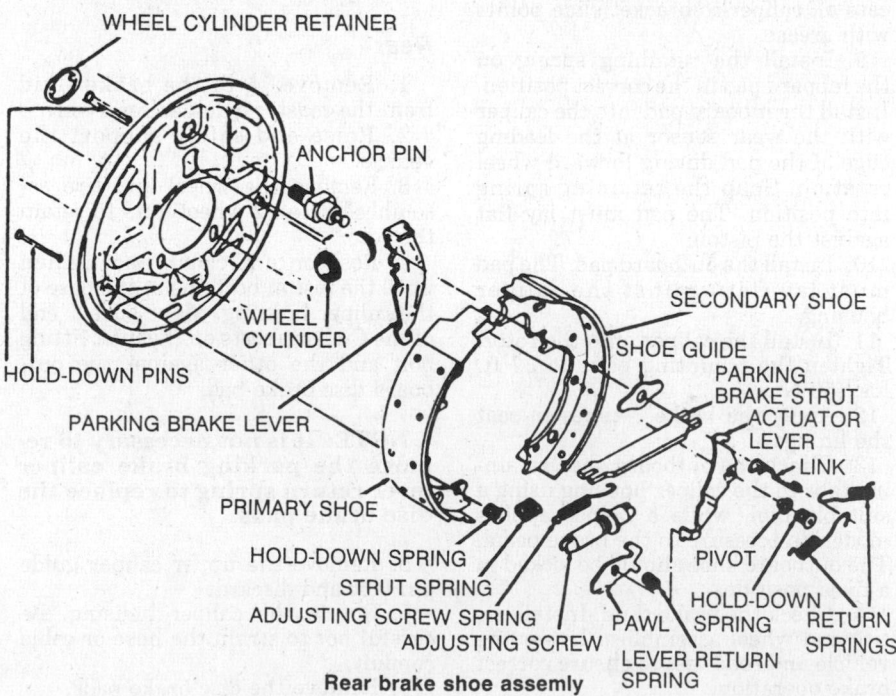

Rear brake shoe assembly

Wheel Cylinder

REMOVAL & INSTALLATION.

1. Raise and support the vehicle safely.

2. Remove the tire and wheel assembly.

3. Remove the brake components, as required.

4. Insert awls or pins, ⅛ in. diameter or less, into the access slots between the wheel cylinder pilot and retainer locking tabs.

5. Bend both tabs away simultaneously until they spring over the abutment shoulder releasing the wheel cylinder. Discard the old retaining clip.

6. Remove the brake line from the wheel cylinder and plug the brake line.

To install:

7. For ease of installation hold the wheel cylinder against the backing plate by inserting a block between the wheel cylinder and the axle shaft flange.

8. Position the wheel cylinder retainers clip so the tabs will be away from and in a horizontal position with the backing plate when installing.

9. Press the new retaining clip over the wheel cylinder abutment and into position using a 1⅛ in. 12-point socket. Make sure the retainer tabs are properly snapped under the abutment shoulder.

10. Install the brake components and the brake line.

11. Install the brake drum; wheel and tire assembly. Bleed and adjust the brake system.

Parking Brake Cable

ADJUSTMENT

The parking brake cable is adjustable only on 1988–89 vehicles. 1990–92 vehicles feature a self-adjusting parking brake.

1988–89

DRUM BRAKES

1. Ensure the rear brakes are prop-

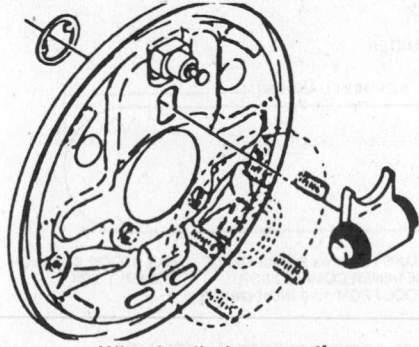

Wheel cylinder mounting

erly adjusted and the hydraulic brake system is functioning properly.

2. Pull the parking brake lever exactly 2 ratchet clicks.

3. Raise and support the rear wheels of the vehicle safely.

4. Tighten the brake cable adjusting nut until the left rear wheel can be turned rearward with both hands but locks when forward rotation is attempted.

5. Release the parking brake lever; both rear wheels must turn freely in either direction without brake drag. Be sure the parking brake cables are not adjusted too tightly causing the brakes to drag.

6. Lower the vehicle.

REAR DISC BRAKES

1. Apply the brake pedal 3 times with a pedal force of approximately 175 lbs. (778 N). Apply and release the parking brake 3 times.

2. Raise and safely support the vehicle.

3. Check the parking brake lever for full release:

   a. Turn the ignition **ON**.

   b. The brake warning light should be **OFF**. If the brake warning light is still **ON** and the parking brake lever is completely released, pull downward on the front parking brake cable to remove slack from the lever assembly.

   c. Turn the ignition switch **OFF**.

4. Remove the rear wheels and tires. Reinstall 2 wheel nuts on each side to retain the brake rotors.

5. Pull the parking lever 4 clicks. The parking brake levers on both calipers should be against the lever stops on the caliper housings. If the levers are not against the stops, check for binding in the rear cables and/or loosen the cables at the equalizer nut until both left and right levers are against their stops.

6. Adjust the equalizer adjusting nut until the parking brake levers on both calipers just begin to move off their stops.

7. Back off the adjuster nut until the levers move back, barely touching their stops.

8. Operate the parking brake lever several times to check adjustment. After cable adjustment, the parking brake lever should travel no more than 14 ratchet clicks. The rear wheels should not turn forward when the parking brake lever is applied 8–16 ratchet clicks.

9. Release the parking brake lever. Both rear wheels must turn freely in both directions. The parking brake levers on both calipers should be resting on their stops.

10. Remove the wheel nuts retaining the rotors. Install the wheel and tire assemblies.

11. Lower the vehicle.

Parking Brake Free-Travel

ADJUSTMENT

1989–92

REAR DISC BRAKES

NOTE: Disc brake pads must be new or parallel to within 0.006 inch (0.15mm). Parking brake adjustment is not valid with heavily tapered pads and may cause caliper/parking brake binding. Replace tapered brake pads. Parking brake free-travel should only be made if the caliper has been taken apart. This adjustment will not correct a condition where the caliper levers will not return to their stops.

1. Have an assistant apply a light brake pedal load, enough to stop the rotor from turning by hand. This takes up all clearances and ensures that components are correctly aligned.

2. Apply light pressure to th caliper lever.

3. Measure the free-travel between the caliper lever and the caliper housing. The free-travel must be 0.0024–0.028 inch (0.6–0.7mm).

4. If the free-travel is incorrect, do the following:

   a. Remove the adjuster screw.

   b. Clean the thread adhesive residue from the threads.

   c. Coat the threads with adhesive.

   d. Screw in the adjuster screw far enough to obtain 0.024–0.028 inch (0.6–0.7mm) free-travel between the caliper lever and the caliper housing.

5. Have an assistant release the brake pedal, then apply the brake pedal firmly 3 times. Recheck the free-tavel and adjust as necessary.

REMOVAL & INSTALLATION

Front Cable

1988–89

1. Raise and safely support the vehicle.

2. Remove the adjusting nut at the equalizer and remove the front cable from the equalizer and bracket.

3. Lower the vehicle.

4. Remove the upper console and lower console rear screws. Lift the rear of the lower console to gain access to the parking brake control.

5. Remove the pin and retainer from the control assembly and front cable.

6. Remove the cable and casing from the control assembly and bracket

and remove the cable and grommet from the vehicle.

7. Installation is the reverse of the removal procedure. Adjust the parking brake.

1990–92

1. Remove the carpet finish moulding.

2. Remove the console assembly.

3. With the parking brake lever in the down position, rotate the arm toward the front of the vehicle until a 3mm metal pen can be inserted into the hole. Insert the metal pin into the hole, locking out the self adjuster.

4. Raise and safely support the vehicle.

5. Disconnect the rear cables from the equalizer.

6. Remove the front cable from the bracket using a fabricated parking brake cable retainer compressor tool.

7. Remove the grommet from the hole.

8. Lower the vehicle.

9. Remove the barrel-shaped button from the adjuster track.

10. Remove the front cable and casing from the control assembly using a fabricated parking brake cable retainer compressor tool.

11. Remove the front cable from the floor pan.

12. Installation is the reverse of removal procedure.

Rear Cable

DRUM BRAKES

1. Raise and safely support the vehicle.

2. Pull the equalizer rearward to gain the necessary cable slack. Insert a spacer to hold the equilizer in place. Remove the left and/or right rear cable from the equalizer.

3. Compress the retainer fingers on the casing and pull the left and/or right rear cable out of the seat belt plate.

4. On the left side, pull the left rear cable through the clip on the axle housing.

5. On the right side, remove the screw and clamp from the right rear cable.

6. Remove the wheel(s) and tire(s).

7. Remove the brake drum(s).

8. Disconnect the left and/or right rear cable from the brake shoe operating lever.

9. Compress the retainer fingers and pull the left and/or right cable from the backing plate.

10. Install in the reverse order of removal. Adjust the parking brake on 1988–89 vehicles.

REAR DISC BRAKES

1. Raise and safely support the vehicle.

2. Pull the equalizer rearward to gain the necessary cable slack. Insert a spacer to hold the equalizer in place. Remove the left and/or right rear cable from the equalizer.

3. Compress the retainer fingers on the casing and pull the left and/or right rear cable out of the seat belt plate.

4. On the left side, pull the left rear cable through the clip on the axle housing.

5. On the right side, remove the screw and clamp from the right rear cable.

6. Remove the wheel(s) and tire(s).

7. Push forward on the caliper lever(s). Remove the left and/or right rear cable from the tang on the caliper lever(s). Release the caliper lever(s).

8. Compress the retainer fingers on the cable casing and pull out of the bracket(s).

9. Installation is the reverse of the removal procedure. Adjust the parking brake on 1988–89 vehicles.

Brake System Bleeding

1. Remove the vacuum reserve by applying the brakes several times, with the engine OFF.

2. Fill the master cylinder reservoir with brake fluid and keep it at least half full of fluid during the bleeding operation.

3. If the master cylinder is known or suspected to have air in the bore, bleed the unit before wheel cylinders or calipers, in the following manner:

a. Disconnect the forward brake line connection at the master cylinder.

b. Allow brake fluid to fill the master cylinder bore until it begins to flow from the forward brake line port at the master cylinder.

c. Connect the forward brake line to the master cylinder and tighten.

d. Have an assistant depress the brake pedal slowly 1 time and hold. Loosen the forward brake line connection at the master cylinder to purge air from the bore. Tighten the connection and have the assistant release the pedal slowly. Wait 15 seconds and repeat the sequence, including the 15 second wait, until all air is removed from the bore. Ensure brake fluid does not contact any painted surface.

e. Repeat the procedure at the rear master cylinder brake line connection.

f. If it is known that the calipers and wheel cylinders do not contain any air, it will not be necessary to bleed them.

4. If it is necessary to bleed all of the wheel cylinder and calipers, follow the proper sequence: Right rear, left rear, right front, left front.

5. Bleed individual wheel cylinders

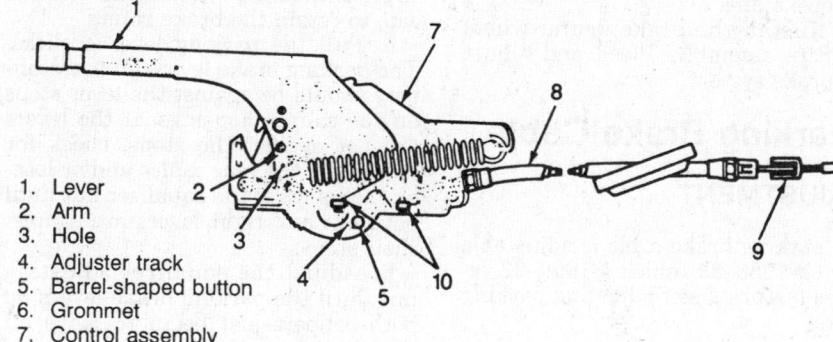

1. Lever
2. Arm
3. Hole
4. Adjuster track
5. Barrel-shaped button
6. Grommet
7. Control assembly
8. Front cable
9. Equalizer
10. Bolt

Parking brake control and front cable assembly—1990–92

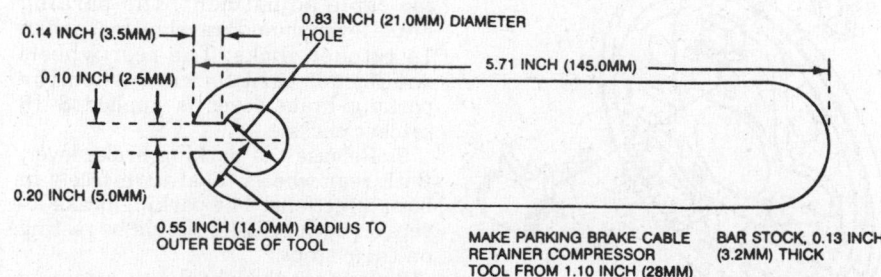

0.14 INCH (3.5MM)
0.10 INCH (2.5MM)
0.20 INCH (5.0MM)
0.83 INCH (21.0MM) DIAMETER HOLE
5.71 INCH (145.0MM)
0.55 INCH (14.0MM) RADIUS TO OUTER EDGE OF TOOL
MAKE PARKING BRAKE CABLE RETAINER COMPRESSOR TOOL FROM 1.10 INCH (28MM)
BAR STOCK, 0.13 INCH (3.2MM) THICK

Fabricated parking brake cable retainer compressor tool—1990–92

or calipers only after all air is removed from the master cylinder.

    a. Place a suitable bleeder wrench over the bleeder valve.

    b. Attach a clear tube over the bleeder valve and allow the tube to hang, submerged in a clear container partially filled with brake fluid.

    c. Have an assistant depress the brake pedal slowly 1 time and hold. Loosen the bleeder valve to purge the air from the cylinder. Tighten the bleeder screw and have the assistant slowly release the pedal. Wait 15 seconds and repeat the sequence, including the 15 second wait, until all air is removed.

    d. It may be necessary to repeat the sequence 10 or more times to remove all of the air.

NOTE: Rapid pumping of the brake pedal pushes the master cylinder secondary piston down the bore in a way that makes it difficult to bleed the rear side of the system.

6. Check the brake pedal for sponginess and the red brake warning light for an indication of unbalanced pressure. Repeat the bleeding procedure to correct either of these conditions.

CHASSIS ELECTRICAL

Air Bag

DISARMING

1. Turn the steering wheel to align the wheels in the straight-ahead position.

2. Turn the ignition switch to the **LOCK** position.

3. Remove the SIR air bag fuse from the fuse block.

4. Remove the left side trim panel and disconnect the yellow 2-way SIR harness wire connector at the base of the steering column.

To enable system:

5. Turn the ignition switch to the **LOCK** position.

6. Reconnect the yellow 2-way connector at the base of the steering column.

7. Reinstall the SIR fuse and the left side trim panel.

8. Turn the ignition switch to the **RUN** position.

9. Verify the SIR indicator light flashes 7–9 times, if not as specified, inspect system for malfunction.

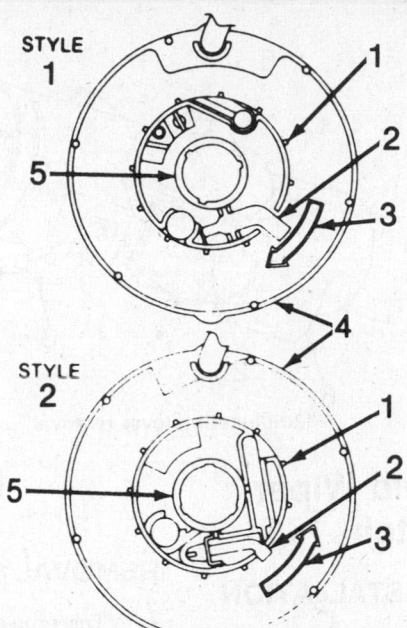

1. Locking tab
2. Spring
3. Hub direction
4. Coil housing
5. Coil hub

Centering the SIR coil asembly

Supplemental Inflatable Restraint (SIR) Coil Assembly

NOTE: After performing repairs on the internals of the steering column the coil assembly must be centered in order to avoid damaging the coil or accidental deployment of the air bag.
There are 2 different styles of coils, 1 rotates clockwise and the other rotates counterclockwise.

ADJUSTMENT

1. With the system properly disarmed, hold the coil assembly with the clear bottom up to see the coil ribbon.

2. While holding the coil assembly, depress the lock spring and rotate the hub in the direction of the arrow until it stops. The coil should now be wound up snug against the center hub.

3. Rotate the coil assembly in the opposite direction approximately 2½ turns and release the lock spring between the locking tabs in front of the arrow.

4. Install the coil assembly onto the steering shaft.

Heater Blower Motor

REMOVAL & INSTALLATION

1. Disconnect the negative battery cable. If necessary, remove the diagonal fender brace at the right rear corner of the engine compartment to gain access to the blower motor.

2. Disconnect the electrical wiring from the blower motor. If equipped with air conditioning, remove the blower relay and bracket as an assembly and swing them aside.

3. Remove the blower motor cooling tube.

4. Remove the blower motor retaining screws.

5. Remove the blower motor and fan as an assembly from the case.

6. Installation is the reverse of the removal procedure.

Windshield Wiper Motor

REMOVAL & INSTALLATION

1. Disconnect the negative battery cable.

2. Remove the left and right wiper arms.

3. Remove the cowl screen on 1988–90 vehicles. On 1991–92 vehicles remove the shroud vent grille.

4. Loosen the transmission drive link to crank arm retaining bolts. Remove the drive link from the motor crank arm.

5. Disconnect the electrical wiring and the washer hoses from the motor assembly.

6. Remove the motor retaining screws. Remove the windshield wiper motor while guiding the crank arm through the hole.

7. Installation is the reverse of the removal procedure. The motor must be in the park position before assembling the crank arm to the drive link.

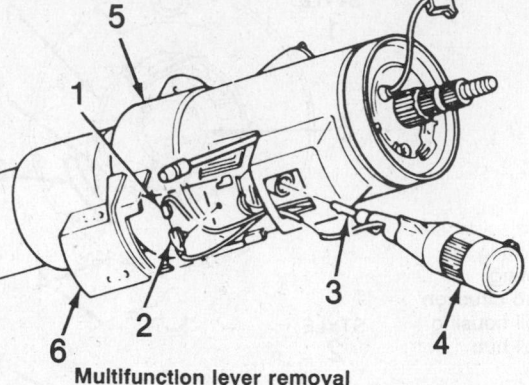

1. Cruise control connector from column
2. Cruise control connector from switch
3. Tang
4. Multifunction lever
5. Steering column
6. Housing cover end cap

Multifunction lever removal

Windshield Wiper Switch

REMOVAL & INSTALLATION

NOTE: If vehicle is equipped with a Supplemental Inflatable Restraint system, make certain to follow the recommended disarming procedure before proceeding.

1. Disconnect the negative battery cable.
2. Remove the column housing end cover by pulling toward front of the vehicle.
3. Disconnect the electrical connector and grommet.
4. Pull the multi-function lever toward the drivers door to remove.
5. Installation is the reverse of the removal procedure.

Instrument Cluster

REMOVAL & INSTALLATION

1. Disconnect the negative battery cable.
2. On 1988–89 Firebird with digital cluster, remove the right and left lower trim plates. Removal of the lower instrument panel covers is not required.
3. On 1990–92 Firebird, remove the instrument panel knee bolster and the instrument panel cluster trim plate.
4. On 1990–92 Camaro, remove the instrument panel knee bolster and the headlight switch knob. Remove the cluster trim plate screws and pull the cluster trim plate forward. Disconnect the electrical connectors and remove the cluster trim plate.
5. Remove the retaining screws from the instrument cluster, pull the cluster back and disconnect the electrical connectors. Remove the instrument cluster.
6. Installation is the reverse of the removal procedure.

Radio

REMOVAL & INSTALLATION

1. Disconnect the negative battery cable.
2. Remove the front trim plate.
3. Remove the screws securing the radio to the console.
4. Pull the radio out from the console and disconnect the electrical connections and antenna lead.
5. Remove the radio assembly.
6. Installation is the reverse of the removal procedure.

Concealed Headlights

MANUAL OPERATION

The concealed headlights used on the Firebird are electrically operated. If an electrical failure involving the head-light actuators should occur, the headlights can be operated manually.

To raise the headlights, rotate the knob, located on the actuator, in a counterclockwise direction until the headlights are fully open. Lower the headlights by turning the actuator knob in a clockwise direction until the headlights are fully closed.

Headlight Switch

REMOVAL & INSTALLATION

1988–89
CAMARO

1. Disconnect the negative battery cable.
2. Remove the insulator screws and nut from under the instrument panel and remove the insulator.
3. Remove the headlight switch knob assembly by depressing the release button on the headlight switch from under the instrument panel.
4. Remove the headlight switch knob trim plate screws and remove the trim plate.
5. Unscrew the retainer attaching the headlight switch to the instrument panel. Disconnect the electrical connector and remove the headlight switch from under the instrument panel.
6. Installation is the reverse of the removel procedure.

FIREBIRD

1. Disconnect the negative battery cable.
2. Remove the right and left lower

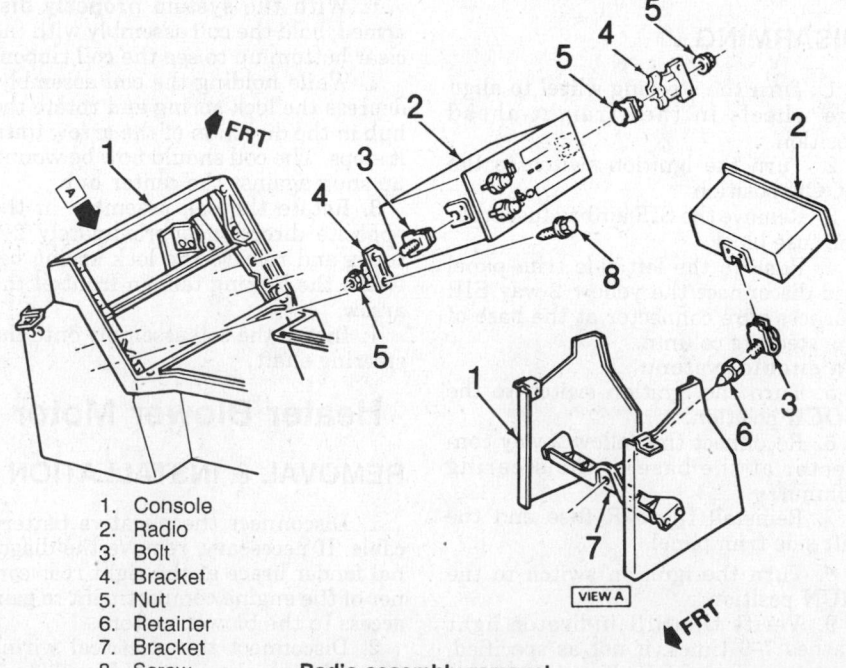

1. Console
2. Radio
3. Bolt
4. Bracket
5. Nut
6. Retainer
7. Bracket
8. Screw

Radio assembly removal

trim plates. Removal of the lower instrument panel covers is not required.

3. Remove the instrument panel cluster trim plate.

4. Remove the 2 switch assembly retaining screws.

5. Depress the side tangs and pull the switch assembly from the instrument panel. Disconnect the electrical connectors and remove the switch assembly.

6. Installation is the reverse of the removal procedure.

1990–92
CAMARO

1. Disconnect the negative battery cable.

2. Remove the instrument panel knee bolster.

3. Remove the switch knob by depressing the release button on the switch from under the instrument panel.

4. Remove the retaining screws from the instrument panel cluster trim plate and pull it forward away from the instrument panel. Disconnect the electrical connectors and remove the cluster trim plate.

5. Remove the retaining nut from the headlight switch and lower the switch out through the bottom of the instrument panel.

6. Disconnect the electrical connectors and remove the switch assembly.

7. Installation is the reverse of the removal procedure.

FIREBIRD

1. Disconnect the negative battery cable.

2. Remove the instrument panel knee bolster.

3. Remove the instrument panel cluster trim plate.

4. Remove the headlight switch retaining screws.

5. Disconnect the electrical connectors and remove the switch assembly.

6. Installation is the reverse of the removal procedure.

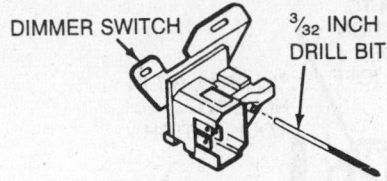

Dimmer switch adjustment

Dimmer Switch

REMOVAL & INSTALLATION

1. Disconnect the negative battery cable.

2. Remove the steering column to instrument panel trim plates.

3. Lower the steering column by loosening the toe plate retaining screws and remove the capsule nuts securing the steering column to the instrument panel. Support the column to prevent it from being damaged.

4. Remove the electrical connector and the retaining nut and remove the dimmer switch.

5. Installation is the reverse of the removal procedure. Adjust the dimmer switch by depressing the mechanism slightly to insert a ³⁄₃₂ inch drill bit. Move the switch to remove the lash and tighten the retaining nut to 35 inch lbs. (4 Nm).

Turn Signal Switch

REMOVAL & INSTALLATION

1. Properly disable the SIR air bag system, if equipped.

2. Disconnect the negative battery cable.

3. Remove the inflator module, if equipped with an air bag.

4. Remove the steering wheel.

5. Ensure the steering wheel is locked in the straight-ahead position and remove the coil assembly retaining ring on air bag equipped vehicles.

6. Pull the coil assembly out and allow it to hang.

7. Using a suitable tool, depress the lock plate to gain access to the snapring. Remove the snapring and remove the lockplate.

8. Remove the turn signal cancelling cam, upper bearing spring and signal switch arm.

9. Remove the hazard warning knob, turn signal lever and steering column wiring protector.

10. Disconnect the turn signal switch connector from the harness connector.

11. Remove the switch retaining screws and remove the turn signal switch.

12. Installation is the reverse of the removal procedure. Make certain to follow the SIR coil recentering procedure.

Ignition Lock

REMOVAL & INSTALLATION

1. Disconnect the negative battery cable.

2. Remove the turn signal switch.

3. Remove the key from the lock cylinder, remove the buzzer switch and reinsert the key in the LOCK position.

4. Remove the lock retaining screw and remove the lock cylinder.

5. If equipped with Vehicle Anti-Theft System (VATS), disconnect the wire connector and gently pull the wire through the steering column housing shroud, steering column housing and lock housing cover.

6. Installation is the reverse of the removal procedure. Place the lock cylinder in the RUN position before installing the buzzer switch.

Ignition Switch

REMOVAL & INSTALLATION

1. Disconnect the negative battery terminal.

1. Nut
2. Retaining ring
3. Coil assembly
4. Wave washer
5. Retaining ring

6. Shaft lock
7. Turn signal cancelling cam
8. Upper bearing spring
9. Screw

10. Screw
11. Signal switch arm
12. Turn signal and hazard warning switch

13. Upper bearing inner race seat
14. Inner race
15. Screw
16. Buzzer switch
17. Buzzer switch retaining clip
18. Lock retaining screw
19. Lock housing cover
20. VATS lock cylinder set

Typical turn signal switch assembly

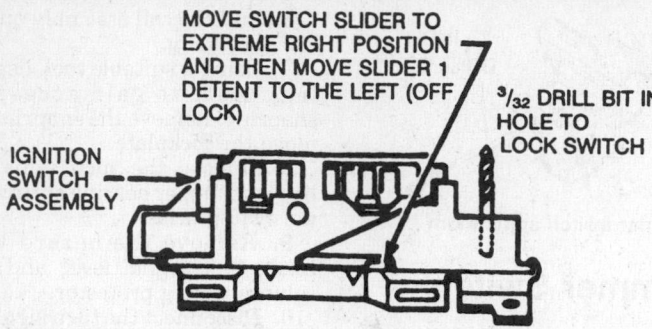

MOVE SWITCH SLIDER TO EXTREME RIGHT POSITION AND THEN MOVE SLIDER 1 DETENT TO THE LEFT (OFF LOCK)

³/₃₂ DRILL BIT IN HOLE TO LOCK SWITCH

IGNITION SWITCH ASSEMBLY

Ignition switch adjustment

2. Remove the column to instrument panel trim plates.

3. Loosen the toe plate screws and remove the capsule nuts to lower the steering column. Support the steering column to avoid damaging the assembly.

4. Disconnect the switch wire connectors.

5. Remove the switch attaching screws and remove the switch.

To install:

6. Position the key in the lock cylinder to the **LOCK** position.

7. Install the switch leaving the attaching screws loose.

8. Adjust the ignition swich by moving the switch slider to the extreme right position and then move the slider 1 detent to the left (OFF LOCK). Tighten the attaching screws to 35 inch lbs. (4 Nm).

9. Reassemble the remaining components of the steering column in reverse of the disassembly procedure.

Stoplight Switch

ADJUSTMENT

1. Insert the switch into the retainer until the switch body seats on the retainer.

2. Pull the brake pedal to the rear against the internal pedal stop. The stoplight switch will move in the retainer giving the proper adjustment.

3. The proper adjustment is obtained when no clicks are heard when the brake pedal is pulled up and the brake lights do not stay **ON**.

REMOVAL & INSTALLATION

1. Disconnect the negative battery cable.

2. Remove the electrical connectors and remove the stoplight switch by pulling it out of the retainer.

3. Install the replacement stoplight switch by inserting it into the retainer. Connect the electrical connectors.

4. Adjust the stoplight switch.

Clutch Switch

REMOVAL & INSTALLATION

1. Disconnect the negative battery cable.

2. Remove the sound insulator on 1988–89 vehicles or the console trim plate on 1990–92 vehicles.

3. Disconnect the clutch switch connector.

4. Remove the switch attaching bolt and remove the clutch switch.

5. Installation is the reverse of the removal procedure.

Neutral Safety Switch

ADJUSTMENT

1. Position the transmission shifter control lever in the **N** position. Align the carrier tang with tang slot on the shift control.

2. Loosen the switch attaching screws.

3. Rotate the switch on the shifter assembly to align the service adjustment hole with the carrier tang hole. Insert a ³/₃₂ in. (2.34mm) diameter gauge pin into the service adjustment hole.

4. Rotate the switch until the gauge pin drops to a depth of ¹⁹/₃₂ in. (15mm).

5. Tighten the attaching screws.

6. Remove the gauge pin.

REMOVAL & INSTALLATION

1. Disconnect the negative battery cable.

2. Remove the console assembly to gain access to the neutral safety switch.

3. Disconnect the electrical connectors.

4. Remove the switch attaching screws and remove the neutral safety switch.

5. Position the transmission control shifter assembly in the **N** notch in the detent plate.

6. Install the switch assembly to the transmission control shifter assembly by inserting the carrier tang into the hole in the shifter lever assembly. Tighten the attaching screws.

7. Move the transmission control shifter assembly out of the **N** position. This will shear the switch internal plastic pin.

8. Assemble the remaining components in the reverse order of their removal.

Fuses, Circuit Breakers and Relays

LOCATION

Fuse Panel

The fuse panel is located on the left side of the vehicle, under the instrument panel assembly. In order to gain access to the fuse panel it may be necessary to first remove the under dash padding.

Circuit Breakers

The circuit breakers are located at the fuse panel.

Relays

All vehicles use a combination of the following electrical relays:

Air Conditoner Blower High Speed Relay—located near the blower module on the air conditioner module.

Burn Off Relay (1988–89)—located behind the ECM.

Power Door Lock Relay—located in the left side shroud in the lower access hole or near the door jamb conduit.

Air Conditioner Compressor Relay—located on the left side engine cowl near the brake booster or on the relay bracket.

Radio Amplifier Relay—located behind the right side of the instrument panel.

Power Antenna Relay (1988–89)—located behind the right side of the instrument panel lower cover near the ECM.

Horn Relay—located in the convenience center, behind the instrument panel to the right of the steering column.

Fuel Pump Relay—located on the left side of the engine cowl on the relay bracket.

Hatch Release Relay—located under the front part of the console or behind the right side of the instument panel.

Cooling Fan Relay (V6 engine)—located on the left side of the firewall.

GM "J" Body
Front Wheel Drive
Buick—Skyhawk **Cadillac**—Cimarron
Chevrolet—Cavalier **Oldsmobile**—Firenza
Pontiac—Sunbird

SPECIFICATIONS

VEHICLE IDENTIFICATION CHART

It is important for servicing and ordering parts to be certain of the vehicle and engine identification. The VIN (vehicle identification number) is a 17 digit number visible through the windshield on the driver's side of the dash and contains the vehicle and engine identification codes. The tenth digit indicates model year and the eighth digit indicates engine code. It can be interpreted as follows:

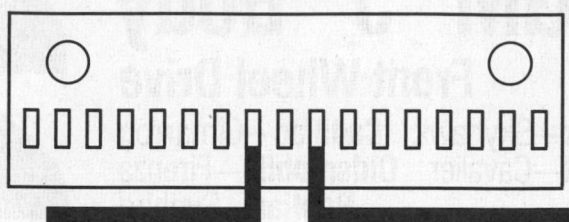

Engine Code

Code	Cu. In.	Liters	Cyl.	Fuel Sys.	Eng. Mfg.
M	121	2.0	4	PFI Turbo	①
1	121	2.0	4	TBI HO	Chevrolet
K	121	2.0	4	TBI	①
G	134	2.2	4	TBI	Chevrolet
W	173	2.8	6	PFI	Chevrolet
T	192	3.1	6	PFI	Chevrolet

Model Year

Code	Year
J	1988
K	1989
L	1990
M	1991
N	1992

HO High Output

PFI Port Fuel Injection

TBI Throttle Body Injection
① Chevrolet-Pontiac-GM of Canada

ENGINE IDENTIFICATION

Year	Model	Engine Displacement cu. in. (liter)	Engine Series Identification (VIN)	No. of Cylinders	Engine Type
1988	Cavalier	121 (2.0)	1	4	OHV
	Cavalier	173 (2.8)	W	6	OHV
	Cimarron	173 (2.8)	W	6	OHV
	Firenza	121 (2.0)	1	4	OHV
	Firenza	121 (2.0)	K	4	OHC
	Sunbird	121 (2.0)	K	4	OHC
	Sunbird	121 (2.0)	M	4	OHC—Turbo
	Skyhawk	121 (2.0)	K	4	OHC
1989	Cavalier	121 (2.0)	1	4	OHV
	Cavalier	173 (2.8)	W	6	OHV
	Sunbird	121 (2.0)	K	4	OHC
	Sunbird	121 (2.0)	M	4	OHC—Turbo
	Skyhawk	121 (2.0)	1	4	OHV
1990	Cavalier	134 (2.2)	G	4	OHV
	Cavalier	192 (3.1)	T	6	OHV
	Sunbird	121 (2.0)	K	4	OHC
	Sunbird	121 (2.0)	M	4	OHC
1991-92	Cavalier	134 (2.2)	G	4	OHV
	Cavalier	192 (3.1)	T	6	OHV
	Sunbird	121 (2.0)	K	4	OHC
	Sunbird	192 (3.1)	T	6	OHV

OHV—Overhead valve

OHC—Overhead cam

OHC-Turbo—Overhead cam with turbocharger

GENERAL ENGINE SPECIFICATIONS

Year	VIN	No. Cylinder Displacement cu. in. (liter)	Fuel System Type	Net Horsepower @ rpm	Net Torque @ rpm (ft. lbs.)	Bore × Stroke (in.)	Compression Ratio	Oil Pressure @ rpm
1988	M	4-121 (2.0)	PFI Turbo	160 @ 5600	160 @ 2800	3.38 × 3.38	8.0:1	65 @ 2500
	1	4-121 (2.0)	TBI (HO)	90 @ 5600	108 @ 3200	3.50 × 3.15	9.0:1	63–77 @ 1200
	K	4-121 (2.0)	TBI	102 @ 5200	130 @ 2800	3.38 × 3.38	8.8:1	45 @ 2000
	W	6-173 (2.8)	PFI	120 @ 4800	155 @ 3600	3.50 × 2.99	8.9:1	50 @ 2400
1989	M	4-121 (2.0)	PFI Turbo	160 @ 5600	160 @ 2800	3.38 × 3.38	8.0:1	65 @ 2500
	1	4-121 (2.0)	TBI (HO)	90 @ 5600	108 @ 3200	3.50 × 3.15	9.0:1	63–77 @ 1200
	K	4-121 (2.0)	TBI	102 @ 5200	130 @ 2800	3.38 × 3.38	8.8:1	45 @ 2000
	W	6-173 (2.8)	PFI	120 @ 4800	155 @ 3600	3.50 × 2.99	8.9:1	50 @ 2400
1990	K	4-121 (2.0)	TBI	102 @ 5200	108 @ 3200	3.39 × 3.39	8.8:1	—
	M	4-121 (2.0)	PFI Turbo	160 @ 5600	160 @ 2800	3.39 × 3.39	8.0:1	—
	G	4-134 (2.2)	TBI	95 @ 5200	120 @ 3200	3.50 × 3.46	9.0:1	63–77 @ 1200
	T	6-192 (3.1)	PFI	140 @ 4500	180 @ 3600	3.50 × 3.31	8.8:1	—
1991–92	K	4-121 (2.0)	TBI	96 @ 4800	118 @ 3600	3.39 × 3.39	8.8:1	—
	G	4-134 (2.2)	TBI	95 @ 5200	120 @ 3200	3.50 × 3.46	9.0:1	63–77 @ 1200
	T	6-192 (3.1)	PFI	140 @ 4500	180 @ 3600	3.50 × 3.31	8.8:1	—

PFI—Port Fuel Injection
TBI—Throttle Body Injection

GASOLINE ENGINE TUNE-UP SPECIFICATIONS

Year	VIN	No. Cylinder Displacement cu. in. (liter)	Spark Plugs Type	Gap (in.)	Ignition Timing (deg.) MT	AT	Compression Pressure (psi)	Fuel Pump (psi)	Idle Speed (rpm) MT	AT	Valve Clearance In.	Ex.
1988	M	4-121 (2.0)	RN9YC4	0.060	①	①	②	25–30	①	①	Hyd.	Hyd.
	1	4-121 (2.0)	RC12LYC	0.035	①	①	②	10–12	①	①	Hyd.	Hyd.
	K	4-121 (2.0)	RN12YC6	0.060	①	①	②	10	①	①	Hyd.	Hyd.
	W	6-173 (2.8)	RS13LYC	0.045	①	①	②	30–37	①	①	Hyd.	Hyd.
1989	M	4-121 (2.0)	RN9YC4	0.060	①	①	②	25–30	①	①	Hyd.	Hyd.
	1	4-121 (2.0)	RC12LYC	0.035	①	①	②	10–12	①	①	Hyd.	Hyd.
	K	4-121 (2.0)	RN12YC6	0.060	①	①	②	10	①	①	Hyd.	Hyd.
	W	6-173 (2.8)	RS13LYC	0.045	①	①	②	30–37	①	①	Hyd.	Hyd.
1990	K	4-121 (2.0)	R44XLS	0.045	①	①	②	9–13	①	①	Hyd.	Hyd.
	M	4-121 (2.0)	R42XLS	0.035	①	①	②	35–38	①	①	Hyd.	Hyd.
	G	4-134 (2.2)	R44LTSM	0.035	①	①	②	9–13	①	①	Hyd.	Hyd.
	T	6-192 (3.1)	R44LTSM	0.045	①	①	②	③	①	①	Hyd.	Hyd.
1991	K	4-121 (2.0)	R44XLS	0.045	①	①	②	9–13	①	①	Hyd.	Hyd.
	G	4-134 (2.2)	R44LTSM	0.035	①	①	②	9–13	①	①	Hyd.	Hyd.
	T	6-192 (3.1)	R44LTSM	0.045	①	①	②	③	①	①	Hyd.	Hyd.
1992		SEE UNDERHOOD VEHICLE EMISSION CONTROL INFORMATION LABEL										

NOTE: The underhood Vehicle Emission Control Information label often reflects tune-up specifications changes made in production. Sticker figures must be used if they disagree with those in this chart.

Hyd—Hydraulic
NA Not available
① See underhood Vehicle Emission Control Information label

② The lowest cylinder compression reading should not be less than 70% of the highest reading and no cylinder should be less than 100 psi.

③ 1–Connect fuel pressure gauge, engine at normal operating temperature.
2–Turn ignition switch on.
3–After approx. 2 seconds; pressure should read 41-47 psi and hold steady.
4–Start engine and idle; pressure should drop 3-10 psi from static pressure.

FIRING ORDERS

NOTE: To avoid confusion, always replace spark plug wires one at a time.

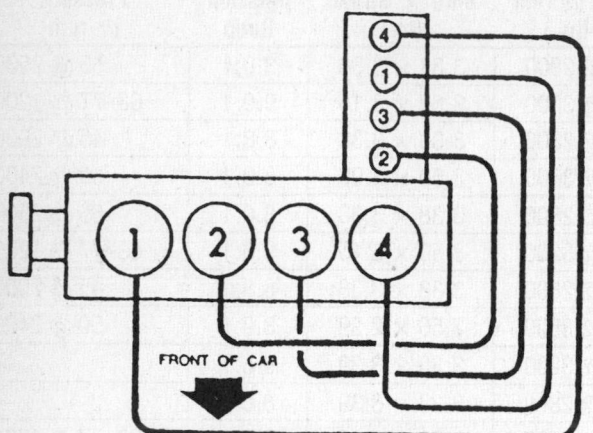

2.0L (VIN 1) and 2.2L Engines
Engine Firing Order: 1–3–4–2
Distributorless Ignition System

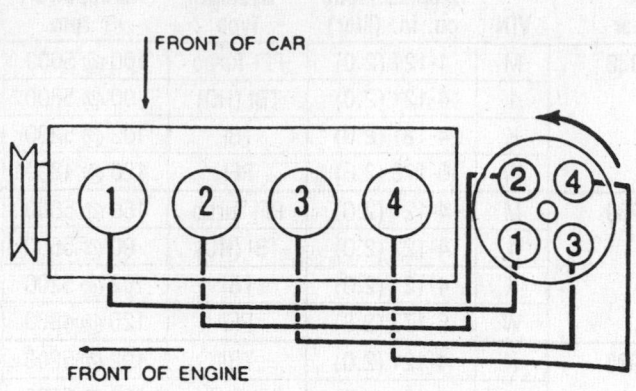

2.0L (VIN K and M) Engines
Engine Firing Order: 1–3–4–2
Distributor Rotation: Counterclockwise

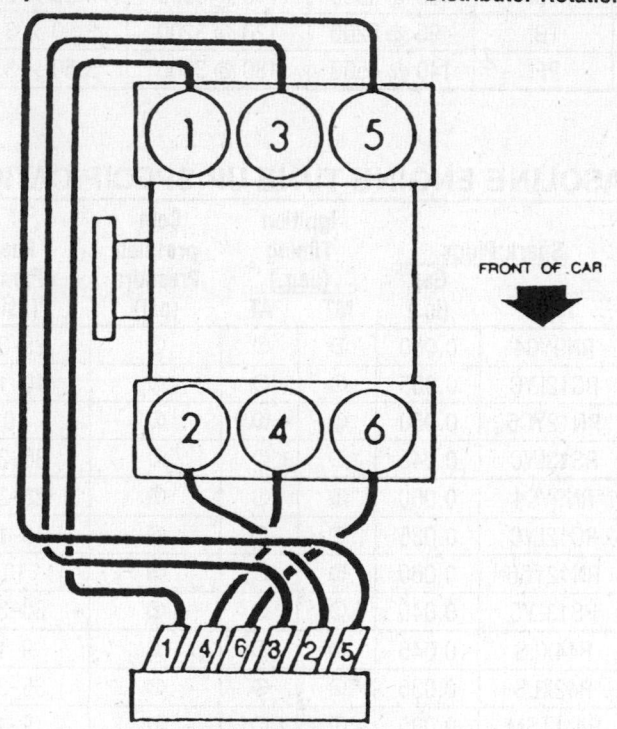

2.8L and 3.1L Engines
Engine Firing Order: 1–2–3–4–5–6
Distributorless Ignition System

CAPACITIES

Year	Model	VIN	No. Cylinder Displacement cu. in. (liter)	Engine Crankcase with Filter	Engine Crankcase without Filter	Transmission (pts.) 4-Spd	Transmission (pts.) 5-Spd	Transmission (pts.) Auto②	Drive Axle (pts.)	Fuel Tank (gal.)	Cooling System (qts.)
1988	Cavalier	1	4-121 (2.0)	4.0①	4.0	6.0	4.0	8.0	—	13.6	9.7
	Cavalier	W	6-173 (2.8)	4.0①	4.0	6.0	4.0	8.0	—	13.6	11.0
	Cimarron	W	6-173 (2.8)	4.0①	4.0	—	4.0	8.0	—	13.6	11.0

CAPACITIES

Year	Model	VIN	No. Cylinder Displacement cu. in. (liter)	Engine Crankcase with Filter	Engine Crankcase without Filter	Transmission (pts.) 4-Spd	Transmission (pts.) 5-Spd	Transmission (pts.) Auto ②	Drive Axle (pts.)	Fuel Tank (gal.)	Cooling System (qts.)
1988	Firenza	1	4-121 (2.0)	4.0①	4.0	—	6.0	8.0	—	13.6	9.7
	Firenza	K	4-121 (2.0)	4.0①	4.0	—	6.0	8.0	—	13.6	8.5
	Sunbird	K	4-121 (2.0)	4.0①	4.0	—	4.0	8.0	—	13.6	8.5
	Sunbird	M	4-121 (2.0)	4.0①	4.0	—	4.0	8.0	—	13.6	8.5
	Skyhawk	K	4-121 (2.0)	4.0①	4.0	—	6.0	8.0	—	13.6	8.5
1989	Cavalier	1	4-121 (2.0)	4.0①	4.0	—	4.0	8.0	—	13.6	9.7
	Cavalier	W	6-173 (2.8)	4.0①	4.0	—	4.0	8.0	—	13.6	11.0
	Sunbird	K	4-121 (2.0)	4.0①	4.0	—	4.0	8.0	—	13.6	8.5
	Sunbird	M	4-121 (2.0)	4.0①	4.0	—	4.0	8.0	—	13.6	8.5
	Skyhawk	1	4-121 (2.0)	4.0①	4.0	—	4.0	8.0	—	13.6	9.7
1990	Cavalier	G	4-134 (2.2)	4.0①	4.0	—	4.0	8.0	—	13.6	8.5
	Cavalier	T	6-192 (3.1)	4.0①	4.0	—	4.0	8.0	—	13.6	11.0
	Sunbird	K	4-121 (2.0)	4.0①	4.0	—	4.0	8.0	—	13.6	8.5
	Sunbird	M	4-121 (2.0)	4.0①	4.0	—	4.0	8.0	—	13.6	8.5
1991–92	Cavalier	G	4-134 (2.2)	4.0①	4.0	—	4.0	8.0	—	13.6	8.5
	Cavalier	T	6-192 (3.1)	4.0①	4.0	—	4.0	8.0	—	13.6	11.0
	Sunbird	K	4-121 (2.0)	4.0①	4.0	—	4.0	8.0	—	13.6	8.5
	Sunbird	T	6-192 (3.1)	4.0①	4.0	—	4.0	8.0	—	13.6	11.0

① When changing the oil filter, additional oil will be needed
② Overhaul capacity, refill capacity is approximately 12 pts.

CAMSHAFT SPECIFICATIONS

All measurements given in inches.

Year	VIN	No. Cylinder Displacement cu. In. (liter)	Journal Diameter 1	Journal Diameter 2	Journal Diameter 3	Journal Diameter 4	Journal Diameter 5	Lobe Lift In.	Lobe Lift Ex.	Bearing Clearance	Camshaft End Play
1988	M	4-121 (2.0)	1.6714–1.6720	1.6812–1.6816	1.6911–1.6917	1.7009–1.7015	1.7108–1.7114	0.2409	0.2409	0.0008	0.0160–0.0640
	1	4-121 (2.0)	1.8670–1.8690	1.8670–1.8690	1.8670–1.8690	1.8670–1.8690	1.8670–1.8690	0.2626	0.2732	0.0010–0.0039	NA
	K	4-121 (2.0)	1.6714–1.6720	1.6812–1.6816	1.6911–1.6917	1.7009–1.7015	1.7108–1.7114	0.2409	0.2409	0.0008	0.0160–0.0640
	W	6-173 (2.8)	1.8678–1.8815	1.8678–1.8815	1.8678–1.8815	1.8678–1.8815	1.8678–1.8815	0.2626	0.2732	0.0010–0.0030	NA
1989	M	4-121 (2.0)	1.6714–1.6720	1.6812–1.6816	1.6911–1.6917	1.7009–1.7015	1.7108–1.7114	0.2409	0.2409	0.0008	0.0160–0.0640
	1	4-121 (2.0)	1.8670–1.8690	1.8670–1.8690	1.8670–1.8690	1.8670–1.8690	1.8670–1.8690	0.2626	0.2626	0.0010–0.0039	NA
	K	4-121 (2.0)	1.6714–1.6720	1.6812–1.6816	1.6911–1.6917	1.7009–1.7015	1.7108–1.7114	0.2409	0.2409	0.0008	0.0160–0.0640
	W	6-173 (2.8)	1.8678–1.8815	1.8678–1.8815	1.8678–1.8815	1.8678–1.8815	1.8678–1.8815	0.2626	0.2732	0.0010–0.0040	NA

CAMSHAFT SPECIFICATIONS

All measurements given in inches.

Year	VIN	No. Cylinder Displacement cu. in. (liter)	Journal Diameter 1	2	3	4	5	Lobe Lift In.	Ex.	Bearing Clearance	Camshaft End Play
1990	K	4-121 (2.0)	1.6706–1.6712	1.6812–1.6818	1.6911–1.6917	1.7009–1.7015	1.7100–1.7106	0.2366	0.2515	0.0011–0.0035	0.0016–0.0063
	M	4-121 (2.0)	1.6714–1.6720	1.6812–1.6816	1.6911–1.6917	1.7009–1.7015	1.7108–1.7114	0.2625	0.2625	0.0011–0.0035	0.0016–0.0063
	G	4-134 (2.2)	1.8670–1.8690	1.8670–1.8690	1.8670–1.8690	1.8670–1.8690	1.8670–1.8690	0.2590	0.2590	0.0010–0.0039	NA
	T	6-192 (3.1)	1.8677–1.8815	1.8677–1.8815	1.8677–1.8815	1.8677–1.8815	1.8677–1.8815	0.2626	0.2732	0.0010–0.0040	NA
1991–92	K	4-121 (2.0)	1.6706–1.6712	1.6812–1.6818	1.6911–1.6917	1.7009–1.7015	1.7100–1.7106	0.2366	0.2515	0.0011–0.0035	0.0016–0.0063
	G	4-134 (2.2)	1.8670–1.8690	1.8670–1.8690	1.8670–1.8690	1.8670–1.8690	1.8670–1.8690	0.2590	0.2590	0.0010–0.0039	NA
	T	6-192 (3.1)	1.8677–1.8815	1.8677–1.8815	1.8677–1.8815	1.8677–1.8815	1.8677–1.8815	0.2626	0.2732	0.0010–0.0040	NA

NA—Not available

CRANKSHAFT AND CONNECTING ROD SPECIFICATIONS

All measurements are given in inches.

Year	VIN	No. Cylinder Displacement cu. in. (liter)	Crankshaft Main Brg. Journal Dia.	Main Brg. Oil Clearance	Shaft End-play	Thrust on No.	Connecting Rod Journal Diameter	Oil Clearance	Side Clearance
1988	M	4-121 (2.0)	①	0.0006–0.0016	0.0030–0.0120	3	1.9278–1.9286	0.0007–0.0024	0.0027–0.0095
	1	4-121 (2.0)	2.4945–2.4954	0.0006–0.0019	0.0020–0.0080	4	1.9983–1.9994	0.0010–0.0031	0.0040–0.0150
	K	4-121 (2.0)	①	0.0006–0.0016	0.0030–0.0120	3	1.9278–1.9286	0.0007–0.0024	0.0027–0.0095
	W	6-173 (2.8)	2.6473–2.6482	0.0016–0.0033	0.0024–0.0083	3	1.9983–1.9994	0.0014–0.0037	0.0063–0.0173
1989	M	4-121 (2.0)	①	0.0006–0.0016	0.0030–0.0120	3	1.9278–1.9286	0.0007–0.0024	0.0027–0.0095
	1	4-121 (2.0)	2.4945–2.4954	0.0006–0.0019	0.0020–0.0080	4	1.9983–1.9994	0.0010–0.0031	0.0040–0.0150
	K	4-121 (2.0)	①	0.0016–0.0016	0.0030–0.0120	3	1.9278–1.9286	0.0007–0.0024	0.0027–0.0095
	W	6-173 (2.8)	2.6473–2.6482	0.0016–0.0033	0.0024–0.0083	3	1.9983–1.9994	0.0014–0.0037	0.0063–0.0173
1990	K	4-121 (2.0)	2.2828–2.2833	0.0006–0.0016	0.0028–0.0118	3	1.9279–1.9287	0.0007–0.0025	0.0028–0.0095
	M	4-121 (2.0)	2.2828–2.2833	0.0006–0.0016	0.0028–0.0118	3	1.9279–1.9287	0.0007–0.0025	0.0028–0.0095
	G	4-134 (2.2)	2.4945–2.4954	0.0006–0.0019	0.002–0.007	4	1.9983–1.9994	0.0009–0.0031	0.0039–0.0149
	T	6-192 (3.1)	2.6473–2.6483	0.0012–0.0030	0.0024–0.0083	3	1.9983–1.9994	0.0011–0.0034	0.0140–0.0270

CRANKSHAFT AND CONNECTING ROD SPECIFICATIONS

All measurements are given in inches.

Year	VIN	No. Cylinder Displacement cu. in. (liter)	Crankshaft				Connecting Rod		
			Main Brg. Journal Dia.	Main Brg. Oil Clearance	Shaft End-play	Thrust on No.	Journal Diameter	Oil Clearance	Side Clearance
1991–92	K	4-121 (2.0)	2.2828–2.2833	0.0006–0.0016	0.0028–0.0118	3	1.9279–1.9287	0.0007–0.0025	0.0028–0.0095
	G	4-134 (2.2)	2.4945–2.4954	0.0006–0.0019	0.002–0.007	4	1.9983–1.9994	0.0009–0.0031	0.0039–0.0149
	T	6-192 (3.1)	2.6473–2.6483	0.0012–0.0030	0.0024–0.0083	3	1.9983–1.9994	0.0011–0.0034	0.0140–0.0270

① Bearings are identified by color:
  Brown 2.2830–2.2832;
  Green 2.2827–2.2830

VALVE SPECIFICATIONS

Year	VIN	No. Cylinder Displacement cu. in. (liter)	Seat Angle (deg.)	Face Angle (deg.)	Spring Test Pressure (lbs.)①	Spring Installed Height (in.)	Stem-to-Guide Clearance (in.)		Stem Diameter (in.)	
							Intake	Exhaust	Intake	Exhaust
1988	M	4-121 (2.0)	45	46	NA	NA	0.0006–0.0020	0.0010–0.0024	NA	NA
	1	4-121 (2.0)	46	45	183 @ 1.33	1.60	0.0011–0.0026	0.0014–0.0030	0.0490–0.0560	0.0630–0.0750
	K	4-121 (2.0)	45	46	NA	NA	0.0006–0.0020	0.0010–0.0024	NA	NA
	W	6-173 (2.8)	46	45	195 @ 1.18	1.57	0.0010–0.0027	0.0010–0.0027	0.0610–0.0730	0.0670–0.0790
1989	M	4-121 (2.0)	45	46	NA	NA	0.0006–0.0020	0.0010–0.0024	NA	NA
	1	4-121 (2.0)	46	45	183 @ 1.33	1.60	0.0011–0.0026	0.0014–0.0030	0.0490–0.0560	0.0630–0.0750
	K	4-121 (2.0)	45	46	NA	NA	0.0006–0.0020	0.0010–0.0024	NA	NA
	W	6-173 (2.8)	46	45	195 @ 1.18	1.57	0.0010–0.0027	0.0010–0.0027	0.0610–0.0730	0.0670–0.0790
1990	K	4-121 (2.0)	45	46	165-197 @ 1.043	NA	0.0006–0.0017	0.0012–0.0024	0.2760–0.2755	0.2753–0.2747
	M	4-121 (2.0)	45	46	165-197 @ 1.043	NA	0.0006–0.0017	0.0012–0.0024	0.2760–0.2755	0.2753–0.2747
	G	4-134 (2.2)	46	45	208-222 @ 1.22	NA	0.0011–0.0026	0.0014–0.0030	NA	NA
	T	6-192 (3.1)	46	45	215 @ 1.291	1.575	0.0010–0.0027	0.0010–0.0027	NA	NA
1991–92	K	4-121 (2.0)	45	46	165-197 @ 1.043	NA	0.0006–0.0017	0.0012–0.0024	0.2760–0.2755	0.2753–0.2747
	G	4-134 (2.2)	46	45	208-222 @ 1.22	NA	0.0011–0.0026	0.0014–0.0031	NA	NA
	T	6-192 (3.1)	46	45	215 @ 1.291	1.575	0.0010–0.0027	0.0010–0.0027	NA	NA

NA—Not available
① Valve open

PISTON AND RING SPECIFICATIONS

All measurements are given in inches.

| Year | VIN | No. Cylinder Displacement cu. in. (liter) | Piston Clearance | Ring Gap | | | Ring Side Clearance | | |
				Top Compression	Bottom Compression	Oil Control	Top Compression	Bottom Compression	Oil Control
1988	M	4-121 (2.0)	0.0012–0.0020	0.0120–0.0200	0.0120–0.0200	0.0160–0.0550	0.0020–0.0030	0.0010–0.0024	—
	1	4-121 (2.0)	0.0098–0.0220	0.0100–0.0200	0.0100–0.0200	0.0100–0.0500	0.0010–0.0030	0.0010–0.0030	0.0006–0.0090
	K	4-121 (2.0)	0.0004–0.0012	0.0120–0.0200	0.0120–0.0200	0.0160–0.0550	0.0020–0.0030	0.0010–0.0024	—
	W	6-173 (2.8)	0.0022–0.0035	0.0100–0.0200	0.0100–0.0200	0.0100–0.0500	0.0020–0.0035	0.0020–0.0035	0.0080 Max
1989	M	4-121 (2.0)	0.0012–0.0020	0.0120–0.0200	0.0120–0.0200	0.0160–0.0550	0.0020–0.0030	0.0010–0.0024	—
	1	4-121 (2.0)	0.0098–0.0220	0.0100–0.0200	0.0100–0.0200	0.0010–0.0500	0.0010–0.0030	0.0010–0.0030	0.0006–0.0090
	K	4-121 (2.0)	0.0004–0.0012	0.0120–0.0200	0.0120–0.0200	0.0160–0.0550	0.0020–0.0030	0.0010–0.0024	—
	W	6-173 (2.8)	0.0022–0.0035	0.0100–0.0200	0.0100–0.0200	0.0100–0.0500	0.0020–0.0035	0.0020–0.0035	0.0080 Max
1990	K	4-121 (2.0)	0.0004–0.0012	0.0098–0.0177	0.0118–0.0197	NA	0.0024–0.0036	0.0019–0.0032	NA
	M	4-121 (2.0)	0.0012–0.0020	0.0098–0.0177	0.0118–0.0197	NA	0.0024–0.0036	0.0019–0.0032	NA
	G	4-134 (2.2)	0.0007–0.0017	0.0100–0.0200	0.0100–0.0200	0.0100–0.0500	0.0019–0.0027	0.0019–0.0027	0.0019–0.0082
	T	6-191 (3.1)	0.00093–0.00222	0.0100–0.0200	0.0200–0.0280	0.0100–0.0300	0.0020–0.0035	0.0020–0.0035	0.008 Max
1991-92	K	4-121 (2.0)	0.0004–0.0012	0.0098–0.0177	0.0118–0.0197	NA	0.0024–0.0036	0.0019–0.0032	NA
	G	4-134 (2.2)	0.0007–0.0017	0.0100–0.0200	0.0100–0.0200	0.0100–0.0500	0.0019–0.0027	0.0019–0.0027	0.0019–0.0082
	T	6-191 (3.1)	0.00093–0.00222	0.0100–0.0200	0.0200–0.0280	0.0100–0.0300	0.0020–0.0035	0.0020–0.0035	0.008 Max

NA—Not available

TORQUE SPECIFICATIONS

All readings in ft. lbs.

| Year | VIN | No. Cylinder Displacement cu. in. (liter) | Cylinder Head Bolts | Main Bearing Bolts | Rod Bearing Bolts | Crankshaft Pulley Bolts | Flywheel Bolts | Manifold | | Spark Plugs |
								Intake	Exhaust	
1988	M	4-121 (2.0)	②	44③	26④	20⑤	48⑥	16	16	15
	1	4-121 (2.0)	①	63–77	34–43	68–89	63①	15–22	6–13	15
	K	4-121 (2.0)	②	44③	26④	20⑤	48⑥	16	16	15
	W	6-173 (2.8)	33⑧	63–83	34–45	66–84	45	18	14–22	15
1989	M	4-121 (2.0)	②	44③	26④	20⑤	48⑥	16	16	15
	1	4-121 (2.0)	⑦	63–77	34–43	68–89	63①	15–22	6–13	15
	K	4-121 (2.0)	②	44③	26④	20⑤	48⑥	16	16	15
	W	6-173 (2.8)	33⑧	63–83	34–45	66–84	45	18	14–22	15

TORQUE SPECIFICATIONS
All readings in ft. lbs.

Year	VIN	No. Cylinder Displacement cu. in. (liter)	Cylinder Head Bolts	Main Bearing Bolts	Rod Bearing Bolts	Crankshaft Pulley Bolts	Flywheel Bolts	Manifold Intake	Manifold Exhaust	Spark Plugs
1990	K	4-121 (2.0)	②	44③	26④	114	⑮	16	⑫	15
	M	4-121 (2.0)	②	44③	26④	114	⑮	18	⑫	15
	G	4-134 (2.2)	⑩	70	38	85	⑬	18	⑪	7–15
	T	6-191 (3.1)	33⑧	73	39	76	52	⑨	18	7–15
1991–92	K	4-121 (2.0)	②	44③	26④	114	⑮	16	⑫	15
	G	4-134 (2.2)	⑭	70	38	77	⑬	18	⑫	11
	T	6-191 (3.1)	33⑧	73	39	76	52	⑨	18	11

CAUTION: Verify the correct original equipment engine is in the vehicle by referring to the VIN engine code before torquing any bolts.

① Auto. Trans.—45–59
② Step 1—18 ft. lbs.
  Step 2—Tighten additional 180 degrees in 3 steps of 60 degrees each
  Step 3—Warm engine—tighten bolts additional 30–50 degree turn
③ Plus additional 45–50 degree turn
④ Plus additional 45 degree turn
⑤ Crankshaft pulley to sprocket bolts
⑥ Plus additional 30 degree turn
⑦ Long bolts—73–83 ft. lbs.
  Short bolts—62–70 ft. lbs.

⑧ Coat thread with sealer an additional 90 degree turn
⑨ Tighten in sequence to 15 ft. lbs., then retighten to 24 ft. lbs.
⑩ Step 1—Tighten all bolts initially to 41 ft. lbs.
  Step 2—Tighten all bolts an additional 45 degrees in sequence
  Step 3—Tighten all bolts an additional 45 degrees in sequence
  Step 4—Tighten the long bolts—8, 4, 1, 5 and 9 an additional 20 degrees and tighten the short bolts—7, 3, 2, 6 and 10 an additional 10 degrees
⑪ Nuts—115 inch lbs.
  Studs—89 inch lbs.

⑫ 115 inch lbs.
⑬ Auto trans-52
  Manual trans-55
⑭ Step 1—Tighten the long bolts—8, 4, 1, 5 and 9 to 46 ft. lbs.
  Tighten the short bolts—7,3,2,6 and 10 to 43 ft. lbs.
  Step 2—Tighten all bolts an additional 90 degrees in sequence.
⑮ Auto trans—48 ft. lbs.—can reuse bolts.
  Man trans—48 ft. lbs.,—plus an additional 30 degrees-must use new bolts.

BRAKE SPECIFICATIONS
All measurements in inches unless noted

Year	Model	Lug Nut Torque (ft. lbs.)	Master Cylinder Bore	Brake Disc Minimum Thickness ①	Brake Disc Maximum Runout	Maximum Brake Drum Diameter ②	Minimum Lining Thickness Front	Minimum Lining Thickness Rear
1988	Cavalier	100	0.940	0.815	0.004	7.929	⅛	⅛
	Sunbird	100	0.940	0.815	0.004	7.929	⅛	⅛
	Firenza	100	0.940	0.815	0.004	7.929	⅛	⅛
	Skyhawk	100	0.940	0.815	0.004	7.929	⅛	⅛
	Cimarron	100	0.940	0.815	0.004	7.929	⅛	⅛
1989	Cavalier	100	0.940	0.815	0.004	7.929	⅛	⅛
	Sunbird	100	0.940	0.815	0.004	7.929	⅛	⅛
	Skyhawk	100	0.940	0.815	0.004	7.929	⅛	⅛
1990	Cavalier	100	0.874	0.815	0.004	7.929	⅛	⅛
	Sunbird	100	0.874	0.815	0.004	7.929	⅛	⅛
1991–92	Cavalier	100	0.874	0.815	0.004	7.929	⅛	⅛
	Sunbird	100	0.874	0.815	0.004	7.929	⅛	⅛

① Discard thickness
② Discard diameter

WHEEL ALIGNMENT

Year	Model	Caster Range (deg.)	Caster Preferred Setting (deg.)	Camber Range (deg.)	Camber Preferred Setting (deg.)	Toe-in (in.)	Steering Axis Inclination (deg.)
1988	Cavalier	13/16N–4 3/16P	1 11/16P	3/16P–1 3/16P ②	13/16P ①	0	13 1/2
	Sunbird	13/16N–4 3/16P	1 11/16P	3/16P–1 3/16P	13/16P	0	13 1/2
	Firenza	13/16N–4 3/16P	1 11/16P	3/16P–1 3/16P	13/16P	0	13 1/2
	Skyhawk	13/16N–4 3/16P	1 11/16P	3/16P–1 3/16P	13/16P	0	13 1/2
	Cimarron	13/16N–4 3/16P	1 11/16P	3/16P–1 3/16P	13/16P	0	13 1/2
1989	Cavalier	11/16P–2 11/16P	1 11/16P	3/16P–1 3/16P ②	13/16P ①	0	13 1/2
	Sunbird	11/16P–2 11/16P	1 11/16P	3/16P–1 3/16P	13/16P	0	13 1/2
	Skyhawk	11/16P–2 11/16P	1 11/16P	3/16P–1 3/16P	13/16P	0	13 1/2
1990	Cavalier	11/16P–2 11/16P	1 11/16P	3/16P–1 3/16P ②	13/16P ①	0	13 1/2
	Sunbird	11/16P–2 11/16P	1 11/16P	3/16P–1 3/16P	13/16P	0	13 1/2
1991-92	Cavalier	11/16P–2 11/16P	1 11/16P	3/16P–1 3/16P ②	13/16P ①	0	13 1/2
	Sunbird	11/16P–2 11/16P	1 11/16P	3/16P–1 3/16P	13/16P	0	13 1/2

① Z-24; 1N-P. Preferred setting is 0 camber
② If vehicle is equipped with P215-60R14 tires setting is 1/8 degree out
P—Positive
N—Negative

ENGINE MECHANICAL

NOTE: Disconnecting the negative battery cable on some vehicles may interfere with the functions of the on board computer systems and may require the computer to undergo a relearning process, once the negative battery cable is reconnected.

Engine Assembly

REMOVAL & INSTALLATION

2.0L (VIN 1) and 2.2LEngines

NOTE: Special tool J-24420 crankshaft pulley hub remover is required. The engine is removed from the top of vehicle.

1. Disconnect the negative battery cable and relieve the fuel system pressure.
2. Drain the cooling system.
3. Remove the air cleaner.
4. Disconnect the accelerator and TV cables.
5. Disconnect the ECM harness at engine.
6. Disconnect the necessary vacuum hoses.
7. Disconnect all of the cooling hoses at engine.
8. Remove the exhaust heat shield.
9. If equipped with air conditioning, remove the adjustment bolt at engine mount.
10. Disconnect engine wiring harness at bulkhead.
11. Remove the windshield washer bottle.
12. Remove the alternator and power steering belt.
13. Disconnect the fuel hoses.
14. Raise and safely support the vehicle. If equipped with air conditioning, remove the air conditioning brace.
15. Remove inner fender splash shield.
16. If equipped with air conditioning, remove the air conditioning compressor.
17. Remove flywheel splash shield.
18. Disconnect and the tag starter wires.
19. Disconnect the front starter brace.
20. Remove the starter.
21. Remove the torque converter bolts.
22. Remove the crankshaft pulley and hub using tool J-24420 or equivalent.
23. Remove the oil filter.
24. Disconnect the engine-to-transaxle bracket.
25. Disconnect the right rear mount.
26. Disconnect exhaust at manifold and center hanger.
27. Disconnect the TV and shift cable.
28. Remove the lower bell housing bolts.
29. Lower the vehicle.
30. Remove the right front engine mount nuts.
31. Remove alternator and adjusting brace.
32. Disconnect the master cylinder and push it aside.
33. Install a suitable lifting device.
34. Remove the right front engine mount bracket.
35. Remove upper bell housing bolts.
36. Remove the power steering pump while lifting engine.
37. Carefully remove the engine from the vehicle.

To install:

38. Install the engine mount alignment bolt M6X1X65 to ensure proper power train alignment.
39. Slowly lower the engine into the vehicle, leaving the lifting device attached.
40. Install the transaxle bracket. Install the mount to the side frame and secure with new bolts.
41. With the engine weight not on the mounts, tighten the transaxle bolts to 48–63 ft. lbs. (65–85 Nm).
42. Tighten the right front mount nuts.
43. Lower the engine weight onto the mounts. Remove the lifting device.
44. Install the power steering pump.
45. Install alternator and adjusting brace.
46. Install the right front engine mount nuts.
47. Connect the master cylinder.
48. Install the lower bell housing bolts.
49. Connect the TV and shift cable.

50. Connect exhaust at manifold and center hanger.

51. Connect the right rear mount.

52. Connect the engine-to-transaxle bracket.

53. Install the oil filter.

54. Install the crankshaft pulley and hub using tool J–24420 or equivalent.

55. Install the torque converter bolts.

56. Install the starter.

57. Connect the front starter brace.

58. Connect the starter wires.

59. Install flywheel splash shield.

60. If equipped with air conditioning, install the air conditioning compressor. Install the air conditioning brace.

61. Install inner fender splash shield.

62. Connect the fuel hoses.

63. Install the alternator and power steering belts.

64. Install the windshield washer bottle.

65. Connect engine wiring harness at bulkhead.

66. If equipped with air conditioning, install the adjustment bolt at the engine mount.

67. Install the exhaust heat shield.

68. Connect all of the cooling hoses at engine.

69. Connect the vacuum hoses.

70. Connect the ECM harness at engine.

71. Connect the accelerator and TV cables.

72. Install the air cleaner.

73. Drain the cooling system.

74. Connect the negative battery cable.

2.0L (VIN K and M) Engines

NOTE: This procedure requires the use of a special powertrain alignment bolt M6X1X65. The engine is removed from the bottom of the vehicle.

1. Disconnect the negative battery cable and relieve fuel pressure.

2. Drain the cooling system into a clean container for reuse.

3. Remove the air cleaner assembly.

4. Remove the cooling fan.

5. Disconnect the engine electrical harness at bulkhead.

6. Disconnect the electrical connector at brake cylinder.

7. Disconnect the air conditioner relay cluster switches.

8. Disconnect the wiper motor electrical connector.

9. Disconnect the cooling fan, relay and ground.

10. Disconnect the ECM harness and pull harness through the bulkhead.

11. Disconnect the temperature switch at the thermostat housing.

12. Disconnect the EFI, MAP sensor and canister vacuum hoses.

13. Disconnect the cables from the throttle bracket and EFI, and shift control at the transaxle.

14. Disconnect the power steering return hose at the pump.

15. Raise and safely support the vehicle.

16. Disconnect the VSS connector or speedometer cable at the transaxle.

17. Disconnect the exhaust pipe at the exhaust manifold and exhaust hangers and swing aside.

18. Disconnect the hoses from the heater core.

19. Disconnect the fuel lines.

20. If equipped with automatic transaxle, disconnect the cooler lines from the radiator.

21. Remove the front wheels.

22. Remove the brake calipers and support using a length of wire.

23. Properly discharge and recover the refrigerant from the air conditioning system.

24. Disconnect the refrigerant lines from the compressor.

NOTE: Cap the refrigerant lines when opening the system to prevent the entry of dirt and moisture and the loss of refrigerant lubricant.

25. Remove the suspension support bolts as follows:

    a. Remove the 2 center bolts on each side.

    b. Remove 1 bolt at each end.

    c. Loosen the remaining bolt.

NOTE: Properly support the vehicle, engine and transaxle during the following steps.

26. If equipped with automatic transaxle, disconnect the lateral strut at the rear of the transaxle.

27. Disconnect the transaxle strut from the front of the transaxle.

28. Support the front of the vehicle with stands under the radiator core support.

29. Reposition the jack to the rear of the cowl with 4 × 4 × 6 in. timber spanning the vehicle width.

30. Raise the vehicle enough to remove the jackstands.

31. Position the dolly under the engine and transaxle with three 4 × 4 × 12 in. blocks as support.

32. Lower the vehicle onto the dolly lightly.

33. Remove the remaining bolt at each end of the right and left front suspension supports.

34. Remove the long mount-to-bracket bolt from the transaxle.

35. Remove the 2 mount-to-bracket bolts from the front engine mount.

36. Remove the 2 mount-to-bracket nuts and reinforcement bracket from the rear engine mount.

37. Carefully scribe the position of the strut on the hub to preserve the camber adjustment. Remove the 2 knuckle-to-strut bolts on each side.

38. Raise the vehicle leaving the engine, transaxle and suspension on the dolly.

39. Separate the engine and transaxle.

To install:

40. With the aid of an assistant, assemble the engine and transaxle. Tighten the engine-to-transaxle bolts to 85 ft. lbs. (75 Nm) for automatic transaxles or 60 ft. lbs. (75 Nm) for manual transaxles. Position the engine and transaxle assembly in the vehicle.

41. Install the 1 long mount-to-bracket bolt at the transaxle. Install the nut and tighten to 80 ft. lbs. (108 Nm).

42. Install the reinforcement bracket and 2 mount-to-bracket nuts at the rear engine mount. Tighten to 18 ft. lbs. (24 Nm).

43. Install the 2 mount-to-bracket bolts at the front engine mount. Tighten to 40 ft. lbs. (54 Nm).

44. Loosely install the bolt at each end of the right and left front suspension supports.

45. Install the knuckle-to-strut bolts. Align the scribe marks made during removal. Tighten the nuts to 133 ft. lbs. (180 Nm).

46. Raise the vehicle and remove the dolly. Using jackstands, remove the 6 in. timber and move the hoist to the front.

47. Install the remaining bolts in the right and left front suspension supports. Tighten to 65 ft. lbs. (88 Nm).

48. Connect the front and rear transaxle struts. Tighten to 50 ft. lbs. (68 Nm).

49. Connect the electrical connectors to the compressor.

50. Install the brake calipers.

51. Install the front wheels.

52. If equipped with automatic transaxle, connect the cooler lines to the radiator.

53. Connect the fuel lines.

54. Connect the hoses to the heater core.

55. Connect the exhaust pipe to the exhaust manifold and hangers.

56. Connect the speedometer cable or VSS connector at the transaxle.

57. Lower the vehicle.

58. Connect the power steering return hose to the pump.

59. Connect the power steering cutoff switch.

60. Connect the transaxle shift cable.

61. Connect the throttle cable to the bracket and EFI unit.

62. Connect the vacuum hoses to the MAP sensor, canister and EFI unit.

63. Connect the electrical connector

to the temperature switch at the thermostat housing.

64. Pull the ECM harness through the bulkhead and connect the ECM connector.

65. Connect the air conditioner relay cluster switches.

66. Connect the wiper motor electrical connector.

67. Connect the cooling fan, relay and ground.

68. Connect the electrical connector to the brake cylinder.

69. Connect engine harness bulkhead connectors.

70. Connect the radiator hose.

71. Replace the compressor fitting O-rings. Lubricate the O-rings with refrigerant oil. Connect the air conditioner refrigerant lines.

72. Connect the negative battery cable.

73. Fill cooling system and check for leaks. Start the engine and allow to come to normal operating temperature. Recheck for leaks. Top-up coolant.

74. Evacuate, recharge and leak test the air conditioning system.

75. Recheck for coolant, fuel, oil and transaxle fluid leaks.

2.8L and 3.1L Engines

NOTE: Always release the fuel pressure before starting repair. The engine is removed from the top of the vehicle.

1. Disconnect the negative battery cable. Drain the cooling system and remove the air cleaner assembly. Mark the bolt location and remove the hood.

2. Remove the air flow sensor. Remove the exhaust crossover heat shield and remove the crossover pipe.

3. Remove the serpentine belt tensioner and belt.

4. Remove the power steering pump mounting bracket. Disconnect the heater pipe at the power steering pump mounting bracket.

5. Disconnect the radiator hoses from the engine.

6. Disconnect the accelerator and throttle valve cable at the throttle valve.

7. Remove the alternator. Tag and disconnect the wiring harness at the engine.

8. Relieve the fuel pressure and disconnect the fuel hose. Disconnect the coolant bypass and the over flow hoses at the engine.

9. Tag and remove the vacuum hoses to the engine.

10. Raise the vehicle and support it safely.

11. Remove the inner fender splash shield. Remove the harmonic balancer.

12. Remove the flywheel cover. Re-

move the starter bolts. Tag and disconnect the electrical connections to the starter. Remove the starter.

13. Disconnect the wires at the oil sending unit.

14. Remove the air conditioning compressor and related brackets.

15. Disconnect the exhaust pipe at the rear of the exhaust manifold.

16. Remove the flexplate-to-torque converter bolts.

17. Remove the transaxle-to-engine bolts. Remove the engine-to-rear mount frame nuts.

18. Disconnect the shift cable bracket at the transaxle. Remove the lower bell housing bolts.

19. Lower the vehicle and disconnect the heater hoses at the engine.

NOTE: It may be necessary to remove the engine hood. Using an awl, scribe marks around the hood hinges to help aid correct hood alignment upon installation.

20. Install a suitable engine lifting device. While supporting the engine and transaxle, remove the upper bell housing bolts.

21. Remove the front mounting bolts.

22. Remove the master cylinder from the booster.

23. Remove the engine assembly from the vehicle.

To install:

24. Install the engine in position in the vehicle.

25. Install the upper transaxle-to-engine bolts.

26. Raise and safely support the vehicle.

27. Install the lower transaxle-to-engine bolts.

28. Reconnect the shift cable bracket to the transaxle.

29. Install the engine mounts, tightening the front mount-to-frame bolts to 61 ft. lbs. (83 Nm) and the engine mount to bracket bolts to 50 ft. lbs. (68 Nm).

30. Install the flywheel to converter bolts.

31. Reconnect the exhaust pipe and install the air conditioning compressor. Install the flywheel cover.

32. Reconnect the coolant hoses and the fuel lines.

33. Install the wiring harness at the engine and install the alternator.

34. Lower the vehicle and install the accessory drive belt.

35. Refill all of the fluids and connect the negative battery cable.

36. Install the hood. Install the air cleaner assembly.

37. Road test the vehicle.

Engine Mounts

REMOVAL & INSTALLATION

Front

1. Disconnect the negative battery cable.

2. Raise the vehicle and support safely.

3. Using a suitable fixture, support the engine and remove the engine mount nuts.

4. Remove the inner fender shield.

5. Remove the engine mount bolts. The manufacturer recommends discarding the engine mount bolts and replacing with new bolts. Note the location and length of each bolt for reassembly.

6. Remove the engine mount.

To install:

7. Install the engine mount.

8. Install the engine mount bolts.

9. Install the inner fender shield.

10. Install the engine mount nuts. Remove the support fixture.

11. Lower the vehicle.

12. Connect the negative battery cable.

Rear

1. Disconnect the negative battery cable.

2. Raise and safely support the vehicle.

3. If equipped with manual transaxle, remove the oil filter in order to gain working clearance.

4. Using a suitable fixture, support the engine and remove the engine mounting nuts.

5. Remove the engine mounting bolts. Remove the engine mount.

To install:

6. Install the engine mount. Install the mounting bolts and nuts.

7. Remove the engine support fixture.

8. If equipped with manual transaxle, install the oil filter.

9. Lower the vehicle.

10. Connect the negative battery cable.

Cylinder Head

REMOVAL & INSTALLATION

2.0L (VIN 1) and 2.2L Engines

NOTE: The engine must be cold before removing the cylinder head. Always release the fuel pressure before starting repair.

1. Disconnect the negative battery cable.

2. Drain the cooling system.

3. Remove the TBI cover. Raise and safely support the vehicle.

4. Remove the exhaust shield. Disconnect the exhaust pipe.

5. Remove the heater hose from the intake manifold. Lower the vehicle.

6. Disconnect the accelerator and TV cable bracket.

7. Lower the vehicle.

8. Tag and disconnect the vacuum lines at the intake manifold and thermostat.

9. Disconnect the accelerator linkage at the TBI unit and remove the linkage bracket.

10. Tag and disconnect all necessary wires. Remove the upper radiator hose at the thermostat.

11. Remove the serpentine belt.

12. Remove the power steering pump and lay it aside.

13. Make sure the fuel system pressure is released and disconnect and plug the fuel lines.

14. Remove the alternator. Remove the alternator brace from the head and remove the upper mounting bracket.

15. Remove the cylinder head cover. Remove the rocker arms and pushrods keeping all parts in order for correct installation.

16. Remove the cylinder head bolts. Remove the cylinder head with the TBI unit, intake and exhaust manifolds still attached.

To install:

17. The gasket surfaces on both the head and the block must be clean of any foreign matter and free of any nicks or heavy scratches. Bolt threads in the block and the bolts must be clean.

18. Place a new cylinder head gasket in position over the dowel pins on the block. Carefully guide the cylinder head into position.

19. Coat the cylinder bolts with sealing compound and install them finger tight.

20. If equipped with the 2.0L engine, tighten the short bolts in sequence to 62–70 ft. lbs. (84–95 Nm) and tighten the long bolts in sequence to 73–83 ft. lbs. (99–113 Nm).

21. If equipped with the 2.2L engine, tighten the bolts in the following sequence:

　a. Tighten all bolts initially to 41 ft. lbs. (56 Nm).

　b. Tighten all bolts 45 degrees in sequence.

　c. Tighten all bolts an additional 45 degrees in sequence.

　d. Tighten the long bolts 8, 4, 1, 5 and 9 an additional 20 degrees.

　e. Tighten the short bolts 7, 3, 2, 6 and 10 an additional 10 degrees.

NOTE: The short bolts, exhaust side, should end up with a total rotation of 100 degrees and the long bolts, intake side, should end up with a total rotation of 110 degrees.

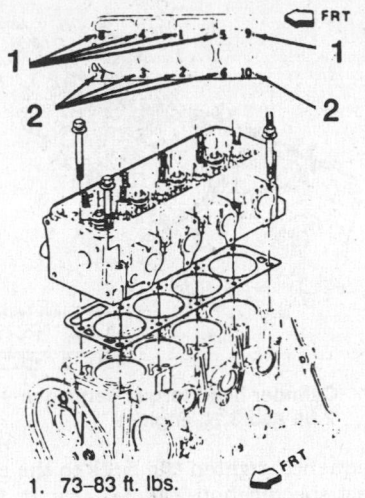

1. 73–83 ft. lbs.
2. 62–70 ft. lbs.

Cylinder head bolt torque sequence— 2.0L (VIN 1) and 2.2L engines

22. Reinstall the alternator. Install the power steering pump and brackets.

23. Reconnect the fuel lines and the hoses. Connect the exhaust pipe to the manifold.

24. Install the valve cover and connect the linkage at the TBI unit. Install the air cleaner and fill all the fluids.

25. Connect the negative battery cable.

26. Start the engine and check for leaks.

2.0L (VIN K and M) Engines

NOTE: Cylinder head gasket replacement is necessary if camshaft carrier/cylinder head bolts are loosened. The head bolts should always be loosened when cold. New head bolts should be used every time camshaft carrier/ cylinder head or gasket are replaced.

1. Disconnect the negative battery cable. Remove the air cleaner and relieve fuel pressure.

2. Drain the cooling system.

3. Remove the alternator and pivot bracket at the camshaft carrier housing.

4. Disconnect the power steering pump and bracket, lay it to one side.

5. Disconnect the ignition coil electrical connections and remove coil.

6. Disconnect the spark plug wires and distributor cap, remove the distributor.

7. Remove the throttle cable from the bracket at intake manifold.

8. Disconnect the throttle cable, downshift cable and TV cable from the EFI assembly.

9. Disconnect the ECM connectors from the EFI assembly.

10. Remove the vacuum brake hose at filter.

11. Disconnect the inlet and return fuel lines at flex joints.

12. Remove the water pump bypass hose at the intake manifold and water pump.

13. Disconnect the ECM harness connectors at intake manifold.

14. Disconnect the heater hose from intake manifold.

15. Disconnect the exhaust pipe at exhaust manifold.

NOTE: On engine Code M, remove the exhaust manifold to turbo connection and oxygen sensor connection.

16. Disconnect the breather hose at camshaft carrier.

17. Remove the upper radiator hose.

18. Disconnect the engine electrical harness and wires from thermostat housing.

19. Remove the timing cover.

20. Remove the timing probe holder.

21. Loosen the water pump retaining bolts and remove timing belt.

22. Loosen the camshaft carrier and cylinder head attaching bolts, a little at a time, in sequence.

23. Remove the camshaft carrier assembly.

24. Remove the cylinder head, intake manifold and exhaust manifold as an assembly.

To install:

25. Install a new cylinder head gasket in position on the block.

26. Apply a continuous bead of sealer to the cam carrier.

27. Install the cylinder head, reassembled with the intake and exhaust manifolds, if removed.

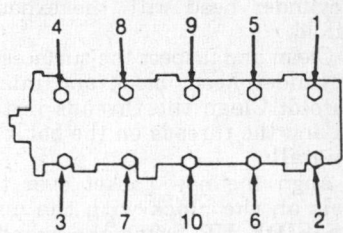

Camshaft carrier/cylinder head bolt loosening sequence—2.0L (VIN K and M) engines

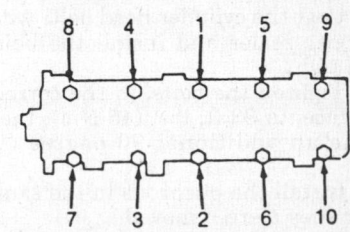

Camshaft carrier/cylinder head bolt torque sequence—2.0L (VIN K and M) engines

28. Install the camshaft carrier on the cylinder head and tighten the bolts, in following sequence, to the correct torque.

    a. Tighten all bolts in sequence to 18 ft. lbs. (24 Nm).

    b. Tighten all bolts an additional 180 degrees in 3 steps of 60 degrees each.

29. Install the timing belt.

30. Reconnect the electrical harness and the breather hose at the camshaft carrier.

31. Connect the exhaust pipe at the manifold and attach the heater hose to the intake manifold.

32. Connect the brake hose at the filter. Connect the throttle and TV cable.

33. Refill the cooling system and connect the negative battery cable.

34. Run the engine, until warm, thermostat open, and tighten all of the the cylinder head/cam carrier bolts an additional 30–50 degrees, in sequence. Check for leaks.

2.8L and 3.1L Engines

LEFT SIDE

1. Relieve the fuel system pressure and disconnect the negative battery cable. Drain the engine coolant into a clean container for reuse. Remove the rocker cover.

2. Remove the intake manifold. Disconnect the exhaust crossover at the right exhaust manifold.

3. Disconnect the oil level indicator tube bracket.

4. Loosen the rocker arms nuts enough to remove the pushrods.

5. Starting with the outer bolts, remove the cylinder head bolts. Remove the cylinder head with the exhaust manifold.

6. Clean and inspect the surfaces of the cylinder head, block and intake manifold. Clean the threads in the block and the threads on the bolts.

To install:

7. Align the new gasket over the dowels on the block with the note **THIS SIDE UP** facing the cylinder head.

8. Install the cylinder head and exhaust manifold crossover assembly on the engine.

9. Coat the cylinder head bolts with a proper sealer and install the bolts hand tight.

10. Tighten the bolts, in the correct sequence, to 33 ft. lbs. (45 Nm), then rotate an additional 90 degree (¼ turn).

11. Install the pushrods in the same order they were removed.

12. Install the rocker arms. Tighten to 18 ft. lbs. (24 Nm).

13. Install the intake manifold using a new gasket and following the correct

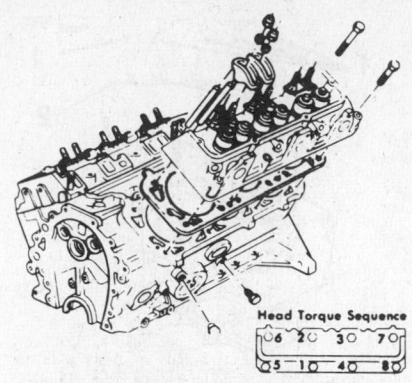

Cylinder head torque sequence — 2.8L and 3.1L engines

sequence, tighten the bolts to the correct specification.

14. Connect the exhaust crossover at the right exhaust manifold.

15. Install the rocker cover.

16. Install the oil level dipstick tube.

17. Connect the negative battery cable.

18. Fill cooling system and check for leaks. Start the engine and allow to come to normal operating temperature. Recheck for leaks. Top-up coolant.

RIGHT SIDE

1. Disconnect the negative battery cable. Drain the engine coolant into a clean container for reuse.

2. Raise and safely support the vehicle. Disconnect the exhaust manifold from the exhaust pipe.

3. Lower the vehicle. Disconnect the exhaust manifold from the cylinder head and remove the manifold.

4. Remove the rocker cover. Remove the intake manifold.

5. Loosen the rocker arms enough so the pushrods can be removed. Note the position of the pushrods for assembly.

6. Starting with the outer bolts, remove the cylinder head bolts and remove the cylinder head.

7. Inspect and clean the surfaces of the cylinder head, engine block and intake manifold.

8. Clean the threads in the engine block and the threads on the cylinder head bolts.

To install:

9. Align the new gasket on the dowels on the engine block with the note **THIS SIDE UP** facing the cylinder head.

10. Install the cylinder head on the engine. Coat the head bolts with a proper sealer. Install and tighten the bolts hand tight.

11. Tighten the bolts, in sequence, to 33 ft. lbs. (45 Nm), then rotate an additional 90 degree (¼ turn).

12. Install the pushrods in the same order as they were removed.

13. Install the rocker arms. The correct rocker arm torque is 18 ft. lbs. (24 Nm).

14. Install the intake manifold using a new gasket. Following the correct sequence, tighten the bolts to the proper specification.

15. Remove the rocker cover.

16. Install the exhaust manifold.

17. Raise and safely support the vehicle.

18. Connect the exhaust manifold to the exhaust pipe.

19. Connect the negative battery cable.

20. Fill cooling system and check for leaks. Start the engine and allow to come to normal operating temperature. Recheck for leaks. Top-up coolant.

Valve Lifters

REMOVAL & INSTALLATION

1. Disconnect the negative battery cable.

2. If equipped with a 2.8L or 3.1L engine, remove the intake manifold.

3. Remove the rocker arm cover.

4. Loosen the rocker arm holding nut and move the rocker arm to the side.

5. Remove the pushrods.

6. Using a suitable tool, remove the valve lifter.

To install:

7. Fill lifter assembly with engine oil and lubricate the bottom of the valve lifter with Molykote® or equivalent. Install the valve lifter.

8. Install the pushrods.

9. Place the rocker arm into position. Tighten the rocker arm nut to 10 ft. lbs. (14 Nm) on the 2.0L engine, 14 ft. lbs. (19 Nm) on the 2.2L engine or 18 ft. lbs. (24 Nm) on the 2.8L and 3.1L engines.

10. Install the rocker arm cover.

11. If equipped with the 2.8L or 3.1L, install the engine. Install the intake manifold.

12. Connect the negative battery cable.

Rocker Arms

REMOVAL & INSTALLATION

2.0L (VIN K and M) Engines

1. Disconnect the negative battery cable. Remove camshaft carrier cover.

2. Hold the valves in place with compressed air, using air adapter tool J–22794 or equivalent, in spark plug hole.

3. Compress the valve springs with special tool J–33302–25 or equivalent.

4. Remove the rocker arms. Keep the rocker arms in order for reassembly.

To install:

5. Compress the valve springs with special tool J–33302–25 or equivalent.

6. If removed, install the valve lash compensators in original positions.

7. Install thrust pieces and rocker arms.

8. Install the camshaft carrier cover.

9. Connect the negative battery cable.

2.0L (VIN 1) and 2.2L Engines

1. Disconnect the negative battery cable. Remove the air cleaner. Remove the rocker cover.

2. Remove the rocker arm nut and ball. Lift the rocker arm off the stud and the pushrods from the engine. Always keep the valve system parts in order.

To install:

3. Coat the rocker arm balls with Molykote® or equivalent.

4. Install the pushrods in the order removed, making sure they seat properly in the lifter.

5. Install the rocker arms, balls and nuts in the order removed. Tighten to 11–18 ft. lbs. (15–24 Nm).

6. Install the rocker cover.

7. Install the air cleaner.

8. Connect the negative battery cable.

2.8L and 3.1L Engines

LEFT SIDE

1. Disconnect the negative battery cable. Disconnect the bracket tube at the rocker cover.

2. Remove the spark plug wire cover. Drain the cooling system into a clean container for reuse and remove the heater hose at the filler neck.

3. Remove the rocker arm cover bolts and remove the rocker cover.

4. Remove the rocker arm nuts and remove the rocker arms. Note the order of removal for installation.

To install:

5. Install the rocker arms in the correct order. Tighten to 14–20 ft. lbs. (19–27 Nm).

6. Install the rocker arm cover. Connect the bracket tube at the rocker cover.

7. Install the spark plug wire cover.

8. Connect the heater hose to the filler neck.

9. Connect the negative battery cable.

10. Fill cooling system and check for leaks. Start the engine and allow to come to normal operating temperature. Recheck for leaks. Top-up coolant.

RIGHT SIDE

1. Disconnect the negative battery cable. Disconnect the brake booster vacuum line at the bracket.

2. Disconnect the cable bracket at the plenum.

3. Disconnect the vacuum line bracket at the cable bracket.

4. Disconnect the lines at the alternator brace stud.

5. Remove the rear alternator brace.

6. Remove the serpentine belt.

7. Remove the alternator and support it out of the way.

8. Remove the PCV valve.

9. Loosen the alternator bracket.

10. Remove the spark plug wires. Remove the rocker cover bolts and remove the rocker cover.

11. Remove the rocker arm nuts and remove the rocker arms. Note the order of removal for installation.

To install:

12. Install the rocker arms in the correct order. Tighten 14–20 ft. lbs. (19–27 Nm).

13. Install the rocker cover and attaching bolts. Install the spark plug wires.

14. Tighten the alternator bracket.

15. Install the PCV valve.

16. Install the alternator.

17. Install the serpentine belt.

18. Install the rear alternator brace.

19. Connect the lines at the alternator brace stud.

20. Connect the vacuum line bracket at the cable bracket.

21. Connect the cable bracket at the plenum.

22. Connect the brake booster vacuum line at the bracket. Connect the negative battery cable.

Intake Manifold

REMOVAL & INSTALLATION

2.0L (VIN K and M) Engines

1. Release the fuel pressure. Disconnect the negative battery terminal from the battery.

2. Remove induction tube and hoses.

3. Disconnect and tag wiring to throttle body, fuel injectors, MAP sensor and wastegate, if equipped.

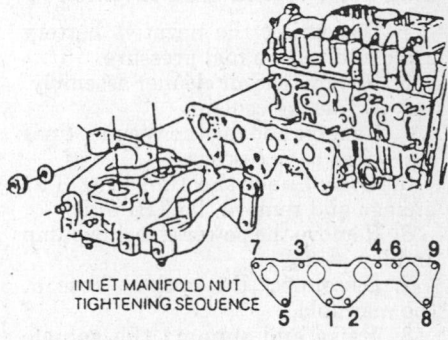

INLET MANIFOLD NUT TIGHTENING SEQUENCE

Intake manifold torque sequence—2.0L (VIN K) engine

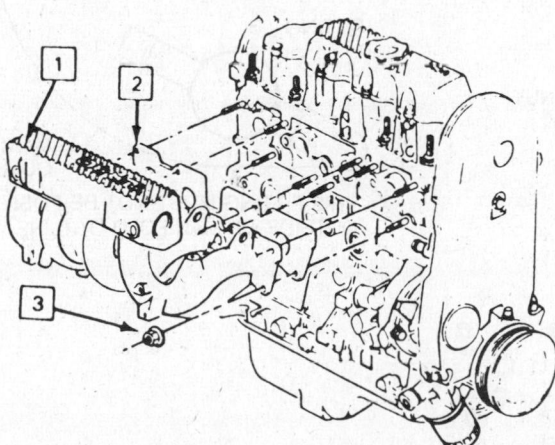

1. Intake manifold
2. Gasket
3. Nut (18 ft. lbs.)

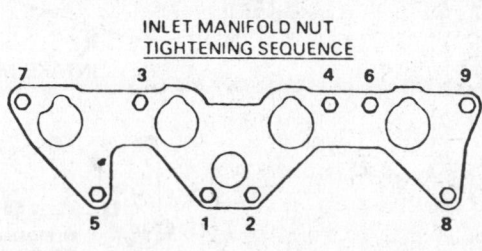

INLET MANIFOLD NUT TIGHTENING SEQUENCE

Intake manifold torque sequence—2.0L (VIN M) engine

4. Disconnect and tag PCV hose and vacuum hoses on the throttle body.

5. Remove the throttle cable and the cruise control cable, if equipped.

6. Remove wiring to the ignition coil and remove the manifold support bracket.

7. Remove the rear bolt from alternator bracket, power steering adjusting bracket and front alternator adjusting bracket.

8. Remove the fuel lines to the fuel rail and regulator outlet.

9. Remove the retaining nuts and washers and intake manifold.

To install:

10. Use a new gasket on the manifold surface and mount the manifold in position.

11. Tighten the bolts to 16 ft. lbs. (22 Nm) for the VIN K engine or 18 ft. lbs. (24 Nm) for the VIN M engine, in the correct sequence.

12. Reconnect the fuel lines. Install the bolt for the power steering adjusting bracket and the alternator.

13. Connect the ignition coil wiring and connect the vacuum hoses.

14. Connect the induction tube and hoses.

15. Reconnect all of the electrical wiring and the battery cable.

2.0L (VIN 1) and 2.2L Engines

1. Disconnect the negative battery cable and relieve fuel pressure.

2. Remove the air cleaner assembly.

3. Drain the coolant.

4. Remove and tag the vacuum lines and wires as necessary.

5. Disconnect the fuel line, TBI linkage and remove the TBI unit.

6. Remove the power steering pump and lay aside.

7. Disconnect the coolant hose at the manifold.

8. Raise and support the vehicle safely.

9. Disconnect the coolant pipe retaining nut, located at the top of the DIS, and move the pipe rearward.

10. Disconnect the accelerator and TV cables and bracket.

11. Remove the lower intake manifold nuts.

12. Lower the vehicle, remove the remaining intake manifold bolts and nuts and remove the manifold.

To install:

13. Use a new gasket on the manifold surface and mount the manifold in position.

14. Install the upper manifold bolts.

15. Raise and safely support the vehicle.

16. Install the lower manifold bolts.

17. Tighten the nuts to 15–22 ft. lbs. (20–30 Nm) in the correct sequence.

18. Connect the accelerator and TV cables and bracket.

19. Move the coolant pipe located at the top of the DIS into position and install the retaining nut.

20. Lower the vehicle.

21. Connect the coolant hose at the manifold.

22. Install the power steering pump.

23. Install the TBI unit, TBI linkage and connect the fuel line.

24. Connect the vacuum lines and wires.

25. Install the air cleaner assembly.

26. Connect the negative battery cable.

27. Fill cooling system and check for leaks. Start the engine and allow to come to normal operating temperature. Recheck for leaks. Top-up coolant.

2.8L and 3.1L Engines

1. Disconnect the negative battery cable and relieve fuel pressure. Remove the air cleaner inlet tube.

2. Disconnect the accelerator cable bracket at the plenum.

3. Disconnect the throttle body and the EGR pipe from the EGR valve. Remove the plenum assembly.

4. Disconnect the fuel line along the fuel rail.

5. Disconnect the serpentine drive belt. Remove the power steering pump mounting bracket.

6. Remove the heater pipe at the power steering pump bracket.

7. Tag and disconnect the wiring at the alternator, remove the alternator.

8. Disconnect the wires from the cold start injector assembly. Remove the injector assembly from the intake manifold.

9. Disconnect the idle air vacuum hose at the throttle body. Disconnect the wires at the injectors.

10. Remove the fuel rail, breather tube and the fuel runners from the engine.

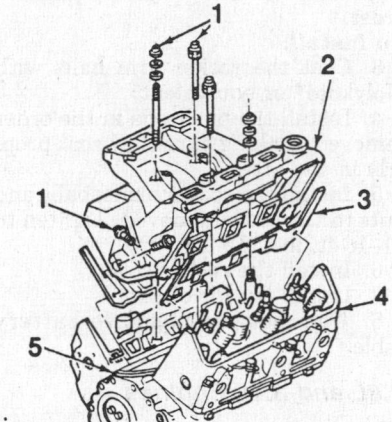

1.
Tighten in proper sequence to 15 ft. lbs. (20Nm), then retighten to 24 ft. lbs. (33Nm)

⑦ ④ ③ ⑥
⑧ ① ② ⑤

2. Intake manifold
3. Gasket
4. Cylinder head
5. Sealer

Intake manifold torque sequence—2.8L and 3.1L engines

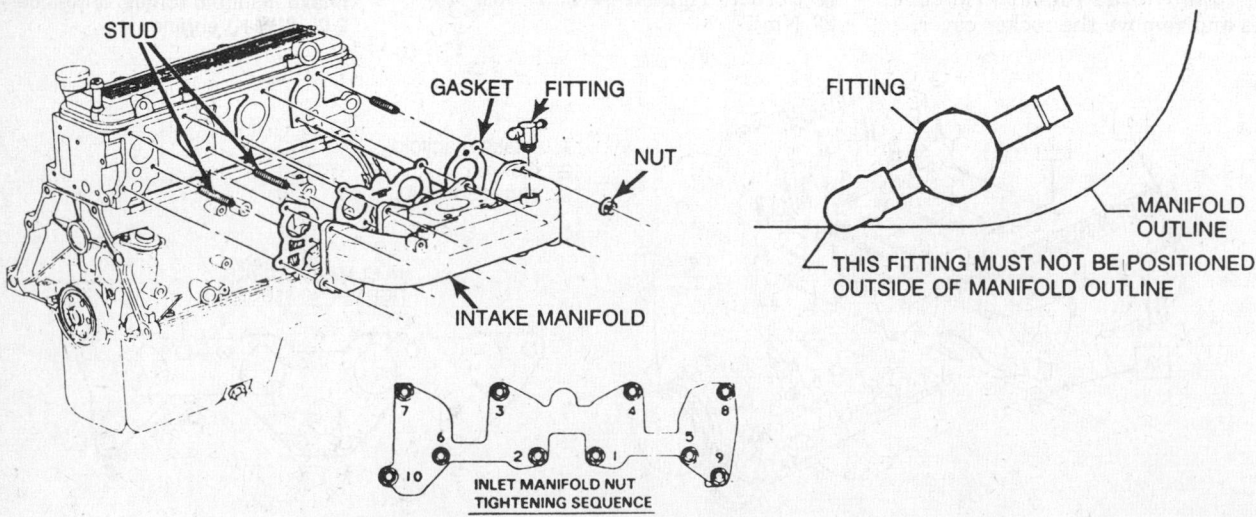

INLET MANIFOLD NUT TIGHTENING SEQUENCE

Intake manifold torque sequence—2.0L (VIN 1) and 2.2L engines

11. Tag and disconnect the coil wires.

12. Remove the rocker arm covers. Drain the cooling system, disconnect the radiator hose at the thermostat housing. Disconnect the heater hose from the thermostat housing and the thermostat wiring.

13. Remove the thermostat assembly housing.

14. Remove the intake manifold bolts and remove the intake manifold from the engine.

To install:

15. Apply a bead of sealant to the points where the manifold meets the block and install new gaskets. The gaskets are marked left and right.

16. Install the intake manifold assembly and tighten the bolts, in sequence, to 15 ft. lbs. (20 Nm), then retighten, in sequence, to 24 ft. lbs. (33 Nm).

17. Install the thermostat housing assembly. Install the rocker arm covers.

18. Reconnect the coil wires. Install the fuel rail, runners and the breather tube.

19. Install the alternator and connect the wiring. Connect the EGR tube to the EGR valve.

20. Install the power steering pump bracket and pump. Install the serpentine belt.

21. Connect the accelerator cable at the plenum and connect the negative battery cable.

22. Install the air cleaner inlet tube.

Exhaust Manifold

REMOVAL & INSTALLATION

2.0L (VIN K and M) Engine

1. Disconnect the negative battery cable.

2. Remove turbocharger induction tube, if equipped.

3. Remove and tag spark plug wires.

4. Remove turbocharger assembly from exhaust manifold, if equipped.

5. Remove exhaust manifold retaining nuts and manifold.

To install:

6. Install the exhaust manifold and retaining nuts. Tighten the exhaust manifold bolts to 16 ft. lbs. (22 Nm) and turbocharger-to-exhaust manifold to 18 ft. lbs. (24 Nm), if equipped.

NOTE: Tighten No. 2 and 3 exhaust manifold retaining nuts prior to No. 1 and 4.

7. Connect the spark plug wires.

8. If equipped, install the turbocharger induction tube.

9. Connect the negative battery cable.

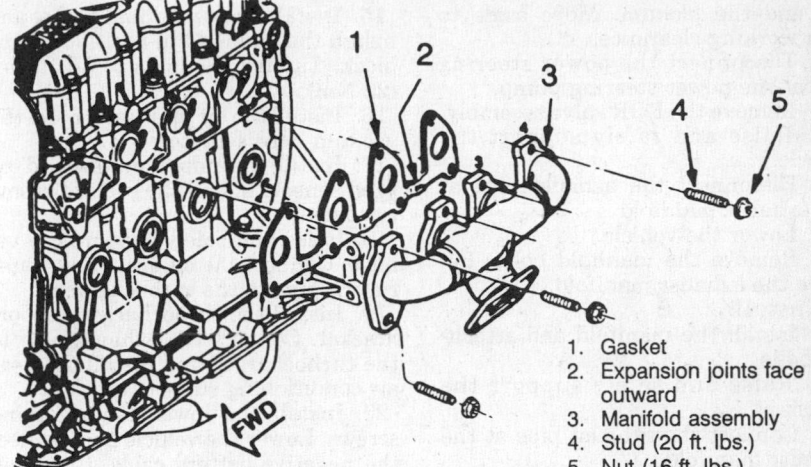

1. Gasket
2. Expansion joints face outward
3. Manifold assembly
4. Stud (20 ft. lbs.)
5. Nut (16 ft. lbs.)

Exhaust manifold torque sequence–2.0L (VIN M) engine

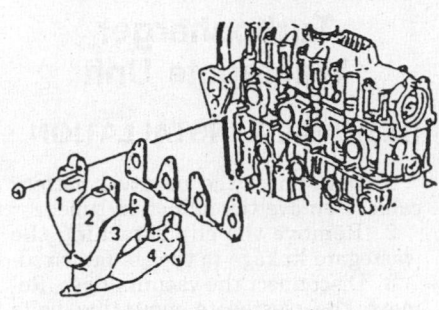

Exhaust manifold installation torque sequence—2.0L (VIN K) engine

2.0L (VIN 1) and 2.2L Engines

1. Disconnect the negative battery cable.

2. Remove the air cleaner. Remove the exhaust manifold shield. Raise and safely support the vehicle.

3. Disconnect the exhaust pipe at the manifold and lower the vehicle.

4. Disconnect the air management-to-check valve hose and remove the bracket. Disconnect the oxygen sensor lead wire.

5. Remove the serpentine belt. Remove the alternator adjusting bolts, loosen the pivot bolt and pivot the alternator upward.

6. Remove the alternator brace and the AIR pipes bracket bolt.

7. Unscrew the mounting bolts and remove the exhaust manifold. The manifold should be removed with the AIR plumbing as an assembly. If the manifold is to be replaced, transfer the plumbing to the new one.

To install:

8. Clean the mating surfaces on the manifold and the head.

9. Install the manifold and tighten the nuts to 6–13 ft. lbs. (8–18 Nm) in the proper sequence.

10. Install the alternator brace and the AIR pipes bracket bolt.

11. Rotate the alternator into posi-

tion and install the pivot bolt. Install the serpentine belt.

12. Connect the oxygen sensor lead wire. Install the air management-to-check valve bracket and connect the hose.

13. Raise and safely support the vehicle. Connect the exhaust pipe to the manifold.

14. Lower the vehicle. Install the exhaust manifold shield. Install the air cleaner.

15. Connect the negative battery cable.

2.8L and 3.1L Engines

LEFT SIDE

1. Disconnect the negative battery cable.

2. Remove the air cleaner assembly.

3. Remove the air flow sensor. Remove the engine heat shield.

4. Remove the crossover pipe at the manifold.

5. Remove the exhaust manifold bolts.

6. Remove the exhaust manifold.

To install:

7. Install the exhaust manifold.

8. Install the exhaust manifold bolts.

9. Install the crossover pipe at the manifold.

10. Install the engine heat shield. Install the air flow sensor.

11. Install the air cleaner assembly.

12. Connect the negative battery cable.

RIGHT SIDE

1. Disconnect the negative battery cable.

2. Remove the air cleaner assembly.

3. Remove the air flow sensor. Remove the engine heat shield.

4. Disconnect the crossover pipe at the manifold.

5. Disconnect the accelerator and throttle valve cable at the throttle le-

ver and the plenum. Move aside to gain working clearance.

6. Disconnect the power steering line at the power steering pump.

7. Remove the EGR valve assembly.

8. Raise and safely support the vehicle.

9. Disconnect the exhaust pipe at the exhaust manifold.

10. Lower the vehicle.

11. Remove the manifold bolts. Remove the exhaust manifold.

To install:

12. Install the manifold and attaching bolts.

13. Raise and safely support the vehicle.

14. Connect the exhaust pipe at the exhaust manifold.

15. Lower the vehicle.

16. Install the EGR valve assembly.

17. Connect the power steering line at the power steering pump.

18. Connect the accelerator and throttle valve cable at the throttle lever and the plenum.

19. Connect the crossover pipe at the manifold.

20. Install the engine heat shield. Install the air flow sensor.

21. Install the air cleaner assembly.

22. Connect the negative battery cable.

Turbocharger

REMOVAL & INSTALLATION

1. Disconnect the negative battery cable.

2. Raise and safely support the vehicle.

3. Remove the Lower fan retaining screws.

4. Disconnect the exhaust pipe at the turbocharger.

5. Remove air conditioning rear support bracket.

6. Remove the turbocharger support bracket from the engine.

7. Disconnect the oil drain and water return pipes at turbo.

8. Lower the vehicle and remove coolant recovery pipe.

9. Remove induction tube, coolant fan and oxygen sensor.

10. Disconnect the oil and water feed pipes.

11. Remove the air intake duct and vacuum hose at the actuator.

12. Remove the exhaust manifold retaining nuts, remove the turbocharger and manifold as an assembly.

13. Remove the turbocharger from exhaust manifold.

To install:

14. Install the turbocharger to the exhaust manifold and tighten the bolts to 18 ft. lbs. (24 Nm).

15. Install a new manifold gasket and install the manifold in position on the block. Tighten the bolts to 16 ft. lbs. (22 Nm).

16. Install the air intake duct and the vacuum hose actuator.

17. Install the induction tube and oxygen sensor. Install the coolant recovery tube.

18. Raise and safely support the vehicle. Connect the oil drain and water return pipe to the turbocharger.

19. Install the turbocharger support bracket. Connect the exhaust pipe to the turbocharger and install the rear air conditioning support bracket.

20. Install the lower fan retaining screws. Lower the vehicle and connect the negative battery cable. Check all fluid levels.

Turbocharger Wastegate Unit

REMOVAL & INSTALLATION

1. Disconnect the negative battery cable. Remove the induction tube.

2. Remove the clip attaching the wastegate linkage to the actuator rod.

3. Disconnect the vacuum hose. Remove the wastegate mounting bolts and remove the wastegate actuator.

To install:

4. Install the wastegate actuator and mounting bolts.

5. Connect the vacuum hose.

6. Install the clip attaching the wastegate linkage to the actuator rod.

7. Install the induction tube.

8. Connect the negative battery cable.

Timing Chain Front Cover

REMOVAL & INSTALLATION

2.0L (VIN 1) and 2.2L Engines

NOTE: The following procedure requires the use of a front cover centering tool J–35468 and crankshaft puller J–24420–B.

1. Disconnect the negative battery cable. Remove the serpentine belt.

2. Although not absolutely necessary, removal of the right front inner fender splash shield will facilitate access to the front cover.

3. Remove the center bolt from the crankshaft pulley and retaining bolts, remove the pulley. Using puller J–24420–B or equivalent, remove hub from the crankshaft.

4. Remove the alternator lower bracket.

5. Remove the oil pan-to-front cover bolts.

6. Remove the front cover-to-block bolts and remove the front cover. If the front cover is difficult to remove, use a rubber mallet to loosen it.

To install:

7. The surfaces of the block and front cover must be clean and free of oil. Apply a ⅛ in. bead of RTV sealant to the cover. The sealant must be wet to the touch when the bolts are torqued down.

NOTE: When applying RTV sealant to the front cover, be sure to keep it out of the bolt holes. When installing hub or pulley note position of key on crankshaft.

8. Position the front cover on the block using a centering tool J–35468 and tighten the screws.

9. Install the oil pan-to-front cover bolts.

10. Install the alternator lower bracket.

11. Using hub installer J–29113 or equivalent, install the crankshaft hub.

12. If removed, install the right front inner fender splash shield.

13. Install the serpentine belt.

14. Connect the negative battery cable.

2.8L and 3.1L Engines

1. Disconnect the negative battery cable.

2. Drain the cooling system and remove the coolant recovery tank from the vehicle.

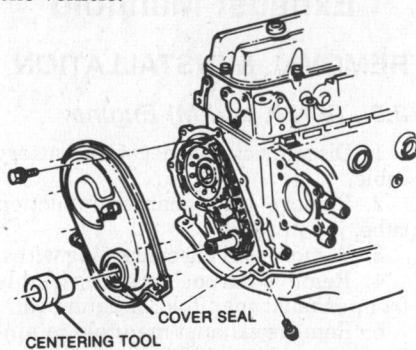

Front cover removal—2.0L (VIN 1) and 2.2L engines

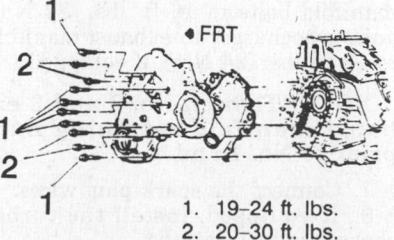

1. 19–24 ft. lbs.
2. 20–30 ft. lbs.

Front cover removal—2.8L and 3.1L engines

3. Disconnect the MAP sensor and EGR sensor solenoids.

4. Remove the serpentine belt and adjusting pulley.

5. Tag and disconnect the heater hose at the power steering bracket.

6. Tag and disconnect the alternator wiring and remove the alternator.

7. Raise the vehicle and support it safely.

8. Remove the inner fender splash shield.

9. Remove the harmonic balancer with tool J–24420 or equivalent puller.

10. Remove the oil pan-to-block bolts and remove the oil pan. Remove the lower cover bolts.

11. Lower the vehicle and disconnect the radiator hoses at the water pump.

12. Remove the heater hose from the thermostat housing.

13. Disconnect the overflow hoses and the canister purge hose.

14. Remove the front cover.

To install:

15. Apply a bead of sealer to the front cover surface.

16. Install a new front cover gasket and front oil seal.

17. Install the front cover and tighten to 20–28 ft. lbs. (27–38 Nm).

18. Raise and safely support the vehicle. Install the oil pan and the lower front cover bolts.

19. Install the crankshaft balancer.

20. Install the inner splash shield and lower the vehicle.

21. Install the radiator hoses and the power steering pump.

22. Install the alternator and the accessory drive belt.

23. Refill the fluids and connect the negative battery cable.

Front Cover Oil Seal

REPLACEMENT

2.0L (VIN 1), 2.2L, 2.8L and 3.1L Engines

1. The oil seal can be replaced with the front cover either on or off the engine.

2. Although not absolutely necessary, removal of the right front inner fender splash shield will facilitate access to the front cover.

3. If the cover is on the engine, remove the crankshaft pulley and hub first.

4. Pry out the seal using a suitable tool, being careful not to distort the seal mating surfaces.

To install:

5. Install the new seal so the lip side, is towards the engine.

6. Press it into place with a seal driver.

7. Install the hub and pulley, if removed.

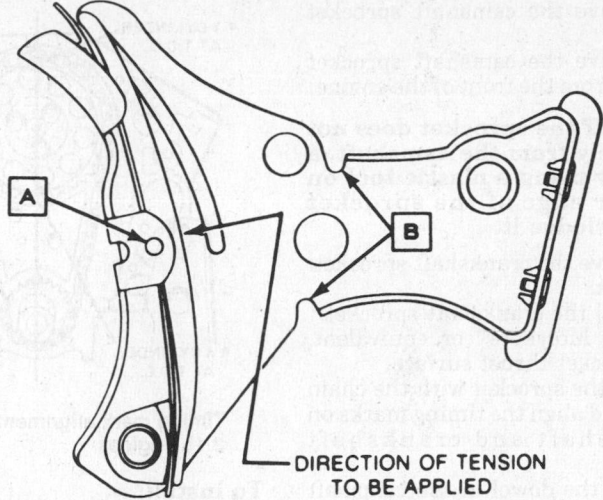

DIRECTION OF TENSION TO BE APPLIED

A INSERT PIN AFTER TENSION HAS BEEN APPLIED

B TABS, USED FOR CAMSHAFT AND CRANKSHAFT ALIGNMENT

Timing chain tensioner—2.0L (VIN 1) and 2.2L engines.

Timing Chain and Sprockets

REMOVAL & INSTALLATION

2.0L (VIN 1) and 2.2L Engines

1. Disconnect the negative battery cable. Remove the front cover.

2. Place the No. 1 piston at **TDC** of the compression stroke so the marks on the camshaft and crankshaft sprockets are in alignment.

3. Loosen the timing chain tensioner nut as far as possible, without actually removing it.

4. Remove the camshaft sprocket bolts and remove the sprocket and chain together. If the sprocket does not slide from the camshaft easily, a light blow with a soft tool at the lower edge of the sprocket will loosen it.

5. Use a gear puller J–22888–20 or equivalent, and remove the crankshaft sprocket.

To install:

6. Press the new crankshaft sprocket onto the crankshaft using crankshaft sprocket installer J–5590 or equivalent.

NOTE: Ensure that the sprocket is fully seated against the crankshaft.

7. Compress the tensioner spring. Insert a cotter pin or nail, into the hole to retain the tensioner.

8. Align the crankshaft and camshaft timing marks with the tabs on the chain tensioner.

9. Install the timing chain over the camshaft sprocket and around the

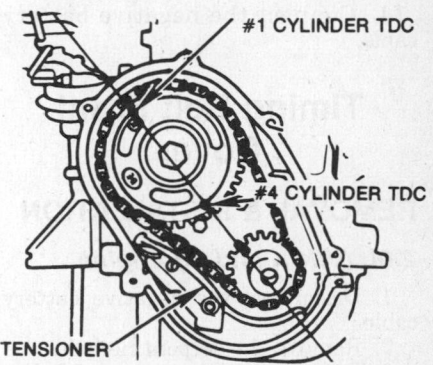

#1 CYLINDER TDC

#4 CYLINDER TDC

TENSIONER

Timing mark alignment—2.0L (VIN 1) and 2.2L engines

crankshaft sprocket. Make sure the marks on the 2 sprockets are in alignment. Lubricate the thrust surface with Molykote® or equivalent.

10. Align the dowel in the camshaft with the dowel hole in the sprocket and install the sprocket onto the camshaft. Use the mounting bolt to draw the sprocket onto the camshaft and tighten to 77 ft. lbs. (105 Nm).

11. Lubricate the timing chain with clean engine oil. Remove the timing chain pin.

12. Install the front cover.

13. Connect the negative battery cable.

2.8L and 3.1L Engines

1. Disconnect the negative battery cable.

2. Remove the front cover.

3. Position the No. 1 piston at **TDC** with the marks on the crankshaft and camshaft sprockets aligned.

4. Remove the camshaft sprocket bolts.

5. Remove the camshaft sprocket and chain from the front of the engine.

NOTE: If the sprocket does not move freely from the camshaft, a light blow using a plastic tool on the lower edge of the sprocket should dislodge it.

6. Remove the crankshaft sprocket.

To install:

7. Install the crankshaft sprocket.

8. Apply Molykote® or equivalent, to the sprocket thrust surface.

9. Hold the sprocket with the chain hanging and align the timing marks on the camshaft and crankshaft sprockets.

10. Align the dowel in the camshaft with the dowel hole in the camshaft sprocket.

11. Draw the camshaft sprocket onto the camshaft using the mounting bolts. Tighten to 18 ft. lbs. (24 Nm).

12. Lubricate the timing chain with engine oil.

13. Install the front cover.

14. Connect the negative battery cable.

Timing Belt Front Cover

REMOVAL & INSTALLATION

2.0L (VIN K and M) Engine

1. Disconnect the negative battery cable.

2. Remove the serpentine belt.

3. Remove the tensioner and bolt.

4. On 1988 vehicles, unsnap the upper cover first. Unsnap the lower the cover.

5. On 1989–92 vehicles, remove the timing belt cover attaching bolts. Remove the cover.

To install:

6. On 1988 vehicles, snap the lower cover into position. Snap the upper cover into position.

7. On 1989–92 vehicles, install the timing belt cover and attaching bolts. Tighten to 89 inch lbs. (11 Nm).

8. Install the serpentine belt. Tighten the timing belt tensioner to 40 ft. lbs. (54 Nm).

9. Connect the negative battery cable.

OIL SEAL REPLACEMENT

2.0L (VIN K and M) Engine

1. Remove the crankshaft sprocket.

2. Remove the crankshaft key and rear thrust washer.

3. Using a suitable prybar, pry out the front oil seal.

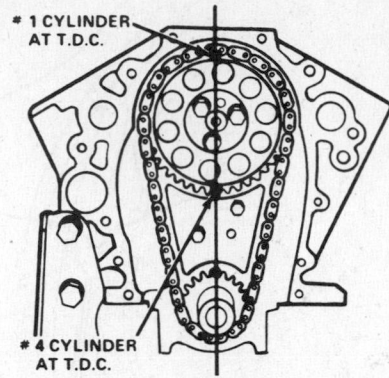

Timing mark alignment—2.8L and 3.1L engines

To install:

4. Place the protective sleeve of special tool set J–33083, seal installer or equivalent, onto the crankshaft.

5. Lubricate the lip of the new seal. Using special tool J–33083, install the seal.

6. Remove the protective sleeve.

7. Install the rear thrust washer and key on the crankshaft.

8. Install the crankshaft sprocket.

Timing Belt and Tensioner

REMOVAL & INSTALLATION

2.0L (VIN K and M) Engine

1. Disconnect the negative battery cable.

2. Remove the serpentine belt and timing belt cover.

3. Loosen the water pump bolts and release tension with tool J–33039 or equivalent.

4. Raise and support the vehicle safely.

5. Remove the crankshaft pulley.

6. Lower the vehicle and remove the timing belt.

To install:

7. Turn the crankshaft and the camshaft gears clockwise to align the timing marks on the gears with the timing marks on the rear cover.

8. Install the timing belt, making sure the portion between the camshaft gear and crankcase gear is in tension.

9. Using tool J–33039 or equivalent, turn the water pump eccentric clockwise until the tensioner contacts the high torque stop. Tighten the water pump screws slightly.

10. Turn the engine by the crankshaft gear bolt 720 degrees to fully seat the belt into the gear teeth.

11. Turn the water pump eccentric counterclockwise until the hole in the tensioner arm is aligned with the hole in the base.

12. Tighten the water pump screws to 18 ft. lbs. (24 Nm) while checking that the tensioner holes remain as adjusted in the prior step.

13. Install the crankshaft pulley, timing belt cover and the serpentine drive belt.

Timing Sprockets

REMOVAL & INSTALLATION

Camshaft Sprocket

1. Disconnect the negative battery cable.

2. Remove the camshaft carrier cover.

3. Remove the timing belt.

4. Hold the camshaft with an open end wrench and remove the sprocket bolt, washer and and sprocket.

To install:

5. Install the sprocket, retaining bolt and washer with the mark on the sprocket lined up with the mark on the rear timing belt cover. Tighten to 77 ft. lbs. (104 Nm) for 2.0L (VIN 1) and 2.2L engines or 34 ft. lbs. (46 Nm) for all others.

6. Install the timing belt.

7. Install the camshaft carrier cover.

8. Connect the negative battery cable.

Crankshaft Sprocket

1. Disconnect the negative battery cable.

2. Remove the timing belt.

3. Remove the crankshaft pulley.

4. Remove the bolt and retaining washer and remove the sprocket.

To install:

5. Install the sprocket over the key on the end of the crankshaft.

6. Install the thrust washer and attaching bolt and tighten to 114 ft. lbs. (155 Nm).

7. Install the crankshaft pulley and timing belt.

8. Connect the negative battery cable.

Camshaft Carrier

REMOVAL & INSTALLATION

2.0L (VIN K and M) Engine

NOTE: Whenever the camshaft carrier bolts are loosened, it is necessary to remove the cylinder head and replace the cylinder head gasket.

1. Disconnect the negative battery cable. Disconnect the crankcase ventilation hose from the camshaft carrier.

2. Mark and remove the distributor.

3. Remove the camshaft sprocket.

4. Loosen the camshaft carrier and cylinder head attaching bolts a little at a time in sequence.

NOTE: Camshaft carrier and cylinder head bolts should be loosened in sequence and only when the engine is cold.

5. Remove the camshaft carrier.
6. Remove the camshaft thrust plate from the rear of the camshaft carrier.
7. Slide the camshaft rearward and remove it from the carrier.
8. Remove the carrier front oil seal.
To install:
9. Install a new carrier front oil seal using tool J–33085.
10. Place the camshaft in the carrier.

NOTE: Take care not to damage the carrier front oil seal when installing the camshaft.

11. Install the camshaft thrust plate and the retaining bolts. Tighten the bolts to 70 inch lbs. (8 Nm).
12. Check the camshaft endplay which should be within 0.016–0.064 in. (0.04–0.16mm).
13. Clean the sealing surfaces on cylinder head and carrier. Apply a continuous 3mm bead of RTV sealer.
14. Install the camshaft carrier on the cylinder head.
15. Install the camshaft carrier and cylinder head attaching bolts.
16. Tighten the bolts a little at a time, in the proper sequence, at cylinder head, to 18 ft. lbs. (24 Nm). Turn each bolt 60 degrees clockwise, in the proper sequence, for 3 times until a 180 degrees rotation is obtained or equivalent, to ½ turn.
17. Install the camshaft sprocket.
18. Install the distributor.
19. Connect the positive crankcase ventilation hose to the camshaft carrier.

NOTE: After remainder of installation is completed, start engine and let it run until the thermostat opens. Tighten all cylinder head bolts an additional 30–50 degrees in the proper sequence.

Camshaft

REMOVAL & INSTALLATION

2.0L (VIN 1) and 2.2L Engines

1. Remove the engine assembly.
2. Remove the intake manifold.
3. Remove the cylinder head cover, pivot the rocker arms to the sides and remove the pushrods, keeping them in order. Remove the valve lifters, keeping them in order.
4. Remove the front cover.
5. Remove the distributor.

6. Remove the fuel pump and its pushrod.
7. Remove the timing chain and sprocket.
8. Carefully pull the camshaft from the block, being sure the camshaft lobes do not contact the bearings.
To install:
9. Lubricate the camshaft journals with clean engine oil. Lubricate the lobes with Molykote® or equivalent. Install the camshaft into the engine, being extremely careful not to contact the bearings with the cam lobes.
10. Install the timing chain and sprocket. Install the fuel pump and pushrod. Install the timing cover. Install the distributor.
11. Install the valve lifters. If a new camshaft has been installed, new lifters should be used to ensure durability of the cam lobes.
12. Install the pushrods and rocker arms and the intake manifold.
13. Install the engine assembly.
14. Install the cylinder head cover.

2.0L (VIN K and M) Engine

1. Disconnect the negative battery cable. Remove the camshaft carrier cover.
2. Hold the valves in place with compressed air, using an air adapter J–22794 or equivalent, in the spark plug hole. Compress the valve springs with a special tool J–33302–25 and remove rocker arms. Keep rocker arms in order for reassembly.
3. Remove the timing belt front cover.
4. Remove the timing belt.
5. Remove the camshaft sprocket.
6. Mark and remove the distributor.
7. Remove the camshaft thrust plate from rear of camshaft carrier.
8. Slide the camshaft rearward and remove it from the carrier.
To install:
9. Install a new camshaft carrier front oil seal using tool J–33085 or equivalent.
10. Place the camshaft in the carrier.

NOTE: Take care not to damage the carrier front oil seal when installing the camshaft.

11. Install the camshaft thrust plate retaining bolts. Tighten bolts to 70 inch lbs. (8 Nm).
12. Check the camshaft endplay, which should be within 0.016–0.064 in.
13. Install the distributor.
14. Install the camshaft sprocket.
15. Install the timing belt.
16. Install the timing belt front cover.
17. Using an air adapter J–22794 or equivalent, in the spark plug hole to hold the valve closed and install valve

train compressing fixture J–33302. Compress valve springs and replace rocker arms.
18. Install the camshaft carrier cover.

2.8L and 3.1L Engines

1. Disconnect the negative battery cable. Remove the engine assembly from the vehicle.
2. Remove the intake manifold.
3. Remove the rocker arm covers. Remove the rocker arm nuts, balls, rocker arms, pushrods and lifters.

NOTE: Always keep valve train parts in order for correct installation.

4. Remove the upper front cover bolts. Remove the lower cover bolts and the front cover.
5. Remove the camshaft sprocket bolts, camshaft sprocket and timing chain.
6. Remove the camshaft by carefully sliding it out the front of the engine. Measure the camshaft bearing journals using a micrometer and replace the camshaft if the journals exceed 0.0009 in. (0.025mm) out of round.
To install:

NOTE: When installing a new camshaft, lubricate the camshaft lobes with GM Engine Oil Supplement (E.O.S.) 1052367 or equivalent.

7. Install the camshaft.
8. Install the timing chain and sprocket.
9. Install the intake manifold.
10. Install the crankcase front cover.
11. Install the lifters, pushrods, rocker arms, balls and rocker arm nuts. Install the valve covers.
12. Install the engine assembly to the vehicle.
13. Connect the negative battery cable.

Piston and Connecting Rod
POSITIONING

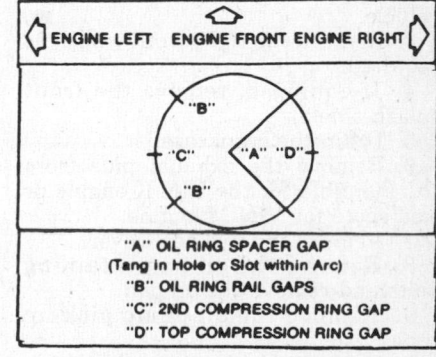

"A" OIL RING SPACER GAP
(Tang in Hole or Slot within Arc)
"B" OIL RING RAIL GAPS
"C" 2ND COMPRESSION RING GAP
"D" TOP COMPRESSION RING GAP

Piston ring gap locations

ENGINE LUBRICATION

Oil Pan

REMOVAL & INSTALLATION

2.0L (VIN 1) and 2.2L Engines

1. Disconnect the negative battery cable.
2. Raise and safely support the vehicle. Drain the crankcase.
3. Remove the air conditioning brace, if equipped.
4. Remove the exhaust shield and disconnect the exhaust pipe at the manifold.
5. Remove the starter motor and position it out of the way.
6. Remove the flywheel cover. Remove the oil pan retaining bolts and remove the oil pan.

To install:

NOTE: Prior to oil pan installation, check the sealing surfaces on the pan, cylinder block and front cover are clean and free of oil. If installing the old oil pan, be sure all old RTV has been removed.

7. Apply a ⅛ in. bead of RTV sealant to the oil pan sealing surface. Use a new oil pan rear seal and install the pan in place. Tighten the bolts to 9–13 ft. lbs. (12–18 Nm).
8. Install the flywheel cover and the starter.
9. Connect the exhaust pipe at the manifold.
10. Install the exhaust shield and install the air conditioning brace.
11. Connect the negative battery cable and run the vehicle to normal operating temperature. Refill and check for leaks.

2.0L (VIN K and M) Engine

1. Disconnect the negative battery cable.
2. Raise and safely support the vehicle.
3. Remove the right front wheel, as required.
4. If equipped, remove the front splash shield.
5. Drain the crankcase.
6. Remove the exhaust pipe from the manifold for the VIN K engine or wastegate for VIN M engine.
7. Remove the flywheel cover.
8. Remove the oil pan retaining bolts and remove the oil pan.
9. Remove the oil pump pick-up tube.
10. Remove the scraper and gasket.

To install:
11. Apply a bead of RTV sealant to the oil pan. Using a new gasket, install the oil scraper.
12. Install the pan and attaching bolts with Loctite®. Tighten the oil pan bolts to 44 inch lbs. (5 Nm).
13. Install the oil pan cover.
14. Install the flywheel cover.
15. Connect the exhaust pipe to the manifold for the VIN K engine or wastegate for the VIN M.
16. Lower the vehicle.
17. If equipped, install the splash shield.
18. If removed, install the right front wheel.
19. Lower the vehicle.
20. Fill the crankcase with oil.
21. Connect the negative battery cable.

2.8L and 3.1L Engines

1988

1. Disconnect the negative battery cable.
2. Raise and support the vehicle safely.
3. Drain the engine oil.
4. Remove the flywheel shield or clutch housing cover and remove the starter.
5. Remove the oil pan bolts and remove the oil pan.

To install:
6. Clean the gasket mating surfaces.
7. Install a new gasket on the oil pan. Apply silicon sealer to the portion of the pan that contacts the rear of the block.
8. Install the oil pan, nuts and retaining bolts.
9. Install the flywheel shield or clutch housing cover and install the starter.
10. Lower the vehicle, fill the crankcase and connect the battery.

1989–92

1. Disconnect the negative battery cable.
2. Remove the serpentine belt and the tensioner.
3. Support the engine with tool J–28467 or equivalent.
4. Raise and safely support the vehicle. Drain the engine oil.
5. Remove the starter shield and the flywheel cover. Remove the starter.
6. Remove the engine to frame mount retaining nuts.
7. Lower the vehicle.
8. Support the engine using tool J–28467–A or equivalent, then raise and support the vehicle safely.
9. Remove the right tire and wheel assembly. Remove the right inner fender splash shield.

10. Remove the oil pan retaining bolts and nuts and remove the oil pan.
To install:
11. Clean the gasket mating surfaces.
12. Install a new gasket on the oil pan. Apply silicon sealer to the portion of the pan that contacts the rear of the block.
13. Install the oil pan retaining nuts. Tighten the nuts to 71 inch lbs. (8 Nm).
14. Install the oil pan retaining bolts. Tighten the rear bolts to 18 ft. lbs. (24 Nm) and the remaining bolts to 71 inch lbs. (8 Nm).
15. Install the right inner fender splash shield.
16. Lower the vehicle and remove the engine support tool.
17. Raise and support the vehicle safely.
18. Install the engine to frame mounting nuts.
19. Install the starter and splash shield. Install the flywheel shield.
20. Lower the vehicle and fill the crankcase with oil, install the belt tensioner and belt and connect the negative battery cable. Run the engine to normal operating temperature and check for leaks.

Oil Pump

REMOVAL & INSTALLATION

2.0L (VIN K and M) Engine

1. Disconnect the negative battery cable. Remove the crankshaft sprocket.
2. Remove the timing belt rear cover.
3. Disconnect the connector at oil pressure switch.
4. Raise and safely support the vehicle. Drain the engine oil and remove the oil pan.
5. Remove the oil filter.
6. Unbolt and remove the oil pick-up tube.
7. Unbolt and remove the oil pump.
To install:
8. Install the pump using a new gasket. Tighten attaching bolts to 5 ft. lbs. (7 Nm).
9. Install the pick-up tube and support with new O-ring.
10. Install the oil pan.
11. Use seal installer J-33083 or equivalent, to install new front oil seal.
12. Install a new oil filter.
13. Connect the oil pressure switch connector.
14. Install the rear timing belt cover.
15. Install the crankshaft sprocket.
16. Connect the negative battery cable.

2.0L (VIN 1) and 2.2L Engines

1. Disconnect the negative battery cable.
2. Raise and safely support the vehicle. Drain the engine oil and remove the engine oil pan.
3. Remove the pump attaching bolts and carefully lower the pump, extension shaft and retainer.

To install:

NOTE: Heat the retainer in hot water prior to assembling the extension shaft.

4. Install extension shaft to the oil pump.

NOTE: To ensure immediate oil pressure on start-up, the oil pump gear cavity should be packed with petroleum jelly.

5. Install the pump to the rear bearing cap and bolt. Tighten to 32 ft. lbs. (43 Nm).
6. Install the oil pan.
7. Lower the vehicle.
8. Fill the crankcase with oil.
9. Connect the negative battery cable.

2.8L and 3.1L Engines

1. Disconnect the negative battery cable.
2. Raise and safely support the vehicle. Drain the engine oil and remove the oil pan.
3. Remove the rear main bearing cap.
4. Remove the oil pump and extension shaft.

To install:

5. Install the oil pump and extension shaft.
6. Engage the drive shaft extension into the drive gear.
7. Install the pump to the rear bearing cap and install the bolt. Tighten to 30 ft. lbs. (41 Nm).
8. Install the oil pan.
9. Lower the vehicle.
10. Fill the crankcase with clean oil.
11. Connect the negative battery cable.

CHECKING

2.0L (VIN 1), 2.2L, 2.8L and 3.1L Engines

1. Drain the oil from the pump and remove the pump cover.
2. Measure the pump gear lash. It should be 0.0037–0.0077 in.
3. Measure the pump gear pocket. It should be as follows:
   a. On pumps with aluminum body the depth should be 1.195–1.198 in. and the diameter should be 1.503–1.506 in.

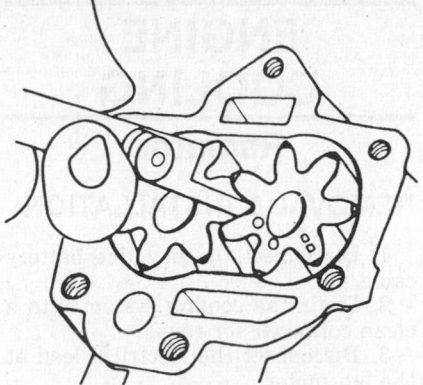

Measuring oil pump gear lash

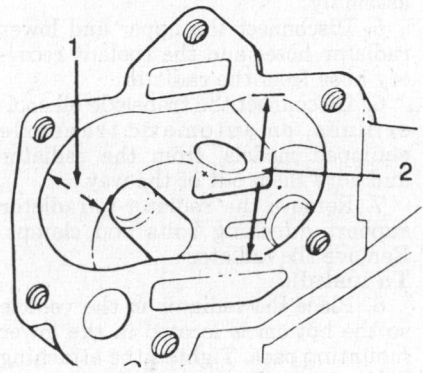

1. Depth of pocket
2. Diameter of pocket

Measuring oil pump gear pocket

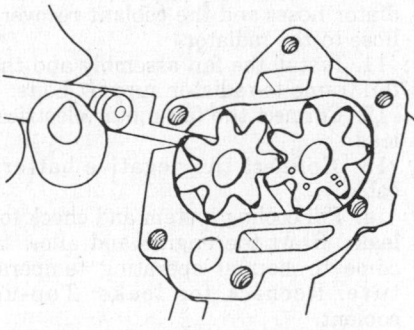

Measuring oil pump gear side clearance

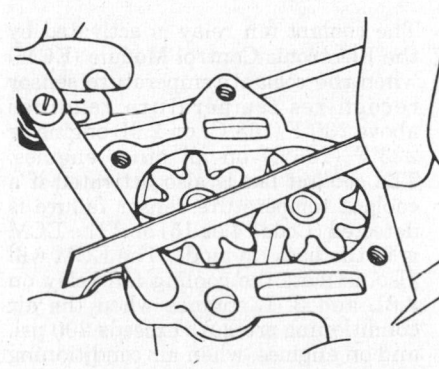

Measuring oil pump end clearance

   b. On pumps with cast iron body the depth should be 1.202–1.205 in. and the diameter should be 1.504–1.506 in.
4. Measure the gear side clearance. It should be 0.003–0.004 in.
5. Measure the gear end clearance. It should be as follows:
   a. On pumps with aluminum body the clearance should be 0.0016–0.0067 in.
   b. On pumps with cast iron body the clearance should be 0.002–0.006 in.
6. Lubricate all internal parts with engine oil during reassembly and install the pump gears.
7. Prime the engine oil galleries by removing the engine oil pump drive unit and rotate the oil pump using a drill motor and appropriate socket and extension.
8. Install the cover and gasket and tighten the pump cover bolts to 89 inch lbs. (10 Nm).

NOTE: Use only original equipment gaskets. The gasket thickness is critical to proper functioning of the pump.

Rear Main Bearing Oil Seal

REMOVAL & INSTALLATION

2.0L (VIN K and M) Engine

NOTE: The rear main bearing oil seal is a 1 piece unit and can be replaced without the removal of the oil pan or crankshaft.

1. Disconnect the negative battery cable. Raise and safely support the vehicle. Remove the transaxle.
2. If equipped with a manual transaxle, remove the pressure plate and clutch disc.
3. Remove the flywheel/flexplate-to-crankshaft bolts and the flywheel/flexplate. Discard the bolts.

NOTE: Flywheel bolts cannot be reused.

4. Using a medium prybar, pry out the old seal. Be careful not to scratch the crankshaft surface.

To install:

5. Clean the block and crankshaft-to-seal mating surfaces.
6. Using the seal installation tool no. J-36227 or equivalent, install the new rear seal into the seal retainer. Lubricate the outside of the seal to aid installation and press the seal in evenly with the tool.
7. If equipped with automatic transaxle, install the flexplate and attaching bolts. Tighten attaching bolts

to 48 ft. lbs. (65 Nm). Flexplate bolts may be reused.

8. If equipped with manual transaxle, install the flywheel using new bolts. Tighten to 48 ft. lbs. (65 Nm), plus and additional 30 degrees.

9. If equipped with manual transaxle, install the pressure plate and disc.

10. Install the transaxle.

11. Lower the vehicle.

12. Connect the negative battery cable.

2.0L (VIN 1) and 2.2L Engines

1. Disconnect the negative battery cable. Raise and safely support the vehicle.

2. Remove the transaxle assembly.

3. Remove the flywheel/flexplate.

4. Remove the seal from the dust lip.

To install:

5. Clean the cylinder block and crankshaft sealing surface.

6. Inspect the crankshaft for damage. Coat the seal and engine mating surface with engine oil.

7. Install the new seal using seal installation tool J–34686 or equivalent.

8. Install the flywheel/flexplate.

9. Install the transaxle assembly.

10. Lower the vehicle.

11. Connect the negative battery cable.

2.8L and 3.1L Engines

NOTE: The rear main bearing oil seal is a 1 piece unit and can be replaced without the removal of the oil pan or crankshaft.

1. Disconnect the negative battery cable. Raise and safely support the vehicle.

2. Remove the transaxle assembly.

3. Remove the flywheel/flexplate.

4. Remove the seal from the dust lip.

NOTE: Care must be exercised during removal so as not to damage the crankshaft outside diameter area.

To install:

5. Clean the cylinder block and crankshaft sealing surface.

6. Inspect the crankshaft for nicks, burrs, scratches, etc.

7. Coat the seal and the engine mating surface with engine oil.

8. Install the new seal, using seal installation tool J–34686 or equivalent.

9. Install the flywheel/flexplate.

10. Install the transaxle assembly.

11. Lower the vehicle.

12. Connect the negative battery cable.

ENGINE COOLING

Radiator

REMOVAL & INSTALLATION

1. Disconnect the negative battery cable.

2. Drain the cooling system into a clean container for reuse.

3. Disconnect the electrical lead at the fan motor.

4. Remove the fan frame-to-radiator support bolts and remove the fan assembly.

5. Disconnect the upper and lower radiator hoses and the coolant recovery hose from the radiator.

6. Disconnect the transaxle oil cooler lines, on automatic transaxle equipped models, from the radiator and wire them out of the way.

7. Remove the radiator-to-radiator support attaching bolts and clamps. Remove the radiator.

To install:

8. Place the radiator in the vehicle so the bottom is located in the lower mounting pads. Tighten the attaching bolts and clamps.

9. In equipped with automatic transaxle, connect the transaxle oil cooler lines and tighten the bolts to 20 ft. lbs. (27 Nm).

10. Connect the upper and lower radiator hoses and the coolant recovery hose to the radiator.

11. Install the fan assembly and the fan frame-to-radiator support bolts.

12. Connect the fan motor electrical lead.

13. Connect the negative battery cable.

14. Fill cooling system and check for leaks. Start the engine and allow to come to normal operating temperature. Recheck for leaks. Top-up coolant.

Electric Cooling Fan

The coolant fan relay is activated by the Electronic Control Module (ECM) when the coolant temperature sensor recognizes temperature readings above 230°F (108°C) on 2.0L engine or 223°F (106°C) on all other engines. The coolant fan is also activated if a coolant temperature sensor failure is detected (Code 14 or 15) or if the ECM is in the back up mode. The ECM will also activate the cooling fan relay on 2.8L and 3.1L engines when the air conditioning pressure exceeds 200 psi. and on engines, when air conditioning is turned on and low pressure switch is closed.

NOTE: The ECM controls the cooling fan by grounding CKT 335 green/yellow wire. Once the ECM turns the fan relay on, it will keep fan on for a minimum of 30 seconds, or until vehicle speed exceeds 70 mph on the 2.8L and 3.1L engine.

TESTING

NOTE: If the fan does not run while connected to the electrical wiring connector, inspect for a defective coolant temperature switch or air conditioning relay, if equipped. Always check body wiring for frayed or loose connections.

1. Disconnect the electrical wiring connector from the electric cooling fan.

2. Using a 14 gauge jumper wire, connect it between the fan and the positive battery terminal; the fan should run.

3. If the fan does not run when connected to the jumper wire, replace the fan assembly.

REMOVAL & INSTALLATION

1. Disconnect the negative battery cable.

2. Tag and disconnect the wiring harness from the fan frame and motor assembly.

3. Remove the fan assembly attaching bolts. Remove the fan and motor assembly from the vehicle.

To install:

4. Install the fan and motor assembly. Install the attaching bolts.

5. Connect the electrical connectors.

6. Connect the negative battery cable.

Heater Core

REMOVAL & INSTALLATION

Without Air Conditioning

1. Disconnect the negative battery cable and drain the cooling system.

2. Remove the heater hoses at the heater core.

3. Remove the heater outlet deflector.

4. Remove the heater core cover retaining screws. Remove the heater core cover.

5. Remove the heater core retaining straps and remove the heater core.

To install:

6. Install the new heater core and retaining straps.

7. Install the heater outlet deflector and heater core cover.

8. Connect the heater hoses to the core.

9. Fill and bleed the cooling system when finished. Check for leaks and the heater operation.

With Air Conditioning

1. Disconnect the negative battery cable and drain the cooling system.

2. Raise and safely support the vehicle.

3. Disconnect the drain tube from the heater case.

4. Remove the rear lateral transaxle support.

5. Remove the heater hoses and evaporator lines from the heater core and evaporator.

6. Lower the vehicle. Remove the right and left hush panels, steering column trim cover, heater outlet duct and glove box.

7. Remove the heater core cover. Pull the cover straight to the rear so it does not damage the drain tube.

8. Remove the heater core clamps and remove the heater core.

To install:

9. Install the heater core and clamps.

10. Install the heater core cover using care not to damage the drain tube.

11. Install the glove box, heater outlet duct, steering column trim cover and hush panels.

12. Raise and support the vehicle safely.

13. Connect the heater hoses and evaporator lines to the heater core and evaporator, connect the drain tube to the case. Install the rear transaxle lateral support.

14. Lower the vehicle, fill the cooling system and connect the negative battery cable.

15. Check the heater operation and bleed the cooling system. Check for leaks.

Water Pump

REMOVAL & INSTALLATION

2.0L (VIN 1), 2.2L, 2.8L and 3.1L Engines

1. Disconnect the negative battery cable.

2. Drain the cooling system into a clean container for reuse.

3. Remove all drive belts.

4. Remove the alternator.

5. Unscrew the water pump pulley mounting bolts and remove the pulley.

To install:

6. Place a ⅛ in. bead of RTV sealant on the water pump sealing surface. While the sealer is still wet, install the pump and tighten the bolts to 15–22 ft. lbs. (20–30 Nm) on 2.0L and 2.2L engines or 6–9 ft. lbs. (8–12 Nm) on 2.8L and 3.1L engines.

7. Install the water pump pulley and the mounting bolts.

8. Install the alternator.

9. Install all drive belts.

10. Connect the negative battery cable.

11. Fill cooling system and check for leaks. Start the engine and allow to come to normal operating temperature. Recheck for leaks. Top-up coolant.

2.0L (VIN M and K) Engines

1. Disconnect negative battery cable.

2. Drain cooling system into a clean container for reuse.

3. Remove timing belt.

4. Remove water pump retaining bolts, water pump and seal ring.

To install:

5. Install a new water pump seal ring, water pump and attaching bolts. Tighten the water pump bolts to 18 ft. lbs. (11 Nm).

6. Install the timing belt.

7. Connect the negative battery cable.

8. Fill cooling system and check for leaks. Start the engine and allow to come to normal operating temperature. Recheck for leaks. Top-up coolant.

Thermostat

REMOVAL & INSTALLATION

2.0L (VIN 1), 2.2L, 2.8L and 3.1L Engines

The thermostat is located inside a housing either on the cylinder head on 2.0L and 2.2L engines or in the thermostat housing on the intake manifold on 2.8L and 3.1L engines. It is not necessary to remove the radiator hose from the thermostat housing when removing the thermostat.

1. Disconnect the negative battery cable.

2. Drain the cooling system and remove the air cleaner.

3. Disconnect the AIR pipe at the upper check valve and the bracket at the water outlet.

4. Disconnect the electrical lead.

5. Remove the 2 retaining bolts from the thermostat housing and lift up the housing with the hose attached. Lift out the thermostat.

To install:

6. Insert the new thermostat, spring end down. Apply a thin bead of silicone sealer to the housing mating surface and install the housing while the sealer is still wet. Tighten the housing retaining bolts to 15–22 ft. lbs. (20–30 Nm) on 2.8L and 3.1L engines or 6–9 ft. lbs. (8–12 Nm) on 2.0L and 2.2L engines.

7. Connect the electrical lead.

8. Connect the AIR pipe at the upper check valve and the bracket at the water outlet.

9. Connect the negative battery cable.

10. Fill cooling system and check for leaks. Start the engine and allow to come to normal operating temperature. Recheck for leaks. Top-up coolant.

2.0L (VIN M and K) Engines

NOTE: The engine must be COLD for this procedure.

1. Disconnect the negative battery cable.

2. Remove the thermostat housing cap.

3. Grasp the handle of the thermostat assembly and gently pull upward.

4. Clean the thermostat housing and O-ring.

To install:

5. Apply a suitable lubricant to the O-ring. Install the thermostat into the housing, pushing down to ensure that the thermostat is firmly seated.

6. Replace the thermostat housing cap.

7. Connect the negative battery cable.

Cooling System Bleeding

After working on the cooling system, even to replace the thermostat, it must be bled. Air trapped in the system will prevent proper filling and leave the radiator coolant level low, causing a risk of overheating.

1. To bleed the system, start with the system cool, the radiator cap off and the radiator filled to about an inch below the filler neck.

2. Start the engine and run it at slightly above normal idle speed. This will insure adequate circulation. If air bubbles appear and the coolant level drops, fill the system with an antifreeze/water mixture to bring the level back to the proper level.

3. Run the engine this way until the thermostat opens. When this happens, coolant will move abruptly across the top of the radiator and the temperature of the radiator will rise.

4. At this point, air is often expelled and the level may drop quite a bit. Keep refilling the system until the level is near the top of the radiator and remains constant.

5. If the vehicle has a coolant recovery tank, fill the radiator up to the fill-

er neck then install the radiator cap and fill recovery tank to correct level.

ENGINE ELECTRICAL

NOTE: Disconnecting the negative battery cable on some vehicles may interfere with the functions of the on board computer systems and may require the computer to undergo a relearning process, once the negative battery cable is reconnected.

Distributor

REMOVAL

2.0L (VIN M and K) Engines

1. Disconnect the negative battery cable.
2. Tag the spark plug wires and remove the wires and ignition coil from the distributor.
3. Disconnect the wiring from the distributor.
4. Remove the 2 distributor hold-down nuts.
5. Mark the tang drive and camshaft for correct reassembly.
6. Remove the distributor.

INSTALLATION

Timing Not Disturbed

1. Align the tang drive according to the previous marking and install the distributor.
2. Tighten the hold-down nuts to 13 ft. lbs. (18 Nm).
3. Connect the wiring to the distributor.
4. Reconnect the cap and spark plug wires and connect the negative battery cable.
5. Check and/or adjust the ignition timing.

Timing Disturbed

1. Remove the No. 1 cylinder spark plug.
2. Place a finger over the spark plug hole while rotating the engine slowly by hand, until compression is felt.
3. Align the timing mark on the crankshaft pulley with the **0** degree mark on the timing scale on the front of the engine. This places the engine at TDC of the compression stroke for No. 1 cylinder.
4. Rotate the distributor shaft until the rotor points to the No. 1 spark plug tower on the distributor cap.

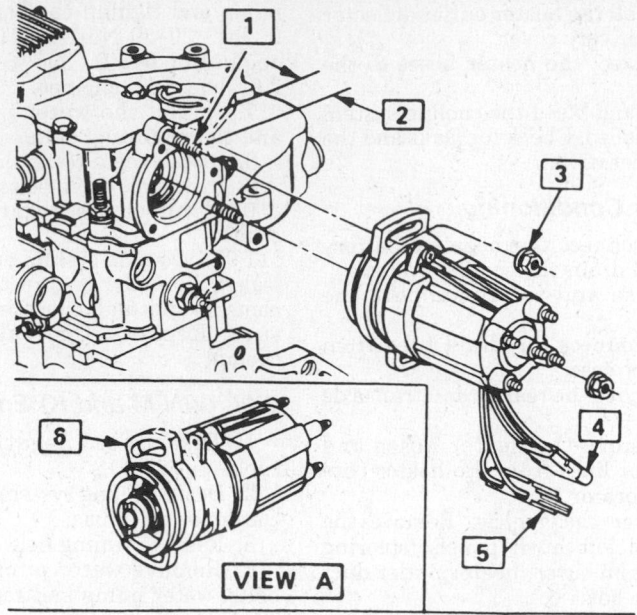

1—STUD	4—E.S.T. CONNECTOR
2—20.5 ±1.0 (BOTH STUDS)	5—COIL CONNECTOR
	6—DISTRIBUTOR ASM.
3—NUT ASM.	

Distributor installation—2.0L (VIN K and M) engines

5. Install the distributor in the engine. Be sure to align the distributor-to-engine matchmarks.
6. Install and tighten the hold-down nuts to 13 ft. lbs. (18 Nm).
7. Connect all wiring to the distributor.
8. Start and run the engine.
9. Check and adjust the ignition timing.

Distributorless Ignition

The Distributorless Ignition System (DIS) is used on the 2.0L (VIN 1), 2.2L (VIN G) and all V6 engines.

REMOVAL & INSTALLATION

DIS Assembly

1. Disconnect the negative battery cable.
2. Disconnect the electrical wires from the DIS assembly.
3. Mark the location of the spark plug wires on the DIS assembly and remove the wires.
4. Remove the DIS assembly mounting bolts and remove the assembly from the block.

NOTE: With the coil pack removed, the coils can each be removed and the ignition module can be removed as well.

To install:
5. Install the DIS assembly on the block.
6. Reconnect the plug wires to their original location.
7. Connect the DIS assembly wiring.
8. Connect the negative battery cable.

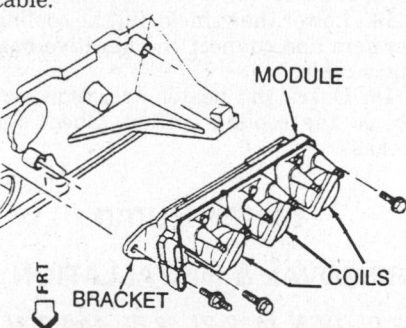

Ignition coil pack (DIS)—2.8L and 3.1L engines

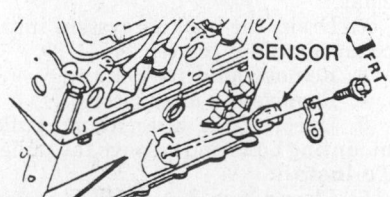

Crankshaft sensor removal— 2.8L and 3.1L engines

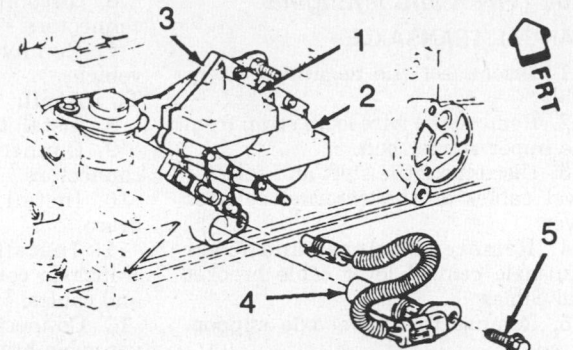

1. 2–3 coil
2. 1–4 coil
3. Module
4. Crank sensor assembly
5. Bolt—tighten to 71 inch lbs.

Ignition coils, module and sensor—2.0L (VIN 1) and 2.2L engines

9. If equipped with the 3.1L engine, perform the idle learn procedure to allow the ECM memory to be updated with the correct IAC valve pintle position and provide for a stable idle speed.

a. Install a Tech 1 scan tool.

b. Turn the ignition to the **ON** position, engine not running.

c. Select **IAC SYSTEM**, then **IDLE LEARN** in the **MISC TEST** mode.

d. Proceed with idle learn as directed by the scan tool.

Crankshaft Sensor

1. Disconnect the negative battery cable.

2. Disconnect the sensor harness plug.

3. Remove the sensor-to-block bolt and remove the sensor from the engine.

4. To install the sensor, position the sensor in the block and install the sensor bolt. Tighten the sensor bolt to 71 inch lbs. (8 Nm).

5. Reconnect the sensor harness plug.

Ignition Timing

ADJUSTMENT

2.0L (VIN 1 and G), 2.8L and 3.1L Engines

The ignition timing on engines with distributorless ignitions, is controlled by the Electronic Control Module (ECM). No adjustments are possible.

2.0L (VIN K and M) Engine
AVERAGING METHOD

1. Refer to the underhood Vehicle Emission Control Information label and follow all of the timing instructions if they differ from below.

2. Warm the engine to normal operating temperature.

3. Place the transmission in **N** or **P**. Apply the parking brake and block wheels.

4. Air conditioning, cooling fan and choke must be **OFF**. Do not remove the air cleaner, except as noted.

5. Ground the ALCL connector under the dash by installing a jumper wire between the **A** and **B** terminals. The Check Engine light should begin flashing.

6. Connect an inductive timing light to the No. 1 spark plug wire lead and record timing.

7. Connect an inductive timing light to the No. 4 spark plug wire lead and record timing.

8. Add the 2 timing numbers and divide by 2 to obtain "average timing".

NOTE: For example: No. 1 timing = 4 degrees and No. 4 timing = 8 degrees; $4 + 8 = 12 \div 2 = 6$ degrees average timing. If a change is necessary, subtract the average timing from the timing specification to determine the amount of timing change to No. 1 cylinder. For example: if the timing specification is 8 degrees and the average timing is 6 degrees, advance the No. 1 cylinder 2 degrees to set the timing.

9. To correct the timing, loosen the distributor hold-down clamp, adjust the distributor and retighten the hold-down bolt.

10. Once the timing is properly set, remove the jumper wire from the ALCL connector.

11. If necessary to clear the ECM memory, disconnect the ECM harness from the positive battery pigtail for 10 seconds with the key in the **OFF** position.

Alternator

PRECAUTIONS

Several precautions must be observed with alternator equipped vehicles to avoid damage to the unit.

• If the battery is removed for any reason, make sure it is reconnected with the correct polarity. Reversing the battery connections may result in damage to the one-way rectifiers.

• When utilizing a booster battery as a starting aid, always connect the positive to positive terminals and the negative terminal from the booster battery to a good engine ground on the vehicle being started.

• Never use a fast charger as a booster to start vehicles.

• Disconnect the battery cables when charging the battery with a fast charger.

• Never attempt to polarize the alternator.

• Do not use test lamps of more than 12 volts when checking diode continuity.

• Do not short across or ground any of the alternator terminals.

• The polarity of the battery, alternator and regulator must be matched and considered before making any electrical connections within the system.

• Never separate the alternator on an open circuit. Make sure all connections within the circuit are clean and tight.

• Disconnect the battery ground terminal when performing any service on electrical components.

• Disconnect the battery if arc welding is to be done on the vehicle.

BELT TENSION ADJUSTMENT

V-Belts

If equipped with 2.0L (VIN K and M) engines, the air conditioner compressor is driven by a separate V-belt. Using a belt tension gauge, adjust the air conditioner belt, to 225 lbs. (1000 N) for a new belt and 115 lbs. (525 N) for a used belt.

Serpentine Belts

Serpentine belts are tensioned by loosening and rotating the belt tensioner. The correct belt tension is indicated on the indicator mark of the belt tensioner. If the indicator mark is not within specification, replace the belt or the tensioner.

NOTE: To remove or install the belt, push and rotate the tensioner. Care should be taken to avoid twisting or bending the tensioner when applying torque.

REMOVAL & INSTALLATION

V-Belt Drive

1. Disconnect the negative battery cable.

2. Disconnect and tag the 2 termi-

nal plug and the battery lead from the rear of the alternator.

3. Loosen the mounting bolts. Push the alternator inwards and slip the drive belt off the pulley.

4. Remove the mounting bolts and remove the alternator.

To install:

5. Position the alternator in its brackets and install the mounting bolts. Do not tighten the bolts.

6. Slip the drive belt over the pulley. Pull outward on the alternator and adjust the belt tension. Tighten the mounting and adjusting bolts.

7. Connect the electrical leads.

8. Connect the negative battery cable.

Serpentine Belt Drive

1. Disconnect the negative battery cable.

2. Disconnect and tag the alternator wiring at the rear of the alternator.

3. Loosen the belt tensioner pivot bolt and rotate the tensioner to remove the belt.

4. Support the alternator and remove the mounting bolts.

5. Remove the alternator from the engine.

To install:

6. Place the alternator in the mounts and install the bolts.

7. Install the serpentine belt and tighten the belt tensioner.

8. Connect the alternator wiring and negative battery cable.

Starter

REMOVAL & INSTALLATION

Except 2.0L (VIN K and M) Engines

1. Disconnect the negative battery cable. Raise and safely support the vehicle.

2. Disconnect and tag the solenoid wires and battery cable at the starter.

3. Remove the rear starter support bracket. Remove the air conditioning compressor support rod, if equipped.

4. Support the starter and remove the 2 starter-to-engine bolts.

5. Remove the starter. Note the location and number of any shims.

To install:

6. Install the starter, replacing the shims, if equipped, in the original location.

7. Tighten the mounting bolts to 25–37 ft. lbs. (34–50 Nm).

8. Install the support bracket and air conditioning compressor rod, if removed.

9. Connect the starter wiring.

10. Connect the negative battery cable and check the starter operation.

2.0L (VIN K and M) Engine
MANUAL TRANSAXLE

1. Disconnect the negative battery cable.

2. Remove the wire loom strap from the upper starter bolt.

3. Disconnect the shift and selector level cables at the external selector lever.

4. Remove the upper and lower transaxle control lever cable bracket and cables.

5. Remove the drive axle support brace.

6. Disconnect the starter electrical connectors.

7. Remove the starter from the vehicle.

To install:

8. Install the starter.

9. Connect the starter electrical connectors.

10. Install the drive axle support brace.

11. Install the upper and lower transaxle control lever cable bracket and cables.

12. Connect the shift and selector lever cable bracket and cable.

13. Install the wire loom strap to the upper starter bolt.

14. Connect the negative battery cable.

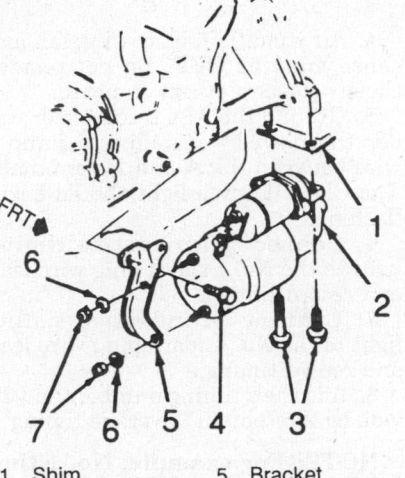

1. Shim
2. Starter
3. Bolt (32 ft. lbs.)
4. Bolt (9 ft. lbs.)
5. Bracket
6. Washer
7. Nut (24 ft. lbs.)

Starter mounting—2.0L (VIN 1) and 2.2L engines

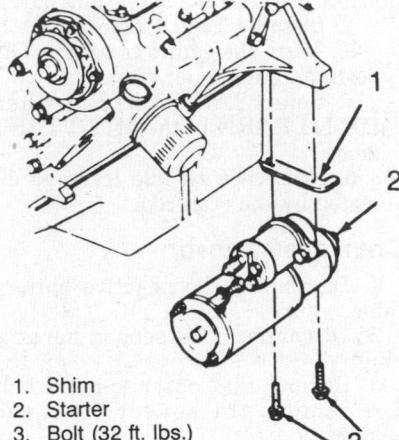

1. Shim
2. Starter
3. Bolt (32 ft. lbs.)

Starter mounting—2.8L and 3.1L engines

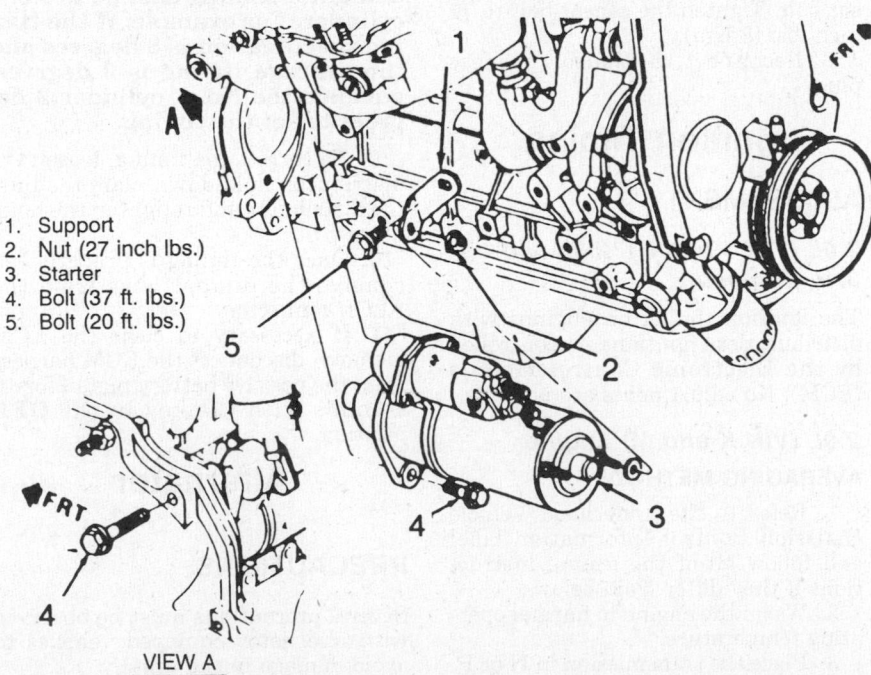

1. Support
2. Nut (27 inch lbs.)
3. Starter
4. Bolt (37 ft. lbs.)
5. Bolt (20 ft. lbs.)

VIEW A

Starter mounting—2.0L (VIN K and M) engine

AUTOMATIC TRANSAXLE

1. Disconnect the negative battery cable.
2. Remove the blower motor.
3. Disconnect the starter motor electrical connectors.
4. Remove the rear starter brace.
5. Remove the wire loom from the upper starter bolt.
6. Remove the upper starter bolt.
7. Remove the transaxle strut.
8. Remove the lower starter bolt.
9. Remove the starter motor from the vehicle through the blower motor opening.

To install:

10. Install the starter motor through the blower motor opening.
11. Install the lower starter bolt.
12. Install the transaxle strut.
13. With the help of an assistant, install the upper starter bolt.
14. Install the wire loom to the upper starter bolt.
15. Install the rear starter brace.
16. Connect the starter motor electrical connectors.
17. Install the blower motor.
18. Connect the negative battery cable.

Emission Warning Lamps

Please refer to "Emission Controls" in the Unit Repair section for system maintenance procedures. Due to the complex nature of modern electronic engine control systems, comprehensive diagnosis and testing procedures fall outside the confines of this repair manual. For complete information on diagnosis, testing and repair procedures concerning all modern engine and emission control systems, please refer to "Chilton's Guide to Fuel Injection and Electronic Engine Controls".

RESETTING

When the Electronic Control Module (ECM) detects a problem, the CHECK ENGINE light will come on and a trouble code will be stored in the ECM. In order to clear the stored trouble code, it is necessary to remove the battery voltage for 10 seconds. This will clear all codes stored in ECM memory. Do this by disconnecting the ECM harness from the positive battery cable with the ignition in the **OFF** position or by removing the ECM fuse.

NOTE: In order to prevent damage to the ECM, the key must be OFF when connecting or disconnecting power to the ECM.

FUEL SYSTEM

Fuel System Service Precautions

Safety is the most important factor when performing not only fuel system maintenance but any type of maintenance. Failure to conduct maintenance and repairs in a safe manner may result in serious personal injury or death. Maintenance and testing of the vehicle's fuel system components can be accomplished safely and effectively by adhering to the following rules and guidelines.

● To avoid the possibility of fire and personal injury, always disconnect the negative battery cable unless the repair or test procedure requires that battery voltage be applied.

● Always relieve the fuel system pressure prior to disconnecting any fuel system component (injector, fuel rail, pressure regulator, etc.), fitting or fuel line connection. Exercise extreme caution whenever relieving fuel system pressure to avoid exposing skin, face and eyes to fuel spray. Please be advised that fuel under pressure may penetrate the skin or any part of the body that it contacts.

● Always place a shop towel or cloth around the fitting or connection prior to loosening to absorb any excess fuel due to spillage. Ensure that all fuel spillage (should it occur) is quickly removed from engine surfaces. Ensure that all fuel soaked cloths or towels are deposited into a suitable waste container.

● Always keep a dry chemical (Class B) fire extinguisher near the work area.

● Do not allow fuel spray or fuel vapors to come into contact with a spark or open flame.

● Always use a backup wrench when loosening and tightening fuel line connection fittings. This will prevent unnecessary stress and torsion to fuel line piping. Always follow the proper torque specifications.

● Always replace worn fuel fitting O-rings with new. Do not substitute fuel hose or equivalent where fuel pipe is installed.

RELIEVING FUEL SYSTEM PRESSURE

The fuel delivery pipe is under high pressure even after the engine is stopped. Direct removal of the fuel line, may result in dangerous fuel spray. Make sure to release the fuel pressure according to the following procedures:

2.0L (VIN 1) and 2.2L Engines

1988

1. Release the fuel vapor pressure in the fuel tank by removing the fuel tank cap and reinstalling it.
2. Remove the fuel pump fuse from the fuse block or from the underhood fuse holder.
3. Start the engine and allow it to run a few seconds until it runs out of fuel.
4. Once the engine is stopped, crank it a few times with the starter for about 3 seconds to dissipate the fuel in the lines.
5. If the fuel pressure can't be released in the above manner because the engine failed to run, disconnect the negative battery cable, cover the union bolt of the fuel line with an sho towel and loosen the union bolt slowly to release the fuel pressure gradually.

1989–92

1. Disconnect the negative battery cable.
2. Release the fuel vapor pressure in the fuel tank by removing the fuel tank cap and reinstalling it.
3. The internal constant bleed feature of the TBI Models 700/220, relieves the fuel pump system pressure when the engine is turned **OFF** and no further pressure relive procedure is required.

2.0L (VIN K) Engine

1. Release the fuel vapor pressure in the fuel tank by removing the fuel tank cap and reinstalling it.
2. Remove the fuel pump fuse from the fuse block.
3. Start the engine and allow it to run a few seconds until it runs out of fuel.
4. Once the engine is stopped, crank it a few times with the starter for about 3 seconds to dissipate the fuel in the lines.
5. If the fuel pressure can't be released in the above manner because the engine failed to run, disconnect the negative battery cable, cover the union bolt of the fuel line with a shop towel and loosen the union bolt slowly to release the fuel pressure gradually.

2.0L (VIN M) Engine

1988

1. Release the fuel vapor pressure in the fuel tank by removing the fuel tank cap and reinstalling it.
2. Disconnect the fuel tank harness connector.
3. Start the engine and allow it to run a few seconds until it runs out of fuel.
4. Once the engine is stopped, crank it a few times with the starter for

about 3 seconds to dissipate the fuel in the lines.

5. If the fuel pressure can't be released in the above manner because the engine failed to run, disconnect the negative battery cable, cover the union bolt of the fuel line with a shop towel and loosen the union bolt slowly to release the fuel pressure gradually.

1989–90

1. Disconnect the negative battery cable.

2. Disconnect the fuel filler cap.

3. Connect gauge J–34730–1 or equivalent, to the fuel pressure connection. Wrap a cloth around the fitting to absorb any fuel leakage.

4. Install the bleed hose into an approved container and open the valve to bleed system pressure.

2.8L and 3.1L Engines

1988

1. Release the fuel vapor pressure in the fuel tank by removing the fuel tank cap and reinstalling it.

2. Disconnect the fuel tank harness connector.

3. Start the engine and allow it to run a few seconds until it runs out of fuel.

4. Once the engine is stopped, crank it a few times with the starter for about 3 seconds to dissipate the fuel in the lines.

5. If the fuel pressure can't be released in the above manner because the engine failed to run, disconnect the negative battery cable, cover the union bolt of the fuel line with a shop towel and loosen the union bolt slowly to release the fuel pressure gradually.

1989–92

1. Disconnect the negative battery cable.

2. Disconnect the fuel filler cap.

3. Connect gauge J–34730–1 or equivalent, to the fuel pressure connection. Wrap a cloth around the fitting to absorb any fuel leakage.

4. Install the bleed hose into an approved container and open the valve to bleed system pressure.

Fuel Filter

REMOVAL & INSTALLATION

The fuel filter is located under the rear of the vehicle near the fuel tank.

1. Relieve the fuel system pressure.

2. Disconnect the negative battery cable.

3. Raise and safely support the vehicle.

4. Disconnect the fuel lines from the filter.

5. Remove the filter retaining bolt and remove the filter from the vehicle.

To install:

6. Install the new filter in position, using new O-ring seals, and connect the fuel lines. Tighten the fuel lines to 20 ft. lbs. (27 Nm).

7. Lower the vehicle.

8. Connect the negative battery cable and run the engine. Check for leaks.

Electric Fuel Pump

PRESSURE TESTING

Throttle Body Injection

1. Relieve the fuel system pressure.

2. Remove the air cleaner and plug the thermal vacuum port on the throttle body unit.

3. Remove the steel fuel line from between the throttle body unit and the fuel filter.

4. Install a fuel pressure gauge with at least a 15 psi capacity between the throttle body and the filter.

5. Start the engine and observe the pressure reading. Pressure should be 9–13 psi. If the pressure is not within these limits, one or more of the following could be at fault:

    a. A short in the system

    b. A clogged fuel filter

    c. A shorted or defective oil pressure switch

    d. Defective fuel pump relay

    e. Defective fuel pump

NOTE: Check each of these components in turn to diagnose the problem before replacing the pump.

6. Follow the cautions at the start of this procedure to depressurize the system. Remove the pressure gauge and install the fuel line. Tighten the nuts to 19–25 ft. lbs. (26–34 Nm).

7. Start the engine and check for leaks.

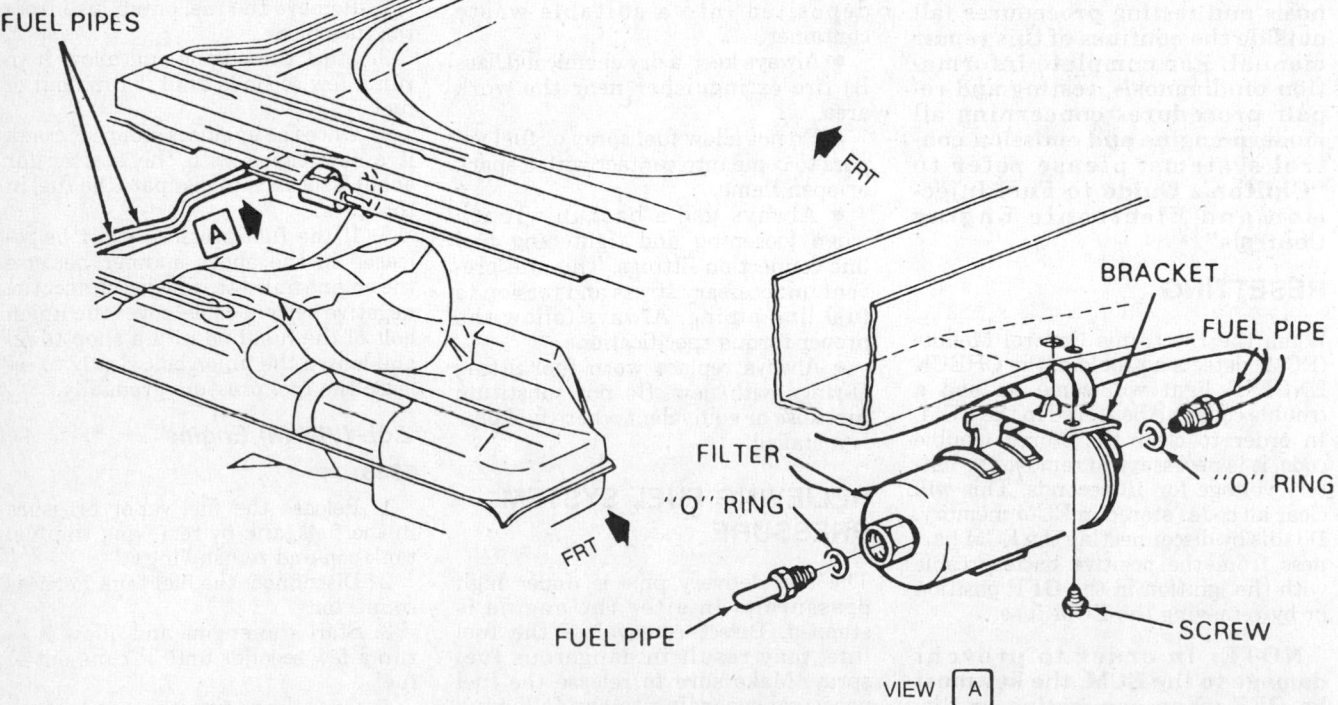

Fuel filter installation

8. Unplug the thermal vacuum port on the throttle body.

Port Fuel Injection

1. Release the fuel system pressure. Wrap a shop towel around fuel pressure connector on the fuel rail to absorb any leakage that may occur when installing gauge.
2. Install a fuel pressure gauge J–34730–1 or equivalent, to pressure connector.
3. With ignition **ON** pump pressure should be as follows:
   a. 40.5–47 psi on 2.8L and 3.1L engines
   b. 35–38 psi on 2.0L (Code M)Engine
4. When engine is idling, pressure should drop 3–10 psi on 2.8L and 3.1L engines or 25–30 psi on 2.0L (Code M) engine.

NOTE: The application of vacuum to the pressure regulator should result in a fuel pressure drop.

5. Remove fuel pressure gauge J–34730–1 or equivalent, from pressure connector.

REMOVAL & INSTALLATION

The electric fuel pump is located in the fuel tank.
1. Relieve the fuel system pressure.
2. Disconnect the battery ground.
3. Raise and safely support the vehicle.
4. Remove the fuel filler cap.
5. Drain the fuel tank.
6. Disconnect the filler neck hose and the vent hose.
7. Remove the fuel tank strap rear support bolts and lower the tank on a jack, just enough, to disconnect the fuel feed line, return and vapor lines from the fuel meter.
8. Remove the tank from the vehicle.
9. Remove the fuel meter/pump assembly by turning the cam lock ring counterclockwise. Lift the assembly from the tank and remove the pump from the meter.
10. Pull the pump up onto the attaching hose while pulling outward from the bottom support. Take care not to damage the rubber insulator and strainer. After the pump is clear of the bottom support pull it out of the rubber connector.
To install:
11. Install the pump to the fuel meter assembly.
12. Install the fuel meter/pump assembly into the tank. Install a new O-ring on the cam lock ring. Install the cam lock ring and tighten by turning clockwise.

13. Place the fuel tank into position.
14. Connect the fuel feed line, return and vapor lines to the fuel meter. Raise the tank and install the fuel tank strap rear support bolts.
15. Connect the filler neck hose and the vent hose.
16. Fill the fuel tank.
17. Install the fuel filler cap.
18. Lower the vehicle.
19. Connect the negative battery cable.

Fuel Injection

IDLE SPEED AND MIXTURE ADJUSTMENT

Idle speed and mixture are controlled by the Electronic Control Module (ECM). No adjustments are necessary.

Fuel Injector

REMOVAL & INSTALLATION

NOTE: Use care in removing injector to prevent damage to the electrical pins on top of the injector. The fuel injectors are an electrical component. Do not immerse in any type of cleaner.

REMOVAL & INSTALLATION

Throttle Body Injection

1. Relieve the fuel pump pressure.
2. Disconnect the negative battery cable.
3. Remove the TBI cover and gasket.
4. Disconnect the electrical connector to fuel injector.
5. Remove the injector retainer.
6. Using a fulcrum, place a suitable tool under the ridge opposite the connector end and carefully pry injector out.
To install:

NOTE: Remove the upper and lower O-rings from injector body and in fuel injector cavity and replace with new O-rings before installing injector.

7. Install the injector.
8. Install the injector retainer.
9. Connect the electrical connector to fuel injector.

NOTE: Be sure the electrical connector end, on the injector is facing in the direction to the cutout in the fuel meter body for the wire grommet to fit properly.

10. Install the TBI cover and gasket.
11. Connect the negative battery cable.

Port Fuel Injection

NOTE: The fuel rail is removed as an assembly, then the injectors can be removed.

1. Disconnect the negative battery cable. Relieve the fuel system pressure.
2. Tag and disconnect the fuel injection electrical connections.
3. Remove the upper intake manifold plenum assembly. Remove the necessary components in order to gain access to the fuel rail retaining bolts.
4. Remove the fuel rail retaining bolts. Remove the fuel rail assembly.
5. Separate the fuel injector from the fuel rail.
To install:
6. Replace the O-rings when installing the injectors.
7. Install the fuel injector to the fuel rail.
8. Install the fuel rail assembly. Install the fuel rail retaining bolts.
9. Install the upper intake manifold plenum assembly.
10. Connect the fuel injector electrical connections.
11. Connect the negative battery cable.

DRIVE AXLE

Halfshaft

REMOVAL & INSTALLATION

NOTE: If equipped with tri-pot joints, care must be exercised not to allow joints to become overextended. Over extending the joint could result in separation of internal components.

1. Raise and safely support the vehicle. Do not support under lower control arms.
2. Remove the front wheels.
3. Remove the hub nut and washer.
4. Remove the caliper bolts and support caliper with a length of wire. Do not let the caliper hang by the brake hose unsupported.
5. Remove the rotor and lower ball joint nut.
6. Remove the stabilizer bolt from lower control arm.

NOTE: Install the halfshaft seal boot protectors J–34754 or equivalent, on the outer drive seal.

7. Install J–28733 or equivalent and press the halfshaft in and away from the hub. The halfshaft should only be pressed in until the press fit between the halfshaft and hub is loose.

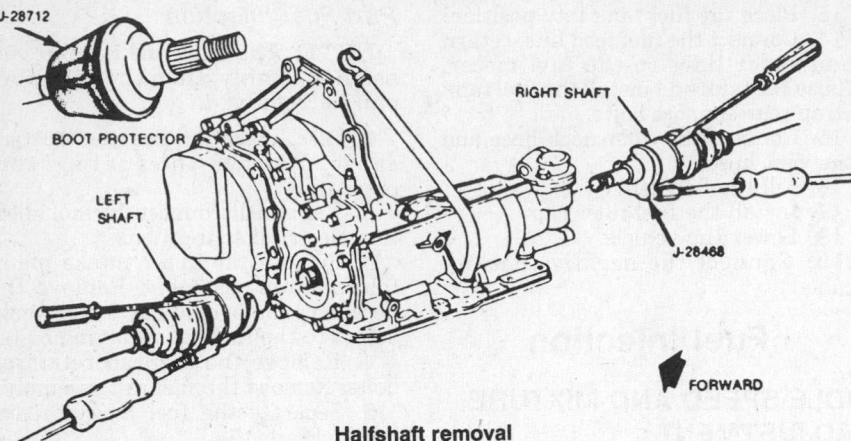

Halfshaft removal

8. Separate and remove the lower ball joint from the steering knuckle.

9. Install J–28468 or equivalent and slide hammer assembly. Remove the halfshaft.

To install:

10. To install the halfshaft, start the splines of the halfshaft into the transaxle and push halfshaft inward until it snaps into place.

11. Verify that the halfshaft is seated into the transaxle by grasping on the housing and pulling outboard.

12. Install the drive axle to hub and bearing assembly.

13. Install the lower ball joint to the steering knuckle. Tighten the ball joint-to-steering knuckle nut to 41 ft. lbs. (56 Nm) with a minimum torque of 50 ft. lbs. (68 Nm). Tighten further to align the next cotter pin hole.

14. Install the washer and new driveshaft nut. Tighten the new axle shaft nut to 74 ft. lbs. (100 Nm).

15. Install the stabilizer bolt to lower control arm.

16. Install the rotor.

17. Install the caliper and attaching bolts.

18. Install the front wheels.

19. Lower the vehicle and apply a final torque to the axle shaft nut of 191 ft. lbs. (259 Nm).

CV-Boot

REMOVAL & INSTALLATION

Outer

1. Remove the halfshaft assembly.

2. Remove the steel deflector ring by using brass drift to tap it off. If the rubber ring is used, slide it off.

3. Cut the seal retaining clamps and lift the boot up to gain access to retaining ring.

4. Using snapring pliers J–8059 or equivalent, spread the retaining ring inside the outer CV-joint and remove joint from shaft.

5. Slide the boot off shaft.

To install:

6. Clean the splines of the shaft and the CV-joint with solvent and repack the joint. Install a new retaining ring inside the joint.

NOTE: When repacking CV-joint, make sure to add grease to axle boot.

7. Install the inner boot clamp, boot, outer boot clamp on shaft.

8. Push the joint assembly onto the shaft until the ring is seated on the shaft.

9. Slide the boot and 2 clamps onto the joint and install the clamps on both the inner and outer part of the boot. Install deflector ring.

10. Install the halfshaft assembly.

Inner

1. Remove the halfshaft assembly.

2. Cut the seal retaining clamps and lift the boot up to gain access to retaining ring for spider assembly.

3. Using snapring pliers J–8059 or equivalent, remove the retaining ring from shaft and remove the spider assembly. Slide the old boot off axle shaft.

To install:

4. Clean the splines of the shaft and the CV-joint with solvent and repack the joint.

NOTE: When repacking CV-joint, make sure to add grease to axle boot.

5. Install the inner boot clamp, boot, outer boot clamp on shaft.

6. Push the tri-pot assembly onto the shaft until the retaining ring is seated on the shaft.

7. Slide the boot and 2 clamps onto the joint and install the clamps on both the inner and outer part of the boot.

8. Install the halfshaft assembly.

NOTE: Be sure the spacer ring is seated in groove on axle at reassembly.

Front Wheel Hub and Bearings

REMOVAL & INSTALLATION

1. Remove the wheel cover, loosen the hub nut, and raise and support the vehicle safely. Remove the front wheel.

2. Install the boot cover protector on 4 cylinder engine with automatic transaxle.

3. Remove the hub nut.

4. Remove the brake caliper and rotor. Support the caliper using a length of wire.

NOTE: Do not allow the brake caliper to hang by the brake hose unsupported.

5. Remove the 3 hub and bearing attaching bolts.

6. Remove the splash shield.

7. Install special tool J–28733 or equivalent, and press the hub and bearing assembly off the halfshaft.

8. Disconnect the stabilizer link bolt at the lower control arm.

9. Separate the ball joint from steering knuckle.

10. Remove the halfshaft from knuckle and support out of the way.

11. Remove the inner knuckle seal using brass drift pin or equivalent.

NOTE: To remove the steering knuckle at this point, remove both strut to knuckle mounting bolts. Before removing the steering knuckle from the strut, be sure to scribe alignment marks between them, so the installation can be easily performed.

To install:

12. Install the hub and bearing assembly to the steering knuckle and install the bolts. Tighten to 70 ft. lbs. (95 Nm).

13. Install the hub and bearing seal.

14. Install the brake rotor and caliper.

15. Install the halfshaft to knuckle.

16. Connect the ball joint to the steering knuckle. Tighten ball joint nut to 42 ft. lbs. (57 Nm).

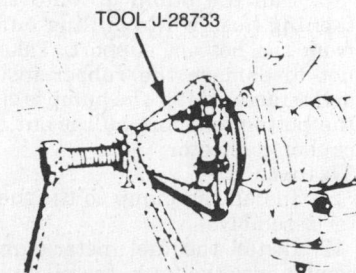

Removing drive axle from hub and bearing assembly

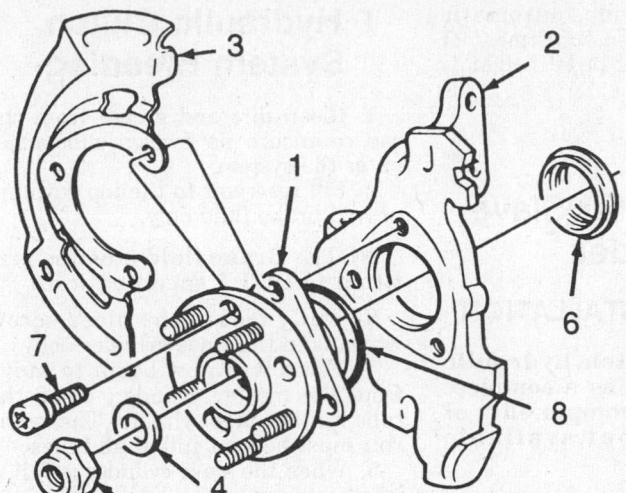

1. Hub and bearing assembly
2. Steering knuckle
3. Shield
4. Washer
5. Hub nut
6. Seal
7. Hub and bearing retaining bolt
8. O-ring

Exploded view of the hub and bearing attachment to the steering knuckle

17. Connect the stabilizer link bolt to the lower control arm. Tighten the steering knuckle-to-strut bolts to 133 ft. lbs. (181 Nm).
18. Install the splash shield.
19. Install the driveshaft nut.
20. Remove the boot cover protector on 4 cylinder engine with automatic transaxle.
21. Install the front wheel.
22. Lower the vehicle.
23. Apply a final torque to the hub and bearing nut to 185 ft. lbs. (251 Nm).
24. Install the wheel cover.

MANUAL TRANSAXLE

For further information on transmissions/transaxles, please refer to "Chilton's Guide to Transmission Repair".

Transaxle Assembly

REMOVAL & INSTALLATION

1. Disconnect the negative battery cable.
2. Install an engine holding bar so one end is supported on the cowl tray over the wiper motor and the other end rests on the radiator support. Use padding and be careful not to damage the paint or body work with the bar. Attach a lifting hook to the engine lift ring and to the bar and raise the engine enough to take the pressure off the motor mounts.

NOTE: If a lifting bar and hook is not available, a chain hoist can be used, however, during the procedure the vehicle must be raised, at which time the chain hoist must be adjusted to keep tension on the engine/transaxle assembly.

3. Remove the heater hose clamp at the transaxle mount bracket. Disconnect the electrical connector and remove the horn assembly.
4. Remove the transaxle mount attaching bolts. Discard the bolts attaching the mount to the side frame; new bolts must be used at installation.
5. Disconnect the clutch master cylinder pushrod from the clutch pedal and disconnect the clutch slave cylinder from the transaxle support bracket and move it aside.
6. Remove the transaxle mount bracket attaching bolts and nuts.
7. Disconnect the ground cables at the transaxle mounting stud.
8. Remove the 4 upper transaxle-to-engine mounting bolts.
9. Raise the vehicle and support it on stands. Remove the left front wheel.
10. Remove the left front inner splash shield. Remove the transaxle strut and bracket.
11. Remove the clutch housing cover bolts.
12. Disconnect the speedometer cable at the transaxle.
13. Disconnect the stabilizer bar at the left suspension support and control arm.
14. Disconnect the ball joint from the steering knuckle.
15. Remove the left suspension support attaching bolts and remove the support and control arm as an assembly.
16. Install boot protectors and disen-gage the halfshafts at the transaxle. Remove the left side shaft from the transaxle.
17. Position a jack under the transaxle case, remove the lower 2 transaxle-to-engine mounting bolts and remove the transaxle by sliding it towards the driver's side, away from the engine. Carefully lower the jack, guiding the right shaft out the transaxle.

To install:

18. Raise the transaxle into position and guide the right halfshaft into its bore as the transaxle is being raised. The right halfshaft can not be readily installed after the transaxle is connected to the engine.
19. Install the transaxle attaching bolts. Tighten the transaxle-to-engine bolts to 55 ft. lbs. (75 Nm) on 1988 vehicles equipped with a Muncie transaxle, 60 ft. lbs. (81 Nm) on 1988 vehicles equipped with an Isuzu transaxle or 85 ft. lbs. (115 Nm) on all 1989–92 vehicles.
20. Install the left side shaft to the transaxle. Remove the boot protectors.
21. Install the control arm as an assembly. Install the left suspension support and attaching bolts.
22. Connect the ball joint to the steering knuckle.
23. Connect the stabilizer bar to the left suspension support and control arm. Tighten the suspension support-to-body attaching bolts to 75 ft. lbs. (102 Nm).
24. Connect the speedometer cable to the transaxle.
25. Install the clutch housing cover and attaching bolts Tighten to 10 ft. lbs. (14 Nm).
26. Install the transaxle strut and bracket. Install the left front inner splash shield.
27. Install the left front wheel. Lower the vehicle.
28. Install the 4 upper transaxle-to-engine mounting bolts.
29. Connect the ground cables to the transaxle mounting stud.
30. Install the transaxle mount bracket attaching bolts and nuts. When installing the bolts attaching the mount-to-transaxle bracket, check the alignment bolt at the engine mount. If excessive effort is required to remove the alignment bolt, realign the powertrain components and tighten the bolts to 40 ft. lbs. (54 Nm) and remove the alignment bolt.
31. Connect the clutch slave cylinder to the transaxle support bracket and connect the clutch master cylinder pushrod to the clutch pedal.
32. Install the transaxle mount attaching bolts. Install new bolts attaching the mount to the side frame. Tighten the transaxle mount-to-side frame to 40 ft. lbs. (54 Nm).
33. Install the heater hose clamp to

the transaxle mount bracket. Install the horn assembly and connect the electrical connector.

34. Remove the engine lifting fixture.

35. Connect the negative battery cable.

CLUTCH

Clutch Assembly

REMOVAL & INSTALLATION

―――――― CAUTION ――――――

The clutch plate contains asbestos, which has been determined to be a cancer causing agent. Never clean the clutch surfaces with compressed air. Avoid inhaling any dust from any clutch surface.

REMOVAL & INSTALLATION

1. Disconnect the negative battery cable. Raise and safely support the vehicle. Remove the transaxle.

2. Mark the pressure plate assembly and the flywheel so they can be assembled in the same position to maintain balance.

3. Loosen the attaching bolts 1 turn at a time until spring tension is relieved.

4. Support the pressure plate and remove the bolts. Remove the pressure plate and the clutch disc.

To install:

5. Inspect the flywheel, pressure plate, clutch disc, release bearing and the clutch fork for wear.

6. Clean the flywheel mating surfaces. Position the clutch disc and pressure plate into the installed position and support with a dummy shaft or clutch aligning tool.

NOTE: Clutch plate must be installed correctly. Clutch plate is marked INSTALL FLYWHEEL SIDE. Always replace clutch and pressure plate as a set.

7. Install the pressure plate-to-flywheel bolts. Tighten in a criss-cross pattern.

8. Lubricate the outside grooves and the inside recess of the release bearing with high temperature grease. Wipe off any excess. Install the release bearing.

9. Install the transaxle.

PEDAL HEIGHT/FREE-PLAY ADJUSTMENT

These vehicles use an hydraulic clutch system which provides automatic clutch adjustment. No adjustment of the clutch linkage or pedal height is required.

Clutch Master/Slave Cylinder

REMOVAL & INSTALLATION

NOTE: The clutch hydraulic system is serviced as a complete unit. Individual components of the system are not available separately.

1. Disconnect the negative battery cable.

2. Remove the left side sound insulator panel.

NOTE: If equipped with a 2.8L or 3.1L engine, remove the air cleaner, mass air flow sensor and air intake duct as an assembly. Disconnect electrical lead at the washer bottle and remove washer bottle from vehicle.

3. Disconnect the master cylinder pushrod from the clutch pedal.

4. Remove the master cylinder-to-cowl brace nuts and remove master cylinder.

5. Remove the slave cylinder retaining nuts at the transaxle and remove slave cylinder. Remove the hydraulic system as a unit from the vehicle.

To install:

6. Install the hydraulic system as a unit from the vehicle. Install the slave cylinder to the transaxle and install the attaching nuts.

7. Install the master cylinder and the master cylinder-to-cowl brace nuts.

8. Connect the master cylinder pushrod to the clutch pedal.

NOTE: If equipped with a 2.8L or 3.1L engine, install the air cleaner, mass air flow sensor and air intake duct as an assembly. Install the washer bottle. Connect the electrical lead to the washer bottle.

9. Install the left side sound insulator panel.

10. Connect the negative battery cable.

11. Bleed the hydraulic system.

NOTE: Do not remove the plastic pushrod retainer from the slave cylinder. The strap will break on the first clutch pedal application.

Hydraulic Clutch System Bleeding

1. Clean dirt and grease from the cap to ensure no foreign substances enter the system.

2. Fill reservoir to the top with approved brake fluid only.

NOTE: Brake fluid must be certified to DOT 3 specification.

3. Fully loosen the bleed screw which is in the slave cylinder body.

4. Fluid will now begin to move from the master cylinder, down the tube, to the slave cylinder. The reservoir must be kept full at all times.

5. When the slave cylinder is full, a steady stream of fluid will come from the slave outlet. At this point, tighten bleed screw.

6. Start the engine, push the clutch pedal to the floor and select reverse gear. There should be no grating of gears. If there is the system still contains air.

AUTOMATIC TRANSAXLE

For further information on transmissions/transaxles, please refer to "Chilton's Guide to Transmission Repair".

Transaxle Assembly

NOTE: By September 1, 1991, Hydra-matic will have changed the name designation of the THM 125C automatic transaxle. The new name designation for this transaxle will be Hydra-matic 3T40. Transaxles built between 1989 and 1990 will serve as transitional years in which a dual system, made up of the old designation and the new designation will be in effect.

REMOVAL & INSTALLATION

1. Disconnect the negative terminal from the battery. Remove the air cleaner, bracket, Mass Air Flow (MAF) sensor and air tube as an assembly.

2. Disconnect the exhaust crossover from the right side manifold and remove the left side exhaust manifold, then, raise and support the manifold/crossover assembly.

3. Disconnect the TV cable from the throttle lever and the transaxle.

4. Remove the vent hose and the shift cable from the transaxle.

5. Remove the fluid level indicator and the filler tube.

6. Using the engine support fixture tool J–28467 or equivalent and the adapter tool J–35953 or equivalent, install them on the engine.

7. Remove the wiring harness-to-transaxle nut.

8. Label and disconnect the wires for the speed sensor, TCC connector and the neutral safety/back up light switch.

9. Remove the upper transaxle-to-engine bolts.

10. Remove the transaxle-to-mount through bolt, the transaxle mount bracket and the mount.

11. Raise and safely support the vehicle.

12. Remove the front wheel assemblies.

13. Disconnect the shift cable bracket from the transaxle.

14. Remove the left side splash shield.

15. Using a modified halfshaft seal protector tool J–34754 or equivalent, install one on each halfshaft to protect the seal from damage and the joint from possible failure.

16. Using care not to damage the halfshaft boots, disconnect the halfshafts from the transaxle.

17. Remove the torsional and lateral strut from the transaxle. Remove the left side stabilizer link pin bolt.

18. Remove the left frame support bolts and move it out of the way.

19. Disconnect the speedometer wire from the transaxle.

20. Remove the transaxle converter cover and matchmark the converter to the flywheel for assembly.

21. Disconnect and plug the transaxle cooler pipes.

22. Remove the transaxle-to-engine support.

23. Using a transmission jack, position and secure it to the transaxle and remove the remaining transaxle-to-engine bolts.

24. Make sure the torque converter does not fall out and remove the transaxle from the vehicle.

NOTE: The transaxle cooler and lines should be flushed any time the transaxle is removed for overhaul or to replace the pump, case or converter.

To install:

25. Put a small amount of grease on the pilot hub of the converter and make sure the converter is properly engaged with the pump.

26. Raise the transaxle to the engine while guiding the right-side halfshaft into the transaxle.

27. Install the lower transaxle mounting bolts, tighten to 55 ft. lbs. (75 Nm) and remove the jack.

28. Align the converter with the marks made previously on the flywheel and install the bolts hand tight.

29. Tighten the converter bolts to 46 ft. lbs. (62 Nm). Retorque the first bolt after the others.

30. Install the starter assembly. Install the left side halfshaft.

31. Install the converter cover, oil cooler lines and cover. Install the subframe assembly. Install the lower engine mount retaining bolts and the transaxle mount nuts.

32. Install the right and left ball joints. Install the power steering rack, heat shield and cooler lines to the frame.

33. Install the right and left inner fender splash shields. Install the tire assemblies.

34. Lower the vehicle. Connect all electrical leads. Install the upper transaxle mount bolts, tighten to 55 ft. lbs. (75 Nm).

35. Attach the crossover pipe to the exhaust manifold. Connect the EGR tube to the crossover.

36. Connect the TV cable and the shift cable. Install the air cleaner and inlet tube.

37. Remove the engine support tool. Connect the negative battery cable.

SHIFT CONTROL CABLE ADJUSTMENT

1. Place the shift lever in the **N**.

NOTE: Neutral can be found by rotating the transaxle selector shaft counterclockwise from P through R to N.

2. Loosely attach the cable to the transaxle shift lever with a nut. Assemble the cable to the cable bracket and to shift lever. Tighten the cable to transaxle shift lever nut.

NOTE: The lever must be held out of P when torquing the nut.

THROTTLE VALVE (TV) CABLE ADJUSTMENT

Setting of the TV cable must be done by rotating the throttle lever at the carburetor or throttle body. Do not use the accelerator pedal to rotate the throttle lever.

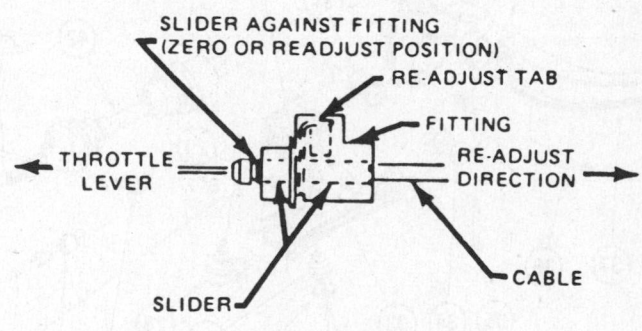

TV cable adjuster

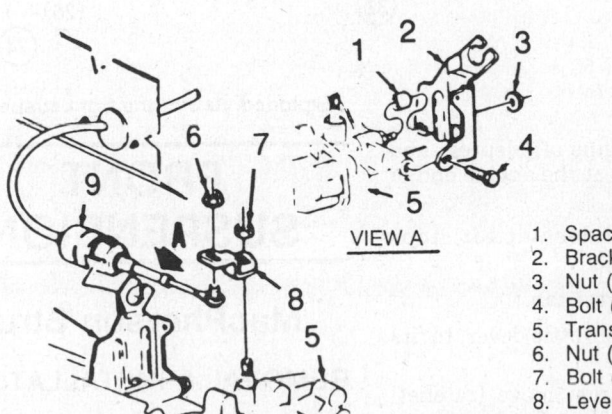

1. Spacer
2. Bracket
3. Nut (20 ft. lbs.)
4. Bolt (20 ft. lbs.)
5. Transaxle assembly
6. Nut (20 ft. lbs.)
7. Bolt (20 ft. lbs.)
8. Lever
9. Cable

Engine compartment shift control cable

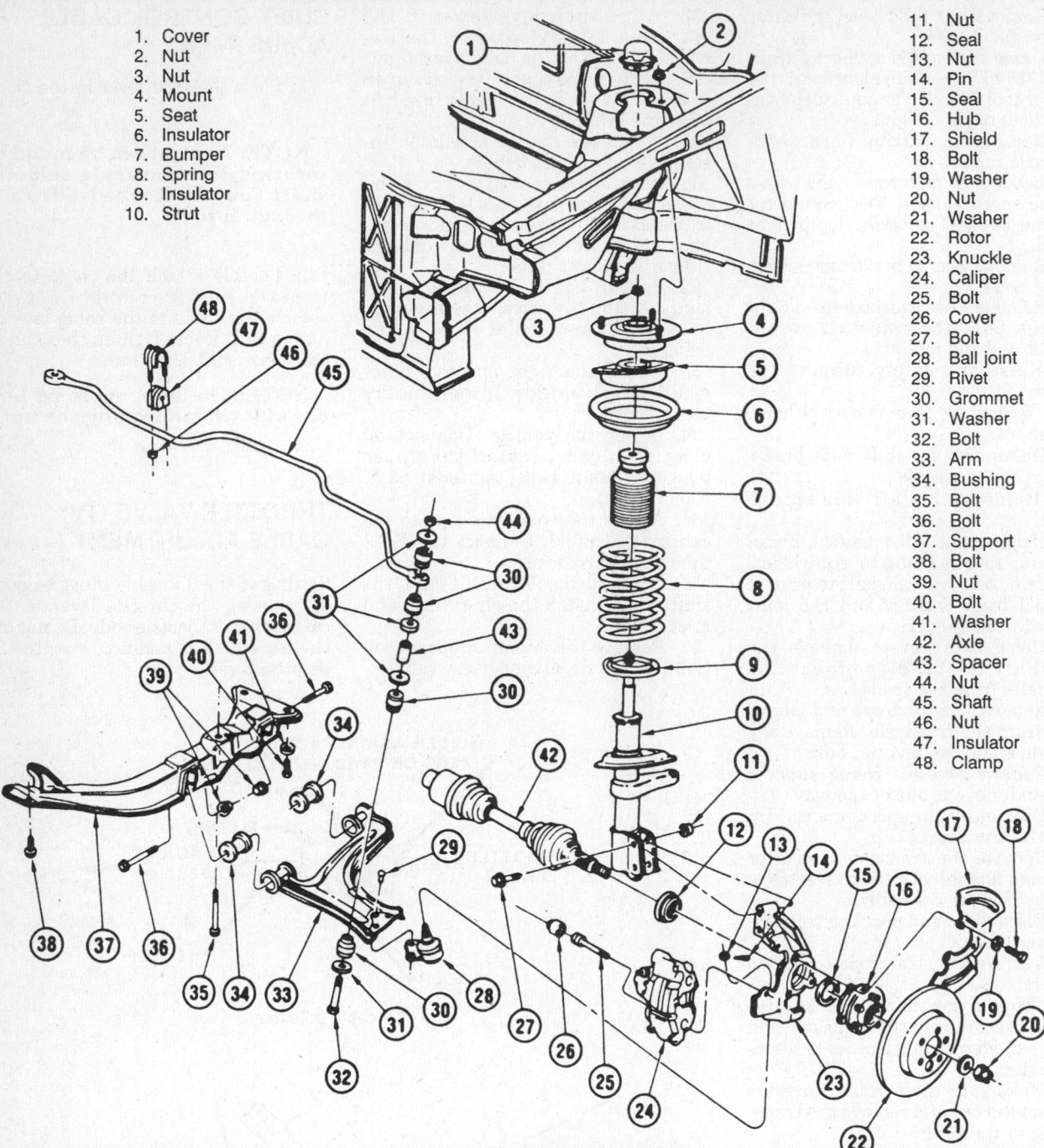

1. Cover
2. Nut
3. Nut
4. Mount
5. Seat
6. Insulator
7. Bumper
8. Spring
9. Insulator
10. Strut

11. Nut
12. Seal
13. Nut
14. Pin
15. Seal
16. Hub
17. Shield
18. Bolt
19. Washer
20. Nut
21. Wsaher
22. Rotor
23. Knuckle
24. Caliper
25. Bolt
26. Cover
27. Bolt
28. Ball joint
29. Rivet
30. Grommet
31. Washer
32. Bolt
33. Arm
34. Bushing
35. Bolt
36. Bolt
37. Support
38. Bolt
39. Nut
40. Bolt
41. Washer
42. Axle
43. Spacer
44. Nut
45. Shaft
46. Nut
47. Insulator
48. Clamp

Exploded view of the front suspension

1. With the engine off, depress and hold the reset tab at the engine end of the TV cable.

2. Move the slider until it stops against the fitting.

3. Release the rest tab.

4. Rotate the throttle lever to its full travel.

5. The slider must move (ratchet) toward the lever when the lever is rotated to its full travel position.

6. Recheck after the engine is hot and road test the vehicle.

FRONT SUSPENSION

MacPherson Strut

REMOVAL & INSTALLATION

NOTE: Before removing front suspension components, their positions should be marked so they may be assembled correctly.

1. Remove the 3 strut-to-body nuts.
2. Raise and safely support the vehicle.
3. Lower the vehicle slightly so the weight rests on jackstands at the frame and not on the control arms.
4. Remove the front wheel and tire assemblies. Remove the tie rod from the strut assembly using tool J–24319 or equivalent.
5. Some vehicles may use a silicone (gray) boot on the inboard axle joint. Use the boot protector tool J–33162 or equivalent, on these boots. All other boots are made from a thermoplastic

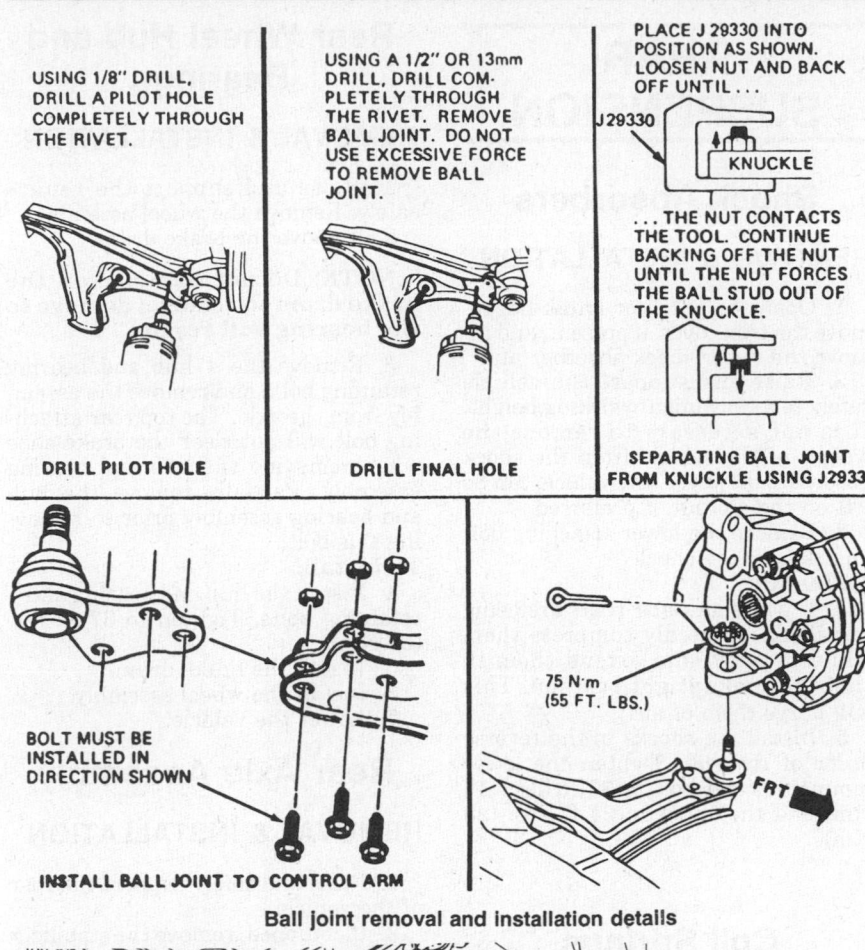

USING 1/8" DRILL, DRILL A PILOT HOLE COMPLETELY THROUGH THE RIVET.

USING A 1/2" OR 13mm DRILL, DRILL COMPLETELY THROUGH THE RIVET. REMOVE BALL JOINT. DO NOT USE EXCESSIVE FORCE TO REMOVE BALL JOINT.

PLACE J 29330 INTO POSITION AS SHOWN. LOOSEN NUT AND BACK OFF UNTIL...

J29330 / KNUCKLE

...THE NUT CONTACTS THE TOOL. CONTINUE BACKING OFF THE NUT UNTIL THE NUT FORCES THE BALL STUD OUT OF THE KNUCKLE.

DRILL PILOT HOLE

DRILL FINAL HOLE

SEPARATING BALL JOINT FROM KNUCKLE USING J29330

BOLT MUST BE INSTALLED IN DIRECTION SHOWN

INSTALL BALL JOINT TO CONTROL ARM

75 N·m (55 FT. LBS.)

FRT

Ball joint removal and installation details

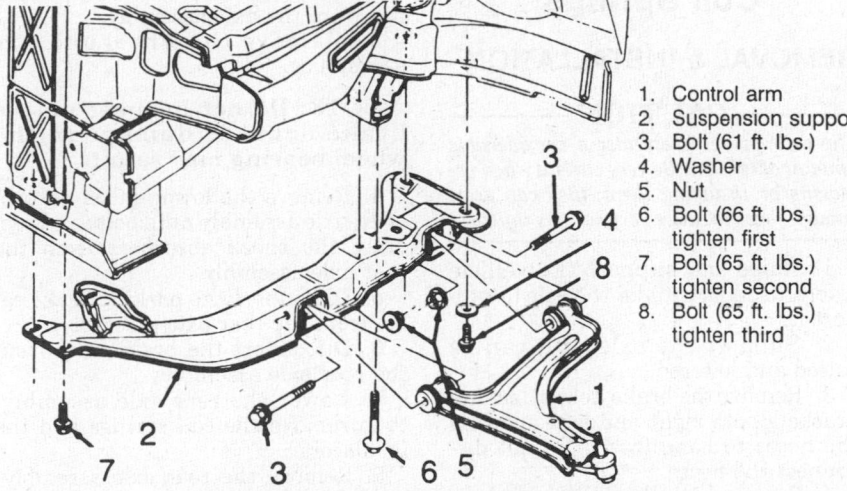

1. Control arm
2. Suspension support
3. Bolt (61 ft. lbs.)
4. Washer
5. Nut
6. Bolt (66 ft. lbs.) tighten first
7. Bolt (65 ft. lbs.) tighten second
8. Bolt (65 ft. lbs.) tighten third

Lower control arm installation torque sequence

material (black) and do not require the use of a boot seal protector.

6. Disconnect the brake line bracket from the strut assembly.

7. Remove the strut-to-steering knuckle bolts.

NOTE: Support steering knuckle to prevent tension from being applied to brake hose.

8. Remove the strut assembly from the vehicle. Care should be taken to avoid chipping or cracking the spring coating when handling the front suspension coil spring assembly.

To install:

9. Install the strut assembly to the vehicle.

10. Install the strut-to-steering knuckle bolts. Tighten to 133 ft. lbs. (184 Nm).

11. Connect the brake line bracket to the strut assembly.

12. If installed, remove the boot protector.

13. Connect the tie rod end to the strut assembly. Install the front wheel and tire assemblies.

14. Lower the vehicle.

15. Install the 3 strut-to-body nuts. Tighten to 18 ft. lbs. (24 Nm).

16. Check and/or adjust the front end alignment.

Lower Ball Joint

INSPECTION

1. Raise and safely support the vehicle allowing the suspension to hang free.

2. Grasp the wheel at the top and bottom, shake it in an "in-and-out" motion. Check for any horizontal movement of the steering knuckle relative to the lower control arm. Replace the ball joint if such movement is noted.

3. If the ball stud is disconnected from the steering knuckle and any looseness is detected or if the ball stud can be twisted in its socket using finger pressure, replace the ball joint.

REMOVAL & INSTALLATION

NOTE: This procedure requires the use of a special tool. The MacPherson strut suspension design does not use an upper ball joint.

1. Raise and support the vehicle safely. Remove the wheel assembly.

2. Use a 1/8 in. drill bit to drill a hole through the center of each of the 3 ball joint rivets.

3. Use a 1/2 in. drill bit to drill completely through the rivet.

4. Use a hammer and punch to remove the rivets. Drive them out from the bottom.

5. Use the special tool J–29330 or a ball joint removal tool, to separate the ball joint from the steering knuckle.

6. Disconnect the stabilizer bar from the lower control arm. Remove the ball joint.

To install:

7. Install the new ball joint into the control arm with the 3 bolts supplied with the replacement joint.

8. Installation of the remaining components is in the reverse order of removal. Use a new cotter pin when installing the castellated nut on the ball joint.

9. Check the toe setting and adjust, as necessary.

Lower Control Arms

REMOVAL & INSTALLATION

1. Raise and support the vehicle safely. Remove the wheel assembly.

2. Disconnect the stabilizer bar from the control arm and/or support.

3. Separate the ball joint from the steering knuckle.

4. Remove the 2 control arm-to-support bolts and remove the control arm.

5. If control arm support bar removal is necessary, unscrew the 6 mounting bolts and remove the support.

To install:

6. If control arm support bar was removed, install the support and the 6 mounting bolts. Tighten the control arm support rail bolts, in sequence.

7. Install the control arm and 2 control arm-to-support bolts.

8. Connect the ball joint to the steering knuckle.

9. Connect the stabilizer bar to the control arm and/or support.

10. Install the wheel. Lower the vehicle.

11. Check the toe and adjust, as necessary.

Stabilizer Bar

REMOVAL & INSTALLATION

1. Raise and support the vehicle safely so the front suspension hang free.

2. Remove the left front wheel and tire.

3. Disconnect the stabilizer from the control arms.

4. Disconnect the stabilizer from the support assemblies.

5. Loosen the front bolts and remove the rear and center bolts from the support assemblies to lower them enough to remove the stabilizer shaft.

6. Remove the stabilizer shaft with grommets and insulators.

To install:

7. Install the stabilizer shaft with grommets and insulators.

8. Install the stabilizer shaft and tighten the front bolts and install the rear and center bolts to the support assemblies.

9. Connect the stabilizer to the support assemblies.

10. Connect the stabilizer to the control arms.

11. Install the left front wheel and tire.

12. Lower the vehicle.

Front Wheel Hub and Bearing

For the front wheel hub and bearing procedure, refer to the Drive Axle section.

REAR SUSPENSION

Shock Absorbers

REMOVAL & INSTALLATION

1. Open the hatch or trunk lid, remove the trim cover, if present, and remove the upper shock absorber nut.

2. Raise and support the vehicle safely to a convenient working height. It is not necessary to remove the weight of the vehicle from the shock absorbers, however, the vehicle can be left on the ground, if preferred.

3. Remove the lower attaching bolt and remove the shock.

To install:

4. If new shock absorbers are being installed, repeatedly compress them while inverted and extend them in their normal upright position. This will purge them of air.

5. Install the shocks in the reverse order of removal. Tighten the lower mount nut and bolt to 35 ft. lbs. (47 Nm) and the upper to 21 ft. lbs. (28 Nm).

Coil Springs

REMOVAL & INSTALLATION

—— CAUTION ——

The coil springs are under a considerable amount of tension. Be very careful when removing or installing them; they can exert enough force to cause very serious injuries.

1. Raise and support the vehicle safely, use a jack under the axle to support it.

2. Support the axle so it can be raised and lowered.

3. Remove the brake hose attaching brackets both right and left, allowing the hoses to hang freely. Do not disconnect the hoses.

4. Remove both shock absorber lower attaching bolts from the axle.

5. Lower the axle. Remove the coil spring and insulator.

To install:

6. Position the spring and insulator on the axle.

7. The leg on the upper coil of the spring must be parallel to the axle, facing the left side of the vehicle.

8. Install the shock absorber bolts. Tighten to 35 ft. lbs. (47 Nm).

9. Install the brake line brackets. Tighten to 8 ft. lbs. (11 Nm).

10. Lower the vehicle.

Rear Wheel Hub and Bearings

REMOVAL & INSTALLATION

1. Raise and support the vehicle safely. Remove the wheel assembly.

2. Remove the brake drum.

NOTE: Do not hammer on the brake drum to remove; damage to the bearing will result.

3. Remove the 4 hub and bearing retaining bolts and remove the assembly from the axle. The top rear attaching bolt will not clear the brake shoe when removing the hub and bearing assembly. Partially remove the hub and bearing assembly prior to removing this bolt.

To install:

4. Install the hub and bearing and 4 retaining bolts. Tighten to 37 ft. lbs. (50 Nm).

5. Install the brake drum.

6. Install the wheel assembly.

7. Lower the vehicle.

Rear Axle Assembly

REMOVAL & INSTALLATION

1. Raise and safely support the rear of the vehicle.

2. If equipped, remove the stabilizer bar from the axle assembly.

3. Remove the wheel and tire assemblies.

NOTE: Do not hammer on the brake drum as damage to the wheel bearing may result.

4. Remove the lower shock absorber-to-axle assembly nuts/bolts and separate the shock absorbers from the rear axle assembly.

5. Disconnect the parking brake cable from the rear axle assembly.

6. Disconnect the brake lines from the rear axle assembly.

7. Lower the rear axle assembly, then remove the coil springs and the insulators.

8. Remove the rear axle assembly-to-chassis bolts and remove the axle assembly.

To install:

9. Install the rear axle assembly and the axle assembly-to-chassis bolts. Tighten to 37 ft. lbs. (50 Nm).

10. Install the coil springs and insulators and raise the rear axle assembly.

11. Connect the brake lines to the rear axle assembly.

12. Connect the parking brake cable to the rear axle assembly.

13. Connect the shock absorbers to the rear axle assembly. Install the low-

er shock absorber-to-axle assembly nuts/bolts.

14. Install the tire and wheel assemblies.

15. If equipped, install the stabilizer bar to the rear axle assembly.

16. Lower the vehicle.

STEERING

Steering Wheel

REMOVAL & INSTALLATION

Standard Steering Wheel

1. Disconnect the negative battery cable.

2. Pull the pad from the wheel. The horn lead is attached to the pad at one end; the other end of the pad has a wire with a spade connector. The horn lead is disconnected by pushing and turning; the spade connector is simply unplugged.

3. Remove the retainer under the pad, if equipped.

4. Remove the steering shaft nut.

5. There should be alignment marks already present on the wheel and shaft. If not, matchmark the parts.

6. Remove the wheel with a puller.

To install:

7. Install the wheel on the shaft, aligning the matchmarks. Install the shaft nut and tighten to 30 ft. lbs. (41 Nm).

8. Install the retainer.

9. Plug in the spade connector, push and turn the horn lead to connect. Install the pad.

10. Connect the negative battery cable.

Sport Steering Wheel

1. Disconnect the negative battery cable.

2. Pry the center cap from the wheel.

3. Remove the retainer, if equipped.

4. Remove the shaft nut.

5. If the wheel and shaft do not have factory alignment marks, matchmark the parts before removal of the wheel.

6. Install a puller and remove the wheel. A horn spring, eyelet and insulator are underneath.

To install:

7. Install the spring, eyelet and insulator into the tower in the column.

8. Align the matchmarks and install the wheel onto the shaft. Install the retaining nut and tighten to 30 ft. lbs. (41 Nm).

9. Install the retainer. Install the center cap. Connect the negative battery cable.

Steering Column

REMOVAL & INSTALLATION

NOTE: Once the steering column is removed from the vehicle, the column is extremely susceptible to damage. Dropping the column assembly on its end could collapse the steering shaft or loosen the plastic injections which maintain column rigidity. If it is necessary to remove the steering wheel, use a standard wheel puller. Under no condition should the end of the shaft be hammered upon, as hammering could loosen or break the plastic injection which maintains column rigidity.

1. Disconnect the negative battery cable.

2. If column repairs are to be made, remove the steering wheel.

3. Remove the sound insulator panels, as necessary, to gain access to the steering column retaining bolts.

4. Remove the nuts and bolts attaching the flexible coupling to the bottom of the steering column. Remove the safety strap and bolt, if equipped.

5. Remove the steering column trim shrouds and column covers.

6. Disconnect all wiring harness connectors. Remove the dust boot mounting screws and column mounting bracket bolts.

7. Remove the shift cable at the actuator and housing holder.

8. Lower the column to clear the mounting bracket and carefully remove from the vehicle.

To install:

9. Install the steering column in the vehicle.

10. Install the shift cable at the actuator and housing holder.

11. Connect all wiring harness connectors. Install the dust boot mounting screws and column mounting bracket bolts.

12. Install the steering column trim shrouds and column covers.

13. Install the nuts and bolts attaching the flexible coupling to the bottom of the steering column. Install safety strap and bolt, if equipped.

14. If removed, install the sound insulator panels.

15. If column repairs were made, install the steering wheel.

NOTE: Some vehicles equipped with tilt steering columns may experience a squeaking noise when turning the steering wheel in a tilted position. This can be caused by insufficient grease in the tilting mechanism.

16. Disconnect the negative battery cable.

Power Steering Rack

REMOVAL & INSTALLATION

1. Disconnect the negative battery cable. Remove the air cleaner.

2. Raise and safely support the vehicle.

3. Remove both front wheel assemblies.

4. Remove the intermediate shaft lower pinch bolt at the steering gear. Remove the intermediate shaft from the stub shaft.

5. Disconnect the electrical lead at the power steering idle switch.

6. Separate the tie rod ends from the knuckle assembly. Remove the rear sub-frame mounting bolts and lower the rear of the sub-frame approximately 4 in.

7. Remove the steering rack heat shield. Disconnect the pressure lines at the steering gear.

8. Remove the rack and pinion mounting bolts, remove the rack and pinion assembly through the left wheel opening.

To install:

9. Install the rack and pinion assembly through the left wheel opening. Tighten the mounting bolts to 59 ft. lbs. (80 Nm). Connect the pressure lines, tighten the fittings to 20 ft. lbs. (27 Nm).

10. Install the rack heat shield, tighten the retaining bolts to 53 inch lbs. (6 Nm). Attach the tie rod ends to the steering knuckle.

11. Connect the electrical lead to the power steering idle switch. Attach the intermediate shaft to the stub shaft, tighten the pinch bolt to 35 ft. lbs. (47 Nm).

12. Install both wheel assemblies. Lower the vehicle.

13. Install the air cleaner. Connect the negative battery cable. Fill and bleed the power steering system.

Power Steering Pump

REMOVAL & INSTALLATION

1. Disconnect the negative battery cable.

2. If equipped with V-belt, loosen the adjusting bolt and pivot bolt on the pump. Remove the pump drive belt.

3. Remove the 3 pump-to-bracket bolts and remove the adjusting bolt.

4. Remove the high pressure fitting from the pump.

5. Disconnect the reservoir-to-pump hose from the pump.

6. Remove the pump.

To install:

7. Install the pump.

8. Connect the reservoir-to-pump hose to the pump.

9. Install the high pressure fitting to the pump.

10. Install the adjusting bolt and the 3 pump-to-bracket bolts.

11. Install the drive belt. If equipped with V-belt, adjust belt tension and tighten the adjusting bolt and pivot bolt on the pump. Remove the pump drive belt.

12. Disconnect the negative battery cable.

13. Bleed the system.

BELT ADJUSTMENT

1. Loosen the adjustment nut and bolt in the slotted bracket. Slightly loosen the pivot bolt.

2. Pull the component outward to increase tension. Push inward to reduce tension. Tighten the adjusting nut, bolt and the pivot bolt.

3. Recheck the drive belt tension which is 135 lbs. on a new belt, 75 lbs. on a used belt and readjust, if necessary.

NOTE: On a serpentine belt the correct tension is indicated on the indicator mark of the belt tensioner. If the indicator mark is not within specification, replace the belt or tensioner.

SYSTEM BLEEDING

1. Raise the front of the vehicle and support safely.

2. With the wheels turned all the way to the left, add power steering fluid to the **COLD** mark on the fluid level indicator.

3. Start the engine and check the fluid level at fast idle. Add fluid, if necessary, to bring the level up to the **COLD** mark.

4. Bleed air from the system by turning the wheels from side-to-side without hitting the stops. Keep the fluid level just above the internal pump casting or at the **COLD** mark.

5. Return the wheels to the center position and continue running the engine for 2–3 minutes.

6. Road test the vehicle to check steering function and recheck the fluid level with the system at its normal operating temperature. Fluid should be at the **HOT** mark.

Tie Rod Ends

REMOVAL & INSTALLATION

1. Loosen both pinch bolts at the outer tie rod.

2. Remove the tie rod end from the strut assembly using a suitable removal tool.

3. Unscrew the outer tie rod end from the tie rod adjuster, counting the number of turns required before they are disconnected.

To install:

4. Install the new tie rod end, screwing it on the same number of turns as counted in Step 3.

5. When the tie rod end is installed, the tie rod adjuster must be centered between the tie rod and the tie rod end, with an equal number of threads exposed on both sides of the adjuster nut. Tighten the pinch bolts to 20 ft. lbs. (27 Nm).

6. Install the tie rod end to the strut assembly and tighten to 50 ft. lbs. (68 Nm). If the cotter pin cannot be installed, tighten the nut up to $\frac{1}{16}$ in. further. Never back off the nut to align the holes for the cotter pin.

7. Check front end alignment.

BRAKES

For all brake system repair and service procedures not detailed below, please refer to "Brakes" in the Unit Repair section.

Master Cylinder

REMOVAL & INSTALLATION

1. Disconnect the electrical connector from the master cylinder.

2. Place a container under the master cylinder to catch the brake fluid. Disconnect the brake tubes from the master cylinder; use a flare nut wrench if one is available. Plug the ends of the tubes.

NOTE: Brake fluid eats paint. Wipe up any spilled fluid immediately and flush the area with clear water.

3. Remove the 2 nuts attaching the master cylinder to the booster or firewall.

To install:

4. Attach the master cylinder to the booster with the nuts. Tighten to 22–30 ft. lbs. (30–41 Nm).

5. Remove the tape from the lines and connect to the master cylinder. Tighten to 10–15 ft. lbs. (14–20 Nm).

Connect the electrical lead.

6. Bleed the brakes.

NOTE: When installing a master cylinder that mounts on an angle, attempts to bleed the system, with the cylinder installed, can allow air to enter the system. To remove air, it is necessary to raise the rear of the vehicle until the master cylinder bore is level.

Proportioning Valve

REMOVAL & INSTALLATION

There is a front and a rear proportioning valve located at the lower left side of the master cylinder.

1. Disconnect the brake lines from the valves. Disconnect the valves from the master cylinder and remove the O-rings.

2. Replace the old O-rings and proportioning valves with new ones and reinstall into the master cylinder.

3. Tighten the proportioning valves to 18–30 ft. lbs. (24–41 Nm).

Power Brake Booster

REMOVAL & INSTALLATION

1. Remove the master cylinder from the booster and set the master cylinder aside. It is not necessary to disconnect the lines from the master cylinder.

2. Disconnect the vacuum booster pushrod from the brake pedal inside the vehicle. It is retained by a bolt. A spring washer is under the bolt head and a flat washer goes on the other side of the pushrod eye, next to the pedal arm.

3. Remove the 4 attaching nuts from inside the vehicle. Remove the booster.

To install:

4. Install the booster on the firewall. Tighten the mounting nuts to 22–33 ft. lbs. (30–45 Nm).

5. Connect the pushrod to the brake pedal.

6. Install the master cylinder. Mounting torque is 22–33 ft. lbs. (30–45 Nm).

Brake Caliper

REMOVAL & INSTALLATION

1. Remove $\frac{2}{3}$ of the brake fluid from the master cylinder.

2. Raise and safely support the vehicle.

3. Remove the wheel and tire and reinstall 2 nuts to retain the rotor.

4. Position a 12 inch adjustable pliers over the inboard brake shoe tab

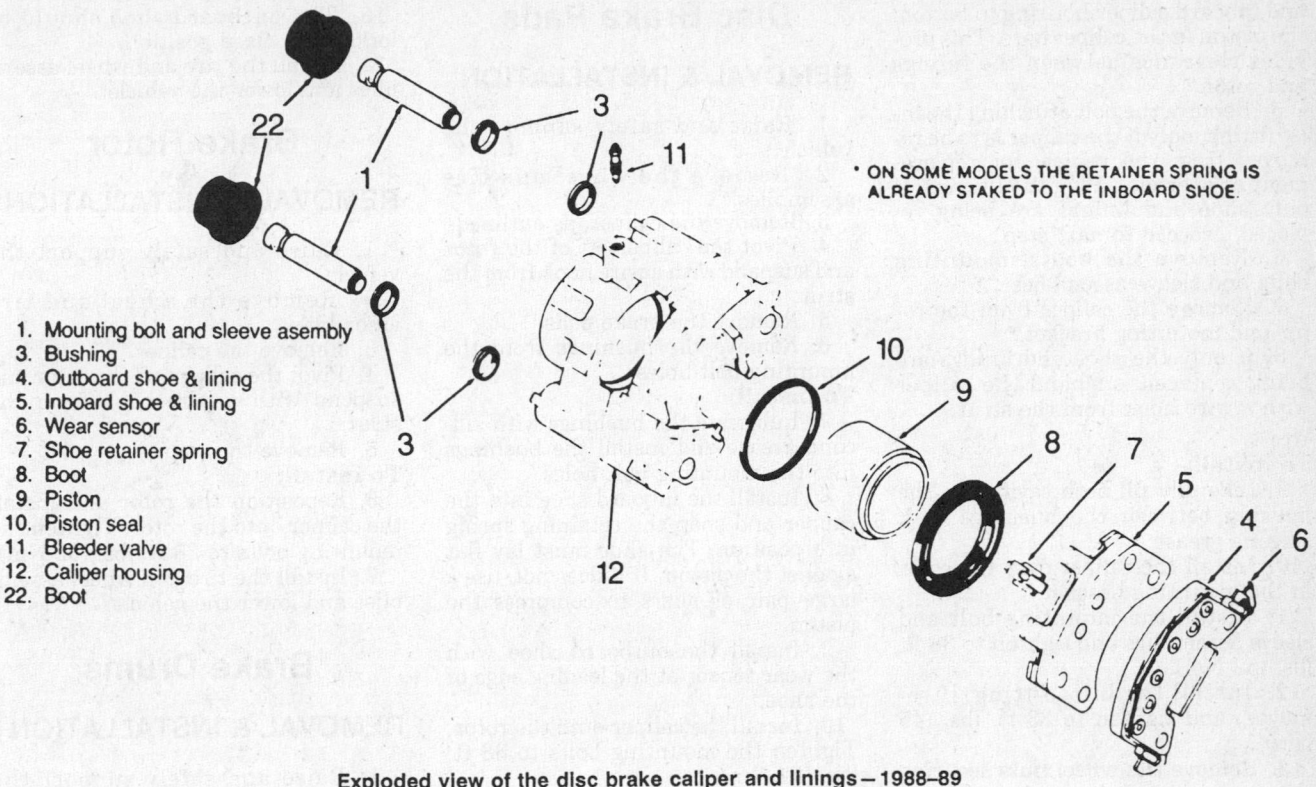

1. Mounting bolt and sleeve asembly
3. Bushing
4. Outboard shoe & lining
5. Inboard shoe & lining
6. Wear sensor
7. Shoe retainer spring
8. Boot
9. Piston
10. Piston seal
11. Bleeder valve
12. Caliper housing
22. Boot

* ON SOME MODELS THE RETAINER SPRING IS
ALREADY STAKED TO THE INBOARD SHOE.

Exploded view of the disc brake caliper and linings — 1988–89

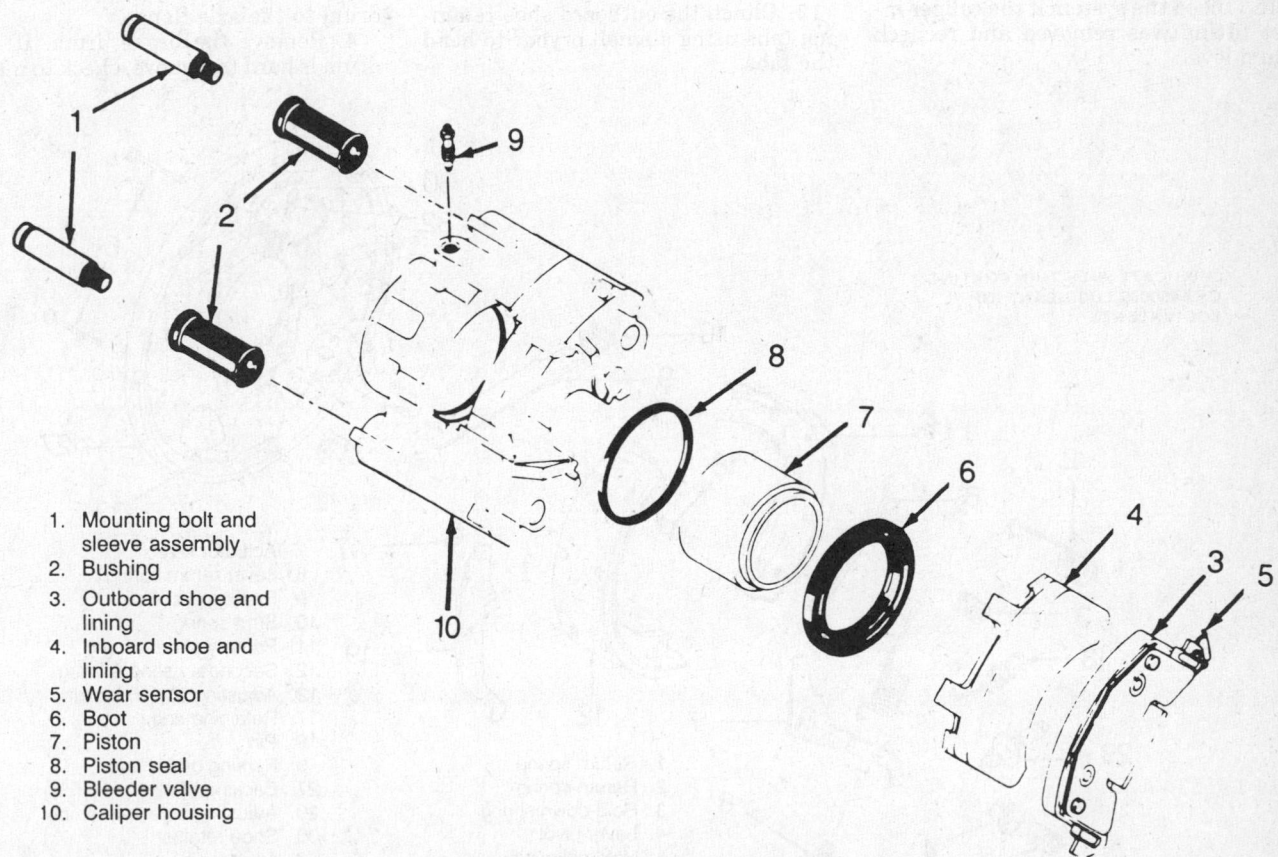

1. Mounting bolt and
   sleeve assembly
2. Bushing
3. Outboard shoe and
   lining
4. Inboard shoe and
   lining
5. Wear sensor
6. Boot
7. Piston
8. Piston seal
9. Bleeder valve
10. Caliper housing

Exploded view of the disc brake caliper and linings — 1990–91

and inboard caliper housing to bottom the piston in the caliper bore. This provides clearance between the linings and rotor.

5. Remove the bolt attaching the inlet fitting, only if the caliper is to be removed from the vehicle for replacement or overhaul. Plug the fittings. If only shoe and linings are being replaced, proceed to next step.

6. Remove the boots, mounting bolts and sleeve assemblies.

7. Remove the caliper from the rotor and mounting bracket.

8. If only the shoe and linings are being replaced, suspend the caliper with a wire hook from the strut.

To install:

9. Liberally fill both cavities in the housing between the bushings with silicone grease.

10. Install the caliper over the rotor in the mounting bracket.

11. Install the mounting bolt and sleeve assemblies and tighten to 38 ft. lbs. (52 Nm).

12. Install the inlet fitting, if removed, and tighten to 33 ft. lbs. (45 Nm).

13. Remove the wheel nuts securing the rotor to the hub.

14. Install the wheel and tire, lower the vehicle and fill the master cylinder.

15. Bleed the system if the caliper inlet fitting was removed and recheck fluid level.

Disc Brake Pads

REMOVAL & INSTALLATION

1. Raise and safely support the vehicle.

2. Remove the wheel and tire assemblies.

3. Remove the caliper, as outlined.

4. Pivot the caliper off of the rotor and suspend with a wire hook from the strut.

5. Remove the brake pads.

6. Remove the bushings from the mounting bolt holes.

To install:

7. Lubricate the bushings with silicone grease and install the bushings into the mounting bolt holes.

8. Install the inboard shoe into the caliper and snap the retaining spring into position. The shoe must lay flat against the piston. If it does not, use a large pair of pliers to compress the piston.

9. Install the outboard shoe with the wear sensor at the leading edge of the shoe.

10. Install the caliper onto the rotor. Tighten the mounting bolts to 38 ft. lbs. (52 Nm).

11. Apply the brake 3 times with approximately 175 lbs. (778 N) of force, this will seat the linings.

12. Clinch the outboard shoe retaining tabs using a small prybar to bend the tabs.

13. The outboard shoe should be locked in a fixed position.

14. Install the tire and wheel assemblies and lower the vehicle.

Brake Rotor

REMOVAL & INSTALLATION

1. Raise and safely support the vehicle.

2. Remove the wheel and tire assemblies.

3. Remove the caliper.

4. Pivot the caliper off the rotor and suspend with a wire hook from the strut.

5. Remove the rotor.

To install:

6. Reposition the rotor and install the caliper onto the rotor. Tighten the mounting bolts to 38 ft. lbs. (52 Nm).

7. Install the tire and wheel assemblies and lower the vehicle.

Brake Drums

REMOVAL & INSTALLATION

1. Raise and safely support the vehicle.

2. Remove the wheel.

3. Mark the relationship of the drum to the axle flange.

4. Remove the brake drum. If the drum is hard to remove, check to make

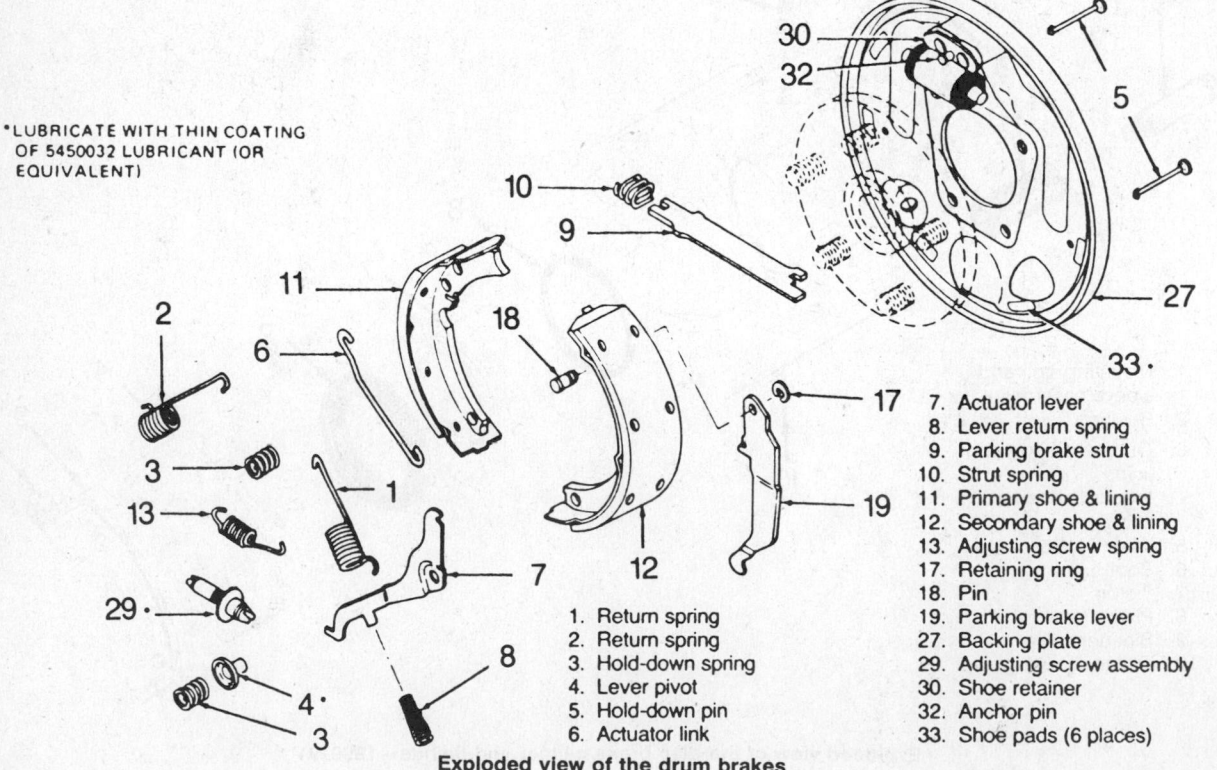

*LUBRICATE WITH THIN COATING OF 5450032 LUBRICANT (OR EQUIVALENT)

1. Return spring
2. Return spring
3. Hold-down spring
4. Lever pivot
5. Hold-down pin
6. Actuator link
7. Actuator lever
8. Lever return spring
9. Parking brake strut
10. Strut spring
11. Primary shoe & lining
12. Secondary shoe & lining
13. Adjusting screw spring
17. Retaining ring
18. Pin
19. Parking brake lever
27. Backing plate
29. Adjusting screw assembly
30. Shoe retainer
32. Anchor pin
33. Shoe pads (6 places)

Exploded view of the drum brakes

sure the parking brake is off and the brake adjuster is not turned all the way out.

To install:

5. Install the brake drum.
6. Install the wheel.
7. Adjust the brakes.
8. Lower the vehicle.

Brake Shoes

REMOVAL & INSTALLATION

1. Raise and safely support the vehicle.
2. Remove the wheel and tire assemblies.
3. Remove the brake drum, if the drum is hard to remove, check to make sure the parking brake is off and the brake adjuster is not turned all the way out.
4. Remove the return springs, using brake spring pliers.
5. Remove the hold-down springs and pins. Remove the lever pivot.
6. Remove the actuator link while lifting up on the actuator lever.
7. Remove the actuator lever and lever return spring.
8. Remove the parking brake strut and strut spring.
9. Remove the brake shoes, after removing the parking brake cable from the shoe.
10. Remove the adjusting screw assembly and spring. Remove the retaining ring, pin and parking brake lever from the secondary shoe.

To install:

11. Install the parking brake lever on

the secondary shoe with the pin and retaining ring.
12. Install the adjusting screw and spring.
13. Install the brake shoe assemblies after installing the parking brake cable on the shoe.
14. Install the parking brake strut and strut spring by spreading the shoes apart.
15. Install the actuator lever and lever return spring.
16. Install the hold-down pins, lever pivot and springs. Install the actuator link on the anchor pin.
17. Install the actuator link into the actuator lever while holding up on the lever.
18. Install the shoe return springs. Install the brake drum. Install the wheel and tire assemblies.
19. Lower the vehicle and apply the brakes repeatedly. Bleed and adjust the brakes, as required.

Wheel Cylinder

REMOVAL & INSTALLATION

1. Raise the rear of the vehicle and support it safely.
2. Remove the rear wheel and brake drum assembly.
3. Disconnect the inlet tube nut and line from the wheel cylinder.
4. Remove the wheel cylinder retainer using 2 awls or pins ⅛ in. diameter or less.
   a. Insert the awls or pins into the access slots between the wheel cylin-

der pilot and the retainer locking tabs.
   b. Bend both tabs away simultaneously.
5. Remove the wheel cylinder.

To install:

6. Position the wheel cylinder and hold it in place using a wooden block placed between the the wheel cylinder and the axle flange.
7. Install a new wheel cylinder retainer over the wheel cylinder abutment using a 1⅛ inch 12 point socket and extension.
8. Reconnect the inlet tube nut and tighten to 12 ft. lbs. (16 Nm).
9. Reinstall the brake drum and bleed the brake system.
10. Install the wheels, lower the vehicle and check for leaks.

Parking Brake Cable

ADJUSTMENT

1. Disconnect the negative battery cable. Raise and safely support the vehicle with both rear wheels off the ground.
2. Pull the parking brake lever exactly 2 ratchet clicks.

NOTE: To prevent damage to the threaded adjusting rod, thoroughly clean and lubricate the threads before turning the adjusting nut.

3. Loosen the equalizer locknut and tighten the adjusting nut until the left rear wheel can just be turned back-

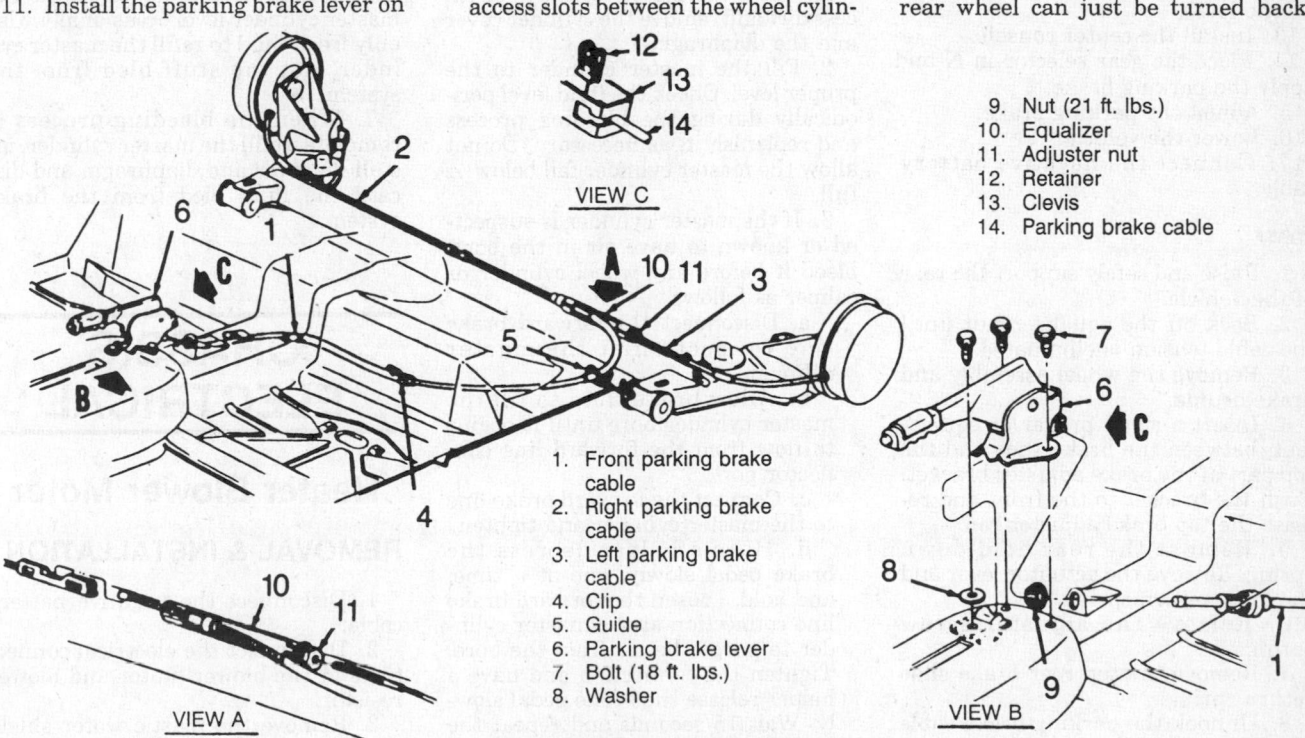

9. Nut (21 ft. lbs.)
10. Equalizer
11. Adjuster nut
12. Retainer
13. Clevis
14. Parking brake cable

1. Front parking brake cable
2. Right parking brake cable
3. Left parking brake cable
4. Clip
5. Guide
6. Parking brake lever
7. Bolt (18 ft. lbs.)
8. Washer

VIEW A

VIEW B

VIEW C

Parking brake lever and cables

ward using 2 hands but is locked in forward rotation.

4. Tighten the locknut.

5. Release the parking brake. Rotate the rear wheels, there should be no drag.

6. Lower the vehicle.

REMOVAL & INSTALLATION

Front

1. Place the gear selector in **N** and apply the parking brake.

2. Remove the center console.

3. Disconnect the parking brake cable from the lever.

4. Remove the cable retaining nut and the bracket securing the front cable to the floor panel.

5. Raise the vehicle and loosen the equalizer nut.

6. Loosen the catalytic converter shield and remove the parking brake cable from the body.

7. Disconnect the cable from the equalizer and remove the cable from the guide and the underbody clips.

To install:

8. Install the cable to the guide and the underbody clips and connect the cable to the equalizer.

9. Install the parking brake cable to the body tighten the catalytic converter shield.

10. Tighten the equalizer nut. Lower the vehicle.

11. Install the cable retaining nut and the bracket securing the front cable to the floor panel.

12. Connect the parking brake cable to the lever.

13. Install the center console.

14. Place the gear selector in **N** and apply the parking brake.

15. Adjust the parking brake.

16. Lower the vehicle.

17. Connect the negative battery cable.

Rear

1. Raise and safely support the rear of the vehicle.

2. Back off the equalizer nut until the cable tension is eliminated.

3. Remove the wheel assembly and brake drums.

4. Insert a small prybar or equivalent, between the brake shoe and the top part of the brake adjuster bracket. Push the bracket to the front and release the top brake adjuster rod.

5. Remove the rear hold-down spring. Remove the actuator lever and the lever return spring.

6. Remove the adjuster screw spring.

7. Remove the top rear brake shoe return spring.

8. Unhook the parking brake cable from the parking brake lever.

9. Depress the conduit fitting retaining tangs and remove the conduit fitting from the backing plate.

10. Remove the cable end button from the connector.

11. Depress the conduit fitting retaining tangs and remove the conduit fitting from the axle bracket.

To install:

12. Install the conduit fitting to the axle bracket.

13. Install the cable end button to the connector.

14. Install the conduit fitting to the backing plate.

15. Hook the parking brake cable to the parking brake lever.

16. Install the top rear brake shoe return spring.

17. Install the adjuster screw spring.

18. Install the actuator lever and the lever return spring. Install the rear hold-down spring.

19. Connect the top brake adjuster rod.

20. Install the wheel assembly and brake drums.

21. Adjust the parking brake.

22. Lower the vehicle.

Brake System Bleeding

The brake system must be bled when any brake line is disconnected or there is air in the system.

NOTE: Never bleed a wheel cylinder when a drum is removed.

1. Clean the master cylinder of excess dirt and remove the cylinder cover and the diaphragm.

2. Fill the master cylinder to the proper level. Check the fluid level periodically during the bleeding process and replenish it, as necessary. Do not allow the master cylinder fall below ½ full.

3. If the master cylinder is suspected or known to have air in the bore, bleed it before any wheel cylinder or caliper as follows:

a. Disconnect the forward brake line connection at the master cylinder.

b. Allow brake fluid to fill the master cylinder bore until it begins to flow from the forward line connector port.

c. Connect the forward brake line to the master cylinder and tighten.

d. Have a helper depress the brake pedal slowly, one at a time, and hold. Loosen the forward brake line connection at the master cylinder to purge the air from the bore. Tighten the connection and have a helper release the brake pedal slowly. Wait 15 seconds and repeat the sequence. Repeat the sequence in-

cluding the 15 second wait until all air is removed from the bore.

e. After all air is removed at the forward connection, repeat the above procedure for the rear connection at the master cylinder.

4. Bleed the individual wheel cylinders or calipers only after all air is removed from the master cylinder.

a. Attach the proper size box end wrench over the bleeder valve.

b. Attach a length of vinyl hose to the bleeder screw of the brake to be bled. Insert the other end of the hose into a clear jar half full of clean brake fluid, so the end of the hose is beneath the level of fluid. The correct sequence for bleeding is to work from the brake farthest from the master cylinder to the one closest; right rear, left rear, right front, left front.

5. Have an assistant depress and release the brake pedal one time and hold. Loosen the bleeder valve to purge the air from the cylinder. Tighten the bleeder screw and slowly release the pedal and wait 15 seconds. Repeat the sequence including the 15 second wait until all air is removed.

NOTE: Make sure an assistant presses the brake pedal to the floor slowly. Rapid pumping of the brake pedal pushes the master cylinder secondary piston down the bore in a way that makes it difficult to bleed the rear side of the system.

6. Repeat this procedure at each of the brakes. Remember to check the master cylinder level occasionally. Use only fresh fluid to refill the master cylinder, not the stuff bled from the system.

7. When the bleeding process is complete, refill the master cylinder, install its cover and diaphragm and discard the fluid bled from the brake system.

CHASSIS ELECTRICAL

Heater Blower Motor

REMOVAL & INSTALLATION

1. Disconnect the negative battery cable.

2. Disconnect the electrical connections at the blower motor and blower resistor.

3. Remove the plastic water shield from the right side of the cowl.

4. Remove the blower motor retaining screws and remove the blower motor and cage.

5. Hold the blower motor cage and remove retaining nut from the blower motor shaft.

6. Remove the blower motor and cage.

To install:

7. Install the cage on the new motor.

8. Check that the retaining nut is on tight, the motor rotates and the fan cage is not interfering with the motor.

9. Install the motor in the heater assembly, connect the wiring and check the motor operation in all speeds.

Windshield Wiper Motor

REMOVAL & INSTALLATION

1. Disconnect the negative battery cable. Loosen, but do not remove the drive link-to-crank arm attaching nuts to detach the drive link from the motor crank arm.

2. Tag and disconnect all electrical leads from the wiper motor.

3. Unscrew the mounting bolts, rotate the motor up, outward and remove.

To install:

4. Guide the crank arm through the opening in the body and install the mounting bolts. Tighten to 48 inch lbs. (5 Nm).

5. Install the drive link to the crank arm with the motor in the park position.

6. Replace the shroud top vent grille and wiper arms.

7. Connect the negative battery cable.

8. Check the operation of the wiper system.

Windshield Wiper Switch

REMOVAL & INSTALLATION

1988–90

1. Disconnect the negative battery cable.

2. Remove the left side insulator panel. If equipped, remove the lower steering column cover.

3. Remove the steering wheel.

4. Remove the shaft lock cover by prying away from steering column housing.

5. Use lock plate compressor tools J–23653 and J–23653–4 to remove the shaft lock retaining ring. Install the tools to the steering column shaft and tighten the nut until the tool slightly depresses the shaft lock.

6. Remove the shaft lock retaining ring.

7. Remove the shaft lock.

8. Remove the cancelling cam assembly.

9. Remove the upper bearing spring.

10. If equipped with column-mounted dimmer control, remove the switch actuator arm assembly.

11. Disconnect the turn signal switch wiring harness connector at the base of the steering column and remove the wire protector. Remove the turn signal switch.

12. Turn the ignition switch to the **RUN** position and remove the key warning buzzer switch by depressing the tangs on the clip.

13. Remove the lock retaining screw. Remove the lock cylinder.

14. Remove the steering column cover attaching screws and remove the cover.

15. Disconnect the wiper switch electrical connector.

16. Remove the screws attaching the column housing to the mast jacket. Note the position of the dimmer switch actuator rod for reassembly. Remove the screws attaching the dimmer switch.

NOTE: Tilt and travel columns have a removable plastic cover on the column housing, providing access to the wiper switch without removing the entire column housing.

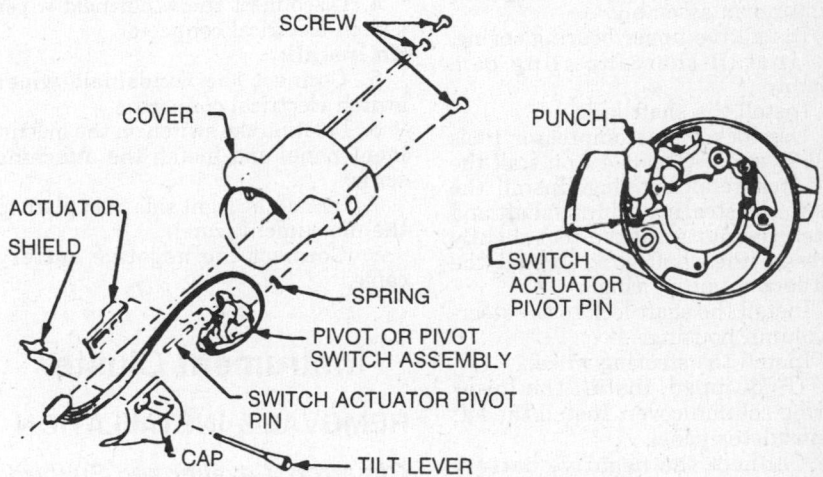

Wiper switch and related parts—adjustable steering column

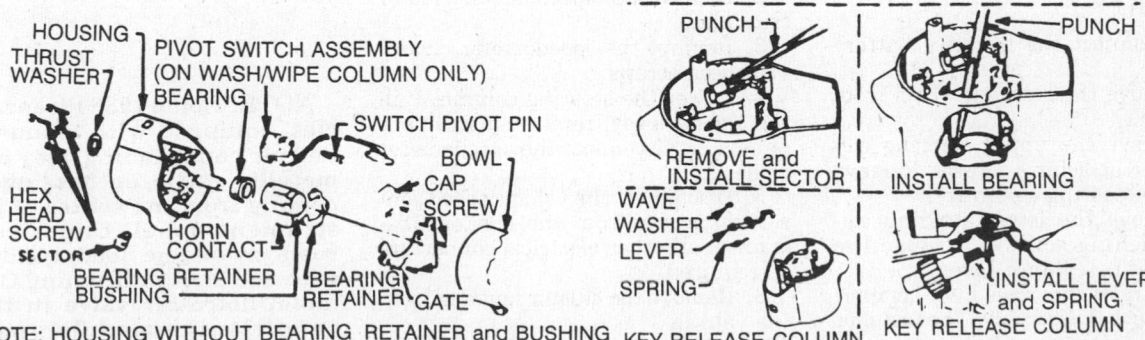

NOTE: HOUSING WITHOUT BEARING RETAINER and BUSHING HAS SPUN-IN BEARING. IF REPAIR IS NECESSARY, COMPLETE HOUSING ASSEMBLY REPLACEMENT IS NECESSARY.

Wiper switch and related parts—standard steering column

17. Remove the screws attaching the ignition switch.

18. Remove the column housing and switches as an assembly.

19. Remove the wiper switch actuator pin and remove the wiper switch.

To install:

20. Place the switch into the housing and install the pivot pin.

NOTE: Ensure that the bearing retainer, horn contact and upper bearing are positioned properly.

21. Position the housing onto the mast jacket and install the attaching screws.

22. Reconnect lower switch wiring connector. Install the wire protector.

23. Install and adjust the dimmer switch.

24. Install and adjust the ignition switch.

25. Install the lock cylinder.

26. Turn the lock to the **RUN** position and install the key warning buzzer switch, using a paper clip to aid installation.

27. Install the turn signal switch and connect the electrical connector at the base of the steering column.

28. If equipped with column-mounted dimmer control, install the switch actuator arm assembly.

29. Install the upper bearing spring.

30. Install the cancelling cam assembly.

31. Install the shaft lock.

32. Use lock plate compressor tools J-23653 and J-23653-4 to install the shaft lock retaining ring. Install the tools to the steering column shaft and tighten the nut until the tool slightly depresses the shaft lock. Install the shaft lock retaining ring.

33. Install the shaft lock to the steering column housing.

34. Install the steering wheel.

35. If equipped, install the lower steering column cover. Install the left side insulator panel.

36. Connect the negative battery cable.

1991–92
CAVALIER

1. Disconnect the negative battery cable.

2. Remove the steering wheel from the column.

3. Remove the upper steering column cover attaching screws. Remove the upper steering column.

4. Remove the lower steering column attaching screws. Remove the lower steering column cover.

5. Separate the rose bud fastener from the jacket assembly. The fastener is integral to the wire harness.

6. Remove the wash/wipe switch attaching screws.

7. Depress the locking tab and remove the wire harness connector from the wash/wipe switch.

To install:

8. Place the wash/wipe switch into position and connect the electrical connector.

9. Install the wash/wipe switch attaching screws.

10. Connect the rose bud fastener to the jacket assembly.

11. Install the lower steering column cover and attaching screws. Tighten to 49 inch lbs. (6 Nm).

12. Install the upper steering column cover and attaching screws. Tighten to 49 inch lbs. (6 Nm).

13. Install the steering wheel onto the steering shaft. Install the hexagonal retaining nut and tighten to 30 ft. lbs. (41 Nm).

14. Connect the negative battery cable.

SUNBIRD

1. Disconnect the negative battery cable.

2. Remove the right side trim plate from the instrument panel.

3. Remove the windshield wiper switch attaching screw.

4. Disconnect the windshield wiper switch electrical connector.

To install:

5. Connect the windshield wiper switch electrical connector.

6. Position the switch on the instrument panel and install the attaching screw.

7. Install the right side trim plate to the instrument panel.

8. Connect the negative battery cable.

Instrument Cluster

REMOVAL & INSTALLATION

Cimarron, Cavalier and Sunbird

1. Disconnect the negative battery cable.

2. Remove the speedometer cluster trim plate.

3. Remove the speedometer cluster attaching screws.

4. Lower the steering column. Pull the cluster away from the instrument panel and disconnect the speedometer cable.

5. Disconnect the vehicle speed sensor connector from the cluster. Disconnect all other electrical connectors as required.

6. Remove the cluster housing from the vehicle.

To install:

7. Install the cluster housing from the vehicle.

8. Connect the vehicle speed sensor connector from the cluster. Connect all other electrical connectors.

9. Connect the speedometer cable. Push the instrument cluster into position on the instrument panel. Raise the steering column.

10. Install the speedometer cluster attaching screws.

11. Install the speedometer cluster trim plate.

12. Connect the negative battery cable.

Skyhawk and Firenza

1. Disconnect the negative battery cable.

2. Remove the steering column trim cover. Remove the left and right hand trim cover.

3. Remove the cluster trim cover.

4. Remove the screws attaching the lens and bezel to the cluster carrier.

5. Lower the steering wheel column by removing the 2 upper steering column attaching bolts.

6. Remove the screws attaching the cluster housing to the cluster carrier. Pull the cluster out slightly from the instrument panel and disconnect the speedometer cable. Disconnect all others connectors.

7. Remove the cluster housing from the vehicle.

To install:

8. Install the cluster housing.

9. Connect the speedometer cable. Connect all others connectors. Install the screws attaching the cluster housing to the cluster carrier.

10. Raise the steering wheel column and install the 2 upper steering column attaching bolts.

11. Install the screws attaching the lens and bezel to the cluster carrier.

12. Install the cluster trim cover.

13. Install the steering column trim cover. Install the left and right hand trim cover.

14. Connect the negative battery cable.

NOTE: Some 1988 Firenza vehicles, equipped with manual control air conditioning, may exhibit metallic rattle or buzzing noise coming from the center of the instrument panel. Under certain road or engine load conditions, with the air conditioning ON, the metal defroster valve in the air conditioning module assembly may vibrate against the case causing the noise. The noise may be heard in all modes except defrost.

Speedometer

REMOVAL & INSTALLATION

1. Disconnect the negative battery cable.
2. Remove speedometer cluster from instrument panel.
3. Remove cluster lens and face plate.
4. Remove screws securing speedometer to cluster assembly and remove speedometer and disconnect the cable.

To install:

5. Install speedometer and connect the cable. Install screws securing speedometer to cluster assembly.
6. Install cluster lens and face plate.
7. Install speedometer cluster from instrument panel.
8. Connect the negative battery cable.

Radio

REMOVAL & INSTALLATION

NOTE: Do not operate the radio with the speaker leads disconnected. Operating the radio without an electrical load will damage the output transistors.

1. Disconnect the negative battery cable.
2. Remove the center instrument panel trim plate.
3. Check the right side of the radio to determine whether a nut or a stud is used for side retention.
4. If a nut is used, remove the hush panel and loosen the nut from below, on vehicles without air conditioning. On vehicles with air conditioning, remove the hush panel, air conditioning duct and air conditioning control head for access to the nut. Do not remove the nut; loosen it just enough to pull the radio out. If a rubber stud is used, go to Step 5.
5. Remove the 2 radio bracket-to-instrument panel attaching screws. Pull the radio forward far enough to disconnect and tag the wiring and antenna. Remove the radio.

To install:

6. Connect the wiring and antenna. Place the radio into position and install the attaching screws.
7. If an attaching nut is used, install the nut and hush panel.
8. Install the center instrument panel trim plate.
9. Connect the negative battery cable.

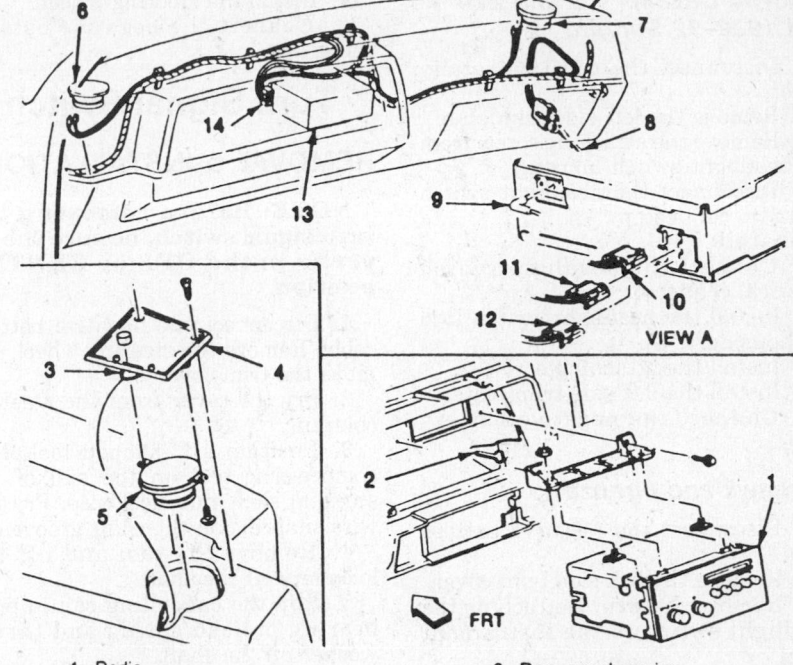

1. Radio
2. Screw on side of radio fits here
3. Retainer
4. Grille
5. Speaker
6. Front speaker
7. Front speaker
8. Rear speaker wire
9. Antenna
10. Rear speakers
11. Front speakers
12. Instrument panel harness
13. Receiver
14. Instrument panel harness

Typical radio installation

Headlight Switch

REMOVAL & INSTALLATION

Cimarron, 1988–90 Base Cavalier and 1988 Sunbird

1. Disconnect the negative battery cable.
2. Pull the knob out fully. Remove the knob from rod by depressing the retaining clip from the underside of the knob.
3. Remove the trim plate.
4. Remove the switch by removing nut, rotating the switch 180 degrees, then tilting forward and pulling it out. Disconnect the wire harness.

To install:

5. Connect the electrical connector. Place the switch into position and install the nut.
6. Install the trim plate.
7. Install the knob.
8. Connect the negative battery cable.

1991–92 Base Cavalier, Cavalier RS and Z24

1. Disconnect the negative battery cable.
2. Remove the steering wheel.
3. Remove the cover attaching screws.
4. Remove the upper steering column cover.
5. Remove the lower steering colum cover screws and remove the lower cover.
6. Separate the rose bud fastener (integral to the wire harness) from the jacket assembly.
7. Remove the switch mounting screws and remove the turn signal switch.
8. Depress the locking tabs and disconnect the wire harness connectors from the turn signal switch assembly.

To install:

9. Connect the wire harness connectors to the turn signal switch assembly.
10. Install the switch and install the attaching screws. Tighten to 49 inch lbs. (6 Nm).
11. Connect the rose bud fastener (integral to the wiring harness) to the jacket assembly.
12. Install the lower steering column cover and attaching screws. Tighten to 49 inch lbs. (6 Nm).
13. Install the upper steering column cover and attaching screws. Tighten to 49 inch lbs. (6 Nm).
14. Install the steering wheel on the steering shaft with the hexagon nut. Tighten to 30 ft. lbs. (41 Nm).
15. Connect the negative battery cable.

1988–90 Cavalier RS and Z24 and 1989–92 Sunbird

1. Disconnect the negative battery cable.
2. Remove the left side trim plate.
3. Remove the attaching screw from the headlight switch housing.
4. Disconnect the headlight switch electrical connector.

To install:

5. Connect the headlight switch electrical connector.
6. Install the headlight switch into the housing.
7. Install the attaching screw.
8. Install the left side trim plate.
9. Connect the negative battery cable.

Skyhawk and Firenza

1. Disconnect the negative battery cable.
2. Remove the left side trim cover.
3. Remove the screws attaching the headlight switch to the instrument panel.
4. Pull the switch rearward in order to release the locking tabs and remove the switch from the vehicle.

To install:

5. Connect the electrical connector.
6. Place the switch into position and install the attaching screws.
7. Install the left side trim panel.
8. Connect the negative battery cable.

Dimmer Switch

REMOVAL & INSTALLATION

1. Disconnect the negative battery cable. Remove the steering wheel. Remove the trim cover.
2. Remove the turn signal switch assembly.
3. Remove the ignition switch stud and screw. Remove the ignition switch.
4. Remove the dimmer switch actuator rod by sliding it from the switch assembly.
5. Remove the dimmer switch bolts and the dimmer switch.

To install:

6. Install the dimmer switch and attaching bolts.
7. Connect the dimmer switch actuator rod.
8. Adjust the dimmer switch by depressing the switch slightly and inserting a $^3/_{32}$ in. drill bit into the adjusting hole. Push the switch up to remove any play and tighten the dimmer switch adjusting screw.
9. Install the turn signal switch assembly.
10. Install the trim cover.

11. Install the steering wheel.
12. Connect the negative battery cable.

Turn Signal Switch

REMOVAL & INSTALLATION

NOTE: Before removing the turn signal switch, be sure the lever is in the OFF or CENTER position.

1. Disconnect the negative battery cable. Remove the steering wheel. Remove the trim cover.
2. Pry the cover from the steering column.
3. Position a U-shaped lockplate compressing tool on the end of the steering shaft nut clockwise. Pry the wire snapring on the shaft groove off.
4. Remove the tool and lift the lockplate off the shaft.
5. Slip the cancelling cam, upper bearing preload spring and thrust washer off the shaft.
6. Remove the turn signal lever.
   a. Make sure the switch is in the center or **OFF** position.
   b. Pull the lever straight out of the turn signal switch.
   c. If equipped with cruise control, attach the connector to mechanic's wire and pull the harness through the column.
7. Remove the hazard flasher button retaining screw and remove the button, spring and knob.
8. Pull the switch connector out of the mast jacket and tape the upper part to facilitate switch removal. Attach a long piece of wire to the turn signal switch connector. When installing the turn signal switch, feed this wire through the column first and then use this wire to pull the switch connector into position. On tilt columns, place the turn signal and shifter housing in **LOW** position and remove the harness cover.
9. Remove the 3 switch mounting screws. Remove the switch by pulling it straight up while guiding the wire harness cover through the column.

To install:

10. Install the replacement switch by working the connector and cover down through the housing and under the bracket. If equipped with tilt steering, the connector is worked down through the housing, under the bracket and then the cover is installed on the harness.
11. Install the switch mounting screws and the connector on the mast jacket bracket. Install the column-to-dash trim plate.
12. Install the flasher knob and turn the signal lever.
13. With the turn signal lever in mid-

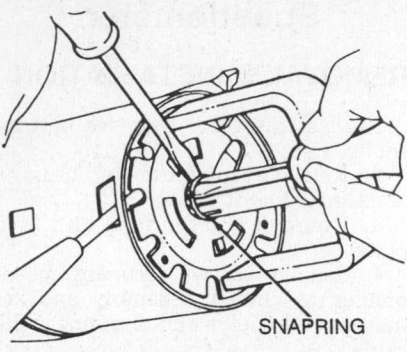

Using the lock plate depressing tool to remove the snapring

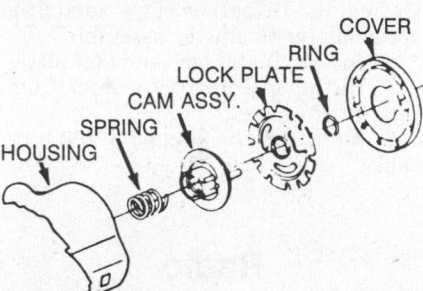

Remove these parts for removal of the turn signal switch

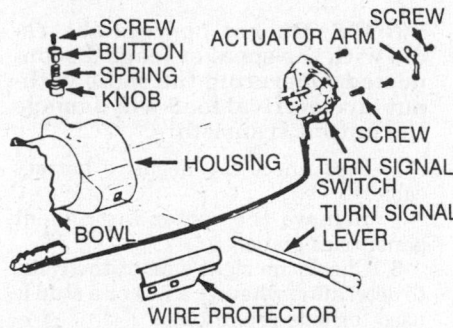

Turn signal switch removal details

dle position and the flasher knob out, slide the thrust washer, upper bearing preload spring and cancelling cam onto the shaft.
14. Position the lock plate on the shaft and press it down until a new snapring can be inserted in the shaft groove. Always use a new snapring when assembling.
15. Install the cover and steering wheel. Connect the negative battery cable.

Ignition Switch

REMOVAL & INSTALLATION

The switch is located inside the channel section of the brake pedal support and is completely inaccessible without first lowering the steering column. The switch is actuated by a rod and rack assembly. A gear on the end of the

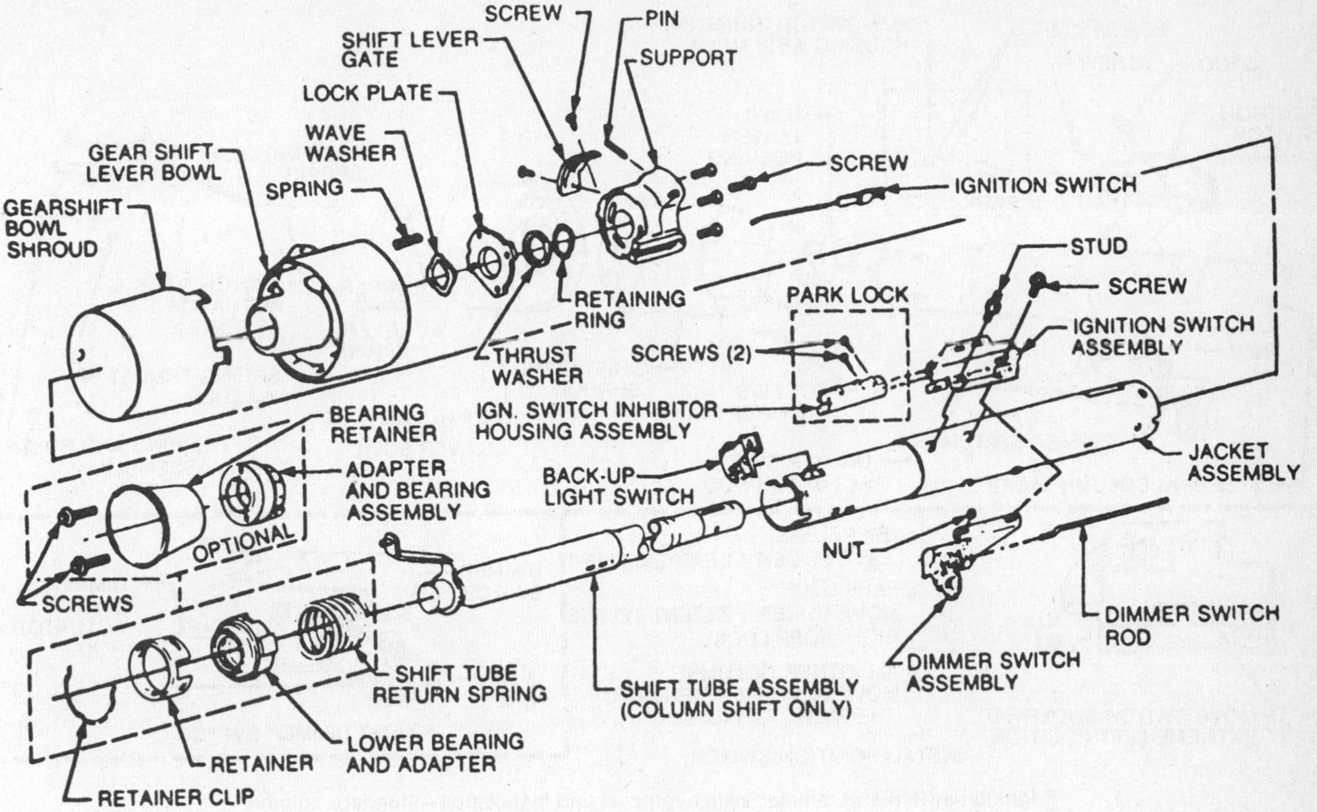

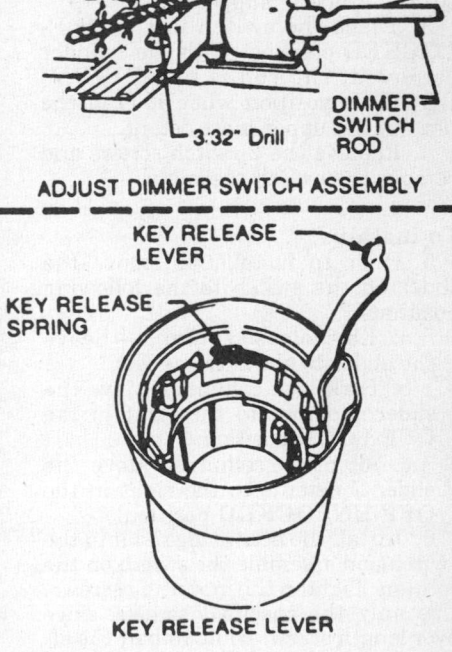

GEARSHIFT BOWL SHROUD

GEAR SHIFT LEVER BOWL

SPRING

WAVE WASHER

LOCK PLATE

SHIFT LEVER GATE

SCREW — PIN

SUPPORT

SCREW

IGNITION SWITCH

RETAINING RING

THRUST WASHER

PARK LOCK

SCREWS (2)

IGN. SWITCH INHIBITOR HOUSING ASSEMBLY

STUD

SCREW

IGNITION SWITCH ASSEMBLY

BEARING RETAINER

ADAPTER AND BEARING ASSEMBLY

OPTIONAL

BACK-UP LIGHT SWITCH

JACKET ASSEMBLY

NUT

SCREWS

DIMMER SWITCH ROD

SHIFT TUBE RETURN SPRING

LOWER BEARING AND ADAPTER

RETAINER

RETAINER CLIP

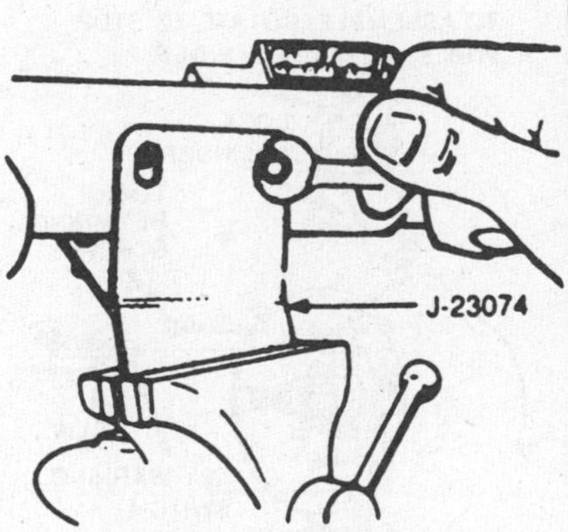

SHIFT TUBE ASSEMBLY (COLUMN SHIFT ONLY)

DIMMER SWITCH ASSEMBLY

J-23074

STEERING COLUMN HOLDING FIXTURE

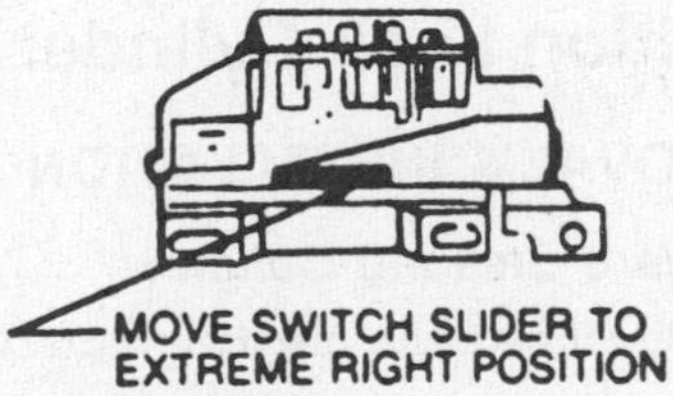

MOVE SWITCH SLIDER TO EXTREME RIGHT POSITION

- KEY RELEASE
  LEAVE SLIDER AT EXTREME RIGHT
- PARK LOCK
  MOVE SLIDER 1 DETENTS TO THE LEFT (OFF LOCK)
- ALL OTHER COLUMNS
  MOVE SLIDER 2 DETENTS TO THE LEFT (OFF UNLOCK)

INSTALL IGNITION SWITCH ASSEMBLY

DIMMER SWITCH ASSEMBLY

3/32" Drill

DIMMER SWITCH ROD

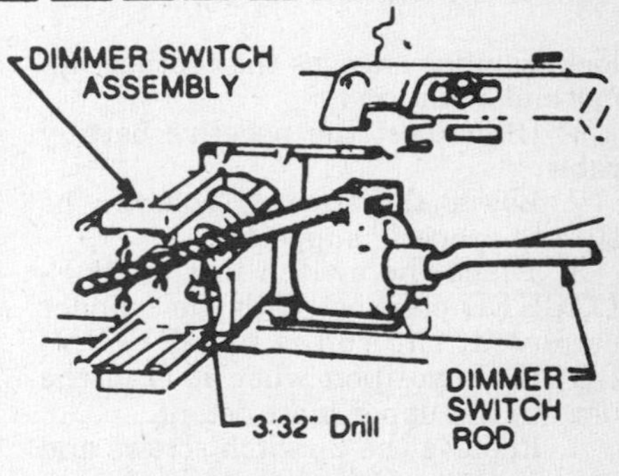

ADJUST DIMMER SWITCH ASSEMBLY

J-23072

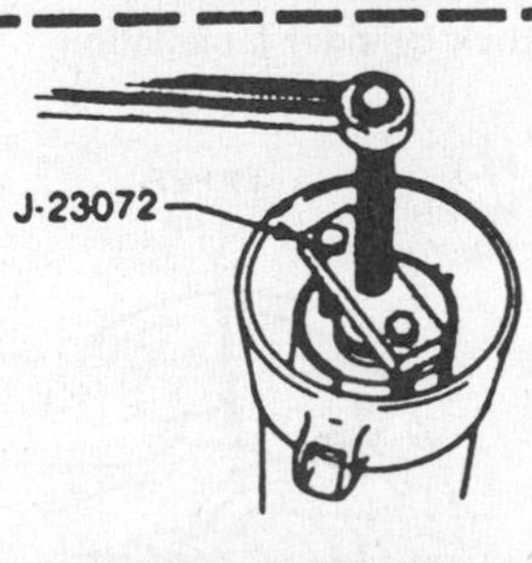

REMOVE SHIFT TUBE ASSEMBLY FROM BOWL

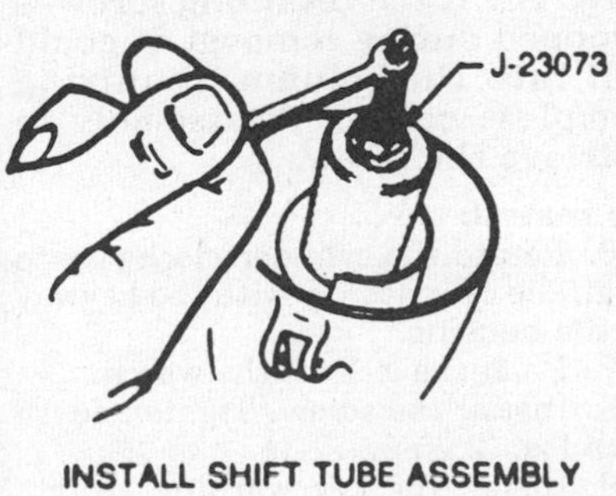

J-23073

INSTALL SHIFT TUBE ASSEMBLY

KEY RELEASE LEVER

KEY RELEASE SPRING

KEY RELEASE LEVER

Ignition switch and dimmer switch removal and installation—adjustable column

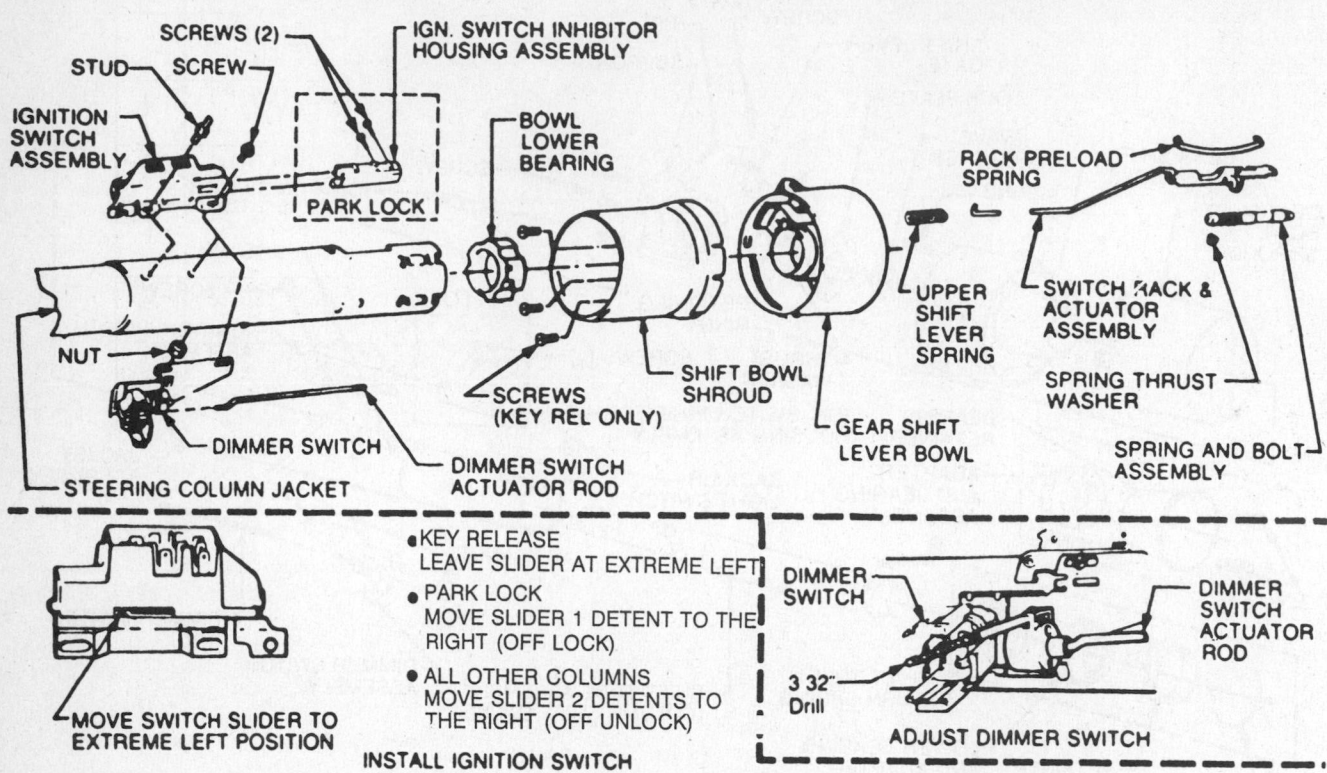

Ignition switch and dimmer switch removal and installation—standard column

lock cylinder engages the toothed upper end of the rod.

1. Disconnect the negative battery cable.

2. Lower the steering column; be sure to properly support it.

3. Place the switch in the **OFF-LOCKED** position. With the cylinder removed, the rod is in **OFF-UN-LOCKED** position when it is in the next to the upper most detent.

4. Remove the 2 switch screws and remove the switch assembly.

To install:

5. Prior to installation, move the slider on the switch to the following positions:

   a. Key release columns—Leave the slider to the extreme left.

   b. Park lock columns—Move the slider 1 detent to the right in the **OFF-LOCK** position.

   c. All other columns—Move the slider 2 detents to the right in the **OFF-UNLOCKED** position.

6. Install the activating rod into the switch and assemble the switch on the column. Tighten the mounting screws. Use only the specified screws, since over length screws could impair the effectiveness of the column to collapse.

7. Install the steering column.

8. Connect the negative battery cable.

Ignition Lock Cylinder

REMOVAL & INSTALLATION

Standard Steering Column

1. Disconnect the negative battery cable.

2. Remove the steering wheel.

3. Turn the ignition key to the **RUN** position.

4. Remove the lock plate, turn signal or combination switch and the key warning buzzer switch. The warning buzzer switch is pulled out with small tool.

5. Remove the lock cylinder retaining screw and lock cylinder.

NOTE: If the retaining screw is dropped during removal, it could fall into the column, requiring complete column disassembly to retrieve the screw.

To install:

6. Rotate the cylinder clockwise to align the cylinder key with the keyway in the housing.

7. Push the lock all the way in.

8. Install the screw. Tighten to 15 inch lbs. (2 Nm).

9. Install the key warning switch. Turn the lock to **RUN** position and push the key warning buzzer switch into place.

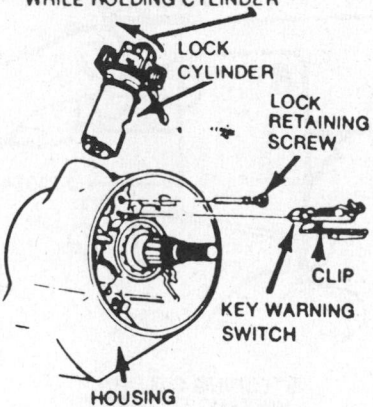

Lock cylinder installation

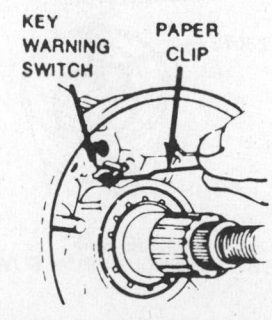

Removing the key warning buzzer

10. Install the turn signal switch and lock plate.

11. Install the steering wheel.

12. Connect the negative battery cable.

Tilt Steering Column

1. Disconnect the negative battery cable. Remove the steering wheel.

2. Remove the rubber sleeve bumper from the steering shaft.

3. Using an appropriate tool, remove the plastic retainer.

4. Using a spring compressor, compress the upper steering shaft spring and remove the C-ring. Release the steering shaft lock plate, the horn contact carrier and the upper steering shaft preload spring.

5. Remove the 4 screws which hold the upper mounting bracket and then remove the bracket.

6. Slide the harness connector out of the bracket on the steering column. Tape the upper part of the harness and connector.

7. Disconnect the hazard button and position the shift bowl, on automatic transmission equipped vehicles in the **P** position. Remove the turn signal lever from the column.

8. If equipped with cruise control, remove the harness protector from the harness. Attach a piece of wire to the switch harness connector. Before removing the turn signal lever, loop a piece of wire and insert it into the turn signal lever opening. Use the wire to pull the cruise control harness out through the opening. Pull the rest of the harness up through and out of the column. Remove the guide wire from the connector and secure the wire to the column. Remove the turn signal lever.

9. Pull the turn signal switch up until the end connector is within the shift bowl. Remove the hazard flasher lever. Allow the switch to hang.

10. Place the ignition key in the **RUN** position.

11. Depress the center of the lock cylinder retaining tab with a suitable tool and then remove the lock cylinder.

To install:

12. Install the lock cylinder.

13. Install the turn signal switch. Install the hazard flasher lever.

14. Install the turn signal lever.

15. If equipped with cruise control, use the guide wire installed during removal to feed the harness into position and connect the electrical connector. Install the harness protector. Install the turn signal lever.

16. If equipped with automatic transmission, position the shift bowl and connect the hazard button.

17. Slide the harness connector into the bracket on the steering column.

18. Install the upper mounting bracket and the 4 attaching screws.

19. Using a spring compressor, compress the upper steering shaft spring and install the C-ring. Release the steering shaft lock plate, the horn contact carrier and the upper steering shaft preload spring.

20. Install the plastic retainer.

21. Install the rubber sleeve bumper to the steering shaft.

22. Install the steering wheel. Connect the negative battery cable.

Stoplight Switch

REMOVAL & INSTALLATION

The stoplight, cruise control and cruise control vacuum switch are all located on the brake pedal mounting bracket and are adjusted in an identical manner.

1. Remove the wiring from switch and remove the switch.

2. To install, insert the retaining tubular clip in bracket on the pedal assembly.

3. With the pedal depressed, insert the switch into the tubular clip until the switch body seats on clip.

NOTE: Audible clicks can be heard as threaded portion of switch is pushed through the clip toward the brake pedal. Vacuum release valve and stoplight switch are self-adjusting.

4. Connect the wiring for the switch.

Clutch Switch

ADJUSTMENT

The clutch start switch is used on vehicles equipped with a manual transaxle. The switch prevents the engine from starting unless the clutch pedal is depressed.

REMOVAL & INSTALLATION

1. Disconnect the negative battery cable.

2. Unbolt the switch from the instrument panel and disconnect the wiring.

To install:

3. Install the switch and the attaching nuts.

4. Connect the clutch switch electrical connector.

5. Connect the negative battery cable.

Neutral Safety System

All column shift automatic transaxle models use a mechanical neutral start system. This system has a mechanical block which prevents cranking the engine when the shift lever is in any position except **P** or **N**.

All floor shift automatic transaxle models use a park lock system. This system uses a flexible cable actuator which is attached at one end to the shift lever and the other end is attached to the column mounted ignition switch where it actuates a locking pin. The locking pin engages an ignition switch sliding contact when the shift lever is in **R**, **N** or **D** and does not allow the ignition switch slider to move to the **LOCK** position. When the shift lever is in **P**, the pin disengages from the slider and allows it to move to the **LOCK** position.

ADJUSTMENT

1. Place the transmission shifter in the **N** notch in the detent plate.

2. Loosen the attaching screws.

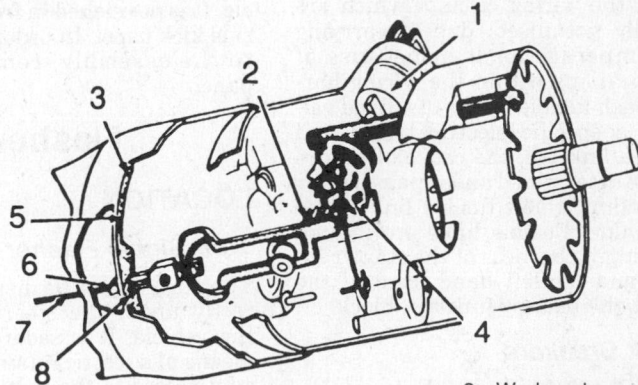

1. Lock cylinder
2. Rack
3. Bowl plate
4. Sector
5. Park position
6. Wedge shape finger
7. Actuator rod assembly
8. Neutral position

Mechanical neutral start system

3. Rotate the switch on the shifter assembly to align the service adjustment hole with the carrier tang hole.

4. Insert a 2.34mm gauge pin or a $\frac{3}{32}$ in. drill bit to a depth of $\frac{5}{8}$ in. (15mm).

5. Tighten the attaching screws to 22 ft. lbs. (30 Nm).

6. Remove the gauge pin.

REMOVAL & INSTALLATION

1. Disconnect the negative battery cable.

2. Disconnect the shift linkage.

3. Disconnect the electrical connector.

4. Remove the attaching bolts.

5. Remove the switch assembly.

To install:

6. Place the shift shaft in the **N** position.

7. Align the flats of the shift shaft with the switch.

8. Assembly the attaching bolts loosely.

9. Insert a 2.34mm diameter gauge pin or $\frac{3}{32}$ in. drill bit into the service adjustment hole and rotate the switch until the pin drops to a depth of $\frac{9}{64}$ in. (9mm).

10. Tighten the attaching bolts to 22 ft. lbs. (30 Nm).

11. Remove the gauge pin.

12. Verify that the engine will only start with the transmission selector in the **P** or **N** position.

Fuses, Circuit Breakers and Relays

LOCATION

Fusible Links

Fusible links are used to prevent major wire harness damage in the event of short circuit or an overload condition in the wiring circuits which are normally not fused, due to carrying high amperage loads or because of their locations within the wiring harness. Each fusible link is of a fixed value for a specific electrical load and should a link fail, the cause of failure must be determined and repaired prior to installing a new fusible link of the same value. Fusible links are located in the engine harness at the starter solenoid and the left hand front of the dash at the battery junction block.

Circuit Breakers

Circuit breakers are used along with the fusible links to protect the various components of the electrical system, such as headlights, the windshield wipers and electric windows. The circuit breakers are located either in the

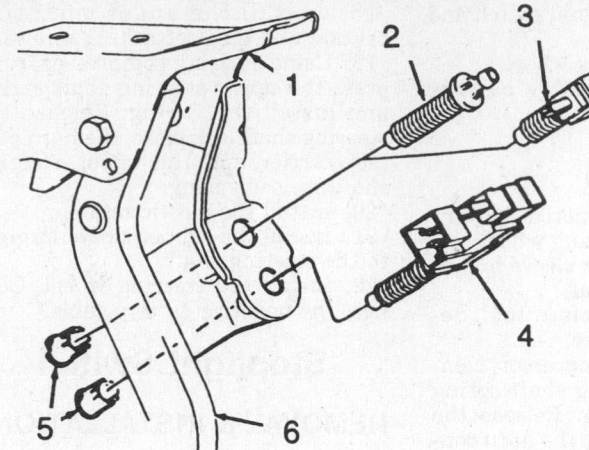

1. Brake pedal bracket
2. Vacuum release valve-manual transaxle
3. Vacuum release valve/switch automatic transaxle
4. Stoplight and cruise control switch
5. Retainer
6. Brake pedal

Cruise control vacuum valve/switch installation

switch or mounted on or near the lower lip of the instrument panel, to the right or left of the steering column.

Fuse Panels

The fuse panel is located on the left side of the vehicle. It is under the instrument panel assembly. In order to gain access to the fuse panel, it may be necessary to first remove the under dash padding.

Convenience Center and Various Relays

The convenience center is located on the underside of the instrument panel near the fuse panel. It provides a central location for various relays, hazard flasher units and buzzers. All units are easily replaced with plug-in modules.

Computer

LOCATION

The Electronic Control Module (ECM) is located on the right side of the vehicle. It is positioned in front of the right side kick panel. In order to gain access to the assembly, remove the trim panel.

Flashers

LOCATION

Turn Signal Flasher

The turn signal flasher is located directly under the steering column of the vehicle. It is secured in place by means of a plastic retainer. In order to gain access to the component, it may be necessary to remove the underdash padding panel.

Hazard Flasher

The hazard flasher is located in the

fuse block. It is positioned on the lower right side corner of the fuse block assembly. In order to gain access to the turn signal flasher, it may be necessary to first remove the under dash padding.

Cruise Control

ADJUSTMENT

Release Switch and Valve

1. Depress the brake pedal and insert the vacuum release valve into the retainer until a click is heard indicating that the valve switch is seated.

2. Allow the brake pedal to travel rearward to the positive stop.

3. The valve switch will be moved through the retainer into the proper position.

NOTE: Audible clicks can be heard as threaded portion of switch is pushed through the clip toward the brake pedal. Vacuum release valve and stoplight switch are self-adjusting.

Servo Cable

1. Install the cable into the engine bracket. Route the cable assembly to the servo bracket.

2. Pull the servo end of the cable towards the servo assembly without moving the throttle lever.

3. Line up the pin in the end of the cable with 1 of the holes in the servo assembly tab.

4. Insert the cable pin into 1 of the 6 holes in the servo bracket. Install the retainer.

NOTE: Do not stretch the cable to make a certain connection as this will prevent the engine from returning to idle. Use the next closest hole.

GM "N" Body
Front Wheel Drive

Buick—Skylark **Pontiac**—Grand Am
Oldsmobile—Calais

SPECIFICATIONS

VEHICLE IDENTIFICATION CHART

It is important for servicing and ordering parts to be certain of the vehicle and engine identification. The VIN (vehicle identification number) is a 17 digit number visible through the windshield on the driver's side of the dash and contains the vehicle and engine identification codes. The tenth digit indicates model year, and the eighth digit indicates engine code. It can be interpreted as follows:

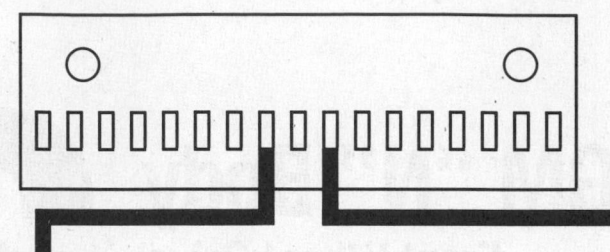

Engine Code						Model Year	
Code	Cu. In.	Liters	Cyl.	Fuel Sys.	Eng. Mfg.	Code	Year
M	122	2.0	4	Turbo	Pontiac	J	1988
A	138	2.3 HO	4	PFI	Oldsmobile	K	1989
D	138	2.3	4	PFI	Oldsmobile	L	1990
U	151	2.5	4	TBI	Pontiac	M	1991
L	181	3.0	6	PFI	Buick	N	1992
N	204	3.3	6	PFI	Buick		

HO High Output
TBI Throttle Body Injection
PFI Port Fuel Injection

ENGINE IDENTIFICATION

Year	Model	Engine Displacement cu. in. (liter)	Engine Series Identification (VIN)	No. of Cylinders	Engine Type
1988	Grand Am	122 (2.0)	M	4	OHC Turbo
	Grand Am	138 (2.3)	D	4	DOHC
	Grand Am	151 (2.5)	U	4	OHV
	Calais	138 (2.3)	D	4	DOHC
	Calais	151 (2.5)	U	4	OHV
	Calais	183 (3.0)	L	6	OHV
	Skylark	138 (2.3)	D	4	DOHC
	Skylark	151 (2.5)	U	4	OHV
	Skylark	181 (3.0)	L	6	OHV
1989	Grand Am	122 (2.0)	M	4	OHC Turbo
	Grand Am	138 (2.3)	D	4	DOHC
	Grand Am	138 (2.3)	A	4	DOHC-HO
	Grand Am	151 (2.5)	U	4	OHV
	Calais	138 (2.3)	D	4	DOHC
	Calais	138 (2.3)	A	4	DOHC-HO
	Calais	151 (2.5)	U	4	OHV
	Calais	204 (3.3)	N	6	OHV
	Skylark	138 (2.3)	D	4	DOHC
	Skylark	151 (2.5)	U	4	OHV
	Skylark	204 (3.3)	N	6	OHV

ENGINE IDENTIFICATION

Year	Model	Engine Displacement cu. in. (liter)	Engine Series Identification (VIN)	No. of Cylinders	Engine Type
1990	Grand Am	138 (2.3)	D	4	DOHC
	Grand Am	138 (2.3)	A	4	DOHC-HO
	Grand Am	151 (2.5)	U	4	OHV
	Calais	138 (2.3)	D	4	DOHC
	Calais	138 (2.3)	A	4	DOHC-HO
	Calais	151 (2.5)	U	4	OHV
	Calais	204 (3.3)	N	6	OHV
	Skylark	138 (2.3)	D	4	DOHC
	Skylark	151 (2.5)	U	4	OHV
	Skylark	204 (3.3)	N	6	OHV
1991–92	Grand Am	138 (2.3)	D	4	DOHC
	Grand Am	138 (2.3)	A	4	DOHC-HO
	Grand Am	151 (2.5)	U	4	OHV
	Calais	138 (2.3)	D	4	DOHC
	Calais	138 (2.3)	A	4	DOHC-HO
	Calais	151 (2.5)	U	4	OHV
	Calais	204 (3.3)	N	6	OHV
	Skylark	138 (2.3)	D	4	DOHC
	Skylark	151 (2.5)	U	4	OHV
	Skylark	204 (3.3)	N	6	OHV

DOHC Double Overhead Cam
OHV Overhead Valve
HO High Output

GENERAL ENGINE SPECIFICATIONS

Year	VIN	No. Cylinder Displacement cu. in. (liter)	Fuel System Type	Net Horsepower @ rpm	Net Torque @ rpm (ft. lbs.)	Bore × Stroke (in.)	Compression Ratio	Oil Pressure @ rpm
1988	M	4-122 (2.0)	Turbo	167 @ 4500	175 @ 4000	3.40 × 3.40	8.0:1	NA
	D	4-138 (2.3)	PFI	150 @ 5200	160 @ 4000	3.62 × 3.35	9.5:1	30 @ 2000
	U	4-151 (2.5)	TBI	98 @ 4300	135 @ 3200	4.00 × 3.00	8.3:1	37 @ 2000
	L	6-181 (3.0)	PFI	125 @ 4900	150 @ 2400	3.80 × 2.70	9.0:1	37 @ 2400
1989	M	4-122 (2.0)	Turbo	167 @ 4500	175 @ 4000	3.40 × 3.40	8.0:1	NA
	A	4-138 (2.3)	PFI	180 @ 6200	160 @ 5200	3.62 × 3.35	10.0:1	30 @ 2000
	D	4-138 (2.3)	PFI	160 @ 6200	155 @ 5200	3.62 × 3.35	9.5:1	30 @ 2000
	U	4-151 (2.5)	TBI	110 @ 5200	135 @ 3200	4.00 × 3.00	8.3:1	37 @ 2000
	N	6-204 (3.3)	PFI	160 @ 5200	185 @ 3200	3.70 × 3.16	9.0:1	45 @ 2000
1990	A	4-138 (2.3)	PFI	180 @ 6200	160 @ 5200	3.62 × 3.35	10.0:1	30 @ 2000
	D	4-138 (2.3)	PFI	160 @ 6200	155 @ 5200	3.62 × 3.35	9.5:1	30 @ 2000
	U	4-151 (2.5)	TBI	110 @ 5200	135 @ 3200	4.00 × 3.00	8.3:1	37 @ 2000
	N	6-204 (3.3)	PFI	160 @ 5200	185 @ 3200	3.70 × 3.16	9.0:1	45 @ 2000

GENERAL ENGINE SPECIFICATIONS

Year	VIN	No. Cylinder Displacement cu. in. (liter)	Fuel System Type	Net Horsepower @ rpm	Net Torque @ rpm (ft. lbs.)	Bore × Stroke (in.)	Compression Ratio	Oil Pressure @ rpm
1991–92	A	4-138 (2.3)	PFI	180 @ 6200	160 @ 5200	3.62 × 3.35	10.0:1	30 @ 2000
	D	4-138 (2.3)	PFI	160 @ 6200	155 @ 5200	3.62 × 3.35	9.5:1	30 @ 2000
	U	4-151 (2.5)	TBI	110 @ 5200	135 @ 3200	4.00 × 3.00	8.3:1	26 @ 800
	N	6-204 (3.3)	PFI	160 @ 5200	185 @ 3200	3.70 × 3.16	9.0:1	60 @ 1850

NA—Not Available
TBI Throttle Body Injection
PFI Port Fuel Injection

TUNE-UP SPECIFICATIONS

Year	VIN	No. Cylinder Displacement cu. in. (liter)	Spark Plugs Type	Gap (in.)	Ignition Timing (deg.) MT	AT	Compression Pressure (psi) ②	Fuel Pump (psi)	Idle Speed (rpm) MT	AT	Valve Clearance In.	Ex.
1988	M	4-122 (2.0)	R42XLS	0.035	①	①	100	35–38	④	④	Hyd.	Hyd.
	D	4-138 (2.3)	FR3LS	0.035	③	③	100	⑤	④	④	Hyd.	Hyd.
	U	4-151 (2.5)	R43TS6	0.060	③	③	100	9–13	④	④	Hyd.	Hyd.
	L	6-181 (3.0)	R44LTS	0.045	—	③	100	⑤	—	④	Hyd.	Hyd.
1989	M	4-122 (2.0)	R42XLS	0.035	①	①	100	35–38	④	④	Hyd.	Hyd.
	A	4-138 (2.3)	FR3LS	0.035	③	③	100	⑤	④	④	Hyd.	Hyd.
	D	4-138 (2.3)	FR3LS	0.035	③	③	100	⑤	④	④	Hyd.	Hyd.
	U	4-151 (2.5)	R43TS6	0.060	③	③	100	9–13	④	④	Hyd.	Hyd.
	N	6-204 (3.3)	R44LTS6	0.060	—	①	100	⑤	—	④	Hyd.	Hyd.
1990	A	4-138 (2.3)	FR3LS	0.035	③	③	100	⑤	④	④	Hyd.	Hyd.
	D	4-138 (2.3)	FR3LS	0.035	③	③	100	⑤	④	④	Hyd.	Hyd.
	U	4-151 (2.5)	R43TS6	0.060	③	③	100	9–13	④	④	Hyd.	Hyd.
	N	6-204 (3.3)	R44LTS6	0.060	—	③	100	⑤	—	④	Hyd.	Hyd.
1991	A	4-138 (2.3)	FR3LS	0.035	③	③	100	⑤	④	④	Hyd.	Hyd.
	D	4-138 (2.3)	FR3LS	0.035	③	③	100	⑤	④	④	Hyd.	Hyd.
	U	4-151 (2.5)	R43TS6	0.060	③	③	100	9–13	④	④	Hyd.	Hyd.
	N	6-204 (3.3)	R45LTS6	0.060	—	③	100	⑤	④	④	Hyd.	Hyd.
1992	SEE UNDERHOOD SPECIFICATIONS STICKER

NOTE: The Underhood Specifications sticker often reflects tune-up specification changes made in production. Sticker figures must be used if they disagree with those in this chart.

Hyd. Hydraulic
① See Underhood Specifications sticker
② Minimum; the lowest reading cylinder should not have less than 70% of the pressure of the highest.

③ Ignition timing is controlled by the ECM and is not adjustable
④ Idle speed is controlled by the ECM and is not adjustable
⑤ 1—Connect fuel pressure gauge, engine at normal operating temperature.

2—Turn ignition switch on.
3—After approx. 2 seconds; pressure should read 41–47 psi and hold steady.
4—Start engine and idle; pressure should drop 3–10 psi from static pressure.

FIRING ORDERS

NOTE: To avoid confusion, always replace spark plug wires one at a time.

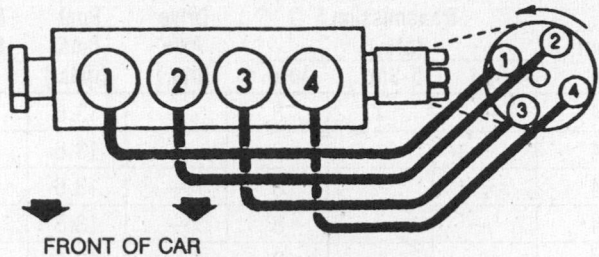

2.0L Engine
Engine Firing Order: 1–3–4–2
Distributor Rotation: Counterclockwise

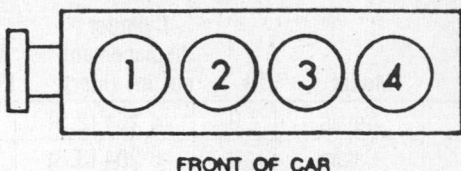

FRONT OF CAR

2.3L Engine
Engine Firing Order: 1–3–4–2
Distributorless Ignition System

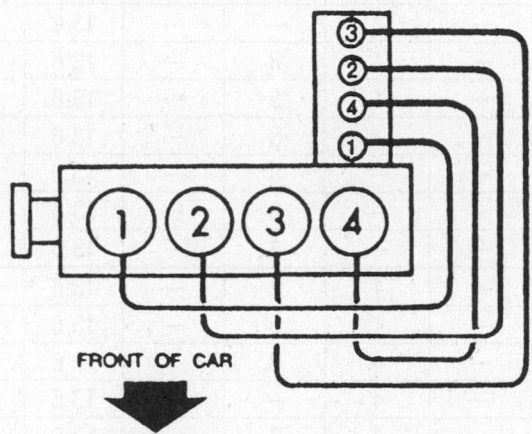

FRONT OF CAR

2.5L Engine
Engine Firing Order: 1–3–4–2
Distributorless Ignition System

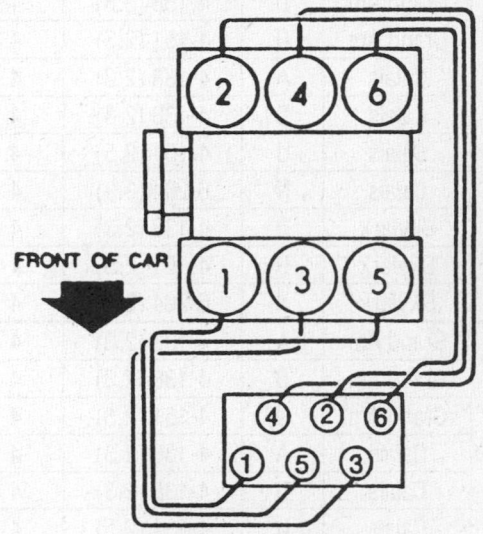

FRONT OF CAR

3.0L and 3.3L Engines
Engine Firing Order: 1–6–5–4–3–2
Distributorless Ignition System

CAPACITIES

Year	Model	VIN	No. Cylinder Displacement cu. in. (liter)	Engine Crankcase (qts.) with Filter ①	without Filter	Transmission (pts.) 4-Spd	5-Spd	Auto. ②	Drive Axle (pts.)	Fuel Tank (gals.)	Cooling System (qts.)
1988	Grand Am	M	4-122 (2.0)	4	4	—	4	8	—	13.6	8
	Grand Am	D	4-138 (2.3)	4	4	—	4	8	—	13.6	8
	Grand Am	U	4-151 (2.5)	4	4	—	5.3	8	—	13.6	8
	Calais	D	4-138 (2.3)	4	4	—	4	8	—	13.6	8
	Calais	U	4-151 (2.5)	4	4	—	5.3	8	—	13.6	8
	Calais	L	6-181 (3.0)	4	4	—	—	8	—	13.6	10
	Skylark	D	4-138 (2.3)	4	4	—	—	8	—	13.6	8
	Skylark	U	4-151 (2.5)	4	4	—	—	8	—	13.6	8
	Skylark	L	6-181 (3.0)	4	4	—	—	8	—	13.6	10
1989	Grand Am	M	4-122 (2.0)	4	4	—	4	8	—	13.6	8
	Grand Am	A	4-138 (2.3)	4	4	—	4	—	—	13.6	8
	Grand Am	D	4-138 (2.3)	4	4	—	4	8	—	13.6	8
	Grand Am	U	4-151 (2.5)	4	4	—	4	8	—	13.6	8
	Calais	A	4-138 (2.3)	4	4	—	4	—	—	13.6	8
	Calais	D	4-138 (2.3)	4	4	—	4	8	—	13.6	8

CAPACITIES

Year	Model	VIN	No. Cylinder Displacement cu. in. (liter)	Engine Crankcase (qts.) with Filter ①	Engine Crankcase (qts.) without Filter	Transmission (pts.) 4-Spd	Transmission (pts.) 5-Spd	Transmission (pts.) Auto. ②	Drive Axle (pts.)	Fuel Tank (gals.)	Cooling System (qts.)
1989	Calais	U	4-151 (2.5)	4	4	—	4	8	—	13.6	8
	Calais	N	6-204 (3.3)	4	4	—	—	8	—	13.6	10
	Skylark	D	4-138 (2.3)	4	4	—	—	8	—	13.6	8
	Skylark	U	4-151 (2.5)	4	4	—	—	8	—	13.6	8
	Skylark	N	6-204 (3.3)	4	4	—	—	8	—	13.6	10
1990	Grand Am	A	4-138 (2.3)	4	4	—	4	—	—	13.6	8
	Grand Am	D	4-138 (2.3)	4	4	—	4	8	—	13.6	8
	Grand Am	U	4-151 (2.5)	4	4	—	4	8	—	13.6	8
	Calais	A	4-138 (2.3)	4	4	—	4	—	—	13.6	8
	Calais	D	4-138 (2.3)	4	4	—	4	8	—	13.6	8
	Calais	U	4-151 (2.5)	4	4	—	4	8	—	13.6	8
	Calais	N	6-204 (3.3)	4	4	—	—	8	—	13.6	10
	Skylark	D	4-138 (2.3)	4	4	—	—	8	—	13.6	8
	Skylark	U	4-151 (2.5)	4	4	—	—	8	—	13.6	8
	Skylark	N	6-204 (3.3)	4	4	—	—	8	—	13.6	10
1991–92	Grand Am	A	4-138 (2.3)	4	4	—	4	8	—	13.6	10.4
	Grand Am	D	4-138 (2.3)	4	4	—	4	8	—	13.6	10.4
	Grand Am	U	4-151 (2.5)	4	4	—	4	8	—	13.6	10.7
	Calais	A	4-138 (2.3)	4	4	—	4	—	—	13.6	10.4
	Calais	D	4-138 (2.3)	4	4	—	4	8	—	13.6	10.4
	Calais	U	4-151 (2.5)	4	4	—	4	8	—	13.6	10.7
	Calais	N	6-204 (3.3)	4	4	—	—	8	—	13.6	12.7
	Skylark	D	4-138 (2.3)	4	4	—	—	8	—	13.6	10.4
	Skylark	U	4-151 (2.5)	4	4	—	—	8	—	13.6	10.7
	Skylark	N	6-204 (3.3)	4	4	—	—	8	—	13.6	12.7

① When changing the oil filter, additional oil may be needed to fill the crankcase.

② Drain and refill capacity shown. Dry capacity is 12 pts.

CAMSHAFT SPECIFICATIONS

All measurements given in inches.

Year	VIN	No. Cylinder Displacement cu. in. (liter)	Journal Diameter 1	Journal Diameter 2	Journal Diameter 3	Journal Diameter 4	Journal Diameter 5	Lobe Lift In.	Lobe Lift Ex.	Bearing Clearance	Camshaft End Play
1988	M	4-122 (2.0)	1.6714–1.6720	1.6812–1.6816	1.6911–1.6917	1.7009–1.7015	1.7108–1.7114	0.2409	NA	0.0008	0.0016–0.0064
	D	4-138 (2.3)	1.3751–1.3760	1.3751–1.3760	1.3751–1.3760	1.3751–1.3760	1.3751–1.3760	0.3400	0.3500	0.0019–0.0043	0.0060–0.0014
	U	4-151 (2.5)	1.8690	1.8690	1.8690	1.8690	1.8690	0.3980	0.3980	0.0007–0.0027	0.0015–0.0050
	L	6-181 (3.0)	1.7850–1.7860	1.7850–1.7860	1.7850–1.7860	1.7850–1.7860	—	0.3580	0.3840	0.0005–0.0025	NA

CAMSHAFT SPECIFICATIONS

All measurements given in inches.

Year	VIN	No. Cylinder Displacement cu. in. (liter)	Journal Diameter 1	2	3	4	5	Lobe Lift In.	Ex.	Bearing Clearance	Camshaft End Play
1989	M	4-122 (2.0)	1.6706–1.6712	1.6812–1.6818	1.6911–1.6917	1.7009–1.7015	1.7100–1.7106	0.2625	0.2625	0.0011–0.0035	0.0016–0.0064
	D	4-138 (2.3)	1.3751–1.3760	1.3751–1.3760	1.3751–1.3760	1.3751–1.3760	1.3751–1.3760	0.3400	0.3500	0.0019–0.0043	0.0060–0.0014
	A	4-138 (2.3)	1.3751–1.3760	1.3751–1.3760	1.3751–1.3760	1.3751–1.3760	1.3751–1.3760	0.4100	0.4100	0.0019–0.0043	0.0060–0.0014–
	U	4-151 (2.5)	1.8690	1.8690	1.8690	1.8690	1.8690	0.2480	0.2480	0.0007–0.0027	0.0014–0.0050
	N	6-204 (3.3)	1.7850–1.7860	1.7850–1.7860	1.7850–1.7860	1.7850–1.7860	—	0.2500	0.2550	0.0005–0.0035	NA
1990	D	4-138 (2.3)	1.5720–1.5728	1.3751–1.3760	1.3751–1.3760	1.3751–1.3760	1.3751–1.3760	0.3400	0.3500	0.0019–0.0043	0.0060–0.0014
	A	4-138 (2.3)	1.5720–1.5728	1.3751–1.3760	1.3751–1.3760	1.3751–1.3760	1.3751–1.3760	0.4100	0.4100	0.0019–0.0043	0.0060–0.0014–
	U	4-151 (2.5)	1.8690	1.8690	1.8690	1.8690	1.8690	0.2480	0.2480	0.0007–0.0027	0.0020–0.0090
	N	6-204 (3.3)	1.7850–1.7860	1.7850–1.7860	1.7850–1.7860	1.7850–1.7860	—	0.2500	0.2550	0.0005–0.0035	NA
1991–92	D	4-138 (2.3)	1.5720–1.5728	1.3751–1.3760	1.3751–1.3760	1.3751–1.3760	1.3751–1.3760	0.3750	0.3750	0.0019–0.0043	0.0009–0.0088
	A	4-138 (2.3)	1.5720–1.5728	1.3751–1.3760	1.3751–1.3760	1.3751–1.3760	1.3751–1.3760	0.4100	0.4100	0.0019–0.0043	0.0009–0.0088–
	U	4-151 (2.5)	1.8690	1.8690	1.8690	1.8690	1.8690	0.2480	0.2480	0.0007–0.0027	0.0020–0.0090
	N	6-204 (3.3)	1.7850–1.7860	1.7850–1.7860	1.7850–1.7860	1.7850–1.7860	—	0.2500	0.2550	0.0005–0.0035	NA

NA—Not Available

CRANKSHAFT AND CONNECTING ROD SPECIFICATIONS

All measurements are given in inches.

Year	VIN	No. Cylinder Displacement cu. in. (liter)	Crankshaft Main Brg. Journal Dia.	Main Brg. Oil Clearance	Shaft End-play	Thrust on No.	Connecting Rod Journal Diameter	Oil Clearance	Side Clearance
1988	M	4-122 (2.0)	①	0.0006–0.0016	0.0030–0.0120	3	1.9278–1.9286	0.0007–0.0024	0.0027–0.0095
	D	4-138 (2.3)	2.0470–2.0474	0.0005–0.0020	0.0034–0.0095	3	1.8887–1.8897	0.0005–0.0025	0.0059–0.0177
	U	4-151 (2.5)	2.3000	0.0005–0.0022	0.0035–0.0085	5	2.0000	0.0005–0.0026	0.0060–0.0220
	L	6-181 (3.0)	2.4988–2.4998	0.0003–0.0018	0.0030–0.0110	2	2.2487–2.2495	0.0003–0.0028	0.0030–0.0150
1989	M	4-122 (2.0)	2.2828–2.2833	0.0006–0.0016	0.0028–0.0118	3	1.9279–1.9287	0.0007–0.0025	0.0028–0.0095
	A	4-138 (2.3)	2.0470–2.0480	0.0005–0.0023	0.0034–0.0095	3	1.8887–1.8897	0.0005–0.0020	0.0059–0.0177

CRANKSHAFT AND CONNECTING ROD SPECIFICATIONS

All measurements are given in inches.

Year	VIN	No. Cylinder Displacement cu. in. (liter)	Crankshaft Main Brg. Journal Dia.	Crankshaft Main Brg. Oil Clearance	Crankshaft Shaft End-play	Crankshaft Thrust on No.	Connecting Rod Journal Diameter	Connecting Rod Oil Clearance	Connecting Rod Side Clearance
		4-138 (2.3)	2.0470–2.0480	0.0005–0.0023	0.0034–0.0095	3	1.8887–1.8897	0.0005–0.0020	0.0059–0.0177
	D	4-151 (2.5)	2.3000	0.0005–0.0020	0.0006–0.0110	5	2.0000	0.0005–0.0030	0.0060–0.0240
	N	6-204 (3.3)	2.4988–2.4998	0.0003–0.0018	0.0030–0.0110	2	2.2487–2.2499	0.0003–0.0026	0.0030–0.0150
1990	A	4-138 (2.3)	2.0470–2.0480	0.0005–0.0023	0.0034–0.0095	3	1.8887–1.8897	0.0005–0.0020	0.0059–0.0177
	D	4-138 (2.3)	2.0470–2.0480	0.0005–0.0023	0.0034–0.0095	3	1.8887–1.8897	0.0005–0.0020	0.0059–0.0177
	U	4-151 (2.5)	2.3000	0.0005–0.0020	0.0006–0.0110	5	2.0000	0.0005–0.0030	0.0060–0.0240
	N	6-204 (3.3)	2.4988–2.4998	0.0003–0.0018	0.0030–0.0110	2	2.2487–2.2499	0.0003–0.0026	0.0030–0.0150
1991–92	A	4-138 (2.3)	2.0470–2.0480	0.0005–0.0023	0.0034–0.0095	3	1.8887–1.8897	0.0005–0.0020	0.0059–0.0177
	D	4-138 (2.3)	2.0470–2.0480	0.0005–0.0023	0.0034–0.0095	3	1.8887–1.8897	0.0005–0.0020	0.0059–0.0177
	U	4-151 (2.5)	2.3000	0.0005–0.0022	0.0059–0.0110	5	2.0000	0.0005–0.0030	0.0060–0.0240
	N	6-204 (3.3)	2.4988–2.4998	0.0003–0.0018	0.0030–0.0110	2	2.2487–2.2499	0.0003–0.0026	0.0030–0.0150

① Brown: 2.2830–2.2833
Green: 2.2827–2.2830

VALVE SPECIFICATIONS

Year	VIN	No. Cylinder Displacement cu. in. (liter)	Seat Angle (deg.)	Face Angle (deg.)	Spring Test Pressure (lbs.)②	Spring Installed Height (in.)	Stem-to-Guide Clearance (in.) Intake	Stem-to-Guide Clearance (in.) Exhaust	Stem Diameter (in.) Intake	Stem Diameter (in.) Exhaust
1988	M	4-122 (2.0)	45	46	NA	NA	0.0006–0.0020	0.0010–0.0024	0.2744–0.2751	0.2739–0.2754
	D	4-138 (2.3)	45	①	159–173 @ 1.043 in.	1.42–1.44	0.0009–0.0027	0.0015–0.0032	0.2744–0.2751	0.2739–0.2754
	U	4-151 (2.5)	46	45	158–170 @ 1.040 in.	1.44	0.0010–0.0026	0.0013–0.0041	0.3130–0.3140	0.3120–0.3130
	L	6-181 (3.0)	46	45	175–195 @ 1.340 in.	1.73	0.0015–0.0035	0.0015–0.0032	0.3401–0.3412	0.3405–0.3412
1989	M	4-122 (2.0)	45	46	165–179 @ 1.043 in.	NA	0.0006–0.0017	0.0010–0.0024	0.2755–0.2760	0.2747–0.2753
	A	4-138 (2.3)	45	①	188–202 @ 1.043 in.	1.42–1.44	0.0009–0.0027	0.0015–0.0032	0.2744–0.2751	0.2740–0.2747
	D	4-138 (2.3)	45	①	159–173 @ 1.043 in.	1.42–1.44	0.0009–0.0027	0.0015–0.0032	0.2744–0.2751	0.2740–0.2747
	U	4-151 (2.5)	46	45	173 @ 1.24 in.	1.68	0.0010–0.0026	0.0013–0.0041	NA	NA
	N	6-204 (3.3)	45	45	200–220 @ 1.315 in.	1.69–1.75	0.0015–0.0035	0.0015–0.0032	NA	NA

VALVE SPECIFICATIONS

Year	VIN	No. Cylinder Displacement cu. in. (liter)	Seat Angle (deg.)	Face Angle (deg.)	Spring Test Pressure (lbs.) ②	Spring Installed Height (in.)	Stem-to-Guide Clearance (in.)		Stem Diameter (in.)	
							Intake	Exhaust	Intake	Exhaust
1990	A	4-138 (2.3)	45	①	193–207 @ 1.043 in.	1.42–1.44	0.0009–0.0027	0.0015–0.0032	0.2744–0.2751	0.2740–0.2747
	D	4-138 (2.3)	45	①	193–207 @ 1.043 in.	1.42–1.44	0.0009–0.0027	0.0015–0.0032	0.2744–0.2751	0.2740–0.2747
	U	4-151 (2.5)	46	45	173 @ 1.24 in.	1.68	0.0010–0.0026	0.0013–0.0041	NA	NA
	N	6-204 (3.3)	45	45	200–220 @ 1.315 in.	1.69–1.75	0.0015–0.0035	0.0015–0.0032	NA	NA
1991–92	A	4-138 (2.3)	45	44	193–207 @ 1.043 in.	0.98–③ 1.00	0.0010–0.0027	0.0015–0.0032	0.2744–0.2751	0.2740–0.2747
	D	4-138 (2.3)	45	44	193–207 @ 1.043 in.	0.98–③ 1.00	0.0010–0.0027	0.0015–0.0032	0.2744–0.2751	0.2740–0.2747
	U	4-151 (2.5)	46	45	173 @ 1.24 in.	1.68	0.0010–0.0026	0.0013–0.0041	NA	NA
	N	6-204 (3.3)	45	45	210 @ 1.315 in.	1.69–1.72	0.0015–0.0035	0.0015–0.0032	NA	NA

NA—Not Available
① Intake: 44°
   Exhaust: 44.5°

② Load—open
③ Measured from top of valve stem to top of camshaft housing mounting surface

PISTON AND RING SPECIFICATIONS

All measurements are given in inches.

Year	VIN	No. Cylinder Displacement cu. in. (liter)	Piston Clearance	Ring Gap			Ring Side Clearance		
				Top Compression	Bottom Compression	Oil Control	Top Compression	Bottom Compression	Oil Control
1988	M	4-122 (2.0)	0.0012–0.0020	0.012–0.020	0.012–0.020	0.016–0.055	0.002–0.003	0.001–0.003	NA
	D	4-138 (2.3)	0.0007–0.0020	0.016–0.025	0.016–0.025	0.016–0.055	0.002–0.004	0.0016–0.0031	NA
	U	4-151 (2.5)	0.0014–0.0022	0.010–0.020	0.010–0.020	0.020–0.060	0.002–0.003	0.001–0.003	0.015–0.055
	L	6-181 (3.0)	0.0010–0.0045	0.010–0.020	0.010–0.022	0.015–0.055	0.001–0.003	0.001–0.003	0.0005–0.0065
1989	M	4-122 (2.0)	0.0012–0.0020	0.010–0.020	0.012–0.020	0.016–0.055	0.002–0.004	0.002–0.003	NA
	A	4-138 (2.3)	0.0007–0.0020	0.014–0.024	0.016–0.026	0.016–0.055	0.002–0.004	0.002–0.003	NA
	D	4-138 (2.3)	0.0007–0.0020	0.014–0.024	0.016–0.026	0.016–0.055	0.002–0.004	0.002–0.003	NA
	U	4-151 (2.5)	0.0014–0.0022	0.010–0.020	0.010–0.020	0.020–0.060	0.002–0.003	0.001–0.003	0.015–0.055
	N	6-204 (3.3)	0.0004–0.0022	0.010–0.025	0.010–0.025	0.010–0.040	0.001–0.003	0.001–0.003	0.001–0.008
1990	A	4-138 (2.3)	0.0007–0.0020	0.014–0.024	0.016–0.026	0.016–0.055	0.003–0.005	0.002–0.003	NA
	D	4-138 (2.3)	0.0007–0.0020	0.014–0.024	0.016–0.026	0.016–0.055	0.002–0.004	0.002–0.003	NA

PISTON AND RING SPECIFICATIONS

All measurements are given in inches.

Year	VIN	No. Cylinder Displacement cu. in. (liter)	Piston Clearance	Ring Gap			Ring Side Clearance		
				Top Compression	Bottom Compression	Oil Control	Top Compression	Bottom Compression	Oil Control
1990	U	4-151 (2.5)	0.0014–0.0022	0.010–0.020	0.010–0.020	0.020–0.060	0.002–0.003	0.001–0.003	0.015–0.055
	N	6-204 (3.3)	0.0004–0.0022	0.010–0.025	0.010–0.025	0.010–0.040	0.001–0.003	0.001–0.003	0.001–0.008
1991–92	A	4-138 (2.3)	0.0007–0.0020	0.014–0.024	0.016–0.026	0.016–0.055	0.003–0.005	0.002–0.003	NA
	D	4-138 (2.3)	0.0007–0.0020	0.014–0.024	0.016–0.026	0.016–0.055	0.002–0.004	0.002–0.003	NA
	U	4-151 (2.5)	0.0014–0.0022	0.010–0.020	0.010–0.020	0.020–0.060	0.002–0.003	0.001–0.003	0.015–0.055
	N	6-204 (3.3)	0.0004–① 0.0022	0.010–0.025	0.010–0.025	0.010–0.040	0.001–0.003	0.001–0.003	0.001–0.008

NA—Not Available
① Measured 1.8 in. (44mm) down from top of piston

TORQUE SPECIFICATIONS

All readings in ft. lbs.

Year	VIN	No. Cylinder Displacement cu. in. (liter)	Cylinder Head Bolts	Main Bearing Bolts	Rod Bearing Bolts	Crankshaft Pulley Bolts	Flywheel Bolts	Manifold		Spark Plugs
								Intake	Exhaust	
1988	M	4-122 (2.0)	①	44②	26②	20	48③	16	16	15
	D	4-138 (2.3)	④	15⑤	15⑥	74⑤	22②	18	27	17
	U	4-151 (2.5)	⑫	70	32	162	⑨	25	⑩	15
	L	6-181 (3.0)	⑪	100	45	219	60	32	37	20
1989	M	4-122 (2.0)	①	44②	26②	20	63②	18	10	15
	A	4-138 (2.3)	④	15⑤	18⑬	74⑤	22②	18	27	17
	D	4-138 (2.3)	④	15⑤	18⑬	74⑤	22②	18	27	17
	U	4-151 (2.5)	⑫	65	29	162	⑨	25	⑩	15
	N	6-204 (3.3)	⑭	90	20②	219	61	7	30	20
1990	A	4-138 (2.3)	⑮	15⑤	18⑬	74⑤	22②	18	⑦	17
	D	4-138 (2.3)	⑮	15⑤	18⑬	74⑤	22②	18	⑦	17
	U	4-151 (2.5)	⑫	65	29	162	⑨	25	⑩	15
	N	6-204 (3.3)	⑭	90	20②	219	61	7	30	20
1991–92	A	4-138 (2.3)	⑮	15⑤	18⑬	74⑤	22②	18	⑦	17
	D	4-138 (2.3)	⑮	15⑤	18⑬	74⑤	22②	18	⑦	17
	U	4-151 (2.5)	⑫	65	29	162	⑨	25	⑩	15
	N	6-204 (3.3)	⑭	26⑯	20②	105⑰	89⑧⑤	89⑧	41	20

① Step 1: 18 ft. lbs.
Step 2: 3 rounds of 60° turns in sequence
Step 3: An additional 30–50° turn after engine warm up
② Plus an additional 40–50° turn
③ Plus an additional 30° turn
④ Short bolts: 26 ft. lbs. plus an additional 80° turn
Long bolts: 26 ft. lbs. plus an additional 90° turn
⑤ Plus an additional 90° turn
⑥ Plus an additional 75° turn

⑦ Nuts: 27 ft. lbs.
Studs: 106 inch lbs.
⑧ Inch lbs.
⑨ Manual transaxle: 69 ft. lbs.
Automatic transaxle: 55 ft. lbs.
⑩ Outer bolts: 26 ft. lbs.
Inner bolts: 37 ft. lbs.
⑪ Step 1: 25 ft. lbs.
Step 2: 2 rounds of 90° turns in sequence, not to exceed 60 ft. lbs.
⑫ Step 1: 18 ft. lbs.
Step 2: 26 ft. lbs., except front bolt/stud

Step 3: Front bolt/stud to 18 ft. lbs.
Step 4: An additional 90° turn
⑬ Plus an additional 80° turn
⑭ Step 1: 35 ft. lbs.
Step 2: An additional 130° turn
Step 3: An additional 30° turn on center 4 bolts
⑮ Short bolts: 26 ft. lbs. plus an additional 100° turn
Long bolts: 26 ft. lbs. plus an additional 110° turn
⑯ Plus an additional 45° turn
⑰ Plus an additional 56° turn

BRAKE SPECIFICATIONS

All measurements in inches unless noted

Year	Model	Lug Nut Torque (ft. lbs.)	Master Cylinder Bore	Brake Disc Minimum Thickness	Brake Disc Maximum Runout	Standard Brake Drum Diameter	Minimum Lining Thickness Front	Minimum Lining Thickness Rear
1988	All	100	①	0.830	0.004	7.879	0.06	0.06
1989	All	100	0.874	0.830	0.004	7.879	0.06	0.06
1990	All	100	0.874	0.830	0.004	7.879	0.06	0.06
1991-92	All	100	0.874	0.786	0.003	7.879	0.06	0.06

① 0.874 in. or 0.937 in.

WHEEL ALIGNMENT

Year	Model		Caster Range (deg.)	Caster Preferred Setting (deg.)	Camber Range (deg.)	Camber Preferred Setting (deg.)	Toe-in (in.)	Steering Axis Inclination (deg.)
1988	Calais	—	11/16P–2 11/16P	1 11/16P	7/32P–1 13/32P	13/16P	0	13 1/2
	Grand Am	—	11/16P–2 11/16P	1 11/16P	7/32P–1 13/32P	13/16P	0	13 1/2
	Skylark	—	11/16P–2 11/16P	1 11/16P	7/32P–1 13/32P	13/16P	0	13 1/2
1989	Calais	front	11/16P–2 11/16P	1 11/16P	1/8P–1 1/2P ②	13/16P ①	0	13 1/2
		rear	—	—	3/4N–1/4P	1/4N	1/4	—
	Grand Am	front	11/16P–2 11/16P	1 11/16P	1/8P–1 1/2P ②	13/16P ①	0	13 1/2
		rear	—	—	3/4N–1/4P	1/4N	1/4	—
	Skylark	front	11/16P–2 11/16P	1 11/16P	1/8P–1 1/2P	13/16P	0	13 1/2
		rear	—	—	3/4N–1/4P	1/4N	1/4	—
1990	Calais	front	11/16P–2 11/16P	1 11/16P	1/8P–1 1/2P ②	13/16P ①	0	13 1/2
		rear	—	—	3/4N–1/4P	1/4N	1/4	—
	Grand Am	front	11/16P–2 11/16P	1 11/16P	1/8P–1 1/2P ②	13/16P ①	0	13 1/2
		rear	—	—	3/4N–1/4P	1/4N	1/4	—
	Skylark	front	11/16P–2 11/16P	1 11/16P	1/8P–1 1/2P ②	13/16P ①	0	13 1/2
		rear	—	—	3/4N–1/4P	1/4N	1/4	—
1991-92	Calais	front	11/16P–2 11/16P	1 11/16P	11/16N–1 11/16P	0	0	13 1/2
		rear	—	—	13/16N–5/16P	1/4N	1/8	—
	Grand Am	front	11/16P–2 11/16P	1 11/16P	11/16N–1 11/16P	0	0	13 1/2
		rear	—	—	13/16N–5/16P	1/4N	1/8	—
	Skylark	front	11/16P–2 11/16P	1 11/16P	11/16N–1 11/16P	0	0	13 1/2
		rear	—	—	3/4N–1/2P	1/4N	1/8	—

N—Negative
P—Positive

① with 16 in. wheels: 0
② with 16 in. wheels: 11/16N–11/16P

ENGINE MECHANICAL

NOTE: Disconnecting the negative battery cable on some vehicles may interfere with the functions of the on board computer systems and may require the computer to undergo a relearning process, once the negative battery cable is reconnected.

Engine Assembly

REMOVAL & INSTALLATION

2.0L and 2.5L Engines

1. Relieve the fuel system pressure.
2. Disconnect both battery cables and ground straps.
3. Drain the cooling system and remove the cooling fan.
4. Remove the air cleaner assembly.
5. Disconnect the ECM connections and feed harness through the bulkhead. Lay the harness across the engine.
6. Label and disconnect the engine wiring harness and all engine-related connectors and lay across the engine.

7. Label and disconnect the radiator hoses and vacuum lines. Disconnect and plug the fuel lines.

8. On 2.5L engine, remove the air conditioning compressor from the engine and lay it aside, without disconnecting the refrigerant lines. Remove the transaxle struts.

9. If equipped with power steering, remove the power steering pump from its mount and lay it aside. Remove the power steering pump bracket from the engine.

10. If equipped with a manual transaxle, disconnect the clutch and transaxle linkage. Remove the throttle cable from the throttle body.

11. If equipped with an automatic transaxle, disconnect the transaxle cooler lines, shifter linkage, downshift cable and throttle cable from the throttle body.

12. Raise and safely support the vehicle.

13. Disconnect all wiring from the transaxle.

14. On 2.0L engine, properly discharge the air conditioning system and remove the compressor. Remove the transaxle strut(s).

15. Disconnect the exhaust pipe from the exhaust manifold and hangers.

16. Disconnect the heater hoses from the heater core tubes and plug them.

17. Remove the front wheels. Remove the calipers and wire them up aside. Remove the brake rotors.

18. Matchmark and remove the knuckle-to-strut bolts.

19. Remove the body-to-cradle bolts at the lower control arms. Loosen the remaining body-to-cradle bolts. Remove a bolt at each cradle side, leaving 1 bolt per corner.

20. Using the proper equipment, support the vehicle under the radiator frame support.

21. Position a jack to the rear of the body pan with a 4 × 4 in. × 6 ft. timber spanning the vehicle.

22. Raise the vehicle enough to remove the support equipment.

23. Position a dolly under the engine/transaxle assembly with 3 blocks of wood for additional support.

24. Lower the vehicle slightly, allowing the engine/transaxle assembly to rest on the dolly.

25. Remove all engine and transaxle mount bolts and brackets. Remove the remaining cradle-to-body bolts.

26. Raise the vehicle, leaving engine and transaxle assembly with the suspension on the dolly.

27. Separate the engine and transaxle.

To install:

28. Assemble the engine and transaxle assembly and position on the dolly.

29. Raise and safely support the ve-hicle. Roll the assembly to the installation position and lower the vehicle over the assembly.

30. Install all engine, transaxle and suspension mounting bolts. Tighten all cradle mounting bolts to 65 ft. lbs. (88 Nm). Connect the wiring to the transaxle.

31. Install the knuckle-to-strut bolts and assemble the brakes.

32. Connect the exhaust pipe to the exhaust manifold and hangers.

33. Connect the heater hoses to the heater core tubes.

34. If equipped with the 2.0L engine, install the air conditioning compressor.

35. Install the wheels and lower the vehicle.

36. If equipped with the 2.5L engine, install the air conditioning compressor.

37. Install the power steering pump and related parts.

38. If equipped with a manual transaxle, connect the clutch and transaxle linkage. Connect the throttle cable to the throttle body.

39. If equipped with an automatic transaxle, connect the transaxle cooler lines, shifter linkage, downshift cable and throttle cable to the throttle body.

40. Connect the radiator hoses, vacuum lines and fuel lines.

41. Connect the engine wiring harness and all engine-related connectors. Feed the ECM connections through the bulkhead and connect.

42. Install the air cleaner assembly.

43. Fill all fluids to their proper levels.

44. Connect the battery cables, start the engine and set the timing, if necessary. Check for leaks.

2.3L Engine

1. Relieve the fuel system pressure.
2. Disconnect both battery cables and ground straps from the front engine mount bracket and the transaxle.
3. Drain the cooling system and remove the cooling fan.
4. Remove the air cleaner duct.
5. Disconnect the heater and radiator hoses from the thermostat housing.
6. Properly discharge the air conditioning system and disconnect the hoses from the compressor.
7. Remove the upper radiator support.
8. Disconnect the 2 vacuum hoses from the front of the engine.
9. Label and disconnect all electrical connectors from engine- and transaxle-mounted devices.
10. Unplug the wires at the starter solenoid.
11. Disconnect the power brake vacuum hose from the throttle body.

12. Disconnect the throttle cable and remove the bracket.

13. Remove the power steering pump bracket and lay the pump aside with the lines attached.

14. Disconnect and plug the fuel lines.

15. If equipped with a manual transaxle, disconnect the shifter cables and the clutch actuator cylinder.

16. If equipped with an automatic transaxle, disconnect the shift and TV cables.

17. Disconnect the transaxle and engine oil cooler pipes, if equipped.

18. Remove the exhaust manifold and heat shield.

19. Remove the lower radiator hose and front engine mount.

20. Install engine support fixture tool J–28467–A.

21. Raise and safely support the vehicle.

22. Remove the wheels, right side splash shield and radiator air deflector.

23. Separate the ball joints from the steering knuckles.

24. Using the proper equipment, support the suspension supports, crossmember and stabilizer shaft. Remove the attaching bolts and remove as an assembly.

25. Disconnect the heater hose from the radiator outlet pipe.

26. Remove the halfshafts from the transaxle.

27. Remove the nut from the transaxle mount through bolt.

28. Remove the nut from the rear engine mount through bolt.

29. Remove the rear engine mount body bracket.

30. Position a suitable support fixture below the engine/transaxle assembly and lower the vehicle so the weight of the engine/transaxle assembly is on the support fixture.

31. Remove the transaxle mount through bolt.

32. Mark the threads on fixture tool J–28467–A so the setting can be duplicated when installing the engine/transaxle assembly. Remove the fixture.

33. Move the engine/transaxle assembly rearward and slowly raise the vehicle from the engine/transaxle assembly.

NOTE: Many of the bell housing bolts are of different lengths; note their locations before removing. It is imperative that these bolts go back in their original locations when assembling the engine and transaxle or engine damage could result.

34. Separate the engine from the transaxle.

To install:

35. Assemble the engine and transaxle. If equipped with an automatic transaxle, thoroughly clean and dry the torque converter bolts and bolt holes, apply thread locking compound to the threads and tighten the bolts to 46 ft. lbs. (63 Nm). If equipped with a manual transaxle, tighten the clutch cover bolts to 22 ft. lbs. (30 Nm).

36. Raise and safely support the vehicle. Position the engine/transaxle assembly and lower the vehicle over the assembly until the transaxle mount is indexed, then install the bolt.

37. Install the engine support fixture and adjust to previously indexed setting. Raise the vehicle off the support fixture.

38. Install the rear mount to body bracket and tighten the bolts to 55 ft. lbs. (75 Nm).

39. Install the rear mount nut and tighten to 55 ft. lbs. (75 Nm).

40. Install the transaxle mount through bolt and tighten the nut to 55 ft. lbs. (75 Nm). Tighten so equal gaps are maintained.

41. Install the halfshafts.

42. Connect the heater hose to the the radiator outlet pipe.

43. Install the suspension supports, crossmember and stabilizer shaft assembly. Tighten the center bolts first, then front, then rear, to 65 ft. lbs. (90 Nm).

44. Install the ball joints and tighten the nuts to a maximum of 50 ft. lbs. (68 Nm).

45. Install the radiator air deflector and splash shield.

46. Install the wheels and lower the vehicle.

47. Install the front engine mount nut and tighten to 41 ft. lbs. (56 Nm). Remove the engine support fixture. Connect the lower radiator hose.

48. Install the exhaust manifold and heat shield.

49. Connect the transaxle and engine oil cooler pipes, if equipped.

50. If equipped with a manual transaxle, connect the shifter cables and the clutch actuator cylinder.

51. If equipped with an automatic transaxle, connect the shift and TV cables.

52. Connect the fuel lines.

53. Install the power steering pump and related parts.

54. Connect the throttle cable and install the bracket.

55. Connect the power brake vacuum hose to the throttle body.

56. Connect the starter wires.

57. Connect all electrical connectors and cables to the proper engine and transaxle-mounted devices.

58. Connect the 2 vacuum hoses at the front of the engine.

59. Install the upper radiator support.

60. Using new seals, connect the air conditioning hoses to the compressor.

61. Connect the heater and radiator hoses at the thermostat housing.

62. Install the air cleaner duct.

63. Fill all fluids to their proper levels.

64. Connect the battery cables, start the engine and check for leaks.

3.0L and 3.3L Engines

1. Disconnect the negative battery cable. Relieve the fuel pressure.

2. Matchmark the hinge-to-hood position and remove the hood.

3. Drain the cooling system. Disconnect and label all electrical connectors from the engine, alternator and fuel injection system, vacuum hoses, and engine ground straps. Remove the alternator.

4. Remove the coolant hoses from the radiator and engine. Remove the radiator and cooling fan assembly.

5. Remove the air intake duct. Disconnect the fuel lines from the fuel rail. Disconnect the throttle, TV and cruise control cables from the throttle body.

6. Raise and safely support the vehicle. Drain the engine oil. Disconnect the exhaust pipe from the exhaust manifold.

7. Remove the air conditioning compressor mounting bolts, and position it aside.

8. Disconnect the heater hoses.

9. Remove the transaxle inspection cover, matchmark the converter to the flexplate and remove the torque converter bolts.

10. Remove the rear engine mount bolts.

11. Remove the lower bell housing bolts. Label and disconnect the starter motor wiring and remove the starter motor from the engine.

12. Lower the vehicle. Remove the power steering pump mounting bolts and set the pump aside.

13. Support the transaxle with a floor jack or equivalent. Attach an engine lifting device to the engine.

14. Remove the upper bell housing bolts.

15. Remove the front engine mount bolts.

16. Lift and remove the engine from the vehicle. If the master cylinder is preventing removal, remove it and plug the brake lines.

To install:

17. Lower the engine into the engine compartment. Align the engine mounts and install the bolts. Tighten the bolts to their proper values:

Front engine mount bracket to block—66 ft. lbs. (90 Nm)

Front engine mount to underbody—54 ft. lbs. (73 Nm)

Front engine mount to engine bracket—15 ft. lbs. (20 Nm)

Rear engine mount to bracket—18 ft. lbs. (24 Nm)

Rear engine mount bracket to underbody—41 ft. lbs. (56 Nm)

Rear engine mount to engine bracket—40 ft. lbs. (54 Nm)

18. Install the upper transaxle-to-engine mounting bolts and tighten to 55 ft. lbs. (75 Nm). Remove the engine lifting fixture from the engine.

19. Raise and safely support the vehicle.

20. Align the converter marks, install the torque converter bolts and tighten to 46 ft. lbs. (63 Nm). Install the transaxle inspection cover.

21. Connect the exhaust pipe to the exhaust manifold. Install the starter motor and connect the wiring.

22. Install the air conditioning compressor. Connect the heater hoses.

23. Lower the vehicle. Install the power steering pump.

24. Install the alternator and belt.

25. Connect all vacuum hoses and electrical connectors to the engine.

26. Connect the fuel lines and all cables to the throttle body. Install the air intake duct.

27. Install the radiator and fan assembly. Connect the fan motor wiring. Connect the radiator hoses and refill the cooling system.

28. Fill all fluids to their proper levels.

29. Connect the battery cables, start the engine and check for leaks.

Engine Mounts
REMOVAL & INSTALLATION

1. Disconnect the negative battery cable.

2. Matchmark the engine mount to its mounting location.

3. Raise and safely support the vehicle, as required. Using the proper equipment, support the weight of the engine.

4. Remove all bolts and nuts that attach the mount to the engine, transaxle or body and remove the mount assembly from the vehicle.

5. Remove the through bolt and separate the insulator from the bracket, as required.

6. The installation is the reverse of the removal procedure. Make sure the matchmarks are aligned before tightening bolts.

Cylinder Head
REMOVAL & INSTALLATION

2.0L Engine

NOTE: Cylinder head gasket re-

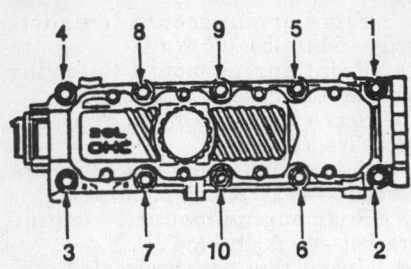

Camshaft carrier/cylinder head bolt torque sequence—2.0L engine

placement is necessary if camshaft carrier/cylinder head bolts are loosened. The head bolts should only be loosened when the engine is cold and should never be reused.

1. Relieve the fuel system pressure. Disconnect the negative battery cable.
2. Drain the coolant. Remove the induction tube.
3. Remove the alternator and bracket.
4. Remove the ignition coil.
5. Matchmark the rotor to the distributor housing and the distributor housing to the cam carrier. Remove the distributor and spark plug wires.
6. Disconnect all cables from the throttle body.
7. Disconnect and tag all electrical connections from the throttle body and intake manifold.
8. Disconnect all vacuum lines and heater hoses.
9. Disconnect and plug the fuel lines.
10. Remove the breather from the camshaft carrier.
11. Remove the upper radiator support.
12. Disconnect the exhaust manifold from the turbocharger and disconnect the oxygen sensor.
13. Label and disconnect wiring at engine harness and thermostat housing.
14. Remove the timing belt.
15. Remove the camshaft carrier/cylinder head bolts in the reverse order of the installation sequence.
16. Remove camshaft carrier, rocker arms and valve lifters.
17. Remove cylinder head and manifolds as an assembly. Remove the head gasket.

To install:
18. Thoroughly clean and dry the mating surfaces and bolt holes. Apply a continuous bead of RTV sealant to the sealing surface of camshaft carrier.
19. Install a new head gasket and position the head on the engine block. Tighten the new head bolts in sequence as follows:

Step 1—Tighten to 18 ft. lbs. (25 Nm).

Step 2—Using a torque angle meter, tighten an additional 60 degrees.

Step 3—Tighten another additional 60 degrees.

Step 4—Tighten a third additional 60 degrees.

Step 5—Tighten and additional 30–50 degrees turn after engine warm up.

20. Install the rear cover and timing belt.
21. Connect all wiring to the engine harness and thermostat housing.
22. Install the exhaust manifold to turbo connection and connect the oxygen sensor.
23. Install the upper radiator support.
24. Install the breather on the camshaft carrier.
25. Connect all vacuum and fuel lines.
26. Connect the heater hoses.
27. Connect all electrical connectors to the throttle body and intake manifold.
28. Connect all cables to the throttle body.
29. Install the distributor and spark plug wires, aligning the matchmarks.
30. Install the ignition coil.
31. Install the alternator and bracket.
32. Fill all fluids to their proper levels.
33. Connect the battery cable, start the engine and check for leaks.
34. Tighten all head bolts another additional 30–50 degrees, in sequence, after full engine warm up.

2.3L Engine

1. Relieve the fuel system pressure. Disconnect the negative battery cable and drain cooling system.
2. Disconnect heater inlet and throttle body heater hoses from water outlet. Disconnect the upper radiator hose from the water outlet.
3. Remove the exhaust manifold.
4. Remove the intake and exhaust camshaft housings.
5. Remove the oil cap and dipstick. Pull oil fill tube upward to unseat from block.
6. Label and disconnect the injector harness electrical connector.
7. Disconnect the throttle body air intake duct. Disconnect the cables and bracket and position aside.
8. Remove the throttle body from the intake manifold.
9. Matchmark and disconnect the vacuum hose from intake manifold.
10. Remove intake manifold bracket to block bolt.
11. Disconnect the coolant sensor connectors.
12. Remove the cylinder head bolts

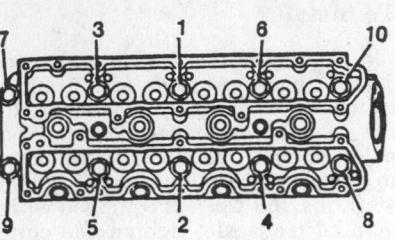

◀ FRONT OF ENGINE

Cylinder head bolt torque sequence—2.3L engine

in reverse order of the installation sequence.
13. Remove the cylinder head and gasket. Inspect the oil flow check valve for freedom of movement.

To install:
14. Thoroughly clean and dry all bolts, bolt holes and mating surfaces. Inspect the head bolts for any damage and replace, if necessary.
15. Install the cylinder head gasket to the cylinder block and carefully position the cylinder head in place.
16. Coat the head bolt threads with clean engine oil and allow the oil to drain off before installing.
17. On 1988–89 engines, tighten the cylinder head bolts in sequence as follows:

Step 1—Tighten all head bolts to 26 ft. lbs. (35 Nm).

Step 2—Using a torque angle meter, tighten the short bolts an additional 80 degrees and the long bolts an additional 90 degrees.

18. On 1990–92 engines, tighten the cylinder head bolts in sequence as follows:

Step 1—Tighten all head bolts to 26 ft. lbs. (35 Nm).

Step 2—Using a torque angle meter, tighten the short bolts an additional 100 degrees and the long bolts an additional 110 degrees.

19. Install the intake manifold bracket.
20. Connect the MAP sensor vacuum hose to the intake manifold.
21. Install the throttle body to the intake manifold.
22. Connect the throttle body air intake duct. Install the throttle cable and bracket.
23. Connect the injector harness electrical connector.
24. Connect the 2 coolant sensor connections.
25. Install the oil cap and dipstick. Install the oil fill tube into the block.
26. Install the exhaust and intake camshaft housings.
27. Install the exhaust manifold.
28. Connect the heater inlet and throttle body heater hoses to the water outlet. Connect the upper radiator hose to the water outlet.

29. Fill all fluids to their proper levels.
30. Connect the battery cable, start the engine and check for leaks.

2.5L Engine

1. Relieve the fuel system pressure.
2. Disconnect the negative battery cable.
3. Drain the coolant and remove the oil dipstick tube.
4. Remove the air cleaner assembly.
5. Raise and safely support the vehicle. Disconnect the exhaust pipe from the manifold.
6. Lower the vehicle.
7. Label and disconnect the electrical wiring and throttle linkage from the throttle body assembly.
8. Disconnect the heater hose from the intake manifold.
9. Remove the ignition coil. Label and disconnect the electrical wiring connectors from the intake manifold and the cylinder head. Remove the alternator.
10. If equipped with a top-mounted air conditioning compressor, remove the compressor and lay it aside.
11. If equipped with power steering, remove the upper bracket from the power steering pump.
12. Remove the radiator hoses from the engine.
13. Remove the valve cover. Label and remove the rocker arms and pushrods.
14. Remove the cylinder head bolts in reverse order of the installation sequence and remove the cylinder head.
To install:
15. Thoroughly clean and dry all bolts, bolt holes and mating surfaces. Inspect the head bolts for any damage and replace if necessary.
16. Install the head gasket to the block and carefully position the cylinder head in place.
17. Tighten the cylinder head bolts in sequence as follows:
    Step 1: Tighten all bolts to 18 ft. lbs. (25 Nm).
    Step 2: Tighten to 26 ft. lbs. (35 Nm), except front bolt/stud.
    Step 3: Tighten front bolt/stud to 18 ft. lbs. (25 Nm).
    Step 4: Using a torque angle meter, tighten all bolts an additional 90 degrees.
18. Install the pushrods and rocker arms in their original positions. Install the valve cover with a new gasket.
19. Install the radiator hoses. Install the power steering pump and upper bracket.
20. Install the air conditioning compressor, if removed.
21. Install the ignition coil and connect the electrical wiring connectors

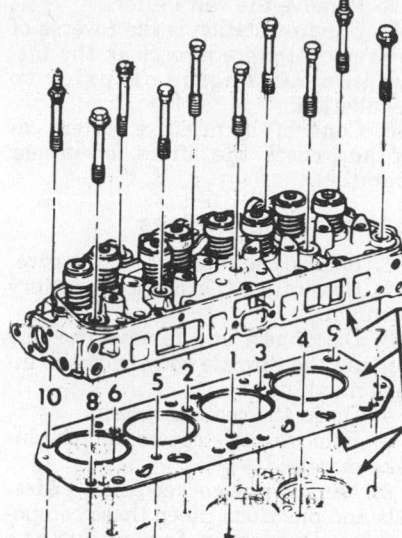

Cylinder head bolt torque sequence— 2.5L engine

to the intake manifold and the cylinder head.
22. Install the alternator.
23. Install the heater hose to the intake manifold.
24. Connect the electrical wiring and throttle linkage to the throttle body assembly.
25. Raise and safely support the vehicle.
26. Install the exhaust pipe to the manifold.
27. Install the air cleaner assembly.
28. Adjust all belt tensions and fill all fluids to their proper levels.
29. Connect the battery cable, start the engine and check for leaks.

3.0L and 3.3L Engines

1. Relieve the fuel system pressure.
2. Disconnect the negative battery cable and drain the coolant.
3. Remove the mass air flow sensor and the air intake duct.
4. Remove C³I ignition module and wiring.
5. Remove the serpentine drive belt, the alternator and bracket.
6. Label and remove all necessary vacuum lines and electrical connections.
7. Remove the fuel lines, the fuel rail and the spark plug wires.
8. Remove the heater/radiator hoses from the throttle body and intake manifold. Remove the cooling fan and the radiator.
9. Remove the intake manifold.
10. Remove the valve covers. Label and remove the rocker arms, pedestals and pushrods.
11. Remove the left side exhaust manifold.
12. Remove the power steering

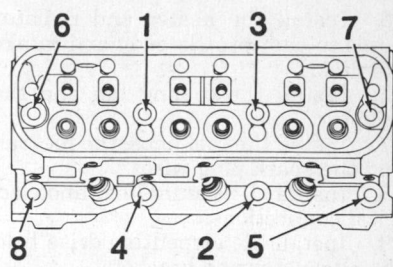

Cylinder head bolt torque sequence— 3.0L and 3.3L engines

pump. Remove the dipstick and dipstick tube.
13. Remove the left side head bolts in reverse order of the installation sequence and lift the left cylinder head from the engine.
14. Raise and safely support the vehicle. Remove the right exhaust manifold-to-engine bolts.
15. Remove the right cylinder head-to-engine bolts in reverse of the installation sequence and lift the right cylinder head from the engine.
To install:
16. Thoroughly clean and dry all bolts, bolt holes and mating surfaces. Inspect the head bolts for any damage and replace if necessary.
17. Install the head gasket to the block and carefully position the cylinder head in place.
18. On the 3.0L engine, tighten the cylinder head bolts, in sequence, as follows:
    Step 1: Tighten to 25 ft. lbs. (34 Nm).
    Step 2: Using a torque angle meter, tighten an additional 90 degrees.
    Step 3: Tighten another additional 90 degrees, to a maximum of 60 ft. lbs. (81 Nm).
19. On the 3.3L engine, tighten the cylinder head bolts, in sequence, as follows:
    Step 1: Tighten to 35 ft. lbs. (47 Nm).
    Step 2: Using a torque angle meter, tighten an additional 130 degrees.
    Step 3: Tighten the 4 center bolts an additional 30 degrees.
20. Install the intake manifold. Raise and safely support the vehicle. Install the exhaust manifold. Lower the vehicle.
21. Install the power steering pump. Install the dipstick and dipstick tube.
22. Install new valve cover gaskets and install the valve covers.
23. Install the rocker arms, pedestals and bolts. Tighten pedestal bolts to 43 ft. lbs. (58 Nm) for the 3.0L engine or 28 ft. lbs. (38 Nm) for the 3.3L engine.
24. Install the intake manifold assembly.

25. Install the heater and radiator hoses to the throttle body and intake manifold.
26. Install the cooling fan and the radiator.
27. Install the fuel lines, the fuel rail and the spark plug wires.
28. Install all vacuum lines and electrical connections.
29. Install the serpentine drive belt, the alternator and bracket.
30. Install the C³I ignition module and wiring.
31. Install the mass air flow sensor and the air intake duct.
32. Fill all fluids to their proper levels.
33. Connect the battery cable, start the engine and check for leaks.

Valve Lifters
REMOVAL & INSTALLATION

2.0L Engine

1. Disconnect the negative battery cable. Remove the camshaft carrier cover.
2. Hold the valves in place with compressed air, using an air adapter in the spark plug hole.
3. Compress the valve springs using a valve spring compressor.
4. Remove rocker arms; keep them in order for reassembly.
5. Remove the lifters.
6. The installation is the reverse of the removal procedure. Soak the lifters in clean engine oil prior to installation.
7. Connect the negative battery cable and check the lifters for proper operation.

2.3L Engine

1. Disconnect the negative battery cable.
2. Remove the camshafts.
3. Remove the lifters from their bores.
4. The installation is the reverse of the removal procedure. Soak the lifters in clean engine oil prior to installation.
5. Connect the negative battery cable and check the lifters for proper operation.

2.5L Engine

1. Relieve the fuel system pressure.
2. Disconnect the negative battery cable.
3. Remove the valve cover and intake manifold.
5. Remove the side pushrod cover.
6. Loosen the rocker arms in pairs and rotate them in order to clear the pushrods.
7. Remove the pushrods, retainer and guide from each cylinder.

8. Remove the valve lifters.
9. The installation is the reverse of the removal procedure. Soak the lifters in clean engine oil prior to installation.
10. Connect the negative battery cable and check the lifters for proper operation.

3.0L and 3.3L Engines

1. Relieve the fuel system pressure.
2. Disconnect the negative battery terminal.
3. Disconnect and remove the fuel rail and the throttle body from the intake manifold.
4. Drain the cooling system.
5. Remove valve covers and the intake manifold.
6. Remove the rocker arms, pedestals and pushrods. Keep these components in order for accurate installation.
7. Remove the valve lifters.
8. The installation is the reverse of the removal procedure. Soak the lifters in clean engine oil prior to installation.
9. Connect the negative battery cable and check the lifters for proper operation.

Rocker Arms
REMOVAL & INSTALLATION

2.0L Engine

1. Disconnect the negative battery cable. Remove the camshaft carrier cover.
2. Hold the valves in place with compressed air, using an air adapter in the spark plug hole.
3. Compress the valve springs using a suitable valve spring compressor.
4. Remove rocker arms. Keep them in order if they are being reused.
5. The installation is the reverse of the removal procedure.
6. Connect the negative battery cable and check for proper operation.

2.5L Engine

1. Relieve the fuel system pressure.
2. Disconnect the negative battery cable.
3. Remove the valve cover and intake manifold.
4. Remove the rocker arm bolts and remove the rocker arms.
5. The installation is the reverse of the removal procedure. Tighten the attaching bolts to 20 ft. lbs. (27 Nm).
6. Connect the negative battery cable and check for proper operation.

3.0L Engine
FRONT HEAD

1. Relieve the fuel system pressure.
2. Disconnect the negative battery cable.

3. Disconnect all electrical components and vacuum hoses which prevent access to the valve cover bolts.
4. Remove the valve cover.
5. Remove the rocker arm pedestal-to-cylinder head bolts, the rocker arm and pedestal assembly.
6. The installation is the reverse of the removal procedure. Tighten the rocker arm pedestal bolts to 45 ft. lbs. (60 Nm).
7. Connect the negative battery cable and check for proper operation.

REAR HEAD

1. Relieve the fuel system pressure.
2. Disconnect the negative battery cable.
3. Remove the C³I ignition coil module. Disconnect the spark plug wires, electrical connectors, EGR solenoid wiring and vacuum hoses.
4. Remove the serpentine belt, alternator wiring and the rear alternator bracket-to-engine bolt. Rotate the alternator toward the front of the vehicle.
5. Remove the power steering pump from the belt tensioner and remove the tensioner assembly.
6. Remove the engine lift bracket and the rear alternator brace.
7. Drain the radiator below heater hose level. Remove the throttle body heater hoses.
8. Remove the valve cover.
9. Remove the rocker arm pedestal-to-cylinder head bolts. Remove the rocker arm and pedestal assembly.
10. The installation is the reverse of the removal procedure. Tighten the rocker arm pedestal bolts to 45 ft. lbs. (60 Nm).
11. Refill the cooling system. Connect the battery cable and check for proper operation.

3.3L Engine
FRONT HEAD

1. Relieve the fuel system pressure.
2. Disconnect the negative battery cable.
3. Disconnect all electrical components and vacuum hoses which prevent access to the valve cover bolts.
4. Remove the serpentine drive belt.
5. Remove the alternator brace bolt and remove the alternator belt.
6. Remove the spark plug wire harness.
7. Remove the valve cover.
8. Remove the rocker arm pedestal-to-cylinder head bolts, the rocker arm and pedestal assembly.
9. The installation is the reverse of the removal procedure. Tighten the rocker arm pedestal bolts to 28 ft. lbs. (38 Nm).
10. Connect the negative battery cable and check for proper operation.

REAR HEAD

1. Relieve the fuel system pressure.
2. Disconnect the negative battery terminal.
3. Remove the serpentine drive belt.
4. Loosen the power steering pump bolts and slide the pump forward.
5. Remove the power steering braces.
6. Remove the spark plug wires from the spark plugs.
7. Remove the valve cover.
8. Remove the rocker arm pedestal-to-cylinder head bolts. Remove the rocker arm and pedestal assembly.
9. The installation is the reverse of the removal procedure. Tighten the rocker arm pedestal bolts to 28 ft. lbs. (38 Nm).
10. Refill the cooling system. Connect the battery cable and check for proper operation.

Intake Manifold

REMOVAL & INSTALLATION

2.0L Engine

1. Relieve the fuel system pressure. Disconnect the negative battery cable.
2. Remove induction tube and hoses.
3. Label and disconnect the wiring to throttle body, fuel injectors, MAP sensor and wastegate.
4. Disconnect the PCV and vacuum hoses on the throttle body.
5. Disconnect the throttle and cruise control cables, if equipped.
6. Remove the fuel return line from the throttle cable support bracket.
7. Disconnect the wiring to the ignition coil.
8. Remove the vacuum hoses from the rear of the manifold.
9. Remove the transaxle fill tube bracket.
10. Remove the manifold support bracket.
11. Remove the heater tube support bracket on the lower side of the manifold.
12. Disconnect the wires from the injectors.
13. Drain and remove the coolant recovery tank.
14. Remove the serpentine drive belt.
15. Remove the rear bolt from alternator bracket, the power steering adjusting bracket and front alternator adjusting bracket.
16. Remove the alternator.
17. Disconnect the fuel lines to fuel rail and regulator outlet.
18. Remove the attaching nuts and washers and remove the intake manifold.

To install:

19. Thoroughly clean and dry the mating surfaces. Install new gaskets and place the intake manifold in position.
20. Tighten the intake manifold attaching nuts to 18 ft. lbs. (24 Nm), starting from the middle and working outward.
21. Connect the fuel return line to the regulator outlet.
22. Install the power steering pump bracket and alternator and power steering pump adjusting brackets.
23. Install the alternator and belt.
24. Install the coolant recovery tank.
25. Connect the injector wiring.
26. Install the heater tube support, manifold support and transaxle fill tube brackets.
27. Connect the vacuum hoses at the rear of the bracket.
28. Install the ignition coil with its bracket.
29. Install the fuel supply line to the throttle cable support bracket.
30. Connect the cables, hoses and connectors to the throttle body.
31. Connect the wiring to the wastegate and MAP sensor.
32. Install the induction tube and hoses.
33. Fill all fluids to their proper levels.
34. Connect the negative battery cable and check for leaks.

2.3L Engine

1. Disconnect the negative battery cable.
2. Remove the coolant fan shroud, vacuum hose and electrical connector from the MAP sensor.
3. Disconnect the throttle body to air cleaner duct.
4. Remove the throttle cable bracket.
5. Remove the power brake vacuum hose, including the attaching bracket to power steering bracket and position it aside.
6. Remove the throttle body from

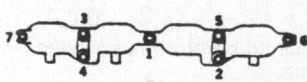

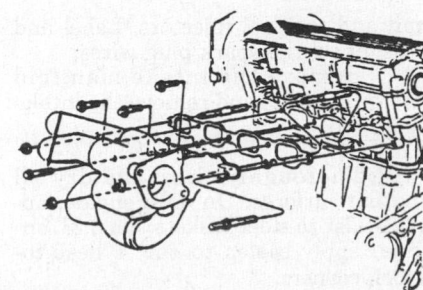

Intake manifold bolt torque sequence— 2.3L engine

the intake manifold with electrical harness, coolant hoses, vacuum hoses and throttle cable attached. Position these components aside.
7. Remove the oil/air separator bolts and hoses. Leave the hoses attached to the separator, disconnect from the oil fill, chain housing and the intake manifold. Remove as an assembly.
8. Remove the oil fill cap and oil level indicator stick.
9. Pull the oil tube fill upward to unseat from block and remove.
10. Disconnect the injector harness connector.
11. Remove the fill tube, rotating as necessary to gain clearance for the oil/air separator nipple between the intake tubes and fuel rail electrical harness.
12. Remove the intake manifold support bracket bolts and nut. Remove the intake manifold attaching nuts and bolts.
13. Remove the intake manifold.

To install:

14. Thoroughly clean and dry the mating surfaces. Install new gaskets and place the intake manifold in position.
15. Tighten the intake manifold bolts/nuts, in sequence, to 18 ft. lbs. (25 Nm). Tighten intake manifold brace and retainers hand tight. Tighten to specifications in the following order:
   a. Nut to stud bolt—18 ft. lbs. (25 Nm).
   b. Bolt to intake manifold—40 ft. lbs. (55 Nm).
   c. Bolt to cylinder block—40 ft. lbs. (55 Nm).
16. Lubricate a new oil fill tube ring seal with engine oil and install tube between No. 1 and 2 intake tubes. Rotate as necessary to gain clearance for oil/air separator nipple on fill tube.
17. Locate the oil fill tube in its cylinder block opening. Align the fill tube so it is approximately in its installed position. Press straight down to seat fill tube and seal into cylinder block.
18. Lubricate the hoses and install the oil/air separator assembly. Install the throttle body to intake manifold using a new gasket.
19. Install the power brake vacuum hose and the attaching bracket to power steering bracket.
20. Install the throttle cable bracket.
21. Connect the throttle body to air cleaner duct.
22. Install the coolant fan shroud, vacuum hose and electrical connector to the MAP sensor.
23. Fill all fluids to their proper levels.
24. Connect the negative battery cable and check for leaks.

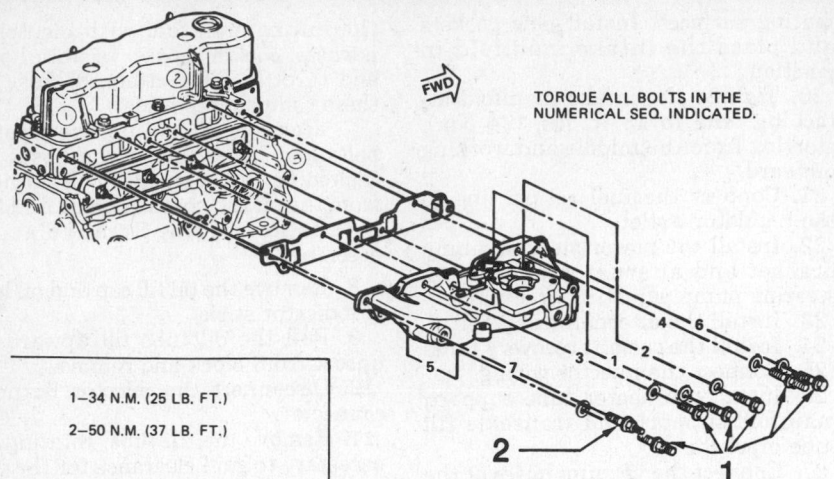

TORQUE ALL BOLTS IN THE
NUMERICAL SEQ. INDICATED.

1—34 N.M. (25 LB. FT.)
2—50 N.M. (37 LB. FT.)

Intake manifold bolt torque sequence—2.5L engine

2.5L Engine

1. Relieve the system fuel pressure. Disconnect the negative battery cable.
2. Drain the coolant. Remove the air cleaner, PCV valve and hose.
3. Disconnect and plug the vacuum lines and fuel lines. Disconnect the wiring and linkages from the throttle body.
4. Disconnect the throttle linkage and bell crank; position the assembly aside for clearance.
5. Disconnect the heater hoses. If equipped with power steering, disconnect and remove the upper power steering pump bracket.
6. Remove the ignition coil.
7. Remove the intake manifold mounting bolts and remove the intake manifold.
8. The installation is the reverse of the removal procedure. Tighten the intake manifold mounting bolts, in sequence, to 25 ft. lbs. (34 Nm).

3.0L and 3.3L Engines

1. Relieve the fuel system pressure.
2. Disconnect the negative battery cable.
3. On 3.0L engine, disconnect the mass air flow sensor. Remove the air intake duct.
4. Remove the serpentine drive belt, alternator and bracket.
5. Remove the C^{3}I ignition module and bracket.
6. Label and remove all the necessary vacuum and electrical wiring connectors.
7. Remove the throttle, cruise control and TV cables from the throttle body assembly.
8. Drain the coolant. Disconnect the heater hoses from the throttle body.
9. Remove the upper radiator hose from the intake manifold.
10. Remove the fuel lines, the fuel

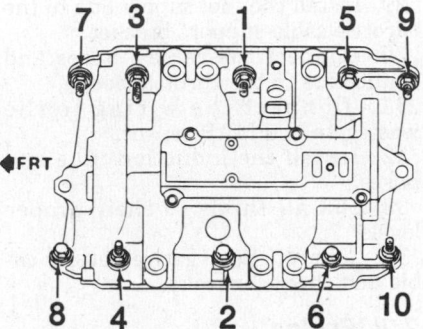

Intake manifold bolt torque sequence—3.0L engine

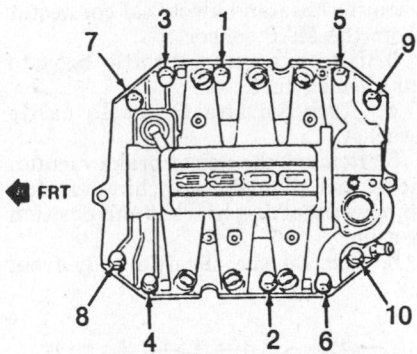

Intake manifold bolt torque sequence—3.3L engine

rail and the fuel injectors. Label and disconnect the spark plug wires.
11. Remove the intake manifold mounting bolts and remove the intake manifold.
To install:
12. Thoroughly clean and dry all mating surfaces. On 3.0L engine, apply sealer to steel gaskets. On 3.3L engine, apply sealer to the 4 head-to-block corners.
13. On 3.0L engine, tighten the intake manifold bolts in sequence to 32 ft. lbs. (44 Nm). Apply thread lock

compound to the threads and tighten to 88 inch lbs. (10 Nm) for 3.3L engine.
14. Install the fuel injectors, rail and lines. Connect the spark plug wires.
15. Connect the heater hoses to the throttle body.
16. Install the upper radiator hose to the intake manifold.
17. Connect the throttle, cruise control and TV cables to the throttle body assembly.
18. Connect all remaining vacuum and electrical wiring connectors.
19. Install the C^{3}I ignition module and bracket.
20. Install the alternator, bracket and serpentine drive belt.
21. Connect the mass air flow sensor, if equipped. Install the air intake duct.
22. Fill all fluids to their proper levels.
23. Connect the negative battery cable and check for leaks.

Exhaust Manifold

REMOVAL & INSTALLATION

2.0L Engine

1. Disconnect the negative battery cable.
2. Raise and safely support the vehicle.
3. Drain the engine coolant.
4. Remove the fan attaching screws.
5. Disconnect the exhaust pipe.
6. Remove the air conditioning compressor rear support bracket.
7. Remove turbocharger support bracket to engine.
8. Disconnect and plug the oil drain pipe at turbocharger.
9. Disconnect water return pipe at turbocharger.
10. Lower vehicle and remove coolant recovery pipe.
11. Remove the air induction tube, coolant fan, oxygen sensor.
12. Disconnect the oil and water feed pipes.
13. Remove air intake duct and vacuum hose at actuator.
14. Remove the exhaust manifold attaching nuts and remove turbocharger and manifold as an assembly.
15. Remove turbocharger from exhaust manifold.
To install:
16. Assemble the turbocharger and exhaust manifold.
17. Clean the exhaust manifold and cylinder head mating surfaces.
18. Install a new gasket and install the manifold/turbocharger assembly to the engine. Tighten the Nos. 2 and 3 manifold runner nuts first, then Nos. 1 and 4, to 18 ft. lbs. (24 Nm).
19. Connect the oil and water feed and return lines.

20. Connect the oxygen sensor.
21. Install the air intake duct and connect the vacuum hose to the actuator.
22. Install the cooling fan.
23. Install the induction tube and coolant recovery tube.
24. Raise and safely support the vehicle.
25. Install the rear turbocharger support bolt.
26. Install the compressor support bracket.
27. Install the oil drain hose.
28. Connect the exhaust pipe.
29. Connect the negative battery cable and check the turbocharger for proper operation and the assembly for leaks.

2.3L Engine

1. Disconnect the negative battery cable and oxygen sensor connector.
2. Remove upper and lower exhaust manifold heat shields.
3. Remove the bolt that attaches the exhaust manifold brace to the manifold.
4. Break loose the manifold to exhaust pipe spring loaded bolts using a 13mm box wrench.
5. Raise and safely support the vehicle.

NOTE: It is necessary to relieve the spring pressure from 1 bolt prior to removing the second bolt. If the spring pressure is not relieved it will cause the exhaust pipe to twist and bind up the bolt as it is removed.

6. Remove the manifold to exhaust pipe bolts from the exhaust pipe flange as follows:

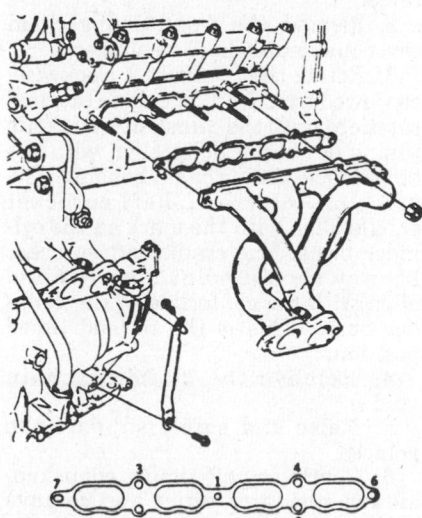

Exhaust manifold bolt torque sequence — 2.3L engine

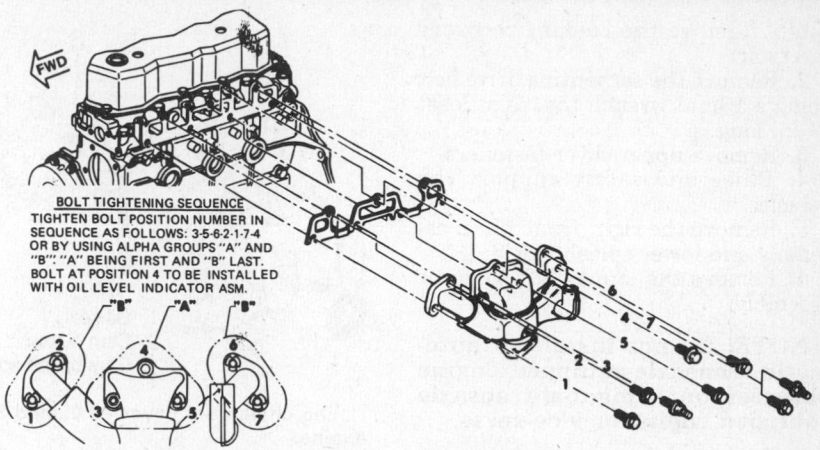

BOLT TIGHTENING SEQUENCE
TIGHTEN BOLT POSITION NUMBER IN SEQUENCE AS FOLLOWS: 3-5-6-2-1-7-4 OR BY USING ALPHA GROUPS "A" AND "B": "A" BEING FIRST AND "B" LAST. BOLT AT POSITION 4 TO BE INSTALLED WITH OIL LEVEL INDICATOR ASM.

Exhaust manifold bolt torque sequence — 2.5L engine

a. Unscrew either bolt clockwise 4 turns.
b. Remove the other bolt.
c. Remove the first bolt.
7. Pull down and back on the exhaust pipe to disengage it from the exhaust manifold bolts.
8. Lower the vehicle.
9. Remove the exhaust manifold mounting bolts and remove the manifold.
10. The installation is the reverse of the removal procedure. Tighten the mounting bolts, in sequence, to 27 ft. lbs. (37 Nm). Install the exhaust pipe flange bolts evenly and gradually to avoid binding.
11. Connect the negative battery cable and check for leaks.

2.5L Engine

1. Disconnect the negative battery cable and oxygen sensor connector. Remove the air cleaner assembly.
2. Remove the upper alternator mount and position the unit to one side.
3. Raise and safely support the vehicle.
4. Disconnect the exhaust pipe-to-exhaust manifold bolts and lower the exhaust pipe.
5. Lower the vehicle.
6. Remove the exhaust manifold mounting bolts and lift the exhaust manifold from the engine.
7. The installation is the reverse of the removal procedure. Tighten the exhaust manifold bolts, in sequence, to 32 ft. lbs. (43 Nm).
8. Connect the negative battery cable and check for leaks.

3.0L and 3.3L Engines
FRONT MANIFOLD

1. Disconnect the negative battery cable.
2. Disconnect air cleaner mounting bolts.
3. Remove the bolts attaching the exhaust crossover pipe to the manifold.
4. Disconnect the spark plug wires.
5. Remove the cooling fan.
6. Remove the mounting bolts and remove the manifold.

NOTE: The oil dipstick tube may have to be removed to provide access to the manifold bolts.

7. The installation is the reverse of the removal procedure. Tighten the mounting bolts to 37 ft. lbs. (50 Nm) for 3.0L engine or 30 ft. lbs. (41 Nm) for 3.3L engine.
8. Connect the negative battery cable and check for leaks.

REAR MANIFOLD

1. Disconnect the negative battery cable.
2. Remove the 2 bolts attaching exhaust pipe to manifold.
3. Disconnect oxygen sensor wire.
4. Disconnect and tag spark plug wires.
5. Remove 2 nuts attaching crossover pipe to manifold.
6. Remove serpentine belt.
7. Remove power steering pump.
8. Remove heater hose from tube, heat shield and C^3I bracket nuts.
9. Remove the bolts attaching manifold to cylinder head.
10. The installation is the reverse of the removal procedure. Tighten the mounting bolts to 37 ft. lbs. (50 Nm) for 3.0L engine or 30 ft. lbs. (41 Nm) for 3.3L engine.
11. Connect the negative battery cable and check for leaks.

Timing Chain Front Cover

REMOVAL & INSTALLATION

2.3L Engine

1. Disconnect the negative battery

cable. Remove the coolant recovery reservoir.

2. Remove the serpentine drive belt using a 13mm wrench that is at least 24 in. long.

3. Remove upper cover fasteners.

4. Raise and safely support the vehicle.

5. Remove the right front wheel assembly and lower splash shield.

6. Remove the crankshaft balancer assembly.

NOTE: Do not install an automatic transaxle-equipped engine balancer on a manual-transaxle equipped engine or vice-versa.

7. Remove lower cover fasteners and lower the vehicle.

8. Remove the front cover.

9. The installation is the reverse of the removal procedure. Tighten the balancer attaching bolt to 74 ft. lbs. (100 Nm).

1990–92 2.5L Engine

1. Disconnect the negative battery cable.

2. Remove the belts. Remove the power steering pump mounting bolts and position it aside.

3. Raise and safely support the vehicle. Remove the inner fender splash shield.

4. Remove the harmonic balancer.

5. Remove the timing case cover-to-engine bolts and the timing case cover.
To install:

6. Thoroughly clean and dry all mating surfaces. Use RTV sealant to seal all mating surfaces.

7. A centering tool fits over the crankshaft seal and is used to correctly position the timing case cover during installation. Install the cover and partially tighten the 2 opposing timing case cover screws.

8. Tighten the remaining cover screws and remove the centering tool from the timing case cover. Tighten to 89 inch lbs. (10 Nm).

9. Install the harmonic balancer and tighten the bolt to 162 ft. lbs. (220 Nm). Install the belts and the power steering pump.

10. Install the splash shield.

11. Connect the negative battery cable and check for leaks.

3.0L and 3.3L Engines

1. Disconnect the negative battery cable.

2. Drain the coolant and the engine oil. Remove the oil filter.

3. Loosen the water pump pulley bolts but do not remove them. Remove the serpentine drive belt and the pulley. Remove the water pump-to-engine bolts and the water pump.

4. Raise and safely support the ve-

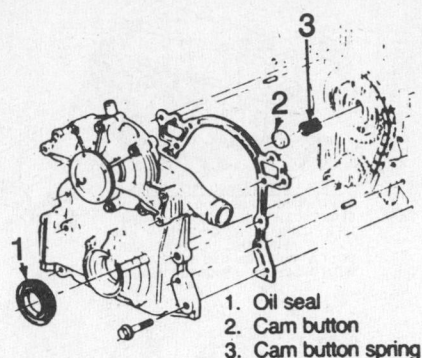

1. Oil seal
2. Cam button
3. Cam button spring

Timing chain front cover — 3.0L and 3.3L engines

hicle. Remove the right front wheel assembly and the right inner fender splash shield.

5. Remove the crankshaft harmonic balancer and the crankshaft sensor.

6. Remove the radiator and heater hoses.

7. Remove the timing case cover-to-engine bolts, the timing case cover and the gasket.

8. Clean the gasket mounting surfaces. Replace the front oil seal.

9. The installation is the reverse of the removal procedure. Coat all timing case cover bolts with thread sealer prior to installation.

10. Fill all fluids to their proper levels.

11. Connect the negative battery cable and check for leaks.

Timing Gear Front Cover

REMOVAL & INSTALLATION

1988–89 2.5L Engine

1. Disconnect the negative battery cable.

2. Remove the belts.

3. Raise and safely support the vehicle. Remove the inner fender splash shield.

4. Remove the harmonic balancer.

5. Remove the cover-to-engine bolts and the timing cover.
To install:

6. Thoroughly clean and dry all mating surfaces. Use RTV sealant to seal all mating surfaces.

7. A centering tool fits over the crankshaft seal and is used to correctly position the timing case cover during installation. Install the cover and partially tighten the 2 opposing timing case cover screws.

8. Tighten the remaining cover screws and remove the centering tool from the timing case cover. Final torque of all screws should be 89 inch lbs. (10 Nm).

9. Install the harmonic balancer and tighten the bolt to 162 ft. lbs. (220 Nm). Install the belts and the power steering pump.

10. Install the splash shield.

11. Connect the negative battery cable and check for leaks.

Front Cover Oil Seal

REPLACEMENT

1. Disconnect the negative battery cable.

2. Remove the front cover.

3. Using a small prybar, pry out the old oil seal.

NOTE: Use care to avoid damage to seal bore or seal contact surfaces.

4. Thoroughly clean and dry the oil seal mounting surface.

5. Use the appropriate installation tool and drive the oil seal into the front cover.

6. Lubricate balancer and seal lip with clean engine oil.

7. The installation is the reverse of the removal procedure.

8. Connect the negative battery cable and check for leaks.

Timing Chain and Sprockets

REMOVAL & INSTALLATION

2.3L Engine

NOTE: It is recommended that the entire procedure be reviewed before attempting to service the timing chain.

1. Disconnect the negative battery cable.

2. Remove the front timing chain cover and crankshaft oil slinger.

3. Rotate the crankshaft clockwise, as viewed from front of engine (normal rotation) until the camshaft sprocket's timing dowel pin holes align with the holes in the timing chain housing. The mark on the crankshaft sprocket should align with the mark on the cylinder block. The crankshaft sprocket keyway should point upwards and align with the centerline of the cylinder bores. This is the normal timed position.

4. Remove the 3 timing chain guides.

5. Raise and safely support the vehicle.

6. Gently pry off timing chain tensioner spring retainer and remove spring.

NOTE: Two styles of tensioner are used. Early production en-

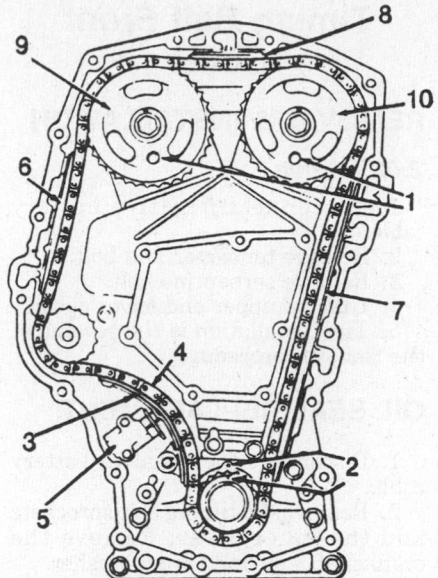

1. Camshaft timing marks
2. Crankshaft timing mark
3. Tensioner shoe assembly
4. Timing chain
5. Tensioner
6. R/H guide
7. L/H guide
8. Upper guide
9. Exhaust camshaft sprocket
10. Intake camshaft sprocket

Timing chain Installaiton—2.3L engine

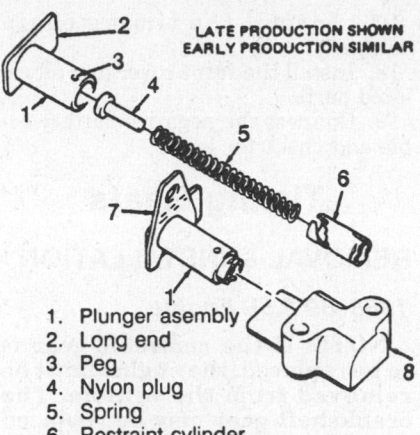

LATE PRODUCTION SHOWN
EARLY PRODUCTION SIMILAR

1. Plunger asembly
2. Long end
3. Peg
4. Nylon plug
5. Spring
6. Restraint cylinder
7. J–36589 anti-release devise
8. Tensioner body

Exploded view of the timing chain tensioner—2.3L engine. Versions may differ slightly with year.

gines will have a spring post and late production ones will not. Both styles are identical in operation and are interchangeable.

7. Remove the timing chain tensioner shoe retainer.

8. Make sure all the slack in the timing chain is above the tensioner assembly; remove the chain tensioner shoe. The timing chain must be disengaged from the wear grooves in the tensioner shoe in order to remove the shoe. Slide a prybar under the timing chain while pulling shoe outward.

9. If difficulty is encountered removing chain tensioner shoe, proceed as follows:

a. Lower the vehicle.

b. Hold the intake camshaft sprocket with a holding tool and remove the sprocket bolt and washer.

c. Remove the washer from the bolt and re-thread the bolt back into the camshaft by hand, the bolt provides a surface to push against.

d. Remove intake camshaft sprocket using a 3-jaw puller in the 3 relief holes in the sprocket. Do not attempt to pry the sprocket off the camshaft or damage to the sprocket or chain housing could occur.

10. Remove the tensioner assembly attaching bolts and the tensioner.

— CAUTION —

The tensioner piston is spring loaded and could fly out causing personal injury.

11. Remove the chain housing to block stud, which is actually the timing chain tensioner shoe pivot.

12. Remove the timing chain.

To install:

13. Tighten intake camshaft sprocket attaching bolt and washer, while holding the sprocket with tool J–36013, if removed.

14. Install the special tool through holes in camshaft sprockets into holes in timing chain housing. This positions the camshafts for correct timing.

15. If the camshafts are out of position and must be rotated more than ⅛ turn in order to install the alignment dowel pins:

a. The crankshaft must be rotated 90 degrees clockwise off of TDC in order to give the valves adequate clearance to open.

b. Once the camshafts are in position and the dowels installed, rotate the crankshaft counterclockwise back to TDC. Do not rotate the crankshaft clockwise to TDC or valve and piston damage could occur.

16. Install the timing chain over the exhaust camshaft sprocket, around the idler sprocket and around the crankshaft sprocket.

17. Remove the alignment dowel pin from the intake camshaft. Using a dowel pin remover tool, rotate the in-take camshaft sprocket counterclockwise enough to slide the timing chain over the intake camshaft sprocket. Release the camshaft sprocket wrench. The length of chain between the 2 camshaft sprockets will tighten. If properly timed, the intake camshaft alignment dowel pin should slide in easily. If the dowel pin does not fully index, the camshafts are not timed correctly and the procedure must be repeated.

18. Leave the alignment dowel pins installed.

19. With slack removed from chain between intake camshaft sprocket and crankshaft sprocket, the timing marks on the crankshaft and the cylinder block should be aligned. If marks are not aligned, move the chain 1 tooth forward or rearward, remove slack and recheck marks.

20. Tighten the chain housing to block stud. The stud is installed under the timing chain. Tighten to 19 ft. lbs. (26 Nm).

21. Reload timing chain tensioner assembly to its 0 position as follows:

a. Assemble restraint cylinder, spring and nylon plug into plunger. Index slot in restraint cylinder with peg in plunger. While rotating the restraint cylinder clockwise, push the restraint cylinder into the plunger until it bottoms. Keep rotating the restraint cylinder clockwise but allow the spring to push it out of the plunger. The pin in the plunger will lock the restraint in the loaded position.

b. Install tool J–36589 or equivalent, onto plunger assembly.

c. Install plunger assembly into tensioner body with the long end toward the crankshaft when installed.

22. Install the tensioner assembly to the chain housing. Recheck plunger assembly installation. It is correctly installed when the long end is toward the crankshaft.

23. Install and tighten timing chain tensioner bolts and tighten to 10 ft. lbs. (14 Nm).

24. Install the tensioner shoe and tensioner shoe retainer. Remove special tool J–36589 and squeeze plunger assembly into the tensioner body to unload the plunger assembly.

25. Lower vehicle and remove the alignment dowel pins. Rotate crankshaft clockwise 2 full rotations. Align crankshaft timing mark with mark on cylinder block and reinstall alignment dowel pins. Alignment dowel pins will slide in easily if engine is timed correctly.

NOTE: If the engine is not correctly timed, severe engine damage could occur.

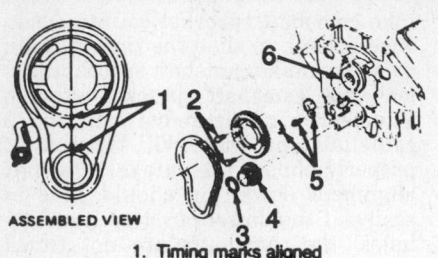

ASSEMBLED VIEW

1. Timing marks aligned
2. 22 ft. lbs. (30 Nm)
3. Seal
4. Crankshaft gear
5. Dampener assembly
6. Camshaft

Timing chain and timing mark alignment —3.0L and 3.3L engines

26. Install 3 timing chain guides and crankshaft oil slinger.

27. Install the timing chain front cover.

28. Connect the negative battery cable and check for leaks.

3.0L, 3.3L and 1990–92 2.5L Engines

1. Disconnect the negative battery cable.

2. Drain the cooling system. Disconnect the cooling hose from the water pump.

3. Raise and safely support the vehicle.

4. Remove the inner fender splash shield.

5. Remove the serpentine drive belt.

6. Remove the crankshaft pulley bolt and slide the pulley from the crankshaft.

7. Remove the front cover.

8. Rotate the crankshaft to align the timing marks on the sprockets. Remove the chain dampener assembly.

9. Remove the camshaft sprocket-to-camshaft bolt(s), remove the camshaft sprocket and chain and thrust bearing.

10. Remove the crankshaft gear by sliding it forward.

11. Clean the gasket mounting surfaces. Inspect the timing chain and the sprockets for damage and/or wear and replace damaged parts.

To install:

12. Position the crankshaft so the No. 1 piston is at TDC of its compression stroke. Install the thrust bearing on 2.5L engine.

13. Temporarily install the gear on the camshaft and position the camshaft so the timing mark on the gear is pointing straight down.

14. Assemble the timing chain to the gears so the timing marks are aligned, mark-to-mark.

15. Install the camshaft sprocket attaching bolt(s).

16. Install the camshaft thrust bearing, if not already done.

17. Install the timing chain dampener.

18. Install the front cover and all related parts.

19. Connect the negative battery cable and check for leaks.

Timing Gears

REMOVAL & INSTALLATION

1988–89 2.5L Engine

NOTE: If the camshaft gear is to be replaced, the engine must be removed from the vehicle. The crankshaft gear may be replaced with the engine in the vehicle.

1. Disconnect the negative battery cable.

2. Raise and safely support the vehicle.

3. Remove the inner fender splash shield.

4. Remove the accessory drive belts. Remove the crankshaft pulley-to-crankshaft pulley bolt and slide the pulley from the crankshaft.

5. If replacing the camshaft gear, perform the following procedures:

   a. Remove the engine from the vehicle and secure it onto a suitable holding fixture.

   b. Remove the camshaft from the engine.

   c. Using an arbor press, press the camshaft gear from the camshaft.

   d. To install the camshaft gear onto the camshaft, press the gear onto the shaft until a thrust clearance of 0.0015–0.0050 in. exists.

6. If removing the crankshaft gear, perform the following procedures:

   a. Remove the front cover-to-engine bolts.

   b. Remove the attaching bolt and slide the crankshaft gear forward off the crankshaft.

7. Clean the gasket mounting surfaces. Inspect the parts for damage and/or wear and replace damaged parts.

8. The installation is the reverse of the removal procedure. Make sure the timing marks are aligned mark-to-mark when installing.

Aligning the timing marks—1988–89 2.5L engine

Timing Belt Front Cover

REMOVAL & INSTALLATION

2.0L Engine

1. Disconnect negative battery cable.

2. Remove tensioner and bolt.

3. Remove serpentine belt.

4. Unsnap upper and lower cover.

5. The installation is the reverse of the removal procedure.

OIL SEAL REPLACEMENT

1. Disconnect the negative battery cable.

2. Remove the timing belt sprockets and the inner cover. Remove the crankshaft key and thrust washer.

3. Using a small prybar, pry out the old oil seal.

NOTE: Use care to avoid damage to seal bore and crankshaft.

4. Thoroughly clean and dry the oil seal mounting surface.

5. Use the appropriate installation tool and drive the oil seal into the front cover.

6. The installation is the reverse of the removal procedure.

7. Connect the negative battery cable and check for leaks.

Timing Belt and Tensioner

ADJUSTMENT

2.0L Engine

1988

1. Disconnect the negative battery cable. Remove the timing belt cover.

2. Adjust the timing belt using tools J–26486–A and J–33039 to adjust the water pump. With the gauge installed, increase the tension to within the band on the gauge will ensure an initial over-tensioning.

3. Crank the engine without starting it about 10 revolutions; a tension loss may occur.

4. Recheck the tension with the gauge. If a tension increase is needed, remove the gauge and adjust the water pump. Repeat until the tension is within specification.

NOTE: Do not increase tension with the gauge installed or the resulting tension will be inaccurate.

5. After the proper tension has been reached, tighten the water pump bolts to 19 ft. lbs. (25 Nm).

6. Install the timing belt cover and all related parts.

7. Connect the negative battery cable and road test the vehicle.

1989

1. Disconnect the negative battery cable. Remove the timing belt cover.

2. Make sure the portion of the belt between the camshaft and crankshaft has no slack.

3. Adjust the timing belt using tool J–33039 to turn the water pump eccentric clockwise until the tensioner contacts the high torque stop. Temporarily tighten the water to prevent movement.

4. Turn the engine 2 revolutions.

5. Turn the water pump eccentric counterclockwise until the hole in the tensioner arm is aligned with the hole in the base.

6. Tighten the water pump bolts to 19 ft. lbs. (25 Nm), making sure the tensioner hole remains aligned.

7. Install the timing belt cover and all related parts.

8. Connect the negative battery cable and road test the vehicle.

REMOVAL & INSTALLATION

2.0L Engine

1988

1. Disconnect the negative battery cable.

2. Remove the timing belt cover.

3. Remove the crankshaft pulley.

4. Remove the coolant reservoir.

5. Loosen the water pump mounting bolts and remove the timing belt.

To install:

6. Position the camshaft so the

mark on its sprocket aligns with the mark on the rear timing belt cover.

7. Position the crankshaft so the mark on the pulley aligns with 10 degrees BTDC on the timing scale.

8. Install the timing belt.

9. Adjust the timing belt using tools J–26486–A and J–33039 to adjust the water pump. Increase the tension—with the gauge installed—to within the band on the gauge will ensure an initial over-tensioning.

10. Crank the engine without starting it about 10 revolutions; a substantial tension loss should occur.

11. Recheck the tension with the gauge. If a tension increase is needed, remove the gauge and adjust the water pump. Repeat until the tension is within specification.

NOTE: Do not increase tension with the gauge installed or the resulting tension will be inaccurate.

12. After the proper tension has been reached, tighten the water pump bolts to 19 ft. lbs. (25 Nm).

13. Install the timing belt cover and all related parts.

14. Install the coolant reservoir.

15. Connect the negative battery cable and road test the vehicle.

1989

1. Disconnect the negative battery cable.

2. Remove the timing belt cover.

3. Remove the crankshaft pulley.

4. Loosen the water pump mounting bolts and relieve the tension using tool J–33039.

5. Remove the timing belt.

To install:

6. Position the camshaft and crank-

shaft so the marks on their sprockets aligns with the marks on the rear cover.

7. Install the timing belt so the portion between the camshaft and crankshaft has no slack.

8. Adjust the timing belt using tool J–33039 to turn the water pump eccentric clockwise until the tensioner contacts the high torque stop. Temporarily tighten the water to prevent movement.

9. Turn the engine 2 revolutions to fully seat the belt into the gear teeth.

10. Turn the water pump eccentric counterclockwise until the hole in the tensioner arm is aligned with the hole in the base.

11. Tighten the water pump bolts to 19 ft. lbs. (25 Nm), making sure the tensioner hole remains aligned as in Step 10.

12. Install the timing belt cover and all related parts.

13. Install the crankshaft pulley.

14. Install the timing belt cover and all related parts.

15. Connect the negative battery cable and road test the vehicle.

Timing Sprockets

REMOVAL & INSTALLATION

1. Disconnect the negative battery cable.

2. If removing the camshaft sprocket, remove the camshaft carrier cover.

3. Remove the timing belt cover.

4. Position the engine so the timing marks are aligned for belt installation.

5. Remove the timing belt.

6. If removing the camshaft sprock-

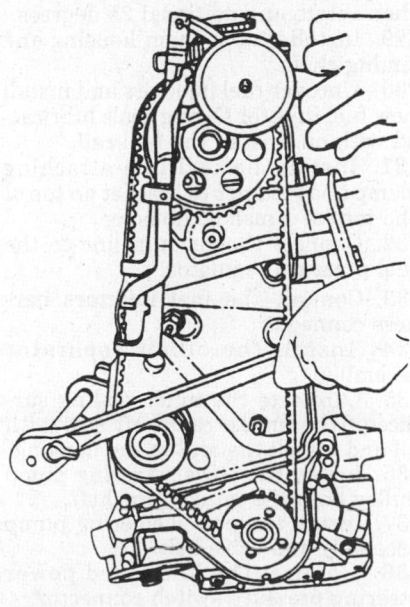

Adjusting the timing belt tension—2.0L engine

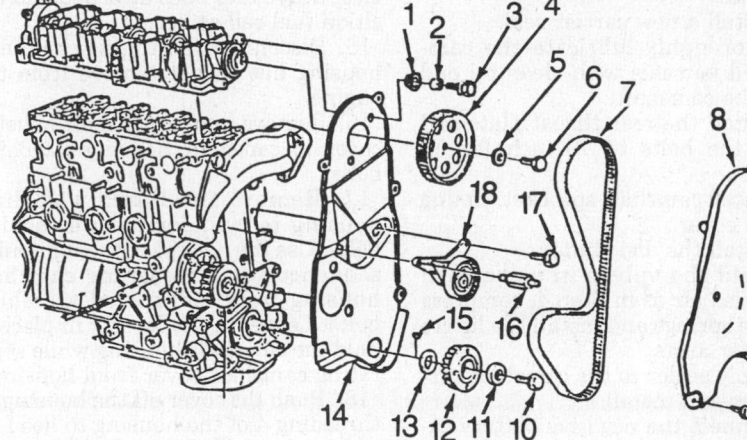

1. Grommet	7. Timing belt	13. Washer
2. Sleeve	8. Front cover	14. Rear cover
3. Bolt	9. Bolt	15. Key
4. Camshaft sprocket	10. Bolt	16. Stud
5. Washer	11. Washer	17. Bolt
6. Bolt	12. Crankshaft sprocket	18. Tensioner

Timing belt and related parts—1989 2.0L engine

et, hold the camshaft with an open-end wrench.

7. Remove the camshaft or crankshaft sprocket attaching bolt, washer and the sprocket.

8. The installation is the reverse of the removal procedure. Tighten the camshaft sprocket bolt to 34 ft. lbs. (45 Nm). Tighten the crankshaft sprocket bolt to 114 ft. lbs. (155 Nm).

9. Connect the negative battery cable and road test the vehicle.

Camshaft

REMOVAL & INSTALLATION

2.0L Engine

1. Relieve the fuel system pressure.
2. Disconnect the negative battery cable.
3. Remove the camshaft carrier cover.
4. Hold the valves in place with compressed air, using air adapters in the spark plug holes.
5. Compress the valve springs with the special valve spring compressing tool.
6. Remove the rocker arms and lifters and keep them in order for reassembly. Hold the camshaft with an open-end wrench and remove the camshaft sprocket. Try to keep the valve timing by using a rubber cord, if possible. If the timing cannot be kept intact, the timing belt will have to be reset.
7. Matchmark and remove the distributor.
8. Remove the camshaft thrust plate from the rear of the carrier.
9. Remove the camshaft by sliding it toward the rear. Remove the front carrier seal.

To install:

10. Install a new carrier seal.
11. Thoroughly lubricate the camshaft and journals with clean oil and install the camshaft.
12. Install the rear thrust plate and tighten the bolts to 70 inch lbs. (8 Nm).
13. Install camshaft sprocket, timing belt and cover.
14. Install the distributor.
15. Hold the valves in place with compressed air as in Step 4, compress the valve springs and install the lifters and rocker arms.
16. Apply sealer to the camshaft carrier cover and install.
17. Connect the negative battery cable and road test the vehicle.

2.3L Engine

INTAKE CAMSHAFT

NOTE: Any time the camshaft housing to cylinder head bolts are loosened or removed, the cam-shaft housing to cylinder head gasket must be replaced.

1. Relieve the fuel system pressure. Disconnect the negative battery cable.
2. Label and disconnect the ignition coil and module assembly electrical connections.
3. Remove 4 ignition coil and module assembly to camshaft housing bolts and remove assembly by pulling straight up. Use a special spark plug boot wire remover tool to remove connector assemblies, if they have stuck to the spark plugs.
4. Remove the idle speed power steering pressure switch connector.
5. Loosen 3 power steering pump pivot bolts and remove drive belt.
6. Disconnect the 2 rear power steering pump bracket to transaxle bolts.
7. Remove the front power steering pump bracket to cylinder block bolt.
8. Disconnect the power steering pump assembly and position aside.
9. Using the special tool, remove the power steering pump drive pulley from the intake camshaft.
10. Remove oil/air separator bolts and hoses. Leave the hoses attached to the separator, disconnect from the oil fill, chain housing and intake manifold. Remove as an assembly.
11. Remove vacuum line from fuel pressure regulator and disconnect the fuel injector harness connector.
12. Disconnect fuel line attaching clamp from bracket on top of intake camshaft housing.
13. Remove fuel rail to camshaft housing attaching bolts.
14. Remove the fuel rail from the cylinder head. Cover injector openings in cylinder head and cover injector nozzles. Leave fuel lines attached and position fuel rail aside.
15. Disconnect the timing chain and housing but do not remove from the engine.
16. Remove intake camshaft housing cover to camshaft housing attaching bolts.
17. Remove the intake camshaft housing to cylinder head attaching bolts. Use the reverse of the tightening sequence when loosening camshaft housing to cylinder head attaching bolts. Leave 2 bolts loosely in place to hold the camshaft housing while separating camshaft cover from housing.
18. Push the cover off the housing by threading 4 of the housing to head attaching bolts into the tapped holes in the cam housing cover. Tighten the bolts in evenly so the cover does not bind on the dowel pins.
19. Remove the 2 loosely installed camshaft housing to head bolts and remove the cover. Discard the gaskets.
20. Note the position of the chain sprocket dowel pin for reassembly. Remove the camshaft carefully; do not damage the camshaft oil seal.
21. Remove intake camshaft oil seal from camshaft and discard seal. This seal must be replaced any time the housing and cover are separated.
22. Remove the camshaft carrier from the cylinder head and remove the gasket.

To install:

23. Thoroughly clean the mating surfaces of the camshaft carrier and the cylinder head, bolts and bolt holes. Install a new gasket and place the housing on the head. Install 1 bolt loosely to hold in place.
24. Install the lifters into their bores. If the camshaft is being replaced, the lifters must also be replaced. Lubricate camshaft lobes, journals and lifters with camshaft and lifter prelube. The camshaft lobes and journals must be adequately lubricated or engine damage could occur upon start up.
25. Install the camshaft in the same position as when removed. The timing chain sprocket dowel pin should be straight up and align with the centerline of the lifter bores.
26. Install new camshaft housing to camshaft housing cover seals into cover; do not use sealer. Make sure the correct color seal is placed in each groove. Install the cover to the housing.
27. Apply thread locking compound to the camshaft housing and cover attaching bolt threads.
28. Install bolts and tighten to 11 ft. lbs. (15 Nm). Rotate the bolts, except the 2 rear bolts that hold the fuel pipe to the camshaft housing, an additional 75 degrees, in sequence. Tighten the excepted bolts to 16 ft. lbs. (15 Nm), then rotate an additional 25 degrees.
29. Install timing chain housing and timing chain.
30. Uncover fuel injectors and install new fuel injector O-ring seals lubricated with oil. Install the fuel rail.
31. Install the fuel line attaching clamp and retainer to bracket on top of the intake camshaft housing.
32. Connect the vacuum line to the fuel pressure regulator.
33. Connect the fuel injectors harness connector.
34. Install the oil/air separator assembly.
35. Lubricate the inner sealing surface of the intake camshaft seal with oil and install the seal to the housing.
36. Install the power steering pump pulley onto the intake camshaft.
37. Install the power steering pump assembly and drive belt.
38. Connect the idle speed power steering pressure switch connector.
39. Clean any loose lubricant that is present on the ignition coil and mod-

ule assembly to camshaft housing bolts. Apply Loctite® 592 or equivalent, onto the ignition coil and module assembly to camshaft housing bolts. Install the bolts and tighten to 13 ft. lbs. (18 Nm).

40. Connect the electrical connectors to ignition coil and module assembly.

41. Connect the negative battery cable and road test the vehicle. Check for leaks.

EXHAUST CAMSHAFT

NOTE: Any time the camshaft housing to cylinder head bolts are loosened or removed the camshaft housing to cylinder head gasket must be replaced.

1. Relieve the fuel system pressure. Disconnect the negative battery cable.

2. Label and disconnect the ignition coil and module assembly electrical connections.

3. Remove 4 ignition coil and module assembly to camshaft housing bolts and remove assembly by pulling straight up. Use a special tool to remove connector assemblies if they have stuck to the spark plugs.

4. Remove the idle speed power steering pressure switch connector.

5. Remove the transaxle fluid level indicator tube assembly from exhaust camshaft cover and position aside.

6. Remove exhaust camshaft cover and gasket.

7. Disconnect the timing chain and housing but do not remove from the engine.

8. Remove exhaust camshaft housing to cylinder head bolts. Use the reverse of the tightening procedure when loosening camshaft housing while separating camshaft cover from housing.

9. Push the cover off the housing by threading 4 of the housing to head attaching bolts into the tapped holes in the camshaft cover. Tighten the bolts in evenly so the cover does not bind on the dowel pins.

10. Remove the 2 loosely installed camshaft housing to cylinder head bolts and remove cover, discard gaskets.

11. Loosely reinstall 1 camshaft housing to cylinder head bolt to retain the housing during camshaft and lifter removal.

12. Note the position of the chain sprocket dowel pin for reassembly. Remove camshaft being careful not to damage the camshaft or journals.

13. Remove the camshaft carrier from the cylinder head and remove the gasket.

To install:

14. Thoroughly clean the mating surfaces of the camshaft carrier and the cylinder head, bolts and bolt holes. Install a new gasket and place the

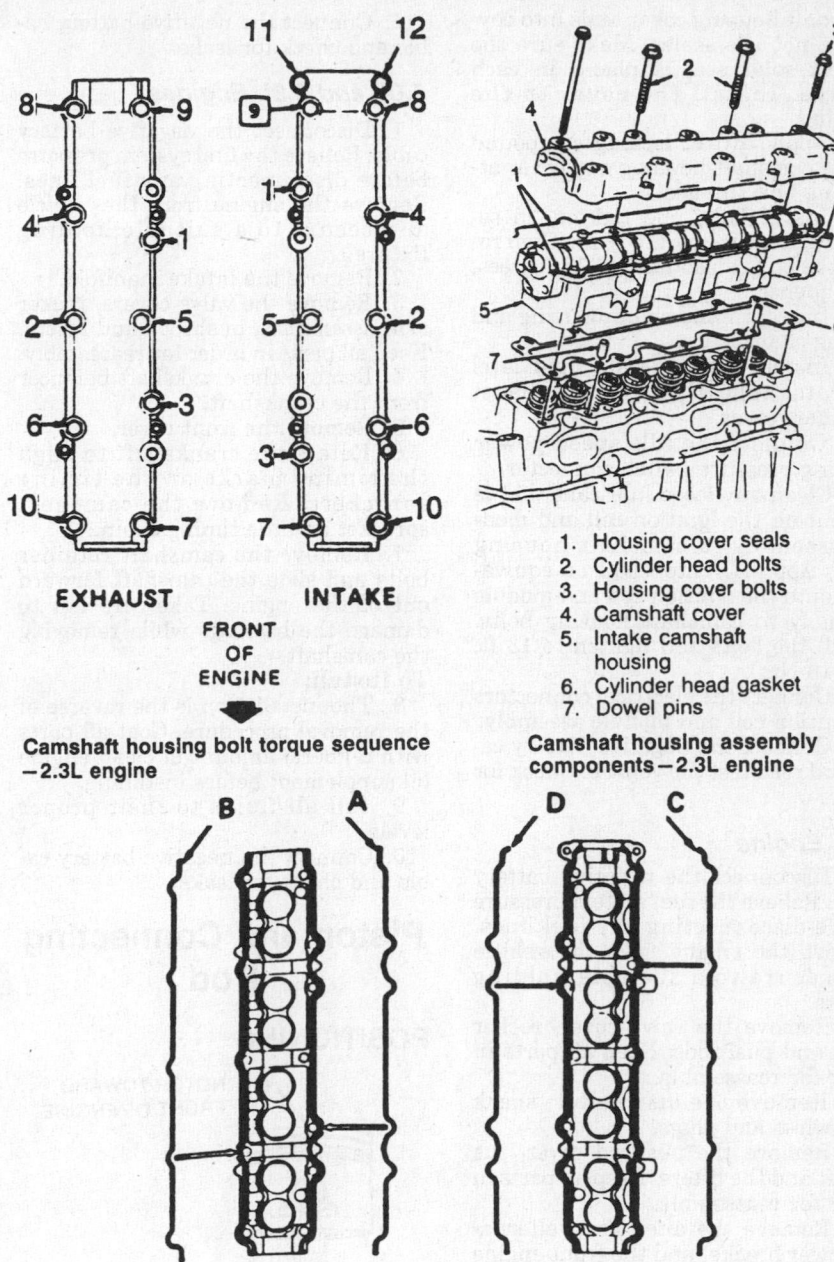

1. Housing cover seals
2. Cylinder head bolts
3. Housing cover bolts
4. Camshaft cover
5. Intake camshaft housing
6. Cylinder head gasket
7. Dowel pins

Camshaft housing bolt torque sequence —2.3L engine

Camshaft housing assembly components—2.3L engine

A. Seal—inner (exhaust—red)
B. Seal—outer (exhaust—red)
C. Seal—outer (intake—blue)
D. Seal—inner (intake—blue)

Camshaft housing cover seal identification—2.3L engine

housing on the head. Install 1 bolt loosely to hold in place.

15. Install the lifters into their bores. If the camshaft is being replaced, the lifters must also be replaced. Lubricate camshaft lobes, journals and lifters with camshaft and lifter prelube. The camshaft lobes and journals must be

adequately lubricated or engine damage could occur upon start up.

16. Install camshaft in same position as when removed. The timing chain sprocket dowel pin should be straight up and align with the centerline of the lifter bores.

17. Install new camshaft housing to

camshaft housing cover seals into cover; do not use sealer. Make sure the correct color seal is placed in each groove. Install the cover to the housing.

18. Apply thread locking compound to the camshaft housing and cover attaching bolt threads.

19. Install bolts and tighten, in sequence, to 11 ft. lbs. (15 Nm). Then rotate the bolts an additional 75 degrees, in sequence.

20. Install timing chain housing and timing chain.

21. Install the transaxle fluid level indicator tube assembly to exhaust camshaft cover.

22. Connect the idle speed power steering pressure switch connector.

23. Clean any loose lubricant that is present on the ignition coil and module assembly to camshaft housing bolts. Apply Loctite® 592 or equivalent, onto the ignition coil and module assembly to camshaft housing bolts. Install the bolts and tighten to 13 ft. lbs. (18 Nm).

24. Connect the electrical connectors to ignition coil and module assembly.

25. Connect the negative battery cable and road test the vehicle. Check for leaks.

2.5L Engine

1. Disconnect the negative battery cable. Relieve the fuel system pressure before disconnecting any fuel lines. Remove the engine from the vehicle and secure to a suitable holding fixture.

2. Remove the valve cover, rocker arms and pushrods. Keep all parts in order for reassembly.

3. Remove the distributor, spark plug wires and plugs.

4. Remove the pushrod cover, the gasket and the lifters. Keep all parts in order for reassembly.

5. Remove the alternator, alternator lower bracket and the front engine mount bracket assembly.

6. Remove the oil pump driveshaft and gear assembly.

7. Remove the crankshaft pulley and front cover Remove the timing chain and gears, if equipped.

8. Remove the 2 camshaft thrust plate screws by working through the holes in the gear.

9. Remove the camshaft, and gear assembly, if gear driven by pulling it through the front of the block. Take care not to damage the bearings while removing the camshaft.

To install:

10. The installation is the reverse of the removal procedure. Coat all parts with a liberal amount of clean engine oil supplement before installing.

11. Fill all fluids to their proper levels.

12. Connect the negative battery cable and check for leaks.

3.0L and 3.3L Engines

1. Disconnect the negative battery cable. Relieve the fuel system pressure before disconnecting any fuel lines. Remove the engine from the vehicle and secure to a suitable holding fixture.

2. Remove the intake manifold.

3. Remove the valve covers, rocker arm assemblies, pushrods and lifters. Keep all parts in order for reassembly.

4. Remove the crankshaft balancer from the crankshaft.

5. Remove the front cover.

6. Rotate the crankshaft to align the timing marks on the timing sprockets. Remove the camshaft sprocket and the timing chain.

7. Remove the camshaft retainer bolts and slide the camshaft forward out of the engine. Take care not to damage the bearings while removing the camshaft.

To install:

8. The installation is the reverse of the removal procedure. Coat all parts with a liberal amount of clean engine oil supplement before installing.

9. Fill all fluids to their proper levels.

10. Connect the negative battery cable and check for leaks.

Piston and Connecting Rod

POSITIONING

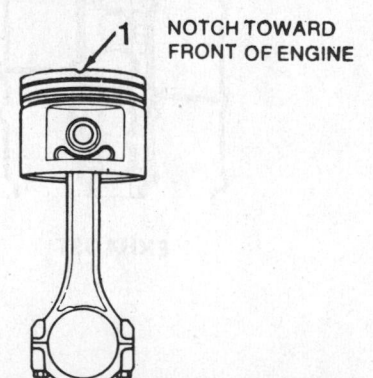

Piston and connecting rod assembly—2.0L and 2.5L engines

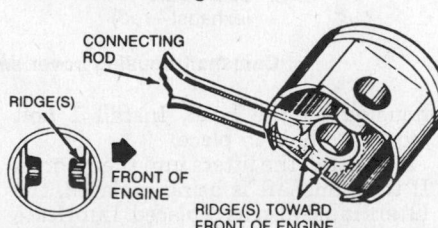

Piston and connecting rod assembly—3.3L engine

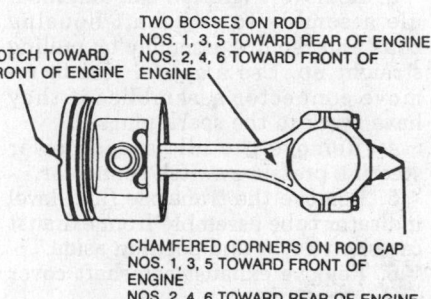

1. Piston
2. Arrow—toward front of engine
3. Connecting rod
4. Oil squirt hole—toward exhaust side
5. Connecting rod bearing
6. Connecting rod cap
7. Cap nuts

Piston and connecting rod assembly—2.3L engine

TWO BOSSES ON ROD
NOS. 1, 3, 5 TOWARD REAR OF ENGINE
NOS. 2, 4, 6 TOWARD FRONT OF ENGINE

NOTCH TOWARD FRONT OF ENGINE

CHAMFERED CORNERS ON ROD CAP
NOS. 1, 3, 5 TOWARD FRONT OF ENGINE
NOS. 2, 4, 6 TOWARD REAR OF ENGINE

Piston and connecting rod assembly—3.0L engine

ENGINE LUBRICATION

Oil Pan

REMOVAL & INSTALLATION

2.0L Engine

1. Disconnect the negative battery cable.

2. Raise and safely support the vehicle. Remove the right front wheel assembly and the splash shield.

3. Drain the engine oil.

4. Remove the exhaust pipe from the turbocharger.

5. Remove the flywheel inspection cover.

6. Remove the oil pan attaching bolts and remove the oil pan, scraper and gasket.

7. The installation is the reverse of the removal procedure. Use a new gasket and apply sealant at the 4 engine block seams. Use thread locking compound on the bolt threads and tighten

to 4 ft. lbs. (6 Nm), starting from the middle and working outward.

8. Fill the crankcase with oil to specification.

9. Connect the negative battery cable and check for leaks.

2.3L Engine

1. Disconnect the negative battery cable. Raise and safely support the vehicle.

2. Remove the flywheel inspection cover.

3. Remove the splash shield-to-suspension support bolt. Remove the exhaust manifold brace, if equipped.

4. Remove the radiator outlet pipe-to-oil pan bolt.

5. Remove the transaxle-to-oil pan nut and stud using a 7mm socket.

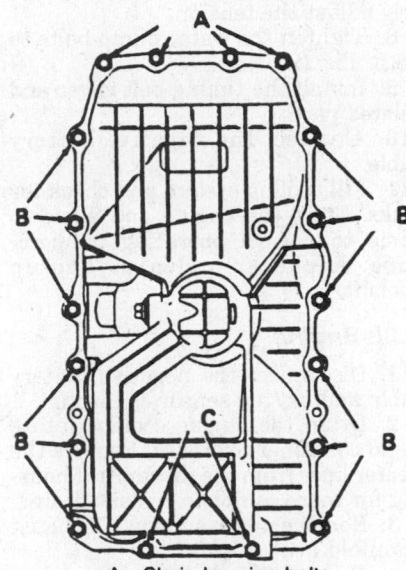

A. Chain housing bolts
B. Block bolts
C. Carrier seal bolts

Oil pan mounting bolts—2.3L engine

6. Gently pry the spacer out from between oil pan and transaxle.

7. Remove the oil pan bolts. Rotate the crankshaft, if necessary, and remove the oil pan and gasket from the engine.

8. Inspect the silicone strips across the top of the aluminum carrier at the oil pan-cylinder block-seal housing 3-way joint. If damaged, these strips must be repaired with silicone sealer. Use only enough sealer to restore the strips to their original dimension; too much sealer could cause leakage.

To install:

9. Thoroughly clean and dry the mating surfaces, bolts and bolt holes. Install the oil pan with a new gasket; do not uses sealer on the gasket. Loosely install the pan bolts.

10. Place the spacer in its approximate installed position but allow clearance to tighten the pan bolt above it.

11. Tighten the pan to block bolts to 17 ft. lbs. (24 Nm) and the remaining bolts to 106 inch lbs. (12 Nm).

12. Install the spacer and stud.

13. Install the oil pan transaxle nut and bolt.

14. Install the slash shield to suspension support.

15. Install the radiator outlet pipe bolt.

16. Install the exhaust manifold brace, if removed.

17. Install the flywheel inspection cover.

18. Fill the crankcase with the proper oil.

19. Connect the negative battery cable and check for leaks.

2.5L Engine

1. Disconnect the negative battery cable.

2. Raise and safely support the vehicle. Drain the engine oil.

3. Disconnect the exhaust pipe and hangers from the exhaust manifold and allow it to swing aside.

4. Disconnect electrical connectors from the starter. Remove the starter-to-engine bolts, the starter and the flywheel housing inspection cover from the engine.

5. Remove the oil pan-to-engine bolts and the oil pan.

To install:

6. Thoroughly clean the mating surfaces, bolts and bolt holes.

7. Apply sealant to the oil pan flange, surrounding all bolt holes. Also, apply sealant to the engine at the front and rear seams.

8. Install the oil pan and tighten the bolts to 20 ft. lbs. (27 Nm) for 1988 vehicles or 89 inch lbs. (10 Nm) for 1989–92 vehicles.

9. Install the flywheel housing cover and exhaust pipe.

10. Fill the crankcase with oil to specification.

11. Connect the negative battery cable and check for leaks.

3.0L and 3.3L Engines

1. Disconnect the negative battery cable.

2. Raise and safely support the vehicle.

3. Drain the engine oil and remove the oil filter.

4. Remove the flywheel cover and the starter.

5. Remove the oil pan, tensioner spring and formed rubber gasket.

6. The installation is the reverse of the removal procedure. Tighten the oil pan-to-engine bolts to 88 inch lbs. (10 Nm) for the 3.0L engine or 124 inch lbs. (14 Nm) for the 3.3L engine.

7. Fill the crankcase with the proper oil.

8. Connect the negative battery cable and check for leaks.

Oil Pump

REMOVAL & INSTALLATION

2.0L Engine

1. Disconnect negative battery cable.

2. Remove the timing belt and crankshaft sprocket.

3. Remove the rear timing belt cover.

4. Disconnect oil pressure sending unit connector.

5. Raise and safely support the vehicle.

6. Drain the engine oil.

7. Remove the oil pan and oil filter.

8. Remove the oil pump mounting bolts and remove the pump and pickup tube.

To install:

9. Prime the pump by pouring fresh oil into the pump intake and turning the driveshaft until oil comes out the pressure port. Repeat a few times until no air bubbles are present.

10. The installation is the reverse of the removal procedure. Use a new gasket and seal and tighten the oil pump bolts to 5 ft. lbs. (7 Nm). Use a new ring for the pickup tube.

11. Fill the crankcase with the proper oil.

12. Connect the negative battery cable, check the oil pressure and check for leaks.

2.3L Engine

1. Disconnect the negative battery cable.

2. Raise and safely support the vehicle.

3. Drain the engine oil and remove the oil pan.

4. Remove the oil pump attaching bolts and nut.

5. Remove the oil pump assembly, shims if equipped, and screen.

To install:

6. With the oil pump assembly off the engine, remove 3 attaching bolts and separate the driven gear cover and screen assembly from the oil pump.

7. Install the oil pump on the block using the original shims, if equipped. Tighten the bolts to 33 ft. lbs. (45 Nm).

8. Mount a dial indicator assembly to measure backlash between oil pump to drive gear.

9. Record oil pump drive to driven gear backlash. Proper backlash is 0.010–0.018 in. When measuring, do not allow the crankshaft to move.

10. If equipped with shims, remove shims to decrease clearance and add

shims to increase clearance. If no shims were present, replace the assembly if proper backlash cannot be obtained.

11. When the proper clearance is reached, rotate crankshaft ½ turn and recheck clearance.

12. Remove oil pump from block, fill the cavity with petroleum jelly and reinstall driven gear cover and screen assembly to pump. Tighten the bolts to 106 inch lbs. (13 Nm).

13. Reinstall the pump assembly to the block. Tighten oil pump-to-block bolts 33 ft. lbs. (45 Nm).

14. Install the oil pan.

15. Fill the crankcase with the proper oil.

16. Connect the negative battery cable, check the oil pressure and check for leaks.

2.5L Engine

1. Disconnect the negative battery cable.

2. Drain the engine oil and remove the oil pan.

3. Remove the oil filter.

4. Remove the oil pump cover assembly.

5. Remove the gerotor pump gears.

------ CAUTION ------

The pressure regulator valve spring is under pressure. Exercise caution when removing the pin or personal injury may result.

6. Remove the pressure regulator pin, spring and valve.

To install:

7. Lubricate all internal parts with clean engine oil and fill all pump cavities with petroleum jelly.

8. Install the pressure regulator valve, spring and secure the pin.

9. Install the gerotor gears.

10. Install the pump cover and tighten the screws to 10 ft. lbs. (14 Nm).

11. Install the oil filter.

12. Install the oil pan.

13. Fill the crankcase with oil to specification.

14. Connect the negative battery cable, check the oil pressure and check for leaks.

3.0L and 3.3L Engines

1. Disconnect the negative battery cable.

2. Remove the timing chain front cover.

3. Raise and safely support the vehicle.

4. Drain the engine oil. Lower the vehicle.

5. Remove the oil filter adapter, the pressure regulator valve and the valve spring.

6. Remove the oil pump cover-to-oil pump screws and remove the cover.

7. Remove the oil pump gears.

To install:

8. Lubricate the oil pump gears with clean engine oil.

9. Pack the pump cavity with petroleum jelly.

10. Install the oil pump cover screws using a new gasket and tighten to 97 inch lbs. (11 Nm).

11. Install the pressure regulator spring and valve.

12. Install the oil filter adaptor using a new gasket. Tighten the oil filter adapter-to-engine bolts to 30 ft. lbs. (41 Nm) for the 3.0L engine or 24 ft. lbs. (33 Nm) for the 3.3L engine.

13. Install the timing chain front cover to the engine.

14. Fill the crankcase with clean engine oil.

ENGINE COOLING

Heater Core
REMOVAL & INSTALLATION

1. Disconnect the negative battery cable.

2. Drain the engine coolant into a clean container for reuse.

3. Raise and safely support the vehicle.

4. Remove the rear lateral transaxle strut mount, if necessary.

5. Remove the drain tube and disconnect the heater hoses from the core tubes. Lower the vehicle.

6. Remove the sound insulators, console extensions and/or steering column filler, as required.

7. Remove the floor or console outlet ductwork and hoses.

8. Remove the heater core cover.

9. Remove the heater core mounting clamps and remove the heater core.

To install:

10. Install the heater core and clamps.

11. Install the heater core cover.

12. Install the outlet hoses and ducts.

13. Install the sound insulators, console extensions and/or steering column filler.

14. Raise and safely support the vehicle. Install the drain tube and connect the heater hoses to the core tubes.

15. Install the rear lateral transaxle strut mount, if removed. Lower the vehicle.

16. Connect the negative battery cable.

17. Fill cooling system and check for leaks. Start the engine and allow to come to normal operating temperature. Recheck for leaks. Top-up coolant.

Water Pump
REMOVAL & INSTALLATION

2.0L Engine

1. Disconnect the negative battery cable.

2. Drain the engine coolant into a clean container for reuse.

3. Remove the timing belt.

4. Remove the water pump attaching bolts, water pump and seal ring.

To install:

5. Thoroughly clean and dry the mounting surfaces, bolts and bolt holes.

6. Using a new sealing ring, install the water pump to the engine and tighten the bolts by hand.

7. Install the timing belt and properly adjust the tension.

8. Tighten the water pump bolts to 18 ft. lbs. (24 Nm).

9. Install the timing belt cover and related parts.

10. Connect the negative battery cable.

11. Fill cooling system and check for leaks. Start the engine and allow to come to normal operating temperature. Recheck for leaks. Top-up coolant.

2.3L Engine

1. Disconnect the negative battery cable and oxygen sensor connector.

2. Drain the engine coolant into a clean container for reuse. Remove the heater hose from the thermostat housing for more complete coolant drain.

3. Remove upper and lower exhaust manifold heat shields.

4. Remove the bolt that attaches the exhaust manifold brace to the manifold.

5. Break loose the manifold to exhaust pipe spring loaded bolts using a 13mm box wrench.

6. Raise and safely support the vehicle.

NOTE: It is necessary to relieve the spring pressure from 1 bolt prior to removing the second bolt. If the spring pressure is not relieved, it will cause the exhaust pipe to twist and bind up the bolt as it is removed.

7. Remove the manifold to exhaust pipe bolts from the exhaust pipe flange as follows:

   a. Unscrew either bolt clockwise 4 turns.

   b. Remove the other bolt.

   c. Remove the first bolt.

8. Pull down and back on the exhaust pipe to disengage it from the exhaust manifold bolts.

9. Remove the radiator outlet pipe from the oil pan and transaxle. If

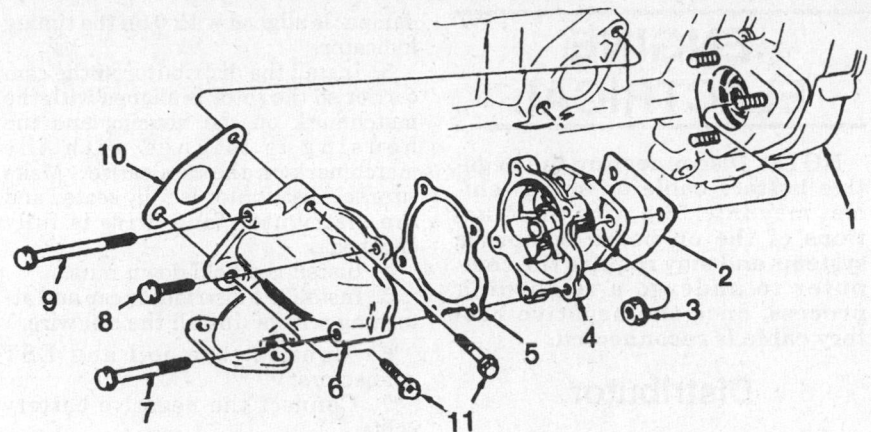

1. Timing chain housing
2. Water pump to timing chain housing gasket
3. Nut
4. water pump
5. Water pump body cover gasket
6. Water pump cover
7. Bolt
8. Bolt
9. Bolt
10. Water pump gasket cover to block gasket
11. water pump cover bolts

Water pump assembly—2.3L engine

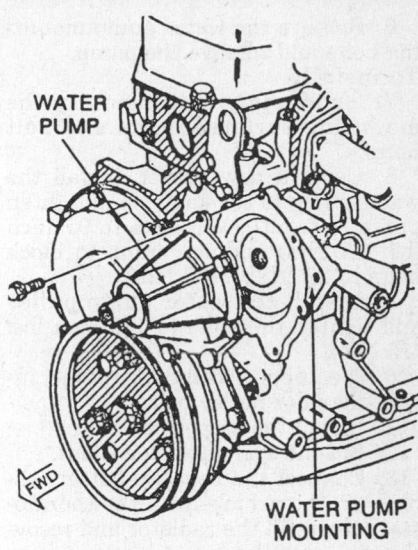

Water pump assembly—2.5L engine

equipped with a manual transaxle, remove the exhaust manifold brace. Leave the lower radiator hose attached and pull down on the outlet pipe to remove it from the water pump.

10. Lower the vehicle.
11. Remove the exhaust manifold, seals and gaskets.
12. Loosen and reposition the rear engine mount and bracket for clearance, as required.
13. Remove the water pump mounting bolts and nuts. Remove the water pump and cover assembly and separate the 2 pieces.

To install:

14. Thoroughly clean and dry all mounting surfaces, bolts and bolt holes. Using a new gasket, install the water pump to the cover and tighten the bolts finger tight.
15. Lubricate the splines of the water pump with clean grease and install the assembly to the engine using new gaskets. Install the mounting bolts and nuts finger tight.
16. Lubricate the radiator outlet pipe O-ring with antifreeze and install to the water pump with the bolts finger tight.
17. With all gaps closed, tighten the bolts, in the following sequence, to the proper values:
   a. Pump assembly to chain housing nuts—19 ft. lbs. (26 Nm).
   b. Pump cover to pump assembly—106 inch lbs. (12 Nm).
   c. Cover to block, bottom bolt first—19 ft. lbs. (26 Nm).
   d. Radiator outlet pipe assembly to pump cover—125 inch lbs. (14 Nm).

18. Install the exhaust manifold.
19. Raise and safely support the vehicle.
20. Install the exhaust pipe flange bolts evenly and gradually to avoid binding.
21. Connect the radiator outlet pipe to the transaxle and oil pan. Install the exhaust manifold brace, if removed. Lower the vehicle.
22. Install the bolt that attaches the exhaust manifold brace to the manifold.
23. Install the heat shields.
24. Connect the oxygen sensor connector.
25. Fill the radiator with coolant until it comes out the heater hose outlet at the thermostat housing. Then connect the heater hose.
26. Connect the negative battery cable, run the vehicle until the thermostat opens, fill the radiator and recovery tank completely.
27. Once the vehicle has cooled, recheck the coolant level.

2.5L Engine

1. Disconnect the negative battery cable.
2. Drain the engine coolant into a clean container for reuse.
3. Remove the drive belts, alternator and air conditioning compressor, as required.
4. Remove the water pump mounting bolts and remove the water pump from the vehicle.

To install:

5. Transfer the water pump pulley to the new pump using the proper pulley removal and installation tools.

6. Thoroughly clean and dry the mounting surfaces, bolts and bolt holes. Place a 1/8 in. bead of RTV sealant on the pump's sealing surface.
7. Install the pump to the engine and coat the bolt threads with sealant as they are installed. Tighten the bolts to 25 ft. lbs. (34 Nm).
8. Install the alternator and/or air conditioning compressor. Install and adjust the drive belts.
9. Connect the negative battery cable.
10. Fill cooling system and check for leaks. Start the engine and allow to come to normal operating temperature. Recheck for leaks. Top-up coolant.

3.0L and 3.3L Engines

1. Disconnect the negative battery cable.
2. Drain the engine coolant into a clean container for reuse.
3. Remove the serpentine belt.
4. If equipped with 3.3L engine, remove the idler pulley bolt.
5. Remove the water pump pulley bolts and remove the pulley. If equipped with 3.0L engine, the long bolt is removed through the access hole in the body side rail.

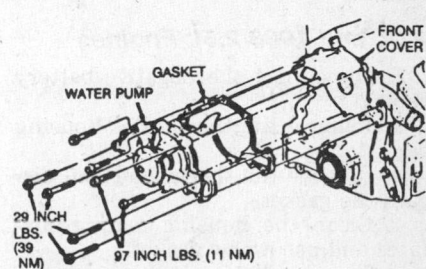

Water pump assembly—3.0L and 3.3L engines

6. Remove the water pump mounting bolts and remove the pump.

To install:

7. Thoroughly clean and dry the mounting surfaces, bolts and bolt holes.

8. Using a new gasket, install the water pump to the engine and tighten the pump to front cover bolts to 97 inch lbs. (11 Nm) and the pump to block bolts to 29 ft. lbs. (39 Nm).

9. Install the water pump pulley and tighten the bolts to 115 inch lbs. (13 Nm).

10. If equipped with 3.3L engine, install the idler pulley bolt.

11. Install the serpentine belt.

12. Fill the system with coolant.

13. Connect the negative battery cable, run the vehicle until the thermostat opens, fill the radiator and recovery tank completely.

14. Once the vehicle has cooled, recheck the coolant level.

Thermostat

REMOVAL & INSTALLATION

Except 2.0L and 1988 2.5L Engines

1. Disconnect the negative battery cable. Drain the coolant down to thermostat level or below.

2. Remove the air cleaner assembly, as required. Disconnect the coolant sensor on 2.3L engine.

3. Disconnect the hose(s) and remove the thermostat housing.

4. Remove the thermostat and discard the gasket.

5. Clean the housing mating surfaces and use a new gasket.

6. The installation is the reverse of the removal procedure.

7. Fill the system with coolant.

8. Connect the negative battery cable, run the vehicle until the thermostat opens, fill the radiator and recovery tank completely.

9. Connect the negative battery cable.

10. Fill cooling system and check for leaks. Start the engine and allow to come to normal operating temperature. Recheck for leaks. Top-up coolant.

2.0L and 1988 2.5L Engines

1. Disconnect the negative battery cable.

2. Remove the thermostat housing cap.

3. Remove the thermostat and discard the gasket.

4. Clean the housing mating surfaces and use a new gasket.

5. The installation is the reverse of the removal procedure.

ENGINE ELECTRICAL

NOTE: **Disconnecting the negative battery cable on some vehicles may interfere with the functions of the on board computer systems and may require the computer to undergo a relearning process, once the negative battery cable is reconnected.**

Distributor

REMOVAL

2.0L Engine

1. Disconnect the negative battery cable.

2. Disconnect the coil and Electronic Spark Timing (EST) connectors.

3. Remove the coil wire. Unscrew the distributor cap hold-down screws and lift off the distributor cap with all ignition wires still connected.

4. Matchmark the rotor to the distributor housing and the distributor housing to the cam carrier.

NOTE: **Do not crank the engine during this procedure. If the engine is cranked, the rotor's matchmark must be disregarded.**

5. Remove the hold-down nuts.

6. Remove the distributor from the engine.

INSTALLATION

Timing Not Disturbed

1. Install a new distributor housing O-ring.

2. Install the distributor in the cam carrier so the rotor is aligned with the matchmark on the housing and the housing is aligned with the matchmark on the cam carrier. Make sure the distributor is fully seated and the distributor tang drive is fully engaged.

3. Install the hold-down nuts.

4. Install the distributor cap and attaching screws. Install the coil wire.

5. Connect the coil and EST connectors.

6. Connect the negative battery cable.

7. Adjust the ignition timing and tighten the hold-down nuts.

Timing Disturbed

1. Install a new distributor housing O-ring.

2. Position the engine so the No. 1 piston is at TDC of the compression stroke and the mark on the vibration damper is aligned with **0** on the timing indicator.

3. Install the distributor in the cam carrier so the rotor is aligned with the matchmark on the housing and the housing is aligned with the matchmark on the cam carrier. Make sure the distributor is fully seated and the distributor tang drive is fully engaged.

4. Install the hold-down nuts.

5. Install the distributor cap and attaching screws. Install the coil wire.

6. Connect the coil and EST connectors.

7. Connect the negative battery cable.

8. Adjust the ignition timing and tighten the hold-down nuts.

Ignition Timing

ADJUSTMENT

NOTE: **All engines except for the 2.0L, are equipped with distributorless ignition systems. Distributorless ignition systems do not give provisions for setting ignition timing; only the timing on the 2.0L engine can be set. Follow all instructions on the Vehicle Emission Control Information label if they are not consistent with these procedures.**

1988–89 Vehicles

1. Start the engine, set the parking brake and run the engine until at normal operating temperature. Keep all lights and accessories off.

2. Connect the red lead of a tachometer to the terminal of the coil labeled **TACH** and connect the black lead to a good ground.

3. If a magnetic timing unit is available, insert the probe into the receptacle near the timing scale.

4. If a magnetic timing unit is not available, connect a conventional power timing light to the No. 1 cylinder spark plug wire.

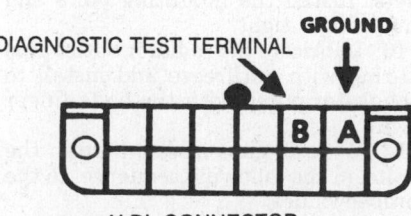

DIAGNOSTIC TEST TERMINAL / **GROUND**

ALDL CONNECTOR

Grounding the ALDL connector

5. With parking brake safely set, place automatic transaxle in **D** or leave manual transaxle in neutral.

6. Ground the ALDL connector under the dash by installing a jumper

wire between the A a... The check engine light shou... flashing.

7. Aim the timing light at the timing scale or read the magnetic timing unit. Record the reading.

8. Repeat Steps 3–6 using the No. 4 spark plug wire. Record the reading.

9. Use the average of the 2 readings to derive an average timing value.

10. Loosen the distributor hold-down nuts so the distributor can be rotated.

11. Using the average timing value, turn the distributor in the proper direction until the specified timing according to the Vehicle Emission Control Information label is reached.

12. Tighten the hold-down nuts and recheck the timing values.

13. Remove the jumper wire from the ALDL connector. To clear the ECM memory, disconnect the ECM harness from the positive battery pigtail for 10 seconds with the key in the OFF position.

Alternator

PRECAUTIONS

Several precautions must be observed with alternator-equipped vehicles to avoid damage to the unit.

• If the battery is removed for any reason, make sure it is reconnected with the correct polarity. Reversing the battery connections may result in damage to the rectifiers.

• When utilizing a booster battery as a starting aid, always connect the positive to positive terminals and the negative terminal from the booster battery to a good engine ground on the vehicle being started.

• Never use a fast charger as a booster to start vehicles.

• Disconnect the battery cables when charging the battery with a fast charger.

• Never attempt to polarize the alternator.

• Do not use a tester of of greater than 12 volts when checking diode continuity.

• Do not short across or ground any of the alternator terminals.

• The polarity of the battery, alternator and regulator must be matched and considered before making any electrical connections within the system.

• Never separate the alternator on an open circuit. Make sure all connections within the circuit are clean and tight.

• Disconnect the battery ground terminal when performing any service on electrical components.

• Disconnect the battery if arc welding is to be done on the vehicle.

• Never disconnect the battery with the engine running.

REMOVAL & INSTALLATION

Except 2.3L Engine

1. Disconnect the negative battery cable.

2. Label and disconnect the wiring from the back of the alternator assembly.

3. On the 2.5L engine, loosen the adjusting bolts and remove the alternator belt. If equipped with a serpentine belt, loosen the serpentine belt tensioner and rotate it counterclockwise to remove the drive belt.

4. Remove the alternator attaching bolts and remove the alternator assembly from the vehicle.

5. The installation is the reverse of the removal procedure.

6. Check and/or adjust the belt tension.

7. Connect the negative battery cable and check the alternator for proper operation.

2.3L Engine

1. Disconnect the negative battery cable.

2. Using a 13mm wrench that is at least 24 in. long, loosen the tensioner pulley bolt, rotate the tensioner counterclockwise and remove the belt from the alternator pulley.

3. Label and disconnect the vacuum lines at the front of engine and remove the attaching bracket.

4. Label and disconnect the injector harness and alternator connectors.

5. Remove the 2 rear alternator mounting bolts.

6. Remove the front alternator bolt and engine harness clip.

NOTE: Care must be taken during removal and installation not to damage the air conditioning hoses.

7. Remove the alternator by manipulating it between the engine lifting eyelet and the air conditioning hoses.

To install:

8. Position the alternator on the engine.

9. Install the front mounting bolt loosely and install the clip.

10. Install the 2 rear mounting bolts and tighten to 37 ft. lbs. (50 Nm).

11. Tighten the front mounting bolt to 20 ft. lbs. (26 Nm).

12. Connect the injector harness and alternator connectors.

13. Connect the vacuum lines and install the bracket.

14. Install the belt.

15. Connect the negative battery cable and check the alternator for proper operation.

Starter

REMOVAL & INSTALLATION

Except 2.3L Engine

1. Disconnect the negative battery cable.

2. Raise and safely support the vehicle. Disconnect the electrical wiring from the starter.

3. Remove the dust cover bolts and pull the dust cover back to gain access to the front starter bolt and remove the front starter bolt.

4. Remove the rear support bracket.

5. Pull the rear dust cover back to gain access to the rear starter bolt and remove the rear bolt.

6. Note the number and location of any shims.

7. Push the dust cover back into place and remove the starter from the vehicle.

8. The installation is the reverse of the removal procedure.

9. Tighten the starter bolts to 30–35 ft. lbs. (41–47 Nm).

2.3L Engine

1988–89 (VIN D)

1. Disconnect the negative battery cable.

2. Remove the air cleaner to throttle body duct.

3. Label and disconnect the TPS, IAC and MAP sensor connectors.

4. Remove vacuum harness assembly from intake and position aside.

5. Remove cooling fan shroud attaching bolts and remove the shroud.

6. Remove upper radiator support.

7. Disconnect the connector from the cooling fan and remove the fan assembly. Do not damage the lock tang on the TPS with the fan bracket.

8. Remove the starter mounting bolts.

9. Tilt the rear of starter towards the radiator, pull the starter out and rotate solenoid towards the radiator to gain access to the electrical connections.

NOTE: If present, do not to damage the crank sensor mounted directly to the rear of the starter.

10. Disconnect the connectors from the solenoid.

11. Move the starter toward the driver's side of the vehicle and remove.

To install:

12. Lower the starter and connect the solenoid connectors.

13. Rotate the starter into installation position, properly install any shims that were removed and install the mounting bolts. Tighten to 74 ft. lbs. (100 Nm).

14. Install the fan, support and shroud.

15. Install the vacuum harness assembly and connect the TPS, IAC and MAP sensor connectors.

16. Install the air cleaner to throttle body duct.

17. Connect the negative battery cable and check the starter for proper operation.

2.3L Engine

1990–92 (VIN D)
1989–92 (VIN A)

1. Disconnect the negative battery cable.

2. Remove the cooling fan assembly.

3. Remove the oil filter, if necessary.

4. Remove the intake manifold brace.

5. Remove the mounting bolts; some engines may have 3 starter mounting bolts. Pull the starter out of the hole and move toward the front of the vehicle.

6. Disconnect the wiring from the solenoid.

7. Remove the starter by lifting it between the intake manifold and the radiator.

To install:

8. Lower the starter between the intake manifold and the radiator and connect the wiring to the solenoid.

9. Rotate the starter into installation position and install the mounting bolts. Tighten to 74 ft. lbs. (100 Nm).

10. Install the intake manifold brace and oil filter.

11. Install the cooling fan assembly.

12. Connect the negative battery cable and check the starter for proper operation.

EMISSION CONTROLS

Please refer to "Emission Controls" in the Unit Repair section for system maintenance procedures. Due to the complex nature of modern electronic engine control systems, comprehensive diagnosis and testing procedures fall outside the confines of this repair manual. For complete information on diagnosis, testing and repair procedures concerning all modern engine and emission control systems, please refer to "Chilton's Guide to Fuel Injection and Electronic Engine Controls".

FUEL SYSTEM

Fuel System Service Precautions

Safety is the most important factor when performing not only fuel system maintenance but any type of maintenance. Failure to conduct maintenance and repairs in a safe manner may result in serious personal injury or death. Maintenance and testing of the vehicle's fuel system components can be accomplished safely and effectively by adhering to the following rules and guidelines.

• To avoid the possibility of fire and personal injury, always disconnect the negative battery cable unless the repair or test procedure requires that battery voltage be applied.

• Always relieve the fuel system pressure prior to disconnecting any fuel system component (injector, fuel rail, pressure regulator, etc.), fitting or fuel line connection. Exercise extreme caution whenever relieving fuel system pressure to avoid exposing skin, face and eyes to fuel spray. Please be advised that fuel under pressure may penetrate the skin or any part of the body that it contacts.

• Always place a shop towel or cloth around the fitting or connection prior to loosening to absorb any excess fuel due to spillage. Ensure that all fuel spillage (should it occur) is quickly removed from engine surfaces. Ensure that all fuel soaked cloths or towels are deposited into a suitable waste container.

• Always keep a dry chemical (Class B) fire extinguisher near the work area.

• Do not allow fuel spray or fuel vapors to come into contact with a spark or open flame.

• Always use a backup wrench when loosening and tightening fuel line connection fittings. This will prevent unnecessary stress and torsion to fuel line piping. Always follow the proper torque specifications.

• Always replace worn fuel fitting O-rings with new. Do not substitute fuel hose or equivalent where fuel pipe is installed.

RELIEVING FUEL SYSTEM PRESSURE

1988 Engines and 2.5L

1. Loosen the fuel filler cap.

2. Remove the fuse marked fuel pump from the fuse block or disconnect the harness connector at the tank.

3. Start the engine and run at idle until it stalls.

4. Crank the engine for an additional 3 seconds to make sure all of the fuel pressure is exhausted from the fuel lines.

5. Turn the ignition switch OFF, disconnect the negative battery cable and reinstall the fuel pump fuse or connect the connector at the tank.

6. Tighten the filler cap.

1989–92 Engines Except 2.5L

1. Disconnect the negative battery cable.

2. Loosen the fuel filler cap.

3. Install a fuel pressure gauge to the fuel pressure connection on the fuel pressure regulator assembly. Wrap a shop towel around the connection to avoid any fuel spray.

4. Install the bleed hose into an approved container and open the valve to bleed the fuel pressure.

5. Drain any residual fuel in the gauge into the container.

6. Tighten the filler cap.

Fuel Filter

The fuel filter is located near the rear of the vehicle, forward of the fuel tank.

REMOVAL & INSTALLATION

1. Relieve the fuel system pressure.

2. Raise and safely support the vehicle.

3. Using a backup wrench, remove the fuel line fittings from the fuel filter.

4. Remove the fuel filter mounting screws and remove the filter from the vehicle.

5. The installation is the reverse of the removal procedure. Replace the O-rings. Tighten the fuel line to filter connectors to 22 ft. lbs. (30 Nm).

Electric Fuel Pump

PRESSURE TESTING

1. Relieve the fuel system pressure.

2. Connect an appropriate fuel pressure gauge to the pressure connection on the fuel pressure regulator assembly, if equipped. If there is no valve, install in-line to the pressure line.

3. Wrap a clean shop towel around the fitting to catch any fuel leakage.

4. Turn the ignition ON and read the pressure on the gauge.

5. If not within specifications, inspect the system for clogs, collapsed hoses, kinks or a faulty pump. The fuel pressure can be measured at different points in the system to locate the problem area.

6. Relieve the fuel system pressure and disconnect the gauge.

REMOVAL & INSTALLATION

1. Relieve the fuel system pressure.

2. Raise and safely support the vehicle.

3. Using the proper approved equipment, drain the fuel tank.

4. Disconnect all wiring and hoses from the tank.

5. Place a transmission jack under the center of the tank and apply slight pressure. Remove the tank straps.

6. Remove the fuel tank from the vehicle.

7. Using a hammer and a brass drift, turn the lock ring counterclockwise to release the pump/sending unit assembly.

8. Disassemble the unit to separate the pump itself from the assembly.

To install:

9. Push the fuel pump onto the attaching hose and install the filter on the end of the pump.

10. Install a new tank seal O-ring to the pump.

11. Install the pump into the tank and install the lock ring with a hammer and brass punch turning the ring clockwise.

12. Install the fuel tank.

13. Connect the negative battery cable, start the engine and check for leaks.

Fuel Injection

IDLE SPEED ADJUSTMENT

The idle speed is controlled by the ECM, which receives data from various sensors and switches within the fuel injection system. Adjustments are preset at the factory and not adjustable.

MANUAL TRANSAXLE

For further information on transmissions/transaxles, please refer to "Chilton's Guide to Transmission Repair".

Transaxle Assembly

REMOVAL & INSTALLATION

1. Disconnect the negative battery cable from the battery and transaxle. Remove air ducts and tubes, etc. to gain access to transaxle mounting bolts.

2. Remove the power steering pump and brackets and position aside, if necessary.

3. Attach an engine support fixture to the engine lift ring and raise the engine enough to take the pressure off the engine mounts.

NOTE: If a lifting bar is not available, a chain hoist can be used. However, during the removal procedure the vehicle must be raised and the chain hoist adjusted to keep tension on the engine/transaxle assembly.

4. Remove the left side steering column opening filler from inside the vehicle.

5. Disconnect the clutch master cylinder pushrod from the clutch pedal.

6. Disconnect the clutch slave cylinder from the transaxle support bracket and move it aside.

7. Remove the transaxle mount-to-transaxle bolts. Discard the bolts attaching the mount to the side frame. New bolts must be used upon installation.

8. Remove the transaxle mount bracket attaching bolts and nuts. Remove the upper transaxle to engine bolts.

9. Remove the transaxle vent tube and disconnect the reverse light switch.

10. Disconnect the shift cables and retaining clips from the transaxle.

11. Raise and safely support the vehicle.

12. Remove the left front wheel assembly.

13. Remove the left front inner splash shield. Drain the transaxle oil.

14. Remove the transaxle strut and bracket, if equipped.

15. Remove the flywheel housing cover bolts.

16. Disconnect the speedometer cable or sensor from the transaxle.

17. If equipped with a 2.3L engine, remove the radiator outlet pipe support bolt from transaxle.

18. Disconnect the stabilizer bar from the left suspension support and control arm.

19. Disconnect the ball joint-to-steering knuckle nut and separate the ball joint from the steering knuckle.

20. Remove the left suspension support attaching bolts, the support and control arm as an assembly.

21. Use boot protectors and disen-

gage the halfshafts from the transaxle. Remove the left halfshaft from the transaxle.

22. Remove engine mount components and remaining transaxle mount bolts, as required.

23. Position a transmission jack under and secure to the transaxle case. Remove the remaining transaxle-to-engine mounting bolts.

24. Remove the transaxle by sliding it toward the driver's side, away from the engine. Carefully lower the jack, guiding the right or intermediate shaft out of the transaxle. Lower the engine to aid the operation, if necessary.

To install:

25. Install the transaxle into position. As the transaxle is being installed, guide the right halfshaft into place. Lower the engine to its installation position.

26. Connect the negative battery cable to the transaxle case.

27. Install engine mount components and remaining transaxle mount bolts. Install the flywheel cover(s).

28. Remove the support jack when the transaxle is securely mounted.

29. Install the left halfshaft.

30. Install the left suspension support.

31. Install the engine mount crossmember nuts, if removed.

32. Connect the stabilizer bar to the left suspension support and control arm.

33. Install the radiator outlet pipe support bolt, if equipped.

34. Connect the speedometer cable or sensor.

35. Install the transaxle bracket and strut, if equipped.

36. Install the splash shield and wheel. Lower the vehicle.

37. Connect the shift cables and install the retaining clips.

38. Install the transaxle vent tube and connect the reverse light switch connector.

39. Install the upper transaxle to engine bolts. Install the transaxle mount bracket attaching bolts and nuts.

40. Install the new transaxle mount-to-transaxle bolts.

41. Connect the clutch slave cylinder to the support bracket.

42. Connect the clutch master cylinder pushrod to the clutch pedal.

43. Install the steering column opening filler panel.

44. Remove the engine support tool.

45. Install the power steering pump and brackets, if they were removed.

46. Install air ducts, etc. that were removed.

47. Fill the transaxle with the proper fluid.

48. Connect the negative battery cable and check the transaxle for proper operation.

CLUTCH

Clutch Assembly

REMOVAL & INSTALLATION

1. Disconnect the negative battery cable. Remove the transaxle.
2. Matchmark the clutch/pressure plate cover and flywheel, if reinstalling old parts. Insert a clutch plate alignment tool into the clutch disc hub.
3. Loosen the flywheel to pressure plate bolts gradually and evenly to avoid warpage.
4. Remove the pressure plate/clutch assembly from the flywheel.
5. Sand the flywheel or replace it, if scored, cracked or heat damaged.
6. Sparingly apply anti-seize compound to the input shaft and clutch disc splines. Install a new release bearing.

To install:

7. Using a clutch disc alignment tool, tighten the pressure plate bolts to center the disc.
8. Tighten the pressure plate/clutch assembly mounting bolts to the flywheel gradually and evenly to 20–25 ft. lbs. (27–34 Nm).
9. Install the transaxle.
10. Connect the negative battery cable and check the clutch and reverse lights for proper operation.

Clutch Master and Slave Cylinders

REMOVAL & INSTALLATION

1. Disconnect the negative battery cable.
2. Remove the steering column opening filler from inside the vehicle.
3. Disconnect the clutch master cylinder pushrod from the clutch pedal.
4. Remove the clutch master cylinder attaching nuts at the front of the dash and disconnect the remote fluid reservoir, if equipped.
5. Remove the actuator cylinder attaching nuts at the transaxle.
6. Remove the hydraulic actuating system as an assembly.

To install:

7. Bleed the system, if necessary.
8. Install the actuator cylinder to the transaxle, aligning the pushrod into the pocket on the lever. Tighten the attaching nuts evenly to prevent damage.

NOTE: New actuators are packaged with plastic straps to retain the pushrod. Do not break the strap off; it will break upon the first clutch application.

9. Install the master cylinder. Tighten the attaching nuts evenly to prevent damage. Connect the remote fluid reservoir, if equipped. If equipped with a bleed screw and bleeding is necessary, bleed the system.
10. Remove the pushrod restrictor from the master cylinder pushrod. Lubricate the bushing on the clutch pedal. Connect the pushrod to the pedal and install the retaining clip. Make sure the cruise control switch is operating properly.

NOTE: When adjusting the cruise control switch, do not use a force of more than 20 lbs. to pull the pedal up, or damage to the master cylinder pushrod retaining ring could result.

11. Install the steering column opening filler from inside the vehicle.
12. Push the clutch pedal down a few times. This will break the plastic straps on the actuator.
13. Connect the negative battery cable and check for proper operation.

ADJUSTMENT

The hydraulic system used provides automatic clutch adjustment, therefore no adjustment to any portion of the system is required.

Hydraulic Clutch System Bleeding

With Bleed Screw

1. Make sure the reservoir is full of DOT 3 fluid and is kept topped off throughout this procedure.
2. Loosen the bleed screw, located on the actuator cylinder body next to the inlet connection.
3. When a steady stream of fluid comes out the bleeder, tighten it to 17 inch lbs. (2 Nm).
4. Refill the fluid reservoir.
5. To check the system, start the engine and wait 10 seconds.
6. Depress the clutch pedal and shift into Reverse. If there is any gear clash, air may still be present.

Without Bleed Screw

1. Remove the actuator cylinder from the transaxle.
2. Loosen the master cylinder attaching nuts to the ends of the studs.
3. Remove the reservoir cap and diaphragm.
4. Depress the actuator cylinder pushrod about ¾ in. into its bore and hold the position.
5. Install the reservoir diaphragm and cap while holding the actuator pushrod.
6. Release the pushrod when the diaphragm and cap are properly installed.
7. With the actuator lower than the master cylinder, hold the actuator vertically with the pushrod end facing the ground.
8. Press the actuator pushrod into its bore with ½ in. strokes. Check the reservoir for bubbles. Continue until no bubbles enter the reservoir.
9. Install the master cylinder and actuator.
10. Refill the fluid reservoir.
11. To check the system, start the engine and wait 10 seconds.
12. Depress the clutch pedal and shift into reverse. If there is any gear clash, air may still be present.

AUTOMATIC TRANSAXLE

For further information on transmissions/transaxles, please refer to "Chilton's Guide to Transmission Repair".

Transaxle Assembly

REMOVAL & INSTALLATION

1. Disconnect the negative battery cable. If necessary, drain the coolant and disconnect the heater core hoses.
2. Remove the air cleaner assembly. If equipped with a 3.0L or 3.3L engine, remove the mass air flow sensor and air intake duct.
3. Disconnect the throttle valve cable from the throttle lever and the transaxle.
4. If equipped with a 2.3L engine, remove the power steering pump and bracket and position it aside.
5. Remove the transaxle dipstick and tube.
6. Install an engine support tool. Insert a ¼ × 2 in. bolt in the hole at the front right motor mount to maintain driveline alignment.
7. Remove the wiring harness-to-transaxle nut. Disconnect the wiring connectors from the speed sensor, TCC connector, neutral safety switch and reverse light switch.
8. Disconnect the shift linkage from the transaxle.
9. Remove the upper 2 transaxle-to-engine bolts and the upper left transaxle mount along with the bracket assembly.
10. Remove the rubber hose from the transaxle vent pipe. Remove the remaining upper engine-to-transaxle bolts.

11. Raise and safely support the vehicle. Remove both front wheels.

12. If equipped with a 2.3L engine, remove both lower ball joints and stabilizer shafts links.

13. Drain the transaxle fluid.

14. Remove the shift linkage bracket from the transaxle.

15. Install a halfshaft boot seal protector on the inner seals.

NOTE: Some vehicles may use a gray silicone boot on the inboard axle joint. Use boot protector tool on these boots. All other boots are made from a black thermo-plastic material and do not require the use of a boot seal protector.

16. Remove both ball joint-to-control arm nuts and separate the ball joints from the control arms.

17. Remove both halfshafts and support them with a cord or wire.

18. Remove the transaxle mounting strut.

19. Remove the left stabilizer bar link pin bolt, left frame bushing clamp nuts and left frame support assembly.

20. Remove the torque converter cover. Matchmark the flexplate and torque converter for installation purposes. Remove the torque converter-to-flexplate bolts.

21. Disconnect and plug the transaxle oil cooler lines.

22. Remove the transaxle-to-engine support bracket and install the transaxle removal jack.

23. Remove the remaining transaxle-to-engine attaching bolts and the transaxle from the vehicle.

To install:

24. Securely mount the transaxle on the jack.

25. Apply a small amount of grease on the torque converter hub and seat in the oil pump.

26. Position the transaxle in the vehicle and install the lower engine to transaxle bolts.

27. Install the transaxle to engine support bracket. Once the transaxle is securely held in place, remove the jack. Connect the cooler lines.

28. Install the torque converter bolts and tighten to specification.

29. Install the torque converter cover.

30. Install the left frame support assembly.

31. Install the left stabilizer shaft frame busing nuts and link pin bolt.

32. Install the transaxle mounting strut.

33. Install the halfshafts. Install the ball joints.

34. Install the shift linkage bracket to the transaxle.

35. Install the wheels and lower the vehicle.

36. Install the upper transaxle to engine bolts.

37. Install the left side transaxle mount.

38. Connect the shift linkage to the transaxle.

39. Connect the wiring connectors to their switches on the transaxle.

40. Remove the 1/4 × 2 in. bolt that was placed in the hole at the front right motor mount to maintain driveline alignment. Remove an engine support tool.

41. Replace the O-ring, lubricate it and install the dipstick tube and dipstick.

42. Install the TV cable and rubber vent tube.

43. Install the air cleaner assembly and air tubes.

44. Connect the heater hoses, if disconnected.

45. Fill all fluids to their proper levels. Adjust cables as required.

46. Connect the negative battery cable and check the transaxle for proper operation and leaks.

FRONT SUSPENSION

MacPherson Strut

REMOVAL & INSTALLATION

1. Remove the 3 mounting nuts from the shock tower under the hood.

2. Raise and safely support the vehicle. Remove the wheel.

1. Strut assembly
2. Steering knuckle
3. Bolts
4. Nuts
5. Suspension support
6. Cover
7. Mounting nut

MacPherson strut assembly

3. The control arms must not be supporting the vehicle's weight.

NOTE: Do not allow the tri-pot joints from becoming overextended or they can get separated and damaged.

4. Matchmark the lower strut mount to the knuckle and remove the strut to knuckle bolts and nuts.

5. While the strut is off of the vehicle, the lower mounting hole may be elongated for alignment purposes. Paint any exposed metal afterward to prevent rusting.

6. The installation is the reverse of the removal procedure. Tighten the upper mounting nuts to 18 ft. lbs. (24 Nm). Do not tighten the lower mounting bolts until the front end alignment has been completed.

7. Perform a front end alignment. Tighten the strut to knuckle nuts to 133 ft. lbs. (180 Nm).

Lower Control Arms

REMOVAL & INSTALLATION

1. Raise and safely support the vehicle. Remove the tire and wheel assembly.

2. Remove the stabilizer shaft.

NOTE: Do not allow the tri-pot joints from becoming overextended or they can get separated and damaged.

3. Remove the ball joint stud attaching nut.

4. Pry the lower control arm from the steering knuckle.

5. Remove the control arm to suspension support bolts and nuts.

6. Remove the control arm from the vehicle.

7. Transfer reusable parts to the new control arm.

8. The installation is the reverse of the removal procedure.

9. Lower the vehicle so the full weight of the vehicle is on the ground.

10. Tighten the nuts to 61 ft. lbs. (83 Nm).

11. Perform a front end alignment.

REAR SUSPENSION

Shock Absorbers

REMOVAL & INSTALLATION

1. Disconnect the negative battery cable.

2. Open the deck lid and remove the trim cover.

3. Remove the upper shock attaching nut. Remove 1 shock at a time if removing both.

4. Raise and safely support the vehicle.

5. Remove the lower mounting bolt.

6. Remove the shock from the vehicle.

7. The installation is the reverse of the removal procedure.

Coil Springs

REMOVAL & INSTALLATION

1. Raise and safely support the vehicle.

2. Using the proper equipment, support the weight of the rear axle. Disconnect the brake lines from the rear axle.

3. Remove the bolts that attach the shock to the lower mounting bracket.

4. Lower the axle and remove the coil spring from the vehicle.

5. The installation is the reverse of the removal procedure.

Rear Wheel Bearings

REMOVAL & INSTALLATION

1. Raise and safely support the vehicle.

2. Remove the wheel assembly.

3. Remove the brake drum.

4. Remove the 4 hub/bearing assembly-to-rear axle assembly nuts/bolts and the hub/bearing assembly from the axle.

NOTE: The top rear attaching bolt will not clear the brake shoe when removing the hub and bearing assembly. Partially remove the hub prior to removing this bolt.

5. The installation is the reverse of the removal procedure. Tighten the hub/bearing assembly-to-rear axle assembly nuts/bolts to 39 ft. lbs. (53 Nm).

ADJUSTMENT

The rear wheel bearing assembly is non-adjustable and is serviced by replacement only.

STEERING

Steering Wheel

REMOVAL & INSTALLATION

1. Disconnect the negative battery cable.

2. Remove the 2 screws that retain the steering pad.

3. Disconnect the horn lead and remove the horn pad.

4. Remove the retainer, nut and dampener, if equipped.

5. Matchmark the steering wheel to the shaft and remove the steering wheel from the vehicle.

6. The installation is the reverse of the removal procedure. Tighten the attaching nut to 30 ft. lbs. (41 Nm).

Power Rack and Pinion Steering Gear

REMOVAL & INSTALLATION

1. Disconnect the negative battery cable. Remove the left side sound insulator.

2. Disconnect the upper pinch bolt on the steering coupling assembly.

3. Disconnect the clamp nuts.

4. Raise and safely support the vehicle. Remove both front wheel assemblies.

5. Remove the clamp nut and the fluid line retainer.

6. Remove the tie rod end-to-steering knuckle cotter pin and castle nut. Using a puller tool, disconnect the tie rod ends from the steering knuckles.

7. Lower the vehicle.

8. Disconnect and plug the fluid lines from the power steering rack.

9. Remove the mounting clamps. Move the steering rack forward and remove the lower pinch bolt on the coupling assembly.

10. Disconnect the coupling from the steering rack.

11. Remove the rack and pinion assembly with the dash seal through the left wheel opening.

To install:

12. If the studs were removed with the mounting clamps, reinstall the studs into the cowl. If the stud is being reused, use Loctite® to secure the threads.

13. Slide the rack and pinion assembly through the left side wheel housing opening and secure the dash seal.

14. Move the assembly forward and install the coupling.

15. Install the lower pinch bolt and tighten to 29 ft. lbs. (40 Nm).

16. Connect the fluid lines.

17. Install the clamp nuts. Tighten the left side clamp first, then tighten the right side. Raise and safely support the vehicle.

18. Connect the tie rod ends to the steering knuckle, tighten the nut to 35 ft. lbs. (47 Nm) and install a new cotter pin. Install the wheels.

19. Install the line retainer and lower the vehicle.

20. Install the upper pinch bolt on the coupling assembly. Tighten to 29 ft. lbs. (40 Nm).

21. Install the sound insulator.

22. Fill the power steering pump with fluid and bleed the system.

23. Connect the negative battery cable and check the rack for proper operation and leaks.

24. Check and adjust front end alignment, as required.

Power Steering Pump

REMOVAL & INSTALLATION

2.3L Engine

1. Disconnect the negative battery cable.

2. Disconnect the pressure and return lines from the pump.

3. Remove the rear bracket to pump bolts.

4. Remove the drive belt and position aside.

5. Remove the rear bracket to transaxle bolts.

6. Remove the front bracket to engine bolt.

7. Remove the pump and bracket as an assembly.

8. Transfer pulley and bracket, as necessary.

9. The installation is the reverse of the removal procedure.

10. Fill the power steering pump with fluid and bleed the system.

11. Connect the negative battery cable and check the pump for proper operation and leaks.

2.5L Engine

1. Disconnect the negative battery cable.

2. Remove the drive belt.

3. Disconnect and plug the pressure tubes from the power steering pump.

4. Remove the front adjustment bracket-to-rear adjustment bracket bolt.

5. Remove the front adjustment bracket-to-engine bolt and spacer.

6. Remove the pump with the front adjustment bracket.

7. If installing a new pump, transfer the pulley and front adjustment bracket to the new pump.

To install:

8. The installation is the reverse of the removal procedure.

9. Adjust the drive belt tension.

10. Fill the power steering pump with fluid and bleed the system.

11. Connect the negative battery cable and check the pump for proper operation and leaks.

2.0L, 3.0L and 3.3L Engines

1. Disconnect the negative battery cable.

2. Remove the serpentine drive belt.

3. Remove the power steering pump-to-engine bolts.

4. Pull the pump forward and disconnect the pressure tubes.

5. Remove the pump and transfer the pulley, as necessary.

6. The installation is the reverse of the removal procedure.

7. Adjust the drive belt tension.

8. Fill the power steering pump with fluid and bleed the system.

9. Connect the negative battery cable and check the pump for proper operation and leaks.

SYSTEM BLEEDING

1. Raise the vehicle so the wheels are off the ground. Turn the wheels all the way to the left. Add power steering fluid to the **COLD** or **FULL COLD** mark on the fluid level indicator.

2. Start the engine and check the fluid level at fast idle. Add fluid, if necessary to bring the level up to the mark.

3. Bleed air from the system by turning the wheels from side-to-side without hitting the stops. Keep the fluid level at the **COLD** or **FULL COLD** mark. Fluid with air in it has a tan appearance.

4. Return the wheels to the center position and continue running the engine for 2–3 minutes.

5. Lower the vehicle and road test to check steering function and recheck the fluid level with the system at its normal operating temperature. Fluid should be at the **HOT** mark when finished.

Tie Rod Ends

REMOVAL & INSTALLATION

Inner Tie Rod

1. Disconnect the negative battery cable. Remove the rack and pinion gear from the vehicle.

2. Remove the lock plate from the inner tie rod bolts.

3. If removing both tie rods, remove both bolts, the bolt support plate and 1 of the tie rod assemblies. Reinstall the removed tie rod's bolt to keep inner parts of the rack aligned. Remove the remaining tie rod.

4. If only removing 1 tie rod, slide the assembly out from between the support plate and the center housing cover washer.

To install:

5. Install the center housing cover washer fitted into the rack and pinion boot.

6. Install the inner tie rod bolts through the holes in the bolt support plate, inner pivot bushing, center

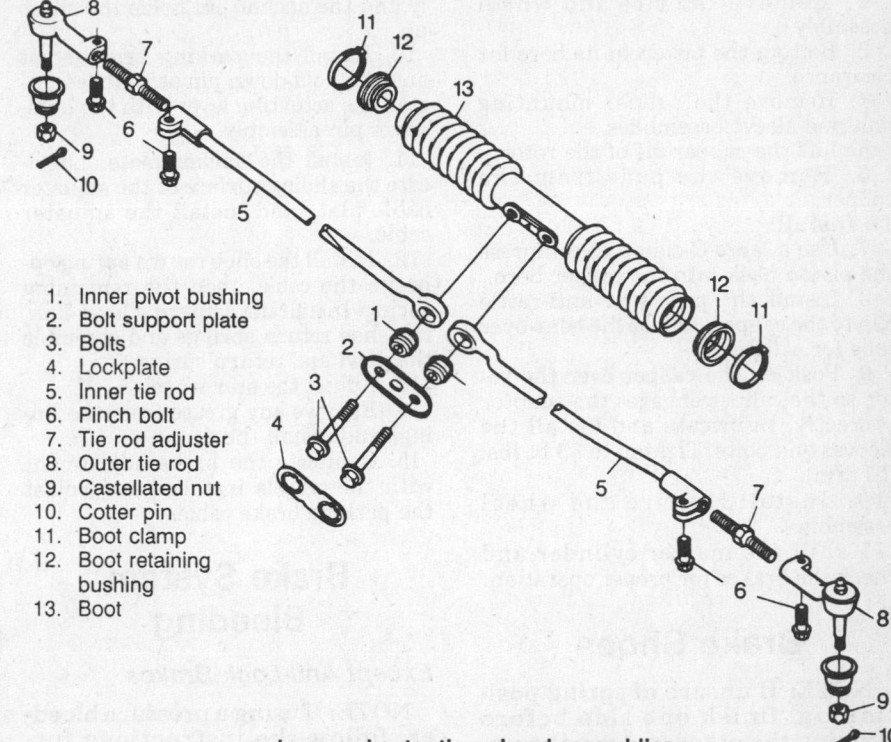

1. Inner pivot bushing
2. Bolt support plate
3. Bolts
4. Lockplate
5. Inner tie rod
6. Pinch bolt
7. Tie rod adjuster
8. Outer tie rod
9. Castellated nut
10. Cotter pin
11. Boot clamp
12. Boot retaining bushing
13. Boot

Inner and outer tie rod end assemblies

housing cover washer, rack housing and into the threaded holes.

7. Tighten the bolts to 65 ft. lbs. (90 Nm).

8. Install a new lock plate with its notches over the bolt flats.

9. Install the rack and pinion gear.

10. Fill the power steering pump with fluid and bleed the system.

11. Connect the negative battery cable and check the rack for proper operation and leaks.

Outer Tie Rod

1. Disconnect the negative battery cable.

2. Remove the cotter pin and the nut from the tie rod ball stud at the steering knuckle.

3. Loosen the pinch bolts.

4. Using the proper tools, separate the tie rod taper from the steering knuckle.

5. Remove the tie rod from the adjuster.

6. The installation is the reverse of the removal procedure.

7. Perform a front end alignment.

BRAKES

For all brake system repair and service procedures not detailed below, please refer to "Brakes" in the Unit Repair section.

Master Cylinder

REMOVAL & INSTALLATION

Except Anti-Lock Brakes

1. Disconnect the negative battery cable. Unplug the fluid level sensor connector.

2. Disconnect and plug the brake lines from the master cylinder.

3. Remove the nuts attaching the master cylinder to the power booster.

4. Remove the master cylinder from the mounting studs.

5. Remove the retaining roll pins and remove the fluid reservoir from the cylinder, if necessary.

To install:

6. Replace the reservoir O-rings and bench bleed the master cylinder.

7. Install to the booster and install the nuts.

8. Install the brake lines to the master cylinder.

9. Fill the reservoir with brake fluid.

10. Connect the negative battery cable and check the brakes for proper operation.

Disc Brake Pads

REMOVAL & INSTALLATION

1. Remove some of the fluid from the master cylinder. Raise and safely support the vehicle.

2. Remove the tire and wheel assembly.

3. Bottom the piston in its bore for clearance.

4. Remove the caliper mounting bolt and sleeve assemblies.

5. Lift the caliper off of the rotor.

6. Remove the pads from the caliper.

To install:

7. Use a large C-clamp to compress the piston back into the caliper bore.

8. Install the pads and anti-rattle clip to the caliper. Adjust the bent-over tabs for a tight fit.

9. Position the caliper over the rotor so the caliper engages the adaptor correctly. Lubricate and install the sleeves and bolts. Tighten to 38 ft. lbs. (51 Nm).

10. Install the tire and wheel assembly.

11. Fill the master cylinder and check the brakes for proper operation.

Brake Shoes

NOTE: If unsure of spring positioning, finish one side before starting the other and use the untouched side as a guide.

REMOVAL & INSTALLATION

1. Remove the wheels and drums. Remove the primary and secondary shoe return springs from the anchor pin but leave them installed in the shoes.

2. Lift on the adjuster lever and remove the adjuster cable. Remove the actuating lever link and pawl return spring.

3. Remove the hold-down pin return springs and cups. Remove the parking brake strut and spring. Remove the actuating lever and pawl.

4. Remove the shoes, held together by the lower spring, while separating the parking brake actuating lever from the shoe with a twisting motion.

5. Lift the wheel cylinder dust boots and inspect for fluid leakage.

6. Thoroughly clean and dry the backing plate.

To install:

7. Remove, clean and dry all parts still on the old shoes. Lubricate the star wheel shaft threads and transfer all the parts to the new shoes in their proper locations.

8. To prepare the backing plate, lubricate the bosses, anchor pin and parking brake actuating lever pivot surface lightly with the brake-compatible lubricant.

9. Spread the shoes apart, engage the parking brake actuating lever and position them on the backing plate so the wheel cylinder pins engage proper-

ly and the anchor pin holds the shoes up.

10. Install the parking brake strut and the hold-down pin assemblies. Install the actuating lever with the hold-down pin assembly.

11. Install the anchor plate. Lubricate the sliding surface of the adjuster cable plate and install the adjuster cable.

12. Install the shoe return spring opposite the cable, then the remaining spring. Install the actuating lever link, the shoe return springs and assemble the pawl and return spring.

13. Adjust the star wheel.

14. Remove any grease from the linings and install the drum.

15. Complete the brake adjustment with the wheels installed and adjust the parking brake cable.

Brake System Bleeding

Except Anti-Lock Brakes

NOTE: If using a pressure bleeder, follow the instructions furnished with the unit and choose the correct adaptor for the application. Do not substitute an adapter that "almost fits" as it will not work and could be dangerous.

MASTER CYLINDER

If the master cylinder is off the vehicle it can be bench bled.

1. Connect 2 short pieces of brake line to the outlet fittings, bend them until the free end is below the fluid level in the master cylinder reservoirs.

2. Fill the reservoir with fresh brake fluid. Pump the piston slowly until no more air bubbles appear in the reservoirs.

3. Disconnect the 2 short lines, refill the master cylinder and securely install the cylinder caps.

4. If the master cylinder is on the vehicle, it can still be bled, using a flare nut wrench.

5. Open the brake lines slightly with the flare nut wrench while pressure is applied to the brake pedal by a helper inside the vehicle.

6. Be sure to tighten the line before the brake pedal is released.

7. Repeat the process with both lines until no air bubbles come out.

CALIPERS AND WHEEL CYLINDERS

1. Fill the master cylinder with fresh brake fluid. Check the level often during the procedure.

2. Starting with the right rear wheel, remove the protective cap from the bleeder, if equipped, and place

where it will not be lost. Clean the bleed screw.

CAUTION

When bleeding the brakes, keep face away from the brake area. Spewing fluid may cause facial and/or visual damage. Do not allow brake fluid to spill on the car's finish; it will remove the paint.

---

3. If the system is empty, the most efficient way to get fluid down to the wheel is to loosen the bleeder about ½–¾ turn, place a finger firmly over the bleeder and have a helper pump the brakes slowly until fluid comes out the bleeder. Once fluid is at the bleeder, close it before the pedal is released inside the vehicle.

NOTE: If the pedal is pumped rapidly, the fluid will churn and create small air bubbles, which are almost impossible to remove from the system. These air bubbles will eventually congregate and a spongy pedal will result.

4. Once fluid has been pumped to the caliper or wheel cylinder, open the bleed screw again, have the helper press the brake pedal to the floor, lock the bleeder and have the helper slowly release the pedal. Wait 15 seconds and repeat the procedure (including the 15 second wait) until no more air comes out of the bleeder upon application of the brake pedal. Remember to close the bleeder before the pedal is released inside the vehicle each time the bleeder is opened. If not, air will be induced into the system.

5. If a helper is not available, connect a small hose to the bleeder, place the end in a container of brake fluid and proceed to pump the pedal from inside the vehicle until no more air comes out the bleeder. The hose will prevent air from entering the system.

6. Repeat the procedure on remaining wheel cylinders in order:
   a. left front
   b. left rear
   c. right front

7. Hydraulic brake systems must be totally flushed if the fluid becomes contaminated with water, dirt or other corrosive chemicals. To flush, bleed the entire system until all fluid has been replaced with the correct type of new fluid.

8. Install the bleeder cap(s), if equipped, on the bleeder to keep dirt out. Always road test the vehicle after brake work of any kind is done.

Anti-Lock Brakes

BRAKE CONTROL ASSEMBLY

NOTE: Only use brake fluid from a sealed container which meets DOT 3 specifications.

1. Clean the area around the master cylinder cap.

2. Check fluid level in master cylinder reservoir and top-up, as necessary. Check fluid level frequently during bleeding procedure.

3. Attach a bleeder hose to the rear bleeder valve on the brake control assembly. Slowly open the bleeder valve.

4. Depress the brake pedal slowly until fluid begins to flow.

5. Close the valve and release the brake pedal.

6. Repeat for the front bleeder valve on the brake control assembly.

NOTE: When fluid flows from both bleeder valves, the brake control assembly is sufficiently full of fluid. However, it may not be completely purged of air. Bleed the individual wheel calipers/cylinders and return to the control assembly to purge the remaining air.

WHEEL CALIPERS/CYLINDERS

NOTE: Prior to bleeding the rear brakes, the rear displacement cylinder must be returned to the top-most position. This can be accomplished using the Tech I Scan tool or T-100 (CAMS), by entering the manual control function and applying the rear motor.

If a Tech I or T-100 are unavailable, bleed the front brakes. Ensure the pedal is firm. Carefully drive the vehicle to a speed above 4 mph to cause the ABS system to initialize. This will return the rear displacement cylinder to the top-most position.

1. Clean the area around the master cylinder cap.

2. Check fluid level in master cylinder reservoir and top-up, as necessary. Check fluid level frequently during bleeding procedure.

3. Raise and safely support the vehicle.

4. Attach a bleeder hose to the bleeder valve of the right rear wheel and submerge the opposite hose in a clean container partially filled with brake fluid.

5. Open the bleeder valve.

6. Slowly depress the brake pedal.

7. Close the bleeder valve and release the brake pedal.

8. Wait 5 seconds.

9. Repeat Steps 5–8 until the pedal begins to feel firm and no air bubbles appear in the bleeder hose.

10. Repeat Steps 5–9, until the pedal is firm and no air bubbles appear in the brake hose, for the remaining wheels in the following order:
    a. left rear
    b. right front
    c. left front.

11. Lower the vehicle.

Anti-Lock Brake System Service

PRECAUTION

Failure to observe the following precautions may result in system damage.

- Before performing electric arc welding on the vehicle, disconnect the Electronic Brake Control Module (EBCM) and the hydraulic modulator connectors.

- When performing painting work on the vehicle, do not expose the Electronic Brake Control Module (EBCM) to temperatures in excess of 185°F (85°C) for longer than 2 hrs. The system may be exposed to temperatures up to 200°F (95°C) for less than 15 min.

- Never disconnect or connect the Electronic Brake Control Module (EBCM) or hydraulic modulator connectors with the ignition switch ON.

- Never disassemble any component of the Anti-Lock Brake System (ABS) which is designated non-serviceable; the component must be replaced as an assembly.

- When filling the master cylinder, always use Delco Supreme 11 brake fluid or equivalent, which meets DOT-3 specifications; petroleum base fluid will destroy the rubber parts.

ABS Hydraulic Modulator Assembly

REMOVAL & INSTALLATION

———— CAUTION ————
To avoid personal injury, use the Tech I Scan tool to relieve the gear tension in the hydraulic modulator. This procedure must be performed prior to removal of the brake control and motor assembly.

1. Disconnect the negative battery cable.

2. Disconnect the 2 solenoid electrical connectors and the fluid level sensor connector.

3. Disconnect the 6-pin and 3-pin motor pack electrical connectors.

4. Wrap a shop towel around the hydraulic brake lines and disconnect the 4 brake lines from the modulator.

NOTE: Cap the disconnected lines to prevent the loss of fluid and the entry of moisture and contaminants.

5. Remove the 2 nuts attaching the ABS hydraulic modulator assembly to the vacuum booster.

6. Remove the ABS hydraulic modulator assembly from the vehicle.

To install:

7. Install the ABS hydraulic modulator assembly to the vehicle. Install the 2 attaching nuts and tighten to 20 ft. lbs. (27 Nm).

8. Connect the 4 brake pipes to the modulator assembly. Tighten to 13 ft. lbs. (17 Nm).

9. Connect the 6-pin and 3-pin electrical connectors and the fluid level sensor connector.

10. Properly bleed the system.

11. Connect the negative battery cable.

CHASSIS ELECTRICAL

Heater Blower Motor

REMOVAL & INSTALLATION

1. Disconnect negative battery cable.

2. Remove the serpentine belt and/or the power steering pressure hose, as required.

3. Disconnect the connector to the blower motor and remove the cooling tube.

4. Remove the attaching screws and remove the blower from the case.

5. If necessary, remove the fan from the blower motor.

6. The installation is the reverse of the removal procedure.

7. Connect the negative battery cable and check the blower motor for proper operation.

Windshield Wiper Motor

REMOVAL & INSTALLATION

1. Disconnect the negative battery cable.

2. Remove the wiper arm assembly(s) and cowl cover, if necessary.

3. Remove the wiper arm drive link from the crank arm.

4. Disconnect the connectors from the motor.

5. Remove the wiper motor attaching bolts.

6. Remove the wiper motor and crank arm by guiding the assembly through the access hole in the upper shroud panel.

7. The installation is the reverse of the removal procedure.

8. Connect the negative battery cable and check the wiper motor for proper operation.

Instrument Cluster

REMOVAL & INSTALLATION
Headlight Switch

REMOVAL & INSTALLATION

1. Disconnect the negative battery cable.
2. Remove the cluster trim, instrument panel trim or headlight switch trim screws, as required.
3. Remove the headlight switch attaching screws.
4. Pull the switch out, unplug the connectors and remove the switch assembly.
5. The installation is the reverse of the removal procedure.
6. Connect the negative battery cable and check the headlight switch for proper operation.

Dimmer Switch

REMOVAL & INSTALLATION

1. Disconnect the negative battery cable.
2. Remove the lower steering column cover.
3. Unplug the switch, located on the lower portion of the steering column.
4. Hold the actuating rod against its upper seat, remove the screw and nut that attaches the switch to the column and remove the switch.
5. The installation is the reverse of the removal procedure. To adjust the switch:
   a. Depress the switch slightly and insert a $3/32$ in. drill.
   b. Force the switch up to remove the lash.
   c. Tighten the screw and nut.
6. Connect the negative battery cable and check the switch for proper operation.

Turn Signal Switch

REMOVAL & INSTALLATION

1. Disconnect the negative battery cable.
2. Remove the lower instrument panel sound insulator, trim pad and steering column trim collar.
3. Straighten the steering wheel so the tires are pointing straight ahead.
4. Remove the steering wheel.
5. Remove the plastic wire protector from under the steering column.
6. Disconnect the turn signal switch connector at the bottom of the column.
7. To disassemble the top of the column:
   a. Remove the shaft lock cover.

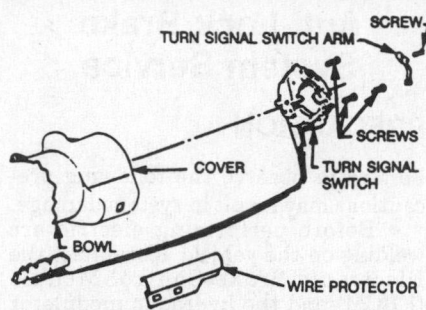

Removing the turn signal switch

   b. If equipped with telescope steering, remove the first set of spacers, bumper, second set of spacers and carrier snapring retainer.
   c. Depress the lockplate with the proper depressing tool and remove the retaining ring from its groove.
   d. Remove the tool, ring, lockplate, cancelling cam and spring.
8. Remove the 3 screws, the turn signal switch and actuator lever.
To install:
9. Install the turn signal switch and lever.
10. To assemble the top end of the column:
   a. Install the spring, cancelling cam, lockplate and retaining ring on the steering shaft.
   b. Depress the plate with the depressing tool and install the ring securely in the groove. Remove the tool slowly.
   c. If equipped with telescope steering, install the carrier snapring retainer, lower set of spacers, bumper and upper set of spacers.
   d. Install the shaft lock cover.
11. Connect the turn signal switch connector and install the wire protector.
12. Install the steering wheel.
13. Install the steering column trim collar, lower instrument panel trim pad and sound insulator.
14. Connect the negative battery cable and check the turn signal switch for proper operation.

Ignition Switch

REMOVAL & INSTALLATION

1. Disconnect the negative battery cable.
2. Remove the left instrument panel insulator.
3. Remove the left instrument panel trim pad and the steering column trim collar.
4. Remove the steering column upper support bracket bolts and remove the support bracket.
5. Lower the steering column and support it safely.

6. Disconnect the wiring from the ignition switch.
7. Remove the ignition switch-to-steering column screws. Remove the ignition from the steering column.
To install:
8. Before installing, place the slider in the proper position (switch viewed with the terminals pointing up), according to the steering column and accessories:
   a. Standard column with key release—extreme left detent.
   b. Standard column with PARK/LOCK—1 detent from extreme left.
   c. All other standard columns—2 detents from extreme left.
   d. Tilt column with key release—extreme right detent.
   e. Tilt column with PARK/LOCK—1 detent from extreme right.
   f. All other tilt columns—2 detents from extreme right.
9. Install the activating rod into the switch and install the switch to the column. Do not use oversized screws as they could impair the collapsibility of the column.
10. Connect the wiring to the ignition switch. Adjust the switch, as required.
11. Install the steering column.
12. Install the steering column trim collar, instrument panel trim pad and insulator.
13. Connect the negative battery cable and check the ignition switch for proper operation.

Fuses, Circuit Breakers and Relays

LOCATION

Fuses and Circuit Breakers

The fuse block, which contains the fuses and also the circuit breakers for power accessories, is located on the lower left side of the instrument panel, behind an access door.

Flashers

LOCATION

Turn Signal

The turn signal flasher is clipped to the instrument panel near the fuse block.

Hazard

On all vehicles except Grand Am, the hazard flasher is in the convenience center, located near the fuse block. On Grand Am, the hazard flasher is clipped to the console front extension bracket.

GM "W" Body
Front Wheel Drive
Buick—Regal **Chevrolet**—Lumina
Oldsmobile—Cutlass Supreme
Pontiac—Grand Prix

SPECIFICATIONS

VEHICLE IDENTIFICATION CHART

It is important for servicing and ordering parts to be certain of the vehicle and engine identification. The VIN (vehicle identification number) is a 17 digit number visible through the windshield on the driver's side of the dash and contains the vehicle and engine identification codes. The tenth digit indicates model year and the eighth digit indicates engine code. It can be interpreted as follows:

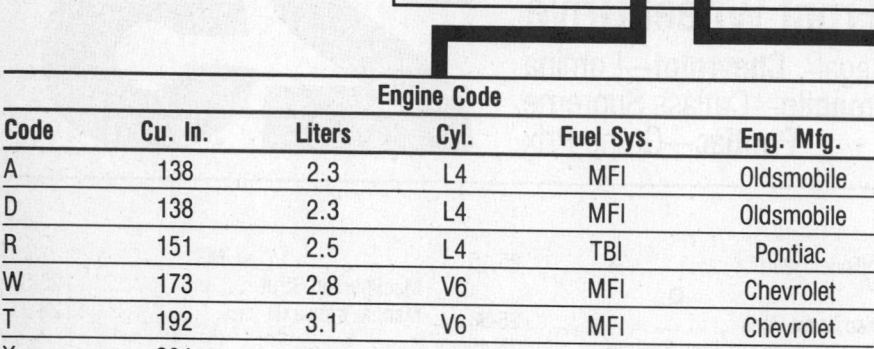

Engine Code						Model Year	
Code	Cu. In.	Liters	Cyl.	Fuel Sys.	Eng. Mfg.	Code	Year
A	138	2.3	L4	MFI	Oldsmobile	J	1988
D	138	2.3	L4	MFI	Oldsmobile	K	1989
R	151	2.5	L4	TBI	Pontiac	L	1990
W	173	2.8	V6	MFI	Chevrolet	M	1991
T	192	3.1	V6	MFI	Chevrolet	N	1992
X	204	3.4	V6	MFI	Chevrolet		
V	231	3.1	V6	MFI	Buick		
L	231	3.8	V6	MFI	Buick		

MFI—Multi Point Fuel Injection
TBI—Throttle Body Injection

ENGINE IDENTIFICATION

Year	Model	Engine Displacement cu. in. (liter)	Engine Series Identification (VIN)	No. of Cylinders	Engine Type
1988	Grand Prix	173 (2.8)	W	6	OHV
	Cutlass Supreme	173 (2.8)	W	6	OHV
	Regal	173 (2.8)	W	6	OHV
1989	Grand Prix	173 (2.8)	W	6	OHV
	Grand Prix	192 (3.1)	T	6	OHV
	Cutlass Supreme	173 (2.8)	W	6	OHV
	Cutlass Supreme	192 (3.1)	T	6	OHV
	Regal	173 (2.8)	W	6	OHV
	Regal	192 (3.1)	T	6	OHV
	Lumina	151 (2.5)	R	4	OHV
	Lumina	192 (3.1)	T	6	OHV
1990	Grand Prix	138 (2.3)	D	4	DOHC
	Grand Prix	192 (3.1)	T	6	OHV
	Grand Prix	192 (3.1)	V	6	OHV-Turbo
	Cutlass Supreme	138 (2.3)	A	4	DOHC
	Cutlass Supreme	138 (2.3)	D	4	DOHC
	Cutlass Supreme	192 (3.1)	T	6	OHV
	Regal	192 (3.1)	T	6	OHV

ENGINE IDENTIFICATION

Year	Model	Engine Displacement cu. in. (liter)	Engine Series Identification (VIN)	No. of Cylinders	Engine Type
1990	Regal	231 (3.8)	L	6	OHV
	Lumina	151 (2.5)	R	4	OHV
	Lumina	192 (3.1)	T	6	OHV
1991-92	Grand Prix	138 (2.3)	D	4	DOHC
	Grand Prix	192 (3.1)	T	6	OHV
	Grand Prix	204 (3.4)	X	6	DOHC
	Cutlass Supreme	138 (2.3)	D	4	DOHC
	Cutlass Supreme	192 (3.1)	T	6	OHV
	Cutlass Supreme	204 (3.4)	X	6	DOHC
	Regal	192 (3.1)	T	6	OHV
	Regal	231 (3.8)	L	6	OHV
	Lumina	151 (2.5)	R	4	OHV
	Lumina	192 (3.1)	T	6	OHV
	Lumina	204 (3.4)	X	6	DOHC

OHV—Overhead valve
DOHC—Double overhead cam

GENERAL ENGINE SPECIFICATIONS

Year	VIN	No. Cylinder Displacement cu. in. (liter)	Fuel System Type	Net Horsepower @ rpm	Net Torque @ rpm (ft. lbs.)	Bore × Stroke (in.)	Compression Ratio	Oil Pressure @ rpm
1988	W	6-173 (2.8)	MFI	125 @ 4500	160 @ 3600	3.500 × 2.990	8.9:1	15 @ 1100
1989	R	4-151 (2.5)	TBI	98 @ 4500	134 @ 2800	4.000 × 3.000	8.3:1	26 @ 800
	W	6-173 (2.8)	MFI	125 @ 4500	160 @ 3600	3.500 × 2.990	8.9:1	15 @ 1100
	T	6-192 (3.1)	MFI	140 @ 4500	185 @ 3600	3.500 × 3.310	8.8:1	15 @ 1100
1990	A	4-138 (2.3) ①	MFI	180 @ 6200	160 @ 5200	3.620 × 3.350	10.0:1	30 @ 2000
	D	4-138 (2.3)	MFI	160 @ 6200	155 @ 5200	3.620 × 3.350	9.5:1	30 @ 2000
	R	4-151 (2.5)	TBI	98 @ 4500	134 @ 2800	4.000 × 3.000	8.3:1	35 @ 2000
	T	6-192 (3.1)	MFI	140 @ 4400	185 @ 3200	3.500 × 3.310	8.8:1	15 @ 1100
	V	6-192 (3.1)	MFI②	205 @ 4800	220 @ 3000	3.500 × 3.310	8.9:1	15 @ 1100
	L	6-231 (3.8)	MFI	165 @ 4800	210 @ 2000	3.800 × 3.400	8.5:1	60 @ 1850
1991	D	4-138 (2.3)	MFI	160 @ 6200	155 @ 5200	3.620 × 3.350	9.5:1	30 @ 2000
	R	4-151 (2.5)	TBI	98 @ 4500	134 @ 2800	4.000 × 3.000	8.3:1	35 @ 2000
	T	6-192 (3.1)	MFI	140 @ 4400	180 @ 3600	3.500 × 3.310	8.8:1	15 @ 1100
	L	6-231 (3.8)	MFI	170 @ 4800	220 @ 3200	3.800 × 3.400	8.5:1	60 @ 1850
	X	6-204 (3.4)	MFI	③	215 @ 4000	3.622 × 3.312	9.25:1	15 @ 1100
1992	R	4-151 (2.5)	TBI	98 @ 4500	134 @ 2800	4.000 × 3.000	8.3:1	35 @ 2000
	T	6-192 (3.1)	MFI	140 @ 4400	180 @ 3600	3.500 × 3.310	8.8:1	15 @ 1100
	L	6-231 (3.8)	MFI	170 @ 4800	220 @ 3200	3.800 × 3.400	8.5:1	60 @ 1850
	X	6-204 (3.4)	MFI	③	215 @ 4000	3.622 × 3.312	9.25:1	15 @ 1100

MFI—Multi Point Fuel Injection
TBI—Throttle Body Injection
① High output (H.O.)
② Turbocharged
③ Manual—210 @ 5200
   Auto—200 @ 5000

ENGINE TUNE-UP SPECIFICATIONS

Year	VIN	No. Cylinder Displacement cu. in. (liter)	Spark Plugs Type	Gap (in.)	Ignition Timing (deg.) MT	AT	Compression Pressure (psi)	Fuel Pump (psi)	Idle Speed (rpm) MT	AT	Valve Clearance In.	Ex.
1988	W	6-173 (2.8)	R43LTSE	0.045	①	①	②	40–47	①	①	Hyd.	Hyd.
1989	R	4-151 (2.5)	R43CTS6	0.060	①	①	②	26–32	①	①	Hyd.	Hyd.
	W	6-173 (2.8)	R43LTSE	0.045	①	①	②	40–47	①	①	Hyd.	Hyd.
	T	6-192 (3.1)	R43LTSE	0.045	①	①	②	40–47	①	①	Hyd.	Hyd.
1990	A	4-138 (2.3)	FR3LS	0.035	①	①	②	40.5–47	①	①	Hyd.	Hyd.
	D	4-138 (2.3)	FR3LS	0.035	①	①	②	40.5–47	①	①	Hyd.	Hyd.
	R	4-151 (2.5)	R43CTS6	0.060	①	①	②	26–32	①	①	Hyd.	Hyd.
	T	6-192 (3.1)	R43LTSE	0.045	①	①	②	40.5–47	①	①	Hyd.	Hyd.
	V	6-192 (3.1)	R42LTS	0.045	①	①	②	40.5–47	①	①	Hyd.	Hyd.
	L	6-231 (3.8)	R44LTS6	0.060	①	①	②	40–47	①	①	Hyd.	Hyd.
1991	D	4-138 (2.3)	FR3LS	0.035	①	①	②	40.5–47	①	①	Hyd.	Hyd.
	R	4-151 (2.5)	R43CTS6	0.060	①	①	②	26–32	①	①	Hyd.	Hyd.
	T	6-192 (3.1)	R43LTSE	0.045	①	①	②	40.5–47	①	①	Hyd.	Hyd.
	L	6-231 (3.8)	R44LTS6	0.060	①	①	②	40–47	①	①	Hyd.	Hyd.
	X	6-204 (3.4)	R42LTSM	0.045	①	①	②	41–47	①	①	Hyd.	Hyd.
1992		SEE UNDERHOOD SPECIFICATION STICKER										

Hyd.—Hydraulic
① Ignition timing and idle speed are controlled by the Electronic Control Module. No adjustment is necessary
② Look for uniformity between cylinders rather than pressure. Lowest reading not less than 70% of the highest. No reading less than 100 psi

FIRING ORDERS

NOTE: To avoid confusion, always replace spark plug wires one at a time.

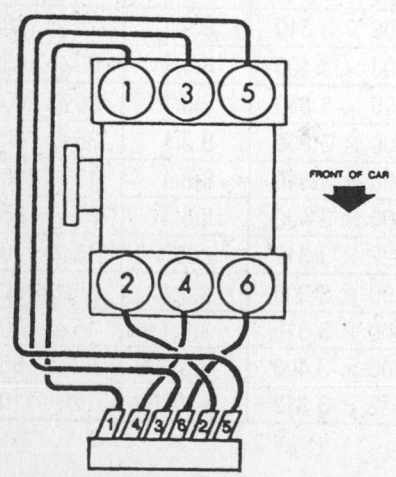

2.8L Engine (1988)
Engine Firing Order: 1–2–3–4–5–6
Distributorless Ignition System

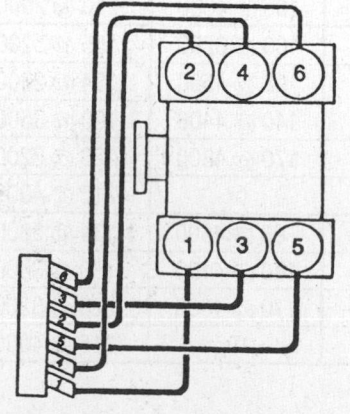

2.8L (1989) and 3.1L Engines
Engine Firing Order: 1–2–3–4–5–6
Distributorless Ignition System

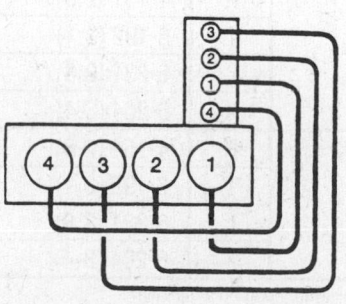

2.5L Engine
Engine Firing Order: 1—3–4–2
Distributorless Ignition System

FIRING ORDERS

NOTE: To avoid confusion, always replace spark plug wires one at a time.

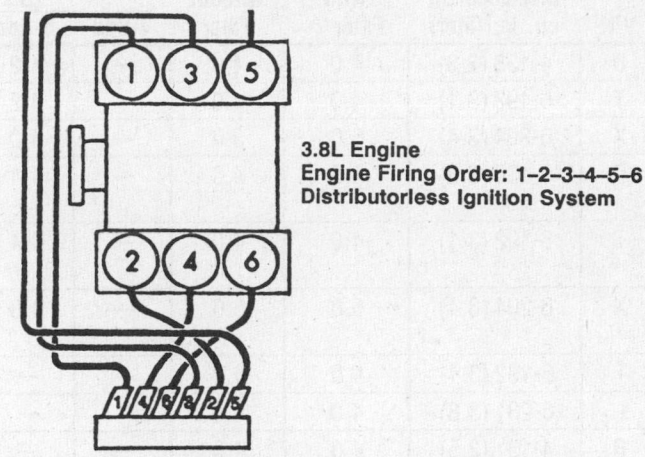

3.8L Engine
Engine Firing Order: 1–2–3–4–5–6
Distributorless Ignition System

CAPACITIES

Year	Model	VIN	No. Cylinder Displacement cu. in. (liter)	Engine Crankcase with Filter ④	Engine Crankcase without Filter	Transmission (pts.) 4-Spd	Transmission (pts.) 5-Spd	Transmission (pts.) Auto.	Drive Axle (pts.)	Fuel Tank (gal.)	Cooling System (qts.)
1988	Grand Prix	W	6-173 (2.8)	4.0	3.8	—	5	16.0①	—	16.0	②
	Cutlass Supreme	W	6-173 (2.8)	4.0	3.8	—	5	16.0①	—	16.0	②
	Regal	W	6-173 (2.8)	4.0	3.8	—	—	16.0①	—	16.0	②
1989	Grand Prix	W	6-173 (2.8)	4.0	3.8	—	5	12.0③	—	16.0	12.6
	Grand Prix	T	6-192 (3.1)	4.0	3.8	—	5	12.0③	—	16.0	12.6
	Cutlass Supreme	W	6-173 (2.8)	4.0	3.8	—	5	12.0③	—	16.0	12.6
	Cutlass Supreme	T	6-192 (3.1)	4.0	3.8	—	5	12.0③	—	16.0	12.6
	Regal	W	6-173 (2.8)	4.0	3.8	—	—	12.0③	—	16.0	12.6
	Regal	T	6-192 (3.1)	4.0	3.8	—	—	12.0③	—	16.0	12.6
	Lumina	R	4-151 (2.5)	4.0	3.8	—	—	12.0③	—	16.0	12.6
	Lumina	T	6-192 (3.1)	4.0	3.8	—	—	12.0③	—	16.0	12.6
1990	Grand Prix	A	4-138 (2.3)	4.0	4.0	—	4.2	⑤	—	16.5	8.9
	Grand Prix	D	4-138 (2.3)	4.0	4.0	—	4.2	⑤	—	16.5	9.2
	Grand Prix	T	6-192 (3.1)	4.0	4.0	—	4.2	⑤	—	16.5	12.5
	Grand Prix	V	6-192 (3.1)	4.0	4.0	—	4.2	⑤	—	16.5	13.2
	Cutlass Supreme	A	4-138 (2.3)	4.0	4.0	—	4.4	⑤	—	16.5	8.9
	Cutlass Supreme	D	4-138 (2.3)	4.0	4.0	—	4.4	⑤	—	16.5	9.2
	Cutlass Supreme	T	6-192 (3.1)	4.0	4.0	—	4.4	⑤	—	16.5	12.5
	Regal	T	6-192 (3.1)	4.0	4.0	—	—	⑤	—	16.5	12.5
	Regal	L	6-201 (3.0)	4.0	4.0	—	—	⑤	—	16.5	11.1
	Lumina	R	4-151 (2.5)	4.0	4.0	—	—	⑤	—	16.0	9.4
	Lumina	T	6-192 (3.1)	4.0	4.0	—	—	⑤	—	16.0	12.6

CAPACITIES

Year	Model	VIN	No. Cylinder Displacement cu. in. (liter)	Engine Crankcase with Filter ④	without Filter	Transmission (pts.) 4-Spd	5-Spd	Auto.	Drive Axle (pts.)	Fuel Tank (gal.)	Cooling System (qts.)
1991-92	Grand Prix	D	4-138 (2.3)	4.0	4.0	—	4.2	⑤	—	16.5	9.2
	Grand Prix	T	6-192 (3.1)	4.0	4.0	—	4.2	⑤	—	16.5	12.5
	Grand Prix	X	6-204 (3.4)	5.0	5.0	—	4.0	⑤	—	16.5	12.7
	Cutlass Supreme	D	4-138 (2.3)	4.0	4.0	—	4.4	⑤	—	16.5	9.2
	Cutlass Supreme	T	6-192 (3.1)	4.0	4.0	—	4.4	⑤	—	16.5	12.5
	Cutlass Supreme	X	6-204 (3.4)	5.0	5.0	—	4.0	⑤	—	16.5	12.7
	Regal	T	6-192 (3.1)	4.0	4.0	—	—	⑤	—	16.5	12.5
	Regal	L	6-231 (3.8)	4.0	4.0	—	—	⑤	—	16.5	11.1
	Lumina	R	4-151 (2.5)	4.0	4.0	—	—	⑤	—	16.0	9.4
	Lumina	T	6-192 (3.1)	4.0	4.0	—	—	⑤	—	16.0	12.6
	Lumina	X	6-204 (3.4)	5.0	5.0	—	4.0	⑤	—	16.5	12.7

① Drain and refill only.
  Complete overhaul—22 pts.
② Without air conditioning—12.3 qts.
  With air conditioning—12.6 qts.
③ Drain and refill only.
  Complete overhaul—16 pts.
④ Add fluid as necessary to bring to appropriate level.
⑤ Hydra-matic 3T40: drain and refill only—14 pts., complete overhaul—18 pts.
  Hydra-matic 4T60: drain and refill only—12 pts., complete overhaul—16 pts.

CAMSHAFT SPECIFICATIONS

All measurements given in inches.

Year	VIN	No. Cylinder Displacement cu. in. (liter)	Journal Diameter 1	2	3	4	5	Lobe Lift In.	Ex.	Bearing Clearance	Camshaft End Play
1988	W	6-173 (2.8)	1.867–1.881	1.867–1.881	1.867–1.881	1.867–1.881	—	0.262	0.273	0.0010–0.0040	—
1989	R	4-151 (2.5)	1.869	1.869	1.869	1.869	—	0.248	0.248	0.0010–0.0030	0.0014–0.0050
	W	6-173 (2.8)	1.867–1.881	1.867–1.881	1.867–1.881	1.867–1.881	—	0.262	0.273	0.0010–0.0040	—
	T	6-192 (3.1)	1.867–1.881	1.867–1.881	1.867–1.881	1.867–1.881	—	0.262	0.273	0.0010–0.0040	—
1990	A	4-138 (2.3)	1.5728–1.5720	1.3751–1.3760	1.3751–1.3760	1.3751–1.3760	1.3751–1.3760	0.410	0.410	0.0019–0.0043	0.0009–0.0088
	D	4-138 (2.3)	1.5728–1.5720	1.3751–1.3760	1.3751–1.3760	1.3751–1.3760	1.3751–1.3760	0.375	0.375	0.0019–0.0043	0.0009–0.0088
	R	4-151 (2.5)	1.869	1.869	1.869	1.869	—	0.248	0.248	0.0007–0.0027	0.0014–0.0050
	T	6-192 (3.1)	1.8677–1.8815	1.8677–1.8815	1.8677–1.8815	1.8677–1.8815	—	0.262	0.273	0.0010–0.0040	—
	V	6-192 (3.1)	1.8677–1.8815	1.8677–1.8815	1.8677–1.8815	1.8677–1.8815	—	0.262	0.273	0.0010 0.0040	—
	L	6-231 (3.8)	1.785–1.786	1.785–1.786	1.785–1.786	1.785–1.786	—	0.250	0.255	0.0005–0.0035	—

CAMSHAFT SPECIFICATIONS

All measurements given in inches.

Year	VIN	No. Cylinder Displacement cu. in. (liter)	Journal Diameter 1	Journal Diameter 2	Journal Diameter 3	Journal Diameter 4	Journal Diameter 5	Lobe Lift In.	Lobe Lift Ex.	Bearing Clearance	Camshaft End Play
1991–92	D	4-138 (2.3)	1.5728–1.5720	1.3751–1.3760	1.3751–1.3760	1.3751–1.3760	1.3751–1.3760	0.375	0.375	0.0019–0.0043	0.0009–0.0088
	R	4-151 (2.5)	1.869	1.869	1.869	1.869	—	0.248	0.248	0.0007–0.0027	0.0014–0.0050
	T	6-192 (3.1)	1.8677–1.8815	1.8677–1.8815	1.8677–1.8815	1.8677–1.8815	—	0.262	0.273	0.0010–0.0040	—
	X	6-204 (3.4)	2.165–2.166	2.165–2.166	2.165–2.166	2.165–2.166	—	0.370	0.370	0.0015–0.0035	—
	L	6-231 (3.8)	1.785–1.786	1.785–1.786	1.785–1.786	1.785–1.786	—	0.250	0.255	0.0005–0.0035	—

CRANKSHAFT AND CONNECTING ROD SPECIFICATIONS

All measurements are given in inches.

Year	VIN	No. Cylinder Displacement cu. in. (liter)	Crankshaft Main Brg. Journal Dia.	Crankshaft Main Brg. Oil Clearance	Crankshaft Shaft End-play	Thrust on No.	Connecting Rod Journal Diameter	Connecting Rod Oil Clearance	Connecting Rod Side Clearance
1988	W	6-173 (2.8)	2.6473–2.6483	0.0016–0.0032	0.0024–0.0083	3	1.9994–1.9983	0.0013–0.0026	0.0060–0.0170
1989	R	4-151 (2.5)	2.3000	0.0005–0.0022	0.0005–0.0180	5	2.0000	0.0005–0.0030	0.0060–0.0240
	W	6-173 (2.8)	2.6473–2.6483	0.0012–0.0027	0.0024–0.0083	3	1.9994–1.9983	0.0014–0.0036	0.0140–0.0270
	T	6-192 (3.1)	2.6473–2.6483	0.0024–0.0027	0.0012–0.0083	3	1.9994–1.9983	0.0014–0.0036	0.0140–0.0270
1990	A	4-138 (2.3)	2.0470–2.0480	0.0005–0.0023	0.0034–0.0095	3	1.8887–1.8897	0.0005–0.0020	0.0054–0.0177
	D	4-138 (2.3)	2.0470–2.0480	0.0005–0.0023	0.0034–0.0095	3	1.8887–1.8897	0.0005–0.0020	0.0054–0.0177
	R	4-151 (2.5)	2.300	0.0005–0.0022	0.0005–0.0180	5	2.0000	0.0005–0.0003	0.0060–0.0240
	T	6-192 (3.1)	2.6473–2.6483	0.0012–0.0030	0.0024–0.0083	3	1.9983–1.9994	0.0016–0.0034	0.0140–0.0270
	V	6-192 (3.1)	2.6473–2.6483	0.0012–0.0030	0.0024–0.0083	3	1.9983–1.9994	0.0011–0.0034	0.0140–0.0270
	L	6-231 (3.8)	2.4988–2.4998	0.0018–0.0030	0.0030–0.0110	3	2.2487–2.2499	0.0003–0.0026	0.0030–0.0150
1991–92	D	4-138 (2.3)	2.0470–2.0480	0.0005–0.0023	0.0034–0.0095	3	1.8887–1.8897	0.0005–0.0020	0.0054–0.0177
	R	4-151 (2.5)	2.300	0.0005–0.0022	0.0005–0.0180	5	2.0000	0.0005–0.0003	0.0060–0.0240
	T	6-192 (3.1)	2.6473–2.6483	0.0012–0.0030	0.0024–0.0083	3	1.9983–1.9994	0.0016–0.0034	0.0140–0.0270
	X	6-204 (3.4)	2.6473–2.6479	0.0013–0.0030	0.0024–0.0083	3	1.9987–1.9994	0.0011–0.0032	0.0140–0.0250
	L	6-231 (3.8)	2.4988–2.4998	0.0018–0.0030	0.0030–0.0110	3	2.2487–2.2499	0.0003–0.0026	0.0030–0.0150

VALVE SPECIFICATIONS

Year	VIN	No. Cylinder Displacement cu. in. (liter)	Seat Angle (deg.)	Face Angle (deg.)	Spring Test Pressure (lbs.)	Spring Installed Height (in.)	Stem-to-Guide Clearance (in.) Intake	Stem-to-Guide Clearance (in.) Exhaust	Stem Diameter (in.) Intake	Stem Diameter (in.) Exhaust
1988	W	6-173 (2.8)	46	45	90 @ 1.70 ①	1.70	0.0010–0.0027	0.0010–0.0027	NA	NA
1989	R	4-151 (2.5)	46	45	75 @ 1.68 ①	1.68	0.0010–0.0026	0.0013–0.0041	NA	NA
	W	6-173 (2.8)	46	45	90 @ 1.70 ①	1.57	0.0010–0.0027	0.0010–0.0027	NA	NA
	T	6-192 (3.1)	46	45	90 @ 1.70 ①	1.57	0.0010–0.0027	0.0010–0.0027	NA	NA
1990	A	4-138 (2.3)	45	②	76 @ 1.43 ①	NA	0.0010–0.0027	0.0015–0.0032	NA	NA
	D	4-138 (2.3)	45	②	76 @ 1.43 ①	NA	0.0010–0.0027	0.0015–0.0032	NA	NA
	R	4-151 (2.5)	45	46	75 @ 1.68	1.68	0.0010–0.0026	0.0013–0.0041	NA	NA
	V	6-192 (3.1)	46	45	90 @ 1.70 ①	1.57	0.0010–0.0027	0.0010–0.0027	NA	NA
	T	6-192 (3.1)	46	45	90 @ 1.70 ①	1.57	0.0010–0.0027	0.0010–0.0027	NA	NA
	L	6-231 (3.8)	46	45	80 @ 1.75 ①	1.70	0.0015–0.0035	0.0015–0.0032	NA	NA
1991-92	D	4-138 (2.3)	45	②	76 @ 1.43 ①	NA	0.0010–0.0027	0.0015–0.0032	NA	NA
	R	4-151 (2.5)	45	46	75 @ 1.68	1.68	0.0010–0.0026	0.0013–0.0041	NA	NA
	T	6-192 (3.1)	46	45	90 @ 1.70 ①	1.57	0.0010–0.0027	0.0010–0.0027	NA	NA
	X	6-204 (3.4)	46	45	75 @ 1.400 ①	1.66	0.0011–0.0026	0.0014–0.0031	NA	NA
	L	6-231 (3.8)	46	45	80 @ 1.75 ①	1.70	0.0015–0.0035	0.0015–0.0032	NA	NA

NA—Not available
① Valve closed
② Intake—44 degrees
   Exhaust—44.5 degrees

PISTON AND RING SPECIFICATIONS

All measurements are given in inches.

Year	VIN	No. Cylinder Displacement cu. in. (liter)	Piston Clearance	Ring Gap Top Compression	Ring Gap Bottom Compression	Ring Gap Oil Control	Ring Side Clearance Top Compression	Ring Side Clearance Bottom Compression	Ring Side Clearance Oil Control
1988	W	6-173 (2.8)	0.0020–0.0030	0.016–0.020	0.010–0.020	0.020–0.055	0.001–0.003	0.001–0.003	0.001–0.008
1989	R	4-151 (2.5)	0.0014–0.0022	0.010–0.020	0.010–0.020	0.020–0.060	0.002–0.003	0.001–0.003	0.015–0.055
	W	6-173 (2.8)	0.0009–0.0022	0.010–0.020	0.010–0.020	0.020–0.055	0.002–0.003	0.002–0.003	0.001–0.008
	T	6-192 (3.1)	0.0009–0.0022	0.010–0.020	0.020–0.028	0.010–0.030	0.002–0.003	0.002–0.003	0.001–0.008

PISTON AND RING SPECIFICATIONS

All measurements are given in inches.

Year	VIN	No. Cylinder Displacement cu. in. (liter)	Piston Clearance	Ring Gap			Ring Side Clearance		
				Top Compression	Bottom Compression	Oil Control	Top Compression	Bottom Compression	Oil Control
1990	A	4-138 (2.3)	0.0007–0.0020	0.013–0.023	0.015–0.025	0.015–0.055	0.002–0.004	0.001–0.003	0.019–0.026
	D	4-138 (2.3)	0.0007–0.0020	0.013–0.023	0.015–0.025	0.015–0.055	0.002–0.003	0.001–0.003	0.019–0.026
	R	4-151 (2.5)	0.0014–0.0022	0.010–0.020	0.010–0.020	0.020–0.060	0.002–0.003	0.001–0.003	0.015–0.055
	V	6-192 (3.1)	0.0009–0.0022	0.010–0.020	0.020–0.028	0.010–0.030	0.002–0.003	0.002–0.003	0.001–0.008
	T	6-192 (3.1)	0.0009–0.0022	0.010–0.020	0.020–0.028	0.010–0.030	0.002–0.003	0.002–0.003	0.001–0.008
	L	6-231 (3.8)	0.0004–0.0022	0.010–0.025	0.010–0.025	0.015–0.055	0.001–0.003	0.001–0.003	0.001–0.008
1991–92	D	4-138 (2.3)	0.0007–0.0020	0.013–0.023	0.015–0.025	0.015–0.055	0.002–0.003	0.001–0.003	0.019–0.026
	R	4-151 (2.5)	0.0014–0.0022	0.010–0.020	0.010–0.020	0.020–0.060	0.002–0.003	0.001–0.003	0.015–0.055
	T	6-192 (3.1)	0.0009–0.0022	0.010–0.020	0.020–0.028	0.010–0.030	0.002–0.003	0.002–0.003	0.001–0.008
	X	6-204 (3.4)	0.0009–0.0023	0.012–0.022	0.019–0.029	0.010–0.030	0.001–0.003	0.001–0.003	0.002–0.008
	L	6-231 (3.8)	0.0004–0.0022	0.010–0.025	0.010–0.025	0.015–0.055	0.001–0.003	0.001–0.003	0.001–0.008

TORQUE SPECIFICATIONS

All readings in ft. lbs.

Year	VIN	No. Cylinder Displacement cu. in. (liter)	Cylinder Head Bolts	Main Bearing Bolts	Rod Bearing Bolts	Crankshaft Pulley Bolts	Flywheel Bolts	Manifold		Spark Plugs
								Intake	Exhaust	
1988	W	6-173 (2.8)	①	72	40	77	46	4	19	18
1989	R	4-151 (2.5)	②	65	29	162	55	25	④	18
	W	6-173 (2.8)	①	70	37	76	46	③	18	18
	T	6-192 (3.1)	①	70	37	76	46	③	18	18
1990	A	4-138 (2.3)	⑤	⑥	⑦	⑧	⑨	18	27	17
	D	4-138 (2.3)	⑤	⑥	⑦	⑧	⑨	18	27	17
	R	4-151 (2.5)	②	65	29	162	55	25	④	18
	V	6-192 (3.1)	①	73	39	76	44	③	18	18
	T	6-192 (3.1)	①	73	39	76	44	③	18	18
	L	6-231 (3.8)	⑩	⑪	⑫	⑬	⑭	⑮	41	20
1991–92	D	4-138 (2.3)	⑤	⑥	⑦	⑧	⑨	18	27	17
	R	4-151 (2.5)	②	65	29	162	55	25	④	18
	T	6-192 (3.1)	①	73	39	76	44	③	18	18

TORQUE SPECIFICATIONS

All readings in ft. lbs.

Year	VIN	No. Cylinder Displacement cu. in. (liter)	Cylinder Head Bolts	Main Bearing Bolts	Rod Bearing Bolts	Crankshaft Pulley Bolts	Flywheel Bolts	Manifold Intake	Manifold Exhaust	Spark Plugs
1991-92	X	6-204 (3.4)	⑯	⑰	39	37	61	18	⑱	11
	L	6-231 (3.8)	⑩	⑪	⑫	⑬	⑭	⑮	41	20

① Torque in 2 steps:
  1st step—33 ft. lbs.
  2nd step—Turn an additional 90 degrees (¼ turn)
② Torque in 3 steps:
  1st step—18 ft. lbs.
  2nd step—Bolts "A" through "J" except "I" to 26 ft. lbs. Tighten bolt "I" to 18 ft. lbs.
  3rd step—Turn an additional 90 degrees (¼ turn)
③ Torque in 2 steps:
  1st step—15 ft. lbs.
  2nd step—24 ft. lbs.
④ Torque inner bolts to 37 ft. lbs. and outer bolts to 26 ft. lbs.

⑤ Torque in 2 steps:
  1st step—Torque all bolts in sequence to 26 ft. lbs.
  2nd step—Torque in sequence bolts number 7 and 9 an additional 100 degrees and the remaining bolts 110 degrees
⑥ 15 ft. lbs. plus an additional 90 degree turn
⑦ 18 ft. lbs. plus an additional 80 degree turn
⑧ 74 ft. lbs. plus an additional 90 degree turn
⑨ 22 ft. lbs. plus an additional 45 degree turn
⑩ Torque in 3 steps:
  1st step—Tighten all bolts in sequence to 35 ft. lbs.
  2nd step—Tighten all bolts in sequence an additional 130 degrees

3rd step—Tighten the center 4 bolts an additional 30 degrees
⑪ 26 ft. lbs. plus an additional 45 degree turn
⑫ 20 ft. lbs. plus an additional 50 degree turn
⑬ 105 ft. lbs. plus an additional 56 degree turn
⑭ 89 inch lbs. plus an additional 90 degree turn
⑮ Intake manifold to cylinder head (lower)—89 inch lbs.
⑯ Torque in 2 steps:
  1st step—Torque all bolts in sequence to 37 ft. lbs.
  2nd step—Turn an additional 90 degrees (¼ turn)
⑰ 37 ft. lbs. plus an additional 75 degree turn
⑱ Torque to 115 inch lbs.

BRAKE SPECIFICATIONS

All measurements in inches unless noted.

Year	Model	Lug Nut Torque (ft. lbs.)	Master Cylinder Bore	Brake Disc Minimum Thickness	Brake Disc Maximum Runout	Standard Brake Drum Diameter	Minimum Lining Thickness Front	Minimum Lining Thickness Rear
1988	Cutlass Supreme	100	0.945	0.972 ①	0.003	—	0.030	0.030
	Grand Prix	100	0.945	0.972 ①	0.003	—	0.030	0.030
	Regal	100	0.945	0.972 ①	0.003	—	0.030	0.030
1989	Cutlass Supreme	100	0.945	0.972 ①	0.004	—	0.030	0.030
	Grand Prix	100	0.945	0.972 ①	0.004	—	0.030	0.030
	Regal	100	0.945	0.972 ①	0.004	—	0.030	0.030
	Lumina	100	0.945	0.972 ①	0.004	—	0.030	0.030
1990	Cutlass Supreme	100	0.945	0.972 ①	0.004	—	0.030	0.030
	Grand Prix	100	0.945	0.972 ①	0.004	—	0.030	0.030
	Regal	100	0.945	0.972 ①	0.004	—	0.030	0.030
	Lumina	100	0.945	0.972 ①	0.004	—	0.030	0.030
1991-92	Cutlass Supreme	100	0.945	0.972 ①	0.004	—	0.030	0.030
	Grand Prix	100	0.945	0.972 ①	0.004	—	0.030	0.030
	Regal	100	0.945	0.972 ①	0.004	—	0.030	0.030
	Lumina	100	0.945	0.972 ①	0.004	—	0.030	0.030

① Rear rotor discard thickness 0.429

| Year | Model | Caster | | Camber | | Toe-in (in.) | Steering Axis Inclination (deg.) |
		Range (deg.)	Preferred Setting (deg.)	Range (deg.)	Preferred Setting (deg.)		
1988	Cutlass Supreme	$1\frac{1}{2}$P–$2\frac{1}{2}$P	2P	$\frac{3}{16}$P–$1\frac{3}{16}$P	$\frac{11}{16}$P	$\frac{3}{32}$N–$\frac{3}{32}$P	NA
	Grand Prix	$1\frac{1}{2}$P–$2\frac{1}{2}$P	2P	$\frac{3}{16}$P–$1\frac{3}{16}$P	$\frac{11}{16}$P	$\frac{3}{32}$N–$\frac{3}{32}$P	NA
	Regal	$1\frac{1}{2}$P–$2\frac{1}{2}$P	2P	$\frac{3}{16}$P–$1\frac{3}{16}$P	$\frac{11}{16}$P	$\frac{3}{32}$N–$\frac{3}{32}$P	NA
1989	Cutlass Supreme	$1\frac{5}{16}$P–$2\frac{5}{16}$P	$1\frac{13}{16}$P	$\frac{3}{16}$P–$1\frac{3}{16}$P	$\frac{11}{16}$P	$\frac{3}{32}$N–$\frac{3}{32}$P	NA
	Grand Prix	$1\frac{5}{16}$P–$2\frac{5}{16}$P	$1\frac{13}{16}$P	$\frac{3}{16}$P–$1\frac{3}{16}$P	$\frac{11}{16}$P	$\frac{3}{32}$N–$\frac{3}{32}$P	NA
	Regal	$1\frac{1}{2}$P–$2\frac{1}{2}$P	2P	$\frac{3}{16}$P–$1\frac{3}{16}$P	$\frac{11}{16}$P	$\frac{3}{32}$N–$\frac{3}{32}$P	NA
	Lumina	$1\frac{1}{2}$P–$2\frac{1}{2}$P	2P	$\frac{3}{16}$P–$1\frac{3}{8}$P	$\frac{11}{16}$P	$\frac{3}{32}$N–$\frac{3}{32}$P	NA
1990	Cutlass Supreme	$1\frac{1}{2}$P–$2\frac{1}{2}$P	2P	$\frac{3}{16}$P–$1\frac{3}{16}$P	$\frac{11}{16}$P	$\frac{3}{32}$N–$\frac{3}{32}$P	NA
	Grand Prix	$1\frac{5}{16}$P–$2\frac{5}{16}$P	$1\frac{13}{16}$P	$\frac{3}{16}$P–$1\frac{3}{16}$P	$\frac{11}{16}$P	$\frac{3}{32}$N–$\frac{3}{32}$P	NA
	Regal	$1\frac{1}{2}$P–$2\frac{1}{2}$P	2P	$\frac{3}{16}$P–$1\frac{3}{16}$P	$\frac{11}{16}$P	$\frac{3}{32}$N–$\frac{3}{32}$P	NA
	Lumina	$1\frac{1}{2}$P–$2\frac{1}{2}$P	2P	$\frac{3}{16}$P–$1\frac{3}{8}$P	$\frac{11}{16}$P	$\frac{3}{32}$N–$\frac{3}{32}$P	NA
1991–92	Cutlass Supreme	$1\frac{1}{2}$P–$2\frac{1}{2}$P	2P	$\frac{3}{16}$P–$1\frac{3}{16}$P	$\frac{11}{16}$P	$\frac{3}{32}$N–$\frac{3}{32}$P	NA
	Grand Prix	$1\frac{5}{16}$P–$2\frac{5}{16}$P	$1\frac{13}{16}$P	$\frac{3}{16}$P–$1\frac{3}{16}$P	$\frac{11}{16}$P	$\frac{3}{32}$N–$\frac{3}{32}$P	NA
	Regal	$1\frac{1}{2}$P–$2\frac{1}{2}$P	2P	$\frac{3}{16}$P–$1\frac{3}{16}$P	$\frac{11}{16}$P	$\frac{3}{32}$N–$\frac{3}{32}$P	NA
	Lumina	$1\frac{1}{2}$P–$2\frac{1}{2}$P	2P	$\frac{3}{16}$P–$1\frac{3}{8}$P	$\frac{11}{16}$P	$\frac{3}{32}$N–$\frac{3}{32}$P	NA

NA—Not adjustable
N—Negative
P—Positive

ENGINE MECHANICAL

NOTE: Disconnecting the negative battery cable on some vehicles may interfere with the functions of the on board computer systems and may require the computer to undergo a relearning process, once the negative battery cable is reconnected.

Engine Assembly

REMOVAL & INSTALLATION

2.3L Engine

1. Disconnect the negative battery cable.
2. Release the fuel system pressure.
3. Mark the hood hinges and remove the hood with an assistant. Drain the engine coolant into a suitable drain pan.
4. Remove the heater hoses at the heater core and thermostat housing.
5. Remove the radiator upper hose.
6. Remove the air cleaner and inlet hose from the vehicle.
7. If equipped with air conditioning, discharge the system.
8. Remove the air conditioning compressor and condenser hose at the compressor, if equipped.
9. Disconnect and label engine the vacuum lines.
10. Disconnect and label the electrical connectors from the alternator, air conditioning compressor, fuel injection harness, starter solenoid, engine ground strap, ignition assembly, coolant sensor, oil pressure sensor, knock sensor, oxygen sensor, Idle Air Control (IAC) valve and Throttle Position Sensor (TPS). The last 2 sensors are located at the throttle body.
11. Disconnect the power brake vacuum hose and throttle cable.
12. Remove the power steering pump and position aside. Do not remove the pump hoses, unless necessary.
13. Release the fuel pressure, if not already done and remove the fuel lines.
14. Remove the engine torque strut mounts.
15. Remove the transaxle fill tube, auto transaxle only.
16. Remove the exhaust heat shield and exhaust pipe-to-manifold.
17. Remove the upper transaxle-to-engine bolts.
18. Raise the vehicle and support it safely.
19. Remove the remaining lower transaxle-to-engine bolts.
20. Remove the exhaust-to-transaxle bracket.
21. Remove the lower radiator hose.
22. Remove the flywheel or converter cover.
23. Scribe a mark on the torque converter and flywheel. Remove the torque converter nuts.
24. Remove the transaxle-to-engine bracket.
25. Lower the vehicle.
26. Install the engine lifting fixture and remove the engine. Place the engine on a workstand.

To install:

NOTE: Make sure all the engine mounting bolts are in their correct location to prevent transaxle and engine damage.

27. Install the engine to a lifting fixture and position the engine in the vehicle. With an assistant, align the engine-to-transaxle.
28. Raise the vehicle and support it safely.
29. Install the transaxle-to-engine bracket and bolts. Torque the engine-to-transaxle bolts to:
   a. Positions No. 2, 3, 4, 5, 6—71 ft. lbs. (96 Nm).
   b. Positions No. 7, 8—41 ft. lbs. (56 Nm).
30. Apply thread locking compound and install the torque converter-to-flywheel bolts. Torque the bolts to 46 ft. lbs. (63 Nm). Install the flywheel cover.
31. At the right side of the vehicle, install the engine mount bolt.
32. Install the lower radiator hose and engine ground wires.
33. Install the air conditioning com-

pressor and condensor hose. Connect the compressor and alternator electrical harnesses.

34. Install the heater hoses at the heater core and throttle body.

35. Install the exhaust-to-transaxle bracket.

36. Lower the vehicle.

37. Install the exhaust pipe-to-manifold and heat shield. Torque the exhaust bolts to 22 ft. lbs. (30 Nm).

38. Install the upper engine mounts.

39. Connect the fuel lines.

40. Install the power steering pump, lines and drive belt.

41. Install the throttle cable and power brake vacuum hose.

42. Connect the electrical connectors to the oxygen sensor, knock sensor, oil pressure sensor, coolant sensor, ignition assembly, TPS sensor, IAC sensor and starter solenoid.

43. Connect all engine vacuum hoses.

44. Install the upper radiator hose and fill the radiator with the specified amount of antifreeze.

45. Refill the engine with the specified amount of engine oil.

46. Evacuate and recharge the air conditioning system.

47. Install the air cleaner and inlet hose.

48. Install the hood assembly with the help of an assistant.

49. Recheck all procedures for completion of repair.

50. Recheck all fluid levels.

51. Connect the negative battery cable. Start the engine and check for fluid leaks.

2.5L Engine

1. Disconnect the negative battery cable.

2. Place a suitable drain pan under the radiator drain valve and drain the engine coolant.

3. Remove the air cleaner assembly. Release fuel system pressure.

4. Mark the hood hinges with a scribe and remove the hood assembly.

5. Mark and remove all engine wiring. Place all the wire assemblies out of the way.

6. Remove the vacuum, heater and radiator hoses labeling for location.

7. Remove the air conditioning compressor from the engine and place to the side with a piece of rope or wire. Do not disconnect the hoses from the compressor.

8. Remove the alternator and bracket.

9. Remove the engine torque strut.

10. Remove the throttle and transaxle linkage.

11. Remove the transaxle-to-engine bolts except the 2 upper bolts.

12. Raise the vehicle and support it safely.

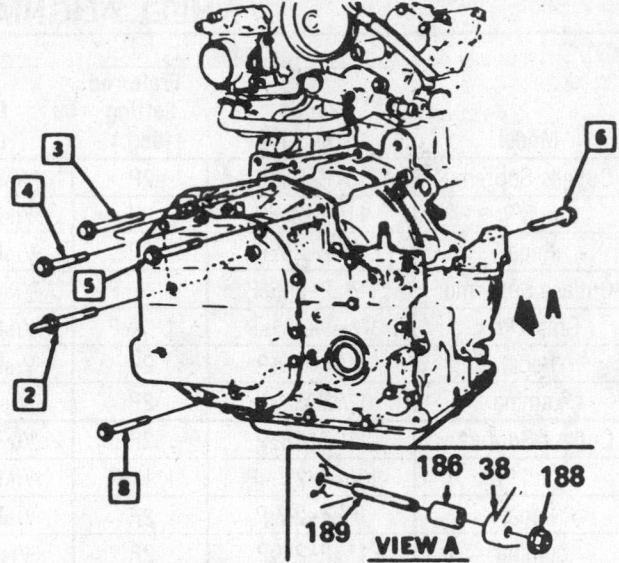

POSITION 2 STUD – 96 N·m (71 LBS. FT.)
POSITION 3 BOLT – 96 N·m (71 LBS. FT.)
POSITIONS 4 AND 6 BOLT – 96 N·m (71 LBS. FT.)
POSITION 5 BOLT – 96 N·m (71 LBS. FT.)
POSITION 8 BOLT – 56 N·m (41 LBS. FT.)
POSITION 7:
 38. OIL PAN
 186. SPACER
 188. NUT – 56 N·m (41 LBS. FT.)
 189. STUD – 12 N·m (106 LBS. IN.)

Engine to automatic transaxle fasteners

13. Remove the engine mount-to-frame bolts.

13. Remove the exhaust pipe from the manifold.

14. Remove the torque converter-to-flywheel bolts.

15. Remove the starter motor.

16. Remove the power steering pump and attach to the inner fender with a piece of rope or wire. Do not disconnect the hoses.

17. Release fuel pressure, if not done prior, and remove the fuel lines at the throttle body assembly.

18. Remove the rear engine support bracket.

19. Support the transaxle assembly with a transaxle holding fixture.

20. Disconnect the transaxle from the engine and support with a jack.

21. Attach an suitable engine lifting device.

22. Remove the engine assembly. Use care not to get under the engine assembly in case of lift failure.

23. Place the engine on a workstand.

To install:

24. Place the engine assembly onto a lifting device.

25. With an assistant, install the engine into the vehicle.

26. Position the engine into the engine mounts and engage the transaxle with the engine.

27. Remove the engine lifting device.

28. Install the torque converter bolts and engine-to-transaxle mounting bolts. Torque the torque converter bolts to 55 ft. lbs. (75 Nm).

29. Remove the transaxle holding fixture.

30. Install the rear support bracket bolts.

31. Install the engine mount nuts and torque to 32 ft. lbs. (43 Nm).

32. Install the rear transaxle mount bracket bolts and torque to 35 ft. lbs. (47 Nm).

33. Install the fuel lines to the throttle body assembly.

34. Install the power steering pump.

35. Install the starter motor assembly.

36. Install the flywheel cover plate.

37. Install the exhaust pipe-to-manifold.

38. Install the engine torque strut.

39. Install the alternator and bracket.

40. Install the air conditioning compressor.

41. Install the heater, radiator and vacuum hoses.

42. Install the throttle and transaxle linkages.

43. Install and reconnect all engine wiring harnesses.

44. Install the hood assembly to its original position with an assistant.

45. Refill the cooling system with engine coolant.

46. Reconnect the negative battery cable.
47. Install the air cleaner assembly.
48. Inspect for proper fluid levels.
49. Recheck every procedure for proper reinstallation.
50. Start the vehicle and check for any fluid leaks.

2.8L, 3.1L and 3.4L Engine
ENGINE REMOVED FROM BOTTOM

1. Remove the air cleaner assembly and ground wire near cleaner bracket.
2. Disconnect the negative battery cable and body ground.
3. Drain the engine coolant into a suitable drain pan. On 3.4L engine, it will be necessary to remove the coolant recovery tank.
4. Remove the battery remote jump start terminal from the body but leave the cables attached.
5. Disconnect the cooling fan electrical connectors and remove fans. Remove wiring harness cover and upper engine wire connectors at right strut tower, if equipped.
6. Remove the transaxle cooler lines at the radiator and the fluid level indicator.
7. Remove the upper and lower radiator hoses.
8. Remove the heater inlet and outlet hoses. On 3.4L engine, disconnect heater hose quick connect at intake manifold.
9. Release the fuel pressure. Disconnect fuel lines at fuel rail.
10. Remove the serpentine belt from the engine.
11. Remove the shift cable linkage and cable from the mounting bracket.
12. Remove the accelerator and cruise control from the throttle linkage, if equipped.
13. Remove the air conditioning pressure switch wire connector.
14. Remove the vacuum check valve from the power brake booster.
15. Remove the canister purge vacuum line at the engine.
16. Remove the torque struts from the engine.
17. Remove all electrical connectors at the right side cowl.
18. Remove the upper bolts securing the wiring harness plastic bracket-to-body side rail.
19. Remove the ECM and fuse block and set on top of the engine. If equipped with convience center, remove center, wiring harness cover, harness clips and low coolant sensor electrical connector.
20. Remove the strut-to-body mounting nuts.
21. Remove the vacuum hose from the vacuum reservoir.
22. Raise the vehicle and support it safely.

23. Remove the front wheel and tire assemblies.
24. Remove both side engine splash shields. On models with Anti-Lock Brakes (ABS), disconnect front brake ABS electrical sensors.
25. Remove the oil from the crankcase and remove the oil filter.
26. Remove the air conditioning compressor and hang from the body with the hoses still connected. On 3.4L engine, the factory recommends to discharge the air conditioner system and remove the lines near the accumulator.
27. Remove the exhaust crossover pipe and converter assembly.

──────── CAUTION ────────
Failure to disconnect the intermediate shaft from the rack and pinion stub shaft can result in damage to steering gear and/or intermediate
shaft. This damage can result in loss of steering control which could cause personal injury.
──────────────────────────

28. Remove the steering gear pinch bolt.
29. Remove the brake hose from the strut.
30. Remove the brake calipers and support to body.
31. Lower the vehicle far enough to place the engine/transaxle table under the frame.
32. Remove the frame bolts.
33. Lower the table with the engine/transaxle attached.
34. Raise the vehicle and remove the engine/transaxle from the vehicle.
35. Separate the engine from the transaxle and place the engine on a workstand.

To install:
36. Attach the transaxle and engine together and tighten. Slowly lower the body onto the drivetrain.
37. Install the strut bolts to the shock towers.
38. Install the frame bolts.
39. Remove the engine/transaxle table from under the vehicle.
40. Install the brake calipers.
41. Install the brake hoses at the struts.

NOTE: When installing the intermediate shaft make sure the shaft is seated prior to pinch bolt installation. If the pinch bolt is inserted into the coupling before shaft installation, the 2 mating shafts may disengage.

42. Install the steering pinch bolt.
43. Install the exhaust crossover pipe and converter assembly.
44. Install the air conditioning compressor and hoses as required.
45. Install the oil filter and splash

shields. Reconnect ABS electrical connector.
46. Install the tire and wheel assemblies and torque the lug nuts to 100 ft. lbs. (136 Nm).
47. Install the wiring harness bracket to body side rail.
48. Lower the vehicle. Add engine oil to correct level.
49. Install the vacuum hose at the vacuum reservoir.
50. Install the ECM and the fuse block.
51. Install the remaining bolts securing the wiring harness bracket to the body side rail.
52. Install the torque struts at engine.
53. Reconnect the canister purge vacuum line and vacuum check valve at the power brake booster.
54. Reconnect the air conditioning pressure switch electrical connector. Recharge air conditioning system as required.
55. Reconnect the accelerator and cruise control cables to the mounting bracket.
56. Install the serpentine belt and fuel lines to fuel rail.
57. Install the heater inlet and outlet hoses. Reconnect heater hose quick connect at intake manifold.
58. Install the radiator upper and lower hoses and coolant recovery tank.
59. Install the transaxle fluid indicator and cooler lines at the radiator.
60. Install radiator fans and reconnect the fan electrical connectors.
61. Install the battery remote jump start terminal to body.
62. Refill all necessary fluids, engine oil, coolant, transaxle fluid.
63. Install the battery cables and all body grounds.
64. Install the air cleaner assembly.
65. Recheck all procedures for proper reinstallation.
66. Turn ignition **ON** for 3 seconds and turn **OFF**. Check for fuel leaks. Repeat this procedure a second time and recheck for fuel leaks.
67. Start the engine and check for any fluid leaks.
68. Test drive vehicle, recheck for fluid leaks.

2.8L, 3.1L AND 3.4L Engine
ENGINE REMOVED FROM TOP

1. Remove the air cleaner and duct assembly.
2. Disconnect the negative battery cable.
3. Mark the hood hinges to ensure proper reinstallation. With an assistant, remove the hood assembly.
4. Mark and remove all necessary engine wiring and place the harnesses out of the way.

5. Remove the throttle, TV and cruise control cables, if equipped, from the throttle body assembly.

6. Release the fuel pressure and remove the fuel lines at engine.

7. Remove the AIR pump and serpentine belt.

8. Position a suitable drain pan under the radiator drain valve and drain the engine coolant. Remove coolant recovery tank.

9. Remove the upper and lower radiator hoses and heater hose quick connect at intake manifold.

10. Remove the air conditioning compressor mounting bolts at the front mounting bracket.

11. Remove the power steering pump and move to the side. Attach to the body with a piece of wire or rope. Do not disconnect the pump hoses.

12. Remove the heater hoses from the engine and move out of the way.

13. Remove the brake booster vacuum hose.

14. Remove the EGR hose from the exhaust manifold. Remove pipe from EGR valve, if equipped.

15. Raise the vehicle and support it safely.

16. Remove the air conditioning compressor from the engine and attach to the body with a piece of rope or wire. The factory recommends removal of the air conditioning manifold from compressor.

17. Remove the right front tire and wheel.

18. Disconnect right ball joint nut and separate from control arm.

19. Remove halfshaft assembly.

20. Remove the flywheel cover, starter motor and torque converter bolts.

21. Remove the transaxle bracket and front engine mount nuts.

22. Remove the exhaust pipe and converter assembly from manifold.

23. Lower the vehicle.

24. Remove the torque struts.

25. Remove the left crossover pipe-to-manifold clamp.

26. Disconnect the bulkhead electrical connector and quick connects near Electronic Control Module (ECM).

27. Remove the right crossover pipe-to-manifold clamp.

28. Support the transaxle with a suitable floor jack or equivalent.

29. Remove the remaining transaxle-to-engine bolts.

30. Attach an engine lifting device and remove the engine from the vehicle. Check for connected wires and hoses as the engine is coming out of the body.

31. Place the engine on a workstand.

To install:

32. With an assistant, install a lifting device onto the engine and position into the vehicle.

33. Remove the lifting device.

34. Install the transaxle-to-engine bolts.

35. Remove the transaxle support.

36. Reconnect the right crossover pipe-to-manifold clamp.

37. Reconnect the bulkhead electrical connector.

38. Reconnect electrical connector at ECM.

39. Install the left crossover pipe-to-manifold clamp.

40. Install the coolant recovery bottle and torque struts.

41. Raise the vehicle and support it safely.

42. Reinstall halfshaft assembly.

43. Reconnect ball joint to control arm.

44. Reinstall tire and wheel. Torque to 100 ft. lbs.

45. Reconnect ABS electrical connector if equipped.

42. Install the crossover pipe and converter assembly.

43. Install the front engine mount retaining nuts and torque to 32 ft. lbs. (43 Nm).

44. Install the transaxle bracket, torque converter bolts and starter motor.

45. Install the flywheel cover.

46. Install the air conditioning compressor to engine.

47. Lower the vehicle.

48. Install the EGR pipe and hose to valve.

49. Reconnect the brake booster vacuum supply, heater hoses and power steering pump.

50. Install the air conditioning compressor front mounting bracket bolts.

51. Install the radiator hoses and fans, serpentine and AIR pump belts. Recharge as required.

52. Reconnect the fuel lines. Install coolant recovery tank.

53. Install the throttle, TV and cruise control linkage to the throttle body.

54. Reconnect all necessary engine electrical and ground wiring.

55. Install the hood assembly with an assistant.

56. Reconnect the battery cables.

57. Turn the ignition **ON** for 3 seconds and then return to **OFF** position. Check for fuel leaks. Repeat this procedure a second time.

58. Install the air cleaner and duct assembly.

59. Recheck all procedures for proper reinstallation and correct if necessary.

60. Refill the engine with engine oil, coolant and transaxle fluid, if needed.

61. Inspect vehicle for fluid leaks before and after starting the engine.

62. Road test the vehicle and recheck for fluid leaks.

3.8L Engine

1. Disconnect the negative battery cable.

2. Remove the air cleaner assembly.

3. Release the fuel system pressure.

4. Disconnect the fuel lines from the rail and mounting brackets.

5. Drain the engine coolant and remove the recovery bottle.

6. Remove the inner fender electrical cover and the fuel injector sight cover.

7. Disconnect the throttle cables from the throttle body and mounting bracket.

8. Remove the rear heat shield from the crossover pipe.

9. Remove the throttle cable mounting bracket and vacuum line as an assembly.

10. Disconnect the exhaust crossover from the manifolds.

11. Disconnect the engine torque strut bolt and strut from the engine.

12. Remove the right side engine cooling fan.

13. Disconnect the vacuum line to the transaxle module.

14. Remove the serpentine belt.

15. Remove the power steering pump and alternator assemblies.

16. Tag and disconnect all electrical connections from the engine.

17. Disconnect the upper and lower radiator, and heater hoses from the engine.

18. Remove the transaxle to engine bolts and ground wire harness.

19. Raise and support the vehicle safely.

20. Remove the right front wheel and inner splash shield.

21. Remove the flywheel cover, scribe a mark on the torque converter and flywheel and remove the flywheel to torque converter bolts.

22. Disconnect the wire harness clamps from the frame near the radiator.

23. Remove the air conditioner compressor from the bracket and lay aside and secure to the frame.

24. Disconnect the wires and remove the starter motor assembly.

25. Safely support the transaxle and remove the transaxle to engine bolt, through the wheel well, using a long extension.

26. Attach a lifting device and remove the engine mount to frame nuts.

27. Drain the engine oil and remove the oil filter.

28. Disconnect the oil cooler pipes from the hose connections.

29. Disconnect the exhaust pipe from the manifold.

30. Lower the vehicle and remove the engine assembly from the vehicle.

To install:

31. With an assistant, install a lifting

device onto the engine and position into the vehicle.

32. Support the transaxle, install the transaxle-to-engine bolts and ground wire harness and torque to 46 ft. lbs.

33. Install the heater and upper and lower radiator hoses to the engine.

34. Install all electrical connections to the engine.

35. Install the alternator, power steering pump and serpentine belt.

36. Install the vacuum line to the transaxle module.

37. Install the engine torque strut and bolt and torque to 41 ft. lbs.

38. Install the exhaust crossover pipe.

39. Install the throttle cable mounting bracket and vacuum lines.

40. Install the heat shield to the crossover pipe and the throttle cables to the throttle body and mounting bracket.

41. Install the inner fender electrical cover and the coolant recovery bottle.

42. Install the fuel hoses to the fuel rail and mounting brackets.

43. Raise and support the vehicle safely.

44. Connect the front exhaust pipe to the manifold.

45. Install the oil filter and oil cooler pipes.

46. Install the engine mount nuts to the frame and torque to 32 ft. lbs.

47. Install the transaxle to engine bolt through the wheel well and torque to 46 ft. lbs.

48. Install the starter motor assembly and connect the electrical connectors.

49. Install the air conditioner compressor to the bracket.

50. Install the wire harness clamps to the frame near the radiator.

51. Align the scribe marks, install the torque converter to flywheel bolts and torque to 46 ft. lbs.

52. Install the flywheel cover and the inner fender splash shield.

53. Install the right front wheel assembly and lower the vehicle.

54. Refill the cooling system and bleed the power steering system.

55. Install the right side cooling fan.

56. Install the fuel injector sight shield and the air cleaner assembly.

57. Connect the negative battery cable and install the hood.

58. Chech and add fluids as required. Test drive vehicle and recheck for leaks and correct levels.

Cylinder Head

REMOVAL & INSTALLATION

2.3L Engine

1. Disconnect the negative battery cable.

2. Drain the cooling system.

3. Remove the heater inlet and throttle body heater hoses from the water inlet.

4. Remove the exhaust manifold.

5. Remove the intake and exhaust camshaft housing.

6. Remove the oil fill cap, tube and retainer. Pull the tube up and out of the block.

7. Disconnect and move the fuel injector harness.

8. Release the fuel system pressure.

9. Remove the throttle body and air inlet tube with the hoses and cables still connected. Position the assembly out of the way.

10. Remove the power brake booster hose and throttle cable bracket.

11. Remove the MAP sensor vacuum hose and all electrical connectors from the intake manifold and cylinder head.

12. Remove the radiator inlet hose and coolant sensor connectors.

13. In the reveres order of installation, remove the cylinder head-to-block retaining bolts.

14. Gently tap the outer edges of the cylinder head with a rubber hammer to dislodge the head gasket. Do not pry a screwdriver between the 2 surfaces.

15. Remove the cylinder head and intake manifold as an assembly.

To install:

16. Clean all gasket mating surfaces with a plastic scraper and solvent. Remove all dirt from the bolts with a wire brush.

17. Clean and inspect the oil flow check valve but do not remove the valve.

18. Check the cylinder head mating surface for flatness using a straight edge and a feeler gauge. Resurface the head, if the warpage exceeds 0.010 inch (0.25mm).

19. Check to see if the dowel pins are installed properly, replace, if necessary.

NOTE: To avoid damage, install new spark plugs after the cylinder head has been installed on the engine. In the mean time, plug the holes to prevent dirt from entering the combustion chamber during reinstallation.

20. Do not use any sealing compounds on the new cylinder head gasket. Match the new gasket with the old one to ensure a perfect match.

21. Install the cylinder head and camshaft housing covers.

22. Torque all bolts to 26 ft. lbs. (35 Nm) plus an additional 100 degrees for bolts No. 7 and 9. Torque an additional 110 degrees for all bolts except No. 7 and 9.

23. Install the throttle body heater hoses, upper radiator hose and intake manifold bracket.

24. Install cylinder head and intake manifold electrical connectors and vacuum hoses.

25. Install the throttle body-to-intake manifold with a new gasket. Install the throttle cable, MAP sensor vacuum hose and air cleaner duct.

26. Lubricate the new oil fill tube O-ring and install the fill tube. Make sure the tube is fully seated in the block.

27. Install and torque the exhaust manifold.

28. Fill the radiator with the specified amount of engine coolant.

29. Recheck all procedures to ensure completion of repair.

30. Connect the negative battery cable, start the engine and check for fluid leaks.

2.5L Engine

1. Disconnect the negative battery cable.

2. Drain the cooling system.

3. Raise and safely support the vehicle.

4. Remove the exhaust pipe and oxygen sensor.

5. Lower the vehicle.

6. Remove the oil level indicator tube and auxiliary ground cable.

7. Remove the air cleaner assembly.

8. Disconnect the EFI electrical connections and vacuum hoses.

9. Release the fuel pressure. Remove the wiring connectors, throttle linkage and fuel lines.

10. Remove the heater hose from the intake manifold.

11. Remove the wiring connectors from the manifold and cylinder head.

1. 26 ft. lbs (35 Nm) plus 110 degrees
2. 26 ft. lbs (35 Nm) plus 100 degrees

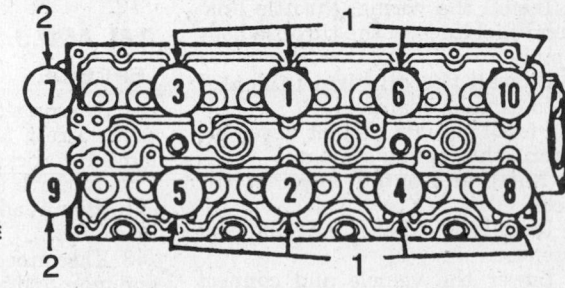

Cylinder head bolt tightening sequence—2.3L engine

12. Remove the vacuum hoses, serpentine belt and alternator bracket.

13. Remove the radiator hoses.

14. Remove the rocker arm cover.

15. Loosen the rocker arm nuts and move the rocker arms to the side enough to remove the pushrods.

16. Mark each pushrod and remove from the engine.

NOTE: Mark each valve component to ensure that they are replaced in the same location as removed.

17. Remove the cylinder head bolts.

18. Tap the sides of the cylinder head with a plastic hammer to dislodge the gasket. Remove the cylinder head with the intake and exhaust manifold still attached.

19. If the cylinder head has to be serviced or replaced, remove the intake manifold, exhaust manifold and remaining hardware.

To install:

20. Before installing, clean the gasket surfaces of the head and block.

21. Check the cylinder head for warpage using a straight edge.

22. Match up the old head gasket with the new one to ensure the holes are exact. Install a new gasket over the dowel pins in the cylinder block.

23. Install the cylinder head in place over the dowel pins.

24. Coat the cylinder head bolt threads with sealing compound and install finger tight.

25. Torque the cylinder head bolts, in sequence, in 3 steps.
   a. Torque all bolts to 18 ft. lbs.
   b. Torque bolts "A" through "J" except "I" to 26 ft. lbs. Torque bolt "I" to 18 ft. lbs.
   c. Turn all bolts an additional 90 degree (¼).

26. Install the pushrods, rocker arms and nuts (or bolts) in the same location as removed. Tighten the nuts (or bolts) to 24 ft. lbs. (32 Nm).

27. Install the rocker arm cover.

28. Install the radiator hoses, alternator bracket and serpentine belt.

29. Connect all intake manifold and cylinder head wiring.

30. Install the vacuum hoses and heater hose at manifold.

31. Install the wiring, throttle linkage and fuel lines to the throttle body assembly.

32. Install the oil level indicator tube-to-exhaust manifold.

33. Install the air cleaner assembly and refill the cooling system.

34. Raise and safely support the vehicle.

35. Install the exhaust pipe and oxygen sensor.

36. Lower the vehicle and connect the negative battery cable.

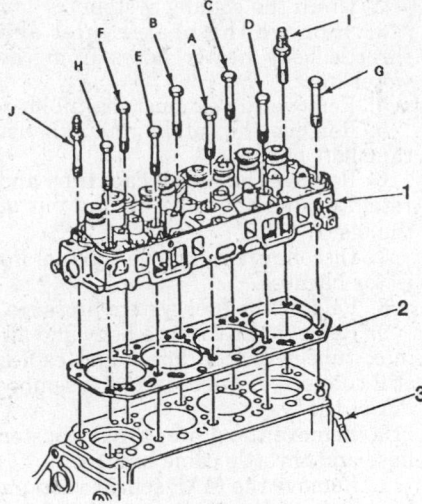

1. Cylinder head
2. Gasket
3. Cylinder block
4. NOTE:Tighten all bolts in proper sequence to 18 ft. lbs. (25 Nm). Tighten bolts A—J (except I) again to 26 ft. lbs. (35 Nm) and bolt I to 18 ft. lbs. (25 Nm). Tighten all the bolts in proper sequence an additional ¼ turn or 90° degrees.

Cylinder head bolt tightening sequence —2.5L engine

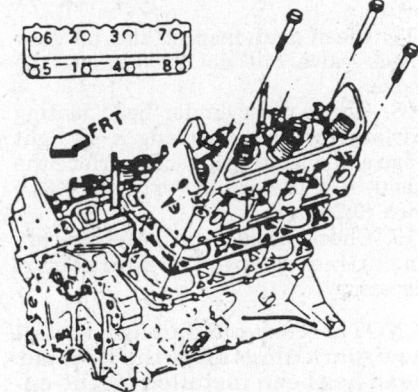

Cylinder head bolt tightening sequence— 2.8L and 3.1L engines

37. Start the engine and check for leaks.

2.8L AND 3.1L Engine

LEFT SIDE

1. Disconnect the negative battery cable. Drain the cooling system. Remove the rocker cover.

2. Remove the intake manifold-to-cylinder head bolts and the intake manifold.

3. Disconnect the exhaust crossover and manifold bolts and remove left exhaust manifold.

4. Disconnect the oil level indicator tube bracket.

5. Loosen the rocker arms nuts, turn the rocker arms and remove the pushrods. Intake and exhaust pushrods are different lengths and are color coded for identification; intake pushrods are marked orange and exhaust pushrods are marked blue in color.

NOTE: Be sure to keep the parts in order for installation purposes.

6. Remove spark plug wires.

7. Remove the cylinder head-to-engine bolts; start with the outer bolts and work toward the center. Remove the cylinder head with the exhaust manifold as an assembly.

To install:

8. Clean the gasket mounting surfaces. Inspect the surfaces of the cylinder head, block and intake manifold for damage and/or warpage. Clean the threaded holes in the block and the cylinder head bolt threads.

9. Use new gaskets, align the new cylinder head gasket over the dowels on the block with the note **THIS SIDE UP** facing the cylinder head.

10. Install the cylinder head and exhaust manifold crossover assembly on the engine.

11. Using GM sealant 1052080 or equivalent, coat the cylinder head bolts and install the bolts hand tight.

12. Using the correct sequence, torque the bolts to 33 ft. lbs. (45 Nm). After all bolts are torqued to 33 ft. lbs. (45 Nm), rotate the torque wrench another 90 degrees or ¼ turn. This will apply the correct torque to the bolts.

13. Install the pushrods in the same order that they were removed. Torque the rocker arm nuts to 14–20 ft. lbs. (19–27 Nm).

14. Install the intake manifold using a new gasket and following the correct sequence, torque the bolts to the correct specification.

15. Install the oil level indicator tube and install the rocker cover. Install the air inlet tube and spark plug wires.

16. Reinstall engine strut bracket and exhaust manifold.

17. Connect the negative battery cable. Refill the cooling system. Start the engine and check for leaks.

RIGHT SIDE

1. Disconnect the negative battery cable. Drain the cooling system. Remove air cleaner assembly. Remove the torque strut at engine.

2. Raise and safely support the vehicle. Remove the exhaust manifold-to-exhaust pipe bolts and separate the pipe from the manifold.

3. Lower the vehicle. Remove coolant recovery tank.

4. Remove the exhaust manifold-to-cylinder head bolts and the manifold.

5. Remove exhaust crossover heat shield and crossover pipe at right exhaust manifold.

6. Remove right side spark plug wires at cylinder head.

7. Remove the rocker arm cover. Remove the intake manifold-to-cylinder head bolts and the intake manifold.

8. Loosen the rocker arms nuts, turn the rocker arms and remove the pushrods. Intake and exhaust pushrods are different lengths and are color coded for identification; intake pushrods are marked orange and exhaust pushrods are marked blue in color.

NOTE: Be sure to keep the components in order for reassembly purposes.

9. Remove the cylinder head-to-engine bolts, starting with the outer bolts and working toward the center and the cylinder head.

To install:
10. Clean the gasket mounting surfaces. Inspect the parts for damage and/or warpage.

11. Clean the engine block's threaded holes and the cylinder head bolt threads.

12. To install, use new gaskets and reverse the removal procedures. Using GM sealant 1052080 or equivalent, coat the cylinder head bolts and install the bolts hand tight.

13. Place the cylinder head gasket on the engine block dowels with the note **THIS SIDE UP** facing the cylinder head.

14. Using the torquing sequence, torque the bolts to 33 ft. lbs. (45 Nm). After all bolts are torqued to 33 ft. lbs. (45 Nm), rotate the torque wrench another 90 degrees or ¼ turn. This will apply the correct torque to the bolts.

15. Install the pushrods in the same order as they were removed. Torque the rocker arm nuts to 14–20 ft. lbs. (19–27 Nm).

16. Follow the torquing sequence, use a new gasket and install the intake manifold.

a. Install spark plug wires to cylinder head.

b. Install right side exhaust crossover pipe, heat shield and manifold.

c. Install exhaust pipe to manifold.

14. Install the oil level indicator tube and install the rocker cover. Install the air inlet tube.

15. Install coolant recovery tank. Refill the cooling system. Start the engine, allow it to reach normal operating temperatures and check for leaks.

3.4L Engine

LEFT SIDE (FRONT)

1. Disconnect the negative battery cable.

2. Drain cooling system. Remove intake manifold.

3. Remove left side cam carrier as follows:

a. Disconnect oil/air breather hose from cam carrier cover. Remove spark plug wires from plugs and remove rear spark plug wire cover.

b. Remove cam carrier cover bolts and lift off cover. Remove gasket and O-rings from cover.

c. Remove secondary timing belt by removing secondary timing belt actuator and tensioner assembly and sliding belt from pulleys.

d. Install 6 sections of fuel line hoses under cam shaft and between lifters. This will hold lifters in the carrier. For this procedure use $\frac{3}{16}$ inch fuel line hose for exhaust valves and $\frac{5}{32}$ inch fuel line hose for the intake valves.

e. Remove exhaust crossover pipe and torque strut.

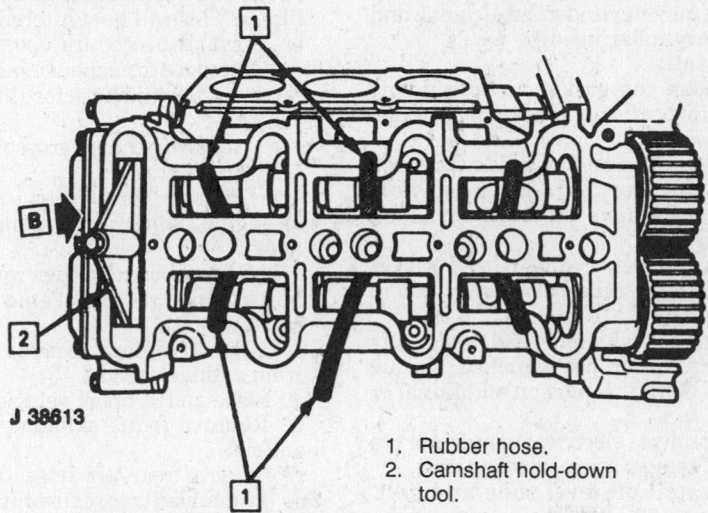

J 38613

1. Rubber hose.
2. Camshaft hold-down tool.

Cam carrier with lifter hold-down hoses in place

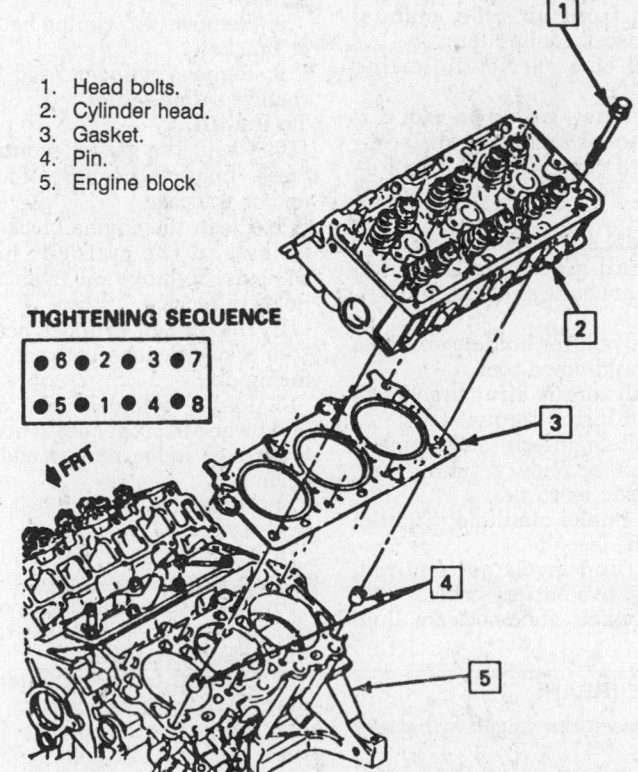

1. Head bolts.
2. Cylinder head.
3. Gasket.
4. Pin.
5. Engine block

TIGHTENING SEQUENCE

| ●6 | ●2 | ●3 | ●7 |
| ●5 | ●1 | ●4 | ●8 |

FRT

Cylinder head and torque sequence on—3.4L engine

f. Remove torque strut bracket at engine.

g. Remove cam carrier mounting bolts and nuts and remove cam carrier.

h. Remove cam carrier gasket from cylinder head.

4. Remove front air hose on manual transaxle only.

5. Remove right cooling fan.

6. Remove exhaust mounting bolts and manifold.

7. Remove oil level indicator tube bolt and tube.

8. Disconnect electrical connector from temperature sending unit.

9. Remove cylinder head bolts and remove cylinder head.

To install:

10. Clean the gasket mounting surfaces. Inspect the parts for damage and/or warpage.

11. Clean the engine block threaded holes and the cylinder head bolt threads. Remove oil from threaded holes in block.

12. Install new cylinder head gasket to block with tabs between cylinders facing up.

13. Install cylinder head and bolts and torque in proper sequence. Torque bolts to 33 ft. lbs. plus an additional ¼ turn.

14. Connect electrical connector to coolant temperature sending unit.

15. Install oil level tube and bolt. Tighten to 89 inch lbs.

16. Install exhaust manifold and nuts. Tighten to 116 inch lbs.

17. Install front air pipe, manual transaxle. Install cooling fan.

18. Install cam carrier following these steps:

a. Install new gasket on cam carrier to cylinder mounting surface.

b. Install cam hold-down tool J–38613 or equivalent, to carrier assembly.

c. Install cam carrier to cylinder head.Install mounting bolts and nuts. Torque bolts and nuts to 18 ft. lbs.

d. Remove lifter hold-down hoses and cam hold-down tool.

e. Install torque strut bracket to engine and install torque strut.

f. Install engine crossover pipe.

g. Install secondary timing belt and cam carrier cover.

19. Install intake manifold. Tighten bolts to 18 ft. lbs.

20. Refill fliud levels as required. Connect negative battery cable.

21. Start vehicle and check for fluid leaks.

RIGHT SIDE (REAR)

1. Disconnect the negative battery cable.

2. Drain cooling system. Remove intake manifold.

3. Remove right side cam carrier as follows:

a. Remove intake plenum and right timing belt cover.

b. Remove right spark plug wires.

c. Remove air/oil separator hose at cam carrier cover.

d. Remove cam carrier cover bolts and lift of cover. Remove gasket and O-rings from cover.

e. Remove secondary timing belt by removing secondary timing belt actuator and tensioner assembly and sliding belt from pulleys.

f. Install 6 sections of fuel line hoses under cam shaft and between lifters. This will hold lifters in carrier. For this procedure use ³/₁₆ inch fuel line hose for exhaust valves and ⁵/₃₂ inch fuel line hose for the intake valves.

g. Remove exhaust crossover pipe and torque strut.

h. Remove torque strut bracket at engine. Remove front engine lift hook.

i. Remove cam carrier mounting bolts and nuts and remove cam carrier.

j. Remove cam carrier gasket from cylinder head.

4. Raise and support vehicle safely.

5. Remove front exhaust pipe at manifold.

6. Remove rear air hose from air pipe on manual transaxle only.

7. Lower vehicle and disconnect electrical connector from oxygen sensor.

8. Remove rear timing belt tensioner bracket.

9. Remove cylinder head bolts and remove cylinder head.

To install:

10. Clean the gasket mounting surfaces. Inspect the parts for damage and/or warpage.

11. Clean the engine block threaded holes and the cylinder head bolt threads. Remove oil from threaded holes in block.

12. Install new cylinder head gasket to block with tabs between cylinders facing up.

13. Install cylinder head and bolts and torque in proper sequence. Torque bolts to 33 ft. lbs. plus an additional ¼ turn.

14. Install rear timing belt tensioner bracket.

15. Connect electrical connector to oxygen sensor.

16. Raise vehicle and support safely.

17. Connect rear air hose to air pipe for manual transaxle.

18. Install front exhaust pipe to manifold.Lower vehicle.

19. Install cam carrier following these steps:

a. Install new gasket on cam carrier to cylinder mounting surface.

b. Install cam hold-down tool J–38613 or equivalent, to carrier assembly.

c. install cam carrier to cylinder head. Install mounting bolts and nuts. Torque bolts and nuts to 18 ft. lbs.

d. Remove lifter hold-down hoses and cam hold-down tool.

e. Install torque strut bracket to engine and install torque strut.

f. Install engine crossover pipe and engine lift hook.

g. Install secondary timing belt and cam carrier cover.

h. Install spark plug wires and cover.

20. Install intake manifold. Torque bolts to 18 ft. lbs.

21. Refill fliud levels as required. Connect negative battery cable.

22. Start vehicle and check for fluid leaks.

3.8L Engine

LEFT SIDE (FRONT)

1. Disconnect the negative battery cable and remove the air cleaner assembly.

2. Drain the cooling system and remove the intake manifold.

3. Remove the valve covers and remove the rocker arm assemblies.

4. Disconnect the torque strut from the bracket at cylinder head.

5. Disconnect the vacuum line from the transaxle.

6. Remove the left exhaust manifold.

7. Disconnect the spark plug wires and remove the spark plugs.

8. Remove the alternator front mount bracket and ignition module with bracket.

9. Remove the cylinder head bolts and remove the cylinder head.

10. Clean all gasket mating surfaces and the cylinder head bolt holes in the block.

To install:

11. Place the cylinder head gasket on the engine block dowels with the note **THIS SIDE UP** facing the cylinder head and the arrow facing the front of the engine.

12. Install the cylinder head bolts and tighten as follows:

a. Tighten the cylinder head bolts, in sequence, to 35 ft. lbs.

b. Rotate each bolt 130 degrees, in sequence.

c. Rotate the center 4 bolts an additional 30 degrees, in sequence.

13. Install the rocker arm assemblies and valve covers.

14. Install the intake and exhaust manifolds.

15. Install the alternator front mount bracket and ignition module

with bracket and torque the bolts to 37 ft. lbs.

16. Install the spark plugs and wires.

17. Install the torque strut to the bracket, at the head and torque to 41 ft. lbs.

18. Fill the cooling system, connect the negative battery cable and install the air cleaner assembly.

RIGHT SIDE (REAR)

1. Disconnect the negative battery cable and remove the air cleaner assembly.

2. Drain the cooling system and disconnect the exhaust crossover pipe.

3. Remove the intake manifold.

4. Raise and support the vehicle safely.

5. Disconnect the front exhaust pipe from the manifold.

6. Remove the valve covers.

7. Remove the belt tensioner pulley.

8. Disconnect the heater hose from the engine.

9. Remove the power steering pump mounting bracket and lay the pump to 1 side.

10. Remove the spark plug wires and remove the spark plugs.

11. Disconnect the exhaust manifold and leave in place.

12. Disconnect the electrical connection from the oxygen sensor.

13. Remove the rocker arm assemblies.

14. Remove the cylinder head bolts and remove the cylinder head.

15. Clean all gasket mating surfaces and the cylinder head bolt holes in the block.

To install:

16. Place the cylinder head gasket on the engine block dowels with the note **THIS SIDE UP** facing the cylinder head and the arrow facing the front of the engine.

17. Install the cylinder head bolts and tighten as follows:

　a. Tighten the cylinder head bolts, in sequence, to 35 ft. lbs.

　b. Rotate each bolt 130 degrees, in sequence.

　c. Rotate the center 4 bolts an additional 30 degrees, in sequence.

18. Connect the electrical connection to the oxygen sensor.

19. Install the exhaust manifold and intake manifold.

20. Install the rocker arm assemblies.

21. Install the valve cover.

22. Install the spark plugs and wires.

23. install the power steering pump bracket and torque the bolts to 37 ft. lbs.

24. Install the belt tensioner pulley.

25. Install the heater hose to the engine.

26. Install the exhaust crossover pipe.

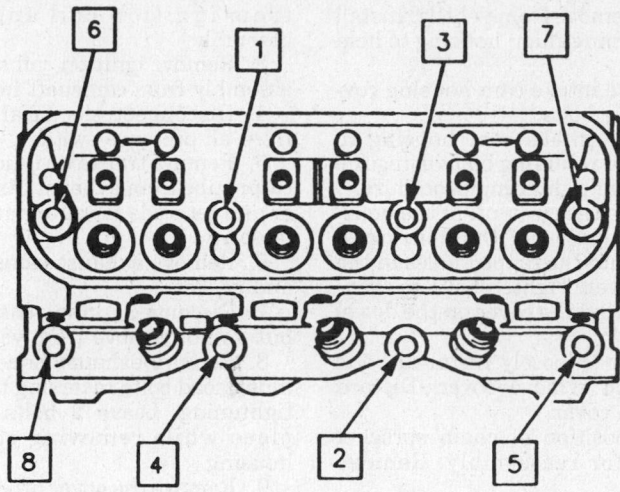

Cylinder head bolt tightening sequence — 3.8L engine

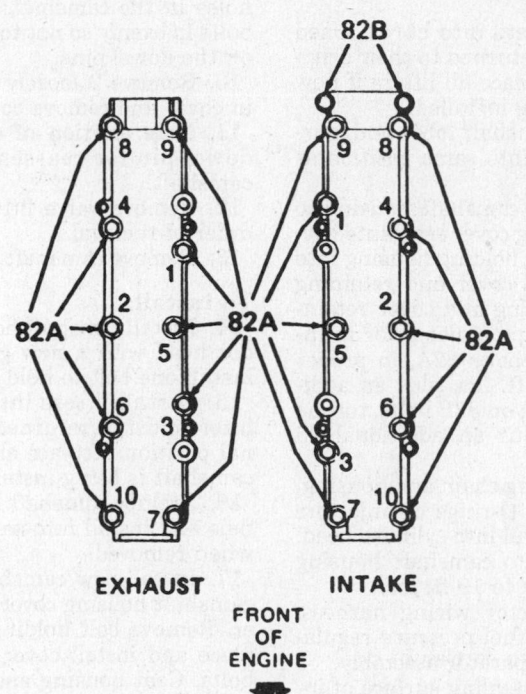

Camshaft housing bolt tightening sequence

27. Raise and support the vehicle safely.

28. Install the front exhaust pipe to the manifold and lower the vehicle.

29. Fill the cooling system, connect the negative battery cable and install the air cleaner assembly.

Valve Lifters

REMOVAL & INSTALLATION

2.3L Engine

INTAKE CAMSHAFT AND LIFTERS

1. Disconnect the negative battery cable.

2. Remove the ignition coil and module assembly electrical connections.

3. Remove the ignition coil and module from engine.

4. Remove idle speed power steering pressure switch connector.

5. Remove power steering drive belt and remove power steering pump as required.

6. Remove oil/air separator hose, fuel harness connector, vacuum hose to fuel regulator and fuel rail as required. Position fuel rail out of the way leaving fuel rail attached to fuel lines.

7. Disconnect timing chain housing

but do not remove from vehicle. Install 2 bolts in timing chain housing to hold into place.

8. Remove intake cam housing cover to housing bolts.

9. Remove intake cam housing to cylinder head retaining bolts using the reverse of the tightening procedure.

10. Remove the cover off of the housing by threading 4 of the housing to head bolts into the tapped holes in the camshaft cover. Tighten bolts in evenly so not to bind the cover on the dowel pins.

11. Remove 2 loosely installed bolts in cover and remove cover. Discard gasket from cover.

12. Note position of chain sprocket dowel pin for reassembly. Remove camshaft.

13 Remove valve lifters keeping in order of removal.

To install:

14. Install lifters into bores. Used lifters must be returned to their original position. Replace all lifters if new camshaft is being installed.

15. Prelube camshaft lobes and journals and install into same position as when removed.

16. Install new camshaft housing to camshaft housing cover seals into cover. Remove bolts holding housing into place and install cover and retaining bolts. Coat housing and cover retaining bolts with pipe sealer prior to installing. Torque bolts 82A, in proper sequence, to 11 ft. lbs. plus an additional 75 degrees; on 82B bolts, torque to 11 ft. lbs. plus an additional 25 degrees.

17. Install timing chain and housing.

18. Install new O-rings on injectors and install fuel rail into cylinder head. Install fuel rail to camshaft housing bolts and tighten to 19 ft. lbs.

19. Install injector wiring harness, vacuum hose to fuel pressure regulator and oil/air separator assembly.

20. Lube inner sealing surface of intake camshaft seal with clean engine oil and install seal into housing using tool J–36009 or equivalent.

21. Install drive pulley onto intake camshaft using tool J–36015 or equivalent.

22. Install power steering pump and drive belt.

23. Install idle speed power steering switch connector.

24. Install ignition module and coil assembly with retainer bolts and reconnect electrical connector.

25. Connect negative battery cable, start engine and check for oil leaks.

EXHAUST CAMSHAFT AND LIFTERS

1. Disconnect the negative battery cable.

2. Disconnect electrical connection

from ignition coil and module assembly.

3. Remove ignition coil and module assembly from camshaft housing.

4. Disconnect electrical connector from oil pressure switch.

5. Remove transaxle fluid level indicator tube from exhaust camshaft cover and set aside for automatic transaxle only.

6. Remove exhaust camshaft cover and gasket.

7. Disconnect timing chain housing but do not remove from vehicle.

8. Remove exhaust housing to cylinder head bolts reversing the order of tightening. Leave 2 bolts loosely in place while removing cover from housing.

9. Remove the cover off of the housing by threading 4 of the housing to head retaining bolts into the tapped holes in the camshaft cover. Tighten bolts in evenly so not to bind the cover on the dowel pins.

10. Remove 2 loosely installed bolts in cover and remove cover.

11. Note position of chain sprocket dowel pin for reassembly. Remove camshaft.

12. Remove valve lifters keeping in order of removal.

13. Remove camshaft housing.

To install:

14. Install camshaft housing to cylinder head with a new gasket. Loosely install one bolt to hold into place.

15. Install lifters into bores. Used lifters must be returned to their original position. Replace all lifters if new camshaft is being installed.

16. Prelube camshaft lobes and journals and install into same position as when removed.

17. Install new camshaft housing to camshaft housing cover seals into cover. Remove bolt holding housing into place and install cover and retaining bolts. Coat housing and cover retaining bolts with pipe sealer prior to installing. Torque bolts, in proper sequence, to 11 ft. lbs. plus an additional 75 degrees.

18. Install timing chain and housing.

19. Install exhaust camshaft housing cover with new gasket in place.

20. Install transaxle level indicator tube to exhaust camshaft cover.

21. Install electrical connection to oil pressure switch.

22. Install ignition coil and module assembly and connect electrical connector.

23. Install negative battery cable and start vehicle. Inspect for leaks.

2.5L Engine

1. Disconnect the negative battery cable.

2. Remove the rocker arm cover.

3. Remove the intake manifold.

4. Remove the pushrod cover.

5. Loosen the rocker arms and move to the side.

6. Mark and remove the pushrods, retainer and lifter guides.

7. Mark and remove the lifters.

NOTE: Mark each valve component location for reassembly.

8. Lubricate all bearing surfaces and lifters with engine oil and install the lifters.

9. Install the lifter guides, retainers and pushrods.

10. Position the rocker arms over the pushrods and tighten the rocker arm nuts to 24 ft. lbs. (32 Nm) with the lifter at the base circle of the camshaft.

11. Install the pushrod cover, intake manifold and rocker arm cover.

12. Connect the negative battery cable.

2.8L and 3.1L Engine

1. Disconnect the negative terminal from the battery.

2. Drain the cooling system.

3. Remove the rocker arm covers and intake manifold.

4. Loosen the rocker arms nuts enough to move the rocker arms to 1 side and remove the pushrods.

5. Remove the lifters from the engine.

6. Using Molykote® or equivalent, coat the base of the new lifters and install them into the engine.

7. Position the pushrods and the rocker arms correctly into their original positions. Torque the rocker arm nuts to 18 ft. lbs. (25 Nm)

8. Install the intake manifold and tighten the intake manifold-to-cylinder head bolts to specification.

9. Install the rocker cover. Connect the negative battery cable.

10. Fill the cooling system.

3.4L Engine

LEFT SIDE (FRONT)

1. Disconnect the negative battery cable.

2. Remove left side cam carrier as follows:

a. Disconnect oil/air breather hose from cam carrier cover. Remove spark plug wires from plugs and remove rear spark plug wire cover.

b. Remove cam carrier cover bolts and lift off cover. Remove gasket and O-rings from cover.

c. Remove secondary timing belt by removing secondary timing belt actuator and tensioner assembly and sliding belt from pulleys.

d. Install 6 sections of fuel line hoses under cam shaft and between lifters. This will hold lifters in the carrier. For this procedure use $\frac{3}{16}$ inch fuel line hose for exhaust valves and $\frac{5}{32}$ inch fuel line hose for the intake valves.

e. Remove exhaust crossover pipe and torque strut.

f. Remove torque strut bracket at engine.

g. Remove cam carrier mounting bolts and nuts and remove cam carrier.

h. Remove cam carrier gasket from cylinder head.

3. Remove the 6 lifter hold-down hoses. Remove the lifters.

To install

4. Lubricate lifters with clean engine oil and install lifters into original position.

5. Install lifter hold-down hoses to cam carrier.

6. Install cam carrier following these steps:

a. Install new gasket on cam carrier to cylinder mounting surface.

b. Install cam hold-down tool J-38613 or equivalent, to carrier assembly.

c. Install cam carrier to cylinder head. Install mounting bolts and nuts. Torque bolts and nuts to 18 ft. lbs.

d. Remove lifter hold-down hoses and cam hold-down tool.

e. Install torque strut bracket to engine and install torque strut.

f. Install engine crossover pipe.

g. Install secondary timing belt and cam carrier cover.

h. Reconnect spark plug cover and wires.

i. Connect breather hose to cam carrier cover.

7. Add fluids as required, reconnect negative battery cable. Start engine and recheck for leaks.

RIGHT SIDE (REAR)

1. Disconnect the negative battery cable. Drain cooling system.

2. Remove right side cam carrier as follows:

a. Remove intake plenum and right timing belt cover.

b. Remove right spark plug wires.

c. Remove air/oil separator hose at cam carrier cover.

d. Remove cam carrier cover bolts and lift of cover. Remove gasket and O-rings from cover.

e. Remove secondary timing belt by removing secondary timing belt actuator and tensioner assembly and sliding belt from pulleys.

f. Install 6 sections of fuel line hoses under cam shaft and between lifters. This will hold lifters in carrier. For this procedure use $\frac{3}{16}$ inch fuel line hose for exhaust valves and $\frac{5}{32}$ inch fuel line hose for the intake valves.

g. Remove exhaust crossover pipe and torque strut.

h. Remove torque strut bracket at engine. Remove front engine lift hook.

i. Remove cam carrier mounting bolts and nuts and remove cam carrier.

j. Remove cam carrier gasket from cylinder head.

3. Remove 6 lifter hold-down hoses.

4. Remove lifters.

To install

5. Lubricate lifters with clean engine oil and install lifters into original position.

6. Install lifter hold-down hoses to cam carrier.

7. Install cam carrier following these steps:

a. Install new gasket on cam carrier to cylinder mounting surface.

b. Install cam hold-down tool J-38613 or equivalent, to carrier assembly.

c. Install cam carrier to cylinder head. Install mounting bolts and nuts. Torque bolts and nuts to 18 ft. lbs.

d. Remove lifter hold-down hoses and cam hold-down tool.

e. Install torque strut bracket to engine and install torque strut.

f. Install engine crossover pipe and engine lift hook.

g. Install secondary timing belt and cam carrier cover.

h. Install spark plug wires and cover.

8. Add fluids as required. Connect negative battery cable. Start engine and check for fluid leaks.

3.8L Engine

1. Disconnect the negative terminal from the battery.

2. Drain the cooling system.

3. Remove the rocker arm covers and intake manifold.

4. Remove the rocker arm assemblies.

5. Remove the guide retainer bolts and retainer.

6. Remove the valve lifter guides and and valve lifters.

To install:

7. Prelube (dip) the valve lifters with oil before installation.

8. Install the lifter guides, guide retainer and bolts and torque to 27 ft. lbs.

9. Install the rocker arm assemblies, intake manifold and valve covers.

10. Fill the cooling system and connect the negative battery cable.

Valve Lash

ADJUSTMENT

All engines use hydraulic valve lifters. No adjustment is necessary.

Rocker Arms

REMOVAL & INSTALLATION

2.5L Engine

1. Disconnect the negative battery cable.

2. Remove the rocker arm cover.

3. Remove the rocker arm bolt and ball.

4. Remove the rocker arm and guide.

NOTE: Mark all valve components so they are reinstalled in their original location.

5. If removed, install the pushrod through the cylinder head and into the lifter seat.

6. Install the guide, rocker arm, ball and bolt. Tighten the rocker arm bolts to 24 ft. lbs. (32 Nm)

7. Install the rocker arm cover.

2.8L and 3.1L Engine

LEFT SIDE

1. Disconnect the negative battery cable. Remove the air cleaner assembly.

2. Remove the ignition wire clamps from coolant tube. Disconnect the bracket tube from the rocker cover.

3. Remove the spark plug wire cover. Drain the cooling system and remove the heater hose from the filler neck. Remove the coolant hose at the coolant pump and the coolant tube.

4. On 2.5L engine, it is necessary to remove the EGR valve.

5. Remove the rocker arm cover-to-cylinder head bolts and the rocker cover.

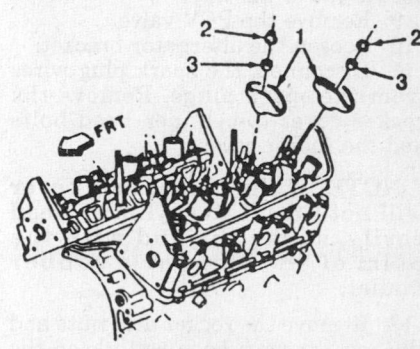

1. Rocker arm
2. 14-20 ft lbs
3. Ball

Rocker arm installation—2.8L and 3.1L engines

NOTE: If the rocker arm cover will not lift off the cylinder head easily, strike the end with the palm of the hand or a rubber mallet.

6. Remove the rocker arm nuts and remove the rocker arms, keep the components in order for installation purposes.

7. Clean the gasket mounting surfaces.

8. To install rocker arms torque the rocker arm nuts to 14–20 ft. lbs. (19–27 Nm).

9. To install new rocker cover gaskets apply a bead of sealant, GM 1052917 or equivalent, to the rocker cover and position on head.

10. Install the spark plug wire cover. Install EGR valve, if removed.

11. Attach the heater hose to the filler neck. Attach the coolant hose at the coolant pump. Fill the cooling system.

12. Install negative battery cable and air cleaner assembly. Start vehicle and check for leaks.

RIGHT SIDE

1. Disconnect the negative battery cable. Disconnect the brake booster vacuum line from the bracket.

2. Disconnect the cable bracket from the plenum. Disconnect throttle, cruise control and transaxle cable from throttle body.

3. Drain cooling system and remove coolant hose at throttle body. Remove coolant recovery tank.

4. Remove serpentine belt. Remove EGR tube at crossover pipe and disconnect crossover pipe from exhaust pipe.

5. Disconnect the vacuum line bracket from the cable bracket.

6. Disconnect the lines from the alternator brace stud.

7. Remove the rear alternator brace and the serpentine drive belt.

8. Remove the alternator and support it out of the way.

9. Remove the PCV valve.

10. Loosen the alternator bracket.

11. Disconnect the spark plug wires from the spark plugs. Remove the rocker cover-to-cylinder head bolts and the rocker cover.

NOTE: If the rocker arm cover will not lift off the cylinder head easily, strike the end with the palm of the hand or a rubber mallet.

12. Remove the rocker arm nuts and the rocker arms; be sure to keep the components in order for installation purposes.

To install:

13. Clean the gasket mounting surfaces.

14. Install rocker assembly and torque nuts to 18 ft. lbs.

15. To install, use new rocker cover gaskets apply a bead of sealant, GM 1052917 or equivalent, to the rocker cover and torque cover bolts to 89 inch lbs.

16. Install the spark plug wire cover and attach the heater hose to the filler neck. Install coolant recovery tank and hose at throttle body. Fill the cooling system.

17. Reconnect exhaust crossover pipe and exhaust pipe.

18. install serpentine belt.

19. Install throttle, transaxle and cruise control cables to throttle body and resecure cable bracket.

20. Refasten all electrical and vacuum connection.

21. Install negative battery cable and air cleaner assembly and start engine. Check for fluid leaks.

3.8L Engine

1. Disconnect the negative battery cable.

2. Remove the valve cover.

3. Remove the rocker arm pedestal retaining bolts and remove the pedestal and rocker arm assembly.

4. Remove the pushrods.

NOTE: Intake and exhaust pushrods are the same length. Store components in order so they can reassembled in the same location.

To install:

5. Install the pushrods and make sure they seat in the lifter.

6. Apply a thread lock compound to the bolt threads before reasssembly.

7. Install the pedestal and rocker arm assemblies and tighten the retaining bolts to 28 ft. lbs.

8. Install the valve covers and connect the negative battery cable.

9. Start engine and check for fliud leaks.

Intake Manifold

REMOVAL & INSTALLATION

2.3L Engine

1. Disconnect the negative battery cable.

2. Remove the coolant fan shroud, vacuum hose and electrical connector from the MAP sensor.

3. Disconnect the throttle body to air cleaner duct.

4. Remove the throttle cable bracket.

5. Remove the power brake vacuum hose, including the retaining bracket to power steering bracket and position it to the side.

6. Remove the throttle body from the intake manifold with electrical harness, coolant hoses, vacuum hoses and throttle cable attached. Position these components aside.

7. Remove the oil/air separator bolts and hoses. Leave the hoses attached to the separator, disconnect from the oil fill, chain housing and the intake manifold. Remove as an assembly.

8. Remove the oil fill cap and oil level indicator stick.

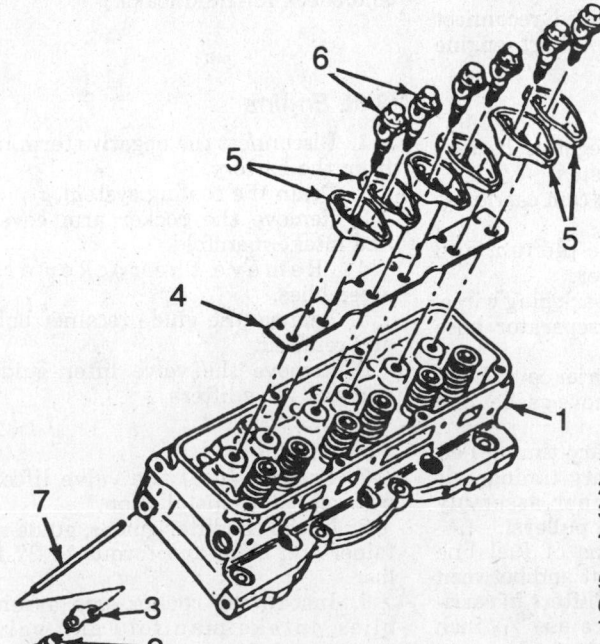

1. Cylinder head
2. Hydraulic valve lifter
3. Lifter guide
4. Valve lifter push rod guide
5. Rocker arms
6. 28 ft. lbs.
7. Pushrod

Rocker arm installation—3.8L engine

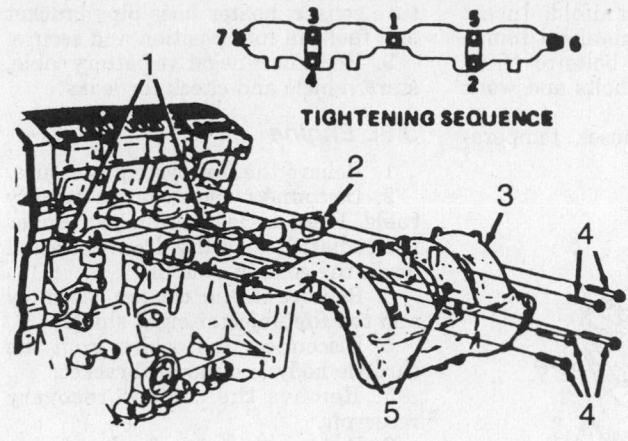

TIGHTENING SEQUENCE

1. Stud
2. Intake manifold gasket
3. Intake manifold
4. Bolt
5. Nut

Intake manifold installation—2.3L engine

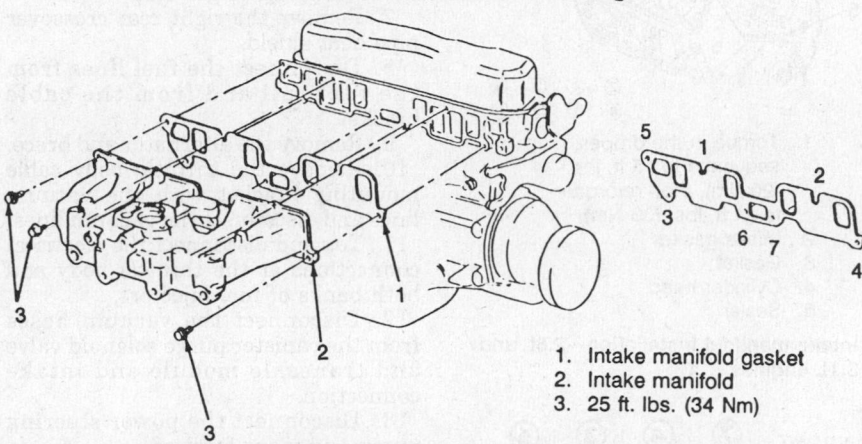

1. Intake manifold gasket
2. Intake manifold
3. 25 ft. lbs. (34 Nm)

Intake manifold installation and torque sequence—2.5L engine

9. Pull the oil tube fill upward to unseat from block and remove.

10. Disconnect the injector harness connector.

11. Remove the fill tube, rotating as necessary to gain clearance for the oil/air separator nipple between the intake tubes and fuel rail electrical harness.

12. Remove the intake manifold support bracket bolts and nut. Remove the intake manifold retaining nuts and bolts.

13. Remove the intake manifold.

To install:

14. Thoroughly clean and dry the mating surfaces. Install new gaskets and place the intake manifold in position.

15. Tighten the intake manifold bolts/nuts, in sequence, to 18 ft. lbs. (25 Nm). Tighten intake manifold brace and retainers hand tight. Tighten to specifications in the following order:

    a. Nut to stud bolt—18 ft. lbs. (25 Nm)

    b. Bolt to intake manifold—40 ft. lbs. (55 Nm)

    c. Bolt to cylinder block—40 ft. lbs. (55 Nm)

16. Lubricate a new oil fill tube ring seal with engine oil and install tube between No. 1 and 2 intake tubes. Rotate as necessary to gain clearance for oil/air separator nipple on fill tube.

17. Locate the oil fill tube in its cylinder block opening. Align the fill tube so it is approximately in its installed position. Press straight down to seat fill tube and seal into cylinder block.

18. Lubricate the hoses and install the oil/air separator assembly.

19. Install throttle body to intake manifold using a new gasket.

20. Install the power brake vacuum hose and the retaining bracket to power steering bracket.

21. Install the throttle cable bracket.

22. Connect the throttle body to air cleaner duct.

23. Install the coolant fan shroud, vacuum hose and electrical connector to the MAP sensor.

24. Fill all fluids to their proper levels.

25. Connect the negative battery cable and check for leaks.

2.5L Engine

1. Disconnect the negative battery cable.

2. Remove the air cleaner assembly.

3. Remove the PCV valve and hose at the throttle body assembly.

4. Drain the engine coolant at the radiator.

5. Release the fuel pressure and remove the fuel lines from the throttle body.

6. Remove the vacuum lines and brake booster hose from the throttle body.

7. Remove all linkage and wiring from the TBI assembly.

8. Remove the power steering pump and position aside.

9. Remove the heater hose.

10. Remove the seven intake manifold retaining bolts and the manifold.

To install:

11. Clean all gasket surfaces on the cylinder head and intake manifold.

12. Install the intake manifold with a new gasket.

13. Install all the retaining bolts and washers hand tight.

14. Tighten the bolts, in proper sequence, to 25 ft. lbs. (34 Nm)

15. Install power steering pump assembly.

16. Install all heater hoses, vacuum hoses, throttle linkages and wiring.

17. Install the fuel lines.

18. Refill the engine coolant.

19. Install the PCV valve and hose to the TBI assembly.

20. Install the air cleaner assembly and connect the negative battery cable. Check for fluid leaks.

2.8L and 3.1L Engine

1. Disconnect the negative battery cable. Drain the cooling system. Relieve fuel system pressure.

2. Disconnect the TV and accelerator cables from the plenum.

3. Disconnect the throttle body-to-plenum bolts and the throttle body. Remove the EGR valve.

4. Remove the plenum-to-intake manifold bolts and the plenum. Disconnect and plug the fuel lines and return pipes at the fuel rail.

5. Remove the serpentine drive belt. Remove the power steering pump-to-bracket bolts and support the pump out of the way; do not disconnect the pressure hoses.

6. Remove the alternator-to-bracket bolts and support the alternator out of the way.

7. Loosen the alternator bracket. From the throttle body, disconnect the idle air vacuum hose.

8. Label and disconnect the electrical connectors from the fuel injectors. Remove the fuel rail.

9. Remove the breather tube. Disconnect the runners.

10. Remove both rocker arm cover-to-cylinder head bolts and the covers.

Remove the radiator hose from the thermostat housing.

11. Label and disconnect the electrical connectors from the coolant temperature sensor and oil pressure sending unit. Remove the coolant sensor.

12. Remove the bypass hose from the filler neck and cylinder head. Remove top radiator hose.

13. Remove the intake manifold-to-cylinder head bolts and the manifold.

14. Loosen the rocker arm nuts, turn them 90 degrees and remove the pushrods; be sure to keep the components in order for installation purposes.

15. Clean all of the gasket mounting surfaces.

To install:

16. Place a bead of RTV sealer or equivalent on each ridge where the intake manifold and block meet. Install the intake manifold gasket in place on the block.

17. Install the pushrods and reposition the rocker arms, tighten the rocker arm nuts to 18 ft. lbs. (25 Nm).

18. Mount the intake manifold on the engine and tighten the bolts to 23 ft. lbs. (29 Nm)

19. Connect the heater inlet pipe to the manifold. Install and connect the coolant sensor.

20. Attach the radiator hoses. Connect the wire at the oil sending switch.

21. Install the rocker covers, tighten the retaining bolts to 90 inch lbs. (10 Nm)

22. Install the runners, breather tube, fuel rail and connect the wires at the fuel injectors.

23. Install the alternator bracket and the alternator. Install the power steering pump.

24. Connect the fuel lines to the fuel rail. Install the EGR valve.

25. Install the plenum and mount the throttle body to the plenum.

26. Connect the accelerator cable and the TV cable.

27. Fill the cooling system. Connect the negative battery cable.

28. Run the engine until it reaches normal operating temperature and check for coolant and oil leaks.

3.4L Engine

1. Relieve fuel system pressure and disconnect negative battery cable.

2. Remove fuel rail.

3. Remove heater hose pipe bracket at thermostat housing. Drain cooling system.

4. Remove temperature sensor from intake manifold. Remove radiator hose from thermostat housing.

5. Remove mounting bolts and manofold.

To install:

6. Clean mating surfaces and install new gaskets.

7. Install intake manifold. Insert rubber isolators into manifold flange and tighten mounting bolts to 18 ft. lbs. Start with center bolts and work in a circular pattern.

8. Install radiator hoses, tempera-

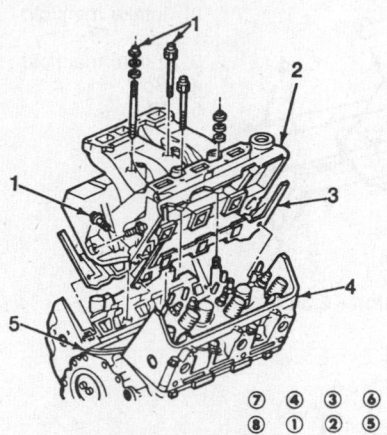

1. Torque in the proper sequence to 15 ft. lbs. (20 Nm), then retorque to 24 ft. lbs. (33 Nm)
2. Intake gasket
3. Gasket
4. Cylinder head
5. Sealer

Intake manifold installation—2.8L and 3.1L engines

FRONT ⑦ ④ ③ ⑥
⑧ ① ② ⑤

Intake manifold torque sequence—2.8L and 3.1 engines

ture sensor, heater hose pipe bracket and fuel rail into position and secure.

9. Reconnect negative battery cable, start vehicle and check for leaks.

3.8L Engine

1. Relieve the fuel system pressure.

2. Disconnect the negative battery cable. Place a clean drain pan under the radiator, open the drain cock and drain the cooling system.

3. Remove the air cleaner assembly and the fuel injector sight shield.

4. Disconnect the cables from the throttle body and mount bracket.

5. Remove the coolant recovery reservoir.

6. Remove the inner fender electrical cover on the right side.

7. Remove the right rear crossover pipe heat shield.

8. Disconnect the fuel lines from the fuel rail and from the cable bracket.

9. Remove the alternator and brace.

10. Remove the throttle body cable mounting bracket with the vacuum lines and disconnect the vacuum lines.

11. Tag and disconnect the electrical connections at the throttle body and both banks of fuel injectors.

12. Disconnect the vacuum hoses from the canister purge solenoid valve and transaxle module and intake connection.

13. Disconnect the power steering pump and move forward.

14. Disconnect the spark plug wires and lay aside.

15. Disconnect the coolant bypass hose from the intake manifold.

16. Disconnect the solenoid valve

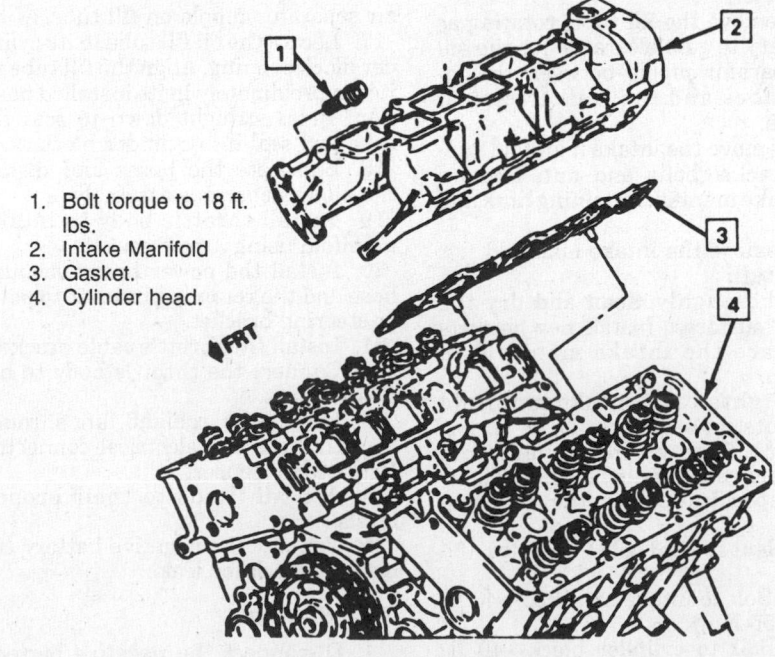

1. Bolt torque to 18 ft. lbs.
2. Intake Manifold
3. Gasket.
4. Cylinder head.

Intake manifold—3.4L engine

mounting bracket and power steering support brace from the intake manifold.

17. Disconnect the heater pipes from the intake and front cover.

18. Disconnect the alternator support brace from the intake.

19. Disconnect the upper radiator hose from the housing.

20. Remove the thermostat housing and thermostat from the intake.

21. Disconnect the electrical connector from the temperature sensor and sensor switch.

22. Remove the intake manifold bolts and manifold as an assembly.

To install:

23. Clean the mating surfaces and install the intake manifold gaskets and seals. Apply sealer to the ends of the of the intake manifold seals.

24. Install the intake manifold and apply thread lock compound to the bolt threads and torque the bolts to 88 inch lbs., twice in sequence.

25. Connect the electrical connector to the temperature sensor and sensor switch.

26. Install the thermostat housing and thermostat with a new gasket.

27. Connect the alternator support brace to the intake.

28. Connect the solenoid valve mounting bracket and power steering support brace to the intake manifold.

29. Connect the heater pipes to the intake and front cover.

30. Connect the coolant bypass hose to the intake manifold.

31. Install the power steering pump support bracket and torque to 37 ft. lbs.

32. Install the spark plug wires on both sides.

33. Install the belt tensioner pulley and tighten to 33 ft. lbs.

34. Install the power steering pump.

35. Connect the vacuum hoses to the canister purge solenoid valve and transaxle module and intake connection.

36. Connect the electrical connections at the throttle body and both banks of fuel injectors.

37. Install the alternator and brace.

38. Connect the throttle body cable mounting bracket with the vacuum lines.

39. Install the right rear crossover pipe heat shield.

40. Install the cables to the throttle body.

41. Connect the fuel lines to the fuel rail and mount bracket.

42. Install the inner fender electrical cover on the right side.

43. Install the coolant recovery reservoir and upper radiator hose. Fill the cooling system.

44. Install the air cleaner assembly and the fuel injector sight shield.

45. Connect the negative battery cable.

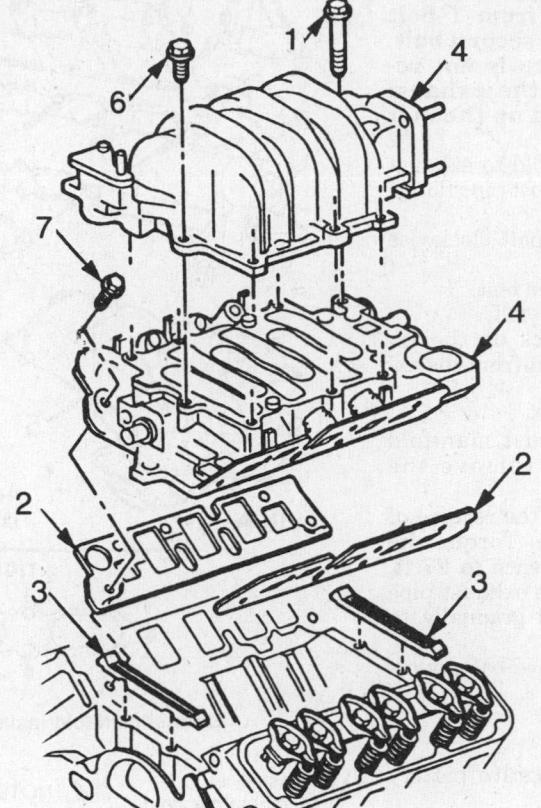

1. 19 ft. lbs.
2. Intake manifold gasket
3. Intake manifold seal
4. Lower intake manifold
5. Upper intake manifold
6. 19 ft. lbs.
7. 88 inch lbs.

Intake manifold installation—3.8L engine

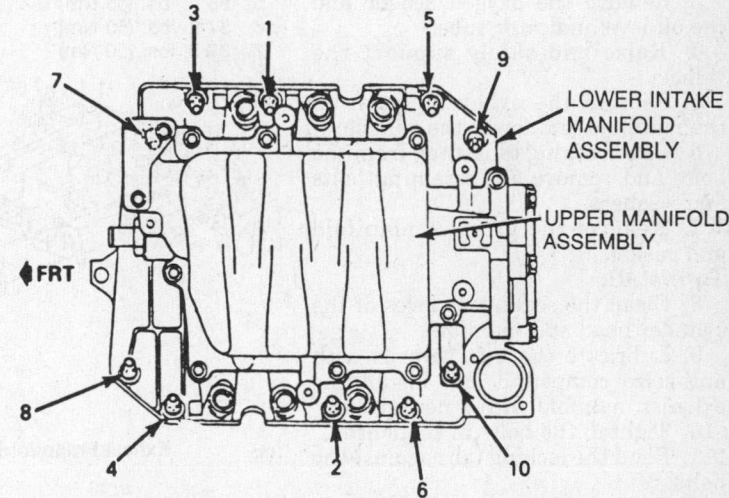

LOWER INTAKE MANIFOLD ASSEMBLY

UPPER MANIFOLD ASSEMBLY

◀FRT

Intake manifold torque sequence—3.8L engine

Exhaust Manifold

REMOVAL & INSTALLATION

2.3L Engine

1. Disconnect the negative battery cable and oxygen sensor connector.

2. Remove upper and lower exhaust manifold heat shields.

3. Remove the bolt that attaches the exhaust manifold brace to the manifold.

4. Break loose the manifold to exhaust pipe spring loaded bolts using a 13mm box wrench.

5. Raise the vehicle and support safely.

NOTE: It is necessary to relieve the spring pressure from 1 bolt prior to removing the second bolt. If the spring pressure is not relieved it will cause the exhaust pipe to twist and bind up the bolt as it is removed.

6. Remove the manifold to exhaust pipe bolts from the exhaust pipe flange as follows:

    a. Unscrew either bolt clockwise 4 turns.

    b. Remove the other bolt.

    c. Remove the first bolt.

7. Pull down and back on the exhaust pipe to disengage it from the exhaust manifold bolts.

8. Lower the vehicle.

9. Remove the exhaust manifold mounting bolts and remove the manifold.

10. The installation is the reverse of the removal procedure. Torque the mounting bolts in sequence to 27 ft. lbs. (37 Nm). Install the exhaust pipe flange bolts evenly and gradually to avoid binding.

11. Connect the negative battery cable and check for leaks.

2.5L Engine

1. Disconnect the negative battery cable.

2. Remove the torque strut bolts at the radiator panel and cylinder head.

3. Remove the oxygen sensor and the oil level indicator tube.

4. Raise and safely support the vehicle.

5. Remove the exhaust pipe from the manifold and lower the vehicle.

6. Bend rocking tabs away from the bolts and remove the retaining bolts and washers.

7. Remove the exhaust manifold and gasket.

To install:

8. Clean the sealing surfaces of the cylinder head and manifold.

9. Lubricate the bolt threads with anti-seize compound and install the exhaust manifold with a new gasket.

10. Tighten the bolts in sequence.

11. Bend the locking tabs against the bolts.

12. Raise and support the vehicle safely.

13. Install the exhaust pipe to the manifold and lower the vehicle.

14. Install the oil level indicator tube, oxygen sensor and torque rod bracket at the cylinder head and radiator support.

15. Connect the negative battery cable.

2.8L and 3.1L Engine

LEFT SIDE

1. Disconnect the negative battery cable.

---

1. Exhaust manifold to cylinder head stud
2. Manifold assembly (VIN D)
3. Nut
4. Gasket
5. Manifold assembly (VIN A)

(HEAT SHIELD REMOVED FOR ILLUSTRATION PURPOSES)

TIGHTENING SEQUENCE

Exhaust manifold installation and torque sequence—2.3L engine

1. Gasket
2. Exhaust manifold
3. Lock
4. 26 ft. lbs. (35 Nm)
5. 26 ft. lbs. (35 Nm)
6. 37 ft. lbs. (50 Nm)
7. 37 ft. lbs. (50 Nm)
8. NOTE: When installing the lock tabs on the exhaust manifold, one tab must be bent against a flat of the hex to prevent rotation.

BOLT TIGHTENING SEQUENCE
TIGHTEN BOLT POSITION NUMBER IN SEQUENCE AS FOLLOWS: 3-5-6-2-1-7-4 OR BY USING ALPHA GROUPS "A" AND "B". "A" BEING FIRST AND "B" LAST. OR SIMULTANEOUS GANG DRIVE.

VIEW A

VIEW B

Exhaust manifold installation and torque sequence—2.5L engine

---

2. Remove the coolant recovery bottle.

3. Relieve the accessory drive belt tension and remove the belt.

4. Remove the air conditioner compressor mounting bolts and support the compressor aside.

5. Remove the right side engine torque strut. Remove the bolts retaining the air conditioner compressor and torque strut mounting bracket, remove the bracket.

6. Remove the heat shield and crossover pipe at the manifold.

7. Remove the exhaust manifold mounting bolts and remove the manifold.

To install:

8. Clean the gasket mounting surfaces.

9. Install the exhaust manifold to the engine, loosely install the mounting bolts.

10. Install the exhaust crossover pipe. Tighten the exhaust manifold bolts to 18 ft. lbs. (25 Nm)

11. Attach the heat shield. Install the air conditioner and torque strut mounting bracket.

12. Install the torque strut. Mount

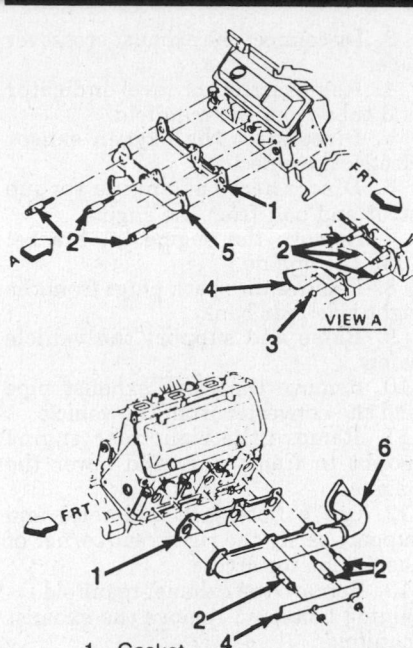

1. Gasket
2. 18 ft. lbs.
3. 90 inch lbs.
4. Heat shield
5. Right exhaust manifold
6. Left exhaust manifold

Exhaust manifold installation—2.8L and 3.1L engines

the air conditioner compressor and install the accessory drive belt.

13. Install the coolant recovery bottle and connect the negative battery cable.

RIGHT SIDE

1. Disconnect the negative battery cable.
2. Raise and safely support the vehicle.
3. Remove the exhaust pipe at the crossover. Lower the vehicle.
4. Remove the coolant recovery bottle and remove the engine torque struts.
5. Pull the engine forward and support it.
6. Remove the air cleaner, breather, mass air flow sensor and heat shield.
7. Remove the crossover at the manifold. Disconnect the accelerator and TV cables.
8. Remove the manifold mounting bolts and remove the manifold. Clean the manifold mounting surfaces.

To install:

9. Install the exhaust manifold, loosely install the mounting bolts.
10. Attach the crossover at the manifold. Tighten the manifold mounting bolts to 18 ft. lbs. (25 Nm).
11. Connect the accelerator and TV cables.
12. Attach the air cleaner, breather and mass air flow sensor.

13. Remove the engine support and allow the engine to roll back into position.
14. Install the coolant recovery bottle and the engine torque struts.
15. Raise and safely support the vehicle. Install the exhaust pipe to the crossover.
16. Lower the vehicle. Connect the negative battery cable.

3.4L Engine

LEFT SIDE (FRONT)

1. Remove air cleaner assembly. Disconnect the negative battery cable.
2. Remove exhaust crossover .
3. Remove the engine torque strut bracket at frame and position out of the way.
4. Remove upper radiator shroud. Remove right side cooling fan.
5. Remove front hose from air pipe for manual transaxle only.
6. Remove exhaust retaining nuts and manifold. Remove old gasket and disgard.
7. To install install gasket, heat shields and manifold.
8. Install manifold nuts and torque to 115 inch lbs.
9. Install cooling fan, radiator shroud and torque strut into position and secure.
10. Install exhaust crossover. Install negative battery cable and air cleaner assembly.

RIGHT SIDE (REAR) WITH AUTOMATIC TRANSAXLE

1. Disconnect the negative battery cable.
2. Remove right side cam carrier as follows:
   a. Remove intake plenum and right timing belt cover.
   b. Remove right spark plug wires.

   c. Remove air/oil separator hose at cam carrier cover.
   d. Remove cam carrier cover bolts and lift of cover. Remove gasket and O-rings from cover.
   e. Remove secondary timing belt by removing secondary timing belt actuator and tensioner assembly and sliding belt from pulleys.
   f. Install 6 sections of fuel line hoses under cam shaft and between lifters. This will hold lifters in carrier. For this procedure use $\frac{3}{16}$ inch fuel line hose for exhaust valves and $\frac{5}{32}$ inch fuel line hose for the intake valves.
   g. Remove exhaust crossover pipe and torque strut.
   h. Remove torque strut bracket at engine. Remove front engine lift hook.
   i. Remove cam carrier mounting bolts and nuts and remove cam carrier.
   j. Remove cam carrier gasket from cylinder head.
3. Remove exhaust manifold to crossover pipe nuts.
4. Raise and safely support vehicle.
5. Remove front exhaust pipe at manifold. Lower vehicle.
6. Remove electrical connector from oxygen sensor.
7. Remove exhaust manifold nuts, heat shield and manifold.

To install:

8. Clean all mating surfaces, install manifold gasket and heat shields.
9. Install exhaust manifold. Torque nuts to 116 inch lbs.
10. Install electrical connector at oxygen sensor.
11. Raise and safely support vehicle.
12. Install exhaust pipe at manifold. Lower vehicle.
13. Install exhaust crossover pipe.

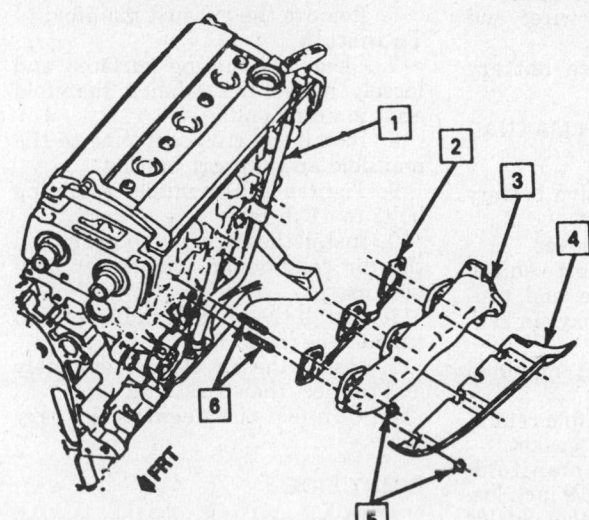

1. Oil level indicator.
2. Gasket.
3. Left exhaust manifold.
4. Heat shield.
5. Nuts torqued to 115 inch lbs.
6. Studs torqued to 13 ft. lbs.

Left side exhaust manifold

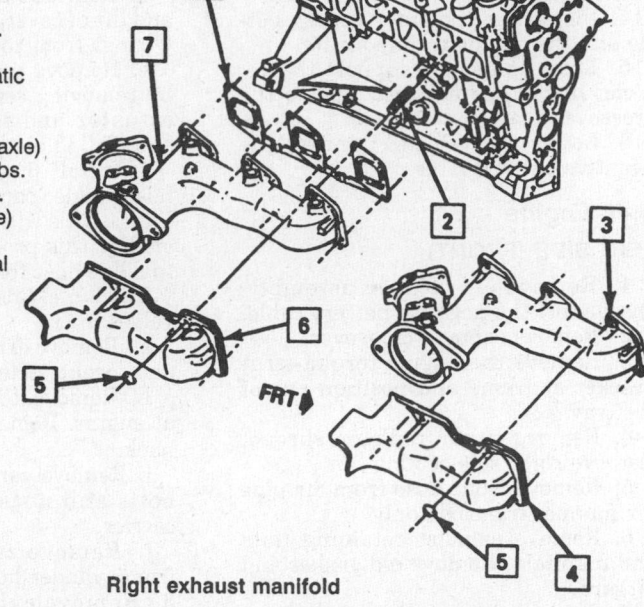

1. Gasket.
2. Stud to 13 ft. lbs.
3. Right exhaust manifold.(automatic transaxle)
4. Right heat shield .(automatic transaxle)
5. Nut to 116 inch lbs.
6. Right heat shield. (manual transaxle)
7. Right exhaust manifold. (manual transaxle)

Right exhaust manifold

14. Install right cam carrier as follows:

a. Install new gasket on cam carrier to cylinder mounting surface.

b. Install cam hold-down tool J–38613 or equivalent, to carrier assembly.

c. Install cam carrier to cylinder head. Install mounting bolts and nuts. Torque bolts and nuts to 18 ft. lbs.

d. Remove lifter hold-down hoses and cam hold-down tool.

e. Install torque strut bracket to engine and install torque strut.

f. Install engine crossover pipe and engine lift hook.

g. Install secondary timing belt and cam carrier cover and gasket.

h. Install spark plug wires and cover.

15. Reconnect negative battery cable.

RIGHT SIDE (REAR) WITH MANUAL TRANSAXLE

1. Disconnect the negative battery cable.

2. Remove exhaust crossover.

3. Raise and safely support vehicle.

4. Remove exhaust pipe and converter assembly. Remove oxygen sensor connector.

5. Remove EGR pipe at manifold and manifold heat shields.

6. Remove exhaust manifold retaining nuts and manifold and gasket.

7. Install gasket and manifold. Torque retaining nuts to 116 inch lbs.

8. Install EGR pipe and heat shields to exhaust manifold.

9. Install electrical connector at oxygen sensor.

10. Install exhaust pipe to manifold and lower vehicle.

11. Install exhaust crossover and negative battery cable.

3.8L Engine

LEFT SIDE

1. Disconnect the negative battery cable.

2. Remove the air cleaner assembly and disconnect the spark plug wires.

3. Disconnect the exhaust crossover pipe.

4. Remove the oil level indicator and tube from the manifold.

5. Disconnect the engine lift bracket and the air conditioner compressor support brace.

6. Remove the exhaust manifold.

To install:

7. Clean the mating surfaces and loosely install the exhaust manifold and retaining bolts.

8. Install the crossover pipe to the manifold and support bracket.

9. Tighten the manifold retaining bolts to 41 ft. lbs.

10. Install the engine lift bracket and the air conditioner compressor support brace.

11. Install the oil level indicator and tube to the manifold.

12. Install the air cleaner assembly and connect the spark plug wires.

13. Connect the negative battery cable.

RIGHT SIDE

1. Disconnect the negative battery cable.

2. Remove the air cleaner assembly and disconnect the spark plug wires.

3. Disconnect the exhaust crossover pipe.

4. Remove the oil level indicator and tube from the manifold.

5. Disconnect the oxygen sensor electrical connector.

6. Disconnect the engine torque strut and bolt from the engine.

7. Remove the engine lift bracket from the engine.

8. Remove the spark plugs from the right side rear bank.

9. Raise and support the vehicle safely.

10. Remove the front exhaust pipe and the converter from the vehicle.

11. Remove the right rear engine mount to frame nuts and lower the engine.

12. Use a floor jack and raise and support safely the right rear corner of the engine for access.

13. Remove the exhaust manifold retaining bolts and remove the exhaust manifold.

To install:

14. Clean the mating surfaces and loosely install the exhaust manifold and retaining bolts.

15. Install the crossover pipe to the manifold and support bracket.

16. Tighten the manifold retaining bolts to 41 ft. lbs.

17. Lower the engine and remove the floor jack.

18. Raise and support the vehicle safely.

19. Install the front exhaust pipe and the converter.

20. Install the right rear engine mount to frame nuts and lower the engine.

21. Tighten the crossover bolts.

22. Install the spark plugs to the right side rear bank.

23. Install the engine lift bracket to the engine.

24. Connect the oxygen sensor electrical connector.

25. Connect the engine torque strut and bolt to the engine and torque to 41 ft. lbs.

26. Install the oil level indicator and tube to the manifold.

27. Install the air cleaner assembly and connect the spark plug wires.

28. Connect the negative battery cable.

Timing Chain/Gear Front Cover

REMOVAL & INSTALLATION

2.3L Engine

1. Disconnect the negative battery cable. Remove the coolant recovery reservoir.

2. Remove the serpentine drive belt

using a 13mm wrench that is at least 24 in. long.

3. Remove upper cover fasteners.

4. Raise the vehicle and support safely.

5. Remove the right front wheel assembly and lower splash shield.

6. Remove the crankshaft balancer assembly.

7. Remove lower cover fasteners and lower the vehicle.

8. Remove the front cover.

9. The installation is the reverse of the removal procedure. Torque the balancer retaining bolt to 74 ft. lbs. (100 Nm) plus an additional 90 degree turn.

2.5L Engine

1. Disconnect the negative battery cable.

2. Remove the torque strut bolt at the cylinder head bracket and move the strut out of the way.

3. Remove the serpentine belt.

4. Install the engine support fixture tool J–28467–A and J–36462.

5. Raise and safely support the vehicle.

6. Remove the right front tire assembly.

7. Disconnect the right lower ball joint from the knuckle.

8. Remove the 2 right frame attaching bolts.

9. Loosen the 2 left frame attaching bolts but do not remove.

10. Lower the vehicle.

11. Lower the engine on the right side. Raise and safely support the vehicle.

12. Remove the engine vibration dampener using a dampener puller.

13. Remove the timing cover retaining bolts and cover.

To install:

14. Clean all gasket mating surfaces with solvent and a gasket scraper.

15. Apply a 3/8 in. wide by $^3/_{16}$ in. thick bead of RTV sealer to the joint at the oil pan and timing cover.

16. Apply a ¼ in. wide by ⅛ in. thick bead of RTV sealer to the timing cover at the block mating surface.

17. Install a new timing cover oil seal using a timing cover seal installer tool J–34995 or equivalent.

18. Install the cover onto the block and install the retaining bolts loosely.

19. Install the timing cover seal installer tool J–34995 to align the timing cover.

20. Tighten the opposing bolts to hold the cover in place.

21. Torque the bolts in sequence and to the proper specification. Remove the timing cover oil seal installer tool.

22. Install the crankshaft vibration dampener and torque the bolt to 162 ft. lbs. (220 Nm)

23. Lower the vehicle.

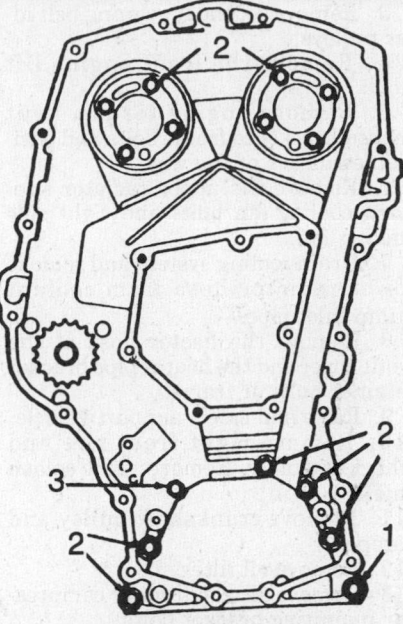

1. Stud end bolt (chain housing to block)
2. Bolt (chain housing to block and cam housing)
3. Stud (timing chain tensioner shoe pivot)

Timing chain cover installation—2.3L engine

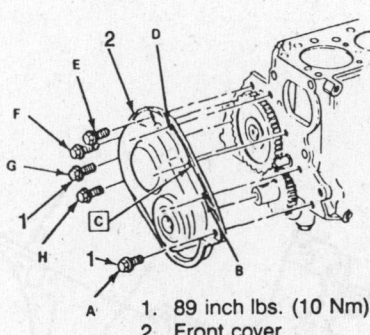

1. 89 inch lbs. (10 Nm)
2. Front cover

Timing case cover assembly—2.5L engine

24. Raise the engine to its proper position using the support fixture.

25. Raise and safely support the vehicle.

26. Raise the frame and install the removed frame bolts. Torque the bolts to 103 ft. lbs. (140 Nm)

27. Install the right ball joint and tighten the nut.

28. Install the right front tire, torque the lug nuts to 100 ft. lbs. (136 Nm) and lower the vehicle.

29. Remove the engine support fixture.

30. Install the torque strut and bolt to the cylinder head bracket.

31. Install the serpentine belt, con-

nect the negative battery cable and check for oil leaks.

2.8L and 3.1L Engine

1. Disconnect the negative terminal from the battery. Drain the cooling system.

2. Remove the serpentine belt and the belt tensioner.

3. Remove the alternator-to-bracket bolts and remove the alternator, with the wires attached, support it out of the way.

4. Remove the power steering pump-to-bracket bolts and support it out of the way. Do not disconnect the pressure hoses.

5. Raise and safely support the vehicle.

6. Remove the right side inner fender splash shield. Remove the flywheel dust cover.

7. Using a crankshaft pulley puller tool. Remove the crankshaft damper.

8. Label and disconnect the starter wires, remove the starter.

9. Drain the engine oil and remove the oil pan. Remove the lower front cover bolts.

10. Lower the vehicle. Disconnect the radiator hose from the water pump.

11. Disconnect the heater coolant hose from the cooling system filler pipe.

12. Remove the bypass and overflow hoses.

13. Remove the water pump pulley. Disconnect the canister purge hose.

14. Remove the spark plug wire shield from the water pump.

15. Remove the upper front cover-to-engine bolts and remove the front cover.

16. Clean front cover mounting surfaces.

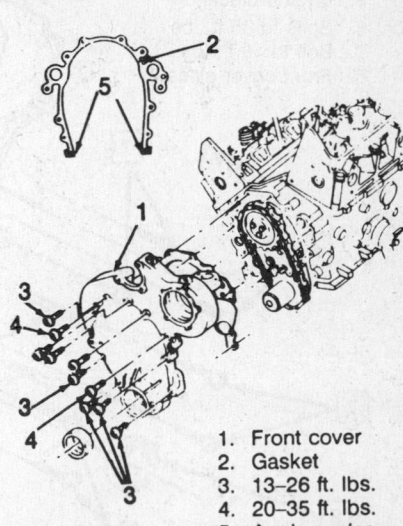

1. Front cover
2. Gasket
3. 13–26 ft. lbs.
4. 20–35 ft. lbs.
5. Apply sealer

Timing case cover assembly—2.8L and 3.1L engines

To install:

17. Apply a thin bead of silicone sealant on the front cover mating surface and using a new gasket, install the front cover on the engine with the top bolts to hold it in place.

18. Raise and safely support the vehicle.

19. Install the oil pan. Install the lower front cover bolts, tighten all of the front cover bolts to 26–35 ft. lbs. (35–48 Nm)

20. Install the serpentine belt and idler pulley. Install the damper on the engine using tool J–29113 or equivalent. Install the starter.

21. Install the inner fender splash shield. Lower the vehicle.

22. Attach the radiator hose too the water pump and attach the heater hoses.

23. Install the power steering pump and the alternator.

24. Attach the spark plug wire shield. Fill the cooling system.

25. Connect the negative battery cable. Check for coolant and oil leaks.

3.4L Engine

1. Disconnect the negative battery cable.

2. Remove secondary timing belt tensioner mounting bracket and gasket by removing tensioner pulley and mounting bracket bolts.

3. Remove secondary timing belt idler pulleys.

4. Remove the front engine lift hook.

5. Remove engine torque strut mount bracket to frame bolts and position strut out of the way.

6. Remove the upper radiator support, cooling fan bolts and right side cooling fan.

7. Drain cooling system and remove lower radiator hose from coolant pump inlet pipe.

8. Remove the heater hose at the front cover and the heater pipe bracket retainer bolts at frame.

9. Raise and safely support vehicle.

10. Remove right front tire and wheel assembly. Remove right splash shield.

11. Remove crankshaft pulley and damper.

12. Remove oil filter.

13. Remove air conditioner compressor mounting bracket bolts.

14. Remove lower front cover bolts.

15. On automatic transaxle vehicles remove the halfshaft.

16. Remove the rear alternator bracket and starter motor. Lower vehicle.

17. Remove the camshaft drive belt sprocket retaining bolt and extract camshaft drive belt sprocket using tool J–38616 or equivalent.

18. Remove upper alternator retaining bolts. Remove the forward light relay center screws and position relay center aside.

19. Remove oil cooler hose at front cover and coolant pump pulley.

20. Remove upper front cover bolts and front cover. Remove the old gasket and clean mating surfaces of front cover and block.

To install:

21. Apply GM sealer 1052080 or equivalent, to lower edges of the sealing surface of the front cover and install. Apply thread sealant to large bolts and tighten cover into place.

22. Install coolant pump pulley. Install oil cooler coolant hose to front cover.

23. Install forward light relay center and upper alternator retaining bolts.

24. Install camshaft drive belt sprocket and retaining bolt.

25. Raise and safely support vehicle.

26. Install starter motor.

27. Reinstall halfshaft and rear alternator bracket.

28. Install lower front cover bolts. Tighten lower cover bolts to 18 ft. lbs. Install air conditioning compressor mounting bolts.

29. Install oil filter, crankshaft damper and crankshaft pulley.

30. Install right side splash shield and wheel assembly. Lower vehicle.

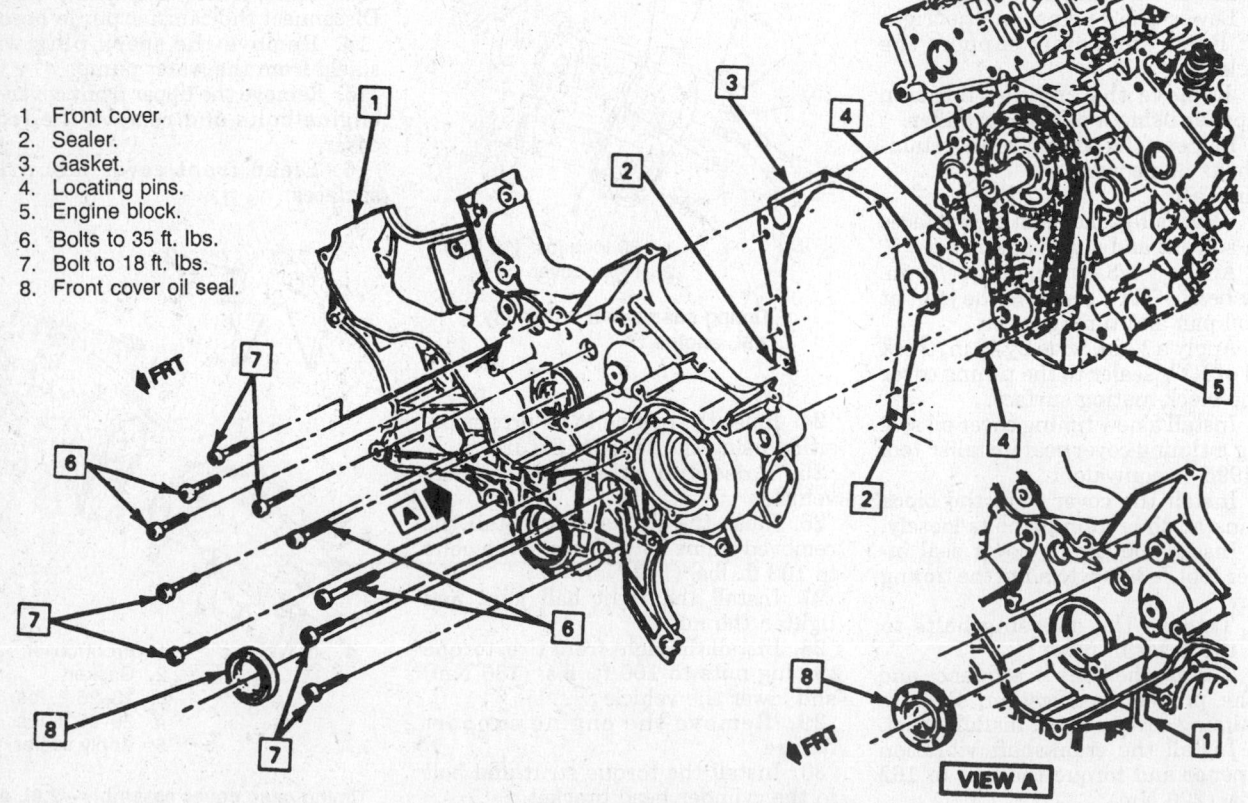

1. Front cover.
2. Sealer.
3. Gasket.
4. Locating pins.
5. Engine block.
6. Bolts to 35 ft. lbs.
7. Bolt to 18 ft. lbs.
8. Front cover oil seal.

FRT

A

VIEW A

Front cover and oil seal 3.4L—engine

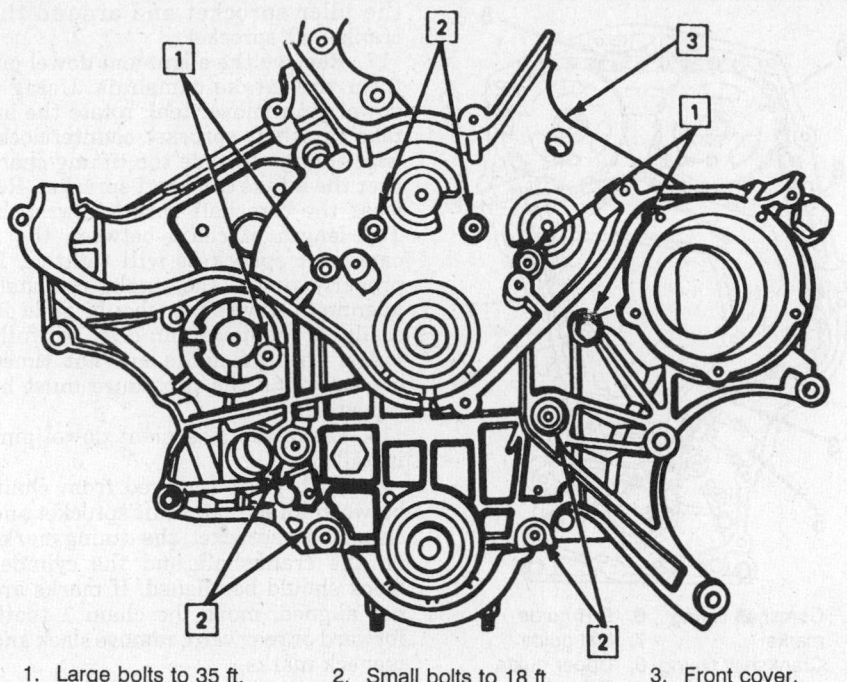

1. Large bolts to 35 ft. lbs.
2. Small bolts to 18 ft. lbs.
3. Front cover.

Front cover bolt locations—3.4L engine

31. Tighten upper front cover small bolts to 18 ft. lbs., front cover large bolts to 35 ft. lbs.

32. Install heater hoses at front cover, lower radiator hose to coolant pump and coolant fan. Add coolant to correct level.

33. Install retainer screws into heater pipe bracket.

34. Install upper radiator support and torque strut to frame bolts.

35. Install front engine lift hook and secondary timing belt idler pulley.

36. Install secondary timing belt tensioner mounting bracket tightening bolts to 37 ft. lbs.

37. Reconnect negative battery cable.

3.8L Engine

1. Disconnect the negative battery cable.

2. Remove the crankshaft balancer.

3. Remove the crankshaft sensor cover.

4. Disconnect the electrical connections at the camshaft, crankshaft and oil pressure sensors.

5. Raise and support the vehicle safely.

6. Drain the engine oil and remove the oil pan to front cover bolts.

7. Remove the oil filter and disconnect the oil cooler pipes from the oil filter adapter housing.

8. Lower the vehicle and drain the cooling system.

9. Remove the alternator and brace.

10. Disconnect the heater hoses and pipe and the bypass hose from the cover.

11. Disconnect the lower radiator hose.

12. Remove the coolant pump pulley.

13. Remove the front cover attaching bolts and cover with the oil filter adapter as an assembly.

14. Remove the oil filter adapter housing.

15. Remove the oil pressure valve, spring and oil pump from the front cover.

16. Remove the coolant pump from the front cover.

17. Pry the oil seal out of the cover using a suitable tool.

To install:

NOTE: The oil pan bolts can be loosened and the pan dropped slightly for front cover clearance. If the oil pan gasket is excessively swollen, the oil pan must be removed and the gasket replaced.

18. Clean the mating surfaces of the front cover and cylinder block with a degreaser.

19. Install the oil filter and adapter housing with the oil pressure valve and spring to the cover. Tighten the bolts to 24 ft. lbs.

20. Install the oil pump assembly to the cover.

21. Use a new gasket, apply sealer to the bolt threads and install the coolant pump to the front cover.

22. Lubricate a new front cover oil seal with clean engine oil and install it to the front cover, using tool J–35354 or equivalent. Use the crankshaft balancer bolt with the tool and tighten the bolt until the seal is seated in the cover. Remove the tool.

23. Install the front cover to the engine and install the upper cover bolts. Tighten the upper cover bolts to 124 inch lbs. (14 Nm)

24. Install the crankshaft sensor and adjust, using tool J–37089 or equivalant.

25. Install the sensor cover and electrical connections.

26. Install the crankshaft balancer.

27. Install the oil cooler lines and the oil filter.

28. Lower the vehicle and install the coolant pump pulley.

29. Install the lower radiator hose, bypass hose and heater hoses.

30. Install the alternator and brace.

31. Add engine coolant, oil and connect the negative battery cable.

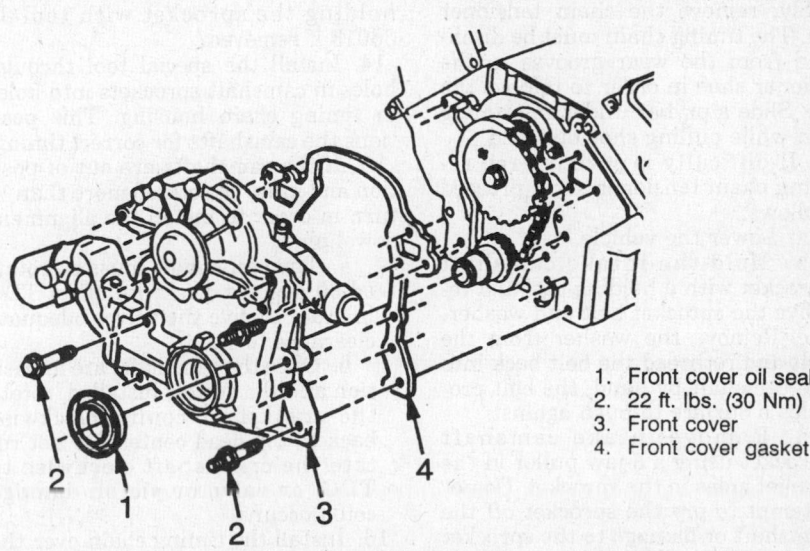

1. Front cover oil seal
2. 22 ft. lbs. (30 Nm)
3. Front cover
4. Front cover gasket

Timing chain cover installation—3.8L engine

Timing Chain and Sprockets

REMOVAL & INSTALLATION

2.3L Engine

NOTE: It is recommended that the entire procedure be reviewed before attempting to service the timing chain.

1. Disconnect the negative battery cable.
2. Remove the front timing chain cover and crankshaft oil slinger.
3. Rotate the crankshaft clockwise, as viewed from front of engine (normal rotation) until the camshaft sprocket's timing dowel pin holes line up with the holes in the timing chain housing. The mark on the crankshaft sprocket should line up with the mark on the cylinder block. The crankshaft sprocket keyway should point upwards and line up with the centerline of the cylinder bores. This is the normal timed position.
4. Remove the 3 timing chain guides.
5. Raise the vehicle and support safely.
6. Gently pry off timing chain tensioner spring retainer and remove spring.

NOTE: Two styles of tensioner are used. Early production engines will have a spring post and late production ones will not. Both styles are identical in operation and are interchangeable.

7. Remove the timing chain tensioner shoe retainer.
8. Make sure all the slack in the timing chain is above the tensioner assembly; remove the chain tensioner shoe. The timing chain must be disengaged from the wear grooves in the tensioner shoe in order to remove the shoe. Slide a prybar under the timing chain while pulling shoe outward.
9. If difficulty is encountered removing chain tensioner shoe, proceed as follows:
    a. Lower the vehicle.
    b. Hold the intake camshaft sprocket with a holding tool and remove the sprocket bolt and washer.
    c. Remove the washer from the bolt and rethread the bolt back into the camshaft by hand, the bolt provides a surface to push against.
    d. Remove intake camshaft sprocket using a 3-jaw puller in the 3 relief holes in the sprocket. Do not attempt to pry the sprocket off the camshaft or damage to the sprocket or chain housing could occur.

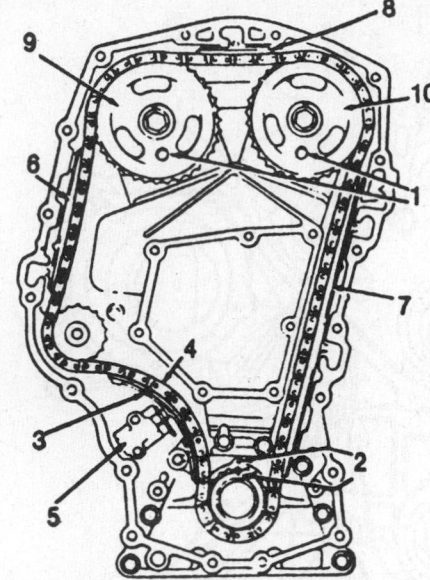

1. Camshaft timing marks
2. Crankshaft timing mark
3. Tensioner shoe assembly
4. Timing chain
5. Tensioner
6. R/H guide
7. L/H guide
8. Upper guide
9. Exhaust camshaft sprocket
10. Intake camshaft sprocket

Timing chain Installation—2.3L engine

10. Remove the tensioner assembly retaining bolts and the tensioner.

— CAUTION —

The tensioner piston is spring loaded and could fly out causing personal injury.

11. Remove the chain housing to block stud (timing chain tensioner shoe pivot).
12. Remove the timing chain.
To install:
13. Tighten intake camshaft sprocket retaining bolt and washer, while holding the sprocket with tool J–36013 if removed.
14. Install the special tool through holes in camshaft sprockets into holes in timing chain housing. This positions the camshafts for correct timing.
15. If the camshafts are out of position and must be rotated more than 1/8 turn in order to install the alignment dowel pins:
    a. The crankshaft must be rotated 90 degrees clockwise off of TDC in order to give the valves adequate clearance to open.
    b. Once the camshafts are in position and the dowels installed, rotate the crankshaft counterclockwise back to top dead center. Do not rotate the crankshaft clockwise to TDC, or valve or piston damage could occur.
16. Install the timing chain over the exhaust camshaft sprocket, around

the idler sprocket and around the crankshaft sprocket.

17. Remove the alignment dowel pin from the intake camshaft. Using a dowel pin remover tool, rotate the intake camshaft sprocket counterclockwise enough to slide the timing chain over the intake camshaft sprocket. Release the camshaft sprocket wrench. The length of chain between the 2 camshaft sprockets will tighten. If properly timed, the intake camshaft alignment dowel pin should slide in easily. If the dowel pin does not fully index, the camshafts are not timed correctly and the procedure must be repeated.
18. Leave the alignment dowel pins installed.
19. With slack removed from chain between intake camshaft sprocket and crankshaft sprocket, the timing marks on the crankshaft and the cylinder block should be aligned. If marks are not aligned, move the chain 1 tooth forward or rearward, remove slack and recheck marks.
20. Tighten the chain housing to block stud (timing chain tensioner shoe pivot). the stud is installed under the timing chain. Tighten to 19 ft. lbs. (26 Nm).
21. Reload timing chain tensioner assembly to its **0** position as follows:
    a. Assemble restraint cylinder, spring and nylon plug into plunger. Index slot in restraint cylinder with peg in plunger. While rotating the restraint cylinder clockwise, push the restraint cylinder into the plunger until it bottoms. Keep rotating the restraint cylinder clockwise but allow the spring to push it out of the plunger. The pin in the plunger will lock the restraint in the loaded position.
    b. Install tool J–36589 or equivalent, onto plunger assembly.
    c. Install plunger assembly into tensioner body with the long end toward the crankshaft when installed.
22. Install the tensioner assembly to the chain housing. Recheck plunger assembly installation. It is correctly installed when the long end is toward the crankshaft.
23. Install and tighten timing chain tensioner bolts and tighten to 10 ft. lbs. (14 Nm)
24. Install the tensioner shoe and tensioner shoe retainer.
25. Remove special tool J–36589 and squeeze plunger assembly into the tensioner body to unload the plunger assembly.
26. Lower vehicle and remove the alignment dowel pins. Rotate crankshaft clockwise 2 full rotations. Align crankshaft timing mark with mark on cylinder block and reinstall alignment dowel pins. Alignment dowel pins will

slide in easily if engine is timed correctly.

NOTE: If the engine is not correctly timed, severe engine damage could occur.

27. Install the 3 timing chain guides and crankshaft oil slinger.
28. Install the timing chain front cover.
29. Connect the negative battery cable and check for leaks.

2.5L Engine

TIMING GEARS

1. Disconnect the negative battery cable.
2. Remove the engine from the vehicle.
3. Remove the damper, front cover and camshaft. Align the timing marks on the crank and cam gears.
4. To remove the camshaft gear, use a arbor press and adapter. Position the thrust plate to avoid damage to the Woodruff key as the gear is removed.
5. Remove the crankshaft gear with a suitable prybar.
6. Support the camshaft in the arbor press using the press adapter. Position the spacer ring, thrust plate and woodruff key over the end of the shaft and press the gear onto the camshaft.
7. Measure the end clearance with a feeler gauge between the cam journal and thrust plate. The measurement should be between 0.0015–0.0050 in. If the measurement is less than 0.0015 in., replace the spacer ring. If the measurement is more than 0.0050 in., replace the thrust plate.
8. Apply assembly lube GM 1052367 or equivalent, to the cam journals and lobes.
9. Install the camshaft into the engine and align the timing marks.
10. Install the front cover, rocker arm cover, damper and install the engine into the engine into the vehicle.
11. Connect the negative battery cable.

2.8L and 3.1L Engine

1. Disconnect the negative battery cable.
2. Remove the front cover assembly.
3. Place the No. 1 piston at TDC with the marks on the crankshaft and the camshaft aligned (No. 4 firing position).
4. Remove the camshaft sprocket and the timing chain.

NOTE: If the camshaft sprocket does not come off easily, a light blow on the lower edge of the sprocket with a rubber mallet should loosen the sprocket.

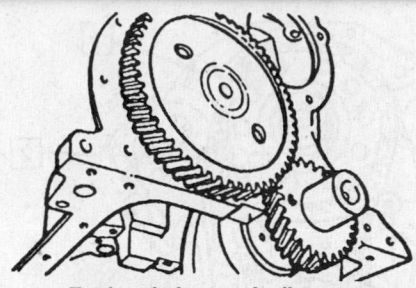

Engine timing mark alignment— 2.5L engine

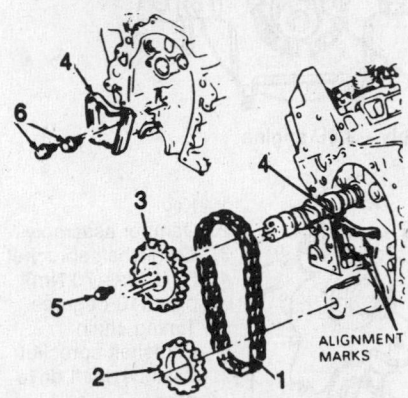

1. Timing chain
2. Crank sprocket
3. Camshaft sprocket
4. Damper
5. 15–20 ft. lbs.
6. 13–18 ft. lbs.

Timing chain and sprockets—2.8L and 3.1L engines

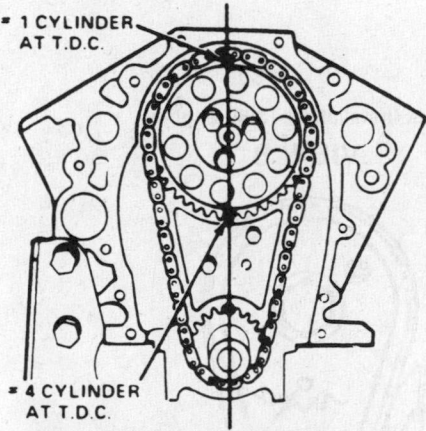

Engine timing mark alignment—2.8L and 3.1L engines

5. Remove the crankshaft sprocket with a suitable prybar.
6. Install the crankshaft sprocket. Apply a coat of Molykote® or equivalent, to the sprocket thrust surface.
7. Hold the camshaft sprocket with the chain hanging down and align the marks on the camshaft and crankshaft sprockets.
8. Align the dowel in the camshaft with the dowel hole in the camshaft sprocket. Install the camshaft sprocket and chain, use the camshaft sprocket bolts to draw the sprocket on to the

camshaft. Tighten the sprocket bolts to 18 ft. lbs. (25 Nm).
9. Lubricate the timing chain with engine oil. Install the front cover assembly.

3.4L Engine

1. Disconnect the negative battery cable.
2. Raise and safely support vehicle.
3. Remove starter motor and flywheel cover.
4. Remove oil pan retaining nuts and bolts. Remove lower frame and powertrain onto transmission table.
5. Remove front cover.
6. Mark intermediate sprocket, chain link, front face of cylinder and crank sprocket for reference.
7. Retract timing chain tensioner shoe by using J–33875 or equivalent, on both sides of the tensioner and pulling on the thru pin in the tensioner arm to retract the spring. While spring is retracted insert a suitable tool.
8. Remove timing chain and crankshaft sprocket using gear puller. If intermediate gear does not slide off easily with timing chain assembly, rotate the crankshaft back and forth to loosen tight fit.

To install:

9. Check to insure that crankshaft key is fully seated and chain tensioner is fully seated and retracted.
10. Install sprockets and chain over shafts maintaining alignment.
11. Make sure the large chamfer and counterbore of crank sprocket are installed towards crank. The intermediate sprocket spline sockets are installed away from the case.
12. Press the crankshaft sprocket on the final 0.31 inch using J–38612 or equivalent. Check to make sure timing was maintained. Remove retaining pin from tensioner.
13. Install front cover.
14. Connect negative battery cable. Start engine and chech for leaks.

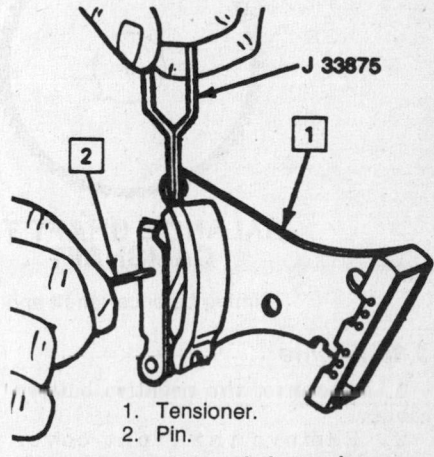

1. Tensioner.
2. Pin.

Retracting timing chain tensioner

25-33

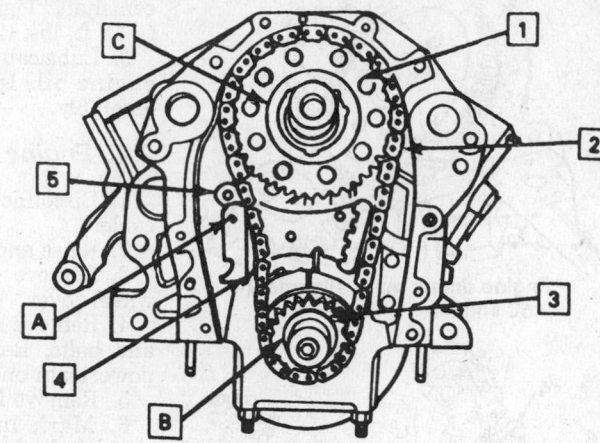

1. Intermediate shaft sprocket.
2. Timing chain.
3. Crankshaft sprocket.
4. Timing chain tensioner.
5. 18 ft. lbs.
A. Spring pin hole.
B. Chamfer and counter bore inward.
C. Sprockets outward.

Timing chain assembly—3.4L engine

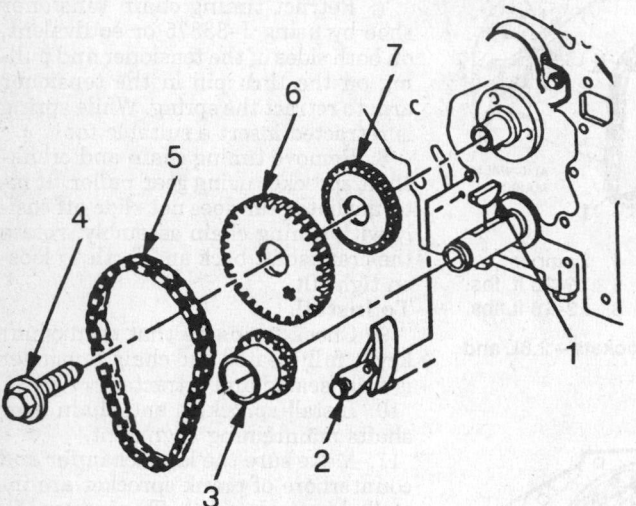

1. Key
2. Damper assembly
3. Crankshaft sprocket
4. 52 ft. lbs. (70 Nm) plus 110 degrees
5. Timing chain
6. Camshaft sprocket
7. Balance shaft drive gear

Timing chain and sprocket installation—3.8L engine

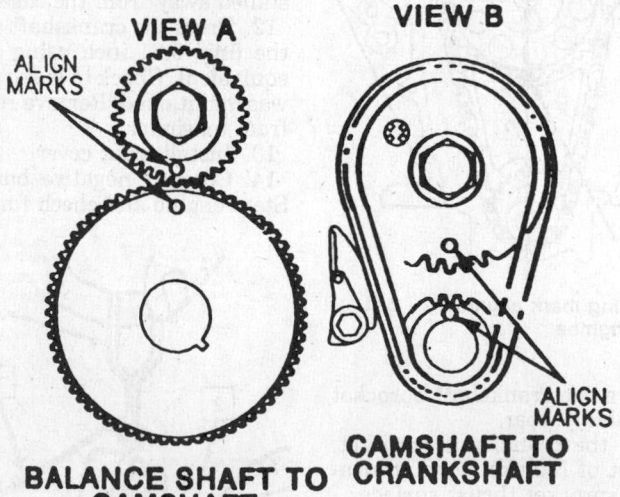

VIEW A
ALIGN MARKS

VIEW B
ALIGN MARKS

BALANCE SHAFT TO CAMSHAFT

CAMSHAFT TO CRANKSHAFT

Timing balancer shaft and camshaft marks—3.8L engines

3.8L Engine

1. Disconnect the negative battery cable.
2. Remove the front cover assembly.
3. Align the timing marks on the sprockets and remove the timing chain damper.
4. Remove the camshaft sprocket bolts, camshaft sprocket and chain.

5. Remove the crankshaft sprocket by applying a light blow on the lower edge of the sprocket with a plastic mallet.

To install:

6. If the pistons have been moved in the engine, do the following:

  a. Turn the crankshaft so the No. 1 piston is at Top Dead Center (TDC).

  b. Turn the camshaft so, with the sprocket temporarily installed, the timing mark is straight down.

7. Assemble the timing chain on the sprockets with the timing marks facing each other.

8. Install the timing chain and sprockets and tighten the camshaft sprocket bolts to 52 ft. lbs. plus an additional 110 degree turn.

9. Install the timing chain damper and tighten the bolt to 14 ft. lbs.

10. Rotate the engine 2 revolutions and make sure the marks are aligned correctly.

11. Install the front cover assembly.

12. Connect the negative battery cable.

Secondary Timing Belt Cover

REMOVAL & INSTALLATION

3.4L Engine

RIGHT SIDE

1. Disconnect negative battery cable.
2. Remove retaining bolts and remove cover.
3. To install position cover and install retaining bolts. Torque bolts to 89 inch lbs.

LEFT SIDE

1. Disconnect the negative battery cable.
2. Remove spark plug wire cover.
3. Remove retaining bolts and cover.
4. To install position cocer and secure with retaining bolts.Torque bolts to 89 inch lbs.
5. Install spark plug wire cover.

CENTER COVER

1. Disconnect the negative battery cable.
2. Disconnect Electronic Control Module (ECM) harness cover.
3. Remove serpentine belt tensioner.
4. Remove right and left side timing belt covers.
5. Remove power steering pipe retaining clip nut at alternator stud.
6. Remove center timing belt cover bolts and remove cover.
7. To install position cover on en-

gine and secure with retainer bolts. Tighten bolts to 89 inch lbs.

8. Reinstall power steering pipe retaining clip nut to alternator stud.

9. Install right and left side covers. Install serpentine belt.

10. Install Electronic Control Module (ECM) harness cover. Reconnect negative battery cable.

Secondary Timing Belt and Tensioner

ADJUSTMENT

Belt tension is set and maintained by fully automatic tensioners. Timing must be reset if sprockets are remove from their shaft.

REMOVAL & INSTALLATION

1. Disconnect the negative battery cable.

2. Remove secondary timing belt actuator.

3. If belt is to be reused mark direction of rotation.

4. Remove tensioner and pulley arm assembly.

5. Remove timing belt by sliding off of pulleys. Do not bend, twist, or kink belt or damage to the belt may occur.

6. Install new belt or old belt taking note of direction of rotation.

7. Install tensioner pulley to mounting base. Tighten bolt to 37 ft. lbs.

8. Rotate the tensioner pulley counterclockwise into the belt using the cast square lug on body and engage ball end of the actuator into socket on pulley arm.

9. Remove tensioner lock pin allowing tensioner shaft to extend and the pulley to move into the belt.

10. Rotate the tensioner pulley counterclockwise applying 12–15 ft. lbs. torque.

11. Rotate the engine clockwise 3 times to seat belt. Align the crankshaft reference marks during final rotation to TDC. Do not allow crankshaft to spring back or reverse direction of rotation.

12. Seat lock ring on the right exhaust and right intake camshaft into the bore by threading in the attaching bolts.

13. Hold sprocket from turning using tool J–38614 or equivalent. Tighten attaching bolt to 81 ft. lbs. taking note of running torque; torque required to turn bolt before seating. Running torque of bolt should be 44–66 ft. lbs. If less torque is required, replace the shim and lock rings and inspect the nose of the camshaft for brinelling. If more torque is required

than replace the shim and lock rings and check the attaching bolt threads for burrs or foreign material.

14. Rotate engine clockwise 1 revolution and realign the balancer marks at TDC. Make sure timing mark on damper lines up with front cover timing mark.

15. Repeat steps 7–13 starting with left intake then left exhaust camshaft.

16. Install secondary timing belt covers and retaining bolts.

17. Reconnect negative battery cable.

Camshaft

REMOVAL & INSTALLATION

2.3L Engine

INTAKE CAMSHAFT

NOTE: Any time the camshaft housing to cylinder head bolts are loosened or removed, the camshaft housing to cylinder head gasket must be replaced.

1. Relieve the fuel system pressure. Disconnect the negative battery cable.

2. Label and disconnect the ignition coil and module assembly electrical connections.

3. Remove 4 ignition coil and module assembly to camshaft housing bolts and remove assembly by pulling straight up. Use a special spark plug boot wire remover tool to remove connector assemblies if they have stuck to the spark plugs.

4. Remove the idle speed power steering pressure switch connector.

5. Loosen 3 power steering pump pivot bolts and remove drive belt.

6. Disconnect the 2 rear power steering pump bracket to transaxle bolts.

7. Remove the front power steering pump bracket to cylinder block bolt.

8. Disconnect the power steering pump assembly and position to the side.

9. Using the special tool, remove the power steering pump drive pulley from the intake camshaft.

10. Remove oil/air separator bolts and hoses. Leave the hoses attached to the separator, disconnect from the oil fill, chain housing and intake manifold. Remove as an assembly.

11. Remove vacuum line from fuel pressure regulator and disconnect the fuel injector harness connector.

12. Disconnect fuel line retaining clamp from bracket on top of intake camshaft housing.

13. Remove fuel rail to camshaft housing retaining bolts.

14. Remove the fuel rail from the cylinder head. Cover injector openings in cylinder head and cover injector noz-

zles. Leave fuel lines attached and position fuel rail aside.

15. Disconnect the timing chain and housing but do not remove from the engine.

16. Remove intake camshaft housing cover to camshaft housing retaining bolts.

17. Remove the intake camshaft housing to cylinder head retaining bolts. Use the reverse of the tightening sequence when loosening camshaft housing to cylinder head retaining bolts. Leave 2 bolts loosely in place to hold the camshaft housing while separating camshaft cover from housing.

18. Push the cover off the housing by threading 4 of the housing to head retaining bolts into the tapped holes in the cam housing cover. Tighten the bolts in evenly so the cover does not bind on the dowel pins.

19. Remove the 2 loosely installed camshaft housing to head bolts and remove the cover. Discard the gaskets.

20. Note the position of the chain sprocket dowel pin for reassembly. Remove the camshaft carefully; do not damage the camshaft oil seal.

21. Remove intake camshaft oil seal from camshaft and discard seal. This seal must be replaced any time the housing and cover are separated.

22. Remove the camshaft carrier from the cylinder head and remove the gasket.

To install:

23. Thoroughly clean the mating surfaces of the camshaft carrier and the cylinder head, bolts and bolt holes. Install a new gasket and place the housing on the head. Install 1 bolt loosely to hold in place.

24. Install the lifters into their bores. If the camshaft is being replaced, the lifters must also be replaced. Lubricate camshaft lobes, journals and lifters with camshaft and lifter prelube. The camshaft lobes and journals must be adequately lubricated or engine damage could occur upon start up.

25. Install the camshaft in the same position as when removed. The timing chain sprocket dowel pin should be straight up and line up with the centerline of the lifter bores.

26. Install new camshaft housing to camshaft housing cover seals into cover; do not use sealer. Make sure the correct color seal is placed in each groove. Install the cover to the housing.

27. Apply thread locking compound to the camshaft housing and cover retaining bolt threads.

28. Install bolts and torque to 11 ft. lbs. Rotate the bolts (except the 2 rear bolts that hold fuel pipe to camshaft housing) an additional 75 degrees in sequence. Rotate the excepted bolts an additional 25 degrees.

29. Install timing chain housing and timing chain.

30. Uncover fuel injectors and install new fuel injector ring seals lubricated with oil. Install the fuel rail.

31. Install the fuel line retaining clamp and retainer to bracket on top of the intake camshaft housing.

32. Connect the vacuum line to the fuel pressure regulator.

33. Connect the fuel injectors harness connector.

34. Install the oil/air separator assembly.

35. Lubricate the inner sealing surface of the intake camshaft seal with oil and install the seal to the housing.

36. Install the power steering pump pulley onto the intake camshaft.

37. Install the power steering pump assembly and drive belt.

38. Connect the idle speed power steering pressure switch connector.

39. Clean any loose lubricant that is present on the ignition coil and module assembly to camshaft housing bolts. Apply Loctite® 592 or equivalent, onto the ignition coil and module assembly to camshaft housing bolts. Install the bolts and torque to 13 ft. lbs. (18 Nm)

40. Connect the electrical connectors to ignition coil and module assembly.

41. Connect the negative battery cable and road test the vehicle. Check for leaks.

Exhaust Camshaft

NOTE: Any time the camshaft housing to cylinder head bolts are loosened or removed the camshaft housing to cylinder head gasket must be replaced.

1. Relieve the fuel system pressure. Disconnect the negative battery cable.

2. Label and disconnect the ignition coil and module assembly electrical connections.

3. Remove 4 ignition coil and module assembly to camshaft housing bolts and remove assembly by pulling straight up. Use a special tool to remove connector assemblies if they have stuck to the spark plugs.

4. Remove the idle speed power steering pressure switch connector.

5. Remove the transaxle fluid level indicator tube assembly from exhaust camshaft cover and position aside.

6. Remove exhaust camshaft cover and gasket.

7. Disconnect the timing chain and housing but do not remove from the engine.

8. Remove exhaust camshaft housing to cylinder head bolts. Use the reverse of the tightening procedure when loosening camshaft housing while separating camshaft cover from housing.

9. Push the cover off the housing by threading 4 of the housing to head retaining bolts into the tapped holes in the camshaft cover. Tighten the bolts evenly so the cover does not bind on the dowel pins.

10. Remove the 2 loosely installed camshaft housing to cylinder head bolts and remove cover, discard gaskets.

11. Loosely install 1 camshaft housing to cylinder head bolt to retain the housing during camshaft and lifter removal.

12. Note the position of the chain sprocket dowel pin for reassembly. Remove camshaft being careful not to damage the camshaft or journals.

13. Remove the camshaft carrier from the cylinder head and remove the gasket.

To install:

14. Thoroughly clean the mating surfaces of the camshaft carrier and the cylinder head, bolts and bolt holes. Install a new gasket and place the housing on the head. Install 1 bolt loosely to hold in place.

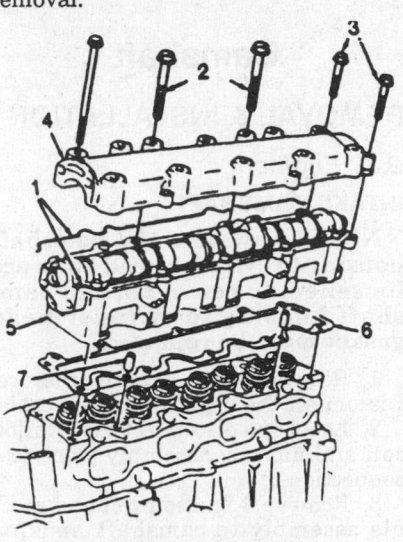

1. Housing cover seals
2. Cylinder head bolts
3. Housing cover bolts
4. Camshaft cover
5. Intake camshaft housing
6. Cylinder head gasket
7. Dowel pins

Camshaft housing assembly—2.3L engine

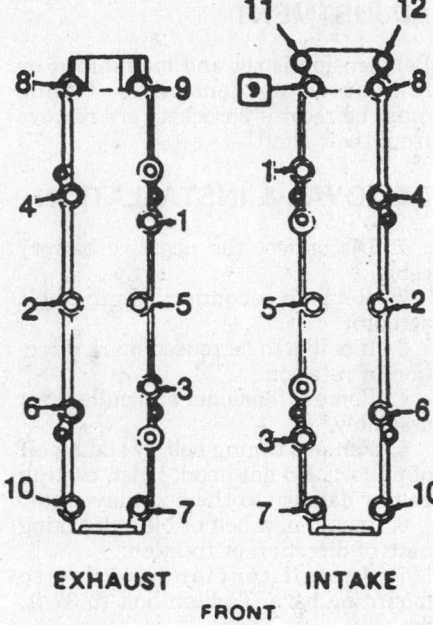

Camshaft housing bolt torque sequence—2.3L engine

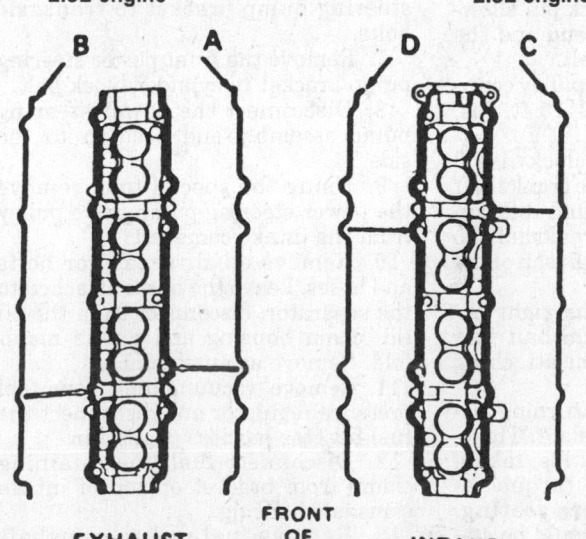

A. Seal – inner (exhaust – red)
B. Seal – outer (exhaust – red)
C. Seal – outer (intake – blue)
D. Seal – inner (intake – blue)

Camshaft cover assembly—2.3L engine

15. Install the lifters into their bores. If the camshaft is being replaced, the lifters must also be replaced. Lubricate camshaft lobes, journals and lifters with camshaft and lifter prelube. The camshaft lobes and journals must be adequately lubricated or engine damage could occur upon start up.

16. Install camshaft in same position as when removed. The timing chain sprocket dowel pin should be straight up and align with the centerline of the lifter bores.

17. Install new camshaft housing to camshaft housing cover seals into cover; do not use sealer. Make sure the correct color seal is placed in each groove. Install the cover to the housing.

18. Apply thread locking compound to the camshaft housing and cover retaining bolt threads.

19. Install bolts and torque in sequence to 11 ft. lbs. Then rotate the bolts an additional 75 degrees, in sequence.

20. Install timing chain housing and timing chain.

21. Install the transaxle fluid level indicator tube assembly to exhaust camshaft cover.

22. Connect the idle speed power steering pressure switch connector.

23. Clean any loose lubricant that is present on the ignition coil and module assembly to camshaft housing bolts. Apply Loctite® 592 or equivalent, onto the ignition coil and module assembly to camshaft housing bolts. Install the bolts and torque to 13 ft. lbs. (18 Nm)

24. Connect the electrical connectors to ignition coil and module assembly.

25. Connect the negative battery cable and road test the vehicle. Check for leaks.

2.5L Engine

NOTE: For the removal of the camshaft, the engine assembly must be removed from the vehicle.

1. Disconnect the negative battery cable.

2. Remove the engine assembly from the vehicle.

3. Remove the rocker arm cover and pushrods.

4. Remove the pushrod cover and valve lifters.

5. Remove the serpentine belt, crankshaft pulleys and vibration dampener.

6. Remove the front cover.

7. Remove the camshaft thrust plate screws.

NOTE: The camshaft journals are the same diameter. Care must be taken when removing the camshaft to avoid damage to the cam bearings.

8. Carefully slide the camshaft and gear through the front of the block.

9. To remove the camshaft gear, use a arbor press and adapter.

10. Old and new camshafts should be cleaned with solvent and compressed air before being installed.

To install:

11. Install the camshaft gear onto the camshaft with an arbor press.

12. Measure the end clearance with a feeler gauge between the cam journal and thrust plate. The measurement should be between 0.0015–0.0050 in. If the measurement is less than 0.0015 in., replace the spacer ring. If the measurement is more than 0.0050 in., replace the thrust plate.

NOTE: Always apply assembly lube, GM Engine Oil Supplement (E.O.S) or equivalent, to the cam journals and lobes. If this procedure is not done, cam damage may result.

13. Carefully install the camshaft into the engine block by rotating and pushing forward until seated.

14. Install the thrust plate screws and torque to 89 inch lbs. (10 Nm).

15. Install the front cover, vibration dampener and serpentine belt.

16. Install the valve lifter and pushrod cover.

17. Install the pushrods and rocker arm cover.

18. Install the engine into the vehicle.

19. Refill all necessary fluids.

20. Start the engine and check for leaks.

2.8L, 3.1L and 3.8L Engine

NOTE: For the removal of the camshaft the engine assembly must be removed from the vehicle.

1. Remove the engine assembly from the vehicle.

2. Remove the rocker covers and remove the valve lifters.

3. Remove the front cover assembly, timing chain and sprockets.

4. Remove the camshaft by sliding it from the block.

To install:

5. Coat the camshaft journals with engine oil. Coat the camshaft lobes with GM Engine Oil Supplement (E.O.S) or equivalent.

6. Slide the camshaft into the block.

7. Install the timing chain and sprockets, making sure to align the timing marks.

8. Install the front cover assembly. Install the valve lifters.

9. Install the engine assembly into

the vehicle. Run the engine and check for leaks.

3.4L Engine

LEFT SIDE

1. Disconnect the negative battery cable.

2. Drain cooling system.

3. Remove left side cam carrier as follows:

   a. Disconnect oil/air breather hose from cam carrier cover. Remove spark plug wires from plugs and remove rear spark plug wire cover.

   b. Remove cam carrier cover bolts and lift off cover. Remove gasket and O-rings from cover.

   c. Remove secondary timing belt by removing secondary timing belt actuator and tensioner assembly and sliding belt from pulleys.

   d. Install 6 sections of fuel line hoses under cam shaft and between lifters. This will hold lifters in the carrier. For this procedure use $^3/_{16}$ inch fuel line hose for exhaust valves and $^5/_{32}$ inch fuel line hose for the intake valves.

   e. Remove exhaust crossover pipe and torque strut.

   f. Remove torque strut bracket at engine.

   g. Remove cam carrier mounting bolts and nuts and remove cam carrier.

   h. Remove cam carrier gasket from cylinder head.

4. Remove the 6 lifter hold-down hoses. Remove the lifters.

5. Install cam hold-down tool J–38613 or equivalent, in place and remove cam sprockets.

6. Remove cam carrier end caps and retainer plate bolts and plate.

7. Remove camshaft hold hold-down tool and carefully remove camshaft out the back of the carrier.

To install:

8. Coat camshaft lobes and journals with clean engine oil and install camshaft into carrier. Install retaining plate and bolts and tighten to 89 inch lbs. Install cam carrier end caps.

9. Install cam sprocket .

10. Adjust cam timing and install cam hold-down tool.

11. Lubricate lifters with clean engine oil and install lifters into original position.

12. Install lifter hold-down hoses to cam carrier.

13. Install cam carrier following these steps:

   a. Install new gasket on cam carrier to cylinder mounting surface.

   b. Install cam hold-down tool J–38613 or equivalent, to carrier assembly.

c. Install cam carrier to cylinder head. Install mounting bolts and nuts. Torque bolts and nuts to 18 ft. lbs.

d. Remove lifter hold-down hoses and cam hold-down tool.

e. Install torque strut bracket to engine and install torque strut.

f. Install engine crossover pipe.

g. Install secondary timing belt and cam carrier cover.

h. Reconnect spark plug cover and wires.

i. Connect breather hose to cam carrier cover.

14. Add fluids as required, reconnect negative battery cable. Start engine and recheck for leaks.

RIGHT SIDE (REAR)

1. Disconnect the negative battery cable. Drain cooling system.

2. Remove right side cam carrier as follows:

a. Remove intake plenum and right timing belt cover.

b. Remove right spark plug wires.

c. Remove air/oil separator hose at cam carrier cover.

d. Remove cam carrier cover bolts and lift of cover. Remove gasket and O-rings from cover.

e. Remove secondary timing belt by removing secondary timing belt actuator and tensioner assembly and sliding belt from pulleys.

f. Install 6 sections of fuel line hoses under cam shaft and between lifters. This will hold lifters in carrier. For this procedure use $\frac{3}{16}$ inch fuel line hose for exhaust valves and $\frac{5}{32}$ inch fuel line hose for the intake valves.

g. Remove exhaust crossover pipe and torque strut.

h. Remove torque strut bracket at engine. Remove front engine lift hook.

i. Remove cam carrier mounting bolts and nuts and remove cam carrier.

j. Remove cam carrier gasket from cylinder head.

3. Remove 6 lifter hold-down hoses.

4. Remove lifters.

5. Install cam hold-down tool J–38613 or equivalent, and remove cam sprocket.

6. Remove cam carrier end caps and retainer plate.

Remove cam hold-down tool and slide cam shaft out rear of carrier.

To install

7. Lubricate cam shaft lobes and journals with clean engine oil and slide into cam carrier. Install retainer plate and bolts and tighten bolts to 89 inch lbs.

8. Install cam carrier end caps and cam sprockets.

9. Install cam shaft carrier hold-down tool and adjust cam timing.

10. Lubricate lifters with clean engine oil and install lifters into original position.

11. Install lifter hold-down hoses to cam carrier.

12. Install cam carrier following these steps:

a. Install new gasket on cam carrier to cylinder mounting surface.

b. Install cam hold-down tool J–38613 or equivalent, to carrier assembly.

c. install cam carrier to cylinder head. Install mounting bolts and nuts. Torque bolts and nuts to 18 ft. lbs.

d. Remove lifter hold-down hoses and cam hold-down tool.

e. Install torque strut bracket to engine and install torque strut.

f. Install engine crossover pipe and engine lift hook.

g. Install secondary timing belt and cam carrier cover.

h. Install spark plug wires and cover.

13. Add fluids as required. Connect negative battery cable. Start engine and check for fluid leaks.

Balance Shaft/ Intermediate Shaft

REMOVAL & INSTALLATION

3.4L Engine

INTERMEDIATE SHAFT

1. Disconnect the negative battery cable.

2. Remove engine from vehicle.

3. Remove right side cylinder head and oil pump drive assembly.

4. Remove the timing chain assembly.

5. Remove thrust plate screws and plate.

6. Remove intermediate using care not to damage journals or bearings.

To install:

7. Lubricate intermediate shaft journals and gear with engine oil. Install shaft, thrust plate and retainer screws. Tighten screws to 89 inch lbs.

8. Replace O-ring after sprocket is installed and install timing chain and gear assembly.

9. Install oil pump drive assembly and cylinder head.

10. Install engine assembly.

3.8L Engine

BALANCE SHAFT

1. Disconnect the negative battery cable. Remove the engine and secure it to a workstand.

2. Remove the flywheel-to-crankshaft bolts and the flywheel.

3. Remove the timing chain cover-to-engine bolts and the cover.

4. Remove the camshaft sprocket-to-camshaft gear bolts, the sprocket, the timing chain and the gear.

5. To remove the balance shaft, perform the following procedures:

a. Remove the balance shaft gear-to-shaft bolt and the gear.

b. Remove the balance shaft retainer-to-engine bolts and the retainer.

c. Using the slide hammer tool, pull the balance shaft from the front of the engine.

To install:

6. If replacing the rear balance shaft bearing, perform the following procedures:

a. Drive the rear plug from the engine.

b. Using the camshaft remover/installer tool, press the rear bearing from the rear of the engine.

c. Dip the new bearing in clean engine oil.

d. Using the balance shaft rear bearing installer tool, press the new rear bearing into the rear of the engine.

e. Install the rear cup plug.

7. Using the balance shaft installer tool, screw it into the balance shaft and install the shaft into the engine; remove the installer tool.

8. Clean the gasket mounting surfaces. Inspect the parts for wear and/or damage; replace the parts, if necessary.

9. Install the balance shaft retainer. Torque the balance shaft retainer-to-engine bolts to 27 ft. lbs.

10. Align the balance shaft gear with the camshaft gear timing marks. Install the balance shaft gear onto the balance shaft. Torque the balance gear-to-balance shaft bolt to 15 ft. lbs, then using a torque angle meter tool, rotate another 35 degrees.

11. Align the marks on the balance shaft gear and the camshaft gear by turning the balance shaft.

12. Turn the crankshaft so the No. 1 piston is at TDC.

13. Install the timing chain and sprocket.

14. Replace the balance shaft front bearing retainer and bolts. Tighten the bolts to 26 ft. lbs.

15. Install the front timing cover and the lifter guide retainer.

16. Install the intake manifold and flywheel assembly. Tighten the flywheel bolts to 89 inch lbs., plus an additional 90 degree turn.

17. Install the engine assembly and connect the negative battery cable. Start the engine and check for leaks.

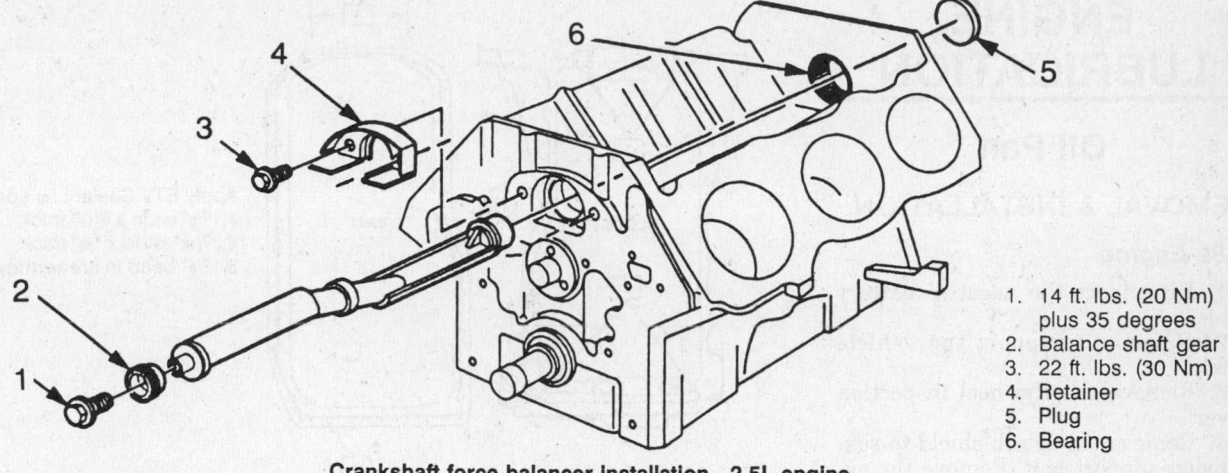

Crankshaft force balancer installation—2.5L engine

1. 14 ft. lbs. (20 Nm) plus 35 degrees
2. Balance shaft gear
3. 22 ft. lbs. (30 Nm)
4. Retainer
5. Plug
6. Bearing

Piston and Connecting Rod

POSITIONING

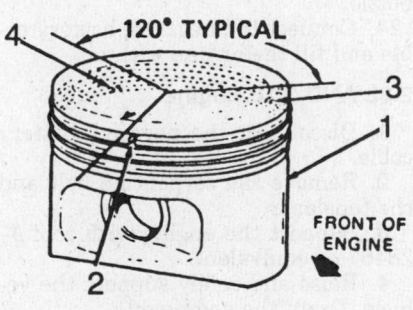

1. Piston
2. Upper compression ring gap
3. Lower compression ring gap
4. Oil ring assembly gap

Piston ring end gap positioning— 2.3L engine

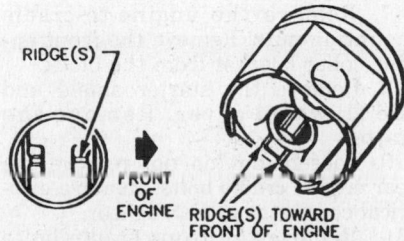

Piston positioning—3.8L engine

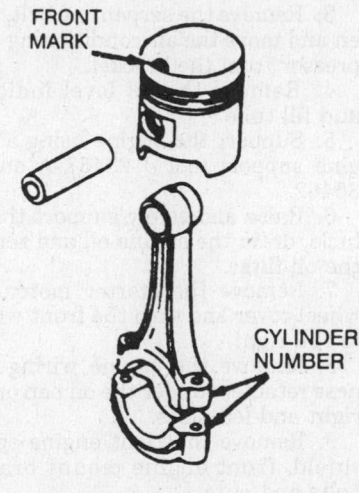

Piston positioning—2.3L, 2.5L, 2.8L and 3.1L engines

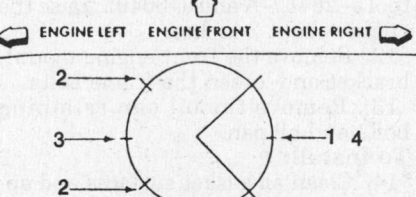

1. Oil ring spacer gap (tang in hole or slot with arc)
2. Oil ring rail gaps
3. 2nd compression ring gap
4. Top compression ring gap

Piston ring end gap positioning—2.5L, 2.8L, 3.1L, 3.4L and 3.8L engines

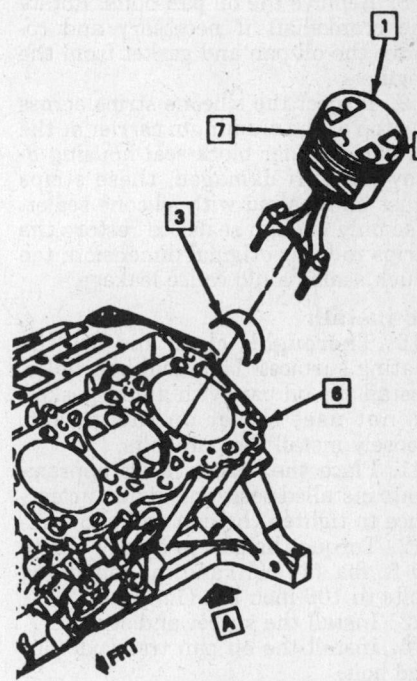

1. Piston and rod.
2. Cap.
3. Bearing.
4. Crankshaft.
5. To 39 ft. lbs.
6. Engine block.
7. Install piston with L in left bank only. Install piston with R in right bank only.
8. Install with stamped arrow pointing towards front of engine.

Piston and connecting rod assembly

ENGINE LUBRICATION

Oil Pan

REMOVAL & INSTALLATION

2.3L Engine

1. Disconnect the negative battery cable.
2. Raise and support the vehicle safely.
3. Remove the flywheel inspection cover.
4. Remove the splash shield-to-suspension support bolt. Remove the exhaust manifold brace, if equipped.
5. Remove the radiator outlet pipe-to-oil pan bolt.
6. Remove the transaxle-to-oil pan nut and stud using a 7mm socket.
7. Gently pry the spacer out from between oil pan and transaxle.
8. Remove the oil pan bolts. Rotate the crankshaft if necessary and remove the oil pan and gasket from the engine.
9. Inspect the silicone strips across the top of the aluminum carrier at the oil pan-cylinder block-seal housing 3-way joint. If damaged, these strips must be repaired with silicone sealer. Use only enough sealer to restore the strips to their original dimension; too much sealer could cause leakage.

To install:

10. Thoroughly clean and dry the mating surfaces, bolts and bolt holes. Install the oil pan with a new gasket; do not uses sealer on the gasket. Loosely install the pan bolts.
11. Place the spacer in its approximate installed position but allow clearance to tighten the pan bolt above it.
12. Torque the pan to block bolts to 17 ft. lbs. (24 Nm) and the remaining bolts to 106 inch lbs. (12 Nm)
13. Install the spacer and stud.
14. Install the oil pan transaxle nut and bolt.
15. Install the slash shield to suspension support.
16. Install the radiator outlet pipe bolt.
17. Install the exhaust manifold brace, if removed.
18. Install the flywheel inspection cover.
19. Fill the crankcase with the proper oil.
20. Connect the negative battery cable and check for leaks.

2.5L Engine

1. Disconnect the negative battery cable.

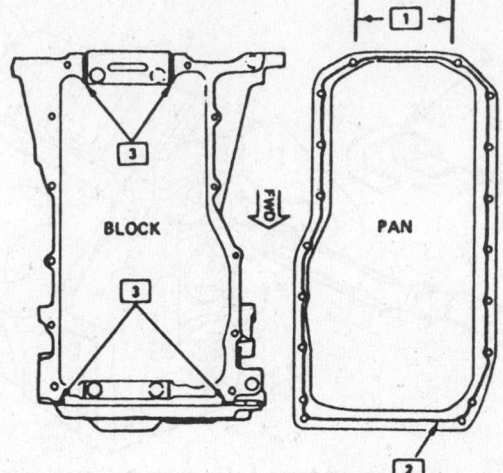

Apply RTV Sealant as specified:
1. 3/8" wide x 3/16" thick
2. 3/16" wide x 1/8" thick
3. 1/8" bead in areas shown

Oil pan sealer locations—2.5L engine

2. Remove the coolant recovery bottle, engine torque strut, air cleaner and the air inlet.
3. Remove the serpentine belt, loosen and move the air conditioning compressor from the bracket.
4. Remove the oil level indicator and fill tube.
5. Support the engine using an engine support tool J–28467–A and J–36462.
6. Raise and safely support the vehicle, drain the engine oil and remove the oil filter.
7. Remove the starter motor, flywheel cover and turn the front wheels to full right.
8. Remove the engine wiring harness retainers under the oil pan on the right and left sides.
9. Remove the right engine splash shield, front engine mount bracket bolts and nuts.
10. Remove the transaxle mount nuts.
11. Using the engine support fixture tool J–28467–A and J–36462, raise the engine about 2 inches.
12. Remove the front engine mount, bracket and loosen the frame bolts.
13. Remove the oil pan retaining bolts and oil pan.

To install:

14. Clean all gasket surfaces and apply RTV sealer to the oil pan and engine surfaces.
15. Install the oil pan and retaining bolts and tighten to 89 inch lbs. (10 Nm)
16. Install the frame bolts and tighten to 103 ft. lbs. (140 Nm)
17. Install the engine mount, bracket, lower the engine into position and install the transaxle mount nuts.
18. Install the engine mount nuts and bracket bolts.
19. Install the engine splash shield,

wiring harness to the oil pan, flywheel cover and the starter motor.
20. Lower the vehicle and remove the engine support fixtures.
21. Install the oil level indicator and tube assembly.
22. Reinstall the air conditioning compressor to its original location and serpentine belt.
23. Install the air inlet, air cleaner, torque strut and coolant recovery bottle.
24. Connect the negative battery cable and fill the engine with oil.

2.8L AND 3.1L Engine

1. Disconnect the negative battery cable.
2. Remove the serpentine belt and the tensioner.
3. Support the engine with tool J–28467 or equivalent.
4. Raise and safely support the vehicle. Drain the engine oil.
5. Remove the right tire and wheel assembly. Remove the right inner fender splash shield.
6. Remove the steering gear pinch bolt. Remove the transaxle mount retaining bolts. Failure to disconnect intermediate shaft from rack and pinion stub shaft can result in damage to the steering gear and/or intermediate shaft. This could cause a loss of steering control which could result in personal injury.
7. Remove the engine-to-cradle mounting nuts. Remove the front engine collar bracket from the block.
8. Remove the starter shield and the flywheel cover. Remove the starter.
9. Loosen, but do not remove the rear engine cradle bolts. Remove electrical connector at DIS sensor.
10. Remove the front cradle bolts and lower front of frame. Remove the

oil pan retaining bolts and nuts. Remove the oil pan.

To install:

11. Clean the gasket mating surfaces.

12. Install a new gasket on the oil pan. Apply silicon sealer to the portion of the pan that contacts the rear of the block.

13. Install the oil pan, nuts and retaining bolts. Tighten rear bolts to 18 ft. lbs. (18–25 Nm),and remaining nuts and bolts to 89 inch lbs.

14. Install the front cradle bolts and tighten the rear cradle bolts.Install DIS connector. Install the starter and splash shield. Install the flywheel shield.

15. Attach the collar bracket to the block, install the engine-to-cradle nuts. Install the transaxle mount nuts.

16. Install the steering pinch bolt. Install the right inner fender splash shield and tire assembly. Lower the vehicle.

17. Remove the engine support tool. Install the serpentine belt and tensioner.

18. Fill the crankcase to the correct level. Connect the negative battery cable. Run the engine to normal operating temperature and check for leaks.

3.4L Engine

1. Disconnect the negative battery cable.

2. Raise and safely support vehicl. Drain engine oil.

3. Remove right front wheel assembly and steering gear heat shield.

4. Remove steering gear retaining bolts and support steering gear to body.

5. Seperate right and left lower ball joints.

6. Disconnect power steering cooler line clamps at frame.

7. Support frame and remove engine mount nuts at frame.

8. Remove frame retaining bolts and remove frame assembly.

9. Remove starter assembly and flywheel cover.

10. Remove oil pan retaining nuts and bolts and remove oil pan.

11. Remove old pan gasket. Clean all mating surfaces.

12. Install new gasket adding sealer to gasket next to rear main bearing cap.Install oil pan and secure rear retaining bolts to 18 ft. lbs. and all other bolts and nuts to 89 inch lbs.

13. Install flywheel cover and starter motor.

14. Install frame assembly and secure all bolts.

15. Install engine mount nuts at frame. Remove frame support.

16. Install power steering cooler lines at frame.

17. Install lower ball joints. Install steering gear to steering gear mounts.

18. Install steering gear retainer bolts and heat shield.

19. Install tire assembly and lower vehicle.

20. Connect negative battery cable and add engine oil.

21. Start vehicle and check for leaks.

3.8L Engine

1. Disconnect the negative battery cable.

2. Disconnect the engine torque strut from the engine.

3. Raise and support the vehicle safely.

4. Disconnect the front exhaust pipe from the manifold.

5. Remove the right front wheel and the inner fender splash shield.

6. Drain the engine oil and remove the oil filter.

7. Disconnect the oil cooler pipes and allow to hang loose for access.

8. Remove both front engine mounts from the frame.

9. Remove the flywheel cover.

10. Raise the engine assembly safely, using a suitable jack and remove the oil pan retaining bolts.

11. Lower the oil pan and disconnect the oil pump screen assembly.

12. Remove the oil pan and pump screen assembly.

To install:

13. Clean the gasket mating surfaces.

14. Use a new oil pan gasket and install the oil pan and screen assembly to the engine.

NOTE: If the rear main bearing cap is being installed, then RTV sealant must be placed on the oil pan gasket tabs that insert into the gasket groove of the outer surface on the rear main bearing cap.

15. Tighten the screen assembly bolts to 115 inch lbs. and the oil pan retaining bolts to 124 inch lbs. Do not overtighten.

16. Lower the engine and install the transaxle converter cover.

17. Install the engine mount nuts to the frame and tighten to 32 ft. lbs.

18. Install the oil cooler pipes and oil filter.

19. Install the inner fender splash shield and wheel assembly.

20. Install the front exhaust pipe to the manifold.

21. Lower the vehicle and install the engine torque strut to the engine.

22. Fill with engine oil and connect the negative battery cable.

Oil Pump

REMOVAL & INSTALLATION

2.3L Engine

1. Disconnect the negative battery cable.

2. Raise and support the vehicle safely.

3. Drain the engine oil and remove the oil pan.

4. Remove the oil pump retaining bolts and nut.

5. Remove the oil pump assembly, shims, if equipped, and screen.

To install:

6. With oil pump assembly off en-

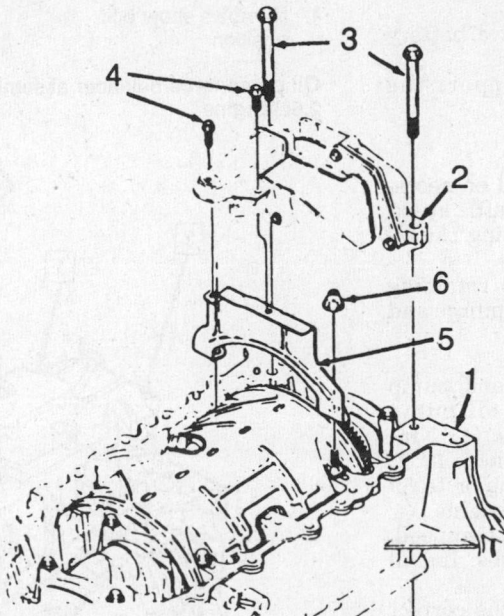

1. Cylinder block
2. Oil pump assembly
3. Oil pump to block bolt
4. Oil pump screen to brace bolt
5. Oil pump to block brace
6. Oil pump brace to block nut

Oil pump installation—2.3L engine

gine, remove 3 retaining bolts and separate the driven gear cover and screen assembly from the oil pump.

7. Install the oil pump on the block using the original shims, if equipped. Tighten the bolts to 33 ft. lbs. (45 Nm).

8. Mount a dial indicator assembly to measure backlash between oil pump to drive gear.

9. Record oil pump drive to driven gear backlash. Proper backlash is 0.010–0.018 in. When measuring, do not allow the crankshaft to move.

10. If equipped with shims, remove shims to decrease clearance and add shims to increase clearance. If no shims were present, replace the assembly if proper backlash cannot be obtained.

11. When the proper clearance is reached, rotate crankshaft ½ turn and recheck clearance.

12. Remove oil pump from block, fill the cavity with petroleum jelly and reinstall driven gear cover and screen assembly to pump. Tighten the bolts to 106 inch lbs. (13 Nm).

13. Reinstall the pump assembly to the block. Torque oil pump-to-block bolts 33 ft. lbs. (45 Nm).

14. Install the oil pan.

15. Fill the crankcase with the proper oil.

16. Connect the negative battery cable, check the oil pressure and check for leaks.

2.5L, 2.8L, 3.1L and 3.4L Engine

NOTE: On the 2.5L engine, the force balancer assembly does not have to be removed to service the oil pump or pressure regulator assemblies.

1. Disconnect the negative battery cable.

2. Raise and safely support the vehicle.

3. Drain the engine oil.

4. Remove the oil pan.

5. On 3.4L engine, it will be necessary to remove the oil pan baffle by extracting the nuts and rotating the oil pick up tube out of the way.

6. Remove the oil pump retaining bolts and remove the oil pump and pump driveshaft.

To install:

7. Install the oil pump and pump driveshaft. Tighten the oil pump mounting bolts to 30 ft. lbs. (41 Nm) for the 2.8L and 3.1L engines, 40 ft. lbs. (54 Nm) for 3.4L engine or to 89 inch lbs. (10 Nm) for 2.5L engine.

8. Install oil pan baffle, if equipped, and tighten nuts to 18 ft. lbs. Install oil pan. Lower the vehicle.

9. Fill the crankcase to the correct level with oil. Run the vehicle and check for leaks.

3.8L Engine

1. Disconnect the negative battery cable.

2. Raise and safely support the vehicle.

3. Drain the engine oil.

4. Remove the front cover assembly.

5. Remove the oil filter adapter, pressure regulator valve and spring.

6. Remove the oil pump cover attaching screws and remove the cover.

7. Remove the oil pump gears.

To install:

8. Lubricate the gears with petroleum jelly and install the gears into the housing.

9. Pack the gear cavity with petroleum jelly after the gears have been installed in the housing.

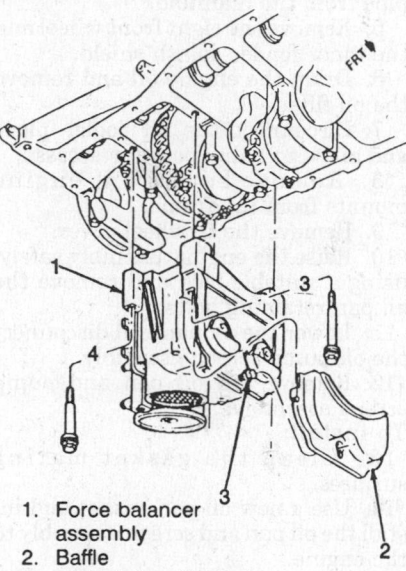

1. Force balancer assembly
2. Baffle
3. 89 inch lbs. (10 Nm)
4. Numbers show bolt position

Oil pump/force balancer assembly—2.5L engine

10. Install the oil pump cover and screws and tighten to 97 inch lbs.

11. Install the oil filter adapter with new gasket, pressure regulator valve and spring.

12. Install the front cover assembly.

13. Fill with clean engine oil and test oil pressure.

NOTE: Running the engine without measurable oil pressure will cause extensive damage.

Rear Main Bearing Oil Seal

REMOVAL & INSTALLATION

2.3L Engine

1. Disconnect the negative battery cable.

2. Remove the transaxle.

3. Remove the flywheel-to-crankshaft bolts and the flywheel.

4. Remove the oil pan-to-seal housing bolts and the block-to-seal housing bolts.

5. Remove the seal housing from the engine.

6. Place 2 blocks of equal thickness on a flat surface and position the seal housing on the 2 blocks. Remove the seal from the housing.

7. The installation is the reverse of the removal procedure. Use new gaskets when installing.

8. Connect the negative battery cable and for leaks.

2.5L Engine

1. Disconnect the negative battery cable.

2. Remove the transaxle assembly.

3. Remove the flywheel.

4. Carefully pry out the seal, using a suitable tool.

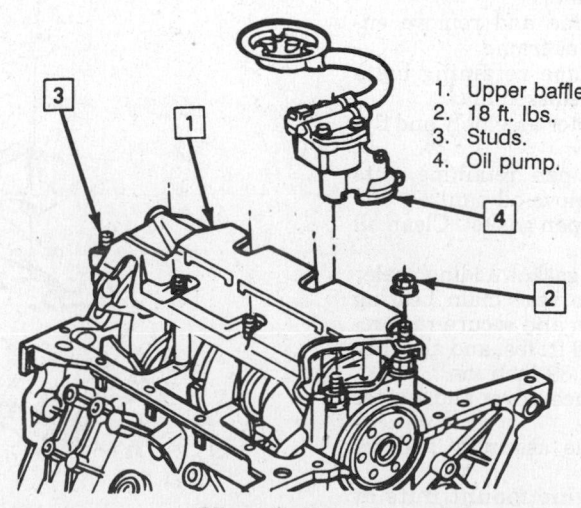

1. Upper baffle.
2. 18 ft. lbs.
3. Studs.
4. Oil pump.

Oil pan baffle—3.4L engine

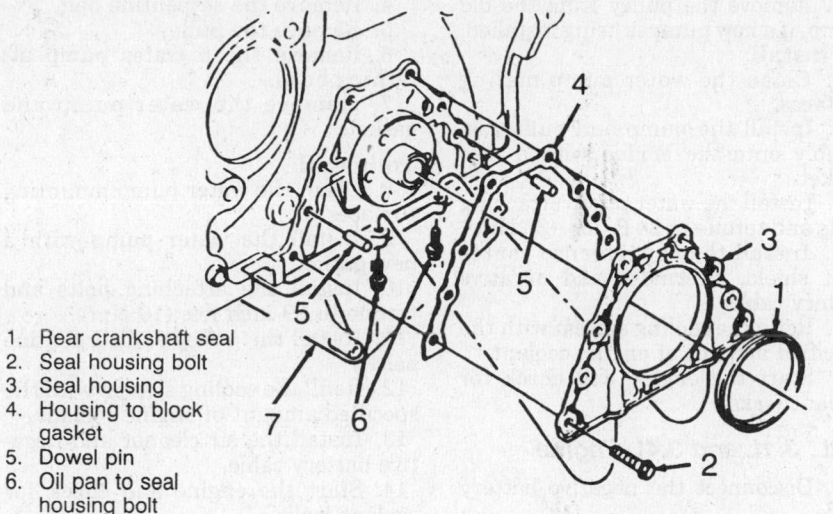

1. Rear crankshaft seal
2. Seal housing bolt
3. Seal housing
4. Housing to block gasket
5. Dowel pin
6. Oil pan to seal housing bolt
7. Oil pan

Rear crankshaft seal installation—2.3L engine

To install:

5. Clean the the block and crankshaft to seal mating surfaces.
6. Apply engine oil to the inside and outside diameter of the new seal.
7. Press the new seal evenly into place, using tool J–34924-A or equivalent.
8. Install the flywheel and transaxle and check for leaks.

2.8L, 3.1L, 3.4L AND 3.8L Engine

NOTE: These engines use a round rear oil seal that requires removal of the transaxle and flywheel.

1. Support the engine with tool J–28467 or equivalent. Raise and safely support the vehicle.
2. Remove the transaxle assembly. Remove the flywheel.
3. Using a small prybar or equivalent, insert it through the dust lip at an angle and pry the old seal from the block.
4. Inspect the seal bore and the crankshaft end for any damage.
5. Coat the inside lip of the seal with engine oil and install it on the seal installation tool J–34686 or equivalent.
6. Align the dowel pin of the tool with the dowel pin of the crankshaft. Install the tool on the crankshaft and turn the wing nut until the tool and seal are fully seated on the crankshaft.
7. Loosen the wing nut and remove the tool. Check the seal to make sure it is properly seated.
8. Install the flywheel and the transaxle.
9. Remove the engine support tool. Run the engine and check for leaks.

ALIGNMENT HOLE   DUST LIP

DOWEL PIN

SEAL

ATTACHING SCREWS   MANDRIL   COLLAR

Rear main seal replacement—2.8L, 3.1L and 3.8L engines

ENGINE COOLING

Radiator

REMOVAL & INSTALLATION

1. Disconnect the negative battery cable.
2. Remove the air cleaner, mounting stud and duct.
3. Drain the engine coolant from the radiator.
4. Remove the coolant recovery bottle.
5. Remove the engine strut brace bolts from the upper tie bar and rotate the struts and brace rearward.

NOTE: To prevent shearing of the rubber bushing, loosen the bolts on the engine strut before swinging the struts.

6. Remove the air intake resonator mounting nut, upper radiator mounting panel bolts and clamps.
7. Disconnect the cooling fan electrical connectors.
8. Remove the upper radiator mounting panel with the fans attached.
9. Remove the upper and lower radiator hoses.
10. Remove low coolant sensor and electrical connector, if used.
11. Remove the automatic transaxle cooler lines from the radiator.
12. Remove the radiator.

To install:

NOTE: If a new radiator is being used, transfer all necessary fittings from the old radiator to the new one.

13. Position the radiator into the lower insulator pads
14. Install the automatic transaxle cooler lines to radiator.
15. Install low coolant sensor and electrical connector.
16. Install the upper and lower radiator hoses and tighten the clamps.
17. Install the upper radiator mounting panel with the fans attached and connect the fan wires.
18. Install the mounting panel bolts and clamps. Torque the bolts to 89 inch lbs. (10 Nm).
19. Install the coolant recovery bottle.
20. Swing the engine strut to the proper position and tighten the bolts.
21. Refill the engine with the specified amount of engine coolant.
22. Install the air cleaner and negative battery cable. Start the engine and check for coolant leaks.

Water Pump

REMOVAL & INSTALLATION

2.3L Engine

1. Disconnect the negative battery cable.
2. Disconnect the upper engine torque strut and rotate the engine rearward.
3. Disconnect and remove the oxygen sensor, if needed.
4. Remove the exhaust heat shield and EGR valve, if equipped.
5. Remove the exhaust pipe from manifold.
6. Remove the exhaust manifold.
7. Partially drain the engine coolant.
8. Remove the coolant return hose and lower coolant pipe from the pump.
9. Remove the pump retaining bolts and pump.

To install:

10. Clean the gasket mating surfaces.

11. Install the pump, retaining bolts and torque to 10 ft. lbs. (25 Nm).

12. Install the lower coolant pipe and torque to 124 inch lbs. (14 Nm).

13. Install the coolant return hose.

14. Install the exhaust manifold and pipe, oxygen sensor, EGR valve and heat shield.

15. Return the engine to its proper position and install the torque strut.

16. Refill the engine with coolant, connect the negative battery cable, start the engine and check for coolant leaks.

2.5L Engine

1. Disconnect the negative battery cable.

2. Remove the alternator.

3. Remove the convenience center heat shield.

4. Drain about a gallon of engine coolant from the radiator. Enough to be below the water pump level.

5. Remove the 4 water pump-to-engine attaching bolts.

6. Remove the water pump and gasket.

7. Remove the pulley from the old pump, if a new pump is being installed.

To install:

8. Clean the water pump mating surfaces.

9. Install the pump and pulley assembly onto the engine with a new gasket.

10. Install the water pump attaching bolts and torque to 24 ft. lbs. (33 Nm).

11. Install the convenience center heat shield, alternator and negative battery cable.

12. Refill the cooling system with the specified amount of engine coolant.

13. Start the engine and check for coolant leaks.

2.8L, 3.1L and 3.4L Engine

1. Disconnect the negative battery cable.

2. Remove the air cleaner assembly.

3. Drain about a gallon of engine coolant from the radiator. The level must be below the water pump level.

4. Remove the serpentine belt.

5. Remove the pulley.

6. Remove the 5 water pump attaching bolts.

7. Remove the water pump and gasket.

To install:

8. Clean the water pump mounting surfaces.

9. Install the water pump with a new gasket.

10. Install the attaching bolts and torque to 89 inch lbs. (10 Nm).

11. Install the pulley and serpentine belt.

12. Refill the cooling system with specified amount of engine coolant.

13. Install the air cleaner and negative battery cable.

14. Start the engine and check for coolant leaks.

3.8L Engine

1. Disconnect the negative battery cable.

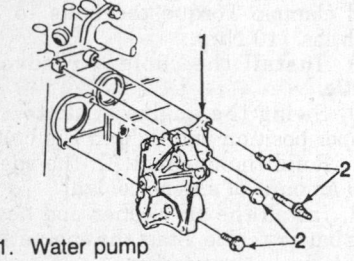

1. Water pump
2. 24 ft. lbs. (33 Nm)

Water pump mounting—2.5L engine

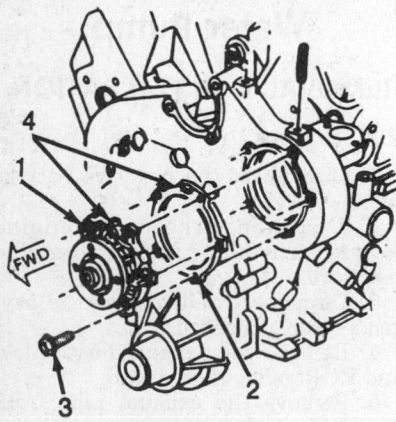

1. Water pump
2. Gasket
3. Mounting bolts
4. Pump locator—must be vertical

Water pump mounting—2.8L and 3.1 engines

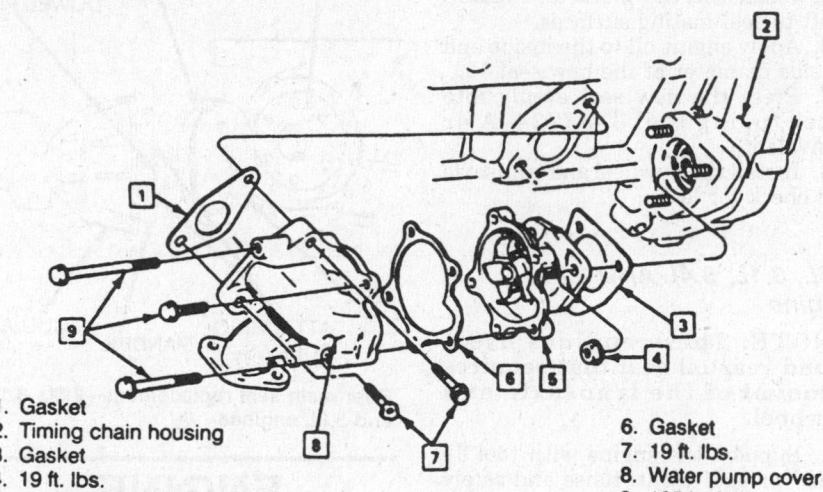

1. Gasket
2. Timing chain housing
3. Gasket
4. 19 ft. lbs.
5. Water pump body

6. Gasket
7. 19 ft. lbs.
8. Water pump cover
9. 125 inch lbs.

Water pump mounting—2.3L engine

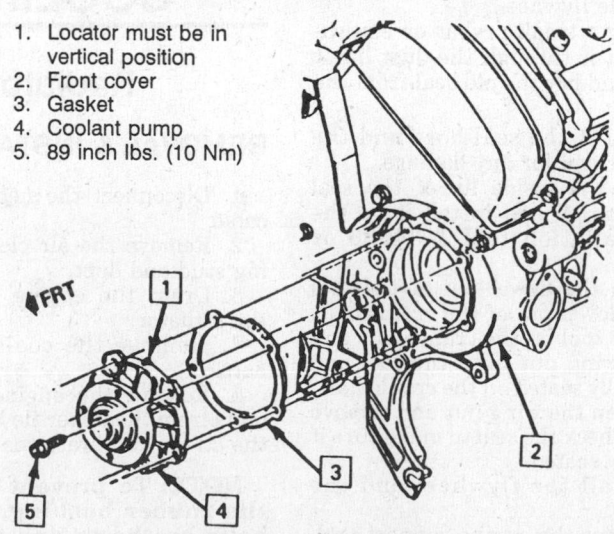

1. Locator must be in vertical position
2. Front cover
3. Gasket
4. Coolant pump
5. 89 inch lbs. (10 Nm)

Coolant pump—3.4L engine

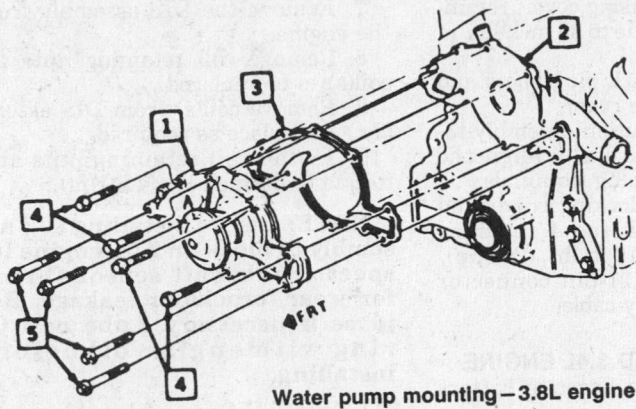

1. Coolant pump
2. Front cover
3. Gasket
4. 13 ft. lbs.
5. 22 ft. lbs.

Water pump mounting—3.8L engine

2. Drain the engine coolant from the radiator.

3. Disconnect the coolant recovery reservoir.

4. Remove the serpentine belt.

NOTE: If more access is needed, remove the inner fender electrical cover.

5. Remove the pulley.

6. Remove the 8 water pump attaching bolts.

7. Remove the water pump and gasket.

To install:

8. Clean the water pump mounting surfaces.

9. Install the water pump with a new gasket.

10. Install the attaching bolts and torque the long bolts to 22 ft. lbs. (30 Nm) and the short bolts to 13 ft. lbs. (18 Nm).

11. Install the pulley and serpentine belt. Tighten the pulley to 115 inch lbs.

12. Reconnect the coolant recovery reservoir.

13. Refill the cooling system with the specified amount of engine coolant.

14. Install the negative battery cable.

15. Start the engine and check for coolant leaks.

Cooling System Bleeding

To insure complete filling of the cooling system, it is necessary to bleed the system.

1. Disconnect the negative battery cable.

2. Park vehicle on level surface.

3. Remove thermostat housing cap and thermostat or open bleed vents:

a. On 2.5L engine, remove the thermostat housing cap and thermostat.

b. On 2.3L and 2.8L engines open bleed valve on thermostat housing 2–3 turns.

c. On 3.1L engine, open the air bleed vents on the thermostat housing and the throttle body return pipe above coolant pump. Open vents 2–3 turns.

d. On 3.4L engine, open the air bleed vents on the thermostat housing and the heater coolant inlet pipe by the master brake cylinder. Open vents 2–3 turns.

e. On 3.8L engine, open air bleed vent on thermostat housing. Open 2–3 turns.

4. Fill cooling system with coolant to base of radiator neck.

5. Reinstall or replace the thermostat and housing and close air vents.

6. Fill coolant reservoir to proper level with ethylene glycol/water mixture.

7. Reconnect negative battery cable. Start vehicle and let engine reach operating temperature adding coolant as needed. Check the cooling system for leaks.

ENGINE ELECTRICAL

NOTE: Disconnecting the negative battery cable on some vehicles may interfere with the functions of the on board computer systems and may require the computer to undergo a relearning process, once the negative battery cable is reconnected.

Distributorless Ignition

REMOVAL & INSTALLATION

Ignition Coil

2.3L ENGINE

1. Disconnect the negative battery cable.

2. Disconnect the 11-pin (IDI) harness connector at the ignition cover.

3. Remove the 4 ignition cover assembly-to-cylinder head bolts.

4. Remove the ignition assembly from the vehicle.

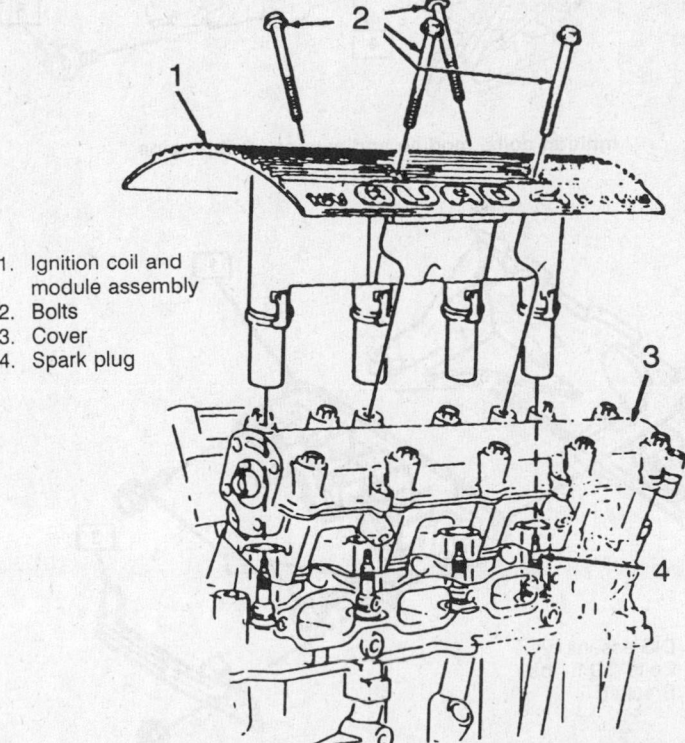

1. Ignition coil and module assembly
2. Bolts
3. Cover
4. Spark plug

Ignition coil and module assembly—2.3L engine

5. Remove the 4 coil housing-to-cover screws.

NOTE: Be careful not to damage the module terminals when pulling the coil assemblies from the module. Pull slowly and carefully away from the ignition assembly.

6. Disconnect the coil harness connectors.

7. Remove the coils, contacts and seals from the cover.

NOTE: If the spark plug boots stick, use a spark plug connector removing tool J–36011 or equivalent, to remove with a twisting motion.

To install:

8. Install the coils-to-cover.
9. Connect the coil harness.
10. Using new seals, install the seals into the housing.
11. Install the contacts-to-housing. use petroleum jelly to retain the contact in place.

12. Install the housing cover, retaining screws and torque to 35 inch lbs. (4 Nm).
13. Install the spark plug boots and retainers-to-ignition cover.
14. Install the ignition assembly-to-cylinder heads. Carefully align the boots to the spark plug terminals.
15. Apply thread locking compound to the bolts. Install the 4 retaining bolts and torque to 19 ft. lbs. (26 Nm).
16. Connect the 11-pin connector and negative battery cable.

2.5L, 2.8L, 3.1L AND 3.4L ENGINE

1. Disconnect the negative battery cable.
2. Raise and safely support the vehicle.
3. Label each spark plug wire for proper installation.
4. Remove the spark plug wires from the ignition coils.
5. Remove the DIS electrical connectors.
6. Remove the 3 DIS assembly to block bolts.

7. Remove the DIS assembly from the engine.
8. Remove coil retaining nuts for coil(s) to be replaced.
9. Remove coil(s) from DIS assembly and replace as required.
10. Install coil retaining nuts and torque to 40 inch lbs.(4.5 Nm).

NOTE: Befofe installing DIS assembly to block on 2.5L engine inspect crankshaft sensor O-ring for wear, cracks or leakage. Replace if necessary. Lube new O-ring with engine oil before installing.

11. Install DIS assembly to block and torque to 20 ft. lbs. (27 Nm).
12. Install the spark plug wires and electrical connectors.
13. Reconnect negative battery cable.

3.8L ENGINE

1. Disconnect negative battery cable.
2. Label and disconnect spark plug wires.
3. Remove 2 screws securing coil to ignition module.
4. Remove coil(s) from module.
5. Install coil using 2 screws, torque to 40 inch lbs.
6. Reconnect spark plug wires and negative battery cable.

Ignition Module

2.3L ENGINE

1. Disconnect the negative battery cable.
2. Remove the electrical harness, 4 retaining bolts and ignition cover assembly from the vehicle.
3. Remove the 4 housing screws, coil housing and coil harness connectors.
4. Remove the 3 module-to-housing cover screws and module.

To install:

5. If replacing the module or a coil, the new unit should come with a package of silicone grease, if not, purchase a tube at your local parts distributor. Spread the grease on the metal face of

1. No. 2-3 ignition coil
2. No. 1-4 ignition coil
3. Ignition module
4. Crankshaft sensor
5. Bolt (20 ft. lbs.)

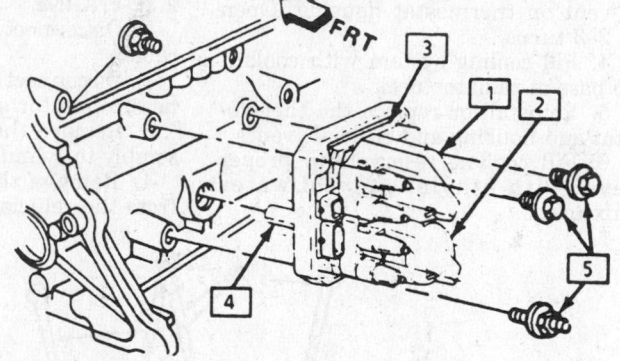

Ignition coils, module and sensor—2.5L engine

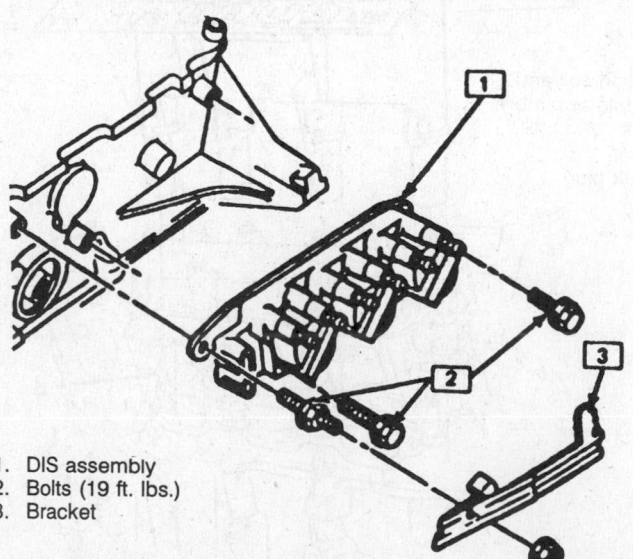

1. DIS assembly
2. Bolts (19 ft. lbs.)
3. Bracket

Coil and module assembly mounting—2.8L and 3.1L engines

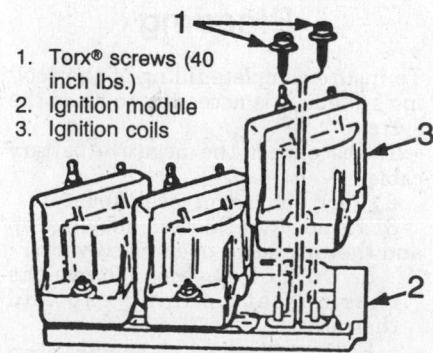

1. Torx® screws (40 inch lbs.)
2. Ignition module
3. Ignition coils

Ignition coil mounting—3.8L engine

the module and on the cover where the module seats. The grease is used for module cooling.

6. Install the module-to-cover, module screws and torque to 35 inch lbs. (4 Nm).

7. Install the coil harness-to-module and housing cover screws. Torque the screws to 35 inch lbs. (4 Nm).

8. Install the spark plug boots and retainers.

9. Install the ignition system-to-cylinder head while carefully aligning the spark plug boots with the terminals.

10. Apply thread locking compound to the bolts. Install the 4 retaining bolts and torque to 19 ft. lbs. (26 Nm).

11. Connect the 11-pin connector and negative battery cable.

2.5L ENGINE

1. Disconnect the negative battery cable.

2. Label each spark plug wire for proper installation.

3. Remove the spark plug and module electrical connectors from the DIS assembly.

4. Remove the 3 DIS assembly-to-engine attaching bolts.

NOTE: Be careful not to damage the crankshaft sensor and module terminals when pulling the DIS assemblies from the engine. Pull slowly and carefully away from the engine.

5. Remove the DIS assembly from the engine.

6. Remove the coils from the DIS assembly.

7. Remove the module from assembly plate.

To install:

8. Carefully engage the sensor to module terminals and install module assembly to plate.

9. Reinstall coils, torque screws to 45 inch lbs. (5 Nm).

10. Install the DIS assembly and torque bolts to 20 ft. lbs. (27 Nm).

11. Reconnect the spark plugs and module electrical connectors to their original positions.

3.1L AND 3.4L ENGINE

1. Disconnect the negative battery cable.

2. Raise the vehicle and support safely.

3. Label each spark plug wire for proper installation.

4. Remove the spark plug and module electrical connectors from the DIS assembly.

5. Remove the 3 DIS assembly attaching bolts.

6. Remove the DIS assembly from the engine.

7. Remove the coils from the DIS as-

sembly. Remove the module from assembly plate.

To install:

8. Install the module to the assembly plate.

9. Install the coil assemblies to the module and torque the screws to 45 inch lbs. (5 Nm).

10. Install the DIS assembly and attaching bolts to the engine and torque to 20 ft. lbs. (27 Nm).

11. Reconnect the spark plug and module electrical connectors to their original positions.

12. Reconnect the negative battery cable.

3.8L ENGINE

1. Disconnect the negative battery cable.

2. Disconnect the 14-way connector at the ignition module.

3. Disconnect and label the spark plug wires.

4. Remove the 6 screws securing the coil assemblies to the ignition module and disconnect the coils from the module.

5. Remove the 3 nuts and washers securing the ignition module assembly to the bracket and remove the module.

To install:

6. Install the coils onto the module and tighten the 6 retaining screws to 40 inch lbs.

7. Install the 3 nuts and washers securing the ignition module assembly to the bracket and tighten to 70 inch lbs.

8. Connect the spark plug wires.

9. Connect the 14-way connector to the ignition module.

10. Reconnect negative battery cable.

Ignition Timing

ADJUSTMENT

Because the reluctor is an integral part of the crankshaft and the crankshaft sensor is mounted in a fixed position, timing adjustment is not possible.

Alternator

PRECAUTIONS

Several precautions must be observed with alternator equipped vehicles to avoid damage to the unit.

• If the battery is removed for any reason, make sure it is reconnected with the correct polarity. Reversing the battery connections may result in damage to the 1-way rectifiers.

• When utilizing a booster battery as a starting aid, always connect the positive to positive terminals and the negative terminal from the booster

battery to a good engine ground on the vehicle being started.

• Never use a fast charger as a booster to start vehicles.

• Disconnect the battery cables when charging the battery with a fast charger.

• Never attempt to polarize the alternator.

• Do not use test lights of more than 12 volts when checking diode continuity.

• Do not short across or ground any of the alternator terminals.

• The polarity of the battery, alternator and regulator must be matched and considered before making any electrical connections within the system.

• Never separate the alternator on an open circuit. Make sure all connections within the circuit are clean and tight.

• Disconnect the battery ground terminal when performing any service on electrical components.

• Disconnect the battery if arc welding is to be done on the vehicle.

BELT TENSION ADJUSTMENT

A single serpentine belt is used to drive all engine mounted components. Drive belt tension is maintained by a spring loaded tensioner.

NOTE: The drive belt tensioner can control the belt tension over a wide range of belt lengths; however, there are limits to the tensioners ability to compensate for various belt lengths. Installing the wrong size belt and using the tensioner outside of its operating range can result in poor tension control and damage to the tensioner, drive belt and driven components.

REMOVAL & INSTALLATION

2.3L Engine

1. Disconnect the negative battery cable.

2. Remove the electrical center fuse block shield.

3. Remove the serpentine belt, by removing the belt guard, and lifting or rotating the tensioner, using a breaker bar.

4. Label and remove the alternator electrical connectors.

5. Remove the alternator brace bolt, rear bolt and front bolt.

NOTE: Use extreme care when removing the alternator, not to damage the air conditioning compressor and condensor hose.

6. Lift the alternator out between the engine lifting eyelet and the air conditioning compressor.

To install:

7. Install the alternator, front, rear and brace retaining bolts.

8. Torque the long bolt to 40 ft. lbs. (54 Nm), the short bolt to 19 ft. lbs. (26 Nm) and the brace bolt to 18 ft. lbs. (25 Nm).

9. Connect the alternator electrical connectors, serpentine belt and fuse block shield.

10. Connect the negative battery cable and check for proper operation.

2.5L Engine

1. Disconnect the negative battery cable.

2. Remove the serpentine belt.

3. Remove the electrical connectors from the back of the alternator.

4. Remove the rear (first), front attaching bolts and heat shield.

5. Remove the alternator assembly carefully making sure all wires are disconnected.

To install:

6. Position the alternator into the mounting bracket.

7. Install the front and rear mounting bolts but do not tighten.

8. Install the heat shield with the rear mounting bolts.

9. Install the electrical connectors and tighten the battery cable nut.

10. Torque the mounting bolts to 18 ft. lbs. (25 Nm).

11. Install the serpentine belt.

12. Reconnect the negative battery cable.

2.8L and 3.1L Engine

1. Disconnect the negative battery cable. Remove the air cleaner assembly.

2. Remove the serpentine belt.

3. Remove the electrical connectors from the back of the alternator.

4. Remove the rear and front attaching bolts, and bolt from brace to alternator.

5. Remove the alternator assembly carefully making sure all wires are disconnected.

NOTE: If alternator brace is removed, studs must be retightened before installation or damage to the brace may result.

To install:

6. Position the alternator into the mounting bracket.

7. Install brace to alternator bolt but do not tighten.

8. Install the front and rear mounting bolts. Torque the mounting bolts as follows:

    a. Long bolt to 35 ft. lbs. (47 Nm)

    b. Short bolt to 18 ft. lbs. (25 Nm)

    c. Bracket bolt to 18 ft. lbs. (25 Nm)

9. Check that tightening of the brace bolts did not bind alternator.

10. Install the electrical connectors and tighten the battery cable nut.

11. Install the serpentine belt.

12. Install the air cleaner and negative battery cable.

3.4L Engine

1. Disconnect the negative battery cable.

2. Remove air cleaner assembly.

3. Remove coolant recovery reservoir and set aside.

4. Remove serpentine belt.

5. Raise and safely support vehicle.

6. Remove power steering pipe retaining clip nut from upper alternator stud and remove alternator stud.

7. Remove right front tire and wheel assembly.

8. Separate lower ball joint from lower control arm.

9. Remove halfshaft from transaxle.

10. Remove right hand engine splash shield.

11. Disconnect connectors and wires from alternator.

12. Remove brace bolt from alternator and loosen brace at engine block.

13. Remove alternator lower mounting bolt and alternator.

To install:

14. Install alternator and loosly install all mounting bolts. If replacement alternator does not fit into mounts, remove adhesive-backed shim from rear of alternator bracket.

15. Tighten alternator lower mounting bolts to 61 ft. lbs. (83 Nm).

16. Install connectors and wires to alternator.

17. Install right hand engine splash shield.

18. Reinstall halfshaft.

19. Install lower ball joint to lower control arm.

20. Reinstall tire and wheel assembly and lower vehicle.

21. Install upper alternator stud and power steering pipe retaining clip nut.

22. Reinstall serpentine belt and coolant recovery reservoir.

23. Reinstall negative battery cable.

24. Install air cleaner assembly.

3.8L Engine

1. Disconnect the negative battery cable.

2. Remove the serpentine belt.

3. Remove the electrical connectors from the back of the alternator.

4. Remove the nut and the positive battery connector from the **BAT** terminal.

5. Remove the alternator mounting bolts and remove the alternator from the vehicle.

To install:

6. Installation is the reverse of removal. Tighten all mounting bolts to 20 ft. lbs. using the following sequence:

    a. Alternator attaching bolt to the direct fire mounting bracket/rear brace.

    b. Alternator attaching bolt to the power steering and tensioner pulley bracket.

    c. Alternator brace bolt to engine.

NOTE: Make sure tightening bolts do not bind alternator.

Starter

REMOVAL & INSTALLATION

2.3L Engine

1. Disconnect the negative battery cable.

2. Remove the air cleaner and inlet hose from the throttle body.

3. Remove and plug the coolant reservoir hose at the radiator filler neck.

4. Remove the coolant reservoir.

5. Remove the intake manifold brace bolts.

6. Place a drain pan under the oil filter and remove the filter.

7. Remove the starter retaining bolts, lower the starter onto the frame member and disconnect the starter electrical connectors.

To install:

8. Position the starter into the vehicle, connect the electrical connectors and torque the retaining bolts to 32 ft. lbs. (43 Nm).

9. Install a new oil filter and add engine oil, as needed.

10. Install the intake manifold brace, coolant reservoir and hoses.

11. Add coolant, if needed.

12. Install the air cleaner and inlet hose.

13. Connect the negative battery cable and check for proper operation.

2.5L Engine

1. Disconnect the negative battery cable.

2. Raise and support the vehicle safely.

3. Remove the flywheel inspection cover bolts and cover.

4. Remove the stud from the starter support bracket.

5. Remove the 2 starter mounting bolts and shim, if equipped.

6. Remove the starter motor. Be careful not to damage the starter wires by letting the starter hang.

7. While holding the starter motor,

disconnect the starter electrical connectors from the starter solenoid.

8. Remove the starter from the rear bracket.

To install:

9. Install the support bracket to the starter.

10. Install the starter adjustment shims, if equipped.

11. Position the starter to the engine mounting flange and torque the bolts to 32 ft. lbs. (43 Nm).

12. Install the bracket-to-engine and torque the stud to 18 ft. lbs. (25 Nm).

13. Install the inspection cover.

14. Lower the vehicle and connect the starter electrical wires. Reconnect the negative battery cable.

2.8L, 3.1L and 3.4L Engine

1. Remove the air cleaner.

2. Disconnect the negative battery cable.

3. Raise the vehicle and support it safely.

4. If equipped with an engine oil cooler, remove the engine oil, oil filter and position the hose next to the starter motor to the side.

5. Remove the nut from the brace at the air conditioning compressor, nut from the brace at the engine and the brace.

6. Remove the flywheel inspection cover.

7. Remove the starter bolts and shims, if equipped. Do not let the starter hang from the starter wires.

8. Remove the starter wires from the solenoid and remove the starter.

To install:

9. While supporting the starter, connect the starter wires at the solenoid.

10. Install the starter motor-to-engine mount with the shims, if equipped, and the mounting bolts. Torque the bolts to 32 ft. lbs. (43 Nm).

11. If equipped with an engine oil cooler, reposition the hose next to the starter motor, install the oil filter and refill the engine with the proper amount of engine oil.

12. Install the flywheel inspection cover and tighten the bolts.

13. Install the starter support brace to the air conditioning compressor and torque the nut to 23 ft. lbs. (31 Nm).

14. Lower the vehicle, reconnect the negative battery cable and install the air cleaner assembly.

3.8L Engine

1. Disconnect the negative battery cable.

2. If necessary, remove the right side cooling fan.

3. Remove the serpentine drive belt.

4. Disconnect the air conditioning

compressor upper support brace and lay the compressor in the fan opening.

5. Raise and support the vehicle safely.

6. Disconnect the engine oil cooler lines at the flex connector.

7. Remove the flywheel inspection cover.

8. Remove the starter motor retaining bolts and remove the starter motor and shims, if used.

9. Disconnect the starter motor wiring and remove the starter from the vehicle.

To install:

10. Position the starter motor and shims, if used, to the engine and tighten the mounting bolts to 32 ft. lbs.

11. Connect the electrical connectors to the starter terminals and tighten the battery nut to 80 inch lbs. and the S terminal nut to 35 inch lbs.

12. Install the flywheel inspection cover. Tighten to 89 inch lbs. (10 Nm).

13. Connect the engine oil cooler lines at the flex connector.

14. Lower the vehicle and install the air conditioner compressor.

15. Install the serpentine drive belt, cooling fan and negative battery cable.

EMISSION CONTROLS

Please refer to "Emission Controls" in the Unit Repair section for system maintenance procedures. Due to the complex nature of modern electronic engine control systems, comprehensive diagnosis and testing procedures fall outside the confines of this repair manual. For complete information on diagnosis, testing and repair procedures concerning all modern engine and emission control systems, please refer to "Chilton's Guide to Fuel Injection and Electronic Engine Controls".

FUEL SYSTEM

Fuel System Service Precautions

Safety is the most important factor when performing not only fuel system maintenance but any type of maintenance. Failure to conduct maintenance and repairs in a safe manner

may result in serious personal injury or death. Maintenance and testing of the vehicle's fuel system components can be accomplished safely and effectively by adhering to the following rules and guidelines.

- To avoid the possibility of fire and personal injury, always disconnect the negative battery cable unless the repair or test procedure requires that battery voltage be applied.

- Always relieve the fuel system pressure prior to disconnecting any fuel system component (injector, fuel rail, pressure regulator, etc.), fitting or fuel line connection. Exercise extreme caution whenever relieving fuel system pressure to avoid exposing skin, face and eyes to fuel spray. Please be advised that fuel under pressure may penetrate the skin or any part of the body that it contacts.

- Always place a shop towel or cloth around the fitting or connection prior to loosening to absorb any excess fuel due to spillage. Ensure that all fuel spillage (should it occur) is quickly removed from engine surfaces. Ensure that all fuel soaked cloths or towels are deposited into a suitable waste container.

- Always keep a dry chemical (Class B) fire extinguisher near the work area.

- Do not allow fuel spray or fuel vapors to come into contact with a spark or open flame.

- Always use a backup wrench when loosening and tightening fuel line connection fittings. This will prevent unnecessary stress and torsion to fuel line piping. Always follow the proper torque specifications.

- Always replace worn fuel fitting O-rings with new. Do not substitute fuel hose or equivalent where fuel is installed.

RELIEVING FUEL SYSTEM PRESSURE

2.5L Engine

1. Remove the fuel filler

2. Remove the fuel pump the fuse block located in the compartment.

3. Start the engine and the engine stops due to the

4. Crank the engine for 3 ensure all pressure is relieved

5. Make sure the negative cable is disconnected.

Except 2.5L Engine

1. Disconnect the negative battery cable. Loosen fuel filler cap.

2. Connect fuel pressure gauge J–34730–1 or equivalent, to the fuel pressure connection.

3. Wrap a shop cloth around the fitting while connecting the gauge to catch any leaking fuel.

4. Install the bleed hose into an approved container and open the valve. Connect the negative battery cable.

5. When the repair to the fuel system is complete check all of the fittings for leaks.

Fuel Filter

REMOVAL & INSTALLATION

1. Disconnect the negative battery cable.

2. Relieve fuel system pressure.

3. Raise and support the vehicle safely.

4. Disconnect the fuel lines from the filter. To reduce fuel spillage, place a shop towel over the fuel lines before disconnecting.

5. Remove the clamp and filter from the vehicle.

6. Loosely install the new filter. Using new O-ring seals, install the fuel lines to the filter. Use a backup wrench to prevent the filter from turning and O-ring damage. Torque the fittings to 16 ft. lbs. (22 Nm).

7. Secure the filter to the vehicle. Tighten fuel filler cap.

8. Reconnect the negative battery cable. Lower the vehicle and start the engine to check for fuel leaks.

9. If equipped with the 3.1L or 3.4L engine, perform the Idle Learn procedure to allow the ECM memory to be updated with the correct IAC valve pintle position for a stable idle speed.

   a. Install a Tech 1 scan tool.

   b. Turn the ignition to the **ON** position, engine not running.

   c. Select **IAC SYSTEM**, then **IDLE LEARN** in the **MISC TEST** mode.

   d. Proceed with idle learn as directed by the scan tool.

Electric Fuel Pump

REMOVAL & INSTALLATION

1. Disconnect the negative battery cable.

2. Drain all fuel from the fuel tank.

3. Raise and safely support the vehicle. Support the fuel tank and remove the retaining straps.

4. Lower the fuel tank slightly and disconnect the fuel lines, hoses and the sending unit electrical connectors.

5. Remove the tank from the vehicle.

6. Remove the sending unit retaining cam using tool J-24187 or equivalent and remove the sending unit assembly from the tank.

7. Replace O-ring and filter on pump assembly and install the unit into the tank.

8. Raise the tank into position and attach all fuel lines, hoses and electrical connectors to the tank.

9. Install the retaining straps. Tighten the tank retaining strap bolts to 26 ft. lbs. (34 Nm).

10. Lower the vehicle and refill the tank. Connect the negative battery cable. Turn the ignition **ON** for 2 seconds, then turn ignition **OFF** and check for leaks. Start vehicle and recheck for leaks.

Fuel Injection

IDLE SPEED ADJUSTMENT

Idle speed and mixture are electronically controlled by the ECM. No adjustments are possible.

DRIVE AXLE

Halfshaft

REMOVAL & INSTALLATION

If equipped with an automatic transaxle, the left halfshaft uses a female spline which installs over a stub shaft protruding from the transaxle. The right halfshaft uses a male spline and interlocks with the transaxle using barrel type snaprings.

If equipped with a manual transaxle, the left halfshaft uses a male spline locking into the transaxle. The right halfshaft axle uses a female spline that installs into the intermediate axle shaft.

1. With the weight of the vehicle on the tires, loosen the hub nut.

2. Raise and safely support the vehicle.

3. Remove the hub nut.

4. Install boot protectors on the boots.

5. Remove the brake caliper with the line attached and safely support it out of the way; do not allow the caliper to hang from the line. Place the ABS sensors out of the way.

6. Remove the brake rotor and caliper mounting bracket.

7. Remove the hub/bearing to strut housing bolts. Pull the hub/bearing out of the strut bracket.

8. Using a halfshaft removal tool J-33008 or equivalent and the extension tool J-29794 or equivalent, remove the halfshafts from the transaxle and support them safely.

9. Pull the axle and hub assembly through the strut housing. Using a spindle remover tool J-28733 or equivalent, remove the halfshaft from the hub and bearing.

To install:

10. Loosely place the halfshaft on the transaxle and in the hub and bearing.

11. Properly position the steering knuckle to the strut bracket and install the bolt. Torque the bolts to 133 ft. lbs. (178 Nm).

12. Install the ABS sensor, brake rotor, caliper bracket and caliper. Place a holding device in the rotor to prevent it from turning.

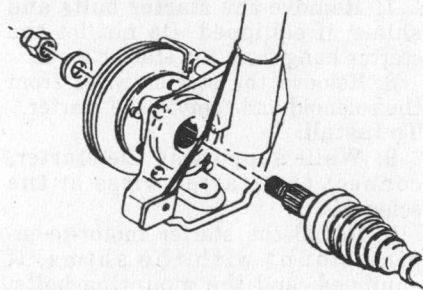

Removing the halfshaft from the hub/knuckle assembly

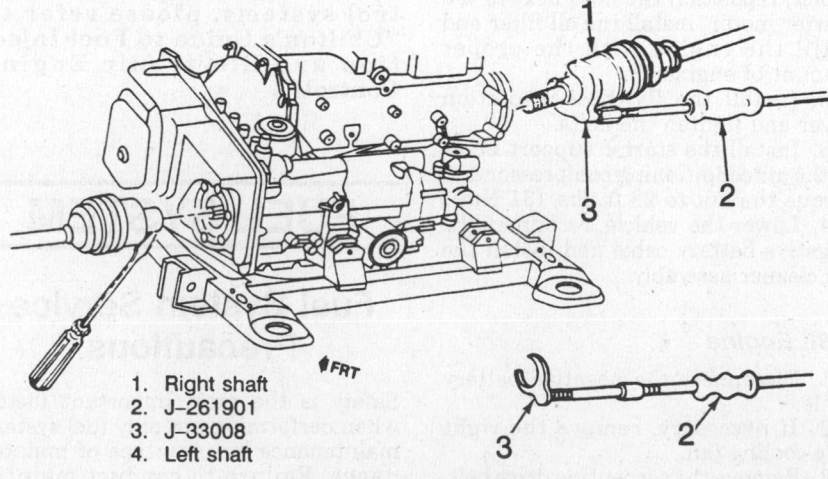

1. Right shaft
2. J-261901
3. J-33008
4. Left shaft

Removing the halfshafts

13. Install the hub nut and washer. Torque the nut to 71 ft. lbs. (95 Nm).

14. Seat the halfshafts into the transaxle.

15. Verify that the shafts are seated by grasping the CV-joint and pulling outwards. Do not grasp the shaft. If the snapring is seated, the halfshaft will remain in place.

16. Remove the boot protectors and lower the vehicle.

17. When the vehicle is lowered with the weight on the wheels, final torque the hub nut to 184 ft. lbs. (250 Nm).

Front Wheel Hub and Bearing

The vehicles are equipped with sealed hub and bearing assemblies. The hub and bearing assemblies are non-serviceable. If the assembly is damaged, the complete unit must be replaced.

REMOVAL & INSTALLATION

1. Disconnect the negative battery cable.

2. Loosen the drive axle shaft nut and washer 1 turn.

3. Raise the vehicle and support it safely.

4. Remove the rear wheel, caliper, bracket and rotor.

5. Remove the halfshaft nut and washer.

6. Loosen the 4 hub/bearing-to-knuckle attaching bolts.

7. Using tool J–28733–A or equivalent, push the halfshaft splines back out of the hub/bearing.

8. Remove the ABS sensor, if equipped, and position out of the way.

9. Protect the halfshaft boots, remove the hub/bearing assembly attaching bolts and remove the hub/bearing assembly.

To install:

10. Install the hub/bearing assembly onto the knuckle. Install the 4 attaching bolts and torque to 52 ft. lbs. (70 Nm).

11. Install the ABS sensor, if equipped.

12. Install the rotor, caliper and bracket.

13. Install the rear wheel and torque the lug nut to 100 ft. lbs. (135 Nm).

14. Lower the vehicle and torque the hub nut to 184 ft. lbs. (250 Nm).

MANUAL TRANSAXLE

For further information on transmissions/transaxles, please refer to "Chilton's Guide to Transmission Repair".

Transaxle Assembly

REMOVAL & INSTALLATION

NOTE: Before performing any maintenance that requires the removal of the slave cylinder, transaxle or clutch housing, the clutch master cylinder pushrod must first be disconnected from the clutch pedal. Failure to disconnect the pushrod will result in permanent damage to the slave cylinder if the clutch pedal is depressed with the slave cylinder disconnected.

1. Disconnect the negative battery cable.

2. Install the engine support tool J–28467 or equivalent.

3. Remove the air cleaner housing and intake tube. Disconnect the clutch slave cylinder from the transaxle.

4. Disconnect the electrical connection at the speed sensor assembly. Disconnect the clutch and shift cables from the transaxle.

5. Remove the exhaust crossover pipe at the left manifold and remove the EGR tube from the crossover.

6. Loosen the crossover-to-right exhaust manifold clamp and move the crossover pipe to gain access to the transaxle bolts for V6 engine.

7. Remove the 2 upper transaxle mounting bolts and remove the 2 upper mounting studs. Leave 1 bottom bolt and stud attached.

8. Disconnect the electrical connection at the backup light switch. Raise and safely support the vehicle.

9. Drain the transaxle fluid. Remove the clutch housing cover. Remove both front tire assemblies.

10. Remove the inner fender splash shields from both side of the vehicle. Disconnect the power steering lines from the frame.

11. Remove the rack and pinion heat shield and remove the rack and pinion from the frame.

12. Disconnect the right and left ball joints. Remove the upper transaxle mount retaining bolts. Remove the lower engine mount retaining nuts.

13. Remove the sub-frame retaining bolts and remove the sub-frame from the vehicle. Remove the starter and support it aside.

14. Remove the right and left halfshafts from the transaxle. Support the halfshafts to the frame with wire to prevent damage to the CV-joints. Support the transaxle and remove the remaining bolt and stud. Remove the transaxle from the vehicle.

To install:

15. Align the transaxle with the engine and install. Install the lower transaxle-to-engine mounting bolt and stud, tightening to 55 ft. lbs. (75 Nm).

16. Install the starter assembly. Install the left and right halfshaft.

17. Install the sub-frame and retaining bolts. Install the lower engine mount retaining nuts.

18. Install the upper transaxle retaining bolts, tightening to 55 ft. lbs. (75 Nm) Install the right and left ball joints to the steering knuckles.

19. Install the rack and pinion, heat shield and lines to the frame. Install the right and left inner fender splash shields.

20. Install the clutch housing cover, tighten the screws to 115 inch lbs. (13 Nm). Lower the vehicle.

21. Attach the crossover pipe to the manifolds and attach the EGR pipe to the crossover.

22. Attach the shift and clutch cables to the transaxle. Connect all of the electrical connectors. Install the air cleaner housing and tube. Remove the engine support tool.

23. Fill the transaxle with fluid. Connect the negative battery cable.

LINKAGE ADJUSTMENT

The shift control and cables are preset at the factory and require no adjustments.

CLUTCH

Clutch Assembly

REMOVAL & INSTALLATION

NOTE: Before any service that requires removal of the slave cylinder, the master cylinder pushrod must be disconnected from the clutch pedal and the connection in the hydraulic lines must be separated using tool J–36221 or equivalent. If not disconnected, permanent damage to the slave cylinder will occur if the clutch pedal is depressed while the system is not resisted by clutch loads.

1. Disconnect the negative terminal from the battery.

2. From inside the vehicle, remove the sound insulator panel.

3. Disconnect the clutch master cylinder pushrod from the clutch pedal and disconnect the quick connect fitting in the hydraulic line. Remove the actuator from the transaxle housing.

4. Remove the transaxle.

5. With the transaxle removed,

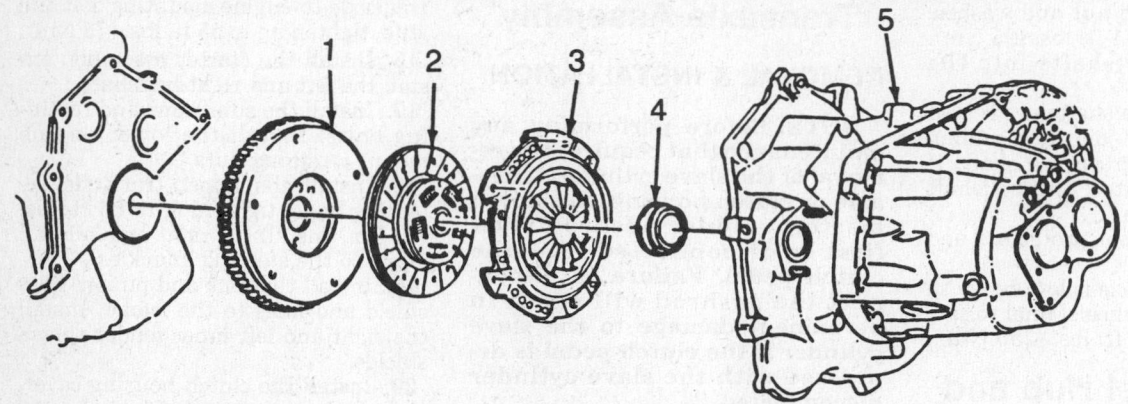

1. Flywheel
2. Driven plate assembly
3. Clutch cover assembly
4. Clutch release bearing
5. Transaxle

Exploded view of the clutch assembly

matchmark the pressure plate and flywheel assembly to insure proper balance during reassembly.

6. Loosen the pressure plate-to-flywheel bolts, a turn at a time, until the spring pressure is removed.

7. Support the pressure plate and remove the bolts.

8. Remove the pressure plate and disc assembly; be sure to note the flywheel side of the clutch disc.

To install:

9. Clean and inspect the clutch assembly, flywheel, release bearing, clutch fork and pivot shaft for signs of wear. Replace any necessary parts.

10. Position the clutch disc and pressure plate in the appropriate position, support the assembly with alignment tool J–29074, J–35822 or equivalent.

NOTE: Make sure the clutch disc is facing the same direction it was removed. If the same pressure plate is being reused, align the marks made during removal and install, install the pressure plate retaining bolts and tighten them gradually and evenly.

11. Remove the alignment tool and torque the pressure plate-to-flywheel bolts to 15 ft. lbs. (21 Nm). Lightly lubricate the clutch fork ends. Fill the recess ends of the release bearing with grease. Lubricate the input shaft with a light coat of grease.

12. Install the transaxle assembly. Install the clutch master cylinder pushrod and install the sound insulator panel.

NOTE: The clutch lever must not be moved towards the flywheel until the transaxle is bolted to the engine. Damage to the transaxle, release bearing and clutch fork could occur if this is not followed.

13. Connect the negative battery cable. Bleed the clutch system and check the clutch operation.

PEDAL HEIGHT/FREE-PLAY ADJUSTMENT

The clutch system is a hydraulic linkage system that provides automatic clutch adjustment and determines the clutch pedal position. No adjustment of clutch linkage or pedal position is required or possible.

Clutch Master Cylinder, Actuator and Reservoir

REMOVAL & INSTALLATION

NOTE: The factory hydraulic system is serviced as a single assembly. Replacement hydraulic assemblies are pre-filled with fluid and do not require bleeding. Individual components of the system are not available separately. Check with an aftermarket part supplier to see if individual components can be purchased separately.

1. Disconnect the negative battery cable.

2. Remove the sound insulator inside the vehicle and disconnect the master cylinder pushrod at the clutch pedal.

3. Remove the left upper secondary cowl panel.

4. Remove the 2 master cylinder reservoir-to-strut tower retaining nuts.

5. Remove the anti-rotation screw located next to the master cylinder flange at the pedal support plate.

6. Using wrench flats on the front end of the master cylinder body, twist the cylinder counterclockwise to release the twist lock attachment-to-plate. Do not torque on the hose connection on top of the cylinder body, damage may occur.

7. Remove the 2 actuator-to-transaxle retaining nuts and actuator assembly.

8. Pull the master cylinder with the pushrod attached forward out of the pedal plate. Lift the reservoir off the strut tower studs and remove the 3 components as a complete assembly.

To install:

9. Install the master cylinder into the opening in the pedal plate and rotate 45 degrees by applying torque on the wrench flats only.

10. Install the anti-rotation screw.

11. Install the fluid reservoir-to-strut tower and torque the retaining nuts to 36 inch lbs. (4 Nm).

12. Install a new pushrod bushing and lubricate before installation.

13. Install the master cylinder pushrod-to-clutch pedal.

14. Install the clutch actuator-to-transaxle.

15. Press the clutch pedal down several times to ensure proper operation. Adjust cruise control switch if equipped.

16. Install the left upper secondary cowl panel, sound insulator and connect the negative battery cable.

Clutch Slave Cylinder

1. Disconnect the negative battery cable.

2. Remove the sound insulator inside the vehicle and disconnect the master cylinder pushrod at the clutch pedal.

3. Remove 2 bolts holding canister to transaxle. Remove 2 actuator retainer nuts and remove actuator from transaxle housing.

4. To install position canister mounting bracket and bolts to transaxle assembly and secure retaining bolts to 28 ft. lbs.

5. Install actuator to housing studs with pushrods centered in pocket of lever in housing. Install actuator retainer nuts to 18 ft. lbs.

6. Install a new pushrod bushing and lubricate before installation.

7. Install the master cylinder pushrod-to-clutch pedal.

8. Install the clutch actuator-to-transaxle.

9. Press the clutch pedal down several times to ensure proper operation. Adjust cruise control switch if equipped.

10. Install the left upper secondary cowl panel, sound insulator and connect the negative battery cable.

Hydraulic Clutch System Bleeding

1. Disconnect the negative battery cable.

2. Disconnect quick connect fittings in clutch hydraulic line. Insert J–36221 or equivalent hydraulic line separator tool and depress plastic sleeve to separate connection.

3. Remove cap and diaphragm and fill reservoir with DOT 3 brake fluid.

4. Remove left hand upper secondary cowl.

5. Remove air from supply hose by squeezing it until no more air bubbles are seen in reservoir.

6. Pump clutch pedal slowly until slight pressure is observed. Hold pressure on pedal and depress internal valve on quick connect fitting.

7. Repeat step 6 until pedal is firm and no bubbles are seen.

8. Reconnect clutch hydraulic line. Refill clutch system and replace reservoir cap. Reconnect battery cable.

AUTOMATIC TRANSAXLE

For further information on transmissions/transaxles, please refer to "Chilton's Guide to Transmission Repair".

Transaxle Assembly

NOTE: On September 1, 1991, Hydra-matic changed the name designations of the THM 125C and THM 440–R4 automatic transaxle. The new name designations are Hydra-matic 3T40 and 4T60. Transaxles built between 1989–1990 will serve as transitional years in which a dual system, made up of the old designation and the new designation will be in effect.

REMOVAL & INSTALLATION

1. Disconnect the negative battery cable. Remove the air cleaner, bracket, Mass Air Flow (MAF) sensor and air tube as an assembly.

2. Disconnect the exhaust crossover from the right side manifold and remove the left side exhaust manifold, then, raise and support the manifold/crossover assembly on V6 engines.

3. Disconnect the TV cable from the throttle lever and the transaxle.

4. Remove the vent hose and the shift cable from the transaxle.

5. Remove the fluid level indicator and the filler tube.

6. Using a engine support fixture tool J–28467 or equivalent and the adapter tool J–35953 or equivalent, install them on the engine.

7. Remove the wiring harness-to-transaxle nut.

8. Label and disconnect the wires for the speed sensor, TCC connector and the neutral safety/backup light switch.

9. Remove the upper transaxle-to-engine bolts.

10. Remove the transaxle-to-mount through bolt, the transaxle mount bracket and the mount.

11. Raise and safely support the vehicle.

12. Remove the front wheel assemblies.

13. Disconnect the shift cable bracket from the transaxle.

14. Remove the left side splash shield.

15. Using a modified halfshaft seal protector tool J–34754 or equivalent, install 1 on each halfshaft to protect the seal from damage and the joint from possible failure. Support the halfshafts to the body to prevent CV-joint damage.

16. Using care not to damage the halfshaft boots, disconnect the halfshafts from the transaxle.

17. Remove the torsional and lateral strut from the transaxle. Remove the left side stabilizer link pin bolt.

18. Remove the left frame support bolts and move it out of the way.

19. Disconnect the speedometer wire from the transaxle. Remove the starter motor.

20. Remove the transaxle converter cover and matchmark the converter to the flywheel for assembly.

21. Disconnect and plug the transaxle cooler pipes.

22. Remove the transaxle-to-engine support.

23. Using a transmission jack, position and secure it to the transaxle and remove the remaining transaxle-to-engine bolts.

24. Make sure the torque converter does not fall out and remove the transaxle from the vehicle.

NOTE: The transaxle cooler and lines should be flushed any time the transaxle is removed for overhaul, or to replace the pump, case or converter.

To install:

25. Put a small amount of grease on the pilot hub of the converter and make sure the converter is properly engaged with the pump.

26. Raise the transaxle to the engine while guiding the right side halfshaft into the transaxle.

27. Install the lower transaxle mounting bolts, tighten to 55 ft. lbs. (75 Nm) and remove the jack.

28. Align the converter with the marks made on the flywheel and install the bolts hand tight.

29. Torque the converter bolts to 46 ft. lbs. (61 Nm). Retorque the first bolt after the others.

30. Install the starter assembly. Install the left side halfshaft.

31. Install the converter cover, oil cooler lines and cover. Install the subframe assembly. Install the lower engine mount retaining bolts and the transaxle mount nuts.

32. Install the right and left ball joints. Install the power steering rack, heat shield and cooler lines to the frame.

33. Install the right and left inner fender splash shields. Install the tire assemblies.

34. Lower the vehicle. Connect all electrical leads. Install the upper transaxle mount bolts, tighten to 55 ft. lbs. (75 Nm).

35. Attach the crossover pipe to the exhaust manifold. Connect the EGR tube to the crossover.

36. Connect the TV cable and the shift cable. Install the air cleaner and inlet tube.

37. Remove the engine support tool. Connect the negative battery cable.

FRONT SUSPENSION

MacPherson Strut/ Knuckle

REMOVAL & INSTALLATION

— CAUTION —

Do not remove the strut cartridge nut without compressing the coil spring first. This procedure must be followed because it keeps the coil spring compressed. Use care to support the strut assembly adequately because the coil spring is under heavy load, if re-

leased too quickly personal injury could result. Never remove the center strut nuts unless the spring is compressed with a Mac-
Pherson strut spring compressor tool J–
26584 or equivalent. The vehicle weight can
be used when the strut assembly is still in
the vehicle and only the strut cartridge is
going to be replaced.

1. Disconnect the negative battery
cable.
2. Loosen the cover plate bolts.
3. Loosen the wheel nuts. Raise and
safely support the vehicle.
4. Remove the wheel assembly. Remove the brake caliper and bracket assembly, hang the caliper aside. Do not
hang the caliper by the brake lines.
5. Remove the brake rotor. Remove
the hub and bearing attaching bolts.
6. Remove the halfshaft. Remove
the tie rod attaching nut. Using tool J–
35917 or equivalent, separate the tie
rod from the steering knuckle.
7. Remove the lower ball joint attaching nut and separate the lower
ball from the lower control arm.
8. Remove the hub and bearing attaching bolts and hub assembly.
9. Remove the cover plate bolts and
remove the strut from the vehicle.
To install:
10. Install the strut mount cover
plate, tighten the nuts after lowering
the vehicle. Install the lower ball joint
and torque to 81 inch lbs. (10 Nm),
plus an additional 120 degrees (2 flats)
turn until the cotter pin hole is lined
up.
11. Install the tie rod and torque to
40 ft. lbs. (54 Nm) to line up the cotter
pin hole.
12. Install the halfshaft and install
the hub and bearing-to-knuckle attaching bolts, tighten to 52 ft. lbs. (70
Nm).
13. Install the brake rotor and caliper assembly.
14. Install the wheel assembly, tight-
en the wheel lug nuts to 100 ft. lbs.
(136 Nm).
15. Lower the vehicle, tighten the
strut cover bolts to 17 ft. lbs. (24 Nm)
and tighten the wheel nuts.
16. Connect the negative battery
cable.

Lower Control Arms

REMOVAL & INSTALLATION

1. Loosen the wheel nuts. Raise and
safely support the vehicle.
2. Remove the wheel assembly.
3. Remove the stabilizer shaft-to-
lower control arm bolts. Remove the
ball joint retaining nut and cotter pin.
4. Using tool J–35917 or equivalent, separate the ball joint from the
control arm.

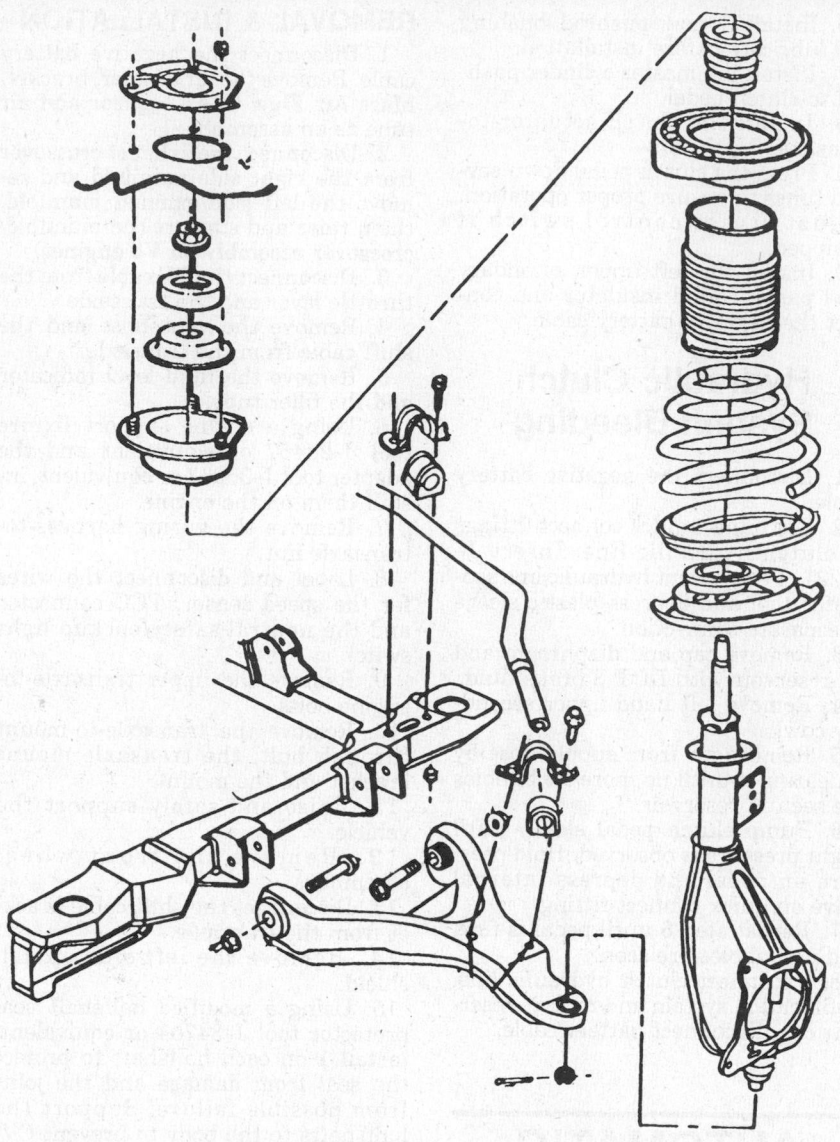

Exploded view of the front suspension

5. Remove the lower control arm-
to-frame attaching nuts and bolts. Remove the lower control arm from the
vehicle.
To install:
6. Install the lower control arm to
the frame and pivot it to the ball joint.
7. Tighten the lower control arm
bolts to 52 ft. lbs. (69 Nm) and the ball
joint nut to 89 inch lbs., plus an additional 120 degrees (2 flats). Install a
new cotter pin.
8. Install the stabilizer shaft to the
lower control arm, tighten the bolts to
35 ft. lbs. (48 Nm).
9. Install the wheel assembly and
lower the vehicle. Tighten the wheel
nuts to 100 ft. lbs. (136 Nm).

Front Wheel Bearings

The vehicles are equipped with sealed

hub and bearing assemblies. The hub
and bearing assemblies are non-serviceable. If the assembly is damaged,
the complete unit must be replaced.
Refer to the "Driveaxle" section for
the procedure.

REAR SUSPENSION

The rear suspension features a lightweight composite fiberglass mono-leaf
transverse spring. Each wheel is
mounted to a tri-link independent suspension system. The 3 links consist of
an inverted U–channel trailing arm
and tubular front and rear rods.

MacPherson Strut

REMOVAL & INSTALLATION

1. Disconnect the negative battery cable.
2. Raise and support the vehicle safely.
3. Remove the rear wheel assembly.
4. Scribe the strut-to-knuckle for proper installation.
5. Remove the auxiliary spring, if equipped.
6. Remove the jack pad.
7. Install a rear leaf spring compressor tool J-35778 or equivalent.
8. Fully compress the spring but do not remove the retention plates or the spring.
9. Remove the 2 strut-to-body bolts.
10. Remove the brake hose from the strut.
11. Remove the strut and auxiliary spring upper bracket from the knuckle.

To install:
12. Position the strut to the body and knuckle bracket.
13. Install the strut-to-body bolts and torque to 34 ft. lbs. (46 Nm).
14. Install the strut-to-knuckle, align the scribe marks and torque the bolts to 133 ft. lbs. (180 Nm).
15. Install the brake hose bracket and remove the spring compressing tool.
16. Install the jack pad and torque the bolts to 18 ft. lbs. (25 Nm).
17. Install the auxiliary spring, if so equipped.
18. Install the wheel and torque the lug nuts to 100 ft. lbs. (136 Nm).
19. Lower the vehicle and connect the negative battery cable.

NOTE: The rear strut assembly is not serviceable. The assembly is replaced as a complete unit.

Knuckle

REMOVAL & INSTALLATION

1. Disconnect the negative battery cable.
2. Raise and support the vehicle safely.
3. Remove the rear wheels and scribe the strut-to-knuckle.
4. Remove the jack pad and install the rear leaf spring compressor tool J-35778 or equivalent.
5. Fully compress the spring but do not remove the spring or retention plates.
6. Remove the auxiliary spring, if equipped. If not equipped, remove the rod-to-knuckle bolt.
7. Remove the front rod-to-knuckle.
8. Remove the brake hose bracket,

caliper and rotor. Do not leave the caliper hang by the brake hose.
9. Remove the hub and bearing assembly, trailing arm and the strut/upper auxiliary spring bracket from the knuckle. Remove the knuckle.

To install:
10. Install the knuckle and position it to the strut/upper auxiliary spring bracket. Hand start the bolts, but do not tighten.
11. Install the front rod and trailing arm-to-knuckle. Hand tighten the bolts.
12. Torque the trailing arm bolt and nut to 192 ft. lbs. (260 Nm).
13. Install the hub/bearing assembly and torque the bolts to 52 ft. lbs. (70 Nm).
14. Install the rotor and caliper.
15. Align the scribe marks to ensure proper alignment. Torque the strut-to-knuckle attaching bolts to 133 ft. lbs. (180 Nm).
16. Remove the rear leaf spring compressor.
17. Install the jack pad, auxiliary spring, if equipped, and rod-to-knuckle bolt. Apply thread locking compound to the knuckle bolts.
18. Torque the rod-to-knuckle bolts to 66 ft. lbs. (90 Nm), plus 120 degrees.
19. Install the rear wheels and torque the lug nuts to 100 ft. lbs. (136 Nm).
20. Check for completion of repair, lower the vehicle and connect the negative battery cable.

Rear Wheel Bearings

The hub and bearing assemblies are sealed units and are non-serviceable. If the assembly is damaged, the complete unit must be replaced.

REMOVAL & INSTALLATION

1. Disconnect the negative battery cable.
2. Raise and support the vehicle safely.
3. Remove the rear wheel, caliper, bracket and rotor.
4. Loosen the 4 hub/bearing-to-knuckle attaching bolts.
5. Remove the hub/bearing assembly.

To install:
6. Install the hub/bearing assembly onto the knuckle. Install the 4 attaching bolts and torque to 52 ft. lbs. (70 Nm).
7. Install the rotor, caliper and bracket.
8. Install the rear wheel and torque the lug nut to 100 ft. lbs. (135 Nm).
9. Lower the vehicle and connect the negative battery cable.

STEERING

Steering Wheel

REMOVAL & INSTALLATION

1. Disconnect the negative battery cable.
2. Remove the screws holding the pad, if equipped. Push down and turn the horn pad and remove retainer.
3. Disconnect the horn electrical lead from the cancelling cam tower.
4. Turn the ignition switch to the **ON** position.
5. Scribe an alignment mark on the steering wheel hub in line with the slash mark on the steering shaft.

NOTE: When removing the steering wheel from a vehicle with redundant accessory control switches on the pad careful and proper use of puller J-1859-03 or equivalent, must be adhered to. Do not screw the bolts of the puller more than 5 turns or contact may be made with the electronic components in the hub.

6. Loosen the steering shaft nut and position the nut at the end of the threads. Install steering wheel puller J-1859-03 or equivalent, and pull the steering wheel free of the shaft. Remove the steering wheel nut and the steering wheel.
7. Align the matchmarks on the wheel hub and shaft and install the steering wheel. Tighten the steering shaft nut to 30 ft. lbs. (41 Nm).
8. Connect the horn electrical lead and install the horn pad.
9. Connect the negative battery cable.

Power Rack

RACK BEARING PRELOAD ADJUSTMENT

1. Disconnect the negative battery cable.
2. Loosen the adjuster plug locknut and turn clockwise until it bottoms in the housing, then back off 50–70 degrees (1 flat).
3. Raise and safely support the vehicle. Center the steering wheel.
4. Tighten the locknut to the adjuster plug. Tighten to 50 ft. lbs. (70 Nm) while holding the adjuster plug stationary. Make sure the steering does not bind. Connect the negative battery cable.

REMOVAL & INSTALLATION

1. Disconnect the negative battery cable. Remove the air cleaner.
2. Raise and safely support the vehicle.
3. Remove both front wheel assemblies.
4. Remove the intermediate shaft lower pinch bolt at the steering gear. Remove the intermediate shaft from the stub shaft.
5. Disconnect the electrical lead at the power steering idle switch.
6. Separate the tie rod ends from the knuckle assembly. Remove the rear sub-frame mounting bolts and lower the rear of the sub-frame approximately 4 in.
7. Remove the steering rack heat shield. Disconnect the pressure lines at the steering rack.
8. Remove the steering rack mounting bolts, remove the rack and pinion through the left wheel opening.

To install:

9. Install the rack and pinion through the left wheel opening. Tighten the mounting bolts to 59 ft. lbs. (81 Nm). Connect the pressure lines, tighten the fittings to 20 ft. lbs. (27 Nm).
10. Install the rack heat shield, tighten the retaining bolts to 53 inch lbs. (6 Nm). Attach the tie rod ends to the steering knuckle.
11. Connect the electrical lead to the power steering idle switch. Attach the intermediate shaft to the stub shaft, tighten the pinch bolt to 35 ft. lbs. (48 Nm).
12. Install both wheel assemblies, tighten lug nuts to 100 ft. lbs. (136 Nm) and lower the vehicle.
13. Install the air cleaner. Connect the negative battery cable. Fill and bleed the power steering system.

Power Steering Pump

REMOVAL & INSTALLATION

2.3L Engine

1. Disconnect the negative battery cable.
2. Remove the air cleaner assembly.
3. Disconnect the left side torque strut from the engine.
4. Separate the throttle cable bracket from the engine torque strut bracket and set aside. Do not remove cables.
5. Remove the engine torque strut bracket.
6. Disconnect the hydraulic pump lines.
7. Remove the rear bracket to pump bolts.
8. Remove the drive belt and lay aside.

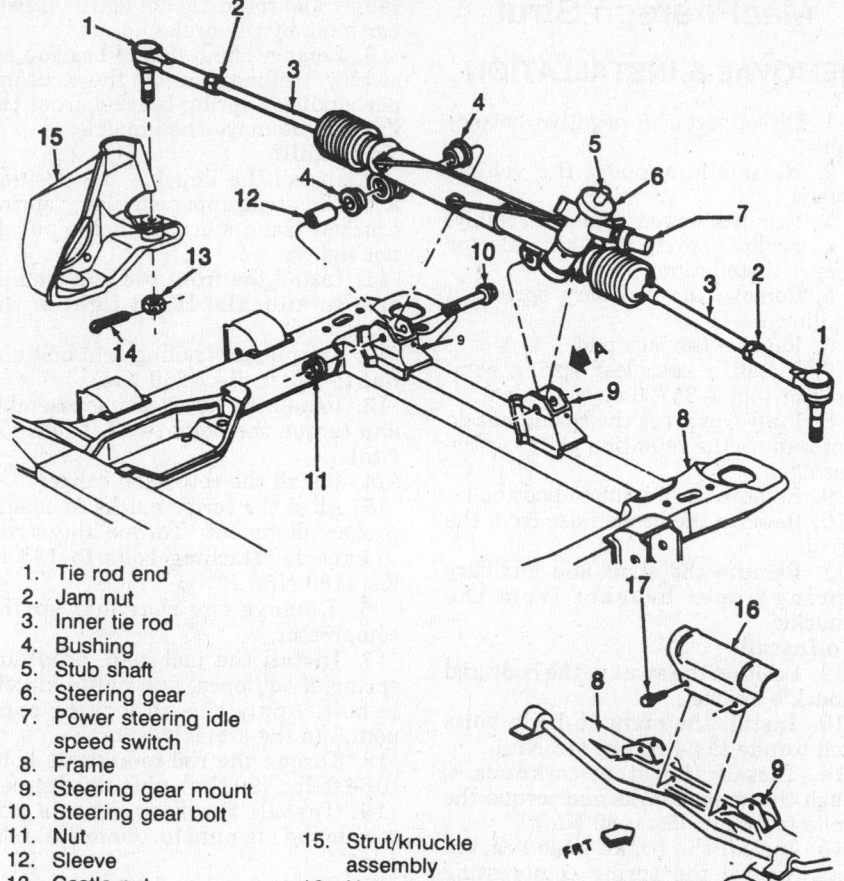

1. Tie rod end
2. Jam nut
3. Inner tie rod
4. Bushing
5. Stub shaft
6. Steering gear
7. Power steering idle speed switch
8. Frame
9. Steering gear mount
10. Steering gear bolt
11. Nut
12. Sleeve
13. Castle nut
14. Cotter pin
15. Strut/knuckle assembly
16. Heat shield
17. Screw

Power rack assembly mounting

9. Remove the rear bracket to transaxle bolts.
10. Remove the front bracket to engine bolt and remove the pump with bracket.
11. Transfer the pulley and bracket, as necessary.

To install:

12. Install the pump, pulley and bracket.
13. Install the front bracket to engine bolt.
14. Install the rear bracket to transaxle bolts.
15. Install the rear bracket to pump bolts.
16. Install the drive belt.
17. Connect the hydraulic pump lines.
18. Install the engine torque strut bracket.
19. Install the throttle cable bracket to the engine torque strut bracket.
20. Connect the left side torque strut to the engine.
21. Install the air cleaner assembly.
22. Fill with fluid and bleed the air from the system.
23. Connect the negative battery cable and check for leaks.

2.5L Engine

1. Disconnect the negative battery cable.
2. Raise and safely support the vehicle.
3. Remove the pressure and return hoses from the pump and drain the fluid.
4. Lower the vehicle, remove the ECM heat shield and serpentine belt.
5. Remove the pump mounting bolts and pump.

To install:

6. Install the pump and tighten the bolts to 20 ft. lbs. (27 Nm).
7. Install the serpentine belt, ECM heat shield. Raise and safely support the vehicle.
8. Install the inlet and outlet hoses and lower the vehicle.
9. Refill the pump with power steering fluid and bleed the system. Connect the negative battery cable.

2.8L, 3.1L and 3.8L Engine

1. Disconnect the negative cable from the battery.
2. Remove the pressure and return

hoses from the pump and drain the system into a suitable container.

3. Cap the fittings at the pump.

4. Remove the serpentine belt.

5. Locate the pump attaching bolts through the pulley and remove the bolts.

6. Remove the pump assembly.

To install:

7. Install the pump and torque the mounting bolts to 18 ft. lbs. (25 Nm) for the 2.8L and 3.1L engines or 20 ft. lbs. (27 Nm), in sequence, top bolt first, bottom bolt second, for the 3.8L engine.

8. Reconnect the hoses to the pump and install the serpentine belt.

9. Refill the power steering pump reservoir and bleed the system. Connect the negative battery cable.

3.4L Engine

1. Disconnect the negative battery cable. remove the coolant recovery reservoir.

2. Remove the serpentine drive belt from the engine. Remove the steering line bracket from the cover.

3. Remove the power steering fluid from the reservoir. Disconnect the lines from the power steering pump. Plug the lines and the connections at the pump housing to prevent dirt from entering.

4. Remove the pump mounting bolts and remove the pump.

5. Installation is the reverse of removal.

System Bleeding

NOTE: Automatic transmission fluid is not compatible with the seals and hoses of the power steering system. Under no circumstances should automatic transmission be used in place of power steering fluid in this system.

1. With the engine turned off, turn the wheels all the way to the left.

2. Fill the reservoir with power steering fluid until the level is at the cold mark on the reservoir.

3. Start and run the engine at fast idle for 15 seconds. Turn the engine off.

4. Recheck the fluid level and fill it to the cold mark.

5. Start the engine and bleed the system by turning the wheels in both directions slowly to the stops.

6. Stop the engine and check the fluid. Fluid that still has air in it will be a light tan color.

7. Repeat this procedure until all of the air is removed from the system.

BRAKES

For all brake system repair and service procedures not detailed below, please refer to "Brakes" in the Unit Repair section.

Master Cylinder

REMOVAL & INSTALLATION

Standard System

NOTE: Always use a proper size flare nut wrench when removing and installing the brake lines. Failure to use the proper wrench may cause damage to the line fittings.

1. Disconnect the negative battery cable and fluid level sensor at the master cylinder.

2. Using a flare nut wrench, remove and plug the brake lines from the master cylinder. Plug the lines to prevent fluid loss and contamination.

3. Remove the 2 master cylinder-to-brake power booster retaining nuts and master cylinder.

To install:

4. Install the master cylinder and torque the retaining nuts to 20 ft. lbs. (27 Nm).

5. Install the brake lines and torque to 15 ft. lbs. (20 Nm), using a flare nut wrench.

6. Connect the fluid level sensor electrical wire.

7. Fill the master cylinder to the proper level with new brake fluid meeting DOT 3 specifications.

8. Bleed the system.

9. Connect the negative battery cable and recheck the fluid level.

10. Do not move the vehicle until a firm brake pedal is felt.

Proportioning Valve

REMOVAL & INSTALLATION

Standard System

The proportioning valves are an integral part of the master cylinder assembly.

Anti-Lock Brake System

There is a remote proportioning valve located in the rear of the vehicle. The valve is not serviceable and must be replaced as a complete unit.

1. Depressurize the ABS brake system as follows:

  a. With the ignition key **OFF**, firmly apply and release the brake pedal a minimum of 40 times.

  b. A noticeable change in the ped-

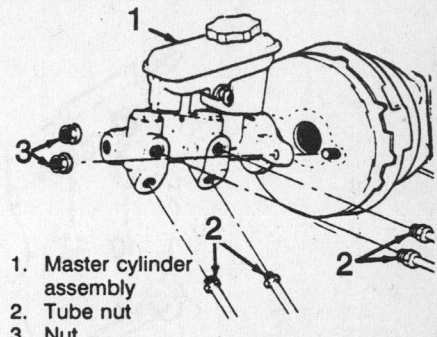

1. Master cylinder assembly
2. Tube nut
3. Nut

Master cylinder mounting

al feel will occur when the accumulator is completely discharged (a hard pedal).

  c. Do not turn the ignition key **ON** after depressurizing the system.

2. Disconnect the negative battery cable.

3. Raise and support the vehicle safely.

4. Disconnect the 3 fluid lines and cap.

5. Remove the proportioner valve.

To install:

6. Position the proportioner valve on the vehicle with the rub pad against the vehicle body.

7. Connect the 3 brake lines and tighten to 13 ft. lbs. (17 Nm).

8. Lower the vehicle, connect the battery cable and bleed the rear brake calipers.

Brake Caliper

REMOVAL & INSTALLATION

Front

1. Remove ⅔ of the brake fluid from the brake reservoir using a syringe or equivalent.

2. Raise and support the vehicle safely.

3. Mark the relationship of the wheel-to-hub and bearing assembly.

4. Remove the tire and wheel. Install 2 lug nuts to retain the rotor.

5. If the caliper is going to be removed, disconnect and plug the brake hose.

6. Remove the caliper mounting bolts and pull the caliper from the mounting bracket and rotor. Support the caliper with wire if not removing.

To install:

7. Inspect the bolt boots and support bushings for cuts or damage, replace if necessary. Inspect the bolts and bushings for corrosion, replace if any corrosion is found. Do not attempt to polish away the corrosion.

8. Install the caliper over the rotor into the mounting bracket. Make sure the bolt boots are in place.

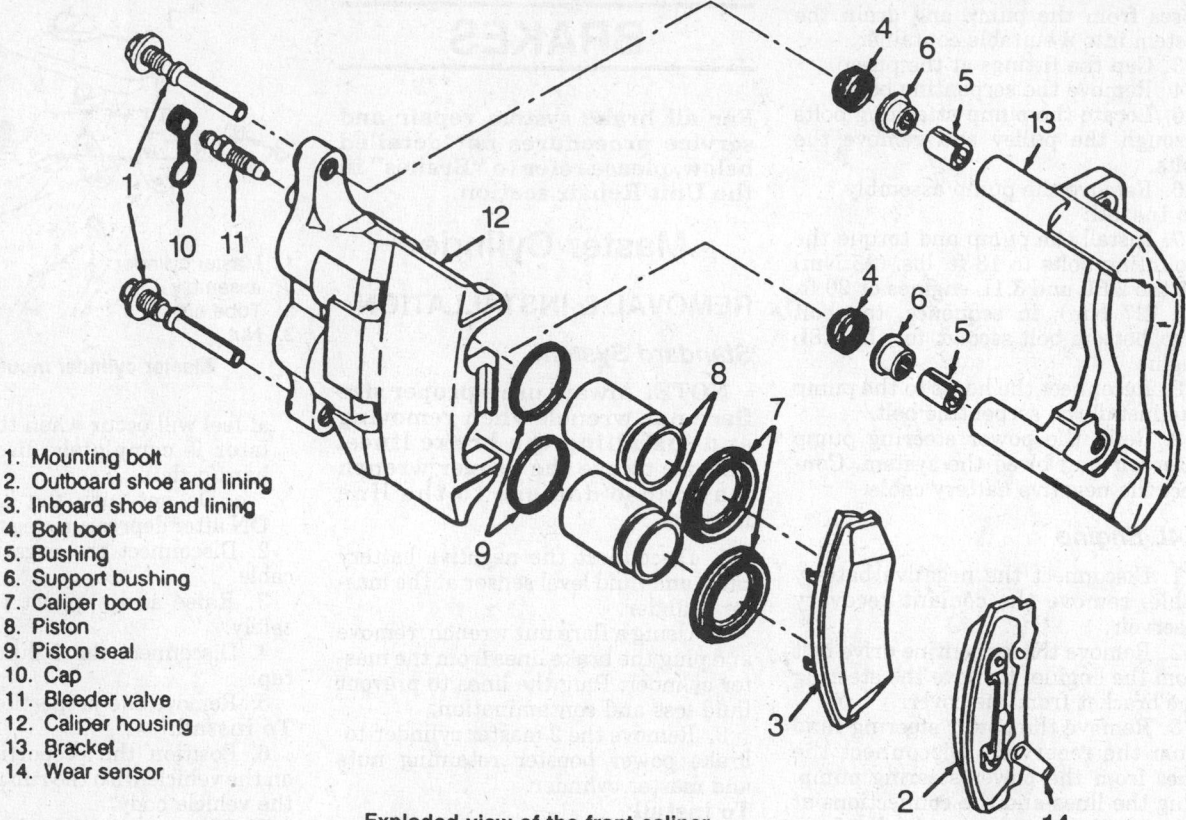

1. Mounting bolt
2. Outboard shoe and lining
3. Inboard shoe and lining
4. Bolt boot
5. Bushing
6. Support bushing
7. Caliper boot
8. Piston
9. Piston seal
10. Cap
11. Bleeder valve
12. Caliper housing
13. Bracket
14. Wear sensor

Exploded view of the front caliper

9. Lubricate the entire shaft of the mounting bolts and cavities with silicone grease.

10. Install the mounting bolts and torque to 79 ft. lbs. (107 Nm).

11. Install the brake hose, using new copper washers and torque to 32 ft. lbs. (44 Nm).

12. Remove the 2 wheel lugs, install the wheels and torque the lug nuts to 100 ft. lbs. (136 Nm).

13. Lower the vehicle.

14. Fill the master cylinder and bleed the calipers.

15. Check for hydraulic leaks. Pump the brake pedal a few times before moving the vehicle.

Rear

1. Remove ⅔ of the brake fluid from the reservoir with a syringe.

2. Raise and support the vehicle safely.

3. Remove the rear wheel assembly and install 2 lug nuts to retain the rotor.

4. Remove the brake shield assembly.

5. Loosen the tension on the parking brake cable at the equalizer.

6. Remove the parking cable and return spring from the lever.

7. Hold the cable lever and remove the lock nut, lever and seal.

8. Push the piston into the caliper bore using 2 adjustable pliers over the inboard pad tabs.

NOTE: Do not allow pliers to contact the actuator screw. Protect the piston so the contact surface does not get damaged.

9. Reinstall the lever seal with the sealing bead against the caliper housing, lever and locknut.

10. Remove and plug the brake hose inlet fitting only if the caliper is going to be removed from the vehicle.

11. Remove the bolt and bracket to gain access to the upper mounting bolt.

12. Remove the caliper mounting bolts, caliper and hang from the suspension with a piece of wire to prevent brake hose damage.

To install:

13. Inspect all brake parts for damage and deterioration. Replace any parts, if necessary.

14. Push the caliper sleeves inward.

15. Install the caliper-to-mounting bracket. Torque the mounting bolts to 92 ft. lbs. (125 Nm).

16. Install the bracket and bolt after the mounting bolts have been torqued.

17. Install the brake hose inlet with new copper washers, if removed. Torque the hose bolt to 32 ft. lbs. (44 Nm).

18. Remove the locknut, lever and seal. Lubricate the lever seal and lever shaft.

19. Install the seal and lever with the lever facing down.

20. Hold the lever back against the stop and torque the lock nut to 35 ft. lbs. (47 Nm).

21. Install the return spring and parking brake cable and adjust.

22. Install the brake shield and rear wheel assembly. Torque the lug nuts to 100 ft. lbs. (136 Nm).

23. Lower the vehicle.

24. Fill the brake reservoir with DOT 3 brake fluid.

25. Bleed the caliper if removed from the vehicle.

26. Inspect the brake system for fluid leaks.

27. Apply the brake pedal 3 times to seat the brake pads before moving the vehicle.

Disc Brake Pads

REMOVAL & INSTALLATION

Front

1. Disconnect the negative battery cable.

2. Raise and support the vehicle safely.

3. Remove the wheel and tire assembly.

4. Remove the 2 caliper mounting bolts, caliper and hang from the suspension with a piece of wire. Do not hang by the brake hose.

5. Using a prybar, lift up the out-

board pad retaining spring so it will clear the center lug.

6. Remove the inboard pad by unsnapping the pad from the pistons.

To install:

7. Remove about 2/3 of the fluid from the brake reservoir with a syringe or equivalent.

8. Bottom the pistons in the caliper bore using a C-clamp and the old inboard brake pad.

9. Install the new inboard brake pad. Make sure both inboard pad tangs are inside the piston cavity.

10. Install the outboard pad by snapping the pad retainer spring over the housing center lug and into the housing slot.

11. Make sure both pads remain free of grease or oil. The wear sensor should be at the trailing edge of the pad during rotation.

12. Install the caliper assembly, wheels assembly and lower the vehicle.

13. Fill the master cylinder to the **FULL** mark and apply the brake pedal 3 times to seat the pads. Connect the negative battery cable.

Rear

1. Raise and support the vehicle safely.

2. Remove the rear wheel assemblies.

3. Remove the rear caliper and hang by the suspension with a piece of wire to prevent brake hose damage.

4. Using a prybar, disengage the buttons on the outboard pad from the holes in the caliper housing.

5. Press in on the edge of the inboard pad and tilt outward to release the pad from the pad retainer.

6. Remove the plug from the end of the caliper piston using a small prybar.

To install:

NOTE: Do not allow pliers to contact the actuator screw. Protect the piston so the contact surface does not get damaged.

7. Bottom the piston into the caliper bore by positioning a twelve inch adjustable pliers over the caliper housing and piston surface.

8. Lubricate a new plug and install it into the end of the piston.

9. Install the inboard brake pad. Engage the pad edge in the retainer tabs closest to the caliper bridge. Press down and snap the tabs at the open side of the caliper. The wear sensor should be at the leading edge of the pad during wheel rotation. The back of the pad must lay flat against the piston. The button on the back of the pad must engage the D-shaped notch in the piston.

NOTE: If the piston will not align or retract into the bore. Turn the piston clockwise using a piston turning tool J–7624 or equivalent.

10. Install the outboard brake pad. Snap the pad retainer spring into the slots in the caliper housing. The back of the pad must lay flat against the caliper.

11. Install the caliper onto the mounting bracket.

12. Apply force at least 3 times to the brake pedal to seat the brake pads before moving the vehicle.

13. Install the rear wheels and torque the lug nuts to 100 ft. lbs. (136 Nm).

14. Lower the vehicle and check for fluid leaks.

Brake Rotor

REMOVAL & INSTALLATION

1. Raise and support the vehicle safely.

2. Remove the wheel and tire assembly.

3. Remove the brake caliper and support with a wire to the body.

4. Slide the rotor off of the hub assembly.

To install:

5. Install the brake rotor over the hub assembly.

6. Install the brake caliper.

7. Install the wheel and tire assembly. Torque the lug nuts to 100 ft. lbs. (136 Nm).

8. Lower the vehicle and pump the brake pedal before moving.

Brake System Bleeding

The brake system must be bled after the hydraulic system has been serviced. Air can enter the system during servicing and has to be removed to prevent poor system performance.

When installing a new master cylinder the time required to bleed the system can be reduced by removing as much air as possible before installing the master cylinder onto the vehicle. This is called bench bleeding the master cylinder. Place the master cylinder in a vise or holding fixture, run tubing from the fluid pipe fittings to the reservoir, fill the cylinder with DOT 3 brake fluid and pump the brake pushrod until most of the air is removed from the master cylinder. Install the master cylinder onto the vehicle and bleed all 4 wheels.

Standard System

1. Fill the master cylinder reservoir with brake fluid and keep the reservoir at least half full during the bleeding operation.

2. If the master cylinder has air in the bore, it must be removed before bleeding the calipers. Bleed the master cylinder as follows:

a. Disconnect the forward brake pipe at the master cylinder.

b. Fill the reservoir until fluid begins to flow from the forward pipe connector port.

c. Reconnect the forward brake pipe and tighten.

d. Depress the brake pedal slowly 1 time and hold. Loosen the forward brake pipe and purge the air from the bore. Tighten the brake pipe, wait 15 seconds and repeat until all air is removed.

e. When the air is removed from the forward brake pipe, repeat the same procedures for the rear brake pipe.

3. Bleed the calipers in the following order: right rear, left front, left rear, right front.

4. Install a box end wrench over the bleeder valve and connect a clear tube onto the valve. Place the other end of the tube into a container of new brake fluid. The end of the tube must be submerged in brake fluid.

5. Depress the brake pedal slowly 1 time and hold. Loosen the bleeder valve to purge the air from the caliper. Close the valve and release the pedal. Wait 15 seconds. Repeat the procedure until all air is removed from the brake fluid. It may take 10 repetitions or more to completely purge the system of air.

6. Do not pump the brake pedal rapidly, this causes the secondary master cylinder piston to push to the end of the bore and make bleeding difficult.

7. After the calipers have been bled, check the brake pedal for sponginess and the **BRAKE** warning lamp for unbalanced pressure.

8. Repeat the bleeding operation if a spongy pedal is felt and fill the reservoir to the **MAX** line.

Anti-Lock Brake System (ABS)

———— **CAUTION** ————

Use only clean DOT 3 brake fluid from a sealed container in the anti-lock brake system. Any other type of fluid may cause severe damage to the internal components causing brake failure and personal injury.

1. Make sure the vehicle ignition is **OFF**.

2. Disconnect the negative battery cable.

3. Depressurize the ABS system as follows.

a. With the ignition key **OFF**,

firmly apply and release the brake pedal a minimum of 40 times.

b. A noticeable change in the pedal feel will occur when the accumulator is completely discharged (a hard pedal).

c. Do not turn the ignition key **ON** after depressurizing the system unless instructed to do so.

4. Clean and remove the reservoir cap.

5. Fill the reservoir with DOT 3 brake fluid.

6. Raise the vehicle and support the vehicle safely.

7. Bleed the right front wheel by attaching a clean hose to the bleeder valve and submerge the other end into a container of partially filled brake fluid.

8. Open the valve and slowly depress the brake pedal.

9. Tap lightly on the brake caliper with a rubber mallet to dislodge the air bubbles.

10. Close the valve and release the brake pedal. Repeat until all air is removed.

11. Repeat steps 7–10 on the left front wheel.

12. Connect the negative battery cable and turn the ignition key to the **RUN** position without starting the vehicle. Allow the pump to run to pressurize the accumulator.

13. Bleed the right rear brake by installing a bleeder hose and container, open the valve, with the ignition **ON** slowly depress the pedal part way until the fluid begins to flow from the bleeder valve and allow the fluid to flow for 15 seconds. Do not fully depress the brake pedal.

14. Close the valve and release the brake pedal.

15. Fill the reservoir with fluid to 1 inch below the FULL mark.

16. Repeat steps 13–16 for the left rear wheel.

17. Lower the vehicle and bleed the Powermaster III isolation valves (at the master cylinder) as follows:

a. Attach a clear hose and container to the Powermaster III inboard bleeder valves.

b. With the ignition in the **ON** position, apply the pedal, slowly open the valve and allow fluid to flow until no air bubbles are seen.

c. Close the valve and repeat the steps to the outboard bleeder valve until no air bubbles are present.

18. Bleed the accumulator as follows.

a. Turn the ignition key to the **OFF** position, depressurize the system and wait 2 minutes.

b. Remove the reservoir cover and check the fluid level. Add if necessary.

c. Install the reservoir cap.

d. Turn the ignition key to the

RUN position but do not start the engine.

e. When the pump has stopped, depress the brake pedal and repeat the **OFF/RUN** procedures 10 times to cycle the solenoids.

19. Apply the brake pedal and note the pedal feel and travel.

20. If the pedal feels firm and smooth without excessive travel, the system is properly bled. Connect the negative battery cable.

Anti-Lock Brake System Service

PRECAUTION

Failure to observe the following precautions may result in system damage.

- The brake system uses a hydraulic accumulator which when fully charged, contains brake fluid at high pressure. Before disconnecting any hydraulic lines, hoses or fittings, be sure that the accumulator is fully depressurized.
- Never disassemble any component of the Anti-Lock Brake System (ABS) which is designated non-servicable; the component must be replaced as an assembly.
- Replace all components included in repair kits used to service the system.
- When filling the master cylinder, always use Delco Supreme 11 brake fluid or equivalent, which meets DOT-3 specifications; petroleum base fluid will destroy the rubber parts.
- Avoid spilling brake fluid on the vehicles painted surfaces, wiring, cables or electrical connectors. Brake fluid will damage paint and electrical connections.

Relieving Anti-Lock Brake System Pressure

─── CAUTION ───

Failure to fully depressurize the system before performing service operations could result in personal injury from a high pressure spray of brake fluid.

1. With the ignition key **OFF**, firmly apply and release the brake pedal a minimum of 40 times.

2. A noticeable change in the pedal feel will occur when the accumulator is completely discharged (a hard pedal).

3. Do not turn the ignition key **ON** after depressurizing the system.

Accumulator

The accumulator is a nitrogen charged pressure vessel which holds the brake fluid under high pressure. The accu-

mulator can not be repaired and must be replaced as an assembly.

REMOVAL & INSTALLATION

1988–89

1. Disconnect the negative battery cable.

2. Depressure the ABS system as follows.

a. With the ignition key **OFF**, firmly apply and release the brake pedal a minimum of 40 times.

b. A noticeable change in the pedal feel will occur when the accumulator is completely discharged (a hard pedal).

c. Do not turn the ignition key **ON** after depressurizing the system unless instructed to do so.

3. Remove the accumulator by turning the hex nut on the end of the accumulator with a 17mm socket. The unit can be removed by sliding out from beneath the ABS unit, towards the left front wheel well.

4. Remove the O-ring seal from the accumulator.

To install:

5. Lightly lubricate the new O-ring seal and install it on the accumulator.

6. Install the accumulator and torque the unit to 23–36 ft. lbs. (31–35 Nm).

7. Bleed the system.

1990–92

1. Disconnect the negative battery cable.

2. Depressurize the ABS system as follows.

a. With the ignition key **OFF**, firmly apply and release the brake pedal a minimum of 40 times.

b. A noticeable change in the pedal feel will occur when the accumulator is completely discharged (a hard pedal).

c. Do not turn the ignition key **ON** after depressurizing the system unless instructed to do so.

3. Remove the air cleaner duct and stud, if used.

4. Disconnect the 30 amp pump motor fusible element **K** from the ABS power center.

5. Remove the accumulator by turning the hex nut on the end of the accumulator with a 17mm socket. The unit can be removed by sliding out from beneath the ABS unit, towards the left front wheel well.

6. Remove the O-ring seal from the accumulator.

To install:

7. Lightly lubricate the new O-ring seal and install it on the accumulator.

8. Install the accumulator and torque the unit to 24 ft. lbs. (33 Nm).

CHASSIS ELECTRICAL

Heater Blower Motor

REMOVAL & INSTALLATION

1. Disconnect the negative terminal from the battery.
2. Disconnect the electrical connections from the blower motor and resistor.
3. Remove the plastic water shield from the right side of the cowl.
4. Remove the blower motor-to-chassis screws and the blower motor.
5. Remove the cage retaining nut and the cage (old style).

NOTE: Some of the new style blower cages are plastic welded to the motor shaft. Use a hot knife to cut a slot in the cage shaft sleeve in 3 places. Cut through the plastic material from the dome to the end of the shaft until the cage splits from the shaft.

To install:
6. Install the cage on the new blower motor with the opening facing away from the motor.
7. Install the blower motor and screws. Install the sound insulator and connect the electrical leads to the motor and resistor.
8. Install the water shield to the cowl. Connect the negative battery cable.

Windshield Wiper Motor

REMOVAL & INSTALLATION

1988–90

1. Disconnect the negative battery cable.
2. Remove wiper module from vehicle, if equipped, as follows:
   a. Remove the washer hose, cap and retaining nut from each wiper arm. Remove the wiper arms from the vehicle.
   b. Remove the screws retaining the cowl cover. Lower the hood partially and remove the cowl cover.
   c. Remove the air inlet panel, underhood light switch, if equipped.

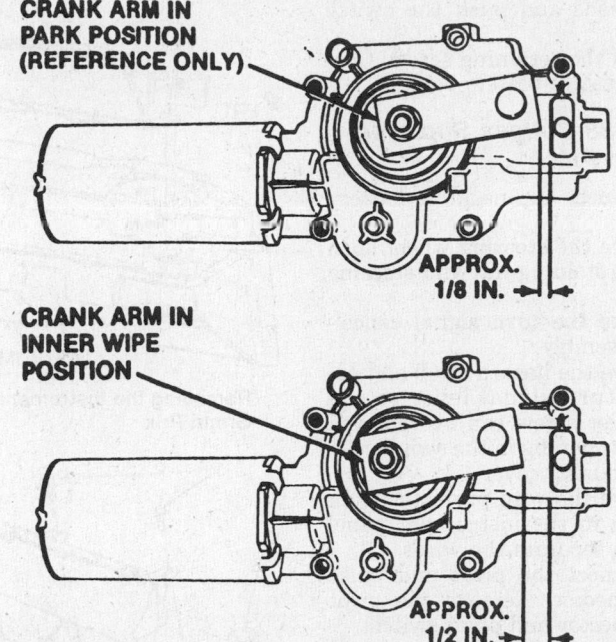

CRANK ARM IN PARK POSITION (REFERENCE ONLY)

APPROX. 1/8 IN.

CRANK ARM IN INNER WIPE POSITION

APPROX. 1/2 IN.

Crank arm in inner wipe and park position

NOTE: Attach holding wire to upper portion of switch before removing retaining nut or switch will fall between panels.

   d. Disconnect 2 wiring harness connectors from motor, and washer hose at firewall.
   e. Position crank arm to inner wipe position, remove 3 screws from bellcrank housing, lower transmission and remove.
3. Remove crank arm from motor.
4. Remove 3 screws retaining the motor and remove the motor.

To install:
5. Attach the motor to the module assembly.
6. Install the crank arm and nut. Tighten to 25–38 ft. lbs. (34–51 Nm).
7. Attach the bellcrank to module assembly and install the wire connectors washer hose, air inlet panel and light switch.
8. Install cowl cover, wiper arms, washer hoses and nuts with protective caps.
9. Connect the negative battery cable.

1991–92

1. Disconnect the negative battery cable.
2. Remove wiper arms.
3. Remove top vent screen shroud.
4. Remove wiper arm drive link from crank arm and electrical connectors.
5. Remove wiper motor attaching bolts.
6. Remove wiper motor while guiding crank arm through hole.

7. Remove crank arm from motor.
To install:
8. Install crank arm to wiper motor.
9. Install wiper motor and connect electrical connectors.
10. Connect wiper arm drive link to crank arm.
11. Reinstall top vent screen shroud and wiper arms.
12. Install negative battery cable.

Windshield Wiper Switch

REMOVAL & INSTALLATION

The wiper/washer switch on the 1988 Grand Prix and Cutlass Supreme are mounted on the instrument panel to the right of the instrument cluster. The wiper switch on the other models are controlled by the multi-function turn signal lever but the actual switch is located in the steering column.

1988 Cutlass Supreme and Grand Prix

1. Disconnect the negative battery cable.
2. Remove the screw retaining the switch panel to the instrument panel.
3. Remove the switch from the instrument panel by pulling the bottom out and releasing the top retaining clips.
4. Disconnect the electrical connector from the switch and remove it from the vehicle.
5. To install the switch, connect the

electrical leads and push the switch into position.

6. Install the retaining screw. Connect the negative battery cable.

Except 1988 Cutlass Supreme and Grand Prix

1. Disconnect the negative battery cable.
2. Remove the steering wheel horn pad, wheel retaining nut and steering wheel.
3. Remove the turn signal cancelling cam assembly.
4. Remove the hazard knob and position the turn signal lever so the housing cover screw can be removed through the opening in the switch. Remove the housing cover.
5. Remove the wire protector from the opening in the instrument panel bracket and separate the wires.
6. Disconnect the pivot and pulse switch connector. Remove the pivot switch connector and pivot switch.

To install:

7. Install the pivot and pulse switch assembly. Install the wiring protector around the instrument panel opening, covering all wires.
8. Install the steering column housing cover and torque the screws to 35 inch lbs. (4 Nm).
9. Install the hazard knob and lubricate the bottom side of the cancelling cam with lithium grease.
10. Install the steering wheel and torque the shaft nut to 30 ft. lbs. (41 Nm).
11. Connect the negative battery cable and check steering column operations.

Instrument Cluster

REMOVAL & INSTALLATION

Cutlass Supreme

1. Disconnect the negative battery cable.
2. Remove the 5 screws retaining the cluster trim plate. Pull the bottom of the trim plate out and remove it from the vehicle.
3. Remove the screws retaining the instrument cluster and remove the cluster from the instrument panel. Disconnect the electrical connectors.

To install:

4. Install the cluster to the instrument panel. Connect the electrical leads.
5. Install the cluster trim panel. Connect the negative battery cable.

Except Cutlass Supreme

1. Disconnect the negative battery cable.

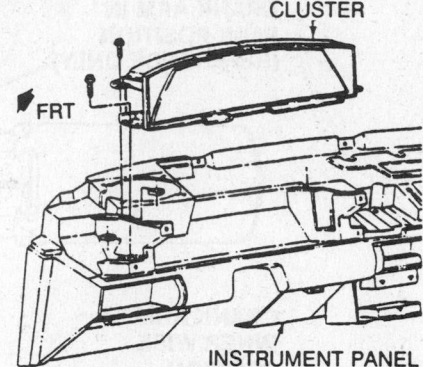

CLUSTER

FRT

INSTRUMENT PANEL

Removing the instrument cluster—Grand Prix

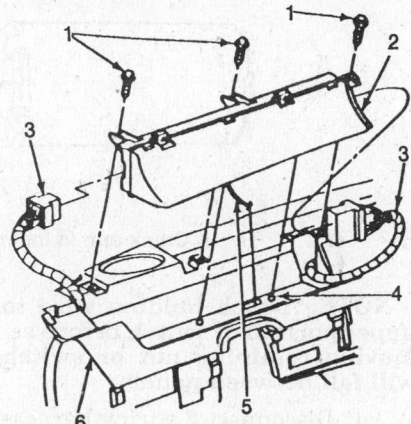

1. Screws	4. Locating tab
2. Cluster	5. PRNDL cable
3. Connector	6. Instrument panel carrier

Removing the instrument cluster—Lumina

2. Remove the instrument panel pad from the vehicle.
3. Remove the cluster trim plate and the 4 screws retaining the instrument cluster. Pull the cluster forward, disconnect the electrical connectors, **PRNDL** cable and remove the cluster from the vehicle.

To install:

4. Install the cluster to the instrument panel. Connect the electrical leads and **PRNDL** cable.
5. Install the upper panel pad.
6. Install the cluster trim panel.
7. Connect the negative battery cable.

Headlight Switch

REMOVAL & INSTALLATION

Cutlass Supreme and 1991–92 Regal

1. Disconnect the negative battery cable.
2. Remove the 4 instrument cluster trim plate retaining screws and plate. Remove the air outlet trim plate.

3. Remove the 2 screws retaining the switch and remove the switch from the instrument panel.
4. Disconnect the electrical connector from the switch and remove the switch.
5. To install the switch, connect the electrical connector and install the switch in the instrument panel.
6. Install the air outlet and cluster trim plates.
7. Connect the negative battery cable.

Grand Prix and 1988–90 Regal

1. Disconnect the negative battery cable.
2. Remove the screw retaining the headlight switch to the instrument panel.
3. Pull the top of the switch out to release the lower retaining clips and remove it from the instrument panel.
4. Disconnect the electrical connector and remove the switch from the vehicle.
5. To install the switch, connect the electrical connector and install the switch in the instrument panel.
6. Connect the negative battery cable.

Lumina

1. Disconnect the negative battery cable.
2. Remove left instrument panel trim plate.
3. Remove retaining screws from switch and extract switch.
4. Disconnect wire connectors.
5. To install switch, connect the electrical connector and install switch into dash.
6. Install screws and trim plate. Reconnect negative battery cable.

Turn Signal Switch

REMOVAL & INSTALLATION

NOTE: Tool J-35689-A or equivalent, is required to remove the terminals from the connector on the turn signal switch.

1. Disconnect the negative battery cable. Remove the steering wheel.
2. Pull the turn signal cancelling cam assembly from the steering shaft.
3. Remove the hazard warning knob-to-steering column screw and the knob.

NOTE: Before removing the turn signal assembly, position the turn signal lever so the turn signal assembly to steering column screws can all be removed.

4. Remove the column housing cover-to-column housing bowl screw and the cover.

NOTE: If equipped with cruise control, disconnect the cruise control electrical connector.

5. Remove the turn signal lever-to-pivot assembly screw and the lever; 1 screw is in the front and the other screw is in the rear.

6. Remove the wiring protector from the opening in the instrument panel bracket and separate from the wires.

7. Using the terminal remover tool J–35689–A or equivalent, disconnect and label the wires **F** and **G** on the connector at the buzzer switch assembly from the turn signal switch electrical harness connector.

8. Remove the turn signal switch-to-steering column screws and the switch.

To install:

9. Install the turn signal switch to the steering column, torque the turn signal switch-to-steering column screws to 35 inch lbs. (4 Nm).

10. Install the electrical connectors and install the turn signal lever to the pivot assembly. Install the hazard flasher knob. Install the cancelling cam. Lubricate bottom side of cancel cam with lithium grease.

11. Install the wiring protector and connect the wiring harness.

12. Install the steering wheel. Connect the negative battery cable.

Clutch Switch

ADJUSTMENT

1. Disconnect the negative battery cable.

2. Remove the lower, left trim panel. Locate the switch on the clutch pedal support. Disconnect connector at switch and check for good connection. If connection is good, proceed with adjustment.

3. Pull back on the clutch pedal and push the switch through the retaining clip noting the clicks.

4. Repeat this procedure until no more clicks can be heard.

5. Reconnect negative battery cable.

Neutral Safety Switch

REMOVAL & INSTALLATION

1. Apply park brake firmly. Place the selector lever in the **N** detent. Disconnect the negative battery cable.

2. Remove the air cleaner, as required.

3. Raise and safely support the vehicle.

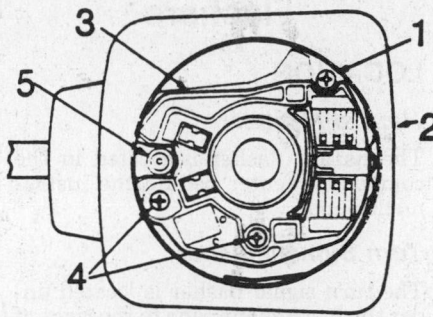

1. Screw
2. Housing cover
3. Turn signal switch
4. Screw
5. Self tapping screw

Turn signal switch mounting

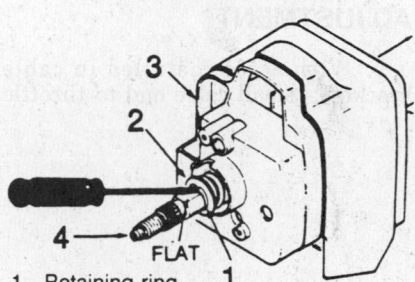

FLAT

1. Retaining ring
2. Thrust washer
3. Turn signal switch housing
4. Steering shaft assembly

Removing the turn signal switch housing

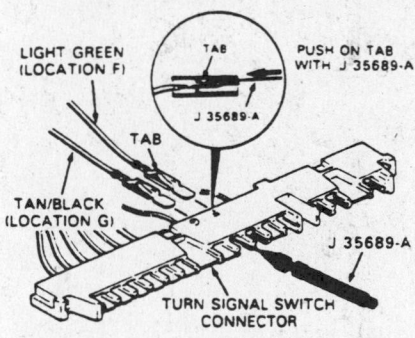

LIGHT GREEN (LOCATION F)

PUSH ON TAB WITH J 35689-A

J 35689-A

TAB

TAN/BLACK (LOCATION G)

J 35689-A

TURN SIGNAL SWITCH CONNECTOR

Removing the terminals from the turn signal switch connector

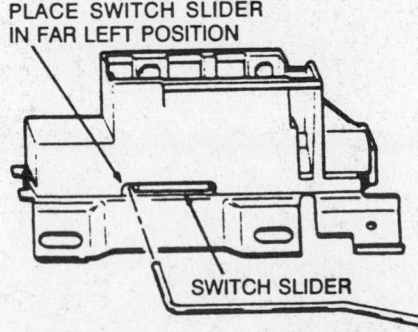

PLACE SWITCH SLIDER IN FAR LEFT POSITION

SWITCH SLIDER

Ignition switch installation position

4. Remove the switch harness.

5. Lower the vehicle.

6. Remove the vacuum lines and electrical connectors from the cruise control servo, if equipped.

7. Remove the shift lever, cruise control servo and switch. Do not disconnect the lever from the cable.

To install:

8. Align the notch on the inner sleeve of the switch with the notch on the switch body.

9. Install the switch and tighten the bolts to 18 ft. lbs. (24 Nm).

10. Install the shift lever and tighten the nut to 15 ft. lbs. (20 Nm).

11. Raise and safely support the vehicle, connect the switch harness and lower the vehicle.

12. Install the cruise control servo, vacuum lines and electrical connectors, if equipped.

13. Install the air cleaner, as required and connect the negative battery cable.

14. After switch has been installed, verify that the engine will not start in any gear other than **P** or **N**. If engine will start in any other position, readjust switch and check again.

Fuses, Circuit Breakers and Relays

LOCATION

Fusible Links

Fusible links are used to protect wiring in circuits that are not normally fused, such as the ignition circuit. In the event of an electrical overload, the fuse link will melt and create an open in the circuit. The fuse link is smaller than the wire it is to protect. The gauge size is marked on the insulation. The replacement fuse link must be the same size as the original link. To replace a damaged fuse link remove the wire section beyond the splice and splice the replacement link into the wiring harness. The fuse links are located at the starter solenoid terminal, the wiring harness near the battery and at the electrical center in the engine compartment.

Circuit Breakers

Circuit breakers are located in the fuse panel and component center located behind the instrument panel.

Fuse Panel

The fuse panel is located on the left side of the instrument panel in Regal, Cutlass Supreme and Lumina. It is located on the right side of the instrument panel in the Grand Prix. In order to gain access to the fuse panel, it may be necessary to first remove the lower trim panel.

Various Relays

The coolant fan, air conditioner compressor, high blower speed and fuel pump relays are located in the engine compartment mounted to the right side of the firewall on the relay bracket. The power door lock relay is located on the left side, behind the instrument panel.

Computer

LOCATION

The Electronic Control Module (ECM) is located in the engine compartment in front of the right strut tower.

Flashers

LOCATION

Hazard Flasher

The hazard flasher is located in the component center behind the instrument panel.

Turn Signal Flasher

The turn signal flasher is located under the instrument pane to the right of the steering column.

Cruise Control

ADJUSTMENT

1. With cable installed in cable brackets, install cable end to throttle linkage. Install cable end over throttle lever pin and secure with retainer. Specific procedures are listed for 2.5L and 3.4L engines.

   a. On 2.5L engines, rotate idler pulley toward cable bracket and install cable to pulley.

   b. On 3.4L engines, rotate idler pulley and insert cable slug into idler pulley slot. Route other end of cable through retainer clip.

2. Pull the servo end of the cable towards the servo as far as possible without moving the throttle.

3. Attach the cable to the servo in the closest alignment holes without moving the throttle.

NOTE: Do not stretch the cable to attach it to the servo. This will not allow the engine to return to idle.

5. Check the system operation and repeat the adjustment as necessary.

Unit Repair Sections

26 Tools and Equipment

In addition to the normal assortment of screwdrivers and pliers, automotive service work requires an investment in wrenches, sockets and the handles needed to drive them, and various measuring tools such as torque wrenches and feeler gauges.

The best approach to gathering the required equipment is to proceed slowly, buying high-quality tools as they are needed. An initial investment should be made in a set of quality wrenches, ranging in size from $\frac{1}{4}$ inch to one inch, if your car has standard bolts, or from 5mm to 19mm if your car has metric fasteners. High quality forged wrenches are available in three styles; open end, box end, and combination open/box end. The combination tools are generally the most desirable as a starter set; the wrenches shown in the illustration are of the combination type.

NOTE: Many later model American cars use both metric and standard nuts and bolts.

The other set of tools inevitably required is a ratchet handle and socket set. This set should have the same size range as your wrench set. The ratchet, extension, and flex drives fro the sockets are available in many sizes; it is advisable to choose a $\frac{3}{8}$ inch drive set initially. One break in the inch/metric sizing war is that metric-sized sockets sold in the U.S. have inch-sized drive ($\frac{1}{4}$, $\frac{3}{8}$, $\frac{1}{2}$, etc.). Sockets are available in six and twelve point versions; six point types are generally cheaper and are a good choice for a first set.

The choice of a drive handle for the sockets should be made with some care. If this is your first set, take the plunge and invest in a flexhead ratchet; it will get into many places otherwise accessible only through a long

chain of universal joints, extensions and adapters. An alternative is a flex handle; such a tool is shown in the illustration, below the ratchet handle. In addition to the range of sockets mentioned, a rubber-lined spark plug socket should be purchased. Spark plugs have either a $\frac{13}{16}$ or a $\frac{5}{8}$ inch hex; get the correct socket for the plugs in your car.

The most important thing to consider when purchasing hand tools is quality. Don't be misled by the low cost of "bargain" tools. Forged wrenches, tempered screwdriver blades, and fine tooth ratchets are a much better investment than their less expensive counterparts. The skinned knuckles and frustration inflicted by poor quality tools make any job an unhappy core. Another consideration is that quality tools sold by reputable firms come with an on-the-spot replacement guarantee; if the tool breaks, you get a new one, no questions asked.

The tools needed for basic maintenance jobs, in addition to those just mentioned, include:

1. Jackstands, for support;
2. Oil filter wrench;
3. Oil filler spout or funnel;
4. Grease gun;
5. Battery hydrometer;
6. Battery post and clamp cleaner;
7. Container for draining oil;
8. Many rags for the inevitable spills.

In addition to these items there are several others which are not absolutely necessary, but handy to have around. These include a transmission funnel and filler tube, a drop (trouble) light on a long cord, an adjustable wrench (crescent wrench), and slip joint pliers.

A more extensive list of tools, suitable for tune-up work, can be drawn

up easily. While the tools involved are slightly more sophisticated, they need not be outrageously expensive. For example, there are several inexpensive tach/dwell meters on the market that are every bit as good for the average mechanic as a $100.00 professional model. The key to these purchases is to make them with an eye towards adaptability and wide range. Using the tach/dwell meter example again, if the model you buy runs up to at least 1,500 rpm on the tachometer scale, the dwell meter works on 4, 6, or 8 cylinder engines, and the tachometer unit is adaptable to both conventional and electronic ignitions, it will serve for a long time on a variety of automobiles. A basic list of tune-up tools could include:

1. A tach/dwell meter;
2. Spark plug gauge and gapping tool;
3. Feeler blades;
4. Timing light.

In this list, the choice of a timing light should be made carefully. A light which works on the DC current supplied by the car battery is the best choice; it should have a xenon tube for brightness. If your car has electronic ignition, the light should have an inductive pick-up (the timing light illustrated has one of these), and since nearly all cars will have electronic ignition in the future, this feature is a reasonable one to look for.

In addition to these basic tools, there are several other tools and gauges you may find useful. These include:

1. A compression gauge. The screw-in type is slower to use, but eliminates the possibility of a faulty reading due to escaping pressure.
2. A manifold vacuum gauge.
3. A test light.
4. An induction meter. This is used

to determine whether or not there is current flowing in a wire, and thus is extremely helpful in electrical troubleshooting.

Finally, you will probably find a torque wrench necessary for all but in the most basic of work. The beam type models are perfectly adequate, although the newer click (break-away) type are more precise. Whichever type you choose, plan on having it recalibrated every once in a while.

Special Tools

Several procedures in this manual refer to special tools needed to make repairs or adjustments. These tools can be purchased from the following companies:

AMC, GM
Special Tool Division
Kent-Moore Corp.
29784 Little Mack
Roseville, MI 48066

Ford
Owatonna Tool Co.
Owatonna, MN 55060

Chrysler
Miller Special Tools
A Division of Utica Tool Co.
32615 Park Lane
Garden City, MI 48135

SPECIAL TEST EQUIPMENT

A variety of diagnostic tools are available to help troubleshoot and repair computerized engine and emission control systems. The most sophisticated of these devices are the console-type engine analyzers that usually occupy a garage service bay, but there are several types of aftermarket electronic testers available that will allow quick circuit tests of the engine control system by plugging directly into a special test connector located in the engine compartment or under the dashboard. Several tool and equipment manufacturers offer simple, hand-held testers that measure various circuit voltage levels on command to check all system components for proper operation. Although these testers usually cost about $300–500, consider that the average computer-controlled carburetor can cost twice as much and the money saved by not replacing perfectly good sensors in an attempt to correct a problem could justify the purchase price of a special diagnostic tester.

These testers can allow quick and easy test measurements while the en-

Aftermarket hand-held testers can make diagnosing computer-controlled systems easier

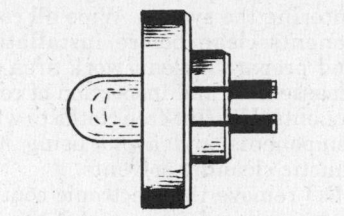

Throttle body fuel injector tester

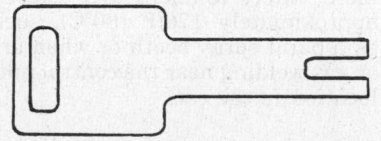

Special key for activating GM on-board diagnosis system. Insert the prongs into the diagnostic test terminals located under the dash

gine is operating or while the car is being driven. In addition, the on-board computer memory can be read to access any stored trouble codes; in effect allowing the computer to tell you where it hurts and aid trouble diagnosis by pinpointing exactly which circuit or component is malfunctioning. In the same manner, repairs can be tested to make sure the problem has been corrected. The biggest advantage these special testers have is their relatively easy hookups that minimize or eliminate the chances of making the wrong connections and getting false voltage readings or damaging the on-board computer.

NOTE: It should be remembered that these testers check voltage levels in circuits; they don't detect mechanical problems or failed components if the circuit voltage falls within the preprogrammed limits stored in the tester PROM unit. Also, most of the

hand-held testes are designed to work only on one or two systems made by a specific manufacturer.

A variety of aftermarket testers are available to help diagnose different computerized engine control systems. Owatonna Tool Company (OTC), for example, markets a device called the OTC Monitor 2000 which plugs directly into the assembly line diagnostic link (ALDL). When the correct manufacturer cartridge is plugged into the unit, the OTC tester makes diagnosis a simple matter of pressing the correct buttons. An adapter is supplied with the tester to allow connection to all types of ALDL links, regardless of the number of pin terminals used.

Servicing Your Car Safely

It is virtually impossible to anticipate all of the hazards involved with automotive maintenance and service, but care and common sense will prevent most accidents. The rules of safety for mechanics range from "don't smoke around gasoline," to "use the proper tool for the job." The trick to avoiding injuries is to develop safe work habits and take every possible precaution.

Any computer-based electronic engine control system is extremely sensitive to electrical voltages and cannot tolerate careless or haphazard testing or service procedures. An inexperienced individual can literally do major damage looking for a minor problem by using the wrong kind of test equipment or connecting test leads or connectors with the ignition switch ON. When selecting test equipment, make sure the manufacturers instructions state that the tester is compatible with whatever type of electronic control system is being serviced. Read all instructions carefully and double check all test points before installing probes or making any connections.

Aftermarket electronic testers are available from a variety of sources, as well as from the manufacturer, but care should be taken that the test equipment being used is designed to diagnose a particular system accurately without damaging the control unit (ECU) or components being tested.

DO'S

• DO keep a fire extinguisher and first aid kit within easy reach.
• DO wear safety glasses or goggles when cutting, drilling, grinding or prying, even if you have 20-20 vi-

sion. If you wear glasses for the sake of vision, they should be made of hardened glass that can serve also as safety glasses, or wear safety goggles over your regular glasses.

• DO shield your eyes whenever you work around the battery. Batteries contain sulphuric acid. In case of contact with the eyes or skin, flush the area with water or a mixture of water and baking soda and get medical attention immediately.

• DO remove the battery cables before charging the battery. Never use a high-output charger on an installed battery or attempt to use any type of "hot shot" (24 volt) starting aid.

• DO use safety stands for any undercar service. Jacks are for raising vehicles; safety stands are for making sure the vehicle stays raised until you want it to come down. Whenever the car is raised, block the wheels remaining on the ground and set the parking brake.

• DO use adequate ventilation when working with any chemicals or hazardous materials. Follow the manufacturer's directions for usage. Brake fluid, anti-freeze, solvents, paints, etc. are all deadly poisons if taken internally. Seal the containers tightly after use and store them safely, out of the reach of children.

• DO use caution when working on clutches or brakes. The asbestos used in the friction material will cause lung cancer if inhaled. Wipe the component with a damp rag to remove dust, and dispose of the rag after use.

• DO disconnect the negative battery cable when working on the electrical system. The secondary ignition system can contain up to 40,000 volts.

• DO properly maintain your tools. Loose hammerheads, mushroomed punches and chisels, frayed or poorly grounded electrical cords, excessively worn screwdrivers, spread open-end wrenches, cracked sockets, slipping ratchets, or faulty droplight sockets can cause accidents.

• DO use the proper size and type of tool for the job being done.

• DO when possible, pull on a wrench handle rather than push on it, and adjust your stance to prevent a fall.

• DO be sure that adjustable wrenches are tightly closed on the nut or bolt and pulled so that the face is on the side of the fixed jaw.

• DO select a wrench or socket that fits the nut or bolt. The wrench or socket should sit straight, not cocked.

• DO strike squarely with a hammer; avoid glancing blows.

• DO set the parking brake and block the drive wheels if the work requires the engine running.

• DO depressurize the fuel system before attempting to disconnect any fuel lines. Although only fuel injection vehicles use a pressurized fuel system, it's a good idea to exercise caution whenever disconnecting any fuel line or hose during service procedures. Take precautions to avoid a fire hazard.

• DO use clean rags and tools when working on an open fuel system and take care to prevent any dirt from entering the system. Wipe all components clean before installation and prepare a clean work area for disassembly and inspection of components. Use lint-free cloths to wipe components and avoid using any caustic cleaning solvents.

• DO remove the electronic control unit (on-board computer) if the vehicle is to be placed in an environment where temperatures exceed approximately 176°F (80°C), such as a paint spray booth or when arc or gas welding near the control unit location in the car.

DON'TS

• DON'T run an engine in a garage or anywhere else without proper ventilation—EVER! Carbon monoxide is poisonous; it takes a long time to leave the human body and you can build up a deadly supply of it in your system by simply breathing in a little every day. You may not realize you are slowly poisoning yourself. Always use power vents, windows, fans or open the garage doors.

• DON'T work around moving parts while wearing a necktie or other loose clothing. Short sleeves are much safer than long, loose sleeves; hard-toed shoes with neoprene soles protect your toes and give a better grip on slippery surfaces. Jewelry such as watches, rings, fancy belt buckles, beads or body adornment of any kind is not safe working around a car. Long hair should be hidden under a hat or cap.

• DON'T use pockets for toolboxes. A fall or bump can drive a screwdriver deep into your body. Even a wiping cloth hanging from the back pocket can wrap around a spinning shaft or fan.

• DON'T smoke when working around gasoline, cleaning solvent or other flammable material.

• DON'T use gasoline to wash your hands; there are excellent soaps available. Gasoline may contain lead, and lead can enter the body through a cut, accumulating in the body until you are very ill. Gasoline also removes all the natural oils from the skin so that bone dry hands will suck up oil and grease.

• DON'T service the air conditioning system unless you are equipped with the necessary tools and training. The refrigerant, R-12, is extremely cold when compressed, and when released into the air will instantly freeze any surface it contacts, including your eyes. Although the refrigerant is normally non-toxic, R-12 becomes a deadly poisonous gas in the presence of an open flame. One good whiff of the vapors from burning refrigerant can be fatal.

• DON'T install or remove battery cables with the key ON or the engine running. Jumper cables should be connected with the key OFF to avoid power surges that can damage electronic control units. Engines equipped with computer controlled systems should avoid both giving and getting jump starts due to the possibility of serious damage to components from arcing in the engine compartment when connections are made with the ignition ON.

• DON'T remove or attach wiring harness connectors with the ignition switch ON, especially to the electronic control unit.

• DON'T drop any components during service procedures and never apply 12 volts directly to any component (like a fuel injector) unless instructed specifically to do so. Some component electrical windings are designed to safely handle only 4 or 5 volts and can be destroyed in seconds if 12 volts are applied directly to the connector.

Air Conditioning Service 27

AIR CONDITIONING SYSTEMS

Automotive air conditioning systems are basic in design and operation, but many different components are used by the vehicle manufacturers to operate and control the systems to their specifications.

Basic System

The basic air conditioning system utilizes the compressor, condenser, evaporator, receiver-drier, expansion valve and a thermostatic or ambient type switch to control evaporator freeze-up. The controls are manually operated and the unit is basic in design. This system is usually installed as an add-on or after-market unit. A sight glass may be used in the system.

GENERAL SERVICING PROCEDURES

The most important aspect of air conditioning service is the maintenance of a pure and adequate charge of refrigerant in the system. A refrigeration system cannot function properly if a significant percentage of the charge is lost. Leaks are common be-

cause the severe vibration encountered in an automobile can easily cause a sufficient cracking or loosening of the air conditioning fittings; as a result, the extreme operating pressures of the system force refrigerant out.

The problem can be understood by considering what happens to the system as it is operated with a continuous leak. Because the expansion valve regulates the flow of refrigerant to the

evaporator, the level of refrigerant there is fairly constant. The receiver-drier stores any excess of refrigerant, and so a loss will first appear there as a reduction in the level of liquid. As this level nears the bottom of the vessel, some refrigerant vapor bubbles will begin to appear in the stream of liquid supplied to the expansion valve. This vapor decreases the capacity of the expansion valve very little as the valve opens to compensate for

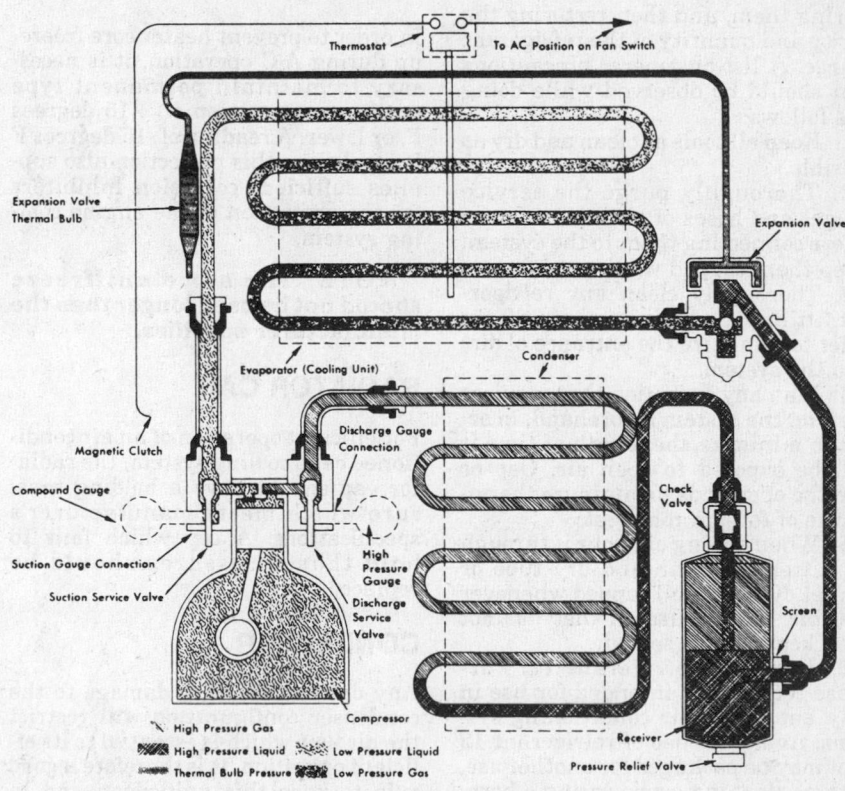

Basic air conditioning system

its presence. As the quantity of liquid in the condenser decreases, the operating pressure will drop there and throughout the high side of the system. As the R-12 continues to be expelled, the pressure available to force the liquid through the expansion valve will continue to decrease, and, eventually, the valve's orifice will prove to be too much of a restriction for adequate flow even with the needle fully withdrawn.

At this point, low side pressure will start to drop, and severe reduction in cooling capacity, marked by freeze-up of the evaporator coil, will result. Eventually, the operating pressure of the evaporator will be lower than the pressure of the atmosphere surrounding it, and air will be drawn into the system wherever there are leaks in the low side.

Because all atmospheric air contains at least some moisture, water will enter the system and mix with the R-12 and the oil. Trace amounts of moisture will cause sludging of the oil, and corrosion of the system. Saturation and clogging of the filter-drier, and freezing of the expansion valve orifice will eventually result. As air fills the system to a greater and greater extent, it will interfere more and more with the normal flows of refrigerant and heat.

From this description, it should be obvious that much of the repairman's time will be spent detecting leaks, repairing them, and then restoring the purity and quantity of the refrigerant charge. A list of general precautions that should be observed while doing this follows:

1. Keep all tools as clean and dry as possible.

2. Thoroughly purge the service gauges and hoses of air and moisture before connecting them to the system. Keep them capped when not in use.

3. Thoroughly clean any refrigerant fitting before disconnecting it, in order to minimize the entrance of dirt into the system.

4. Plan any operation that requires opening the system beforehand, in order to minimize the length of time it will be exposed to open air. Cap or seal the open ends to minimize the entrance of foreign material.

5. When adding oil, pour it through an extremely clean and dry tube or funnel. Keep the oil capped whenever possible. Do not use oil that has not been kept tightly sealed.

6. Use only refrigerant 12. Purchase refrigerant intended for use in only automatic air conditioning systems. Avoid the use of refrigerant 12 that may be packaged for another use, such as cleaning, or powering a horn, as it is impure.

7. Completely evacuate any system that has been opened to replace a component, or that has leaked sufficiently to draw in moisture and air. This requires evacuating air and moisture with a good vacuum pump for at least one hour.

If a system has been open for a considerable length of time it may be advisable to evacuate the system for up to 12 hours (overnight).

8. Use a wrench on both halves of a fitting that is to be disconnected, so as to avoid placing torque on any of the refrigerant lines.

9. When overhauling a compressor, pour some of the oil into a clean glass and inspect it. If there is evidence of dirt or metal particles, or both, flush all refrigerant components with clean refrigerant before evacuating and recharging the system. In addition, if metal particles are present, the compressor should be replaced.

10. Schrader valves may leak only when under full operating pressure. Therefore, if leakage is suspected but cannot be located, operate the system with a full charge of refrigerant and look for leaks from all Schrader valves. Replace any faulty valves.

Additional Preventive Maintenance Checks

ANTIFREEZE

In order to prevent heater core freeze-up during A/C operation, it is necessary to maintain permanent type antifreeze protection of +15 degrees F, or lower. A reading of -15 degrees F is ideal since this protection also supplies sufficient corrosion inhibitors for the protection of the engine cooling system.

NOTE: The same antifreeze should not be used longer than the manufacturer specifies.

RADIATOR CAP

For efficient operation of an air conditioned car's cooling system, the radiator cap should have a holding pressure which meets manufacturer's specifications. A cap which fails to hold these pressures should be replaced.

CONDENSER

Any obstruction of or damage to the condenser configuration will restrict the air flow which is essential to its efficient operation. It is therefore a good rule to keep this unit clean and in proper physical shape.

NOTE: Bug screens are regarded as obstructions.

CONDENSATION DRAIN TUBE

This single molded drain tube expels the condensation, which accumulates on the bottom of the evaporator housing, into the engine compartment. If this tube is obstructed, the air conditioning performance can be restricted and condensation buildup can spill over onto the vehicle's floor.

Safety Precautions

Because of the importance of the necessary safety precautions that must be exercised when working with air conditioning systems and R-12 refrigerant, a recap of the safety precautions are outlined.

1. Avoid contact with a charged refrigeration system, even when working on another part of the air conditioning system or vehicle. If a heavy tool comes into contact with a section of copper tubing or a heat exchanger, it can easily cause the relatively soft material to rupture.

2. When it is necessary to apply force to a fitting which contains refrigerant, as when checking that all system couplings are securely tightened, use a wrench on both parts of the fitting involved, if possible. This will avoid putting torque on refrigerant tubing. (It is advisable, when possible, to use tube or line wrenches when tightening these flare nut fittings.)

3. Do not attempt to discharge the system by merely loosening a fitting, or removing the service valve caps and cracking these valves. Precise control is possible only when using the service gauges. Place a rag under the open end of the center charging hose while discharging the system to catch any drops of liquid that might escape. Wear protective gloves when connecting or disconnecting service gauge hoses.

4. Discharge the system only in a well ventilated area, as high concentrations of the gas can exclude oxygen and act as an anaesthetic. When leak testing or soldering, this is particularly important, as toxic gas is formed when R-12 contacts any flame.

5. Never start a system without first verifying that both service valves are back-seated, if equipped, and that all fittings throughout the system are snugly connected.

6. Avoid applying heat to any refrigerant line or storage vessel. Charging may be aided by using wa-

ter heated to less than 125° to warm the refrigerant container. Never allow a refrigerant storage container to sit out in the sun, or near any other source of heat, such as a radiator.

7. Always wear goggles when working on a system to protect the eyes. If refrigerant contacts the eyes, it is advisable in all cases to see a physician as soon as possible.

8. Frostbite from liquid refrigerant should be treated by first gradually warming the area with cool water, and then gently applying petroleum jelly. A physician should be consulted.

9. Always keep refrigerant drum fittings capped when not in use. Avoid sudden shock to the drum, which might occur from dropping it, or from banging a heavy tool against it. Never carry a drum in the passenger compartment of a car.

10. Always completely discharge the system before painting the vehicle (if the paint is to be baked on), or before welding anywhere near refrigerant lines.

AIR CONDITIONING TOOLS AND GAUGES

Test Gauges

Most of the service work performed in air conditioning requires the use of a set of two gauges, one for the high (head) pressure side of the system, the other for the low (suction) side.

The low side gauge records both pressure and vacuum. Vacuum readings are calibrated from 0 to 30 inches and the pressure graduations read from 0 to no less than 60 psi.

The high side gauge measures pressure from 0 to at least 600 psi. Both gauges are threaded into a manifold that contains two hand shut-off valves. Proper manipulation of these valves and the use of the attached test hoses allow the user to perform the following services:

1. Test high and low side pressures.
2. Remove air, moisture, and contaminated refrigerant.
3. Purge the system (of refrigerant).
4. Charge the system (with refrigerant).

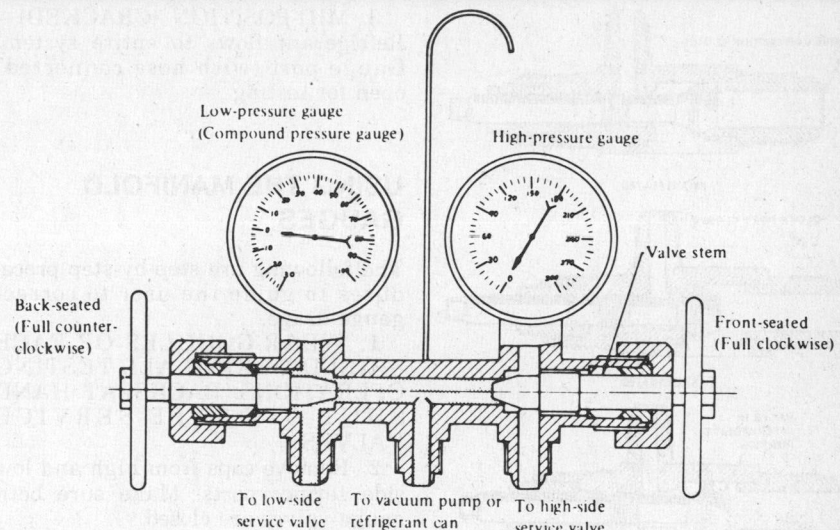

Typical manifold gauge set

NOTE: Chrysler Corp. requires the use of a third gauge on those units that have an evaporator pressure regulator (EPR) valve mounted on the suction side of the compressor.

The manifold valves are designed so they have no direct effect on gauge readings, but serve only to provide for, or cut off, flow of refrigerant through the manifold. During all testing and hook-up operations, the valves are kept in a closed position to avoid disturbing the refrigeration system. The valves are opened only to purge the system of refrigerant or to charge it.

When purging the system, the center hose is uncapped at the lower end, and both valves are cracked open slightly. This allows refrigerant pressure to force the entire contents of the system out through the center hose. During charging, the valve on the high side of the manifold is closed, and the valve on the low side is cracked open. Under these conditions, the low pressure in the evaporator will draw refrigerant from the relatively warm refrigerant storage container into the system.

Service Valves

For the user to diagnose an air conditioning system he or she must gain "entrance" to the system in order to observe the pressures. There are two types of terminals for this purpose, the hand shut off type and the familiar Schrader valve.

The Schrader valve is similar to a tire valve stem and the process of connecting the test hoses is the same as threading a hand pump outlet hose to a bicycle tire. As the test hose is threaded to the service port the valve

core is depressed, allowing the refrigerant to enter the test hose outlet. Removal of the test hose automatically closes the system.

Extreme caution must be observed when removing test hoses from the Schrader valves as some refrigerant will normally escape, usually under high pressure. (Observe safety precautions.)

Some systems have hand shut-off valves (the stem can be rotated with a special ratcheting box wrench) that can be positioned in the following three ways:

1. FRONT SEATED—Rotated to full clockwise position.
   a. Refrigerant will not flow to

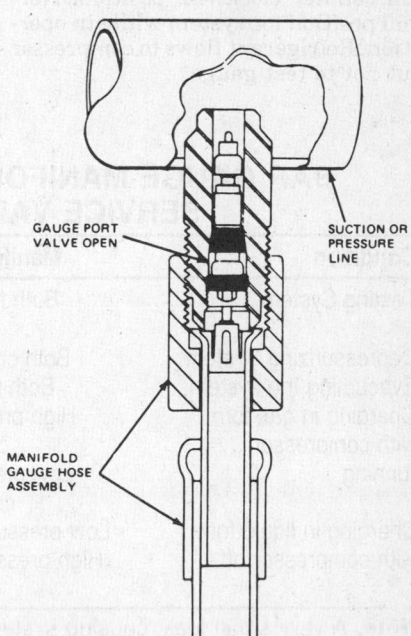

Manifold gauge hose connected to a Schraeder type service port

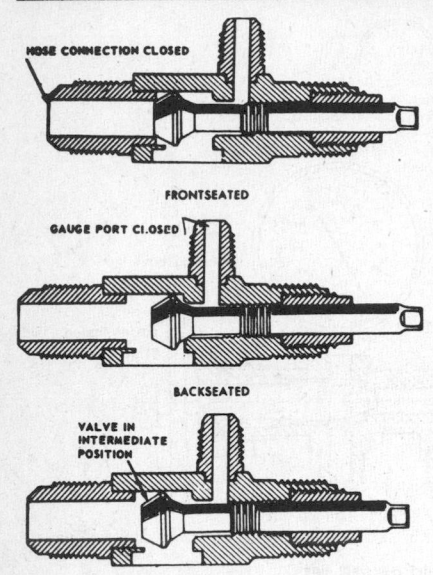

Manual service valve positions

compressor, but will reach test gauge port. COMPRESSOR WILL BE DAMAGED IF SYSTEM IS TURNED ON IN THIS POSITION.

b. The compressor is now isolated and ready for service. However, care must be exercised when removing service valves from the compressor as a residue of refrigerant may still be present within the compressor. Therefore, remove service valves slowly observing all safety precautions.

2. BACK SEATED — Rotated to full counter clockwise position. Normal position for system while in operation. Refrigerant flows to compressor but not to test gauge.

3. MID-POSITION (CRACKED) — Refrigerant flows to entire system. Gauge port (with hose connected) open for testing.

USING THE MANIFOLD GAUGES

The following are step-by-step procedures to guide the user to correct gauge usage.

1. WEAR GOGGLES OR FACE SHIELD DURING ALL TESTING OPERATIONS. BACKSEAT HAND SHUT-OFF TYPE SERVICE VALVES.

2. Remove caps from high and low side service ports. Make sure both gauge valves are closed.

3. Connect low side test hose to service valve that leads to the evaporator (located between the evaporator outlet and the compressor).

4. Attach high side test hose to service valve that leads to the condenser.

5. Mid-position hand shutoff type service valves.

6. Start engine and allow for warm-up. All testing and charging of the system should be done after engine and system have reached normal operation temperatures (except when using certain charging stations).

7. Adjust air conditioner controls to maximum cold.

8. Observe gauge readings.

When the gauges are not being used it is a good idea to:

a. Keep both hand valves in the closed position.

b. Attach both ends of the high and low service hoses to the manifold, if extra outlets are present on the manifold, or plug them if not.

Also, keep the center charging hose attached to an empty refrigerant can. This extra precaution will reduce the possibility of moisture entering the gauges. If air and moisture have gotten into the gauges, purge the hoses by supplying refrigerant under pressure to the center hose with both gauge valves open and all openings unplugged.

DISCHARGING, EVACUATING AND CHARGING

Discharging the System

CAUTION

Perform operation in a well-ventilated area.

When it is necessary to remove (purge) the refrigerant pressurized in the system, follow this procedure:

1. Operate air conditioner for at least 10 minutes.

2. Attach gauges, shut off engine and air conditioner.

3. Place a container or rag at the outlet of the center charging hose on the gauge. The refrigerant will be discharged there and this precaution will avoid its uncontrolled exposure.

4. Open low side hand valve on gauge slightly.

5. Open high side hand valve slightly.

NOTE: Too rapid a purging process will be identified by the appearance of an oily foam. If this occurs, close the hand valves a little more until this condition stops.

6. Close both hand valves on the gauge set when the pressures read 0 and all the refrigerant has left the system.

Evacuating the System

Before charging any system it is necessary to purge the refrigerant and draw out the trapped moisture with a suitable vacuum pump. Failure to do so will result in ineffective charging and possible damage to the system.

Use this hook-up for the proper evacuation procedure:

1. Connect both service gauge hoses to the high and low service outlets.

BAR GAUGE MANIFOLD AND COMPRESSOR SERVICE VALVE SETTINGS

Condition	Manifold Valves	Compressor Valves
Testing System	Both fully closed	Both cracked off backseat
Depressurizing System	Both cracked open	Both at mid position
Evacuating the system	Both wide open	Both at mid position
Charging in gas form with compressor running	High pressure valve closed	High pressure valve cracked off backseat
	Low pressure valve cracked	Low pressure valve at mid position
Charging in liquid form with compressor off	Low pressure valve closed	Both valves mid positioned
	High pressure valve wide open	

Note: A very small leak, causing system discharge about every two weeks, can be caused by a leaky Schrader type service valve. Check these valves with extra care when testing for a small leak.

2. Open high and low side hand valves on gauge manifold.

3. Open both service valves a slight amount (from back seated position), allow refrigerant to discharge from system.

4. Install center charging hose of gauge set to vacuum pump.

5. Operate vacuum pump for at least one hour. (If the system has been subjected to open conditions for a prolonged period of time it may be necessary to "pump the system down" overnight. Refer to "System Sweep" procedure.)

NOTE: If low pressure gauge does not show at least 28" hg. within 5 minutes, check the system for a leak or loose gauge connectors.

6. Close hand valves on gauge manifold.

7. Shut off pump.

8. Observe low pressure gauge to determine if vacuum is holding. A vacuum drop may indicate a leak.

System Sweep

An efficient vacuum pump can remove all the air contained in a contaminated air conditioning system very quickly, because of its vapor state. Moisture, however, is far more difficult to remove because the vacuum must force the liquid to evaporate before it will be able to remove it from the system. If a system has become severely contaminated, as, for example, it might become after all the charge was lost in conjunction with vehicle

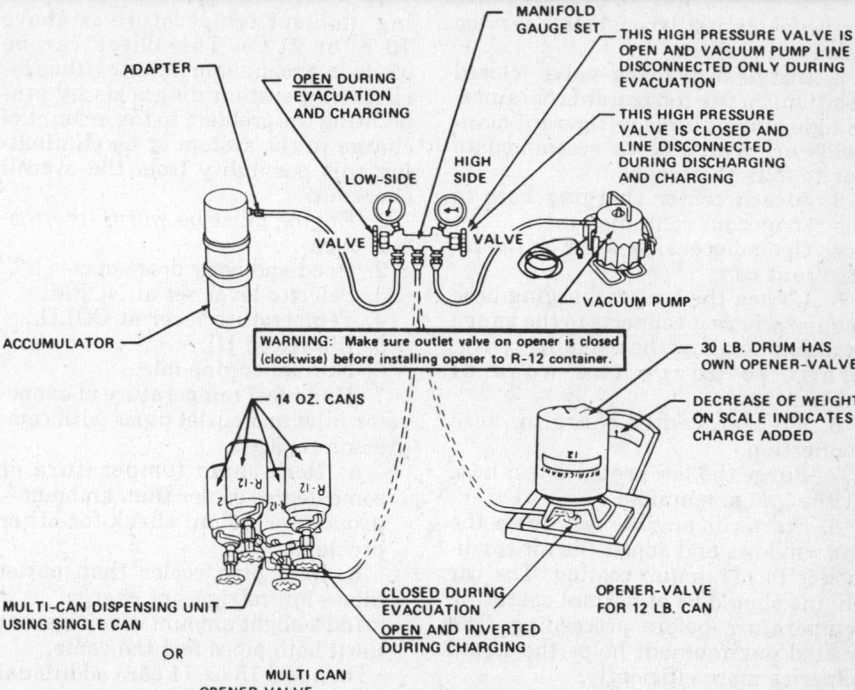

Typical gauge connections for discharge, evacuation and charging the system

accident damage, moisture removal is extremely time consuming. A vacuum pump could remove all of the moisture only if it were operated for 12 hours or more.

Under these conditions, sweeping the system with refrigerant will speed the process of moisture removal considerably. To sweep, follow the following procedure:

1. Connect vacuum pump to

gauges, operate it until vacuum ceases to increase, then continue operation for ten more minutes.

2. Charge system with 50% of its rated refrigerant capacity.

3. Operate system at fast idle for ten minutes.

4. Discharge the system.

5. Repeat twice the process of charging to 50% capacity, running the system for ten minutes, and discharging it, for a total of three sweeps.

6. Replace drier.

7. Pump system down as in Step 1.

8. Charge system.

Charging the System

——— CAUTION ———

Never attempt to charge the system by opening the high pressure gauge control while the compressor is operating. The compressor accumulating pressure can burst the refrigerant container, causing sever personal injuries.

BASIC SYSTEM

In this procedure the refrigerant enters the suction side of the system as a vapor while the compressor is running. Before proceeding, the system should be in a partial vacuum after adequate evacuation. Both hand valves on the gauge manifold should be closed.

1. Attach both test hoses to their respective service valve ports. Mid-

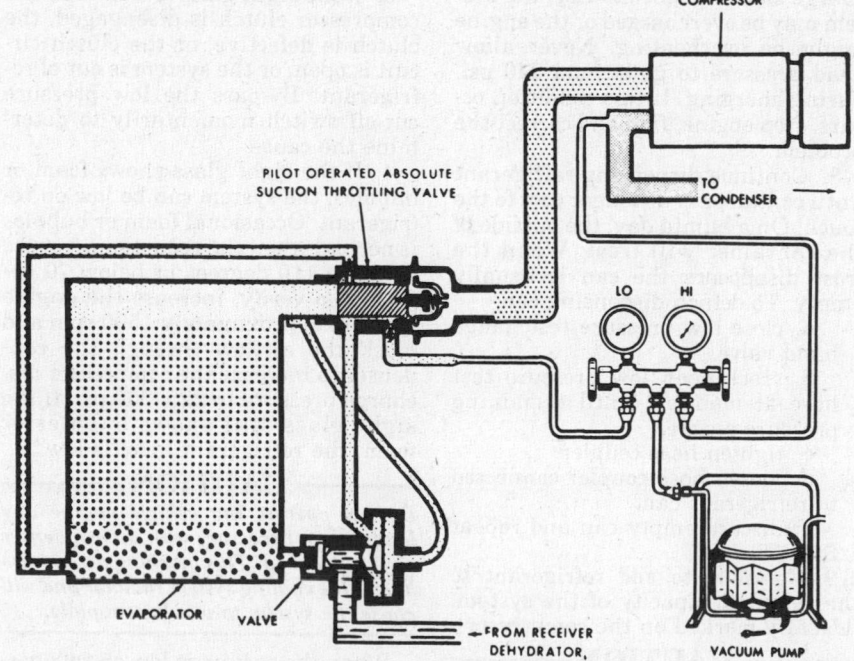

Schematic for evacuating the system

position manually operated service valves, if present.

2. Install dispensing valve (closed position) on the refrigerant container. (Single and multiple refrigerant manifolds are available to accommodate one to four 15 oz. cans.)

3. Attach center charging hose to the refrigerant container valve.

4. Open dispensing valve on the refrigerant can.

5. Loosen the center charging hose coupler where it connects to the gauge manifold to allow the escaping refrigerant to purge the hose of contaminants.

6. Tighten center charging hose connection.

7. Purge the low pressure test hose at the gauge manifold.

8. Start car engine, roll down the car windows and adjust the air conditioner to maximum cooling. The car engine should be at normal operating temperature before proceeding. The heated environment helps the liquid vaporize more efficiently.

9. Crack open the low side hand valve on the manifold. Manipulate the valve so that the refrigerant that enters the system does not cause the low side pressure to exceed 40 psi. Too sudden a surge may permit the entrance of unwanted liquid to the compressor. Since liquids cannot be compressed, the compressor will suffer damage if compelled to attempt it. If the suction side of the system remains in a vacuum the system is blocked. Locate and correct the condition before proceeding any further.

NOTE: Placing the refrigerant can in a container of warm water (no hotter than 125°F) will speed the charging process. Slight agitation of the can is helpful too, but be careful not to turn the can upside down.

Some manufacturers allow for a partial charging of the A/C system in the form of a liquid (can inverted and compressor off) by opening the high side gauge valve only, and putting the high side compressor service valve in the middle position (if so equipped). The remainder of the refrigerant is then added in the form of a gas in the normal manner, through the suction side only.

SYSTEMS WITHOUT SIGHT GLASS, EXCEPT CCOT SYSTEM

The following procedure can be used to quickly determine whether or not an air conditioning system has the proper charge of refrigerant (provid-ing ambient temperature is above 70°F, or 21°C). This check can be made in a manner of minutes, thus facilitating system diagnosis by pinpointing the problem to the amount of charge in the system or by eliminating this possibility from the overall checkout.

1. Engine must be warm (thermostat open).

2. Hood and body doors open.

3. Selector lever set at NORM.

4. Temperature lever at COLD.

5. Blower on HI.

6. Normal engine idle.

7. Hand-feel temperature of evaporator inlet and outlet pipes with compressor engaged.

   a. Both same temperature or some degree cooler than ambient—proper condition: check for other problems.

   b. Inlet pipe cooler than outlet pipe—low refrigerant charge.
   • Add a slight amount of refrigerant until both pipes feel the same.
   • Then add 15 oz. (1 can) additional refrigerant.

   c. Inlet pipe has frost accumulation—outlet pipe warmer: proceed as in Step b above.

If during the charging process the head pressure exceeds 200 psi, place an electric fan in front of the car and direct the turbulent air to the condenser. If no fan is available, repeatedly pour cool water over the top of the condenser. These cooling actions may be necessary on an extremely warm day to help dissipate the heat emitted by the engine during idle.

If this fails and pressure on the discharge side continues to rise, the system may be overcharged or the engine might be overheating. Never allow head pressure to go beyond 240 psi. during charging. If this condition occurs, stop engine, find and correct the problem.

8. Continue dispensing refrigerant until container is no longer cool to the touch. On a humid day, the outside of the container will frost. When the frost disappears the can is usually empty. To detach dispensing can:

   a. close low pressure test gauge hand valve.

   b. crack open low pressure test hose at manifold until remaining pressure escapes.

   c. tighten hose coupler.

   d. loosen hose coupler connected to refrigerant can.

   e. discard empty can and repeat Steps 2–8.

9. Continue to add refrigerant to the required capacity of the system. (Usually marked on the compressor).

— CAUTION —
DO NOT OVERCHARGE. This condition is usually indicated by an abnormally high side pressure reading and a noisy compressor resulting in ineffective cooling and damage to the system.

SYSTEMS WITH A SIGHT GLASS

The air conditioning systems that use a sight glass as a means to check the refrigerant level, should be carefully checked to avoid under or over charging. The gauge set should be attached to the system for verification of pressures.

To check the system with the sight glass, clean the glass and start the vehicle engine. Operate the air conditioning controls on maximum for approximately five minutes to stabilize the system. The room temperature should be above 70 degrees. Check the sight glass for one of the following conditions:

1. If the sight glass is clear, the compressor clutch is engaged, the compressor discharge line is warm and the compressor inlet line is cool, the system has a full charge of refrigerant.

2. If the sight glass is clear, the compressor clutch is engaged and there is no significant temperature difference between the compressor inlet and discharge lines, the system is empty or nearly empty. By having the gauge set attached to the system a measurement can be taken. If the gauge reads less than 25 psi, the low pressure cutoff protection switch has failed.

3. If the sight glass is clear and the compressor clutch is disengaged, the clutch is defective, or the clutch circuit is open, or the system is out of refrigerant. By-pass the low pressure cut-off switch momentarily to determine the cause.

4. If the sight glass shows foam or bubbles, the system can be low on refrigerant. Occasional foam or bubbles is normal when the room temperature is above 110 degrees or below 70 degrees. To verify, increase the engine speed to approximately 1500 rpm and block the airflow through the condenser to increase the compressor discharge pressure to 225–250 psi. If the sight glass still shows bubbles or foam, the refrigerant level is low.

— CAUTION —
Do not operate the vehicle engine any longer than necessary with the condenser airflow blocked. This blocking action also blocks the cooling system radiator and will cause the system to overheat rapidly.

When the system is low on refrigerant, a leak is present or the system was not properly charged. Use a leak detector and locate the problem area and repair. If no leakage is found, charge the system to its capacity. (Refer to the refrigerant capacity chart at the end of this section).

—————— **CAUTION** ——————

It is not advisable to add refrigerant to a system utilizing the suction throttling valve and a sight glass, because the amount of refrigerant required to remove the foam or bubbles will result in an overcharge and potentially damaged system components.

CCOT SYSTEM

When charging the CCOT system, attach only the low pressure line to the low pressure gauge port, located on the accumulator. Do not attach the high pressure line to any service port or allow it to remain attached to the vacuum pump after evacuation. Be sure both the high and the low pressure control valves are closed on the gauge set. To complete the charging of the system, follow the outline supplied.

1. Start the engine and allow to run at idle, with the cooling system at normal operating temperature.

2. Attach the center gauge hose to a single or multi-can dispenser.

3. With the multi-can dispenser inverted, allow one pound or the contents of one or two 14 oz. cans to enter the system through the low pressure side by opening the gauge low pressure control valve.

4. Close the low pressure gauge control valve and turn the A/C system on to engage the compressor. Place the blower motor in its high mode.

5. Open the low pressure gauge control valve and draw the remaining charge into the system. Refer to the capacity chart at the end of this section for the individual vehicle or system capacity.

6. Close the low pressure gauge control valve and the refrigerant source valve, on the multi-can dispenser. Remove the low pressure hose from the accumulator quickly to avoid

Check item \ Amount of refrigerant	Almost no refrigerant	Insufficient	Suitable	Too much refrigerant
Temperature of high pressure and low pressure lines.	Almost no difference between high pressure and low pressure side temperature.	High pressure side is warm and low pressure side is fairly cold.	High pressure side is hot and low pressure side is cold.	High pressure side is abnormally hot.
State in sight glass.	Bubbles flow continuously. Bubbles will disappear and something like mist will flow when refrigerant is nearly gone.	The bubbles are seen at intervals of 1 - 2 seconds.	Almost transparent. Bubbles may appear when engine speed is raised and lowered.\n\nNo clear difference exists betwen these two conditions.	No bubbles can be seen.
Pressure of system.	High pressure side is abnormally low.	Both pressure on high and low pressure sides are slightly low.	Both pressures on high and low pressure sides are normal.	Both pressures on high and low pressure sides are abnormally high.
Repair.	Stop compressor immediately and conduct an overall check.	Check for gas leakage, repair as required, replenish and charge system.		Discharge refrigerant from service valve of low pressure side.

Using a sight glass to determine the relative refrigerant charge

loss of refrigerant through the Schrader valve.

7. Install the protective cap on the gauge port and check the system for leakage.

8. Test the system for proper operation.

Leak Testing the System

There are several methods of detecting leaks in an air conditioning system; among them, the two most popular are (1) halide leak-detection or the "open flame method," and (2) electronic leak-detection.

The halide leak detection is a torch like device which produces a yellow-green color when refrigerant is introduced into the flame at the burner. A purple or violet color indicates the presence of large amounts of refriger-

ant at the burner.

An electronic leak detector is a small portable electronic device with an extended probe. With the unit activated the probe is passed along those components of the system which contain refrigerant. If a leak is detected, the unit will sound an alarm signal or activate a display signal depending on the manufacturer's design. It is advisable to follow the manufacturer's instructions as the design and function of the detection may vary significantly.

—————— **CAUTION** ——————

Caution should be taken to operate either type of detector in well ventilated areas, so as to reduce the chance of personal injury, which may result from coming in contact with poisonous gases produced when R-12 is exposed to flame or electric spark.

REFRIGERANT FLUSHING INFORMATION CHART

Refrigerant	Vaporizes °C(°F)①	Approximate Closed Container Pressure① kPa (psi)②					Adaptability
		15.57°C (60°F)	21.13°C (70°F)	26.69°C (80°F)	32.25°C (90°F)	37.81°C (100°F)	
R-12	−29.80 (−21.6)	393 (57)	483 (70)	579 (84)	689 (100)	807 (117)	Self Propelling
F-114	3.56 (38.4)	55.16 (8)	89.63 (13)	131 (19)	172 (25)	221 (32)	
F-11③	23.74 (74.7)	27 (8 in Hg)	10 (3 in Hg)	7 (1)	34 (5)	62 (9)	
F-113	47.59 (117.6)	74 (22 in Hg)	64 (19 in Hg)	54 (16 in Hg)	44 (13 in Hg)	27 (8 in Hg)	Pump Required

①At sea level atmospheric pressure.
②kPa (psi) unless otherwise noted.
③F-11 is also available in pressurized containers. This makes it suitable for usage when special flushing equipment is not available. However, it is more toxic than R-12 and F-114.

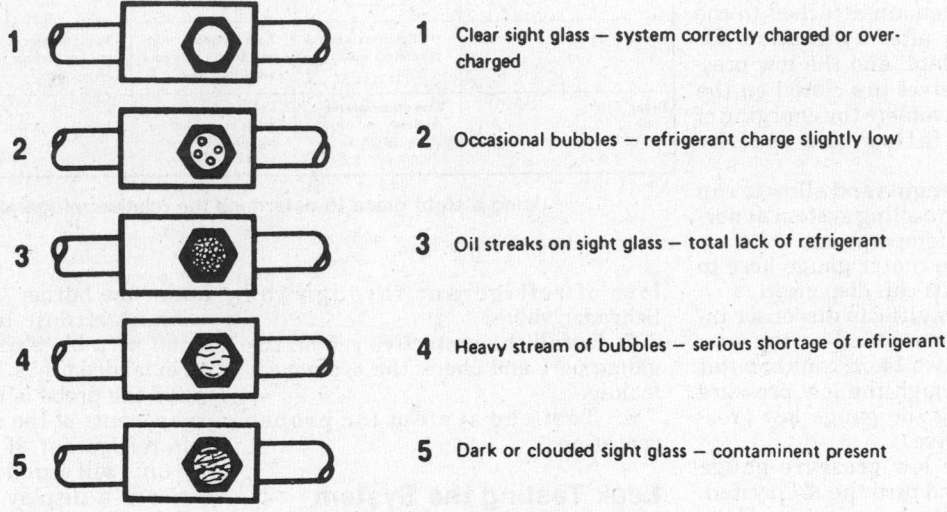

1 Clear sight glass — system correctly charged or over-charged

2 Occasional bubbles — refrigerant charge slightly low

3 Oil streaks on sight glass — total lack of refrigerant

4 Heavy stream of bubbles — serious shortage of refrigerant

5 Dark or clouded sight glass — contaminent present

Sight glass examination of refrigerant flow

Carburetor Service

FORD MOTORCRAFT CARBURETORS

Model 7200 VV

Since the design of the 7200 VV (variable venturi) carburetor differs considerably from the other carburetors in the Ford lineup, an explanation in the theory and operation is presented here.

In exterior appearance, the variable venturi carburetor is similar to conventional carburetors and like a conventional carburetor, it uses a normal float and fuel bowl system. However, the similarity ends there. In place of a normal choke plate and fixed area venturis, the 7200 VV carburetor has a pair of small oblong castings in the top of the upper carburetor body where the choke plate would normally be located. These castings slide back and forth across the top of the carburetor in response to air/fuel demands. Their movement is controlled by a spring-loaded diaphragm valve regulated by a vacuum signal taken below the venturis in the throttle bores. As the throttle is opened, the strength of the vacuum signal increases, opening the venturis and allowing more air to enter the carburetor.

Fuel is admitted into the venturi area by means of tapered metering rods that fit into the main jets. These rods are attached to the venturis and the venturis open or close in response to air demand. The fuel needed to maintain the proper mixture increases or decreases as the metering rods slide in the jets. In comparison to a conventional carburetor with fixed venturis and a variable air supply, this system provides much more precise control of the fuel/air supply during all modes of operation. Because of the variable venturi principle, there are fewer fuel metering systems and fuel passages. The only auxiliary fuel metering systems required are an idle trim, accelerator pump (similar to a conventional carburetor), starting enrichment and cold running enrichment.

NOTE: Adjustment, assembly and disassembly of this carburetor require special tools for some of the operations. Do not attempt any operations on this carburetor without first obtaining special tools needed for that particular operation. Special tools needed for the following adjustments are identified in the procedure.

FLOAT LEVEL

Adjustment

1. Remove and invert the upper part of the carburetor, with the gasket in place.

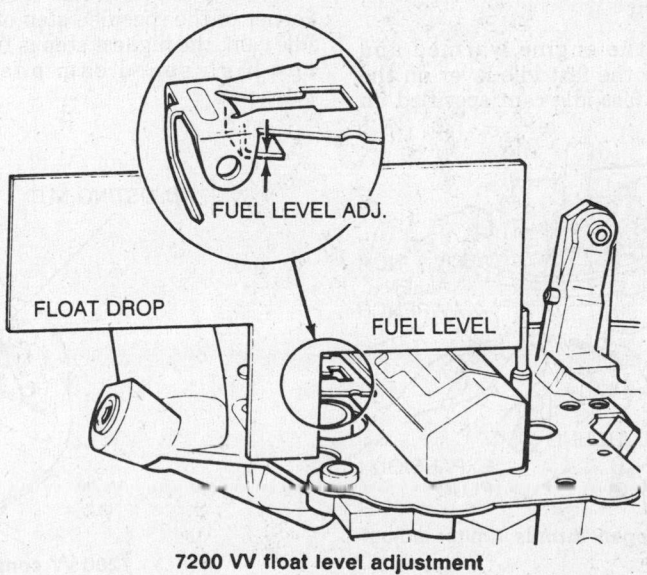

FUEL LEVEL ADJ.

FLOAT DROP

FUEL LEVEL

7200 VV float level adjustment

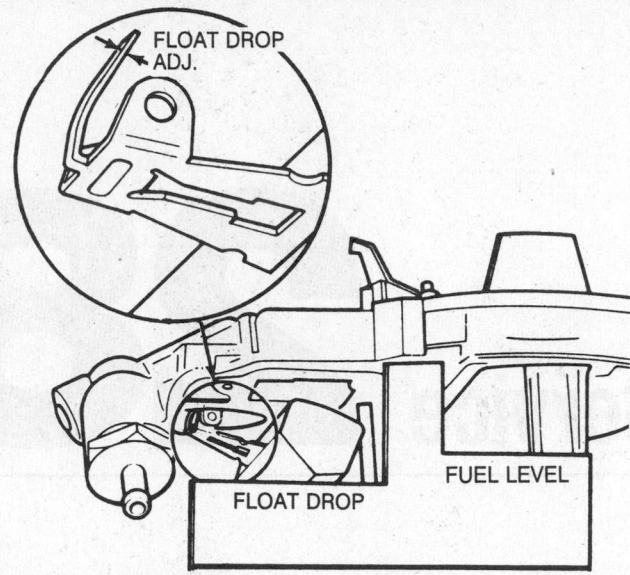

7200 VV float drop adjustment

3. If the adjustment is being made with the carburetor removed, hold the throttle lightly closed with a rubber band.

4. Turn the stator cap tool clockwise until the lever contacts the fast idle cam adjusting screw.

5. Turn the fast idle cam adjusting screw until the index mark on the cap lines up with the specified mark on the casting.

6. Remove the stator cap tool. Install the choke coil cap and set to the specified housing mark.

COLD ENRICHMENT METERING ROD

Adjustment

A dial indicator and a stator cap tool T77L–9848–A or equivalent, are required for this adjustment.

1. Remove the choke coil cap.

2. Attach a weight to the choke coil mechanism to seat the cold enrichment rod.

3. Install and zero a dial indicator with the tip on top of the enrichment rod. Raise and release the weight to verify zero on the dial indicator.

4. With the stator cap at the index position, the dial indicator should read the specified dimension on the specification tag. If needed, turn the adjusting nut to correct.

5. Install the choke cap at the correct setting.

VENTURI VALVE LIMITER

Adjustment

1. Remove the carburetor. Take off the venturi valve cover and the 2 rollers.

2. Use a center punch to loosen the expansion plug at the rear of the carburetor main body on the throttle side and remove the expansion plug.

3. Use an Allen wrench to remove

2. Measure the vertical distance between the carburetor body, outside the gasket and the bottom of the float.

3. To adjust, bend the float operating lever that contacts the needle valve. Make sure that the float remains parallel to the gasket surface.

FLOAT DROP

Adjustment

1. Remove and hold the upper part of the carburetor upright.

2. Measure the vertical distance between the carburetor body, outside the gasket and the bottom of the float.

3. Adjust by bending the stop tab on the float lever that contacts the hinge pin.

FAST IDLE SPEED

Adjustment

1. With the engine warmed and idling, place the fast idle lever on the step of the fast idle cam specified on

the engine compartment sticker or in the specifications chart. Disconnect and plug the EGR vacuum line.

2. Make sure the high speed cam positioner lever is disengaged.

3. Turn the fast idle speed screw to adjust to the specified speed.

FAST IDLE CAM

Adjustment

Use of a stator cap special tool T77L–9848–A or equivalent, is required for this procedure. It fits over the choke thermostatic lever when the choke cap is removed.

1. Remove the choke coil cap. The top rivets will have to be drilled out; the bottom rivet will have to be driven out from the rear. New rivets must be used upon installation.

2. Place the fast idle lever in the corner of the specified step of the fast idle cam, the highest step is first, with the high speed cam positioner retracted.

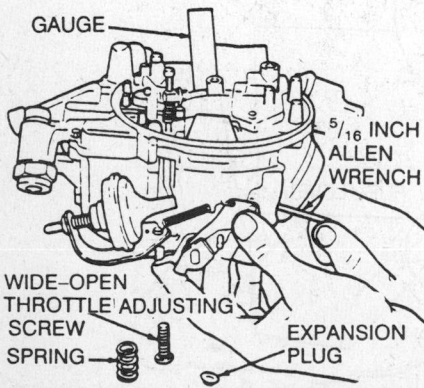

7200–Wide open throttle limiter adjustment

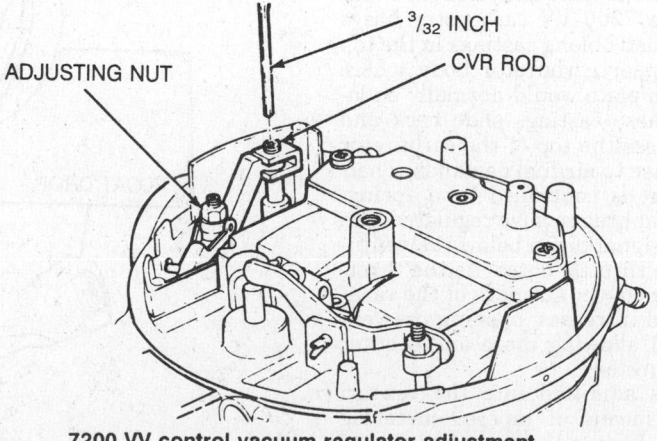

7200 VV control vacuum regulator adjustment

Year	Model	Float Level (in.)	Float Drop (in.)	Fast Idle Cam Setting/Step	Cold Enrichment Metering Rod (in.)	Control Vacuum (in. H₂0)	Venturi Valve Limiter (in.)	Choke Cap Setting (notches)
1988	E7AE-AA	1.010–1.070	1.430–1.490	0.360/2nd step	③	②	①	Index
	E8AE-AA	1.010–1.070	1.430–1.490	0.360/2nd step	③	②	①	Index
1989-90	ALL	1.010–1.070	1.430–1.490	0.360/2nd step	③	②	①	Index

① Maximum opening: 0.99/1.01
  Wide open on throttle: 0.39/.41
② See text
③ Maximum opening: 0.99/1.01
  Wide open on throttle: 0.74/.76

the venturi valve wide open stop screw.

4. Hold the throttle wide open.

5. Apply a light closing pressure on the venturi valve and check the gap between the valve and the air horn wall. To adjust, move the venturi valve to the wide open position and insert an Allen wrench into the stop screw hole. Turn clockwise to increase the gap. Remove the wrench and check the gap again.

6. Replace the wide open stop screw and turn it clockwise until it contacts the valve.

7. Push the venturi valve wide open and check the gap. Turn the stop screw to bring the gap to specifications.

8. Reassemble the carburetor with a new expansion plug.

CONTROL VACUUM REGULATOR (CVR)

Adjustment

The cold enrichment metering rod adjustment must be checked and set before making this adjustment.

1. After adjusting the cold enrichment metering rod, leave the dial indicator in place but remove the stator cap. Do not re-zero the dial indicator.

2. Press down on the CVR rod until it bottoms on its seat. Measure this amount of travel with the dial indicator.

3. If the adjustment is incorrect, hold the ⅜ in. CVR adjusting nut with a box wrench to prevent it from turning. Use a ³/₃₂ in. Allen wrench to turn the CVR rod; turning counter-clockwise will increase the travel and vice versa.

HOLLEY CARBURETORS

Model 2280 and 6280

FLOAT LEVEL

Adjustment

1. Remove the carburetor air horn.

2. Invert the carburetor body taking care to catch the pump intake check ball so that only the weight of the floats is forcing the needle against the seat. Hold a finger against the hinge pin retainer to fully seat the float in the float pin cradle.

3. Lay a straight edge across the float bowl. The toe of each float should be as per specifications from the straight edge. If necessary, bend the float tang to adjust.

ACCELERATOR PUMP STROKE

Measurement

MODEL 2280

1. Remove the bowl vent cover plate and vent valve lever spring. Take care to avoid loosening the vent valve retainer.

2. Make sure that the accelerator pump connector rod is in the inner hole of the pump operating lever and the throttle is at curb idle.

3. Place a straight edge on the bowl vent cover surface of the air horn, over the accelerator pump lever.

4. The lever surface should be flush

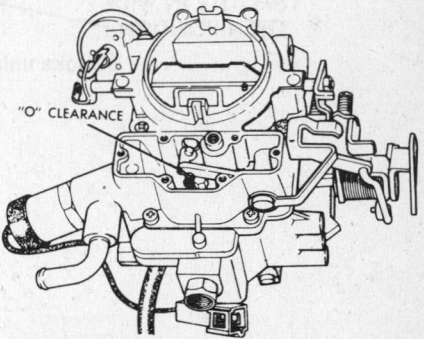

Accelerator pump stroke adjustment—model 6280

with the air horn. If not, adjust it by bending the pump connector rod at the 90 degree bend.

NOTE: If this adjustment is changed, both the bowl vent and the mechanical power valve adjustments must be reset.

MODEL 6280

1. Remove the bowl vent cover plate and gasket.

2. With all pump links and levers installed, adjust the accelerator pump cap nut for zero clearance between the pump lever and the cap nut. Check that the wide open throttle can be reached without binding.

3. Install the gasket and the bowl vent cover plate.

CHOKE UNLOADER

Adjustment

1. Hold the throttle valves in the wide open position.

2. Lightly press a finger against the

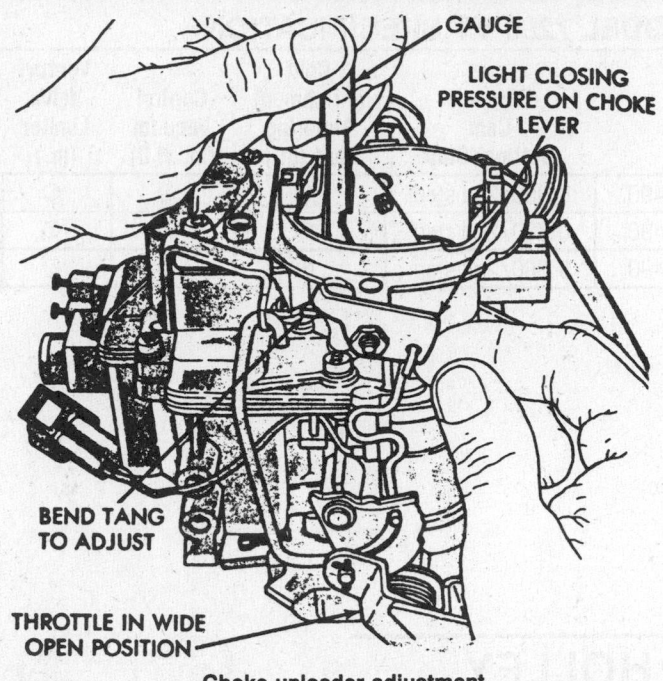

Choke unloader adjustment

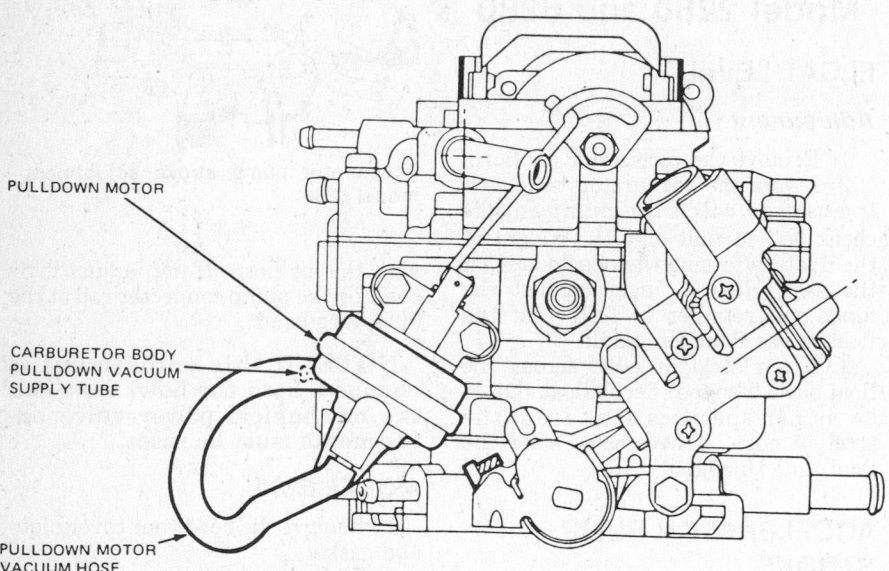

Fast Idle adjustment—measure the choke clearance

2. Disconnect the vacuum hose from the carburetor and connect it to an auxiliary vacuum source with a length of hose. Apply at least 15 in. Hg.

3. Completely compress the choke lever spring in the diaphragm stem without distorting the linkage.

4. Insert the specified gauge between the top of the choke valve and the air horn wall.

5. Adjust by bending the diaphragm link. Check for free movement. Replace the vacuum hose.

FAST IDLE CAM POSITION

Adjustment

1. Position the adjusting screw on the second highest step of the fast idle cam.

2. Move the choke towards the closed position with light finger pressure.

3. Insert the specified gauge between the choke valve and the air horn wall.

4. Adjust by opening or closing the U-bend in the fast idle connector link.

MECHANICAL POWER VALVE

Adjustment

MODEL 2280

1. Remove the bowl vent cover plate, vent valve lever, spring and retainer. Remove the lever pivot pin.

2. Hold the throttle in the wide open position.

3. Using a $5/64$ in. Allen wrench, press the mechanical power valve adjustment screw down and release it to determine if clearance exists. Turn the screw clockwise until clear is zero.

4. Adjust by turning the screw a turn counterclockwise.

5. Install all parts.

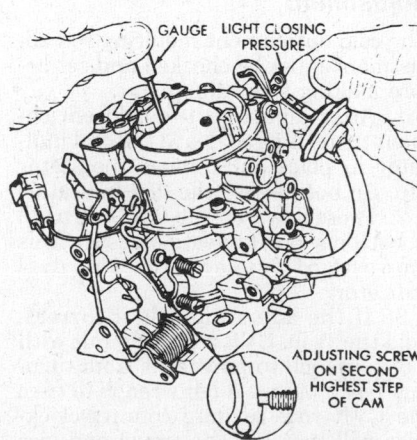

Fast Idle cam position adjustment

control lever to move the choke valve toward the closed position.

3. Insert the specified gauge between the top of the choke valve and the air horn wall.

4. Adjust, if necessary, by bending the tang on the accelerator pump lever.

CHOKE VACUUM KICK

Adjustment

1. Open the throttle, close the choke, then close the throttle to trap the fast idle cam at the closed choke position.

HOLLEY MODEL 2280/6280
Chrysler Corporation

Year	Carb. Part No.	Float Level (in.)	Accelerator Pump Adjustment (in.)	Fast Idle (rpm)	Choke Unloader Clearance (in.)	Vacuum Kick (in.)	Fast Idle Cam Position (in.)	Choke
1988	R-40276A	$9/32$	0.180	①	0.280	0.130	0.060	Fixed
	R-40354A	$9/32$	0.050	①	0.200	0.140	0.052	Fixed
1989	R-40354A	$9/32$	0.180	①	0.280	0.130	0.060	Fixed

① Refer to underhood sticker

ROCHESTER CARBURETORS

Quadrajet

The Rochester Quadrajet carburetor is a 2 stage, 4-barrel downdraft carburetor. It has been built in many variations designated as 4MC, M4MC, M4MCA, M4ME, M4MEA, E4MC and E4ME.

The first M in the identification indicates that the carburetor is of a modified primary metering (open loop) design, while the first E indicates electronically controlled. The C has an integral hot air choke, while the E has an electric choke.

The primary side of the carburetor is equipped with 2 primary bores and a triple venturi with plain tube nozzles. During off idle and part throttle operation, the fuel is metered through tapered metering rods operating in specially designed jets positioned by a manifold vacuum responsive piston.

The secondary side of the carburetor contains 2 secondary bores. An air valve is used on the secondary side for metering control and supplements the primary bore. The secondary air valve operates tapered metering rods which regulate the fuel in constant proportion to the air being supplied.

FAST IDLE SPEED

Adjustment

1. Position the fast idle lever on the high step of the fast idle cam.

Quadrajet fast idle adjustment

2. Be sure that the choke is wide open and the engine warm. Plug the EGR vacuum hose. Disconnect the vacuum hose to the front vacuum break unit, if there are 2.

3. Make a preliminary adjustment by turning the fast idle screw out until the throttle valves are closed, then turning it in the specified number of turns after it contacts the lever.

4. Use the fast idle screw to adjust the fast idle to the speed and under the conditions, specified on the engine compartment sticker or in the specifications chart.

FAST IDLE CAM CHOKE ROD

Adjustment

1. Adjust the fast idle and place the cam follower on the highest step of the fast idle cam against the shoulder of the high step.

2. Close the choke valve by exerting counter-clockwise pressure on the external choke lever. Remove the coil assembly from the choke housing and push upon the choke coil lever. On models with a fixed (riveted) choke cover, push up on the vacuum brake lever tang and hold in position with a rubber band.

3. Insert a gauge of the proper size between the upper edge of the choke valve and the inside air horn wall.

4. To adjust the valve, bend the tang on the fast idle cam. Be sure that the tang rests against the cam after bending.

PRIMARY (FRONT) VACUUM BREAK

Adjustment

A choke valve measuring gauge J–26701 or equivalent, is used to measure angle (degrees instead of inches).

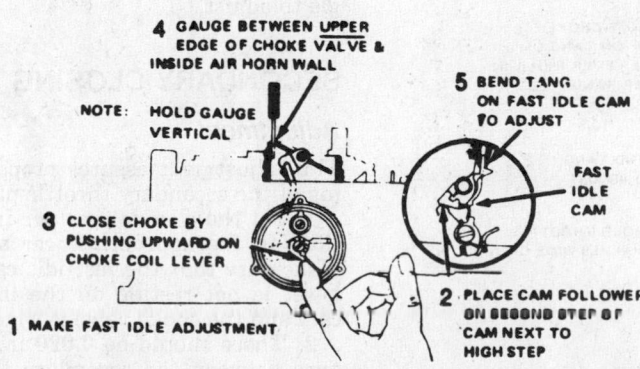

Quadrajet choke rod (fast idle cam) adjustment—typical

1. ATTACH RUBBER BAND TO GREEN TANG OF INTERMEDIATE CHOKE SHAFT

2. OPEN THROTTLE TO ALLOW CHOKE VALVE TO CLOSE

3. SET UP ANGLE GAGE AND SET TO SPECIFICATION

4. RETRACT VACUUM BREAK PLUNGER USING VACUUM SOURCE. AT LEAST 18" HG. PLUG AIR BLEED HOLES WHERE APPLICABLE

ON QUADRAJETS, AIR VALVE ROD MUST NOT RESTRICT PLUNGER FROM RETRACTING FULLY. IF NECESSARY, BEND ROD (SEE ARROW) TO PERMIT FULL PLUNGER TRAVEL. FINAL ROD CLEARANCE MUST BE SET AFTER VACUUM BREAK SETTING HAS BEEN MADE.

5. WITH AT LEAST 18" HG STILL APPLIED, ADJUST SCREW TO CENTER BUBBLE

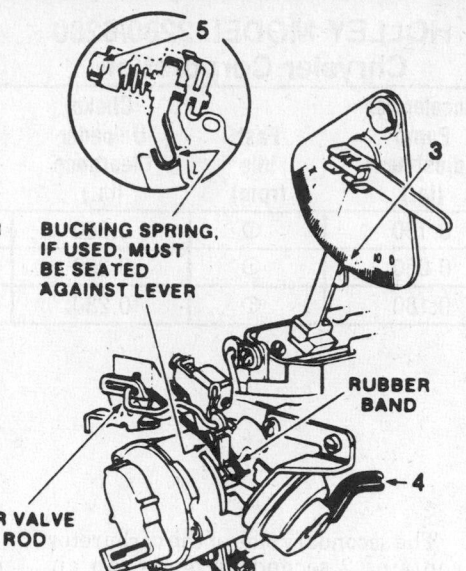

BUCKING SPRING, IF USED, MUST BE SEATED AGAINST LEVER

RUBBER BAND

AIR VALVE ROD

Quadrajet front vacuum break adjustment

1. ATTACH RUBBER BAND TO GREEN TANG OF INTERMEDIATE CHOKE SHAFT.

2. OPEN THROTTLE TO ALLOW CHOKE VALVE TO CLOSE.

3. SET UP ANGLE GAGE AND SET ANGLE TO SPECIFICATION.

RETRACT VACUUM BREAK PLUNGER. USING VACUUM SOURCE. AT LEAST 18" HG. PLUG AIR BLEED HOLES WHERE APPLICABLE

4A. ON QUADRAJETS. AIR VALVE ROD MUST NOT RESTRICT PLUNGER FROM RETRACTING FULLY. IF NECESSARY BEND ROD HERE TO PERMIT FULL PLUNGER TRAVEL. WHERE APPLICABLE. PLUNGER STEM MUST BE EXTENDED FULLY TO COMPRESS PLUNGER BUCKING SPRING.

5. TO CENTER BUBBLE. EITHER:
A. ADJUST WITH 1/8" HEX WRENCH (VACUUM STILL APPLIED)

-OR-

B. SUPPORT AT "S" AND BEND VACUUM BREAK ROD (VACUUM STILL APPLIED)

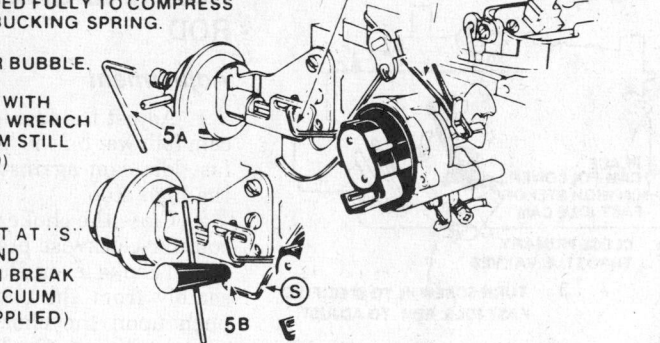

Quadrajet rear vacuum break adjustment—typical

4. GAUGE BETWEEN UPPER EDGE OF CHOKE VALVE AND AIR HORN WALL (SEE NOTE*)

3. ON WARM ENGINE. CLOSE CHOKE VALVE BY PUSHING UP ON TANG ON INTERMEDIATE CHOKE LEVER (HOLD IN POSITION WITH RUBBER BAND)

5. BEND TANG TO ADJUST

2. HOLD THROTTLE VALVES WIDE OPEN

1. INSTALL CHOKE THERMOSTATIC COVER AND COIL ASSEMBLY IN HOUSING ALIGN INDEX MARK WITH SPECIFIED POINT ON HOUSING

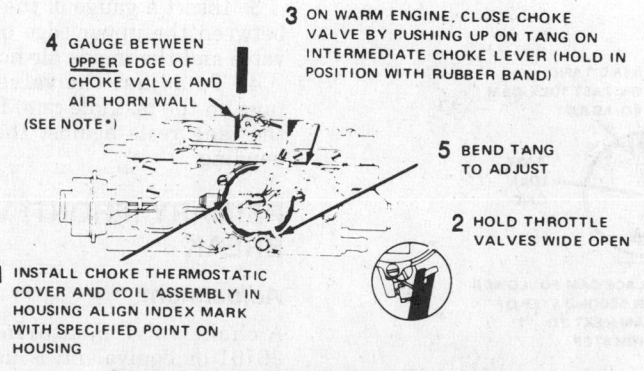

Quadrajet unloader adjustment—typical

SECONDARY (REAR) VACUUM BRAKE

Adjustment

A choke valve measuring gauge J-26701 or equivalent, is used to measure the angle (degrees instead of inches).

CHOKE LINK

Adjustment

Refer to the illustration for E4MC fast idle cam adjustment.

CHOKE UNLOADER

Adjustment

1. Push up on the vacuum break lever to close the choke valve and fully open the throttle valves.
2. Measure the distance from the upper edge of the choke valve to the air horn wall.
3. To adjust, bend the tang on the fast idle lever.

CHOKE COIL LEVER

Adjustment

MC AND ME CARBURETORS

1. Remove the choke cover and thermostatic coil from the choke housing. On models with a fixed (riveted) choke cover, the rivets must be drilled out. A choke stat kit is necessary for assembly. Place the fast idle cam follower on the high step.
2. Push up on the coil tang (counter-clockwise) until the choke valve is closed. The top of the choke rod should be at the bottom of the slot in the choke valve lever.
3. Insert a 0.120 in. drill bit in the hole in the choke housing.
4. The lower edge of the choke coil lever should just contact the side of the plug gauge.
5. Bend the choke rod at the top angle to adjust.

SECONDARY CLOSING

Adjustment

This adjustment assures proper closing of the secondary throttle plates.

1. Set the slow idle as per instructions in the appropriate car section. Make sure that the fast idle cam follower is not resting on the fast idle cam and the choke valve is wide open.
2. There should be 0.020 in. clearance between the secondary throttle actuating rod and the front of the slot

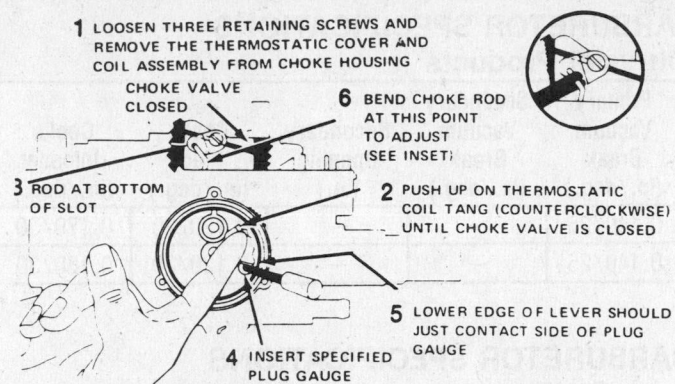

1 LOOSEN THREE RETAINING SCREWS AND REMOVE THE THERMOSTATIC COVER AND COIL ASSEMBLY FROM CHOKE HOUSING

CHOKE VALVE CLOSED

6 BEND CHOKE ROD AT THIS POINT TO ADJUST (SEE INSET)

3 ROD AT BOTTOM OF SLOT

2 PUSH UP ON THERMOSTATIC COIL TANG (COUNTERCLOCKWISE) UNTIL CHOKE VALVE IS CLOSED

5 LOWER EDGE OF LEVER SHOULD JUST CONTACT SIDE OF PLUG GAUGE

4 INSERT SPECIFIED PLUG GAUGE

Quadrajet choke coil lever adjustment—typical

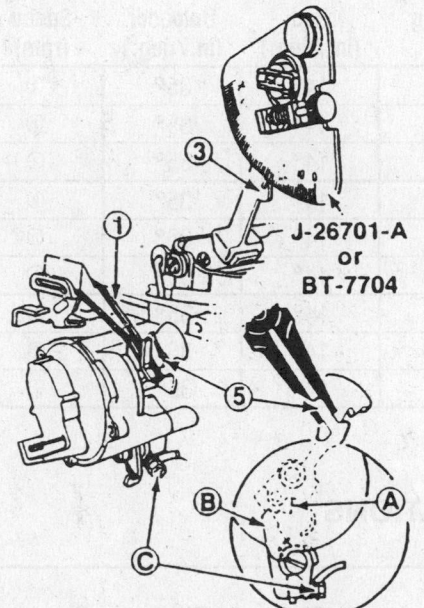

J-26701-A or BT-7704

① Attach rubber band to Vacuum Break Lever of Intermediate Choke Shaft.

② Open Throttle to allow Choke Valve to close.

③ Set up Angle Gage and set to specification.

④ Place Fast Idle Cam Ⓐon second step against Cam Follower Lever Ⓑ, with Lever contacting rise of High Step. If Lever does not contact Cam, turn Fast Idle Adjusting Screw Ⓒin additional turn(s).

⑤ Adjust, if bubble is not recentered, by bending Fast Idle Cam Kick Lever with pliers.

Feedback Quadrajet—Fast Idle cam adjustment

on the secondary throttle lever with the closing tang on the throttle lever resting against the actuating lever.

3. Bend the secondary closing tang on the primary throttle actuating rod or lever to adjust.

SECONDARY OPENING

Adjustment

1. Open the primary throttle valves until the actuating link contacts the upper tang on the secondary lever.

2. With the 2 point linkage, the bottom of the link should be in the center of the secondary lever slot.

3. With the 3 point linkage, there should be 0.070 in. clearance between the link and the middle tang.

4. Bend the upper tang on the secondary lever to adjust as necessary.

FLOAT LEVEL

Adjustment

With the air horn assembly removed, measure the distance from the air horn gasket surface (gasket removed) to the top of the float at the toe ($^{3}/_{16}$ in. back from the toe).

NOTE: Make sure the retaining pin is firmly held in place and that the tang of the float is lightly held against the needle and seat assembly.

On carburetors without the computer controlled systems, remove the float and bend the float arm to adjust. For (E4MC and E4ME) the computer controlled systems carburetors, use the following steps:

1. Remove air horn and gasket.

2. Remove solenoid plunger, metering rods and float bowl insert.

NOTE: If necessary to remove solenoid lean mixture adjusting screw count and record the number of turns it takes to lightly bottom the screw and return to the exact position when reassembling.

3. Attach tool J–34817 or equivalent to float bowl.

4. Place tool J–34817–3 or equivalent in base with contact pin resting on outer edge of float lever.

5. With tool J–9789–90 or equivalent, measure the distance from the top of the casting to top of the float, at a point $^{3}/_{16}$ in. from large end of float.

6. If more than $^{2}/_{32}$ in. from specification, use tool J–34817–15 or equivalent to bend lever up or down.

7. Recheck float alignment.

8. Install the parts, turning the mixture solenoid screw in until it is lightly bottomed, then unscrewing it the exact number of turns counted earlier.

ACCELERATOR PUMP

Adjustment

The accelerator pump is not adjustable on computer controlled carburetors (E4MC and E4ME).

1. Close the primary throttle valves by backing out the slow idle screw and making sure that the fast idle cam follower is off the steps of the fast idle cam.

2. Bend the secondary throttle closing tang away from the primary throttle lever, if necessary, to insure that the primary throttle valves are fully closed.

3. With the pump in the appropriate hole in the pump lever, measure from the top of the choke valve wall to the top of the pump stem.

4. To adjust, bend the pump lever.

5. After adjusting, readjust the secondary throttle tang and the slow idle screw.

AIR VALVE SPRING

Adjustment

To adjust the air valve spring windup, loosen the Allen head lockscrew and turn the adjusting screw counterclockwise to remove all spring tension. With the air valve closed, turn the adjusting screw clockwise the specified number of turns after the torsion spring contacts the pin on the shaft. Hold the adjusting screw in this position and tighten the lockscrew.

QUADRAJET CARBURETOR SPECIFICATIONS
Chrysler Products

Year	Carburetor Identification	Float Level (in.)	Air Valve Spring (turn)	Pump Rod (in.)	Primary Vacuum Break (in./deg.)	Secondary Vacuum Break (in./deg.)	Secondary Opening (in.)	Choke Rod (in./deg.)	Choke Unloader (in./deg.)	Fast Idle Speed (rpm)
1988	17085433	14/32	7/8	—	0.140/25	—	—	0.120/20	0.179/30	①
1989	17085433	14/32	7/8	—	0.140/25	—	—	0.120/20	0.180/30	①

① Refer to the underhood sticker

QUADRAJET CARBURETOR SPECIFICATIONS
Cadillac

Year	Carburetor Identification	Float Level (in.)	Air Valve Spring (turn)	Pump Rod (in.)	Primary Vacuum Break (deg.)	Secondary Vacuum Break (deg.)	Secondary Opening (in.)	Choke Rod (in./deg.)	Choke Unloader (in./deg.)	Fast Idle Speed (rpm)
1988	17086008	11/32	1/2	Fixed	25°	43°	①	14°	35°	②
	17086009	14/32	1/2	Fixed	25°	43°	①	14°	35°	②
	17088115	11/32	1/2	Fixed	25°	43°	①	14°	35°	②
1989	17086008	11/32	1/2	Fixed	25°	43°	①	14°	35°	②
	17086009	14/32	1/2	Fixed	25°	43°	①	14°	35°	②
	17088115	11/32	1/2	Fixed	25°	43°	①	14°	35°	②
1990	17086008	11/32	1/2	Fixed	25°	43°	①	14°	35°	②
	17086009	14/32	1/2	Fixed	25°	43°	①	14°	35°	②
	17088115	11/32	1/2	Fixed	25°	43°	①	14°	35°	②

① No measurement necessary on two point linkage
② See underhood decal

QUADRAJET CARBURETOR SPECIFICATIONS
Buick

Year	Carburetor Identification	Float Level (in.)	Air Valve Spring (turn)	Pump Rod (in.)	Primary Vacuum Break (deg.)	Secondary Vacuum Break (deg.)	Secondary Opening (in.)	Choke Rod (deg.)	Choke Unloader (deg.)	Fast Idle Speed (rpm)
1988	17086008	11/32	1/2	Fixed	25°	43°	①	14°	35°	②
	17088115	11/32	1/2	Fixed	25°	43°	①	14°	35°	②
1989	17088115	11/32	1/2	Fixed	25°	43°	①	14°	35°	②
1990	17088115	11/32	1/2	Fixed	25°	43°	①	14°	35°	②

① No measurement necessary on two point linkage
② See underhood decal

QUADRAJET CARBURETOR SPECIFICATIONS
Chevrolet

Year	Carburetor Identification	Float Level (in.)	Air Valve Spring (turn)	Pump Rod (in.)	Primary Vacuum (deg./in.)	Secondary Vacuum (deg./in.)	Secondary Opening (in.)	Choke Rod (deg./in.)	Choke Unloader (deg./in.)	Fast Idle Speed (rpm)
1988	17087306	11/32	7/8	Fixed	27°	—	①	20°	32°	②
	17087129	11/32	7/8	Fixed	27°	—	①	20°	32°	②
	17087132	11/32	7/8	Fixed	27°	—	①	20°	32°	②
1989	17088115	11/32	1/2	Fixed	25°	43°	①	14°	35°	②
1990	17088115	11/32	1/2	Fixed	25°	43°	①	14°	35°	②

① No measurement necessary on two point linkage
② See underhood decal

QUADRAJET CARBURETOR SPECIFICATIONS
Oldsmobile

Year	Carburetor Identification	Float Level (in.)	Air Valve Spring (turn)	Pump Rod (in.)	Primary Vacuum Break (in./deg.)	Secondary Vacuum Break (in./deg.)	Secondary Opening (in.)	Choke Rod (in./deg.)	Choke Unloader (in./deg.)	Fast Idle Speed (rpm)
1988	17086008	11/32	1/2	Fixed	25°	43°	①	14°	35°	②
	17088115	11/32	1/2	Fixed	25°	43°	①	14°	35°	②
1989	17088115	11/32	1/2	Fixed	25°	43°	①	14°	35°	②
1990	17088115	11/32	1/2	Fixed	25°	43°	①	14°	35°	②

① No measurement necessary on two point linkage
② See underhood decal

QUADRAJET CARBURETOR SPECIFICATIONS
Pontiac

Year	Carburetor Identification	Float Level (in.)	Air Valve Spring (turn)	Pump Rod (in.)	Primary Vacuum Break (in./deg.)	Secondary Vacuum Break (in./deg.)	Secondary Opening (in.)	Choke Rod (in./deg.)	Choke Unloader (in./deg.)	Fast Idle Speed (rpm)
1988	17086008	11/32	1/2	Fixed	25°	43°	—	14°	35°	②
	17088115	11/32	1/2	Fixed	25°	43°	—	14°	35°	②
1989	17088115	11/32	1/2	Fixed	25°	43°	①	14°	35°	②

① No measurement necessary on two point linkage; see text
② See underhood decal

QUADRAJET CARBURETOR SPECIFICATIONS
All Canadian Models

Year	Carburetor Identification	Float Level (in.)	Air Valve Spring (turn)	Pump Rod (in.)	Primary Vacuum Break (deg./in.)	Secondary Vacuum Break (deg./in.)	Secondary Opening (deg./in.)	Choke Rod (deg./in.)	Choke Unloader (deg./in.)	Fast Idle Speed (rpm)
1988	17086008	11/32	1/2	Fixed	25/0.142	43°	①	14°	35°	②
	17088115	11/32	1/2	Fixed	25/0.142	43°	①	14°	35°	②
	17087211	11/32	1/2	9/32	21/0.117	—	①	14°	28/0.164	②
1989	17086008	11/32	1/2	Fixed	25/0.142	43°	①	14°	35°	②
	17088115	11/32	1/2	Fixed	25/0.142	43°	①	14°	35°	②
	17087211	11/32	1/2	9/32	21/0.117	—	①	14°	28/0.164	②
1990	17086008	11/32	1/2	Fixed	25/0.142	43°	①	14°	35°	②
	17088115	11/32	1/2	Fixed	25/0.142	43°	①	14°	35°	②
	17087211	11/32	1/2	9/32	21/0.117	—	①	14°	28/0.164	②

① No measurement necessary on two point linkage
② See underhood decal

AISAN CARBURETORS

Festiva 2 Barrel

The Festiva uses a 2 barrel electronically controlled feedback carburetor made by Aisan. To set the idle mixture on this vehicle requires the use of an exhaust gas analyzer. Before condemning this carburetor, make certain fuel pressure is correct. All ignition and electronic controls and must also be functioning properly. If removing the base plate note the location of the hollow attaching screw, the hole provides vacuum to the power valve. The following adjustments can be made with the carburetor on the vehicle except the throttle plate adjustments.

FLOAT LEVEL

Adjustment

1. Hold air horn upright with float hanging free.
2. Measure the distance from the gasket to the bottom edge of the float. Float drop should be 1.850–1.929 in. (47–49mm).
3. Bend tap on hinge if not within specifications.
4. Invert the air horn to adjust float level.
5. Measure the distance from the gasket to the top edge of the float. Float level should be 0.327–0.366 in. (8.3–9.3mm).

CHOKE BREAKER

Adjustment

1. Set choke plate to fully closed position.
2. Disconnect vacuum hose at the pulldown and apply 16 in. Hg of vacuum.
3. Hold choke plate closed as far as possible without forcing it.
4. Set fast idle cam on fourth step, if ambient temperature is below 86°F or third step if above 86°F.
5. Check choke gap with $^5/_{16}$ drill. Adjust by bending breaker adjuster tab.

CHOKE UNLOADER

Adjustment

1. Hold the choke plate closed as far

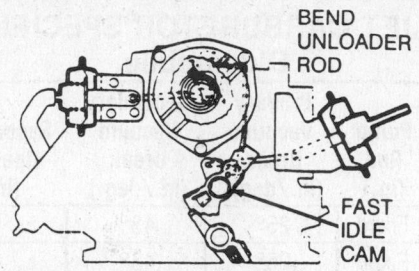

Festiva choke unloader adjustment

as possible without forcing it.
2. Distance between choke plate and air horn should be 0.059–0.076 in. (drill size $^1/_{16}$) while holding throttle wide open.
3. Adjust clearance by bending rod.

CHOKE PLATE CLEARANCE

Adjustment

1. Set fast idle on third step of cam.
2. Distance between choke plate and air horn should be 0.024–0.037 in. (drill size $^1/_{32}$).
3. Adjust clearance by bending the tap.

SECONDARY THROTTLE PLATE

Adjustment

1. With carburetor assembly removed from engine, slowly open throttle while watching the secondary plate.
2. When the secondary plate just starts to open the clearance to the throttle wall should be 0.0372 in. ($^3/_8$ drill size).
3. Adjust clearance by bending tap at secondary shaft.

THROTTLE OPENING

Adjustment

1. With carburetor assembly removed from engine, position idle cam against the third step.
2. The distance between the throttle and venturi wall should be 0.009–0.014 in. (0.25–0.36mm).
3. Adjust to specifications using the fast idle cam screw.

IDLE MIXTURE

Adjustment

NOTE: To perform the idle mixture adjustment an emission analyzer must be used to identify CO concentration.

1. Run engine to reach normal temperature.
2. Insert probe into the secondary air hose and plug hose to prevent leaking past the probe lead.
3. Adjust mixture screw until analyzer shows CO concentration of 1.5–2.5 percent.

CURB IDLE

Adjustment

1. Run engine to reach normal temperature, place transaxle in **N** and set parking brake.
2. Adjust idle to 700–760 rpm using idle adjusting screw.

FAST IDLE BREAKER

Adjustment

1. Run engine to reach normal temperature.
2. Set fast idle cam on 2nd step.
3. Turn fast idle cam breaker adjusting screw to obtain an engine speed of 1650–2150 rpm.

FAST IDLE ADJUSTMENT

Adjustment

1. Disconnect and plug vacuum hose at the fast idle cam servo.
2. Set fast idle cam on 2nd step.
3. Adjust engine speed to 1650–2150 rpm using fast idle screw.

ELECTRICAL LOAD IDLE-UP

Adjustment

1. Run engine to reach normal temperature.
2. Disconnect brown electrical connector at electrical vacuum solenoid.
3. Increase engine speed to 2000 rpm and let return to idle.
4. Adjust servo nut to obtain at idle of 750–850 rpm.

AIR CONDITIONING IDLE-UP

Adjustment

1. Run engine to reach normal temperature.
2. Disconnect orange electrical connector at air conditioning vacuum solenoid.
3. Increase engine speed to 2000 rpm and let return to idle.
4. Adjust A/C idle-up screw to obtain 1200–1300 rpm.

Turbocharging **29**

Theory

The internal combustion engine can be thought of as an air pump. The action of the pistons moving down or up in their cylinders when the intake or exhaust valves are open alternately draws air and fuel into the engine or expels burnt gases into the atmosphere. The amount of air and fuel pulled into the engine (known as an engine's volumetric efficiency) is governed by the drawing efficiency of the piston as it descends in its cylinder, and by the scavenging effect of the exiting exhaust gases, which act to pull additional air/fuel mixture in through the open intake valves during valve overlap periods. The more air and fuel each cylinder pulls in, the more power the engine will produce.

Theoretically, a normally aspirated engine should be able to draw in an amount of air and fuel equal to its displacement (e.g. a 350 cu in. engine should draw in 350 cu in. of air and fuel). In practice, however, only about 80% of the displacement capacity is drawn through because of flow restrictions, the slight pressure drop through the carburetor, and the inability of the exhaust stroke to drive out all of the burnt gases.

There are several ways to increase an engine's drawing power (volumetric efficiency). These include increasing valve overlap, increasing engine bore and/or stroke, supercharging the engine, or (the most popular approach) turbocharging.

In effect, the turbocharger is an air pump which crams more air/fuel mixture into the cylinders than they could possibly draw in by themselves.

In doing so, the turbocharger increases the engine's volumetric efficiency past its normal 80%, which proportionately increases engine horsepower and torque output.

Perhaps the most advantageous aspect of the turbocharger is that it does not require usable engine horsepower to operate. By comparison, say a car is climbing a steep hill and the driver decides to turn on the air conditioner. The moment the air conditioner is turned on, a power drain on the engine can usually be felt. That's because some of the power that was being used to drive the car up the hill is now being used to turn the air conditioner compressor. A turbocharger, on the other hand, does not drain power from the engine to operate because it uses the free energy of the exhaust gases as they are blown out of the en-

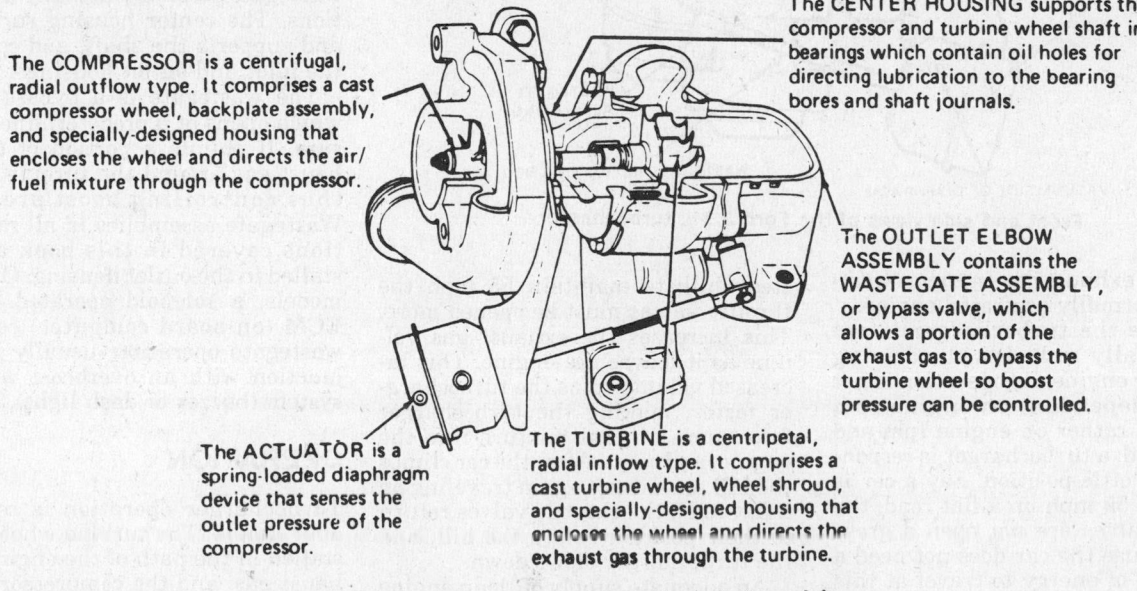

The **COMPRESSOR** is a centrifugal, radial outflow type. It comprises a cast compressor wheel, backplate assembly, and specially-designed housing that encloses the wheel and directs the air/fuel mixture through the compressor.

The **CENTER HOUSING** supports the compressor and turbine wheel shaft in bearings which contain oil holes for directing lubrication to the bearing bores and shaft journals.

The **OUTLET ELBOW ASSEMBLY** contains the **WASTEGATE ASSEMBLY**, or bypass valve, which allows a portion of the exhaust gas to bypass the turbine wheel so boost pressure can be controlled.

The **ACTUATOR** is a spring-loaded diaphragm device that senses the outlet pressure of the compressor.

The **TURBINE** is a centripetal, radial inflow type. It comprises a cast turbine wheel, wheel shroud, and specially-designed housing that encloses the wheel and directs the exhaust gas through the turbine.

Turbocharger components, typical of all models

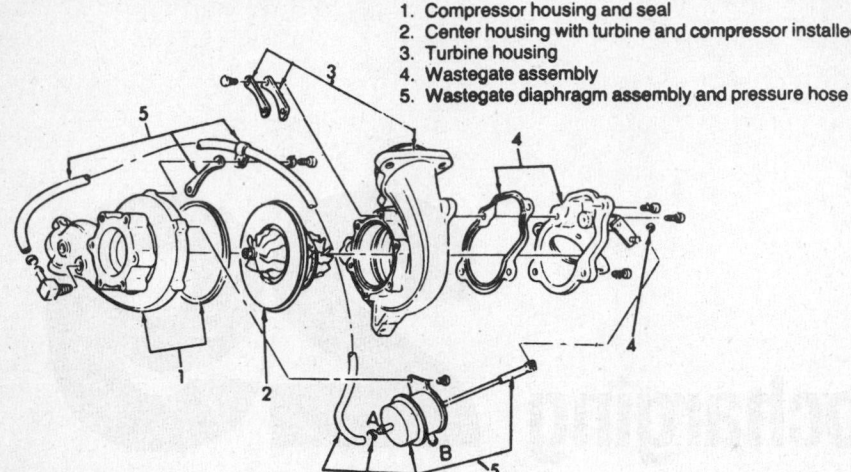

1. Compressor housing and seal
2. Center housing with turbine and compressor installed
3. Turbine housing
4. Wastegate assembly
5. Wastegate diaphragm assembly and pressure hose

Typical GM 3.8L (231 cu in.) engine turbocharger. "A" is pressure side of wastegate diaphragm, "B" is vacuum side

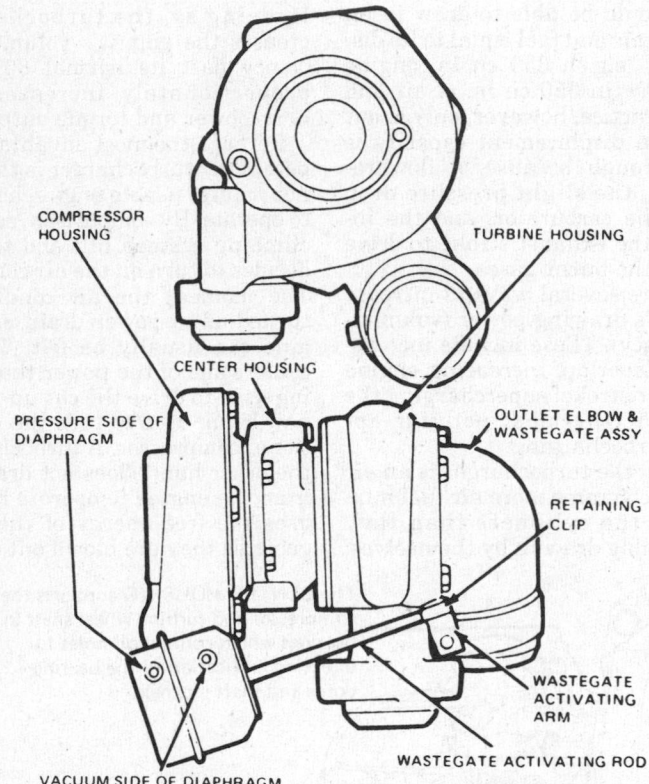

COMPRESSOR HOUSING

TURBINE HOUSING

CENTER HOUSING

PRESSURE SIDE OF DIAPHRAGM

OUTLET ELBOW & WASTEGATE ASSY

RETAINING CLIP

WASTEGATE ACTIVATING ARM

WASTEGATE ACTIVATING ROD

VACUUM SIDE OF DIAPHRAGM

Front and side views of the Ford 2.3 L turbocharger

gine. This exhaust gas energy is wasted on a normally aspirated engine.

Because the turbocharger is not mechanically linked to the driving part of the engine, its operation is not directly dependent on engine rpm alone, but rather on engine rpm and engine load: a turbocharger is responsive to throttle position. Say a car is driving at 55 mph on a flat road: the throttle valves are not open a great deal, because the car does not need a great deal of energy to travel at this speed. Soon the car starts to climb a steep hill: to maintain 55 mph the throttle valves must be opened more. This increases the exhaust gas volume as it leaves the engine. This increased volume spins the turbocharger faster, making the turbocharger force more air/fuel mixture into the engine, and so on. After the car climbs the hill and is once again traveling on a flat road, the throttle valves return to their position before the hill, and the turbocharger slows down.

An adequate supply of clean engine oil is essential for cooling and lubrica-tion and to maintain the turbocharger bearing assembly. The turbocharger wheels routinely operate at 130,000–140,000 rpm during boost and any interruption in the oil supply to the bearing assembly can result in major turbocharger damage. Contamination of the engine oil can also cause serious damage. Any time a basic engine bearing (main, connecting rod or camshaft) is replaced due to damage, the oil and oil filter must be changed and the turbocharger flushed with clean engine oil to remove any contamination. In addition, any time the turbocharger is removed for service or as part of another procedure, the oil and oil filter should be changed. When first starting the engine after removing the turbocharger, fill the turbocharger oil passage with clean engine oil and crank the engine a few times to allow oil pressure to build up. It's also a good idea to allow the engine to idle for one minute before shutting it off, especially when running at freeway speeds for long periods of time, to prevent the possibility of turbocharger bearing damage due to sudden oil starvation.

COMPONENTS

The turbocharger unit consists of two vaned wheels (compressor and turbine) connected by a common axle (shaft), and a housing which can be sub-divided into three sections: inlet (or compressor), center, and outlet (or turbine). The inlet housing surrounds the compressor wheel, and connects to the air intake and the intake manifold. The outlet housing surrounds the turbine wheel, and connects to the exhaust system; it also houses the wastegate assembly in many installations. The center housing surrounds and supports the shaft, and connects the inlet and outlet housings.

The wastegate is a bypass valve, which opens at a predetermined pressure. It shunts a portion of the exhaust gas around the turbine wheel, thus controlling boost pressure. Wastegate assemblies in all installations covered in this book are installed in the outlet housing. On some models, a solenoid operated by the ECM (on-board computer) controls wastegate operation, usually in conjunction with an overboost warning system (buzzer or dash light).

OPERATION

Turbocharger operation is remarkably simple. The turbine wheel is installed in the path of the engine's exhaust gas, and the compressor wheel is installed in the intake path. Exhaust gas is directed through the turbine housing, causing the turbine wheel to spin. This spinning motion is transferred by the connecting shaft to the compressor wheel. As the compressor wheel spins, it packs the intake charge into a dense mass, which is fed into the engine. Combustion converts the charge into exhaust. The exhaust charge is directed through the turbine housing, where it spins the turbine wheel, and then out through the turbine housing discharge into the exhaust system.

Thus, turbocharger operation is self-perpetuating. However, unchecked turbocharger operation will increase compressor pressure (called boost pressure) beyond the design limits of the engine, and will seriously damage internal engine components. Boost pressure is controlled by the wastegate. When boost pressure rises to a predetermined value, the wastegate opens, bypassing exhaust flow around the turbine.

Greater volumetric efficiency is a benefit of the turbocharging process, but increased cylinder pressure is a drawback, because it raises the engine's octane requirement. The two are inseparable, so a method must be devised to compensate for the increased octane requirement to avoid detonation (spark knock). Water injection, alcohol injection, low boost pressures, charge intercoolers, ignition spark retardation, and alcohol fuels have all been used to control detonation, with varying degrees of success.

Ford controls detonation by limiting boost and by spark retardation. Wastegate operation begins at five psi, and enough exhaust gas is routed around the turbine to limit boost to a maximum of six psi. The electronic ignition system has been modified in the turbocharged engine to include two spark retardation points. When boost pressure reaches approximately one-half to one psi, a switch in the intake manifold sends a signal to the ignition module, which retards ignition timing six degrees. A second manifold switch sends its signal when boost reaches four psi, resulting in an additional six degrees of retard.

The General Motors system of detonation control is slightly different. Boost is limited to a maximum of approximately six psi. In addition, a detonation sensor is installed in the engine block (V6) or intake manifold (V8). Vibrations caused by detonation are transmitted to the sensor, which sends a signal to the Electronic Spark Control (ESC) module. The module processes this signal, and sends a command signal to the HEI distribu-

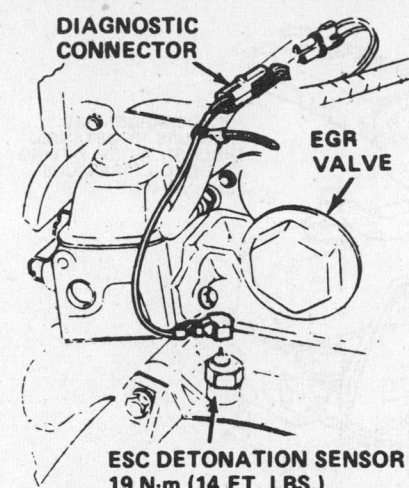

Buick 231 V6 (3.8 L) detonation sensor installation

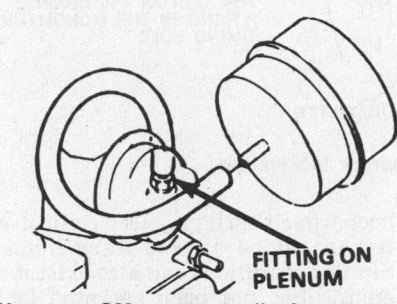

New type GM wastegate diaphragm uses plenum vacuum only-

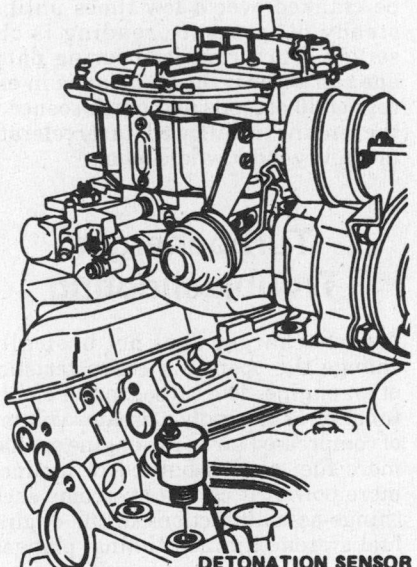

Pontiac turbocharged V8 detonation sensor location

tor to retard timing. Timing retard ranges up to 22° on V6s, or 15° on V8s.

LUBRICATION

The turbocharger shaft spins in bearings lubricated by engine oil. Turbine speeds routinely reach 120,000–140,000 rpm, making an adequate and well-filtered oil supply critical for proper operation. Any interruption or contamination of the oil supply will result in engine damage as well. Ford cautions that accelerating the engine to top rpm immediately after starting can result in engine and turbocharger damage (due to the lack of oil pressure). Immediately shutting down the engine after it has been operated at high rpm for an extended period can also result in turbocharger damage, since oil pressure will be shut off, but the turbine will continue to spin for a few moments. Shutting the throttle abruptly when the engine is at high speed can also cause extensive damage, but for a different reason: sudden closed throttle operation causes the mixture to become very lean, resulting in detonation, high engine temperature, and consequent damage.

General Motors recommends the following procedure before starting the engine when changing the oil and filter, or performing any operation which results in oil drainage or loss:

1. Disconnect the ignition switch connector (pink wire) from the HEI distributor module.
2. Crank the engine several times until the oil light goes out. Do not crank the engine for more than thirty seconds at a time to avoid starter damage.
3. Reconnect the pink wire. Start the engine.

Turbocharger Maintenance

Proper maintenance is important, particularly regarding air and oil filtration, to maximize the service life and performance of the turbocharger. Experience has shown that the main cause of turbocharger failure is due to oil lag, restriction or lack of oil flow and dirt in the oil. The second principle cause of failure is foreign objects entering the compressor and/or turbine wheels.

AIR INTAKE SYSTEM

Dust or sand entering the turbocharger compressor housing from a leaky air inlet system can seriously erode the compressor wheel blades and will result in deterioration of turbocharger and engine performance. The wearing away of the blades, if uneven, can induce shaft motion which will pound out the turbocharger shaft bearings. Ingestion of sand or dust will also cause excessive wear on engine parts, such as pistons, rings, valves, etc.

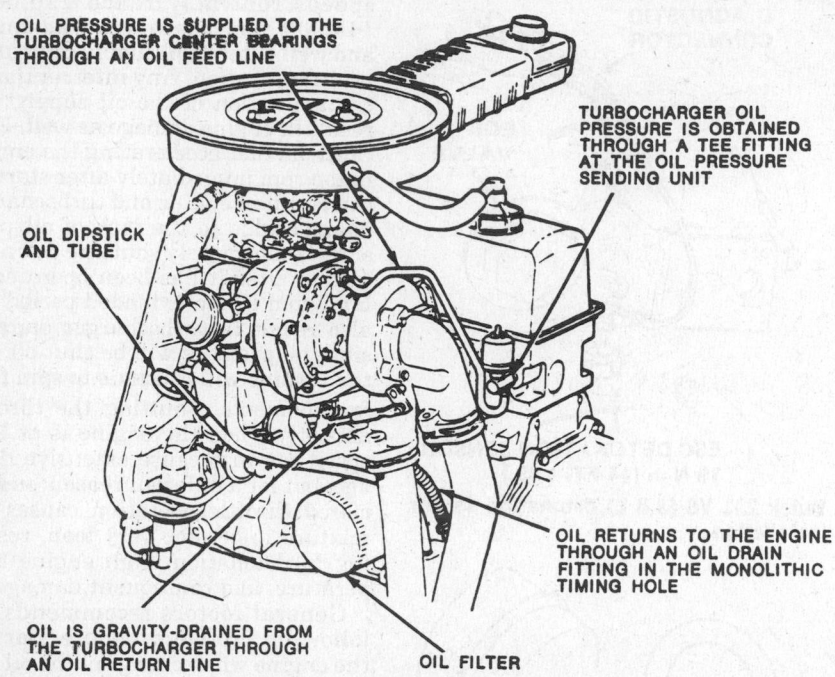

Ford 2.3 L turbocharger lubrication

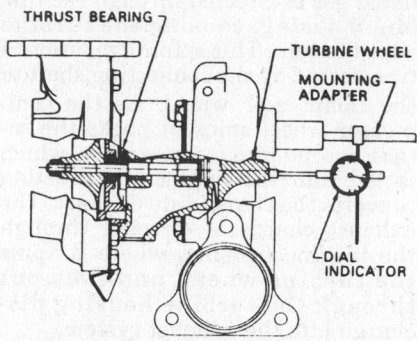

Thrust bearing clearance measurement

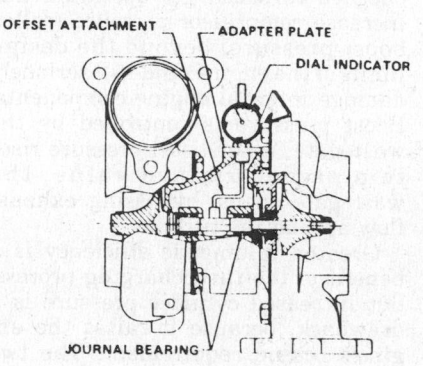

Journal bearing clearance measurement

Plugged or restricted air cleaner systems (due to neglected air filter changes) will reduce air pressure and volume at the compressor air inlet and cause the turbocharger to lose performance. The restricted air cleaner and the resultant air pressure drop between cleaner and turbocharger can, during engine idle periods, cause oil pullover at the compressor end of the turbocharger and result in an oil leak at the seal.

LUBRICATION SYSTEM

Dirt or foreign material, when introduced into the turbocharger bearing system by the lube oil, causes wear on the center housing bearing bore surfaces. Contaminents act as abrasives and will eventually cause the shaft hub and either or both wheels to rub on the housings, causing the rotating assembly to turn slower. Engine power loss, excessive smoke, excessive noise and appearance of oil at either or both ends of the turbocharger could be noted. Contaminated and dirty oil problems can be eliminated by regular oil and filter changes.

A turbocharger should never be operated under engine load conditions with less than 30 psi oil pressure. The turbocharger is much more sensitive to a limited oil supply than an engine, due to the high rotational speed of the shaft and relatively small area of the bearing surfaces. Oil pressure and flow lag during engine starting can have a detrimental effect on the tur-

bocharger bearings, most critical after an engine oil and filter change. Similar conditions can also exist if an engine has not been operated for a long period of time, since engine lube systems tend to bleed down. Before allowing the engine to start, it should be cranked over a few times until a steady oil pressure reading is observed. Turbocharger bearing damage can occur if the oil delay is in excess of 30 seconds and much sooner if the engine is allowed to accelerate much beyond low idle rpm.

Turbocharger Troubleshooting

A turbocharger does not basically change the operating characteristics of an engine. The turbocharger's only function is to supply a greater volume of compressed air to the engine so that more fuel can be burned to produce more power. It cannot overcome such things as malfunctions in the engine fuel system, ignition timing, plugged air cleaner elements, etc. If a turbocharged engine system has malfunctioned and the turbocharger has been inspected and determined to be functioning normally, proceed with troubleshooting as though the engine were naturally aspirated (non-turbocharged). Simply replacing a good turbocharger with another will not correct engine deficiencies. Always inspect and asses turbocharger condi-

tion before removing it from the engine as follows:

1. Remove the inlet and exhaust ducts from the turbocharger.

2. Inspect both turbocharger wheels for blade damage caused by foreign material entering the turbocharger. The wheels can be visually checked by simply looking through the compressor housing inlet opening while holding the the throttle blade open. A light is necessary when examining the turbine wheel blade tips since they are positioned inside the turbine housing. Look between the turbine wheel blades from the exhaust outlet end of the turbine housing.

3. Inspect the outer blade tip edges on both wheels adjacent to their respective housing bores and check for wheel rub.

4. Rotate the shaft wheel assembly by hand and feel for drag or binding conditions. Push the shaft to one side, rotate it and feel for rub. It should turn smoothly.

5. Lift both ends of the shaft up and down at the same time and feel for excessive journal bearing clearance. If clearance is normal, very little shaft movement will be detected. Actual shaft end play can be measured with a dial indicator without removing the turbocharger from the engine.

6. If the shaft assembly rotates

freely and no wheel damage, binding or rub has been noted, it can be assumed that the turbocharger is not in need of service.

——— **CAUTION** ———
Operation of the turbocharger without all normally installed inlet ducts and filters connected can result in personal injury and equipment damage from foreign objects entering the turbocharger.

TESTING WASTEGATE OPERATION

As noted before, the wastegate is a safety valve for the engine. If the wastegate sticks shut, boost pressure will build until the air/fuel mixture charge becomes too powerful for the mechanical components (pistons, bearings, etc.) and causes engine damage.

If the wastegate sticks open, little or no boost will be received from the turbocharger, which translates into mediocre engine performance. The simplest wastegate test is to remove the pressure hose at the wastegate diaphragm unit, connect a pressure pump (such as the type used for cooling system testing) and apply pressure. At the specified opening pressure (7 psi for Ford, 8.5–9.5 for GM), the link between the wastegate and its diaphragm unit will just move (about .015 in). The movement is not great, but it should be easy to see.

If the wastegate does not move, try to operate the linkage by hand. It should move under moderate hand pressure. If it moves, the problem is probably in the diaphragm unit (broken diaphragm). To test the diaphragm, remove the vacuum hose from the diaphragm, hook up a manual vacuum pump and apply 25 in. Hg of vacuum to the diaphragm unit. If the vacuum drops below 18 in. Hg within one minute, replace the diaphragm unit.

NOTE: Some 1981 and later GM turbos have a new type of diaphragm which opens the wastegate during idle and part throttle, when there's no boost, to reduce engine backpressure and improve fuel economy. To test this type of unit, apply about 20 in. Hg of vacuum to the diaphragm unit: the wastegate link should move slightly. This unit operates solely with plenum vacuum and can be identified by the absence of a boost pressure signal line on the diaphragm unit.

TESTING OPERATION OF GM DETONATION SENSOR

Connect a tachometer and timing light to the engine, run the engine at 1800–2500 rpm and tap on the intake manifold next to the detonation sensor.

NOTE: Be careful to keep all wires, clothing and tools away from moving engine parts.

Rap continuously, quickly and moderately hard. This should trigger the detonation sensor. When it triggers, engine speed should drop at least 200 rpm and timing should retard at least 4°, probably more.

TURBOCHARGER TROUBLESHOOTING

Problem	Cause	How To Check	Solution
No boost	Gasket leak, hole in exhaust system	Temporarily block tailpipe with engine running. Any exhaust leaks in the system will be heard.	Repair leaks (usually at gasket surfaces)
	Dirty air filter	Remove air filter and check	Replace or clean filter
	Blocked air intake	Visually inspect for blockage	Clear intake
	Worn valves or rings	Compression test engine	Repair
	Throttle valves not opening completely	Manually operate throttle linkage, check valve movement	Adjust linkage, repair carburetor
	Exhaust blockage	Check catalytic converter for melted and blocked catalyst, check muffler and exhaust pipes for debris	Replace catalytic converter, repair exhaust system
	Wastegate stuck open	Test wastegate operation	Repair or replace wastegate assembly
Fuel odor under boost	Leak at compressor or intake manifold	Look for fuel stains at fittings	Tighten fittings or replace gaskets
Ignition miss at high speed, under load	Spark plug gap too large	Remove spark plugs, measure gap	Reduce gap
	Faulty coil	Test Coil	Replace
Ignition miss (often)	Excessive resistance in ignition cables	Check cable resistance (see Tune-Up Unit Repair section)	Replace cables as necessary
Oil leaks into turbine	Blocked oil return hose	Remove hose and check for blockage or crimps	Repair or replace hose

TURBOCHARGER TROUBLESHOOTING

Problem	Cause	How To Check	Solution
Detonation	Fuel octane rating too low	Check octane rating of fuel used against that recommended by manufacturer (consult owner's manual)	Switch to higher octane unleaded fuel
	Faulty sensor	Check G.M. as instructed here; have Ford system checked by qualified technician	Replace as necessary
	Faulty ignition retard unit	Refer to qualified technician	Repair or replace as necessary
	Engine overheating	Check coolant level, debris clogged radiator, no coolant circulation, blocked thermostat	Repair or replace as necessary
Poor idle	Air leak between compressor and carburetor	Listen at joints for hissing sound while the engine idles	Repair

Emission Controls 30

GENERAL INFORMATION

The earth's atmosphere, at or near sea level, consists of approximately 78% nitrogen, 21% oxygen and 1% other gases. If it were possible to remain in this state, 100% clean air would result. However, many varied causes allow other gases and particulates to mix with clean air, causing the air to become unclean or polluted. Some of these pollutants are visible while others are invisible, with each having the capability of causing distress to the eyes, ears, throat, skin and respiratory system. These pollutants can also cause damage to the environment and to the many man made objects that are exposed to the elements. To better understand the causes of air pollution, pollutants can be categorized into 3 separate types, natural, industrial and automotive.

Natural Polution

This type of pollution has been present on earth before man appeared and is still a factor to be considered when discussing air pollution, although it causes only a small percentage of the present overall pollution problem existing in our country. It is the direct result of decaying organic matter, windborn smoke and particulates from such natural events as forest fires, volcanic ash, sand and dust which can spread over a large area of the countryside.

Industrial Polution

This type of pollution is caused primarily by industrial processes which are the burning of coal, oil and natural gas. The by-product of which in turn produces smoke and fumes. This type of polution occurs most severely during still, damp and cool weather. Working with Federal, State and Local mandated rules, regulations and by carefully monitoring the emissions, industries have greatly reduced the amount of pollutant emitted from their industrial sources.

Automotive Polution

This type of air pollution is the automotive emissions. The emissions from the internal combustion engine were not an appreciable problem years ago because of the small number of registered vehicles and the nation's small highway system. However, during the early 1950's, the trend of the American people was to move from the cities to the surrounding suburbs. This caused an immediate problem in the transportation area because the majority of the suburbs were not afforded mass transit conveniences. This lack of transportation created an attractive market for the automobile, which resulted in a dramatic increase in the number of vehicles produced and sold, along with a marked increase in highway construction between the cities and the suburbs. Multi-car families emerged with much emphasis placed on the individual vehicle per family member. As the increase in vehicle ownership and usage occurred, so did the pollutant levels in and around the cities. It was noted that a fog and smoke type haze was being formed and at times, remained in suspension over the cities and did not quickly dissipate. At first this smog, was thought to result from industrial pollution, but it was determined that the automobile emission was largely to blame.

CATEGORIZING VEHICLE EMISSIONS AND CONTROLS

To recognize the sources and methods used to control vehicle emission, 3 major categories have been established. They are crankcase emissions and controls, fuel evaporative emissions and controls and exhaust emissions and controls.

Regardless of the manufacturer, or

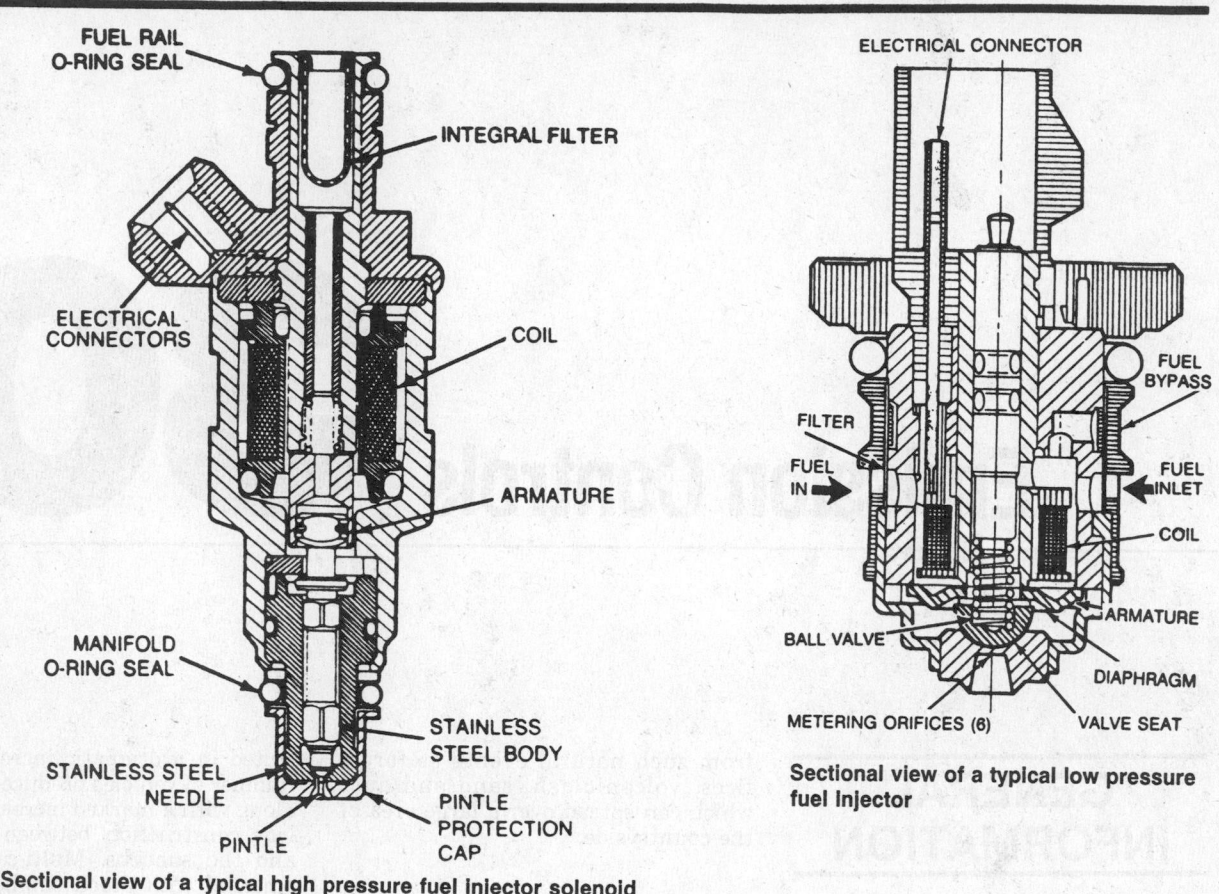

FUEL RAIL O-RING SEAL

INTEGRAL FILTER

ELECTRICAL CONNECTORS

COIL

ARMATURE

MANIFOLD O-RING SEAL

STAINLESS STEEL NEEDLE

PINTLE

STAINLESS STEEL BODY

PINTLE PROTECTION CAP

Sectional view of a typical high pressure fuel injector solenoid

ELECTRICAL CONNECTOR

FILTER

FUEL IN

FUEL BYPASS

FUEL INLET

COIL

ARMATURE

BALL VALVE

DIAPHRAGM

VALVE SEAT

METERING ORIFICES (6)

Sectional view of a typical low pressure fuel injector

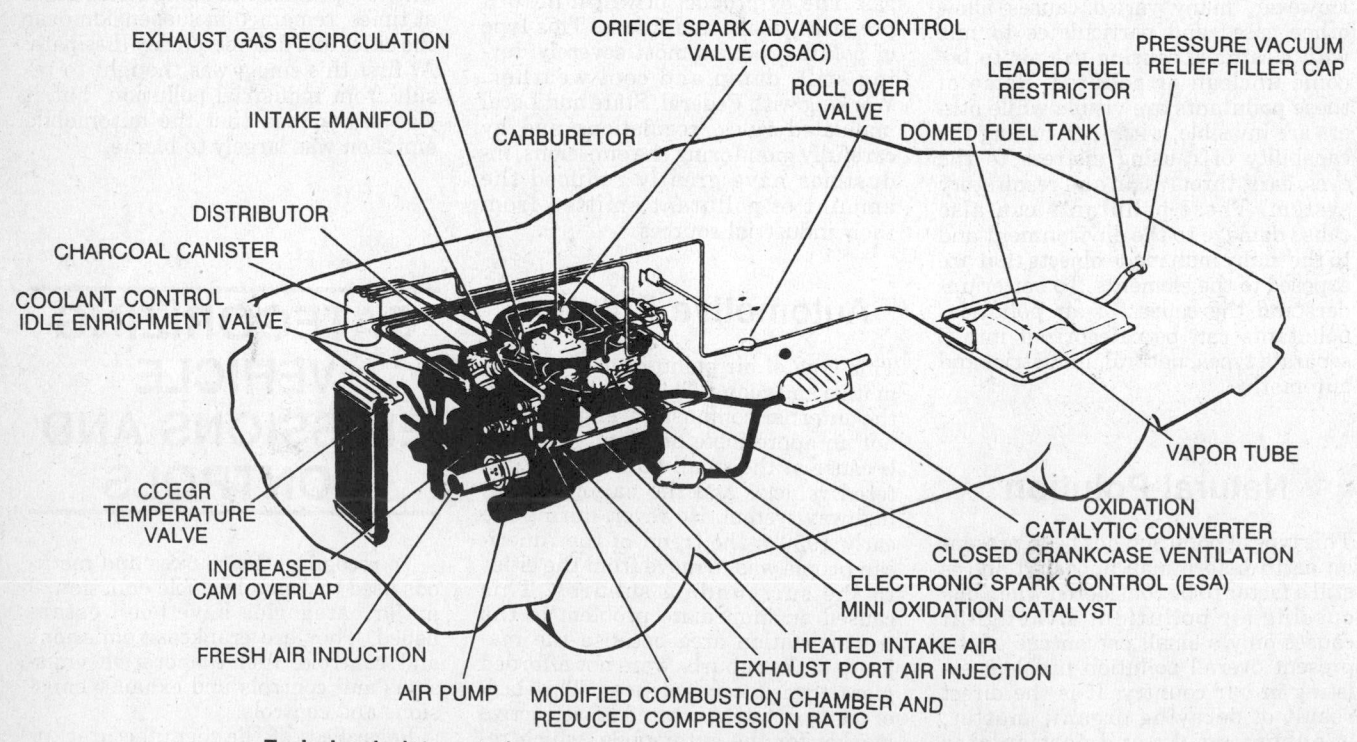

EXHAUST GAS RECIRCULATION

ORIFICE SPARK ADVANCE CONTROL VALVE (OSAC)

PRESSURE VACUUM RELIEF FILLER CAP

INTAKE MANIFOLD

ROLL OVER VALVE

LEADED FUEL RESTRICTOR

CARBURETOR

DOMED FUEL TANK

DISTRIBUTOR

CHARCOAL CANISTER

COOLANT CONTROL IDLE ENRICHMENT VALVE

CCEGR TEMPERATURE VALVE

INCREASED CAM OVERLAP

FRESH AIR INDUCTION

AIR PUMP

MODIFIED COMBUSTION CHAMBER AND REDUCED COMPRESSION RATIO

HEATED INTAKE AIR EXHAUST PORT AIR INJECTION

MINI OXIDATION CATALYST

ELECTRONIC SPARK CONTROL (ESA)

CLOSED CRANKCASE VENTILATION

OXIDATION CATALYTIC CONVERTER

VAPOR TUBE

Typical emission control system component schematic—carbureted vehicles

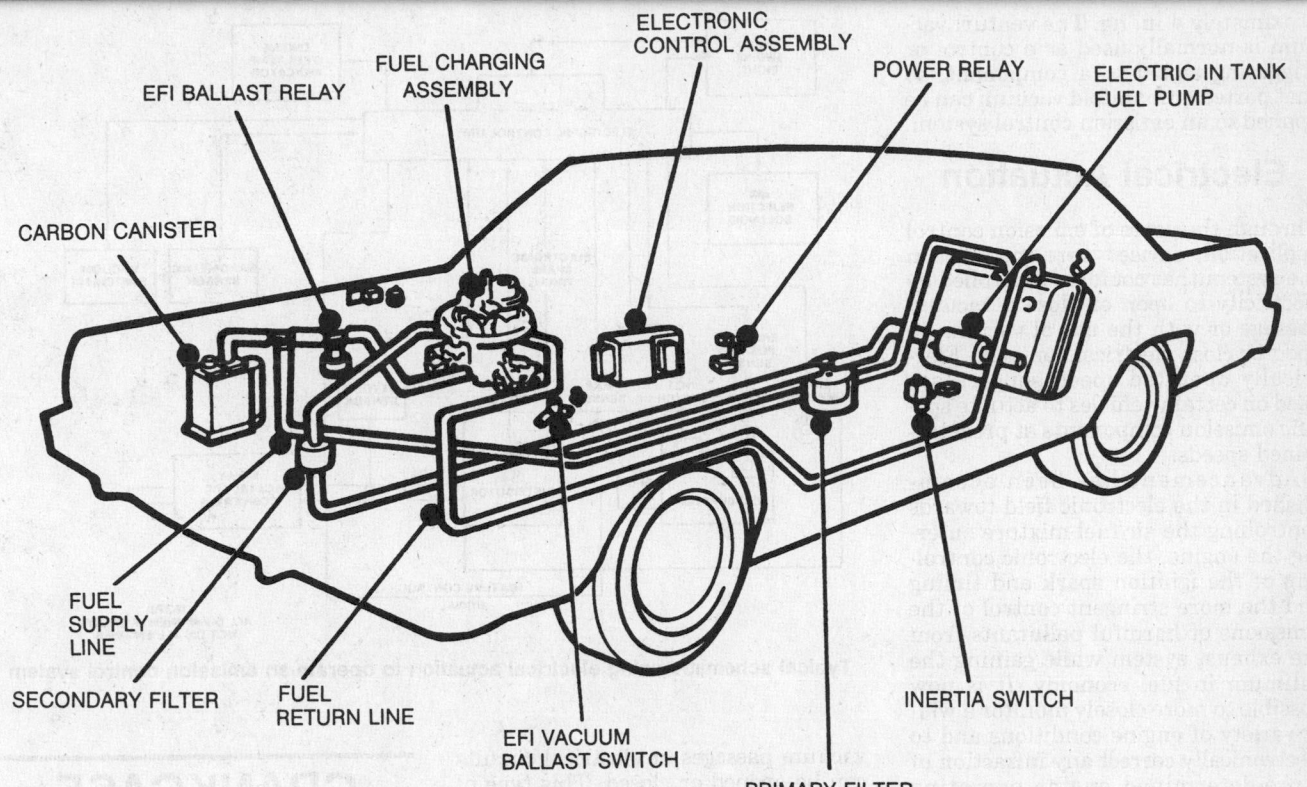

Typical emission control system component schematic—fuel injected vehicle

type of emission control device that is used, a means of actuation must be applied to the device in order for it to operate at a specific time or temperature during either the vehicle operation modes or during the combustion process. The actuating methods commonly used are vacuum, electrical, temperature and mechanical.

Vacuum Sources

The most common method of component actuation is by engine vacuum and has been used since the conception of emission controls. Three major sources of vacuum are obtained from the engine are manifold vacuum, ported vacuum and venturi vacuum. However, with the increased use of smaller engines, the demand for engine vacuum could not be totally supplied by the engine. Vacuum pumps were added to the engine and assisted in supplying the necessary required vacuum needs.

MANIFOLD VACUUM

The engine could be considered a large vacuum pump by having a constant negative pressure developed within the intake manifold as the engine is operated. The tap for this type vacuum is taken from the below the throttle plates or directly from the intake man-

ifold. This source of vacuum will vary in strength between 17–22 in. hg. at idle, to approximately 0 in. hg. at wide open throttle. With the engine in a deceleration mode, the manifold vacuum will be at its highest, which is above idle specifications. Manifold vacuum, normally in conjunction with a vacuum reservoir or amplifier to insure an adequate vacuum volume, is used to actuate the emission control components rapidly.

PORTED VACUUM

The ported vacuum tap is located directly above the throttle plates. When the throttle plates are closed, no ported vacuum signal is present, but as the throttle plates are opened, the vacuum tap is exposed and senses the vacuum below the throttle plates. The ported vacuum signal will vary from 0 in. hg. to approximately 14 in. hg., depending upon the throttle plate opening and the manner in which the port and tap are designed. It should be remembered that ported vacuum is not present at times when the throttle plates are closed or at wide open throttle. Ported vacuum is used in both control and actuation of various systems and components, depending upon the operational needs of the component.

VENTURI VACUUM

Such as the name implies, the venture vacuum tape is located in the venturi chamber of the carburetor throat and is depending upon the velocity of air flowing through the venturi chamber. An example would be as the throttle plates are opened and the velocity of the air flow increases, so would the venturi vacuum signal. The venturi vacuum varies from 0 in. hg. to ap-

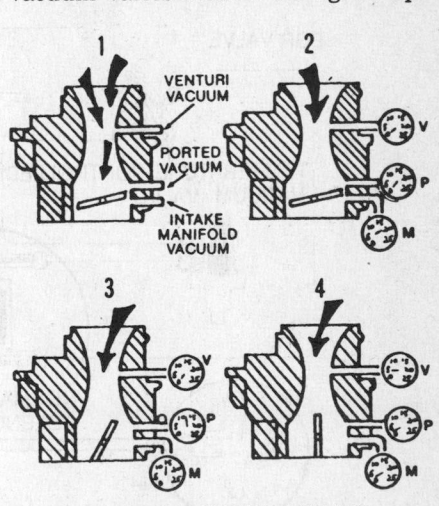

1 - SOURCES OF DIFFERENT VACUUMS
2 - IDLE
3 - OFF IDLE
4 - WIDE OPEN THROTTLE (WOT)

Engine vacuum sources

proximately 4 in. hg. The venturi vacuum is normally used as a control or triggering signal to a component, so that ported or manifold vacuum can be applied to an emission control system.

Electrical Actuation

Through the years of emission control application, devices were installed in the system that could be controlled by electricity to open or close a vacuum passage or with the use of vacuum to open or close electrical contacts. Electrically operated speed sensors are used on certain vehicles to actuate specific emission components at predetermined speeds.

Advancement has been accomplished in the electronic field towards controlling the air/fuel mixture entering the engine, the electronic controlling of the ignition spark and timing and the more stringent control of the emissions of harmful pollutants from the exhaust system while gaining the optimum in fuel economy. It is now possible to more closely monitor a wider variety of engine conditions and to electronically correct any infraction of a pre-determined engine operating mode, through the vehicle computer. Most computer systems are programmed to store and release malfunctioning or system defects information by electronically probing its many circuits.

Temperature Actuation

Various temperature switches are used on the engine to sense coolant temperature change and to react at specific temperature points so that

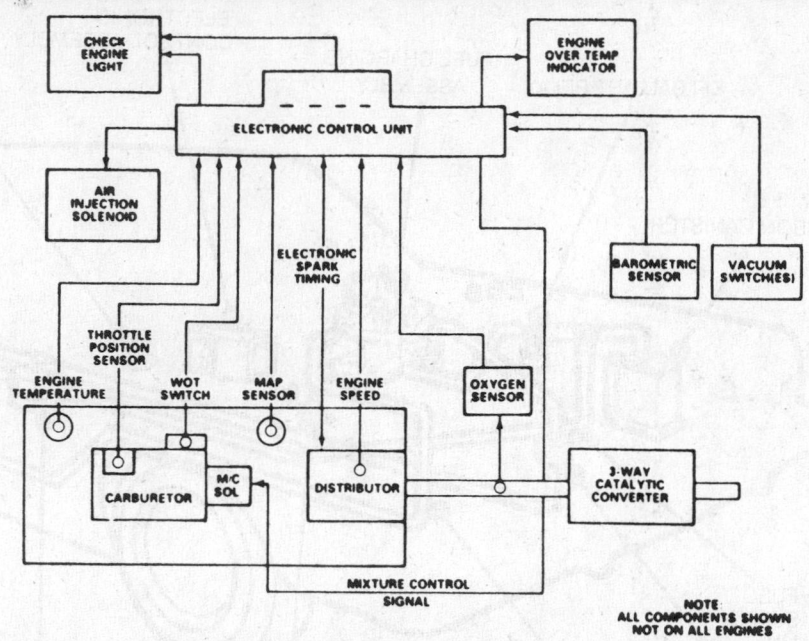

Typical schematic using electrical actuation to operate an emission control system

vacuum passages or electrical circuits can be opened or closed. This type of switch can control numerous vacuum or electrical circuits from a single supply source.

Mechanical Actuation

Mechanical switches are used to control the opening or closing of vacuum passages or the operation of electrical switches through the use of linkages, transmission shift rails or through manual operation.

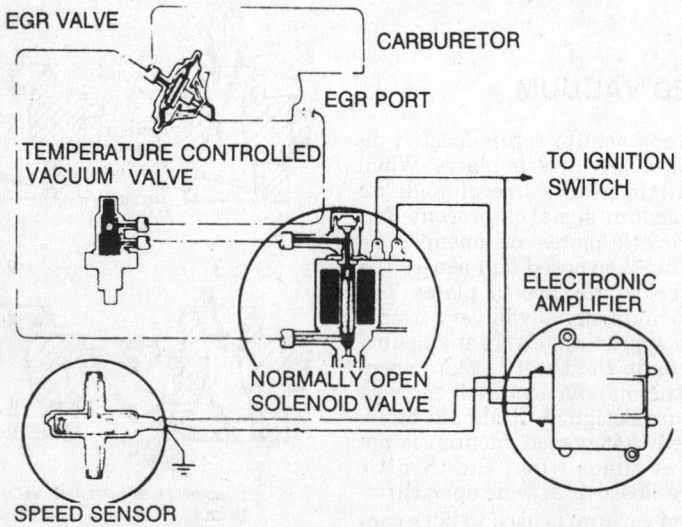

Typical schematic using temperature, electronics and manual actuation to operate an emission control system

CRANKCASE EMISSION CONTROL SYSTEM

System Description

Crankcase emissions are responsible for approximately 20% of all harmful automotive pollutants before any emission controls were installed on the vehicles. Crankcase emissions are the result of compression gasses being forced past the piston rings on both the compression and power strokes, resulting in an accumulation of gases, known as blowby gases, in the engine crankcase. These blowby gases become mixed with vapors from the agitated lubricating oil and must be relieved from the crankcase area to prevent internal engine pressures from building up.

Prior to the early 60's a road draft tube was used to ventilate the crankcase, which allowed the pollutants to be emitted into the atmosphere. With the installation of a regulating valve and necessary plumbing, the road draft tube was eliminated and the gases routed to the air intake area, to be drawn into the engine to be reburned with the air/fuel mixture. At first, engine vacuum was used as the controlling factor to draw the crankcase gases into the engine, but was found that the vacuum source varied at the wrong times. Different systems were experimented with, some with flow control valves while others merely direct the gases to the air cleaner assembly. Other systems had open breather caps to allow fresh air to enter the system while others had sealed breather caps with the fresh air supply being tapped from the air cleaner snorkel.

By 1968, all vehicles manufactured in the United States were equipped with a closed crankcase ventilation system, which did not allow any of the blowby gases and oil vapors to escape into the atmosphere. It is a closed system which utilizes a flow regulating valve called the positive crankcase ventilation valve, PCV valve, or may use a restrictor orifice in place of the PCV valve.

System Operation

The PCV valve is constructed and calibrated to perform the task of metering the gases from the crankcase as required and is matched to engine operation in the following manner. When the engine is idling, only a small amount of air and fuel is needed for combustion, resulting in a small amount of blowby gases being produced because the compression and power strokes are not occuring as frequently as at higher speeds. The PCV valve reacts to this lack of blowby gases and tends to restrict the flow into the induction system. As the engine speed increases, the compressions and power strokes occur more often, along with the addition of more fuel and air needed for combustion. This results in the formation of more blowby gases and the need to purge the crankcase of them. The PCV valve reacts to this increase in blowby gases by allowing more of the gases to be drawn into the air/fuel mixture. The PCV valve is constructed and calibrated in such a manner as to prevent engine backfires from entering the crankcase to avoid detonation of the accumulated blowby gases.

In the closed crankcase ventilation system, the fresh air intake that is located in the air cleaner or snorkel, has a dual role. Not only is it a source of fresh air for the crankcase ventilation system, but it doubles as an overload release of blowby gases into the system air stream should the PCV valve fail to control the build up of blowby gases. This happens rather than allowing the excess gases to escape into the atmosphere. With the use of this closed system the hydrocarbon (HC) emissions produced in the crankcase are prevented from entering the atmosphere.

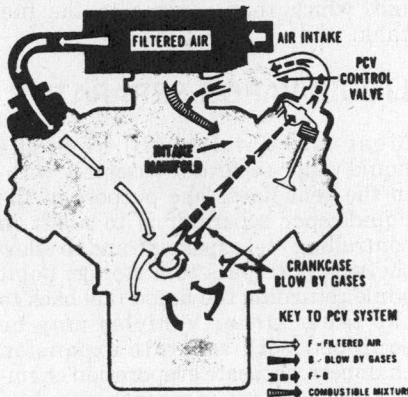

Typical closed crankcase ventilation control system

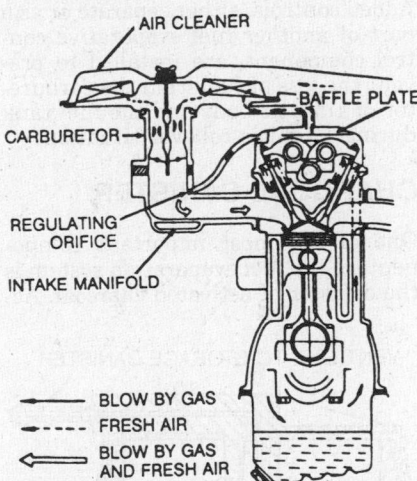

Positive crankcase ventilation system using a regulating orifice rather that a PCV valve

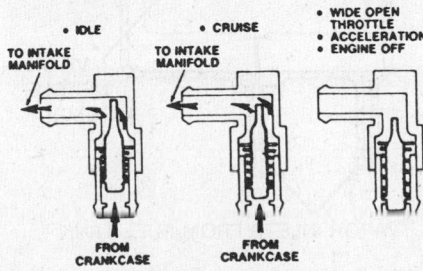

PCV valve operation

FUEL EVAPORATIVE SYSTEM

System Description

Fuel evaporation vapors are found to account for approximately 20% of the total automotive emission problem and is more severe as the temperature increases. The sources of the hydrocarbon vapor emissions were the fuel tank and carburetor bowl, both which were vented into the atmosphere. Another problem was the overfilling of the fuel tank, which under changes of temperatures or by having the vehicle parked on an incline, would spill gasoline from the tank. A means of trapping the vapor emission and preventing gasoline leakage was a major undertaking.

One of the early systems used, was the engine crankcase to store the fuel vapors when the engine was not running. When the engine was started, the vapors were purged from the crankcase by the positive crankcase ventilation system. Certain drawbacks were noted in this system, some of which were the dilution of engine lubricating oils with gasoline, an over rich air/fuel mixture during the purge cycle and the danger of gasoline vapor detonation within the crankcase during engine start up.

To prevent fuel loss from the tank due to expansion, an expansion dome has been manufactured into the top of the fuel tank and the fill pipes have been redesigned to prevent filling the fuel tank above a desired level. Certain vehicles use added plumbing to increase the area volume needed, should the fuel expand. This added plumbing is normally part of the vapor control system with necessary valves to control both vapors and liquids included. After much experimenting and testing, a general system was designed that could control both vapor and liquid emissions by sealing the fuel system from the atmosphere. Although each vehicle manufacturer has designed their own vapor control system, similar components are used, resulting in systems that are basically the same in the manner or vapor collection and storage. However, the manner in which the vapors are purged may vary greatly.

System Components

SEALED FUEL TANK CAP

The first step in sealing the fuel system was to replace the vented fuel tank filler cap with a sealed cap. The venting of the fuel tank is accomplished through another component of the system which controls the vapor emission by storage. To prevent damage to the tank should excessive internal or external pressure exist due to this closed system, a pressure/vacuum relief valve is incorporated in the sealing cap. A tank pressure of ½-1 psi.

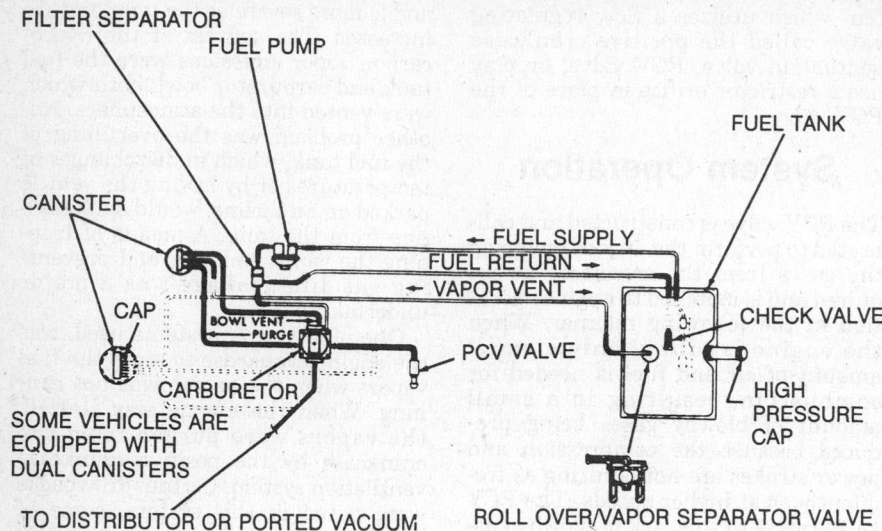

Typical evaporative control system schematic

FILTER SEPARATOR
FUEL PUMP
CANISTER
CAP
BOWL VENT PURGE
CARBURETOR
SOME VEHICLES ARE EQUIPPED WITH DUAL CANISTERS
TO DISTRIBUTOR OR PORTED VACUUM
FUEL TANK
← FUEL SUPPLY
FUEL RETURN →
← VAPOR VENT
CHECK VALVE
HIGH PRESSURE CAP
PCV VALVE
ROLL OVER/VAPOR SEPARATOR VALVE

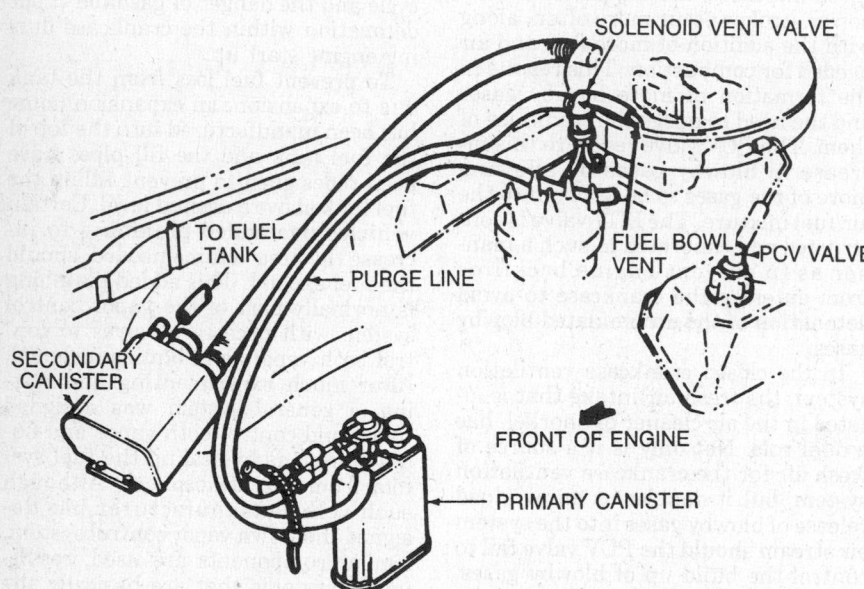

Typical box type evaporative control system schematic

SOLENOID VENT VALVE
TO FUEL TANK
PURGE LINE
SECONDARY CANISTER
FUEL BOWL VENT
PCV VALVE
FRONT OF ENGINE
PRIMARY CANISTER

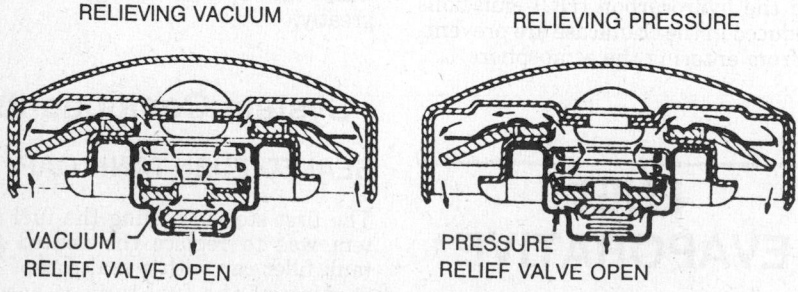

Sealed fuel tank cap operation

RELIEVING VACUUM
RELIEVING PRESSURE
VACUUM RELIEF VALVE OPEN
PRESSURE RELIEF VALVE OPEN

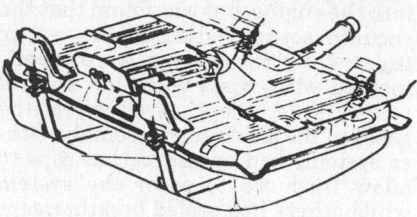

Typical emission related fuel tank

can exist in the tank and is controlled by the relief valve.

FUEL TANK

The fuel tank has been redesigned to provide approximately 10% of the total tank volume for expansion space, should the fuel expand due to temperature changes. An overfill protector is provided by the filler neck to assure the expansion space is maintained during the tank filling. The tank vapor venting is controlled by having a vent tap at or near the expansion chamber dome. A foam type filter or a vapor separator is used to allow the vapors to pass, but prevents the passage of liquid, which then returns to the fuel tank.

LIQUID/VAPOR SEPARATOR

Most all vehicles will have the liquid/vapor separator assembly within the vent lines. The purpose of the liquid/vapor separator is to assist in controlling fuel expansion and to allow the vapors to pass to a storage point while returning the liquid fuel back to the tank. Other vehicles may be equipped with separate expansion chambers, separate evaporation chambers and with one or more check valves in the lines. These components are usually located near the fuel tank. Added controls, either separate or as a part of another fuel evaporative control component, are installed to prevent the loss of fuel from the carburetor or throttle body and the fuel tank during a vehicle rollover situation.

CHARCOAL CANISTER

One of the most important components of the fuel evaporation system is the canister of activated charcoal. Ac-

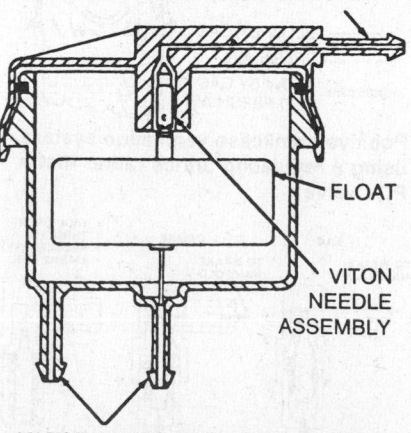

VENT LINE TO STORAGE CANISTER
FLOAT
VITON NEEDLE ASSEMBLY
VAPOR INLETS FROM FUEL TANK

Typical liquid check valve assembly

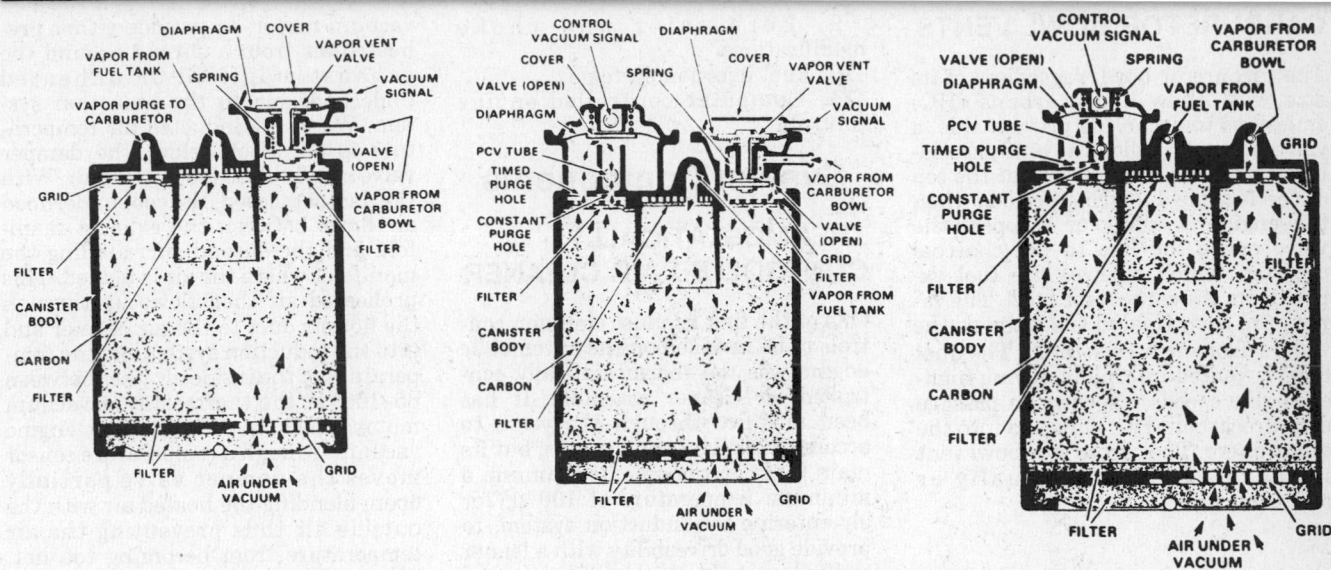

Typical vapor canisters used in the evaporative emission control system

tivated charcoal has the capabilities to absorb and store fuel vapors. The charcoal can be purged with fresh air and cleaned of its vapors, resulting in its reuse many times. When the vehicle is not running, the fuel vapors are routed to the canister where they are absorbed and stored by the activated charcoal. When the engine is started, either engine vacuum or the air flow through the air cleaner, draws the vapors from the charcoal by allowing a metered amount of air to pass over and through the charcoal and then routed into the engine's induction system.

Three different types of purge methods are used to cleanse the vapors from the canister. The first method is the variable purge which draws the va-por laden air into the air cleaner from the canister. The variable air flow through the canister is dependent upon the air flow through the air cleaner and into the engine. The second method is called the demand purge system which connects the canister to the fuel metering system. As the throttle is opened, the engine vacuum draws the cleansing air through the canister and into the engine air flow. This type of system only operates when the throttle plates are open, therefore preventing the vapors from entering the fuel sytem metering device when the engine is idling. The third system is a combination of a constant and demand purge. One purge line is routed to the PCV line which provides a continual purge, but by having a restriction in the passage, the flow rate is controlled. Another passage is routed to the fuel metering device. However, as the throttle plates are opened, the engine vacuum acts upon the restriction in the PCV line, opening the restriction and allowing more vapors and air to flow through the PCV system and into the intake air/fuel flow.

With the increased use of fuel injection and electronics, the vapors are purged from the canister at specific engine rpm and temperatures, so as not to upset the controlled air/fuel mixture and cause increased emissions from the combustion chambers.

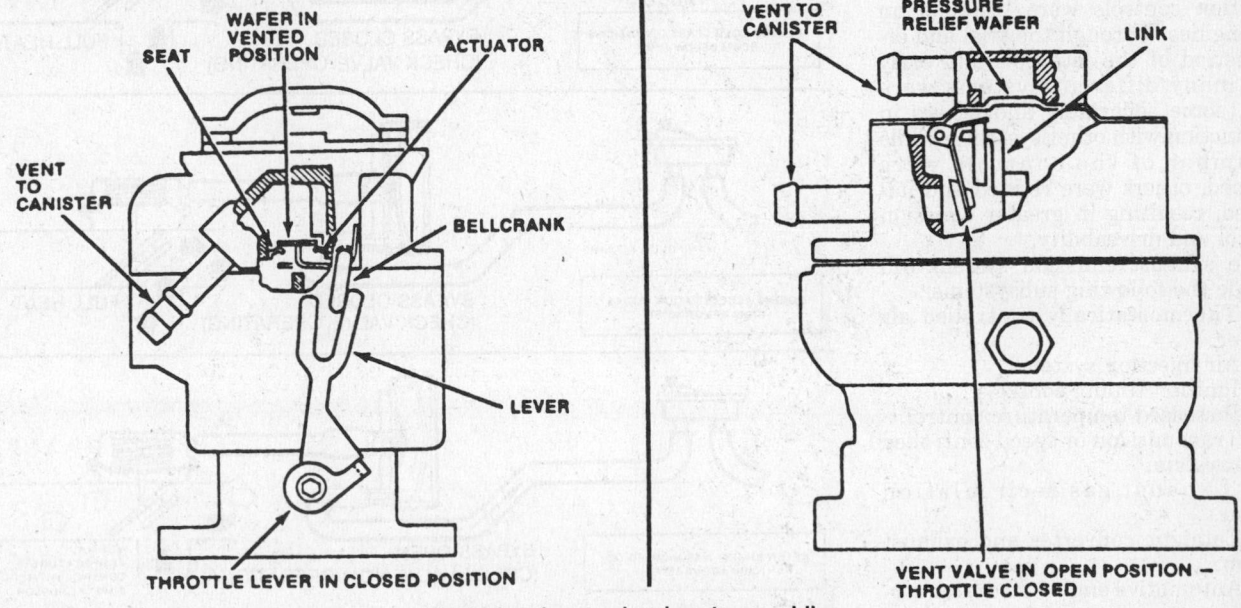

Typical carburetor bowl vent assemblies

CARBURETOR BOWL VENTS

The carburetor bowl, regardless of its size, will allow hydrocarbons (HC) emissions to occur. To prevent this, a valve normally called the anti-percolation valve, has been located at the top of the fuel bowl. During periods when the engine is idling or stopped, the valve opens the line to the charcoal canister, which receives the fuel vapors from the carburetor bowl. The vapors are then treated the same as the vapors from the fuel tank. The fuel bowl is vented internally during higher engine speeds, allowing no passage of hydrocarbon (HC) emission into the atmosphere. The carburetor bowl vent is operated either manually or electronically.

EXHAUST EMISSION CONTROL SYSTEM

System Description

The exhaust emission control system encompasses the automotive engine from the entrance of air into the engine's induction system until the exhaust byproduct of the combustion process emerges from the tail pipe. The engine exhaust was found to be responsible for approximately 60% of all automobile emissions before any pollution controls were installed on the engines. Through the trial and error period of the late 60's and early 70's many different systems were used, some separately and others in conjunction with other systems. While a number of the controls were dropped, others were refined and improved, resulting in greater emission control and driveability.

The exhaust emission system will include the following subsystems.

1. Thermostatically controlled air cleaner.
2. Air injection systems.
3. Ignition timing controls.
4. Increased temperature control.
5. Transmission or speed controlled spark system.
6. Exhaust gas recirculation system.
7. Catalytic converter and exhaust system.
8. Automotive engine feed back carburetor systems.

9. Carburetor and choke modifications.
10. Fuel injection systems.
11. Computer controlled engine controls.

System Components

THERMOSTATICALLY CONTROLLED AIR CLEANER

One of the first exhaust emission controls to be installed on the automobile engine was the thermostatically controlled air cleaner assembly. It has been modified through the years to accomodate many applications, but its main function remains, to maintain a minimum temperature of 100° F for air entering the induction system, to provide good driveability with a leaner air/fuel mixture which will help to reduce the exhaust emission of hydrocarbons (HC) and carbon monoxide (CO).

A damper valve, located in the snorkel or nozzle of the air cleaner housing, is operated by a thermostat or a vacuum motor, to provide either preheated air from a shroud around the exhaust manifold or unheated underhood air to the induction system. When the inducted air temperature is 85° F or below, the damper valve is fully closed to outside air. With the damper closed, the cool underhood air flows between the exhaust manifold and the shroud surrounding the manifold, where the air is heated. This preheated air then flows up through the hot air duct to the air cleaner and into the induction system. As the temperature of the inside air rises between 85–100° F, the thermostat or vacuum motor which is controlled by engine vacuum through a temperature sensor moves the damper valve partially open, blending the heated air with the outside air thus preventing the air temperature from becoming too hot. When the temperature in the air cleaner rises above 130° F, the thermostat or temperature sensor controlled vacuum motor opens the damper valve, assisted by a damper door spring, to outside air.

Regardless of the name given to this

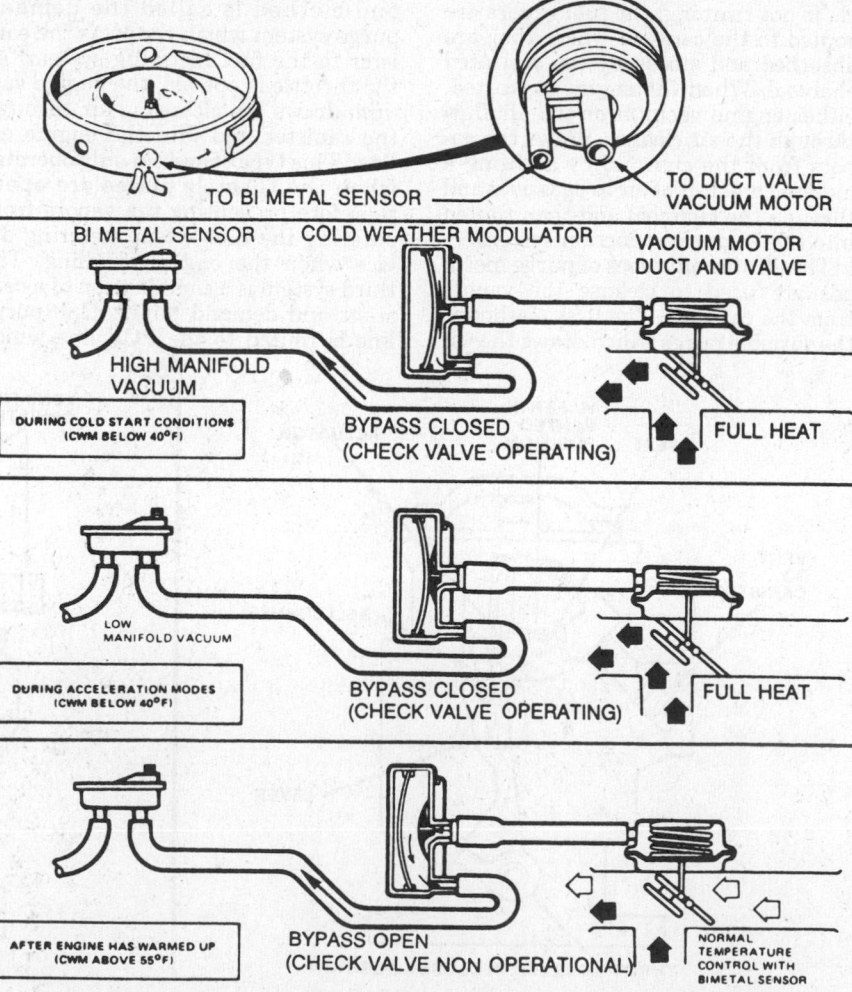

Operating sequences of the vacuum sensor operated air control valve

system by the vehicle manufacturer, its basic function and operation has remained the same throughout its application years. It is extremely important in the control of engine warm-up emissions, vehicle driveability and prevention of fuel icing, that all components of the thermostatically controlled air cleaner assembly be connected and operating.

AIR INJECTION SYSTEM

One of the major problems of the internal combustion engine is the fact that complete burning of the air/fuel mixture does not occur within the combustion chambers. This unburned mixture is swept from the combustion chamber along with the burned exhaust gases and emitted through the exhaust system and into the atmosphere. To prevent excessive emission of the unburned and burned gases containing large portions of hydrocarbons (HC) and carbon monoxide (CO), a means was devised to further burn the gases as they were forced from the cylinders by injecting fresh, oxygen laden air into the heated exhaust gas stream.

The addition of oxygen to the exhaust manifold and the exhaust pipes reducs the amount of unburned gases in the emitted exhaust. This after burner type system remains in use on many engine families. The vehicle manufacturers have modified, added and reduced the control components of the original system through the years, but the basic system to inject fresh air into the heated exhaust stream, still remains.

Two different systems are used on present day engines, one system using a belt driven air pump and the other system using the positive and negative exhaust system pressure pulsation to draw in fresh air.

Air Pump System

The major components of the air pump systems are.
1. Air pump which supplies filtered low pressure air to the system, normally 2–5 psi.
2. Diverter valve which diverts air pump output air to the atmosphere during deceleration to prevent backfire. A pressure relief valve is incorporated to protect the system.
3. Check valve which prevents hot exhaust gases from entering the system.
4. Air manifold which distributes the fresh air to each exhaust port of other area of the exhaust system.
5. Air nozzle which injects the air into the exhaust system.
6. Manifold vacuum signal line which senses manifold vacuum to acti-

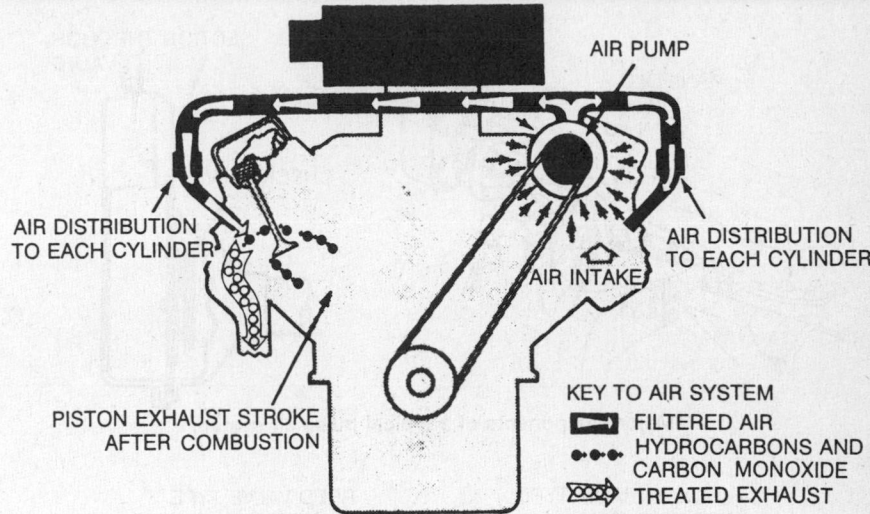

Typical air injection system flow schematic

KEY TO AIR SYSTEM
- FILTERED AIR
- HYDROCARBONS AND CARBON MONOXIDE
- TREATED EXHAUST

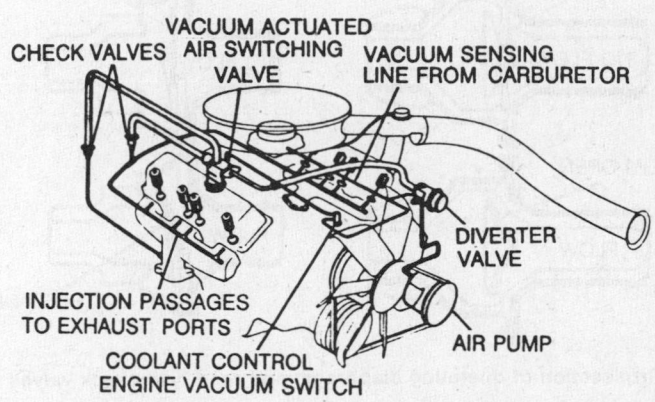

Major components of a typical air injection system

vate the diverter valve.

During the engine's normal operating condition, the air pump is supplying fresh air to the diverter valve which passes the air on to the air nozzles for injection into the exhaust stream. As the air is mixed with the hot exhaust gases, further combustion takes place in the exhaust system. During periods of deceleration, high manifold vacuum and a rich mixture is present in the combustion chambers. If fresh air was to be injected into the exhaust system during this condition, backfiring would occur. The diverter valve senses the high intake manifold vacuum and vents the fresh air from the air pump to the atmosphere. A switch valve is used by some manufacturers to redirect the fresh air from the exhaust valve port area to another location down stream in the exhaust system when the engine has warmed up to normal operating temperature, in order to avoid increasing the oxides of nitrogen (NOx) emissions while relying upon the heat of the exhaust to further the burning of the hydrocar-

bons (HC) at the exhaust valve ports. Since the addition of the three-way catalytic converters and oxygen sensors to the exhaust system, the air injection system has been modified and electronically controlled to critically monitor the entrance of fresh air into the exhaust stream during different engine operational modes and at different locations.

Pulse Air Injection System

The pulse air injection system does not use an air pump, but relies on the positive high pressure and negative low pressure pulses of the exhaust flow from the engine to operate one or more one-way check valves, which allows filtered fresh air to enter the exhaust stream at periods of negative low pressure and to prevent the leakage of exhaust gases back through the inlet air tubing. The fresh air induction through the one-way valve is normally accomplished at idle or slightly above.

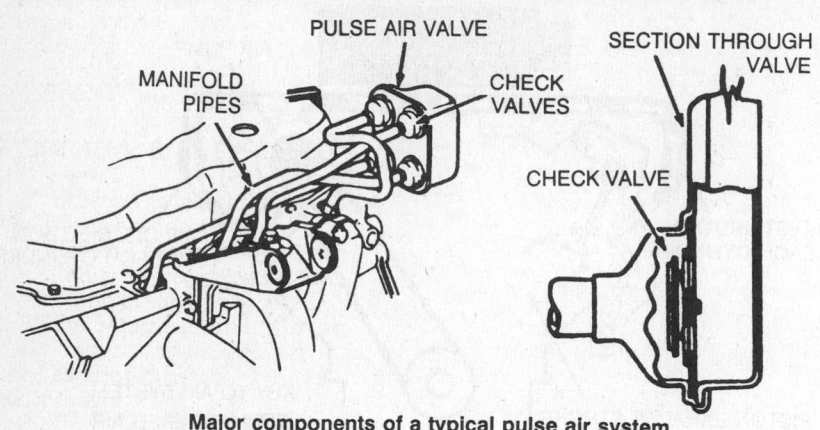

Major components of a typical pulse air system

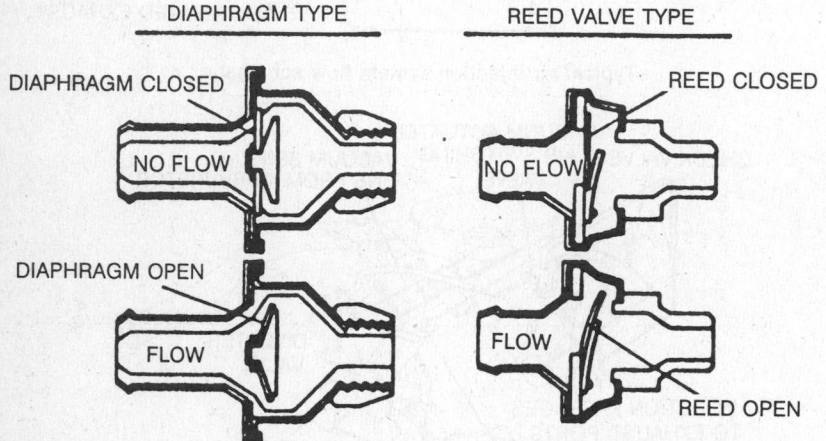

Crossection of operating diaphragm and reed type check valves

IGNITION TIMING CONTROLS

The purpose of the timing or spark controls is to fire the air/fuel mixture in the combustion chamber at a specific time in order to derive the most power from the mixture, while burning it as completely as possible to rid it of excess hydrocarbons (HC) and still maintain a combustion chamber temperature that prevents excess formations of nitrogen oxides (NOx).

Electronic Ignition

To increase the ignition system's durability to over 50,000 miles of engine operation and to eliminate adjustment or replacement of the contact point sets, and electronically controlled ignition system was introduced as standard equipment on American made automobiles, beginning with the 1975 model year.

The distributor primary circuit was changed from a breaker plate and cam assembly in the distributor, to a magnetic signal generating system which detects the distributor shaft position and sends electrical pulses to an electronic control module, which takes the place of the mechanically operated point set, in the off-on switch of the primary current to the ignition coil. The armature and pick-up assembly has no effect on the dwell period which is controlled by the control module. Therefore, dwell never needs to be adjusted. With a dwell that remains constant, we do not need to continually read just the ignition timing to compensate for mechanical wear in the ignition system.

With the use of the electronics in the controlling of the distributor operations, on-board vehicle computers were added to control the many operations of the electronic components, such as the fuel delivery and spark timing for optimum engine performance. Many electronic sensors have been added to the systems to inform the computer(s) that adjustments may be necessary to maintain the vehicle's electronic performance. An example of the quickness of the computer to regulate and direct changes to the electronic components, is that changes are continually being made by an electronic component between 10–50 times per second to maintain the engine in its optimum performance.

Ignition Coil

The ignition coil is the component of the ignition system that must produce voltage of enough intensity and strength to cross a predetermined spark plug gap to ignite the air/fuel mixture in the combustion chamber, under any and all engine operation conditions. To achieve this responsibility, the ignition coil must increase the primary voltage form an average of 12 volts to a secondary voltage, through induction, as high as 30,000 volts.

Distribution of Spark

To route the secondary voltage to the spark plugs, heavily insulate wiring, distributor caps and rotors are used. With the use of electronic ignition sys-

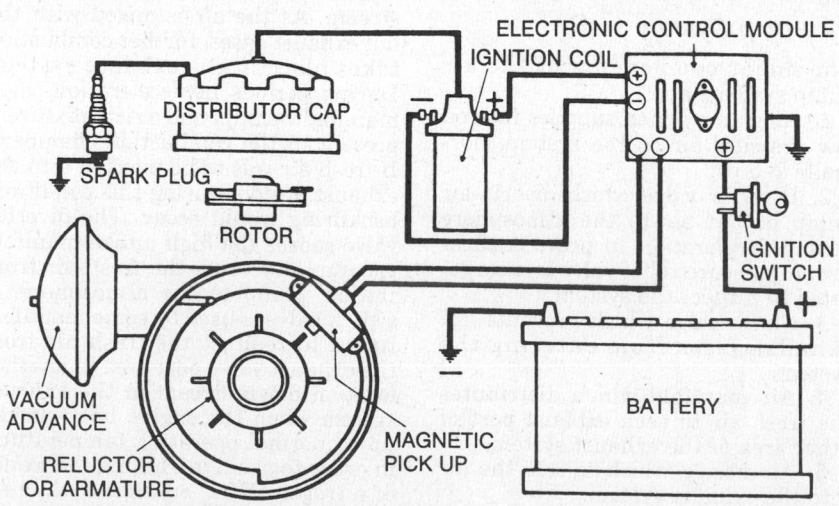

Typical electronic ignition system electrical schematic

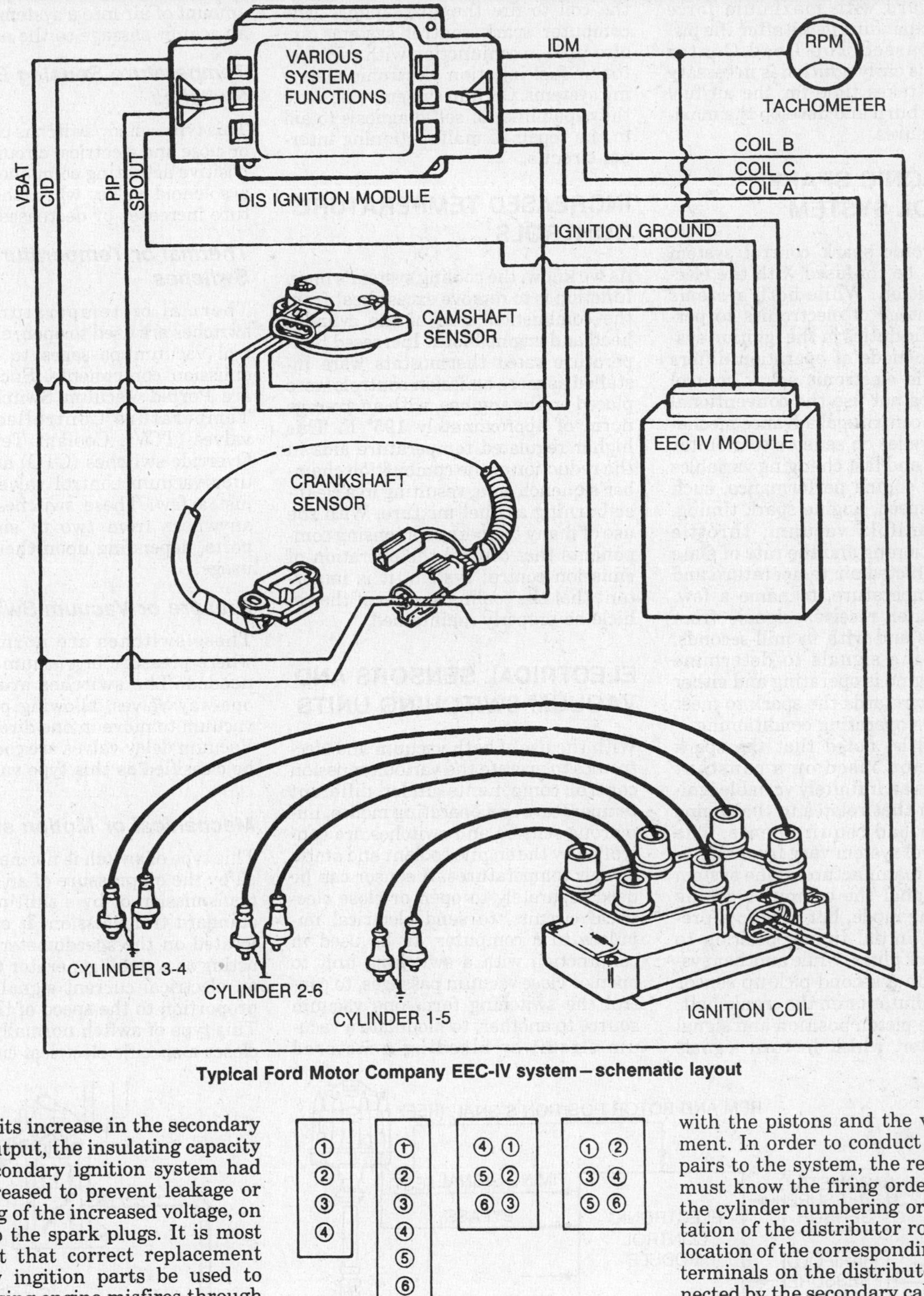

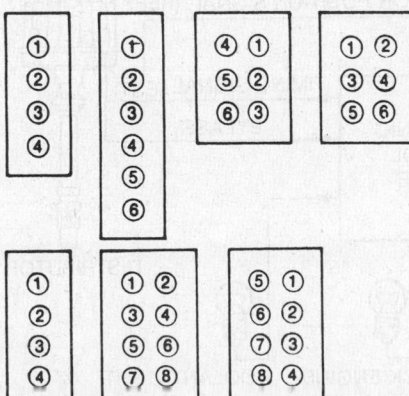

Typical Ford Motor Company EEC-IV system—schematic layout

Labels in the schematic:
VARIOUS SYSTEM FUNCTIONS
IDM
RPM
TACHOMETER
COIL B
COIL C
COIL A
DIS IGNITION MODULE
IGNITION GROUND
VBAT
CID
PIP
SPOUT
CAMSHAFT SENSOR
CRANKSHAFT SENSOR
EEC IV MODULE
CYLINDER 3-4
CYLINDER 2-6
CYLINDER 1-5
IGNITION COIL

tems and its increase in the secondary voltage output, the insulating capacity of the secondary ignition system had to be increased to prevent leakage or crossfiring of the increased voltage, on its way to the spark plugs. It is most important that correct replacement secondary ingition parts be used to avoid causing engine misfires through the loss of secondary current due to the use of inferior replacement parts.

Firing Sequence

To have the ignition system ignite the air/fuel mixture in the proper cylinder at the right time, a firing order sequence must be established by the engine manufacturer, in proper time

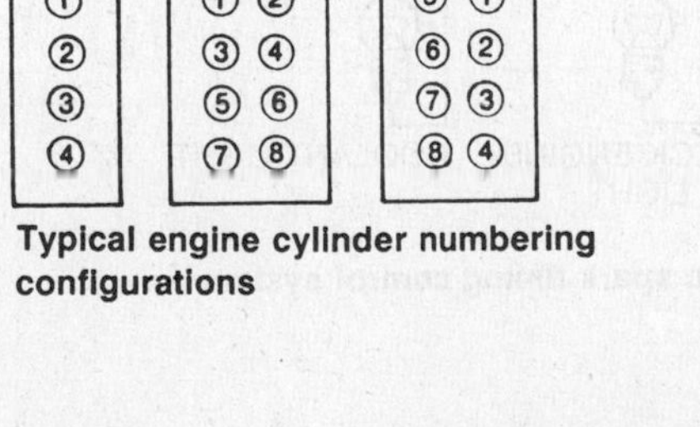

Typical engine cylinder numbering configurations

with the pistons and the valve movement. In order to conduct tests or repairs to the system, the repair person must know the firing order sequence, the cylinder numbering order, the rotation of the distributor rotor and the location of the corresponding electrical terminals on the distributor cap, connected by the secondary cables to their respective spark plugs, which are screwed into the cylinder's combustion chamber, beginning with the No.1 cylinder.

Spark Timing

The air/fuel mixture must be ignited before the piston reaches Top Dead Center (TDC) to have the piston driven downward with maximum force from the expanding gases after the piston has passed Top Dead Center (TDC). This early ignition is necessary because it takes time for the air/fuel mixture to burn and develop the maximum pressures.

ELECTRONIC SPARK CONTROL SYSTEM

The electronic spark control system should not be confused with the electronic ignition. While both systems take advantage of electronics to perform specific duties in the ignition systems, their mode of operation differs greatly. The electronic spark control system does not use the conventional vacuum or centrifugal advance mechanism, but relies on sensors to monitor the critical and fast changing variables that affect engine performance, such as engine speed, engine spark timing, intake manifold vacuum, throttle plate postitioning and the rate of plate change, inducted air temperature and coolant temperature, to name a few. The computer receives signals from the sensors and with in mill-seconds, computes the signals to determine how the engine is operating and either advances or retards the spark to meet the engine's operating conditioning.

It should be noted that the spark timing is not based on a constant curve, but is an infinitely variable ignition system that relates to the engine speed and load requirements. The spark control system vary from manufacturer to manufacturer. One system may not signal the computer during the cranking mode, but rely on a predetermined initial timing position to fire the spark plugs, while another system relies on a second pick-up sensor in the distributor or on the crankshaft, to sense the piston position and signal the computer, which in turn signals

the coil to fire the spark plugs. The computer spark control systems are also used in conjunction with carburetor or fuel injection electronic metering systems. Certain systems also have the capabilities of self-diagnosis to aid in the repair of malfunctioning internal circuits.

INCREASED TEMPERATURE CONTROLS

As we know, the cooling system's main function is to remove excess heat from the combustion area of the cylinder head and engine block. Increased temperature rated thermostats were installed as more emission controls were placed on the engines, with an average norm of approximately 195° F. This higher regulated temperature aids in the reduction of the combustion chamber's quench area, resulting in a cleaner burning air/fuel mixture. With the use of many temperature sensing components that control the operation of emission control system, it is important that the cooling system of the vehicle be properly maintained.

ELECTRICAL SENSORS AND VACUUM SWITCHING UNITS

With the use of both vacuum and electronics to operate the various emission control components under different temperature and operating modes, numerous sensors and switches are controlled by the engine coolant and ambient air temperatures. A sensor can be used separately to open or close electrical circuits, to send electrical impulses to a computer, to be used in conjunction with a switching unit to open or close vacuum passages, to control the switching form one vacuum source to another, to modulate a vacuum circuit by bleeding a metered

amount of air into a system or opening a vacuum passage to the atmosphere.

Temperature Sensing Electrical Switches

This type sensor/switch is used to open or close and electrical circuit to a more positive actuating component, such as a solenoid valve, when the temperature increases or decreases.

Thermal or Temperature Switches

Thermal or temperature sensing switches are used to open, close or control vacuum passages to the varied emission components. Such switches are Ported Vacuum Switches (PVS), Temperature Controlled Vacuum valves (TCV), Coolant Temperature Override switches (CTO) and Distributor vacuum control valves to name just a few. These switches can have anywhere from two to six different ports, depending upon their intended usage.

Presure or Vacuum Switches

These switches are normally used where pressure or vacuum sensing is needed. The switches are normally one-way valves, allowing pressure or vacuum to move in one direction only. Vacuum delay valves are considered to be classified as this type valve.

Mechanical or Motion switches

This type of switch is normally operated by the oil pressure of an automatic transmission or by a shifting rail of a standard transmission. It can also be located on the speedometer cable, reacting as a small generator to produce an electrical current signal, in direct proportion to the speed of the vehicle. This type of switch normally opens or closes a specific electrical circuit.

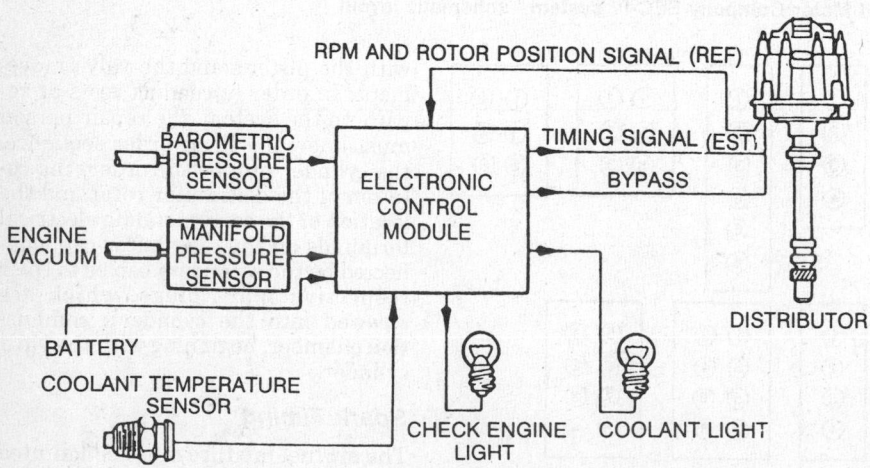

Typical first generation electronic spark timing control system

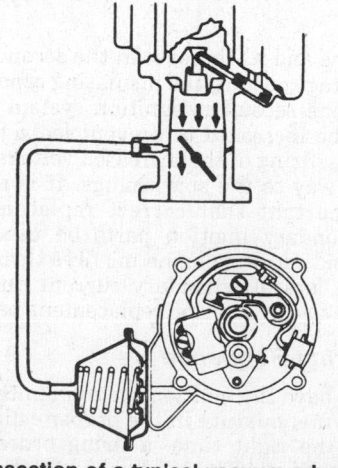

Crossection of a typical vacuum advance mechanism

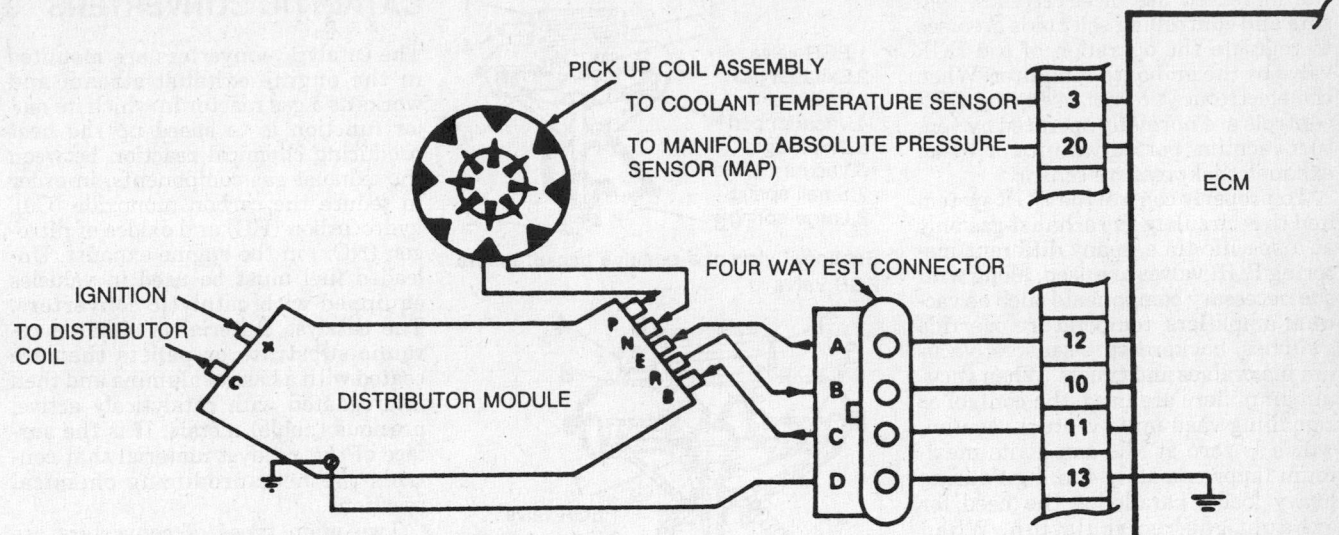

PICK UP COIL ASSEMBLY
TO COOLANT TEMPERATURE SENSOR
TO MANIFOLD ABSOLUTE PRESSURE SENSOR (MAP)
ECM
IGNITION
TO DISTRIBUTOR COIL
FOUR WAY EST CONNECTOR
DISTRIBUTOR MODULE

Typical usage of vacuum and temperature sensors to control EST

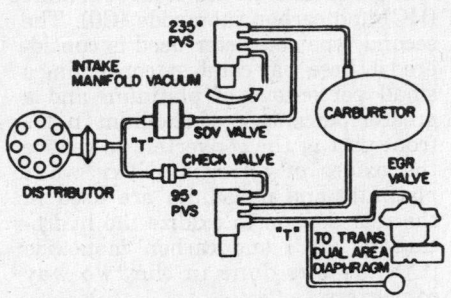

Typical temperature sensing valves used to control vacuum actuated components

EXHAUST GAS RECIRCULATION SYSTEM (EGR)

In their attempt to reduce hydrocarbons (HC) emission, the manufacturers increased the combustion chamber temperatures to more thoroughly burn the air/fuel mixture. With the increase in temperature and pressure, another pollutant was created, oxides of nitrogen (N0x).

At temperatures below 2500° F, nitrogen remains an inert gas, but with combustion temperatures reaching as high as 4500° F, the nitrogen combines with the oxygen in the air/fuel mixture, resulting in the formation of oxides of nitrogen (NOx) a harmful pollutant. Through experimentation, it was found that a portion of the exhaust gases could be redirected into the combustion chamber, along with the inducted air/fuel mixture, resulting in the temperature of the burning process being lowered, causing a reduction in the emission of the oxides of nitrogen (NOx). The amount of the recirculated exhaust gases has to be carefully controlled. If too much exhaust gas is supplied at the wrong time, the engine may stall and be very rough at idle. If not enough exhaust gas is recirculated, the oxides of nitrogen (NOx) will not be reduced.

EGR Valves

Varied types of EGR valves are used with different control components, so that the proper amount of recirculated gas is directed into the air/fuel mixture at a specific time. The EGR valves are vacuum operated, either by intake manifold of by ported vacuum. With

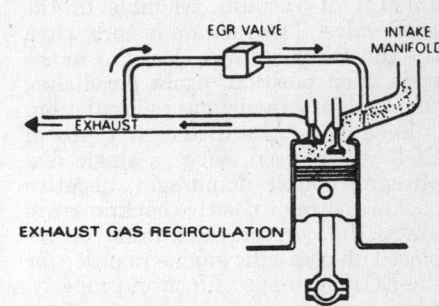

Typical EGR system schematic

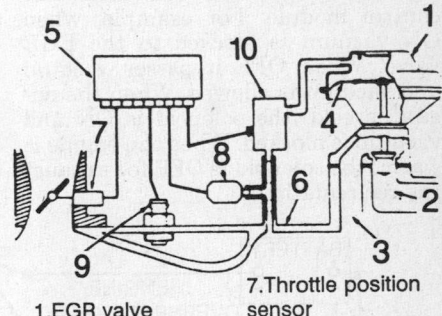

1. EGR valve
2. Exhaust gas
3. Intake air
4. Diaphragm
5. Electronic control module
6. Manifold vacuum
7. Throttle position sensor
8. Manifold pressure sensor
9. coolant temperature sensor
10. EGR control solenoid

Typical operation of a vacuum solenoid controlled EGR valve

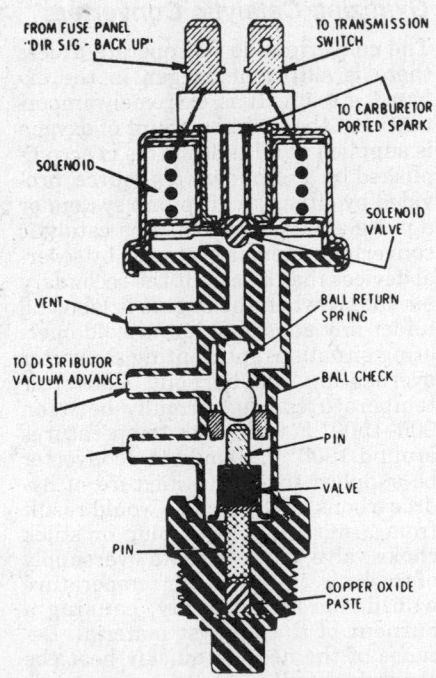

FROM FUSE PANEL "DIR SIG - BACK UP"
TO TRANSMISSION SWITCH
TO CARBURETOR PORTED SPARK
SOLENOID
SOLENOID VALVE
VENT
BALL RETURN SPRING
TO DISTRIBUTOR VACUUM ADVANCE
BALL CHECK
PIN
VALVE
PIN
COPPER OXIDE PASTE

Crossection of a typical vacuum control valve for distributor vacuum advance operation

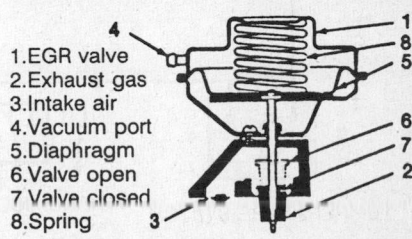

1. EGR valve
2. Exhaust gas
3. Intake air
4. Vacuum port
5. Diaphragm
6. Valve open
7. Valve closed
8. Spring

Sectional view of a ported EGR valve

the increased use of electronics, sensors and controlling solenoids are used to regulate the operation of the EGR valve by the on-board computer. When the electronics are not used, the EGR controls are normally operated by venturi vacuum, ported vacuum or by an exhaust backpressure sensor.

To properly control the EGR system and to recirculate the exhaust gas only at a specific time, many different metering EGR valves are used, along with the necessary components such as vacuum amplifiers, temperature override switches, backpressure sensors, vacuum bias valves and timers. When vacuum amplifiers are used, the control or signalling vacuum is venturi vacuum, which is zero at idle and at its maximum (approximately 4 in. hg.) during heavy loads, paralleling the need for exhaust gas recirculation. When ported vacuum is used, the position of the throttle plate regulates the amount of vacuum available to the EGR valve. The vacuum is zero when the throttle plates are closed or in the wide open position, again paralleling the need for exhaust gas recirculation.

Because of the different types of EGR valves used, such as single diaphragm, double diaphragm, negative backpressure or positive backpressure units, the correct valve must be replaced on a specific engine in order for the EGR system to function properly. Some new models control the EGR valve by a signal from the electronic control module. For example, when ON vacuum is blocked to the EGR valve. When OFF it passes vacuum and vacuum is allowed. When the engine is cold, the solenoid is ON and vacuum is blocked. When the engine is warm, the solenoid is OFF for exhaust gas recirculation.

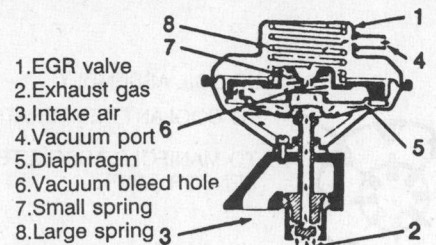

1. EGR valve
2. Exhaust gas
3. Intake air
4. Vacuum port
5. Diaphragm
6. Vacuum bleed hole
7. Small spring
8. Large spring

Sectional view of a positive backpressure EGR valve

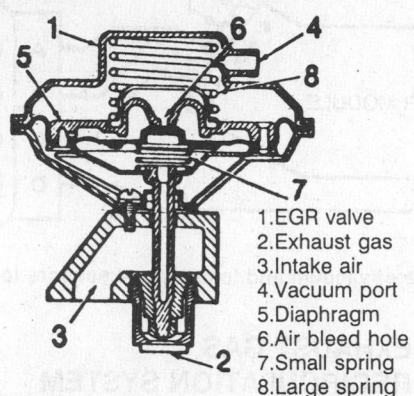

1. EGR valve
2. Exhaust gas
3. Intake air
4. Vacuum port
5. Diaphragm
6. Air bleed hole
7. Small spring
8. Large spring

Sectional vies of a negative backpressure EGR valve

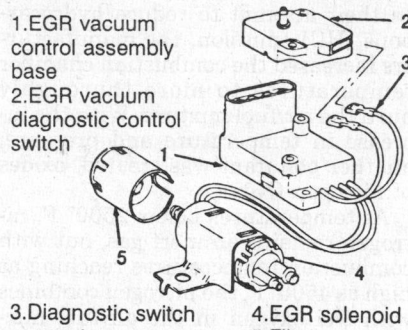

1. EGR vacuum control assembly base
2. EGR vacuum diagnostic control switch
3. Diagnostic switch connectors
4. EGR solenoid
5. Filter

Typical General Motors EGR control solenoid

CATALYTIC CONVERTERS

The catalytic converters are mounted in the engine exhaust stream and works as a gas reactor in which its major function is to speed up the heat producing chemical reaction between the exhaust gas components, in order to reduce the carbon monoxide (CO), hydrocarbon (HC) and oxides of nitrogen (NOx) in the engine exhaust. Unleaded fuel must be used in vehicles equipped with catalytic converters. The catalyst material is either a ceramic substrate or pellets that are coated with a base of alumina and then impregnated with catalyticaly active, precious (noble) metals. It is the surface of the catalyst material that controls the heat producing chemical reaction.

Two main types of converters are used. The first type contains, platinum and palladium to effectively catalyze the oxidation of the hydrocarbons (HC) and carbon monoxide (CO). The second type converter used is considered a three-way catalyst, containing a small percentage of platinum and a greater percentage of rhodium in the front part of the converters to reduce the oxides of nitrogen (NOx), while platinum and palladium are used in the rear section to oxidize the hydrocarbons (HC) and carbon monoxide (CO), as was done in the two way converters.

Oxidizing Catalytic Converters

The converters do not operate unless there is sufficient oxygen in the exhaust stream. It is extremely important that the proper amount of oxygen is supplied at all times. This is accomplished by a secondary air source, provided by either an air pump system or a pulse air type system. The catalytic converter system is protected by several devices that block out the secondary air supply when the engine is laboring under any abnormal hot or cold operating situations, preventing converter overheating and burnout. Converter temperatures are normally between 900–1500° F with peak temperatures around 1800° F. Should the converter be supplied too rich a mixture of hydrocarbons (HC), such as would result from a misfiring spark plug or stuck choke valve, along with an oversupply of fresh air, the converter temperature would increase sharply, causing a burnout of the catalyst material. Because of the need to quickly heat the converters units, smaller or mini converters are placed in the exhaust stream before the main converter to preheat the exhaust gas.

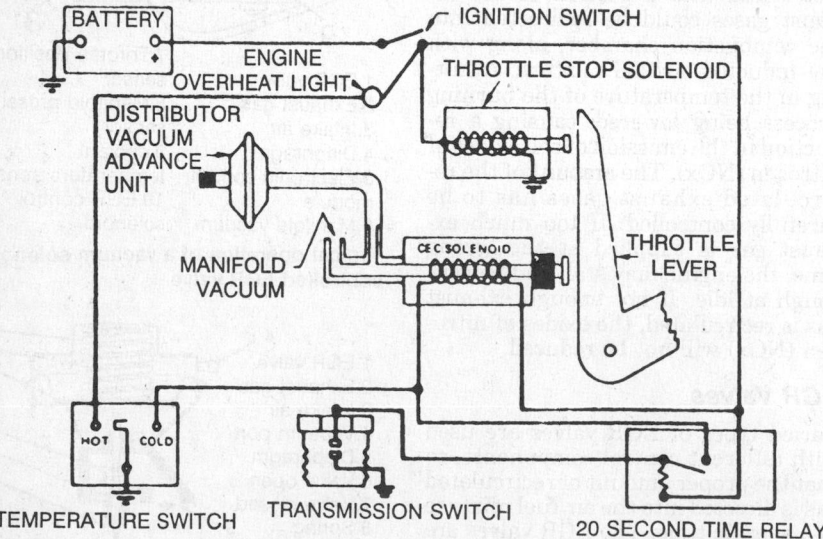

EGR system using various types of emission sensing and actuating controls

BATTERY
ENGINE OVERHEAT LIGHT
IGNITION SWITCH
THROTTLE STOP SOLENOID
DISTRIBUTOR VACUUM ADVANCE UNIT
MANIFOLD VACUUM
CEC SOLENOID
THROTTLE LEVER
HOT COLD
TEMPERATURE SWITCH
TRANSMISSION SWITCH
20 SECOND TIME RELAY

Three-way Catalytic Converters

The three-way catalytic converters use a combination of catalyst which produce two different chemical reactions, oxidation and reduction. By adding fresh air to the unburned hydrocarbons (HC) and carbon monoxide (CO) within the converter, the oxidizing of combustion process takes place. Just the reverse process is required to lower the oxides of nitrogen (NOx) emissions. The oxides of nitrogen (NOx) already contains excessive oxygen and the process of separating the excess oxygen from the nitrogen is called a reducing reaction. This reducing or reduction process is done in the front section of the converter while the oxidizing process is accomplished in the section. A fresh air connector is located on the center of the converter shell, to add fresh air from the air systems as required. To enable the three-way con-

verter to operate properly, the engine's air/fuel ratio must be held within a tight range, called a Stoichiometric range. This is accomplished with the use of the latest computer controlled electronic engine components. Different control components are used by the vehicle manufacturers to prevent converter damage and/or burnout.

THERMAL REACTOR SYSTEM

The thermal reactor is installed in place of the exhaust manifold. It is much heavier and heat resistant. Its purpose is to collect the exhaust gases in a common area, to keep their temperature higher for a longer period of time, thus allowing further oxidation or burning of the exhaust gas and secondary air mix burned emissions.

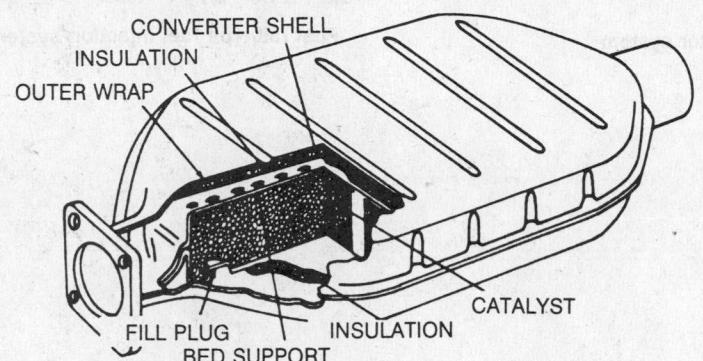

CONVERTER SHELL
INSULATION
OUTER WRAP
CATALYST
FILL PLUG
BED SUPPORT
INSULATION

INLET GAS   CATALYTIC PELLET COMPOUND   OUTLET GAS

Pellet type catalytic converter

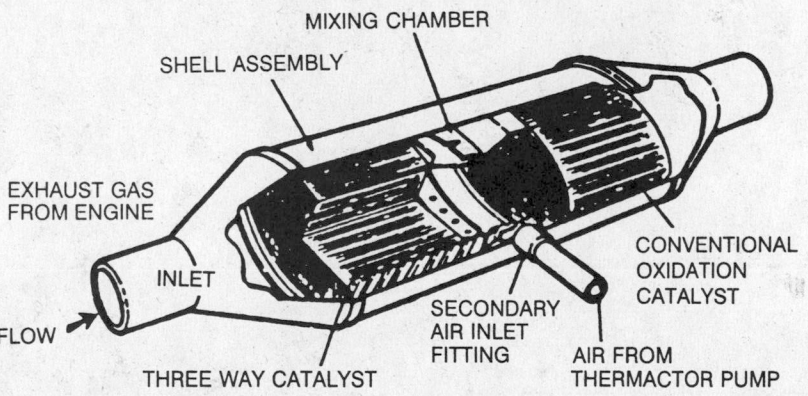

MIXING CHAMBER
SHELL ASSEMBLY
EXHAUST GAS FROM ENGINE
INLET
FLOW
THREE WAY CATALYST
SECONDARY AIR INLET FITTING
AIR FROM THERMACTOR PUMP
CONVENTIONAL OXIDATION CATALYST

Typical conversion of exhaust emissions from a three way catalytic converter

FUEL INJECTION SYSTEMS

A number of different fuel injection systems are used on today's vehicles to improve the emissions, miles per gallon and the driveability demands on the automobile. The injection systems range from a metering of the inducted air velocity to the use of computer or microprocessor units to operate the systems, allowing the proper amount of fuel to mix with the inducted air. Sensors are used to signal the control unit that changes in air/fuel mixture is needed as the driving mode changes. The fuel injectors can be located near the intake valve ports on some systems and this system is know as Multiport Fuel Injection. The fuel is injected into the ports from injector groups of two, three or four injectors. A sequential type of fuel injection is also used, known as SFI and is based on the firing order of the engine. The throttle body injection system, known as TBI, uses an injector unit(s), mounted in the throttle body to inject fuel as needed into the intake manifold. An electronic control module is used, in conjunction with sensors to control the system.

Engine Condition

A definate relationship exists between emission controls and engine tune-up. An engine that is out of tune after many miles of operation, can result in failure of the emission controls to properly perform their jobs. Malfunctioning emission controls can cause poor engine performance, so we can understand that proper engine operation is dependent upon thorough testing and servicing of all components related to both performance and emission controls. Because the engine is responsible for approximately 80% of the emissions and its condition reflects on its operational capacities, a knowledge of its basic operation is necessary when attempting to service it. We know that compression, ignition and fuel distribution are the basic needs of the engine in order to start and run. However, how these three basic needs are coordinated through the action of the crankshaft, camshaft, cylinder components and distributor must be understood, in order for the engine to operate under all speeds and loads. If a malfunctioning cylinder or other internal problems are suspected, a compression test, a cylinder balance test or other mechanical tests should be done to determine the engine's internal condition, prior to any attempted tune-up on the engine.

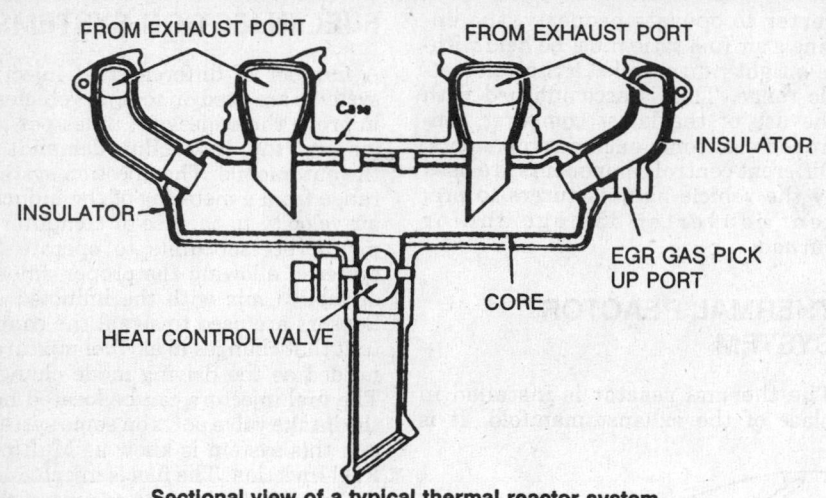

Sectional view of a typical thermal reactor system

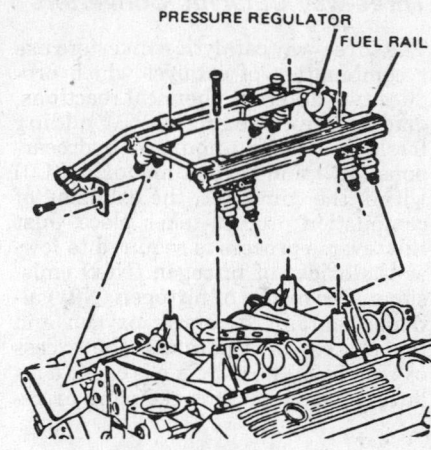

Fuel rail type fuel injection system

Electronic Engine Controls

ENGINE ELECTRONICS

In the ladder part of the 1960's, Robert Bosch introduce the first true electronically controlled engine with an on-board computer. Today, almost every car produced has some kind of electronic control. The once mechanically controlled engine functions are all but extinct.

The first system, Bosch D-Jetronic, is comprised of electrically energized fuel injectors in which the injection time is controlled by an electronic control unit (ECU). The early system delivered a basic quantity of fuel and varied from this point depending upon engine load, engine speed and engine temperature.

Since the early days of ECU, the controls have become more complex, with a much greater amount of computer memory and even the ability to learn.

In this section the topics will include different types of electronically controlled fuel induction, spark control, the sensors and switches that provide the ECU with information, other non-engine related controls that the ECU might supply and some ECU self-diagnostics.

The most common fuel induction system with an ECU is electronic fuel injection. In this system fuel can be delivered many different ways. One of which is the single point injection (SPI) were one or two injectors are mounted on a throttle body assembly. Fuel is delivered constantly through the injector(s), but in varying quanti-ties. The SPI system very much resembles a carbureted system. SPI is more commonly known as throttle body injection (TBI). Another fuel injection system is multi-point injection (MPI). This system supplies one injector for each cylinder, usually positioned in the intake manifold, just above the intake valve. In MPI, fuel can be injected in two ways. One is to energize a group of injectors, thus atomizing fuel in the intake manifold and storing it for a short time until the intake valve opens. The second way is to sequentially energize each cylinder's injector as the intake valve is opened. This injection is the more efficient, effective and more complex system.

Another fuel induction system utilizing an ECU is the feedback carburetor (FBC). A conventional carburetor is still used but it has a more precise air/fuel mixture control which is achieved through an integral mixture control solenoid. The solenoid is energized on and off by the ECU to maintain mixture demand. The ECU calculates air/fuel mixture demand changes by the data it receives through remote sensors. The most important sensor (and makes the system possible) is an oxygen (O_2) sensor (which will be discussed later in this section). The ECU monitors the exhaust gases for rich/lean conditions by way of the O_2 sensor and, in turn, controls the air/fuel mixture by increasing or decreasing the duty cycles (on and off) to the mixture control solenoid for an optimum 14.7:1 air/fuel ratio.

ECU Self-Diagnostics

The ECU can detect a malfuction or abnornality in the sensors or in the ECU itself and display a warning light on the instrument panel when it does. When this occurs, the ECU stores a trouble code for future system diagnosis. If the problem is sever enough to where it inhibits closed loop operation, the ECU will assume a backup system. This fail-safe circuit is pre-programmed into the ECU for minimal driveability operation so the vehicle can be driven to a nearby service facility. The trouble codes are usually a two digit numbers identified by the number of diagnostic LED or check engine light flashes. The trouble codes assist the service technician in isolating a faulty circuit or component within the system.

Electronic Data Sensors

The engine control system consists of various data sensors. Although data sensor names and applications vary from system to system, the most common input sensors/switches are:

- oxygen (O_2) sensor
- coolant temperature sensor
- manifold air pressure (MAP) sensor
- vehicle speed sensor (VSS)
- throttle position sensor (TPS)
- engine speed reference or distributor reference (rpm)
- air flow sensor
- air intake temperature sensor
- crankshaft sensor
- detonation (knock) sensor
- throttle body temperature sensor
- throttle idle switch
- transmission or drive switch
- a/c compressor clutch switch

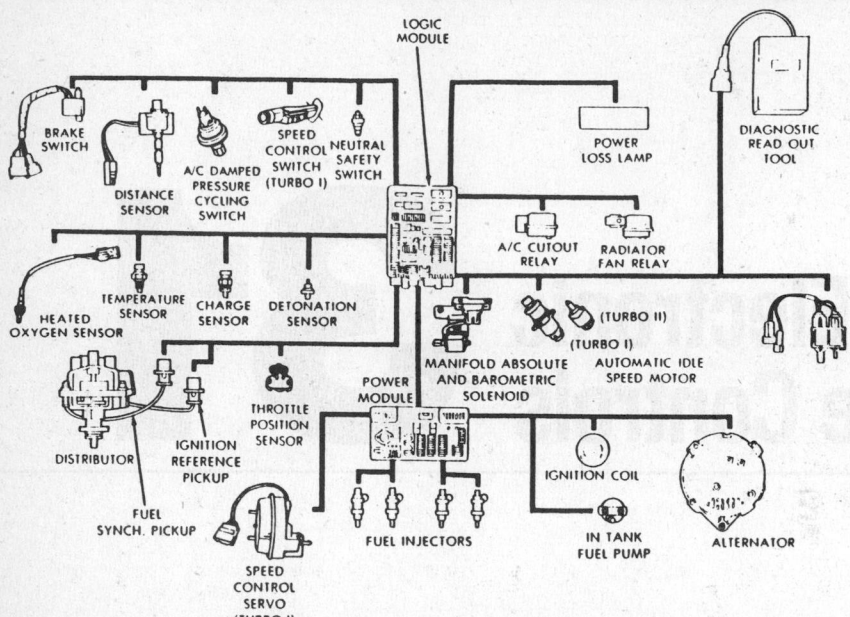

Electronic engine control components

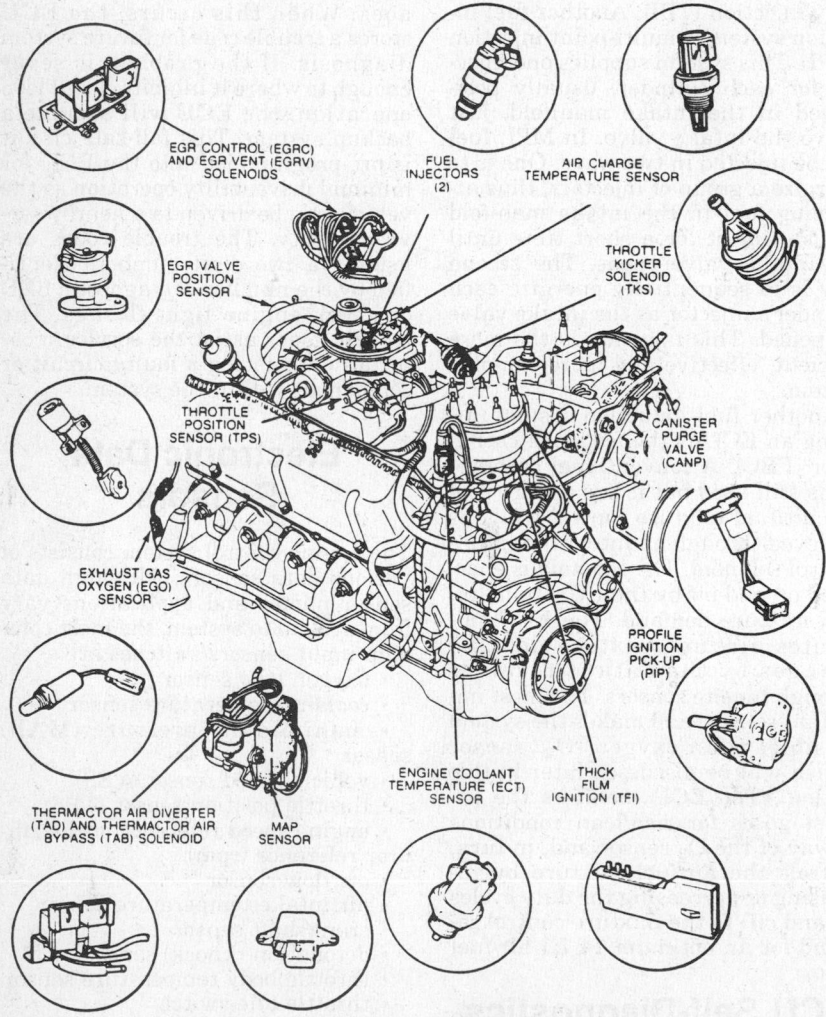

Electronic engine control component locations

- power steering pump switch
- altitude or barometric pressure sensor
- wide open throttle switch

Electronically Controlled Devices

Some of the output devices that the ECU may control vary from system to system, but the most common output or ECU controlled devices are:

- fuel injector(s)
- air/fuel mixture solenoid
- fuel pump relay
- a/c compressor clutch relay
- idle air control (IAC) valve
- idle speed control (ISC) motor
- ignition spark/timing
- canister purge solenoid
- torque converter clutch solenoid (automatic transmission)
- air management system (air induction)
- idle-up or throttle kicker solenoid
- alternator field control (charging system)
- turbocharger boost wastegate
- cooling fan relay

Component Description

THROTTLE BODY

The throttle body, in most fuel injected systems, is usually an alumunum housing that consists of one or two throttle blades which are attached to a throttle shaft. The housing has a throttle position sensor (TPS) sensor, idle air control motor and, in some cases, throttle body temperature sensor. On SPI systems, the housing also has an injector(s) and (in some cases) a fuel pressure regulator. The throttle body throttle blade controls the amount of air that enters the engine as well as the amount of vacuum.

ELECTRONIC CONTROL UNIT (ECU)

The ECU monitors and controls all engine control functions. The ECU consists of input and output devices, a central processing unit, a power supply and various memory banks. The input and output devices of the ECU convert electrical signals received by the data sensors and switches to the digital signal that are used by the central processing unit. The central processing unit receives digital signals that are used to perform all mathematical computations and logic

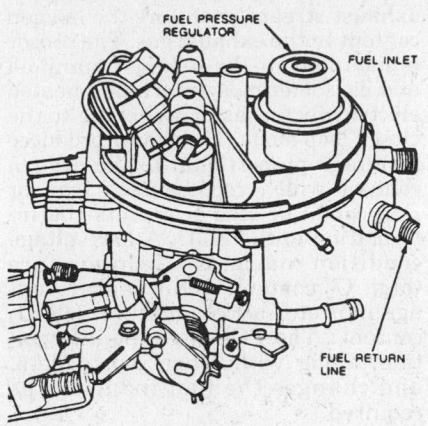

Throttle body – TBI

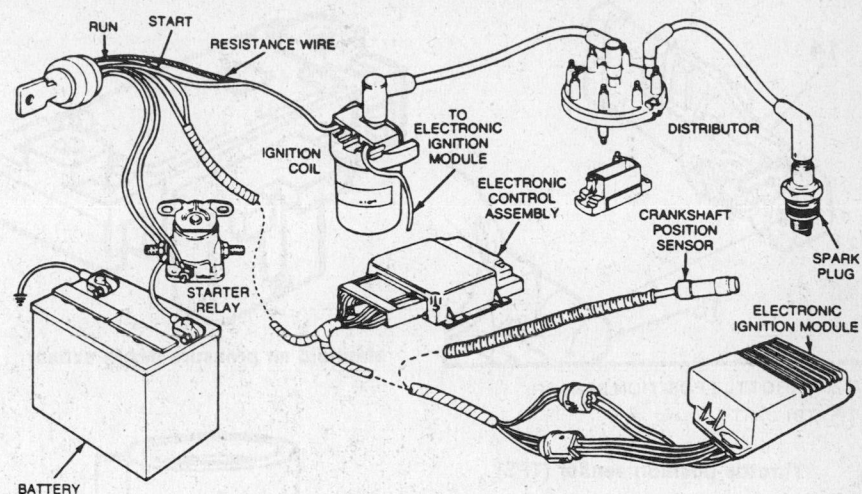

Electronic ignition system using ECU

functions necessary to deliver proper air/fuel mixture. The central processing unit is also responsible for calculating spark timing information. The main source of power that allows the ECU to function is generated from the battery of the vehicle and transported through the ignition system. The memory bank of the ECU is programmed with exact information that is used by the ECU during the open loop mode. This data is also used when a sensor of other component fails, allowing the vehicle to be driven to a repair facility.

CALIBRATION ASSSEMBLY OR PROM (PROGRAMMABLE READ ONLY MEMORY)

Some vehicle manufactures use one ECU for several different model vehicles. This interchangeable ECU is possible through the use of a calibration assembly or prom. Information about the vehicle's engine, transmission, body and drive axle ratio are programmed and permanently stored into the assembly. If the battery supply should become disconnected from the ECU, the data stored into the assembly is not lost.

ELECTRONIC SPARK CONTROL (ESC)

The vehicles equipped with an ESC have the ability to change the ignition timing under any and all operating conditions. Data from various remote sensors (coolant temperature, throttle position, rpm, etc.) is transmitted to the ESC. The ESC computes the information and triggers the ignition spark at precisely the right instant. Some ESC systems (ie.,turbocharged engines) use a detonation (knock) sensor which senses pre-ignition and

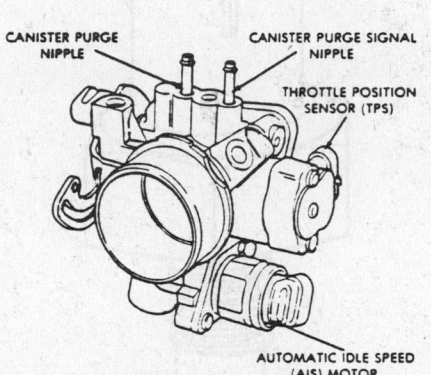

Throttle body – MFI

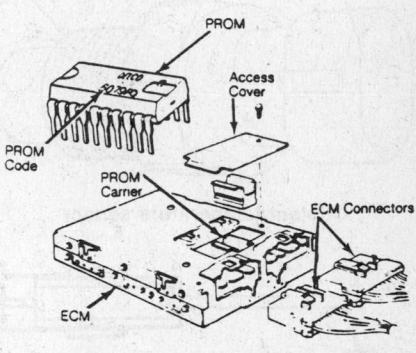

Electronic control unit prom

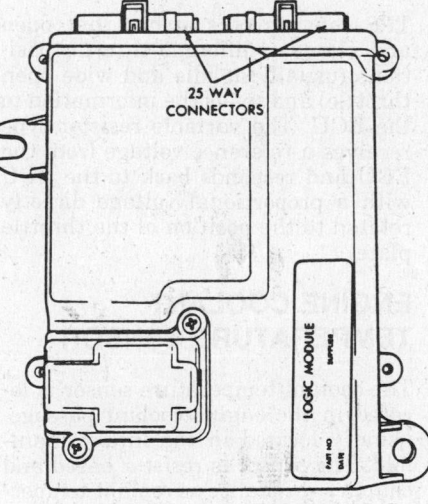

Electronic control unit

transmits the information to the ESC. The ESC modifies spark advance and boost pressure in order to eliminate knock.

MASS AIR FLOW SENSOR

The mass air flow (MAF) sensor is

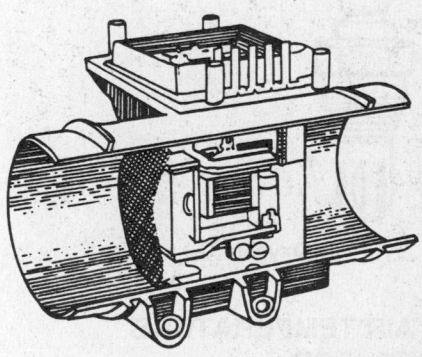

Mass air flow (MAF) sensor

only incorporated in some Multi-point fuel injection systems. The MAF sensor is a very complex device which measures the air mass of the engine intake. Because the air mass is always changing with temperature, humidity and altitude, the fuel delivery rate must be adjusted to compensate for these changes so that a precise fuel mixture can be maintained.

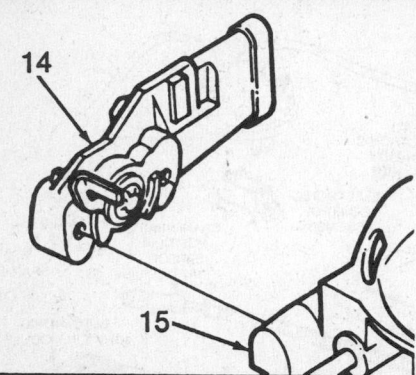

14 THROTTLE POSITION SENSOR
15 TBI UNIT

Throttle position sensor (TPS)

Coolant temperature sensor

Oxygen (O₂) sensor

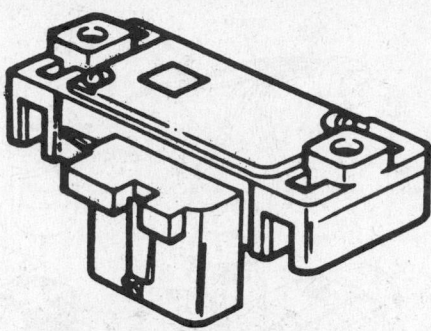

Manifold air pressure (MAP) sensor

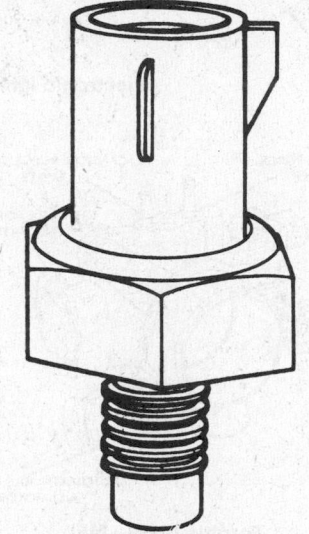

Detonation (knock) sensor

exhaust stream, monitors the oxygen content in the exhaust gas. The sensor is mounted in the exhaust manifold and is sometimes internally heated electrically for faster switching to the closed loop mode. The sensor produces a voltage proportional to the oxygen content which represents a lean or rich condition and transmits the information to the ECU. A low voltage condition indicates a lean mixture (high O_2 content) and a higher voltage indicates a rich mixture (low O_2 content). The ECU uses the information, along with other sensor data, and changes the fuel induction as required.

CYLINDER HEAD TEMPERATURE SENSOR

The cylinder head temperature sensor monitors the temperature of the cylinder head and transmits the information to the ECU. The sensor is located in the cylinder head and is a temperature sensitive resistive unit known as a thermistor.

VEHICLE SPEED SENSOR (VSS)

The VSS provides vehicle speed data to the ECU in the form of pulse signals. There are many different types of VSS, some using a reed switch installed in the speed meter unit and others using a optical type. In the optical type a light emitting diode (LED) is used to transmit light and photo diode receives the light. A shutter device, which is usually in-line with the speedometer cable, allows the LED light to reach the photo diode in vehicle speed related pulses. The reed switch type relies on a reed switch that opens and closes by way of a rotating magnet. The magnet rotates proportionally with the vehicle speed.

MANIFOLD AIR PRESSURE (MAP) SENSOR

The MAP sensor is a device that monitors manifold absolute pressure. The sensor is mounted remotely and senses vacuum through a connecting hose. The MAP sensor has a reference voltage from the ECU and transmits remaining voltage to the ECU to calculate engine load. The ECU uses this data along with other data to determine fuel demands.

DETONATION (KNOCK) SENSOR

The detonation sensor generates a

TPS consists of switches that open and close at different throttle positions (usually at idle and wide open throttle) and sends the information to the ECU. The variable resistor type receives a reference voltage from the ECU and responds back to the ECU with a proportional voltage directly related to the position of the throttle plate.

ENGINE COOLANT TEMPERATURE SENSOR

The coolant temperature sensor is located in the engine coolant passage, usually located in the intake manifold. The sensor is resistor based and changes resistance as coolant temperature changes. The sensor uses a reference voltage and the output voltage is sent to the ECU. The ECU calculates engine warm up and provides an optimum fuel enrichment when the engine is cold.

OXYGEN (O₂) SENSOR

The O_2 sensor, which is placed in the

AIR TEMPERATURE SENSOR

The air temperature sensor is located in the air stream of the air flow meter. The sensor supplies incoming air temperature information to the ECU. The ECU uses this data, along with other data, to regulate fuel injection rate.

THROTTLE POSITION SENSOR (TPS)

The TPS can be either a switch (or a combination of switches) or a variable resistor which is much more accurate in throttle position. The switch type

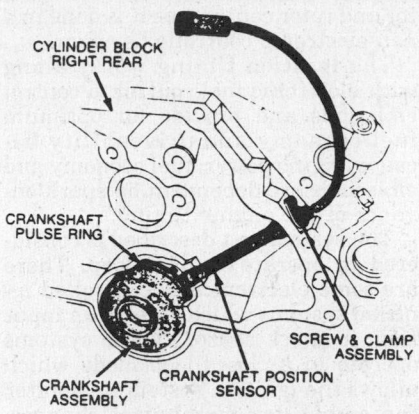

Crankshaft position sensor

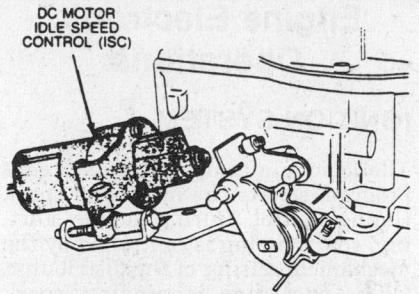

Idle speed control (ISC) motor

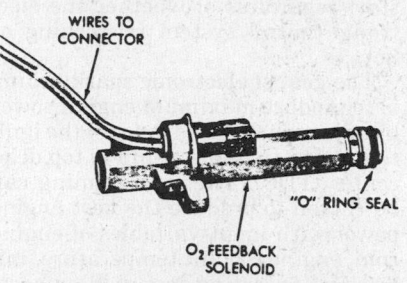

Air/fuel mixture solenoid

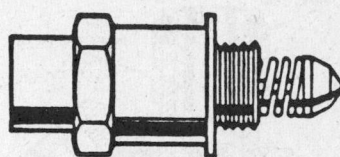

DUAL TAPER VALVE

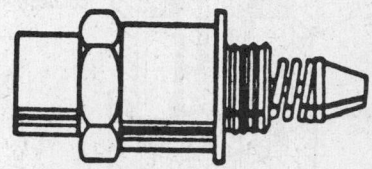

BLUNT PINTLE

Idle air control (IAC) valves

signal when pre-ignition (knock) occurs in one or more combustion chambers. The sensor is made of a material that is sensitive to oscillation that the engine knock produces and sends signals to the ECU. The ECU, in turn, delays the ignition signal which retards the ignition timing and continues to do this until the engine knock ceases.

CRANKSHAFT (REFERENCE MARK) SENSOR

The crankshaft sensor may be located at either the rear of the engine, at the flywheel or at the front of the engine, near the crankshaft pulley. The sensor detects crankshaft position in relation to top dead center and transmits the signals to ECU.

IDLE SPEED CONTROL (ISC) MOTOR

The ISC is sometimes included on a feedback carburetor system and mounted to the side of the carburetor. The motor driven ISC would maintain a steady idle by way of the ECU. When an added load is put on the engine (air conditioning or when vehicle is in drive) the ECU could increase the idle via the ISC by extending a plunger which would open the throttle valve.

AIR/FUEL MIXTURE SOLENOID

The air/fuel mixture solenoid on feedback carburetor operates in conjuntion with the fixed metering jets and/or the manually adjustable idle speed mixture screw. The ECU energizes and de-energizes the solenoid in the closed loop mode. The solenoid usually controls a fixed air bleed and/or fuel discharge port.

IDLE AIR CONTROL (IAC)

The IAC in a fuel injection system controls the air flow around the throttle plate by extending and retracting a bypass valve in the bypass port. The ECU controls the valve by sending voltage pulses called counts or steps to increase or decrease the bypass air flow, thus increasing and decreasing the idle speed.

FUEL INJECTOR

Throttle Body Type

The fuel injector is an electric solenoid controlled by the ECU. The ECU controls the injector by varying voltage pulse widths. When electrical current is supplied to the injector a spring loaded ball is lifted from its seat. This allows fuel to flow through spray orifices and deflects off the sharp edge of the injector nozzle. This action causes the fuel to form a 45° cone shaped spray pattern before entering the air stream in the throttle body.

Multiport Type

The fuel injector is an electric solenoid controlled by the ECU. The ECU controls the injector by varying voltage pulse widths. When electrical current is supplied to the injector, the armature and pintle move a short distance against a spring, opening a small orifice. Fuel is supplied to the inlet of the injector by the fuel pump, then passes through the injector,

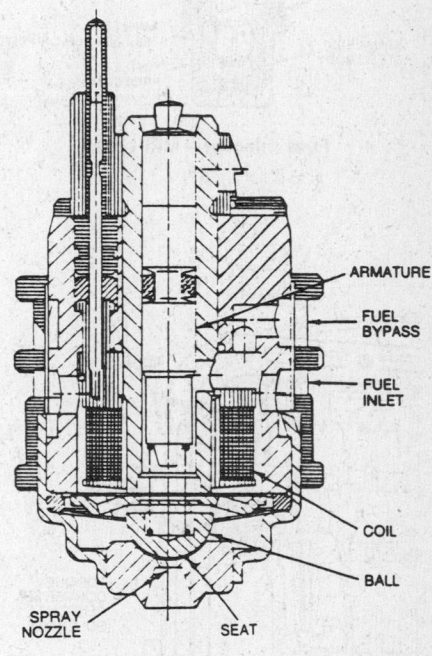

Fuel injector—TBI type

around the pintle and out the orifice. Since the fuel is under high pressure, a fine spray is developed in the shape of a hollow cone. The injector, through this spraying action, atomizes the fuel and distributes it into the air entering the combustion chamber.

TORQUE CONVERTER CLUTCH (TCC) SOLENOID

The TCC solenoid is used on some automatic transmission, which allows for better fuel economy. When certain engine and vehicle speeds have been met, the ECU energizes the solenoid. This allows transmission fluid to flow into passages in the torque converter,

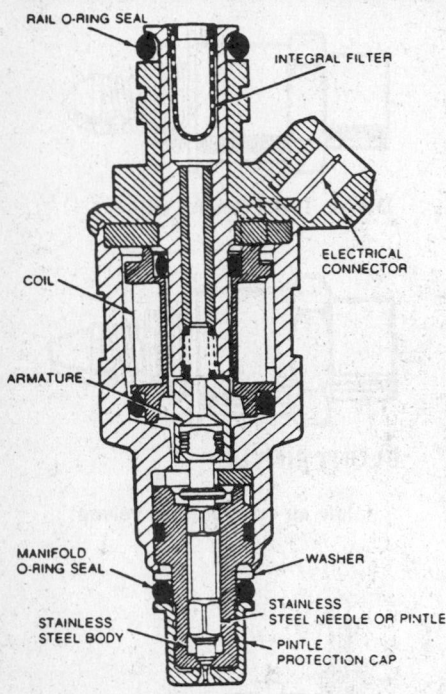

Fuel Injector – MFI type

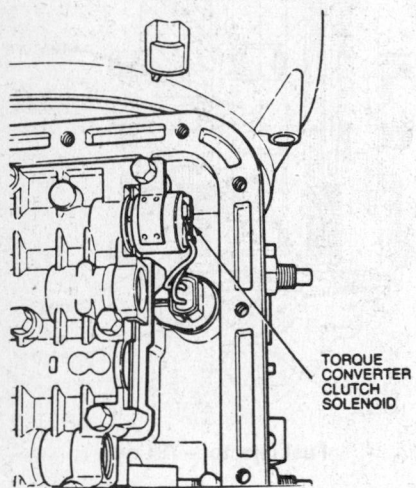

Torque converter clutch (TCC) solenoid

which causes the converter to lock up. This lockup is similar to a direct connection made possible in a manual transmission.

FUEL PUMP RELAY

The fuel is supplied under pressure, usually by an electric fuel pump. The ECU controls the fuel pump relay, which controls the fuel pump operation. When the ignition is switch ON, the fuel pump relay is energized and the fuel pump is activated. The pump primes the fuel system with fuel to a pre-determined pressure.

Engine Electronic Operations

IGNITION SYSTEM

The logic in a computerized system's program selects the method of spark timing control. During engine starting, spark timing is controlled by the mechanical setting of the distributor. Once the engine is running, spark timing is turned over to the ECU. This scheme ensures that the car will start regardless of whether the electronic control system is working or not.

The goal of electronic spark timing is to produce maximum engine power by adjustment the advance of the ignition firing in relationship to top dead center (TDC). The spark timing can be chosen to produce the best engine power with input variables of engine rpm, engine coolant temperature, initial and operating manifold or barometric pressure.

The total spark advance is determined by computing the information received from the various engine sensors which affect spark timing. The processor will then adjust the timing according to information that has been calibrated in it. The processor has programmed into it specific information on:

Warm-Up Spark Advance – this is used when the engine is cold, since a greater amount of advance is required while the engine warms up.

Special Spark Advance – to improve fuel economy during steady driving conditions.

Spark Advance Due to Barometric Pressure – this is used when barometric pressure exceeds a preset calibrated amount.

All of this information is then added together and the initial mechanical advance (if equipped) is subtracted to determine the final spark advance.

The processor receives a timing pulse from a sensor which indicates crankshaft position for top dead center and engines rpm. The processor makes a decision based upon this information and the information that was calibrated into it. at that time, the computer sends a pulse to the ignition actuator circuit which opens the ignition coil primary circuit to generate a secondary voltage pulse to fire the spark plugs. In some cases, the circuitry to open the primary of the coil may be in the computerized controller. The spark selection is performed mechanically by the distribu-

tor and rotor contacts as it is done in a non-electronic controlled system.

The ignition timing works along with electronic fuel control to control emissions and provide for optimum fuel economy and driveability because engine power, fuel economy and emissions are dependent on spark advance of the engine timing.

The system just described is considered to operate in open-loop. There are some electronically controlled ignition systems which receive an input from a knock sensor. These systems operate in a closed-loop mode which allows the ignition system to monitor the engine for mechanical changes, such as engine knock.

Engine knock is a condition where the air/fuel mixture in the cylinder does not burn normally. the pressure rise during this burning is so rapid compared to normal combustion that it is accompanied by an audible "knock".

Through some low level knock is acceptable, it is important to avoid excessive knock. To control engine knock, a knock sensor is installed in the engine or intake manifold. This helps to detect excessive engine knock.

The knock sensor is a tuned accelerometer and produces an output voltage depending on the amount of engine vibration occurring in a certain frequency band. When the processor receives a signal from the knock sensor, it retards the spark advance until the knocking stops and then starts increasing it again. This cycle is repeated as long as engine knock occurs.

FUEL CONTROL

In order for the processor to control fuel, it requires a sensor or sensors to monitor the state of the engine, and one or more actuators to do the actual controlling. The sensors measure: exhaust gas oxygen, manifold or barometric absolute pressure, engine rpm and speed, inlet air and coolant temperatures. Actuators are energized to control the air/fuel ratio.

The primary purpose of this control system is to maintain air/fuel ratio at or near 14.7:1 ratio. This is accomplished in two modes (during normal engine operation) open and closed loop. The electronic fuel control system can operate in closed loop only when certain conditions are satisfied. Open loop mode is employed whenever these conditions are not satisfied. However, for either mode, the exhaust emissions will satisfy federal requirements if the average air/fuel ratio is held within the tolerance limits.

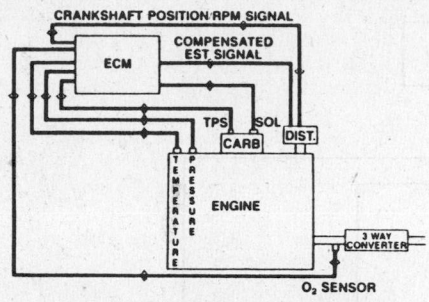

Electronic spark timing system

In addition to open and closed loop control modes, a practical fuel control system has other operating modes depending on engine conditions. These handle such conditions as starting, rapid acceleration or heavy load, sudden deceleration, idling, etc.

An automotive engine has various operating modes as the operating conditions change. Preprogrammed into the processor, control logic determines the operating mode from the engine conditions that exist. From these engine conditions, the system determines which operating modes are to be performed.

There are seven different engine operating modes which affect fuel control: engine crank, engine warmup, open loop, closed loop control, hard acceleration, deceleration and idle. The program for mode control logic determines the engine operating mode by reading various sensors.

When the ignition switch is initially switched on, the mode control logic automatically selects an engine-start control scheme which provides the low air/fuel ratio required for starting the engine. Once the engine rpm rises above the cranking value, the controller identifies the engine-started mode and passes control to the program for the engine warm-up mode. This operating mode keeps the air/fuel ratio low to prevent engine stall during cool weather until engine coolant temperature rises above a preset value.

When the coolant temperature rises, the mode control logic directs the system to operate in the open loop control mode until a certain time has elapsed and the exhaust gas sensor warms up enough to provide accurate readings. This condition is detected by monitoring the exhaust gas sensor's output for voltage readings above a certain minimum air/fuel mixture voltage set point. when the sensor has indicated a rich mixture a certain number of times (depending on calibration), and after the engine has been in open loop for a specific time, the control mode logic selects the closed loop mode for the system.

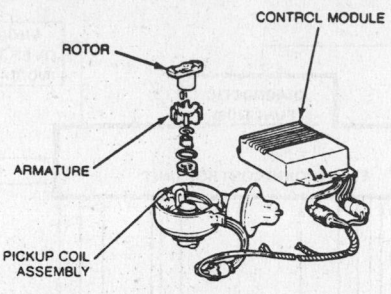

Solid state ignition system

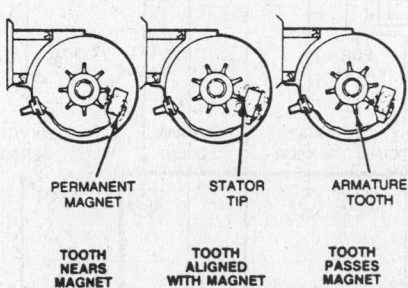

Distributor pick-up coil and armature assembly

The engine remains in the closed loop mode until either the exhaust gas sensor cools and fails to switch (from rich to lean) for a certain length of time, or a hard acceleration or deceleration occurs. If the sensor cools, the control mode logic selects the open loop mode again.

During hard acceleration of heavy engine loads, the control mode logic chooses a scheme which provides a rich air/fuel mixture for the duration of the acceleration or heavy load. This scheme provides maximum power, but poor emissions control and poor fuel economy. After the need for enrichment has passed, control is returned to either open or closed loop depending on the control mode logic selection conditions that exist at that time.

During periods of deceleration, the air/fuel ratio is increased to reduce emissions of HC and CO due to unburned fuel. When idle conditions are present, control mode logic passes system control to the idle speed control mode. In this mode, the engine speed is controlled to reduce engine roughness and stalling which might occur because the idle load has changed due to air conditioner compressor operation, alternator operation, or gearshift positioning from PARK or NEUTRAL to DRIVE.

Engine Crank

While the engine is being cranked,

the fuel control system must provide an intake air/fuel ratio anywhere from 2:1 to 12:1, depending on engine temperature. Low temperatures affect the carburetor's ability to atomize or mix the incoming air and fuel. At low temperature, the fuel tends to form into large droplets. The larger fuel droplets tend to increase the apparent air/fuel ratio because the amount of usable fuel in the air is reduced, therefore, the system must provide a decreased air/fuel ratio to provide the engine with a more combustible air/fuel mixture. The engine temperature is read by the processor through an analog to digital converter from a temperature sensor in the engine water coolant passage. The processor's calibration determines what the proper air/fuel ratio must be at that temperature. The air/fuel is determined and controlled as in the open loop mode.

Engine Warm-up

While the engine is warming up, an enriched air/fuel ratio is still needed to keep it running smoothly, but the required air/fuel ratio changes as the temperature increases. Therefore, the fuel control system will stay in the open loop mode, but the air/fuel ratio commands continue to be altered due to the temperature changes. The emphasis in this control mode is on rapid and smooth engine warm-up. Fuel economy and emission control are still a secondary concern. The controller determines the warm-up time period based on the coolant temperature when the warm-up mode was selected. Naturally, an initially cold engine requires a longer warm-up time than a warm engine. The time allowed by the controller timer is chosen according to the calibration of the processor.

OPEN LOOP CONTROL

Open loop fuel control is used when the engine has not reached a preset operating condition. This condition is sensed by various sensors located in and around the engine, and include engine coolant temperature, air charge temperature, engine time on, etc. After all these preset conditions are met, the system will go into closed loop. During certain operating conditions, such as a wide open throttle condition the system will go back into open loop.

CLOSED LOOP CONTROL

Closed loop fuel control is selected when the engine is warm and the exhaust gas oxygen sensor exceeds its minimum operating temperature. The intake air/fuel ratio is controlled in a closed loop by measuring the ex-

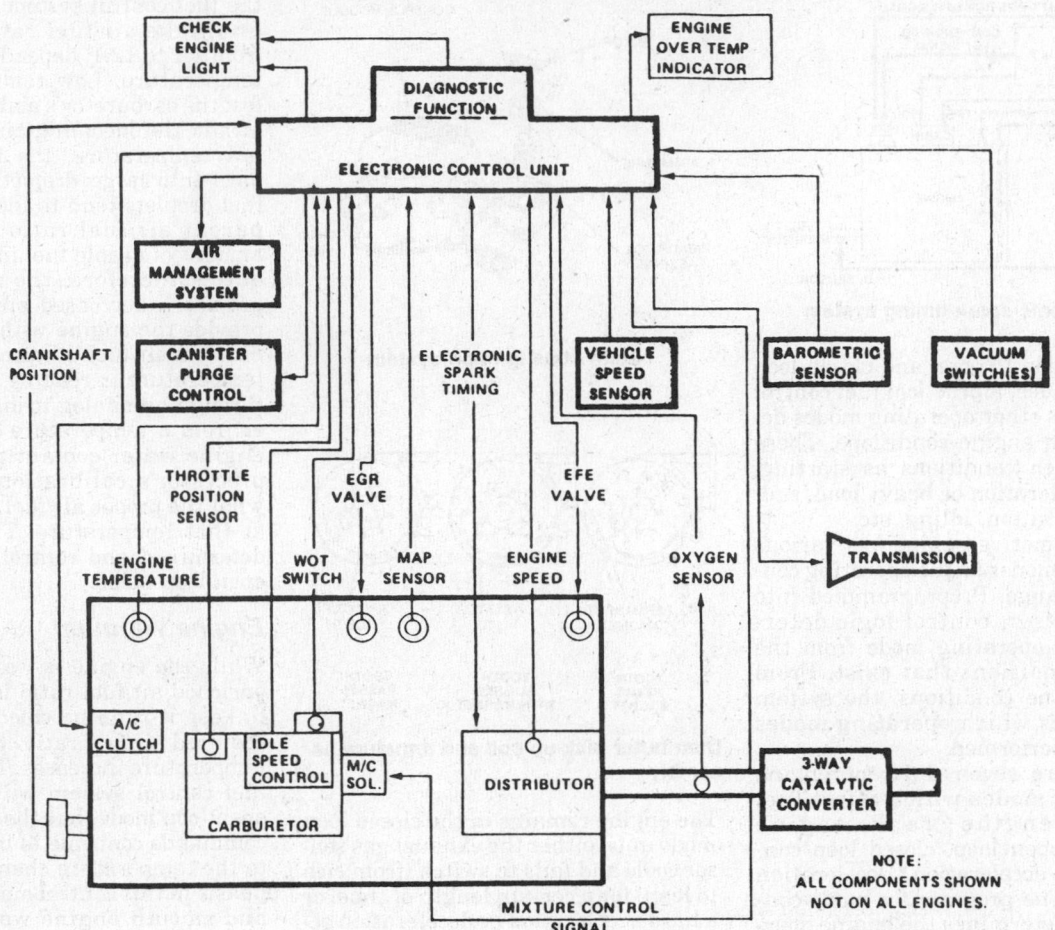

Typical electronic feedback carburetor system

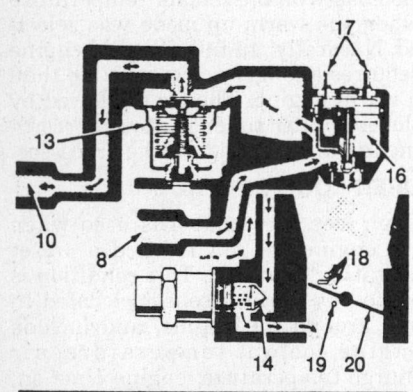

8 FUEL SUPPLY
10 FUEL RETURN
13 PRESSURE REGULATOR (PART OF FUEL METER COVER)
14 IDLE AIR CONTROL (IAC) VALVE (SHOWN OPEN)
16 FUEL INJECTOR
17 FUEL INJECTOR TERMINALS
18 PORTED VACUUM SOURCES*
19 MANIFOLD VACUUM SOURCE*
20 THROTTLE VALVE

*May Be Different on some Models.

Throttle body injection air and fuel flow

haust gas at the exhaust manifold and altering the input fuel flow rate or the air entering the main metering systems (depending on the type of fuel system used).

ACCELERATION ENRICHMENT (OPEN LOOP)

During periods of heavy engine load, such as wide open acceleration, fuel control is adjusted to provide an enriched ratio to maximize engine power while neglecting fuel economy and emission.

The computer detects this condition by reading the throttle position sensor voltage or the MAP sensor. Low intake manifold vacuum or throttle position corresponds to heavy engine loads. The fuel control system controller responds by increasing the amount of fuel to enter the intake manifold or to decrease the amount of air n the main metering system. This enrichment allows the engine to operate with a power greater than that allowed when emissions and fuel economy are controlled within specifications.

DECELERATION AND IDLE SPEED CONTROL (OPEN LOOP)

During periods of light engine load and high rpm, such as during closed throttle deceleration, coasting or engine idle, the engine requires a very lean air/fuel ratio to reduce excess emissions of HC and CO. Deceleration is indicated by a sudden increase in manifold vacuum and throttle position, indicating a closed throttle. When these conditions are detected by the processor, it computes a change in the amount of fuel required or amount of air entering the main or idle speed passages (depending on type of fuel system used). On certain engine engine applications which electronic fuel injection, the fuel may even be turned completely off during closed throttle deceleration.

Idle speed control is used to prevent engine stall during idle. The goal is to allow the engine to idle at as low an rpm as possible, yet keep the engine from running rough and stalling when power takeoff accessories such as air conditioning compressors are turned on.

Engine Rebuilding

This section describes, in detail, the procedures involved in rebuilding a typical engine. The procedures are basically identical to those used in rebuilding engines of nearly all design and configurations.

The section is divided into two parts. The first, Cylinder Head Reconditioning, assumes that the cylinder head is removed from the engine, all manifolds are removed, and the cylinder head is on a workbench. The camshaft should be removed from overhead cam cylinder heads. The second section, Cylinder Block Reconditioning, covers the block, pistons, connecting rods and crankshaft. It is assumed that the engine is mounted on a work stand, and the cylinder head and all accessories are removed.

Procedures are identified as follows:

Unmarked—Basic procedures that must be performed in order to successfully complete the rebuilding process.

Starred(*)—Procedures that should be performed to ensure maximum performance and engine life.

Double starred (**)—Procedures that may be performed to increase engine performance and reliability.

In many cases, a choice of methods is also provided. Methods are identified in the same manner as procedures. The choice of method for a procedure is at the discretion of the user.

The tools required for the basic rebuilding procedure should, with minor exceptions, be those included in a mechanic's tool kit. An accurate torque wrench, and a dial indicator (reading in thousandths) mounted on a universal base should be available. Special tools, where required, all are readily available from the major tool suppliers. The services of a competent automotive machine shop must also be readily available.

When assembling the engine, any parts that will be in frictional contact must be pre-lubricated, to provide protection on initial start-up. Any product specifically formulated for this purpose may be used. NOTE: *Do not use engine oil.* Where semi-permanent (locked but removable) installation of bolts or nuts is desired, threads should be cleaned and coated with Loctite® or a similar product (non-hardening).

Aluminum has become increasingly popular for use in engines, due to its low weight and excellent heat transfer characteristics. The following precautions must be observed when handling aluminum engine parts:

—Never hot-tank aluminum parts.

—Remove all aluminum parts (identification tags, etc.) from engine parts before hot-tanking (otherwise they will be removed during the process).

—Always coat threads lightly with engine oil or anti-seize compounds before installation, to prevent seizure.

—Never over-torque bolts or spark plugs in aluminum threads. Should stripping occur, threads can be restored using any of a number of thread repair kits available (see next section).

Magnaflux and Zyglo are inspection techniques used to locate material flaws, such as stress cracks. Magnafluxing coats the part with fine magnetic particles, and subjects the part to a magnetic field. Cracks cause breaks in the magnetic field, which are outlined by the particles. Since Magnaflux is a magnetic process, it is applicable only to ferrous materials. The Zyglo process coats the material with a fluorescent dye penetrant, and then subjects it to blacklight inspection, under which cracks glow brightly. Parts made of any material may be tested using Zyglo. While Magnaflux and Zyglo are excellent for general inspection, and locating hidden defects, specific checks of suspected cracks may be made at lower cost and more readily using spot check dye. The dye is sprayed onto the suspected area, wiped off, and the area is then sprayed with a developer. Cracks then will show up brightly. Spot check dyes will only indicate surface cracks; therefore, structural cracks below the surface may escape detection. When questionable, the part should be tested using Magnaflux or Zyglo.

REPAIRING DAMAGED THREADS

Several methods of repairing damaged threads are available. Heli-Coil® (shown here), Keenserts® and Microdot® are among the most widely used. All involve basically the same principle—drilling out stripped threads, tapping the hole and installing a pre-wound insert— making welding, plugging and oversize fasteners unnecessary.

Two types of thread repair inserts are usually supplied—a standard type for most Inch Coarse, Inch Fine, Metric Coarse and Metric Fine thread sizes and a spark plug type to fit most spark plug port sizes. Consult the individual manufacturer's catalog to determine exact applications. Typical thread repair kits will contain a selection of prewound threaded inserts, a tap (corresponding to the outside diameter threads of the insert) and an installation tool. Most manufacturers also supply blister-packed thread repair inserts separately and a master kit with a variety of taps and inserts plus installation tools.

Before effecting a repair to a threaded hole, remove any snapped, broken or damaged bolts or studs. Penetrating oil can be used to free frozen threads; the offending item can be removed with locking pliers or with a screw or stud extractor. After the hole is clear, the thread can be repaired as follows.

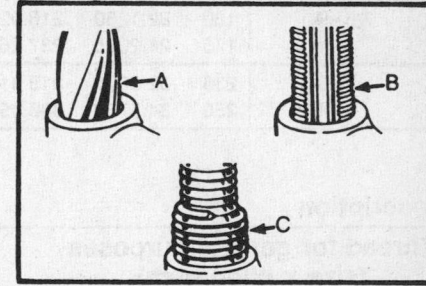

A. Drill out the damaged threads with the specified drill. Drill completely through the hole or to the bottom of a blind hole.

B. With the tap supplied tap the hole to receive the threaded insert. Keep the tap well oiled and back it out frequently to avoid clogging the threads.

C. Screw the threaded insert onto the installation tool until the tang engages the slot. Screw the insert into the tapped hole until it is ¼–½ turn below the top surface. After installation, break the tang off with a hammer and punch.

32 ENGINE REBUILDING

STANDARD TORQUE SPECIFICATIONS AND CAPSCREW MARKINGS

Newton-Meter has been designated as the world standard for measuring torque and will gradually replace the foot-pound and kilogram-meter torque measuring standard. Torquing tools are still being manufactured with foot-pounds and kilogram-meter scales, along with the new Newton-Meter standard. To assist the repairman, foot-pounds, kilogram-meter and Newton-Meter are listed in the following charts, and should be followed as applicable.

U.S. BOLTS

SAE Grade Number	1 or 2			5			6 or 7			8		
Capscrew Head Markings Manufacturer's marks may vary. Three-line markings on heads below indicate SAE Grade 5. 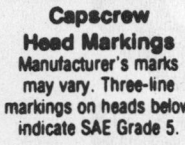												
Usage	Used Frequently			Used Frequently			Used at Times			Used at Times		
Quality of Material	Indeterminate			Minimum Commercial			Medium Commercial			Best Commercial		
Capacity Body Size	Torque			Torque			Torque			Torque		
(inches)–(thread)	Ft-Lb	kgm	Nm	Ft-Lb	kgm	Nm	Ft-Lb	kgm	Nm	Ft-Lb	kgm	Nm
1/4–20	5	0.6915	6.7791	8	1.1064	10.8465	10	1.3630	13.5582	12	1.6596	16.2698
–28	6	0.8298	8.1349	10	1.3830	13.5582				14	1.9362	18.9815
5/16–18	11	1.5213	14.9140	17	2.3511	23.0489	19	2.6277	25.7605	24	3.3192	32.5396
–24	13	1.7979	17.6256	19	2.6277	25.7605				27	3.7341	36.6071
3/8–16	18	2.4894	24.4047	31	4.2873	42.0304	34	4.7022	46.0978	44	6.0852	59.6560
–24	20	2.7660	27.1164	35	4.8405	47.4536				49	6.7767	66.4351
7/16–14	28	3.8132	37.9629	49	6.7767	66.4351	55	7.6065	74.5700	70	9.6810	94.9073
–20	30	4.1490	40.6745	55	7.6065	74.5700				78	10.7874	105.7538
1/2–13	39	5.3937	52.8769	75	10.3725	101.6863	85	11.7555	115.2445	105	14.5215	142.3609
–20	41	5.6703	55.5885	85	11.7555	115.2445				120	16.5860	162.6960
9/16–12	51	7.0533	69.1467	110	15.2130	149.1380	120	16.5960	162.6960	155	21.4365	210.1490
–18	55	7.6065	74.5700	120	16.5960	162.6960				170	23.5110	230.4860
5/8–11	83	11.4789	112.5329	150	20.7450	203.3700	167	23.0961	226.4186	210	29.0430	284.7180
–18	95	13.1385	128.8027	170	23.5110	230.4860				240	33.1920	325.3920
3/4–10	105	14.5215	142.3609	270	37.3410	366.0660	280	38.7240	379.6240	375	51.8625	508.4250
–16	115	15.9045	155.9170	295	40.7985	399.9610				420	58.0860	568.4360
7/8–9	160	22.1280	216.9280	395	54.6285	535.5410	440	60.8520	596.5520	605	83.6715	820.2590
–14	175	24.2025	237.2650	435	60.1605	589.7730				675	93.3525	915.1650
1–8	236	32.5005	318.6130	590	81.5970	799.9220	660	91.2780	894.8280	910	125.8530	1233.7780
–14	250	34.5750	338.9500	660	91.2780	849.8280				990	136.9170	1342.2420

METRIC BOLTS

Description	Torque ft-lbs. (Nm)			
Thread for general purposes (size x pitch (mm))	Head Mark 4		Head Mark 7	
6 x 1.0	2.2 to 2.9	(3.0 to 3.9)	3.6 to 5.8	(4.9 to 7.8)
8 x 1.25	5.8 to 8.7	(7.9 to 12)	9.4 to 14	(13 to 19)
10 x 1.25	12 to 17	(16 to 23)	20 to 29	(27 to 39)
12 x 1.25	21 to 32	(29 to 43)	35 to 53	(47 to 72)
14 x 1.5	35 to 52	(48 to 70)	57 to 85	(77 to 110)
16 x 1.5	51 to 77	(67 to 100)	90 to 120	(130 to 160)
18 x 1.5	74 tc 110	(100 to 150)	130 to 170	(180 to 230)
20 x 1.5	110 to 140	(150 to 190)	190 to 240	(160 to 320)
22 x 1.5	150 to 190	(200 to 260)	250 to 320	(340 to 430)
24 x 1.5	190 to 240	(260 to 320)	310 to 410	(420 to 550)

CAUTION: Bolts threaded into aluminum require much less torque

CYLINDER HEAD RECONDITIONING

Procedure	Method
Identify the valves:	Invert the cylinder head, and number the valve faces front to rear, using a permanent felt-tip marker.
Remove the rocker arms (OHV engines only):	Remove the rocker arms with shaft(s) or balls and nuts. Wire the sets of rockers, balls and nuts together, and identify according to the corresponding valve.
Remove the camshaft (OHC engines only):	See the engine service procedures earlier in this book for details concerning specific engines.
Remove the valves and springs:	Using an appropriate valve spring compressor (depending on the configuration of the cylinder head), compress the valve springs. Lift out the keepers with needlenose pliers, release the compressor, and remove the valve, spring, and spring retainer.
Remove glow plugs and fuel injectors (Diesel engines only):	Label and remove all fuel injectors and glow plugs from the head. Glow plugs unscrew. See the appropriate car section for injector removal. Inspect glow plugs for bulges, cracks or signs of melting. Clean injector tips with a steel brush, then inspect for evidence of melting.
**Remove pre-combustion chamber inserts (Diesel engines only):	**Remove the pre-combustion chambers using a hammer and a thin, blunt brass drift, inserted through the injector hole (or glow plug hole, whichever is more convenient). If chamber is to be reused, carefully remove all carbon from it. NOTE: *Remove chamber only if being replaced, if a glow plug tip has broken off and must be removed, or if chamber is obviously damaged or loose.*

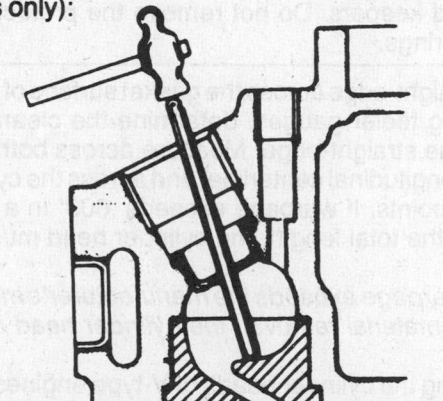

Removing pre-combustion chamber with a drift (© G.M. Corp.)

Check the valve stem-to-guide clearance:	Clean the valve stem with lacquer thinner or a similar solvent to remove all gum and varnish. Clean the valve guides using solvent and an expanding wire-type valve guide cleaner. Mount a dial indicator so that the stem is at 90° to the valve stem, as close to the valve guide as possible. Move the valve off its seat, and measure the valve guide-to-stem clearance by rocking the stem back and forth to actuate the dial indicator. Measure the valve stems using a micrometer, and compare to specifications, to determine whether stem or guide wear is responsible for excessive clearance.

DIAL INDICATOR

VALVE STEM

Checking the valve stem-to-guide clearance

CYLINDER HEAD RECONDITIONING

Procedure	Method

De-carbon the cylinder head and valves:

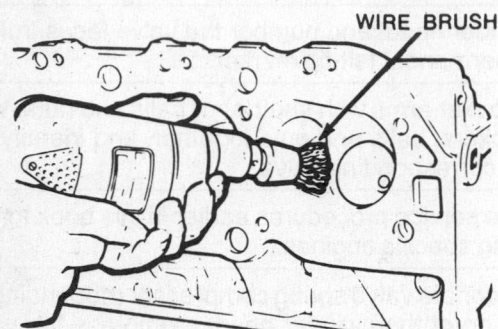

WIRE BRUSH

Removing carbon from the cylinder head

Chip carbon away from the valve heads, combustion chambers, and ports, using a chisel made of hardwood. Remove the remaining deposits with a stiff wire brush.
NOTE: *Ensure that the deposits are actually removed, rather than burnished.*

Hot-tank the cylinder head (cast iron heads only):
CAUTION: *Do not hot-tank aluminum parts.*

Have the cylinder head hot-tanked to remove grease, corrosion, and scale from the water passages.
NOTE: *In the case of overhead cam cylinder heads, consult the operator to determine whether the camshaft bearings will be damaged by the caustic solution.*

Degrease the remaining cylinder head parts:

Using solvent (i.e., Gunk), clean the rockers, rocker shaft(s) (where applicable), rocker balls and nuts, springs, spring retainers, and keepers. Do not remove the protective coating from the springs.

Check the cylinder head for warpage:

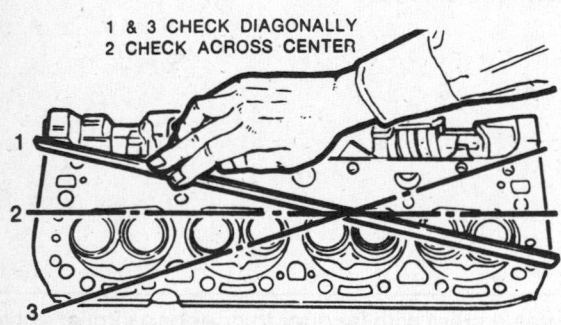

1 & 3 CHECK DIAGONALLY
2 CHECK ACROSS CENTER

Checking cylinder head for warpage

Place a straight-edge across the gasket surface of the cylinder head. Using feeler gauges, determine the clearance at the center of the straight-edge. Measure across both diagonals, along the longitudinal centerline, and across the cylinder head at several points. If warpage exceeds .003′ in a 6′ span, or .006′ over the total length, the cylinder head must be resurfaced.
NOTE: *If warpage exceeds the manufacturer's maximum tolerance for material removal, the cylinder head must be replaced.*
When milling the cylinder heads of V-type engines, the intake manifold mounting position is altered, and must be corrected by milling the manifold flange a proportionate amount.

**Porting and gasket matching:

**Coat the manifold flanges of the cylinder head with Prussian blue dye. Glue intake and exhaust gaskets to the cylinder head in their installed position using rubber cement and scribe the outline of the ports on the manifold flanges. Remove the gaskets. Using a small cutter in a hand-held power tool gradually taper the walls of the port out to the scribed outline of the gasket. Further enlargement of the ports should include the removal of sharp edges and radiusing of sharp corners. Do not alter the valve guides.
NOTE: *The most efficient port configuration is determined only by extensive testing. Therefore, it is best to consult someone experienced with the head in question to determine the optimum alterations.*

Procedure	Method

*Knurling the valve guides:

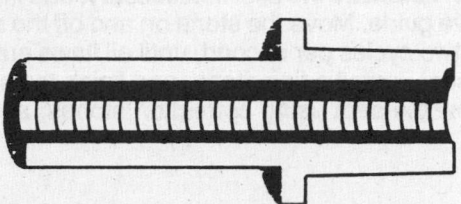

Cut-away view of a knurled valve guide

*Valve guides which are not excessively worn or distorted may, in some cases, be knurled rather than replaced. Knurling is a process in which metal is displaced and raised, thereby reducing clearance. Knurling also provides excellent oil control. The possibility of knurling rather than replacing valve guides should be discussed with a machinist.

Replacing the valve guides:
NOTE: *Valve guides should only be replaced if damaged or if an oversize valve stem is not available.*

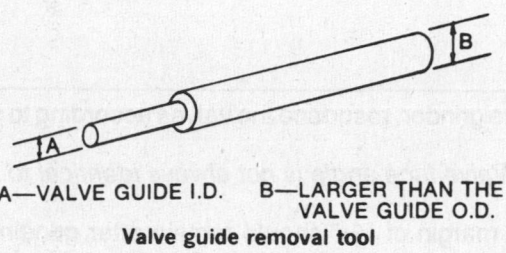

A—VALVE GUIDE I.D. B—LARGER THAN THE VALVE GUIDE O.D.
Valve guide removal tool

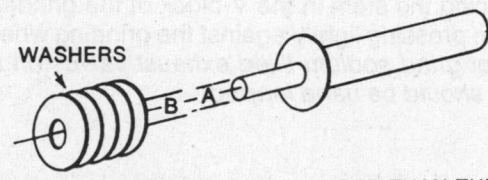

WASHERS

A—VALVE GUIDE I.D. B—LARGER THAN THE VALVE GUIDE O.D.

Valve guide installation tool (with washers used for installation)

Depending on the type of cylinder head, valve guides may be pressed, hammered, or shrunk in. In cases where the guides are shrunk into the head, replacement should be left to an equipped machine shop. In other cases, the guides are replaced as follows: Press or tap the valve guides out of the head using a stepped drift (see illustration). Determine the height above the boss that the guide must extend, and obtain a stack of washers, their I.D. similar to the guide's O.D., of that height. Place the stack of washers on the guide, and insert the guide into the boss.
NOTE: *Valve guides are often tapered or beveled for installation.*
Using the stepped installation tool (see illustration), press or tap the guides into position. Ream the guides according to the size of the valve stem.

Replacing valve seat inserts:

Replacement of valve seat inserts which are worn beyond resurfacing or broken, if feasible, must be done by a machine shop.

Resurfacing the valve seats using reamers:

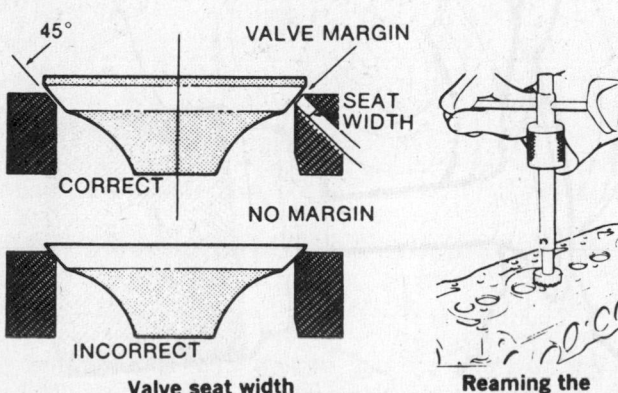

45° VALVE MARGIN SEAT WIDTH
CORRECT
NO MARGIN
INCORRECT
Valve seat width and centering

Reaming the valve seat

Select a reamer of the correct seat angle, slightly larger than the diameter of the valve seat, and assemble it with a pilot of the correct size. Install the pilot into the valve guide, and using steady pressure, turn the reamer clockwise.
CAUTION: *Do not turn the reamer counterclockwise.*
Remove only as much material as necessary to clean the seat. Check the concentricity of the seat (see below). If the dye method is not used, coat the valve face with Prussian blue dye, install and rotate it on the valve seat. Using the dye marked area as a centering guide, center and narrow the valve seat to specifications with correction cutters.
NOTE: *When no specifications are available, minimum seat width for exhaust valves should be 5/64", intake valves 1/16".*
After making correction cuts, check the position of the valve seat on the valve face using Prussian blue dye.
NOTE: *Do not cut induction hardened seats; they must be ground.*

CYLINDER HEAD RECONDITIONING

Procedure	Method

*Resurfacing the valve seats using a grinder:

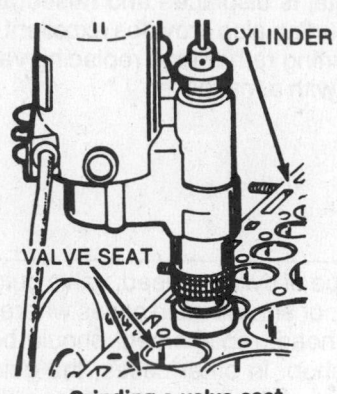

CYLINDER

VALVE SEAT

Grinding a valve seat

*Select a pilot of the correct size, and a coarse stone of the correct seat angle. Lubricate the pilot if necessary, and install the tool in the valve guide. Move the stone on and off the seat at approximately two cycles per second, until all flaws are removed from the seat. Install a fine stone, and finish the seat. Center and narrow the seat using correction stones, as described above.

Resurfacing (grinding) the valve face:

FOR DIMENSIONS, REFER TO SPECIFICATIONS

CHECK FOR BENT STEM

DIAMETER

VALVE FACE ANGLE

1/32" MINIMUM

THIS LINE PARALLEL WITH VALVE HEAD

Critical valve dimensions

Using a valve grinder, resurface the valves according to specifications.
CAUTION: *Valve face angle is not always identical to valve seat angle.*
A minimum margin of 1/32" should remain after grinding the valve. The valve stem top should also be squared and resurfaced, by placing the stem in the V-block of the grinder, and turning it while pressing lightly against the grinding wheel.
NOTE: *Do not grind sodium filled exhaust valves on a machine. These should be hand lapped.*

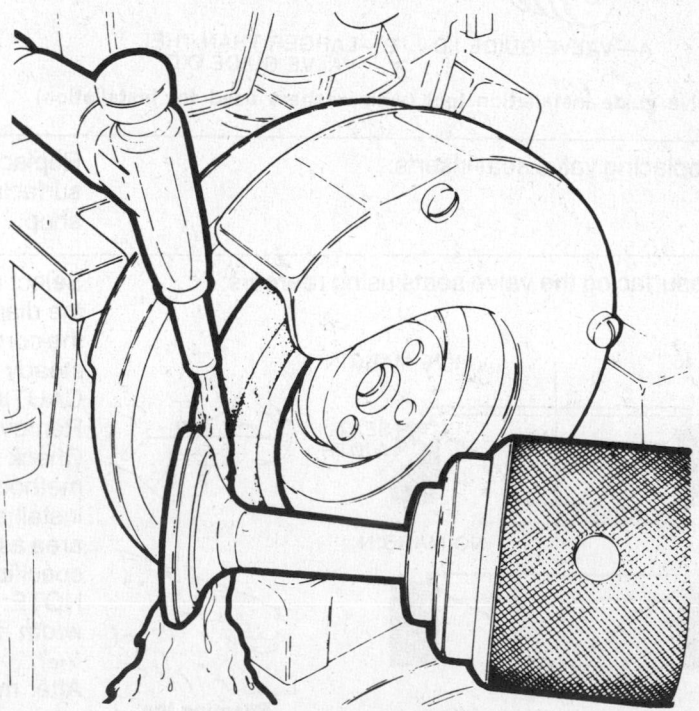

Valve grinding by machine

CYLINDER HEAD RECONDITIONING

Procedure	Method

Checking the valve seat concentricity:

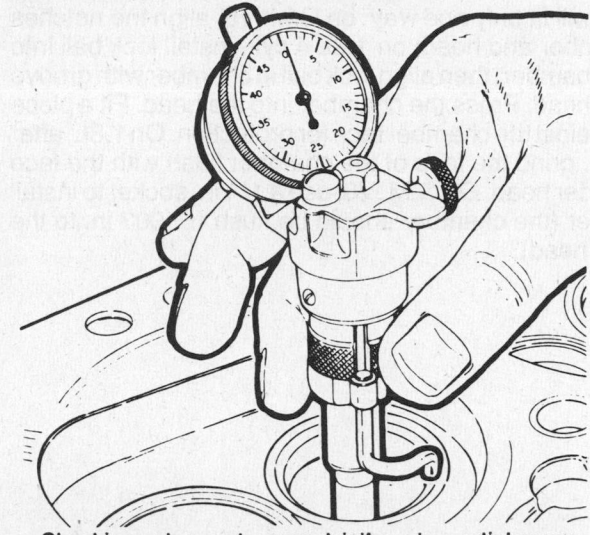

Checking valve seat concentricity using a dial gauge

Coat the valve face with Prussian blue dye, install the valve, and rotate it on the valve seat. If the entire seat becomes coated, and the valve is known to be concentric, the seat is concentric.

*Install the dial gauge pilot into the guide, and rest the arm on the valve seat. Zero the gauge, and rotate the arm around the seat. Run-out should not exceed .002″.

---

*Lapping the valves:
NOTE: *Valve lapping is done to ensure efficient sealing of resurfaced valves and seats.*

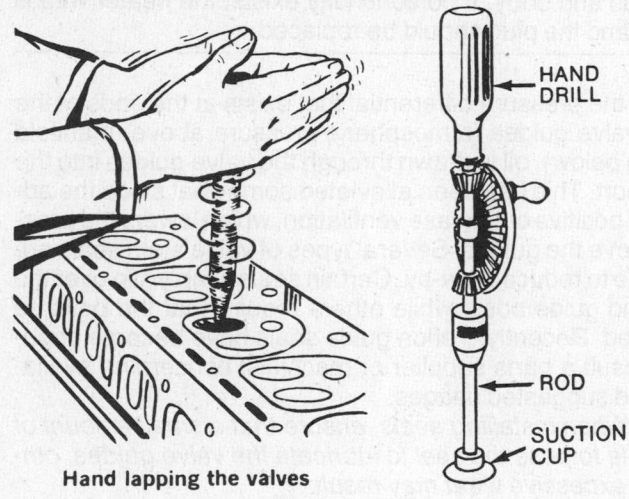

Hand lapping the valves

HAND DRILL

ROD

SUCTION CUP

Home made mechanical valve lapping tool

*Invert the cylinder head, lightly lubricate the valve stems, and install the valves in the head as numbered. Coat valve seats with fine grinding compound, and attach the lapping tool suction cup to a valve head.
NOTE: *Moisten the suction cup.*
Rotate the tool between the palms, changing position and lifting the tool often to prevent grooving. Lap the valve until a smooth, polished seat is evident. Remove the valve and tool, and rinse away all traces of grinding compound.
**Fasten a suction cup to a piece of drill rod, and mount the rod in a hand drill. Proceed as above, using the hand drill as a lapping tool.
CAUTION: *Due to the higher speeds involved when using the hand drill, care must be exercised to avoid grooving the seat.* Lift the tool and change direction of rotation often.

Check the valve springs:

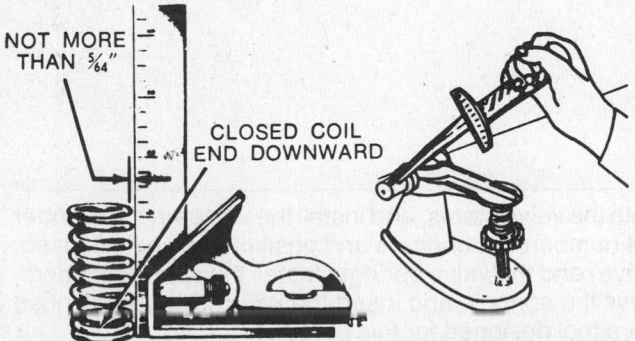

NOT MORE THAN ⁵⁄₆₄″

CLOSED COIL END DOWNWARD

Checking valve spring free length and squareness

Measuring valve spring test pressure

Place the spring on a flat surface next to a square. Measure the height of the spring, and rotate it against the edge of the square to measure distortion. If spring height varies (by comparison) by more than $1/16″$ or if distortion exceeds $1/16″$, replace the spring.
**In addition to evaluating the spring as above, test the spring pressure at the installed and compressed (installed height minus valve lift) height using a valve spring tester. Springs used on small displacement engines (up to 3 liters) should be ∓ 1 lb. of all other springs in either position. A tolerance of ∓ 5 lbs. is permissible on larger engines.

CYLINDER HEAD RECONDITIONING

Procedure	Method

Install pre-combustion chambers (Diesel engines only)

Pre-combustion chambers are press-fit into the head. The chambers will fit only one way: on G.M. V8, align the notches in the chamber and head; on 1.8L 4 cyl., install lock ball into groove in chamber, then align lock ball in chamber with groove in cylinder head. Press the chamber into the head. Fit a piece of metal against the chamber face for protection. On 1.8L, after installation, grind the face of the chamber flush with the face of the cylinder head. On G.M. V8, use a 1¼ in. socket to install the chamber (the chamber should be flush ± .003 in. to the face of the head).

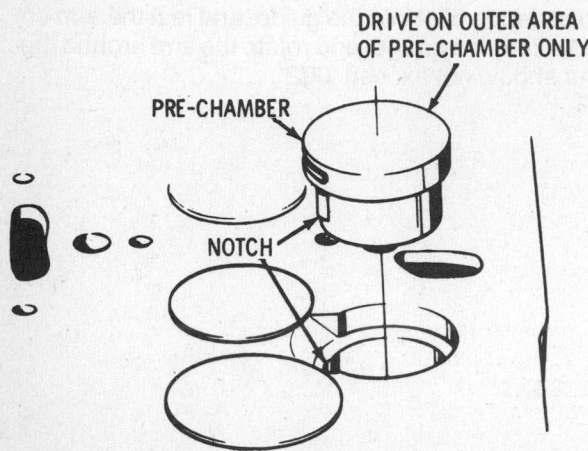

DRIVE ON OUTER AREA OF PRE-CHAMBER ONLY

PRE-CHAMBER

NOTCH

Align the notches to install the pre-combustion chamber (© G.M. Corp.)

Install fuel injectors and glow plugs (Diesel engines)

Before installing glow plugs, check for continuity across plug terminals and body. If no continuity exists, the heater wire is broken and the plug should be replaced.

---

*Install valve stem seals:

*Due to the pressure differential that exists at the ends of the intake valve guides (atmospheric pressure above, manifold vacuum below), oil is drawn through the valve guides into the intake port. This has been alleviated somewhat since the addition of positive crankcase ventilation, which lowers the pressure above the guides. Several types of valve stem seals are available to reduce blow-by. Certain seals simply slip over the stem and guide boss, while others require that the boss be machined. Recently, Teflon guide seals have become popular. Consult a parts supplier or machinist concerning availability and suggested usages.

NOTE: *When installing seals, ensure that a small amount of oil is able to pass the seal to lubricate the valve guides; otherwise, excessive wear may result.*

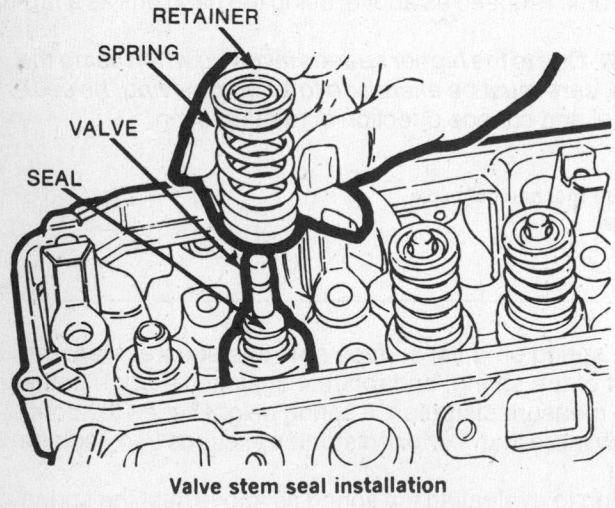

RETAINER

SPRING

VALVE

SEAL

Valve stem seal installation

Install the valves:

Lubricate the valve stems, and install the valves in the cylinder head as numbered. Lubricate and position the seals (if used, see above) and the valve springs. Install the spring retainers, compress the springs, and insert the keys using needlenose pliers or a tool designed for this purpose.

NOTE: *Retain the keys with wheel bearing grease during installation.*

Procedure	Method
Check valve spring installed height:	Measure the distance between the spring pad and the lower edge of the spring retainer, and compare to specifications. If the installed height is incorrect, add shim washers between the spring pad and the spring. CAUTION: *Use only washers designed for this purpose.*

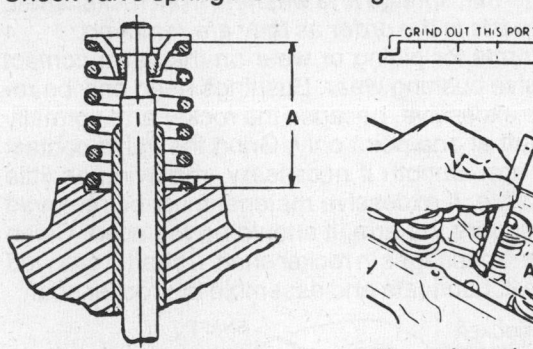

Valve spring installed
height dimension

Measuring valve spring
installed height

Install the camshaft (OHC engines only) and check end play:	See the engine service procedures earlier in this book for details concerning specific engines.
Inspect the rocker arms, balls, studs, and nuts (OHV engines only):	Visually inspect the rocker arms, balls, studs, and nuts for cracks, galling, burning, scoring or wear. If all parts are intact, liberally lubricate the rocker arms and balls, and install them on the cylinder head. If wear is noted on a rocker arm at the point of valve contact, grind it smooth and square, removing as little material as possible. Replace the rocker arm if excessively worn. If a rocker stud shows signs of wear, it must be replaced (see below). If a rocker nut shows stress cracks, replace it. If an exhaust ball is galled or burned, substitute the intake ball from the same cylinder (if it is intact), and install a new intake ball. NOTE: *Avoid using new rocker balls on exhaust valves.*

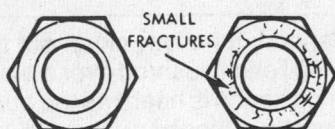

Stress cracks in the rocker nuts

Replacing rocker studs (OHV engines only):	In order to remove a threaded stud, lock two nuts on the stud, and unscrew the stud using the lower nut. Coat the lower threads of the new stud with Loctite®, and install. Two alternative methods are available for replacing pressed in studs. Remove the damaged stud using a stack of washers and a nut (see illustration). In the first, the boss is reamed .005–.006″ oversize, and an oversize stud pressed in. Control the stud extension over the boss using washers, in the same manner as valve guides. Before installing the stud, coat it with white lead and grease. To retain the stud more positively drill a hole through the stud and boss, and install a roll pin. In the second method, the boss is tapped, and a threaded stud installed. Retain the stud using Loctite® Stud and Bearing Mount.

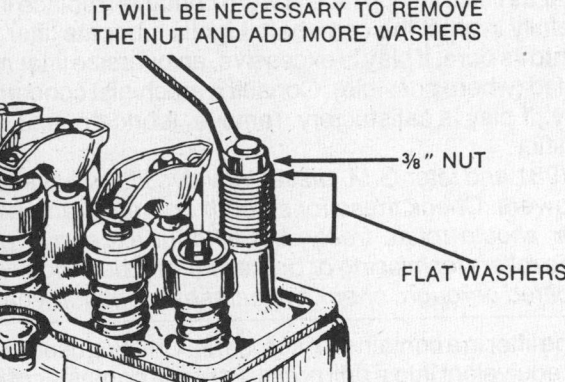

AS STUB BEGINS TO PULL UP,
IT WILL BE NECESSARY TO REMOVE
THE NUT AND ADD MORE WASHERS

⅜″ NUT

FLAT WASHERS

Extracting a pressed-in rocker stud

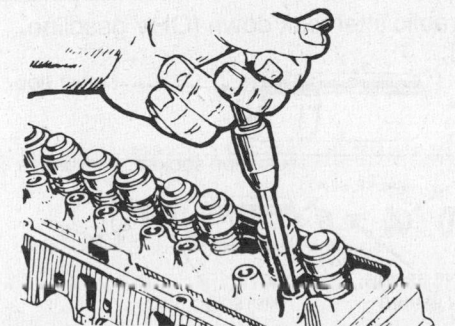

Reaming the stud bore for oversize rocker studs

CYLINDER HEAD RECONDITIONING

Procedure	Method

Inspect the rocker shaft(s) and rocker arms (OHV engines only):

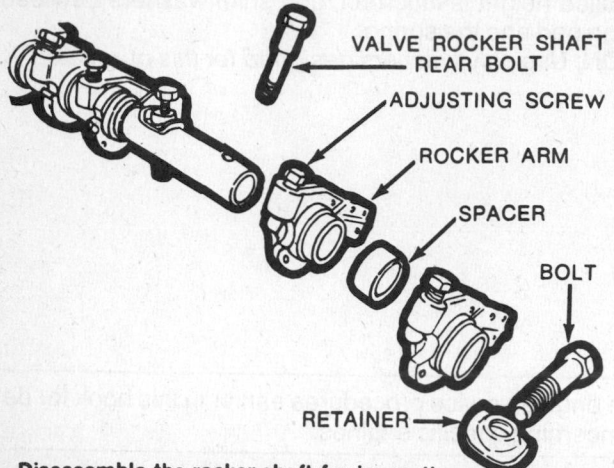

VALVE ROCKER SHAFT REAR BOLT

ADJUSTING SCREW

ROCKER ARM

SPACER

BOLT

RETAINER

Disassemble the rocker shaft for inspection

Remove rocker arms, springs and washers from rocker shaft. NOTE: *Lay out parts in the order as they are removed.* Inspect rocker arms for pitting or wear on the valve contact point, or excessive bushing wear. Bushings need only be replaced if wear is excessive, because the rocker arm normally contacts the shaft at one point only. Grind the valve contact point of rocker arm smooth if necessary, removing as little material as possible. If excessive material must be removed to smooth and square the arm, it should be replaced. Clean out all oil holes and passages in rocker shaft. If shaft is grooved or worn, replace it. Lubricate and assemble the rocker shaft.

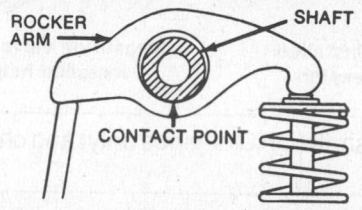

ROCKER ARM   SHAFT

CONTACT POINT

Rocker arm-to-rocker shaft contact area

Inspect the camshaft bushings and the camshaft (OHC engines):

See next section.

Inspect the pushrods (OHV engines only):

Remove the pushrods, and, if hollow, clean out the oil passages using fine wire. Roll each pushrod over a piece of clean glass. If a distinct clicking sound is heard as the pushrod rolls, the rod is bent, and must be replaced.

*The length of all pushrods must be equal. Measure the length of the pushrods, compare to specifications, and replace as necessary.

Inspect the valve lifters (OHV engines only):

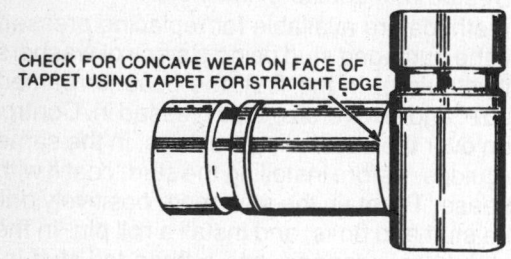

CHECK FOR CONCAVE WEAR ON FACE OF TAPPET USING TAPPET FOR STRAIGHT EDGE

Checking the lifter face

Remove lifters from their bores, and remove gum and varnish, using solvent. Clean walls of lifter bores. Check lifters for concave wear as illustrated. If face is worn concave, replace lifter, and carefully inspect the camshaft. Lightly lubricate lifter and insert it into its bore. If play is excessive, an oversize lifter must be installed (where possible). Consult a machinist concerning feasibility. If play is satisfactory, remove, lubricate, and reinstall the lifter.
NOTE: *1981 and later G.M. diesel V8 valve lifters have roller cam followers. Check these for smooth operation and wear. The roller should rotate freely, but without excessive play. Check the rollers for missing or broken needle bearings. If the roller is pitted or rough, check the camshaft lobe for wear.*

---

*Testing hydraulic lifter leak down (OHV gasoline engines only):

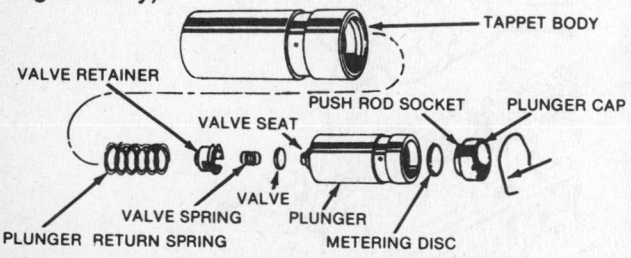

TAPPET BODY

VALVE RETAINER

PUSH ROD SOCKET   PLUNGER CAP

VALVE SEAT

VALVE

PLUNGER RETURN SPRING

VALVE SPRING   PLUNGER

METERING DISC

Typical exploded view of hydraulic valve lifter

Submerge lifter in a container of kerosene. Chuck a used pushrod or its equivalent into a drill press. Position container of kerosene so pushrod acts on the lifter plunger. Pump lifter with the drill press, until resistance increases. Pump several more times to bleed any air out of lifter. Apply very firm, constant pressure to the lifter, and observe rate at which fluid bleeds out of lifter. If the fluid bleeds very quickly (less than 15 seconds), lifter is defective. If the time exceeds 60 seconds, lifter is sticking. In either case, recondition or replace lifter. If lifter is operating properly (leak down time 15–60 seconds), lubricate and install it.

CYLINDER HEAD RECONDITIONING

Procedure	Method
Bleed the hydraulic lifters (diesel engines only):	After the cylinder heads are installed on G.M. V8 diesels, the valve lifters must be bled down before the crankshaft is turned. Failure to bleed down the lifters will cause damage to the valve train. See diesel engine rocker arm replacement procedure in Oldsmobile 88, 98, etc. car section for procedures. NOTE: *When installing new lifters, prime by working the lifter plunger while submerged in clean kerosene or diesel fuel.*

CYLINDER BLOCK RECONDITIONING

Procedure	Method
Checking the main bearing clearance: 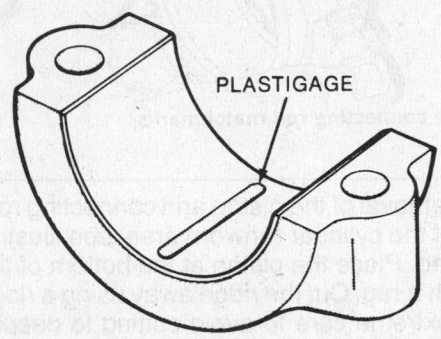 **Plastigage® installed on the lower bearing shell** 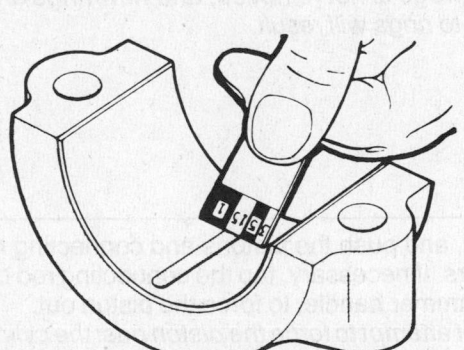 **Measuring Plastigage® to determine bearing clearance**	Invert engine, and remove cap from the bearing to be checked. Using a clean, dry rag, thoroughly clean all oil from crankshaft journal and bearing insert. NOTE: *Plastigage is soluble in oil; therefore, oil on the journal or bearing could result in erroneous readings.* Place a piece of Plastigage along the full length of journal, reinstall cap, and torque to specifications. Remove bearing cap, and determine bearing clearance by comparing width of Plastigage to the scale on Plastigage envelope. Journal taper is determined by comparing width of the Plastigage strip near its ends. Rotate crankshaft 90° and retest, to determine journal eccentricity. NOTE: *Do not rotate crankshaft with Plastigage installed.* If bearing insert and journal appear intact, and are within tolerances, no further main bearing service is required. If bearing or journal appear defective, cause of failure should be determined before replacement. *Remove crankshaft from block (see below). Measure the main bearing journals at each end twice (90° apart) using a micrometer, to determine diameter, journal taper and eccentricity. If journals are within tolerances, reinstall bearing caps at their specified torque. Using a telescope gauge and micrometer, measure bearing I.D. parallel to piston axis and at 30° on each side of piston axis. Subtract journal O.D. from bearing I.D. to determine oil clearance. If crankshaft journals appear defective, or do no meet tolerances, there is no need to measure bearings; for the crankshaft will require grinding and/or undersize bearings will be required. If bearing appears defective, cause for failure should be determined prior to replacement.
Checking the connecting rod bearing clearance:	Connecting rod bearing clearance is checked in the same manner as main bearing clearance, using Plastigage. Before removing the crankshaft, connecting rod side clearance also should be measured and recorded. *Checking connecting rod bearing clearance, using a micrometer, is identical to checking main bearing clearance. If no other service is required, the piston and rod assemblies need not be removed.

CYLINDER BLOCK RECONDITIONING

Procedure	Method
Removing the crankshaft:	Using a punch, mark the corresponding main bearing caps and saddles according to position (i.e., one punch on the front main cap and saddle, two on the second, three on the third, etc.). Using number stamps, identify the corresponding connecting rods and caps, according to cylinder (if no numbers are present). Remove the main and connecting rod caps, and place sleeves of plastic tubing over the connecting rod bolts, to protect the journals as the crankshaft is removed. Lift the crankshaft out of the block.

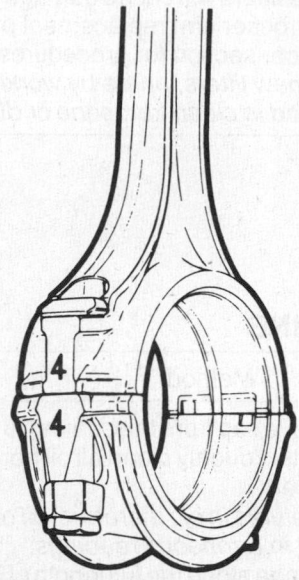

Connecting rod matched to cylinder with a number stamp

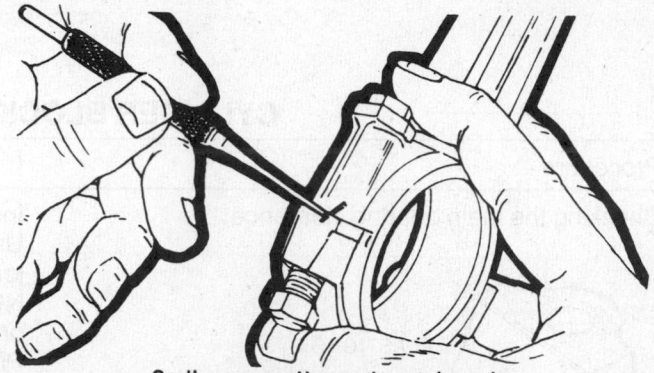

Scribe connecting rod matchmarks

Remove the ridge from the top of the cylinder:	In order to facilitate removal of the piston and connecting rod, the ridge at the top of the cylinder (unworn area; see illustration) must be removed. Place the piston at the bottom of the bore, and cover it with a rag. Cut the ridge away using a ridge reamer, exercising extreme care to avoid cutting to deeply. Remove the rag, and remove cuttings that remain on the piston. CAUTION: *If the ridge is not removed, and new rings are installed, damage to rings will result.*

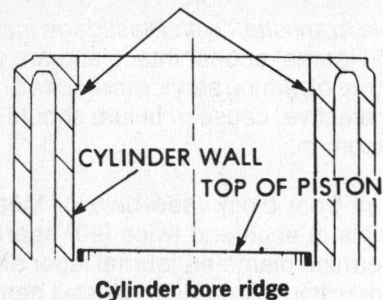

RIDGE CAUSED BY CYLINDER WEAR

CYLINDER WALL

TOP OF PISTON

Cylinder bore ridge

Removing the piston and connecting rod:	Invert the engine, and push the pistons and connecting rods out of the cylinders. If necessary, tap the connecting rod boss with a wooden hammer handle, to force the piston out. CAUTION: *Do not attempt to force the piston past the cylinder ridge* (see above).

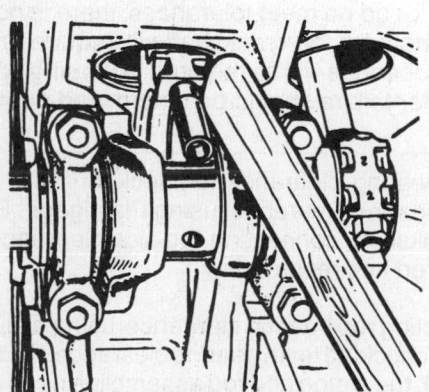

Removing the piston

CYLINDER BLOCK RECONDITIONING

Procedure	Method
Service the crankshaft:	Ensure that all oil holes and passages in the crankshaft are open and free of sludge. If necessary, have the crankshaft ground to the largest possible undersize. **Have the crankshaft Magnafluxed, to locate stress cracks. Consult a machinist concerning additional service procedures, such as surface hardening (e.g., nitriding, Tuftriding) to improve wear characteristics, cross drilling and chamfering the oil holes to improve lubrication, and balancing.
Removing freeze plugs:	Drill a small hole in the middle of the freeze plugs. Thread a large sheet metal screw into the hole and remove the plug with a slide hammer.
Remove the oil gallery plugs:	Threaded plugs should be removed using an appropriate (usually square) wrench. To remove soft, pressed in plugs, drill a hole in the plug, and thread in a sheet metal screw. Pull the plug out by the screw using pliers.
Hot-tank the block: NOTE: *Do not hot-tank aluminum parts.*	Have the block hot-tanked to remove grease, corrosion, and scale from the water jackets. NOTE: *Consult the operator to determine whether the camshaft bearings will be damaged during the hot-tank process.*

Check the block for cracks:	Visually inspect the block for cracks or chips. The most common locations are as follows:
	   Adjacent to freeze plugs.
	   Between the cylinders and water jackets.
	   Adjacent to the main bearing saddles.
	   At the extreme bottom of the cylinders.
	Check only suspected cracks using spot check dye (see introduction). If a crack is located, consult a machinist concerning possible repairs.
	
	**Magnaflux the block to locate hidden cracks. If cracks are located, consult a machinist about feasibility of repair.

Install the oil gallery plugs and freeze plugs:	Coat freeze plugs with sealer and tap into position using a piece of pipe, slightly smaller than the plug, as a driver. To ensure retention, stake the edges of the plugs. Coat threaded oil gallery plugs with sealer and install. Drive replacement soft plugs into block using a large drift as a driver. *Rather than reinstalling lead plugs, drill and tap the holes, and install threaded plugs.
*Check the deck height:	*The deck height is the distance from the crankshaft centerline to the block deck. To measure, invert the engine, and install the crankshaft, retaining it with the center main cap. Measure the distance from the crankshaft journal to the block deck, parallel to the cylinder centerline. Measure the diameter of the end (front and rear) main journals, parallel to the centerline of the cylinders, divide the diameter in half, and subtract it from the previous measurement. The results of the front and rear measurements should be identical. If the difference exceeds .005", the deck height should be corrected. NOTE: *Block deck height and warpage should be corrected at the same time.*

CYLINDER BLOCK RECONDITIONING

Procedure	Method
Check the block deck for warpage:	Using a straightedge and feeler gauges, check the block deck for warpage in the same manner that the cylinder head is checked (see Cylinder Head Reconditioning). If warpage exceeds specifications, have the deck resurfaced. NOTE: *In certain cases a specification for total material removal (Cylinder head and block deck) is provided. This specification must not be exceeded.*

Procedure	Method
Check the bore diameter and surface:  **Measuring the cylinder bore with a dial gauge**	Visually inspect the cylinder bores for roughness, scoring, or scuffing. If evident, the cylinder bore must be bored or honed oversize to eliminate imperfections, and the smallest possible oversize piston used. The new pistons should be given to the machinist with the block, so that the cylinders can be bored or honed exactly to the piston size (plus clearance). If no flaws are evident, measure the bore diameter using a telescope gauge and micrometer, or dial guage, parallel and perpendicular to the engine centerline, at the top (below the ridge) and bottom of the bore. Subtract the bottom measurements from the top to determine taper, and the parallel to the centerline measurements from the perpendicular measurements to determine eccentricity. If the measurements are not within specifications, the cylinder must be bored or honed, and an oversize piston installed. If the measurements are within specifications the cylinder may be used as is, with only finish honing (see below). NOTE: *Prior to boring, check the block deck warpage, height and bearing alignment.* CAUTION: *The 4 cyl. 140 G.M. engine cylinder walls are impregnated with silicone. Boring or honing can be done only by a shop with the proper equipment.*

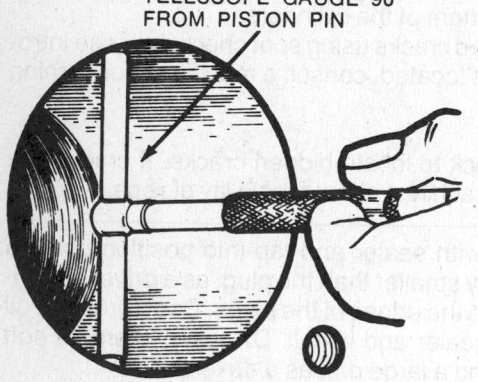

Measuring cylinder bore with a telescope gauge

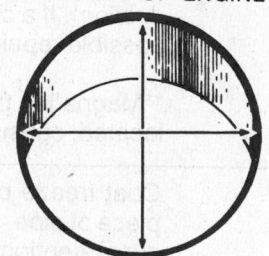

A—AT RIGHT ANGLE TO CENTERLINE OF ENGINE
B—PARALLEL TO CENTERLINE OF ENGINE

Cylinder bore measuring points

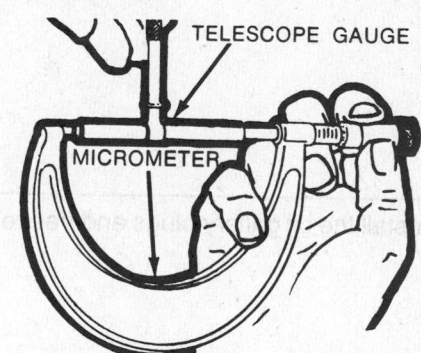

Determining cylinder bore by measuring telescope gauge with a micrometer

Procedure	Method
Check the cylinder block bearing alignment: 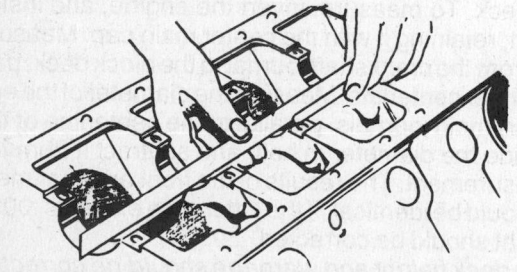 **Checking main bearing saddle alignment**	Remove the upper bearing inserts. Place a straightedge in the bearing saddles along the centerline of the crankshaft. If clearance exists between the straightedge and the center saddle, the block must be alignbored.

Procedure	Method

Clean and inspect the pistons and connecting rods:

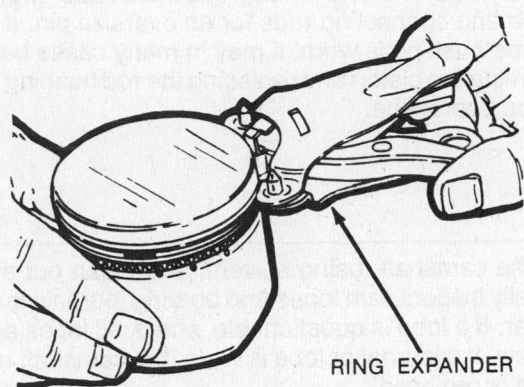

Removing the piston rings

RING EXPANDER

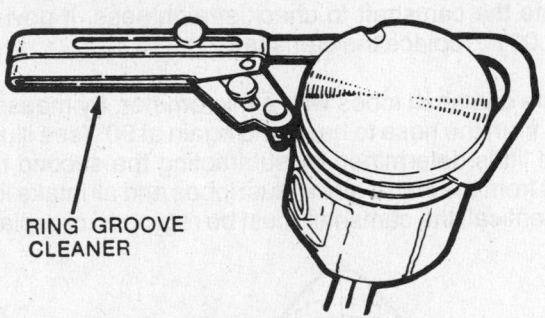

Cleaning the piston ring grooves

RING GROOVE CLEANER

Using a ring expander, remove the rings from the piston. Remove the retaining rings (if so equipped) and remove piston pin.

NOTE: *If the piston pin must be pressed out, determine the proper method and use the proper tools; otherwise the piston will distort.*

Clean the ring grooves using an appropriate tool, exercising care to avoid cutting too deeply. Thoroughly clean all carbon and varnish from the piston with solvent.

CAUTION: *Do not use a wire brush or caustic solvent on pistons.*

Inspect the pistons for scuffing, scoring, cracks, pitting, or excessive ring groove wear. If wear is evident, the piston must be replaced. Check the connecting rod length by measuring the rod from the inside of the large end to the inside of the small end using calipers (see illustration). All connecting rods should be equal length. Replace any rod that differs from the others in the engine.

*Have the connecting rod alignment checked in an alignment fixture by a machinist. Replace any twisted or bent rods.

*Magnaflux the connecting rods to locate stress cracks. If cracks are found, replace the connecting rod.

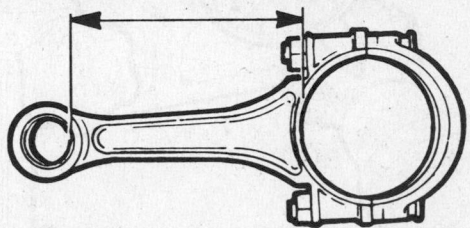

Check the connecting rod length (arrow)

Fit the pistons to the cylinders:

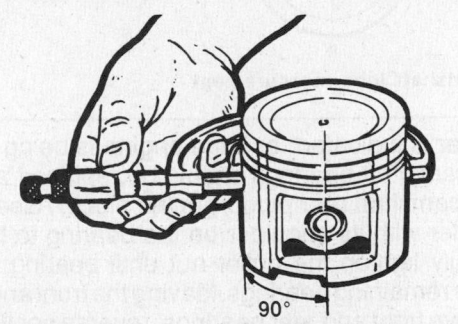

90°

Measuring the piston prior to fitting

Using a telescope gauge and micrometer, or a dial gauge, measure the cylinder bore diameter perpendicular to the piston pin, 2½° below the deck. Measure the piston perpendicular to its pin on the skirt. The difference between the two measurements is the piston clearance. If the clearance is within specifications or slightly below (after boring or honing), finish honing is all that is required. If the clearance is excessive, try to obtain a slightly larger piston to bring clearance within specifications. Where this is not possible, obtain the first oversize piston, and hone (or if necessary, bore) the cylinder to size.

Assemble the pistons and connecting rods:

Inspect piston pin, connecting rod small end bushing, and piston bore for galling, scoring, or excessive wear. If evident, replace defective part(s). Measure the I.D. of the piston boss and connecting rod small end, and the O.D. of the piston pin. If within specifications, assemble piston pin and rod.

CAUTION: *If piston pin must be pressed in, determine the proper method and use the proper tools; otherwise the piston will distort.*

CYLINDER BLOCK RECONDITIONING

Procedure	Method

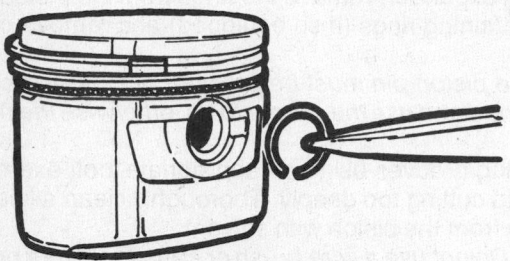

Installing piston pin lock rings

Install the lock rings; ensure that they seat properly. If the parts are not within specifications, determine the service method for the type of engine. In some cases, piston and pin are serviced as an assembly when either is defective. Others specify reaming the piston and connecting rods for an oversize pin. If the connecting rod bushing is worn, it may in many cases be replaced. Reaming the piston and replacing the rod bushing are machine shop operations.

Clean and inspect the camshaft:

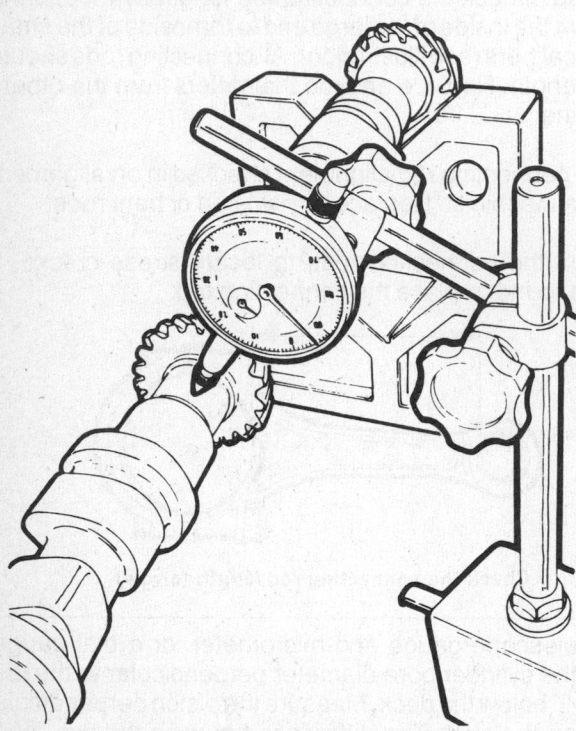

Checking the camshaft for straightness

Degrease the camshaft, using solvent, and clean out all oil holes. Visually inspect cam lobes and bearing journals for excessive wear. If a lobe is questionable, check all lobes as indicated below. If a journal or lobe is worn, the camshaft must be reground or replaced.

NOTE: *If a journal is worn, there is a good chance that the bushings are worn.*

If lobes and journals appear intact, place the front and rear journals in V-blocks, and rest a dial indicator on the center journal. Rotate the camshaft to check straightness. If deviation exceeds .001°, replace the camshaft.

*Check the camshaft lobes with a micrometer, by measuring the lobes from the nose to base and again at 90° (see illustration). The lift is determined by subtracting the second measurement from the first. If all exhaust lobes and all intake lobes are not identical, the camshaft must be reground or replaced.

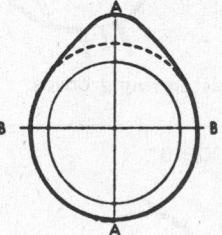

Camshaft lobe measurement

Replace the camshaft bearings (OHV engines only):

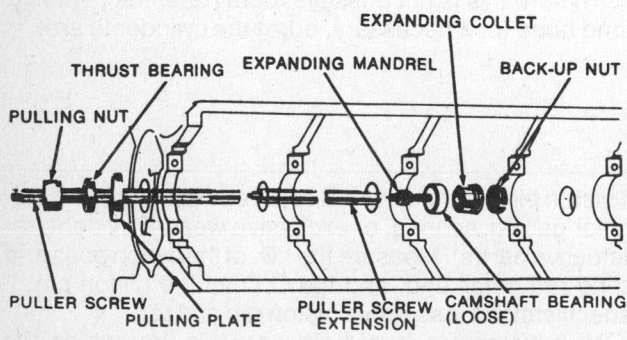

Camshaft removal and installation tool (typical)

If excessive wear is indicated, or if the engine is being completely rebuilt, camshaft bearings should be replaced as follows: Drive the camshaft rear plug from the block. Assemble the removal puller with its shoulder on the bearing to be removed. Gradually tighten the puller nut until bearing is removed. Remove remaining bearings, leaving the front and rear for last. To remove front and rear bearings, reverse position of the tool, so as to pull the bearings in toward the center of the block. Leave the tool in this position, pilot the new front and rear bearings on the installer, and pull them into position: Return the tool to its original position and pull remaining bearings into postion.

NOTE: *Ensure that oil holes align when installing bearings.*

Replace camshaft rear plug, and stake it into position to aid retention.

CYLINDER BLOCK RECONDITIONING

Procedure	Method

Finish hone the cylinders:

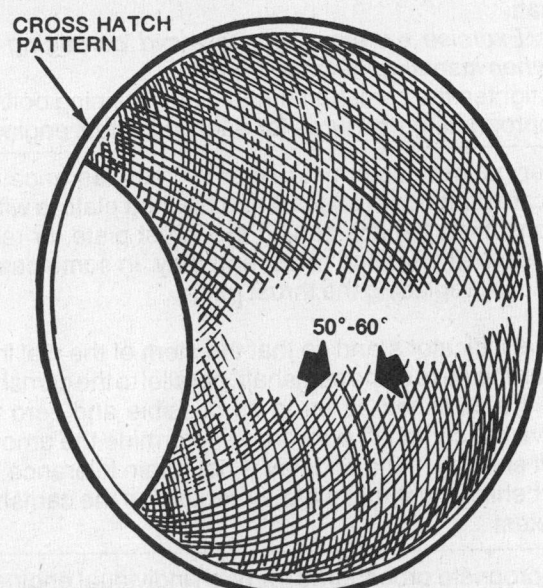

CROSS HATCH PATTERN

50°-60°

Chuck a flexible drive hone into a power drill, and insert it into the cylinder. Start the hone, and move it up and down the cylinder at a rate which will produce approximately a 60° cross-hatch pattern (see illustration).
NOTE: *Do not extend the hone below the cylinder bore.*
After developing the pattern, remove the hone and recheck piston fit. Wash the cylinders with a detergent and water solution to remove abrasive dust, dry, and wipe several times with a rag soaked in engine oil.

Check piston ring end-gap:

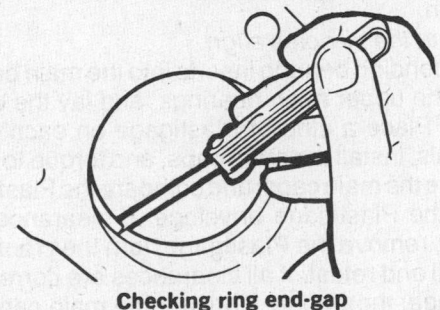

Checking ring end-gap

Compress the piston rings to be used in a cylinder, one at a time, into that cylinder, and press them approximately 1″ below the deck with an inverted piston. Using feeler gauges, measure the ring end-gap, and compare to specifications. Pull the ring out of the cylinder and file the ends with a fine file to obtain proper clearance.
CAUTION: *If inadequate ring end-gap is utilized, ring breakage will result.*

Install the piston rings:

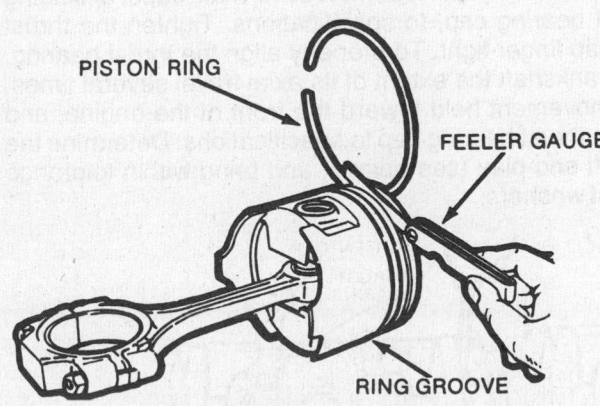

PISTON RING

FEELER GAUGE

RING GROOVE

Checking ring side clearance

Inspect the ring grooves in the piston for excessive wear or taper. If necessary, recut the groove(s) for use with an overwidth ring or a standard ring and spacer. If the groove is worn uniformly, overwidth rings, or standard rings and spacers may be installed without recutting. Roll the outside of the ring around the groove to check for burrs or deposits. If any are found, remove with a fine file. Hold the ring in the groove, and measure side clearance. If necessary, correct as indicated above.
NOTE: *Always install any additional spacers above the piston ring.*
The ring groove must be deep enough to allow the ring to seat below the lands (see illustration). In many cases, a "go-no-go" depth gauge will be provided with the piston rings. Shallow grooves may be corrected by recutting, while deep grooves require some type of filler or expander behind the piston. Consult the piston ring supplier concerning the suggested method. Install the rings on the piston, lowest ring first, using a ring expander.
NOTE: *Position the ring markings as specified by the manufacturer (see car section).*

CYLINDER BLOCK RECONDITIONING

Procedure	Method
Install the camshaft (OHV engines only):	Liberally lubricate the camshaft lobes and journals, and install the camshaft. CAUTION: *Exercise extreme care to avoid damaging the bearings when inserting the camshaft.* Install and tighten the camshaft thrust plate retaining bolts. See the appropriate procedures for each individual engine.

Check camshaft end-play (OHV engines only):

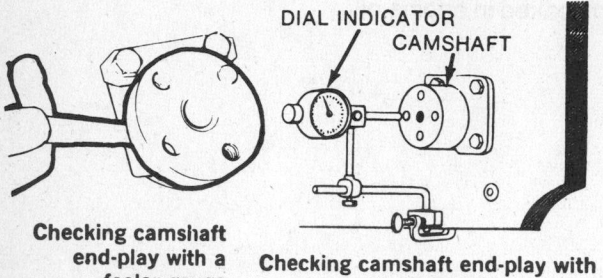

DIAL INDICATOR
CAMSHAFT

Checking camshaft end-play with a feeler gauge

Checking camshaft end-play with a dial indicator

Using feeler gauges, determine whether the clearance between the camshaft boss (or gear) and backing plate is within specifications. Install shims behind the thrust plate, or reposition the camshaft gear and retest end-play. In some cases, adjustment is by replacing the thrust plate.

*Mount a dial indicator stand so that the stem of the dial indicator rests on the nose of the camshaft, parallel to the camshaft axis. Push the camshaft as far in as possible and zero the gauge. Move the camshaft outward to determine the amount of camshaft endplay. If the endplay is not within tolerance, install shims behind the thrust plate, or reposition the camshaft gear and retest.

Install the rear main seal (where applicable):	See the appropriate procedures for each individual engine.

Install the crankshaft:

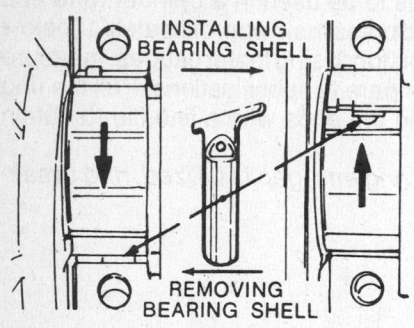

INSTALLING BEARING SHELL

REMOVING BEARING SHELL

Removal and installation of upper bearing insert using a roll-out pin

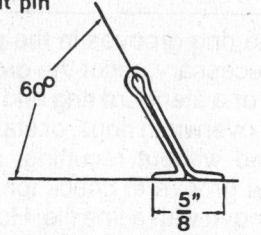

60°

$\frac{5}{8}$"

Home-made bearing roll-out pin

Thoroughly clean the main bearing saddles and caps. Place the upper halves of the bearing inserts on the saddles and press into position.
NOTE: *Ensure that the oil holes align.*
Press the corresponding bearing inserts into the main bearing caps. Lubricate the upper main bearings, and lay the crankshaft in position. Place a strip of Plastigage on each of the crankshaft journals, install the main caps, and torque to specifications. Remove the main caps, and compare the Plastigage to the scale on the Plastigage envelope. If clearances are within tolerances, remove the Plastigage, turn the crankshaft 90°, wipe off all oil and retest. If all clearances are correct, remove all Plastigage, thoroughly lubricate the main caps and bearing journals, and install the main caps. If clearances are not within tolerance, the upper bearing inserts may be removed, without removing the crankshaft, using a bearing roll out pin (see illustration). Roll in a bearing that will provide proper clearance, and retest. Torque all main caps, excluding the thrust bearing cap, to specifications. Tighten the thrust bearing cap finger tight. To properly align the thrust bearing, pry the crankshaft the extent of its axial travel several times, the last movement held toward the front of the engine, and torque the thrust bearing cap to specifications. Determine the crankshaft end-play (see below), and bring within tolerance with thrust washers.

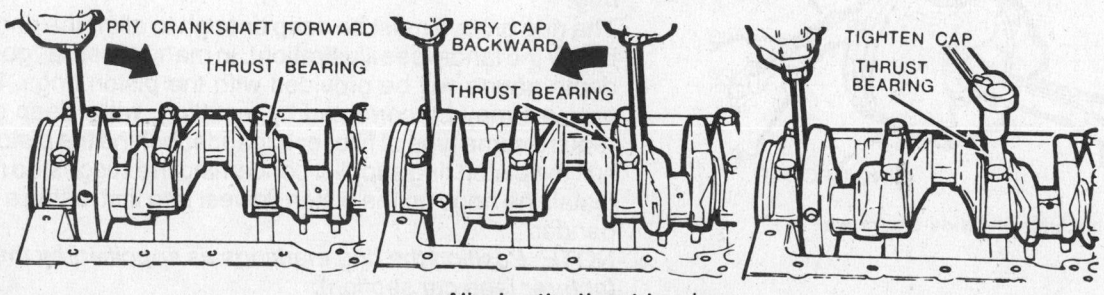

PRY CRANKSHAFT FORWARD
THRUST BEARING

PRY CAP BACKWARD
THRUST BEARING

TIGHTEN CAP
THRUST BEARING

Aligning the thrust bearing

Procedure

Method

Measure crankshaft end-play:

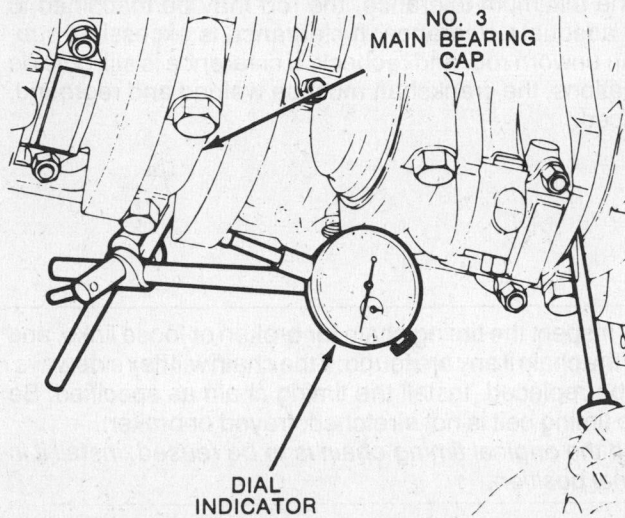

NO. 3
MAIN BEARING
CAP

DIAL
INDICATOR

Checking crankshaft end-play with a dial indicator

Mount a dial indicator stand on the front of the block, with the dial indicator stem resting on the nose of the crankshaft, parallel to the crankshaft axis. Pry the crankshaft the extent of its travel rearward, and zero the indicator. Pry the crankshaft forward and record crankshaft end-play.

NOTE: *Crankshaft end-play also may be measured at the thrust bearing, using feeler gauges* (see illustration).

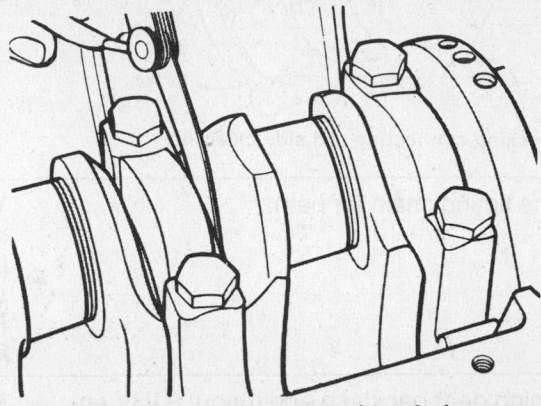

Checking crankshaft end-play with a feeler gauge

Install the pistons:

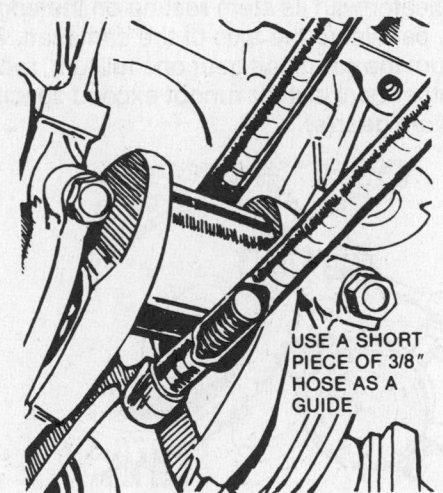

USE A SHORT
PIECE OF 3/8"
HOSE AS A
GUIDE

Tubing used to protect crankshaft journals and cylinder walls during piston installation

Press the upper connecting rod bearing halves into the connecting rods, and the lower halves into the connecting rod caps. Position the piston ring gaps according to specifications (see car section), and lubricate the pistons. Install a ring compressor on a piston, and press two long (8") pieces of plastic tubing over the rod bolts. Using the tubes as a guide, press the pistons into the bores and onto the crankshaft with a wooden hammer handle. After seating the rod on the crankshaft journal, remove the tubes and install the cap finger tight. Install the remaining pistons in the same manner. Invert the engine and check the bearing clearance at two points (90° apart) on each journal with Plastigage.

NOTE: *Do not turn the crankshaft with Plastigage installed.* If clearance is within tolerances, remove *all* Plastigage, thoroughly lubricate the journals, and torque the rod caps to specifications. If clearance is not within specifications, install different thickness bearing inserts and recheck.

CAUTION: *Never shim or file the connecting rods or caps.* Always install plastic tube sleeves over the rod bolts when the caps are not installed, to protect the crankshaft journals.

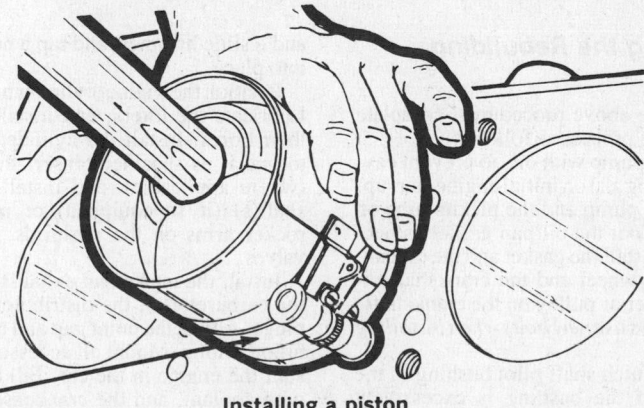

RING COMPRESSOR

Installing a piston

CYLINDER BLOCK RECONDITIONING

Procedure	Method
Check connecting rod side clearance: 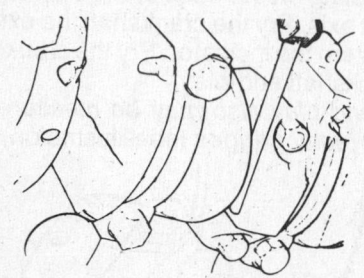 **Checking connecting rod side clearance**	Determine the clearance between the sides of the connecting rods and the crankshaft, using feeler gauges. If clearance is below the minimum tolerance, the rod may be machined to provide adequate clearance. If clearance is excessive, substitute an unworn rod, and recheck. If clearance is still outside specifications, the crankshaft must be welded and reground, or replaced.
Inspect the timing chain (or belt):	Visually inspect the timing chain for broken or loose links, and replace the chain if any are found. If the chain will flex sideways, it must be replaced. Install the timing chain as specified. Be sure the timing belt is not stretched, frayed or broken. NOTE: *If the original timing chain is to be reused, install it in its original position.*
Check timing gear backlash and runout (OHV engines): 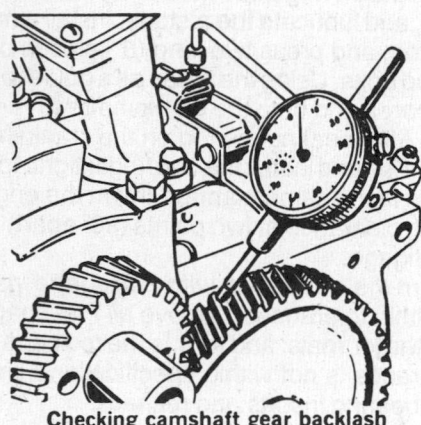 **Checking camshaft gear backlash**	Mount a dial indicator with its stem resting on a tooth of the camshaft gear (as illustrated). Rotate the gear until all slack is removed, and zero the indicator. Rotate the gear in the opposite direction until slack is removed, and record gear backlash. Mount the indicator with its stem resting on the edge of the camshaft gear, parallel to the axis of the camshaft. Zero the indicator, and turn the camshaft gear one full turn, recording the runout. If either backlash or runout exceed specifications, replace the worn gear(s).  **Checking camshaft gear runout**

Completing the Rebuilding Process

Following the above procedures, complete the rebuilding process as follows:

Fill the oil pump with oil, to prevent cavitating (sucking air) on initial engine start up. Install the oil pump and the pickup tube on the engine. Coat the oil pan gasket as necessary, and install the gasket and the oil pan. Mount the flywheel and the crankshaft vibration damper or pulley on the crankshaft. NOTE: *Always use new bolts when installing the flywheel.*
Inspect the clutch shaft pilot bushing in the crankshaft. If the bushing is excessively worn, remove it with an expanding puller and a slide hammer, and tap a new bushing into place.

Position the engine, cylinder head side up. Lubricate the lifters, and install them into their bores. Install the cylinder head, and torque it as specified. Insert the pushrods (where applicable), and install the rocker shaft(s) (if so equipped) or position the rocker arms on the pushrods. Adjust the valves.

Install the intake and exhaust manifolds, the carburetor(s), the distributor and spark plugs. Adjust the point gap and the static ignition timing. Mount all accessories and install the engine in the car. Fill the radiator with coolant, and the crankcase with high quality engine oil.

Break-in Procedure

Start the engine, and allow it to run at low speed for a few minutes, while checking for leaks. Stop the engine, check the oil level, and fill as necessary. Restart the engine, and fill the cooling system to capacity. Check the point dwell angle and adjust the ignition timing and the valves. Run the engine at low to medium speed (800–2500 rpm) for approximately ½ hour, and retorque the cylinder head bolts. Road test the car, and check again for leaks.

Follow the manufacturer's recommended engine break-in procedure and maintenance schedule for new engines.

U-Joint, CV-Joint Overhaul 33

UNIVERSAL JOINTS

U-Joint is mechanic's jargon for universal joint. U-Joints should not be confused with U-bolts, which are U-shaped bolts used to connect U-joints to the differential pinion flange.

Universal joints provide flexibility between the driveshaft and axle housing to accommodate changes in the angle between them. Changes of length are accommodated by the sliding splined yoke between the driveshaft and transmission. The engine and transmission are mounted rigidly on the car frame. The angles between the transmission, driveshaft and axle change constantly as the car responds to various road conditions.

To give flexibility and still transmit power as smoothly as possible, several types of universal joints are used. The most common type of universal joint is the cross and yoke type. Yokes are used on the ends of the driveshaft with the yoke arms opposite each other. Another yoke is used opposite the driveshaft and when placed together, both yokes engage a center member, or cross, with four arms spaced 90° apart. The U-joint cross is alternately referred to as a spider, and the arms are called trunnions. A bearing cup (or cap) is used on each arm of the cross to accommodate movement as the driveshaft rotates. The bearings used are needle bearings.

A conventional universal joint will cause the driveshaft to speed up and slow down through each revolution and cause a corresponding change in the velocity of drive shaft. This change in speed causes natural vibrations to occur through the driveline, necessitating a third type of universal joint: The constant velocity joint. A rolling ball moves in a curved groove, located between two yoke-and-cross universal joints, connected to each other by a coupling yoke. The result is a uniform motion as the driveshaft rotates, avoiding the fluctuations in driveshaft speed. This type of joint is found in cars with sharp driveline angles, or where the extra measure of isolation is desirable.

Cross And Yoke U-Joint

OVERHAUL

There are two types of cross and yoke U-joints. One type retains the cross within the yoke with C-shaped snap rings. This type is found on all American Motors, Chrysler, and Ford Cars. GM cars generally use the second type of joint, which is held together by injection molded plastic retainer rings. The second type cannot be reassembled with the same parts, once disassembled. However, repair kits are available.

Snapring Type

1. Remove the driveshaft. For the correct procedure, see the car section for the model you are working on.
2. If the front yoke is to be disassembled, matchmark the driveshaft and sliding splined yoke (transmission yoke) so that driveline balance is preserved upon reassembly. Remove the snap rings which retain the bearing caps.

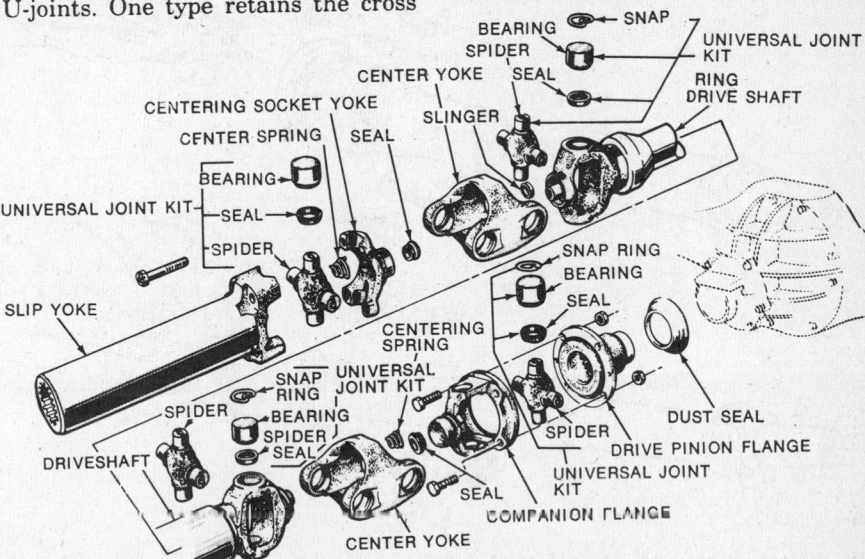

Typical driveshaft with cardan type U—joints

33-1

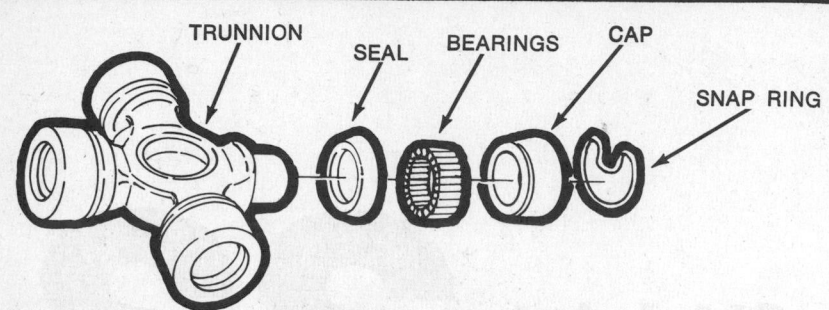

Snap ring type universal joint

3. Select two sockets, one small enough to pass through the yoke holes for the bearing caps, the other large enough to receive the bearing cap.

4. Using a vise or a press, position the small and large sockets on either side of the U-joint. Press in on the smaller socket so that it presses the opposite bearing cap out of the yoke and into the larger socket. If the cap does not come all the way out, grasp it with a pair of pliers and work it out.

5. Reverse the position of the sockets so that the smaller socket presses on the cross. Press the other bearing cap out of the yoke.

6. Repeat the procedure on the other bearings.

7. To install, grease the bearing caps and needles throughly if they are not pregreased. Start a new bearing cap into one side of the yoke. Position the cross in the yoke.

8. Select two sockets small enough to pass through the yoke holes. Put the sockets against the cross and the cap, and press the bearing cap ¼ inch below the surface of the yoke. If there is a sudden increase in the force needed to press the cap into place, or if the cross starts to bind, the bearings are cocked, They must be removed and restarted in the yoke. Failure to do so will greatly reduce the life of the bearing.

9. Install a new snap ring.

10. Start a new bearing into the opposite side. Place a socket on it and press in until the opposite bearing contacts the snap ring.

11. Install a new snap ring. It may be necessary to grind the facing surface of the snap ring slightly to permit easier installation.

12. Install the other bearings in the same manner.

13. Check the joint for free movement. If binding exists, smack the yoke ears with a brass or plastic faced hammer to seat the bearing needles. Do not strike the bearings, and support the shaft firmly. Do not install the driveshaft until free movement exists at all joints.

Plastic Retainer Type

Remove and install the bearing caps and trunnion (cross) as described for the snap-ring type universal joints. On an original universal joint, however, the bearing caps will be secured in the yokes with injected plastic. The plastic will shear when the bearing caps are pressed. Service snap-rings are installed in the groove on the inside (of yoke) of the installed caps.

NOTE: The plastic which retains the bearing will be sheared when the bearing cup is pressed out. Be sure to remove the remains of the plastic retainer from the ears of the yoke. It is easier to remove the remains if a small pin or punch is first driven through the injection holes in the yoke. Failure to remove all of the plastic remains may prevent the bearing cups from being pressed into place and the bearing retainers from being properly seated.

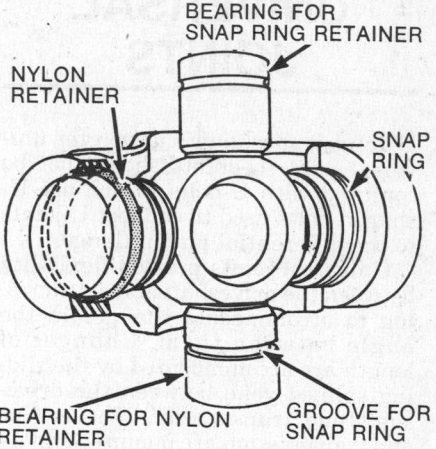

U-joint locking methods

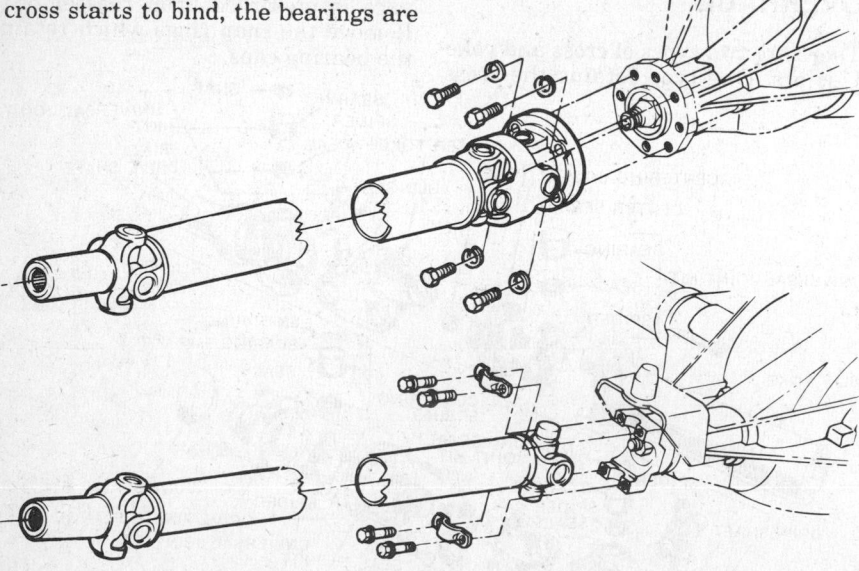

The driveshaft may be retained to the differential pinion by a flange (top) or by U-bolts or straps (bottom)

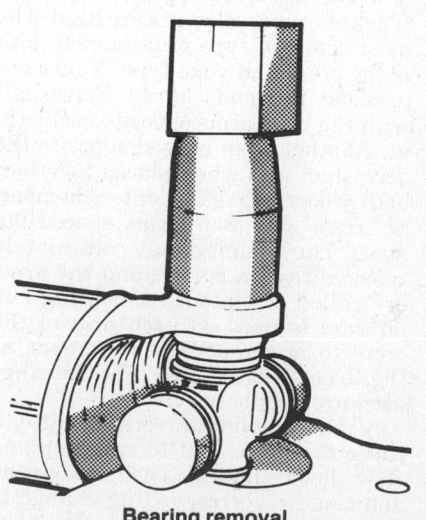

Bearing removal

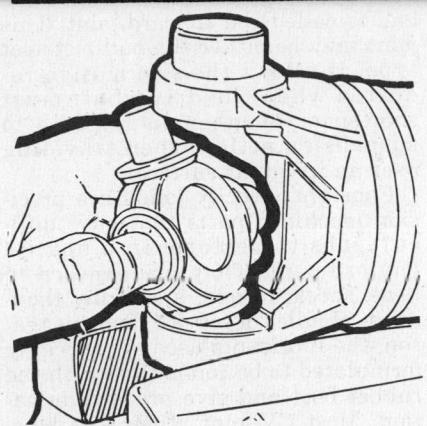

Press a bearing cap into the yoke, then install the cross

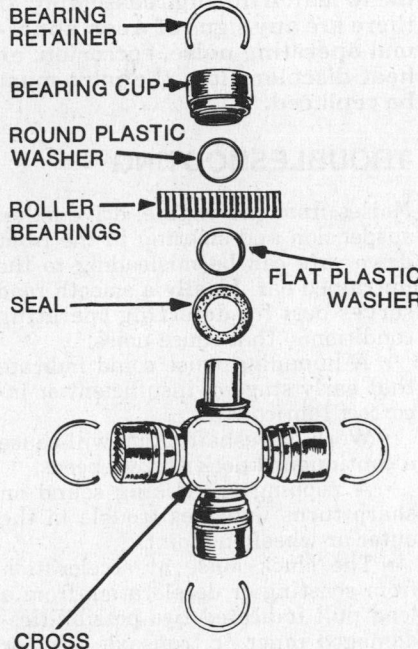

BEARING RETAINER

BEARING CUP

ROUND PLASTIC WASHER

ROLLER BEARINGS

SEAL

FLAT PLASTIC WASHER

CROSS

Plastic retainer U-joint repair kit components

Cardan Type U-Joint

OVERHAUL

Ford and Chrysler products with Cardan type U-joints use snap rings to retain the bearing cups in the yokes. Most GM cars have plastic retainers. Be sure to obtain the correct rebuilding kit.

1. Use a punch to mark the coupling yoke and the adjoining yokes before disassembly, to ensure proper reassembly and driveline balance.

2. It is easiest to remove the bearings from the coupling yoke first. Follow the order indicated in the illustration.

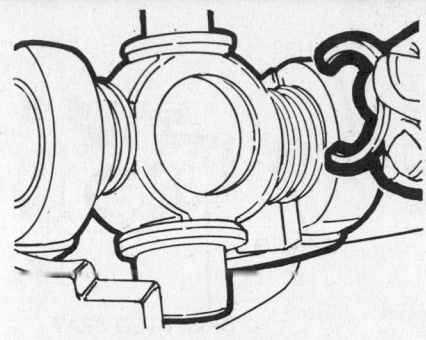

Service snap rings are installed inside the yoke

3. Support the driveshaft horizontally on a press stand, or on the workbench if a vise is being used.

4. If snap rings are used to retain the bearing cups, remove them. Place the rear ear of the coupling yoke over a socket large enough to receive the cup. Place a smaller socket, or a cross press made for the purpose, over the opposite cup. Press the bearing cup out of the coupling yoke ear. If the cup is not completely removed, insert a spacer and complete the operation, or grasp the cup with a pair of slip joint pliers and work it out. If the cups are retained by plastic, this will shear the retainers. Remove any bits of plastic.

5. Rotate the driveshaft and repeat the operation on the opposite cup.

6. Disengage the trunnions of the spider, still attached to the flanged yoke, from the coupling yoke, and pull the flanged yoke and spider from the center ball on the ball support tube yoke.

NOTE: The joint between the shaft and coupling yoke can be

serviced without disassembly of the joint between the coupling yoke and flanged yoke.

7. Pry the seal from the ball cavity, remove the washers, spring and three seats. Examine the ball stud seat and the ball stud for scores or wear. Worn parts can be replaced with a kit. Clean the ball seat cavity and fill it with grease. Install the spring, washer, ball seats, and spacer (washer) over the ball.

8. To assemble, insert one bearing cup part way into one ear of the ball support tube yoke and turn this cup to the bottom.

9. Insert the spider (cross) into the tube so that the trunnion (arm) seats freely in the cup.

10. Install the opposite cup part way, making sure that both cups are straight.

11. Press the cups into position, making sure that both cups squarely engage the spider. Back off if there is a sudden increase in resistance, indicating that a cup is cocked or a needle bearing is out of place.

12. As soon as one bearing retainer groove clears the yoke, stop and install the retainer (plastic retainer models). On models with snap rings, press the cups into place, then install the snap rings over the cups.

13. If difficulty is encountered installing the plastic retainers or the snap rings, smack the yoke sharply with a hammer to spring the ears slightly.

14. Install one bearing cup part way into the ear of the coupling yoke. Make sure that the alignment marks are matched, then engaged the coupling yoke over the spider and press

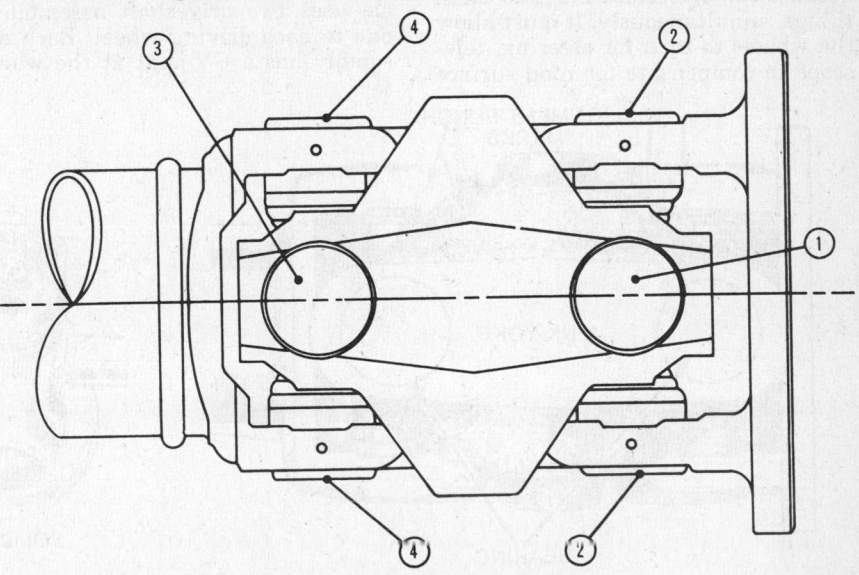

Cardan joint disassembly sequence

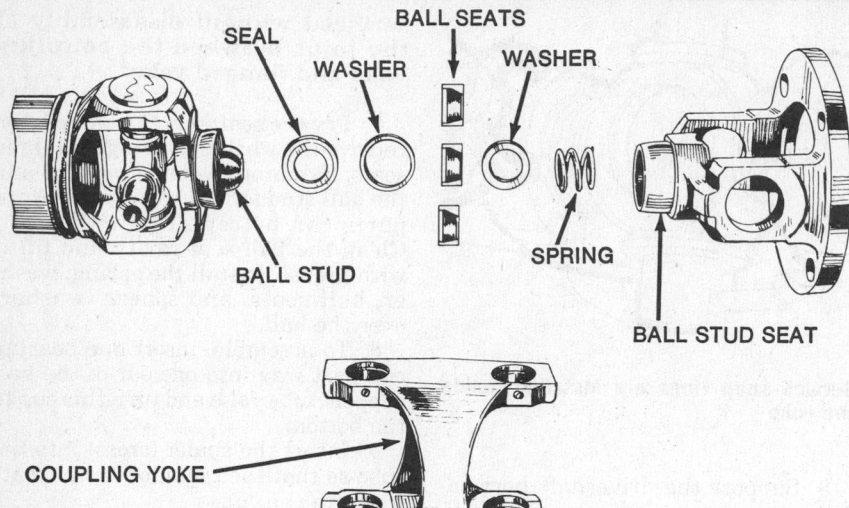

Cardan type joint

COUPLING YOKE

in the cups, installing the retainers or snap rings as before.

15. Install the cups and spider into the flanged yoke as with the previous yoke.

NOTE: The flange yoke should snap over center to the right or left and up or down by the pressure of the ball seat spring.

CONSTANT VELOCITY JOINTS

Front wheel drive vehicles present several unique problems to engineers because the driveshaft must do three things, simultaneously. It must allow the wheels to turn for steering, telescope to compensate for road surface vibrations, and it must transmit torque continuously without vibration.

To compensate for these three factors a two-joint driveshaft allows the front wheels to perform these functions. This driveshaft mates disc type straight groove ball joint design with the bell type Rzeppa CV universal joint.

The Rzeppa joint on the outboard end of each driveshaft provides steering ability by allowing drive wheels to steer up to 43° while transmitting all available torque to the wheels. The inboard joint allows telescoping (up to 1½ in.) through the rolling actions of balls in straight grooves and operates at angles up to 20°. The combined action of these two ball type U-joints eliminates vibration.

The typical front wheel drive vehicle uses two driveshaft assemblies-one to each driving wheel. Each assembly has a CV-joint at the wheel end is called the inboard joint. This joint may be either the ball or tripot type. It allows the slip motion required when the driveshaft must shorten or lengthen in response to suspension action when traveling over an irregular surface.

Constant velocity joints are precision machined parts that have difficult jobs to perform in a hostile enviornment. They are exposed to heat, shock, torque, and many thousands of miles of service. For this reason, the lubricants used are specially formulated to be compatible with the rubber boot and give proper lubrication. Most CV-joint repair kits have this special lubricant included.

NOTE: Wear patterns in a used ball or tripot CV-joint are impossible to match during reassembly. If there are any signs of wear, abnormal operating noise, corrosion, or heat discoloration, the joint must be replaced.

TROUBLESHOOTING

Noises from the engine, drive axles, suspension and steering in the front drive cars can be misleading to the untrained ear. Ideally a smooth road serves best for detecting operating condition(s) that cause noise.

• A humming noise could indicate that early stage of insufficient or incorrect lubricant.

• Worn driveshaft joints will cause a continuous knock at low speeds.

• A popping or clicking sound on sharp turns indicates trouble in the outer or wheel end joint.

• The cluck noise at acceleration from coasting or deceleration from a load pull indicated two possibilities-damaged inner or transaxle joint or differential problem(s).

• An inner joint will create a vibra-

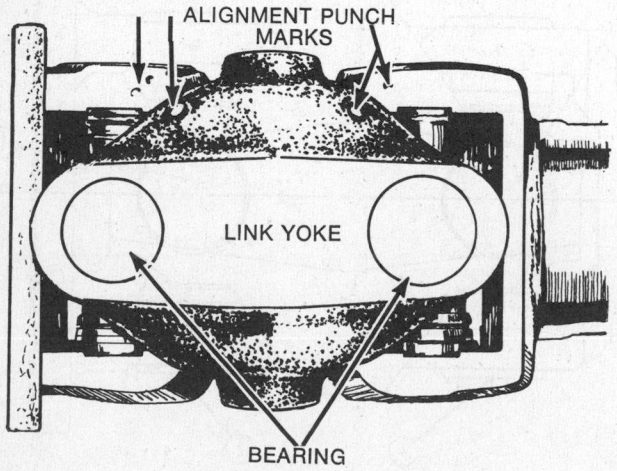

Match marks for double cardan joint

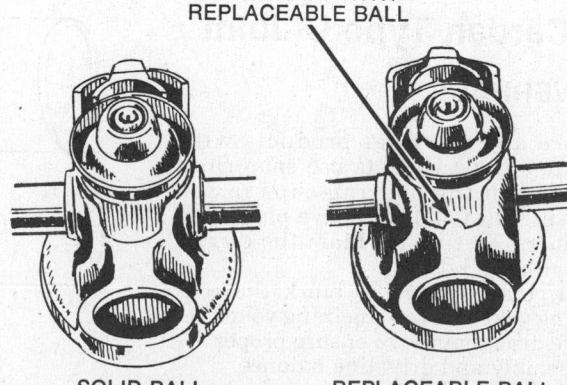

Solid and replaceable U-Joint balls

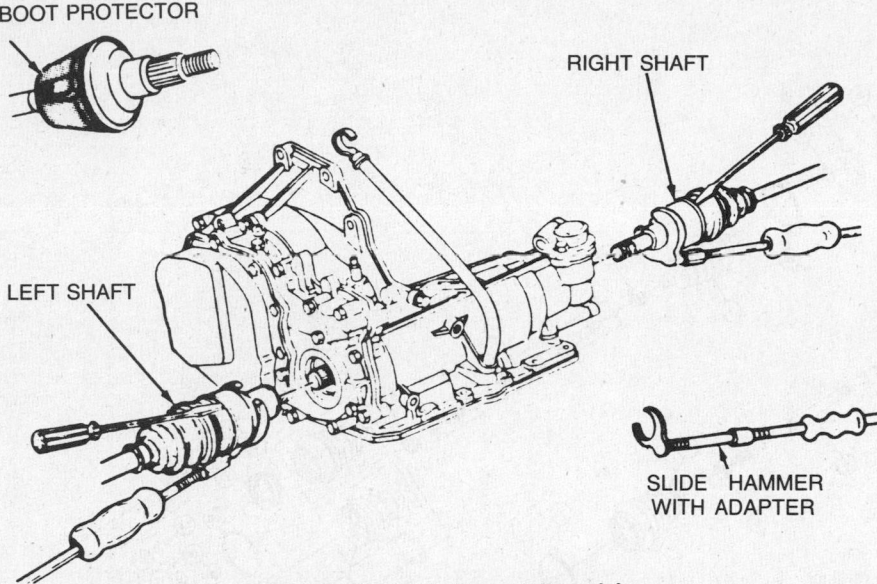

BOOT PROTECTOR

RIGHT SHAFT

LEFT SHAFT

SLIDE HAMMER
WITH ADAPTER

Removing axle shafts on GM models

tion during acceleration due to plunging action hanging up and releasing repeatedly. Probable cause would be foreign particles or lack of lubrication, or improper assembly.

• Remember that tires, suspension, engine, and exhaust system are all up front to add their noises.

• Make a check with front wheels elevated off ground. Spin the wheels by hand to determine if wheel bearing could be noisy or if out of round tires are causing vibration. Many wheel bearings are prelubed and sealed at the factory.

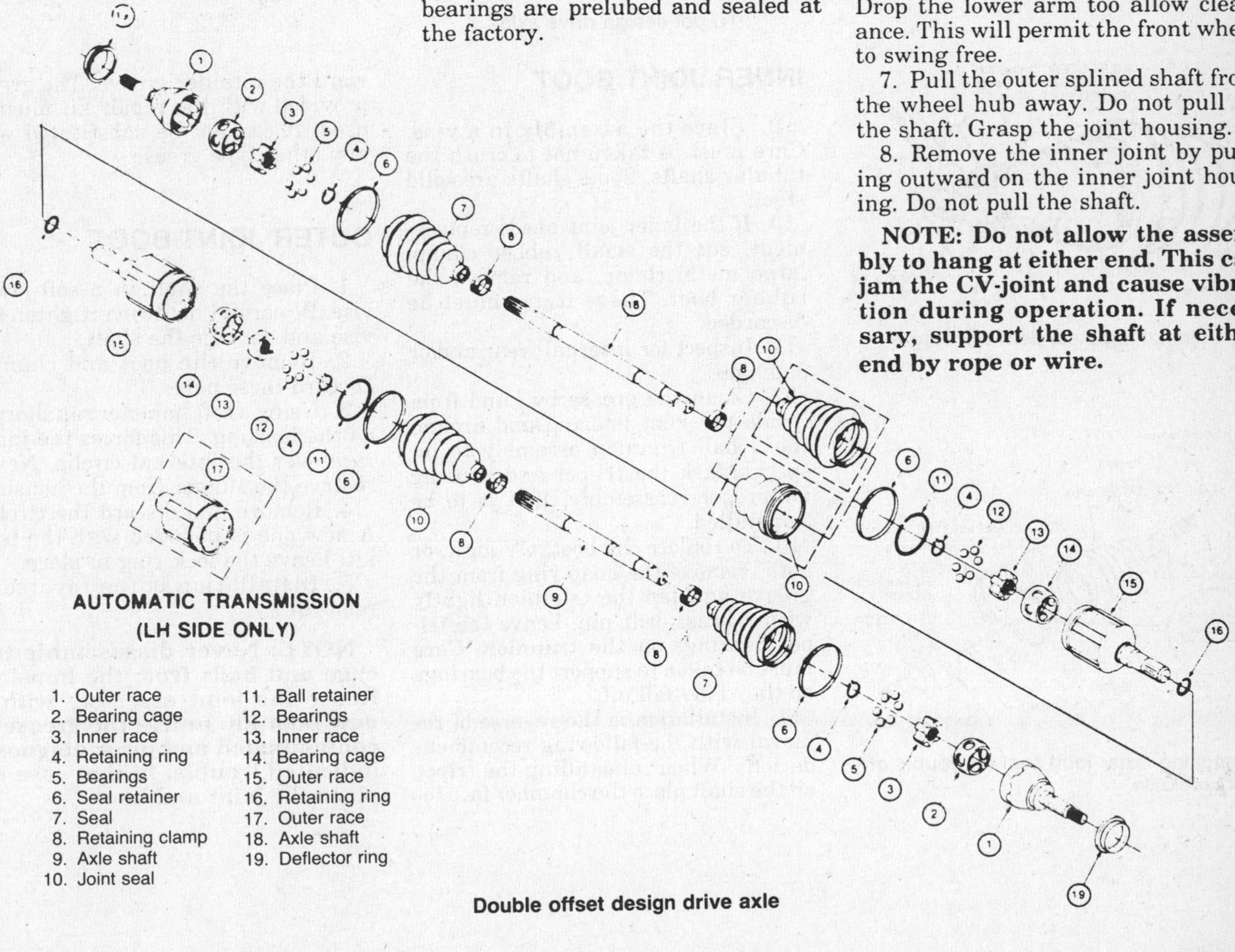

— CAUTION —

Personal injury can occur from spinning wheels by engine power. Spinning a wheel at excess speed may cause damage to CV-joints that could be operating at angles too steep when wheels are allowed to hang. Overspeeding might also cause damage to tires and the differential.

SHAFT REMOVAL

1. Remove the hub nut and discard it.

2. Drain the lubricant from the transaxle. Remove the differential cover (Chrysler only).

3. The speedometer pinion gear assembly must be removed before the right driveshaft can be removed (automatic transaxles only).

4. Rotate the driveshaft to view the circlip.

5. Compress the circlip tangs with needle nose pliers as you pry into the side gear. This compresses the circlip in position for shaft removal later. Keep an awl between the differential pinion shaft and the end face of the shaft to prevent circlip reentry to the groove.

NOTE: This applies to Chrysler cars only.

6. Remove the ball joint clamp bolt. Drop the lower arm too allow clearance. This will permit the front wheel to swing free.

7. Pull the outer splined shaft from the wheel hub away. Do not pull on the shaft. Grasp the joint housing.

8. Remove the inner joint by pulling outward on the inner joint housing. Do not pull the shaft.

NOTE: Do not allow the assembly to hang at either end. This can jam the CV-joint and cause vibration during operation. If necessary, support the shaft at either end by rope or wire.

**AUTOMATIC TRANSMISSION
(LH SIDE ONLY)**

1. Outer race
2. Bearing cage
3. Inner race
4. Retaining ring
5. Bearings
6. Seal retainer
7. Seal
8. Retaining clamp
9. Axle shaft
10. Joint seal
11. Ball retainer
12. Bearings
13. Inner race
14. Bearing cage
15. Outer race
16. Retaining ring
17. Outer race
18. Axle shaft
19. Deflector ring

Double offset design drive axle

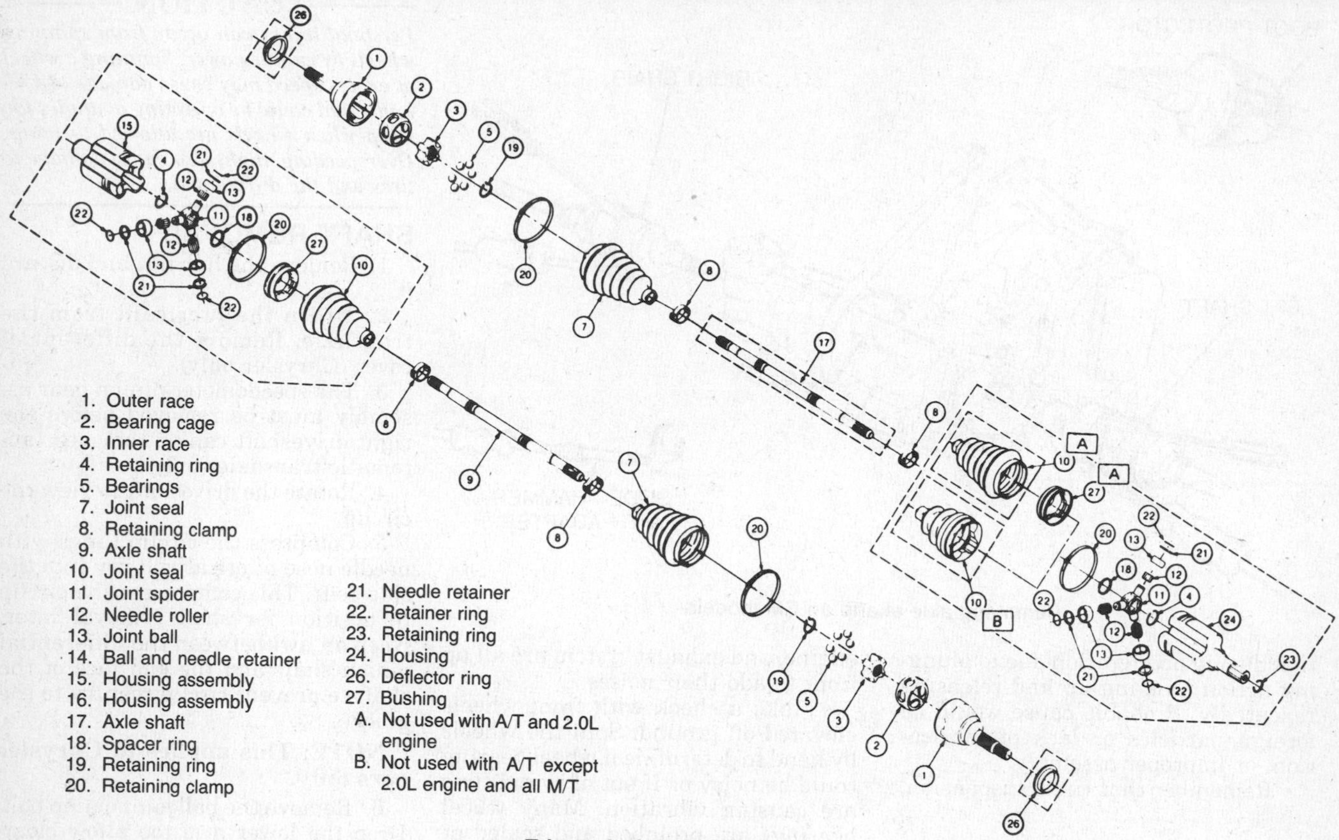

1. Outer race
2. Bearing cage
3. Inner race
4. Retaining ring
5. Bearings
7. Joint seal
8. Retaining clamp
9. Axle shaft
10. Joint seal
11. Joint spider
12. Needle roller
13. Joint ball
14. Ball and needle retainer
15. Housing assembly
16. Housing assembly
17. Axle shaft
18. Spacer ring
19. Retaining ring
20. Retaining clamp

21. Needle retainer
22. Retainer ring
23. Retaining ring
24. Housing
26. Deflector ring
27. Bushing
A. Not used with A/T and 2.0L engine
B. Not used with A/T except 2.0L engine and all M/T

Tri-pot design drive axle

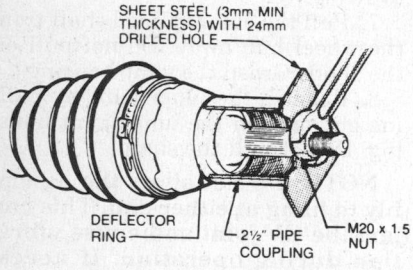

SHEET STEEL (3mm MIN THICKNESS) WITH 24mm DRILLED HOLE

DEFLECTOR RING

2½" PIPE COUPLING

M20 x 1.5 NUT

Installing steel deflector ring

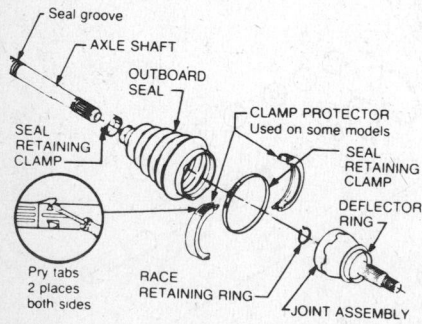

Seal groove
AXLE SHAFT
OUTBOARD SEAL
SEAL RETAINING CLAMP
CLAMP PROTECTOR Used on some models
SEAL RETAINING CLAMP
DEFLECTOR RING
Pry tabs 2 places both sides
RACE RETAINING RING
JOINT ASSEMBLY

Removing outer joint seal on double off-set type axle

INNER JOINT/BOOT

9. Place the assembly in a vise. Care must be taken not to crush the tubular shafts. Some shafts are solid steel.

10. If the inner joint needs replacement, cut the small rubber clamp, large metal clamp, and remove the rubber boot. These items must be discarded.

11. Inspect for internal wear and/or damage.

12. Clean the grease by hand from inside the joint housing and around the 3 ball trunnion assembly to inspect. Mark the tri-pot and housing for proper reassembly, If it is to be reinstalled.

13. To replace the boot, CV-joint, or both, remove the snap ring from the groove and tap the trunnion lightly with a brass drift pin. Leave the tri-pot bearings on the trunnion. Care must be taken to support the bearings as they may fall off.

14. Installation is the reverse of removal with the following recommendations. When reinstalling the tripot on the shaft place the chamber face to-ward the retainer groove. The grease provided with the repair kit must be used. It can not be substituted with any other type grease.

OUTER JOINT/BOOT

1. Place the shaft in a soft-jawed vise. Be careful not to overtighten the vise and damage the shaft.

2. Remove the boot and clamps. Discard these parts.

3. Using a soft hammer rap sharply on the housing. This forces the inner race over the internal circlip. Never remove the slinger from the housing.

4. Remove and discard the circlip. A new one is included with the boot kit. Leave the lock ring in place.

5. Installation is the reverse of removal.

NOTE: Never disassemble the cage and balls from the housing. Reuse the joint assembly with a new boot kit, unless the grease is contaminated and prior diagnosis indicated trouble. In that case replace the joint and boot.

Strut Overhaul **34**

STRUT SERVICE AND REPAIR

MacPherson struts are appearing on the front (and rear) wheels of more and more cars. The strut design takes up less room in the engine compartment, compared to a conventional upper and lower arm with shock absorber arrangement. The trend toward smaller, lighter and more efficient packaging mandates the use of a strut suspension to permit more room for engine accessories and front wheel drive components.

Strut Suspension Design

In a conventional front suspension, the wheel is attached to a spindle, which is in turn, connected to upper and lower control arms through upper and lower ball joints. A coil spring between the control arms (sometimes on top of the upper arm) supports the weight of the vehicle and a shock absorber controls rebound and dampens oscillations.

In a strut type suspension, the strut performs a shock dampening function, like a shock absorber, but unlike a conventional shock absorber, the strut is a structural part of the vehicle's suspension.

The strut assembly usually contains a spring seat to retain the coil spring that supports the vehicle's weight. The shock absorber is built into the body of the strut housing. The strut is normally attached at the bottom to the lower control arm and at the top to the car body. The upper mount usually features a bearing that permits the

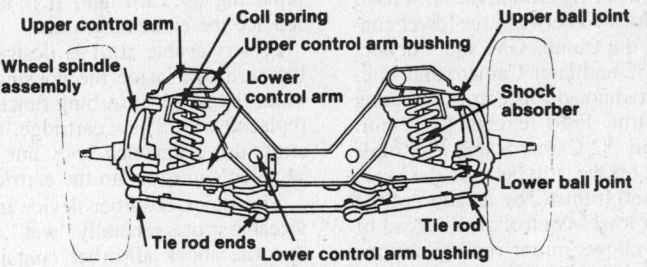

Conventional upper and lower arm suspension

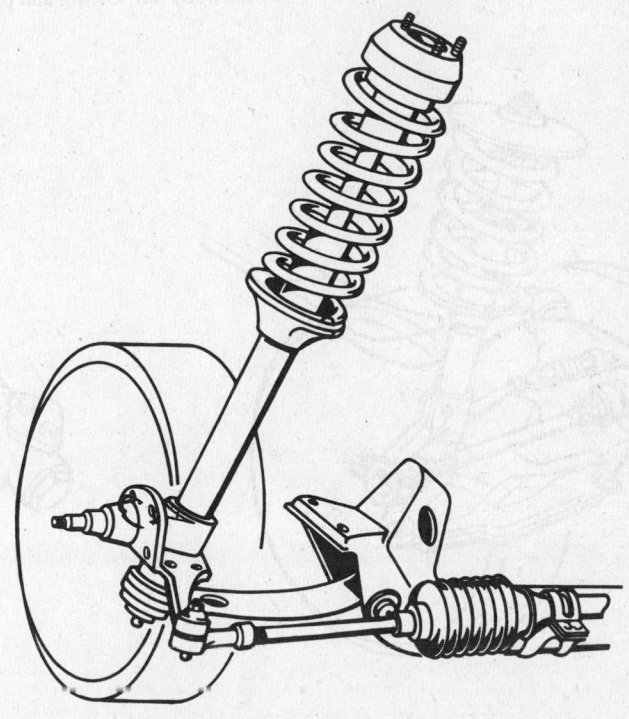

Strut with concentric coil spring (rear wheel drive)

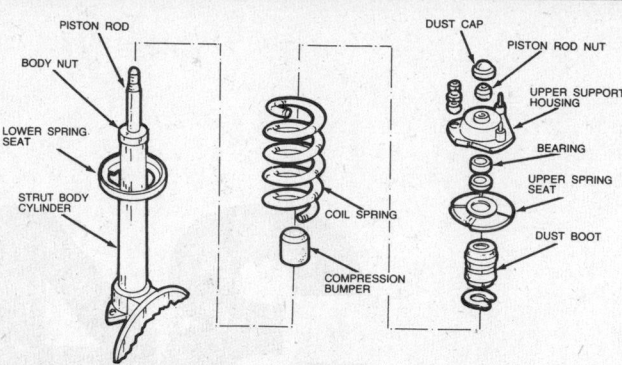

Exploded view of a typical strut

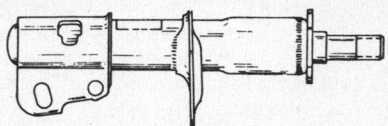

A sealed strut has no body nut and is serviceable by replacement

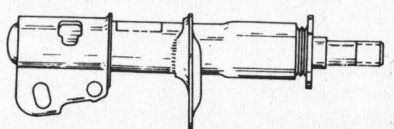

Serviceable struts have a removeable body nut to allow replacement of the strut cartridge

coil spring to rotate as the wheels turn for smoother steering. The entire design eliminates the need for the upper control arm, upper ball joint and many of the conventional suspension bushings. The lower ball joint is no longer a load carrying unit, because it is isolated from the weight of the vehicle.

Domestic struts have taken 2 forms—a concentric coil spring around the strut itself and a spring located between the lower control arm and the frame. GM and Chrysler (except for '82 and later Camaro and Firebird) use the traditional concentric coil spring around the strut. Ford (except the Escort and Lynx) and '82 Camaros and Fire-birds use the spring off the strut between the lower control arm and frame. The location of the spring on the lower control arm instead of on the strut, allows minor road vibrations to be absorbed through the chassis rather than be fed back to the driver through the steering system.

Serviceability

Struts fall into 2 broad categories—serviceable and sealed units. A sealed strut is designed so that the top closure of the strut assembly is permanently sealed. There is no access to the shock absorber cartridge inside the strut housing and no means of replacing the cartridge. It is necessary to replace the entire strut unit.

A serviceable strut is designed so that the cartridge inside the housing, that provides the shock absorbing function, can be replaced with a new cartridge. Serviceable struts use a threaded body nut in place of a sealed cap to retain the cartridge.

The shock absorber device inside a serviceable strut is generally "wet". This means that the shock absorber contains oil that contacts and lubricates the inner wall of the strut body. The oil is sealed inside the strut by the body nut, O-ring and piston rod seal.

Servicing a "wet" strut with the equivalent components involves a thorough cleaning of the inside of the strut body, absolute cleanliness and great care in reassembly.

Cartridge inserts were developed to simplify servicing "wet" struts. The insert is a factory sealed replacement for the strut shock absorber. The replacement cartridge is simply substituted for the original shock absorber cartridge and retained with the body nut, avoiding the near laboratory-like conditions required to service a "wet" strut with "wet" service components.

Most OEM domestic struts are serviced by replacement of the entire unit. There is no strut cartridge to replace. Exceptions to this general rule are the struts used on GM front wheel drive J-cars and A-cars, which feature an internally threaded housing, accessible by removing the OEM cap from the housing. Once the old cartridge is removed, a new cartridge can be threaded

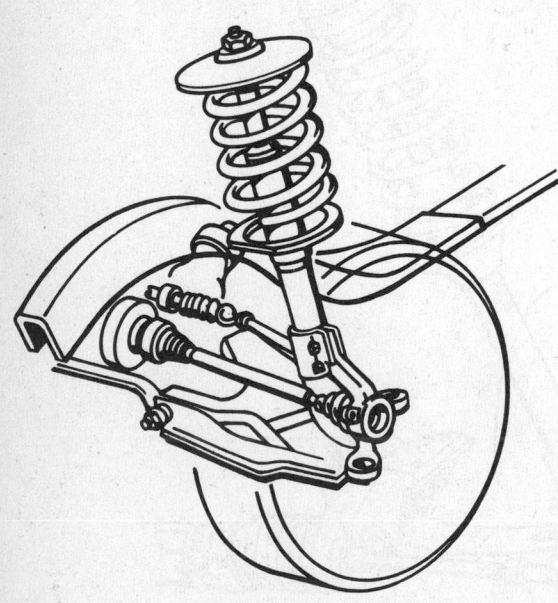

Strut with concentric coil spring (front wheel drive)

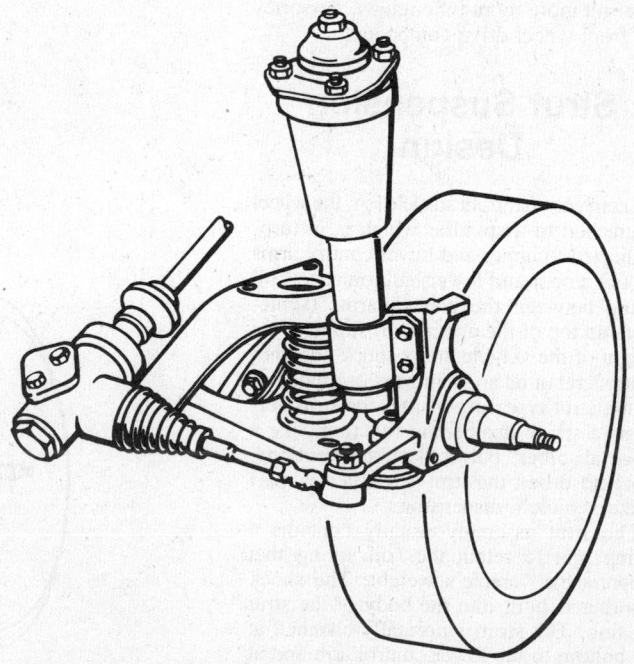

Modified MacPherson strut design with coil spring on the lower arm

into the housing.

Sealed, OEM units can also be serviced by replacement with an aftermarket unit, that will permit future servicing by cartridge replacement.

WHEEL ALIGNMENT

It is not always necessary to re-align the wheels after struts are serviced. If care is taken matchmarking affected components and in reassembling, alignment may be unaffected. However, if wheels were not in proper alignment prior to service, or if the entire strut assembly was replaced, a wheel alignment check should be made. Generally, only camber is adjustable, and then only within a narrow range.

Do not attempt to bend components to correct wheel alignment.

Since the majority of OEM struts are serviced by replacement, most manufacturers recommend wheel alignment following strut replacement.

Tools

Without the right tools, a strut job will take longer than necessary and can be dangerous.

A normal selection of hand tools such as open end and box wrenches, sockets, pliers, screwdrivers and hammers are necessary to work on struts. Extensions and universal joints will help reach tight spots. Be sure to have both metric and inch-sized wrenches on hand. Two big time-savers are "crowsfeet" and ratcheting box wrenches in assorted sizes. Torx fasteners are also showing up more and more in chassis fasteners.

In addition to the normal handtools, some sort of spanner is necessary to remove the body nut on serviceable struts. Sometimes a pipe wrench can be used successfully.

Strut and cartridge replacement requires a spring compressor.

Makeshift tools for compressing coil springs—threaded rod, chains, wire or other methods—should never be used. The coil spring is under tremendous compression and can fly off causing personal injury and damage to equipment. Use only a good quality spring compressor such as described below.

Economy, or manual, spring compressors are the least expensive but more time consuming to use. Angle hooks grasp the spring coils and must be compressed with a wrench. For those who service struts infrequently, this is probably the wisest investment for purchase.

Other manual spring compressors (jaws type) are faster to operate, have a more positive gripping action and can be used on or off the car. These types are probably not cost effective for the do-it-yourselfer, but can be rented from auto supply stores for single-time use.

For volume work, compressors that are pneumatically or hydraulically operated are

MAINTAINING WHEEL ALIGNMENT

The location and method of adjusting wheel alignment determines the components that must be match-marked to maintain wheel alignment. There are 4 basic methods of adjusting wheel alignment. Almost all cars use one of these or a slight variation.

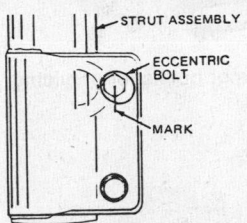

Mark the eccentric (camber adjusting bolt) relative to the clevis mounting bracket.

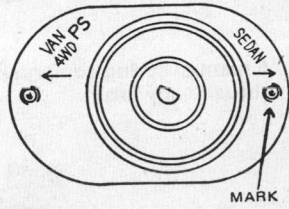

Mark the mounting stud that faces the front of the vehicle. This type of bracket is reversible for varying applications.

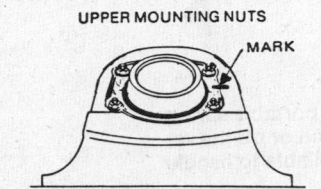

Mark the upper support housing relative to the inner fender before removing the strut from the upper mount.

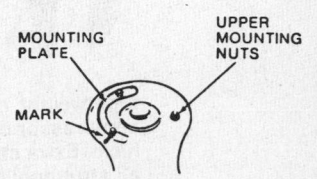

Mark the location of the mounting plate relative to the location on the inner fender.

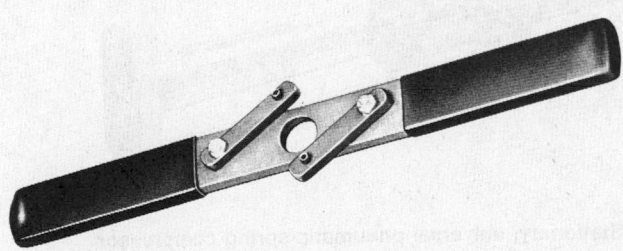

A simple spanner wrench designed for use with body nuts equipped with recessed lugs. A pipe wrench is a frequent substitute

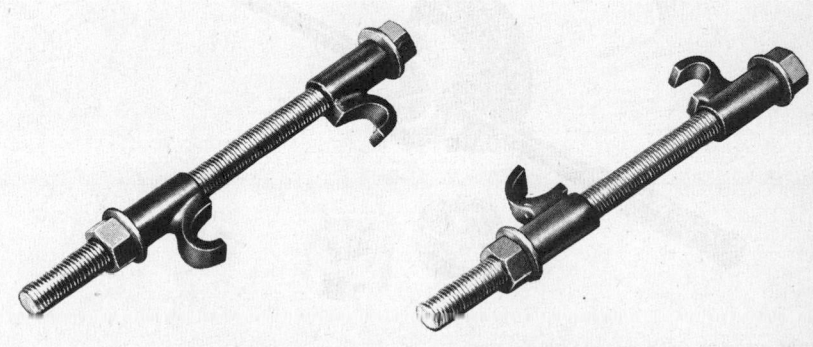

An economical manual spring compressor

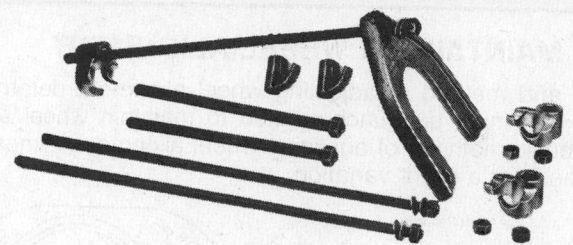

A manual spring compressor with plates or hooks for servicing virtually any strut

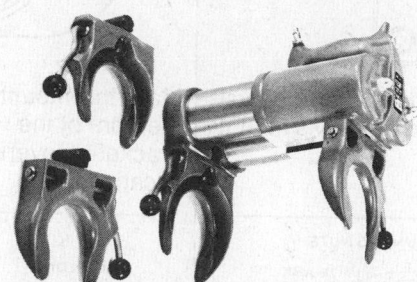

Lightweight, air operated, portable spring compressor can be used on or off the vehicle. Extra shoes are available to handle all strut applications

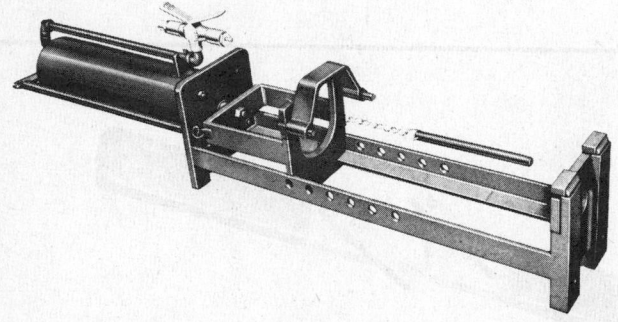

Stationary, universal pneumatic spring compressor

Spanner wrench with adaptor inserts for various applications of body nuts. This type of spanner can be used with a torque owrench for retorqueing the body nut

best. Air operated compressors are suitable for all types of struts (through use of adaptors), are lightweight and can be used on or off the vehicle. Bench mounted hydraulically operated units are probably the safest, but are also the most expensive and require that the strut be removed from the vehicle, which means separating brake lines and other connections which can be time consuming.

There are also universal kits that fit all struts in either the manual or air operated types.

Regardless of what type of spring compressor you're using, GM front wheel drive A-, J-, and X-cars as well as Chrysler Corp. Omni, Horizon and K-cars, require the use of a special spring compressor with self-leveling plates to grasp the spring seats as the spring is compressed. Likewise, the portable, pneumatic units have extra wide shoe sets suitable for these cars. The shoes are also epoxy coated to avoid scratching the coated springs on these models.

GM front wheel drive A-, J- and X-cars also make use of a camber assist tool, that makes camber adjustment a one man job.

A tube cutter is necessary on GM J-cars to cut the welded top from the strut housing for cartridge replacement.

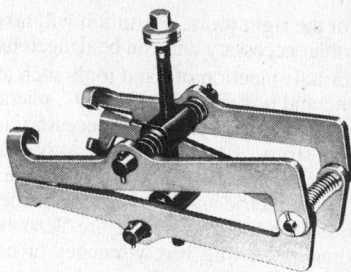

"Jaws" type spring compressor

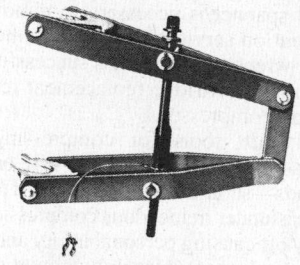

Spring compressor for GM and Chrysler product applications

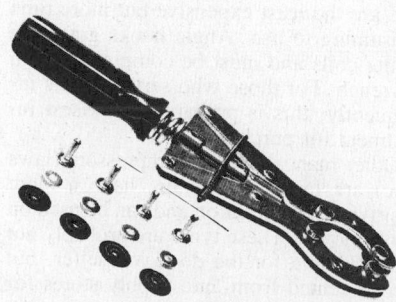

A tube cutter allows opening of the GM J-car struts for cartridge replacement

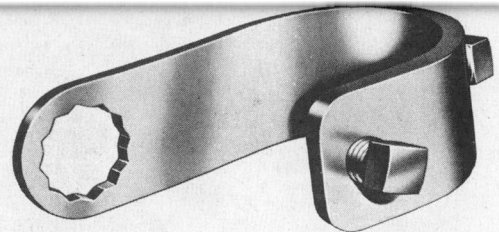

A camber assist tool makes GM cars a one-man job

Mark the position of the attachments that control wheel alignment. See Maintaining Wheel Alignment earlier in this section

Repair Tips

1. Make sure you have all the tools you'll need. NEVER IMPROVISE A SPRING COMPRESSOR.

2. Normally both front struts should be repaired or replaced at the same time.

3. The easiest way to work on most struts is to remove the entire unit from the vehicle, unless you have access to an air operated spring compressor. Some struts, however, can, and should, be repaired while installed on the vehicle.

4. Always read the instructions packaged with any replacement parts. In particular, note whether the body nut is supplied new or re-used.

5. Mark the position(s) of any bearing plate nuts or cam bolts to assure proper alignment after installation.

6. Be sure to protect the rubber boot on the drive axle of front wheel drive cars.

7. If necessary to remove the brake caliper, do not let the caliper hang by the brake hose. Suspend the caliper from a wire hook or rope.

8. Be careful in clamping a strut in a vise. Special fixtures are available to hold struts in a vise, but are not necessary if care is used to be sure the housing is not crushed or dented. A block of soft wood on either side of the housing will prevent most damage.

9. Use a spring compressor to relieve tension from the spring. Be sure to clean and lubricate the screw threads, particularly on hand operated (manual) spring compressors.

Some springs have a special coating that should not be scuffed.

10. If you are replacing the strut cartridge, clean the inside of the strut housing and the body nut threads before replacing the oil and installing a new cartridge.

11. Be sure to use OEM quality fasteners any time a fastener is replaced.

STRUT OVERHAUL (OFF-CAR)

Following is a typical overhaul procedure of a serviceable MacPherson strut, after having removed the strut from the vehicle. The vehicle should be firmly supported. If it is necessary, to separate the brake line from the strut for strut removal, the brakes will have to be bled after reinstallation. See the manufacturer's car section for specific MacPherson strut removal and installation procedures.

Photos Courtesy Gabriel Div., Maremont Corp.

Step 1. Examine the strut assembly for damage, dented strut body, spring seat, broken or missing strut mounting parts. Any of these will require replacement of the complete assembly. Also inspect other suspension components for wear or damage

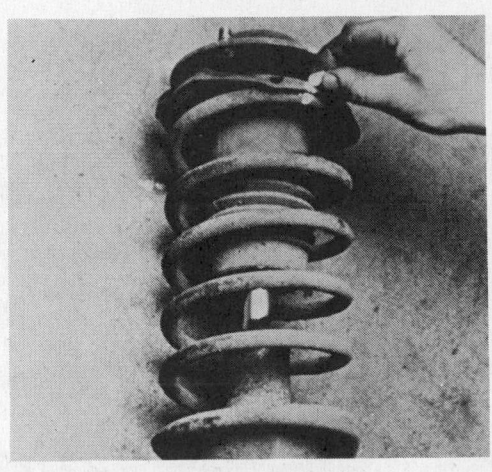

Step 2. Matchmark the upper end of the coil spring and bearing plate to avoid confusion during reassembly

Step 3. To make servicing easier, clamp the strut in a strut vise. The strut vise is designed to clamp the strut tight without damage to strut cylinder. It is very handy for strut work and can be used in your shop vise or mounted to any bench

Step 4. Before using the manual spring compressor, lubricate both sides of the thrust washers and the threads with a light coat of grease

Step 5. Install the compressor hooks on opposite sides of the coil spring with the hooks attached to the upper-most and lower-most spring coils. To avoid possible slippage, use tape or small hose clamps on either side of the compressor hooks

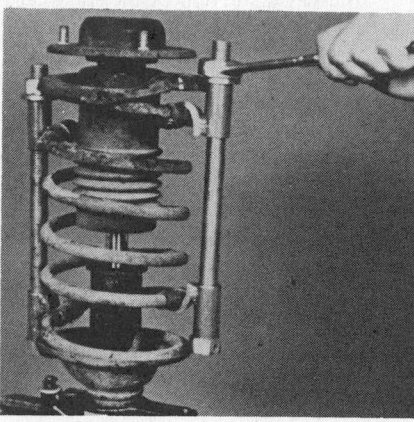

Step 6. Alternately tighten the bolts a few turns at a time until all tension is removed from the spring seat

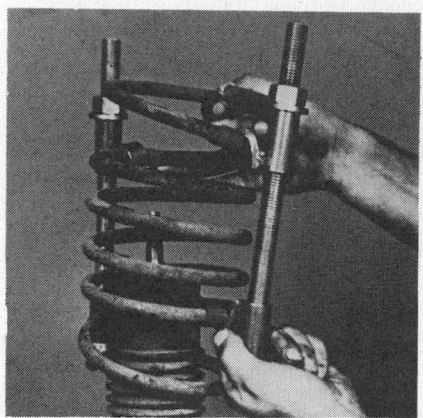

Step 7. Remove the piston rod nut and disassemble the upper mounting parts, keeping them in order for reassembly. Remove the coil spring. There is no need to remove the compressor from the coil spring

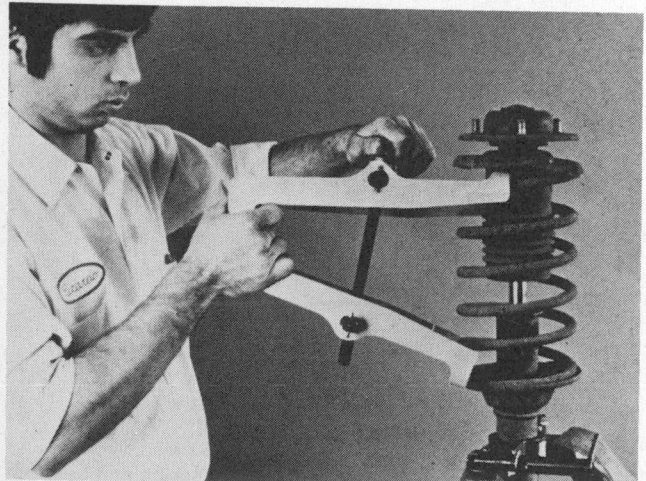

Step 8. An alternative to the manual compressor is the "jaws" type. Turn the load screw to open or close the compressor until the maximum number of spring coils can be engaged

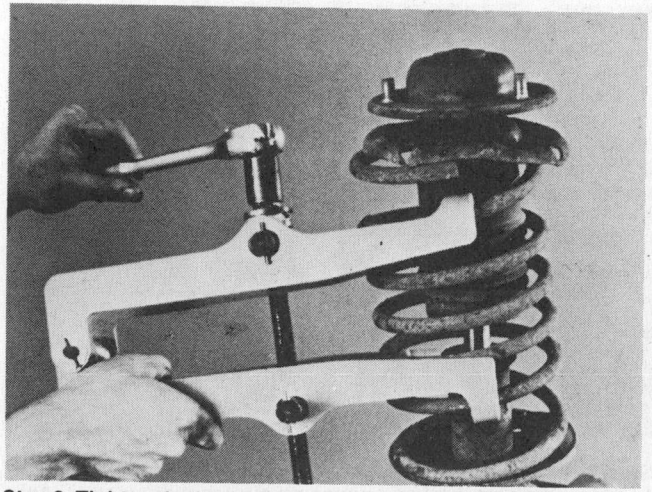

Step 9. Tighten the load screw until the coil spring is loose from the spring seats. There is no need to compress the spring any further

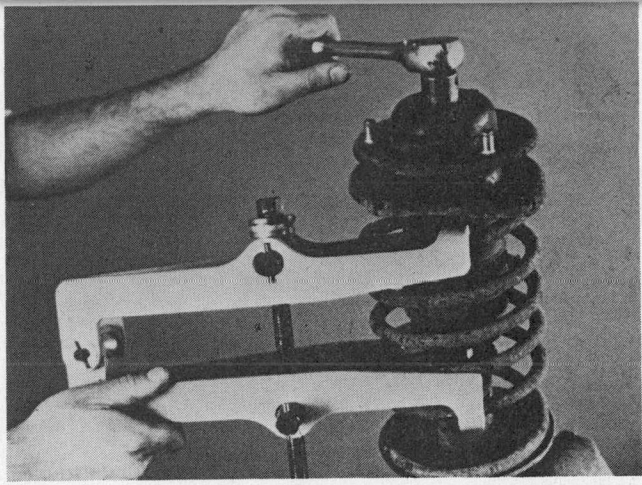

Step 10. Remove the piston rod nut and disassemble the upper mounting parts

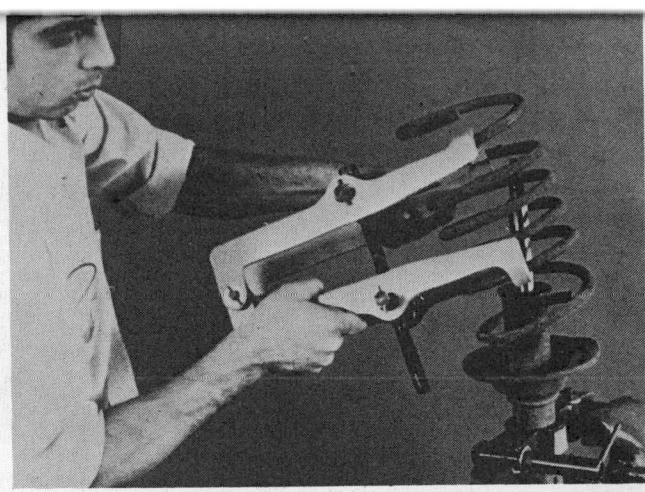

Step 11. Like the manual compressor, there is no need to remove the compressor from the coil spring. Remove the coil spring and compressor

Step 12. Keep the upper mounting parts in order of their removal. They'll be re-assembled in reverse order

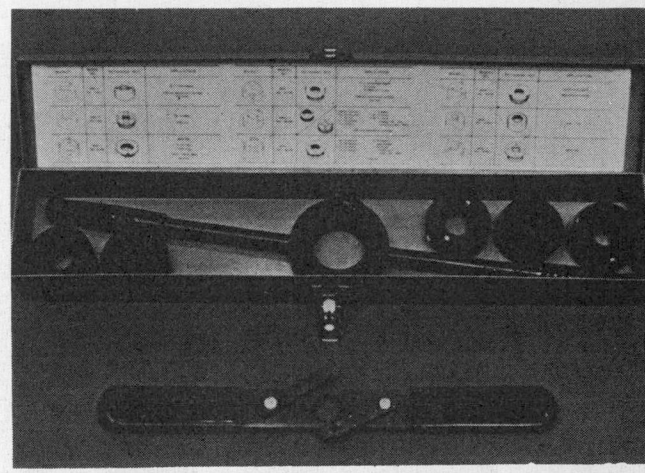

Step 13. A spanner wrench is necessary to remove body nuts, although a pipe wrench will do the job

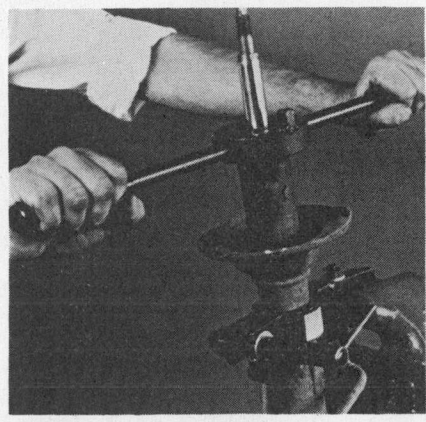

Step 14. Use the spanner wrench or pipe wrench to loosen the body nut

Step 15. Remove the body nut and discard if a new body nut came with the replacement cartridge. If not, save the body nut

Step 16. Use a scribe or suitable tool to remove the O-ring from the top of the housing

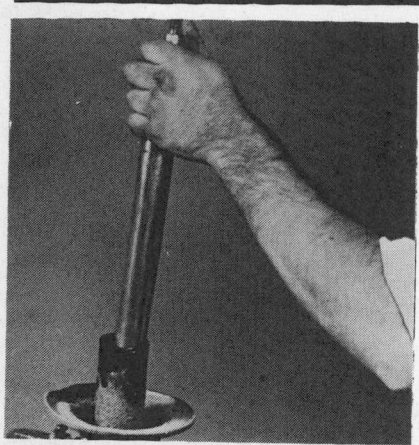

Step 17. Grasp the piston rod and pull cartridge out of the housing. Remove it slowly to avoid splashing oil. Be sure all pieces come out of the housing

Step 18. Pour all of the strut fluid into a suitable container, clean the inside of the strut cylinder, and inspect the cylinder for dents and to insure that all loose parts have been removed from inside of strut body

Step 19. Refill the cylinder with one ounce (a shot glass) of the original oil or fresh oil. The oil helps dissipate internal cartridge heat during operation and results in a cooler running, longer lasting unit. Do not put too much oil in—otherwise the oil may leak at the body nut after it expands when heated

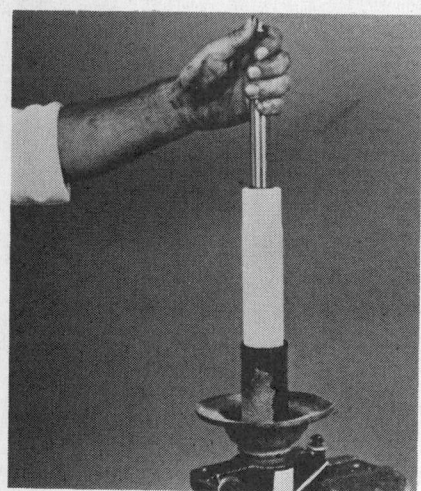

Step 20. Insert the new replacement cartridge into the strut body

Step 21. Push the piston rod *all* the way down, to avoid damage to the piston rod if the spaner wrench slips, and start the body nut by hand. Be sure it is not cross-threaded

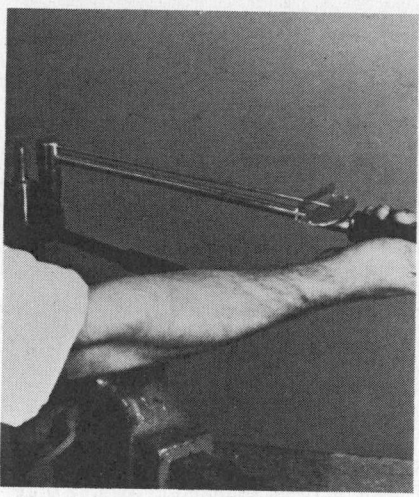

Step 22. Tighten the body nut securely

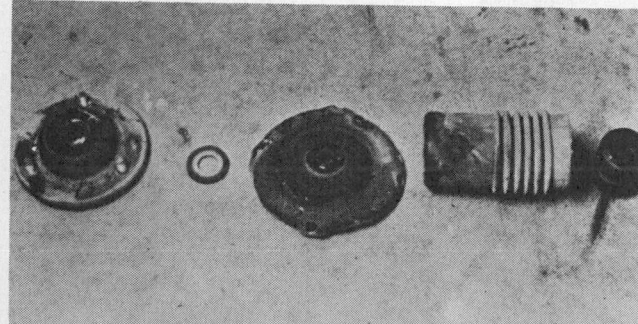

Step 23. Inspect the loose parts prior to re–assembly. Note the chalk mark location for proper seating of the upper spring seat

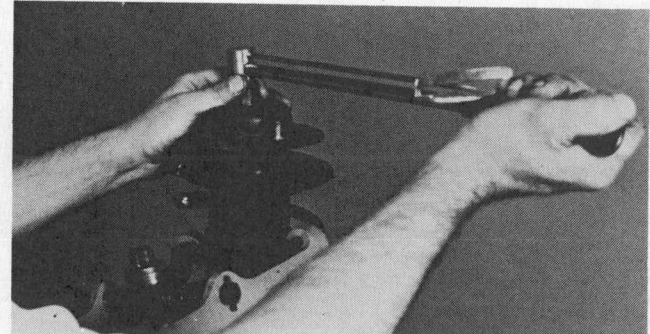

Step 24. Re–assemble the coil spring and upper mounting parts in reverse order. Tighten the piston rod nut and remove the spring compressor. Install the dust cap. Install the strut in the vehicle. See the car section for details

STRUT OVERHAUL

Most domestic car OEM MacPherson struts are sealed units and not repairable. The exceptions are GM front wheel drive A- and J-cars, which use replaceable cartridges. All other cars must use aftermarket struts to be serviceable at a future date. The following procedures cover disassembly of the strut, installation of a serviceable strut, reassembly and cartridge replacement on GM front wheel drive A- and J-models. Consult the applicable manufacturer's car section for removal and installation procedures.

Photos Courtesy Gabriel Div., Maremont Corp.

Step 1. Most domestic cars are serviced initially by replacing the entire strut rather than by using a replacement cartridge. This is necessary because the original equipment struts are sealed shut and cannot be serviced with a replacement cartridge. After-market struts are designed with serviceable threaded body nuts which means they can be serviced in the future by installing a replacement cartridge, using normal cartridge service methods, rather than by replacing the entire strut

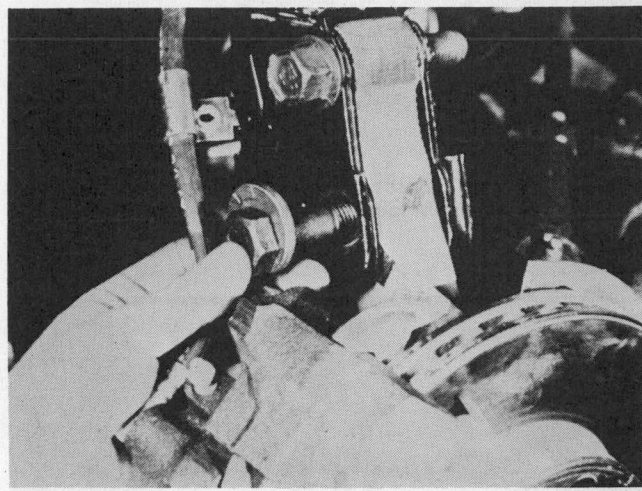

Step 2. An X-car is shown, but the lower mount on the Citation is typical of many vehicles. They all have two bolt clevis mounts and the position of the strut determines the camber adjustment. This means that if you are replacing a sealed strut, front end alignment is necessary because the original alignment is eliminated when you change the strut. If the car has a serviceable strut, you can retain the alignment by marking the position of the mounting bolt relative to the strut. GM has made a running change on the lower mount of their X-Car. The earlier type had an eccentric bolt for camber adjustment. Camber on the latest type is adjusted by pushing or pulling on the wheel with the bolts loosened slightly, but the eccentric can be installed on later cars

Step 3. A special type spring compressor is required for the GM cars and Chrysler K and L cars. A compressor should be used that does not damage the protective coating on the coil spring. Virtually any compressor can be used on other car lines/models

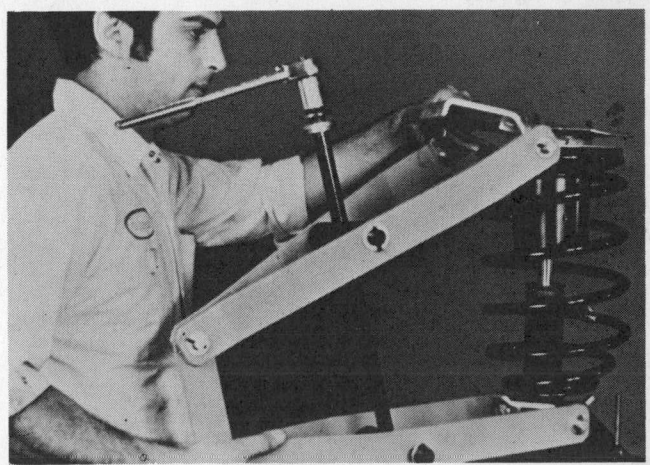

Step 4. Secure the strut in the strut vise; turn the load screw counter-clockwise until the lower plate can be fitted under the lower spring seat and the upper plate can be fitted between the upper spring seat and support housing

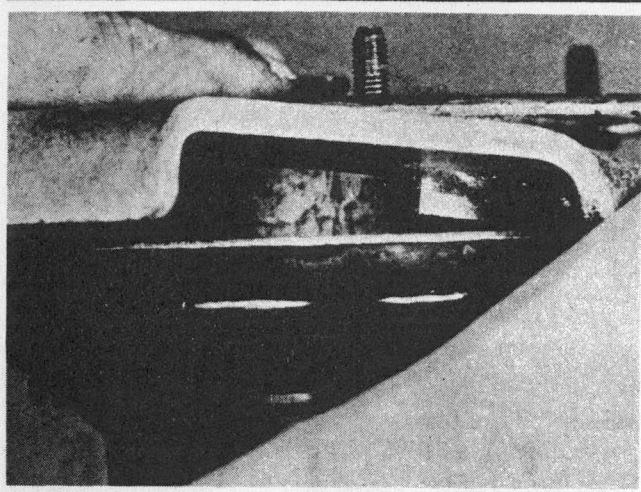

Step 5. Make sure that the crescent shaped bars on the upper compression plate are located inside the upper spring seat

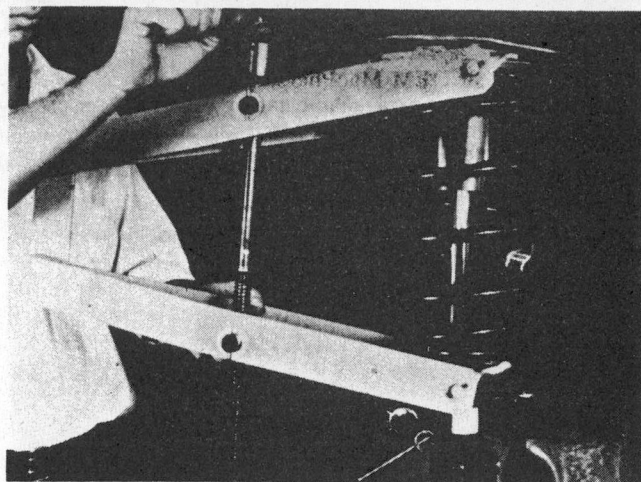

Step 6. Turn the load screw clockwise enough to tighten the compression plates on the spring seats. Stop and make sure that the coil spring will not arch, and that the pivot points are aligned with the center-line of the coil spring

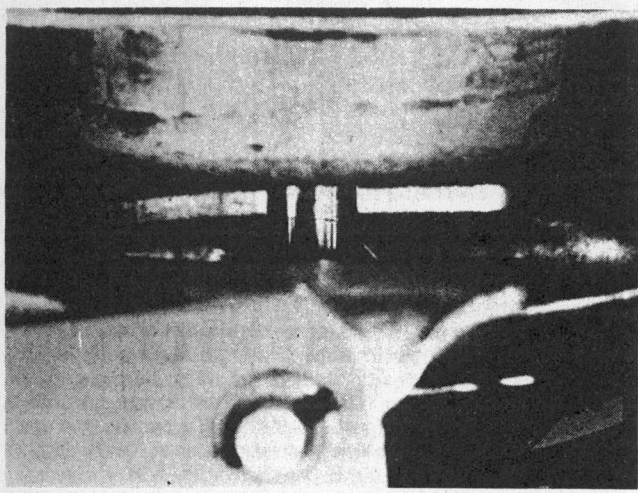

Step 7. Continue to tighten the load screw until the upper support housing can be pulled up to expose about ½ inch of piston rod. This assures that the spring load has been removed from upper spring seat

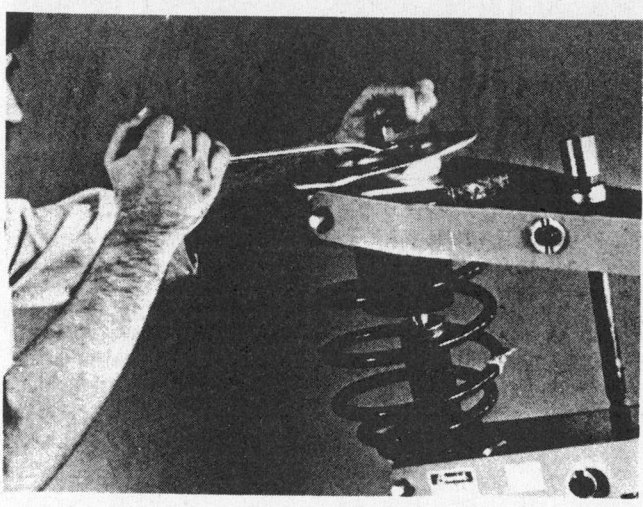

Step 8. Remove the piston rod nut with the aid of a wrench to keep the piston rod from turning and remove upper support housing

Step 9. Turn the load screw counter-clockwise until the spring tension is completely relieved. Remove the compressor, coil spring and upper support housing from the strut.

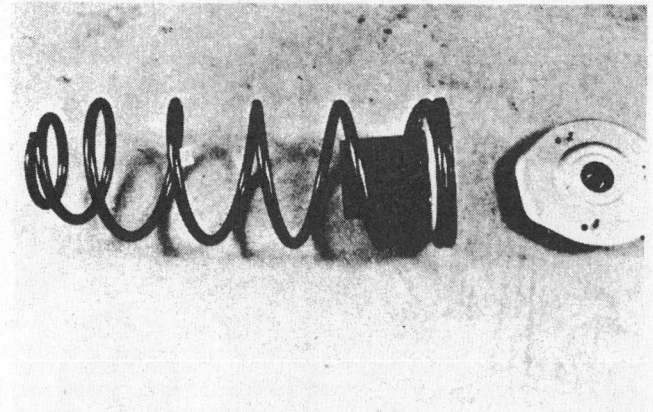

Step 10 Assemble the upper mounting parts in order of their removal. They'll be re-assembled in reverse order

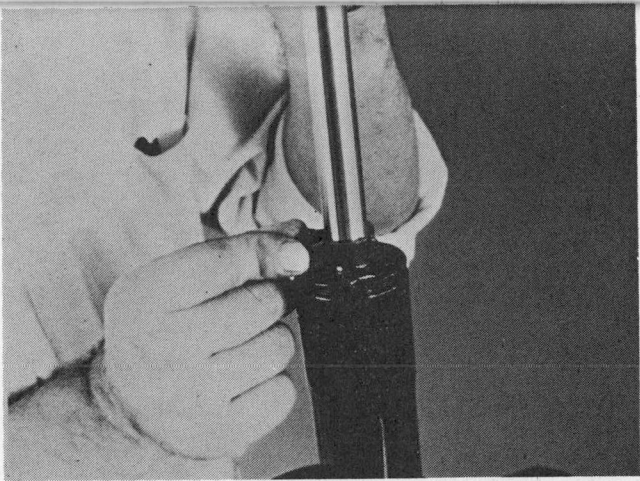

Step 11. Place the new strut in the vise and extend piston rod fully and install clip (spring type clothes pin will do) as shown. This keeps the piston rod extended while assembling the spring and upper mounting parts.

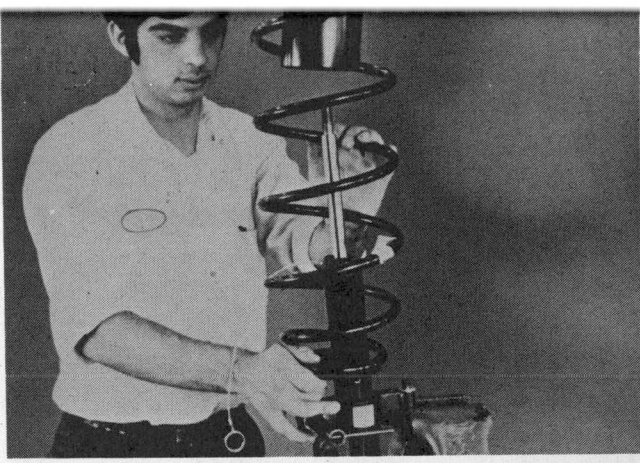

Step 12. Install the coil spring and upper spring seat on the new strut

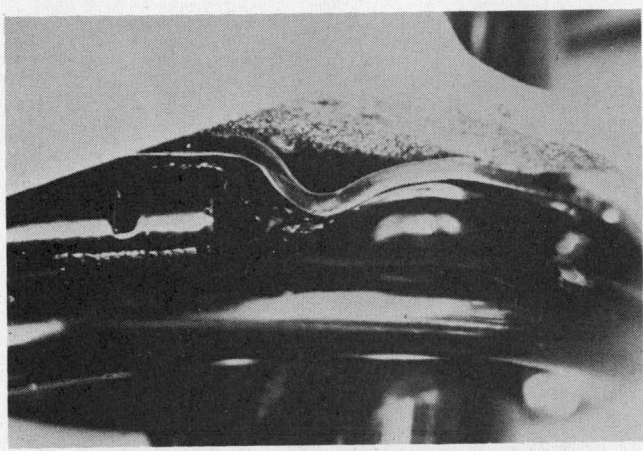

Step 13. Make sure that the spring helix is aligned with the lower spring seat

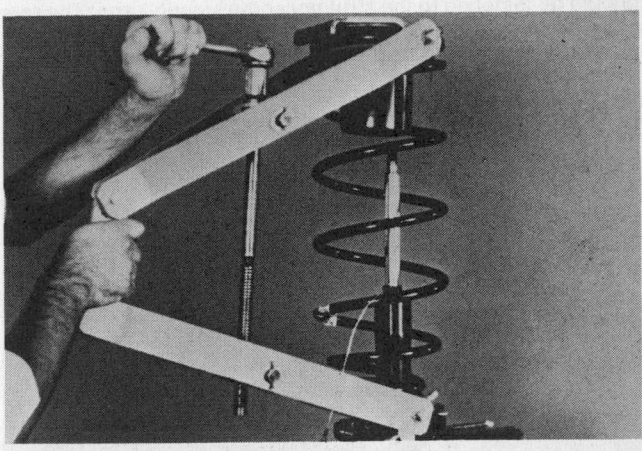

Step 14. Locate upper and lower compression plates on spring seat

Step 15. Make sure that the crescent shaped bars on the upper compression plate are located on the upper spring seat as shown. Turn the load screw clockwise enough to tighten the compression plates on the upper and lower spring seats. Stop. Again, to assure that the coil spring will not arch, make sure that the pivot points are aligned with the centerline of the coil spring. Then continue turning the load screw clockwise until about 1½ inches of piston rod is showing above the upper spring seat

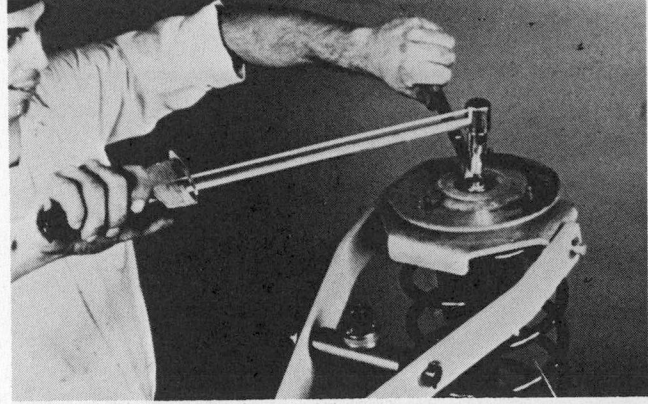

Step 16. Install upper support housing on piston rod. Tighten the piston rod nut and remove the compressor from the strut and the strut from the vise. Install the strut. See the car section for details

GM J- AND A-CARS ONLY

Step 1. Place the strut assembly in a vise, and compress the coil spring. Remove the piston rod, upper support housing, spring seat and coil spring. If the universal pneumatic spring compressor is used, an adaptor provided with the compressor should be fastened to the strut under the steering arm. The ears of the adaptor should be aligned with steering arm. The adaptor provides a square seating surface for the strut while it is being compressed

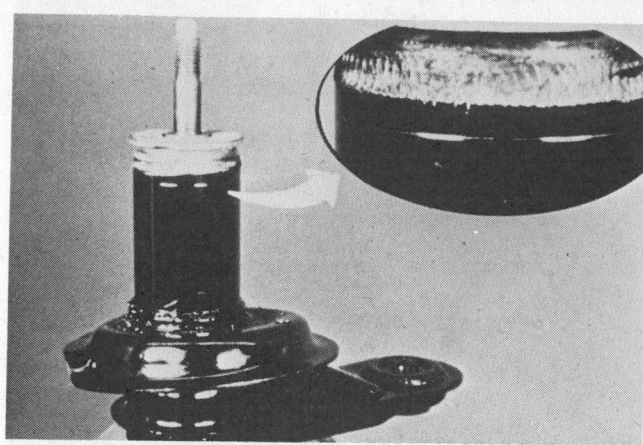

Step 2. J and A–car struts have a welded upper closure, but the strut is designed so the damping mechanism can be replaced with a cartridge insert. Just below the spin weld there is a cut-line scribed in the strut body

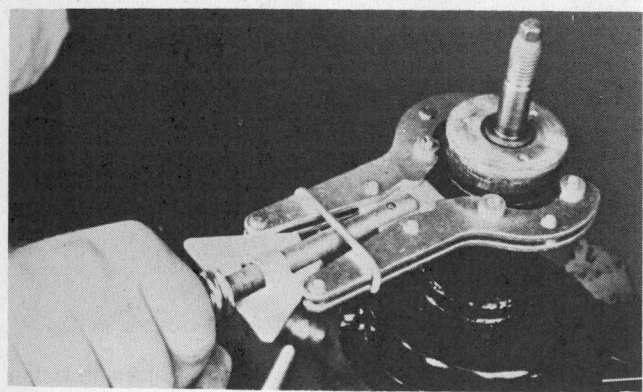

Step 3. Using a pipe cutter, cut open the strut body at the scribed line. (Note: *It is important that the cut be made on the cut-line*)

Step 4. Remove the cartridge and oil from the strut. Note the threads on the inside of the strut. Deburr the top of the strut body if necessary

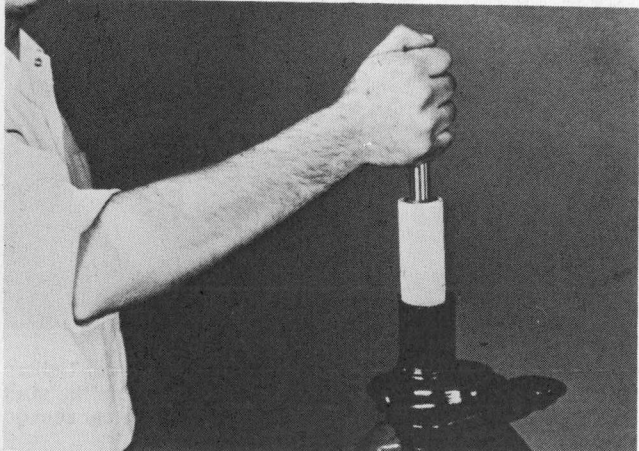

Step 5. Pour about one ounce of oil into the strut body and insert the replacement cartridge. Push the piston rod down and start the body nut by hand. Tighten the nut securely

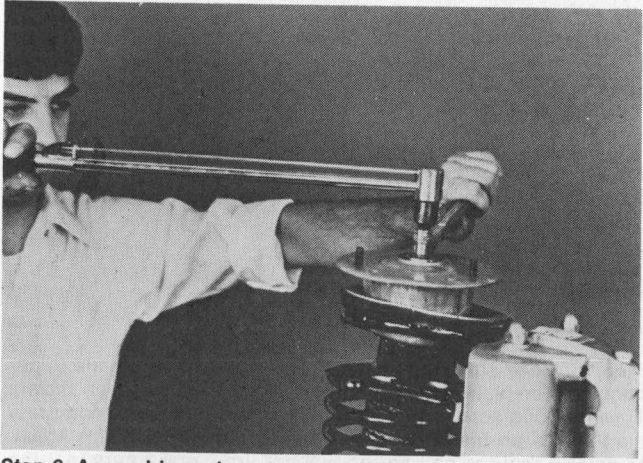

Step 6. Assemble spring, upper spring seat, and upper support housing on the strut, and tighten the new piston rod nut. The renewed strut is now ready to install on the vehicle. Release the spring tension

Problems with MacPherson struts generally fall into 3 main categories: suspension, tire wear and steering. In general, the symptoms encountered are not significantly different from those encountered on conventional suspensions.

Suspension

Sag

Vehicle "sag" is a visible tilt of the car from one side to the other or one end to the other while parked on a level surface.

Weak or damaged strut springs could cause this condition and should be repaired immediately.

Sag will also cause steering and tire wear problems to be more pronounced and vehicle instability on rough roads. Front wheel alignment will not solve the problem.

Weak strut springs increase vehicle sag. See "Tire Cupping".

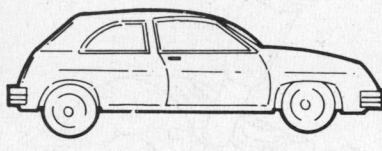

Cartridge Leaks

Strut cartridge leaks (not seepage) indicate the need for cartridge or strut replacement. Be sure the leakage is coming from the strut, and not from elsewhere on the vehicle.

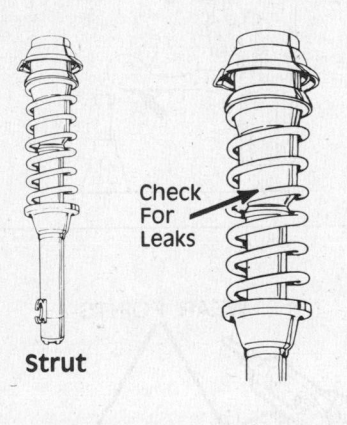

Check For Leaks

Strut

Abnormal Tire Wear

Wear on One Side

One sided tire wear indicates incorrect camber. Check the causes in the accompanying illustration and be sure the wheel alignment is correct.

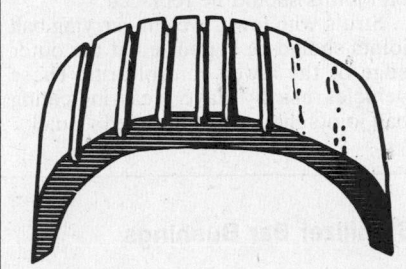

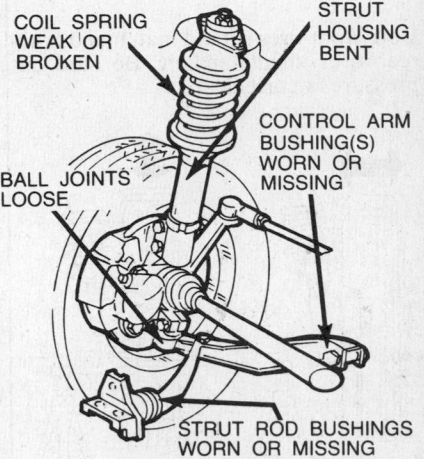

COIL SPRING WEAK OR BROKEN

STRUT HOUSING BENT

CONTROL ARM BUSHING(S) WORN OR MISSING

BALL JOINTS LOOSE

STRUT ROD BUSHINGS WORN OR MISSING

Tire "Cupping"

Cupped tires indicate any or all of the following problems.

1. A weak strut cartridge can be verified by bouncing each corner of the car vigorously and letting go. The car should not bounce more than once, if the shock absorber cartridges are good.

2. Weak strut springs allow sag to increase with only a slight amount of downward pressure. A visual inspection will reveal any broken springs or shiny spots.

3. Check for loose or worn wheel bearings with the weight of the car off of the wheel.

4. Check the wheel balance.

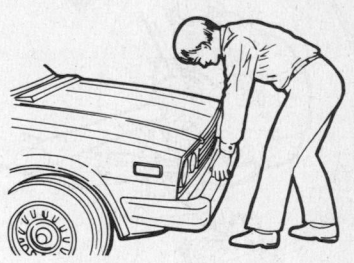

Tread Edge Wear

Wear along tread edges (feathering) indicates a suspension or steering system problem.

1. Strut rod bushings are worn or missing.

2. Tie rod end wear can be determined by grabbing the tie rod end firmly and forcing it up, down or sideways to check for lost motion.

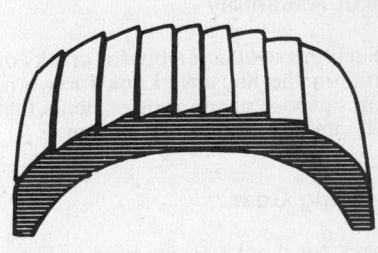

MACPHERSON STRUT PROBLEM DIAGNOSIS

Problems with MacPherson struts generally fall into 3 main categories: suspension, tire wear and steering. In general, the symptoms encountered are not significantly different from those encountered on conventional suspensions.

Tires

Both front tires should match and both rear tires should match. Be sure air pressure is correct.

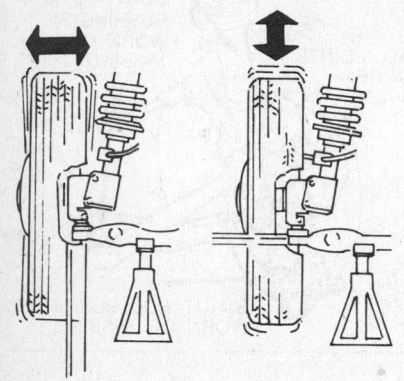

Strut Rod Bushings

Grasp the strut rod and shake it. Any noticeable play indicates excessive wear and need for parts replacement.

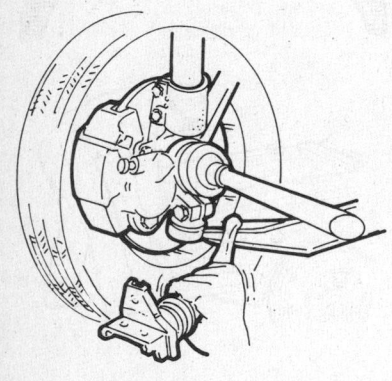

Steering

Ball Joints

Support the car under the frame or crossmember so that the jack does not interfere with the control arm. Rock the tire in and out and up and down. Excessive movement means that both ball joints should be replaced.

Struts with lower weight-carrying ball joints should be supported at the outer edge of the lower control arm. These vehicles usually have wear indicating ball joints that can be checked visually.

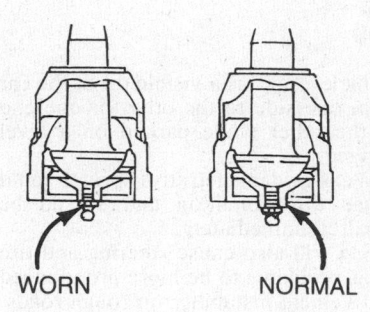

WORN   NORMAL

Stabilizer Bar Bushings

Check for worn bushings or lost motion with the vehicle level and the weight evenly distributed on all wheels.

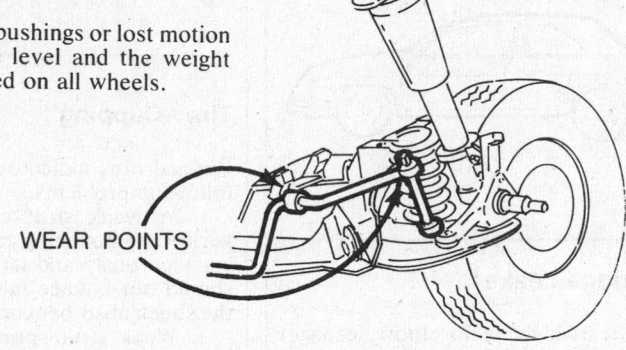

WEAR POINTS

Control Arm Bushings

Support the car under the frame or body and remove the weight from the wheel and control arm. Check for free-play in the bushings at the pivot point, using a pry bar.

NOTE: Some control arm bushings are serviceable only by replacing the entire arm.

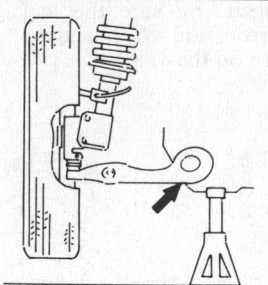

Strut Assembly

Check the strut assembly for cracks or dents in the housing. Look for worn, bent or loose piston rods or dents that will inhibit piston rod movement.

Steering Gear

Check for worn steering gear or loose or worn mounting bolts and bushings.

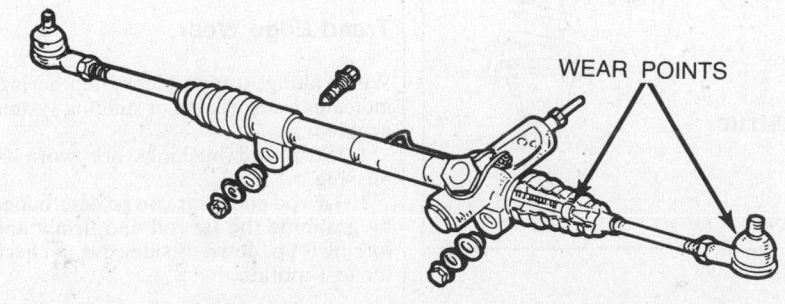

WEAR POINTS

Brakes 35

Hydraulic Brake Component Service

BASIC OPERATING PRINCIPLES

The hydraulic brake system transports the power required to force the frictional surfaces of the braking system together from the pedal to the individual brake units at each wheel. A hydraulic system is used for 2 reasons. First, fluid under pressure can be carried to all parts of an automobile by small hoses (some of which are flexible) without taking up a significant amount of room or posing routing problems. Second, a great mechanical advantage can be given to the brake pedal end of the system and the foot pressure required to actuate the brakes can be reduced by making the surface area of the master cylinder pistons smaller than that of any of the pistons in the wheel cylinders or calipers.

The master cylinder consists of a double reservoir and piston assembly as well as other springs, fittings, etc. Double (dual) master cylinders are designed to separate the wheels from the others into a pair of hydraulic systems. The standard approach has been have separate circuits for the front and rear wheels. Newer models may have a diagonally split system; i.e. a front wheel and the opposite side rear wheel are in a separate circuit from the other front and rear wheel.

Steel lines carry the brake fluid to a point on the vehicles frame near each wheel. A flexible hose usually carries the fluid to the disc caliper or wheel cylinder. The flexible line allows for suspension and steering movement.

The rear wheel cylinders contain 2 pistons, 1 at either end, which push outward in opposite directions. Most brake calipers contain a single piston, however in some cases they may contain more.

All pistons employ some type of seal, usually made of rubber, to minimize fluid leakage. A rubber dust boot seals the outer end of the cylinder against dust and dirt. The boot fits around the outer end of the piston on disc brake calipers and around the brake actuating rod on the wheel cylinders.

The hydraulic system operates as follows: When at rest, the entire system, from the piston(s) in the master cylinder to those in the wheel cylinders or calipers, is full of brake fluid. Upon application of the brake pedal, fluid trapped in front of the master cylinder piston(s) is forced through the lines to the wheel cylinders and calipers. Here, it forces the pistons outward, in the case of drum brakes and inward toward the disc, in the case of disc brakes. The motion of the pistons is opposed by return springs mounted outside the cylinders in drum brakes and by internal springs or spring seals, in disc brakes.

Upon release of the brake pedal, a spring located inside the master cylinder immediately returns the master cylinder pistons to the normal position. The pistons contain check valves and the master cylinder has compensating ports drilled in it. These are uncovered as the pistons reach their normal position. The piston check valves allow fluid to flow toward the wheel cylinders or calipers as the pistons withdraw. Then, as the rubber boot/seal or return springs force the brake pads or shoes into the released position, the excess fluid returns to the reservoir through the compensating ports.

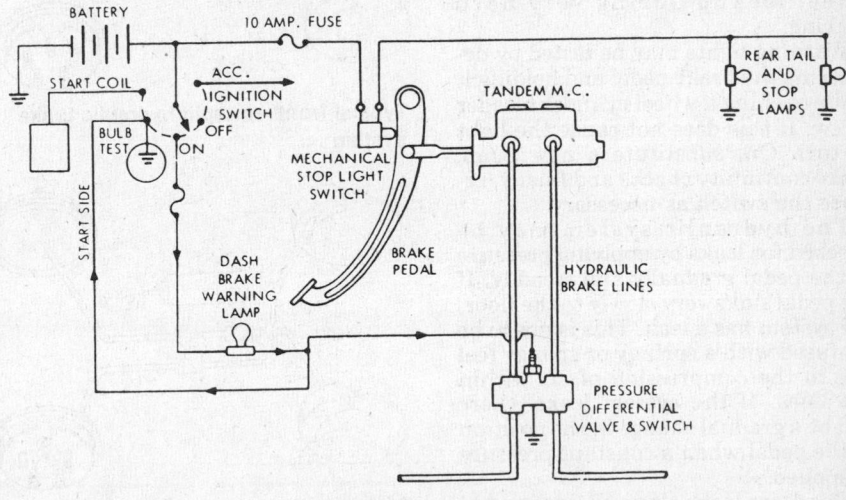

Typical dual brake system

The dual master cylinder has 2 pistons, located 1 behind the other. The primary piston is actuated directly by mechanical linkage from the brake pedal. The secondary piston is actuated by fluid trapped between the 2 pistons. If a leak develops in front of the secondary piston, it moves forward until it bottoms against the front of the master cylinder. The fluid trapped between the pistons will operate opposite sides of the split system. If the other side of the system develops a leak, the primary piston will move forward until direct contact with the secondary piston takes place and it will force the secondary piston to actuate the other side of the split system. In either case the brake pedal drops closer to the floor board and less braking power is available.

The brake system uses a switch to warn the driver when only half of the brake system is operational. This switch is usually located in a valve body which is mounted on the firewall or the frame below the master cylinder. A hydraulic piston receives pressure from both circuits, each circuit's pressure being applied to opposite end of the piston. When the pressures are in balance, the piston remains stationary. When a circuit has a leak, however, the greater pressure in that circuit during brake application will push the piston to 1 side, closing the switch and activating the brake warning light.

In disc brake systems, this valve body contains a metering valve and, in some cases, a proportioning valve or valves. The metering valve keeps pressure from traveling to the disc brakes on the front wheels until the brake shoes on the rear wheels have contacted the drums, ensuring that the front brakes will never be used alone. The proportioning valve controls the pressure to the rear brakes to avoid rear wheel lockup during very hard braking.

Warning lights may be tested by depressing the brake pedal and holding it while opening a wheel cylinder bleeder screw. If this does not cause the light to turn On, substitute a new lamp, make continuity checks and finally, replace the switch as necessary.

The hydraulic system may be checked for leaks by applying pressure to the pedal gradually and steadily. If the pedal sinks very slowly to the floor, the system has a leak. This is not to be confused with a springy or spongy feel due to the compression of air within the lines. If the system leaks, there will be a gradual change in the position of the pedal when a constant pressure is applied.

Check for leaks along all lines and at wheel cylinders or calipers. If no external leaks are apparent, the problem is inside the master cylinder.

DISC BRAKES

Disc brake systems utilize a disc (rotor) with brake pads positioned on either side of it. Braking effect is achieved in a manner similar to the way you would squeeze a spinning phonograph record between your fingers. The disc (rotor) is a casting which may be equipped with cooling fins between the 2 braking surfaces. The fins (if equipped) enable air to circulate between the braking surfaces making them less sensitive to heat buildup and more resistant to fade. Dirt and water do not affect braking action since contaminants are thrown off by the centrifugal action of the rotor or scraped off by the pads. Also, the equal clamping action of the brake pads tends to ensure uniform, straightline stops. Disc brakes are inherently self-adjusting.

DRUM BRAKES (REAR)

Drum brakes employ 2 brake shoes mounted on a stationary backing plate. These shoes are positioned inside a circular drum which rotates with the wheel assembly. The shoes are held in place by springs, this allows them to slide toward the drums (when they are applied) while keeping the linings and drums in alignment. The shoes are actuated by a wheel cylinder

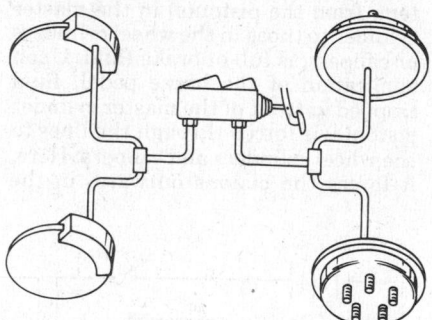

Typical front/rear split hydraulic brake system

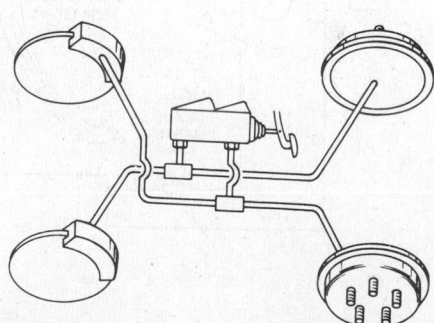

Typical diagonally split hydraulic brake system

which is mounted at the top of the backing plate. When the brakes are applied, hydraulic pressure forces the wheel cylinder's actuating links outward. Since these links bear directly against the top of the brake shoes, the tops of the shoes are then forced against the inner side of the drum. This action forces the bottoms of the 2 shoes to contact the brake drum by rotating the entire assembly slightly (known as servo action). When pressure within the wheel cylinder is relaxed, return springs pull the shoes back away from the drum.

Rear drum brakes are (in most cases) designed to self-adjust themselves during application. Motion causes both shoes to rotate very slightly with the drum, rocking an adjusting lever, thereby causing rotation of the adjusting screw or lever.

POWER BRAKE SYSTEM

Power brakes operate just as standard brake systems except in the actuation of the master cylinder pistons. A vacuum diaphragm is located on the front of the master cylinder and assists the driver in applying the brakes, reducing both the effort and travel he must put into moving the brake pedal.

The vacuum diaphragm housing is connected to the intake manifold by a vacuum hose. A check valve is placed at the point where the hose enters the diaphragm housing, so that during periods of low manifold vacuum brake assist vacuum will not be lost.

Depressing the brake pedal closes off the vacuum source and allows atmospheric pressure to enter on 1 side of the diaphragm. This causes the master cylinder pistons to move and apply the brakes. When the brake pedal is released, vacuum is applied to both sides of the diaphragm and the return springs return the diaphragm and the master cylinder pistons to the released position. If the vacuum fails, the brake pedal rod will butt against the end of the master cylinder actuating rod and direct mechanical application will occur as the pedal is depressed.

MASTER CYLINDERS

——— CAUTION ———
The master cylinder unit is a highly calibrated unit specifically designed for the vehicle it is on. Although cylinders may look alike there are many differences in calibration. If replacement is necessary, make sure the replacement unit is the correct cylinder for the vehicle.

NOTE: Some GM vehicles are equipped with "Quick Take-Up"

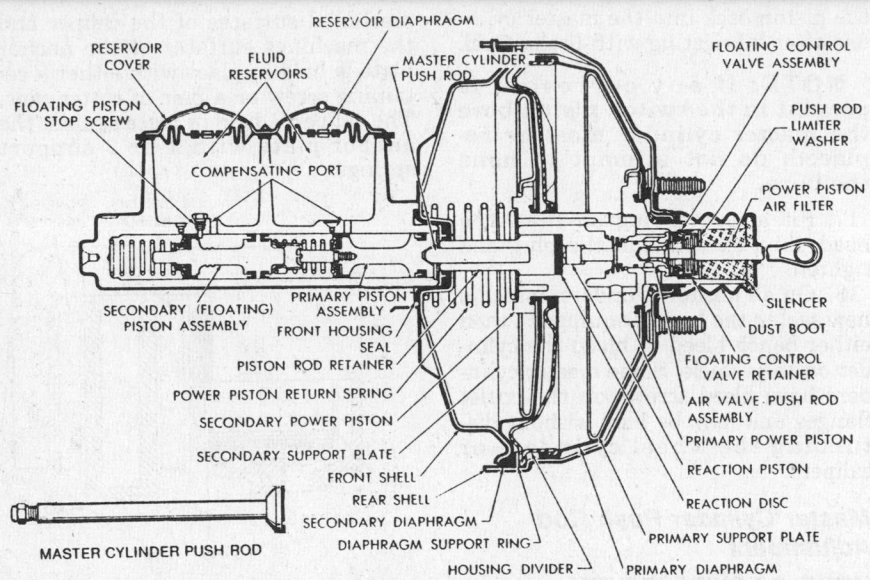

Typical dual master cylinder

MASTER CYLINDER PUSH ROD

(Labels in the diagram:)
RESERVOIR COVER
FLUID RESERVOIRS
RESERVOIR DIAPHRAGM
MASTER CYLINDER PUSH ROD
FLOATING CONTROL VALVE ASSEMBLY
FLOATING PISTON STOP SCREW
COMPENSATING PORT
PUSH ROD LIMITER WASHER
POWER PISTON AIR FILTER
SECONDARY (FLOATING) PISTON ASSEMBLY
PRIMARY PISTON ASSEMBLY
SILENCER
FRONT HOUSING SEAL
DUST BOOT
PISTON ROD RETAINER
FLOATING CONTROL VALVE RETAINER
POWER PISTON RETURN SPRING
AIR VALVE-PUSH ROD ASSEMBLY
SECONDARY POWER PISTON
PRIMARY POWER PISTON
SECONDARY SUPPORT PLATE
REACTION PISTON
FRONT SHELL
REACTION DISC
REAR SHELL
SECONDARY DIAPHRAGM
PRIMARY SUPPORT PLATE
DIAPHRAGM SUPPORT RING
HOUSING DIVIDER
PRIMARY DIAPHRAGM

master cylinders which provide a large volume of fluid to the brakes at low pressure when the brake pedal is initially applied. This large volume of fluid is needed because self retracting piston seals are used on the caliper pistons. The piston seals pull the pistons into the calipers after the brakes are released, thereby preventing the brake pads from causing a drag on the rotors.

The "Quick Take-Up" master cylinder has a hydraulically operated brake warning light switch incorporated in the master cylinder body. The piston is accessible by removing the large plug at the front of the master cylinder body. Only remove the plug when overhauling the cylinder, as brake fluid will escape.

Overhaul procedures on these master cylinders are basically the same as those on conventional master cylinders.

Servicing Master Cylinders

NOTE: Plastic reservoirs need to be removed only for the following reasons: Reservoir is damaged or the rubber grommet(s) be-

tween the reservoir and bore is leaking. Removal of stop pin from Chrysler style plastic reservoir master cylinder to allow removal of pistons. Pin is located underneath front reservoir nipple. Service "Quick Take-up" valve on GM quick take-up master cylin-

ders. The reservoir should be removed by first clamping the cylinder flange in a vice. Next remove the reservoir for the Chrysler style. Grasp the reservoir base on the end and pull away from the body. GM reservoirs must be removed by prying between the reservoir and casting with a pry bar. Grommets can be reused if they are in good condition. Whether or not the reservoir is removed, it and the cover or caps should be thoroughly cleaned.

1. Remove the cylinder from the vehicle and drain the brake fluid.
2. Mount the cylinder in a vise so that the outlets are up and remove the rubber boot seal from the hub.
3. Remove the stop pin or screw from the bottom of the front reservoir, if present.
4. Remove the snapring from the front of the bore and the primary piston assembly.
5. Remove the secondary piston assembly using compressed air or a piece of wire.
6. Clean the metal parts in brake fluid and discard the rubber parts.
7. Inspect the bore for damage or wear, then check the pistons for damage and proper clearance in the bore.

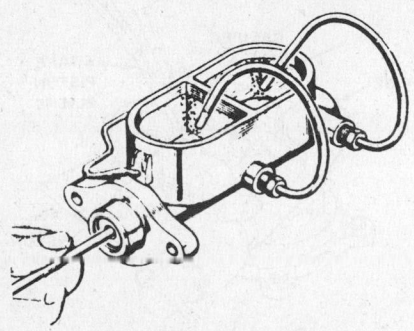

Pre-bleeding master cylinder

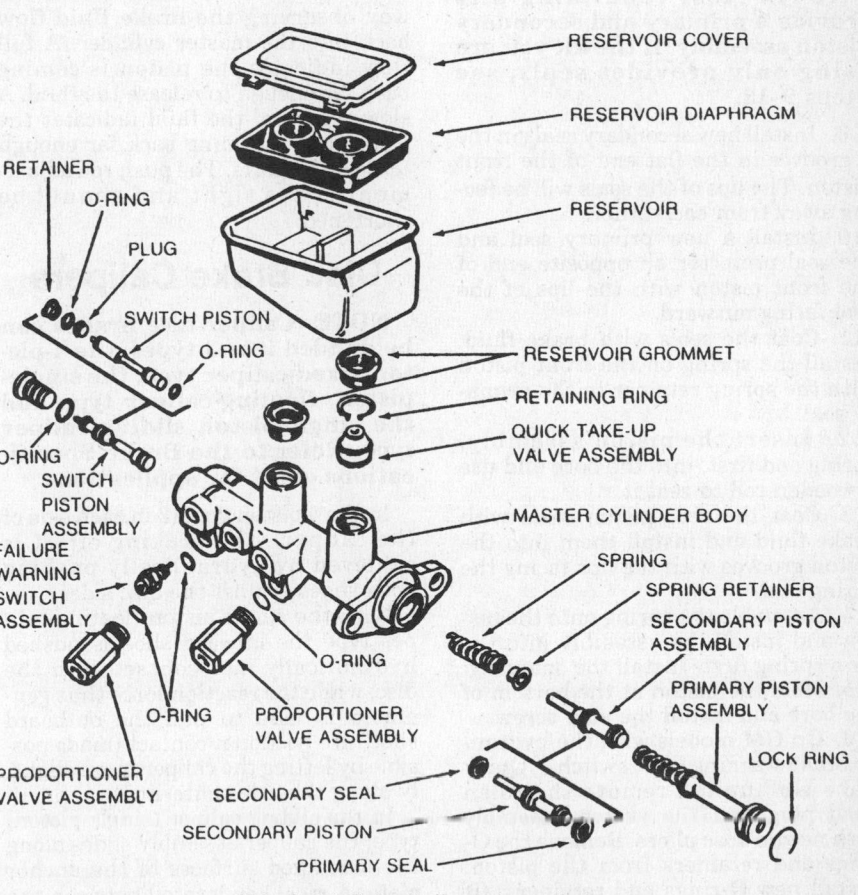

GM "Quick Take Up" master cylinder

(Labels in the diagram:)
RETAINER
O-RING
PLUG
SWITCH PISTON
O-RING
O-RING
SWITCH PISTON ASSEMBLY
FAILURE WARNING SWITCH ASSEMBLY
RESERVOIR COVER
RESERVOIR DIAPHRAGM
RESERVOIR
RESERVOIR GROMMET
RETAINING RING
QUICK TAKE-UP VALVE ASSEMBLY
MASTER CYLINDER BODY
SPRING
SPRING RETAINER
SECONDARY PISTON ASSEMBLY
PRIMARY PISTON ASSEMBLY
LOCK RING
O-RING
O-RING
PROPORTIONER VALVE ASSEMBLY
PROPORTIONER VALVE ASSEMBLY
SECONDARY SEAL
SECONDARY PISTON
PRIMARY SEAL

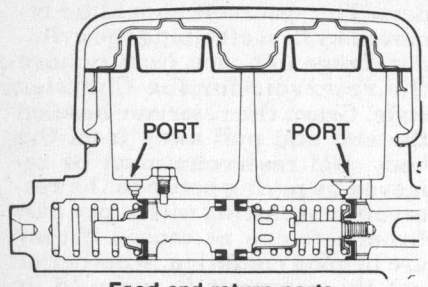
Feed and return ports

CAUTION

Aluminum cylinder bores cannot be honed. The cylinder must be replaced if the bore is pitted or scored.

---

8. If the bore is only slightly scored or pitted it may be honed. Always use hones that are in good condition and completely clean the cylinder with brake fluid when the honing is completed. If any sign of wear or corrosion is apparent on "Quick Take-Up" master cylinder bores, the master cylinder must be replaced; it cannot be honed. If any evidence of contamination exists in the master cylinder the entire hydraulic system should be flushed and refilled with clean brake fluid. Blow out the passages with compressed air.

NOTE: Most rebuilding kits provide a primary and secondary piston assembly. If the kit you are using only provides seals, see Steps 9–13.

9. Install new secondary seals in the 2 grooves in the flat end of the front piston. The lips of the seals will be facing away from each other.
10. Install a new primary seal and the seal protector on opposite end of the front piston with the lips of the seal facing outward.
11. Coat the seals with brake fluid. Install the spring on the front piston with the spring retainer in the primary seal.
12. Insert the piston assembly, spring end first, into the bore and use a wooden rod to seat it.
13. Coat the rear piston seals with brake fluid and install them into the piston grooves with the lips facing spring end.
14. Assemble the spring onto the piston and install the assembly into the bore spring first. Install the snapring.
15. Hold the piston at the bottom of the bore and install the stop screw.
16. On GM models with the hydraulic brake warning light switch ("Quick Take-Up" units), remove the Allen head plug and the switch assembly with needle nose pliers. Remove the O-rings and retainers from the piston. Install new O-rings and retainers, fit

---

the piston back into the master cylinder after lubricating with brake fluid.

NOTE: If any corrosion is present in the switch piston bore the master cylinder must be replaced: do not attempt to hone the bore.

17. Fit a new O-ring on the Allen head plug, then install the plug and tighten.
18. On all master cylinders, install a new seal in the hub (if equipped), then either bench bleed or bleed the cylinder on the vehicle. Some master cylinders have bleed screws on the outlet flanges and may be bled without disturbing the wheel cylinders or calipers.

Master Cylinder Push Rod Adjustment

MODELS EQUIPPED WITH ADJUSTABLE PUSH ROD

After assembly of the master cylinder to the power section, the piston cup in the hydraulic cylinder should just clear the compensating port hole when the brake pedal is full released. If the push rod is too long, it will hold the piston over the port. A push rod that is too short, will give too much loose travel (excessive pedal play). Apply the brakes and release the pedal all the way observing the brake fluid flow back into the master cylinder. A full flow indicates the piston is coming back far enough to release the fluid. A slow return of the fluid indicates the piston is not coming back far enough to clear the ports. The push rod adjustment is too tight and should be shortened.

Disc Brake Calipers

NOTE: Caliper disc brakes can be divided into 3 types: the 4-piston, fixed-caliper type; the single-piston, floating-caliper type and the single-piston sliding-caliper type. Refer to the Brake Specifications chart for applications.

In the 4 piston type (2 in each side of the caliper), the braking effect is achieved by hydraulically pushing both shoes against the disc sides.

With the single piston floating-caliper type the inboard shoe is pushed hydraulically into contact with the disc, while the reaction force thus generated is used to pull the outboard shoe into frictional contact (made possible by letting the caliper move slightly along the axle centerline).

In the sliding caliper (single piston) type, the caliper assembly slides along the machined surfaces of the anchor plate. A steel key located between the

---

machined surfaces of the caliper and the machines surfaces of the anchor plate is held in place with either a retaining screw or a pair of cotter pins. The caliper is held in place against the anchor plate with 1 or 2 support springs.

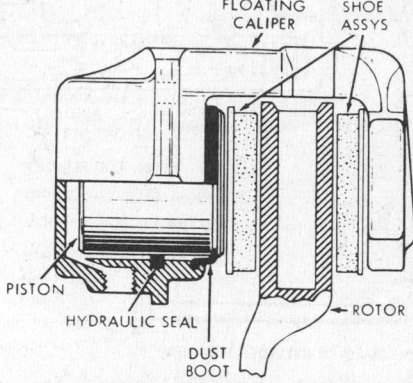
Floating caliper disc brake

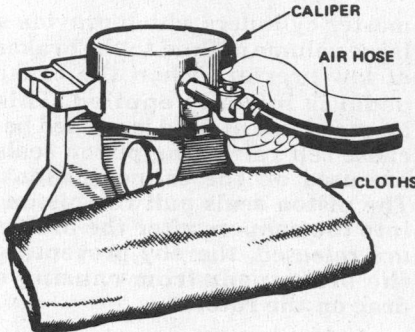
Removing piston pneumatically

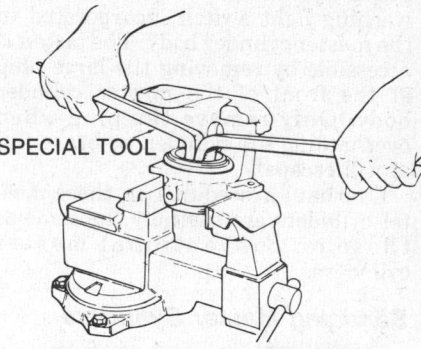
Removing hollow end piston

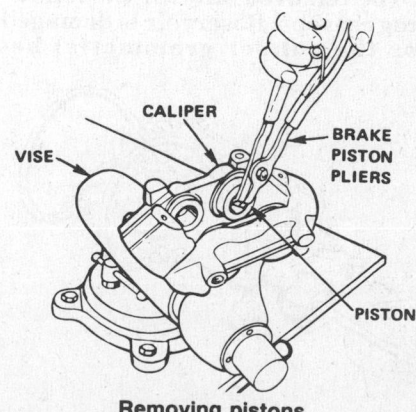
Removing pistons

SERVICING THE CALIPER ASSEMBLY

NOTE: The following is a general caliper service procedure. Before proceeding, check under the individual disc brake section for your vehicle (Delco Moraine, Bendix, etc.) for any special servicing procedures.

1. Raise and support the front of the vehicle on jackstands, then remove the front wheels.

2. Working on a side at a time only, disconnect the hydraulic inlet line from the caliper and plug the end. Remove the caliper mounting bolts or pins and the shims (if used), then slide the caliper off the disc.

3. Remove the disc pads from the caliper or mounting adapter. If the old ones are to be reused, make them so that they can be reinstalled in their original positions.

4. Open the caliper bleed screw and drain the fluid. Clean the outside of the caliper and mount it in a vise with padded jaws.

——— CAUTION ———

When cleaning any brake components, use only brake fluid or denatured (Isopropyl) alcohol. Never use a mineral-based solvent, such as gasoline or paint thinner, since it will swell and quickly deteriorate the rubber parts. Alcohol must NOT be allowed to remain in any component that will hold brake fluid, this would lower the boiling point of the brake fluid.

5. Remove the bridge bolts (fixed type), separate the caliper halves and remove the 2 O-ring seals from the transfer holes.

6. Pry the lip on (each) piston dust boot from its groove, then remove the piston assemblies and spring(s) from the bore(s). If necessary, air pressure may be used to force the pistons(s) out of the bore(s), using care to prevent the piston from popping out of control.

7. Remove the boot(s) and seal(s) from the piston(s), then clean the piston(s) in brake fluid. Blow out the caliper passages with an air hose.

8. Inspect the cylinder bore(s) for scoring, pitting or corrosion. Corrosion is a pitted or rough condition not to be confused with staining. Light rough spots may be removed by rotating crocus cloth, using finger pressure, in the bores. DO NOT polish with an in and out motion or use any other abrasive.

9. If the piston(s) are pitted, scored or worn, they must be replaced. A corroded or deeply scored caliper should also be replaced.

10. Check the clearance of the piston(s) in the bores using a feeler gauge. Clearance should be 0.002–0.006 in. If there is excessive clearance the caliper must be replaced.

11. Replace all rubber parts and lubricate with brake fluid. Install the seals (or square cut rings) and boots in the grooves in each piston. The seal should be installed in the groove closest to the closed end of the piston with the seal lips facing the closed end. The lip on the boot should be facing the seal.

12. Lubricate the piston and bore with brake fluid. Position the piston return spring (if equipped), large coil first, in the piston bore.

13. Install the piston in the bore, taking great care to avoid damaging the seal lip as it passes the edge of the cylinder bore.

14. Compress the lip on the dust boot into the groove in the caliper. Be sure the boot is full seated in the groove, as poor sealing will allow contaminants to ruin the bore.

15. On fixed calipers: Position the O-rings in the cavities around the caliper transfer holes and fit the caliper halves together. Install the bridge bolts (lubricated with brake fluid) and be sure to torque to specification.

16. Install the disc pads in the caliper or adapter and remount the caliper on the hub. Connect the brake line to the caliper and bleed the brakes. Replace the wheels. Recheck the brake fluid level, check the brake pedal travel and road test the vehicle.

OVERHAUL TIPS

Field reports indicate that 2 factors determine whether to replace or rebuild calipers: Can the piston or pistons be removed? Will the bleed screw break off when removal is attempted? (Rebuilders will not accept a caliper with a broken bleed screw.) Since there is no way to predict how a bleed screw will react, follow this procedure to attempt removal.

1. Insert a drill shank into the bleed screw hole (snug fit).

2. Tap the screw on all sides.

3. Using a 6-point wrench, apply pressure gently while working the drill up and down slightly.

4. If the drill starts to bind, the screw is beginning to collapse and cannot be removed intact.

5. Heating the caliper is another successful, but time consuming, bleed screw removal technique. Remove the caliper from the vehicle. Heat the caliper. Shrink the bleed screw by applying dry ice and attempt to remove.

DISC BRAKE BLEEDER SCREW REPLACEMENT

1. Using the existing hole in bleed screw for a pilot, drill a ¼ in. hole completely through existing bleeder.

2. Increase the hole size to $^{7}/_{16}$ in.

3. Tap the hole using a ¼ in. (18-national pipe thread) ½ in. deep (full thread).

4. Install the bleeder repair kit.

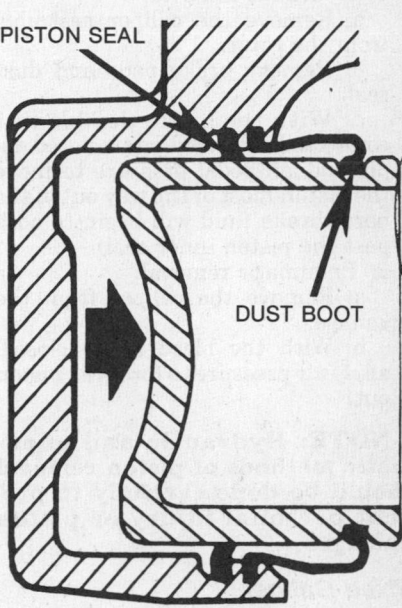

Brake applied

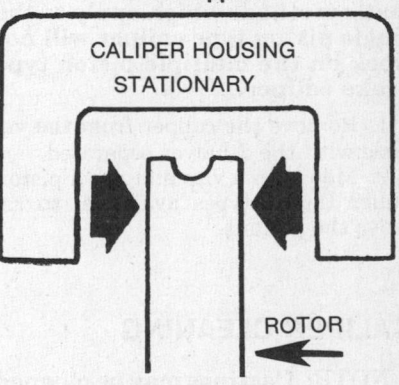

Fixed caliper type

Replacing disc brake bleeder screw

5. Test for leaks and full brake pedal pressure.

FROZEN PISTONS

Sliding or Floating Caliper

1. Hydraulic removal:
   a. Remove the caliper assembly from the rotor.
   b. Remove brake pads and dust seal.
   c. With the brake flexible line connected and bleed screw closed apply enough pedal pressure to move the piston most of the way out of the bore (brake fluid will begin to ooze past the piston inner seal).
2. Pneumatic removal:
   a. Remove the caliper from the vehicle.
   b. With the bleed screw closed, apply air pressure to force the piston out.

NOTE: Hydraulic and pneumatic methods of piston removal should be done carefully to prevent personal injury or piston damage.

Fixed Caliper

NOTE: The hydraulic or pneumatic methods which apply to the single piston type caliper will not work on the multiple piston type brake caliper.

1. Remove the caliper from the vehicle with the 2 halves separated.
2. Mount in a vise and use a piston puller (many types available) to remove the pistons.

CALIPER CLEANING

NOTE: Castings may be cleaned with any type cleaning fluid after all the rubber seals have been removed.

It is important that all traces of cleaning fluid be completely removed from the caliper casting. Rubber components are compatible with alcohol and/or brake fluid. Use a lint free wiping cloth to clean the caliper and parts. Black stains on the pistons or walls, caused by the seals, will not do harm; however, extreme cleanliness is essential. Blow out the passages with compressed air. A fine grade of crocus cloth may be used to correct minor imperfections in the cylinder bore. Slide crocus cloth with finger pressure in a circular rather than a lengthwise motion. DO NOT use any form of abrasive on a plated piston. Discard a piston which is pitted or has signs of plating wear.

REBUILDING CALIPERS

NOTE: If a fine stone honing of a caliper bore is necessary it should be done with skill and caution. Some vehicles can develop 800 psi hydraulic pressure on severe application so the honing must never exceed 0.003 in. Also

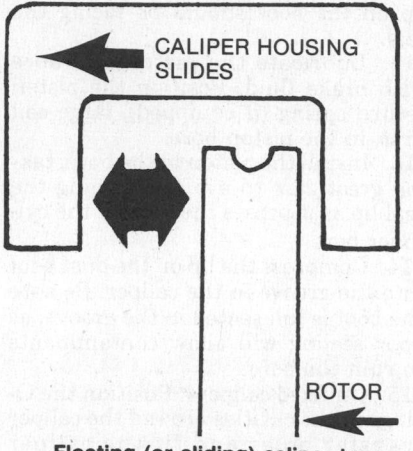

Floating (or sliding) caliper type

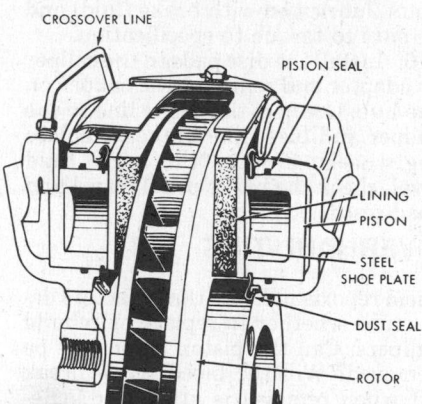

Fixed caliper disc brake

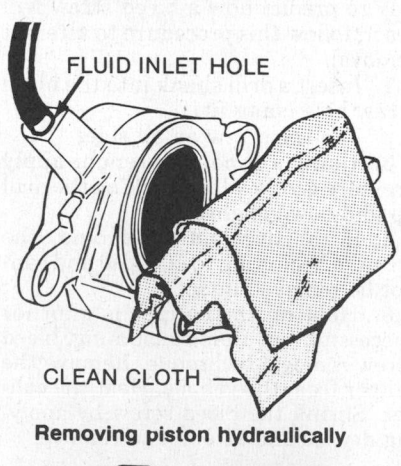

Removing piston hydraulically

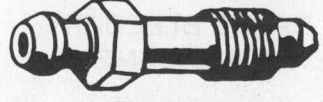

Bleed screw

the dust seal groove must be free of rust or nicks so that a perfect mating surface is possible on the piston and casting.

Installing Stroking Type Seals and Boots

Stretch the boot and seal over the piston and seat them. The seal lip on the Bendix and Delco styles, faces toward hydraulic pressure; boot lips face toward the brake shoe. Locate the return spring (if used) in the cylinder and carefully start the piston into the cylinder to avoid nicking the seal. Alignment tools are available for inserting the lip cup seals. Fully depress the piston into the bore in order to fasten the boot lip to the caliper housing. On the Delco types, use a wooden drift or a special seating tool to seat the boot ring in the caliper counterbore. It must be flush or below the caliper machined surface.

Installing Fixed Position (Rectangular Ring) Seals and Boots

Insert a rectangular ring seal into bore and at any location, push the ring into the seal groove. From this area, with a finger, gently work around the bore until the ring is seated in this channel. Be sure the ring does not twist or roll in the groove. When the boot lip is retained inside the cylinder bore, insert the boot in the same manner. Then work the inside of the boot over the pressure end of the piston, stretching the boot with a small plastic tool and pressing the piston through the seal, straight in, until it bottoms. The inside of the boot should slide on the piston and come to rest in the boot groove. If the boot lip is retained outside of the cylinder bore, first stretch boot over the piston and seat it in its groove, then press the piston through the seal. Fully depress the piston to 50–100 lbs. in order to fasten the boot lip in place. On the Delco-Moraine types, use a wooden drift or a special seating tool to seat the metal boot ring in the caliper counterbore below the face of the caliper.

Installing Fixed-Caliper Bridge Bolts

If the caliper contains internal fluid crossover passages, be sure to install new O-ring seals at the joints. Mate the caliper halves and install high tensile strength bridge bolts. Never replace the bridge bolts with ordinary standard hardware bolts.

Wheel Cylinders

Wheel cylinders contain a pair of op-

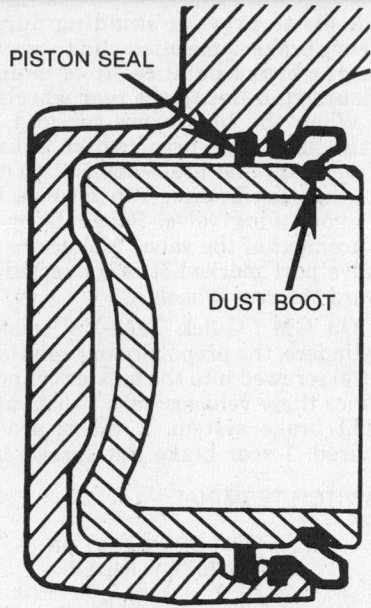

Brake released

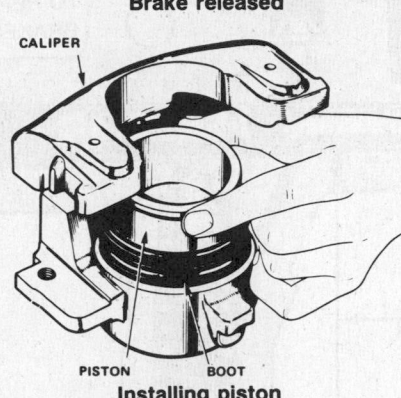

Installing piston

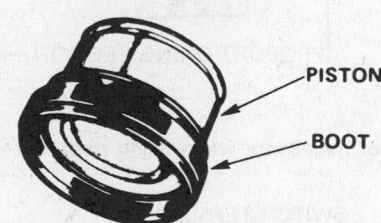

Assembling boot on piston

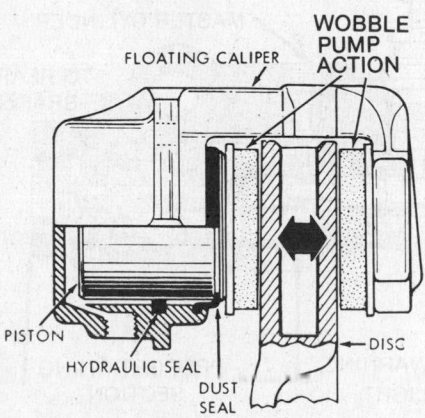

Wobble pump action

posed pistons fitted with rubber cups, compression spring and sometimes expander washers to keep the cups tight against the pistons.

SERVICING

1. Raise and support the vehicle on jackstands. Remove the wheel and drum assemblies from the side to be serviced.

2. Remove the brake shoes, then clean the backing plate and the wheel cylinder. Rebuilding can be done on the vehicle, depending on the design of the brake backing plate. If the backing plate is recessed to the point that it is impossible to get a hone into the cylinder, the cylinder has to be removed.

3. To remove the cylinder; disconnect the brake line from the rear of the cylinder, remove the mounting bolts or retainers and the cylinders.

NOTE: On some models, the wheel cylinder is contained by a retaining ring. In order to remove the rear wheel cylinders, remove the wheel cylinder retainer. Insert 2 pin punches or equivalent tools into the access slots and bend both tabs at the same time thereby releasing the cylinder. Use a new retainer when reinstalling the wheel cylinder. The new retainer can be driven on using a 1 $\frac{1}{8}$ in. socket with an extension bar.

4. Remove the rubber boots (dust covers) from the ends of the cylinder. Remove the pistons, the piston cups (expanders, if equipped) and the spring from the inside of the cylinder. Remove the bleeder screw and make sure it is not clogged.

5. Discard all of the parts that the rebuilding kit will replace.

6. Examine the inside of the cylinder. If it is severely rusted, pitted or scratched install a new or rebuilt cylinder.

7. If the condition of the cylinder in-

dicates that it can be rebuilt, hone the bore. Light honing will provide a new surface on the inside of the cylinder which promotes better cup sealing.

8. Wash out the cylinder with brake

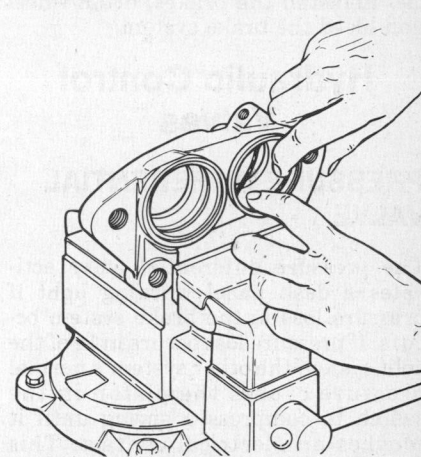

Installing fixed position rectangular ring seal (seal lip toward pressure side)

Wheel cylinder components

Honing cylinder bore

fluid after honing. Reassemble the cylinder using the new parts provided in the kit. When assembling the cylinder dip all parts in brake fluid.

9. Install the cylinder on the vehicle. Reinstall the brakes, drum/wheel and bleed the brake system.

Hydraulic Control Valves

PRESSURE DIFFERENTIAL VALVE

The pressure differential valve activates a dash panel warning light if pressure loss in the brake system occurs. If pressure loss occurs in ½ of the split system the other system's normal pressure causes the piston in the switch to compress a spring until it touches an electrical contact. This turns the warning lamp on the dash panel to light, thus warning the driver of possible brake failure.

On some vehicles, the spring balance piston automatically resets as the brake pedal is released warning the driver only upon brake application. On other vehicles, the light remains on until manually cancelled.

Valves may be located separately, as part of a combination valve, or incorporated into the master cylinder.

Resetting Valves

On some vehicles, the valve piston(s) remain off center after failure until necessary repairs are made. The valve will automatically reset itself (after repairs) when pressure is equal on both sides of the system.

If the light does not go out, bleed the brake system that is opposite the failed system. If front brakes failed, bleed the rear brakes, this should force the light control piston toward center.

If this fails, remove the terminal switch. If brake fluid is present in the electrical area, the seals are gone, replace the complete valve assembly.

METERING VALVE

The metering valve's function is to improve braking balance between the front disc and rear drum brakes, especially during light brake application.

The metering valve prevents the application of the front disc brakes until the rear brakes overcome the return spring pressure. Thus, when the front disc pads contact the rotor, the rear shoes will contact the brake drum at the same time.

Inspect the metering valve each time the brakes are serviced. A slight amount of moisture inside the boot does not indicate a defective valve,

however, fluid leakage indicates a damaged or worn valve. If fluid leakage is present the valve must be replaced.

The metering valve can be checked very simply. With the vehicle stopped, gently apply the brakes. At about 1 in. of travel a very small change in pedal effort (like a small bump) will be felt if the valve is operating properly. Metering valves are not serviceable and must be replaced (if defective).

PROPORTIONING VALVE

The proportioning (pressure control) valve is used, on some vehicles, to reduce the hydraulic pressure to the rear wheels to prevent skidding during heavy brake application and to provide better brake balance. It is usually mounted in line to the rear wheels.

When the brakes are serviced the valve should be inspected for leakage. Premature rear brake application during lighting braking can mean a bad proportioning valve. Repair is by replacement of the valve. Make sure the valve port marked **R** is connected toward the rear wheels.

On GM "Quick Take-Up" master cylinders, the proportioning valve(s) is (are) screwed into the master cylinder. Since these vehicles have a diagonally split brake system, 2 valves are required. 1 rear brake line screws into

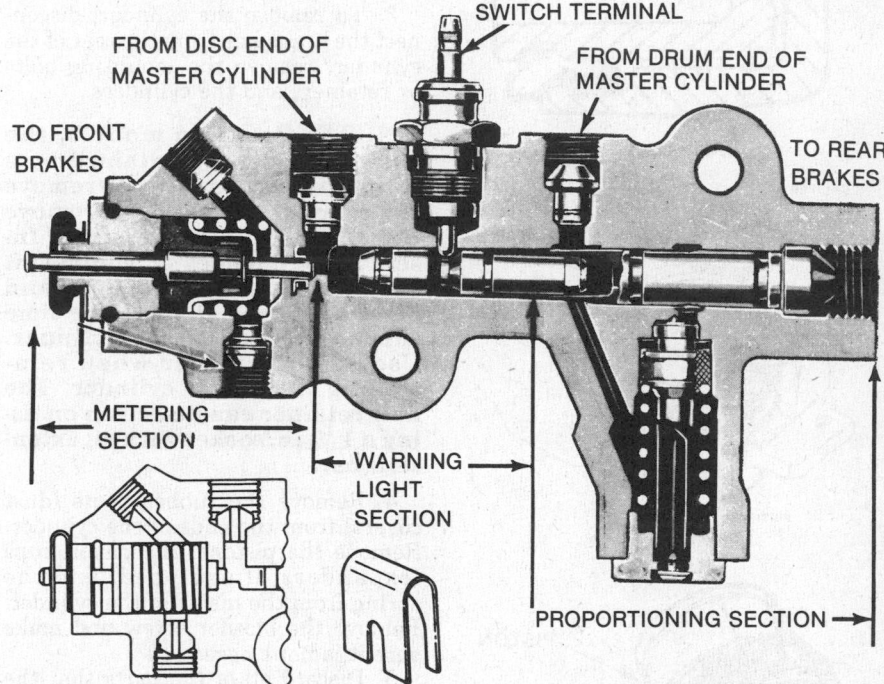

Hold valve out .060 in pressure bleed only-not necessary when using pedal bleed method

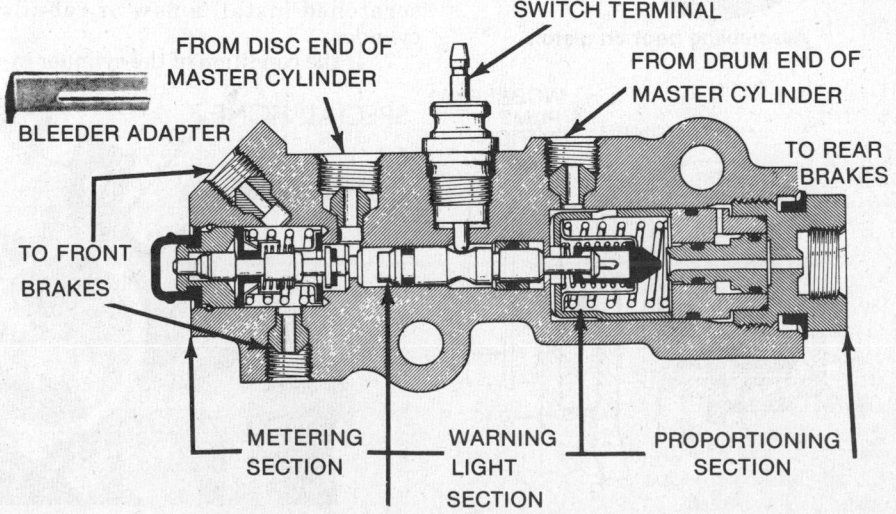

Push valve in when pressure bleeding-not necessary when using pedal bleed method

each valve. The early type valves (GM front wheel drive) were steel and silver colored, an occasional "clunking" noise was encountered on some early models, but does not affect brake efficiency. Replacement valves are now made of aluminum. Never mix an aluminum valve with a steel valve, always use 2 aluminum valves.

COMBINATION VALVE

The combination valve may perform 2 or 3 functions. They are: metering, proportioning and brake failure warning.

Variations of the two-way combination valve are: proportioning and brake failure warning or metering and brake failure warning.

A three-way combination valve directs the brake fluid to the appropriate wheel, performs necessary valving and contains a brake failure warning.

The combination valve is usually mounted under the hood close to the master cylinder, where the brake lines can easily be connected and routed to the front or rear wheels.

The combination valve is non-serviceable and must be replaced if malfunctioning.

Brake Bleeding

The hydraulic brake system must be free of air to operate properly. Air can enter the system when hydraulic parts are disconnected for servicing or replacement, or when the fluid level in the master cylinder reservoirs is very low. Air in the system will give the brake pedal a spongy feeling upon application.

The quickest and easiest of the 2 ways for system bleeding is the pressure method but special equipment is needed to externally pressurize the hydraulic system. The other, more commonly used method of brake bleeding is done manually.

BLEEDING SEQUENCE

Bleeding may be required at only 1 or 2 wheels or at the master cylinder, depending upon what point the system was opened to air. If after bleeding the cylinder caliper that was rebuilt or replaced and the pedal still has a spongy feeling upon application, it will be necessary to bleed the entire system. Bleed the system in the following order:

1. **Master cylinder**: If the cylinder is not equipped with bleeder screws, open the brake line(s) to the wheels slightly while pressure is applied to the brake pedal. Be sure to tighten the line before the brake pedal is released. The

procedure for bench bleeding the master cylinder is in the following section.

2. **Power Brake Booster**: If the unit is equipped with bleeder screws, it should be bled after the master cylinder. The vehicle's engine should be off and the brake pedal applied several times to exhaust any vacuum in the booster. If the unit is equipped with 2 bleeder screws, always bleed the higher bleeder screw first.

3. **Combination Valve**: If equipped with a bleeder screw.

4. **Front/Back Split Systems**: Start with the wheel farthest away from the master cylinder, usually the right-rear wheel. Bleed the other rear wheel then the right-front and left-front.

NOTE: If you are unsuccessful in bleeding the front wheels, it may be necessary to deactivate the metering valve. This is accomplished by either pushing in, or pulling out a button or stem on the valve. The valve may be held by hand, with a special tool or taped, it should remain deactivated while the front brakes are bled.

5. **Diagonally Split System**: Start with the right-rear then the left-front. The left-rear then the right-front.

6. **Rear Disc Brakes**: If the vehicle is equipped with rear disc brakes and the calipers have 2 bleeder screws, bleed the inner first then the outer.

NOTE: DO NOT allow brake fluid to spill on the vehicles finish, it will remove the paint. Flush the area with water.

MANUAL BLEEDING

1. Clean the bleed screw at each wheel.

2. Start with the wheel farthest from the master cylinder (right-rear).

3. Attach a small rubber hose to the bleed screw and place the end in a clear container of brake fluid.

4. Fill the master cylinder with brake fluid. (Check often during bleeding). Have an assistant slowly pump up the brake pedal and hold pressure.

5. Open the bleed screw about one-quarter turn, press the brake pedal to the floor, close the bleed screw and slowly release the pedal. Continue until no more air bubbles are forced from the cylinder on application of the brake pedal.

6. Repeat procedure on remaining wheel cylinders and calipers, still working from cylinder/caliper farthest from the master cylinder.

NOTE: Master cylinders equipped with bleed screws may

be bled independently. When bleeding the Bendix-type dual master cylinder it is necessary to solidly cap 1 reservoir section while bleeding the other to prevent pressure loss through the cap vent hole.

—— CAUTION ——
The bleeder valve at the wheel cylinder must be closed at the end of each stroke and before the brake pedal is released, to insure that no air can enter the system. It is also important that the brake pedal be returned to the full up position so the piston in the master cylinder moves back enough to clear the bypass outlets.

PRESSURE BLEEDING DISC BRAKES

Pressure bleeding disc brakes will close the metering valve and the front brakes will not bleed. For this reason it is necessary to manually hold the metering valve open during pressure bleeding. Never use a block or clamp to hold the valve open and never force the valve stem beyond its normal position.

Of the 2 different types of valves used, the most common type requires the valve stem to be held in while bleeding the brakes, while the second type requires the valve stem to be held out (0.060 in. minimum travel). Determine the type of visual inspection.

—— CAUTION ——
Special adapters are required when pressure bleeding cylinders with plastic reservoirs. Pressure bleeding equipment should be diaphragm type; placing a diaphragm between the pressurized air supply and the brake fluid. This prevents moisture and other contaminants from entering the hydraulic system.

NOTE: Front disc/rear drum equipped vehicles use a metering valve which closes off pressure to the front brakes under certain conditions. These systems contain manual release actuators, which must be engaged to pressure bleed the front brakes.

1. Connect the tank hydraulic hose and adapter to the master cylinder.

2. Close the hydraulic valve on the bleeder equipment.

3. Apply air pressure to the bleeder equipment following the equipment manufacturer's recommendations for correct air pressure.

4. Open the valve to bleed air out of the pressure hose to the master cylinder. Never bleed this system using the secondary piston stopscrew on the bottom of many master cylinders.

5. Open the hydraulic valve and bleed each wheel cylinder or caliper.

Bleed the rear brake system first when bleeding both front and rear systems.

FLUSHING HYDRAULIC BRAKE SYSTEMS

Hydraulic brake systems must be totally flushed if the fluid becomes contaminated with water, dirt or other corrosive chemicals. To flush, simply bleed the entire system until all of the fluid has been replaced with the correct type of new fluid.

BENCH BLEEDING MASTER CYLINDER

Bench bleeding the master cylinder before installing it on the vehicle reduces the possibility of getting air into the lines.

1. Connect 2 short pieces of brake line to the outlet fittings, bend them until the free end is below the fluid level in the master cylinder reservoirs.

2. Fill the reservoirs with fresh brake fluid. Pump the piston until no more air bubbles appear in the reservoir(s).

3. Disconnect the 2 short lines, refill the master cylinder and securely install the cylinder cap(s).

4. Install the master cylinder on the vehicle. Attach the lines but do not completely tighten them. Force any air that might have been trapped in the connection by slowly depressing the brake pedal. Tighten the lines before releasing the brake pedal.

GM QUICK TAKE-UP SYSTEM BLEEDING

Bleed the master cylinder as follows: disconnect the left-front brake line from the master cylinder. Fill the cylinder with fluid until it flows from the opened port. Connect the line and tighten the fitting. Apply the brake pedal slowly 1 time and keep it applied. Loosen the same brake line fitting to allow any air to escape. Retighten the fitting and release the brake pedal slowly. Wait 15 seconds and repeat the procedure until all of the air is expelled. Bleed the right-front connection in the same manner. Bleed the cylinders and calipers after you are sure all the air is out of the master cylinder.

— CAUTION —

Rapid pumping will move the secondary piston down the bore and make it difficult to bleed the system. Always apply slow pedal pressure.

Power Brakes

VACUUM OPERATED BOOSTER

Power brakes operate just as standard brake systems except in the actuation of the master cylinder pistons. A vacuum diaphragm is located on the front of the master cylinder to assist in applying the brakes, reducing both the effort and travel needed to move the brake pedal.

The vacuum diaphragm housing is connected to the intake manifold by a vacuum hose. A check valve is placed at the point where the hose enters the diaphragm housing, so that during periods of low manifold vacuum brake assist vacuum will not be lost.

Depressing the brake pedal closes off the vacuum source and allows atmospheric pressure to enter on 1 side of the diaphragm. This causes the master cylinder pistons to move and apply the brakes. When the brake pedal is released, vacuum is applied to both sides of the diaphragm, the return springs return the diaphragm and master cylinder pistons to the released position. If the vacuum fails, the brake pedal rod will butt against the end of the master cylinder actuating rod and direct mechanical application will occur as the pedal is depressed.

The hydraulic and mechanical problems that apply to conventional brake systems also apply to power brakes should be checked if the tests and chart below do not reveal the problem. Tests for a system vacuum leak as described below:

1. Operate the engine at idle with the transmission in Neutral without touching the brake pedal for at least 1 minute.

2. Turn **OFF** the engine and wait 1 minute.

3. Test for the presence of assist vacuum by depressing the brake pedal and releasing it several times. Light application will produce less and less pedal travel, if vacuum was present. If there is no vacuum, air is leaking into the system somewhere.

4. Test the system operation as follows:

   a. Pump the brake pedal (with engine off) until the supply vacuum is entirely gone.

   b. Put light, steady pressure on the pedal. Start the engine and operate it at idle with the transmission in neutral.

   c. If the system is operating, the brake pedal should fall toward the floor when constant pressure is maintained on the pedal.

NOTE: Power brake systems may be tested for hydraulic leaks just as ordinary systems are tested, except that the engine should be idling with the transmission in neutral throughout the test.

POWER BRAKE BOOSTER TROUBLESHOOTING

NOTE: The following items are in addition to those listed in the "General Troubleshooting" section. Check those items first.

Hard Pedal

1. Faulty vacuum check valve
2. Vacuum hose kinked, collapsed, plugged leaky or improperly connected
3. Internal leak in unit
4. Damaged vacuum cylinder
5. Damaged valve plunger
6. Broken or faulty springs
7. Broken plunger stem

Grabbing Brakes

1. Damaged vacuum cylinder
2. Faulty vacuum check valve
3. Vacuum hose leaky or improperly connected
4. Broken plunger stem

Pedal Goes to Floor

Generally, when this problem occurs, it is not caused by the power brake booster. In rare cases, a broken plunger stem may be at fault.

OVERHAUL

Most power brake boosters are serviced by replacement only. In many cases, repair parts are not available. A good many special tools are required for rebuilding these units. For these reasons, it would be most practical to replace a failed booster with a new or remanufactured unit.

Hydro-Boost Hydro-Boost II

Hydro-Boost differs from conventional power brake systems, in that it operates from the power steering pump fluid pressure rather than intake manifold vacuum.

The Hydro-Boost unit contains a spool valve with an open center which controls the strength of pump pressure when braking occurs. A lever assembly controls the valve's position. A boost piston provides the force necessary to operate the conventional master cylinder on the front of the booster.

A reserve of at least 2 assisted brake applications is supplied by an accumulator which is spring loaded on earlier and pneumatic on later models. The accumulator is an integral part of the Hydro-Boost II unit. The brakes can be applied manually if the reserve system is depleted.

All system checks, tests and troubleshooting procedure are the same for the 2 systems.

HYDRO-BOOST SYSTEM CHECKS

1. A defective Hydro-Boost cannot cause any of the following conditions: Noisy brakes, fading pedal or pulling brakes. If any of these occur, check elsewhere in the brake system.
2. Check the fluid level in the master cylinder. It should be within ¼ in. of the top; if not, add only DOT-3 or DOT-4 brake fluid until the correct level is reached.
3. Check the fluid level in the power steering pump. The engines should be at normal running temperature and stopped. The level should register on the pump dipstick. Add power steering fluid to bring the reservoir level up to the correct level. Low fluid level will result in both poor steering and stopping ability.

── CAUTION ──
The brake hydraulic system uses brake fluid only, while the power steering and Hydro-Boost systems use power steering fluid only. Don't mix the two.

4. Check the power steering pump belt tension and inspect all of the power steering/Hydro-Boost hoses for kinks or leaks.
5. Check and adjust the engine idle speed, as necessary.
6. Check the power steering pump fluid for bubbles. If air bubbles are present in the fluid, bleed the system. Fill the power steering pump reservoir to specifications with the engine at normal operating temperature. With the engine running, rotate the steering wheel through its normal travel 3–4 times, without holding the wheel against the stops. Check the fluid level again.
7. If the problem still exists, go on to the Hydro-Boost test sections and troubleshooting chart.

HYDRO-BOOST TESTS

Functional Test

1. Check the brake system for leaks or low fluid level. Correct as necessary.
2. Place the transmission in Neutral and stop the engine. Apply the brakes 4–5 times to empty the accumulator.
3. Keep the pedal depressed with moderate (25–40 lbs.) pressure and start the engine.
4. The brake pedal should fall slightly and then push back up against your foot. If no movement is felt, the Hydro-Boost system is not working.

Accumulator Leak Test

1. Run the engine at normal idle. Turn the steering wheel against either stop; hold it there for no longer than 5 seconds. Center the steering wheel and stop the engine.
2. Keep applying the brakes until a ''hard'' pedal is obtained. There should be a minimum of 2 power (1 on Hydro-Boost II) assisted brake applications when pedal pressure of 20–25 lbs. is applied.
3. Start the engine and allow it to idle. Rotate the steering wheel against the stop. Listen for a light ''hissing'' sound; this is the accumulator being charged. Center the steering wheel and stop the engine.
4. Wait 1 hour and apply the brakes without starting the engine. As in Step 2, there should be at least 2 (1 on Hydro-Boost II) stops with power assist. If not, the accumulator is defective and must be replaced.

Hydro-Boost System Bleeding

NOTE: The system should be bled whenever the booster is removed and installed.

1. Fill the power steering pump until the fluid level is at the base of the pump reservoir neck. Disconnect the battery lead from the distributor.

NOTE: On diesel engines remove the electrical lead to the fuel solenoid terminal on the injection pump before cranking the engine.

2. Raise the front of the vehicle, turn the wheels all the way to the left and crank the engine for a few seconds.
3. Check the steering pump fluid level. If necessary, add fluid to the **ADD** mark on the dipstick.
4. Lower the vehicle, connect the battery lead and start the engine. Check the fluid level and add fluid to the **ADD** mark if necessary. With the engine running, turn the wheels from side-to-side to bleed air from the system. Make sure that the fluid level stays above the internal pump casting.
5. The Hydro-Boost system should now be fully bled. If the fluid is foaming after bleeding, stop the engine, let the system set for 1 hour. Add fluid to the **ADD** mark if necessary, then with the engine running, turn the wheels from side-to-side to bleed air from the system. Repeat this step if necessary.
6. The preceding procedures should be effective in removing excess air from the system, however, sometimes air may still remain trapped. When this happens the booster may make a ''gulping'' noise when the brake is applied. Lightly pumping the brake pedal with the engine running should cause this noise to disappear. After the noise stops, check the pump fluid level and add as necessary.

HYDRO-BOOST TROUBLESHOOTING

High Pedal and Steering Effort (Idle)

1. Loose/broken power steering pump belt
2. Low power steering fluid level
3. Leaking hoses or fittings
4. Low idle speed
5. Hose restriction
6. Defective power steering pump

High Pedal Effort (Idle)

1. Binding pedal/linkage
2. Fluid contamination
3. Defective Hydro-Boost unit

Poor Pedal Return

1. Binding pedal linkage
2. Restricted booster return line
3. Internal return system restriction

Pedal Chatter/Pulsation

1. Power steering/pump drivebelt slipping
2. Low power steering fluid level
3. Defective power steering pump
4. Defective Hydro-Boost unit

Brakes Oversensitive

1. Binding pedal/linkage
2. Defective Hydro-Boost unit

Noise

1. Low power steering fluid level
2. Air in the power steering fluid
3. Loose power steering pump drivebelt
4. Hose restrictions

OVERHAUL

Ford Motor Company services the Hydro-Boost unit with a replacement new or rebuilt unit only. No provisions are made for overhaul of the unit. GM Hydro-Boost units may be overhauled by qualified mechanics.

──── CAUTION ────
DO NOT attempt to interchange the parts between the Hydro-Boost units of different makes of vehicles, because of pressure differentials and differences of the tolerances of the internal parts. Pressure could exceed the normal accumulator release pressure of 1400 psi and injury or damage could result.

DISC BRAKES

Disc Brake Rotors

RUNOUT

Manufacturers differ widely on permissible runout but too much can sometimes be felt as a pulsation at the brake pedal. A wobble pump effect is created when a rotor is not perfectly smooth and the pad hits the high spots forcing fluid back into the master cylinder. This alternating pressure causes a pulsating feeling which can be felt at the pedal when the brakes are applied.

To check the actual runout of the rotor, perform the following procedures:

1. Tighten the wheel spindle nut to a snug bearing adjustment, end-play removed.

2. Fasten a dial indicator on the suspension at a convenient place so that the indicator stylus contacts the rotor face approximately 1 in. from its outer edge.

3. Set the dial at zero. Check the total indicator reading while turning the rotor 1 full revolution. If the rotor is warped beyond the runout specification, it is likely that it can be successfully remachined.

Lateral Runout: A wobbly movement of the rotor from side-to-side as it rotates. Excessive lateral runout causes the rotor faces to knock back the disc pads and can result in chatter, excessive pedal travel, pumping or fighting pedal and vibration during the braking action.

Parallelism (lack of): Refers to the amount of variation in the thickness of the rotor. Excessive variation can cause pedal vibration or fight, front end vibrations and possible "grab" during the braking action; a condition comparable to an "out-of-round brake drum." Check parallelism with a micrometer. "Mike" the thickness at 8 or more equally spaced points, equally distant from the outer edge of the rotor, preferably at midpoints of the braking surface. Parallelism then is the amount of variation between maximum and minimum measurements.

Surface or Micro-inch finish, flatness, smoothness: Different from parallelism, these terms refer to the degree of perfection of the flat surface on each side of the rotor; that is, the minute hills, valleys and swirls inherent in machining the surface. In a visual inspection, the remachined surface should have a find ground polish with, at most, only a faint trace of non-directional swirls.

SERVICING THE DISC ROTOR

Disc Replacement

1. Raise and support the vehicle on jackstands, then remove the wheel/tire assembly.

2. Remove the caliper. Secure the caliper out of the way suspended by wire, DO NOT allow the caliper to hang by the brake hose.

3. Remove the wheel bearing nut from the spindle and the outer wheel bearing from the hub.

4. Remove the hub and disc assembly from the spindle.

5. To install, reverse the removal procedures.

NOTE: The disc is removable from the hub on the Eldorado, Toronado and Corvette (rear only). To separate the rear disc and hub on a (1982-87) Corvette the 3 hub-to-disc attaching rivets must be drilled out. This can be done with the hub and rotor mounted on the vehicle. It is not necessary to install new rivets when the disc is installed.

DRUM BRAKES

— CAUTION —

The asbestos dust thrown off from the brake linings or disc pads may be dangerous to your health if inhaled. Never use compressed air or your own breath to blow the dust from the brake assembly. Use an aerosol brake cleaner, damp rag or a vacuum cleaner with an approved asbestos filter. Dispose of the rag or cleaner bag properly. Do not move a vehicle until a firm brake pedal is obtained.

Brake Drums

BRAKE DRUM TYPES

The FULL-CAST drum has a cast iron web (back) of $3/16-1/4$ in. thickness (passenger vehicle sizes) whereas the COMPOSITE drum has a steel web approximately ⅛ in. thick. These 2 types of drums, with few exceptions are not interchangeable.

BRAKE DRUM DEPTH

Rest a straight edge across the drum diameter on the open side. The actual drum depth then is the measurement at a right angle from the straight edge to that part of the web which mates against the hub mounting flange.

ALUMINUM DRUMS

When replaced by other types, aluminum drums must be replaced in pairs.

METALLIC BRAKES

Drums designed for use with standard brake linings should not be used with metallic brakes.

REMOVING TIGHT DRUMS

Difficulty removing a brake drum can be caused by shoes which are expanded beyond the drum's inner diameter or shoes which have cut into and ridged the drum. In either case back off the adjuster to obtain sufficient clearance for removal.

BRAKE DRUM INSPECTION

The condition of the brake drum surface is just as important as the surface to the brake lining. All drum surfaces should be clean, smooth, free from hard spots, heat checks, score marks and foreign matter embedded in the drum surface. They should not be out of round, bell-mouthed or barrel shaped. It is recommended that all drums be first checked with a drum micrometer to see if they are within oversize limits. If the drum is within safe limits, even though the surface appears smooth, it should be turned not only to assure a true drum surface but also to remove any possible contamination in the surface from previous brake linings, road dusts, etc. Too much metal removed from a drum is unsafe and may result in:

1. Brake fade due to the thin drum being unable to absorb the heat generated.

2. Poor and erratic brake due to distortion of drums.

3. Noise due to possible vibration caused by thin drums.

4. A cracked or broken drum on a severe or very hard brake application.

NOTE: Brake drum run-out should not exceed 0.005 in. Drums turned to more than 0.060 in. oversize are unsafe and should be replaced with new drums, except for some heavy ribbed drums which have an 0.080 in. limit. It is recommended that the diameters of the left and right drums on every axle be within 0.010 in. of each other. In order to avoid erratic brake action when replacing drums, it is always good to replace the drums on both wheels at the same time. If the drums are true, smooth up any slight scores by polishing with fine emery cloth. If deep scores or grooves are present, which cannot be removed by this method, then the drum must be turned.

Mechanics Data 36

SI METRIC TABLES

The following tables are given in SI (International System) metric units. SI units replace both customary (English) and the older gavimetric units. The use of SI units as a new worldwide standard was set by the International Committee of Weights and Measures in 1960. SI has since been adopted by most countries as their national standard.

These tables are general conversion tables which will allow you to convert customary units, which appear in the text, into SI units.

The following are a list of SI units and the customary units, used in this book, which they replace:

To measure:	Use SI units:	Which replace (customary units):
mass	kilograms (kg)	pounds (lbs)
temperature	Celsius (°C)	Fahrenheit (°F)
length	millimeters (mm)	inches (in.)
force	newtons (N)	pounds force (lbs)
capacities	liters (l)	pints/quarts/gallons (pts/qts/gals)
torque	newton-meters (N·m)	foot pounds (ft lbs)
pressure	kilopascals (kPa)	pounds per square inch (psi)
volume	cubic centimeters (cm³)	cubic inches (cu in.)
power	kilowatts (kW)	horsepower (hp)

If you have had any prior experience with the metric system, you may have noticed units in this chart which are not familiar to you. This is because, in some cases, SI units differ from the older gravimetric units which they replace. For example, newtons (N) replace kilograms (kg) as a force unit, kilopascals (kPa) replace atmospheres or bars as a unit of pressure, and, although the units are the same, the name Celsius replaces centigrade for temperature measurement.

If you are not using the SI tables, have a look at them anyway; you will be seeing a lot more of them in the future.

ENGLISH TO METRIC CONVERSION: MASS (WEIGHT)

Current mass measurement is expressed in pounds and ounces (lbs. & ozs.). The metric unit of mass (or weight) is the kilogram (kg). Even although this table does not show conversion of masses (weights) larger than 15 lbs, it is easy to calculate larger units by following the data immediately below.

To convert ounces (oz.) to grams (g): multiply th number of ozs. by 28
To convert grams (g) to ounces (oz.): multiply the number of grams by .035

To convert pounds (lbs.) to kilograms (kg): multiply the number of lbs. by .45
To convert kilograms (kg) to pounds (lbs.): multiply the number of kilograms by 2.2

lbs	kg	lbs	kg	oz	kg	oz	kg
0.1	0.04	0.9	0.41	0.1	0.003	0.9	0.024
0.2	0.09	1	0.4	0.2	0.005	1	0.03
0.3	0.14	2	0.9	0.3	0.008	2	0.06
0.4	0.18	3	1.4	0.4	0.011	3	0.08
0.5	0.23	4	1.8	0.5	0.014	4	0.11
0.6	0.27	5	2.3	0.6	0.017	5	0.14
0.7	0.32	10	4.5	0.7	0.020	10	0.28
0.8	0.36	15	6.8	0.8	0.023	15	0.42

ENGLISH TO METRIC CONVERSION: TEMPERATURE

To convert Fahrenheit (°F) to Celsius (°C): take number of °F and subtract 32; multiply result by 5; divide result by 9

To convert Celsius (°C) to Fahrenheit (°F): take number of °C and multiply by 9; divide result by 5; add 32 to total

Fahrenheit (F)	Celsius (C)			Fahrenheit (F)	Celsius (C)			Fahrenheit (F)	Celsius (C)		
°F	°C	°C	°F	°F	°C	°C	°F	°F	°C	°C	°F
−40	−40	−38	−36.4	80	26.7	18	64.4	215	101.7	80	176
−35	−37.2	−36	−32.8	85	29.4	20	68	220	104.4	85	185
−30	−34.4	−34	−29.2	90	32.2	22	71.6	225	107.2	90	194
−25	−31.7	−32	−25.6	95	35.0	24	75.2	230	110.0	95	202
−20	−28.9	−30	−22	100	37.8	26	78.8	235	112.8	100	212
−15	−26.1	−28	−18.4	105	40.6	28	82.4	240	115.6	105	221
−10	−23.3	−26	−14.8	110	43.3	30	86	245	118.3	110	230
−5	−20.6	−24	−11.2	115	46.1	32	89.6	250	121.1	115	239
0	−17.8	−22	−7.6	120	48.9	34	93.2	255	123.9	120	248
1	−17.2	−20	−4	125	51.7	36	96.8	260	126.6	125	257
2	−16.7	−18	−0.4	130	54.4	38	100.4	265	129.4	130	266
3	−16.1	−16	3.2	135	57.2	40	104	270	132.2	135	275
4	−15.6	−14	6.8	140	60.0	42	107.6	275	135.0	140	284
5	−15.0	−12	10.4	145	62.8	44	112.2	280	137.8	145	293
10	−12.2	−10	14	150	65.6	46	114.8	285	140.6	150	302
15	−9.4	−8	17.6	155	68.3	48	118.4	290	143.3	155	311
20	−6.7	−6	21.2	160	71.1	50	122	295	146.1	160	320
25	−3.9	−4	24.8	165	73.9	52	125.6	300	148.9	165	329
30	−1.1	−2	28.4	170	76.7	54	129.2	305	151.7	170	338
35	1.7	0	32	175	79.4	56	132.8	310	154.4	175	347
40	4.4	2	35.6	180	82.2	58	136.4	315	157.2	180	356
45	7.2	4	39.2	185	85.0	60	140	320	160.0	185	365
50	10.0	6	42.8	190	87.8	62	143.6	325	162.8	190	374
55	12.8	8	46.4	195	90.6	64	147.2	330	165.6	195	383
60	15.6	10	50	200	93.3	66	150.8	335	168.3	200	392
65	18.3	12	53.6	205	96.1	68	154.4	340	171.1	205	401
70	21.1	14	57.2	210	98.9	70	158	345	173.9	210	410
75	23.9	16	60.8	212	100.0	75	167	350	176.7	215	414

ENGLISH TO METRIC CONVERSION: LENGTH

To convert inches (ins.) to millimeters (mm): multiply number of inches by 25.4

To convert millimeters (mm) to inches (ins.): multiply number of millimeters by .04

Inches	Decimals	Milli-meters	Inches to millimeters inches	mm	Inches	Decimals	Milli-meters	Inches to millimeters inches	mm
1/64	0.051625	0.3969	0.0001	0.00254	33/64	0.515625	13.0969	0.6	15.24
1/32	0.03125	0.7937	0.0002	0.00508	17/32	0.53125	13.4937	0.7	17.78
3/64	0.046875	1.1906	0.0003	0.00762	35/64	0.546875	13.8906	0.8	20.32
1/16	0.0625	1.5875	0.0004	0.01016	9/16	0.5625	14.2875	0.9	22.86
5/64	0.078125	1.9844	0.0005	0.01270	37/64	0.578125	14.6844	1	25.4
3/32	0.09375	2.3812	0.0006	0.01524	19/32	0.59375	15.0812	2	50.8
7/64	0.109375	2.7781	0.0007	0.01778	39/64	0.609375	15.4781	3	76.2
1/8	0.125	3.1750	0.0008	0.02032	5/8	0.625	15.8750	4	101.6
9/64	0.140625	3.5719	0.0009	0.02286	41/64	0.640625	16.2719	5	127.0
5/32	0.15625	3.9687	0.001	0.0254	21/32	0.65625	16.6687	6	152.4
11/64	0.171875	4.3656	0.002	0.0508	43/64	0.671875	17.0656	7	177.8
3/16	0.1875	4.7625	0.003	0.0762	11/16	0.6875	17.4625	8	203.2
13/64	0.203125	5.1594	0.004	0.1016	45/64	0.703125	17.8594	9	228.6
7/32	0.21875	5.5562	0.005	0.1270	23/32	0.71875	18.2562	10	254.0
15/64	0.234375	5.9531	0.006	0.1524	47/64	0.734375	18.6531	11	279.4
1/4	0.25	6.3500	0.007	0.1778	3/4	0.75	19.0500	12	304.8
17/64	0.265625	6.7469	0.008	0.2032	49/64	0.765625	19.4469	13	330.2
9/32	0.28125	7.1437	0.009	0.2286	25/32	0.78125	19.8437	14	355.6
19/64	0.296875	7.5406	0.01	0.254	51/64	0.796875	20.2406	15	381.0
5/16	0.3125	7.9375	0.02	0.508	13/16	0.8125	20.6375	16	406.4
21/64	0.328125	8.3344	0.03	0.762	53/64	0.828125	21.0344	17	431.8
11/32	0.34375	8.7312	0.04	1.016	27/32	0.84375	21.4312	18	457.2
23/64	0.359375	9.1281	0.05	1.270	55/64	0.859375	21.8281	19	482.6
3/8	0.375	9.5250	0.06	1.524	7/8	0.875	22.2250	20	508.0
25/64	0.390625	9.9219	0.07	1.778	57/64	0.890625	22.6219	21	533.4
13/32	0.40625	10.3187	0.08	2.032	29/32	0.90625	23.0187	22	558.8
27/64	0.421875	10.7156	0.09	2.286	59/64	0.921875	23.4156	23	584.2
7/16	0.4375	11.1125	0.1	2.54	15/16	0.9375	23.8125	24	609.6
29/64	0.453125	11.5094	0.2	5.08	61/64	0.953125	24.2094	25	635.0
15/32	0.46875	11.9062	0.3	7.62	31/32	0.96875	24.6062	26	660.4
31/64	0.484375	12.3031	0.4	10.16	63/64	0.984375	25.0031	27	690.6
1/2	0.5	12.7000	0.5	12.70					

ENGLISH TO METRIC CONVERSION: TORQUE

To convert foot-pounds (ft. lbs.) to Newton-meters: multiply the number of ft. lbs. by 1.3

To convert inch-pounds (in. lbs.) to Newton-meters: multiply the number of in. lbs. by .11

in lbs	N·m	in lbs	N·m	in lbs	N·m	in lbs	N·m	in lbs	N·m
0.1	0.01	1	0.11	10	1.13	19	2.15	28	3.16
0.2	0.02	2	0.23	11	1.24	20	2.26	29	3.28
0.3	0.03	3	0.34	12	1.36	21	2.37	30	3.39
0.4	0.04	4	0.45	13	1.47	22	2.49	31	3.50
0.5	0.06	5	0.56	14	1.58	23	2.60	32	3.62
0.6	0.07	6	0.68	15	1.70	24	2.71	33	3.73
0.7	0.08	7	0.78	16	1.81	25	2.82	34	3.84
0.8	0.09	8	0.90	17	1.92	26	2.94	35	3.95
0.9	0.10	9	1.02	18	2.03	27	3.05	36	4.07

ENGLISH TO METRIC CONVERSION: TORQUE

Torque is now expressed as either foot-pounds (ft./lbs.) or inch-pounds (in./lbs.). The metric measurement unit for torque is the Newton-meter (Nm). This unit—the Nm—will be used for all SI metric torque references, both the present ft./lbs. and in./lbs.

ft lbs	N-m	ft lbs	N-m	ft lbs	N-m	ft lbs	N-m
0.1	0.1	33	44.7	74	100.3	115	155.9
0.2	0.3	34	46.1	75	101.7	116	157.3
0.3	0.4	35	47.4	76	103.0	117	158.6
0.4	0.5	36	48.8	77	104.4	118	160.0
0.5	0.7	37	50.7	78	105.8	119	161.3
0.6	0.8	38	51.5	79	107.1	120	162.7
0.7	1.0	39	52.9	80	108.5	121	164.0
0.8	1.1	40	54.2	81	109.8	122	165.4
0.9	1.2	41	55.6	82	111.2	123	166.8
1	1.3	42	56.9	83	112.5	124	168.1
2	2.7	43	58.3	84	113.9	125	169.5
3	4.1	44	59.7	85	115.2	126	170.8
4	5.4	45	61.0	86	116.6	127	172.2
5	6.8	46	62.4	87	118.0	128	173.5
6	8.1	47	63.7	88	119.3	129	174.9
7	9.5	48	65.1	89	120.7	130	176.2
8	10.8	49	66.4	90	122.0	131	177.6
9	12.2	50	67.8	91	123.4	132	179.0
10	13.6	51	69.2	92	124.7	133	180.3
11	14.9	52	70.5	93	126.1	134	181.7
12	16.3	53	71.9	94	127.4	135	183.0
13	17.6	54	73.2	95	128.8	136	184.4
14	18.9	55	74.6	96	130.2	137	185.7
15	20.3	56	75.9	97	131.5	138	187.1
16	21.7	57	77.3	98	132.9	139	188.5
17	23.0	58	78.6	99	134.2	140	189.8
18	24.4	59	80.0	100	135.6	141	191.2
19	25.8	60	81.4	101	136.9	142	192.5
20	27.1	61	82.7	102	138.3	143	193.9
21	28.5	62	84.1	103	139.6	144	195.2
22	29.8	63	85.4	104	141.0	145	196.6
23	31.2	64	86.8	105	142.4	146	198.0
24	32.5	65	88.1	106	143.7	147	199.3
25	33.9	66	89.5	107	145.1	148	200.7
26	35.2	67	90.8	108	146.4	149	202.0
27	36.6	68	92.2	109	147.8	150	203.4
28	38.0	69	93.6	110	149.1	151	204.7
29	39.3	70	94.9	111	150.5	152	206.1
30	40.7	71	96.3	112	151.8	153	207.4
31	42.0	72	97.6	113	153.2	154	208.8
32	43.4	73	99.0	114	154.6	155	210.2